Women
and
Health

Women
and
Health

Edited by

Marlene B. Goldman

Department of Epidemiology
Harvard School of Public Health
Boston, Massachusetts

Maureen C. Hatch

Department of Community and Preventive Medicine
Division of Epidemiology
Mt. Sinai School of Medicine
New York, New York

ACADEMIC PRESS
A Harcourt Science and Technology Company

San Diego   San Francisco   New York   Boston   London   Sydney   Tokyo

Cover photograph: Bharati Chaudhuri/Superstock.

This book is printed on acid-free paper.

Copyright © 2000 by ACADEMIC PRESS

All Rights Reserved.
No part of this publication may be reproduced or transmitted in any form or by any
means, electronic or mechanical, including photocopy, recording, or any information
storage and retrieval system, without permission in writing from the publisher.

Requests for permission to make copies of any part of the work should be mailed to:
Permissions Department, Harcourt Inc., 6277 Sea Harbor Drive,
Orlando, Florida 32887-6777

Academic Press
A Harcourt Science and Technology Company
525 B Street, Suite 1900, San Diego, California 92101-4495, U.S.A.
http://www.apnet.com

Academic Press
24-28 Oval Road, London NW1 7DX, UK
http://www.hbuk.co.uk/ap/

Library of Congress Catalog Card Number: 99-61535

International Standard Book Number: 0-12-288145-1

PRINTED IN THE UNITED STATES OF AMERICA
99 00 01 02 03 04 MM 9 8 7 6 5 4 3 2 1

I thank my parents, David and Charlotte Goldman, and grandparents, Ida and Edward Hurwitch, in memoriam, for their unconditional love and faith in me; Heidi Ellenbecker, Michelle Flaherty, and Edward Flaherty, my inspirations—in the hope that their generation will reap the benefits of books like this one—and most of all, I dedicate this book to Michael Ellenbecker, whose strength and sustained optimism help me stay the course.

M. B. G.

I dedicate this book to my husband, Richard Hatch, our sons, Tony and Peter, and my good friend Sue Lehmann. I also acknowledge Satch, Gus, and Jack, who helped round out the days while the book was coming together.

M. C. H.

Contents

Contributors

Numbers in parentheses indicate the pages on which the authors' contributions begin.

Ruth H. Allen (607), Extramural Division of Cancer Control and Population Sciences, National Cancer Institute, Bethesda, Maryland 20892

Hani K. Atrash (160), Pregnancy and Infant Health Branch, Division of Reproductive Health, National Centers for Disease Control, Atlanta, Georgia 30341

Donna Day Baird (126), Epidemiology Branch, National Institute of Environmental Health Sciences, Research Triangle Park, North Carolina 27709

Carol M. Baldwin (1129), Department of Medicine, Respiratory Sciences Center, College of Medicine, University of Arizona, Tucson, Arizona 85723

Robert L. Barbieri (196), Department of Obstetrics and Gynecology and Reproductive Biology, Brigham and Women's Hospital, Boston, Massachusetts 02115

Richard Beasley (724), Wellington Asthma Research Group (WARG), Department of Medicine, Wellington School of Medicine, Wellington South, New Zealand

Iris R. Bell (1129), Tucson Veterans Affairs Medical Center, Tucson, Arizona 85723

Gertrud S. Berkowitz (81), Departments of Community and Preventive Medicine and Obstetrics and Gynecology and Reproductive Sciences, The Mount Sinai Hospital, Mount Sinai School of Medicine, New York, New York 10029

Leslie Bernstein (871), Department of Preventative Medicine, Norris Comprehensive Cancer Center, University of Southern California, Los Angeles, California 90033

F. Xavier Bosch (932), Catalan Institute of Oncology, L'Hospitalet del Llobregat, Barcelona, Spain

Judith Bradford (64), Virginia Commonwealth University, Center for Public Policy, Richmond, Virginia 23284

Naomi Breslau (1024), Departments of Psychiatry, Biostatistics, and Epidemiology, Henry Ford Health Systems, Detroit, Michigan 48202

Robin L. Brey (797), Department of Medicine, Division of Neurology, University of Texas Health Sciences Center, San Antonio, Texas 78204

Louise A. Brinton (855), Environmental Epidemiology Branch, National Cancer Institute, Bethesda, Maryland 20892

Evelyn J. Bromet (987), Department of Psychiatry and Behavioral Science, State University of New York at Stony Brook, Stony Brook, New York 11794

Deborah Brooks-Nelson (369), Center for Clinical Epidemiology and Biostatistics, University of Pennsylvania, Philadelphia, Pennsylvania 19104

Joelle M. Brown (311), Departments of Biostatistics and Epidemiology, University of California, Berkeley, California 94720

Dedra Buchwald (1059), Department of Medicine, University of Washington School of Medicine, Seattle, Washington 98117

Diana S. M. Buist (757), Center for Health Studies, Group Health Cooperative of Puget Sound, University of Washington, Seattle, Washington 98101

Gale Burstein (273), Division of Adolescent and School Health Center for Disease Control and Prevention, Johns Hopkins University School of Medicine, Hyattsville, Maryland 20782

Willard Cates, Jr. (381), Family Health International, Research Triangle Park, North Carolina 27709

Jane A. Cauley (1182), University of Pittsburgh, Pittsburgh, Pennsylvania 15261

Connie L. Celum (302), Department of Medicine, University of Washington, and King County Department of Public Health, Harborview STD Clinic, Seattle, Washington 98104

David C. Christiani (634), Harvard School of Public Health, Harvard Medical School, Massachusetts General Hospital, Boston, Massachusetts 02115

Carolyn M. Clancy (50), Center for Outcomes and Effectiveness Research, Agency for Health Care Policy and Research, Department of Health and Human Services, Rockville, Maryland 20852

Karen Scott Collins (55), The Commonwealth Fund, New York, New York 10021

Linda S. Cook (916), Department of Community Health Science, University of Calgary, Alberta, Canada; and Fred Hutchinson Cancer Research Center, Seattle, Washington 98104

Karen J. Cruickshanks (1212), Department of Ophthalmology and Visual Sciences, University of Wisconsin, Madison, Wisconsin 53705

Janet R. Daling (81), Division of Public Health Sciences, University of Washington, Fred Hutchinson Cancer Research Center, Seattle, Washington 98104

Michelle E. Danielson (1182), University of Pittsburgh, Pittsburgh, Pennsylvania 15261

Karen Davis (55), The Commonwealth Fund, New York, New York 10021

Susan S. Devesa (863), Descriptive Studies Section, Biostatistics Branch, Division of Cancer, Epidemiology, and Genetics, National Cancer Institute, Rockville, Maryland 20852

Kay Dickersin (253), Department of Community Health, Brown University, Providence, Rhode Island 02912

William R. Downs (529), Center for the Study of Adolescence, University of Northern Iowa, Cedar Falls, Iowa 50614

Mark Drangsholt (1120), Department of Oral Medicine, Dental Public Health Sciences and Epidemiology, University of Washington, Seattle, Washington 98195

Carin E. Dugowson (674), Bone and Joint Center, University of Washington School of Medicine, Seattle, Washington 98104

Rhobert W. Evans (839), Department of Epidemiology, University of Pittsburgh, Pittsburgh, Pennsylvania 15261

Mercedes Fernandez (1129), Tucson Veterans Affairs Medical Center, Tucson, Arizona 85723

Jodi A. Flaws (625), Department of Epidemiology and Preventive Medicine, University of Maryland School of Medicine, Baltimore, Maryland 21201

Katherine M. Flegal (830), National Center for Health Statistics, Centers for Disease Control and Prevention, Hyattsville, Maryland 20782

Betsy Foxman (361), Department of Epidemiology, University of Michigan School of Public Health, Ann Arbor, Michigan 48109

William D. Fraser (182), Department of Obstetrics and Gynecology, Laval University, Quebec City, Canada G1L 3L5

Lynn P. Freedman (428), Center for Population and Family Health, Joseph L. Mailman School of Public Health, Columbia University, New York, New York 10032

Nancy H. Fultz (1202), Institute of Social Research, University of Michigan, Ann Arbor, Michigan 48109

Samuel Gandy (1227), New York University at N.S. Kline Institute, Orangeburg, New York 10962

Edward Giovannucci (962), Channing Laboratory, Department of Medicine, Harvard Medical School, Departments of Nutrition and Epidemiology, Harvard School of Public Health, Boston, Massachusetts 02115

Karen Glanz (977), Cancer Research Center of Hawaii, University of Hawaii, Honolulu, Hawaii 96813

Marlene B. Goldman (5, 196), Department of Epidemiology, Harvard School of Public Health, Boston, Massachusetts 02115

David A. Grainger (215), Division of Reproductive Endocrinology, University of Kansas School of Medicine, Wichita, Kansas 67214

Jack Guralnik (1143, 1147), Epidemiology, Demography, and Biometry Program, National Institute of Aging, Bethesda, Maryland 20892

Katherine A. Halmi (1032), Department of Psychiatry, Weill Medical College of Cornell University Medical College, Eating Disorder Program, New York Presbyterian Hospital, Westchester Division, White Plains, New York 10605

Sandra W. Hamelsky (1084), Department of Neurology, Albert Einstein College of Medicine, Bronx, New York 10461

Siobán D. Harlow (99), Department of Epidemiology, School of Public Health and Population Studies Center, University of Michigan, Ann Arbor, Michigan 48109

Patricia Hartge (907), Division of Cancer Epidemiology and Genetics, National Cancer Institute, National Institutes of Health, Bethesda, Maryland 20892

Maureen C. Hatch (5, 37), Division of Epidemiology, Department of Community and Preventive Medicine, Mt. Sinai School of Medicine, New York, New York 10029

Donna J. Hawley (1068), School of Nursing, Wichita State University, Wichita, Kansas 67260

Suzanne G. Haynes (25, 37), Office on Women's Health, Department of Health and Human Services, Washington, DC 20201

Robin Herbert (474), Department of Community and Preventive Medicine, Division of Environmental and Occupational Medicine, Mt. Sinai School of Medicine, New York, New York 10029

A. Regula Herzog (1202), Institute for Social Research, Institute of Gerontology, Department of Epidemiology, University of Michigan, Ann Arbor, Michigan, 48104

Shirley Y. Hill (1042), Department of Psychiatry, University of Pittsburgh School of Medicine, Pittsburgh, Pennsylvania 15213

Anne N. Hirshfield (625), Department of Anatomy and Neurobiology, University of Maryland School of Medicine, Baltimore, Maryland 21201

Marc C. Hochberg (1192), Department of Medicine and Epidemiology and Preventive Medicine, Division of Rheumatology and Clinical Immunology, University of Maryland School of Medicine, Baltimore, Maryland 21201

Michael Hodgson (503), Centers for Disease Control, Washington, DC 20016

Carol J. Rowland Hogue (15), Rollins School of Public Health, Emory University, Women and Children's Center, Atlanta, Georgia 30322

Victoria L. Holt (226), Department of Epidemiology, University of Washington, Fred Hutchinson Cancer Research Center, Seattle, Washington 98109

Richard Holubkov (771), LHAS Women's Heart Center, Division of Cardiology and Department of Epidemiology, Graduate School of Public Health, University of Pittsburgh Medical Center, Pittsburgh, Pennsylvania 15261

Corinne G. Husten (563), Office on Smoking and Health, Centers for Disease Control and Prevention, Atlanta, Georgia 30341

Noreen A. Hynes (285), Division of Infectious Diseases, Johns Hopkins University School of Medicine, Baltimore, Maryland 21205

Susan Izett (404), Research, Action, and Information Network for the Bodily Integrity of Women (RAINBO), West Coxsackie, New York 12192

Naomi Jay (324), University of California, San Francisco, California 94143

Janna Jenkins (226), Fred Hutchinson Cancer Research Center, Division of Public Health Sciences, Seattle, Washington 98101

Mehdi Kamarei (1110), Division of Urology, University of California, San Diego, San Diego, California 92103

Mary L. Kamb (336), Division of HIV/AIDS Prevention and Surveillance and Epidemiology, National Center for HIV, STD, and TB Prevention, Centers for Disease Control and Prevention, Atlanta, Georgia 30333

Quarraisha Abdool Karim (420), Medical Research Council, Southern African Fogarty, Durban, 4013 South Africa

Jennifer Kelsey (1251), Department of Health, Research, and Policy, Division of Epidemiology, Stanford University School of Medicine, Stanford, California 94305

Karla Kerlikowske (895), Departments of Medicine and Epidemiology and Biostatistics, General Internal Medicine, Veterans Affairs Medical Center, University of San Francisco, San Francisco, California 94121

Ronald C. Kessler (997), Department of Health Care Policy, Harvard Medical School, Boston, Massachusetts 02115

Samia J. Khoury (686), Center for Neurologic Diseases, Brigham and Women's Hospital, Boston, Massachusetts 02115

Mona Kimball-Dunn (724), Wellington Asthma Research Group (WARG), Department of Medicine, Wellington School of Medicine, Wellington South, New Zealand

Steven J. Kittner (797), Departments of Neurology and Epidemiology and Preventive Medicine, University of Maryland School of Medicine, Baltimore, Maryland 21201

Natasha A. Koloski (1098), University of Sydney, Nepeon Hospital, New South Wales, Australia

Peter Kopp (655), Division of Endocrinology, Metabolism and Molecular Medicine, Northwestern University, Chicago, Illinois 60611

Michael S. Kramer (182), Departments of Epidemiology and Biostatistics and of Pediatrics, McGill University Faculty of Medicine, Montreal, Quebec, H3A IA2 Canada

Andrea Z. LaCroix (757), Department of Epidemiology, Women's Health Initiative Clinical Coordinating Center, Fred Hutchinson Cancer Research Center, School of Public Health and Community Medicine, University of Washington, Seattle, Washington 98104

Gail Schoen LeMaire (253), Department of Epidemiology and Preventive Medicine, University of Maryland School of Medicine, Baltimore, Maryland 21201

Linda LeResche (1120), Department of Oral Medicine, University of Washington, Seattle, Washington 98195

Margaret Lethbridge-Cejku (1192), University of Maryland School of Medicine, Baltimore, Maryland 21201

Suzanne Leveille (1147), Epidemiology, Demography and Biometry Program, National Institute on Aging, Bethesda, Maryland 20892

Marja-Liisa Lindbohm (463), Department of Epidemiology and Biostatistics, Finnish Institute of Occupational Health, Helsinki, 00250 Finland

Richard B. Lipton (1084), Department of Neurology, Albert Einstein College of Medicine, Bronx, New York 10461

Paulo A. Lotufo (819), Division of Preventive Medicine, Brigham and Women's Hospital, Harvard Medical School, Boston, Massachusetts 02215; and University of Sao Paulo School of Medicine, Sao Paulo, Brazil

Andrea Lucas (514), Joseph L. Mailman School of Public Health, Columbia University, New York, New York 10032

Deborah Maine (395), Center for Population and Family Health, Joseph L. Mailman School of Public Health, Columbia University, New York, New York 10032

Ann-Marie Malarcher (563), Office on Smoking and Health, Centers for Disease Control and Prevention, Atlanta, Georgia 30341

JoAnn E. Manson (819), Division of Preventive Medicine, Brigham and Women's Hospital, Harvard Medical School, Harvard School of Public Health, Boston, Massachusetts 02215

Susan M. Manzi (704), Department of Medicine and Epidemiology, Division of Rheumatology, University of Pittsburgh, Pittsburgh, Pennsylvania 15213

Jeanne M. Marrazzo (302), Department of Medicine, University of Washington, Seattle–King County Department of Public Health, Harborview STD Clinic, Seattle, Washington 98104

Lynn M. Marshall (240), Department of Epidemiology, School of Public Health and Community Medicine, University of Washington, Seattle, Washington 98109

Richard Mayeux (1227), Gertrude H. Sergievsky Center, Columbia University, New York, New York 10032

Therese McGinn (395), Center for Population and Family Health, Joseph L. Mailman School of Public Health, Columbia University, New York, New York 10032

Anne McTiernan (1169), Cancer Prevention Research Program, Fred Hutchinson Cancer Research Center, Seattle, Washington 98109

Elaine Meilahn (782), Department of Epidemiology and Population Studies, London School of Hygiene and Tropical Medicine, London WCIE 7HT, United Kingdom

Kathleen Ries Merikangas (1010), Departments of Epidemiology and Public Health and Psychiatry, Yale University School of Medicine, New Haven, Connecticut 06510

Karen Messing (455), CINBIOSE Department of Biological Sciences, University of Quebec at Montreal, Montreal, Quebec H3B 3H5, Canada

Brenda A. Miller (529), Research on Urban Social Work Practice, School of Social Work, University of Buffalo, Buffalo, New York 14203

Daniel R. Mishell, Jr. (138), University of Southern California, Los Angeles, California 90033

Stacey A. Missmer (196), Department of Epidemiology, Harvard School of Public Health, Boston, Massachusetts 02115

Robert Mittendorf (171), Department of Obstetrics and Gynecology, Chicago Lying-in Hospital, University of Chicago, Chicago, Illinois 60631

Manoj Monga (1110), Division of Urology, University of California, San Diego, San Diego, California 92103

Joseph F. Mortola[†] (114), Division of Reproduction and Endocrinology, Cook County Hospital, Chicago, Illinois 60612

Anna-Barbara Moscicki (324), Department of Adolescent Medicine, University of California, San Francisco, San Francisco, California 94143

Nancy E. Moss (541), Pacific Institute for Women's Health, Los Angeles, California 94303

Eileen V. Moy (634), Occupational Health Program, Harvard School of Public Health, Boston, Massachusetts 02115

Heidi Mueller (789), Department of Psychology, State University of New York at Stony Brook, Stony Brook, New York 11794

Nubia Muñoz (932), International Agency for Research on Cancer, Lyon, France

Roberta B. Ness (369, 753), Departments of Epidemiology and Obstetrics and Gynecology, University of Pittsburgh, Pittsburgh, Pennsylvania 15261

Beth Newman (884), Queensland University of Technology School of Public Health, Kelvin Grove, Queensland 4059, Australia

Katherine M. Newton (757), Center for Health Studies, Group Health Cooperative of Puget Sound, Seattle, Washington 98101

Michael J. Olek (686), Center for Neurological Diseases, Brigham and Women's Hospital, Harvard Medical School, Boston, Massachusetts 02115

Nancy S. Padian (269, 381), Center for Family Planning and Reproductive Epidemiology, Department of Obstetrics and Gynecology and Reproductive Sciences, University of California, San Francisco, San Francisco, California 94143

Ann L. Parke (740), University of Connecticut Health Center, Division of Rheumatology, Farmington, Connecticut 06030

Sheila Hill Parker (578), Arizona Prevention Center, University of Arizona, Tucson, Arizona 85724

C. Lowell Parsons (1110), Division of Urology, University of California, San Diego, San Diego, California 92103

Neil Pearce (724), Wellington Asthma Research Group (WARG), Department of Medicine, Wellington School of Medicine, Wellington South, New Zealand

Elizabeth A. Platz (962), Department of Nutrition, Harvard School of Public Health, Boston, Massachusetts 02115

Rachel A. Pollock (1010), Departments of Epidemiology and Public Health and Psychiatry, Yale University School of Medicine, New Haven, Connecticut 06510

Laura Punnett (474), Department of Work Environment, University of Massachusetts Lowell, Lowell, Massachusetts 01854

Rosalind Ramsey-Goldman (651, 704), Department of Medicine, Division of Arthritis and Connective Tissue Diseases, Northwestern University, Chicago, Illinois 60611

Tom Rea (1059), Department of Medicine, University of Washington School of Medicine, Seattle, Washington 98104

Steven E. Reis (771), Division of Cardiology, LHAS Women's Heart Center, University of Pittsburgh Medical Center, Pittsburgh, Pennsylvania 15213

[†]Deceased.

Anne M. Rompalo (273, 285), Johns Hopkins University School of Medicine, Baltimore, Maryland 21287

Lynn Rosenberg (811), Slone Epidemiology Unit, Boston University School of Public Health, Brookline, Massachusetts 02446

Gary S. Rubin (1212), Lions Vision Center, Wilmer Eye Institute, Johns Hopkins University, Baltimore, Maryland 21210

Marcia Russell (589), Research Institute on Addictions, Buffalo, New York 14203

Mary Sabolsi (819), Division of Preventive Medicine, Brigham and Women's Hospital, Harvard Medical School, Boston, Massachusetts 02215

Audrey F. Saftlas (160), Department of Preventive Medicine and Environmental Health, University of Iowa, College of Medicine, Iowa City, Iowa 52242

Mark Schiffman (942), Division of Cancer Epidemiology and Genetics, National Cancer Institute, Bethesda, Maryland 20852

Cathy Schoen (55), The Commonwealth Fund, New York, New York 10021

Theresa O. Scholl (85), Department of Obstetrics and Gynecology, University of Medicine and Dentistry of New Jersey, Stratford, New Jersey 08084

Karen B. Schmaling (1055), Department of Psychiatry and Behavioral Sciences, University of Washington, Seattle, Washington 98122

Gary E. R. Schwartz (1129), Tucson Veterans Affairs Medical Center, Tucson, Arizona 85723

Stephen M. Schwartz (240), Program in Epidemiology, Division of Public Health Sciences, Fred Hutchinson Cancer Research Center, Seattle, Washington 98109

Jane R. Schwebke (352), University of Alabama at Birmingham, Birmingham, Alabama 35294

Jean C. Scott (1192), Departments of Medicine and Epidemiology and Preventive Medicine, University of Maryland School of Medicine, Baltimore, Maryland 21201

Barbara Seaman (27), National Women's Health Network, New York, New York 10023

Teresa E. Seeman (1238), Department of Medicine, Division of Geriatrics, University of California Los Angeles, Los Angeles, California 90033

Mary V. Seeman (989), Center for Addictions and Mental Health, Clarke Institute of Psychiatry, University of Toronto, Toronto, Ontario, MST 1RS Canada

Fady I. Sharara (625), Department of Reproductive Endocrinology and Infertility, University of Maryland School of Medicine, Baltimore, Maryland 21201

Donna Shoupe (138), University of Southern California, Los Angeles, California 90033

Ellen K. Silbergeld (559, 601, 625), Department of Epidemiology and Preventive Medicine, University of Maryland School of Medicine, Baltimore, Maryland 21201

Debra T. Silverman (492), Division of Cancer Epidemiology and Genetics, National Cancer Institute, Bethesda, Maryland 20892

Diane Solomon (942), Division of Cancer Prevention and Control, National Cancer Institute, Bethesda, Maryland 20892

Glorian Sorensen (523), Harvard School of Public Health and Center for Community Based Research, Dana Farber Cancer Institute, Boston, Massachusetts 02115

MaryFran R. Sowers (1155), Department of Epidemiology, University of Michigan School of Public Health, Ann Arbor, Michigan 48103

Darcy V. Spicer (871), Department of Medicine, University of Southern California School of Medicine, Norris Comprehensive Cancer Center, Los Angeles, California 90033

Zena Stein (391, 420), HIV Center and New York State Psychiatric Institute, Joseph L. Mailman School of Public Health, Columbia University, New York, New York 10032

Jeanne Mager Stellman (514), Joseph L. Mailman School of Public Health, Columbia University, New York, New York 10032

Walter F. Stewart (1084), Department of Epidemiology, The Johns Hopkins University School of Hygiene, Baltimore, Maryland 21205

Eileen Storey (503), Division of Occupational and Environmental Medicine, University of Connecticut Health Center, Farmington, Connecticut 06030

Beverly I. Strassmann (126), Department of Anthropology, University of Michigan, Ann Arbor, Michigan 48109

Nicholas J. Talley (1098), University of Sydney, Nepean Hospital, Penrith, New South Wales 2750, Australia

Helena Taskinen (463), Department of Occupational Medicine, Finnish Institute of Occupational Health, Fin 00250 Helsinki, Finland

Maria Testa (589), Research Institute on Addictions, Buffalo, New York 14203

Bruce L. Tjaden (215), Division of Reproductive Endocrinology, University of Kansas School of Medicine, Wichita, Kansas 67214

Nahid Toubia (404), Research Action Information Network for Bodily Integrity of Women (RAINBO), New York, New York 10010

Rebecca Troisi (907), Division of Cancer Epidemiology and Genetics, National Cancer Institute, Bethesda, Maryland 20892

Debra Umberson (553), Department of Sociology and Population Research Center, University of Texas, Austin, Austin, Texas 78712

Jennifer B. Unger (1238), Institute for Health Promotion and Disease Prevention Research, University of Southern California School of Medicine, Los Angeles, California 90033

Giske Ursin (871), Department of Preventive Medicine, Norris Comprehensive Cancer Center, University of Southern California, Los Angeles, California 90033

Thomas M. Vogt (977), Kaiser Permanente Center for Health Research, Cancer Research Center of Hawaii, Honolulu, Hawaii 96813

Anna Wald (311), Department of Medicine and Epidemiology, University of Washington, Seattle, Washington 98122

Jane Walstedt (447), Women's Bureau, Department of Labor, Washington, DC 20210

Mary H. Ward (492), Division of Cancer and Genetics and Epidemiology, National Cancer Institute, Bethesda, Maryland 20892

Gerdi Weidner (789), Department of Psychology, State University of New York at Stony Brook, Stony Brook, New York 11794

Noel S. Weiss (916), Department of Epidemiology, University of Washington, Seattle, Washington 98195

Jocelyn C. White (64), Center for Public Policy, Virginia Commonwealth University, Richmond, Virginia 23284

Kristi Williams (553), Department of Sociology and Population Research Center, University of Texas, Austin, Texas 78712

Michelle A. Williams (171), Department of Epidemiology, University of Washington, Swedish Medicine Center for Perinatal Studies, Seattle, Washington 98122

Sharon C. Wilsnack (589), Department of Neuroscience, University of North Dakota School of Medicine and Health Sciences, Grand Forks, North Dakota 58203

Phyllis A. Wingo (1169), American Cancer Society, Atlanta, Georgia 30329

Frederick Wolfe (1068), University of Kansas School of Medicine, Wichita, and Wichita Arthritis Research and Clinical Centers, Wichita, Kansas 67214

Susan F. Wood (27), Office on Women's Health, Department of Health and Human Services, Washington, DC 20201

Pascale M. Wortley (336), Division of HIV/AIDS Prevention—Surveillance and Epidemiology, National Center for HIV, STD, and TB Prevention, Centers for Disease Control and Prevention, Atlanta, Georgia 30333

Anna H. Wu (949), Department of Preventive Medicine, Norris Comprehensive Cancer Center, University of Southern California School of Medicine, Los Angeles, California 90033

Shelia Hoar Zahm (441, 492), Division of Cancer Epidemiology and Genetics, National Cancer Institute, Rockville, Maryland 20892

Preface

Women are diverse individuals striving to maintain their health within a complex world of cultural, psychological, social, and biological influences. In this light, the content of this book was designed to address the behavioral and societal, as well as the biological, determinants of the health and well-being of women and girls throughout their life span. We have known for many years that diseases may not affect women in the same way as they affect men and that knowledge gained from studying men is not necessarily applicable to understanding disease occurrence, diagnosis, or treatment in women, yet no comprehensive reference text has explicitly taken this into account. Our intent is that the information presented here will improve the quality of thinking about women's health and in so doing will improve the quality of women's lives.

The evolution of women's health as a discipline is reflected in the breadth and depth of the chapters collected here. We have defined women's health in broad terms and have chosen not to limit the scope of the book to those conditions uniquely or predominantly affecting women, but to recognize that an understanding of the roles of sex and gender is integral to all aspects of excellent health care and medical research. The preparation of this book required a shift in focus for many of the contributors—from thinking of their discipline as gender-neutral to realizing the importance of examining gender-related differences. Frequently, a potential contributor in an area not historically considered "women's health" would say "I've never considered my research in that way. In fact, we usually control for gender to remove its confounding effects." Then after the chapter was done, these contributors would thank us for asking them to address their research in a new and productive way. The development of a gendered-perspective—for the improvement of the health of women and men—has been the goal of all of us who have brought this book into existence. Creating "gender awareness" parallels the evolution of the definition of women's health from concerns about reproduction and "traditionally female" ailments to encompass the broader vision represented here. Concern for women's health is not about sexism; it is about unnecessary pain, stress, and mutilation due to lack of prevention strategies and delayed diagnosis and treatment because research on the ways diseases affect women has not been done or disseminated.

For examples of how pervasive traditional concepts of male-focused medicine can be this late in the 20th century, we need only look at instances where knowledge gained from only one sex (usually male) is thought to be sufficient for application to both. We are reminded of the famous, but unfortunate, marathoner who was suspended after failing a drug test—a test of testosterone hormone levels that was developed for use in male athletes and never validated in women. Her misfortune was that she was a woman. Another example is the initial case definitions for AIDS-related complex and AIDS that were based on signs and symptoms in men. Failure to construct case definitions that include disease characteristics present in women that may not occur in men has put women at a disadvantage in obtaining quality health care. It took 10 years and the realization that some women were presenting with cervical cancer, pelvic inflammatory disease, or candidiasis rather than Kaposi's sarcoma or pneumocystis penumonia for the case definition to be changed and for many women to be correctly diagnosed, recognized, and counted in the AIDS epidemic. It is incumbent upon us to challenge those who cling to traditional models by demanding that new drugs be tested in diverse populations that include both sexes, that research studies and clinical trials include women and men, that screening tests be validated in the populations in which they are to be implemented, and that diagnostic and treatment procedures be applied without bias by gender.

While the rationale for this volume was a recognition that a textbook that provided the medical and scientific community with a synthesis of the latest research results on a comprehensive range of diseases and conditions that affect women was needed, we also intend that the extensive material presented here will assist consumers of health care. It is our conviction that the very best clinical care can be obtained only when patients have access to the cutting-edge knowledge that research medicine has to offer.

As the book evolved from concept to reality, many individuals provided guidance, professional assistance, and encouragement. We acknowledge several of them here. The idea for this volume developed from conversations with Dr. Dimitrios Trichopoulos, to whom we owe our gratitude for his foresight. Dr. Jennifer Kelsey generously shared her publishing experience and helped us frame our initial proposal. The section editors, Drs. Trudy S. Berkowitz, Louise A. Brinton, Evelyn J. Bromet, Janet R. Daling, Jack M. Guralnik, Suzanne G. Haynes, Roberta B. Ness, Nancy S. Padian, Rosalind Ramsey-Goldman, Karen B. Schmaling, Ellen K. Silbergeld, Glorian Sorensen, Zena A. Stein, and Shelia Hoar Zahm, were essential to the development of the book's content and mission, assisted in the selection of contributors, and reviewed the submitted chapters. We particularly acknowledge the chapter authors for their commitment to scholarship, flexibility in accommodating multiple revisions, and grace in honoring the deadlines that the timely production of a book of this size necessitated. We are most appreciative of the efforts of one of our colleagues, a contributor to the Reproductive Health Section, Dr. Joseph Mortola, who made a remarkable effort to complete his chapter shortly before his untimely death.

At Academic Press, Charlotte Brabants, then Acquisitions Editor, was the catalyst for the book. Without Charlotte's repeated encouragement during the book's early development, it is unlikely that it would have reached fruition. Tari Paschall expertly saw the assembly of the manuscript through from beginning to end.

Hazel Emery, Rachelle Ferrari, Destiny Irons, Alvara McBean, and Stacey Missmer cheerfully and efficiently managed the hundreds of clerical and research tasks that kept us on schedule.

We thank our friends and colleagues Drs. Trudy Berkowitz, Neta Crawford, Shelia Zahm, and Li Zemp for their constructive criticism and support.

While we have attempted to be comprehensive within time and space constraints, we are aware of health conditions that had to be omitted. We hope that we will be able to include them in future editions. If the information you seek is not here, please let us know.

Marlene B. Goldman
Maureen C. Hatch

Part I

WOMEN, HEALTH, AND MEDICINE

Section 1

WHY WOMEN'S HEALTH?

1

An Overview of Women and Health

MARLENE B. GOLDMAN* AND MAUREEN C. HATCH†

*Department of Epidemiology, Harvard School of Public Health, Boston, Massachusetts; †Division of Epidemiology, Department of
Community and Preventive Medicine, Mount Sinai School of Medicine, New York, New York

I. Why Women's Health?

This introductory chapter discusses what is meant by women's health—both what we mean by it and what others have meant—and why we believe a book like this is needed. We next present an overall picture of women's health using epidemiologic evidence derived from major health indicators. In the final section, we make suggestions about where to go next. The chapters that follow this one explore the influences of sex and gender on the occurrence of a variety of diseases and conditions, both inside and outside the United States, and will address possible reasons for similarities and differences between disease occurrence in women and men.

As women's health has gained recognition as a discipline, there have been legislative and organizational attempts to define its scope, research and clinical priorities, and fundamental tenets [1–5]. Since the 1970s, a number of definitions have been proposed (Figure 1.1). Early definitions focused on women's

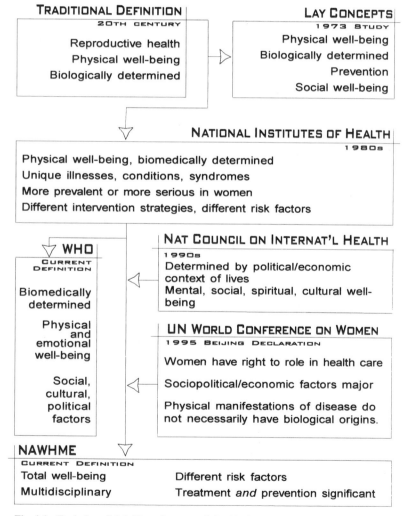

Fig. 1.1 Evolution of definitions for women's health. Sources: [1,6–10]. NAWHME, National Academy on Women's Health Medical Education.

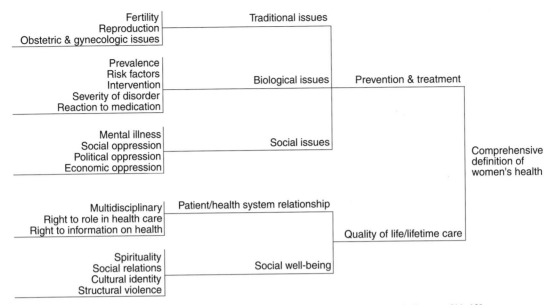

Fig. 1.2 Components of a comprehensive definition of women's health. Sources: [11–13].

reproductive health and were based on the biomedical model [6], while more recent ones recognize that a variety of factors influences a woman's health during her life span [1,7–10]. These later definitions encompass all diseases and disorders that affect women, include an awareness of the impact of social, cultural, economic, and political influences, and emphasize prevention as well as treatment.

For example, the definition formulated by the National Academy on Women's Health Medical Education [8] states that

- women's health is devoted to facilitating the
  - preservation of wellness and
  - prevention of illness in women

and includes

- screening, diagnosis, and management of conditions that
  - are unique to women,
  - are more common in women,
  - are more serious in women,
  - or have manifestations, risk factors, or interventions that are different in women.

It also

- recognizes the importance of the study of gender differences,
- recognizes multidisciplinary team approaches,
- includes the values and knowledge of women and their own experience of health and illness,
- recognizes the diversity of women's health needs over the life cycle and how these needs reflect differences in race, class, ethnicity, culture, sexual preference, and levels of education and access to medical care, and
- includes the empowerment of women to be informed participants in their own health care.

In Figure 1.2 we summarize the major components that contribute to a comprehensive model of women's health.

Along with the developing awareness of health (and health care) issues of particular concern to women has come the realization that the training of health care professionals must incorporate these new dimensions in order to address gender bias and inequalities in health care delivery [14,15].

The American College of Women's Health Physicians (www.acwh.org) defines the practice of women's health care as

- A sex- and gender-informed practice centered on the whole woman in the diverse contexts of her life, grounded in an interdisciplinary sex- and gender-informed biopsychosocial science [16].

Necessary changes to the training of health care professionals are to require courses in women's health and the role of gender in medical and public health school curricula, to create a separate woman's health medical specialty, and to identify and remove barriers to successful biomedical careers for women [16–20].

II. Why This Book?

The field of women's health is at the point where the need exists for a comprehensive reference that brings together authoritative, state-of-the-art reports—written from an epidemiologic perspective—on a broad range of topics that include social, environmental, and occupational determinants of health as well as various diseases and disorders.

In the book we have taken a life span approach, with sections that follow a woman's life course from her entry into the reproductive years, to the chronic diseases of midlife, including mental disorders, and then to the special conditions associated with aging. The virtue of this chronologic perspective is that it groups together health events that occur against a specific age-related backdrop of biologic development and social roles.

The topics covered in the book's fourteen sections include important and familiar concerns such as heart disease and cancer. Also included are conditions unique to women that receive scant attention from the research community. Examples are domestic violence and rape, premenstrual syndrome, and maternal morbidity. Several sections deal with conditions that are more common among women. These include autoimmune disorders and a group we have called poorly understood conditions—for example, chronic fatigue syndrome and migraine. The conditions discussed in the section on aging, such as urinary incontinence and sensory impairment, affect more women than men, if only because women comprise the majority of the aging population. Specific social factors influencing the health of women, and women in various social contexts—at work, in the environment, in the global setting—are also covered in this comprehensive book. While the book is international in scope, the overview that follows is based primarily on U.S. statistics.

III. Overview of the Health of Women

Women constitute 51% of the U.S. and 50% of the world's population [21,22]. We make three-fourths of all health care decisions in American households and spend almost two of every three health care dollars, approximately $500 billion annually [23]. More than 61% of physician visits are made by women and 59% of prescription drugs are purchased by women, yet health problems common or unique to women are often overlooked when research priorities are established [24]. Only recently has the scientific community made a commitment to include women in federally funded research studies [25–28]. In the medical community, efforts have now begun to address the needs of women in the clinical setting [29,30].

A familiar paradox of women's health is that women live longer than men but have poorer health. Worldwide, women usually live longer than men in similar socioeconomic circumstances. In most of the developed countries, the difference between female and male life expectancy is about 6.5 years [31]. In the United States, life expectancy at birth is 79.1 years for women and 73.1 years for men [32] (Figure 1.3). For both sexes, life expectancy at birth has increased since 1940, although the increase has been greater for women than men.

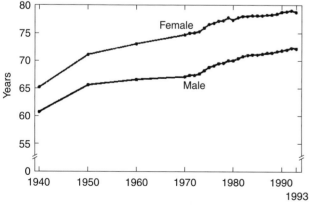

Fig. 1.3 Life expectancy at birth by sex: United States, 1940–1993. Source: Centers for Disease Control and Prevention, National Center for Health Statistics, National Vital Statistics System [33].

Women have not always lived longer than men. The longer life expectancy of women first became apparent in the late nineteenth century. Declines in fertility and maternal mortality combined with greater nutrition are credited with reducing premature mortality in women [31].

As a result of their longer life expectancy, women make up more of the elderly population. In the United States, 75% of nursing home residents over the age of 75 are women. By 2030, one in every four women will be 65 years of age or older. There will also be increased diversity among women. Women of color, currently 27%, will make up 40% of all women in the United States by 2030 [21,34].

Historically considered a disease of men, heart disease is the leading cause of death in both U.S. men and women, followed by cancer and cerebrovascular disease [35] (Figure 1.4). Worldwide, ischemic heart disease is the leading cause of death, followed by cerebrovascular disease and lower respiratory infections. In developing regions, lower respiratory infections cause slightly more deaths than heart disease today, although noncommunicable diseases are expected to account for 70% of all deaths in developing countries by 2020 [36].

In the United States between 1970 and 1979, women and men experienced similar rates of decline in mortality from heart disease; after 1979 the decline was less rapid for women than for men. The decrease since 1979 is attributed to a decline in deaths from ischemic heart disease [33] (Figure 1.5). Death rates for heart disease are higher for African-American women than white women. In 1995, the age-adjusted death rate for women aged 45 and older was 544.5 per 100,000 African-American women and 499.6 per 100,000 white women [33,34]. In contrast, rates for American Indian or Alaskan Native women, Asian or Pacific Islander women, Hispanic women, and women in Western and Southern Europe, Scandinavia, and Japan are lower than for non-Hispanic white American women.

Heart disease develops, on average, ten years later in women than men and may go undetected and untreated until it has become severe. According to the American Heart Association, 42% of women who have had a myocardial infarction die within one year compared to 24% of men, underscoring the importance of designing studies that address heart disease risk factors, treatment, and prevention in women [23,37].

Known cardiovascular risk factors include smoking, hypertension, higher levels of total serum cholesterol, lower levels of high density lipoprotein cholesterol, diabetes, adiposity, and a sedentary lifestyle. Between 1965 and 1990 the prevalence of smoking in the United States declined more among men than among women, although between 1990 and 1993 smoking prevalence actually increased in young women 18 to 24 years of age. U.S. statistics from the late 1990s show that 23% of women are smokers [38]. Worldwide, use of tobacco is projected to cause more premature death and disability than any single disease [36].

The prevalence of hypertension increases with age. Men under 50 years of age are more likely to be hypertensive, but after age 70, women are more likely to be affected. Hypertension is more common in African-American women (1988–1991 age-adjusted prevalence = 31%) than non-Hispanic white women (21%) or Mexican-American women (22%). Younger women are more likely to treat their hypertension with either medication

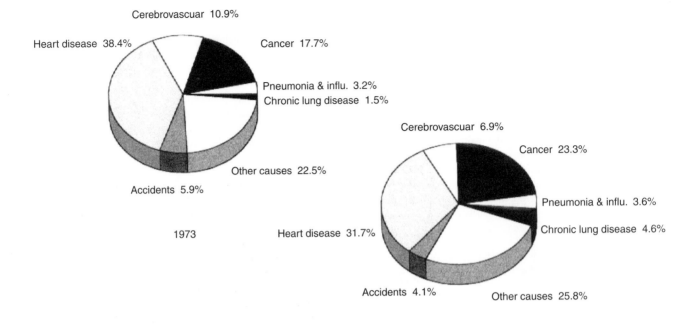

Fig. 1.4 Leading causes of death in the United States, 1973 versus 1996, men and women combined. Source: SEER Cancer Statistics Review, 1973–1996, National Center Institute [35].

or nonpharmacologic means than are older women or men of any age [33].

The prevalence of overweight has increased dramatically in the United States over the last decade. When overweight is defined as a body mass index (BMI) greater than 27.3 kg/m², 34% of all U.S. women were reported to be overweight by the National Health and Nutrition Examination Survey [33]. Using the new cutoff of a BMI > 25 kg/m², 51% of U.S. women are either overweight or obese [39]. A related finding is that among women 25 years and older, the prevalence of a sedentary lifestyle (defined by the National Health Interview Survey as no self-reported leisure time physical activity during the past two weeks) was 30%. The likelihood of reporting a sedentary lifestyle increased with decreasing level of education, a pattern also observed with cigarette smoking.

When the data are stratified by age at death, cancer is the leading cause of death in U.S. women younger than age 75, followed by heart disease and stroke. The most common cancers are breast, lung, and colorectal; lung, breast, and colorectal are the first, second, and third leading causes of death from cancer [40] (Figure 1.6). Time trends in cancer incidence and mortality between 1973 and 1996 are shown in Figure 1.7 [35]. The age-adjusted death rate for lung cancer in women almost tripled, compared to a modest increase in lung cancer deaths in men during the same time period. Lung cancer overtook breast cancer as the leading cause of cancer death in 1987 (Figure 1.8). There is recent evidence that women may be more susceptible to the harmful effects of tobacco-related carcinogens than men [42].

With the exception of lung cancer, the incidence rates for most common cancers in women have remained relatively sta-

ble. Breast cancer incidence did rise 25% between 1973 and 1996, with the majority of the increase between 1980 and 1987. The rise is usually attributed to earlier diagnosis and greater use of mammographic screening. Since 1987 breast cancer rates have been stable [33] (Figure 1.9).

In U.S. women aged 15 to 54, breast cancer is the leading cause of cancer death. Although the age-adjusted breast cancer incidence rate for white women is higher than that for African-American women, black women are almost 30% more likely to die from their breast cancer than white women [33] (Figure 1.10).

Reproductive health and sexuality remain important sources of morbidity and reasons for seeking health care services for

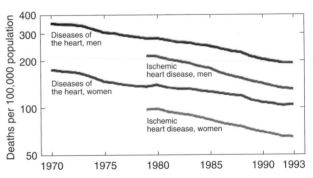

Fig. 1.5 Death rates for heart disease and ischemic heart disease by sex: United States, 1970–1993. Source: Centers for Disease Control and Prevention, National Center for Health Statistics, National Vital Statistics System [33].

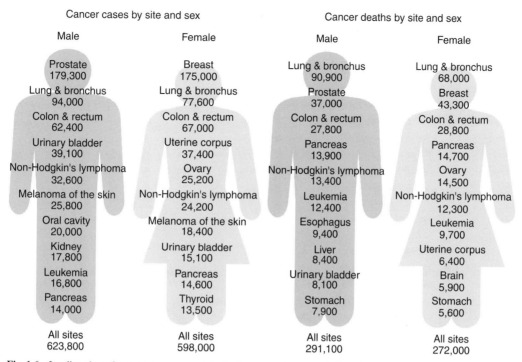

Fig. 1.6 Leading sites of new cancer cases and deaths for men and women, 1999 estimates, excluding basal and squamous cell skin cancer and carcinomas in situ except urinary bladder. Source: American Cancer Society, 1999 Cancer Facts and Figures [40].

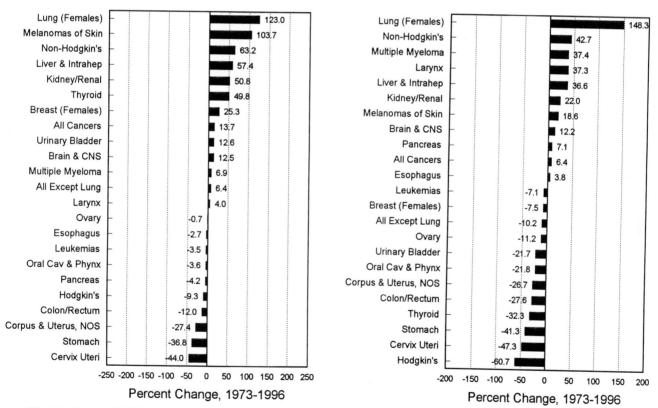

Fig. 1.7 Trends in SEER incidence rates (left) and U.S. mortality rates (right) by primary cancer site for women, 1973–1996. Source: SEER Cancer Statistics Review 1973–1996, National Cancer Institute [35].

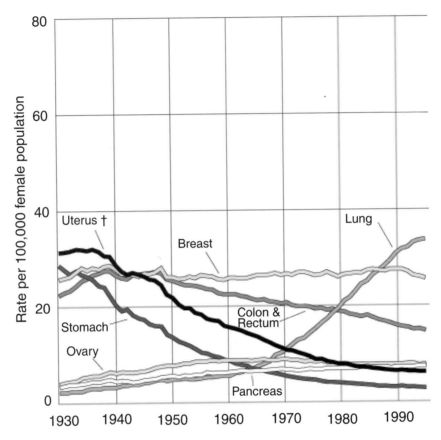

Fig. 1.8 Age-adjusted cancer death rates by site, females, United States, 1930–1994. Rates are per 100,000 women and are age-adjusted to the 1970 U.S. standard population. Due to changes in *International Classification of Diseases* coding, numerator information has changed over time. Rates for cancers of the uterus, ovary, lung, and colon and rectum are affected by these coding changes. Uterine cancer death rates are for cervix and endometrium combined. Source: Vital Statistics of the United States, 1997, American Cancer Society, Surveillance Research, Cancer Facts and Figures, 1998 [41].

women. In *The Global Burden of Disease,* Murray and Lopez [36] report that, while the burden of reproductive ill health is almost entirely confined to developing countries, it is so great that, even worldwide, maternal conditions account for three of the ten leading causes of disease burden in reproductive age women. (The term "disease burden" captures the impact of both premature death and disability.) The three conditions are obstructed labor, chlamydia, and maternal sepsis. In developing countries, half of the ten leading causes of disease burden are related to reproductive health and include unsafe abortion and chlamydia infection. Henderson [43] states that it is "clearly impossible to separate non-reproductive from reproductive research issues in considering women's health because the interplay of steroidal and non-steroidal hormones is an almost certain influence on the growth, function, durability, and immunocompetence of cells and systems in women of most, if not all, ages." Thus, decisions that women make with regard to childbearing, contraception, and replacement hormone use, for example, have an impact on both their immediate and long-term health and well-being.

Compared to men, women are at greater risk of many musculoskeletal disorders, including arthritis and osteoporosis.

In the National Health Interview Survey, 28% of women aged 45 to 64 years reported having arthritis [33]. The proportion increased with age: 49% of women aged 65 to 74 and 57% of those 75 and older were affected. At each age group women were more likely to be affected than men (Figure 1.11). The risk of developing arthritis has been associated with a BMI greater than 29 kg/m² [44].

Half of all U.S. women aged 50 and older have moderately reduced bone density or osteopenia and 20% have severely reduced bone density or osteoporosis—a preventable condition [33]. One out of two women aged 80 and older have osteoporosis (Figure 1.12). Osteoporosis increases the risk of falls and hip fracture, a source of serious morbidity and mortality in women [45]. In the United States, the incidence of hip fracture is equal to the combined risks of having breast, endometrial, and ovarian cancer.

Although the rates of psychiatric disorder are similar in men and women, the types of disorders differ. Depression and anxiety affect women almost twice as often as men. In both developing and developed countries, unipolar major depression was reported as the leading cause of disease burden in women [36]. In the U.S. alone, 25% of American women report being depressed at

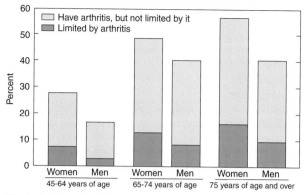

Fig. 1.11 Proportion of persons with arthritis and proportion with limitations of activities due to arthritis by age and sex: United States, 1993–1994. Source: Centers for Disease Control and Prevention, National Center for Health Statistics, National Health Interview Survey [33].

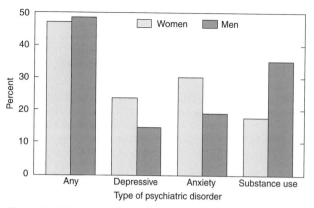

Fig. 1.13 Lifetime prevalence of psychiatric disorders among persons 15 to 54 years of age by sex: United States, 1990–1992. Source: University of Michigan, Institute for Social Research/Survey Research Center, National Comorbidity Survey [33].

some time in their lives (Figure 1.13). Positron emission tomography (PET) imaging of human brains has shown a lower rate of serotonin synthesis in women than in men—a possible biological difference that may relate to the higher incidence of depression reported in women [46]. Senile dementia affects 50% of those over age 85—a population of primarily women [47].

What then is the relationship between women's mortality advantage and their morbidity disadvantage? Why do women suffer from more chronic, disabling illnesses such as arthritis, osteoporosis, and affective disorders? In the 1991 U.S. National Health Interview Survey, nearly 15% of women aged 45 to 64 reported some disability [33] (Figure 1.14). The proportions increased to 24 and 41% for women aged 65 to 74 years and 75 years or older, respectively. Within each age group, the total proportion disabled was nearly twice as high for women as men.

Verbrugge and Wingard [49] postulated that differences in inherent risk, acquired risk, illness and prevention orientations,

and/or health and death reporting behaviors might account for women's poorer health and men's shorter lives. Men's higher mortality at every age (including prenatally) suggests that men may be disadvantaged by their inherent vulnerability to disease. Second, the risks associated with lifestyle, occupational and recreational activities, and management of stress differ between men and women. For example, men's higher smoking rates and alcohol use and more hazardous occupations place them at higher risk of death. Women are less likely to die from leisure activities and sports and report lower alcohol intake. On the other hand, Verbrugge reported that having numerous and more

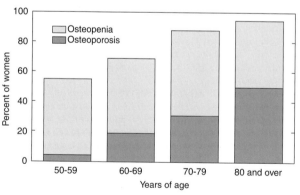

Fig. 1.12 Prevalence of reduced hip bone density among women 50 years of age and older by age and severity: United States, 1988–1991. Osteopenia is defined as a bone mineral density 1–2.5 standard deviations below the mean of white, non-Hispanic women 20–29 years of age as measured in NHANES III (Phase I); osteoporosis is defined as a bone mineral density value of more than 2.5 standard deviations below the mean of young white, non-Hispanic women (WHO expert panel). Source: Centers for Disease Control and Prevention, National Center for Health Statistics, National Health and Nutrition Examination Survey III (Phase I) [33].

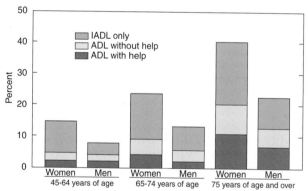

Fig. 1.14 Disability status among noninstitutionalized persons 45 years of age and over by sex and age: United States, 1991. Three mutually exclusive categories of disability status are presented: IADL = persons who have difficulty with one or more instrumental activities of daily living and no difficulty with any activities of daily living; ADL without help = persons who have difficulty with one or more activities of daily living but do not receive help; and ADL with help = persons with the greatest degree of disability, including persons who have difficulty and receive help with the activities of daily living. A disability was defined as any restriction resulting from an impairment of ability to perform an activity in the manner or within the range considered normal for a human being [48]. An impairment was defined as any loss or abnormality of psychological, physiological, or anatomical structure or function. Source: Centers for Disease Control and Prevention, National Center for Health Statistics, National Health Interview Survey [33].

fulfilling roles was associated with better health [50]. Women may be at risk of poorer health because they have fewer and less satisfying roles or because they experience greater between-role conflict.

Third, perceptions of illness and willingness to take preventive or curative actions may differ between men and women, with women leading longer lives because they are more willing to seek help for their health problems and to participate in preventive actions such as screening for asymptomatic disease. Last, health reporting behaviors or reporting bias may account for differential rates of morbidity detected or cause of death reported on death certificates.

We refer the reader to the subsequent sections of this book for a detailed review of the issues we have raised here.

IV. Future Directions in the Study of Women's Health

What do the trends described in this chapter tell us about women's health, and what are some important next steps?

First, it continues to be true that "women get sick and men die," although the mortality advantage has been somewhat reduced in recent years, reflecting decreased death rates for heart disease and cancer among men but not women. These changes have been attributed to women's increasing exposure to risk factors such as smoking. Increased exposure to stress has also been perceived to be important. Are there enough gender-sensitive studies that focus on stress or on change in risk-related behavior? What about studies investigating the biologic underpinnings for differential susceptibility among women to smoking and alcohol? While some research has been done, there are no firm answers as yet.

Second, the data presented point to a variability in disease risk among subgroups of women, based on race/ethnicity and income. Here again explanations may lie in the social sphere— differences in exposure and effect modifiers—or in the biologic—genetic differences affecting metabolism of exposures, for example. Descriptive research is only just beginning to yield results, and explanatory studies are few.

Third, women's overall morbidity is greatly influenced by the incidence of a variety of diseases and disorders that preferentially affect them. These have not received much attention from the research community, and there are few testable hypotheses to explain the sex ratio, for instance, of autoimmune diseases.

It is not sufficient to repeat studies previously undertaken in men in populations of women. There is the need to reframe research questions with an understanding of the influences on health that are unique to each gender. The evidence in this book and elsewhere represents a challenge to the traditional biomedical model in which the 70 kg white male was meant to represent all of us. The limitations and in some cases the perils of ignoring gender in research have become clear [43,51,52].

Here is a list of recommendations that are applicable to a range of women's health topics rather than specific to particular exposures, such as hormone replacement therapy, or particular diseases, such as ovarian cancer.

• We should compile more systematic descriptions of sex similarities and differentials in health and new testable hypotheses about what might explain them—e.g., why an established risk factor appears to affect one sex more or less adversely than another.

• Emphasis in future studies should be placed on exposures common to women and on mechanisms thought to be critical in creating sex differentials.

• Efforts to study hard-to-reach populations of women should continue.

• Among women, we need more systematic investigation of subgroup differences and, again, what might explain them.

• Consideration should also be given to the whole context of a woman's life, using psychological and social, as well as biological, variables.

• Further development and application of women-sensitive research methods are essential if studies are to be successful and informative.

To many people right now, women's health research has a limited meaning. As important as breast cancer is—in reality and even more so in the perception of most women—there are other conditions affecting women's health that rightly deserve attention [53]. Although a long-term, nation-wide randomized controlled trial of hormone replacement, diet, and calcium supplementation may be very important, not all studies with a gender perspective need be designed to duplicate the scope and cost of the Women's Health Initiative in order to contribute to the field [54]. What seems essential is to look at women's health separately from that of men—not simply combining the data or adjusting for sex in the analysis—and to use a gender-specific approach. This implies studying women in context, considering cultural, behavioral, and psychosocial factors; using appropriately validated data collection instruments, as well as case definitions and tests; and incorporating gender biology whenever possible.

The chapters that follow are intended to stimulate new ideas about research worth pursuing and the methodologic issues that must be addressed as women's health enters a new era.

References

1. Pinn, V. W. (1997). "Office of Research on Women's Health Overview." National Institutes of Health. *www4.od.nih.gov/orwh/overview.html*.
2. LaRosa, J. H., and Alexander, L. L. (1997). "Women's Health Research." Report prepared for the U.S. Public Health Service's Office on Women's Health. U.S. Department of Health and Human Services.
3. Chesney, M. A., and Ozer, E. M. (1995). Women and health: In search of a paradigm. *Women's Health: Research on Gender, Behavior, and Policy* **1,** 3–26.
4. Johnson, K., and Hoffman, E. (1994). Women's health and curriculum transformation: The role of medical specialization. *In* "Reframing Women's Health" (Alice J. Dan, ed.) pp. 27–39. Sage Publications, Thousand Oaks, CA.
5. Ruzek, S. B. (1996). Continuing or emerging gaps in knowledge about women's health. Research agendas focusing on diverse pathways to health. Testimony for Office of Research on Women's Health, September 25, 1996. Session 1.
6. Ostergard, D. R., Gunning, J. E., and Marshall, J. R. (1975). Training and function of a women's health-care specialist, a physicians assistant, or nurse practitioner in obstetrics and gynecology. *Am. J. Obstet. Gynecol.* **121,** 1029–1037.
7. National Council on International Health (now the Global Health Council). (1999). "Overview: Women's Reproductive Health Initia-

tive." *www.nich.org/women.html* and *www.globalhealthcouncil.org.*

8. National Academy on Women's Health Medical Education. (1994). "Definition of Women's Health." Minutes of Second Meeting, MCP-Hahnemann School of Medicine. *www.auhs.edu/institutes/iwh/nawhme/defin.html.*

9. Beijing Declaration and Platform for Action. (1995). "Fourth World Conference on Women." September 15, 1995. A/CONF. 177/20 October 17, 1995.

10. World Health Organization. (1998). "Mission statement. Women's Health and Development Programme." *www.who.org/frh-whd/index.html.*

11. Cohen, M. (1998). Towards a framework for women's health. *Patient Education Counseling* **33,** 187–196.

12. Fathalla, M. F. (1997). Global trends in women's health. *Int. J. Gynaecol. Obstet.* **58,** 5011.

13. Birn, A. E. (1999). Skirting the issue: women and international health in historical perspective. *Am. J. Public Health* **89,** 399–407.

14. Ruiz, M. T., and Verbrugge, L. M. (1997). A two way view of gender bias in medicine. *J. Epidemiol. Comm. Health* **51,** 106–109.

15. Schwartz. L. M., Fisher, E. S., Tosteson, N. A., Woloshin, S., Chang, C. H., Virnig, B. A., Plohman, J., and Wright, B. (1997). Treatment and health outcomes of women and men in a cohort with coronary artery disease. *Arch. Intern. Med.* **157,** 1545–1551.

16. Hoffman, E., Maraldo, P., Coons, H. L., and Johnson, K. (1997). The women-centered health care team: Integrating perspectives from managed care, women's health, and the health professional workforce. *Womens Health Issues* **7,** 362–374.

17. U.S. Department of Health and Human Services. Health Resources and Human Services Administration. (1997). "Women's Health in the Medical School Curriculum: Report of a Survey and Recommendations." HRSA-A-OEA-96-1.

18. Johnson, K., and Hoffman, E. (1993). Women's health: Designing and implementing an interdisciplinary specialty. *Womens Health Issues* **3,** 115–120.

19. Hoffman, E., and Johnson, K. (1996). Women's health and health reform: Who will deliver primary care to women? *Yale J. Biol. Med.* **68,** 201–206.

20. U.S. Department of Health and Human Services. National Institutes of Health. Office of Research on Women's Health. (1995). "Women in Biomedical Careers: Dynamics of Change. Strategies for the 21st Century. Full Report of the Workshop." Bethesda, MD. June 11–12, 1992. NIH Publication No. 95-3565.

21. U.S. Bureau of the Census. Population Division. (1999). "United States Population Estimates, by Age, Sex, Race, and Hispanic Origin, 1990 to 1997." *www.census.gov/population/estimates/nation.*

22. United Nations. (1992 statistics projected to 1995). "Sex and Age Distribution of the World's Population." *www.un.org.*

23. Society for the Advancement of Women's Health Research. (1999). *www.womens-health.org/factsheet.html.*

24. U.S. Department of Health and Human Services. Office of Research on Women's Health. (1992). "Report of the National Institutes of Health: Opportunities for Research on Women's Health." September 4–6, 1991, Hunt Valley, Maryland, NIH Publication No. 92-3457.

25. LaRosa, J. H., Seto, B., Caban, C. E., and Hayunga, E. G. (1995). Inclusion of women and minorities in clinical research. *Appl. Clin. Trials* **4,** 31–38.

26. NIH guidelines on the inclusion of women and minorities as subjects in clinical research. (1994). *Federal Register* **59,** 14508–14513.

27. Macklin, R. (1994). Reversing the presumption: The IOM report on women in health research. *JAMWA* **49,** 113–116, 121.

28. U.S. Department of Health and Human Services. Office of the Director. (1995). "Recruitment and Retention of Women in Clinical Studies." NIH Publication No. 95-3756.

29. Eidsness, L., and Wilson, A. L. (1994). Women voicing their autonomy: the changing picture of women's health care. *S. Dakota J. Med.* **47,** 227–229.

30. Broom, D. H. (1998). By women, for women: The continuing appeal of women's health centres. *Women Health* **28,** 5–22.

31. Doyal, L. (1995). "What Makes Women Sick." Rutgers University Press, New Brunswick, New Jersey.

32. Peters, K. D., Kochanek, K. D., and Murphy, S. L. (1998). Deaths: final data for 1996. National vital statistics reports; Vol. 47 No. 9, p. 22. National Center for Health Statistics, Hyattsville, MD.

33. National Center for Health Statistics. (1996). "Health, United States, 1995." Public Health Service, Hyattsville, MD. *www.cdc.gov/nchswww/products/pubs/pubd/hus/hus.htm.*

34. U.S. Department of Health and Human Services. Office of Research on Women's Health. (1998). "Women of Color Health Data Book." NIH Publication No. 98-4247.

35. Ries, L. A. G., Kosary, C. L., Hankey, B. F., Miller, B. A., and Edwards, B. K. (eds.). (1999). "SEER Cancer Statistics Review, 1973–1996." National Cancer Institute, Bethesda, MD. *www-seeer.ims.nci.nih.gov.*

36. Murray, C. J. L., and Lopez, A. D. (eds). (1996). "The Global Burden of Disease." Harvard University Press, Cambridge, MA.

37. American Heart Association. Biostatistical Fact Sheet, 1999. *www.americanheart.org/statistics/biostats/biowo.htm.*

38. Pamuk, E., Makuc, D., Heck, K., Reuben, C., and Lochner, K. (1998). Socioeconomic Status and Health Chartbook. Health, United States, 1998. National Center for Health Statistics, Hyattsville, MD. *www.cdc.gov/nchswww/products/pubs/pubd/hus/hus.htm.*

39. National Heart, Lung, and Blood Institute. National Institutes of Health. (1998). "Clinical guidelines on the identification, evaluation, and treatment of overweight and obesity in adults. The evidence report." Bethesda, MD. NIH Publication No. 98-4803.

40. American Cancer Society. (1999). "1999 Cancer Facts and Figures." *www.cancer.org/statistics/cff99/data.*

41. American Cancer Society. (1998). "1998 Cancer Facts and Figures." *www.cancer.org/statistics/cff98/graphicaldata.html.*

42. Zang, E. A., and Wynder, E. L. (1996). Differences in lung cancer risk between men and women: Examination of the evidence. *J. Natl. Cancer Inst.* **88,** 183–192.

43. Henderson, M. M. (1994). Priorities for research to meet women's health needs. *Ann. Epidemiol.* **4,** 172–173.

44. Sahyoun, N. R., Hochberg, M. C., Helmick, C. G., Harris, T., and Pamuk, E. R. (1999). Body mass index, weight change, and incidence of self-reported physician-diagnosed arthritis among women. *Am. J. Public Health* **89,** 391–394.

45. Hochberg, M. C., Williamson, J., Skinner, E. A., Guralnik, J., Kasper, J. D., and Fried, L. P. (1998). The prevalence and impact of self-reported hip fracture in elderly community-dwelling women: The Women's Health and Aging Study. *Osteoporosis Int.* **8,** 385–389.

46. Nishizawa, S., Benkelfat, C., Young, S. N., Leyton, M., Mzengeza, S., de Montigny, C., Blier, P., and Diksic, M. (1997). Differences between males and females in rates of serotonin synthesis in human brain. *Proc. Natl. Acad. Sci. U.S.A.* **94,** 5308–5313.

47. Bachman, D. L., Wolf, P. A., Linn, R., Knoefel, J. E., Cobb, J., Belanger, A., D'Agostino, R. B., and White, L. R. (1992). Prevalence of dementia and probable senile dementia of Alzheimer type in the Framingham Study. *Neurology* **42,** 115–119.

48. World Health Organization. (1980). "International Classification of Impairments, Disabilities, and Handicaps." World Health Organization, Geneva.

49. Verbrugge, L. M., and Wingard, D. L. (1987). Sex differentials in health and mortality. *Women Health* **12,** 103–145.

50. Verbrugge, L. M. (1986). Role burdens and physical health of women and men. *Women Health* **11,** 47–77.

51. Kunkel, S. R., and Atchley, R. C. (1996). Why gender matters: being female is not the same as not being male. *Am. J. Prev. Med.* **12,** 294–295.

52. Krieger, N., and Zierler, S. (1995). Accounting for the health of women. *Curr. Issues Public Health* **1,** 251–256.

53. Walters, V. (1992). Women's views of their main health problems. *Canadian J. Public Health* **83,** 371–374.

54. Rossouw, J. E., Finnegan, L. P., Harlan, W. R., Pinn, V. W., Clifford, C., McGowan, J. A. (1995). The evolution of the Women's Health Initiative: Perspectives from the NIH. *JAMWA* **50,** 50–55.

2

Gender, Race, and Class: From Epidemiologic Association to Etiologic Hypotheses

CAROL J. ROWLAND HOGUE
Rollins School of Public Health
Emory University
Atlanta, Georgia

I. Introduction

The goal of this chapter is to provide the beginnings of an approach to causal hypothesis formation that can be used for assessing the epidemiologic and related literature on a specific health problem (*e.g.,* cardiovascular disease or breast cancer), identifying gaps in knowledge about the impact of group identities on the health problem, and then designing epidemiologic studies that can fill in one or more gaps. The approach suggested in this chapter is a conceptual work-in-progress, encompassing several untested or only partially tested assumptions. The chapter first discusses some of the key assumptions, outlines the steps to hypothesis formation, and mentions some analytic approaches that may prove useful for hypothesis testing. These ideas then are illustrated with a few current controversies in women's health epidemiology—how behavioral risk factors (smoking and being sedentary) can be modified and why African-American women have high rates of preterm delivery. A biopsychosocial theory of disease causation may be used to develop and test hypotheses about these issues.

II. Assumptions

A. Biologic and Cultural Basis for between-Gender Health Differentials

This discussion arises from the assumption that between-gender health differences can be caused by biological differences between males and females, by physiological differences that are a consequence of culturally constructed genders, or by some combination of these sets of factors [1,2]. In the past, epidemiologists, recognizing that health status varies between men and women, have generally limited their studies to one gender or controlled for gender while examining other potential etiologic factors. Much of this research was based on the assumption that differences between men's and women's health status were due entirely to biological differences in the sexes [3]. As women's voices began to be heard in the research community, alternative hypotheses were proposed that were based on the impact of differential access to power associated with historical and ongoing sexism [1,3]. To test such hypotheses requires theorizing about how relationships between groups defined by gender (*i.e.,* "culture-bound conventions about appropriate roles and behaviors for, as well as relations between, women and men" [1, p. 89]) affect the health of individuals in the subordinate group (as well as in the dominant group, which is not the focus of this book). This means, literally, how social relationships enter the body. In the classic epidemiologic framework of agent, host, and environment, the agent is sexism ("an oppressive system of gender relations, justified by ideology, premised on the subordination of women by men" [1, p. 89]), or more specifically, some aspect of sexism; the host is the individual woman, and the environment is the context of her life. For example, sexism may be hypothesized to increase the individual's exposure to occupational injury through her increased chance of working in a high-risk job (such as clerk in an all-night shop) or her susceptibility to becoming ill, given exposure to the disease, due to working longer hours in the combination of paid and unpaid labor.

B. Social Construction of Race/Ethnicity/Class

The environment, or context, of the woman's life includes, importantly, her simultaneous membership in a social group (her social class and position; Moss, Chapter 43) and in a racial/ethnic group. Most health conditions vary by social class and condition [1,2,4–6], and those investigated for gender differences within social class or social position often find that gender is an effect modifier for class/disease associations [1–3,6–9]. Within gender and class, further effect modification commonly occurs by race and ethnicity [1,2,6]. As overall health has improved during the twentieth century, racial disparities have actually worsened, especially between African-Americans and whites [5]. Within genders, all-cause mortality generally increases linearly on a log scale with decreasing income (with exceptions among the elderly population). However, for a given income, blacks have higher mortality than whites, and higher income is not as protective for blacks as for whites [6]. These complex and consistent associations of health with race, class, and gender dictate that studies of women's health be done within the context of race and class.

But how are we to conceptualize race and class? A second assumption is that gender, race, ethnicity, and class are primarily social constructs rather than biological phenomena. A socially constructed group is one that reflects "social and ideological conventions, not natural distinctions" [1, p. 85]. A full discussion of social class is given in Chapter 43 and will not be repeated here. While some have claimed that social class is biologically determined, there is very little evidence that poverty is genetically predetermined or biologically defined [Moss, Chapter 43]. Ethnic groups share a common ancestry, culture, and history. Although there are often (but not necessarily) phenotypic similarities within ethnic groups, "the term is typically used to highlight cultural and social characteristics instead of biological ones" [5, p. 325].

There is consensus between social scientists and epidemiologists on the social origins of social class and ethnicity, but epidemiologists have traditionally viewed racial categories as biological realities while social scientists have for some time rejected the biological validity of racial categories [1,2,4,5]. Their reasons for claiming that race is a social construct are given below.

- The origins of classifying individuals into races predate genetic studies and were used from the beginning to establish superiority and justify exploitation. Racial definitions vary from culture to culture by how the dominant group views itself and others.
- Genetic studies of phenotypical races have revealed that the external characteristics used to distinguish between races do not correlate well with other genetic characteristics. Except for small, isolated groups (*e.g.,* Inuits), an estimated 95% of genetic variation is among individuals within racial groups rather than between racial groups.

Given the social construction of race, how are we to study its effect on women's health? While Williams [5] argues strenuously that epidemiologists will gain the greatest etiologic benefit from studies that focus on what causes health disparities between races from the perspective of the social construction of race, he does not rule out the need for genetic studies of differences within and between racial groups, especially for gaining understanding of gene–environment interactions. Williams [5] offers a useful causal framework:

"[The framework] indicates that race is a complex multidimensional construct reflecting the confluence of biological factors and geographical origins, culture, economic, political and legal factors, as well as racism. All of these forces are interrelated, and they may combine both additively and interactively to affect health status and the utilization of medical care. Race, an analytic variable of interest to many epidemiologists, is one of several social-status categories created by these large-scale societal forces and institutions. Social-status categories reflect, in part, differential exposure to risk factors and resources that ultimately affect health through biological pathways" [5, p. 327].

Within the framework, basic causes (Krieger's "spiders" spinning the web of causation [10]) are the societal forces that define race, gender, and class and result in racism, sexism, and classism. These forces create groups with distinct social statuses, as well as disparities in socioeconomic status, and in social position within race based on gender, age, marital status, etc. One of the social status correlates of racism is residential segregation, which has been associated with adverse health effects, including overall mortality and infant mortality [11]. Social status differentials create "surface" causes of health differentials: health practices, stress, psychosocial resources, and medical care. These surface causes are mediated through biological processes (central nervous system, endocrine, metabolic, immune, and cardiovascular) that culminate in health status differentials. Williams' framework fits within the general biopsychosocial model of disease causation related to social status differentials, and can be used to develop tests of the theory.

C. Increased Complexity in Epidemiologic Study Design and Analysis

Digging below the surface of between-gender or within-gender differences will require epidemiologic methodologies that reflect the complexities of studying a variety of factors simultaneously [9]. These complexities occur at both the design and analysis stages.

In broadening the scope of epidemiologic investigation, it is important to develop a theoretical construct of how the complex relationships among groups distinguished by gender, race, and class cause ill health. Using an analogy to Snow's understanding of cholera, "taking the pump handle off" to prevent a disease caused or exacerbated by social inequity requires an understanding that "cholera" is caused by "something" in the "water system" rather than "bad air," theorizing how that something enters the water system, and then stopping the flow of that something before it can "infect" individuals [12]. If social inequity is the disease agent, how does it affect human health and how can humans be protected from its negative health impact? Is there a coherent specific theory of disease causation that can be used to posit tests of circumstances that would invalidate it?

Perhaps a theory is emerging, in the broad sense that membership in a social group must be related to health in a meaningful way; that is, there must be a pathway or pathways for the individual's group identity to enter her body and produce physiological changes. One of those ways is through differential exposure to known risk factors, by dint of geographical clustering (*e.g.,* by residence or occupation) or through individual actions made in reaction to group membership that place the individual at greater disease risk (*e.g.,* smoking or sedentary life style), or through relations with members of another group (*e.g.,* domestic violence). Why the group member is exposed to one or another of these types of exposures may vary. For example, residential choices may be limited by external forces. The lure of smoking may be increased through targeted advertisement. Domestic violence may be a greater risk in the absence of mediating community forces. The specific exposures and how they come to be differentially distributed by group membership may vary according to which health condition is under investigation. Nevertheless, this theory predicts that, for diseases with a multifactorial cause, lesser access to control over one's circumstances will increase the individual's exposure to known etiologic factors. Depending on the health condition, the epidemiologist needs to specify, in advance, how known factors are likely to vary (or not vary) with class/race/ethnicity/gender and why they should be expected to behave in this way [1,2,4,5]. To posit these associations, it may be very helpful to enlist collaborators from the groups being investigated as well as to examine anthropological, psychological, and sociological studies for clues [1,2,4,5].

A corollary to the theory of disease causality through differential exposures is that, while all exposures ultimately affect disease causation within the individual, not all exposures that affect an individual are "individual-level" risk factors (*i.e.,* measured by examining the individual). Some, such as residential area or cultural norms regarding male/female relationships, are more accurately described at the group community level [Moss, Chapter 43]. The investigator must specify the appropriate measurement level a priori, along with the proposed causal pathway.

Regardless of the disease or health condition under study, a full examination of the distribution of known risk factors cannot account for all the variation in disease status between and among socially defined groups. A complementary theory is that very membership in a socially constructed subordinate group creates perceived stressful circumstances that affect members' health through biological pathways of the nervous, immune, and endocrine systems and related biologic processes. One expression of this biopsychosocial theory is presented by Evans and colleagues [13]. Building on the seminal work of Selye [14] Cassel [15], Syme [16] and Rose [17], and evidence from animal research, they have advanced a general theory that "humans, like monkeys, baboons, rats, and dogs, may respond to a stressful environment that they cannot control with physiological changes that are harmful to their health" [18, p. 183]. They further posit that "biological processes that underlie the correlation between socioeconomic status and health differ among individuals as a result of differences in initial genetic endowment, . . . in previously 'learned' physiological responses, . . . and in the contemporaneous balance between external stresses and coping resources" [18, p. 184]. Their treatise focuses on social class, but its approach pertains equally well to other socially constructed groups and to the combined socially subordinated group of poor minority women. Thus, it provides a unifying theory of the process whereby social inequity enters the body and increases the individual's susceptibility to disease.

Consistent and coherent findings from research on the biological processes resulting from stress lead to the following postulates [18].

• Similar stresses induce different effects, depending on the individual's perception of the event, on the individual's coping styles (inherent characteristics, *e.g.,* resilience, anger expression) and skills (learned strategies, *e.g.,* meditation, exercise), on the availability of mediating supports (*e.g.,* family, intimate friends, financial resources), and on the context in which the stressful event occurs.

• Perceived control of the situation is positively related to better health outcomes, through direct physiological mediation and through indirect means (avoidance of health-harming behaviors).

• Some stress leads to positive results. However, stress becomes negative when the individual is unable to respond optimally to the stress. This can occur because of

   • interruption of the normal neural system development at critical points in early life, leading to reduced "plasticity" of neurons and lessened ability to "turn off" the fight-or-flight reactions to a stressful situation

   • prolonged exposure to stressful situations, leading to protracted periods of circulating cortisol, which also results in reduced capacity to learn and to turn off the stress reaction.

• Levels of stress early in life may have an impact later in life, not only through reduced plasticity of neural cells but also due to triggering previously "learned" physiological responses.

• Varying stress levels over the life course and across cohorts born at different times requires consideration of time-dependent associations.

• Nurturing environments may be protective of high stress levels throughout the life cycle.

• Social position affects one's ability to respond optimally to stress. Individuals in subordinate positions are subject to prolonged stress and chronic elevations in cortisol and reduced immune function. Individuals in dominant positions generally respond to stress with more effective physiological functioning, but when confronted with uncertainty, they exhibit chronic elevations in cortisol and reduced immune function.

• Some individuals are inherently hyperreactive to stress, but the negative health effects of hyperreactivity are expressed only in a stressful environment and may be buffered by intensive nurturing.

> "Since most common diseases today are likely due to interaction between the genotype and environmental factors, it is not surprising that most studies show a clear relationship between illness and social class. But the connection to genetic predisposition is masked in these studies. The important point is that genes determine *who* may get sick within a class, but environmental factors determine the *frequency* of sickness among susceptibles. Progress in understanding how illness is distributed in the population will be made by studies that control for one set of influences (genetic or environmental), while investigating the other." [authors' italics] [19, p. 143].

Like any theory, the complete model cannot be tested in any one study, but rather correlates or postulates of the theory will "predict circumstances that would invalidate it, and further, that it is just these circumstances that the investigator most profitably pursues" [12, p. 1551]. Much work must be done before the utility of this approach is adequately investigated. Several researchers have applied aspects of this biopsychosocial model to advance research hypotheses for specific health conditions, including preterm delivery among African-American women [20] and cardiovascular disease [21], as well as numerous examples in the following chapters. In attempting to apply the model, researchers have identified pragmatic problems that can hamper effective theory testing.

One of these problems is the dearth of standardized measures of stress, especially for minority women [20]. The epidemiologic investigation of race/ethnicity, class, gender, and health requires a thorough understanding of cultures that can be gained through partnering with social scientists familiar with the groups [1,5]. Through combinations of qualitative and quantitative research techniques, community members may be involved in the research process and, among other benefits, provide valuable input into the development of appropriate measures of perceived stress and strain [22,23].

Another problem is inconsistent definitions of group membership (especially race/ethnicity and class/position). The quality of epidemiologic studies into the joint effects of race, class, and gender is affected by the definition and precision of these variables. Measurement issues for class variables are discussed in Chapter 43. Especially for women, how social position is measured affects the degree of association between social status and health [2,24]. Measurement issues for race variables include differential classification depending on the source of information, change in self-reported identity over time (especially among Native Americans), classification of persons of mixed racial parentage, and census undercount of racial minorities [4]. Some persons will be classified as white, African-American, or another race, depending on who classifies them (*i.e.*, self-report, investigator observation, proxy report, or record abstraction). This distinction is more than a question of accurate classification. Self-identification is most relevant when examining the aspects of race and racism from the individual's perceptions. However, investigator observation might be more relevant when examining the aspects of race and racism from the societal perspective. For example, skin color seems to be a marker for social position among African-Americans, as well as an indicator of the degree of individual discrimination experienced from whites [4].

Apart from the theoretical construct of which individuals should be placed into which classifications, there are statistical concerns when using two or more data sets with different sources of information on race. For example, birth certificates now classify infants on the basis of their mother's race (which may be self-reported or extracted from medical records) whereas death certificates classify decedents on the basis of the medical examiner's opinion or a proxy report. Infant mortality rates vary depending on whether the infant's race is defined as race at birth or race at death [25].

A major issue in using large national data sets is the lack of relevant information on social status. For example, birth certificates include maternal and paternal educational status but no information on income and wealth. Moreover, much of the smaller scale research to date has failed to define groups by race/ethnicity, gender, and class simultaneously, which leads to reduced ability of the existing literature to answer targeted, theory-based questions [1,2].

For future studies, the epidemiologist must specify clearly and as completely as possible the presumptive causal model for gender, race/ethnicity, and class differences among women [26]. New understandings of gendered health differences will come from studies that collect enough information to test fully articulated hypotheses and that use relevant analytic techniques.

Relevant variables can include not only individual-level variables—the traditional definition of epidemiologic risk factors—but also group-level variables that define the social and/or cultural context of women's lives [Moss, Chapter 43]. Variables can also be time-defined, as social definitions are modified over time. For example, gender has undergone a tremendous change in definition as a result of the nineteenth century women's movement, women's involvement in the political process, and women's increased access to power and resources [3]. Thus, cohorts experience gender differently as they grow, develop, and age.

Complex study designs require analytic techniques that allow for the layering of effects. In addition, it is important that in using a framework such as that proposed by Williams that variables be analyzed with the causal framework in mind. It is be-

yond the scope of this chapter to compare and contrast the numerous modeling strategies available and being developed for this work. However, an example may be helpful in illustrating the benefits of a multilevel modeling approach. If one is studying deaths due to cervical cancer and posits that social status differentials are related to access to health care, the health care access variable should be correlated to social status. Williams' framework assigns social status a causal role in health care access. Therefore, assuming correlation occurs in the study, a single model with both social status and health care access entered will "overcontrol" because of the causal relationship between these variables. A two-stage analytic model might be employed to estimate first the odds ratio for cervical cancer deaths associated with poor access to health care and then to use that estimate in the second stage to estimate any additional effects of social status differentials that are not captured by the pathway of health care access.

III. Examples

The examples have been chosen to illustrate how the biopsychosocial theory may be used to develop etiologic hypotheses. In these examples, I am focusing more on factors related to disease causation and less on health differentials that occur as a result of gendered differences in health care—the topic of other chapters.

A. Smoking

Cigarette smoking, a personal habit that is greatly influenced by social and cultural forces [27–34], is a direct health risk. It has been hypothesized that people smoke because they feel it helps them cope with the stresses of everyday life [28,30]. According to the biopsychosocial theory, members of subordinate groups will include a higher frequency of smokers than members of dominant groups. There should be a dose-response relationship between the degree of subordination and the frequency of smoking. However, because smoking is a result of personal volition—choosing to smoke—smoking initiation may be mediated by cultural forces defining its acceptability for members of the group as well as the availability of positive mediating factors for group-related stress [28,30–32,34].

Women (subordinate group) smoke less than men [27,29,31]. Among women, smoking status varies substantially. Poor women smoke more than near poor women, who smoke more than middle income women, who smoke more than high income women [7,8]; this dose-response relationship is predicted by the model. Within these social classes defined by income, white non-Hispanic women (dominant group) smoke more than black non-Hispanic women, who smoke more than Hispanic women [7,8]. Even these observations do not fully describe the complex interrelationship between race/ethnicity for women smokers. Figure 2.1 shows smoking prevalence during pregnancy (not controlling for social class). These women are most likely to be heavily addicted to smoking, as they continue to smoke while pregnant. They can be grouped into three categories: heavy smokers, including white women and American Indian or Alaska Native women (quite dissimilar with respect to class); moderate smokers, including black, non-Hispanic and Puerto

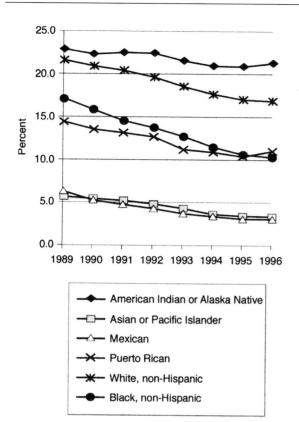

Fig. 2.1   Percentage of mothers who smoked cigarettes during pregnancy, according to mother's race, and Hispanic origin, selected states, 1989–1996. Source: National Center for Health Statistics, Health, United States, 1998 with Socioeconomic Status and Health Chartbook. Hyattsville, Maryland.

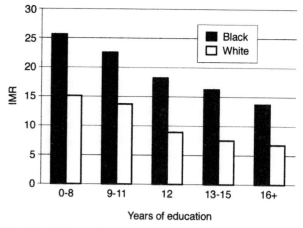

Fig. 2.2   Infant mortality risk per 1000 live births by race and maternal education, single-delivery infants born during 1980. Source: C. Hogue *et al.* (1987) Overview of the national infant mortality surveillance (NIMS) project—design, methods, results. *Public Health Reports* **102,** 126–138.

Rican women; and light smokers, including Asian or Pacific Islander women and Mexican women (who vary greatly by class). Similarities in smoking prevalence between race/ethnicity groups undoubtedly arise from different etiologies within each racial/ethnic group. What is striking is the wide gap in women's experiences, suggesting that different groups have solved the problem of smoking more or less successfully, and by perhaps different means. An in depth examination of the etiology of smoking initiation should unearth important findings for smoking prevention.

Despite the apparent dissimilarities between these observations and a simple oppressed group/stress model, the smoking patterns of women (and men) defined by their gender, race/ethnicity, and class are consistent with the fuller group stress support model described by Evans *et al.* [13]. Membership in a group defined socially by race/ethnicity, gender, and class includes cultural norms that can protect group members from choosing to smoke [30–32,34]. Understanding what these norms are for each group, how they may differ across groups, and how they may be adapted to groups in which norms encourage smoking is fertile ground for developing potentially effective interventions to encourage nonsmoking behavior.

The observation that African-American women who do choose to smoke are less likely to stop smoking offers the opportunity

to test Geronimus' weathering hypothesis [30]. That is, for those who start to smoke, smoking is viewed as a buffer against life's strains. Because African-American women have more strains than white women, it is easier for them to become dependent on this stress-reducer. The key is to choose alternative stress-reduction strategies and to not start to smoke. Following the theoretical framework of Evans and colleagues, it would appear that the postulates regarding early life stresses and nurturing should be the focus of further investigation, because smoking usually begins in adolescence. This part of the theory involves behavioral (rather than physiological) response to stress. What individual- and group-level characteristics increase the risk that a young adolescent will start to smoke? Do these vary by race/ethnicity/gender/class groups? Can vulnerable individuals be protected through targeted "nurturing," and how should that nurturing be defined? These are just some of the questions that emerge from examining the theoretical model. While some research has begun to address these types of issues [31–34], much more needs to be accomplished before we understand the causes of smoking initiation and ways to prevent it.

B. Preterm Delivery Among African-American Women

Since it was first identified as a public health problem, infant mortality among African-American babies has been double that among white babies [35], with a trend in recent years towards even greater racial disparity. This black-white gap was long viewed as an indication of socioeconomic disparities, but that view was modified during the 1980s when data became available to assess the impact of infant mortality by both socioeconomic status (measured by maternal education) and race (Fig. 2.2). Although infant mortality did decline with increasing maternal education, the slope was steeper for white babies than for black babies. Additionally, at each educational level, black infant mortality was higher than white infant mortality. That meant that the black-white gap in infant mortality was actually higher among babies born to college-educated women than among babies born to women with fewer than nine years of education [35].

The black-white infant mortality gap for babies born to college-educated women could not be explained by differences in known risk factors, including maternal age, parity, marital status, and timing of first prenatal visit [37]. The vast majority of them sought prenatal care in their first trimester, were married at the time of delivery, and most were giving birth to their first or second child, at what is considered optimal childbearing age. Adjustment for these factors reduced the crude odds ratio by only 5% (from 1.9 to 1.8). Most of the excess infant mortality could be attributed to a threefold risk of birth weight less than 1500 g—a very low birthweight baby who is most likely to have been born preterm and at very high risk of dying before his first birthday.

Because well-educated African-American women have, on the average, lower incomes and less wealth [1,2], these findings do not entirely rule out the role of social class in their excess preterm delivery. Due to their high educational attainment, women in this group who have fewer economic resources might be under psychosocial stress associated with social status incongruity and frustrated expectations. Such status incongruity could result from institutional manifestations of racism and sexism [1,2,20,38]. It was also hypothesized that well-educated, African-American women experience and perceive individualized racism through increased exposure to the dominant culture in their work and home environments [22,23].

Several theories have been proposed to account for variations in poor pregnancy outcomes [38,39]. The biopsychosocial theory of disease causation among socially subordinate populations is only one of these theories. It is included here to continue the illustration of the utility of designing research based on a coherent theory. While the focus is on well-educated, African-American women, the theory would apply to any subordinate population of pregnant women, perhaps with some indication of the degree of subordination relative to the general population. A rationale for focusing on just one well-defined segment of the population is that the definition of stressors is specific to the kinds of subordination experienced and must come from the population perceiving subordination-associated stress. Following the outline presented in the previous section, some of these hypotheses are listed below.

1. The stresses, coping styles, coping skills, and mediating supports of well-educated African-American women can be determined through grounded theory (an anthropological research strategy) with a combination of qualitative and quantitative studies [22,23].

2. Relative social position (income, wealth, marital status) may be indirect mediators of psychosocial stress in this population.

3. Early neural system development might lead to increased risk of preterm delivery among these women [40,41]. Direct tests of early stressful situations and subsequent risk of preterm delivery are needed.

4. Higher circulating levels of cortisol—reflecting prolonged exposure to stressful situations—as well as maternal corticotrophin releasing factor and catecholamine levels [42] may identify women who have been chronically stressed and are at greater risk of adverse health effects, such as bacterial vaginosis, when exposed to acute stress in pregnancy.

5. Age may mediate the effect of the stress/preterm delivery relationship, either owing to the cohort effect of external stressors or to individual factors such as increasing maturity or decreasing health from weathering.

6. Women within stronger nurturing environments may be less likely to experience stress-related health effects; if so, the mediating mechanism(s)—e.g., less perceived stress, greater sense of control, greater assets—should be determined.

7. Women who experience loss in status should have increased stress-related risk of low birthweight deliveries. This effect was seen among women who had previously given birth to an infant weighing ⩾2500 g and who, between that birth and the subsequent birth, had a loss of social status [43].

8. Hyperreactivity to stress is hypothesized to place some well-educated African-American women at greater risk of stress-related preterm delivery, but only when they are exposed to acute stress at critical times during pregnancy.

Preterm delivery is only one of many adverse health states that well-educated African-American women experience. The biopsychosocial theory maintains that any disease or health condition affected by chronic stimulation of the central nervous system and related immune and endocrine systems should be associated with chronic stressors. Thus, this model, if useful in predicting preterm delivery, may also prove useful in predicting other health conditions among these women.

C. The Obesity Epidemic

Obesity has risen alarmingly since the 1970s (Fig. 2.3). Approximately one-third of white women and more than one-half of Hispanic and African-American women are obese [7]. There is a strong inverse correlation between socio-economic status and obesity among women, but researchers have not established the causal direction for this association. That is, we do not know whether poorer women are obese because of their environment and impoverished access to resources or whether obese women are poorer because they are socially disadvantaged due to their obesity [45,46]. The causal arrow may go in both directions, with poverty increasing the risk of obesity and obesity increasing the risk of becoming poorer, regardless of initial social status.

Much of the socio-economic gradient in obesity cannot be explained with known risk factors for obesity [46]. The proximate causes of the obesity epidemic are assumed to include eating more and more high-fat foods coupled with an increasingly sedentary lifestyle. However, there is little evidence that exercise differs greatly by social class. Strenuous physical exercise, a behavior assumed to mitigate the effects of contemporary, high-fat diets [47], is self-reported in less than 20% of adult women, irrespective of race/ethnicity (Fig. 2.4) [48]. Although some surveys have reported slight differences in physical activity by race/ethnicity [49,50], more of the social class difference in obesity appears to be related to diet than to exercise. Yet even controlling for diet, strong social class differences in obesity remain [46].

To the extent that obesity causes social disadvantage, it is important to protect children and adolescents from becoming

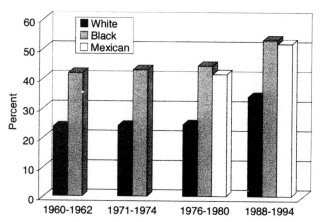

Fig. 2.3 Age-adjusted percentage of overweight women 20 years of age and over according to race and Hispanic origin: United States, average annual, 1960–1962, 1971–1974, 1976–1980, and 1988–1994. Source: National Center for Health Statistics, Health, United States, 1998 with Socioeconomic Status and Health Chartbook. Hyattsville, Maryland.

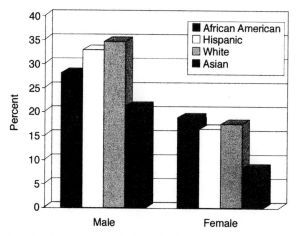

Fig. 2.4 Percentage of adults (18 and older) who report that they exercise hard at least four times per week, according to sex, race/ethnicity, and gender, 1994. Source: Hogue, C. J. R. (1999). Avoiding smoking, eating well, and exercising: Health promotion among minority populations. *In* Minority Health in America: Findings and Policy Implications from the Commonwealth Fund Minority Health Survey." (Hogue, C. J. R., Hargraves, M. A., and Collins, K. S., Eds.). Johns Hopkins University Press, Baltimore (in press).

obese. This protection will help them to achieve their maximum potential educational and social development. Preventing childhood and adolescent obesity is also important because obese adults have problems losing weight and maintaining weight loss, and these problems get worse with increasing age. The postulates of the biopsychosocial model might serve as a useful research agenda for explaining obesity differentials and developing effective interventions for children and adolescents. For example, there is some evidence that reducing stress through family therapy prevents obesity in children [51]. Overeating can be a symptom of psychological stress and distress. Among poor and minority girls, the stress associated with discrimination and subordinated status early in life may predispose them to become obese, thereby setting in motion a cycle of continuing discrimination related to obesity as well as subordinated class status. Here are just some examples of the types of questions that might be tested, based on the biopsychosocial postulates.

1. Would teaching children coping skills, e.g., to express anger appropriately, affect their diet and weight gain?

2. Since perceived control of the situation is negatively related to obesity among children [52], would interventions to strengthen locus of control among at-risk youth reduce overeating and enhance weight loss among obese children and adolescents?

3. Are children whose normal neural system development at critical points in early life has been interrupted more likely to become obese, irrespective of genetic predisposition to obesity? If so, what screening tools can be developed to identify children early enough to intervene successfully? For example, is higher circulating cortisol among young children a predictor of adult obesity? Are children who are hyperreactive to experimental stressors more likely to become obese?

4. What predicts vigorous, cotinuing physical activity among pre-adolescent and adolescent girls? Despite similarities in physical activity by race/ethnicity and class, do factors that predict physical activity vary by race/ethnicity and class? Can interven-

tions be developed to train young girls to use vigorous exercise as an outlet for anger and frustrations?

IV. Summary and Conclusions

Until the 20th century, the average life expectancy for men was greater than that for women, owing to women's very high risk of dying in childbirth. Thanks to vast improvements in maternal survival, women's average life expectancy now exceeds men's in all but the most primitive societies. Despite major progress towards elimination of health-related gendered discrimination, women continue to be disadvantaged with respect to men in the control they exercise over their environment. Minority women are "doubly disadvantaged," and poor minority women are "triply disadvantaged." Women's health status is consistently found to be associated with gender, race/ethnicity, and social class. In general, poorer and minority women suffer poorer health, with a dose-response relationship between degree of disadvantage (whether measured by wealth or skin color) and extent of health problems. There are exceptions that challenge us to look beyond simplistic explanations for these associations and seek to develop causal theories that can both further science and, eventually, lead to better health for all.

Some would argue that the only way to eliminate these health disparities is to eliminate social disparities. While social equality is an important public health goal, it is not likely to be achieved soon. In fact, trends in income distribution in the United States have been in the opposite direction. Others would argue that because social inequities seem firmly imbedded in modern culture, continuing to document the health costs associated with social disparities will do nothing to improve women's health. It may be useful to return to our infectious disease analogy. Cholera, like most infectious disease agents, has not been eliminated from the environment. Instead, drinking water is protected. Moreover, when safe drinking water is unavailable, the scourge of cholera reemerges. For some other infectious agents, means to immunize susceptible individuals have been devised. In short, with a thorough understanding of how a noxious agent interacts within a given environment to make a susceptible person ill, it should be possible to intervene along the causal pathway to prevent the illness.

This chapter includes an outline of one theory of how social disparity enters the body and impairs its ability to withstand disease agents, as well as some examples of how that theory can be used to develop hypotheses of disease causation related to sexism and racism. This is not meant as an endorsement of the theory as the only possible mechanism of effect; rather, its inclusion is meant to stimulate thought about what would constitute a useful coherent theory. The goal of epidemiology devolves into defining the noxious agents, the environmental factors contributing to transmission, human susceptibilities, and how to intervene effectively. As epidemiology expands its perspective to include sexism, racism, and a classed society among the list of noxious agents, epidemiologists need to define these agents very specifically, develop and test theoretical models about how they adversely affect health, and recommend actions to protect the health of all.

References

1. Krieger, N., Rowley, D. L., Herman, A. A., Avery, B., and Phillips, M. T. (1993). Racism, sexism, and social class: Implications for studies of health, disease, and well-being. *Am. J. Prev. Med.* **9** (Suppl.), 82S–122S.
2. Krieger, N., Williams, D. R., and Moss, N. E. (1997). Measuring social class in U.S. public health research: Concepts, methodologies, and guidelines. *Annu. Rev. Public Health* **18**, 341–378.
3. Weisman, C. S. (1997). Changing definitions of women's health: Implications for health care and policy. *Maternal Child Health J.* **1**, 179–189.
4. Williams, D. (1996). Race/ethnicity and socioeconomic status: Measurement and methodological issues. *Int. J. Health Serv.* **26**, 483–505.
5. Williams, D. R. (1997). Race and health: Basic questions, emerging directions. *Ann. Epidemiol.* **7**, 322–333.
6. Kaufman, J. S., Long, A. E., Liao, Y., Cooper, R. S., and McGee, D. L. (1998). The relation between income and mortality in U.S. blacks and whites. *Epidemiology* **9**, 147–155.
7. National Center for Health Statistics (1998). "Health United States, 1998 with Socioeconomic Status and Health Chartbook." National Center for Health Statistics, Hyattsville, MD.
8. Pamuk, E., Makuc, D., Heck, K., Reuben, C., and Lochner, K. (1998). Socioeconomic status and health chartbook. National Center for Health Statistics, Hyattsville, MD.
9. Dressel, P., Minkler, M., and Yen, I. (1997). Gender, race, class, and aging: Advances and opportunities. *Int. J. Health Serv.* **27**, 579–600.
10. Krieger, N. (1994). Epidemiology and the web of causation: Has anyone seen the spider? *Soc. Sci. Med.* **39**, 887–903.
11. Polednak, A. P. (1996). Segregation, discrimination, and mortality in U.S. blacks. *Ethnicity Dis.* **6**, 99–108.
12. Paneth, N., Vinten-Johansen, P., Brody, H., and Rip, M. (1998). A rivalry of foulness: Official and unofficial investigations of the London cholera epidemic of 1854. *Am. J. Public Health* **88**, 1545–1553.
13. Evans, R. G., Morris, L., and Marmor, T. R., eds. (1994). "Why Are Some People Healthy and Others Not?" de Gruyter, New York.
14. Selye, H. (1976). "The Stress of Life," Rev. ed. McGraw-Hill, New York.
15. Cassel, J. C. (1976). The contribution of the social environment to host resistance. *Am. J. Epidemiol.* **104**, 107–123.
16. Syme, S. L. (1991). Control and health: A personal perspective. *Advances* **7**, 16–27.
17. Rose, G. (1985). Sick individuals and sick populations. *Int. J. Epidemiol.* **14**, 32–38.
18. Evans, R. G., Hodge, M., and Pless, I. B. (1994). If not genetics, then what? Biological pathways and population health. *In* "Why Are Some People Healthy and Others Not?" (R. G. Evans, M. L. Barer, and T. R. Marmor, eds.), pp. 161–188. de Gruyter, New York.
19. Baird, P. A. (1994). The role of genetics in population health. *In* "Why Are Some People Healthy and Others Not?" (R. G. Evans, M. L. Barer, and T. R. Marmor, eds.), pp. 133–160. de Gruyter, New York.
20. Rowley, D. L., Hogue, C. J. R., Blackmore, C. A., Ferre, C. D., Hatfield-Timajchy, K., Branch, P., and Atrash, H. K. (1993). Preterm delivery among African-American women: A research strategy. *Am. J. Prev. Med.* **9** (Suppl.), 1S–6S.
21. Marmot, M. G., and Mustard, J. F. (1994). Coronary heart disease from a population perspective. *In* "Why Are Some People Healthy and Others Not?" (R. G. Evans, M. L. Barer, and T. R. Marmor, eds.), pp. 189–216. de Gruyter, New York.
22. Jackson, F. M. (1998). Stress, depression, and reproductive health in African American women. Paper presented at the Maternal, Infant, and Child Health (MICH) Workshop, Atlanta, GA.
23. Hogue, C. J. R., and Jackson, F. M. (1999). Developing measures of stress. Paper presented at the annual meeting of the Southeastern Psychological Association, Savannah, GA.
24. Gregorio, D. I., Walsh, S. J., and Paturzo, D. (1997). The effects of occupation-based social position on mortality in a large American cohort. *Am. J. Public Health* **87**, 1472–1475.
25. Hahn, R. A., Mulinare, J., and Teutsch, S. M. (1992). Inconsistencies in coding of race and ethnicity between birth and death in U.S. infants. A new look at infant mortality, 1983 through 1985 *J. Am. Med. Assoc.* **267**, 259–263.
26. Hogue, C. J. R. (1997). Getting to the why. *Epidemiology* **8**, 230.
27. Fiore, M. C., Novotny, T. E., Pierce, J. P., Hatziandreu, E. J., Patel, K. K. M., and Davis, R. M. (1989). Trends in cigarette smoking in the United States. *J. Am. Med. Assoc.* **261**, 49–55.
28. Feigelman, W., and Gorman, B. (1989). Toward explaining the higher incidence of cigarette smoking among black Americans. *J. Psychoac. Drugs* **21**, 299–305.
29. Escobedo, L. G., Anda, R. F., Smith, P. F., Remington, P. L., and Mast, E. E. (1990). Sociodemographic characteristics of cigarette smoking initiation in the United States. *J. Am. Med. Assoc.* **264**, 1550–1555.
30. Geronimus, A. T. (1992). The weathering hypothesis and the health of African-American women and infants: Evidence and speculations. *Ethnicity Dis.* **2**, 207–221.
31. Feigelman, W., and Lee, J. (1995). Probing the paradoxical pattern of cigarette smoking among African-Americans: Low teenage consumption and high adult use. *J. Drug Educ.* **25**, 307–320.
32. Siegel, D., and Faigeles, B. (1996). Smoking and socioeconomic status in a population-based inner city sample of African-Americans, Latinos and whites. *J. Cardiovasc. Risk* **3**, 295–300.
33. Manley, A. F. (1997). Cardiovascular implications of smoking: The Surgeon General's point of view. *J. Health Care Poor Underserved* **8**, 303–309.
34. Gritz, E., Prokhorov, A. V., Hudmon, K. S., Chamberlain, R. M., Taylor, W. C., DiClemente, C. C., Johnston, D. A. Hu, S., Jones, L. A., Jones, M. M., Rosenblum, C. K., Ayars, C. L., and Amos, C. I. (1998). Cigarette smoking in a multiethnic population of youth: Methods and baseline findings. *Prev. Med.* **27**, 365–384.
35. Hargraves, M. A. H. (1992). The social construction of infant mortality—From grass roots to medicalization. Ph.D. Thesis, University of Texas Health Science Center, Houston.

36. Hogue, C. J. R., Buehler, J. W., Strauss, L. T., and Smith, J. C. (1987). Overview of the National Infant Mortality Surveillance (NIMS) project—design, methods, results. *Public Health Rep.* **102,** 126–138.

37. Schoendorf, K., Hogue, C. J. R., Kleinman, J. C., and Rowley, D. (1992). Mortality among infants of black as compared with white college-educated parents. *N. Engl. J. Med.* **326,** 556–563.

38. Rutter, D. R., and Quine, L. (1990). Inequalities in pregnancy outcome: A review of psychosocial and behavioural mediators. *Soc. Sci. Med.* **30,** 553–568.

39. Kogan, M. D., and Alexander, G. R. (1998). Social and behavioral factors in preterm birth. *Prenat. Neonat. Med* **3,** 29–31.

40. Polednak, A. P., and King, G. (1998). Birth weight of U.S. biracial (black-white) infants: Regional differences. *Ethnicity Dis.* **8,** 340–349.

41. Hogue, C. J. R., and Hargraves, M. A. (1993). Class, race, and infant mortality in the United States. *Am. J. Public Health* **83,** 9–12.

42. Petraglia, F., Hatch, M., Lapinski, R., Stomati, M., Reis, F. M., and Berkowitz, T. (1999). Maternal corticotrophin-releasing factor, catecholamine levels and psychosocial stress in pregnant women. Manuscript in preparation.

43. Basso, O., Olsen, J., Johansen, A. M. T., and Christensen, K. (1997). Change in social status and risk of low birth weight in Denmark: Population-based Cohort Study. *Br. Med. J.* **315,** 1997.

44. Sorensen, T. I. (1995). Socio-economic aspects of obesity: causes or effects? *International Journal of Obesity & Related Metabolic Disorders* **19,** Suppl. 6, S6–8.

45. Stunkard, A. J. (1996). Socioeconomic status and obesity. *Ciba Foundation Symposium.* **201,** 174–182.

46. Jeffery, R. W., and French, S. A. (1996). Socioeconomic status and weight control practices among 20- to 45-year-old women. *Am. J. Public Health* **86,** 1005–1010.

47. Hill, J. O., and Peters, J. C. (1998). Environmental contributions to the obesity epidemic. *Science* **280,** 1371–1373.

48. Hogue, C. J. R. (1999). Avoiding smoking, eating well, and exercising: Health promotion among minority populations. *In* "Minority Health in America: Findings and Policy Implications from the Commonwealth Fund Minority Health Survey" (C. J. R. Hogue, M. A. Hargraves, and K. S. Collins, eds.). Johns Hopkins University Press, Baltimore, MD (in press).

49. Burke, G. L., Savage, P. J., Manolio, T. A., Sprafka, J. M., Wagenknecht, L. E., Sidney, S., Perkins, L. L., Liu, K., and Jacobs, D. R., Jr. (1992). Correlates of obesity in young black and white women: The CARDIA study. *Am. J. Public Health* **82,** 1621–1625.

50. Boardley, D. J., Sargent, R. G., Coker, A. L., Hussey, J. R., and Sharpe, P. T. The relationship between diet, activity, and other factors, and postpartum weight change by race. *Obstetrics Gynecology* **86,** 834–838.

51. Glenny, A. M., O'Meara, S., Melville, A., Sheldon, T. A., and Wilson, C. (1997). The treatment and prevention of obesity: A systematic review. *International Journal of Obesity & Related Metabolic Disorders* **21,** 715–737.

52. Broman, C. L. (1995). Leisure-time physical activity in an African-American population. *Journal of Behavioral Medicine* **18,** 341–353.

Section 2

THE ROLE OF WOMEN IN HEALTH CARE AND RESEARCH

Suzanne G. Haynes

Office on Women's Health
Department of Health and Human Services
Washington, DC

The history of women's health research in the United States is best told by women's health advocates who have been at the heart of the movement. The story told by Seaman and Wood in Chapter 3 is an empowering account of advances made for women's health by pioneers from different professions. Throughout our history, three important lessons have been learned:

1. Women have rallied into action as a result of adverse effects from drugs or surgical treatments (*e.g.,* diethylstilbestrol (DES), the early oral contraceptive pills, and radical mastectomies).

2. The heart of the women's health movement was born out of inadequacies in reproductive health and breast cancer research.

3. Most of the mistakes made by the medical establishment regarding women's health have resulted from inadequate drug testing, the failure to listen to patients, and the abandonment of the precautionary principle. As adopted by the United States in the 1992 Rios agreement, the precautionary principle states:

When an activity raises threats of harm to human (women's) health or the environment, precautionary measures should be taken even if some cause and effect relationships are not fully established scientifically. In this context, the proponent of an activity rather than the public should bear the burden of proof [1].

If we apply the precautionary principle in every research study we undertake, the terrible mistakes of the past, like the DES story, will not occur again. It is shocking today to think that "third" generation daughters of women who were prescribed DES from the 1950s to the 1970s (for the prevention of miscarriage) might be affected with deleterious cancer or reproductive abnormaliies. Unfortunately, early animal studies suggest that this possibility may exist [2]. It is even more shocking that, like the tobacco industry, many of the pharmaceutical companies responsible for mistakes that have harmed women's health have not admitted these errors, much less apologized for them. Women's health cannot bear a repetition of these mistakes.

It is no wonder that some mistakes were made in the development or application of drug treatments for women. Most of the research until the 1990s was conducted in men, or in both sexes with the assumption that women were really "little men." When Maureen Hatch and I set out to review the state-of-the art methods in women's health research in Chapter 4, we were shocked to see how scant the published literature was before 1990. After the sentinnel 1990 GAO report noting the lack of inclusion of women in NIH funded grants, at least two major conferences were convened to discuss research methods for women's health [3]. In 1993, Dr. Vivian Pinn, the Associate Director for Research on Women's Health at the NIH, convened the first conference on the recruitment and retention of women in clinical studies [4]. The important lessons learned at that conference hold true today, along with scores of new methodologic discoveries published during the 1990s. I recommend the NIH report to all new graduate students who are training to become researchers in women's health. In 1997, The Michigan Initiative for Women's Health hosted a workshop on methods and measures for women's reproductive and long-term health. Important recommendations from that workshop were made regarding multidisciplinary, qualitative, and environmental contaminant research [5]. A comprehensive review of the gaps in women's health research, the Agenda for Research on Women's Health for the Twenty-First Century, will be published in 1999 by the Office of Research on Women's Health at the NIH [6]. We hope our review of the key methodological issues for women's health research will be useful in designing new and innovative research for women's health during the first decade of the twenty-first century, and for decades beyond that. Our daughters, granddaughters, and great grandaughters deserve only the best state-of-the-art research for their health and the health of the nation.

When looking at the health of women from a broader perspective, it is important to recognize the surprising challenges that women face when entering and using the health care system. In Chapter 5, Clancey nicely describes the "gender bias" that women face when receiving medical services. Multiple examples exist to demonstrate that women receive fewer expensive, high-technology services than men (*e.g.,* diagnostic and therapeutic interventions for heart disease, etc.) It goes without saying that medical school training in 1985 did not recognize gender differences in the presentation of symptoms for heart disease in men and women, nor the diagnostic procedures that we now know are essentially different for women as compared to men [7]. And who would have guessed that the gender of your physician makes a significant difference in the type of care you receive? Clancey cites several studies that show that women physicians are better communicators and are more likely to provide Pap smears, clinical breast exams, and mammography to their female patients than male physicians. Are male physicians embarrassed to discuss or perform exams on women's reproductive organs? The medical community needs to encourage male physicians to overcome these barriers so that women can receive the full benefit from medical care, regardless of the gender of the provider. Otherwise, women may self-select to receive their care from women providers. Researchers should ask about the gender of health care providers in future studies so that this effect can be examined and explained.

One of the greatest hopes and perhaps greatest disappointments for the improvement of women's health in the twentieth century has been the proliferation of managed care organizations. In Chapter 6, Davis *et al.* have provided enlightening and hard-to-find data on women enrolled in managed care plans. In 1997, almost 70% of privately insured women were enrolled in managed care plans. The dissatisfaction, worry, and problems with specialty care received under these plans as compared to fee-for-service plans is alarming. Equally alarming is the 14% of women who have no insurance at all. The conduct of research, especially clinical trials, often assumes that a woman has usual medical care coverage. The increasing refusal of managed care organizations to cover usual medical care expenditures if a woman is enrolled in a clinical trial poses grave danger for the testing of effective treatments for serious diseases such as breast cancer. What good is new scientific knowledge from women's health research if it is not put into practice because of

payment restrictions? As scientists, we can no longer ignore the policies and politics of the health care system—our research is influenced by and dependent on the huge changes that are occurring in the delivery of health care to women in this country.

Finally, we were pleased to have Bradford and White present an overview in Chapter 7 of the well-received Institute of Medicine (IOM) report on lesbian health [8]. The methodologic discussion and references are important readings for any researcher. Being one of the original IOM panel members, Bradford provides a unique perspective on the steps that must be undertaken to move lesbian health research forward. Of all the topics in women's health that have been studied in the 1990s, lesbian health research has suffered most because of methodological challenges. When I was asked in 1993 to review the literature on the risk factors for and rates of breast cancer in the lesbian population, there were no representative samples to turn to. Our national surveillance systems for cancer and mortality provide no information on sexual orientation. Since that time, questions on sexual orientation and behavior have been included in the Women's Health Initiative, the Nurses Health study, and representative surveys for alcohol and drug use. Definition of the population is a major challenge outlined in the report, along with the need for representative samples. For scientists who are interested in research for which few pathways have been paved, this chapter is crucial reading. The frontiers of knowledge gained from this area of research will make significant contributions to research in other hard-to-reach populations. The Office of Research on Women's Health at the NIH has designated lesbians as a priority for funding under the Women's Health Research Enhancement Awards Program for fiscal year 1999.

References

1. Horan, R. (1998). The new public health of risk and radical engagement. *Lancet* **352,** 251.
2. Newbold, R. R., Hanson, R. B., Jefferson, W. N., Bullock, B. C., Haseman, J., and McLachlan, J. A. (1998). Increased tumors but uncompromised fertility in the female descendants of mice exposed developmentally to diethylstilbestrol carcinogens. **19,** 1655–1633.
3. Bass, M., and Howes, J. (1992). Women's health: And the making of a powerful new public issue commentary. *Women's Health Issues* **2,** 3–5.
4. U.S. Department of Health and Human Services, National Institute of Health, Office of Research on Women's Health (1995). "Recruitment and Retention of Women in Clinical Studies," NIH Publ. No. 95-3756. USDHHS, Bethesda, MD.
5. Harlow, S. O., Bainbridge, K., Howard, D., Myntti, C., Potter, L., Sussman, N., Olphen, J., Williamson, N., and Young, E. (1999). Methods and measures: Emerging strategies in women's health research. *J. Women's Health* **8,** 139–147.
6. U.S. Department of Health and Human Services, Public Health Service, National Institute of Health (1999). "Agenda for Research on Women's Health for the 21st Century. A Report of the Task Force on the NIH Women's Heart Research Agenda for the 21st Century," Vol. 2, NIH Publ. No. 99-4386. USDHHS, Bethesda, MD.
7. Douglas, P. S., and Ginsbury, G. S. (1996). The evaluation of chest pain in women. *N. Engl. J. Med.* **334,** 1311–1315.
8. Solarz, A. L. (1999). "Lesbian Health: Current Assessment and Directions to the Future." Institute of Medicine, National Academy Press, Washington, DC.

3
Role of Advocacy Groups in Research on Women's Health

BARBARA SEAMAN* AND SUSAN F. WOOD†
*National Women's Health Network, New York, New York; †Office on Women's Health, Department of Health and Human Services, Washington, D.C.

"I can use a 'no' as well as a 'yes.' I just use it differently."
Doris Haire, President, American Foundation for Maternal and Child Health; Past President, International Childbirth Education Association (ICEA)
National Women's Health Network (NWHN). Author of *The Cultural Warping of Childbirth*.
Statement made in 1998 after 47 years of activism.

I. Introduction

Progress in women's health measured at the end of the twentieth century began in the early years of the century and is due to advances in public health, increases in scientific and clinical knowledge, and improved access to care. However, much of the overall improvement in health would not have been targeted to women without the involvement of advocates who worked to ensure that research and policies focused on the health of women. Dating back at least as far as 1913—when a news headline announced that "rich women begin a war on cancer"—medical research, policies, and practices in the twentieth century United States have been profoundly influenced by three distinct waves of women's health advocates: the first spearheaded by progressive wealthy women; the second by feminist activists; the third by professionals, including doctors, scientists, legislators, lawyers, and corporate executives. Within each wave, key individuals and the organizations they founded or led identified their priority issues and advocated to make them priorities for the nation. They worked not only to bring women's health to the top of the agenda but also to bring women's voices and input to the development of research initiatives and policies. This chapter reviews some of the key accomplishments of several of these advocates who laid the foundation for future progress in women's health.

II. The First Wave: Progressive Ladies

Elsie Mead, daughter of a New York gynecologist, and a friend of John D. Rockefeller, whom she enlisted in her cause, epitomizes the wealthy women of her day who took up various diseases as their "charity." Mead helped found, and vigorously chaired, the finance committee of the American Society for the Control of Cancer (ASCC), whose goal—by placing articles in the popular and medical press—was to substitute "a message of hope and early detection" for the fear, denial, secrecy, and despair attached to cancer diagnoses. The organization declined while Mead was in France for the Red Cross during World War I but, on her return, she wrote 2000 letters to people in the New York Social Register and got nearly 690 recruits of whom many paid considerably more than the $5 membership fee [1–3].

The ASCC languished during World War II but was rescued by Mary Woodard Lasker, a philanthropist, advocate, and art collector, who would eventually come to be recognized by insiders as "the most powerful person in modern medicine." She reorganized the ASCC as the American Cancer Society, expanded its budget from $102,000 to $14,000,000 in just five years (1943–1948), and established a research program which, over time, would demonstrate the effectiveness of the Pap smear and find cures for childhood leukemia. Lasker modestly described herself as "a self-employed health lobbyist." In fact she was a consummate power broker who exercised unparalleled control over many aspects of health policy, generating tremendous publicity for research into cancer and other diseases, and flogging Congress to boost the budget of the National Institutes of Health from $2.8 million in 1945 to $11 billion in 1994—the year she died at age 93.

Lasker's success, according to her colleague, the lobbyist Mike Gorman, "was due to a high class kind of subversion, very high class." For example, she "made her way into the Kennedy White House by writing Jackie the first check—for $10,000—to redecorate the White House." She obtained the then experimental drug L-dopa for the powerful Senator Lister Hill, whose wife had Parkinson's disease. Like Elsie Mead, Lasker courted influential journalists and leaned on her friends in the media, including Ann Landers, to support her advocacy campaigns. As Landers (herself no shrinking violet) recalls Lasker, "She was intimidating; everything about her was so strong" [4–9].

Ironically, Lasker's husband, Albert, an advertising genius, had created the slogan "Reach for a Lucky instead of a sweet" which persuaded millions of women to take up cigarette smoking. However, tobacco industry money led to the endowment of the annual Albert Lasker Medical Research Awards, originated in 1946, which have gone to support the work of 52 eventual Nobel prize winners.

III. Transitional Advocates/Activists of the 1950s and Early 1960s

In the era following World War II, reproductive experimentation on women by medical doctors was unchecked such that faculty at prestigious institutions, including Harvard, the University of Chicago, and Tulane, fed hormones to healthy pregnant patients, often under the guise that they were vitamins to "grow bigger and better babies." The major hormone used,

Diethylstilbestrol (DES), was later found to be carcinogenic to both the mothers and their offspring. However, these same years also brought the first eruptions of what would later become a militant and sweeping feminist health movement that would profoundly alter the power relationship between male physicians and their once compliant female patients and would transform medicine into a more welcoming profession for women.

Like Mead and Lasker, Doris Haire and Terese Lasser were associated with wealth. Haire's husband, John, managed money for the Vanderbilts, while Lasser's husband, J.K., wrote the perennial best-seller Your Income Tax. Each woman was "radicalized" by a hospital experience.

In 1951, at age 26, Haire, part-Cherokee, born in Oklahoma from the working class, delivered a healthy daughter in Pittsfield, MA. Wide awake and with her husband at her side, Haire got permission to follow the precepts of British natural-childbirth proponent Dr. Grantley Dick-Reed because, and only because, Anne Vanderbilt was principal patroness of the hospital and demanded compliance of the reluctant staff. Joyous, yet feeling "miserably over-privileged," Haire pondered why one had to be under Vanderbilt "protection" to be involved in decisions about one's own delivery and birth. Haire soon consecrated her life to study and activism, compiling a record as president of the International Childbirth Education Association (ICEA), the National Women's Health Network (NWHN), the American Foundation for Maternal and Child Health, and author of The Cultural Warping of Childbirth (1972), which has inspired, and provided the basis for, many subsequent books and articles and actions restoring birth rights to mothers. It is said that when Doris Haire marched down the halls at the Food and Drug Administration (FDA) the walls trembled, for in overcoming the opposition of both the medical and pharmaceutical industries, she has gotten that agency to withdraw approval for questionable drugs such as oxytocin in the induction of labor. Valuing results more than recognition, Haire has exemplified the activist adage "If you don't demand credit for things you can push them through" [10,11].

Terese "Ted" Lasser wrote, when recalling her Halsted radical mastectomy performed following her biopsy at New York's Memorial hospital in 1952 while she was still unconscious and without prior discussion, "You awake to find yourself wrapped in bandages from midriff to neck—bound like a mummy in surgical gauze, somewhere deep inside you a switch is thrown and your mind goes blank. You do not know what to think, you do not want to guess, you do not want to know."

With good reason, the Halsted radical mastectomy has been called "the greatest standardized surgical error of the twentieth century." Introduced in the 1880s by William Halsted, a surgeon at Johns Hopkins Hospital, it was debilitating, even crippling, and based on Halstead's unproven belief that breast cancer was a local disease that could be fully, if brutally, excised before it spread.

Soignée, energetic, and imperious, Lasser was unaccustomed to being patronized or hustled and was furious at the doctor's failure to state her options prior to her surgery. Later, neither her surgeon nor anyone else at Memorial hospital would give her the specific information she demanded on treatments or exercises to regain the use of her arm, when or how to resume sexual relations, or even how to shop for a "falsie." Determined not to

become a cripple, Lasser herself developed an innovative program of stretch exercises through which she regained her strength; better yet, she concluded, would have been to start exercising right after surgery.

A compulsion came over Lasser to teach what she knew. Founding Reach to Recovery (R2R), Lasser adopted a practice of slipping into the hospital rooms of new "mastectomees," exhorting them to arise from their beds and crawl the fingers of their affected arms up the wall. Lasser personally counseled thousands of patients. Most who had the stamina to do the work showed results that seemed "miraculous" to their doctors, who were ultimately convinced.

Lasser maintained a pretense that her calls were made at the requests of the patients' surgeons or families. In truth, in the early years she was often "escorted out the front door of Memorial Hospital when she was found visiting patients at random and without the consent of the responsible surgeon." In 1969, R2R merged with the American Cancer Society, becoming tame and traditional, activists say, such that they now find it hard to appreciate Lasser's courage or the ground that she first broke [12–14].

In 1960, the widespread use of the drug thalidomide in nearly twenty countries set off an epidemic of limbless babies born to mothers whose doctors prescribed a "mild sedative." Evidence of the link was discovered and published in West Germany in November 1961 but was not reported here and was obscured by the U.S. manufacturer, Merrell, who had furnished nearly 1100 doctors with samples of the drug and who followed-up with warning letters to only 10% of them. At the FDA, medical officer Dr. Frances Oldham Kelsey, who had resisted pressure to approve thalidomide, was also not informed by her superiors that they had received the West German reports. In 1962, after the Washington Post broke the story, Dr. Kelsey received the Presidents Award for Distinguished Federal Civilian Service, but the revelations had come too late for Sherri Finkbine, a thalidomide-exposed pregnant woman whose local hospital refused her request for an abortion following that exposure and whose harrowing well-publicized odyssey took her to Sweden for the procedure. By unflinchingly arguing her case in public, Finkbine aroused new sentiment for abortion law reform, which came to fruition in Roe vs Wade eleven years later [15].

Another key figure in women's reproductive rights was Margaret Sanger, whose long years as an advocate left a mixed legacy. Folk heroine and ever enlarging mythic figure, Sanger published The Woman Rebel in 1914, opened the first U.S. birth control clinic in 1916, and founded the American Birth Control League (later to become Planned Parenthood) in 1921. Her great works notwithstanding, Sanger became an elitist, and in the early 1950s as Lasser and Haire drew their lines in the sand on the rights of women patients, Sanger was writing fund-raising letters to help Gregory Pincus develop the first "universal contraceptive," namely the birth control pill. In a letter to Katharine McCormick, heiress to a farm-machinery fortune, Sanger states: "I consider that the world and almost our civilization for the next twenty-five years is going to depend upon a simple, cheap, safe contraceptive to be used in poverty stricken slums and jungles, and among the most ignorant people . . . I believe that now, immediately, there should be national sterilization for certain dysgenic types of our population who are being encouraged to breed and would die out were the government not feeding them." In the 1920s, Sanger had thrown her lot in with physician advocates of birth control and seems also to have been influenced by "eugenics thinking which contributed to the transformation of the birth control movement, beginning in the 1920s, from a women's rights focus to a more conservative emphasis on family planning, understood as an appeal to responsible motherhood and population control" [16,17].

IV. Phase Two: Militant Activism

Post World War II feminist advocacy, to combat sex discrimination in all spheres of life (social, political, economic, and psychological), surfaced with the forming of the National Organization of Women (NOW) in 1966. Most of its twenty-nine founders were middle-aged, middle class, Caucasian, and distinguished in areas such as academia and government service. Even Betty Friedan, author of the controversial best-seller The Feminine Mystique, was a member of the magazine journalist "establishment". At first, the public response to NOW was limited.

However, some of the issues Friedan and other traditional feminists brought up sparked recognition in radical young women who had been toiling in the activist movements of the 1960s: civil rights, antiwar, prostudent, socialist, and who had therefore had a training, somewhat unique to late twentieth century women, in civil disobedience, demonstrating, organizing, and making headline news.

By 1969, social control, financial exploitation, and excessive medicalization of women through health care (especially obstetrics/gynecology and psychiatry) were major topics in their CR (consciousness raising) discussions. In May, Bread and Roses, a grassroots, socialist, and feminist group held a gathering for several hundred women from the Boston area, at which social worker Nancy Miriam Hawley led a workshop called "Women and their Bodies." Participants hoped to come up with a list of good ob-gyn doctors, but, recalls Jane Pincus, "We realized we didn't know what questions to ask to find out if they were good or not." Hawley and Pincus began a study group at which they distributed copies of their research. A year and one half later, the forerunner of the Boston Women's Health Book Collective self-published its work still called "Women and their Bodies" after Hawley's workshop, in a 138-page newsprint edition of findings, through the New England Free Press. Demand grew and the name changed to Our Bodies, Ourselves. They kept printing and expanding the "underground" booklet and had distributed 250,000 copies by 1973 when Simon and Schuster brought out the trade edition. At this point it became an international bestseller, described as "the most powerful revolutionary document since Das Kapital because it induced women to seize the means of Reproduction" [11,18].

Also in 1969, in the nation's capital, Alice Wolfson and other members of Washington, D.C. Women's Liberation were analyzing the "body issues," including specifically their disenchantment with the birth control pill, promoted as a great advance for women but apparently causing many more side effects (some lethal) than manufacturers or most prescribing doctors would acknowledge. They read a review by Victor Cohn in the Washington Post of a book by one of the authors of this

chapter (Seaman), The Doctors' Case against the Pill, and then they learned that Senate hearings, based on the book and chaired by Wisconsin Senator Gaylord Nelson, were to open on January 14, 1970. At issue was not primarily the safety of the pill, for British studies had established the blood-clotting associated deaths, but rather informed consent.

D.C. Women's Liberation, which was also a "feeder" for the radical feminist newspaper Off Our Backs, proved to be another history-bearing group. The very young observer/demonstrators whom Alice Wolfson rounded up to attend the Senate Pill Hearings included such future feminist heavyweights as Charlotte Bunch and Marilyn Salzman Webb. They arrived in "straight-lady" clothes and sat demurely through the first day of testimony, but, on the morning of the second day, as Dr. Roy Hertz, then Associate Medical Director of the Population council at Rockefeller University, concluded his talk, which included pessimistic analysis of increased cancer risks among pill users, what the record describes as "a disturbance among several women in the audience" occurred. The Wolfson women peppered Hertz and Nelson with insistent questions such as "Why are no patients testifying?" and "Why isn't there a pill for men?" The world press was in attendance and was fascinated by the disruptions, which continued to occur throughout the nine days of hearings, which concluded March 4. [19].

Just one of the scientific experts whose testimony was interrupted was unafraid to declare his support in public. On January 23, Dr. Phillip Corfman, Director of the Center for Population Research at NIH and a member of the FDA's Advisory Committee on the Pill, stated after the demonstrators were dragged away, as usual, by the guards, "Incidentally, some of the questions placed by the people who interrupted our hearing were quite important" [19, p. 6395]. Corfman would soon prove to be the essential "insider" who, for example, notified advocates when the FDA's Pill Advisory Committee meetings were scheduled so that Alice Wolfson could appear demanding the right to observe because "it's our bodies you're talking about."

Wolfson's demonstrations, with the brilliant strategies and repetitions of deceptively simple questions, were to the women's health movement as the Boston Tea Party was to the American Revolution. As Science Magazine pointed out in its special issues on "Women's Health Research" (August 11, 1995), "the dissent helped to launch a political movement focusing on women's health. By 1975, nearly 2000 women's self-help medical projects were scattered across the United States, many of them groups of volunteers without an institution" [11,20–23].

Although the NWHN did not formally incorporate until 1975, it grew out of the activist association of Seaman and Wolfson, along with three others who joined with them in the early 1970s. These founders included: Belita Cowan, publisher of Herself Newspaper in Ann Arbor, who exposed seriously flawed research on DES as a morning-after contraceptive that had been published in *The Journal of the American Medical Association;* Mary Howell, first woman Dean at Harvard Medical School, who helped force medical schools to abandon their quotas against women through her underground broadside Why Would a Girl Go into Medicine?; and Phyllis Chesler, who held mental health practitioners to a less sexist standard through her classic book Women and Madness. Other key activists were Sherry Lebowitz for DES-Action; Helen Rodriguez Trias, campaigning to curb sterilization abuse of vulnerable women; Byllye Avery, focusing on childbirth and special health problems of black women; Judy Norsigian for Our Bodies, Ourselves and a fourteen year NWHN Board member; Rose Kushner, who chaired the NWHN Breast Cancer Task Force; Barbara Ehrenreich, coauthor of Witches, Midwives, and Nurses; Anne Kasper, who started the NWHN Clearinghouse; Denise Fuge, the feminist "mole" at Sloan Kettering; Chicago's feminist psychiatrist, Anne Seiden; Doris Haire; Philadelphia health activist JoAnne Fischer Wolf; and a young Dayton, Ohio widower named James Luggen.

Little progress is likely to be made by reformers unless yet more militant advocates appear to threaten at the gates. The radicals of Bread and Roses, D.C. Women's Liberation, Redstockings, and other small groups made NOW seem both more moderate and more relevant, and therefore more mainstream. However, the most original and daring women's health revolutionary was Carol Downer who initiated self-help gynecology on April 7, 1971 at a Los Angeles bookstore when she jumped on a table, inserted a speculum into her vagina, and invited the other women to observe her cervix. The extent of Downer's radical imagination—bringing menstrual extraction as well as self-examination to laywomen—made those activists involved with the NWHN to appear "respectable." Cynthia Pearson, who became the Executive Director of the NWHN in the 1990s, received her early activist training at Downer's feminist health centers [11, p. 53, also pp. 1–2,54–58,60,145f,155,167–169, 173,195,203,214; 24–34].

Throughout the 1970s, organizations were created that focused on issues of particular concerns to minority women. The issue of sterilization abuse was real, and the activism by women of color led to changes in sterilization regulations at both the state and federal level. Organizations such as the Committee to End Sterilization Abuse, co-founded by Helen Rodriguez Trias, led the fight. Health disparities, then as now, were apparent, and women of color faced higher mortality and morbidity from diseases such as breast cancer, hypertension, and lupus. They also faced higher infant mortality rates. The National Black Women's Health Project, founded by Byllye Avery in 1981 with the support of the NWHN, stands as the model for other organizations focused on improving minority women's health. Other organizations targeting the health needs of their own constituency formed subsequently, including the National Latina Health Organization and the Native American Women's Health and Education Resource Center.

By the early 1980s, activists had made enormous progress in institutionalizing informed consent for both research and treatment, succeeded in getting regulations to restrain sterilization abuse against minority women, and secured the ownership of medical records by patients. They had also opened the FDA to regular consumer participation and had established the custom that patients share an equal voice with scientists and physicians at Congressional hearings. Advocates had popularized natural childbirth, revived midwifery, and reclaimed labor and delivery as a family event. Women were allowed to more freely enroll in medical schools so that female enrollment had tripled [35]. Advocates had effectively challenged the use of radical mastectomies and created an atmosphere where patients could become

partners in selecting their treatment plans. Throughout this period, activists played a central role in the legalization and continued availability of abortion, often through underground actions.

V. The Third Wave: Professional Advocates

In the mid 1980s, the third wave of women's health activism began. Although still not proportional to their overall numbers, women were now becoming established as health professionals, scientists, and policymakers, and therefore advocates began to make their presence known from the inside of government and the halls of Congress. This wave of activism began with a focus on research on women's health and got its momentum from the issue of lack of inclusion of women as research subjects in clinical studies. Advocates from the scientific community partnered with women members of Congress to push through a legislative agenda on women's health research.

A key leader in this movement was Florence Haseltine who, outside of her role as a National Institutes of Health (NIH) scientist, founded along with others the Society for the Advancement of Women's Health Research. Initially, her concern was the lack of research in the areas of contraception and infertility, which had become extremely limited due to the political debates over abortion and to the fears by the pharmaceutical industry of potential lawsuits. The Society was at that time a small group of women scientists who based their activities out of the women-owned political consulting firm Bass and Howes, who in 1989 broadened their focus to include other questions of research ranging from breast cancer and osteoporosis to women in clinical trials.

At that time, they began their discussions with the Congressional Caucus for Women's Issues which was cochaired by Representatives Patricia Schroeder and Olympia Snowe. Together they worked with Representative Henry Waxman, chair of the House Subcommittee on Health and the Environment, to develop a strategy to bring the issue of research on women's health to the forefront. Key congressional staffers—Lesley Primmer with the Caucus, Andrea Camp from Rep. Schroeder's office, and Ruth Katz from the Subcommittee—developed a request for an investigation by the General Accounting Office (GAO) to examine the implementation by the NIH of its 1986 policy to include women in clinical research studies. Florence Haseltine was later to refer to this investigation as "Lesley's crowbar" because it broke open the issue of women's health research for the public to see.

This policy had been established in 1986 based on a report in 1985 by the Public Health Service's Task Force on Women's Health, which had been led by Ruth Kirschstein, the only woman Institute Director at the NIH. The Task Force report recommended including more women in clinical studies as well as expanding the research portfolio on women's health. The now infamous "GAO report" released in June 1990 found that the policy adopted by the NIH had scarcely been implemented, and, indeed, the NIH could not answer the question regarding how many women were actually included in NIH-funded studies.

Unlike most GAO reports, this one did not gather dust on a shelf but instead served to catalyze action by the public, by Congress, and by the NIH. An enormous amount of press coverage, which had been coordinated by the Society for the Advancement of Women's Health, followed the congressional hearing held in June of 1990. The Congresswomen and Senator Barbara Mikulski introduced their first Women's Health Equity Act, which called for increased focus on women's health through research, services, and prevention activities. Although the research portions of that legislation did not become law until 1993, NIH responded immediately by establishing the Office of Research on Women's Health at the NIH, initially led by Ruth Kirschstein, and by issuing new guidelines requiring the inclusion of women and minorities in all NIH funded clinical trials.

In the spring of 1991, Bernadine Healy was confirmed as the first woman Director of the NIH and the NIH launched the Women's Health Initiative (WHI) a multiphase, multiyear research study focusing on the major causes of death and disability in older women. The WHI includes the largest clinical trial in the history of the NIH and will assess the impact of dietary fat, hormone replacement therapy, and calcium and vitamin D supplementation on heart disease and cancer in women over the age of 50. However, Healy and the WHI were also challenged by women's health advocates. The NWHN worked diligently to have the study modified to exclude women with an intact uterus from the part of the trial that included estrogen replacement therapy (ERT). ERT alone is known to cause endometrial cancer and advocates succeeded in convincing the NIH that the study was putting women at unacceptable risk. Questions about the thoroughness of the informed consent form with regard to other cancer risks were also raised by the NWHN, leading to revisions in the consent form. However, these changes in the study did not come easily. As noted in Science by Charles Mann [38] "Bernadine Healy, . . . ,rejected their concerns, arguing that plans to monitor the test subjects with yearly biopsies would protect female subjects." Only in January 1995, after high rates of cellular abnormalities were seen in a related study, "the original plans were quietly dropped."

Meanwhile, other advocates were working on specific women's health issues, the most prominent and most successful being the focus on breast cancer. The National Breast Cancer Coalition, made up of grassroots organizations from around the country, began to make its collective voice heard in Washington and at the NIH. Led by Fran Visco, and once again working with Bass and Howes, funding levels for breast cancer research began to grow. From less than a hundred million dollars in 1990 to over five hundred million in 1998, the level of funding for breast cancer currently more fully reflects the level of concern felt by women around the nation. However, much of this funding did not come by way of the usual channels.

In the early 1990s, funding for areas such as biomedical research were limited by ceilings established by legislation. Therefore, any increase in funding had to be offset by decreases in other areas. To get around this limit, the National Breast Cancer Coalition and others focused on the Defense Department budget. By getting funding from there, not only would there be greater opportunities for increased funding, but advocates also believed that this would create the opportunity for change and innovation in the research process—by allowing consumers and advocates to have a voice in priority setting. Breast cancer advocates were successful in implementing this strategy by working not just with the women in Congress but also with Senator Tom Harkin, chair of the Senate Appropriations Subcommittee on Labor, Health, and Human Services and Education. Senator Harkin had lost two sisters to breast cancer and thus became a champion within the Senate. His amendments to the appropriations bills in 1992 led to the creation of a major breast cancer research program within the Defense Department which broke new ground in the level of consumer and advocate involvement in the funding process. This program now serves as a model for linking the scientific and advocacy communities.

By this time, the "mainstreaming" of women's health had moved to the highest levels. During the campaign for the presidency in 1992, women's health had been identified by the Clinton campaign as a top priority. One of President Clinton's first acts after being sworn in in 1993 was to sign the NIH legislation that mandated the appropriate inclusion of women and minorities in clinical research trials and called for more research on breast and cervical cancer, osteoporosis, contraception, and infertility, and permanently established the NIH Office of Research on Women's Health [40]. He also signed executive orders lifting restrictions on access to information about abortion services and began expansion of the federally funded family planning program.

The National Breast Cancer Coalition did not stop at increasing funding. In October of 1993, they delivered 2.6 million signatures to the White House calling for the establishment of a National Action Plan on Breast Cancer with the goal of eliminating breast cancer. By the end of that year, Secretary of Health and Human Services, Donna Shalala, had convened a conference to identify the critical issues and to establish the Action Plan [39]. Advocates have played the leading role in both the creation and the agenda for this Plan, which has focused on catalyzing efforts in previously overlooked or underaddressed areas, particularly in the research arena. Many of the issues, such as ensuring appropriate informed consent, questions about discrimination based on hereditary susceptibility to breast cancer, and increased advocacy input to the development of research priorities reflect the earlier issues raised by the advocates of the 1960s and 1970s. Understanding the etiology of breast cancer through increased research continues to be a high priority for breast cancer advocates, who want to take the knowledge gained and target it towards prevention and cure.

The Department of Health and Human Services itself has brought women's health advocacy "inside." Through the establishment of additional women's health offices at the Food and Drug Administration, the Substance Abuse and Mental Health Services Administration, and the Centers for Disease Control and Prevention, as well as the Office on Women's Health within the Office of the Secretary, women's health advocates have increased involvement in the design and implementation of programs and policies, including the research agenda.

Beginning in 1991 with the conference convened by the NIH Office of Research on Women's Health in Hunt Valley, Maryland, and subsequently in 1997 through a conference series known as "Beyond Hunt Valley," the NIH has assessed current research needs in women's health with a view towards ensuring that the research funded by the NIH addresses the key areas identified. The research agenda that has been developed by the

NIH Office of Research on Women's Health, now led by Vivian Pinn, drew from women's health advocates as well as from health professionals and biomedical research scientists.

Another trend throughout the nineties was that new issue areas, which previously had had little research, moved to the forefront due to the initiatives of advocates. Areas such as menopause, silicone breast implants, TMJ, autoimmune disorders, interstitial cystitis, ovarian cancer, and the long-term effects of DES, among others, became the focus of both Congressional concern and government-funded research. Each one of these conditions has advocates and organizations focused on their specific topic, and in most cases focused on increasing research in the area. Although some in the scientific community have raised concerns about targeting research funding to specific diseases and about involving nonscientists in priority setting, patient advocates—modeling after early women's health activists, AIDS activists, and newer organizations such as the Breast Cancer Coalition—have been successful in having their voices heard.

In retrospect, women's health advocates have had a profound impact, particularly in the areas of informed consent and biomedical research. In 1970, Alice Wolfson and the NWHN in formation opened up the FDA to consumer observers, which led, in turn, to consumer representatives on FDA panels and through the 1970s and 1980s to increasing demands for direct consumer participation in an expanding range of regulatory decisions and taxpayer-funded research. In the 1990s, advocates brought about changes not only in the topic areas under study but also in the study designs to ensure that women's health was addressed. Questions about racial disparities in health status are moving to the front burner. In 1998, the Institute of Medicine published its report Scientific Opportunities and Public Needs which endorsed increased consumer participation in priority setting at NIH [7]. However, before their recommendations are implemented, there are crucial issues that require review and discussion, as there is far more at stake than just how much money will be allocated to research on which disease. Issues include ensuring that consumers have a voice in the design and monitoring of studies, improving the clarity and information provided in informed consent forms, and ensuring that clinical research on vulnerable populations is carried out ethically and with appropriate access to care. The future of women's health research will depend on advocates to continue to serve both as watchdogs and also as sources of new ideas and issues to be addressed.

VI. Selected Key Organizations

National Women's Health Network
514 10th Street, NW Suite 400
Washington, DC 20004
(202) 347-1140 Office
628-7814 Clearinghouse
347-1168 Fax

National Black Women's Health Project
1211 Connecticut Avenue NW Suite 310
Washington, DC 20036
(202) 835-0117
833-8790 Fax

Boston Women's Health Book Collective
240A Elm Street
Somerville, MA 02144
(617) 625-0277
625-0294 Fax

VII. Selected Bibliography

A. Overview Histories

1. Books

Ruzek, S. B., "The Women's Health Movement; Feminist Alternatives to Medical Control. Praeger Special Studies, New York, 1978.

Ruzek, S. B., Olesen, V. L., and Clark, A. E., eds., "Women's Health; Complexities and Differences." Columbus, Ohio State University Press, Columbus, 1997.

Weiss, K., "Women's Health Care; A Guide to Alternatives." Reston Publishing, Reston, VA, 1984.

Weisman, C. S. "Women's Health Care; Activist Traditions and Institutional Change." Johns Hopkins University Press, Baltimore, MD, 1998.

Worcester, N. and Whatley, M. H., eds., "Women's Health; Readings on Social, Economic and Political Issues." Kendall/Hunt Publishing, Dubuque, IA, 1988.

B. Periodical

Science Magazine, "Women's Health Research," Spec. Issue, August 11, 1995, pp. 766–801.

B. Film, and Selected Publications Provoking Action/ Change in Concepts of Women's Healthrights 1969–1999

Documentary Film

Taking our Bodies Back: The Women's Health Movement by Cambridge Documentary Films c/o Margaret Lazarus PO BOX 390385, Cambridge, MA, 02139 Fax: 617-484-0754 Phone: 617 484-0754. 1974

Selected Articles and Books, 1967–1999
1967
Peters, M., and Vera, M. D., Carcinoma of the breast, stage II. *Journal of the American Medical Association,* April 10.
1969
–Seaman, B. "The Doctors' Case against the Pill." Wyden/Avon, New York.

(revised and updated: Doubleday Dolphin, New York, 1980; 25th Anniversary edition, updated, new material, with original text restored: Alameda, Hunter House, 1995).

–(1972) Free and Female New York: Coward McCann/Fawcett.

–(1977) Seaman, Barbara and Gideon Seaman Women and the Crisis in Sex Hormones. New York: Rawson/Bantam.

Freidson, Eliot "Client Control and Medical Practice" *American Journal of Sociology* (January) 374–82.

KNOW is founded by Pittsburgh NOW to reprint and distribute feminist articles and tracts (including many early health movement materials) at cost.
1970
Cisler, Lucinda. "Unfinished Business: Birth Control and Women's Liberation." In R. Morgan, ed. Sisterhood Is Powerful, pp. 245–89 New York: Vintage Books.

Firestone, Shulamith. *The Dialectic of Sex.* New York: William Morrow.

Boston Women's Health (Book) Collective. *Women and Their Bodies.* Sommerville, Mass. New England Free Press

–(1971) name changed to *Our Bodies, Ourselves.*

–(1973) *Our Bodies, Ourselves.* New York: Simon and Schuster. First Trade Edition

–(1984) Retitled *The New Our Bodies, Ourselves.*

–(1992) Updated and Expanded for the ''90s.

–(1996) 25th Anniversary Edition

Herbst, A. L., and R. E. Scully. "Adenocarcinoma of the Vagina in Adolescence: A Report of Seven Cases including Six Clear-Cell Carcinomas (So-Called Mesonephromas)." *Cancer* 25 (April): 745–57.

Bunker, John. "Surgical Manpower: A Comparison of Operations and Surgeons in the United States and in England and Wales." *New England Journal of Medicine* 282 (January): 135–44.

Wolfson, Alice. "Caution: Health Care May Be Hazardous to Your Health." *Up From Under* 1 (May/June) 5–10.

1971

Bart, Pauline. "Depression in Middle-Aged Women (AKA Portnoy's Mother's Complaint) in V. Gornick and B. Moran, eds., *Woman in Sexist Society; Studies in Power and Powerlessness,* pp. 99–117, New York: Basic Books/Signet.

Weisstein, Naomi. *Psychology Constructs the Female, or the Fantasy Life of the Male Psychologist.* Sommerville, Mass: New England Free Press

1972

Chesler, Phyllis. *Women and Madness.* Garden City, N.Y: Doubleday/Avon

(1997) New York: Four Walls Eight Windows. 25th Anniversary Edition.

Downer, Carol "Covert Discrimination Against Women as Medical Patients." Address to the American Psychological Association, September. Honolulu. Mimeographed.

Ehrenreich, Barbara and Deirdre English. *Witches, Midwives and Nurses; A History of Women Healers.* Glass Mountain Pamphlet no. 1. Old Westbury, N.Y.: The Feminist Press.

–(1973) *Complaints and Disorders; The Sexual Politics of Sickness.* Glass Mountain Pamphlet, no. 2. Old Westbury, N.Y.: The Feminist Press

–(1989) *For Her Own Good; One Hundred Fifty Years of Experts Advice to Women.* New York: Doubleday Anchor.

Frankfort, Ellen. *Vaginal Politics* New York: Quadrangle Books.

Haire, Doris. *The Cultural Warping of Childbirth; Special Report of the President of the International Childbirth Education Association.* ICEA News. Special Issue.

Tanzer, Deborah with Jean Libman Block *Why Natural Childbirth?* New York. Schocken.

Lang, Raven. *The Birth Book.* Ben Lomond, Calif.: Genesis Press

Dejanikus, Tacie. Various articles in *Off Our Backs,* 1972, 73, 74. Topics include "Super-Coil Controversy" and "Dalkon Shield Exposed." (Also see articles by Frances Chapman.) Dejanikus was health editor of *Off Our Backs,* —in Washington DC,—the first national feminist newspaper. Other feminist periodicals providing reliable, consistent expose health reporting

in the 1970s were *Her-Self* in Ann Arbor, edited by Belita Cowan, *Ms. Magazine* (founded in New York in 1972), where Nina Finkelstein was Health Editor, and *Majority Report,* a newspaper, also in New York. *The Second Wave, Journal of Female Liberation* and *Up From Under* also featured health topics. National Women's Health Network continues to publish *The Network News,* a bi-monthly, while the Black Woman's Health Project has published *Vital Signs,* a magazine. OBOS formerly distributed regular packets of health and medical news, and is set to resume sending "consumer and feminist critiques of current controversies in women's health" to reporters. (1999.)

Seaman's *Free and Female* and Chesler's *Women and Madness* are both excerpted in *Ms.* Surprisingly, Seaman's chapter "How to Liberate Yourself From Your Gynecologist" is also excerpted in *New York Magazine,* under the title "Do Gynecologists Exploit Women?"

Reverby, Susan "Health: Women's Work" *Health/PAC Bulletin* 40 15–20.

–"The Sorcerer's Apprentice" *Health/PAC Bulletin* 46 10–16

Zola, Irving Kenneth, "Medicine as an Institution of Social Control." *The Sociological Review* 20 (November) 487–504.

1973

Hirsch, Jeanne. "Some Unorthodox Methods of Birth Control." *The Monthly Extract* (May/June) 6–7

Howell, Mary. *Why Would a Girl Go Into Medicine? A Guide For Women.* Old Westbury, N.Y: The Feminist Press. Published under the pseudonmy Margaret Campbell, M.D.

Klemesrud, Judy. "Why Women are Losing Faith in Their Doctors." *McCalls* (June) 76–77.

Ramsey, Judith. "The Modern Woman's Health Guide to Her Own Body. *Family Circle* (July) 113–120

Naismith, Grace. "How Safe is Do-It-Yourself Gynecology? *Family Health Magazine* (February) 24–25

(The above three titles- Klemesrud, Ramsey, Naismith—illustrate that by 1973- radical health-feminist discussions had invaded traditional women's magazines.)

Scully, Diana, and Pauline Bart. "A Funny Thing Happened on the Way to the Orifice: Women in Gynecology Textbooks." *American Journal of Sociology* (January): 1045–50.

Feminist Women's Health Center Report. 1973. Los Angeles: FWHC. Also see Reports for 1974, 75, 77.

Rennie, Susan and Kirsten Grimsted. *The New Woman's Survival Catalog.* New York: Coward, McCann.

Lennane, K. J., and R. J. Lennane. "Alleged Psychogenic Disorders in Women—a Possible Manifestation of Sexual Prejudice." *New England Journal of Medicine* (February) 288–292.

1974

Borman, Nancy "Harvey Karman: Savior or Charlatan? *Majority Report* (January) 6.

Dodson, Betty. *Liberating Masturbation; A Meditation on Self Love.* New York: Bodysex Designs

Howell, Mary. "What Medical Schools Teach About Women." *The New England Journal of Medicine* (August) 304–307

1975

Brownmiller, Susan. *Against Our Will; Men, Women and Rape.* New York: Simon and Schuster.

Gordon, Linda. *Woman's Body, Woman's Right; A Social History of Birth Control in America.* New York: Penguin.

Hite, Shere. *The Hite Report.* New York: Macmillan

Kushner, Rose. *Breast Cancer: A Personal History and An Investigative Report.* New York: Harcourt Brace Jovanovich.

Lorber, Judith "Good Patients and Problem Patients: Conformity and Deviance in a General Hospital." *Journal of Health and Social Behavior* 16:213–25

Seaman, Barbara "Pelvic Autonomy: Four Proposals." *Social Policy* 6 (September/October) 43–47.

Lacy, Luise. *Lunaception; A Feminine Odyssey Into Fertility and Contraception.* New York: Coward McCann

1976

Katz Rothman, Barbara "In Which A Sensible Woman Persuades Her Doctor, Her Family and Her Friends to Help Her Give Birth at Home. *Ms Magazine* December. pp 25–32.

Eagan, Andrea. "Breast cancer: Facts a Woman Needs to Know." *Healthright.* 2 No. 3

(1977) "The Home Birth Movement." *Healthright* 3 No. 2.

Seaman, Barbara. "How Late Can You Wait to Have A Baby?" *Ms Magazine* January.

Marieskind, Helen. "Gynecological Services and the Women's Movement: A Study of Self-Help Clinics and other Modes of Delivery." Ph.D. dissertation, School of Public Health, University of California, Los Angeles.

Barker-Benfield, G. J. *The Horrors of the Half-Known Life.* New York: Harper & Row.

1977

Corea, Gena. *The Hidden Malpractice; How American Medicine Treats Women as Patients and Professionals.* New York: William Morrow.

(1985) *The Mother Machine; Reproductive Technology from Artificial Insemination to Artificial Wombs.* New York: Harper & Row

Dreifus, Claudia. (Editor) *Seizing Our Bodies: The Politics of Women's Health.* Nwq York: Vintage Books.

Reitz, Rosetta. *Menopause, a Positive Approach.* New York: Penguin.

Wertz, Richard and Dorothy C. Wertz. *Lying-In; A History of Childbirth in America.* New York: Macmillan, The Free Press.

Rich, Adrienne. *Of Woman Born; Motherhood as Experience and Institution.* New York: Norton..

1978

Ohrbach, Susie. *Fat Is a Feminist Issue; A Self-Help Guide for Compulsive Eaters.* New York: Paddington/Berkley.

1979

Gage. Susann *When Birth Control Fails: How to Abort Ourselves Safely.* Hollywood, California: Speculum Press/Self-Health Circle, Inc.

1981

Howell, Elizabeth and Marjorie Bayes (Editors) *Women and Mental Health.* New York: Basic Books.

Federation of Feminist Women's Health Centers. *How to Stay Out of the Gynecologist's Office.* Culver City, Calif: Peace Press.

1983

Rothman, Barbara Katz. *In Labor; Women and Power in the Birthplace.* New York: Norton 1982/ Reissued 1993

————*The Tentative Pregnancy; Prenatal Diagnosis and the Future of Motherhood.* New York: Viking 1986/Reissued Norton 1993

Meyers, Robert *D.E.S. The Bitter Pill.* New York: Seaview/ Putnam.

1984

Lorber, Judith *Women Physicians; Careers, Status and Power.* New York and London: Tavistock.

–(1997) *Gender and the Social Construction of Illness.* Thousand Oaks, Ca: Sage.

Arditti, Rita, Renate Duelli Klein and Shelley Minden (Editors) *Test-Tube Women; What Future for Motherhood?* London: Pandora Press

1987

Payer, Lynn. *How to Avoid a Hysterectomy.* New York: Pantheon.

1989

National Women's Health Network. *Abortion Then and Now; Creative Responses to Restricted Access.* Washington, DC.

————*Taking Hormones and Women's Health; Choices, Risks and Benefits.* Various updated editions through 1999.

1990

White, Evelyn C. (Editor) *The Black Women's Health Book; Speaking for Ourselves* Seattle: Seal Press.

Banzhaf, Marion, Tracy Morgan and Karen Ramspacher "Reproductive Rights and AIDS: The Connections" in *Women, AIDS and Activism* (The ACT UP/NY Women and AIDS Group) Boston: South End Press. 199–209.

Hubbard, Ruth. *The Politics of Women's Biology.* New Jersey: Rutgers University Press

1991

Klein, Renate, Janice G. Raymond, and Lynette J. Dumble. RU 486: *Misconceptions, Myths, and Morals.* Australia: Spinifex Press.

1992

Chalker, Rebecca and Carol Downer. *A Woman's Book of Choices; Abortion, Menstrual Extraction, RU-486.* New York: Four Walls Eight Windows.

Cody, Pat and Fred. *The Life and Times of Cody's A Berkeley Bookstore 1956–1977.* San Francisco: Chronicle Books

Herman, Judith. *Trauma and Recovery.* New York: Basic Books.

1993

O'Leary Cobb, Janine *Understanding Menopause; Answers and Advice for Women in the Prime of Life.* New York: Plume

Institute of Medicine, *An Assessment of the NIH Women's Health Initiative.* Washington, D.C.: National Academy Press.

1994

Batt, Sharon. *Patient No More; The Politics of Breast Cancer.* Charlottetown, P.E. I. Canada: Gynergy Books.

Coney, Sandra *The Menopause Industry; How the Medical Establishment Exploits Women.* Alameda, Ca: Hunter House.

Grant, Linda *Sexing the Millenium; Women and the Sexual Revolution.* New York: Grove Press

Hite, Shere *The Hite Report on the Family; Growing Up Under Patriarchy.* New York: Grove Press,

Nuland, Sherwin B. *How We Die; Reflections on Life's Final Chapter.* New York: Alfred A. Knopf

1996

Baker, Christina Looper and Christina Baker Kline. *The Conversation Begins; Mothers and Daughters Talk about Living Feminism.* New York; Bantam.

(See sections on health activist mothers such as Patsy Mink, Helen Rodriguez-Trias, Eleanor Smeal, Maryann Napoli, Barbara Ehrenreich, Barbara Seaman, - and their daughters.).

1997

Webb, Marilyn *The Good Death; The New American Search to Reshape the End of Life.* New York: Bantam.

Adler, Margot *Heretic's Heart; A Journey Through Spirit and Revolution.* Boston; Beacon Press.

Love, Susan, with Karen Lindsey. *Dr. Susan Love's Hormone Book.* New York: Random House

————— *Dr. Susan Love's Breast Book.* New York. Random House.

1998

Institute of Medicine, *Scientific Opportunities and Public Needs; Improving Priority Setting and Public Input at the National Institute of Health.* Washington, D.C.: National Academy Press.

Avery, Byllye *An Alter of Words; Wisdom, Comfort and Inspiration for African American Women.* New York: Broadway Books.

(First entry is "Activism.")

Watkins, Eliozabeth Siegel. *On the Pill; A Social History of Oral Contraceptives 1950–1970.* Baltimore: Johns Hopkins University Press.

1999

Institute of Medicine. *The Unequal Burden of Cancer; An Assessment of NIH Research and Programs for Ethnic Minorities and the Medically Underserved.* Washington, D.C. National Academy Press.

Special category; some books that recognize issues of corporate censorship in health care, and may help readers to distinguish legitimate grass roots groups from astro-turf imitations.

Szockyi and James G. Fox. *Corporate Victimization of Women,* especially see Finley, Lucinda M. "The Pharmaceutical industry and Women's Reproductive Health p. 59–110. Boston: Northeastern University Press. 1996

Stauber, John and Sheldon Rampton. *Toxic Sludge Is Good for You; Lies, Damn Lies and the Public Relations Industry.* Monroe, Me: Common Courage Press 1995

Flanders, Laura *Real Majority, Media Minority; The Cost of Sidelining Women in Reporting.* Monroe, Me.: Common Courage Press: 1997

Phillips, Peter and Project Censored. *Censored 1999; The News That Didn't Make the News; The Year's Top 25 Censored Stories.* New York: Seven Stories Press.

Published annually by Seven Stories Press. Editions available for 1998, 97, 96.

Special category B; Cautionary tales of what happened to Norma McCorvey AKA Jane Roe:

McCorvey, Norma with Andy Meisler. *I Am Roe; My Life, Roe v. Wade, and Freedom of Choice.* New York: HarperCollins 1994.

McCorvey, Norma with Gary Thomas. *Won by Love; Norma McCorvey, Jane Roe of Roe v. Wade, Speaks Out for the Unborn as She Shares Her New Conviction for Life.* Publisher: Thomas Nelson 1998

Also see: Weddington, Sarah *Choice; By the Lawyer Who Won Roe v. Wade.* New York; Penguin. 1992.

References

1. Ross, W. S. "Crusade—The Official History of the American Cancer Society." Arbor House, New York.
2. Patterson, J. (1987). "The Dread Disease: Cancer and Modern American Culture.
3. Batt, S. (1994). "Patient No More: The Politics of Breast Cancer."
4. Magazine (1976). Medical Dimensions, March.
5. Moss, R. W. (1989). The Cancer Industry." Paragon House, New York.
6. Rettig, R. A. (1977). "Cancer Crusade: The Story of the National Cancer Act of 1971." Princeton University Press, Princeton, NJ.
7. Institute of Medicine (1998). "Scientific Opportunities and Public Needs." National Academy Press, Washington, DC.
8. U.S. News Online "Mary and her "little Lambs" launch a war."
9. Albert and Mary Lasker Foundation Online "Mission Statement," "About the Lasker Medical Research Awards."
10. Mothering Magazine (1998). Living treasures, Doris Haire. *Mothering Mag.,* July–August. (Publications available from American Foundation for Maternal and Child Health, 439 E. 51 St., New York, N.Y. 10022)
11. Ruzek, S. B. (1978). "The Women's Health Movement, Feminist Alternatives to Medical Control." Praeger, New York.
12. Seaman, S. S. (1965). "Always a Woman; What Every Woman Should Know About Breast Surgery." Argonaut Books, Larchmont, NY.
13. Lasser, T., and Clarke, W. K. (1972). "Reach to Recovery." Simon & Schuster, New York.
14. Seaman, B. (1997). Beyond the Halsted radical. *On the Issues Mag.* Fall.
15. Seaman, B. (1998). Thalidomide. *In* "The Reader's Companion to U.S. Women's History" (W. Mankiller, G. Mink, M. Navarro, B. Smith, and G. Steinem, eds.). Houghten Mifflin, Boston.
16. Weisman, C. S. (1998). "Women's Health Care; Activist Traditions and Institutional Change," Hopkins University Press, Baltimore, MD. pp. 62–63.
17. Seaman, B., and Seaman, G. (1977). "Women and the Crisis in Sex Hormones." Rawson, New York; Bantam Books, New York, 1978, p. 79.
18. Rimer, S. (1997). Women's health; A special section. *New York Times,* Sunday, June 22, p. 27.
19. Anonymous (1970). "Competitive Problems in the Drug Industry," Part 15, p. 6053. U.S. Gov. Printing Office, Washington, DC.
20. Mann, C. (1995). *Science* **269,** 766–770.
21. Watkins, E. S. (1998). Oral contraceptives and informed consent. *In* "On the Pill; A Social History of Contraceptives, 1950–1970," pp. 103–131. Johns Hopkins University Press, Baltimore, MD.
22. Wolfson, A. J. (1998). Clenched fist, open heart. *In* "The Feminist Project Memoir, Voices from Women's Liberation" (R. B. DuPlessis and A. Snitow, eds.), pp. 268–283. Crown Publishers (Three Rivers Press), New York.
23. Seaman, B. (1995). "The Doctor's Case against the Pill," 25th Anniversary updated ed., pp. 222–226, Hunter House, Alameda, CA.
24. MS Magazine (1996), "Snapshot," Carol Downer.
25. Campbell, M. A. (a pseudonym for Mary Howell) (1973). "Why Would a Girl go into Medicine?" The Feminist Press, Old Westbury, NY.
26. Saxon, W. (1998). Mary Howell, a leader in medicine, dies at 65. Lead obituary. *N.Y. Times,* February 6.
27. Figures on Medical School admissions, American Medical Association compilations.
28. (1970). "Our Bodies, Ourselves." New England Free Press.
29. (1973). "Our Bodies, Ourselves." Simon & Schuster, New York.
30. (1976). "Our Bodies, Ourselves," rev. ed.

31. (1979). "Our Bodies, Ourselves," rev. ed.

32. (1984). "The New Our Bodies Ourselves."

33. (1992). "The New Our Bodies Ourselves," rev., updated.

34. (1998). "Our Bodies Ourselves for the New Century."

35. U.S. Department of Health and Human Services, Public Health Service, Health Resources and Services Administration, Council on Graduate Medical Education (1995). "Fifth Report: Women and Medicine, Part II. USDHHS, PHS, Washington, DC.

36. U.S. Department of Health and Human Services (1985). "Women's Health: Report of the Public Health Service Task Force on Women's Health Issues," Vol. II, Publ. No. (PHS) 85-50206. USDHHS, Washington, DC.

37. U.S. General Accounting Office (1990). "Statement of Mark V. Nadel," Associate Director National and Public Health Issues Human Resources Division before the Subcommittee on Health and the Environment Committee on Energy and Commerce, House of Representatives, GAO/T-HRD-90-38. GAO, Washington, DC.

38. Mann, C. (1995). *Science* **269,** 770.

39. National Institutes of Health (1993). "Secretary's Conference to Establish a National Action Plan on Breast Cancer: Proceedings." NIH, Washington, DC.

40. National Institutes of Health, Office of Research on Women's Health, Office of the Director (1991). "Report of the National Institutes of Health: Opportunities for Research on Women's Health." NIH, Washington, DC.

4

State-of-the-Art Methods for Women's Health Research

SUZANNE G. HAYNES* AND MAUREEN HATCH†

*Office on Women's Health, Department of Health and Human Services, Washington, DC; †Division of Epidemiology, Department of Community Medicine, Mt. Sinai School of Medicine, New York, New York

I. Introduction

It is still shocking for most contemporary scientists to realize that the study of women's health only has come to the forefront in the 1990s. Prior to this time, women were excluded from most research unless it centered around questions related to hormones, reproduction, or childbearing. Several reasons have been put forward to explain this situation [1,2]. In the first part of the twentieth century, there was little collection and recording of data on the health of Americans overall. The low status of women resulted in insufficient attention to women's health. After World War II, with new emphasis and priority focused on scientific research, women continued to be excluded from clinical studies because of cost and complexity issues [2]. The inclusion of women, it was argued, would raise the cost of research in large population studies such as those used by epidemiologists and statisticians for clinical trials and prospective cohort studies. Women were thought to make the study designs more complex because of the potential effect of hormonal fluctuations or pregnancy on outcomes [2]. Major resistance to the inclusion of women in research came at the unfortunate time in the United States's history when it was discovered that thalidomide and diethylstilbestrol had immediate and delayed teratogenic effects on the offspring of childbearing women [1].

The turning point for women's health research came in 1990 with the now infamous report by the General Accounting Office on the failure of the National Institutes of Health (NIH) to include women in research despite policies established in the 1980s requiring such inclusion [3]. With the creation of the Office of Research on Women's Health in 1990, and the passage of the NIH Revitalization Act of 1993 requiring the inclusion of women and minorities in research supported by Federal funds, the funding for women's health research has risen exponentially [4,5]. Two years later the Centers for Disease Control and Prevention (CDC) passed similar guidelines, and recently the Food and Drug Administration (FDA) passed a rule to allow FDA to place a clinical hold on any Investigational New Drug Application (IND) that excludes women with reproductive potential who have a life threatening disease but are otherwise eligible for drug testing.

Since the surge in funding for women's health research is very recent, it is not surprising that there has been little attention to methodologic differences in conducting research on women until the last few years. In 1995, the NIH sponsored the first workshop ever to address the issues of the recruitment and retention of women in clinical studies [6]. This chapter will summarize the problems that have been encountered in recruiting women into research studies, particularly special populations of women, such as minority, low socioeconomic status, and older

women. We will also document success stories and strategies that have worked to recruit these special populations of women to studies. Given the long awaited appearance of women in the history of research, we must ensure that all women are included and that we do not err in a similar fashion to our ancestors by overlooking some populations of women because they are hard to recruit.

In addition to the issue of recruitment, we will discuss challenges to retention in studies that are unique to women. The whole issue of family and physician support for research has been ignored until recently. The reliability and validity of questionnaire responses, including questionnaire or interview format and the type of questions women prefer, has been virtually ignored in epidemiologic research. Social desirability in reporting will be discussed in this light. In discussing questionnaire design issues, we draw on the reproductive health, dietary, and psychological literature. Finally, the behavior of women when answering questions of a sensitive nature, such as alcohol consumption, abortion, sexual partners, or illicit drug use will be addressed.

Moving from questionnaire design, we will review issues, both biologic and nonbiologic, that need to be considered in research on women's health, including the importance of a lifespan approach that takes into account hormones and hormonal fluctuations and body size and composition, as well as family responsibilities. We hope this chapter will assist researchers in developing different approaches and designs for including women in research.

II. Problems Related to Recruitment

A. Minorities, Lower Socioeconomic Status, and Older Women

Prior to the 1990s, little attention was given in the scientific literature to the recruitment of women into research studies. The NIH Revitalization Act of 1993, which mandated the inclusion of women in clinical trials in NIH-supported clinical research, immediately spurred the recruitment of women into research studies [5]. As investigators began recruiting women into clinical trials and observational studies, it became immediately clear that there were no apparent problems recruiting women, per se, into studies. In a report by Hayunga et al [7], 51.8% of participants in NIH extramural studies and 49.0% of participants in NIH intramural studies conducted in fiscal year (FY) 1994 (October 1994–September 1995) were women. However, concerns about recruiting minority, lower socioeconomic status, and older women have been raised.

Although several studies have documented low participation rates of minorities in clinical trials, problems regarding the recruitment of minority women into trials became painfully obvious during the initial stages of the Women's Health Initiative (WHI) [8]. The "Tuskegee" effect has often been cited as a reason for the nonparticipation of blacks in health studies [9], and this held true in the WHI trial as well. The WHI is a multicentered clinical trial to determine the efficacy of low-fat diets, hormonal replacement therapy, and Vitamin D/calcium supplementation to prevent several diseases in women aged 50–79 years. A short 7–10 minute survey of early nonrespondents in one of the WHI clinics revealed that almost one-third of black women agreed that scientists could not be trusted compared to 4% of white women [8]. Only 28% of black women felt that clinical research in the United States was ethical as compared to 47% of white women, and 37% preferred to be treated by a black physician/scientist [8]. These findings suggest that black women would be more likely to participate in studies focused exclusively on blacks. However, black women were underrepresented in the African-American Study of Kidney Disease and Hypertension, where only 25% of the clinic-recruited participants were women, although twice this proportion were initially contacted [10]. Likewise, low participation (20%) was initially reported when patient logs were used to recruit low-income black women for a trial of an education intervention to promote screening for cancer [11]. Higher participation (50%) was achieved when black women were recruited from public housing projects, churches, and the National Black Women's Health Project [11].

The participation of Hispanic, Asian-Pacific Islander, and American Indian/Alaskan native women in NIH funded research is even lower than participation by black women. The participation for these groups of women in NIH extramural studies is 9.1, 3.6, 1.1, and 23.4% respectively. Participation in NIH intramural studies is even lower for these four race/ethnicity groups: 2.7, 2.50, .2, and 10.9%, respectively [7]. Strategies to include Hispanic women in NIH clinical research have been described by Caban [12] and depend on the inclusion of Hispanics in all phases of research projects as well as empowering communities to establish better means of encouraging their members to participate in clinical research studies. Minority investigators are often scarce on research teams, a factor that can be the fatal flaw for a research study.

A question might be raised as to whether race, per se, or low socioeconomic status (SES) explains the lower participation among minority women in research. In two breast cancer risk counseling trials targeting women with a family history of breast cancer, a lower level of education (high school or less) was significantly associated with lower participation rates (42% and 36%, respectively) [13,14]. Race was not a significant factor for participation in either trial. In the Coronary Artery Risk Development in Young Adults study (Cardia), using typical random digit dial techniques initially resulted in lower participation rates (under 50%) for lower SES (high school or less) women of both races (white and black) aged 18–30 years [15]. These rates were improved after sampling in more targeted census tracts, followed by a community-wide publicity campaign that had the support of community leaders.

After pilot testing a convenience sample of low-income, uninsured or underinsured women over 40 years of age, a successful approach for recruiting low income African-American women into a breast cancer screening program was developed by Zabora et al. [16]. The key to success in recruiting these women was door-to-door contact by a community services coordinator. Door-to-door contact provided the opportunity for the investigators to respond immediately to questions in an informal and nonthreatening manner. It also reinforced the importance of screening and served as the initial teaching opportunity. Working with community representatives and ministerial contacts also contributed to the recruitment of these women.

The unique predominance of women among the older population (65+ years) is one demographic that has often been ignored in study designs. In the past, most clinical trials and observational studies stopped recruitment at age 65. Given that 15% of women fall in this older age group, what do we know about recruiting older women into research? Participation has generally been quite good for older women in several major clinical trials. In the Beta-Carotene and Retinol Efficiency Trial (CARET), drop-out rates were identical (24%) for women as compared to men in the age range 65–69 years [17]. In the Cholesterol Reduction in Seniors Program (CRISP), major efforts were made to successfully recruit women aged 65 and older, with the result that 71% of the participants were women [18]. Likewise, older women aged 60–80 were just as likely as men to be recruited in the Trial of Nonpharmacologic Intervention in the Elderly (TONE) trial of hypertensives [19].

The reasons for success in recruiting older (60+ years) women into trials were assessed in the Systolic Hypertension in the Elderly Program (SHEP) [20]. The most important reasons older women gave for joining SHEP were the desire to contribute to science, improve the health of others, improve their own health, obtain free medical care, and have someone to talk to. Transportation was more likely to be a problem for women (18%) compared to men (13%) [21].

The response rates for survey questionnaires given to older women are generally in the 70% or greater range, as seen for the SHEP study (81%) and other community samples such as one in Baltimore (68%) that interviewed women 65 and older [20,21]. There is little evidence in the literature that older respondents report factual information less validly than younger respondents [22], so that studies including older women yield both high participation rates and valid responses. Thus, women's health researchers need to design studies that include women 65 years and older. Otherwise, we will miss the opportunity to study a large segment of our population of women.

In sum, we are calling for inclusiveness in health research around gender, race, class, and aging. The challenges to increasing inclusiveness have been explored by Dressel et al. [23]. The authors encourage us not to use the "add-and-stir" approach, where researchers simply add overlooked samples or groups to their data collection without appreciating that the questions that frame their research may not be responsive to the realities of these new groups. Similarly, tokenism toward minority and older women (i.e. limited inclusion in research projects, conferences, or journal issues) should be avoided in favor of the inclusive mainstreaming of these issues in all women's health research studies. Dressel's treatise is a thoughtful challenge to researchers who are serious about the recruitment of a diverse mix of women into research studies.

B. Hard-to-Reach Populations

In order to recruit representatives from the total universe of women with health concerns, several hard-to-reach populations of women need to be considered by researchers. These groups are often missed by typical recruitment strategies. They include drug users, HIV-positive women, lesbians, and victims of domestic violence. Unlike the demographically defined populations, the barriers to recruit these hard-to-reach women include social stigmatization, legal prohibitions, fear of employment or insurance discrimination, and reprisals. Efforts to reach these women are now in the formative stage, challenging methodologists to discover new and creative means for recruitment. Some of the obstacles are societal, which suggest that our solutions may have to move outside of the usual study design techniques, "out of the box," and into the community at large.

An excellent example of this challenge is presented in a review on the recruitment and retention of adolescent women in drug treatment research [24]. Few randomized clinical trials have been attempted in the field of drug abuse treatment, much less for adolescent women aged 14–19 years who are abusing drugs. However, it is this group who should be targeted because most drug abuse problems begin in adolescence. In the past, recruitment was hampered by motivation problems, rigid selection criteria, and institutional constraints. In addition, parental approval was often not attainable. Because adolescent drug users use so-called gateway drugs (alcohol, marijuana, or tobacco) that are not thought to be addictive, as compared to heavy drugs, participation by adolescents in drug treatment programs has been extremely low (*i.e.,* under 5%). In the Positive Adolescent Life Skills (PALS) study, recruiting teenage girls from dozens of sources yielded a recruitment of almost 30% of those who were eligible. Participants in the trial (skills training vs no skills training) scored higher in mental health, peer relations, educational status, vocational status, and aggressive behavior, and delinquency than did nonparticipants [24]. Of note, participants were more likely to be referred by noninstitutional sources (other participants, self-referral, parent, or guardian) than nonparticipants. Overall, participants were younger, included more Latinas and fewer whites and other ethnic groups, and were less likely to drop out of school. The lesson learned from the trial was that informal or noninstitutional sources of referrals were important as a means for recruiting this hard-to-reach population.

Likewise, word of mouth was the most frequent response given by enrolled women in the Human Immunodeficiency Virus Epidemiology Research (HER) Study when asked "How did you find out about this study?" [25]. The 66% enrollment rate in this prospective follow-up study of HIV-infected women and demographically matched uninfected women was reasonable given the stigma associated with HIV-positive status. Even better was the retention of 88% of participants at the third six-month follow-up. In this study, lower retention was associated with currently injecting drugs, not having dependent children, and not being infected with HIV at enrollment.

The Institute of Medicine issued a landmark report on lesbian health and recommendations for the future [26], which will be described in depth by Bradford and White in Chapter 7. It is important to point out that the lesbian population, like other hard-to-reach groups, is difficult to identify, constitutes a mod-

est percentage of women (under 10%), and has less access to medical care and medical research [26].

III. Unique Role of Social Support in Women's Participation in Research

The role of family and physician support for the decision to join a clinical trial is not often considered in study design. However, for women it may have special importance. In this section, we will use examples of problems encountered in recruiting women to three tamoxifen trials to illustrate the role that family and personal physicians have in the decision to join or not to join a study.

It has been common practice to use health maintenance organizations (HMO) or cancer registry lists to recruit women to clinical trials of tamoxifen. Using the Wisconsin Cancer Reporting System, women's physicians were asked to refer node-negative breast cancer patients to join a toxicity trial of tamoxifen [27]. The total yield of patients who ended up in the study after physician contact was only 5.8%. This was due to a large nonresponse from the patients' physicians (61.5%) and the fact that half of the women contacted did not meet the eligibility criteria or were unwilling to participate.

The experience from an HMO recruitment effort in one of the clinics in the National Cancer Institute's Breast Cancer Prevention Trial (BCPT) of tamoxifen yielded only slightly higher numbers of women patients for study [28]. In this clinic, 63% of the names generated met the eligibility criteria and participated in a screening interview, and of these, 45% expressed interest in the trial, which means that only 28% of the women contacted were eligible and interested in the trial. The most significant factor associated with interest in participation and/or willingness to take a pill was anticipated family support. The fact that family support was found to be the only significant facilitator of participation supports a growing body of literature that social support systems are associated with health behaviors, particularly for women and older persons. In the decision-making process to join a trial, there must be enough social support to overcome the usual barriers of side effects, increased visits to the doctor, and thoughts about cancer.

This observation was also found by another BCPT clinic [29]. Lack of participation in that site was related to not being able to take hormone replacement therapy (HRT), concerns about side effects, the possibility of getting a placebo, costs associated with the trial, and lack of conviction that significant others would be reassured if the respondent was taking tamoxifen. Both BCPT examples suggest that recruitment rates in clinical trials could be improved by bolstering family and physician support. This is especially true in the Hispanic family, as referred to earlier in this chapter. This aspect, which is often ignored by researchers, may be just as important to a researcher as addressing concerns about drug toxicity and side effects.

The importance of family member's influence on the decision to participate in research was actually tested by Muncie *et al.* [30]. In this study, 315 mentally competent persons aged 65 years or older (77% women) were asked about their decision to participate in several hypothetical studies. Their designated proxies, if they were to become mentally incompetent, were separately asked what choice they would make for their charges.

The correlation of the proxies' decisions with those of the subjects themselves was low and not different from random. Proxies' decisions were more highly associated with their own decision to participate rather than with their charge's decision to participate. Protectiveness by proxies was highest for a new vaccine, followed by stroke drugs, drugs for urinary tract infection, blood drawing, cancer drugs, external catheter, food intake, and decubitus cream, in that order.

IV. Resolving Recruitment Problems

Where to approach a targeted population and how best to do it has long been a concern of epidemiologists, health educators, and health providers. A number of reports in the literature compare the results obtained using different recruitment methods. Some of these studies were undertaken specifically to evaluate the effectiveness of various approaches. Supplementing this methodologic research are observations from several successful studies of women that are currently in progress or have been recently completed. Topics discussed below are: the means of recruitment; the approach to potential enrollees; considerations such as trust and perceived benefit to participants; barriers to participation (of paramount importance in studying women); and incentives to participate.

A. Means of Recruitment

1. Occupation

Recruiting women into studies through their workplace or professional affiliation has had substantial success. Examples of occupational cohorts that have been studied include female radium-dial painters [31], female radiologists [32], and female anesthesiologists [33], to name a few. Perhaps the best known contemporary cohort of women selected on the basis of occupation is the Nurses' Health Study. This longitudinal investigation enrolled 121,964 subjects through a mailing to 172,413 female registered nurses in eleven large states. While the initial response rate was 71% [34], follow-up has been extremely successful, with over 90% of participants responding to each of the cycles since 1988 [35]. Recently, the California Teachers Study assembled a cohort of 135,000 women using a list maintained by the teachers' retirement fund [Dr. Peggy Reynolds, personal communication].

2. Health Services

Female users of health services are another group that has been recruited into studies, with recruitment usually taking place directly at the service site or via membership in a HMO. This has sometimes worked well but recruitment of women through their physicians has not always had a high yield, as noted previously. The rate of physician refusals on behalf of patients varies depending on the nature of the research and the nature of the population of patients (i.e., diagnosis, age, prognosis).

3. General Population

Community-based recruitment presents greater challenges than recruitment based in hospitals or work sites. Recruitment approaches aimed at women from the general population have included standard mail or telephone methods. Several studies have made use of lists maintained by the Motor Vehicle Bureau or the Health Care Financing Administration (HCFA) to identify and approach potential subjects by mail and/or telephone. For example, the study population for the Women's Health and Aging Study was drawn from HCFA's Medicare enrollment file [36]. Participation rates are not always optimal with this strategy, however. Recruitment of a population-based sample using the random digit dialing method of telephone contact has yielded response rates of around 65% [37], low enough to cause concern about the representativeness of the subjects enrolled. In addition, the telephone approach can lead to bias in areas where phone coverage is not high. For instance, among low-income elderly women in the South, "not having a phone" was a strong predictor of nonparticipation in cervical cancer screening [38]. Although it is now common in the U.S. for rates of telephone ownership to be high, some elderly women who have telephones may feel uncomfortable using them. Mail surveys, however, typically yield lower response rates than telephone surveys. A mail survey sent to female college graduates of four predominantly black Atlanta institutions resulted in a response rate of 45% [39]. The response rate to a single mailing sent to white women by the Iowa Women's Health Study was 40%. These rates can be increased by multiple mailings and incentives.

Apart from mail or telephone, recruitment for community studies or health-related programs has taken place in people's homes, on the sidewalk in a busy area, or in a local institution such as a church or community center. Home-based sampling is more expensive than telephone recruitment but may also be more effective. One study in an Arizona retirement community found substantially higher participation rates when the first contact was a home visit (81%) compared to a telephone call (55%) [40]. In Melbourne, Australia, a study of mid-life to elderly individuals also found the direct "doorstep" approach to be much more effective than the telephone method of recruitment (participation rates of 76% and 47%, respectively) [41]. These studies would suggest that, in older populations, recruitment is enhanced by home visits to the potential participant.

Other population subgroups may also respond to an approach that takes the study to their doorstep. For instance, one study of a high-risk population [42] found that bringing sexually transmitted disease (STD) screening to homeless shelters, soup kitchens, and drug treatment centers led to higher participation, even compared with street-based outreach and referral of potential participants to a community center for screening. Similarly, a study of women enrolled in the Women, Infants, and Children (WIC) food supplementation program found that a personal escort system was over five times more effective in getting women vaccinated against measles than was the use of a vaccination voucher [43].

Another indication that the doorstep approach produces higher participation than recruitment strategies based on phone or mail comes from the experience of the National Health and Examination Survey (NHANES) [44]. Once a household is targeted for possible inclusion, a NHANES staff member visits the home to determine whether someone eligible resides there and is willing to become a participant. This approach has produced participation rates as high as 85% [44]. Clearly, however, the strategy requires many more resources to implement than a mail or telephone recruitment campaign.

Other enrollment strategies for community studies include recruiting on the street or through local institutions. Soliciting subjects from a sidewalk table will essentially produce a self-selected sample, making it important to consider the effect of "volunteerism" in interpreting study results. Recruitment from a local institution can potentially be more systematic because a list of all members' names will often be available. Nonetheless, in this case the study base is composed of people who are "joiners" and may therefore be unrepresentative of the community at large. Among the community institutions that have participated in health studies are senior citizens centers, churches, and schools.

"Snowball sampling" is an approach that has been developed for studying populations such as the homeless, who may be hard for investigators to reach but who know where to find one another. The idea is to ask subjects to recruit other potential participants from among their acquaintances, who would then be asked to do the same [45]. The sample assembled with the aid of informants is intended to be typical of the target population but can end up with serious omissions. Informants who draw from their own network may produce a biased list of other homeless women. For example, homeless women who are problem drinkers may nominate other problem drinkers for study rather than a sample more representative of the whole population of homeless women.

It has been suggested that users of the World Wide Web could serve as a potential study population. People would recruit themselves on the basis of electronic advertisements [46]. While this sounds farfetched at present, it may prove to be a feasible way of recruiting volunteer subjects from certain age and educational groups in the future, understanding that respondents will be biased toward younger, more educated women.

B. Approach to Participants

Several studies of women have used mass media to inform potential subjects about the proposed research and to stimulate interest during the precontact stage. In addition, researchers can use the media to enlist volunteers directly. To be effective, media strategy must clearly be shaped with the audience in mind. For example, if Native American women were the target population, tribal newsletters, newspapers, or radios could be used to publicize the study and its potential benefits for improving the health of tribal women. An example of successful media outreach to a Spanish-speaking audience is the radio program "Cuidando Su Salud" in Washington, DC, which discusses the value of screening programs and other health topics [47].

Among the types of approaches suggested to be effective in eliciting participation—whether in a medical service program, a community intervention, or a research project—are personal invitations from a woman's physician or nurse [48] or from community volunteers [41]. It is important that the study team include culture and gender relevant members, even at the highest level [49]. The Black Women's Health Study, for instance, has both black and white women scientists as leading investigators [50]. The most effective approach to potential participants will vary according to the specific situation. For instance, prenatal patients are almost always best approached through their obstetric care provider.

Common to all of these successful methods is the sense of a personal endorsement of the study and encouragement to participate by someone important or relevant to the target population. It has been suggested that, in addition to invitations from service providers or from culturally compatible field staff or neighbors, community leaders should be asked to make an approach to potential subjects or at least to help publicize the study during a prerecruitment period. Clearly, involving community leaders is no easy matter. Pitfalls include identifying the wrong individuals as leaders or failing to win support from the true leaders.

An important element in avoiding these pitfalls appears to be familiarity with the subject's environment. Recommended strategies to increase knowledge of the sociocultural environment include one-on-one interviews, focus groups, and ethnographic research in the target population [51]. Results of one such survey among low-income Mexican-American women concerning knowledge and attitudes related to cancer prevention identified the need for educational materials that would be sensitive to their fears and fatalistic beliefs about external forces [52]. Among the barriers to care for Asian/Pacific Islander American women are cultural norms encouraging modesty about their bodies, as well as confusion regarding Western medical practices [53]. To overcome such barriers, the authors proposed using culturally compatible personnel to conduct outreach and education activities. Another study in a Puerto Rican population [54] found that incorporating Puerto Rican folktales into the treatment modality improved utilization statistics, producing an attendance rate of over 80% at group sessions.

Qualitative methods such as focus groups can be important to understanding women's lives and women's needs. They can potentially help in identifying research questions, developing recruitment strategies, designing data collection tools, and interpreting results. While use of these methods is increasing, the integration of qualitative material into a quantitative framework needs further attention. For instance, there are no guidelines about how much weight to give to information gathered in small focus groups, which often comprise volunteers. A development by Federal agencies and other researchers is the formation of cognitive laboratories [55]. The purpose of these laboratories is to reduce response error in survey data by using various cognitive techniques to study sensitive questions and make recommendations for survey design when such questions are administered.

An approach that is increasingly recognized as valuable in longitudinal studies of hard-to-reach populations is frequent contact with the study subject, especially in-person contact, as well as access to the participant's network [56,57]. Frequent contacts presumably serve to increase the subject's sense of connection with and commitment to the study team, while access to the individuals central to the subject's social network helps to ensure continuing contact and retention in the study. Because name changes present unique difficulties for the follow-up of women subjects, social networks for women can help surmount these problems.

C. Trust and Perceived Benefit

Involvement with community groups is an important concern for some populations of women who favor a collaborative model of research. Some examples of those who tend to prefer a collaborative model are breast cancer advocates and African-American women [58]. Preference for this model over the standard model in some cases reflects a distrust of medical research. This distrust could be based on personal experiences with the medical system or on historical instances involving abuse of minorities (the "Tuskegee effect") [9], prisoners, or the mentally retarded. Working out agreements as to the nature and extent of community or target group involvement in research has proved to be a serious challenge.

The concept of having multiple partners sponsor the proposed research—the research institution together with local community, service, or business groups—is intended to build trust and support. This is a relatively new idea and as yet there is not much experience with it. Elements that appear to be essential to its success are: initiating the partnership early in the planning stage of the study; drawing on a broad base of organizations within the community; and being prepared to make a long-term commitment to the community partners. The report on recruitment and retention of women in clinical studies by the Office of Research on Women's Health gives the example of Heart, Body and Soul, a project that provides health education and prevention services in east Baltimore, Maryland [59,60]. Local acceptance of the program grew out of a partnership between Johns Hopkins University, as the local medical institution, and a group of religious leaders who call themselves Clergy United for Renewal in east Baltimore.

D. Overcoming Barriers

Providing childcare and transportation to women is now recognized as an important means of overcoming practical barriers to participation in research. In addition, offering flexible hours for telephone interviews and home or clinic visits has been recommended as a way of addressing practical obstacles and may be particularly important for working women with families. For women living in rural areas, time and distance may make clinic visits unappealing despite extended hours and free transportation; for these women, at-home visits may be essential.

In addition to practical matters related to domestic and work responsibilities, cultural influences on women may impinge. In some Hispanic groups, for instance, the husband is the sole decision maker and agreement to participate depends on his approval. The San Antonio Women's Health Initiative has developed a model to address this issue [61]. As discussed earlier, support from family members appears to be key to the successful recruitment of women into research studies. One study that invited African-American women to participate in a physical activity intervention found that the absence of other African-Americans in the exercise environment was discomforting to them [62]. Overcoming cultural barriers will require adopting some of the measures discussed earlier, including ethnographic research in the target community; the development of study materials that are appropriate in terms of culture and language; inclusion of target group members on the study team, especially as lead investigators; and sensitivity training for staff. Forging partnerships with relevant community groups should also help in overcoming cultural barriers.

Confidentiality concerns related to discrimination or legal problems may arise among HIV-infected women, lesbian women, or drug-abusing pregnant women and are another barrier to participation. One approach to this problem is to have women enroll in the study using a pseudonym to protect their identity. Another approach would involve giving, in addition to the usual confidentiality assurance, a specific assurance that data will be kept confidential from whomever the subjects are particularly concerned about. For instance, to facilitate studies of worker cohorts, the National Institute of Occupational Safety and Health (NIOSH) has issued a special form assuring subjects that none of the information they provide will be turned over to their employer.

E. Incentives

There are probably more empiric data on the efficacy of various incentives than on any other topic related to study participation. Types of incentives that have been offered include financial compensation (cash or gift certificates); results of clinical tests that were part of the study protocol (e.g., bone density, cholesterol level, hormone concentrations); lectures or informational materials on a health-related topic (usually the topic under study); or gifts (e.g., a cookbook, movie tickets, or a calendar on women's health). While all of these represent an appeal to self-interest, the specific incentive offered should be tailored to the target population.

Test results and informational materials appeal to a participant's perception of the project as salient to her concerns. Lack of salience appears to be a disincentive to participate. For instance, the absence of a family history of cancer has been associated with refusal to participate in Pap smear screening programs [39]. Likewise, participation in a breast cancer risk factor counseling trial for first-degree relatives was moderated by time since the relative's diagnosis [13]. Women were more likely to participate if they were approached within two months of their relative's diagnosis of breast cancer, particularly women with more than a high school education [13].

Monetary incentives are a form of payment for the time and effort involved in participation. Payment, and particularly prepayment, have been found to be effective in securing participation of low-wage employees [63]. Amounts offered as compensation vary according to the cost of living in the study area and the demands imposed by the study protocol. A blood draw may be compensated by $5–10 in one area but by as much as $25 in another. Regimens that demand frequent trips to the clinic for procedures or that require the participant to collect bodily fluids for days or weeks will usually be compensated with amounts of $100 or more. There is concern, especially among institutional review boards, that what is offered to potential participants be fair compensation for their time and effort but not so excessive as to appear coercive.

Participation in health studies or health programs has been declining over time. To boost response rates and, hence, the validity of study results, recruitment strategies must take account of women's culture and community and should be piloted carefully prior to the start of the study.

V. Women as Survey Respondents

The reliability and validity of responses to surveys has only recently been considered by gender. Theoretically, we would expect the quality of women's responses to be superior to that of men because of women's primary role of caretaker and medical gatekeeper for their families. Because women often act as representatives for their children with physicians, accurate recall of symptoms and events is important. Alternatively, responses on socially undesirable behaviors are likely to be problematic for women. In this section we will address the following issues on data quality: information about outcomes; normal risk factors, such as exercise and diet; sensitive information such as alcohol consumption and sexual behaviors; medicine use, particularly HRT; and the use of proxy responses among older women.

First, one of the most ambitious undertakings to foster methodologic knowledge about the implications of various measures and design decisions for studies of outcomes is the Inter-PORT Workgroup on Outcomes Assessment [64]. The best questions to ask, the way data collection modalities and proxies effect study results, and the comparative advantage of different study designs on outcomes are being determined in this working group funded by the Agency for Health Care Policy and Research. Women are included in most of the experiments, with topics ranging from acute myocardial infarction to hip fractures. The results from this group will hopefully improve the way we design women's health surveys. Although the results from the Outcomes Assessment Workgroup are not published yet, there are clues noted in this paper as to what they will find for women.

Second, when reporting life style habits, such as physical exercise, Warneke et al. [65] found that women are more likely than men to exhibit the social desirability trait based on the Marlow Crowne Social Desirability Scale. African-American and Mexican-American respondents were more likely to demonstrate the social desirability trait than non-Hispanic whites. On the other hand, women were less likely to be acquiescent (tend to agree) than men. No sex differences were found in extreme response styles.

The reporting of dietary intake has been problematic in population-based studies since the 1980s. It is not surprising then that several studies document the poor reporting of dietary components by women. The long-term reliability of a food frequency questionnaire in the New York University Women's Health Study ranged from 0.31 to .64 for nutrients or food groups in volunteer responders, and it was even worse in nonresponders [66]. Correlation of responses from two binge eating questionnaires in obese women were only modest (kappa of 0.45), even though this behavior along with other eating disorders is a high priority topic for women's health [67]. Reliability estimates are much higher for food frequency reporting on specific nutrients, such as calcium, when data are classified into quartiles [68].

Racial/ethnic and socioeconomic differences have also been reported in food frequency questionnaires and four-day food records [69]. Among women in the Women's Health Trial Feasibility Study in minority populations, the correlation between the food frequency questionnaires and the food records was lower for blacks than for either whites or Hispanics, who did not differ. Of interest, the correlations were lower for women with fewer years of education as compared with women at high levels of education (0.19 for less than 12 years to 0.42 for women with ≥ 16 years of education) [69]. Telephone vs face-to-face interviews among five major ethnic groups of women in Hawaii yielded relatively high interclass correlation coefficients among women, ranging from 0.61 for vitamin C to 0.74 for saturated fat [70].

Third, differences have been observed on the reporting of culturally sensitive issues, such as alcohol consumption and sexual behaviors. Women, particularly Mexican-American women, reported that talking about alcohol consumption with an interviewer from their own culture would be uncomfortable. Sex differences in response styles have been found in adult women in two Swedish studies [71,72]. Using two different methods for the measurement of alcohol consumption, women were more likely to respond differently to the two approaches than men [71]. The reliability of responses on two self-reported scoring tests on alcoholism was also poorer in women than men, particularly the CAGE instrument [72]. The CAGE questionnaire, which is commonly employed in screening for problem drinking and alcohol dependence, has been found to be predictive for men but not for women college students [73]. There is uniform agreement among investigators that further development of female-oriented alcohol questions and sex-related modes of answering alcohol questions is needed.

One of the most fascinating examples of recall bias related to sensitive sexual/reproductive health reports by women was illustrated in the debate on the relationship between abortion and breast cancer. In over 20 studies that have been published on this issue, small or nonsignificant relative risks (usually under 1.5) have been reported between induced abortion and breast cancer [74]. However, contradictory conclusions have been drawn from the same literature, some authors saying the evidence is inadequate to infer a relationship [75], others calling it a significant independent risk factor and a public health tragedy [76]. Powerful evidence that recall bias may explain the findings was presented by Rookus and van Leeuwen [77] from the Netherlands, where relative risks were low in liberal areas of the country and high in conservative areas of the county. This suggested that women living in liberal areas of the country (i.e., areas where abortion was socially accepted) were more likely to give honest reports on abortion history. Likewise, women in conservative areas (i.e., areas where abortion was not socially accepted) were more likely not to reveal their true abortion history. The authors provided some evidence of this phenomenon in regard to another reproductive matter: oral contraceptives. The methodologic solution for overcoming this bias has to be verification of the abortion procedure or prescriptions with objective medical records.

Fourth, the reporting of medicines has also been a source of concern for studies of women. In follow-up interviews of over 2500 women aged 45–55, a large proportion of women had used nonprescription drugs (92%) and nonprescription drugs (47%) [78]. After three years of follow-up, 43% of the women had switched categories of use, defined as nonuse, nonprescription use, prescription use, and mixed use [78].

One of the most widely researched drug used by women has been postmenopausal estrogens or HRT. In a Postmenopausal Estrogen-Progestin Intervention Trial (PEPI) clinic, the accuracy of a single self-report questionnaire about postmenopausal estrogen use was determined over a three-month period, comparing it to a history obtained by the Hormone Use Interview [79]. Overall, there was excellent (95%) agreement between self-reported estrogen use on the two instruments. Using the interview as the standard, the self-report misclassified 2.3% of the women as false positives and 6.3% as false negatives. Four factors were associated with discordant reporting: route of administration, recency of hormone use, duration of hormone use, and race [79].

In preventive medicine, numerous examples exist where questionnaire design improvements are sorely needed. Problems in self-reported frequency of Pap smears and mammograms have been reported. Among women reporting a Pap smear within the last three years, only 61.2% were verified with pathology reports [80]. The reliability of mammography utilization in low-income women was good for "ever had a mammogram" (0.82) but lower for information about timing, mammogram in last year (0.65), date of last mammogram (0.54), and number of mammograms (0.72) [81]. The definition of perimenopause has also challenged researchers. One solution was proposed in a prospective study of women going through menopause in Massachusetts that defined perimenopause as 3–11 months of amenorrhea and increased menstrual irregularity for those without amenorrhea [82]. Special considerations need to be taken in studying menopause in racial/ethnic groups, as illustrated by a review of the menopausal experience of African-American women [83]. Not surprisingly, women seem to have excellent recall of past histories of hysterectomy and tubal sterilization as compared to medical reports (96% agreement) [84].

Fifth, contrary to expectation, the Baltimore Study of women 65 years and older found excellent agreement between older women and their proxies [21]. The agreement was substantial (kappa over 0.6) for five of nine chronic conditions, six of seven physical tasks of daily living, and seven of seven instrumental tasks of daily living [21]. Slight to moderate agreement was observed for eleven of eleven health symptoms (kappa between 0.24 and 0.59). Proxies were best at reporting observable areas of health and functioning and less accurate in reporting health symptoms for women that are not observable. The same investigative team found good agreement for summary activities of daily living (instrumental functioning and physical functioning) [85]. More bias was observed for proxies in rating cognitive functioning and CES-D depression symptoms [86]. Similar findings on the poor agreement in depression indices were published in earlier work by this team [87]. Thus, more subjective questions included in affective functional scales are harder to observe than objective items included in cognitive scales. In one study, about two-thirds of proxies gave accurate information on the date of death, and 87% and 90% accurately remembered the year and month, respectively [88].

Kelsey *et al.* [89] have also suggested that investigators should phrase questions for proxy respondents so as to refer to explicitly defined and distinct behaviors and activities, ask about points of fact rather than opinions, and limit the number of response options. Kelsey *et al.* also recommend limiting the length of questionnaires for older persons to 45 minutes, simplifying dietary questionnaires, using closed ended questions with predetermined response categories, and using words that are familiar to older women, such as "change of life" instead of menopause [89]. Many older women will also digress during interviews, which has to be handled tactfully by interviewers.

VI. A Core Questionnaire for Women's Health Research

In 1995, the National Action Plan on Breast Cancer (NAPBC) Etiology Working Group initiated an ambitious project to develop, validate, and disseminate two core questionnaires (self-administered and interviewer-administered) and six detailed modular questionnaires for use in breast cancer case-control studies [90]. The six modules contain detailed questions about potential risk factors in the following topic areas: sociodemographic and cultural factors, medical and reproductive history, personal behaviors and lifestyle, food and nutrition, environmental exposures, and occupational history. Scores of scientists initially screened hundreds of questions based on a review of the existing literature and national survey instruments. The best items were chosen using the best available psychometric, reliability, and validity data. After this process was completed, the Institute of Survey Research at Temple University convened 21 focus groups representing a variety of age, racial, ethnic, religious, and socioeconomic groups. Less than half of the groups were breast cancer patients or survivors because it was equally important to test the instruments in women who did not have breast cancer. The age of the focus group participants ranged from the mid-twenties to 82 years old. Focus groups were convened around the country, including Baltimore, San Francisco, Washington, DC, and the Midwest.

The focus groups helped the scientists revise the questionnaires by establishing construct validity and by helping to reduce systematic error and measurement error in the instruments. The compilation of items is comprehensive and could be used for women's health studies in general, not just breast cancer studies. These items are currently available for review at the NAPBC website (www.NAPBC.org). This is the most comprehensive set of questions compiled for epidemiologic studies in women's health, particularly breast cancer. The lessons learned from the focus groups are quite instructive.

Women's recall of detailed information was problematic [90]. For example, women indicated that they would not be able to recall detailed information about prescribed medication use, such as dosage, dates, or the names of prescribing physicians. However, they did feel they could report whether or not they had "ever" taken a medication and possibly for what type of condition. In terms of calendars, women felt that they were more likely to remember their age at the time of an event rather than the date of an event. Likewise, photographs of pills in addition to lists of names of pills were requested by women to refresh their memories of usage. This approach is commonly used in the NHANES [44] interviews. Most importantly, being told before their interview that they would be asked medication questions would have allowed them to think about their medication history. This request also applied to prenatal, family, personal, medical, and environmental exposure histories.

Women in the focus groups also suggested several additions to the Core Questionnaire. First, they wanted to see questions on alternative medical treatments. An entire section on these treatments was added. Second, when answering questions on exposure to radiation, women wanted to add more screening procedures, like back and spinal X-rays and flouroscopy for shoe fittings, as well as treatments with infrared/heat lamps for tonsillitis. Third, women felt that autoimmune diseases should be assessed. Fourth, the women in the focus groups indicated that entire family histories could best be captured by a family tree. There was a recognition that deaths of siblings and adoptions may be kept secret by women. Finally, lists of brands of pest control products used for personal use were requested [90].

Differences in reporting information by age cohorts were also noted. Older women were thought not to know about the details of pelvic procedures and surgeries. For example, they did not believe most women would know about whether oophorectomies were total or partial [90]. Family history of cancer, or for that matter many other causes of death of family members, may not be known by older women.

Honesty in answering sensitive questions was also discussed [90]. As noted earlier, honesty about abortions (induced or spontaneous) was an issue, depending on the social or political environment. Women in the focus groups laughed about their honesty in reporting weight and height. They suggested that diagrams of body shapes in questionnaires should include more shapes, with shorter legs and larger breasts. The wealth of information collected from these focus groups is worth reviewing before undertaking research on women.

VII. Gender-Specific Issues in Research

Apart from the issues of recruiting women into research studies and developing valid data collection instruments, there are both biologic and social considerations that need to be addressed in research on women's health.

A. Biological Factors

Many of the observed sex differences in morbidity and mortality are determined by biologic factors, some examples of which are given below.

1. Steroid Hormones and Immunologic Function

In general, women have enhanced immunoreactivity compared to males [91], a phenomenon that is largely estrogen-driven [92]. As a result of the estrogen influence, both pregnancy and the menstrual cycle have an impact on autoimmune conditions such as lupus erythematosus, rheumatoid arthritis, and multiple sclerosis [93]. Steroid hormones can also affect acquisition and severity of certain bacterial, parasitic, and viral infections [94]. Malaria increases in severity during pregnancy and chlamydia tends to present during the early part of the menstrual cycle.

Menstruation has been described as a cyclic autoimmune process regulated by estrogen, which dominates the preovulatory or follicular part of the cycle, and progesterone, which rises in the postovulatory or luteal stage of the cycle [95]. For many research questions, it may be important to know a woman's hormonal milieu. One may need to know the stage of a women's cycle at the time of a clinical visit or intervention because of the potential effect that hormones may have on the variables being measured. Likewise, one may need to schedule a visit to avoid the expected time of menses because it may interfere with a clinical procedure. There are any number of reasons for gathering menstrual cycle data. Day or stage of the menstrual cycle can be used as a surrogate measure of a woman's hormone profile. Such data are far easier to obtain than the samples of serum, urine, or saliva needed to make actual measurements. Cycle day or stage can give a fair indication of the estrogen to progesterone ratio, if not the absolute hormone levels.

2. Measurement of Steroid Hormones

Measuring hormone concentrations requires a valid and feasible sample collection scheme standardized across subjects. For premenopausal women, collection must be standardized to cycle day and time, usually requiring serial samples. Investigators have to identify a research laboratory that is equipped to perform the required assays and can provide data on the sensitivity and precision of each test, along with reference ranges for women in the relevant age group. In choosing which laboratory will perform the hormone measurements, data on interlaboratory comparisons of the sex steroid assays are valuable.

3. Life Span

The importance of cyclic hormone levels has led to the suggestion that a life span approach be taken in studying women. Puberty, the reproductive years, perimenopause, and postmenopause are all points in a woman's life span where susceptibility to exposure may be different due to changes in the hormonal milieu at critical periods. In the life span approach, data must be collected on a subject's total exposures to date as well as on exposures at specific points in the life course.

Women have more body fat than men and may store more fat-soluble toxic material even when exposed to the same amounts as men. Subsequent changes in body weight resulting from pregnancy or from weight loss at older ages or changes in bone density at menopause can trigger mobilization of stored toxicants and lead to increases in circulating levels. Plasma dioxin concentrations have been found to be elevated in women as compared to men (exposed at the same level) even after adjustment for a number of factors such as body mass index [96]. The authors suggest that body fat, hormones, and metabolic differences are possible explanations.

4. Susceptibility Factors

One of the most consistent observations of gender differences in susceptibility is the two- to threefold higher relative risks for lung cancer related to smoking among women as compared to men with equivalent exposures [97]. In one case-control study of lung cancer, which compared a history of 40 pack-years relative to lifelong nonsmoking, the odds ratio for women was 27.9 while the odds ratio for men was 9.6 [98].

The gender difference in odds ratios could in part reflect enhanced susceptibility among females to the effects of toxic exposures. The biologic factor that might account for this has not as yet been identified. Hormones have been proposed to play a role, by deregulating growth and differentiation via receptor binding [99]. Because metabolizing genes are not sex-linked, there should be no gender difference in frequency, although there could be gender differences in the basal expression of genetic polymorphisms. Hormones could interact to affect inducibility of detoxifying enzymes. Some argue that neither expression nor induction of metabolizing enzymes is likely to account for observed gender differences in adverse effects of exposure [100]; instead, an as yet unidentified difference in sensitivity to toxic compounds is the crux of the matter. Clearly, it is extremely important to find the reason for the apparent increased sensitivity in women. Until this variation in susceptibility is understood, it will continue to be important to compare results in women with those obtained in men, where possible, so that we can better appreciate the true extent of these differences.

Certain genetic traits among women, like the BRCA gene mutations, greatly increase susceptibility to, in this case, breast and ovarian cancer. A great deal of attention is currently being devoted to genetic testing, mainly of high-risk populations. One issue that is important for women is the potential impact on health insurance. Cassel points out that testing results can be used to increase insurance rates or exclude people from coverage altogether, and that women as a group tend to be less well covered than men or totally uninsured [101].

B. Nonbiological Factors

In addition to gender-based biology, the sociocultural environment and economic status of women are important considerations in women's health. We briefly discuss examples of this topic which is developed at greater lengths in Section 6 of this book.

Gender differences in death rates sometimes arise not from any true difference in disease severity but as a result of culturally determined delays in diagnosis or failures to provide women with appropriate health services. A recent U.S. study based on national practice patterns found that women with myocardial infarction (MI) received less treatment than men with MI, after adjustment for confounding factors [102].

Access to health care problems should decline as women are increasingly included in clinical research studies, thus forming an empiric base from which recommendations can be made about diagnostic and treatment approaches appropriate for female patients. In the interval, the influence of selective mortality differences needs to be kept in mind when interpreting gender differences in disease.

Segregation of employment by gender is another phenomenon that can lead to nonbiological gender differences. Women engage in different work activities than men and so have different exposures and health risks. In the U.S., the localization of female workers is in two areas of employment: clerical work and services. Clerical work involves exposures to stress, video display terminals (VDT), repetitive motion, and indoor air pollution, among other potential hazards. Exposures in the services sector, which includes women working as teachers, nurses, and in restaurants, include stress, hepatitis, acquired immunodeficiency syndrome (AIDS), tuberculosis, and other infectious diseases. With the exception of VDT use, and perhaps stress, many exposures that are common in female-dominated industries have not received much attention.

Family responsibilities are another sociocultural phenomenon that influence women's health. For most women workers, the end of the work day only marks the beginning of the so-called "second shift." One expert in the area of work stress and coronary heart disease has argued that women's domestic responsibilities could prevent nighttime recovery from job-related increases in blood pressure, leading to potentially serious chronic elevations [103]. Workplace injury is another health outcome that may be affected by a woman worker's domestic responsibilities. One study of female aerospace workers found that women with young children were at increased risk of on-the-job injury compared to women without young children. This may reflect fatigue or a preoccupation with family responsibilities [104].

Studies of women should incorporate data on social factors (household and childcare responsibilities, amount of leisure time, hours of sleep) as well as biological factors (body size, hormones, susceptibility genes). In addition, some variables that may affect a woman's health and well-being operate at the community level, beyond the household and the workplace (*e.g.,* the threat of violence within their neighborhoods or their workplaces, manifestations of poverty). Collecting information at the level of the community should also be considered in research on women. Statistical methods now exist for integrating such ecologic data with data collected at the individual level.

VIII. Summary

During the 1990s, a wealth of evidence suggests that the methods used to recruit, survey, and determine biological measures in women are unique. This chapter has presented successful strategies for recruiting women, with particular focus on minority, lower socioeconomic status, older, and hard-to-reach women. The success of the word of mouth or snowball sample for recruiting women suggests that this older sampling technique deserves some revisiting from methodologists. We were surprised to find that family support for participation in research is very important to women. It was also enlightening to see that community approaches to recruitment have been far more successful than recruitment through physicians or medical institutions.

Based on our review, there is little question that in-person interviews are likely to yield more valid responses among women. The societal effect on womens' responses to sensitive questions related to reproductive health, substance abuse, or body weight opens the door for a whole new era of research on questionnaire approaches and/or external validation.

The use of a life-span approach in study design for both biologic and nonbiologic conditions brings a brand new dimension to research. The key to women's increased sensitivity to toxic exposures may lie in that delicate balance of hormonal fluctuations that take place over a woman's lifetime. The challenges we face when taking women's hormonal profiles into account are complex. In sum, when conducting women's health research, it is safe to say that all the simple studies have been done and that the best scientific minds are needed to solve the complex issues before us in the next century of research on women's health.

References

1. McCarthy, C. R. (1994). Historical background of clinical trials involving women and minorities. *Acad. Med.* **69,** 695–698.
2. Bush, J. K. (1994). The industry perspective on the inclusion of women in clinical trials. *Acad. Med.* **69,** 708–715.
3. Nadel, M. V. (1990). "National Institutes of Health: Problems in Implementing Policy on Women in Study Populations." U.S. General Accounting Office, Washington, DC.
4. LaRosa, J. H., Seto, B., Caban, C. E., and Hayunga, E. G. (1995). Inclusion of women and minorities in clinical research. *Appl. Clin. Trials* **4,** 31–38.
5. National Institutes of Health (1994). NIH guidelines on the inclusion of women and minorities as subjects in clinical research. *Fed. Regis.* **59,** 14508–14513.

6. U.S. Department of Health and Human Services (1995). "Recruitment and Retention of Women in Clinical Studies," NIH Publ. No. 95-3756, pp. 1–104. USDHHS, Washington, DC.

7. Hayunga, E. G., Costello, M. D., and Pinn, V. W. (1997). Demographics of study populations. *Appl. Clin. Trials* **6**, 41–45.

8. Mouton, C. P., Harris, S., Rovi, S., Solozano, P., and Johnson, M. S. (1997). Barriers to black women's participation in cancer clinical trials. *J. Natl. Med. Assoc.* **11**, 721–727.

9. Brawley, O. W. (1998). The study of untreated syphilis in the Negro male. *Int. J. Radiat. Oncol. Biol. Phys.* **40**, 5–8.

10. Whelton, P. K., Lee, J. Y., Kusek, J. W., Charleston, J., DeBarge, J., Douglas, M., Faulkner, M., Green, P. G., Jones, C. A., Kiefer, S., Kirk, K. A., Levell, B., Norris, K., Powers, S. N., Retta, T. M., Smith, D. E., and Ward, H. (1996). Recruitment experience in the African American study of kidney disease and hypertension (AASK), pilot study. *Controlled Clin. Trials* **17**, 17S–33S.

11. Blumenthal, D. S., Sung, J., Coates, R., Williams, J., and Liff, J. (1995). Recruitment and retention of subjects for a longitudinal cancer prevention study in an inner city black community. *Health Serv. Res.* **30**, 197–205.

12. Caban, C. E. (1995). Hispanic research: Implications of the National Institutes of Health Guidelines on inclusion of women and minorities in clinical research. *NCI Monogr.* **18**, 165–169.

13. Rimer, B. K., Schildkraut, J. M., Lerman, C., Lin, T. H., and Audrain, S. (1996). Participation in a women's breast cancer risk counseling trial. *Cancer (Philadelphia)* **77**, 2348–2355.

14. Lerman, C., Rimer, B. K., Daly, M., Lustbader, E., Sands, C., Balsham, A., and Marney, A. (1994). Recruiting high risk women into a breast cancer health promotion trial. *Cancer Epidemiol., Biomarkers Prev.* **3**, 271–276.

15. Orden, S. R., Dyer, A. R., and Liu, K. (1990). Recruiting young adults in an urban setting: The Chicago CARDIA Experience. *Am. J. Prev. Med.* **6**, 176–182.

16. Zabora, J. R., Morrison, C., Olsen, S. J., and Ashley, B. (1997). Recruitment of underserved women for breast cancer detection programs. *Cancer Pract.* **5**, 297–303.

17. Thornquist, M. D., Patrick, D. L., and Omenn, G. S. (1991). Participation and adherence among older men and women recruited to the Beta-Carotene and Retinol Efficiency Trial (CARET). *Gerontologist* **31**, 593–597.

18. LaRosa, J. C., Applegate, W., Crouse, J. R., 3rd, Hunninghake, D. B., Grimm, R., Knopp, R., Eckfeldt, J. H., Davis, C. E., and Gordon, D. J. (1994). Cholesterol lowering in the elderly: Results of the Cholesterol Reduction in Seniors Program (CRISP) Pilot Study. *Arch. Intern. Med.* **154**, 529–539.

19. Whelton, P. K., Babnson, J., Appel, L. J., Charleston, J., Cosgrove, N., Espeland, M. A., Folmar, S., Hoagland, D., Krieger, S., Lacy, C., Lichtermann, L., Oates-Williams, F., Tayback, M., and Wilson, A. C. (1997). Recruitment in the trial of nonpharmacologic intervention in the elderly (TONE). *J. Am. Geriatr. Soc.* **45**, 185–193.

20. Schron, E. B., Wassertheil-Smoller, S., and Pressel, S. (1997). Clinical trial participant satisfaction: Survey of SHEP enrollees. SHEP Co-operative research group: Systolic hypertension in the elderly program. *J. Am. Geriatr. Soc.* **45**, 934–938.

21. Magaziner, J., Bassett, S. S., Hebel, J. R., and Gruha-Baldini, A. (1996). Use of proxies to measure health and functional status in epidemiologic studies of community-dwelling women aged 65 years and older. *Am. J. Epidemiol.* **143**, 283–292.

22. Herzog, A. R., and Dielman, L. (1985). Age differences in response accuracy for factual survey questions. *J. Gerontol.* **40**, 350–357.

23. Dressel, P., Minkler, M., and Yen, I. (1997). Gender race, class, and aging: Advances and opportunities. *Int. J. Health Serv.* **27**, 579–600.

24. Palinkas, L. A., Atkins, C. J., Noel, P., and Miller, C. (1996). Recruitment and retention of adolescent women in drug treatment research. *NIDA Res. Monogr.* **166**, 87–109.

25. Smith, D. K., Warren, D. L., Vlahov, D., Shuman, P., Stein, M. D., Greenberg, B. L., and Holmberg, S. D. (1997). Design and baseline participant characteristics of the human Immunodeficiency Virus Epidemiology Research (HER) Study: A Perspective cohort study of human immunodeficiency virus infection in U.S. women. *Am. J. Epidemiol.* **146**, 459–469.

26. Solarz, A. L. (1999). "Lesbian Health: Current Assessment and Direction for the Future." Institute of Medicine, Washington, DC.

27. Newcomb, P. A., Love, R. R., Phillips, J. L., and Buckmaster, B. J. (1990). Using a population based cancer registry for recruitment in a pilot cancer control study. *Prev. Med.* **19**, 61–65.

28. Daly, M., Seay, J., Balshem, A., Lerman, C., and Engstrom, P. (1992). Feasibility of telephone survey to recruit health maintenance organization members into a tamoxifen chemoprevention trial. *Cancer Epidemiol., Biomarkers Prev.* **1**, 413–416.

29. Yeaman-Kinney, A., Vernon, S. W., Frankowski, R. F., Weber, D. M., Bitsura, J. M., and Vogel, V. G. (1995). Factors related to enrollment in the breast cancer prevention trial at a comprehensive cancer center during the first year of recruitment. *Cancer (Philadelphia)* **76**, 46–56.

30. Muncie, H. L., Magaziner, J., Hebel, J. R., and Warren, J. W. (1997). Proxies decisions about clinical research participation for their charges. *J. Am. Geriatr. Soc.* **45**, 929–933.

31. Stebbings, J. H., Lucas H. F., and Stehney, A. F. (1984). Mortality from cancers of major sites in female radium dial workers. *Am. J. Ind. Med.* **5**, 435–459.

32. Knill-Jones, R. P., Moir, D. B., and Rodriguez, L. V. (1972). Anesthetic practice and pregnancy: Controlled survey of women anesthetists in the United Kingdom. *Lancet* **1**, 1326–1328.

33. Matanowski, G., Seltser, R., Sartwell, P., Diamond, E. H., and Elliott, E. A. (1975). The current mortality rates of radiologists and other physician specialist; specific causes of death. *Am. J. Epidemiol.* **101**, 199–210.

34. Willett, W. C., Hennekens, C. H., Bain, C., Rosner, B., and Speizer, F. E. (1981). Cigarette smoking and non-fatal myocardial infarction in women. *Am. J. Epidemiol.* **113**, 575–582.

35. Colditz, G. A., Manson, J. E., and Hankinson, S. E. (1927). The Nurses' Health Study: 20-year contribution to the understanding of health among women. *J. Women's Health* **6**, 49–62.

36. Guralnik, J. M., Fried, L. P., Simonsick, E. M., Kasper, J. D., and Lafferty, M. E., eds. (1995). "The Women's Health and Aging Study: Health and Social Characteristics of Older Women with Disabilities," NIH Publ. No. 95-4009. National Institute of Aging, Bethesda, MD.

37. Olson, S. H., Kelsey, J. L., Pearson, T. A., and Levin, B. (1992). Evaluation of random digit dialing as a method of control selection in case-control studies. *Am. J. Epidemiol.* **135**, 210–222.

38. Weinrich, S., Coker, A. L., Weinrich, M., Eleeazer, G. P., and Green, F. L. (1995). Predictors of pap smear screening in socioeconomically disadvantaged elderly women. *J. Am. Geriatr. Soc.* **43**, 267–270.

39. McGrady, G. A., Sung, J. F., Rowley, D. L., and Hogue, C. J. (1992). Preterm delivery and low birth weight among first-born infants of black and white college students. *Am. J. Epidemiol.* **136**, 266–276.

40. Cartmel, B., and Moon, T. E. (1992). Study of strategies for the recruitment of elders including the use of community volunteers. *J. Am. Geriatr. Soc.* **40**, 173–177.

41. Livingston, P. M., Guest, C. S., Bateman, A., Woodcock, N., and Taylor, H. R. (1994). Cost-effectiveness of recruitment methods

in a population-based epidemiological study: the Melbourne Visual Impairment Project. *Aust. J. Public Health* **18**, 314–318.

42. Geringer, W. M., and Hinton, M. (1993). Three models to promote syphilis screening and treatment in a high risk population. *J. Commun. Health* **18**, 137–151.

43. Birkhead, G. S., LeBaron, C. W., Parsons, P., Grabau, J. C., Barr-Gale, L., Fuhrman, J., Brooks, S., Rosenthal, J., Hadler, S. C., and Morse, D. L. (1995). The immunications of children enrolled in the Special Supplemental Food Program for Women, Infants, and Children (WIC). The Impact of different strategies. *JAMA, J. Am. Med. Assoc.* **274**, 312–316.

44. Household Adult Data File Documentation (1994). "The Third National Health and Nutrition Examination Survey (1988–1994)," Vital Health Stat., Ser. 1, No. 32. U.S. Dept. of Health and Human Services, Hyattsville, MD.

45. Kish, L. (1965). "Survey Sampling," p. 408. John Wiley, New York.

46. Rothman, K., and Walker, A. M. (1997). Epidemiology and the Internet. *Epidemiology* **8**, 124–125.

47. Kaufman, B. D., and Rodriguez-Trias, H. (1995). Participant and community issues in the recruitment and retention of women in clinical studies. *In* "Recruitment and Retention of Women in Clinical Studies" (Office of Research on Women's Health), NIH Publ. No. 95-3756, pp. 17–23. USOHHS, Washington, DC.

48. White, G. E., McAvoy, B. R., and Gleisner, S. (1993). Increasing the uptake of cervical smears: Strategies implemented among general practitioners in Auckland. *N.Z. Med. J.* **106**, 357–360.

49. Sherraden, M. S., and Barrera, R. E. (1995). Qualitative research with an understudied population: In-depth interviews with women of Mexican descent. *Hispanic J. Behav. Sci.* **17**, 452–470.

50. Rosenberg, L., Adams-Campbell, L., and Palmer, J. R. (1995). The Black Women's Health Study: A follow-up study for causes and preventions of illness. *J. Am. Med. Women's Assoc.* **50**, 56–58.

51. Masse, L. C., Ainsworth, B. E., Tortolero, S., Levin, S., Fulton, J. E., Henderson, K. A., and Mayo, K. (1998). Measuring physical activity in midlife, older, and minority women: Issues from an expert panel. *J. Women's Health* **7**, 57–67.

52. Suarez L., Nichols, D., Roche, R.-A., and Simpson, D. M. (1997). Knowledge, behavior, and fears concerning breast and cervical cancer among older low-income Mexican-American women. *Am. J. Prev. Med.* **13**, 137–142.

53. True, R.-H., and Guillermo, T. (1996). Asian/Pacific Islander American women. *In* "Race, Gender, and Health" (M. Bayne-Smith, (ed.), pp. 94–120. Sage, Publ. Thousand Oaks, CA.

54. Malgady, R. G. (1984). Mental health evaluation and psychotherapeutic treatment of hispanic children. *In* Society for the Study of Social Problems (SSSP). Sociological Abstracts, Inc.

55. Willis, G. B. (1997). The use of psychological laboratory to study sensitive survey topics. *NIDA Res. Monogr.* **167**, 416–438.

56. Sullivan, C. M., Rumptz, M. H., Campbell, R., Eby, K. K., and Davidson, W. S., II (1996). Retaining participants in longitudinal community research: A comprehensive protocol. *J. Appl. Behav. Sci.* **32**, 262–276.

57. Conover, S., Berkman, A., Gheith, A., Jahiel, R., Stanley, D., Geller, P. A., and Susser, E. (1997). Methods of successful follow-up of elusive urban populations: An ethnographic approach with homeless men. *Bull. N. Y. Acad. Med.* [2] **74**, 90–108.

58. Hatch, J., Moss, N., Saran, A., Presley-Cantrell L., Mallory C., and Dressler, W. W. (1993). Community research: Partnership in black communities. *Am. J. Prev. Med.* **9**, 27–31.

59. Kuller, L. H. (1995). Current experiences in Women's Health Research (1995). *In* "Recruitment and Retention of Women in Clinical Studies" (Office of Research on Women's Health), NIH Publ. No. 95-3756, pp. 25–28. USDHHS, Washington, DC.

60. Kuller, L. H. (1995). Conclusions and recommendations. *In* "Recruitment and Retention of Women in Clinical Studies" (Office of Research on Women's Health), NIH Publ. No. 95-3756, pp. 29–31. USDHHS, Washington, DC.

61. Steinle, S. (1997). Women still face many hurdles to clinical trial participation. *J. Natl. Cancer Inst.* **89**, 545–546.

62. Carter-Nolan, P. L., Adams-Campbell, L. L., and Williams, J. (1996). Recruitment strategies for black women at risk for noninsulin-dependent diabetes mellitus into exercise protocols: a qualitative assessment. *J. Natl. Med. Assoc.* **88**, 558–562.

63. Schweitzer, M., and Asch, D. A. (1995). Timing payments to subjects of mail surveys: Cost-effectiveness and bias. *J. Clin. Epidemiol.* **48**, 1325–1329.

64. Fowler, F. J., Cleary, P. D., Magaziner, J., Patrick, D. L., and Benjamin, K. L. (1994). Methodological issues in measuring patient-reported outcomes: The agenda of the working group on outcome assessment. *Med. Care* **32**, 565–576.

65. Warnecke, R. B., Johnson, T. P., Chavez, N., Sudman, S., O'Rourke, D. P., Lacey, L., and Horm, J. (1997). Improving question wording in surveys of cultural diverse populations. *Ann. Epidemiol.* **7**, 334–342.

66. Riboli, E., Toniolo, P., Kaaks, R., Shore, R. E., Casagrande, C., and Pasternack, B. S. (1997). Reproductibility of a food frequency questionnaire used in the New York University Women's Health Study: Effect of self-selection by study subjects. *Eur. J. Clin. Nutr.* **51**, 437–442.

67. Gladis, M. M., Wadden, T. A., Foster, G. D., Vogt, R. A., and Wingate, B. I. (1998). A comparison of two approaches to the assessment of binge eating in obesity. *Int. J. Eat. Disord.* **23**, 17–26.

68. Wilson, P., and Horwarth, C. (1996). Validation of a short food frequency questionnaire for assessment of dietary calcium intake in women. *Eur. J. Clin. Nutr.* **50**, 220–228.

69. Kristal, A. R., Feng, Z., Coates, R. J., Oberman, A., and George, V. (1997). Association of race/ethnicity, education, and dietary intervention with the validity and reliability of a food frequency questionnaire: The Women's Health Trial Feasibility Study in Minority Populations. *Am. J. Epidemiol.* **146**, 856–869.

70. Lyu, L. C., Hankin, J. H., Liu, L. Q., Wilkens, L. R., Lee, J. H., Goodman, M. T., and Kolonel, L. N. (1998). Telephone vs. face-to-face interview for quantitative food frequency assessment. *Int. J. Eat. Disord.* **23**, 44–48.

71. Romelsjo, A., Leifman, H., and Nyshem, S. (1995). A comparative study of two methods for the measurement of alcohol consumption in the general population. *Int. J. Epidemiol.* **24**, 929–936.

72. Osterling, A., Beyland, M., Nilsson, L., and Kristenson, H. (1993). Sex differences in response style to two self-report screening tests on alcoholism. *Scand. J. Soc. Med.* **21**, 83–89.

73. O'Hara, T., and Tran, T. V. (1997). Predicting problem drinking in college students: Gender differences and the CAGE questionnaire. *Addict. Behav.* **22**, 13–21.

74. Weed, D. L., and Kramer, B. S. (1996). Induced abortion, bias, and breast cancer: Why epidemiology hasn't reached its limits. *J. Natl. Cancer. Inst.* **88**, 1698–1699.

75. Michels, K. B., and Willet, W. C. (1996). Does induced or spontaneous abortion effect the risk of breast cancer? *Epidemiology* **7**, 521–528.

76. Brind, J., Churchhill, V. M., Severs, W. B., and Summy-Long, J. (1996). A comprehensive review and meta-analysis for breast cancer. *J. Epidemiol. Commun. Health* **50**, 481–496.

77. Rookuss, M. A., and van Leeuen, F. E. (1996). Induced abortion and risk for breast cancer: Reporting (recall) bias in a Dutch case-control study. *J. Natl. Cancer Inst.* **88**, 1759–1764.

78. Hemminki, E., Ferdock, M. J., Rahkonen, O., and McKinley, S. S. (1989). Clustering and consisting of use of medicines among mid-aged women. *Med. Care* **27,** 859–868.

79. Greendale, G. A., James, M. K., Espeland, M. A., and Barrett-Connor, E. (1997). Can we measure postmenopausal estrogen/progestin use: The Postmenopausal Estrogen/Progesten Interventions Trial. The PEPI investigators. *Am. J. Epidemiol.* **146,** 763–770.

80. Bowman, J. A., Sanson-Fisher, R., and Redman, S. (1997). The accuracy of self-reported pap smear utilization. *Soc. Sci. Med.* **44,** 969–976.

81. Vacek, P. M., Mickey, R. M., and Warden, J. K. (1997). Reliability of self-reported breast screening information in a survey of lower income women. *Prev. Med.* **26,** 287–291.

82. Brambilla, D. J., McKinlay, S. M., and Johannes, C. B. (1994). Defining the perimenopause for application in epidemiologic investigations. *Am. J. Epidemiol.* **140,** 1091–1095.

83. Rousseau, M. E., and McCool, W. F. (1997). The menopausal experience of African American women: Overview and suggestions for research. *Health Care Women Int.* **18,** 233–250.

84. Green, A., Purdie, D., Green, L., Dick, M. L., Bain, C., and Siskind, V. (1997). Validity of self-reported hysterectomy and tubal sterilization. The Survey of Women's Health Study Group. *Aust. N.Z.J. Public Health* **21,** 337–340.

85. Magaziner, J., Zimmerman, S. I., Gruber-Baldini, A. L., Hebel, R., and Fox, K. M. (1997). Proxy reporting in five areas of functional status. *Am. J. Epidemiol.* **146,** 418–428.

86. Bassett, S. S., Magaziner, J., and Hebell, J. R. (1990). Reliability of proxy response in mental health indices for aged, community-dwelling women. *Psychol. Aging* **5,** 127–132.

87. Zimmerman, S. I., and Magaziner, J. (1994). Methodological issues in measuring the functional status of cognitive impaired nursing home residents: The use of proxies and performance-based measures. *Alzheimer Dis. Assoc. Disord.* **8,** S281–S290.

88. Fultz, N. H., and Herzog, A. R. (1995). Quality of survey informants report about death: Verification of dates through a record check. *Gerontologist* **35,** 553–555.

89. Kelsey, J. L., O'Brien, L. A., Grisso, J. A., and Hoffman, S. (1989). Reviews and commentary: Issues in carrying out epidemiologic research in the elderly. *Am. J. Epidemiol.* **130,** 857–866.

90. Werner, E. (1997). "Focus Group Research for the Breast Cancer Core Questionnaire Project: Final Report." PHS Office on Women's Health, National Action Plan on Breast Cancer, U.S. Dept. of Health and Human Services, Washington, DC.

91. Cannon, J. G., and St. Pierre, G. A. (1997). Gender differences in host defense mechanisms. *J. Psychiatr. Res.* **31,** 99–113.

92. Legato, M. J. (1997). Gender-specific physiology: How real is it? How important is it? *Int. J. Fertil.* **42,** 19–29.

93. Inman, R. D. (1978). Immunologic sex differences and the female preponderance in systemic lupus erythematosus. *Arthritis Rheum.* **21,** 849–852.

94. Styrt, B., and Sugarman, B. (1991). Estrogens and infection. *Rev. Infect. Dis.* **13,** 1139–1150.

95. Hertz, R. (1988). Hypothesis: Menstruation is a steroid-regulated, cyclic, autoimmune process. *Am. J. Obstet. Gynecol.* **155,** 374–375.

96. Landi, M. T., Consonni, D., Patterson, D. G., Jr., Needham, L. L., Lucier, G., Brambilla, P., Cazzaniga, M. A., Mocarelli, P., Pesatori, A. C., Bertazzi, P. A., and Caporaso, N. E. (1998). 2,3,7,8-Tetrachlorodibenzo-p-dioxin plasma levels in Seveso 20 years after the accident. *Environ. Health Perspect.* **106,** 273–277.

97. Zang, E. A., and Wynder, E. L. (1996). Differences in lung cancer risk between men and women: Examination of the evidence. *J. Natl Cancer Inst.* **88,** 183–192.

98. Risch, H. A., Howe, G. R., Jain, M., Burch, D. J., Holoway, E. J., and Miller, A. B. (1993). Are female smokers at higher risk for lung cancer than male smokers? A case-control analysis by histologic type. *Am J. Epidemiol.* **138,** 281–293.

99. Perera, F. P. (1997). Environment and cancer: Who are susceptible? *Science* **278,** 1068–1073.

100. Wood, N. F., Bingham, E., Boekelheide, K., Faustman, D., Safe, S. H., Wegman, D. H., Hinshaw, A. S., Setlow, V. P., Lawson, C. E., Depugh, L. A., and Tinker, J. (1998). *In* "Gender Differences in Susceptibility to Environmental Factors. A Priority Assessment." Appendix B: Gender differences in metabolism and susceptibility to environmental exposures, pp. 53–58. (V. P. Setlow, C. E. Lawson, and N. F. Woods, eds.). National Academy Press, Washington, DC.

101. Cassel, C. K. (1997). Policy implications of the human genome project for women. *Women's Health Issues* **7,** 225–229.

102. Chandra, N. C., Ziegelstein, R. C., Rogers, W. J., Tiefenbrunn, A. J., Gore, J. M., French, W. J., and Rubison, M. (1998). Observations of the treatment of women in the United States with myocardial infarction. *Arch. Intern. Med.* **158,** 981–988.

103. Frankenheuser, M. (1989). A biopsychosocial approach to work life issues. *Int. J. Health Serv.* **19,** 747–758.

104. Wohl, A., Morgenstern, H., and Kraus, J. (1995). Occupational injury in female aerospace workers. *Epidemiology* **6,** 110–114.

5

Gender Issues in Women's Health Care[1]

CAROLYN M. CLANCY

Center for Outcomes and Effectiveness Research
Agency for Health Care Policy and Research
Department of Health and Human Services
Rockville, Maryland

During the 1960s women began to express their discontent with the traditional hierarchical model of physician–patient interaction [1]. A growing research capacity to examine patterns of health service use on a large scale [2], the demands of women and legislators for an enhanced focus on women's health concerns in biomedical research, a professional evolution in health care that recognizes the central importance of patients' preferences in clinical decision making [3], increased recognition of the central importance of women's voices in family health care decisions [4], and the changing demographics of the medical profession have converged to establish a new focus on the use of health services associated with patient gender.

As a result, the federal research enterprise has established policies that require the inclusion of women in biomedical research, and professional organizations have identified specific curricular approaches to address women's health problems. Studies that examine multiple dimensions of women's experiences with health care systems and providers also have been published. In short, a dynamic focus on women's use of and experiences with health care now provides a framework for assessing professional education, public policy, and measuring and improving quality of care. This chapter provides an overview of major themes and controversies with respect to women's experiences with health care, reviews some key findings related to the impact of a substantially increased proportion of women physicians on women's health services, and identifies important future challenges for all stakeholders in the health care system.

I. Overview of Women's Use of Services

Throughout the life span, women use health services more frequently than do men, even after excluding reproductive services [5], so it is not surprising that much of the reaction to managed care has focused on women's health concerns. A coherent explanation for this phenomenon, well documented by researchers since the 1970s has not yet been fully articulated. Through adolescence there are few differences in the source of care for boys and girls, though one study suggests that even as young children, girls are more likely than boys to seek care for a similar problem. Adolescence is characterized by infrequent use of physicians' services. Analyses of the National Ambulatory Medical Care Survey, a nationally representative survey of office practice, indicate that the most common reason for ado-

lescents to see physicians is for prenatal care. Bartman's study of adolescents' access to care found that while approximately sixteen million of twenty-five million adolescents nationwide experienced symptoms, only one-third sought medical care [6].

Once women enter the reproductive years, they encounter a health care system that has historically isolated reproductive services from other health services. For adult women this has meant that women face a choice that may have important implications for their care. They can seek care from a generalist (family physician or internist), an obstetrician–gynecologist, or a nurse practitioner or nurse midwife.[2] In a nationally representative survey, 90.6% of adult women identified a generalist or obstetrician–gynecologist as their usual source of care [7]. While seeing a separate clinician for pregnancy-related needs may be manageable, women have gynecological needs throughout their lives. The separation of reproductive from ''all other'' health needs has often meant that women have had to cope with a health-care system that has not been organized to provide woman-centered care. A substantial number of women have chosen to obtain care from more than one physician, a response that may result in redundant effort (*e.g.,* two physicians reminding a woman to receive a mammogram) or gaps created by problems such as domestic violence, sexual harassment, and eating disorders that do not fit neatly within the domain of any specialty [1].

During the middle and later years of women's lives, women are more likely to identify either a family physician or internist as the usual source of care. Only 5.1% of women between the ages of 45 and 64 and 1.3% of women over 65 report that an obstetrician–gynecologist provides care for most problems. As the U.S. population ages, it should be emphasized that aging is becoming a women's health issue. Fifty-six percent of the U.S. population between the ages of 65 and 69 are women. For people age 85 and older, women comprise 69%, and 80% of centenarians are women.

Although women live longer and have a longer active life expectancy than men, women spend more years disabled because, among those with severe disability, women survive longer than men. After adjusting for age, women age 65 or older are twice as likely as men to report limitations of instrumental activities of daily living (IADL) such as shopping or handling finances, 1.5 times as likely to report one or two limitations of activities of daily living (ADL) such as bathing or dressing, and

[1]The opinions in this chapter are the author's and do not represent official policy of the Agency for Health Care Policy and Research or the Department of Health and Human Services.

[2]At this writing there are no public data sources that consistently and reliably include the contributions made by advance practice nurses, including nurse practitioners and nurse midwives, or physician assistants.

twice as likely to report three to six ADL limitations. Elderly women bear a considerable burden of chronic illness. For example, 59% report arthritis, more than half report hypertension, 43% report visual impairment, and 25% report cardiac disease. Eighty-two percent of women between the ages of 65 and 85 report at least one chronic disease, a half report more than one, and a quarter report three or more chronic conditions [8]. This increased occurrence of chronic conditions presents new challenges to health care systems and providers for both women and men, but the disproportionate burdens of chronic disease in later life fall on women [9,10].

II. Which Services Are Appropriate?

Access to care is only the first step in delivering services that meet women's needs. Increasing awareness of the phenomenon of practice variations, including the work of Wennberg [11], Chassin [12], and others, has demonstrated that substantial variations in medical practice exist that cannot be explained by observed differences in patient disease severity or access to care. For women's health, two distinct categories of variations are notable: variations in services provided to women compared with those provided to men, and variations in services for conditions unique to women (*e.g.,* reproductive services, breast diseases).

The term "gender bias" has been used to indicate differential receipt of services associated with gender. Multiple studies have demonstrated that compared with men, women are less likely to receive expensive, high-technology services. In particular, studies of access to dialysis and transplantation, diagnosis of lung cancer, and access to specific diagnostic and therapeutic interventions for heart disease suggest that women receive less aggressive care [13]. Determining the cause of these differences to plan strategies for improvement is a flourishing area of inquiry. For example, comparative use of services has been most extensively studied for patients with suspected or confirmed coronary artery disease. While numerous studies demonstrate that women receive fewer invasive diagnostic and therapeutic procedures than men, Bickell and colleagues found that women receive a substantially higher proportion of services that are considered appropriate, suggesting that some men may be overtreated [14]. The challenge in interpreting these studies lies in assessing which variations reflect differences in disease condition (biology), differences in patient preferences, or a societal bias that is shared by physicians and other health providers. On one hand, the difference is consistently in the same direction: women receive fewer services. On the other hand, developing specific strategies to identify the cause of the observed discrepancies— or unwanted variation—requires consideration of the multiple points in the process of care when clinical decisions may be made differently for women than for men as well as an understanding of gender-related differences in access to all services.

Practice variations also are ubiquitous when patterns of care for conditions unique to women are examined. Rates of hysterectomy for noncancerous uterine conditions, of Cesarian sections (either for a first or subsequent birth), and of the use of breast-sparing surgery for early-stage breast cancer suggest that too often in health care, "geography is destiny;" where a woman lives often predicts which treatments are recommended

[15–17]. Plausible explanations for observed variations in either domain (*i.e.,* conditions affecting both women and men or those unique to women) include clinical uncertainty, variations in availability of services such as radiation oncology, clinical differences in disease presentation, the role of socioeconomic factors, and the role of patient preferences in clinical decision making. Since the early 1990s, the existence of practice variations as a reflection of an inadequate knowledge base to inform clinical decision making has stimulated development of the field of outcomes research. Outcomes research focuses on linking the process of care to outcomes—or end results—that individual patients experience and care about [18].

In the case of coronary artery disease, women on average are older than men at first presentation, so differential use of services may reflect the presence of other conditions that may influence the decision-making process. Identifying appropriate strategies to inform clinical practice has been an important focus of efforts to assess and improve quality of care. Some studies of treatment patterns for coronary artery disease have suggested that biologic differences between women and men may be important, such as the observation that women who undergo coronary artery bypass surgery may be less likely to experience a technically excellent result due to smaller arteries [19].

Within the domain of conditions or procedures unique to women, increased rates of cesarian sections in the U.S. since the 1980s also have raised considerable concern that women may be exposed to avoidable surgical risks. Research supported by the Agency for Health Care Policy and Research has demonstrated that demographic changes (*e.g.,* the increased proportion of women deferring children until they are older) account for only about one-quarter of the observed increase [Childbirth PORT (Patient Outcomes Research Team), UCLA and RAND, Santa Monica, CA, unpublished data]. Other factors may include increased liability concerns and patient preferences for repeat Cesarian sections. Economic incentives appear to be a less important determinant of Cesarian section rates [20].

III. Women's Interactions with the Health Care System

The process of care leading to the clinical decision of interest is critical to consider. Horner's analysis of differential service delivery associated with race identified multiple possible etiologies of observed differences in treatment patterns such as clinical differences in disease, economic factors, patients' roles in decision making including beliefs about disease susceptibility and differences in seeking care, and the importance of the physician–patient decision-making process [21]. Before 1980, women were often underrepresented in clinical trials, so the knowledge base to guide clinical decisions is underdeveloped in many areas. Other studies derived from analyses of medical claims or hospital discharges can describe whether a service was provided but provide few insights regarding why it was not performed. When a woman first presents with chest pain or other suspicious symptoms, she may present to a primary care clinician or seek care directly from a cardiologist. Failure to receive a procedure can thus indicate differences in primary care physicians' propensity to refer women for further evaluation, cardiologists' willingness to perform further evaluation, a woman's decision to seek care or not, a woman's preference to select a less aggressive approach, or some combination of these factors. Most large datasets do not allow one to understand *why* observed variations occurred. In addition, ambulatory care data that permit linkage of initial clinical presentation with subsequent use of services beyond the initial visit remain underdeveloped compared with hospital discharge and medical claims data.

An additional wrinkle for women relates to the historic fragmentation of services in the U.S. health care system. The traditional separation of reproductive and other health services for women has meant that women seek primary care from multiple types of clinicians, and up to one-third seek primary care from two or more different physicians. The Commonwealth Fund Survey of Men and Women in 1993 found that compared with men, women are more likely to have a regular doctor (90% vs 85%), and also that throughout the life span, women report seeing more physicians than men do [22]. Thus while women may be perceived to be enthusiastic users of health services, they also confront multiple opportunities for variation in clinical decision making.

The influence of patient socioeconomic factors on clinical decision making, and the implications for women's use of services, is also an essential aspect of assessing women's experiences with care [23]. Across the life span, socioeconomic status (SES), whether measured by income, education, or race, is a powerful predictor of health status, and women are more likely than men to be of lower SES. The impact of SES is mediated both through direct and through indirect pathways. SES is an important nonmedical determinant of health, and SES also is known to affect patients' experiences with health care [7].

Insurance status is likely to influence which providers a woman can see as well as coverage for specific services. Overall, women are less likely than men to be without insurance, but the pattern of insurance coverage is strikingly different. In particular, women are much more likely to be covered by public insurance, particularly Medicaid, and are overrepresented among the ranks of the service workers and part-time workers, so they are much less likely than men to have private insurance through work [1]. In addition, women are substantially more likely than men to be *underinsured*, meaning that they have less generous benefits [24]. In short, the existence and generosity of insurance coverage is closely linked with employment and societal status, which may place women at risk of undertreatment. In addition to insurance status, it is well known that income and educational status are associated with differences in clinical decision making. Changes in welfare eligibility, historically linked with eligibility for Medicaid, are likely to present substantial challenges for poor women, the impact of which is largely unknown.

IV. New Developments

Changes in medical education and health care delivery offer great promise for ameliorating some of these structural challenges. Since 1969, the specialty of family medicine has trained physicians who can provide both general and gynecologic care, and internal medicine training programs are now required to include training in primary care gynecology. Since 1995, residency programs in obstetrics–gynecology include a nonreproductive primary care component. Managed care organizations have emphasized the importance of primary care, and many also permit women to select a gynecologist as a primary care provider. Efforts to assess the quality of care provided by health plans, particularly the National Committee on Quality Assurance's accreditation system, have made evidence-based clinical prevention a centerpiece of performance monitoring [25].

Perhaps the transformation of greatest interest to women's health is the changing demographics of the medical profession. In 1970, 7.7% of actively practicing allopathic physicians were women; by 1996 this had increased to 22.0% and is projected to increase to 29.4% by 2010 [26,27]. Statistics for osetopathic physicians reflect a comparable trend: in 1994, 17.0% of practicing osteopathic physicians were women compared with 12.8% in 1989. Since the 1970s, moreover, the number and percentage of women applicants, entrants, and graduates underwent a dramatic increase from 11.1% in 1970 to nearly equal numbers of men and women starting medical school in 1999. The specialty with the highest number of women is internal medicine; those with the highest proportions include pediatrics, psychiatry, obstetrics and gynecology. Overall, about 60% of women practice in five specialties: family practice, internal medicine, obstetrics–gynecology, pediatrics, and psychiatry. Based on residencies selected, this clustering of women in primary care specialties is likely to continue. The absolute numbers of women entering surgical fields, especially general surgery, have increased, but the overall percentages remain low. Women also are underrepresented in internal medicine specialties such as cardiology; one report found that just 5% of practicing adult cardiologists and 10% of trainees are women [28]. Women infrequently occupy positions of leadership within the profession, but multiple professional organizations including the American Medical Association (currently led by the first woman president) are actively addressing this issue. Entrance of women of color and underrepresented ethnic minorities into the profession has increased only slightly, lagging far behind that of Caucasian women.

The impact of an increased number and proportion of women in medicine on clinical care delivery has been particularly notable in the areas of clinical prevention and in patient communication and satisfaction. Both the age and gender of the physician are clearly related to the delivery of preventive care: younger physicians and women physicians are more likely to comply with evidence-based recommendations for clinical preventive services. Independent of physician age, women physicians are more likely to provide Pap smears, clinical breast examinations, and recommend mammography to their women patients [29]. These positive findings persist even after adjusting for patients' attitudes toward prevention [30]. Women's preferences for female physicians and the fact that female physicians have stronger beliefs about the value of prevention both contribute to the finding that female patients of women physicians are significantly more likely to receive appropriate screening for breast and cervical cancer [31].

Several studies also have found that women physicians have communication styles that have positive effects on patient satisfaction. Women physicians spend more time with patients, converse more with patients, and appear to listen more closely. Patients of women physicians report greater willingness to reveal difficult problems, such as personal violence related to family or sexual abuse [32]. As an increasing number of patients become familiar with obtaining medical information from the Internet and other sources, these findings suggest that the behaviors of women physicians can provide a model for their male colleagues.

V. Future Directions and Challenges

In response to women's health advocates, including legislators and women health professionals, the profession and the health care system have made important changes in training and health care delivery that can enhance the quality of care provided to women. In addition, demands by health care purchasers and consumers for evidence of performance in multiple domains, including both technical and interpersonal domains, will promote continued improvements.

Now and in the future, three issues will require continued efforts of all stakeholders to improve women's health. First is the relationship between socioeconomic status and health. Gaps in income between women and men for comparable positions have narrowed somewhat but remain substantial. Coupled with the dramatic growth in single parent families overwhelmingly led by women, it is likely that women will continue to have lower incomes than men. To the extent that within a market-oriented health care system generosity of insurance coverage is linked with economic status, the ultimate impact of current quality improvement efforts will require close scrutiny to assure that the benefits are fairly distributed.

Second is the critical need for coordination of care. The emergence and popularity of multidisciplinary women's health centers is a direct reflection of women's preferences for a comprehensive approach to care that minimizes opportunities for redundancy and provides comprehensive care [33]. Some have also called for establishment of a women's health specialty whose members would be broadly trained and serve as a resource to the multiple disciplines that provide care for women, similar to the multidisciplinary model developed for geriatrics.

Third is the broad issue of women's interactions with the health care system. Efforts to expand the current knowledge base for clinical decisions by including women in clinical trials will increase the availability of scientific evidence for practice. Effective communication between patients and health professionals is essential for accurate diagnosis and management, as well as for addressing psychosocial concerns and routine preventive health care needs. A focus on assessing and incorporating patients' preferences into clinical decision making is important for all patients, and is increasingly recognized as an important component of training and practice.[7] The increased availability of medical information to consumers through the Internet (*e.g.,* the new Department of health and Human Services website, http://www.4woman.gov, or the website from the Agency for Health Care Policy and Research, http://www.ahcpr.gov) and other sources can encourage women to be "comanagers" of their health and health care—but this will require physicians who are ready to be partners. Available evidence indicates that patients who are actively involved in managing their care have superior outcomes [34], but we have only begun to articulate sustained partnerships between physicians and patients as an important goal [35]. Translating that goal into reality will require continued education in communication skills, particularly

for physicians and patients who do not share a common gender or race/ethnicity.

Finally, an increased number and proportion of elderly women, and the challenges inherent in providing care to individuals with one or more chronic conditions, indicate that an important unfinished agenda for improving women's health lies in developing and implementing interventions that enhance or maintain functional status and independence. As health care systems struggle to make a long overdue transition from a primary focus on the treatment of illness in individuals to a broader focus that encompasses the promotion of health and well being in populations, the challenge lies in aligning organizational and financial structures and incentives with health needs. Women, who suffer from a higher burden of illness and disability than men, and who have unique health needs across the life span, stand to benefit most from this transition [7].

References

1. Clancy, C. M., and Massion, C. T. (1992). American women's health care: A patchwork quilt with gaps. *JAMA,* **268,** 1918–1920.
2. Clancy, C. M., and Eisenberg, J. M. *J. Am. Med. Assoc.* (1998). Outcomes research at the Agency for Health Care Policy and Research. *Dis. Manag. Clin. Outcomes* **3,** 72–80.
3. Laine, C. and Davidoff, F. (1996). Patient-centered medicine. A professional evolution. *J. Am. Med. Assoc.* **275,** 152–156.
4. Agency for Health Care Policy and Research (1996). "Survey of Consumers' Experiences with Health Care" (unpublished analyses). Kaiser Family Foundation.
5. Verbrugge, L. M. and Steiner, R. P. (1981). Physician treatment of men and women patients: Sex bias or appropriate care? *Med. Care* **19,** 609–632.
6. Bartman, B. A., Moy, E. and D'Angelo, L. J. (1997). Access to ambulatory care for adolescents: The role of a usual source of care. *J. Health Care Poor Underserved* **8,** 214–226.
7. Bartman, B. A., Clancy, C. M., Moy, E., and Langenberg, P. (1996). Cost differences among women's primary care physicians. *Health Aff (Millwood)* **15**(4), 77–82.
8. Bierman, A. S., and Clancy, C. M. (1999). Women's health, chronic disease, and disease management: New words and old music? *Women's Health Issues* (in press).
9. Wagner, E. H. (1997). Managed care and chronic illness: Health services research needs. *Health Serv. Res.* **32,** 702–714.
10. Wagner, E. H., Austin, B. T., and von Korff, M. (1996). Organizing care for patients with chronic illness. *Milbank Q.* **74,** 511–544.
11. Wennberg, J. E., and Gittelson, A. (1982). Variations in medical care among small areas. *Sci. Am.* **246,** 120–133.
12. Chassin, M. R., Kosecoff, J., Park, R. E. *et al.* (1979). Does inappropriate use explain geographic variations in the use of health care services? A study of three procedures. *J. Am. Med. Assoc.* **258,** 2533–2537.
13. Council on Ethical and Judicial Affairs, American Medical Association (1991). Gender disparities in clinical decision making. *J. Am. Med. Assoc.* **266,** 559–562.
14. Bickell, N. A., Califf, R. M., Pryor, D. B., Glower, D. D., Mark, D. B., Lee, K. L., and Pieper, K. S. (1992). Referral patterns for coronary artery disease treatment: Gender bias or good clinical judgment? *Ann. Intern. Med.* **116,** 791–797.
15. New York Times (1997). Health care's weird geography (editorial). *New York Times,* October 25.
16. Dartmouth Medical School (1996). "The Dartmouth Atlas of Health Care." American Hospital Publishing, Chicago.
17. Johantgen, M. E., Clinton, J. J., Levy, H., Harris, D. R., and Coffey, R. M. (1995). Treating early-stage breast cancer: hospital charac-

teristics associated with breast-conserving surgery. *Am. J. Public Health* **85,** 1432–1434.

18. Clancy, C. M., and Eisenberg, J. M. (1998). Outcomes research: Measuring the end results of health care. *Science* **282,** 245–246.

19. O'Connor, N. J., Morton, J. R., Birkmeyer, J. D., Olmstead, E. M., and O'Connor, G. T. (1996). Effect of coronary artery diameter in patients undergoing coronary bypass surgery: Northern New England Cardiovascular Disease Study Group. *Circulation* **93,** 652–655.

20. Keeler, E. B., and Brodie, M. (1993). Economic incentives in the choice between vaginal delivery and cesarean section. *Milbank Q.* **71,** 365–404.

21. Horner, R. D., Oddone, E. Z., and Matchar, D. B. (1995). Theories explaining racial differences in the utilization od diagnostic and therapeutic procedures for cerebrovascular disease. *Milbank Q.* **73,** 443–462.

22. Weisman, C. (1998). "Women's Health Care. Activist Traditions and Institutional Change," p. 250. Johns Hopkins University Press, Baltimore, MD.

23. Eisenberg, J. M. (1979). Sociologic influences on decision-making by clinicians. *Ann. Intern. Med.* **90,** 957–964.

24. Short, P. F., and Banthin, J. S. (1995). New estimates of the under-insured younger than 65 years. *J. Am. Med. Assoc.* **274,** 1302–1306.

25. Iglehart, J. K. (1996). The National Committee for Quality Assurance. *N. Engl. J. Med.* **335,** 995–999.

26. Council on Graduate Medical Education (1995). "Women in Medicine," U.S. Department of Health and Human Services, Rockville, MD.

27. American Medical Association (1997/1998). "Physician Characteristics and Distribution in the U.S." AMA, Chicago.

28. American College of Cardiology (1998). The ACC professional life survey: Career decisions of women and men in cardiology. A report of the committee on women in cardiology. *J. Am. Coll. Cardiol.* **32,** 827–835.

29. Lurie, N., Slater, J., McGovern, P., Ekstrum, J., Quam, I., and Margolis, K. (1993). Preventive care for women. Does the sex of the physician matter? *N. Engl. J. Med.* **329,** 478–482.

30. Franks, P., and Clancy, C. M. (1993). Physician gender bias in clinical decisionmaking: Screening for cancer in primary care. *Med. Care* **31,** 213–218.

31. Lurie, N., Margolis, K. L., McGovern, P. G., Mink, P. J., and Slater, J. S. (1997). Why do patients of female physicians have higher rates of breast and cervical cancer screening? *J. Gen. Intern. Med.* **12,** 34–43.

32. Roter, D., Lipkin, M., Jr., and Korsgaard, A. (1991). Sex differences in patients' and physicians' communication during primary care medical visits. *Med. Care* **29,** 1083–1093.

33. Weisman, C. S., Curbow, B., and Khoury, A. J. (1995). The National Survey of Women's Health Centers: Current models of women-centered care. *Women's Health Issues* **5,** 103–117.

34. Kaplan, S. H., Greenfield, S., and Ware, J. E., Jr. (1989). Assessing the effects of physician-patient interactions on the outcomes of chronic disease. *Med. Care* **27**(3 Suppl.), S110–S127.

35. Leopold, N., Cooper, J., and Clancy, C. (1996). Sustained partnership in primary care. *J. Fam. Pract.* **42,** 129–137.

6

Women's Health and Managed Care

KAREN DAVIS, KAREN SCOTT COLLINS, AND CATHY SCHOEN

The Commonwealth Fund
New York, New York

I. Introduction

Women have a special stake in understanding and helping inform policy on the transformation of the U.S. health care system by the growth of managed care. Not only are women more likely to need regular access to primary and preventive care, they often play a key role as decision makers about health care for family members as well as themselves. Women are typically the primary care givers when a child, spouse, or parent is ill or incapacitated.

Women's concerns with the evolution of managed care cut across private and public health insurance boundaries. As a result of low incomes, a higher proportion of women than men under the age of 65 depend on Medicaid for health insurance, and women comprise the majority of the Medicare population and of those reliant on Medicare and Medicaid's support of long-term care services. As a result, women are directly affected by both public policies and employer decisions to restructure health insurance programs.

This chapter reviews what is currently known about the effectiveness of managed care in addressing women's health concerns and presents new information, drawn from the 1997 Kaiser/Commonwealth National Survey of Health Insurance, about women's experiences obtaining health care in the evolving health care marketplace. As Medicare and Medicaid place reliance on managed care as a cost containment strategy, information on quality of plans, quality standards, and the monitoring of plan performance becomes increasingly important. Holding plans accountable for delivering high quality care while protecting the financial security of women and their families remains the challenge in front of the nation as markets and industry mature.

II. Background

Managed care is an increasingly important source of women's health care. Sixty-nine percent of privately insured women are enrolled in health maintenance organizations (HMOs) [1]. Managed care plans can be attractive to women because they typically cover preventive services, have low out-of-pocket costs, and permit coordination of primary and obstetrical/gynecological care.

To best serve women's health care needs, plans should also allow women opportunities to discuss their care and participate in decisions about that care, actively promote preventive services, and provide full access to reproductive health care and to mental health services, when needed. Women's health physician Karen Carlson writes that while managed care plans have the potential to emphasize primary care and improve coordination

of care for women, the new system can also threaten quality of care by imposing time constraints on doctor–patient interactions, for example [2].

The data on experiences with managed care support this mixed picture. A study by Collins and Simon compares the experiences of women enrolled in managed care plans with women who are privately insured and women with Medicaid coverage [3]. Collins and Simon's analysis looked at early signs of managed care's effect on women with private health insurance, using data from 2500 participants in the Commonwealth Fund 1993 Survey of Women's Health [3]. Among privately insured women aged 18–64, women in HMOs are more likely to receive regular preventive care. Three out of four women enrolled in HMOs receive preventive services such as Pap tests and clinical breast examinations, compared with two out of three privately insured women cared for through other arrangements. Women outside HMOs are more likely to cite the high cost of care as a reason for not getting preventive services. Lack of physician counseling is an important reason for both groups [3].

Coverage, whether under an HMO or other private insurance, is not a guarantee of access to care, however. Collins and Simon note that 14% of women in HMOs and 10% of insured women not enrolled in HMOs report that they failed to get needed medical care in the previous year. Reasons for not getting needed care vary; cost is the primary reason for women not in HMOs, while for HMO members, inability to get an appointment is a major reason [3].

Furthermore, women in HMOs are more likely than women not in HMOs to rate their physician as fair or poor for amount of time spent with them (23% vs 16%) and for explaining symptoms (20% vs 13%). A higher percentage of women in HMOs than of women not in HMOs (34% vs 28%) say their doctors "talk down" to them [3].

Analysis of the Commonwealth Fund's 1994 Survey of Patient Experiences with Managed Care [4] also identifies concerns with provider choice and access but greater satisfaction with costs and benefits covered. Seventy percent of managed care enrollees had been in their plan for less than five years. Women rated managed care plans fair–poor more often than fee-for-service on several dimensions of access, including convenience of office location (14% vs 4%), waiting time for appointments (31% vs 10%), waiting time for emergency care (20% vs 8%), and access to specialty care (24% vs 7%). In terms of affordability, however, managed care plans were less likely to be rated fair or poor than fee-for-service on out-of-pocket costs (19% vs 34%) and on coverage for preventive services (13% vs 29%) [4].

The Survey results point to greater problems for women who are at risk either due to low income or poorer health status. Low income women enrollees (*i.e.,* women with family incomes less than $25,000) reported significantly higher dissatisfaction with dimensions of access than did higher income women. Women enrollees reporting fair or poor health also reported greater dissatisfaction with access to services and with the range of services covered than did those in good or excellent health [4].

Some of women's dissatisfaction with access to care and time with their physician is also reflected in a survey of physicians and managed care [5]. Women physicians report a high level of concern over the amount of time they are able to spend with patients, with dissatisfaction increasing as the extent of managed care patients in the practice increases. Forty-eight percent of women physicians with 50% or more managed care patients were dissatisfied with the amount of time spent with patients, compared with 17% of women physicians who have 10% or fewer patients in managed care.

A 1996 review of managed care coverage and costs for women's services supports the important role of managed care for women in providing affordable coverage for preventive services [6]. In 1994, 86% of HMOs covered mammograms for women over 50 and 78% covered Pap tests. However, cost-sharing and intervals for receiving the screening tests vary by type of managed care plan. Data on health plan performance from the National Committee for Quality Assurance show that 70% of women aged 52–69 enrolled in plans reporting performance measures had received a mammogram in the past two years, although the screening rates for individual plans varied from 30 to 90% [7]. Cervical cancer screening showed similar variation, with 70% of women aged 15–24 receiving at least one Pap test in the past three years; rates for individual plans ranged from 25 to 100% [7].

Focused planning of women's health services has included assessment of current clinical services, surveys of women enrollees' preferences in health care, and a review of women's health programs at other plans [8]. Some managed care plans are making a special effort to implement model women's health programs. The Commonwealth Fund commissioned a case study of best practices in the delivery of clinical preventive services to women in managed care organizations by Mathematica Research Policy, Inc. [9]. Six HMOs and one medical group practice were included in the study. The sites included Comprehensive Health Services, Inc./The Wellness Plan, Group Health Cooperative of Puget Sound, Harvard Community Health Plan, Kaiser Foundation Health Plan of Southern California, Keystone Health Plan East, Park Nicollet Clinic (a medical group), and U.S. Healthcare. Interventions used by the sites to improve the delivery of breast and cervical cancer screening included patient reminders, provider reminders, performance feedback to providers, financial incentive for providers, and office staff reminders and procedures. Successful programs had the following characteristics:

- Careful selection of preventive services to be included, careful design of the intervention whether a stand-alone program or part of a comprehensive program, and identification of high-risk women

- Engaging patients and providers in early stages of program development
- Collecting accurate and timely data that can be used for a variety of purposes, including monitoring performance and meeting external reporting requirements
- Monitoring and evaluating programs and integrating the data collection process within the health care services delivery system

Different types of managed care plans result in different experiences for women. Tightly organized managed care plans have the potential to address certain women's health problems effectively. Training programs, for example, can help physicians recognize and treat neglected problems such as depression or domestic violence. Managed care plans with good information systems can also notify women when they are due to receive periodic preventive services.

III. 1997 National Survey

A new national survey of health insurance provides an up-to-date profile of women and their experiences obtaining needed health care in the rapidly evolving health care marketplace. The Kaiser/Commonwealth National Survey of Health Insurance, conducted by Louis Harris and Associates, surveyed 4001 adults ages 18 and over, including 2081 women, between November 1996 and March 1997. The total sample includes 3761 adults interviewed by telephone and 240 adults interviewed in person because they did not have telephones in their homes. The data were weighted to the March 1996 Current Population Survey for accurate representation of Americans by sex, race, age, education, and health insurance.

The survey results indicate that women are more likely than men to be dependent on Medicare and Medicaid for health insurance coverage; 28% of women compared to 19% of men are covered by these public programs (see Fig. 6.1). Women are twice as likely as men to be covered by Medicaid (9% vs 4%), reflecting their poorer economic status and the categorical restrictions targeting Medicaid coverage on single parent families headed by women. Women are also somewhat more likely to be covered by Medicare (19% vs 15%) due to their longer life expectancy. By contrast, men are more likely to be covered under private health insurance plans provided by employers. They are also somewhat more likely than women to be uninsured (18% vs 14%).

Women's health insurance coverage is often unstable. Thirty-six percent of women have been in their plan for less than two years. Women report that most of the recent change in health plans was a result of a change in job, loss of job, or their employer changing health plans. These reports indicate that most plan change is involuntary; only 16% of those changing plans in the past two years said they did so because the new plan offered better benefits, lower costs, or other attractive features.

Women are more likely than men to report difficulties either obtaining health care services or paying their medical bills. Overall, 20% of women compared with 15% of men report that there was a time in the last year when they did not get needed care or failed to fill a prescription because of the cost of doing so (see Fig. 6.2 and Table 6.1). When they do obtain care,

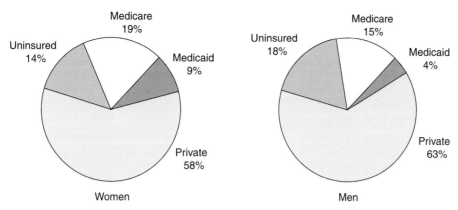

Fig. 6.1 Sources of health insurance coverage for adult women and men, 1997. Source: [1].

women report somewhat greater difficulty in paying their medical bills; 18% of women and 14% of men report problems paying medical bills. Together, 28% of women and 22% of men experience either a problem obtaining needed care or prescription drugs or paying medical bills.

Problems obtaining medical care or paying medical bills are particularly severe for uninsured women and women who have had a recent period without insurance. Indeed, currently insured women with a recent time uninsured report rates of access and cost difficulties similar to currently uninsured women. Even when insured, low-income women also face serious difficulties, reflecting the lower quality of their health insurance coverage and the burden of premiums and uncovered medical expenses. As shown in Table 6.2, nearly one-third (31%) of currently or recently uninsured women aged 18–64 report a time they did not get needed care. This percentage is nearly four times higher than among women who had been insured for the past two years (*i.e.,* 8% of low-income women (family incomes less than 200% of poverty) and 7% of higher income women (family incomes more than 200% of poverty).

Uninsured women and low-income insured women aged 18–64 are also less likely to have a regular doctor than are insured higher income women. Uninsured women are defined to include those currently uninsured and those who have been uninsured within the past two years. Studies indicate that having a regular doctor is key to receiving regular preventive care and seeking care for early diagnosis and treatment of a medical problem. Forty percent of uninsured women or women with a recent lapse in coverage do not have a regular doctor, nor do 22% of low-income insured women, as compared to 12% of insured nonpoor women.

Rates of receiving preventive care are disturbingly low across women of all incomes and insurance status, with the worst screening rates seen among uninsured and poor women. Even though regular mammograms for older women and Pap smears for all women are part of the recommended guidelines for preventive care, 38% of all older women (aged 50–64) did not get a mammogram in the last year, and 36% of all women (aged 18–64) did not get a Pap smear. Preventive care is particularly at risk among uninsured and low-income women. Over half of uninsured women aged 50–64 (54%) and insured poor or near

poor women (52%) failed to get an annual mammogram. Similarly, almost half of uninsured women (45%) and insured low-income women (40%) failed to get an annual Pap smear. In addition, even one-fourth of insured higher income women also said they did not have a mammogram (26%) or Pap smear (27%) in the past year.

The cost of care can be a major barrier to obtaining needed care. Even when women do go to see a physician for a medical problem, they may not be able to afford prescription medications, and, as a consequence, they may fail to follow through with recommended treatment. Thirty percent of uninsured women aged 18–64 reported that they failed to fill a prescription due to cost, as did 17% of insured low-income women, compared to 4% of insured higher income women.

Combining the failure to either get needed care or failure to fill a prescription, almost half of uninsured women (45%) reported inadequate care, as compared to 21% of insured low-income women and 10% of insured high income women. Clearly, insurance coverage is a major factor in obtaining needed care. However, even with insurance coverage, the extent of that coverage and any out-of-pocket expenses facing women can affect whether they are able to obtain the care they need.

In fact, financial burdens from paying medical bills are a major problem for women, who may be pursued by collection

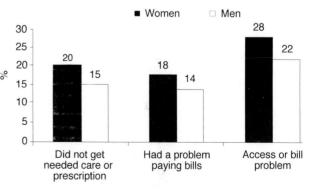

Fig. 6.2 Problems obtaining health care or paying bills, women and men, 1997. Source: [1].

Table 6.1
Women and Men Care Experiences: Access, Cost, Primary Care, and Worries

	Total	Men	Women
Access problems			
Did not get needed care	11	10	12
Did not fill prescription	10	7	13
Difficult to get medical care (extremely, very, or somewhat)	17	17	18
Did not get needed care or did not fill prescription	18	15	20
Cost experiences			
Problems paying medical bills	16	14	18
Had to significantly change way of life because of medical bills	7	5	8
Contacted by a collection agency because of medical bills	10	7	12
Out-of-pocket expenses	54	57	51
$500 to $1000	14	15	14
$1001 or more	12	13	11
Bill or access problem (not get needed care or not fill prescription)	25	22	28
Physician care and preventive services			
No regular doctor	27	34	20
Been with current doctor for 5 years or more	39	37	40
No physician visit in the past year	19	29	11
No mammogram, women age 50 or older	NA	NA	42
No Pap smear	NA	NA	40
No prostate exam, men age 50 or older	NA	43	NA
No routine physical exam or check-up	43	52	35
Satisfaction with services (percent very or somewhat dissatisfied)			
Dissatisfied with health care received by self or family (all respondents)	14	15	13
Insured respondents satisfaction and experiences with health plans			
Dissatisfied with choice of doctors in current plan (insured respondents)	10	10	10
Dissatisfied with health insurance (insured respondents)	14	15	13
Insured respondents major or minor problems			
Plan not covering treatment thought necessary	17	16	17
Plan delaying care while waiting for approval	10	10	10
Having to deal with rules that were complex	24	25	23
Plan delaying payment for services	19	18	20
Worries (percent who worry a lot or a great deal about)			
Worry about being denied a medical procedure	28	25	30
Worry about not being able to get specialty care	38	33	42
Number of respondents	4001	1922	2079

Source: [1].

agencies or have to change their way of life in order to pay medical bills. Thirty-eight percent of uninsured women aged 18–64 reported problems paying medical bills during the past year, as compared to 19% of insured low-income women and 10% of higher income insured women. One-fifth (21%) of uninsured women had to change their way of life significantly due to medical bills and 22% had been contacted by a collection agency to pay these bills.

The twin problems of being able to get needed health care and struggles to pay for such care affect a third of all women. Among uninsured women, 57% reported either failing to get needed care, failing to fill a prescription, or difficulty paying medical bills in the past year, contrasted with 31% of insured low-income women and 15% of insured high income women.

IV. Managed Care

Managed care is now the dominant form of health insurance for women. Among women with insurance, three-fifths (59%) report managed care enrollment. Among privately insured adult women, the proportion rises to 69% (see Fig. 6.3). Public programs such as Medicare and Medicaid have a much lower percentage of beneficiaries enrolled in managed care, but the numbers have increased rapidly in recent years and are expected to grow considerably over the next five years. When surveyed, 55% of poor and near poor insured women described their health plans as managed care, as did 30% of women aged 65 and over.

Managed care plans vary considerably in their characteristics. Common to all managed care plans is the requirement that pa-

Table 6.2
Gaps in Women's Insurance Coverage, Poverty, and Care Experiences—Women Ages 18 to 64

	Total women 18–64	Insured women 18–64	Uninsured or recent gap 18–64	<200% of poverty		>200% of poverty	
				Insured 18–64	Uninsured or recent gap 18–64	Insured 18–64	Uninsured or recent gap 18–64
Access problems							
Did not get needed care	14	7	31	8	34	7	24
Did not fill prescription due to cost	15	9	30	17	36	4	17
Difficult to get medical care	20	11	43	17	49	7	29
Did not get needed care or did not fill prescription	23	14	45	21	51	10	30
Cost experiences							
Problems paying medical bills	21	13	38	19	41	10	32
Had to significantly change way of life to pay medical bills	10	5	21	9	24	3	11
Contacted by a collection agency because of medical bills	14	10	22	16	26	7	15
Out-of-pocket expenses							
$500 to $1000	15	15	14	12	14	18	16
$1001 or more	12	9	17	10	14	9	26
Bill or access problem	32	21	57	31	62	15	46
Physician care and preventive services							
No regular doctor	32	16	40	22	41	12	35
No mammogram (women over 50)	38	35	54	52	57	26	44
No Pap smear	36	32	45	40	45	27	43
No routine physical exam or check up	37	32	47	31	47	33	44
Worries							
Worry about being denied a needed medical procedure	32	25	49	35	53	20	42
Worry about not being able to get specialty services	46	38	63	50	66	33	57
Number of respondents	1708	1201	506	396	349	726	138

Note. Insured women include those who were insured at the time of the survey and had no time uninsured in the past two years. Uninsured women include those uninsured at the time of the survey and those who had been uninsured during the past two years although they were insured at the time of the survey. Source: [1].

tients seek care from an approved network of physicians, hospitals, and other health care providers who have contracts with the plan, setting the terms of payment. Some managed care plans have a point-of-service option (POS) that permits beneficiaries to see out-of-network providers at an extra cost.

HMOs also typically require that patients have a primary care physician who must authorize any specialized services (the "gatekeeping" function). The advantage of such a requirement is that every patient has a regular doctor familiar with all the care they are receiving and who is responsible for seeing that patients get regular preventive services. A possible disadvantage is that financial incentives to limit specialty care may pose a barrier to obtaining more advanced treatment in the case of serious medical problems.

Contrasts between experiences reported by women insured by managed care plans and those insured by traditional health insurance plans indicate that managed care plans help improve access to preventive care services yet perform surprisingly un-

evenly on other access, care, and satisfaction measures. For example, it is somewhat surprising that women in managed care plans are no more likely to have a regular doctor than are women enrolled in traditional fee-for-service (FFS) plans; 13% of

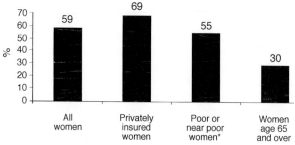

Fig. 6.3 Women enrolled in managed care, 1997. *, Income less than 200% of poverty. Source: [1].

Table 6.3

Women's Access and Care Experiences by Type of Plan: Comparisons of Managed Care and FFS

	Total women (n = 2079)	Uninsured women (n = 198)	MC women (n = 961)	FFS women (n = 697)
Access problems				
Did not get needed care in the past year	12	35	7	5
Did not fill prescription due to costs in the past year	13	32	8	8
Difficult to get medical care	18	55	11	8
Did not get needed care or did not fill prescription	20	49	13	11
Cost experiences				
Problems paying medical bills in the past year	18	40	12	10
Had to significantly change way of life because of medical bills	8	24	4	5
Contacted by a collection agency because of medical bills	12	24	10	6
Out-of-pocket expenses				
$500 to $1000	14	14	14	13
$1001 or more	11	17	8	10
Bill or access problem (not get care or not fill prescription)	28	61	20	17
Physician care and preventive services				
No regular doctor	20	46	13	14
Been with current doctor for 5 years or more	40	26	39	55
No visit to a physician in the past year	11	27	7	9
No mammogram in the past year (women 50+)	42	62	35	44
No Pap smear	40	56	32	47
No routine physical exam or check up	35	58	32	29
Satisfaction with services (percent somewhat or very dissatisfied)				
Dissatisfied with health care received by self or family	13	22	12	7
Insured respondents satisfaction and experiences with health plans				
Dissatisfied with choice of doctors in current plan	10	NA	13	2
Dissatisfied with health insurance	13	NA	14	8
Major or minor problems				
Plan not covering treatment you thought necessary	17	NA	19	13
Plan delaying care while you waited for approval	10	NA	12	6
Having to deal with rules that were complex	23	NA	28	13
Plan delaying payment for services	20	NA	21	16
Had to change doctor when joined current plan (changed plan in last two years)	30	NA	33	5
Worry about being denied a medical procedure	30	59	27	20
Worry about not being able to get specialty care	42	69	38	30

Note. Insured women includes only those with no time uninsured in the past two years. Source: [1].

women enrolled in managed care plans do not have a regular doctor compared with 14% of women enrolled in FFS plans (see Table 6.3). Because managed care plans require women to obtain care from a specified network of physicians, women in managed care plans are much more likely to have changed doctors when they joined a managed care plan and, as a result, to have been with the same doctor for a shorter period of time. Over half of women (55%) in FFS plans have been with the same doctor for five years or more compared with 39% of women in managed care plans.

One concern about managed care is the possibility that instability in plan coverage may undermine continuity of care and satisfaction with care. Among women who had changed plans in the last two years, one-third of managed care patients had changed doctors as compared with 5% of women in FFS plans. The higher turnover in doctors is often a reflection of change in health plan coverage. About one-third of managed care enrollees had been in the current plan less than two years, and over three-fourths of these had done so involuntarily. Length of time in health plans appears to affect satisfaction. In general, women enrolled in their health plans for two years or more were more satisfied with care and insurance than those in plans for less than two years. Greater familiarity with managed care plans appears to reduce the problems encountered with plan rules and coverage decisions.

Managed care seems to do a better job than traditional FFS plans of seeing that women get preventive care. One major incentive for managed care plans to provide preventive care is that their performance in seeing that women obtain mammograms and Pap smears is reported to the National Committee for Quality Assurance (NCQA) as part of HEDIS (Health Employer Data Information Systems), and these data are made publicly available.

The principal concern about managed care is that the pressure to control costs will come at the sacrifice of quality care. Patient dissatisfaction with care and worries about being denied care have led to serious consideration of the need for national quality standards and better information about plan performance on quality indicators. The Kaiser/Commonwealth National Survey of Health Insurance found that, overall, satisfaction rates with insurance were high. However, there was some evidence of greater dissatisfaction with coverage and care among managed care enrollees than among women enrolled in FFS plans. For example, 14% of women in managed care were somewhat or very dissatisfied with their health insurance compared with 8% of women enrolled in FFS plans. Higher rates of dissatisfaction also extend to choice of doctors available in the plan; 13% of women in managed care plans were dissatisfied with their choice of doctor compared with only 2% of women in FFS plans. One in eight women enrolled in managed care (12%) was dissatisfied with health care services overall compared with 7% of women in FFS plans. The disparity between those enrolled in managed care and those enrolled in FFS is disturbing.

Differences between women enrolled in managed care and FFS plans also emerged in reports of plans not covering treatment thought to be necessary and problems with plans delaying treatment. Almost one-fifth of women in managed care (19%) reported major or minor problems with plans not covering treatment they thought was necessary compared with 13% of women in FFS plans. Similarly, 21% of women in managed care reported problems with plans delaying payment for treatment compared with 16% of women in FFS plans.

Experiences with difficulties in obtaining needed care are reported in the media and are shared with family and friends. Concern about the potential for being denied care, therefore, is widespread. One in three women (30%) is worried she will be denied a medical procedure when it is needed. This concern is particularly high among uninsured women (59%) and is somewhat higher among women enrolled in managed care (27%) than among women enrolled in FFS plans (20%). Similarly, two-fifths of women (42%) are worried they will not be able to get specialty care when they need it. Again, worries about access to specialists are particularly high among uninsured women (69%) and are higher among women enrolled in managed care (38%) than among women enrolled in FFS plans (30%).

One hope of managed care is that it will provide better financial protection of medical care needs and help close existing gaps in health insurance coverage. In the survey, however, a similar proportion of women insured by managed care plans and FFS plans reported problems paying medical bills and estimated out-of-pocket costs exceeding $500 annually (See Table 6.3).

Further analyses of the Kaiser/Commonwealth 1997 National Survey of Health Insurance and other surveys are now underway to explore the experiences of women with health problems. Research from earlier surveys suggests that dissatisfaction with managed care is particularly high for patients with a serious illness or injury in the family or where the patient is in fair or poor health [10].

While managed care is often considered a single form of coverage, in fact it takes on many faces. Many different organizational forms exist. Prior research suggests that patients report fewer difficulties with nonprofit managed care plans than for-profit plans and fewer problems with those HMOs organized on a group or staff model basis [11]. Further research will investigate the importance of these attributes in women's experiences with obtaining needed care.

V. Medicaid and Managed Care

Managed care has special implications for low-income women and their families. Managed care can be an attractive system of care for low-income families. The services HMOs provide are comprehensive and out-of-pocket costs are often minimal; if any fees are charged beyond the capitated premium, they are usually very low. Disturbingly, however, both consumer surveys and health outcome studies show that low-income Americans may fare less well in managed care plans than in more traditional plans [12–14].

Despite these warning signs, and if current trends continue, five years from today the vast majority of Medicaid beneficiaries will receive their health care through managed care plans. State governments have moved aggressively in recent years to enroll Medicaid beneficiaries into managed care. Several states, such as Kentucky, Tennessee, and Oregon, have obtained federal waivers to convert their entire Medicaid programs to managed care. From 1991 to 1995, the proportion of Medicaid beneficiaries enrolled in managed care more than tripled, from 10 to 40% [15]. In 1995, 13.3 million Medicaid beneficiaries were enrolled in managed care plans.

Some states have been more successful than others in making the transition to Medicaid managed care. To help determine why, the Henry J. Kaiser Family Foundation and The Commonwealth Fund have supported in-depth case studies of selected states' efforts [16]. Factors that correlated with success in making a smooth transition included gradual implementation, education of beneficiaries and providers about the new system's rules and procedures, broad use of managed care by working populations in advance of the Medicaid initiative, and state assistance to "safety net" providers, such as public hospitals and community health centers, to boost their ability to compete. On the other hand, states that failed to put adequate time and staffing into matching beneficiaries with appropriate plans, setting fair payment rates, and protecting the quality of care tended to experience problems in moving people into managed care. Low capitation rates were a particular problem, sometimes leading to poor care or threatening the financial viability of the plan.

Low-income families, disabled and frail elderly people, and other Medicaid enrollees pose special challenges for managed care, especially for plans that have mainly served younger, healthier, working families in the past. Even with insurance, many poor, near-poor, and minority patients lack resources, such as telephones and transportation, that make it possible to receive regular health care. Managed care can add to the burden by imposing complex preapproval and referral procedures, limited locations or hours of operation, and early hospital discharge. In addition, established managed care plans have not developed expertise in meeting the unique and complex health and social needs of Medicaid beneficiaries and the disenfranchised. Unless health care systems build outreach, education, and support services and develop links to the public health and social services sectors and traditional safety net providers, these less fortunate groups may be unable to comply with health plan rules and medical care instructions.

To help monitor the impact of managed care on Medicaid beneficiaries, the Commonwealth Fund and the Henry J. Kaiser Family Foundation have cosponsored a set of state-level surveys of patients. Initial analysis of the *Kaiser/Commonwealth State Low Income Surveys* finds that Medicaid managed care enrollees are more likely to rate their overall health care services as fair or poor than beneficiaries remaining in traditional Medicaid (21% vs 14%) and much more likely to rate their physicians as fair or poor (26% versus 9%). However, considerable variation exists across states, plans, and beneficiaries. In some states, managed care enrollees are more likely to receive preventive care than those covered by traditional Medicaid. In other states, Medicaid managed care enrollees continue to receive very episodic care, often from emergency rooms.

Surveys show a high correlation between patients' ability to choose their managed care provider and their satisfaction with that provider. For example, one study found that when New York City Medicaid beneficiaries voluntarily enrolled in managed care plans, they were more likely to give higher satisfaction ratings to their plans than people enrolled in traditional Medicaid FFS plans [17].

Low-income families experience frequent changes in insurance status, moving from Medicaid coverage to no coverage to employer coverage and back again to Medicaid, depending on their employment status. Nationally, among women aged 18–64 receiving Medicaid, 28% have been in the program for less than a year [18]. Many women lose coverage—and, under managed care, access to their previous doctor—after they give birth, when they get a job, or when their earnings increase. Two-thirds of those who lose Medicaid coverage become uninsured.

Frequent changes in insurance status not only undermine beneficiaries' ability to see a particular physician, but they also make it more difficult for them to get continuous care. In addition, frequent turnover in patients means that managed care plans have less incentive to provide preventive care. Steps to assure continuation of Medicaid coverage as changes in employment, earnings, health status, and marital status occur are necessary to improve the health care of low income families.

VI. Medicare and Managed Care

For the 38 million elderly and disabled people enrolled in the Medicare program, the shift from traditional FFS plans into managed care has the potential to bring great benefits. Managed care, with its integrated networks of physicians, is capable of providing the coordinated primary and specialty care that best serves patients with chronic illnesses [19].

In managed care, the burden of coordinating care is often moved from the patient to a health care team whose members plan treatment and share medical records. Capitated payment systems can offer flexibility to arrange and pay for services as needed while also freeing providers from traditional insurance's narrow definitions of covered benefits. In addition, because Medicare patients are likely to stay in their plans for extended periods, plans have a clear incentive to invest in cost-effective treatments.

In practice, however, competing incentives have led many plans to put their energies into attracting healthier patients, because most public and private purchasers pay on the basis of the average cost of care, with only minor adjustments for age, sex, or disability. Current market dynamics seem to be rewarding plans that avoid risk and penalizing those that gain reputations for outstanding care of seriously or chronically ill patients. Especially troubling is recent evidence that health outcomes are worse over a four-year period for chronically ill elderly people and low-income people with serious health problems cared for in HMOs than for those in FFS care [14].

Other surveys show a wide variety of opinions about the quality of health care, suggesting substantial variations from one plan to the next. Surveys of Medicare enrollees by the Health Care Financing Administration, for instance, show that HMO enrollees are more likely than FFS enrollees to give either very high ratings or very low ratings to the quality of their care [20]. These responses reinforce the need for setting quality standards and providing better information about plans so that beneficiaries can make informed choices [21].

VII. The Research Agenda

Women can become more effective purchasers of health care services by learning more about managed care plans' performance. The Commonwealth Fund, along with other foundations and public agencies, has supported the National Committee for Quality Assurance (NCQA) in its efforts to develop and collect data on quality indicators, including women's health indicators, and make the data available to employers and consumer groups. An NCQA Measurement Advisory Panel (MAPS) is considering the development of a fuller set of women's health measures.

These efforts will provide employers, government agencies, and consumers with comparative data about managed care plans so that they can make informed choices. Already underway are case studies, surveys of vulnerable populations, and development of new methods to adjust payment systems to reward managed care plans that attract and care for people with serious mental or chronic illnesses. Further research, however, to develop suitable measures of women's health, analyze variable performance across plans, and identify best practices in improving women's health care is essential.

While improved information for making health care decisions is an important ingredient, there is also a need for government, whether at the federal or state level, to establish the rules of the game that will assure minimum quality standards, systems to monitor access and quality, public disclosure of information on performance, patient rights and protections, guarantees of physician independence in medical decision making, appropriate financial incentives, and fair marketing practices. We will all benefit from quality standards that make the health care system as a whole more responsive to patients and that increase accountability of plans for the health of women and their families. Beneficiary education to understand the choices available and assist women in selecting plans best for their families' needs is particularly important for Medicaid and Medicare, with their concentration of low-income, chronically ill, and frail elderly beneficiaries. Most fundamentally, all Americans, including the poor and the uninsured, must have access to quality health care.

The nation is still in the early stages of its experiment with managed care. Health care paradigms will continue to shift, but the risks and rewards of a market-driven system are beginning to reveal themselves with greater clarity. Women have a common interest in shaping initiatives to meet the health care needs of the nation and in realizing the promise of better health care for themselves and their families.

References

1. Louis Harris and Associates (1997). "Kaiser/Commonwealth National Survey of Health Insurance." Louis Harris and Associates, New York.
2. Carlson, K. J. (1997). Primary care for women under managed care: Clinical issues. *Women's Health Issues,* November/December, 349–361.
3. Collins, K. S., and Simon, L. J. (1996). Women's health and managed care: Promises and challenges. *Women's Health Issues,* January/February, 39–44.
4. Wyn, R., Collins, K. S., and Brown, E. R. (1997). Women and managed care: Satisfaction with provider choice, access to care, plan costs, and coverage. *J. Am. Med. Women's Assoc.,* Spring, 60–64.
5. Collins, K. S., Schoen, C. A., and Khoransanzadeh, F. (1997). Practice satisfaction and experiences for women physicians in an era of managed care. *J. Am. Med. Women's Assoc.,* Spring, 52–56, 64.
6. Bernstein, A. B. (1996). Women's Health in HMOs: What we know and what we need to find out. *Women's Health Issues,* January/February, 51–59.
7. National Committee for Quality Assurance (1997). "The State of Managed Care Quality." Washington, DC.
8. Warner, C. K. (1996). Planning for women's health care services. *Women's Health Issues,* January/February, 60–63.
9. Heiser, N., and St. Peter, R. (1997). "Improving the Delivery of Clinical Preventive Services to Women in Managed Care Organizations: A Case Study Analysis." The Commonwealth Fund, New York.
10. Davis, K. (1997). "Managed Care and Patients at Risk," Annual Report 1996. The Commonwealth Fund, New York.
11. Schoen, C., and Davidson, P. (1996). Image and reality. *Bull. N. Y. Acad. Med.* **73,** 506–531.
12. Davis, K., Collins, K. S., Schoen, C., and Morris, C. (1995). Choice matters: Enrollees' views of their health plans. *Health Affairs* **14** (Summer), 99–112.
13. Ware, J. E., Jr., *et al.* (1986). Comparison and health outcomes at a health maintenance organization with those of fee-for-service care. *Lancet* **1,** 1017–1022.
14. Ware, J. E., Jr., *et al.* (1996). Differences in four-year health outcomes for elderly and poor, chronically ill patients treated in HMO and fee-for-service systems. *JAMA, J. Am. Med. Assoc.* **276,** 1039–1047.
15. Health Care Financing Administration (1996). "Medicaid Managed Care Enrollment Report." Washington, DC.
16. Gold, M., Sparer, M., and Chu, K. (1996). Medicaid managed care: Lessons from five states. *Health Affairs* **15** (Fall), 153–166.
17. Sisk, J. *et al.* (1996). Evaluation of Medicaid managed care. *JAMA, J. Am. Med. Assoc.* **276,** 50–55.
18. Short, P. F. (1996). "Medicaid's Role in Insuring Low Income Women." The Commonwealth Fund, New York.
19. Jones, S. B. (1996). "Why Not the Best for the Chronically Ill? Research Agenda Brief." George Washington University Center for Health Policy Research, Washington, DC.
20. Health Care Financing Administration. (1996). "Profiles of Medicare," p. 93 (based on data from the Medicare Current Beneficiary Survey, 1993). Washington, DC.
21. Davis, K. (1996). Rules needed to ensure efficiency and quality in Medicare managed care. *Internist* **37,** 9–12.

7

Lesbian Health Research

JUDITH BRADFORD* AND JOCELYN C. WHITE†

*Survey and Evaluation Research Laboratory, Virginia Commonwealth University, Richmond, Virginia; †Department of Medicine, Legacy Good Samaritan Hospital, Portland, Oregon

I. Importance of Studying Lesbian Health

Lesbians are found throughout all areas of the world and among all subgroups of the general population, in all racial and ethnic groups, all occupations and professions, all ages, and all socioeconomic strata. Health care organizations and providers who attempt to meet the needs of lesbians typically do so without the benefit of formal training and professional experience with lesbian clients. A paucity of literature about the specific health care needs of lesbians can result in a perception that lesbians' needs are no different than those of other women and at a minimum results in a health workforce uninformed about how to identify and treat these individuals. Official recognition of the importance of studying the health status and health care needs of lesbians was granted in 1997, when the National Institutes of Health (NIH) Office of Research on Women's Health funded a workshop study on Lesbian Health Research Priorities to be conducted by the Institute of Medicine (IOM). Additional funding for the study was contributed by the Centers for Disease Control and Prevention (CDC) through the NIH. By convening the Institute of Medicine Committee on Lesbian Health Research Priorities, the need for research-based knowledge about lesbian health was acknowledged publicly. The IOM Study Committee identified several reasons why attention should be paid to lesbian health, based on its deliberations from June through December 1997 and on expert testimony given at its October 1997 two-day invitational workshop [1].

Studying lesbian health should result in knowledge that will be useful for improving the health status and health care of lesbians, and knowledge of how the health of lesbians differs from that of other women may provide insights to improve the health of all women. Uncovering the reality, or lack of reality, behind various myths about lesbians and countering misconceptions about their health risks may lead to improved health outcomes for these women. Identification of health risks faced by lesbians and how their risks may differ from those of other women will assist lesbians and health providers in their efforts to prevent unwanted conditions and promote healthy lives for this underserved population.

II. Definition of the Population

Lesbians have been defined in various ways, including "a woman who calls herself a lesbian," "a woman-identified-woman," and "a woman who has sex with other women (WSW) [1]." There is no one right way to define lesbianism or lesbian, which presents researchers with the challenge to de-velop operational definitions for each new study. Overall, it seems best to adopt a flexible approach to defining the population, allowing considerable latitude for participants to identify themselves in ways that feel both authentic and safe, while making it possible for researchers to recruit samples that satisfy study demands and protect the interests of participants. After considerable discussion, the IOM Committee decided to focus for its report on two groups of women who fit the most generally accepted definitions of "lesbian": women who have sex with and/or primary emotional partnerships with women. Other groups of women who some would want to include, such as bisexual women, were not considered lesbian for the purpose of the IOM study because there has not been enough research about them to determine whether their needs were similar to, or quite different from, those of lesbians.

That flexibility is of the utmost importance when classifying particular individuals as lesbians is made clear by the research of Laumann and colleagues at the National Opinion Research Center (NORC) when they conducted the "National Sex Study" earlier in the 1990s [2]. Consistent with prevailing conceptions of sexual orientation as including behavioral, affective, and cognitive dimensions, these researchers asked a series of questions to explore the sexual experiences of a random sample of adults 18 and over throughout the United States. Results validated the multidimensional conceptions of sexual orientation that a number of other researchers had put forward [3–5].

Women in the NORC sample exhibited differing patterns of same-sex sexual behavior, attraction/desire, and identity, in combinations that varied among respondents. Variations reported by women in the Laumann *et al.* data set are shown in Figure 7.1, where it can be seen that among 150 women (8.5% of 1749 sampled) who fit at least one definition of same-sex sexuality as measured for this study, a majority (58.7%) reported desire only (no same-sex behavior and heterosexual identity). Everyone who self-identified as bisexual or homosexual reported either same-sex desire without same-sex behavior (0.7%) or both same-sex desire and same-sex behavior (15%). One in four of these 150 women self-identified as heterosexual but reported same-sex behavior without current desire (12.7%) or current same-sex desire and behavior (12.7%).

Researchers who wish to study lesbians must select an operational definition of eligible participants for each new project. Using the Laumann distribution as a way to frame this challenge, a researcher could broadly define "lesbians" as all women who report any level of response to any of the three dimensions measured (current desire, current or past same-sex behavior, current identity as homosexual or bisexual). Alternatively, the researcher could define lesbians as narrowly as women who

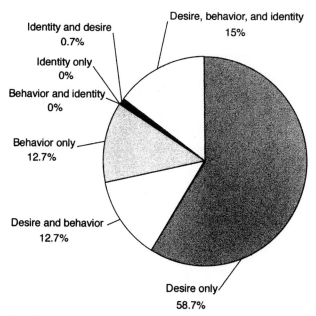

Fig. 7.1 Interrelation of the different aspects of same-sex sexuality (current desire, current or past same-sex behavior, current identity as homosexual or bisexual) for 150 women (8.6% of the total 1749) who report any adult same-sex sexuality. Source: [2].

identify as homosexual or as somewhere in-between these two measurement extremes.

A. Population Estimates for Lesbians

Without a generally accepted definition of lesbian, there can be no agreement on the number of lesbians or their prevalence within the general population of women. Estimating the number and prevalence of lesbians is further complicated by the preference of many women who have same-sex desire and/or behavior, or who self-identify as lesbian or bisexual, to keep this information hidden from others. Despite the 1973 reclassification of homosexuality by the American Psychiatric Association as neither an illness nor a pathological condition, negative attitudes about gays and lesbians continue to be held by the general public and many states have laws that explicitly target gays and lesbians [6–8]. Same-sex marriage is not legal anywhere in this country and is specifically banned in 25 states. Although a handful of cities and states have passed ordinances or laws to allow domestic partner benefits, leadership in this area has come from business rather than from government. Despite growing public support for the right of lesbians to adopt children [6], in some states efforts are underway to prevent same-sex couples from adopting and from serving as foster parents. These attitudes and practices provide very powerful incentives for women who might otherwise identify as lesbians to hide aspects of their lives from others—to "closet" themselves in order to enjoy the greatest possible independence and access to societal benefits enjoyed by heterosexuals.

Anticipated social discrimination (with consequences such as loss of employment or restriction of housing options) is perhaps the primary reason why many lesbians choose not to disclose

their sexual orientation, but demographic and cultural characteristics also affect their lifestyle and disclosure patterns. In the Laumann sample, proportions of women who could be counted as lesbian varied on the basis of age, marital status, education, religion, race/ethnicity, and place of residence. Variations in the proportion of respondents who could be considered lesbian were based on how this was measured; from 1.4% who self-identified as homosexual or bisexual, 3.8% who reported one or more same-sex partners since puberty, to 4.3% who reported one or more specific forms of same-sex behavior (such as oral sex, sex for pay, etc.), to 7.5% who reported that they were attracted to women or found sex with women to be appealing (Table 7.1).

The proportions of women who could be considered lesbians also varied by age (younger women were more likely to self-identify as homosexual or bisexual), marital status (never married women were more likely to report desire or attraction to women), and education (higher proportions of the most educated women fit one or more operational definitions of lesbian). Religion, race/ethnicity, and place of residence also were associated with varying "rates" of lesbianism.

These results are quite consistent with those of other studies conducted with nonprobability samples of self-identified lesbians, such as the 1984–1985 National Lesbian Health Care Survey (NLHCS) [9]. Among this sample of 1925 self-identified lesbians, only a small proportion lived in the same town or city where they had been born, and most lived in metropolitan areas with gay communities or in other areas where lesbian-only events and organizations could be found. Ninety-two percent of NLHCS respondents had been affiliated with a religious community while they were growing up, but only 33% had a religious community at the time of the survey, and most of those who did had shifted from the traditional groups of their youth to groups known for their acceptance of gays and lesbians. Similar to women in the Laumann *et al.* study, respondents' age was also related to disclosure about sexual orientation. Younger lesbians in the NLHCS sample were much more likely to be open about their interests in same-sex sexuality than were their older counterparts.

B. Becoming a Lesbian

No one really knows why certain individuals become homosexuals, although studies have suggested a biological basis and that environmental factors may also be involved [10]. Most experts believe that sexual orientation is not a conscious choice. Some studies have also challenged the common misconception that sexual abuse or family dysfunction can cause children to become gay or lesbian as they grow up [11–13]. The question of etiology is best framed at this time from a developmental perspective, recognizing that a proportion of all women will self-identify as lesbians, partner with other women, feel attraction and desire to other women, and/or engage in sexual behavior with other women. This proportion ranges from 2 to 10% or more [4], depending on a mix of sociodemographic and other characteristics.

The process of a girl or woman's discovery that she may be homosexual can begin quite early in life and is often associated with notable emotional distress [11]. A study of 358 adult women found that recall of age of first sexual or romantic attraction

Table 5.1

Percentage of Women Reporting Various Types of Same-Sex (SS) Sexuality by Selected Social and Demographic Variables[a]

	Any SS partners since puberty[b]	SS activity since puberty[c]	Desire SS partner[d]	Identify as homosexual or bisexual
Total	3.8	4.3	7.5	1.4
Age				
18–29	2.9	4.2	6.7	1.6
30–39	5.0	5.4	9.2	1.8
40–49	4.5	4.6	8.3	1.3
50–59	2.1	1.9	4.6	0.4
Martial status				
Never married	5.6	5.9	10.4	3.7
Married	2.6	2.8	5.2	0.1
Divorced, widowed, separated	4.1	5.5	9.6	1.9
Education				
Less than high school	3.3	1.8	3.3	0.
High school graduate	1.8	2.3	5.3	0.4
Some college or vocational	3.9	5.1	7.3	1.2
College graduate	6.7	7.3	12.8	3.6
Religion				
None	9.9	11.3	15.8	4.6
Type 1 Protestant	2.1	2.0	5.2	0.5
Type 2 Protestant	2.9	3.3	5.5	0.3
Catholic	3.4	4.2	8.4	1.7
Jewish	6.9	12.5	10.3	3.4
Other	18.9	14.7	16.2	5.4
Race or ethnicity				
White	4.0	4.7	7.8	1.7
Black	3.5	2.8	7.0	0.6
Hispanic	3.8	3.5	7.6	1.1
Place of residence				
Top 12 central cities	6.5	4.6	9.7	2.6
Next 88 cities	5.7	7.7	7.8	1.6
Suburbs of top 12 cities	5.7	4.1	9.0	1.9
Suburbs of next 88 cities	3.3	4.8	9.8	1.6
Other urban areas	2.7	3.4	6.9	1.1
Rural areas	2.1	2.2	2.1	0.0

[a]Based on data gathered in the National Health and Social Life Survey (N=1749 adult women).

[b]Any SS partners since puberty (having had a same-sex partner at any time since puberty).

[c]SS activity since puberty (having engaged in specific sexual behaviors with another woman at any time since puberty).

[d]Desire SS partner (attracted to women or found sex with women to be appealing).

and self-acknowledgment of sexual orientation did not vary based on whether participants self-identified as heterosexual, bisexual, or homosexual [14]. In a study of 194 lesbian, gay, and bisexual youth 21 and older, respondents reported that their first awareness of sexual orientation (and themselves as different from others) occurred at about age 10 [15]. On average, about six years passed before these youths disclosed their sexual orientation to another person.

A woman becomes a lesbian through a developmental process of two dimensions: first, coming out to herself by acknowledging her same-sex feelings and accepting herself as a nonheterosexual, and second, coming out to others. This process has been described in a number of ways over time and can be generically

discussed as a shift in core identity that takes place in four stages: (1) awareness of homosexual feelings, (2) testing and exploration, (3) identity acceptance, and (4) identity integration and disclosure to others [11,16–19]. Because prevailing social attitudes and the anticipated response of family and friends influence the experience of coming out, each lesbian must consider the advantages and disadvantages of disclosing to others, and some perform a conscious weighing of costs and benefits for each situation in which they consider self-disclosure.

In the NLHCS, respondents were asked to disclose the percentage of persons within four groups to whom they had come out. Although 88% had disclosed their lesbianism to all the gay people they knew, much smaller percentages were out to everyone in the other three groups: 28% were out to all of their straight friends, 27% to all family members, and 17% to all of their coworkers [9]. Patterns of outness varied on the basis of sociodemographic characteristics and place of residence, as did proportions of women respondents to the Laumann *et al.* study [2] who fit the various definitions of lesbian, conducted 10 years later. Frequency of attendance at religious events (among NLHCS respondents) was negatively associated with all measures of outness, as community of religious affiliation or having no affiliation at all were associated with varying rates of lesbianism estimated by the Laumann *et al.* measures of homosexuality.

In 1963, the sociologist Erving Goffman described the two-world nature of life for persons with stigmatized conditions [20]. Integrating the psychological demands of public and private life, while maintaining necessary boundaries to protect both of these, presents significant challenge for all human beings. For those with a stigmatizing condition, such as mental illness or physical disability, the need to master this challenge is particularly poignant. Telling others may result in stereotypical views of who you are, while not telling others means hiding any potentially observable signs of the condition.

So it goes with lesbianism, or at least such is the fear of many women who exhibit feelings or behaviors associated with this ''condition.'' Their concern is valid and must be taken into consideration by researchers who relate to lesbians. Understanding what lies behind this concern and how it affects the health and access to health care of lesbians is necessary before an effective research strategy to address their unique needs can be created.

III. Historical Evolution and Trends

Systematic research about lesbians was not initiated until the 1950s, when the first studies focused on etiology, seeking to identify factors that would cause a woman to be a lesbian. Tracing the development of studies about lesbians over four decades, Tully [21] notes that a second phase of studies during the 1960s and 1970s compared clinical and nonclinical samples of lesbian and heterosexual women to determine if lesbianism was a form of psychopathology. During the end of this period, researchers (many of whom were themselves lesbian) began to conceptualize lesbianism as a psychologically healthy way of life and to focus explicitly on the psychosocial aspects of life experienced by women who self-identified as lesbian.

Beginning with the 1980s, researchers studying lesbians began to explore psychosocial concerns to understand lesbians' life span development, and in 1984/1985 a higher level of interest emerged through design and implementation of the NLHCS and several other substantial health survey efforts that took place during that time period [22]. The NLHCS made use of a systematic approach to questionnaire design and data collection methodology, seeking to identify the health needs and health care utilization patterns of a broadly representative sample of lesbians. The researchers and their organizational sponsor, the National Lesbian and Gay Health Foundation, clearly stated the intent to study lesbians and their health care needs in order to understand how to improve health care delivery to this population. The conceptual linkage from research about lesbians to an

effective response from health care providers was made. Since then, others have focused their efforts on lesbian health and a body of knowledge has begun to develop [1].

A. How Early Surveys Were Done

In her 1995 review of lesbian health literature for practicing physicians, Dr. Kate O'Hanlan provides a framework for understanding the approaches taken in early lesbian health surveys, as well as a practical way to assess the relevance of their findings. In addition to the NLHCS, O'Hanlan acknowledges the contributions to basic understanding about lesbian health of surveys conducted by the Michigan Department of Public Health, the Los Angeles Gay and Lesbian Community Services Center, and the San Francisco Department of Public Health [22]. Of particular significance to an early understanding about lesbian diversity was the Black Women's Relationship Project, conducted in 1984 and directing its focus to the influence of multiple levels of discrimination on the relationships and community interactions of African-American lesbians [23]. During this same period of time, the Fenway Community Health Center in Boston conducted its own national survey to develop a comprehensive understanding of the sexual practices and other health-related concerns of lesbians who could be in need of care. In its very early efforts to integrate primary research and health care development, Fenway served as a benchmark for other gay and lesbian community health centers that were opening doors to care for populations previously unable to find knowledgeable, caring providers.

Early lesbian health surveys were subject to the same limitations that researchers face for contemporary studies, such as lack of common definitions, difficulty in recruiting participants, inadequate funding, and general lack of interest from the scientific community. Nevertheless, these early surveys and others that are not as widely known were successful in building a foundation upon which the serious study of lesbians' health status and health care needs can be pursued. Although these studies have never received broad acknowledgement from the mainstream scientific community, the research results have been widely disseminated throughout the gay and lesbian community. Professional and lay members of this community assessed their relevance and felt confident enough in their quality to create a body of accepted knowledge from which to build a self-generated system of health education, clinical care, and supportive services.

B. Barriers Encountered by Researchers

Universities and other research settings are often viewed by the general public as very liberal environments, accepting of virtually all difference, yet this has not been the reality for lesbian employees. Tierney [24] described a paradox between the emphasis on academic freedom and the constraints on permissible fields of study, which do not include research on gay and lesbian issues. In a study conducted at Oberlin College, well known for its progressivism, students and employees were found on the one hand to be acquainted with lesbian, gay, or bisexual individuals (40% knew well at least one such individual), while on the other hand nearly one in five of lesbian, gay, and bisexual employees had been socially ostracized, verbally insulted, and threatened with exposure [25]. Over 80% of all student respondents, regardless of reported sexual orientation, had heard stereotypical or derogatory remarks or seen homophobic graffiti on campus. Studies have been conducted on other campuses with similar results, including Yale University [26], Pennsylvania State University [27], the University of Massachusetts [28], and the University of California at Santa Cruz [29].

Anticipating an increased interest in research about lesbians, in 1996 An Uncommon Legacy Foundation funded a mail survey of lesbian researchers to assess their interest in participating in a network designed to provide them with technical assistance and peer support. A questionnaire was designed to gather information about the training and resource needs faced by those who wanted to do research about lesbians. As the survey mailing list was being developed, 430 lesbian researchers expressed interest in joining such a network. Nearly 300 (in 41 states and 14 countries) completed questionnaires, providing data about their educational and research backgrounds, experiences related to being a lesbian and doing lesbian research, areas of expertise and research skills, mentoring and technical assistance needs, and recommendations for prioritizing lesbian research network activities [30]. Results were generally consistent with studies about university acceptance of gays and lesbians and provided evidence that to be productive researchers, lesbians may well be in need of assistance.

A majority of respondents had at least 10 years of experience as researchers, yet fewer than a third had what they described as a very supportive work environment in which to do research about lesbians. One in four described working situations that were not supportive, more than a third (37%) reported that doing lesbian research had negatively affected their employability, and 29% reported that it had negatively affected their professional development. The impact of being a lesbian had been experienced as even more negative than doing research about lesbians. Nineteen percent thought conducting research about lesbians had limited their chances for employment, and a greater proportion (28%) thought being a lesbian had this consequence.

Despite the negative conditions under which many respondents were carrying out research about lesbians, a majority who did (59%) thought there had been a positive impact from their work (e.g., "It helped me realize the importance of being out if change is to occur." "All of the research we do in the area of lesbian issues will open doors for lesbians.") However, not even half of the lesbian researchers who participated in this study had actually done research about lesbians themselves, although a very large majority wanted to. Survey results indicate that a cadre of trained researchers may be available and ready to participate on research teams, if appropriate opportunities are created. Although most respondents were in academic positions, 13% worked for community-based organizations and 10% worked in health care settings, providing further evidence that well-rounded teams with researchers from various perspectives are realistic to envision.

IV. Specific Health Conditions of Concern to Lesbians

A general lack of interest in research about lesbian health, coupled with inadequate institutional support for researchers who would like to pursue work in this area, has resulted in a dearth of scientific knowledge from which to develop clinical and programmatic interventions. In addition, existing studies have not been conducted with sufficient rigor to provide evidence about whether lesbians may be at higher or lower risk of certain health problems relative to heterosexual women or women in general. Nevertheless, from several decades of research conducted primarily by the lesbian research community, convergent results from numerous studies have created a fairly clear consensus of what the salient areas for research should be. These areas include a variety of physical and mental health concerns and incorporate a necessary focus on health issues faced by WSWs and the unique aspects of lesbian life brought about by societal stigmatization.

A. Cancer Risks and Screening

Breast cancer is thought to be theoretically more common in lesbians than in other women because of the presumed confluence of risk factors such as nulliparity, heavy alcohol use, high body mass index, and cigarette use common in lesbians [22]. Only one study to date has directly compared populations of lesbians with other women and found differences in risk factors for breast cancer. Roberts and colleagues [31] reported that lesbians in this retrospective chart review at one clinic had a higher body mass index and lower rates of smoking, pregnancy, and oral contraceptive use than heterosexual women. Prospective studies in larger populations are needed to elucidate breast cancer risks in lesbians.

Cervical cancer appears less common among lesbians than among bisexual or heterosexual women, as suggested by lower rates of dyplasia and abnormal Pap smears [32,33], but it does occur in women who are sexually active only with women [34].

Ovarian cancer has been reported to occur more frequently in women who have not used oral contraception and those who have not given birth. Lesbians are more likely to be nulliparous and to not be current or recent users of oral contraceptives. They often appear to have a higher BMI and to smoke more than heterosexual women [22]. Endometrial cancer is also more common in nulliparous women. Thus, many lesbians can be expected to have these risk factors for cancer, and physicians should follow current screening guidelines.

Lesbians over 40 report smoking and drinking alcohol [9,22, 35–38] more often than their heterosexual counterparts. No data have been reported on lung or head and neck cancers in lesbians.

Because no population-based studies of cancer rates in lesbians have been reported, clinicians should screen patients based on individual risk factors using standard screening guidelines for women.

B. Cardiovascular Risks

Cardiovascular diseases (heart, stroke, and artheroschlerosis) represent the leading causes of death for women in general. Risk factors for heart disease include cigarette smoking, hypertension, high blood cholesterol, excessive weight, use of oral contraceptives, and physical inactivity [39]. Nonrandom studies have indicated that lesbians may smoke more and have a higher BMI and thus may be at higher risk for cardiovascular disease. Because lesbians appear to seek health care less often, they are less likely to receive blood pressure and cholesterol screening, further compounding their risk. Based on these probable risks, and because there are no population-based studies on lesbians and cardiovascular disease, additional research is needed to elucidate lesbians' risks in this area.

C. Mental Health and Psychosocial Issues

Lesbian adolescents are particularly vulnerable to the emotional distress of coming out, and this distress often confounds their developmental tasks [40]. Parental acceptance during this process, especially from the mother, may be the primary determinant of the development of healthy self-esteem in adolescent lesbians [41]. In one report that found that gay and lesbian youth are two to three times more likely to attempt suicide and may account for 30% of completed youth suicides, signs of confusion about sexual orientation in adolescents included depression, diminished school performance, alcohol and substance abuse, acting out, and suicidal ideation [42,43]. It should be noted that data on gay youth are difficult to collect, and studies on youth suicide are controversial. Additional studies are needed to clarify degree and nature of risk. At this time, it is critical that primary care professionals screen adolescents for these signs and consider confusion about sexual orientation in the differential diagnosis of depression and substance abuse.

Domestic violence has been reported in lesbian relationships. In one study, more than a third of lesbians 22 to 52 years of age reported experiencing battery by a partner; alcohol and drug use was involved in 64% of these incidents [44]. Early research suggested that interpersonal power imbalances were the underlying source of lesbian battery, and more recent studies suggest that lesbian battery may be characterized by mutual abuse [45,46]. One partner acts as the batterer while the other engages in emotional abuse. In these cases the power imbalance cycles between the two partners.

Lesbians may be victims of other forms of violence as well. According to a study for the U.S. Department of Justice, lesbians and gay men may be the most victimized groups in the nation [47]. The number of hate or bias crimes against lesbians, including verbal abuse, threats of violence, property damage, physical violence, and murder, increases each year [27,48]. Lesbians at universities report being victims of sexual assault twice as frequently as heterosexual women [49]. Perpetrators of hate crimes often include family members and community authorities [48]. Many gay and lesbian adolescents have been forced out of their homes because of abuse related to their sexual orientation [12,50–52]. Although the actual number of lesbian and gay runaways and "throwaways" is not known, local reports are very disturbing. For example, in Seattle, 40% of homeless youth are estimated to be lesbian or gay [53].

D. Sexually Transmitted Diseases

No known gynecological problems are unique to WSWs or lesbians and none occur more often in lesbians than in bisexual or heterosexual women. Actually, sexually transmitted diseases appear to be less common in women who identify as lesbian and in women who are sexually active only with women than in either heterosexual women or gay men. This may be due in part

to a relative epidemiologic isolation of this group from men and the lack of penile–vaginal intercourse. Lesbian sexual practices include kissing, breast stimulation, manual and oral stimulation of the genitals and anus, friction of the clitoris against the partner's body, and penetration of the vagina and anus with fingers and devices.

Human papillomavirus and bacterial vaginosis have been shown to be transmissible between women and do occur in lesbians [34,54]. Candidiasis and *Trichomonas vaginalis* infection do occur in lesbians and appear to be transmissible between women [55]. Women sexually active only with women appear to have a lower incidence of syphilis and gonorrhea than any other group except those who have never been sexually active. Infections with chlamydia or herpes virus disease appear to be less common in lesbians who have been sexually active exclusively with women, but all are theoretically transmissible [32,33,55]. Hepatitis A, amebiasis, shigellosis, and helminthism also have a low prevalence in these women. Hepatitis B and C occur only when other risk factors are present [56,57].

E. Human Immunodeficiency Virus (HIV)

The number of women who are HIV positive who report having sexual contact with women is low, and more than 90% of lesbians with the acquired immunodeficiency syndrome (AIDS) are injection drug users [58]. To date, HIV may have been transmitted between women as a result only of sexual contact in as many as nine cases, but this has not been proven by rigorous methods [59–63]. Exposure to menstrual and traumatic bleeding was probably the source of transmission in these cases. HIV has been cultured from cervical and vaginal secretions and cervical biopsy specimens taken throughout the menstrual cycle, however [64–69]. Therefore, it may theoretically be transmitted by infected women who are not bleeding.

Lesbians and WSWs who seek to become pregnant frequently use donor insemination. These women who undergo artificial insemination with either fresh semen from donors in the community or frozen semen from sperm banks are also at risk for HIV infection [70]. Sperm banks routinely test donors for HIV infection at the time of donation and six months later before releasing the specimen for use. Because of delays in seroconversion, however, it is possible for lesbians to be exposed to HIV with frozen semen or fresh semen from a seronegative donor. Fresh semen donors, often gay men, may also be tested at the time of donation and six months later. The delay in seroconversion is a risk in these cases also, as is the chance that new exposure to the virus may have occurred in the intervening six months. Clinicians discourage this practice.

F. Substance Abuse

Epidemiologic studies on alcohol and other drug abuse rarely ask about sexual orientation. Studies of the lesbian population have only occasionally surveyed alcohol and drug use, and often with methodologic flaws [71]. Early studies of alcohol use in lesbians recruited participants from bars. McKirnan and Peterson reported that lesbians were less likely to abstain from alcohol than heterosexual women, more likely to be moderate users, and equally likely to be heavy drinkers [35,36]. Interestingly, lesbians reported having alcohol problems almost twice as often as heterosexual women. These studies also reported less decline in alcohol use with age for lesbians than for heterosexual women. Skinner also reported this lower decline in alcohol use with age and found higher rates of drinking among lesbians than among women in a geographically matched sample [37]. Bradford and Ryan also found that lesbians' alcohol use did not decline with age but appeared to increase with age [9].

Using random phone surveys in the bay area, Bloomfield, by contrast, reported that there were no significant differences in levels of drinking and bar-going behavior between lesbians and bisexual women and their heterosexual counterparts [72]. The proportion of women who reported being in alcohol recovery was greater for lesbians and bisexual women than for the heterosexual women.

Most of the research on risk factors for alcohol abuse has been based on studies of gay men and women in the general population. Suggested risk factors include reliance on gay bars for socialization and discrimination related to homophobic attitudes. In addition, Hughes suggests age factors, employment factors, having multiple social roles, being in a cohabitating but unmarried relationship, depression, stress, having a partner who drinks, and perhaps a history of childhood sexual abuse [71]. Domestic violence is also strongly associated with heavy use of alcohol.

Even fewer studies have examined drug abuse in lesbians. McKirnan and Peterson [35] found that rates of marijuana and cocaine use were higher among lesbians, and gender differences in use were smaller than those found in the general population. Similarly, Skinner and Otis [38] found few gender differences, but lesbians reported higher rates of use of marijuana and cocaine than did gay men in some age groups. In addition, lesbians in that study were more likely to report smoking than were their gay male counterparts.

V. Barriers to Accessing Health Care

In a frank 1977 discussion of physicians' allegiance to the social attitudes and institutions they represent, Thomas Szasz admonished his readers to require experts to disclose the agents and values they represent as a protection against abuse of the power inherent in their specialized knowledge and skills [73]. Two decades later, his advice is as applicable for health researchers as it is for the providers whose effectiveness is based not just on their own values and those of the institutions they work for, but also on the quality of research from which interventions are developed. These concerns hold true for understudied and underserved populations in general, including lesbians.

Lesbians experience barriers to accessing health care that other women experience, but in addition, there are several barriers they face specifically because they are lesbian. First and most challenging are negative attitudes toward lesbians that many providers hold. Forty percent of physicians in one study were sometimes or often uncomfortable providing care to lesbian or gay patients [74]. Many lesbians report that their doctors are not sensitive to or knowledgeable about their particular health risks and needs and so do not disclose pertinent information [75,76]. Varying numbers of lesbians across the country

report being unable to come out to their primary care provider. In one Michigan survey [77], 61% of lesbians reported that they were unable to disclose their sexual orientation to their physician. In contrast, in Oregon, 90% of lesbian respondents had disclosed their sexual orientation to their health care providers, and of these, 92% had raised the issue themselves with the provider [78,79]. The degree to which a lesbian discloses sexual orientation to a provider appears to have an effect on health risks and screenings.

In the Oregon study, lesbians who had come out to their primary care provider were more likely to seek health and preventive care, have ever had a Pap test, be a nonsmoker, and be comfortable discussing difficult issues with the provider. Likewise, difficulty communicating with the primary care provider was associated with delay in seeking health care. The communication style of the provider was rated by respondents as the most important provider characteristic in determining ease of discussing difficult issues. The study suggests that lesbians appear to be motivated to disclose sexual orientation to their primary care providers, but provider attitudes and communication skills may pose a significant barrier that affects not only patient satisfaction but also health status.

Physicians are not alone among health providers in their capability to discriminate against and even to cause harm to their patients. In a random sample survey of Virginia mental health providers, respondents acknowledged having lesbians in their practices yet having none or very little training about the special needs of these clients [80]. A majority of practitioners in this study gave inadequate or inappropriate definitions of lesbianism and many stated that they did not think the concerns of lesbian clients were any different from those of heterosexual women. Findings were comparable to those of the California study of physicians [74].

In part because of the attitudes of health care providers, lesbians as a group face financial, structural, and personal and cultural barriers to accessing competent, sensitive health care services. When a lesbian encounters any of these barriers, she may not receive the screening and prevention services she needs and she may also experience delays in receiving care for acute conditions.

Reacting to poor access to the general health care system, for two decades lesbian communities have been building elements of their own system. The lesbian health movement was initiated in the early 1970s as lesbians began providing health services for other lesbians at women's health clinics, often during the evening outside of normal clinic hours. These "lesbian health nights" were the first known efforts to provide safe and supportive health care to this population of women [81]. The movement grew steadily throughout the 1980s, when it was sidetracked for a number of years as many lesbian providers and activists shifted their priorities to the AIDS epidemic.

By the 1990s, however, lesbians throughout the country began reinvigorating lesbian health services, culminating in a contemporary national response which ranges from Miami to Anchorage and from Hartford to San Diego [81]. As this response grows, awareness of the barriers lesbians face in gaining access to quality health care will continue to increase, and the general inadequacy of data-based knowledge on which improvements can be designed and implemented will be improved.

A. Financial

Lesbians as a group have a lower socioeconomic status despite higher educational level than their heterosexual counterparts. Many are self-employed in businesses or work as artists or crafts workers or work part time [9]. As such, many have few or no health insurance benefits, nor do they have access to health care benefits through their partner's employer. Lesbians are at a disadvantage compared to married heterosexual women because of the prohibition against spousal benefits for unmarried partners [82,83]. While in many localities low-cost or free health and screening services are available to women who are seeking birth control, lesbians who do not have a need for birth control find it hard to locate affordable health care services.

An unpublished analysis of NLHCS data provided evidence that subgroups of lesbians may be at particularly high risk of negative health consequences due to lack of insurance coverage. Within the NLHCS sample, lack of health insurance was significantly correlated with being younger, unemployed, in school, of lower income, and African-American. Mental health issues were more prominent among uninsured respondents, who reported significantly higher levels of anxiety and suicidal ideation. Uninsured respondents were also more likely to have experienced physical and/or sexual abuse and reported much greater concern about sometimes feeling unable to meet their routine responsibilities. Certain physical health conditions were also more prominent, including ulcers and other intestinal disorders, substance abuse, and eating disorders. There was a statistically significant correlation for this sample between not having health insurance and believing that being lesbian had affected their access to health care [84].

B. Structural

Throughout the health care system there exists an institutionalized assumption of heterosexuality. This assumption is evident from the moment a lesbian patient sits down to fill out questionnaires to the time she leaves with health education brochures in hand. Health care intake forms routinely assume a patient is single or married and leave no room for discussion of a significant other and what relationship that person may have to the patient. In the Michigan Lesbian Health Survey, 9% of respondents reported that their health providers had not allowed their female partners to stay with them during treatment or see them in a treatment facility; 9% also said that providers had not included their partners in discussion about treatment [85].

Intake forms covering sexual history rarely include the option for giving information on same-gender sexual partners. Researchers are taught throughout training to ask questions that assume heterosexuality while taking a history. For example, physicians routinely ask "are you married, single, widowed, or divorced?" "What kind of birth control do you use?" In hospitals, emergency rooms, and intensive care units, visitation policies routinely allow the next of kin to visit the patient, usually excluding partners of lesbian patients. Likewise, policies and laws regarding surrogate decision making, that is who can make decisions with the health care team if the patient is unable to communicate, relate only to legal next of kin, unless a durable power of attorney for health care is in place.

The educational system over the years has failed to educate researchers regarding the unique aspects of lesbian health. Clinicians with an inadequate knowledge base about lesbian health may well fail to meet standards of care when treating or conducting research with their lesbian patients.

C. Personal and Cultural

Many lesbians find that the most challenging barrier to accessing health care is the memory of having come out to a clinician and receiving a negative or inappropriate response. In the NLHCS sample, 27% of respondents said their current providers had assumed they were heterosexual, 16% felt they could not come out to their providers, and 11% said providers had "forced" birth control on them. One in seven of these lesbians (14%) had found their ob-gyn providers hard to talk to [9]. Barriers such as these cause many lesbians to avoid seeking care, or if in care, to fail to disclose their sexual orientation and family structure. Because of this, many lesbians do not receive proper counseling, screening, and psychosocial or family support services from their health care providers.

Clinicians may also fail to accurately diagnose and treat patients without all of the medically relevant information. There is a dearth of science-based information available to providers, many of whom may simply be unaware of the presence of lesbians in their practices. Not knowing that lesbians are presenting for care, and without education about their special health care needs, providers cannot deliver sensitive care. The generation and dissemination of research findings about this population is directly related to solving these problems.

Many clinicians experience discomfort in working with lesbian patients because they may believe that being a lesbian is immoral or is evidence of physiological or psychiatric disturbance. Others may simply believe they do not have a repertoire of questions in the social and sexual histories that will not be offensive to lesbians. They may also worry that failing to assume heterosexuality may offend their heterosexual patients. Still other clinicians assume heterosexuality in structuring their questions and counseling and therefore fail to communicate openness to information regarding lesbian issues. Among physicians interviewed for a cancer screening project conducted by the Mautner Project for Lesbians with Cancer, approximately half stated that they assumed lesbians were in their practices but did not see any reason to respond to this in a direct way [86]. These providers expressed an eagerness to learn more about the needs of lesbians and stated emphatically that they would make changes when they had information about what steps would be appropriate. Additional studies need to be undertaken and results disseminated to providers upon which to base guidelines for lesbian-sensitive services.

Finally, lesbians themselves, when faced with an uncomfortable interaction with a clinician, may lack the skills or self-efficacy to defend against negative experiences. They may feel unable to change physicians or to leave the room in an uncomfortable situation or to speak openly with a clinician about their discomfort. Much of this stems from a history of discrimination and the power imbalance traditionally experienced between clinician and patient. While some referral services and directories of gay- and lesbian-sensitive clinicians exist, many lesbians do not know how to find these providers. The primary care physician panel structure of many managed care insurance companies also limits choice of physicians. Many lesbians may lack the skills or feel uncomfortable interviewing physicians regarding communication skills and knowledge base in lesbian health care.

Clearly further research is needed in understanding and overcoming lesbians' barriers to accessing health care. We also need more information regarding the skills that lesbians successfully use to overcome these barriers and how these skills may be taught to others.

VI. Best Practices for Research about Lesbians

Conducting lesbian health research presents many challenges that must be successfully met before studies can generate reliable and valid data. These challenges begin with the formidable task of defining the study population and include all of the difficulties inherent in doing research about rare and underserved groups. Very specific obstacles must be overcome in order to gain access to the study population and to ensure that knowledgeable, unbiased research teams are assembled. Special risks are undertaken by lesbian participants in research studies, and researchers must take care to ameliorate these. Ensuring confidentiality is of the utmost concern [87,88].

A. General Methodological Concerns

A complex set of methodological concerns must be addressed in each individual study and across multiple studies in order to compare findings from various settings and time periods. These concerns include the following characteristics of past studies about lesbian health: (1) inconsistencies in the way sexual orientation is defined, (2) the lack of standard measures of sexual orientation, (3) the use of small, nonprobability samples, (4) the lack of appropriate control or comparison groups, and (5) the lack of longitudinal data about lesbian development [1]. Until these concerns have been successfully met, it will continue to be difficult to compare results across lesbian health studies, to generalize from study findings to the larger population of lesbians, to assess the health of lesbians relative to other groups of women, and to understand lesbian development and its implications for defining and measuring lesbian sexual orientation.

Attempting to capture sufficient information to identify homosexuals in the National Health and Social Life Survey, Laumann and colleagues developed multiple measures and reported a variety of rates of attraction/desire, behavior, and identity [2]. The variety of measures used and range of percentages generated from the data provide one clear example of how expert researchers face the complexities inherent in conducting research about lesbians. There are no agreed-upon standard questions for assessing whether or not a study participant is a lesbian, and over time researchers have used an array of questions to do this. Despite researchers' awareness of the multidimensional nature of sexual orientation (which Laumann and colleagues attempted to measure), Sell and Petrulio reported in 1996 that self-reported lesbian, homosexual, or bisexual identity was by far the most common method used to categorize lesbians in published articles about public health research [3]. Very clear and specific decisions about definition and measurement must

be made for each study about lesbians if findings are to be useful for understanding this population, yet researchers clearly find this difficult to accomplish. This difficulty results in a lack of confidence in study results depending on accurate assessment of participants' sexual orientation.

Developing representative samples of lesbians, with appropriate comparison or control groups of nonlesbians, is even more difficult because this challenge cannot be met until the study population has been defined. Researchers with sufficient financial resources can take an approach such as that used by Laumann and colleagues, with multiple measures and a sample size large enough to produce some cases in each of the desired cells. Survey research that uses probability sampling, such as the Laumann study, can provide a strong basis for making comparisons about differences among the sample on a wide variety of health and social conditions. However, the validity of survey research results is limited by the degree to which respondents are correctly classified, and in the case of lesbians (and gay men) where definition is fluid, study findings remain open to challenge. Challenge may come from other researchers on methodological grounds and it may also come from spokespersons for the study population on the very basic issue of face validity.

B. Clinical Research Issues

Randomized, controlled clinical trials do not play a significant role in current research about lesbians, and it will be difficult to find resources for such studies until sufficient evidence has been amassed to indicate that lesbians have differential health conditions compared to other women [1]. Notable exceptions are the Women's Health Initiative, in which a question about sexual orientation was added after study initiation and in which some lesbians have self-identified, and a study funded by the National Institute for Allergy and Infectious Diseases (NIAID) that includes a component examining the transmission of genital herpes simplex virus between herpes-discordant lesbian partners [1].

Despite the paucity of clinical research specifically about lesbians or including identified lesbians in the sample, researchers who wish to conduct clinical trials with women participants (not specifically lesbians) will need to recognize that some study participants may be lesbians or WSWs. In the past, lesbians have not been eager to participate in research because they feared negative consequences of disclosure of their sexual orientation and because they did not trust that researchers would always use health information about lesbians for good purposes. We now know that research on lesbian health is important to advancing our understanding the health needs of lesbians and all women. For this reason, it is important to assess the sexual orientation of all clinical trial participants and to provide ethical safeguards and protections for this sensitive information.

1. Assessing Sexual Orientation

All research studies conducted with women participants need to include items that assess sexual orientation. Current recommendations state that even though the subject of the clinical trial is not necessarily related to sexual orientation, it is appropriate to include items that measure the three domains of affectional preference, sexual behavior, and identity. It is important to understand the sexual orientation characteristics of clinical trial/research study participants for two reasons. First, methodological errors may occur in the absence of adequate information about sexual orientation and are most likely to occur in trials concerning sexually transmitted diseases, HIV, cancer risk, reproductive medicine, cardiovascular disease, and the behavioral sciences. Second, failure to recognize that participants are lesbians or WSWs may lead to clinical errors during the course of the trial. For example, WSWs may be treated with oral contraceptives when they do not need birth control, thus incurring unnecessary side effect risks. In another example, researchers may instruct participants in STD prevention but may fail to provide adequate information on transmission prevention between women, placing participants at an increased risk of infection. In fairness to and to provide adequate protection for study participants who place their trust in researchers, all research studies should include an assessment of sexual orientation.

2. Ethical Safeguards

Clinical research is governed by guidelines stated in several ethical codes. The Department of Health and Human Services (DHHS) requires that proposed research be reviewed by an institutional review board (IRB). This IRB must evaluate the risks and benefits to participants of the research procedures and recommend approval or disapproval of the project. The DHHS also requires that participants receive this information about risks and benefits so that they or their guardians may provide informed consent to participating in the research.

A discussion of the risks of participating in a trial or other health study should include information about the risks of disclosure of sexual orientation or the safeguards in place for keeping this sensitive information confidential. Different from financial information but similar to information on injection drug use, release of information about a participant's sexual orientation information could place her at risk of loss of employment, loss of child custody or inability to become a parent through alternative insemination or adoption, dismissal from military service, or rejection by family members. Lesbians are extremely attuned to protecting the confidentiality of their own sexual orientation. Researchers who study women's health should be well informed about the risks of sexual orientation disclosure for lesbian participants and forthright about safeguards to protect their confidentiality.

Privacy and confidentiality safeguards are often delineated in writing in a prominent position within the informed consent document. Researchers wishing to work closely with the lesbian community, those who expect to have many lesbian participants due to the nature of their studies, and those who will need to recruit samples of WSW all need to focus on protection of confidentiality as a primary concern. Methods for protecting participants from being identified should be considered, such as the use of encrypted identifiers to replace names in a data base or the option of using code names to identify participants and/or their significant others from whom data may also be gathered.

Many clinical trials involve spouses or family members, and it is critical for researchers to treat these individuals with full acceptance and respect because these individuals provide significant support to participants and may thus have an impact on outcomes. The informed consent discussion often includes spouses because the study may have repercussions that affect them and other members of the family. It is thus appropriate to treat the partners of lesbian study participants in the same way heterosexual spouses would be treated. Partners often have power of attorney for financial matters and durable powers of attorney for health care. They are often a primary source of support for participants and a valuable source of clinical information. For researchers working closely with the lesbian community or hoping to recruit large numbers of lesbians to the study, it is helpful to state up front that partners and other identified family members are welcome to participate in the process just as legally married spouses would be.

C. Role of the Lesbian Community

The community–research linkage is a very important component of lesbian health research, as it is for other hard-to-reach, understudied populations [89–91]. Through this linkage, important research questions are formulated, recruitment of research participants is facilitated, researchers gain knowledge about the community in which participants live, and feedback to participants is enhanced. Too often in the past, researchers have presented themselves as elite members of the scientific community, charged to study the lives of other individuals who are referred to and thought of as "subjects." If members of the study population perceive the research as irrelevant to them or insensitive in its inception, trust between researcher and study participants is damaged and unnecessary constraints are placed on the feasibility and quality of the work.

Input from the lesbian community can be sought in a variety of ways, such as a series of steps taken by the research team for the NLHCS [92]. Even before questionnaire development began, discussion groups were convened in various areas of the country and at national meetings to determine what topics were most important to include. The questionnaire was developed through an iterative process in which successive versions were completed and reviewed in regional focus groups. Once finalized, questionnaires were distributed through a multilevel network of lesbian organizations, groups, and individuals (various gay men's groups also assisted), and notices about the survey were posted on bulletin boards and printed in newsletters of gay and lesbian and women's organizations and businesses, wherever these could be located.

Over 42% of the questionnaires were completed and returned, comprising a total of 1925 surveys representing every state in the country. Questionnaires were coded by first and second level distributors and included as variables the respondent's current zip code and the source from which she received the questionnaire. Thus, the "migration" of survey forms could be tracked through as many as three points, illustrated by a completed questionnaire that was distributed in Miami at a Latina discussion group and later returned from Alaska, where the respondent had received it through a local social group.

NLHCS researchers frequently meet lesbians who responded to the survey and who remember the details of how and when they learned about the study and received their questionnaires. A sense of community ownership was engendered as a result of the inclusivity of the study's methodology, and the benefits were obvious. An unusually high response rate was achieved for a study of this kind, and the sense of community ownership made it possible for results to stimulate ongoing discussion and to provide a base from which other research has been developed. Data were stored in a prestigious data archive, the Inter-university Consortium for Political and Social Research (ICPSR) at the University of Michigan, thus becoming available for secondary research by students and other researchers. No fewer than six dissertations and theses have been based on analysis of the NLHCS data set, the first one on lesbian health to be made available in this way.

Experience with the NLHCS is one example of how researchers and community members alike can benefit from attention to community involvement in study development and implementation. Potential benefits for the researcher include: (1) gaining support and cooperation, (2) more sensitive, unbiased studies, (3) appropriateness and feasibility of study design, (4) understanding of participants' concerns, fears, and special needs, resulting in improved sampling, (5) improved measurement—asking the right questions the right way, (6) more appropriate interactions with participants, assuring enhanced reliability and validity of study results, and (7) more in-depth, on-target analysis and interpretation.

There are clear benefits for the lesbian community as well, including: (1) methodological expertise to improve studies and acceptability of results; (2) justification for increased resources to conduct health care and education programs, (3) information which may assist lesbians to be healthier, (4) higher quality research, and (5) opportunities to pilot test model programs for health interventions.

Historically, much of the available research about lesbians has been conducted by community-based (lesbian) researchers and graduate students. The existence of this work and its significance underscore the importance of developing a research infrastructure that places high value on these linkages and includes processes to support them. At every step of the research process, research team members must respect and value each other's areas of expertise. Academic researchers, whether or not they are lesbian, do not have superior knowledge simply because they work in elite institutions, and lesbian community researchers do not have superior moral authority simply because they do not work in elite institutions.

Much like research conducted on racial minorities, an ideal approach to doing research about lesbians includes development of a research team, with individuals from the study population and experienced lesbian researchers as full members of the team [94]. Necessary ingredients include: (1) establishing a relationship of trust and ability to work together in advance of the study, (2) belief that the effort will be justified through use of the results, (3) intelligent planning and implementation—community members and researchers working together throughout all aspects of the study, from planning to analysis and information dissemination, (4) acceptance of resource limitations and willingness to develop consensus priorities, (5) careful protection of participants' confidentiality, and (6) demonstrated benefit to the community—researchers giving back.

In keeping with these principles, the Institute of Medicine prior to initiating meetings of the study committee formed an ad hoc public liaison group [1]. The primary purpose of creating this group was to facilitate involvement of a larger number of people with expertise in lesbian health and to provide a way for members of the community to have input into the study. Formation of the public liaison group was a visible sign of the IOM's recognition of the expertise and knowledge of the issues held by members of the lesbian health community and reflected a concern about the political importance of having the community involved.

D. Creating an Infrastructure to Support Lesbian Health Research

Fortunately, recent years have seen the emergence of what can become an effective infrastructure to support lesbian research. As interest in research about lesbian health continues to build, various public and private organizations are beginning to respond. Nearly a decade after the first federal funding for a lesbian health study was awarded by the National Institute of Mental Health (NIMH) to support data analysis for the NLHCS, supplemental funding has been provided since 1994 for NIH-funded researchers to include lesbians and bisexual women in ongoing studies, and sexual behavior questions were included in the NIH Women's Health Initiative, a multisite longitudinal study of women's health initiated in 1996. At least two studies specifically about lesbian health have been funded through NIH competitive grants during the first six months of 1998. An ongoing multisite study of breast cancer in lesbians was initiated through the Centers for Disease Control and Prevention (CDC) in 1994 and a corollary study about how to effectively educate providers to screen and treat lesbians for breast and cervical cancer is now being conducted under the auspices of the Mautner Project for Lesbians with Cancer in Washington, D.C. In conjunction with the IOM Study, these initiatives demonstrate a meaningful response from the federal research agencies.

Private foundations, such as the Astraea Foundation, the Lesbian Health Fund (LHF), and An Uncommon Legacy Foundation, annually provide small grants to fund research about lesbians and offer significant opportunities for inexperienced researchers to develop pilot data and for more experienced researchers to develop new areas of study. These organizations have been invaluable in supporting development of a cadre of researchers prepared to gain substantial funding as federal agencies become more receptive to studies about lesbian health. The LHF's requirement that grantees submit at least one article of study results for publication demonstrates how a thoughtful research sponsor can push grantees to leverage the benefits of small grant dollars. Each research article published in a refereed journal increases the visibility of the field as well as the individual reputation of the researcher.

The initiation of new journals dedicated to lesbian health studies or to lesbian health within the broader context of general lesbian studies addresses another critical component of the emerging lesbian health research infrastructure. Quarterly publication of the *Journal of the Gay and Lesbian Medical Association* (*JGLMA*) was initiated in 1997, providing a multidisciplinary, peer-reviewed journal devoted to the study of the health of lesbian, gay male, bisexual, and transgendered populations. *JGLMA* places a priority on reporting the results of hypothesis-driven research performed according to the principles of the scientific method [93].

The *Journal of Lesbian Studies,* also initiated in the spring of 1997, is an international journal devoted to descriptive, theoretical, empirical, applied, and multicultural perspectives on the topic of lesbian studies. It encourages submissions from all academic areas as well as personal accounts. The journal has a feminist perspective and authors are encouraged to avoid overly technical language. An issue has been dedicated to lesbian health, and articles relevant to lesbians' health and health care needs are regularly included [94].

Although without a previous focus on lesbian health, *Women's Health Issues* (published by the Jacobs Institute for Women's Health) recently invited researchers who have studied lesbian health to consider submitting articles for review. The *Journal of Women's Health,* published by the Society for the Advancement of Women's Health Research, publishes high quality clinical research about women's health. Articles about lesbian health are typically included.

Local and regional efforts are also underway for integrating lesbian health services research and care. At the Fenway Community Health Center, administrators and health care providers are strengthening a longstanding focus on lesbian health research to complement the Center's comprehensive lesbian health care programs. Fenway sponsored a weekend research forum on lesbian and bisexual women's health in February 1998 and drew participants from throughout the Northeast [95]. At the Mautner Project for Lesbians with Cancer, CDC funding for educating providers to effectively screen for and treat cancer among lesbians stimulated efforts to conduct rigorous needs assessment and program evaluation. These health clinics and care programs have recognized emerging opportunities for clinical and applied research and are preparing to respond through staff development and skills building, the intentional creation of networks with the community, and an emergent national network of researchers committed to the study of lesbians' health and health care needs.

A group of lesbian researchers meet regularly in Seattle to discuss ongoing studies and develop ideas for new ones [K. Stine, University of Washington, personal communication, 1997]. In Richmond, university-based lesbian researchers are working with a community group of African-American lesbians to develop an ongoing research program. Through a series of focus groups conducted during 1997–1998 in Dallas, Richmond, Chicago, Portland, and Seattle for an LHF-funded study, researchers learned that these efforts are not isolated or without realistic possibility for future work. Lesbians are very interested in participating in studies when their safety is protected and potential results are perceived to be meaningful [96]. It seems very likely that when an infrastructure to support lesbian health research has been created, study participants will come forward to work with researchers who apply thoughtful, informed approaches based in a genuine understanding of the lesbian community and its needs.

VII. The Future of Lesbian Health Research

Why don't we have more knowledge about lesbian health on which to base effective interventions? Answers range from issues of research methodology (an emerging population, difficult to sample, with imprecise definition) to less satisfying and more enduring sociopolitical concerns that stem from negative social sanctions for homosexuals, provider reluctance to respond to

their needs, and a nearly complete lack of support for training researchers to enter this field of inquiry [97]. Unlike the surge of research about gay men that was stimulated by the AIDS epidemic, research about lesbians has not been stimulated by a frightening health epidemic or another perceived threat to society at large. Although change in social attitudes and health care system response have been and will continue to be slower than many would like, both the scientific and practitioner communities have long understood these concerns. Pragmatic understanding about what can be done to move forward is often less easy to come by, particularly when the challenges are both broad and diverse.

Major limitations that will need to be confronted and overcome include inadequate funding for studies, lack of opportunities to support the training and mentorship of future researchers, inadequate mechanisms for experienced researchers to develop shared research programs, and a poorly developed infrastructure to support the emerging field of lesbian health research. Each of these can be remedied, and recommendations of the IOM Study Committee may provide a sound direction for the future [1]. These recommendations include:

1. Public and private funding to support research on lesbian health needs to be increased in order to enhance knowledge about risks to health and protective factors, to improve methodologies for gathering information about lesbian health, to increase understanding of the diversity of the lesbian population, and to improve lesbians' access to mental and physical health care services.

2. Methodological research needs to be funded and conducted to improve measurement of the various dimensions of lesbian sexual orientation.

3. Pilot studies are needed to test the feasibility of including questions about sexual orientation on data collection forms in relevant studies in the behavioral and biomedical sciences to capture the full range of female experience and to increase knowledge about associations between sexual orientation and health status.

4. Researchers studying lesbian health should consider the full range of racial, ethnic, and socioeconomic diversity among lesbians when designing studies on lesbian health; strive to include members of the lesbian study population in the development and conduct of research; and give special attention to the confidentiality and privacy of the study population.

5. A large-scale probability survey should be funded to determine the range of expression of sexual orientation among all women and the prevalence of various risk and protective factors for health, stratified by sexual orientation.

6. Conferences should be held on an ongoing basis to disseminate information about the conduct and results of research on lesbian health, including the protection of human subjects.

7. Federal agencies, including the National Institutes of Health and the Centers for Disease Control and Prevention, foundations, health professional associations, and academic institutions should develop and support mechanisms for broadly disseminating information and knowledge about lesbian health to health care providers, researchers, and the public.

8. The committee encourages development of strategies to train researchers in conducting lesbian health research at both the predoctoral and the postdoctoral levels.

Whether lesbian or not, researchers who would like to conduct studies about lesbian health can expect increased opportunities and support. A blueprint has been provided, and those who are fortunate enough to be able to respond can expect expanded opportunities in the near future. Building on the strong foundation created by lesbian researchers and health advocates, those researchers who take time to develop solid relationships with the lesbian community and understand the unique history and needs of this population can be expected to provide leadership for an emerging national program of lesbian health research.

References

1. Solarz, A., ed. (1999). "Lesbian Health: Current Assessment and Directions for the Future." Committee on Lesbian Health Research Priorities, National Academy Press, Washington, DC.

2. Laumann, E. O., Gagnon, J. H., Michael, R. T., and Michaels, S. (1994). "The Social Organization of Sexuality: Sexual Practices in the United States." University of Chicago Press, Chicago.

3. Sell, R. L., and Petrulio, C. (1996). Sampling homosexuals, bisexuals, gays, and lesbians for public health research: A review of the literature from 1990 to 1992. *J. Homosex.* **30**(4), 31–47.

4. Gonsiorek, J. C., and Weinrich, J. D. (1991). The definition and scope of sexual orientation. *In* "Homosexuality: Research Implications for Public Policy" (J. C. Gonsiorek and J. D. Weinrich, eds.). Sage Publ., Newbury Park, CA.

5. Greene, B. (1994b). Lesbian and gay sexual orientations. *In* "Lesbian and Gay Psychology: Theory, Research and Clinical Applications" (B. Greene and G. M. Herek, eds.), Vol. 1. Sage Publ., Thousand Oaks, CA.

6. Bradford, J. B., Honnold, J. A., and Ryan, C. C. (1997). Disclosure of sexual orientation in survey research on women. *J. Gay Lesbian Med. Assoc.* **1**(3), 169–177.

7. Dejowski, E. F. (1992). Public endorsement of restrictions on three aspects of free expression by homosexuals: Socio-demographic and trends analysis 1973–1988. *J. Homosex.* **23**(4), 1–19.

8. Ryan, C., and Bogard, R., (1998). "What Every Lesbian and Gay American Needs to Know about Health Care Reform." HRCF Foundation and the Human Rights Campaign Fund, Washington, DC.

9. Bradford, J. B., and Ryan, C. C. (1988). "National Lesbian Health Care Survey: Final Report." National Lesbian and Gay Health Foundation, Washington, DC.

10. Savin-Williams, R. C. (1988). Theoretical perspectives accounting for adolescent homosexuality. *J. Adolesc. Health* **9**, 95.

11. Ryan, C. C., and Futterman, D. (1997). Lesbian and gay youth: Care and counseling. *In* "Adolescent Medicine: State of the Art Reviews," Vol. 8, Part 2. Hanley & Belfus, Philadelphia.

12. American Academy of Pediatrics (1993). Homosexuality and adolescence. *Pediatrics* **92**, 631.

13. Bailey, J. M., Pillard, R. C., Kitsinger, C., and Wilkinson, S. (1997). Sexual orientation: Is it determined by biology? *In* "Women, Men and Gender: Ongoing Debates" (M. R. Walsh, ed.). Yale University Press, New Haven, CT.

14. Pattatuci, A. M., and Hamer, D. H. (1995). Development and familiarity of sexual orientation in females. *Behav. Genet.* **25**(5), 407–420.

15. D'Augelli, A. R., and Hershberger, S. L. (1993). Lesbian, gay and bisexual youth in community settings: Personal challenges and mental health problems. *Am. J. Commun. Psychol.* **21**(4), 421–448.

16. Perrin, E. C. (1996). Pediatricians and gay and lesbian youth. *Pediatr. Rev.* **17**(9), 311–318.

17. Sullivan, T. R. (1994). Obstacles to effective child welfare service with gay and lesbian youths. *Child Welfare* **73**(4), 291–304.

18. Troiden, R. R. (1988). Homosexual identity development. *J. Adolesc. Health Care* **9**, 105–113.

19. Troiden, R. R. (1989). The formation of homosexual identities. *J. Homosex.* **17**, 43–73.

20. Goffman, E. (1963). "Stigma: Notes on the Management of Spoiled Identity." Prentice-Hall, Englewood Cliffs, NJ.

21. Tully, C. T. (1995). In sickness and in health: Forty years of research on lesbians. *In* "Lesbian Social Services: Research Issues" (C. T. Tully, ed.). Harrington Park Press/Haworth Press, New York.

22. O'Hanlan, K. A. (1995). Lesbian health and homophobia: Perspectives for the treating obstetrician/gynecologist. *Curr. Probl. Obstet., Gynecol. Fertil.* **18**(4), 97–133.

23. Mays, V. M., and Cochran, S. D. (1988). The black women's relationships project: A national survey of black lesbians. *In* "A Sourcebook of Gay/Lesbian Health Care" (M. Shernoff and W. A. Scott, eds.), 2nd ed. National Lesbian and Gay Health Foundation, Washington, DC.

24. Tierney, W. G. (1993). Academic freedom and the parameters of knowledge. *Harv. Educ. Rev.* **63**, 143–160.

25. Norris, W. P. (1992). Liberal attitudes and homophobic acts: The paradoxes of homosexual experience in a liberal institution. *In* "Coming out of the Classroom Closet." Haworth Press, New York.

26. Herek, G. M. (1993). Documenting prejudice against lesbians and gay men on campus: The Yale Sexual Orientation Study. *J. Homosex.* **25**, 15–30.

27. D'Augelli, A. R. (1989). Lesbians' and gay mens' experiences of discrimination and harassment in a university community. *Am. J. Commun. Psychol.* **17**, 317–321.

28. Yeskel, F. (1985). The consequences of being gay. Unpublished report from the Program for Gay, Lesbian and Bisexual Concerns, University of Massachusetts, Amherst.

29. Nelson, R., and Baker, H. (1990). The educational climate for gay, lesbian, and bisexual students at the University of California at Santa Cruz. Unpublished report from the Gay, Lesbian, Bisexual Community Concerns Advisory Committee, University of California, Santa Cruz.

30. Ryan, C. C., and Bradford, J. B. (1997). "Lesbian Research Network: Final Report—Year 1." Survey and Evaluation Research Laboratory, Richmond, VA.

31. Roberts, S. A., Dibble, S. L., Scanlon, J. L., Paul, S. M., and Davids, H. (1998). Differences in risk factors for breast cancer: Lesbians and heterosexual women. *J. Gay Lesbian Med. Assoc.* **2**(3), 93–102.

32. Johnson, S. R., Smith, S. M., and Guenther, S. M. (1987). Comparison of gynecologic health care problems between lesbians and bisexual women. *J. Reprod. Med.* **32**(11), 805–811.

33. Robertson, P., and Schacter, J. (1981). Failure to identify veneral disease in a lesbian population. *Sex. Transm. Dis.* **8**(2), 75–76.

34. Marrazzo, J. M., Koutsky, L. A., Stine, K., Kuypers, J., Grubert, T. A., Galloway, D. A., Kiviat, N. B., and Handsfield, H. H. (1999). Genital human papillomavirus infection in women who have sex with women. *J. Infect. Dis.* (in press).

35. McKirnan, D. J., and Peterson, P. L. (1989). Alcohol and drug use among homosexual men and women: Epidemiology and population characteristics. *Addict. Behav.* **14**, 545–563.

36. McKirnan, D. J., and Peterson, P. L. (1989). Psychosocial and cultural factors in alcohol and drug abuse: An analysis of a homosexual community. *Addict. Behav.* **14**, 555–563.

37. Skinner, W. F. (1994). The prevalence and demographic predictors of illicit and licit drug use among lesbians and gay men. *Am. J. Public Health* **84**, 1307–1310.

38. Skinner, W. F., and Otis, M. (1992). Drug use among lesbians and gay people: Findings, research, design, insights, and policy issues from the Trilogy Project. *In* "Proceedings of the Research Sympo-

sium on Alcohol and Other Drug Problem Prevention Among Lesbians and Gay Men," pp. 34–60. Evaluation, Management and Training Group, Sacramento, CA.

39. National Heart Lung and Blood Institute (1997). Facts about heart disease and women: Are you at risk? [Updated August 1996] [WWW document]. URL gopher://fido.nhlbi.nih.gov:70/00/ . . . ealth/ cardio/other/gp/hdwmnrsk.txt (accessed December 1, 1997).

40. Schneider, M. (1989). Sappho was a right-on adolescent: Growing up lesbian. *J. Homosex.* **17**(1–2); 111–30.

41. Savin-Williams, R. C. (1989). Coming out to parents and self esteem among gay and lesbian youths. *J. Homosex.* **18**(1–2), 1–35.

42. Anstett, R., Kiernan, M., and Brown, R. (1987). The gay-lesbian patient and the family physician. *J. Fam. Pract.* **25**(4), 339–344.

43. Feinleib, M. R. (1989). "Report of the Secretary's Task Force on Youth Suicide." U.S. Department of Health and Human Services, Rockville, MD.

44. Schilit, R., Lie, G., and Montagne, M. (1990). Substance use as a correlate of violence in intimate lesbian relationship. *J. Homosex.* **19**(3), 151–165.

45. Renzetti, C. M. (1992). "Violent Betrayal: Partner Abuse in Lesbian Relationships." Sage Publ., New York.

46. Ristock, J. L. (1997). Understanding abuse in lesbian relationships. *Gay Lesbian Med. Assoc. Symp.*

47. National Gay and Lesbian Task Force (1986). "Anti-gay Violence: Causes, Consequences, Responses." National Gay and Lesbian Task Force, Washington, DC.

48. Herek, G. H. (1989). Hate crimes against lesbians and gay men: Issues for research and policy. *Am. Psychol.* **44**(6), 948–955.

49. Duncan, D. F. (1990). Prevalence of sexual assault victimization among heterosexual and gay/lesbian university students. *Psychol. Rep.* **66**, 65–66.

50. Bidwell, R. J. (1992). Sexual orientation and gender identity. *In* "Comprehensive Adolescent Health Care" (S. B. Friedman, M. Fisher, and S. K. Schonberg, eds.). Quality Medical Publishing, St. Louis, MO.

51. Gonsiorek, J. C. (1988). Mental health issues of gay and lesbian adolescents. *J. Adolesc. Health Care* **9**, 114.

52. Hunter, J. (1992). Violence against lesbian and gay male youths. *In* "Hate Crimes: Confronting Violence against Lesbians and Gay Men" (G. M. Herek and K. T. Berrill, eds.). Sage Publ., Newbury Park, CA.

53. Kruks, G. P. (1991). Gay and lesbian homeless/street youth: Special issues and concerns. *J. Adolesc. Health* **12**, 515.

54. Berger, B. J., Kolton, S., Zenilman, J. M., Cummings, M. C., Feldman, J., and McCormack, W. M. (1995) Bacterial vaginosis in lesbians: A sexually transmitted disease. *Clin. Infect. Dis.* **21**(6), 1402–1405.

55. Degen, K., and Waitkevicz, H. J. (1982). Lesbian health issues. *Br. J. Sex. Med.,* May, pp. 40–47.

56. Walter, M. H., and Rector, W. G. (1986) Sexual transmission of hepatitis A in lesbians. *JAMA, J. Am. Med. Assoc.* **56**, 594.

57. William, D. C. (1981). Hepatitis and other sexually transmitted diseases in gay men and lesbians. *Sex. Transm. Dis.* **8**(4), 330–332.

58. Chu, S. Y., Hammet, T. A., and Buehler, J. W. (1992). Update: Epidemiology of reported cases of AIDS in women who have sex with other women, United States, 1980–1991. *AIDS* **6**, 518–519.

59. Sabatini, M. T., Patel, K., and Hirschman, R. (1984). Kaposi's sarcoma and T-cell lymphoma in an immunodeficient woman: A case report. *AIDS Res.* **1**, 135–137.

60. Marmor, M., Weiss, L. R., Lyden, M., *et al.* (1986). Possible female-to-female transmission of human immunodeficiency virus. *Ann. Intern. Med.* **105**, 969.

61. Monzon, D. T., and Capellan, J. M. B. (1987). Female-to-female transmission of HIV. *Lancet* **2**, 40–41.

62. Perry, S., Jacobsberg, L., and Fogel, K. (1989). Orogenital transmission of human immunodeficiency virus (HIV). *Ann. Intern. Med.* **111**, 951–952.

63. Rich, J. D., Buck, A., Tuomala, R. E., *et al.* (1993). Transmission of human immunodeficiency virus infection presumed to have occurred via female homosexual contact. *Clin. Infect. Dis.* **17**, 1003–1005.

64. Henin, Y., Mandelbrot, L., Henrion, R., *et al.* (1993). Virus excretion in the cervicovaginal secretions of pregnant and nonpregnant HIV-infected women. *J. Acquired Immune Defic. Syndr.* **1**, 72–75.

65. Zorr, B., Schafer, A. P., Dilger, I., *et al.* (1994). HIV-1 detection in endocervical swabs and mode of HIV-1 infection. *Lancet* **8901**, 852.

66. Clemetson, D. B., Moss, G. B., Willerford, D. M., *et al.* (1993). Detection of HIV DNA in cervical and vaginal secretions: Prevalence and correlates among women in Nairobi, Kenya. *JAMA, J. Am. Med. Assoc.* **269**, 2860–2864.

67. Vogt, M. W., Witt, D. J., Craven, D. E., *et al.* (1986). Isolation of HTLV-III/LAV from cervical secretions of women at risk for AIDS. *Lancet* **8480**, 525–527.

68. Vogt, M. W., Witt, D. J., Craven, D. E., *et al.* (1987). Isolation patterns of the human immunodeficiency virus from cervical secretions during the menstrual cycle of women at risk for the acquired immunodeficiency syndrome. *Ann. Intern. Med.* **3**, 380–382.

69. Wofsy, D. B., Cohen, J. B., Hauer, L. B., *et al.* (1986). Isolation of AIDS-associated retrovirus from genital secretions of women with antibodies to the virus. *Lancet* **8480**, 527–529.

70. Araneta, M. R. G., Mascola, L., Eller, A., O'Neill, L., Ginsberg, M. M., *et al.* (1995). HIV transmission through donor artificial insemination. *JAMA, J. Am. Med. Assoc.* **273**(1), 854–858.

71. Hughes, T. L., and Wilsnack, S. C. (1997). Use of alcohol among lesbians research and clinical implications. *Am. J. Orthopsychiatry* **1**, 20–36.

72. Bloomfield, K. A. (1993). A comparison of alcohol consumption between lesbians and heterosexual women in an urban population. *Drug Alcohol Depend.* **33**, 257–269.

73. Szasz, T. (1977). "The Theology of Medicine: The Political-Philosophical Foundations of Medical Ethics." Harper & Row, New York.

74. Mathews, W. C., Boose, M. W., Turner, J. D., *et al.* (1986). Physicians' attitudes toward homosexuality—survey of a California County Medical Society. *West. J. Med.* **144**, 106.

75. Smith, E. M., Johnson, S. R., and Guenther, S. M. (1985). Health care attitudes and experiences during gynecologic care among lesbians and bisexuals. *Am. J. Public Health* **75**, 1085.

76. Trippet, S. E., and Bain, J. (1992). Reasons American lesbians fail to seek traditional health care. *Health Care Women Int.* **13**, 145.

77. Bybee, D. (1990). "Michigan Lesbian Health Survey," A Report to the Michigan organization for human rights and the Michigan Department of Public Health. Michigan Department of Health and Human Services, Lansing.

78. White, J. C., and Dull, V. T. (1997). Health risk factors and health seeking behavior in lesbians. *J. Women's Health* **6**(1), 103–112.

79. White, J. C., and Dull, V. T. (1998). Room for improvement: Communication between lesbians and primary care providers. *J. Lesbian Stud.* **2**(1), 95–110.

80. Ryan, C. C., Bradford, J. B., and Honnold, J. A. (1999). Social workers' and counselors' understanding of lesbian needs. *J. Gay Lesbian Soc. Serv.* **9**, 1–26.

81. Plumb, M. (1998). "A Call for a Progressive Research Agenda," Paper presented at the Lesbian and Bisexual Women's Health Research Forum. Fenway Community Health Center, Boston.

82. Denenberg, R. (1995). Report on lesbian health. *Women's Health Issues* **5**(2), 181–191.

83. Stevens, P. E. (1995). Structural and interpersonal impact of heterosexual assumptions on lesbian health care clients. *Nurs. Res.* **44**(1), 25–30.

84. Bradford, J. B., and Honnold, J. A. (1994). Unpublished analysis of data from the National Lesbian Health Care Survey.

85. Bybee, D., and Roeder, V. (1990). "Michigan Lesbian Health Survey: Results Relevant to AIDS," A Report to the Michigan Organization for Human Rights and the Michigan Department of Public Health. Michigan Department of Health and Human Services, Lansing.

86. Bradford, J. B., and Dye, L. Physicians' readiness for providing cancer screening to lesbians. Unpublished manuscript.

87. National Research Council (1989). "AIDS: Sexual Behavior and Intravenous Drug Use." National Academy Press, Washington, DC.

88. National Research Council (1997). "Evaluating Genetic Diversity." National Academy Press, Washington, DC.

89. Centers for Disease Control and Prevention (1996). "Inclusion of Women and Racial and Ethnic Minorities in Research," Manual Guide, General Administration CDC-80 Transmittal Notice 96.2. CDC, Atlanta, GA.

90. O'Hanlan, K. A. (1995). Recruitment and retention of lesbians in health research trials. *In* "Recruitment and Retention of Women in Clinical Studies 1995," NIH Publ. No. 95-3756; pp. 101–104. National Institutes of Health, Bethesda, MD.

91. Dressler, W. W. (1993). Commentary on "Community research: Partnership in black communities." *Am. J. Prev. Med.* **9**(Supple.), 32–24.

92. Bradford, J. B., and Ryan, C. C. (1987). "National Lesbian Health Care Survey: Mental Health Implications," NIMH Contract No. 86MO19832201D. U.S. Department of Health and Human Services, Rockville, MD.

93. Townsend, M. H., and White, J. C. (1997). *J. Gay Lesbian Med Assoc.* **1**(1).

94. Rothblum, E., ed. (1997). "Journal of Lesbian Studies," Vol. 1, Part 1. Haworth Press, New York.

95. Newsletter, (Spring) (1997). Fenway Community Health Center, Boston.

96. Bradford, J. B., Honnold, J. A., White, J. C., and Ryan, C. C. (1998). Refining survey research methods to enhance participation from lesbians. Unpublished manuscript.

97. Bradford, J. B. (1996). Overview of methodology. *Lesbian Health Res. Inst.,* Seattle, WA, Opening discussion.

Part II

SEXUAL AND REPRODUCTIVE HEALTH

Section 3

REPRODUCTIVE HEALTH

Gertrud S. Berkowitz

Departments of Community and Preventive Medicine
and Obstetrics, Gynecology and Reproductive Science
The Mount Sinai Hospital
Mount Sinai School of Medicine
New York, New York

Janet R. Daling

University of Washington
Member, Division of Public Health Sciences
Fred Hutchinson Cancer Research Center
Seattle, Washington

Women's reproductive years span the time from puberty to menopause. Hormones are clearly intimately connected with reproductive success but other factors such as physical health, genetic make-up, environmental exposures, diet, and access to health care can directly or indirectly impact this process. Sexual maturation is accompanied by a surge of endogenous hormones as well as physical growth and the development of secondary sexual characteristics. The childbearing years, generally classified as ages 15–44, are characterized by cyclical fluctuations of hormone levels except during pregnancy and lactation. Pregnancy is associated with marked increases in estrogen and progesterone levels as well as other profound metabolic and physiological changes. The cessation of ovarian secretion and menstrual cycling signals the onset of menopause. In total, it has been estimated that women spend 35.9 years between menarche and menopause [1].

The reproductive history of women can be described as a series of stages or events that have both biological and social implications [1]. Data from the 1982 and 1988 National Surveys of Family Growth [2,3] were used to calculate the timing of these reproductive stages and to evaluate any changes in their occurrence over time [1]. The median ages for these reproductive stages for 1988 are illustrated in Figure 1 [4]. The median age at menarche in 1988 was 12.5 years, which represents a slight drop from 12.7 years in 1982. Trends over a longer time period have shown a marked decline in the age at menarche from 17 years in 1800 and 14 years in 1900 [5].

Women are also initiating sexual activity at an earlier age: the median age at first intercourse declined from 18.0 years in 1982 to 17.4 years in 1988. Median age at first marriage, in contrast, increased from 23.3 years to 24.3 years. Similarly, the median ages at first birth increased from 25.0 years to 26.0 years, as did the median age at completion of desired family size (from 29.1 years in 1982 to 30.0 years in 1988). Only a slight shift was apparent in the median age at sterility (tubal ligation or a partner's vasectomy) or noncontraceptive infertility. The median for this reproductive stage transition was 35.6 years in 1982 and 35.7 years in 1988.

The timing of these stages tends to differ across racial/ethnic groups and poverty status [1]. Age at menarche was youngest for Hispanic girls (12.1 years), intermediate for African-American girls (12.4 years), and oldest for white girls (12.5 years). Other studies suggest that the gap between black and white girls is larger [6]. The difference in the proportion of girls with early menarche (<12 years) is even more striking when black (21.3%) and white girls (11.6%) are compared [6].

Median ages at first intercourse and at first pregnancy were lower for African-American women as compared to Hispanic and white women. Median age at cohabitation or marriage was substantially later for African-American (24.0 years) than Hispanic (21.8 years) or white women (21.8 years). Because lower-income women were also younger at first intercourse and first birth than higher-income women, the racial/ethnic differences may reflect disparities in income level. Thus, further studies of

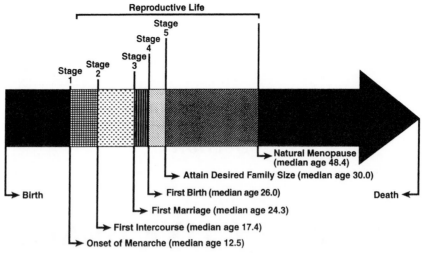

Fig. 1   The timing of a woman's reproductive stages. Reprinted with permission from [4].

racial/ethnic groups are needed that control for poverty status as well as for other potential confounders such as education and lifestyle behavior.

According to estimates based on the two National Surveys of Family Growth [2,3], 14% of the reproductive years are spent sexually inactive in the postmenarcheal years. Another 24% are spent being sexually active before the first birth. Attainment of desired family size only occupies 11% of the total time from the first birth. In contrast, approximately 51% of a woman's reproductive life is spent in the period from attainment of desired family size to menopause [1].

Social changes and advances in technology and knowledge during the latter half of the twentieth century have led to significant improvements, at least in industrialized countries, in a woman's ability to control her reproductive process and to achieve conception, in the safety of childbearing, and in the reduction of gynecologic morbidity. An increasing proportion of women is choosing to remain childless, to marry at a later age, or to postpone childbearing.

The fertility rate in 1996 (65.3 live births per 1000 women aged 15–44 years) was the lowest since 1976 when it was 65.0 live births per 1000 women aged 15–44 [7]. The proportion of first births that occurs among women aged 30 years and older has increased dramatically. In 1996 this proportion was 22% as compared to 5% in 1975 [7].

These trends presumably reflect the changing role of women in society, including expanded educational, professional, and, at least to some extent, political attainments. Despite these unprecedented changes in women's lives, barriers to health care access and reproductive choices still exist in this country and especially in the developing world. Many of the constraints on women's reproductive lives are rooted in poverty, ignorance, or cultural beliefs and traditions, but lack of sensitization to women's reproductive concerns as well as their educational and career aspirations has also played an important role.

Puberty, as it heralds fertility, represents a time of celebration in many societies. However, this can also be a time of emotional and psychological upheaval. As the age of sexual maturity has decreased over time, this enables an earlier age at pregnancy. Although the rate of teenage pregnancy has recently appeared to decline in the United States, the rate is still substantially higher than in other industrialized countries. These trends as well as the characteristics and sequelae of early sexual maturation and adolescent pregnancy are discussed in Chapter 8.

Menstrual cycling pervades much of a women's reproductive life. Historically, menstruation has been associated with myths and superstitions that still exist in religions and in many developing societies. Menstruating women were regarded as unclean and were often isolated or relegated to separate quarters. Perhaps because of these beliefs, the impact of menstrual dysfunction on women's health and well-being has received relatively little attention in the epidemiology literature. The existing knowledge regarding the prevalence of and risk factors for menstrual dysfunction is reviewed in Chapter 9.

Premenstrual syndrome (PMS) represents one of the more disruptive of the menstrual disorders. The physical and mental attributes of this syndrome can interfere with daily life, although previous beliefs that PMS can be used to explain violence or poor cognitive functioning have been largely discounted. The etiology of PMS is not well understood and little is known about its epidemiologic risk factors. However, as is described in Chapter 10, there is a greater understanding of the pathophysiology of PMS and this in turn has led to improved treatment modalities.

Women's ability to conceive, or fecundity, has traditionally been regarded as the most important function of women. As women have gained more basic rights during the twentieth century, childbearing has become less of a central issue, at least for some women. Nevertheless, an understanding of women's reproductive capacity is important not only for individual women but also for populations as a whole. Estimates of fecundability and factors associated with its variability are the subjects of Chapter 11. Infertility (Chapter 16) affects 10–15% of married couples. Infertility is generally defined as not conceiving while having unprotected intercourse for 12 months or more. The major types of infertility include ovulatory dysfunction, tubal infertility, endometriosis-related infertility, and infertility due to a male factor [8]. In many instances the proximal cause of the failure to conceive is not determined. Delay in having children until older age, as well as other lifestyle choices, may have a role in the apparent frequency of infertility.

Increasingly, women are seeking medical assistance to achieve conception. The birth of the first "test-tube" baby in the early 1980s and subsequent expansion and improvement of assisted reproductive technologies are giving infertile women the possibility to conceive. Chapter 17 gives an overview of the remarkable development of reproductive technologies to assist what appears to be a growing population of women who cannot conceive.

The development of oral contraceptives during the 1950s and 1960s and, since then, of such other contraceptives as long-acting injectable contraceptive steroids, has given women more control over their reproductive process than ever before. As described in Chapter 12, in 1995, 64% of women 15–44 years of age were contracepting, whereas only 5% of women who were sexually active and who did not wish to conceive were using no method of contraception. Contraceptive choice is always changing as new methods are researched and marketed, often in response to health concerns raised by the medical research community about existing methods. The prevalence of HIV infection has led to a resurgence of barrier methods either as the primary method or in conjunction with a more effective method that inhibits ovulation or disrupts implantation. The long-acting injectable contraceptives that were introduced in the U.S. in the mid-1990s, after having been used internationally for some time, are just becoming the method of choice by young women. Women who have completed their families are selecting more permanent methods such as tubal ligation or vasectomy of the partner. With the exception of oral contraceptives, sterilization was the most frequently used method in 1995. Finally, postcoital contraception has been approved by the FDA; however, in contrast to Great Britain, where the majority of women of reproductive age are aware of its availability, awareness of, and hence use of, this method in the U.S. is lagging.

The legalization of induced abortions in this country and elsewhere around the world has similarly left women with a much greater choice over childbearing decisions. The history of abortion in the U.S. and worldwide is described in Chapter 13. The woman who is undergoing an abortion is more likely to be single, 18–24 years of age, and to have an income less than $15,000 per year. In the United States in 1995, over one-half of all abortions were performed for a pregnancy that was conceived while the woman was using some means of contraception. Research is needed on factors related to contraceptive failure and on the tailoring of effective contraception to characteristics of each woman. Although abortion services are widespread, access to these services is still restrictive for many women seeking them. Abortion facilities are more likely to be located in urban areas and are declining in number. Federal funds have not been available for abortion services since 1977, which further restricts availability. The approval of Mifepristone (RU-486) by the Food and Drug Administration should increase the availability of abortion, as it does not involve a surgical procedure and can be administered by physicians under more private circumstances.

While childbearing is considerably safer today than in the past, the reduction in complications during the antepartum, intrapartum, and postpartum course appears to have received less attention than the efforts to help women achieve pregnancy. Maternal morbidity is considerable, as Chapter 14 relates. It should be noted that pregnancy outcome is not included in this section as it impacts primarily the neonate rather than the mother. A large proportion of pregnancy complications results from underlying medical conditions. Nevertheless, many complications, such as gestational diabetes or preeclampsia, are unique to pregnancy. Given the large number of pregnancy complications, this chapter is not meant to be an exhaustive review of the epidemiology of morbid conditions. Rather, it focuses on those conditions that arise during pregnancy, worsen or improve during pregnancy, or deteriorate after pregnancy.

Chapter 15 considers the impact of selected procedures or complications that occur during labor and delivery. As this chapter illustrates, the introduction or increasing use of some procedures, such as fetal heart rate monitoring, cesarean section deliveries, and episiotomy, may not be beneficial to women's health and may actually lead to added health costs as well and further morbidity.

Two conditions that affect a woman's health, reproductive capacity, and quality of life are endometriosis and leiomyomata and are described in Chapters 18 and 19, respectively. The etiology of these diseases remains uncertain, and few modifiable risk factors have been identified. The rate of leiomyomata is increased two- to threefold among black women as compared to whites, but the reason for this excess among blacks is not known. Because both endometriosis and leiomyomata can occur without noticeable symptoms, research into factors related to these diseases has been hindered by the difficulty in identifying a control group free of disease. Both diseases are characterized by excessive menstrual bleeding and pelvic pain and are associated with reduced fertility. Among the treatments for endometriosis and leiomyomata is hysterectomy. As described in Chapter 20, hysterectomy is the most common nonobstetric surgery performed in the United States on reproductive-aged women. In a national survey conducted in 1988–1994, over 20% of women aged 17 and older had experienced a hysterectomy. The peak prevalence age was 60–69, where 41.7% of women reported having had a hysterectomy. The diagnosis of uterine leiomyomata is the most frequent reason for the procedure, followed by endometriosis.

Menopause, which represents the end of a woman's reproductive life, is discussed in Chapter 92 under the Aging Section. The risks and benefits of estrogen replacement therapy are addressed in Chapter 93.

In summary, while the improvement in the diagnosis and treatment of reproductive disorders has been revolutionary for women's physical and mental wellbeing, further progress is needed to reduce teenage pregnancy; elucidate the epidemiology of menstrual disorders; lessen the frequency and severity of premenstrual syndrome; identify the factors associated with fertility and infertility; continue the advances in the effectiveness and safety of assisted reproduction; provide safe and effective contraceptives and access to legally induced abortions; minimize morbidity and mortality during the prenatal, intrapartum, and postpartum course; and add a better understanding of the mechanisms underlying gynecologic disorders, such as endometriosis and leiomyomata, and the factors leading to hysterectomy.

The role of genetics and inclusion of markers of genetic susceptibility in the evaluation of reproductive events and related conditions is an area of research in need of expansion. Research areas where genetics may have a role include age at menarche and menopause, menstrual characteristics, endometriosis, and leiomyomata. Research into effective methods of contraception also needs to be continued. The medical–legal climate has greatly impacted the enthusiasm of private industry to initiate research into new products. Further research is also needed on familial and particularly transgenerational effects on reproduction. There is growing evidence that intrauterine exposures may have adverse effects on subsequent pregnancy-related events, but whether these effects are related to hormone levels or other factors is not known. Early life events undoubtedly play a role in reproductive processes but the nature of this role is still unclear. Finally, concern over rising health care costs mandates research on the development of techniques that are not only effective but less costly for the diagnosis and treatment of various reproductive disorders.

References

1. Forrest, J. D. (1993). Timing of reproductive life stages. *Obstet. Gynecol.* **82,** 105–111.
2. Bachrach, C. A. (1984). Contraceptive practice among American women, 1973–1982. *Fam. Plann. Perspect.* **16,** 253–259.
3. Mosher, W. D., and Bachrach, C. A. (1996) Understanding U.S. fertility: Continuity and change in the National Survey of Family Growth, 1988–1995. *Fam. Plann. Perspect.* **28,** 4–12.
4. Cates, W., Jr. (1996). Contraception, unintended pregnancies, and sexually transmitted diseases: Why isn't a simple solution possible? *Am. J. Epidemiol.* **143,** 311–318.
5. Eveleth, P. B., and Tanner, J. M. (1990). "Worldwide Variation in Human Growth," 2nd ed. Cambridge University Press, New York.
6. MacMahon, B. (1973). "Age at Menarche," U.S. DHEW Publ. No. (HRA) 74-1615, NCHS Ser. 11, No. 133. National Center for Health Statistics, Rockville, MD.
7. Ventura, S. J., Martin, J. A., Curtin, S. C., and Matthews, T. S. (1998). "Report of Final Natality Statistics, 1996," Mon. Vital Stat. Rep., Vol. 46, No. 11, Suppl. National Center for Health Statistics, Hyattsville, MD.
8. Mueller, B. A., and Daling, J. R. (1989). Epidemiology of infertility. *In* "Controversies in Reproductive Endocrinology and Infertility" (M. R. Soules, ed.), pp. 1–13. Elsevier, New York.

8

Puberty and Adolescent Pregnancy

THERESA O. SCHOLL
Department of Obstetrics and Gynecology
University of Medicine and Dentistry of New Jersey
Stratford, New Jersey

I. Introduction

Puberty is a time of life that is characterized by maturation of the hypothalamic-pituitary-gonadal axis, accelerated physical growth, and the development of the secondary sexual characteristics. The underlying mechanism that gives rise to these changes is not known. Physical modifications, however, are part of the complex alteration that takes place. The growth spurt modifies the size and shape of almost every bodily dimension— the face grows and markedly alters, the heart, lungs, viscera, and reproductive tract all expand in size, weight, and volume. The skeleton (long bones, vertebrae), musculature, and, in females, the fat mass also increase. In the end, a girl will be transformed physically in a woman-adult in the size, shape, and composition of her body, with mature secondary sexual characteristics, and she is likely to be sexually active and capable of reproduction. The ages at which the growth spurt commences, breasts begin to bud (thelarche), pubic hair first appears (pubarche), and menarche (first menstrual bleeding) starts vary by nearly a third of the life span of a prepubertal girl.

II. Pubertal Spurt

The growth spurt that occurs at puberty involves all skeletal and muscular dimensions and, with a few exceptions (brain, thymus), almost every other organ and system of the body [1]. During the pubertal, or adolescent, spurt, weight is gained at about the same rate as in infancy and childhood whereas stature velocity (rate of gain in stature) is somewhat lower. It is estimated that adolescent linear growth accounts for 15–20% of adult stature, growth in the mass and volume of the skeleton accounts for 50% of adult skeletal mass, and growth in weight accounts for about 50% of ideal adult weight [1,2].

During adolescence rapid changes in the rate of growth occur at different times within cohorts of the same age, sex, and ethnicity. Unlike other times of life, the adolescent years are marked by rapid acceleration of the rate of growth (growth velocity) until a maximum is reached (peak velocity). Peak velocity is followed by a slowing in the rate of growth (deceleration) until eventually maturity is reached and growth ceases (Fig. 8.1) [32].

III. Linear Growth

On average, the spurt in linear growth is less intense and occurs about two years earlier in girls then it does in boys [1]. The start of the height spurt (age at takeoff) is marked by an abrupt increase in height (stature) velocity. In healthy, well-nourished girls from Europe and the United States this occurs earlier (mean age 8.5–10.3 years) than is generally realized and the range is wide (6–13 years) [3]. Peak height velocity falls closer to the age of twelve (range 10–14 years), averaging nearly 8 cm/year (range 5–10 cm/year); linear growth slows down about two years after peak velocity has been reached [1,3] (Fig. 8.1).

In reality, the linear spurt is a composite of segment lengths and skeletal breaths, each of which has a peak velocity that contributes to, but may not be coincident with, peak height velocity. When examined by fine intervals, growth is not a continuous process but seems to be marked by periods of stasis (no growth) and saltation (rapid growth) and the duration and definition of each is debated [4–6].

The growth spurt proceeds upward from the foot and involves the lower extremities first. Within extremity, distal spurts proceed before proximal; for example, the hand spurts before forearm and forearm before upper arm. The spurt in leg length occurs before the spurt in the spine (estimated as sitting height). Peak velocities for width of the shoulders and the hips occur at approximately the same time as that of sitting height. Growth of the tibia (lower leg) persists longer than that for any other dimension except for growth of the spine [7]. Consequently, a girl early in the throes of puberty will have relatively longer legs with a shorter trunk, but ultimately the trunk should make the greater contribution to her adolescent growth.

The final phase of growth occurs on the downside of the adolescent spurt after menarche has been reached (Fig. 8.1). Linear growth continues after menarche, albeit at a slower rate than before. Roche and Davila [8] estimated the median age of adult stature attainment at 17.3 years, meaning that 50% of the young women in their study continued to grow after that time. By the age of 21 years the proportion still growing had fallen to 10%. The median amount of growth in stature between menarche and the age when growth was completed amounted to 7.4 cm (10th percentile 4.3 cm, 90th percentile 10.7 cm) in their sample. Frisch [9], Berkey *et al.* [10], and Zacharias and Rand [11] described similar amounts of postmenarcheal growth.

Sensitive measures (the Knee Height Measuring Device) suggest that about 50% of adolescent women are still growing at six years postmenarche [12]. More traditional measures show comparable growth in stature and sitting height for young Puerto Rican and white women, but less linear growth for blacks [12].

Late adolescent growth in stature (spinal growth) proceeds in young women even after skeletal maturity is attained (epiphyseal closure of the hand and wrist) and averages 2.3 cm after the skeletal age of 18 years (skeletal maturity by the Greulich-Pyle

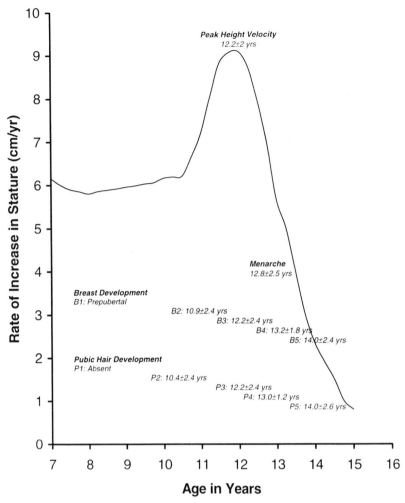

Fig. 8.1 Chronological ages are given with *95% confidence intervals* for the Tanner Breast Stages (B2–B5), Tanner Pubic Hair Stages (P2–P5), peak height velocity, and menarche. Data for peak height velocity, breast development, and pubic hair development are from Sizonenko (116). Data for menarche are from MacMahon (32). Confidence intervals for menarche were calculated by predicting age at menarche for the 2.5th and 97.5th percentiles. From Coupey, S. M., and Saunders, D. S. (1985). Physical maturation. *In* Pediatric and Adolescent Obstetrics and Gynecology (J. P. Lavery and J. S. Sanfilippo, Eds.). Springer-Verlag, New York. Originally from Tanner, J. M. (1962). "Growth at Adolescence," 2nd ed. Blackwood Scientific, Oxford.

method) [13]. Pelvic growth continues throughout late adolescence. At menarche, the pelvis is a "late bloomer," small and less mature compared to stature, because it has reached a lesser proportion of its adult dimensions. After menarche, as statural growth slows down, the pelvis continues to increase in size [14].

IV. Body Composition

Body weight generally reaches peak velocity within a year of peak height velocity. The weight spurt, peaking at gains of 7–9 kg/year for females in Tanner's classic studies, includes skeletal tissue gained as part of the height spurt, increases in the weight of the internal organs, and gains in muscle mass and fat [1,3]. Fat mass increases during childhood and throughout adoles-

cence in both males and females, however, a sex difference in fatness emerges around the time of puberty. Fat increases rapidly in females during later childhood and throughout the pubertal spurt, at approximately twice the rate found in males. During their period of maximal growth, females gain an estimated 2.8 kg of fat (1.4 kg/year) compared to 1.5 kg (0.7 kg/year) in males [15]. By late adolescence young women have accumulated a fat mass approximately twice as large as the fat mass of young men.

On the other hand, at puberty males have a spurt in fat free mass (lean tissue) that is approximately two times larger (14.3 kg fat free mass at 7.2 kg/year) than the spurt in females (7.1 kg at 3.5 kg/year), establishing yet another sex difference at adolescence [1,3]. Because fat mass increases more rapidly in adolescent females and fat free mass increases more rapidly in adolescent males, fat contributes a greater proportion to the body weight of females (percent body fat) than it does in males; this is also termed "relative fatness." During peak height velocity, the relative fat mass increases at approximately 0.9%/year in females while decreasing by 0.5%/year in males [3,15]. Changes in fat mass also are reflected in skinfolds (subcutaneous fat) which increase in females and decrease in males during puberty and the linear growth spurt [15].

Substantial changes in the deposition of fat (internal versus subcutaneous) and in the pattern of the subcutaneous fat depots (trunk versus limb skinfolds) occur in young women as linear growth decelerates. Changes in the ratio of subcutaneous fat to the total fat mass (subcutaneous plus internal) suggest that fat accumulates as menarche approaches and most of the accumulation is internal rather than subcutaneous [15]. At menarche and afterward there is change in the placement of fat depots on the body and proportionately more subcutaneous fat accumulates on the trunk compared to the extremities [15]. Following menarche, young women continue to gain both subcutaneous fat and total fat mass. Obese adolescent girls have a more central pattern of fat with deposits localized primarily on the trunk (upper back) [16]. However, even with normal weight at puberty there is an increased deposition of fat on the trunk compared to the limbs as adolescence progresses [17]. Excessive weight gain during late adolescence exacerbates the normal pattern of subcutaneous fat deposition leading to even larger gains on central body (trunk) sites [17].

V. Maturation

Although pubertal events cluster together and follow a fairly uniform sequence, chronological age is not the best indicator of when these changes will begin or end. A twelve year old girl, for instance, may be prepubertal and physically childlike at the start of her growth spurt, with some development of breasts and pubic hair, or she may have adult secondary sexual characteristics, be menstruating, and be nearly full grown. As a result, pubertal events in adolescence are better described by the concept of biological maturation.

A. Indicators

There are several indicators used to mark the advancement of biological maturity and the progression from child to adult. Measures based on the development of the secondary sexual characteristics and menarche are most useful during puberty whereas skeletal or bone age can be used to assess biological maturity over the span from infancy to adolescence. The assessment of skeletal age is, however, an invasive procedure requiring X-ray exposure of the hand and wrist. Sexual maturation can be assessed by observation or by self report; age at menarche may be obtained prospectively (in longitudinal series), retrospectively, or estimated via the status quo (yes/no) method for a population.

1. Skeletal Age

Skeletal age is a measure of the distance the skeleton has covered in the course of its development. Traditionally, skeletal age is assessed in the hand and wrist (left side), so chosen because of the large number of bones (N = 30) and epiphyses (N = 21) present *in an extremity that is small enough* to be X-rayed in total and effectively shielded from the rest of the body, but varied enough to be informative about the status of the entire skeleton [1]. The maturity of other skeletal areas (knee, hip, elbow, foot, and shoulder) has been assessed but is not routinely used.

Bones of the hand and wrist progress from cartilage to bone via centers of ossification. Ossification of the diaphysis (shaft) of the long bones (ulna and radius) and short bones (metacarpals and phalanges) of the hand and wrist is nearly complete *in utero*. Ossification in the epiphyses of the long bones of the hand and wrist and in the carpals or round bones of the hand (capitate, hamate triquetral, lunate, navicular, trapezium, trapezoid sesamoid, and pisiform) appears at or around birth. The carpals enlarge and change shape in a way that is unique to each bone as cartilage ossifies and the adult shapes and contours are assumed. Long and short bones also show characteristic shape changes in their epiphyses and in the corresponding diaphyses as the growth plate approaches closure. At maturity, epiphyses and diaphyses fuse.

Individual bones have specific stages that occur in a predictable sequence. These stages form the basis for the assessment of skeletal or bone age using a global method (Greulich-Pyle) [18] that assesses the status of the entire hand and wrist. A bone-specific approach (Tanner Whitehouse II) [19] examines and grades the maturity of 20 selected bones in the hand and wrist and assigns a letter grade to each (later converted to a score) based on bone-specific maturity indicators. A third method, useful for population comparisons, involves the age at which centers of ossification of the hand and wrist are first visible [20].

Both the Greulich-Pyle and Tanner Whitehouse II methods are interpretable as a skeletal age for a subject's chronological age and sex. At birth girls are more skeletally mature than boys by a few weeks; this difference enlarges during childhood to a difference of several months. At puberty, girls are advanced by about two years in comparison to boys [1] and thus, on the average, experience adolescent changes at an earlier point in time. A young woman with a skeletal age of 12.5 but a chronological age of 11.0 is said to be skeletally advanced relative to her chronological age or early maturing whereas in the reverse situation (skeletal age of 11.0, chronological age of 12.5) she would be considered to be skeletally delayed or late maturing. A young woman who has a skeletal age and a chronological age within one year of each other (skeletal age of 11.5, chronological age of 12.0) is considered an average maturer with a skeletal age that is chronologically more or less "on-target."

2. Secondary Sexual Characteristics

One of the first signs of impending puberty in a childlike girl without breasts (Tanner stage B1) or pubic hair (Tanner stage P1) is the budding of her breasts (Tanner stage B2) accompanied by an increase in the diameter of the areola (Fig. 8.2). The breasts and areola will continue to enlarge but have not yet separated (Tanner stage B3). Later on, the areola and the papilla will mound above the level of the breast (Tanner stage B4). As the breast becomes adult, only the papilla will project (Tanner stage B5).

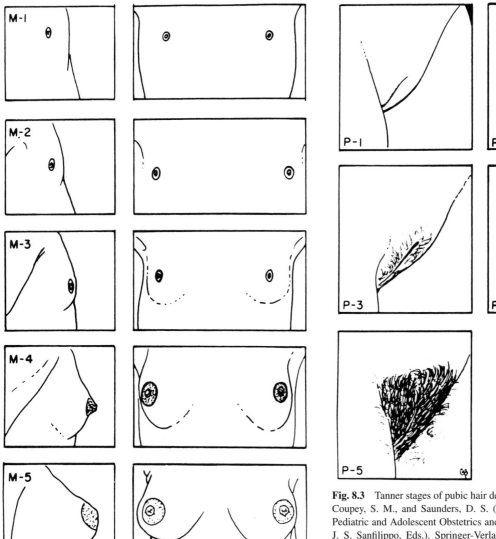

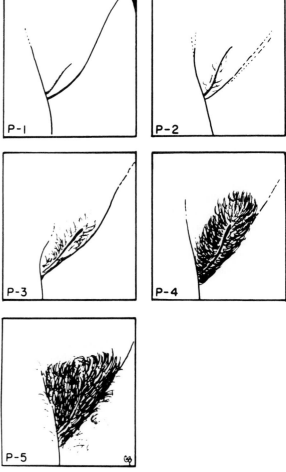

Fig. 8.3 Tanner stages of pubic hair development in the female. From Coupey, S. M., and Saunders, D. S. (1985). Physical maturation. *In* Pediatric and Adolescent Obstetrics and Gynecology (J. P. Lavery and J. S. Sanfilippo, Eds.). Springer-Verlag, New York. Originally from Tanner, J. M. (1962). "Growth at Adolescence," 2nd ed. Blackwood Scientific, Oxford.

Fig. 8.2 Tanner stages of breast development in the female. From Coupey, S. M., and Saunders, D. S. (1985). Physical maturation. *In* Pediatric and Adolescent Obstetrics and Gynecology (J. P. Lavery and J. S. Sanfilippo, Eds.). Springer-Verlag, New York. Originally from Tanner, J. M. (1962). "Growth at Adolescence," 2nd ed. Blackwood Scientific, Oxford.

Likewise, a sprinkling of fine pubic hair appears on the labia majora (Tanner stage P2) (Fig. 8.3). Subsequently, pubic hair becomes thicker, darker, and more wiry (Tanner stage P3) and forms an inverted triangle covering the perineum and mons veneris (Tanner stage P4). Ultimately there is spread to the thighs per the adult pattern (Tanner stage P5).

Development of breasts and pubic hair, while correlated, actually exhibit considerable independence. At Tanner stage B3, for example, while many also have reached stage P3 (33%), others are at stages P1 (22%), P2 (28%), P4 (16%), or P5 (1%) [21]. Similarly, pooled data on British and Swiss participants in longitudinal studies [3] suggest that most young women reach

menarche at Tanner stage B4 and P4. However, other breast stages from B2 to B5 and pubic hair stages from P1 to P5 also are represented at menarche [3].

The onset of breast development at puberty is paralleled by concomitant uterine growth in length, fundal diameter, and endometrial thickness. Growth of the fundus exceeds growth at the cervix, resulting in a pear shaped organ, which also serves to mark the onset and progress of the pubertal process [22].

Data on sexual maturity may be obtained by physical examination or by self report. Duke and colleagues [23] reported agreement between ratings performed by a single investigator and adolescent self ratings of breast stage (Kappa = 0.81) and pubic hair stage (Kappa = 0.91) in 43 white females. Neinstein [24] studied self assessments in a small (N = 22 females) triethnic sample of teenagers (white, black, Hispanic) and found agreement of 77% (r = 0.87) for exact breast stage and 73% (r = 0.86) for exact pubic hair stage between female subjects and the pediatrician rater; these correlations were consistent for white, black, and Hispanic young women. These two studies, both of small sample size, suggest that older adolescents may be able to gauge the development of their secondary sexual characteristics fairly accurately. The issue, however, requires further study, particularly for ethnic minorities and younger teenagers.

Age at menarche signals the onset of reproductive capacity and is the best indicator of the timing of maturation [20]. While it is not as precise an indicator of timing of maturation as peak height velocity, skeletal age, or pubertal staging, it is the only indicator that does not require assessment or longitudinal monitoring through adolescence and it can be reliably collected retrospectively.

Bergsten-Brucefors [25] reported a correlation (r = 0.81) between recalled and recorded menarcheal dates in Swedish girls (N = 339) four years after the event. A total of 63% were able to recall the date of the event within ± 3 months. Likewise, of the 101 members of a cohort of girls (N = 657) who had experienced menarche, 59% of respondents at the third annual contact recalled their menarche with an exact date and year (average duration of recall was 430 days) [26]. Age at menarche can be recalled fairly reliably decades after the event, although the exact date may not be well remembered. Damon and Bajema [27] recontacted 143 white women whose ages at menarche had been recorded some 39 years earlier and found that 86% were able to recall their menarche within one year of the actual age. After 34 years, 59% of the 160 women followed as part of the Menstrual and Reproductive History Study [28] were able to repeat their previously reported age at menarche precisely to the year, while a total of 90% were accurate to within one year. Mean age at menarche based on recall (12.67 ± 1.1 years) did not differ from the prospectively determined mean age (12.68 ± 0.09 years). Recollections of 50 year old participants (N = 91) with prospectively determined age of menarche showed a significant correlation (r = 0.67); 84% were able to remember their menarche to within 1 year of the actual age [29]. Data on recall of menarche in minority women, for whom the event may have a different cultural context, are sorely needed.

Like skeletal age, the division of age at menarche signifying early (<12 years), average (12.0–13.9 years), and late (≥14 years) maturation is well established [1]. The average age at menarche in the United States and Europe is the basis for this division. Early maturers, classified on the basis of age at menarche, show consistently advanced skeletal age throughout adolescence and pass more quickly through breast and pubic hair stages [1,3]. The relationship of menarche to peak height velocity is fairly well described [1,3]. Early maturers (menarche up to 11.9 years) have been shown to reach peak height velocity about 8 months before menarche (age 10.3). Average maturers (menarche age 12.0–13.9) reach peak height velocity about a year later (age 11.5) and late maturers (menarche ≥ 14 years) reach peak height velocity the latest of all. Early maturers also have a more intense growth spurt than average or later maturing girls and a greater amount of postmenarcheal growth [1].

The correlations between skeletal age, sexual maturation, and menarcheal age are positive and on the order of 0.6–0.9. This means that a young woman who is advanced in one indicator of maturity is likely to be advanced in the others. Intercorrelations

grow stronger as adolescence proceeds and an underlying factor unidentified but probably hormonal discriminates adolescents who are early, average, or late in the timing of adolescent events. However, no single indicator by itself provides a complete description of the tempo of maturation at puberty [3].

Secondary sexual characteristics may be used in combination with age at menarche to gauge the timing of pubertal events [30]. Tanner stage B2 (breast bud) is often the first sign of puberty in females and precedes peak height velocity in stature by about one year (Fig. 8.1). The correlation between age at peak height velocity and age at menarche is moderate to high (r = 0.7–0.9). On average, menarche occurs 1.3 years after peak height velocity (range 0.01–2.5 years) for average maturers. The proportion of girls experiencing menarche before peak height velocity is estimated at less than one percent [3,31].

Menarche usually begins on the downside of the growth spurt a little more than a year after peak height velocity has passed [1] (Fig. 8.1). Young women who experience menarche at the age of 12.8, the United States median [32], are likely to have reached peak height velocity sometime around 11.5 years. A girl who has neither attained stage B2 nor experienced menarche is probably prepubescent and has not yet started her pubertal spurt in linear growth. One who has attained stage B2 (at least) but not yet menstruated is likely to be close to peak height velocity. A young woman who has attained stage B2 (or higher) and is menstruating, however, probably has passed peak height velocity and completed much of her adolescent growth spurt [30].

B. Factors Influencing Maturation

1. Ethnicity

There are significant ethnic differences in the timing of maturation in girls, including skeletal age, sexual maturation, and age at menarche. Young black women from the United States have an earlier average age at menarche than whites. While white girls in North America and Europe attain menarche at an average of 12.8–13.2 years (range 9–16 years), black girls in the United States have an average age at menarche of 12.5 years [1,3,32]. This is paralleled by an approximately two-fold increased likelihood of early menarche (<12 years) when black (21.3%) and white (11.6%) girls are compared [32]. African girls living in Africa, however, experience menarche relatively late (mean 13.0–13.4 years), even when living under good economic circumstances [20]. A correlated event, skeletal ossification, although accelerated at birth among African infants [1], tends not to persist in that environment.

Few reliable data are available for Hispanic groups in the United States, but national statistics from Cuba indicate an average age at menarche of 12.9 years for Cuban girls, similar to that for U.S. whites [33]. The only published study citing an average age at menarche for Puerto Rican girls indicated an average age closer to that of blacks in the same sample [34]. According to Eveleth and Tanner's compilation [20], affluent Argentinean, Brazilian, Chilean, and Venezuelan girls of European descent have early mean menarcheal ages (12.0–12.6 years). Rural Mexican girls from the state of Oaxaca menstruate late (mean 14.3 years) whereas urban or middle-class girls from Mexico City seem to have timing comparable to that of white U.S. and European girls (mean age 12.3–12.8). Although the estimate is likely to have been biased upward by the partial exclusion of early maturing girls, Mexican-American girls included in Hispanic NHANES (National Health and Nutrition Examination Survey) experienced Tanner stage B4 (the breast stage at which many undergo menarche) at the mean age of 13.8 years [35]. In contrast, Mexican-American females included in the Ten State Nutrition Study, are said to have experienced menarche at the mean age of 12.7 years [3].

Asian girls living in California are reported to experience menarche at an average of 13.2 years [20], while Japanese girls in Brazil are slightly younger (mean 12.9 years) when they begin to menstruate [20]. Affluent girls from Hong-Kong and Singapore begin to menstruate a few months sooner (mean age at menarche 12.4–12.7 years). The latest recorded ages at menarche are experienced by the Bundi from highland New Guinea (mean 18.0 years) [20].

The ethnic difference in age at menarche between white and black girls from the United States is reflected in two correlated events—advanced skeletal ossification and earlier development of the secondary sexual characteristics. Black U.S. infants are precocious in the appearance of the ossification centers of the hand and wrist during the first year of life and up until the age of 7 years [36,37]. Skeletal advancement occurs for black children of both sexes. Advancement, however, is more pronounced for females than for males; black girls are more than half of a standard deviation unit (SDU) ahead of white girls despite a large gulf in socioeconomic status. Roche [38] reported that mean skeletal ages (Greulich-Pyle method) of black females aged 6–11 years were usually advanced for their chronological age and were ahead of white girls. Except at age 8, where there was no difference, advancement in skeletal age varied from +0.7 months to +2.7 months for black females during the interval in question.

Differences in the timing of the appearance of secondary sexual characteristics also reflect advanced maturation of black females. Harlan *et al.* [39] compared Tanner staging for breast and pubic hair in white and black females aged 12–17 included in NHANES III. Because girls under the age of 12 years were not examined, information on Tanner Stages 4 and 5 are the most meaningful for ethnic comparison. At the chronological age of 12, for example, a smaller proportion of white girls (14.2%) compared to black girls (30.5%) were adult (Tanner stage B5) in breast development. This was also true for pubic hair, where the proportion of adult (Tanner stage P5) girls among blacks (28.2%) was increased three-fold compared to white girls (9.2%).

Hermann-Giddens and colleagues [40] reported data on more than 17,000 girls aged 3–12 years using data from pediatric practices in 34 states and Puerto Rico. Confirming prior reports [41,42], black girls were more likely than white girls to have developed breasts and/or pubic hair (Tanner stage 2) at each chronological age. At 7 years, the typical age for first grade, there was a threefold difference in the proportion of each ethnic group at Tanner stage B2 or greater (5% of white and 15.4% of black 7-year-old girls). By age 8, prevalence was increased to 10.5% among white and 37.8% among black girls. Ethnic differences in the prevalence of girls at Tanner stage P2 (or better) were also substantial and consistent with earlier sexual maturity of black females. Median age at menarche, estimated by the status quo (yes/no) method and probit analysis, was at the ex-

pected age (12.88 years) for whites but was earlier than expected (12.16 years) for black girls.

These data, however, were gathered from physician office practices and raise the question of whether early maturing children, especially those who are black, were more likely to have been brought in for examination because of parental concerns about precocious development. In addition, the survey is confined exclusively to early and average maturers and excluded later maturing children (*i.e.,* those who experience menarche after the age of 12 years). Approximately 27% of white girls and 26% of black girls are expected to begin menarche between the ages of 13 and 17 years [32]. Exclusion of these individuals would have the additional effect of biasing downward the average ages at which Tanner stage 2 and Tanner stage 3 were attained as well as the average age at menarche.

2. Nutrition and Nutritional Status

There has been a long-term or secular increase in childhood growth that is accompanied by a more rapid rate of maturation and a decline in the age at menarche. In Europe, during the past 150 years, the age of menarche has fallen by three months per decade. Data from the United States also provide evidence for a secular decline in the age at menarche that is slightly less (about two months per decade), falling from age 14.7 at the turn of the nineteenth century to age 12.8 at the end of the twentieth century [20,43]. There is debate about whether the secular trend continues or has ended in the developed world. There are even examples of reversals of the trend during periods of deprivation, for example in post-World War I Germany [44]. Long-term changes in nutrition and public health are believed to be the causes of the secular change as the trend has paralleled improvements in sanitation, food availability, and health care, and declines in the prevalence of childhood infectious diseases and clinically detectable nutritional deficiency disorders such as scurvy and pellagra [43,44].

The nutritional hypothesis for secular change centers around body size and composition. Earlier menarche is associated with increased weight and fatness [1]. Frisch proposed that a "critical weight" was the trigger for menarche [45]. As a result of more rapid growth, children attained the "critical weight" at an earlier chronological age. At present, there is little support for the hypothesis that "weight" or "fat" is causally related to menarche. However, the etiologic factor is probably correlated in some way with pubertal growth, body weight, and composition. For example, it is well known that children from regions of the world where chronic malnutrition is endemic have reduced weights and heights and attain menarche relatively late in their adolescence. In small, rural Egyptian villages, the average age at menarche is 13.9 years, and in India (Hyperdabad) poor rural girls menstruate at a mean age of 14.6 years [20]. Data on poorly nourished Alabama girls demonstrated that rate of skeletal maturation and the age at epiphyseal fusion of the long bones were delayed in comparison to well-nourished controls [46]. Onset of menarche was 24 months later for the undernourished (14.4 years) compared to the well-nourished (12.4 years) girls. Because both growth and maturation were delayed to the same extent, adult stature was not substantially affected (<1 cm) by undernutrition. Barbadian girls diagnosed with moderate to severe protein-energy malnutrition in the first year of life experienced delays in the appearance of the secondary sexual characteristics; both breasts and pubic hair were said to be affected [47]. For example, at the age of 10 years, 25% of girls with clinically severe malnutrition had attained Tanner stage P2 in comparison to 46% of girls without such a history; by the age of 15, 40% of the cases had experienced menarche compared to 100% of controls [47].

Poor children from the developing world show some "catch-up" in weight and linear growth when they are adopted. Korean children adopted by U.S. families at an early age are closer to U.S. norms for weight and height than those adopted later in life [48]. However, other studies have documented that when children from the subcontinent of India are adopted into Swedish homes, sexual maturation and age at menarche are more accelerated than linear growth [49]. Adopted girls reached menarche at an average of 11.6 years, at least one year in advance of members of the Indian privileged classes and 2–3 years before the lower social classes into which these young women had been born. Earlier menarche was, in turn, associated with reduced adult stature for the children. They were barely taller as adults (1 cm) than less privileged girls still living in India. Thus, while poor nutrition tends to delay growth and maturation, remediation in contrast tended to speed up sexual development. A greatly improved standard of living, which includes both nutrition and health care, seems to greatly accelerate childhood growth, but not necessarily adult stature, and may bring about a very early onset of pubertal growth and sexual maturity.

Several studies have suggested that dietary factors including vegetarianism [50], and the intake of dietary fiber [51] or fat [52] affect age at menarche. In animal studies, fat intake increases rate of weight gain, which consequently decreases age at first estrus [53]. Because intake can influence weight through the accumulation of fat and lean tissue, diet may influence age at menarche either directly (dietary related hormonal changes) or indirectly (body composition).

The early maturer in any ethnic group is distinguished by a number of biological characteristics. Perhaps the most obvious of these are differences in growth and body composition. Prior to achieving menarche, early maturing girls are taller and have a higher weight and body mass index and larger skinfolds than later maturing girls [1,52,54]. Supporting the hormonal hypothesis, Merzenich [52] found an independent effect of dietary fat on age at menarche. There was a two-fold-increased risk of early menarche when fat intake (energy adjusted) was high (highest quartile); adjustment for percent body fat did not diminish the association.

Mosian and colleagues [54] examined the correlates of early menarche using prospectively collected data from fifth-grade French-Canadian girls. Energy and carbohydrate and fat intake all were increased among the earlier maturing girls (age at menarche 9.5–12.5 years). Consistent with their higher intake, girls were taller and heavier and had a higher body mass index (weight/height²), larger skinfolds, and greater energy expenditure compared to controls. After adjusting for weight, no effect of dietary intake on earlier menarche was evident in this population, supporting the hypothesis that the influence of nutrition on menarche was indirect (body composition). Maclure and

colleagues [55] found no effect of macronutrient intake on age at menarche but did detect a strong association between menarche and body composition; they interpreted these data as consistent with the hypothesis that nutrition influences menarche via growth and development.

3. Physical Activity

Early maturers may have increased energy expenditure that is correlated with their larger body size and increased fatness [54], although when expressed per kg of lean body mass their energy expenditure may actually be below that of later maturers [56]. Early maturers also appear to be less physically active than girls who mature later on and one anticipated consequence of reduced activity would be increased fat accrual. Consequently, energy balance (the difference between energy intake and energy expenditure) may be an important factor in the timing of maturation.

A number of reports confirm the fact that girls participating in leisure time physical activity such as dancing or sports programs begin to menstruate at a later age compared to nonparticipants. Moisan et al. reported a two-fold decrease in risk of earlier menarche for physically active French-Canadian girls [54]. Similarly, more hours per week spent in school and leisure time sports correlate with later onset of menarche, after controlling for the dietary fat intake and the higher fat mass that often are associated with earlier maturation [52]. Malina and Bouchard [3] have summarized data showing, for example, a later average age at menarche for runners and jumpers (13.5–14.0 years), gymnasts (13.7–15.0 years), and for women engaged in most other sports as well. Ballerinas also have a delayed age at menarche. In the series reported by Warren [57], ballet dancers (N = 15) averaged 15.4 years at menarche versus 12.5 years for age-matched controls. Dancers also were delayed in breast but not pubic hair development; pubertal progression was linked to periods of light activity. Episodes of amenorrhea, experienced by 85% of dancers who had achieved menarche, correlated with increased activity. Like dancers, women athletes, particularly those with early onset of sports training, also exhibit later menarche. Athletes who began training before menarche had an onset at 15.1 years versus 12.8 years in women who initiated training after they had started to menstruate [58].

These findings have raised the question of whether physical activity delays menarche or if late maturers are more likely to choose intensive training in sports or dance. Malina argues that ballet and some sports (gymnastics, distance running) self select for the linear physique associated with later maturation [59]. On the other hand, the relationship between training start and age at menarche would offer support for a causal relationship between intense activity and delayed menarche if potential confounding factors (nutrition, ethnicity, heredity) had been taken into account. For example, Frisch et al. [58] documented, but did not control, the lower macronutrient intakes of the athletes who began their training before menarche. In comparison to a postmenarcheal sample, energy (−22.5%), protein (−22.8%), and fat (−31.5%) all were significantly reduced among young women engaged in early preparation for sports.

Malina [60] examined age at menarche in athletes and their mothers and sisters reasoning that if self selection was involved, family members, along with the athletes, would exhibit later age at menarche. While no nonathletes, along with their mothers and female sibling controls, were studied, Malina demonstrated that the mother–daughter correlation (r = 0.25) and sister–sister correlation (r = 0.44) in menarche were comparable to correlations in the published literature. Interestingly, among white families, mother–daughter correlations tended to be higher when both mother and daughter were athletes (r = 0.44) than when only the daughter was athletic (r = 0.18). In the total sample, as well as among the white women, there was a reverse secular trend. Mean menarcheal age was later for athletes, following sisters by about half a year and mothers by about three-quarters of a year.

It has been pointed out that athletes who train strenuously have body composition (fat and lean body mass) and menarcheal age very much like that of anorectic young women [61]. Thus, it seems possible that some athletes, especially those involved in sports where they need to "make weight" or where there is an optimal performance weight or an emphasis on thinness, may be more prone to restrict intake or to develop an eating disorder [59]. At one end of the spectrum, dietary restriction, coupled with a demanding schedule of training, could potentially give rise to changes in weight and body composition, delayed menarche, and other menstrual problems.

4. Genetic, Family, and Other Factors

Aside from overall ethnic differences, both genetic and environmental factors (nutrition, socioeconomic status) affect the timing of maturation in females. Clear evidence for genetic factors in the timing of maturation comes from studies of twins, where it has been shown that monozygotic twins achieve menarche at nearly the same age (within 2–3 months). Dizygotic twins, in contrast, differ in age at menarche by an average of nine months [62]. Timing of maturation also is transgenerational as in some studies there is a fairly sizable correlation (r = 0.40) in age at menarche between mothers and daughters [1]. Furthermore, daughters of migrants to Australia from various regions of Europe maintain their relative timing of menarche despite moving to a new environment [63].

The timing of adolescent maturation may be altered by environmental and socioeconomic factors. In the developing world, where there is a large gulf between rich and poor, children from higher social classes are taller, heavier, and mature earlier than lower class children, and urban girls mature sooner than rural ones [1,20]. However, in developed countries differences between the social classes are less apparent. In addition, adolescent girls from the lower social classes are fatter, whereas adolescents from a higher socioeconomic status have a physique that is more linear, which in many cases is the intended effect of diet and exercise. These differences in physique may attenuate any differences in maturation associated with social class. For example, in the United States family income is not correlated with menarcheal age [32]; in Sweden neither income nor paternal occupation predicts age at menarche [64]. However, in Poland and in the United Kingdom, parental education and/or occupation still remain weakly correlated with menarche. The children of better educated white-collar workers or professional fathers menstruate slightly earlier, that is, by about half a year in Poland and approximately 2 months in England [3].

Family size also influences age at menarche whereas birth order does not appear to do so. As the number of children in the family increases, so does age when menstruation starts. For example, among athletes, family size is correlated with increased menarcheal age by 0.17 years among white athletes and by 0.21 years among black athletes [65]. However, because it is uncertain whether or not athletes come from families with more children (no nonathlete control group), it is unclear whether or not family size confounds the association of athletics with later menarche or if it is an independent risk factor. Summarized data [3] suggest a delay of menarche by 0.11 to 0.22 additional years per additional child in the family, although factors such as nutritional status and ethnicity are not always controlled. Other factors being equal, however, an estimated delay of 0.12 years/child implies that a U.S. girl from a sibship of five would be expected to experience menarche more than 6 months after a girl with no siblings.

Stress is another sociobiological risk factor thought to accelerate maturation. Girls growing up in difficult environments—where parents separate or live in conflict—have menarche that is earlier by approximately four months. Evolutionary psychologists have hypothesized that stress leads to premature flight from the home which is facilitated by early maturation [66]. Tests of this theory have so far been negative. Neither manifestations of childhood stress (bed wetting, nightmares) nor contextual stressors (father's presence in home, parental education and work status) predict age at menarche [67]. Stress also has not been found to lower metabolic rate and increase storage of fat thus leading to an earlier age at menarche [68].

C. Sequelae of Maturation

Early maturers (menarche at ages 10–11) have been shown to reach peak height velocity about 8 months before menarche whereas average maturers (menarche at age 12) reach peak height velocity about a year later (relative to chronologic age) [1]. Early maturers classified on the basis of age at menarche show advanced skeletal age and pass more quickly through breast and pubic hair stages [69] but this has not been confirmed by others [69a]. Physical changes in size and shape occur at an earlier chronological age, in early maturers compared to other girls.

1. Sexual Debut

Hormonal changes at adolescence lead to increased libido in adolescent women. Pubertal hormones also lead to the development of the secondary sexual characteristics that attract the sexes to each other. The timing of sexual attraction presumably occurs sooner for young women who mature early. Testosterone concentration is correlated with the timing of noncoital sexual activity as well as with age at sexual debut for U.S. black and white females [70]. While concentration of testosterone approximately doubles during the pubertal period, the largest increase occurs just before menarche. Testosterone levels are said to be higher and may thus increase sooner in early maturing girls [70].

Early maturers are more likely to engage in sexual intercourse as young teenagers. Soefer et al. [71] utilized data from two national probability surveys (1971, 1976) to examine the relationship between early menarche and onset of sexual activity in white and black women aged 15–19 years. The probability of engaging in premenarcheal sexual intercourse was uncommon; this occurred in fewer than 5% of the blacks or whites surveyed. Early maturers experienced premarital intercourse sooner than average maturers and both early and average maturers experienced intercourse prior to the later maturers. For example, among black young women relative risk comparisons are most strikingly different at ages 12 and 13. A 13 year old black female in the early menarche group (<12 years) was 4.4 times more likely in 1971 and 3.3 times more likely in 1976, to have experienced premarital intercourse than a black female in the late menarche group (≥14 years). By age 17, the relative risks were all extremely close to unity. A 13 year old white female in the early menarche group was 6.3 times more likely in 1971 and 5.0 times more likely in 1976 to have experienced premarital intercourse than a white female in the late menarche group. Udry also reported early sex (<16 years) in girls with early menarche (<12 years) [71a]. Zabin [72] confirmed these finding in black girls from Baltimore in 1981. While reporting a higher rate of sexual activity in the 1980s, she also found substantially increased risks of early coitus for early maturing women. For example, at the age of 15, the cumulative probability of first intercourse was 55% when menarche was experienced at age 11 or younger versus 32% for those experiencing menarche by age 14 or later. Others have confirmed these associations, [73] extending them to Hispanic U.S. teenagers [74] and young women from developing countries [75,76].

2. Adolescent Fecundity

Early maturers may present a different hormonal profile than later maturers and be more fecund at an earlier chronological age. Maturation is orchestrated by neuroendocrine factors and modified by autocrine and paracrine events in the ovary. During the first few months after menarche, the hypothalamic-pituitary-ovarian axis is immature and the developing follicle secretes estrogen alone; positive feedback to trigger ovulation develops later on. This is the period of so-called "adolescent sterility," although subfecundity is probably a more appropriate term.

Both age at menarche and gynecological age (years since menarche) correlate with the probability of ovulation in teenagers. Six months after menarche, 40% of the cycles in early maturing girls (<12 years at menarche) are ovulatory as opposed to less than 20% in average (12.0–12.9 at menarche) and later maturing (≥13 years at menarche) girls [77]. As the number of years since menarche increases, the proportion of ovulatory cycles also increases from 15% in the first postmenarcheal year to 80% by the sixth year after menarche [78]. The timetable also seems to depend on age at menarche. At 1.5 years after menarche, 80% of the cycles for early maturers are ovulatory compared to 30% in the average maturers and 20% for the late maturers. The amount of time from menarche to the point where half of all cycles are ovulatory takes 1 year for early maturers, 3 years for average maturers, and 4.5 years for late maturers [77]. Early maturers are thus more fecund at an earlier chronological age and an earlier gynecological age than are later maturing girls.

In ovulatory cycles, progesterone opposes estrogen. Cycles where estrogen is unopposed by progesterone are typically irregular with variable menstrual flow. Ovulation is correlated with

various indicators of pubertal development including breast development and pubic hair development. At Tanner stage B1 none of the teenagers (aged 13–16 years) studied by Vihko and Apter [78] had ovulatory cycles. The proportions increased markedly with development. At Tanner stages B2–3, 38% had ovulatory cycles; this proportion increased to 44% at Tanner stage B4 and 57% by Tanner stage B5. Likewise, there were no teenagers with ovulatory cycles at Tanner stages P1 or P2 but 35% of girls at Tanner stages P3–4 and 60% at Tanner stage P5 had ovulatory cycles.

3. Adolescent Pregnancy

Low-income women with early menarche are more than twice as likely to have had intercourse by age 16 and are almost twice as likely to have conceived and terminated the pregnancy or given birth by age 18 [71a]. Thus, earlier maturation appears to increase the risk of adolescent pregnancy.

Each year in the U.S., nearly 1 million young women aged 19 and under conceive; of these, approximately half will give birth [78b]. The adolescent childbearing rate has declined from the very high levels of 1960 (about 90/1000) to 69.5/1000 in 1970 and 54.1/1000 in 1980, but the rate rose again to 61.3/1000 in 1990. In the 1990s there have been some year to year fluctuations, with the zenith (63.5/1000) reached in 1991. In 1999, the adolescent childbearing rate (55.6/1000) is slightly higher than the rates in effect between the years 1976 (54.0/1000) and 1988 (54.3/1000) [78b]. The rate of adolescent childbearing has fallen since the 1991 peak with approximately a 5% decrease between the two most recent reports (1995 and 1996); parallel declines have occurred in the adolescent pregnancy rate. [78b].

It is well recognized that women raised under favorable social and economic circumstances are more likely to terminate a pregnancy that occurs at an early age or which is unintended, whereas rates of adolescent childbearing are highest in disadvantaged minority women.

That teenagers who give birth are more likely to have reached menarche before the age of 12 has been noted in many studies of pregnancy outcome [79,80]. While there are a few reports in the literature on conception before menarche [81], there are many more instances of teenagers who conceive at a low gynecological age (i.e., within two or three years of menarche [82,83]).

a. GYNECOLOGIC AGE. In 1977, Zlatnik and Burmeister [84] coined the term "gynecologic age" to distinguish gravidas (pregnant women) who were physiologically immature from those who were not. Their rationale originated in the well-known differences in the onset and timing of adolescent development. Just as there are endocronologic differences between adolescents and adults in the menstrual cycle, there also appear to be differences between biologically mature and immature teenagers. These differences are indexed by gynecological age.

Adolescent menstrual cycles are characterized by a longer follicular phase and a shorter luteal phase, than adult menstrual cycles. In the adolescent cycle, ovulation occurs approximately five days later than in adults [85]. The dominant follicle is selected late and remains smaller throughout the follicular phase. Follicle stimulating hormone (FSH) increases more slowly early in the cycle, luteal progesterone concentrations are lower, and deficient luteal phases are more common in adolescents than in adults. During the first several years following menarche there is a gradual decrease in the duration of the follicular phase [86]. Among pregnant adolescents as well, the duration of the follicular phase of the menstrual cycle which preceded conception is thought to have been longer for the gynecologically immature [87] and thus may have implications for the (shorter) duration of gestation.

Young maternal age is sometimes associated with an increased risk of preterm labor and/or delivery. However, the literature on this subject is divided, with some studies reporting an increased risk of preterm labor/delivery [88a] and others finding no adverse effect on gestational duration [89]. Discrepancies may well be a result of the underlying prevalence of low gynecological age. In studies examining biologic immaturity, increased risk of preterm delivery appears to be specific to low gynecological age and not to chronological age per se [82,83]. Low gynecological age is an indicator of immaturity practically unique to the teenage years. Maternal age has a weak negative association with an increased risk of preterm delivery that amounts to a 3% decline in risk per year of increasing age. Low gynecological age shows a much stronger relationship, with a 20% decline in risk/year after adjustment for chronologic age [83]. After subdividing preterm delivery according to etiology (idiopathic preterm labor, premature rupture of the membranes, and medical or obstetric indications and complications), risk of idiopathic preterm labor and preterm delivery preceded by preterm labor are increased two-fold or more for young adolescents (<16 years) of low gynecological age (<2 years) [90].

Gynecological age is a good indicator of immaturity among young persons who are not far removed from menarche, however, there are situations where it may be an insensitive indicator of risk. For example, under conditions where chronic and clinical malnutrition is prevalent, menarche often is delayed until the completion of breast and pubic hair stages. In this situation, low gynecological age may be neither a sensitive nor a specific indicator of biologic immaturity [91].

b. MATERNAL GROWTH DURING PREGNANCY. Adolescent growth and teenage pregnancy often coincide. The results of one study indicated that 63% of black teenagers (aged 12–18) had postpartum hand-wrist radiographs showing open epiphyses, thus denoting their skeletal immaturity [92]. Consequently, many teenagers retain the potential to grow while pregnant. However, when serial measures of stature are taken, the continued growth of many pregnant adolescents is not clinically apparent due to a diminution in maternal stature with pregnancy [93]. Vertebral compression secondary to postural lordosis and gestational weight gain leaves the mistaken impression that growth has ceased [94]. Using sensitive methods (Knee Height Measuring Device) and a body segment (lower leg) less susceptible to "shrinkage" to index growth, about 50% of teenagers continue to grow while pregnant [93,95].

During pregnancy, maternal growth is associated with larger weight gains, increased fat stores, and greater postpartum weight retention [95–97]. Despite these changes, which should be associated with increased fetal size, growing gravidas experience a reduction in infant birth weight (150–200 g) [98]. Competition for nutrients between the still-growing mother and her fetus

is suggested by marked reductions in placental blood flow, as measured by Doppler Ultrasound, and transfer of micronutrients from the still-growing mother to her fetus. Ferritin and folate levels from cord bloods of growing teenagers fall below levels of teenagers who do not grow; the difference amounts to a reduction of more than 20% in the concentration for each nutrient [99]. The systole/diastole (S/D) ratio, an indicator of vascular resistance, is measured by umbilical artery Doppler Ultrasound. Nearly twice as many of the growing gravidas had S/D ratios of three or greater, indicating a high resistance circuit, which also suggests reduced blood flow [97].

In an animal model of maternal-fetal competition [100], still-growing rat dams, mated at vaginal opening and fed *ad libitum,* had larger gestational weight gains compared to mature pregnant dams. The adolescent dams delivered significantly smaller pups with retarded skeletal ossification and increased mortality. Adolescent ewes, implanted with ova from superovulated adults inseminated by a single sire, were fed a diet formulated to insure low or rapid maternal growth [100a]. Maternal weight was greatly increased among the super-fed young ewes with concomitant effects on body weight (+2 kg at embryo transfer, +22 kg at day 95 gestation), but differences in maternal linear growth are unknown. Rapid growth in the adolescent ewe was associated with decreased lamb birth weight and placental weight and increased risk of spontaneous abortion during pregnancy and pup mortality after birth. Thus, across species, the drive to tissue synthesis in the still-growing young mother appears to take precedence over the nutrient requirements of the fetus and uterus.

Competition between mother and offspring may occur during lactation as well as during pregnancy. Teenagers who breastfeed have significantly lower breast milk volumes in comparison to mature breastfeeding women [101]. Rat pups cross-fostered to adolescent dams had lower rates of growth than when cross-fostered to adults, implying that the quantity or quality of breast milk is affected [100]. Young super-fed and rapidly growing ewes have reduced colostrum at delivery [100a]. Hypothetically, production of the colostrum (and possibly growth of the mammary gland) may have been compromised by reduced secretion of placental hormones during pregnancy.

One reason for the increased gestational weight gain is that growing gravidas continue to accrue fat during the third trimester when fat stores are typically mobilized to support fetal growth [96]. In younger mothers, maternal stores are not mobilized but instead are conserved. Additional weight and fat accrued during pregnancy are retained by young still-growing mothers in the postpartum period. On average, growing teenagers retain more than 40% of their gestational weight gain at 4–6 weeks postpartum compared to 14% for same aged nongrowing gravidas and mature women.

c. OTHER RISK FACTORS AND COMPLICATIONS. Behavioral risk factors such as maternal smoking, drinking, and drug use appear to be less common among teenage gravidas than among mature women, particularly when the young women are ethnic minorities [80,102]. The prevalence of most complications (placental abruption, placenta previa, premature rupture of the membranes, urinary tract infections, and anemia) and characteristics of labor (duration by stage, laceration and episiotomy, use of amniotomy or pitocin augmentation) are similar between adolescents and mature women. At present, Cesarean section appears to be less prevalent among teenagers than among mature women; reports of increased Cesarean delivery now stem predominantly from the developing world [80,102].

Hypertensive disorders of pregnancy have been reported to be more common in young teenagers and in teenagers from the developing world [102]. However, many published studies do not control for parity, thus confounding risk associated with nulliparity with any additional risk due to young maternal age. Studies with adequate controls suggest that even very young gravidas (<16 years) do not experience increased risk [80]. Further, there has been a secular decline in maternal anemia and pregnancy-induced hypertension among teenage gravidas in comparison to the risk sustained by more mature women. In the past, but not at present, these risks were found in excess among teenagers [102]. Thus, despite earlier claims to the contrary, pregnancy, labor, and delivery do not now appear to impose excessive medical or obstetric risks to teenage women.

d. Long-term Sequelae. Early age at menarche (<12 years) is a well-recognized risk factor for breast cancer. Risk declines by about 10% for each year that menarche is delayed [103]. Menarche at an early age is associated with an early onset of ovulation and regular menstrual cycling [77]. The longer cumulative exposure to estrogen and progesterone and possibly the higher exposure to estradiol as a young adult are thought to influence risk via increased breast epithelial proliferation [104]. Breast cancer risk is offset by the greater likelihood that early maturing women will experience adolescent pregnancy and childbearing. Women with an early age at first full-term pregnancy (<20 years) have a diminished risk of breast cancer compared to women with later childbearing onset and the earlier the pregnancy the lower the risk [103,104].

Relative to later maturing females, girls who achieve menarche early tend to be heavier and to have increased body mass indices and fatness throughout childhood and puberty which persist into adulthood [1,3,105,106]. Patterns of body fat distribution established in adolescence also persist, so that heavier earlier maturers show a more "centripetal" fat distribution (greater trunk to limb fat ratios) and an increased risk of overweight and obesity in adolescence and throughout adulthood [105,107].

Excessive rates of gestational weight gain are more common in young teenagers and teenagers who grow while pregnant [106] and may abet the increased risk of obesity and the central body fat distribution associated with early maturation. There is increased potential for central fat deposition when large weight gains occur at adolescence, during pregnancy, and in the nonpregnant state as well. For example, in the postpartum period, teenagers with excessive gestational weight gains are more likely to have retained more weight overall and to have a higher postpartum BMI and a greater risk of being overweight [108,109]. Deposition of fat at central body sites also is increased in young mothers when gestational weight gain is large or excessive [110].

Adolescence is a critical period for the development of obesity [111]; only a fraction (20%) of obese adolescent girls return to a more normal weight within 10 years [112]. With the increased postpartum weight retention associated with adolescent pregnancy, the transient high insulin levels of adolescence [113]

may persist and be exacerbated if maternal weight increases with age or subsequent pregnancies. Long-term follow-up studies suggest that adolescent obesity may increase the risk of chronic disease in later life, including noninsulin dependent diabetes mellitus, coronary heart disease (CHD), and arthritis [114]. Case-control studies show associations between teenage pregnancy and high parity or gravidity and CHD [115]. This relationship could, in theory, be mediated via persistence of the insulin resistance of puberty into adult life.

References

1. Tanner, J. M. (1962). "Growth at Adolescence," 2nd ed. Blackwell, Oxford.
2. Garn, S. M., and Wagner, B. (1969). The adolescent growth of the skeletal mass and its implications to mineral requirements. *In* "Adolescent Nutrition and Growth" (F. P. Heald, ed.), pp. 139–161. Appleton-Century-Crofts, New York.
3. Malina, R. M., and Bouchard, C. (1991). "Growth, Maturation and Physical Activity." Human Kinetics Book, Champaign, IL.
3a. Sizonenko, P. C. (1987). Normal sexual maturation. *Pediatrics* **14,** 191–201.
4. Lampl, M., and Johnson, M. L. (1993). A case study of daily growth during adolescence: A single spurt or changes in the dynamics of saltatory growth? *Ann. Hum. Biol.* **20,** 595–603.
5. Thalange, N. K. S., Foster, P. J., Gill, M. S., Price, D. A., and Clayton, P. E. (1996). Model of normal prepubertal growth. *Arch. Dis. Child.* **75,** 427–431.
6. Lampl, M., Veldhuis, J. D., and Johnson, M. L. (1992). Saltation and status: A model of human growth. *Science* **258,** 801–803.
7. Cameron, N., Tanner, J. M., and Whithouse, R. H. (1982). A longitudinal analysis of the growth of limb segments in adolescence. *Am. Hum. Biol.* **9,** 211–220.
8. Roche, A. F., and Davila, G. H. (1972). Late adolescent growth in stature. *Pediatrics* **50,** 874–880.
9. Frisch, R. E. (1976). Fatness of girls from menarche to age 18 years, with a nomogram. *Hum. Biol.* **48,** 353–359.
10. Berkey, C. S., Dockery, D. W., Wang, X., Wypij, D., and Ferris, B. G., Jr. (1993). Longitudinal height velocity standards for U.S. adolescence. *Stat. Med.* **12,** 403–414.
11. Zacharias, L., and Rand, W. M. (1983). Adolescent growth in height and its relation to menarche in contemporary American girls. *Ann. Hum. Biol.* **10,** 209–222.
12. Cronk, C. E., Schall, J. I., Hediger, M. L., and Scholl, T. O. (1996). Growth of postmenarcheal girls from three ethnic groups. *Am. J. Hum. Biol.* **8,** 31–42.
13. Garn, S. M., Rohmann, C. G., and Apfelbaum, B. (1961). Complete epiphyseal union of the hand. *Am. J. Phys. Anthropol.* **19,** 365–372.
14. Moerman, M. L. (1982). Growth of the birth canal in adolescent girls. *Am. J. Obstet. Gynecol.* **43,** 528–532.
15. Malina, R. M., and Bouchard, C. (1988). Subcutaneous fat distribution during growth. *In* "Fat Distribution During Growth and Later Health Outcomes" (C. Bouchard and F. E. Johnston, eds.), pp. 63–48. Liss, New York.
16. Deutsch, M. I., Mueller, W. H., and Malina, R. M. (1985). Androgyny in fat patterning is associated with obesity in adolescents and young adults. *Ann. Hum. Biol.* **12,** 275–286.
17. Hediger, M. L., Scholl, T. O., Schall, J. I., and Cronk, C. E. (1995). One year changes in weight and fatness in girls during late adolescence. *Pediatrics* **96,** 253–258.
18. Gruelich, W. W., and Pyle, S. I. (1959). "Radiographic Atlas of Skeletal Development of the Hand and Wrist," 2nd ed. Stanford University Press, Stanford, CA.

19. Tanner, J. M., Whitehouse, R. H., Marshall, W. A., Healy, M. J. K., and Goldstein, H. (1975). "Assessment of Skeletal Maturity and Prediction of Adult Height." Academic Press, New York.

20. Eveleth, P. B., and Tanner, J. M. (1990). "Worldwide Variation in Human Growth." Cambridge University Press, Cambridge, UK.

21. Marshall, W. A., and Tanner, J. M. (1969). Variations in pattern of pubertal changes in girls. Arch. Dis. Child. 44, 291–303.

22. Bridges, N. A., Cooke, A., Healy, M. J. R., Hindmarsh, P. C., and Brook, C. G. D. (1996). Growth of the uterus. Arch. Dis. Child. 75, 330–331.

23. Duke, P. M., Litt, I. F., and Gross, R. T. (1980). Adolescents' self-assessment of sexual maturation. Pediatrics 66, 918–920.

24. Neinstein, L. S. (1982). Adolescent self-assessment of sexual maturation. Clin. Pediatr. 12, 482–501.

25. Bergsten-Brucefors, A. (1976). A note on the accuracy of recalled age at menarche. Ann. Hum. Biol. 3, 71–73.

26. Koo, M. M., and Rohan, T. E. (1997). Accuracy of short-term recall of age at menarche. Ann. Hum. Biol. 24, 61–64.

27. Damon, A., and Bajema, C. J. (1974). Age at menarche: accuracy of recall after thirty-nine years. Hum. Biol. 46, 381–384.

28. Bean, J. A., Leeper, J. D., Wallace, R. B., Sherman, B. M., and Jagger, H. (1979). Variations in the reporting of menstrual histories. Am. J. Epidemiol. 109, 181–185.

29. Casey, V. A., Dwyer, J. T., Coleman, K. A., Krall, E. A., Gardner, J., and Valadian, I. (1991). Accuracy of recall by middle-aged participants in a longitudinal study of their body size and indices of maturation earlier in life. Ann. Hum. Biol. 18, 155–166.

30. World Health Organization (1995). "Physical Status: The Use and Interpretation of Anthropometry," W.H.O. Tech. Rep. Ser. No. 854. WHO, Geneva.

31. Marshall, W. A., and Tanner, J. M. (1986). Puberty in human growth a comprehensive trentise. In "Postnatal Growth, Neurobiology" (F. Falkner and J. M. Tanner, eds.), Vol. 2, pp. 171–203. Plenum, New York.

32. MacMahon, B. (1973). "Age at Menarche: United States," DHEW Publ. No. (HRA) 74-1615, NCHS Ser. 11, No. 133. National Center for Health Statistics, Rockville, MD.

33. Jordan, J. R. (1979). "Desarollo Humano en Cuba. Ediciones Cientifico-Technica." Ministeno de Cultura, La Habana, Cuba.

34. Litt, I. F., and Cohen, M. I. (1973). Age of menarche: A changing pattern and its relationship to ethnic origin and delinquency. J. Pediatr. 82, 288–289.

35. Villarreal, S. F., Martorell, R., and Mendoza, F. (1989). Sexual maturation of Mexican-American adolescents. Am. J. Hum. Biol. 1, 87–95.

36. Kelly, H. J., and Reynolds, L. (1947). Appearance and growth of ossification centers and increases in the body dimensions of white and negro infants. Am. J. Roentgenol. Radium Ther. 57, 477–516.

37. Garn, S. M., Sandusky, S. T., Nagy, J. M., and McCann, M. B. (1972). Advanced skeletal development in low-income negro children. J. Pediatr. 80, 965–969.

38. Roche, A. F. (1974). "Skeletal Maturity of Children 6–11 Years," DHEW Publ. No. (HRA) 75-1622, NCHS Ser. 11, No. 140. National Center for Health Statistics, Rockville, MD.

39. Harlan, W. R., Harlan, E. A., and Grillo, G. P. (1980). Secondary sex characteristics of girls 12 to 17 years of age: United States Health Examination Survey. J. Pediatr. 96, 1074–1078.

40. Herman-Giddens, M. E., Slora, E. J., Wasserman, R. C., Bourdony, C. J., Bhapkar, M. V., Koch, G. G., and Hasemeier, C. M. (1997). Secondary sexual characteristics and menses in young girls seen in office practice: A study from the pediatric research office settings network. Pediatrics 99, 505–512.

41. Morrison, J. A., Barton, B., Biro, F. M., Sprecher, D. L., Falkner, F., and Obarzanek, E. (1994). Sexual maturation and obesity in 9- and 10-year old black and white girls: The National Heart, Lung, and Blood Institute Growth and Health Study. J. Pediatr. 124, 889–895.

42. Richards, R. J., Svec, F., Bao, W., Srinivasan, S. R., and Berenson, G. S. (1992). Steroid hormones during puberty: Racial (black-white) differences in androstenedione and estradiol: The Bogalusa Heart Study. J. Clin. Endocrinol. Metab. 75, 624–631.

43. Wyshak, G., and Frisch, R. E. (1982). Evidence for a secular trend in age of menarche. N. Engl. J. Med. 306, 1033–1035.

44. Garn, S. M. (1987). The secular trend in size and maturational timing and its implications for nutritional assessment. J. Nutr. 117, 817–823.

45. Frisch, R. E. (1994). The right weight: Body fat, menarche and fertility. Proc. Nutr. Soc. 53, 113–129.

46. Dreizen, S., Spirakis, C. N., and Stone, R. E. (1967). A comparison of skeletal growth and maturation in under-nourished and well-nourished girls before and after menarche. J. Pediatr. 70, 256–264.

47. Galler, J. R., Ramsey, F., and Solimano, G. (1985). A follow-up study of the effects of early malnutrition on subsequent development-I. Physical growth and sexual maturation during adolescence. Pediatr. Res. 79, 518–523.

48. Winick, M., Meyer, K. K., and Harris, R. C. (1975). Malnutrition and environmental enrichment by early adoption. Science 190, 1173–1175.

49. Proos, L. A. (1993). Anthropometry in adolescence—secular trends, adoption, ethnic and environmental differences. Horm. Res. 39(Suppl. 3), 18–24.

50. Sanchez, A., Kissinger, D. G., and Phillips, R. I. (1981). A hypothesis on the etiological role of diet on age of menarche. Med. Hypotheses 7, 1339–1345.

51. Hughes, R. E., and Jones, E. (1985). Intake of dietary fiber and the age of menarche. Am. Hum. Biol. 4, 325–332.

52. Merzenich, H., Boeing, H., and Wahrendorf, J. (1993). Dietary fat and sports activity as determinants for age at menarche. Am. J. Epidemiol. 138, 217–224.

53. Frisch, R. E., Hegsted, D. M., and Yoshinaga, K. (1975). Body weight and food intake at early estrus of rats on high-fat diet. Proc. Natl. Acad. Sci. U.S.A. 72, 4172–4176.

54. Moisan, J., Meyer, F., and Gingras, S. (1990). A nested case-control study of the correlates of early menarche. Am. J. Epidemiol. 132, 953–961.

55. Maclure, M., Travis, L. B., Willett, W., and MacMahon, B. (1991). A prospective cohort study of nutrient intake and age at menarche. Am. J. Clin. Nutr. 54, 649–656.

56. Post, G. B., and Kemper, H. C. G. (1993). Nutrient intake and biological maturation during adolescence: The Amsterdam Growth and Health Longitudinal Study. Eur. J. Clin. Nutr. 47, 400–408.

57. Warren, M. P. (1980). The effects of exercise on pubertal progression and reproductive function in girls. J. Clin. Endrocrinol. Metab. 51, 1150–1157.

58. Frisch, R. E., Gotz-Welbergen, A. V., McArthur, J. W., Albright, T., Witschi, J., Bullen, B., Birnholz, J., Reed, R. B., and Hermann, H. (1981). Delayed menarche and amenorrhea of college athletes in relation to age of onset of training. JAMA; J. Am. Med. Assoc. 246, 1559–1563.

59. Malina, R. M. (1994). Physical growth and biological maturation of young athletes. Exercise Sport Sci. Rev. 22, 389–433.

60. Malina, R. M., Ryan, R. C., and Bonci, C. M. (1994). Age at menarche in athletes and their mothers and sisters. Ann. Hum. Biol. 21, 417–422.

61. Bale, P., Doust, J., and Dawson, D. (1996). Gymnasts, distance runners, anorexics body composition and menstrual status. J. Sports Med. Phys. Fitness 36, 49–53.

62. Fischbein, S. (1977). Onset of puberty in MZ and DZ twins. *Acta Genet. Med. Gemellol.* **26**, 151–158.

63. Eveleth, P. B. (1996). Timing of menarche: Secular trend and population differences. *In* "Life Cycle and Biological Development" (J. B. Lancaster and B. A. Hamburg, eds.), pp. 39–52. Aldine, New York.

64. Lindgren, G. (1976). Height, weight and menarche in Swedish urban school children in relation to socio-economic status and regional factors. *Ann. Hum. Biol.* **3**, 501–528.

65. Malina, R. M., Katzmarzyk, P. T., Bonci, C. M., Ryan, R. C., and Wellens, R. E. (1997). Family size and age at menarche in athletes. *Med. Sci. Sports Exercise* **29**, 99–106.

66. Belsky, J., Steinberg, L., and Draper, P. (1991). Childhood experience, interpersonal development and reproductive strategy: An evolutionary theory of socialization. *Child Dev.* **62**, 647–670.

67. Campbell, B. C., and Udry, J. R. (1995). Stress and age at menarche of mothers and daughters. *J. Biosoc. Sci.* **27**, 127–134.

68. Moffitt, T. E., Casp, A., Belsky, J., and Silva, P. A. (1992). Childhood experience and the onset of menarche: A test of a sociobiological model. *Child Dev.* **63**, 47–58.

69. Apter, D., and Vihko, R. (1985). Premenarcheal changes in relation to age at menarche. *Clin. Endocrinol.* **22**, 753–760.

69a. Largo, R. H., and Prader, A. (1983). Pubertal development in Swiss girls. *Helv. Paediatr. Acta* **38**, 229–243.

70. Halpern, C. T., Udry, J. R., and Suchindran, C. (1997). Testosterone predicts initiation of coitus in adolescent females. *Psychosom. Med.* **59**, 161–171.

71. Soefer, E. F., Scholl, T. O., Sobel, E., Tanfer, K., and Levy, D. B. (1985). Menarche: Target age for reinforcing sex education for adolescents. *J. Adolesc. Health Care* **6**, 383–386.

71a. Udry, J. R. (1979). Age at menarche, first intercourse and at first pregnancy. *J. Biosoc. Sci.* **11**, 433–441.

72. Zabin, L. S., Smith, E. A., Hirsch, M. B., and Hardy, J. B. (1986). Ages of physical maturation and first intercourse in black teenage males and females. *Demography* **23**, 595–605.

73. Phinney, V. G., Jensen, L. C., Olsen, J. A., and Cundich, B. (1990). The relationship between early development and psychosexual behaviors in adolescent females. *Adolescence* **25**, 321–332.

74. Adolph, C., Ramos, D. E., Linton, K. L., and Grimes, D. A. (1995). Pregnancy among Hispanic teenagers: Is good parental communication a deterrent? *Contraception* **51**, 303–306.

75. Udry, J. R., and Cliquet, R. L. (1982). A cross-cultural examination of the relationship between ages at menarche, marriage, and first birth. *Demography* **19**, 53–63.

76. Buga, G. A. B., Amoko, D. H. A., and Ncayiyana, D. J. (1996). Sexual behaviour, contraceptive practice and reproductive health among school adolescents in rural Transkei. *S. Afr. Med. J.* **86**, 523–527.

77. Apter, D., and Vihko, R. (1983). Early menarche, a risk factor for breast cancer, indicates early onset of ovulatory cycles. *J. Clin. Endocrinol. Metab.* **57**, 82–86.

78. Vihko, R., and Apter, D. (1984). Endocrine characteristics of adolescent menstrual cycles: Impact of early menarche. *J. Steroid Biochem.* **20**, 231–236.

78b. Ventura, S. J., Martin, J. A., Curtin, S. C., and Matthews, T. S. (1997). "Report of Final Natality Statistics, 1995," Mon. Vital Stat. Rep., Vol. 45, No. 11, Suppl. 2. National Center for Health Statistics, Hyattsville, MD.

79. Scholl, T. O., Decker, E., Karp, R. J., Greene, G., and De Sales, M. (1984). Early adolescent pregnancy outcome in young adolescents and mature women. *J. Adolesc. Health Care* **5**, 167–171.

80. Perry, R. L., Mannino, B., Hediger, M. L., and Scholl, T. O. (1996). Pregnancy in early adolescence: Are there obstetric risks? *J. Maternal-Fetal Med.* **5**, 333–339.

81. Bender, S. (1989). Premenarcheal pregnancy. *Br. Med. J.* **1**, 760.

82. Scholl, T. O., Hediger, M. L., Salmon, R. W., Belsky, D. H., and Ances, I. G. (1989). Association between low gynaecological age and preterm birth. *Paediatr. Perinatal Epidemiol.* **3**, 357–366.

83. Scholl, T. O., Hediger, M. L., Huang, J., Johnson, F. E., Smith, W., and Ances, I. G. (1992). Young maternal age and parity. *Ann. Epidemiol.* **2**, 565–575.

84. Zlatnik, F. J., and Burmeister, L. F. (1977). Low "gynecologic age' an obstetric risk factor. *Am. J. Obstet. Gynecol.* **128**, 183–186.

85. Apter, D. (1996). Hormonal events during female puberty in relation to breast cancer risk. *Eur. J. Cancer Prev.* **5**, 476–482.

86. Apter, D., Viinikka, L., and Vihko, R. (1978). Hormonal pattern of adolescent menstrual cycles. *J. Clin. Endocrinol. Metab.* **47**, 944–954.

87. Stevens-Simon, C., and McAnarney, E. R. (1995). Further evidence of reproductive immaturity among gynecologically young pregnant adolescents. *Fertil. Steril.* **64**, 1109–1112.

88. Brown, H. L., Fan, Y. D., and Gonsoulin, W. J. (1991). Obstetric complications in young teenagers. *South. Med. J.* **84**, 46–48.

88a. Fraser, A. M., Brockert, J. E., and Ward, R. H. (1995). Association of young maternal age with adverse reproductive outcomes. *N. Engl. J. Med.* **333**, 1113–1117.

89. Lubarsky, S. L., Schiff, E., Friedman, S. A., Mercer, B. M., and Sibai, B. M. (1994). Obstetric characteristics among multiparas under age 15. *Obstet. Gynecol.* **84**, 365–368.

90. Hediger, M. L., Scholl, T. O., Schall, J. I., and Krueger, P. M. (1997). Young maternal age and preterm labor. *Ann. Epidemiol.* **7**, 400–406.

91. Frisancho, H. R., Matos, J., and Bolletino, L. A. (1984). Role of gynecological age and growth maturity status in fetal maturation and prenatal growth of infants born to young still-growing adolescent mothers. *Hum. Biol.* **56**, 583–593.

92. Stevens-Simon, C., and McAnarney, E. R. (1993). Skeletal maturity and growth of adolescent mothers. *J. Adolesc. Health Care* **13**, 428–432.

93. Scholl, T. O., Hediger, M. L., Ances, I. G., and Cronk, C. E. (1988). Growth during early teenage pregnancies. *Lancet* **1**, 701–2.

94. Sukanich, A. C., Rogers, K. D., and McDonald, H. M. (1986). Physical maturity and outcome of pregnancy in primiparas younger than 16 years of age. *Pediatrics* **78**, 31–6.

95. Scholl, T. O., Hediger, M. L., Cronk, C. E., and Schall, J. I. (1993). Maternal growth during pregnancy and lactation. *Horm. Res.* **39**(Suppl. 3), 59–67.

96. Scholl, T. O., Hediger, M. L., Schall, J. I., Khoo, C. S., and Fischer, R. L. (1994). Maternal growth during pregnancy and the competition for nutrients. *Am. J. Clin. Nutr.* **60**, 183–188.

97. Scholl, T. O., Hediger, M. L., and Schall, J. I. (1997). Maternal growth and fetal growth: Pregnancy course and outcome in the Camden Study. *Ann. N.Y. Acad. Sci.* **817**, 292–301.

98. Scholl, T. O., Hediger, M. L., and Ances, I. G. (1990). Maternal growth during pregnancy and decreased infant birth weight. *Am. J. Clin. Nutr.* **51**, 790–793.

99. Scholl, T. O., Hediger, M. L., Schall, J. I., Mead, J. P., and Fischer, R. L. (1995). Reduced micronutrient levels in the cord blood of growing teenage gravida. *JAMA; J. Am. Med. Assoc.* **274**, 26–27.

100. Hashizume, K., Ohashi, K., and Hamajima, F. (1991). Adolescent pregnancy and growth of progeny in rats. *Physiol. Behav.* **49**, 367–371.

100a. Wallace, J. M., Aitken, R. P., and Cheyne, M. A. (1996). Nutrient partitioning and fetal growth in rapidly growing adolescent ewes. *J. Reprod. Fertil.* **107**, 183–190.

101. Motil, K. J., Kertz, B., Thotath, U., and Chery, M. (1997). Lactational performance of adolescent mothers shows preliminary

difference from that of adult women. *J. Adolesc. Health Care* **20,** 442–446.

102. Scholl, T. O., Hediger, M. L., and Belsky, D. H. (1994). Prenatal care and maternal health during adolescent pregnancy: A review and meta-analysis. *J. Adolesc. Health Care* **15,** 444–456.

103. Kelsey, J. L., Gammon, M. D., and John, E. M. (1993). Reproductive factors and breast cancer. *Epidemiol. Rev.* **15,** 36–47.

104. Kelsey, J. L., and Bernstein, L. (1996). Epidemiology and prevention of breast cancer. *Annu. Rev. Public Health* **17,** 47–67.

105. Wellens, R., Malina, R. M., Roche, A. F., Chumlea, W. C., Guo, S., and Siervogel, R. M. (1992). Body size and fatness in young adults in relation to age at menarche. *Am. J. Hum. Biol.* **4,** 783–787.

106. Hediger, M. L., Scholl, T. O., and Schall, J. I. (1997). Implications of the Camden Study of adolescent pregnancy: Interactions among maternal growth, nutritional status and body composition. *Ann. N.Y. Acad. Sci.* **817,** 281–291.

107. Garn, S. M., LaVelle, M., Rosenberg, K. R., and Hawthorne, V. M. (1986). Maturational timing as a factor in female fatness and obesity. *Am. J. Clin. Nutr.* **43,** 879–883.

108. Scholl, T. O., Hediger, M. L., Schall, J. I., Ances, I. G., and Smith, W. K. (1995). Gestational weight gain, pregnancy outcome, and postpartum weight retention. *Obstet. Gynecol.* **86,** 423–427.

109. Segel, J. S., and McAnarney, E. R. (1994). Adolescent pregnancy and subsequent obesity in African-American girls. *J. Adolesc. Health Care* **15,** 491–494.

110. Scholl, T. O., Hediger, M. L., and Schall, J. I. (1996). Excessive gestational weight gain and chronic disease risk. *Am. J. Hum. Biol.* **8,** 735–741.

111. Dietz, W. H. (1994). Critical periods in childhood for the development of obesity. *Am. J. Clin. Nutr.* **59,** 955–959.

112. Garn, S. M. (1985). Continuities and changes in fatness from infancy through adulthood. *Curr. Probl. Pediatr.* **15,** 1–47.

113. Amiel, S. A., Caprio, S., Sherwin, R. S., Plewe, G., Haymond, M. W., and Tamborlane, W. V. (1991). Insulin resistance of puberty: A defect restricted to peripheral glucose metabolism. *J. Clin. Endocrinol. Metab.* **72,** 277–282.

114. Must, A., Jacques, P. F., Dallal, G. E., Bejema, C. J., and Dietz, W. H. (1992). Long-term morbidity and motality of overweight adolescents: A follow-up of the Harvard Growth Study of 1922 to 1935. *N. Engl. J. Med.* **327,** 1350–1355.

115. Palmer, J. R., Rosenberg, L., and Shapiro, S. (1992). Reproductive factors and risk of myocardial infarction. *Am. J. Epidemiol.* **136,** 408–416.

116. Sizonenko, P. C. (1987). Normal sexual maturation. *Pediatrics* **14,** 191–201.

9

Menstruation and Menstrual Disorders
The Epidemiology of Menstruation and Menstrual Dysfunction

SIOBÁN D. HARLOW
Department of Epidemiology
School of Public Health
University of Michigan
Ann Arbor, Michigan

I. Introduction

Periodic bleeding, in its presence or absence, is an integral part of a woman's experience throughout her reproductive life. In addition to potential fertility problems and the concern provoked by unexplained alterations in bleeding patterns, considerable morbidity is directly attributable to menstrual dysfunction. Menstrual disturbances are a common complaint both in industrialized and in nonindustrialized countries throughout the reproductive lifespan [1–4]. Menstrual characteristics (*e.g.,* cycle length) that place a woman at increased risk of developing chronic disease may also be considered pathological, but current data suggest that specific characteristics may increase the risk of one disease while decreasing the risk of another. Our understanding of the relationship between menstrual cycle characteristics and women's long-term risk of chronic disease remains limited [1]. Despite the importance of menstruation in women's lives, information on the incidence and prevalence of menstrual dysfunction is scarce and our knowledge of risk factors for menstrual dysfunction is confined predominantly to studies of the impact of weight, physical activity, and stress on the risk of amenorrhea [1].

This chapter will focus on the epidemiology of menstruation and menstrual disorders. First it will describe normal variation in menstrual cycle length, in the duration and amount of menstrual bleeding, and in the probability of ovulation across the reproductive lifespan. It also will address what is known about how these patterns differ by geographic region, ethnicity, and socioeconomic status. Next it will define menstrual dysfunction and describe what is known about the population frequency of and risk factors for menstrual dysfunction. The chapter concludes with a discussion of methodological issues in the measurement of menstrual function and in the design and analysis of studies of menstrual function with a focus on directions for future research.

II. Normal Variation in Menstrual Cycles across the Lifespan

A. *Menstrual Cycle Length*

Information on the length and variability of the menstrual cycle across the reproductive lifespan comes predominantly from four seminal menstrual diary studies. Matsumoto *et al.* [5] analyzed two years of menstrual records from 701 Japanese women aged 13–52. Treloar *et al.* [6] followed 2700 white US women aged 10–56 for up to 29 years. Chiazze *et al.* [7] observed bleeding patterns in 2316 US and Canadian women aged 15–44 for two years. Vollman [8] observed 691 Swiss women aged 11–58 for one to 39 years. A series of retrospective and short menstrual diary studies conducted by the World Health Organization (WHO) [9–11] provide some comparative data on women from industrialized and developing countries. A scattering of other small studies provide data on menstrual function in, for example, adolescent Swiss [12] and Nigerian [3,13] girls and Indian women [14]. For the most part these studies corroborate previous findings on age-related variation in menstrual cycle length. With few exceptions, they are not sufficiently detailed to enable evaluation of regional differences in menstrual cycle characteristics.

Table 9.1 provides data on the cross-sectional population distributions of menstrual cycle lengths by age from menarche to menopause obtained from each of the four large menstrual diary studies. Population variability in menstrual cycle length is greatest immediately after menarche and shortly before menopause, with each transition lasting approximately 2–5 years. Both transitions are characterized by an increased frequency of both very long and very short cycles and, consequently, by an increased range of cycle lengths. Cycle length from age 20 to age 40 exhibits considerably less variability and the population mean cycle length shortens from 30.1 to 27.3 days across these two decades [6].

These data describe the gross changes in bleeding patterns a woman may expect throughout reproductive life. However, they provide less information about how menstrual cycle length changes for an individual woman over her reproductive lifespan. Analyses of menstrual diaries collected by Treloar *et al.* suggest that the decline in mean cycle length between age 20 and 40 is initiated in many women by the sudden occurrence of some shorter cycle lengths (*e.g.,* a 24 or 25 day cycle for a woman who had previously experienced cycles ranging from 27 to 29 days) which over time become more frequent and then become the normative cycle length. However, among other women this decline is initiated by a reduction in the frequency of periodic long cycles (*e.g.,* of cycles longer than 30 or 35 days). Similarly, approximately 50% of women will have a menstrual cycle of 120 days or longer for the first time within one year of their final menstrual period, 20% of women will have a menstrual cycle of 120 days or longer for the first time from two to four years before their final menstrual period, and

Table 9.1
Menstrual Cycle Length (Days) by Age, Results of Four Studies

	Treloar et al. [6]				Vollman [8]			Chiazze et al. [7]		Matsumoto et al. [5]		
	Person-year mean	SD[a]	Median	Range (5–95%)	Mean	Median	Range (5–95%)	Mean (15–45 day cycles)	SD	Mean	SD	Range (10–90%)
Postmenarche												
Year 1	36.9	11.2	29.1	18.3–83.1	35.0	29.0	17.8–76.5					
Year 2	34.1	8.7	29.1	18.4–63.5	31.2	28.7	19.9–57.5					
Year 5	31.2	4.4	28.2	21.7–40.4	30.1	27.8	19.9–48.5					
Age (years)												
20	30.1	3.9	27.8	22.1–38.4	29.0	27.9	19.7–39.2					
20–24								29.1	4.6	31.0	5.7	26–38
25	29.8	3.5	27.8	22.7–37.1	30.7	28.2	23.6–43.5					
25–29								28.5	3.6	31.3	7.5	26–37
30	29.3	3.2	27.2	22.5–35.4	29.6	27.9	23.1–39.4					
30–34								28.0	3.5	30.1	5.5	25–36
35	28.2	2.7	26.7	22.3–33.4	29.1	27.9	22.8–36.4					
35–39								27.3	3.4	29.4	4.8	25–35
40	27.3	2.8	26.2	21.8–32.0	27.3	26.7	22.0–33.6					
40–44								26.9	3.7			
45					28.3	26.7	20.1–39.8					
Premenopause												
5 years	28.4	6.4	25.5	17.8–38.8								
2 years	43.5	19.5	26.6	15.4–80.0								
1 year	57.1	35.5	27.9	14.9–∞								

[a]SD, standard deviation.
Reprinted with permission from Harlow and Ephross [1].

approximately one-third of women will never experience a cycle as long as 120 days prior to their final menstrual period.

Data from the only two studies that provide quantitative estimates of probabilities of change in cycle length from one cycle to the next suggest that for 17–19 year old girls [5] and for women aged 15–45 [15], the majority of menstrual cycles will fall in the midrange of cycle lengths (approximately 26–34 days) regardless of the length of the previous cycle. Long cycles and short cycles do tend to recur in a woman, but the probability of such recurrence is relatively low (approximately 15–25%). Annual variation in menstrual cycle length within individual women tends to be underestimated. A population-based study from Denmark reported that almost 30% of women aged 15–44 experienced more than 14 days variation in cycle length in the past year [16]. The proportion of women who experienced this large amount of variation ranged from 47% in women aged 15–19 to a low of 11.9% among women aged 35–39 and the proportion rose again to 18.9% among women aged 40–44 years old [16].

Only a few articles provide population data on length of the follicular or luteal phase. Differences in methods used to assess ovulatory status, in rules used to define the start of the luteal phase, and in eligibility criteria create some disparities in phase length estimates. One study estimated day of ovulation by the peak day of cervical mucus among women participating in natural family planning trials in five countries [10]. Three studies examined daily plasma hormonal profiles from volunteer samples of Swedish and English women with regular (undefined), ovulatory menstrual cycles [17–19]. Vollman [8] and Matsu-moto et al. [5] determined day of ovulation by basal body temperature (BBT). In the three studies that excluded long cycles, estimates of follicular phase length varied from approximately 10 to 23 days (mean = 13–15 days) and luteal phase length varied from 8 to 17 days (mean = 13–14 days). Inclusion of longer menstrual cycles yielded longer estimates for the follicular phase (mean length = 17–18 days) [5,8]. Follicular and luteal phase length are positively correlated with cycle length in ovulatory menstrual cycles (r = 0.55 and 0.62, respectively) and negatively correlated with each other (r = −0.31) [10]. Mean length of the follicular phase declines with age, from about 14 days at age 18–24 to about 10.5 days at age 45–60 [19], although one report observed an increase after age 37 from 14.6 days to 15.9 days [20]. Lenton et al. suggest that short luteal phases (<11 days) are more frequent both in younger and in older women [18], but Vollman's data [8] suggest that once reproductive maturity is attained luteal phase length remains relatively constant through menopause.

B. Duration of Bleeding and Amount of Flow

Fewer data are available on the population distribution of the amount and duration of menstrual bleeding. Of the large population-based studies, only Matsumoto et al. [5] reported on bleeding length and subjective amount of flow. Duration of bleeding after ovulatory cycles ranged from 2 to 12 days, with 80% of bleeding episodes lasting 3–6 days (mode = 5, mean = 4.6, SD = 1.3). The heaviest flow was usually reported on the

Table 9.2
Probability of Anovulation: Findings from Four Studies

Author	Ovulation measurement method	Cycle length (days)	Age (years)	Percent anovulatory cycles
By cycle length				
Vollman [8]	BBT[a]	16–20	all ages[b]	24–44%
		21–24		6–22%
		25–30		2–4%
		31–44		5–19%
		>45		17–41%
Matsumoto et al. [5]	BBT	<25	20–39[b]	17%
		25–38		3%
		>38		18%
By age				
Doring [23]	BBT	all cycles[c]	12–17	43–60%
			18–20	20%
			21–25	13%
			26–40	3–7%
			41–50	12–15%
Vollman [8]	BBT	all cycles[c]	1 year postmenarche	56%
			29	1%
			40–45	34%
Metcalf [24]	weekly urine	all cycles[c]	10–14	52%
			20–24	28%
			30–39	2%
			>50	34%
By age and cycle lenggth				
Metcalf [24]	weekly urine	20–35[c]	10–24	36%
			40–55	11%
		36–49[c]	10–24	38%
			40–55	68%
		≥50[c]	10–24	57%
			40–55	73%

[a]BBT = basal body temperature.
[b]For all cycle length categories shown.
[c]For all age categories shown.
Reprinted with permission from Harlow and Ephross [1].

second day. Long menstrual periods (>8 days) were more common in anovulatory cycles. Women over the age of 35 reported bleeding about half a day less and report lighter flow than women aged 20–24. Older women also have been reported, in other studies, to experience fewer bleeds longer than seven days [21].

Information on actual blood loss comes predominantly from a Swedish population-based study [22]. Mean volume of blood loss per bleeding episode was 43.4 ml (with 80% of episodes ranging between 10 and 84 ml). Among women who considered their bleeding to be normal, mean blood loss was 38.5 ml. The mean for women with subjectively normal blood loss who also had adequate iron levels was 33.2 ml. Fifteen-year-old girls bled, on average, 1–2 ml less and 50-year-old women bled about 6 ml more than women aged 20–45. Women approaching menopause also were more likely to experience very heavy bleeding than younger women (90th percentile = 133 ml in women aged 50 versus 86–88 ml for women aged 30–45).

C. Ovulation and Anovulation

Population data on the probability of ovulation across the reproductive lifespan come principally from four studies. In their longitudinal investigations, Vollman [8] and Matsumoto et al. [5] obtained basal body temperatures on subsamples of women. Doring [23] charted BBTs and Metcalf [24] assayed urine for pregnanediol in age-stratified samples of German women and New Zealand women of European descent, respectively (Table 9.2). Anovulation is associated with menstrual cycle length. Short and long cycles are 10–30% more likely to be anovulatory than are cycles of 25–35 days. Thus, the probability of anovulation is greatest during the postmenarcheal and premenopausal years when long and short cycles are also more common. From about age 25 to 39, approximately 2–7% of cycles are anovulatory. In contrast, 50–60% of cycles in 10–14 year old girls and about 34% of cycles in women over age 50 are anovulatory. Much of this age-related change appears to be associated with concurrent changes in variability in menstrual cycle length. Among women aged 40–55, 95% of those with no recent change in menstrual cycle length ovulated consistently, compared with only 34% of those who reported a recent history of oligomenorrhea (cycles longer than 35 days) [25].

D. Regional, Ethnic, and Socioeconomic Differences in Menstrual Cycle Parameters

Marked geographic differences have been reported in the timing of reproductive maturation and, to a lesser extent, senescence [26,27]. However, data on regional differences in menstrual cycle characteristics are more limited. Although one would not expect major differences in the nature of ovarian or menstrual function across population groups, subtle alterations in the timing of follicular growth, in the timing of age-related changes in ovarian function, in the probability of ovulation across the reproductive life cycle, or in the efficiency of endometrial function may exist [10,28–30]. No differences were noted in the timing of age-related changes in ovulatory frequency between the Lese of Zaire and European women, but the Lese ovulated much less frequently [30]. One multicenter study conducted by the WHO reported differences in the length of the follicular and luteal phases in Mexican women compared with women from other countries (13.6 versus 15.0 days and 14.5 versus 13.5 days, respectively) [10]. The WHO data also suggest that Mexican and Latin American women may have shorter bleeding episodes (mean = 4 days), whereas European women may have longer episodes (mean = 5.9 days) than women in other regions [9,11,31,32]. Other studies suggest that Chinese women may have both longer and heavier bleeds [11,32].

One study of postmenarcheal girls in the United States suggests that menstrual characteristics of European-American girls may differ substantially from those of African-American girls. The European-American girls had longer average duration of menstrual bleeding but fewer episodes of heavy bleeding, a longer mean cycle length, increased between-subject standard deviation in menstrual cycle length, and an increased probability of having cycles longer than 45 days than their African-American counterparts [33,33a].

Two studies provide evidence that menstrual function varies by socioeconomic status. A population-based study of Danish women aged 15–44 found nearly a tripling of the odds of having greater than 14 days variation in menstrual cycle length in women of low versus high social class [16]. A study of South Indian women reported that women with less than a high school education were more than twice as likely to have a cycle longer than 36 days than women with at least a high school education [14].

Regional and socioeconomic differences in menstrual function are of particular interest because such differences can offer clues as to how social and environmental conditions affect menstrual function. Evaluation of such differences is also critical to understanding how risks of hormone-related diseases and response to hormonal interventions might differ between populations. The above data, though highly provocative, are clearly insufficient to evaluate regional, ethnic, or socioeconomic differences in menstrual patterns. Furthermore, the biologic mechanisms that underlie these observed differences have yet to be explored.

III. Menstrual Disorders

A. Conceptualization of Menstrual Dysfunction

Historically, menstrual dysfunction was defined primarily in terms of disruptions in bleeding patterns, such as menorraghia (heavy or prolonged bleeding), oligomenorrhea (infrequent menstruation), polymenorrhea (frequent menstruation), or amenorrhea (cessation of menses). Definitions based on ovarian function, such as anovulation and luteal deficiency, are now also common. Functional disturbances may or may not manifest as alterations in bleeding patterns. A third, entirely separate construct of menstrual dysfunction stems from a consideration of pain (dysmenorrhea) and other symptomatology (premenstrual syndrome) associated with the onset of menses. Bleeding disturbances may be pathological if they are associated with significant blood loss, disrupt activities of daily living, indicate ovarian alterations incompatible with conception, or signal underlying pathology such as cancer. Functional disturbances may be considered pathological to the extent that they interfere with the capacity to reproduce or provoke bleeding disturbances. In the absence of a desire to become pregnant, however, the pathological nature of anovulation or luteal phase inadequacy is unclear.

Ellison has proposed a theory of the ecology of ovarian function which provides a unified framework for understanding dysfunction [30]. He suggests that ovarian function shows a graded response continuum to important ecological, behavioral, and constitutional variables. The mildest form of ovarian suppression is reflected in a lower profile of progesterone secretion resulting in an inadequate luteal phase, followed by follicular suppression which, if severe enough, can lead to ovulatory failure without noticeable changes in menstrual patterns, or, eventually, to cessation of menstrual function. This theory provides a conceptual basis for the integration of that group of menstrual dysfunctions manifested by disruptions in cycle length and by disruptions of ovarian function. This model of a graded response to environmental insult has also helped to synthesize much of the extant literature on risk factors for menstrual dysfunction. However, as this model primarily addresses menstrual dysfunction in the context of fertility, it remains somewhat less clear whether the model can incorporate menstrual dysfunction characterized by heavy or profuse bleeding or associated symptomatology. This section describes menstrual cycle length disorders (the frequency of anovulation has already been discussed). Most of the literature on risk factors for menstrual dysfunction is focused in this area. The literature on abnormal uterine bleeding and dysmenorrhea is then considered. Premenstrual syndrome is discussed in Chapter 11.

B. Cycle Length Disorders

Amenorrhea is defined as the cessation of menses. Primary amenorrhea refers to failure to achieve menarche by age 16. Secondary amenorrhea refers to cessation of menses after menarche. Although clinically the criteria for amenorrhea is absence of menses for 6 months or for three times the length of the previous menstrual cycle [34], in research a definition of no menses for 90 days is most commonly used. *Oligomenorrhea* is defined as the absence of menses for shorter intervals or as unevenly spaced menses and is generally operationalized as menstrual cycles of length 35–90 days. *Polymenorrhea* is defined as frequent menses, that is generally operationalized as menstrual cycles shorter than 21 days.

Few studies have been conducted on the population distribution of secondary amenorrhea, oligomenorrhea, or polymenorrhea. In a random sample of Swedish women aged 18–45 years, Pettersson *et al.* [35] reported a one-year incidence rate of 3.3%, a one-year prevalence of 4.4%, and a point prevalence of 1.8% for amenorrhea of more than three months duration based on questionnaire data. A population-based study of Danish women aged 15–44 obtained women's menstrual calendars for the previous year and reported on prevalence of oligomenorrhea and polymenorrhea. Only 0.5% and 0.9% of regularly cycling women reported a usual cycle length shorter than 21 days or longer than 35 days, respectively, but 18.6% of women reported experiencing at least one short cycle and 29.5% of women reported experiencing at least one long cycle in the previous year [16]. A questionnaire-based study of US college women [36] reported a point prevalence of 11.3% for oligomenorrhea (cycle lengths of 35–90 days) and 2.6% for amenorrhea of more than 90 days, with the higher point prevalences reflecting the young age of this population. In a study of Indian women, 20.3% complained of amenorrhea [2], while 4.7% were found to have amenorrhea and 22.4% were found to have oligomenorrhea (no operational definitions provided) on examination. Amenorrhea is common in women who are breastfeeding. The duration of postpartum amenorrhea depends both on the frequency and intensity of suckling as well as on the nutritional status of the mother. Ultrasound studies suggest that 17–23% of women volunteers who have never sought treatment for menstrual disturbances, 26% of patients with amenorrhea and 87% of patients with oligomenorrhea have polycystic ovaries [37–40].

Luteal phase defect is defined as a lag of more than two days in histologic development of the endometrium compared to day of the menstrual cycle, presumably due to inadequate progesterone secretion or action [34]. Although an endometrial biopsy is the gold standard, most studies of the population frequency of inadequate luteal phase have utilized measures of serum progesterone or length of the luteal phase. Although clinically a short luteal phase is often defined as less than eight days, Lenton *et al.* [18] suggest that 100% of luteal phases shorter than 10 days, 74% of 10-day phases, 22% of 11-day phases, and 2% of 12-day phases are abnormal. A common definition of inadequate luteal function is progesterone levels that do not reach 16 nmol/l (5 ng/mL) for at least 5 days [17] or progesterone levels of less than 10–12 ng/mL one week prior to the onset of menstruation [34].

Although 3.5% of women who are evaluated for infertility [41] and 23–38% of women with recurrent spontaneous abortions [34] are reported to have luteal phase defects, the prevalence of this condition in the general population is less clear. Doring [23] reported that 37% of menstrual cycles in women aged 18–20 years as compared with only 9% of cycles in women aged 35–39 years had temperature elevations lasting less than 10 days. Vollman observed that 15% of cycles in adult women had luteal phases shorter than 11 days [42]. Using a definition of eight days, 3.2% of women in the multinational WHO study of participants in natural family planning programs

[10] and 2.1% of women in a smaller clinical study [18] had short luteal phases. Landgren *et al.* [17] determined that 6% of women demonstrated insufficient progesterone production; however, their definition was based on having a progesterone value below the 95th percentile.

C. Risk Factors for Amenorrhea, Oligomenorrhea, and Anovulation

Knowledge about determinants of variability in menstrual function is largely confined to studies measuring the effect of weight, physical activity, and stress on menstrual cycle length, the probability of ovulation, and adequacy of luteal function [1]. Recently attention also has been focused on the importance of the combined presence of these risk factors (*e.g.*, dietary behaviors and stress in athletes).

1. Weight

Both absolute weight and change in weight appear to influence menstrual function. Acute and moderate weight loss lead to alterations in ovarian function, with the degree of ovarian suppression depending on the severity and duration of weight loss. Extreme environmental or pathological conditions, such as famine and anorexia nervosa, which result in starvation and severe weight loss, are known to provoke amenorrhea [43,44]. Weight-loss associated amenorrhea also has been reported in women without other evidence of overt physical or mental illness who present at 20–30% below ideal weight or with a loss of more than 10% of premorbid weight. Menstruation generally resumes after a mean weight gain of about 4 kg or to within 5–15% of ideal weight [43,45]. During periods of seasonal food shortages among Lese women of the Ituri Forest in Zaire, the probability of ovulation and peak progesterone production in ovulatory cycles were correlated with baseline weight-for-height and with amount of weight lost during the study period [30]. Moderate weight loss among normal-weight Boston women resulted in similar patterns of ovarian suppression [33]. The prevalence of inadequate luteal phase has been shown to increase when women lose just 5% of their initial weight [46] or lose an average of 1 kg [47]. Weight stabilization precipitates a return towards normal ovarian function [30].

Studies of female athletes [43,47–52], US college students [36], and the general population in Scandinavia [53] have demonstrated that individuals with highly variable cycles or amenorrhea tend to weigh less, to have lower percent body fat, and to report more weight loss than women with normal menstrual cycles. In the only population-based case-control study of amenorrhea, women reported recent weight loss of ≥5 kg and recent dieting more frequently than did age-matched controls [53].

Considerably less data are available on women who are heavy, although in the United States overweight is more prevalent than underweight. Obesity has been associated with amenorrhea and polycystic ovarian disease [54,55] and a high waist-to-hip ratio with oligomenorrhea [56]. One study of US college women found that heavier women had the highest probability of having long menstrual cycles [57]. Symons *et al.* [58] have demonstrated a nonlinear relationship between cycle length and BMI, fat mass, and lean mass in women aged 24–45 years old (mean age 38 years). Women with the longest mean cycle lengths had greater BMI, body fat mass, and body lean mass, but women at the lowest decile of BMI and body fat mass also had long mean cycle lengths. In contrast, another study of women aged 29–31 years reported that being in the lowest quartile of weight-for-height was associated with a longer mean cycle length, increased cycle variability, and an increased probability of having cycles longer than 41 days [58a].

Weight influences menstrual function through multiple mechanisms. Reproductive function may depend on nutritional status, with either direct caloric energy balance or food composition being the relevant nutritional factor [30,43,46,59–61]. Dietary restraint (e.g., restriction of calories or avoidance of high calorie foods) independent of actual weight loss has been shown to alter both ovarian function [62] and menstrual cycle length [57], and it is now understood that aberrant eating behaviors in the absence of frank psychiatric illness also alter menstrual function. Body fat may directly influence endocrine function because fat tissue is a reservoir for steroid hormones and is a site of estrogen production [54,63,64], and obesity appears to influence sex-hormone-binding-globulin concentration and estrogen metabolism [54,65]. Both weight and reproductive function may be affected by a third factor such as underlying disease (i.e., anorexia nervosa or milder forms of eating disorders or endocrine disorders) or a concomitant environmental exposure (i.e., stress)).

2. Physical Activity

High and moderate levels of physical activity can alter menstrual function. Although the frequency of menstrual disturbance among athletes varies considerably across studies, women athletes, particularly ballet dancers and runners, have a higher frequency of amenorrhea, anovulation, and luteal phase defects than do nonathletes [48,52,66]. Among women aged 29–31 years, daily activity in vigorous sports was associated with an increase in cycle variability and an increased probability of having long cycles [58a]. Recreational exercise has been shown to increase mean menstrual cycle length, the probability of having a long menstrual cycle, and the probability of anovulation in school girls [57,66a]. Data on training intensity are inconsistent, but the reported frequency of amenorrhea is higher in studies of competitive athletes than in studies of recreational runners. Apparent discrepancies in the literature are largely attributable to divergent definitions of dysfunction or to failure to control for age. In general, amenorrheic runners weigh less, have lower weight-to-height ratios, have less body fat, and report more weight loss than other runners or nonrunner controls [47,50, 57,67–69]. Evidence that halting training without weight gain restores normal menses suggests, however, that physical training also has an independent effect [43]. Prospective studies support this hypothesis [66,70,71]. In one study [70], training provoked abnormal luteal function and anovulation in all but five of 53 observed menstrual cycles. Anovulation was most frequent in women fed a diet that ensured weight loss but also occurred in women who maintained their pretraining weight.

The probable hormonal pathway through which physical activity induces menstrual alterations is inhibition of gonadotropin-releasing hormone (GnRH) and gonadotropin activity accompanied by a decline in serum estrogen levels. Cumming hypothesized that these hormonal alterations reflect a physio-logic response to stress that, in the presence of additional risk factors, may be sufficient to affect menstrual function [59]. Amenorrheic athletes are significantly more likely to perceive running as stressful [51] and to report higher levels of stress [67,72]. Higher mean serum [50,73] and urinary [74] cortisol levels also have been demonstrated in amenorrheic compared with eumenorrheic runners. However, because moderate levels of aerobic activity also alter ovarian function and are less consistent with a stress model [57,66a,71], alternative mechanisms such as negative energy balance remain plausible [30,72]. Amenorrheic athletes have been found to score higher than eumenorrheic athletes on the Eating Attitudes Test (EAT), a measure of aberrant eating behaviors [74a], with scores on the EAT being inversely correlated with fat intake and simple carbohydrate intake [75]. Studies of ballet dancers have also found a strong association between presence of amenorrhea and anorexia nervosa [76]. Women who experience menstrual alterations after intense physical activity may also be more susceptible to menstrual dysfunction. Amenorrheic athletes tend to have a later menarche and to report a history of highly variable or long menstrual cycles [51,52,77].

3. Stress

Although psychological stress is generally acknowledged to affect menstruation, studies of stress and menstrual function consist mainly of studies of major life changes, of catastrophic events such as incarceration and war, and of girls leaving home [44,78,79]. Studies of more common events such as leaving home to go to school [79–81], to the military [78], or to work [82] suggest that separation from home and family [44] increases the probability of amenorrhea. The designs of most of these studies are seriously flawed, lacking comparison groups or comparing menstrual histories with prospective observations of cycle length. However, Metcalf and Mackenzie [82] have shown that 5–8 years after menarche, 72% of girls living at home ovulated as compared with only 40% of those who lived away from home. In older women, separation from family had only a minimal effect.

Data from population-based studies are less consistent. In cross-sectional surveys, history of perceived emotional upset, history of death or separation from close relatives, and self-perceived excessive physical or mental workload did not differ between women with and without amenorrhea [36], while psychological state and life events were unrelated to excessive bleeding [83]. On the other hand, a population-based case-control study in Scandinavia reported that stressful life events and consumption of tranquilizers were more common among amenorrheic women [53]. A prospective study of the acute effects of stress during a given menstrual cycle [57] found that major life events associated with loss and separation were too infrequent to explain cycle alterations in this population of young women but that events associated with gain or newness such as starting a new job, starting college, or starting a new relationship increased the probability of having a long cycle.

As is true for physical activity, the mechanism through which stress influences menstrual function remains uncertain. Central psychogenic disturbances may be mediated through the hypothalamus via changes in prolactin or the endogenous opiates [43]. Alternatively, a systemic physiologic response to stress, such as elevation in basal cortisol levels, may provoke alterations in hypothalamic response. One study demonstrated reduced luteinizing hormone (LH) pulse frequency in amenorrheic women reporting a history of stressful life events and reduced LH pulse amplitude in women meeting clinical criteria for anxiety or depressive disorders [84]. Women with functional hypothalamic amenorrhea have also been shown to have higher cortisol concentrations and a blunted response to administration of corticotropin-releasing hormone (CRH) than women with other forms of anovulation or eumenorrheic women [85,86].

4. Diet

In addition to caloric intake and the psychogenic effects of dietary restraint already noted, some studies suggest that diet composition may also influence menstrual function. Vegetarian diets have been associated with anovulation, decreased pituitary responsiveness, shorter follicular phase, and infrequent cycling (<10 menstrual cycles per year) [60,87]. In an experimental study, women had slightly longer menstrual cycles and slightly longer bleeding periods while consuming a low-fat (20%) as compared with a high-fat (40%) diet [88]. Amenorrheic runners report consuming fewer calories, less red meat, and less fat than menstruating runners [89]. Subclinical eating disorders accompanied by severe restriction in dietary fat are present in normal weight, nonathletic women with functional amenorrhea [90].

5. Occupational and Environmental Exposures

Few studies have examined the impact of occupational or chemical exposures on menstruation, despite suggestive evidence that a wide range of environmental factors may influence menstrual function. For example, the implications of the data on physical activity and menstrual function for women whose occupations require heavy physical labor remain unclear. Rural Indian women who perform heavy agricultural work do have a longer mean cycle length than urban women (32 versus 31 days) [14], and, in this same population, having an active versus inactive occupation has been associated with the risk of having a long cycle [91]. A study of Nepalese women who work predominantly in agricultural and pastoral occupations found that progesterone levels were depressed during seasons of increased workload when weight loss also occured [92]. In a study of Vietnamese factory workers, women who worked in textile factories were more likely than other factory workers to report having irregular (undefined) menstruation (p = 0.008), and irregular menstruation was associated with exposure to noise and high intensity of work [93].

The potential for environmental chemicals to act as ovarian toxins and to alter menstrual function has been clearly demonstrated for a few pharmacologic exposures, specifically exposure to antineoplastic agents. Cytotoxic drugs have been shown to induce ovarian failure, including loss of follicles, anovulation, oligomenorrhea, and amenorrhea [94–96], although return of normal menses after treatment does occur [97]. Low dose occupational exposure to antineoplastic drugs has also been associated with an increased odds of menstrual dysfunction (broadly defined as amenorrhea, short or long cycles, and short or long duration of menstrual flow), particularly among women aged 30–44 [98]. Neuroleptics also have been reported to provoke amenorrhea [99,100].

In addition to ovarian toxicity, however, it is also possible that environmental chemicals may influence menstrual function through other mechanisms. Tobacco serves as a good model of the broad range of effects chemicals may have on menstrual function. Tobacco smoke has been shown to alter estrogen metabolism [101], is associated with elevated follicular phase plasma estrogen and progesterone levels [102], and is a risk factor for infertility [103] and for earlier onset of menopause [104]. Preliminary studies suggest that smoking may be a risk factor for dysmenorrhea, infrequent menstrual cycles, and heavy or prolonged bleeding [105]. Studies of occupational populations have also observed associations between smoking and irregular cycles (>6 day variation in cycle length) or long cycles (>32 days) [106], and longer mean menstrual cycle length [107]. A few investigators have also pointed to the potential implications of chemical induction of liver metabolism and early reports have suggested the potential for pesticides such as DDT and solvents to alter endocrine metabolism [108–111]. Studies in rhesus monkeys have found that exposure to lead is associated with longer and more variable menstrual cycles and shorter menstrual flow [112].

Nonetheless, only a few studies have examined the effect of occupational or environmental chemicals on menstrual function. Anecdotal evidence from as early as the 1940s suggested that xylene exposure may produce menorraghia [109]. A report from Singapore found no association between toluene exposure and dysfunctional uterine bleeding (defined as cycle irregularity and prolonged or heavy bleeding) [113]. However, a case series from Korea suggested a potentially strong association between amenorrhea and exposure to solvents containing 2-bromopropane [114]. A small study in Colorado reported menstrual alterations in farm laborers [115], but this finding has not been followed up. In a large comprehensive study of women who worked in the semiconductor industry, Gold et al. [107] found that women who worked in the thin film and ion implantation processes had a longer mean cycle length (3.2 days), a greater probability of having a cycle longer than 35 days, and increased variance compared to women who worked in nonfabrication jobs. Women who worked in photolithography processes had an increased cycle variance and increased probability of having cycles shorter than 24 days. In a study of menstrual function in cannery and slaughterhouse workers, slaughterhouse workers, who had more variable work schedules and worked in temperatures below 18 degrees centigrade, reported an increased frequency of irregular cycles (>6 days variability in cycle length) and amenorrhea as compared to cannery workers [106]. Finally, consumption of polychlorinated biphenol (PCB) contaminated fish has been associated with having a shorter menstrual cycle in a population-based study of families of licensed anglers (1.1 days for women reporting consumption of >1 fish meal per month or 1.0 days for women with a moderate/high estimated PCB index) [116].

6. Menstrual Synchrony: Social and Environmental Signals

Variations in the length of the day have long been known to regulate and synchronize the timing of estrus in animals that are seasonal breeders [117]. These photoperiod effects are now known to be mediated through melatonin excretion from the pineal gland [118]. Alterations in melatonin secretion have been noted in women with hypothalmic amenorrhea [119], and patterns of melatonin secretion appear to differ between athletes with and without amenorrhea [120]. The question of whether women who work or live together start cycling together [121] generates considerable interest; however, studies of this phenomenon have produced inconsistent results [122,123] and are often methodologically weak. More rigorous studies that have addressed many of the methodological shortcomings of previous work have demonstrated clear shifts towards synchrony among roommates who are close friends [124]. Both animal and human studies suggest that social interactions can modify endocrine function [125], although a simple theory of social proximity probably is not sufficient. Pheromones (substances secreted by one individual that influence the behavior of another individual which are frequently perceived through olfaction) involved in same-sex [126] or opposite-sex contact [127] may reduce variability in cycle length and synchronize the onset of menses. Others argue that a range of common environmental signals are relevant to the establishment of menstrual and reproductive synchrony [122].

7. Endocrine Disturbances

Overt or subclinical endocrine diseases, such as diabetes and hyper- and hypothyroidism, often are associated with menstrual disturbances. Diabetics are much more likely to have cycle length disturbances than nondiabetics. A Danish study found the six month prevalence of amenorrhea (absence of menses for three months) to be 8.2% in diabetic patients as compared with 2.8% in nondiabetic controls and the six month prevalence of oligomenorrhea (cycle length of 36 days to 3 months) to be 10.6% and 4.8%, respectively [128]. Studies among the Pima Indians support an association among diabetes, hyperinsulinemia, and an increased prevalence of amenorrhea and oligomenorrhea [129]. Polycystic ovarian disease is associated with obesity, insulin resistance, and oligomenorrhea [130,131]. Occurrence of oligo/amenorrhea among women with polycystic ovary disease appears to be linked to the simultaneous occurrence of insulin insensitivity [132] and to obesity [38]. Hyperthyroidism is most commonly associated with oligomenorrhea that may progress to amenorrhea, with as many as 21.5–64.7% of patients affected. Hypothyroidism is associated mainly with polymenorrhea and menorraghia [133].

D. Bleeding Disorders

A range of terms is used to describe excessive, prolonged, or too frequent bleeding. The terminology depends on the amount of information available on (1) the actual bleeding pattern, (2) the underlying ovarian function, and (3) the presence of organic pathology. In the absence of precise information on one or more of these parameters, the term abnormal uterine bleeding (AUB) is commonly used to describe bleeding complaints. As noted previously, traditional definitions based on bleeding patterns include menorrhagia (regular, normal intervals, excessive flow and duration), metrorrhagia (irregular intervals, excessive flow and duration), and polymenorrhea (intervals less than 21 days) [34]. The term dysfunctional uterine bleeding (DUB) has been used to describe various bleeding disruptions in anovulatory cycles that are not attributable to other organic pathology [34].

The terms ovulatory DUB and anovulatory DUB are increasingly used to describe excessive bleeding associated with predictable menstrual cycles and bleeding that is irregular in both timing and amount, respectively. Here we will focus on definitions of heavy or prolonged menstrual bleeding and factors that affect the duration or amount of menstrual flow.

Excessive blood loss is defined as blood loss of more than 80 ml per menses [22]. As noted previously, anovulation can provoke abnormal bleeding. In the presence of normal ovulation, conditions such as coagulation disorders, endometriosis, fibroids, uterine infections, and possibly prostaglandin imbalances may cause heavy bleeding [134]. Neither the incidence nor prevalence of abnormal uterine bleeding has been well established, but it is estimated that 9–14% of reproductive age women have blood loss exceeding 80 ml per cycle [4] and that 20% of women will report excessive bleeding during their reproductive life [22]. Prolonged menstruation often is defined as bleeding for more than 7 or 8 days, but one study suggests that a cut point of 10 days distinguishes abnormal and normal bleeding most precisely. An English study of medical records in one practice estimated the yearly incidence of menorrhagia to be seven cases per 1000 women, with a median age at presentation of 35–39 years [136]. In another study, 5% of Indian village women complained of profuse periods, but a significantly larger proportion, 15.2%, were diagnosed on clinical examination as having menorraghia [2]. A population-based survey in Australia [137] also found that about 5% of women reported prolonged (>7 days) menstrual bleeding. A study of Nigerian teenagers reported that 12.1% of girls had menorraghia (≥80 ml of blood loss) [3]. DUB is thought to increase during the menopausal transition, but there are no studies of the cumulative incidence of DUB during the perimenopause. One cross-sectional study found no difference in the frequency of menorrhagia between those classified as early or late perimenopause [138].

Studies of factors influencing the duration of menstrual bleeding have found that prenatal exposure to DES [139] shortens the duration while being sedentary increases bleed duration [58a,140]. Body mass index and dieting [33,58a,140] appear to affect both bleed duration and bleed amount, although the direction of their effects seems to depend on other population characteristics. Perceived stress may decrease bleed duration but increase the probability of having a heavy bleeding episode [33]. Women with dysfunctional uterine bleeding are more likely to report recent stressful life changes [141] and experimental data suggest that blood coagulation factor activity may decrease in response to prolonged stress [142]. Some data suggest that prolonged bleeding is more common in smokers [58a,105,137] and that heavy bleeding is more common in obese women [143]. Although most studies of athletes focus exclusively on cycle length disorders, one Nigerian study has reported an increased risk of menorrhagia (defined as profuse menstrual flow due to no demonstrable anatomical or pathological lesion at least five times in the previous year) among athletes as compared to nonathletes (14% versus 8% respectively, p = 0.002) [144]. Case reports also suggest an association between menorraghia and mild hypothyroidism [145]. A case series from England suggests that up to 17% of patients referred for treatment of menorrhagia may have an inherited bleeding disorder [146]. Intra-uterine devices are known to augment

blood flow substantially, except for devices containing progesterone which reduce the amount of flow [105]. Use of oral anticoagulants, including aspirin, may increase menstrual blood loss
[147] and the role of frequent or daily use of aspirin may warrant further study. In contrast, nonsteroidal anti-inflammatory
medication (NSAID) reduces amount and duration of flow and
is now considered appropriate treatment for the medical management of AUB. Coulter *et al.* have reviewed the relative efficacy of a range of NSAIDs in reducing blood flow and suggest
that mefaminic acid is the most efficacious [148].

E. Dysmenorrhea

Although the term dysmenorrhea is sometimes used to refer
to a broad panorama of symptomatology, it is appropriately defined as abdominal pain, cramping, or backache associated with
menstrual bleeding. Related gastrointestinal symptomatology,
such as nausea or diarrhea, may also occur. In cross-sectional
surveys, 30–60% of women of reproductive age report experiencing menstrual pain, although the proportion of women who
report severe pain sufficient to interfere with daily activity is
considerably lower, ranging from 7–15% [2,13,81,105,137,
149–154]. Variation in prevalence estimates between studies is
largely attributable to differences in the age of study populations and to differences in definitions of dysmenorrhea (*e.g.*, any
versus moderate pain and ever versus usual experience).

Young adult women, aged 17–24, report a higher prevalence
of dysmenorrhea, with estimates ranging from 67–72% [13,
151–154]. As would be expected, given that dysmenorrhea is
associated with having ovulatory cycles, the prevalence of dysmenorrhea during adolescence increases with gynecologic age
[152]. In a Finnish study, 54% of girls aged 10–20 reported at
least occasional pain, while 13% reported that they always experienced pain. In the first year postmenarche, only 7% of girls
reported having cramps, as opposed to 25% in the fifth to ninth
years [153]. Data from US and Swedish adolescents are similar
[151,152,154], whereas studies from Nigeria [13] and Turkey
[155] report slightly higher prevalences (72% and 78%, respectively). In a one year prospective menstrual diary study of US
college women aged 17–19 years old, 60% reported at least one
episode of severe dysmenorrhea, but only 13% reported severe
pain in more than half of their menstrual periods [156]. Women
who experienced a severe pain episode had a 62% probability
of experiencing moderate to severe pain during their subsequent
menstrual period. The median duration of pain was two days.

Studies in adult women are less consistent and often focus on
subgroups of working women. In India, 15% of women complained of dysmenorrhea while 57% were found to experience
some level of pain on examination [2]. In studies from Australia
[137], Singapore [149], and Finland [153], 29–75% of women
reported menstrual pain, with the prevalence largely dependent
on how the question was phrased. Although dysmenorrhea is
often touted as the greatest cause of lost work time for women,
the data disprove the assertion. A comprehensive study conducted prior to the introduction of NSAIDs determined that dysmenorrhea accounted for only 3.7% and 2.5% of absences
among female factory and office workers, respectively [157],
with the greatest cause of lost work time for US women actually
being influenza [109].

In addition to age, increased prevalence, severity, and duration of dysmenorrhea is correlated with earlier age at menarche
[149,151,156], longer and heavier menstrual flow [151,152,
156], and nulliparity [149,151]. The prevalence of menstrual
cramps also increases with cycle length. Women who use oral
contraceptives report less severe pain. Several studies found a
higher prevalence and increased severity of dysmenorrhea in
smokers [105,137,150,156,158]. Most studies have found no
effect of physical exercise [81,144,150,151,156], although one
study of Nigerian athletes reported a lower prevalence of dysmenorrhea compared to sedentary women [159]. One prospective study reported that obesity also increased the prevalence
and duration of menstrual pain [156]. Specific working conditions may increase the risk of dysmenorrhea. Exposure to cold
and physical workload in canneries and slaughterhouses significantly increased both the prevalence of dysmenorrhea and the
likelihood of taking sick leave [160,160a].

IV. Contraception and Menstrual Function

Hormonal contraception is designed to alter ovarian function
and thereby prevent ovulation and/or the establishment of a
pregnancy by overriding the endogenous menstrual cycle. Although many preparations are designed to mimic menstrual
periodicity, it should be acknowledged that these bleeding episodes are not actually menstrual periods. The periodic bleeding
is artificial and both the intended periodic bleeding and unintended "intermenstrual" bleeding represent treatment effects.
Oral contraceptives reduce both duration and amount of bleeding but are associated with an increased likelihood of intermenstrual bleeding [105,161,162]. One study of monophasic oral
contraceptives found that 30–50% of women reported intermenstrual bleeding after 6 months of use [163]. A second study
[164] reported that breakthrough bleeding/spotting occurred in
11% of cycles during the first six months of use and in 6% of
cycles thereafter. Amenorrhea (no withdrawal bleeding after one
pill cycle) occurred in 0.5–1.8% of treated cycles. Studies of
triphasic preparations [165] have found that 33–63% of women
who use triphasic pills report at least one episode of intermenstrual bleeding in the first six months of use, with a per cycle
frequency of 25% in the first cycle of use and 8% in the sixth
cycle of use. Oral contraceptives also are associated with decreased frequency of menstrual cramps [161,162,166], with the
mechanism of action being related to inhibition of uterine activity [167] and decreased synthesis of prostaglandins and leukotrienes [168].

Long-acting progestin-only preparations also alter bleeding
patterns. Women using long-acting injectables tend to report
infrequent but prolonged bleeding episodes, with over 40% of
women reporting amenorrhea by 12 months of use [169].
Amenorrhea is reported by 57% of depot medroxyprogesterone
acetate users after one year and by 68% after two years [170].
Subdermal implants are associated with significant alterations
in the rhythm and periodicity of bleeding; most commonly an
increase in the frequency and duration of bleeding is seen
[171,172]. Two-thirds of implant users experience intermenstrual bleeding in the first year [170] but bleeding patterns tend
to stabilize after the first year of use [171]. Even in the third
year 9–15% of users will report more than four bleeding/spotting

episodes and 19–25% will report at least one episode of more than eight days duration in any given 90 day period [172]. Ten to 25% of users report amenorrhea [170]. Lighter weight women are most likely to experience oligo/amenorrhea while women weighing >60 kg are most likely to experience menometrorrhagia, that is, excessive bleeding both during menses and between menstrual cycles, [172].

Women who use intrauterine devices (IUD) are more likely to report prolonged and heavy periods than are nonusers (22.7% versus 5.3–14.2% and 16.2% versus 5.0–13.2%, respectively) [105]. Studies on the effect of tubal sterilization on subsequent menstrual function are inconsistent. The more rigorous studies which included pre- and postevaluation of bleeding characteristics and a nonsterilized comparison group found no differences between sterilized and nonsterilized women in menstrual cycle length, duration of menstrual flow, or in the occurrence of intermenstrual bleeding within one year after sterilization [173]. The prevalence of dysmenorrhea was found to be much more likely to increase over the one year period in women who had been sterilized than in the comparison group. However, the possibility of long term impact of tubal sterilization on menstrual bleeding has not been ruled out [174].

V. Instruments and Issues in the Measurement of Menstrual Function

Two of the major barriers to furthering our understanding in this area are the difficulties of measuring menstrual function and of defining dysfunction. The few papers published on the reliability of menstrual data suggest that women have difficulty recalling their menstrual history and that prospective recording of bleeding experience using menstrual calendars or diaries is needed to obtain precise and valid data on menstrual bleeding patterns [9,175]. Specifically, women have difficulty remembering the date of the first day of their last menstrual period, demonstrate strong digit preference for 28 and 30 day cycles, and have difficulty reporting an average or usual cycle length when their cycles tend to vary in length. Duration of bleeding is recalled with more accuracy than cycle length, but quantification of bleeding amount is particularly problematic.

Relatively simple menstrual calendars have been used extensively in a wide range of populations across a wide range of literacy levels. The WHO has demonstrated the usefulness of a basic calendar in many multicenter studies comparing bleeding patterns in women using various contraceptive methods [9]. More detailed menstrual diaries have also been employed in epidemiologic studies [15,33,33a,107,139]. Menstrual calendars tend to measure timing and duration of bleeding fairly well, but methods to assess the amount of bleeding are still under development. Counting the number of pads and/or tampons used per day is problematic because women use pads for reasons other than bleeding (e.g., for incontinence) and tend to think in terms of the frequency with which they change their sanitary protection. Higham et al. have developed a pictorial blood loss assessment scale [176], while others have used a subjective scale with definitions given to provide common reference points for amount of bleeding.

Another problem of measuring blood loss stems from the fact that women have no common understanding of or criterion on which to grade the severity of their own blood loss. Mean blood loss for women who report heavy bleeding is greater than mean loss for women who report light or moderate bleeding, however, many women who experience considerable blood loss do not report excessive bleeding [22,177]. Hallberg et al. found that 37% of women with >80 ml of blood loss reported their bleeding as moderate [22]. Conversely, 14% of those who lost <20 ml of blood considered their bleeding to be heavy. Data from postpartum women suggest that self report can describe withinwoman changes fairly accurately and that lighter than normal bleeds predict anovulation [178]. Two instruments have been developed and validated to evaluate menorraghia, the pictorial blood loss assessment chart mentioned above [176] and a 13-item self-administered questionnaire [179]. Nonetheless, unpredictability and sudden heavy flow cause problems in the social management of bleeding and may be as important as total blood loss during a bleeding episode [180]. Further research is needed regarding the precise bleeding parameters that engender a complaint of abnormal bleeding as well as the social contexts in which such complaints arise.

A lack of clarity as to what distinguishes normal and abnormal function also persists for cycle length disorders. Definitions of normal cycles span as little as 8 days (24–32 days) to as many as 15 days (20–35 or 21–36 days). The term "irregular," although frequently employed by health professionals and researchers, has no common meaning for women in the community. In research, "irregular" has been used to refer to long cycles (e.g., >37 or 38 days), long and short cycles (e.g., <21 or >35 days or <20 or >50 days), cycles that vary in length (>5 days), and inability to predict menses. For amenorrhea and oligomenorrhea, definitions of cycles >90 days and cycles of 35–90 days are generally accepted and should be used more universally. Development of a better conceptualization of the boundaries of normal function, of the domain of "irregularity," and of what constitutes the most meaningful cutpoints for short and long cycles are still needed

Finally, much previous research has focused on individual risk factors without measuring other important potential confounders. For example, most studies of menstrual function in ballet dancers have considered weight and physical activity, but not aberrant eating behaviors.

VI. Directions for Future Research

As illustrated by this review, epidemiologic data on the menstrual cycle are inadequate. Population data on menstrual cycle length and blood loss lack the detail on within-woman variability necessary to enable women and clinicians to anticipate specific bleeding changes that are likely to occur at different life stages, to differentiate potentially pathological alterations from short-term aberrations, and to identify bleeding patterns that may be risk factors for the development of chronic disease. The absence of data that characterize bleeding changes as women approach and pass through menopause is of particular concern given the high frequency of physician visits for abnormal uterine bleeding and the prevalence of hysterectomies after age 35. The fact that women have little comparative information on how much bleeding constitutes "too much" suggests that objective criteria still need to be developed by which women can self rate their daily blood loss, and more education is needed to inform women of what constitutes menstrual dysfunction.

Basic research is needed to define population patterns of menstrual dysfunction and to identify potentially modifiable risk factors. Fairly extensive data are available on the effects of low weight and physical activity on menstrual cycle length and the probability of ovulation. More focus should be placed on examining the influence of weight at the upper end of the spectrum, of recreational activity in gynecologically mature women, and of hard physical activity in the context of women's daily work life, as well as on the interaction of low weight and physical activity in nonindustrialized countries where women typically are occupied in nonaerobic but energy-intensive tasks. Research needs in the area of chemical exposures, stress, and social factors are more fundamental. Further evaluation of work-related stress, multiple social roles, violence and discrimination would be informative. Additionally, more attention should be paid to new risk factors, particularly factors such as diet and noise that are amenable to public health intervention. Given the widespread use of pesticides and the increasing participation of women in industrial production throughout the world, investigation of the impact of chemical exposures on menstrual function and on contraceptive efficacy also is necessary. Finally, much of our understanding of menstrual function across the reproductive lifespan derives from studies of highly educated women of European descent from industrialized countries. Information is needed about change in menstrual function with age from other populations, as well as about host and environmental factors that influence menstrual function in these populations.

References

1. Harlow, S. D., and Ephross, S. A. (1995). Epidemiology of menstruation and its relevance to women's health. *Epidemiol. Rev.* **17**, 265–286.
2. Bang, R. A., Bang, A. T., Baitule, M., Choudhary, Y., Sarsnukaddam, S., and Talle, O. (1989). High prevalence of gyneacological diseases in rural Indian women. *Lancet* **1**, 85–88.
3. Barr, F. S. M., Brabin, L., Agbaje, S., Buseri, F., Ikimalo, J., and Briggs, N. D. (1998). Menstrual health in adolescents: The hidden reproductive health issue. *2nd Eur. Congr. Trop. Med.*, Liverpool, 1998, Abstract.
4. Van Eijkeren, M. A., Christiaens, G. C. M. L., Sixma, J. J., and Haspels, A. A. (1989). Menorrhagia: A review. *Obstet. Gynecol. Surv.* **44**, 421–429.
5. Matsumoto, S., Nogami, Y., Ohkuri, S. (1962). Statistical studies on menstruation: A criticism on the definition of normal menstruation. *Gunma J. Med. Sci.* **11**, 294–318.
6. Treloar, A. E., Boynton, R. E., Behn, B. G., and Brown, B. W. (1967). Variation of the human menstrual cycle through reproductive life. *Int. J. Fertil.* **12**, 77–126.
7. Chiazze, L., Brayer, F. T., Macisco, J. J., Parker, M. P., and Duffy, B. J. (1968). The length and variability of the human menstrual cycle. *J. Am. Med. Assoc.* **203**, 377–380.
8. Vollman, R. F. (1977). "The Menstrual Cycle." Saunders, Philadelphia.
9. World Health Organization Task Force on Psychosocial Research in Family Planning, Special Programme of Research, Development and Research Training in Human Reproduction (1981) Women's bleeding patterns: Ability to recall and predict menstrual events. *Stud. Fam. Plan.* **12**, 17–27.
10. World Health Organization: (1983). Task force on methods for the determination of the fertile period. A prospective multicentre trial of the ovulation method of natural family planning. III. Characteristics of the menstrual cycle and of the fertile phase. *Fertil. Steril.* **40**, 773–778.
11. World Health Organization Task Force on Adolescent Reproductive Health (1986). World Health Organization Multicenter Study on menstrual and ovulatory patterns in adolescent girls. *J. Adolesc. Health Care* **7**, 236–244.
12. Flug, D., Largo, R. H., and Prader, A. (1984). Menstrual patterns in adolescent Swiss girls: A longitudinal study. *Ann. Hum. Biol.* **11**, 495–508.
13. Thomas, K. D., Okonofua, F. E., and Chiboka, O. (1990). A study of the menstrual patterns of adolescents in Ile-Ife, Nigeria. *Int. J. Gynecol. Obstet.* **33**, 1–4.
14. Jeyaseelan, L., and Rao, P. S. S. (1993). Correlates of menstrual cycle length in South Indian women: A prospective study. *Hum. Biol.* **65**, 627–634.
15. Harlow, S. D., and Zeger, S. L. (1991). An application of longitudinal methods to the analysis of menstrual diary data. *J. Clin. Epidemiol.* **44**, 1015–1025.
16. Munster, K., Schmidt, L., and Helm, P. (1992). Length and variation in the menstrual cycle—a cross-sectional study from a Danish county. *Reprod. Sci.* **99**, 422–429.
17. Landgren, B. M., Unden, A. L., and Diczfalusy, E. (1980). Hormonal profile of the cycle in 68 normally menstruating women. *Acta Endocrinol. (Copenhagen)* **94**, 89–98.
18. Lenton, E. A., Landgren, B. M., and Sexton, L. (1984). Normal variation in the length of the luteal phase of the menstrual cycle: Identification of the short luteal phase. *Br. J. Obstet. Gynaecol.* **91**, 685–689.
19. Lenton, E. A., Landgren, B. M., and Sexton, L. (1984). Normal variation in the length of the follicular phase of the menstrual cycle: Effect of chronological age. *Br. J. Obstet. Gynaecol.* **91**, 681–684.
20. Fitzgerald, C. T., Seif, M. W., Killick, S. R., and Elstein, M. (1994). Age-related changes in the female reproductive cycle. *Br. J. Obstet. Gynaecol.* **101**, 229–233.
21. Collett, M. E., Wertenberge, G. E., Fiske, V. M., (1954). The effect of age upon the pattern of the menstrual cycle. *Fertil. Steril.* **5**, 437–448.
22. Hallberg, L., Hogdahl, A. M., Nilsson, L., and Rybo, G. (1966). Menstrual blood loss—a population study. Variation at different ages and attempts to define normality. *Acta Obstet. Gynecol. Scand.* **45**, 320–351.
23. Doring, G. K. (1969). The incidence of anovular cycles in women. *J. Reprod. Fertil. Suppl.* **6**, 77–81.
24. Metcalf, M. G. (1983). Incidence of ovulation from the menarche to the menopause: Observations of 622 New Zealand women. *N. Z. Med. J.* **96**, 645–648.
25. Metcalf, M. G. (1979). The incidence of ovulatory cycles in women approaching the menopause. *J. Biosoc. Sci.* **11**, 39–48.
26. Gray, R. H., and Doyle, P. E. (1983). The epidemiology of conception and fertility. *In* "Obstetrical Epidemiology" (S. L. Barron and A. M. Thomson, Eds.) pp. 25–60. Academic Press, London.
27. Boulet, M. J., Oddens, B. J., Lehert, P., Vemer, H. M., and Visser, A. (1994). Climacteric and menopause in seven south-east Asia countries. *Maturitas* **19**, 157–176.
28. Baanders-Van Halewijn, E. A., and de Waard, F. (1968). Menstrual cycles shortly after menarche in European and Bantu girls. *Hum. Biol.* **40**, 314–322.
29. Odujinrin, O. M. T., and Ekunwe, E. O. (1991). Epidemiologic survey of menstrual patterns amongst adolescents in Nigeria. *West Afr. J. Med.* **10**, 244–249.
30. Ellison, P. T., and Cabot, T. D. (1990). Human ovarian function and reproductive ecology: New hypotheses. *Am. Anthropol.* **92**, 933–952.

31. Belsey, E. M., and Perogoudov, S. (1988). Task force on long-acting systemic agents for fertility regulation. Determinants of menstrual bleeding patterns among women using natural and hormonal methods of contraception: I. Regional variations. *Contraception* **38**, 227–242.

32. Ji, G., Ma, L. Y., Zeng, S., Fan, H. M., and Han, L. H. (1981). Menstrual blood loss in healthy Chinese women. *Contraception* **23**, 591–601.

33. Harlow, S. D., and Campbell, B. (1996). Ethnic differences in the duration and amount of menstrual bleeding during the postmenarcheal period. *Am. J. Epidemiol.* **144**, 980–988.

33a. Harlow, S. D., Campbell, B., Lin, X., and Raz, J. (1997). Ethnic differences in the length of the menstrual cycle during the postmenarcheal period. *Am. J. Epidemiol.* **146**, 572–580.

34. Speroff, L., Glass, R. H., and Kase, N. G. (1994). "Clinical Gynecologic Endocrinology and Infertility," 5th ed. Williams & Wilkins, Baltimore, MD.

35. Pettersson, F., Fries, H., and Nillus, S. J. (1973). Epidemiology of secondary amenorrhea I. Incidence and prevalence rates. *Am. J. Obstet. Gynecol.* **117**, 80–86.

36. Bachmann, G., and Kemmann, E. (1982). Prevalence of oligomenorrhea and amenorrhea in a college population. *Am. J. Obstet. Gynecol.* **144**, 98–102.

37. Adams, J., Polson, D. W., and Franks, S. (1986). Prevalence of polycystic ovaries in women with anovulation and idiopathic hirsutism. *Br. Med. J.* **293**, 355–359.

38. Botsis, D., Kassanos, D., Pyrgiotis, E., and Zourlas, P. A. (1995). Sonographic incidence of polycystic ovaries in a gynecological population. *Ultrasound Obstet. Gynecol.* **6**, 182–185.

39. Farquhar, C. M., Birdsall, M., Manning, P., Mitchell, J. M., and France, J. T. (1994). The prevalence of polycystic ovaries on ultrasound scanning in a population of randomly selected women. *Aust. N.Z. J. Obstet. Gynaecol.* **34**, 67–72.

40. Polson, D. W., Adams, J., Wadsworth, J., and Franks, S. (1988) Polycystic ovaries—A common finding in normal women. *Lancet* **1**, 870–872.

41. Jones, G. S. (1976). The luteal phase defect. *Fertil. Steril.* **27**, 351–356.

42. Vollman, R. F. (1968). The length of the premenstrual phase by age of women. *Int. Congr. Ser.—Exerpta Med.* **133**, 1171–1175.

43. Warren, M. P. (1983). Effects of undernutrition on reproductive function in the human. *Endoc. Rev.* **4**, 363–377.

44. Drew, F. L. (1961). The epidemiology of secondary amenorrhea. *J. Chronic Dis.* **14**, 396–407.

45. Nakamura, Y., Yoshimura, Y., Oda, T., Katayama E., Kamei, K., Tanabe, K., and Iizuka, R. (1985). Clinical and endocrine studies of patients with amenorrhea associated with weight loss. *Clin. Endocrinol.* **23**, 643–651.

46. Schweiger, U., Schwingenschloegel, M., Laessle, R., Schweiger, M., Pfister, H., Pirke, K-M. and Hoehl, C. (1987). Diet induced menstrual irregularities: Effects of age and weight loss. *Fertil. Steril.* **48**, 746–751.

47. Pirke, K. M., Laessle, R. G., Schweiger, U., Broocks, A., Strowitzki T., Huber, B., Tuschl, R. J., and Middendorf, R. (1989). Dieting causes menstrual irregularities in normal weight young women through impairment of episodic luteinizing hormone secretion. *Fertil. Steril.* **51**, 263–268.

48. Abraham, S. F., Beumont, P. J., Fraser, I. S., and Llewellyn-Jones, D. (1982). Body weight, exercise and menstrual status among ballet dancers in training. *Br. J. Obstet. Gynaecol.* **89**, 507–510.

49. Baker, E. R., Mathur, R. S., Kirk, R. F., and Williamson, H. O. (1981). Female runners and secondary amenorrhea: Correlation with age, parity, milage, and plasma hormonal and sex-hormone-binding-globulin concentrations. *Fertil. Steril.* **36**, 183–187.

50. Glass, A. R., Deuster, P. A., Kyle, S. B., Yahiro, J. A., Vigersky, R. A., and Schoomaker, E. B. (1987). Amenorrhea in olympic marathon runners. *Fertil. Steril.* **48**, 740–745.

51. Schwartz, B., Cumming, D. C., Riordan, E., Selye, M., Yen, S. S. C., and Rebar, R. W. (1981). Exercise associated amenorrhea: A distinct entity? *Am. J. Obstet. Gynecol.* **141**, 662–670.

52. Speroff, L., and Redwine, D. B. (1979) Exercise and menstrual dysfunction. *Phys. Sportsmed.* **8**, 42–52.

53. Fries, H., Nillius, S. J., and Pettersson, F. (1974). Epidemiology of secondary amenorrhea: II. A retrospective evaluation of etiology with special regard to psychogenic factors and weight loss. *Am. J. Obstet. Gynecol.* **118**, 473–479.

54. Friedman, C. I., and Kim, M. H. (1985). Obesity and its effect on reproductive function. *Clin. Obstet. Gynecol.* **28**, 645–663.

55. Harlass, F. E., Plymate, S. R., Fariss, B. L., and Belts, R. P. (1984). Weight loss is associated with correction of gonadotropin and sex steroid abnormalities in the obese anovulatory female. *Fertil. Steril.* **42**, 649–652.

56. Hartz, A. J., Ripley, D. C., and Rimm, A. A. (1984). The association of girth measurements with disease in 32,856 women. *Am. J. Epidemiol.* **119**, 71–80.

57. Harlow, S. D., and Matanoski, G. M. (1991). The association between weight, physical activity and stress and variation in the length of the menstrual cycle. *Am. J. Epidemiol.* **133**, 38–49.

58. Symons, J. P., Sowers, M. R., and Harlow, S. D. (1997). Relationship of body composition measures and menstrual cycle length. *Ann. Hum. Biol.* **24**, 107–116.

58a. Cooper, G. S., Sandler, D. P., Whelan, E. A., and Smith, K. R. (1996). Association of physical and behavioral characteristics with menstrual cycle patterns in women age 29–31 years. *Epidemiol.* **7**, 624–628.

59. Cumming, D. C., Wheeler, G. D., and Harber, V. J. (1994). Physical activity, nutrition and reproduction. *Ann. N.Y. Acad. Sci.* **709**, 55–76.

60. Pirke, K. M., Schweiger, U., Laessle, R., Dickhaut, B., Schweiger, M., and Waechtler, M. (1986). Dieting influences the menstrual cycle: Vegetarian versus nonvegetarian diet. *Fertil. Steril.* **46**, 1083–1088.

61. Ellison, P. T., Lipson, S. F., O'Rourke, M. T., Bentley, G. R., Harrigan, A. M., Panter-Brick, C., and Vitzhum, V. J. (1993). Population variation in ovarian function. *Lancet* **342**, 433–434.

62. Schweiger, U., Tuschl, R., Platte, P., Broocks A., Laessle, R. G., and Pirke, K-M. (1992). Everyday eating behavior and menstrual function in young women. *Fertil. Steril.* **57**, 771–775.

63. Deslypere, J. P., Verdonek, L., and Vermeulen, A. (1985). Fat tissue: A steroid reservoir and site of endocrine metabolism. *J. Clin. Endocrinol. Metab.* **61**, 564–570.

64. Longcope, C., Baker, R., and Johnston, C. C. (1986). Androgen and estrogen metabolism: Relationship to obesity. *Metabol. Clin. Exp.* **35**, 235–237.

65. Schneider, J., Bradlow, H. L., Strain, G., Levin, J., Anderson, K., and Fishman, J. (1983). Effects of obesity on estradiol metabolism: Decreased formation of nonuterotropic metabolites. *J. Clin. Endocrinol. Metab.* **6**, 973–978.

66. Dale, E., Gerlach, D. H., and Wilhite, A. L. (1979). Menstrual dysfunction in distance runners. *Obstet. Gynecol.* **54**, 47–53.

66a. Bernstein, L., Ross, R. K., Lobo, R. A., Hanisch, R., Krailo, M. D., and Henderson, B. E. (1987). The effects of moderate physical activity on menstrual cycle patterns in adolescence: Implications for breast cancer prevention. *Br. J. Cancer* **55**, 681–685.

67. Galle, P. C., Freeman, E. W., Galle, M. G., Huggins, G. R., and Sondheimer, S. J. (1983). Physiologic and psychologic profiles in a survey of women runners. *Fertil. Steril.* **39**, 633–639.

68. Ouellette, M. D., MacVicar, M. G., and Harlan, J. (1986). Relationship between percent body fat and menstrual patterns in athletes and nonathletes. *Nurs. Res.* **35,** 330–333.

69. Carlberg, K. A., Buckman, M. T., Peake, G. T., and Riedesel, M. L. (1983). Body composition of oligo/amenorrheic athletes. *Med. Sci. Sports Exercise* **15,** 215–217.

70. Bullen, B. A., Skrinar, G. S., Beitins, I. Z., von Mering, G., Turnbull, B. A., and McArthur, J. W. (1985). Induction of menstrual disorders by strenuous exercise in untrained women. *N. Engl. J. Med.* **312,** 1349–1353.

71. Ellison, P. T., and Lager, C. (1986). Moderate recreational running is associated with lowered salivary progesterone profiles in women. *Am. J. Obstet. Gynecol.* **154,** 1000–1003.

72. Schweiger, U., Herrman, F., Laessle, R., Riedel, W., Schweiger, M., and Pirke, K-M. (1988). Caloric intake, stress, and menstrual function in athletes. *Fertil. Steril.* **49,** 447–450.

73. Ding, J. H., Sheckter, C. B., Drinkwater, B. L., Soules, M. R., and Bremner, W. J. (1988). High serum cortisol levels in exercise-associated amenorrhea. *Ann. Intern. Med.* **108,** 530–534.

74. Loucks, A. B., Mortola, J. F., Girton, L., and Yen, S. C. C. (1989). Alterations in the hypothalamic-pituitary-ovarian and the hypothalamic-pituitary-adrenal axes in athletic women. *J. Clin. Endocrinol. Metab.* **68,** 402–411.

74a. Garner, D. M., and Garfinkel, P. E. (1979). The eating attitudes test: An index of symptoms of anorexia nervosa. *Psychol. Med.* **9,** 273–279.

75. Perry, A. C., Crane, L. S., Applegate, B., Marquez-Sterling, S., Signorile, J. F., and Miller, P. C. (1996). Nutrient intake and psychological and physiological assessment in eumenorrheic and amenorrheic female athletes: A preliminary study. *Int. J. Sports Nutr.* **6,** 3–13.

76. Brooks-Gunn, J., Warren, M. P., and Hamilton, L. H. (1987) The relation of eating problems and amenorrhea in ballet dancers. *Med. Sci. Sports Exercise* **19,** 41–44.

77. Malina, R. M., Spirduso, W. W., Tate, C., and Bayor, A. M. (1978). Age at menarche and selected menstrual characteristics in athletes at different competitive levels and in different sports. *Med. Sci. Sports* **10,** 218–222.

78. Boehm, F. H., and Salerno, N. J. (1973). Menstrual abnormalities in the wave recruit population. *Mil. Med.* **138,** 30–31.

79. Osofsky, H. J., and Fisher, S. (1967). Psychological correlates of the development of amenorrhea in a stress situation. *Psychosom. Med.* **29,** 15–23.

80. Matsumoto, S., Tamada, T., and Konuma, S. (1979). Endocrinological analysis of environmental menstrual disorders. *Int. J. Fertil.* **24,** 233–239.

81. Wilson, C., Emans, S. J., Mansfield, J., Podolsky, C., and Grace, E. (1984). The relationships of calculated percent body fat, sports participation, age, and place of residence on menstrual patterns in healthy adolescent girls at an independent New England high school. *J. Adolesc. Health Care* **5,** 248–253.

82. Metcalf, M. G., and Mackenzie, J. A. (1980). Incidence of ovulation in young women. *J. Biosoc. Sci.* **12,** 345–352.

83. Gath, D., Osborn, M., Bungay, G., Iles, S., Day, A., Bond, A., and Passingham, C. (1987). Psychiatric disorder and gynaecological symptoms in middle age women: A community survey. *Br. Med. J.* **294,** 213–218.

84. Facchinetti, F., Fava, M., and Fioroni, L. (1993). Stressful life events and affective disorders inhibit pulsatile LH secretion in hypothalamic amenorrhea. *Psychoneuroendocrinology* **18,** 397–404.

85. Berga, S. L., Daniels, T. L., and Giles, D. E. (1997). Women with functional hypothalamic amenorrhea but not other forms of anovulation display amplified cortisol concentrations. *Fertil. Steril.* **67,** 1024–1030.

86. Biller, B. M. K., Federoff, H. J., Koenig, J. I., and Klibanski, A. (1990). Abnormal cortisol secretion and responses to corticotropin-releasing hormone in women with hypothalamic amenorrhea. *J. Clin. Endocrinol. Metab.* **70,** 311–317.

87. Pedersen, A. B., Bartholomew, M. J., Dolence, L. A., Aljadir, L. P., Netteburg, K. L., and Loyd, T. (1991). Menstrual differences due to vegetarian and non-vegetarian diets. *Am. J. Clin. Nutr.* **53,** 879–885.

88. Jones, D. Y., Judd, J. T., Taylor, P. R., Campbell, W. S., and Nair, P. P. (1987). Influence of dietary fat on menstrual cycle and menses length. *Hum. Nutr. Clin. Nutr.* **41C,** 341–345.

89. Kaiserauer, S., Snyder, A. C., Sleeper, M., and Zierath, J. (1989). Nutritional, physiological, and menstrual status of distance runners. *Med. Sci. Sports Exercise* **21,** 120–125.

90. Laughlin, G. A., Dominguez, C. E., and Yen, S. S. (1998). Nutritional and endocrine-metabolic aberrations in women with functional hypothalamic amenorrhea. *J. Clin. Endocrinol. Metab.* **83,** 25–32.

91. Jeyaseelan, L., and Rao, P. S. S. (1995) Effect of occupation on menstrual cycle length: Causal model. *Hum. Biol.* **67,** 283–290.

92. Panter-Brick, C., Lotstein, D. S., and Ellison, P. T. (1993). Seasonality of reproductive function and weight loss in rural Nepali women. *Hum. Reprod.* **8,** 684–690.

93. Matsuda, S., Luong, N. A., Hoai, N. V., Thung, D. H., Trinh, L. V., Cong, N. T., Hien, H. M., Dat, P. H., and Tri, D. D. (1997) A study of complaints of fatigue by workers employed in Vietnamese factories with newly imported technology. *Ind. Health* **35,** 16–28.

94. Chapman, R. M., Sutcliffe, S. B., and Malpas, J. S. (1979). Cytotoxic-induced ovarian failure in women with Hodgkin's disease: I. Hormone function. *JAMA, J. Am. Med. Assoc.* **242,** 1877–1881.

95. Sobrinho, L. G., Levine, R. A., and Deconti, R. C. (1971). Amenorrhea in patients with Hodgkin's disease treated with antineoplastic agents. *Am. J. Obstet. Gynecol.* **109,** 135–139.

96. Warne, G. L., Fairley, K. F., Hobbs, J. B., and Martin, F. I. R. (1973). Cyclophosphamide-induced ovarian failure. *N. Engl. J. Med.* **289,** 1159–1162.

97. Gershenson, D. M. (1988). Menstrual and reproductive function after treatment with combination chemotherapy for malignant ovarian germ cell tumors. *J. Clin. Oncol.* **6,** 270–275.

98. Shortridge, L. A., Lemasters, G. K., Valanis, B., and Hertzberg, V. (1995). Menstrual cycles in nurses handling antineoplastic drugs. *Cancer Nurs.* **18,** 439–444.

99. Meltzer, H. Y., and Fang, V. S. (1976). The effect of neuroleptics on serum prolactin in schizophrenic patients. *Arch. Gen. Psychiatry* **33,** 279–286.

100. Gingell, K. H., Darley, J. S., Lengua, C. A., and Baddela, P. (1993). Menstrual changes with antipsychotic drugs (letter). *Br. J. Psychiatry* **162,** 127.

101. Michnovicz, J. J., Hershcopf, R. J., Naganuma, H., Bradlow, H. L., and Fishman, J. (1986) Increased 2-Hydroxylation of estradiol as a possible mechanism for the anti-estrogenic effect of cigarette smoking. *N. Engl. J. Med.* **315,** 1305–1309.

102. Zumoff, B., Miller, L., Levit, C. D., Miller, E. H., Heinz, U., Kalin, M., Denman, H., Jandorek, R., and Rosenfeld, R. S. (1990). The effect of smoking on serum progesterone, estradiol, and luteinizing hormone levels over a menstrual cycle in normal women. *Steroids* **55,** 507–511.

103. Baird, D. D., and Wilcox, A. J. (1985). Cigarette smoking associated with delayed conception. *JAMA, J. Am. Med. Assoc.* **253,** 2979–2983.

104. Baron, J. A. (1984). Smoking and estrogen related disease. *Am. J. Epidemiol.* **119,** 9–22.

105. Brown, S., Vessey, M., and Stratton, I. (1988). The influence of method of contraception and cigarette smoking on menstrual patterns. *Br. J. Obstet. Gynaecol.* **95,** 905–910.

106. Messing, K., Saurel-Cubizolles, M-J., Bourgine, M., and Kaminski, M. (1992). Menstrual-cycle characteristics and work conditions of workers in poultry slaughterhouses and canneries. *Scand. J. Work Environ. Health* **18,** 302–309.

107. Gold, E. B., Eskenazi, B., Hammond, S. K., Lasley, B. L., Samuels, S. J., O'Neill Rasor, M., Hines, S. J., Overstreet, J. W., and Schenker, M. B. (1995). Prospectively assessed menstrual cycle characteristics in female water-fabrication and nonfabrication semiconductor employees. *Am. J. Ind. Med.* **28,** 799–815.

108. Martucci, C., and Fishman, J. (1993). P450 enzymes of estrogen metabolism. *Pharmac. Ther.* **5,** 237–257.

109. Harlow, S. D. (1986). Function and dysfunction: An historical critique of the literature on menstruation and work. *Health Care Women Int.* **7,** 39–50.

110. Guengerich, F. P. (1988). Oxidation of 17α-Ethynylestradiol by human liver cytochrome P-450. *Mol. Pharmacol.* **33,** 500–508.

111. Ungvary, G., Varga, B., Horvath, E., Tatrai, E., and Folly, G. (1981). Study on the role of maternal sex steroid production and metabolism in the embryotoxicity of para-xylene. *Toxicology* **19,** 263–268.

112. Franks, P. A., Laughlin, N. K., Dierschke, D. J., Bowman, R. E., and Meller, P. A. (1989). Effects of lead on luteal function in rhesus monkeys. *Biol. Reprod.* **41,** 1055–1062.

113. Ng, T. P., Foo, S. C., and Yoong, T. (1992). Menstrual function in workers exposed to toluene. *Br. J. Ind. Med.* **49,** 799–803.

114. Kim, Y., Jung, K., Hwang, T., Jung, G., Kim, H., Park, J., Kim, J., Park, J., Park, D., Park, S., Choi, K., and Moon, Y. (1996). Hematopoietic and reproductive hazards of Korean electronic workers exposed to solvents containing 2-bromopropane. *Scand. J. Work Environ. Health* **22,** 387–391.

115. Chase, H. P., Barnett, S. E., Welch, N., Briese, F. W., and Krassner, M. L. (1973). Pesticides and U.S. farm labor families. *Rocky Mt. Med. J.* **70,** 27–31.

116. Mendola, P., Buck, G. M., Sever, L. E., Zielezny, M., and Vena, J. E. (1997) Consumption of PCB-contaminated freshwater fish and shortened menstrual cycle length. *Am. J. Epidemiol.* **146,** 995–960.

117. Karsch, F. J., Bittman, E. L., Foster, D. L., Goodman, R. L., Legan, S. J., and Robinson, S. E. (1984). Neuroendocrine basis of seasonal reproduction. *Recent Prog. Horm. Res.* **40,** 185–232.

118. Karsch, F. J., Bittman, E. L., Robinson, J. E., Yellon, S. M., Wayne, N. L., Olster, D. H., and Kaynard, A. H. (1986). Melatonin and photorefractoriness: Loss of response to the melatonin signal leads to seasonal reproductive transitions in the ewe. *Biol. Reprod.* **34,** 265–274.

119. Berga, S., Mortola, J., and Yen, S. S. (1988). Amplification of nocturnal melatonin secretion in women with functional hypothalamic amenorrhea. *J. Clin. Endocrinol. Metab.* **66,** 242–244.

120. Laughlin, G. A., Loucks, A. B., and Yen, S. S. (1991). Marked augmentation of nocturnal melatonin secretion in amenorrheic athletes, but not in cycling athletes: Unaltered by opioidergic or dopaminergic blockade. *J. Clin. Endocrinol. Metab.* **73,** 1321–1326.

121. Graham, C. A., and McGrew, W. C. (1980). Menstrual synchrony in female undergraduates living on a coeducational campus. *Psychoneuroendocrinology* **5,** 245–252.

122. Little, B. B., Guzick, D. S., Malina, R. M., and Rocha Ferreira, M. D. (1989). Environmental influences cause menstrual synchrony, not pheromones. *Am. J. Hum. Biol.* **1,** 53–57.

123. Wilson, H. C., Kiefhaber, S. H., and Gravel, V. (1991). Two studies of menstrual synchrony: Negative results. *Psychoneuroendocrinology* **16,** 353–359.

124. Weller, L., Weller, A., and Avinir, O. (1995). Menstrual synchrony: Only in roommates who are close friends? *Physiol. Behav.* **58,** 883–889.

125. Faulkes, C. G., Abbot, D. H., and Jarvis, J. U. (1990). Social suppression of ovarian cyclicity and wild colonies of naked mole-rats, *Heterocephalus glaber. J. Reprod. Fertil.* **88,** 559–568.

126. Preti, G., Cutler, W. B., Garcia, C. R., Huggins, G. R., and Lawley, H. J. (1986). Human axillary secretions influence women's menstrual cycles: The role of donor extract of females. *Horm. Behav.* **20,** 474–482.

127. Cutler, W. B., Preti, G., Krieger, A., Huggins, G. R., Garcia, C. R., and Lawley, H. J. (1986). Human axillary secretions influence women's menstrual cycles: The role of donor extract from men. *Horm. Behav.* **20,** 463–473.

128. Kjaer, K., Hagen, C., Sando, S. H., and Eshoj, O. (1992). Epidemiology of menarche and menstrual disturbances in an unselected group of women with insulin-dependent diabetes mellitus compared to controls. *J. Clin. Endocrinol. Metab.* **75,** 524–529.

129. Weiss, D. J., Charles, M. A., Dunaif, A., Prior, D. E., Lillioja, S., Knowler, W. C., and Herman, W. H. (1994). Hyperinsulinemia is associated with menstrual irregularity and altered serum androgens in Pima Indian women. *Metab. Clin. Exp.* **43,** 803–807.

130. Redmond, G. (1996). Diabetes and women's health. *Semin. Reprod. Endocrinol.* **14,** 35–43.

131. Pettigrew, R., and Hamilton-Fairley, D. (1997). Obesity and female reproductive function. *Br. Med. Bull.* **53,** 341–358.

132. Robinson, S., Kiddy, D., Gelding, S. V., Willis, D., Niththyananthan, R., Bush, A., Johnston, D. G., and Franks, S. (1993). The relationship of insulin insensitivity to menstrual pattern in women with hyperandrogenism and polycystic ovaries. *Clin. Endrocrinol. (Oxford)* **39,** 351–355.

133. Koutras, D. A. (1997) Disturbances of menstruation in thyroid disease. *Ann. N.Y. Acad. Sci.* **816,** 280–284.

134. Smith, S. K., Abel, M. H., Kelly, R. W., and Baird, D. T. (1981). A role for prostacyclin (PGI(2) in excessive menstrual bleeding. *Lancet* **1,** 522–524.

136. Stott, P. C. (1983). The outcome of menorrhagia: A retrospective case control study. *J. Coll. Gen. Pract.* **33,** 715–720.

137. Wood, C. (1972) Gynaecological survey in a metropolitan area of Melbourne. *Aust. N. Z. J. Obstet. Gynaecol.* **12,** 147–156.

138. Ballinger, C. B., Browning, M. C. K., and Smith, A. H. W. (1987). Hormone profiles and psychological symptoms in perimenopausal women. *Maturitas* **9,** 235–251.

139. Hornsby, P. P., Wilcox, A. J., Weinberg, C. R., and Herbst A. L. (1994). Effects on the menstrual cycle of in utero exposure to diethylstilbestrol. *Am. J. Obstet. Gynecol.* **170,** 709–715.

140. Harlow, S. D., and Campbell, B. C. (1994). Host factors that influence the duration of menstrual bleeding. *Epidemiology* **5,** 352–355.

141. Tudiver, F. (1983). Dysfunctional uterine bleeding and prior life stress. *J. Fam. Pract.* **17,** 999–1003.

142. Palmblad, J., Blomback, M., Egberg, N., Froberg, J., Karlsson, C-G., and Levi, L. (1977) Experimentally induced stress in man: Effects on blood coagulation and fibrinolysis. *J. Psychosom. Res.* **21,** 87–92.

143. Hartz, A. J., Baboriak, P. N., Wong, A., Katayama, K. P., and Rimm, A. H. (1979). The association of obesity with infertility and related menstrual abnormalities in women. *Int. J. Obes.* **3,** 57–73.

144. Toriola, A. L., and Mathur, D. N. (1986). Menstrual dysfunction in Nigerian athletes. *Br. J. Obstet. Gynaecol.* **93,** 979–985.

145. Wilansky, D. L., and Greisman, B. (1989). Early hypothyroidism in patients with menorrhagia. *Am. J. Obstet. Gynecol.* **160,** 673–577.

146. Kadir, R. A., Economides, D. L., Sabin, C. A., Owens, D., and Lee, C. A. (1998). Frequency of inherited bleeding disorders in women with menorrhagia. *Lancet* **351,** 485–489.

147. van Eijkeren, M. A., Christiaens, G. C. M. L., Haspels, A. A., and Sixma, J. J. (1990). Measured menstrual blood loss in women with a bleeding disorder or using oral anticoagulant therapy. *Am. J. Obstet. Gynecol.* **162,** 1261–1263.

148. Coulter, A., Kelland, J., Peto, V., and Rees, M. C. (1995). Treating menorrhagia in primary care. An overview of drug trials and a survey of prescribing practice. *Int. J. Technol. Assess. Health Care* **11,** 456–471.

149. Ng, T. P., Tan, N. C. K., and Wansaicheong, G. K. L. (1992). A prevalence study of dysmenorrhoea in female residents aged 15–54 years in Clementi Town, Singapore. *Ann. Acad. Med. Singapore* **21,** 323–327.

150. Pullon, S., Reinken, J., Sparrow, M. (1988). Prevalence of dysmenorrhoea in Wellington women. *N. Z. Med. J.* **101,** 52–54.

151. Sundell, G., Milsom, I., and Andersch, B. (1990). Factors influencing the prevalence and severity of dysmenorrhoea in young women. *Br. J. Obstet. Gynaecol.* **97,** 588–594.

152. Klein, J. R., and Litt, I. F. (1981). Epidemiology of adolescent dysmenorrhea. *Pediatrics* **68,** 661–664.

153. Widholm, O. (1979). Dysmenorrhea during adolescence. *Acta Obstet. Gynecol. Scand.* **87** (Suppl.); 61–66.

154. Svanberg, L., and Ulmsten, U. (1981). The incidence of primary dysmenorrhea in teenagers. *Arch. Gynecol.* **230,** 173–177.

155. Vicdan, K., Kukner, S., Dabakoglu, T., Ergin, T., Keles, G., and Gokmen, O. (1996). Demographic and epidemiologic features of female adolescents in Turkey. *J. Adolesc. Health* **18,** 54–58.

156. Harlow, S. D., and Park, M. (1996). A longitudinal study of risk factors for the occurrence, duration and severity of menstrual cramps in a cohort of college women. *Br. J. Obstet. Gynaecol.* **103,** 1134–1142.

157. Svennerud, S. (1959). Dysmenorrhea and absenteeism: Some gynecologic and socio-medical aspects. *Acta Obstet. Gynecol. Scand.* **38** (Suppl. 2).

158. Parazzini, F., Tozzi, L., Mezzopane, R., Luchini, L., Marchini, M., and Fedele, L. (1994). Cigarette smoking, alcohol consumption, and risk of primary dysmenorrhea. *Epidemiology* **5,** 469–472.

159. Okonofua, F. E., Balogun, J. A., Ayangade, S. O., and Fawole, J. O. (1990). Exercise and menstrual function in Nigerian university women. *Afr. J. Med. Sci.* **19,** 185–190.

160. Mergler, D., and Vezina, N. (1985). Dysmenorrhea and cold exposure. *J. Reprod. Med.* **30,** 106–111.

160a. Messing, K., Saurel-Cubizolles, M.-J., Bourgine, M., and Kaminski, M. (1993) Factors associated with dysmenorrhea among workers in French poultry slaughterhouses and canneries. *JOM, J. Occup. Med.* **35,** 493–500.

161. Larsson, G., Milsom, I., Lindstedt, G., and Rybo, G. (1992). The influence of a low-dose combined oral contraceptive on menstrual blood loss and iron status. *Contraception* **46,** 327–334.

162. Milman, N., Rosdahl, N., Lyhne, N., Jorgensen, T., and Graudal, N. (1993). Iron status in Danish women aged 35–65 years. Relation to menstruation and method of contraception. *Acta Obstet. Gynecol. Scand.* **72,** 601–605.

163. Saleh, W. A., Burkman, R. T., Zacur, H. A., Kimball, A. W., Kwiterovich, P., and Bell, W. K. (1993). A randomized trial of three oral contraceptives: Comparison of bleeding patterns by contraceptive types and steroid levels. *Am. J. Obstet. Gynecol.* **168,** 1740–1747.

164. Corson, S. L. (1990). Efficacy and clinical profile of a new oral contraceptive containing norgestimate. *Acta Obstet. Gynecol. Scand., Suppl.* **152,** 25–31.

165. Droegemueller, W., Katta, L. R., Bright, T. G., and Bowes, W. A., Jr. (1989). Triphasic randomized clinical trial: Comparative frequency of intermenstrual bleeding. *Am. J. Obstet. Gynecol.* **161,** 1407–1411.

166. Milsom, I., Sundell, G., and Andersch, B. (1990). The influence of different combined oral contraceptives on the prevalence and severity of dysmenorrhea. *Contraception* **42,** 497–507.

167. Hauksson, A., Ekstrom, P., Juchnicka, E., Laudanski, T., and Akerlund, M. (1989). The influence of a combined oral contraceptive on uterine activity and reactivity to agonists in primary dysmenorrhea. *Acta Obstet. Gynecol. Scand.* **68,** 31–34.

168. Bieglmayer, C. Hofer, G., Kainz, C., Reinthaller, A., Kopp, B., and Janisch, H. (1995). Concentrations of various arachidonic acid metabolites in menstrual fluid are associated with menstrual pain and are influenced by hormonal contraceptives. *Gynecol. Endocrinol.* **9,** 307–312.

169. Belsey, E. M. (1988). Vaginal bleeding patterns among women using one natural and eight hormonal methods of contraception. *Contraception* **38,** 181–206.

170. Burkman, R. T., Jr. (1994) The role of hormonal contraceptives: Noncontraceptive effects of hormonal contraceptives: Bone mass, sexually transmitted disease and pelvic inflammatory disease, cardiovascular disease, menstrual function, and future fertility. *Am. J. Obstet. Gynecol.,* suppl. **170,** 1569–1575.

171. Sivin, I., Diaz, S. Holma, P., Alvarez-Sanchez, F., and Robertson, D. N. (1983) A four-year clinical study of NORPLANT implants. *Stud. Fam. Plann.* **14,** 184–191.

172. Sivin, I. (1994). Contraception with NORPLANT implants. *Hum. Reprod.* **9,** 1818–1826.

173. Rulin, M. C., Davidson, A. R., Philliber, S. G., Graves, W. L., and Cushman, L. F. (1989). Changes in menstrual symptoms among sterilized and comparison women: A prospective study. *Obstet. Gynecol.* **74,** 149–154.

174. Wilcox, L. S., Martinez-Schnell, B., Peterson, H. B., Ware, J. H. and Hughes, J. M. (1992). Menstrual function after tubal sterilization. *Am. J. Epidemiol.* **135,** 1368–1381.

175. Bean, J. A., Leper, J. D., Wallace, R. B., Sherman, B. M., and Jagger, H. (1979). Variation in the reporting of menstrual histories. *Am. J. Epidemiol.* **109:** 181–185.

176. Higham, J. M., Obrien, P. M. S., and Shaw R. W. (1990) Assessment of menstrual blood loss using a pictorial chart. *Obstet. Gynecol. (N.Y.)* **97,** 734–739.

177. Fraser, I. S., McCarron, G., and Markham, R. (1984). A preliminary study of factors influencing perception of menstrual blood loss volume. *Am. J. Obstet. Gynecol.* **149,** 788–793.

178. Campbell, O. M. R., and Gray, R. H. (1993). Characteristics and determinants of postpartum ovarian function in women in the United States. *Am. J. Obstet. Gynecol.* **169,** 55–60.

179. Ruta, D. A., Garratt, A. M., Chadha, Y. C., Flett, G. M., and Hall, M. H. (1995). Assessment of patients with menorrhagia: how valid is a structured clinical history as a measure of health status? *Qual. Life Res.* **4:** 33–40.

180. Geller, S. E., Harlow, S. D., and Bernstein, S. J. (1999) Menstrual bleeding characteristics and their relationship to functional status and attitudes towards menstruation. *J. Women Health.* **8:**533–540.

10

Premenstrual Syndrome

JOSEPH F. MORTOLA[1]
Division of Reproductive Endocrinology
Cook County Hospital
Chicago, Illinois

I. Definition

Premenstrual syndrome (PMS) is a clinical disorder characterized by the repeated occurrence of disabling behavioral and physical symptoms during the luteal phase of the menstrual cycle [1]. The timing of the symptoms is more crucial to the definition of the disorder than the symptoms themselves because the symptoms often overlap those of other physical and psychiatric disorders. The World Health Organization characterizes PMS as a disorder of the genitourinary system, a designation that emphasizes the pivotal role of the ovary and its hormones in the pathophysiology of the disorder [2]. Nonetheless, the clinical presentation of PMS typically includes multisystem complaints, with those of the central nervous, gastrointestinal, and circulatory systems being most common [3]. The term Luteal Phase Dysphoric Disorder has been suggested by the American Psychiatric Association as a replacement for the term PMS, although the Association has rejected the incorporation of this entity into the standard psychiatric nomenclature at present [4]. The term premenstrual syndrome is preferred because it reflects a broad symptom constellation rather than the single symptom of "dysphoria." The term premenstrual syndrome should be used when, as in the vast majority of cases, both behavioral and physical symptoms are present [3].

II. Prevalence

Premenstrual symptoms are found in up to 80% of reproductive age women. Such symptoms are so prevalent that they are considered in the gynecologic literature to be a sign of normal ovarian function. The term "premenstrual molimina" is the appropriate designation for this symptom constellation, and it should be viewed as a normal, although often bothersome, physiologic event.

Unfortunately, confusion regarding the distinction between premenstrual molimina and premenstrual syndrome has led to completely erroneous reports of prevalence estimates of PMS. This has precipitated an understandable movement to abandon the term PMS. However, at least within the United States, the number of scientific papers that has been published with appropriately diagnosed PMS patients has been sufficient to reclarify the term PMS as used in clinical practice and to validate its continued use.

The social implications of these overestimates of PMS have been deleterious to those women who truly suffer from the disorder. As a result of inclusion of women who experience lesser severity of irritability and mood swings, there has been a tendency to minimize the importance of the disorder. At the extreme, the consequences of this minimization have led to PMS becoming the subject of situation comedies and greeting cards that are designed to be humorous.

The prevalence estimate of the disorder, using strict diagnostic criteria [5], is 2.5–5% of women [6–8]. As such, it is among the most common disorders in reproductive age women. It is therefore responsible for millions of dollars annually in lost productivity [1] and for untold costs to society and individual lives which result from inability to parent and maintain normal social and economic relationships during the premenstrual time period.

The clinical diagnosis of PMS requires that the symptoms occur both reproducibly in the luteal phase of the menstrual cycle and are of sufficient severity to cause impairment in social or economic performance. The prevalence estimates of premenstrual syndrome, as defined by this severity criterion, are dramatically lower than the prevalence of premenstrual symptoms.

III. Epidemiology

Data on the worldwide prevalence of PMS using strict diagnostic criteria are difficult to obtain due to the inconsistency with which these criteria are applied in cross-cultural studies. As early as 1972, cross-cultural studies were performed to assess the prevalence of premenstrual symptoms. Using retrospective surveys of Caucasian-American, Apache, Greek, Japanese, Nigerian, and Turkish women, Janiger et al. [9] determined that both the prevalence and severity of symptoms were highest among Turkish and Nigerian women. In contrast, Japanese women had the lowest prevalence of symptoms. American Caucasians showed an intermediate pattern of symptom prevalence and severity. Cultural differences have also been observed in the relative prevalence of some premenstrual symptoms as compared to others. In a comparison of women from the United States, Italy, and Bahrain [10], at least 30% of women from all cultures experienced premenstrual symptoms. However, irritability and mood swings were more common among American women, whereas swelling and breast tenderness were experienced by more Italian women. In a study that examined the prevalence of premenstrual symptoms in an inner-city population in the United States, the frequency of symptoms was similar among African-Americans, Hispanics, and Caucasians [11]. However, Hispanics differed from African-Americans and Caucasians in that they reported less premenstrual irritability, were less likely to recognize PMS as a distinct entity, and had a greater tendency to consider premenstrual symptoms normal. Overall, there do appear to be ethnic and cultural differences in the occurrence of premenstrual symptoms. However, these differences

[1]Deceased.

do not seem to be large when one considers the overall number and severity of symptoms reported by different populations.

In addition to ethnic factors, age appears to predict the prevalence of premenstrual syndrome to some degree. Of course, PMS is restricted to reproductive age women (approximately 13–51 years). Within this age group, however, there is some evidence to suggest that women are most vulnerable to PMS in their late twenties to mid thirties [12]; however, other studies have failed to find this association [13].

Increasingly, there has been interest in the possible familial association of premenstrual syndrome. Anecdotally, patients with PMS report a high prevalence of affected relatives [14], and severe symptoms in mothers may indicate an increased predisposition to PMS in their daughters [15]. Data from twin studies suggest heritability. Monozygotic twins have been found to be twice as likely as dizygotic twins to have similar PMS severity scores [16]. Overall, an estimated heritability of 30–40% has been estimated [17].

Several other risk factors have been suggested for PMS, including the use of oral contraceptive pills, dysmenorrhea, menstrual cycle characteristics, and personality variables. At this time, there is no firm evidence to support these associations.

IV. Historical Perspective

The earliest recorded reference to PMS appears in the writings of Hippocrates wherein he described the symptoms as "headache" and "a sense of heaviness" [18]. Later, Pliny described the marked behavioral changes [19]:

> On approach of a woman in this state, grass withers away, garden plants are parched up, new wine become sour, and the fruit will fall from the tree beneath which she sits.

Although the reports of these episodes of PMS date back to the earliest writings of Western Medicine, it was not until 1931 that the first reports in modern scientific literature appeared. Frank is credited with the nosologic designation of PMS when he characterized the condition as a state of "indescribable tension" [20]. After 60 years of research, the pathophysiology of the syndrome is just becoming elucidated.

V. Diagnostic Criteria

A. Symptoms

Although more than 150 symptoms have been ascribed to PMS [21], it is now apparent that the symptom constellation is rather specific and well defined. The repeated occurrence of either irritability or depression and fatigue during the luteal phase of the cycle accompanied by bloated sensations in the abdomen or extremities, breast tenderness, or headache is seen in almost all patients (Table 10.1). In addition, a select group of other symptoms occurs with sufficient frequency to merit their inclusion in the syndrome. The behavioral symptoms include labile mood with alternating sadness and anger (81%), oversensitivity (69%), crying spells (65%), social withdrawal (65%), forgetfulness (56%), and difficulty concentrating (47%). The common physical symptoms include acne (71%) and gastrointestinal upset (48%). Appetite changes and food cravings are

Table 10.1
University of California, San Diego Diagnostic Criteria for Premenstrual Syndrome

1. The presence by self-report of at least one of the following somatic *and* affective symptoms during the five days prior to menses in each of the three prior menstrual cycles:

Affective	*Somatic*
Depression	Breast tenderness
Angry outbursts	Abdominal bloating
Irritability	Headache
Anxiety	Swelling
Confusion	
Social withdrawal	

2. Relief of the above symptoms within 4 days of the onset of menses, without recurrence until at least cycle day 12.

3. The symptoms are present in the absence of any pharmacologic therapy, hormone ingestion, drug or alcohol use.

4. The symptoms occur reproducibly during two cycles of prospective recording.

5. Identifiable dysfunction in social or economic performance by one of the following criteria:

   Marital or relationship discord confirmed by partner
   Difficulties in parenting
   Poor work or school performance, attendance/tardiness
   Increased social isolation
   Legal difficulties
   Suicidal ideation
   Seeking medical attention for a somatic symptom(s)

From Mortola *et al.* [5].

seen in 70% of women with PMS. Vasomotor flushes (18%), heart palpitations (13%), and dizziness (13%) are less commonly observed.

B. Temporal Patterns

None of the symptoms of PMS is unique to the syndrome. What is diagnostic of the disorder is the marked fluctuation of symptoms with the menstrual cycle. During the time from the fourth day after the onset of menses until at least cycle day 12, symptoms, if they occur at all, are sporadic and no more frequent than those seen in the general population. This criterion is applicable to the vast majority of reproductive age women, although women with cycles that are typically shorter than 26 days in length may have the onset of symptoms slightly earlier than day 12. Using prospective recording of symptoms, the pattern of symptom severity in women with PMS has been defined. Mean symptom severity increases gradually throughout the luteal phase, reaches a peak just prior to the onset of menses, and subsequently declines rapidly over the first four days following the onset of menses.

C. Level of Impairment

Diagnosis of PMS also requires assessment of symptom severity. Therefore, a measurement of the *degree of impairment* should be ascertained. With respect to the physical symptoms of breast tenderness and bloating, for instance, their occurrence is so common as to be considered normal. Such symptoms should be more accurately termed premenstrual molimina. In contrast, premenstrual headaches may be so severe as to prohibit work. Because there is no "gold standard" by which to compare the severity of a self-reported symptom in one individual with that in another, more objective external manifestations of the disorder are required. This permits differentiation of PMS as a distinct syndrome at one end of the continuum between women who discern no physical and emotional differences during the course of the menstrual cycle and those who become incapacitated during the luteal phase. The assessment is most verifiable if objective criteria for impairment are used based on *impairment of social or economic performance*. Identifiable disruption in performance can be ascertained by (1) marital or relationship discord confirmed by the partner, (2) difficulties in parenting children manifested by behavioral disturbance in the child, (3) poor work or school performance, (4) increased social isolation, (5) legal difficulties, (6) expressed suicidal ideation, or (7) seeking medical attention for a somatic symptom.

D. Spontaneous Cyclicity

In addition to the application of criteria for symptom timing and degree of impairment, the diagnosis of PMS can only be made in the presence of spontaneous menstrual cycles. Frequently, side-effects from use of oral contraceptive pills may mimic the symptoms of PMS. In such patients, accurate diagnosis depends on discontinuation of the oral contraceptive and assessment of symptoms in the absence of any pharmacologic intervention. In addition, self-prescribed substances such as alcohol or marijuana may obscure the ability to assess the progression of symptoms during the course of the menstrual cycle.

VI. Symptom Recording

A. Prospective Inventories

Because of the inability to diagnose PMS based on the symptoms themselves as opposed to the *timing* of the symptoms, the importance of prospective recording has become apparent. Although many of the scales used most frequently in research to measure PMS symptoms are retrospective in nature [22,23], they have not had acceptance in clinical practice. Little information is afforded beyond that which can be learned during the clinical interview. Differences between prospective inventories and retrospective measures have been demonstrated [24].

B. Calendar of Premenstrual Experiences (COPE)

Several prospective rating scales are available [3,24–27]. One such scale, the Calendar of Premenstrual Experiences (COPE), has been tested in both PMS populations and asymptomatic controls [3]. It provides a valid, reliable, and simple to use measure (Fig. 10.1). A follicular phase score is rapidly tabulated by summing the total points obtained on this inventory on days 3–9 of the menstrual cycle, and a luteal phase score is calculated by summing the total points on the last seven days of the cycle. The diagnosis of PMS is based on the following criteria: (1) the luteal phase score is at least twice that of the follicular phase score, (2) the luteal phase score is at least 42, and (3) the follicular phase score is less than 40. A follicular phase score of greater than 40 should alert the clinician that the patient has a disorder other than PMS because, in well-selected populations, women with PMS alone do not achieve elevated follicular phase scores on this inventory.

C. Visual Analogue Scales and Research Tools

In addition to prospective diaries, such as the Calendar of Premenstrual Experiences, other tools have been used for research purposes. A large literature supports the advantages of visual analogue scales as accurate indicators of self-reported mood states. Visual analogue scales typically use 10 cm lines, the extremes of which are opposites, such as happy–sad. These are more cumbersome for use in clinical practice, but they can be important adjuncts for research purposes.

VII. Differential Diagnosis

A. Disorders in the Differential

The differential diagnosis of premenstrual syndrome includes a large number of medical and psychiatric disorders. In preparation of strict diagnostic criteria [5], 263 women presenting for the complaint of PMS were evaluated to elucidate the differential diagnosis. 10.2% were found to be experiencing early menopausal symptoms, 20.5% admitted having no symptom-free interval or were found to have no such interval on prospective recording, 11% had affective or personality disorders, 10.6% were using hormonal contraceptives, 5.3% had eating disorders, 3.8% were alcohol or other substance abusers, and 8.4% were complaining of symptoms which could also be attributed to a previously diagnosed medical disorder such as diabetes or hypothyroidism. 16.6% of subjects were noted to have menstrual irregularities with cycles less than 26 days or more than 34 days in length. Thus, more than 85% of patients with PMS either had another diagnosis which could account for symptoms or were noted to have another complaint which required correction before an accurate diagnosis of PMS could be made.

B. Comorbidity

Several reports have indicated a high incidence of affective disorder in patients with premenstrual syndrome. In addition, premenstrual exacerbations of anxiety disorders have been observed [28–30]. For both research purposes and clinical practice it is best to reserve the term premenstrual syndrome for those cases in which underlying psychiatric disorder is excluded. This distinction is made easily by prospective symptom recording on a daily symptom inventory. Those cases in which premenstrual symptoms occur concomitantly with such psychiatric disorders should be classified in such a way as to specify the psychiatric diagnosis (*e.g.,* recurrent major depression with premenstrual exacerbation). Such cases should be managed first with a view toward the diagnosis. In contrast, those individuals presenting with PMS in the absence of other disorders are most appropriately

CALENDAR OF PREMENSTRUAL EXPERIENCES

Name _____ Month/Year _____ Age _____ Unit ⧧ _____

Begin your calendar on the *first* day of your menstrual cycle. Enter the calendar date below the cycle day. Day **1** is your *first* day of bleeding. Shade the box above the cycle day if you have bleeding. ■ Put an **X** for spotting. ☒

If more than one symptom is listed in a category, i.e., nausea, diarrhea, constipation, you do not need to experience all of these. Rate the most disturbing of the symptoms on the 1-3 scale.

Weight: Weigh yourself before breakfast. Record weight in the box below date.
Symptoms. Indicate the severity of your symptoms by using the scale below. Rate each symptom at about the same time each evening.

    0 = **None** (symptom not present)   2 = **Moderate** (interferes with normal activities)
    1 = **Mild** (noticeable but not troublesome)   3 = **Severe** (intolerable, unable to perform normal activities)

Other Symptoms: If there are other symptoms you experience, list and indicate severity.
Medications: List any medications taken. Put an **X** on the corresponding day(s).

Bleeding																																								
Cycle Day	1	2	3	4	5	6	7	8	9	10	11	12	13	14	15	16	17	18	19	20	21	22	23	24	25	26	27	28	29	30	31	32	33	34	35	36	37	38	39	40
Date																																								
Weight																																								
SYMPTOMS																																								
Acne																																								
Bloatedness																																								
Breast tenderness																																								
Dizziness																																								
Fatigue																																								
Headache																																								
Hot flashes																																								
Nausea, diarrhea, constipation																																								
Palpitations																																								
Swelling (hands, ankles, breast)																																								
Angry outbursts, arguments, violent tendencies																																								
Anxiety, tension, nervousness																																								
Confusion, difficulty concentrating																																								
Crying easily																																								
Depression																																								
Food cravings (sweets, salts)																																								
Forgetfulness																																								
Irritability																																								
Increased appetite																																								
Mood swings																																								
Overly sensitive																																								
Wish to be alone																																								
Other Symptoms																																								
1. _____																																								
2. _____																																								
Medications																																								
1. _____																																								
2. _____																																								

Fig. 10.1 The Calendar of Premenstrual Experiences. From Mortola *et al.* [3].

managed by therapy directed at premenstrual syndrome (see Treatment section).

C. Biochemical Markers

Based on the similarity of some of the mood symptoms of PMS and those observed in depression, the biochemical markers of endogenous depression have been examined in PMS patients. The most consistent neuroendocrine marker of depression is alteration of the secretory episodes of serum cortisol. Using frequent sampling techniques, serum cortisol levels assayed at 20 minute intervals for a 24-hour period revealed no differences in any cortisol secretory parameter between women with PMS and asymptomatic women. In contrast, women with endogenous depression showed increased cortisol secretion as compared with both normal controls and women with PMS. Another neuroendocrine marker for depression, the thyroid-stimulating hormone (TSH) response to thyrotropin-releasing hormone (TRH), has also been demonstrated to be similar in women with PMS and asymptomatic women [31]. It has been suggested that serum circadian melatonin profiles were altered in women with PMS [32], as they have been shown to be altered in depression. However, the PMS population studied included some subjects who had evidence of depression. When women with evidence of depression were excluded, no difference in serum melatonin levels was found.

VIII. Lack of Evidence for Psychogenic Etiology

Development of treatment strategies for PMS has been hampered by incomplete understanding of the pathophysiology of the disorder. Several hypotheses on the etiology of PMS have been advanced, including the possibility that PMS is a stress-induced disorder, a variant of affective disorder, or a personality disorder. Compelling evidence is available to exclude these psychogenic postulates on the etiology of PMS. These include the findings that concurrent psychosocial stress accounts for less than 12% of the unique variance in PMS mood and physical symptoms in women with PMS over the course of the menstrual cycle [33]. It also has been shown that PMS can be easily distinguished from affective disorder both by biochemical methods, particularly serum cortisol secretory parameters, and by psychometric indices [5]. Use of the most widely validated personality inventory, the Minnesota Multiphasic Personality Inventory, has also demonstrated that women with PMS in whom psychiatric disorders are excluded do not score differently than controls.

IX. Pathophysiology

A. Ovarian Steroids

PMS has been widely hypothesized to be due to differences in ovarian steroid levels. Attempts to discern differences in serum ovarian steroid levels in women with PMS, however, have been contradictory. While some investigators have implicated a decreased level of luteal phase progesterone in women with PMS [34–36], others have failed to replicate these results [37–39]. Similar studies reporting that estradiol (E_2) or estradiol/progesterone (E_2/P_4) ratios are increased in subjects with PMS as compared to controls [34,35,40] have not been confirmed by

other investigators [37,41]. The differences in these results can be explained by a variety of factors including the use of inadequate sampling frequencies, lack of appropriate subject selection criteria, and the absence of sufficiently sophisticated symptom measurement techniques. In clinical practice, measurements of estradiol or progesterone are not useful in confirming the diagnosis of PMS.

B. Steroid Metabolites

It has been hypothesized that differences in ovarian steroid metabolism may be present in women with PMS [42]. One study compared first morning urinary levels of the estrogen metabolite estrone-1-glucuronide (E_1G) and the progesterone metabolite pregnanediol glucuronide (PdG) in samples obtained daily throughout several consecutive menstrual cycles in women with PMS and in controls [43]. PMS women in this study were shown to have similar menstrual cycle lengths, including both similar follicular phase lengths and similar luteal phase lengths. No differences in either E_1G or PdG values were seen on any cycle day. The ratio of E_1G/PdG also was similar on all days throughout the cycle, as was the ratio of total E_1G to total PdG produced during the luteal phase.

Despite inability to detect differences in urinary steroid metabolites in women with PMS, there is a clear association between premenstrual symptoms and menstrual cycle phase [44]. In both the follicular and luteal phase, however, the correlation between E_1G levels and either mood or physical symptoms is close to zero. Moreover, no relationship exists between either physical symptoms or mood and the E_1G/PdG ratio throughout the menstrual cycle or the slope of the late luteal phase PdG decline. In contrast, correlation has been found between PdG levels and both physical symptoms and mood. These correlations increase steadily as symptoms are lagged one, two, three, and four days behind PdG levels. After a 4-day lag period, the physical symptoms are highly correlated with PdG levels. These correlations disappear abruptly after a 5-day lag, when correlation coefficients approach zero. Therefore, at least on a correlational basis, it appears that progesterone may influence symptoms, but its effects are maximal only after a 4-day lag. This may account for worsening symptom severity in the late luteal phase when progesterone levels are actually declining.

Although differences in urinary steroid metabolites have not been detected, it has been shown that women with PMS have higher serum levels of the progesterone metabolite allopregnenolone than do controls [45]. This metabolite of progesterone has been shown to interact with the benzodiazepine receptor complex in a manner similar to that of barbiturates. Thus it is possible that women with PMS may have differences in progesterone metabolism that may contribute to the disorder.

It also has been suggested that estrogen, classically considered a hormone that enhances well-being, may promote PMS symptoms [46].

C. Neurotransmitters

Substantial indirect evidence supports the concept that the occurrence of the behavioral symptoms of PMS, which, in the majority of patients, are the most disabling, are the result of ovarian steroid influences on central neurotransmitters. Currently, there is evidence to support opioidergic, adrenergic, serotonergic, and γ-amino butyric acid (GABA) involvement in PMS symptoms. Of these, the most evidence has been accumulated to suggest a pivotal role for serotonin. Nonetheless, a combination of neurotransmitter–enzyme steroid interactions undoubtedly occurs.

1. Opiates

Endogenous opiate activity in the central nervous system is increased both by estrogen and by progesterone. As a result of investigations in both animals and humans, it is apparent that endogenous opioid activity increases in the estrogen-dominated follicular phase of the cycle and reaches even higher levels in the high estrogen, high progesterone environment of the early to midfollicular phase. Subsequently, in the late luteal phase, endogenous opiate levels abruptly fall, creating a mini-opiate withdrawal. The ability of a relatively short-term exposure to opiates (such as that which occurs during the menstrual cycle) to produce physical dependence has been demonstrated in animals [47]. In the human, decreased levels of β-endorphin have been reported in the peripheral circulation of women with PMS [48], although the extent to which these peripheral levels reflect central opioidergic activity is uncertain.

2. Adrenergic System

The overlapping distribution of opiate and α-adrenergic neuronal systems, along with evidence that administration of the opiate antagonist, naloxone, can increase firing rates of noradrenergic neurons, suggests that there may be an interaction between these symptoms. This is supported by several experimental models in animals [49–51]. In humans, the α-adrenergic agent clonidine, when given to opiate addicts undergoing withdrawal, results in a rapid and dramatic reversal of objective signs and subjective symptoms of the withdrawal [52].

Taken together, opiate and α-adrenergic systems may explain many of the constellation of behavioral symptoms observed in women with PMS. Increased opiate activity in the midluteal phase may stimulate appetite, resulting in the episodes of binge-eating seen in PMS [53]. Fatigue and depression at this time may result from diminished release of norepinephrine occasioned by opiate inhibition of the catecholamine system [54]. Acute withdrawal of opiates as menses approaches may lead to rebound hyperactivity of noradrenergic neurons due to acquired receptor sensitivity [55] and the increased number of receptors, resulting in irritability, anxiety, tension, and aggression.

3. GABA

Given that the most commonly observed symptom of PMS is fatigue, the steroid fluctuations during the menstrual cycle have been examined with respect to inhibitory neurotransmitters. Of these, gamma-amino-butyric acid (GABA) is the most studied. The GABA receptor complex constitutes the major site of action of benzodiazepines [56], where they serve to promote binding of GABA agonists to receptors [57]. It is now apparent that progesterone and its metabolites bind to the GABA receptor complex at a site distinct from the GABA binding site and consequently influence the GABA effect [58]. In addition, both estrogen and progesterone may serve to regulate GABA activity by attenuation of the number of GABA receptors or modulation of GABA release [59]. This finding has led to exploration of the role of GABA active agents in PMS (See Treatment section).

4. Serotonin

There is a large body of evidence to suggest that serotonin (5-HT) is a neurotransmitter of considerable importance in PMS symptomatology. In the rat, serial injection of progesterone results in increased 5-HT uptake in several areas of the brain [60] as well as increased 5-HT turnover [61]. During the normal estrus cycle, 5-HT receptors in the median forebrain undergo cyclic fluctuations, being unregulated following ovulation [62]. In women, decreased 5-HT uptake by platelets has been reported in the premenstrual phase of the cycle following steroid withdrawal [63] and has been correlated with the severity of some PMS symptoms [64]. Differences in platelet 5-HT uptake mechanisms also have been noted in women with PMS [65]. Similar alterations have been demonstrated in whole blood 5-HT [66]. In addition, the 5-HT metabolite 5-HIAA has been measured in urine throughout the menstrual cycle and found to be highest at midluteal phase and lower in the late luteal phase [67].

In rats, m-chlorophenylpiperazine (m-CCP), a postsynaptic serotonin agonist, produces profound anorexia [68], as does the serotonin agonist fenfluramine, when injected into the medial hypothalamus [69]. In humans, cyproheptadine, a serotonin receptor antagonist, increases appetite in anorexia nervosa and potentiates hyperphagia in bulimia [70]. In addition to producing anorexia, m-CCP has been shown in mice [71] and rhesus monkeys [72] to decrease locomotor activity. In humans, decreased cerebrospinal fluid levels of 5-HIAA are associated with depression [73]. Thus, decreased serotonergic activity, as is found in the late luteal phase of the menstrual cycle, may be implicated in many of the symptoms of PMS, including depression, increased psychomotor activity, and increased appetite.

D. Dietary and Nutritional Factors

A role for dietary and nutritional factors in the etiology of PMS has been widely postulated. Implicated in this regard have been vitamin B_6 deficiency [74], magnesium deficiency [75–77], and an alteration in carbohydrate metabolism [78]. The vitamin B_6 hypothesis has been largely excluded based on the preponderance of evidence that vitamin B_6 does not ameliorate PMS symptoms [79]. Moreover, definitive evidence for alterations in carbohydrate metabolism has not been found. The possibility that magnesium deficiency or calcium deficiency has a central role in the pathophysiology of PMS has been largely dismissed but not completely excluded.

X. Treatment

A. Lifestyle Alterations

In the absence of an understanding of the pathophysiology of premenstrual syndrome, nonspecific attempts at alleviating symptomatology achieved popularity. These included dietary restrictions, including the elimination of caffeine, chocolate, and sweet and salty foods [80]. There is little evidence to demonstrate the efficacy of these treatments. While anecdotal reports have purported benefits of these treatments, their efficacy has not been proven. Some data suggest that women engaged in aerobic exercise have a decreased severity of PMS symptoms as compared to nonexercising women with PMS. There also have been uncontrolled studies showing a benefit to exercise. Because of the reported benefits of this treatment, an exercise program with its overall health benefits appears a reasonable prescription for women with a low degree of symptom severity.

B. Serotonin Reuptake Inhibitors

Serotonin reuptake inhibitors have become the first line of therapy for PMS. The serotonin reuptake inhibitors include fluoxetine, sertraline, and paroxetine. A related compound, venlafaxine, shares many of the properties of the serotonin reuptake inhibitors. The best studied serotonin receptor inhibitor in premenstrual syndrome is fluoxetine. In double-blind placebo-controlled studies, its efficacy has been unequivocally demonstrated [81,82]. Given data that serotonin levels are altered by ovarian steroids and that serotonin metabolism may differ in women with PMS, its use in the syndrome is supported by a rationale based on pathophysiology. The dose of 20 mg daily has been established to be effective, although smaller doses show clinical efficacy in some patients. It is best taken in the morning. Approximately 15% of women are unable to tolerate the medication primarily because of jitteriness, nausea, or headache, and less commonly because of sedation. In those who experience sedation, a nighttime regimen may be considered. Of those who are able to tolerate the medication, approximately 75% experience a reduction of symptoms greater than 50%.

It has been demonstrated that the beneficial effects of fluoxetine are not due to the generalized antidepressant effects [83]. When women treated with fluoxetine were compared to women treated with the tricyclic antidepressant imipramine, a significant response was noted in 70% of subjects with fluoxetine as compared to only 25% of those treated with imipramine. This is observed in depressed patients where the response to imipramine is similar to that of fluoxetine. A uniformly high response to fluoxetine was noted in women presenting either with anxious or with depressed symptoms of PMS and those presenting with a premenstrual appetite or sleep disturbance.

C. Benzodiazepines

At least two double-blind studies have demonstrated the efficacy of alprazolam in the treatment of premenstrual syndrome [84,85]. As such, it clearly deserves a prominent place in the treatment options to be considered. The usual dose range is 0.25 mg four times a day during the luteal phase of the menstrual cycle. Occasionally, higher doses of 0.5 mg up to four times a day are required. Although efficacy has been demonstrated at these doses, clinically, many patients report significant improvement when the medication is taken during the luteal phase on an as needed basis.

The side effect of greatest concern with alprazolam is its addictive potential. This has prompted a number of clinicians to substitute other benzodiazepines for alprazolam in the treatment of PMS. While there is sound theoretical rationale to posit that other benzodiazepines may have efficacy similar to alprazolam based on their biochemical similarity, this has not been demonstrated in controlled studies. Moreover, while alprazolam may be more addictive than some other benzodiazepines, all agents in this class carry a substantial risk of addiction. For this reason, the use of benzodiazepines in PMS should be carefully restricted to luteal phase administration in the reliable patient. Addiction to alprazolam has not been reported when restricted to use during this prescribed time interval.

D. Gonadotropin-Releasing Hormone Analogues

Based on the association of PMS symptoms with progesterone in the luteal phase of the cycle, an ideal treatment for PMS would be one in which the progesterone levels could be decreased or eliminated. This led to the hypothesis that PMS would be improved by administration of gonadotropin-releasing hormone (GnRH) agonists. The use of GnRH agonists for this indication depends on their ability to cause pituitary desensitization to native GnRH. This desensitization is thought to be the result of internalization of the GnRH receptor [86,87]. Depending on the potency of the agonist, the desensitization phase (termed downregulation) requires 7–21 days. Once downregulation has been established, it persists for as long as the agonist is administered. During downregulation, luteinizing hormone (LH) and follicle-stimulating hormone (FSH) secretion by the pituitary is substantially reduced. As a result, there is insufficient stimulation of the ovary for normal sex steroid production. Circulating estrogen levels are therefore in the postmenopausal range, and progesterone levels are similarly low.

The first report of the efficacy of GnRH agonist treatment for PMS was by Muse et al. in 1984 [88]. These investigators were able to show a more than 75% reduction in PMS symptoms during the second month of GnRH-agonist treatment, once downregulation of pituitary receptors had been achieved. This work, using the GnRH analogue D-His[6] Pro[9] NEt-GnRH, was subsequently replicated by other investigators with similarly excellent results using other GnRH agonists [89,90]. As a result, the use of GnRH analogues constituted the first clearly successful treatment for PMS. Since that time, their effectiveness has been confirmed in several other studies. The use of GnRH analogues as a first line therapy for PMS has been supplanted by serotonin reuptake inhibitors or a benzodiazepine. The use of GnRH agonists is now generally reserved for patients who fail to respond to other therapies. The primary reasons for not using GnRH analogues for premenstrual syndrome include cost, the discomfort resulting from the menopausal symptoms induced, and long-term health risks associated with estrogen deprivation in women. Although cost still mitigates against the use of GnRH agonists as a first line therapy, the ill effects of estrogen deprivation have been largely eliminated by add-back therapy. The dose used is 0.625 mg of conjugated equine estrogens (CEE) or the equivalent. This dose has been shown to be effective for both relief of hot flashes and prevention of osteoporosis.

E. GnRH Analogues and Estrogen/Progestin "Add-back"

The primary concerns with administration of GnRH analogues for periods longer than six months are the likelihood that the patient will develop osteoporosis [91] and the potential for increased cardiovascular disease [92], both of which result from long-term estrogen deprivation.

The results of one study showing beneficial effects of GnRH-agonist plus estrogen/progestin therapy represented an advance in the treatment of PMS in that they suggested that low-dose steroid hormone replacement may maintain the beneficial effects, and circumvent the unacceptable consequences of the use of GnRH-agonists [46]. Administration of estrogen alone in combination with a GnRH-agonist would be expected to expose patients to the risks of unopposed estrogen that are observed in postmenopausal women on this regimen. Thus, 10 mg medroxyprogesterone acetate (MPA) for 10 days is added. This dose has been shown to obviate the adverse endometrial effects of unopposed estrogen replacement. Combined estrogen/progestin replacement therapy thus offers a safe and effective method of obtaining the beneficial effects of GnRH-agonists on PMS.

Further study is required to document a sustained beneficial effect of this regimen over repeated treatment cycles. In addition, study of other hormone replacement regimens such as continuous daily administration of CEE and MPA, are warranted in the PMS population receiving GnRH-agonists. At present, however, the use of GnRH-agonists in combination with standard cyclic postmenopausal hormone replacement therapy provides a promising, safe, long-term hormonal treatment alternative in PMS.

F. Danazol

Several reports have indicated the efficacy of danazol (Danacrine) in premenstrual syndrome [93]. This therapy has been shown to be particularly efficacious in the treatment of premenstrual migraines [94]. Danazol is a derivative of the synthetic androgen 17-alpha-ethinyl testosterone [95]. As such, it possesses significant androgenic properties. Administration of danazol results in amenorrhea in the majority of women on the medication.

The side effect profile of danazol is considerable, due both to its androgenic activity and to its anti-estrogen properties. Acne and weight gain are commonly reported. Decreased breast size and fluid retention are particularly disturbing complaints for many women. More rarely, overtly masculinizing side effects are noted. These include deepening of the voice and clitoromegaly [96].

The anti-estrogenic side effects, while better tolerated by most women than the androgen effects, are at times quite bothersome. These are the same side effects as those observed when taking GnRH analogues and include hot flashes, vaginal dryness, and emotional lability. Although the osteoporosis that accompanies GnRH agonist therapy is less of a concern with danazol, the effects on lipid profiles are more worrisome. This is due to the combined adverse effects of hypoestrogenism and hyperandrogenism. For this reason, the use of danazol should be accompanied by monitoring of lipid profiles.

Taken together, approximately 80% of women on danazol will experience side effects. Based on the risk/benefit ratio, danazol is a relatively poor choice for most patients with PMS. Although its success equals the excellent results obtained with GnRH agonists, it has considerably more side effects.

XI. Unproven Treatment

A. Vitamin B6

At least one placebo-controlled study of vitamin B6 in PMS has shown efficacy [97], although other studies have failed to replicate these results [98]. The efficacy of vitamin B6 therapy in the syndrome is therefore controversial. At high doses (>600 mg/day), vitamin B6 therapy has been associated with peripheral neuropathy [99,100]. In the event this therapy is tried in PMS, patients must be cautioned against excess dosages.

B. Diuretics

1. Spironolactone

Of the diuretics that have been used in treating premenstrual syndrome, spironolactone has achieved the greatest popularity. This is largely the result of specific properties of this agent that are uniquely suited to hormonally-based disorders. Spironolactone is an aldosterone inhibitor. As such, it also shares considerable structural similarity with steroid hormones such as estrogen and progesterone. This property, together with its diuretic effects, suggested that spironolactone might be an effective agent in treating PMS [101]. It was hypothesized that the sex steroid-like properties of the compound may alleviate the mood changes in PMS while the diuretic action may improve the physical symptoms of breast tenderness, water retention, and weight gain. Overall, the results with spironolactone in treating the broad spectrum of symptoms that accompany PMS have been disappointing, and it appears to be a mildly helpful agent at best. Nonetheless, because of its anti-steroidal properties, it has remained the diuretic most often chosen in treatment of the water-retention symptoms of PMS. Many of the agents that are most effective in treating overall PMS symptoms, particularly the mood and appetite disturbances, are less effective in relieving the water retention symptoms of bloating and breast tenderness. In these women spironolactone, limited to the luteal phase of the cycle, provides a relatively safe, although unproven, adjunctive therapy.

2. Other Diuretics

Diuretics other than spironolactone have also been tried in treating PMS, with anecdotal success in alleviating the water retention symptoms. The most commonly used class of diuretics after spironolactone is the thiazides. Thiazide diuretics often are combined with an antikaliuretic agent such as triamterene. Even with these preparations, however, hypokalemia remains the major concern. Not uncommonly, diuretics become the drugs of abuse in women who become highly concerned about their weight. "Water pills" are generally viewed by the public as benign agents and overuse is therefore common. Particularly in women with PMS who are concerned about bloated sensations, cautions about the overuse of diuretics should be stressed.

C. Oral Contraceptive Pills

The success of treating premenstrual syndrome with oral contraceptive agents has not been consistent. For the most part, the side effects of oral contraceptive agents, including mood effects (particularly depression), water retention, and appetite changes, are precisely those that women with PMS experience [102,103]. There appears to be a small percentage of women for whom oral contraceptives provide a preferable hormonal milieu to their endogenous estrogen and progesterone. In general, however, these agents have not been established as effective in the treatment of PMS.

XII. Ineffective Treatment

Until the mid 1980s, progesterone given in the form of vaginal or rectal suppositories was widely prescribed for PMS [104]. This was based on uncontrolled studies. Progesterone now has been shown to be no more effective than a placebo in treating PMS symptoms [105,106]. Moreover, there is evidence that both the physical and emotional symptoms of PMS may be progesterone induced [44]. Thus administration of progesterone commonly results in increased breast tenderness, bloating of the abdomen and extremities, and emotional lability. The use of progesterone in treatment of PMS therefore cannot be advocated.

XIII. Future Research

As recently as 1985 the available information on PMS consisted almost entirely of a plethora of unsubstantiated anecdotal reports and poorly designed studies about its prevalence, pathophysiology, and treatment. This information has been dramatically altered by more than 100 well-controlled studies providing insights into pathophysiology and treatment. The data from the 1990s have revolutionized medical understanding of the disorder and have raised important questions for future investigation.

One area of important research concerns the relative contribution of estrogen. As mentioned earlier, evidence suggests that estrogen, previously regarded as a relatively universal mood elevator, may, in some women, actually underlie the symptomatology of PMS. Because of the myriad of changes induced by estrogen, as well as progestin, on neurotransmitters and peripheral enzymes, use of agents that block these steroid effects are more likely to be effective than agents that influence single neurotransmitter systems. The development of such agonists and antagonists is therefore an important area of investigation.

The complex interaction of ovarian steroids with neuronal systems continues to be an active area of investigation. Further research into the genetic and environmental factors that create the propensity for the development of PMS is required. In the more distant future, research may provide methods of identifying high risk women and offer pharmacologic insights into the properties of compounds that will be optimally effective in treating PMS.

XIV. Summary and Conclusions

Premenstrual syndrome affects approximately 2.5–5% of women. Previous reports of higher prevalence rates were based on women who suffered symptoms rather than the well-defined entity of PMS. Some reports suggest that there are more similarities than differences in the racial or ethnic prevalence of PMS. The pathophysiology of PMS is now understood to be a complex interaction between fluctuations in ovarian steroids and central neurotransmitters as well as peripheral effects of these hormones. Highly effective treatments for the disorder have been developed that include agents that alter central neurotransmitters and circulating ovarian steroid levels. Future research will provide more specific treatments and aid in an understanding of the factors that influence the diathesis toward PMS.

References

1. Reid, R. L., and Yen, S. S. C. (1981). Premenstrual syndrome. *Am. J. Obstet. Gynecol.* **139,** 85–103.
2. World Health Organization (1992). "International Classification of Diseases," 10th ed. WHO, Geneva.
3. Mortola, J. F., Girton, L., Beck, L., and Yen, S. S. C. (1990). Diagnosis of premenstrual syndrome by a simple prospective, and reliable instrument: The calendar of premenstrual experiences. *Obstet. Gynecol.* **76,** 302–307.
4. American Psychiatric Association, Committee on Nomenclature and Statistics (1993). "Diagnostic and Statistical Manual of Mental Disorders," 4th ed. American Psychiatric Association Press, Washington, DC.
5. Mortola, J. F., Girton, L., and Yen, S. S. C. (1989). Depressive episodes in premenstrual syndrome. *Am. J. Obstet. Gynecol.* **161,** 1682–1687.
6. Steiner, M., and Wilkins, A. (1996). Diagnosis and assessment of premenstrual dysphoria. *Psychiatr. Ann.* **26,** 571–575.
7. Rivera-Tovar, A. D., and Frank, E. (1990). Late luteal phase dysphoric disorder in young women. *Am. J. Psychiatry* **147,** 1634–1636.
8. Ramcharan, S., Love, E. J., Fick, G. H., and Jaffe, E. L. (1992). The epidemiology of premenstrual symptoms in a population based sample of 2650 urban women. *J. Clin. Epidemiol.* **45,** 377–392.
9. Janiger, O., Riffenburgh, R., and Kwersh, R. (1972). Cross-cultural study of premenstrual symptoms. *Psychosomatics* **13,** 226–235.
10. Dan, A. J., and Monagle, L. (1994). Sociocultural influences on women's experiences of premenstrual symptoms. *In* "Premenstrual Dysphorias: Myths and Realities" (J. H. Gold and S. K. Severino, eds.), pp. 61–79. American Psychiatric Press, Washington, DC.
11. Bason, S., Nani, J. M., Tennery, S., and Mortola, J. F. (1998). A cross-sectional comparison of menstrual cycle-related symptoms among ethnically diverse women. *Abstr. 45th Annu. Meet. Soc. Gynecol. Invest.,* Atlanta, GA.
12. Freeman, E. W., Rickels, K., Schweizer, E., and Ting, T. (1995). Relationships between age and symptom severity in women seeking medical treatment for premenstrual syndrome. *Psychol. Med.* **25,** 309–315.
13. Johnson, S. R. (1987). The epidemiology and social impact of premenstrual symptoms. *Clin. Obstet. Gynecol.* **30,** 367–376.
14. Pearlstein, T. B. (1995). Hormones and depression: What are the facts about premenstrual syndrome, menopause and hormone replacement therapy. *Am. J. Obstet. Gynecol.* **173,** 646–653.
15. Freeman, E. W., Sondheimer, S. J., and Rickels, K. (1988). Effects of medical history factors on symptom severity in women meeting criteria for premenstrual syndrome. *Obstet. Gynecol.* **72,** 236–239.
16. Condon, J. T. (1993). The premenstrual syndrome: A twin study. *Br. J. Psychiatry* **162,** 481–486.
17. Kendler, K. S., Silberg, J. L., Neale, M. C., Kessler, R. C., Heath, A. C., and Eaves, L. J. (1992). Genetic and environmental factors in the aetiology of menstrual, premenstrual and neurotic symptoms: A population-based twin study. *Psychol. Med.* **22,** 85–100.
18. Simon, H. (1978). "Mind and Madness in Ancient Greece," p. 273. Cornell University Press, Ithaca, NY.
19. Dennerstein, L., and Burrows, G. D. (1979). Affect and the menstrual cycle. *J. Affective Disord.* **1,** 77–92.
20. Frank, R. T. (1931). The hormonal causes of premenstrual tension. *Arch. Neurol. Psychiatry* **26,** 1053–1057.

21. Hamilton, J. A., Parry, B. L., Alagna, S., Blumenthal, S., and Herz, E. (1984). Premenstrual mood changes: A guide to evaluation and treatment. *Psychiatr. Ann.* **14,** 426–435.

22. Moos, R. H. (1968). The development of a menstrual distress questionnaire. *Psychosom. Med.* **30,** 853–867.

23. Stout, A. L., and Steege, J. (1985). Psychological assessment of women seeking treatment for premenstrual syndrome. *J. Psychosom. Res.* **29,** 621–629.

24. Endicott, J., and Halbreich, U. (1982). Retrospective report of premenstrual depressive changes: Factors affecting confirmation by daily ratings. *Psychopharmacol. Bull.* **18,** 109–112.

25. Sampsom, G. A., and Prescott, P. (1981). The assessment of the symptoms of premenstrual syndrome and their response to therapy. *Br. J. Psychiatry* **138,** 399–405.

26. Steiner, M., Haskett, R. F., and Carroll, B. J. (1980). Premenstrual tension syndrome: The development of diagnostic criteria and new rating scales. *Acta Psychiatr. Scand.* **62,** 229–234.

27. Taylor, J. W. (1974). The timing of menstruation-related symptoms assessed by a daily rating scale. *Acta Psychiatr. Scand.* **60,** 87–105.

28. Endicott, J., Halbreich, U., Schacht, S., and Nee, J (1981). Premenstrual changes and affective disorders. *Psychosom. Med.* **43,** 519–529.

29. Kashiwagi, T., McClure, J. H., and Wetzel, R. D. (1981). Premenstrual affective syndrome and psychiatric disorder. *Dis. Nerv. Syst.* **37,** 116–119.

30. DeJong, R., Rubinow, D. R., Roy-Byrne, P., Hobin, M. D., Grover, G. N., and Post, R. M. (1985). Premenstrual mood disorder and psychiatric illness. *Am. J. Psychiatry* **142,** 1359–1363.

31. Casper, R. F., Patel-Christopher, N., and Powell, A. M. (1989). Thyrotropin and prolactin responses to thyrotropin releasing hormone in premenstrual syndrome. *J. Clin. Endocrinol. Metab.* **68,** 608–612.

32. Parry, B. L., Berga, S. R., Kripke, D. F., Klauber, M. R., Laughlin, G. A., Yen, S. S. C., and Gillin, J. C. (1990). Altered wave form of plasma nocturnal melatonin secretion in premenstrual depression. *Arch. Gen. Psychiatry* **47,** 1139–1146.

33. Beck, L. E., Gevirtz, R., and Mortola, J. F. (1990). The predictive role of psychosocial stress on symptom severity in premenstrual syndrome. *Psychosom. Med.* **52,** 536–543.

34. Abraham, G. E., Elsner, C. W., and Lucas, L. A. (1978). Hormonal and behavioral changes during the menstrual cycle. *Senologia* **3,** 33.

35. Backstrom, T., and Carstensen, H. (1974). Estrogen and progesterone in plasma in relation to premenstrual tension. *J. Steroid Biochem.* **5,** 257–260.

36. Munday, M. R., Brush, M. U., and Taylor, R. W. (1981). Correlation between progesterone, estradiol and aldosterone levels in premenstrual syndrome. *Clin. Endocrinol. (Oxford)* **14,** 1–11.

37. Haspels, A. A. (1981). A double-blind, placebo controlled multicenter study of the efficacy of hydrogesterone (Buphaston). *In* "Premenstrual Syndrome" (P. A. Van Keep and W. H. Utian, eds.). MTP Press, Lancaster, England.

38. O'Brian, P. M., Selby, C., and Symonds, E. M. (1981) Progesterone, fluids and electrolytes in premenstrual syndrome. *Br. Med. J.* **1,** 1161–1166.

39. Taylor, J. W. (1979). Plasma progesterone, oestradiol-17-beta and premenstrual symptoms. *Acta Psychiatr. Scand.* **60,** 76–86.

40. Hammarbäck, S., Damber, J. E., and Backström, T. (1989). Relationship between symptom severity and hormone changes in women with premenstrual syndrome. *J. Clin. Endocrinol. Metab.* **68,** 125–129.

41. Andersch, B., and Abrahamson, L. (1979). Hormone profile in premenstrual tension: Effects of bromocriptine and diuretics. *Clin. Endocrinol. (Oxford)* **11,** 657–662.

42. MacDonald, P. C., Dombroski, R. A., and Casey, M. L. (1991). Recurrent secretion of progesterone in large amounts: An endocrine/metabolic disorder unique to young women? *Endocr. Rev.* **12,** 372–401.

43. Mortola, J. F., Laughlin, G., and Lepine, L. (1992). Comparison of daily ovarian steroid metabolites in premenstrual syndrome and asymptomatic women. *39th Annu. Meet. Soc. Gynecol. Invest.,* San Antonio, TX.

44. Mortola, J. F., Beck, L., Girton, L., and Fischer, U. (1989). The temporal delay between ovarian steroid fluctuations and symptoms in premenstrual syndrome. *Proc. 71st Annu. Meet. Endocr. Soci.,* Washington, DC.

45. Rapkin, A., Morgan, M., Brann, D., Wiedneier, V., Somone, D., and Mahesh, V. (1997). Anxiolytic steroid levels in premenstrual syndrome. *Abstr. 44th Annu. Meet. Soc. Gynecol. Invest.,* San Diego, CA.

46. Mortola, J. F., Girton, L., and Fischer, U. (1991). Successful treatment of severe premenstrual syndrome by combined use of gonadotropin-releasing hormone agonist and estrogen/progestin. *J. Clin. Endocrinol. Metab.* **72,** 252A–252F.

47. Wei, E., and Loh, H. (1976). Physical dependence on opiate-like peptides. *Science* **193,** 1262–1264.

48. Chuong, C. J., Coulam, C. B., Koo, P. C., Gergstralh, E. J., and Go, V. L. W. (1985). Neuropeptide levels in premenstrual syndrome. *Fertil. Steril.* **44,** 760–766.

49. Cedarbaum, J. M., and Aghajanian, G. K. (1976). Noradrenergic neurons of the locus coeruleus: Inhibition by epinephrine and activation by the alpha-antagonist piperoxane. *Brain Res.* **112,** 413–419.

50. Young, W. S., and Kuhar, M. J. (1980). Noradrenergic receptors: Light microscopic autoradiographic localization. *Proc. Natl. Acad. Sci. U.S.A.* **77,** 1696–1700.

51. Young, S. W., and Kuhar, M. J. (1979). A new method for receptor and autoradiography: [³H] opioid receptors in rat brain. *J. Brain Res.* **179,** 255–270.

52. Gold, M. S., Redmond, D. E., Jr., and Kleber, H. D. (1979). Noradrenergic hyperactivity in opiate withdrawal supported by clonidine reversal of opiate withdrawal. *Am. J. Psychiatry* **136,** 100–102.

53. Morley, J. E. (1980). The neuroendocrine control of appetite: The role of endogenous opiates, cholecystokinin, TRH, gamma aminobutyric-acid and the diazepam receptor. *Life Sci.* **27,** 355–361.

54. Hamburg, M., and Tallman, J. F. (1981). Chronic morphine administration increases the apparent number of α_2-adrenergic receptors in rat brain. *Nature (London)* **291,** 493–496.

55. Schwartz, J. C., Pollard, H., Llorens, C., Malfroy, G., Gros, C., Pradelles, P. H., and Dray, F. (1978). Endorphins and endorphin receptors in striatum: Relationships with dopaminergic neurons. *Adv. Biochem. Psychopharmacol.* **19,** 245–257.

56. Tallman, J. F., Thomas, J. W., and Gallager, D. W. (1978). GABAergic modulation of benzodiazepine binding site sensitivity. *Nature (London)* **274,** 283–287.

57. Gallagher, D. W. (1978). Benzodiazepines: Potentiation of a GABA inhibitory response in the dorsal raphe nucleus. *Eur. J. Pharmacol.* **49,** 133–139.

58. Majewska, M. D., Harrison, N. L., Schwartz, R. D., Barker, J. L., and Paul, S. M. (1982). Steroid hormone metabolites are barbiturate like modulators of the GABA receptor. *Science* **232,** 1004–1007.

59. Lasaga, M., Duvilanski, B. H., Seilicovich, A., Afiones, S., and Dubeljuk, L. (1988). Effect of sex steroids on GABA receptors in the rat hypothalamus and anterior pituitary gland. *Eur. J. Pharmacol.* **155,** 163–167.

60. Hackman, E., Wirz-Justice, A., and Lichtensteiner, M. (1973). Uptake of dopamine and serotonin in the rat brain due to progesterone decline. *Psychopharmacologia.* **4,** 183–190.

61. Ladisich, W. (1977). Influence of progesterone on serotonin metabolism: A possible causal factor for mood changes. *Psychoneuroendocrinology* **2**, 257–262.

62. Biegon, A., Bercovitz, H., and Samuel, D. (1980). Serotonin receptor concentration during the estrous cycle of the rat. *Brain Res.* **187**, 221–225.

63. Wirz-Justice, A., and Chappius Arntt, E. (1976). Sex differences in chlorimipramine inhibition of serotonin uptake in human platelets. *Eur. J. Pharmacol.* **40**, 21–32.

64. Taylor, D., Matthew, R. J., and Ho, B. T. (1982). Serotonin levels and platelet uptake during premenstrual tension. Serotonin in biological psychiatry. *15th Annu. Symp. Tex. Res. Inst. Ment. Sci.*, Hamilton.

65. Ashby, C. R., Carr, L. A., Cook, C. L., Steptoe, M. M., and Franks, D. D. (1988). Alteration of platelet serotinin mechanisms and monoamine oxidase activity in premenstrual syndrome. *Biol. Psychiatry* **24**, 225–231.

66. Rapkin, A. J., Edulmuth, E., Chang, L. C., and Goldman, L. M. (1987). Whole blood serotonin in premenstrual syndrome. *Obstet. Gynecol.* **70**, 533–539.

67. Leudo de Tejado, A., Carrero, Q. F. B., Estela, C. A., and Contera, P. (1978). Climination urina de acido 5-hidroxiindolico durante el cicle menstrual humano [Urinary excretion of 5-hydroxyindolacetic acid during the human menstrual cycle]. *Cenicol. Obstet. Mex.* **44**, 85–81.

68. Samanin, R., Caccia, S., Benoulti, C., Borsini, F., Borroni, E., Invernizzi, R., Pataccini, R., and Monnini, T. (1980). Further studies on the mechanism of serotonin-dependent anorexia in rats. *Psychopharmacology* **68**, 99–104.

69. Hoebel, B. G., and Liebowitz, S. F. (1981). Brain monoamines in the modulation of self-stimulation, feeding and body weight. *In* "Brain, Behavior and Bodily Disease" (H. Weiner, M. F. Hofer, and A. J. Stunrand, eds.), pp. 263–268. Raven Press, New York.

70. Halmi, K. A., Eckert, E., La Du, T. J., and Cohen, J. (1986). Anorexia nervosa: Treatment efficacy of cyproheptadine and amitryptiline. *Arch. Gen. Psychiatry* **43**, 177–184.

71. Vetulani, J., Sansone, M., Bednarczyk, B., and Hano, J. (1982). Different effects of 3-chlorophenylpiperazine on locomotor activity and acquisition of conditioned avoidance response in different strains of mice. *Naunyn Schmiedeberg's Arch. Pharmacol.* **319**, 271–274.

72. Aloi, J. A., Insel, T. R., Mueller, E. A., and Murphy, D. L. (1984). Neuroendocrine and behavioral effects of m-chlorophenylpiperazine administration in rhesus monkeys. *Life Sci.* **34**, 1325–1331.

73. Asberg, M., Bertilsson, L., Tuck, D., Cronholm, B., and Sjöquist, F. (1973). Indolamine metabolites in the cerebrospinal fluid of depressed patients before and during treatment with nortriptyline. *Clin. Pharmacol. Ther.* **14**, 277–286.

74. Steward, A. (1991). Vitamin B6 in the treatment of the premenstrual syndrome: Review. *Br. J. Obstet. Gynaecol.* **98**, 329–334.

75. Facchinetti, F., Borella, P., Sances, G., Fioroni, L., Nappi, R. E., and Genazzani, A. R. (1991). Oral magnesium successfully relieves premenstrual mood changes. *Obstet. Gynecol.* **78**, 177–181.

76. Sherwood, R. A., Rocks, B. F., Steward, A., and Saxton, R. S. (1986). Magnesium and the premenstrual syndrome. *Ann. Clin. Biochem.* **23**, 667–670.

77. Facchinetti, F., Borella, P., Valentini, M., Fioroni, L., and Genazzani, A. R. (1988). Premenstrual increase of intracellular magnesium levels in women with ovulatory, asymptomatic menstrual cycles. *Gynecol. Endocrinol.* **2**, 249–256.

78. Van den Vange, N., Klogsterbuer, H. J., and Haspels, A. A. (1987). Effect of seven low-dose combined oral contraceptive preparations on carbohydrate metabolism. *Am. J. Obstet. Gynecol.* **156**, 918–923.

79. Kleijnen, J., Ter Riet, G., and Knipschild, P. (1991). Vitamin B6 in the treatment of the premenstrual syndrome: Review reply. *Br. J. Obstet. Gynaecol.* **98**, 329–330.

80. Rossignol, A. M., and Bonnlander, H. (1991). Prevalence and severity of the premenstrual syndrome: Effects of foods and beverages that are sweet or high in sugar content. *J. Reprod. Med.* **36**, 131–136.

81. Stone, A. B., Perlstein, T. B., and Brown, W. A. (1990). Fluoxetine in the treatment of premenstrual syndrome. *Psychopharmacol. Bull.* **26**, 331–335.

82. Wood, S. H., Mortola, J. F., Chen, Y. F., Moosazadeh, F., and Yen, S. S. C. (1992). Treatment of premenstrual syndrome with fluoxetine: A double blind placebo controlled crossover study. *Obstet. Gynecol.* **80**, 339–344.

83. Mortola, J. F., and Moossazadeh, F. (1991). A randomized trial of fluoxetine and imipramine in the treatment of premenstrual syndrome. *Proc. 38th Annu. Meet. Soc. Gynecol. Invest.*, San Diego, CA.

84. Smith, S., Rinehart, J. S., Ruddick, V. E., and Schiff, I. (1987). Treatment of premenstrual syndrome with alprazolam: Results of a double-blind, placebo-controlled, randomized crossover clinical trial. *Obstet. Gynecol.* **70**, 37–43.

85. Harrison, W. H., Endicott, J., Rabkin, J. G., Nee, J. C., and Sandberg, D. (1987). Treatment of premenstrual dysphoria with alprazolam and placebo. *Psychopharmacol. Bull.* **23**, 150–153.

86. Hazum, E., Cuatrecasas, P., Marion, J., and Conn, P. M. (1980). Receptor-mediated internalization of gonadotropin releasing hormone by pituitary gonadotrophes. *Proc. Natl. Acad. Sci. U.S.A.* **77**, 6692–6673.

87. Morel, G., Dihl, F., Aubert, M. L., and Dubois, P. M. (1987). Binding and internalization of native gonadoliberin (GnRH) by anterior pituitary gonadotrophs of the rat. A quantitative autoradiographic study after cryoultramicrotomy. *Cell Tissue Res.* **248**, 541–550.

88. Muse, K. N., Cetel, N. S., Futterman, L. A., and Yen, S. S. C. (1984). The premenstrual syndrome: Effects of medical ovariectomy. *N. Engl. J. Med.* **311**, 1345–1349.

89. Hammarback, S., and Backström, T. (1988). Induced anovulation as treatment of premenstrual tension syndrome. A double-blind cross-over study with GnRH-agonist versus placebo. *Acta Obstet. Gynecol. Scand.* **67**, 159–166.

90. Bancroft, J., Boyle, H., Warner, P., and Fraser, H. M. (1987). The use of an LHRH antagonist, buserelin, in the long-term management of premenstrual syndromes. *Clin Endocrinol. (Oxford)* **27**, 171–182.

91. Johansen, J. S., Riis, B. J., Hassager, C., Moen, M., Jacobsen, J., and Christiansen, C. (1988). The effect of a gonadotropin-releasing hormone agonist analog (nafarelin) on bone metabolism. *J. Clin. Endocrinol. Metab.* **67**, 701–706.

92. Stampfer, M. J., Willett, W. C., Colditz, G. A., Rosner, B., Speizer, F. E., and Hennekens, C. H. (1985). A prospective study of postmenopausal estrogen therapy and coronary heat disease. *N. Engl. J. Med.* **313**, 1044–1049.

93. Casson, P., Hahn, P. M., Van Vugt, D. A., and Reid, R. L. (1990). Lasting response to ovariectomy in severe intractable premenstrual syndrome. *Am. J. Obstet. Gynecol.* **162**, 99–105.

94. Carlton, G. J., and Burnett, J. W. (1984). Danazol and migraine. *N. Engl. J. Med.* **310**, 721–727.

95. Dmowski, W. P. (1979). Endocrine properties and clinical application of danazol. *Fertil. Steril.* **31**, 237–244.

96. Warole, P. G., Whitehead, M. I., and Mills, R. P. (1983). Nonreversible and wide ranging vocal changes after treatment with danazol. *Br. Med. J.* **287**, 946–951.

97. Abraham, G. E., and Hargrove, J. T. (1980). Effect of vitamin B6 on premenstrual symptomatology in women with premenstrual

tension syndromes: A double blind crossover trial. *Infertility* **3**, 155–161.

98. Kleijnen, J., Ter Riet, G., and Knipschild, P. (1990). Vitamin B6 in the treatment of the premenstrual syndrome—A review. *Br. J. Obstet. Gynaecol.* **97**, 847–852.

99. Schaumberg, H., Kaplan, J., Windebank, A., Vick, N., Rasmus, J., Pleasure, D., and Brown, M. J. (1983). Sensory neuropathy from pyridoxine abuse. *N. Engl. J. Med.* **309**, 445–452.

100. Parry, G. J., and Bredesen, D. E. (1985). Sensory neuropathy with low dose pyridoxine. *Neurology* **35**, 1466–1468.

101. O'Brien, P. M., Craven, D., Selby, C., and Symonds, E. M. (1979). Treatment of premenstrual syndrome by spironolactone. *Br. J. Obstet. Gynaecol.* **86**, 142–149.

102. Forrest, A. R. W. (1979). Cyclical variations in mood in normal women taking oral contraceptives. *Br. Med. J.* **1**, 1403–1410.

103. Cullberg, J. (1972). Mood changes and menstrual symptoms with different gestagen/estrogen combinations. A double blind comparison with placebo. *Acta Psychiatr. Scand.* **236**(Suppl.), 1–15.

104. Greene, R., and Dalton, K. (1953). The premenstrual syndrome. *Br. Med. J.* **1**, 1007–1011.

105. Dennerstein, L., Spencer-Gardner, C., Gotts, G., Brown, J. B., Smith, M. A., and Burrows, G. D. (1985) Progesterone and the premenstrual syndrome: A double-blind crossover trial. *Br. Med. J.* **290**, 1617–1621.

106. Rapkin, A., Chang, L. C., and Reading, A. E. (1987). Premenstrual syndrome: A double blind placebo controlled study of treatment with progesterone vaginal suppositories. *J. Obstet. Gynecol.* **7**, 217–222.

11

Women's Fecundability and Factors Affecting It

DONNA DAY BAIRD* AND BEVERLY I. STRASSMANN†

*Epidemiology Branch, National Institute of Environmental Health Sciences, Research Triangle Park, North Carolina; †Department
of Anthropology, University of Michigan, Ann Arbor, Michigan

I. Introduction

World population first exceeded a billion people in the early 1800s and it took approximately a century for the next billion increase. In 1999, our population exceeds six billion. Another billion increase is projected in just over a decade [1]. The rapid increase results from high fertility populations where average numbers of live births typically range from five to eight [2]. Other parts of the world have undergone demographic transition in association with industrialization and hover at or below replacement reproductive rates. The transition from high fertility to low fertility is influenced by complex social changes thought to be unrelated to basic biological capacity to reproduce.

Reproduction is a relatively rare event for women in industrialized countries. Only 6.5% of US women of reproductive age (15–44) gave birth in 1995 [3], and one out of six women aged 40–45 have never had a child [4]. Although most girls grow up assuming that they will be able to have children when and if they choose to do so, an estimated 10–15% of live births require more than a year to conceive [5], suggesting that these couples may be experiencing some fertility problems.

This chapter focuses on variability in biological capacity to reproduce. How variable are different populations? What accounts for variability among women within a population? Do women with abundant food have greater capacity for reproduction?

Terminology for describing fertility and fertility problems is not uniform across disciplines. We will follow Leridon [6]. *Fertility* refers to number of live births, a focus of demographic research. *Fecundity* denotes the biological capacity to reproduce, a focus of medical research. Fecundity is inherently difficult to measure; it requires successful interaction of several complex biological processes. Women may be fecund but choose to contracept and not demonstrate fertility. Conversely, they can be fertile despite impaired fecundity by utilizing specialized infertility treatments such as *in vitro* fertilization. *Fecundability*, the probability of conceiving in a given time interval, provides a measurement tool for the study of fecundity. It usually is measured as the probability of conceiving in any given menstrual cycle (or month) among couples who are sexually active and doing nothing to prevent pregnancy. The probability of conceiving is a function of the fecundity of the male and female partners but also varies with frequency and timing of sexual intercourse. As for any probability, it cannot be assessed for an individual couple but must be estimated for a group. If human conceptions could be identified at time of fertilization, we could measure *total fecundability*. Instead, most data provide estimates of *effective fecundability* (the probability of conceiving a pregnancy that survives to birth) or *apparent fecundability* (the probability of having a clinically recognizable conception).

This chapter explores the variability among couples in their ability to conceive, as measured by fecundability. Unless specifically stated otherwise, fecundability will refer to apparent fecundability, the probability of conceiving a clinically recognized pregnancy in any given menstrual cycle (or month). The broader questions about social and economic determinants of fertility and family size are beyond our scope. We start by summarizing fecundability estimates from both contracepting and noncontracepting populations. We then consider the major biological processes required for successful pregnancy and begin to quantify how failure of these processes contributes to reducing fecundability. The largest section summarizes research on factors affecting women's fecundability. Finally, we propose directions for future research.

II. Estimates of Fecundability

The majority of estimates of fecundability come from natural fertility populations (populations in which contraception is not used to limit family size). Today, natural fertility is most likely to be found among rural populations of developing countries and among conservative religious sects such as the Hutterites and Amish of North America [7,8]. There are possible theoretical as well as practical advantages to studying natural fertility populations. Natural fertility is thought to be the reproductive pattern of the vast majority of our evolutionary past, so natural fertility populations may be particularly suitable for exploring the evolved mechanisms that underlie differences in female fecundability [9,10]. Practically, the lack of contraceptive use can simplify data collection. Waiting-time data for calculating fecundability estimates can be conveniently collected for first birth intervals (time from entry into a sexual union, *e.g.*, marriage, to the date of first conception, imputed from date of live birth). When sexual union begins at marriage, existing marriage and birth records can be used to estimate fecundability retrospectively.

Wood *et al.*'s study of the Pennsylvania Amish provides one of the best examples of first birth interval studies [11]. They used the carefully-kept marriage and birth records of the Amish community to establish waiting times. Nearly all women married before age 30, so a fecundability estimate was calculated for women aged 18–29. The estimated effective fecundability for the study sample of 271 women was 0.25 (the probability of becoming pregnant in any given menstrual cycle was 25%). Similar methods were used in Taiwanese and Sri Lankan samples where effective fecundability varied from 0.16 to 0.30 for 25–29 year olds [11]. Prior estimates of fecundability (based on clinically recognized conceptions) in the first birth interval were summarized by Wood [12]. Populations were from the US, Taiwan, Peru, Brazil,

Mexico, France (including historical data), Tunis, and Quebec. Fecundability estimates ranged from 0.14 to 0.31, with corresponding conception waits of 10 months to 5.2 months.

Estimates of fecundability at the time of marriage often are limited to young women and tend to be elevated by high coital frequencies commonly seen with the onset of marriage [13,14]. Few studies have provided fecundability estimates for women across the reproductive life span because they require accurate information on the length of lactational amenorrhea. Studies in contracepting populations require added information on birth control usage. These concerns can be addressed best with prospective studies in which individual women are followed to collect accurate waiting-time data.

John *et al.* [15,16] conducted the first prospective study of fecundability in a natural fertility population, the rural Bangladesh of Matlab Thana. Family planning in this population was minimal. Women were sought for interviews once a month. Nonetheless, absences of the women from home on their interview days resulted in gaps of two to four months in the records. A sample of 403 married women aged 14–49 participated. Fecundability was 0.19 for nonbreastfeeding women and less than 0.07 for breastfeeding women. However, even with the prospectively-collected data, concern about accuracy of postpartum amenorrhea information led Leridon [14] to question these fecundability estimates.

Strassmann and Warner [10] studied the waiting time to conception in a Dogon village of 460 people in Mali, West Africa. The total fertility rate of the postreproductive women in this village was 8.6 births, and none of the women in any cohort reported that they had ever used contraception. During menses Dogon women spend five nights sleeping at a menstrual hut, which made it possible to monitor female reproductive status prospectively without interviews. By censusing the women present at the menstrual huts in the study village every night for two years, Strassmann and Warner were able to prospectively monitor the time from a woman's first postpartum menstruation to the onset of her last menstruation before a subsequent pregnancy. Urinary steroid hormone profiles for 93 women in two villages showed that, over a 10 week period, women in the principal study village went to the menstrual huts during 87.5% of all menses and did not go to the huts at other times [17]. Thus, menstrual hut visitation provided a reasonably reliable indication of menstruation.

The Dogon sample included 50 women aged 15–41 with prospectively observed conception waits. Fecundability was estimated at 0.11 with covariates assigned mean values for the population (covariates included age, time since marriage, gravidity, and lactation). This fecundability estimate corresponds to a conception wait of 8.3 months.

In contracepting populations prospective studies that can enroll women at the time they stop using contraception in order to conceive provide the most accurate waiting-time data. Though such studies have been done, none were done primarily to measure fecundability and none present fecundability estimates based on statistical modelling of the entire distribution of waiting times. However, first cycle conception rates provide a good estimate of mean fecundability for a population (described in Leridon [6]), and these data have been published. We describe three prospective studies that reported these data.

Tietze [18] reported data on 611 US women who had their IUDs removed in order to become pregnant. Ages ranged from 17 to 42, with median age of about 25. The apparent fecundability (estimated by first cycle conception rates for clinically recognized pregnancies) was 0.33. Most of the women had been pregnant before, so this estimate represents fecundability for couples of proven fertility. If women with no prior pregnancies had been included, the estimate of mean fecundability probably would have been lower.

The second study, the North Carolina Early Pregnancy Study, enrolled 221 volunteers at the time they began trying to conceive [19]. Follow-up continued for six months or through the eighth gestational week for those conceiving during the study. Women ranged in age from 21 to 42, with 80% between the ages of 26 and 35; 35% were nulligravid. The apparent fecundability was 0.24 (estimated by first cycle conception rates for clinically recognized pregnancies). This is probably a low estimate because some women stopped contraception well into their first cycle, so the opportunity to become pregnant during that first cycle was reduced for these women. On the other hand, women with known fertility problems were excluded from participation, so we would expect this factor to bias the estimate upward.

The third study [20], conducted in Denmark, enrolled 411 couples at the time they began trying to conceive a first pregnancy. Couples were recruited from four trade unions: metal workers, office workers, nurses, and daycare workers. Most participants (92%) were in their twenties. Couples were followed until clinically pregnant or through six menstrual cycles, whichever came first. Fecundability as estimated by the first cycle conception rate for clinical pregnancies was 0.16.

The fecundability estimates presented here are highly variable (0.11 to 0.33), and variation exists both among the contracepting samples and among the natural fertility samples (Table 11.1). Some of this variation may have methodologic explanations. In natural fertility populations, accurate estimates depend on ascertainment of when a couple begins having sexual intercourse and when lactational amenorrhea ends. Reporting accuracy may vary with study design as well as with education and other characteristics of participants.

In contracepting populations, waiting times can be measured accurately in prospective studies of women stopping contraception in order to conceive, but these studies are based solely on women planning a pregnancy. This is a select group. Some women never attempt to conceive because they do not want children. Other women become pregnant even though they are not intending to. In the United States about half of pregnancies are unplanned [21]. Though nearly half of these unplanned pregnancies were to women not using contraception, the others were conceived during months when birth control had been used. This latter group of women might be expected to have high mean fecundability because they became pregnant while using some form of birth control around the time they conceived. Therefore, fecundability estimates based only on women trying to conceive are likely to be lower than true fecundability in the population. Given the methodologic issues and the limited number of studies, the degree to which populations differ in their true fecundability is not known.

Fecundability varies within populations as well. Regardless of the mean fecundability for the population and whether it is a

Table 11.1
Estimates of Fecundability in Selected Populations

References	Population	Design	N	Estimate	Estimation method
[11]	Amish, US natural fertility	Retrospective	271	0.25[a]	Model distribution of time from marriage to imputed conception
[11]	Taiwan, 1967 natural fertility	Retrospective	445	0.21[a] 25–29 year olds	Model distribution of time from marriage to imputed conception
[11]	Taiwan, 1973 natural fertility	Retrospective	471	0.30[a] 25–29 year olds	Model distribution of time from marriage to imputed conception
[11]	Sri Lanka, 1987 natural fertility	Retrospective	655	0.16[a] 25–29 year olds	Model distribution of time from marriage to imputed conception
[15,16]	Matlab Thana, Bangladesh natural fertility	Prospective	403	0.19[a] nonnursing	Model distribution of waiting times
[10]	Dogon, Mali natural fertility	Prospective	50	0.11[a]	Model distribution of waiting times
[18]	US women having IUDs removed to conceive	Prospective	611	0.33[b]	First cycle conception rate
[19]	North Carolina, US women stopping contraception to conceive	Prospective	211	0.24[b]	First cycle conception rate
[20]	Danish couples stopping contraception to attempt first pregnancy	Prospective	430	0.16[b]	First cycle conception rate

[a]Effective fecundability.
[b]Apparent fecundability.

natural fertility population or a contracepting population, nearly all studies show that some couples are more fecund than others. This heterogeneity is expressed in the gradual decrease in conception rates during successive months of trying. Using Tietze's data [18] as an example (Fig. 11.1), the conception rate in the first month of trying to conceive was 0.33, declining to 0.21 by

month three and to 0.12 by month nine. This pattern is produced because the most fecund couples become pregnant quickly. Those still trying in the latter months are those couples who tend to have low fecundability. Identifying and quantifying the impact of factors that can account for this variability in fecundability is an active area of current research. Age and frequency of intercourse are obvious examples, but many other factors may also play a role.

III. Biological Processes Reflected in Measures of Fecundability

Figure 11.2 illustrates the reproductive failure or loss inherently measured in apparent and effective fecundability, using midrange estimates from the literature [21a]. For 100 women beginning the first menstrual cycle in which they attempt to conceive, 20–25% will start a viable pregnancy. The other 75 to 80 cycles represent some form of reproductive failure. The vast majority of women aged 25–39 who are not using hormonal contraceptives ovulate each cycle [22,23], so anovulation cannot account for much of this early reproductive failure. The rate of fertilization is unknown, but *in vitro* fertilization rates are high [24]. If sexual intercourse is timely, *in vivo* fertilization rates may also be high. There may be considerable loss prior to implantation, but no data are available.

Very early pregnancy testing can be used to estimate pregnancy loss after implantation, before normal clinical detection of pregnancy. Highly sensitive tests for the pregnancy hormone, human chorionic gonadotropin (hCG), can detect pregnancies around the time of implantation using daily first morning urine

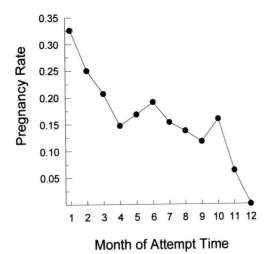

Fig. 11.1 Decline in fecundability seen in a population over months of attempt time. As the more fecund couples conceive and drop out of the pool of waiting couples, the sample becomes more and more selected for low fecundability couples. Data are from 611 women having IUDs removed in order to conceive [18].

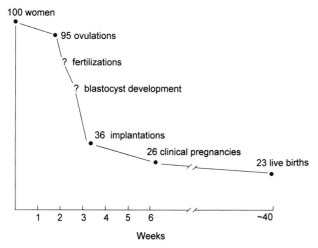

Fig. 11.2 Time line of reproductive failure. Of 100 sexually active women who start a noncontracepting menstrual cycle, approximately 95 will ovulate. An unknown number will have eggs fertilized and blastocysts develop. About 36 women will show urinary hCG evidence of pregnancy at around the time of implantation, but only about 26 of these pregnancies will survive long enough to be clinically detectable at about six weeks after the start of the last menstrual period. Compared to the earlier weeks, relatively little loss occurs after clinical detection. See text for references (figure adapted from Figure 1 in [21a]).

specimens from women trying to conceive. Early pregnancy losses can be identified by a rise and fall of hCG. The North Carolina Early Pregnancy Study, which used a very sensitive hCG assay, defined chemically detected pregnancies by hCG of 0.025 ng/ml or greater for at least 3 consecutive days, levels that exceeded those found in a group of women with tubal ligations [19]. In that study, 22% of chemically detected pregnancies were unrecognized clinically and another 2% were recognized clinically but lost within the same time period (*i.e.,* within six weeks of the start of the last menstrual period) [25]. Thus, very early postimplantation pregnancy loss was estimated to be 24%.

Two more studies tend to corroborate this estimate of early loss. Both followed somewhat similar protocols but used less sensitive assays for hCG (the North Carolina study used a polyclonal antibody to detect the intact molecule [26], and the antibody is no longer available). Bonde *et al.* [20] identified pregnancy by hCG $\geq$ 1 IU/L (0.076 ng/ml) in the first 10 days of the menstrual cycle following the conception cycle and reported a loss rate of 17%. Zinaman *et al.* [27] identified pregnancy by 3 consecutive days of hCG > 0.15 ng/ml and reported a loss rate of 13%. If these two sets of criteria for detecting pregnancy are applied to the North Carolina data, the loss rates in that study would be 17% and 14%, respectively, suggesting that early loss rates in the three studies are remarkably similar.

Projecting these data onto the timeline of loss in Figure 11.2, we estimate that of the 100 women entering an attempt cycle, only 36 would have chemically detectable implantations. The combination of failure to fertilize, preimplantation loss, and loss at the time of implantation accounts for the majority of reproductive loss. By comparison, loss after clinical recognition is relatively small. Only about 3% of the original 100 cycles re-

sults in a clinically identified pregnancy that is lost spontaneously. This 3% represents a 10–15% spontaneous abortion rate [28,29] and a stillbirth rate (loss after 26 weeks) of less than 2% among clinical pregnancies [30].

Reproductive loss at the different stages involves alterations in diverse biological processes. Failure to ovulate can arise from impaired gamete production/maturation or from hormonal imbalance. Fertilization requires timely oocyte release, sexual intercourse, and transport of gametes through the female reproductive tract. Survival to clinical recognition requires successful blastocyst formation, uterine receptivity, and implantation. Failure in any of the biological processes will result in an apparent nonconceptive menstrual cycle. Factors that interfere with any of these biological processes will reduce fecundability.

IV. Factors Affecting Fecundability

Factors affecting fecundability may operate through any of the biological pathways necessary for successful development of a conceptus. Frequency and timing of sexual intercourse is obviously important. Though fecundability clearly depends on effects on both male and female partners, we focus the remainder of the section on factors affecting female fecundity. Many factors have been identified, but the mechanisms by which they influence fecundability may not be known. Such factors include age, reproductive tract infections, previous methods of birth control, health-related behaviors such as cigarette smoking and exercise, and occupational exposures. Past exposures with long-term adverse effects may be as important as current conditions. Even a woman's exposures during her prenatal development could influence future fecundability by affecting her reproductive tract development and lifetime supply of ova.

A. Frequency and Timing of Sexual Intercourse

Sexual intercourse must occur close enough to the time when the egg is released from the ovary for fertilization to occur. Two studies provide detailed data on the length of the fertile window and day-specific estimates of conception probabilities relative to the day of ovulation. A third large study is in progress with preliminary results published. The first study enrolled 241 couples who used natural family planning [31]. The data were reanalyzed by Schwartz *et al.* [32] who reported a 7-day fertile window ending on day of basal body temperature rise, with highest probabilities of conception on the middle three days. A 6-day fertile window ending on estimated day of ovulation was reported by Wilcox *et al.* [33]. Pregnancies were identified by assay of daily urine specimens and included pregnancies that were lost very early. Conception probabilities for the last three days of the 6-day window were about twice those of the earlier three days. However, the conceptions occurring on the day of ovulation were at high risk of very early loss [34]. The probability of conceiving a surviving pregnancy was highest on the two days before ovulation (Fig. 11.3). This pattern is consistent with the earlier study [32] and preliminary data from the ongoing European Multicenter Study [35].

Couples who have frequent intercourse will be more likely to have sex on days when the probability of conception is high, and these couples should have higher fecundability. This is

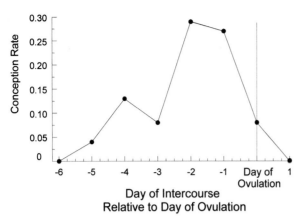

Fig. 11.3 The probability of clinical pregnancy following intercourse on a given day relative to ovulation. The estimated probability of conception is 0 outside of the 6-day window ending on the day of ovulation. Data are from the North Carolina Early Pregnancy Study (figure adapted from Figure 1 of [34]).

Table 11.2

Effective Fecundability by Age among the Dogon of Mali

Age	Fecundability
15	0.03
20	0.10
25	0.18
30	0.18
35	0.11
40	0.03

Note. N = 50 [10].

supported by data from natural fertility populations that show higher fecundability associated with recent marriage, an indirect measure of frequency of intercourse [10]. Higher fecundability also has been associated with frequent intercourse in studies based on interview data from women in the US [36,37]. Concern has been raised about possible adverse effects of too frequent intercourse (short abstinence time). Though a short abstinence time is associated with lower sperm count and sperm concentration [37a], even daily intercourse had no measureable adverse impact on fecundability in a study of couples with no known fertility problems [33,37b].

For couples having difficulty conceiving, purposeful timing of intercourse could be even more effective than increasing frequency, but some methods for timing intercourse are not optimal [38]. The basal body temperature shift comes too late. Urinary luteinizing hormone (LH) provides a signal close to ovulation [39], but gives no information about earlier fertile days. Couples using LH kits who wait for the urinary LH signal often will miss the two most fertile days. However, earlier days can be identified by cervical mucus characteristics [40]. Couples who have intercourse frequently after cervical mucus becomes receptive will tend to have intercourse on those days with the highest probabilities of conception.

B. Age

When age-specific fecundability data are available, women in their twenties usually show the highest fecundability. Table 11.2 shows this pattern in data from the Dogon. Effective fecundability had an inverse U-shaped relationship with the woman's age, peaking at 0.19 for ages 26–29 years [10]. The lower fecundability of adolescents is consistent with known menstrual cycle irregularities (including anovulation) in the years immediately following menarche [22,23]. The drop-off in fecundability with later ages is consistent with the decline in coital frequency usually found among older women [41]. However, women's fecundity also declines. A large French study of

women undergoing artificial insemination, whose husbands were sterile, found that the number of insemination cycles required for pregnancy increased with age [42]. Women in their twenties had the highest fecundability. It was somewhat lower for women in their early 30s, and those over 35 had a more marked reduction in fecundability. A similar study in the Netherlands has reported the same pattern [43].

The biological changes that underlie reduced fecundability in older women are not well understood. Several factors probably operate together. Ellison [44] summarizes three hormonal studies showing reduced levels of salivary progesterone in older cycling women. Others have considered the relative importance of egg quality and uterine receptivity [45]. Egg quality declines in older women, as demonstrated by higher pregnancy rates in older women having oocyte donation compared with other older women. However, the uterus also is implicated because among donor egg recipients, older women have lower pregnancy rates than younger women [45a]. Further research is needed to investigate other aspects of the reproductive process in older women. For example, do older women have fewer preovulatory days of fertile cervical mucus?

C. Lactation

Lactation suppresses ovulation and results in amenorrhea. Even after menses has resumed, breastfeeding appears to reduce fecundability. For example, in Assam, India, fecundability was 41% lower among cycling women who were breastfeeding compared to nonbreastfeeding women [46], and a strong effect also was seen in the Matlab of Bangladesh [15,16]. However, reproductive hormones are reported to return to normal cyclicity in the first couple of menstrual cycles after lactational amenorrhea [47,48], and fecundability also has been reported to rise rapidly during those first few menstrual cycles [47]. Thus, if women continue to breastfeed long after resumption of menses, this prolonged lactation may not reduce fecundability.

Leridon [14] points out how challenging it is to evaluate fecundability in postpartum women who nurse. Collecting sufficiently detailed time-dependent data on resumption of menses, nursing habits, and the inherently correlated factors of nutrition and health is very difficult. Small prospective studies in different ecologic settings that can collect detailed hormonal and nursing data along with information on sexual intercourse may reveal more about lactation and subfecundity than large, less-detailed fecundability studies.

D. Pelvic Infection and Other Medical Conditions

Pelvic infections are a common cause of reduced fecundability and sterility [49]. Pathogens can ascend the female reproductive tract after vaginal entry and cause infection and subsequent scarring of the oviducts. Though tubal damage is the most studied sequelae of pelvic infections, chronic inflammation of the cervix and uterus might also reduce fecundability by interfering with gamete transport and implantation.

The sexually transmitted organisms, *Chlamydia trachomatis* and *Neisseria gonorrhoea* are the two widely recognized pathogens that cause pelvic inflammation [49]. Both are found throughout the world and are probably responsible for dramatically reduced fertility in sub-Saharan Africa [50]. In the United States, asymptomatic infection is thought to be even more common than diagnosed infection, and adverse effects on fecundability are not limited to symptomatic cases [49,51]. Trichomoniasis also has been associated with reduced fecundability [52].

Sexually transmitted pathogens may not be the only important microbes. A history of Cesarean-section [53] or a ruptured appendix [54] has been associated with reduced fecundability, and spread of infection to oviducts is a potential mechanism. Bacterial vaginosis and mycoplasmas also have been associated with pelvic disease [55,56]. It is not clear whether bacterial vaginosis directly reduces fecundability or predisposes the reproductive tract to infection by sexually transmitted organisms [57].

Other factors also may influence fecundability by affecting susceptibility to infection. The adverse effects on fertility associated with prior use of IUDs, especially the Dalkon Shield, is attributed to increased risk of pelvic infection [58,59]. On the other hand, the combination of barrier and chemical contraceptive methods (*e.g.,* diaphragm with spermicide) has been found to protect against tubal damage [60], presumably by reducing infection. Vaginal douching also may reduce fecundability by increasing susceptibility to infection [61]. Several studies show douching to be a risk factor for pelvic inflammatory disease (reviewed in Zhang *et al.* [62]), and supporting laboratory work shows that douching can reduce the prevalence of vaginal lactobacillus, a bacterium that protects against pathogens [63].

Several other medical conditions in women are associated with reduced fecundability, the most common being thyroid disease [64], endometriosis [65], polycystic ovary syndrome [66], and uterine fibroids [67]. Shared HLA-DR serotypes between male and female partners also may reduce fecundability [68].

E. Nutrition and Exercise

A woman's nutritional status and energy balance are expected to influence fecundability [69], but data are limited. Extreme cases of malnutrition, as in anorexia, result in anovulation and amenorrhea [70]. Similarly, intense physical training, as for ballet dancers or marathon runners, can result in anovulation and amenorrhea [71].

Effects on fecundability of more moderate nutrient deficits or exercise regimens are unclear, but menstrual cycle hormones are known to be affected. Bullen *et al.* [72] reported reduced luteal progesterone levels in untrained women after starting an exercise program. Salivary progesterone data from four separate studies (Boston, Poland, Zaire, and Nepal) showed similar patterns associated with energetic stress (reviewed by Ellison [44]). The Boston women were on a voluntary exercise program, the Polish women did seasonal agricultural work with no mechanized farm equipment, and the women in Zaire and Nepal belonged to subsistence populations with highly seasonal food resources. The latter two populations also were known to have seasonal changes in birth rates consistent with low fecundability in seasons of high energetic stress.

Despite hormonal evidence, actual links between hormones and fecundability in these populations have not been demonstrated. Higher luteal progesterone is associated with higher fecundability in some, but not all, of the few available studies, and these were all conducted in the US (summarized in Baird *et al.,* [73]).

John *et al.* initiated a study in Bangladesh designed to assess fecundability changes with chronic undernutrition [15,16]. They measured women's height and collected weight data monthly to determine changes in body mass. However, this measure of nutrition had no significant effect on fecundability. The nonsignificant results led the investigators to conclude that seasonality of conceptions in Bangladesh may have more to do with changes in coital frequency than to changes in nutrition. Physiologic studies in laboratory animals suggest that sexual behavior may be more sensitive than ovarian function to increased energetic demand [74].

Nutritional balance is more than just getting enough calories. Body fat may not be the best marker; specific nutrients may be more important. This is supported by findings from a clinical trial of folate supplementation designed to evaluate the vitamin's effects on birth defects. After randomization, the folate-supplemented group had significantly higher pregnancy rates [75]. These findings suggest the need for investigation of more nutritional markers when studying fecundability.

F. Obesity and Body Weight Distribution

Obese women tend to take longer to conceive than women of normal weight [76–78], probably because of endocrine imbalances (reviewed in Harlow and Ephross [79]). A history of obesity in adolescence is associated with reduced fecundity regardless of adult body mass [78,80]. The distribution of fat in the body also has been found to be associated with fecundability [81]. A 0.1 unit increase in waist-hip ratio (*e.g.,* the difference between a 26-inch waist and a 30-inch waist for women with 36-inch hips) was associated with a 30% decrease in fecundability. Women with higher waist-hip ratios have been reported to have low pH of the endocervical mucus and higher androgen levels [82].

To what extent elevated androgen levels and subclinical polycystic ovarian disease may be an underlying mechanism for obesity-related subfecundity is unknown. A 13-year follow-up study of adolescent girls showed that androgen levels tended to track over time (*i.e.,* higher adolescent levels were predictive of higher levels later in life) [83]. Furthermore, those with high androgen levels were less likely to have become pregnant during the 13-year follow-up, and this could not be explained by other factors such as differences in sexual behavior or use of oral contraceptives. Further longitudinal research on androgen levels, the relationship with fecundability, and early-life factors that might cause elevation in androgen levels is needed.

G. Oral Contraceptives and Other Medication

Oral contraceptives were introduced in the 1960s, and by 1970 Wolfers [84] had shown that pill users had a temporary reduction in fertility during the first few months after discontinuing pill use. The effect appears to be mediated by endocrine effects [85]. The short-term subfecundity was substantiated by Harlap and Baras [86] who found a 30% reduction in the first month after discontinuation. The temporary reduction is followed by cycles during which pill users appear to have higher fecundability than the other women still attempting pregnancy. The effect was more marked in older women who had used the pill for longer durations. Pill users did not show more long conception waits (>12 months). The absence of long-term adverse fecundity effects of pill use has been substantiated in the Nurses Health Study [87]. A recent study also found temporary subfecundity after stopping the pill [20], suggesting that this effect occurs even with the new low-dose medication.

Little research has been done to evaluate effects of other common medications. This is a difficult area of research because women who take a medication have some underlying condition that might account for any observed decrease in fecundability. A case-control study of self-reported prescription medication use and ovulatory infertility found associations with thyroid preparations, antidepressants, tranquilizers, and asthma medication [88], but no adjustments could be made for underlying disease.

Laboratory animal evidence suggests that analgesic medications such as aspirin, indomethacin, and the nonsteroidal anti-inflammatory drugs like ibuprofen may reduce fecundability, but human evidence is still very limited. These drugs block prostaglandin production by inhibiting the cyclooxygenase enzymes. The reproductive function of these enzymes has been studied in knockout mice that lack the gene for one or the other of the two cyclooxygenase enzymes, COX-I or COX-II. COX-II appears to be necessary for implantation and also for release of the egg from the ovary during ovulation [89]. Thus, the cyclooxygenase inhibiting drugs might be expected to result in reduced fecundability in women. Small experimental studies of egg release measured by ultrasound show retention of eggs in women given indomethacin [90,91]. Ibuprofen has not been studied, but may show similar adverse effects when taken around the time of ovulation.

H. Prenatal Factors

A woman's lifetime supply of eggs is produced during fetal life, peaking at around six months gestation [66]. Prenatal exposures might reduce fecundability during adulthood by limiting egg numbers or affecting egg quality. Prenatal exposures can also affect reproductive tract development. Diethylstilbestrol (DES) provides the best documented example in humans. Prenatally exposed women are at higher risk for many reproductive problems, including reduced fecundability [92].

Reduced fecundability associated with reduced oocyte numbers has been demonstrated in laboratory animals exposed prenatally to benzo(a)pyrene, a component of cigarette smoke [93]. In women, prenatal exposure to cigarette smoke has been linked to reduced fecundability in one study [94], but not in others [95,96].

Repositories of data (and sometimes serum samples) from past studies of large cohorts of pregnant women present resources for research on prenatal effects on fecundability (e.g., the California Child Health and Development Study and the Collaborative Perinatal Study). The daughters born to the participants are now adults whose fecundability can be studied [96]. Data on prenatal exposures are available from interviews conducted with their mothers during pregnancy.

I. Lifestyle Factors

Cigarette smoking was first linked with reduced fecundability in a Danish study in 1983 [97], and the relationship was investigated in several studies during the subsequent decade with most studies finding adverse effects (reviewed by Baird [98]). The association has been most convincingly confirmed in a multicenter study based on data from nine areas in Europe [99]. The researchers reported that nulliparous women smoking >10 cigarettes/day had significantly reduced fecundability. Results were fairly consistent across study populations, making it quite unlikely that the effect could be explained by bias.

Alcohol drinking also was linked to reduced fecundability in the 1983 Danish study [97]. Though very heavy drinking has clear adverse effects on fecundability [100,101], studies of moderate alcohol consumption have shown little consistency [102]. However, a report from the Danish prospective study of fecundability found adverse effects of even moderate alcohol consumption (1–5 drinks/week) [102a]. This study collected detailed data on alcoholic beverage consumption during each menstrual cycle of trying, thus providing more accurate data than were available in any of the previous studies. It may be that prior studies failed to find effects because of excessive misclassification of alcohol intake.

Caffeine and reduced fecundability was first reported in 1988 [103] for a subset of study participants who presumably provided the best data on caffeine intake. Results from further studies have been inconsistent, showing adverse, positive, and no measurable effect (reviewed in Dlugosz and Bracken [104] and Jensen et al. [105]). Furthermore, adverse caffeine effects are often found only in subgroups of study populations. For example, Olsen [106] found an effect only for smokers with very high caffeine intake, whereas Stanton and Gray [107] found an association only in nonsmokers. Results from the European Study of Infertility and Subfecundity, with data from five locations, show a small reduction in fecundability (10% reduction) only at very high caffeine levels, the effect being somewhat stronger in smokers [108]. A prospective study with cycle-specific caffeine data showed a weak (nonsignificant) decrease in fecundability only in nonsmokers [105]. The effect that is reported also is not consistent across various types of caffeinated beverages [109]. Because coffee drinking is related to other health-related behaviors like smoking and stress, the small effects seen in most studies could be explained by poorly measured associations with other factors. At this point, no clear conclusions can be drawn. An intervention study with detailed prospectively recorded caffeine data may be the only way to clarify the issue.

Psychological stress has been associated with infertility problems, but because psychological stress has been measured after fertility problems are recognized, it may well be a result, rather than a cause, of reduced fecundability [110]. One prospective study measured psychological stress before couples had ever tried to conceive [111]. They reported reduced fecundability in stressed women whose menstrual cycles were long, but not in the majority of stressed women, and suggested those with long cycles may represent women who were particularly susceptible to stress. An intervention study designed to evaluate the effects of a stress-reduction program for infertility patients is now being conducted in the United States. Only more prospective studies can provide meaningful information on this important issue.

J. Occupational Exposures

The discovery of reduced fertility and sterility in males exposed to the fumigant dibromochloropropane [112] focused attention on occupational effects on fecundability. The initial search was directed at male exposure and its effects on semen characteristics, but subsequent studies have included or even focused on occupational effects on females. The first was a Danish study that examined a broad range of occupations and chemical and physical exposures in both men and women [113]. In the 1990s came reports of adverse effects for women associated with nitrous oxide [37], video display terminals [114], mercury vapor [115], and pharmacists working with antibiotics [116] (see reviews in Gold and Tomich [117] and Baird *et al.* [118]). Other studies have implicated employment as a hairdresser [119], shift work [120,121], employment requiring high energy demand or long hours [122,123], pesticides [124,125], lead [126], and solvents [125,127–129].

However, other studies find little evidence for occupational exposures adversely affecting women's fecundability (*e.g.,* Spinelli *et al.* [130]), and results for specific exposures often show apparently differing results. The lack of consistent findings is not surprising. Unless a study is focused on a group of exposed women, the number of women with a given occupational exposure often will be too few to estimate effects. Also, exposures usually have been assessed by self report of past employment situations so that exposure levels may not be comparable across studies. Future studies will need to focus more on enrolling large numbers of exposed women and quantifying exposure more precisely.

V. Summary and Directions for Future Research

The description of fecundability began with demographers and has only recently been addressed by epidemiologists and anthropologists. The available data suggest that fecundability may vary widely among populations. Differences between natural fertility populations and contracepting populations appear to be much smaller than variability within each group. Some differences in fecundability probably reflect different study designs and methodology, but known determinants of fecundability also differ among populations. No study has systematically tried to compare fecundability estimates among populations by adjusting for differences in age, frequency of intercourse, and prevalence of *Chlamydia* and gonorrhea.

Methodology for anthropologic and epidemiologic investigation of fecundability is still being developed, and issues of validity and bias have been addressed [21a,131–134]. Prospective studies can provide accurate waiting-time data but involve select populations with limited generalizability. Retrospective studies also have their own methodologic challenges. Though research supports the general validity of recalled time-to-pregnancy data [135,136], studies based on recall data involve pregnancy attempts that began at various times in the past. Variation over time in family planning practices, desired family size, and prevalence of exposures can be difficult to adjust for in such studies.

Concern has been raised about whether fecundity has declined with the possible increase of reproductive toxicants in the environment [137,138]. However, fecundability in developing countries shows increases over time [139]. Given the current limitations in methodology, it is easier to study the effects on fecundability of specific exposures than to compare fecundability over time.

Perhaps the most efficient research strategy is to incorporate multiple approaches. Studies of fecundability in the general population (such as the European Infertility and Subfecundity Studies) can provide data on prevalent lifestyle factors like cigarette smoking and can provide insight into methodologic issues of measurement and selection bias. Any exposure that is only experienced by a small portion of the population (which includes most occupational exposures) must be studied in specially identified groups known to have high prevalence of exposure (*e.g.,* selection of microelectronics workers for study of glycol ethers).

Combining fecundability studies with laboratory research designed to elucidate possible biological mechanisms can be very helpful. For example, data showing adverse effects of vaginal douching on fecundability are very limited, but laboratory data showing that douching can shift the vaginal ecology toward pathogen susceptibility makes the fecundability data more persuasive. Similarly, if folate supplementation were found to enhance follicular growth, support more rapid midcycle luteinization, or reduce very early pregnancy loss, the limited data on fecundability effects would be more convincing. Future research that involves collaboration among biologists, toxicologists, epidemiologists, and anthropologists is likely to have the most impact.

Acknowledgments

We thank Robert McConnaughey for graphical displays, Andrew Rowland and Allen Wilcox who provided useful comments on an earlier draft of the manuscript, and NIEHS library staff who assisted in locating literature.

References

1. Famighette, R., ed. (1996). "The World Almanac and Book of Facts, 1997." World Almanac Books, Mahwah, NJ.
2. Wood, J. (1994). "Dynamics of Human Reproduction: Biology, Biometry, Demography." de Gruyter, New York.
3. Abma, J., Chandra, A., Mosher, W., Peterson, L., and Piccinino, L. (1997). Fertility, family planning, and women's health: New

data from the 1995 National Survey of Family Growth. National Center for Health Statistics. *Vital Health Stat.* **23**(19), 1–114.

4. National Center for Health Statistics (1997). "Health and Injury Chartbook." U.S. National Center for Health Statistics, Hyattsville, MD.

5. Greenhall, E., and Vessey, M. (1990). The prevalence of subfertility: A review of the current confusion and a report of two new studies. *Fertil. Steril.* **54**, 978–983.

6. Leridon, H. (1977). "Human Fertility: The Basic Components." University of Chicago Press, Chicago.

7. Campbell, K. L., and Wood, J. W. (1988). Fertility in traditional societies. *In* "Natural Human Fertility: Social and Biological Determinants" (P. Diggory, M. Potts, and S. Teper, eds.), pp. 39–69. Macmillan, London.

8. Wood, J. (1990). Fertility in anthropological populations. *Ann. Rev. Anthropol.* **19**, 211–242.

9. Strassmann, B. I. (1997). The biology of menstruation in *Homo sapiens:* Total lifetime menses, fecundity, and nonsynchrony in a natural fertility population. *Curr. Anthropol.* **38**, 123–129.

10. Strassmann, B. I., and Warner, J. (1998). Predictors of fecundability and conception waits among the Dogon of Mali. *Am. J. Phys. Anthropol.* **105**, 167–184.

11. Wood, J. W., Holman, D. J., Yashin, A. I., Peterson, R. J., Weinstein, M., and Chang, M.-C. (1994). A multistate model of fecundability and sterility. *Demography* **31**, 403–426.

12. Wood, J. W. (1989). Fecundity and natural fertility in humans. *Oxford Rev. Reprod. Biol.* **11**, 61–109.

13. Goldman, N., Westoff, C. F., and Paul, L. E. (1985). Estimation of fecundability from survey data. *Stud. Fam. Plann.* **16**, 252–259.

14. Leridon, H. (1993). Fecundability and postpartum sterility: An insuperable interaction? *In* "Biomedical and Demographic Determinants of Reproduction" (R. Gray, H. Leridon, and A. Spira, eds.), pp. 345–358. Clarendon Press, Oxford.

15. John, A. M., Menken, J. A., and Chowdhury, A. K. M. A. (1987). The effects of breastfeeding and nutrition on fecundability in rural Bangladesh: A hazards-model analysis. *Popul. Stud.* **41**, 433–446.

16. John, A. M. (1988). Lactation and the waiting time to conception: An application of hazard models. *Hum. Biol.* **60**, 873–888.

17. Strassmann, B. I. (1996). Menstrual hut visits by Dogon women: A hormonal test distinguishes deceit from honest signaling. *Behav. Ecol.* **7**, 304–315.

18. Tietze, C. (1968). Fertility after discontinuation of intrauterine and oral contraception. *Int. J. Fertil.* **13**, 385–389.

19. Wilcox, A. J., Weinberg, C. R., O'Connor, J. F., Baird, D. D., Schlatterer, J. P., Canfield, R. E., Armstrong, E. G., and Nisula, B. C. (1988). Incidence of early loss of pregnancy. *N. Engl. J. Med.* **319**, 189–194.

20. Bonde, J. P. E., Hjollund, N. H. I., Jensen, T. K., Ernst, E., Kolstad, H., Henriksen, T. B., Giwercman, A., Skakkebaek, N. E., Andersson, A.-M., and Olsen, J. (1998). A follow-up study of environmental and biologic determinants of fertility among 430 Danish first-pregnancy planners: Design and methods. *Reprod. Toxicol.* **12**, 19–27.

21. Henshaw, S. K. (1998). Unintended pregnancy in the United States. *Fam. Plann. Perspect.* **30**, 24–29, 46.

21a. Baird, D. D., Wilcox, A. J., and Weinberg, C. R. (1986). Use of time to pregnancy to study environmental exposures. *Am. J. Epidemiol.* **124**, 470–480.

22. Doring, G. K. (1969). The incidence of anovular cycles in women. *J. Reprod. Fertil. Suppl.* **6**, 77–81.

23. Metcalf, M. G. (1983). Incidence of ovulation from the menarche to the menopause: Observations of 622 New Zealand women. *N. Z. Med. J.* **96**, 645–648.

24. Edwards, R. G. (1985). Current status of human conception *in vitro. Proc. R. Soc. London, Ser. B* **223**, 417–448.

25. Weinberg, C. R., Hertz-Picciotto, I., Baird, D. D., and Wilcox, A. J. (1992). Efficiency and bias in studies of early pregnancy loss. *Epidemiology* **3**, 17–22.

26. Armstrong, E. G., Ehrlich, P. H., Birken, S. *et al.* (1984). Use of a highly sensitive and specific immunoradiometric assay for detection of human chorionic gonadotropin in urine of normal, nonpregnant, and pregnant individuals. *J. Clin. Endocrinol. Metab.* **59**, 867–874.

27. Zinaman, M. J., O'Connor, J., Clegg, E. D., Selevan, S. G., and Brown, C. C. (1996). Estimates of human fertility and pregnancy loss. *Fertil. Steril.* **65**, 503–509.

28. Harlap, S., Shiono, P. H., and Ramcharan, S. (1980). A life table of spontaneous abortions and the effects of age, parity and other variables. *In* "Human Embryonic and Fetal Death" (I. H. Porter and E. B. Hook, eds.), pp. 145–164. Academic Press, New York.

29. Goldhaber, M. K., and Fireman, B. H. (1991). The fetal life table revisited: Spontaneous abortion rates in three Kaiser Permanente cohorts. *Epidemiology* **2**, 33–39.

30. Feldman, G. B. (1992). Prospective risk of stillbirth. *Obstetr. Gynecol.* **79**, 547–553.

31. Barrett, J. C., and Marshall, J. (1969). The risk of conception on different days of the menstrual cycle. *Popul. Stud.* **23**, 455–461.

32. Schwartz, D., MacDonald, P. D. M., and Heuchel, V. (1980). Fecundability, coital frequency and the viability of ova. *Popul. Stud.* **34**, 397–400.

33. Wilcox, A. J., Weinberg, C. R., and Baird, D. D. (1995). Timing of sexual intercourse in relation to ovulation: Effects on the probability of conception, survival of the pregnancy, and sex of the baby. *N. Engl. J. Med.* **333**, 1517–1521.

34. Wilcox, A. J., Weinberg, C. R., and Baird, D. D. (1998). Postovulatory ageing of the human oocyte and embryo failure. *Hum. Reprod.* **13**, 394–397.

35. Masarotto, G., and Romualdi, C. (1997). Probability of conception on different days of the menstrual cycle: An ongoing exercise. *Adv. Contraception* **13**, 105–115.

36. Baird, D. D., and Wilcox, A. J. (1985). Cigarette smoking associated with delayed conception. *J. Am. Med. Assoc.* **253**, 2979–2983.

37. Rowland, A. S., Baird, D. D., Weinberg, C. R., Shore, D. L., Shy, C. M., and Wilcox, A. J. (1992). Reduced fertility among women employed as dental assistants exposed to high levels of nitrous oxide. *N. Engl. J. Med.* **327**, 993–997.

37a. Blackwell, J. M., and Zaneveld, L. J. D. (1992). Effect of abstinence on sperm acrosin, hypoosmotic swelling, and other semen variables. *Fertil. Steril.* **58**, 798–802.

37b. Zhou, H., and Weinberg, C. R. (1996). Modeling conception as an aggregated Bernoulli outcome with latent variables via the EM algorithm. *Biometrics* **52**, 945–954.

38. Agarwal, S. K., and Haney, A. F. (1994). Does recommending timed intercourse really help the infertile couple? *Obstet. Gynecol.* **84**, 307–310.

39. Collins, W. P., Branch, C. M., Collins, P. O., and Sallam, H. M. (1983). Biochemical indices of the fertile period in women. *Int. J. Fertil.* **26**, 196–202.

40. Katz, D. F., Slade, D. A., and Nakajima, S. T. (1997). Analysis of pre-ovulatory changes in cervical mucus hydration and sperm penetrability. *Adv. Contraception* **13**, 143–151.

41. Frank, O., Bianchi, P. G., and Campana, A. (1994). The end of fertility: Age, fecundity and fecundability in women. *J. Biosoc. Sci.* **26**, 349–368.

42. Federation CECOS, Schwartz, D., and Mayaux, M. J. (1982). Female fecundity as a function of age. *N. Engl. J. Med.* **306**, 404–406.

43. Van Noord-Zaadstra, B. M., Looman, C. W. N., Alsbach, H., Habbema, J. D. F., te Velde, E. R., and Karbaat, J. (1991). Delay-

ing childbearing: Effect of age on fecundity and outcome of pregnancy. *Br. Med. J.* **302,** 1361–1365.

44. Ellison, P. T. (1994). Salivary steroids and natural variation in human ovarian function. *Ann. N. Y. Acad. Sci.* **709,** 287–298.

45. Pellicer, A., Simon, C., and Remohi, J. (1995). Effects of aging on the female reproductive system. *Hum. Reprod.* **10**(Suppl. 2), 77–83.

45a. Meldrum, D. R. (1993). Female reproductive aging—ovarian and uterine factors. *Fertil. Steril.* **59,** 1–5.

46. Singh, K. K., Suchindran, C. M., and Singh, K. (1993). Effects of breast feeding after resumption of menstruation on waiting time to next conception. *Hum. Biol.* **65,** 71–86.

47. Diaz, S., Miranda, P., Cardenas, H., Salvatierra, A. M., Brandeis, A., and Croxatto, H. B. (1992). *Fertil. Steril.* **58,** 498–503.

48. Campbell, O. M. R., and Gray, R. H. (1993). Characteristics and determinants of postpartum ovarian function in women in the United States. *Am. J. Obstet. Gynecol.* **169,** 55–60.

49. Cates, W., Jr., Wasserheit, J. N., and Marchbanks, P. A. (1994). Pelvic inflammatory disease and tubal infertility: The preventable conditions. *Ann. N. Y. Acad. Sci.* **709,** 179–195.

50. Cates, W., Farley, T. M. M., and Rowe, P. J. (1985). Worldwide patterns of infertility: Is Africa different? *Lancet* **2,** 596–598.

51. Cates, W., Jr., Joesoef, M. R., and Goldman, M. B. (1993). Atypical pelvic inflammatory disease: Can we identify clinical predictors? *Am. J. Obstet. Gynecol.* **169,** 341–346.

52. Grodstein, F., Goldman, M. B., and Cramer, D. W. (1993). Relation of tubal infertility to history of sexually transmitted diseases. *Am. J. Epidemiol.* **137,** 577–584.

53. Hemminki, E., Graubard, B. I., Hoffman, H. J., Mosher, W. D., and Fetterly, K. (1985). Cesarean section and subsequent fertility: Results from the 1982 National Survey of Family Growth. *Fertil. Steril.* **43,** 520–528.

54. Mueller, B. A., Daling, J. R., Moore D. E., Weiss, N. S., Spadoni, L. R., Stadel, B. V., and Soules, M. R. (1986). Appendectomy and the risk of tubal infertility. *N. Engl. J. Med.* **315,** 1506–1508.

55. Sweet, R. L. (1995). Role of bacterial vaginosis in pelvic inflammatory disease. *Clin. Infect. Dis.* **20**(Suppl. 2), S271–S275.

56. Koch, A., Bilina, A., Teodorowicz, L., and Stary, A. (1997). *Mycoplasma hominis* and *Ureaplasma urealyticum* in patients with sexually transmitted diseases. *Wien. Klin. Wochenschr.* **109,** 584–589.

57. Hill, G. B. (1993). The microbiology of bacterial vaginosis. *Am. J. Obstet. Gynecol.* **169,** 450–454.

58. Daling, J. R., Weiss, N. S., Metch, B. J., Chow, W. H., Soderström, R. M., Moore, D. E., Spadoni, L. R., and Stadel, B. V. (1985). Primary tubal infertility in relation to the use of an intrauterine device. *N. Engl. J. Med.* **312,** 937–941.

59. Cramer, D. W., Schiff, I., Schoenbaum, S. C., Gibson, M, Belisle, S., Albrecht, B., Stillman, R. J., Berger, M. J., Wilson, E., Stadel, B. V., and Seibel, M. (1985). Tubal infertility and the intrauterine device. *N. Engl. J. Med.* **312,** 941–947.

60. Cramer, D. W., Goldman, M. B., Schiff, I., Belisle, S., Albrecht, B., Stadel, B., Gibson, M., Wilson, E., Stillman, R., and Thompson, I. (1987). The relationship of tubal infertility to barrier method and oral contraceptive use. *J. Am. Med. Assoc.* **257,** 2446–2450.

61. Baird, D. D., Weinberg, C. R., Voigt, L. F., and Daling, J. R. (1996). Vaginal douching and reduced fertility. *Am. J. Public Health* **86,** 844–850.

62. Zhang, J., Thomas, A. G., and Leybovich, E. (1997). Vaginal douching and adverse health effects: A meta-analysis. *Am. J. Public Health* **87,** 1207–1211.

63. Hillier, S. L. (1991). The vagina as an ecosystem. *9th Int. Meet. Int. Soc. STD Res.,* Banff, Alberta, Canada, 1991.

64. Koutras, D. A. (1997). Disturbances of menstruation in thyroid disease. *Ann. N. Y. Acad. Sci.* **816,** 280–284.

65. Moen, M. H., and Schei, B. (1997). Epidemiology of endometriosis in a Norwegian county. *Acta Obstet. Gynecol. Scand.* **76,** 559–562.

66. Speroff, L., Glass, R. H., and Kase, N. G., eds. (1994). "Clinical Gynecologic Endocrinology and Infertility," 5th ed. Williams & Wilkins, Baltimore, MD.

67. Farhi, J., Ashkenazi, J., Feldberg, D., Dicker, D., Orvieto, R., and Ben-Rafael, Z. (1995). Effect of uterine leiomyomata on the results of in-vitro fertilization treatment. *Hum. Reprod.* **10,** 2576–2578.

68. Ober, C., Elias, S., Kostyu, D. D., and Hauck, W. W. (1992). Decreased fecundability in Hutterite couples sharing HLA-DR. *Am. J. Hum. Genet.* **50,** 6–14.

69. Warren, M. P. (1983). Effects of undernutrition on reproductive function in the human. *Endocr. Rev.* **4,** 363–377.

70. Copeland, P. M., Sacks, N. R., and Herzog, D. B. (1995). Longitudinal follow-up of amenorrhea in eating disorders. *Psychosom. Med.* **57,** 121–126.

71. Henley, K., and Vaitukaitis, J. L. (1988). Exercise-induced menstrual dysfunction. *Ann. Rev. Med.* **39,** 443–451.

72. Bullen, B. A., Skrinar, G. S., Beitins, I. Z., von Mering, G., Turnbull, B. A., and McArthur, J. W. (1985). Induction of menstrual disorders by strenuous exercise in untrained women. *N. Engl. J. Med.* **312,** 1349–1353.

73. Baird, D. D., Wilcox, A. J., Weinberg, C. R., Kamel, F., McConnaughey, D. R., Musey, P. I., and Collins, D. C. (1997). Preimplantation hormonal differences between the conception and non-conception menstrual cycles of 32 normal women. *Hum. Reprod.* **12,** 2607–2613.

74. Wade, G. N., Schneider, J. E., and Li, H-Y. (1996). Control of fertility by metabolic cues. *Am. J. Physiol.* **270,** E1–E19.

75. Czeizel, A. E., Dudas, I., and Metneki, J. (1994). Pregnancy outcomes in a randomised controlled trial of periconceptional multivitamin supplementation: Final report. *Arch. Gynecol. Obstet.* **255,** 131–139.

76. Green, B. B., Weiss, N. S., and Daling, J. R. (1988). Risk of ovulatory infertility in relation to body weight. *Fertil. Steril.* **50,** 721–726.

77. Grodstein, F., Goldman M. B., and Cramer, D. W. (1994). Body mass index and ovulatory infertility. *Epidemiology* **5,** 247–250.

78. Lake, J. K., Power, C., and Cole, T. J. (1997). Women's reproductive health: The role of body mass index in early and adult life. *Int. J. Obes. Relat. Metab. Disord.* **21,** 432–438.

79. Harlow, S. D., and Ephross, S. A. (1995). Epidemiology of menstruation and its relevance to women's health. *Epidemiol. Rev.* **17,** 265–286.

80. Rich-Edwards, J. W., Goldman, M. B., Willett, W. C., Hunter, D. J., Stampfer, M. J., Colditz, G. A., and Manson, J. E. (1994). Adolescent body mass index and infertility caused by ovulatory disorder. *Am. J. Obstet. Gynecol.* **171,** 171–177.

81. Zaadstra, B. M., Seidell, J. C., Van Noord, P. A. H., te Velde, E. R., Habbema, J. D. F., Vrieswijk, B., and Karbaat, J. (1993). Fat and female fecundity: Prospective study of effect of body fat distribution on conception rates. *Br. Med. J.* **306,** 484–487.

82. Jenkins, J. M., Brook, P. F., Sargeant, S., and Cooke, I. D. (1995). Endocervical mucus pH is inversely related to serum androgen levels and waist to hip ratio. *Fertil. Steril.* **63,** 1005–1008.

83. Apter, D., and Vihko, R. (1990). Endocrine determinants of fertility: Serum androgen concentrations during follow-up of adolescents into the third decade of life. *J. Clin. Endocrinol. Metab.* **71,** 970–974.

84. Wolfers, D. (1970). The probability of conception after discontinuance of oral contraceptives: A note on "Oral contraception, coital frequency, and the time required to conceive" by Westoff, Bumpass, and Ryder. *Soc. Biol.* **17,** 57.

85. Rice-Wray, E., Correu, S., Gorodovsky, J., Esquivel, J., and Gold-zieher, J. W. (1967). Return of ovulation after discontinuance of oral contraceptives. *Fertil. Steril.* **18,** 212–218.

86. Harlap, S., and Baras, M. (1984). Conception-waits in fertile women after stopping oral contraceptives. *Int. J. Fertil.* **29,** 73–80.

87. Chasan-Taber, L., Willett, W. C., Stampfer, M. J., Spiegelman, D., Rosner, B. A., Hunter, D. J., Colditz, G. A., and Manson, J. E. (1997). Oral contraceptives and ovulatory causes of delayed fertility. *Am. J. Epidemiol.* **146,** 258–265.

88. Grodstein, F., Goldman M. B., Ryan, L., and Cramer, D. W. (1993). Self-reported use of pharmaceuticals and primary ovulatory infertility. *Epidemiology* **4,** 151–156.

89. Lim, H., Paria, B. C., Das, S. K., Dinchuk, J. E., Langenbach, R., Trzaskos, J. M., and Dey, S. K. (1997). Multiple female reproductive failures in cyclooxygenase 2-deficient mice. *Cell (Cambridge, Mass.)* **91,** 197–208.

90. Athanasiou, S., Crayford, T. J. B., Bourne, T. H., Hagström, H.-G., Khalid, A., Campbell, S., Okokon, E. V., and Collins, W. P. (1996). Effects of indomethacin on follicular structure, vascularity, and function over the periovulatory period in women. *Fertil. Steril.* **65,** 556–560.

91. Killick, S., and Elstein, M. (1987). Pharmacologic production of luteinized unruptured follicles by prostaglandin synthetase inhibitors. *Fertil. Steril.* **47,** 773–777.

92. Senekjian, E. K., Potkul, R. K., Frey, K., and Herbst, A. L. (1988). Infertility among daughters either exposed or not exposed to diethylstilbestrol. *Am. J. Obstet. Gynecol.* **158,** 493–498.

93. MacKenzie, K. M., and Angevine, D. M. (1981). Infertility in mice exposed in utero to benzo(a)pyrene. *Biol. Reprod.* **24,** 183–191.

94. Weinberg, C. R., Wilcox, A. J., and Baird D. D. (1989). Reduced fecundability in women with prenatal exposure to cigarette smoking. *Am. J. Epidemiol.* **129,** 1072–1078.

95. Baird, D. D., and Wilcox, A. J. (1986). Future fertility after prenatal exposure to cigarette smoke. *Fertil. Steril.* **46,** 368–372.

96. Schwingl, P. (1992). Prenatal smoking exposure in relation to female adult fecundability. Doctoral Dissertation, University of North Carolina, Chapel Hill.

97. Olsen, J., Rachootin, P., Schiodt, A. V., and Damsbo, N. (1983). Tobacco use, alcohol consumption and infertility. *Int. J. Epidemiol.* **12,** 179–184.

98. Baird, D. D. (1992). Evidence for reduced fecundity in female smokers. *In* "Effects of Smoking on the Fetus, Neonate, and Child," pp. 5–22. Oxford University Press, Oxford.

99. Bolumar, F., Olsen, J., Boldsen, J., and the European Study Group on Infertility and Subfecundity (1996). Smoking reduces fecundity: A European multicenter study on infertility and subfecundity. *Am. J. Epidemiol.* **143,** 578–587.

100. Wilsnack, S. C., Klassen, A. D., and Wilsnack, R. W. (1984). Drinking and reproductive dysfunction among women in a 1981 national survey. *Alcohol.: Clin. Exp. Res.* **8,** 451–458.

101. Mello, N. K., Mendelson, J. H., King, N. W., Bree, M. P., Skupny, A., and Ellingboe, J. (1988). Alcohol self-administration by female macaque monkeys: A model for study of alcohol dependence, hyperprolactinemia and amenorrhea. *J. Stud. Alcohol* **49,** 551–560.

102. Olsen, J., Bolumar, F., Boldsen, J., Bisanti, L., and the European Study Group on Infertility and Subfecundity (1997). Does moderate alcohol intake reduce fecundability? A European multicenter study on infertility and subfecundity. *Alcohol.: Clin. Exp. Res.* **21,** 206–212.

102a. Jensen, T. K., Hjollund, N. H. I., Henriksen, T. B., Scheike, T., Kolsad, H., Giwercman, A., Ernst, E., Bonde, J. P., Skakkebaek, N. E., and Olsen, J. (1998). Does moderate alcohol consumption affect fertility? Follow up study among couples planning first pregnancy. *Br. Med. J.* **317,** 505–510.

103. Wilcox, A. J., Weinberg, C. R., and Baird, D. D. (1988). Caffeinated beverages and decreased fertility. *Lancet* **2,** 1453–1456.

104. Dlugosz, L., and Bracken, M. B. (1992). Reproductive effects of caffeine: A review and theoretical analysis. *Epidemiol. Rev.* **14,** 83–100.

105. Jensen, T. K., Henriksen, T. B., Hjollund, N. H. I., Scheike, T., Kolstad, H., Giwercman, A., Ernst, E., Bonde, J. P., Skakkebaek, N. E., and Olsen, J. (1998). Caffeine intake and fecundability: A follow-up study among 430 Danish couples planning their first pregnancy. *Reprod. Toxicol.* **12,** 289–295.

106. Olsen, J. (1991). Cigarette smoking, tea and coffee drinking, and subfecundity. *Am. J. Epidemiol.* **133,** 734–739.

107. Stanton, C. K., and Gray, R. H. (1995). Effects of caffeine consumption on delayed conception. *Am. J. Epidemiol.* **142,** 1322–1329.

108. Bolumar, F., Olsen, J., Rebagliato, M., Bisanti, L., and the European Study Group on Infertility and Subfecundity (1997). Caffeine intake and delayed conception: A European Multicenter Study on Infertility and Subfecundity. *Am. J. Epidemiol.* **145,** 324–334.

109. Wilcox, A. J., and Weinberg, C. R. (1991). Tea and fertility (letter). *Lancet* **337,** 1159–1160.

110. Edelmann, R. J., Connolly, K. J., Cooke, I. D., and Robson, J. (1991). Psychogenic infertility: Some findings. *J. Psychosom. Obstet. Gynecol.* **12,** 163–168.

111. Henriksen, T. B. (1999). General psychosocial and work-related stress and reduced fertility. *Scand. J. Work Environ. Health* **25** (Suppl. 1): 38–39.

112. Whorton, D., Krauss, R. M., Marshall, S., and Milby, T. H. (1977). Infertility in male pesticide workers. *Lancet* **2,** 1259–1261.

113. Rachootin, P., and Olsen, J. (1983). The risk of infertility and delayed conception associated with exposures in the Danish workplace. *J. Occup. Med.* **25,** 394–402.

114. Brandt, L. P. A., and Nielsen, C. V. (1992). Fecundity and the use of video display terminals. *Scand. J. Work Environ. Health* **18,** 298–301.

115. Rowland, A. S., Baird, D. D., Weinberg, C. R., Shore, D. L., Shy, C. M., and Wilcox, A. J. (1994). The effect of occupational exposure to mercury vapour on the fertility of female dental assistants. *Occup. Environ. Med.* **51,** 28–34.

116. Schaumburg, I., and Olsen J. (1989). Time to pregnancy among Danish phamacy assistants. *Scand. J. Work Environ. Health* **15,** 222–226.

117. Gold, E. B., and Tomich, E. (1994). Occupational hazards to fertility and pregnancy outcome. *Occup. Med.: State Art Rev.* **9,** 435–469.

118. Baird, D. D., Rowland, A. S., and Weinberg, C. R. (1994). Relative fecundability as an outcome measure in epidemiologic studies of infertility. *Infertil. Reprod. Med. Clin. North Am.* **5,** 309–320.

119. Kersemaekers, W. M., Roeleveld, N., and Zielhuis, G. A. (1997). Reproductive disorders among hairdressers. *Epidemiology* **8,** 396–401.

120. Bisanti, L., Olsen, J., Basso, O., Thonneau, P., Karmaus, W., and The European Study Group on Infertility and Subfecundity (1996). Shift work and subfecundity: A European multicenter study. *J. Occup. Environ. Med.* **38,** 352–358.

121. Ahlborg, G., Jr., Axelsson, G., and Bodin, L. (1996). Shift work, nitrous oxide exposure and subfertility among Swedish midwives. *Int. J. Epidemiol.* **25,** 783–790.

122. Florack, E. I. M., Zielhuis, G. A., and Rolland, R. (1994). The influence of occupational physical activity on the menstrual cycle and fecundability. *Epidemiology* **5,** 14–18.

123. Tuntiseranee, P., Olsen, J., Geater, A., and Kor-anantakul, O. (1998). Are long working hours and shiftwork risk factors for subfecundity? A study among couples from southern Thailand. *Occup. Environ. Med.* **55,** 99–105.

124. Abell, A., Juul, S., and Bonde, J. P. (1997). Time to pregnancy in female greenhouse workers. *Int. Symp. Environ., Lifestyle, Fertil.,* Aarhus, Denmark, 1997.

125. Smith, E. M., Hammonds-Ehlers, M., Clark, M. K., Kirchner, H. L., and Fuortes, L. (1997). Occupational exposures and risk of female infertility. *J. Occup. Environ. Med.* **39,** 138–147.

126. Sallmen, M., Anttila, A., Lindbohm, M.-L., Kyyronen, P., Taskinen, H., and Hemminki, K. (1995). Time to pregnancy among women occupationally exposed to lead. *J. Occup. Environ. Med.* **37,** 931–934.

127. Sallmen, M., Lindbohm, M.-L., Kyyronen, P., Nykyri, E., Anttila, A., and Taskinen, H. (1995). Reduced fertility among women exposed to organic solvents. *Am. J. Ind. Med.* **27,** 699–713.

128. Correa, A., Gray, R. H., Cohen, R., Rothman, N, Shah, F., Seacat, H., and Corn, M. (1996). Ethylene glycol ethers and risks of spontaneous abortion and subfertility. *Am. J. Epidemiol.* **143,** 707–717.

129. Eskenazi, B., Gold, E. B., Samuels, S. J., Wight, S., Lasley, B. L., Hammond, S. K., O'Neill-Rasor, M., and Schenker, M. B. (1995). Prospective assessment of fecundability of female semiconductor workers. *Am. J. Ind. Med.* **28,** 817–831.

130. Spinelli, A., Figa-Talamanca, I., and Osborn, J. (1997). Time to pregnancy and occupation in a group of Italian women. *Int. J. Epidemiol.* **26,** 601–609.

131. Olsen, J., and Skov, T. (1993). Design options and methodological fallacies in the studies of reproductive failures. *Environ. Health Perspect.* **101**(Suppl. 2), 145–152.

132. Baird, D. D., Weinberg, C. R., Schwingl, P., and Wilcox, A. J. (1994). Selection bias associated with contraceptive practice in time-to-pregnancy studies. *Ann. N. Y. Acad. Sci.* **709,** 156–164.

133. Weinberg, C. R., Baird, D. D., and Rowland, A. S. (1993). Pitfalls inherent in retrospective time-to-event studies: The example of time to pregnancy. *Stat. Med.* **12,** 867–879.

134. Weinberg, C. R., Baird, D. D., and Wilcox, A. J. (1994). Sources of bias in studies of time to pregnancy. *Stat. Med.* **13,** 671–681.

135. Zielhuis, G. A., Hulscher, M. E. J. L., and Florack, E. I. M. (1992). Validity and reliability of a questionnaire on fecundability. *Int. J. Epidemiol.* **21,** 1151–1156.

136. Joffe M., Villard, L., Li, Z., Plowman, R., and Vessey, M. (1995). A time to pregnancy questionnaire designed for long-term recall: Validity in Oxford, England. *J. Epidemiol. Commun. Health* **49,** 314–349.

137. Feichtinger, W. (1991). Environmental factors and fertility. *Hum. Reprod.* **6,** 1170–1175.

138. Olsen, J. (1994). Is human fecundity declining—and does occupational exposures play a role in such a decline if it exists? *Scand. J. Work Environ. Health* **20** (Spec. Issue), 72–77.

139. Kallan, J., and Udry, J. R. (1986). The determinants of effective fecundability based on the first birth interval. *Demography* **23,** 53–66.

12

Contraception

DONNA SHOUPE AND DANIEL R. MISHELL, JR.
University of Southern California
Los Angeles, California

I. Contraceptive Use in the United States

Data accumulated from three national surveys conducted in 1982, 1988, and 1995 provide useful information about the contraceptive choices of American women [1]. It is encouraging to observe that overall contraceptive usage has increased from 56 to 64% in women of reproductive age (15–44 years old). Only 5% of sexually active women who did not wish to conceive were using no method of contraception [1]. However, it is of some concern that 30% of the contraceptive users were using methods with high typical failure rates. Because many sexually active women in the US do not use the most effective methods of contraception, there is a high incidence of unintended pregnancy and subsequent elective abortion in contrast to lower rates in both Western Europe and Canada. Unintended pregnancy is endemic in the United States, with nearly half (49%) of the 5.4 million pregnancies in 1994 unintended.

Between 1988 and 1995, the percentage of oral contraceptive users decreased from 31 to 27% while condom use increased from 15 to 20%. The drop in pill use was most marked among black women younger than 25 years, falling from 75 to 32%. Pill use increased among women aged 35–39 and increased sixfold among those aged 40–44. The increase in condom use was higher among blacks and Hispanics than among whites. In 1995, the prevalence of condom use was about 20% and roughly equal in all three groups (Table 12.1). Reliance on the intrauterine device (IUD) in Hispanic women dropped while use of the

diaphragm in college-educated white women also fell. The decline in pill and diaphragm use and the increased use of condoms may reflect the increased concern about HIV and other sexually transmitted diseases (STD) in reproductive age couples. Overall, contraceptive use in 1995 reached an all time high of 64.2%. Compared to 1988, the proportion of women using contraceptives increased in all age groups among Hispanic, white, and black women [1] (Table 12.2).

Factors influencing contraceptive choice are age, marital status, education, income, and fertility intentions. In 1995, never-married women younger than 30 years, with one or more years of college education, predominantly selected the pill while formerly married women in their thirties and forties commonly selected tubal sterilization. In response to the growing concern about STDs, almost 8% of women presently use two methods, one of which is condoms [1].

II. Contraceptive Effectiveness

Extended use effectiveness reflects the effectiveness of a method over extended periods of time using actuarial statistical techniques including life table analysis. Most contraceptive studies are performed in controlled clinical trials where frequent contact with supportive clinic personnel may result in lower failure rates and higher continuation rates than may actually occur in field use.

Table 12.1

Percentage Distribution of Contraceptive Users Aged 15–44, by Current Method, According to Race and Ethnicity, 1982–1995

Method	Hispanic 1982 (N = 245)	Hispanic 1988 (N = 342)	Hispanic 1995 (N = 977)	Non-Hispanic white 1982 (N = 2.231)	Non-Hispanic white 1988 (N = 3.142)	Non-Hispanic white 1995 (N = 4.352)	Non-Hispanic black 1982 (N = 1.688)	Non-Hispanic black 1988 (N = 1.572)	Non-Hispanic black 1995 (N = 1.606)
Female sterilization	23	32	37	22	26	25	30	38	40
Male sterilization	5	4	4	13	14	14	2	1	2
Pill	30	33	23	26	30	29	38	38	24
Implant	na	na	2	na	na	1	na	na	2
Injectable	na	na	5	na	na	2	na	na	5
IUD	19	5	2	6	2	1	9	3	1
Diaphragm	5	2	1	9	7	2	3	2	1
Male condom	7	14	21	13	15	20	6	10	20
Other	11	10	6	11	7	7	12	8	5
Total	100	100	100	100	100	100	100	100	100
No. in 000s:	2,224	2,799	3,957	23,666	26,800	28,120	3,520	4,208	5,098

Note. na = not applicable. In subsequent tables, totals for 1995 include methods not included in the other methods' category; the emergency contraceptive pill, the female condom, the cervical cap, and the suppository. From [1].

Table 12.2

Percentage of U.S. Women Aged 15–44 Currently Using
Contraceptives, and Percentage Sexually Active in the Prior Three
Months but not Using a Method, by Selected Characteristics.
National Survey of Family Growth, 1982–1995

Characteristic	1982 (N = 7.969)	1988 (N = 8.450)	1995 (N = 10.847)
Using a method	55.7	60.3	64.2
Age			
15–19	24.2	32.1	29.3
20–24	55.8	59.0	63.5
25–29	65.9	64.5	69.2
30–34	67.6	68.0	72.8
35–39	61.9	70.2	73.1
40–44	61.2	66.6	71.4
Race ethnicity			
Hispanic	50.6	50.4	59.0
Non-Hispanic white	57.3	52.9	66.1
Non-Hispanic black	51.6	56.8	62.1
Sexually active, not using	7.4	6.7	5.2
Age			
15–19	9.9	7.5	7.1
20–24	8.2	8.1	6.0
25–29	7.1	7.5	4.7
30–34	6.8	5.4	4.4
35–39	6.4	4.6	4.3
40–44	6.6	5.5	5.1
Race ethnicity			
Hispanic	6.5	5.6	5.6
Non-Hispanic white	6.2	5.5	5.0
Non-Hispanic black	13.6	10.3	7.0
No. in 000s:	54,099	57,900	60,201

Table 12.3

Percentage of Women Experiencing a Contraceptive Failure
during the First Year of Typical Use and the First Year of Perfect
Use and the Percentage Continuing Use at the End of the First
Year, United States

Method	% of women experiencing an accidental pregnancy within the first year of use		% of women continuing use at one year
	Typical use	Perfect use	
Chance	85	85	
Spermicides	21	6	43
Periodic abstinence	20		67
Calendar		9	
Ovulation method		3	
Symptothermal		2	
Postovulation		1	
Withdrawal	19	4	
Cap			
Parous women	36	26	45
Nulliparous women	18	9	58
Sponge			
Parous women	36	20	45
Nulliparous women	18	9	58
Diaphragm	18	6	58
Condom			
Female (reality)	21	5	56
Male	12	3	63
Pill	3		72
Progestin only		0.5	
Combined		0.1	
IUD			
Progesterone T	2.0	1.5	81
Copper T380A	0.8	0.6	78
LNg 20	0.1	0.1	81
Depo-Provera	0.3	0.3	70
Norplant (6 capsules)	0.09	0.09	85
Female sterilization	0.4	0.4	100
Male sterilization	0.15	0.10	100

From Trussell J, Hatcher RA, Cates W, et al: American Health Consultants, Contraceptive Technology Update 17(2):13, 1995.

Contraceptive failure occurs more often in couples seeking to delay a wanted birth as compared with those seeking to prevent any more births. The woman's age has a strong negative correlation with failure, as does socioeconomic status and level of education. Failure rates are lower among married women compared to unmarried women and are usually higher during the first year of use than in subsequent years. Table 12.3 lists typical use and perfect use failure rates within the first year of use of various contraceptives and one-year continuation rates [2].

The actual use failure rate for five years of use for certain methods of long acting contraceptives has been reported [3–5]. The cumulative failure rate for five years use of Norplant® in clinical trials is 1.1% [3]. In a large World Health Organization (WHO) study, the cumulative failure rates of the Copper T 380 IUD were 1.0, 1.4, and 1.6% after three, five, and seven years of use [4]. The percentages are comparable to those for tubal sterilization, which are reported as 1.3% after five years and 1.9% ten years after the procedure [5].

III. Contraceptive Cost

All contraceptive methods reduce health care costs. An economic model to compare costs of 15 contraceptive methods is

shown in Figure 12.1 [6]. The model projects the five-year cost of using each method by adding the direct cost of each method with costs associated with method failures and side effects. Due to the high costs of unintended pregnancies, use of any of the 15 methods studied was less costly than use of no method. The most cost effective methods were the IUD, vasectomy, and progestin implant or injection.

IV. Contraceptive Types

A. Spermicides: Foams, Creams, Suppositories

All spermicidal products contain a surfactant, usually nonoxynol 9, which immobilizes or kills sperm on contact. The effectiveness of these agents increases with increasing age of

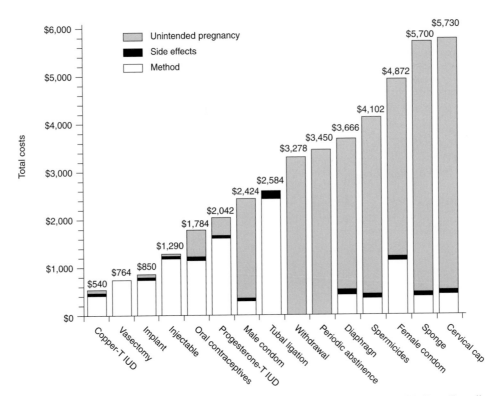

Fig. 12.1 Five-year costs associated with contraceptive methods in the managed payment model. (From Trussell, J., Leveque, J. A., and Koenig, J. D. *Am. J. Public Health* 85(49):494, 1995, [6].)

the woman and is comparable to that of the diaphragm in all age and income groups. In 1994, the manufacturer of the intravaginal sponge discontinued its production and it is no longer available. Although a few studies linked the use of spermicides at the time of conception to an increased risk of congenital malformations, several well-performed studies showed no increased risk of either congenital malformations [7,8] or karyotypic abnormalities [9] in pregnancies of women who conceived while using spermicides.

B. Barrier Techniques

1. Diaphragm

A diaphragm must be fitted by a health professional. The largest size diaphragm that does not cause discomfort should be used. For the most effective protection, the diaphragm should be left in place for eight hours after intercourse. If repeated intercourse occurs, additional spermicide should be used. The number of urinary tract infections is higher in women who use diaphragms than in nonusers, probably because pressure of the diaphragm on the urethra may result in partial obstruction of urinary outflow [10]. The diaphragm should not be left in place for more than 24 hours as ulceration of the vaginal epithelium may occur with prolonged use.

2. Cervical Cap

The cervical cap is a cup-shaped plastic or rubber device that is designed to fit around the cervix. It has been used for decades in Europe, especially in Britain, but was only approved for use

in the US in 1988. The cervical cap is manufactured in four sizes, can be left in the vagina for up to 48 hours, and is more comfortable than the diaphragm. Spermicide should always be placed inside the cap before use. The cap requires more training than the diaphragm, both for the provider in order to fit the cap correctly and for the user to place it correctly.

In a large clinical trial, one-year pregnancy rates were 17% for the cap and diaphragm [11]. Because of concern about a possible adverse effect on cervical tissue, the cervical cap should be used only in women with normal cervical cytology.

3. Male Condom

The condom is one of the oldest and most widely used forms of contraception (used by over 40 million couples worldwide) and was used as early as the 1500s for protection against the spread of venereal infection. The male condom is currently the most effective method of contraception to prevent transmission of sexually transmitted diseases for both men [12] and women [13]. Epidemiologic evidence is encouraging that the condom in actual use is protective against transmission of HIV [14,15]. There is evidence that condom use has a protective and even therapeutic effect on cervical-cell abnormalities by limiting transmission of agents associated with cervical carcinoma [16] and by reducing the risk of pelvic inflammatory disease (PID) in women users by 40% [13]. In one study, after ten years of condom or diaphragm use, the relative risk of women developing cervical dysplasia was 0.2 [17].

The majority of condoms today are made of vulcanized latex rubber and come in a great variety of shapes, sizes, textures, and colors. Because of vast improvements in quality control, product defects of present-day condoms that could lead to failure are rare. First-year failure rates range between 0.4 and 6% in motivated populations, and between 8 and 31.9% in young, inexperienced, or inconsistent users [1,18,19]. Future advances in condom technology may be towards ultra-thin condoms, a greater variety of spermicidally lubricated condoms, and more stringent quality control.

4. Female Condom

The female condom was approved in the US in 1994. It consists of a soft, loose-fitting sheath and two flexible polyurethane rings. One ring is placed inside the vagina and serves as an internal anchor to hold the sheath in the proper position within the vagina. The outer ring remains outside the vagina, thus providing external protection to the vulva and penis. The condom is prelubricated and is intended for one time use.

In comparison with the male condom, the female condom may be inserted anytime prior to sexual activity and can be left in place longer. Similar to latex, polyurethane does not allow virus transmission and it is hoped that the outer ring may provide the female with even greater protection against STD transmission than does the male condom. Additionally, polyurethane is stronger than latex and is less likely to rupture. In a multicenter clinical trial, the pregnancy rate after six months of use was 12.4%, although it was only 2.6% in those with perfect use [20]. At the end of six months, about one-third of the women had discontinued use of the method.

C. Periodic Abstinence/Natural Family Planning

As defined by the WHO, natural methods of family planning include methods for planning or avoiding pregnancies by observing the natural signs and symptoms of the fertile and infertile phases of the menstrual cycle. Many couples are motivated to use these methods in order to avoid technology, to allow them to take autonomous control of their fertility, and because of religious or cultural reasons.

There are several techniques used to identify the female fertile period. The oldest is the calendar rhythm method which relies on three assumptions: (1) the human ova is capable of being fertilized for only 10–24 hours after ovulation; (2) spermatozoa retain their fertilizing ability for only 48 hours; and (3) ovulation occurs 12–16 days prior to the onset of the next menses. The first day of the fertile period is established by subtracting 18–20 days from the length of the shortest cycle, while the last day of the fertile phase is determined by subtracting 11 days from the longest cycle. This method often results in long periods of abstinence. It is not surprising that this method is rarely used as a single index of fertility.

In the temperature method, couples must refrain from intercourse from the onset of the cycle until the morning of the third day of a sustained elevated basal body temperature. In the cervical-mucus method, the first fertile day of the cycle is the first day when either the sensation of moistness or mucus occurs. The last fertile day is the fourth day after peak symptoms. Changes in the cervix opening (softening or shift upward in the pelvis) before ovulation are also used to indicate the start of the fertile period. The symptothermal methods use changes in cervical mucus or calendar calculations to determine the onset of the fertile period and changes in mucus or basal temperature to

estimate the end. Simple, self-administered tests that measure urinary estrogen and pregnanediol glucuronide are now available. Such tests are performed for about 12 days each month and can usually reduce the days of abstinence required to about seven.

These techniques are used alone or in combination. Method-related failures are usually low and range from 1 to 7% [21–23]. However, as with many other contraceptives, the user failure rates are substantially higher than the method-related failure rates and range from 6.6 to 39% [24,25]. Generally, the many varieties of the symptothermal methods are the most commonly used and are associated with low failure rates [26].

D. Oral Contraceptives (OCs)

OCs are reliable, reversible, and are the most thoroughly studied pharmacological agent of the twentieth century. Initially marketed in the US in 1960, OCs quickly became the most widely used method of contraception. Most OC formulations combine an estrogen and a progestin. The major action of the progestin component is to inhibit ovulation and produce other contraceptive actions such as thickening of the cervical mucus. The major effect of the estrogen component is to maintain the endometrium and prevent unscheduled bleeding. Initially marketed OC formulations contained 150 μg of estrogen (mestranol) and 9.85 mg of progestin (norethynodrel). Many recently introduced OCs contain only 20 μg of estrogen and 100 to 150 μg of progestin.

1. Pharmacology/Selectivity

There are three major types of OC formulations: combination phasic, fixed-dose combination, and daily progestin (minipill). The phasic formulations were designed to lower the total monthly steroid dose. All currently marketed formulations are made from synthetic steroids and contain no natural estrogen or progestin. Progestins are either a derivative of 19-nortestosterone or 17α-acetoxyprogesterone. The acetoxyprogesterones, including medroxyprogesterone acetate and megestrol acetate, are not used in current OCs. The 19-nortestosterone progestins are of two major types: estranes (Fig. 12.2) and gonanes (Fig. 12.3). The gonanes have greater progestational activity per unit weight than the estranes and thus a smaller amount of gonane is used in OC formulations. The parent compound of the gonanes is DL-norgestrel, which consists of two isomers, dextro and levo. Only the levo form is biologically active. Levonorgestrel is 10–20 times more potent than the estrane progestin norethinedrone [27,28]. Three less androgenic derivatives of levonorgestrel, namely desogestrel, norgestimate, and gestodene, have been marketed in Europe for many years. OCs with desogestrel and norgestimate, but not gestodene, have been marketed in the US since 1992.

All current OC formulations with less than 50 μg of estrogen contain ethinyl estradiol. Formulations of OCs containing $\geqslant$50 μg of estrogen (either ethinyl estradiol or mestranol as shown in Fig. 12.4) are termed first generation OCs. Those with 20–35 μg of ethinyl estradiol are second generation OCs, unless they contain any of the three newest progestins. The progestins in third generation OCs, desogestrel, norgestimate, or gestodene, have similar or greater progestogenic potency compared to other gonane progestins but less androgenic activity per unit weight. Selectivity is the comparison of a progestin's progestational

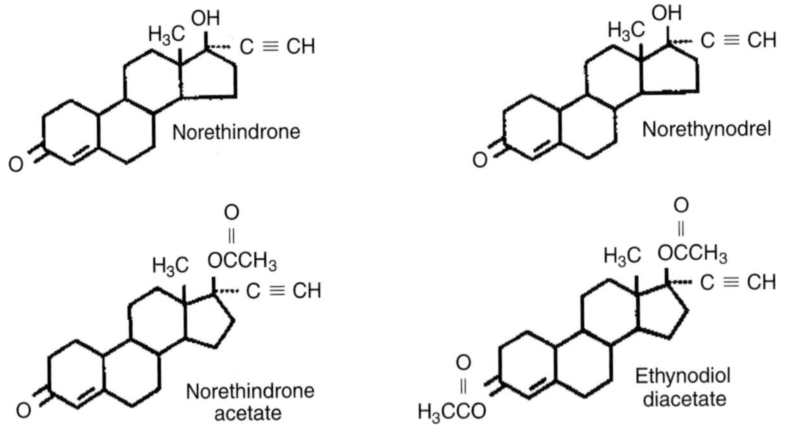

Fig. 12.2 Chemical structures of the estrane progestins used in oral contraceptives.

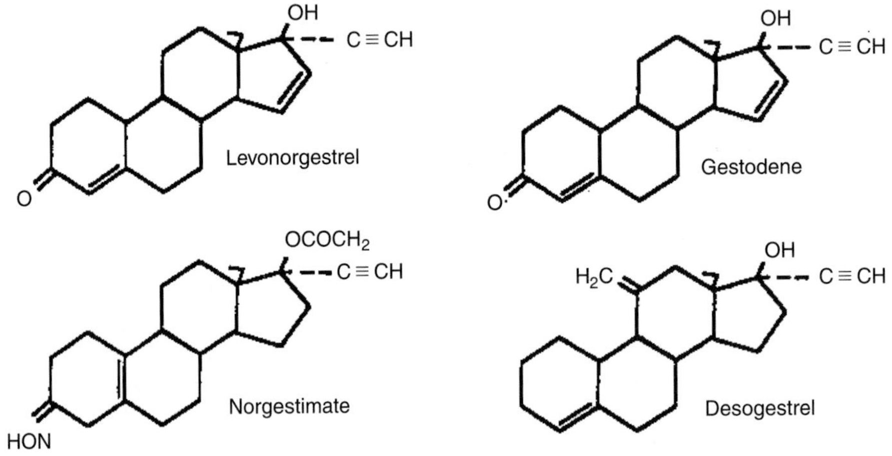

Fig. 12.3 Chemical structure of the gonane progestins used in oral contraceptives.

potency to its androgenic potency. Ethinyl estradiol is about 1.7 times as potent as mestranol and 100 times more potent than an equivalent weight of conjugated estrogen [29].

2. Mechanism of Action

The combination estrogen plus progestin OCs are the most effective formulations, as these preparations consistently inhibit the midcycle gonadotropin surge and thus prevent ovulation.

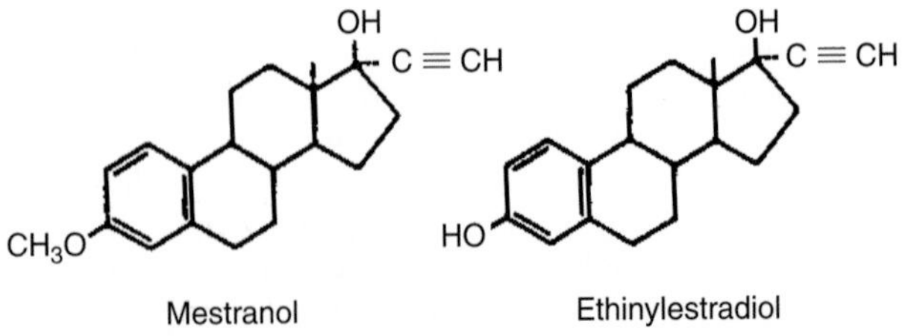

Fig. 12.4 Structures of the two estrogens used in combination oral contraceptives.

Several studies report that most women on OCs have a marked suppression of the release of luteinizing hormone (LH) and follicle-stimulating hormone (FSH) after infusion of gonadotropin releasing hormone (GnRH) indicating a direct inhibitory effect on the pituitary as well as on the hypothalamus. The amount of suppression is unrelated to the age of the woman or the duration of steroid use but is related to the potency of the OC formulation [30]. In comparison, progestin-only pills have a lower dose of progestin and do not consistently inhibit ovulation [31].

Both combination and progestin-only OCs act on many other aspects of the reproductive process including: (1) altering the cervical mucus, making it viscid and thick, thus retarding sperm penetration; (2) decreasing motility of the oviduct and uterus, impairing ova and sperm transport; (3) suppressing the endometrium so that glandular production of glycogen is decreased and therefore allowing less available energy for the blastocyst to survive in the uterine cavity; and (4) altering ovarian responsiveness to gonadotropin stimulation. With current OCs, neither ovarian steroidogenesis nor gonadotropin secretion is completely abolished. Women on OCs have circulating levels of endogenous estradiol similar to those found in the early follicular phase of a normal cycle [32].

3. Effectiveness

No significant differences in clinical effectiveness have been demonstrated among the various combination OCs. (Table 12.3) As long as no tablets are omitted (perfect use), the annual pregnancy rate for combination OCs is less than 0.2%. Among OC users in the United States, about 3% become pregnant during the first year of typical use [33]. During the first year of typical use of the minipill, the percentage of women becoming pregnant ranges from 1.1 to 13.2% [34]. In lactating women, the minipill is nearly 100% effective [35]. Failure rates occur more frequently when one to two pills are missed at the end of the pill-free period when compared to cycles where pills are missed during a cycle.

4. Side Effects

Most women taking new formulation OCs do not experience any of the estrogen-mediated side effects such as nausea (a central nervous system effect), breast tenderness, fluid retention, melasma, and leukorrhea. OCs may cause minor, clinically insignificant changes in circulating vitamin levels, such as decreases in B complex and ascorbic acid [36] and increases in vitamin A [37]. Older, high dose OCs are reported to accelerate the appearance of symptoms of gallbladder disease, related to an estrogenic effect on the gallbladder whereby cholesterol concentration is increased [38–40].

While all new low-dose pills tend to have a beneficial impact on acne and oily skin, some progestins produce androgenic effects. The androgenic side effects that are related to the progestin component of the pill include increased appetite and weight gain, acne, oily skin [41], adverse effect on lipids, diabetogenic effect, pruritis, depression, mood changes, nervousness, and fatigue [36]. The new progestins, as well as use of combination pills with a low dose of progestin, generally minimize these androgenic effects.

Both the estrogen and progestin component of the pill may contribute to problems such as hypertension, headaches, and breast tenderness. Failure to have withdrawal bleeding or amenorrhea is due to a progestin dominant pill and can be alleviated by switching to a more estrogenic formulation.

a. LIVER PROTEIN EFFECTS: SHBG AND CLOTTING CHANGES. A major hepatic effect of androgenic progestins is a dose-dependent suppression of sex hormone binding globulin (SHBG). Because estrogens increase SHBG, measurement of SHBG is a good way to determine the relative estrogen/androgenic balance of different formulations (Fig. 12.5). The greatest increases in SHBG occur after ingestion of combination pills with desogestrel, gestodene, cyproterone acetate [42] and norgestimate, suggesting that these formulations are most useful for treating women with symptoms of hyperandrogenism such as acne, hirsutism, or android obesity.

Orally ingested synthetic estrogens in OCs cause a dose-related increase in hepatic protein production. The increases in proteins leading to increased clotting activity (Factor V, VIII, X, and fibrinogen) are countered by changes in factors that increase antithrombosis action [43] (Fig. 12.6) Another globulin that is

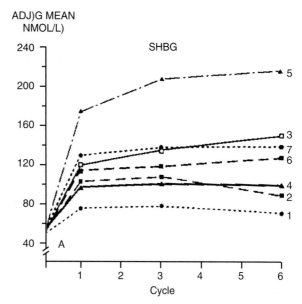

Fig. 12.5 Effects of monophasic levonorgestrel (1), monophasic cyproterone acetate (2), monophasic desogestrel (3), triphasic levonorgestrel (4), monophasic cyproterone acetate (5), triphasic gestodene (6), and biphasic desogestrel (7) on plasma levels of SHBG in 70 healthy women. (From van der Vange, M. A., Blankenstein, M. A., Kloosterboer, H. J., *et al. Contraception* 41:345, 1990, [42].)

increased is angiotensinogen, which when converted to angiotensin may increase blood pressure [44]. While angiotensinogen levels in users taking OCs that contain 30–35 μg of ethinyl estradiol are lower than those in users who ingest higher dose pills, a slight but significant increase in mean blood pressure may occur [45]. The circulating levels of these globulins are directly correlated with the amount of estrogen in the pill and lower dose pills appear to have fewer deleterious effects.

The most serious complications associated with OC use are related to circulatory diseases due to coagulation changes. Cardiovascular disease is most likely to occur in pill users who smoke or who are overweight, diabetic, hypertensive, or over age 50. Epidemiologic studies have shown that the incidence of both arterial and venous thrombosis are directly related to the dose of estrogen [45a] (Table 12.4). The risk of serious cardiovascular disease attributable to the use of new low-dose pills is low. The incidence of hospitalizations for serious cardiovascular

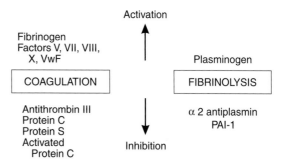

Fig. 12.6 Factors involved in coagulation and fibrinolysis.

Table 12.4

Ratio of Observed to Expected Embolism and Thrombosis in Relation to Type and Dose of Estrogen in Combined Oral Contraceptives

| | Estrogen | | | | | |
| | Mestranol dose (μg) | | | | Ethinyl estradiol dose (μg) | |
	150	100	75–80	50	100	50
Fatal pulmonary embolism[a]	2.8	1.5	0.9	0.6	1.0	0.5
Nonfatal pulmonary embolism[b]	2.3	1.2	0.9	1.0	1.8	0.7
Cerebral thrombosis[c]	3.2	1.2	0.7	0.5	—	0.8
Coronary thrombosis[a]	2.6	1.2	0.3	1.1	1.8	0.9

Modified from Mann, J. I.: *Am. J. Obstet. Gynecol.* **142**: 752, 1982.
[a]Linear trend test: $P < 0.05$.
[b]Linear trend test: $P < 0.01$.
[c]Linear trend test: $P < 0.001$.

complications associated with, or not avoided by, use of low-dose combination OCs is one myocardial infarction, three strokes, and 11 venous thromboses/emboli per 100,000 OC users [46].

The background rate of venous thrombosis/embolism (VTE) in nonpregnant women of reproductive age (not using OCs) is about 0.8 per 10,000 woman-years. A large observational study found that the incidence of VTE among users of OCs with 20–50 μg ethinyl estradiol was 3 per 10,000 woman-years. While this was about four times the background rate, it was still one-half the rate of 6 per 10,000 woman-years associated with pregnancy [47]. Women with an inherited coagulation disorder (such as protein C, S, Leiden factor, or antithrombin III deficiency or activated Protein C resistance) are at higher risk. In one study, the overall risk of developing thrombosis increased from a baseline rate of 6 to 30 per 10,000 woman-years in OC users [48]. Screening for these coagulation deficiencies is recommended in women with a strong family or personal history of thrombotic events.

Four observational studies reported that the risk of VTE in users of third generation OCs was increased compared to that of second generation OCs. While these studies raised concern, the results relating to duration of use were inconsistent and the odds ratio for myocardial infarction (MI) was significantly lower in women using third generation products compared to those using

Table 12.5

Risk of Idiopathic DVT and OC use

Study	Comparison	RR (CI)
WHO	LNG: DSG,GTD	2.6 (1.4–2.8)
UKR	LNG: DSG,GTD	2.2 (1.1–4.0)
Transnational	LNG: DSG,GTD	1.5 (1.1–2.2)
Leiden	LNG: DSG	2.2 (0.9–5.4)
Transnational Reanalysis	LNG: DSG,GTD	1.0
Odds Ratio of MI Transnational Study		
3rd generation OCs versus no use		0.8 CI (.3–2.3)
3rd generation OCs versus 2nd generation OCs		0.3 CI (0.1–0.9)

Note. LNG, levonorgestrel; DSG, desogestrel; GTD, gestodene.

second generation pills. In a reanalysis of the Transnational study, after the data were analyzed according to duration of OC use, the risk of VTE for first-time users was dependent on duration of prior OC use and essentially identical for second and third generation pills (Table 12.5).

b. CARBOHYDRATE METABOLISM. Carbohydrate metabolism is not clinically affected in women on current low-dose OCs, although older high-dose pills had a profound effect on glucose and insulin levels. Conflicting data exist on the effect of the estrogen component; however, it probably acts synergistically with progestin to impair glucose tolerance [49]. Generally, the higher the dose and potency of the progestin, the greater the magnitude of impaired glucose metabolism. Due to the minor effects on glucose, insulin, and glucagon levels, low-dose pills are safe to use in women with a history of gestational diabetes [50] and in many insulin-dependent diabetics. In the Nurses Health Study, although type 2 diabetes developed in more than 2000 women, the risk was not increased among current OC users (relative risk (RR) = 0.86, 95% confidence interval (CI) = 0.46–1.61) [51].

c. LIPIDS AND LONG TERM RISK OF ATHEROSCLEROSIS. Estrogens in OCs increase HDL-cholesterol, total cholesterol, and triglyceride levels and decrease LDL-cholesterol levels. The progestin component causes decreases in HDL-, total cholesterol, and triglyceride levels and increases LDL-cholesterol. While older formulations with progestin-dominant formulations had an overall adverse lipid effect, current low-dose formulations report little change in HDL- and LDL- or total cholesterol, although triglyceride levels are increased. Formulations with a high estrogen to progestin ratio (*i.e.,* those containing ≤0.5 mg norethindrone or new progestins) increase HDL- and triglycerides and decrease LDL-cholesterol [52,53] (Fig. 12.7) [53a].

In spite of any changes in lipids, there is no increased risk of atherosclerosis in former users of old or new formulation OCs [54,55]. In fact, data continue to accumulate showing that OCs have a long-term protective effect on vascular disease [56], although there is an increased short-term risk of thrombosis associated with current OC use. Evidence that myocardial infarction

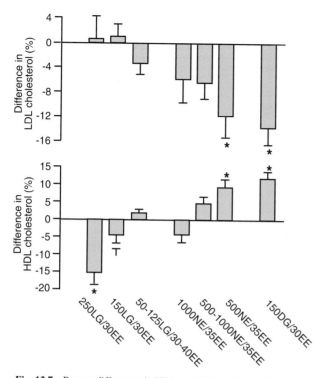

Fig. 12.7 Percent differences in HDL and LDL cholesterol levels and in the incremental area for insulin in response to the OGTT between women taking one of seven combination oral contraceptives and those not taking oral contraceptives. The T bars indicate 1 SD. The asterisk (p < 0.001) and dagger (p < 0.01) indicate significant differences between users and nonusers in the mean values for the principal metabolic variables. (Modified from Godsland, I. F., Crook, D., Simpson, R., *et al. N. Engl. J. Med. 323:*1375, 1990 [53a].)

in current pill users is due to thrombosis and not atherosclerosis is supported by an angiographic study of young women diagnosed with an MI, where evidence of coronary atherosclerosis was present in 79% of non-OC-users but only 36% of OC users [57]. Again, data support a long-term protective effect rather than an increase in risk.

d. Neoplastic Effects.

Breast Cancer Risk Because estrogen stimulates the growth of breast tissue, there are concerns that OCs may either initiate or promote breast cancer among users. In 1991, a comprehensive meta-analysis of all published epidemiologic studies reported that a summary relative risk in OCs users was 1.0 (95% CI = 1.0–1.1) (Table 12.6) [58]. Eight case-control studies and one cohort study investigated the relative risk of breast cancer in OC users who did and did not have a family history of breast cancer. None of the studies showed a significant difference in either group [36].

In 1996, an international group reanalyzed the entire worldwide epidemiologic data on OCs and breast cancer [59]. Analysis was performed on data from 54 studies representing 25 countries and more than 53,000 women with breast cancer and over 100,000 controls. Women who used OCs had a slightly increased risk of developing breast cancer (RR = 1.24, 95%

CI = 1.15–1.33) [59]. The risk declined steadily after stopping use of OCs so that ten years after use, the relative risk was 1.01 (95% CI = 0.96–1.05). It is of interest that the breast cancers diagnosed in women taking OCs were less advanced clinically. The risk of having breast cancer that had spread beyond the breast compared to a localized tumor was significantly reduced (RR = 0.88, 95% CI = 0.81–0.95) in current OC users compared to nonusers.

Overall, the vast body of data concerning OCs is reassuring. It is likely that contraceptive steroids promote the growth and increase the chance of early diagnosis of existing cancers, thus avoiding the usual pattern where it takes many years before breast cancer is clinically detected. Use of the lowest dose OC is advocated.

Cervical Cancer The epidemiologic data regarding the risk of invasive cervical cancer in OC users are conflicting. Confounding factors such as the number of sexual partners, age at first intercourse, exposure to human papillomavirus (HPV), use of barrier contraceptives and spermicides, cigarette smoking, and cytologic screening frequency may account for the results in some studies. Schlesselman's review of 14 studies of over 3800 women with invasive cervical cancer reported a significantly increased risk with increased duration of use (Fig. 12.8) [60]. Other studies reported that long-term use increased the risk 1.5- to 2.5-fold [61,62]. In contrast to these findings, other well-controlled studies indicated no change in the risk of cervical intraepithelial neoplasia (CIN) with OC use [61,63]. A population-based case-control study reported that OC use and cigarette smoking influenced the ability of HPV infection to cause invasive cervical cancer [64]. The authors suggest that OC use may only be important in the etiology of invasive squamous cell cervical tumors if the use occurs at a critical time in the development of a woman's reproductive tract, that is at ages ≤17 years.

Endometrial Cancer Thirteen of fifteen case-control and cohort studies reported that OCs protect against endometrial cancer, the third most common cancer among US women [65]. Women who used OCs for at least one year had a 50% lower risk of developing endometrial cancer between ages 40 and 55 compared to nonusers. A significant trend of decreasing risk with increasing duration of combined OC use has been observed (Fig. 12.9) [60]. The protective effect is highest in nulliparous women and occurs with combination formulations with both high and low doses of progestins [66].

Ovarian Cancer Eighteen of 20 published reports find a reduction in the risk of ovarian cancer among OC users, specifically the most common type, epithelial ovarian cancer. (Fig. 12.10) [67]. The summary relative risk for development of ovarian cancer among ever users of OCs was 0.64, a 36% reduction. OCs reduce the risk of the four major histologic types of epithelial ovarian cancer (serous, mucinous, endometrioid, and clear-cell) and the risk of both invasive ovarian cancers and those with low malignant potential. The magnitude of the decrease in risk is directly related to the duration of use, ranging from a 40% reduction with four years of use to a 53% reduction after eight years and a 60% reduction after 12 years. The protective effect begins within 10 years of first use and continues for at least 20 years after OC use ends. There is a similar level of protection with low-dose monophasic formulations as well as

Table 12.6

Relative Risks of Breast Cancer in Women in Developed Countries Who Ever Used Oral Contraceptives:
Case-Control Studies of Women of All Ages at Risk of Exposure

Source, Year	Upper age limit of cases	Cases/Controls		RR estimate (95% CI)[a]
		Users	Nonusers	
Henderson *et al.,* 1974	64	59/69	248/238	0.7 [0.5–1.2]
Pallenbarger *et al.,* 1977	50	226/398	226/474	1.1 [0.9–1.4]
Sartwell *et al.,* 1977	74	22/34	262/333	0.9 (0.5–1.5)
Awnihar *et al.,* 1979	64	30/65	160/315	0.9 [0.6–1.5]
Vessey *et al.,* 1981	74	30/141	300/1207	0.9 (0.6–1.3)
Harris *et al.,* 1982	54	36/189	73/279	1.0 (0.6–1.4)
Vessey *et al.,* 1983	50	537/554	639/622	1.0 (0.8–1.2)
Rosenberg *et al.,* 1984	59	397/2558	794/2468	0.9 (0.8–1.1)
Paganini *et al.,* 1985	79	15/23	353/351	0.7 (0.4–1.4)
CASH 1986	54	2743/2802	1870/1774	1.0 (0.9–1.1)
Paul *et al.,* 1986	54	310/708	123/189	0.9 (0.7–1.3)
La Vecchia *et al.,* 1986	60	104/178	672/1104	1.1 (0.8–1.5)
Awnihar *et al.,* 1988	54	162/467	372/1522	1.6 (1.3–2.1)
Stanford *et al.,* 1989	>60	481/515	1541/1668	1.0 (0.9–1.2)
WHO 1990	62	438/1496	716/1888	1.1 (0.9–1.3)
Summary RR[b]	—	—	—	1.0 [1.0–1.1]

From Thomas, D. B. Contraception **43:**597, 1991 [58].

[a]Indicates confidence intervals estimated from published data.

[b]P-value of chi-square test for heterogeneity = 0.08.

higher dose agents [68]. As with endometrial cancer, the protective effect occurs only in women of low parity (≤4 births) who are at greatest risk for this type of cancer.

Liver Adenoma and Cancer   The development of benign hepatocellular adenoma is a rare occurrence associated with prolonged use of high-dose OC formulations, particularly those containing mestranol. Although two British studies reported an increased risk of liver cancer among users, the number of patients was small and there were many confounding factors [69]. Data from a large multicenter epidemiologic study coordinated

by the WHO found no increased risk of liver cancer in OC users, even with increasing duration of use [70].

Pituitary Adenoma   OCs mask the symptoms of a prolactinoma, namely amenorrhea and galactorrhea. When OC use is discontinued, these symptoms appear, suggesting a causal relationship. However, data from three studies indicate that the incidence of pituitary adenoma is not higher in pill users compared to matched controls [71].

Malignant Melanoma   Several epidemiologic studies assessing the relationship between OC use and the development of

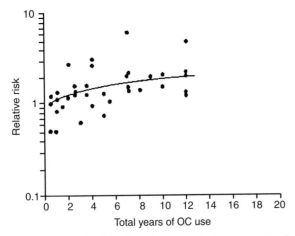

Fig. 12.8   Relative risk of cervical cancer by total years of oral contraceptive use. (From Schlesselman, J. J. *Obstet. Gynecol.* 85: 793, 1995, [60].)

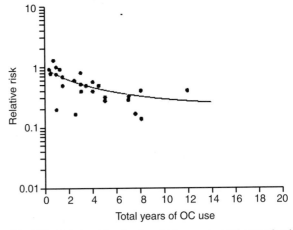

Fig. 12.9   Relative risks of endometrial cancer by total years of oral contraceptive use. (From Schlesselman, J. J. *Obstet. Gynecol.* 85: 793, 1995, [60].)

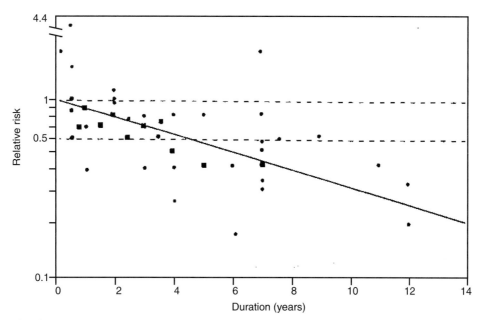

Fig. 12.10   Relative risk of ovarian cancer associated with different durations of oral contraceptive use; findings of 15 studies. Study categories, indicating category weights ranging from smallest (weight in bottom 25% of range) to largest (weight in top 25% of range): red circles = 1 (smallest); red squares = 2; black circles = 3; black squares = 4 (largest). (From Hankinson, S. E., Colditz, G. A., Hunter, D. J., and Rosner, B. *Obstet. Gynecol.* 80:708, 1992, [67].)

malignant melanoma report ambiguous results. A report from the Royal College of General Practitioners' Oral Contraceptive Study and the Oxford/Family Planning Association Study, involving over 450,000 woman-years of observation, concluded that oral contraceptive use is probably not associated with an increased risk of melanoma [72].

5. Contraindications for Use of OCs

It is safe to prescribe OCs to the majority of reproductive aged women. Absolute contraindications include a history of thromboembolism, atherosclerosis, stroke, or any systemic disease with associated vascular disease (such as lupus or diabetes with retinopathy or nephropathy). Cigarette smoking in women over age 35, uncontrolled hypertension, pregnancy, pituitary macroadenoma, migraine headaches with localizing signs, acute liver disease, and cancer of the breast or endometrium are also contraindications. Women with functional heart disease should not use OCs as fluid retention may lead to heart failure. Hypertension in older or obese women is a relative contraindication.

The following are not contraindications to pill use: asymptomatic mitral valve prolapse, age over 40, past history of liver disease, family history of breast cancer, controlled diabetes, and prolactin secreting pituitary microadenoma. OCs should be stopped in women who develop more severe headaches, fainting, temporary loss of vision or speech, or paresthesias while taking OCs.

6. Drug Interactions

Some drugs interfere with the action of OCs by inducing liver enzymes that convert the steroids to more polar and less biologically active metabolites. These drugs include barbiturates, sulfonamides, cyclophosphamide, and rifampin. There may be a higher incidence of OC failure in women using these drugs. The clinical data concerning failures during treatment with certain antibiotics (*e.g.* penicillin, ampicillin, sulfonamides, and tetracycline), analgesics, and barbiturates (e.g. phenytoin) are less clear.

7. Noncontraceptive Health Benefits

In addition to being one of the most effective methods of contraception, OCs provide many health benefits.

a. Reproductive Effects.   OC use appears to have a long-term beneficial effect on fertility due to a decreased risk of salpingitis, ectopic pregnancy, and ovarian cyst formation. However, there is a short-term negative impact due to a variable delay in the return of ovulation after stopping OCs. For two years after stopping pill use, fertility is lower than in women discontinuing barrier methods (Fig. 12.11) [73,74].

b. Benefits from Antiestrogenic Action of Progestins. Both natural and synthetic progestins inhibit the proliferative effect of estrogen, the so-called antiestrogenic effect. Progestins decrease the number of estrogen receptors and also stimulate the activity of the enzyme estradiol 17-β-dehydrogenase within the endometrial cell. This enzyme converts the more potent estradiol to the less potent estrone, reducing estrogenic action within the cell. The progestin action limits the height of the endometrium compared to an ovulatory cycle, resulting in a reduction in the amount of blood loss at the time of endometrial shedding. In a normal ovulatory cycle the mean blood loss is about 35 ml compared to 20–25 ml for OC users. As a result, OC users are

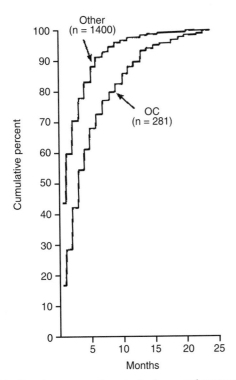

Fig. 12.11 Cumulative conception rates for former oral contraceptive and other contraceptive users, Yale-New Haven Hospital 1980 to 1982. (From Bracken, M. B., Hellenbrand, K. G., and Holford, T. R. *Fertil. Steril.* 53:21, 1990, [74].)

Table 12.7

Rate Ratio Estimates for Functional Ovarian Cysts Comparing Each Oral Contraceptive Category with No Oral Contraception

	Rate ratio[a]	95% Confidence interval
No prescription	1.00	Reference category
Active prescription:		
Multiphasic	0.91	0.30–2.31
≤35 μg estrogen	0.52	0.17–1.33
>35 μg estrogen	0.24	0.01–1.34

From Lanes, A. F., Birmann, B., Walter, A. M., and Singer, S. *Am. J. Obstet. Gynecol.* 166:956, 1992 [79].

[a]Rate ratios standardized to age distribution of index (i.e., "exposed") category.

order linked to incessant ovulation, ovarian cancer, is significantly reduced in users (see above).

d. OTHER BENEFITS. Several European studies reported that the risk of development of rheumatoid arthritis in OC users was only half that of controls [79] while other studies have not found this protection [80]. Another benefit is protection against salpingitis, commonly referred to as PID. The risk is reduced by 50% and may relate to a decreased duration of menstrual flow or to thickened cervical mucus preventing bacteria from ascending to the upper genital tract. OCs reduce the risk of ectopic pregnancy by more than 90% in current users [41]. There are several studies indicating that OCs reduce bone loss in perimenopausal women [81], particularly those with oligomenorrhea, reduce the risk of fibroids [82] and improve bleeding control [83].

E. Long-Acting Injectable Contraceptives

Because most of the long-acting steroid formulations contain only a progestin, without an estrogen, the endometrium is not maintained and uterine bleeding occurs at irregular and unpredictable intervals. The three types of injectable steroid formulations include depo-medroxyprogesterone acetate (DMPA), norethindrone enanthate, and several once-a-month injections of combinations of progestins and estrogens. Injectable contraceptives are a popular method of contraception worldwide although only DMPA is approved for use in the US.

1. Depot Formulations of MPA

DMPA is a 17-acetoxy progesterone and is the only progestin used for contraception that is not a 19-nortestosterone derivative. The acetoxyprogestins are structurally related to progesterone and do not have androgenic activity. MPA was commonly used in OC formulations until US regulatory approval ended in response to studies reporting that MPA increased the risk of mammary cancer in beagle dogs. It was later learned that beagles uniquely metabolize 17-acetoxy progestins to estrogen thus stimulating mammary hyperplasia. In humans DMPA is not metabolized to estrogen. In 1992, after epidemiologic studies reported that DMPA does not increase the risk of breast cancer in

about half as likely to develop iron deficiency anemia and are less likely to suffer menorrhagia, irregular menstruation, or intermenstrual bleeding. Because these disorders are frequently treated by curettage or hysterectomy, users require these procedures less frequently. Additionally, adenocarcinoma of the endometrium is significantly less likely in OC users [65].

Estrogen exerts a proliferative effect on breast tissue, which contains estrogen receptors. Progestins probably inhibit the synthesis of estrogen receptors in the breast, thus exerting an antiestrogenic action on the breast. Several studies have shown that OCs reduce the incidence of benign breast disease, indicating that this reduction is directly related to the amount of progestin. Studies report that current OC users have an 85% reduction in fibroadenomas and a 50% reduction in chronic cystic disease and nonbiopsied breast lumps compared with nonusers [75].

c. BENEFITS FROM INHIBITION OF OVULATION. Disorders such as dysmenorrhea and premenstrual tension occur less frequently in anovulatory cycles. Only about 30% of users previously suffering from dysmenorrhea report no improvement after starting OCs [76]. Low-dose, combined OCs virtually eliminate cyclic PMS symptoms for most women [77]. When ovulation is inhibited, functional cysts are less likely to develop [78] (Table 12.7). The incidence of functional cysts is reduced by 80–90% in women using OCs [75]. Users of multiphasic OCs also have protection from cysts, although to a lesser extent. Another dis-

humans, the Food and Drug Administration (FDA) approved DMPA as a contraceptive agent.

In a large WHO trial, the pregnancy rate at one year was only 0.1% and the cumulative rate at two years was 0.4%. Inhibition of ovulation is the primary mechanism of action, although DMPA also causes the endometrium to become thin, thus preventing implantation. In addition, DMPA makes the cervical mucus thick and viscous, making it unlikely that sperm are able to reach the oviduct.

a. PHARMACOKINETICS.   DMPA can be detected in the systemic circulation 30 minutes after injection, and levels rise steadily to effective blood levels (>0.5 ng/ml) within 24 hours. MPA levels plateau at 1.0–1.5 ng/ml for about three months, decline to about 0.2 ng/ml during the fifth month, and are often detectable for up to 7–9 months. Estradiol levels remain below 100 pg/ml during the first four months. Return of follicular activity precedes the return of luteal activity by 2–3 months. Return of ovulation may not occur in some users for 7–9 months when MPA levels fall to <0.01 ng/ml.

b. OVULATORY SUPPRESSION.   Mishell *et al.* reported that during the first two months after injection, although the midcycle LH/FSH peak is suppressed, LH and FSH are still secreted in a pulsatile manner [84]. During this time, estradiol levels varied from 5 to 100 pg/ml (mean 42 pg/ml) among women who used DMPA for 4–5 years, although none had hypoestrogenic complaints.

c. RETURN OF FERTILITY.   Because of the lag time in clearing DMPA from the circulation, resumption of ovulation may be delayed for as long as one year after the last injection. After this delay, there is normal fecundability [85] (Fig. 12.12) The delay in return to fertility is not related to the number of injections but increases with increasing weight, most likely because the drug is absorbed into adipose tissue and is not rapidly cleared.

d. ENDOMETRIAL CHANGES.   Endometrial biopsies taken at intervals of 1.5, 3, 6, 9, and 12 months after the first DMPA injection in a group of women receiving DMPA every three months showed about half the biopsies were proliferative at six weeks. After the second injection, <10% of the biopsies were proliferative. The majority of specimens showed a quiescent type of endometrium, but after one year of DMPA, about 40% were characterized as atrophic.

e. ADVERSE EFFECTS.

Clinical   The major side effect of DMPA is complete disruption of the menstrual cycle. In the three months after the first injection, about 30% of women are amenorrheic and another 30% have irregular bleeding and spotting occurring more than 11 days per month. The bleeding is usually light and does not cause anemia. By the end of two years, about 70% of women treated with DMPA are amenorrheic [86]. After discontinuation of DMPA, about half resume regular cycles within six months and three-fourths have regular menses within one year.

In five cross-sectional studies, users of DMPA weighed more than women not using hormonal contraceptives [36]. Several longitudinal studies have indicated that DMPA users gain be-

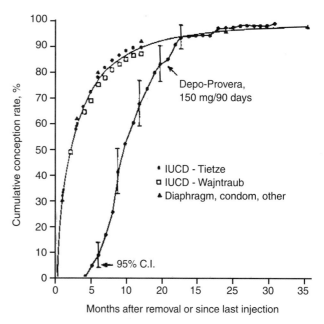

Fig. 12.12   Cumulative conception rates of women who discontinued a contraceptive method to become pregnant. (From Schwallie, P. C., and Assenzo, J. R. *Contraception* 10(2):181, 1974, [85].)

tween 1.5 and 4 kg per year [87] which is higher than the typical weight gain of about 1 kg per year.

The product labeling lists depression and mood changes as side effects of DMPA. Several studies, however, indicate that the incidence of depression and mood change is less than 5%. No clinical trials with a comparison group have been performed to determine whether a causal relation between use of DMPA and development of depression exists. Although development of headaches is the most frequent medical event reported by DMPA users and is a common reason for discontinuation, there are no comparative studies indicating that DMPA increases the incidence or severity of either tension or migraine headaches.

Metabolic   DMPA does not increase liver globulin production; therefore, there are no alterations in blood clotting factors or angiotensinogen levels associated with its use. DMPA is not associated with an increased incidence of hypertension or thromboembolism. A WHO study reported that blood pressure measurements were unchanged in DMPA users after two years of injections [88].

There is little or no change in mean triglyceride and total cholesterol levels with DMPA use. In all seven studies in which HDL cholesterol level was measured, levels were lower among DMPA users. Of five studies measuring LDL cholesterol, three noted an increase among users [36]. There are no studies reporting an increased incidence of cardiovascular events among current or former long-term DMPA users and there is no evidence that DMPA is associated with an acceleration of atherosclerosis.

Bone Loss   In one cross-sectional study using DEXA bone scanning, 30 long-term DMPA users had a reduction in lumbar spine and femoral bone mineral density (BMD) compared to 30 nonusers [89]. A follow-up report indicated that after stopping DMPA, there was normalization of BMD [89]. Data suggest that loss of BMD with DMPA can be minimized or avoided with adequate calcium intake among users. Other long-term studies have not shown a decrease in BMD with DMPA use and, to date, no studies have been able to demonstrate an increased risk of fracture in current or former users [36].

Cancer Risk Approval of DMPA for contraceptive use in the US was delayed for many years due to a concern about an increased risk of cervical, breast, and endometrial cancer. Two large case-control studies indicated that the relative risk of developing breast cancer among DMPA users was not increased [90,91]. In those who had used the drug for more than five years or more than 14 years earlier, the risk of developing breast cancer was also not increased (RR = 1.0, 95% CI 0.7–1.5 and RR = 0.89, 95% CI 0.63–1.3, respectively). However, among women under age 35 who had started use within the last five years, there was a significantly increased risk of breast cancer (RR = 2.0 95% CI = 1.5–2.8), similar to that found with use of OCs. Overall, DMPA does not appear to change the incidence of developing breast cancer.

A WHO case-control study found the risk of developing endometrial cancer to be significantly reduced among DMPA users (RR = 0.21, 95% CI = 0.06–0.79). This reduction in risk persisted for at least eight years after stopping use [92].

In a WHO case-control study, the risk of developing ovarian cancer among DMPA users was unchanged (RR = 1.1, 95% CI = 0.6–1.8) [93]. These findings do not demonstrate a protective effect similar to that observed with OCs despite inhibition of ovulation with both agents. The lack of a protective effect observed with DMPA was probably due to the fact that in the countries studied, DMPA was given only to multiparous women (women at low risk of developing epithelial ovarian cancer).

In several large case-control studies, the risk of developing invasive cancer of the cervix was not significantly increased (RR = 1.1, 95% CI = 1.0–1.3) with either short- or long-term use. The overall risk of developing cancer *in situ* was slightly increased in the WHO study (RR = 1.4, 95% CI = 1.2–1.7), but not in the Costa Rica study site (RR = 1.0, 95% CI = 0.6–1.8) [94] or in two New Zealand studies [95,96].

f. CONTRACEPTIVE HEALTH BENEFITS. There is good epidemiologic evidence that use of DMPA reduces the risk of developing iron deficiency anemia, PID, and endometrial cancer and has beneficial effect in women with sickle cell disease (Table 12.8).

Table 12.8
Noncontraceptive Health Benefits of Contraceptive Use of DMPA

Definite
  Salpingitis
  Endometrial cancer
  Iron deficiency anemia
  Sickle cell problems
Probable
  Ovarian cysts
  Dysmenorrhea
  Endometriosis
  Epileptic seizure
  Vaginal candidiasis
  Premenstrual syndrome

DMPA also reduces seizure frequency in women with epilepsy and probably reduces the incidence of primary dysmenorrhea, ovulation pain, and functional ovarian cysts. DMPA reduces the symptoms of endometriosis, and in two small studies it reduced the incidence of vaginal candidiasis [36].

2. Norethindrone Enanthate (NET-EN)

NET-EN is an injectable progestin that is approved for contraceptive use in more than 40 countries but not in the US. Because of a shorter duration of action, it is recommended that NET-EN (200 mg in an oily suspension) be given every 60 days for at least the first six months of use and not less often than every 12 weeks thereafter.

3. Progestin-Estrogen (Monthly) Injectable Formulations

Because of the bleeding problems associated with progestin-only formulations, several combined progestin-estradiol ester injectables have been developed. They are designed for once-a-month administration and are often associated with regular withdrawal bleeding. Four formulations are currently the most popular. A formulation containing 17α-hydroxyprogesterone caproate 250 mg and estradiol valerate 5 mg is used in China. A combination of dihydroxyprogesterone acetophenide 150 mg plus estradiol enanthate 10 mg is widely used in Mexico and Latin America. The WHO has developed two new formulations that are used by several national family planning programs. A combination of MPA 25 mg and estradiol cypionate is marketed as Cyclofem (or Cycloprovera) and a combination containing NET-EN 50 mg and estradiol valerate 5 mg is marketed as Mesigyna. These preparations offer better bleeding control than progestin-only products. About 85% of cycles are regular and amenorrhea rates are low. The results of five clinical trials with Mesigyna and Cyclofem demonstrate 12 month pregnancy rates of ≤0.4% and 0.2%, respectively [97].

F. Subdermal Implants

Subdermal implants of polydimethylsiloxane (Silastic) capsules containing levonorgestrel (levo) were developed by the Population Council and approved in the US in 1990. To date, Norplant has been used by more than 5½ million women and is currently approved by regulatory agencies of 60 countries. The rate of steroid delivery is directly proportional to the surface area of the capsules, and the duration of action depends on the amount of steroid within the capsules. There are six cylindrical capsules, 3.4 cm long and 2.4 mm in outer diameter, each filled with 36 mg of crystalline levo.

After insertion, blood levels of levo rise rapidly to reach levels between 1000 and 2000 pg/ml in 24 hours. After the first month, levo levels remain relatively constant during the first year [98] (Fig. 12.13). After five years, mean levo levels are 170–350 pg/ml. Heavier women have lower circulating levels than thin women. The daily release rate of levo averages 50 pg/ml during the first year and drops to 30 pg/ml after eight years.

1. Mechanism of Action

It is difficult to determine the incidence of ovulatory cycles in Norplant users. In one report where serum progesterone levels > 3 ng/ml were used as evidence of ovulation, one-third of cycles were presumably ovulatory. However, progesterone levels

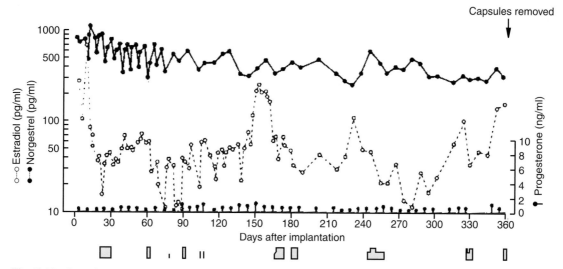

Fig. 12.13 Serum levels of estradiol, progesterone, and *d*-norgestrel in a subject with six polysiloxanne capsules, each containing 33.9 mg of *d*-norgestrel, implanted on day 0. Hatched bars represent uterine bleeding. (From Moore, D. E., Roy, S., Stanczyk, F. Z., *et al. Contraception* 17:315, 1978, [98].)

in users were much lower than those in controls, indicating a high incidence of luteal deficiency and/or anovulatory luteinized follicles. Daily scanning of Norplant users with regular cycles and elevated luteal-phase progesterone levels demonstrated that only one-third of the cycles had ovarian ultrasonic findings consistent with ovulation [99]. Because only half of the cycles in users are associated with regular menses, these data suggest that less than 20% of users' cycles are ovulatory and a high percentage of these have deficient progesterone production.

The low levels of circulating levo in users do not completely suppress gonadotropins, and ovarian follicular activity results in periodic peaks of estradiol. Circulating levo suppresses the positive feedback effect of estradiol on LH release and therefore, ovulation, especially during the first two years of use, rarely occurs. Additionally, these low levels of levo prevent the normal midcycle thinning of cervical mucus and normal sperm penetration does not occur [100].

2. Effectiveness

Some of the early studies reported higher failure rates after two years of Norplant use in women who were heavy (1.7% per year in women weighing more than 70 kg) [101]. However, in these studies the capsules were made of uncommonly dense tubing and serum levels of levo were lower than levels seen with regular tubing. The relationship of efficacy to body weight has disappeared in recent clinical trials. Annual pregnancy rates for the first five years of current use are 0.2 per 100 women of all body weights yielding a cumulative five year pregnancy rate of 1.1% [92].

3. Ectopic Pregnancies

As with all progestin-only methods, when pregnancies occur with Norplant, a high percentage (17%) are ectopic. Because of its high rate of effectiveness, however, the overall frequency of ectopic pregnancy in users (1.3 per 1000 woman-years) is reduced compared with the proportion of pregnancies that are

ectopic among women in the US who are not using contraception at the time of conception (1.0–1.5%) [3].

4. Side Effects

a. CLINICAL. Mean estradiol levels in users are similar to levels in women with regular ovulatory cycles. About half of users have periodic, irregular peaks of estradiol (up to 400 pg/ml), 30% have fluctuating levels (above 400 pg/ml), and about 10% have low levels (below 75 pg/ml) [102]. Following falls in estradiol, endometrial sloughing and uterine bleeding or spotting usually occurs.

Bleeding episodes tend to be more prolonged and irregular during the first year of use, after which regular patterns are more common. During the first year of use, about 35% of cycles are regular, 55% irregular, and 10% amenorrheic. (Fig. 12.14). By the fifth year, about two-thirds of cycles are regular and one-third are irregular. The mean number of days of bleeding declines steadily from about 54 days in the first year to 44.1 days in the fifth year [103]. Despite the increased number of spotting and bleeding days, hemoglobin concentrations rise in implant users, even in those who stop the method because of bleeding, because of the decreased average amount of menstrual blood loss (<25 ml per month).

Other problems include local infection, irritation, or painful reaction at the insertion site. Occasionally, expulsion of the capsule occurs, usually in association with infection. The incidence of insertion-site infection is less than 1% [3]. Headache is the most frequent medical condition causing removal of the implants (30%). Weight gain is another common reason for removal in US studies, whereas weight loss is reported in the Dominican Republic. Other medical problems include acne, mastalgia, mood changes, anxiety, depression, and nervousness. Because ovarian follicular development without subsequent ovulation is common, adnexal enlargement due to persistent unruptured follicles occurs eight times more frequently in users compared to normally cycling women. These enlarged follicles,

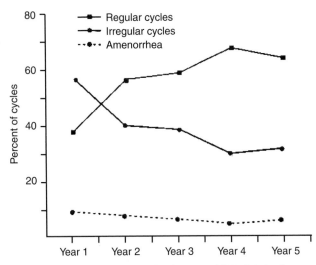

Fig. 12.14 Bleeding patterns calculated on a monthly basis in implant users during 5 years of use. (From Shoupe, D., Mishell, D. R. Jr., Bopp, B. L., and Fielding, M. *Obstet. Gynecol.* 77:256, 1991, [103].)

which may reach 5–7 cm in diameter, regress spontaneously. Additionally, users infected with genital herpes simplex virus complain of more frequent outbreaks after insertion [3].

b. METABOLIC.   Studies of carbohydrate metabolism, serum chemistries, liver function, serum cortisol levels, thyroid function, immunoglobulins, and blood coagulation reveal only minimal changes in users. In most studies, levels of triglycerides, total cholesterol, and LDL-cholesterol decline slightly during use, whereas HDL cholesterol declines slightly or increases [104]. There is little change in the cholesterol/HDL cholesterol ratio, indicating that Norplant should not enhance the development of atherosclerosis [36].

5. Insertion and Removal

Insertion is performed in an outpatient setting and the entire procedure takes about five minutes. After infiltration of the skin with local anesthesia, a small (3 mm) incision is made with a scalpel, usually in the upper inner arm. Capsules are implanted into the subcutaneous tissue in a radial pattern. Removal of the implants is a more difficult procedure than insertion, in part because fibrous tissue develops around each capsule. Difficult removals are generally associated with capsules that have been placed too deeply. Return to ovulation is similar to that in women discontinuing other hormonal contraceptives, reaching 50% at three months and 86% at one year. Continuation rates range from 76 to 99% at one year and from 33 to 78% at five years. These rates are similar to those observed with IUDs and are due in part to the fact that the user must return to the clinic for removal.

6. Implants on the Horizon

A new implant consisting of two solid rods called Norplant-2® has been approved by the US FDA. These rods are easier to insert and remove than six capsules. Two 4-cm rods, containing a homologous mixture of silastic and crystalline levo, have the same total release rate of 50 µg/day as Norplant's six 3.4-cm capsules. Single implants with other progestins such as desogestrel (Implanon) are undergoing clinical trials. Implanon implants have a duration of action of three years and are easy to insert and remove. Side effects are low and efficacy rates remain high [105].

G. Emergency Contraception

Various compounds have been used for postcoital contraception, or for what is commonly referred to as the "morning after pill." It is estimated that use of emergency contraception could reduce the number of unintended pregnancies each year by 1.7 million and the number of abortions by 800,000 [106]. Emergency use of OCs reduces the risk of pregnancy after unprotected intercourse by at least 74% [99] (Table 12.9) [107]. Emergency contraceptive pills are ordinary OCs (2–4 pills in one dose) taken within 72 hours of unprotected coitus. A second dose is taken 12 hours later. The FDA has approved Lo/Ovral, Levlen, Nordette, Triphasil, and Trilevelen for use as emergency contraceptive pills. The side effects, including nausea, vomiting, breast soreness, and menstrual irregularities, tend to reduce compliance.

Another postcoital option is insertion of a copper 380 T IUD for up to 7–10 days after unprotected sex. This is used in women who are candidates for an IUD and plan to continue its use. Mifepristone (RU 486) is also reported as an effective postcoital contraceptive when used as a 600 mg dose [108].

H. Intrauterine Devices (IUDs)

The main benefits of IUDs are a high level of effectiveness, a lack of associated systemic metabolic changes, and the need for a single visit for insertion. Despite these advantages, only 1% of reproductive age women in the US use the IUD compared to 14–30% in most European countries and Canada.

1. Effectiveness

First-year failure rates with the copper (Cu) T 380A IUD are less than 1% and are 2–3% for the progesterone-releasing IUD [4] Expulsion and pregnancy rates are related to the skill of the clinician inserting the device. The cumulative pregnancy rate after seven years use of the Cu T 380A is only 1.6%. The incidence of all major adverse events with IUDs, including pregnancy, expulsion, or removal for bleeding and/or pain, steadily decreases with increasing age (Table 12.10) [4]. The Cu T 380A is currently approved for 10 years of use.

2. Types of IUDs

The IUDs of the 1960s were made of polyethylene impregnated with barium sulfate to make them radiographic. In order to diminish the frequency of side effects, including bleeding and pain, smaller plastic devices covered with copper were developed. In the 1980s, devices bearing a larger amount of copper, including sleeves on the horizontal arm, such as the Cu T 380 A, the Cu T 220 C, and the multiload Cu 250 and Cu 375, were introduced. The multiload Cu 375 is widely used in Europe.

Adding a reservoir of progesterone to the vertical arm also increases the effectiveness of the T-shaped devices. The Proges-

Table 12.9
Observed and Expected Pregnancies According to Various Methods of Postcoital Contraception

Treatment	Number of studies	Total number of patients considered	Number of observed pregnancies	Pregnancy rate (%)
Ethinyl estradiol—high dosage	4	3168	19	0.6
Other estrogens—high dosage	2	975	11	1.1
Ethinyl estradiol plus *dl*-norgestrel	11	3802	69	1.8
Danazol	3	998	20	2.0
IUD	9	879	1	0.1

From Fasoli, M., Parazzini, F., Cecchetti, G., *et al. Contraception* 39:459, 1989 [67].
    For heterogeneity = 35.31; $P < 0.001$.

tasert IUD releases 65 mg of progesterone into the endometrial cavity daily that prevents pregnancy by a local action without a measurable increase in peripheral serum progesterone levels. Because of this progestational effect on the endometrium, the amount of uterine bleeding is reduced and this IUD can be used to treat menorrhagia. The Progestasert needs to be replaced every year, as the reservoir of progesterone is depleted after 18 months.

A T-shaped IUD containing a reservoir of levo on the vertical arm is currently marketed in Europe. A large comparative trial of the Cu T 380A and the levo-IUD reported similar effectiveness and continuation rates. The levo-IUD has a duration of action of at least five years. This IUD reduces menstrual blood loss and can be used therapeutically.

3. Mechanisms of Action

The IUD's main mechanism of contraceptive action is spermicidal by a local sterile inflammatory reaction produced by the presence of a foreign body in the uterus. There is a 1000% increase in the number of leukocytes in washings of the human endometrial cavity after insertion of an IUD [109]. Leukocytes cause phagocytosis of sperm and the tissue breakdown products of leukocytes are toxic to both sperm and blastocyst. The amount of inflammatory reaction and thus contraceptive effectiveness is directly related to the size and composition of the IUD. Copper markedly increases the extent of the inflammatory reaction and also impedes sperm transport and viability. Addi-

tionally, copper impedes sperm transport through and viability in the cervical mucus. Very few, if any, sperm reach the oviducts and the ovum usually does not become fertilized.

Sperm transport from the cervix to the oviduct in the first 24 hours after coitus is markedly impaired in women wearing IUDs. Additionally, in a study where women had unprotected sexual intercourse shortly before ovulation, normally cleaving fertilized ova were found in the tubal flushings in about half the women not wearing IUDs, whereas none were found in users [110]. Further evidence that the IUD acts as a spermicide rather than an abortifacient is data showing that while intrauterine pregnancy rates gradually increase with duration of use, the ectopic rates remain low and constant [111]. The progesterone-IUD has a higher ectopic pregnancy rate than the copper IUD and may act by slowing tubal transport of the embryo or preventing implantation of the fertilized ovum [36]. The progesterone-IUD adds to the endometrial action of the IUD, causing the endometrium to become decidualized with atrophy of the glands. In addition, progesterone thickens the cervical mucus, making sperm penetration difficult. The levo-IUD also partially inhibits ovarian follicular development and ovulation.

4. Fertility after Removal

Upon removal of the IUD, the inflammatory reaction rapidly disappears. Resumption of fertility is prompt and occurs at the same rate as resumption of fertility following discontinuation of barrier methods [85,112] (Fig. 12.12) Concern regarding the association of PID and IUD use can now be minimized with use of safe and effective IUDs placed into women with low risk of exposure to STDs.

5. Adverse Effects

In general, the first year of IUD use is associated with a 1% pregnancy rate, a 10% expulsion rate, and a 15% rate of removal for medical reasons, mainly bleeding and pain. The incidence of these events diminishes steadily in subsequent years.

6. Uterine Bleeding

A common reason for removal of the Cu T involves prolonged menses or intermenstrual bleeding (Table 12.11) [113]. This heavy bleeding may be due to an increased local release of prostaglandins in response to the presence of a foreign body. The deposition of calcium salts on IUDs is irritating to the endometrium and can cause excessive bleeding. Prostaglandins

Table 12.10
Cumulative Discontinuation Rate for Copper T 380A IUD

Event	Years since insertion		
	3	5	7
Pregnancies	1.0	1.4	1.6
Expulsions	7.0	8.2	8.6
Medical removals	14.6	20.8	25.8
Nonmedical removals	13.8	25.6	34.4
Loss to follow-up	10.2	15.5	22.1
All discontinuations	32.2	46.7	56.3
Woman-months	38,571	56,010	67,885

Modified from World Health Organization (WHO): Contraception 42:141. 1990 [4].

Table 12.11
First-Year Event Rates per 100 Women for Copper T 380A

Event	1 Year parous
Pregnancy	0.5
Expulsion	2.3
Bleeding/pain	3.4
Infection	0.3
Other medical	0.5
Planning pregnancy	0.6
Other personal	0.7

From Rosenberg, M. J., Foldesy, R., Mishell, D. R., Jr., *et al. Contraception* 53:197, 1996 [13].

stimulate uterine contractions and these may prolong the duration of bleeding and cramps. The amount of blood loss each month is significantly increased in women wearing inert and copper IUDs (50 ml) compared to nonusers (35 ml). With use of the progestin containing IUDs, the amount of blood loss is significantly reduced to ≤25 ml/cycle [114].

A sensitive indicator of iron stores is the serum ferritin level, and levels less than 4 µg/l require oral iron supplementation to prevent anemia. In a study, women wearing the Cu T IUD had no significant change in serum ferritin levels after 3, 6, and 12 months [115]. Excessive bleeding that continues can be treated with prostaglandin synthetase inhibitors. Mefenamic acid (500 mg three times/day) significantly reduces blood loss in IUD users [116].

7. Perforation

Although uncommon, one potentially serious complication is perforation of the uterine fundus. Perforation occurs at the time of insertion, although a partial perforation can turn into a full perforation if uterine contractions push the IUD through the wall into the peritoneal cavity. The incidence of perforation is generally related to the shape of the device as well as to the experience of the insertor and is best prevented by straightening the uterine axis with a tenaculum and probing the cavity with a uterine sound prior to insertion. In a large multiclinic trial, perforation rates for the Cu 7 were about 1/1000 insertions, but only about 1/3000 for the Cu T [117].

Perforation of the cervix has been reported to range from about 1/600 to 1/1000 insertions [84]. A plastic ball has been added to the distal vertical arm of the Cu T to reduce the rate of cervical perforation.

8. Complications Related to Pregnancy

If pregnancy occurs with an IUD in place, implantation takes place at a site distant from the device itself, so the device is always extra-amniotic. Although there is a paucity of published data, there is no increased incidence of congenital anomalies.

In studies of all types of IUDs left *in situ* during pregnancy, the incidence of fetal death was not increased, but the frequency of spontaneous abortion was about 55%, approximately three times greater than would occur normally [118]. However, the frequency of spontaneous abortion was only 20% if the device was removed or spontaneously expelled. Thus, the IUD should be removed if the appendage is visible. Several reports indicate that during early gestation, it is possible to safely remove an IUD without a visible appendage using sonographic guidance [119].

If the IUD is not removed during early gestation, the risk of septic abortion is also increased. This is due, at least in part, to the fact that IUDs increase the risk of spontaneous abortion (50%) and 2% of all spontaneous abortions are septic. Serious problems with septic abortions were linked to the Dalkon shield IUD (no longer used) and its multifilament tail. Bacteria entered into the spaces between the filaments of the tail and as the IUD was pulled into the uterine cavity as gestation advanced, it caused severe and sometimes fatal infections [120].

9. Ectopic Pregnancies

The ratio of ectopic pregnancies to total pregnancies in women wearing the Cu T IUD is 39 per 1000 pregnancies, which is three times greater than the 14.1 per 1000 pregnancies in the general population. However, because the Cu T so effectively prevents all pregnancies, the estimated ectopic pregnancy rate is only 0.2–0.4 per 1000 woman-years [121]. The estimated ectopic pregnancy rate among sexually active US women using no contraceptive has been estimated to be between 1.5–4.5 per 1000 woman-years, thus women using a Cu T have a substantially reduced risk of having an ectopic pregnancy [121]. The progesterone-IUD has a pregnancy rate of about 30 per 1000 woman-years and an ectopic pregnancy rate of 7.5 per 1000 woman-years. Thus, about one in four pregnancies occurring with this device will be ectopic.

The two types of IUDs currently marketed in the US have differing effects on the risk of ectopic pregnancy. The Cu T lowers the risk while in situ, and the progesterone-IUD increases the risk [122].

Users of an IUD are more likely than users of other types of contraception, but less likely than noncontraceptors, to have a tubal pregnancy while the IUD is in situ [123]. Tubal pregnancy is more likely to occur among IUD users than among women using oral contraceptives (RR = 3.8, 95% CI = 1.5–1.9) or barrier contraceptives (RR = 3.6, 95% CI = 1.6–8.1) [124]. However, past use of an intrauterine device for an extended time is associated with an increased risk of tubal pregnancy [123]. Women who have used an IUD, including copper-containing devices, for three or more years are reported to be twice as likely as women who have never used an IUD to have a tubal pregnancy (adjusted RR = 2.5, 95% CI = 1.5–4.3).

10. Prematurity

The rate of prematurity is reported to be four times greater when the Cu T is left in place during pregnancy compared to when it is removed. A higher incidence of prematurity was also noted in a British study that reported that 13.6% of infants conceived during IUD use weighed less than 2800 g at birth compared to 3% of infants conceived during use of other methods [125].

11. Pelvic Inflammatory Disease (PID)

There were many studies published in the 1970s that suggested that IUD use increased the risk of PID. In 1966, a study was published where cultures were made of endometrial tissue obtained at various intervals after insertion of a loop IUD. During the first 24 hours, the normally sterile endometrial cavity was infected with bacteria. After 24 hours, however, 80% of the cavities were sterile [126]. These findings indicate that development of PID more than one month after insertion of the IUD is due to infection with a sexually transmitted pathogen and not to the presence of the device.

Confirming these findings were the results of a large multicenter study of 24,000 IUD users. The PID rate was highest in the first three weeks after insertion and then remained low for the next eight years at 0.5 per 100 woman-years. Other studies confirm that the risk of PID in Cu 7 and loop users was increased only during the first four months of use. In a Centers for Disease Control and Prevention study of a group of IUD users with only one sexual partner during the last six months, the only increased risk of PID was in previously married or never married women. The authors postulated that the partners of such women had an increased risk of transmitting a pathogen responsible for PID. In married or cohabiting women who had used an IUD for more than four months, the risk of developing PID was 1.0 [127].

There is evidence that IUD users may have an increased risk for colonizing Actinomyces in the upper genital tract. If Actinomyces are identified on a routine Pap smear, appropriate antimicrobial therapy should be started. The IUD does not have to be removed in asymptomatic users. The levo-IUD is reported to reduce the risk of pelvic infections [128].

12. Infertility

The use of the Dalkon shield and possibly of plastic IUDs, other than those that contain copper, can lead to infertility in nulligravid women [129,130]. The risk of primary tubal infertility in women who had ever used an IUD was reported to be 3.7 times higher compared to women who had never used one (95% CI = 2.2–6.1). The increased relative risk associated with use was highest for the Dalkon Shield (RR = 6.0, 95% CI = 2.7–13.3). There was also a significantly increased risk among users of copper-containing IUDs (RR = 3.9, 95% CI = 2.0–7.4) [129].

In a similar study, the adjusted risk of primary tubal infertility associated with any IUD use before a first live birth was 2.0 (95% CI = 1.5–2.6) relative to nonuse. Users of copper-containing IUDs had a relative risk of 1.6 (95% CI = 1.7–5.2). Women who reported having only one sexual partner had no increased risk of primary tubal infertility associated with IUD use. The adjusted risk of secondary tubal infertility associated with use of a copper IUD after a first live birth was not statistically significant [130].

13. Contraindications

Absolute contraindications for IUD insertion include: (1) pregnancy or suspicion of pregnancy; (2) acute PID; (3) postpartum endometritis or infected abortion within three months; (4) known or suspected uterine or cervical malignancy; (5) genital bleeding of unknown etiology; (6) untreated acute cervicitis; and (7) a previously inserted IUD that has not been removed. There are few data available to support the assertions that Wilson's disease (allergy to copper) and genital actinomycosis are contraindications. Relative contraindications are: (1) distortion

of the uterine cavity; (2) history of PID; (3) untreated vaginitis; (4) patient or her partner with multiple sexual partners; (5) conditions associated with increased susceptibility to infections; and (6) women with previous endocarditis, rheumatic heart disease, or a prosthetic heart valve. Women with type 1 or 2 diabetes mellitus may use an IUD.

The increased risk of impairment of future fertility from PID developing in the first month after IUD insertion, as well as the possibility of ectopic pregnancy in the event of contraceptive failure, must be considered when deciding whether to use an IUD in a nulliparous woman, especially if she, or her partner, has multiple sexual partners. OCs, diaphragms, and condoms reduce the risk of developing salpingitis in women infected with gonorrhea and should be a first line option for nulliparous women or women at high risk for STDs. Silent PID, which is associated with subsequent infertility and increased risk of ectopic pregnancy, is often not identified or treated [131,132].

14. Overall Safety

The IUD is not associated with an increased incidence of endometrial or cervical carcinoma. The IUD is a particularly useful method of contraception for women who have completed their families and do not wish permanent sterilization. An analysis reported that after five years of use, the IUD was the most cost-effective method of contraception [6] (Fig. 12.1) Studies indicate that users have a high level of satisfaction with their IUDs [36].

I. Sterilization

In 1995, sterilization of one member of a couple was the most widely used method of preventing pregnancy in the US, with the exception of OCs. The popularity of this method was greatest if: (1) the wife was over age 30; (2) the couple had been married for more than 10 years; and (3) the couple desired no further children. Sterilization should be considered permanent, although reanastomosis after tubal ligation or vasectomy is possible.

Voluntary sterilization is legal in all 50 states. The decision should be made solely by the patient in consultation with the physician. More than one counselor is useful if the woman seeking the procedure is younger than 25 years of age. The rationale for such careful scrutiny is that younger candidates tend to change their minds more often, their attitudes are less fixed, and they face a longer period of reproductive life during which divorce, remarriage, or death among their children can occur. In the US, approximately 7000 women request reversal each year. About 2 per 1000 sterilized women will eventually have a reversal [133].

1. Female Sterilization Techniques

Laparoscopic sterilization can be achieved with unipolar and bipolar electrosurgery, ultrasound destruction, or by using mechanical means (clips, silastic rings, sutures). Discomfort is minimal and a return to full activity in one to two days is common. Tubal ligation by minilaparotomy is the most common female sterilization technique worldwide, although it is generally used for postpartum patients in the US. Local anesthesia can be used while making a small 3–5 cm incision. The simple Pomeroy-type tubal ligation is the most commonly used method.

The most effective, least destructive method of tubal occlusion is the most desirable in younger patients because ovarian dysfunction and adhesion formation are diminished. The effective laparoscopic band techniques or the modified Pomeroy technique (also called partial salpingectomy) should be used in women younger than 25 years old. Reversal of these procedures is followed by pregnancy rates of around 75%.

2. Effectiveness

Female sterilization by surgical removal of a portion of the oviduct, or mechanically occluding a portion of the lumen by clips, bands, or electrocoagulation, was previously reported to be the most effective form of pregnancy prevention [134]. However, the results of a long-term study indicate that pregnancies continue to occur for many years after the sterilization procedure. The failure rate increased from 0.55 per 100 women at 1 year to 1.31 at 5 years and 1.85 at 10 years after sterilization [135]. For women 18–27 years of age, 2.8% became pregnant 5–10 years after bipolar coagulation of the fallopian tubes. There are several reversible methods of contraception with a failure rate similar to those of tubal sterilization. For example, cumulative five-year failure rates with Norplant are 1.1% and with the copper T 380 IUD are 1.2%, compared to 1.3% with female sterilization.

Additionally, among women who become pregnant after female sterilization, about one-third have ectopic pregnancies compared with one-fourth with Norplant and 5% with the Copper T 380 IUD. Because the Copper T 380 IUD has an effective life span of ten years with a failure rate comparable to female sterilization and a lower ectopic pregnancy rate, as well as being less expensive, it should be considered as an alternative to tubal sterilization.

References

1. Piccinio, L. J. and Mosher, W. D. (1998). Trends in contraception use in the United States 1982–1995. *Fam. Plann. Perspec.* **30**(1), 4–10.
2. Trussell, J., Hatcher, R. A., Cates, W. *et al.* (1995). American Health Consultants, Contraceptive technology update. **17**(2), 13.
3. Sivin, I. (1994). Contraception with Norplant® implants. *Hum. Reprod.* **9**, 1818–1826.
4. World Health Organization (1990). The TCU220, multiload 250 and Nova T IUDs at 3, 5 and 7 years of use. Results from three randomized multicentre trials. *Contraception* **42**, 141–158.
5. Peterson, H. B., Xia, Z., Hughes, J. M. *et al.* (1996). The risk of pregnancy after tubal sterilization: Findings from the U.S. Collaborative Review of Sterilization. *Am. J. Obstet. Gynecol.* **174**, 1161–1168.
6. Trussell, J., Leveque, J. A., and Koenig, J. D. (1995). *Am. J. Public Health* **85**(49), 494.
7. Bracken, M. B., and Vita, K. (1983). Frequency of non-hormonal contraception around conception and association with congenital malformations in offspring. *Am. J. Epidemiol.* **117**, 281.
8. Louik, C., Mitchell, A. A., Werler, M. M. *et al.* (1987). Maternal exposure to spermicides in relation to certain birth defects. *N. Engl. J. Med.* **317**, 474.
9. Strobino, B., Kline, J., Lai, A. *et al.* (1986). Vaginal spermicides and spontaneous abortion of known karyotype. *Am. J. Epidemiol.* **123**, 432.
10. Fihn, S. D., Latham, R. H., Roberts, P., *et al.* (1986). Association between diaphragm use and urinary tract infection. *JAMA, J. Am. Med. Assoc.* **254**, 240.
11. Klitsch, M. (1988). FDA approval ends cervical cap's marathon. *Fam. Plann. Perspect.* **20**, 137.
12. Barlow, D. (1977). The condom and gonorrhoea. *Lancet* **2**, 811.
13. Kelaghan, J., Rubin, G. L., Ory, H. W., and Layde, P. M. (1982). Barrier-method contraceptives and pelvic inflammatory disease. *JAMA, J. Am. Med. Assoc.* **248**, 184.
14. Conant, M., Hardy, D., Sernatinger, J. *et al.* (1986). Condoms prevent transmission of AIDS-associated retrovirus. *JAMA, J. Am. Med. Assoc.* **255**, 1706.
15. Fischl, M., Dickinson, G. M., Scott, G. B. *et al.* (1987). Evaluation of heterosexual partners, children and household contacts of adults with AIDs. *JAMA, J. Am. Med. Assoc.* **257**, 640.
16. Richardson, A. C., and Lyon, J. B. (1981). The effect of condom use on squamous cell cervical intraepithelial neoplasia. *Am. J. Obstet. Gynecol.* **140**, 909.
17. Harris, R. W. C., Brinton, L. A., Cowdell, R. H. *et al* (1980). Characteristics of women with dyusplaisa or carcinoma in situ of the cervix uteri. *Br. J. Cancer* **42**, 359.
18. Vessey, M., Lawless, M., and Yeates, D. (1982). Efficacy of different contraceptive methods. *Lancet* **1**, 841.
19. John, A. P. K. (1973). Contraception in a practice community. *J. R. Coll. Gen. Practitioners* **23**, 665.
20. Trussel, J., Sturgen, K., Stricker, J., and Dominik, R. (1994). Comparative contraceptive efficacy of the female condom and other barrier methods. *Fam. Plann. Perspect.* **26**, 66–72.
21. Marshall, J. (1968). A field trial of the basal body temperature method of regulation births. *Lancet* **2**, 8–10.
22. World Health Organization (1981). task force on methods for the determination of the fertile period. Special programme of research, development and research training in human reproduction. A prospective multi-center trial of the ovulation method of natural family planning. 2. Effectiveness phase. *Fertil. Steril.* **36**, 591–598.
23. Perez, A., Zabala, A., Larrain, *et al.* (1983). The clinical efficiency of the ovulation method (Billings). *Rev. Chil. Obstet. Ginecol.* **48**,(2), 97.
24. Rice, F. J., Lanctot, C. A., and Garcia-Devesa, C. (1981). Effectiveness of the sympto-thermal method of natural family planning: An International Study. *Int. J. Fertil.* **26**, 222–230.
25. Wade, M. E., McCarthy, P., Abernathy, J. R. *et al.* (1979). A randomized prospective study of the use effectiveness of two methods of natural family planning: An Interim Report. *Am. J. Obstet. Gynecol.* **134**, 628–631.
26. Marshall, J. (1985). A prospective trial of the mucothermic method of natural family planning. *Int. Rev. Nat. Fam. Plann.* **9**, 139–143.
27. Ferin, J. (1972). Orally active progestational compounds. Human studies: Effects on the utero-vaginal tract. *In* "International Encyclopedia of Pharmacology and Therapeutics," Vol. 2. Pergamon, Oxford.
28. Dorlinger L. (1985). Relative potency of progestins used in oral contraceptives. *Contraception* **31**, 557.
29. Goldqieher, J. W., Tazewell, S., de la Pean, D., and A. (1980). Plasma levels and pharmacokinetics of ethinyl estrogens in various populations. *Contraception* **21**, 17.
30. Scott, J. A., Letzky, O. A., Brenner, P. F. *et al.* (1978). Comparison of the effects of contraceptive steroid formulations containing two doses of estrogen on pituitary function. *Fertil. Steril.* **30**, 141.
31. Mishell, D. R., Jr., Kletzky, O. A., Brenner, P. F. *et al.* (1978). The effect of contraceptive steroids on hypothalamic-pituitary function. **130**, 817.
32. Mishell, D. R., Jr., Thorneycroft, I. H., Nakamura, R. M. *et al.* (1972). Serum estradiol in women ingesting combination oral contraceptive steroids. *Am. J. Obstet. Gynecol.* **114**, 923.

33. Trussell, J., Hatcher, R. A., Cates, W., Stewart, F. H., and Kost, K. (1990). Contraceptive failure in the United States: An update. *Stud. Fam. Plann.* **21**(1), 51–54.

34. Sheth, A., Jain, U., Sharma, S. *et al.* (1982). A randomized double-blind study of two combined and two progestogen-only oral contraceptives. *Contraception* **25**(3), 243–252.

35. Shaaban, M. M. (1991). Contraception with progestogens and progesterone during lactation. *J. Steroid Biochem. Mol. Biol.* **40** (4–6), 705–710.

36. Mishell, D. R. (1997). Family planning. *In* "Comprehensive Gynecology" (Mishell, Herbst, Stenchever, and Droegemueller, eds.), 3rd ed. Mosby, St. Louis, MO.

37. Gal, I., Parkinson, C., and Craft, I. (1971). Effect of oral contraceptives on human plasma vitamin A levels. *Br. Med. J.* **2**, 436.

38. Drife, J. (1989). Complications of combined oral contraception. *In* "Contraception Science and Practice" (M. Filshie and J. Guillebaud, eds.). Butterworth, London.

39. Layde, P. M., Vessey, M. P., and Yeates, D. (1982). Risk factors for gall-bladder disease: A Cohort Study of Young Women Attending Family Planning Clinics. *J. Epidemiol. Commun. Health* **36**, 274.

40. Boston Collaborative Drug Surveillance Project (1973). Oral contraceptives and venous thromboembolic disease, surgically confirmed gallbladder disease and breast tumours. *Lancet* **1**, 1399.

41. Kols, A., Rinehart, W., Piotrow, P. *et al.* (1982). Oral contraceptives in the 1980s. *Popul. Rep.* **10**, A191–A222.

42. Van der Vange, N., Blankenstein, M. A., Kloosterboer, H. J. *et al.* (1990). Effects of seven low-dose combined oral contraceptives on sex hormone binding globulin, corticosteroid binding globulin, total and free testosterone. *Contraception* **41**, 345.

43. Meade, T. W., (1982). Oral contraceptives, clotting factors, and thrombosis. *Am. J. Obstet. Gynecol.* **142**, 758.

44. Wilson, E. S., Cruickshank, J., McMaster, M. *et al.* (1984). A prospective controlled study of the effect on blood pressure of contraceptive preparations containing different types of dosages and progestogen. *Br. J. Obstet. Gynaecol.* **91**, 1254.

45. Khaw, K.-T., and Peart, W. S. (1982). Blood pressure and contraceptive use. *Br. Med. J.* **285**, 403.

45a. Mann, J. I. (1982). *Am. J. Obstet. Gynecol.* **142**, 752.

46. Harlap, S., Kost, K., and Forrest, J. D. (1991). "Preventing Pregnancy, Protecting Health: A New Look at Birth Control Choices in the United States." The Alan Guttmacher Institute, New York and Washington, DC.

47. Farmer, R. D. T., and Preston, (1995). The risk of venous thrombosis associated with low oestrogen oral contraceptives. *J. Obstet. Gynecol.* **15**, 195.

48. Vandenbroucke, J. P., Koster, T., Briet, E. *et al.* (1994). Increased risk of venous thrombosis in oral-contraceptive users who are carriers of Factor V. Leiden mutation. *Lancet* **344**, 1453.

49. Wynn, V., and Goldsland, I. F. (1986). Effects of oral contraceptives on carbohydrate metabolism. *J. Reprod. Med.* **31**(Suppl. 9), 892–897.

50. Shoupe, D., and Bopp, B. (1991). Contraceptive options for the gestational diabetic woman. *Int. J. Fertil., Suppl.* **2**, 80–86.

51. Rimm, E. B., Manson, J. E., Stampfer, M. J., *et al.* (1992). Oral contraceptive use and the risk of type 2 (non-insulin-dependent) diabetes mellitus in a large prospective study of women. *Diabetologia* **35**, 967–972.

52. Petersen, Skouby, S. O., and Pedersen, R. G. (1991). Desogestrel and gestodene in oral contraceptives: 12 months' assessment of carbohydrate and lipoprotein metabolism. *Obstet. Gynecol.* **78**, 666.

53. Speroff, L., DeCherney, A., and the Advisory Board for the New Progestins (1993). Evaluation of a new generation of oral contraceptives. *Obstet. Gynecol.* **81**, 1034.

53a. Goldsland, I. F., Crook, D., Simpson, R. *et al.* (1990). *N. Engl. J. Med.* **323**, 1375.

54. Layde, P. M., Ory, H. W., and Schlesselman, J. J. (1982). the risk of myocardial infarction in former users of oral contraceptives. *Fam. Plann. Perspect.* **14**, 78.

55. Stampfer, M. J., Willett, W. C., Colditz, G. A. *et al.* (1988). A prospective study of past use of oral contraceptive agents and risk of cardiovascular diseases. *N. Engl. J. Med.* **319**, 1313.

56. Adams, M. R., Clarkson, T. B., Kortinik, D. R. *et al.* (1987). Contraceptive steroids and coronary artery atherosclerosis in cynomologus macaques. *Fertil. Steril.* **47**, 1010.

57. Engel, H.-J., Engel, E., and Lichtlen, P. R. (1983). Coronary atherosclerosis and myocardial infarction in young women—role of oral contraceptives. *Eur. Heart J.* **4**, 1.

58. Thomas, D. B. (1991). Oral contraceptives and breast cancer. Review of the epidemiologic literature. *Contraception* **43**, 597–642.

59. Collaborative Group on Hormonal Factors in Breast Cancer (1996). Breast cancer and hormonal contraceptives: Collaborative reanalyzes of individual data on 53,297 women with breast cancer and 100,239 women without breast cancer from 54 epidemiological studies. *Lancet* **347**, 1713–1727.

60. Schlesselman, J. J. (1995). Net effect of oral contraceptive use on the risk of cancer in women in the United States. *Obstet. Gynecol.* **85**, 793–801.

61. Kjaer, S. K., Engholm, G., Dahl, C. *et al.* (1993). Case-control study of risk factors for cervical squamous-cell neoplasia in Denmark. III. Role of oral contraceptive use. *Cancer Causes Control* **4**, 513–519.

62. Ursin, G., Peters, R. K., Henderson, B. E. *et al.* (1994). Oral contraceptive use and adenocarcinoma of cervix. *Lancet* **334**, 1390.

63. Coker, A. L., McCann, M. F., Hulka, B. S., and Walter, L. A. (1992). Oral contraceptive use and cervical intraepilelial neoplasia. *J. Clin. Epidemiol.* **45**, 1111.

64. Daling, J. R., Madeleine, M. M., McKnight, Carter, J. J., Wipf, G. C., Ashley, R., Schwartz, *et al.* (1996). The relationship of Human papillomavirus-related cervical tumors to cigarette smoking, oral contraceptive use, and prior herpes simples virus type 2 infection. *Cancer Epidem., Biomarkers Prev.* **5**, 541–548.

65. Centers of Disease Control (1987). Combination oral contraceptives use and risk of endometrial cancer. *JAMA, J. Am. Med. Assoc.* **257**, 976.

66. Voigt, L. F., Deng, O., and Weiss, N. S. (1994). Recency, duration, and progestin content of oral contraceptives in relation to the incidence of endometrial cancer. *Cancer Causes Control* **5**, 227–233.

67. Hankinson, S. E., Colditz, G. A., Hunter, D. J. *et al.* (1992). A quantitative assessment of oral contraceptive use and risk of ovarian cancer. *Obstet. Gynecol.* **80**, 708.

68. Rosenberg, L., Palmer, J. R., Lesko, S. M. *et al.* (1990). Oral contraceptive use and the risk of myocardial infarction. *Am. J. Epidemiol.* **131**, 1009.

69. Forman, D., Vincent, T. J., and Doll, R. (1986). Cancer of the liver and the use of oral contraceptives. *Br. Med. J.* **292**, 1357.

70. World Health Organization (1989). Combined oral contraceptives and liver cancer. *Int. J. Cancer* **43**, 254.

71. Pituitary Adenoma Study Group (1983). Pituitary adenomas and oral contraceptives: A multicenter case-control study. *Fertil. Steril.* **39**, 753.

72. Hannaford, P. C., Villard-Mackintosh, L., Vessey, M. P., and Kay, C. R. (1991). Oral contraceptives and malignant melanoma. *Br. J. Cancer* **63**, 430.

73. Vessey, M. P., Wright, N. H., McPherson, K. *et al.* (1978). Fertility after stopping different methods of contraception. *Br. Med. J.* **1**, 265.

74. Bracken, M. B., Hellenbrand, K. G., and Holford, T. R. (1990). *Fertil. Steril.* **53**, 21.

75. Ory, H. W., Forrest, J. D., and Lincoln, R. (1983). "Making Choices: Evaluating the Health Risks and Benefits of Birth Control Methods." The Allan Guttmacher Institute, New York.

76. Robinson, J. C., Plichta, B. A., Weismann, C. S., Nathanson, C. A., and Ensminger, M. (1992). Dysmenorrhea and use of oral contraceptives in adolescent women attending a family planning clinic. *Am. J. Obstet. Gynecol.* **166**(2), 578–583.

77. (1994). Menstrual problems and common gynecologic concerns. *In* "Contraceptive Technology" (R. A. Hatcher, J. Trussell, Steward, F. H. Stewart, D. Kowal, F. Guest, W. Cates, and Policar, eds.). 16th rev. ed. Irvington Publishers, New York.

78. Lanes, A. F., Birmann, B., Walter, A. M., and Singer, S. (1992). Oral contraceptive type and functional ovarian cysts. *Am. J. Obstet. Gynecol.* **166**, 956.

79. Hazes, J. M. W., Dijkmans, B. A. C., Vanderbroucke, J. P. *et al.* (1990). Reduction of the risk of rheumatoid arthritis among women who take oral contraceptives. *Arthritis Rheum.* **33**, 173.

80. Del Junco, D. J., Annegers, J. F., Luthra, H. S., *et al.* (1985). Do oral contraceptives prevent rheumatoid arthritis? *JAMA, J. Am. Med. Assoc.* **254**, 1938–1941.

81. Gambacciani, M., Spinetti, A., Toponeco, F. *et al.* (1993). Longitudinal evaluation of perimenopausal vertebral bone loss: Effects of a low-dose oral contraceptive preparation on bone mineral density and metabolism. *Obstet. Gynecol.* **83**, 392–396.

82. Ross, R. K., Pike, M. C., Vessey, M. P. *et al.* (1986). Risk factors for uterine fibroids: Reduced risk associated with oral contraceptives. *Br. Med. J.* **293**, 359–362.

83. Guillebaud, J. (1991). "The Pill and Other Hormones for Contraception," fourth ed. Oxford University Press, Oxford.

84. Mishell, D. R., Kharma, K. M., Thorneycroft, I. H., and Nakamura, R. M., (1972). Estrogenic activity in women receiving an injectable progestogen for contraception. *Am. J. Obstet. Gynecol.* **113**, 372.

85. Schwallie, P. C., and Assenzo, J. R. (1974). Contraceptive use—efficacy study utilizing medroxyprogesterone acetate administered as an intramuscular injection once every 90 days. *Contraception* **10**(2), 181.

86. Schwallie, P. C., and Assenzo, J. R. (1973). The effect of depo-medroxyprogesterone acetate on pituitary and ovarian function, and the return of fertility following its discontinuation: A review. *Fertil. Steril.* **24**, 331.

87. Moore, L. L., Valuck, R., McDougall, C. *et al.* (1995). A comparative study of one-year weight gain among users of medroxyprogesterone acetate, levonorgestrel implants, and oral contraceptives. *Contraception* **52**, 215–220.

88. World Health Organization Expanded Programme of Research, Development and Research Training in Human Reproduction Task Force on Long-Acting System Agents for the Regulation of Fertility (1983). Multinational comparative clinical evaluation of two long-acting injectable contraceptive steroids, norethisterone enanthate and medroxyprogesterone acetate. Final report. *Contraception* **18**, 1.

89. Cundy, T., Cornish, J., Evans, M. C. *et al.* (1994). recovery of bone density in women who stop using medroxyprogesterone acetate. *Br. Med. J.* **308**, 247–248.

90. Skegg, D. C., Noonan, E. A., Paul, C. *et al.* (1995). Depot medroxyprogesterone acetate and breast cancer: A pooled analysis of the World Health Organization and New Zealand studies. *JAMA, J. Am. Med. Assoc.* **273**, 799–804.

91. WHO Collaborative Study of Neoplasia and Steroid Contraceptives (1991). Breast cancer and depot-medroxyprogesterone acetate: A Multinational Study. *Lancet* **338**, 833–838.

92. WHO Collaborative Study of Neoplasia and Steroid Contraceptive (1991). DMPA and risk of endometrial cancer. *Int. J. Cancer* **49**, 186–190.

93. WHO Collaborative Study of Neoplasia and Steroid Contraceptives (1991). DMPA and risk of epithlelial ovarian cancer. *Int. J. Cancer* **49**, 191–195.

94. Thomas, D. B., and Ray, R. M. (1995). Depot-medroxyprogesterone acetate (DMPA) and risk of invasive adenocarcinomas and adenosquamous carcinomas of the uterine cervix. *Contraception* **52**, 307–312.

95. New Zealand Contraception and Health Study Group (1994). Risk of cervical dysplasia in users of oral contraceptives, intrauterine devices or depot medroxyprogesterone acetate. *Contraception* **50**, 431.

96. New Zealand Contraception and Health Study Group (1994). History of long-term use of depot-medroxyprogesterone acetate in patients with cervical dysplasia: Case-control Analysis Nested in a Cohort Study. *Contraception* **50**:443–449.

97. World Health Organization (1989). Task Force of long-acting systemic agents for fertility regulation. Special Programmer of Research, Development and Research Training in Human Reproduction; a multicentre Phase III comparative study of two-hormonal contraceptive preparations given once-a-month by intramuscular injection: II. The comparison of bleeding patterns. *Contraception* **40**, 531.

98. Moore, D. E., Roy, S., Stanczyk, F. Z. *et al.* (1978). Bleeding and serum d-norgestrel, estradiol, and progesterone patterns in women using d-norgestrel subdermal polysiloxane capsules for contraception. *Contraception* **17**, 315.

99. Shoupe, D., Horenstein, J., Mishell, D. R., Jr. *et al.* (1991). Characteristics of ovarian follicular development in Norplant users. *Fertil. Steril.* **55**, 766–770.

100. Brache, V., Faundes, A., Johansson, E. *et al.* (1985). Anovulation, inadequate luteal phase and poor sperm penetration in cervical mucus during prolonged use of Norplant implants. *Contraception* **31**, 261.

101. Sivin, I. (1988). International experience with Norplant and Norplant II. *Contraception* **19**(2), 81–94.

102. Brache, V., Alvarez-Sanchez, Faundes, A. *et al.* (1990). Ovarian endocrine function through five years of continuous treatment with Norplant subdermal contraceptive implants. *Contraception* **41**, 169.

103. Shoupe, D., Mishell, D. R., Bopp, B. I., and Fielding, M. (1991). The significance of bleeding patterns in Norplant implant users. *Obstet. Gynecol.* **77**, 256.

104. Population Countil (1990). "Norplant Levonorgestrel Implants: A summary of Scientific Data," Monograph. Population Council, New York.

105. Davies, G. C., Li, X. F., and Newton, J. R. (1993). Release characteristics, ovarian activity and menstrual bleeding pattern with a single contraceptive implant releasing 3-keodesogestrel. *Contraception* **47**(3), 251–261.

106. Van Look Paul, F. A., and Stewart, F. (1998). Emergency contraception. *In* "Contraceptive Technology" (R. A. Hatcher, J. Trussell, F. H. Stewart, W. Cates, G. K. Stewart, F. Guest, and D. Kowal, eds.). Ardent Media, New York.

107. Fasoli, M., F. Parazzini, G. Cecchetti *et al.* (1989). *Contraception* **39**, 459.

108. Glasier, A., Thong, K. J., Dewar, M., Mackie, M., and Baird, D. T. (1992). Mifepristone compared with high-dose estrogen and progestogen for emergency postcoital contraception. *N. Engl. J. Med.* **327**, 1041–1044.

109. Hatcher, R. A., Trussell, J., Stewart, F. *et al.* (1994). "Contraceptive Technology," 16th ed., pp. 347–377. Irvington Publishers, New York.

110. Moyer, D. L., and Mishell, D. R., Jr. (1971). Reactions of human endometrium to the intrauterine foreign day. II. Long term effects on the endometrial histology and cytology. *Am. J. Obstet. Gynecol.* **111,** 66.

111. Tredway, D. R., Umezaki, C. U., and Mishell, D. R., Jr. (1975). Effect of intrauterine devices on sperm transport in the human being. *Am. J. Obstet. Gynecol.* **123,** 734.

112. Vessey, M. P., Lawless, M., McPherson, K. *et al.* (1983). Fertility after stopping use of intrauterine contraceptive device. *Br. Med. J.* **286,** 106.

113. Rosenberg, M. J., Foldesy, R., Mishell, D. R., Jr. *et al.* (1996). *Contraception* **53,** 197.

114. Rybo, G. (1977). The IUD and endometrial bleeding. *J. Reprod. Med.* **20,** 715.

115. Milsom, I., Andersson, K., Jonasson, K. *et al.* (1995). The influence of the Gyne-T 380S IUD on menstrual blood loss and iron status. *Contraception* **52,** 175–179.

116. Anderson, A. B. M., Haynes, P. J., Guillebaud, J. *et al.* (1976). Reduction of menstrual blood loss by prostaglandin synthetase inhibitor. *Lancet* **1,** 774.

117. Sivin, I., and Stern, J. (1979). Long-acting more effective copper T IUDs: A summary of U.S. experience, 1970–75. *Stud. Fam. Plann.* **10,** 276–281.

118. Tatum, H. J., Schmidt, F. H., and Jain, A. K. (1976). Management and outcome of pregnancies associated with the copper T intrauterine contraceptive device. *Am. J. Obstet. Gynecol.* **126,** 869.

119. Shalev, E., Edelstein, S., Engelhard, J. *et al.* (1987). Ultrasonically controlled retrieval of an intrauterine contraceptive device in early pregnancy. *J. Clin. Ultrasound* **15,** 525.

120. Tatum, H. J., Schmidt, F. J., Phillips, D. M., *et al.* (1975). The Dalkon shield controversy, structural and bacteriologic studies of IUD tails. *JAMA, J. Am. Med. Assoc.* **4,** 253.

121. Sivin, I. (1991). Dose- and age-dependent ectopic pregnancy risks with intrauterine contraception. *Obstet. Gynecol.* **78,** 291–298.

122. Ory, H. W., and the Women's Health Study (1981). Ectopic pregnancy and intrauterine contraceptive devices, new perspectives. *Obstet. Gynecol.* **57,** 137–144.

123. Rossing, M. A., Daling, J. R., and Weiss, N. S. (1993). Past use of an intrauterine device and risk of tubal pregnancy. *Epidemiology* **4,** 245–251.

124. Rossing, M. A., Daling, J. R., and Voigt, L. F. (1993). Current use of an intrauterine device and risk of tubal pregnancy. *Epidemiology* **4,** 252–258.

125. Vessey, M. P., Johnson, B., Doll, R. *et al.* (1984). Outcome of pregnancy in women using an intrauterine device. *Lancet* **1,** 495.

126. Mishell, D. R., Jr., Bell, J. H., Good, R. G. *et al.* (1996). The intrauterine device: A bacteriologic study of the endometrial cavity. *Am. J. Obstet. Gynecol.* **96,** 119.

127. Lee, N. C., and Rubin, G. I. (1988). The intrauterine device and pelvic inflammatory disease revisited: New results from the women's health study. *Obstet. Gynecol.* **71,** 1.

128. Toivonen, J., Luukkainen, T., and Alloven, H. (1991). Protective effect of intrauterine release of levonorgestrel on pelvic infection: Three years—comparative experience of levonorgestrel and copper-releasing intrauterine devices. *Obstet. Gynecol.* **77,** 261.

129. Daling, J. R., Weiss, N. S., Voigt, L. F., McKnight, B., and Moore, D. E. (1992). The intrauterine device and primary tubal infertility. *N. Engl. J. Med.* **326,** 203–204.

130. Cramer, D. W., Schiff, I., Schoenbaum, S. C. *et al.* (1985). Tubal infertility and the intrauterine device. *N. Engl. J. Med.* **312,** 914–917.

131. Cates, W., Wasserheit, J. N., and Marchbanks, P. A. (1994). Pelvic inflammatory disease and tubal infertility: The preventable conditions. *Ann. N.Y. Acad. Sci.* **709,** 179–195.

132. Cates, W., Joesoef, M. R., and Goldman, M. B. (1993). A typical pelvic inflammatory disease: Can we identify clinical predictors? *Am. J. Obstet. Gynecol.* **169,** 341–346.

133. Wilcox, L. S., Chu, S. Y., and Peterson, H. B. (1990). Characteristics of women who considered or obtained tubal reanastomosis: Results from a prospective study of tubal sterilization. *Obstet. Gynecol.* **75,** 661.

134. Trussel, J., Hatcher, R. A., Cates, W. J. R., Stewart, F. H., and Kost, K. (1990). Contraceptive failure in the United States: An update. *Study Fam. Plann.* **21,** 51.

135. Peterson, H. B., Zia, Z., Hughes, J. M. *et al.* (1996). The risk of pregnancy after tubal sterilization: Findings from the U.S. collaborative review of sterilization. *Am. J. Obstet. Gynecol.* **174,** 1161–1170.

13

Induced Abortion

HANI K. ATRASH* AND AUDREY F. SAFTLAS[†]

*Division of Reproductive Health, National Center for Chronic Disease Prevention and Health Promotion, Centers for Disease Control and Prevention, Atlanta, Georgia; [†]Department of Epidemiology and Public Health, Yale School of Medicine, New Haven, Connecticut

I. History and Background

Induced abortion has been a controversial issue throughout the ages; anthropologists have found evidence of its existence in every known culture [1,2]. The earliest records of recognized abortifacients, which are presumed to have been written more than 4500 years ago, are found in ancient Chinese texts [3]. As early as 384 B.C.E., philosophers in ancient Greece had accepted abortion as a permissible act to end an unwanted pregnancy or to control population [3]. Reflecting a variety of religious, social, and political forces, laws and regulations permitting or restricting abortion have been enacted throughout the centuries [3,4].

In 1990, the Alan Guttmacher Institute (AGI) reported on the status of abortion laws around the world. Of countries with populations of 1 million or more, 53 (containing 25% of the world's population) had restrictive abortion laws that prohibit abortion except when the woman's life is endangered if the pregnancy is carried to term [5]. Another 42 countries (comprising 12% of the world's population) had statutes authorizing abortion on broader medical grounds—to avert a threat to the woman's general health—and sometimes for genetic or judicial indications (such as rape or incest), but not for social indications alone or on request. Another 14 countries (comprising 23% of the world's population) allowed abortions for social or socio-medical reasons. A total of 23 countries permitted abortion at the request of the woman; these countries comprise 40% of the world's population and include some of the most populous countries (China, the United States, half of Western Europe, and countries previously constituting the Soviet Union) [5].

In 1987, an estimated 26 to 31 million legal abortions and 10 to 22 million illegal abortions were performed worldwide [5]. Even in countries where abortion is legal and available on request, millions of unsafe, illegal, clandestine abortions are still performed [5,6].

In the United States, abortion is the most frequently performed gynecologic procedure; an estimated 1.5 million legal induced abortions are performed each year [5]. Focusing on the United States, this chapter provides an overview of abortion services, the reasons women seek induced abortions, the characteristics of women who obtain abortions, and the morbidity and mortality associated with the procedure.

II. Abortion Services

A. *Providers and Facilities*

Legal induced abortions in the United States are provided in various health care settings, including hospitals, freestanding surgical centers, abortion clinics, and private physicians' offices. Legalization of abortion following the U.S. Supreme Court's 1973 *Roe vs. Wade* decision led to an immediate increase in the number of health care facilities providing abortion services; by 1982, this number had reached 2908 [7]. Since then, the number of facilities has declined to 2618 in 1987, 2582 in 1988, and 2380 in 1992 [7,8]. This decline occurred mostly among hospitals and in nonmetropolitan counties. In 1977, 1654 hospitals offered abortion services; this number fell to 1040 in 1988 and to 855 in 1992 [7,8]. The traditional concentration of abortion facilities in urban areas has become more pronounced in the 1990s. In 1988, 90% of all abortion facilities and providers were in metropolitan counties [7]. In 1992, 8% of women having an abortion in nonhospital facilities traveled more than 100 miles for abortion services and an additional 16% traveled 50–100 miles [9]. After abortion was legalized, the rising demand for abortion procedures led to the creation of a new institution—the freestanding abortion clinic. In 1973, more than 50% of all abortions were performed in hospitals; this percentage declined to 10% in 1988 and 7% in 1992 [7,8]. Meanwhile, the percentage of abortions performed in freestanding clinics increased from 46% in 1973 to 86% in 1988 and to 89% in 1992 [7,8]. This change in the percentage of procedures performed at hospitals does not reflect an actual decline in the number of abortions performed at hospitals; as the number of procedures rose throughout the 1970s and 1980s, most of these additional procedures were performed at clinics.

B. *Barriers*

Although abortion has been legal and available "on request" since 1973 in the U.S., access to abortion services is still problematic for many women because of barriers related to distance, cost, legislative restrictions, provider training, violence and harassment, and other barriers.

1. *Geographic Barriers*

Because clinics are concentrated in metropolitan counties, women seeking abortions outside of metropolitan areas may face logistical barriers to access. As stated previously, it is estimated that 8% of women seeking abortions at clinics must travel more than 100 miles, and 16% must travel 50–100 miles [9,10].

2. *Financial Barriers*

The most obvious barrier to service is the cost of the abortion procedure and related special services such as administration of Rh immune globulin, general anesthesia, or human immune deficiency (HIV) testing [10]. Since 1977, Congress prohibited the use of federal funds to pay for abortions for Medicaid recipients, except in cases where a full-term pregnancy endangers the life of the mother. As of April 1994, only 15 states continue voluntarily or under court order to fund abortions that are necessary to protect the woman's physical or mental health (California, Hawaii, Maryland, North Carolina, New York, and Washington provide funds voluntarily; Alaska, Connecticut, Idaho, Massachusetts, Minnesota, New Jersey, Oregon, Vermont, and West Virginia are required to by court order) [11]. In 1994, 37.3% of the United States population lived in these 15 states [12]; however, these states provided 52% of all abortions performed in 1994 [13].

3. Legislative Restrictions

Legislation that totally excludes abortion from federal Medicaid coverage limits access to abortion services for poor women. On the other hand, many states have enacted legislation limiting access to abortion services for all women. Such legislation includes waiting periods, counseling requirements that involve more than one visit to the provider, banning late-term abortions, and, for minors, parental notification and consent requirements [9]. In 1996, three states attempted to impose clinic requirements designed to make abortion more difficult to obtain, 11 states were enforcing state-scripted compulsory counseling coupled with waiting periods, and 26 states were enforcing parental involvement laws for minors seeking abortion services [14].

4. Provider Training

Access to services has been hindered by the decline in the number of hospitals that perform abortion procedures. Because fewer hospitals perform abortions, fewer residency training programs (mostly hospital-based) in family practice and obstetrics and gynecology teach abortion techniques. Thus, upon completion of their training, fewer physicians are familiar with and competent to perform the basic procedures [15–17]. A 1991 Columbia University poll reported that only 53% of graduating obstetrics and gynecology residents had ever performed a first-trimester abortion, and 47% had performed a midtrimester dilatation and evacuation [18]. Three surveys conducted during the last two decades reported major shifts in emphasis on abortion training. Investigators collected information from directors of obstetrics and gynecology training programs in 1976, 1987, and 1994 [15,19,20]. In 1976, first-trimester abortion training was required in 26.3% of programs; it was optional in 66.2% and was not offered in 7.2% of programs [19]. By 1991–1992, first-trimester abortion training was required in only 12.4% of programs; it was optional in 57.9% and was not offered in 29.6% of programs [20]. Second trimester abortion training was required in 22.5% of programs in 1976 and 6.9% in 1991–1992; 16% of programs did not offer this training in 1976 and 34.8% did not offer it in 1991–1992 [19,20]. In a survey of chief residents and program directors, the chief residents reported that residents received less abortion training than was estimated by program directors [21]. Busy residents with multiple competing responsibilities are unlikely to take advantage of elective activities, particularly knowing that written and oral board examinations do not include questions on abortion [22].

5. Violence and Harassment

Abortion providers and women seeking abortion have been subjected to harassment and repeated attacks. In 1985, of 1250 abortion facilities that provided services to most women seeking an abortion in the United States, 73% were the target of at least one illegal activity; 6% were subjected solely to picketing [23]. Between 1977 and 1988, there were 3.7 incidents per 100 abortion providers per year, and 7.2 incidents per 100 nonhospital abortion facilities per year [24]. From 1993 to 1998, abortion providers and facilities were subjected to nine shootings and bombings resulting in six deaths (two physicians, two guards, and two clinic workers) and 15 injuries [25].

6. Other Barriers

Other barriers that affect women's ability to obtain abortion include the limited services offered by providers. For example, only 43% of abortion facilities provide abortions beyond 12 weeks gestation, and approximately one-third of nonhospital abortion facilities do not serve women who are HIV positive [10].

III. Why Women Seek Abortions

A. Unplanned Pregnancy

Although a small proportion of planned pregnancies end in induced abortion because of maternal and fetal health problems or changes in the woman's circumstances [26,27], it is estimated that most pregnancies ending in induced abortion were unplanned. In surveys of nationally representative samples of U.S. abortion patients in 1987 and 1994–1995, AGI reported that more than half of all abortion patients (51% in 1987 and 58% in 1994–1995) used some means of contraception during the month in which conception occurred, and a substantial proportion of those who were not using contraception had stopped doing so only a few months before becoming pregnant [28,29]. Poor, unemployed, less educated, and Hispanic women were least likely to be using a contraceptive at the time of conception [28,29]. In 1987, the percentage of women who used contraception increased with age (from 39.5% of women less than 18 years old to 58.8% of women aged 30 or older), but by 1994–1995, these age differences had almost disappeared [28,29].

B. Other Factors

Most women who responded to AGI's 1987 survey reported that multiple factors contributed to their decision to obtain an abortion. The majority of respondents (76%) reported that having a baby would interfere with work, school, or other responsibilities (16% reported that this was their main reason for having an abortion). Of the respondents, 68% reported that they could not afford a child (21% indicated that this was their main reason for having an abortion); 51% said they did not want to be a single parent or that they had problems in their relationship with the father (12% indicated that this was their main reason for having an abortion). Other reported reasons for obtaining abortion included: not ready for responsibility (31% reported that this was a factor and 21% said it was their main reason), did not want others to know about the pregnancy (31% reported

Table 13.1

Number of Abortions, Abortion Ratio, Abortion Rate, Number of Abortion-Related
Deaths, and Abortion Mortality Rate in the United States, Selected Years[a]

	1972	1980	1990	1995
Number of Abortions	586,760	1,297,606	1,429,577	1,210,883
Abortion Ratio[b]	180	359	345	311
Abortion Rate[c]	13	25	24	20
Abortion-related Deaths				
Legal	24	9	5	NA[d]
Illegal	39	1	0	NA
Abortion Mortality Rate[e]	4.1	0.7	0.3	NA

[a]Sources: [13,32,33].

[b]Number of legal abortions per 1000 live births.

[c]Number of legal abortions per 1000 women aged 15 to 44 years.

[d]NA = not available.

[e]Legal abortion deaths per 100,000 abortions.

this reason, 1% said it was their main reason), not mature enough or too young to have a child (30% reported this reason, 11% said it was their main reason), had all the children she wanted or children had grown up (26% reported this reason, 8% said it was their main reason), husband or partner wanted her to have an abortion (23% reported this reason, 1% said it was their main reason), fetal health problems (13% reported this reason, 3% said it was their main reason), maternal health problems (7% reported this reason, 3% said it was their main reason), woman's parents wanted her to have an abortion (7% reported this reason, a negligible number said it was their main reason), woman was the victim of rape or incest (1% reported this reason and 1% said it was their main reason), and other reasons [27].

Women having abortions at 16 or more weeks of gestation reported several factors that contributed to their delay in seeking abortion services. Each woman could report more than one contributing factor. The most common reason given was that they did not realize they were pregnant or they did not know soon enough the actual gestation of their pregnancy (71%). Other reasons reported were that they had trouble arranging the abortion, usually because they needed time to raise money (48%), they feared telling a partner or parents (33%), and they needed additional time to decide to have an abortion (24%) [27].

IV. Characteristics of Women Obtaining Abortions

Since 1969, the Centers for Disease Control and Prevention (CDC) monitored and published the numbers and characteristics of women obtaining legal abortions in the United States. In compiling abortion statistics, the CDC relies to a great extent on reports from central health agencies in 52 reporting areas (the 50 states, the District of Columbia, and New York City). In 1994, the latest year for which final abortion statistics are available, 47 central health agencies reported abortion statistics to the CDC; data for the remaining five areas were provided by hospitals and other medical facilities [13]. Because state abortion reporting laws vary and because some providers may not report or may underreport the number of abortions to central health agencies, the number of abortions reported to the CDC is less than the actual number of abortions performed in the United

States. AGI is the only organization other than the CDC that collects national abortion statistics. AGI's numbers are based on a direct survey of U.S. abortion providers [30]. AGI has consistently reported an average of 20% more abortions than the CDC [31]. Although AGI does not routinely collect information on the characteristics of women who obtain abortions, the organization conducted a survey in 1994–1995 detailing a broad range of characteristics of abortion patients, such as socioeconomic status, religious affiliation, residence, childbearing intention, and use of contraception prior to pregnancy [29].

A. Numbers, Ratios, and Rates

From 1969 through 1995, 29,269,812 legal abortions were reported to the CDC [13,32,33]. In 1969, 22,670 abortions were reported to the CDC, and the national abortion ratio (number of abortions per 1000 live births) was 6.3. In 1972, just before the *Roe vs. Wade* decision, 586,760 legal abortions were reported to the CDC and the national abortion ratio was 180 (Table 13.1) [32]. The number of abortions increased steadily, peaking at 1,429,577 in 1990 (Table 13.1). The abortion ratio increased from 180 in 1972 to 364 in 1984, and then decreased to 311 in 1995 (Table 13.1 and Fig. 13.1) [33]. In 1994, abortion ratios were highest for women in the youngest (younger than 15 years and 15 to 19 years) and oldest (40 years and older) age groups. For women less than 15 years of age, the abortion ratio was 704, for women aged 15 to 19 years it was 415, and for women aged 40 years and older, it was 412 abortions per 1000 live births. The ratio was lowest for women aged 30 to 34 years (172 abortions per 1000 live births) [13]. The abortion ratio was 538 for black women, 217 for white women, and 290 for Hispanic women. The abortion ratio was almost nine times higher for unmarried women (689) than for married women (79) [13].

The abortion rate (number of abortions per 1000 females of reproductive age) increased from 13 in 1972 to a peak of 25 in 1980 and then decreased to 20 in 1995 [13,33]. In 1994, abortion rates were highest for women aged 20 to 24 years (39 abortions per 1000 women) and lowest for women in the youngest and oldest age groups (for women less than 15 years of age and for women aged 40 years and above, the rate was two abor-

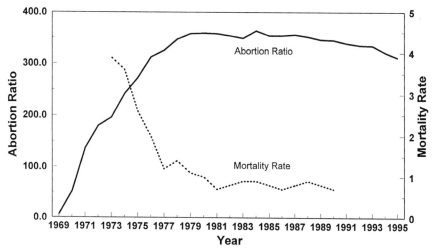

Fig. 13.1 Abortion ratios and abortion mortality rates in the United States. Abortions per 1000 live births. Abortion deaths per 100,000 abortions; three-year moving average.

tions per 1000 women). In comparison, AGI's 1990 world review of induced abortion reported abortion rates for other countries ranging from 5.3 in the Netherlands (1986) to 112 in countries of the Soviet Union (1987) [5]. More recent data are available from AGI but have not been published (Fig. 13.2).

B. General Characteristics

Between 1973 and 1995, women obtaining abortions were mostly young, white, unmarried, and of low parity, and most of them had their abortions performed by suction curettage at 12 weeks of gestation or earlier (Table 13.2) [13,32]. The characteristics of women obtaining abortions have changed over time; a higher proportion of women seeking abortion are 25 years or older, unmarried, nonwhite, and have had one or more live

births (Table 13.2). In 1994, 54% of women who obtained an abortion were doing so for the first time; 27% had obtained one previous abortion, and 18% had obtained two or more previous induced abortions [13]. The proportion of women obtaining abortions who had the procedure performed at 8 weeks gestation or earlier increased with age. Only 34.5% of women younger than 15 years of age and 44.4% of women aged 15 to 19 obtained an abortion at or before 8 weeks of gestation; 57.4% of women aged 25 to 29 and 62.7% of women aged 40 and older had the procedure performed early in pregnancy. Black women were less likely to obtain early abortions than white women (47.4% and 56.7%, respectively) [13].

AGI's 1994–1995 survey of U.S. abortion patients provides additional information about the distribution of abortion patients by age, race, marital status, and number of previous live

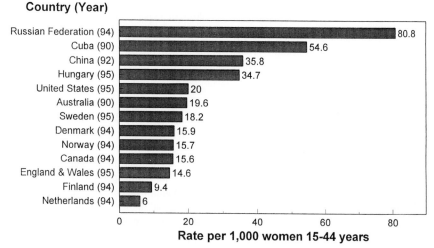

Fig. 13.2 Legal abortion rates for selected countries with accurate abortion statistics. Data for the United States from [28]; data for other countries from personal communication with Dr. Stanley Henshaw, Alan Guttmacher Institute.

Table 13.2

Characteristics of US Women Who Obtain Abortions (Percent Distribution) in the United States, Selected Years[a]

Characteristic	1972	1980	1990	1995
Age (years)				
≤19	32.6	29.2	22.4	20.1
20–24	32.5	35.5	33.2	32.5
≥25	34.9	35.3	44.4	47.4
Race				
White	77.0	69.9	64.8	59.5
Black	23.0[b]	30.1[b]	31.8	35.0
Other	—	—	3.4	5.5
Marital Status				
Married	29.7	23.1	21.7	20.3
Unmarried	70.3	76.9	78.3	79.7
Number of Live Births				
0	49.4	58.4	49.2	45.2
1	18.2	19.4	24.4	26.5
2	13.3	13.7	16.9	18.0
≥3	19.1	8.5	9.5	10.3
Type of Procedure				
Curettage	88.6	95.5	98.8	98.9
Suction	65.2	89.8	96.0	96.6
Sharp	23.4	5.7	2.8	2.3
Intrauterine instillation	10.4	3.1	0.8	0.5
Other[c]	1.1	1.4	0.4	0.6
Gestation (weeks)				
≤8	34.0	51.7	51.6	54.0
9–10	30.7	26.2	25.3	23.1
11–12	17.5	12.2	11.7	10.9
13–15	8.4	5.1	6.4	6.3
16–20	8.2	3.9	4.0	4.3
≥21	1.3	0.9	1.0	1.4

[a]Sources: Refs. 13,32,33.
[b]Reported as black and other races.
[c]Includes hysterectomy and hysterotomy.

births that is similar to information reported by the CDC for 1995 [29]. AGI found that women who live with a partner outside marriage or have no religious identification were 3.5 to 4 times more likely to have an abortion than women in the general population. Nonwhites, women aged 18 to 24, Hispanics, separated and never-married women, and women with an annual income of less than $15,000 were 1.6 to 2.2 times more likely than the general population to have an abortion. Residents of metropolitan areas and women who had already had a live birth also were more likely to have an abortion. Catholics were as likely to have an abortion as the general population, whereas Protestants were only 69% as likely and Evangelical or born-again Christians were only 39% as likely [29].

C. Gestational Age and Type of Procedure

In 1995, a larger percentage of women who obtained abortions did so during the first trimester (88%) than in 1973 (83.4%). The greatest change was in the percentage of abortions performed at or before 8 weeks gestation (34% in 1972, 54% in

1995) (Table 13.2). Between 1972 and 1995, the procedures used to perform abortions also changed. Curettage was used for 88.6% of all abortions in 1972 and 98.9% of all abortions in 1995. The proportion of instillation (labor-induction) procedures for pregnancy termination decreased from 10.4% in 1972 to 0.5% in 1995; by 1995, the use of hysterectomy, hysterotomy, or other procedures (such as vaginal suppositories and intramuscular injections) for pregnancy termination was negligible (Table 13.2). This shift in type of procedure can be explained by the higher proportion of abortions performed earlier in pregnancy and the lower complication rates associated with dilatation and evacuation performed later in pregnancy [34].

Certain medications for the induction of abortion (medical abortion), including antimetabolites (methotrexate) and steroid analogues (mifepristone), have entered the spotlight. Methotrexate is an antimetabolite that has been available for many years for the treatment of neoplastic diseases such as choriocarcinoma, for severe psoriasis, and for adult rheumatoid arthritis. Its mechanism of action is the inhibition of dehydrofolic acid reductase, interfering with DNA synthesis, repair, and cellular replication [35]. As a result, actively proliferating tissues, such as certain malignant cells and fetal and trophoblastic cells, are particularly susceptible to the drug. There have been a number of reports on the success of this drug in the medical treatment of early unruptured ectopic pregnancy [36–39]. Moreover, there is good evidence that the anticellular effects of the medication in seemingly low doses offer the possibility of medical induction of abortion in the future, with minimal risk. In combination with misoprostol (a prostaglandin analogue), methotrexate has been found to be a safe and effective method for termination of early pregnancy [36].

Mifepristone (RU-486) is a derivative of norethindrone, a progestin frequently used in oral contraceptives. Investigators have conducted numerous clinical studies of RU-486 in 15 countries; they have found that the drug, when used alone, is very effective in inducing an abortion if administered orally shortly after conception. However, although successful induction of abortion occurred in more than 90% of women whose last normal menstrual period occurred within the previous 6 weeks, the effectiveness in slightly later pregnancies fell far short of expectations. When RU-486 is used in combination with one of the prostaglandin analogues, it is 95% to 99% successful in inducing early abortion [40–44]. Its effectiveness, particularly if administered with a prostaglandin analogue, is comparable to the use of suction curettage before 7 weeks of gestation [45]. The most important advantage of using RU-486 is that it is not a surgical procedure. The participants in the 1995 Population Council mifepristone/misoprostol clinical trial were queried about their experience. The majority reported that they chose medical abortion because they wanted to avoid surgery and believed that the method was safer than surgical abortion. Almost half of the participants believed that mifepristone was a more natural method and a third said that they believed this method had the least risk of infection and that it could be used early in their pregnancy [46]. Nearly three-fourths of the participants (73%) stated that they were very satisfied, and 16% reported that they were somewhat satisfied with their experience of medical abortion; 94% said they would recommend medical abortion to a friend, and 87% said they would choose medical abortion if they had to terminate another pregnancy [46]. Medical abortion is not yet generally available to American women [46].

V. Complications of Induced Abortion

A. Abortion Mortality

1. Rates and Trends

The CDC has monitored abortion-related mortality since 1972. From 1972 through 1991, the CDC identified and investigated 287 deaths related to legal abortions [13]. The annual number of deaths related to legal abortions decreased from 24 deaths in 1972 to five in 1990 (Table 13.1, Fig. 13.1) [13]. The mortality rate decreased from 4.1 per 100,000 abortions in 1972 to 0.3 per 100,000 abortions in 1990.

2. Risk Factors

In a detailed analysis of deaths occurring between 1972 and 1987, investigators found that the main contributors to an increased risk of abortion-related death were age, race, parity, gestational age, and type of procedure [47]. Analysis of deaths through 1991 confirmed previous findings and revealed no substantial changes [Hani K. Atrash, unpublished data from the CDC].

From 1972 through 1987, women in their teens had the lowest risk of abortion-related death (1.0 deaths per 100,000 procedures). The risk of death increased with age and was highest for women aged 40 years or older (3.1 deaths per 100,000 procedures) [47]. Black and other minority women had 2.5 times the risk of death from legal abortion than did white women. The risk of death increased with parity; however, only women who had given birth three or more times had a statistically significant increased risk of abortion-related death (2.8 times higher than nulliparous women) [47]. The risk of abortion-related death increased markedly with gestational age. For abortions performed during the first 8 weeks of gestation, the rate of abortion-related death was 0.4 deaths per 100,000 procedures compared to 2.9 per 100,000 abortions at 13–15 weeks, 9.3 per 100,000 abortions at 16–20 weeks, and 12.0 per 100,000 abortions performed at 21 weeks of gestation or later [47].

Abortions performed by suction or sharp curettage during the first trimester were associated with the lowest mortality rate (0.5 per 100,000 procedures); the risk of death was significantly higher for abortions performed by dilatation and evacuation after 12 weeks of gestation (6.8 times higher for abortions performed by curettage), for instillation procedures (13.0 times higher), and for hysterectomy/hysterotomy procedures (95 times higher) [47].

3. Causes of Death

Embolism, infection, hemorrhage, and complications resulting from anesthesia contributed to abortion-related mortality; together, these causes accounted for 82% of all deaths related to legal abortions between 1972 and 1987. However, the leading causes of abortion-related death have changed over time; during the 1970s, the two leading causes were infection and embolism but during the 1980s, the leading cause was complications resulting from anesthesia [47].

B. Abortion Morbidity

1. Complication Rates

Because of a lack of consensus about what constitutes a "complication," abortion-related morbidity is very difficult to measure. Complications can range from very minor complications such as headache and nausea to death. Because there are no surveillance data on abortion-related morbidity, all information is derived from published study reports. In fact, most of what we know today about abortion morbidity is based on findings from three national prospective studies conducted by the CDC and the Population Council between 1971 and 1978 (the Joint Program for the Study of Abortion [JPSA]). Between 73,000 and 84,000 women were involved in each of the three phases of the study. The investigators defined a major complication as one or more of the following 15 complications: death, cardiac arrest, convulsion, endotoxic shock, fever for 3 or more days, hemorrhage necessitating blood transfusion, hypernatremia (the presence of an abnormally high level of sodium in the blood), injury to the bladder, ureter, or intestines, pelvic infection with 2 or more days of fever and a peak temperature of at least 40° C or with hospitalization for 11 days or more, pneumonia, psychiatric hospitalization for 11 days or more, pulmonary embolism or infarction, thrombophlebitis, major surgical treatment of complications, and wound disruption after hysterotomy or hysterectomy [34,48]. In summarizing their findings from the JPSA studies, the CDC researchers reported that the rate of major abortion complication was less than 1% [49]. The risk of developing major complications from legal abortion decreased significantly during the 1970s, from 1.0% to 0.29% [34,48,50]. In a report of 170,000 first-trimester abortions performed in three freestanding clinics in New York City from 1971 to 1987, Hakim-Elahi and colleagues [51] reported a total complication rate of 0.9% and a major complication rate of 0.07%. Data compiled by the National Abortion Federation for 1988 on 68,828 patients indicate that about 0.5% of patients required hospitalization because of abortion-related complications [52]. Preliminary analysis of data collected by the CDC from one metropolitan abortion clinic found a total complication rate of 1.59% and a major complication rate of 0.08% for 1973 to 1988 [Audrey Saftlas and Hani Atrash, unpublished data from the CDC]. The higher overall complication rate reported in the CDC study is probably related to the better follow up and ascertainment of patients with complications.

2. Risk Factors

The key risk factors for abortion morbidity are gestational age and type of procedure. Complication rates are lowest among women who obtain abortions before 8 weeks of gestation; complication rates are twice as high for abortions performed at 13 to 14 weeks and four to ten times higher for procedures performed after 15 weeks of gestation. Complication rates associated with curettage procedures are lower than complication rates associated with other procedures; the risk of complications is six times higher for abortions performed using labor-induction procedures and more than ten times higher for abortions performed by hysterectomy or hysterotomy [48,49,53]. For midtrimester abortions, the rate of complication is highest for intra-amniotic instillation of prostaglandin ($PGF_{2\alpha}$), lower for saline instillation, and lowest for dilatation and evacuation [54]. The risk of abortion complications associated with late gestational age and the risk associated with type of procedure are not independent, because suction curettage alone—the safest procedure—cannot be performed effectively after approximately 14 weeks of gestation.

The risk of complications increases among previously pregnant women. Women who have previously had one or more induced abortions have more than a 50% increased risk of complications before 12 weeks of gestation but no increased risk after 12 weeks. Women with one or more previous deliveries have a 34% increased risk of complications across all gestational ages. No increased risk of complications has been observed among women with a history of spontaneous abortion [49].

General anesthesia is associated with higher rates of uterine hemorrhage, uterine perforation, intra-abdominal hemorrhage, and cervical trauma; on the other hand, local anesthesia is associated with higher rates of fever and convulsions [55]. In a study of 10,000 abortions performed in Italy between 1982 and 1986, investigators found that the overall risk of complications following abortion was nearly doubled if a general anesthetic was used rather than a local anesthetic [56]. In this same study, the use of general anesthesia increased the risk of hemorrhage 4.6 times, the risk of injury 1.3 times, and the risk of other complications 1.6 times [56].

Higher rates of cervical injury and uterine perforation were found when using rigid dilators for cervical dilatation, and when the abortion was performed by a resident in training [57,58].

3. Abortion Complications

Complications of abortion can be categorized as early, delayed, and late complications depending upon their time of occurrence. Early complications occur during or within 48 hours of the abortion procedure, delayed complications occur between 48 hours and 2 weeks after the procedure, and late complications occur more than 2 weeks after the procedure.

a. EARLY COMPLICATIONS.

i. Hemorrhage Because of a lack of consensus regarding the definition of hemorrhage (definitions range from blood loss of 100 ml to blood loss of 1000 ml) and clinicians' imprecision in estimating volumes of blood loss, the reported incidence of hemorrhage varies from 0.05 to 4.9 per 100 abortions [57,59]. The most objective index of significant blood loss is blood transfusion. The reported rate of transfusion associated with suction curettage is 0.06 transfusions per 100 abortions [59]. The incidence of bleeding varies depending on the method of abortion: 0.26% for dilatation and evacuation, 0.32% for urea-prostaglandin, and 1.72% for saline instillation [60,61]. The risk increases with gestational age and with the use of general anesthesia, especially with agents that produce uterine relaxation [55,56].

ii. Uterine Perforation The incidence of uterine perforation has been consistently reported at 0.2 per 100 abortions [59,62]. However, the real incidence may be much higher because of asymptomatic, unsuspected perforations. In a study of women who underwent first-trimester abortions at the same time they underwent laparoscopic sterilization (a procedure that allowed the operator to detect perforations visually), Kaali and colleagues reported a 2% incidence of perforation [63]. Lack of

operator skill, advancing gestational age, and multiparity have been identified as risk factors for uterine perforation [56].

iii. Cervical Lacerations Reported rates of cervical injury range from 0.01% to 1.6% [51,58,59]. The lowest rates were noted for suction curettage (0.18% to 0.96%), and higher rates were reported for dilatation and evacuation at 13 to 20 weeks of gestation (1.16%). The cervical injury rate for labor-induction procedures was 0.55%. Women who previously have received an abortion have a reduced risk of cervical injury. Women under the age of 17 years have an increased risk.

b. DELAYED COMPLICATIONS.

i. Incomplete Abortion Retained products of conception is one of the most important causes of abortion morbidity and may result in infection, bleeding, or both. The incidence of this complication after dilatation and evacuation is reportedly less than 1%, whereas the incidence following instillation procedures may be as high as 36% [34,51,64].

ii. Infection Infection rates vary by abortion procedures. The rate of infection is lowest after suction curettage (0.75%), higher following dilatation and evacuation (1.5%), and highest following instillation (5%) [60,61]. Women are at increased risk of postabortion infection if the procedure is performed later in pregnancy or if local rather than general anesthesia is used for suction curettage [65]. Although retained products of conception is the most common cause of infection, endometritis or pelvic peritonitis may occur in the absence of retained tissue, particularly among women who have a preexisting gonococcal or *Chlamydia* infection.

iii. Other Complications Other complications of abortion include venous thrombophlebitis, pulmonary embolism, severe coagulopathies such as disseminated intravascular coagulation, anesthesia complications, hypernatremia, water intoxication, pregnancy left intact, and delivery of a live-born fetus. Sufficient population-based data are not available to assess the incidence of these conditions.

c. LATE COMPLICATIONS.

i. Secondary Infertility Several prospective studies have concluded that induced abortion has no significant effect on secondary infertility [65–70]. In fact, two studies [66,70] reported that fertility was greater among women following their abortions, as indicated by significantly shorter interpregnancy intervals. However, because women who experience an unwanted pregnancy are likely to be among the more fertile group of women [70], shorter interpregnancy intervals among this group are not surprising.

ii. Ectopic Pregnancy Studies have concluded that undergoing one or more induced abortions does not increase a woman's risk of ectopic pregnancy [71,72]. Studies that reported an association between induced abortion and ectopic pregnancy involved small numbers of ectopic pregnancy cases, failed to control for important risk factors, or were conducted in countries where abortion was illegal [73–79]. Chung and colleagues reported a significant increase in the incidence of ectopic pregnancy following abortions complicated by infection or retained products of conception [80].

iii. Spontaneous Abortion Because the ascertainment of first-trimester spontaneous abortions is very difficult, most of our knowledge about the incidence of spontaneous abortion following induced abortion pertains to the second trimester. In a review of the literature, Atrash and Hogue concluded that suction curettage, commonly used to terminate pregnancies during the first trimester, carried no increased risk of spontaneous abortion; on the other hand, dilatation and evacuation during the midtrimester was associated with a significantly increased risk of future spontaneous abortion [73].

iv. Preterm Delivery Most studies have found no increased risk of preterm delivery following an induced abortion, regardless of the type of procedure or gestational age. The risk of preterm delivery among women whose first pregnancy ended in induced abortion is no higher than the risk for nulliparous women with no history of abortion [81–86]. However, the risk of preterm delivery for women whose first pregnancy ended in induced abortion is higher than for women who have had one full-term birth and no prior abortive outcomes (this slightly increased risk is not statistically significant) [81,83–87]. This finding suggests that induced abortion does not confer the well-known protective effect of a term delivery [73].

v. Low Birth Weight Overall, studies have shown no significantly increased risk of low birth weight following abortions performed with vacuum aspiration. However, dilatation and evacuation has been associated with an increased risk of low birth weight [73,74,86]. The reported relative risk of low birth weight was higher when women with abortions induced by vacuum aspiration were compared with women who had one previous pregnancy ending in a live birth than when they were compared with women who had one prior pregnancy ending in an abortive outcome [73,81,83–89]. This observation suggests that women whose first pregnancy was terminated by induced abortion have a higher risk of low birth weight than women who carried their pregnancy to term, which suggests that induced abortion does not reduce the well-known risk of low birth weight for firstborn offspring.

vi. Psychological Effects Some investigators have reported that a large number of women experience emotional problems following an abortion. An expert panel appointed by the American Psychological Association to review the literature on psychological effects of induced abortion concluded that severe psychological reactions following abortion are rare and they parallel those following other normal life stresses [90]. Some individual women may experience severe distress or psychopathology, but it is not clear whether these are causally linked to abortion [91]. After reviewing more than 250 studies of the psychological outcomes of abortion, then Surgeon General C. Everett Koop concluded in 1989 that most research in this area has serious flaws and could not support the contention that abortion is dangerous to women's mental health [92].

vii. Breast Cancer Investigators have focused a great deal of attention on the potential association between induced abortion and breast cancer. This research has been based on the hypothesis that an interrupted pregnancy leads to breast cell proliferation without the protective effect of breast cell differentiation that occurs later in pregnancy [93]. In 1981, a case-control interview study reported a 2.4-fold increased risk of breast cancer in young women with a history of induced abortion [94]. This provocative finding sparked numerous subsequent case-control studies that have yielded inconsistent findings [95–

98]. Most studies reporting a positive association have found the increased risk to be limited primarily to certain subgroups of women (nulliparous women, young women, and women whose pregnancies were terminated before first birth). A meta-analysis of 23 observational studies of induced abortion and breast cancer risk by Brind *et al.* reported an overall odds ratio of 1.3 and concluded that induced abortion is a risk factor [99]. This meta-analysis was limited in that it did not impose any methodologic standards or quality criteria on studies for inclusion in the analysis and it did not stratify by study design. In contrast, a qualitative review of 32 studies on the risk of breast cancer following spontaneous or induced abortion reported that the existing scientific evidence was insufficient to reach a conclusion about this relationship [100].

Case-control studies, which rely on women's self-reports of past abortions, are vulnerable to recall bias. Because abortion is a highly sensitive and emotionally charged issue, women with breast cancer are likely to be more motivated (or less reluctant) than healthy control subjects to disclose a past elective abortion. Three case-control studies from Sweden, Holland, and the United States have suggested that much of the increased risk of breast cancer associated with induced abortion could be explained by differential recall of abortion history among breast cancer patients and healthy subjects [97,98,101]. Newcomb *et al.* found that the overall risk of breast cancer following abortion was elevated by 23%; however, the increased risk among women who reported having an abortion since its legalization in 1973 was 12% compared to an increased risk of 35% among those who reported abortions prior to legalization [97]. In a case-control study, Daling and colleagues did not find a difference in risk of breast cancer associated with a history of induced abortion prior to legalization compared with the overall risk estimate [95]. Daling *et al.* [95] have also challenged the analysis of the Swedish study [100], which concluded that a spurious 50% increase in risk could be attributed to recall bias. Based on a reanalysis of the Swedish data, Daling *et al.* have asserted that reporting differences between case and control patients account for only a 16% difference in breast cancer risk [95].

In 1997, a Danish study reported that induced abortion was not associated with an increased risk of breast cancer (relative risk = 1.00; 95% confidence interval = 0.94–1.06) [102]. This conclusion was based on analysis of a population-based cohort study of 1.5 million women born between April 1, 1935 and March 31, 1978 in Denmark where reporting of induced abortions has been mandatory since 1939. This study obtained information on the number and dates of abortions from Denmark's National Registry of Induced Abortions and utilized linkage of the Abortion Registry with the Danish Cancer Registry to identify all new cases of breast cancer. Because of the prospective cohort design of the study, the large number of study subjects, the use of record linkage to ascertain abortion history in women from an entire nation, and the statistical adjustment for reproductive variables, the overall conclusion of this study provides strong evidence in support of a null association [103].

viii. Other Late Complications Investigators have studied placenta previa as a possible late sequela of induced abortion. One case-control study reported a ten-fold increased risk of placenta previa [104]. However, in a later study that accounted for the potential confounding risk factors of gravidity and age in a predominantly black population, investigators found the relative risk to be 1.1 and not statistically significant [105]. In a prospective cohort study, Frank *et al.* [106] conducted a 10-year followup of 6418 women who had induced abortions and 8059 women who had unplanned pregnancies but did not terminate the pregnancies. Pregnancies occurred among 729 women in the case group and 1754 women in the control group. Prior induced abortion had no material effect on the rate of pregnancy-related morbidity (including hemorrhage before labor, mental illness, preeclampsia, hemorrhage in labor, fetopelvic disproportion, trauma at delivery, forceps delivery, cesarean section, and fetal distress). However, the incidence of anemia during pregnancy was significantly reduced among women with a history of induced abortion, and the incidence of urinary tract infection was significantly increased [106].

References

1. McFarland, D. R. (1993). Induced abortion: An historical overview. *Gynecol. Health* **7**, 25–30.
2. Sheeran, P. J. (1987). "Women, Society, the State and Abortion: A Structuralist Analysis." Praeger, New York.
3. Mankekar, K. (1973). "Abortion: A Social Dilemma." Vikas Publishing House, Delhi, India.
4. Hall, R. E. (1971). "A Doctor's Guide to Having an Abortion." New American Library, New York.
5. Henshaw, S. K., and Morrow, E. (1990). "Induced Abortion, A World Review, 1990 Supplement." Alan Guttmacher Institute, New York.
6. Binkin, N., Gold, J., and Cates, W., Jr. (1982). Illegal abortions: Why are they still occurring? *Fam. Plann. Perspect.* **14**, 163–166.
7. Henshaw, S. K., and Van Vort, J. (1990). Abortion services in the United States, 1987 and 1988. *Fam. Plann. Perspect.* **22**, 102–109.
8. Henshaw, S. K., and Van Vort, J. (1994). Abortion services in the United States, 1991 and 1992. *Fam. Plann. Perspect.* **26**, 100–106, 112.
9. Henshaw, S. K. (1995). Factors hindering access to abortion services. *Fam. Plann. Perspect.* **27**, 54–59, 87.
10. Henshaw, S. K. (1991). Accessibility of abortion services in the United States. *Fam. Plann. Perspect.* **23**, 246–253.
11. Kolbert, K., and Miller, A. (1994). Government in the examining room: Restrictions on the provision of abortion. *J. Am. Med. Women's Assoc.* **49**, 153–155.
12. U.S. Bureau of the Census (1997). "Statistical Abstract of the United States: 1997," 117th ed. U.S. Bureau of the Census, Washington, DC.
13. Koonin, L. M., Smith, J. C., Ramick, M., Strauss, L. T., and Hopkins, F. W. (1997). Abortion surveillance—United States, 1993 and 1994. *Morbid. Mortal. Wkly. Rep.* **46**(SS-46), 37–98.
14. Sollom, T. (1997). State actions on reproductive health issues in 1996. *Fam. Plann. Perspect.* **29**, 35–40.
15. Darney, P. D., Landy, U., MacPherson, S., and Sweet, R. L. (1987). Abortion training in U.S. obstetrics and gynecology residency programs. *Fam. Plann. Perspect.* **19**, 158–162.
16. Grimes, D. A. (1992). Clinicians who provide abortions: The thinning ranks. *Obstet. Gynecol.* **80**, 719–723.
17. Henshaw, S. K., Forrest, J. D., and Van Vort, J. (1987). Abortion services in the United states, 1984 and 1985. *Fam. Plann. Perspect.* **19**, 63–70.
18. Westfall, J. M., Kallail, K. J., and Walling, A. D. (1991). Abortion attitudes and practices of family and general practice physicians. *J. Fam. Pract.* **33**, 47–51.
19. Lindheim, B., and Cotterill, M. (1978). Training in induced abortion by obstetrics and gynecology residency programs. *Fam. Plann. Perspect.* **10**, 24–28.
20. MacKay, H. T., and MacKay, A. P. (1995). Abortion training in obstetrics and gynecology residency programs in the United States, 1991–1992. *Fam. Plann. Perspect.* **27**, 112–115.
21. Westhoff, C., Marks, F., and Rosenfield, A. (1993). Residency training in contraception, sterilization, and abortion. *Obstet. Gynecol.* **81**, 311–314.
22. Westhoff, C. (1994). Abortion training in residency programs. *J. Am. Med. Women's Assoc.* **49**, 150–152.
23. Forrest, J. D., and Henshaw, S. K. (1987). The harassment of U.S. abortion providers. *Fam. Plann. Perspect.* **19**, 9–13.
24. Grimes, D. A., Forrest, J. D., Kirkman, A. L., and Radford, B. (1991). An epidemic of antiabortion violence in the United States. *Am. J. Obstet. Gynecol.* **165**, 1263–1268.
25. Scruggs, K. (1998). Birmingham clinic bombing: Too early to establish Atlanta link. *Atlanta Journal-Constitution,* January 30, p. A-11.
26. Henshaw, S. K. (1998). Unintended pregnancy in the United States. *Fam. Plann. Perspect.* **30**, 24–29.
27. Torres, A., and Forrest, J. D. (1988). Why do women have abortions? *Fam. Plann. Perspect.* **20**, 169–176.
28. Henshaw, S. K., and Silverman, J. (1988). The characteristics and prior contraceptive use of U.S. abortion patients. *Fam. Plann. Perspect.* **20**, 158–168.
29. Henshaw, S. K., and Kost, K. (1996). Abortion patients in 1994–1995: Characteristics and contraceptive use. *Fam. Plann. Perspect.* **28**, 140–147.
30. Henshaw, S. K., and Van Vort, J., eds. (1988). "Abortion Services in the United States, Each State and Metropolitan Area, 1984–1985." Alan Guttmacher Institute, New York.
31. Atrash, H. K., Lawson H., and Smith, J. C. (1990). Legal abortion in the United States: Trends and mortality. *Contemp. Obstet. Gynecol.* **35**, 58–69.
32. Centers for Disease Control (CDC) (1983). "Abortion Surveillance, 1979–1980." CDC, Atlanta, GA.
33. Centers for Disease Control and Prevention (CDC) (1997). Abortion surveillance: Preliminary analysis—United States, 1995. *Morbid. Mortal. Wkly. Rep.* **46**, 1133–1137.
34. Grimes, D. A., Schulz, K. F., Cates, W., Jr., and Tyler, C. W., Jr. (1977). Midtrimester abortion by dilatation and evacuation. *N. Engl. J. Med.* **296**, 1141–1145.
35. Hammer, R. A., and Milad, M. P. (1998). Overview of the management of tubal pregnancy. *In* "Gynecology and Obstetrics" (J. J. Sciarra, ed.), Vol. 1, Chapter 68, pp. 1–14. Lippincott-Raven, Philadelphia.
36. Hausknect, U. D. (1995). Methotrexate and misoprostol to terminate early pregnancy. *N. Engl. J. Med.* **333**, 537–540.
37. Stovall, T. G., and Ling, F. W. (1993). Single-dose methotrexate: An expanded clinical trial. *Am. J. Obstet. Gynecol.* **168**, 1759–1765.
38. Fernandez, H., Benifla, J. L., Lelaidier, C., Baton, C., and Frydman, R. (1993). Methotrexate treatment of ectopic pregnancy: 100 cases treated by primary transvaginal injection under sonographic control. *Fertil. Steril.* **59**, 773–777.
39. Ichinoe, K., Wake, N., Shinkai, N., Shiina, Y., Miyazaki, T., and Tanaka, T. (1987). Nonsurgical therapy to preserve oviduct function in patients with tubal pregnancies. *Am. J. Obstet. Gynecol.* **156**, 484–487.
40. Avrech, O. M., Golan, A., Weinraub, Z., Bukovsky, I., and Caspi, E. (1991). Mifepristone (RU486) alone or in combination with a prostaglandin analogue for termination of early pregnancy: A review. *Fertil. Steril.* **56**, 385–393.

41. Swahn, M. L., Gottlieb, C., Green, K., and Bygdeman, M. (1990). Oral administration of RU 486 and 9-methylene PGE2 for termination of early pregnancy. *Contraception* **41**, 461–473.

42. Rodger, M. W., Logan, A. F., and Baird, D. T. (1989). Induction of early abortion with mifepristone (RU486) and two different doses of prostaglandin pessary (gemeprost). *Contraception* **39**, 497–502.

43. Wu, S., Gao, J., Wu, Y., Wu, M., Fan, H., Yao, G., Zheng, S., Wang, P., Du, M., Huang, Z., Huang, J., Zhu, G., Lei, Z., Chen, X., Peng, D., Song, L., Wu, X., Huang, S., Xia, J., and Zhang, J. (1992). Clinical trial on termination of early pregnancy with RU486 in combination with prostaglandin. *Contraception* **46**, 203–210.

44. Aubeny, E. (1991). RU486 combined with PG analogs in voluntary termination of pregnancy. *Adv. Contraception* **7**, 339–343.

45. Kaunitz, A. M., Rovira, E. Z., Grimes, D. A., and Schulz, K. F. (1985). Abortions that fail. *Obstet. Gynecol.* **66**, 533–537.

46. Beckman, L. J., and Harvey, S. M. (1997). Experience and acceptability of medical abortion with mifepristone and misoprostol among U.S. women. *Women's Health Issues* **7**, 253–262.

47. Lawson, H. W., Frye, A., Atrash, H. K., Smith, J. C., Shulman, H. B., and Ramick, M. (1994). Abortion mortality, United States, 1972–1987. *Am. J. Obstet. Gynecol.* **171**, 1365–1372.

48. Tietze, C., and Lewit, S. (1972). Joint program for the study of abortion (JPSA): Early medical complications of legal abortion. *Stud. Fam. Plann.* **3**, 97–122.

49. Buehler, J. W., Schulz, K. F., Grimes, D. A., and Hogue, C. J. R. (1985). The risk of serious complications from induced abortion: Do personal characteristics make a difference? *Am. J. Obstet. Gynecol.* **153**, 14–20.

50. Cates, W. Jr., and Grimes, D. A. (1981). Morbidity and mortality of abortion in the United States. *In* "Abortion and Sterilization: Medical and Social Aspects" (S. E. Hodgson, ed.), pp. 155–180. Academic Press, San Diego, CA.

51. Hakim-Elahi, E., Tovell, H. M. M., and Burnhill, M. S. (1990). Complications of first-trimester abortion: A report of 170,000 cases. *Obstet. Gynecol.* **76**, 129–135.

52. Gold, R. B. (1990). "Abortion and Women's Health: A Turning Point for America." Alan Guttmacher Institute, New York.

53. Grimes, D. A., Schulz, K. F., Cates, W., Jr., and Tyler, C. W. (1977). The Joint Program for the Study of Abortion/CDC: A preliminary analysis. *In* "Abortion in the Seventies" (W. Hern and B. Andrikopoulos, eds.), pp. 41–46. National Abortion Federation, New York.

54. Grimes, D. A., Schulz, K. F., Cates, W., Jr., and Tyler, C. W., Jr. (1977). Methods of midtrimester abortion: Which is safest? *Int. J. Gynaecol. Obstet.* **15**, 184–188.

55. Grimes, D. A., Schulz, K. F., Cates, W., Jr., and Tyler, C. W., Jr. (1979). Local versus general anesthesia: Which is safer for performing suction curettage abortions? *Am. J. Obstet. Gynecol.* **135**, 1030–1035.

56. Osborn, J. F., Arisi, E., Spinelli, A., and Stazi, M. A. (1990). General anesthesia, a risk factor for complication following induced abortion? *Eur. J. Epidemiol.* **6**, 416–422.

57. Grimes, D. A., Schulz, K. F., and Cates, W., Jr. (1984). Prevention of perforation during curettage abortion. *JAMA, J. Am. Med. Assoc.* **251**, 2108–2111.

58. Schulz, K. F., Grimes, D. A., and Cates, W., Jr. (1983). Measures to prevent cervical injury during suction curettage abortion. *Lancet* **1**, 1182–1185.

59. Cates, W. Jr., Schulz, K. F., Grimes, D. A., and Tyler, C. W., Jr. (1979). Short-term complications of uterine evacuation techniques for abortion at 12 weeks' gestation or earlier. *In* "Pregnancy Termination Procedures, Safety and New Developments" (G. I. Zatuchni, J. J. Sciarra, and J. J. Speidel, eds.), p. 127. Harper & Row, Hagerstown, MD.

60. Kafrissen, M. E., Schulz, K. F., Grimes, D. A., and Cates, W., Jr. (1984). Midtrimester abortion, intrauterine instillation of hyper osmolar urea and prostaglandin F_{2a} versus dilatation and evacuation. *JAMA, J. Am. Med. Assoc.* **251**, 916–919.

61. Binkin, N. J., Schulz, K. F., Grimes, D. A., and Cates, W., Jr. (1983). Urea-prostaglandin versus hypertonic saline for instillation abortion. *Am. J. Obstet. Gynecol.* **146**, 947–952.

62. Lindell, G., and Flam, F. (1995). Management of uterine perforations in connection with legal abortions. *Acta Obstet. Gynecol. Scand.* **74**, 373–375.

63. Kaali, S. G., Szigetvari, I. A., and Bartfai, G. S. (1989). The frequency and management of uterine perforations during first-trimester abortions. *Am. J. Obstet. Gynecol.* **161**, 406–408.

64. Burkman, R. T., and King, T. M. (1984). Second trimester termination of pregnancy. *In* "Clinical and Diagnostic Procedures in Obstetrics and Gynecology. Part B: Gynecology" (E. M. Symonds and F. P. Zuspan, eds.), p. 241. Dekker, New York.

65. World Health Organization Task Force on Sequelae of Abortion, Special Programme of Research, Development and Research Training in Human Reproduction (1984). Secondary infertility following induced abortion. *Stud. Fam. Plann.* **15**, 291–295.

66. Stubblefield, P. G., Monson, R. R., Schoenbaum, S. C., Wolfson, C. E., Cookson, D. J., and Ryan, K. J. (1984). Fertility after induced abortion: A prospective follow-up study. *Obstet. Gynecol.* **63**, 186–193.

67. Daling, J. R., Weiss, N. S., Voigt, L., Spadoni, L. R., Soderström, R., Moore, D. E., and Stadel, B. V. (1985). Tubal infertility in relation to prior induced abortion. *Fertil. Steril.* **43**, 389–394.

68. Cramer, D. W., Schiff, I., Schoenbaum, S. C., Gibson, M., Belisle, S., Albrecht, B., Stillman, R. J., Berger, M. J., Wilson, E., and Stadel, B. V. (1985). Tubal infertility and the intrauterine device. *N. Engl. J. Med.* **312**, 941–947.

69. Frank, P. I., McNamee, R., Hannaford, P. C., Kay, C. R., and Hirsch, S. (1993). The effect of induced abortion on subsequent fertility. *Br. J. Obstet. Gynaecol.* **100**, 575–580.

70. Hogue, C. J., Schoenfelder, J. R., Gesler, W. M., and Shachtman, R. H. (1978). The interactive effects of induced abortion, interpregnancy interval, and contraceptive use on subsequent pregnancy outcome. *Am. J. Epidemiol.* **107**, 15–26.

71. Atrash, H. K., Strauss, L. T., Kendrick, J. S., Skjeldestad, F. E., and Ahn, Y. W. (1997). The relation between induced abortion and ectopic pregnancy. *Obstet. Gynecol.* **89**, 512–518.

72. Skjeldestad, F. E., Gargiullo, P. M., and Kendrick, J. S. (1997). Multiple induced abortions as risk factor for ectopic pregnancy. *Acta Obstet. Gynecol. Scand.* **76**, 691–696.

73. Atrash, H. K., and Hogue, C. J. R. (1990). The effect of pregnancy termination on future reproduction. *Bailliere's Clin. Obstet. Gynecol.* **4**, 391–405.

74. Hogue, C. J. R., Cates, W., Jr., and Tietze, C. (1982). The effects of induced abortion on subsequent reproduction. *Epidemiol. Rev.* **4**, 66–94.

75. Sawazaki, C., and Tanaka, S. (1966). The relationship between artificial abortion and extrauterine pregnancy. *In* "Harmful Effects of Induced Abortion" (Y. Koya, ed.), pp. 49–63. Family Planning Association of Japan, Tokyo.

76. Shinagawa, S., and Nagayama, M. (1969). Cervical pregnancy as a possible sequela of induced abortion. Report of 19 cases. *Am. J. Obstet. Gynecol.* **105**, 282–284.

77. Panayotou, P. P., Kaskarelis, D. B., Miettinen, O. S., Trichopoulos, D. B., and Kalaudidi, A. K. (1972). Induced abortion and ectopic pregnancy. *Am. J. Obstet. Gynecol.* **114**, 507–510.

78. Dziewulska, W. (1973). Abortion in the past versus the fate of the subsequent pregnancy. *Ginekol. Pol.* **44**, 1003–1011.

79. Hren, M., Tomazevic, T., and Seigel, D. (1974). Ectopic pregnancy. *In* "The Ljubljana Abortion Study, 1971–1973" (L. Andolsck, ed.), pp. 34–38. Center for Population Research, National Institutes of Health, Bethesda, MD.

80. Chung, C. S., Smith, R. G., Steinhoff, P. G., and Mi, M. P. (1982). Induced abortion and ectopic pregnancy in subsequent pregnancies. *Am. J. Epidemiol.* **115**, 879–887.

81. Chung, C. S., Steinhoff, P. G., and Smith, R. G. (1981). "Effects of Induced Abortion on Subsequent Reproductive Function and Pregnancy Outcome," Final rep., Contract No. (N01-HD-62801). University of Hawaii, Honolulu.

82. Daling, J. R., and Emanuel, I. (1977). Induced abortion and subsequent outcome of pregnancy in a series of American women. *N. Engl. J. Med.* **297**, 1241–1245.

83. Frank, P. I. (1985). Sequelae of induced abortion. *Ciba Found. Symp.* **115**, 67–82.

84. Lerner, R. C., and Varma, A. O. (1981). "Prospective Study of the Outcome of Pregnancy Subsequent to Previous Induced Abortion," Final rep., Contract No. (NO1-HD-62803). Downstate Medical Center, State University of New York.

85. Logrillo, V., Quickenton, P., and Thériault, G. D. (1980). "Effect of Induced Abortion on Subsequent Reproductive Function," Final report to National Institute of Child Health and Human Development. New York State Health Department, Albany.

86. World Health Organization Task Force on the Sequelae of Abortion (1979). Gestation, birthweight, and spontaneous abortion in pregnancy after induced abortion. *Lancet* **1**, 142–145.

87. Meirik, O., and Bergström, R. (1983). Outcome of delivery subsequent to vacuum aspiration abortion in nulliparous women. *Acta Obstet. Gynecol. Scand.* **63**, 45–50.

88. Obel, E. (1979). Pregnancy complications following legally induced abortion. *Acta Obstet. Gynecol. Scand.* **58**, 485–490.

89. Pickering, R. M., and Forbes, J. F. (1985). Risks of preterm delivery and small-for-gestational age infants following abortion: A population study. *Br. J. Obstet. Gynaecol.* **92**, 1106–1112.

90. Adler, N. E., David, H. P., Major, B. N., Roth, S. H., Russo, S. H., and Wyatt, G. E. (1992). Psychological factors in abortion: A review. *Am. Psychol.* **47**, 1194–1204.

91. Dagg, P. K. B. (1991). The psychological sequelae of therapeutic abortion—Denied and completed. *Am. J. Psychiatry* **148**, 578–585.

92. Hearings before the Human Resources and Intergovernmental Relations Subcommittee of the Committee on Government Operations, House of Representatives, 101st Congress, 1st Session (1989). Testimony of C. Everett Koop, MD, ScD, Surgeon General.

93. Russo, J., and Russo, I. H. (1980). Susceptibility of the mammary gland to carcinogenesis. II. Pregnancy interruption as a risk factor in tumor incidence. *Am. J. Pathol.* **100**, 497–512.

94. Pike, M. C., Henderson, B. E., Casagrande, J. T., Rosario, I., and Gray, G. E. (1981). Oral contraceptive use and early abortion as risk factors for breast cancer in young women. *Br. J. Cancer* **43**, 72–76.

95. Daling, J. R., Brinton, L. A., Voigt, L. F., Weiss, N. S., Coates, R. J., Malone, K. E., Schoenberg, J. B., and Gammon, M. (1996). Risk of breast cancer among white women following induced abortion. *Am. J. Epidemiol.* **144**, 373–380.

96. Daling, J. R., Malone, K. E., Voigt, L. F., White, E., and Weiss, N. S. (1994). Risk of breast cancer among young women: Relationship to induced abortion. *J. Natl. Cancer Inst.* **86**, 1584–1592.

97. Newcomb, P. A., Storer, B. E., Longnecker, M. P., Mittendorf, R., Greenberg, E. R., and Willett, W. C. (1996). Pregnancy termination in relation to risk of breast cancer. *JAMA, J. Am. Med. Assoc.* **275**, 283–287.

98. Rookus, M. A., and Van Leeuwen, F. E. (1996). Induced abortion and risk for breast cancer: Reporting (recall) bias in a Dutch case-control study. *J. Natl. Cancer Inst.* **88**, 1759–1764.

99. Brind, J., Chinchilli, V. M., Severs, W. B., and Summy-Long, J. (1996). Induced abortion as an independent risk factor for breast cancer: A comprehensive review and meta-analysis. *J. Epidemiol. Commun. Health* **50**, 481–496.

100. Wingo, P. A., Newsome, K., Marks, J. S., Calle, E. E., and Parker, S. L. (1997). The risk of breast cancer following spontaneous or induced abortion. *Cancer Causes Control* **8**, 93–108.

101. Lindefors-Harris, B. M., Eklund, G., Adami, H. O., and Meirik, O. (1991). Response bias in a case-control study utilizing comparative data concerning legal abortions from two independent Swedish studies. *Am. J. Epidemiol.* **134**, 1003–1008.

102. Melbye, M., Wohlfahrt, J., Olsen, J. H., Frisch, M., Westergaard, T., Helwig-Larsen, K., and Andersen, P. K. (1997). Induced abortion and the risk of breast cancer. *N. Engl. J. Med.* **336**, 81–85.

103. Hartge, P. (1997). Abortion, breast cancer, and epidemiology. *N. Engl. J. Med.* **336**, 127–128.

104. Barrett, J. M., Boehm, F. H., and Killam, A. P. (1981). Induced abortion: A risk factor for placenta previa. *Am. J. Obstet. Gynecol.* **141**, 769–772.

105. Grimes, D. A., and Techman, T. (1984). Legal abortion and placenta previa. *Am. J. Obstet. Gynecol.* **149**, 501–504.

106. Frank, P. I., Kay, C. R., Scott, L. M., Hannaford, P. C., and Haran, D. (1987). Pregnancy following induced abortion: Maternal morbidity, congenital abnormalities and neonatal death: Royal College of General Practitioners/Royal College of Obstetricians and Gynaecologists Joint Study. *Br. J. Obstet. Gynaecol.* **94**, 836–842.

14

Maternal Morbidity

MICHELLE A. WILLIAMS* AND ROBERT MITTENDORF†

*Department of Epidemiology, University of Washington, Seattle, Washington; †Department of Obstetrics and Gynecology, Chicago Lying-in Hospital, MC-2050, University of Chicago, Chicago, Illinois

I. Introduction and Background

Pregnancy is characterized by profound metabolic and other physiologic alterations involving virtually every organ system. Alterations include changes in circulating concentrations of steroid and adrenocorticotropic hormones, changes in circulating plasma lipids, hemodynamic changes such as plasma volume, red cell mass, and blood viscosity, and anatomical changes such as a flaring of the lower ribs. These metabolic alterations, when overlaid upon chronic medical illnesses predating pregnancy, may contribute to a worsening, improvement, or no change in the woman's health. Among apparently healthy women, the metabolic challenges of pregnancy may unmask susceptibility for metabolic chronic disorders. Available clinical and epidemiologic data support the hypothesis that common medical complications of pregnancy, including gestational diabetes mellitus (GDM), pregnancy-induced hypertension, and pregnancy-associated thyroid disorders, may represent, at least for some women, their underlying risk for developing related chronic medical conditions as they age.

This chapter is not intended to be a comprehensive review of maternal medical complications during pregnancy. Rather, we offer a brief description of selected physiologic changes that occur in medically uncomplicated pregnancies. We then provide a summary of the clinical and epidemiologic characteristics of selected chronic medical conditions that are relatively prevalent among reproductive age women, and whose clinical course may either be exacerbated, set into remission, or both as a result the physiologic demands of pregnancy. The selected chronic medical conditions to be discussed include multiple sclerosis (MS), systemic lupus erythematosus (SLE), thyroid disorders, and asthma. A review of maternal medical conditions arising in apparently healthy women and the possible association of these conditions with subsequent chronic metabolic disorders also will be discussed.

II. Physiological Changes in Normal Pregnancy

Several alterations in immunologic status occur during pregnancy, and among them is a shift away from cell-mediated immunity towards increased humoral immunity [1,2]. The shift, from a predominance of type 1, or proinflammatory (helper T cells), to type 2, or anti-inflammatory (helper T cells), is considered to have an important role in allowing maternal tolerance of the fetus. An inversion of this shift during the latter half of the third trimester is thought to contribute to the initiation of labor.

Although details of the nature and direction of the influence of altered immunologic status in pregnancy are still highly debated, there is growing consensus that pregnancy-associated immunosuppressive cytokines may account for modifications in the behavior of autoimmune disorders. Immunologic alterations during pregnancy may explain the variable course of autoimmune disorders in pregnancy.

Every aspect of lipid metabolism is influenced by pregnancy [3]. Maternal serum or plasma cholesterol and triglyceride concentrations increase (1.5- and 2–3-fold, respectively) during pregnancy, with the major increase occurring in the second and third trimesters [4]. In addition to absolute changes in plasma lipid levels, changes in lipoprotein composition [4–6] also are noted in uncomplicated pregnancy. Some lipoproteins, such as high density lipoproteins (HDL) and low density lipoproteins (LDL), become more triglyceride enriched during pregnancy. Very low density lipoprotein (VLDL) increases its cholesterol and triglyceride proportionately and can be up to 5-fold of the ratio seen in nonpregnant women [7,8].

Many of the observed alterations in lipid metabolism during pregnancy are attributed to changes in hormonal status, although few investigators have specifically examined the effects of hormones on lipids and lipoprotein changes during pregnancy [9]. Estrogen and progesterone concentrations increase to values 16- and 7-fold, respectively, by 30 weeks gestation [10]. In general, studies assessing the effects of exogenous hormones suggest that the effects that estrogens have on lipids and lipoproteins are in the opposite direction from those of progestagens [11]. The continuous elevation of triglyceride levels in all lipoprotein fractions is thought to be a result of increasing estrogen concentrations [5,7,12]. Elevations in high density lipoproteins (HDL) and associated lipoprotein subfractions throughout the first two trimesters of pregnancy are thought to be related to rises in estrogen levels [13], whereas the decline in HDL-cholesterol that occurs after 24 weeks gestation is thought to be associated with increasing insulin concentrations [10]. This combination of hormonal and carbohydrate metabolic changes in uncomplicated pregnancies makes pregnancy a diabetogenic state. Specifically, uncomplicated pregnancy is characterized by progressive insulin resistance, hyperinsulinemia, and a deterioration of glucose tolerance in the third trimester.

As summarized by Knopp *et al.* [5], the apparent physiologic hyperlipidemia of pregnancy is significant for several reasons. First, it is possible that the increase in plasma triglycerides may enhance the availability of essential and nonessential fatty acids for placental transfer to the fetus, and, as such, is a physiologic adaptation that would serve to favor fetal nutrition and growth. Second, the elevations in cholesterol may provide the excess needed for placental progesterone synthesis and transplacental cholesterol transfer to the fetus. Third, the plasma triglyceride elevation may be the barometer of a general metabolic adaptation by the mother to augment nutrient flow to the fetus. Fourth, the hyperlipidemia may stress maternal lipid homeostasis to an extent that subclinical or mild hyperlipidemia becomes clinically detectable, analogous to the prediabetes recognized in women who develop gestational diabetes [14,15]. Finally, the hyperlipidemia, which in many ways resembles some dyslipidemic syndromes in nonpregnant individuals, namely increased triglycerides and LDL and decreased HDL, could itself function as an arteriosclerosis risk factor.

Pregnancy represents a considerable challenge to maternal energy balance, particularly in the third trimester, which is the period of rapid fetal growth. Leptin, the product of the *ob* gene, is produced primarily by adipose tissue and has been implicated in the regulation of energy balance via its central actions on food intake and energy expenditure [16]. In humans, circulating leptin concentrations correlate with adiposity [17,18] and decrease after weight loss [18,19], fasting [19], or caloric restriction [19]. Some studies, however, suggest that during pregnancy, leptin is not derived solely from adipocytes. Maternal serum leptin concentrations have been shown to increase during pregnancy to values greater than those predicted for the degree of adiposity [20–22]. Emerging evidence from *in vivo* and *in vitro* studies indicate that the placenta, an important endocrine organ, is capable of synthesizing and secreting leptin into maternal and fetal circulation and may contribute to pregnancy-associated elevations in circulating leptin concentrations. Placentally-derived leptin also is thought to be a source of leptin detected in amniotic fluid and in arterial and venous cord blood [23]. While the processes that control the synthesis and secretion of leptin in maternal and fetal circulation during pregnancy are incompletely understood, two independent reports indicate that the placenta is an important source of leptin in human pregnancies. Indirect immunofluorescence analyses of first and third trimester placental tissue revealed strong staining for immunoreactive leptin in first trimester chorionic villi and third trimester syncytiotrophoblasts, respectively [20]. These results are in agreement with those of Senaris and colleagues [23] who demonstrated that leptin is synthesized in syncytiotrophoblast cells harvested from delivered placentas. Studies involving animal models also demonstrate a 2.7-fold increase in uterine leptin receptor mRNA levels concurrent with a 1.8-fold increase in leptin concentration during gestation [24]. Thus, the pregnant uterus likely makes a significant contribution to leptin levels in maternal blood. Longitudinal studies are needed to further elucidate the processes whereby leptin may play a role in regulating maternal fat accumulation and mobilization as well as fetal growth.

A large number of hemodynamic changes are noted throughout the 40 weeks of a normal uncomplicated pregnancy. These include an increase in plasma volume from the time of conception to approximately 32 weeks, peaking at 40–50% above the nonpregnant baseline [25]. This occurs despite the increase in red blood cell mass which is on the order of 20–35% above baseline. The greater increase in plasma volume accounts for pregnancy-associated anemia and decrease in blood viscosity. The modest decrease in systolic blood pressure and a more pronounced decrease in diastolic blood pressure is secondary to a decrease in systemic vascular resistance which occurs in uncomplicated pregnancies. Heart rate is increased by 10–20%

during pregnancy [26]. Cardiac output increases late in the first trimester, peaking at 40% above baseline in the early second trimester, and may increase further during labor (depending on pain control and anesthesia) [27]. Pregnancy also is associated with marked increases in circulating fibrinogen and coagulation factors [28].

Alterations in renal hemodynamics during pregnancy include increased glomerular filtration rates and increased renal plasma flow. Consequently, creatinine clearance is increased and serum creatinine and urea nitrogen decrease. Alterations in tubular function and dilation of the renal collecting duct systems also are noted in uncomplicated pregnancies [28].

Much of what is known about homocyst(e)ine metabolism may be attributed to the work of Steegers-Theunissen *et al.* [29–31] and others [32] who have been investigating the relationship between decreased maternal circulatory levels of folate and other B vitamins in relation to the pathogenesis of neural tube defects. Homocyst(e)ine concentrations in pregnancy are attenuated by approximately 50% from the concentrations measured during the nonpregnant state [32]. The substantial elevations in steroid hormones during pregnancy are thought to account for the reductions in circulating homocyst(e)ine concentrations [33]. This thesis is supported by the observations that postmenopausal women, as compared with premenopausal women, have higher circulating concentrations of homocyst(e)ine [34,35]. Fetal uptake of homocyst(e)ine from maternal circulation also has been suggested as another potential mechanism whereby circulatory concentrations in pregnant women may be reduced [36].

Uncomplicated pregnancy is associated with a 20% increase in oxygen consumption and a 15% increase in maternal metabolic rate [37]. This demand is met by an impressive array of anatomic and physiologic changes. Anatomic changes in the chests of pregnant women include a flaring of the lower ribs [27]; subcostal angle increases as the transverse diameter of the chest increases (approximately 2 cm), and there is a diaphragmatic elevation of approximately 4 cm so as to compensate for the enlarged uterus, particularly in the third trimester. Placentally-derived hormones may also stimulate alterations in respiration during pregnancy, resulting in compensatory respiratory alkalosis. Briefly, progesterone, which peaks during the third trimester, may stimulate the respiratory centers of the brain to produce hyperventilation. Resting minute ventilation may increase by as much as 40–50% over baseline. Hyperventilation may cause a decrease in partial pressure of arterial carbon dioxide, resulting in respiratory alkalosis, decreased bicarbonate levels, and changes in arterial pH.

In summary, pregnancy is associated with significant changes in organ function and physiologic levels from the nonpregnant state. An understanding of the natural history and determinants of these alterations (*i.e.,* variations in magnitude and timing) as well as how these changes differ for women with medically complicated and uncomplicated pregnancies may provide greater specificity whereby clinically heterogeneous medical complications of pregnancy (such as pregnancy-induced hypertensive disorders) may be classified and managed clinically. Additionally, a breaking-down of the sum (the syndrome) to its parts may allow assessment of how underlying genetic and nongenetic determinants of plasma homocyst(e)ine or plasma triglyceride concentrations, for example, contribute to variations noted in uncomplicated and complicated pregnancies. Finally, in phenotypically diverse pregnancy-associated disorders such as preeclampsia, classification of preeclampsia according to metabolic disturbances associated with the clinical diagnosis may assist in defining homogeneous case populations. Such improvements may enhance the resolution of etiological studies of the disorder and accelerate the processes for identifying appropriate and efficient prevention strategies.

III. Autoimmune Disorders and Pregnancy

Autoimmune disorders, relatively common among women, are generally diagnosed during the childbearing years. The diseases are heterogeneous with respect to their clinical spectrum, the presence of antibodies, and the type of tissue involved in the pathology. The disorders may be broadly classified into two groups. The first group includes multisystem disorders such as multiple sclerosis (MS) and systemic lupus erythematosus (SLE). The second group comprises tissue-specific disorders including autoimmune thyroid diseases such as Graves' disease, Hashimoto's disease, postpartum thyroiditis, and autoimmune thrombocytopenic purpura.

The study of autoimmune disease in the context of pregnancy has provided important insights into the nature of the disease process and the relevance of circulating autoantibodies to pathological effects during pregnant and nonpregnant states [38]. Below we summarize briefly what is known about how pregnancy impacts the clinical course of selected disorders and we summarize the impact these disorders have on maternal health during pregnancy. Additionally, where possible, we describe what little is known about risk factors or "triggers" that may contribute to the worsening or improvement of these selected disorders.

A. Multiple Sclerosis

Multiple sclerosis (MS), an organ-specific autoimmune disease of the central nervous system, affects an estimated 300,000 Americans. The disorder is diagnosed at least twice as frequently among reproductive age women as compared with their male counterparts of the same age. The clinical characteristics of MS reflect its multifocal involvement. Its course may be subacute, with relapses followed by remission, or chronic and progressive. Pathophysiologically, MS is characterized by a diverse combination of inflammation, demyelination, and axonal damage in the central nervous system [39].

Results from several small studies suggest that exacerbation of the condition is decreased during pregnancy and increased during the postpartum period [40–42]. The Pregnancy in Multiple Sclerosis (PRIMS) study, a European multicenter prospective study of 254 women with MS during 269 pregnancies, compared the course of MS during pregnancy with the course during the year before conception. They reported that women experienced a 70% reduction in the frequency of relapse during the third trimester, followed by a 70% increase in the frequency of exacerbation during the first three months postpartum [43]. This observation is similar to some, though not all, previous reports [40–42,44].

Immunologic alterations during pregnancy may explain the variable course of autoimmune disorders in pregnant women with MS. The pattern of disease remission and relapse may be mediated by shifts in the balance of proinflammatory and anti-inflammatory cytokines. For instance, anti-inflammatory cytokines such as interleukin-10 (IL-10), known to downregulate the production of other proinflammatory cytokines such as tumor necrosis factor-alpha (TNF-α), may contribute to an immunodepression that could explain the maternal tolerance of the fetus and the spontaneous remission of MS during the third trimester. Inversion of the cytokine balance concomitant with labor and delivery may account for the postpartum exacerbation of MS and other T-cell mediated autoimmune disorders including rheumatoid arthritis [45]. Following this thesis, it would stand to reason that β-cell mediated autoimmune disorders such as SLE would tend to worsen during pregnancy.

B. Systemic Lupus Erythematosus

Systemic lupus erythematosus (SLE), characterized by a broad spectrum of immunologic abnormalities that appear to be related to defects in regulatory components of the immune response, has received much attention by reproductive immunologists [28,46]. The major pathological features of SLE are the presence of autoantibodies which, either by themselves or by forming immune complexes, mediate inflammatory processes and tissue damage resulting in the clinical features of the disease. Clinical features include arthritis, splenomegaly, thrombocytopenia, immune complex nephritis, and skin lesions.

In other β-cell mediated autoimmune disorders there is a tendency towards exacerbation of the clinical spectrum of SLE in the puerperium, and to a lesser extent in the first half of pregnancy [38]. This pattern is, however, not seen in all populations studied. For instance, Petri *et al.* did not detect any particular pattern in the timing of SLE flares during pregnancy [47]. SLE disease activity at conception appears to be predictive of its activity during pregnancy. The more active the disease is at conception, the more likely an exacerbation is to occur during pregnancy [48]. In patients with lupus nephritis, there is a tendency towards increasing severity of disease during pregnancy, though it is unclear whether this pattern is a direct result of pregnancy or the inevitable course of the disorder [28].

Lupus flares, defined as a change of 1.0 in the Physician's Global Assessment when using a 0–3 visual analogue scale [49], are noted to be more common during pregnancy [49–51] and are particularly more frequent among pregnant African-American women [49]. Petri *et al.*, in their followup study of patients with SLE, reported that over 60% experienced flares during pregnancy [47]. Investigators also have noted differences in the distribution of flares by organ system; flares involving the musculoskeletal system are less common during pregnancy [45] whereas flares involving the renal and hematological systems are increased during pregnancy.

Pregnant women with SLE are at an increased risk of delivering preterm and stillborn infants [28,47]. The frequency of fetal wastage is noted to be as high as 50% in patients with active lupus nephritis [28]. The disorder is associated with two-fold to nearly seven-fold increases in the risk of developing pregnancy-induced hypertension (including preeclampsia), urinary tract infections, diabetes mellitus, and hyperglycemia [47] as compared with controls. Results from the Johns Hopkins cohort study suggest that 7% of pregnant women with SLE experience clinically devastating complications including uterine rupture, bilateral retinal detachment (during labor), severe retinopathy, stroke, hemolysis, liver enzyme, and low platelet counts (HELLP syndrome), and deep vein thrombosis [47].

C. Autoimmune Thyroid Disease

As reviewed by Diehl [52], thyroid dysfunction is the second most common endocrine disorder (after diabetes) among women of childbearing age. Autoimmune thyroid diseases comprise two broad categories, Graves' and autoimmune thyroiditis (Hashimoto's disease). Together these two variants form a continuous spectrum of immunopathologic processes [38]. As with other autoimmune disorders, thyroid disease is more likely to be diagnosed among women than men and is likely to be diagnosed during the childbearing years. Physiologic and metabolic changes secondary to pregnancy are thought to stimulate either the remission or exacerbation of certain thyroid disorders. Most of the pregnancy-induced changes in thyroid physiology are considered to be stimulated by elevations in pregnancy estrogens [53]. Hyperestrogenemia may cause increased synthesis and release of thyroxine-binding globulin (TBG). An increase in TBG may subsequently lead to a reduction in triiodothyronine renin uptake (T_3 RU) and increases in serum concentrations of thyroxine (T_4) and triiodothyronine (T_3) [53].

1. Hyperthyroidism/Graves' Disease

Thyrotoxicosis, seen in one per 2000 pregnancies and primarily caused by Graves' disease, is an organ-specific autoimmune process associated with thyroid-stimulating antibody (TSAb) activity. Pregnancy is thought to affect the natural history of Graves' disease. For instance, it has been noted that the diagnosis tends to be made within a year of delivery [54]. Further, data indicating that women with inactive Graves' disease may experience transient elevations in free T_4 during the late first trimester [53] suggest that the diseased thyroid gland may be stimulated by human chorionic gonadotropin (hCG). This thesis is supported by the observation that women with trophoblastic disease, a condition characterized by high plasma concentrations of hCG, are at an increased risk of developing Graves' disease [52]. Also supporting the thesis is the observation that hyperemesis gravidarum is associated with the severity of hyperthyroidism in pregnancy [55]. Specifically, hyperemesis gravidarum is a self-limiting pregnancy-associated disorder that usually resolves during the second half of the first trimester of pregnancy when maternal hCG serum concentrations begin to decline.

Women with poor metabolic control of hyperthyroidism during pregnancy or women who remain untreated throughout pregnancy are noted to be at an increased risk of preeclampsia, congestive heart failure, and adverse perinatal outcomes [56,57]. Although the clinical course of Graves' disease is likely to be altered during pregnancy, investigators note that the condition is no more difficult to control during pregnancy [58].

2. Thyroid Storm

Poor control of, or untreated, hyperthyroidism during pregnancy may result in a serious medical complication called thyroid storm. The condition may be life-threatening and is characterized by altered mental status, hyperthermia, diarrhea, and congestive heart failure [52,56,59]. Although the crisis may occur during the antepartum, intrapartum, or postpartum period, it is most often associated with a precipitating factor such as infection, labor, or cesarean section delivery [52]. Untreated hyperthyroidism during pregnancy is associated with a ten-fold increase in risk of thyroid storm, including heart failure [56].

3. Postpartum Thyroiditis

The immediate postpartum period is characterized by an increase in thyroid stimulation [60]. Although the cause of this thyroid hyperstimulation is unknown, it is thought that the rebounding of the maternal immune system during the puerperium may be a contributing factor. Postpartum thyroiditis occurs in approximately 10% of pregnant women [38]. Carriage of the HLA-DR4 antigen and a personal or family history of an autoimmune thyroid disorder are known to be risk factors. Additionally, women who develop postpartum thyroiditis have higher antithyroid antibody titers than euthyroid controls [52]. Available evidence suggests that postpartum thyroiditis may be a harbinger of clinical Graves' disease [38].

4. Hypothyroidism/Hashimoto's Disease

As reviewed by Diehl [52], hypothyroidism in pregnancy, like hyperthyroidism, is associated with autoimmune disease of the thyroid. In the case of hypothyroidism, however, the autoimmune process destroys the gland rather than stimulating it [52]. Hashimoto's disease, the most common cause of hypothyroidism, is detected in one per 3000 pregnancies. Overt hypothyroidism is rarely noted to be a complicating disorder of pregnancy, as it is associated with reduced fecundity and infertility [53]. Among those women with hypothyroidism who conceive, their pregnancies are associated with an increased frequency of placental abruption, preeclampsia, and heart failure [53]. The frequency of the disorder is reported to be increased up to three-fold among women with type-1 diabetes [61].

A better understanding of the biological mechanisms underlying pregnancy-related increases and decreases in autoimmune activity could lead to new therapeutic strategies for controlling these disorders during pregnancy and during the postpartum period, when maternal morbidity may interfere with the ability of a mother to provide adequate care for her newborn.

IV. Other Disorders during Pregnancy

A. Asthma

Nearly 20% of reproductive age women suffer from some form of allergic disorder that affects pulmonary function [27,62]. The incidence of asthma and mortality from asthma have increased by over 40% since the 1980s. Briefly, asthma is a heterogeneous chronic lung disease. The disorder is characterized by recurrent bouts of wheezing and dyspnea resulting in airway obstruction and is the most common respiratory crisis encountered during pregnancy. Investigators have reported that 4% of all pregnancies are complicated by an asthma crisis [63,64]. The airways of asthmatics are hyperresponsive to stimuli such as allergens, viral infections, air pollutants, and cold air. The hypersensitivity is manifested by broncospasm, mucosal edema, and mucus plugging that results in air trapping and hyperinflation of the lungs [27]. The cause of asthma is unknown, although certain pathophysiologic mechanisms have been implicated.

The effect of pregnancy on the course of asthma is variable. Asthma has been noted to improve, worsen, or remain unchanged during pregnancy [62]. A review of more than 1000 pregnant asthmatics in nine studies found worsening of asthma in 22%, improvements in 29%, and no change in the remaining 49% [65]. Experiences during a previous pregnancy may be somewhat predictive of experience in subsequent pregnancies. Schatz and colleagues noted that in approximately 60% of women, the clinical course of asthma is similar in successive pregnancies [66]. There is some evidence to support an association between the severity of asthma prior to pregnancy and the likelihood of deterioration during pregnancy. Gluck and Gluck noted that women with severe asthma prior to conception were more likely to experience worsening of asthma during pregnancy as compared with women with mild disease [65]. This finding was confirmed by White et al. [67]. Beecroft et al. [68] noted that the course of asthma during pregnancy was influenced by fetal gender. Women carrying female fetuses were more likely to experience moderate to severe deterioration as compared with those carrying male fetuses.

Methodologic limitations related to the subjective nature of measuring breathlessness and heightened maternal awareness and concern about disease control during pregnancy—in addition to undermedication secondary to maternal concerns about possible fetal toxicity—are important limitations of many of these studies. Mechanisms involved in changes of the course of asthma during pregnancy have not been defined. The peaking of plasma progesterone concentrations during the third trimester, coincident with improvements in the course of asthma in the third trimester, is considered a hint for a causal model. Some researchers [69–71], though not all [62], note that the frequency of preeclampsia is increased among pregnant asthmatics. Women with severe asthma and those with poor asthma control during pregnancy have been noted to be at increased risk of delivering by cesarean section and of experiencing postpartum hemorrhage.

B. Diabetes Mellitus

Approximately 4–6% of all pregnancies in the United States are complicated by diabetes mellitus. The vast majority of these pregnancies (88%) are classified as gestational diabetes (GDM), with the remaining 8% and 4%, respectively, representing cases of type-2 and type-1 diabetes mellitus [72]. For the purposes of this chapter, we shall limit our discussion to the experiences of women apparently free of diabetes mellitus prior to pregnancy but who are diagnosed as developing the disorder during pregnancy.

Women who develop gestational diabetes mellitus (GDM) are thought to have a compromised physiologic capacity to adapt to the metabolic challenges of late pregnancy. The third trimester is characterized by profound metabolic stresses on maternal lipid and glucose homeostasis favoring the transfer of nutrients to the fetus [5]. Maternal metabolic changes occurring in the third trimester include marked insulin resistance, hyperinsulinemia, a progressive worsening of glucose intolerance [5], and hyperlipidemia (particularly hypertriglyceridemia [4]). The endocrinologic challenges of pregnancy and the variation in maternal physiologic adaptation to these challenges have led several investigators to speculate that pregnancy serves to unmask a predisposition to glucose metabolic disorders in some women [5]. Women with a history of GDM have an increased risk of developing type-2 diabetes and impaired glucose tolerance later in life [14,15]. For instance, O'Sullivan and colleagues reported that approximately 50% of women with a history of GDM develop type-2 diabetes within 25 years of the affected pregnancy [15].

Relatively few risk factors have been identified for GDM [73–76]. Several of the consistently reported GDM risk factors are common to type-2 diabetes. Women with a high prepregnancy body mass index are noted to have an increased risk of GDM [74,76]. There is a progressive increase in risk of GDM above a prepregnancy body mass index of 22 kg/m^2 [74]. A body mass index of 27.3 kg/m^2, which corresponds to a prepregnancy weight that is 120% of ideal body weight, is associated with a two-fold increased risk of GDM as compared to a BMI of <20.0 kg/m^2 [74]. Advanced maternal age, a family history of diabetes mellitus, and a prior history of GDM are three well-recognized risk factors of GDM [74,76,76a,76b]. Women with a positive family history of diabetes mellitus, as compared with women without such a history, have been noted to experience a 1.5 to two-fold increase in risk of GDM [74,76,76b]. Cigarette smoking has not been identified consistently as a risk factor for GDM [74,76]. Available data suggest that the magnitude of any possible association between maternal smoking during pregnancy and GDM may be modest [74]. Asian, Hispanic, and Native American women, as compared with Nonhispanic White women, have an increased risk of GDM [74–76]. African-American women have been reported to have an increased risk of GDM, as compared with Nonhispanic Whites, by some [74,75], though not all [76] investigators.

Although the role of genetic factors in susceptibility to GDM is suggested by observations of increased risk of diabetes mellitus in first degree relatives of gestational diabetics [76b], genes conferring susceptibility to the disorder are presently unknown. Variation in one or more candidate genes that either directly or indirectly regulate carbohydrate metabolism, including the insulin gene variable number of tandem repeats, insulin-like growth factor II, insulin receptor, and glucose transporter genes, are thought to contribute to the glucose intolerance of GDM [76c], though results from early studies have been inconsistent [76c–76e].

Given the strength and consistency of the reported inverse association between size at birth and risk of impaired glucose tolerance and type-2 diabetes in later life [76f–76j], we and others [76k] hypothesized that intrauterine fetal growth retardation was associated with an increased risk of developing GDM. Using linked birth certificates from the state of Pennsylvania for a cohort of women born in 1974, Plante [76k] reported that White women who were small-for-gestational-age at birth, as compared with women who were appropriate weight for gestational age, had a relative risk of 4.3- (95% CI 2.6–7.0) of having their pregnancy complicated by GDM. Results from an ongoing study of women from Washington state are in general agreement with these findings [76l]. Our results and those of Plante [76k] are consistent with the growing body of evidence documenting the importance of birth weight as a determinant of impaired glucose tolerance and diabetes in adulthood. Moreover, these results serve to illustrate the potential benefits that may be appreciated from considering maternal medical complications of pregnancy within the larger context of women's health over their entire life span, since women who first experience their diabetes during pregnancy are likely candidates for developing type-2 diabetes later in life [15].

C. Preeclampsia

Hypertensive disorders, including pregnancy-induced hypertension (with and without proteinuria), and pregnancy-induced hypertension superimposed on chronic hypertension, complicate approximately 12% of pregnancies in the United States. The majority of women with hypertensive disorders of pregnancy have preeclampsia (defined as hypertension diagnosed after 20 weeks gestation with proteinuria), making it the most common medical disease during pregnancy [77,78] as well as a leading cause of maternal mortality and premature delivery worldwide [79].

The cardinal features of preeclampsia are hypertension, proteinuria, and edema. When seizures or coma further complicate these pregnancies, the condition is generally referred to as eclampsia [80]. Elevated hypertension without concomitant proteinuria is generally referred to as gestational hypertension. Hereinafter, preeclampsia will be used to refer to proteinuric pregnancy-induced hypertension. Pregnancy-induced hypertension without proteinuria will be referred to as gestational hypertension. Pathologic changes associated with preeclampsia include maladaptation of spiral arteries of the placental bed [79], hypertriglyceridemia [4,81,82], hypercholesterolemia [4,81,82], excessive lipid peroxidation [83], endothelial cell dysfunction [84], sympathetic nervous system overreactivity [85], plasma elevations of proinflammatory cytokines [86–89], an imbalance in thromboxane and prostacyclin in favor of vasoconstriction [90,91], glucose intolerance or hyperinsulinemia [92], hyperuricemia [92], hypovolemia [93], and decreased angiotensin II [93]. Histologic studies of placental arteries delivered from preeclamptic women show fibrin and complement deposition and the involvement of foam cells in atheromatous lesions resembling those noted in renal allograph rejection [79]. Few longitudinal studies have been conducted to: (1) determine whether these changes are a cause or a consequence of preeclampsia; (2) determine the occurrence of these events relative to each other; and (3) determine the extent to which, if any, these events are associated with maternal behavioral characteristics, including dietary habits. Almost nothing is known about the relation of maternal behavioral characteristics to preeclampsia and the aforementioned pathophysiologic features.

Several clinical and epidemiologic studies have been conducted to identify risk factors for hypertensive disorders of pregnancy. Previously reported risk factors include nulliparity [80], primigravidity [94], changed paternity [95,96], previous preeclamptic pregnancy [97], family history of preeclampsia [98–100], African-American race [97,101,102], young maternal age (<20 years) [103], advanced maternal age (≥ 35 years) [80], high body mass index [101,104], and limited exposure to seminal antigen [105]. One report suggests that high milk consumption during pregnancy may be associated with an increased risk of preeclampsia, though this finding has yet to confirmed [101].

Not all epidemiologic studies of hypertension in pregnancy are able to distinguish between nonproteinuric and proteinuric hypertension. Several are inherently limited by the data bases available to the investigators. For example, discharge diagnosis data and birth certificate files seldom distinguish between the different categories of hypertensive disorders [102,106]. Results of the few studies that have attempted to distinguish between preeclampsia and gestational hypertension suggest that, in general, observed associations are stronger for preeclampsia [107–111]. Furthermore, studies that have evaluated perinatal outcomes of women with the two syndromes indicate that the presence of proteinuria confers increased risks of poor fetal outcomes, including a two-fold increase in fetal mortality [112]. It appears, therefore, that inclusion of nonproteinuric hypertensives in a study of preeclampsia may result in a diminution of estimated associations.

Several of the aforementioned risk factors of preeclampsia are also predictive of coronary heart disease (CHD) in nonpregnant individuals. Additionally, many of the protective factors for preeclampsia are also protective of CHD. It may be worth noting that much of the overlap in risk factors is concentrated among behavioral and metabolic characteristics which are known to be related to lipid and carbohydrate metabolism.

Few studies of the genetic control of susceptibility to preeclampsia have been published. Available data from the UK and the U.S. suggest that the disorder has a familial tendency, although the genetic basis of the susceptibility is as yet unclear [113]. In a study of probands with a history of toxemia (preeclampsia) and women who had normotensive pregnancies, 29% of probands had themselves been born of a toxemic pregnancy, as compared to 17% of controls [99]. This observation was later confirmed in a study of eclampsia pedigrees [113]. Chesley and Cooper [98] reported that the frequency of preeclampsia and eclampsia in the sisters and daughters of preeclamptics was higher than that in the general population. Preeclampsia also has been reported to be more common in the daughters of women who have had the disease than in daughters-in-laws [114].

The recurrence risk of preeclampsia is very high. Women with a previous history of preeclampsia are approximately seven-times more likely to have a subsequent pregnancy complicated by the disorder, as compared to women with no history of preeclampsia [98,100,101]. This suggests that there is a subgroup of women who may be predisposed to developing this very dangerous complication of pregnancy.

Accumulating epidemiologic and clinical evidence supports the thesis that pregnancy represents a metabolic challenge to those individuals who may be susceptible to metabolic disorders. Consider, for instance, the following case reports of severe lipid alterations in pregnant women genetically predisposed to hyperlipidemia. Ma *et al.* [115,116] reported that mutations in the human lipoprotein lipase (LPL) gene are associated with pregnancy-induced chylomicronemia and that the condition appears to be aggravated in the presence of a single apolipoprotein E2 allele. The receptor binding activity of E2 is only 2% of that of the E3 allele [117]. This work, although focused on an extremely rare pregnancy complication, serves as one poignant example of the potential importance of systematically evaluating the individuals' genetic predisposition to abnormal lipid metabolism in the context of pregnancy [115,116]. Indeed, the observations of Ma *et al.* [115] lend support to the thesis that pregnancy-associated hyperlipidemia may stress maternal lipid homeostasis to an extent that subclinical or mild hyperlipidemia becomes clinically detectable [5,7].

Our own studies of preeclampsia include a case-control pilot study to evaluate the relationship between maternal ApoE phenotype and risk of preeclampsia [118]. Fifty-seven consecutive women with preeclampsia and 56 normotensive pregnant women delivering appropriate-for-gestational-age term infants comprised the case and control groups, respectively. ApoE phenotypes were determined by isoelectric focusing. White women with at least one E2 allele experienced a 12.9-fold increased risk of preeclampsia (95% CI 3.2–51.8) as compared to women with the E3/E3 phenotype. Preeclampsia risk also was positively associated with the E4 allele (OR = 3.8, 95% CI 1.3–11.2). The associations remained after adjusting for possible confounding by maternal prepregnancy weight, diabetes, maternal age, and parity (OR = 11.0, 95% CI 1.5–82.5 and OR = 3.2, 95% CI 0.9–10.9 for E2 and E4, respectively). These pilot data further suggest that ApoE alleles, established genetic risk markers for dyslipidemia in men and in nonpregnant women, may also be associated with preeclampsia. These preliminary findings were supported by Nagey *et al.* [119] who reported that the frequency of the ApoE E2 allele was statistically significantly higher among preeclampsia cases as compared with pregnant controls (16.6% vs 12.9%, p < 0.001). There was no difference in the frequency distribution of the ApoE E4 allele among cases and controls. Similar studies of the relationship between point mutations in candidate genes, specific metabolic disturbances, and preeclampsia risk are now appearing in the emerging literature [120–122]. For instance, investigators have reported that disturbances in the renin-angiotensin system and the likely role variants in the angiotensinogen (AGT) gene may be important in the pathogenesis of preeclampsia [122].

V. Historical Evolution/Trends and Future Direction

Since the 1950s, infant mortality has been substantially reduced mostly due to general improvements in women's health and to substantial technical advances in neonatology. Improvements have been made in diagnosing and treating medical disorders which complicate pregnancy. However, comparatively little is known about the environmental and genetic determinants of many of the physiologic changes of pregnancy and whether manipulation of these changes by modifying dietary intake and physical activity or by using pharmacological therapeutics may impact the occurrence of medical complications during pregnancy. It is important to note that inferences from most of the existing data are limited by small samples sizes, variable exe-

cution of clinical screening protocols (as is likely the case for GDM), and lack of objective measures of complex heterogeneous outcomes with variable clinical manifestations. The generalizabililiy of these numerous studies also is limited because virtually all of the available literature is from studies of women of European ancestry and residents of developed countries.

What are the public health and clinical implications of the aforementioned limitations? To respond, we must consider several important demographic trends. First, an increasing proportion of women are electing to delay childbearing. Second, improvements in medical reproductive technology will result in a steady increase in the proportion of women with metabolic disorders in antepartum clinics. Third, as the first generation of neonatal intensive care infants, particularly those surviving very preterm delivery, progress through the reproductive years, it is likely that those who are fecund will have their significant medical histories influence the outcome of their pregnancies. As an example, there is suggestive evidence that individuals with a history of low birth weight are at increased risk for a range of chronic disorders and adverse pregnancy events [123–125]. These points, when taken together, serve to underscore the urgent need for basic and epidemiologic research that provides the empirical basis upon which appropriate risk assessment and clinical management protocols can be developed.

Acknowledgments

This work was supported in part by awards from the National Institutes of Health: HDR01-32562 and HDR01-34888.

References

1. Wegmann, T. G., Lin, H., Guilbert, L., and Mosman, T. R. (1993). Bidirectional cytokine interactions in the maternal-fetal relationship: Is successful pregnancy a TH2 phenomenon? *Immunol. Today;* **14,** 353–356.
2. Sargent, I. L., Arenas, J., and Redman, C. W. (1987). Maternal cell-mediated sensitization to paternal HLA may occur, but is not a regular event in normal human pregnancy. *J. Reprod. Immunol.* **10,** 111–120.
3. Perro, R., (1990). "Nutrition and Metabolism in Pregnancy: Mother and Fetus," pp. 41–65. Oxford University Press, New York.
4. Potter, J. M., and Nestel, P. J., (1979). The hyperlipidemia of pregnancy in normal and complicated pregnancies. *Am. J. Obstet. Gynecol.* **15,** 165–170.
5. Knopp, R. H., Bergelin, R. O., Wahl, P. W., Walden, C. E., Chapman, M., and Irvine, S. (1982). Population-based lipoprotein lipid reference values for pregnant women compared to nonpregnant women classified by sex hormone status. *Am. J. Obstet. Gynecol.* **143,** 626–637.
6. Maseki, M., Nishigka, I., Hagihara, M., Tomoda, Y., and Yagi, K. (1981). Lipid peroxide levels and lipid content of serum lipoprotein fractions of pregnant subjects with or without preeclampsia. *Clin. Chim. Acta* **115,** 155–161.
7. Knopp, R. H., Bergelin, R. O., Wahl, P. W., and Walden, C. E. (1985). Effects of pregnancy, postpartum lactation and oral contraceptive use on the lipoprotein cholesterol/triglyceride ratio. *Metab. Clin. Exp.* **34,** 893–899.
8. Montes, A., Walden, C. E., Knopp, R. H., Cheung, M., Chapman, M. B., and Albers, J. J. (1984). Physiologic and supraphysiologic increases in lipoprotein lipids and apoproteins in late pregnancy and postpartum: Possible markers for the diagnosis of 'prelipemia'. *Arteriosclerosis* **4,** 407–417.

9. Al, M., Houwelingen, A. C., Badart-Smook, A., and Honstra, G. (1995). Some aspects of neonatal essential fatty acid status are altered by linoleic supplementation of women during pregnancy. *J. Nutr.* **125,** 2822–2830.

10. Desoye, G., Schweditsch, M., Pfieffer, K. P., Zechner, R., and Kostner, G. M. (1987). Correlation of hormones with lipid and lipoprotein levels during normal pregnancy and postpartum. *J. Clin. Endocrinol. Metab.* **64,** 704–712.

11. Miller, V. T. (1990). Dyslipoproteinemia in women. *Endocrinol. Metab. Clin. North Am.* **19,** 381–399.

12. Knopp, R. H., Warth, M. R., Charles, D., Childs, M., Li, J. R., Mabuchi, H., and Van Allen, M. I. (1986). Lipoprotein metabolism in pregnancy, fat transport to the fetus, and the effects of diabetes. *Biol. Neonate* **50,** 297–317.

13. Fahraeus, L., Larsson-Cohn, U., and Wallentin, L. (1985). Plasma lipoproteins including high density lipoprotein subfractions during normal pregnancy. *Obstet. Gynecol.* **66,** 468–472.

14. Peters, R. K., Kjos, S. L., Xiang, A., and Buchanan, T. A. (1996). Long-term diabetogenic effect of single pregnancy in women with previous gestational diabetes mellitus. *Lancet* **347,** 227–230.

15. O'Sullivan, J. B. (1984). Long term follow-up of gestational diabetes. *In* "Early Diabetes" R. A. Camerini Davalos, and H. S. Cole, eds., pp. 1009–1027. Academic Press, New York.

16. Schwartz, M. W., Seeley, R. J., Campfield, L. A., Burn, P., and Baskin, D. G. (1996). Identification of targets of leptin action in rat hypothalamus. *J. Clin. Invest.* **98,** 1101–1106.

17. Schwartz, M. W., Peskind, E., Raskind, M., Boyko, E. J., and Porte, D. (1996). Cerebrospinal fluid leptin levels: Relationship to plasma levels and to adiposity in humans. *Nat. Med.* **2,** 589–593.

18. Havel, P. J., Kasim-Karakas, S., Mueller, W. M., Johnson, P. R., Gingerich, R. L., and Stern, J. S. (1996). Relationship of plasma leptin to plasma insulin and adiposity in normal weight and overweight women: Effects of dietary fat content and sustained weight loss. *J. Clin. Endocrinol. Metab.* **81,** 4406–4413.

19. Weigle, D. S., Duell, P. B., Connor, W. E., Steiner, R. A., Soules, M. R., and Kuipjer, J. L. (1997). Effect of fasting, refeeding, and dietary fat restriction on plasma leptin levels. *J. Clin. Endocrinol. Metab.* **82,** 561–565.

20. Masusaki, H., Ogawa, Y., Sagawa, N., Hosoda, K., Matsumoto, T., Mise, H., Nishimura, H., Yoshimasa, Y., Tanaka, I., Mori, T., and Nakao, K. (1997). Nonadipose tissue production of leptin: Leptin as a novel placental-derived hormone in humans. *Nat. Med.* **3,** 1029–1033.

21. Schubring, C., Kiess, W., Englaro, P., Rascher, W., Dotsch, J., Hanitsch, S., Attanasio, A., and Blum, W. F. (1997). Levels of leptin in maternal serum, amniotic fluid, and arterial and venous cord blood: Relation to neonatal and placental weight. *J. Clin. Endocrinol. Metab.* **82,** 1480–1483.

22. Butte, N. F., Hopkinson, J. M., and Nicolson, M. A. (1997). Leptin in human reproduction: Serum leptin levels in pregnant and lactating women. *J. Clin. Endocrinol. Metab.* **82,** 585–589.

23. Senaris, R., Garcia-Caballero, T., Casabiell, X., Gallego, R., Castro, R., Considine, R. V., Dieguez, C., and Casanueva, F. F. (1997). Synthesis of leptin in human placenta. *Endocrinology* (*Baltimore*) **138,** 4501–4505.

24. Chien, E. K., Hara, M., Rouard, M., Yano, H., Phillippe, M., Polonsky, K. S., and Bell, G. I. (1997). Increase in serum leptin and uterine leptin receptor messenger RNA levels during pregnancy in rats. *Biochem. Biophys. Res. Commun.* **237,** 476–480.

25. Metcalf, J., McAnulty, J. H., and Ueland, K. (1981). Cardiovascular physiology. *Clin. Obstet. Gynecol.* **24,** 693–710.

26. Katz, R., Karliner, J. S., and Resnik, R. (1978). Effects of a natural volume overload state (pregnancy) on left ventricular performance in normal human subjects. *Circulation* **58,** 434–441.

27. Danzell, J. D. (1998). Pregnancy and pre-existing heart disease. *J. La. State Med. Soc.* **150,** 97–102.

28. Mason, E., Rosene-Montella, K., and Powrie, R. (1998). Medical problems during pregnancy. *Med. Clin. North Am.* **82,** 249–269.

29. Steegers-Theunissen, R. P. M., Boers, G. H. J., Blom, H. J., Nijhuis, J. G., Thomas, C. M., Borm, G. F., and Eskes, T. K. (1995). Neural tube defects and elevated homocysteine levels in amniotic fluid. *Am. J. Obstet. Gynecol.* **172,** 1436–1441.

30. Steegers-Theunissen, R. P. M., Boers, G. H. J., Trijbels, J. M. F., and Eskes, T. K. (1992). Hyperhomocysteinemia and recurrent spontaneous abortion or abruptio placentae. *Lancet* **1,** 1122–1123.

31. Steegers-Theunissen, R. P. M., Boers, G. H. J., Trijbels, J. M. F., Finkelstein, J. D., Blom, H. J., Thomas, C. M., Borm, G. F., Wouters, M. G., and Eskes, T. K. (1994). Maternal hyperhomocysteinemia: A risk factor for neural tube defects. *Metab., Clin. Exp.* **43,** 1475–1480.

32. Kang, S. S., Wong, P. W. K., Zhou, J., and Cook, Y. (1986). Total homocyst(e)ine in plasma and amniotic fluid of pregnant women. *Metab., Clin. Exp.* **35,** 889–891.

33. Moghadasian, M. H., McManus, B. M., and Frohlich, J. J. (1997). Homocysteine and coronary artery disease. *Arch. Intern. Med.* **157,** 2299–2308.

34. Rasmussen, K., Moller, J., Lyngback, M., Pedersen, A. M. H., and Dybkjaer, L. (1996). Age- and gender-specific reference intervals for total homocysteine and methylmalonic acid in plasma before and after vitamin supplementation. *Clin. Chem.* (*Winston-Salem, N.C.*) **42,** 630–636.

35. Wouters, M. G., Moorrees, M. T., van der Mooren, M. J., Blom, H. J., Boers, G. H., Schellekens, L. A., Thomas, C. M., and Eskes, T. K. (1995). Plasma homocysteine and menopausal status. *Eur. J. Clin. Invest.* **25,** 801–805.

36. Malinow, M. R., Duell, P. B., Hess, D. L., Anderson, P. H., Kruger, W. D., Phillipson, B. E., Gluckman, R. A., Block, P. C., and Upson, B. M. (1998). Reduction of plasma homocyst(e)ine levels by breakfast cereal fortified with folic acid in patients with coronary heart disease. *N. Engl. J. Med.* **338,** 1009–1015.

37. Nelson-Piercy, C., and Moore, G. J. (1996). Asthma in pregnancy. *Br. J. Hosp. Med.* **55,** 115–117.

38. Jones, W. R. (1994). Autoimmune disease and pregnancy. *Aust. N. Z. J. Obstet. Gynecol.* **34,** 251–258.

39. Trapp, B. D., Peterson, J., Ransohoff, R. M., Rudick, R., Mork, S., and Bo, L. (1998). Axonal transection in the lesions of multiple sclerosis. *N. Engl. J. Med.* **338,** 278–285.

40. Ghezzi, A., and Caputo, D. (1981). Pregnancy: A factor influencing the course of multiple sclerosis? *Eur. Neurol.* **20,** 115–117.

41. Nelson, L. M., Franklin, G. M., and Jones, M. C. (1988). Risk of multiple sclerosis exacerbation during pregnancy and breastfeeding. *JAMA, J. Am. Med. Assoc.* **259,** 3441–3443.

42. Korn-Lubetzki, I., Kahana, E., Cooper, G., and Abramsky, O. (1984). Activity of multiple sclerosis during pregnancy and puerperium. *Ann. Neurol.* **16,** 229–231.

43. Confavreux, C., Hutchinson, M., Hours, M. M., Cortinovis-Tourniaire, P., and Moreau, T. (1998). Rate of pregnancy-related relapse in multiple sclerosis. *N. Engl. J. Med.* **339,** 285–291.

44. Damek, D. M., and Shuster, E. A. (1997). Pregnancy and multiple sclerosis. *Mayo Clin. Proc.* **72,** 977–989.

45. Hench, P. S. (1938). The ameliorating effect of pregnancy on chronic atrophic (infectious rheumatoid) arthritis, fibrositis, and intermittent hydrarthrosis. *Proc. Staff Meet. Mayo Clin.* **13,** 161–167.

46. Nelson, J. L. (1996). Autoimmune disease in pregnancy: Relationships with HLA. *In* "HLA and the Maternal-Fetal Relationship" (J. S. Hunt, ed.), pp. 158–188. R. G. Landes Co., Austin, TX.

47. Perti, M. (1997). Hopkins Lupus Pregnancy Center: 1987–1996. *Rheum. Dis. Clin. North Am.* **23,** 1–13.

48. Dombroski, R. A. (1989). Autoimmune disease in pregnancy. *Med. Clin. North Am.* **73**, 605–621.

49. Perti, M., Howard, D., and Repke, J. (1991). Frequency of lupus flares in pregnancy: The Hopkins Lupus Pregnancy Center experience. *Arthritis Rheum.* **34**, 1538–1545.

50. Zulman, J. I., Talal, N., Hoffman, G. S., and Epstein, W. V. (1980). Problems associated with the management of pregnancies in patients with systemic lupus erythematosus. *J. Rheumatol.* **7**, 37–49.

51. Lima, F., Buchanan, N. M. M., Khamasthta, M. A., Kerslake, S., and Hughes, G. R. (1995). Obstetric outcome in systemic lupus erythematosus. *Semin. Arthritis Rheum.* **25**, 184–192.

52. Diehl, K. (1998). Thyroid dysfunction in pregnancy. *J. Perinatal. Neonatal Nurs.* **11**, 1–12.

53. American College of Obstetricians and Gynecologists Technical Bulletin (1993). Thyroid disease in pregnancy. *Int. J. Gynaecol. Obstet.* **43**, 82–88.

54. Jansson, R., Dahlberg, P. A., Winsa, B., Meirik, O., Safwenberg, J., and Karlsson, A. (1987). The postpartum period constitutes an important risk for the development of clinical Graves' disease in young women. *Acta Endocrinol. (Copenhagen)* **116**, 321–325.

55. Goodwin, T. M., Montoro, M., and Mestman, J. H. (1992). Transient hyperthyroidism and hyperemesis gravidarum: Clinical aspects. *Am. J. Obstet. Gynecol.* **167**, 648–652.

56. Davis, L., Lucas, M., Hankins, G. D. V., Roark, M. L., and Cummingham, F. G. (1989). Thyrotoxicosis complicating pregnancy. *Am. J. Obstet. Gynecol.* **160**, 63–70.

57. Millar, L., Wing, D., Leung, A., Koonings, P. P., Montoro, M. N., and Mestman, J. H. (1994). Low birth weight and preeclampsia in pregnancies complicated by hyperthyroidism. *Obstet. Gynecol.* **84**, 946–949.

58. Burrow, G. N. (1985). The management of thyrotoxicosis in pregnancy. *N. Engl. J. Med.* **313**, 562–565.

59. Prihoda, J., and Davis, L. (1919). Metabolic emergencies in obstetrics. *Obstet. Gynecol. Clin. North Am.* **18**, 304–306.

60. Lazarus, J., and Othman, S. (1991). Thyroid disease in relation to pregnancy. *Clin. Endocrinol. (Oxford)* **34**, 91–98.

61. Mastman, J. H., Goodwin, T. M., and Montoro, M. M. (1995). Thyroid disorders of pregnancy. *Endocrinol. Metab. Clin. North Am.* **24**, 41–71.

62. Schatz, M., and Zeiger, R. S. (1997). Asthma and allergy in pregnancy. *Clin. Perinatol.* **24**, 407–432.

63. Mabie, W. C. (1996). Asthma in pregnancy. *Clin. Obstet. Gynecol.* **39**, 56–69.

64. McFadden, E. R., and Gilbert, I. A. (1992). Asthma. *N. Engl. J. Med.* **327**, 1928–1937.

65. Gluck, J. C., and Gluck, P. A. (1976). The effects of pregnancy on asthma. A prospective study. *Ann. Allergy* **37**, 164–168.

66. Schatz, M., Harden, K., Forsythe, A., Chilingar, L., Hoffman, C., Sperling, W., and Zeiger, R. S. (1988). The course asthma during pregnancy, postpartum, and with successive pregnancies: A prospective analysis. *J. Allergy Clin. Immunol.* **81**, 509–517.

67. White, R. J., Coutts, I. I., Gibbs, C. J., and MacIntyre, C. (1989). A prospective study of asthma during pregnancy and the puerperium. *Respir. Med.* **83**, 103–106.

68. Beecroft, N., Cochrane, G. M., and Milburn, H. J. (1998). Effects of sex on fetus on asthma during pregnancy: Blind prospective study. *Br. Med. J.* **317**, 856–857.

69. Bahna, S. L., and Bjerkedal, T. (1972). The course and outcome of pregnancy in women with bronchial asthma. *Acta Allergol.* **27**, 397–406.

70. Stenius-Aarniala, B., Piirila, P., and Teramo, K. (1988). Asthma and pregnancy: A prospective study of 198 pregnancies. *Thorax* **43**, 12–18.

71. Lehrer, S., Stone, J., Lapinski, R., Lockwood, C. J., Schachter, B. S., Berkowitz, R., and Berkowitz, G. S. Association between pregnancy-induced hypertension and asthma. *Am. J. Obstet. Gynecol.* **168**, 1463–1466.

72. Engelson, M. M., Herman, W. H., Smith, P. J., German, R. R., and Aubert, R. E. (1995). The epidemiology of diabetes and pregnancy in the U.S. *Diabetes Care* **18**, 1029–1033.

73. McLellan, J. A. S., Barrow, B. A., Levy, J. C., Hammersley, M. S., Hattersley, A. T., Gillmer, M. D. G., and Turner, R. C. (1995). Prevalence of diabetes mellitus and impaired glucose tolerance in parents of women with gestational diabetes. *Diabetologia* **38**, 693–698.

74. Solomon, C. G., Willett, W. C., Carey, V. J., Rich-Edwards, J., Hunter, D. J., Colditz, G. A., Stampfer, M. J., Speizer, F. E., Spiegelman, D., and Manson, J. E. (1997). A prospective study of pregravid determinants of gestational diabetes mellitus. *JAMA, J. Am. Med. Assoc.* **278**: 1078–1083.

75. Dooley, S. L., Mertzger, B. E., Cho, N., and Liu, K. (1991). The influence of demographic and phenotypic heterogeneity in the prevalence of gestational diabetes mellitus. *Int. J. Gynaecol. Obstet.* **35**, 13–18.

76. Berkowitz, G. S., Lapinski, R. H., Wein, R., and Lee, D. (1992). Race ethnicity and other risk factors for gestational diabetes. *Am. J. Epidemiol.* **135**, 965–973.

76a. Coustan, D. R., Nelson, C., Carpenter, M. W., Carr, S. R., Rotondo, L., and Widness, J. A. (1989). Maternal age and screening for gestational diabetes: A population-based study. *Obstet. Gynecol.* **73**, 557–560.

76b. Martin, A. O., Simpson, J. L., Ober, C., and Freinkel, N. (1985). Frequency of diabetes mellitus in mothers of probands with gestational diabetes: Possible maternal influence on the predisposition to gestational diabetes. *Am. J. Obstet. Gynecol.* **151**, 471–475.

76c. Ober, C., Xiang, K.-S., Thisted, R. A., Indovina, K. A., Wason, C. J., and Dooley, S. (1989). Increased risk of gestational diabetes mellitus associated with insulin receptor and insulin-like growth factor II restriction fragment length polymorphisms. *Gene. Epidemiol.* **6**, 559–569.

76d. Allan, C. J., Argyropoulos, G., Bowker, M., Zhu, J., Lin, P.-M., Stiver, K., Golichowski, A., and Garvey, W. T. (1997). Gestational diabetes mellitus and gene mutations which affect insulin secretion. *Diabetes Res. Clin. Pract.* **36**, 135–141.

76e. Lapolla, A., Betterle, C., Sanzari, M., Zanchetta, R., Pfeifer, E., Businaro, A., Fagiolo, U., Plebani, M., Marini, S., Photiou, E., and Fedele, D. (1996). An immunological and genetic study of patients with gestational diabetes mellitus. *Acta Diabetol. Lat.* **33**, 139–144.

76f. Yajnik, C. S., Fall, C. H., Vaidya, U., Pandit, A. N., Bavdekar, A., Bhat, D. S., Osmond, C., Hales, C. N., Barker, D. J. (1995). Fetal growth and glucose and insulin metabolism in four-year-old Indian children. *Diabetes Med.* **12**, 330–336.

76g. Forrester, T. E., Wilks, R. J., Bennett, F. I., Simeon, D., Allen, M., Chung, A. P., and Scott, P. (1996). Fetal growth and cardiovascular risk factors in Jamaican schoolchildren. *Br. Med. J.* **312**, 156–160.

76h. Hofman, P. L., Cutfield, W. S., Robinson, E. M., Bergman, R. N., Menon, R. K., Sperling, M. A., and Gluckman, P. D. (1997). Insulin resistance in short children with intrauterine growth retardation. *J. Clin. Endocrinol. Metab.* **82**, 330–336.

76i. Leger, J., Levy-Marchal, C., Bloch, J., Pinet, A., Chevenne, D., Porquet, D., Collin, D., and Czarnichow, P. (1997). Reduced final height and indications of insulin resistance in 20 year olds born small for gestational age: Regional cohort study. *Br. Med. J.* **315**, 341–347.

76j. McCance, D. R., Pettitt, D. J., Hanson, R. L., Jacobson, L. T. H., Knowler, W. C., and Bennett, P. H. (1994). Birthweight and non-insulin dependent diabetes: Thrifty genotype, thrifty phenotype or surviving baby genotype? *Br. Med. J.* **308**, 942–945.

76k. Plante, L. A. (1998). Small size at birth and later diabetic pregnancy. *Obstet. Gynecol.* **92**, 781–784.

76l. Williams, M. A., Emanuel, I., Kimpo, C., Leisenring, W. M., and Hale, C. D. (1999). A population-based cohort study of the relation between maternal birthweight and risk of gestational diabetes mellitus in four racial/ethnic groups. *Paediatr. Perinatal Epidemiol.* (In press).

77. Sibai, B. M., Caritis, S. N., Thom, E., Klebanoff, M., McNellis, D., Rocco, L., Paul, R. H., Romero, R., Witter, F., and Rosen, M. (1993). Prevention of preeclampsia with low-dose aspirin in healthy, nulliparous pregnant women. *N. Engl. J. Med.* **329**, 1213–1218.

78. Zuspan, F. P., and Samuels, P. (1993). Preventing preeclampsia. *N. Engl. J. Med.* **329**, 1265–1266.

79. Roberts, J. M., and Redman, C. W. (1993). Pre-eclampsia: More than pregnancy-induced hypertension. *Lancet* **341**, 1447–1451.

80. Pritchard, J. A., MacDonald, P. C., and Gant, N. F. (1985). "Williams Obstetrics," 17th ed. Appleton-Century-Crofts, Norwalk, CT.

81. Maseki, M., Nishigka, I., Hagihara, M., Tomoda, Y., and Yagi, K. (1981). Lipid peroxide levels and lipid content of serum lipoprotein fractions of pregnant subjects with or without preeclampsia. *Clin. Chim. Acta* **115**, 155–161.

82. Mikhail, M. S., Basu, J., Palan, P. R., Furgiuele, J., Romney, S. L., and Anaegbunam, A. (1995). Lipid profile in women with preeclampsia: Relationship between plasma triglyceride levels and severity of preeclampsia. *J. Assoc. Acad. Minority Physicians* **6**, 43–45.

83. Hubel, C. A., Roberts, J. M., Taylor, R. N., Musci, T. J., Rogers, G. M., and McLaughlin, M. K. (1989). Lipid peroxidation in pregnancy: New perspectives on preeclampsia. *Am. J. Obstet. Gynecol.* **161**, 1025–1034.

84. Roberts, J. M., Taylor, R. N., Musci, T. J., Rodgers, G. M., Hubel, C. A., and McLaughlin, M. K. (1989). Preeclampsia: An endothelial cell disorder. *Am. J. Obstet. Gynecol.* **161**, 1200–1204.

85. Schobel, H. P., Fischer, T., Heuszer, K., Geiger, H., and Schmieder, R. E. (1996). Preeclampsia—a state of sympathetic overactivity. *N. Engl. J. Med.* **335**, 1480–1485.

86. Kupferminc, M. J., Peaceman, A. M., Wigton, T. R., Rehnberg, B. A., and Socol, M. L. (1994). Tumor necrosis factor-alpha is elevated in plasma and amniotic fluid of patients with severe preeclampsia. *Am. J. Obstet. Gynecol.* **170**, 1752–1759.

87. Greer, I. A., Lyall, F., Perera, T., Boswell, F., and Macara, L. M. (1994). Increased concentration of cytokines interleukin-6 and interleukin-1 receptor antagonist in plasma of women with preeclampsia: A mechanism for endothelial dysfunction? *Obstet. Gynecol.* **84**, 937–940.

88. Williams, M. A., Farrand, A., Mittendorf, R., Sorensen, T. K., Zingheim, R. W., O'Reilly, G. O., King, I. B., Zebelman, A. M., and Luthy, D. A. (1999). Maternal second-trimester serum tumor necrosis factor-α soluble receptor p55 (sTNFR55) and subsequent risk of preeclampsia. *Am. J. Epidemiol.* **149**, 323–329.

89. Williams, M. A., Mahomed, K., Farrand, A., Woelk, G. B., Mudzamiri, S., Madzime, S., King, I. B., and McDonald, G. B. (1998). Plasma tumor necrosis factor-α soluble receptor p55 (sTNFp55) concentrations in eclamptic, preeclamptic and normotensive pregnant Zimbabwean women. *J. Reprod. Immunol.* **40**, 159–173.

90. Walsh, S. W., Michael, J. B., and Allen, N. H. (1985). Placental prostacyclin production in normal and toxemic pregnancies. *Am. J. Obstet. Gynecol.* **151**, 110–115.

91. FitzGerald, D. J., Entman, S. S., Mulloy, K., and FitzGerald, G. A. (1987). Decreased prostacyclin biosynthesis preceding the clinical manifestation of pregnancy-induced hypertension. *Circulation.* **75**, 956–963.

92. Kaaja, R., Tikkanen, M. J., Vinnikka, L., and Ylikorkala, O. (1995). Serum lipoproteins, insulin, and urinary prostanoid metabolites in normal and hypertensive pregnant women. *Obstet. Gynecol.* **85**, 353–356.

93. Gant, N. F., Whalley, P. J., Everett, R. B., Worley, R. J., and MacDonald, P. C. (1987). Control of vascular reactivity in pregnancy. *Am. J. Kidney Dis.* **9**, 303–307.

94. Campbell, D., MacGillivray, I., and Carr-Hill, P. (1985). Preeclampsia in second pregnancy. *Br. J. Obstet. Gynaecol.* **92**, 131–140.

95. Fenney, J., and Scott, J. (1983). Preeclampsia in pregnancies from donor inseminations. *J. Reprod. Immunol.* **5**, 329–338.

96. Ikedife, D. (1980). Eclampsia in multipara. *Br. Med. J.* **5**, 985–986.

97. Eskenazi, B., Fenster, L., and Sidney, S. (1991). A multivariate analysis of risk factors for preeclampsia. *JAMA, J. Am. Med. Assoc.* **266**, 237–241.

98. Chesley, L. C., and Cooper, D. W. (1986). Genetics of hypertension in pregnancy. *Br. J. Obstet. Gynaecol.* **93**, 898–908.

99. Cooper, D. W., and Liston, W. A. (1979). Genetic control of severe preeclampsia. *J. Med. Genet.* **16**, 409.

100. Mahomed, K., Williams, M. A., Woelk, G. B., Jenkins-Woelk, L., Mudzamiri, S., Madzime, S., and Sorensen, T. K. (1998). Risk factors for preeclampsia among Zimbabwean women: Recurrence risk and familial tendency towards hypertension. *J. Obstet. Gynaecol.* **18**, 218–222.

101. Richardson, B. E., and Baird, D. D. (1995). A study of milk and calcium supplement intake and subsequent preeclampsia in a cohort of pregnant women. *Am. J. Epidemiol.* **141**, 667–673.

102. Savitz, D. A., and Zhang, J. (1992). Pregnancy-induced hypertension in North Carolina, 1988 and 1989. *Am. J. Public Health* **82**, 675–679.

103. Saftlas, A. F., Olson, D. R., Franks, A. L., Atrash, H. K., and Pokras, R. (1990). Epidemiology of preeclampsia and eclampsia in the United States, 1979–1986. *Am. J. Obstet. Gynecol.* **163**, 460–465.

104. Mahomed, K., Williams, M. A., Woelk, G. B., Jenkins-Woelk, L., Mudzamiri, S., Longstaff, S., and Sorensen, T. K. (1998). Risk factors for preeclampsia among Zimbabwean women: Maternal arm circumference and other anthropometric measures of obesity. *Paediat. Perinatal Epidemiol.* **12**, 253–262.

105. Klonoff-Cohen, H. S., Savitz, D. A., Cefalo, R. C., and McCann, M. F. (1989). Am epidemiologic study of contraception and preeclampsia. *JAMA, J. Am. Med. Assoc.* **262**, 3143–3147.

106. Velentgas, P. Benga-De, E., and Williams, M. A. (1994). Chronic hypertension, pregnancy-induced hypertension and low birthweight. *Epidemiology* **5**, 345–348.

107. Belizan, J. M., Villar, J., Gonzalez, L., Campodonico, L., and Bergel, E. (1991). Calcium supplementation to prevent hypertensive disorders of pregnancy. *N. Engl. J. Med.* **325**, 1439–1440.

108. Marcoux, S., Brisson, J., and Fabia, J. (1991). Calcium intake from dairy products and supplements and the risks of preeclampsia and gestational hypertension. *Am. J. Epidemiol.* **133**, 1266–1272.

109. Marcoux, S., Brisson, J., and Fabia, J. (1989). The effect of leisure time physical activity on the risk of preeclampsia and gestational hypertension. *J. Epidemiol. Commun. Health* **43**, 147–152.

110. Marcoux, S., Brisson, J., and Fabia, J. (1989). The effect of cigarette smoking on the risk of preeclampsia and gestational hypertension. *Am. J. Epidemiol.* **130**, 950–957.

111. Marcoux, S., Berube, S., Brisson, J., and Fabia, J. (1992). History of migraine and risk of pregnancy-induced hypertension. *Epidemiology* **3**, 53–56.

112. MacGillivray, I. (1983). "Pre-eclampsia: The Hypertensive Disease of Pregnancy." Saunders, London.

113. Cooper, D. W., Hill, J. A., Chesley, L. C., and Bryans, C. L. (1988). Genetic control of susceptibility to eclampsia and miscarriage. *Br. J. Obstet. Gynaecol.* **95,** 644–653.

114. Sutherland, A., Cooper, D. W., Howie, P. W., Liston, W. A., and MacGillivary, I. (1981). The incidence of severe preeclampsia amongst mothers and mothers-in-law of preeclamptics and controls. *Br. J. Obstet. Gynaecol.* **88,** 785–791.

115. Ma, Y., Ooi, T. C., Lui, M.-S., Zhang, H., McPherson, R., Edwards, A. L., Forsythe, I. J., Frohlich, J., Brunzell, J. D., and Hayden, M. R. (1994). High frequency of mutations in the human lipoprotein lipase gene in pregnancy-induced chylomicronemia: Possible association with apolipoprotein E2 isoform. *J. Lipid Res.* **35,** 1066–1075.

116. Ma, Y., Liu, M.-S., Ginzinger, D., Frolich, J., Brunzell, J. D., and Hayden, M. R. (1993). Gene-environment interaction in the conversion of a mild-to-severe phenotype in a patient homozygous for a ser (172) to cys mutation in the lipoprotein lipase gene. *J. Clin. Invest.* **91,** 1953–1958.

117. Chamberlain, J. C., and Galton, D. J. (1991). Atherosclerosis: The genetic analysis of a multifactorial disease. *NATO ASI Ser., Ser. A* 240.

118. Williams, M. A., Zingheim, R. W., King, I. B., Sorensen, T. K., Marcovina, S. M., and Luthy, D. A. (1996). Apolipoprotein E (Apo E) phenotypes and risk of preeclampsia. *Am. J. Obstet. Gynecol.* **174** (Part 2), 447 (abstr.).

119. Nagy, B., Rigo, J. Jr., Fintor, L., Karadi, I., and Toth, T. (1998). Apolipoprotein E alleles in women with severe preeclampsia. *J. Clin. Pathol.* **51,** 324–325.

120. Sohda, S., Arinami, T., Hamada, H., Yamada, N., Hamaguchi, H., and Kubo, T. (1997). Methylenetetrahydrofolate reductase polymorphism and pre-eclampsia. *J. Med. Genet.* **34,** 525–526.

121. Chen, G., Wilson, R., Wang, S. H., Zheng, H. Z., Walker, J. J., and McKillop, J. H. (1996). Tumour necrosis factor-alpha (TNF-α) gene polymorphism and expression of pre-eclampsia. *Clin. Exp. Immunol.* **104,** 154–159.

122. Ward, K., Jeunemaitre, X., Helin, C., Nelson, L., Nanikawa, C., Farrington, P. F., Ogasawara, M., Suzumori, K., Tomoda, S., Berrebi, S., Sasaki, M., Corvol, P., Lifton, R. P., and Lalouel, J.-M. (1993). A molecular variant of angiotensinogen associated with preeclampsia. *Nat. Gene.* **4,** 59–61.

123. Emanuel, I. (1997). Invited commentary: An assessment of maternal intergenerational factors in pregnancy outcomes. *Am. J. Epidemiol.* **146,** 820–825.

124. Emanuel, I., Leisenring, W., Williams, M. A., Kimpo, C., O'Brien, W., and Hale, C. B. (1999). The Washington State intergenerational study of birth outcomes: Methodology and some comparisons of maternal birthweight and infant birthweight and gestation in four ethnic groups. *Paediatr. Perinatal Epidemiol.* **13,** 352–371.

125. Kuh, D., and Ben-Shlomo, Y. (1997). "A Life Course Approach to Chronic Disease Epidemiology." Oxford University Press, New York.

15
Labor and Delivery

WILLIAM D. FRASER* AND MICHAEL S. KRAMER[†]
*Department of Obstetrics and Gynecology, Laval University, Québec City, Canada; [†]Department of Epidemiology and Biostatistics and of Pediatrics, McGill University, Montréal, Canada

I. Introduction

No other physiological event, with the possible exception of her own birth, has as great a potential to benefit or to harm a woman's life as labor and delivery. Consequently, health care planners, educators, and providers hold the responsibility to ensure that the risk of mishap during this period is minimized. They must regularly assess relevant indicators of maternal and perinatal health to identify problems. Using this information, they must develop and implement training, policies, and programs to ensure effective approaches to prevention and to care. These programs must be based on the best available scientific evidence.

In an attempt to reduce peripartum maternal mortality and morbidity, most developed countries have shifted overwhelmingly toward hospital births. For example, it is estimated that in the UK in 1927, 15% of births occurred in institutions. In 1992, 97.2% of births occurred in hospitals, 1.4% in isolated general practitioner units, and 1.3% at home [1]. Unfortunately, the routine application of surgical interventions and new obstetrical technologies to normal, low-risk women without the prior assessment of their effects has led to significant iatrogenic morbidity. Examples of routine interventions leading to increased maternal morbidity include electronic fetal monitoring [2] and epidural analgesia [3,4], both of which are associated with an increase in operative delivery, and routine episiotomy, which results in an increased risk of perineal injury [5,6].

A document developed at the 1995 World Health Organization-Pan American Health Organization conference in Forteleza, Brazil recommended that health ministries and planning authorities both in developed and in developing countries establish explicit policies promoting the appropriate use of technologies for perinatal care [7]. Such policies should be designed to ensure access to emergency services for women with complications of labor and delivery. Policies also should be designed to limit the application of technologies to groups that are likely to benefit from their use. Technologies should be introduced into practice use only after careful evaluation has shown the benefits to outweigh the risks.

Both in developed and in developing countries, improvement in the quality of perinatal services depends on a cyclic and iterative process. This process includes: (1) developing consensus on definitions for perinatal morbidity events; (2) ascertaining the types and distributions of morbid events, along with an evaluation of the causes. Factors in the chain of causality may be biological, social, and those related to the organization and delivery of care; (3) conducting research on approaches that are designed to reduce morbidity. This research may be primary, that is conducted on cohorts of patients, or secondary, that is

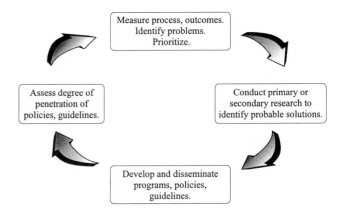

Fig. 15.1 Cycle of improvement of obstetrical services.

based on systematic review of existing research; (4) developing and disseminating policies and guidelines; (5) assessing the degree of penetration of the policies and guidelines, and (6) re-evaluating the distribution of morbidity events. A graphic representation of this cyclic process in shown in Figure 15.1. This chapter is structured around this simple model. In the next section, we briefly discuss how complications of labor and delivery can be ascertained. Section III reviews available information on the incidence and causes of these complications, and Section IV reviews the evidence on the effectiveness of interventions to prevent them. Section V discusses approaches to effecting change in clinical practice and in the organization of health care services bearing on labor and delivery. It is followed by a brief Conclusion (Section VI).

II. Ascertaining Complications of Labor and Delivery

The Working Group on Perinatal Audit of the European Association of Perinatal Medicine has proposed routine monitoring of the rates of operative delivery in normal, nulliparous women [8]. While cesarean section is an indicator of process, it is also a quasi-indicator of outcome. Other labor and delivery complications proposed for routine assessment include post-cesarean infection and damage to the anal sphincter at delivery. Middle- to long-term complications proposed for routine audit include urinary stress incontinence, fecal incontinence, and postpartum dyspareunia.

Special surveys are required to assess the prevalence of complications that result from labor but only become symptomatic after the immediate postpartum period. These include bladder and bowel disorders, psychological problems including "baby blues" and postpartum depression, and problems related to breastfeeding. Indicators of postpartum pelvic floor function include targeted questionnaires, clinical pelvic-squeeze scoring [9], intravaginal manometry using an air-filled balloon [9], transvaginal bladder sonograph [9], anal sphincter endosonography [10], and pudendal nerve latency [11]. An intact perineum at delivery does not exclude sphincter damage [10]. The incidence of sphincter damage increases when episiotomy is associated with perineal tear. One study showed that persistent pudendal nerve latency was associated with prolonged second stage and a large baby, but the correlation with clinical symptoms was poor [11].

Obstetric fistula is a serious morbidity that presents special problems with respect to ascertainment. As symptoms may appear only following vaginal delivery at home or following discharge from the hospital, facility-based data underestimate the prevalence of this problem. Because of the social stigma attached to the disorder, women may be reluctant to present for investigation and treatment [12].

III. Incidence and Causes of Labor and Delivery Complications

In most developed countries, there has been a marked reduction in maternal mortality related to such peripartum factors as hemorrhage and anesthesia-related causes. However, postpartum hemorrhage remains the leading cause of maternal mortality in France [13]. Other labor- and delivery-related causes (amniotic fluid embolus, pulmonary embolus) continue to play a role [14]. In developing countries, particularly when deaths occur outside of the hospital, causes of death are more difficult to assess. In a 1982 population-based study of maternal mortality in Addis Ababa, Ethiopia, maternal mortality was estimated as 566 per 100,000 live births. Approximately half of the deaths were complications of abortions, while the other half were due to peripartum causes (uterine rupture, sepsis, anesthesia-related causes, hemorrhage, and hypertensive disorders) [15].

Both in developed and in developing countries, among the most frequent complications of labor and delivery are infection, hemorrhage with or without transfusion of blood products, and injury to the anorectal area or the urinary tract. Perineal injuries involving the anal sphincter or anorectal mucosa can produce long-term morbidity [16], including fecal and urinary incontinence. Obstructed labor in settings where obstetrical referral services are not available can result in serious injury to the urogenital tract, including rectovaginal and vesico-vaginal fistulae. Further maternal complications of trauma to the urogenital region include pelvic inflammatory disease with its secondary infertility, renal failure, hydroureter, rectal atresia, vaginal stenosis, and foot-drop [12]. Fetal death is a common complication of these events. In addition to the physical handicap, women with obstetric fistulae are frequently ostracized by their families and communities and suffer deepening poverty as the consequence of social isolation [12]. In an analysis of factors contributing to vesico-vaginal fistulae in Nigeria, prolonged labor was present in 83% of cases and spontaneous delivery in 58%.

IV. Effectiveness of Interventions to Prevent Complications of Labor and Delivery

Obstetrics was among the first of the medical disciplines to maintain systematic records of adverse events and to attempt to assess their causes [17]. However, only recently has a concerted effort been made to place obstetric practice on the footing of carefully documented evidence. The 1980s witnessed a revolution in the assessment of the effectiveness of health technologies. Perinatology was the first health discipline to establish a database of systematic reviews (meta-analyses) of controlled clinical trials concerning the effectiveness of a wide range of interventions [18,19]. This electronic database made research findings rapidly available to health care providers and consumers. This model of technology transfer has been expanded to a wide range of health disciplines through the activities of the international Cochrane Collaboration [19]. The following summaries of the effectiveness of a number of approaches to labor and delivery rely, whenever possible, on the evidence generated through systematic reviews of controlled clinical trials; many of these reviews have been performed by members of the Pregnancy and Childbirth Review Group of the Cochrane Collaboration and are published electronically in the Cochrane Library.

While randomized clinical trials remain the gold standard for the evaluation of obstetrical interventions, certain characteristics of trials limit the information that they can provide. It is difficult or impossible to achieve blinding of treatment allocation for certain interventions. Once the treatment group is revealed, the preference of the obstetrician, the midwife, or the mother for one or another of the treatments may influence the probability of certain outcomes of interest. When care providers or consumers have strong beliefs concerning the effectiveness of certain approaches to care, it may be difficult to obtain acceptance of randomization, and even if the intervention is randomized, compliance may be poor [20]. Cluster randomization can be useful in assessing the impact of programs requiring complex educational interventions to ensure compliance. Cluster randomization allows implementation of the intervention while reducing the likelihood of contamination [21].

A. Postpartum Hemorrhage

Postpartum hemorrhage (PPH) is the major cause of maternal mortality and morbidity worldwide [22], and the risk of PPH is increased by anemia. The majority of cases of PPH can be prevented through the routine application by trained obstetrical personnel of "active management of the third stage" (AMTS). This approach consists of a package of interventions including administration of prophylactic uterotonic agents with or immediately after delivery of the baby, early cord clamping and cutting, and controlled traction to deliver the placenta [22]. AMTS has been compared with expectant management in several randomized, controlled trials and the results have been summarized in a meta-analysis [22]. AMTS is associated with statistically and clinically significant reductions in a number of indicators of blood loss including PPH > 500 ml, PPH > 1000 ml, maternal hemoglobin < 90 g/l, and blood transfusion. Benefits were observed both among unselected populations and among women who were selected as being at low risk for PPH. The optimal uterotonic agent remains to be defined. Ergometrine is a more effective uterotonic than oxytocin alone but is associated with a number of undesirable side effects such as hypertension, nausea, and vomiting [23]. Misoprostol, a prostaglandin that can be administered orally or rectally, may offer advantages in the developing country context because it does not require refrigeration and is easy to administer [24]. A small trial comparing prophylactic misoprostol to Syntometrine (an oxytocin-ergometrine combination) found the two groups similar with respect to blood loss and postpartum hemoglobin concentration. However, less diastolic hypertension was noted with misoprostol [25]. A WHO-sponsored multicenter trial of misoprostol in the developing country context is currently in progress. The safety of the selective use of misoprotol by traditional birth attendants (TBAs) to prevent PPH in the primary care setting will require careful assessment. The prevention of death due to obstetric hemorrhage that is unresponsive to medical treatment requires ready access to a surgical team and transfusion service.

B. Sepsis

Factors predisposing a woman to maternal sepsis include prolonged labor, premature and prolonged rupture of the membranes, cesarean section, and instrumental vaginal delivery. Microbial colonization of the amniotic fluid is present at cesarean section in 66% of women with ruptured membranes, in 23% of women in labor with intact membranes, and in 13% of women not in labor with intact membranes [26]. The routine use of prophylactic antibiotics at the time of cesarean section is effective in reducing postoperative infectious morbidity both among populations that are at high risk of wound infection [27] and also among those that are at low risk [28,29]. A French study noted that while 71.7% of emergency obstetrical hysterectomies were performed for uncontrolled hemorrhage, 25% were for the indication of infection [30]. Maternal mortality associated with the procedure was 20%.

Basic hygiene is difficult to maintain for home deliveries both in rural settings and in urban slums in developing countries. Providing TBAs with clean delivery kits, risk screening and referral, tetanus toxoid immunization, transportation to first referral services, and early and aggressive antibiotic treatment are essential components of sepsis prevention programs. A large study carried out in a developing country setting used historical controls to test the hypothesis that a manual cleansing of the birth canal with a 0.25% chlorhexidine solution at the time of vaginal examination reduced infectious morbidity. The study intervention included wiping the baby with chlorhexidine. Maternal admissions due to postpartum infection and neonatal mortality due to infectious causes were both reduced by approximately two-thirds during the intervention period [31].

C. Eclampsia

A review of cases of eclampsia carried out in three teaching hospitals in Melbourne, Australia, noted 5 maternal deaths and 17 perinatal deaths in 90 cases of eclampsia [32]. The definitive treatment of preeclampsia/eclampsia syndrome (PEE) is deliv-

ery, and the labor and delivery period is a high-risk period for eclampsia. While comprehensive antenatal care with early detection and intervention is the keystone in effective treatment of PEE, appropriate management during the intrapartum period is critical to secondary prevention of maternal mortality and morbidity due to this disorder. The two pillars of morbidity prevention in severe preeclampsia are antihypertensive drugs and magnesium sulfate to prevent seizures. The use of antihypertensives in this setting, although not assessed in trials, is widely accepted. Hydralazine and labetalol are the two drugs of choice [33]. A landmark trial has clearly demonstrated magnesium sulfate to be superior to diazepam or diphenylhydantoin in the prevention of eclampsia recurrence [34]. Trends toward reductions in maternal mortality were seen with magnesium sulfate. A published trial demonstrated that magnesium sulfate was more effective than a placebo in preventing the first eclamptic convulsion. One woman out of 345 women who received magnesium sulphate developed eclampsia, while in the placebo group, 11 out of 340 women developed eclampsia (RR = 0.09, 95%CI 0.01–0.69) [35]. A Canadian study found that 3.9% of women with preeclampsia experienced convulsions when managed with a placebo [36]. The use of magnesium sulfate appears to be increasing in some regions following the results of the Collaborative Trial [37]. While concerns have been voiced in some circles that magnesium sulfate may prolong labor [38] one trial found no evidence of an effect of the medication on labor duration [39]. In this study, treatment with the drug was associated with a fourfold increase in the risk of postpartum hemorrhage. However, the Collaborative Eclampsia Trial [34] found no evidence for an effect of magnesium sulfate on blood loss >500 ml. Lindheimer commented "The results (of the Collaborative Trial of Eclampsia) should end decades of controversy on the efficacy of magnesium sulfate . . . using a simple treatment protocol which is applicable in underdeveloped nations" [40].

D. Cesarean Delivery

The most dramatic change in obstetrical practice witnessed since the 1970s has been the rise in cesarean births, for which dystocia (abnormal labor progress), repeat cesarean, and fetal distress are the main contributors [41,42]. In the U.S., 20–25% of all births are now by cesarean [43]. Certain developing countries also have witnessed disturbing increases in cesarean rates [44,45]. The data, however, suggest that cesarean rates in certain regions are beginning to plateau and or even to decrease [46,47]. Table 15.1 shows the decrease in cesarean sections in the Province of Québec, Canada.

Both in developed and in developing countries, maternal morbidity associated with cesarean section is high and includes febrile morbidity (24–27%), blood loss > 1000 ml (4–8%), urinary tract infections (3%), and wound dehiscence (0–10%) [48,49]. Maternal postpartum depression also appears to be increased [50].

Certain risk factors for cesarean section are related to the woman and her pregnancy, such as short stature, advanced maternal age, increasing birth weight of the baby, gestational age >41 weeks, excess pregnancy weight gain, and excess prepregnancy weight [51]. In one region of the U.S. [52], a shift in the childbearing population toward later childbearing, lower parity, and larger birthweight from the early 1970s to the late 1980s was estimated to be responsible for an increase in the cesarean rate by 18% over what it would have been if these variables had remained stable over the time period.

Marked variations in cesarean rates have been observed between regions and even among hospitals in the same region [53]. Several nonmedical factors have been shown to be key determinants of cesarean section and include payment method [54] and the practice pattern of the obstetrician. Goyert *et al.* [55] noted that the obstetrician's individual cesarean rate was second only to parity as a predictor of cesarean section risk.

Variations in cesarean section rates among obstetrical units and among regions are partly explained by differences in policies and practices concerning relative indications for cesarean, such as breech presentation and repeat cesarean, and partly by differences in patient risk profile. The "standard" nullipara has been proposed as a unit for interhospital comparisons and for audits that focus on intervention rates [56]. The standard nullipara is defined as a woman without known major risk factors for cesarean; she is of normal height (>155 cm), aged 20–34 years, has a single term baby in cephalic presentation, and is without medical complications. An audit of 15 obstetrical units in the North West Thames Region of the UK found cesarean rates in standard nulliparae ranging from 5 to 12.5% [57].

1. Dystocia

Dystocia is the most frequent indication for cesarean section in nulliparae both in developed and in developing countries. In the case of true cephalopelvic disproportion, cesarean section can be lifesaving both for mother and for baby. In some regions of developing countries where transportation to obstetrical services is not readily available, obstructed labor is by far the most frequent obstetrical complication and the one leading to the highest maternal mortality ratio [58,59]. Female genital mutilation is a documented risk factor for maternal mortality due to dystocia-related causes [60].

No physical, radiographic, or physiological indicator or group of indicators has been identified that is sufficiently sensitive and specific to identify, either before labor or early in labor, women who will require cesarean section for dystocia. The absence of good evidence, or even a strong consensus, concerning the diagnosis and treatment of dystocia opens the door to marked variation in rates of medical intervention (augmentation of labor) and cesarean section for this indication. Active management of labor (AML) is a structured clinical protocol for intrapartum care designed to standardize the diagnosis and treatment of dystocia. Interest in this protocol has been high because it represents an integrated management strategy for in-hospital intrapartum care and it is one of the few such programs to focus on the problem of dystocia. Based on observational data (some with historical controls) [61–63], AML has been proposed as a solution to the problem of high cesarean rates. The protocol includes several components: prenatal education, selective admission to the labor ward, and continuous intrapartum professional support. Decision making regarding labor augmentation (amniotomy and oxytocin) is based on strict interpretation of a partograph or labor graph. Epidural analgesia is used selectively and midforceps are proscribed [64].

During the 1990s, several randomized trials have assessed the efficacy of AML and its components, mostly in developed countries. A single trial comparing the overall AML "package" to usual care [65] found no evidence of an effect on cesarean section rates. As an isolated intervention, early amniotomy prevents delays in labor progress that are severe enough to qualify as dystocia [66]. However, amniotomy increases the hourly rate of fetal heart rate decelerations [67] and, depending on the context of care, may lead to an increased cesarean rate for the indication of fetal distress [66,67]. In a systematic review of trials of early amniotomy, we noted a statistically nonsignificant trend toward an increase in the global rate of cesarean section in association with amniotomy [68]. The WHO Reproductive Health Library is an electronic publication dedicated to disseminating high-quality evidence concerning the effectiveness of interventions [69]. It also proposes evidence-based guidelines that are relevant to care in settings with limited resources. The systematic review of the effects of amniotomy in shortening spontaneous labor notes that in some developing countries, a trend toward a more interventionist approach to membrane management has been observed. In contexts where resources are scarce and where cesarean-related morbidity and mortality are often higher than in settings with adequate resources, a conservative approach to membrane management is advocated [70].

In a separate systematic review, we assessed trials that compared a policy of routine early labor augmentation using both early amniotomy and early oxytocin to more conservative forms of care [71]. This meta-analysis stratified trials according to whether they enrolled women with normal progress in labor (primary prevention trials; n = 7) or were restricted to women with established dystocia (secondary prevention trials; n = 3). In both strata, early augmentation had a significant labor-shortening effect. There was little evidence for a beneficial effect of early augmentation on cesarean section risk in the primary prevention trials [pooled odds ratio (OR) = 0.9, 95% CI 0.7–1.1]. In the secondary prevention trials, a large but nonsignificant-trend toward a reduction in cesarean risk was observed with earlier intervention (OR = 0.6, 95% CI 0.2–1.4); the small number of women in these trials led to the wide confidence interval. Of the three trials [72–74] that reported on women's perception of pain associated with AML, two studies [73,74] found significantly more maternal discomfort with early intervention. One trial [72] found no evidence of such an effect. We concluded that early augmentation is without benefit for women with mild delays in labor progress.

The partograph, a graphic representation of labor progress, has been assessed in underresourced settings as a decision aid for care during labor. While working in southern Africa in the early 1970s, Philpott introduced the partograph to distinguish women who required transfer for in-hospital obstetrical care from those who could be permitted to labor and deliver in peripheral midwifery units [75,76]. Women who crossed the so-called "alert line" were candidates for transfer and those who crossed the "action line" were considered as requiring medical intervention (Fig. 15.2). In 1993, a WHO technical working group within the Safe Motherhood Initiative approved a modified

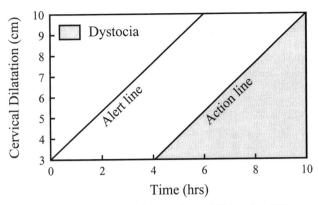

Fig. 15.2 Labor partograph [adapted from [75], Fig. 4, p. 595].

version of the Philpott partograph and developed a user's guide for its dissemination [77]. A multicenter study of implementation of the partograph was conducted in four pairs of hospitals in Indonesia, Malaysia, and Thailand [78]. The investigators noted a nearly two-thirds reduction in labor augmentation, a reduction in forceps use of nearly one-third, and a similar reduction in cesarean section after implementation of the partograph. Labors lasting longer than 18 hours and postpartum sepsis also were significantly reduced. Thus, in the developing country context, the use of the WHO partograph along with an established treatment algorithm appears to be a useful triage tool and decision aid both for in-hospital and for planned home deliveries. For women who plan to deliver outside of the hospital, the partograph can identify the subgroup of women who would benefit from timely transfer and thereby reduce untreated dystocia and its consequences. The use of the partograph requires basic literacy skills. Special adaptations may be necessary for its use by TBAs with poor literacy skills. To date, there have be no published trials evaluating the effect of the WHO partograph in developed country settings.

2. Vaginal Birth after Cesarean Section

Repeat cesarean section was a major contributor to increasing cesarean section rates observed in the 1970s and 1980s [41,42]. There have been no randomized controlled trials of elective repeat cesarean section versus attempted vaginal delivery. However, based on observational data, a consensus developed in the early 1980s that attempted vaginal delivery for women with a single previous low transverse cesarean section was associated with a lower risk of complications than elective repeat cesarean section [79,80]. Following the U.S. and Canadian consensus conferences on cesarean birth in the early 1980s, professional bodies developed guidelines promoting vaginal birth after cesarean (VBAC) [81,82]. VBAC has been an important factor in the reduction in cesarean rates noted in certain regions [43]. An observational study has reported that a trial of labor was not associated with a reduction in major delivery complications compared with a policy of routine elective cesarean section [83]. However, the rational for excluding cesarean delivery, itself a major maternal morbidity event, from the primary outcome was questioned in the accompanying editorial [84].

The pregnant woman herself plays an important role in the VBAC versus repeat cesarean decision. Between 55 and 90% of women identify themselves as the primary decision maker in the choice [85]. We conducted a randomized controlled trial of a prenatal education program designed to promote VBAC [86]. Using a 10-cm visual analog scale, we asked women (n = 1275) in the second and early third trimesters to indicate their degree of agreement with the statement "I plan to attempt a vaginal delivery in this pregnancy." We then dichotomized responses as indicating low motivation (0–5.0 cm) or high motivation (5.1–10.0 cm) for VBAC. Twenty-nine percent of women had low motivation by this definition. Women were then stratified by their degree of motivation and randomized to receive a personalized two-hour prenatal educational intervention provided by a research nurse. We found no evidence for an effect of the educational intervention on the method of delivery in either stratum. However, women who prior to randomization indicated a low degree of motivation for vaginal delivery were more than three times as likely to have a scheduled cesarean section as women with high motivation (46.5 versus 12.5%). That the attitudes of women in the "low VBAC motivation" category were refractory to the educational program suggests that poorly understood personal and cultural factors were more important in influencing health-related decisions than was the information concerning the probability of successful vaginal delivery. These results raise the ethical issue of whether health care providers should perform nonmedically-indicated interventions solely on the basis of the woman's beliefs and preferences. It also suggests the need for qualitative research to gain an understanding of women's perceptions of their own needs prior to designing "educational" programs.

With the increasing rates of cesarean section in the developing world, the question of the safety of attempted vaginal delivery for these women in their subsequent pregnancy becomes more urgent. In developing countries, primary cesarean section is frequently performed as an emergency procedure in women whose labors are complicated by dystocia (of a severity not found in developed countries), serious infection, or other conditions that could have an impact on wound healing. Furthermore, transfusion and anesthesia services may be less readily available if a trial of labor results in uterine rupture. Given these factors, it is possible that complication rates from trial of labor in developing countries could be higher than those observed in developed countries. From a social and cultural perspective, cesarean section is viewed as a denial of normal motherhood in many parts of the developing world. Moreover, its costs are high and it may be viewed as limiting fertility. Where hospitals do not offer a trial of labor, some women might prefer an unsupervised labor at home with its attendant risks to a scheduled cesarean section. To evaluate the safety and effectiveness of a policy of trial of labor for women with a previous cesarean section delivering in sub-Saharan Africa, Boulvain et al. conducted a meta-analysis of 17 published reports in these settings [87]. These studies included data on 4500 women with previous cesarean section. The probability of vaginal delivery among these women was 69% (95% CI 63–75%), a value similar to that observed in women attempting VBAC in developed countries. Maternal mortality among all women with a previous cesarean section was 190 per 100,000, a value higher than that found in developed countries but lower than the average for sub-Saharan Africa. The rate of uterine rupture was 2.1%, which was slightly higher than that observed in developed countries. Thus, trial of labor would appear to be both effective and reasonably safe in hospitals that are equipped with emergency cesarean services. However, absolute contraindications to attempts at vaginal delivery remain, including a vertical uterine scar, previous vesico-vaginal or recto-vaginal fistula, previous ruptured uterus, and severe cephalo-pelvic disproportion.

3. Electronic Fetal Monitoring and Cesarean Risk

Electronic fetal monitoring (EFM) has largely replaced intermittent auscultation (IA) as the primary method of intrapartum fetal assessment in most developed countries. This technology was introduced into obstetrical practice prior to careful evaluation of its benefits and risks. Several trials comparing EFM to IA were conducted in the context where decisions regarding intervention for fetal indications were based solely on EFM or IA findings. These early studies documented an increase in the rate of operative delivery associated with EFM, including cesarean section, without evidence of any important neonatal benefits [88]. The Dublin EFM trial [89], which is the largest monitoring trial to date and involved more than 13,000 women, found EFM to decrease the risk of neonatal convulsions compared with IA. However, the prevalence of cerebral palsy at 4 years was similar in the two groups [90]. Women who were in the EFM group had slightly more operative vaginal deliveries but the rate of cesarean in the two groups was similar. The use of fetal scalp pH as an adjunct to EFM in this trial (to reduce the false positives associated with monitoring) is a likely explanation of the low rate of intervention for fetal indications in the experimental group.

Given the uncertain benefits of EFM and the increase in maternal morbidity associated with the procedure in specific contexts of care, the Society of Obstetrician-Gynecologists of Canada 1995 guidelines for intrapartum fetal assessment proposed IA as the standard method of fetal assessment for women with uncomplicated labors [91]. These guidelines imply significant changes in clinical practice for Canadian obstetricians, midwives, and nurses who have depended on EFM since the 1970s. The extent to which these guidelines will be implemented remains to be determined.

E. Complications of Forceps Delivery

It has been traditionally held that a duration of the second stage of labor (*i.e.,* between full cervical dilation and delivery) over 2 hours places the baby at increased risk of adverse outcome [92]. Other data suggest that with careful monitoring of fetal status, a prolonged second stage may not place the baby at increased risk [93–95]. The American College of Obstetrics and Gynecology's Committee on Obstetrics has defined a prolonged second stage for nulliparous women with epidural analgesia as greater than 3 hours [96]. Forceps often are used to hasten delivery when the second stage of labor is prolonged.

Rates of operative vaginal delivery approaching 30% have been reported in North American hospitals [51,97,98]. Opera-tive vaginal delivery, particularly midpelvic procedures and those involving rotation of the fetal head, are associated with increased maternal and perinatal morbidity. Maternal injury, defined as a 3rd and 4th degree perineal tear or a vaginal laceration, occurs in 36.5% of midforceps deliveries compared with 13% of outlet forceps [99]. Low forceps with rotation >45° has been associated with a maternal trauma rate greater than that of midforceps (44%) [99]. Thus, both midforceps and low forceps with rotation >45° are associated with a significant risk of maternal injury.

Randomized, controlled trials [100] have demonstrated that when assisted vaginal delivery is required, the ventouse should be the instrument of choice because it is less likely to cause vaginal or perineal injury than forceps. While cephalohematoma and retinal hemorrhages are more frequent with the ventouse, other cranial and facial injuries are more frequent with forceps.

F. Complications of Epidural Anesthesia

Epidural analgesia (EA) has been implicated as a risk factor for operative vaginal delivery and possibly also for cesarean section [3,4]. Clinical trials comparing EA to other forms of analgesia have shown EA to be associated with malpositions of the fetal head (OR = 5.42, 95% CI 1.33–22.04) [101]. The wider diameter of the fetal head in the OP (occipito-posterior) position can impede descent [102,103]. The OP position is associated with an increased risk of operative delivery and particularly with procedures requiring rotation of the fetal head. A meta-analysis comparing EA to other forms of pain relief in labor demonstrated that if EA was maintained in the second stage, there was a marked increase in operative vaginal delivery (OR = 2.59, 95% CI 1.29–5.19) [101]. Others have estimated that EA increases the risk of rotational forceps 20-fold [102].

Second stage duration is increased with EA [104], in part as the result of motor block. While several observational studies have found EA to be associated with an increased cesarean section risk [3,105], the few randomized trials that have compared epidural with nonepidural analgesia have yielded divergent results [4,106–108]. It appears that if epidural is indeed a risk factor for cesarean section, the effect likely depends on a number of context-specific factors, including the degree of motor block produced by the epidural agent and the threshold for intervention of the delivery obstetrician.

The routine administration of oxytocin in the second stage of labor has been advocated as a possible approach to the reduction of second stage operative deliveries for women with EA. This approach is based on two observations: the absence of the normal physiologic increase in circulating oxytocin [109] and the reduction in uterine contractility [110] among women in second stage with EA. A single randomized clinical trial comparing routine oxytocin administration to an attempt to avoid oxytocin in the second stage for women with EA has been published [111]. An increase in the rate of spontaneous delivery (50 versus 40%) and a reduction in the rate of nonrotational forceps and vacuum (31 versus 47%) was observed in the group receiving oxytocin. However, a statistically significant increase in the frequency of rotational forceps was noted in the group treated with oxytocin (18 versus 9%). Because of the potential of rotational forceps to cause adverse maternal and neonatal sequelae, the routine augmentation of second stage of labor with oxytocin cannot be recommended at present. Further studies are required to address this question.

A number of clinical trials have assessed variations in the method of epidural administration, including the types, timing, and concentrations of medications used. It has been suggested that simply by terminating the epidural at the end of the first stage, second-stage operative interventions can be reduced. However, the two trials evaluating this approach have yielded divergent results [112,113]. Pain rapidly increased to preepidural levels among women deprived of continuous infusion [114].

The addition of narcotics (fentanyl, sufentanil) to the epidural solution provides the same degree of sensory blockade at lower concentrations of local anaesthetic agent, thus reducing motor blockade [115–118]. One randomized trial [119] found that women receiving a sufentanil-bupivacaine (0.125%) combination had a statistically significant decrease in the incidence of instrumental delivery (24 versus 36%) compared to those receiving bupivacaine alone.

Delayed pushing is an approach that can reduce operative vaginal delivery for women with epidural anesthesia. Women are advised to avoid voluntary expulsive efforts until an irresistible urge to push is felt or until the presenting part has descended to the perineum under the force of uterine contractions. Its goal is to conserve the woman's physical and psychological energies until the presenting part has descended beyond the midpelvis. We completed a multicenter trial of a policy of delayed pushing for women with epidural anesthesia [120]. Women in the delayed pushing group had a median waiting period of 110 minutes after full dilatation, compared with 5 minutes among those in the early pushing group. The primary outcome for the trial was "difficult delivery," defined as any of the following methods of delivery: cesarean section, midpelvic delivery, or low pelvic delivery with rotation of the fetal head more than 45 degrees. Delayed pushing reduced the risk of difficult delivery (RR = 0.79, 95% CI 0.66–0.95), mainly owing to a reduction in mid-pelvic procedures. While there was an increase in abnormal umbilical cord blood pH results among babies in the delayed pushing group, there was no clinical evidence of a difference in the frequency of adverse neonatal events in the two groups. The reduction in difficult delivery was most marked when the fetal head was in the transverse to posterior position at full dilatation [120].

G. Perineal Injury

As early as 1741, Ould reported on surgical incision of the perineum to prevent severe perineal tears [121]. Beneficial effects alleged to be associated with episiotomy included reduction in the likelihood of severe perineal tears, reduction in urinary and fecal incontinence, improved sexual function, and easier repair and healing. Although episiotomy has become one of the most commonly performed surgical procedures in the world, it was introduced into practice without scientific evidence of its effectiveness [122]. During the 1980s and early 1990s, a series of clinical trials compared the effects of liberal versus restrictive use of episiotomy on the occurrence of a number of postpartum morbidity indicators, including severe

Table 15.1

Trends in Cesarean Section and Episiotomy in Quebec from 1982 to 1997

Year	Episiotomy (%)	Cesarean section (%)
1982–1983	72.2	17.0
1986–1987	66.9	19.0
1991–1992	56.7	17.4
1992–1993	52.5	17.0
1993–1994	48.0	16.9
1994–1995	43.0	16.7
1995–1996	38.6	16.4
1996–1997	36.1	16.7

perineal injury. The restrictive use of episiotomy is associated with a lower risk of posterior perineal trauma, reduced requirement for perineal suturing, and reduction in healing complications at 7 days. No differences have been observed in severe vaginal or perineal trauma, dyspareunia, or urinary incontinence [5].

In Canada, Graham and Graham noted a decline in episiotomies from 66.8% in 1981–1982 to 37.7% in 1993–1994 [123]. This reduction was observed both for spontaneous and for operative vaginal deliveries. Graham suggests that this trend may reflect the convergence of obstetrical practice to evidence-based practice guidelines. Trends of episiotomy rates in Quebec Province are shown in Table 15.1.

Research indicates that postpartum urinary and fecal incontinence have higher prevalences than previously assumed [124]. When 906 women were interviewed 10 months after delivery, they reported a prevalence of fecal incontinence of 6.6% [125]. Sultan et al. [11] performed anal endoscopy and neurophysiological testing in 202 women before and 150 women six weeks after delivery. Postpartum occult sphincter defects were observed in 35% of primiparous women and 44% of multiparous women. Occult sphincter damage was observed more frequently among women with episiotomy than among women with spontaneous second-degree tears. There was a strong correlation between the presence of endoscopically identified occult tears and defecatory symptoms.

The results of two trials indicate that prenatal perineal massage may be an effective strategy to reduce perineal injury at delivery [126,127]. As yet, no data are available concerning the longer-term effects of perineal massage on sexual function or urinary or anal symptoms.

H. The Effects of Ambulation, Posture, and Position in Labor

The woman may assume one of several postures during both the first and second stages of labor. These include vertical positions (standing, walking, sitting, squatting, and kneeling), various reclining positions (with back support provided by a labor coach, a wedge, or an adjustable chair), and recumbent positions (supine or lateral tilt). Women may wish to intermittently vary their position throughout labor. In most primitive societies, women give birth in a vertical posture [128]. Mariceau, a seventeenth century obstetrician, was among the first to popularize the use of the bed for delivery. The recumbent position has been the obstetrician's preferred position for delivery of women throughout most of the twentieth century; the advent of routine electronic fetal monitoring has further reduced the use of other postures.

Maternal position during the first stage of labor may influence uterine activity. Both standing and lateral-recumbent positions are associated with contractions of greater intensity but lower frequency than those observed in women in the dorsal recumbent or sitting positions [129–131]. The dorsal recumbent position can have an adverse effect on maternal hemodynamic status, as well as lead to fetal heart rate abnormalities.

Trials of the effects of ambulation on the duration of the first stage of labor have yielded variable results, with some showing a labor-shortening effect [132–136] and other studies showing no evidence of an effect [137–140]. A meta-analysis of five trials found ambulation to reduce medical intervention for pain relief in labor (OR = 0.63, 95% CI 0.46–0.86) [141]. A systematic review of the effect of maternal posture on the second stage of labor was unable to detect an effect of posture on the duration of the second stage, nor was there consistent evidence of an effect on the risk of operative delivery [142]. A single trial has demonstrated a marked reduction in operative vaginal delivery with the squatting position as compared with the semirecumbent posture [143]. Three trials [143–145] have assessed the risk of postpartum blood loss >500 ml with the upright versus the recumbent position. The results, which are summarized in a meta-analysis, show an increased risk of PPH associated with upright postures [142]. In summary, women should be encouraged to move freely about in labor, if that is their desire. Although the advantages of upright posture are most evident during the first stage of labor, women should be encouraged to exercise choice over their position throughout labor.

I. The Effects of Primary-Care Provider and Place of Delivery

The presence during labor of a trained health care professional is a key element in reducing maternal morbidity, including operative delivery rates [146]. In some developing countries, it is not possible in the short to medium term to ensure the presence of trained midwives or physicians at all or even at the majority of deliveries. Initiatives have been developed to train TBAs in safe practices during labor, to attend to the woman's comfort, to diagnose and to treat simple conditions, and, when possible, to arrange for timely transfer in the case of complications [147]. Intrapartum support, even when provided by a nonprofessional trained "doula," can reduce perinatal morbidity. A meta-analysis of randomized trials found augmented intrapartum social support to be associated with a reduction in the requirement for epidural anesthesia and in operative vaginal delivery, cesarean section, and episiotomy. Maternal postpartum anxiety and depression also were less frequent among women receiving support [146].

In the absence of an efficient method of transport, delivery at a place distant from emergency medical services poses a major threat to women's safety. Often, women may not have the information necessary to allow them to exercise wise choices.

Delivering at home, even when the residence is distant from emergency obstetrical services, may be more appealing to the women than leaving their home and family to travel to a hospital or other birthing facility. Especially once labor has commenced, women may have little power to influence decisions regarding transfer for care. Decisions concerning transfer to the hospital often are taken by relatives who may harbor their own anxieties about contact with health services, including fear of the costs of care [148].

In developed countries, controversy remains concerning the optimum type of health professional to lead intrapartum care for low-risk women. Based on a literature review, Enkin *et al.* suggests that intrapartum care is best led by a midwife [149]. A number of trials have addressed this issue. We performed a meta-analysis of six trials [150–155] comparing selected low-risk women who were randomized during the antepartum period to receive either midwifery-led or obstetrician-led care. The analysis was by intention to treat, in that midwifery clients who required transfer to medical care either during the antepartum or the intrapartum periods remained in the midwifery group for the purpose of analysis. As shown in Table 15.2, a high degree of statistical heterogeneity was noted both for epidural analgesia and for cesarean section. Based on the random-effects model of DerSimonian and Laird [156], midwife-led care was associated with a reduction in the use of epidural anesthesia and in the risk of instrumental vaginal delivery.

We conducted a cohort study in Quebec, Canada comparing midwifery care provided in birthing centers to that provided by obstetricians and family doctors (N = 1922). Study participants were matched on a number of sociodemographic and obstetrical variables. Analysis was by intention to treat: women who were booked for midwifery care were analyzed in that group, even though some required transfer to medical care during the antepartum or intrapartum periods. We observed marked reductions in obstetrical interventions among women in the midwifery group, including induction of labor (RR = 0.23, 95% CI 0.18–0.31), episiotomy (RR = 0.19, 95% CI 0.15–0.24), primary cesarean section (RR = 0.54, 95% CI 0.40–0.75), and intrapartum electronic monitoring (RR = 0.24, 95% CI 0.21–0.27). Severe perineal injury in the form of third- and fourth-degree tears also was reduced (RR = 0.28, 95% CI 0.16–0.49). However, we noted that neonatal respiratory depression requiring bag and mask resuscitation was more frequent in the midwifery care group than in the group receiving obstetric care. Our data suggest that intrapartum care provided by a midwife substantially decreases maternal morbidity. On the other hand, sole reliance by the midwife on intermittent auscultation may lead to a failure to diagnose fetal distress during labor and an increase in neonatal respiratory depression and requirement for assisted ventilation. A large study reporting on 11,814 births from the National Birthing Center Cohort in the U.S. also suggests that midwifery care is associated with a reduction in maternal morbidity compared with physician care [157].

In developed countries, controversy continues as to whether home delivery places the woman and her fetus/newborn at excess risk of morbidity. After reviewing the epidemiologic pitfalls surrounding this question, Campbell and Macfarlane [1] concluded it unlikely that unbiased data will be available to bring resolution to this question in the near future.

Table 15.2

Meta-analysis of Randomized Trials of Midwifery Care versus Traditional Medical Care for Several Labor and Delivery Outcomes [150–155]

Outcome measures	No. of trials	Relative risk (95% CI Random)	Heterogeneity X^2 (p value)
Epidural	6	0.81 (0.68–0.96)	14.24 (0.014)
Instrumental vaginal delivery	6	0.86 (0.75–0.99)	0.74 (0.98)
Cesarean section	6	0.94 (0.76–1.18)	9.24 (0.10)

V. Effecting Change in Clinical Practice and in the Organization of Health Services

In order for effective approaches to care to have an impact on women's health, a structured network linking those involved in health technology assessment to health planners and providers must be in place. Regional and national surveys can be useful in assessing the degree to which polices have been put into practice. For instance, a 1995 national survey contacted all 572 Canadian hospitals where obstetrics was practiced [158]. Ninety-nine percent of hospitals encouraged the woman's partner to be involved in the birth experience. Eleven and sixteen percent of hospitals persisted in maintaining policies requiring suppositories/enemas and partial shave preparations, respectively, despite evidence that such procedures are not beneficial [159,160]. Almost two-thirds of hospitals required at least initial electronic fetal monitoring on admission to the labor ward.

Lomas *et al.* [161] have shown that knowledge of perinatal care guidelines is not equivalent to a change in practice. Diffusion of innovation is a form of social change, the rapidity of which depends on a number of factors, including the nature of the innovation, the communication channels that are available, and the social structure into which the change is introduced [162]. Research is required to assess the effectiveness of strategies to implement guidelines. While in one study, the use of opinion leaders was more effective than the simple dissemination of guidelines in promoting VBAC [163], other studies have yielded less encouraging results [164]. Knowledge is not a sufficient cause for the adoption of a practice innovation, but it is likely a necessary component in the causal network of change.

The first step in the implementation of change is to develop an analysis of the magnitude of the problem and the contributing causes. However, when even vital registration data are unavailable because of resource limitations, alternative methods must be developed. Facility-based audit, tracing the route taken by the injured or deceased women prior to arrival in hospital, and assessment of physical, sociocultural, and economic barriers may be useful. While these are essentially qualitative methodologies, the assessment of process indicators can be of assistance in defining the mix of interventions that are most likely to reduce maternal mortality and morbidity [165].

VI. Conclusion

Marked differences exist between developed and developing nations in the frequency and types of labor and delivery complications. Problems that are faced in organizing intrapartum services vary correspondingly. However, certain unifying principles cut across contexts and are key to the development of effective solutions. Two such principles are: (1) respect of women's perceptions and preferences should be central in planning services, and (2) perinatal technologies should be appropriate to each woman's needs. The Forteleza statement [7] reflects a broadening trend to reassess the underlying principles that serve as the basis for perinatal services planning and to affirm the central role of the woman and her family in planning and decision making during this period. For example, the Canadian Institute of Child Health has proposed "Family-Centred Principles" as the basis for planning perinatal services in Canada [166]. Similarly, the 1993 UK Department of Health Report of the Expert Maternity Working Group entitled "Changing Childbirth" emphasized the importance of women's exercising choices regarding their care and their need for accurate information to inform these choices [167]. Continuity is viewed as an essential element in women-centered maternity services. Choice, or in many cases lack of choice, is also the central issue for laboring women in developing countries. The vulnerability of a woman during her pregnancy, labor, and delivery frequently mirrors the subordinate status to which she is subjected during the previous phases of her life.

The implementation of women-centered labor and delivery services requires strong administrative structures at the national, regional, and local levels with leadership that is committed to providing choices to women. Leaders must be prepared to set priorities and to implement policies and programs that are consistent with care which is centered on women's needs. Effective leadership is essential in developing an organization that is responsive to the unique physical and psychosocial needs of women during the perinatal period. Quality assurance programs can have an important impact on complications of labor and delivery when there is collaboration between various levels of government, health planners, and health providers [168,169]. Skilled leadership is an essential tool in reducing these important sources of maternal morbidity and mortality.

References

1. Campbell, R., and Macfarlane, A. (1994). "Where to be Born?" National Perinatal Epidemiology Unit, Oxford.
2. Thacker, S. B., Stroup, D. F., and Peterson, H. B. (1995). Efficacy and safety of intrapartum electronic fetal monitoring: An update. *Obstet. Gynecol.* **86,** 613–620.
3. Morton, S., Williams, M., Keeler, E., Gambone, J., and Kahn, K. (1994). Effect of epidural analgesia for labor on the cesarean delivery rate. *Obstet. Gynecol.* **83,** 1045–1052.

4. Thorp, J., Hu, D., Albin, R., McNitt, J., Meyer, B., Cohen, G., and Yeast, J. (1993). The effect of intrapartum epidural analgesia on nulliparous labor: A randomized controlled prospective trial. *Am. J. Obstet. Gynecol.* **169,** 851–858.

5. Carolli, G., Belizan, J., and Stamp, G. (1999). Episiotomy policies in vaginal births (Cochrane Review). *In* "The Cochrane Library, Issue 1." Oxford, Update Software.

6. Labrecque, M., Baillargeon, L., Dallaire, M., Tremblay, A., Pineault, J.-J., and Gingras, S. (1997). Association between median episiotomy and severe perineal lacerations in primiparous women. *Can. Med. Assoc. J.* **156,** 797–802.

7. World Health Organization (1985). Appropriate technology for birth. *Lancet* **2** 436–437.

8. European Association of Perinatal Medicine Working Party (1996). Perinatal audit: A report for the European Association of Perinatal Medicine. *Prenat. Neonat. Med.* **1,** 160–194.

9. Peschers, U. M., Schaer, G. N., DeLancey, J. O. L., and Schuessler, B. (1997). Levator ani function before and after childbirth. *Br. J. Obstet. Gynaecol.* **104,** 1004–1008.

10. Frudinger, A., Bartram, C. I., Spencer, J. A. D., and Kamm, M. A. (1997). Perineal examination as a predictor of underlying external anal sphincter damage. *Br. J. Obstet. Gynaecol.* **104,** 1009–1013.

11. Sultan, A. H., Kamm, M. A., Hudson, C. N., Thomas, J. M., and Bartram, C. I. (1993). Anal sphincter disruption during vaginal delivery. *N. Engl. J. Med.* **329,** 1905–1911.

12. Arrowsmith, S., Hamlin, E. C., and Wall, L. L. (1996). Obstructed labor injury complex: Obstetric fistula formation and the multifaceted morbidity of maternal birth trauma in the developing world. *Obstet. Gynecol. Surv.* **51,** 568–574.

13. Bouvier-Colle, M. H., Varnoux, N., and Bréart, G. (1995). Maternal deaths and substandard care: The results of a confidential survey in France. Medical Experts Committee. *Eur. J. Obstet. Gynecol. Reprod. Biol.* **58,** 3–7.

14. Steinberg, W. M., and Farine, D. (1985). Maternal mortality in Ontario from 1970 to 1980. *Obstet. Gynecol.* **66,** 510–512.

15. Kwast, B. E., Rochat, R. W., and Kidane-Mariam, W. (1986). Maternal mortality in Addis Ababa, Ethiopia. *Stud. Fam. Plann.* **17,** 288–300.

16. Sultan, A. H., and Kamm, M. A. (1997). Faecal incontinence after childbirth. *Br. J. Obstet. Gynaecol.* **104,** 979–982.

17. Simon, J. (1858). "General Board of Health Papers Relating to the Sanitary State of England." HM Stationery Office, London.

18. Chalmers, I., Heatherington, J., Newdick, M., *et al.* (1986). The Oxford Database of Perinatal Trials: Developing a register of published reports of controlled trials. *Controlled Clin. Trials* **7,** 306–24.

19. Chalmers, I., Enkin, M., Keirse, M. J. N. C., Eds. (1989). "Effective Care in Pregnancy and Childbirth." Oxford University Press, Oxford.

20. Klein, M., Kaczorowski, J., Robbins, J., Gauthier, R., Jorgensen, S., and Joshi, A. (1995). Physicians' beliefs and behaviour during a randomized controlled trial of episiotomy: Consequences for women in their care. *Can. Med. Assoc. J.* **153,** 769–779.

21. Donner, A., and Klar, N. (1994). Methods for comparing event rates in intervention studies when the unit of allocation is a cluster. *Am. J. Epidemiol.* **140,** 279–289.

22. Prendiville, W. J., Elbourne, D., and McDonald, S. (1999). Active versus expectant management of the third stage of labour (Cochrane Review). *In* "The Cochrane Library, Issue 2." Oxford Update Software.

23. McDonald, S., Prendiville, W. J., and Elbourne, D. (1999). Prophylactic syntrometrine vs oxytocin in the third stage of labour (Cochrane Review). *In* "The Cochrane Library, Issue 2." Oxford Update Software.

24. Walder, J. (1997). Misoprostol: Preventing postpartum haemorrhage. *Mod. Midwife* **7,** 23–27.

25. Bamigboye, A. A., Merrell, D. A., Hofmeyr, G. J., and Mitchell, R. (1998). Randomized comparison of rectal misoprostol with Syntometrine for management of third stage of labor. *Acta Obstet. Gynecol. Scand.* **77,** 178–181.

26. Keski-Nisula, L., Kirkinen, P., Katila, M., Ollikainen, M., and Saarikoski, S. (1997). Cesarean delivery. Microbial colonization in amniotic fluid. *J. Reprod. Med.* **42,** 91–98.

27. Duff, P. (1987). Prophylactic antibiotics for cesarean delivery: A simple cost-effective strategy for prevention of postoperative morbidity. *Am. J. Obstet. Gynecol.* **157,** 794–798.

28. Bibi, M., Megdiche, H., Ghanem, H., Sfaxi, I., Nouira, M., Essaidi, H., Chaieb, A., Slama, A., and Khairi, H. (1994). L'antibioprophylaxie dans les césariennes a priori sans "haut risque infectieux". Expérience d'une maternité Tunisienne. *J. Gynecol. Obstet. Biol. Reprod.* **23,** 451–455.

29. Mahomed, K. (1988). A double blind randomized controlled trial on the use of prophylactic antibiotics in patients undergoing elective caesarean section. *Br. J. Obstet. Gynaecol.* **95,** 689–692.

30. Diouf, A., Faye, E., Moreira, P., Guisse, A., Sangare, M., Cisse, C., and Diadhiou, F. (1998). Emergency obstetrical hysterectomy. *Contraception Fertil. Sex.* **26,** 167–172.

31. Taha, T. E., Biggar, R. J., Broadhead, R. L., Mtimavalye, L. A., Justesen, A. B., Liomba, G. N., Chiphangwi, J. D., and Miotti, P. G. (1997). Effect of cleansing the birth canal with antiseptic solution on maternal and newborn morbidity and mortality in Malawi: Clinical trial. *Br. Med. J.* **315,** 216–219.

32. Cincotta, R., and Ross, A. (1996). A review of eclampsia in Melbourne: 1978–1992. *Aust. N. Z. J. Obstet. Gynaecol.* **36,** 264–267.

33. Rey, E., LeLorier, J., Burgess, E., Lange, I. R., and Leduc, L. (1997). Pharmacologic treatment of hypertensive disorders in pregnancy. *Can. Med. Assoc. J.* **157,** 1245–1254.

34. The Eclampsia Collaborative Group (1995). Which anticonvulsant for women with eclampsia? Evidence from the Collaborative Eclampsia Trial. *Lancet* **345,** 1455–1463.

35. Coetzee, E. J., Dommisse, J., and Anthony, J. (1998). A randomised controlled trial of intravenous magnesium sulphate versus placebo in the management of women with severe pre-eclampsia. *Br. J. Obstet. Gynaecol.* **105,** 300–303.

36. Burrows, R. F., and Burrows, E. A. (1995). The feasibility of a control population for a randomized control trial of seizure prophylaxis in the hypertensive disorders of pregnancy. *Am. J. Obstet. Gynecol.* **173,** 929–935.

37. Smith, G., and McEwan, H. (1997). Use of magnesium sulphate in Scottish obstetric units. *Br. J. Obstet. Gynaecol.* **104,** 155–156.

38. Chen, F., Chang, S., and Chu, K. (1995). Expectant management in severe preeclampsia: Does magnesium sulfate prevent the development of eclampsia? *Acta Obstet. Gynecol. Scand.* **74,** 852–853.

39. Witlin, A., Freidman, S., and Sibai, B. (1997). The effect of magnesium sulfate therapy on the duration of labor in women with mild preeclampsia at term: A randomized, double-blind, placebo-controlled trial. *Am. J. Obstet. Gynecol.* **176,** 623–627.

40. Lindheimer, M. D. (1996). Pre-eclampsia-eclampsia, 1996: Preventable? Have disputes on its treatment been resolved? *Curr. Opin. Nephrol. Hypertens.* **5,** 452–458.

41. Shiono, P. H., McNellis, D., and Rhoads, G. G. (1987). Reasons for the rising cesarean delivery rates: 1978–1984. *Obstet. Gynecol.* **96,** 696–700.

42. Taffel, S. M., Placek, P. J., and Liss, T. (1987). Trends in the United States cesarean section rate and reasons for the 1980–85 rise. *Am. J. Public Health* **77,** 955–959.

43. Taffel, S. M., Placek, P. J., Moien, M., and Kosary, C. L. (1991). 1989 U.S. cesarean section rate steadies—VBAC rate rises to nearly one in five. *Birth* **18,** 73–77.

44. Murray, S. F., and Serani, P. F. (1997). Cesarean birth trends in Chile, 1986 to 1994. *Birth* **24,** 258–263.

45. Onsrud, L., and Onsrud, M. (1996). Increasing use of cesarean section, even in developing countries. *Tidsskr. Nor. Laegeforen.* **116,** 67–71.

46. Clarke, S. C., and Taffel, S. (1995). Changes in cesarean delivery in the United States, 1988 and 1993. *Birth* **22,** 63–67.

47. Notzon, F. C., Cnattingius, S., Bergsjo, P., Cole, S., Taffel, S., Irgens, L., and Daltveit, A. K. (1994). Cesarean section delivery in the 1980s: International comparison by indication. *Am. J. Obstet. Gynecol.* **170,** 495–504.

48. van Ham, M. A., van Dongen, P. W., and Mulder, J. (1997). Maternal consequences of caesarean section. A retrospective study of intra-operative and postoperative maternal complications of caesarean section during a 10-year period. *Eur. J. Obstet. Gynecol.* **74,** 1–6.

49. Ali, Y. (1995). Analysis of caesarean delivery in Jimma Hospital, south-western Ethiopia. *East Afr. Med. J.* **72,** 60–63.

50. Garel, M., Lelong, N., and Kaminski, M. (1987). Psychological consequences of caesarean childbirth in primiparas. *J. Psychosom. Obstet. Gynaecol.* **6,** 197–209.

51. Turcot, L., Marcoux, S., Fraser, W. D., and The Canadian Early Amniotomy Study Group (1997). Multivarate analysis of risk factors for operative delivery in nulliparous women. *Am. J. Obstet. Gynecol.* **176,** 395–402.

52. Parrish, K. M., Holt, V. L., Easterling, T. R., Connell, F. A., and LoGerfo, J. P. (1994). Effect of changes in maternal age, parity, and birth weight distribution on primary cesarean delivery rates. *JAMA, J. Am. Med. Assoc.* **271,** 443–447.

53. Anderson, G., and Lomas, J. (1984). Determinants of the increasing cesarean birth rate. *N. Engl. J. Med.* **311,** 887–892.

54. Haynes de Regt, R., Minkoff, H. L., Feldman, J., and Schwarz, R. H. (1986). Relation of private or clinic care to the cesarean birth rate. *N. Engl. J. Med.* **315,** 619–624.

55. Goyert, G. L., Bottoms, S. F., Treadwell, M. C., and Nehra, P. C. (1989). The physician factor in cesarean birth rates. *N. Engl. J. Med.* **320,** 706–709.

56. Paterson, C. M., Chapple, J. C., Beard, R. W., Joffe, M., Steer, P. J., and Wright, C. S. W. (1991). Evaluating the quality of the maternity services—a discussion paper. *Br. J. Obstet. Gynaecol.* **98,** 1073–1078.

57. Cleary, R., Beard, R. W., Chapple, J., Coles, J., Griffin, M., Joffe, M., and Welch, A. (1996). The standard primipara as a basis for inter-unit comparisons of maternity care. *Br. J. Obstet. Gynaecol.* **103,** 223–229.

58. Wollast, E., Renard, F., and Vandenbussche, P. (1993). Research report: Detecting maternal morbidity and mortality by traditional birth attendants in Burkina Faso. *Health Policy Plann.* **8,** 161–168.

59. Khan, S., and Roohi, M. (1995). Obstructed labour: The preventable factors. *J. Pak. Med. Assoc.* **45,** 261–263.

60. Campbell, M., and Abu Sham, Z. (1995). Sudan: Situational analysis of maternal health in Bara District, North Kordofan. *World Health Stat. Q.* **48,** 60–66.

61. O'Driscoll, K., Jackson, R. J. A., and Gallagher, J. T. (1969). Prevention of prolonged labour. *Br. Med. J.* **24,** 477–480.

62. Akoury, H., MacDonald, F., Brodie, G., Caddick, R., Chaudhry, N., and Frize, M. (1991). Oxytocin augmentation of labour and perinatal outcome in nulliparas. *Obstet. Gynecol.* **78,** 227–230.

63. Turner, M. J., Brassil, M., and Gordon, H. (1988). Active management of labor associated with a decrease in the cesarean section rate in nulliparas. *Obstet. Gynecol.* **71,** 150–154.

64. O'Driscoll, K., and Meagher, D. (1980). "Active Management of Labour." Saunders, Philadelphia.

65. Frigoletto, F. D., Lieberman, E., Lang, J. M., Cohen, A., Barss, V., Ringer, S., and Datta, S. (1995). A clinical trial of active management of labor. *N. Engl. J. Med.* **333,** 745–750.

66. Fraser, W., Marcoux, S., Moutquin, J. M., Christen, A., and The Canadian Early Amniotomy Study Group (1993). Effect of early amniotomy on the risk of dystocia in nulliparous women. *N. Engl. J. Med.* **328,** 1145–1149.

67. Goffinet, F., Fraser, W., Marcoux, S., Bréart, G., Moutquin, J.-M., Daris, M., and The Canadian Early Amniotomy Study Group (1997). Early amniotomy increases the frequency of fetal heart rate abnormalities. *Br. J. Obstet. Gynaecol.* **104,** 548–553.

68. Brisson-Carrol, G., Fraser, W., Bréart, G., Krauss, I., and Thornton, J. (1996). The effect of routine early amniotomy on spontaneous labor: A meta-analysis. *Obstet. Gynecol.* **87,** 891–896.

69. World Health Organization (1997). "Amniotomy to Shorten Spontaneous Labour." WHO Reproductive Health Library, Washington, DC.

70. Fraser, W., Krauss, I., Brisson-Carrol, G., Thornton, J., and Bréart, G. (1995). Amniotomy to shorten spontaneous labour. *In* "Pregnancy and Childbirth Module of the Cochrane Database of Systematic Reviews [updated 01 September 1997]. Available in The Cochrane Library [database on disk and CDROM]. "The Cochrane Collaboration, Issue 4" (J. Neilson, C. Crowther, E. Hodnett, and G. Hofmeyr, eds.). Update Software, Oxford (updated quarterly).

71. Fraser, W., Venditelli, F., Krauss, I., and Bréart, G. (1998). Effects of early augmentation of labour with amniotomy and oxytocin in nulliparous women: A meta-analysis. *Br. J. Obstet. Gynaecol.* **105,** 189–194.

72. Bréart, G., Mlika-Cabane, N., Kaminski, M., Alexander, S., Herruzo-Nalda, A., Mandruzzato, P., Thornton, J., and Trakas, D. (1992). Evaluation of different policies for the management of labour. *Early Hum. Dev.* **29,** 309–312.

73. Labrecque, M., Brisson-Carrol, G., Fraser, W., and Plourde, D. (1994). Evaluation of obstetrical labour augmentation with oxytocin during the latent phase: A pilot study. Unpublished paper.

74. Hemminki, E., Lenck, M., Saarikoski, S., and Henriksson, L. (1985). Ambulation vs oxytocin in protracted labour: a pilot study. *Eur. J. Obstet. Gynecol. Reprod. Biol.* **20,** 199–208.

75. Philpott, R. H., and Castle, W. M. (1972). Cervicographs in the management of labour in primigravidae: II. the action line and treatment of abnormal labour. *J. Obstet. Gynaecol. Br. Commonw.* **79,** 599–602.

76. Philpott, R. H., and Castle, W. M. (1972). Cervicographs in the management of labour in primigravidae: I. the alert line for detecting abnormal labour. *J. Obstet. Gynaecol. Br. Commonw.* **79,** 592–598.

77. World Health Organization (1993). "The Partograph. A Managerial Tool for the Prevention of Prolonged Labour. Section I: The Principle and Strategy. Section II: A User's Manual." WHO, Geneva.

78. Kwast, B. E. (1994). World Health Organization partograph in management of labour. *Lancet* **343,** 1399–1404.

79. Boucher, M., Tahilramaney, M. P., Eglinton, G. S., and Phelan, J. P. (1984). Maternal morbidity as related to trial of labor after previous cesarean delivery: A quantitative analysis. *J. Reprod. Med.* **29,** 12–16.

80. Shy, K. K., Logerto, J. P., and Karp, L. E. (1981). Evaluation of elective repeat cesarean section as a standard of care: An application of decision analysis. *Am. J. Obstet. Gynecol.* **139,** 123–129.

81. American College of Obstetricians and Gynecologists Committee on Obstetrics: Maternal and Fetal Medicine (1984). "Guidelines for Vaginal Delivery after a Previous Cesarean Birth." ACOG, Washington, DC.

82. Society of Obstetrician-Gynecologists of Canada (1986). Indications for cesarean section: Final statement of the panel of the

National Consensus Conference on Aspects of Cesarean Birth. *Can. Med. Assoc. J.* **134,** 1348–1352.

83. McMahon, M. J., Luther, E. R., Bowes, W. A., and Olshan, A. F. (1996). Comparison of a trial of labor with an elective second cesarean section. *N. Engl. J. Med.* **335,** 689–995.

84. Paul, R. H. (1996). Toward fewer cesarean sections—the role of a trial of labor. *N. Engl. J. Med.* **335,** 735–736.

85. McClain, C. S. (1987). Patient decision making: The case of delivery method after a previous cesarean section. *Cult. Med. Psychiatry* **11,** 495–508.

86. Fraser, W., Maunsell, E., Hodnett, E., Moutquin, J.-M., and The Childbirth Alternatives Postcesarean Group (1997). Randomized controlled trial of a prenatal vaginal birth after cesarean section education and support program. *Am. J. Obstet. Gynecol.* **176,** 419–425.

87. Boulvain, M., Fraser, W. D., Brisson-Carroll, G., Faron, G., and Wollast, E. (1997). Trial of labour after caesarean section in sub-Saharan Africa: A meta-analysis. *Br. J. Obstet. Gynaecol.* **104,** 1385–1390.

88. Thacker, S. (1987). The efficacy of intrapartum electronic fetal monitoring. *Am. J. Obstet. Gynecol.* **156,** 24–30.

89. MacDonald, D., Grant, A., Sheridan-Pereira, M., Boylan, P., and Chalmers, I. (1985). The Dublin randomized controlled trial of intrapartum fetal heart rate monitoring. *Am. J. Obstet. Gynecol.* **152,** 524–539.

90. Grant, A., O'Brien, N., Joy, M. T., Hennessy, E., and MacDonald, D. (1989). Cerebral palsy among children born during the Dublin randomized trial of intrapartum monitoring. *Lancet* **2,** 1233–1235.

91. Society of Obstetricians and Gynaecologists of Canada (1995). Fetal health surveillance in labour. *J SOGC* **17,** 859–900.

92. Hellman, L. M., and Prystowsky, M. (1952). The duration of the second stage of labor. *Am. J. Obstet. Gynecol.* **63,** 1223–1233.

93. Reynolds, F. (1989). Epidural analgesia in obstetrics: Pros and cons for mother and baby. *Br. Med. J.* **299,** 751–752.

94. Moon, J., Smith, C., and Rayburn, W. (1990). Perinatal outcome after a prolonged second stage of labor. *J. Reprod. Med.* **35,** 229–231.

95. Cohen, W. (1977). Influence of the duration of second stage labor on perinatal outcome and puerperal morbidity. *Obstet. Gynecol.* **49,** 266–269.

96. Maresh, M., Choong, K., and Beard, R. (1983). Delayed pushing with lumbar epidural analgesia in labour. *Br. J. Obstet. Gynaecol.* **90,** 623–627.

97. Guillemette, J., and Fraser, W. (1992). Differences between obstetricians in caesarean section rates and the management of labour. *Br. J. Obstet. Gynaecol.* **99,** 105–108.

98. Cyr, R., Usher, R., and McLean, F. (1984). Changing patterns of birth asphyxia and trauma over 20 years. *Am. J. Obstet. Gynecol.* **148,** 490–498.

99. Hagadorn-Freathy, A., Yeomans, E., and Hankins, G. (1991). Validation of the 1988 ACOG forceps classification system. *Obstet. Gynecol.* **77,** 356–360.

100. Johanson, R. B., and Menon, V. J. (1998). Vacuum extraction vs forceps delivery (Cochrane Review). *In* "The Cochrane Library, Issue 2." Update Software, Oxford (updated quarterly).

101. Howell, C. J. (1998). Epidural vs non-epidural analgesia in labour. (Cochrane Review). *In* "The Cochrane Library, Issue 2." Update Software, Oxford (updated quarterly).

102. Studd, J., Crawford, J., Duignan, N., Rowbotham, C., and Hughes, A. (1980). The effect of lumbar epidural analgesia on the rate of cervical dilatation and the outcome of labour of spontaneous onset. *Br. J. Obstet. Gynaecol.* **87,** 1015–1021.

103. Kaminski, H., Stafl, A., and Aiman, J. (1987). The effect of epidural analgesia on the frequency of instrumental obstetric delivery. *Obstet. Gynecol.* **69,** 770–773.

104. Thorp, J. A., Parisi, V. M., Boylan, P. C., and Johnston, D. A. (1989). The effect of continuous epidural analgesia on cesarean section for dystocia in nulliparous women. *Am. J. Obstet. Gynecol.* **161,** 670–675.

105. Thorp, J., Hu, D., Albin, R., McNitt, J., Meyer, B., Cohen, G., and Yeast, J. (1993). The effect of intrapartum epidural analgesia on nulliparous labor: A randomized, controlled, prospective trial. *Am. J. Obstet. Gynecol.* **169,** 851–858.

106. Philipsen, T., and Jensen, N. H. (1990). Maternal opinion about analgesia in labour and delivery. A comparison of epidural blockade and intramuscular pethidine. *Eur. J. Obstet. Gynecol. Reprod. Biol.* **34,** 205–210.

107. Ramin, S. M., Gambling, D. R., Lucas, M. J., Sharma, S. K., Sidawi, J. E., and Leveno, K. J. (1995). Randomized trial of epidural versus intravenous analgesia during labor. *Obstet. Gynecol.* **86,** 783–789.

108. Sharma, S. K., Sidawi, J. E., Ramin, S. M., Lucas, M. J., Leveno, K. J., and Cunningham, F. G. (1997). Cesarean delivery: A randomized trial of epidural versus patient-controlled meperidine analgesia during labor. *Anesthesiology* **87,** 487–494.

109. Goodfellow, C., Hull, M., Swaab, D., Dogterom, J., and Buijs, R. (1983). Oxytocin deficiency at delivery with epidural analgesia. *Br. J. Obstet. Gynaecol.* **90,** 214–219.

110. Bates, R., Helm, C., Duncan, A., and Edmonds, D. (1985). Uterine activity in the second stage of labour and the effect of epidural analgesia. *Br. J. Obstet. Gynaecol.* **92,** 1246–1250.

111. Saunders, N. J., Spiby, H., Gilbert, L., Fraser, R. B., Hall, J. M., Mutton, P. M., Jackson, A., and Edmonds, D. K. (1989). Oxytocin infusion during second stage of labour in primiparous women using epidural analgesia: A randomised double blind placebo controlled trial. *Br. Med. J.* **299,** 1423–1426.

112. Chestnut, D., Vandewalker, G., Owen, C., Bates, J., and Choi, W. (1987). The influence of continuous epidural bupivacaine analgesia on the second stage of labor and method of delivery in nulliparous women. *Anesthesiology* **66,** 774–780.

113. Phillips, K., and Thomas, T. (1983). Second stage of labour with or without extradural analgesia. *Anaesthesia* **38,** 972–976.

114. Chestnut, D. H., Laszewski, L. J., Pollack, K. L., Bates, J. N., Manago, N. K., and Choi, W. W. (1987). Continuous epidural infusion of 0.0626% bupivacaine-0.0002% fentenyl during the second stage of labour. *Anesthesiology* **72,** 613–618.

115. Chestnut, D., Owen, C., Bates, J., Ostman, L., Choi, W., and Geiger, M. (1988). Continuous infusion epidural analgesia during labor: A randomized, double-blind comparison of 0.0625% bupivacaine/0.0002% fentanyl versus 0.125% bupivacaine. *Anesthesiology* **68,** 754–759.

116. Youngstrom, P., Eastwood, D., Patel, H., Bhatia, R., Cowan, R., and Sutheimer, C. (1984). Epidural fentanyl and bupivacaine in labor: double blind Study. *Anesthesiology* **61,** 3A.

117. Westmore, M. (1990). Epidural opioids in obstetrics—A review. *Anaesth. Intensive Care* **18,** 292–312.

118. Celleno, D., and Capogna, G. (1988). Epidural fentanyl plus bupivacaine 0.125 per cent for labour: analgesic effects. *Can. J. Anaesth.* **35,** 375–378.

119. Vertommen, J., Vandermeulen, E., Van Aken, H., Vaes, L., Soetens, M., Van Steenburge, A., Mourisse, P., Willaert, J., Noorduin, H., Devilieger, H., and Van Assche, A. (1991). The effects of the addition of sufentanil to 0.125% bupivicaine on the quality of analagesia during labor and on the incidence of instrumental deliveries. *Anesthesiology* **74,** 809–814.

120. Fraser, W. D., Marcoux, S., Douglas, J., Goulet, C., Krauss, I., and the PEOPLE Study Group (1997). *Acta. Obstet. Gynecol. Scan,* **76** (Suppl. 167:1), 45.

121. Ould, F. (1741). "A Treatise of Midwifery," pp. 145–146. Buckland, London.

122. Lede, R. L., Belizan, J. M., and Carroli, G. (1996). Is routine use of episiotomy justified? *Am. J. Obstet. Gynecol.* **174,** 1399–1402.

123. Graham, I. D., and Graham, D. F. (1997). Episiotomy counts: Trends and prevalence in Canada, 1981/1982 to 1993/1994. *Birth* **24,** 141–147.

124. Sultan, A. H. (1997). Anal incontinence after childbirth. *Curr. Opini. Obstet. Gynecol.* **9,** 320–324.

125. MacArthur, C., Bick, D. E., and Keighley, M. R. B. (1997). Faecal incontinence after childbirth. *Br. J. Obstet. Gynaecol.* **104,** 46–50.

126. Shipman, M. K., Boniface, D. R., Tefft, M. E., and McCloghty, F. (1997). Antenatal perineal massage and subsequent perineal outcomes: A randomised controlled trial. *Br. J. Obstet. Gynaecol.* **104,** 787–791.

127. Labrecque, M., Eason, E., Marcoux, S., Lemieux, F., Pineault, J.-J., Feldman, P., and Laperrière, L. (1999). Randomized controlled trial of prevention of perineal trauma by perineal massage during pregnancy. *Am. J. Obstet. Gynecol.* **180,** 593–600.

128. Householder, M. (1974). A historical perspective on the obstetric chair. *Surg. Gynecol. Obstet.* **139,** 423.

129. Caldeyro-Barcia, R., Noriega Guerra, L., and Cibils, L. (1960). Effect of position changes on the intensity and frequency of uterine contractions during labour. *Am. J. Obstet. Gynecol.* **80,** 284–290.

130. Mendez-Bauer, C., Arroyo, J., Garcia Ramos, C., Menendez, A., Lavilla, M., Izquierdo, F., Villa Elizaga, I., and Zamarriego, J. (1975). Effects of standing position on spontaneous uterine contractility and other aspects of labour. *J. Perinat. Med.* **3,** 89–100.

131. Roberts, J., Mendez Bauer, C., and Wodell, D. (1983). The effects of maternal position on uterine contractility and efficiency. *Birth* **4,** 243–249.

132. Caldeyro-Barcia, R. (1979). The influence of maternal position on time of spontaneous rupture of the membranes progress of labour and fetal head compression. *Birth Fam. J.* **6,** 7–15.

133. Diaz, A., Schwarcz, R., Fescina, R., and Caldeyro-Barcia, R. (1980). Vertical position during the first stage of the course of labour and neonatal outcome. *Eur. J. Gynecol. Reprod. Biol.* **11,** 1–7.

134. Flynn, A., Kelly, J., Hollins, G., and Lynch, P. (1978). Ambulation in labour. *Br. Med. J.* **2,** 591–593.

135. Liu, Y. C. (1974). Posture in labor: A comparison of 30 women who lay in a semi-upright position with 30 women who lay flat, in terms of their uterine contractions and duration of labor, and the Apgar scores of their infants. *Am. J. Nurs.* **74,** 2203–2205.

136. Mitre, I. (1974). The influence of maternal position on duration of the active phase of labour. *Int. J. Obstet. Gynecol.* **12,** 181–183.

137. Calvert, J., Newcombe, R., and Hibbard, B. (1982). An assessment of radiotelemetry in the monitoring of labour. *Br. J. Obstet. Gynaecol.* **89,** 285–291.

138. Chan, D. (1963). Positions during labour. *Br. Med. J.* **1,** 100–102.

139. McManus, T., and Calder, A. (1978). Upright position and the efficiency of labour. *Lancet* **1,** 72–74.

140. Williams, R., Thom, M., and Studd, J. (1980). A study of the benefits and acceptability of ambulation in spontaneous labour. *Br. J. Obstet. Gynaecol.* **87,** 122–126.

141. Roberts, J. (1989). Maternal position during the first stage of labor. *In* "Effective Care in Pregnancy and Childbirth" (I. Chalmers, M. Enkin, and M. Keirse, eds.), pp. 883–892. Oxford University Press, Oxford.

142. Sleep, J., Roberts, J., and Chalmers, I. (1989). Care during the second stage of labour. *In* "Effective Care in Pregnancy and Childbirth" (I. Chalmers, M. Enkin, and M. Keirse, eds.), Vol. 2, Part 8, pp. 1129–1144. Oxford University Press, Oxford.

143. Gardosi, J., Hutson, N., and Lynch, C. B. (1989). Randomised controlled trial of squatting in the second stage of labour. *Lancet* **2** 74–77.

144. Stewart, P., and Spiby, H. (1989). A randomized study of the sitting position for delivery using a newly designed obstetric chair. *Br. J. Obstet. Gynaecol.* **96,** 327–333.

145. Crowley, P., Elbourne, D., Ashurst, H., Garcia, J., Murphy, D., and Duignan, N. (1991). Delivery in an obstetric birth chair: A randomized controlled trial. *Br. J. Obstet. Gynaecol.* **98,** 667–674.

146. Hodnett, E. D. (1994). Support from caregivers during childbirth (Cochrane Review). *In* "The Cochrane Library, Issue 2." Update Software, Oxford (updated quarterly).

147. Weston, L. (1986). Reducing maternal deaths in developing countries. *Population, Health and Nutrition Technical Note 86-19, World Bank,* pp. 1–69.

148. Kwast, B. E., Roads to maternal death case histories including comments on preventive strategies. Personal communication.

149. Enkin, M., Keirse, M., Renfrew, M., and Neilson, J. (1995). "A Guide to Effective Care in Pregnancy and Childbirth." Oxford University Press, Oxford.

150. Flint, C., Poulengeris, P., and Grant, A. (1989). The "Know Your Midwife' scheme—a randomised trial of continuity of care by a team of midwives. *Midwifery* **5,** 11–16.

151. MacVicar, J., Dobbie, G., Owen-Johnstone, L., Jagger, C., Hopkins, M., and Kennedy, J. (1993). Simulated home delivery in hospital: A randomised controlled trial. *Br. J. Obstet. Gynaecol.* **100,** 316–323.

152. Rowley, M. J., Hensley, M. J., Brinsmead, M. W., and Wlodarczyk, J. H. (1995). Continuity of care by a midwife team versus routine care during pregnancy and birth: A randomised trial. *Med. J. Aust.* **163,** 289–293.

153. Harvey, S., Jarrel, J., Brant, R., Stainton, C., and Rach, D. (1996). A randomized, controlled trial of nurse-midwifery care. *Birth* **23,** 128–135.

154. Turnbull, D., Holmes, A., Shields, N., Cheyne, H., Twaddle, S., Gilmour, W. H., McGinley, M., Reid, M., Johnstone, I., Geer, I., McIlwaine, G., and Lunan, C. B. (1996). Randomised, controlled trial of efficacy of midwife-managed care. *Lancet* **348,** 213–218.

155. Waldenström, U., Nilsson, C.-A., and Winbladh, B. (1997). The Stockholm Birth Centre Trial: Maternal and infant outcome. *Br. J. Obstet. Gynaecol.* **104,** 410–418.

156. DerSimonian, R., and Laird, N. (1986). Meta-analysis in clinical trials. *Controlled Clin. Trials* **7,** 177–188.

157. Rooks, J. P., Weatherby, N. L., and Ernst, E. K. M. (1992). The national Birth Center Study. Part III. Intrapartum and immediate postpartum and neonatal care, outcomes, and client satisfaction. *J. Nurse-Midwifery* **37,** 361–397.

158. Levitt, C., Hanvey, A. D., Chance, G., and Kaczorowski, J. (1995). "Survey of Routine Maternity Care and Practices in Canadian Hospitals." Health Canada and Canadian Institute of Child Health, Ottawa.

159. Kantor, H., Rember, R., Tabio, P., and Buchanon, R. (1965). Value of shaving the pudendal area in delivery preparation. *Obstet. Gynecol.* **25,** 509–512.

160. Romney, M., and Gordon, H. (1981). Is your enema really necessary? *Br. Med. J.* **282,** 1269–1271.

161. Lomas, J., Anderson, G., Enkin, M., Vayda, E., Roberts, R., and MacKinnon, B. (1988). The role of evidence in the consensus process. Results from a Canadian consensus exercise. *JAMA, J. Am. Med. Assoc.* **259,** 3001–3005.

162. Rogers, E. M. (1995). Lessons for guidelines from the diffusion of innovations. *J. Qual. Improvement* **21,** 324–328.

163. Lomas, J., Enkin, M., Anderson, G., Hannah, W., Vayda, E., and Singer, J. (1991). Opinion leaders vs audit and feedback to implement practice guidelines: Delivery after previous cesarean section. *JAMA, J. Am. Med. Assoc.* **265,** 2202–2207.

164. Thomson, M., Oxman, A., Haynes, R., Davis, D., Freemantle, N., and Harvey, E. (1998). Local opinion leaders to improve health

professional practice and health care outcomes. (Cochrane Review).). *In* "The Cochrane Library, Issue 2." Update Software, Oxford (updated quarterly).

165. AbouZahr, C., Wardlaw, T., Stanton, C., and Hill, K. (1996). Maternal mortality. *World Health Stat. Q.* **49,** 77–87.

166. Health Canada (In press). Family centered maternity and newborn care: National guidelines (2nd ed). Government of Canada, Ottawa.

167. Department of Health. (1993). "Changing Childbirth. Part I. Report of the Expert Maternity Group." HM Stationery Office, London.

168. Omaswa, F., Burnham, G., Baingana, G., Mwebesa, H., and Morrow, R., (1997). Introducing quality management into primary health care services in Uganda. *Bull. W.H.O.* **75,** 155–161.

169. Gates, P. E. (1995). Think globally, act locally: An approach to implementation of clinical practice guidelines. *J. Qual. Improv.* **21,** 71–85.

16

Infertility

MARLENE B. GOLDMAN,* STACEY A. MISSMER,* AND ROBERT L. BARBIERI†

*Department of Epidemiology, Harvard School of Public Health, Boston, Massachusetts; †Department of Obstetrics, Gynecology and Reproductive Biology, Brigham and Women's Hospital, Boston, Massachusetts

I. Introduction and Background

A. Definition of Infertility

In epidemiologic research infertility is often defined as the inability to conceive after twelve months of unprotected sexual intercourse [1]. Twelve months is derived from the biological and clinical observations that about 90% of noncontracepting couples of normal fertility will conceive within a year [2]. Although biologically supported, this definition is arbitrary and other definitions extend the trying time to twenty-four months or shorten it to six months, depending on the age group and population under study.

Related terms used by clinicians, epidemiologists, and demographers are: *infecundity*—a synonym for infertility; *sterility*—the complete absence of reproductive capability; *fertility*—the demonstrated capacity to conceive; *fecundity*—the physiological ability to reproduce, whether or not it has been demonstrated; and *fecundability*—the monthly probability of conception. "Normal" fecundability in a healthy young couple is approximately 0.25, that is, a 25% probability of conception per month.

Infertility is further classified as primary or secondary and by pathologic type because the etiologic factors for each may differ. Primary infertility describes a couple who has attempted, but never achieved, conception. Secondary infertility arises after having conceived at least once, regardless of the outcome, but being unable to conceive subsequently. After clinical assessment, a couple's infertility may be attributed to one or more of the following pathologies: ovulatory dysfunction, tubal occlusion or adhesions, endometriosis, uterine and/or cervical factors, or male factor. Infertility is often a multifactorial condition with more than one contributing cause. For example, a woman may have both a tubal factor and endometriosis or a woman may have ovulatory dysfunction and her partner may have oligospermia. Couples whose fertility evaluation identifies no abnormalities but who are unable to conceive are said to have idiopathic, or unexplained, infertility.

B. Prevalence of Infertility

In industrialized nations, 10–15% of couples are estimated to experience either primary or secondary infertility. Half of these couples will not succeed in having as many children as they desire [3]. Another 1% of women will suffer recurrent pregnancy losses and have no live-born children [4]. In global estimates of the number of infertile men and women, the World Health Organization (WHO) reported that 8% of couples, or approximately 50–80 million people worldwide, experienced some form of infertility [5]. Sub-Saharan African countries are reported to have the highest prevalence, with estimates ranging from 30 to 50% of women [6–8]. A study of Somali women found that 7% of the married women over age 45 were primarily infertile and an additional 20% were subfertile, that is they had fewer children than they wished [9]. In a population-based survey in Norway, Sundby and Schei reported primary infertility in 2.6% of women aged 40–42 and subfertility in another 7.7% [10]. They concluded that up to 10% of Norwegian women experienced infertility at some time during their reproductive years.

In the United States, the percentage of infertile women increased from 8.4% in the 1982 and 1988 cycles of the National Survey of Family Growth (NSFG) to 10.2% in the 1995 cycle [11]. This is equivalent to 6.2 million infertile women in the United States in 1995. Stephen and Chandra project that between 5 and 6.3 million U.S. women will be infertile in the year 2000, increasing to 5.4–7.7 million in 2025 [11]. Percentages of infertile women increased uniformly for all women of reproductive age but were highest for women who were aged 35–44. The greater increase in older age groups reflects the fact that more women delayed childbearing until an age when their fertility declined. The authors attribute the increase in the number of infertile women that was seen in all age groups to three factors: (1) heightened public awareness of infertility which may have increased the likelihood of self-report; (2) an increased prevalence of unrecognized pelvic inflammatory disease secondary to gonorrhea or chlamydia infection; and (3) a decline in hysterectomy rates that led to the classification of women as infertile who in earlier years would have been classified as surgically sterile.

The distribution of specific types of infertility varies in different populations and from study to study, depending on methods for diagnosing, classifying, and treating patients. In addition, there are differences in the extent to which each couple undergoes evaluation and in the selection of patients for treatment, as well as differences in length of patient follow-up.

A WHO task force studied patients from thirty-three medical centers in twenty-five countries including Eastern and Western Europe, Canada, Australia, Scandinavia, Africa, Asia, Latin America, and the Mediterranean [12]. Among the developed countries, ovulatory disorders were diagnosed in 33%, tubal occlusion or adhesions in 36%, and endometriosis in 6% of the infertile women. In 40% of women, no demonstrable cause was found. (Percentages add to more than 100% because patients could have more than one diagnosis.) In men, varicocele accounted for 11% of male factor infertility, with infection or sperm disorders accounting for another 28%. In 49% of the cases of male infertility there was no demonstrable cause and in 12% the cause was not available. Among women, the distribution of causes was similar for developing countries in Asia, Latin America, and the Eastern Mediterranean, with the exception of endometriosis, which was higher in Asia (10%). However, the pattern differed in Africa where over 85% of the infertile women had diagnoses such as tubal occlusion or adhesions that were attributable to infection. For men, the percentages of infertility due to varicocele and infection were higher in Africa and Latin America.

In a review of 21 published reports that included more than 14,000 infertile couples, Collins reported that the primary diagnoses in the infertile couples were disorders of oocyte production including anovulation or oligoovulation (27%), disorders of sperm production such as low sperm concentration or low sperm motility (25%), tubal defects (22%), endometriosis (5%), uterine, cervical and other factors (4%), and unexplained infertility (17%) [13]. Male factor contributes to infertility 50% of the time, as the primary diagnosis in 25% of couples and as a contributing factor in an additional 25% [14].

C. Emotional Impact

Because infertility is unanticipated, often unexplained, and may continue for an indeterminate length of time, a couple's difficulty conceiving creates a stressful situation that may tax existing coping mechanisms [15]. During the process of infertility diagnosis and treatment, many couples experience emotions that range from surprise to denial to anger [16–18]. Feelings of guilt can affect the couple's relationship, especially if one partner feels at fault due to past experiences such as choice of contraception, induced abortion, occurrence of sexually transmitted infection, or sexual practices [16–18].

The socialized desire to have children transcends sex, age, race, religion, and social class [15]. While cultures exert considerable societal pressure on couples to reproduce, the emotional burden is imposed differently upon men and women. Women, who have been socialized to view motherhood as their primary adult role, may find it difficult to move past the belief that a "family" includes children [17,19–21]. One review reported that while 50% of infertile women considered infertility to be the most distressing experience of their lives, only 15% of infertile men did [18]. Whiteford and Gonzalez [15] suggest that "the stigma attached to infertility rests not on the perceptions of a physical deformity but on the sense of having broken a group norm." They propose that the perception of a woman's other accomplishments is affected, transforming biological infertility into socially defined inadequacy. Whether it is the man or the woman who is physically responsible for the infertility, it is the woman's identity that is affected.

Draye et al. [19] surveyed 26 couples receiving infertility treatment and found that women were more likely than men to report having decreased self-esteem (65% vs 22%), to feel that they were being punished (49% vs 12%), and as if life's accomplishments were less important without children (46% vs 15%). A study of 185 couples recruited through infertility specialists found that, at least for women who were seeking medical treatment for their infertility, the importance of children, the number of tests received, and attributions of responsibility to the physician were significant predictors of increased stress [20]. In contrast, for men the number of physicians seen, future treatment costs, and household income predicted stress.

Infertility treatment may last for as little as six months or continue for years, with couples experiencing cycles of hope, frustration, and disappointment each month [22,23]. When all thoughts and actions revolve around conceiving, the onset of menstruation triggers a crushing sense of failure [24]. Litt et al. [25] found that 17% of women experienced emotions severe enough to be defined as clinical depression one to two weeks after treatment failure.

Couples undergoing in vitro fertilization (IVF) cycles experience significant stress. In one study, the female partner of the couple reported increased stress associated with ovarian stimulation [26]. In addition, stress levels increased markedly when couples were informed that a pregnancy test obtained two weeks after embryo transfer was negative, but not when they were told that the pregnancy test was positive. The failure to conceive in an IVF cycle is a loss for which the couple must grieve.

For couples found to be irreversibly infertile, the grieving process may be compounded by a lack of social support systems and the struggle to express feelings regarding the inadequately recognized loss of a potential rather than an actual person. Couples may not feel entitled to grieve and may have drifted away from friends and family with children. Pressure from physicians or family members to adopt may prompt a couple to begin adoption proceedings prematurely, without bringing closure to the intense physical and emotional process of dealing with their infertility and before they are prepared to embark on the equally costly and emotional process of adoption [16]. Couples who adopt a child may continue to express feelings of loss and distress associated with their inability to have a biological child [17].

D. Access to and Use of Medical Services

Wilcox and Mosher reported that the number of U.S. couples who sought medical advice or treatment for infertility increased by 25% from 1982 to 1988 [27,28]. The relative share of infertility-related visits among all physician visits increased from 69.5/100,000 visits in 1968 to 101.9/100,000 visits in 1980 (Table 16.1) [4,29–30]. As knowledge of treatment options has increased, couples are more likely to seek medical advice [29]. In addition, the number of physicians who are trained to provide these services has increased dramatically, improving access to treatment and the frequency of referral [29]. However, costs and insurance coverage have a direct effect on the use of services. In 1993 in France, the average couple paid 7% of the cost of IVF treatment, whereas couples in the United States were responsible for 85% of the cost. Consequently, the number of IVF cycles initiated per capita in France was five times that in the United States [31].

When couples are careful about the timing of pregnancies and choose to delay childbearing, the interval of desired reproduction is condensed. The result is increased pressure to conceive in a shorter amount of time, lowering the threshold for seeking intervention [29]. A study of use of services in five European countries found that the percentage of women who sought consultation after just 6 to 11 months of intercourse

Table 16.1

Factors Contributing to the Increasing Request for Infertility Services

Increasing age-specific infertility rates?
Condensing of childbearing into a shorter interval
Aging of the baby-boom generation
Delayed childbearing (greater biologic infertility
    due to aging or lengthened exposure to external
    risk factors)
Delayed conception due to oral contraceptive use

    → Greater absolute number of
       infertile couples

Heightened expectations and awareness of treatment
    options
Increasing financial ability to obtain health care

    → Increased proportion of couples
       seeking medical care

Decline in the demand for obstetric services
    (baby bust)
Greater demand from private patients
More sophisticated diagnosis and treatment

    → Increased number of service
       providers

    ⇒ Increased request
       for services

Greater acceptance of medical intervention
Heightened expectations of fertility control
Profamily movement (backlash against the sexual
    revolution)
Movement of the family planning industry toward
    more comprehensive care

    → More conducive social milieu

Adapted with permission from [29] (*JAMA*, 1983, **250**, 2327–2331) and [29a].

without contraception ranged from 15% in Poland to 36% in Denmark [32].

Even so, the proportion of infertile couples who seek medical advice or treatment remains relatively low, with estimates of 31–48% for women who had been attempting to conceive for greater than one year [30,33–35]. In an analysis of Danish women who had unprotected intercourse for greater than two years, the proportion of women who used health care services for infertility remained below 50% [36]. Ten years later, a study conducted in Scotland found that a higher proportion—81% of women who had attempted to conceive for greater than two years—sought medical services from either a general practitioner or a hospital [37].

Age cohort has been found to predict use of infertility services. Women of younger age cohorts are more likely to seek medical advice or treatment than are women of older cohorts [27,30,33]. In a study by Gunnell and Ewings, 59% of women currently between the ages of 36 and 40 had sought care compared to only 33% of women now aged 46 to 50 [33]. Templeton and colleagues in Scotland reported similar findings: 89% of younger women (now ages 36 to 40) had sought care compared to 69% of those who were now older (ages 46 to 50) [37]. There are, however, two Danish studies that found no significant relationship between age cohort and use of services [34,36].

Couples with primary infertility are more likely to seek medical intervention than couples with secondary infertility. Gunnell and Ewings found that 50% of women with primary infertility sought help from a general practitioner, whereas only 34% of those with secondary infertility did [33]. In the United

States, an analysis of data from the 1982 NSFG found that 51% of women with primary infertility had ever sought services as compared to 22% of those with secondary infertility [30]. In these data, white race, education beyond high school, prior treatment for pelvic inflammatory disease, younger age at menarche, high level of income, and having a husband with a skilled, non-manual occupation were predictive of service-seeking among those with secondary infertility, but not among those with primary infertility. Prior use of oral contraception, shorter duration of marriage, and younger age were predictive of service-seeking whether infertility was primary or secondary [30].

In addition to age cohort and classification of infertility, researchers have evaluated the relationship between the use of health care and race, income, education, occupation, marital status, parity, geographic region, and reproductive pathology. In an analysis of the 1988 NSFG, Wilcox and Mosher found that women who were white, had a history of endometriosis, were college graduates, had a high income, or were currently married were more likely to seek infertility services [27]. However, the same survey found that married women who had not conceived after one year of unprotected intercourse were more likely to be older, black, have a high school education, a lower income, and less skilled occupations. Black women were half as likely and low-income women were one-third as likely to use specialized services such as ovulation induction drugs, artificial insemination, IVF, or surgery.

Not surprisingly, low socioeconomic status has consistently been found to be a deterrent to seeking infertility services [27,30,33,38]. As a result, if economic inequality continues to widen, the proportional demand for infertility services may decrease, regardless of improved prognoses or geographic expansion of health care availability [31].

Few studies have addressed the influence of religion, sexual orientation, or cultural mores on the use of medical care for infertility [39]. Research on barriers to access is needed to elucidate those aspects of medical care—be it current treatment protocols, geographic availability of services, financial impediments, personal preference, or other reasons—that prevent more than 50% of affected couples from seeking care.

II. Clinical Presentation

Successful initiation of pregnancy requires the ovulation of a mature oocyte, production of competent sperm, proximity of sperm and oocyte in the reproductive tract, fertilization of the oocyte, transport of the conceptus into the uterus, and implantation of the embryo into a properly prepared endometrium. Diseases that result in dysfunction in any one of these complex biological steps can cause infertility. For example, the azoospermia of Klinefelter's syndrome or the anovulation resulting from hyperprolactinemia clearly cause infertility. For conditions such as endometriosis or oligospermia, however, the data suggest that there is decreased fecundability but cannot prove a cause and effect relationship.

Our limited understanding of the physiological mechanisms of pregnancy make it difficult to develop a scientific catalog of the diseases that cause infertility. Most tabulations of medical conditions that cause infertility are subjective and rely on assumptions that are based on clinical judgment rather than on scientific evidence. For example, controversy persists as to whether uterine leiomyomata and minimal or mild endometriosis are causes of infertility.

A. Disorders of Oocyte Production

Disorders of oocyte production are the most common causes of female infertility. Adult onset anovulation can result from hypothalamic underproduction of gonadotropin-releasing hormone (GnRH) (40% of cases), pituitary prolactin-secreting tumors (20%), polycystic ovary syndrome (30%), or premature ovarian failure (10%). Hypothalamic underproduction of GnRH is caused by abnormalities of body mass or nutrition, stress, or excess exercise. Prolactin-secreting pituitary tumors, or prolactinomas, are usually small (<10 mm in diameter) and monoclonal. Polycystic ovary syndrome consists of chronic anovulation, hyperandrogenism as evidenced by hirsutism, acne, and/or an elevated free testosterone concentration, and hypersecretion of luteinizing hormone (LH). The number of oocytes a woman has decreases with age (Fig. 16.1) [39a]. A woman's number of oocytes and follicles is fixed *in utero* and declines exponentially from the second trimester of gestation. Premature ovarian failure is the loss of all oocytes before age 40. The etiology of premature ovarian failure has not been fully elucidated, but familial and genetic patterns clearly exist.

Women who have monthly menses and report moliminal symptoms such as breast tenderness or dysmenorrhea are almost always ovulatory. Anovulation is commonly associated with

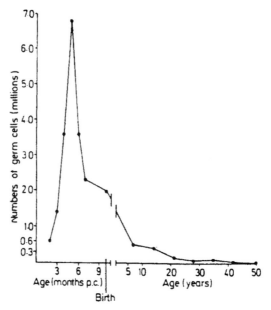

Fig. 16.1 Changes in the total number of oocytes (and follicles) in human ovaries before and after birth. The number of germ cells in the ovary peaks *in utero* during the second trimester. From [39a], with permission.

amenorrhea (absence of menses) or oligomenorrhea (menstrual cycle lengths > 35 days). A simple laboratory technique for detecting ovulation is basal body temperature recording. The basal oral temperature obtained in the morning before rising from bed is less than 98°F prior to ovulation. After ovulation, progesterone produced by the ovary appears to raise the hypothalamic set point for basal temperature by roughly 0.6°F. Therefore, the normal luteal phase is associated with a temperature above 98°F for about 10 to 14 days.

In addition to basal body temperature, a serum progesterone level greater than 3 ng/ml is diagnostic of ovulation. A midluteal phase progesterone concentration above 10 ng/ml is associated with a greater per cycle pregnancy rate than cycles with lower midluteal progesterone concentrations [40] (Table 16.2).

B. Tubal Factor Infertility

The second most prevalent cause of infertility in women is blockage or adhesions of the fallopian tubes. Two commonly used tests of tubal function and patency are hysterosalpingography (HSG) and laparoscopy. Hysterosalpingography uses few resources, generates data concerning the shape and size of the uterine cavity, and may increase fecundability in the ensuing months by altering the peritoneal environment or by opening tubes blocked by small particulate matter. However, hysterosalpingography cannot be used to diagnose concomitant endometriosis and often does not detect ovarian adhesions.

C. Endometriosis

Endometriosis is the presence of tissue that resembles endometrial glands and/or stroma outside of the uterus. The most

Table 16.2

Initial Laboratory Tests for Investigation of the Causes of Infertility

Primary Tests for Infertility

Documentation of ovulation
  Midluteal progesterone > 10 ng/ml
  Day 3 serum FSH concentration < 10 mIU/ml

Normal parameters for semen analysis

Volume	2–6 ml
Concentration	>20 million/ml
Motility	>40%
Morphology	>40% normal forms

Documentation of tubal patency
  Hysterosalpingogram

Secondary Tests for Infertility

Laparoscopy
  Identify and treat endometriosis
  Evaluate pelvic adhesions and anatomy

Postcoital test

Endometrial biopsy

---

common sites of endometriotic lesions are the peritoneal surfaces of the cul-de-sac of Douglas, the peritoneum covering the bladder, the ovarian surfaces, the stroma of the ovary, the bowel serosa, the fallopian tubes, and the appendix. Women with endometriosis usually present with pelvic pain, a pelvic mass (endometrioma), or infertility. See Chapter 18 for an in-depth discussion of endometriosis.

The American Society of Reproductive Medicine has published a surgical staging system for endometriosis that divides the disease into four stages. Women with advanced endometriosis (stages III and IV) often have major pelvic adhesions that involve the ovarian surface and distort the fallopian tubes. Laboratory and clinical evidence suggest that advanced endometriosis is associated with infertility [41–43]. The relationship between early stage endometriosis (stages I and II) and infertility is controversial. Marcoux and colleagues reported that the surgical treatment of stage I and II endometriosis improved fecundability in infertile women [44]. The investigators randomized 341 women with stage I or II endometriosis to either a diagnostic laparoscopy only or a diagnostic laparoscopy combined with surgical resection or ablation of endometriotic lesions. During 36 weeks of postoperative follow-up, the fecundability for pregnancies carried beyond 20 weeks was 2.4% in the group that had only a diagnostic laparoscopy and 4.7% in the group that had surgical resection or ablation (Rate Ratio (RR) = 1.9, 95% Confidence Interval (CI) 1.2–3.1). The 36-week cumulative probabilities of pregnancies carried beyond 20 weeks were 17.7% in the diagnostic laparoscopy group and 30.7% in the laparoscopic surgery group (RR = 1.7, 95% CI 1.2–2.6). This was the first large clinical trial to test the effectiveness of surgical treatment in enhancing fecundability in early stage endometriosis. The results suggest that early stage endometriosis causes infertility and that surgical treatment will increase the probability of pregnancy.

D. Uterine and Cervical Causes of Infertility

Uterine and cervical abnormalities have been reported in 4% of women with infertility [45]. Examples of malformations include a bicornate or septate uterus and the cervical ectopy and cervicovaginal hoods reported in women exposed to diethylstilbestrol (DES) *in utero* [46,47]. Uterine leiomyomata, also known as fibroids or myomas, are the most common pelvic tumors of women and have been associated with infertility [48]. Myomas are monoclonal, benign smooth muscle tumors of the uterus that often demonstrate chromosomal rearrangements. A gene at chromosome 12q15, HMGI-C, is mutated in many cases [49,50]. There are few reports that explore the effects of uterine leiomyomata on fecundability. In one study of the effects of leiomyomata on pregnancy during IVF treatment, the investigators reported that women with leiomyomata had a 22% pregnancy rate per transfer and women with tubal factor infertility had a 25% pregnancy rate per transfer, not a statistically significant difference. The women with leiomyomata were divided into two groups based on the results of an HSG. The women with uterine leiomyomata and a normal uterine cavity had a pregnancy rate of 30% per transfer. The women with uterine leiomyomata and an abnormal uterine cavity had a 9% pregnancy rate per transfer. This difference was statistically significant and suggests that uterine leiomyomata that distort the uterine cavity may be associated with a decrease in fecundability [51].

The cervix is an active participant in transporting sperm from the vagina to the upper reproductive tract. Cervical mucus has physiochemical properties that facilitate sperm transport. Cervical factor infertility is assessed by the postcoital test performed during the late follicular phase of the menstrual cycle. The infertile couple is asked to have sexual intercourse and the female partner is seen soon after. A small amount of cervical mucus is obtained by the use of oval forceps or a syringe. The glycoproteins in the mucus support the property of spinnbarkeit, or stretchability of the mucus. A portion of the specimen is allowed to dry on a glass slide to test for "ferning" because the high salt concentration of cervical mucus in the late follicular phase will cause the mucus to develop a "fern" pattern when it dries. Another aliquot is placed on a glass slide, overlaid with a coverslip and examined under the high power objective for the presence of motile sperm.

Two problems with the postcoital test are that there is a lack of agreement concerning the definition of a normal test and that the test has poor predictive value for pregnancy. Some authorities believe that a normal test requires more than 20 sperm per high power field [52,53]. Other authorities conclude that the presence of a single sperm is indicative of a normal test result [54]. In addition, the inter- and intraobserver variation of the test is high [55].

E. Disorders of Sperm Production: The Male Factor

The processes of spermatogenesis and sperm transport and maturation through the male reproductive tract are complex. We will not describe male factor diagnosis in detail, other than to outline the diagnostic steps. The initial approach in the evaluation of sperm production is a semen analysis which, in its basic form, measures volume, concentration, motility, and morphology (Table 16.2). The semen analysis has significant variability that can be introduced by laboratory techniques or by the large variation between ejaculates in the same man. A combination of concentration and motility, the "motile sperm density" (sperm concentration multiplied by percent motility) appears to predict fecundability. Two independent studies suggest that a total motile sperm density below 5 million motile sperm per milliliter is associated with a decrease in fecundability [56,57].

A varicocele is a dilatation of the pampiniform plexus of the scrotal veins that may decrease semen quality by increasing testicular temperature. Approximately 11% of men with a normal semen analysis have a clinically detected varicocele, while approximately 25% of men with an abnormal semen analysis have a varicocele [58]. Most studies indicate that the treatment of varicocele improves semen parameters [59]. However, the effect of varicocele treatment on fecundability is controversial. In one study, the investigators randomized men with varicoceles to observation or to treatment [59]. In the men who did not receive treatment, the fecundability in the 12 months following randomization was 1%. In the men who had the varicoceles treated by high spermatic vein ligation, the fecundability was significantly higher, 4%; however, other investigators have not replicated these findings [60].

F. Unexplained Infertility

"Unexplained" infertility is a diagnosis reached by exclusion. The following tests must have been performed: serum progesterone level or endometrial biopsy to demonstrate ovulation, semen analysis to assess male factor, hysterosalpingogram or other test to demonstrate a normal uterine cavity and patency of the fallopian tubes, and laparoscopy to demonstrate the absence of endometriosis and pelvic adhesions. Because many fertility specialists question the clinical utility of laparoscopy in women with endometriosis, it has been proposed that "idiopathic" infertility be diagnosed when a laparoscopy has not been performed, but all the other fertility tests listed above have been performed and are normal.

Many fertility specialists have proposed that normal ovarian reserve must be demonstrated prior to diagnosing unexplained infertility. Ovarian reserve can be evaluated by measuring cycle day 3 follicle stimulating hormone (FSH) levels or by a clomiphene challenge test (Table 16.3).

III. Methodological Issues in Conducting Epidemiologic Research

A. Study Designs and Sources of Data on Infertility

Information on the distribution and determinants of infertility has been collected using both descriptive and analytic observational study designs. An in-depth review of the advantages and disadvantages of each type of study design that can be used to conduct infertility research has been published and will be reviewed briefly here [61].

A number of descriptive studies have provided scientists with an understanding of the burden of infertility in populations and have identified areas for further study [5,6,9–12]. These studies

are most useful for suggesting hypotheses that are then tested by analytic study designs. Examples of other descriptive studies of infertility are case series such as those that first described the effects of DES [62] and reported septic abortion and pelvic inflammatory disease (PID) after intrauterine device use [63], incidence statistics such as those published by Houston et al. [64] on endometriosis, and registry data such as the IVF-ET registry [65]. The National Survey of Family Growth [66] provides cross-sectional data on patterns and trends of infertility in married couples and correlational studies provide international insights into trends in sperm counts [67].

The analytic study designs, such as case-control and follow-up studies, include formal comparison groups and permit hypothesis testing. A particular type of follow-up study that has been developed to study infertility is the time-to-pregnancy design [68]. This type of study looks for relationships between waiting times to conception and various environmental, lifestyle, or medical factors. Time-to-pregnancy designs can be conducted either retrospectively or prospectively. In the prospective design, women (or couples) who are planning to conceive are asked about their exposure histories and then are followed until they conceive or until the closing date of the study. The fecundability of those who are exposed to the agent under study is compared to that of nonexposed persons [69]. This comparison of conception probabilities is called the *fecundability ratio* [70]. A time-to-pregnancy study can be carried out retrospectively by enrolling women who recently gave birth and asking them if they planned their pregnancies and, if so, how many months it took them to conceive [71]. They are then asked to remember their exposures prior to their pregnancy.

The time-to-pregnancy study is a sensitive design for detecting exposures that cause conception delay, particularly if conceptions are detected by biochemical testing [72]. However, these studies do not provide information on women who don't conceive within the study time period (often twelve months) or who never conceive. Another disadvantage is that women or couples who plan their pregnancies may have different lifestyle habits and exposures than nonplanners [70,73]. In addition, these studies may systematically exclude the most fertile couples, because they are most likely to conceive without planning. This exclusion may result in an overrepresentation of subfertile couples in the study population.

Case-control studies have provided much of the epidemiologic knowledge that has been learned about risk factors for infertility [74,75]. Advantages of case-control study designs are that excellent information on infertility diagnosis and pathology can be collected from medical records and that large numbers of patients with similar types of infertility can be assembled. Additionally, information on many prior exposures may be obtained via personal interview, mail questionnaire, or examination of medical records. However, information collected from infertile patients may be subject to recall bias and the selection of appropriate control subjects may be difficult [76–79]. Often, in order to ensure that controls do not have infertility, subjects with demonstrated fertility are chosen. These parous subjects may recall their prior lifestyle, sexual, and reproductive histories differently than infertile subjects.

Prospective follow-up studies are not subject to recall bias because the study subjects are enrolled based on their exposure

Table 16.3
The Diagnosis, Differential Diagnosis, and Treatment of Common Causes of Infertility

Cause of infertility	History	Physical exam	Laboratory tests	Differential diagnosis	Treatment
Disorders of oocyte production					
Anovulation	Amenorrhea, oligo-menorrhea, absence of moliminal symptoms	Evidence for hypo-estrogenism, *e.g.,* scant cervical mucus	Basal body temp. chart, serum proges-terone, FSH, prolac-tin, testosterone, TSH	Hypogonadotropic hypogonadism, polycystic ovary syndrome, hyperprolactinemia	Clomiphene, gonado-tropin injections, pul-satile GnRH, dopamine agonist, IVF
Aging oocyte or depleted oocyte reserve	Age >35 years, his-tory of cigarette smoking, history of bilateral ovarian sur-gery, familial history of early menopause		Elevated cycle day 3 FSH, abnormal clom-iphene challenge test	Premature ovarian failure, depleted ovarian oocyte reserve	Empirical ovarian stimulation, IVF, oocyte donation
Tubal factor	History of ruptured appendix, previous pelvic surgery, pelvic inflammatory disease, or chlamydia or gonorrhea	Frozen pelvis, adnexal mass	Hysterosalpingo-gram, test for pres-ence of chlamydia and gonorrhea	Various infections	Antibiotics, surgical repair of the pelvis, IVF
Endometriosis	Severe dysmenorrhea or dyspareunia	Uterosacral ligament nodularity, cervical stenosis, lateral dis-placement of the cervix	Laparoscopy		Surgical excision of endometriosis le-sions, empirical ovar-ian stimulation, IVF
Uterine leiomyomata	History of menorrhagia	Pelvic mass	Sonography		Myomectomy, empir-ical ovarian stimula-tion, IVF
Male factor	No history of previ-ous paternity	Small testes, eunu-choid habitus, varicocele	Semen analysis	Varicocele, Klinefel-ter's, vas deferens oc-clusion, Kallman's syndrome	Varicocele repair, ICSI, donor sperm
Unexplained infertility			Standard infertility evaluation is normal	Rule out endome-triosis, depleted ovar-ian oocyte pool	Empirical ovarian stimulation, IVF

histories and followed to assess the occurrence of fertility prob-lems [80]. This study design is particularly useful in the occu-pational setting where large groups of exposed individuals can be identified and monitored and exposures can be documented. However, when studying relatively rare outcomes, there may be few study subjects with any one type of fertility problem.

In contrast to observational study designs, experimental stud-ies such as clinical trials are used to assess the efficacy of new treatments or to compare success rates after IVF therapy [44].

B. Sample Selection

Patients who seek medical care at an infertility center or physician's office often form the basis for selection of cases in case-control studies and are those who identify themselves as infertile in some follow-up study designs. However, infertility patients are a small subset of all infertile people and are not representative of all individuals or couples with infertility. Such

referral bias must be considered when interpreting prevalence statistics or the results of clinical and epidemiologic studies.

Detection bias is also a possibility because couples may have more than one type of infertility identified, depending on the nature and extent of their infertility evaluation. More fertile cou-ples will achieve pregnancy earlier in the diagnosis and treat-ment process, while less fertile ones will continue on to the next stage of diagnostic tests and treatments. Furthermore, treatment is often empirical and may begin before evaluation is complete. At any point in infertility evaluation or treatment, spontaneous pregnancies can occur, removing the more fertile couples from observation.

Because the definition of infertility describes the absence of an event (in fact, some scientists describe it as a symptom rather than as a disease entity), there is the potential for misclassifica-tion of disease. In some studies, particularly time-to-pregnancy designs, a heterogeneous group of conditions that manifest as the lack of conception are studied as if they were a single entity.

This violates the epidemiologic principle that all affected individuals have the same disease, presumably of similar etiology. To avoid such misclassification, studies of infertile patients should attempt to precisely characterize any underlying pathology. Furthermore, conditions may coexist, with more than one cause contributing to a couple's infertility. Investigators deal with this nonspecificity by classifying patients in different ways during the data analysis. For example, patients may be classified by *any* diagnosis of a condition, by the doctor's assessment of the *most significant* cause of their infertility, or by restricting case populations to patients who have *only* one diagnosis.

Rather than defining infertility as the absence of conception, the time-to-pregnancy design measures the number of menstrual cycles from entry into the study until a pregnancy is detected. These studies often use the terms "subfertility" or "delay to conception" to describe their outcomes. With the development of sensitive radioimmunoassays the concept of a biochemical pregnancy has emerged. These laboratory tests are able to measure levels of human chorionic gonadotropin (hCG) in a woman's urine after implantation of the fertilized ovum [81,82]. These early pregnancies may be unrecognized by the woman or her clinician and result in early unrecognized pregnancy losses. Baird *et al.* have hypothesized that about 80% of all conceptions are lost prior to recognition of the pregnancy, although others place the estimate lower [68]. Before the use of hCG testing, a couple with an early unrecognized pregnancy loss would be perceived as "infertile." We now have the ability to identify these early biochemical conceptions and subsequent early fetal losses, allowing us to identify and study reproductive hazards that act after implantation but prior to the first missed menses. These women, however, are not "infertile" in the way that most studies use the term, because all must conceive to be able to measure their "trying time" to conception.

The selection of the controls in case-control studies of infertility is often more of an art than a science. In some infertility studies, parous controls (with no prior history of infertility diagnosis or therapy) are selected on the theory that they comply with the first rule of control selection—that the controls do not have the disease under study. Therefore, controls who have delivered at least one live born child are said to have demonstrated fertility (at least with the partner who fathered the child). While using parous controls is common practice, there are clearly many other factors to consider. Recall bias is likely to be maximized when parous controls' responses are compared to those of infertile women. If women who are multiparous are selected as controls, they must be asked to recall exposures prior to their first conception in order to approximate the stage in life during which the infertile couple was trying to conceive. In studying secondary infertility, there is the problem of assessing only those exposures that have occurred since the last pregnancy. An index date can be calculated, with exposure histories sought for the time prior to the estimated date of conception of the controls' index pregnancy. Other studies limit the choice of controls to primiparous women for primarily infertile cases and match on number of known prior conceptions for women who are secondarily infertile.

In follow-up studies, the selection of nonexposed subjects can be problematic, particularly in occupational studies. Often no truly nonexposed subjects are available and "lesser exposed"

subjects are enrolled. In addition, if the exposed subjects are employed outside the home, the comparison subjects must also be in the workforce in order to avoid confounding by employment, because employed women have fewer children and less favorable reproductive histories than unemployed women [83,84]. This phenomenon has been characterized as the "unhealthy pregnant worker effect" [84].

C. Exposure Assessment

Accurate assessment of the types of exposures that may be sustained by study subjects has been a weak area in infertility research. For example, in industrial and environmental settings, few studies have incorporated on-site exposure measurements to minimize misclassification of exposure variables. Rigorous exposure assessment would be particularly valuable in investigations of occupational hazards, water pollutants, physical activity, and environmental causes of infertility.

Recording the timing of exposure is critical. Savitz and Harlow have pointed out that better and more specific information on the nature and timing of exposures prior to, during, and after conception is needed to help clarify the reproductive health endpoint that is observed [85].

Epidemiologic research relies heavily on exposure, and sometimes, disease histories that are obtained by self-report during personal interview or by mail questionnaire. As previously mentioned, case-control and retrospective follow-up studies, including retrospective time-to-pregnancy designs, are subject to information bias because the responses that study subjects provide may be influenced by the fact that they are infertile. Validation of exposure histories by medical record review or other objective means is desirable, although not always possible. Questionnaires that use well-designed, closed-ended questions minimize the opportunity for bias and misclassification [86,87]. Studies of infertility seek information that is personal, not only about medical, reproductive, and social histories, but also about sexual and contraceptive practices, induced abortions, and prescription and recreational drug use. Investigators need to understand the impact of their questions and design and administer them with sensitivity so as to maximize cooperation and minimize information bias. This may mean creating a confidential interview setting that provides an atmosphere conducive to answering extremely personal questions. Constructing a calendar of reproductive and life events is helpful for establishing the timing of events. Pictures of contraceptive devices and pill packs are useful as memory aids. Menstrual calendars that record menstrual symptoms, bleeding characteristics and days, contraceptive use, and timing of intercourse have been developed and may soon be available in electronic form.

IV. Etiologic Factors

A number of epidemiologic investigations have been conducted during the 1990s to identify determinants of the major types of infertility (Table 16.4). The studies conducted include both case-control and follow-up designs that have focused on constitutional factors (*e.g.,* body mass index, menstrual characteristics), reproductive histories (*e.g.,* choice of contraceptive, history of sexually transmitted diseases), and lifestyle practices

Table 16.4

Summary of Epidemiologic Studies Investigating Risk Factors for the Major Types of Primary Infertility

References	Exposure under study	Study design	Study population	Results
PRIMARY OVULATORY INFERTILITY				
[80]	Body mass index at age 18	Nested case-control	2,527 married nulliparous nurses with self-reported ovulatory disorder 46,718 married parous nurses with no history of infertility Source: 116,678 registered nurses in 14 US states (NHS II)	BMI 22–23.9: RR = 1.1, 95% CI 1.0–1.2 BMI 24–25.9: RR = 1.3, 95% CI 1.2–1.6 BMI 26–27.9: RR = 1.7, 95% CI 1.4–2.1 BMI 28–29.9: RR = 2.4, 95% CI 1.8–3.1 BMI 30–31.9: RR = 2.7, 95% CI 1.9–3.8 BMI ≥ 32: RR = 2.7, 95% CI 2.0–3.7 Compared to referent BMI of 20–21.9
[89]	Body mass index	Multicenter case-control	597 cases with ovulatory infertility 1,695 primiparous controls Source: 1,880 patients and 4,023 parous controls from 7 hospitals/infertility clinics in the U.S. & Canada between 1981 & 1983	BMI < 17: OR = 1.6, 95% CI 0.7–3.9 BMI 17–19.9: OR = 0.7, 95% CI 0.6–0.9 BMI 25–26.9: OR = 1.2, 95% CI 0.8–1.9 BMI ≥ 27: OR = 3.1, 95% CI 2.2–4.4 Compared to referent BMI of 20–24.9
[90]	Body mass index	Population-based case-control	204 primary ovulatory infertility cases 204 matched parous controls Source: 1,005 infertile women aged 20–39 who were evaluated between 1979 & 1981 and resided in King County, WA and 1,009 primigravid controls matched on age, race, and census tract	Percentage of ideal body weight: ≤ 85%: RR = 4.7, 95% CI 1.5–14.7 ≥ 120%: RR = 2.1, 95% CI 1.0–4.3 Compared to referent women who were 86% to 119% of their ideal body weight
[91]	Exercise	Population-based case-control	187 primary ovulatory infertility cases 187 matched parous controls Source: 1,005 infertile women aged 20–39 who were evaluated between 1979 & 1981 and resided in King County, WA and 1,009 primigravid controls matched on age, race, and census tract	Average min/day of vigorous exercise: < 60 min/day: RR = 0.6, 90% CI 0.4–0.9 ≥ 60 min/day: RR = 1.9, 90% CI 0.6–5.1 Compared to non-exercisers
[92]	Pharmaceuticals	Multicenter case-control study	597 primary ovulatory infertility cases 3,833 parous controls Source: 1,880 patients and 4,023 parous controls from 7 hospitals/infertility clinics in the US & Canada between 1981 & 1983	Drug use of ≥ 6 months compared to never use Thyroid replacement hormone: OR = 2.3, 95% CI 1.5–3.5 Antidepressants: OR = 2.9, 95% CI 0.9–8.3 Tranquilizers (current use): OR = 3.2, 95% CI 1.1–8.5 Asthma medication before age 21: OR = 2.5, 95% CI 1.0–5.9
[93]	Recreational drug use	Population-based case-control	150 married women with primary ovulatory infertility 150 married women who had given birth 1 yr after the infertile women had started trying to conceive Source: 1,005 infertile women aged 20–39 who were evaluated between 1979 & 1981 and resided in King County, WA and 1,009 primigravid controls matched on age, race, and census tract	Ever smoked marijuana: RR = 1.7, 95% CI 1.0–3.0 Marijuana use within 1 yr of trying to conceive: RR = 2.1, 95% CI 1.1–4.0 Compared to never-users
[94]	Alcohol drinking	Multicenter case-control	431 primary ovulatory infertility cases 3,833 parous controls Source: 1,880 patients and 4,023 parous controls from 7 hospitals/infertility clinics in the US & Canada between 1981 & 1983	Moderate drinkers (≤100g of alcohol/wk): OR = 1.3, 95% CI 1.0–1.7 Heavier drinkers (>100g alcohol/wk): OR = 1.6, 95% CI 1.1–2.3 Compared to nondrinkers
[95]	Occupational exposures	Case-referent	124 ovulatory infertility cases 216 parous women Source: 281 infertility clinic patients recruited 1991–92 who resided in Iowa or Illinois and 216 fertile women recruited from postpartum units of adjacent hospitals	Solvents: OR = 1.8, 95% CI 1.0–3.0 Dusts: OR = 3.0, 95% CI 1.2–7.5 Pesticides: OR = 3.8, 95% CI 1.3–11.4
[96]	Agricultural work history	Case-control	124 ovulatory infertility cases 216 parous women Source: 281 infertility clinic patients recruited 1991–92 who resided in Iowa or Illinois and 216 fertile women recruited from postpartum units of adjacent hospitals	Agricultural industry: OR = 10.2, 95% CI 3.2–32.6 Agricultural occupation: OR = 16.1, 95% CI 3.5–75.0 Lived on a farm: OR = 1.9, 95% CI 1.1–3.2
PRIMARY TUBAL INFERTILITY				
[74]	Intrauterine device use	Population-based case-control	159 tubal infertility cases 159 matched primiparous controls Source: 1,005 infertile women aged 20–39 who were evaluated between 1979 & 1981 and resided in King County, WA and 1,009 primigravid controls matched on age, race, and census tract	Ever use of an IUD: RR = 2.6, 95% CI 1.3–5.2 Dalkon Shield only: RR = 11.3, 95% CI 1.4–95.0 Lippes Loop/Saf-T-Coil only: RR = 4.4, 95% CI 0.5–41.8 Copper IUD only: RR = 1.3, 95% CI 0.6–3.0
[75]	Intrauterine device use	Multicenter case-control	283 primary tubal infertility cases 3,833 parous controls Source: 1,880 patients and 4,023 parous controls from 7 hospitals/infertility clinics in the US & Canada between 1981 & 1983	Ever use of an IUD: RR = 2.0, 95% CI 1.5–2.6 Dalkon Shield only: RR = 3.3, 95% CI 1.7–6.1 Lippes Loop/Saf-T-Coil only: RR = 2.9, 95% CI 1.7–5.2 Copper IUD only: RR = 1.6, 95% CI 1.1–2.4

(continued)

Table 16.4

(continued)

References	Exposure under study	Study design	Study population	Results
PRIMARY TUBAL INFERTILITY *(continued)*				
[97]	Appendectomy	Population-based case-control	158 primary tubal infertility cases 501 parous controls Source: 1,005 infertile women aged 20–39 who were evaluated between 1979 & 1981 and resided in King County, WA and 1,009 primigravid controls matched on age, race, and census tract	Previous appendectomy with rupture: RR = 4.8, 95% CI 1.5–14.9 Compared to no appendectomy
[98]	Prior pelvic inflammatory disease	Population-based case-control	162 primary tubal infertility cases 518 primiparous controls Source: 1,005 infertile women aged 20–39 who were evaluated between 1979 & 1981 and resided in King County, WA and 1,009 primigravid controls matched on age, race, and census tract	History of physician-diagnosed PID: RR = 5.9, 99% CI 3.1–11.4 compared to no history of PID Overt vs Silent PID: Use of OCs for >3 yr: RR = 0.5, 95% CI 0.3–0.8 for silent PID RR = 0.9, 95% CI 0.3–2.5 for overt PID, compared to short-term and nonusers Use of an IUD: RR = 2.4, 95% CI 1.3–4.3 for silent PID And RR = 9.7, 95% CI 3.7–25.5 for overt PID, compared to nonusers of an IUD
[99]	History of sexually trans-mitted diseases	Multicenter case-control	283 primary tubal infertility cases 3,833 parous controls Source: 1,880 patients and 4,023 parous controls from 7 hospitals/infertility clinics in the US & Canada between 1981 & 1983	History of gonorrhea: OR = 2.4, 95% CI 1.3–4.4 History of trichomoniasis: OR = 1.9, 95% CI 1.3–2.8 Compared to women with no history of these infections Risk increased with increasing number of episodes
[93]	Recreational drug use	Population-based case-control	84 married women with primary tubal infertility 84 married women who had given birth 1 yr after the infertile women had started trying to conceive Source: 1,005 infertile women aged 20–39 who were evaluated between 1979 & 1981 and resided in King County, WA and 1,009 primigravid controls matched on age, race, and census tract	Cocaine use: RR = 11.1, 95% CI 1.7–70.8 Compared to never users
[102]	Caffeine intake	Multicenter case-control	230 primary tubal infertility cases 3,833 parous controls Source: 1,880 patients and 4,023 parous controls from 7 hospitals/infertility clinics in the US & Canada between 1981 & 1983	Greater than 7g of caffeine per month: RR = 1.5, 95% CI 1.1–2.0, compared to women who consumed 3g or less per month
[104]	Smoking	Multicenter case-control	283 primary tubal infertility cases 1,264 parous controls Source: 1,880 patients and 4,023 parous controls from 7 hospitals/infertility clinics in the US & Canada between 1981 & 1983	Current smoking: RR = 1.6, 95% CI 1.1–2.1, compared to nonsmokers
[105]	Use of barrier methods or oral contraceptives	Multicenter case-control	283 primary tubal infertility cases 3,833 parous controls Source: 1,880 patients and 4,023 parous controls from 7 hospitals/infertility clinics in the US & Canada between 1981 & 1983	Ever use of barrier methods: RR = 0.6, 95% CI 0.5–0.8 Ever use of oral contraceptives: RR = 1.2, 95% CI 0.8–1.6 Compared to nonuse
[95]	Occupational exposures	Case-referent	84 tubal infertility cases 216 parous women Source: 281 infertility clinic patients recruited 1991–92 who resided in Iowa or Illinois and 216 fertile women recruited from postpartum units of adjacent hospitals	Solvents: OR = 2.0, 95% CI 1.1–3.5 Dusts: OR = 2.9, 95% CI 1.1–7.9
[96]	Agricultural work history	Case-control	84 tubal infertility cases 216 parous women Source: 281 infertility clinic patients recruited 1991–92 who resided in Iowa or Illinois and 216 fertile women recruited from postpartum units of adjacent hospitals	Agricultural industry: OR = 3.9, 95% CI 0.9–17.9 Agricultural occupation: OR = 6.3, 95% CI 1.0–38.2 Lived on a farm: OR = 1.7, 95% CI 0.9–3.3
INFERTILITY DUE TO ENDOMETRIOSIS				
[106]	Menstrual characteristics, smoking, exercise	Multicenter case-control	268 women with primary infertility due to endometriosis 3,794 parous women Source: 1,880 patients and 4,023 parous controls from 7 hospitals/infertility clinics in the US & Canada between 1981 & 1983	Average cycle length ≤ 27 days: RR = 2.1, 95% CI 1.5–2.9 compared to 28–34 days Duration of flow ≥ 8 days: RR = 2.4, 95% CI 1.4–4.0 compared to ≤ 7 days Mild (RR = 1.7, 95% CI 1.1–2.6), moderate (RR = 3.4, 95% CI 2.2–5.2), and severe (RR = 6.7, 95% CI 4.4–10.2) menstrual pain compared to none Exercise regularly: RR = 0.6, 95% CI 0.4–0.8
[94]	Alcohol drinking	Multicenter case-control	180 women with primary infertility due to endometriosis 3,833 parous controls Source: 1,880 patients and 4,023 parous controls from 7 hospitals/infertility clinics in the US & Canada between 1981 & 1983	Moderate drinkers (≤ 100g of alcohol/wk): OR = 1.6, 95% CI 1.1–2.3 Heavier drinkers (> 100g alcohol/wk): OR = 1.5, 95% CI 0.8–2.7 Compared to nondrinkers
[95]	Occupational exposures	Case-referent	36 women with infertility due to endometriosis 216 parous women Source: 281 infertility clinic patients recruited 1991–92 who resided in Iowa or Illinois and 216 fertile women recruited from postpartum units of adjacent hospitals	Video display terminal exposure: OR = 3.7, 95% CI 1.5–9.1

Note. Abbreviations: BMI = Body mass index, CI = Confidence interval, IUD = Intrauterine device, OC = Oral contraceptives, OR = Odds ratio, PID = Pelvic inflammatory disease, RR = Relative risk.

(*e.g.*, smoking, coffee drinking, alcohol use). Notably, two studies, published simultaneously, quantified the risk of primary tubal infertility after use of an intrauterine device (IUD) [74,75]. These two studies included large populations of infertile patients and have provided data for a number of subsequent epidemiologic investigations of risk factors for infertility [88]. Many other likely determinants of infertility, including diet, occupational exposures, and immunologic and genetic factors remain largely unexplored.

Many other studies have examined the influence of lifestyle factors such as smoking, alcohol use, caffeine intake, and environmental exposures on fecundability or waiting times to conception. This literature is described in Chapter 11.

A. Ovulatory Infertility

Anovulation, oligoovulation, and subfertility are commonly observed in women above or below their ideal body weight as measured by the body mass index (BMI = weight/height^2). Rich-Edwards *et al.*, using a nested case-control design within the Nurses Health Study II cohort, reported that women with a BMI $\geq$ 24.0 were at increased risk of ovulatory infertility compared to women with a BMI of 20.0 to 21.9 [80]. Similar results were reported by Grodstein *et al.* [89] who found that the risk of primary ovulatory infertility was highest in obese women and slightly increased in moderately overweight and underweight women (BMI < 17.0, odds ratio (OR) = 1.6, 95% CI 0.7–3.9; BMI = 25.0–26.9, OR = 1.2, 95% CI 0.8–1.9; and BMI $\geq$ 27.0, OR = 3.1, 95% CI 2.2–4.4, all compared to women with BMI = 20.0–24.9). After adjustment for age and exercise, the odds ratio decreased to 2.4 (95% CI 1.7–3.3) for women with BMI $\geq$ 27.0. Risks were highest for women with polycystic ovary syndrome. Green *et al.* [90] observed an increased risk of ovulatory infertility in nulligravid women who were $\leq$85% or $\geq$120% of their ideal body weight ($\leq$85%, RR = 4.7, 95% CI 1.5–14.7; $\geq$120%, RR = 2.1, 95% CI 1.0–4.3), compared to women who were 86 to 119% of their ideal body weight.

To elucidate whether women who participated in vigorous exercise had an increased risk of primary ovulatory infertility, Green *et al.* [91] reviewed the exercise histories of 187 women with primary ovulatory infertility. Vigorous exercise was defined as aerobic activity such as running, dancing, or aerobic exercise that required approximately 6 KCAL/minute for a woman of average weight. Women who participated in vigorous exercise for $\geq$ 1 hour per day had twice the risk of ovulatory infertility (RR = 1.9, 90% CI 0.6–5.1) compared to women who did not exercise. When the case group was further refined to include only those 137 women with primary ovulatory infertility who had no evidence of concomitant tubal disease, the relative risk increased to 6.2 (90% CI 1.0–39.8), but this result was based on only seven cases. No increased risk was noted for women who had secondary ovulatory infertility.

Little is known about the effects on fertility of popular prescription and over-the-counter medications. Grodstein *et al.* [92] reported elevated risks of primary ovulatory infertility for women who used thyroid replacement hormones for 6 months or more (OR = 2.3, 95% CI 1.5–3.5). Information on drug use was obtained by self-report and the authors were unable to separate any effects of the underlying thyroid condition from the

drugs themselves. Such confounding by indication could explain the association if, for example, hypothyroidism, even when treated, produced amenorrhea and a resultant infertility diagnosis. With respect to other prescription medications the authors reported that use of antidepressants for 6 months or more was associated with a tripling of risk (OR = 2.9, 95% CI 0.9–8.3). For women reporting use for 2 years or more, the risk rose to 4.3 (95% CI 0.7–26.0) but was based on only two cases. Increases were also noted for current and extended use of tranquilizers (use for > 2 years: OR = 2.9, 95% CI 0.8–11.0) and asthma medication first used before age 21 (OR = 2.5, 95% CI 1.0–5.9). When type of asthma medication was examined, increased risk was related to use of beta-agonists rather than theophylline. Risks were also doubled for use of codeine, steroids, and acetaminophen for 6 months or more.

When recreational drug use was examined (marijuana, LSD, speed, and cocaine), smoking marijuana within one year of trying to conceive was reported to increase the risk of ovulatory infertility (RR = 2.1, 95% CI 1.1–4.0) [93]. Because the history of drug use was obtained by self-report, recall or response bias could account for these results if controls were more likely than cases to have underreported their drug use.

Moderate use of alcohol has been reported to increase the risk of ovulatory infertility (OR = 1.3, 95% CI 1.0–1.7 for moderate drinkers and 1.6, 95% CI 1.1–2.3 for heavier drinkers) compared to nondrinkers [94]. Moderate drinking was defined as one drink or less per day.

Few occupational hazards have been assessed for their relationship to specific types of infertility, but two reports from the same study population have identified elevated risks of ovulatory infertility for women who worked in the agricultural industry or who were occupationally exposed to solvents, dusts, or pesticides [95,96].

B. Tubal Infertility

Tubal disease is identified in approximately 22% of the female partners of infertile couples [13]. Pelvic inflammatory disease, sexually transmitted diseases, appendicitis, septic abortion, previous tubal surgery, and use of an IUD resulting in pelvic infection are major contributors to tubal disease [74,75, 97–99]. The prevalence of tubal infertility has been reported to be 12%, 23%, and 54% after one, two, or three episodes of PID [100]. Subclinical pelvic infections with *Chlamydia trachomatis* may be another cause of tubal infertility [101].

In a population-based case-control study of 159 primary tubal infertility cases, Daling *et al.* [74] reported that use of any IUD tripled the risk of tubal infertility (OR = 2.6, 95% CI 1.3–5.2), compared to no IUD use. Increased risks were noted for ever use of all IUD types but were greatest for the Dalkon Shield (OR = 6.8, 95% CI 1.8–25.2) and lowest for copper IUDs (OR = 1.9, 95% CI 0.9–4.0). In a multicenter case-control study of 283 nulliparous women with tubal infertility and 3833 parous control women, Cramer *et al.* [75] reported a doubling of risk related to IUD use (OR = 2.0, 95% CI 1.5–2.6) compared to no IUD use. Again, users of the Dalkon Shield were at highest risk (OR = 3.3, 95% CI 1.7–6.1). Risks for other IUDs were also elevated, although less so for copper IUDs. Interestingly, risk was limited to women who had more than one sexual partner.

The authors concluded that IUDs were not recommended for nulliparous women who wished to preserve their fertility, with the possible exception of women with one sexual partner (who are, therefore, at low risk for sexually transmitted diseases). Tubal infertility had previously been recognized as a sequela of PID, but prior to these two studies, the link between tubal infertility and use of the IUD had not been documented.

The influence of a variety of lifestyle factors and personal habits on the occurrence of tubal infertility has been assessed. Cocaine use (RR = 11.1, 95% CI 1.7–70.8) [93], consumption of caffeinated beverages (RR = 1.5, 95% CI 1.1–2.0 for women who consumed the equivalent of more than two cups of coffee or four cans of cola per day, compared to those who consumed less) [102], vaginal douching [103], and current smoking (RR = 1.6, 95% CI 1.1–2.1, compared to women who had never smoked) [104] have all been related to increased risks of tubal infertility.

The protective effect of barrier methods of contraception, especially those methods that combine a mechanical and chemical barrier, such as diaphragms, cervical caps, and condoms plus spermicides, in reducing the risk of tubal infertility has been demonstrated [105].

In a study of infertility and agricultural work history, scientists reported increased risks of tubal infertility associated with working in an agricultural occupation (OR = 6.3, 95% CI 1.0–38.2), or being occupationally exposed to solvents (OR = 2.0, 95% CI 1.1–3.5) or dusts (OR = 2.9, 95% CI 1.1–7.9) [95,96].

C. Infertility Due to Endometriosis

The risk of infertility due to endometriosis has been related to menstrual cycle characteristics [106–109]. Women with shorter cycle lengths (≤27 days), longer duration of flow (≥8 days), dysmenorrhea, and perhaps heavier menstrual flow as evidenced by use of both napkins and tampons have been reported to be at more than twice the risk of infertility due to endometriosis [106,107]. A similar pattern of frequent cycles of longer, heavier menstrual flow accompanied by severe menstrual cramps has been reported for women with endometriosis that is not necessarily related to infertility [108].

A significant increase in the risk of endometriosis-related infertility has been reported in women who consumed 5.1–7.0 grams of caffeinated beverages per month (RR = 1.9, 95% CI 1.2–2.9) or more than 7 grams/month (RR = 1.6, 95% CI 1.1–2.4), compared to women who consumed 3 grams or less per month [102]. Seven grams of caffeine per month is roughly equivalent to more than two cups of coffee or four cans of caffeinated soda per day.

An association with alcohol use has also been reported. The odds ratio for endometriosis was 1.6 (95% CI 1.1–2.3) for moderate drinkers and 1.5 (95% CI 0.8–2.7) for heavier drinkers, compared to women who did not drink [94]. Moderate drinkers were defined as those who consumed 100 grams or less of beer, wine, or liquor per week (about one drink or less per day) and those drinking more than 100 grams per week were defined as heavier drinkers.

Two other lifestyle factors, cigarette smoking and physical exercise, have been reported to protect against infertility attributable to endometriosis [106]. While the adverse effects of smoking outweigh any potential protective effect, the mecha-

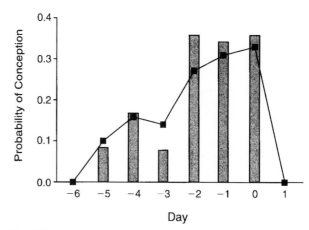

Fig. 16.2 Probability of conception on specific days near the day of ovulation. The bars represent data reported by 129 women who had sexual intercourse on only one day in the six-day interval ending on the day of ovulation (day 0). The solid line shows the probability of conception based on a statistical analysis of data from 625 cycles. From [109a]. Wilcox, A. J., Weinberg, C. R., and Baird, D. D. (1995). Timing of sexual intercourse in relation to ovulation. *N. Engl. J. Med.* **333,** 1517–1521. Copyright 1995 Massachusetts Medical Society. All rights reserved.

nism by which it exerts a protective effect is of interest and warrants further study. Smokers are also known to reach menopause an average of two years earlier than nonsmokers and a protective effect against endometrial cancer has been reported.

Cramer *et al.* [106] found that the risk of endometriosis was significantly reduced in women who exercised (RR = 0.6, 95% CI 0.4–0.8) compared to women who did not exercise regularly. Physical exercise affects endogenous estrogen levels, with less dysmenorrhea and lighter flow reported by those who exercise regularly.

V. Clinical Intervention

A. Diagnosis

There is no widely accepted standardized approach to establish a diagnosis of infertility. One approach is to quickly screen for diseases, such as anovulation, azoospermia, or tubal occlusion that are associated with a fecundability of 0.00 (Table 16.3). Anovulation can be excluded by a progesterone level higher than 3.0 ng/ml or by a biphasic basal body temperature chart. Azoospermia can be diagnosed by a semen analysis that demonstrates the absence of sperm. Tubal occlusion can be diagnosed by performing a hysterosalpingogram or laparoscopy. If these tests are normal and the female partner is younger than 35 years of age, the couple can be asked to time intercourse to a "fertile time" of the cycle with the use of an ovulation predictor test that measures the level of luteinizing hormone in the urine. The day of ovulation and the four or five preceding days appear to be the most fertile days of the menstrual cycle (Fig. 16.2) [109a]. One day after ovulation, the probability of achieving pregnancy markedly decreases.

If pregnancy does not occur after an interval of timed intercourse in a couple with normal semen analysis, hysterosalpingogram, and luteal phase progesterone level, then the couple

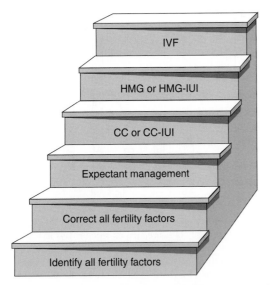

Fig. 16.3 Staircase approach to empirical infertility treatment using empirical ovarian stimulation. For women over 35 years old, the first three steps in the algorithm should be rapidly completed. In women less than 30 years old, more time can be spent on the first three steps in the staircase.

can enter an empirical ovarian stimulation program (Fig. 16.3). If the female partner of the infertile couple is 35 years of age or older, the clinician may want to assess oocyte reserve. This can be performed by measuring a basal FSH on day 3 of the menstrual cycle or by performing a clomiphene challenge test. During menses, FSH secretion is suppressed by inhibin B, a product of the small ovarian follicles, and to a lesser degree by estradiol. If the residual pool of follicles is reduced, FSH on cycle day 3 will be abnormally increased. An elevated serum FSH on cycle day 3 is a biochemical marker of a depleted follicular pool and is associated with decreased fecundability [110]. The clomiphene challenge test appears to be a more sensitive method than a basal day 3 FSH test for detecting depleted oocyte reserve. The clomiphene challenge test is performed by administering 100 mg of clomiphene citrate daily during cycle days 5 to 9. FSH levels are measured on cycle days 3 and 10. An elevated FSH on either cycle day is associated with diminished ovarian follicle number. In some series, for every 100 women with an elevated day 10 FSH on a clomiphene challenge test, only 40 had an elevated day 3 FSH. The clomiphene challenge test is probably effective in identifying women with diminished follicle reserve because it blocks the negative feedback effect of estradiol on FSH, leaving only inhibin B to suppress FSH production. In women with diminished follicle reserve, inhibin B concentrations are low and are incapable of suppressing FSH production without the assistance of estradiol. In women with normal follicle reserve, inhibin B concentration is capable of keeping FSH suppressed without the assistance of estradiol. Once a diminished ovarian follicle pool has been identified it is often "too late." In this situation, fertility treatments that rely on the female partner's oocytes have low success rates and oocyte donation is often required to achieve pregnancy [65,111].

B. Treatment Plans

1. Anovulation

Anovulatory women below their ideal BMI often have low GnRH production. Even without an increase in BMI, pulsatile GnRH treatment can induce ovulation and pregnancy [112]. However, inducing ovulation in a woman with an abnormally low BMI results in an increased risk of delivering a low birth weight infant [112]. Anovulatory women above their ideal BMI often have polycystic ovary syndrome (PCOS). For these women, weight loss is associated with the resumption of ovulation and pregnancy [113,114]. For example Pasquali and colleagues [113] reported that obese women with anovulation and PCOS who were treated with a 1000 calorie/day diet lost an average of 10 kg after six months. Weight loss was associated with ovulation and pregnancy in many of the women who were infertile.

a. CLOMIPHENE. Clomiphene is a nonsteroidal estrogen agonist/antagonist approved for the treatment of anovulation in 1967. For clinical purposes, the WHO classification recognizes two types of anovulation. WHO group I includes those women with low levels of endogenous gonadotropins and minimal endogenous estrogen production. These women often have low GnRH production due to low BMI, intensive exercise, or stress. Women who fail to menstruate after a progestin challenge (medroxyprogesterone acetate 10 mg orally per day for 5 days) are typically in WHO group I [115]. WHO group II includes those women with normal or elevated gonadotropin levels and significant endogenous estrogen production. Women with PCOS comprise a significant percent of all women in WHO group II.

Clomiphene is most effective in inducing ovulation in women in WHO group II. Women in WHO group I rarely respond to clomiphene and typically require pulsatile GnRH treatment or gonadotropin treatment. The approved dosages of clomiphene are 50 or 100 mg daily for a maximum of 5 days per cycle. After spontaneous menses, or the induction of menses with a progestin withdrawal, clomiphene is started on cycle day 3, 4, or 5 at 50 mg for 5 days. Starting clomiphene on either cycle day 3 or 5 does not influence the per cycle pregnancy rate [116]. In properly selected women, approximately 50% will ovulate at the 50 mg daily dosage. Another 25% will ovulate if the dosage is increased to 100 mg daily [117]. Anovulatory women in WHO group II have a baseline fecundability of 0.00. During the first three to six cycles of clomiphene treatment, their fecundability rises to 0.08–0.25. The age of the female partner influences the effectiveness of clomiphene treatment (Fig. 16.4) [117a].

b. GONADOTROPIN INDUCTION OF OVULATION. In 1958, Gemzell and associates [118] reported the efficacy of pituitary extracts of FSH to induce ovulation. In 1962, Lunenfeld and colleagues [119] demonstrated the efficacy of extracts of LH and FSH from menopausal urine in the induction of ovulation in women in WHO group I. Preparations of recombinant FSH (Gonal-F, Follistim) also have been released for clinical use. Recombinant FSH is produced using genetically engineered Chinese hamster ovary cell lines and is purified from the culture medium by immunochromatography.

FSH is the primary hormone responsible for follicular recruitment and growth in the human. Experiments in nonhuman

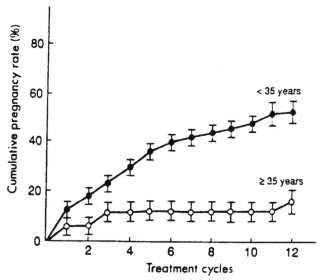

Fig. 16.4 Cumulative pregnancy rates following gonadotropin treatment for anovulatory women who did not respond to clomiphene induction of ovulation (WHO Group II). Solid circles represent the cumulative pregnancy rate in women less than 35 years old. Open circles represent the cumulative pregnancy rate in women greater than 35 years old. From [117a], with permission.

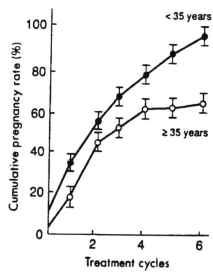

Fig. 16.5 Cumulative pregnancy rates for hypogonadotropic anovulatory women (WHO Group I) treated with gonadotropins. Solid circles represent the cumulative pregnancy rate in women less than 35 years old. Open circles represent the cumulative pregnancy rate in women greater than 35 years old. From [117a], with permission.

primates in which LH is reduced to extremely low levels demonstrate that the administration of FSH stimulates follicular recruitment and growth but that estradiol is abnormally suppressed [120]. It appears that most anovulatory women secrete enough LH so that no exogenous LH is required to achieve successful follicular growth and adequate estradiol production. For most women in WHO group I, administration of exogenous hCG is required to effect final maturation of the follicle and trigger ovulation.

In a large group of WHO group I women treated with urinary menopausal gonadotropins, Lunenfeld and Insler reported a six cycle cumulative pregnancy rate of 91%, which reflects a per cycle fecundability in the range of 0.33 (Fig. 16.5) [117a]. The fecundability observed in gonadotropin treatment cycles depends on the age of the female partner and the underlying cause of the anovulation.

c. PULSATILE GnRH TO INDUCE OVULATION.   In anovulatory women in WHO group I, administration of pulsatile GnRH with a programmable computerized pump is effective in the induction of ovulation. The advantage of GnRH induction of ovulation is that there is a reduced need for cycle monitoring and a reduced risk of multiple gestation due to the presence of intact feedback interaction between the ovary and the pituitary. The main disadvantage of pulsatile GnRH treatment is that it requires the woman to wear an infusion pump attached to an intravenous or subcutaneous catheter for up to 21 days. Santoro and colleagues [121] proposed eight criteria for identifying women most likely to safely achieve ovulation with pulsatile GnRH: (1) primary or secondary amenorrhea for 6 months; (2) absence of hirsutism; (3) weight not below 90% of ideal weight; (4) no excessive exercise or stress; (5) normal serum prolactin, thyrotropin (TSH), DHEA-S, and testosterone; (6) low gonadotropin

concentration; (7) no evidence of a pituitary or hypothalamic tumor; and (8) no recent hormone treatment. GnRH is delivered at a dose of 75 to 100 ng/kg every 90 minutes by an intravenous catheter [121].

2. Tubal Factor Infertility

In the 1970s, the major approach to the treatment of tubal factor infertility was pelvic reconstructive surgery, which was a modestly effective treatment. Currently, IVF is far more successful than surgery in the treatment of tubal factor infertility. The benefits and risks related to IVF are discussed in Chapter 17. Using advanced embryo culture media and day 3 embryo transfer, pregnancy rates in the range of 50% per treatment cycle have been observed in women under 35 years of age. In one randomized study of treatment for tubal factor infertility, IVF was documented to be more effective than no treatment in achieving pregnancy [122].

3. Male Factor Infertility

Approximately 10% of men with azoospermia or severe oligospermia have abnormal karyotypes, suggesting that a stem cell mutation is responsible for the disorder of sperm production. For men with azoospermia, the most common treatment is insemination of the female partner with donor sperm. For men with severe oligospermia, treatment with donor sperm or harvesting of sperm for intracytoplasmic sperm injection (ICSI) are the most commonly used infertility treatments. ICSI involves the direct injection of a single sperm into the oocyte cytoplasm. Only one live sperm is required for successful ICSI treatment.

VI. Adverse Effects of Infertility Therapy

The medical side effects of infertility treatment must be considered. While the majority of couples undergoing infertility evaluation are treated with conventional therapies, the advanced fertility procedures, especially IVF, have generated considerable controversy concerning their medical consequences.

Ovarian stimulation with gonadotropin is associated with an approximately 1% risk of ovarian hyperstimulation syndrome (OHSS). In IVF cycles, risk factors for OHSS appear to be peak estradiol concentrations greater than 2000 pg/ml, more than 15 follicles larger than 12 mm in diameter, and the establishment of a successful pregnancy [123]. OHSS that develops during IVF cycles typically resolves with conservative management. Harvesting of granulosa cells at the time of oocyte retrieval may reduce the chance of developing OHSS in an IVF cycle [124]. Oocyte retrieval can be associated with pelvic bleeding or pelvic infection. Both complications are rare, occurring in less than 1 in 500 cases.

Several studies have suggested that drugs used to induce ovulation may be associated with an increased risk of ovarian cancer [125,126]. Complicating factors are that a reasonably strong association appears to exist between ovarian cancer and nulliparity per se, whether due to infertility or choice, and that detection bias may occur due to visualization of the ovaries during infertility therapy or subsequent follow-up. The presence of bias can be assessed by examining the interval between treatment and subsequent ovarian cancer diagnosis and by looking for a relationship between the number of treated cycles and the occurrence of cancer. Additional research that addresses this question is underway.

The major medical risk associated with IVF is the potential for high order multiple gestations and the accompanying increased risks of pregnancy complications including prematurity and low birth weight [127]. The advent of blastocyst transfer will minimize the number of embryos that are transferred during each procedure while improving the chances of conception.

VII. Cost of Fertility Therapy

Some authorities have suggested that the cost of an IVF live birth ranges from $44,000 to $211,940 [128]. This analysis assumes a live birth rate in the range of 10% per cycle, far below that actually achieved at most centers. The Brigham and Women's Hospital IVF program analyzed the actual charges and live births for calendar year 1993 [129]. Couples were assigned to one of three groups based on their clinical characteristics. Group A consisted of couples where the woman was younger than 32 years of age and the male partner had a normal semen analysis. Group B consisted of women under 40 years of age with a male partner who had an abnormal semen analysis. Group C consisted of women 40 to 42 years of age with a male partner who had a normal semen analysis. The live birth rates in the three groups were 35%, 24%, and 19%. For group A, the cost of a live birth was $23,000. For group B, the cost of a live birth was $34,000. For group C, in the first cycle of IVF, the cost of a live birth was $43,000. Because the live birth rate was "low" in Group C, many of these couples had multiple cycles of IVF. When the analysis was restricted to couples in Group C who completed three cycles of IVF the cost of a live birth was $75,000.

These studies demonstrate that the cost of a live birth from IVF is directly related to the clinical characteristics of the couple. Currently, the cost of adoption in the United States is in the range of $12,000 to $30,000. The cost of detecting one abnormal fetus by alpha-fetoprotein screening is in the range of $40,000. These comparisons support the use of IVF for the treatment of infertility that has not responded to less intensive treatments.

As the cost of IVF decreases and the success rate increases there will be an increase in the use of IVF. If the cost of IVF decreases to $3000 per cycle and the success rate increases to 50% per cycle, the cost of the evaluation of the infertile couple might be limited to those tests that cost less than several thousand dollars. Non-IVF treatment would be limited to those interventions where the total cost of the treatment was less than approximately $4000 per live birth. This trend might reduce the use of classic surgical interventions for the diagnosis and treatment of infertility.

VIII. Summary and Directions for Future Research, Diagnosis, and Treatment

Infertility affects 10–15% of married couples in industrialized nations and has been reported to affect more than 30% of women in sub-Saharan Africa. Infertility has long been understudied, partly because of the difficulties of identifying the population of infertile women and in defining what is an inherently heterogeneous group of conditions joined only by a single manifestation—the absence of conception. Epidemiologic studies have identified risk factors for several major types of infertility. Ovulatory dysfunction has been associated with extremes of body weight, polycystic ovary syndrome, and excessive exercise. An increased risk of tubal infertility has been associated with histories of pelvic inflammatory disease, sexually transmitted diseases, intrauterine device use, and lifestyle and dietary factors such as cigarette smoking, alcohol drinking, and caffeine intake. The association of infertility with moderate and advanced endometriosis is related to pelvic adhesions, but the relationship of minimal and mild endometriosis to infertility remains controversial. Advances in diagnostic techniques and treatments for infertility and a more open social climate have resulted in an increase in the number of couples seeking medical assistance to help them achieve conception. Epidemiologic studies of infertility face methodological challenges including the definition of infertility, the choice of study design, and the selection of appropriate comparison subjects.

Continued research into the causes of infertility and any consequences of infertility therapy are a high priority. Future studies must recognize the multifactorial nature of infertility and strive to establish specificity in the disease definition. Research on the role of dietary components and micronutrients, occupational exposures, and environmental hazards is needed. In addition, investigators must realize that the preconceptual exposures of both the woman and her partner need to be considered, and techniques for exposure assessment, including industrial hygiene and toxicological assessments, must be improved.

Studies on the physiology of fertility that result from *in vitro* fertilization techniques will continue to increase our understanding of influences on oocyte quality (in addition to a woman's age and a history of smoking) and have the potential to improve the prognostic value of sperm assessment. Advances in infertility therapy including expansion of services to those currently underserved and techniques to improve success rates while minimizing the likelihood of multiple births need to be developed and evaluated.

References

1. Marchbanks, P. A., Peterson, H. B., Rubin, G. L., and Wingo, P. A. (1989). Research on infertility: Definition makes a difference. *Am. J. Epidemiol.* **130,** 259–267.

2. Cramer, D. W., Walker, A. M., and Schiff, I. (1979). Statistical methods in evaluating the outcome of infertility therapy. *Fertil. Steril.* **32,** 80–86.

3. Healy, D. L., Trounson, A. O., and Andersen, A. N. (1994). Female infertility: Causes and treatment. *Lancet* **343,** 1539–1544.

4. Hoxsey, R., and Rinehart, J. S. (1997). Infertility and subsequent pregnancy. *Clin. Perinatol.* **24,** 321–342.

5. World Health Organization (1987). Infections, pregnancies, and infertility: Perspectives on prevention. *Fertil. Steril.* **47,** 964–968.

6. Sciarra, J. (1994). Infertility: An international health problem. *Int. J. Gynecol. Obstet.* **46,** 155–163.

7. Belsey, M. A. (1976). The epidemiology of infertility: A review with particular reference to sub-Saharan Africa. *Bull. W.H.O.* **54,** 319–341.

8. Population Information Program, Johns Hopkins University (1983). Infertility and sexually transmitted disease: A public health challenge. *Popul. Rep. Ser. L* **4,** L-113–L-151.

9. Omar, M. M. (1994). Fertility, infertility and child survival of Somali women. *Scand. J. Soc. Med.* **22,** 194–200.

10. Sundby, J., and Schei, B. (1996). Infertility and subfertility in Norwegian women aged 40–42. *Acta Obstet. Gynecol. Scand.* **75,** 832–837.

11. Stephen, E. H., and Chandra, A. (1997). Updated projections of infertility in the United States: 1995–2025. *Fertil. Steril.* **70,** 30–34.

12. Cates, W., Farley, T. M. M., and Rowe, P. J. (1985). Worldwide patterns of infertility: Is Africa different? *Lancet* **1,** 596–598.

13. Collins, J. A. (1995). Unexplained infertility. *In* "Infertility: Evaluation and Treatment" (W. R. Keye, R. J. Chang, R. W. Rebar, and M. R. Soules, eds.), pp. 249–262. Saunders, Philadelphia.

14. Howards, S. S. (1995). Treatment of male infertility. *N. Engl. J. Med.* **332,** 312–317.

15. Whiteford, L. M., and Gonzalez, L. (1995). Stigma: The hidden burden of infertility. *Soc. Sci. Med.* **40,** 27–36.

16. Menning, B. E. (1980). The emotional needs of infertile couples. *Fertil. Steril.* **34,** 313–319.

17. Valentine, D. P. (1986). Psychological impact of infertility: Identifying issues and needs. *Soc. Work Health Care* **11,** 61–69.

18. Robinson, G. E., and Stewart, D. E. (1996). The psychological impact of infertility and new reproductive technologies. *Harv. Rev. Psychiatry* **4,** 168–172.

19. Draye, M. A., Fugate Woods, N., and Mitchell, E. (1988). Coping with infertility in couples: Gender differences. *Health Care Women Int.* **9,** 163–175.

20. Abbey, A., Hallman, L. J., and Andrews, F. M. (1992). Psychosocial, treatment, and demographic predictors of the stress associated with infertility. *Fertil. Steril.* **57,** 122–128.

21. Cook, R. (1993). The relationship between sex role and emotional functioning in patients undergoing assisted conception. *J. Psychosom. Obstet. Gynaecol.* **14,** 31–40.

22. Becker, G., and Nachtigall, R. D. (1994). 'Born to be a mother': The cultural construction of risk in infertility treatment in the U.S. *Soc. Sci. Med.* **39,** 507–518.

23. Boivin, J., Takefman, J. E., Tulandi, T., and Brender, W. (1995). Reactions to infertility based on extent of treatment failure. *Fertil. Steril.* **63,** 801–807.

24. Mahlstedt, P. P. (1985). The psychological component of infertility. *Fertil. Steril.* **43,** 335–346.

25. Litt, M. D., Tennan, H., Affleck, G., and Klock, S. (1992). Coping and cognitive factors in adaptation to in vitro fertilization failure. *J. Behav. Med.* **15,** 171–187.

26. Boivin, J., and Takefman, J. E. (1995). Stress level across stages of in vitro fertilization in subsequently pregnant and non-pregnant women. *Fertil. Steril.* **64,** 802–810.

27. Wilcox, L. S., and Mosher, W. D. (1993). Use of infertility services in the United States. *Obstet. Gynecol.* **82,** 122–127.

28. Mosher, W. D., and Pratt, W. F. (1990). "Fecundity and Infertility in the United States, 1965–1988," DHHS Publ. No. (PHS)91-1250, Adv. Data No. 192. National Center for Health Statistics, Hyattsville, MD.

29. Aral, S. O., and Cates, W. (1983). The increasing concern with infertility: Why now? *JAMA, J. Am. Med. Assoc.* **250,** 2327–2331.

29a. Office of Technology Assessment (1988). "Infertility: Medical and Social Choices," Report Brief. Office of Technology Assessment, Washington, DC.

30. Hirsch, M. B., and Mosher, W. D. (1987). Characteristics of infertile women in the United States and their use of infertility services. *Fertil. Steril.* **47,** 618–625.

31. Phipps, W. R. (1996). The future of infertility services. *Fertil. Steril.* **66,** 202–204.

32. Olsen, J., Kuppers-Chinnow, M., and Spinelli, A. (1996). Seeking medical help for subfecundity: A study based upon surveys in five European countries. *Fertil. Steril.* **66,** 95–100.

33. Gunnell, D. J., and Ewings, P. (1994). Infertility prevalence, needs assessment, and purchasing. *J. Public Health Med.* **16,** 29–35.

34. Schmidt, L., Munster, K., and Helm, P. (1995). Infertility and the seeking of infertility treatment in a representative population. *Br. J. Obstet. Gynaecol.* **102,** 978–984.

35. Buckett, W., and Bentick, B. (1997). The epidemiology of infertility in a rural population. *Acta Obstet. Gynecol. Scand.* **76,** 233–237.

36. Rachootin, P., and Olsen, J. (1981). Social selection in seeking medical care for reduced fecundity among women in Denmark. *J. Epidemiol. Commun. Health* **35,** 262–264.

37. Templeton, A., Fraser, C., and Thompson, B. (1991). Infertility—epidemiology and referral practice. *Hum. Reprod.* **6,** 1391–1394.

38. Henshaw, S. K., and Orr, M. T. (1987). The need and unmet need for infertility services in the United States. *Fam. Plann. Perspect.* **19,** 180–186.

39. Schenker, J. G. (1997). Infertility evaluation and treatment according to Jewish law. *Eur. J. Obstet. Gynecol. Reprod. Biol.* **71,** 113–121.

39a. Baker, T. G. (1971). Radiosensitivity of mammalian oocytes with particular reference to the human female. *Am. J. Obstet. Gynecol.* **110,** 746–761.

40. Hull, M. G., Savage, P. E., Bromham, D. R., Ismail, A. A., and Morris, A. F. (1982). The value of a single serum progesterone measurement in the midluteal phase as a criterion of a potentially fertile cycle derived from treated and untreated conception cycles. *Fertil. Steril.* **37,** 355–360.

41. Schenken, R. S., Asch, R. H., and Williams, R. F. (1984). Etiology of infertility in monkeys with endometriosis: Luteinized unruptured follicles, luteal phase defects, pelvic adhesions and spontaneous abortions. *Fertil. Steril.* **41,** 122–127.

42. Olive, D. L., and Lee, K. L. (1986). Analysis of sequential treatment protocols for endometriosis-associated infertility. *Am. J. Obstet. Gynecol.* **154,** 613–619.

43. Garcia, C. R., and David, S. S. (1977). Pelvic endometriosis: Infertility and pelvic pain. *Am. J. Obstet. Gynecol.* **129,** 740–746.

44. Marcoux, S., Maheux, R., and Berube, S. (1997). The Canadian Collaborative Group on Endometriosis. Laparoscopic surgery in infertile women with minimal or mild endometriosis. *N. Engl. J. Med.* **337,** 217–222.

45. Acién, P. (1997). Incidence of Müllerian defects in fertile and infertile women. *Hum. Reprod.* **12,** 1372–1376.

46. Antonioli, D. A., Burke, L., and Friedman, E. A. (1980). Natural history of diethylstilbestrol-associated genital tract lesions: Cervical ectopy and cervicovaginal hood. *Am. J. Obstet. Gynecol.* **137,** 847–853.

47. Jefferies, J. A., Robboy, S. J., O'Brien, P. C., Bergstralh, E. J., Labarthe, D. R., Barnes, A. B., Noller, K. L., Hatab, P. A., Kaufman, R. H., and Townsend, D. E. (1984). Structural anomalies of the cervix and vagina in women enrolled in the Diethylstilbestrol Adenosis (DESAD) Project. *Am. J. Obstet. Gynecol.* **148,** 59–66.

48. Marshall, L. M., Spiegelman, D., Goldman, M. B., Manson, J. E., Colditz, G. A., Barbieri, R. L., Stampfer, M. J., and Hunter, D. J. (1998). A prospective study of reproductive factors and oral contraceptive use in relation to the risk of uterine leiomyomata. *Fertil. Steril.* **70,** 432–439.

49. Rein, M. S., Freidman, A. J., Barbieri, R. L., Pavelka, K., Fletcher, J. A., and Morton, C. C. (1991). Heterogeneous cytogenetic abnormalities are associated with uterine leiomyomata. *Obstet. Gynecol.* **77,** 923–926.

50. Schoenmakers, E. F., Wanschura, S., Mols, R., Bullerdiek, J., and Van den Berghe, H. (1995). Recurrent rearrangements in the high mobility group protein gene, HMGI-C, in benign mesenchymal tumours. *Nat. Genet.* **10,** 436–444.

51. Farhi, J., Ashkenazi, J., Feldberg, D., Dicker, D., Orvieto, R., and Ben Rafael, Z. (1995). Effect of uterine leiomyomata on the results of in-vitro fertilization treatment. *Hum. Reprod.* **10,** 2576–2578.

52. Jette, N. T., and Glass, R. H. (1972). Prognostic value of the postcoital test. *Fertil. Steril.* **23,** 29–32.

53. Oei, S. G., Bloemenkamp, K. W., Helmerhorst, F. M., Naaktgeboren, N., and Keirse, M. J. (1996). Evaluation of the postcoital test for assessment of 'cervical factor' infertility. *Eur. J. Obstet. Gynecol. Reprod. Biol.* **64,** 217–220.

54. Kovacs, G. T., Newman, G. B., and Henson, G. L. (1978). The postcoital test: What is normal? *Br. Med. J.* **1,** 818.

55. Glatstein, I. Z., Best, C. L., Palumbo, A., Sleeper, L. A., Friedman, A. J., and Hornstein, M. D. (1995). The reproducibility of the postcoital test: A prospective study. *Obstet. Gynecol.* **85,** 396–400.

56. Dunphy, B. C., Neal, L. M., and Cooke, I. D. (1989). The clinical value of conventional semen analysis. *Fertil. Steril.* **51,** 324–329.

57. Peng, H. Q., Collins, J. A., Wilson, E. H., and Wrixon, W. (1987). Receiver-operating characteristics curves for semen analysis variables: Methods for evaluating diagnostic tests of male gamete function. *Gamete Res.* **17,** 229–236.

58. World Health Organization (1992). The influence of varicocele on parameters of fertility in a large group of men presenting to infertility clinics. *Fertil. Steril.* **57,** 1289–1293.

59. Madgar, I., Weissenberg, R., Lunenfeld, B., Karasik, A., and Goldwasser, B. (1995). Controlled trial of high spermatic vein ligation for varicocele in infertile men. *Fertil. Steril.* **63,** 120–124.

60. Breznk, R., Vlaisavljevic, V., and Borko, E. (1993). Treatment of varicocele and male fertility. *Arch. Androl.* **30,** 157–160.

61. Goldman, M. B. (1994). Types of epidemiologic studies used to investigate infertility. *Infertil. Reprod. Med. Clin. North Am.* **5,** 239–258.

62. Herbst, A. L., Ulfelder, H., and Poskanzer, D. C. (1971). Adenocarcinoma of the vagina. Association of maternal stilbestrol therapy with tumor appearance in young women. *N. Engl. J. Med.* **284,** 878–881.

63. Christian, C. D. (1974). Maternal deaths associated with an intrauterine device. *Am. J. Obstet. Gynecol.* **119,** 441–444.

64. Houston, D. E., Noller, K. L., Melton, L. J., Selwyn, B. J., and Hardy, R. J. (1987). Incidence of pelvic endometriosis in Rochester, Minnesota, 1970–1979. *Am. J. Epidemiol.* **125,** 959–969.

65. Society for Assisted Reproductive Technology and the American Society for Reproductive Medicine (1998). Assisted reproductive technology in the United States and Canada: 1995 results generated from the American Society for Reproductive Medicine/Society for Assisted Reproductive Technology Registry. *Fertil. Steril.* **69,** 389–398.

66. Abma, J., Chandra, A., Mosher, W., Peterson, L., and Piccinino, L. (1997). "Fertility, Family Planning, and Women's Health: New data from the 1995 National Survey of Family Growth," Vital Health Stat. Ser. 23, No. 19, DHHS Publ. No. (PHS) 97-1995. National Center for Health Statistics. Hyattsville, MD.

67. Sharpe, R. M., and Skakkebaek, N. E. (1993). Are oestrogens involved in falling sperm counts and disorders of the male reproductive tract? *Lancet* **341,** 1392–1395.

68. Baird, D. D., Wilcox, A. J., and Weinberg, C. R. (1986). Use of time to pregnancy to study environmental exposures. *Am. J. Epidemiol.* **124,** 470–480.

69. Heckman, J. J., and Walker, J. R. (1990). Estimating fecundability from data on waiting times to first conception. *J. Am. Stat. Assoc.* **85,** 283–294.

70. Weinberg, C. R., Baird, D. D., and Wilcox, A. J. (1994). Sources of bias in studies of time to pregnancy. *Stat. Med.* **13,** 671–681.

71. Weinberg, C. R., Baird, D. D., and Rowland, A. S. (1993). Pitfalls inherent in retrospective time-to-event studies: The example of time-to-pregnancy. *Stat. Med.* **12,** 867–879.

72. Wilcox, A. J., Weinberg, C. R., O'Connor, J. F., Baird, D. D., Schlatterer, J. P., Canfield, R. E., Armstrong, E. G., and Nisula, B. C. (1988). Incidence of early loss of pregnancy. *N. Engl. J. Med.* **319,** 189–194.

73. Baird, D. D., Weinberg, C. R., Schwingl, P., and Wilcox, A. J. (1994). Selection bias associated with contraceptive practice in time-to-pregnancy studies. *Ann. N. Y. Acad. Sci.* **709,** 156–164.

74. Daling, J. R., Weiss, N. S., Metch, B. J., Soderström, R. M., Moore, D. E., Spadoni, L. R., and Stadel, B. V. (1985). Primary tubal infertility in relation to the use of an intrauterine device. *N. Engl. J. Med.* **312,** 937–941.

75. Cramer, D. W., Schiff, I., Schoenbaum, S. C., Gibson, M., Belisle, S., Albrecht, B., Stillman, R. J., and Berger, M. J. (1985). Tubal infertility and the intrauterine device. *N. Engl. J. Med.* **312,** 941–947.

76. Neutra, R. R., Swan, S. H., Hertz-Picciotto, I., Windham, G. C., Wrensch, M., Shaw, G. M., Fenster, L., and Deane, M. (1992). Potential sources of bias and confounding in environmental epidemiologic studies of pregnancy outcomes. *Epidemiology* **3,** 134–142.

77. Bryant, H. E., Visser, N., and Love, E. J. (1989). Records, recall loss, and recall bias in pregnancy: A comparison of interview and medical records data of pregnant and postnatal women. *Am. J. Public Health* **79,** 78–80.

78. Werler, M. M., Pober, B. R., Nelson, K., and Holmes, L. B. (1989). Reporting accuracy among mothers of malformed and nonmalformed infants. *Am. J. Epidemiol.* **129,** 415–421.

79. Weiss, N. S., Daling, J. R., and Chow, W. H. (1985). Control definition in case-control studies of ectopic pregnancy. *Am. J. Public Health* **75,** 67–68.

80. Rich-Edwards, J. W., Goldman, M. B., Willett, W. C., Hunter, D. J., Stampfer, M. J., Colditz, G. A., and Manson, J. E. (1994). Adolescent body mass index and infertility caused by ovulatory disorder. *Am. J. Obstet. Gynecol.* **171,** 171–179.

81. Canfield, R. E., O'Connor, J. F., and Wilcox, A. J. (1988). Measuring human chorionic gonadotropin for detection of early pregnancy loss. *Reprod. Toxicol.* **2,** 199–203.

82. Lasley, B. L., Lohstroh, R., Kuo, A., Gold, E. B., Eskenazi, B., Samuels, S. J., Stewart, D. R., and Overstreet, J. W. (1995). Laboratory methods for evaluating early pregnancy loss in an industry-based population. *Am. J. Ind. Med.* **28,** 771–781.

83. Lemasters, G. K., and Pinney, S. M. (1989). Employment status as a confounder when assessing occupational exposures and spontaneous abortion. *J. Clin. Epidemiol.* **42,** 975–981.

84. Savitz, D. A., Whelan, E. A., Rowland, A. S., and Kleckner, R. C. (1990). Maternal employment and reproductive risk factors. *Am. J. Epidemiol.* **132,** 933–945.

85. Savitz, D. A., and Harlow, S. D. (1991). Selection of reproductive health end points for environmental risk assessment. *Environ. Health Perspect.* **90,** 159–164.

86. Mitchell, A. A., Cottler, L. B., and Shapiro, S. (1986). Effect of questionnaire design on recall of drug exposure in pregnancy. *Am. J. Epidemiol.* **123,** 670–676.

87. Eskenazi, B., and Pearson, K. (1988). Validation of a self-administered questionnaire for assessing occupational and environmental exposures of pregnant women. *Am. J. Epidemiol.* **128,** 1117–1129.

88. Buck, G. M., Sever, L. E., Batt, R. E., and Mendola, P. (1997). Life-style factors and female infertility. *Epidemiology* **8,** 435–441.

89. Grodstein, F., Goldman, M. B., and Cramer, D. W. (1994). Body mass index and ovulatory infertility. *Epidemiology* **5,** 247–250.

90. Green, B. B., Weiss, N. S., and Daling, J. R. (1988). Risk of ovulatory infertility in relation to body weight. *Fertil. Steril.* **50,** 721–726.

91. Green, B. B., Daling, J. R., Weiss, N. S., Liff, J. M., and Koepsell, T. (1986). Exercise as a risk factor for infertility with ovulatory dysfunction. *Am. J. Public Health* **76,** 1432–1436.

92. Grodstein, F., Goldman, M. B., Ryan, L., and Cramer, D. W. (1993). Self-reported use of pharmaceuticals and primary ovulatory infertility. *Epidemiology* **4,** 151–156.

93. Mueller, B. A., Daling, J. R., Weiss, N. S., and Moore, D. E. (1990). Recreational drug use and the risk of primary infertility. *Epidemiology* **1,** 195–200.

94. Grodstein, F., Goldman, M. B., and Cramer D. W. (1994). Infertility in women and moderate alcohol use. *Am. J. Public Health* **84,** 1429–1432.

95. Smith, E. M., Hammonds-Ehlers, M., Clark, M. K., Kirchner, H. L., and Fuortes, L. (1997). Occupational exposures and risk of female infertility. *J. Occup. Environ. Med.* **39,** 138–147.

96. Fuortes, L., Clark, M. K., Kirchner, H. L., and Smith, E. M. (1997). Association between female infertility and agricultural work history. *Am. J. Ind. Med.* **31,** 445–451.

97. Mueller, B. A., Daling, J. R., Moore, D. E., Weiss, N. S., Spadoni, L. R., Stadel, B. V., and Soules, M. R. (1986). Appendectomy and the risk of tubal infertility. *N. Engl. J. Med.* **315,** 1506–1508.

98. Mueller, B. A., Luz-Jimenez, M., Daling, J. R., Moore, D. E., McKnight, B., and Weiss, N. S. (1992). Risk factors for tubal infertility. Influence of history of prior pelvic inflammatory disease. *Sex. Transm. Dis.* **19,** 28–34.

99. Grodstein, F., Goldman, M. B., and Cramer, D. W. (1993). Relation of tubal infertility to history of sexually transmitted diseases. *Am. J. Epidemiol.* **137,** 577–584.

100. Westrom, L. (1980). Incidence, prevalence and trends of acute pelvic inflammatory disease and its consequences in industrialized countries. *Am. J. Obstet. Gynecol.* **138,** 880–892.

101. Patton, D. L., Askienazy-Elbhar, M., Henry-Suchet, J., Campbell, L. A., Cappuccio, A., Tannous, W., Wang, S. P., and Kuo, C. C. (1994). Detection of *Chlamydia trachomatis* in fallopian tube tissue in women with postinfectious tubal infertility. *Am. J. Obstet. Gynecol.* **171,** 95–101.

102. Grodstein, F., Goldman, M. B., Ryan, L., and Cramer, D. W. (1993). Relation of female infertility to consumption of caffeinated beverages. *Am. J. Epidemiol.* **137,** 1353–1360.

103. Baird, D. D., Weinberg, C. R., Voigt, L. F., and Daling, J. R. (1996). Vaginal douching and reduced fertility. *Am. J. Public Health* **86,** 844–850.

104. Phipps, W. R., Cramer, D. W., Schiff, I., Belisle, S., Stillman, R., Albrecht, B., Stadel, B., Gibson, M., Berger, M. J., and Wilson, E. (1987). The association between smoking and female infertility as influenced by cause of the infertility. *Fertil. Steril.* **48,** 377–382.

105. Cramer, D. W., Goldman, M. B., Schiff, I., Belisle, S., Albrecht, B., Stadel, B., Gibson, M., Wilson, E., Stillman, R. J., and Thompson, I. (1987). The relationship of tubal infertility to barrier method and oral contraceptive use. *JAMA, J. Am. Med. Assoc.* **257,** 2446–2450.

106. Cramer, D. W., Wilson, E., Stillman, R. J., Berger, M. J., Belisle, S., Schiff, I., Albrecht, B., Gibson, M., Stadel, B. V., and Schoenbaum, S. C. (1986). The relation of endometriosis to menstrual characteristics, smoking, and exercise. *JAMA, J. Am. Med. Assoc.* **255,** 1904–1908.

107. Goldman, M. B., and Cramer, D. W. (1990). The epidemiology of endometriosis. *In* "Current Concepts in Endometriosis" (D. R. Chadha and V. C. Buttram, eds.), pp. 15–31. Liss, New York.

108. Darrow S. L., Vena, J. E., Batt, R. E., Zielezny, M. A., Michalek, A. M., and Selman, S. (1993). Menstrual cycle characteristics and the risk of endometriosis. *Epidemiology* **4,** 135–142.

109. Darrow, S. L., Selman, S., Batt, R. E., Zielezny, M. A., and Vena, J. E. (1994). Sexual activity, contraception, and reproductive factors in predicting endometriosis. *Am. J. Epidemiol.* **140,** 500–509.

109a. Wilcox, A. J., Weinberg, C. R., and Baird, D. D. (1995). Timing of sexual intercourse in relation to ovulation. *N. Engl. J. Med.* **333,** 1517–1521.

110. Scott, R. T., and Hormann, G. E. (1995). Prognostic assessment of ovarian reserve. *Fertil. Steril.* **63,** 1–11.

111. Warburton, D. (1987). Reproductive loss: How much is preventable. *N. Engl. J. Med.* **316,** 158–160.

112. Abraham, S., Mira, M., and Llewellyn-Jones, D. (1990). Should ovulation be induced in women recovering from an eating disorder or who are compulsive exercisers. *Fertil. Steril.* **53,** 566–568.

113. Pasquali, R., Antenucci, D., and Casimirri, F. (1989). Clinical and hormonal characteristics of obese and amenorrheic hyperandrogenic women before and after weight loss. *J. Clin. Endocrinol. Metab.* **68,** 173–178.

114. Clark, A. M., Ledger, W., Galletly, C., Tomlinson, L., Blaney, F., Wang, X., and Norman, R. J. (1995). Weight loss results in significant improvement in pregnancy and ovulation rates in anovulatory obese women. *Hum. Reprod.* **10,** 2705–2712.

115. Hull, M. G., Knuth, U. A., Murray, M. A., and Jacobs, H. S. (1979). The practical value of the progestogen challenge test, serum estradiol estimation or clinical examination in assessment of the estrogen state and response to clomiphene in amenorrhea. *Br. J. Obstet. Gynaecol.* **86,** 799–805.

116. Wu, C. H., and Winkel, C. A. (1989). The effect of therapy initiation day on clomiphene citrate therapy. *Fertil. Steril.* **52,** 564–568.

117. Gysler, M., March, C. M., Mishell, D. R., and Bailey, E. J. (1982). A decade's experience with an individualized clomiphene treatment regimen including its effect on the postcoital test. *Fertil. Steril.* **37,** 161–167.

117a. Lunenfeld, B., and Insler, V. (1995). Human gonadotropins. *In* Reproductive Medicine and Surgery'' (E. E. Wallach and H. A. Zacur, eds.), p. 617. Mosby, St. Louis, Mo.

118. Gemzell, C. A., Diczfalusy, E., and Tillinger, K. G. (1958). Clinical effect of human pituitary follicle stimulating hormone. *J. Clin. Endocrinol. Metab.* **18,** 1333–1339.

119. Lunenfeld, B., Sulimovici, S., Rabau, E., and Eshkol, A. (1962). L'induction de l'ovulation dans les amenorrhea hypophysaires par un traitement combaine de gonadotropine urinaires menopausiques et de gonadotropines chorionique. *C. R. Soc. Fr. Gynecol.* **35,** 346–351.

120. Karnitis, V. J., Townson, D. H., Friedman, C. I., and Danforth, D. R. (1994). Recombinant human follicle stimulating hormone

stimulates multiple follicular growth, but minimal estrogen production in gonadotropin releasing hormone antagonist treated monkeys: Examining the role of luteinizing hormone in follicular development and steroidogenesis. *J. Clin. Endocrinol. Metab.* **79,** 91–97.

121. Santoro, N., Wierman, M. E., Filicori, M., Waldstreicher, J., and Crowley, W. F. (1986). Intravenous administration of pulsatile gonadotropin releasing hormone in hypothalamic amenorrhea. Effects of dosage. *J. Clin. Endocrinol. Metab.* **62,** 109–116.

122. Soliman, S., Daya, S., Collins, J., and Jarrell, J. (1993). A randomized trial of in vitro fertilization versus conventional treatment for infertility. *Fertil. Steril.* **59,** 1239–1244.

123. Forman, R. G., Frydman, R., Egan, D., Ross, C., and Barlow, D. H. (1990). Severe ovarian hyperstimulation syndrome using agonists of gonadotropin releasing hormone for in vitro fertilization: A European series and a proposal for prevention. *Fertil. Steril.* **53,** 502–509.

124. Gonen, Y., Powell, W. A., and Casper, R. F. (1991). Effect of follicular aspiration on hormonal parameters in patients undergoing ovarian stimulation. *Hum. Reprod.* **6,** 356–358.

125. Spirtas, R., Kaufman, S. C., and Alexander, N. J. (1993). Fertility drugs and ovarian cancer: Red alert or red herring? *Fertil. Steril.* **59,** 291–293.

126. Whittemore, A. S., Harris, R., Itnyre, J., and the Collaborative Ovarian Cancer Group (1992). Characteristics relating to ovarian cancer risk: Collaborative analysis of 12 U.S. case-control studies. II. Invasive epithelial ovarian cancers in white women. *Am. J. Epidemiol.* **136,** 1184–1203.

127. Callahan, T. L., Hall, J. E., Ettner, S. L., Christiansen, C. L., Greene, M. F., and Crowley, W. F. (1994). The economic impact of multiple-gestation pregnancies and the contribution of assisted-reproduction techniques to their incidence. *N. Engl. J. Med.* **331,** 244–249.

128. Neumann, P. J., Gharib, S. D., and Weinstein, M. C. (1994). The cost of a successful delivery with in vitro fertilization. *N. Engl. J. Med.* **331,** 239–243.

129. Trad, F. S., Hornstein, M. D., and Barbieri, R. L. (1995). In vitro fertilization: A cost effective alternative for infertile couples? *J. Assist. Reprod. Genet.* **12,** 418–421.

17

Assisted Reproductive Technologies

DAVID A. GRAINGER AND BRUCE L. TJADEN

Division of Reproductive Endocrinology
University of Kansas School of Medicine—Wichita
Wichita, Kansas

I. Introduction

It has been twenty years since the birth of Louise Brown, the first successful *in vitro* fertilization (IVF) pregnancy. The introduction of this assisted reproductive technology (ART) by Steptoe and Edwards [1] revolutionized the approach to the couple with infertility. Currently, IVF represents the final common pathway in the treatment of infertility, regardless of etiology. With improving success rates, compelling medical and economic arguments can be made for the introduction of this efficacious modality earlier in the treatment algorithm. The intent of this chapter is for the reader to better understand the population of patients to which ART is applied, to review the process of ART, to identify factors affecting success or failure of the technique, and to examine potential complications of the procedure. Success in ART can be assessed using several different measurements including the pregnancy per cycle rate, the live birth per cycle rate, the live birth per egg retrieval rate, and the live birth per embryo transfer rate.

II. Epidemiology of Infertility

Infertility is a common condition worldwide with important medical, social, and economic implications. While the prevalence of the disease has been stable over the 1990s, the demand for services has exploded [2–4]. The traditional definition of infertility is the inability to conceive after one year of unprotected intercourse. However, because infertility increases with maternal age, a newly married 39 year-old woman may (should) not be asked to attempt unprotected intercourse for one year prior to any diagnostic testing. The epidemiology of infertility, including risk factor and prevalence, is presented in Chapter 16 and will be reviewed only briefly here.

Data from the 1976, 1982, and 1988 National Surveys of Family Growth indicate a fairly stable prevalence of infertility in the United States. Exclusion of sterilized individuals reveals a prevalence of infertility of 13.3% in 1965 and 13.7% in 1988 [5]. While these estimates of the prevalence of infertility are representative, they do not reflect the use of infertility services. There are many barriers to accessibility of these services: inability to pay (in the United States, widespread restrictions apply to reimbursement for infertility related services), lack of adequate information regarding effective therapies, geographic barriers, and social barriers (or lack of interest in seeking infertility services). A sample of women with primary infertility in the United States revealed that only 50% had ever sought medical services for their disorder [6]. The National Survey of Family Growth compared 1982 data to 1995 data and confirmed a slight increase in the proportion of women ages 15–44 reporting impaired fecundity (8 to 10%) [7]. However, the proportion of women with impaired fecundity that sought infertility services did not change (44%). Women seeking fertility services tended to be older and of higher socioeconomic status. The increase in impaired fecundity occurred across education, race, income, and ethnicity subgroups.

III. Assisted Reproductive Technologies (ART)

The fertilization of a human egg outside of the body, the culturing of the embryo(s) in the laboratory, and the subsequent transfer of embryo(s) to the uterus has resulted in the birth of tens of thousands of children worldwide. Registries in the United States and abroad reflect an explosion both in the number of programs offering ART and in the utilization of these technologies since the 1980s. The growth in utilization has led to increased governmental oversight in the United States. The Fertility Clinic Success Rate and Certification Act of 1992 mandates clinic-specific reporting of success rates to the Secretary of Health and Human Services through the Centers for Disease Control and Prevention (CDC). The law also requires the Secretary to annually publish and distribute to the States and public the clinic-specific success rates. The law also requires IVF laboratory oversight and certification. Implementation of this law has led to a joint and productive effort between the Society for Assisted Reproductive Technologies (SART) and the CDC. While the process continues to be refined, clinic-specific reports are now available on the CDC web page [8]. There has been a tremendous growth in the number of IVF retrievals (see Fig. 17.1) and programs in the United States offering these highly specialized services. In 1990, 96 programs reported 8725 IVF retrievals. In 1996, 300 programs reported 44,647 IVF retrievals. It is estimated that in the United States in 1998, over 60,000 IVF retrievals were performed by close to 400 programs [SART, personal communication].

IVF was originally envisioned as a solution for severe tubal factor infertility. Fertilization, embryo culture, and transport to the uterus are normal functions of the human fallopian tube. The laboratory serves as a surrogate for this process. Initially, IVF was performed in a "natural" cycle—that is one in which no medications for ovarian stimulation were employed. This resulted in the recovery of a single oocyte, with the transfer (hopefully) of a single embryo. This process has been modified over the years to include ovarian stimulation (initially with clomiphene citrate, later with human menopausal gonadotropins, and

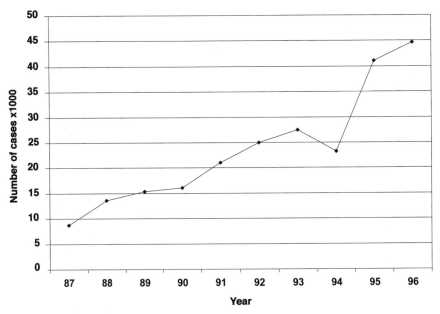

Fig. 17.1   Number of IVF retrievals in the United States: 1987–1996.

currently with recombinant gonadotropins), such that multiple oocytes may be retrieved. This results in the fertilization of more than one oocyte and the potential transfer of more than one embryo. Furthermore, retrieval techniques have evolved from laparotomy, to laparoscopy, to ultrasound guided transabdominal oocyte retrieval, to transurethral oocyte retrieval, to the current practice of transvaginal ultrasound guided oocyte retrieval. This procedure does not require general anesthesia and can be easily accomplished in an outpatient setting. It additionally has the advantage of ovarian accessibility, as many of the patients undergoing IVF have severe pelvic adhesive disease that would make laparoscopic retrieval virtually impossible.

The use of controlled ovarian hyperstimulation with markedly elevated estrogen levels resulted in premature ovulation and loss of the cycle in 20–30% of cases [9]. The introduction of gonadotropin releasing hormone analogs has greatly reduced the cancellation rate of IVF cycles due to premature ovulation. Thus, it is easy to appreciate that the medically controlled cycle is far from "natural"; some have felt that the resulting endometrial environment is so nonconducive to pregnancy that all embryos should be frozen and replaced in a subsequent natural cycle. The use of supplemental progesterone in the luteal phase of the cycle, however, appears to provide an adequate environment for implantation. In most programs, fresh embryo transfers result in pregnancy rates that are twice as good as those for frozen embryo transfers. Endometrial maturation and the process of implantation remain areas of intense investigation.

Along with the ability to reliably stimulate the ovary and recover multiple oocytes came the attendant problem of embryo transfer. How many embryos should be replaced in the uterus? It became clear that the probability of pregnancy was related to the number of embryos transferred to the uterus (Fig. 17.2).

However, the probability of multiple pregnancies also increased with increasing number of embryos transferred. The incidence of multiple pregnancy following IVF remains a major concern. First, any multiple pregnancy—even twins—increases the risk for both mother and babies. The primary risk is prematurity, but maternal risks of hypertension, gestational diabetes, and pre-eclampsia are also increased [10]. This risk increases with the increasing number of fetuses. The high order multiple pregnancy also places a tremendous burden on the health care system with regard to utilization of services. At Brigham and Women's Hospital in Boston, it was estimated that a triplet pregnancy would cost the health care system in excess of $109,000 [11]. The Cleveland Clinic estimated the cost of a triplet or quadruplet pregnancy to be $340,000 [12]. In countries where IVF is a covered treatment, the payor (usually the government) will restrict the number of embryos to be transferred. In a system that does not cover the expense of the treatment (ranging from $8,000–12,000), patients and doctors accept a higher risk of multiple pregnancy in order to maximize the chance of pregnancy. Improvements in laboratory techniques have resulted in increased implantation rates (the likelihood of a single embryo to implant). This has allowed programs to transfer fewer embryos and maintain acceptable pregnancy rates. Ideally, a single embryo would be transferred to the uterus; with improvements in blastocyst culture and transfer, this may become a reality. We would then have come full circle—Louise Brown resulted from a single embryo transfer.

IV. General Steps Involved with Assisted Reproductive Technologies

Current strategies used in IVF protocols involve manipulation of the menstrual cycle in order to achieve the maturation of

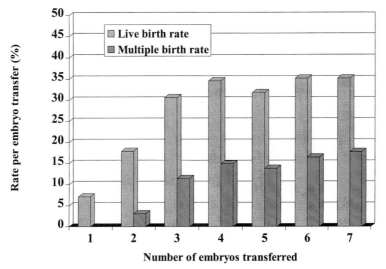

Fig. 17.2 Live birth and multiple birth rates among women younger than age 35, by number of embryos transferred, 1995.

more than one oocyte. Figure 17.3 diagrams a typical stimulation protocol. In general, there are six phases to an ART cycle:

1. Downregulation of the ovaries, frequently with some combination of oral contraceptive pills, gonadotropin releasing hormone (GnRH) analog, and progestins.

2. Stimulation of the ovaries with exogenous gonadotropins (follicle stimulating hormone (FSH) and luteinizing hormone (LH)).

3. Oocyte retrieval using a transvaginal ultrasound guided approach.

4. Fertilization, embryo culture, and any micromanipulation procedures to be performed.

5. Embryo transfer, which involves the placement of embryos into the uterus through the cervix (or laparoscopically into the fallopian tube with ZIFT).

6. Luteal support with exogenous progesterone, human chorionic gonadotropin (HCG), or both.

The specifics of each stimulation cycle depend on patient and program characteristics. Likewise, the specifics of laboratory conditions for culture including media used, protein source, days of culture, and techniques of micromanipulation may differ from program to program. The oocytes (Fig. 17.4; see color insert) are identified from the follicular fluid that is aspirated from the follicle and are transferred to an incubator at 37°C and 5% CO_2. Mature oocytes are inseminated approximately 6–8 hours after collection. Sixteen hours later the inseminated oocytes are checked for fertilization (Fig. 17.5; see color insert—the presence of male and female pronuclei). This is also the only chance to observe the oocytes for polyspermic fertilization (Fig. 17.6; see color insert), as polyspermic embryos may initiate cleavage in an apparently normal fashion. Two to three days after retrieval, the embryos are transferred to the uterus (Fig. 17.7; see color insert). This involves selection of the highest grade of embryos using morphologic criteria, loading the

embryos in a transfer catheter, and placing them 1–2 cm from the uterine fundus. The use of abdominal ultrasound may help in guiding the transfer catheter to the appropriate intrauterine location. The embryos are slowly transferred in a small volume of transfer media (15–25 μl). The catheter is slowly removed and checked under a dissecting microscope for any retained embryos. Retained embryos are retransferred to the uterus using a similar technique.

In the case of gamete intrafallopian transfer (GIFT), oocytes are collected in the fashion described above. Three to four mature oocytes and 500,000 sperm are transferred to the ampullary portion of the fallopian tube via laparoscopy. Transcervical tubal canalization techniques have been described; they are not applied in a widespread fashion at this time. Zygote intrafallopian transfer (ZIFT) involves the tubal transfer of zygotes (fertilized oocytes), usually one day after insemination. Like GIFT, the procedure requires a laparoscopy for the tubal placement of the zygotes. Both GIFT and ZIFT require the presence of at least one normal tube. With improvements in laboratory embryo culture techniques, most programs do not see dramatic improvements in pregnancy rates for GIFT or ZIFT over IVF. IVF has the obvious advantage of avoiding laparoscopy.

Micromanipulation techniques include: (1) intracytoplasmic sperm injection (ICSI—Fig. 17.8; see color insert); (2) assisted zona hatching; and (3) embryo biopsy for preimplantation genetic diagnosis (PGD). The first two of these techniques are used in over two-thirds of the IVF laboratories in the United States. PGD remains highly specialized and is performed routinely in only a few centers in the United States.

V. Indications for Assisted Reproductive Technologies

The selection of ART for establishing pregnancy depends on patient characteristics, economic feasibility, and provider characteristics. In the mid-1980s, IVF was "experimental"—a term

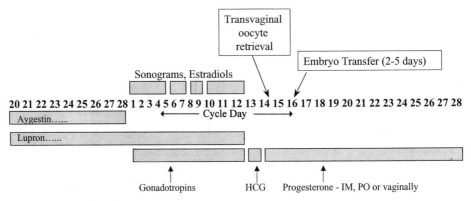

Fig. 17.3 General IVF protocols by cycle day.

insurance providers clung to into the late 1980s and early 1990s. Success rates were low, the treatment was expensive, and IVF remained an uncovered benefit. There has been a gradual change in attitude towards the technology with improvements in pregnancy and delivery rates. It is not unreasonable to expect continued improvements in IVF pregnancy rates. Many programs now have consistent pregnancy rates in excess of 50% per transfer, limiting the number of embryos transferred to two or three [8]. This success—even with a decreased number of embryos transferred—must be evaluated in parallel with success with other treatments and with the increased risk of multiple pregnancy with ART.

A. Absolute Indications

For some couples, the only chance of having their own biologic children is via IVF. These patients include women with absent fallopian tubes and men with severe male factor infertility. Success with IVF requires at minimum a uterus, one ovary (or piece thereof), and as many sperm as oocytes. With ICSI, successful pregnancies have been established using testicular biopsies and round spermatids as the sperm source. Additionally, women with premature ovarian failure utilizing donor oocytes require the use of ART.

B. Relative Indications

Women with severe pelvic adhesive disease, damaged (but patent) fallopian tubes, severe endometriosis, and some recalcitrant ovulation disorders are all examples of patients with relative indications for IVF. In the absence of a surgically correctable lesion in the male, most clinicians will treat moderate male factor infertility with ovulation induction and timed intrauterine inseminations. If a fallopian tube(s) is open and a reasonable amount of sperm is present in the ejaculate, there is always a chance of pregnancy.

C. IVF as a Final Common Treatment Pathway

The general treatment of the infertile couple involves a diagnostic phase and a treatment phase. The diagnostic phase, in general, consists of a semen analysis, hysterosalpingogram to evaluate tubal patency, endometrial biopsy to evaluate the luteal phase, postcoital test to evaluate cervical mucus, tests for ovulation including LH predictor kits or serum progesterone, and evaluation of peritoneal factors with laparoscopy. The treatment phase may include operative procedures through the laparoscope (adhesiolysis, vaporization of endometriosis, distal tubal surgery). Any specific abnormalities identified in the diagnostic phase are addressed. The treatment then follows a fairly predictable algorithm: controlled ovarian hyperstimulation (usually with clomiphene initially, followed by injectable gonadotropins) combined with intrauterine inseminations. This therapy may be attempted for six to nine months. Failure generally results in the recommendation for IVF.

There is an accumulation of data supporting IVF as a cost-effective procedure for many infertility diagnoses [13]. Algorithms and practice guidelines for the cost-effective management of the infertile couple have been validated and published [14]. However, policy decisions regarding allocation of funds to cover these services are lacking. Thus, much of the care received by infertile couples is not covered by insurance, limiting access to effective therapies to those who can afford them.

VI. Patient Selection for ART

The discussion in Section V reviewed the indications for ART. Which patients are actually utilizing these resources? Examination of the SART/CDC registry data reveals the age distribution of female partners using ART and use of ART resources by primary diagnosis (Figs. 17.9, 17.10). Notably lacking in this report is utilization of resources based on socioeconomic factors. As future data are collected and validated, this information will become available. Only 44% of infertile women ever avail themselves of infertility services; it is not known what percentage do not attempt ART because of inability to pay [7]. The current reimbursement system creates a hierarchy of care: ART for those who can afford it, infertility surgery for those with insurance that covers surgery but not ART, and minimal treatment for the rest.

It is possible to select patients for success. Programs (frequently "money-back" guarantee programs) will screen entrants

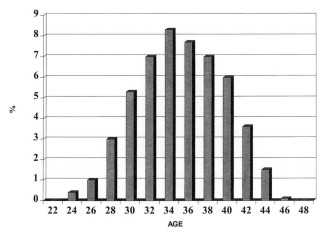

Fig. 17.9 ART user in the United States by age, 1995.

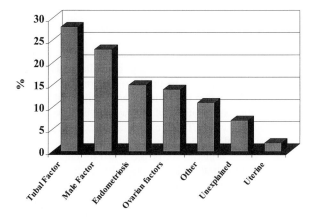

Fig. 17.10 Primary diagnosis for ART procedures.

for known predictive factors for failure (age, cycle day 3 FSH, previous failed attempts using ART). Only those patients lacking poor prognostic factors are allowed in the program. This effectively limits treatment to those patients most likely to have a live birth.

Other factors not included in the clinic-specific reports are number of previous controlled ovarian hyperstimulation (COH) cycles, number of previous inseminations, and number of previous ART cycles. By failing all of these treatments, a different population of patients may be selected as compared to those with an identical length of infertility (again, emphasizing potential differences in the patient populations between programs' success rates). Analyzing these types of data requires multivariate analyses. Several attempts have been made to use such statistical methods for predicting success; most agree that each program will require individual analysis, as the clinical and laboratory teams play such a crucial role in success [15].

VII. Defining Success in ART

The patients define success as taking home a baby. Indeed, live birth rates are clearly the most important outcome with ART. The following definitions are used by SART and the CDC [8]:

Pregnancy per cycle rate—this refers to the percentage of ART cycles started that result in a pregnancy (defined as the presence of an intrauterine gestational sac). This is the highest percentage, as some of these pregnancies will be lost through miscarriage, therapeutic abortion, or stillbirth.

Live birth per cycle rate—this is the percentage of cycles started that result in the delivery of one or more live infants.

Live birth per egg retrieval rate—this rate will exclude patients that had a cancelled cycle (perhaps because of a poor response to medications).

Live birth per embryo transfer rate—this rate includes only those cycles in which an embryo or eggs and sperm are transferred back into the woman. This rate will exclude those cycles in which no eggs were retrieved, no fertilization of eggs occurred, or when the embryos formed were deemed to be abnormal and were not transferred.*

Figure 17.11 shows the ART outcomes for 1995 and 1996 in the United States using these definitions. Live birth (LB) per

*Please note that while these statistics are technically proportions of the form a/(a + b) they are commonly called rates. Clear definition of the denominator (*i.e.,* a + b) is essential to understand ART success measures.

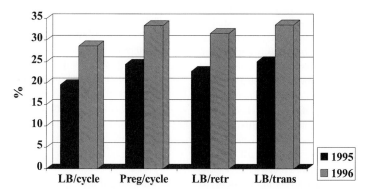

Fig. 17.11 1995 and 1996 ART pregnancy success rates for women <35, using fresh embryos from nondonor eggs.

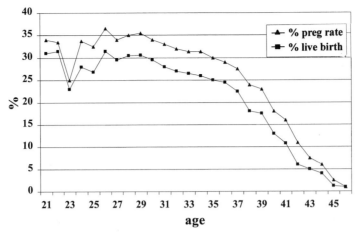

Fig. 17.12   Pregnancy and live birth rates for fresh, nondonor ART cycles by age of woman, 1996

cycle initiated is the rate that most couples are interested in. It is imperative that patients understand clearly the rates being discussed when considering ART. While live birth per cycle initiated is important, the live birth rate per oocyte retrieval may be more reflective of the quality of the ART laboratory. This presumes that the program is not canceling an inordinately large number of cycle starts (another question that must be asked).

VIII.   Predicting Success with ART: Is It Possible?

What are my chances of having a baby? This is the most frequently asked question when consulting with patients about ART. The outcome is binary: they either take home a baby or they do not. It is the placement of a probability on this binary outcome that is difficult. There are obvious deterrents to success using ART. Female age has long been considered a predictor of success. Until ICSI became widely utilized, severe male factor infertility had a poor prognosis. The ability to stimulate the ovaries, recover multiple oocytes, and transfer multiple embryos affect success. What are the quality of the embryos transferred? In spite of our most intellectual estimates of probability, the practitioners of ART are occasionally surprised at the outcome. By using the IVF registry data, we can begin to see some of the predictors of success [8].

Female age is a clear marker of decreased fecundity. This observation arises from several types of longitudinal studies [16,17] and is confirmed by evaluating the ART national registry (Fig. 17.12). This figure clearly demonstrates two important features of aging: (1) the pregnancy rates decline with age, and (2) the live birth rates decline disproportionately with age—that is the miscarriage rates are higher in the older population. Biologic markers of aging such as cycle day 3 FSH levels, and perhaps inhibin, appear to have a very powerful negative predictive value. Most programs will screen women over the age of 35 with cycle day 3 FSH. If this value exceeds 13 mIU/ml, the chance of pregnancy is reduced, perhaps as much as 50% [18]. Pregnancy rates with cycle day 3 FSH levels in excess of 20 mIU/ml are close to zero. The impact of aging is clearly seen when donor oocyte data are examined. Pregnancy rates using

donor oocytes do not vary across age groups for either fresh or frozen transfers (Fig. 17.13).

Figure 17.14 shows the live birth rate per cycle as a function of infertility diagnosis. This is the primary diagnosis recorded in the chart (presumed to be the most important condition requiring ART). There is often more than one diagnosis per couple; currently, combinations of diagnoses cannot be analyzed using the registry data. The live birth rates across diagnoses are fairly comparable, with the lowest rate identified for "uterine factor." Note that uterine factor accounted for only 2% of the cases of ART for 1996. It does not appear from the national data that primary diagnosis affects success with ART.

The number of embryos transferred has a direct impact on the pregnancy rate and live birth rate per transfer. Reviewing Figure 17.2 will clearly show the increasing live birth rate with increasing number of embryos transferred up to three embryos. There was no apparent increase in pregnancy rates by increasing the number of embryos transferred past three; however, there was a slightly higher multiple birth rate.

Embryo quality is related to implantation rates and pregnancy rates [19]. Several categorical scales are available for morphologic grading of the embryo. Embryo quality may relate to many factors including patient age, underlying ovulatory disturbances, and laboratory conditions.

Having a history of a previous live birth positively affects the outcome of the ART cycle. It appears that women who have had a previous live birth, regardless of age category, have better outcomes than those who have had no previous live births.

The 1995 data did not show any correlation of success to the size of the clinic performing the ART procedure (as measured by number of procedures per year). However, the 1996 data did show a tendency towards increasing success with larger programs [8].

IX.   Complications of ART

Most would now agree that the use of ART is medically indicated for the treatment of infertility. The use of ART is elective, as the couple could choose to remain childless. Therefore, it

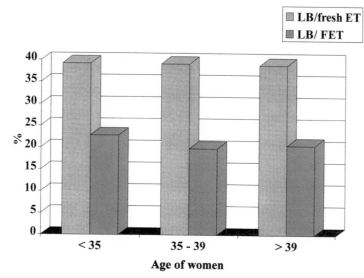

Fig. 17.13 Live births per transfer utilizing donor oocytes, by age for fresh embryos transfers and frozen embryo transfer, 1996.

is incumbent on the practitioner of ART to adequately inform his or her patients about the risks of the process. Fortunately, complications arising from the use of ART are rare. These risk and complications may relate to one of three categories: (1) controlled ovarian hyperstimulation (COH); (2) the surgical retrieval of oocytes (or sperm, for the male); or (3) cycle outcome-related.

A. Stimulation-Related Complications

Risks from controlled ovarian hyperstimulation may be immediate (ovarian hyperstimulation syndrome, or OHSS) or delayed, such as the potentially increased risk of ovarian cancer.

1. Ovarian Hyperstimulation Syndrome (OHSS)

Ovarian hyperstimulation syndrome is a condition of ovarian enlargement and ascites that occasionally follows COH [20–24]. The syndrome can range from mild, consisting of ovarian enlargement less than 5 cm, to severe, with ovarian enlargement up to 12 cm, massive ascites, and severe fluid and electrolyte disturbances. The incidence of OHSS ranges from 0.1 to 2.0% for patients undergoing COH. With surgical aspiration of all follicles in ART, the incidence of OHSS is probably at the low end of this range. Practitioners and patients must remember that the disorder can be life threatening and that it is iatrogenic.

The pathogenesis of OHSS has not been completely elucidated but involves most likely the ovarian renin–angiotensin cascade (with angiotensin II the most likely candidate for increased mesothelial permeability) [20]. Additionally, there is an increased production of histamines, prostaglandin mediators, and some cytokines. Third spacing of fluid is the hallmark of the disorder, and it is the attendant hemoconcentration that alerts the clinician to a possible impending disaster.

The risk factors for OHSS include predominately polycystic ovary syndrome (PCOS), conceptual cycles, and exogenous

or endogenous HCG stimulation in the luteal phase. Other risk factors include high serum estradiols, multiple immature and intermediate follicles, and possibly young age and lean body mass.

Many strategies have been employed for the prevention of OHSS; most are unsuccessful. In the high-risk patient (lean PCOS with previous stimulations resulting in high estradiols), long suppression with GnRH agonists have been used along with decreased stimulation with gonadotropins. Additionally, care is taken to avoid the use of exogenous HCG for luteal phase support. The clinician must monitor these cycles carefully, and the clinician and patient must be willing to cancel a cycle. As the severe forms of OHSS are almost always in conception cycles, cryopreservation of all embryos—averting the risk of endogenous HCG production—may eliminate OHSS or at least reduce the severity.

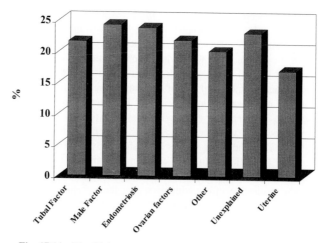

Fig. 17.14 Live birth rates among users of fresh, nondonor ART cycles, by primary diagnosis, 1996.

As the underlying pathogenesis of OHSS is not known, the treatment is empiric in nature. For mild OHSS, observation with daily weights and increased fluid intake is usually sufficient. For moderate OHSS, careful monitoring of hematocrit is necessary; with the hematocrit remaining below 45%, rest and electrolyte-containing fluids such as sport drinks are helpful. Increasing hematocrit, increasing ascites, and decreasing urine outputs are indications of severe OHSS. Hematocrits greater than 50% warrant hospitalization with IV hydration. Fluid shifts into third spaces result in ascites and occasional pleural effusions, electrolyte imbalance, thromboembolic phenomena, and potentially renal and hepatic damage. The treatment of severe OHSS involves volume expansion with crystalloids, recognizing that much of the infused fluid will escape into third spaces. Plasma expanders such as human serum albumin are used to maintain osmotic pressure. Tense, symptomatic ascites should be treated by paracentesis. Indications for paracentesis include the need for symptomatic relief (with respiratory compromise), oliguria, rising creatinine, and hemoconcentration unresponsive to medical therapy. Paracentesis is performed with ultrasound guidance, secondary to the massive enlargement of the ovaries. The most serious complications of OHSS are renal failure, thromboembolic events, and adult respiratory distress syndrome (ARDS). Critical OHSS may necessitate pregnancy interruption as a life-saving measure.

2. Ovulation-Inducing Drugs and Ovarian Cancer

The use of fertility medications has been linked to possible increases in ovarian cancer. Many articles have been published confirming or refuting this hypothesis. The biologic plausibility is twofold: (1) persistent uninterrupted ovulation may be associated with increased risk of ovarian cancer—increasing sites of ovulation per cycle may increase this risk further; and (2) increased gonadotropin stimulation to the epithelium of the ovary may increase the risk.

A series of articles published by Whittemore et al. evaluated risk factors associated with epithelial ovarian cancer [25–30]. These articles combined 12 previously published U.S. case-control studies that included 2197 cases and 8893 controls throughout a period of time from 1956–1986. Trends of decreasing risk were identified for increasing number of pregnancies (regardless of outcome), increased duration of breast-feeding, and oral contraceptive use. A history of tubal ligation odds ratio (OR) = 0.15, 95% confidence interval (CI) = 0.03–0.88) or hysterectomy with ovarian conservation (OR = 0.66, 95% CI = 0.50–0.86) were also associated with reduced risk. There was an increased risk of ovarian cancer associated with fertility drug use and long periods of unprotected intercourse without contraception (although this study did not show an increased risk when a clinical diagnosis of infertility was made). This and other studies have shown trends toward increased protection with each additional pregnancy (14–22% per additional pregnancy); however, there seems to be a consistent protective effect with even one term pregnancy that may be more important than benefits imparted by additional pregnancies.

Infertility has also been associated with an increased risk of ovarian cancer in a case-cohort study by Rossing et al. [31] The study was designed to evaluate the risk of ovarian tumors associated with infertility and whether the risk was associated with infertility or its treatment. The age-standardized incidence ratio for an invasive epithelial ovarian tumor was 1.5 (95% CI = 0.4–3.7). However, the risk of a borderline tumor was substantially higher than expected (age-standardized incidence ratio of 3.3 (95% CI = 1.1–7.8)). The overall standardized incidence ratio was 2.5 (95% CI = 1.3–4.5). Tumors were diagnosed an average of 6.9 years after enrollment. Thus, the increased incidence ratio was not thought to be due to increased surveillance. Overall cancer risks for this cohort were not different from those of the general population.

The results of this study demonstrated a substantially elevated risk for women taking clomiphene for greater than 12 months (relative risk (RR) = 11.1, 95% CI = 1.5–82.3). No increase in risk was seen with the use of clomiphene for less than 12 cycles. Infertile women without exposure to clomiphene (or any other ovulation inducing medications) had an age-standardized incidence ratio of 1.4 (95% CI = 0.2–5.0), indicating little if any increase in risk. The conclusion of the study was that prolonged use of clomiphene may increase the risk of ovarian tumors. Additional studies were recommended to test this hypothesis.

Shushan et al. published a case-control study performed in Israel between January 1, 1990 and September 1, 1993 [32]. Two hundred women with histologically proven ovarian cancer or borderline tumor were enrolled in the study. Controls (408) were obtained from random digit dialing in the same area code as the cases. Twenty-four women with epithelial ovarian cancer (12%) and 29 healthy controls (7.1%) reported ever having used fertility drugs (adjusted OR = 1.31, 95% CI = 0.63–2.74). The risk of ovarian cancer associated with human menopausal gonadotropin alone or in combination with clomiphene citrate was elevated but nonsignificant (adjusted OR = 1.42, 95% CI = 0.65–3.12). Human menopausal gonadotropin use was associated with a particularly increased risk for women with borderline ovarian tumors (adjusted OR = 9.38, 95% CI = 1.66–52.08). This study suggested that the use of ovulation-inducing medications, particularly human menopausal gonadotropins, may increase the risk of epithelial ovarian tumors.

A pilot study of risk perception at two fertility clinics indicated that only 10% of infertile women would not take fertility drugs because of the potentially increased risk of ovarian cancer. Fifty percent of those surveyed would accept a 2–4% increase in risk and 40% would accept a maximum lifetime risk of greater than 4% [33]. A review summarizes the literature on ovarian cancer risks and fertility drug exposure [34].

B. Retrieval Related Complications

The stimulated ovaries are, in general, quite close to the vaginal fornix. This allows for transvaginal ultrasound guided access. The blood supply to the vagina may occasionally be transgressed; the upper vagina is always inspected at the end of the procedure, and if bleeding occurs, pressure to the area is usually sufficient. If needed, a small suture may be placed for hemostasis. More significant bleeding may be encountered with the hypogastric artery and vein, which lie posteriorly and lateral to the stimulated ovary. The vein, on cross section, has many ultrasound characteristics of a follicle. Thus, if entered, the needle must be carefully removed without lateral motion of the transducer. Lacerations of this vessel may necessitate laparotomy for repair—an extremely rare complication. Laparoscopy for GIFT or ZIFT often reveals a fairly large amount of intraperitoneal blood and hemorrhage into the ovarian follicles. Occasionally hematomas over the bladder reflection or lateral sidewalls are seen; if stable, these require no further intervention.

The vagina is not sterile. Most programs will have the patients douche with povidone iodine solution on the evening prior to retrieval. At the time of oocyte retrieval, the vagina is cleansed with sterile saline. Antibiotic prophylaxis is often employed. Fortunately, serious pelvic infections are rare following transvaginal oocyte retrieval. Infections can occur, however, and the sequelae may be serious. Pelvic abscesses requiring drainage, usually by laparotomy, have been reported. Myometrial abscess formation, if extensive, may require hysterectomy.

C. Outcome-Related Complications

The results of an IVF cycle are binary: the patient is pregnant or she is not pregnant. Not conceiving with IVF has psychological implications. Regardless of the extent of counseling regarding success, individual patients believe that the process will be successful for them. It is "other people" for whom the technology is not successful. For this reason, mental health professionals are an integral part of most IVF teams.

A positive pregnancy test following IVF brings a set of potentially negative outcomes. Patients with less than optimal outcomes may have a variety of clinical presentations. They may have a chemical pregnancy (in which no intrauterine gestational sac is identified), a nonviable clinical pregnancy that ends in first trimester miscarriage, an ectopic (tubal) pregnancy, a therapeutic abortion (generally for genetic abnormalities), a second

trimester loss, a premature delivery, or a stillbirth. Figure 17.15 summarizes the frequency of these outcomes using the 1996 registry data.

XI. Conclusions and Future Directions

The use of ART has expanded dramatically over the 1990s. This increase in utilization has brought medical and ethical concerns regarding multiple gestations and cryopreservation of embryos. Risks of the therapies appear low; the long-term risks of ovulation-inducing medications are not completely clear at this time.

Aging of the female partner presents special concerns and challenges when using ART. The data clearly demonstrate a decrease in the probability of having a healthy live born infant with increasing maternal age. Novel techniques to circumvent this oocyte abnormality are currently under investigation. It appears that the increase in embryonic aneuploidy with increasing maternal age may relate to cytoskeletal abnormalities within the oocyte. Thus, transfer of "younger" oocyte cytoplasm to the "older" oocyte via micromanipulation has been attempted, as has transfer of the germinal vesicle containing the maternal genetic material to a "younger" oocyte recipient [35,36]. The number of patients undergoing these techniques is quite limited at this time. There are ethical concerns about splitting the female genetic contribution (nuclear and mitochondrial DNA), the impact on oocyte donors, and alterations of the germline of the resulting children [37].

The transfer of blastocysts instead of 4–8 cell embryos is gaining popularity. This technique involves five days of laboratory culture using special growth culture media. The process naturally selects embryos that are more likely to implant. A prospective randomized comparison revealed increased implantation rates with the transfer of blastocysts [38]. This technology allows for a reduction in the number of embryos transferred. Ideally, a single embryo could be transferred.

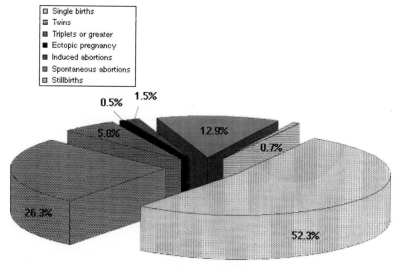

Fig. 17.15   Live births per transfer utilizing donor oocytes, by age for fresh embryo transfers and frozen embryo transfer, 1996.

The risks of blastocyst transfer include failure to achieve a transfer (approximately 7–10% of patients will have no embryos to transfer) and the increased risk of monozygotic twinning.

The use of preimplantation genetic diagnosis (PGD) may also expand in the coming years [39]. This technique may allow for the pretransfer identification of aneuploid embryos in older patients. These defects are known to be a major cause of first trimester loss and may also be a cause of earlier failures of implantation. In addition to screening for abnormalities in chromosome number, single gene defects may be identified. These include autosomal mutations (cystic fibrosis, Duchenne's muscular dystrophy, Tay-Sachs disease, Marfan Syndrome, and others) and X-linked mutations (hemophilia, retinitis pigmentosa, X-linked mental retardation, and others). Using PGD requires knowledge about the specific autosomal mutation in each family that is affected, as many mutations may result in a single phenotype. For X-linked diseases, determining the sex of the embryos and transferring only female embryos will result in healthy offspring. However, misdiagnoses can and have occurred with PGD.

Last, the medical and psychological sequelae of infertility should drive reimbursement policy. Adequate reimbursement is lacking in most states (13 now cover IVF). Hopefully, increasing awareness of the cost-effectiveness of IVF will result in insurance coverage for this process, removing the financial barriers that prohibit equal usage of effective infertility services. As we can see from the National Surveys of Family Growth [5], the disease does not discriminate among socioeconomic classes.

References

1. Steptoe, P. C., and Edwards, R. G. (1978). Birth after reimplantation of a human embryo. *Lancet* **2**, 366.
2. Gray, R. H. (1990). Epidemiology of infertility. *Curr. Opin. Obstet. Gynecol.* **2**, 154–158.
3. Templeton, A., Fraser, C., and Thompson, B. (1991). Infertility—Epidemiology and referral practice. *Hum. Reprod.* **101**, 1391–1394.
4. Arral, S. O., and Cates, W. (1983). The increasing concern with infertility. Why now? *JAMA, J. Am. Med. Assoc.* **250**, 2327.
5. Mosher, W. D., and Pratt, W. F. "Fecundity and Infertility in the United States, 1965–88." National Center for Health Statistics, Hyatsville, MD.
6. Hirsch, M., and Mosher, W. (1987). Characteristics of infertile women in the United States and their use of infertility services. *Fertil. Steril.* **47**, 618–625.
7. Chandra, A., and Stephen, F. H. (1998). Impaired fecundity in the United States: 1982–1995. *Fam. Plann. Perspect.* **30**, 34–42.
8. CDC web page address for 1996 Clinic Specific Report: *www.cdc. gov/nccdphp/drh/art96/*
9. Meldrum, D. R. (1992). Ovulation induction protocols. *Arch. Pathol. Lab. Med.* **116**, 406–409.
10. Wolff, K. M., McMahon, M. J., Kuller, J. A., Walmer, D. K., and Meyer, W. R. (1997). Advanced maternal age and perinatal outcome: Oocyte recipiency versus natural conception. *Obstet. Gynecol.* **89**, 519–523.
11. Callahan, T. L., Hall, J. E., Ettner, S. L., Christiansen, C. L., Greene, M. R., and Crowley, W. F., Jr. (1994). The economic impact of multiple-gestation pregnancies and the contribution of assisted-reproductive techniques to their incidence. *N. Engl. J. Med.* **331**, 244–249.
12. Goldfarb, J. M., Austin, C., Lisbona, H., Peskin, B., and Clapp, M. (1996). Cost-effectiveness of in vitro fertilization. *Obstet. Gynecol.* **87**, 18–21.
13. Van Voorhis, B. J., Stovall, D. W., Allen, B. G. S., and Syrop, C. H. (1998). Cost-effective treatment of the infertile couple. *Fertil. Steril.* **70**, 995–1005.
14. Bates, G. W., and Bates, S. R. (1996). The economics of infertility: Developing an infertility managed-care plan. *Am. J. Obstet. Gynecol.* **174**, 1200–1207.
15. Wheeler, C. A., Cole, B. F., Frishman, G. N., Seifer, D. B., Lovegreen, S. B., and Hackett, R. J. (1998). Predicting probabilities of pregnancy and multiple gestation from in vitro fertilization—a new model. *Obstet. Gynecol.* **91**, 696–700.
16. Menken, J., Trussell, J., Larsen, U. (1986). Age and infertility. *Science* **233**, 1389–1394.
17. Federation CECOS, Schwartz, D., and Mayaux, M. J. (1982). Female fecundity as a function of age. *N. Engl. J. Med.* **306**, 404–406.
18. Lindheim, S. R., Sauer, M. V., Francis, M. M., Macaso, T. M., Lobo, R. A., and Paulson, R. J. (1996). The significance of elevated early follicular-phase follicle stimulation hormone (FSH) levels: Observations in unstimulated in vitro fertilization cycles. *J. Assist. Reprod. Genet.* **13**, 49–52.
19. Shulman, A., Ben-Non, I., Ghetler, Y., Kaneti, H., Shilon, M., and Beyth, Y. (1993). Relationship between embryo morphology and implantation rate after in vitro fertilization treatment in conception cycles. *Fertil. Steril.* **60**, 123–126.
20. Navot, D., Bergh, P. A., and Laufer, N. (1992). Ovarian hyperstimulation syndrome in novel reproductive technologies: Prevention and treatment. *Fertil. Steril.* **58**, 249–260.
21. Navot, D., Margalioth, E. J., Laufer, N., Birkenfeld, A., Relou, A., Rosler, A., and Schenker, J. G. (1987). Direct correlation between plasma renin activity and severity of ovarian hyperstimulation syndrome. *Fertil. Steril.* **48**, 57–61.
22. Blankenstein, J., Shalev, J., Saadon, T. *et al.* (1987). Ovarian hyperstimulation syndrome: Prediction by number and size of preovulatory ovarian follicles. *Fertil. Steril.* **47**, 597–602.
23. Smitz, J., Camus, M., Devroey, P., Erard, P., Wisanto, A., and van Steirteghem, A. C. (1990). Incidence of severe ovarian hyperstimulation syndrome after GnRH agonist/hMG superovulation for in vitro fertilization. *Hum. Reprod.* **5**, 933–937.
24. Rizk, B., and Aboulghar, M. A. (1991). Modern management of ovarian hyperstimulation syndrome. *Hum. Reprod.* **6**, 1082–1087.
25. Whittemore, A. S., Harris, R., Itnyre, J., Halpern, J., and the Collaborative Ovarian Cancer Group (1992). Characteristics relating to ovarian cancer risk: Collaborative analysis of 12 U.S. case-control studies. I. Methods. *Am. J. Epidemiol.* **136**, 1175–1183.
26. Whittemore, A. S., Harris, R., Itnyre, J., and the Collaborative Ovarian Cancer Group (1992). Characteristics relating to ovarian cancer risk: Collaborative analysis of 12 U.S. case-control studies. II. Invasive epithelial ovarian cancers in women. *Am. J. Epidemiol.* **136**, 1184–1203.
27. Harris, R., Whittemore, A. S., Itnyre, J., and the Collaborative Ovarian Cancer Group (1992). Characteristics relating to ovarian cancer risk: Collaborative analysis of 12 U.S. case-control studies. III. Epithelial tumors of low malignant potential in white women. *Am. J. Epidemiol.* **136**, 1204–1211.
28. Whittemore, A. S., Harris, R., Itnyre, J., and the Collaborative Ovarian Cancer Group (1992). Characteristics relating to ovarian cancer risk: Collaborative analysis of 12 U.S. case-control studies. IV. The pathogenesis of epithelial ovarian cancer. *Am. J. Epidemiol.* **136**, 1212–1220.

29. Wu, M. L., Whittemore, A. S., Paffenbarger, R. S., Sarles, D. L., Kampert, J. B., Grosser, S., Jung, D. L., Balloon, S., Hendrickson, M., and Mohle-Boetano, J. (1988). Personal and environmental characteristics related to epithelial ovarian cancers. I. Reproductive and menstrual events and oral contraceptive use. *Am. J. Epidemiol.* **128,** 1216–1227.

30. Whittemore, A. S., Wu, M. L., Paffenbarger, R. S., Sarles, D. L., Kampert, J. B., Grosser, S., Jung, D. L., Balloon, S., and Hendrickson, M. (1988). Personal and environmental characteristics related to epithelial ovarian cancers. II. Exposures to talcum powder, tobacco, alcohol, and coffee. *Am. J. Epidemiol.* **128,** 1228–1240.

31. Rossing, M. A., Daling, J. R., Weiss, N. S., Moore, D. E., and Self, S. G. (1994). Ovarian tumors in a cohort study of infertile women. *N. Engl. J. Med.* **331,** 771–776.

32. Shushan, A., Paltiel, O., Iscovich, J., Elchalal, U., Peretz, T., and Schenker, J. G. (1996). Human menopausal gonadotropins and the risk of epithelial ovarian cancer. *Fertil. Steril.* **65,** 13–18.

33. Rosen, B., Irvine, J., Ritvo, P. *et al.* The feasibility of assessing women's perceptions of the risks and benefits of fertility drug therapy in relation to ovarian cancer risk. *Fertil. Steril.* **68,** 90–94.

34. Glud, E., Kruger Kjaer, S., Troisi, R., and Brinton, L. A. (1998). Fertility drugs and ovarian cancer. *Epidemiol. Rev.* **20,** 237–257.

35. Cohen, J., Scott, R. T., Schimmel, T., Levron, J., and Willadsen, S. (1997). Birth of an infant after transfer of anucleate donor oocyte cytoplasm into recipient eggs. *Lancet* **350,** 186–187.

36. Takeuchi, T., Ergun, B., Huang, T. H., David, O. K., Veeck, L. L., Rosenwaks, Z. *et al.* Preliminary experience of nuclear transplantation in human oocytes.

37. Robertson, J. A. (1999). Reconstituting eggs: The ethics of cytoplasm donation. *Fertil. Steril.* **71,** 219–221.

38. Gardner, D. K., Schoolcraft, W. B., Wagley, L., Schlenker, T., Stevens, J., and Hesla, J. (1998). A prospective randomized trial of blastocyst culture and transfer in in-vitro fertilization. *Hum. Reprod.* **13,** 3434–3440.

39. Verlinsky, Y., Handyside, A., Grifo, J. *et al.* (1994). Preimplantation diagnosis of genetic and chromosomal disorders. *J. Assist. Reprod. Genet.* **11,** 236–243.

18

Endometriosis

VICTORIA L. HOLT* AND JANNA JENKINS†

*University of Washington, Department of Epidemiology, School of Public Health and Community Medicine, Seattle, Washington;
†Fred Hutchinson Cancer Research Center, Division of Public Health Sciences, Seattle, Washington

I. Description and Impact of Endometriosis

Endometriosis is a relatively common abnormality that, in many women, may be a serious and chronic gynecologic disease. It is responsible for substantial morbidity and health care costs among women of reproductive age. While endometriosis has been identified as such for almost 75 years, its etiology remains uncertain and few modifiable risk factors have been identified.

Endometriosis is defined as the presence of functioning endometrial glands and stroma outside of the uterus, usually within the peritoneal cavity. Diagnosis typically is made by surgical examination that reveals pink-red vascularized or brown, blue, or black pigmented lesions adherent to pelvic structures, and the pelvic cavity may also contain evidence of past hemorrhage and inflammatory response in the form of scarring (adhesions). The disease, confirmed through microscopic examination of excised tissue, is found on the ovary, posterior broad ligament, anterior and posterior cul-de-sacs, and uterosacral ligament; it is found less commonly on the fallopian tubes, bladder, colon, small bowel, and ureters [1,2]. Adenomyosis, the ectopic implantation of endometrial glands and stroma into the uterine myometrium, has in the past been considered "internal endometriosis." However, adenomyosis is now seen by most researchers and clinicians as an etiologically and epidemiologically distinct condition, and typically is excluded from studies of endometriosis [3].

While endometriosis can exist asymptomatically, symptoms usually associated with the disease include premenstrual spotting, heavy menstrual bleeding, and painful menstrual cramps. In addition to menstrual dysfunction, adhesions associated with endometriosis may cause dyspareunia or chronic and severe pelvic pain throughout the menstrual cycle. The impact of these symptoms on the health of affected women is considerable. In an analysis of 1984–1992 U.S. National Health Interview Survey data, Kjerulff *et al.* found that half of respondents who reported having endometriosis had at least one day of bed rest due to this condition in the previous 12 months, with a mean of 17.8 days per affected woman [4].

There has long been an assumed causal relationship between endometriosis and infertility, based on the plausibility of pelvic adhesions interfering with tubal patency, toxic alterations seen in peritoneal fluid of women with endometriosis, and the high frequency of an endometriosis diagnosis in women with infertility not ascribed to other causes [5–9]. Epidemiologic studies typically have compared women presenting for laparoscopic diagnosis of unexplained infertility with those undergoing tubal sterilization and have found endometriosis to be three to ten times more common among the infertile women [10–14]. The interpretation of these studies should be cautious, however, because there is the possibility of less rigorous pelvic examination or disease notation among women undergoing sterilization procedures than those undergoing infertility diagnosis, and all of the studies were cross-sectional. Animal studies do support this hypothesized association: surgically induced endometriosis in rats, rabbits, and monkeys has been found to lead to a 50–67% reduction in subsequent fecundability [15–17]. Two small clinical studies provide additional information in support of the association between endometriosis and infertility. In these studies, all women undergoing artificial insemination with donor sperm were routinely examined laparoscopically before insemination. In a study of 98 regularly ovulating women with subfertile partners, Jansen found that the seven women with minimal endometriosis had 70% lower fecundability in subsequent insemination attempts than did laparoscopically normal women [18]. In the other study, among 90 women with azoospermic partners, the 12 with grade 1 endometriosis and no other infertility factors had 50% lower postinsemination fecundability than did laparoscopically-normal women without any infertility factors, but this difference was not statistically significant [18,19].

Record-linkage studies of Swedish women report increased cancer risk among women with endometriosis. Brinton *et al.* found that a history of hospitalization for endometriosis was associated with a significantly increased risk of breast cancer (RR = 1.3), nonHodgkin's lymphoma (RR = 1.8), and ovarian cancer (RR = 1.9), and women with a history of ovarian endometriosis had an ovarian cancer relative risk of 3.1 [20]. Another study found a three-fold increase in breast cancer risk among Swedish women with hysterectomy performed only because of endometriosis, compared with the background population, but the interpretation of the results of both of these studies is limited by the researchers' inability to control for important confounders such as endometriosis treatment [21].

II. Theories of Endometriosis Pathogenesis

While the pathogenesis of endometriosis is not entirely clear, several mechanisms have been proposed to explain the distribution, maintenance, and growth of endometrial tissue outside of the endometrium. One early theory posited that endometriosis develops from metaplasia of the celomic epithelium. The female reproductive tract lining contains portions of mullerian ducts and mesenchymal tissue that retain their ability for multi-potential growth. Therefore, it was thought that metaplastic transformation into endometrial tissue could occur, perhaps with a substance such as menstrual debris or under the influence of estrogen and progesterone. While this theory still has adherents, there are few scientific data to confirm it. Endometriosis does not typically behave as celomic metaplasia would dictate: it rarely occurs in men, does not occur in all sites where celomic membranes are present, and does not occur with increasing frequency with age [22,23].

The most widely accepted pathogenic theory was first proposed by Sampson in the 1920s [24]. He hypothesized that endometriosis is a result of the transportation of menstrual blood containing endometrial tissue back from the uterus through the fallopian tubes into the peritoneal cavity, where it then implants on pelvic structures. This theory is supported by the distribution of lesions within the pelvic cavity and the results of numerous clinical and epidemiologic studies; however, research indicates that most women have retrograde menstruation to some extent and yet the majority do not develop endometriosis [1,25–28]. It is plausible that other factors, such as excessive hormonal stimulation or altered immune function, may be necessary or facilitative for initiation of disease.

The appearance of endometriosis only after menarche and the diminution of disease after menopause suggest that a woman's cyclic estrogen and progestin levels may play a strong role in disease etiology and maintenance [29]. This hypothesis is further supported by observation that medical interventions that decrease the production of estrogen usually decrease the extent of disease, and oophorectomy usually cures it [30]. Additionally, as detailed more fully later in the chapter, indicators of increased endogenous estrogen levels have been associated with increased endometriosis risk, whereas factors that may decrease estrogen production or alter estrogen metabolism have been found to decrease risk.

There has been interest in the association between endometriosis and the immune system, but it is unclear if altered immune status is a cause of or a response to the disease. It has been posited that a healthy immune system would prevent the implantation of ectopic endometrium cells, and therefore endometriosis represents a deficiency in immune function [31,32]. There is evidence for the association between decreased immune system function and endometriosis. Women with endometriosis have been shown to have depressed cell-mediated immunity, with decreased T cell and natural killer cell responsiveness [31,33,34]. However, the opposite is true of the humoral immune responsiveness in affected women, in whom high serum levels of general autoantibodies and specific antiendometrial antibodies are seen [33–35,35–37]. This increased humoral immunity has been hypothesized variously to be a consequence of dealing with an excessive amount of refluxed endometrial tissue or a genetically predisposed response [38].

III. Disease Ascertainment

While the diagnostic signs of adnexal or cul-de-sac scarring, thickening, or nodularity may be discernible via physical examination, this procedure has a very low sensitivity as a screening tool for mild endometriosis [39]. Ultrasound or MRI examinations fairly reliably diagnose cystic ovarian endometriosis (endometriomas), but these modalities do not discover the majority of endometriotic lesions [40–43]. As a consequence of the limitations of these noninvasive methods, the current American Fertility Society (now the American Society for Reproductive

Medicine) and the American College of Obstetricians and Gynecologists standard for the diagnosis of endometriosis is direct visualization of lesions during a surgical laparoscopy or laparotomy procedure, with histological confirmation if possible [44,45]. In hospital discharge diagnosis summaries, the lesions are categorized using the International Classification of Diseases Ninth Revision (ICD-9) diagnostic codes 617.0–617.9, although 617.0 denotes both adenomyosis and external uterine endometriosis.

The American Society for Reproductive Medicine initially proposed a classification system for endometriosis severity in 1979, with revisions in 1985 and 1997 [44,46,47]. This system, designed to provide a standard for the recording of pathologic findings, categorizes disease severity in stages from I (minimal) to IV (severe) based on weighted points for disease extent and location. The latest revision attempts to produce scores that predict the probability of pregnancy following treatment by incorporating the morphological appearance of lesions into the system. No information is available about the usefulness of this revision, but the correspondence between symptoms and stage of disease defined using previous versions has generally been poor [48–51].

IV. Endometriosis Incidence/Prevalence and Age Distribution

The possibility of asymptomatic disease and the necessity for laparoscopic ascertainment make endometriosis incidence rates difficult to determine because some women with the condition may remain undiagnosed [22,30,52–54]. The most widely used estimate of endometriosis incidence was based on a review of 1970–1979 medical records of one Minnesota community's entire health care system [55]. In this study, the annual incidence of surgically confirmed endometriosis was found to be 1.6/1000 white women 15–49 years of age. Additional information is available from the National Center for Health Statistics National Hospital Discharge Survey (NHDS), an annual sample of all U.S. nonfederal short stay hospitalizations. Since 1983, the rates of hospitalization with a first-listed or any-listed diagnosis of endometriosis have decreased by one-third among 15–44 year old women, from 2.1/1000 (first-listed) and 5.3/1000 (any-listed) in 1983 to 1.3/1000 and 3.4/1000 in 1995. (Fig. 18.1) These rates do not reflect the true incidence of endometriosis for two reasons: (1) incident cases diagnosed by outpatient laparoscopic procedures, which have been used increasingly in the past decade, are not included in these data, and (2) previously diagnosed cases who are hospitalized for definitive treatment such as hysterectomy are included. Additionally, the ICD-9 code 617.0 (adenomyosis and uterine endometriosis) comprises almost half of the endometriosis diagnoses among 15–44 year olds and three-fourths of diagnoses among 45–64 year olds. Therefore, the rates of hospitalization for true endometriosis may be at least 50% lower than the apparent rates.

Endometriosis prevalence also is difficult to estimate. Houston's annual incidence rate of 1.6/1000 15–49 year old women would lead to a 3% prevalence in this age group if the disease duration was 20 years. Somewhat lower self-reported disease prevalences have been obtained in two population-based studies, one in the U.S. and one in Norway. The U.S. National Health Interview Survey (NHIS) consists of in-home interviews of a

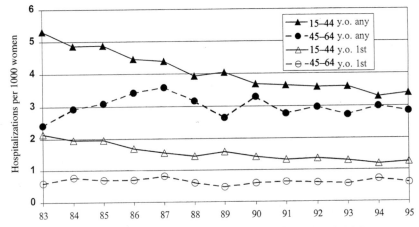

Fig. 18.1   Rates of endometriosis/adenomyosis hospitalization (ICD9 617.0–617.9) by age group and year, and whether diagnosis was first-listed or any-listed, US women, 1983–1995. Source: U.S. Vital and Health Statistics, Series 13 reports.

representative sample of the U.S. civilian noninstitutionalized population. In NHIS surveys from 1984–1994 combined, 0.7% of 15–44 year old women interviewed stated they had had endometriosis in the prior 12 months [4]. In the Norwegian study, 2% of the 40–42 year old women in one county reported that they had had endometriosis [56]. Another source of information about endometriosis prevalence is reports of examination of fertile women during tubal ligation procedures. In several studies, the prevalence of undiagnosed endometriosis found at tubal ligation generally ranged from 2–7%. These cases were mainly

asymptomatic and minimal or mild in nature, as shown in Table 18.1 [11–14,52,54,56a–60].

The diagnosis of endometriosis usually is limited to the menstrual or reproductive years, and the risk of endometriosis has been shown to be positively associated with age even within this period. Houston found that the peak incidence occurred in women between the ages of 35 and 44, with the lowest age-specific rates for women 15–19 years of age [55]. Increasing rates of first endometriosis diagnosis until age 45 also were reported by Vessey in the follow-up of a cohort of family plan-

Table 18.1

Prevalence and Severity Distribution of Endometriosis Found at Tubal Ligation

Reference	Location	N	Mean age (yrs.)	Prevalence (%)	Severity
[11]	U.S.	296	NA	1.4	NA
[10]	U.S.	43	32.2	4.7	NA
[13]	U.S.	200	29.5	2.0	100% mild
[56a]	U.S.	50	35.0	15.0	NA
[57]	U.K.	75	NA	42.7	100% minimal or mild
[12]	U.S.	251	33.6	5.2	77% minimal 23% mild
[54,58]	U.S.	566	30.1	7.4	67% minimal 24% mild 10% moderate
[58a]	Netherlands	200	35.3	2.5	20% mild 80% moderate
[14]	U.K.	598	NA	6.2	81% minimal or mild 19% moderate
[52,59]	Norway	206	37.0	20.0	81% minimal 5% mild 10% moderate 5% severe
[60]	U.S.	3384	NA	3.7	91% minimal 5% mild 4% moderate

ning clinic attendees in England [61]. Similar results can be seen in the 1988–1990 NHDS data: among 15–44 year old women, the endometriosis/adenomyosis hospitalization rates were highest (6.1/1000 women) in those aged 40–44 years [62]. Because 65% of all endometriosis/adenomyosis hospitalizations were associated with hysterectomy, they are more likely to reflect advanced disease than initial diagnosis, however.

V. Risk Factors for Endometriosis

In case-control studies of endometriosis, the source populations used for subject selection may greatly affect study results and their interpretation. Case groups often have been partially or entirely composed of infertility clinic patients without another known cause of infertility. This group may not be representative of all endometriosis cases, as asymptomatic infertile women may be diagnosed only because of laparoscopic investigation of their infertility problems, whereas similarly diseased fertile women may remain undiagnosed. Additionally, infertile women seeking infertility treatment are not demographically representative of all infertile women, so even this portion of endometriosis may not be fully assessed using infertility clinic cases [63]. Another commonly used case source consists of women with endometriosis diagnosed at time of tubal ligation. Because of their proven fertility and the generally mild nature of the disease diagnosed in this manner, these women also are unlikely to be representative of all women with endometriosis. Control group selection may be problematic as well. In an effort to ensure that controls are free of undiagnosed endometriosis, many investigators have selected women who have undergone a laparoscopic procedure for another reason, such as another gynecologic disease or tubal sterilization, or who have infertility due to another cause. These control groups may be atypical, particularly in terms of reproductive history. A few investigators have attempted to address this problem by using two control groups, and contrasting results obtained, for instance, using friend controls with those obtained using gynecology clinic controls [64,65]. Friend controls may more closely resemble the entire source population for the cases, but the absence of disease is unconfirmed in them.

A. Demographic Factors

1. Race/Ethnicity

While the clinical observation that endometriosis is more common among white women has been repeated for several decades, little supportive epidemiologic evidence is available. The results of two Texas studies of endometriosis prevalence at tubal sterilization are mixed. One study found that the prevalence of endometriosis was similar among white, black, and Hispanic women, while a larger study reported that, compared to white and Hispanic women, endometriosis was 30% less common among black women and eight times more common among Asian women [54,60]. The U.S. NHDS provides population-based information on endometriosis hospitalization rates by race, and one analysis of 1988–1990 data found that white women were hospitalized with endometriosis or adenomyosis at rates 40% higher than those of nonwhite women (3.4/1000 vs 2.4/1000) [62]. Because of the small sample size, nonwhite race

was not further categorized in this analysis, so the extent of a possible excess risk among Asian women remains to be verified.

2. Socioeconomic Status

The clinical impression has persisted that the risk of endometriosis is higher among affluent women, but this association is particularly susceptible to selection bias and results of epidemiologic studies are not consistent. Using case groups of infertility or gynecology patients with endometriosis and control groups of other infertile women, other gynecology patients, or women with livebirths, studies in North America, Lebanon, and Malaysia have found positive associations between family income or social class and endometriosis risk [66–68]. Similar studies in Spain and Scotland have failed to find such associations [69,70].

B. Menstrual Characteristics

The most consistent risk factors identified for endometriosis are characteristics related to increased menstruation, and this evidence has contributed to the acceptance of the transport theory of menstrual retroflux. A modest positive association generally has been found between early menarche and endometriosis risk in studies using a variety of sources for cases and controls, with menarche before age 12 usually associated with an increase in endometriosis risk of 20–40% [56,65,69,71–74] (Table 18.2) Short cycle length also has been associated fairly consistently with increased likelihood of endometriosis; women with cycles under 28 days in length have nearly twice the disease risk of other women [56,60,64,65,69,71,72,74].

Other aspects of menstrual history that have been associated with an increased risk of endometriosis include regular menstrual cycles and long duration of flow during each menstrual period [64,65,71,73,74]. Heavy menstrual flow and moderate to severe menstrual pain also have been associated with increased risk, but the interpretation of these findings should be cautious because these factors may be symptoms of existing disease [64,65,71]. While menstrual characteristics have been associated with endometriosis in studies in several settings, control group choice may still affect these associations, as shown by Signorello et al. in a study of 50 women with infertility-associated endometriosis and separate control groups of fertile and infertile women [65]. These authors found that several associations between menstrual factors and endometriosis seen using the fertile control group moderated in analyses using infertile controls, perhaps due to their unusual hormonal profiles. The relationships between other menstrually-associated risk factors and endometriosis are less clear. Two researchers report no association with current tampon use whereas another found an odds ratio (OR) of 3.6 associated with 14 or more years of tampon use among women under 30, and one study found that infertile patients with endometriosis have intercourse during menses more frequently than do women with infertility due to other causes [64,71,75,76].

C. Reproductive History

1. Spontaneous Abortion

Despite reports of unusually high rates of prior spontaneous abortion among women diagnosed with endometriosis, this

Table 18.2
Risk of Endometriosis in Relation to Menstrual Factors

Reference	Location	Subject source (N)		Exposure	Odds ratio	95% CI
		Cases	Controls			
Age at menarche						
[71]	U.S.	Infert C1 (268)	Ob Pts (3794)	≤11 yrs	1.2[a]	0.9–1.7
				12–15 yrs	1.0[a]	(ref.)
				≥16 yrs	0.8	0.4–1.7
[72]	Italy	Hosp Pts (114)	Hosp Pts (1127)	≤12 yrs	1.0	(ref.)
				13–14 yrs	1.1[b]	NA
				≥15 yrs	0.4[b]	NA
[73]	Italy	Hosp Pts (376)	Hosp Pts (522)	≤11 yrs	1.0	(ref.)
				12–13 yrs	0.8[c]	NA
				≥14 yrs	0.9[c]	NA
[69]	Spain	Infert C1 (174)	Infert C1 (174)	≤12 yrs	1.2[d]	0.8–2.0
				≥13 yrs	1.0	(ref.)
[65]	U.S.	Infert Hosp (50)	Steril Hosp (89)	≤11 yrs	1.4/2.2[e]	0.5–4.6/0.5–9.2
			Infert Hosp (47)	12 yrs	1.0	(ref.)
				≥13 yrs	1.1/1.0[e]	0.4–2.8/0.4–2.6
[56]	Norway	Self-report survey (79)	Self-report survey (3955)	≤12 yrs	1.8[d]	1.2–2.9
				≥13 yrs	1.0	(ref.)
[74]	Malaysia	Hosp Pts (305)	Hosp Pts (305)	≤11 yrs	0.9[d]	0.4–1.5
				≥12 yrs	1.0	(ref.)
Cycle length						
[71]	U.S.	Infert Pts (268)	Ob Pts (3794)	≤27 days	2.1[a]	1.5–2.9
				28–34 days	1.0	(ref.)
				≥35 days	0.5[a]	0.3–1.1
[72]	Italy	Hosp Pts (114)	Hosp Pts (1127)	≤30 days	1.0	(ref.)
				≥31 days	0.7[b]	0.3–1.7
[64]	U.S.	Gyn Clin (104)	Friends (100)	≤27 days	0.9/1.3[f]	0.3–2.4/0.5–3.5
			Gyn Clin (98)		2.0/2.4[g]	0.6–6.1/0.8–7.2
				≥28 days	1.0	(ref.)
[69]	Spain	Infert Pts (174)	Infert Pts (174)	≤27 days	1.8[d]	1.0–3.1
				≥28 days	1.0	(ref.)
[60]	U.S.	Steril Pts (126)	Steril Pts (504)	≤29 days	1.0	(ref.)
				≥30 days	1.7[h]	1.1–2.5
[74]	Malaysia	Hosp Pts (305)	Hosp Pts (305)	≤27 days	1.8[i]	1.6–2.6
				≥28 days	1.0	(ref.)
[56]	Norway	Self-report survey (79)	Self-report survey (3955)	"frequent periods"	1.7[d]	1.1–2.8
[65]	U.S.	Infert Pts (50)	Steril Pts (89)	≤27 days	1.9/0.7[e]	0.7–5.1/0.2–2.2
			Infert Pts (47)	28 days	1.0	(ref.)
				≥29 days	1.3/0.6[e]	0.5–3.4/0.2–1.6

[a]Adjusted for center, age, years from menarche to index date, education, religion.
[b]Adjusted for age, education.
[c]Adjusted for age.
[d]Unadjusted.
[e]Adjusted for age, education, height, weight.
[f]Among women <30, adjusted for religion, gravidity, BMI.
[g]Among women ≥30, adjusted for religion, gravidity, BMI.
[h]Adjusted for age, parity.
[i]Adjusted for parity, age first birth, social class.

association remains inconclusive [77,78]. Two small studies of differing populations have found an increased risk of endometriosis among women with a history of spontaneous abortion. (Table 18.3) One of these studies assessed parous women at

tubal sterilization and the other used cases who were patients at a reproductive and endometriosis clinic, the majority of whom presented because of infertility [54,79]. The latter study found that women whose first pregnancy had been a spontaneous abor-

Table 18.3
Risk of Endometriosis in Relation to Reproductive History

Reference	Location	Subject source (N)		Exposure	Odds ratio	95% CI
		Cases	Controls			
Spontaneous abortion						
[54]	U.S.	Steril Pts (42)	Steril Pts (524)	≥ 1	1.6[a]	0.7–3.5
[79]	U.S.	Gyn Clin (104)	Friends (100) Gyn Clin (98)	Spont. abort. 1st preg.	6.4/3.6[a]	1.7–28.6/1.0–16.6
[72]	Italy	Hosp Pts (114)	Hosp Pts (1127)	≥ 1	1.0[b]	0.5–1.7
[73]	Italy	Hosp Pts (376)	Hosp Pts (522)	≥ 1	0.7[c]	0.4–1.2
Induced abortion						
[72]	Italy	Hosp Pts (114)	Hosp Pts (1127)	≥ 1	0.5[b]	0.2–1.0
[79]	U.S.	Gyn Clin (104)	Friends (100) Gyn Clin (98)	Ind. abort. 1st preg.	3.9/1.0[a]	1.3–12.0/0.4–2.7
[73]	Italy	Hosp Pts (376)	Hosp Pts (522)	≥ 1	0.4[c]	0.2–0.8
Parity						
[72]	Italy	Hosp Pts (114)	Hosp Pts (1127)	0	1.0	(ref.)
				1–2	0.6[d]	NA
				≥ 3	0.3[d]	NA
[61]	U.K.	Cohort (142 with endometriosis)		0	1.0	(ref.)
				1	1.2[e]	0.6–2.8
				2	0.9[e]	0.5–1.8
				3	0.8[e]	0.4–1.7
				≥ 4	0.7[e]	0.3–1.8
[73]	Italy	Hosp Pts (376)	Hosp Pts (522)	0	1.0	(ref.)
				1	0.4[c]	NA
				≥ 2	0.2[c]	NA
[56]	Norway	Self-report survey (79)	Self-report survey (3955)	0	3.7[f]	1.9–7.2
				≥ 1	1.0	(ref.)

[a]Unadjusted odds ratio and 95% confidence interval calculated from data presented in article.
[b]Adjusted for age, education, gravidity.
[c]Adjusted for age.
[d]Adjusted for age, education, age first birth.
[e]RR adjusted for age.
[f]Unadjusted.

tion had three to six times the risk of endometriosis compared to women with another first pregnancy outcome, depending on the control group used (controls from the same clinic with other diagnoses or friend controls). In contrast, two hospital-based Italian studies found no association between spontaneous abortion and endometriosis. In one of these studies, cases were women with endometriomas, and the other included those with all types of endometriosis, 25% of whom were diagnosed because of infertility [72,73]. Both studies used control groups consisting of women hospitalized with fractures, low back pain, or abdominal surgery. The relationship between spontaneous abortion and endometriosis may be complex. Altered immune status has been associated with both conditions. Therefore, even in the presence of a strong association it may be inappropriate to conclude that spontaneous abortion is an independent risk factor for endometriosis.

2. Induced Abortion

The association between induced abortion and endometriosis also is uncertain. Darrow *et al.* found that in comparison to

friend controls, the first pregnancies of women with endometriosis were much more likely to have ended in induced abortion [79] (Table 18.3). In contrast, the two hospital-based Italian studies both found a decrease in endometriosis risk associated with a history of induced abortion [72,73]. However, women who had never been pregnant were included in these analyses, a choice that would spuriously decrease the risk associated with induced abortion if more cases than controls were nulligravid.

3. Gravidity and Parity

Increasing parity may be hypothesized to decrease endometriosis risk by decreasing the total number of menstrual cycles, thereby decreasing the peritoneal cavity's exposure to retrofluxed menstrual fluid. In support of this theory, an inverse dose-response association has been found in two Italian hospital-based studies [72,73]. As shown in Table 18.3, two other studies confirm the relationship between parity and endometriosis. A self-report survey of all 40–42 year old women in one Norwegian county found nulliparous women to have almost four times the endometriosis risk of parous women, and a study of endometriosis diagnosed at tubal ligation reported that women with one birth were 2.5 times as likely to have endometriosis as those with two or more births [56,60]. A prospective cohort study conducted among women at 17 family planning clinics in England had less conclusive findings. Compared with nulliparous women, the relative risk of endometriosis among women with one birth was 1.2, but this risk decreased linearly to 0.7 for those with four births or more [61].

The associations generally seen between nulliparity and endometriosis have led to speculation that endometriosis is a consequence of voluntarily deferred childbearing. An alternative explanation to this view is that nulliparity or reduced parity is an early sign of endometriosis, rather than a modifiable risk factor. This hypothesis is consistent with the results of research by Darrow et al., who found that women with endometriosis were no more likely than other women to delay seeking pregnancy, but they were much more likely to have problems becoming pregnant [79].

D. Contraceptive History

1. Oral Contraceptive Use

Oral contraceptive (OC) use may be hypothesized to decrease endometriosis risk due to its direct effect on the ovary. Women using OCs have no estradiol preovulatory surge, but merely have low and constant endogenous estradiol levels [80,81]. In the absence of sufficient estrogen, there is no proliferation of endometrial tissue and menstrual flow is lessened, potentially decreasing retroflux of menstrual blood containing endometrial tissue [59]. Epidemiologic evidence is somewhat contradictory

concerning this association, however, as seen in Table 18.4. While Strathy et al. found OC ever use to be associated with a sharply reduced risk of endometriosis in a 1982 study of infertile cases and tubal ligation controls, another study using cases presenting for pain or infertility and controls hospitalized for other reasons found an OR of 2.1 with OC ever use, and others have found only a modest positive association [13,14,69,72]. These apparently discrepant results may be explained by differences in the sources used for subjects or by the differing effects of former and current OC use. Current use generally has been found to be modestly protective, while former or "usual" use has been associated with a 40–80% elevation in risk [58–61, 64,73]. These findings have been attributed to an increased likelihood that women with menstrual symptoms of endometriosis will be prescribed OCs, with disease suppression during current use and subsequent diagnosis after OC discontinuation [61,82].

2. Intrauterine Devices (IUDs)

If the transport theory of retrofluxed menstruation is correct, IUD use may be expected to increase endometriosis risk because it increases menstrual cramping, menstrual flow, endometrium irritation, and endometrial tissue shedding. Again, however, there are contradictory results from relevant epidemiologic studies. Among women presenting for tubal ligation, two studies found that those with endometriosis more often had a history of IUD use or lengthy IUD use, whereas another found no association with ever use [58–60]. Studies of symptomatic women have found variously that IUD use does not affect risk, that there is decreased risk associated with current use, increased risk associated with former use, and increased risk associated with longer duration of use [14,64,67,71].

Table 18.4
Risk of Endometriosis in Relation to Use of Oral Contraceptives

Reference	Location	Subject source (N)		Exposure	Odds ratio	95% CI
		Cases	Controls			
[13]	U.S.	Infert/Fert (25)	Infert/Fert (275)	Ever use	0.1[a]	0.0–0.5
[72]	Italy	Hosp Pts (114)	Hosp Pts (1127)	Ever use	2.1[b]	1.2–3.6
[14]	U.S.	Steril Pts (227)	Steril Pts (1315)	Ever use	1.3[a]	1.0–1.8
[67]	Spain	Infert C1 (174)	Infert C1 (174)	Ever use	1.3[c]	0.8–2.2
[54]	U.S.	Steril Pts (42)	Steril Pts (524)	Current use	0.7[a]	0.3–1.4
[59]	Norway	Steril Pts (42)	Steril Pts (164)	Current use	1.0[a]	0.4–2.5
[60]	U.S.	Steril Pts (126)	Steril Pts (504)	Current use	0.4[d]	0.2–0.7
[61]	U.K.	Cohort (142 with endometriosis)		Current use	0.4[e]	0.2–0.7
				Former use	1.8[e]	1.0–3.1
[73]	Italy	Hosp Pts (377)	Hosp Pts (522)	Current use	0.9[f]	0.5–1.9
				Former use	1.7[f]	1.3–2.4
[79]	U.S.	Gyn Clin (104)	Friends (100) Gyn Clin (98)	OC is usual method	1.4/0.7[a]	0.8–2.5/0.4–1.3

[a]Unadjusted odds ratio and 95% confidence interval calculated from data presented in article.

[b]Adjusted for age, education.

[c]Unadjusted.

[d]Adjusted for age, parity, duration IUD use.

[e]Adjusted for age, parity.

[f]Adjusted for age, education, marital status, parity.

3. Tubal Sterilization

Sterilization may be hypothesized to lower endometriosis risk due to the occlusion of the route for retroflux of menstrual products, or perhaps due to the decreased estrogen production and excretion that occur following tubal ligation [83,84]. Little epidemiologic research is available on this exposure, although some case reports have found endometriosis to be relatively common after sterilization, especially among women with interval sterilizations using electrocautery [85–90].

E. Gynecologic History

Investigations of the associations between other aspects of gynecologic health and endometriosis have provided further evidence in favor of the pathogenic theory of retrograde menstruation. Olive and Henderson found that women with mullerian anomalies of uterine anatomy had higher than expected rates of endometriosis, and further, those whose disease resulted in patent fallopian tubes, functioning endometrium, and outflow obstruction had over twice the risk of comparable women without outflow obstruction [91]. Further support of the mechanical transportation theory comes from case reports describing endometriosis growing in incision scars from uterine and fallopian tube surgeries and cesarean section deliveries, but no epidemiologic studies are available to confirm or quantify these associations [92]. While gynecologic surgery and PID can be hypothesized to increase risk by providing a location for attachment of the transported endometrial tissue, the interpretation of studies of these associations should be cautious. Moen's population-based Norwegian survey [56] found that a history of cervical conization, infertility surgery, ovarian surgery, hysterectomy, or pelvic inflammatory disease (PID) treatment was more common among women with self-reported endometriosis, but some of these procedures are plausibly a consequence of endometriosis symptoms rather than risk factors for the disease. A prospective cohort study of British women in one regional health authority area did find that women with PID had a subsequent endometriosis diagnosis 5.5 times as frequently as those without PID [93]. Differential observation is a possible explanation for these results, however, as women with chronic sequelae of PID may be more likely than others to undergo laparoscopy.

F. Other Personal Factors

1. Physical Activity and Characteristics

Because endometriosis risk has been posited to be influenced by a woman's levels of circulating estrogen, associations with several body characteristics and behaviors that may impact estrogen levels have been investigated. Strenuous exercise has been shown to decrease estradiol levels in normally-menstruating women, leading to the hypothesis that exercise might lower endometriosis risk [94]. This hypothesis has generally been borne out in epidemiologic studies, which have found that women who engage in strenuous regular exercise have an endometriosis risk 10–60% lower than other women [56,65,71].

Weight and body mass index (BMI = weight (kg)/height $(m)^2$) have been investigated by many researchers, with mixed results. Cramer et al. [71] found that women weighing 64 kg and over had a 30% reduction in endometriosis risk, compared with 56–63 kg women, and Signorello et al. [65] reported a 20–30% decrease per 10 kg increase in weight, depending on the control group used. The latter study found a similar decrease in risk per 5 (kg/m^2) increase in BMI. However, Darrow et al. found that endometriosis cases had lower BMI only in comparison with infertile controls, Moen found no association between endometriosis and BMI in a self-report survey study of women in one county in Norway, and a study of women at tubal ligation found an OR of 1.2 for BMI over 30 [56,60,64]. Weight and BMI are not related to plasma estradiol levels in premenopausal women, so the associations seen may reflect other factors relevant to case and control group composition, such as socioeconomic status [95].

Body fat distribution is a good marker of endogenous estrogen levels in premenopausal women, as women with a higher ratio of androgens to estrogens appear to concentrate body fat centrally, and those with a higher estrogen to androgen ratio have more peripherally-accumulating fat [96]. One study of gynecology clinic cases and friend controls found no association between endometriosis risk and weight, height, or BMI, but reported endometriosis risk to be substantially increased in women under aged 30 with a low waist-to-hip (OR = 6.2) or waist-to-thigh ratio (OR = 2.4) [97]. Height has been found by other researchers to be positively related to endometriosis risk. Cramer et al. [71] reported an OR of 1.4 for women over 166 cm compared with those ≤160 cm, and Signorello et al. [65] found an OR of 2.8–3.0 per 10 cm height increase. This association may be explained by the higher follicular-phase plasma-estradiol levels that have been noted among taller women [95].

2. Diet and Substance Use

Cigarette smoking, which is associated with an androgenic distribution of fat in women, increases sex hormone binding globulin and may therefore decrease available estradiol. The proposed resultant decrease in endometriosis risk has generally been reported, with ORs ranging from 0.3 to 0.8 [56,60,64, 69,71]. Exceptionally, Darrow et al. [64] found that the OR for high lifetime cigarette use was decreased among women under 30, but increased (OR = 1.4) for those 30 or older, and Signorello et al. [65] found no associations between smoking and endometriosis.

Alcohol is a phytoestrogen, and moderate alcohol consumption has been shown to increase plasma and urinary estrogen levels in premenopausal women, so it is plausible that alcohol use may increase endometriosis risk [98]. Accordingly, Signorello et al. [65] found that any alcohol use doubled endometriosis risk. It is possible that alcohol use may be a marker for undiagnosed disease, however, as 31% of endometriosis patients in one study reported increased alcohol consumption when experiencing gynecological symptoms [99]. Available epidemiologic evidence does not confirm this explanation for the association. Grodstein et al. found an increase in endometriosis-associated infertility with alcohol use (RR = 1.5 for moderate or heavy drinking) that did not change when limiting the analysis to women without dysmenorrhea [100].

Associations between diet and endometriosis have been reported or proposed. One epidemiologic study using a comparison group of women delivering a liveborn child found that ingestion of 5–7 g/month of caffeine (around 1.5–2.0 cups per day) was associated with a nearly twofold risk of endometriosis-associated infertility compared with ingestion of under 3 g/month [101]. Fish oil containing eicosapentaenoic acid (EPA) and docosahexaenoic acid (DHA) has been found in animal studies to decrease endometriosis risk, but no epidemiologic data are available currently [102].

G. Environmental Exposures

The known suppressant effects of irradiation and polychlorinated biphenyl (PCB) or dioxin exposure on the immune system and the results of animal research suggest that these factors may increase the risk of endometriosis in women [103–106]. Two epidemiologic studies have reported on these risk factors, with somewhat differing results. Mayani *et al.* found discernable dioxin levels in the blood of Israeli women with endometriosis and infertility six times as often as in women with tubal infertility [107]. In contrast, a small Canadian case-control study found that women with endometriosis did not have significantly higher current PCB or chlorinated pesticide blood levels than other women [108]. These authors did not measure dioxin directly, but noted that they expected there would be a strong correlation between dioxin and other organochlorines.

VI. Biologic Markers for Susceptibility

There is evidence that endometriosis may have a genetic basis. Women with endometriosis are more likely than other women to have first-degree relatives with endometriosis, and there is a high concordance for the disease among monozygotic twin sisters [109–113]. The age of onset of symptoms is similar in sister-pairs rather than the year of onset, implying a genetic inheritance rather than an entirely environmental influence [114]. Characteristics of endometriosis appear hereditary as well. Compared with endometriosis patients without a family history of the disease, patients with a family history have been found to be more likely to have severe disease [32,115].

The mechanism for inheritance of endometriosis is unclear, and research results do not suggest a simple recessive or dominant trait. Coxhead posits a polygenic mechanism involving an interaction of a number of genes rather than a single mutation, or a multifactorial mode of inheritance, with the likelihood of an interaction between environmental factors and genetic expression [110]. The biological marker human leukocyte antigen (HLA) has not been shown to be useful in predicting endometriosis risk [116–118]. However, research has suggested possible relationships between endometriosis and two separate genetic mutations (glutathione S-transferase M1 deletion and the N314D mutation of the galactose-1-phosphate uridyl transferase gene) [119,120].

VII. Endometriosis Treatment

Medical and surgical treatment are used individually and together in attempts to accomplish two main goals: pain alleviation and fertility improvement. Medical treatment consists of the pharmaceutical creation of a chronic hypoestrogenic state comparable to that of postmenopause, thereby causing atrophy of the endometriotic implants. Surgical treatment, often administered during diagnostic laparoscopy, involves laser or electrocautery destruction of visible adhesions and endometriotic implants. Large endometriomas or extensive adhesions may necessitate a larger abdominal incision (laparotomy), and women with severe symptoms and long-standing disease may undergo the definitive radical surgical treatment of oophorectomy and hysterectomy [121–123].

A. Medical Treatment

Any of several medications may be administered in a 3–6 month long course. The synthetic steroid danazol, an orally-administered weak androgen, has been used since the 1970s to treat endometriosis, acting by inhibition of ovarian steroidogenesis and direct suppression of endometriotic tissue growth. Danazol has androgenic side effects, including weight gain, mood swings, oily hair and skin, constipation, increased LDL cholesterol, and decreased HDL cholesterol [124]. The more recently introduced GnRH agonists (buserelin, leuprolide acetate, nafarelin, gosereline) are administered via intranasal spray or injected in a long-acting depot and create a protracted state of decreased gonadotropin secretion [30]. Their side effects are the signs and symptoms of hypoestrogenism: hot flashes, vaginal dryness, decreased libido, and decreased bone density [125,126]. Progestins (norethindrone, medroxyprogesterone acetate) are administered in high doses for endometriosis treatment, alone or as part of "add-back" therapy accompanying GnRH agonists in order to counteract GnRH agonists' detrimental effects on bone density [30,127]. Other anti-estrogens (tamoxifen) or continuously-administered high dose oral contraceptives currently are less commonly prescribed for women with active disease.

While not effective at treating large endometriomas, pharmaceutical treatment of other pelvic endometriosis provides effective pain relief by the end of a 6-month course of treatment for the majority of patients, and danazol and GnRH agonists are equally effective [127–134]. After successful pharmaceutical treatment, symptoms recur within 5 years in 37–74% of women, and recurrence may depend on the severity of initial disease [133,135–137]. However, in spite of an atrophic effect on endometrial implants, pharmaceutical treatment has not been shown to improve fertility. In a meta-analysis of seven randomized trials and cohort studies, Hughes *et al.* calculated a common OR of 0.9 for the likelihood of pregnancy after medical treatment with danazol or progestins, compared with placebo or no treatment [138]. These authors also found that GnRH agonist treatment was no more likely to improve fertility than was danazol.

B. Surgical Treatment

Surgical treatment provides initial pelvic pain relief in 60–75% of patients, and in these women the 5-year recurrence of endometriosis signs or symptoms ranges from 19 to 40% [139–141]. The impact of surgical treatment on fertility is less clear. Hughes' meta-analysis of one quasi-random and five cohort studies comparing laparoscopic laser or cautery treatment with danazol or no treatment found that surgery may improve fertility (pooled OR = 2.7), but the individual ORs ranged from 0.8 to

7.6, calling into question the advisability of calculating a pooled risk estimate [138]. These authors found similar associations with conservative laparotomy (pooled OR of 1.6 compared with danazol or no treatment, individual ORs ranging from 0.3 to 4.4). One randomized trial confirms a significant laparoscopic treatment effect on less severe endometriosis. Marcoux *et al.* found that 31% of infertile women with mild or minimal endometriosis who had surgical removal of all visible implants and lysis of adhesions via cautery and/or laser became pregnant in the subsequent 36 weeks, compared with 18% of similar women who underwent diagnostic laparoscopy only (adjusted RR = 1.9), but this pregnancy rate was still only one-third that of fertile women [142].

VIII. Outlook

Epidemiologic research has thus far identified relatively few conclusive risk factors for endometriosis. Characteristics associated with menstruation, particularly those that create the potential for increased amounts of retrofluxed menstrual fluid (early menarche, short menstrual cycles, heavy menstrual flow, and long flow duration) have provided the most consistent positive associations. Additionally, emerging evidence indicates that factors associated with lower circulating estrogen levels (exercise, smoking, short height, and androgenic body fat distribution) may decrease endometriosis risk. The impact of a woman's contraceptive and reproductive history on her risk of endometriosis is less clear, and the associations seen may be at least partially a function of the researchers' choice of populations from which case and control subjects are drawn.

Past epidemiologic research on this disease primarily has been clinically or hospital based, with cases obtained from infertility, gynecology, or endometriosis clinics or from cohorts of women undergoing tubal ligation. The former group is likely to overrepresent women with infertility, while the latter underrepresents infertile women and those with symptomatic disease. Some clinicians and researchers question whether mild, asymptomatic endometriosis is a disease at all or just a temporary phase of an ongoing process of cytolysis of recently implanted endometrial fragments from normal retrograde menstruation. As this physiological state may be present in all women at some point, one viewpoint is that endometriosis only becomes a disease when the implantations cause symptoms or local destruction or both [143,144]. This perspective has led to recommendations that research should focus on determining factors associated with severe or symptomatic disease only. Unfortunately, symptoms and severity (usually classified using American Fertility Society staging criteria) do not correlate well in endometriosis, and it is unclear whether infertility should be considered a disease symptom [49].

Control selection also deserves careful consideration in the study of this disease. Past research has used as controls a variety of women: infertility clinic patients with other infertility causes, patients hospitalized with other conditions, tubal ligation patients, obstetrical patients, and friends of cases. While the first three of these sources yield women who have been examined surgically and are free of endometriosis, the unrepresentative nature of these groups may outweigh the bias that would be obtained with the use of a more representative control group

including a small proportion of undiagnosed cases. Population-based studies with randomly-selected controls may allow more accurate etiologic investigations, particularly of the associations between endometriosis and contraceptive and reproductive history. Future epidemiologic research also should incorporate emerging noninvasive techniques for disease ascertainment, as well as new methods of exposure assessment. Promising genetic markers are under investigation, and our ability to measure the impact of environmental and other exposures accurately is improving with the use of biological assays of chemical or hormone levels. These advances should enhance our ability to identify etiologic relationships in the study of this enigmatic disease.

References

1. Jenkins, S., Olive, D. L., and Haney, A. F. (1986). Endometriosis; pathogenetic implications of the anatomic distribution. *Obstet. Gynecol.* **67**, 335–338.
2. Adamson, G. D. (1990). Diagnosis and clinical presentation of endometriosis. *Am. J. Obstet. Gynecol.* **162**, 568–569.
3. Lu, P. Y., and Ory, S. J. (1995). Endometriosis: Current management. *Mayo Clin. Proc.* **70**, 453–463.
4. Kjerulff, K. H., Erickson, B. A., and Langenberg, P. W. (1996). Chronic gynecological conditions reported by U.S. women: Findings from the National Health Interview Survey, 1984 to 1992. *Am. J. Public Health* **86**, 195–199.
5. Weed, J. C., and Arquembourg, P. C. (1980). Endometriosis: Can it produce an autoimmune response resulting in infertility? *Clin. Obstet. Gynecol.* **23**, 885–893.
6. Goldman, M. B., and Cramer, D. W. (1990). The epidemiology of endometriosis. *Prog. Clin. Biol. Res.* **323**, 15–31.
7. Ronnberg, L. (1990). Endometriosis and infertility. *Ann. Med.* **22**, 91–96.
8. Davis, K. M., and Rock, J. A. (1992). Endometriosis. *Curr. Opin. Obstet. Gynecol.* **4**, 229–237.
9. Mangtani, P., and Booth, M. (1993). Epidemiology of endometriosis. *J. Epidemiol. Commun. Health* **47**, 84–88.
10. Drake, T. S., and Grunert, G. M. (1980). The unsuspected pelvic factor in the infertility investigation. *Fertil. Steril.* **34**, 27–31.
11. Hasson, H. M. (1976). Incidence of endometriosis in diagnostic laparoscopy. *J. Reprod. Med.* **16**, 135–138.
12. Verkauf, B. S. (1987). The incidence, symptoms and signs of endometriosis in fertile and infertile women. *J. Fla. Med. Assoc.* **74**, 671–674.
13. Strathy, J. H., Molgaard, C. A., Coulam, C. B., and Melton, L. J. (1982). Endometriosis and infertility: A laparoscopic study of endometriosis among fertile and infertile women. *Fertil. Steril.* **38**, 667–672.
14. Mahmood, T. A., and Templeton, A. (1991). Prevalence and genesis of endometriosis. *Hum. Reprod.* **6**, 544–549.
15. Schenken, R. S., and Asch, R. H. (1980). Surgical induction of endometriosis in the rabbit: Effects on fertility and concentrations of peritoneal fluid prostaglandins. *Fertil. Steril.* **34**, 581–587.
16. Schenken, R. S., Asch, R. H., Williams, R. F., and Hodgen, D. G. (1984). Etiology of infertility in monkeys with endometriosis: Luteinized unruptured follicles, luteal phase defects, pelvic adhesions, and spontaneous abortions. *Fertil. Steril.* **41**, 122–130.
17. Vernon, M. W., and Wilson, E. A. (1985). Studies on the surgical induction of endometriosis in the rat. *Fertil. Steril.* **44**, 684–694.
18. Chauhan, M., Barratt, C. L., Cooke, S. M., and Cooke, I. D. (1989). Differences in the fertility of donor insemination recipients—a study to provide prognostic guidelines as to its success and outcome. *Fertil. Steril.* **51**, 815–819.

19. Jansen, R. P. (1986). Minimal endometriosis and reduced fecundability: Prospective evidence from an artificial insemination by donor program. *Fertil. Steril.* **46**, 141–143.

20. Brinton, L. A., Gridley, G., Persson, I., Baron, J., and Bergqvist, A. (1997). Cancer risk after a hospital discharge diagnosis of endometriosis. *Am. J. Obstet. Gynecol.* **176**, 572–579.

21. Schairer, C., Persson, I., Falkeborn, M., Naessen, T., Troisi, R., and Brinton, L. A. (1997). Breast cancer risk associated with gynecologic surgery and indications for such surgery. *Int. J. Cancer* **70**, 150–154.

22. Haney, A. F. (1990). Etiology and histogenesis of endometriosis. *Prog. Clin. Biol. Res.* **323**, 1–14.

23. Oral, E., and Arici, A. (1997). Pathogenesis of endometriosis. *Obstet. Gynecol. Clin. North Am.* **24**, 219–232.

24. Sampson, J. A. (1940). The development of the implantation theory for the origin of peritoneal endometriosis. *Am. J. Obstet. Gynecol.* **40**, 549–557.

25. Halme, J., Hammond, M. G., Hulka, J. F., Raj, S. G., and Talbert, L. M. (1984). Retrograde menstruation in healthy women and in patients with endometriosis. *Obstet. Gynecol.* **64**, 151–154.

26. Koninckx, P. R., Ide, P., and Vandenbroucke, W. (1980). New aspects of the pathophysiology of endometriosis and associated infertility. *J. Reprod. Med.* **24**, 257–260.

27. Badawy, S. Z., Cuenza, V., Marshall, L., Munchback, R., Rinas, A. C., and Coble, D. A. (1984). Cellular components in peritoneal fluid in infertile patients with and without endometriosis. *Fertil. Steril.* **42**, 704–708.

28. Bartosik, D., Jacobs, S. L., and Kelly, L. J. (1986). Endometrial tissue in peritoneal fluid. *Fertil. Steril.* **46**, 796–800.

29. Houston, D. E., Noller, K. L., and Melton, L. J. (1988). The epidemiology of pelvic endometriosis. *Clin. Obstet. Gynecol.* **31**, 787–800.

30. Barbieri, R. L. (1990). Etiology and epidemiology of endometriosis. *Am. J. Obstet. Gynecol.* **162**, 565–567.

31. Steele, R. W., Dmowski, W. P., and Marmer, D. J. (1984). Immunologic aspects of human endometriosis. *Am. J. Reprod. Immunol.* **6**, 33–36.

32. Dmowski, W. P., Braun, D., and Gebel, H. (1990). Endometriosis: Genetic and immunologic aspects. *Prog. Clin. Biol. Res.* **323**, 99–122.

33. Dmowski, W. P., Steele, R. W., and Baker, G. F. (1981). Deficient cellular immunity in endometriosis. *Am. J. Obstet. Gynecol.* **141**, 377–383.

34. Dmowski, W. P., Gebel, H. M., and Braun, D. P. (1994). The role of cell-mediated immunity in pathogenesis of endometriosis. *Acta Obstet. Gynecol. Scand., Suppl.* **159**, 7–14.

35. Mathur, S. P., Holt, V. L., Lee, J. H., Jiang, H., and Rust, P. F. (1998). Levels of antibodies to transferrin and alpha 2-HS glycoprotein in women with and without endometriosis. *Am. J. Reprod. Immunol.* **40**, 69–73.

36. Gleisher, N., El-Roely, A., Confino, E., and Friberg, J. (1987). Is endoemtriosis an autoimmune disease? *Obstet. Gynecol.* **70**, 115–122.

37. Fernandez-Shaw, S., Hicks, B. R., Yudkin, P. L., Kennedy, S., Barlow, D. H., and Starkey, P. M. (1993). Anti-endometrial and anti-endothelial auto-antibodies in women with endometriosis. *Hum. Reprod.* **8**, 310–315.

38. Garza, D., Mathur, S., Dowd, M. M., Smith, L. F., and Williamson, H. O. (1991). Antigenic differences between the endometrium of women with and without endometriosis. *J. Reprod. Med.* **36**, 177–182.

39. Koninckx, P. R., Meuleman, C., Oosterlynck, D., and Cornillie, F. J. (1996). Diagnosis of deep endometriosis by clinical examination during menstruation and plasma CA-125 concentration. *Fertil. Steril.* **65**, 280–287.

40. Volpi, E., De Grandis, T., Zuccaro, G., La Vista, A., and Sismondi, P. (1995). Role of transvaginal sonography in the detection of endometriomata. *J. Clin. Ultrasound* **23**, 163–167.

41. Ha, H. K., Lim, Y. T., Suh, T. S., Song, H. H., and Kim, S. J. (1994). Diagnosis of pelvic endometriosis: Fat-suppressed T1-weighted vs conventional MR images. *Am. J. Radiol.* **163**, 127–131.

42. Sugimura, K., Okizuka, H., Imaoka, I., Kaju, Y., Takahashi, K., Kitao, M., and Ishida, T. (1993). Pelvic endometriosis: Detection and diagnosis with chemical shift MR imaging. *Radiology* **188**, 435–438.

43. Togashi, K., Nishimura, K., Kimura, I., Tsuda, Y., Yamashita, K., Shibata, T., Nakano, Y., Konishi, J., Konishi, I., and Mori, T. (1991). Endometrial cysts: Diagnosis with MR imaging. *Radiology* **180**, 73–78.

44. American Fertility Society (1985). Revised American Fertility Society classification of endometriosis. *Fertil. Steril.* **43**, 351–352.

45. Hornstein, M. D., Gleason, R. E., Orav, J., Haas, S. T., Friedman, A. J., Rein, M. S., Hill, J. A., and Barbieri, R. L. (1993). The reproducibility of the revised American Fertility Society classification of endometriosis. *Fertil. Steril.* **59**, 1015–1021.

46. American Fertility Society (1979). Classification of endometriosis. *Fertil. Steril.* **32**, 633–634.

47. American Society for Reproductive Medicine (1997). Revised American Society for Reproductive Medicine classification of endometriosis: 1996. *Fertil. Steril.* **67**, 817–821.

48. Marana, R., Muzii, L., Caruana, P., Sell-Acqua, S., and Mancuso, S. (1991). Evaluation of the correlation between endometriosis extent, age of the patients and associated symptomatology. *Acta Eur. Fertil.* **22**, 209–212.

49. Jansen, R. P. (1993). Endometriosis symptoms and the limitations of pathology-based classification of severity. *Int. J. Gynaecol. Obstet.* **40** (Suppl.), S3–S7.

50. Palmisano, G. P., Adamson, G. D., and Lamp, E. J. (1993). Can staging systems for endometriosis based on anatomic location and lesion type predict pregnancy rates? *Int. J. Fertil.* **38**, 241–249.

51. Guzick, D. S., Silliman, N. P., Adamson, G. D., Buttram, V. C., Canis, M., Malinak, L. R., and Schenken, R. S. (1997). Prediction of pregnancy in infertile women based on the American Society for Reproductive Medicine's revised classification of endometriosis. *Fertil. Steril.* **67**, 822–829.

52. Moen, M. H. (1987). Endometriosis in women at interval sterilization. *Acta Obstet. Gynecol. Scand.* **66**, 451–454.

53. Houston, D. E. (1984). Evidence for the risk of pelvic endometriosis by age, race and socioeconomic status. *Epidemiol. Rev.* **6**, 167–191.

54. Kirshon, B., Poindexter, A. N., and Fast, J. (1989). Endometriosis in multiparous women. *J. Reprod. Med.* **34**, 215–217.

55. Houston, D. E., Noller, K. L., Melton, L. J., Selwyn, B. J., and Hardy, R. J. (1987). Incidence of pelvic endometriosis in Rochester, Minnesota, 1970–1979. *Am. J. Epidemiol.* **125**, 959–969.

56. Moen, M. H., and Schei, B. (1997). Epidemiology of endometriosis in a Norwegian county. *Acta Obstet. Gynecol. Scand.* **76**, 559–562.

56a. Kresch, A. J., Seifer, D. B., Sachs, L. B., and Barrese, I. (1984). Laparoscopy in 100 women with chronic pelvic pain. *Obstet. Gynecol.* **64**, 672–674.

57. Liu, D. T., and Hitchcock, A. (1986). Endometriosis: Its association with retrograde menstruation, dysmenorrhoea and tubal pathology. *Br. J. Obstet. Gynaecol.* **93**, 859–862.

58. Kirshon, B., and Poindexter, A. N. (1988). Contraception: A risk factor for endometriosis. *Obstet. Gynecol.* **71**, 829–831.

58a. Trimbos, J. B., Trimbos-Kemper, G. C., Peters, A. A., van der Does, C. D., and van Hall, E. V. (1990). Findings in 200 consecutive asymptomatic women having a laparoscopic sterilization. *Arch. Gynecol. Obstet.* **247**, 121–124.

59. Moen, M. H. (1991). Is a long period without childbirth a risk factor for developing endometriosis? *Hum. Reprod.* **6,** 1404–1407.

60. Sangi-Haghpeykar, H., and Poindexter, A. N. (1995). Epidemiology of endometriosis among parous women. *Obstet. Gynecol.* **85,** 983–992.

61. Vessey, M. P., Villard-Mackintosh, L., and Painter, R. (1993). Epidemiology of endometriosis in women attending family planning clinics. *Br. Med. J.* **306,** 182–184.

62. Velebil, P., Wingo, P. A., Xia, Z., Wilcox, L. S., and Peterson, H. B. (1995). Rate of hospitalization for gynecologic disorder among reproductive-age women in the United States. *Obstet. Gynecol.* **86,** 764–769.

63. Wilcox, L. S., and Mosher, W. D. (1993). Use of infertility services in the United States. *Obstet. Gynecol.* **82,** 122–127.

64. Darrow, S. L., Vena, J. E., Batt, R. E., Zielezny, M. A., Michalek, A. M., and Selman, S. (1993). Menstrual cycle characteristics and the risk of endometriosis. *Epidemiology* **4,** 135–142.

65. Signorello, L. B., Harlow, B. L., Cramer, D. W., Spiegelman, D., and Hill, J. A. (1997). Epidemiologic determinants of endometriosis: A hospital-based case-control study. *Ann. Epidemiol.* **7,** 267–274.

66. Beral, V., Rolfs, R., Joesoef, M. R., Aral, S., and Cramer, D. W. (1994). Primary infertility: Characteristics of women in North America according to pathological findings. *J. Epidemiol. Commun. Health.* **48,** 576–579.

67. Makhlouf Obermeyer, C., Armenian, H. K., and Azoury, R. (1986). Endometriosis in Lebanon: A case-control study. *Am. J. Epidemiol.* **124,** 762–767.

68. Arumugam, K., and Welluppilai, S. (1993). Endometriosis and social class: An Asian experience. *Asia-Oceania J. Obstet. Gynaecol.* **19,** 231–234.

69. Matorras, R., Rodiquez, F., Piuoan, J. I., Ramon, O., Gutierrez de Teran, G., and Rodriguez-Escudero, F. (1995). Epidemiology of endometriosis in infertile women. *Fertil. Steril.* **63,** 34–38.

70. Arumugam, K., and Templeton, A. A. (1990). Endometriosis and social class (letter). *Med. J. Aust.* **153,** 567.

71. Cramer, D. W., Wilson, E., Stillman, R. J., Berger, M. J., Belisle, S., Schiff, I., Albrecht, B., Gibson, M., Stadel, B. V., and Schoenbaum, S. C. (1986). The relation of endometriosis to menstrual characteristics, smoking, and exercise. *JAMA, J. Am. Med. Assoc.* **255,** 1904–1908.

72. Parazzini, F., La Vecchia, C., Franceschi, S., Negri, E., and Cecchetti, G. (1989). Risk factors for endometrioid, mucinous and serous benign ovarian cysts. *Int. J. Epidemiol.* **18,** 108–112.

73. Parazzini, F., Ferraroni, M., Fedele, L., Bocciolone, L., Rubessa, S., and Riccardi, A. (1995). Pelvic endometriosis: Reproductive and menstrual risk factors at different stages in Lombardy, northern Italy. *J. Epidemiol. Commun. Health* **49,** 61–64.

74. Arumugam, K., and Lim, J. M. (1997). Menstrual characteristics associated with endometriosis. *Br. J. Obstet. Gynaecol.* **104,** 948–950.

75. Lamb, K. T., and Berg, N. (1985). Tampon use in women with endometriosis. *J. Commun. Health* **10,** 215–225.

76. Filer, R. B., and Wu, C. H. (1989). Coitus during menses: Its effect on endometriosis and pelvic inflammatory disease. *J. Reprod. Med.* **34,** 887–890.

77. Wheeler, J. M., Johnston, B. M., and Malinak, L. R. (1983). The relationship of endometriosis to spontaneous abortion. *Fertil. Steril.* **39,** 656–660.

78. Metzger, D. A., Olive, D. L., Stohs, G. F., and Franklin, R. R. (1986). Association of endometriosis and spontaneous abortion: Effect of control group selection. *Fertil. Steril.* **45,** 18–22.

79. Darrow, S. L., Selman, S., Batt, R. E., Zielezny, M. A., and Vena, J. E. (1994). Sexual activity, contraception, and reproductive factors in predicting endometriosis. *Am. J. Epidemiol.* **140,** 500–509.

80. Mishell, D. R., Kletzky, O. A., Brenner, P. F., Roy, S., and Nicoloff, J. (1977). The effect of contraceptive steroids on hypothalamic-pituitary function. *Am. J. Obstet. Gynecol.* **128,** 60–74.

81. Mishell, D. R., Thorneycroft, I. H., Nakamura, R. M., Nagata, Y., and Stone, S. C. (1972). Serum estradiol in women ingesting combination oral contraceptive steroids. *Am. J. Obstet. Gynecol.* **114,** 924–928.

82. Guillebaud, J. (1993). Epidemiology of endometriosis (letter). *Br. Med. J.* **306,** 931.

83. Thomas, E. J. (1990). Preventive, symptomatic and expectant management of endometriosis. *Prog. Clin. Biol. Res.* **323,** 197–208.

84. Cattanach, J. F., and Milne, B. J. (1988). Post-tubal sterilization problems correlated with ovarian steroidogenesis. *Contraception* **38,** 541–550.

85. Denton, G. W., Schofield, J. B., and Gallagher, P. (1990). Uncommon complications of laparoscopic sterilisation. *Ann. R. Coll. Surg. Engl.* **72,** 210–211.

86. Stock, R. J. (1982). Postsalpingectomy endometriosis: A reassessment. *Obstet. Gynecol.* **60,** 560–570.

87. McCausland, A. (1982). Endosalpingosis (endosalpingoblastosis) following laparoscopic tubal coagulation as an etiologic factor of ectopic pregnancy. *Am. J. Obstet. Gynecol.* **143,** 12–24.

88. Fakih, H. N., Tamura, R., Kesselman, A., and DeCherney, A. H. (1985). Endometriosis after tubal ligation. *J. Reprod. Med.* **30,** 939–941.

89. Donnez, J., Casanas-Roux, F., Ferin, J., and Thomas, K. (1984). Tubal polyps, epithelial inclusions, and endometriosis after tubal sterilization. *Fertil. Steril.* **41,** 564–568.

90. Rock, J. A., Parmley, T. H., King, T. M., Laufe, L. E., and Su, B. C. (1981). Endometriosis and the development of tuboperitoneal fistulas after tubal ligation. *Fertil. Steril.* **35,** 16–20.

91. Olive, D. L., and Henderson, D. Y. (1987). Endometriosis and mullerian anomalies. *Obstet. Gynecol.* **69,** 412–415.

92. Brenner, C., and Wohlgemuth, S. (1990). Scar endometriosis. *Surg., Gynecol. Obstet.* **170,** 538–540.

93. Buchan, H., Vessey, M., Goldacre, M., and Fairweather, J. (1993). Morbidity following pelvic inflammatory disease. *Br. J. Obstet. Gynaecol.* **100,** 558–562.

94. Boyden, T. W., Pamenter, R. W., Stanforth, P. R., Rotkis, T. C., and Wilmore, J. H. (1983). Sex steroids and endurance running in women. *Fertil. Steril.* **39,** 629–632.

95. Dorgan, J. F., Reichman, M. E., Judd, J. T., and Brown, C. (1995). The relation of body size to plasma levels of estrogens and androgens in premenopausal women. *Cancer Causes Control* **6,** 3–8.

96. Rebuffe-Scrive, M. (1988). Steroid hormones and distribution of adipose tissue. *Acta Med. Scand., Suppl.* **723,** 143–146.

97. McCann, S. E., Freudenheim, J. L., Darrow, S. L., Batt, R. E., and Zielezny, M. A. (1993). Endometriosis and body fat distribution. *Obstet. Gynceol.* **82,** 545–549.

98. Reichman, M. E., Judd, J. T., Longscope, C., Schatzkin, A., Clevidence, B. A., Nair, P. P., Campbell, W. S., and Taylor, P. R. (1993). Effects of alcohol consumption on plasma and urinary hormone concentration in premenopausal women. *J. Natl. Cancer Inst.* **85,** 722–727.

99. Perper, M. M., Breitkopf, L. J., Breitstein, R., Cody, R. P., and Manowitz (1993). MAST scores, alcohol consumption, and gynecological symptoms in endometriosis patients. *Alcohol.: Clin. Exp. Res.* **17,** 272–278.

100. Grodstein, F., Goldman, M. B., and Cramer, D. W. (1994). Infertility in women and moderate alcohol use. *Am. J. Public Health* **84,** 1429–1432.

101. Grodstein, F., Goldman, M. B., Ryan, L., and Cramer, D. W. (1993). Relation of female infertility to consumption of caffeinated beverages. *Am. J. Epidemiol.* **137,** 1353–1360.

102. Covens, A. L., Christopher, P., and Casper, R. F. (1988). The effect of dietary supplementation with fish oil fatty acids on surgically induced endometriosis in the rabbit. *Fertil. Steril.* **49,** 698–703.

103. Rier, S. E., Martin, D. C., Bowman, R. E., Dmowski, W. P., and Becker, J. L. (1993). Endometriosis in rhesus monkeys (*Macaca mulatta*) following chronic exposure to 2,3,7,8-tetrachlorodibenzo-p-dioxin. *Fundam. Appl. Toxicol.* **21,** 433–441.

104. Fanton, J. W., and Golden, J. G. (1991). Radiation-induced endometriosis in *Macaca mulatta. Radiat. Res.* **126,** 141–146.

105. Eskenazi, B., and Kimmel, G. (1995). Workshop on perinatal exposure to dioxin-like compounds. II: Reproductive effects. *Environ. Health Perspect., Suppl.* **103,** 143–145.

106. Ahlborg, U. G., Lipworth, L., Titus-Ernstoff, L., Hsieh, C. C., Hanberg, A., Baron, J., Trichopoulos, D., and Adami, H. O. (1995). Organochlorine compounds in relation to breast cancer, endometrial cancer, and endometriosis: An assessment of the biological and epidemiological evidence. *Crit. Rev. Toxicol.* **25,** 463–531.

107. Mayani, A., Barel, S., Soback, S., and Almagor, M. (1997). Dioxin concentrations in women with endometriosis. *Hum. Reprod.* **12,** 373–375.

108. Lebel, G., Dodin, S., and Dewailly, E. (1998). Organochlorine exposure and the risk of endometriosis. *Fertil. Steril.* **69,** 221–228.

109. Simpson, J. L., Elias, S., Malinak, L. R., and Buttram, V. C. (1980). Heritable aspects of endometriosis. I. Genetic studies. *Am. J. Obstet. Gynecol.* **137,** 327–331.

110. Coxhead, D., and Thomas, E. J. (1993). Familial inheritance of endometriosis in a British population: A case control study. *J. Obstet. Gynecol.* **13,** 42–44.

111. Moen, M. H., and Magnus, P. (1993). The familial risk of endometriosis. *Acta Obstet. Gynecol. Scand.* **72,** 560–564.

112. Moen, M. H. (1994). Endometriosis in monozygotic twins. *Acta Obstet. Gynecol. Scand.* **73,** 59–62.

113. Hadfield, R. M., Mardon, H. J., Barlow, D. H., and Kennedy, S. H. (1997). Endometriosis in monozygotic twins. *Fertil. Steril.* **68,** 941–942.

114. Kennedy, S., Hadfield, R., Mardon, H., and Barlow, D. (1996). Age of onset of pain symptoms in non-twin sisters concordant for endometriosis. *Hum. Reprod.* **11,** 403–405.

115. Malinak, L. R., Buttram, V. C., Elias, S., and Simpson, J. L. (1980). Heritable aspects of endometriosis. II: Clinical characteristics of familial endometriosis. *Am. J. Obstet. Gynecol.* **137,** 332–337.

116. Moen, M., Bratlie, A., and Moen, T. (1984). Distribution of HLA-antigens among patients with endometriosis. *Acta Obstet. Gynecol. Scand.* **123,** 25–27.

117. Simpson, J. L., Malinak, L. R., Elias, S., Carson, S., and Radvany, R. A. (1984). HLA association in endometriosis. *Am. J. Obstet. Gynecol.* **148,** 395–397.

118. Maxwell, C., Kilpatrick, D. C., Haining, R., and Smith, S. K. (1989). No HLA-DR specificity is associated with endometriosis. *Tissue Antigens* **34,** 145–147.

119. Baranova, H., Bothorishvilli, R., Canis, M., Albuisson, E., Perriot, S., Glowaczower, E., Bruhat, M. A., Baranov, V., and Malet, P. (1997). Glutathione S-transferase M1 gene polymorphism and susceptibility to endometriosis in a French population. *Mol. Hum. Reprod.* **3,** 775–780.

120. Cramer, D. W., Hornstein, M. D., Ng, W. G., and Barbieri, R. L. (1996). Endometriosis associated with the N314D mutation of galactose-1-phosphate uridyl transferase (GALT). *Mol. Hum. Reprod.* **2,** 149–152.

121. Barbieri, R. L., and Kistner, R. W. (1986). Endometriosis. *In* "Gynecology Principles and Practice" (R. W. Kistner, ed.), pp. 393–414. Year Book Medical Publishers, Chicago.

122. Cunningham, F. G. (1989). "Williams Obstetrics," 18th ed. Appleton & Lange, Norwalk, CT.

123. Adamson, G. D., and Nelson, H. P. (1997). Surgical treatment of endometriosis. *Obstet. Gynecol. Clin. North Am.* **24,** 375–409.

124. Packard, C. J., and Shepherd, J. (1994). Action of danazol on plasma lipids and lipoprotein metabolism. *Acta Obstet. Gynecol. Scand., Suppl.* **159,** 35–40.

125. Dawood, M. Y. (1994). Hormonal therapies for endometriosis: Implications for bone metabolism. *Acta Obstet. Gynecol. Scand.* **159,** 22–34.

126. Orwoll, E. S., Yuzpe, A. A., Burry, K. A., Heinrichs, L., Buttram, V. C., and Hornstein, M. D. (1994). Nafarelin therapy in endometriosis: Long-term effects on bone mineral density. *Am. J. Obstet. Gynecol.* **171,** 1221–1225.

127. Hornstein, M. D., Surrey, E. S., Weisberg, G. W., and Casino, L. A. (1998). Leuprolide acetate depot and hormonal add-back in endometriosis: A 12-month study. *Obstet. Gynecol.* **91,** 16–24.

128. Schenken, R. S. (1990). Gonadotropin-releasing hormone analogs in the treatment of endometriomas. *Am. J. Obstet. Gynecol.* **162,** 579–581.

129. Rolland, R., and van der Heijden, P. F. (1990). Nafarelin versus danazol in the treatment of endometriosis. *Am. J. Obstet. Gynecol.* **162,** 586–588.

130. Henzl, M. R., Corson, S. L., Moghissi, K., Buttram, V. C., Berqvist, C., and Jacobson, J. (1988). Administration of nasal nafarelin as compared with oral danazol for endometriosis. *N. Engl. J. Med.* **318,** 485–489.

131. Shaw, R. W. (1990). Nafarelin in the treatment of pelvic pain caused by endometriosis. *Am. J. Obstet. Gynecol.* **162,** 574–576.

132. Telimaa, S., Puolakka, J., Ronnberg, L., and Kauppila, A. (1987). Placebo-controlled comparison of danazol and high-dose medroxyprogesterone acetate in the treatment of endometriosis. *Gynecol. Endocrinol.* **1,** 13–23.

133. Buttram, V. C., Reiter, R. C., and Ward, S. (1985). Treatment of endometriosis with danazol: Report of a 6-year prospective study. *Fertil. Steril.* **43,** 353–360.

134. Kettel, L. M., Hummel, W. P. (1997). Modern medical management of endometriosis. *Obstet. Gynecol. Clin. North Am.* **24,** 361–373.

135. Barbieri, R. L., Evans, S., and Kistner, R. W. (1982). Danazol in the treatment of endometriosis: Analysis of 100 cases with a 4-year follow-up. *Fertil. Steril.* **37,** 737–746.

136. Hornstein, M. D., Yuzpe, A. A., Burry, K. A., Heinrichs, L. R., Buttram, V. L., and Orwoll, E. S. (1995). Prospective randomized double-blind trial of 3 versus 6 months of nafarelin therapy for endometriosis associated pelvic pain. *Fertil. Steril.* **63,** 955–962.

137. Waller, K. G., and Shaw, R. W. (1993). Gonadotropin-releasing hormone analogues for the treatment of endometriosis: Long-term follow-up. *Fertil. Steril.* **59,** 511–515.

138. Hughes, E. G., Fedorkow, D. M., and Collins, J. A. (1993). A quantitative overview of controlled trials in endometriosis-associated infertility. *Fertil. Steril.* **59,** 963–970.

139. Sutton, C. J., Ewen, S. P., Whitelaw, N., and Haines, P. (1994). Prospective, randomized, double-blind, controlled trial of laser laparoscopy in the treatment of pelvic pain associated with minimal, mild, and moderate endometriosis. *Fertil. Steril.* **62,** 696–700.

140. Redwine, D. B. (1991). Conservative laparoscopic excision of endometriosis by sharp dissection: Life table analysis of reoperation and persistent or recurrent disease. *Fertil. Steril.* **56,** 628–634.

141. Sutton, C., Hill, D. (1990). Laser laparoscopy in the treatment of endometriosis: A 5-year study. *Br. J. Obstet. Gynaecol.* **97,** 181–185.

142. Marcoux, S., Maheux, R., and Berube, S. (1997). Laparoscopic surgery in infertile women with minimal or mild endometriosis. *N. Engl. J. Med.* **337,** 217–222.

143. Koninckx, P. R., Oosterlynck, D., D'Hooghe, T., and Muelemen, C. (1994). Deeply infiltrating endometriosis is a disease whereas mild endometriosis could be considered a non-disease. *Ann. N. Y. Acad. Sci.* **734,** 333–341.

144. Thomas, E. J. (1995). Endometriosis, 1995-confusion or sense? *Int. J. Gynecol. Obstet.* **48,** 149–155.

19

Uterine Leiomyomata

STEPHEN M. SCHWARTZ* AND LYNN M. MARSHALL†

*Program in Epidemiology, Division of Public Health Sciences, Fred Hutchinson Cancer Research Center, Seattle, Washington;
†Department of Epidemiology, School of Public Health and Community Medicine, University of Washington, Seattle, Washington

I. Introduction

Uterine leiomyomata, commonly called "fibroids," are benign neoplasms of uterine smooth muscle [1,2]. These tumors frequently cause menstrual abnormalities, pelvic pain, and other symptoms that seriously affect a woman's quality of life. Symptomatic uterine leiomyomata often require major surgery and increased medical utilization; in the U.S. they account for 33% of all hysterectomies [3]. Despite their substantial impact on gynecologic morbidity, relatively little is known about the etiology of these neoplasms or the characteristics of women who are at increased risk for developing them. Moreover, there is limited information on the clinical outcomes of women diagnosed with uterine leiomyomata.

This chapter begins with a summary of the histologic and molecular characteristics of uterine leiomyomata, followed by issues in the clinical presentation, diagnosis, and ascertainment of patients with uterine leiomyomata relevant to the conduct of epidemiologic research. We present prevailing hypotheses as to the role of steroid hormones in uterine leiomyomata pathogenesis, focusing on the evidence from epidemiologic studies. Following a summary of the clinical epidemiology, we conclude with recommendations for further research.

II. Anatomic, Histopathologic, and Molecular Characteristics of Uterine Leiomyomata

The pathologic characteristics of uterine leiomyomata have been reviewed extensively [4,5]. Briefly, these tumors appear as a whorled mix of smooth muscle cells and extracellular fibrous tissue comprising collagen, proteoglycan, and fibronectin [5]. Uterine leiomyomata may be intramural (completely within the myometrium), submucosal (partially or entirely within the endometrium or uterine cavity), or subserosal (partially or entirely outside of the uterus) [6]. All cells from a single uterine leiomyoma have the same glucose-6-phosphate dehydrogenase isoform and inactivation pattern of other genes located on the X chromosome [7–9]. Multiple uterine leiomyomata developing within a single uterus have different patterns of X chromosome inactivation [8,9], indicating that they are clonal growths that develop independently of one another.

Up to 50% of uterine leiomyomata contain cytogenetic abnormalities, three-quarters of which involve 6p, 12q, or 7q [10]. Tumors with chromosomal abnormalities tend to be larger and exhibit increased proliferative activity compared to tumors without abnormalities [11–13]. The development of uterine leiomyomata does not appear to involve dysregulated cell cycling or increased genetic or epigenetic alterations in cellular oncogenes or tumor suppressor genes that are characteristic of malignant neoplasms, including uterine leiomyosarcomas [14–19]. Studies focusing on regions commonly involved in cytogenetic abnormalities, however, have identified the High Mobility Group (HMG) protein genes as possible loci within the 6p and 12q regions that are disrupted in uterine leiomyomata [20–22]. HMG proteins alter the conformation of DNA to facilitate the binding of regulatory proteins during the gene transcription process, and inhibition of HMGs suppresses cell transformation *in vitro* [23]. A variety of mitogens such as epidermal growth factor and insulin-like growth factors 1 and 2 (IGF-1, IGF-2) and growth factor receptors are overexpressed in uterine leiomyomata relative to normal myometrial tissue [24]. Taken together, the available data suggest that an underlying molecular defect in uterine leiomyomata may be abnormal expression of HMG proteins leading to aberrant transcription of growth factors that contribute to the proliferation of smooth muscle cells, accumulation of extracellular material such as collagen, or both.

III. Issues in the Design of Epidemiologic Studies of Uterine Leiomyomata

A. *Symptomatology and Methods of Diagnosis*

Excessive menstrual bleeding and pelvic pain are the primary symptoms associated with uterine leiomyomata. Up to 76% of patients report at least one of these two symptoms [6,25,26]. Other symptoms include infertility, increased urinary frequency or incontinence, constipation, abdominal bloating, and fatigue (likely due to anemia resulting from heavy or chronic bleeding) [26]. The spectrum and severity of symptoms is thought to depend on the size, location, and number of leiomyomata present in the uterus [6]. A large proportion of uterine leiomyomata, however, are diagnosed in the absence of symptoms. For example, in a sample of women from a cohort of female nurses who reported a diagnosis of uterine leiomyomata confirmed by ultrasound or hysterectomy, 33% reported no symptoms [27]. Among women treated surgically for uterine leiomyomata, a smaller proportion (12%) were asymptomatic [26].

Initial diagnoses usually are made at pelvic examination [28]. However, because the clinical signs and symptoms typical of uterine leiomyomata are also associated with other gynecologic conditions, pelvic examinations do not provide definitive diagnoses [28]. In addition, women with symptoms of uterine leiomyomata may not be diagnosed at pelvic exam if they have submucos or intramural tumors. For this reason, abdominal [29] or transvaginal [30] ultrasound examinations provide reasonably sensitive, noninvasive confirmation of a suspected uterine leiomyomata diagnosis. Only a subset of women with a diagnosis confirmed by ultrasound will ultimately proceed to surgery and histologic confirmation (the gold standard). The decision to manage uterine leiomyomata surgically depends on several factors, most importantly the development and severity of symptoms (See Clinical Epidemiology of Uterine Leiomyomata). Operative and pathology reports from hysterectomy or myomectomy (surgical removal of individual uterine leiomyomata) potentially provide more precise information regarding the size, location, and number of tumors than ultrasound procedures.

B. Methodologic Issues

The major methodologic issues to be considered in designing valid studies of uterine leiomyomata are the definition and ascertainment of cases. Interpretation of the results of epidemiologic research often hinges on the strengths and limitations of different approaches to the choice of cases, and by extension, the definition and selection of noncases.

Although histologic evidence of uterine leiomyomata is the gold standard for determining case status, epidemiologic studies that include only women with histologically-confirmed tumors have limited etiologic interpretation [31]. For example, such studies may preferentially identify etiologic factors operating relatively late in the growth and development of the tumors. These risk factors could be very important both from clinical and from public health perspectives because they may provide avenues to preventing the development of uterine leiomyomata requiring hysterectomy. However, studies relying solely on surgically diagnosed cases may be particularly susceptible to bias. Histologically-confirmed cases may be more likely than cases confirmed only by ultrasound to have characteristics associated with unsuccessful conservative management, patient preferences for surgical management, or both. For example, a woman may be more likely to choose hysterectomy as therapy if she has completed childbearing. Moreover, there may be an extended interval between the initial diagnosis of uterine leiomyomata (e.g., by ultrasound) and surgical treatment, creating challenges in measuring the timing of exposures in relation to the development of the tumor. Epidemiologic studies of uterine leiomyomata can reduce the impact of these biases by including cases newly-diagnosed by ultrasound, whether or not the tumor is confirmed histologically.

Information on new uterine leiomyomata diagnoses may be obtained from self-reported history, from hospital or clinical records, or from automated hospital discharge and ambulatory care databases. Original diagnostic imaging, surgical, and pathology reports should be reviewed to examine the validity of the ascertainment source. For example, self-reported diagnoses were accurate in 93% of women who agreed to release their medical records in a sample of respondents from a large cohort of female nurses [27]. With such data, the sensitivity and specificity of the ascertainment source can be computed and the effect on the study results estimated.

Whatever source of information is used to identify cases, important limitations in case ascertainment should be appreciated. If the study population includes a substantial proportion of women who infrequently use, or have limited access to, gynecologic care, asymptomatic cases or cases with mild symptoms may be underascertained. Even within populations with relatively high and uniform access to gynecologic care, asymptomatic cases may be missed if tumors are not palpated or the uterus is not enlarged, and more sensitive diagnostic procedures are not performed subsequently. A woman's risk-factor history may also influence the extent to which uterine leiomyomata are diagnosed in the absence of symptoms. For example, women who have pelvic exams performed in conjunction with prescriptions for hormonal birth control would be expected to have an increased opportunity to have asymptomatic uterine leiomyomata detected.

Some uterine leiomyomata cases will be identified incidentally during examinations prompted by gynecologic symptoms caused by other conditions. Incidental diagnoses are difficult to avoid, because usually it will not be possible to determine with confidence whether the symptoms are due to the uterine leiomyomata or to other pathologic conditions. However, if a characteristic is not related to uterine leiomyomata, but is related to a condition that often coexists with these tumors, a spurious association with uterine leiomyomata could result. To reduce the potential impact of incidental diagnoses, cases should not be selected preferentially from specialty clinics that treat gynecologic, urologic, or infertility conditions.

Potential biases arising from ascertainment of asymptomatic cases or potentially incidental diagnoses may be addressed analytically if the appropriate data are available. Therefore, a study should collect information regarding the presence, nature, and duration of symptoms leading to diagnosis, such as presence of coexisting conditions, methods of diagnosis, and pharmacologic and surgical treatments. In addition, a history of gynecologic examinations preceding the diagnosis of the cases (or a similar date for the noncases) should be assessed.

Clearly defining the source population from which the cases will be drawn provides the basis for establishing the criteria by which the comparison population, the noncases, will be selected. In general, at any particular point during the case accrual period, noncases would be women who have an intact uterus and have not been diagnosed with uterine leiomyomata. In cohort studies, such as the Nurses' Health Study II [27,32] and the Oxford Family Planning Study [33], rates of uterine leiomyomata diagnoses are compared between the exposed and nonexposed groups of participants. In case-control studies, controls are selected to represent the distribution of exposure in the source population. In studies conducted within a managed care organization, controls would be selected from among women who are members of that organization during the case accrual period. In hospital- or clinic-based studies, it is frequently difficult to define the source population because it is made up of women who, had they had uterine leiomyomata that would come to diagnosis, would have been diagnosed at the particular hospital(s) or clinic(s) from which the cases are enrolled.

Whatever the source population, if a proportion of the cases are diagnosed in the absence of symptoms, a similar proportion of controls should also be selected from among women who received the same types of gynecologic care, such as a pelvic exam, that lead to the diagnosis of asymptomatic uterine leiomyomata. The apparent high frequency of uterine leiomyomata in women of reproductive age raises the likelihood that tumors will be present, yet undiagnosed, in a substantial proportion of women who are selected as controls. In general, unidentified cases among the control group will impair a study's ability to identify risk factors, particularly those that operate early in the pathogenesis of the tumors. This limitation applies in varying degrees to epidemiologic studies of many "benign" reproductive health conditions as well as some malignant tumors (*e.g.,* prostate cancer) and thus is not unique to uterine leiomyomata.

Some investigators recommend that epidemiologic studies include as controls only women who have had their uteri removed and examined histologically in order to exclude the presence of uterine leiomyomata [34]. An alternative approach might be to choose controls who have no evidence of uterine leiomyomata based on ultrasound examinations. Because hysterectomy and gynecologic ultrasound are performed to treat or diagnosis gynecologic conditions, controls identified in such ways almost certainly will yield spurious positive or negative associations with uterine leiomyomata when studying characteristics related to other conditions for which these procedures are performed [35,36]. In some research settings it may be possible to use noninvasive diagnostic methods in a way that is less likely to produce biased results, for example by recruiting controls into a protocol that includes screening ultrasound examinations to detect asymptomatic, clinically occult, uterine leiomyomata. Exposures could then be compared not only between cases (women with clinically apparent uterine leiomyomata) and all controls, but also between cases and those controls whose uteri are free of occult uterine leiomyomata, and between controls with and without occult uterine leiomyomata.

IV. Frequency of Occurrence and Demographic Patterns

A. Incidence

Rates of uterine leiomyomata diagnoses in U.S. populations are based on data from the National Hospital Discharge Survey (NHDS) [3,37], the National Health and Nutrition Examination Survey (NHANES) Epidemiologic Follow-up Study [38], and the Nurses' Health Study II (NHS II) cohort [27]. Rates vary according to case definition (Table 19.1), ranging from 12.8 per 1000 person-years for all diagnoses to approximately 2.0 per 1000 person-years for hysterectomy-confirmed cases. Rates of hysterectomy for uterine leiomyomata from the NHDS and from the NHS II are similar [3,27] and lower than rates from the NHANES [38].

Little is known about rates of diagnoses in populations outside the U.S. Based on data from the Oxford Family Planning Study (OFPS) in the U.K., the rate of hysterectomy-confirmed uterine leiomyomata diagnoses among white women is similar to that in the U.S. [33]. Formal international comparisons of uterine leiomyomata diagnoses are not possible because rates have been reported only in the U.S. and the U.K.

1. Age and Race

Rates of uterine leiomyomata increase with age through the reproductive years [3,27,33,37] and, at least for hysterectomy-confirmed cases, decline in the postmenopausal years [3,33]. The age-specific rates of hysterectomy-confirmed uterine leio-

Table 19.1

Diagnosis Rates of Uterine Leiomyomata per 1000 Person-Years among US Women

Diagnosis type	Source	Race		
		All	Whites	Black
Pelvic examination, ultrasound, or hysterectomy	[27]	12.8	12.5	37.9
Ultrasound or hysterectomy	[27]	9.2	8.9	30.6
Hospitalization[a]	[37]	3.0	2.4	5.3[b]
Hysterectomy	[3]	1.9	1.6	3.8
	[27]	2.0	2.0	4.5
	[38]	3.6	n.d.[c]	n.d.

[a]Hospitalization for gynecologic conditions unrelated to pregnancy among women ages 15–44.
[b]Rate for nonwhite women.
[c]Not determined.

myomata appear to be similar among women in the U.S. and in the U.K. (Fig. 19.1). Compared to rates of uterine leiomyomata confirmed by hysterectomy, rates of diagnoses confirmed by other surgical procedures or ultrasound are 2 to 16 times greater at each age [27,37].

Age-adjusted uterine leiomyomata rates among black women are two to three times greater than rates among whites [3,27,38]. The higher rates among blacks are evident at all ages [27,37], and rates peak earlier among blacks (35–39 years) than among whites (40–44 years) [27]. Several other findings are consistent with the black-white differences in diagnosis rates. Compared to whites, black women are twice as likely to report having physician-diagnosed uterine leiomyomata in the prior 12 months [26]. Among women undergoing hysterectomy for any reason, 58–65% of blacks have uterine leiomyomata as the primary indication compared to 29–45% among whites [3,39,40]. Among women undergoing hysterectomy specifically for uterine leiomyomata, the mean ages at first diagnosis and at hysterectomy are lower among blacks than whites [26,39]. Interestingly, reports from Nigeria suggest an earlier peak age at diagnosis for African women (30–39 years) [25] compared to European women (40–49 years) [41,42] that parallels the difference in age at diagnosis between U.S. blacks and whites.

The reason for the more frequent and earlier diagnoses of uterine leiomyomata among blacks is unclear. The excess does not seem to be attributable to differences in the types and severity of symptoms, the use of health care, or in the prevalence of putative risk factors [26,27]. In the NHS II cohort, the two- to three-fold elevated rates among blacks persisted after adjustment for several suspected risk factors and after restricting the analysis to women who reported having a recent physical examination or a Papanicolaou smear [27]. That black women have more uterine leiomyomata, larger tumors, and heavier uterine weight compared to whites [26,27] suggests a biological basis for the racial difference.

Few data exist on the frequency of uterine leiomyomata diagnoses among other racial or ethnic groups. Observations from

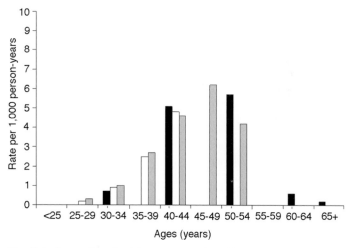

Fig. 19.1 Rates of uterine leiomyomata confirmed by hysterectomy, by age. ■, National Hospital Discharge Survey, 1988–1991 [3]—U.S. women, age ⩾15 years (rates are plotted at the midpoint of the 10-year age intervals 25–34, 35–44, 45–54, and 55–64); □, Nurses Health Study II, 1989–1993 [27]—U.S. female nurses, ages 25–44; ■, Oxford Family Planning Study, 1968–1985 [33]—UK married women, ages 25–57.

the NHS II cohort suggest that Hispanic and Asian women experience rates similar to those of whites [27].

2. Time Trends

NHDS data indicate that the average annual rate of hysterectomy for uterine leiomyomata from 1980 through 1990 was relatively constant at 1.9 per 1000 women per year [3,43]. Similar findings were reported in a cohort of reproductive-age women in the United Kingdom during a comparable time period [44]. Secular trends of uterine leiomyomata diagnoses in different racial groups have not been reported.

B. Mortality

Uterine leiomyomata influence a woman's risk of dying to the extent that surgical management results in procedure-related mortality. Compared to women undergoing hysterectomy but in whom uterine leiomyomata are not diagnosed, women diagnosed with these tumors have significantly lower in-hospital mortality [39]. There are no studies directly comparing mortality among women receiving hysterectomy for uterine leiomyomata to the mortality in similarly-aged women not undergoing hysterectomy.

V. Etiologic Hypotheses and Risk Factors

A. Etiologic Hypotheses

The development of a uterine leiomyoma probably involves an initiation phase during which a myometrial cell acquires a

somatic genetic or epigenetic change conferring increased sensitivity to a variety of growth factors [45]. An expansion and growth phase likely follows, during which multiplication of the altered cell into a clonal lesion is accompanied by stimulation of a local extracellular environment that provides structure and increased mass to the tumor [45]. Factors involved in the initiation of uterine leiomyomata have not been extensively explored, but animal models suggest the possibility that expression of viral genes (and thus possibly human counterparts) could be important [46]. That estrogen plays a prominent role in promoting uterine leiomyoma growth has been considered since early in the twentieth century [47,48]. Evidence from clinical and laboratory-based studies has strengthened the argument for a role for estrogen but also has raised the possibility that progesterone may be an important growth enhancing factor.

1. Estrogen

Compared to normal myometrium, smooth muscle cells in uterine leiomyomata exhibit increased expression of a wide variety of steroid hormone receptors, growth factors, and growth factor receptors, most of which are regulated by, or interact with, estrogen [24,49]. Uterine leiomyomata also show increased capacity to convert estradiol to catechol estrogens suspected of enhancing tumorigenesis [50]. Thus, uterine leiomyomata appear to have an increased sensitivity, and possibly a generalized altered response, to a woman's hormonal milieu. The hypersensitive phenotype is similar to the pregnant uterus and suggests that the growth of uterine leiomyomata occurs under stimulation from normal cycling levels of estrogen. Whether the development of altered sensitivity to growth factors precedes or follows the occurrence of cytogenetic changes [10] has not been directly studied. Increased growth could allow expansion of cell populations, increasing the probability that one or more cells would develop a molecular change that further enhances tumor growth potential. Alternatively, somatic events could lead to the loss of normal control of the expression of growth factors and/or their receptors, followed by clonal expansion of cells with growth advantages. Also, the extent to which growth factors act via deposition of collagen versus proliferation of smooth muscle tissue is not clear.

A combination of animal and human experimental studies provide more direct evidence of the role of estrogen in uterine leiomyomata growth and development. Eker rats spontaneously develop uterine leiomyomata that are histologically identical to those arising in humans [51,52]. Growth of cell lines derived from these tumors is enhanced by estrogen and inhibited by estrogen antagonists such as tamoxifen [53,54]. Similar findings have been reported in other animal models [46,55]. Clinical studies of women with uterine leiomyomata treated with gonadotropin-releasing hormone (GnRH) agonists provide parallel data from humans. GnRH agonists are highly potent, synthetic analogues of luteinizing hormone releasing-hormone that create a temporary hypoestrogenic state by reducing biologically active gonadotropin secretions of the pituitary gland. As discussed more fully later in this chapter, GnRH agonist administration markedly reduces uterine leiomyomata size [56].

2. Progesterone

The possibility that progesterone promotes the development of uterine leiomyomata is suggested by several lines of evidence [57]. Molecular and histopathologic studies show that: (1) among premenopausal women, the proliferative activity and expression of the anti-apoptotic proto-oncogene *bcl*-2 in uterine leiomyomata is highest in the secretory phase [58–60]; (2) among premenopausal women, the proliferative activity of uterine leiomyoma appears to be higher in women using medroxyprogesterone acetate (MPA) compared to women using combined oral contraceptives or women who are not taking any hormonal contraception [61]; and (3) among postmenopausal women, the proliferative activity of uterine leiomyomata is higher in women receiving combined estrogen and progesterone replacement therapy compared to women receiving estrogen replacement alone [59]. A growth-promoting role of progesterone also is supported by most studies of women treated for uterine leiomyomata with progestins or progesterone inhibitors: (1) postmenopausal women with small, asymptomatic uterine leiomyomata receiving estrogen replacement therapy (ERT) plus 5 mg MPA showed roughly a 50% increase in myoma volume compared to women receiving ERT plus 2.5 mg MPA [62]; (2) RU-486 causes myoma shrinkage in a dose-response fashion [63,64]; and (3) GnRH agonist-induced regression of leiomyomata is impaired when administered concurrently with norethindrone [65] or MPA [66]. However, not all studies have observed a reduced GnRH agonist effect when MPA is included in the regimen [67,68].

B. Risk Factors

In this section we summarize the results of epidemiologic studies of risk factors for uterine leiomyomata, focusing on putative markers of exposure to endogenous and exogenous steroid hormones and their relevance to hypothesized roles of estrogen and progesterone as described previously. The majority of the evidence regarding hormonal exposures comes from nine studies (Table 19.2). In most studies, cases were women who had undergone hysterectomy or other surgery and whose diagnosis of uterine leiomyomata was confirmed histologically. However, in some studies the designation of case status was based on a lower level of diagnostic certainty. In all studies the comparison population consisted of women among whom uterine leiomyomata had not been clinically diagnosed.

1. Markers of Endogenous Hormone Levels

a. MENARCHE AND MENOPAUSE. In three studies, two of surgically-confirmed uterine leiomyomata and one of uterine leiomyomata diagnosed by ultrasound or surgery, risk decreased with increasing age at menarche [69–71]. No association with age at menarche was observed in one study of surgically-confirmed uterine leiomyomata [41], however, or in a study of self-reported history [72]. In contrast, regardless of the method of case ascertainment or diagnostic confirmation, studies consistently show that postmenopausal women are at 70–90% reduced risk [33,41,70,72].

The infrequent clinical diagnosis of uterine leiomyomata following menopause provides strong evidence that steroid hormones play an important role in the growth of the tumors. The epidemiologic data are supported by pathologic studies of hysterectomy specimens that find a reduction both in the size and in the number of leiomyomata in postmenopausal compared to premenopausal women [34]. Women with early onset of menses will, on the average, have increased lifetime exposure to ovulatory cycles. Because mitotic activity in the myometrium appears to be greatest during the luteal phase of the menstrual cycle [73], a longer history of cycling might be expected to be associated with increased risk. On this basis it would also be predicted that late age at menopause would be associated with increased risk of uterine leiomyomata, but no studies have examined this possibility.

b. PREGNANCY AND CHILDBEARING HISTORY. The risk of uterine leiomyomata confirmed by surgery alone, or by either surgery or ultrasound, has been reported to be 20–50% lower [69,71,74] among women who have given birth compared to nulliparous women. Furthermore, the risk appears to decrease with increasing parity [33,69,71,74]. In several studies, women who have had four or five births were at 70–80% lower risk than nulliparous women [33,69,74]. Incomplete pregnancies due to miscarriages appear to be unrelated to risk [38,74], although in one study a nearly two-fold increase in risk was observed for three or more abortions (with no distinction made between induced and spontaneous terminations) [69]. In contrast, women with a history of two or more induced abortions were at 40% lower risk compared to women who never had an induced abortion [41]. A history of infertility also has been reported to be a risk factor for uterine leiomyomata [71,74].

Table 19.2

Characteristics of Published Epidemiologic Studies of Uterine Leiomyomata

Study type	Source	Location	Enrollment date	End of follow-up	Age at enrollment	Basis of case ascertainment	Number of cases	Person-years (p-y) or # of controls
Cohort	[88]	U.S.	1968–1972	1977	18–54	Hospital discharge	unknown	107,165[a]
	[38]	U.S.	1971–1975	1992	25–49	Hospital discharge	unknown	unknown
	[27]	U.S.	1989	1993	25–42	Self report[b]	2967[c]	327,065
	[89]	U.K.	1968–1969	1971 (?)	15–45	Physician report	168	87,000[d]
	[33]	U.K.	1968–1974	1985	29–35	Pathology report	538	198,653
Case-Control	[70]	U.S.	1985–1987	—	40–64	Pathology report	345	749
	[41,74,81,87]	Italy	1986–1993	—	≤58	Pathology report[e]	621[f]	1,051
	[69]	Thailand	1991–1993	—	≤60	Pathology report	901	2,709
	[72]	U.S.	1980–1982	—	20–54	Self report[g]	201	1,503

[a]Person-time in the entire study population. Person-time used to calculate rates of uterine leiomyomata included only women with intact uteri but was not reported in the publication.

[b]In a sample of cases, 93% of the self-reported diagnoses were confirmed in the medical record.

[c]667 cases were confirmed by hysterectomy, and 2300 were confirmed by ultrasound.

[d]Estimated from information provided in the published manuscript.

[e]Cases were limited to women with uterine leiomyomata (1) causing persistent menorrhagia or (2) >10 cm size regardless of symptoms.

[f]The numbers of cases and controls interviewed through 1993. Earlier reports were based on smaller numbers of cases and controls.

[g]The self report of uterine leiomyomata diagnosis was not validated.

With one exception [70], studies have not observed any association between age at first term birth and risk of surgically-confirmed uterine leiomyomata [33,38,69,74]. However, in two studies age at last birth was independently associated, with relative risks (RR) of surgically-confirmed uterine leiomyomata ranging from 0.4 [33] to 0.6 [69] among women last giving birth at age 35 years or older compared to women who were younger at their last birth. The possible importance of jointly considering the timing of both first and last births was examined in a large cohort study [71]. Compared to nulliparous women, the risk was reduced similarly (58–65%) regardless of age at first birth (under 25 years of age vs 25 years or older) among women who last gave birth in the past 3 years. However, as the time since last birth increased the risks among women with early and with late age at first birth both approached the risk among nulliparous women, and only the risk among women 25 years of age or older at first birth was reduced (approximately 25%) compared to nulliparous women. A history of breastfeeding has not been shown to be independently related to risk after adjustment for parity [69,72].

The inverse associations with parity and timing of either first or last term birth could, to some extent, be artifactual. Both relations would be observed if the true underlying association was with infertility and/or its treatment, or to the extent that uterine leiomyomata impair a woman's ability to conceive or carry a pregnancy to term. In the Nurse's Health Study II, however, parous women were at lower risk of uterine leiomyomata regardless of infertility history, and associations with the timing of the first and last birth were not altered when women with a history of infertility were excluded [71]. While these findings suggest that infertility and parity are independently associated

with uterine leiomyomata, they do not address the possibility that infertility results from, rather than precedes, the development of uterine leiomyomata.

Imperfect knowledge regarding myometrial physiology during pregnancy makes it difficult to interpret the associations observed for childbearing characteristics with respect to specific biological models. Altered endocrine profiles following a first or second pregnancy [75–77], especially if initiated late in reproductive life, could underlie the inverse relations of uterine leiomyomata with parity and late age at last birth. Similarly, pregnancy may lead to a reduction in estrogen receptor levels in myometrial tissue [73]. If uterine leiomyomata growth results from estrogen hypersensitivity [49], one or more pregnancies, or pregnancies occurring late in reproductive life, could act to reduce the sensitivity of these tumors to hormonal stimuli.

Alternatively, childbearing may counteract the development of uterine leiomyomata through nonhormonal mechanisms. For example, myometrial hypertrophy during pregnancy could inhibit growth of small clones of transformed myometrial cells, and reductions in collagen content and smooth muscle cell cytoplasm in the postpartum could eliminate or drastically reduce the size of minute uterine leiomyomata. The postpartum also could represent an intersection of hormonal and nonhormonal pathways, as loss of smooth muscle cell cytoplasm could diminish estrogen receptor levels and thus, at least temporarily, decrease the estrogen responsiveness of existing uterine leiomyomata.

c. BODY SIZE.   In two cohort studies [32,33] a linear relationship between body mass index (BMI) and risk was observed, with the heaviest women at two- to threefold greater risk than the leanest women [32,33]. Data from case-control studies show a modest elevation (30–50%) in risk for BMI of 25–29 kg/m² compared to ≤24 kg/m², but no increased risk among the heavier women [41,69]. Weaker associations were observed when uterine leiomyomata confirmed by ultrasound were included, with the heaviest women at 20–30% greater risk than lean women [32]. In one case-control study, BMI was unrelated to self-reported history of uterine leiomyomata [72].

Increasing weight gain during adulthood is associated with increasing risk of uterine leiomyomata [32]. In comparison to stable weight since age 18 (+/− 3.0 kg), a weight gain of 15 kg or more since age 18 was associated with an elevated risk of uterine leiomyomata confirmed by ultrasound or hysterectomy (RR = 1.5) and with risk of hysterectomy-confirmed uterine leiomyomata (RR = 2.0) [32]. Neither BMI at age 18 nor height were related to risk of uterine leiomyomata confirmed by ultrasound or hysterectomy [32].

Body size influences endogenous steroid hormone levels through a variety of pathways which differ according to a woman's menopausal status. Among obese postmenopausal women, excess estrogens result from the aromatization of androstenedione in peripheral adipose tissue [78]. In premenopausal women, obesity is associated with lower progesterone levels [76,79,80]. Therefore, obesity may be considered a marker of excess endogenous estrogen unopposed, or inadequately opposed, by progesterone. The association seen with adult BMI, but not with BMI at age 18, and the stronger associations with surgically-confirmed uterine leiomyomata, suggest that estrogen acts by promoting the growth of existing tumors.

d. CIGARETTE SMOKING.   Four studies suggest an inverse relationship between risk of surgically-confirmed uterine leiomyomata and cigarette smoking [33,69,70,81]; the risk ranged from 20–50% lower among smokers depending on the definition of smoking (*e.g.* current or ever). A fifth study, limited to premenopausal women, found no association with cigarette smoking [32]. It is unclear whether the inverse association depends on the amount smoked. Two studies showed an inverse trend [33,70], whereas one showed a reduction in risk regardless of amount smoked [81]. The risk among former cigarette smokers does not appear to differ from that of women who have never smoked cigarettes [32,81].

Cigarette smoking is associated with a reduced risk of endometrial cancer, a disease in which excess or unopposed estrogens clearly are etiologically important [82]. Thus, it is tempting to interpret the inverse association between cigarette smoking and uterine leiomyomata as supporting a role for unopposed estrogen. The great majority of uterine leiomyomata, however, develop in premenopausal women, and the sum of evidence suggests little difference in blood or urine estrogen levels between premenopausal smokers and nonsmokers [80,83–86]. Rather than affecting estrogen levels per se, cigarette smoking may exert its effect on uterine leiomyomata by altering levels of steroid receptors or hormone metabolizing enzymes in these tumors.

2. Use of Exogenous Hormones

a. ORAL CONTRACEPTIVES.   Although the relationship between oral contraceptive (OC) use and uterine leiomyomata has been studied extensively, no clear patterns have emerged. The risk of uterine leiomyomata among women who have used OCs has been observed to be reduced [33,69], similar [70,72,87], and increased [88] relative to never users. In two studies, current OC use was associated with a 20–60% lower risk, whereas past use was unrelated to risk [71,89]. Results regarding duration of use are similarly inconsistent [33,70,71,87,88]. Risk does not appear to be associated with time since most recent use of OCs [33,87], but one study observed an increased risk associated with initiating use of OCs prior to 17 years of age (RR = 1.29, 95% Confidence Interval [CI] = 1.05–1.51) [71].

Data on the risk of leiomyomata in relation to specific hormonal formulations of OCs are limited to one study in which, for each of nineteen different combinations of estrogen and progestin dose, the ratio of months of use among cases and controls was computed [33]. These results suggested that use of OCs containing progestins with estrogenic properties was more common in cases than in controls [31].

The inconsistent patterns of risk with OC use may be due, at least in part, to methodologic limitations. Some studies included controls who had their uteri removed, and who would not be at risk for uterine leiomyomata and have had less opportunity to use OCs. Other potential sources of bias may also influence the findings. For example, one cohort study [89] showed that 40% of the women who stopped taking OCs did so because of side effects or "intercurrent morbidity," and Ross *et al.* [33] found that women frequently quit using OCs after their initial uterine leiomyomata diagnosis. Ramcharan *et al.* [88] reported that past OC users were generally less healthy than current or never users. Because the surgical diagnosis in many cases could have followed the initial clinical diagnosis by months or years, an inverse association between OCs and uterine leiomyomata could be explained by reduced use of these medications following the initial diagnosis. Alternatively, positive associations could similarly arise if OCs tended to be prescribed to treat the bleeding that brings many cases to clinical attention. Interpreting these data is made more difficult by the fact that the hormonal formulations of OCs have changed dramatically over the period in which the published studies were conducted. However, this cannot be considered a complete explanation for the divergent results because even studies conducted during the same time periods lacked consistency.

b. DEPOT MEDROXYPROGESTERONE ACETATE (DMPA).   A study in Thailand showed a strong inverse association (odds ratio[OR] = 0.4, 95% CI = 0.3–0.5) between a history of DMPA use and the risk of surgically-confirmed uterine leiomyomata [69]. The risk declined with increasing duration of use, such that use for more than 5 years reduced the risk of uterine leiomyomata by 90%. In addition, the inverse association weakened with increasing time since last use. These findings are consistent with those for childbearing and menstrual history in suggesting that progestins might inhibit, rather than promote, the development of uterine leiomyomata.

c. ESTROGEN REPLACEMENT THERAPY.   The hypothesis that unopposed or excess estrogens are important in the etiology of uterine leiomyomata predicts that use of estrogen replacement therapy would increase risk. In one study, women who were current estrogen users were at six-fold increased risk of uterine leiomyomata requiring hospitalization compared to never users [88]. The risk of surgically-confirmed uterine leiomyomata was 1.9 times greater (95% CI = 1.5–2.5) among women who had used estrogen replacement therapy [70]. Nearly all of this increase was due to women with 8 or more years of use (OR = 2.3, 95% CI = 1.1–9.7), whereas there was no excess risk among women who had used estrogen replacement therapy for fewer than 4 years (OR = 1.2; 95% CI = 0.7–1.9) or for 4–7 years (OR = 1.3, 95% CI = 0.5–3.2). As the cases in these studies were diagnosed prior to the era in which progestins were commonly added to hormone replacement therapy regimens, the association with combined therapy could not be determined.

3. Other Potential Risk Factors

a. DEMOGRAPHIC CHARACTERISTICS.   Women diagnosed with uterine leiomyomata are more likely to have been married [27,41], have more years of education [41,69], and are more likely to have an occupation categorized as "professional" [69]. These associations may reflect greater access to medical care and/or may be correlated with one or more putative risk factors described above.

b. USE OF NONHORMONAL CONTRACEPTIVE METHODS.   The risk of surgically-confirmed uterine leiomyomata has not been associated with use of the intrauterine device [33,41,69]. In one study, a modest elevation in risk (OR = 1.7, 95% CI = 1.4–2.0) was observed among women who had a history of tubal ligation [69].

c. FAMILIAL AGGREGATION AND INHERITED SUSCEPTIBILITY.   Although gynecologic texts state that family history of uterine leiomyomata is an important risk factor [28], only one study has evaluated this hypothesis [69]. A family history of uterine leiomyomata was associated with a 3.5-fold excess risk in a study conducted among Thai women [69]. The interpretation of this finding is limited by a lack of information on how "family history" was defined in that investigation. In addition, the association may be due, in whole or in part, to greater knowledge among cases of female family members affected by uterine leiomyomata.

Familial aggregation of disease may be due to inherited susceptibility, shared reproductive, lifestyle, and environmental factors, or a combination of genetic and nongenetic factors. Only one study has examined germline DNA variation in relation to the risk of uterine leiomyomata [90]. These investigators found an IGF-2 gene AvaII restriction fragment length polymorphism to be more common among women undergoing hysterectomy for uterine leiomyomata (82%) than in a comparison population (62%) [90]. Although the functional relevance of this polymorphism is uncertain, this finding is notable because IGF-2 appears to be an important growth factor expressed in uterine leiomyomata.

d. PHYSICAL ACTIVITY.   In a study of former college athletes and nonathletes surveyed about their history of benign gynecologic diseases, nonathletes were more likely to report a history of benign uterine tumors (almost all of which were likely to have been uterine leiomyomata) [91]. Increased levels of physical activity appear to be related to a reduction in estrogen lev-

els, and thus these data would be consistent with a role for estrogen in promoting the development of uterine leiomyomata.

e. IONIZING RADIATION.   In a follow-up of women who survived the atomic bombings in Hiroshima and Nagasaki, the risk of uterine leiomyomata increased with increasing estimated uterine radiation dose [92]. There were insufficient data collected in this study to refute the possibility that the association was due to a dose-related increase in gynecologic surveillance. To the extent that ionizing radiation could contribute to somatic chromosomal damage at loci involved in the initiation or growth of uterine leiomyomata, however, these findings may also represent a true etiologic relation in this cohort.

C. Summary of Uterine Leiomyomata Etiology and Risk Factors

The evidence from the relatively small number of epidemiologic studies conducted to date generally supports the hypothesis that ovarian hormones are important in the etiology of uterine leiomyomata. Elevated risks associated with estrogen replacement therapy and increasing body mass index during adulthood, but not during late adolescence, suggest in particular that unopposed estrogen promotes the growth of uterine leiomyomata. In contrast to molecular and clinical studies, epidemiologic evidence does not support the notion that progesterone promotes uterine leiomyomata. Indeed, the lower risk among DMPA users and possibly among women with higher parity and late age at menarche suggest that progesterone may inhibit uterine leiomyomata growth. It is notable that many of the epidemiologic associations observed have been reported for uterine leiomyosarcoma [93–95], suggesting that benign and malignant neoplasms of the uterine myometrium may share a similar etiology.

VI. Clinical Epidemiology of Uterine Leiomyomata

Women diagnosed with uterine leiomyomata are managed with surgical, pharmacologic, and expectant approaches [96,97]. The specific indications for different management approaches are based largely on the type and extent of a woman's symptoms and her desire to retain the ability to bear children. Hysterectomy provides definitive therapy for women with symptomatic uterine leiomyomata [96]. The high costs associated with hysterectomy, however, have stimulated development of more conservative approaches. Clinical epidemiologic issues of interest therefore include delineating predictors of the clinical course of uterine leiomyomata and the outcomes of conservative therapies.

Few data have been reported on the frequency and predictors of different surgical or pharmacologic interventions among women diagnosed with uterine leiomyomata. In the Maine Women's Health Study, which included only women whose uteri were ≥8 weeks gestational size, 25% of patients had surgery within one year of diagnosis [98]. In a study including patients with a broader range of uterine sizes, approximately 26% had surgery within 5 years [99]. The risk of progressing to hysterectomy in that study was directly associated with abnormal bleeding as an initial presenting symptom (RR = 1.7, 95% CI = 1.1–2.6), a history of prior surgery for pelvic adhesions

(RR = 4.8, 95% CI = 2.6–8.9) or gallbladder removal (RR = 2.4, 95% CI = 1.1–5.3), and inversely related to recommended therapeutic use of oral contraceptives and/or progestins (RR = 0.5, 95% CI = 0.3–0.8) [99]. The 3-year cumulative hysterectomy rate was extremely high (83%) among the subgroup of women who were >35 years of age, who had uteri >8 weeks size, and who had at least one of the other characteristics found to be related to increased risk of hysterectomy in that study. Interestingly, there was little difference in the risk of hysterectomy related to uterine leiomyomata risk factors identified in epidemiologic studies, such as African-American race, absence of cigarette smoking, and nulliparity.

Myomectomy is the primary surgical alternative to hysterectomy [96]. Evidence suggests that the rate of hospitalization for uterine leiomyomata treated with myomectomy has increased over time and is highest for young, black women [100]. Studies have reported cumulative recurrence ranging from 4–50% [101,102]. The variation in recurrence among studies is likely to be due to differences in length of follow-up, ethnic composition of populations, and methods of ascertaining and diagnosing recurrences. For example, one study found a 5-year recurrence of 7% based on clinical examination for uterine leiomyomata ascertainment, whereas the frequency increased to 51% when ultrasonography was used [101]. Predictors of recurrence have not been examined in rigorously designed studies, but available data suggest that the risk of recurrence is lower among women with only a single myoma removed and among women who bear at least one child following myomectomy [101,102]. There are conflicting data on whether use of GnRH agonists prior to myomectomy is associated with an increased risk of recurrence [103,104]. Whether women with multiple leiomyomata removed have host or environmental exposures that predispose them to develop new tumors, or are more likely to harbor undiagnosed, microscopic uterine leiomyomata at the time of surgery, is unknown. That uterine leiomyomata often recur in parts of the uterus that differ from the location of the tumor removed by myomectomy [102] supports the latter hypothesis. The inverse association with a subsequent birth could be due to infertility or early pregnancy loss caused by recurrent leiomyomata, including growth of small leiomyomata that were undetected at the time of myomectomy. Alternatively, hormonal changes associated with pregnancy could inhibit the growth of leiomyomata. Other characteristics of women, such as body size and weight, use of oral contraceptives, and physical activity, have not been studied in relation to recurrence risk.

GnRH agonists are the primary pharmacologic agents used to reduce both symptoms and tumor size among women with uterine leiomyomata [56,105]. Suppression of uterine leiomyomata depends on continued use of GnRH agonists, and prominent bone loss results from the ensuing long-term hypoestrogenic milieu [68,106,107]. This complication limits the use of GnRH agonists to short-term regimens, such as those designed to shrink tumors prior to myomectomy, or among late perimenopausal women whose tumors are more likely to shrink spontaneously following cessation of ovarian function [108,109]. There is a possibility that GnRH agonists could be used for extended periods when coupled with low-doses of estrogen replacement therapy or estrogen–progestin replacement therapy [65,110].

VII. Summary and Directions for Future Research

Uterine leiomyomata are a common cause of reproductive health problems and medical care utilization in women. Estrogen and progesterone are almost certainly involved in their pathogenesis, but the mechanisms through which these hormones act are not well defined at the molecular or epidemiologic levels. With the possible exception of obesity, no modifiable risk factors have been identified. Thus, avenues toward the primary prevention of uterine leiomyomata are limited.

Clarifying the etiology of uterine leiomyomata and associated risk factors will require further research at the population and molecular levels. Epidemiologic studies that bridge these areas by incorporating biologic measures of exposure and disease markers will be essential to develop stronger inferences as to specific risk factors and mechanisms. Included among such molecular epidemiologic investigations would be studies of blood levels of steroid hormones and possibly also of environmental contaminants and other compounds suspected of exerting hormonal effects [111]. It would also be valuable to conduct studies in which cases are stratified according to ultrasonographic, histopathologic, or molecular characteristics to identify subsets of uterine leiomyomata that are more likely to be associated with particular exposures. The contribution of germline polymorphisms in genes coding for factors involved in the synthesis and metabolism of steroid hormones, steroid hormone receptors, and collagen needs to be examined. Such studies may identify genetic factors that, although having low penetrance, may contribute to the development of a large proportion of uterine leiomyomata in the population either alone or through interactions with nongenetic factors. Population-based, family investigations also should be conducted to quantify familial aggregation, determine patterns of inheritance, find candidate loci of major genes through linkage analysis, and eventually identify the genes and associated mutations that might contribute to uterine leiomyomata occurrence within families [112].

Etiologic hypotheses other than those directly related to steroid hormones also need to be explored. For example, similarities between smooth muscle cell proliferation in uterine leiomyomata and atheromas, and reported associations between hypertension and surgically-diagnosed uterine leiomyomata, suggest the potential for shared etiologic mechanisms with coronary heart disease, such as insulin resistance [113–115]. The occurrence of nonuterine leiomyomata in patients immunocompromised due to HIV infection [116] suggests possible roles for infectious agents and/or immunological factors. Such hypotheses could be tested within epidemiologic studies by testing both leiomyomata tissue from cases and blood samples from cases and controls. There have been no published epidemiologic studies of diet or nutritional factors, but the relatively consistent association with obesity suggests that dietary fat may contribute to the development of uterine leiomyomata. In theory, promising advances also might result from studies that seek to explain behavioral, social, and biologic factors that underlie the differences in the frequency, size, and number of uterine leiomyomata among women of different racial and ethnic origins. For example, factors involved in the production of keloids, which are benign, collagen-rich skin tumors that occur more frequently among U.S. blacks than whites [117], may be important in the etiology of uterine leiomyomata.

Finally, studies focused on medical utilization and clinical outcomes are needed to identify characteristics of patients and their providers that influence the use of different treatments for uterine leiomyomata at different stages of development. In addition, studies should attempt to determine the extent to which patient characteristics, ultrasonographic features, and physiologic measures such as hormone levels are related to the progression of uterine leiomyomata from clinical diagnosis to surgical therapy, particularly hysterectomy. For example, one study suggested that increased tumor vascularity, as measured by Doppler ultrasonography, is strongly associated with increased myoma growth [118]. Similar studies, augmented by analyses of molecular features of excised leiomyomata, may help define predictors of recurrence risk following myomectomy. The results of these studies could provide patients and their health care providers with information that will enhance the management of uterine leiomyomata.

References

1. Ferenczy, A., Richart, R. M., and Okagaki, T. (1971). A comparative ultrastructural study of leiomyosarcoma, cellular leiomyoma, and leiomyoma of the uterus. *Cancer (Philadelphia)* **28**, 1004–1018.

2. Konishi, I., Fujii, S., Ban, C., Okuda, Y., Okamura, H., and Tojo, S. (1983). Ultrastructural study of minute uterine leiomyomas. *Int. J. Gynecol. Pathol.* **2**, 113–120.

3. Wilcox, L. S., Koonin, L. M., Pokras, R., Strauss, L. T., Xia, Z., and Peterson, H. B. (1994). Hysterectomy in the United States, 1988–1990. *Obstet. Gynecol.* **83**, 549–555.

4. Koutsilieris, M. (1992). Pathophysiology of uterine leiomyomas. *Biochem. Cell Biol.* **70**, 273–278.

5. Scully, R. E. (1992). Pathology of leiomyomas. *Semin. Reprod. Endocrinol.* **10**, 325–331.

6. Buttram, V. C., Jr., and Reiter, R. C. (1981). Uterine leiomyomata: Etiology, symptomatology, and management. *Fertil. Steril.* **36**, 433–445.

7. Townsend, D. E., Sparkes, R. S., Baluda, M. C., and McClelland, G. (1970). Unicellular histogenesis of uterine leiomyomas as determined by electrophoresis of glucose-6-phosphate dehydrogenase. *Am. J. Obstet. Gynecol.* **107**, 1176–1179.

8. Mashal, R. D., Schoenberg Fejzo, M. L., Friedman, A. J., Mitchner, N., Nowak, R. A., Rein, M. S., Morton, C. C., and Sklar, J. (1994). Analysis of androgen receptor DNA reveals the independent clonal origins of uterine leiomyomata and the secondary nature of cytogenetic aberrations in the development of leiomyomata. *Genes Chromosomes Cancer* **11**, 1–6.

9. Hashimoto, K., Azuma, C., Kamiura, S., Kimura, T., Nobunaga, T., Kanai, T., Sawada, M., Noguchi, S., and Saji, F. (1995). Clonal determination of uterine leiomyomas by analyzing differential inactivation of the X-chromosome-linked phosphoglycerokinase gene. *Gynecol. Obstet. Invest.* **40**, 204–208.

10. Nilbert, M., and Heim, S. (1990). Uterine leiomyoma cytogenetics. *Genes Chromosomes Cancer* **2**, 3–13.

11. Rein, M. S., Walters, F., Weremowicz, S., and Morton, C. (1995). Cytogenetic abnormalities in uterine leiomyomata are associated with leiomyomata site. *J. Soc. Gynecol. Invest.* **2**, 379 (abstr.).

12. Brosens, I., Johannisson, E., Dal Cin, P., Deprest, J., and Van den Berghe, H. (1996). Analysis of the karyotype and desoxyribonucleic acid content of uterine myomas in premenopausal, menopausal, and gonadotropin-releasing hormone agonist-treated females. *Fertil. Steril.* **66**, 376–379.

13. Brosens, I., Deprest, J., Dal Cin, P. D., and Van den Berghe, H. (1998). Clinical significance of cytogenetic abnormalities in uterine myomas. *Fertil. Steril.* **69**, 232–235.

14. Chou, C. Y., Huang, S. C., Tsai, Y. C., Hsu, K. F., and Huang, K. E. (1997). Uterine leiomyosarcoma has deregulated cell proliferation, but not increased microvessel density compared with uterine leiomyoma. *Gynecol. Oncol.* **65**, 225–231.

15. Arheden, K., Nilbert, M., Heim, S., Mandahl, N., and Mitelman, F. (1989). No amplification or rearrangement of INT1, GLI, or COL2A1 in uterine leiomyomas with t(12;14)(q14-15; q23-24). *Cancer Genet. Cytogenet.* **39**, 195–201.

16. Fotiou, S. K., Tserkezoglou, A. J., Mahera, H., Konstandinidou, A. E., Agnantis, N. J., Pandis, N., and Bardi, G. (1992). Chromosome aberrations and expression of ras and myc oncogenes in leiomyomas and a leiomyosarcoma of the uterus. *Eur. J. Gynaecol. Oncol.* **13**, 340–345.

17. De Vos, S., Wilczynski, S. P., Fleischhacker, M., and Koeffler, P. (1994). p53 alterations in uterine leiomyosarcomas versus leiomyomas. *Gynecol. Oncol.* **54**, 205–208.

18. Gloudemans, T., Pospiech, I., Van Der Ven, L. T., Lips, C. J., Schneid, H., Den Otter, W., and Sussenbach, J. S. (1992). Expression and CpG methylation of the insulin-like growth factor II gene in human smooth muscle tumors. *Cancer Res.* **52**, 6516–6521.

19. Vu, T. H., Yballe, C., Boonyanit, S., and Hoffman, A. R. (1995). Insulin-like growth factor II in uterine smooth-muscle tumors: maintenance of genomic imprinting in leiomyomata and loss of imprinting in leiomyosarcomata. *J. Clin. Endocrinol. Metab.* **80**, 1670–1676.

20. Schoenmakers, E. F. P. M., Wanschura, S., Mols, R., Bullerdiek, J., Van den Berghe, H., and Van de Ven, W. J. M. (1995). Recurrent rearrangements in the high mobility group protein gene, HMGI-C, in benign mesenchymal tumours. *Nat. Genet.* **10**, 436–444.

21. Kazmierczak, B., Bol, S., Wanschura, S., Bartnitzke, S., and Bullerdiek, J. (1996). PAC clone containing the HMGI(Y) gene spans the breakpoint of a 6p21 translocation in a uterine leiomyoma cell line. *Genes Chromosomes Cancer* **17**, 191–193.

22. Schoenberg Fejzo, M., Ashar, H. R., Krauter, K. S., Powell, W. L., Rein, M. S., Weremowicz, S., Yoon, S. J., Kucherlapati, R. S., Chada, K., and Morton, C. C. (1996). Translocation breakpoints upstream of the HMGIC gene in uterine leiomyomata suggest dysregulation of this gene by a mechanism different from that in lipomas. *Genes Chromosomes Cancer* **17**, 1–6.

23. Bustin, M., and Reeves, R. (1996). High-mobility-group chromosomal proteins: Architectural components that facilitate chromatin function. *Prog. Nucleic Acid Res. Mol. Biol.* **54**, 35–100.

24. Andersen, J. (1996). Growth factors and cytokines in uterine leiomyomas. *Semin. Reprod. Endocrinol.* **14**, 269–282.

25. Emembolu, J. O. (1987). Uterine fibromyomata: Presentation and management in northern Nigeria. *Int. J. Gynaecol. Obstet.* **25**, 413–416.

26. Kjerulff, K. H., Langenberg, P., Seidman, J. D., Stolley, P. D., and Guzinski, G. M. (1996). Uterine leiomyomas. Racial differences in severity, symptoms and age at diagnosis. *J. Reprod. Med.* **41**, 483–490.

27. Marshall, L. M., Spiegelman, D., Barbieri, R. L., Goldman, M. B., Manson, J. E., Colditz, G. A., Willett, W. C., and Hunter, D. J. (1997). Variation in the incidence of uterine leiomyoma among premenopausal women by age and race. *Obstet. Gynecol.* **90**, 967–973.

28. Hillard, P. A. (1996). Benign diseases of the female reproductive tract: Symptoms and signs *In* "Novak's Gynecology" (J. S. Berek, ed.), 12th ed., pp. 331–397. Williams & Wilkins, Baltimore, MD.

29. Loutradis, D., Antsaklis, A., Creatsas, G., Hatzakis, A., Kanakas, N., Gougoulakis, A., Michalas, S., and Aravantinos, D. (1990). The validity of gynecological ultrasonography. *Gynecol. Obstet. Invest.* **29**, 47–50.

30. Fedele, L., Bianchi, S., Dorta, M., Brioschi, D., Zanotti, F., and Vercellini, P. (1991). Transvaginal ultrasonography versus hysteroscopy in the diagnosis of uterine submucous myomas. *Obstet. Gynecol.* **77**, 745–748.

31. Cramer, D. W. (1992). Epidemiology of myomas. *Semin. Reprod. Endocrinol.* **10,** 320–324.

32. Marshall, L. M., Spiegelman, D., Manson, J. E., Goldman, M. B., Barbieri, R. L., Stampfer, M. J., Willett, W. C., and Hunter, D. J. (1998). Risk of uterine leiomyomata among premenopausal women in relation to body size and cigarette smoking. *Epidemiology* **9,** 511–517.

33. Ross, R. K., Pike, M. C., Vessey, M. P., Bull, D., Yeates, D., and Casagrande, J. T. (1986). Risk factors for uterine fibroids: Reduced risk associated with oral contraceptives. *Br. Med. J.* **293,** 359–362.

34. Cramer, S. F., and Patel, A. (1990). The frequency of uterine leiomyomas. *Am. J. Clin. Pathol.* **94,** 435–438.

35. Rothman, K. J., and Greenland, S. (1998). "Modern Epidemiology," 2nd ed. Lippincott-Raven, Philadelphia.

36. Walker, A. M. (1991). "Observation and Inference." Epidemiology Resources Inc., Chestnut Hill, MA.

37. Velebil, P., Wingo, P. A., and Xia, Z. (1995). Rate of hospitalization for gynecologic disorders among reproductive-age women in the United States. *Obstet. Gynecol.* **86,** 764–769.

38. Brett, K. M., Marsh, J. V., and Madans, J. H. (1997). Epidemiology of hysterectomy in the United States: Demographic and reproductive factors in a nationally representative sample. *J. Women's Health* **6,** 309–316.

39. Kjerulff, K., Langenberg, P., and Guzinski, G. (1993). The socioeconomic correlates of hysterectomies in the United States. *Am. J. Public Health* **83,** 106–108.

40. Meilahn, E. N., Matthews, K. A., Egeland, G., and Kelsey, S. F. (1989). Characteristics of women with hysterectomy. *Maturitas* **11,** 319–329.

41. Parazzini, F., LaVecchia, C., Negri, E., Cecchetti, G., and Fedele, L. (1988). Epidemiologic characteristics of women with uterine fibroids: A case-control study. *Obstet. Gynecol.* **72,** 853–857.

42. Lindegard, B. (1990). Breast cancer among women from Gothenburg with regard to age, mortality and coexisting benign breast disease or leiomyoma uteri. *Oncology* **47,** 369–375.

43. Pokras, R., and Hufnagel, V. G. (1988). Hysterectomy in the United States, 1965–84. *Am. J. Public Health* **78,** 852–853.

44. Vessey, M. P., Villard-Mackintosh, L., McPherson, K., Coulter, A., and Yeates, D. (1992). The epidemiology of hysterectomy: Findings in a large cohort study. *Br. J. Obstet. Gynaecol.* **99,** 402–407.

45. Barbieri, R. L., and Andersen, J. (1992). Uterine leiomyomas: The somatic mutation theory. *Semin. Reprod. Endocrinol.* **10,** 301–309.

46. Romagnolo, B., Molina, T., Leroy, G., Blin, C., Porteux, A., Thomasset, M., Vandewalle, A., Kahn, A., and Perret, C. (1996). Estradiol-dependent uterine leiomyomas in transgenic mice. *J. Clin. Invest.* **98,** 777–784.

47. Witherspoon, J. T. (1933). A possible cause of uterine fibroids. *Endocrinology (Baltimore)* **17,** 703–708.

48. Witherspoon, J. T., and Butler, V. W. (1934). The etiology of uterine fibroids. *Surg., Gynecol. Obstet.* **58,** 57–61.

49. Andersen, J., and Barbieri, R. L. (1995). Abnormal gene expression in uterine leiomyomas. *J. Soc. Gynecol. Invest.* **2,** 663–672.

50. Liehr, J. G., Ricci, M. J., Jefcoate, C. R., Hannigan, E. V., Hokanson, J. A., and Zhu, B. T. (1995). 4-Hydroxylation of estradiol by human uterine myometrium and myoma microsomes: Implications for the mechanism of uterine tumorigenesis. *Proc. Natl. Acad. Sci. U.S.A.* **92,** 9220–9224.

51. Everitt, J. I., Wolf, D. C., Howe, S. R., Goldsworthy, T. L., and Walker, C. (1995). Rodent model of reproductive tract leiomyomata. Clinical and pathological features. *Am. J. Pathol.* **146,** 1556–1567.

52. Howe, S. R., Gottardis, M. M., Everitt, J. I., Goldsworthy, T. L., Wolf, D. C., and Walker, C. (1995). Rodent model of reproductive tract leiomyomata. Establishment and characterization of tumor-derived cell lines. *Am. J. Pathol.* **146,** 1568–1579.

53. Howe, S. R., Gottardis, M. M., Everitt, J. I., and Walker, C. (1995). Estrogen stimulation and tamoxifen inhibition of leiomyoma cell growth in vitro and in vivo. *Endocrinology (Baltimore)* **136,** 4996–5003.

54. Fuchs-Young, R., Howe, S., Hale, L., Miles, R., and Walker, C. (1996). Inhibition of estrogen-stimulated growth of uterine leiomyomas by selective estrogen receptor modulators. *Mol. Carcinog.* **17,** 151–159.

55. Blin, C., L'Horset, F., Romagnolo, B., Colnot, S., Lambert, M., Thomasset, M., Kahn, A., Vanderwalle, A., and Perret, C. (1996). Functional and growth properties of a myometrial cell line derived from transgenic mice: Effects of estradiol and antiestrogens. *Endocrinology (Baltimore)* **137,** 2246–2253.

56. Lemay, A., and Maheux, R. (1996). GnRH agonists in the management of uterine leiomyoma. *Infertil. Reprod. Med. Clin. North Am.* **7,** 33–55.

57. Rein, M. S., Barbieri, R. L., and Friedman, A. J. (1995). Progesterone: A critical role in the pathogenesis of uterine myomas. *Am. J. Obstet. Gynecol.* **172,** 14–18.

58. Kawaguchi, K., Fujii, S., Konishi, I., Nanbu, Y., Nonogaki, H., and Mori, T. (1989). Mitotic activity in uterine leiomyomas during the menstrual cycle. *Am. J. Obstet. Gynecol.* **160,** 637–641.

59. Lamminen, S., Rantala, I., Helin, H., Rorarius, M., and Tuimala, R. (1992). Proliferative activity of human uterine leiomyoma cells as measured by automatic image analysis. *Gynecol. Obstet. Invest.* **34,** 111–114.

60. Matsuo, H., Maruo, T., and Samoto, T. (1997). Increased expression of Bcl-2 protein in human uterine leiomyoma and its up-regulation by progesterone. *J. Clin. Endocrinol. Metab.* **82,** 293–299.

61. Tiltman, A. J. (1985). The effect of progestins on the mitotic activity of uterine fibromyomas. *Int. J. Gynecol. Pathol.* **4,** 89–96.

62. Sener, A. B., Seckin, N. C., Ozmen, S., Gokmen, O., Dogu, N., and Ekici, E. (1996). The effects of hormone replacement therapy on uterine fibroids in postmenopausal women. *Fertil. Steril.* **65,** 354–357.

63. Murphy, A. A., Morales, A. J., Kettel, L. M., and Yen, S. S. (1995). Regression of uterine leiomyomata to the antiprogesterone RU486: Dose-response effect. *Fertil. Steril.* **64,** 187–190.

64. Murphy, A. A. (1996). RU 486 in the treatment of leiomyomata uteri. *Infertil. Reprod. Med. Clin. North Am.* **7,** 57–68.

65. Friedman, A. J., Daly, M., Juneau-Norcross, M., Gleason, R., Rein, M. S., and LeBoff, M. (1994). Long-term medical therapy for leiomyomata uteri: A prospective, randomized study of leuprolide acetate depot plus either oestrogen-progestin or progestin 'add-back' for 2 years. *Hum. Reprod.* **9,** 1618–1625.

66. Carr, B. R., Marshburn, P. B., Weatherall, P. T., Bradshaw, K. D., Breslau, N. A., Byrd, W., Roark, M., and Steinkampf, M. P. (1993). An evaluation of the effect of gonadotropin-releasing hormone analogs and medroxyprogesterone acetate on uterine leiomyomata volume by magnetic resonance imaging: A prospective, randomized, double blind, placebo-controlled, crossover trial. *J. Clin. Endocrinol. Metab.* **76,** 1217–1223.

67. Benagiano, G., Morini, A., Aleandri, V., Piccinno, F., Primiero, F. M., Abbondante, G., and Elkind-Hirsch, K. (1990). Sequential Gn-RH superagonist and medroxyprogesterone acetate treatment of uterine leiomyomata. *Int. J. Gynaecol. Obstet.* **33,** 333–343.

68. Caird, L. E., West, C. P., Lumsden, M. A., Hannan, W. J., and Gow, S. M. (1997). Medroxyprogesterone acetate with Zoladex™ for long-term treatment of fibroids: Effects on bone density and patient acceptability. *Hum. Reprod.* **12,** 436–440.

69. Lumbiganon, P., Rugpao, S., Phandhu-fung, S., Laopaiboon, M., Vudhikamraksa, N., and Werawatakul, Y. (1996). Protective effect

of depot-medroxyprogesterone acetate on surgically treated uterine leiomyomas: A multicentre case—control study. *Br. J. Obstet. Gynaecol.* **103**, 909–914.

70. Romieu, I., Walker, A. M., and Jick, S. (1991). Determinants of uterine fibroids. *Post Market Surveill.* **5**, 119–133.

71. Marshall, L. M., Spiegelman, D., Goldman, M. B., Manson, J. E., Colditz, G. A., Barbieri, R. L., Stampfer, M. J., and Hunter, D. J. (1998). A prospective study of reproductive factors and oral contraceptive use in relation to the risk of uterine leiomyomata. *Fertil. Steril.* **70**, 432–439.

72. Samadi, A. R., Lee, N. C., Flanders, W. D., Boring, J. R. 3rd, and Parris, E. B. (1996). Risk factors for self-reported uterine fibroids: A case-control study. *Am. J. Public Health* **86**, 858–862.

73. Kawaguchi, K., Fujii, S., Konishi, I., Iwai, T., Nanbu, Y., Nonogaki, H., Ishikawa, Y., and Mori, T. (1991). Immunohistochemical analysis of oestrogen receptors, progesterone receptors and Ki-67 in leiomyoma and myometrium during the menstrual cycle and pregnancy. *Virchows Arch. A: Pathol. Anat. Histopathol.* **419**, 309–315.

74. Parazzini, F., Negri, E., La Vecchia, C., Chatenoud, L., Ricci, E., and Guarnerio, P. (1996). Reproductive factors and risk of uterine fibroids. *Epidemiology* **7**, 440–442.

75. Bernstein, L., Pike, M. C., Ross, R. K., Judd, H. L., Brown, J. B., and Henderson, B. E. (1985). Estrogen and sex hormone-binding globulin levels in nulliparous and parous women. *J. Natl. Cancer Inst.* **74**, 741–745.

76. Dorgan, J. F., Reichman, M. E., Judd, J. T., Brown, C., Longcope, C., Schatzkin, A., Campbell, W. S., Franz, C., Kahle, L., and Taylor, P. R. (1995). Relationships of age and reproductive characteristics with plasma estrogens and androgens in premenopausal women. *Cancer Epidemiol. Biomarkers Prev.* **4**, 381–386.

77. Musey, P. I., Collins, D. C., Bradlow, H. L., Gould, K. G., and Preedy, J. R. (1987). Effect of diet on oxidation of 17 beta-estradiol in vivo. *J. Clin. Endocrinol. Metab.* **65**, 792–795.

78. Enriori, C. L., and Reforzo-Membrives, J. (1984). Peripheral aromatization as a risk factor for breast and endometrial cancer in postmenopausal women: A review. *Gynecol. Oncol.* **17**, 1–21.

79. Sherman, B. M., and Korenman, S. G. (1974). Measurement of serum LH, FSH, estradiol and progesterone in disorders of the human menstrual cycle: The inadequate luteal phase. *J. Clin. Endocrinol. Metab.* **39**, 145–149.

80. Westhoff, C., Gentile, G., Lee, J., Zacur, H., and Heilbig, D. (1996). Predictors of ovarian steroid secretion in reproductive-age women. *Am. J. Epidemiol.* **144**, 381–388.

81. Parazzini, F., Negri, E., La Vecchia, C., Rabaiotti, M., Luchini, L., Villa, A., and Fedele, L. (1996). Uterine myomas and smoking. Results from an Italian study. *J. Reprod. Med.* **41**, 316–320.

82. Weiss, N. S. (1990). Cigarette smoking and the incidence of endometrial cancer. *In* "Smoking and Hormone-Related Disorders" (N. Wald and J. Baron, eds.), pp. 145–153. Oxford University Press, Oxford.

83. MacMahon, B., Trichopoulos, D., Cole, P., and Brown, J. (1982). Cigarette smoking and urinary estrogens. *N. Engl. J. Med.* **307**, 1062–1065.

84. Longcope, C., and Johnston, C. C., Jr (1988). Androgen and estrogen dynamics in pre- and postmenopausal women: A comparison between smokers and nonsmokers. *J. Clin. Endocrinol. Metab.* **67**, 379–383.

85. Zumoff, B., Miller, L., Levit, C. D., Miller, E. H., Heinz, U., Kalin, M., Denman, H., Jandorek, R., and Rosenfeld, R. S. (1990). The effect of smoking on serum progesterone, estradiol, and luteinizing hormone levels over a menstrual cycle in normal women. *Steroids* **55**, 507–511.

86. Daniel, M., Martin, A. D., and Faiman, C. (1992). Sex hormones and adipose tissue distribution in premenopausal cigarette smokers. *Int. J. Obes. Relat. Metab. Disord.* **16**, 245–254.

87. Parazzini, F., Negri, E., LaVecchia, C., Fedele, L., Rabaiotti, M., and Luchini, L. (1992). Oral contraceptive use and risk of uterine fibroids. *Obstet. Gynecol.* **79**, 430–433.

88. Ramcharan, S., Pelligrin, F. A., Ray, R., and Hsu, J-P. (1981). The Walnut Creek Contraceptive Drug Study. A prospective study of the side effects of oral contraceptives. NIH Publ. No. 81-564. *Cent. Popul. Res. Monogr.* **3**, 69–74.

89. Royal College of General Practitioners (1974). "Oral Contraceptives and Health." Pittman Medical, London.

90. Gloudemans, T., Pospiech, I., Van der Ven, L. T., Lips, C. J., Den Otter, W., and Sussenbach, J. S. (1993). An avaII restriction fragment length polymorphism in the insulin-like growth factor II gene and the occurrence of smooth muscle tumors. *Cancer Res.* **53**, 5754–5758.

91. Wyshak, G., Frisch, R. E., Albright, N. L., Albright, T. E., and Schiff, I. (1986). Lower prevalence of benign diseases of the breast and benign tumours of the reproductive system among former college athletes compared to non-athletes. *Br. J. Cancer* **54**, 841–845.

92. Wong, F. L., Yamada, M., Sasaki, H., Kodama, K., Akiba, S., Shiimaoka, K., and Hosoda, Y. (1993). Noncancer disease incidence in the atomic bomb survivors: 1958–1986. *Radiat. Res.* **135**, 418–430.

93. Harlow, B. L., Weiss, N. S., and Lofton, S. (1986). The epidemiology of sarcomas of the uterus. *J. Natl. Cancer Inst.* **76**, 399–402.

94. Schwartz, S. M., Weiss, N. S., Daling, J. R., Newcomb, P. A., Liff, J. M., Gammon, M. D., Thompson, W. D., Watt, J. D., Armstrong, B. K., Weyer, P., Isaacson, P., and Ek, M. (1991). Incidence of histologic types of uterine sarcoma in relation to menstrual and reproductive history. *Int. J. Cancer* **49**, 362–367.

95. Schwartz, S. M., Weiss, N. S., Daling, J. R., Gammon, M. D., Liff, J. M., Watt, J., Lynch, C. F., Newcomb, P., Armstrong, B. K., and Thompson, W. D. (1996). Exogenous sex hormone use, correlates of endogenous hormone levels, and the incidence of histologic types of sarcoma of the uterus. *Cancer (Philadelphia)* **77**, 717–724.

96. Committee on Technical Bulletins of the American College of Obstetricians and Gynecologists (1994). Uterine leiomyomata. *Int. J. Gynaecol. Obstet.* **46**, 73–82.

97. Davies, A., and Magos, A. L. (1997). Indications and alternatives to hysterectomy. *Bailliere's Clin. Obstet. Gynaecol.* **11**, 61–75.

98. Carlson, K. J., Miller, B. A., and Fowler, F. J., Jr (1994). The Maine Women's Health Study: II. Outcomes of nonsurgical management of leiomyomas, abnormal bleeding, and chronic pelvic pain. *Obstet. Gynecol.* **83**, 566–572.

99. Weber, A. M., Mitchinson, A. R., Gidwani, G. P., Mascha, E., and Walters, M. D. (1997). Uterine myomas and factors associated with hysterectomy in premenopausal women. *Am. J. Obstet. Gynecol.* **176**, 1213–1219.

100. Wilcox, L., Lepine, L., and Kieke, B. (1995). Demographic differences among women receiving hysterectomy or myomectomy for uterine leiomyoma. *Am. J. Epidemiol.* **141**, S49 (abstr.).

101. Candiani, G. B., Fedele, L., Parazzini, F., and Villa, L. (1991). Risk of recurrence after myomectomy. *Br. J. Obstet. Gynaecol.* **98**, 385–389.

102. Fedele, L., Parazzini, F., Luchini, L., Mezzopane, R., Tozzi, L., and Villa, L. (1995). Recurrence of fibroids after myomectomy: A transvaginal ultrasonographic study. *Hum. Reprod.* **10**, 1795–1796.

103. Fedele, L., Vercellini, P., Bianchi, S., Brioschi, D., and Dorta, M. (1990). Treatment with GnRH agonists before myomectomy and the risk of short-term myoma recurrence. *Br. J. Obstet. Gynaecol.* **97**, 393–396.

104. Friedman, A. J., Daly, M., Juneau-Norcross, M., Fine, C., and Rein, M. S. (1992). Recurrence of myomas after myomectomy in women pre-treated with leuprolide acetate depot or placebo. *Fertil. Steril.* **58**, 205–208.

105. Friedman, A. J., Hoffman, D. I., Comite, F., Browneller, R. W., and Miller, J. D. (1991). Treatment of leiomyomata uteri with leuprolide acetate depot: A double-blind, placebo-controlled, multicenter study. *Obstet. Gynecol.* **77,** 720–725.

106. Matta, W. H., Shaw, R. W., Hesp, R., and Katz, D. (1987). Hypogonadism induced by luteinising hormone releasing hormone agonist analogues: Effects on bone density in premenopausal women. *Br. Med. J.* **294,** 1523–1524.

107. Nencioni, T., Penotti, M., Barbieri-Carones, M., Ortolani, S., Trevisan, C., and Polvani, F. (1991). Gonadotropin releasing hormone agonist therapy and its effect on bone mass. *Gynecol. Endocrinol.* **5,** 49–56.

108. Davis, K. M., and Schlaff, W. D. (1995). Medical management of uterine fibromyomata. *Obstet. Gynecol. Clin. North Am.* **22,** 727–738.

109. Vavala, V., Lanzone, A., Monaco, A., Scribanti, A., Guida, C., and Mancuso, S. (1997). Postoperative GnRH analog treatment for the prevention of recurrences of uterine myomas after myomectomy. A pilot study. *Gynecol. Obstet. Invest.* **43,** 251–254.

110. Thomas, E. J. (1996). Add-back therapy for long-term use in dysfunctional uterine bleeding and uterine fibroids. *Br. J. Obstet. Gynaecol.* **103,** 18–21.

111. Safe, S. H. (1994). Dietary and environmental estrogens and antiestrogens and their possible role in human disease. *Environ. Sci. Pollut. Res.* **1,** 29–33.

112. Zhao, L. P., Hsu, L., Davidov, O., Potter, J., Elston, R. C., and Prentice, R. L. (1997). Population-based family study designs: An interdisciplinary research framework for genetic epidemiology. *Genet. Epidemiol.* **14,** 365–388.

113. Moss, N. S., and Benditt, E. P. (1975). Human atherosclerotic plaque cells and leiomyoma cells. *Am. J. Pathol.* **78,** 175–187.

114. Lovell, H. G., Miall, W. E., and Stewart, D. B. (1966). Arterial blood pressure in Jamaican women with and without uterine fibroids. *West Indian Med. J.* **15,** 45–51.

115. Shikora, S. A., Niloff, J. M., Bistrian, B. R., Forse, A., and Blackburn, G. L. (1991). Relationship between obesity and uterine leiomyomata. *Nutrition* **7,** 251–255.

116. Chadwick, E. G., Connor, E. J., Hanson, I. C. G., Joshi, V. V., Abu-Farsakh, H., Yogev, R., McSherry, G., McClain, K., and Murphy, S. B. (1990). Tumors of smooth-muscle origin in HIV-infected children. *JAMA, J. Am. Med. Assoc.* **263,** 3182–3184.

117. Datubo-Brown, D. D. (1990). Keloids: A review of the literature. *Br. J. Plast. Surg.* **43,** 70–77.

118. Tsuda, H., Kawabata, M., Nakamoto, O., and Yamamoto, K. (1998). Clinical predictors in the natural history of uterine leiomyoma: Preliminary study. *J. Ultrasound Med.* **17,** 17–20.

20

Hysterectomy

KAY DICKERSIN* AND GAIL SCHOEN LEMAIRE†

*Department of Community Health, Brown University, Providence, Rhode Island; †Women's Health Research and Resources, Towson, Maryland

I. Introduction

Hysterectomy, the surgical removal of the uterus, is the most common nonobstetric surgery performed on reproductive-aged women in the United States (U.S.) [1]. It is considered an accepted form of treatment for both malignant and nonmalignant gynecologic conditions [2,3].

In 1946, concerns were first raised as to the wisdom of performing hysterectomy for nonmalignant conditions simply because it was technically possible and the mortality rate was acceptably low [4]. Controversy continues on this topic [5–9] because it is major surgery with associated mortality and morbidity, and medical or more conservative surgery may be just as effective in many cases [10]. In addition, it is viewed in a unique context because of its relationship to a woman's fertility and, thus, femininity.

Rates of hysterectomy vary by age, education, region, marital status, and insurance coverage (see Tables 20.1–20.8), implying that the decision to recommend hysterectomy is, at least to some extent, discretionary. The most common underlying indications for hysterectomy, excluding malignant and pregnancy-related conditions, are leiomyoma, endometriosis, uterine prolapse, and dysfunctional uterine bleeding (see Tables 20.10 and 20.11); it is largely the symptoms of these conditions that lead women to seek care. Typically, however, surgery has been reserved for serious and life-threatening conditions. Thus, much of the discussion on the appropriateness of hysterectomy has focused on the relative weights that should be assigned to a woman's suffering, lost work time, surgery-related morbidity and mortality, and cost [2].

The debate on the appropriateness of hysterectomy emerged and continues for numerous reasons. These include the increasing role of the gynecologist in the primary health care of women, changes in the typical specialty of the physician performing hysterectomy (*i.e.,* surgeons versus gynecologists), the growing dependence in the U.S. on technical solutions to disease, the changing nature of the physician–patient relationship, the economics of health care, and the women's health movement [10]. The issue of the appropriateness of hysterectomy is of interest to consumers, health care providers, third-party payers, and governmental agencies [11,12].

This chapter focuses on issues related to hysterectomy that is performed for nonmalignant and nonpregnancy related gynecologic conditions. Our purpose is to provide the reader with an overview of the epidemiology of hysterectomy in the U.S., including identification of data sources and presentation of current data on the incidence and prevalence of the procedure. Risk factors, including age, race, education, and geographic variation, as well as morbidity and mortality associated with the pro-

cedure, are discussed. While much of the data are derived from previously published, primarily government sources, some new analyses are also presented. We have not systematically reviewed (*i.e.,* systematically identified and summarized all relevant literature) the literature on the epidemiology of hysterectomy or the procedure's effectiveness for nonmalignant conditions. Rather, our goal is to introduce the reader to certain aspects of hysterectomy and its sequelae, including topics already mentioned, as well as those related to patient-centered outcomes and emerging alternatives to the procedure.

II. History

Hysterectomy is an ancient procedure, reportedly mentioned as early as 2000 BC for the treatment of uterine prolapse [13]. Uterine prolapse, the displacement of the uterus and cervix downward through the vaginal canal, is usually due to a weakening of the pelvic musculature as a consequence of multiple pregnancies [14]. Soranus is credited with performing the first vaginal hysterectomy (the surgical removal of the uterus through the vagina) in the second century AD for the removal of a gangrenous prolapsed uterus [13]. From the fifth century BC to the mid-fourteenth century, removing the uterus was a last resort for treating prolapse; practitioners preferred to treat the condition using nonsurgical means such as mechanical support of the uterus [15–17]. Uterine prolapse was first formally recorded as an indication for vaginal hysterectomy in the early nineteenth century [15–17].

Before the twentieth century, the majority of patients undergoing vaginal hysterectomy died. Anecdotal accounts suggested that the mortality rate for this procedure in 1830 approached 90% [15]. The removal of the uterus using intra-abdominal surgery (abdominal hysterectomy) was viewed as nearly impossible in the early nineteenth century given the absence of anesthesia and other factors [18]. The first successful, albeit inadvertent, abdominal hysterectomy occurred in 1843 when A. M. Heath, reportedly planning to remove an ovary, instead removed the patient's uterus after finding uterine leiomyomas (fibroids) [13]. In the mid-nineteenth century, Walter Burnham in the U.S. attempted abdominal hysterectomy in 15 patients, 12 of whom died following surgery [18]. During the late nineteenth century, significant technological advances, including enhanced instrumentation, anesthesia, and antisepsis, led to a decrease in mortality rates associated with both abdominal and vaginal surgical approaches [2,15]. By the middle of the twentieth century, an "acceptable" mortality rate had been achieved (25/1000 women having the procedure for vaginal hysterectomy and 30/1000 women having the procedure for the abdominal approach), leading to both an increase in the use of the surgery and in the number of trained specialists able to perform the procedure in the U.S. and internationally [15]. In the last half of the twentieth century, scientific advances, including antibiotics, intravenous therapy, and blood transfusion [2,15], further contributed to a decrease in mortality associated with hysterectomy in the U.S. to approximately 1/1000 procedures [15].

III. Epidemiology of Hysterectomy

A. Surveillance in the United States

1. Sources of Data

Incidence of hysterectomy is defined as the number of new cases of hysterectomy in a population at risk within a defined time period (usually one year). The most reliable source of epidemiologic data on the incidence of hysterectomy in the U.S. is the Centers for Disease Control and Prevention's (CDC) National Hospital Discharge Survey (NHDS), which is analyzed regularly as part of their Hysterectomy Surveillance System. The NHDS, conducted annually since 1965, is a national probability sample of patient discharges from non-Federal U.S. short-stay hospitals in the 50 states and the District of Columbia. It is the only surveillance system based on national, population-based inpatient data [1]. Persons at risk for hysterectomy are defined as female, civilian U.S. residents aged 15 and older. Population estimates of female civilians that are used to compute incidence rates are obtained using data from the U.S. Bureau of the Census for each year [1]. NHDS estimates of hysterectomy incidence retain women who have previously had a hysterectomy in the population of women at risk, therefore, rates may be underestimates.

Sampling for the NHDS was redesigned in 1988 to facilitate comparisons between NHDS and other CDC surveys, such as the National Health Interview Survey. The NHDS collects information on patient demographics (age, race, and sex), geographic region of the hospital, admission and discharge dates (from which length of stay is calculated), diagnoses, procedures performed, and expected sources of payment [19]. The redesign of the NHDS in 1988 resulted in a decrease in the estimated numbers and rates of hysterectomy in the U.S. from 1988 onward, making it more difficult to analyze trends over the entire 14-year surveillance period from 1980 to 1993 [20]. Surveillance data from the NHDS are influenced both by the incidence and prevalence of gynecologic conditions and by changes in patterns of gynecologic care [1].

NHDS data should be interpreted cautiously for several reasons. First, information on clinical factors such as number of pregnancies, which affect hysterectomy rates, is not collected in the survey. Second, indications for hysterectomy are not validated by medical record review [1]. Third, a large proportion of NHDS data on race are missing; from 1982 to 1989, 9–11% of discharged cases are missing race information and from 1990 to 1992, 16–20% are missing [1]. Because eliminating cases with unknown race might underestimate hysterectomy rates for certain races, missing race data have been imputed according to the discharge distribution among women with known race [1].

Another source of hysterectomy incidence data, length of stay, and cost is the National Inpatient Sample (NIS), a component of the Agency for Health Care Policy and Research's

(AHCPR'S) Healthcare Cost and Utilization Project (HCUP-3), conducted since 1988. Estimates of national inpatient utilization are based on a stratified sample of hospitals. Stratification is on five characteristics (ownership/control, bed size, teaching status, urban/rural status, and geographic region) [21]; the sample is based on 6,538,976 discharges from 913 hospitals in 17 states. The NIS contains data on hysterectomy discharges for patients aged 15 and older, mean total charges and payment source, in-hospital complications, mean length of stay, and discharge status [21].

Prevalence of hysterectomy is defined as the number of existing cases of hysterectomy in a defined population at a particular point in time. Several sources of data on the prevalence of hysterectomy in the U.S. are available. The National Health and Nutrition Examination Survey (NHANES) uses self-report and physical examination to collect data on demographic characteristics, medical history, nutrition, and health information from a representative civilian sample. NHANES I was administered from 1971 to 1975 and included persons 25 to 74 years of age. Although no hysterectomy-related data were collected at baseline, these data were collected on women participants [22] who were subsequently followed prospectively by the NHANES I Epidemiologic Follow-Up Study (NHEFS). The NHEFS is a national, longitudinal study designed to examine the relationships between baseline data collected in NHANES I and morbidity, mortality, and other data collected at four follow-up time points (1982–1984, 1986, 1987, and 1992) [22]. The 1992 NHEFS included data collection on all living members of the original NHANES I cohort [22], of which 8596 were women [23]. Respondents were asked whether they had their womb, and if not, their age when it was removed.

NHANES III, administered from 1988 to 1994, was the first NHANES survey to collect baseline information on hysterectomy [oral communication from S. Wiser to G. Lemaire, February 9, 1998]. The "Household" data collection form was administered to approximately 33,994 persons two months of age and older during the six-year period. A subset of respondents completed the "Examination" part of the survey, yielding hysterectomy status data for 9354 women aged 17 and older who were asked whether or not they still had their uterus, and if not, their age when it was removed [24].

Hysterectomy prevalence data are also available from the Behavioral Risk Factor Surveillance System (BRFSS), a nationwide state-based survey, initiated in 1984, of a random sample of civilian, noninstitutionalized adults aged 18 years and older; it is used to measure progress toward national and state health objectives [25]. Each month, states select a random sample of adults who are interviewed about risk factors for chronic disease and other leading causes of death [E-mail communication from C. Leutzinger to G. Lemaire, March 4, 1998]. Individual state sample size is based on the number of respondents needed to make reliable estimates for selected subgroups and on available state funding. For 1996, the total nationwide BRFSS sample size was 120,354 [E-mail communication from C. Leutzinger to G. Lemaire, March 30, 1998]. A single survey question asks respondents whether or not they have had a hysterectomy [26].

2. Incidence

From 1965 to 1981, hysterectomy was the most common surgery in the U.S. for women of reproductive age. Since 1981,

Table 20.1

Incidence[a] of Hysterectomy by United States Geographic Region and Year[b]

Year	Overall	Region Northeast	Midwest	South	West
1972	8.3	6.7	7.9	9.6	8.9
1975	8.8	6.6	9.0	9.9	9.6
1980	7.1	5.3	7.5	8.7	6.4
1981	7.3	4.7	7.2	8.7	7.9
1982	6.9	4.7	7.1	8.5	6.6
1983	7.1	5.4	6.8	8.5	6.9
1984	6.9	4.8	6.6	8.3	7.2
1985	6.9	4.3	6.6	8.3	7.8
1986	6.6	4.4	6.8	7.6	7.0
1987	6.6	4.1	6.5	7.4	8.1
1988	5.8	4.1	5.9	7.4	4.8
1989	5.4	3.8	5.9	6.3	4.7
1990	5.8	4.0	6.1	7.4	4.8
1991	5.4	3.7	5.1	6.7	5.1
1992	5.6	4.0	5.3	6.7	5.8
1993	5.4	4.2	5.1	6.7	4.6
1994	5.3	4.2	5.4	6.4	4.5
1995	5.5	4.2	4.8	7.0	5.0

[a]Incidence = number of hysterectomy discharge diagnoses per 1000 women in population; women ≥ 15 years of age.

[b]Source: National Center for Health Statistics, National Hospital Discharge Survey (1972 to 1995).

Table 20.2

Prevalence of Hysterectomy by Age and Survey Year[a]

Age group	1990 No. in sample	1990 % with hyst.	1995 No. in sample	1995 % with hyst.
18–24	1,936	1.3	5,456	0.7
25–29	2,042	2.3	56,818	1.4
30–34	2,461	5.4	7,022	4.7
35–39	2,316	12.1	6,546	16.6
40–44	1,929	21.6	6,546	16.6
45–54	2,625	34.9	9,962	31.7
≥55	6,860	40.5	21,356	42.8
Total	20,169	22.8	63,454	22.9

[a]Source: Behavioral Risk Factor Surveillance System (1990, 1995); women ≥ 18 years of age.

the total number of cesarean sections each year has exceeded the number of hysterectomies [27]. Data from the NHDS indicate that the number of hysterectomies performed in the U.S. in women 15 years of age and older increased from 426,000 in 1965 to 724,000 in 1975, the year with the greatest number of hysterectomies to date [28]. Incidence increased concomitantly from an estimated 6.1 per 1000 women in 1965 to 8.6 per 1000 women in 1975. Incidence has declined since then and has remained fairly stable since the late 1980s (see Table 20.1). The most recent estimates are from 1995 when an estimated 583,000 hysterectomies [1,20] (5.5 per 1000 women) were performed. Thus, between 1980 and 1995, an estimated 9.7 million U.S. women underwent hysterectomy (excludes hysterectomy for advanced pelvic cancer) [1,29].

The lower than projected incidence of hysterectomy [28,30] observed in recent years can be partially explained by the NHDS 1988 redesign. Other possible explanations for this decline include: adoption of quality assurance activities and second opinion programs for elective hysterectomy [1,30–33]; the increased availability of alternate medical and surgical treatments [2,34]; increased popular opinion that hysterectomy is invasive and undesirable [10]; and delayed childbearing resulting in a desire to preserve fertility [1].

3. Prevalence

For 1988 to 1994, the estimated prevalence of hysterectomy in the NHANES III [24] sample of 9354 women aged 17 to 90 years at time of survey (mean age = 46.5 years; SD = 20.4) was 201 per 1000 women (20.1%). In the 1990 BRFSS sample of 20,454 women aged 18 to 99 (mean age = 46.5 years; SD =

18.4) at time of survey asked whether they had had a hysterectomy, the estimated prevalence was 229 per 1000 (22.9%) (see Table 20.2) [26]. For the 1995 BRFSS sample of 64,385 women aged 18 to 99 (mean age = 47.5 years, SD = 18.2) at time of survey, the prevalence was similar, 228 per 1000 (22.8%) [26]. Thus, both BRFSS samples (1990 and 1995) yielded a higher prevalence than did the NHANES III survey which was conducted during approximately the same time period, 1988 to 1994. Further work is needed to identify factors that may have contributed to the variation in prevalence observed between the two surveys.

IV. Risk Factors for Hysterectomy

A. Age, Race, Education, and Geographic Variation

For the period 1988 to 1993, women 40 to 44 years of age had the highest incidence rate of hysterectomy (12.9 per 1000 women) and women 15 to 24 had the lowest rate (0.4 per 1000 women) (see Table 20.3) [1].

As would be expected, the prevalence of hysterectomy rises with age. In the NHANES III sample, prevalence of hysterectomy was highest for women 60 to 69 at the time of survey (41.7%) and similar for women 50 to 59 (39.6%) and 70 to 79 years of age (38.8%) (see Table 20.4). Prevalence was about the same for whites and blacks, although prevalence in blacks was generally somewhat higher among women 50–59 and lower in women older than 80 years. No hysterectomies were reported for women less than 19 years of age and the prevalence was low, 0.7% and 6.6%, among women ages 20 to 29 and 30 to 39, respectively [24].

The 1990 and 1995, BRFSS [26] used different age groupings, but data appear similar to those from NHANES III, with hysterectomy prevalence rising with age from approximately 1% among women ages 18 to 24 years, to 5% among women ages 30 to 34, and to 40% among women ages 55 and older (see Table 20.2). Prevalence was somewhat lower in 1995 compared to 1990 for younger ages, although the prevalence overall was similar. For both years, the prevalence was slightly higher in whites compared to blacks (see Table 20.5).

Table 20.3

Estimated Number of Hysterectomies, Percentage of Total, and Incidence Rates, by Age and Race[a]

Age (years)	White			Black			Other[b]			Total		
	No. with hyst.	%	Rate[c]	No. with hyst.	%	Rate	No. with hyst.	%	Rate	No. with hyst.	%	Rate
15–24	39,974	1.4	0.5	4,762	1.1	—[d]	301	.3	—	45,037	1.3	0.4
25–29	186,467	6.6	3.7	23,775	5.6	2.8	3,817	3.5	1.4	214,059	6.4	3.5
30–34	325,994	11.6	6.0	63,915	15.0	7.4	7,886	7.2	2.7	397,795	11.9	6.0
35–39	479,853	17.0	9.5	107,953	25.4	13.9	16,972	15.5	6.2	604,779	18.1	9.9
40–44	572,278	20.3	12.5	99,744	23.5	15.7	30,868	28.1	13.2	702,890	21.0	12.9
45–54	664,734	23.6	9.8	88,484	20.8	10.1	32,411	29.5	11.0	785,629	23.4	9.9
≥ 55	546,642	19.4	3.4	36,676	8.6	2.3	17,453	15.9	4.3	600,772	17.9	3.3
Totals	2,815,943	100.0	5.5	425,309	100.0	5.9	109,708	100.0	4.8	3,350,961	100.0	5.5

[a]Source: National Center for Health Statistics, National Hospital Discharge Survey (1988–1993); women ≥ 15 years of age.
[b]Includes Asian, Pacific Islander, American Indian, Alaskan Native, and other races.
[c]Rates are number of hysterectomy discharges per 1000 women in population ≥ 15 years of age, by age and race.
[d]Numbers too small for meaningful estimate.

Overall, hysterectomy incidence rates did not differ meaningfully between whites and blacks [1], although the rates for black women were slightly higher than for white women overall and for all but the two youngest age groups (15 to 24 and 25 to 29 years). The slightly higher incidence reported by NHDS for 1988 to 1993 for black women (5.9 per 1000 compared with 5.5 per 1000 for whites) (see Table 20.3) was not reflected in the NHANES and BRFSS prevalence data, where a smaller overall proportion of blacks compared with whites reported hysterectomy (see Tables 20.4 and 20.5). The higher incidence in blacks was seen in the 30 to 54 year age range, whereas lower incidences for black women were observed in the younger and older age ranges (see Table 20.3). This could be related to methodological differences in the surveys. These data may also reflect real differences in indications for surgery, access to and availability of alternative treatments, reimbursement factors, and other health care system and societal factors.

In general, women with less education have a higher prevalence of hysterectomy than women with more education. Data obtained from the 1992 NHANES I National Epidemiologic Follow-up Study indicated that women who had some high school education were most likely to have had a hysterectomy (27.8%), whereas those with greater than a high school education had the lowest rates of hysterectomy (16.6%) [23]. Increased prevalence of hysterectomy was also associated with lower education levels both in BRFSS and in NHANES III (see Tables 20.5 and 20.6). Data from the BRFSS indicated that overall prevalence was lowest for women with four or more years of college; prevalence was 16.2% in 1990 and 14.3% in 1995 (see Table 20.5). In 1995, 22.3% of those who never attended school, 39.3% of those completing grades 1–8, and 34.0% of those completing grades 9–11 reported having had a hysterectomy. In 1995, white women had generally higher prevalence of hysterectomy than black women and those of "other" race, except among college graduates. The higher prevalence among white women may reflect increased knowledge about health or confidence in the health care system, increased access to or use of care and reimbursement, or a systematic sampling

difference. Prevalence among those who did not complete high school was higher in the BRFSS survey than in NHANES III. Data from NHANES III (1988–1994) [24] indicated that the prevalence for black women was higher than for whites only for those with less than a ninth grade education, and this group had the highest prevalence overall (28.5%) (see Table 20.6).

NHANES III provided data on the association between other demographic variables and hysterectomy. Women who were separated, widowed, or divorced were more likely to report having had a hysterectomy compared to married or never married women. Prevalence was similar across income levels but was higher for insured women (see Table 20.7), perhaps reflecting their easier access to the health care system or a tendency of physicians to recommend surgery more often in women with insurance coverage.

There is considerable geographic variation in hysterectomy incidence rates [3,35,36]. These patterns have remained stable since 1970 when U.S. surveillance of the procedure was initiated [1]. For 1995, NHDS data indicated that rates of hysterec-

Table 20.4

Prevalence of Hysterectomy by Race and Age at Survey[a]

Age group	White		Black		Total	
	No. in sample	% with hyst.	No. in sample	% with hyst.	No. in sample	% with hyst.
20–29	1124	0.6	641	0.8	1765	0.7
30–39	1077	5.7	656	8.1	1733	6.6
40–49	822	22.1	457	27.6	1279	24.1
50–59	676	35.2	278	50.4	954	39.6
60–69	818	40.5	295	45.1	1113	41.7
70–79	768	38.9	171	38.0	939	38.8
≥80	654	31.5	83	19.3	737	30.1
Total	5939	22.3	2581	20.8	8520	21.8

[a]Source: NHANES III (1988–1994); women ≥ 17 years of age.

Table 20.5
Prevalence of Hysterectomy by Race and Education[a]

| | 1990 | | | | | | | | 1995 | | | | | | | |
| | White | | Black | | Other | | Total | | White | | Black | | Other | | Total | |
Education	No.[b]	%[c]	No.	%	No.	%	No.	%	No.	%	No.	%	No.	%	No.	%
Never attended/ kindergarten	0	0	0	0	0	0	0	0	652	23.3	102	21.6	122	17.2	876	22.3
Grades 1–8	1086	37.3	203	37.4	151	29.1	1440	36.5	2225	41.8	391	39.9	393	24.4	3009	39.3
Grades 9–11	1675	33.7	293	32.1	181	19.9	2149	32.3	4203	35.8	825	32.8	581	22.0	5609	34.0
Grade 12	6161	25.8	672	17.9	397	17.9	7230	24.6	18,056	27.1	2012	18.8	1634	17.1	21,702	25.6
College 1–3 years	3869	18.6	462	14.9	300	16.0	4631	18.1	14,772	21.8	1458	15.8	1435	14.4	17,665	20.7
College 4+ years	4172	16.3	323	17.3	286	13.3	4781	16.2	12,698	14.4	947	15.0	1129	12.1	14,774	14.3
Total	16,963	23.3	1953	21.3	1315	18.0	20,231	22.8	52,606	23.8	5735	20.9	5294	16.4	63,635	22.9

[a]Source: Behavioral Risk Factor Surveillance System (1990, 1995); women ≥ 18 years of age.
[b]No. = Number in sample.
[c]Percentage with hysterectomy.

tomy were highest in the South (7.0/1000 women) and lowest in the Northeast (4.2/1000 women) (see Table 20.1). Rates were 5.0/1000 in the West and 4.8/1000 in the Midwest [20]. Age at hysterectomy appears to vary by geographic region. For example, for the period 1988 to 1993, the average age at hysterectomy for women in the South was 41.6 years, younger than for women in the other geographic regions. In the Northeast, the average age was 47.7, in the Midwest it was 44.5, and in the West it was 44.0 years [1].

Observed geographic variation in hysterectomy rates and the associated underlying conditions leading to hysterectomy have been an issue of continuing concern for many years [37]. The Rand Corporation has published two reports: the first, "Hysterectomy: A Review of the Literature Regarding Risks and Com-

plications" [38], and a subsequent revision, "Hysterectomy: A Review of the Literature on Indications, Effectiveness, and Risks" [9] (hereafter referred to collectively as the "Rand Report"). They assembled the scientific evidence addressing these and related questions. The reports concluded that there is no evidence that links differences in indications with differences in incidence by geographic area. These data contribute to the continuing research and debate related to the appropriateness of the surgery [9].

NHANES III prevalence data for 1988 to 1994 showed a somewhat different geographic pattern from incidence rates derived from the NHDS data, although both incidence and preva-

Table 20.6
Prevalence of Hysterectomy by Race and Education[a]

| | White | | Black | | Total[b] | |
Education	No. in sample	% with hyst.	No. in sample	% with hyst.	No. in sample	% with hyst.
< High school (n = 1946)	1528	22.6	418	28.5	1946	23.8
Some high school (n = 1678)	1077	22.9	601	19.0	1678	21.5
High school graduate (n = 2987)	1965	22.5	1022	17.4	2987	20.8
> High school (n = 2421)	1683	16.8	738	16.9	2421	16.9

[a]Source: NHANES III (1988–1994); women ≥ 17 years of age.
[b]Excludes 272 women of other races.

Table 20.7
Prevalence of Hysterectomy by Marital Status, Income, and Insurance Status[a]

Characteristic	No. in sample	% with hyst.
Total women surveyed	9354	20.2
Marital status (n = 9333)		
Married or living as married	4795	21.0
Separated, divorced, or widowed	2717	29.1
Never married	1821	4.8
Income (n = 8378)[b]		
<$10,000	1864	20.3
$10,000 to $19,999	2348	21.4
$20,000 to $34,999	1937	20.8
$35,000 to $49,999	1132	17.4
≥$50,000	1097	18.4
Insurance status (n = 9003)		
Insured	7524	22.5
Uninsured	1479	11.0

[a]Source: NHANES III (1988–1994); women ≥ 17 years of age.
[b]10.4% of sample missing data on income.

Table 20.8

Prevalence of Hysterectomy by Race and Region[a]

Region	White		Black		Total[b]	
	No. in sample	% with hysterectomy	No. in sample	% with hysterectomy	No. in sample	% with hysterectomy
Northeast	776	20.4	559	13.1	1335	17.3
Midwest	1290	21.3	541	19.8	1831	20.9
South	2431	24.8	1491	21.5	3922	23.6
West	1792	16.1	202	18.3	1994	16.3
Total	**6289**	**21.1**	**2793**	**19.3**	**9082**	**20.5**

[a]Source: NHANES III (1988–1994), women ≥ 17 years of age.
[b]Excludes 272 women of other races.

lence of hysterectomy were highest in the South (see Tables 20.1 and 20.8). In NHANES III, the highest prevalence was in the South (23.6%) and the lowest prevalence was in the West (16.3%). Except for the West, reported prevalence was consistently higher for whites, with black women in the Northeast having the lowest prevalence of any group (13.1%).

V. International Surveillance

Concern about the appropriateness of hysterectomy for non-malignant conditions extends beyond the U.S. Although the extent and method of hysterectomy surveillance may vary across countries, making international comparisons difficult, there appears to be substantial variation in rates both among and within industrialized countries such as Sweden, the United Kingdom, and Denmark [23,35,39–42]. The rate of hysterectomy in the U.S. is the highest (5.5/1000 women) among countries with published data [43a,43b] (see Table 20.9), and Sweden ranks lowest, at 1.95/1000 women. So few data are available that these rankings may not be reliable. If real, the higher rate in the U.S. may be related to a number of factors, including the fee-for-service payment method in the U.S. health care system; the failure to develop, test, use, and make available alternative treatments; attitudes regarding acceptable indications for performing the procedure; and other factors which may impact on access to health care [9]. In addition, provider factors such as physician autonomy and characteristics such as training, sex, and fear of malpractice litigation may also impact on variation in hysterectomy rates [9]. Patient-related factors such as race, education,

Table 20.9

Reported Annual Hysterectomy Incidence Rates by Year and Region

Country	Year	Rate/1000 women
United States [20]	1995	5.5
New South Wales, Australia [43a]	1994–1995	5.3
Finland [66]	1991	4.7
England [43b]	1994–1995	2.9
Denmark [67]	1984–1986	2.8
Sweden [40]	1991	2.0

parity, religious beliefs, attitudes toward gynecologic surgery, and perception of the value and importance of the uterus may also play a role [9,10].

VI. Hysterectomy Indications and Associated Risk Factors

The number of hysterectomies and incidence rates in the United States are presented by reason for surgery or indication (*i.e.*, discharge diagnosis) and race for the years 1988 to 1993 in Table 20.10. The discharge diagnoses most frequently associated with hysterectomy were uterine leiomyoma or "fibroids" (33.4% of all hysterectomies), endometriosis (18.9%), and uterine prolapse (16.4%). Uterine leiomyoma has been the most frequently reported primary discharge diagnosis for hysterectomy since 1965. Over one million hysterectomies were performed for this indication between 1988 and 1993. Endometriosis became the second most frequently reported hysterectomy-related diagnosis starting with the period 1982 to 1984, supplanting uterine prolapse [43]. Endometriosis and uterine prolapse each accounted for more than one-half million hysterectomies from 1988 to 1993. Cervical dysplasia and menstrual disturbances combined were responsible for almost 16% of hysterectomies, while about 10% were performed for cancer [1]. White women tend to dominate this distribution, with leiomyomas accounting for 28.7%, endometriosis 20.5%, and prolapse 18.2% of hysterectomies in this population. The vast majority of hysterectomies in black women were for leiomyoma (61.5%), with endometriosis and prolapse accounting for only 9.3% and 5.2%, respectively. For women of "other" races, leiomyoma was the primary indication for hysterectomy in 44.9% of cases [1].

Indications for hysterectomy also differ by age (see Table 20.11). For women 15 to 24 and 25 to 29 years of age, hysterectomy has been relatively rare and has been most often attributed to "other" conditions such as cervical dysplasia and menstrual disturbances. Endometriosis has been the most frequent indication for ages 30 to 34, and leiomyoma for ages 35 to 54 years. Prolapse has been the most frequent indication for hysterectomy in women 55 and older [1]. These differences by age may be related to true incidence and prevalence differences in the underlying conditions, differences in the severity of illness at the time surgery is performed, differences in attitudes regarding the need to maintain reproductive capacity, and other factors.

Table 20.10

Numbers of Hysterectomies, Percentage of Total, and Incidence Rates by Primary Discharge Diagnosis and Race[a]

	White			Black			Other[b]			Total		
Diagnosis	No. with hyst.	%	Rate[c]	No. with hyst.	%	Rate	No. with hyst.	%	Rate	No. with hyst.	%	Rate
Uterine leiomyoma	807,598	28.7	1.6	260,783	61.5	3.6	49,190	44.9	2.2	1,117,571	33.4	1.8
Endometriosis	577,846	20.5	1.1	39,434	9.3	0.5	14,377	13.1	0.6	631,657	18.9	1.0
Uterine prolapse	513,049	18.2	1.0	21.911	5.2	0.3	13,697	12.5	0.6	548,657	16.4	0.9
Other[d]	459,316	16.3	0.9	63,362	14.9	0.9	10,803	9.9	0.5	533,481	15.9	0.9
Cancer	294,892	10.5	0.6	28,179	6.6	0.4	17,177	15.7	0.8	340,248	10.2	0.6
Endometrial hyperplasia	164,592	5.8	0.3	10,582	2.5	0.1	4,173	3.8	—[e]	179,347	5.4	0.3
Total hysterectomies	**2,817,293**	**100.0**	**5.5**	**424,251**	**100.0**	**5.9**	**109,417**	**100.0**	**4.8**	**3,350,961**	**—**	**5.5**

[a]Source: National Center for Health Statistics, National Hospital Discharge Survey (1988–1993); women ⩾ 15 years of age.

[b]Includes Asian, Pacific Islander, American Indian, Alaskan Native, and other races.

[c]Rates are number of hysterectomy discharges per 1000 women in population ⩾ 15 years of age, by diagnosis and race.

[d]Includes cervical dysplasia and menstrual disturbances.

[e]Numbers are too small for meaningful estimate.

Although the indications for hysterectomy are typically presented as medical conditions, it is most often the symptoms associated with the nonmalignant indications and not the conditions themselves that lead women to seek medical or surgical treatment. Common symptoms of endometriosis include subfertility, pelvic pain, dysmenorrhea, and dyspareunia [14]. Menorrhagia is the most common symptom associated with (but not necessarily caused by) leiomyomas, although pelvic pain may occur, as well as urinary tract symptoms such as frequency and ureteral obstruction. Uterine prolapse is associated with a feeling of pressure in the vagina and a sensation of heaviness, and frequently, urinary stress incontinence [14].

There are some data indicating that women who have had a tubal ligation are at increased risk of hysterectomy due to "posttubal sterilization syndrome." This syndrome is characterized by a wide array of symptoms that have been noted following sterilization and which include abnormal bleeding, pain, changes in sexual behavior and mood, and menstrual symptoms [44]. While some studies have reported a positive association between previous sterilization and hysterectomy [45,46], evidence is inconclusive and additional data are needed to confirm or refute this association. If there is a real increased risk for hysterectomy among women who have been sterilized, this could be associated with their greater willingness to relinquish the uterus, or it may reflect a biologic effect of tubal sterilization [46].

VII. Types and Surgical Route of Hysterectomy

There are three general types of hysterectomy for nonmalignant, nonpregnancy related conditions: total hysterectomy, subtotal hysterectomy, and hysterectomy with adnexectomy (see Table 20.12) [2]. Hysterectomy may be accomplished by vari-

Table 20.11

Number of Hysterectomies by Primary Diagnosis and Age at Discharge[a]

	15–24		25–29		30–34		35–39		40–44		45–54		>55		
Diagnosis	No.[b]	%	No.	%	No.	%	No.	%	No.	%	No.	%	No.	%	Total
Uterine leiomyoma	881	2.0	17,456	8.2	72,732	18.3	202,990	33.6	341,741	48.6	415,760	52.9	66,010	11.0	1,117,571
Endometriosis	7,288	16.2	64,883	30.3	115,553	29.0	160,615	26.6	149,801	21.3	108,052	13.8	25,465	4.2	631,657
Uterine prolapse	7,442	16.5	31,506	14.7	62,028	15.6	78,515	13.0	70,553	10.0	89,858	11.4	208,755	34.7	548,657
Other[c]	23,669	52.6	79,501	37.1	106,178	26.7	103,212	17.1	86,532	12.3	65,465	8.3	68,925	11.5	533,482
Cancer	4,809	10.7	18,408	8.6	29,555	7.4	40,291	6.7	31,063	4.4	46,234	5.9	169,888	28.3	340,248
Endometrial hyperplasia	948	2.1	2,305	1.1	11,749	3.0	19,156	3.2	23,200	3.3	60,260	7.7	61,729	10.3	179,347
Total No. of hysterectomies	**45,037**	**100.0**	**214,059**	**100.0**	**397,795**	**100.0**	**604,779**	**100.0**	**702,890**	**100.0**	**785,629**	**100.0**	**600,772**	**100.0**	**3,350,961**

[a]Source: National Center for Health Statistics, National Hospital Discharge Survey (1988–1993); women ⩾ 15 years of age.

[b]No. = Number of hysterectomy discharges.

[c]Includes cervical dysplasia and menstrual disturbances.

Table 20.12
Types of Hysterectomy and Tissue Removed

Type of hysterectomy	Tissue removed
Total hysterectomy	Corpus and cervix
Subtotal hysterectomy	Corpus only (cervix is retained)
Hysterectomy with adnexectomy (Unilateral or Bilateral)	Fallopian tube (salpingectomy) Ovary (oophorectomy) Fallopian tube and ovary (salpingo-oophorectomy)

ous surgical routes or approaches: abdominal, vaginal, or a combination of laparoscopic surgery and vaginal hysterectomy known as laparoscopic assisted vaginal hysterectomy (LAVH) [14]. Abdominal hysterectomy involves a horizontal or vertical incision in the abdomen through which the uterus is removed. In vaginal hysterectomy, the uterus is removed through the vagina with the aid of surgical instruments. LAVH, first introduced in 1989 [47], permits visualization of and surgical access to the abdomino-pelvic cavity by way of a laparoscope inserted through the umbilicus during a vaginal hysterectomy [14]. This approach combines the less invasive vaginal route with a means to view the interior of the pelvis without the need for a large incision. Specific criteria to determine hysterectomy route have not been standardized, although contraindications to the various routes have been suggested [48].

Vaginal hysterectomy has advantages when compared to abdominal hysterectomy because it is less invasive, results in less morbidity, and is associated with comparatively rapid recovery [11,14]. Nevertheless, 74% of all hysterectomies were performed abdominally and 28% vaginally between 1980 and 1993 [1], demonstrating little change from the previous decade [1,49].

During 1980 to 1993, there was an increase in the percentage of vaginal hysterectomies that were accompanied by laparoscopy. While laparoscopy was associated with less than 1% of all vaginal hysterectomies during the 1980s, this percentage rose from 1.4% in 1990 to 15.3% and 14.2% in 1992 and 1993, respectively [1]. The relatively small proportion of vaginal compared to abdominal hysterectomies may be due to several factors, including surgical training, inaccessibility of the ovaries, and the relatively large percentage of hysterectomies performed for uterine leiomyoma (the enlarged uterus resulting from this condition may make its removal through the vagina difficult).

VIII. Prophylactic Bilateral Oophorectomy

Prophylactic bilateral salpingo-oophorectomy (BSO) is the removal of healthy fallopian tubes and ovaries to decrease a women's risk of ovarian cancer; BSO is the most frequent concomitant procedure performed in conjunction with abdominal hysterectomy [1]. The decision to perform BSO is associated with surgical route, diagnosis, and patient age. It is not possible, from available data, to differentiate prophylactic BSO performed with hysterectomy for a uterine condition from BSO for an ovarian condition performed with a ''prophylactic'' hysterectomy. According to Lepine [1], for 1988 to 1993, BSO was

performed with 66% of hysterectomies for cancer or endometrial hyperplasia, with 50% for leiomyoma or endometriosis, and with 20% of hysterectomies for uterine prolapse.

From 1965 to 1984, the proportion of hysterectomies accompanied by BSO increased from 25 to 41% [29], and from 1988 to 1993, this percentage rose from 47 to 52% [1]. Concomitant BSO increased with age at hysterectomy until ages 45 to 54, with 18% of women aged 15 to 24 and 76% of women 45 to 54 undergoing BSO, and 62% of women 55 years and older undergoing the combined procedure [1]. In 1993, 63% of women who had an abdominal hysterectomy had BSO, compared to 18% of those who had the vaginal approach.

IX. Length of Stay

Based on NHDS data, the average length of stay (LOS) for women undergoing hysterectomy decreased from 12.2 days in 1965 to 7.2 days in 1984 [29] and 5.7 days in 1987. In 1992, the average LOS for women having hysterectomy for abnormal uterine bleeding was 3.7 days [43]. Using discharge data from the Maryland Health Services Cost Review Commission, Kjerulff [50] found a similar decrease in LOS from a mean of 5.7 days in 1986 to a mean of 4.8 days in 1989 among 34,441 Maryland women having hysterectomy. Abdominal hysterectomy with BSO was associated with the longest LOS (mean of 5.2 days) and vaginal hysterectomy with unilateral oophorectomy with the shortest LOS (mean of 4.0 days). Weber, using Ohio discharge data, found that from 1988 to 1994, mean length of stay decreased from 5.0 days to 3.0 days for abdominal hysterectomy, 4.0 to 2.0 days for vaginal hysterectomy, and 5.0 to 2.0 days for laparoscopic assisted vaginal hysterectomy [51].

Based on 1993 data for 518,000 abdominal and vaginal hysterectomy-related discharges for women aged 15 and older from the Nationwide Inpatient Sample of the Healthcare Cost and Utilization Project (HCUP-3), the LOS for women with private insurance, Medicaid, Medicare, and self-payment sources were 3.6, 4.2, 5.4, and 4.4 days respectively [21].

X. Cost of Hysterectomy

In the early 1990s, the estimated annual direct costs of hospitalization for hysterectomy totaled more than $5 billion [34]. Both direct and indirect costs should be considered in any assessment of the economic impact of hysterectomy, however. Direct costs for hysterectomy generally include those related to hospital stay, operating room, and physician expenses, whereas indirect costs represent the economic impact of time lost from work and daily activities during recovery and convalescence [52]. In several U.S. studies published since 1992, hospital charges alone ranged from $3954 to $7000 for abdominal hysterectomy; $3116 to $5343 for vaginal hysterectomy; and $4914 to $11,931 for LAVH [53–56]. LAVH, while more conservative than abdominal hysterectomy in terms of the size of incision and requiring a shorter postoperative recovery time, usually results in higher direct costs primarily because of greater supply costs [53–56] and increased operating time, including that required for training [53,57]. Indirect costs depend on recovery time, annual income, and value placed on work in the home and may be lower for vaginal compared with abdominal hysterectomy because the vaginal route is associated with shorter hospitalization, recovery, and convalescence [52].

XI. Patient Centered Outcomes of Hysterectomy

Few studies have examined the effects of hysterectomy on long-term outcomes relating to patient quality of life [3]. There is evidence, however, suggesting that hysterectomy for nonmalignant conditions effectively relieves symptoms that lead women to seek surgical treatment (*e.g.,* bleeding, pelvic or abdominal pain, dysmenorrhea, fatigue, depression, and anxiety) [39,58–60].

A. Psychosocial Issues and Sexuality

Although studies designed to evaluate the effects of hysterectomy on psychological and sexual function should consider preexisting function, indication for surgery, type of hysterectomy performed, and whether or not oophorectomy is performed simultaneously, few studies have met these criteria [9]. Considerable variation has been found among studies regarding measures of psychiatric symptoms, follow-up time, and use of comparison or control groups, thus making it difficult to draw firm conclusions about the impact of hysterectomy on psychological symptoms and sexual function. The Rand Report [9,38] summarized studies of psychological complications of hysterectomy, primarily depression, from the 1940s forward. Early studies that suggested an increase in psychiatric morbidity following hysterectomy were hampered by possible biases in patient recall and lack of prehysterectomy assessment of psychological symptoms [61,62]. Subsequent studies that assessed preoperative symptoms suggested that psychiatric morbidity was higher among women undergoing hysterectomy than in the general population, both before and after surgery. Postsurgical psychiatric morbidity, primarily anxiety and depression, was strongly associated with presurgical psychological functioning [62]. The Maine Women's Health Study found new postsurgical psychological symptoms following hysterectomy in a small proportion of the 418 participants [59]. These symptoms included depression (8%), lack of interest in sex (7%), and anxiety (6%) [59]. Given that more than one-half million hysterectomies are performed in the U.S. annually [20], these rates of psychiatric sequellae represent a women's health problem of potentially sizable magnitude (*e.g.,* 40,000 cases of hysterectomy-associated depression each year). If there is a true association between hysterectomy and psychological symptoms, there are multiple possible explanations. For example, for some women, the perceived importance of the uterus may be high and its removal may be related to postoperative negative psychosexual effects [2,63].

The few studies that have examined sexual function have had methodological limitations. Two uncontrolled cohort studies reported decreased sexual function following hysterectomy. As noted earlier, lack of interest in sex was reported as a new problem at 6 and 12 months after hysterectomy by 7% of 418 women participating in the Maine Women's Hlth Study [59]. In another study, 21% of 104 women interviewed one month before and one year after subtotal hysterectomy reported a deterioration in sexuality (including frequency of desire, coital frequency, and orgasm frequency and multiplicity) with preoperative sexual activity positively associated with postoperative activity [64]. A third study, involving 366 women, found that surgery improved the quality of sexual activity three months postoperatively, but had little influence on its frequency [65]. Among the 1057 women participating in The Maryland Women's Health Study who had a hysterectomy and responded to the relevant questions at 6, 12, 18, and 24 months after surgery, most reported that desire, frequency and intensity of orgasm, pain during sex, and frequency of sexual activity improved from baseline at all interview points after hysterectomy. Improvement in some areas of sexual function appeared greatest at the earlier follow-up points, however. For example, the number of days that women desired sex "less than one day per month" or "not at all" declined from 11.8% before the procedure to 5.8%, 5.9%, and 5.7% at 6, 12, and 18 months, respectively, then increased to 7.1% at 24 months. Likewise, the frequency of sexual activity increased from 6.2 times during the month prior to hysterectomy to 7.9 times at 6 months, but then declined to 7.4, 7.5, and 6.7 times at 12, 18, and 24 months respectively. The most frequent new problems reported 12 months after surgery included weak orgasms (7.9%), vaginal dryness (7.3%), and no sexual activity (7.1%) [45]. Bachmann [2], in her extensive literature review, concluded that psychological and sexual problems following hysterectomy are the result of physical, psychological, social, and cultural factors such as declines in ovarian hormones following surgery, preoperative sexual problems, and misconceptions about the surgery and its effects on femininity.

Harris, in his review of 78 articles, concluded that BSO concurrent with hysterectomy in premenopausal women is generally associated with psychosexual dysfunction (*i.e.,* psychological and sexual problems) [63]. He also noted that preoperative sexual function is the best predictor of postoperative sexual function and that well-being is enhanced by preoperative counseling [63].

XII. Risks and Complications

A. Mortality

The Rand Report [9,38] provides a systematic review of the relevant literature published between 1970 and 1993 concerning hysterectomy as a treatment for nonmalignant gynecologic conditions. Because this review is comprehensive and relatively recent, its findings will be only briefly summarized here. A limitation to any review is that comparisons of findings across studies are difficult due to variations in study design and outcome definitions.

Postoperative mortality is the major risk associated with hysterectomy, although there is no agreement on the precise definition of this outcome. Some authors have suggested that postoperative mortality (the time interval since hysterectomy during which mortality was considered attributable to the surgery) be defined as deaths up to 42 days postoperatively, to reflect more accurately those deaths occurring as a result of hysterectomy [11,66]. Others have defined hysterectomy mortality as only those deaths occurring during the hospitalization for hysterectomy, which clearly misses some surgically-related deaths [50,51,67]. Comparison of mortality findings across studies is also difficult because of differences in populations studied, the setting in which hysterectomy was performed, patient factors such as age, previous diagnoses, and health status, and the reason for surgery (*i.e.,* a nonmalignant or malignant condition) [9,38].

The Rand Report estimated the mortality rate for hysterectomy for nonmalignant conditions at approximately 1 per 1000 women from the 1950s forward, with risk increasing with age [9,38]. There is no current evidence as to whether surgical route or indication for the procedure influence the risk of mortality. Mortality rates reported since 1985 for hysterectomy for nonmalignant conditions are relatively low (0.4/1000 to 1.6/1000) [11,50,51,67], whereas rates for pregnancy and cancer-related hysterectomy are higher (2.9/1000 to 17.2/1000) [51,68,69]. Data for 1979 and 1980 from the Commission on Professional and Hospital Activities Professional Activity Study indicated that the mortality rate among black women undergoing hysterectomy (2.1/1000 women), regardless of indication or surgical approach, was almost double that of white women (1.1/1000 women) [68]. In a study that included more than 53,000 Maryland women and that used the Maryland Health Services Cost Review Commission hospital discharge data for 1986 to 1991, Kjerulff [69] found that black women having a hysterectomy were over three times as likely to die as white women, after adjusting for diagnosis, surgical route, age, comorbid conditions, payment source, and hospital characteristics. The overall mortality rate for black women was 2.5/1000 compared to 1.7/ 1000 for white women. For noncancerous conditions the rate was 1.3/1000 for black women and 0.6/1000 for white women, and for malignant conditions it was 17.5/1000 for black women compared to 9.7/1000 for white women [69].

Most studies reporting mortality do not specify the cause of hysterectomy-related death. Where it is specified, causes have included cancer [70], cardiopulmonary arrest (2 of 3322 hysterectomies) [71], venous thrombosis, and an overdose of phenobarbital (1 each in a series of 1851 hysterectomies) [11].

B. Postoperative Morbidity

Hysterectomy is associated with both short and long-term morbidity. Short-term complications (*i.e.,* those occurring within the first 21 days after surgery) include urinary tract infection, wound infection, bleeding, damage to organs (*e.g.,* ureter, bladder, bowel), and reoperation [9,38]. Long-term effects include possible ovarian pathology as well as increased risk of coronary heart disease. Hysterectomy-associated morbidity leads to longer hospital stay and recovery time, of importance both to women and to society.

The following paragraphs provide summary data on postoperative morbidity. Comparison of postoperative complication rates for abdominal and vaginal hysterectomy using data obtained in nonrandomized studies is not reliable, in large part because of underlying patient differences in indications and pathology associated with these routes. Thus, additional research is needed to identify differences in morbidity, mortality, and

outcome between the two standard techniques (abdominal and vaginal) and the newer LAVH [9,38].

1. Febrile Morbidity and Infection

Febrile morbidity (oral temperature of 38°C or 100.4°F or greater) [14], urinary tract infection, and wound infection are the most common causes of posthysterectomy morbidity [9,38]. Rand's systematic review of data from studies conducted through early 1988 [9,38] found that prophylactic antibiotic treatment before surgery is beneficial in preventing posthysterectomy infection, even though small individual studies of antibiotic prophylaxis have demonstrated no significant differences between antibiotic and placebo or between various antibiotics. The meta-analysis indicated that among patients with vaginal hysterectomy, those receiving antibiotics had an average febrile morbidity of about 13% compared to 37% in those not receiving prophylaxis. For abdominal hysterectomy, antibiotic prophylaxis did not reduce the incidence of wound infection (2% for both antibiotic and placebo groups) and had no effect on febrile morbidity; 25% of patients in both the antibiotic and placebo groups had fever. Pelvic infection, however, was on average 7% in the antibiotic group and 18% in the placebo group for abdominal surgery and 7% and 28%, respectively, for vaginal surgery. For both abdominal and vaginal surgical routes, urinary tract infections were less frequent with prophylactic antibiotics. For abdominal hysterectomy, the summary rates of urinary tract infection were 8.4% in the antibiotic group and 14.5% in the placebo group; for vaginal hysterectomy, the summary rates were 5.2% and 13.9%, respectively [38]. The Rand analysis confirmed previous reports of the comparable effectiveness of single-dose versus multiple-dose prophylactic antibiotic regimens for febrile morbidity and pelvic infection and showed somewhat less effectiveness for the single dose regimen for the prevention of urinary tract infection with abdominal hysterectomy. Analyses also demonstrated little difference in effectiveness among a wide range of broad spectrum antibiotics [9,38].

2. Bleeding

Postoperative bleeding after hysterectomy results from damage to blood vessels or organs and may result in reoperation, transfusion, shock, or death [14]. Reoperation puts a woman at the same surgical and postoperative risks as the first surgery and lengthens her total LOS and rehabilitation time. Transfusion increases the risk of transmission of blood-borne disease (*e.g.,* hepatitis B and C) and transfusion reactions. Studies published between 1962 and 1993 estimated that 2–13% of patients having hysterectomy received blood transfusions [9]. Studies published between 1993 and 1997 found similar rates: 1.6–28.2% of women underwent postoperative transfusion, and many of them were likely to be autologous [57,65,72–76].

3. Damage to Organs

Intraoperative injury to organs such as the bladder, ureters, and bowel during hysterectomy may be due to factors such as poor surgical technique or assistance, inadequate lighting, equipment, anatomical abnormalities, or underlying disease (*e.g.,* malignancies) [14]. These injuries are most often repaired during the surgery and may or may not have long-term consequences. The urinary bladder, more commonly injured during vaginal than during abdominal hysterectomy, is the most frequently injured urinary tract organ [9]. Injury to the ureter is more common in abdominal hysterectomy and is more serious than bladder damage because it is frequently not identified until after surgery [9]. Hysterectomy is also associated with an intraoperative risk of bowel injury due to the close proximity of the uterus to the intestines [9].

Urinary and bowel dysfunction after hysterectomy have been reported and discussed in the literature [34,77]. It has been hypothesized that dysfunction in these areas may be related to anatomical alterations and/or disturbances in innervation caused by the surgery [64,78]. In a review of 41 articles on this issue, Thakar and colleagues [78] concluded that a relationship between hysterectomy and subsequent bladder and bowel function has yet to be established and that data from additional well-designed studies are needed to inform surgical practice.

4. Long-Term Morbidity and Risks Associated with Hysterectomy

In addition to the patient-centered outcomes discussed earlier, long-term effects associated with hysterectomy may include ovarian pathology and dysfunction [79–81], increased coronary heart disease (CHD) [82–86], and a decreased risk of ovarian cancer [87–89]. Evidence of long-term morbidity associated with hysterectomy is inconclusive. Interpretation of morbidity findings must take into account the type of surgery, route, and concurrent surgeries (such as BSO), study design (for example, whether data were collected retrospectively or prospectively, and whether control groups were used) [9], and patient factors such as age, previous diagnosis, and health status.

5. Ovarian Function and Coronary Heart Disease Risk Following Hysterectomy

As discussed earlier, BSO is commonly performed at the time of hysterectomy. Concern has been voiced regarding the long-term effects associated with the routine removal of apparently healthy ovaries to prevent cancer, a relatively rare occurrence [90]. Prophylactic oophorectomy increases women's risk for coronary heart disease by inducing surgical menopause. Conservation of at least one ovary has been recommended [79], as has the retention of both ovaries in women of any age [90], and the retention of both ovaries in women 45 years of age or younger [34,85,86,90,91].

Although it is still widely held that hysterectomy does not adversely affect the function of retained ovaries, some authors have suggested that ovarian function may deteriorate following hysterectomy [79,80]. Decreased ovarian function results in lower estradiol levels which are associated with increased risk of cardiovascular disease [82–86]. The premise that hysterectomy may result in ovarian failure and subsequent earlier menopause in some women is supported by the findings of Kritz-Silverstein and colleagues [86] indicating that women who retained their ovaries posthysterectomy were younger at time of menopause than women who had not undergone hysterectomy (mean = 46.5 versus 49.9 years). A study by Siddle and

Table 20.13

Alternatives to Hysterectomy for Benign Gynecologic Conditions

Condition	Surgical treatment	Medical treatment
Uterine leiomyoma	Myomectomy Endometrial resection	Hormone suppression
Endometriosis	Conservative surgery (removal of disease)	Hormone suppression Nonsteroidal anti-inflammatory agents
Uterine prolapse	Repair and restoration Support	Pessary Estrogen cream Kegel exercises
Dysfunctional uterine bleeding	Endometrial ablation Endometrial resection	Hormone suppression Nonsteroidal anti-inflammatory agents

colleagues [81] supported these findings. Additional research is needed to understand better the relationship between hysterectomy and subsequent ovarian function.

Some studies have reported a reduced risk of ovarian cancer following hysterectomy in women who have retained their ovaries [87–89]. Possible explanations for an inverse association include: hormonal dysfunction prior to hysterectomy which results in uterine pathology and subsequent hysterectomy, while at the same time reducing ovarian cancer risk; uterine pathology requiring treatment (such as hormones) prior to hysterectomy which decreases ovarian cancer risk; hormonal, mechanical, or circulatory changes resulting from the surgical procedure; and a "screening" effect resulting from the opportunity to visually examine the ovaries during hysterectomy [89].

XIII. Alternatives to Hysterectomy

Less invasive medical and surgical alternatives are now available for many of the nonmalignant conditions for which hysterectomy has typically been the treatment of choice [34]. Alternative treatments, specific to the indication for which hysterectomy has been performed, continue to evolve and are listed in Table 20.13. Bachmann has suggested that hysterectomy be avoided in cases without clear pathology, when the woman is a poor candidate for, or at risk from, surgery, desires to keep her uterus, or has a lifestyle that does not permit adequate time for postoperative recovery.

The availability of alternative treatments in place of hysterectomy for nonmalignant conditions such as menorrhagia, leiomyoma, and endometriosis has had a negligible effect on hysterectomy rates in the U.S. [49]. Although additional randomized trials are needed, medical and surgical treatment alternatives that allow women to keep their uterus show some promise for the treatment of menorrhagia [92,93].

There is currently very little good evidence regarding the efficacy of medical treatments for menorrhagia. Oral contraceptives and GnRH agonists can be useful in reducing bleeding for some women, but the side effects can lead women to discontinue treatment and, in the case of GnRH agonists, long-term treatment can be expensive [94–96]. Antifibrinolytics, such as tranexamic acid, used extensively in Scandinavia, have been shown to be promising treatments, however, and bear further investigation [97,98]. While not generally used for this indication in the U.S., progesterone/progestagen releasing intrauterine devices have been tested in the context of randomized trials and have been used in Europe to treat abnormal bleeding [14,99].

In randomized studies that have compared hysterectomy with endometrial ablation or resection (the removal or destruction of the endometrium or uterine lining), women have experienced high levels of satisfaction with both types of surgery. One study has indicated that the reoperation rate for resection/ablation appears to be about 20% within three years [72]. These studies are limited however, in that most followed women for insufficient time to estimate the incidence of additional surgery in the resection/ablation group, and losses to follow-up were relatively high [72,92,93,100]. Randomized studies evaluating both hysteroscopic and nonhysteroscopic endometrial ablation were in progress in 1998 [101].

There is some evidence that leiomyoma may be treated conservatively with myomectomy (surgical removal of leiomyomas) in women who wish to maintain their fertility, with hysterectomy reserved for women who no longer desire fertility [14]. Endometriosis may be treated medically with hormones or with conservative surgery (resection or ablation of endometrial implants). In some instances, however, definitive surgery may be necessary to relieve permanently pain and other symptoms associated with leiomyomas and endometriosis [14].

XIV. Implications for Practice

When hysterectomy is recommended as a possible treatment, women should be advised of what is known and not known about its effectiveness and the risks associated with the surgery. Where there are alternatives to hysterectomy for a given indication and there is reliable evidence regarding effectiveness, these alternatives should be made available. Similarities and differences among treatments should be explained to women using information obtained from randomized trials whenever possible. Evidence suggests that postoperative outcomes are enhanced by educating women about the nature of the surgery and its effects [63]. While individual patient and physician factors are important in determining the need and appropriateness of hysterectomy, today's health care system also requires consideration of other issues including insurance reimbursement, quality assurance and second opinion programs, and geographic variations in practice [2].

XV. Implications for Research

Further research is needed to inform practice regarding the effectiveness of hysterectomy, especially in terms of short- and long-term outcomes. In 1994, at the urging of Congress, the Agency for Health Care Policy and Research (AHCPR) held a conference that specifically examined hysterectomy and other treatment options for the treatment of nonmalignant uterine conditions [3]. The conference demonstrated the growing interest in hysterectomy on the part of policymakers and others, especially with regard to medical and treatment alternatives. Randomized trials have been undertaken to compare surgical variations of

hysterectomy, as well as hysterectomy and other surgical procedures for nonmalignant conditions, in terms of patient outcomes and cost. Further research is needed on patient quality of life and psychological outcomes following hysterectomy, and the effects of BSO and hormone replacement therapy should be examined in this context. Additional exploration and collection of data on factors related to race and ethnicity that may be associated with hysterectomy differences (e.g., income, education, access to care), assessment of black and other minority women's experiences with hysterectomy, and examination of increased mortality rates among black women and other minorities are needed. Also needed are efforts to determine factors affecting practice variation across national and international regions.

Because hysterectomy is associated with morbidity, mortality, and high cost, and because in nonmalignant conditions it is most often an elective procedure performed to relieve symptoms for which other treatments may exist, more conservative and potentially safer alternatives should be investigated in the context of randomized trials to provide evidence on which to base medical practice. Systematic reviews that compare hysterectomy with alternative interventions, such as those currently being performed by the Cochrane Collaboration [102,103], should be performed and continuously updated.

Acknowledgments

We thank the following individuals for their valuable assistance in the preparation of this chapter: Eric Manheimer, M.S., for review of data and assisting in preparation of the manuscript; Lisa Fredman, Ph.D., for analysis of prevalence data and critically reviewing the manuscript; Kim Mitchell, M.S., for assisting in the assembly of the literature, performing analysis of prevalence data, and critically reviewing the manuscript; Robert Pokras, M.A., for assisting with interpretation of previously published analyses and critically reviewing the manuscript; Jane Scott, Sc.D., for providing relevant literature and critically reviewing the manuscript; Angela Coulter, Ph.D., for providing reference information and current hysterectomy incidence data for England and critically reviewing the manuscript; and Marcie Richardson, M.D., for critically reviewing the manuscript. Drs. Dickersin and Lemaire were supported in part by Agency for Health Care Policy and Research grant #5 U01 HS09506, *Surgical Treatments Outcomes Project for Dysfunctional Uterine Bleeding.*

References

1. Lepine, L. A., Hillis, S. D., Marchbanks, P. A., Koonin, L. M., Morrow, B., Kieke, B. A., and Wilcox, L. S. (1997). Hysterectomy surveillance—United States, 1980–1993. *MMWR CDC Surveill. Summ.* **46,** 1–15.
2. Bachmann, G. A. (1990). Hysterectomy. A critical review. *J. Reprod. Med.* **35,** 839–862.
3. Agency for Health Care Policy and Research (1995). "Conference Summary. Treatment of Common Non-Cancerous Uterine Conditions: Issues for Research." AHCPR Publ. No. 95-0067. Agency for Health Care Policy and Research, Rockville, MD.
4. Miller, N. (1946). Hysterectomy: Therapeutic necessity or surgical racket? *Am. J. Obstet. Gynecol.* **51,** 804.
5. Bunker, J. P. (1976). Public-health rounds at the Harvard School of Public Health. Elective hysterectomy: Pro and con. *N. Engl. J. Med.* **295,** 264–268.
6. Dicker, R. C., Scally, M. J., Greenspan, J. R., Layde, P. M., Ory, H. W., Maze, J. M., and Smith, J. C. (1982). Hysterectomy among women of reproductive age. Trends in the United States, 1970–1978. *JAMA, J. Am. Med. Assoc.* **248,** 323–327.

7. Easterday, C. L., Grimes, D. A., and Riggs, J. A. (1983). Hysterectomy in the United States. *Obstet. Gynecol.* **62**, 203–212.

8. Wennberg, J. E., Bunker, J. P., and Barnes, B. (1980). The need for assessing the outcome of common medical practices. *Annu. Rev. Public Health* **1**, 277–295.

9. Bernstein, S. J., Fiske, M. E., McGlynn, E. A., and Gifford, D. S. (1997). "Hysterectomy: A Review of the Literature on Indications, Effectiveness, and Risks." Rand Corporation, Santa Monica, CA.

10. Kasper, A. S. (1985). Hysterectomy as social process. *Women's Health* **10**, 109–127.

11. Dicker, R. C., Greenspan, J. R., Strauss, L. T., Cowart, M. R., Scally, M. J., Peterson, H. B., DeStefano, F., Rubin, G. L., and Ory, H. W. (1982). Complications of abdominal and vaginal hysterectomy among women of reproductive age in the United States. The Collaborative Review of Sterilization. *Am. J. Obstet. Gynecol.* **144**, 841–848.

12. Bernstein, S. J., McGlynn, E. A., Siu, A. L., Roth, C. P., Sherwood, M. J., Keesey, J. W., Kosecoff, J., Hicks, N. R., and Brook, R. H. (1993). The appropriateness of hysterectomy. A comparison of care in seven health plans. Health Maintenance Organization Quality of Care Consortium. *JAMA, J. Am. Med. Assoc.* **269**, 2398–2402.

13. Leonardo, R. A. (1994). "History of Gynecology." Froben Press, New York.

14. Stovall, T. (1996). Hysterectomy. *In* "Novak's Gynecology" (J. Berek, ed.), 12th ed., pp. 727–767. Williams & Wilkins, Baltimore, MD.

15. Benrubi, G. I. (1988). History of hysterectomy. *J. Fla. Med. Assoc.* **75**, 533–538.

16. McDonald, T. W. (1993). Hysterectomy-indications, types and alternatives. *In* "Textbook of Gynecology" (I. J. Copeland, ed.), pp. 779–801. Saunders, Philadelphia.

17. Emge, L. A., and Durfee, R. B. (1966). Pelvic organ prolapse: Four thousand years of treatment. *Clin. Obstet. Gynecol.* **9**, 997–1032.

18. Ricci J. V. (1945). "One Hundred Years of Gynecology—1800–1900." Blakiston, Philadelphia.

19. Graves, E. J. (1995). National Hospital Discharge Survey: Annual summary, 1993. *Vital Health Stat.* **13**, 1–63.

20. Graves, E. J., and Owings, M. F. (1997). "1995 Summary: National Hospital Discharge Survey. Advance Data From Vital and Health Statistics," No. 291. National Center for Health Statistics, Hyattsville, MD.

21. Elixhauser, A., Johantgen, M., and Andrews, R. (1997). "Descriptive Statistics by Insurance Status for Most Frequent Hospital Diagnoses and Procedures," AHCPR Publ. No. 97-0009, Healthcare Cost and Utilization Project (HCUP-3), Res. Note 5. Agency for Health Care Policy and Research, Rockville, MD.

22. National Center for Health Statistics (1997). "National Health Interview Survey." National Center for Health Statistics, Centers for Disease Control and Prevention, Hyattsville, MD.

23. Brett, K. M., Marsh, J. V., and Madans, J. H. (1997). Epidemiology of hysterectomy in the United States: Demographic and reproductive factors in a nationally representative sample. *J. Women's Health* **6**, 309–316.

24. National Center for Health Statistics (1996). "Third National Health and Nutrition Examination Survey, 1988–1994" [CD-ROM]. National Center for Health Statistics, Centers for Disease Control and Prevention, Hyattsville, MD.

25. Powell-Griner, E., Anderson, J. E., and Murphy, W. (1997). State- and sex-specific prevalence of selected characteristics—behavioral risk factor surveillance system, 1994 and 1995. *MMWR CDC Surveill. Summ.* **46**, 1–31.

26. National Center for Chronic Disease Prevention and Health Promotion "Behavioral Risk Factor Surveillance System, Survey Data 1984–1995" [CD-ROM]. National Center for Chronic Disease Prevention and Health Promotion, Centers for Disease Control and Prevention, Atlanta, GA.

27. Pokras, R., and Hufnagel, V. G. (1988). Hysterectomy in the United States, 1965–84. *Am. J. Public Health* **78**, 852–853.

28. Pokras, R. (1989). Hysterectomy: Past, present and future. *Stat. Bull. Metrop. Insur. Co.* **70**, 12–21.

29. Pokras, R., and Hufnagel, V. G. (1987). Hysterectomies in the United States. *Vital Health Stat.* **13**, 1–32.

30. Gambone, J. C., and Reiter, R. C. (1990). Nonsurgical management of chronic pelvic pain: A multidisciplinary approach. *Clin. Obstet. Gynecol.* **33**, 205–211.

31. Finkel, M. L., and Finkel, D. J. (1990). The effect of a second opinion program on hysterectomy performance. *Med. Care* **28**, 776–783.

32. McCarthy, E. G., and Finkel, M. L. (1980). Second consultant opinion for elective gynecologic surgery. *Obstet. Gynecol.* **56**, 403–410.

33. Dyck, F. J., Murphy, F. A., Murphy, J. K., Road, D. A., Boyd, M. S., Osborne, E., De Vlieger, D., Korchinski, B., Ripley, C., Bromley, A. T., and Innes, P. B. (1977). Effect of surveillance on the number of hysterectomies in the province of Saskatchewan. *N. Engl. J. Med.* **296**, 1326–1328.

34. Carlson, K. J., Nichols, D. H., and Schiff, I. (1993). Indications for hysterectomy. *N. Engl. J. Med.* **328**, 856–860.

35. Roos, N. P. (1984). Hysterectomy: Variations in rates across small areas and across physicians' practices. *Am. J. Public Health* **74**, 327–335.

36. Walker, A. M., and Jick, H. (1979). Temporal and regional variation in hysterectomy rates in the United States, 1970–1975. *Am. J. Epidemiol.* **110**, 41–46.

37. Wennberg, J. E., Barnes, B. A., and Zubkoff, M. (1982). Professional uncertainty and the problem of supplier-induced demand. *Soc. Sci. Med.* **16**, 811–824.

38. Bernstein, S. J., McGlynn, E. A., Fiske, M. E., Kamberg, C. J., Siu, A. L., and Brook, R. H. (1993). "Hysterectomy: A Review of the Literature Regarding Risks and Complications." Rand Corporation, Santa Monica, CA.

39. Coulter, A., McPherson, K., and Vessey, M. (1988). Do British women undergo too many or too few hysterectomies? *Soc. Sci. Med.* **27**, 987–994.

40. Hagenfeldt, K., Brorsson, B., and Bernstein, S. J. (1995). "Hysterectomy. Ratings of Appropriateness," SBU-Rep. No. 125E. Swedish Council on Technology Assessment in Health Care, Stockholm.

41. Schofield, M. J., Hennrikus, D. J., Redman, S., and Sanson-Fisher, R. W. (1991). Prevalence and characteristics of women who have had a hysterectomy in a community survey. *Aust. N. Z. J. Obstet. Gynaecol.* **31**, 153–158.

42. Hall, R. E., and Cohen, M. M. (1994). Variations in hysterectomy rates in Ontario: Does the indication matter? *Can. Med. Assoc. J.* **151**, 1713–1719.

43. Pokras, R. (1997). Hysterectomy rates. Unpublished report of NHDS Data.

43a. Yusuf, F., and Siedlecky, S. (1997). Hysterectomy and endometrial ablation in New South Wales, 1981 to 1994–1995. *Aust. N. Z. J. Obstet. Gynaecol.* **37**, 210–216.

43b. Government Statistical Service (1996). "Hospital Episode Statistics (1994–1995)." Government Statistical Service, London Department of Health, London.

44. Gentile, G. P., Kaufman, S. C., and Helbig, D. W. (1998). Is there any evidence for a post-tubal sterilization syndrome? *Fertil. Steril.* **69**, 179–186.

45. Kjerulff, K. H., Guzinski, G. M., Langenberg, P., and Stolley, P. D. (1997). Effectiveness and outcomes of hysterectomy. Unpublished report.

46. Goldhaber, M. K., Armstrong, M. A., Golditch, I. M., Sheehe, P. R., Petitti, D. B., and Friedman, G. D. (1993). Long-term risk of hysterectomy among 80,007 sterilized and comparison women at Kaiser Permanente, 1971–1987. *Am. J. Epidemiol.* **138,** 508–521.

47. Reich, H., DeCaprio, J., and McGlynn, F. (1989). Laparoscopic hysterectomy. *J. Gynecol. Surg.* **5,** 213–216.

48. American College of Obstetricians and Gynecologists (1994). "Quality Assessment and Improvement in Obstetrics and Gynecology." American College of Obstetricians and Gynecologists, Washington, DC.

49. Wilcox, L. S., Koonin, L. M., Pokras, R., Strauss, L. T., Xia, Z., and Peterson, H. B. (1994). Hysterectomy in the United States, 1988–1990. *Obstet. Gynecol.* **83,** 549–555.

50. Kjerulff, K. H., Guzinski, G. M., Langenberg, P. W., Stolley, P. D., and Kazandjian, V. A. (1992). Hysterectomy: An examination of a common surgical procedure. *J. Women's Health* **1,** 141–147.

51. Weber, A. M., and Lee, J. C. (1996). Use of alternative techniques of hysterectomy in Ohio, 1988–1994. *N. Engl. J. Med.* **335,** 483–489.

52. Brumsted, J. R., Blackman, J. A., Badger, G. J., and Riddick, D. H. (1996). Hysteroscopy versus hysterectomy for the treatment of abnormal uterine bleeding: A comparison of cost. *Fertil. Steril.* **65,** 310–316.

53. Harris, M. B., and Olive, D. L. (1994). Changing hysterectomy patterns after introduction of laparoscopically assisted vaginal hysterectomy. *Am. J. Obstet. Gynecol.* **171,** 340–343.

54. Summitt, R. L. Jr, Stovall, T. G., Lipscomb, G. H., and Ling, F. W. (1992). Randomized comparison of laparoscopy-assisted vaginal hysterectomy with standard vaginal hysterectomy in an outpatient setting. *Obstet. Gynecol.* **80,** 895–901.

55. Nezhat, C., Bess, O., Admon, D., Nezhat, C. H., and Nezhat, F. (1994). Hospital cost comparison between abdominal, vaginal, and laparoscopy-assisted vaginal hysterectomies. *Obstet. Gynecol.* **83,** 713–716.

56. Dorsey, J. H., Holtz, P. M., Griffiths, R. I., McGrath, M. M., and Steinberg, E. P. (1996). Costs and charges associated with three alternative techniques of hysterectomy. *N. Engl. J. Med.* **335,** 476–482.

57. Yuen, P. M., and Rogers, M. S. (1996). Is laparoscopically-assisted vaginal hysterectomy associated with low operative morbidity? *Aust. N. Z. J. Obstet. Gynaecol.* **36,** 39–43.

58. Schofield, M. J., Bennett, A., Redman, S., Walters, W. A., and Sanson-Fisher, R. W. (1991). Self-reported long-term outcomes of hysterectomy. *Br. J. Obstet. Gynaecol.* **98,** 1129–1136.

59. Carlson, K. J., Miller, B. A., and Fowler, F. J., Jr. (1994). The Maine Women's Health Study: I. Outcomes of hysterectomy. *Obstet. Gynecol.* **83,** 556–565.

60. Kjerulff, K. H., Langenberg, P., Seidman, J. D., Stolley, P. D., and Guzinski, G. M. (1996). Uterine leiomyomas. Racial differences in severity, symptoms and age at diagnosis. *J. Reprod. Med.* **41,** 483–490.

61. Stockman, A. F. (1995). Gynecologic surgery. *In* "Psychological Aspects of Women's Reproductive Health" (M. W. O'Hara, R. C. Reiter, S. R. Johnson, A. Milburn, and J. Engeldinger, eds.), pp. 81–95. Springer, New York.

62. Gath, D., Cooper, P., and Day, A. (1982). Hysterectomy and psychiatric disorder: I. Levels of psychiatric morbidity before and after hysterectomy. *Br. J. Psychiatry* **140,** 335–342.

63. Harris, W. J. (1997). Complications of hysterectomy. *Clin. Obstet. Gynecol.* **40,** 928–938.

64. Helstrom, L., Lundberg, P. O., Sorbom, D., and Backstrom, T. (1993). Sexuality after hysterectomy: A factor analysis of women's sexual lives before and after subtotal hysterectomy. *Obstet. Gynecol.* **81,** 357–362.

65. Clarke, A., Black, N., Rowe, P., Mott, S., and Howle, K. (1995). Indications for and outcome of total abdominal hysterectomy for benign disease: A prospective cohort study. *Br. J. Obstet. Gynaecol.* **102,** 611–620.

66. Virtanen, H. S., and Makinen, J. I. (1995). Mortality after gynaecologic operations in Finland, 1986–1991. *Br. J. Obstet. Gynaecol.* **102,** 54–57.

67. Loft, A., Andersen, T. F., Bronnum-Hansen, H., Roepstorff, C., and Madsen, M. (1991). Early postoperative mortality following hysterectomy. A Danish population based study, 1977–1981. *Br. J. Obstet. Gynaecol.* **98,** 147–154.

68. Wingo, P. A., Huezo, C. M., Rubin, G. L., Ory, H. W., and Peterson, H. B. (1985). The mortality risk associated with hysterectomy. *Am. J. Obstet. Gynecol.* **152,** 803–808.

69. Kjerulff, K. H., Guzinski, G. M., Langenberg, P. W., Stolley, P. D., Moye, N. E., and Kazandjian, V. A. (1993). Hysterectomy and race. *Obstet. Gynecol.* **82,** 757–764.

70. Kjerulff, K., Langenberg, P., and Guzinski, G. (1993). The socioeconomic correlates of hysterectomies in the United States. *Am. J. Public Health* **83,** 106–108.

71. Boyd, M. E., and Groome, P. A. (1993). The morbidity of abdominal hysterectomy. *Can. J. Surg.* **36,** 155–159.

72. O'Connor, H., Broadbent, J. A., Magos, A. L., and McPherson, K. (1997). Medical Research Council randomised trial of endometrial resection versus hysterectomy in management of menorrhagia. *Lancet* **349,** 897–901.

73. Cooper, M. J., Cario, G., Lam, A., Carlton, M., Vaughan, G., and Hammill, P. (1996). Complications of 174 laparoscopic hysterectomies. *Aust. N. Z. J. Obstet. Gynaecol.* **36,** 36–38.

74. Kovac, S. R. (1995). Guidelines to determine the route of hysterectomy. *Obstet. Gynecol.* **85,** 18–23.

75. Pinion, S. B., Parkin, D. E., Abramovich, D. R., Naji, A., Alexander, D. A., Russell, I. T., and Kitchener, H. C. (1994). Randomised trial of hysterectomy, endometrial laser ablation, and transcervical endometrial resection for dysfunctional uterine bleeding. *Br. Med. J.* **309,** 979–983.

76. Dwyer, N., Hutton, J., and Stirrat, G. M. (1993). Randomised controlled trial comparing endometrial resection with abdominal hysterectomy for the surgical treatment of menorrhagia. *Br. J. Obstet. Gynaecol.* **100,** 237–243.

77. Nathorst-Boos, J., Fuchs, T., and von Schoultz, B. (1992). Consumer's attitude to hysterectomy. The experience of 678 women. *Acta Obstet. Gynecol. Scand.* **71,** 230–234.

78. Thakar, R., Manyonda, I., Stanton, S. L., Clarkson, P., and Robinson, G. (1997). Bladder, bowel and sexual function after hysterectomy for benign conditions. *Br. J. Obstet. Gynaecol.* **104,** 983–987.

79. Beksac, M. S., Kisnisci, H. A., Cakar, A. N., and Beksac, M. (1983). The endocrinological evaluation of bilateral and unilateral oophorectomy in premenopausal women. *Int. J. Fertil.* **28,** 219–224.

80. Machin, D., and Williams, J. D. (1988). The effect of hysterectomy on the age at ovarian failure. *Fertil. Steril.* **49,** 378–380.

81. Siddle, N., Sarrel, P., and Whitehead, M. (1987). The effect of hysterectomy on the age at ovarian failure: identification of a subgroup of women with premature loss of ovarian function and literature review. *Fertil. Steril.* **47,** 94–100.

82. Centerwall, B. S. (1981). Premenopausal hysterectomy and cardiovascular disease. *Am. J. Obstet. Gynecol.* **139,** 58–61.

83. Speroff, T., Dawson, N. V., Speroff, L., and Haber, R. J. (1991). A risk-benefit analysis of elective bilateral oophorectomy: Effect of changes in compliance with estrogen therapy on outcome. *Am. J. Obstet. Gynecol.* **164,** 165–174.

84. Rosenberg, L., Hennekens, C. H., Rosner, B., Bélanger, C., Rothman, K. J., and Speizer, F. E. (1981). Early menopause and the risk of myocardial infarction. *Am. J. Obstet. Gynecol.* **139,** 47–51.

85. Stoney, C. M., Owens, J. F., Guzick, D. S., and Matthews, K. A. (1997). A natural experiment on the effects of ovarian hormones on cardiovascular risk factors and stress reactivity: Bilateral salpingo oophorectomy versus hysterectomy only. *Health Psychol.* **16,** 349–358.

86. Kritz-Silverstein, D., Barrett-Connor, E., and Wingard, D. L. (1997). Hysterectomy, oophorectomy, and heart disease risk factors in older women. *Am. J. Public Health* **87,** 676–680.

87. Irwin, K. L., Weiss, N. S., Lee, N. C., and Peterson, H. B. (1991). Tubal sterilization, hysterectomy, and the subsequent occurrence of epithelial ovarian cancer. *Am. J. Epidemiol.* **134,** 362–369.

88. Parazzini, F., Negri, E., La Vecchia, C., Luchini, L., and Mezzopane, R. (1993). Hysterectomy, oophorectomy, and subsequent ovarian cancer risk. *Obstet. Gynecol.* **81,** 363–366.

89. Weiss, N. S., and Harlow, B. L. (1986). Why does hysterectomy without bilateral oophorectomy influence the subsequent incidence of ovarian cancer? *Am. J. Epidemiol.* **124,** 856–858.

90. Bukovsky, I., Halperin, R., Schneider, D., Golan, A., Hertzianu, I., and Herman, A. (1995). Ovarian function following abdominal hysterectomy with and without unilateral oophorectomy. *Eur. J. Obstet. Gynecol. Reprod. Biol.* **58,** 29–32.

91. Everson, S. A., Matthews, K. A., Guzick, D. S., Wing, R. R., and Kuller, L. H. (1995). Effects of surgical menopause on psychological characteristics and lipid levels: The Healthy Women Study. *Health Psychol.* **14,** 435–443.

92. Alexander, D. A., Naji, A. A., Pinion, S. B., Mollison, J., Kitchener, H. C., Parkin, D. E., Abramovich, D. R., and Russell, I. T. (1996). Randomised trial comparing hysterectomy with endometrial ablation for dysfunctional uterine bleeding: Psychiatric and psychosocial aspects. *Br. Med. J.* **312,** 280–284.

93. Gannon, M. J., Holt, E. M., Fairbank, J., Fitzgerald, M., Milne, M. A., Crystal, A. M., and Greenhalf, J. O. (1991). A randomised trial comparing endometrial resection and abdominal hysterectomy for the treatment of menorrhagia. *Br. Med. J.* **303,** 1362–1364.

94. Sculpher, M. J., Bryan, S., Dwyer, N., Hutton, J., and Stirrat, G. M. (1993). An economic evaluation of transcervical endometrial resection versus abdominal hysterectomy for the treatment of menorrhagia. *Br. J. Obstet. Gynaecol.* **100,** 244–252.

95. Noble, A. D. (1985). Management of menorrhagia. *Br. Med. J.* **291,** 296–297.

96. Crosignani, P. G., Vercellini, P., Mosconi, P., Oldani, S., Cortesi, I., and De Giorgi, O. (1997). Levonorgestrel-releasing intrauterine device versus hysteroscopic endometrial resection in the treatment of dysfunctional uterine bleeding. *Obstet. Gynecol.* **90,** 257–263.

97. Bonnar, J., and Sheppard, B. (1996). Treatment of menorrhagia during menstruation: Randomised controlled trial of ethamsylate, mefenamic acid, and tranexamic acid. *Br. Med. J.* **313,** 579–582.

98. Preston, J. T., Cameron, I. T., Adams, E. J., and Smith, S. K. (1995). Comparative study of tranexamic acid and norethisterone in the treatment of ovulatory menorrhagia. *Br. J. Obstet. Gynaecol.* **102,** 401–406.

99. Cooke, I. and Rees, M. (1998). Progesterone/progestagen releasing IUCDs vs either placebo or any other medication for heavy menstrual bleeding (Cochrane Review). *In* "The Cochrane Library, Issue 3." Update Software, Oxford.

100. Sculpher, M. J., Dwyer, N., Byford, S., and Stirrat, G. M. (1996). Randomised trial comparing hysterectomy and transcervical endometrial resection: Effect on health related quality of life and costs two years after surgery. *Br. J. Obstet. Gynaecol.* **103,** 142–149.

101. Weber, A. M., and Munro, M. G. (1988). Endometrial ablation versus hysterectomy: STOP-DUB. *Medscape Women's Health* **3,** 1–6.

102. Dickersin, K., and Manheimer, E. (1998). The Cochrane Collaboration: Evaluation of health care and services using systematic reviews of the results of randomized controlled trials. *Clin. Obstet. Gynecol.* **41,** 315–331.

103. The Cochrane Library [database on disk and CD-ROM] (1999). "The Cochrane Collaboration." Update Software, Oxford (updated quarterly).

Section 4

SEXUALLY TRANSMITTED DISEASES

Nancy S. Padian

Center for Reproductive Health Research and Policy,
Department of Obstetrics, Gynecology and Reproductive Sciences,
University of California at San Francisco
San Francisco, California

In the United States, from 1973 through 1992, more than 150,000 women in the United States died of causes related to sexually transmitted diseases (STDs) (including human immunodeficiency virus [HIV]) [1]. In developing countries, STDs are the second largest health burden (after maternal causes) for women as measured by disability-adjusted life years [2]. Among all women, adolescents are particularly at risk [3]. For example, in the U.S., while overall rates of gonorrhea have declined in the general population, in the first half of the 1990s, rates increased among female adolescents of all races [3]. Similar patterns have been observed for other STDs. Such trends may in part be attributed to the fact that compared to men, women have increased susceptibility to STDs and HIV, a larger proportion of asymptomatic infections resulting in delayed health care-seeking behavior, and more frequent and severe sequelae associated with initial infection.

I. Increased Susceptibility to STD Infections among Women

Throughout the world, women bear a rate and burden of STDs that is disproportionate compared to men. Much of this overrepresentation can be attributed to increased biological and behavioral susceptibility for many sexually transmitted infections [4,5]. Biologically, young women have an increased proportion of columnar epithelium lining their cervix that is particularly susceptible to infection from chlamydia and gonorrhea. With maturity, columnar epithelium is replaced by squamous epithelium that is less penetrable to infection. Other biological factors include vaginal ecology and the presence of resident flora and concomitant pH [6,7]. As females progress through adolescence to adulthood, developmental changes also involve the replacement of endogenous vaginal flora, predominantly characterized by enteric organisms and a basic pH, to an environment predominated by lactobacilli and an acidic environment that is hostile to many STD pathogens [8]. Another change observed post-adolescence is an increased thickening of cervical mucus that is less permeable to STD pathogens and may also harbor immunological defenses such as antibodies and white blood cells [9–11]. Together, these factors help explain the increased risk for STD acquisition among young women.

Permutations of these developmental changes may be seen in adult women. For example, although the vagina is normally acidic, factors such as bacterial vaginosis may alter a healthy vaginal ecosystem and cause a reduction in the kinds of lactobacilli that produce hydrogen peroxide, resulting in a net increase in pH, an environment more favorable to STD-associated pathogens [12,13]. Gender differences in anatomy represent another biological risk. During intercourse, the increased exposure time of semen in the vaginal vault as compared to the contact of female secretions with the male genital tract is a simple contributor that also might account for increased susceptibility of STDs among women compared to men. In addition, although adult women have a thickened cervical mucus lining, the mucus plug at the cervix, which protects the upper genital tract, becomes more permeable during menstruation and may account for high rates of upper genital tract infection during the first week of menses [14,15]. Finally, pregnancy can also alter susceptibility

269

to STDs and has been associated with progression of human papilloma virus (HPV) and more frequent and severe recurrence of genital herpes [16–19].

Many of the behavioral factors that increase women's susceptibility to infection can be attributed to a gender imbalance; compared to women, men make more decisions about sexual activity and whether or not to use protection [5]. This is particularly critical with regard to condom use, which offers protection against acquisition of many STDs. Women may be able to negotiate their use, but whether they are actually used or not ultimately depends on the man [20]. This problem may be further magnified in certain societies where passivity and subordination are considered the norm for women [21], making it even more difficult for women to assert their needs.

Socioeconomic and cultural factors may also increase a woman's behavioral risk. In addition to their own health needs, women often have to balance the needs of their family, including health care for other family members as well as basic needs such as shelter, food, and care [22]. Finally, issues such as sexual violence as well as age and economic disparities resulting from the pairing of younger women with older men may further increase the chances of involuntary unprotected intercourse [23,24].

II. Asymptomatic STD Infections among Women

In addition to increased susceptibility, women are more likely to be asymptomatically infected with an STD compared to men. For example, 30–80% of women infected with gonorrhea may be asymptomatic compared to approximately 5% of men [25]. Likewise, as many as 85% of women with chlamydial infection may be asymptomatic compared to 40% of infected men [26]. When STD symptoms do become manifest, among women they are less likely to be attributed to STDs and may be ignored [27] (e.g., a discharge may be considered a normal occurrence). Both aysmptomatic and undiagnosed infections are associated with reduced health-care seeking behavior [4,27]. For example, a delay in seeking medical care after an initial cervical chlamydial infection is a clearly established risk of pelvic inflammatory disease (PID) [28]. Because the symptoms of initial chlamydial infection may be nonexistent, women may not seek medical care until infection has spread to the upper genital tract resulting in PID. Likewise, unattended HPV infection may be associated with higher rates of cervical cancer.

III. Gender-Associated Sequelae of STDs

A delay in seeking health care after initial infection has been associated with severe long-term outcomes. Women infected with gonorrhea or chlamydia are at risk for upper genital tract infection and PID. If left untreated or inadequately treated, upper tract infection can lead to infertility or ectopic pregnancy [29,30]. Certain types of HPV are associated not only with the development of cervical cancer, which in the U.S. ranks seventh of all cancers in women [31], but also with cancer of the vulva, vagina, and anus [32]. STDs during pregnancy present additional risks both for the mother and the infant, including preterm delivery, premature membrane rupture, puerperal sepsis and postpartum infection infertility, low birth rate, and other adverse outcomes for the neonate [33].

IV. Section Overview

In this section we explore the epidemiology and clinical manifestations (including diagnosis and treatment options) for the most common STDs that affect women. In Chapter 21, Rompalo and Burstein discuss one of the bacterial STDs causing cervicitis, *Chlamydia trachomatis,* and Hynes and Rompalo discuss the other most common bacterial STD, *Neisseria gonorrhea,* in Chapter 22. Marrazo and Celum review current knowledge about syphilis (Chapter 23) which, although easily treatable, has experienced recent outbreaks in the United States, Eastern Europe, Africa, and Asia. Wald and Brown review the epidemiology, natural history, and public health significance of one of the most prevalent viral STDs, genital herpes, (Chapter 24), while Jay and Moscicki discuss another viral pathogen, Human Papilloma Virus (Chapter 25), which is arguably the most ubiquitous STD and can result in STD-associated genital cancer, cancer of the cervix.

In some respects, HIV represents a common denominator because it interacts with a range of other STDs. Infection with chlamydia, gonorrhea, and trichomoniasis and other STDs almost certainly increases susceptibility to HIV, while infection with HIV likely increases the probability of complications stemming from genital herpes and HPV. Kamb and Wortley review current knowledge with regard to HIV and AIDS among women primarily in the United States, but consider the global context of disease burden as well (Chapter 26).

In Chapter 27, Schwebke reviews the three most common vaginal infections: vulvovaginal candidiaisis, trichomoniasis, and bacterial vaginosis. Of these, only trichomoniasis has been proven to be sexually transmitted, yet both other outcomes have been associated with sexual activity. In particular, bacterial vaginosis, sexually transmitted or not, has been associated with increased risk for other STDs including HIV. Foxman also considers urinary tract infection (UTI) (Chapter 28). Here too, while perhaps not classically considered a STD, UTI has been associated with sexual activity (particularly duration of the most recent sexual partnership) as well as with method of contraception (oral contraceptives, diaphragms, and cervical caps seem to increase risk).

PID and its attendant sequelae, including infertility and ectopic pregnancy, is one of the most frequently observed long-term sequelae associated with initial infection from bacterial STDs, and is reviewed in Chapter 29. Finally, noting the interaction between fertility and STDs, and the fact that both pregnancy and infection result from the same exposure, in Chapter 30 Cates and Padian discuss the contraceptive "trade-off" between contraceptive choices. In the absence of a perfect method, women must choose from among methods that prevent either pregnancy or STDs, but not both. Methods such as the oral contraceptive pill and sterilization that may be highly effective in preventing pregnancy but have little effect in protecting against STD acquisition, or, in the case of hormonal contraception, may actually increase risk. Similarly, methods such as the male condom may be highly effective against bacterial STDs and HIV, but less so for pregnancy.

V. New Directions

Sexually transmitted diseases are indeed the hidden epidemic [1], but more so for women than men because women have more infection and more severe consequences. New directions include quick, noninvasive diagnostic techniques that rely on more sensitive nucleic acid amplification techniques from urine specimens [34]. However, these tests are expensive and results are not immediate. The need for rapid, inexpensive diagnostic tests that can be used in field-based settings remains paramount to increasing accessibility to diagnosis and treatment [35]. Likewise, there is an urgent need for development of female-controlled topical microbicides (which do not require complicity from men) that are effective against STDs and HIV but that do not affect fertility [5,20], enabling women to protect themselves against infection while retaining control over fertility decisions. Finally, integrating health care services so that primary health care, family planning, and STD diagnosis and treatment are all offered under the same roof should increase access and use of STD services and facilitate appropriate control and care [35].

References

1. Institute of Medicine (1997). "The Hidden Epidemic: Confronting Sexually Transmitted Diseases." National Academy Press, Washington, DC.

2. World Bank (1993). "World Development Report, 1993: Investing in Health." Oxford University Press, New York.

3. CDC, DSTDP (Division of STD Prevention) (1995). "Sexually Transmitted Disease Surveillance 1994." U.S. Department of Health and Human Services, Public Health Service, Centers for Disease Control and Prevention, Atlanta, GA.

4. Cates, W., Jr. (1990). Epidemiology and control of sexually transmitted diseases in adolescents. *In* "AIDS and Other Sexually Transmitted Diseases" (M. Schydlower and M. A. Shafer, eds.), pp. 409–427. Hardy and Belfus, Philadelphia.

5. Elias, C. J., and Heise, L. (1994). Challenges for the development of female-controlled vaginal microbicides. *AIDS* **8**, 1–9.

6. Cruickshank, R. *et al.* (1934). The biology of the vagina in the human subject. II. The bacterial flora and secretion of the vagina at various age periods and their relation to glycogen in the vaginal epithelium. *Obstet. Gynaecol. Br. Emp.* **41**, 208–226.

7. Hammerschlag, M. R. *et al.* (1978). Microbiology of the vagina in children: Normal and potentially pathogenic organisms. *Pediatrics* **62**, 57–62.

8. Klebanoff, S. J., and Coombs, R. W. (1991). Virucidal effect of lactobacillus acidophilus on human immunodeficiency virus type 1: Possible role in heterosexual transmission. *J. Exp. Med.* **174** (7), 289–292.

9. Platz-Christensen, J. J. *et al.* (1993). Endotoxin and interleukin-1 alpha in the cervical mucus and vaginal fluid of pregnant women with bacterial vaginosis. *Am. J. Obstet. Gynecol.* **169**(5), 1161–1166.

10. Madile, B. M. (1976). The cervical epithelium from fetal age to adolescence. *Obstet. Gynecol.* **47**, 536–539.

11. Vickery, H. E. *et al.* (1968). The cervix and its secretion in mammals. *Physiol. Rev.* **48**, 135–154.

12. Klebanoff, S. J., Hillier, S. L., Eschenbach, D. A., and Waltersdorph, A. M. (1991). Control of the microbial flora of the vagina by H202-generating lactobacilli. *J. Infect. Dis.* **164**(7), 94–100.

13. Hillier, S. (1998). The vaginal microbial ecosystem and resistance to HIV. *AIDS Res. Hum. Retroviruses* **14**(1), S17–S21.

14. Eschenbach, D. A. *et al.* (1977). Orthogenesis of acute pelvic inflammatory disease: Role of contraception and other risk factors. *Am. J. Obstet. Gynecol.* **128**, 838–850.

15. Sweet, R. *et al.* (1986). The occurrence of chalmydial and gonococcal salpingitis during the menstrual cycle. *JAMA, J. Am. Med. Assoc.* **255**, 2062–6064.

16. Rando, R. F. *et al.* (1989). Increased frequency of detection of human papillomavirus deoxyribonucleic acid in exfoliated cervical cells during pregnancy. *Am. J. Obstet. Gynecol.* **161**, 50–55.

17. Schneider, A. *et al.* (1987). Increased prevalence of human papillomavirus in the lower genital tract of pregnant women. *Int. J. Cancer* **40**, 198–201.

18. Brown, Z. A. *et al.* (1985). Genital herpes in pregnancy: Risk factors associated with recurrence and asymptomatic shedding. *Am. J. Obstet. Gynecol.* **153**, 24–30.

19. Vontver, L. A. *et al.* (1982). Recurrent genital herpes simplex virus infection in pregnancy: Infant outcome and frequency as asymptomatic recurrences, *Am. J. Obstet. Gynecol.* **143**, 75–84.

20. Rosenberg, M. J., and Gollub, E. (1992). Commentary: Methods women can use that may prevent sexually transmitted diseases including HIV. *Am. J. Public Health* **82**, 1473–1478.

21. Amaro, H. (1988). Considerations for prevention of HIV infection among Hispanic women. *Psychol. Women Q.* **12**, 429–433.

22. Mays, V. M. and Cochran, S. D. (1988). Issues in the perception of AIDS risk and risk reduction activities by black and Hispanic/Latina women. *Am. Psychol.* **43**, 949–987.

23. Finkelhor, D. *et al.* (1986). "A Sourcebook on Child Sexual Abuse." Sage Publ., Beverly Hills, CA.

24. O'Leary, A., and Jemmott, L. S. (1995). Future directions. *In* "Women at Risk: Issues in the Primary Prevention of AIDS" (A. O'Leary and L. S. Jemmott, eds.), pp. 257–259. Plenum, New York.

25. Hook, E. W., and Handsfield, H. H. (1990). Gonococcal infections in the adult. *In* "Sexually Transmitted Diseases" (K. K. Holmes, P.-A. Mardh, P. F. Sparling, P. J. Weisner, W. Cates, and S. Lemon, eds.), 2nd ed., pp. 149–65. McGraw-Hill, New York.

26. Fish, A. N., Fairweather, D. V., Oriel, J. S., and Ridgeway, G. L. (1989). Chlamydia trachomatis infection in a gynecology clinic population: Identification of high-risk groups and the value of contact tracing. *Eur. J. Obstet. Gynecol. Reprod. Biol.* **31**, 67–74.

27. Aral, S. O. and Guinan, M. E. (1984). Women and sexually transmitted diseases. *In* "Sexually Transmitted Diseases" (K. K. Holmes, P.-A. Mardh, F. Sparling, and P. Weisner, eds.), 1st ed., pp. 85–89. McGraw-Hill, New York.

28. Hillis, D., Joesoef, R., Marchbanks, P. *et al.* (1993). Delayed care of pelvic inflammatory disease as a risk factor for impaired fertility. *Am. J. Obstet. Gynecol.* **168**, 1503–1509.

29. Westrom, L. (1980). Incidence, prevalence and trends of acute pelvic inflammatory disease and its consequences in industrialized countries. *Am. J. Obstet. Gynecol.* **138**, 880.

30. Platt, R. *et al.* (1983). Risk of acquiring gonorrhea and prevalence of abnormal adnexal finding among women recently exposed to gonorrhea. *JAMA, J. Am. Med. Assoc.* **150**, 3205–3209.

31. Koutsky, L. *et al.* (1999). Genital HPV infection. *In* "Sexually Transmitted Diseases" (K. K. Holmes, P. F. Sparling, P. Mardh, *et al.*, eds.), 3rd edi., pp. 117–127. McGraw Hill, New York.

32. American Cancer Society (1996). "Cancer Facts and Figures—1996." American Cancer Society, Atlanta, GA.

33. Brunham, R. C., Holmes, K. and Embree, J. (1990). Sexually transmitted diseases in pregnancy. *In* "Sexually Transmitted Diseases" (K. K. Holmes, P.-A. Mardh, P. F. Sparling, P. J. Weisner, W. Cates, and S. Lemon, eds.), 2nd ed., pp. 771–801. McGraw-Hill, New York.

34. Ehrhardt, A., Bolan, G., and Wasserheit, J. (1999). Gender perspectives and STDs. *In* "Sexually Transmitted Diseases" (K. K. Holmes, P. F. Sparling, P.-A., Mardh, eds.), 3rd ed., pp. 117–127. McGraw Hill, New York.

35. McDermott, J., Bangser, M., Ngugi, E., and Sandvold, I. (1993). Infection: Social and medical realities. *In* "The Health of Women: A Global Perspective" (M. Koblinsky, J. Timyan, and J. Gay, eds.), pp. 91–105. Westview Press, Boulder, CO.

21

Chlamydia

GALE BURSTEIN AND ANNE ROMPALO
Division of Infectious Disease
Johns Hopkins School of Medicine
Baltimore, Maryland

I. Introduction

Chlamydia trachomatis, originally regarded as innocuous in the early 1970s, is now considered the most common preventable cause of tubal factor infertility [1–5]. *C. trachomatis* genital infection can result in a variety of clinical syndromes but is usually asymptomatic. Most cervical infections are identified by screening both symptomatic and asymptomatic women [6–8]. Sexually active adolescent women are at highest risk for chlamydia infection and for repeated infections, which further increase their risk of sequelae [6,9–16]. Since 1993, the Centers for Disease Control and Prevention (CDC) has recommended screening all sexually active women under the age of 20 years for chlamydia genital infection whenever undergoing a pelvic exam [1]. Traditionally, chlamydia laboratory tests were time consuming, expensive, and of low sensitivity. Since 1995, nucleic acid amplification tests such as polymerase chain reaction (PCR) and ligase chain reaction (LCR) have been developed that are more sensitive and specific than previously available chlamydia tests [17–22]. In addition, these tests can be performed using either a cervical, self-administered vaginal swab or a urine sample, alleviating the need for a pelvic exam [8,23]. The magnitude of morbidity, the young age group at risk of infection, and the convenience of new highly sensitive tests have resulted in more aggressive public health approaches to preventing the sequelae of chlamydial infection.

II. Epidemiology

A. Adolescents and Chlamydia

Chlamydia is the most common reportable bacterial sexually transmitted disease (STD) in the United States, with an esti-

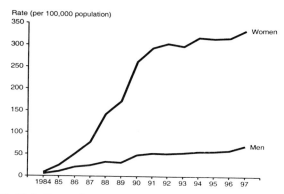

Fig. 21.1 Reported rates of *Chlamydia* infection, United States, 1984–1997. From [26].

mated 3 million new cases per year [24,25]. Rates of reported chlamydia infection have risen each year since it became a notifiable disease in 1986 (Fig. 21.1) [26]. The vast majority of reported cases are among females because control programs only target women. In 1997, the reported U.S. chlamydia rate for females was 316.0 per 100,000 population [26]. Adolescent females bear the brunt of chlamydial morbidity in both numbers infected and sequelae risk (Table 21.1). The CDC's 1997 national surveillance data report that females 15–19 years old maintain the highest chlamydia rates of 2044 per 100,000 population [26]. In a 1996–1997 national sample of 13,204 female U.S. Army recruits aged 17–39 years screened for chlamydia genital infection, the highest proportion of positive chlamydia tests was detected among the 17 year olds [27] (Fig. 21.2). Across the country, local data have identified genital chlamydia

Table 21.1

Ages of Females with Highest Chlamydia Prevalence Detected in Large Screening Programs

	Peak age (years)	Peak rate (%)	n	Test	Ages screened (years)
[27,U.S.]	17	12.2%	13,204	LCR[a]	17–39
[28,Baltimore]	14	27.5%	20.938	LCR[a]	12–60
[29,Seattle]	15–19	13.0%	1,804	Culture	15–34
[12,Wisconsin]	10–14*	41.9%*	38,866	Culture, EIA,[b] or DFA[c]	10–44

[a]LCR = ligase chain reaction.
[b]EIA = enzyme immunoassay.
[c]DFA = direct fluorescent antibody.
*Recurrent infection.

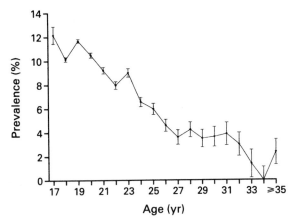

Fig. 21.2 Mean (+SE) age-specific prevalence of chlamydial infection among 13,204 female Army recruits, according to ligase chain reaction assays of urine specimens. From [27].

infections to be an emerging health problem among adolescent females. Among 18,617 visits made by sexually active females between the ages of 12 and 60 to Baltimore city clinics, 14 year olds had the highest proportion of positive chlamydia tests (27.5%) [28]. In Seattle primary care clinics, 13% of females aged 15–19 years were diagnosed with chlamydia infection as compared to 2% of females 25 years or older [29].

Adolescents are also very susceptible to recurrent chlamydia infections. In Wisconsin family planning clinics, 42% of 717 screened 10–14 year old females were diagnosed with recurrent chlamydia infections within 1–6 years [12]. In the U.S. Pacific Northwest family planning clinics, 18% of 3298 females 15–19 years old were diagnosed with a recurrent chlamydia infection within a three year period [13]. In an Indiana adolescent clinic, 38% of 177 sexually active adolescent females with a positive chlamydia culture followed by a sterile test of cure were found to have a recurrent infection at a subsequent visit [30]. The majority of the 68 recurrent infections (54%) were documented within nine months of the initial chlamydia infection. In an Oklahoma adolescent clinic, 12.5% of 319 sexually active adolescent females screened for chlamydia by direct fluorescent antibody test (DFA) had a recurrent infection [31]. The mean time between first and second positive chlamydia tests was approximately five months. In Baltimore clinics, the median time to a repeat positive chlamydia DNA amplification test for sexually active adolescent females was six months, with 25% incident cases occurring within four months during 9760 person months of observation [6]. These frequencies to diagnosis of recurrent infection are approximately twice that of the current recommendations for annual primary care chlamydia screening of adolescent females considered to be at risk [32–34].

B. Other Demographic Determinants of Infection

Studies utilizing less sensitive chlamydia tests than are currently available identified black race and low socioeconomic status as demographic independent predictors of infection [29,35]. However, other studies utilizing predominately white suburban and predominately white rural adolescent samples have detected chlamydia rates of greater than 10% [36,37]. In addition, Marrazzo *et al.* found that race was not a predictor for chlamydia infection among 4968 adolescent females screened with nucleic acid amplification tests in the U.S. Pacific Northwest [38].

C. Biologic and Behavioral Considerations

Various biologic determinants may account for increased chlamydia susceptibility for adolescent females. During childhood, the cervix is lined with columnar epithelium. During adolescence and into adulthood, the cervical columnar epithelium is gradually replaced by squamous epithelium in its entirety [39,40]. However, while the cervical columnar epithelium persists, the adolescent is more susceptible to chlamydia transmission because *C. trachomatis* infects columnar, not squamous, epithelium. In addition, *C. trachomatis* infection imparts a partial protective immunity [41–43]. Therefore, an older woman who has been sexually active for a period of time is likely to have partial immunity against chlamydia transmission from a prior infection, whereas a younger sexually inexperienced adolescent who has not yet been exposed to chlamydia is more susceptible to infection.

Although younger adolescents are progressing in physical development, cognitive abilities of this age group are still concrete. Young adolescents are still not able to think abstractly, to conceptualize, or to develop a plan for the future [44,45]. These limitations explain why many young adolescents do not consider possible adverse outcomes of risky behaviors and are not able to plan ahead for condom use if sexually active.

As adolescents mature, they develop the ability to think abstractly [45,46]. However, abstract thinking leads to a belief in their own "personal fable" where they presume that "nothing bad will happen to *me* . . . those things only happen to *other people*" [45]. Although most adolescents in this age group understand that engaging in unprotected sex can lead to deleterious consequences, they believe their omnipotence and immortality keep them in control of all situations. While these assumptions seem unrealistic to adults, most adolescents probably have not yet encountered any personal life experience to validate the probability of a bad outcome to a risky behavior (*i.e.*, acquiring a STD as a result in engaging in unprotected sex).

Peers become the major social influence during this period of development. Intense conformity with peer values can impel adolescents to engage in risky sexual behaviors against their better judgement [47]. Peer pressure may often be the influence for initiation of sexual activity. Despite concerns about human immunodeficiency virus (HIV), the average age of sexual debut in the United States has steadily decreased [40].

Early and mid-adolescents engage in serial monogamous relationships, a succession of short duration relationships, each involving only one partner [40]. Even though the adolescent is exposed to multiple partners over a period of time, these relationships are perceived as low risk because each is monogamous. Therefore, a need for protective behaviors, such as condom use, is less commonly perceived.

III. Transmission

Chlamydia genital infection is sexually transmitted. Although previous studies using less sensitive tests suggested that transmission was more effective from men to women compared to women to men, data using more sensitive DNA amplification tests demonstrate that transmission rates between genders are comparable [48–50]. Quinn and colleagues tested genital specimens for chlamydia by culture and PCR in 494 sexual couples presenting to STD clinics [50]. Positive chlamydia cultures in female partners of infected males (57%) were significantly more common than positive cultures in male partners of infected females (42%). However, when specimens were analyzed by PCR, no significant difference in the probability of transmission was found by gender. Seventy percent of female partners of infected males and 68% of male partners of infected females had positive chlamydia PCR test results. The presence of STD symptoms did not affect risk of transmission. This study supports the public health opinion that a large reservoir of asymptomatic chlamydia infected males exists, and that these males have viable chlamydia which can be easily transmitted to their sexual partners.

IV. Natural History of Infection

Although chlamydia causes a great deal of reproductive tract inflammation, most genital infections are asymptomatic. No symptoms were reported in 68% of 1070 clinic visits with a chlamydia diagnosis that were made by sexually active adolescent females attending school-based clinics [6]. Only 23% of clinic visits in this sample were for STD symptoms (discharge, dysuria, abdominal pain) and only 19% were for an STD contact. None of the 1219 female U.S. Army recruits found to have a positive chlamydia test reported reproductive health symptoms at their physical exam [27]. Seventy percent (112) of the 372 adolescent and young adult females diagnosed with chlamydia genital infection were asymptomatic in a Seattle-based community screening program [38].

Chlamydia infection in women can persist for a varying duration of time if left untreated. In a study of college females with a positive cervical chlamydia culture, four of seven (57%) of females who denied subsequent antibiotic use were found to have persistent infection 16–17 months after the initial exam [51]. Three of these four women denied any sexual activity 11–15 months prior to the second chlamydia test. Only one of the seven was found to have abnormal vaginal discharge on exam. The potential long duration and the high proportion of asymptomatic infections demonstrate the importance of screening asymptomatic females for chlamydia infection.

V. Screening Strategies

Because most chlamydia infections in women are asymptomatic and universal screening can be a costly endeavor, researchers have attempted to formulate screening criteria to identify those who are most at risk of infection as candidates for chlamydia testing. Risk factors such as sex behaviors, demographic factors such as age, and a positive chlamydia test as outcome have been analyzed. The older literature using a positive chlamydia cell culture as the outcome described demographic factors such as age younger than 25 years, risk behaviors such as a new partner or more than one partner in the past three months, or inconsistent condom use as independent predictors of infection [10,29]. These studies found that by selectively

screening women with ≥2 independent predictors of infection, the proportion of females screened would be substantially lowered (44–65%) and most infections (>85%) would still be identified. Selective chlamydia screening of females has been adopted by public health policymakers such as the CDC [1]. Data using DNA amplification chlamydia tests as outcome have identified young age as the single strongest predictor of chlamydia infection. By PCR testing in 18,617 clinic visits made by sexually active females 12–60 years old in Baltimore, 80% of the 1969 chlamydia infections would have been detected by selective screening of females younger than 25 years regardless of prior STDs, condom use, or multiple partner risks. Tests would only have been performed at 50% of clinic visits, resulting in great cost savings, but 80% of infections would have been detected [28]. If age younger than 30 years had been used as a selective screening criterion in this sample, 91% of infections would have been detected by chlamydia screening at 70% of the clinic visits. None of the behavioral or historical independent predictors of infection could identify a population with the majority of the infections. In the sample of 13,204 female U.S. Army recruits, 95% of chlamydia infections could be identified by LCR screening of 88% of the females if age younger than 25 years was used as a selective screening criterion [27]. Age-based universal chlamydia screening with DNA amplification tests, therefore, has been found to be a cost effective chlamydia control strategy in several populations [17,52].

Given the high risks for chlamydia infection, health providers caring for adolescents need to ask if they are looking hard enough for chlamydia. The CDC recommends chlamydia screening of all sexually active females under the age of 20 whenever undergoing a pelvic exam [1]. Results of the 1997 Youth Risk Behavior Surveillance, a national high school-based survey, demonstrated that 48.4% of high school students had engaged in sexual intercourse [52]. Despite a high prevalence of sexual activity among high school adolescents, only 37% of Los Angeles County high school students surveyed in 1992 reported discussing condom use for vaginal intercourse and only 15% reported discussing their sex life with physicians [53]. Unfortunately, only 65% of adolescents in this sample trusted physician confidentiality regarding sexual activity and only 44% about STDs. Less than half of the pediatricians in the New York City area surveyed in 1993 reported having vaginal speculums available and over half did not feel comfortable managing STDs [54]. Adolescent practitioners surveyed in California reported that fewer than half screened their patients for chlamydia even though two-thirds of clinicians routinely performed Pap smears [55]. A study looking at services rendered to adolescents enrolled in Massachusetts Health Maintenance Organizations (HMOs) found that a large discrepancy existed between the proportion of females screened for STDs and the proportion of females estimated to be sexually active [56]. Only 11% of the enrolled 15–19 year old females had any STD testing despite an estimated sexual activity rate of 53% for females in this age group [56]. In a national survey, 43.6% of 564 primary care providers claimed they would perform more chlamydia screening of adolescent females if DNA amplification tests were available to them [G. R. Burstein, unpublished]. Current guidelines for delivery of adolescent primary care services recommend yearly chlamydia screening for those adolescent females considered to be at risk [32–35]. Because primary care is the point of entry

Table 21.2

Findings at Physical Exam among Sexually Active Females Aged 12–60 Years Tested for *Chlamydia trachomatis* Infection[a] by Polymerase Chain Reaction in Baltimore City Clinics, January, 1994–September, 1996

Exam finding	Total sample, % n = 15,193	Chlamydia positive, % 1719	AOR (95% CI)[b]
Mucopus	26	38	1.5 (1.2–1.7)
Ectopy	8	16	1.5 (1.3–1.8)
Friability	12	21	1.7 (1.4–2.0)
Normal	60	42	0.9 (0.7–1.1)

[a]*Chlamydia trachomatis* positive by polymerase chain reaction chlamydia cervix or urine test.

[b]AOR, adjusted odds ratio; CI, confidence interval.

Table 21.3

Microbiologic Isolation of *Chlamydia trachomtis* in Females with Laparoscopically Verified Pelvic Inflammatory Disease

Study (location)	Specimen site	Test	% CT positive n (%)
[65] (Norway)	Cervix	Culture	26/56 (46)
[65] (Norway)	Fallopian tubes	Culture	5/31 (16)
[66] (Sweden)	Cervix	Culture	19/53 (36)
[66] (Sweden)	Fallopian tubes	Culture	6/20 (30)
[67] (Finland)	Genital tract	Culture	13/26 (50)
[67] (Finland)	Upper genital tract	Culture	9/26 (35)
[68] (U.S.)	Cervix and/or abdomen	Culture	8/50 (16)
[69] (U.S.)	Cervix	Culture	13/84 (15)
[70] (France)	Cervix	Culture	4/17 (24)
[70] (France)	Tubes and/or peritoneum	Culture	16 (38)

into the health care system for most adolescents, it is crucial that adolescent primary care providers screen their patients for chlamydia.

VI. Clinical Syndromes

A. Mucopurulent Cervicitis

Classic mucopurulent cervicitis (MPC) is a clinical syndrome characterized by a purulent or mucopurulent endocervical exudate [26]. Although most cases of chlamydia cervicitis are asymptomatic, *C. trachomatis* is the most common bacterial cause of MPC for which testing is routinely performed [26]. Females with MPC may complain of irregular vaginal discharge, irregular vaginal bleeding, and/or dyspareunia. Findings on pelvic exam include purulent or mucopurulent endocervical exudate visible in the endocervical canal (Fig. 21.3; see color insert) or on an endocervical swab specimen (positive swab test) (Fig. 21.4; see color insert), an erythematous or edematous area of ectopy, and/or easily induced cervical bleeding on exam denoting friability [57–61].

The greater sensitivity of the new chlamydia DNA amplification tests compared to cell culture may account for the greater proportion of chlamydia infected women without signs of MPC on exam. DNA amplification tests can detect those women with low numbers of organisms who are less likely to show signs of infection. Of the 1719 positive chlamydia PCR tests in our Baltimore sample of sexually active females aged 12–60 years where pelvic exam findings were available, only 38% were found to have mucopus, 21% friability, and 16% cervical ectopy [28] (Table 21.2). Forty-two percent of females with a positive chlamydia test had a normal exam documented. A study of 1583 women attending Seattle college health and STD clinics found that only 67% of 153 females with a positive chlamydia cervical culture had signs of cervicitis on exam [60]. As the sensitivity of chlamydia tests improves, the utility of MPC as a predictor of chlamydia infection may decline.

Microscopic examination of a cervical Gram stain has been used as a laboratory test to support the diagnosis of MPC [58–61]. Recommendations for numbers of polymorphonuclear leukocytes (PMNs) per high-powered field (HPF) in an endocervi-

cal mucous specimen as a significant finding are inconsistent and range from >10 to >30. Currently, the CDC does not recommend using number of PMNs in a cervical Gram stain as a diagnostic criterion because it has not been standardized, it has a low positive predictive value, and it is not available in many clinical settings [52].

B. Urethritis

Chlamydia trachomatis has been identified as a causative agent for acute urethral syndrome, defined as acute dysuria and frequent urination in women whose voided urine was sterile or contained $<10^5$ organisms per milliliter [57,62–64]. Stamm isolated *C. trachomatis* by cell culture in cervix and/or urethral specimens in 7 of 16 (44%) women with urethral syndrome, sterile bladder urine, and pyuria [62]. Evidence of a recent chlamydia infection with a fourfold change in *C. trachomatis* antibody titer was found in an additional 3 (19%) women. Panja isolated *C. trachomatis* in the urethra by cell culture in 12 of 25 (48%) women with dysuria, urinary frequency, and pyuria [63].

Similar to the epidemiology of cervical chlamydia infections, urethral infections in women can also be asymptomatic. In a study of women screened for *C. trachomatis* by cell culture with cervical and urethral samples in a Denver STD clinic, 11% of 751 urethral cultures were positive [9]. No significant association with signs and symptoms was found. Of the 79 positive urethral cultures, 22 (28%) were negative in the cervix, which makes urine-based DNA amplification testing more appealing to diagnose chlamydia genitourinary infections.

C. Salpingitis/Pelvic Inflammatory Disease

Salpingitis, also referred to as pelvic inflammatory disease (PID), is inflammation of the fallopian tubes. *C. trachomatis* infection may ascend from the cervix through the uterus to the fallopian tubes resulting in salpingitis, the most serious complication of chlamydia cervical infection. *C. trachomatis* is a major etiologic agent in PID [65–70] (Table 21.3). The risk of a chlamydia infection ascending into the pelvis increases with the duration of untreated cervicitis [71]. Screening asymptomatic

Table 21.4

Center for Disease Control and Prevention Recommended Criteria for Diagnosis of Pelvic Inflammatory Disease

Minimum criteria

   Lower abdominal tenderness

   Adnexal tenderness

   Cervical motion tenderness

Additional criteria

   Oral temperature > 100°F (>38.3OC)

   Abnormal cervical or vaginal discharge

   Elevated erythrocyte sedimentation rate

   Elevated C-reactive protein

   Laboratory documentation of cervical infection with *N. gonorrhoeae* or *C. trachomatis*

Definitive criteria

   Histopathological evidence of endometritis endometrial biopsy

   Transvaginal sonography or other imaging techniques show thickened fluid filled tubes with or without free pelvic fluid or tuboovarian complex

   Laparoscopic abnormalities consistent with pelvic inflammatory disease

females for chlamydia cervicitis has been shown to decrease the incidence of PID in a primary care setting [72].

Symptomatic salpingitis usually presents with complaints of lower abdominal pain. Affected individuals may also complain of vaginal discharge, abnormal vaginal bleeding, dyspareunia, dysuria, nausea, or fever. However, evaluations of women with tubal factor infertility affirm that a significant proportion of chlamydia salpingitis is asymptomatic or has minimal symptoms, a condition referred to as atypical or "silent" PID [16]. Laparoscopy has demonstrated severe fallopian tube inflammatory changes due to chlamydia salpingitis in women with relatively mild symptoms found by clinical exam [65,70]. The magnitude and severity of atypical or "silent" PID is evident by the fact that the majority of women with tubal factor infertility and evidence of prior chlamydia infection have no recall of previous PID [3,70,72,73].

In current practice, PID is a clinical diagnosis based on criteria recommended by the CDC [52] (Table 21.4). However, diagnosis based on clinical exam has been shown to be imprecise. The positive predictive value for *symptomatic* PID is 65–90% in comparison with laparoscopy [16]. Most of the cases of atypical PID, which is estimated to comprise 60% of all PID cases, are missed until consultation is sought for sequelae [16]. Our technically advanced medical capabilities are still in need of satisfactory diagnostic methods for one of the most serious reproductive health infections in women.

VII. Complications

Acute perihepatitis, inflammation of the liver capsule and adjacent peritoneum, is known as the "Fitz-Hugh-Curtis Syndrome" when associated with salpingitis. Adhesions form between the liver capsule and adjacent parietal peritoneum forming dense "violin string" adhesions between the liver capsule and

the abdominal wall [16] (Fig. 21.5; see color insert). Acute onset of severe pleuritic right upper quadrant (RUQ) abdominal pain is the classic presenting symptom. Abdominal exam reveals RUQ tenderness and guarding. Occasionally a friction rub may be auscultated over the liver. Liver enzymes are normal because the inflammation does not involve the liver parenchyma.

Although the "Fitz-Hugh-Curtis Syndrome" was initially described as a complication of gonorrheal PID, a higher proportion has since been described in association with chlamydial salpingitis. In a study of 23 females with perihepatitis and salpingitis, 20 (83%) had immunologic evidence of a recent chlamydia infection by either the presence of IgM antibody, at least a fourfold change of IgM and/or IgG antibody, and/or IgG antibody titer ≥1:1,024 [74]. *C. trachomatis* was isolated from the cervix in three of ten (30%) patients in whom cultures were obtained. In another study, 7 of 18 (39%) females with laparoscopically confirmed perihepatitis associated with salpingitis had *C. trachomatis* isolated in their cervical culture [65].

VIII. Sequelae

Inflammation in the fallopian tubes as a result of chlamydia infection can result in significant scarring and obstruction, regardless of presence of pelvic symptoms. Reproductive health can be severely and permanently impaired, with an increased risk of chronic pelvic pain, ectopic pregnancy, and tubal factor infertility.

A. Ectopic Pregnancy

The emerging chlamydia epidemic has contributed significantly to the increased rates of ectopic pregnancies in the U.S. since the 1980s [60,73,75–78]. Other contributing factors include increased maternal age, use of an intrauterine device, douching, and improved detection [16,75]. Ectopic pregnancy is a major cause of maternal morbidity and mortality in the developed world [76,77]. In 1992 there were an estimated 108,000 ectopic pregnancies in the United States at a rate of 19.7 per 1000 reported pregnancies [76].

Researchers have found that a greater proportion of women with ectopic pregnancies have serologic evidence of a prior chlamydia infection as compared to fertile women of similar age (Table 21.5). Many of these women were never diagnosed with a STD. In a case control study, 47% of 43 women with ectopic pregnancies had chlamydial IgG antibody titers ≥1:32 as compared to 22% of 49 pregnant controls [60]. Only 10% of the females with ectopic pregnancies recalled ever having a pelvic infection. A matched-pair case-control study examining the association of past exposure to *C. trachomatis* and ectopic pregnancy found that 71% of 306 women with ectopic pregnancies had chlamydial IgG antibody titers at 1:64 versus 39% of 266 pregnant controls [75]. The adjusted matched-pair odds ratio for a chlamydia antibody titer of ≥1:64 associated with an ectopic pregnancy was 3.1 (95% confidence interval (CI) = 2.2–4.3). The authors estimated that almost 50% of ectopic pregnancies in the U.S. can be attributed to a prior chlamydia infection. A chlamydia control program would have a great potential impact on the incidence of ectopic pregnancies.

Table 21.5

Serologic Evidence of Prior *C. trachomatis* Infection in Case Control Studies of Women Evaluated for Ectopic Pregnancy or Tubal Factor Infertility

Study	Pathology	*Chlamydia* antibody titers	Cases n (%)	Controls n (%)	p value
[68] (Canada)	INF	IgG $\geq$ 1:32	13/18 (72)	17/119 (14)	<0.005
[72] (U.S.)	INF	IgG or IgM $\geq$ 1:8	74/123 (60)	13/49 (27)	<0.001
[70] (France)	INF	NS	23/40 (58)	5/28 (18)	<0.05
[4] (Canada)	INF	IgM $\geq$ 1:800/IgG or >1:400	34/43 (79)	29/77 (29)	<0.001
[76] (Sweden)	EP	IgG $\geq$ 1:16	73/112 (65)	25/156 (16)[a]	<0.0001
[78] (U.S.)	EP	IgG $\geq$ 1:128	41/60 (68)	12/60 (20)	<0.0001
[75] (U.S.)	EP	IgG $\geq$ 1:64	218/306 (71)	103/266 (39)	<0.05
[68] (Canada)	EP	IgG $\geq$ 1:32	20/43 (47)	11/49 (22)	0.02

[a]Excluding current cervicitis or salpingitis.

B. Infertility

C. trachomatis infection is the most important preventable cause of PID and subsequent tubal factor infertility in the U.S. [1–5,15,70,73]. The fallopian tube fibrosis as a result of inflammation from chlamydial salpingitis can be severe enough to cause complete bilateral tubal obstruction (BTO), often without symptoms. In a sample of 187 women evaluated for infertility, 72 (39%) had antibody titers to *C. trachomatis* $\geq$ 1:16; of which 34 (64%) had no history of PID [73]. Laproscopic evaluations were performed in 51 women in the sample with a history of PID. There was a marked association between tubal inflammatory injury and chlamydia antibody titers $\geq$1:16. Women with more severe grades of adnexal adhesions had a higher prevalence of chlamydia antibody titers (p = 0.01). In a multisite study, 71% of 45 women with infertility due to BTO had serologic evidence of a prior chlamydia infection [3]. Less than half (44%) with BTO and chlamydia antibodies gave a history of previous PID symptoms.

C. Neonatal Transmission

Perinatal exposure to a mother's chlamydia-infected cervix can result in neonatal infection. The risk of transmission is about 50% and neonatal ocular prophylaxis does not prevent perinatal transmission of *C. trachomatis* from mother to newborn [79]. Approximately 70% of infant infections are in the nasopharynx, but perinatal infection may also occur in the rectum and vagina [79,80]. An untreated infection may persist in the infant for more than one year. Twenty-two infants who were not appropriately treated for a perinatally acquired *C. trachomatis* infection were serially tested for recurrent/persistent infection at well-child care visits [80]. The median ages at the time of the last positive culture for a recurrent/persistent infection was 5–6 months. One infant was still infected at 28.5 months of age. Chlamydia is the most frequently identified cause of infectious neonatal conjunctivitis and conjunctivitis is the most frequent clinical syndrome of neonatal chlamydia infection [79]. The incubation period is 5–15 days after delivery. Presentation can vary from minimal discharge to severe conjunctivitis. Although less common, a perinatally acquired chlamydia infection may also result in pneumonitis at 1–3 months of age.

IX. Risk of HIV Transmission

The exudative bacterial STDs have been implicated as important facilitators in HIV transmission. Studies in Africa and the U.S. have demonstrated that infection with bacterial STDs is associated with an increased risk of HIV seroconversion [81,82]. Odds ratio estimates for risk of HIV infection have ranged from 3.0 to 6.0 with a concurrent chlamydia infection [83]. In areas with a high STD burden in the U.S., heterosexual contacts and young women account for an increasing number and percentage of adults with the acquired immunodeficiency syndrome (AIDS) [84]. Public health authorities believe that a large number of new HIV infections are contracted during adolescence and that STDs, such as chlamydia, play a role in that transmission.

X. Microbiology

Chlamydia trachomatis is an obligate intracellular parasite. It undergoes a replicative cycle inside its host's cells which results in cell death. The following description illustrates the developmental growth cycle of chlamydiae inside the host cell. The infectious particle, called the elementary body, first attaches to the host cell. After entering the host cell, this elementary body becomes metabolically active, divides, and forms a larger reticulate body, which is sometimes called the initial body. At this stage, chlamydia synthesizes its own RNA, DNA, and protein, using the host cell's precursors. Glycogen inclusions form that can be stained and detected with iodine. At this active stage the reticulate body divides by binary fission, and some particles reorganize into the smaller infectious elementary bodies. Eventually, the elementary bodies predominate and the host cell expands and ruptures, releasing infectious elementary bodies that may infect other cells or be passed on to other hosts through infected secretions.

Traditionally, cell culture was considered the gold standard for *C. trachomatis* diagnosis [85]. The organism can be recovered from infected patients by culture 48–72 hours after inoculation. Because *C. trachomatis* is an obligate intracellular parasite, it requires a living host cell to support its growth. Maintaining living cell lines requires stringent laboratory procedures and technical expertise. In addition, specimen collec-

Table 21.6

Sensitivity of Available Chlamydial Diagnostics According to Specimen Type and Patient Gender

Population	Specimen	Sensitivity					
		DFA	EIA	Probe	Culture	PCR	LCR
Women	Cervix	70–90%	60–85%	60–75%	50–80%	70–95%	85–95%
	Urine	60–70%	40–50%	?	<30%	90–95%	90–95%
Men	Urethra	70–85%	70–80%	70–80%	50–80%	85–95%	85–95%
	Urine	76–60%	45–85%	?	<20%	85–95%	85–95%
Infant	Conjunctivae	>90%	>90%	?	70–90%	>90%	>90%

tion techniques are important. Specimens must be collected with cotton-tipped swabs that do not have wooden sticks. Swabs with wooden sticks were found to contain inhibitors to chlamydia cell growth. Specimens must be collected from the endocervical and urethral canal so as to contain adequate numbers of infected columnar epithelial cells. For this reason, some clinicians use cytobrushes to collect endocervical specimens [86]. After collection, the specimen must be placed in specific transport media, refrigerated immediately, and transported to the laboratory for cell culture inoculation within 24 hours. Specimens should also be transported "cold" for best results.

With the development and commercial availability of nonculture tests, diagnosis of chlamydia infections became technically easier [87]. Antigen detection methods of chlamydia lipopolysaccharide using DFA or enzyme immunoassay (EIA) techniques were among the first available tests. Specimens of potentially infected secretions are collected in the same manner as for culture, but transport and handling do not require refrigeration and maintenance of the cold chain. Results can be available within one or two days. Several on-site rapid antigen detection tests have been developed that provide results immediately. Sensitivity, however, is an issue with all of these tests, especially the rapid tests. Table 21.6 summarizes several studies comparing the sensitivity of various *C. trachomatis* diagnostic methods. The DFA tests, depending of the type and adequacy of the specimen and the experience of laboratory personnel, have sensitivities ranging between 60 and 90% [88]. EIA tests sensitivities range from 40 to 85% for genital specimens. The performance of rapid diagnostic tests based on antigen detection is even more disappointing and their use is discouraged until newer methods are developed [88,89].

Tests to detect, but not amplify, chlamydia nucleic acid were developed next. Such tests include GenProbe [86]. Sensitivities of these tests are within the range of DFA and EIA tests, but these tests cannot be used reliably with urine specimens [90].

The development of automated methods for the detection of amplified *C. trachomatis* DNA or RNA dramatically increased the sensitivities of chlamydia detection for all specimens, particularly in urine, while maintaining specificities of >99% [87,88]. As compared to culture or nonamplified detection methods that require between 10^4 to 10^8 infectious organisms to register a positive result, these amplification tests have been reported to detect as few as one to ten elementary bodies. These tests are therefore considered very sensitive and have redefined the gold standard for chlamydia infection. LCR has been reported to de-

tect 15 to 40 more infected individuals compared to cell culture and is consistently more sensitive when using endocervical specimens than is culture [91]. PCR has been reported to detect 40% more urethral chlamydial infection than culture, but endocervical performance has varied in several published reports [19,92–96]. Inhibitors in endocervical mucus may be responsible for this variability, but further study is needed. A newly developed transcription-mediated amplification (TMA) method amplifies chlamydial ribosomal RNA and appears to perform as well as PCR or LCR.

The options for diagnostic methods to detect chlamydia infections are many. The choice of which method is appropriate will depend on availability of proper specimen transport, proximity and experience of laboratory facilities, the patient population's acceptance of various specimen collection techniques, and cost.

XI. Management

The first line treatment for an uncomplicated *C. trachomatis* infection in women is a one week course of doxycycline or a single dose of azithromycin [52] (Table 21.7). Although two

Table 21.7

Centers for Disease Control and Prevention 1998 Recommendations for Treatment of Uncomplicated Cervical Chlamydia Infection

Recommended regimens:
  Azithromycin 1 g orally in a single dose
  OR
  Doxycycline 100 mg orally twice a day for 7 days
Alternative regimens:
  Erythromycin base 500 mg orally four times a day for 7 days
  Erythromycin ethyl succinate 800 mg orally four times a day for 7 days
  Ofloxacin 300 mg orally twice a day for 7 days
Recommended regimens during pregnancy:
  Erythromycin base 500 mg orally four times a day for 14 days
  Erythromycin ethyl succinate 800 mg orally four times a day for 7 days
  Erythromycin ethyl succinate 400 mg orally four times a day for 14 days
  Azithromycin 1 g orally in a single dose

published reports comparing the use-effectiveness of doxycycline versus azithromycin found them to be comparable regardless of reported compliance with doxycycline, the CDC recommends single dose therapies for those suspected to be noncompliant or difficult to reach for follow-up, such as adolescents [52,97,98]. Single-dose therapy may be cost prohibitive for STD control programs with limited resources, such as STD clinics, as the cost differential between azithromycin and doxycycline therapies is substantial ($15–9 vs $1–2). Erythromycin is an alternative therapy, but gastrointestinal side effects make this a less desirable treatment. Ofloxacin is more expensive than doxycycline, offers no advantage for compliance since it is a one week course of twice daily dosed medication, and is not recommended for individuals <16 years of age.

Recommended regimens during pregnancy are less than satisfactory. Erythromycin may be particularly difficult to tolerate during pregnancy due to its gastrointestinal side effects. Amoxicillin is not highly efficacious for the treatment of *C. trachomatis* infection. Doxycycline and ofloxacin are contraindicated for use during pregnancy. Azithromycin has not been well studied during pregnancy and lactation. A test of cure is recommended three weeks after completion of therapy because some of the recommended regimens are less efficacious and compliance may be difficult with side effects during pregnancy.

All sex contacts within 60 days preceding the onset of symptoms or chlamydia diagnosis should be referred for STD testing and treatment [52]. Reinfection is a major contributor to high chlamydia rates, especially among adolescents [6,30,31]. The individual treated with a single-dose antibiotic, therefore, must understand that she is *treated* but not immediately *cured*. She should be counseled to abstain from sexual intercourse until seven days after she and her partner(s) have both completed treatment [52]. Treatment of sex partners is an important cost-effective component of STD control [99]. Providers may take advantage of partner notification (PN) services offered by some health departments to assist with locating, contacting, and ensuring treatment of sex partners. Most health departments do not have the resources available for chlamydia-related PN services and the onus of PN falls on the provider.

XII. STD Control

Chlamydia control programs have targeted screening for asymptomatic females based on selective demographic, clinical, and behavioral criteria. A state wide chlamydia control program in Wisconsin was able to decrease chlamydia prevalence rates by 53% in family planning clinics by selective screening based on clinical and behavioral criteria [100]. The Chlamydia Project in the Pacific Northwest successfully decreased family planning clinic chlamydia prevalence rates by 58% [12]. However, the decline in prevalence came to a halt in the mid 1990s and a new endemic rate was established despite continued screening of women [26] (Fig. 21.6).

The question arises as to how to further decrease chlamydia rates and to ultimately eradicate disease. Although innovative programs to screen populations at risk for chlamydia have been instituted in correctional facilities, school clinics, and community outreach settings, a broad population-based approach needs

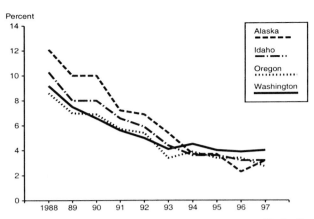

Fig. 21.6 *Chlamydia* positivity rates among women tested in family planning clinics by age group in the U.S. Pacific Northwest, 1988–1987. From [26].

to be taken to address this public health problem. Possible strategies include screening men, age-based screening, and increased frequency of screening.

A. Screening Males

In Denver nonclinical facility and field settings, 6.6% of 486 adolescent males screened for chlamydia with urine PCR tested positive [101]. Only one of the males admitted to urethritis symptoms. In Baltimore city school clinics, 7.6% of 251 adolescent male urine DNA amplification tests were positive [102]. Over 80% of visits were for asymptomatic primary care related reasons. Currently, programmatic support for chlamydia screening of asymptomatic men does not exist. However, a reservoir of asymptomatic male chlamydia infection exists that propagates disease in the population. Therefore, screening programs among young sexually active men should be considered in future program planning.

B. Age-Based Screening

In both Baltimore city clinics and in a national sample of female army recruits, age was the strongest predictor of a positive urine DNA amplification chlamydia test [6,27]. Either most chlamydia infections would have been missed or tests would have been performed at almost every clinic visit if other behavioral or demographic criteria would have been used as screening criteria.

C. Increased Frequency of Screening

Although the CDC recommends annual chlamydia screening for sexually active adolescent females, increasing current annual recommendations for chlamydia screening to twice a year has been advocated as an approach to chlamydia control by some experts. The CDC and primary health advisory panels advocate annual chlamydia screening of females at risk which includes females younger than 20 years old [1,32–34]. Longitudinal testing of adolescent females yields repeat positive chlamydia tests within six months in >50% of adolescent females with recurrent infections [6,30,31]. Shortening the duration of untreated chlamydia decreases infection prevalence rates within a population [103].

XIII. Cost Effectiveness

Although chlamydia screening can incur large costs, decreasing the incidence of chlamydia sequelae, such as PID, ectopic pregnancy, infertility, and neonatal transmission, can ultimately decrease total health care expenditures. Howell and colleagues found that annual urine or cervical DNA amplification chlamydia screening of females younger than 30 years in family planning clinics can be cost effective if prevalence is >2.4% [17,52]. Genc and Mardh determined that chlamydia cervical DNA amplification screening of asymptomatic females combined with single-dose treatment of positive patients to be the most cost-effective strategy when prevalence is ≥6% [104]. Shafer *et al.* concluded that urine-based DNA amplification screening was the most cost-effective strategy to detect chlamydial and gonococcal genital infection in asymptomatic sexually active adolescent females [105]. The authors recommended dismissing the pelvic exam for urine-based testing.

XIV. Conclusions

Chlamydia is a silent yet potentially devastating disease for women. Costs incurred on a societal level include large health care expenditures for screening and management of sequelae. Costs incurred on a personal level include loss of fertility, self-esteem, and sexual health. Chlamydia screening is finally a non-invasive procedure with highly sensitive and specific tests. Wherever we look for chlamydia we find it, especially among adolescents. The onus is on health care providers to take the initiative to screen and treat females for chlamydia infection.

References

1. Centers for Disease Control and Prevention (1993). Recommendations for the prevention and management of *Chlamydia trachomatis* infections. *Morbid. Mortal. Wkly. Rep.* **42,** RR-12.
2. Brunham, R. C., Maclean, I. W., Binns, B., and Peeling, R. W. (1985). *Chlamydia trachomatis:* Its role in tubal infertility. *J. Infect. Dis.* **152,** 1275–1282.
3. World Organization Task Force on the Prevention and Management of Infertility (1995). Tubal infertility: Serologic relationship to past chlamydial and gonococcal infection. *Sex. Transm. Dis.* **22,** 71–76.
4. Sellors, J. W., Mahony, J. B., Chernesky, M. A., and Rath, D. J. (1988). Tubal factor infertility: An association with prior chlamydial infection and asymptomatic salpingitis. *Fertil. Steril.* **49,** 451–457.
5. Kelver, M. E., and Nagamani, M. (1989). Chlamydial serology in women with tubal infertility. *Int. J. Fertil.* **34,** 42–45.
6. Burstein, G. R., Gaydos, C. A., Diener-West, M., Howell, M. R., Zenilman, J. M., and Quinn, T. C. (1998). Incident *Chlamydia trachomatis* infections among inner city adolescent females *JAMA, J. Am. Med. Assoc.* **280,** 521–526.
7. Centers for Disease Control and Prevention (1997). *Chlamydia trachomatis* genital infections—United States, 1995. *Morbid. Mortal. Wkly. Rep.* **46,** 193–198.
8. Schachter, J., Shafer, M. A., Young, M., and Ott, M. (1995). Routine pelvic examinations in asymptomatic young women. *N. Engl. J. Med.* **335,** 1847–1848.
9. Magder, L. S., Harrison, H. R., Ehret, J. M., Anderson, T. S., and Judson, F. N. (1988). Factors related to genital *Chlamydia trachomatis* and its diagnosis by culture in a sexually transmitted disease clinic. *Am. J. Epidemiol.* **128,** 298–308.
10. Handsfield, H. H., Jasman, L. L., Roberts, P. L., Hanson, V. W., Kathenbeutel, R. L., and Stamm, W. E. (1986). Criteria for selective screening for *Chlamydia trachomatis* infection in women attending family planning clinics. *JAMA, J. Am. Med. Assoc.* **255,** 1730–1734.
11. Marlotte, C. K., Wiesmeier, E., and Gelineau, K. J. (1990). Screening for chlamydia cervicitis in a sexually active population. *Am. J. Public Health* **80,** 469–471.
12. Hillis, S. D., Black, K., Newhall, J., Walsh, C., and Groseclose, S. L. (1995). New opportunities for chlamydia prevention: Applications of science to public health practice. *Sex. Transm. Dis.* **22,** 197–202.
13. Mosure, D. J., Berman, S., Kleinbaum, D., and Halloran, M. E. (1996). Predictors of *Chlamydia trachomatis* among female adolescents: A longitudinal analysis. *Am. J. Epidemiol.* **144,** 997–1003.
14. Oh, M. K., Cloud, G. A., Fleenor, M., Sturdevant, M. S., Nesmith, J. D., and Feinstein, R. A. (1996). Risk for gonococcal and chlamydial cervicitis in adolescent females: Incidence and recurrence in a prospective cohort study. *J. Adolesc. Health* **18,** 270–275.
15. Scholes, D., Stergachis, A., Heidrich, F. E., Andrilla, H., Holmes, K. K., and Stamm, W. E. (1996). Prevention of pelvic inflammatory disease by screening for cervical chlamydial infection. *N. Engl. J. Med.* **334,** 1362–1366.
16. Westrom, L., and Eschenbach, D. A. (1999). Pelvic inflammatory disease. *In* "Sexually Transmitted Diseases" (K. K. Holmes, P. F. Sparling, P. A. Mardh *et al.,* eds.), 3rd ed., pp. 783–809. McGraw-Hill, New York.
17. Howell, M. R., Quinn, T. C., and Gaydos, C. A. (1998). Screening for *Chlamydia trachomatis* in asymptomatic women attending family planning clinics. *Ann. Intern. Med.* **128,** 277–284.
18. Courtney, J., Ellen, J., and Bolan, G. (1997). Meta-analysis of the sensitivity and specificity of chlamydia diagnostic tests in females. *12th Meet. Int. Soc. Sex. Transm. Dis. Res., 1997,* Abst. No. O263.
19. Quinn, T. C., Welsh, L., Lentz, A. *et al.* (1996). Diagnosis by AMPLICOR PCR of *Chlamydia trachomatis* infection in urine samples from women and men attending sexually transmitted disease clinics. *J. Clin. Microbiol.* **34,** 1401–1406.
20. Hook, E. W., Smith, K., Mullen, C. *et al.* (1997). Diagnosis of genitourinary *Chlamydia trachomatis* infections by using the ligase chain reaction on patient-obtained vaginal swabs. *J. Clin. Microbiol.* **35,** 2133–2135.
21. Schachter, J., Moncada, J., Whidden, R. *et al.* (1995). Noninvasive tests for diagnosis of *Chlamydia trachomatis* infection: Application of ligase chain reaction of first-catch urine specimens in women. *J. Infect. Dis.* **172,** 1411–1414.
22. Lee, H. H., Chernesky, M. A., Schachter, J. *et al.* (1995). Diagnosis of *Chlamydia trachomatis* genitourinary infection in women by ligase chain reaction assay of urine. *Lancet* **345,** 213–216.
23. Orr, D. (1997). Urine based diagnosis of sexually transmitted infections using amplified DNA techniques: A shift in paradigms? *J. Adolesc. Health* **20,** 3–5.
24. Centers for Disease Control and Prevention (1996). The leading nationally notifiable infectious diseases, United States, 1995. *Morbid. Mortal. Wkly. Rep.* **45,** 883–884.
25. Alexander, L. A., Cates, J. R., Herndon, N., and Ratcliff, J. M. (1998). "Sexually Transmitted Diseases in America." Kaiser Family Foundation, Menlo Park, CA.

26. Division of STD Prevention (1998). "Sexually Transmitted Disease Surveillance, 1997." U.S. Department of Health and Human Services, Public Health Service, Centers for Disease Control and Prevention, Atlanta, GA.

27. Gaydos, C. A., Howell, M. R., Pare, B. *et al.* (1998). *Chlamydia trachomatis* infections in female military recruits. *N. Engl. J. Med.* **339,** 739–744.

28. Burstein, G. R., Gaydos, C. A., Diener-West, M. *et al.* (1997). Predictors of *C. trachomatis* infection by polymerase chain reaction among females presenting to U.S. inner city clinics. *Poster Presentation, Int. Conf. Sex. Transm. Dis.* Seville, Spain, *1997.*

29. Stergachis, A., Scholes, D., Heidrich, F. E., Sherer, D. M., Holmes, K. K., and Stamm, W. E. (1993). Selective screening for *Chlamydia trachomatis* infection in a primary care population of women. *Am. J. Epidemiol.* **138,** 143–153.

30. Blythe, M. J., Katz, B. P., Orr, D. P., Caine, V. A., and Jones, R. B. (1988). Historical and clinical factors associated with *Chlamydia trachomatis* genitourinary infection in female adolescents. *J. Pediatr.* **112,** 1000–1004.

31. Fortenberry, J. D., and Evans, D. L. (1989). Routine screening for genital chlamydia trachomatis in adolescent females. *Sex. Transm. Dis.* **16,** 168–172.

32. American Academy of Pediatrics (1995). Committee on practice and ambulatory medicine. Recommendations for pediatric and preventive health care. *Pediatrics* **96,** 373–374.

33. Elster, A., and Kuznets, N. (1994). "AMA Guidelines for Adolescent Preventive Services (GAPS): Recommendations and Rationale." Williams & Wilkins, Baltimore, MD.

34. U.S. Preventive Services Task Force (1996). "Guide to Clinical Preventive Services," 2nd ed. Williams & Wilkins, Baltimore, MD.

35. Mosure, D. J., Berman, S., Fine, D., DeLisle, S., Cates, W., Jr., and Boring, J. R. (1997). Genital chlamydia infections in sexually active female adolescents: Do we really need to screen everyone? *J. Adolesc. Health* **20,** 6–13.

36. Fisher, M., Swenson, P. D., Risucci, D., and Kaplan, M. H. (1987). *Chlamydia trachomatis* in suburban adolescents. *J. Pediatr.* **111,** 617–620.

37. Winter, L., Goldy, S., and Baer, C. (1990). Prevalence and epidemiologic correlates of *Chlamydia trachomatis* in rural and urban populations. *Sex. Transm. Dis.* **17,** 30–36.

38. Marrazzo, J. M., White, C. L., Krekeler, B. *et al.* (1997). Community-based screening for *Chlamydia trachomatis* with a ligase chain reaction assay. *Ann. Intern. Med.* **127,** 796–803.

39. Beach, R. K. (1992). Female genitalia: Examination and findings. *In* "Comprehensive Adolescent Health Care" (S. B. Friedman, M. Fisher, and S. K. Schonberg, eds.), pp. 956–989. Quality Medical Publishing, St. Louis, MO.

40. Berman, S. M., and Hein, K. (1999). Adolescents and STDs. *In* "Sexually Transmitted Diseases" (K. K. Homes, P. F. Sparling, P. A. Mardh, *et al.,* eds.), 3rd ed., pp. 129–142. McGraw-Hill, New York.

41. Knight, S. C., Iqball, S., Woods, C., Stagg, A., Ward, M. E., and Tuffrey, M. (1995). A peptide of *Chlamydia trachomatis* shown to be a primary T-cell epitope *in vitro* induces cell-mediated immunity *in vivo*. *Immunology* **85,** 8–15.

42. Schachter, J., Cles, L. D., Ray, R. M., and Hesse, F. E. (1983). Is there immunity to chlamydial infection of the human genital tract? *Sex. Transm. Dis.* **10,** 123–125.

43. Katz, B. P., Bateiger, B. E., and Jones, R. B. (1987). Effect of prior sexually transmitted disease on the isolation of *Chlamydia trachomatis. Sex. Transm. Dis.* **14,** 160–164.

44. Felice, M. E. (1987). 11–13 Years: Early adolescence—the age of rapid changes. *In* "Encounters with Children" (S. D. Dixon and M. T. Stein, eds.), pp. 319–328. Year Book Medical Publishers, Chicago.

45. Elkind, D. (1992). Cognitive development. *In* "Comprehensive Adolescent Health Care" (S. B. Friedman, M. Fisher, and S. K. Schonberg, eds.), pp. 24–26. Quality Medical Publishing, St. Louis, MO.

46. Felice, M. E. (1987). 14–16 years: Mid-adolescence—the dating game. *In* "Encounters with Children" (S. D. Dixon and M. T. Stein, eds.), pp. 331–341. Year Book Medical Publishers, Chicago.

47. Hamburg, B. A. (1992). Psychosocial development. *In* "Comprehensive Adolescent Health Care" (S. B. Friedman, M. Fisher, and S. K. Schonberg, eds.), pp. 27–38. Quality Medical Publishing, St. Louis. MO.

48. Worm, A. M., and Peterson, C. S. (1987). Transmission of chlamydia infections to sexual partners. *Genitourin. Med.* **63,** 19–21.

49. Ramstedt, K., Forssman, L., Giesecke, J., and Johannisson, G. (1991). Epidemiologic characteristics of two different population of women with *Chlamydia trachomatis* infections and their male partners. *Sex. Transm. Dis.* **18,** 205–210.

50. Quinn, T. C., Gaydos, C. A., Shepherd, *et al.* (1996). Epidemiologic and microbiologic correlates of Chlamydia trachomatis infection in sexual partnerships. *JAMA, J. Am. Med. Assoc.* **276,** 1737–1742.

51. McCormack, W. M., Alpert, S., McComb, D. E., Nichols, R. L., Semine, D. Z., and Zinner, S. H. (1979). Fifteen-month follow-up of women infected with *Chlamydia trachomatis. N. Engl. J. Med.* **300,** 123–125.

52. Centers for Disease Control and Prevention (1998). Youth risk behavior surveillance—United States, 1997. *Morbid. Mortal. Wkly. Rep.* **47,** SS-3.

53. Schuster, M. A., Bell, R. M., Petersin, L. P., and Kanouse, D. E. (1996). Communication between adolescents and physicians about sexual history. *Arch. Pediatr. Adolesc. Med.* **150,** 906–913.

54. Fisher, M., Golden, N. H., Bergeson, R. *et al.* (1996). Update on adolescent health care in pediatric practice. *J. Adolesc. Health* **19,** 394–400.

55. Montes, J. M., Glaser, M. K., and Schuyler, J. G. (1996). An assessment of general knowledge, screening and counseling practices and barriers to screening for chlamydia infections among practitioners of adolescent health care in California. *Abstr., Sex. Transm. Dis. Prev. Conf.*, Tampa, FL., *1996.*

56. Thrall, J. S., McCloskey, L., Spivak, H., Ettner, S. L., Tighe, J. E., and Emans, S. J. (1998). Performance of Massachusetts HMOs in providing pap smear and sexually transmitted disease screening to adolescent females. *J. Adolesc. Health* **22,** 184–189.

57. Sweet, R. L., Schachter, J., and Landers, D. V. (1983). Chlamydial infection in obstetrics and gynecology. *Clin. Obstet. Gynecol.* **26,** 143–163.

58. Brunham, R. C., Paavonen, J., Stevens, C. E. *et al.* (1984). Mucopurulent cervicitis—the ignored counterpart in women of urethritis in men. *N. Engl. J. Med.* **311,** 1–6.

59. Harrison, H. R., Costin, M., Meder, J. B. *et al.* (1985). Cervical *Chlamydia trachomatis* infection in university women: Relationship to history, contraception, ectopy, and cervicitis. *Am. J. Obstet. Gynecol.* **153,** 244–251.

60. Critchlow, C. W., Wolner-Hanssen, P., Eschenbach, D. A. *et al.* (1995). Determinants of cervical ectopia and of cervicitis: Age, oral contraception, specific cervical infection, smoking, and douching. *Am. J. Obstet. Gynecol.* **173,** 534–543.

61. Holmes, K. K., and Stamm, W. E. (1999). Lower genital tract infection syndromes in women. *In* "Sexually Transmitted Diseases" (K. K. Homes, P. F. Sparling, P. A. Mardh, *et al.,* eds.), 3rd ed., pp. 761–782. McGraw-Hill, New York.

62. Stamm, W. E., Wagner, K. F., Amsel, R. *et al.* (1980). Causes of the acute urethral syndrome in women. *N. Engl. J. Med.* **303**, 409–415.

63. Panja, S. K. (1983). Urethral syndrome in women attending a clinic for sexually transmitted diseases. *Br. J. Vener. Dis.* **59**, 179–181.

64. Stamm, W. E. (1999). *Chlamydia trachomatis* infections in the adult. *In* "Sexually Transmitted Diseases" (K. K. Homes, P. F. Sparling, P. A. Mardh *et al.*, eds.), 3rd ed., pp. 407–422. McGraw-Hill, New York.

65. H. Gjonnaess, K. Dalaker, G. Anestad, P. A. Mardh, G. Kvile, and T. Bergan, (1982). Pelvic inflammatory disease: Etiologic emphasis on chlamydial infection. *Obstet. Gynecol.* **59**, 550–555.

66. Mardh, P. A., Ripa, T., Svensson, L., and Westrom, L. (1977). Chlamydia trachomatis infection in patients with acute salpingitis. *N. Engl. J. Med.* **296**, 1377–1379.

67. Paavonen, J., Teisala, K., Heinonen, P. K. *et al.* (1987). Microbiological and histobiological findings in acute pelvic inflammatory disease. *Br. J. Obstet. Gynaecol.* **94**, 454–460.

68. Brunham, R. C., Binns, B., McDowell, J., and Paraskevas, M. (1986). *Chlamydia trachomatis* infection in women with ectopic pregnancy. *Obstet. Gynecol.* **67**, 722–726.

69. Soper, D. E., Brockwell, N. J., Dalton, H. P., and Johnson, D. (1994). Observations concerning the microbial etiology of acute salpingitis. *Am. J. Obstet. Gynecol.* **170**, 1008–1017.

70. Henry-Suchet, J., Utzmann, C., De Brux, J. *et al.* (1987). Microbiologic study of chronic inflammation associated with tubal factor infertility: Role of *Chlamydia trachomatis. Fertil. Steril.* **47**, 274–277.

71. Aral, S. O., and Wasserheit, J. N. (1998). Social and behavioral correlates of pelvic inflammatory disease. *Sex. Transm. Dis.* **25**, 378–385.

72. Jones, R. B., Ardery, B. P., Hui, S. L. *et al.* (1982). Correlation between serum antichlamydial antibodies and tubal factor as a cause of infertility. *Fertil. Steril.* **38**, 553–557.

73. Gump, D. W., Gibson, M., and Ashikaga, T. (1983). Evidence of prior pelvic inflammatory disease and its relationship to *Chlamydia trachomatis* antibody and intrauterine contraceptive device use in infertile women. *Am. J. Obstet. Gynecol.* **146**, 153–159.

74. Wang, S. P., Eschenbach, D. A., Holmes, K. K., Wager, G., and Grayston, J. T. (1980). *Chlamydia trachomatis* infection in Fitz-Hugh-Curtis syndrome. *Am. J. Obstet. Gynecol.* **138**, 1034–1038.

75. Chow, J. M., Yonekura, L., Richwald, G. A., Greenland, S., Sweet, R. L., and Schachter, J. (1990). The association between *Chlamydia trachomatis* and ectopic pregnancy. *JAMA, J. Am. Med. Assoc.* **263**, 3164–3167.

76. Svensson, L., Mardh, P. A., Ahlgren, M., and Nordenskjold, F. (1985). Ectopic pregnancy and antibodies to *Chlamydia trachomatis. Fertil. Steril.* **44**, 313–317.

77. Hartford, S. L., Silva, P. D., DiZerga, G. S., Schachter, J., and Yonekura, M. L. (1987). Serologic evidence of prior chlamydia infection in patients with tubal ectopic pregnancy and contralateral tubal disease. *Fertil. Steril.* **47**, 118–121.

78. Walters, M. D., Eddy, C. A., Gibbs, R. S., Schachter, J., Holden, A. E. C., and Pauerstein, C. J. (1988). Antibodies to *Chlamydia trachomatis* and risk for tubal pregnancy. *Am. J. Obstet. Gynecol.* **159**, 942–946.

79. Hammerschlag, M. R. (1996). Chlamydia. *In* "Nelson's Textbook of Pediatrics" (R. E. Berhman, R. M. Kliegman, and A. M. Arvin, eds.), 3rd ed., pp. 827–831. Saunders, Philadelphia.

80. Bell, T. A., Stamm, W. E., Wang, S. P., Kuo, C. C., Holmes, K. K., and Grayston, T. (1992). Chronic *Chlamydia trachomatis* infections in infants. *JAMA, J. Am. Med. Assoc.* **267**, 400–412.

81. Kassler, W. J., Zenilman, J. M., Erickson, B., Fox, R., Peterman, T. A., and Hook, E. W. (1994). Seroconversion in patients attending sexually transmitted disease clinics. *AIDS* **8**, 351–355.

82. Laga, M., Mamadou, D. O., and Buve, A. (1994). Inter-relationship of sexually transmitted diseases and HIV: Where are we now? *AIDS* **8**(Suppl. 1), S119–S124.

83. Wasserheit, J. N. (1992). Epidemiological synergy: Interrelationships between human immunodeficiency virus infection and other sexually transmitted diseases. *Sex. Transm. Dis.* **19**, 61–77.

84. Lindegren, M. L., Hanson, C., and Miller, K. (1994). Epidemiology of human immunodeficiency virus infections in adolescents, United States. *J. Pediatr. Infect. Dis.* **13**, 525–535.

85. Schacter, J. *et al.* Chlamydial infections. *N. Engl. J. Med.* **289**, 428.

86. Schacter, J. *et al.* (1995). Chlamydia. *In* "Manual of Medical Microbiology" (P. R. Murray *et al.*, eds.), 6th ed., pp. 669–677.

87. Stamm, W. E. (1990). Laboratory diagnosis of hlamydial infection. *In* "Chlamydial Infections. Proceedings of the 7th International Symposium on Human Chlamydial Infections" (W. R. Bowie *et al.*, eds.), pp. 459–470. Cambridge University Press, Cambridge, UK.

88. Marrazzo, J. M. *et al.* (1999). New approaches to the diagnosis, treatment, and prevention of chlamydial infections. *Curr. Top. Infect. Dis.*

89. Kluytmans, J. A. *et al.* (1994). Evaluation of Clearview and Magic Lite tests, polymerase chain reaction, and cell culture for detection of *Chlamydia trachomatis* in urogenital specimens. *J. Clin. Microbiol.* **31**, 3204–3210.

90. Schwebke, J. R. *et al.* (1991). Use of urine enzyme immunoassay as a diagnostic tool for *Chlamydia trachomatis* urethritis in men. *J. Clin. Microbiol.* **29**, 2446–2449.

91. Rumpianese, F. *et al.* (1996). Detection of *Chlamydia trachomatis* by a ligase chain reaction amplification method. *Sex. Transm. Dis.* **23**, 177–180.

92. Mahoney, J. *et al.* (1992). Confirmatory polymerase chain reaction testing for *Chlamydia trachomatis* in first-void urine from asymptomatic and symptomatic men. *J. Clin. Microbiol.* **30**, 2241–2245.

93. Bauwen, J. E. *et al.* (1993). Diagnosis of *Chlamydia trachomatis* urethritis in men by polymerase chain reaction assay of first-catch urine. *J. Clin. Microbiol.* **31**, 3013–3016.

94. Toye, B. *et al.* (1996). Diagnosis of *Chlamydia trachomatis* infections in asymptomatic men and women by PCR assay. *J. Clin. Microbiol.* **34**, 1395–1400.

95. Bianchi, A. *et al.* (1994). An evaluation of the polymerase chain reaction (Amplicor) *Chlamydia trachomatis* in male urine and female urogenital specimens. *Sex. Transm. Dis.* **21**, 196–200.

96. Pasternack, R. *et al.* (1996). Detection of CT infections in women by Amplicor PCR: Comparison of diagnostic performance with urine and cervical specimens. *J. Clin. Microbiol.* **34**, 995–998.

97. Thorpe, E. M., Stamm, W. E., Hook, E. W. *et al.* (1996). Chlamydial cervicitis and urethritis: Single dose treatment compared with doxycycline for seven days in community based practices. *Genitourin. Med.* **72**, 93–97.

98. Hillis, S. D., Coles, F. B., Litchfield, B. *et al.* (1998). Doxycycline and azithromycin for prevention of chlamydial persistence or recurrence one month after treatment in women. *Sex. Transm. Dis.* **25**, 5–11.

99. Howell, M. R., Quinn, T. C., Brathwaite, W., and Gaydos, C. A. (1998). Screening women for *Chlamydia trachomatis* in family planning clinics: The cost-effectiveness of DNA amplification assays. *Sex. Transm. Dis.* **25**, 108–117.

100. Addiss, D. G., Baughn, M. L., Ludka, D., Pfister, J., and Davis, J. (1999). Decreased prevalence of *Chlamydia trachomatis* infection

associated with a selective screening program in family planning clinics. In press.

101. Rietmeijer, C. A., Yamaguchi, K. J., Oritz, C. G. *et al.* (1997). Feasibility and yield of screening urine for *Chlamydia trachomatis* by polymerase chain reaction among high-risk male youth in field-based and other nonclinic setting. *Sex. Transm. Dis.* **24,** 429–435.

102. Burstein, G. R., Waterfield, G., Hauptman, P., Quinn, T. C., Gaydos, C. A., and Joffe, A. (1992). Chlamydia screening by urine based DNA amplification in adolescent males attending school based health centers. *J. Adolesc. Health* **24,** 120 (abstr.).

103. Brunham, R. C., and Plummer, F. A. (1990). A general model of sexually transmitted disease epidemiology and its implication for control. *Med. Clin. North Am.* **74,** 1339–1352.

104. Genc, M., and Mardh, P. A. (1996). A cost-effectiveness analysis of screening and treatment for *Chlamydia trachomatis* infection in asymptomatic women. *Ann. Intern. Med.* **124,** 1–7.

105. Shafer, M. A. B., Pantell, R. H., and Schachter, J. (1999). Is the routine pelvic examination needed with the advent of urine-based screening for sexually transmitted diseases? *Arch. Pediatr. Adolesc. Med.* **153,** 119–125.

22

Gonococcal Infection in Women

NOREEN A. HYNES*,† AND ANNE M. ROMPALO*

*Division of Infectious Diseases, Johns Hopkins University School of Medicine, †STD Program, Baltimore City Health Department, Baltimore, Maryland; *Centers for Disease Control and Prevention, Atlanta, Georgia

I. Introduction

Sexually transmitted diseases (STDs), including gonorrhea, represent a burgeoning but silent epidemic in the United States notable for the enormous burden placed upon the economy and the health and well-being of the population. In 1997, there were 324,901 cases of gonorrhea reported in the United States, approximately 50% of which were in women [1]. It is estimated that at least an equal number of unreported cases occur due to failure to screen asymptomatic women for infection, use of syndromic treatment algorithms without use of laboratory confirmation of suspected infection, and failure to report identified cases. The estimated total cost (direct and indirect) of gonorrhea infection, including related adverse sequelae such as pelvic inflammatory disease (PID), is over $750 million per year [2]. Because women are more likely to suffer complications of infection than are men, over 85% of the estimated costs are probably incurred by women. Furthermore, each year between 1 and 15 women die from disseminated gonococcal infection and 200 to 300 women die of pelvic inflammatory disease caused by gonorrhea or chlamydia [3].

Up to 50% of women with gonococcal infection are asymptomatic [4]. This critical epidemiologic feature, silent infection, leads to delayed diagnosis which, in turn, is associated with an increased period of communicability and increased risk of complications, including PID. Furthermore, the risk of acquiring human immunodeficiency virus (HIV) infection from an HIV-infected person is increased among persons coinfected with gonorrhea [5] and, in the setting of asymptomatic gonorrhea, may increase the duration of increased risk. In addition, HIV-infected persons coinfected with gonorrhea may be more likely to transmit HIV to an uninfected partner due to increased HIV concentration in their genital secretions.

II. Microbiology

Neisseria gonorrhoeae, the causative agent of gonococcal infection, is a uniquely human pathogen belonging to the Family Nesseriaceae. The organism is Gram-negative on stain. It has fastidious growth requirements including 3–5% CO_2 (candle jar environment), incubation temperature ranging from 35–37°C, and needs selective growth media to optimize recovery of the organism. The gonococcus divides by binary fission every 20 to 30 minutes. The organism produces very high levels of catalase, the enzyme that hydrolyzes peroxide to water and oxygen. This reaction may represent an important survival mechanism for gonorrhea, particularly in women. Peroxide, which the gonococcus hydrolyzes, is produced by the normal vaginal resident organism, lactobacillus, and is critical to maintaining an acidic vaginal pH (<4.5) hostile to most pathogens. In addition to catalase, *N. gonorrhoeae,* like other *Neisseria,* produces oxidase. In certain clinical settings, testing for the presence of oxidase alone is sufficient for making the diagnosis of *N. gonorrhoeae.* The most definitive diagnosis based on culture techniques includes sugar fermentation testing to differentiate *Neisseria* species. The gonococcus only ferments glucose whereas other family members have different fermentation patterns.

Various structural components of the gonococcal cell wall have been identified and functionally characterized (Table 22.1). The gonococcus, which infects noncornified epithelial tissue including that found in the urogenital tract, rectum, oropharynx, and conjunctivae, attaches to nonciliated epithelial cells via pili [6]. The main outer membrane protein in the gonococcus is por, formerly known as protein I (PI). This porin protein serves as a channel through which low molecular weight aqueous solutes pass, triggering ingestion by the host cell [7]. Por has been shown to exhibit minimal genetic variation when compared with the other major functional components of the organism [7a]. Immunoreactivity to por defines the gonococcal serotype of the isolate which can be further subclassified into serovars based on reactivity to monoclonal antibodies [7b]. The presence of anti-por antibodies in serum and in genital secretions have both cidal and opsonic properties. In some women (commercial sex workers and those with gonococcal PID), the presence of this antibody may be protective against infection with the same por serotype and serovar [8]. Organisms of the I-A protein type are more often associated with resistance to the normal bactericidal effect of nonimmune human serum and are associated with disseminated infection [7b,9]. Protein II, a group of five opacity proteins, functions to assist the organism to adhere to epithelial cells, whereas the conserved Protein III serves as a target for blocking antibody [7b,10]. Lipooligosaccharide (LOS) plays an important role in pathogenesis, including the induction of the intense inflammatory response noted in gonococcal urethritis and most of the tissue damage noted in salpingitis [10]. Evasion of host defenses is through multiple mechanisms including antigenic variation of surface proteins, blocking antibody, and cleavage of mucosal IgA_1 by a specifically elaborated protease [7b].

III. Epidemiology of Gonococcal Infections in Women

Large scale screening for gonorrhea began in the late 1970s. This enhanced screening resulted in an increase in the number of reported cases (Fig. 22.1) which was followed by a steady decline. Although the gonorrhea rates for women in more than 20 states have reached the revised Healthy People 2000 target

Table 22.1

Characteristics and Functions of the Major Proteins of *Neisseria gonorrhoeae*

Cell wall component	Description	Function(s)	References
Pili	Hairlike filamentous appendages extending from the cell surface	1. Important virulence factor 2. Piliated organisms associated with classic gonococcal urethritis 3. Mediates adherence of organism to host cell, especially to microvilli of noncilliate columnar cells	[6]

Por (formerly Protein I)	1. The most prominent outer membrane protein of the organism
	2. A porin
	3. Two subclasses
	  a. (IA: >23 serovars
	  b. IB: >31 serovars
	4. Always expressed
		1. Triggers endocytosis of the organism by host cell
		2. May inhibit neutrophil function
		3. May moderate attachment
			[7,7a,7b]

Protein II	1. Also called Opa protein 2. Heat modifiable 3. Not always expressed 4. Associated with environmental growth conditions	1. Cell-to-cell adhesion 2. May be a virulence factor	[8]
Protein III	1. Also called Pmp 2. Invariant among strains 3. Poorly immunologic	Blocks cidal antibodies directed at other surface proteins	[9]
Lipooligosaccharide (LOS)	Purported key in pathogenesis	Mediates most of the toxic damage that occurs through release of tumor necrosis factor-alpha	[10]

of 100 infections per 100,000 women [1,11], the national rate remains above this goal (Fig. 22.2). The gonorrhea rate among U.S. women in 1997 was 119.3 per 100,000 women.

In 1997, the 15- to 19-year-old age group had the highest gonorrhea rate among all women. The reported rate among these adolescent females of 718 per 100,000 (Fig. 22.3) was almost twice the rate seen in males in the same age group [1]. The gonorrhea rate among 10 to 14 year old girls has decreased from 82.3 per 100,000 in 1994 to 53.8 per 100,000 in 1997.

In the developed world, the incidence of gonorrhea usually is highest in areas characterized by high concentrations of persons living in poverty and with poor access to health care [12,13].

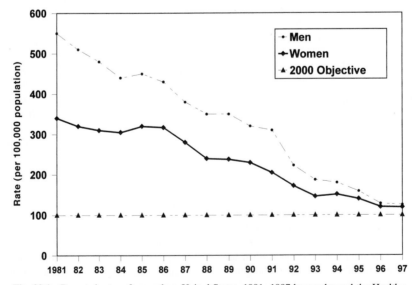

Fig. 22.1 Reported rates of gonorrhea: United States, 1981–1997 by gender and the Healthy People year 2000 objective. Source: [1].

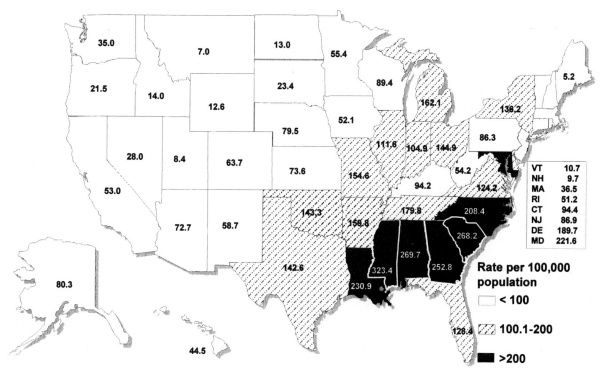

Fig. 22.2 Rates of gonorrhea among women, by state, 1997. Source: [1].

This observation underscores the fact that intrinsic to the maintenance of transmission within any area is the disease transmission efficiency between sexual partners, the rate of sexual partner turnover, and the duration of infectivity [14,15]. The risk of a woman acquiring gonorrhea during a single act of sexual intercourse with an infected male partner ranges from 60 to 90%

[16], whereas transmission from an infected female to a male is approximately 20 to 30% [17,18].

Risk factors for gonococcal infection include previous treatment for an STD (a purported surrogate marker for sexual behaviors that place one at risk for acquiring an STD) [19], a new sexual partner [20], and removal of vaginal secretions prior to

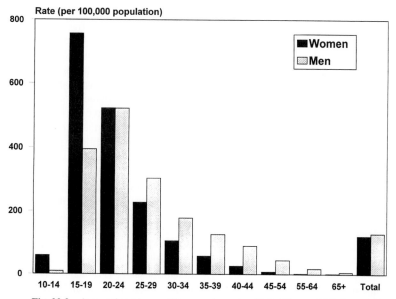

Fig. 22.3 Age- and gender-specific gonorrhea rates: United States, 1997. Source: [1].

Table 22.2
Clinical Spectrum of Gonococcal Infection in Women

Direct mucosal infection	Extension of local infection	Distant/disseminated infection
Asymptomatic	PID	Arthritis
Cervicitis	Perihepatitis	Dermatitis
Urethritis		Tenosynovitis
Bartholinitis		Perihepatitis
Skenitis		Endocarditis
Pharyngitis		Meningitis
Proctitis		
Conjunctivitis		

engaging in vaginal intercourse (dry sex) [21,22]. Additional risk markers for gonorrhea in females include urban residence [23], adolescence [24], or being unmarried [25].

IV. Clinical Spectrum of Infection in Women

Gonococcal infections, often asymptomatic in women, are a major cause of PID, tubal infertility, ectopic pregnancy, and chronic pelvic pain [26]. Routine screening, therefore, is a critical factor in identifying infections. Infection with *N. gonorrhoeae* can result in a spectrum of clinical manifestations in women ranging from asymptomatic infection to life-threatening gonococcal sepsis (Table 22.2). The majority of infections are limited to the mucosal site of inoculation. However, local extension of infection can occur, as can bacteremic dissemination.

A. Asymptomatic Infection

Among women attending urban, public clinics, approximately 50% of gonococcal infections in females are asymptomatic among those screened [4]. From the epidemiologic perspective, asymptomatic persons are a key factor in maintaining endemic/epidemic activity in a community because in the absence of symptoms they are less likely to defer or curtail sexual activity than those with symptomatic infection. Colonization does not occur in gonococcal infection. Therefore, laboratory-confirmed gonorrhea represents true infection and must be treated regardless of the presence or absence of symptoms. Although the cervix is the most commonly infected site of asymptomatic infection in women, any other mucosal surface, such as the rectum or the pharynx, may be involved. Infections that go undetected are of major importance in the development of PID and its sequelae [27]. Additionally, asymptomatic mucosal infection is very common in disseminated gonococcal infection (DGI) [28]. Therefore, screening of sexually active females should be a standard part of outpatient reproductive health care in areas with gonorrhea circulating in the community. In areas with low rates of gonorrhea, routine screening may be limited to women at the highest risk of infection. Any woman who meets at least one of the following criteria is at the highest risk for acquiring gonococcal infection and should be screened for gonorrhea at the time of presentation for health care, regardless of presenting complaint: (1) has multiple sexual partners, (2) has a sexual partner with more than one partner, (3) has entered into a new sexual partnership in the last 30 days, or (4) reports that a partner has been recently diagnosed with gonorrhea. Those in the latter group should be treated presumptively at the time of presentation for care because 40 to 50% of women in this category are found to have a laboratory-confirmed infection [29].

B. Clinical Syndromes Associated with Local Mucosal Infection

1. Cervicitis

Inflammation of the endocervix or the ectocervix, known as cervicitis, can have an infectious or noninfectious etiology. Cervicitis is a clinical sign found in 24–40% of female patients attending STD clinics [4] and 30–35% of patients seen in the private ambulatory care setting [30]. Because the uterine cervix is the primary site of acute, uncomplicated gonococcal infection in women, it is a major cause of cervicitis. The incubation period for gonococcal cervicitis is thought to be about 10 days and is less clear than that for men with gonococcal urethritis, which is 4 to 7 days [31]. Symptoms associated with gonococcal cervicitis are nonspecific and may include dysuria, vaginal discharge, genital pruritis, vaginal bleeding or spotting (especially postcoital), and lower abdominal pain. These nonspecific complaints are also noted by women with urinary tract infections, vaginal infections, and nongonococcal cervicitis. Therefore, full assessment of these reported symptoms includes a urinalysis (with culture if indicated) and an evaluation of vaginal secretions for candidiasis, trichomoniasis, and bacterial vaginosis, and collection of cervical secretions for the laboratory identification of *N. gonorrhoeae* and *Chlamydia trachomatis*. Additionally, a Papanicolaou smear should be collected from any sexually active woman who has not been screened in the past 12 months.

2. Urethritis

Cultures of specimens collected from the urethra of women with proven gonococcal cervicitis are positive for *N. gonorrhoeae* in 40–60% of these cases [4]. This may explain the common complaint of dysuria in women with gonococcal cervicitis. In women who have had a hysterectomy, the urethra is the primary site for gonococcal infection as the vaginal epithelium is relatively resistant to infection by *N. gonorrhoeae* due to cornification of the vaginal epithelium at puberty. Women are usually unaware of urethral discharge but may have urinary frequency in addition to dysuria. Internal dysuria, defined as discomfort perceived at the commencement of micturition, is often associated with urethritis. External dysuria, characterized by discomfort when the urinary stream contacts the external genitalia, is seen commonly in vulvovaginitis associated with candidiasis or genital herpes simplex virus infections. Gonococcal-associated pyuria is usually present in urethritis and routine urine cultures are negative unless coinfected with a common urinary pathogen. Patients with gonococcal or chlamydia urethritis alone may be misdiagnosed with a urinary tract infection (UTI). These patients may respond initially to standard trimethoprim-sulfamethoxazole UTI therapy due to this agent's activity against some gonococci [32]. Recurrence of a culture-negative UTI suggests the possibility of sexually transmitted urethritis.

3. Bartholinitis and Bartholin's Gland Abscess

The greater vestibular glands of Bartholin are located in the superficial perineal space, one on each side and posterior to the vestibular bulbs that connect to the clitoris. These 1-cm structures, which are not usually palpable, secrete lubricating mucus into 2-cm-long ducts lined with noncornified columnar epithelium. The ducts empty into the grooves at the junction of the posterior and middle thirds of the inner surface of the labia minora. Inflammation of the duct can lead to obstruction of the meatus with an associated perineal cellulitis known as bartholinitis. Infection of the Bartholin's gland itself can result in abscess formation. *N. gonorrhoeae* accounts for 20% of all infections of these glands and their associated ducts. Although gonococcal bartholinitis is usually asymptomatic, in 20% of infected women the gland is palpable or tender. When cellulitis occurs it is accompanied by pain, swelling, redness, and at times, purulent discharge [33]. Therefore, examination of the external genitalia in women should include attempted palpation of the glands. If discharge is found, it should be cultured for gonorrhea. Fever or leukocytosis is rarely noted with these local infections. Gonococcal Bartholin gland abscesses often undergo spontaneous external rupture with no associated sequelae. Early supportive therapy including Sitz baths may be helpful. However, failure to respond by the third day following treatment for acute uncomplicated gonorrhea warrants incision and drainage [34].

4. Skenitis

Inflammation of the paraurethral glands (ducts) of Skene can occur in gonoccocal infections. These small and usually invisible structures open on either side of the urethral orifice and are the female analog of the male prostate gland. When infected, the patient may complain of dysuria and purulent material may be expressed at times from the external gland orifice. Skenitis, like bartholinitis, usually occurs in conjunction with cervicitis or urethritis but can be seen as an isolated infection. Treatment is that for acute uncomplicated gonorrhea. In the rare case of abscess formation, surgical drainage is required.

5. Pharyngitis

N. gonorrhoeae can cause both symptomatic and asymptomatic pharyngeal infection. Approximately 10 to 20% of heterosexual women who have gonococcal cervicitis or urethritis are found to have oropharyngeal infection with *N. gonorrhoeae*, which is usually asymptomatic [35]. Pharyngeal infection appears to be more efficiently acquired by fellatio than cunnilingus based on studies among heterosexual men and women and men who have sex with men. No studies of pharyngeal gonorrhea among women who have sex with women have been reported, although gonorrhea, in general, has been noted in an inner-city STD clinic population to occur less frequently among lesbians (2%) than among heterosexual women (7%) [36]. Transmission by kissing and other nonsexual-related modes is rare [37,38].

N. gonorrhoeae needs to be considered in the differential diagnosis of any case of acute pharyngitis as it has been found to be the causative agent in about 1% of cases of symptomatic pharyngeal infection [39–41]. Although a pharyngeal exudate may be seen in some symptomatic cases, the Gram stain has no role in the diagnosis due to the common occurrence of the Gram-negative diplococci in the oropharynx, particularly *N. meningitidis* [42].

Pharyngeal specimens collected that identify *Neisseria* spp. on culture must be confirmed by biochemical or immunologic testing due to common colonization of the oropharynx by nongonococcal *Neisseria* species. It is important to note that pharyngeal infection with *N. gonorrhoeae* is less responsive to some antimicrobial treatments; therefore, presumptive treatment of gonorrhea in women who have had orogenital sexual exposure needs to be with an agent active at this location. Additionally, pharyngeal infection has been linked to systemic (disseminated) infection [43].

6. Anorectal Infection

Approximately 40% of women with uncomplicated gonococcal infections have positive rectal cultures for *N. gonorrhoeae*, but only 5% of women have this as the only site of isolation [44]. Organisms isolated from the rectum are more likely to bear a genetic locus (the *mtr* locus) which confers a survival advantage in the relatively hydrophobic environment of the rectum [45]. Rectal infections are usually asymptomatic [46]. Occasionally acute proctitis is noted and is characterized by purulent discharge, rectal bleeding, tenesmus, or anal pruritis. Symptoms may develop 5 to 7 days after exposure [46]. Findings on anoscopy are nonspecific and include mucopurulent exudate and inflammatory rectal mucosa [47].

7. Conjunctivitis

Although gonococcal conjunctivitis was commonly seen in the past among infants born to infected and untreated women, ocular infection continues to be occasionally seen in adults as a result of autoinoculation. Asymptomatic infection is thought to occur rarely. Gonococcal conjunctivitis requires prompt recognition and treatment to avert sight-threatening complications such as corneal ulceration, corneal perforation, and endophthalmitis [48–50].

C. Infection Associated with Local Extension

1. Pelvic Inflammatory Disease

PID is a clinical syndrome resulting from the ascending spread of microorganisms from the vagina or endocervix to the endometrium, fallopian tubes, and/or contiguous structures. It is the most common complication of gonococcal cervicitis, occurring in 10 to 20% of untreated infections [51] and up to 9% of cases in the year following treatment [52]. This translates to approximately one million women in the United States experiencing an episode of PID each year, requiring an estimated 2.5 million out-patient visits, more than 275,000 hospitalizations, and more than 100,000 surgical procedures [53]. Risk factors for PID include the adolescent age group [54], multiple sex partners [12], cigarette smoking [55], and, possibly, vaginal douching [56].

PID has a broad clinical spectrum (Table 22.3), from silent to chronic disease. The clinical features of acute PID present a wide variation in symptoms and signs and no single diagnostic tool consistently provides a sensitive diagnosis of acute PID.

Table 22.3

Clinical Spectrum of Pelvic Inflammatory Disease

Acute PID
Silent PID
Atypical PID
PID residual syndrome (chronic PID)
Postpartum/postabortal PID

Table 22.4

Clinical Diagnostic Criteria for Acute Pelvic Inflammatory Disease

Minimum criteria (all 3 must be present)
 Lower abdominal tenderness
 Adnexal tenderness
 Cervical motion tenderness
Additional criteria
 Oral temperature >101°F (>38.3°C)
 Abnormal cervical or vaginal discharge
 Elevated erythrocyte sedimentation rate
 Elevated C-reactive protein
 Laboratory documentation of cervical infection with *N. gonorrhoeae*
  or *C. trachomatis*
Definitive criteria
 Histopathologic evidence of endometritis on endometrial biopsy
 Transvaginal sonography or other imaging techniques showing
  thickened fluid-filled tubes with or without free pelvic fluid or
  tubo-ovarian complex
 Laparoscopic abnormalities consistent with PID

Source: [56a].

Laparoscopy is considered the diagnostic "gold standard" for identifying intra-abdominal PID-associated inflammation. However, it is not readily available in most clinical settings and its use is hard to justify when symptoms are mild or vague. Furthermore, it fails to diagnose endometrial PID or tubal PID when inflammatory changes are minimal. Similarly, clinical diagnosis is imprecise for symptomatic cases. Nevertheless, three minimum clinical criteria have been established to permit the empiric treatment: (1) lower abdominal tenderness, (2) adnexal tenderness, and (3) cervical motion tenderness (Table 22.4) [56a]. These currently applied criteria result in a diagnosis of acute PID with suboptimal sensitivity and specificity [57–59]. The application of the published "additional" criteria (Table 22.4) may enhance the specificity of the diagnosis. Notably, a history of tubal ligation does not negate a woman's risk for developing acute PID and needs to be considered in any woman presenting with a constellation of signs and symptoms consistent with this diagnosis [60,61]. Treatment should be initiated as soon as the presumptive diagnosis is made because prevention of long-term sequelae is linked to immediate initiation of therapy [62].

The proportion of PID attributable to gonorrhea depends on the rate of gonococcal infection in the community. In inner cities where gonorrhea is often the most common cause of bacterial cervicitis in women, *N. gonorrhoeae* may be the main cause of PID. Acute gonococcal PID is more likely to be associated with an abrupt febrile onset and findings consistent with peritoneal irritation than other causes of PID [63].

Complications of PID include chronic pelvic pain, tubal infertility, and ectopic pregnancy. The risk for complication following PID increases with each subsequent episode [64], with the severity of infection (when scored laparoscopically) and with delay in seeking care when symptomatic (Table 22.5). These data point to a substantial societal impact of PID and its complications. The economic burden exceeded $4.2 billion in 1990 and is projected to exceed $15 billion by 2000 [65].

2. Perihepatitis

Acute perihepatitis, also known as Fitz-Hugh-Curtis syndrome, is most commonly viewed as an extrapelvic manifestation of PID. It represents the spread, usually of *N. gonorrhoeae* or *C. trachomatis,* through the peritoneal cavity from the fallopian tubes to the Glisson's capsule surrounding the liver [66]. However, some cases probably occur by lymphatic or hematologic spread as the diagnosis has been reported in men [67,68]. Patients usually present with a complaint of right upper quadrant pain and have hepatic tenderness on physical examination, often with signs of peritoneal irritation. Overt clinical PID is often present but may be absent. Therefore, perihepatitis due to gonorrhea or chlamydia should be considered in any young, sexually active female with risk factors for gonorrhea and/or PID who has isolated right upper quadrant pain and/or hepatic tenderness. When unaccompanied by clinical PID it may be mistaken for acute cholecystitis or viral hepatitis.

D. Infection Associated with Bacteremic Spread to Distant Sites

Disseminated gonococcal infection (DGI) encompasses a spectrum of conditions including arthritis, dermatitis, tenosynovitis,

Table 22.5

Complications of Pelvic Inflammatory Disease among Women

	Percentage of women with complication per episode of PID			
Complication	No PID	1st Episode	2nd Episodes	>2nd Episodes
Chronic pelvic pain	7	12	30	53
Tubal factor infertility	2	16	37	66
Ectopic pregnancy	1	6	12	22

Source: [64].

perihepatitis, endocarditis, meningitis, osteomyelitis, adult respiratory distress syndrome, and sepsis. It is the result of bacteremic spread of *N. gonorrhoeae* and occurs in 0.5–3.0% of infected patients [43,69]. It is usually preceded by a local asymptomatic mucosal infection. Bacteremia begins 7 to 30 days after this initial infection and, in women, approximately half have the onset of symptoms within 7 days of a menstrual period [70]. The gonococcal organisms associated with DGI are more likely to have unique nutritional and growth characteristics, have an overrepresentation of certain protein IA serovars, and resist the bactericidal action of nonimmune human serum [71,72]. Risk factors for DGI include pharyngeal or rectal infection, female sex, possibly pregnancy, and terminal serum complement protein deficiency. Approximately 5% of persons who develop DGI have a homozygous deficiency in one of the terminal components of the complement system (C5, C6, C7, or C8) which predisposes them to disseminated Neisserial infections. Persons who present with a second episode of any disseminated Neisserial infection should be screened for this disorder.

The most common form of DGI is the arthritis–dermatitis syndrome [73–75]. The patient initially complains of polyarthriagias, which may be migratory, involving the elbows, knees, and more distal joints. In most cases, physical examination reveals objective signs of arthritis or tenosynovitis in two or more joints. Fever and systemic toxicity are common but usually mild. Asymmetric pauciarticular joint involvement is a distinguishing feature of this syndrome. In up to 75% of DGI patients dermatitis is also seen [73,75], characterized by discrete papules and pustules, often with a hemorrhagic component. Between 5 and 50 lesions are seen, usually on the extremities. Occasionally, necrotic lesions or hemorrhagic bullae are noted. Blood cultures at this time often yield the organism, whereas synovial fluid cultures are usually sterile. If DGI is not diagnosed and treated during the presentation of the arthritis–dermatitis syndrome, the dermatitis and the polyarthropathy will usually resolve. However, the arthritis usually continues and progresses to septic arthritis in one or more of the involved joints if treatment is not initiated. Arthrocentesis usually reveals purulent synovial fluid that is often culture positive for *N. gonorrhoaea*. Blood cultures at this stage are usually negative. Occasionally, a patient will develop gonococcal septic arthritis without previous dermatitis or arthralgias [73].

When DGI is suspected, all potential sites of primary infection need to be cultured, even in the absence of localized symptoms, because sites of local infection are more likely to yield the organism than are blood or synovial fluid cultures [73,75]. A DGI case is classified as "definite" if the gonococcus is isolated from a normally sterile site, such as blood or joint fluid. The diagnosis is considered "presumptive" in the setting of a clinical syndrome consistent with DGI and a gonococcal isolate recovered only from a mucosal site.

Gonococcal infective endocarditis is an infrequent, but very serious, manifestation of DGI occuring in up to 2% of patients with disseminated infection [76]. It is usually associated with the arthritis–dermatitis syndrome but has been seen in isolation. The aortic valve is the most commonly affected site [76]. The destruction of the valve is reportedly more rapid than with viridans streptococci but slower than seen in *Staphylococcus aureus* endocarditis. The other manifestations of DGI are rare but

should be considered in the sexually active, at-risk woman who presents with meningitis [77], osteomyelitis [78], sepsis [79], or adult respiratory distress syndrome [80].

E. Special Considerations: Gonococcal Infections in Pregnant or HIV-Infected Women

1. Gonococcal Infection in Pregnant Women

A test for *N. gonorrhoeae* should be performed at the first prenatal visit among woman at risk for infection and those women living in an area where the prevalence of gonococcal infection is high. A repeat test should be performed during the third trimester for women who are at continued risk for infection.

The clinical manifestations of gonorrhea are unchanged in pregnancy when compared with the nonpregnant state, except that PID is uncommon after the first trimester [81]. This observation is felt to be secondary to the complete obstruction of the uterine cavity by the placenta, fetus, and amniotic fluid thereby preventing ascending spread of the organism. Pregnant women with gonococcal infection have higher rates of spontaneous abortion, premature rupture of the fetal membranes, premature labor, and perinatal infant mortality [81]. It is unclear whether these increases are directly due to the gonococcus or to some other, as yet unidentified, associated risk factor.

Data suggest that pregnant women may be at increased risk for DGI [43,82]. It has been postulated that in later pregnancy, when sexual intercourse may become less comfortable due to expanding abdominal girth, orogenital contact may become more frequent between sexual partners. Therefore, at this time, women may be at increased risk of acquiring pharyngeal gonococcal infection. In addition, oropharyngeal infection, as previously noted, has been associated with a higher risk of DGI than has cervical infection alone [43].

2. Gonococcal Infection in HIV-Infected Women

There are no data to suggest that infection with HIV alters the clinical manifestations of local mucosal gonococcal infection in women [83]. However, HIV has been associated with more severe clinical manifestations of gonococcal PID, although the treatment strategy is unchanged [84]. There is evidence to suggest that HIV-infected women with cervical gonococcal infection transmit HIV more efficiently to male partners than HIV-infected women who are not coinfected with gonorrhea [85]. Also, women with gonorrhea appear to be at increased risk of HIV acquisition than women without gonorrhea when sexually exposed to an HIV-infected partner [5].

V. Diagnosis of Gonococcal Infections

A. Gram Staining of Collected Secretions

The Gram-stained smear for clinical specimens has long been used to assist in the rapid diagnosis of gonococcal infections. The smear is considered to be positive for the presumptive diagnosis of infection if Gram-negative intracellular diplococci (GNID) are seen within polymorphonuclear leukocytes, equivocal if the organisms are seen extracellularly only, and negative if no organisms are seen. Nongonococcal *Neisseria* species can be seen, particularly in women, and may be morphologically

Table 22.6

Sensitivity and Specificity of Gram Stain for the Identification of Gram-Negative Intracellular Diplococci in Various Clinical Settings

Gender	Clinical setting	Percent sensitivity	Percent specificity	References
Female	Uncomplicated endocervical infection	50–70	95–100	[86]
	Pelvic inflammatory disease	60–70	95–100	[87,88]
	Symptomatic urethritis	50–60	95–100	[87,89]
Male	Symptomatic urethritis	90–95	95–100	[88,90]
	Asymptomatic urethritis	50–70	95–100	[90,91]
Either	Proctitis (blind swab specimen collection procedure)	40–60	95–100+	[92,93]
	Proctitis (anoscopically collected specimen)	70–80	95–100+	[93]

Note. +, specificity for finding a gram-negative intracellular diplococcus which may be either the gonococcus or the meningococcus.

indistinguishable from *N. gonorrhoeae.* The sensitivity of the smear is much higher in men than in women (Table 22.6) [86–93] and in anoscopically collected rectal specimens than in those obtained by the blind swab technique.

Although the sensitivity of the Gram stain in gonococcal cervicitis is only 40–60%, the specificity exceeds 90%. Finding 20 to 30 polymorophonuclear neutrophils (PMNs) per oil immersion field in the densest portion of the Gram-stained specimen correlates with the presence of either *N. gonorrhoeae* or *C. trachomatis,* but the sensitivity and positive predictive value are too low for a definitive diagnosis [94]. However, a Gram stain with GNID mandates treatment for both gonorrhea and chlamydia. It is important to note that although mucopurulent cervicitis (MPC) can be caused by *N. gonorrhoeae* or *C. trachomatis,* in most cases neither organism can be isolated [88].

B. Culture

Isolation of *N. gonorrhoeae* remains the gold standard for the diagnosis of gonococcal infection. However, it is possible that this standard may be supplanted in the future by the use of genetic techniques that are becoming increasingly more sensitive and specific than culture. The most sensitive culture results are obtained when a nonselective chocolate agar medium is used along with a selective medium such as Modified Thayer-Martin (MTM), New York City medium, GC-Lect, or Martin Lewis [7]. However, specimens from normally sterile sites do not require selective media. The purpose of the selective media, all of which contain several antibiotics, is to inhibit the growth of normal genitourinary, pharyngeal, or rectal flora. All gonorrhea culture plates should be stored at 4°C and allowed to reach room temperature (25°C) before use. In the STD clinic setting, where a high volume of cultures for *N. gonorrhoeae* identification are collected, any medium can be used with good results by placing the inoculated plates in a humidified candle extinction jar that can be stored in a 35–37°C incubator [95]. In a clinical setting where fewer cultures for gonorrhea are collected, the John E. Martin Biological Environmental Chamber (JEMBEC) plate is commonly used which permits specimen shipping, even over long distances, in a carbon-dioxide enhanced and humid environment [96].

Cultures are incubated for 48 to 72 hours and a presumptive diagnosis is made on the basis of the colonial morphology on the culture plate, a positive oxidase reaction, and a Gram-stained smear of a single colony revealing Gram-negative diplococci. As *Kingella denitrificans,* for example, can grow on selective media and appear as a Gram-negative diplococcus, a confirmatory test is subsequently applied. For pharyngeal specimens, a second confirmatory test is used to increase the reliability of the finding.

C. Antigen and DNA Direct Detection Methods

The new antigen and DNA detection methods have the notable advantages of few, if any, transport problems. They can be rapidly performed (compared with culture) and are easily automated, thereby permitting efficient processing of large specimen volumes.

1. Nonamplification Methods

There are three major commercially available nonamplification tests: Pace-2 (Gen-Probe, San Diego, CA), Gonozyme (Abbott Laboratories, Chicago, IL), and Gonostat (Biotech Diagnostics, Baton Rouge, LA). Although each type of test may be nearly comparable to the Gram stain as a screening test in men with symptomatic gonococcal urethritis and may be slightly more sensitive than Gram staining for endocervical specimens, overall they are equal or less sensitive than culture and less specific than culture [97]. Nonetheless, these methods offer an important diagnostic alternative in areas where on-site microscopy for Gram staining is not available and the prevalence of infection is low, or where specimens must be shipped a great distance for processing.

2. Amplification Methods

There are two amplification methods currently available for diagnosis of *N. gonorrhoeae* infection: polymerase chain reaction (PCR) and ligase chain reaction (LCR). Both types of assays involve *in vitro* nucleic acid amplification that exponentially amplifies targeted DNA sequences. In numerous studies, PCR and LCR have been found to be as sensitive or more sensitive than culture for the identification of *N. gonorrhoeae* from urethral, endocervical, pharyngeal, rectal, urine, and patient-collected vaginal specimens [98–100]. Furthermore, simultaneous detection of *N. gonorrhoeae* and *Chlamydia trachomatis* using either swab or urine specimens offers increased sensitivity of diagnosis compared with culture and increased efficiency in specimen collection and processing [88,101–103]. The associated ease of specimen collection in nonclinical settings offers the future option for wide-scale screening of asymptomatic women and subsequent treatment of some groups of women who would not otherwise have been identified and treated. Such screening in nontraditional settings, when coupled with treatment of infected persons with single dose, directly observed oral therapy, should reduce the long-term sequelae of untreated infection, including PID, ectopic pregnancy, and tubal factor infertility.

Because of the serious social and medicolegal consequences of misdiagnosing gonorrhea or misidentifying strains of *N. gonorrhoeae*, the use of strict clinical criteria for reporting diagnoses of gonorrhea is suggested [96]. Three levels of diagnosis (Table 22.7) are defined on the basis of clinical findings or the results of laboratory diagnostic tests. A definitive diagnosis of gonorrhea is required when there are medicolegal issues associated with the case. However, the majority of cases diagnosed for surveillance and reporting purposes are presumptive clinical cases [104].

VI. Specimen Collection

On speculum examination, a swab test should be performed. When performing the swab test, a large cotton-tipped swab is inserted into the endocervix and examined in good light. A normal examination reveals a white cotton tip or a tip covered with clear cervical mucus whereas a positive swab test is noted when the swab tip turns yellow and indicates that mucopurulent cervicitis is present. The swab test also removes adherent cervicovaginal secretions from the os required prior to specimen collection for Gram stain and/or culture. Cervicitis, particularly MPC, may often be accompanied by a friable, edematous, and erythematous cervical os. Despite these findings, cervicitis and MPC are often asymptomatic. The presence of cervical ectopy, often mistaken for cervicitis, represents endocervical epithelium on the visible exocervix and is not cervicitis and is not pathologic. An endocervical specimen for Gram stain and culture should be collected using a sterile, cotton-tipped, nonwooden shafted swab inserted 1 to 2 cm into the external os and rotated gently for 20 to 30 seconds [86].

A urethral specimen may be indicated if the woman has had a hysterectomy or has a culture-negative UTI. A specimen should be collected from the urethral meatus if there is discharge present. If discharge is not present the urethra should be gently stripped by placing the gloved index finger into the vagina, applying gentle pressure to the roof of the vaginal vault and moving the finger forward toward the meatal opening to express the discharge. If no discharge is expressed from the meatus, a calcium alginate swab on a metal shaft should be inserted approximately 3.5 cm (1.5 inches) into the urethra and removed while rotating the shaft. The swab should be rolled onto a clean glass slide and then inoculated on selective media or appropriate transport media. The urethra has cuboidal epithelial cells that should be present in sheets on a normal smear. The distal urethra is colonized by normal bacterial flora, usually Gram-positive cocci and Gram-positive rods. Therefore, they are not considered to be clinically significant when seen on Gram-stained smear of urethral secretions. Recent voiding, within 2 hours, will reduce the number of white cells seen on Gram stain. Therefore, even the presence of a small number of PMNs on a postmicturition urethral smear strongly suggests the presence of urethritis, particularly if there was spontaneous or expressed meatal discharge. Although the sensitivity of finding GNID in male urethral

Table 22.7

Clinical Diagnostic Categories of Infection with *Neisseria gonorrhea*

Diagnostic category	Case definition
Suggestive diagnosis	Presence of a mucopurulent endocervical or urethral exudate on physical examination Sexual exposure to a person infected with *N. gonorrhoeae*
Presumptive diagnosis	Any 2 of the following 3 must be present: Typical gram-negative intracellular diplococci on microscopic examination of a smear or urethral exudate (men) of endocervical secretions (women) Growth of *N. gonorrhoeae,* from the urethra (men) or endocervix (women), on culture medium, and demonstration of typical colonial morphology, positive oxidase reaction, and typical gram-negative morphology Detection of *N. gonorrhoeae* by a nonculture laboratory test (Antigen detection test (*e.g.,* Gonozyme [Abbott]), direct specimen nucleic acid probe test (*e.g.,* Pace II [GenProbe]), nucleic acid amplifcation test (*e.g.,* LCR [Abbott]).
Definitive diagnosis	Isolation of *N. gonorrhoeae* from sites of exposure (*e.g.,* urethra, endocervix, throat, rectum) by culture (usually a selective medium) and demonstrating typical colonial morphology, positive oxidase reaction, and typical gram-negative morphology Confirmation of isolates by biochemical, enzymatic, serologic, or nucleic acid testing (*e.g.,* carbohydrate utilization, rapid enzyme substrate tests, serologic methods such as coagglutination or fluorescent antibody tests supplemented with additional tests that will ensure accurate identification of isolates) or a DNA probe culture confirmation technique

Source: [96].

smear is 95%, the sensitivity is only 50% for female urethral specimens [87]. Nevertheless, the finding of GNID on Gram stain mandates treatment.

Urine specimens are most likely to be collected for nucleic acid amplified testing. The first 10 to 20 ml of voided urine for culture is rarely used as it requires immediate processing because urine is toxic to the organism, thereby having marked impact on culture sensitivity.

In women who report anorectal contact and who have no symptoms suggestive of proctitis, a rectal specimen for culture can be collected by blindly passing a swab 2 to 3 cm into the anal canal, but it should be discarded if gross fecal contamination is noted upon removal of the swab. If symptoms of proctitis are present, then anosocopy should be undertaken and the specimen should be collected by direct visualization to increase sensitivity of the Gram-stained smear and thereby facilitate more rapid diagnosis [92,93].

Pharyngeal specimens, collected for culture only, are collected by swabbing the posterior pharynx, faucial pillars, and tonsils, if present.

Patients with suspected DGI should have blood cultures collected. The chance of recovery of the organism is greatest early in the course of disease [82]. The laboratory should be alerted that DGI is the suspected diagnosis to permit the addition of 1% gelatin to overcome the inhibitory growth effects of the anticoagulant sodium polyanetholsulfonate present in some commercial blood culture bottles [105]. Joint aspirates should be rapidly transported to the laboratory for culture and should not be inoculated into blood culture bottles.

VII. Treatment of Women with Gonococcal Infections

A. Dual Therapy for Gonococcal and Chlamydial Infections

Patients who are infected with *N. gonorrhoeae* are often coinfected with *C. trachomatis,* with studies demonstrating coinfection rates among women ranging from 35 to 50% [89,106]. Routine dual treatment of both infections without testing for chlamydia has been recommended by the Centers for Disease Control and Prevention (CDC) since 1985 [107] in the setting of presumptive gonococcal infection (Table 22.7) or when a person presents as the sexual contact of a person with a presumptive or definite diagnosis of gonorrhea from any site. Such an approach was found to be cost effective in populations where chlamydial infection accompanies 20–40% of gonococcal infections [88,89,105,108]. The cost effectiveness calculation is based on the cost of therapy for chlamydia, using doxycycline, compared with the cost of testing for chlamydia [109–111]. Additionally, because most gonococci in the United States are sensitive to both major agents used to treat chlamydia (doxycycline and azithromycin), cotreatment may have the salutary effect of hindering the development of antimicrobial resistance to *N. gonorrhoeae.*

B. Treatment Options for Gonococcal Infection in Women

The ideal agent for the treatment of a gonococcal infection is one with the following characteristics: (1) effective in killing the organism at the site(s) of infection, (2) ease of administration, (3) given in a manner to insure adherence, (4) can be given

for the shortest duration possible, (5) inexpensive, (6) low side effect profile, (7) high patient acceptability, and (8) efficacious against coexisting *C. trachomatis* and incubating syphilis, the other major causes of treatable STDs. The degree to which the selected agent meets the individual criteria to be an "ideal" agent is related to the spectrum of activity of the agent, type (local, extension, disseminated) and location of infection, host factors such as pregnancy, age, and allergy history, and the cost constraints on treatment.

Failure to cure a gonococcal infection has public health implications beyond the individual. For example, patients with asymptomatic infection often remain sexually active, thereby serving as a continuing source of transmission in the community. Regimens should have efficacies approaching 100%, and treatments with efficacies less than 95% should be avoided [106,112]. When an agent with suboptimal efficacy must be used, then a test of cure is warranted [29].

1. Uncomplicated Gonococcal Infection of the Cervix, Urethra, or Rectum

The treatment of uncomplicated gonococcal infection of the cervix, urethra, or rectum in adults is single dose cephalosporin or fluoroquinolone therapy (Table 22.8). Ceftriaxone (250 mg intramuscularly), ciprofloxacin (500 mg orally), and cefixime (400 mg orally) have been demonstrated to eradicate infection in males with symptomatic urethritis from urine in 4 hours and from mucosa and semen in 24 hours [113]. No similar study of time to organism eradication has been conducted in women. However, the recommended dose of oral ciprofloxacin has been shown to effect cure in all women with pharyngeal or cervical infection and the recommended dose of intramuscular ceftriaxone cures 99% of cervical infections and all pharyngeal infections in women [114,115].

Ceftriaxone and cefixime therapies have the advantage of probable activity against incubating syphilis [112,116]. Fluoroquinolone therapy should not be used for pregnant women due to potential teratogenicity [117] or for females under the age of 18 years due to concerns about cartilage toxicity [115]. Spectinomycin, which is effective in treating both genital and rectal gonorrhea, is reserved for the treatment of pregnant women with a history of rapid-onset (*i.e.,* IgE-mediated) reactions to penicillin or a documented cephalosporin allergy [112,115,116]. Neither fluoroquinolones nor spectinomycin inhibit *Treponema pallidum.*

2. Uncomplicated Gonococcal Infection of the Pharynx

Uncomplicated gonococcal infection of the pharynx is much more difficult to treat than infections of urogenital or anorectal areas. Ciprofloxacin, ceftriaxone, and ofloxacin have been shown to eradicate pharyngeal infection in at least 95% of cases [112,115,118]. Cefixime has not demonstrated this level of efficacy and therefore should not be used in pharyngeal infection. Similarly, spectinomycin is not optimally efficacious against pharyngeal gonococcal infection. Therefore, penicillin-allergic pregnant women who have pharyngeal infection who receive this agent should have a repeat throat culture collected 3 to 5 days after therapy to confirm cure.

3. Pelvic Inflammatory Disease

PID, a major complication of gonococcal infection in women, requires prompt treatment to decrease the occurrence of long-

Table 22.8
Recommended and Alternative Regimens for the Treatment of *Neisseria gonorrhoeae* Infections

Clinical presentation	Recommended regimen	Alternative regimens
Uncomplicated infection of the cervix, urethra, or rectum	Cefixime 400 mg PO in a single dose OR Cetriaxone 125 mg IM in a single dose OR Ciprofloxacin 500 mg PO in a single dose OR Ofloxacin 400 mg PO in a single dose PLUS Azithromycin 1 gm PO in a single dose OR Doxycycline 100 mg PO twice daily for 7 days	Spectinomycin 2 gm IM in a single dose OR Ceftizoxime 500 mg IM in a single dose OR Cefotaxime 500 mg IM in a single dose OR Cefotetan 1 gm IM in a single dose OR Cefoxitin 2 gm IM in a single dose with probenicid 1 gm PO OR Enoxacin 400 mg PO in a single dose OR Lomefloxacin 400 mg PO in a single dose OR Norfloxacin 800 mg PO in a single dose PLUS An antichlamydial agent
Uncomplicated infection of the pharynx	Ceftriaxone 125 mg IM in a single dose OR Ciprofloxacin 500 mg PO in a single dose OR Ofloxacin 400 mg PO in a single dose PLUS Antichlamydial therapy	As above
Allergy, Intolerance, Adverse Reaction	Spectinomycin 2 gm IM in a single dose	Not applicable
Pregnancy	Ceftriaxone 125 IM as a single dose PLUS Erythromycin base + 500 mg PO four times a day for 7 days OR Amoxicillin 500 mg PO three times a day for 7 days	Ceftizoxime 500 mg PO as a single dose OR Cefotaxime 500 mg IM in a single dose OR Cefotetan 1 gm IM in a single dose PLUS Erythromycin base 250 mg PO fours times a day for 14 days OR Erythromycin ethylsuccinate 800 mg PO four times a day for 14 days OR Azithromycin 1 gm PO in a single dose

(continued)

term sequelae. The treatment of this condition is empiric and provides broad coverage for the most likely pathogens, including *N. gonorrhoeae, C. trachomatis,* anaerobes, streptococci, and Gram-negative facultative bacteria. For hospitalized patients, initial intravenous therapy is indicated (Table 22.8) and should be given for at least 48 hours after the patient demonstrates substantial clinical improvement. At that time, the patient may be switched to oral medications to complete therapy. The preferred outpatient regimen for PID is ofloxacin combined with metronidazole for 14 days because it optimally covers the most likely pathogens. However, this regimen is costly, and therapy consisting of a single dose of parenteral third-generation cephalosporin combined with doxycycline for 14 days is also recommended and commonly used. Data on the use of azithromycin in the treatment of PID are insufficient to recommend

its use at this time. Immunodeficient women, including those with HIV infection and a low CD4+ cell count, those taking immunosuppressive therapy, or those having any other immunodeficiency, should receive initial parenteral therapy in the hospital [119]. PID, although uncommon in pregnant women, can occur in the first trimester of pregnancy and is associated with high rates of maternal morbidity, fetal wastage, and preterm delivery [120]. Therefore, pregnant women who have suspected PID should be hospitalized and receive parenteral therapy.

4. Disseminated Gonococcal Infection

DGI is best treated initially in the hospital with parenteral ceftriaxone. For all forms of DGI other than meningitis and endocarditis, parenteral therapy should be continued for 24 to 48 hours after clinical improvement is noted and then can be

Table 22.8

(continued)

Clinical presentation	Recommended regimen	Alternative regimens
Pregnancy with cephalosporin allergy/ intolerance	Spectinomycin 2 gm IM in a single dose	Not applicable
HIV infection	Same treatment as those without HIV infection	
Gonococcal conjunctivitis	Ceftriaxone 1 gm IM in a single dose PLUS Lavage the eye with saline solution one time	Unknown

Disseminated gonococcal infection (excluding meningitis or endocarditis)	Ceftriaxone 1 gm IM or IV every 24 hours for 24–48 h after improvement begins
	    THEN
	Cefixime 400 mg PO twice a day to complete 1 wk of Rx
	    OR
	Ciprofloxacin 500 mg PO twice a day to complete 1 wk of Rx
	    OR
	Ofloxacin 400 mg PO twice a day to complete 1 wk of Rx
	    PLUS
	Antichlamydial therapy
		Cefotaxime 1 gm IV every 8 hours for 24–48 h after improvement begins
		    OR
		Ceftizoxime 1 Gm IV every 8 hours for 24–48 h after improvement begins
		    THEN
		As noted under recommended regimen

DGI (excluding meningitis or endocarditis) in person with allergy to β-lactam drugs	Ciprofloxacin 500 mg IV every 12 hours for 24–48 h
	    OR
	Ofloxacin 400 mg IV every 12 h for 24–48 h
	    OR
	Spectinomycin 2 gm IM every 12 hours for 24–48 h
	    THEN
	Ciprofloxacin 500 mg PO twice a day to complete 1 wk of Rx
	    OR
	Ofloxacin 400 mg PO twice a day to complete 1 wk of Rx
	    PLUS
	Antichlamydia therapy
		Not applicable

Gonococcal meningitis	Ceftriaxone 1–2 gm IV every 12 hours for 10 to 14 days PLUS Antichlamydial therapy	
Gonococcal endocarditis	Cetriaxone 1–2 gm IV every 12 hours for at least 4 weeks PLUS Antichlamydial therapy	

Source: [56a].

changed to twice daily oral therapy with cefixime, ciprofloxacin, or ofloxacin to complete 1 week of therapy (Table 22.8). Gonococcal meningitis requires twice daily intravenous therapy with ceftriaxone for 10 to 14 days. In gonococcal endocarditis, intravenous therapy should continue for at least 4 weeks. Expert consultation should be sought to assist in the management of patients with gonococcal meningitis or endocarditis.

C. *N. gonorrhoeae Antimicrobial Resistance*

Antimicrobial resistance is an important consideration in the selection of the best agents with which to treat gonococcal infections. The CDC sponsors the Gonococcal Isolate Surveillance Project (GISP), a collaborative endeavor functioning since 1987, to monitor antimicrobial resistance in *Neisseria gonorrhoeae* in the United States [121]. In addition to the CDC, five regional laboratories and selected STD clinics participate in the project. In GISP, *N. gonorrhoeae* isolates are collected from the first 20 men with urethral gonorrhea attending STD clinics each month in approximately 25 cities in the United States. The susceptibilities of these isolates to antimicrobial agents, including broad-spectrum cephalosporins and fluoroquinolones currently recommended for the treatment of uncomplicated gonorrhea, are determined at regional laboratories. Susceptibilities are interpreted according to criteria recommended by the National Committee for Clinical Laboratory Standards. In 1996, all isolates were susceptible to spectinomycin, ceftriaxone and, cefixime (minimum inhibitory concentration (MIC) ≤ 0.25 mg/L).

In 1996 approximately 29% of all gonococcal isolates collected in the GISP were resistant [1] to penicillin, tetracycline, or both (Figure 22.4). Although the percentage of isolates that were penicillinase-producing *N. gonorrhoeae* (PPNG) decreased by 9% between 1991 and 1996, chromosomally mediated penicillin resistance has increased 3% during that same period.

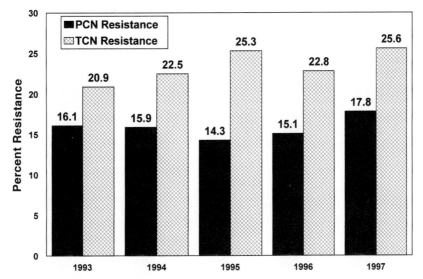

Fig. 22.4 Gonococcal Isolate Surveillance Project (GISP): Trends in combined plasmid- and chromosomally mediated resistance to penicillin and tetracycline, 1993–1997.

Chromosomally mediated tetracycline resistance has remained relatively stable over the 4-year-period beginning in 1993.

Fluroquinolone resistance has been found to be widespread in many parts of the world, particularly in Asia [122] and Africa [123]. Although the problem remains minor and geographically confined in the United States, in 1991 clusters were noted in isolated areas. In 1996 only 5 in 1000 (0.5%) gonococcal isolates demonstrated decreased susceptibility or resistance to ciprofloxacin (MIC $\geq$ 0.125 mg/l) compared with 0.3% (17/5238) of the isolates tested in 1991 and 1.3% (67/4996) of isolates tested in 1994. Most isolates with decreased sensitivities were found in Cleveland, OH (50% of isolates), Pacific coastal states, and Albuquerque, NM [124,125]. In 1996, only two ciprofloxacin-resistant isolates (MIC $\geq$ 1.0 mg/ml) were identified. Both were from Seattle, WA. Two ciprofloxacin-resistant isolates were identified in San Diego, CA in 1998 [126].

VIII. Patient Follow-Up After Therapy

Patients treated for uncomplicated infection of the cervix, urethra, rectum, or pharynx do not require a test of cure if a recommended regimen has been used [119]. Penicillin-allergic pregnant women with pharyngeal infection who are treated with spectinomycin require a test of cure [62]. In a patient who has not had sexual contact since treatment and has persistent symptoms, a culture for *N. gonorrhoeae* should be again collected and any isolate grown should be tested for antimicrobial sensitivities. However, the most common cause of infection following adequate treatment is reinfection, often due to failure to treat all of the woman's sex partners [118].

Women who have been treated for PID should have considerable clinical improvement within 3 days of initiation of therapy. Patients receiving outpatient therapy should be reexamined after 3 days of therapy and if there is no improvement noted, hospitalization is indicated for parenteral therapy and further evaluation. Hospitalized patients who do not demonstrate sub-

stantial clinical improvement within 3 days usually require further evaluation for other conditions. An ultrasound at this time may reveal a tuboovarian abscess or ectopic pregnancy, both of which require surgical consultation. Rescreening for *N. gonorrhoeae* and *C. trachomatis* 4 to 6 weeks after the completion of therapy for PID is recommended by some experts [62].

IX. Public Health Aspects of Gonococcal Infections

A. Management of Sex Partners

Identification and treatment of all exposed sexual partners of a person with a gonococcal infection, regardless of clinical presentation, is key to the control of this STD. Each patient should be advised to refer all of her sexual partners for the 60-day period before the onset of symptoms for evaluation and treatment. If the woman has had no partner in the last 60 days, then her most recent partner should be referred for treatment. Partners, regardless of symptom history, should be treated for both gonorrhea and chlamydia at the time of evaluation rather than waiting for laboratory test results. This strategy interrupts transmission and insures treatment, particularly among groups who are unlikely to return in a timely manner for test results. The patient should be instructed to refrain from sexual contact until therapy is completed and she no longer has symptoms and all of her sex partners have been treated and no longer have symptoms.

B. Prevention and Control

Preventing the spread of STDs, including gonorrhea, requires that those at risk for transmitting infection or acquiring infection modify their current sexual behaviors. Health care providers need to take complete sexual histories and provide patients with prevention messages and reasonable risk reduction strategies.

Every woman should be counseled that she and any new partner should be tested for STDs, including HIV, before beginning sexual contact. Thereafter, regardless of pregnancy risk, a new latex condom should be used with each act of sexual intercourse. The male condom has been shown to provide an effective mechanical barrier to gonorrhea and other STDs, including HIV [127]. On the other hand, no clinical data exist on the efficacy of the female condom in protecting against any STD except recurrent trichomoniasis [128]. Vaginal spermicides (used alone without condoms), the diaphragm, and the cervical sponge (not available in the United States) have been shown to reduce the risk of acquiring cervical gonococcal infection [129–131].

C. Reporting

Gonorrhea, chlamydia, and syphilis are reportable under the law in all jurisdictions in the United States [132]. Timely reporting is important for examining trends in morbidity as well as targeting resources for intervention and control. Questions regarding the mechanisms and requirements of reporting should be directed to your state or local health department's STD control program.

References

1. Division of STD Prevention (1998). "Sexually Transmitted Disease Surveillance, 1997." U.S. Department of Health and Human Services, Public Health Service. Centers for Disease Control and Prevention, Atlanta, GA.
2. Eng, T. R., and Butler, T. (1997). "The Hidden Epidemic: Confronting Sexually Transmitted Diseases." National Academy Press, Washington, DC.
3. Ebrahim, S. H., Peterman, T. A., Zaiki, A. A., and Kamb, M. L. (1997). Mortality related to sexually transmitted diseases in U.S. women, 1983 through 1992. Am. J. Public Health 87, 938–944.
4. McCormack, W. M., Strumacher, R. J., Johnson, K., and Donner, A. (1977). Clinical spectrum of gonococcal infection in women. Lancet 1, 1182–1185.
5. Wasserheit, J. N. (1992). Epidemiological synergy. Interrelationship between human immunodeficiency virus infection and other sexually transmitted diseases. Sex. Transm. Dis. 19, 61–77.
6. Boslego, J. W., Tramont, E. C., and Chung, R. C. (1991). Efficacy trial of a parenteral pilus vaccine in men. Vaccine 9, 154–162.
7. Levi, M. H. (1997). Current concepts in the laboratory diagnosis of gonorrhea. In "Sexually Transmitted Diseases: Epidemiology, Pathology, Diagnosis, and Treatment" (K. A. Borchart and M. A. Noble, eds.). CRC Press, New York.
7a. Swanson, J., Belland, R. J., and Hill, S. A. (1992). Neisserial surface variation: How and why? Curr. Opin. Genet. Dev. 2, 805–811.
7b. Knapp, J. S., Tam, M. R., Nowinski, R. C., Holmes, K. K., and Sandström, E. G. (1984). Serological classification of Neisseria gonorrhoeae with use of monoclonal antibodies to gonococcal outer membrane protein I. J. Infect. Dis. 50, 44–48.
8. Plummer, F. A., Simonsen, J. N., Chubb, H., Slaney, L., Kimata, J., Bosire, M., Ndinya-Achola, J. O., and Ngugi, E. N. (1989). Epidemiologic evidence for the development of serovar-specific immunity after gonococcal infection. J. Clin. Invest. 83, 1472–1476.
9. Brunham, R. C., Plummer, F., Slaney, L., Rand, F., and DeWitt, W. (1985). Correlation of auxotype and protein I with expression of disease due to Neisseria gonorrhoeae. J. Infect. Dis. 152, 339–343.
10. Salyers, A. A., and Whitt, D. D. (1994). "Bacterial Pathogenesis: A Molecular Approach." ASM Press, Washington, DC.
11. U.S. Department of Health and Human Services (1995). "Healthy People 2000: Midcourse Review and 1995 Revisions." USDHHS, Public Health Service, U.S. Govt. Printing Office, Washington, DC.
12. Rothenberg, R. B. (1983). Geography of gonorrhea: Empirical demonstration of core group transmission. Am. J. Epidemiol. 117, 688–693.
13. Zenilman, J. M., Whittington, W. L., Frazier, D., Rice, R. J., and Knapp, J. S. (1988). Penicillinase-producing Neisseria gonorrhoeae in Dade County, Florida: Evidence of core-group transmitters and the impact of illicit antibiotics. Sex. Transm. Dis. 15, 158–163.
14. Becker, K. M., Glass, G. E., Brathwaite, W. S., and Zenilman, J. M. (1998). Geographic epidemiology of gonorrhea in Baltimore, Maryland using a geographic information system. Am. J. Epidemiol. 147, 709–716.
15. Rothenberg, R. B., and Potterat, J. J. (1988). Temporal and social aspects of gonorrhea transmission: The force of infectivity. Sex. Transm. Dis. 15, 88–92.
16. Thin, R. N. T., Williams, I. A., and Nicol, C. S. (1971). Direct and delayed methods of immunofluorescent diagnosis of gonorrhea in women. Br. J. Vener. Dis. 47, 27–30.
17. Hooper, R. R., Reynolds, G. H., Jones, O. G., Zaidi, A., Wiesner, P. J., Latimer, K. P., Lester, A., Campbell, A. F., Harrison, W. O., Karney, W. W., and Holmes, K. K. (1978). Cohort study of venereal disease: I. The risk of gonorrhea transmission from infected women to men. Am. J. Epidemiol. 108, 136–144.
18. Holmes, K. K., Johnson, D. W., and Trostle, J. H. (1970). An estimate of the risk of men acquiring gonorrhea by sexual contact with females. Am. J. Epidemiol. 91, 170–174.
19. Oh, M. K., Cloud, G. A., Fleenor, M., Sturdevant, M. S., Nesmith, H. D., and Feinstein, R. A. (1996). Risk for gonococcal and chlamydial cervicitis in adolescent females: Incidence and recurrence in a prospective study. J. Adolesc. Health 18, 270–275.
20. Upchurch, D. M., Brady, W. E., Reichart, C. A., and Hook, E. W., III (1990). Behavioral contributions to acquisition of gonorrhea in patients attending an inner city sexually transmitted disease clinic. J. Infect. Dis. 161, 938–941.
21. Civic, D., and Wilson, D. (1996). Dry sex in Zimbabwe and implication for condom use. Soc. Sci. Med. 42, 91–98.
22. Brown, J. E., Ayowa, O. B., and Brown, R. C. (1993). Dry and tight: Sexual practices and potential AIDS risk in Zaire. Soc. Sci. Med. 37, 989–994.
23. Thomas, J. C., Schoenbach, V. J., Weiner, D. H., Parker, E. A., and Earp, J. A. (1996). Rural gonorrhea in the southwestern United States: A neglected epidemic. Am. J. Epidemiol. 143, 269–277.
24. Louv, W. C., Austin, H., Perlman, J., and Alexander, W. J. (1989). Oral contraceptive use and the risk of chlamydial and gonococcal infections. Am. J. Obstet. Gynecol. 160, 396–402.
25. Mertz, K. J., Hadgu, A., Dorian, K. J., Berman, S. M., Mosure, D. J., and Levine, W. C. (1997). Screening women for gonorrhea: Demographic screening criteria for general clinical use. Am. J. Public Health 87, 1535–1538.
26. Jonnson, M., Wadell, G., Gustavsson, A., Evander, M., Edlund, K., Boden, E., Rylander, E., and Karlsson, R. (1995). The silent suffering women—a population based study on the association between symptoms and past and present infections of the lower genital tract. Genitourin. Med. 71, 158–162.
27. Jossens, M. O., Eskenazi, B., Schacter, J., and Sweet, R. L. (1996). Risk factors for pelvic inflammatory disease. A Case-control Study. Sex. Transm. Dis. 23, 239–247.
28. Ross, J. D. (1996). Systemic gonococcal infection. Genitourin. Med. 72, 404–407.
29. Carne, C. A. (1997). Epidemiologic treatment and tests of cure in gonococcal infection: Evidence of value. Genitourin. Med. 73, 12–15.
30. Carey, V. (1996). Cervicitis. In "Manual of Medical Practice." (R. E. Rankel, ed.). Saunders, Philadelphia.

31. Platt, R., Rice, P. A., and McCormack, W. M. (1983). Risk of acquiring gonorrhea and prevalence of abnormal adnexal findings among women recently exposed to gonorrhea. *JAMA, J. Am. Med. Assoc.* **250,** 3205–3209.

32. Bushby, S. R. M. (1973). Trimethoprim-sulfamethoxazole: In vitro microbiologic aspects. *J. Infect. Dis.* **128,** S442–S462.

33. Rees, E. (1967). Gonococcal bartholonitis. *Br. J. Vener. Dis.* **43,** 1506.

34. Azzan, B. B. (1978). Bartholin's cyst and abscess. A review of 53 cases. *Br. J. Clin. Pract.* **32,** 101.

35. Rodin, P., Monteiro, G. E., and Scrimgeour, G. (1972). Gonococcal pharyngitis. *Br. J. Vener. Dis.* **48,** 182–183.

36. Skinner, C. J., Stokes, J., Kirlew, Y., Kavanaugh, J., and Forster, G. E. (1996). A case-controlled study of sexual health needs of lesbians. *Genitourin. Med.* **72**(4), 277–280.

37. Willmott, F. E. (1974). Transfer of gonococcal pharyngitis by kissing? *Br. J. Vener. Dis.* **50,** 317–318.

38. Jewett, J. F. (1979). Nonsexual acquisition of genital gonococcal infection. *N. Engl. J. Med.* **301,** 1347.

39. Odegaard, K., and Gundersen, T. (1973). Gonococcal pharyngeal infection. *Br. J. Vener. Dis.* **49,** 350–352.

40. Ratnatunga, C. S. (1972). Gonococcal pharyngitis. *Br. J. Vener. Dis.* **48,** 184–186.

41. Monto, A. S., and Ullman, B. M. (1974). Acute respiratory illness in an American community. The Tecumseh Study. *JAMA, J. Am. Med. Assoc.* **227,** 164–169.

42. Tice, R. W., Jr., and Rodriguez, V. L. (1981). Pharyngeal gonorrhea. *JAMA, J. Am. Med. Assoc.* **246,** 2717–2719.

43. Handsfield, H. H. (1975). Disseminated gonococcal infection. *Clin. Obstet. Gynecol.* **18,** 131–142.

44. Klein, E. J., Fisher, L. S., Chow, A. W., and Guze, L. B. (1977). Anorectal gonococcal infection. *Ann. Intern. Med.* **86,** 340–346.

45. Morse, S. A., Lysko, P. G., McFarland, L., Knapp, J. S., Sandström, E., Critchlow, C., and Holmes, K. K. (1982). Gonococcal strains from homosexual men have outer membranes with reduced permeability to hydrophobic molecules. *Infect. Immun.* **37,** 432.

46. Barlow, D., and Phillips, I. (1978). Gonorrhea in women. Diagnostic, clinical, and laboratory aspects. *Lancet* **1,** 761–764.

47. Quinn, T. C., Corey, L., Chaffee, R. G., Schuffler, M. D., Brancato, F. P., and Holmes, K. K. (1981). The etiology of anorectal infections in homosexual men. *Am. J. Med.* **71,** 395–406.

48. Kestelyn, P., Bogaerts, J., and Meheus, A. (1987). Gonococcal keratoconjunctivitis in African adults. *Sex. Transm. Dis.* **14,** 191–194.

49. Ullman, S., Roussel, T. J., Culbertson, W. W., Rorster, R. K., Alfonso, E., Mendlelsohn, A. D., Heidemann, D. G., and Holland, S. P. (1987). *Neisseria gonorrhoeae* keratoconjunctivitis. *Ophthalmology* **94,** 525–531.

50. Deschenes, J., Seamone, C., and Baines, M. (1990). The ocular manifestations of sexually transmitted diseases. *Can. J. Ophthalmol.* **25,** 177–185.

51. Kottmann, L. M. (1995). Pelvic inflammatory disease: Clinical overview. *J. Obstet. Gynecol. Neonatal Nurs.* **24,** 759–767.

52. Rothman, K. J., Lanza, L., Lal, A., Peskin, E. G., and Dreyer, N. A. (1996). Incidence of pelvic inflammatory disease among women treated for gonorrhea or chlamydia. *Pharmacoepidemiol. Drug Saf.* **5,** 409–414.

53. Rolfs, R. T., Galaid, E. T., and Aaidi, A. A. (1990). Epidemiology of pelvic inflammatory disease: Trends in hospitalizations and office visits, 1979–1988. *J. Meet. Cent. Dis. Control Natl. Inst. Health Pelvic Inflamm. Dis. Prev. Manag. Res. 1990s,* Bethesda, MD.

54. Bell, T. A., and Holmes, K. K. (1984). Age-specific risks of syphilis, gonorrhea, and hospitalized pelvic inflammatory disease in sexually experienced U.S. women. *Sex. Transm. Dis.* **11,** 291–295.

55. Marchbanks, P. A., Lee, N. C., and Peterson, B. C. (1990). Cigarette smoking as a risk factor for pelvic inflammatory disease. *Am. J. Obstet. Gynecol.* **162,** 639–644.

56. Wølner-Hanssen, P., Eschenbach, D. A., Paavonen, J., Stevens, C. E., Kaviat, N. B., Critchlow, C., DeRouen, T., Koutsky, L., and Holmes, K. K. (1990). Association between vaginal douching and acute PID. *JAMA, J. Am. Med. Assoc.* **263,** 1936–1941.

56a. Centers for Disease Control and Prevention (1998). "Guidelines for Treatment of Sexually Transmitted Diseases." U.S. Department of Health and Human Services, U.S. Public Health Service, Atlanta, GA.

57. Kahn, J. G., Walker, C. K., Washington, A. E., Landers, D. V., and Sweet, R. L. (1991). Diagnosing pelvic inflammatory disease. A comprehensive analysis and considerations for developing a new model. *JAMA, J. Am. Med. Assoc.* **266,** 2596–2604.

58. Korn, A. P., Hessol, N., Padian, N., Bolan, G., Muzsnai, D., Donegan, E., Jonte, J., Schacter, J., and Landers, D. V. (1995). Commonly used diagnostic criteria for pelvic inflammatory disease have poor sensitivity for plasma cell endometritis. *Sex. Transm. Dis.* **22,** 335–341.

59. Jacobson, L., and Weström, L. (1969). Objectivized diagnosis of acute pelvic inflammatory disease: Diagnostic and prognostic value of routine laparoscopy. *Am. J. Obstet. Gynecol.* **105,** 1088–1098.

60. Abbuhl, S. B., Muskin, E. B., and Shofer, F. S. (1997). Pelvic inflammatory disease in patients with bilateral tubal ligation. *Am. J. Emerg. Med.* **15,** 271–274.

61. Green, M. M., Vicario, S. J., Sanfilippo, J. S., and Lochhead, S. A. (1991). Acute pelvic inflammatory disease after surgical sterilization. *Ann. Emerg. Med.* **20,** 344–347.

62. Munday, P. E. (1997). Clinical aspects of pelvic inflammatory disease. *Hum. Reprod.* **12**(11 Suppl.), 121–126.

63. Holmes, K. K., Eschenbach, D. A., and Knapp, J. S. (1980). Salpingitis: Overview of etiology and epidemiology. *Am. J. Obstet. Gynecol.* **138,** 893–900.

64. Weström, L. (1980). Impact of sexually transmitted diseases on human reproduction: Swedish studies of infertility and ectopic pregnancy. *In* "Sexually Transmitted Diseases Status Report," NIAID Study Group. NIH Publications, Washington, DC.

65. Washington, A. E., and Katz, P. (1991). Cost of and payment for pelvic inflammatory disease. Trends and projections, 1983 through 2000. *JAMA, J. Am. Med. Assoc.* **266,** 2565–2569.

66. Lopez-Zeno, J. A., Keith, L. G., and Berger, G. S. (1985). The Fitz-Hugh-Curtis syndrome revisited. Changing perspectives after half a century. *J. Reprod. Med.* **30,** 567–582.

67. Kimball, M. W., and Knee, S. (1970). Gonococcal perihepatitis in the male. The Fitz-Hugh-Curtis Syndrome. *N. Engl. J. Med.* **282,** 1082–1084.

68. Francis, T. I., and Osoba, A. O. (1972). Gonococcal hepatitis (Fitz-Hugh-Curtis Syndrome) in a male patient. *Br. J. Vener. Dis.* **48,** 187–188.

69. Masi, M. T., and Eisenstein, B. I. (1981). Disseminated gonococcal infection (DGI) and gonococcal arthritis (GCA): II. Clinical manifestations, diagnosis, complications, treatment and prevention. *Semin. Arthritis Rheum.* **10,** 173–197.

70. Holmes, K. K., Counts, G. W., and Beaty, H. N. (1971). Disseminated gonococcal infection. *Ann. Intern. Med.* **74,** 979–993.

71. Knapp, J. S., and Holmes, K. K. (1975). Disseminated gonococcal infection caused by *Neisseria gonorrhoeae* with unique nutritional requirements. *J. Infect. Dis.* **132,** 204–208.

72. Bohnhoff, M., Morello, J. A., and Lerner, S. A. (1986). Auxotypes, penicillin susceptibility and serogroups of *Neisseria gonorrhoeae* form disseminated and uncomplicated infections. *J. Infect. Dis.* **154,** 225–230.

73. Rompalo, A. M., Hook, E. W., III, Roberts, P. L., Ramsey, P. G., Handsfield, H. H., and Holmes, K. K. (1987). The acute arthritis-

dermatitis syndrome: The changing importance of *Neisseria gon-orrhoeae* and *Neisseria meningitidis. Arch. Intern. Med.* **147,** 281–283.

74. Manicourt, D. H., and Orloff, S. (1982). Gonococcal arthritis-dermatitis syndrome: Study of serum and synovial fluid immune complex levels. *Arthritis Rheum.* **25,** 574–578.

75. Holmes, K. K., Weisner, P. J., and Pedersen, A. H. B. (1971). The gonococcal arthritis-dermatitis syndrome. *Ann. Intern. Med.* **75,** 470–471.

76. Cartwright, I., Petty, M., Reyes, M., and Illuminati, L. (1996). Gonococcal endocarditis. *Sex. Transm. Dis.* **23,** 181–183.

77. Cash, J. M., and Erzurum, S. C. (1988). Gonococcal meningitis: Case report and review of the literature. *S. D. J. Med.* **41,** 5–7.

78. Gantz, N. M., McCormack, W. M., Laughlin, L. W., Shauffer, I. A., and Cohen, A. S. (1976). Gonococcal osteomyelitis. An unusual complication of gonococcal arthritis. *JAMA, J. Am. Med. Assoc.* **236,** 2431–2432.

79. Markham, J. D., Vilseck, J. R., Jr., and O'Donohue, W. J., Jr. (1976). Acute respiratory distress syndrome associated with gon-ococcal septicemia. *Chest* **70,** 667–670.

80. Walters, D. G., and Goldstein, R. A. (1980). Adult respiratory distress syndrome and gonococcemia. *Chest* **77,** 434–436.

81. Gilstrap, L. C., and Faro, S. (1997). Sexually transmitted diseases: I. Gonorrhea and chlamydia. *in* "Infections in Pregnancy," 2nd ed. Wiley-Liss, New York.

82. O'Brien, J. A., Goldenberg, D. L., and Rice, P. A. (1983). Dissem-inated gonococcal infection: A prospective analysis of 49 patients and review of pathophysiology and immune mechanisms. *Medi-cine (Baltimore)* **62,** 395–406.

83. McNeeley, S. G. (1989). Gonococcal infections in women. *Obstet. Gynecol. Clin. North Am.* **16,** 467–468.

84. Kamenga, M. C., DeCock, K. M., St. Louis, M. E., Toure, C. K., Zakaria, S., N'gbichi, J. M., Ghys, P. D., Holmes, K. K., Eschen-bach, D. A., and Gayle, H. D. (1995). The impact of human im-munodeficiency virus infection on pelvic inflammatory disease: A case-control study in Abidjan, Ivory Coast. *Am. J. Obstet. Gyne-col.* **172,** 919–925.

85. Levine, W. C., Talkington, D. F., Mitchell, S., Farshy, C. E., Zaidi, A. A., Lewis, J. S., Tambe, P., Bhoomkar, A., and Pope, V. (1998). Increase in endocervical CD4 lymphocytes among women with nonulcerative sexually transmitted diseases. *J. Infect. Dis.* **177,** 167–174.

86. Zenilman, J. M. (1993). Gonorrhea: Clinical and public health issues. *Hosp. Pract.* **28**(2A), 29–50.

87. Barlow, D., and Philips, I. (1978). Gonorrhea in women: Diagnos-tic, clinical, and laboratory aspects. *Lancet* **1,** 761–764.

88. Rothenberg, R. B., Simon, R., Chipperfield, E., and Catterall, R. D. (1976). Efficacy of selected diagnostic tests for sexually transmit-ted diseases. *JAMA, J. Am. Med. Assoc.* **235,** 49–51.

89. Thin, R. N., and Shaw, E. J. (1979). Diagnosis of gonorrhea in women. *Br. J. Vener. Dis.* **55,** 10–13.

90. Jacobs, N. F., and Kraus, S. J. (1975). Gonococcal and nongono-coccal urethritis in men: Clinical and laboaotory differentiation. *Ann. Intern. Med.* **82,** 7–12.

91. Hansfield, H. H., Lipman, T. O., Harnisch, J. P., Tronca, E., and Holmes, K. K. (1974). Asymptomatic gonorrhea in men: Diagno-sis, natural course, prevalence, and significance. *N. Engl. J. Med.* **290,** 117–123.

92. Deheragoda, P. (1977). Diagnosis of rectal gonorrhea by blind anorectal swabs compared with direct vision swabs taken via a proctoscope. *Br. J. Vener. Dis.* **53,** 311–313.

93. William, D. C., Felman, Y. M., and Riccardi, N. B. (1981). The utility of anoscopy in the rapid diagnosis of symptomatic anorectal gonorrhea in men. *Sex. Transm. Dis.* **8,** 16–17.

94. Brunham, R. C., Paavonen, J., Stevens, C. E., Kiviat, N., Kuo, C. C., Critchlow, C. W., and Holmes, K. K. (1984). Mucopurulent cervicitis—the ignored counterpart in women of urethritis in men. *N. Engl. J. Med.* **311,** 1–6.

95. Martin, J. E., and Jackson, R. L. (1975). A biological environmen-tal chamber for the culture of *Neisseria gonorrhoeae. Sex. Transm. Dis.* **2,** 28–30.

96. Centers for Disease Control and Prevention (1998). The gonorrhea page. Available at: *www.cdc.gov/ncidod/dastlr/gcdir/Neldent/ Ngon.html*

97. DiDomenico, N., Link, H., Knobel, R., Caratsch, T., Weschler, W., Loewy, Z. G., and Rosenstraus, M. (1996). COBAS AMPLICOR: Fully automated RNA and DNA amplification and detection sys-tem for routine diagnostic PCR. *Clin. Chem. (Winston-Salem, N.C.)* **42,** 1915–1923.

98. Hook, E. W., III, Ching, S. F., Stephens, J., Hardy, K. F., Smith, K. R., and Lee, H. H. (1997). Diagnosis of *Neisseria gonorrhoeae* infections in women by using the ligase chain reaction on patient-obtained vaginal swabs. *J. Clin. Microbiol.* **35,** 2129–2132.

99. Smith, K. R., Ching, S., Lee, H., Ohhashi, Y., Hu, H. Y., and Fisher, H. C., III (1995). Evaluation of ligase chain reaction for use with urine for identification of *Neisseria gonorrhoeae* in females attending a sexually transmitted disease clinic. *J. Clin. Microbiol.* **33,** 455–457.

100. Ching, S., Lee, H., Hook, E. W., III, and Zenilman, J. (1995). Ligase chain reaction for detection of *Neisseria gonorrhoeae* in urogenital swabs. *J. Clin. Microbiol.* **33,** 3111–3114.

101. Stary, A., Ching, S. F., Teodorowicz, L., and Lee, H. (1997). Com-parison of ligase chain reaction and culture for detection of *Neis-seria gonorrhoeae* in genital and extragenital specimens. *J. Clin. Microbiol.* **35,** 239–242.

102. Buimer, M., van Doornum, G. J., Ching, S., Peerbooms, P. G., Plier, P. K., Ram, D., and Lee, H. H. (1996). Detection of *Chla-mydia trachomatis* and *Neisseria gonorrhoeae* by ligase chain re-action-based assays with clinical specimens from various sites: Implications for diagnostic testing and screening. *J. Clin. Micro-biol.* **34,** 2395–2400.

103. Carroll, K. C., Aldeen, W. E., Morrison, M., Anderson, R., Lee, D., and Mottice, S. (1998). Evaluation of the Abbott Lcx ligase chain reaction assay for detection of *Chlamydia trachomatis* and *Neisseria gonorrhoeae* in urine and genital swab specimens form a sexually transmitted disease clinic population. *J. Clin. Micro-biol.* **36,** 1630–1633.

104. Wharton, M., Chorba, T. L., Vogt, R. L., Morse, D. L., and Beuhler, J. W. (1990). Case definitions for public health surveil-lance. *Morbid. Mortal. Wkly. Rep.* **39**(RR-13), 11.

105. Evalgelista, A. T., Beilstein, H. R., and Cumitech, R. A. (1993). "Laboratory Diagnosis of Gonorrhea." American Society of Microbiology, Washington, DC.

106. Batteiger, B. E., and Jones, R. B. (1987). Chlamydial infections. *Infect. Dis. Clin. North Am.* **1,** 55–81.

107. Centers for Disease Control (1985). 1985 STD treatment guide-lines. *Morbid. Mortal. Wkly. Rep.* **34**(Suppl. 4S), 1–35.

108. Nettleman, M. D., Jones, R. B., Roberts, S. D., Katz, B. P., Washington, A. E., Dittus, R. S., and Quinn, T. S. (1986). Cost-effectiveness of culturing for *Chlamydia trachomatis:* A study in a clinic for sexually transmitted diseases. *Ann. Intern. Med.* **105,** 189–196.

109. Washington, A. E., Browner, W. S., and Korenbrot, C. C. (1987). Cost effectiveness of combination therapy for endocervical gon-orrhea: Considering co-infection with *Chlamydia trachomatis. JAMA, J. Am. Med. Assoc.* **257,** 2056–2060.

110. Braunstein, H. (1987). Gonorrhea as a marker of chlamydia infec-tion. *JAMA, J. Am. Med. Assoc.* **258,** 1330–1331.

111. Begley, C. E., McGill, L., and Smith, P. B. (1989). The incremental cost of screening, diagnosis, and treatment of gonorrhea and chlamydia in a family planning clinic. *Sex. Transm. Dis.* **16,** 63–67.

112. Moran, J. S., and Levine, W. C. (1995). Drugs of choice for the treatment of uncomplicated gonorrhea infections. *Clin. Infect. Dis., Suppl.* **1,** S47–S65.

113. Haizlip, J., Isbey, S. F., Hamilton, H. A., Jerse, A. E., Leone, P. A., Davis, R. H., and Cohen, M. S. (1995). Time required for elimination of *Neisseria gonorrhoeae* from the urogenital tract in men with symptomatic urethritis: Comparison of oral and intramuscular single-dose therapy. *Sex. Transm. Dis.* **22,** 145–148.

114. Hook, E. W., III, Jones, R. B., Martin, D. H., Bolan, G. A., Mrocykowsk, T. F., Neuman, T. M., Haag, J. J., and Echols, R. (1993). Comparison of ciprofloxacin and ceftriaxone as single-dose therapy for uncomplicated gonorrhea in women. *Antimicrob. Agents Chemother.* **24,** 784–786.

115. Echols, R. M., Heyd, A., O'Keefe, B. J., and Schact, P. (1994). Single dose ciprofloxacin for the treatment of uncomplicated gonorrhea: A worldwide summary. *Sex. Transm. Dis.* **21,** 345–352.

116. Bowie, W. R. (1994). Antibiotics and sexually transmitted diseases. *Infect. Dis. Clin. North Am.* **8,** 841–857.

117. Ison, C. A. (1996). Antimicrobial agents and gonorrhea: Therapeutic choice, resistance, and susceptibility testing. *Genitourin. Med.* **72,** 253–257.

118. Schact, P., Arcieri, G., Branolte, J., Ruick, H., Chysky, V., Griffith, E., Greunwaldt, G., Hullman, R., Konopka, C. A., and O'Brien, B. (1988). Worldwide clinical data on efficacy and safety of ciprofloxacin. *Infection* **16**(Suppl. 1), S29–S43.

119. Centers for Disease Control and Prevention (1998). Guidelines for the treatment of sexually transmitted diseases. *Morbid. Mortal. Wkly. Rep.* **7**(RR-1), 1–116.

120. Blanchard, A. C., Pastorek, J. G., II, and Weeks, T. (1987). Pelvic inflammatory disease during pregnancy. *South. Med. J.* **80,** 1363–1365.

121. Division of STD Prevention (1994). "Sexually Transmitted Disease Surveillance, 1994." U.S. Department of Health and Human Services, Public Health Service, Centers for Disease Control and Prevention, Atlanta, GA.

122. Knapp, J. S., Fox, K. K., Trees, D. L., and Whittington, W. L. (1997). Fluoroquinolone resistance in *Neisseria gonorrhoeae. Emerg. Infect. Dis.* **3,** 33–39.

123. Van Dyck, E., Crabbé, F., Nzila, N., Boegaerts, J., Munyabikali, J. P., Ghys, P., Diallo, M., and Laga, M. (1997). Increasing resistance of *Neisseria gonorrhoeae* in west and central Africa. Consequence on therapy of gonococcal infection. *Sex. Transm. Dis.* **24,** 32–37.

124. Kilmarx, P. H., Knapp, J. S., Xia, M., St. Louis, M. E., Sayers, D., Doyle, L. J., Roberts, M. C., and Whittington, W. L. (1998). Intercity spread of gonococci with decreased susceptibility to fluoroquinolones: A unique focus in the United States. *J. Infect. Dis.* **177,** 677–682.

125. Centers for Disease Control and Prevention (1995). Fluoroquinolone resistance in *Neisseria gonorrhoeae*—Colorado and Washington. *Morbid. Mortal. Wkly. Rep.* **44,** 761–764.

126. Centers for Disease Control and Prevention (1998). Fluoroquinolone resistant *Neisseria gonorrhoeae*—San Diego, California, 1997. *Morbid. Mortal. Wkly. Rep.* **47,** 405–408.

127. Centers for Disease Control and Prevention (1988). Condoms for prevention of sexually transmitted diseases. *Morbid. Mortal. Wkly. Rep.* **37,** 133–137.

128. Campbell, P. (1993). Efficacy of the female condom. *Lancet* **341,** 1155.

129. Louv, W. C., Austin, H., Alexander, W. J., Stangno, S., and Cheeks, J. (1988). A clinical trial of nonoxynol-9 for preventing gonococcal and chlamydial infections. *J. Infect. Dis.* **158,** 518–523.

130. Rosenberg, M. J., Rojanapithayakorn, W., Feldblum, P. J., and Higgins, J. E. (1987). Effect of the contraceptive sponge on chamydial infection, gonorrhea, and candidiasis: A comparative clinical trial. *JAMA, J. Am. Med. Assoc.* **257,** 2308–2312.

131. Kelaghan, J., Rubin, G. L., Ory, H. W., and Layde, P. M. (1982). Barrier-method contraceptives and pelvic inflammatory disease. *JAMA, J. Am. Med. Assoc.* **248,** 184–187.

132. Niskar, A. S., and Koo, D. (1998). Differences in notifiable infectious disease morbidity among adult women—United States, 1992–1994. *J. Women's Health* **7,** 451–458.

23

Syphilis in Women

JEANNE M. MARRAZZO AND CONNIE L. CELUM
Department of Medicine, University of Washington
Seattle-King County Department of Public Health
Seattle, Washington

I. Introduction and History

References to clinical syndromes consistent with syphilis as a disease of humans exist in biblical and ancient Chinese records. The disease was so common in late fifteenth century Europe that it earned the sobriquet "the Great Pox," aimed largely at distinguishing it from the concurrent scourge of smallpox [1]. The bacterial etiology of syphilis was discovered in 1905 by Schaudinn and Hoffmann, who named the corkscrew-shaped organism *Treponema pallidum.* By that time, the common clinical syndromes of syphilis were well-characterized and included genital ulcers (called "hard" chancres to distinguish them from the "soft" chancres of chancroid), rash, cardiovascular and neurologic complications, and numerous congenital abnormalities including spontaneous abortion, saddle nose deformity, and characteristic abnormalities of the teeth and bones (see Section VIII). In the U.S., syphilis was a leading cause of blindness and dementia during the first half of the twentieth century. The discovery in the 1940s that penicillin, a widely available and affordable antibiotic, was highly active against *T. pallidum* heralded the possibility for eventual control of the disease [2]. This hope was heightened when inexpensive and reasonably accurate blood screening tests for syphilis—syphilis serologies—became widely available in the 1940s. However, few infectious conditions illustrate so well the challenge of controlling a sexually transmitted disease (STD) whose viability at the population level depends in large part on complex socioeconomic factors.

The disease induced by *T. pallidum* is often characterized by clinical features that potentiate its perpetuation in the human population, with no physical signs or symptoms during much of its clinical course and establishment of latency in sites relatively protected from immunologic control (see Section IV). This allows for an infected person to unknowingly transmit *T. pallidum* to a sex partner or, in the case of congenital transmission, from mother to fetus. It also reduces the likelihood that an infected person will recognize a need to seek health care on the basis of the infection [3]. These factors allow for the maintenance of an asymptomatic reservoir of infected individuals in the population.

Equally important, while serologic screening aids in the detection of asymptomatic infection (see Section V), its value depends on its use in the correct settings (for example, persons at risk for sexual acquisition of *T. pallidum* and pregnant women) and on adequate clinical assessment and therapy of positive test results. Individuals without access to, or the ability to maintain, appropriate health care constitute the largest segment of the population with syphilis today. These features partly address why syphilis preferentially afflicts the socioeconomically disenfranchised, particularly populations experiencing social disruption such as war, epidemics of illicit drug use, or severe economic crises. Such factors exacerbate the biologically formidable ability of the organism to maintain an asymptomatic state in the human host and to ensure its transmission to susceptible individuals.

Of great additional concern is the well-established interaction between syphilis and infection with human immunodeficiency virus (HIV). The presence of genital ulcers caused by *T. pallidum* significantly enhances a person's ability to transmit and to acquire HIV. This is likely due to the epithelial breakdown defined by ulcers as well as the recruitment of white blood cells to the site of the ulcer, which provide a susceptible target for, and source of, HIV. One prospective study demonstrated over a threefold increase in the likelihood of HIV seroconversion in female injection drug users (all in a methadone maintenance program) who developed early syphilis at any point during the study relative to those who did not [4–7]. The presence of genital ulcers has also been shown to increase the risk of HIV acquisition in commercial sex workers [8]. The dynamic between syphilis and HIV infection is critical in areas where the incidences of both diseases are high, notably Southeast Asia and many parts of Africa, both of which are experiencing alarming epidemics of HIV infection. It has also highlighted the role that the control of bacterial STDs, including syphilis, gonorrhea, chancroid, and chlamydia, has in control of the HIV pandemic.

II. Epidemiology of Syphilis

A. The Global Picture

The World Health Organization (WHO) estimated that during 1995, 12 million cases of syphilis occurred among adults worldwide. The highest burden lay in South and Southeast Asia, with cases estimated at 5.8 million, followed by sub-Saharan Africa, at 3.5 million [9]. Considerable incidence of syphilis was also estimated to occur in Latin America and in the Caribbean, with 1.3 million cases. By comparison, the number of cases estimated for Western Europe was 200,000, and for North America, 140,000. The global picture is one of a simmering endemic in many parts of the world, with foci of epidemics occurring as biological, social, and economic infrastructures are eroded.

The incidence of syphilis, in general, peaks predictably in settings experiencing the consequences of social and economic

Copyright © 2000 by Academic Press.
All rights of reproduction in any form reserved.

upheaval: poverty, epidemics of illicit drug use, and commercial sex work [10]. In particular, the incidence of congenital syphilis is a sensitive marker of these disruptions because it reflects not only the epidemiology of early syphilis (defined as primary or secondary and early latent syphilis; see Section IV) in women but also the quality of prenatal care [11,12]. Because women with early syphilis are the most likely to transmit the disease to their infants, congenital syphilis generally follows trends similar to that for early syphilis, with approximately a one-year lag time (Fig. 23.1). Foci of disease continue to be identified, mostly in urban areas or those too remote and underserved to ensure adequate prenatal screening [11,13,14]. The human and economic costs of congenital syphilis are devastating: in the U.S. alone, the average annual cost of treating infants with congenital syphilis was estimated in 1995 to be $18.4 million [15,16].

Currently, the newly independent states of the former Soviet Union, as well as other areas of Eastern Europe, are experiencing a rapidly growing syphilis epidemic of major proportions [17]. From 1985 to 1996, the reported incidence of syphilis among women in the Russian Federation rose from 8.7 cases/100,000 persons to 252 cases/100,000. The increase was highest—126-fold—in women aged 15–17 years, with an 80-fold increase among women aged 18–19 years. In 1997, there were 50 cases of congenital syphilis in the city of Moscow alone. Explanations for this disturbing trend probably lie in the abrupt changes in the socioeconomic milieu wrought by political and market reforms and wars. Increases in unemployment and migration have accompanied declines in the standard of living and economic instability, along with severe compromise of the health care delivery infrastructure. A burgeoning demand for drugs and commercial sex may also have contributed. With the incidence of HIV transmission also increasing in this region, measures to stem this epidemic of syphilis are critically needed.

B. The United States

Reporting for syphilis began in the U.S. in 1941. The highest number of reported cases of syphilis in the twentieth century occurred in the U.S. during World War II, after which the incidence declined steadily. An exception to this trend occurred in the 1980s, when an increase in early and congenital syphilis coincided with outbreaks of crack cocaine use in many urban areas. From 1990 to 1997, the incidence of syphilis in the U.S. declined by 84%, probably due to intensified community- and clinic-based control efforts, concomitant HIV prevention activities (such as promotion of condom use), and a decline in the use of crack cocaine. With this decline, the disease became more geographically localized (Fig. 23.2) [18]. In 1997, 50% of all early syphilis occurred in only 31 counties (representing 1% of all U.S. counties); 85% of all cases occurred in 186 U.S. counties (13% of all U.S. counties) [19]. In 1997, early syphilis occurred at a rate of 3.2 cases per 100,000 population, with a ratio of 40: 1 for non-Hispanic blacks relative to whites. The syphilis rate for the southern U.S. was over twice the rate in the whole U.S. (6.6 cases per 100,000 persons). Rates for African-Americans and Hispanics continue to be higher than those for whites [18]. While the reasons for this are complex, underlying differences in socioeconomic status, access to health care, and education about health likely play a critical role. Other explanations put forward have included the maintenance of the infection within discrete sexual networks and public vs private provider attitudes regarding the diagnosis and reporting of STDs, which may be affected by the race of both patients and providers. One example of the complex history of the interactions between race and STD in the U.S. is the legacy of the Tuskegee Study, in which black men with syphilis were denied penicillin therapy after it had become available in order to observe the natural history of the

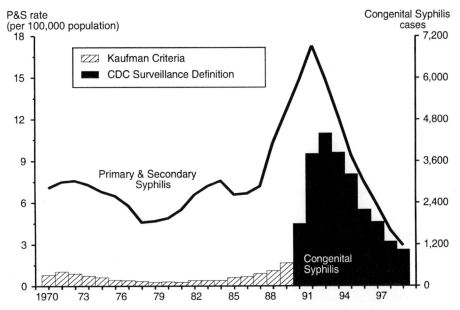

Fig. 23.1 Age- and gender-specific rates of primary and secondary syphilis (United States, 1997). Note: The surveillance case definition for congenital syphilis changed in 1988 (see Appendix).

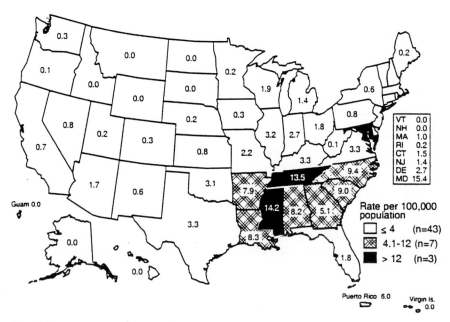

Fig. 23.2 Primary and secondary syphilis—rates for women by state (United States and outlying areas), 1997. The total rate of primary and secondary syphilis for women in the United States and outlying areas (including Guam, Puerto Rico, and Virgin Islands) was 2.9 per 100,000 population. The Healthy People year 2000 objective is 4.0 per 100,000 population.

disease [20]. The U.S. government offered a presidential apology for the study in 1997, but significant mistrust on the part of many communities, and a regard of syphilis as a disease of the truly disenfranchised, remain; such attitudes may affect not only infected persons' attitudes about seeking health care, but also providers' assumptions about preexisting STD risk and reporting of infections.

The case definition for congenital syphilis in the U.S. was liberalized in 1988 to include presumptive cases (mother with untreated or partially treated syphilis at the time of delivery; positive treponemal test, or radiologic, clinical, or cerebrospinal fluid (CSF) evidence of syphilis in the infant). The change prompted a marked increase in reported cases, but several trends are evident even accounting for this change [21]. Reported cases of congenital syphilis peaked in 1991 at 107.3 cases per 100,000 live births, and subsequently declined by 72% to 30.4 in 1996 [18] (Fig. 23.1). In the U.S., 1181 cases of congenital syphilis occurred in 1996; only four states (California, Illinois, New York, and Tennessee) had case reports totaling over 100. Corresponding with the epidemiology of early syphilis in the U.S. in 1996, the rate of congenital syphilis among African-Americans was 127.8 per 100,000 live births, 36.4 among Hispanics, and 3.2 among whites [18]. Urban outbreaks of early syphilis continue to occur in the U.S., generally in association with substance abuse, HIV infection, and failure of prenatal screening [19,22,23].

1. The Future of Syphilis Control in the U.S.

While the decline in reported cases of syphilis is encouraging, the direct economic cost of syphilis in the U.S. remains high: an estimated $79.4 million in 1994 [24]. Based on the highly localized epidemiology of syphilis in the U.S., and because humans

are the only host for *T. pallidum,* several authorities have proposed the possibility of eliminating the disease [25,26]. The success of such an effort would depend in part on overcoming the negative historical associations between syphilis and disenfranchised groups seen as susceptible to the disease. Furthermore, improvements in surveillance capabilities and prenatal screening, control of local outbreaks, and management of sex partners of persons with syphilis are critical [11]. Clearly, these changes will require a strengthening of the public health infrastructure and more effective outreach, screening, and prevention efforts in affected communities, and an acknowledgment that STD control in general is a high fiscal and philosophical priority.

III. Considerations in Transmission of Syphilis

Among adults, syphilis is virtually always sexually transmitted, requiring direct contact of infected mucosa (as with a chancre) or skin (as with the infectious rash of secondary syphilis) to a susceptible epithelial surface (generally, genital or oral mucosa). Anecdotal cases of syphilis spread by "household contact" (for example, drinking from shared vessels) are sometimes cited, but confirmation of such modes of transmission is lacking. A person's infectiousness is strongly modified by the stage of the disease and peaks when the level of treponemes is highest (see Section IV). This feature also has implications for the infectiousness of persons with early latent infection, in whom the 4-year period after acquisition confers the highest risk of transmission to sex partners.

Congenital transmission, from mother to fetus, accounts for the other major mode of transmission of *T. pallidum* [27]. As with sexual transmission, the risk of fetal infection depends largely on the maternal stage of syphilis and is highest when spirochetemia is most intense (*i.e.,* during early syphilis, and during the first 4 years after acquisition). The rate of transmission to the fetus can be remarkably high. Premature birth and congenital syphilis occur in up to 50% of pregnant women with early syphilis, 20–40% with early latent, and 9–10% with late latent syphilis [28]. The fetus can be infected at any time during pregnancy, but risk increases as the pregnancy advances. Antibiotic treatment of the mother during pregnancy is highly effective in treating fetal infection and averting its consequences.

IV. Natural History and Clinical Manifestations of Syphilis

Approximately 30% of individuals exposed to an individual during the infectious stages of syphilis within the prior 30 days will acquire syphilis, based on the placebo arm of placebo-controlled studies of preventive antibiotic therapy [29]. The infectiousness per contact of *T. pallidum* has not been carefully ascertained, however. Once infected, local replication of trepo-

nemes occurs and lymphocytes are recruited to the site. An indurated, painless lesion, called a chancre, develops at the site of inoculation, either genital, perianal, or oral (Fig. 23.3). The extragenital chancres can be atypical; perianal chancres can be painful and atypical in appearance. This stage of syphilis with a chancre is called the primary stage and occurs an average of 3 weeks after acquisition (range 10–90 days). If untreated, the primary lesion resolves in several weeks and may be unnoticed by the patient because of its painless nature.

Secondary syphilis, which represents hematogenous and lymphatic dissemination of *T. pallidum,* typically occurs several weeks to months after primary syphilis but can overlap the primary stage. Clinical manifestations of secondary syphilis are highly variable, leading to the term "the great imitator." Typically, a cutaneous or mucosal rash develops that often involves the palms and soles of the feet (Fig. 23.4). Sometimes it is accompanied by systemic features, such as headache, low grade fever, generalized lymphadenopathy, and sore throat. A small

Fig. 23.3 A healing syphilitic chancre on the left labia with surrounding edema during the primary syphilis stage.

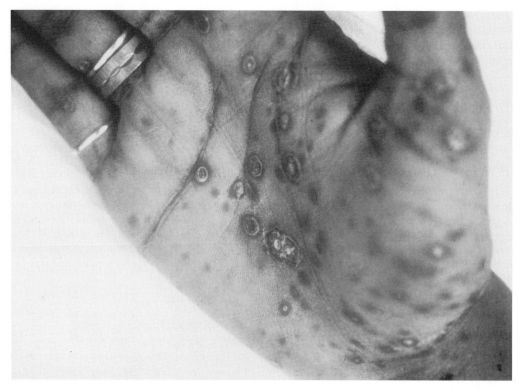

Fig. 23.4   Hyperpigmented rash on the palms during the secondary stage of syphilis.

proportion of patients (*e.g.,* 10%) will develop patchy hair loss, hepatitis, or renal involvement (*e.g.,* nephrotic syndrome). The clinical manifestations of secondary syphilis can be mild or non-specific, and thus, may be either unnoticed by the patient or attributed to the flu or a viral illness.

After the secondary stage, people enter a latent stage that is divided into early and late latent based on the likelihood of spontaneous mucocutaneous relapses. Approximately 25% of patients with untreated secondary syphilis will relapse, usually in the first year but as long as 5 years later [30]. Early latent syphilis is also regarded as a potentially infectious stage, as evidenced by the recommendations to empirically treat contacts of early latent as well as primary and secondary syphilis. The U.S. Public Health Service defines early latent syphilis as one year from onset of infection, which is determined by prior se-rologic tests and time of exposure to an early syphilis index case, but other groups as in Russia define a longer interval such as two years.

Seventy percent of people who enter latent syphilis will re-main latent for life without further clinical manifestations, and the remaining 30% develop tertiary syphilis. The proportion who develop tertiary syphilis today is likely to be lower than that from published natural history studies in the early 1900s before antibiotics became widely available. The manifestations of tertiary syphilis include cutaneous and bony gummas, now rarely seen, but described as occurring from 2 to 40 years after initial infection. Cardiovascular complications, in which the aorta is involved with a vasculitis sometimes leading to an ascending aortic aneurysm, manifest 20 to 30 years after infec-

tion. Neurologic involvement may occur early, such as during secondary syphilis, or in the first 1–2 years after infection, pre-senting either without symptoms, aseptic meningitis, or several years later as a stroke due to meningovascular involvement. Late neurologic complications of syphilis result from parenchy-mal damage without pathologic evidence of spirochetes and in-clude tabes dorsalis, in which the posterior columns of the spinal cord are affected, leading to sensory loss in the feet and a wide-based gait. Late neurologic complications also include demen-tia, typically presenting several decades after infection.

The Oslo natural history study (1891–1910) reported that untreated men, when examined approximately 30 years after infection, had about a twofold higher risk of developing neurol-ogic and cardiovascular complications of syphilis than women [31]. Neurosyphilis occurred in 9.2% of men and 5.1% of women with syphilis, and cardiovascular syphilis developed in 14.9% of men compared to 8% of women. While most of these late complications of untreated syphilis are now rarely seen, occasionally early meningeal and meningovascular manifesta-tions are still observed. The specific indications for lumbar puncture in different stages of syphilis remain somewhat con-troversial. Whereas abnormalities in the cerebrospinal fluid are common (approximately 40%) in early syphilis [31], and the standard treatments for early syphilis do not achieve treponem-icidal levels in the spinal fluid, the risk of developing sympto-matic neurosyphilis is low overall (6% in the Oslo study). However, persons with the most abnormal cerebrospinal fluid parameters have been shown to be at higher risk for developing neurosyphilis. Because higher doses of shorter-acting penicil-

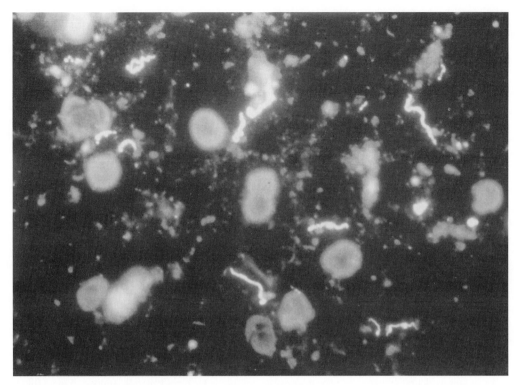

Fig. 23.5 *Treponema pallidum* demonstrated by monoclonal antibody fluorescent stain of exudate from a chancre.

lins are used to treat neurosyphilis, most experts agree that a lumbar puncture should be performed in any patient with syphilis who has new or unexplained neurologic findings. Lumbar puncture should be considered in anyone who is HIV-infected, due to case reports of neurologic relapse in HIV-infected persons who received adequate treatment for their stage of syphilis [32]. The yield of lumbar punctures in persons with HIV-negative latent syphilis is very low [33,34].

V. Diagnostic Methods

Syphilis can be diagnosed through direct identification of *T. pallidum* in specimens taken from lesions in primary, secondary, or congenital syphilis, or indirectly through serologic testing. Direct examination of specimens for treponemes is performed by darkfield microscopy of fresh specimens. Darkfield microscopy must be performed immediately by someone experienced in identification of the small corkscrew-shaped organisms that are slightly larger than a red blood cell and have characteristic motility. Many laboratories do not have darkfield microscopes available and instead perform fluorescence microscopy on acetone-fixed exudate that is incubated with fluorescein-labeled anti-*T. pallidum* globulin (Fig. 23.5).

Serologic diagnosis of syphilis relies on a combination of nontreponemal tests, which detect antibodies against lipoidal antigens, and treponemal tests, which detect specific antibodies against *T. pallidum*. The most commonly used nontreponemal tests are the rapid plasma reagin (RPR) and the Venereal Disease Research Laboratory (VDRL) test. The VDRL and RPR are used for screening, and positive tests are confirmed by specific trepo-

nemal tests: either the microhemagglutination assay (MHA-TP) or the fluorescent treponemal absorption test (FTA-ABS). The nontreponemal tests can be quantitated through serial dilutions of reactive sera, which provide useful information in staging syphilis and a method to assess response to treatment. The levels of antibody rise rapidly in the first year after infection, reach low levels in late syphilis, and spontaneously revert to nonreactive in about 25% of untreated patients [35]. Typical VDRL titers are 1:8 in primary syphilis, 1:128 in secondary syphilis, and 1:4 in late latent syphilis. The sensitivity of the nontreponemal and treponemal tests in different stages of syphilis is shown in Table 23.1 [36]. The FTA-ABS and MHA-TP often become reactive before the VDRL or RPR in primary syphilis, and approximately 75–85% of these will remain reactive after treatment for the patient's lifetime [37].

VI. Treatment

T. pallidum is a slowly replicating organism that can be sequestered in lymph nodes and the central nervous system. A basic tenet in treating syphilis is to achieve adequate concentrations of effective antibiotics over a long enough interval to kill *T. pallidum* in mucosal and tissue sites. The mainstay of treatment has been penicillin for all stages of syphilis; *T. pallidum* is exquisitely sensitive to penicillin and there has been no evidence of development of penicillin resistance. Benzathine penicillin provides low concentrations of penicillin for several weeks after intramuscular administration. For early syphilis (primary, secondary, and early latent), the standard treatment is a single dose of 2.4 million units of benzathine penicillin

Table 23.1
Sensitivity of Serologic Tests during Stages of Untreated Syphilis

	Stage of disease			
	Primary	Secondary	Latent	Late
VDRL	59–87%	100%	73–91%	37–94%
FTA-ABS	86–100	99–100	96–99	96–100
MHA-TP	64–87	96–100	96–100	94–100

Source: [36].

(administered as 1.2 million units intramuscularly into each buttock). Persons with early syphilis often develop symptoms of fever, malaise, musculoskeletal pain, and headache (called the Jarisch-Herxheimer reaction) within 24 hours after treatment due to the release of antigen. The Jarisch-Herxheimer reaction can be prophylaxed or treated with anti-inflammatory drugs. For latent syphilis of greater than one year's duration and syphilis of unknown duration, the dose of benzathine penicillin is 2.4 million units weekly for three weeks for a total dose of 7.2 million units. For patients with a history of allergy to penicillin, the alternative regimen is oral doxycycline, 15 days for early syphilis and 30 days for late latent syphilis. Due to poor compliance with a two to four week oral regimen and poor penetration of doxycycline into the spinal fluid, experts recommend a lumbar puncture prior to the use of doxycycline to assess whether an individual has asymptomatic neurosyphilis.

The advantage of benzathine penicillin is compliance with a single dose, but its main limitation is the low concentration of penicillin in the cerebrospinal fluid. Alternative regimens that achieve higher antibiotic concentrations in the cerebrospinal fluid are all multidose and are associated with lower adherence. As a result, when these regimens are used, diagnostic lumbar punctures are helpful for identifying persons with neurosyphilis who require more intensive antibiotic regimens. An oral antibiotic regimen that achieves higher levels in the cerebrospinal fluid is amoxicillin (up to 6 grams daily) plus probenecid which reduces renal excretion of penicillin. The preferred regimens for treating neurosyphilis are parenteral administration of higher doses of shorter-acting penicillins, such as intramuscular procaine penicillin plus probenecid or intravenous penicillin G, for 10–14 days. Once-daily intramuscular or intravenous ceftriaxone, a third generation cephalosporin, is being studied as a possible alternative for the treatment of neurosyphilis [38].

The regimens described above are highly effective in treating syphilis, but treatment failures do occur in up to 10% of cases. In order to distinguish treatment failure from reinfection through sexual contact with an untreated partner, patients should have close serologic follow-up with nontreponemal tests in the first year after treatment (*e.g.,* the RPR or VDRL, one month and then every three months for the first year and at 24 months after treatment). For patients treated for primary and secondary syphilis, the titers should decline approximately fourfold after three months and eightfold after six months. The VDRL usually becomes nonreactive one year after treatment for primary syphilis and two years after secondary syphilis, whereas in the majority of treated patients, the FTA-ABS and MHA-TP remain reactive for life.

A key component to the treatment and control of syphilis is management of sexual partners. The recommendation is to clinically and serologically evaluate sexual partners of patients with early syphilis (*e.g.,* primary and secondary syphilis as well as those with VDRL or RPR titers $\geqslant$1:32). Partners of patients with early syphilis who have had contact within the prior 90 days are treated empirically with 2.4 million units of benzathine penicillin. Sexual partners with contact longer than the prior 90 days are tested serologically and treated on the basis of their test results, or empirically if follow-up cannot be assured. Sexual partners of persons with late latent syphilis are treated on the basis of their serologic and clinical evaluation, and empiric treatment is not generally recommended.

VII. Effects of HIV on Syphilis

Given the overlapping epidemiology of syphilis and HIV in many parts of the world, there has been considerable interest in the effect of HIV on the serologic and clinical course of syphilis and response of treatment. Case reports in the medical literature have described atypical and severe cutaneous manifestations of syphilis in HIV-infected persons, such as malignant or ulcerating secondary syphilis, as well as higher rates of overlapping primary and secondary syphilis and delayed serologic response to treatment [39,40]. Small series of HIV-infected patients with syphilis have not generally supported more fulminant manifestations or rapid progression of early syphilis in HIV-infected persons, although HIV-infected patients may have slightly higher titers in early syphilis [6,41]. However, some case reports have also documented lower titers among HIV-infected persons with early syphilis; given the rising titer of antibodies with increasing time after infection and the difficulty in accurately timing onset of infection, these serologic findings may not represent a clinically significant difference.

Due to the concern about possibly higher rates of neurologic relapse among HIV-infected persons, a prospective multisite randomized controlled study was conducted of standard therapy (2.4 million units of benzathine penicillin) versus enhanced antibiotic therapy (six grams of amoxicillin every day plus probenicid). The study attempted to address whether HIV-infected patients with early syphilis had higher rates of treatment failure, whether outcome was improved among HIV-infected persons who received enhanced treatment, and whether central nervous system involvement during early syphilis was clinically important [42]. The study was not able to enroll the sample size goal of 1000, so it may have been underpowered to observe a difference in treatment outcome between the 430 HIV-negative and 101 HIV-infected persons enrolled with early syphilis. This study did find that there were minor differences between HIV-infected and HIV-negative persons in the clinical manifestations of early syphilis; HIV-infected persons more commonly had multiple chancres and Jarisch-Herxheimer reactions after treatment. CSF involvement was found in 21% of the total study population and was not significantly higher among the HIV-infected persons. Syphilis in the CSF, either demonstrated through a positive PCR or rabbit inoculation test, did not affect treatment outcome. Although the HIV-infected persons had a slower serologic response than the HIV-negative patients, there was only one treatment failure overall. Most experts do not recommend altering the treatment for syphilis in HIV-infected persons but do recommend close serologic follow-up to monitor their response to therapy. Given the limited power of the study to demonstrate a difference in asymptomatic neurosyphilis between HIV-infected and negative persons, many experts still advise lumbar punctures for HIV-positive persons with syphilis.

VIII. Syphilis and Pregnancy

Rates of congenital syphilis parallel rates of early syphilis in women of reproductive age; thus, cases of congenital syphilis increased significantly in the U.S. during the late 1980s and in Russia in the 1990s during their respective epidemics of early syphilis among young adults. Although a mother can transmit syphilis to the fetus from as early as 9 weeks of gestation to 35 weeks, the risk of congenital syphilis is highest if the mother is diagnosed in the second half of pregnancy. The mother's stage of infection is also an important factor in determining the likelihood of transmission to the fetus; the mother is more infectious in secondary and early latent stages but can still transmit syphilis as long as eight years after infection. One report found congenital syphilis in over half of infants born to mothers with unknown duration of syphilis who were treated with either single dose or three doses of benzathine penicillin, highlighting the potential for transmission in later stages of syphilis and the need for earlier detection to prevent transmission to the infant [12,43].

The manifestations of congenital syphilis can be devastating; untreated seropositive mothers have up to a 12-fold increased risk of stillbirth and miscarriage [11]. Infants who are born with congenital syphilis either have active (approximately 30%) or latent disease. Early congenital syphilis manifests as rhinitis, osteochondritis, periostitis, hepatitis, mucocutaneous rash, and neurosyphilis. However, most congenital syphilis is asymptomatic. In these situations, the clinician must evaluate both the mother and infant by staging their infections based on the mother's titer, clinical symptoms prior to treatment, her serologic response to treatment, physical examination of the infant, and pathologic and dark field examination of the placenta.

The diagnosis of congenital syphilis in asymptomatic infants born to seropositive mothers provides a challenge to the clinician due to the passage of maternal antibodies. A number of assays have been developed to try to discriminate passive transfer of maternal antibodies from active infection, such as IgM enzyme immunoassays and Western blots. Unfortunately, no single assay is sensitive enough to rely on it for the diagnosis of congenital syphilis [44]. If the infant has an abnormal examination or a nontreponemal titer greater than the mother's, a more extensive evaluation including lumbar puncture, long bone X-rays, and liver function tests, is recommended. To better capture the full spectrum of congenital syphilis for surveillance, reporting criteria for congenital syphilis were expanded in 1988 to include presumptive cases, specifically infants born to mothers with untreated or partially treated syphilis at delivery, infants with higher nontreponemal titers than their mothers, and consistent clinical findings.

Congenital syphilis is a preventable disease; penicillin crosses the placenta and is highly effective in treating the infected fetus [45]. A high priority in public health is to prevent congenital syphilis, but this requires multiple approaches. A central component is early detection of syphilis in pregnant women through antenatal screening. Serologic screening for syphilis is routinely performed in the first trimester, and in high prevalence locales and for high-risk women, it is repeated in the third trimester and sometimes at delivery. However, a significant proportion of congenital syphilis cases occur because women present late in pregnancy, diagnosis is delayed, and treatment is not initiated or completed before delivery. Thus, prevention of congenital syphilis will require effective prevention of syphilis in adults, increased serologic screening for syphilis among high-risk populations (such as sex workers, and injection drug and cocaine users), access to and early serologic screening during prenatal care, improved clinical recognition and diagnostic skills by obstetricians, and use of effective therapies to treat the mother's stage of syphilis. The consensus is to only use penicillin to treat pregnant mothers with syphilis because doxycycline is contraindicated in pregnancy and erythromycin does not cross the placenta. Azithromycin is being studied for the treatment of syphilis in pregnancy and appears promising in small trials [46], but the data are not sufficient to widely recommend it at this time. Women with a penicillin allergy can be tested by penicillin skin tests and desensitized, if necessary.

IX. Summary

Syphilis is an important sexually transmitted infection for women and their offspring, given the high morbidity and premature deaths associated with congenital syphilis. Social disruption, drug use (particularly crack cocaine), and economic disenfranchisement have fueled recent syphilis epidemics in the United States, Russia, eastern Europe, Africa, and Asia. Syphilis is easily transmitted to sexual partners and the fetus during the early infectious stages. The clinical diagnosis of syphilis can be difficult due to the highly variable manifestations of secondary syphilis and the lack of symptoms or clinical findings during the latent stage. Serologic screening for syphilis is a critical component in identifying infected women and preventing congenital transmission through treatment of seropositive mothers. Treatment of syphilis with penicillin is highly effective, even in the setting of HIV infection.

References

1. Parran, T. (1937). "Shadow on the Land." Reynal and Hitchcock, New York.
2. Sparling, P. F. (1999). Natural history of syphilis. *In* "Sexually Transmitted Diseases" (K. K. Holmes, P.-A Mårdh, P. F. Sparling, *et al.*, eds.), 3rd ed. McGraw-Hill, New York.
3. Garnett, G. P., Aral, S. O., Hoyle, D. V., Cates, W., and Anderson, R. M. (1997). The natural history of syphilis: Implications for the transmission dynamics and control of infection. *Sex. Transm. Dis.* **24,** 185–200.
4. Torian, L. V., Weisfuse, I. B., Makki, H. A., Benson, D. A., DiCamillo, L. M., and Toribio, F. E. (1995). Increasing HIV-1 seroprevalence associated with genital ulcer disease, New York City, 1990–1992. *AIDS* **9,** 177–81.
5. Williams, M. L., Elwood, W. N., Weatherby, N. L., Bowen, A. M., Zhao, Z., Saunders, L. A., and Montoya, I. D. (1996). An assessment of the risks of syphilis and HIV infection among a sample of not-in-treatment drug users in Houston, Texas. *AIDS Care* **8,** 671–682.
6. Gourevitch, M. N., Hartel, D., Schoenbaum, E. E., Selwyn, P. A., Davenny, K., Friedland, G. H., and Klein, R. S. (1996). A prospec-

tive study of syphilis and HIV infection among injection drug users receiving methadone in the Bronx, N. Y. *Am. J. Public Health* **86,** 1112–1115.

7. Otten, M. W., Jr., Zaidi, A. A., Peterman, T. A., Rolfs, R. T., and Witte, J. J. (1994). High rate of HIV seroconversion among patients attending urban sexually transmitted disease clinics. *AIDS* **8,** 549–553.

8. Plummer, F. A., Simonsen, J. N., Cameron, D. W., Ndinya-Achola, J. O., Kreiss, J. K., Gakinya, M. N., Waiyaki, P., Cheang, M., Piot, P., Ronald, A. R., and Ngugi, E. N. (1991). Cofactors in male-female sexual transmission of human immunodeficiency virus type 1. *J. Infect. Dis.* **163,** 233–239.

9. World Health Organization Office of HIV/AIDS and Sexually Transmitted Diseases (1998). An overview of selected curable STDs. ASD online at *http://www.who.int/asd*

10. Kilmarx, P. H., Zaidi, A. A., Thomas, J. C., Nakashima, A. K., St. Louis, M. E., Flock, M. L., and Peterman, T. A. (1997). Sociodemographic factors and the variation in syphilis rates among U.S. counties, 1984 through 1993: An ecological analysis. *Am. J. Public Health* **87,** 1937–1943.

11. Wilkinson, D., Sach, M., and Connolly, C. (1997). Epidemiology of syphilis in pregnancy in rural South Africa: Opportunities for control. *Trop. Med. Intern. Health* **2,** 57–62.

12. McFarlin, B. L., Bottoms, S. F., Dock, B. S., and Isada, N. B. (1994). Epidemic syphilis: Maternal factors associated with congenital infection. *Am. J. Obstet. Gynecol.* **170,** 535–540.

13. Humphrey, M. D., and Bradford, D. L. (year). Congenital syphilis: Still a reality in 1996. *Med. J. Aust.* **165,** 382–385.

14. How, J. H., and Bowditch, J. D. (year). Syphilis in pregnancy: Experience from a rural aboriginal community. *Aust. N. Z. J. Obstet. Gynaecol.* **35,** 229–230.

15. Institute of Medicine (1996). "The Hidden Epidemic: Confronting Sexually Transmitted Diseases." National Academy Press, Washington, DC.

16. de Lissovoy, G., Zenilman, J., Nelson, K. E., Ahmed, F., and Celentano, D. D. (1995). The cost of a preventable disease: Estimated U.S. national medical expenditures for congenital syphilis, 1990. *Public Health Rep.* **110,** 403–409.

17. Tichonova, L., Borisenko, K., Ward H., Meheus, A., Gromyko, A., and Renton, A. (1997). Epidemics of syphilis in the Russian Federation: trends, Origins, and priorities for control. *Lancet* **350,** 210–213.

18. Centers for Disease Control and Prevention (1997). "Sexually Transmitted Disease Surveillance 1996." U.S. Department of Health and Human Services Public Health Service, Atlanta, GA.

19. Centers for Disease Control and Prevention (1998). Primary and secondary syphilis—United States, 1997. *Morbid. Mortal. Wkly. Rep.* **47,** 493–497.

20. Brawley, O. W. (1998). The study of untreated syphilis in the negro male. *Int. J. Radiat. Oncol. Biol. Phys.* **40,** 1–2.

21. Centers for Disease Control and Prevention (1988). Guidelines for the prevention and control of congenital syphilis. *Morbid. Mortal. Wkly. Rep.* **37**(S-1), 1–13.

22. Klass, P. E., Brown, E. R., and Pelton, S. I. (1994). The incidence of prenatal syphilis at the Boston City Hospital: A comparison across four decades. *Pediatrics* **94,** 24–28.

23. Murph, J. R. (1994). Rubella and syphilis: Continuing causes of congenital infection in the 1990s. *Semin. Pediatr. Neurol.* **1,** 143–000.

24. Bateman, D. A., Ciaran, S. P., Joyce, T., and Heagerty, M. C. (1997). The hospital cost of congenital syphilis. *J. Pediatr.* **130,** 752–758.

25. St. Louis, M. E., and Wasserheit, J. N. (1998). Elimination of syphilis in the United States. *Science* **281,** 353–354.

26. Hook, E. W., III (1998). Is elimination of endemic syphilis transmission a realistic goal for the USA? *Lancet* **351**(S3), 19–21.

27. Radoff, J. D., Sánchez, P. J., Schulz, K. F., and Murphy, F. K. (1999). Congenital syphilis. *In* "Sexually Transmitted Diseases" (K. K. Holmes, P-A Mřdh, P. F. Sparling, *et al.,* eds.), 3rd ed. McGraw-Hill, New York.

28. Fiumara, N. J. (1984). Review of congenital syphilis. *Sex. Transm. Dis.* **11**(1), 49–50.

29. Schroeter, A. L. (1971). Therapy for incubating syphilis: Effectiveness of gonorrhea treatment. *JAMA, J. Am. Med. Assoc.* **218,** 711–000.

30. Geistland, T. (1955). The Oslo study of untreated syphilis: An epidemiologic investigation of the natural course of syphilitic infection based on a restudy of the Boeck-Brunsgaard material. *Acta Derm.-Venereol.* **35S,** 356–368.

31. Lukehart, S. A., Hook, E. W., Baker-Zander, S. A. *et al.* (1988). Invasion of the central nervous system by *Treponema pallidum:* Implications for diagnosis and therapy. *Ann. Intern. Med.* **109,** 855–862.

32. Hook, E. W., and Marra, C. W. (1992). Acquired syphilis in adults. *N. Engl. J. Med.* **326,** 1060–1069.

33. Carey, L. A., Glesby, M. J., Mundy, L. M. *et al.* (1995). Lumbar puncture for evaluation of latent syphilis in hospitalized patients. High prevalence of cerebrospinal fluid abnormalities unrelated to syphilis. *Arch. Intern. Med.* **155,** 1657–1662.

34. Wiesel, J., Rose, D. N., Silver, A. L. *et al.* (1985). Lumbar puncture in asymptomatic late syphilis. *Arch. Intern. Med.* **145,** 465–468.

35. Hart, G. (1986). Syphilis tests in diagnostic and therapeutic decision making. *Ann. Intern. Med.* **104,** 368–376.

36. Jaffe, H. W., and Musher, D. W. (0000). Management of the reactive syphilis serology. *In* "Sexually Transmitted Diseases" (K. K. Holmes *et al.,* eds.), 2nd ed. McGraw-Hill, New York.

37. Romanowski, B., Sutherland, R., Fick, G. H. *et al.* (1991). Serologic response to treatment of infectious syphilis. *Ann. Intern. Med.* **114,** 1005–1009.

38. Centers for Disease Control and Prevention (1998). "STD Treatment Guidelines." CDC, Atlanta, GA.

39. Yinnon, A. M., Coury-Doniger, P., Polito, R., and Reichman, R. C. (1996). Serologic response to treatment of syphilis in patients with HIV infection. *Arch. Intern. Med.* **156,** 321–325.

40. Schofer, H., Imhof, M., Thoma-Greber, E., Brockmeyer, N. H., Hartmann, M. *et al.* (1996). Active syphilis in HIV infection: A multicentre Retrospective Survey. The German AIDS Study Group (GASG). *Genitourin. Med.* **72,** 176–181.

41. Goeman, J., Kivuvu, M., Nzila, N., Behets, F., Edidi, B. *et al.* (1995). Similar serological response to conventional therapy for syphilis among HIV-positive and HIV-negative women. *Genitourin. Med.* **71,** 275–279.

42. Rolfs, R. T., Joesoef, M. R., Hendershot, E. F. *et al.* (1997). A randomized trial of enhanced therapy for early syphilis in patients with and without human immunodeficiency virus infection. *N. Engl. J. Med.* **337,** 307–314.

43. Donders, G. G., Desmyter, J., Hooft, P., and Dewet, G. H. (1997). Apparent failure of one injection of benzathine penicillin G for syphilis during pregnancy in human immunodeficiency virus-seronegative African women. *Sex. Transm. Dis.* **24,** 94–101.

44. Stoll, B. J., Lee, F. K., Larsen, S. *et al.* (1993). Clinical and serologic evaluation of neonates for congenital syphilis: A continuing diagnostic dilemma. *J. Infect. Dis.* **167,** 1093–1099.

45. Radcliffe, M., Meyer, M., Roditi, D., and Malan, A. (1997). Single dose benzathine penicillin in infants at risk of congenital syphilis—results of a randomised study. *S. Afr. Med. J.* **87,** 62–65.

46. Mashkilleyson, A. L., Gomberg, M. A., Mashkilleyson, N., Kutin, S. A. (1996). Treatment of syphilis with azithromycin. *Int. J. STD AIDS* **7S,** 13–15.

24

Genital Herpes

ANNA WALD* AND JOELLE M. BROWN†

*Departments of Medicine and Epidemiology, University of Washington, Seattle, Washington; †Department of Biostatistics and Epidemiology, University of California, Berkeley, California

I. Introduction

Infections with herpes simplex virus (HSV) are ubiquitous in human populations. Initially identified as the cause of fever blisters, or cold sores, HSV was recognized as an infection of the genitals in the nineteenth century [1]. However, the clinical manifestations and sequelae of genital herpes were not described until the 1970s and the epidemiologic features of this infection were not delineated until the development of accurate serologic assays in the 1980s. Thus, despite its long coexistence with humans, the recognition of the morbidity associated with HSV has occurred relatively recently.

Two types of HSV exist: HSV-1 and HSV-2 [2,3]. HSV-1 is the cause of cold sores and is often acquired in childhood from intimate (nonsexual) contact with family members. Conversely, HSV-2 is the predominant cause of genital herpes and is almost always transmitted through sexual contact. However, HSV-1 can also cause genital herpes and the frequency of isolation of HSV-1 from genital herpes lesions has been increasing.

Herpes simplex viruses cause infections that persist for the life of the host, are incurable, and likely can be transmitted for many years after infection [4,5]. The effect of HSV infection on health and psychosexual well-being varies tremendously. The most severe consequences of the infection occur in the neonate, where fatal outcome is not uncommon even with appropriate therapy. Persons with immunosuppression, including HIV-related, also can experience significant morbidity due to HSV infections. Although most persons are unaware of having HSV infections, those who are aware have troublesome recurrent genital ulcerations. The recent association of genital ulcers in general, and HSV-2 in particular, with transmission and acquisition of HIV infection has heightened the importance of preventing genital herpes infections. Unfortunately, the continuing spread of HSV-2 infection suggests that progress in prevention will be slow.

As with many other sexually transmitted diseases (STDs), genital herpes infections affect women disproportionately in several ways. First, the infection is more prevalent among women, regardless of the level of sexual activity. Second, the disease tends to be more clinically severe among women, who are more likely to develop systemic illness and neurologic complications during the initial infection. Finally, due to the threat of potential complications during pregnancy and infection of the neonate, women are more affected by the consequences of genital herpes. This chapter will review current knowledge of the epidemiology, natural history, and public health importance of genital herpes infections.

II. Epidemiology

A. *Prevalence and Incidence*

Genital herpes is the most prevalent STD in the United States. Despite the decreasing numbers of bacterial STDs since the 1980s, infections with HSV-2, the most common cause of genital herpes, have been increasing. Estimates of genital herpes in the population are based on a number of new clinical diagnoses and serologic assays. The Centers for Disease Control and Prevention (CDC), based on the number of office visits to sentinel physicians for new cases of genital herpes, estimate that about 210,000 new infections are diagnosed per year [6]. However, as most infected persons are not aware of having genital herpes, the number likely underestimates the incidence of genital herpes as much as five- to tenfold.

More reliable estimates of HSV-2 prevalence are obtained from serologic surveys that employ type-specific assays for HSV-2 antibodies. Currently, the best available data derive from the National Health and Nutrition Examination Survey (NHANES) III, a random sample of noninstitutionalized civilians in the U.S. [7]. This survey, conducted between 1988 and 1994, showed that 22% of persons aged 12 or older in the U.S. have HSV-2 infection. As shown in Figure 24.1, this infection is more frequent among women compared to men, and among African-Americans compared to whites. A comparison of this 1988–1994 survey with the NHANES survey conducted 12 years previously shows a 30% increase in HSV-2 prevalence. Of particular concern is the

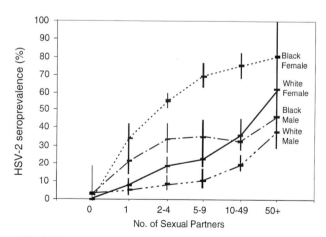

Fig. 24.1 Seroprevalence of HSV-2 according to the number of sexual partners in men and women, whites and African-Americans. These data reflect findings from National Health and Nutrition Survey III (adapted from [7]).

Table 24.1

HSV-2 Seroprevalence and the Percentage of Patients with HSV-2 Infection Who Have a History of Diagnosed Genital Herpes in Selected Populations of Women Worldwide

Study setting	Number of women tested	Percentage with HSV-2 infection	Percentage with history of genital herpes among persons seropositive for HSV-2
GENERAL POPULATION—United States			
NHANES[a][7]	6687	25.6	9.1
Family Practice, Seattle WA [11]	610	29	26
Family Planning Clinic, Pittsburgh, PA [10]	4527	21.6	12.6
University students, South Carolina [113]	1093	4.3	0
San Francisco neighborhood survey, CA [12]	611	41	13
Women's Clinic, Albuquerque, NM [114]	587	31	11.1
Lesbians, Seattle, WA	133	13	41
[J. Marrazzo, personal communication]			
GENERAL POPULATION—Other Countries			
Blood Donors, London, UK [14]	1325	7.6	19
Pregnant women, Sao Paulo, Brazil [115]	655	39	8.2
Household survey of women, Costa Rica [116]	766	39.4	1.1
Pregnant women, Sydney, Australia [15]	229	14.5	0
Pregnant women, Stockholm, Sweden [117]	1000	33	NA
HIGH RISK SETTINGS			
STD Clinic, Seattle WA, U.S. [9]	776	43	22
STD Clinic, Australia	27	55	42[b]
STD Clinic, London, UK [118]	347	24.5	30.1
STD Clinic, Goteborg, Sweden [119]	475	26	30
STD Clinic, heterosexuals, Italy [120]	783	35	14
Prostitutes, Senegal	272	80	NA
Nairobi prostitutes, Kenya	115	61	NA
Prostitutes, HIV+ Kinshasa, Zaire	181	95	NA
Prostitutes, HIV− Kinshasa, Zaire	187	75	NA

From [12].
[a]From National Health and Nutrition Examination Survey III.
[b]Separate % for each gender not available.
NA, not available.

fivefold increase in HSV-2 prevalence in white teenagers, indicating a shift toward HSV-2 infection at an earlier age.

Similar to the recent NHANES findings, smaller surveys conducted in STD clinics, reproductive health clinics, and among the general population have found that women are more likely than men to be infected with HSV-2 (Table 24.1) [8–12]. Not surprisingly, infection rates appear highest in high-risk settings, specifically among persons who attend STD clinics and among women who self-identify as prostitutes.

The studies that measured HSV-2 infection in women from lower risk settings found between 13 and 41% of women to be infected with HSV-2. Of these seropositive women, only between 0 and 41% reported a history of genital herpes. For example, one study among women attending a Family Planning Clinic in the U.S. found that 22% of women were infected with HSV-2; however, only 13% of these HSV-2 seropositive women were aware of their herpes infection [10]. These findings suggest that the majority of women infected with HSV-2 are unaware of their infection.

Only selected estimates of HSV-2 seroprevalence are available from other parts of the world. HSV-2 infection appears less prevalent in Canada, Western Europe, and Australia than in the U.S. [13,14]. Limited data from Equatorial Africa, Southeast Asia (Thailand), and the Caribbean (Haiti) suggest that infection rates are very high [15–17]. HSV-2 seroprevalence among sexually active adults in developing countries ranges from 60 to 90%. While these figures may partly reflect the focus on high-risk populations, they also suggest that HSV-2 infections are widespread throughout the world.

The incidence of HSV is difficult to estimate as most infections are acquired subclinically. A unique study of Swedish girls followed prospectively for 15 years showed rates of HSV-1 acquisition of 2.9%, 3.5%, and 1.5% per year for 14–19, 19–23, and 24–30 year olds, respectively [18]. Rates of HSV-2 acquisition were 0.5%, 2.4%, and 2.3% per year in the same age groups. In an STD clinic in Seattle, the rate of HSV-2 acquisition among women was 3% per year in the late 1980s [Anna Wald, unpublished observation].

B. Correlates of HSV-2 Infection

The prevalence of HSV-2 increases with age, poverty, and less education. Consistently, the prevalence of infection is approximately 1.5 times higher among women than men. The reasons for this gender discrepancy are likely to encompass anatomy (differences in genital epithelium in men and women) and behavior (younger women are more likely to have sex with older men, thus increasing the probability of exposure to the virus). In the United States, African-Americans have a higher prevalence of HSV-2 infection than whites [7]. The rates of HSV-2 infection remain high among African-Americans even after adjustment for sexual behavior and are likely to reflect the high background prevalence of HSV-2 among sexual partners in the African-American community. Other behavioral risk factors associated with HSV-2 seropositivity include the lifetime number of sexual partners and young age of sexual debut. In the NHANES study, cocaine use was also associated with HSV-2 seropositivity.

Because the risk of HSV-2 infection correlates well with the lifetime number of sexual partners, HSV-2 antibody has been proposed as a serologic marker for sexual activity in populations [14]. This correlation may be accurate for some populations, but it may not be accurate at a very high and very low prevalence of HSV-2. For example, the risk of HSV-2 infection levels off in African-American women who report two or more lifetime sexual partners whereas it rises steeply for white women (see Fig. 24.1). At the other extreme, the seroprevalence is so low (3%) among New Zealand youth aged 21 years that the association between sexual activity and the risk of HSV-2 seropositivity is not detected [19]. In addition, characteristics of sexual partners may be an important risk factor for HSV-2 infection. Selection of sexual partners has not been studied extensively in relation to HSV-2 antibody status. One study noted that HSV-2 seropositive persons are more likely to report choosing sex partners who belonged to higher risk demographic groups for STDs than HSV-2 seronegative persons [20].

C. Transmission of HSV-2

As with all STDs, sexual encounter with a person with genital herpes does not necessarily result in transmission of HSV-2. Gender and previous infection with HSV-1 appear to modify the risk of HSV-2 infection. Seronegative women are at greatest risk for acquisition of HSV-2 infection [21,22]. Other factors that may influence HSV-2 transmission include frequency of recurrences in the potential source partner, intercourse during active lesions, consistent use of barrier contraceptives, and chronic use of antiviral agents in the partner with herpes. In about 25% of couples who believe they are discordant for genital herpes, serologic tests show that both have HSV-2 infection. This suggests that the infection was transmitted unknowingly or possibly acquired from prior partners [23]. In couples discordant for HSV-2 antibody, the mean rate of transmission is approximately 12% per year [21]. However, the efficiency of transmission appears to vary widely, as some persons acquire HSV-2 after a single exposure while other couples do not transmit despite years of unprotected sexual activity [24]. This variability in transmission rates is poorly understood.

D. HSV-2 as a Risk Factor for HIV Acquisition and Transmission

HSVs are the most common cause of genital ulcers worldwide: this has been well documented in developed countries and has been shown in the developing world [17,25,26]. The presence of these genital ulcers has been suggested as a potential risk factor for HIV acquisition since the onset of the HIV epidemic [27,28]. Numerous epidemiologic studies have supported the association of genital ulcers in general, and genital herpes in particular, with HIV infection [16,29–31]. For example, a prospective cohort [31] examining the risk of HIV acquisition among men with and without genital ulcers found that men with genital ulcers had three times the risk for HIV seroconversion. In addition, both case-control [16,27,28,30,32] and nested case-control [33,34] studies have shown that HSV-2 infection is associated with acquisition of HIV. The magnitude of the estimates of relative risk for HIV infection has varied from 1.2 to 8.5 [35]. The elevated risk has been found in male-to-male, male-to-female, and female-to-male transmission.

Two biologic mechanisms have been postulated that suggest how genital herpes infections may be a risk factor for HIV infection. First, acquisition of HIV may be facilitated by the mucosal disruption that accompanies genital herpes infections. Second, as herpetic ulcerations are associated with influx of CD4-bearing lymphocytes (due to the local inflammatory response), a larger number of target cells for HIV attachment and entry are present in the genital tract of persons with genital herpes infections [36,37]. Both the mucosal disruption and the presence of increased numbers of activated CD4 cells suggest that individuals with genital herpes infection will be at increased risk for acquisition of HIV.

The assessment of HSV-2 as a risk factor for HIV transmission (as opposed to HIV acquisition) is even less well understood, as people who transmit HIV infection have not been as intensively studied as those who acquire HIV infection. One report of efficient transmission of HIV describes a man with a history of genital herpes transmitting HIV to multiple sex partners [38]. Another study, of couples, found a relative risk of 1.9 for HIV seropositivity in women whose husbands had genital herpes [29]. It has been postulated that this efficiency of HIV transmission among persons with HSV-2 infection is due to migration of activated lymphocytes to genital herpes lesions. Among persons with HIV infection, this migration can result in increased local HIV replication on mucosal surfaces. This postulation is supported by the fact that HIV has been cultured from genital ulcers in African studies [40] and detection of HIV virions has been reported in over 95% of genital herpes episodes in HIV-infected men [41]. Because many HIV-infected persons reactivate HSV frequently [42], local bursts of HIV shedding from genital lesions may provide an explanation as to how genital ulcers facilitate HIV transmission.

E. Genital HSV-1 Infection

Studies of persons infected with HSV-1 or with HSV-2 at the same time in the oral and genital area have shown that the severity of recurrent infection depends on the interaction of the viral type with the anatomic site of infection. Thus, the rate of recurrence is highest for genital HSV-2 infection, then oral HSV-1 infection, then genital HSV-1 infection, and, finally, lowest for oral HSV-2 infection [43]. Despite the fact that HSV-1 is an infrequent cause of recurrent genital herpes, it is increasingly common as the cause of a first episode of genital herpes. In Great Britain, HSV-1 is more frequently isolated than HSV-2 from persons with first-episode genital HSV infection [44–46]. Other places with a high proportion of genital HSV-1 from which data are available include Singapore and Japan [47,48]. In Seattle, HSV-1 accounts for approximately 30% of first-episode genital herpes infections [49].

The reason for the increasing frequency of HSV-1 as the cause of first-episode genital herpes is not clear. One explanation frequently invoked is the increasing practice of oral-genital contact. For example, a case-control study identified oral sex as a risk factor for genital herpes acquisition, most of which was HSV-1 [50]. Another explanation for the increasing frequency of HSV-1 as the cause of first-episode genital herpes is the decreasing rates of HSV-1 acquisition in childhood. As a result, more young people reach adolescence seronegative, and their first encounter with HSV-1 may be during oral sex.

III. Clinical Epidemiology

A. Spectrum of Disease among HSV-2 Seropositive Persons

Among HSV-2 seropositive persons, only small fractions have been diagnosed with genital herpes. For example, in the NHANES survey, only 9.1% of women and 9.2% of men with HSV-2 antibody reported a history of genital herpes [7]. However, 12.2% of whites knew that they had the infection compared with 3.7% of African-Americans. While there is no evidence that the clinical expression of HSV-2 depends on race or ethnic background, access to health care is likely to play a role in recognition of infection [8]. Alternatively, the rates of concurrent HSV-1 infection are higher in African-American persons, and prior HSV-1 infection may make HSV-2 infections less clinically apparent [52]. The biology of HSV-2 in persons unaware of the infection has not been well defined. However, epidemiologic studies show that persons without a history of genital herpes are the major source of new infections, and preliminary data suggest that the virus reactivates in virtually all infected persons [24,53,54].

Many people with genital herpes experience troublesome, recurrent, genital ulcerations for many years after infection. Typically, recurrences are associated with mild physical discomfort; however, some patients experience severe or frequent recurrences that clearly impair health. In addition to the physical discomfort of recurrent disease, genital herpes causes significant amounts of psychosocial dysfunction [55,56]. Several studies have shown that genital herpes is associated with depression, isolation, and perception of social stigma. For many patients, these feelings are exacerbated during recurrences. Comparison of patients with genital herpes to patients with gonorrhea or dermatologic conditions shows that the psychosocial dysfunction is worse in persons with genital herpes [57]. For most patients with genital herpes, fear of transmitting the infection to sexual partners is the greatest concern. New relationships are often difficult, as persons with herpes need to disclose their infection to others and fear rejection [58].

B. Clinical Manifestations

The clinical and virologic course of genital HSV infection is influenced by the immune status of the host. The primary genital herpes infection, defined as infection occurring in persons who are seronegative to both HSV-1 and HSV-2, is the most severe infection [52,59]. These primary infections are associated with prolonged clinical symptoms and viral shedding (2–3 weeks). Nonprimary initial infection, defined as infection with a new HSV type in a person with preexisting antibody to heterologous virus, tends to be less severe. Recurrent genital herpes, caused by reactivation of latent virus, is associated with shortest duration of lesions and viral shedding (2–8 days).

Although an individual's immune status influences the clinical manifestations of genital herpes infections, a large overlap exists between the clinical severity of primary, nonprimary, and recurrent infections. Because of this overlap in clinical symptoms, virologic and serologic laboratory evaluation is necessary to establish the diagnosis in an individual patient. For example, about 25% of patients who present with a clinical first episode of genital herpes already have antibody to the type of virus that is isolated from the genital area, thus indicating previous infection. These clinical first episodes, therefore, represent the first recognized recurrence of previously acquired genital herpes infection.

Primary and nonprimary initial genital herpes are often accompanied by systemic symptoms in addition to genital ulcerations. Neurologic complications, fever, malaise, myalgias, and headache are common, especially among women. Dysuria, both internal and external, may be present in the early stages of infection, and aseptic meningitis is reported among 25% of women and may require hospitalization [60]. The genital symptoms will often start with nonspecific itching and tingling in the vulvar area but rapidly progress to lesions. Early lesions characteristic of primary genital herpes are vesicular, although the vesicles may rupture rapidly in women and never be noticed. Subsequent lesions are multiple, bilateral, shallow ulcers with an erythematous base. These lesions may coalesce forming large areas of ulcerations on the vulva. The cervix is involved in 80% of episodes of primary herpes and occasionally is the only clinically recognized site of infection. Herpetic cervicitis can be severe with large areas of necrosis of the ectocervical epithelium. Such women may present with vaginal bleeding or discharge as their sole presenting complaint. In patients not receiving antiviral therapy, new crops of lesions may appear in the initial one to two weeks of infection. The ulcers heal completely over two to three weeks and are usually not associated with scarring.

The clinical illness has a similar course in men. While lesions tend to be more numerous among men than women, they are generally associated with less pain [61]. Vesicular lesions tend to persist longer and occasionally dry and scab without the ulcerative stage. Systemic and neurologic manifestations tend to be less frequent, except for *MSM* who may present with herpetic proctitis [62]. This syndrome is often associated with autonomic nerve paresis and patients may experience urinary retention, constipation, and sexual dysfunction. The autonomic syndromes are

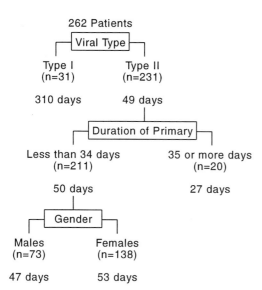

262 Patients

Viral Type

Type I   Type II
(n=31)   (n=231)

310 days   49 days

Duration of Primary

Less than 34 days   35 or more days
(n=211)   (n=20)

50 days   27 days

Gender

Males   Females
(n=73)   (n=138)

47 days   53 days

Fig. 24.2 Natural history of recurrent genital HSV infection after virologically and serologically documented first episode disease. Using discriminant analysis, predictors of median time to first recurrence were determined in a cohort of 262 patients. As shown, type of virus—HSV-1 vs HSV-2—was the strongest predictor of subsequent rate of recurrences (adapted from [64]).

almost always transient and tend not to recur during subsequent episodes of reactivation.

The course of recurrent genital herpes infections is similar to that of primary and nonprimary infections but is milder and shorter in both genders [59]. For many people a recurrent episode starts with a premonitory complex of symptoms (referred to as prodrome) prior to the development of genital lesions [63]. Prodromal symptoms commonly include itching and tingling in the genital area and neuralgia distant to the site of lesions (*i.e.*, in the buttock or thigh area). These prodromal symptoms are often more distressing than the discomfort of the lesions. The average duration of a recurrent episode tends to fall between four and seven days, although the length of the episode varies widely among patients and within patients from one episode to another. Recurrent episodes are often associated with the development of only a few lesions in a well-defined area. New lesion formation is less frequent than during primary episodes. While local genital pain is common among women during recurrent disease, systemic illness and neurologic complications are uncommon. In some patients, recurrences may occur outside of the genital area (most commonly occurring on the buttocks and the leg) [64].

The natural history of genital herpes in persons who present with serologically documented first episode of genital herpes has been well characterized. Almost all patients (89%) with newly acquired genital HSV-2 infection have at least one recurrence in the first year of infection [65]. Thirty-eight percent have ≥6 recurrences and 20% have >10 recurrences in the initial year. Persons who have a prolonged primary episode (>35 days) have a shorter median time to first recurrence than those with shorter primary episodes (27 vs 50 days) (Figure 24.2).

Surprisingly, men appear to have a shorter time to first recurrence than women (47 vs 53 days). Thus, the median number of recurrences is four for women and five for men during the first year of HSV-2 infection. The rate of recurrence declines slowly with time in most patients. However, a subset will experience an increase in recurrence rates, sometimes years after the initial infection.

C. Atypical Genital Herpes

The signs and symptoms described above appear to affect only a minority of HSV-2 infected persons. However, careful studies of HSV-2 seropositive persons show that most have genital symptoms and signs but they tend to be mild or atypical and thus may go unrecognized. For example, following an individualized education session, 50% of HSV-2 seropositive women who did not have a history of genital herpes recognized typical recurrences [66]. Of note, recurrent genital herpes was the most common genitourinary complaint in this group of women during six months of follow-up. Undiagnosed HSV-2 infection can be the cause of genital ulcerations or rashes of atypical appearance or in an atypical location [9]. Atypical appearances include genital fissures, papules, or superficial cervical ulcerations. Lesions that occur in the perianal area are often attributed to hemorrhoids or anal fissures. Not infrequently patients with genital herpes are diagnosed with recurrent yeast infections or "jock itch," urinary tract infections, eczema, or contact dermatitis. These diagnoses are often supported by apparent response to therapy, which may appear effective as healing occurs naturally in genital herpes. The availability of type-specific serologies will facilitate diagnosis of HSV-2 infection in persons with atypical presentations.

D. Viral Shedding

HSV can be isolated from the genital area during initial and recurrent infection, as well as intermittently between clinically recognized episodes. During the primary and the recurrent episodes of genital infection, viral shedding tends to last a shorter time than the lesions. In primary infection, the virus can be isolated for approximately two weeks, and in recurrent episodes, the virus can be isolated for two to four days. Studies have documented that HSV can be present in the genital area also during times between clinically apparent recurrences. Viral shedding in the absence of genital lesions has been termed asymptomatic or subclinical shedding. Epidemiologically, subclinical shedding has been implicated in most cases of sexual as well as perinatal HSV transmission and is likely to be the key concept in control strategies for genital herpes [24,67–70]. While sex during recurrent genital herpes may be associated with a higher risk of transmission, asymptomatic transmission is implicated in most cases of HSV shedding, as safer sex is less likely to be practiced in the absence of lesions.

Asymptomatic shedding has been studied by instructing patients to obtain swabs on a daily basis from genital secretions [71]. Because viral shedding is an infrequent event, studies conducted for a short period of time tend to underestimate the proportion of persons who shed HSV asymptomatically as well as the rate of shedding. Asymptomatic shedding is most common near the time of acquisition of genital herpes [68,72]. In the first two years of HSV-2 infection, the virus can be isolated from the genital area on 10% of all days and 6% of days without lesions (Fig. 24.3) [73]. Subsequently, the shedding rate falls, albeit slowly. Overall, the mean subclinical shedding rate in women is approximately 2.4% of days [72]. However, large variation exists so that some women may shed very infrequently and others may shed virus up to 25% of days.

In women, the most common sites of asymptomatic shedding are the perianal area and the vulva, followed by the cervix. Men also shed virus asymptomatically from the genital area, predominantly from the penile skin and the perianal area [74]. In both men and women, shedding is more likely during the week prior to or the week following a clinical recurrence of HSV. The mean subclinical shedding rate is 2.4% of the days, but in the week prior to a recurrence the shedding rate is as high as 10% of the days. While prodrome may account for some of the shedding during that time period, the symptoms that define prodrome are nonspecific in many patients and may not dissuade them from engaging in sexual activities.

E. Genital HSV-1 Infections

Genital HSV-1 infection has been rising in many areas of the developed world and in some places it accounts for 50% of initial genital herpes. The initial presentation of primary herpes is indistinguishable between HSV-1 and HSV-2 genital infections. However, unlike genital HSV-2 infection, genital HSV-1 recurs infrequently [43,65], thus determining the virus type has prognostic significance. As shown in Figure 24.2, the median time to first recurrence after acquisition of genital HSV-1 infection is 310 days. In addition, HSV-1 is infrequently shed asymptomatically. Thus, patients with genital HSV-1 infection may be reassured that the course of infection is likely to be mild. It is not known how often genital HSV-1 infection is acquired from oral vs genital contact.

IV. Neonatal Herpes

For pregnant women with genital herpes infections, the fear of transmitting the infection to their newborn is a major concern. Infection of the newborn with HSV is a rare but devastating disease. The incidence of neonatal herpes varies geographically, with the Pacific Northwest and Scandinavian countries appearing to have the highest rates in the world. At the University of Washington Medical Center, neonatal herpes occurs in approximately 1 of 2200 births. Not surprisingly, the incidence of neonatal herpes is lower on the European continent (*i.e.,* 1 of 35,000 deliveries in the Netherlands), as the prevalence of HSV-2 is also lower [75]. Interestingly, the incidence of neonatal herpes is also low among populations heavily infected with HSV-2. This low incidence may be due to the fact that in these populations, herpes infections occur at a young age, soon after initiation of sexual activity, and are therefore well established by the time pregnancy occurs. Most infected newborns acquire the infection during birth from contact with infected genital secretions. About 15% of infections are acquired after birth, often from oral HSV-1 from close family members [76]. Despite

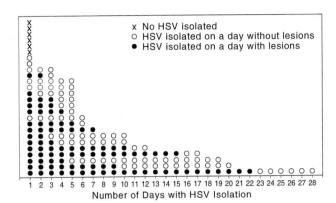

Fig. 24.3 Variability in the rate of shedding of HSV-2 in the genital tract. Each horizontal row represents a woman who obtained swabs of genital secretions on a daily basis for about 70 consecutive days. Days with positive cultures are marked by solid circles if genital lesions were present on the day of the culture and by empty circles if no genital lesions were noted.

availability of antiviral therapy, neonatal herpes infections have a high mortality, and developmental and cognitive impairment persist in most of the survivors [53].

Studies have elucidated the risk factors for HSV transmission during birth and suggest a potential for prevention. First, most episodes of neonatal transmission occur from women without a history of genital herpes. Second, several investigators have found that acquisition of HSV infection during late pregnancy is associated with high risk of neonatal disease. In a cohort of 15,923 women evaluated at delivery, HSV was isolated from 56 women with serologic results available for 52 women [77]. Of the 34 women with HSV who had preexisting antibodies to the same type of virus, one transmitted HSV to her infant. In contrast, of 18 women with virus, but who lacked antibodies to that type of virus, six transmitted the infection to the newborn. Thus, recurrent genital HSV is associated with low risk of neonatal herpes, even when viral shedding occurs at delivery [78]. In contrast, acquisition of genital HSV near the time of labor is associated with extremely high risk (30–50%) of neonatal herpes. Finally, the risk of transmission is also increased by the use of fetal scalp electrodes, as the skin puncture facilitates HSV infection [77]. These findings suggest that prevention of neonatal herpes will require prevention of acquisition of genital HSV in late pregnancy and avoidance of unnecessary use of invasive monitoring during birth.

Interestingly, acquisition of genital herpes earlier during pregnancy does not appear to increase the risk of neonatal herpes. In a study of 7046 women at risk for HSV-2 acquisition, the HSV-2 seroconversion rate, adjusted for duration of pregnancy, was 3.7% for HSV-1 seronegative women and 1.7% for HSV-1 seropositive women [79]. Adverse outcomes during pregnancy were not increased in women who seroconverted during pregnancy and none transmitted the infection to their infants. Thus, passively acquired antibody appears to protect the infant from acquiring herpes.

Current guidelines in the United States call for delivery by Cesarean section (CS) (abdominal delivery) if a lesion is noted in the genital area at the time of labor [80]. Unfortunately, morbidity increases in women with genital herpes in whom CS is performed to prevent transmission of HSV to neonates. In fact, a decision analysis has shown that CS for women with current genital herpes at delivery results in maternal morbidity, mortality, and potential costs that likely exceed the benefit to the newborn [81]. Few data are available to show that this practice decreases the rate of HSV acquisition to the neonate. In some countries, the practice of CS for recurrent genital herpes at term has been abandoned without apparent increase in the incidence of neonatal HSV infections [75]. Furthermore, CS for genital lesions also ignores the potential risk of transmission from asymptomatic shedding at the time of labor. Thus, other methods for preventing neonatal HSV should be investigated.

Use of antiviral agents, such as acyclovir, during pregnancy has been examined as a method to prevent neonatal HSV. Three studies, including two small randomized clinical trials, have investigated the use of acyclovir near term to prevent recurrences during labor [82,83]. The data are inconclusive, and further studies are needed to establish the usefulness as well as the safety of antiviral therapy in this setting.

While acyclovir appears safe in pregnancy [85,86], concern remains about the potential for toxicity of antiviral agents during gestation. A registry that has monitored the outcome of pregnancy during which women took acyclovir has shown no increase in fetal abnormalities compared to the background rate and no consistent pattern to the abnormalities observed [86]. While these data are reassuring, the number of women is too small to definitively assure safety of acyclovir during the first trimester of pregnancy. Towards the end of pregnancy, acyclovir could potentially cause fetal nephrotoxicity and some experts do not feel that sufficient safety data are available to recommend its widespread use.

V. Diagnosis

Because most infected persons do not have the typical signs and symptoms of herpes infections and are unaware of their infection, the clinical diagnosis of genital herpes is insensitive. However, in persons with characteristic genital ulcerations in parts of the world where other causes of genital ulcer disease—namely chancroid and syphilis—are uncommon, clinical diagnosis is usually reliable. Even when other causes of genital ulcers are uncommon, it is quite useful to obtain a viral culture. First, counseling a patient who just acquired a chronic STD is easier if laboratory confirmation of the diagnosis is available. Second, as previously described, typing of the HSV isolate has prognostic implications. Unfortunately, even when an individual presents with a primary episode of genital herpes, HSV is isolated in only about 80% of specimens [87]. Failure to isolate HSV may result from improper collection of specimens and delay in delivery and processing of the specimen. The rate of HSV isolation is even lower in recurrent disease, where only about 50% of episodes yield virus. To optimize recovery of virus, the patient should be seen early during the episode when the lesion is likely to be vesicular or ulcerative, as viral replication is highest during this time.

In addition to viral cultures, several antigen detection tests are available that are comparable in sensitivity to viral culture, but most do not identify virus type [88]. Several studies have also evaluated HSV DNA polymerase chain reaction (PCR) for viral diagnosis [26,89,90]. Not surprisingly, the PCR assay is more sensitive than viral culture for detection of HSV in genital lesions. However, this test is not commercially available and the role of PCR in the diagnosis of genital ulcer disease has not been determined.

Serology has been a mainstay of diagnosis of other chronic viral infections, such as HIV, cytomegalovirus, and hepatitis B. Serology can also be used for diagnosis of HSV infection because infected persons make antibodies that appear to persist for the life of the host. The cross-reactivity between HSV-1 and HSV-2 has made type-specific serologic tests for this infection difficult to develop. The current assays do not distinguish HSV-1 from HSV-2 infection accurately, despite their manufacturers' claims to the contrary [91]. This is especially true when diagnosing HSV-2 in persons with prior HSV-1 infection.

In contrast to the current generation of assays, Western blot can reliably distinguish between HSV-1 and HSV-2 infection and is the gold standard for HSV serologies [92]. However, Western blot is expensive, labor intensive, and requires reading by highly trained personnel. As such, it has not been adapted for commercial use. Newer type-specific serologic tests have been developed and are likely to be commercially available in 1999 [93,94]. Most of these type-specific tests rely on the antigenic differences between glycoprotein G in HSV-1 and HSV-2, and they approximate the accuracy of the Western blot in comparative trials. These type-specific tests may be useful in STD clinics to diagnose atypical genital complaints and to define the susceptibility of partners of patients with genital herpes. These tests may also be used as a screening tool to identify HSV-2 infected persons unaware of their disease or women at risk for acquiring HSV during pregnancy, especially HSV seronegative women with HSV-2 seropositive partners.

VI. Therapy

Although there is no cure for HSV infections, effective antiviral therapy is available to palliate the symptoms of genital HSV infection. Unfortunately, the current therapy does not eradicate latent virus, and treatment of symptomatic episodes does not influence the natural history of subsequent viral reactivation [97]. As the therapy is expensive and not curative, it remains underutilized.

Acyclovir, one commonly used antiviral agent, has been used for genital herpes therapy since the 1980s and is available in oral, intravenous, and topical preparations [98]. It is a nucleoside analogue, which requires phosphorylation by viral thymidine kinase for activity. Thus, acyclovir is only active in cells that are infected with virus; this characteristic likely accounts for acyclovir's unusually safe profile.

Acyclovir decreases viral shedding, duration of lesions, and symptoms when used during first-episode genital herpes infections [99]. In large clinical trials, acyclovir prevented development of neurologic complications associated with primary genital herpes. However, when used for recurrent genital herpes, acyclovir decreased viral shedding and shortened the duration of lesions by one to two days. Thus, the benefit of episodic therapy is marginal for individuals with recurrent lesions.

A more effective way to treat recurrent genital HSV infection is with suppressive therapy, described as chronic daily dosing of acyclovir. When taken daily, acyclovir prevents 70% of recurrences [100,101], and many individuals will experience no recurrences of genital herpes. If a recurrence does occur while an individual is taking acyclovir, the episode is typically mild and short-lasting.

Despite the potential benefits of suppressive therapy with acyclovir, it tends to be used only for patients with frequent recurrences. First, suppressive therapy requires patient compliance and it is costly. Second, clinical trials have only demonstrated benefit with suppressive therapy in patients with at least six recurrences a year. However, the therapy is also likely to be effective in patients with fewer than six recurrences a year. Regardless of the number of recurrences per year, some patients often experience psychological distress during recurrences and are, therefore, good candidates for suppressive therapy. For example, one study showed improvement in psychosocial functioning in patients receiving daily acyclovir [102]. Nonetheless, many clinicians will not offer suppressive therapy to patients with fewer than six recurrences a year.

Drug resistance and safety have been cited as potential barriers to more frequent usage of suppressive therapy. However, in immunocompetent patients, chronic use of acyclovir has not been associated with the emergence of acyclovir-resistant strains of herpes virus [103]. Acyclovir-resistance has been described in persons severely immunocompromised as a result of HIV infection, malignancy, chemotherapy, or transplant [108,109]. Clinically, acyclovir-resistant herpes occurs infrequently but it is associated with significant morbidity. The safety of acyclovir has been assessed in long-term (up to ten year) studies, and no clinical or laboratory abnormalities have been associated with long-term use of acyclovir.

For doses and regimens of antiviral medications currently recommended, the reader is referred to the CDC STD treatment guidelines, available in hard copy and at *www.cdc.gov/rchstp/dstd* [110].

VII. Prevention

Universal immunization with an effective vaccine against HSV-2 infection, delivered in infancy, is likely to be the most effective strategy for control of genital herpes. Such strategy has proven effective against another chronic viral infection, specifically the hepatitis B virus infection. Unfortunately, an effective vaccine appears unlikely in the next few years. A vaccine for genital herpes has been of interest for many decades, and several candidate vaccines have been tested [111]. Promising results have been obtained with the recombinant glycoproteins D and B preparation [112]. This preparation stimulated production of neutralizing antibodies equal to or exceeding that seen in natural infection. However, the results of efficacy clinical trials have been disappointing, as the vaccine failed to protect against acquisition of HSV-2 in susceptible adults [113].

Other candidate vaccines are also under evaluation. Another glycoprotein vaccine is in efficacy trials and results should be available in 1999. Other constructs, including an HSV DNA vaccine and genetically attenuated viruses that are unable to establish latency, are also being studied [114–116]. In the meantime, the failure of the previous vaccines has shed light on the complexities of the immune response to HSV and the determinants of protection.

As with other STDs, counseling to prevent spread of infection to sexual partners is an essential part of management of genital herpes [110]. Unfortunately, it is not clear what messages should be conveyed. Currently, the use of condoms has been encouraged for prevention of genital herpes transmission despite minimal data supporting their efficacy in preventing HSV transmission. Given the wide anatomic area of HSV reactivation in the genital tract, condoms are probably less protective against HSV than STDs characterized by discharge. Although it is reasonable to recommend condoms for sexually active adults, given that they protect against acquisition of other STDs and HIV, compliance with condom use is poor, especially in long-term monogamous couples. In our experience of 107 couples discordant for HSV-2 infection, only four used condoms consistently for prevention of HSV transmission and another 15 used condoms as the sole means of birth control. Other methods that cover a greater part of the genital mucosa, such as microbicides and female condoms, may offer better protection than male condoms; data to demonstrate their efficacy for prevention of HSV-2 infection are lacking.

In our clinic, STD counseling is tailored to patients' individual needs. Persons with new sexual partners are counseled to tell their partners that they have genital herpes. Then each couple can make a decision about how to avoid transmission. Patients are also advised to avoid sexual intercourse during recurrences, as the risk of transmission is greater at that time.

In the future, patients with genital herpes who are concerned about transmitting the virus to a partner may be given the option of suppressive therapy. The studies showing that antiviral drugs abrogate asymptomatic shedding suggest that they may be effective in reducing the risk of sexual transmission of genital herpes [73]. An ongoing trial is evaluating the ability of valaciclovir to interrupt sexual transmission in monogamous couples discordant for HSV-2. In this study, the partner with genital herpes receives daily suppressive therapy. If the clinical trial demonstrates a benefit from antiviral therapy, it will become an option for some couples concerned about transmission of HSV. However, such intervention is unlikely to have a widespread public health benefit.

Implementation of any prevention program, short of universal vaccination, requires identification of persons with HSV-2 infection. This screening is best achieved with type-specific serologies. However, it is unclear if an asymptomatic person who tested positive for HSV-2 will change their sexual behavior, decreasing the likelihood of transmission to partners. Knowledge of HIV serostatus has been shown to decrease high-risk behavior in some persons [117]. While currently there is medical benefit to early identification of HIV infection, such intervention was not available in the mid-1980s when voluntary HIV testing was introduced. Currently, HSV is in a similar situation because the infection can be identified in an asymptomatic person, but effective interventions have yet to be proven and made available. Thus, in a traditional sense of "screening," HSV-2 does not qualify. However, knowledge of infection is likely to be associated with some behavioral change.

Widespread testing may evoke objections from physicians, as the time required for counseling an asymptomatic HSV-2 seropositive person is not trivial. Therefore, similar to HIV testing and counseling, alternative venues for HSV-2 testing and counseling need to be implemented. While there is some concern about the psychological impact on an individual who unexpectedly tests positive for HSV-2 infection, learning that one has HSV-2 infection after transmitting it to a partner is also very stressful. Currently, the practice at STD clinics is to reassure the patient that he or she is free of infection when the bacterial syndromes are excluded. However, HSV-2 is likely to be among the most common STDs in this setting and yet testing for it is currently not offered. Patient demand is likely to result in wider use of serologic testing, once the test becomes commercially available.

Serologic tests are also likely to play a role in management of pregnant women. In areas where the rate of bacterial STDs has declined, neonatal HSV has become the most common cause of neonatal morbidity and mortality. Identifying women at risk through serologic testing and counseling is one way to decrease the poor neonatal outcomes associated with HSV. Women who are identified as HSV-2 seropositive can be reassured that the risk of transmission to the newborn is minimal, and they may be candidates for suppressive therapy near the end of pregnancy. Women at risk for acquisition of genital herpes during pregnancy can be identified and counseled to avoid sexual exposure near the end of pregnancy (genital–genital and oral–genital contact). If antiviral therapy is effective at reducing the transmission of genital herpes, then it offers a potential alternative strategy for prevention of transmission of HSV-2 during pregnancy.

VIII. Resources for Persons with Genital Herpes

Numerous resources are available that provide peer support for persons with genital herpes. As the pathophysiology of the disease is complex and the individual course highly variable, many patients find that their health providers are unable to provide them with up-to-date information about natural history and therapeutic options. Under the sponsorship of the American Social Health Association (ASHA), the Herpes Resource Center was initiated in 1979. The Center publishes a quarterly newsletter, *the helper,* that provides research news relevant to herpes therapies and addresses many common questions. ASHA oversees local support groups for people with herpes. Several lay books are available that explain genital herpes. Such books include "Managing Herpes" by C. Ebel and "The Truth About Herpes" by S. Sacks [118,119]. The Internet has several sites devoted to genital herpes that provide medical information and chat rooms that allow exchange of experiences. Examples include *www.cafeherpes.com, www.viridae.com, www.advicecenter.com,* and *www.ashastd.org.*

The proliferation of the above resources is consistent with the findings in surveys that show that patients with herpes are often dissatisfied with the care they receive from their physicians [55]. While to many patients, having acquired herpes is a life-changing experience, most physicians perceive it as a trivial disorder. This difference in perception has resulted in the creation of alternative sources of information and support for people with herpes.

References

1. Hutfield, D. C. (1996). History of herpes genitalis. *Br. J. Vener. Dis.* **42**, 263.

2. Nahmias, A. J., and Dowdle, W. R. (1968). Antigenic and biologic differences in *Herpesvirus hominis. Prog. Med. Virol.* **10,** 110.

3. Nahmias, A. J., Josey, W. E., Naib, Z. M., Luce, C. F., and Guest, B. A. (1970). Antibodies to herpesvirus hominis types 1 and 2 in humans: Patients with genital infections. *Am. J. Epidemiol.* **92,** 547–552.

4. Baringer, J., and Swoveland, P. (1973). Recovery of herpessimplex virus from human trigeminal ganglions. *N. Engl. J. Med.* **288,** 648–650.

5. Baringer, J. (1974). Recovery of herpes simplex virus from human sacral ganglions. *N. Engl. J. Med.* **291,** 828–830.

6. Centers for Disease Control and Prevention (1996). "Sexually Transmitted Disease Surveillance." CDC, Atlanta, GA.

7. Fleming, D., McQuillan, G., Johnson, R., Nahmias, A., Aral, S., Lee, F., and St. Louis, M. (1997). Herpes simplex virus type 2 in the United States, 1976 to 1994. *N. Engl. J. Med.* **337,** 1105–1111.

8. Koutsky, L. A., Ashley, R. L., Holmes, K. K., Stevens, C. E., Critchlow, C. W., Kiviat, N., Lipinski, C. M., Wolner-Hanssen, P., and Corey, L. (1990). The frequency of unrecognized type 2 herpes simplex virus infection among women: Implications for the control of genital herpes. *Sex. Transm. Dis.* **17,** 90–94.

9. Koutsky, L., Stevens, C., Holmes, K., Ashley, R., Kivat, N., Critchlow, C., and Corey, L. (1992). Underdiagnosis of genital herpes by current clinical and viral-isolation procedures. *N. Engl. J. Med.* **326,** 1539–1553.

10. Breinig, M. K., Kingsley, L. A., Armstrong, J. A., Freeman, D. J., and Ho, M. (1990). Epidemiology of genital herpes in Pittsburgh: Serologic, sexual and racial correlates of apparent and inapparent herpes simplex infections. *J. Infect. Dis.* **161,** 299–305.

11. Oliver, L., Wald, A., Mihee, K., Zeh, J., Selke, S., Ashley, R., and Corey, L. (1995). Seroprevalence of herpes simplex virus infections in a family medicine clinic. *Arch. Fam. Med.* **4,** 228–232.

12. Siegel, D., Golden, E., Washington, A., Morse, S., Fullilove, M., Catania, J., Marin, B., and Hulley, S. (1992). Prevalence and correlates of herpes simplex infections: The population-based AIDS in Multiethnic Neighborhoods Study. *JAMA, J. Am. Med. Assoc.* **268,** 1700–1708.

13. Sacks, S. L. (1995). Genital HSV infection and treatment. (S. L. Sacks, S. E. Strauss, R. J. Whitley, and P. D. Griffiths, eds.). *In* "Clinical Management of Herpes Viruses." IOS Press, Amsterdam.

14. Cowan, F., Johnson, A., Ashley, R., Corey, L., and Mindel, A. (1994). Antibody to herpes simplex virus type 2 as serological marker of sexual lifestyle in populations. *Br. Med. J.* **309,** 1325–1329.

15. Quinn, T. C., Piot, P., McCormick, J. B., Feinsod, F. M., Taelman, H., Kapita, B., Stevens, W., and Fauci, A. S. (1987). Serologic and immunologic studies in patients with AIDS in North America and Africa. The potential role of infectious agents as cofactors in human immunodeficiency virus infection. *JAMA, J. Am. Med. Assoc.* **257,** 2617–2621.

16. Boulos, R., Ruff, A. J., Nahmias, A., Holt, E., Harrison, L., Magder, L., Wiktor, S. Z., Quinn, T. C., Margolis, H., and Halsey, N. A. (1992). Herpes simplex virus type 2 infection, syphilis, and hepatitis B virus infection in Haitian women with human immunodeficiency virus type 1 and human T lymphotropic virus type 1 infections. *J. Infect. Dis.* **166,** 418–420.

17. Beyrer, C., Jitwatcharanan, K., Natpratan, C., Kaewvichit, R., Nelson, K. E., Chen, C. Y., Weiss, J. B., and Morse, S. A. (1998). Molecular methods for the diagnosis of genital ulcer disease in a sexually transmitted disease clinic population in northern Thailand: Predominance of herpes simplex virus infection. *J. Infect. Dis.* **178,** 243–246.

18. Christenson, B., Bottiger, M. Svensson, A., and Jeansson, S. (1992). A 15 year surveillance study of antibodies to herpes simplex virus types 1 and 2 in a cohort of young girls. *J. Infect.* **25,** 147–154.

19. Eberhart-Phillips, J., Dickson, N., Paul, C., Fawcett, J., Holland, D., Taylor, J., and Cunningham, A. (1998). STI 1997/173—Herpes simplex type 2 infection in a cohort aged 21 years. *Sex. Transm. Infect.* **74,** 216–218.

20. Catania, J. A., Binson, D., and Stone, V. (1996). Relationship of sexual mixing across age and ethnic groups to herpes simplex virus-2 among unmarried heterosexual adults with multiple sexual partners. *Health Psychol.* **15,** 362–370.

21. Mertz, G. J., Benedetti, J., Ashley, R., Selke, S. A., and Corey, L. (1992). Risk factors for the sexual transmission of genital herpes. *Ann. Intern. Med.* **116,** 197–202.

22. Bryson, Y. J., Dillon, M., Bernstein, D. I., Radolf, J., Zakowski, P., and Garratty, E. (1993). Risk of acquisition of genital herpes simplex virus type 2 in sex partners of persons with genital herpes: A Prospective Couple Study. *J. Infect. Dis.* **167,** 942–946.

23. Mertz, G. J., Coombs, R. W., Ashley, R., Jourden, J., Remington, M., Winter, C., Fahnlander, A., Guinan, M., Ducey, H., and Corey, L. (1988). Transmission of genital herpes in couples with one symptomatic and one asymptomatic partner: A Prospective Study. *J. Infect. Dis.* **157,** 1169–1177.

24. Mertz, G. J., Schmidt, O., Jourden, J. L., Guinan, M. E., Remington, M. L., Fahnlander, A., Winter, C., Holmes, K. K., and Corey, L. (1985). Frequency of acquisition of first-episode genital infection with herpes simplex virus from symptomatic and asymptomatic source contacts. *Sex. Transm. Dis.* **12,** 33–39.

25. Bogaerts, J., Ricart, C. A., van Dyck, E., and Piot, P. (1989). The etiology of genital ulceration in Rwanda. *Sex. Transm. Dis.* **16,** 123–126.

26. Morse, S. A., Trees, D. L., Htun, Y., Radebe, F., Orie, K. A., Dangor, Y., Beck-Sague, C. M., Schmid, S., Fehler, G., Weiss, J. B., and Ballard, R. C. (1997). Comparison of clinical diagnosis and standard laboratory and molecular methods for the diagnosis of genital ulcer disease in Lesotho: Association with human immunodeficiency virus infection. *J. Infect. Dis.* **175,** 001–007.

27. Stamm, W. E., Handsfield, H. H., Rompalo, A. M., Ashley, R. L., Roberts, P. L., and Corey, L. (1988). Association between genital ulcer disease and acquisition of HIV infection in homosexual men. *JAMA, J. Am. Med. Assoc.* **260,** 1429–1433.

28. Greenblatt, R. M., Lukehart, S. A., Plummer, F. A., Quinn, T. C., Critchlow, C. W., Ashley, R. L., D'Costa, L. J., Ndinya-Achola, J. O., Corey, L., Ronald, A. R., and Holmes, K. K. (1988). Genital ulceration as a risk factor for human immunodeficiency virus infection. *AIDS* **2,** 47–50.

29. Latif, A., Katzenstein, D., Bassett, M., Houston, S., Emmanuel, J., and Marowa, E. (1989). Genital ulcers and transmission of HIV among couples in Zimbabwe. *AIDS* **3,** 519–523.

30. Hook, E., Cannon, R., Nahmias, A., Lee, F., Campbell, J., C. H. Glasser, D., and Quinn, T. (1992). Herpes simplex virus infection as a risk factor for human immunodeficiency virus infection in heterosexuals. *J. Infect. Dis.* **165,** 251–255.

31. Holmberg, S. D., Stewart, J. A., Gerber, R., Byers, R. H., Lee, F. K., O'Malley, P. M., and Nahmias, A. J. (1988). Prior herpes simplex virus type 2 infection as a risk factor for HIV infection. *JAMA, J. Am. Med. Assoc.* **259,** 1048–1050.

32. Gwanzura, L., McFarland, W., Alexander, D., Burke, R. L., and Katzenstein, D. (1998). Association between human immunodeficiency virus and herpes simplex virus type 2 seropositivity among male factory workers in Zimbabwe. *J. Infect. Dis.* **177,** 481–484.

33. Keet, I. P. M., Lee, F. K., van Griensven, G. J. P., Lange, J. M. A., Nahmias, A., and Coutinho, R. A. (1990). Herpes simplex virus type 2 and other genital ulcerative infections as a risk factor for HIV-1 acquisition. *Genitourin. Med.* **66,** 330–333.

34. Telzak, E. E., Chiasson, A., Bevier, P. J., Stoneburner, R. L., Castro, K. G., and Jaffe, H. W. (1993), HIV-1 seroconversion in pa-

tients with and without genital ulcer disease: A Prospective Study. *Ann. Intern. Med.* **119,** 1181–1186.

35. Dickerson, M. C., Johnston, J., Delea, T. E., White, A., and Andrews, E. (1996). The causal role for genital ulcer disease as a risk factor for transmission of human immunodeficiency virus: An application of the Bradford Hill criteria. *Sex. Transm. Dis.* **23,** 429–440.

36. Cunningham, A. L., Turner, R. R., Miller, A. C., Para, M. F., and Merigan, T. C. (1985). Evolution of recurrent herpes simplex lesions: An Immunohistologic Study. *J. Clin. Invest.* **75,** 226–233.

37. Koelle, D. M., Abbo, H., Peck, A., Ziegweld, K., and Corey, L. (1994). Direct recovery of herpes simplex virus (HSV)—specific T lymphocyte clones from recurrent genital HSV-2 lesions. *J. Infect. Dis.* **169,** 956–961.

38. Clumneck, N., Taelman, H., Hermans, P., Piot, P., Schoumacher, M., and DeWit, S. (1989). A cluster of HIV infection among heterosexual people without apparent risk factors. *N. Engl. J. Med.* **321,** 1460–1462.

39. Cameron, D. W., Simonsen, J. N., D'Costa, L. J., Ronald, A. R., Maitha, G. M., Gakinya, M. N., Cheang, M., Ndinya-Achola, J. O., Piot, P., Burnham, R. C., and Plummer, F. A. (1989). Female to male transmission of human immunodeficiency virus type 1: Risk factors for seroconversion in men. *Lancet* 403–407.

40. Kreiss, J. K., Coombs, R., Plummer, F., Holmes, K. K., Nikora, B., Cameron, W., Ngugi, E., Ndinya-Achola, J. O., and Corey, L. (1989). Isolation of human immunodeficiency virus from genital ulcers in Nairobi prostitutes. *J. Infect. Dis.* **160,** 380–384.

41. Schacker, T., Ryncarz, A., Goddard, J., Shaughnessy, M., and Corey, L. (1998). Frequent recovery of HIV from genital herpes simplex virus lesions in HIV infected persons. *JAMA, J. Am. Med. Assoc.* **280,** 61–66.

42. Schacker, T., Zeh, J., Hu, H., Hill, J., and Corey, L. (1998). Frequency of symptomic and asymptomatic HSV-2 reactivations among HIV-infected men. *J. Infect. Dis.* **178,** 1616–1622.

43. Lafferty, W. E., Coombs, R. W., Benedetti, J., Critchlow, C., and Corey, L. (1987). Recurrences after oral and genital herpes simplex virus infection: Influence of anatomic site and viral type. *N. Engl. J. Med.* **316,** 1444–1449.

44. Rodgers, C. A., and O'Mahony, C. (1995). High prevalence of herpes simplex virus type 1 in female anogenital herpes simplex. *Int. J. STD AIDS* **6,** 144–146.

45. Ross, J. D. C., Smith, I. W., and Elton, R. A. (1993). The epidemiology of herpes simplex types 1 and 2 infection of the genital tract in Edinburgh 1978–1991. *Genitourin. Med.* **69,** 381–383.

46. Tayal, S. C., and Pattman, R. S. (1994). High prevalence of herpes simplex virus type 1 in female anogenital herpes simplex in Newcastle upon Tyne 1983–92. *Int. J. STD AIDS* **5,** 359–361.

47. Cheong, W. K., Thirumoorthy, T., Doraisingham, S., and Ling, A. E. (1990). Clinical and laboratory study of first episode genital herpes in Singapore. *Int. J. STD AIDS* **1,** 195–198.

48. Hashido, M., Lee, F. K., Nahmias, A. J., Tsugami, H., Isomura, S., Nagata, Y., Sonoda, S., and Kawana, T. (1998). An epidemiologic study of herpes simplex virus type 1 and 2 infection in Japan based on type-specific serological assays. *Epidemiol. Infect.* **120,** 179–186.

49. Wald, A., Benedetti, J., Davis, G., Remington, M., Winter, C., and Corey, L. (1994). A randomized, double-blind, comparative trial comparing high and standard dose oral acyclovir for first-episode genital herpes infections. *Antimicrob. Agents Chemother.* **38,** 174–176.

50. Berry, J., Crowley, T., Homer, P., and Harvey, I. (1997). Orogenital sex as risk factor for the development of primary genital herpes; a case-control study. *12th Meet. Int. Soc. Sex. Transm. Dis. Res.,* Seville, Spain, *1997.*

51. Lafferty, W., Wald, A., and Celum, C. (1997). Anogenital HSV type and sexual orientation. *12th Meet. Int. Soc. Sex. Transm. Dis. Res.*, Seville, Spain, *1997*.

52. Corey, L., Adams, H. G., Brown, Z. A., and Holmes, K. K. (1983). Clinical course of genital herpes simplex virus infections in men and women. *Ann. Intern. Med.* **48,** 973.

53. Whitley, R. J., Alford, C. A., Hirsch, M. S., Schooley, R. T., Luby, J. P., Aoki, F. Y., Hanley, D., Nahmias, A. J., and Soong, S. (1986). Vidarabine versus acyclovir therapy in herpes simplex encephalitis. *N. Engl. J. Med.* **314,** 144–149.

54. Wald, A., Kim, M., Catlett, L., Selke, S., Ashley, R., and Corey, L. (1996). Genital HSV-2 shedding in women with HSV-2 antibodies but without a history of genital herpes. *Programs Abstr., 36th Intersci. Conf. Antimicrob. Agents Chemother.*, New Orleans, *1996*.

55. Catotti, D., Clarke, P., and Catoe, K. (1993). Herpes revisited: Still a cause of concern. *Sex. Transm. Dis.* **20,** 77–80.

56. Bierman, S. (1985). Recurrent genital herpes simplex infection: A trivial disorder. *Arch. Dermatol.* **121,** 513–517.

57. Carney, O., Ross, E., Bunker, C., Ikkos, G., and Mindel, A. (1994). A prospective study of the psychological impact on patients with a first episode of genital herpes. *Genitourin. Med.* **70,** 40–45.

58. Limandri, B. J. (1989). Disclosure of stigmatizing conditions: The discloser's perspective. *Arch. Psychiatr. Nurs.* **3,** 69–78.

59. Corey, L., and Spear, P. G. (1986). Infections with herpes simplex viruses (part 1). *N. Engl. J. Med.* **314,** 686–691.

60. Bernstein, D., Lovett, M., and Bryson, Y. (1984). Serologic analysis of first episode nonprimary genital herpes simplex virus infection. *Am. J. Med.* **77,** 1055–1060.

61. Sacks, S. L., Tyrrell, L. D., Lawee, D., Schlech, W. I., Gill, M. J., Aoki, F. Y., Martel, A. Y., Singer, J., and the Canadian Cooperative Study Group (1991). Randomized, double-blind, placebo-controlled, clinic-initiated, Canadian multicenter trial of topical edoxudine 3.0% cream in the treatment of recurrent genital herpes. *J. Infect. Dis.* **164,** 665–672.

62. Quinn, T. C., Corey, L., Chaffee, R. G., Schuffler, M. D., Brancato, F. P., and Holmes, K. K. (1981). The etiology of anorectal infection in homosexual men. *Am. J. Med.* **71,** 395–406.

63. Sacks, S. L. (1984). Frequency and duration of patient-observed recurrent genital herpes simplex virus infection: Characterization of the nonlesional prodrome. *J. Infect. Dis.* **150,** 873–877.

64. Benedetti, J. K., Zeh, J., Selke, S., and Corey, L. (1995). Frequency and reactivation of nongenital lesions among patients with genital herpes simplex virus. *Am. J. Med.* **98,** 237–242.

65. Benedetti, J., Corey, L., and Ashley, R. (1994). Recurrence rates in genital herpes after symptomatic first-episode infection. *Ann. Intern. Med.* **121,** 847–854.

66. Langenberg, A., Benedetti, J., Jenkins, J., Ashley, R., Winter, C., and Corey, L. (1989). Development of clinically recognizable genital lesions among women previously identified as having "asymptomatic" HSV-2 infection. *Ann. Intern. Med.* **110,** 882–887.

67. Barton, S. E., Davis, J. M., Moss, V. W., Tyms, A. S., and Munday, P. E. (1987). Asymptomatic shedding and subsequent transmission of genital herpes simplex virus. *Genitourin. Med.* **63,** 102–105.

68. Koelle, D. M., Benedetti, J., Langenberg, A., and Corey, L. (1992). Asymptomatic reactivation of herpes simplex virus in women after first episode of genital herpes. *Ann. Intern. Med.* **116,** 433–437.

69. Rooney, J. F., Felser, J. M., Ostrove, J. M., and Straus, S. E. (1986). Acquisition of genital herpes from an asymptomatic sexual partner. *N. Engl. J. Med.* **314,** 1561–1564.

70. Yeager, A. S. (1984). Genital herpes simplex infections: Effect of asymptomatic shedding and latency on management of infections in pregnant women and neonates. *J. Invest. Dermatol.* **83,** 53s–56s.

71. Brock, B. V., Selke, S., Benedetti, J., Douglas, J. M., and Corey, L. (1990). Frequency of asymptomatic shedding of herpes simplex virus in women with genital herpes. *JAMA, J. Am. Med. Assoc.* **263,** 418–420.

72. Wald, A., Zeh, J., Selke, S., Ashley, R. L., and Corey, L. (1995). Virologic characteristics of subclinical and symptomatic genital herpes infections. *N. Engl. J. Med.* **333,** 770–775.

73. Wald, A., Zeh, J., Barnum, G., Davis, L. G., and Corey, L. (1996). Suppression of subclinical shedding of herpes simplex virus type 2 with acyclovir. *Ann. Intern. Med.* **124,** 8–15.

74. Wald, A., Taylor, L., Warren, T., Remington, M., Selke, S., Zeh, J., and Corey, L. (1995). Subclinical shedding of herpes simplex virus in men with genital herpes. *Programs Abstr., 35th Intersci. Conf. Antimicrob. Chemother.*, San Francisco, *1995*.

75. van Everdingen, J. J., Peeters, M. F., and ten Have, P. (1993). Neonatal herpes policy in The Netherlands. Five years after a consensus conference. *J. Perinat. Med.* **21,** 371–375.

76. Sullivan-Bolyai, J., Hull, H. F., Wilson, C., and Corey, L. (1983). Neonatal herpes simplex virus infection in King County, Washington: Increasing incidence and epidemiological correlates. *JAMA, J. Am. Med. Assoc.* **250,** 3059–3062.

77. Brown, Z. A., Benedetti, J., Ashley, R., Burchett, S., Selke, S., Berry, S., Vontver, L. A., and Corey, L. (1991). Neonatal herpes simplex virus infection in relation to asymptomatic maternal infection at the time of labor. *N. Engl. J. Med.* **324,** 1247–1252.

78. Prober, C. G., Sullender, W. M., Yasukawa, L. L., Au, D. S., Yeager, A. S., and Arvin, A. M. (1987). Low risk of herpes simplex virus infections in neonates exposed to the virus at the time of vaginal delivery to mothers with recurrent genital HSV infections. *N. Engl. J. Med.* **316,** 240–244.

79. Brown, Z. A., Selke, S. A., Zeh, J., Kopelman, J., Maslow, A., Ashley, R. L., Watts, D. H., Berry, S., Herd, M., and Corey, L. (1997). Acquisition of herpes simplex virus during pregnancy. *N. Engl. J. Med.* **337,** 509–515.

80. Prober, C. G., Corey, L., Brown, Z. A., Hensleigh, P. A., Frenkel, L. M., Bryson, Y. J., Whitley, R. J., and Arvin, A. M. (1992). The management of pregnancies complicated by genital infections with herpes simplex virus. *Clin. Infect. Dis.* **15,** 1031–1038.

81. Randolph, A. G., Washington, A. E., and Prober, C. G. (1993). Cesarean delivery for women presenting with genital herpes lesions. *JAMA, J. Am. Med. Assoc.* **270,** 77–82.

82. Scott, L. L., Sanchez, P. J., Jackson, G. L., Zeray, F., and Wendel, G. D., Jr. (1996). Acyclovir suppression to prevent cesarean delivery after first-episode genital herpes. *Obstet. Gynecol.* **87,** 69–73.

83. Smith, J. R., Cowan, F. M., and Munday, P. (1998). The management of herpes simplex virus infection in pregnancy. *Br. J. Obstet. Gynaecol.* **105,** 255–268.

84. Randolph, A. G., Hartshorn, R. M., and Washington, A. E. (1996). Acyclovir prophylaxis in late pregnancy to prevent neonatal herpes: A cost-effectiveness analysis. *Obstet. Gynecol.* **88,** 603–609.

85. Frenkel, L. M., Brown, Z. A., Bryson, Y. J., Corey, L., Unadkat, J. D., Hensleigh, P. A., Arvin, A. M., Prober, C. G., and Connor, J. D. (1991). Pharmacokinetics of acyclovir in the term human pregnancy and neonate. *Am. J. Obstet. Gynecol.* **164,** 569–576.

86. Andrews, E. B., Yankaskas, B. C., Cordero, J. F., Schoeffler, K., Hampp, S., and the Acyclovir in Pregnancy Registry Advisory Committee (1992) Acyclovir in pregnancy registry: Six years' experience. *Obstet. Gynecol.* **79,** 7–13.

87. Ashley, R. (1993). Laboratory techniques in the diagnosis of herpes simplex infection. *Genitourin. Med.* **69,** 174–83.

88. Cone, R. W., Swenson, P. D., Hobson, A. C., Remington, M., and Corey, L. (1993). Herpes simplex virus detection from genital lesions: A comparative study using antigen detection (HerpChek) and culture. *J. Clin. Microbiol.* **31,** 1774–1776.

89. Safrin, S., Shaw, H., Bolan, G., Cuan, J., and Chang, C. S. (1997). Comparison of virus culture and the polymerase chain reaction for diagnosis of mucocutaneous herpes simplex virus infection. *Sex. Transm. Dis.* **24,** 176–180.

90. Cone, R. W., Hobson, A. C., Palmer, J., Remington, M. L., and Corey, L. (1991). Extended duration of herpes simplex DNA in genital lesions detected by the polymerase chain reaction. *J. Infect. Dis.* **164,** 757–760.

91. Ashley, R., Cent, A., Maggs, V., and Corey, L. (1991). Inability of enzyme immunoassays to discriminate between infections with herpes simplex virus type 1 and 2. *Ann. Intern. Med.* **115,** 520–526.

92. Ashley, R. L., Militoni, J., Lee, F., Nahmias, A., and Corey, L. (1988). Comparison of Western blot (Immunoblot) and G-specific immunodot enzyme assay for detecting antibodies to herpes simplex virus types 1 and 2 in human sera. *J. Clin. Microbiol.* **26,** 662–667.

93. Ashley, R., Wu, L., Pickering, J., Tu, M., and Schnorenberg, L. (1998). Premarket evaluation of a commercial glycoprotein-G based enzyme immunoassay for herpes simplex virus type-specific antibodies. *J. Clin. Microbiol.* **36,** 294–295.

94. Ashley, R., and Eagleton, M. (1999). Evaluation of a novel point-of-care test for antibodies to herpes simplex virus type 2. *Sex. Transm. Infect.* (in press).

95. Fairley, I., and Monteiro, E. (1997). Patient attitudes to type specific serologic tests in the diagnosis of genital herpes. *Genitourin. Med.* **73,** 259–262.

96. Ashley, R., and Corey, L. (1997). HSV type specific antibody tests: Patients are ready, are clinicians? *Genitourin. Med.* **73,** 235–236.

97. Corey, L., Mindel, A., Fife, K. H., Sutherland, S., Benedetti, J., and Adler, M. W. (1985). Risk of recurrence after treatment of first episode genital herpes with intravenous acyclovir. *Sex. Transm. Dis.* **12,** 215–218.

98. Whitley, R. J., and Gnann, J. W., Jr. (1992). Acyclovir: A decade later. *N. Engl. J. Med.* **327,** 782–789.

99. Corey, L., Benedetti, J., Critchlow, C., Mertz, G., Douglas, J., Fife, K., Fahnlander, A., Remington, M. L., Winter, C., and Dragavon, J. (1983). Treatment of primary first-episode genital herpes simplex virus infections with acyclovir: Results of topical, intravenous and oral therapy. *J. Antimicrob. Chemother.* **12** (Suppl. B), 79–88.

100. Mindel, A., Carney, O., Freris, M., Faherty, A., Patou, G., and Williams, P. (1988). Dosage and safety of long term suppressive acyclovir therapy for recurrent genital herpes. *Lancet* **1,** 926–928.

101. Goldberg, L. H., Kaufman, R., Kurtz, T., Conant, M. A., Eron, L. I., and Batenhorst, R. I. (1993). Long-term suppression of recurrent genital herpes with acyclovir: A 5-year benchmark. *Arch. Dermatol.* **129,** 582–587.

102. Carney, O., Ross, E., Ikkos, G., and Mindel, A. (1993). The effect of suppressive oral acyclovir on the psychological morbidity associated with recurrent genital herpes. *Genitourin. Med.* **69,** 457–459.

103. Fife, K. H., Crumpacker, C. S., Mertz, G. J., Hill, E. L., Boone, G. S., and the Acyclovir Study Group (1994). Recurrence and resistance patterns of herpes simplex virus following cessation of ≥ 6 years of chronic suppression with acyclovir. *J. Infect. Dis.* **169,** 1338–1341.

104. Reitano, M., Tyring, S., Lang, W., Thoming, C., Worm, A. M., Borelli, S., Chambers, L. O., Robinson, J. M., Corey, L., and the International Valaciclovir HSV Study Group (1998). Valaciclovir for the suppression of recurrent genital herpes simplex virus infection: A Large-scale Dose Range-finding Study. *J. Infect. Dis.* **178,** 603–610.

105. Diaz-Mitoma, F., Sibbald, R. G., Shafran, S. D., Boon, R., Saltzman, R. L., for the Collaborative Famciclovir Genital Herpes Research Group (1998). Oral famciclovir for the suppression of

recurrent genital herpes: A Randomized Controlled Trial. *JAMA, J. Am. Med. Assoc.* **280,** 887–892.

106. Meyers, J. D., Wade, J. C., Mitchell, C. D., Saral, R., Lietman, P. S., Durack, D. T., Levin, M. J., Segreti, A. C., and Balfour, H. H. (1982). Multicenter collaborative trial of intravenous acyclovir for treatment of mucocutaneous herpes simplex virus infection in the immunocompromised host. *Am. J. Med.* **73,** 229–235.

107. Conant, M. A. (1988). Prophylactic and suppressive treatment with acyclovir and the management of herpes in patients with acquired immunodeficiency syndrome. *J. Am. Acad. Dermatol.* **18,** 186–188.

108. Crumpacker, C. S., Schnipper, L. E., Marlowe, S. I., Kowalsky, P. N., Hershey, B. J., and Levin, M. J. (1982). Resistance to antiviral drugs of herpes simplex virus isolated from a patient treated with acyclovir. *N. Engl. J. Med.* **306,** 343–346.

109. Pottage, J. C., and Kessler, H. A. (1995). Herpes simplex virus resistance to acyclovir: Clinical relevance. *Infect. Agents Dis.* **4,** 115–124.

110. Centers for Disease Control and Prevention (1998). CDC 1998 guidelines for treatment of sexually transmitted diseases. *Morbid. Mortal. Wkly. Rep.* **47,** 1–118.

111. Burke, R. L. (1993). Current developments in herpes simplex virus vaccines. *Virology* **4,** 187–197.

112. Langenberg, A. G., Burke, R. L., Adair, S. F., Sekulovich, R., Tigges, M., Dekker, C. L., and Corey, L. (1995). A recombinant glycoprotein vaccine for herpes simplex virus type 2: Safety and immunogenicity. *Ann. Intern. Med.* **12,** 889–898.

113. Corey, L., Ashley, R., Sekulovich, R., Izu, A., Douglas, J., Handsfield, H. H., Warren, T., Marr, L., Tyring, S., DiCarlo, R., Adimora, A., Leone, P., Dekker, C., Burke, R. L., and Langenberg, A. (1997). Lack of efficacy of a vaccine containing recombinant gD2 and gB2 antigens in MF59 adjuvant for the prevention of genital HSV-2 acquisition. *Programs Abstr., 37th Intersci. Conf. Antimicrob. Agents Chemother.,* Toronto, Canada, *1997.*

114. Bourne, N., Stanberry, L. R., Bernstein, D. I., and Dew, D. (1996). DNA immunization against experimental herpes simplex virus infection. *J. Infect. Dis.* **173,** 800–807.

115. Boursnell, M. E., Entwisle, C., Blakeley, D., Roberts, C., Duncan, I. A., Chisholm, S. E., Martin, G. M., Jennings, R., Ni-Challana'in, D., Sobek, I., Inglisk, S. C., and McLean, C. S. (1997). A genetically inactivated herpes simplex virus type 2 (HSV-2) vaccine provides effective protection against primary and recurrent HSV-2 disease. *J. Infect. Dis.* **175,** 16–25.

116. Kriesel, J. D., Spruance, S. L., Daynes, R. A., and Araneo, B. A. (1996). Nucleic acid vaccine encoding gD2 protects mice from herpes simplex virus type 2 disease. *J. Infect. Dis.* **173,** 536–541.

117. Higgins, D. L., Galavotti, C., O'Reilly, K. R., Schnell, D. J., Moore, M., Rugg, D. L., and Johnson, R. (1991). Evidence for the effects of HIV antibody counseling and testing on risk behaviors. *JAMA, J. Am. Med. Assoc.* **266,** 2419–2429.

118. Ebel, C. (1998). "Managing Herpes: How to Live and Love with a Chronic STD." American Social Health Association, Research Triangle Park, NC.

119. Sacks, S. L. (1997). "The Truth about Herpes." Gordon Soules Book Publishers Ltd., West Vancouver, Canada.

120. Corey, L., and Wald, A. (1999). Genital herpes. *In* "Sexually Transmitted Diseases" (K. Holmes, ed.), 3rd ed. McGraw-Hill, New York.

121. Gibson, J., Carlton, A., Alexander, G., Lee, F., Potts, W., and Nahmias, A. (1990). A cross-sectional study of herpes simplex virus types 1 and 2 in college students: Occurrence and determinants of infection. *J. Infect. Dis.* **162,** 306–312.

122. Becker, T., Blount, J., and ME, G. (1985). Genital herpes infections in private practice in the United States, 1966 to 1981. *JAMA, J. Am. Med. Assoc.* **253,** 1601–1603.

123. Weinberg, A., Canto, C., Pannuti, C., Kwang, W., Garcia, S., and Zugaib, M. (1993). Herpes simplex virus type 2 infection in pregnancy: Asymptomatic viral excretion at delivery and seroepidemiologic survey of two socioeconomically distinct population in Sao Paulo, Brazil. *Rev. Inst. Med. Trop. Sao Paulo* **35,** 285–290.

124. Oberle, M. W., Rosero-Bixby, L., Lee, F. K., Sanz chez-Braverman, M., Nahmias, A. J., and Guinan, M. E. (1989). Herpes simplex virus type 2 antibodies: High prevalence in monogamous women in Costa Rica. *Am. J. Trop. Med. Hyg.* **41,** 224–218.

125. Cunningham, A., Lee, F., Ho, D., Field, P., Law, C., Packham, D., McCrossin, I., Sjögren-Jansson, E., and Nahmias, A. (1993). Herpes simplex virus type 2 antibody in patients attending antenatal or STD clinics. *Med. J. Aust.* **158,** 525–528.

126. Forsgren, M. F., Skoog, E., Jeansson, S., Olofsson, S., and Giesecke, J. (1994). Prevalence of antibodies to herpes simplex virus in pregnant women in Stockholm in 1969, 1983 and 1989: Implications for STD epidemiology. *Int. J. STD AIDS* **5,** 113–116.

127. Pasquini, P., Mele, A., Franco, E., Ippolito, G., and Svennerholm, B. (1988). Prevalence of herpes virus type 2 antibodies in selected population groups in Italy. *Eur. J. Clin. Microbiol. Infect. Dis.* **7,** 54–57.

128. Cowan, F., Johnson, A., Ashley, R., Corey, L., and Mindel, A. (1996). Relationship between antibodies to herpes simplex virus (HSV) and symptoms of HSV infection. *J. Infect. Dis.* **174,** 470–475.

129. Bassett, I., Donovan, B., Bodsworth, N., Field, P., Ho, D., Jeansson, S., and Cunningham, A. (1994). Herpes simplex virus type 2 infection of heterosexual men attending a sexual health centre. *Med. J. Aust.* **160,** 697–700.

130. Lowhagen, G., Jansen, E., Nordenfelt, E., and Lycke, E. (1990). Epidemiology of genital herpes infections in Sweden. *Acta Derm-Venereol.* **70,** 330–334.

131. Mele, A., Franco, E., Caprilli, F., Gentili, G., Capitanio, B., Crescimbeni, E., Di Napoli, A., Zaratti, L., Conti, S., Corona, R., Rezza, G., Pana, A., and Pasquini, P. (1988). Genital herpes infection in outpatients attending a sexually transmitted disease clinic in Italy. *Eur. J. Epidemiol.* **4,** 386–388.

25

Human Papilloma Virus Infection in Women

NAOMI JAY AND ANNA-BARBARA MOSCICKI
University of California
San Francisco, California

I. Introduction

Human papilloma virus (HPV) infections were recognized as early as ancient Greece and Rome. Writings from ancient medical records described exophytic fig-like lesions on the genitalia consistent with HPV-induced condyloma acuminata (genital warts). While the recognition of their viral association was unknown, the authors of these logs recognized the association of these lesions with sexual activity [1]. To date, condyloma acuminatum remains the most common clinically recognized manifestation of HPV infection. However, the major morbidity (as well as societal costs) related to HPV infection is its association with the development of cervical and other genital tract squamous precancerous lesions and cancers. The association between HPV and genital cancers was suspected as early as 1974 [2]. Now strong epidemiology and laboratory-based research have confirmed this important relationship. Although the annual incidence of cervical cancer in the United States is relatively low, less than 10 per 100,000 women, there are approximately 50 cases of precancerous lesions found for every new cancer case, resulting in large economic cost [3]. In contrast, in less industrialized areas such as Africa or Latin America, the lifetime risk for developing invasive cervical cancer is as high as 10,000 per 100,000 women [3]. Eighty percent of the 425,000 new annual cases that occur worldwide are in developing countries. Worldwide it is the fifth most common cancer, and it is the second most common cancer in women [4].

II. Diagnosis of HPV Infection

The spectrum of disease associated with HPV infection is wide-ranging—from the detection of DNA with no evidence of clinical disease to invasive cancer. Knowledge of the diagnostic methodology, including gross visual inspection, HPV DNA identification, cytology, colposcopy, and biopsy, is important in understanding the spectrum of HPV disease. Although standard clinical visualization is adequate for diagnosis of the clinically overt condyloma acuminata, this accounts for less than 1% of all HPV infection. Molecular DNA testing affords the most sensitive level of testing but its clinical utility is limited because most women with HPV have no sequelae or clinically relevant disease. Tests based on amplification techniques, such as the polymerase chain reaction (PCR), are considered the gold standard for DNA detection. PCR-type tests are more sensitive than most other types such as dot blot, in situ hybridization, and Southern blot hybridization. The only FDA approved and commercially available test is Hybrid Capture (HC) (Digene Diagnostics, Silver Spring, MD). Its analytical sensitivity is 10^5

copies of HPV, compared to PCR which can detect as few as 10–100 copies of HPV genomes [5]. HC currently offers a probe that detects a mixture of low risk types (6,11,42,43,44) and a probe that detects a mixture of high risk types (16,18, 31,33,35,45,51,52,56). Although this test offers a quantitative result (*i.e.,* how much virus), it is unable to distinguish high viral load of a single type versus a multiple type infection within the mixtures. A second generation Hybrid Capture Microplate (HCM) reportedly detects 1000 copies of HPV genomes and its sensitivity and specificity are currently being evaluated [6].

The most frequently used clinical technique to screen for HPV infection is cytology, or the Papanicolaou (Pap) smear. The Pap smear is prepared from exfoliated cells collected from the surface of the cervix (or other area) which are smeared onto a slide. The smear is routinely stained using the Papanicolaou technique (hence the *Pap* Smear). The Pap smear is then evaluated under microscopic magnification by the cytopathologist for the presence of changes induced by HPV on the cellular level, using specific cytologic criteria (see below). Diagnosis made using a Pap smear is considered preliminary, because the sensitivity and specificity of a single Pap smear is estimated to be around 40% and 90%, respectively [7]. The gold standard for diagnosis of HPV-related SIL (squamous intra-epithelial lesions) requires histologic verification by biopsy which is routinely obtained under the guidance of a colposcopic examination. Colposcopic examination provides the clinician with a magnified view of the tissue, and the application of 3% acetic acid prior to the examination results in certain epithelial and vascular changes in areas with SIL, which directs the clinician to the point of biopsy. The biopsied material is then sent for histologic evaluation by a pathologist. Although colposcopy is the standard adjunct to cytology, colposcopy has extremely poor specificity in the diagnosis of HPV in the face of normal cytology and should never be used as a screening tool.

Based on the above methodological tools, HPV infection frequently is classified into the following categories: latent, condyloma acuminata, and subclinical or precancerous. Latent infection appears to be the most frequent category of HPV infection. Latency in this case is defined by the presence of HPV DNA, ascertained by molecular testing, without epithelial signs of clinical infection by gross visualization, colposcopy, cytology or histology. Condyloma acuminata are exophytic lesions (Fig. 25.1) commonly referred to as genital warts. Although the least prevalent of the categories, condyloma acuminata are the most common clinically recognized HPV-related disease. Subclinical infections are referred to when HPV-associated lesions appear invisible to the naked eye but can be viewed with the aid of a colposcope and acetic acid. However, because colposcopic changes

Table 25.1
Bethesda Classification System for Cytology Evaluation Compared with Other Systems

Normal	Atypical cells of undetermined significance (ASCUS)	Squamous intraepithelial lesion (SIL)		Cancer
		Low grade (LSIL)	High grade (HSIL)	
	Atypia	Cervical intraepithelial neoplasia (CIN)		
		Condyloma	CIN I	CIN II	CIN III	
			Dysplasia		
		Mild mild	Mild	Moderate	Severe	
					Carcinoma *in situ*	

are frequently nonspecific, subclinical infections are more commonly determined either by cytology or by histologic verification of the colposcopic lesion. The histologic interpretation of these subclinical colposcopic lesions is that they are precancerous lesions, termed SIL. The colposcopic changes associated with subclinical lesions include well-demarcated acetowhitening, granular or smooth surfaces, and vascular changes (Figs. 25.2, 25.3). Colposcopic examination can be used for identification of subclinical lesions of the entire anogenital tract, but with the same precautions as described above for cervical lesions.

The classification of precancerous lesions has undergone several transformations that reflect advances in our understanding of the pathophysiology of HPV infection [8]. The Bethesda system for rating cytology, now widely accepted in the United States, reflects a two-tier schemata in which noncancerous lesions are classified into low (L) or high (H) grade squamous intraepithelial lesions (SIL) (Table 25.1). LSIL includes the following terms which are often used interchangeably and/or have been used historically: HPV effect, flat condyloma, mild dysplasia, or cervical intraepithelial neoplasia (CIN) I. LSILs are changes characteristic of productive or actively replicating HPV infection and are considered benign. Production of certain HPV proteins is most likely the direct cause of the cytopathologic effect. Consequently, the cytologic changes described in LSIL are frequently termed "HPV-effect." The term HSIL encompasses the terms moderate or severe dysplasia, CIN II, CIN III, as well as carcinoma in situ (CIS). In contrast to LSIL, HSIL is thought to be a true precursor to squamous cell cancers. The recognition that LSIL and HSIL are characterized differently was based on studies that examined regression and progression rates of LSIL and HSIL. In a meta-analysis of natural history studies, LSIL was found to regress in up to 65% of cases and HSIL in 32–43% of cases [9]. Progression from LSIL (defined as CIN I) to HSIL (defined as CIN III) occurred in 11% of cases; progression from HSIL (defined as CIN II) to CIN III occurred in 22% of cases [9]. Progression to cancer has been estimated to occur in less than 1% of LSIL and in 5–12% of HSIL [9,10]. In Europe, the Bethesda system is not widely accepted and the CIN classification is more frequently used (see Table 25.1). This is based on the belief that CIN III is a distinct lesion from CIN II and is considered the precursor lesion most likely to progress to squamous cell cancer (SCC).

Classification of squamous intraepithelial lesions in other areas of the genital tract are similar and are referred to as vaginal, vulva, and anal SIL. The natural histories of these SILs are less well understood than that of cervical SIL. Although HPV is found in association with many vulvar, vaginal, and anal cancers, the probability that these SILs will progress to invasive cancer in these genital areas is currently unknown.

III. Virology

Papilloma viruses (PV) affect several animals such as cows, deer, and rabbits, as well as humans, but they are species specific. That is, for example, humans cannot get rabbit PV and vise versa. Human papilloma viruses are small, double-stranded DNA viruses containing approximately 7900 basepairs. Each of the approximately 80 different types of HPV that has been identified shows a particular tropism for specific tissue types. Approximately 30 of these have been detected in the area of the genital tract [11]. Table 25.2 demonstrates the tissue distribution of common HPV types. The different HPV types are based on

Table 25.2
HPV Types and Tissue Distribution

HPV type	Epithelial site	Type of wart/disease
1	Soles of feet	Plantar, palmar
2, 4, 7	Hands	Common, plantar, butcher's
3, 10	Arms	Flat, juvenile
26–29, 34	Forehead, arms, trunk	Common, Bowen's disease
5, 8, 9, 12, 14, 15, 17, 19–25, 36, 37, 38	Forehead, arms, trunk	Flat, macular, skin cancer in patients with epidermodysplasia verruciformis
6, 11	Anogenital tract, oral	Condyloma, laryngeal papillomatosis, conjuntival, Buschke-Lowenstein, SIL
16, 18, 31, 33, 35, 39, 41–45, 51–56	Anogenital tract	Bowenoid papulosis, SIL anogenital cancers
13, 32	Oral cavity	Epithelial hyperplasia
30, 40	Larynx	Laryngeal cancer

lack of DNA homology. To be considered a new viral type the homology must be less than 90% of the DNA sequence of the HPV E6, E7, and L1 proteins (see section on Mechanisms of Oncogenicity) and overall DNA homology of a new HPV type is less than 50% with other types [12].

Genital HPV types have been categorized as "low risk," "high risk," and "intermediate risk" according to their potential oncogenic risk. The "low risk" types are associated with condyloma acuminata. Over 90% of such lesions contain HPV 6 or 11. These "low risk" types are also associated with low grade squamous intraepithelial lesions (LSIL) but are rarely detected in isolation in high grade squamous intraepithelial lesions (HSIL) and squamous cell cancers (SCC) [13]. The "high risk" HPV types 16,18, 33, 39,45, 52, and 56 are most commonly found in HSIL and SCC. They are also commonly found in latent infection, benign condyloma, and LSIL. Of all the HPV genital types, HPV type 16 is the most prevalent HPV found in women with cervical HSIL as well as SCC [5]. HPV type 16 is also the predominant HPV type found in anal, oral, penile, vulvar, and vaginal cancers. However, the interpretation of "high risk" should not necessarily imply "risk for progression" because the high risk types such as HPV 16 are also the most common types found in women with latent infection. HPV types 18, 45, and 56, on the other hand, are found less frequently in HSIL than in cancers. In one study of 2627 women, these HPV types were found in 26.8% of cancers but only accounted for 6.5% of HPV types found in HSIL [14]. The term "intermediate risk" type is sometimes used for HPV types 31,33,39,52, and 58 that are associated with SCC but less frequently·than the "high risk" types.

These associations of "low," "intermediate," and "high" risk types are not clear cut and the actual "risk" of developing disease may vary by type within each risk group [14]. For example some have suggested that HPV 18 may progress more rapidly than HPV 16 [15]. Categorization of HPV types into "like-groups" as opposed to single types by phylogenetic trees based on DNA homology is a recent development that may provide better characterization of these types and their importance [16]. "HPV-16-like types" include types 31,33,35,52, and 58. "HPV-18-like types" include 45 and 39.

IV. Mechanism of Oncogenicity

Advances in the understanding of the molecular biology of HPV have elucidated its role in the formation of SIL and cancers. Several pathways have been identified that are thought to promote tumorigenesis. The HPV genome is divided into three regions: an early region (E), late region (L), and the upstream regulatory region (URR) which is also called the long control region (LCR). The E region proteins are associated with cell replication and cell transformation, which are thought to be important oncogenic functions. The L region, consisting of L1 and L2, is required for viral assembly. The LCR located between the E and L regions is important for regulation of viral gene transcription (making viral proteins). Of the early proteins, E6 and E7 are thought to be the most important viral proteins in the development of malignancy. They are capable of inducing cellular transformation by binding to tumor suppressor genes p53 and Rb. These genes are responsible for the normal regulation of cellular growth, repair, and replication. Under normal physi-

ological conditions, DNA damage occurs frequently during cell replication (*i.e.,* accidents of nature). Cells are normally programmed to stop replicating for repair through the p53 and Rb functions. When p53 and Rb functions are interfered with through binding with the HPV E6 and E7 proteins, abnormal cell replication may result in cancerous changes.

Normally, E6 and E7 are regulated by other proteins from the E region and the LCR. Integration of HPV DNA into host cells, which is thought to be a key event in cancer formation, results in a loss of the E6 and E7 regulatory proteins and hence increased transcription of E6 and E7 occurs. Subsequently, viral integration may enhance E6 and E7 function, resulting in cell dysregulation. Differences in the biologic activity of these proteins have been found in "high" versus "low" risk HPV types. Higher expression of E7 has been found in tissues infected with "high risk" HPV types compared to those with "low risk" types [17]. E6 proteins of the "high risk" HPV types bind with p53 more efficiently than in "low risk" HPV types [18]. There are also differences in the higher transforming efficiency of HPV 18 versus HPV 16 *in vitro* [19]. However, because most HPV infections, including those caused by "high risk" types, do not progress to invasive cancer, other factors must be involved in the development of cancer. For example elements in the LCR are responsive to regulatory growth factors, such as transforming growth factors (TGFs) beta 1 and 2, which have been shown to inhibit proliferation of normal epithelial cells, but are also thought to function as regulators for HPV gene expression in infected cells [20]. Last, and certainly least understood, the host immune response to initial HPV infection is most likely critical in eliminating infection or inducing latency.

Most of the data regarding the molecular biology of HPV are derived from studies of HPV types 16 and, to a lesser extent, 18, as well as the "low risk" types 6 and 11. It is important to be cognizant of this when generalizing the behavior of HPV. Inconsistencies may be attributable to differences in the molecular biology of strains that have not been well studied.

V. Transmission of Infection

The majority, if not all, of genital HPV infections are sexually transmitted in adolescents and adults. Women who report never having had vaginal intercourse rarely have HPV isolated from the cervix or vagina [21,22]. Infection appears to begin with skin to skin contact with the virus entering at a site of epithelial disruption, such as small tears or sores, where the virus has close proximity to the basal epithelial layer. The cervical and anal columnar epithelium may be more vulnerable to disruption because columnar epithelium is quite thin compared to well-keratinized (thickened) squamous epithelium.

Relatively high transmission rates (approximately 30–60%) have been reported after exposure to condyloma acuminata in partners [23–25]. The length of time between sexual exposure and evidence of external genital warts (EGW) is approximately 3–6 months. Although not substantiated, it is postulated that viral load is correlated with the size of condyloma and with transmission probabilities. That is, individuals with extensive condyloma are more likely to transmit HPV than individuals with relatively small condyloma. The rate of transmission from SIL or latent infection is unknown. However, because of the

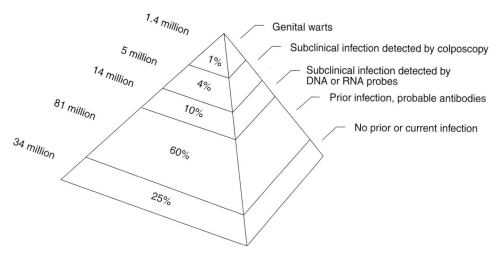

Fig. 25.4 Estimated prevalence of genital HPV infection from Koutsky. Reprinted from *American Journal of Medicine,* **102**(5a), Koutsky, L. A., Epidemiology of genital human papillomavirus infection, 3–8, 1997, with permission from Excerpta Medica, Inc.

ubiquitous nature of this virus, it is thought to be equally high if not higher than that of condyloma.

Evidence for nonsexual transmission of HPV is more controversial. Several studies have described external genital condyloma in toddlers and infants without any evidence of child abuse [26,27]. In such cases it was explained that transmission may occur digitally through autoinoculation or by others because up to 25% of genital warts in children are HPV type 2, or common skin warts [28,29]. The other 75% are due to genital HPV types, most commonly types 6 and 11, and were explained by possible persistence of vertical transmission (mother to infant). However, several studies have shown that persistence of perinatal transmission of HPV is extremely rare in children [30]. When persistence did occur, the HPV types were those not commonly found in condyloma from children, such as HPV types 16 and 18. Last, condyloma from children who have been known to be sexually abused contain predominantly HPV types 6 and 11, followed by HPV types 16 and 18, supporting the role of sexual transmission in condyloma found in children.

Vertical transmission of HPV from mother to infant during childbirth has been associated with the appearance of juvenile-onset recurrent laryngeal papillomatosis (RLP). The incidence of this rare disease is estimated to be 2000 per 4 million births annually. The case for vertical transmission is supported by the fact that children of mothers with a history of external genital warts (EGW) are more likely to have RLP than children of women without EGW [31]. Although RLP is found more often in children delivered vaginally than by cesarean section [32], the rare incidence of this disease does not warrant routine c-sections in women with HPV infections. Other STDs such as *Chlamydia trachomatis, Neisseria gonorrhoeae,* or HIV have also been known to be vertically transmitted and most of these infections result in nongenital infections as well.

Other modes of transmission most likely do not exist. The role of fomite transmission, for example from soap, towel, or hot tub, has not been substantiated and is probably a rare occurrence. There have been isolated reports of HPV detection in

amniotic fluid, with one documented case report of an infant delivered by cesarean section who was born with condylomata acuminata. [33,34].

VI. Prevalence of HPV

HPV is the most common sexually transmitted disease in the U.S.. However, the prevalence of infection varies widely depending on the method of detection, as described earlier. The lifetime risk of having an HPV infection detected by molecular testing has been estimated to be as high as 79% [35]. In contrast, EGW are diagnosed much less frequently, with best estimates of EGWs in cohorts of sexually active women ranging from 1.5 to 2% [36,37]. True prevalence is not known because EGW is not a Centers for Disease Control and Prevention reportable STD. Not surprisingly, higher rates, ranging from 4–13%, are found in cohorts of women attending sexually transmitted disease (STD) clinics [36–39]. The prevalence of subclinical HPV using the Pap smear range from 4–24%, with the lower rates found in more general populations and higher rates found in high-risk behavior groups [35,37,40,41].

Using PCR techniques for HPV DNA detection, prevalence ranges from 2.8–57% [42–48]. These rates are found to vary by age and by number of genital sites tested (*i.e.,* vagina, vulva, cervix, and anus). Several studies have shown that higher rates are found in younger women when compared to older women [49–51]. Unfortunately, most studies have compared women from different clinical settings with different risk behaviors. Therefore, it is not clear whether the difference in prevalence observed is due to an age-related biologic phenomena such as immunologic immaturity or due to differences in sexual behavior. Prevalence using these different methods of detection is summarized in Figure 25.4 [52].

Differences in the prevalence of all HPV disease have been shown to be age-related. Young age is associated with higher rates of HPV DNA and LSIL and rates are noted to decrease after the age of 30 [49–51]. The incidence of condyloma acuminata also has been shown to be age-related, with decreases in prevalence seen after 25 years of age [40,53]. In contrast, older age is a greater risk for cervical cancer, with the peak incidence of >15.3 per 100,000 in women aged 35 years and older and the peak incidence of ≤ 8.1 per 100,000 in women aged 29 years and younger (SEER Program, 1987–1991). Although not usually addressed, racial or ethnic differences in HPV disease have not been found in studies that have analyzed this information [54].

The highest prevalence for HPV has been described in HIV-infected populations. HIV-seropositive adolescents and adults have frequencies of abnormal cervical cytology that are as high as 63% and these frequencies have been uniformly elevated compared to HIV-seronegative controls, even when controlling for sexual behavior [41,54–56]. Elevated rates of condyloma and HPV DNA are also found in HIV-seropositive women in cervical as well as anal specimens [57–59]. Although there is little documentation showing elevated rates of cervical cancer among HIV-seropositive women, an estimated observed to expected incidence ratio of 2.28 was found, based on the AIDS surveillance data in New York City. This resulted in the inclusion of cervical cancer as an AIDS-defining disease in 1993 [41].

In the United States, the incidence of cervical cancer has been shown to differ by race. Data from the Surveillance, Epidemiology, and End Results (SEER) program for 1987–1991 showed that African-Americans have a twofold increased incidence of cervical cancer compared to whites (7.8 per 100,000 white women compared to 14.0 per 100,000 black women).

The prevalence of HPV-associated extra-cervical disease is less well studied. Although EGW include the vulvar and perianal areas, the actual incidence of vulvar and anal HPV or SIL is unknown. The interpretation of HPV detection and SIL in these areas is highly debatable because both cancers appear to be less frequently associated with HPV than cervical cancers. HPV is found in only 50% of vulvar cancers and 85% of anal SCC, which suggests that the natural history of HPV infection in these areas is most likely different than it is in cervical infections. However, developmental and histologic similarities between the anal canal and the cervix have made it possible to adapt similar diagnostic methodologies for anal canal HPV infection and SIL. Although the initial observations regarding anal HPV were on gay males, data support the notion that HPV also infects the anal canal in women. Studies that have examined anal HPV infection by molecular testing in nonimmunocompromised women have found HPV infections in the range of 2–60% [56,57]. These studies cannot be generalized because most of the studies have engaged selected risk groups. The few studies that have examined older, monogamous women have found little to no evidence of anal HPV infection, underscoring the sexual nature of these infections [60,61]. Anal SIL has also been shown to exist in women. Using the Pap smear for diagnosis, studies have found that only 4–8% of nonimmunocompromised women show evidence of SIL in the anal canal [57,62]. The incidence of anal cancer is rare; it is estimated to be less than 1 per 100,000 women. However, it should be noted that there is a 2:1 female to male preponderance in the incidence of anal cancer.

VII. Risk Factors for HPV Acquisition and Disease

Evidence supporting or refuting the importance of various risk factors for HPV infection has been somewhat confusing, because studies frequently use noncomparable outcomes (*e.g.,* interchanging SIL with HPV detection). Even in DNA detection studies, the methodology for DNA detection has varied, making any comparisons difficult. Most studies agree, however, that the predominant risk factor for HPV infection is sexual activity [48,51,63]. While the majority of studies defining these risks were cross-sectional, the relationship with sexual activity has been substantiated in several longitudinal incidence studies. Ho *et al.* [45] reported that the significant risks for incident HPV infection were multiple partners in the past 6 months, as well as the reported number of lifetime partners of the male partner. The association with recent, but not past, sexual behavior of the individual in this and other studies underscores the transient nature of HPV infection. In comparison, nonsexual transmission plays virtually no role in cervical infections of children, adolescents, and adult women. Although sexual activity is certainly the single most important factor in anal and vulvar HPV infection, less is understood regarding the role of nonsexual transmissions to these areas. It is certainly plausible that cervical shedding in vaginal discharges may contaminate both the vulva and anus. However, the relationship between this type of exposure to HPV and actual infection or development of disease is highly questionable.

Young age (less than 25 years old) has been frequently sited as a risk factor for HPV infection. The strongest argument for age is its association with high risk sexual behavior. Young women, specifically adolescents, have continually been shown to engage in higher risk sexual behavior, including poor partner choices, multiple sexual partners, and frequent sexual intercourse compared to older women [64]. In addition, partners of young women tend to have a greater number of sexual partners. The contribution of partner behavior has been difficult to adjust for in analysis of epidemiologic studies. Franco *et al.* [65] reported that sexual behavior accounted for the majority of age-related differences in detecting both low risk and high risk HPV types in a Brazilian cohort. Other studies reported that the risk of age in HPV infection is independent of sexual activity, suggesting that there is a biologic risk associated with young age [49–51]. Shew and colleagues [66] reported that adolescents who had recently begun their menses were more likely to have an HPV infection than adolescents who were several years postmenarche, suggesting a vulnerability to HPV near the onset of menarche. However, if age were a risk, it would be expected that age at first intercourse would be associated with latent HPV detection as well. Several studies have shown that age at first intercourse is not significant after adjusting for lifetime number of partners [67–69].

Similar to HPV DNA detection studies, sexual behavior appears to be a significant risk factor for the development of HPV-associated disease, or SIL. Risk factors for SIL include early age at first intercourse, number of sex partners, number of sex partners of the male partner, and coinfection with other sexually transmitted diseases. Unfortunately, these risk factors have not shown consistent significance from study to study. Inconsistencies may be attributed to confounding factors, as discussed previously for HPV DNA detection studies, or these sexual risks may simply reflect the likelihood of exposure to HPV infection. Because all LSIL is HPV infection, but not all HPV is SIL, these distinctions become murky.

In contrast to the data on risk factors for HPV DNA detection, lifetime sexual behaviors, as compared to recent sexual activity and age at first intercourse, appear to play a significant role in the development of SIL and SCC. Several studies have shown that lifetime number of sexual partners is related to slightly elevated odds ratios (1.5) for the development of SIL and SCC [70,71]. Similarly, studies have shown that women who are exposed to HPV at a young age may be at particularly high risk for the development of SIL and SCC. Age at first intercourse was strongly associated with the risk for invasive cervical cancer. The risk for women who reported first intercourse before the age of 18 years was five times higher than that of women reporting age at first intercourse after age 22 [70–72]. Similar findings regarding SIL were reported by Wright [73]. Mean age at diagnosis of LSIL was 22.5 years and of HSIL was 25 years in the group reporting age at first intercourse between 10 and 14 years, compared to 34 years and 37 years for LSIL and HSIL, respectively, in the group reporting initiation of intercourse at 20 years or older.

The use of oral contraceptives (OC) has been sited in several studies as a risk factor for HPV infection, HSIL, and SCC. Vandenvelde (1992) [74] reported that HPV DNA positivity was associated with OCs but only with those OCs containing simultaneous and continuous potent estrogen and highly active progesterone, suggesting this relationship may in fact be due to the specific progesterone contained in the preparations. Most studies examining OCs and HPV detection found no correlation after controlling for sexual behavior [45,48,75]. On the other hand, HSIL and SCC seem to have a stronger correlation with prolonged OC use, suggesting that these are cofactors important in the development of HPV-related disease, but not in its acquisition. Moscicki et al. [76] and Ho et al. [45] also found no correlation between OCs and persistence of HPV infection.

The endogenous hormonal influence associated with pregnancy has long been considered a risk factor for HPV-associated disease. Most studies have documented that HPV DNA detection and condylomas are not more common in pregnant women [77,78]. However, several observations suggest that condylomas occurring during pregnancy are more resistant to therapy. Although rapid progression of SIL during pregnancy had been reported in the past, research does not support this finding. It is noted that colposcopy during pregnancy can be quite misleading because of the distortion that occurs to the cervical architecture. Another interesting association with pregnancy is the observation of an association between high parity (>7) and cervical cancer [70,71]. It has been speculated that the multiple traumas associated with birth or chronic hormonal influences may play a role. On the other hand, these findings have been found to be more consistent in third world countries and the chance of high-risk partners (partners who engage in sex with prostitutes) is much higher in multiparous women [79].

Although cigarette smoking has not been seen to play a role in HPV DNA detection or persistence, its role in SIL and SCC is more convincing. Certainly, tobacco has well known carcinogenic properties. Interestingly, the tobacco metabolites, nicotine and cotinine, are measurable in the cervical secretions of smokers. In addition, studies have shown that these metabolites are correlated with immune function in the cervical epithelium. Barton et al. [80] found that the number of Langerhans cells

(LC), antigen presenting cells common to the genital tract epithelium, was decreased in women with high levels of nicotine in cervical mucous as compared to women with lower levels. Others have found that decreased numbers of LC were common in SIL, suggesting a mechanism for nicotine in the development of SIL. However, epidemiology studies examining cigarette smoking have been inconsistent. Several studies have shown an increased risk for women smokers for LSIL, HSIL, and SCC [81–83]. Olsen et al. [84] reported a 66-fold increased risk for HSIL in women who were both smokers and had persistent HPV 16 DNA, compared to women without HSIL. Duration of smoking also appears to be important [85]. On the other hand, a few studies found that after adjustment for other factors, including sexual behavior, the effect of smoking was lost [68,86].

The interest between nutrition and cancer has extended to cervical cancer as well. Most of the studies in this area have been linked to clinical trials, limiting our ability to generalize this information to the natural history of HPV. Folic acid, retinoids, and antioxidants, such as beta carotene, have generated the most interest. Butterworth et al. [87] reported that deficiencies in folate and carotenoids were associated with development of SIL. Liu et al. [88] found that higher plasma levels of zinc and retinol were associated with regression of SIL. However, disappointing results have been reported in studies using nutritional supplements for treatment of SIL [88].

One of the most important, yet least understood, factors in the development of SIL is the host immune response. Although serum antibody response to HPV is well documented, it does not appear that the humoral (or antibody) response is important in control of HPV infection or in development of disease. Women with SIL, SCC, or merely HPV detection are more likely to have serum antibodies to HPV than women without disease or detection. This shows that antibody response is a relatively good measure of HPV exposure, but that it is not important for the prevention of disease development. In contrast, strong evidence suggests that cell-mediated immunity plays a critical role in the development of disease. The strongest evidence lies in the epidemiologic observations of patients with defects in cell-mediated immunity, such as organ transplant recipients, congenital immunodeficiencies, or persons with HIV. These persons are noted to have higher rates of condyloma, HPV detection, and anogenital cancers. A 100-fold increased risk of cervical, vulvar, and anal cancers has been observed in renal transplant patients compared to control groups. Numerous studies also have documented increased incidences of cervical and anal HPV DNA detection and SIL in women with HIV compared to noninfected controls. This association has been thought to be predominantly driven by the immunosuppression related to the loss of CD4 cells. Women with CD4 counts less than 200 cells/ml seem to be consistently at higher risk for developing SIL [41]. However, some studies suggest that other immune mechanisms may be involved in HIV infected women, including abnormal cytokine production by local HIV infected lymphocytes and HIV-1 enhanced expression of E6 and E7 [89]. In contrast to the observed higher rates of SIL, HPV, and condyloma, increased rates of cervical, vulvar, and anal cancer among HIV-infected women have not been reported. This discordance may be explained by several factors. First, the time required for development of cancer may be longer than the current life span of HIV-positive women. However, it is hypothesized that the incidence of these slow-developing squamous anogenital cancers may increase along with the improved prognosis in life span expected with the advent of highly active anti-retroviral therapy (HAART). The effectiveness of screening programs and identification and treatment of precancerous lesions may also have had an effect on the incidence of cancer.

Although cell-mediated immune (CMI) function is very difficult to study, several studies support the role of CMI in HPV expression in nonimmunocompromised women. Nakagawa *et al.* [90] showed that women with SIL were less likely to have a CMI response to HPV 16 than women infected with HPV but with no evidence of SIL. Other studies in mice have supported the role of CMI in mediating HPV expression.

Although studies have found that HPV infection is transient in most young women, persistent HPV infection appears critical to the development of HSIL and SCC. Persistence of HPV has been associated with a 14- to 40-fold increased risk of developing HSIL compared to women with transient or no infection. The range of risk appears to increase with increasing time of persistence, is highest in women attending STD clinics, is strongest for persistence of "high risk" HPV types, and is lowest in young women [43,45,65]. In one study, Ho et al. [45] reported that the increased risk for the development of SIL with persistent high risk HPV types was 37-fold, whereas the risk associated with persistent low risk HPV types was only 7-fold. Moscicki *et al.* [76] showed a slightly lower risk (14-fold) in the development of HSIL among adolescents and young women with high risk HPV type than that reported by Ho *et al.* [45], underscoring the risk associated with "time" of persistence, since the age group in Moscicki *et al.*'s study was younger. While these risks are high, it should be emphasized that the majority of women in these studies remain SIL-free despite HPV infection or persistence, underscoring the importance of other cofactors in the development of disease. The risk of HPV persistence has been difficult to define because most women have transient infections and repeated HPV testing in women has shown that current tests have a certain variability in their ability to pick up latent, or low-level, infections. Given these limitations, the few studies examining long-term persistence have found few epidemiologic correlations or risks associated with HPV persistence itself.

VIII. Clinical Application of HPV Screening

Screening for cervical cancer was introduced into clinical practice in the late 1950s with the advent of Pap smears. The premise of screening is that cervical cancer is preceded by a long precancerous stage. Early detection allows for treatment and elimination of the precursor lesion and thereby reduces the risk and incidence of cancer. Its effectiveness in reducing the incidence of cervical cancer is not disputed; rates of cervical cancer based on data from the SEER program were 50/100,000 women before the advent of Pap smear screening and are currently 8/100,000 women in screened populations in the United States [91]. Half of cervical cancers in the United States are found in women who have never had a Pap smear and an additional 10% are found in women who have not had a Pap smear in 5 years [92].

Although Pap smears have dramatically reduced cervical cancer rates, there are also well known limitations associated with Pap smear screening. Sensitivity of PAP smears averages 80% [93,94]. Approximately 4500 women develop cervical cancer annually in the United States who have been screened at some point [95]. Laboratory quality control as well as problems with sampling techniques contribute to the false negative rate. These problems have prompted clinicians to use alternative methods of cytologic screening, such as liquid-based cytology and adjunct testing using molecular HPV DNA testing.

Traditionally, the cervical cells collected by a brush or swab are smeared (frequently unevenly) onto a glass slide and fixed at the clinical sites and then sent to a laboratory for Pap smear staining. Usually, only a small portion of squamous cells (estimated at around 10%) make it onto the slide; most remain on the brush or swab which is discarded. In addition mucus, blood, and debris adhere to the slide. Liquid-based cytology uses the cells collected in the conventional manner but the cells are shaken off into a liquid media (PreservCyt Cytyc Corporation, Marlborough, MA). Studies have shown that more cells do make it off the brush/swab into the media compared to the traditional smearing technique. The media is sent to the laboratory where the cells are processed, using ThinPrep, the company's processor. The processor filters the cells, deleting the debris, mucus and blood, and spins the cells, laying them onto a glass slide in an orderly single layer. Although the cells are easier to read, only 5% of the sample is applied to the glass slide. Some studies have shown that ThinPrep can result in a larger number of samples classified as inadequate because of cells missing from the transformation zone. Several studies are underway to determine cost-effective strategies in Pap smear screening using ThinPrep.

Two computer automated cytology systems, Papnet and AutoPap 300QC, have received F.D.A. approval in the United States for assistance in quality control. These computerized systems are designed to read the Pap smear using neural network technology which allows the computer to recognize normal and abnormal cells. Koss *et al.* [96] reported that Papnet recognized 97% of known abnormal slides and was able to detect an additional 30% of slides thought to be false-negative in women who subsequently developed HSIL or SCC [97]. The reported sensitivity of AutoPap was 88.4% for LSIL, 95.1% for HSIL, and 96.4% for SCC [98]. While these techniques have currently been approved only for quality control, tests are ongoing to determine their utility as a primary screening instrument. Their efficacy and cost-benefit in routine practice have not yet been established.

A larger controversy regards the utility of HPV DNA testing in clinical practice. Several studies have evaluated HPV DNA testing as an adjunct to the management of women with atypical squamous cells of undetermined significance (ASCUS) or LSIL Pap smear. Current standards of practice vary regarding ASCUS—some clinical centers repeat the Pap smear in 3–6 months, whereas other centers refer the woman for a colposcopy examination because it is estimated that 20–30% of these women have HSIL when evaluated. Although an LSIL Pap smear generally elicits a colposcopy referral, some clinical protocols repeat the Pap smear before referral. However, treatment of LSIL is frequently deferred unless the lesion persists or pro-

Table 25.3
Common Treatment Modalities for External Genital Warts

Treatment	Mechanism of action	Mode of application	Side effects
Podophylin 10–25% resin in tincture of benzoine	Causes wart necrosis	Clinician applied in office, every 1–2 weeks	May cause ulceration, local irritation and discomfort
TCA/BCA 85–95% solution	Destroys warts by chemical coagulation	Clinician applied in office, every 1–2 weeks	May cause ulcerations, local irritation and burning
Cryotherapy liquid nitrogen	Cryolysis—induces tissue sloughing	Clinician applied in office with cryoprobe or applicator	May cause ulcerations, local irritation and discomfort, infection
Podofilox 0.5% gel or solution	Causes wart necrosis	Patient applied at home, 3 consecutive days weekly for 4 weeks	May cause ulcerations, erosions, local irritation and discomfort
Imiquimod 5% cream	Induces local interferon α and cytokines	Patient applied to wart at bedtime and alternate days, 3x weekly for up to 16 weeks	May cause burning and inflammation, erosions

gresses. Because most women with ASCUS or LSIL do not require treatment, a test that would better triage those in need of colposcopy or treatment could be beneficial. HPV DNA studies cited previously have shown that presence and persistence of "high risk" HPV DNA types are risk factors for progression of SIL to HSIL or SCC. The rationale for adjunct testing, therefore, is that women with indeterminate Pap smears (*e.g.*, ASCUS) or LSIL would be referred for further testing only if they were also HPV DNA positive for "high risk" types. Data from a study evaluating HPV "reflex" testing in which HPV DNA testing is automatically done on ASCUS Pap smears showed sensitivity levels of 93% for detecting HSIL [99]. Sensitivity of HPV DNA testing in combination with cytology ranged from 95–100% in women with ASCUS or LSIL who were at high risk or over 25 years of age [100,101]. Unfortunately, because of the high prevalence of HPV infection specifically in young women, the specificity of DNA testing is quite low (30–44%) [99]. Such low specificity may result in additional costs related to over-referrals for colposcopy examinations. On the other hand, the addition of routine HPV DNA testing in women over 35 years of age may prove more specific because these women identified with HPV are most likely reflective of persistent infection and are at high risk for eventual development of HSIL. HPV DNA testing as a primary screening method may have more clinical utility in developing countries where cytology screening programs are costly and women may only receive one Pap smear in their lifetime.

IX. HPV Vaccines

Advances in our understanding of HPV molecular structure, immune response, and in recombinant molecular DNA technology have led to the development of vaccines for the prevention of infection and for treatment of HPV-1 related diseases. Prophylactic vaccines are aimed at preventing LSIL or HSIL by inducing anti-HPV immune responses. Several animal vaccines have shown promising results in prevention of infection to bovine papilloma virus, canine oral papilloma virus, and cottontail rabbit papilloma virus [102–104]. Therapeutic vaccines, based on immunologically important epitopes, target existent HSIL or

SCC for treatment. Protection against some E6- and E7-positive tumors has been found in animal studies [105]. Gene therapy, whereby potentially beneficial genetic material is transferred to a cell, is another type of therapeutic vaccine that has been used in treatment of other cancers. Several human HPV vaccine trials are in progress for patients with late stage cervical cancer as well as large-scale prevention studies. Because of the time lag between infection and development of HSIL, and even more so of SCC, it will be several years before any conclusions can be derived from the large scale prevention trials.

X. Treatment

Treatment protocols for genital warts are fairly standard, with both physician applied and patient-applied therapies giving similar success rates (approximately 50–70%). (Table 25.3). Treatment of genital warts can be a prolonged, frustrating process for some women, requiring multiple office visits over several months. Consequently, self-applied therapies are gaining popularity. Although the length of treatment to cure is similar, taking several weeks to months, self-application therapies do not require office visits. Aggressive therapy with laser or excision is expensive but may be necessary in recalcitrant cases. Treatment of immunocompromised women is specifically difficult because recalcitrant cases and recurrence rates are extremely high, despite aggressive therapy.

Treatment of cervical SIL depends on a number of factors including grade of SIL, size of lesion, age, and immune status of the woman. As discussed previously, LSIL is best observed, since spontaneous regression is common. However, persistence for one year or longer or progression to HSIL are criteria for intervention. The goal of treatment is ablation or eradication of the lesion. Several ablative therapies are available. Cryotherapy is the least expensive and continues to be the mainstay of treatment for most cervical SIL. New techniques, such as loop electrosurgical excision procedure (LEEP), are now standard for many types of lesions. Although more expensive, LEEP removes a fairly large area of tissue so that the pathologist can ensure that the borders of the lesions do not contain cancerous tissue. Other techniques include laser ablation or cold-knife conization. Both are now used less frequently than LEEP therapy. While these ablative techniques are standard treatment for vaginal and vulvar HSIL, there are no current standards of practice regarding treatment of anal HSIL. Nonablative therapies including the application of chemotherapeutic agents such as 5 fluoro-uracil have been used for extensive genital tract lesions. Research using systemic drug treatments are ongoing as well [106].

XI. Counseling

The wide range of clinical expression associated with HPV has resulted in confusion both for the clinician and for the woman faced with the diagnosis. Longitudinal studies of HPV have expanded our understanding of its natural history, but the many questions that remain continue to be a source of frustration for patients in search of information. Although the importance of the role of sexual activity in transmission is not disputed, the rate of transmission from latent infection remains unknown. This "unknown" exposure to HPV frequently is viewed with anxiousness regarding questions of partner fidelity. Certainly, if not approached sensitively and honestly, the diagnosis can introduce a source of conflict in a current or future relationship.

The availability of HPV DNA testing may not be of direct benefit to the patient, who now has to contend with the diagnosis of an infectious STD that may or may not have clinical sequelae, that may or may not regress, or that may or may not progress to a cancerous lesion. Although the benefits of testing were discussed earlier, it must be emphasized that most women will neither have persistent infection nor progress to HSIL or SCC. The negative psychological impact of the diagnosis of HPV is unknown. A study of college women who were asked to respond to the question of being hypothetically positive for HPV stated that this diagnosis was associated with feelings of guilt, anger, being anxious, and feeling dirty [107]. These negative feelings may result in sexual dysfunction or introduce higher risk behaviors [108]. The process of disclosing a diagnosis should be accompanied by educational counseling that emphasizes the ubiquitous nature of the infection, the low likelihood of progression, and the current limitations of medical knowledge. This also represents an opportune time to discuss sexual risk and safe sex behaviors.

XII. Conclusions

Research gains in our understanding of the molecular function and natural history of HPV infection have improved our ability to evaluate, diagnose, and appropriately follow-up women with HPV infection and those at risk for HPV-associated disease. There is, however, a disconnection between scientific knowledge, which could lead to increased monitoring for the sequelae related to HPV infection, and a health care environment in which cost-containment is often paramount in practice decisions. On the other hand, underevaluation related to cost-containment may be offset by a medico-legal atmosphere that dictates overevaluation to protect the clinician from the possibility of having missed disease. It seems likely that HPV DNA testing will be incorporated into clinical practice in the future

but the benefits of testing will need to be better understood before community standards are revised. Research efforts will also need to address the psychological impact of diagnosing latent infection before HPV testing becomes part of routine clinical practice. Efforts into new treatment modalities and prevention through vaccines will likely change the course and epidemiology of HPV in the future.

References

1. Bafverstedt, B. (1967). Condylomata Acuminata—past and present. *Acta Derm.-Vernereol.* **47**, 376–381.
2. Hausen, Hz. (1974). Attempts to detect virus specific DNA in human tumors. *Int. J. Cancer* **13**, 650–656.
3. Franco, E. L. F. (1996). Epidemiology of anogenital warts and cancer. *Obset. Gynecol. Clin. North Am.* **23**, 597–623.
4. Bosch, F., Sanjose, S., Castellsaugue, X., and Munoz, N. (1997). Geographical and social patterns of cervical cancer incidence. *In* "New Developments in Cervical Cancer Screening and Prevention," (E. L. Franco and J. Monsonego, eds.), pp. 23–33. Blackwell, Oxford.
5. IARC (1995). "IARC monographs on the Evaluation of Carcinogenic Risks to Humans: Human Papillomaviruses," Vol. 64. World Health Organization, Lyon.
6. Lorincz, A. (1997). Methods of DNA hybridization and their clinical applicability to human papillomavirus detection. *In* "New Developments in Cervical Cancer Screening and Prevention" (E. L. Franco and J. Monsonego, eds.), pp. 325–337. Blackwell, Oxford.
7. Shulman, J., Leyton, M., and Hamilton, R. (1974). The Papanicolaou smear: An insensitive case-finding procedure. *Am. J. Obstet. Gynecol.* **120**, 446–451.
8. Wright, T. C., Kurman, R. J., and Ferenczy, A. (1994). Precancerous lesions of the cervix. *In* "Blaustein's Pathology of the Female Genital Tract" (R. J. Kurman, ed.), 4th ed., pp. 229–277. Springer-Verlag, New York.
9. Ostor, A. G. (1993). Natural history of cervical intraepithelial neoplasia: A critical review. International *J. Gynecol. Pathol.* **12**, 186–192.
10. Syrjanen, K. (1997). Biological behaviour of cervical intraepithelial neoplasia. *In* "New Developments in Cervical Cancer Screening and Prevention" (E. Franco and J. Monsonego, eds.), pp. 93–108. Blackwell, Oxford.
11. Chan, S. Y., Delius, H., Halpern, A. L., and Bernard, H. U. (1995). Analysis of genomic sequences of 95 papillomavirus types: uniting typing, phylogeny, and taxonomy. *J. Virol.* **69**, 3074–3083.
12. de Villiers, E. M. (1994). Human pathogenic papillomavirus types: An update. *Curr. Top. Microbiol. Immunol.* **186**, 1–12.
13. Southern, S., and Herrington, C. (1998). Molecular events in uterine cervical cancer. *Sex. Transm. Infect.* **74**, 101–109.
14. Lorincz, A. T., Reid, R., Jenson, A. B., Greenberg, M. D., Lancaster, W., and Kurman, R. J. (1992). Human papillomavirus infection of the cervix: Relative risk associations of 15 common anogenital types. *Obstet. Gynecol.* **79**, 328–337.
15. Cullen, A. P., Reid, R., Campion, M., and Lorincz, A. T. (1991). Analysis of the physical state of different human papillomavirus DNAs in intraepithelial and invasive cervical neoplasm. *J. Virol.* **65**, 606–612.
16. Van Ranst, M., Tachezy, R., and Burk, R. D. (1996). Human papillomavirus: A never ending story? *In* "Papillomavirus Reviews: Current Research on Papillomaviruses" (C. Lacey, ed.), pp. 1–20. Leeds University Press, Leeds.
17. Smotkin, D., Prokoph, H., and Wettstein, F. O. (1989). Oncogenic and nononcogenic human genital papillomaviruses generate the E7 mRNA by different mechanisms. *J. Virol.* **63**, 1441–1447.

18. Werness, B. A., Levine, A. J., and Howley, P. M. (1990). Association of human papillomavirus types 16 and 18 E6 proteins with p53. *Science* **248**, 76–79.

19. Romanczuk, H., and Howley, P. M. (1992). Disruption of either the E1 or the E2 regulatory gene of human papillomavirus type 16 increases viral immortalization capacity. *Proc. Natl. Acad. Sci. U.S.A.* **89**, 3159–3163.

20. Woodworth, C. D., Notario, V., and DiPaolo, J. A. (1990). Transforming growth factors beta 1 and 2 transcriptionally regulate human papillomavirus (HPV) type 16 early gene expression in HPV-immortalized human genital epithelial cells. *J. Virol.* **64**, 4767–4775.

21. Fairley, C. K., Chen, S., Tabrizi, S. N., Leeton, K., Quinn, M. A., and Garland, S. M. (1992). The absence of genital human papillomavirus DNA in virginal women. *Int. J. STD AIDS* **3**, 414–417.

22. Rylander, E., Ruusuvaara, L., Almstromer, M. W., Evander, M., and Wadell, G. (1994). The absence of vaginal human papillomavirus 16 DNA in women who have not experienced sexual intercourse. *Obstet. Gynecol.* **83**, 735–737.

23. Barrett, T. J., Silbar, J. D., and McGlinley, J. P. (1954). Genital warts—a venereal diseas. *JAMA, J. Am. Med. Assoc.* **154**, 333–334.

24. Oriel, J. D. (1971). Natural history of genital warts. *Br. J. Vener. Dis.* **47**, 1–13.

25. Barrasso, R., Debrux, J., Croissant, O., and Orth, G. (1987). High prevalence of papillomavirus-associated penile intraepithelial neoplasia in sexual partners of women with cervical intraepithelial neoplasia. *N. Engl. J. Med.* **317**, 916–923.

26. Ingram, D. L., Everett, V. D., Lyna, P. R., White, S. T., and Rockwell, L. A. (1992). Epidemiology of adult sexually transmitted disease agents in children being evaluated for sexual abuse. *Pediat. Infect. Dis. J.* **11**, 945–950.

27. Handley, J., Dinsmore, W., Maw, R., Corbette, R., Burrows, D., Bharucha, H., Swann, A., and Bingham, A. (1993). Anogenital warts in prepubertal children; sexual abuse or not? *Int. J. STD AIDS* **4**, 271–279.

28. Nuovo, G. J., Lastarria, D. A., Smith, S., Lerner, J., Comite, S. L., and Eliezri, Y. D. (1991). Human papillomavirus segregation patterns in genital and nongenital warts in prepubertal children and adults. *Am. J. Clin. Pathol.* **95**, 467–474.

29. Obalek, S., Jablonska, S., Favre, M., Walczak, L., and Orth, G. (1990). Condylomata acuminata in children: Frequent association with human papillomaviruses responsible for cutaneous warts. *J. Am. Acad. Dermatol.* **23**, 205–213.

30. Watts, D. H., Koutsky, L. A., Holmes, K. K., Goldman, D., Kuypers, J., Kiviat, N. B., and Galloway D. A. (1998). Low risk of perinatal transmissin of human papillomavirus: Results from a prospective cohort study. *Am. J. Obstet. Gynecol.* **178**, 363–373.

31. Quick, C. A., Watts, S. L., Kryzyzek, R. A., and Faras, A. J. (1980). Relationship between condylomata and laryngeal papillomata. Clinical and molecular virological evidence. *Ann. Otol. Rhinol. Laryngol.* **89**, 467–471.

32. Shah, K., Kashima, H., Polk, B. F., Shah, F., Abbey, H., and Abramson, A. (1986). Rarity of cesarean delivery in cases of juvenile-onset respiratory papillomatosis. *Obstet. Gynecol.* **68**, 795–799.

33. Sedlacek, T. V., Lindheim, S., Eder, C., Hasty, L., Woodland, M., Ludomirsky, A., and Rando, R. F. (1989). Mechanism for human papillomavirus transmission at birth. *Am. J. Obstet. Gynecol.* **161**, 55–59.

34. Rogo, K. O., and Nyansera, P. N. (1989). Congenital condylomata acuminata with meconium staining of amniotic fluid and fetal hydrocephalus: Case report. *East Afr. Med. J.* **66**, 411–413.

35. Syrjanen, K., Hakama, M., Saarikoski, S., Vayrynen, M., Yliskoski, M., Syrjanen, S., Kataja, V., and Castren, O. (1990). Prevalence, incidence, and estimated life-time risk of cervical human papillomavirus infections in a nonselected Finnish female population. *Sex. Transm. Dis.* **17**, 15–19.

36. Koutsky, L., Galloway, D. A., and Holmes, K. K. (1988). Epidemiology of genital human papillomaviruses. *Epidemiol. Rev.* **10**, 122–163.

37. Kiviat, N. B., Koutsky, L. A., Paavonen, J. A., Galloway, D. A., Critchlow, C. W., Beckmann, A. M., McDougall, J. K., Peterson, M. L., Stevens, C. E., and Lipinski, C. M. (1989). Prevalence of genital papillomavirus infection among women attending a college student health clinic or a sexually transmitted disease clinic. *J. Infect. Dis.* **159**, 293–302.

38. Lacey, C. J. N., and Mulcahy, F. M. (1986). Koilocyte frequencey and prevalence of cervical human papillomavirus infection. *Lancet* 557–558.

39. Martinez, J., Smithe, R., Farmer, M., Resau, J., Alger, L., Daiel, R., Gupta, J., Shah, K., and Naghashfar, Z. (1988). High prevalence of genital tract papillomavirus infection in female adolescents. *Pediatrics* **92**, 604–608.

40. Meisels, A. (1991). Cytologic diagnosis of human papillomavirus: Influence of age and pregnancy stage. *Acta Cytol.* **36**, 480–482.

41. Wright, T. C. (1997). Papillomavirus infection and neoplasia in women infected with human immunodeficiency virus. *In* "New Developments in Cervical Cancer Screening and Prevention" (E. L. Franco and J. Monsonego, eds.), pp. 131–143. Blackwell, Oxford.

42. Bauer, H. M., Ting, Y., Greer, C. E., Chambers, J. C., Tashiro, C. J., Chimera, J., Reingold, A., and Manos, M. M. (1991). Genital human papillomavirus infection in female university students as determined by a PCR-based method [see comments]. *JAMA, J. Am. Med. Assoc.* **265**, 472–477.

43. Hildesheim, A., Schiffman, M. H., Gravitt, P. E., Glass, A. G., Greer, C. E., Zhang, T., Scott, D. R., Bush, B. B., Lawler, P., Sherman, M. E., Kurman, R. J., and Manos, M. M. (1994). Persistence of type-specific human papillomavirus infection among cytologically normal women. *J. Infect. Dis.* **169**, 235–240.

44. Kotloff, K. L., Wasserman, S. S., Russ, K., Shapiro, S., Daniel, R., Brown, W., Frost, A., Tabara, S. O., and Shah, K. (1998). Detection of genital human papillomavirus and associated cytological abnormalities among college women. *Sex. Transm. Dis.* **25**, 243–250.

45. Ho, G. Y. F., Bierman, R., Beardsley, L., Chang, C. J., and Burk, R. D. (1998). Natural history of cervicovaginal papillomavirus infection in young women. *N. Engl. J. Med.* **338**, 423–428.

46. Moscicki, A. B. (1996). Genital HPV infections in children and adolescents. *Obstet. Gynecol. Clin. North Am.* **23**, 675–697.

47. Melkert, P. W., Hopman, E., van den Brule, A. J., Risse, E. K., van Diest, P. J., Bleker, O. P., Helmerhorst, T., M. E. S., Meijer, C. J., and Walboomers, J. M. (1993). Prevalence of HPV in cyto-morphologically normal cervical smears as determined by the polymerase chain reaction, is age-dependent. *Int. J. Cancer* **53**, 919–923.

48. Moscicki, A. B., Palefsky, J., Gonzales, J., and Schoolnik, G. K. (1990). Human papillomavirus infection in sexually active adolescent females: Prevalence and risk factors. *Pediatr. Res.* **28**, 507–513.

49. Burk, R. D., Kelly, P., Feldman, J., Bromberg, J., Vermund, S. H., DeHovitz, J. A., and Landesman, S. H. (1996). Declining prevalence of cervicovaginal human papillomavirus infection with age is independent of other risk factors. *Sex. Transm. Dis.* **23**, 333–341.

50. Wheeler, C. M., Parmenter, C. A., Hunt, W. C., Becker, T. M., Greer, C. E., Hildesheim, A., and Manos, M. M. (1993). Determinants of genital human papillomavirus infection among cytologically normal women attending the University of New Mexico student health center. *Sex. Transm. Dis.* **20**, 286–289.

51. Bauer, H. M., Hildesheim, A., Schiffman, M. H., Glass, A. G., Rush, B. B., Scott, D. R., Cadell, D. M., Kurman, R. J., and Manos, M. M. (1993). Determinants of genital human papillomavirus infection in low-risk women in Portland, Oregon. *Sex. Transm. Dis.* **20,** 274–278.

52. Koutsky, L. A. (1997). Epidemiology of genital human papillomavirus infection. *Am. J. Med.* **102**(5A), 3–8.

53. Becker, T. M., Stone, K. M., and Alexander, E. R. (1987). Genital human papillomavirus infection. *Obstet. Gynecol. Clin. North Amer.* **14,** 389–396.

54. Vermund, S. H., Holland, C., Wilson, C. M., Crowley-Nowick, P., and Moscicki, A. B. (1998). HPV infection and HIV status in adolescent girls: The reaching for excellence in adolescent care and health (REACH) project. *12th World AIDS Conf.,* Geneva, *1998,* 142.

55. Calore, E. E., Cavaliere, M. J., and Calore, N. M. P. (1998). Squamous intraepithelial lesions in cervical smears of human immunodeficiency virus-seropositive adolescents. *Diagn. Cytopathol.* **18,** 91–92.

56. Hillemanns, P., Ellerbrock, T. V., McPhillips, S., Dole, P., Alperstein, S., Johnson, D., Sun, X. W., Chiasson, M. A., and Wright, T. C. (1996). Prevalence of anal human papillomavirus infection and anal cytologic abnormalities in HIV-seropositive women. *AIDS* **10,** 1641–1647.

57. Williams, A. B., Darragh, T. M., Vranizan, K., Ochia, C., Moss, A. R., and Palefsky, J. M. (1994). Anal and cervical human papillomavirus infection and risk of anal and cervical epithelial abnormalities in human immunodeficiency virus-infected women. *Obstet. Gynecol.* **83,** 205–211.

58. Chiasson, M. A., Ellerbrock, M. D., Bush, T. J., Sun, X. W., and Wright, T. C. (1997). Increased prevalence of vulvovaginal condyloma and vulvar intraepithelial neoplasia in women infected with the human immunodeficiency virus. *Obstet. Gynecol.* **89,** 690–694.

59. Sun, X. W., Kuhn, L., Ellerbrock, T. V., Chiasson, M. A., Bush, T. J., and Wright, T. C. (1997). Human papillomavirus infection in women infected with the human immunodeficiency virus. *N. Engl. J. Med.* **337,** 1343–1349.

60. Ogunbiyi, O. A., Scholefield, J. H., Robertson, G., Smith, J. H., Sharp, F., and Rogers, K. (1994). Anal human papillomavirus infection and squamous neoplasia in patients with invasive vulvar cancer. *Obstet. Gynecol.* **83,** 212–216.

61. Ogunbiyi, O. A., Scholefield, J. H., Raftery, A. T., Smith, J. H., Duffy, S., Sharp, F., and Rogers, K. (1994). Prevalence of anal human papillomavirus infection and intraepithelial neoplasia in renal allograft recipients. *Br. J. Surg.* **81,** 365–367.

62. Moscicki, A. B., Palefsky, J., Darragh, T. M., Hills, N., and Shiboski, S. (1997). Abnormal anal cytology in young heterosexual women. *Hum. Papillomavirus Virus, 16th Congr.,* Siena, Italy, *1997,* 505.

63. Schiffman, M. H., Bauer, H. M., Hoover, R. N., Glass, A. G., Cadell, D. M., Rush, B. B., Scott, D. R., Sherman, M. E., Kurman, R. J., and Wacholder, S. (1993). Epidemiologic evidence showing that human papillomavirus infection causes most cervical intraepithelial neoplasia. *J. Natl. Cancer Inst.* **85,** 958–964.

64. Moscicki, A. B., Millstein, S. G., Broering, J., and Irwin, C. E. (1993). Risks of human immunodeficiency virus infection among adolescents attending three diverse clinics. *J. Pediatr.* **122,** 813–820.

65. Franco, E. L., Villa, L. L., Richardson, H., Rohan, T. E., and Ferenczy, A. (1997). Epidemiology of cervical human papillomavirus infection. *In* "New Developments in Cervical Cancer Screening and Prevention" (E. L. Franco and J. Monsonego, eds.). Blackwell, Oxford, 14–22.

66. Shew, M. L., Fortenberry, J. D., Miles, P., and Amortegui, A. J. (1994). Interval between menarche and first sexual intercourse,

related to risk of human papillomavirus infection. *J. Pediatr.* **125,** 662–666.

67. Kjaer, S. K., de Villiers, E. M., Caglayan, H., Svare, E., Haugaard, B. J., and Engholm, G. (1993). Human papillomavirus, herpes simplex virus and other potential risk factors for cervical cancer in a high-risk area (Greenland) and a low-risk area (Denmark)— a second look. *Br. J. Cancer* **67,** 830–837.

68. Ley, C., Bauer, H. M., Reingold, A., Schiffman, M. H., Chambers, J. C., Tashiro, C. J., and Manos, M. M. (1991). Determinants of genital human papillomavirus infection in young women. *J. Natl. Cancer Inst.* **83,** 997–1003.

69. Villa, L. L., and Franco, E. L. (1989). Epidemiologic correlates of cervical neoplasia and risk of human papillomavirus infection in asymptomatic women in Brazil. *J. Natl. Cancer Inst.* **81,** 332–340.

70. Stone, K. M., Zaidi, A., Rosero-Bixby, L., Oberle, M. W., Reynolds, G., Larsen, S., Nahmias, A. J., Lee, F. K., Schachter, J., and Guinan, M. E. (1995). Sexual behavior, sexually transmitted diseases, and risk of cervical cancer. *Epidemiology* **6,** 409–414.

71. Brinton, L. A., Herrero, R. W. C. R., de Britton, R. C., Gaitan, E., and Tenorio, F. (1993). Risk factors for cervical cancer by histology. *Gynecol. Oncol.* **51,** 301–306.

72. La Vecchia, C., Franceschi, S., Decarli, A., Fasoli, M., Gentile, A., Parazzini, F., and Regallo, M. (1986). Sexual factors, venereal diseases and the risk of intraepithelial and invasive cervical neoplasia. *Cancer (Philadelphia)* **58,** 935–941.

73. Wright, V. C., and Rioplelle, M. A. (1984). Age at beginning of coitus versus chronologic age as a basis for Papanicolau smear screening: An analysis of 747 cases of preinvasive disease. *Am. J. Obstet. Gynecol.* **149,** 824–830.

74. Vandenvelde, C., and Van Beers, D. (1992). Risk factors inducing the persistence of high risk genital papillomaviruses in the normal cervix. *J. Med. Virol.* **38,** 226–232.

75. Smith, E. M., Johnson, S. R., Jiang, D., Zaleski, S., Lynch, C. F., Brundage, S., Anderson, R. D., and Turek, L. P. (1991). The association between pregnancy and human papilloma virus prevalence. *Cancer Detect. Prev.* **15,** 397–402.

76. Moscicki, A. B., Shiboski, S., Broering, J., Powell, K., Clayton, L., Jay, N., Darragh, T., Brescia, R., Kanowitz, S., Miller, S. B., Stone, J., Hanson, E., and Palefsky, J. (1998). The natural history of human papillomavirus infection as measured by repeated DNA testing in adolescent and young women. *J. Pediat.* **132,** 277–284.

77. Tenti, P., Zappatore, R., Migliora, P., Spinillo, A., Maccarini, U., and De Benedittis, M. (1997). Latent human papillomavirus infection in pregnant women at term: A case-control study. *J. Infect. Dis.* **176,** 277–280.

78. de Roda Husman, A. M., Walboomers, J. M., Hopman, E. O.P. B., Helmerhorst, T. M., Rozendaal, L., Voorhorst, F. J., and Meijer, C. J. (1995). HPV prevalence in cytomorphologically normal cervical scrapes of pregnant women as determined by PCR: The age-related pattern. *J. Med. Virol.* **46,** 97–102.

79. Brinton, L. A., Reeves, W. C., Brenes, M. M., Herrero, R., Gaitan, E., Tenorio, F., de Britton, R. C., Garcia, M., and Rawls, W. E. (1989). The male factor in the etiology of cervical cancer among sexually monogamous women. *Int. J. Cancer* **44,** 199–203.

80. Barton, S. E., Hollingworth, A., Maddox, P. H., Edwards, R., Cuzick, J., McCance, D. J., Jenkins, D., and Singer, A. (1989). Possible cofactors in the etiology of cervical intraepithelial neoplasia. An immunopathologic study. *J. Reprod. Med.* **34,** 613–616.

81. Burger, M. P. M., Hollema, H., Gouw, A. S. H., Pieters, W. J. L. M., and Quint, W. G. V. (1993). Cigarette smoking and human papillomavirus in patients with reported cervical cytological abnormality. *Br. Med. J.* **306,** 749–752.

82. Becker, T. M., Wheeler, C. M., McGough, N. S., Parmenter, C. A., Stidley, C. A., Jamison, S. F., and Jordan, S. W. (1994). Cigarette smoking and other risk factors for cervical dysplasia in southwest-

ern Hispanic and non-Hispanic white women. *Cancer Epidemiol., Biomarkers Prev.* **3,** 113–119.

83. Feldman, J. G., Chirgwin, K., Dehovitz, J. A., and Minkoff, H. (1997). The association of smoking and risk of condyloma acuminatum in women. *Obstet. Gynecol.* **89,** 346–350.

84. Olsen, A. O., Dillner, J., Skrondal, A., and Magnus, P. (1998). Combined effect of smoking and human papillomavirus type 16 infection in cervical carcinogenesis. *Epidemiology* **9,** 346–349.

85. Barton, S. E., Maddox, P. H., Jenkins, D., Edwards, R., and Cuzick, J. (1988). Effect of cigarette smoking on cervical epithelial immunity: A mechanism for neoplastic change? *Lancet* **2,** 652–654.

86. Koutsky, L. A., Holmes, K. K., Critchlow, C. W., Stevens, C. E., Paavonen, J., Beckmann, A. M., DeRouen, T. A., Galloway, D. A., Vernon, D., and Kiviat, N. B. (1992). A cohort study of the risk of cervical intraepithelial neoplasia grade 2 or 3 in relation to papillomavirus infection. *N. Engl. J. Med.* **327,** 1272–1278.

87. Butterworth, C. J., Hatch, K. D., Macaluso, M., Cole, P., Sauberlich, H. E., Soong, S. J., Borst, M., and Baker, V. V. (1992). Folate deficiency and cervical dysplasia. JAMA, *J. Am. Med. Assoc.* **267,** 528–533.

88. Liu, T., Soong, S., Alvarez, R. D., and Butterworth, C. E. (1995). A longitudinal analysis of human papillomavirus 16 infection, nutritional status, and cervical dysplasia progression. *Cancer Epidemiol., Biomarkers Prev.* **4,** 373–380.

89. Vernon, S. D., Hart, C. E., Reeves, W. C., and Icenogle, J. P. (1993). The HIV-1 tat protein enhances E2-dependent human papillomavirus 16 transcription. *Virus Res.* **27,** 133–145.

90. Nakagawa, M., Stites, D. P., Farhat, S., Sisler, J. R., Moss, B., Kong, F., Moscicki, A. B., and Palefsky, J. M. (1997). Cytotoxic T lymphocyte responses to E6 and E7 proteins of human papillomavirus type 16: Relationship to cervical intraepithelial neoplasia. *J. Infect. Dis.* **175,** 927–931.

91. Palefsky, J. M. (1996). Anogenital neoplasia in HIV-positive women and men. *HIV Adv. Res. Ther.* **6,** 10–17.

92. Summary of the NIH consensus development conference on cervical cancer. *Oncology* **11,** 672–674.

93. Soost, H.-J., Lange, H. J., Lechmacher, W., and Ruffing-Kullmann, B. (1991). The validation of cervical cytology: Sensitivity, specificity and predictive values. *Acta Cytol.* **35,** 8–13.

94. Gay, J., Donaldson, L., and Goellner, J. (1985). False-negative results in cervical cytologic studies. *Acta Cytol.* **29,** 1043–1046.

95. Cox, J. T. (1996). Clinical role of HPV testing. *Obstet. Gynecol. Clin. North Am.* **23,** 811–851.

96. Koss, L. G. (1997). Automation in cervicovaginal cytology: System requirements and benefits. *In* "New Developments in Cervical Cancer Screening and Prevention" (E. L. Franco and J. Monsonego, eds.). Blackwell, Oxford, 274–278.

97. Koss, L. G., Lin, E., Schreiber, K., Elgert, P., and Mango, L. (1994). Evaluation of the PAPNET cytologic screening system

for quality control of cervical smears. *Am. J. Clin. Pathol.* **101,** 220–229.

98. Richart, R. M., Patten, S. E., and Lee, L. J. (1997). Automated screening using the Autopap 300 device. *In* Franco "New Developments in Cervical Cancer Screening and Prevention" (E. J. Franco and J. Monsonego, eds.). Blackwell, Oxford, 279–283.

99. Wright, T. C., Lorincz, A., Ferris, D. G., Richart, R. M., Ferenczy, A., Mielzynska, I., and Borgatta, L. (1998). Reflex human papillomavirus deoxyribonucleic acid testing in women with abnormal Papanicolaou smears. *Am. J. Obstet. Gynecol.* **178,** 962–966.

100. Wright, T. C., Sun, X. W., and Koulos, J. (1995). Comparison of management algorithms for the evaluation of women with low-grade cytologic abnormalities. *Obstet. Gynecol.* **85,** 202–210.

101. Cox, J. T., Lorincz, A. T., Schiffman, M. H., Sherman, M. E., Cullen, A., and Kurman, R. J. (1995). HPV testing by hybrid capture appears to be useful in triaging women with a cytologic diagnosis of ASCUS. *Am. J. Obstet. Gynecol.* **172,** 946–954.

102. Roden, R., Hubbert, N., Kirnbauer, R., Christensen, N., Lowy, D., and Schiller, J. (1996). Assessment of the serological relatedness of genital human papillomaviruses by hemagglutination inhibition. *J. Virol.* **70,** 5875–5883.

103. Suzich, J., Ghim, S., Palmer-Hill, F., White, W., Tamura, J., Bell, J., Newsome, J. A., Jenson, A. B., and Schlegel, R. (1995). Systemic immunization with papillomavirus L1 protein completely prevents the development of viral mucosal papillomas. *Proc. Nat. Acad. Sci. U.S.A.* **92,** 11553–11557.

104. Breitburd, F., Kirnbauer, R., Hubbert, N., Nonenmacher, B., Trin-Dinh-Desmarquet, C., and Orth, G. (1995). Immunization with virus-like particles from cottontail rabbit papillomavirus can protect against experimental CRPV infection. *J. Virol.* **69,** 3959–3963.

105. Krul, M., Tijhaar, E., Kleije, J., Van Loon, A., Schipper H., Geerse, L., Vanderkolk, M., Steerenberg, P. A., Modi, F. R., and Den Otter, W. (1996). Induction of an antibody response in mice against human papillomavirus type 16 after immunization with HPV recombinant Salmonella strains. *Cancer Immunol. Immunother.* **43,** 44–48.

106. Mitchell, M. F., Hittelman, W. K., Lotan, R., Nishioka, K., Tortolero-Luna, G., Richards-Kortum, R., Wharton, J. T., and Hong, W. K. (1995). Chemoprevention trials and surrogate end point biomarkers in the cervix. *Cancer (Philadelphia)* **76,** 1956–1977.

107. Rameriz, J. E., Ramos, D. M., Clayton, L. *et al.* (1997). Genital human papillomavirus infections: Knowledge, perception of risk, and actual risk in a nonclinic population of young women. *J. Women's Health* **6,** 113–121.

108. Brookes, J. L., Haywood, S., and Green, J. (1993). Adjustment to the psychological and social sequelae of recurrent genital herpes simplex infection. *Genitourin. Med.* **69,** 384–387.

26

Human Immunodeficiency Virus and AIDS in Women

MARY L. KAMB AND PASCALE M. WORTLEY
Prevention Services Research Branch and HIV/AIDS Surveillance Branch,
Division of HIV/AIDS Prevention—Surveillance and Epidemiology, National Center for HIV, STD, and TB Prevention, Centers for
Disease Control and Prevention, Atlanta, Georgia

I. Introduction

The acquired immunodeficiency syndrome (AIDS) was first described in 1981 among a cluster of gay and bisexual men in California [1]. Although the first AIDS case in a woman was described later that year [2], in the United States and other industrialized nations AIDS and its casual agent, human immunodeficiency virus (HIV), have occurred predominantly among men. In the United States, gay men and injection drug users were affected earliest by AIDS. However, the distribution of AIDS has changed over time as an increasingly larger proportion of disease is related to heterosexual contact. In the 1990s, AIDS has been increasing in frequency in young adults and inner-city indigent populations and has become a leading cause of death among reproductive-aged women [3–6]. Since the emergence of AIDS, one constant has been the considerably disproportionate number of U.S. AIDS cases among minorities, particularly minority women.

As AIDS has become an increasingly significant problem for disadvantaged communities in wealthy nations, it has become an enormous problem for the world's poorest nations. Globally, because as many as 90% of new infections are acquired through heterosexual contact, AIDS affects nearly as many women as men [7]. Since 1981, an estimated 31 million people worldwide, including 12 million women and more than 3 million children, have been infected with HIV [7]. In sub-Saharan Africa, some population-based studies indicate that infection rates for women substantially exceed those for men, a finding probably related to women's greater susceptibility to HIV through heterosexual sex [8]. Contributing to this has been women's relative lack of control over many conditions of their lives, including those under which they have sex.

While AIDS continues a relentless expansion through the poorer nations of the world, in wealthier nations the advent of highly active antiretroviral therapies has led to precipitous declines in AIDS cases and even greater declines in AIDS-related mortality in adults and infants [9,10]. Despite these successes, delayed diagnosis of HIV has been a problem for some, including women, who have historically been less likely to seek, or less able to obtain, health care early [11]. Drugs are not universally available. Even in the United States, the most marginalized members of society have difficulty gaining access to the information and services of well-conducted HIV programs [7]. Of particular concern, considering the remarkable achievements of new therapeutic regimens, is that new HIV infections in the United States appear to be continuing to occur at a substantial rate for some groups. Although data describing HIV incidence are sparse, available information suggests that HIV incidence in

the United States is declining both in men and in women infected through injection drug use, but not in women infected through heterosexual contact [12–14]. Long-term analyses suggest that in most of the world, the HIV epidemic is not receding, it is merely shifting its focus to new vulnerable populations [7].

In this chapter we review current knowledge about HIV and AIDS in women, focusing on the United States but also examining the global contexts of AIDS. We examine how AIDS is defined and monitored in various settings, how women acquire infection, and the distribution of AIDS among women in the United States and in other areas of the world. We consider some of the viral, host, and other factors that determine how and why HIV transmission occurs in women, as well as social, cultural, and environmental factors that may indirectly promote or impede HIV infection. Reviewing the natural history of HIV in women, we discuss the contribution of factors, such as access to and availability of health care, that may affect women and men differently. Finally, we present strategies for preventing HIV, some proven effective and others potentially effective, as they apply to women.

II. Defining AIDS and Monitoring the Epidemic

On June 5, 1981, a report in the *Morbidity and Mortality Weekly Report* (*MMWR*) described five cases of *Pneumocystis carinii* (PCP) pneumonia in men, all of whom reported sex with men [1]. Within weeks, similar cases were reported in cities across the United States and by year's end, 189 cases had been reported to the Centers for Disease Control and Prevention (CDC) [2]. Of these cases, 97% were in men, most of whom were hemophiliac, gay, or bisexual. In addition to PCP, other rare opportunistic infections (*e.g., Cryptosporidium* and *Mycobacterium avium* and *M. intracellulare*) were reported with increasing frequency. These events marked the emergence of a newly recognized disease entity which was given the name acquired immunodeficiency syndrome (AIDS). In 1984, the causal agent—human immunodeficiency virus (HIV)—was recognized, and in 1985 an antibody test that detected presence of the virus was first licensed.

As knowledge about AIDS has increased, the case definition has changed. This is particularly true in the United States, where four distinct changes in the AIDS case definition have been made since 1982. Although these changes have led to sporadic artifacts in trends for men and women, they have resulted in better estimations of AIDS cases in women and in some disadvantaged populations, including injection drug users and racial and ethnic minorities. AIDS has been a reportable condition in

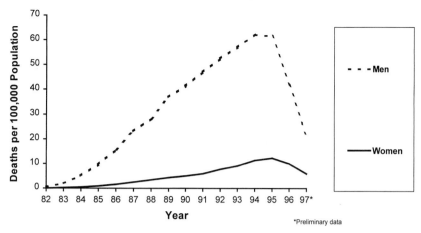

Fig. 26.1 Trends in mortality from HIV infection among men and women 25–44 years old, USA, 1982–1997 (Data source: Richard Selick, National Vital Statistics, CDC.)

all 50 U.S. states since 1982; as of 1997, HIV is reportable in just over half of the states. Until recently, CDC has reported trends in the HIV epidemic on the basis of the number of adult and adolescent women and men with AIDS each year (AIDS incidence). AIDS incidence was considered a reliable, albeit delayed, picture of trends in the epidemic. However, as new treatment advances have greatly prolonged the time between initial infection and the development of AIDS-defining conditions (Fig. 26.1), AIDS incidence is less reliable for tracking the HIV epidemic in the United States and HIV case surveillance is playing an increasingly important role.

The quality of HIV and AIDS surveillance data differs considerably throughout the world. Most industrialized nations have active surveillance systems based on AIDS or HIV reporting, similar to those in United States. Many poorer nations have established sentinel surveillance systems along guidelines from the World Health Organization. Sentinel surveillance is less costly and time intensive than the active surveillance of cases but can still be used to identify affected populations in order to focus resources and prevention efforts most effectively. Among higher-risk populations, such as commercial sex workers or long-distance truck drivers, sentinel surveillance can be useful in determining where HIV may be the bridge from higher-risk to lower-risk persons, whereas in antenatal women it may be used to estimate the extent of HIV infection in the sexually active general population. Sentinel surveillance may lead to biased prevalence estimates if the persons sampled are not representative of the population of interest. For example, if fertility is affected by HIV, antenatal surveys may overestimate HIV infection in younger women and underestimate infection in older women.

III. HIV Transmission: Modes of Exposure

Since 1981, it has become increasingly clear that HIV is transmitted in limited ways, primarily through intimate sexual contact, but also parenterally through the use of contaminated needles or syringes or the transfusion of infected blood components or clotting factors and through mother-to-child (perinatal)

transmission. Although HIV has been isolated from a number of body fluids, including blood, saliva, semen, urine, cerebrospinal fluid, and even sweat, relatively few AIDS cases have been related to exposures other than the primary modes listed. Furthermore, in the United States, the heat treatment of factor concentrates, initiated in 1984, and HIV antibody screening of all blood and plasma donations, instituted nationally in 1985, have virtually eliminated HIV transmission through transfusion [15]. Transfusion remains an important route of HIV infection in areas of the world where blood and blood products are not routinely screened for HIV.

In the United States, almost all AIDS cases in women have been related to injection drug use (IDU) or heterosexual contact with an HIV-infected male sex partner. Heroin was once the injection drug of choice, but cocaine, amphetamines, and other drugs increasingly have been used intravenously and may be injected more frequently and involve more needle sharing than heroin [16,17]. Most IDU-related AIDS among U.S. women has been concentrated in the large cities along the northeastern seaboard. New AIDS cases related to drug injection peaked in 1995; since then the number has decreased, in part because of the increasing use of antiretroviral therapy [4], but probably also because of decreasing IDU-related infections [13,18]. Many experts believe that community-based programs (*e.g.,* outreach, needle exchange, drug abuse treatment), changing drug paraphernalia laws, and networking have contributed to reduced needle and syringe sharing and thus to fewer new HIV infections. Other factors have also been postulated (*e.g.,* the availability of purer forms of heroin which can be snorted rather than injected) [19,20].

In 1994, heterosexual contact surpassed IDU as the predominant route of transmission to U.S. women with a diagnosis of AIDS [6]. According to a birth cohort analysis of women born between 1930 and 1974, AIDS incidence from 1983 through 1995 for women who injected drugs was highest in the 1950–1959 cohort (*i.e.,* women who were in their twenties in the 1980s) [13] (Fig. 26.2a). AIDS incidence decreased consistently with each younger cohort, and by the early 1990s new AIDS cases in injecting women had stabilized in most cohorts [13]. In

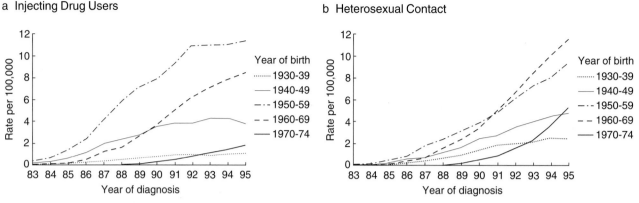

Fig. 26.2 (a and b) AIDS incidence rates per 100,000 women by birth cohort and mode of transmission. (*Data were adjusted for reporting delay, unreported risk, and the 1993 expansion in the AIDS surveillance case definition.) (Data source: P. M. Wortley, M.D., M.P.H., based on national AIDS surveillance, CDC.)

contrast, AIDS incidence among women infected through heterosexual contact increased in the most recent birth cohorts (*i.e.,* women born in the 1950s, 1960s, and 1970s), and by 1995 AIDS incidence was highest in the younger birth cohorts (Fig. 26.2b). Among women diagnosed with AIDS in 1997, IDU accounted for 38% and heterosexual contact for 58% of cases [5]. Among the women with AIDS related to heterosexual contact, a third (34%) had a male sex partner who injected drugs, which corroborates that IDU remains a major force behind heterosexual HIV transmission in U.S. women. The majority (54%) were reported to have an HIV-infected partner, many of whom had likely injected drugs (but this information was not in the medical record, the source of information for the surveillance report). Of women with heterosexual contact, 10% reported sex with a bisexual man and 6% reported sex with a man who had hemophilia or who had been transfused with blood products [5]. While this suggests that bisexual men have not played a prominent role in HIV transmission to women, often people are unaware of their sex partners' behaviors. A substantial proportion of female partners of bisexual men, even those in long-term relationships, do not know their partner is bisexual [21].

An uncommon mode of HIV transmission, at least in the United States, is occupational exposure. As of June 1995, 46 health care workers in the United States, of whom most (78%) were women, had documented seroconversion related to occupational exposure. Most were either laboratory technicians (39%) or nurses (35%), and the majority (83%) had percutaneous exposure to blood (R. Metler, RN, MSPH, CDC, personal communication). An additional 97 health care workers who had no other identifiable risk factors were probably infected through occupational exposures, but seroconversion was not documented at the time of exposure. Occupational exposure is believed to play a much larger role in HIV transmission in some other nations.

Rare modes of HIV transmission to women include female-to-female sexual transmission and artificial insemination. Most HIV infections among women who report sex with women are related to other behavioral risks, such as IDU or sex with a man [22]. An investigation of transmission risk among women with

HIV/AIDS from nine U.S. states found that 65 of 1122 (5.8%) women reported sex with a female partner. Most of the 65 were bisexual: 55 (85%) also reported sex with men [23]. For 64 of the 65 women, sex with a man or another risk behavior may have contributed to their HIV infection, and many women reported numerous risks (*e.g.,* more than half reported IDU). For one woman, sex with a woman was the most compelling risk behavior reported. This study highlights how important it is that health care providers ask about behavior and not assume behaviors on the basis of sexual identity. HIV infection by means of intravaginal insemination with donor semen is rare [24]. Most case reports are related to unscreened semen. There are methods of processing semen from HIV-infected donors to reduce infectiousness, but at least one woman who was inseminated with the processed sperm of her HIV-infected husband became HIV-1 infected [25]. HIV screening of semen for artificial insemination is recommended but is not mandatory in the United States [25,26]. Although commercial sperm banks routinely screen semen for HIV, many physicians who perform inseminations do not require HIV testing for donors [27].

Maternal-to-child transmission has accounted for approximately 8000 AIDS cases in U.S.–born infants [5]. Worldwide, this transmission mode is an extensive source of HIV infection and a leading cause of death in children under 5 years of age. Infants may become infected *in utero,* at the time of delivery, or after birth through breast feeding. Prospective studies from the U.S. and Europe suggest a 13–40% overall rate of perinatal transmission; studies in Africa suggest slightly higher transmission, probably reflecting prevalence of breast feeding and other health factors [28]. In 1994, a large clinical trial (ACTG Study 076) found that zidovudine therapy during pregnancy, labor, and delivery for HIV-infected women reduced HIV transmission to their babies by as much as two-thirds [10]. Since that trial, antiretroviral therapy has become the standard of care for HIV-infected pregnant women in industrialized nations, and vertically acquired HIV infection has plummeted [5]. In some European countries, a combination of antiretroviral therapy and caesarian section has virtually eliminated neonatal HIV [29]. However, for many poor nations antiretroviral therapy is too costly to consider for HIV-infected pregnant women, and the rates of AIDS and the AIDS-related deaths of children remain unaffected by the new therapeutic strategies [30].

In most industrialized nations, HIV transmission patterns have been similar to those in the United States; early AIDS cases in women have been concentrated among drug injectors. In many countries including Western Europe, Canada, and Australia, AIDS cases in women related to IDU and heterosexual contact have decreased with the advent of antiretroviral therapies [31]. In some Eastern European nations, AIDS incidence has increased, reflecting IDU, heterosexually acquired infection fueled by high sexually transmitted disease (STD) prevalence, and use of unsterile injection equipment in some hospital settings [27]. Worldwide, more than 90% of HIV infections have been acquired heterosexually [7]. Through 1996, sub-Saharan Africa, closely followed by Asia, accounted for most of the cumulative total of HIV infections in adults. New HIV epidemics are exploding throughout Asia, most notably in Southeast Asia and in India. While sub-Saharan Africa still accounts for two-thirds of all HIV infections, some estimates suggest that in the future Asia will surpass Africa in the cumulative total of persons with HIV infection [7,32–34].

In many parts of the world, including Asia and sub-Saharan Africa, the pattern of HIV spread historically has been related to introduction of HIV into groups with high rates of partner change, such as women practicing commercial sex work or trading sex for necessities, and traveling men (*e.g.,* long distance drug drivers, seamen, or military recruits). These men have provided a bridge to lower-risk women, leading to the spread of HIV infection into the sexually active general population. As is the case for many STDs, a common theme throughout the world is that women tend to become infected at a younger age than men, with numbers of new infections peaking between ages 15 and 25 years for women and between 25 and 35 years for men [8,35]. In most nations, HIV infection rates are higher in larger urban areas, and rapid urbanization and other population movements are believed to have contributed to the spread of HIV to rural areas [36,37].

IV. Distribution of HIV/AIDS among Women

By 1997 in the United States, just under 100,000 women with AIDS had been reported to CDC, of whom 13% were diagnosed during 1997 [5]. The 13,000 cases in women in 1997 accounted for 22% of all adult AIDS cases diagnosed that year. As of 1995, an estimated 107,000 to 160,000 women of reproductive age were living with HIV infection, of whom one-third to one-half had not yet developed an AIDS-defining opportunistic illness [13,38]. By 1995, HIV infection became the third leading cause of death among women 25 to 44 years of age, and the leading cause of death among black women, accounting for 22% of deaths in that age group [4]. The use of antiretroviral therapies has slowed the progression to AIDS and has led to a decrease in new AIDS cases in all risk groups (Fig. 26.1). However, the decreases in AIDS incidence between 1996 and 1997 were greater for men than for women, reflecting greater decreases among men who report sex with men and drug injectors than among those men infected heterosexually [5].

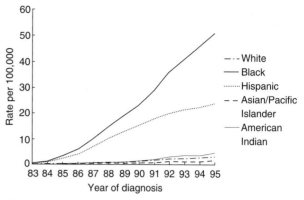

Fig. 26.3 Acquired immunodeficiency syndrome (AIDS) incidence in women by race/ethnicity, 1983–1995. (*Data were adjusted for by reporting delay, unreported risk, and the 1993 expansion in the AIDS surveillance case definition). (Data source: P.M. Wortley, M.D., M.P.H., based on national AIDS surveillance, CDC.)

Many HIV-infected U.S. women are young. In 1997, 80% of the women with AIDS were 13–44 years old; half were less than 35 years old, and 7% were 13–24 years old. However, of women reported with HIV in 1997, 92% were 13–44 years old; 70% were less than 35 years old, and 24% were 13–24 years old [5]. Given the long incubation time of HIV, a great number of new infections probably occur in women still in their teens or early twenties. As is true for other parts of the world, women in the United States tend to become infected at earlier ages than men. For example, HIV incidence among women aged 16–21 years in the Job Corps, a national training program for disadvantaged and out-of-school youth, has typically been 40% higher than for young men of the same age (2.8 per 1000 vs 2.3 per 1000, respectively) [39]. Among the 1997 AIDS cases, the percentage of cases attributed to heterosexual transmission for younger women was twice that for older women [5], and the greatest increases in AIDS incidence rates have been among very young women who were infected heterosexually (Fig. 26.2) [13].

Since the first cases in women were diagnosed, AIDS in the United States has been concentrated among black and Hispanic women. The proportion of women with AIDS who were black or Hispanic was 75% in 1987 and 76% in 1997 [5]. However, race/ethnicity-specific incidence rates better illustrate the gap between groups (Fig. 26.3). In 1997, AIDS cases per 100,000 women were much higher among black (58.8) and Hispanic (21.5) women than among American Indian (4.7), white (3.0), or Asian/Pacific Islander (1.5) women [5]. HIV seroprevalence from a national, population-based sample of women who gave birth to infants during 1989–1994 was consistent with AIDS incidence data. State-specific HIV seroprevalence rates among black women were 8 to 30 times higher than those among white women [14]. Rates among Hispanic women fell between those for black and white women in the northeastern and midwestern states but were similar to those for white women in California and Texas, reflecting differences in risk according to Hispanic origin (Puerto Ricans were at highest risk) [40]. Differences in AIDS incidence by race and ethnicity are probably multifactorial.

High AIDS rates among minority women likely reflect the concentration of minority women who live in communities where injection drug use or crack cocaine use is high, male sex partners are more likely to be HIV infected, and where outbreaks of other STDs (which may enhance HIV transmission) have occurred recently [40–42]. Although such intervening factors have been repeatedly specified in epidemiologic studies as risks for HIV and AIDS, the roots of the disparity in rates of AIDS among minority populations are probably related to more intangible factors, such as distrust of the institutions of medicine and public health, poverty or economic stress, poor race relations, absence of community cohesion, or lack of meaningful jobs [43–45].

V. Factors Associated with HIV Transmission

HIV is an infectious disease almost invariably requiring intimate contact with an infected person, either a sex partner or a needle-sharing partner. Consequently, the behaviors of a woman's partners may be more important than her own behaviors in determining her HIV risk. A partner's risks are particularly important when a woman cannot control the conditions under which she has sex. In addition, HIV prevalence in her community is critical in determining her risk for infection. In communities where HIV prevalence is very low, no behavioral or biologic risk factor is likely to substantially increase an individual's HIV risk, because exposure is unlikely [46]. In contrast, in urban settings in sub-Saharan Africa, antenatal sentinel surveillance suggests that one-third to one-half of sexually active women of reproductive age are HIV-infected [47]. In such settings, where the likelihood of exposure to HIV is high, even modest risk factors can enhance susceptibility and virtually everyone must consider ways to minimize risks. Behavioral, biologic, social, and environmental factors are all believed to enhance women's susceptibility to HIV infection. These factors are highly interrelated and difficult to discuss independently.

A. *Probability of Sexual Transmission*

Although parenteral HIV transmission is highly effective (the likelihood of infection after transfusion with infected blood is estimated at 90%) [48], HIV is less effectively transmitted sexually. Estimates of efficacy of sexual transmission range from as high as 3% among "efficient" transmitters in a cohort of homosexual men to as low as 1 in 2000 sex acts with an HIV-infected partner based on transmission in cohort studies of heterosexual HIV-discordant couples [49–52]. HIV infectivity seems to differ widely for different individuals. Some infected persons are much more efficient transmitters than others, and some individuals appear to be more susceptible to HIV than others. In addition, the amount of virus present (see Section V.B) is likely a critical determinant in transmission.

Other factors being equal, unprotected anal receptive sex with an HIV-infected partner seems to be the most efficient way of acquiring infection sexually. Women who had anal sex with their HIV-infected male partners were two to four times more likely to be infected than women who had only vaginal sex [46]. For vaginal sex, cohort studies of HIV-discordant couples were consistent in suggesting that male-to-female HIV transmission

Table 26.1

Per Act Probability of HIV Transmission and Relative Risk of HIV Transmission from HIV-Infected Partner to Uninfected Partner, by Type of Sex Act

Type of sex for HIV-uninfected partner	Probability of transmission	Per act RR of transmission
Oral Insertive[a]	unknown	1 (referent)
Oral Receptive[a]	unknown	2
Vaginal Insertive[b]	0.0003–0.0009	10
Vaginal Receptive[c]	0.0005–0.0015	20
Anal Insertive[a]	unknown	15
Anal Receptive[d]	0.008–0.032	100

Source: J. M. Maher and B. Varghese, personal communication.
[a]Relative risk (RR) based on best guess after reviewing literature.
[b][49, 50, 54].
[c][50].
[d][50, 51, 53, 59].

is two to three times more efficient than female-to-male transmission [49,50,52–56], as is true of other infectious agents that are concentrated in semen. However, these studies may have underestimated transmission risk. Factors related to the couple (*e.g.,* length of relationship, knowledge of HIV status) and to the length and type of study may have led to infrequent HIV transmission in these discordant partners, and some statistical analyses may also have underestimated risk [56]. One cohort study suggested that female-to-male transmission can be fairly common in stable discordant couples (as common as 5 in 100 sex acts) [57].

Oral sex seems to be less risky than anal or vaginal sex, but few people report only oral sex. There are reports of HIV infection after fellatio with an HIV-infected male partner and after cunnilingus with an HIV-infected female partner [46,56,58,59]. From published transmission studies, it is possible to estimate the per act risk of transmission and from this to develop a risk continuum for types of sex acts (Table 26.1). These estimates suggest that for women, oral receptive sex carries the lowest risk. Vaginal receptive sex, with a risk of transmission of 5–15 per 1000 sex acts, is about 10 times as risky as oral receptive sex [49,50,54,58,59]. Anal receptive sex is about five times as risky as vaginal receptive sex (or 50 times as risky as oral receptive sex). However, because total risk involves both likelihood of transmission related to type of exposure *and* number of exposures, multiple episodes of oral sex with an HIV-infected partner could afford risk as great as or greater than that from a single episode of unprotected anal or vaginal sex.

B. *Viral Factors*

Viral as well as host characteristics are believed to affect one's susceptibility to HIV and the transmissibility of HIV. Viral type seems to be important in transmission [48,60]. HIV type-1 (HIV-1) is responsible for most AIDS cases worldwide and is more efficiently transmitted than HIV type-2 (HIV-2), which is geographically limited mostly to West Africa [7]. In addition, some researchers postulate that certain HIV subtypes may be more transmissible than others, and some have preferable routes of entry [61]. The amount of virus present is probably a key determinant of whether HIV infectiousness leads to heterosexual and perinatal transmission [62,63]. Several studies reported maternal viral load at delivery to be a strong predictor of maternal-to-child transmission [63]. HIV-infected persons in advanced states of illness, when concentration of virus is likely to be high, are more likely to transmit HIV to sex partners than are persons who are asymptomatic [49,50,62]. Although definitive evidence is lacking, modelers have postulated that HIV infectivity may be high during early ("primary") infection, before the development of antibodies to HIV [64,65], thus sexual contact with a recently infected partner may greatly increase the risk for HIV transmission.

Given that the semen of HIV-infected men contains abundant HIV-infected lymphocytes and macrophages as well as cell-free virus, an issue of import for prevention strategies is whether HIV enters the female genital tract inside a cell or as free virus [66,67]. The initial step in HIV infection may be the attachment of the virus to a target cell through CD4, the receptor for HIV. Vaginal epithelial cells lack CD4 cells, suggesting that women become infected from free virus primarily when they have lesions or inflammation of their vaginal epithelia, which attract T cells. This hypothesis is supported by findings of enhanced HIV transmission in people infected with STDs [68]. Others postulate that HIV inside cells is more transmissible to the female genital tract epithelial cells and that HIV is able to infect CD4-negative cells [69]. A proposed mechanism is that cell-associated virus is responsible for transmission across intact epithelial surfaces but that cell-free virus can infect mucosa when present in high concentrations or when epithelial surfaces are disrupted, for example with ulcerative STDs [67–70]. The exact point of virus entry into the female genital tract is of interest in predicting effectiveness of some barrier methods [66,67]. Generally, the cervix has been considered the probable primary point of entry. However, it has been demonstrated that hysterectomized macaques can be infected with Simian immunodeficiency virus (SIV), indicating that the cervix is not necessary for SIV transmission [70] and that disrupted vaginal epithelial cells allow HIV entry [68].

C. Host Factors

Several other biologic factors have been proposed as enhancing women's susceptibility to HIV infection, including menstrual hygiene, intercourse during menses, traumatic sex (reports of vaginal bleeding during intercourse), and use of hormonal contraceptives [49,53,55,71,72]. However, a major difficulty in assessing the contribution of any single hypothesized cofactor is that few of the studies are large enough to assess the potential effects of other relevant variables (e.g., number of sex contacts, use of condoms, presence of asymptomatic STDs, or infectiousness of an HIV-infected partner). Results from cross-sectional and longitudinal studies have suggested that hormonal contraceptives (whether oral or implanted) may enhance the risk for the acquisition and transmission of HIV. Because steroid hormones are used widely by sexually active women worldwide, this is particularly concerning but is difficult to untangle. Possible mechanisms by which steroid hormones enhance HIV transmis-

sion are increased viral shedding in HIV-infected women or increased cervical ectopy (the exposed vascular endocervical epithelium and cervical transitional zone that is also found among many women during puberty and pregnancy) in uninfected women. Most reports have come from developing world settings, and not all studies in these settings have demonstrated the association. In contrast, a long-term European cohort study evaluating HIV transmission found a 60% reduction in risk for HIV infection associated with oral contraceptive use [73]. Douching and "dry sex" have also been reported to enhance HIV transmission, possibly through trauma which allows the direct introduction of HIV or by some other mechanism that enhances susceptibility to HIV, such as increasing vaginal pH [74]. Cell-free HIV and cell-associated HIV-infected lymphocytes are known to be rapidly inactivated by low pH (the usual situation in a healthy vagina) [74]. Blood, semen, and some contraceptives also increase vaginal pH. Another postulated link with HIV is female circumcision, possibly by enhancing susceptibility to HIV or through contaminated instruments [75]. Several prospective and many cross-sectional studies have suggested a modest-to-moderate association between lack of circumcision in men and HIV infection [76–78].

There is fairly strong biologic and epidemiologic evidence that STDs can enhance HIV transmission and acquisition [71, 79–87]. Some studies of HIV-infected persons have found increased viral shedding in the presence of ulcerative and non-ulcerative STDs, and other studies have noted decreased HIV shedding with STD treatment [84–86]. In prospective studies, associations have been observed between HIV acquisition and prior infection with specific STDs, particularly ulcerative but also nonulcerative infections [71,79–83]. The most compelling evidence that HIV acquisition is enhanced by other STDs is based on the results of a randomized community trial conducted in the Mwanza region of Tanzania. In communities that received enhanced syndromic management of symptomatic STD cases, HIV incidence decreased by 40% compared with communities that received usual STD management [87]. However, STD control is unlikely to prevent HIV in every setting, as was demonstrated by the results of a randomized community trial in Uganda's Rakai District [88]. That study found no HIV prevention benefit to mass STD treatment compared with a placebo intervention delivered in a similar manner. Why STD treatment prevented new HIV infections in Mwanza but not in Rakai has been the subject of considerable debate. However, the two interventions differed in target populations, intervention approach, and duration of delivery, and it is possible that characteristics of the two rural settings themselves may have differed.

Sexual behaviors that have been associated with HIV infection in women (and in men) include the number of sex partners and the number of unprotected sex acts (i.e., when a barrier, such as the male latex condom, is not used). Used consistently and correctly, the male latex condom prohibits the transmission of bacterial and viral infections [52,89,90]. However, several studies have suggested that condoms are often used inconsistently and sometimes incorrectly, which may lead to their breaking or slipping off [91]. A few studies have even suggested that condom use may not be a very meaningful marker for "safe" behavior, because people probably use condoms selectively (i.e., with riskier partners but not with partners considered safe)

[92,93]. Other mechanical barriers against HIV and STDs include the diaphragm and the cervical cap (both of which cover only the cervix), and the female condom. The diaphragm protects against gonorrhea and chlamydia, but its utility against viral agents including HIV is unproven [94]. The female condom, although not well studied, has the advantage of covering the vaginal epithelium as well as the cervix and theoretically should be as effective against HIV and other STDs as the male condom. Chemical barriers, such as vaginal microbicides, may also be effective against HIV [66,95]. Two prospective studies of the microbicidal agent nonoxynol-9 found it modestly (up to 25%) protective against gonorrhea and chlamydia [96–98]. However, as yet no study has found it effective against HIV.

Among drug injectors, sharing used syringes is the usual means of transmission of HIV and other blood-borne viruses. The equipment and water used to prepare drugs for injection can become contaminated with the blood remaining in used syringes. For injecting women, risky sex practices may be as closely related to HIV infection as unsafe drug behaviors [99,100]. The use of other illicit drugs, particularly smokable freebase cocaine (crack) and other forms of cocaine, has been linked to heterosexual and IDU-related HIV transmission [42,101–103]. Crack was introduced into the United States in 1985 and because of its low cost and high addictive potential, it gained widespread and rapid popularity in inner cities. Among crack addicts, exchange of drugs for sex with multiple partners led to intersecting epidemics of syphilis and HIV in the early 1990s, each infection enhancing the transmission of the other [42]. The syphilis and crack epidemics were probably influential in fueling the heterosexual HIV epidemic in some urban U.S. cities [42]. Crack use spread quickly to regions outside of large cities, contributing to ongoing syphilis outbreaks and to high rates of heterosexually transmitted HIV in some rural areas [102]. Cocaine has also been linked to unsafe drug use practices, such as more injections, sharing of injection equipment, frequenting of shooting galleries, and "booting" (washing out any drug left in the syringe with the user's own blood while the needle and syringe are still in the vein) [17]. Other substances (*e.g.,* alcohol and marijuana) have also been linked to risky sex and needle sharing practices, such as sex with anonymous partners, inconsistent or incorrect condom use, and sharing drug equipment [37].

D. Social and Environmental Factors

Aside from individual characteristics, social and environmental factors may impact HIV and other diseases. Central among these factors is the social support for women to control their sexual health or the conditions under which they have sex. Many women are unable to negotiate whether or when they have sex or whether condoms or other barriers can be used [37,104,105]. In some cultures, sex is linked to male rights and fertility, fostering a norm of silence about sexuality issues. Breaking that silence involves risks because it challenges accepted social patterns [106]. Related to this is a society's willingness to discuss sexual practices openly or to allow formal education about sex. If young women are encouraged to remain uninformed about sexual practices and their sexual health, they will be unable to recognize STD symptoms or obtain effective treatment [75]. The practice of older men marrying young women may increase women's risk of HIV in areas where HIV prevalence is high and older men are more likely to be infected [37,75]. In some settings it is customary or acceptable for men to have multiple sex partners or to have sex with commercial sex workers before marrying or during marriage. All of these practices can increase a woman's risk for HIV or other STDs even if she herself is behaviorally low-risk. While many of these factors pertain to traditional settings, some, including support for women to negotiate condom use with their male partners, apply across the spectrum of sexually active women, both in industrialized and poor nations, to young and old, and to sex workers and married women alike.

Environmental factors may affect a woman's risk for sexually transmitted infections, including HIV [45,75,106]. Economic and political factors that determine the use of health care funding dictate the number of health care providers per 100,000 population, thus the availability of physicians trained in HIV-related illness. Interstate routes can serve as conduits for illegal drugs and may also encourage commercial sex work, alter sexual mixing patterns, or both [36,37]. Economic and political disruptions related to wars, national disasters, or famine have resulted in sudden poverty, population migrations, and abrupt increases in the number of sex workers [107]. In many nations, laws prevent women from owning property. These laws, coupled with social factors favoring low education for women, assure that women are economically dependent on men, further undermining their power to make decisions about their lives, health, and sexual practices [8]. Even in wealthy nations, environmental factors can affect disease transmission in important ways. In the United States, environmental factors linked to high prevalence of syphilis in some communities include availability of health services, distribution of power and influence in a community, employment opportunities and median wage for unskilled jobs, intra-agency coordination, intercommunity dynamics, and availability of outreach services [45].

VI. Natural History, Clinical Manifestations, and Treatment of HIV Infection and AIDS

Although early observations suggested that HIV-infected women may progress to AIDS and death faster than men, these could not take into account other relevant factors, such as entry to treatment and transmission mode [108,109]. Prospective studies indicate that, after other factors are accounted for, the overall clinical course of HIV infection in women and men is probably quite similar [110–113]. Several factors, however, may result in HIV-infected women not receiving appropriate care and early preventive interventions. Compared with HIV-infected men, infected women are more likely to be minority, to have low incomes, and to live in communities with limited services, long waits, and poor continuity of care [114,115]. Health care providers may have a lower threshold of suspicion about HIV-related disease in their women patients [115]. This may account for past findings that, compared with men, women were diagnosed later in the course of HIV infection and more often because of illness [11]. Women are often the primary care givers for other family members; they must balance their own need for health care against other family needs [114]. Women who inject drugs or who have an HIV-infected or a high risk sex partner

Table 26.2
Contraception for the HIV-Infected Woman by Method, Potential Benefits, and Potential Drawbacks

Method	Potential benefits	Potential drawbacks
Surgical sterilization	• Highly effective • Low maintenance	• Permanent—only useful for women who desire no more children • No STD protection • No HIV protection for sex partner
Male, female condom	• Effective if used consistently and correctly • STD protection • HIV protection for sex partner	• Male condom requires partner cooperation; partner cooperation helpful with female condom
Steroid hormones (*Depo-Provera, Norplant, Oral Contraceptives*)	• Highly effective • Low maintenance (particularly *Norplant* and *Depo-Provera*)	• Unclear interaction of steroids and immune function • Possible increased shedding of virus from cervix • No protection from STDs • No HIV protection for sex partner • (Oral contraceptives) Interaction with some antibiotics, antiretrovirals, other drugs
Diaphragm, cap, spermicides	• Some STD protection	• Requires good technique • Vulvovaginal irritation increases vulnerability to urinary track infections for some users
IUD	• Highly effective • Low maintenance	• Contraindicated due to risk of pelvic inflammatory disease • No STD protection • No HIV protection for sex partner • Increased days of bleeding, possible anemia

Note. Adapted from Felicia Guest, MPH, CHES, Contraceptive Technology, 1998.

commonly face severe disincentives for seeking treatment, such as state laws or local statutes or ordinances that permit authorities to remove dependent children from women who seek substance abuse treatment, prosecute pregnant women who admit illicit drug use, or deny pregnant women entry to drug treatment programs [114].

After diagnosis, HIV-infected women seem to be treated less aggressively than their male counterparts. Two studies of health care found that, after other differences were controlled, HIV-infected women were less likely than men to be treated with zidovudine [116,117]. Although gender differences in antiretroviral treatment have equalized with time, urban, university-based HIV clinics have reported that women, particularly minority women, referred to the clinics were less likely than men to have received PCP prophylaxis [110,118,119]. HIV-infected women who are receiving care tend to receive fewer services than men [120]. Historically, access to clinical trials has been much more limited for women [109], which was probably influenced by demographic and geographic factors as well as by concerns about pregnancy and contraception.

A. Reproductive Health and Pregnancy

Sexually active HIV-infected women can transmit HIV to their sex partners and they need to be counseled about ways to avoid this (*e.g.,* nonpenetrative sex, oral sex, consistent use of condoms). Infected women should also be counseled about the advantages and disadvantages of different types of contraception (Table 26.2). Nonetheless, many women first learn they are HIV infected when they are tested during prenatal care [11]. Furthermore, the pregnant women at greatest risk may never receive prenatal care or HIV testing [121–123]. These findings, coupled with the availability of safe and effective therapies, have led many clinicians to push toward routinely encouraging all pregnant women to accept HIV counseling and screening (current U.S. practice) [124,125]. Once women know they are HIV infected, their fertility rates are lower than demographically similar, uninfected women [126]. But once pregnant, HIV-infected women are no more likely to choose to terminate the pregnancy than are uninfected women [127]. Pregnancy poses serious concerns, the foremost of which is the risk of HIV transmission to a fetus or infant, even with antiretroviral therapy. In addition, it is clear that substantial transmission occurs during breastfeeding [128]. The availability and the accessibility of safe water supplies for formula feeding is critical. Even if shortcourse antiretroviral therapy were available in many poor countries, any reduced neonatal transmission gained by these drugs could be almost entirely offset by transmission during breastfeeding.

Aside from preventing perinatal transmission, the effect of pregnancy on the course of HIV disease in the woman must also be considered. Pregnancy itself is associated with some degree of immune compromise (manifested by lower CD4 counts), and early case studies suggested that pregnancy might accelerate the progression of HIV-related disease. However, larger longitudinal cohorts of HIV-infected women have found no evidence of that and suggest, rather, that the effect of pregnancy on the course of HIV infection may be minor for many women [129,130].

B. Clinical Manifestations of HIV

The prevalence of most AIDS-defining conditions is similar for men and for women after differences in race or ethnicity and mode of transmission are accounted for. However, the early clinical manifestations of HIV may differ significantly [131]. For women, gynecologic symptoms such as vaginal candidiasis, pelvic inflammatory disease, cervical dysplasia, menstrual abnormalities, or genital herpes or warts unresponsive to standard therapy are often the first sign of HIV infection. Vaginal candidiasis ('yeast infection'), a common gynecologic disorder in women with normal immune systems, is even more common in immune compromised women. Compared with uninfected women, HIV-infected women are more likely to be symptomatic and have frequent, longer, and more severe episodes [132–134]. Increasing severity of vaginal candidiasis has been linked to declining immune competence, and recurrent vaginal candidiasis often precedes oral and esophageal candidiasis. Initial treatment with topical antifungal medications is suggested, as cure rates for immune competent HIV-infected women should be similar to that of uninfected women [135]. In severe HIV disease, treatment failure and recurrence are not unusual, and some women require oral imidazole treatment and long-term suppression [136–137].

Sexually active HIV-infected women are also at risk for other STDs, some of which may be particularly aggressive in immunocompromised patients. For example, pelvic inflammatory disease (PID), bacterial infection of the upper reproductive tract, has been linked to HIV infection and seems to follow a more aggressive course in HIV-infected women, which is manifested by a tendency toward more frequent tuboovarian abscesses and need for surgical intervention [138–140]. PID is often caused by *Neisseria gonorrhea* or *Chlamydia trachomatis,* but it may also be caused by nonsexually transmitted bacteria that are part of normal vaginal flora, including anaerobes and Gram negative bacteria. Although PID in HIV-infected and uninfected women is likely to involve similar microorganisms [139,140], aggressive management with intravenous antibiotics is recommended because the course is more severe in HIV-infected women [135]. Screening for bacterial STDs, including gonorrhea and chlamydia, is not a routine part of HIV management, but when screening has been done, coinfection was not uncommon, even among women in long-term care [141].

Since 1993, cervical cancer in the presence of HIV-infection has been considered an AIDS-defining disease in the United States. Certain subtypes of human papilloma virus, a sexually transmitted infection, are implicated as the cause of most cervical cancers [142]. According to several studies, cervical intraepithelial lesions, the precursors of cervical cancer, are more common and more aggressive among HIV-infected women than among uninfected women [143–145]. While few well-controlled studies exist, current evidence suggests that compared with uninfected women, HIV-infected women—particularly those with advanced disease—are at high risk for cervical cancer. Guidelines recommend biannual Pap smear screening during the first year after diagnosis of HIV infection and, if the results are normal, annual Pap smears thereafter. Colposcopy is advised if the Pap smear shows any cellular atypia, including persistent inflammation [136,137,142].

Menstrual irregularities and abnormalities, including amenorrhea, are commonly reported by HIV-infected women, especially in late-stage disease [146]. With the exception of amenorrhea associated with weight loss and wasting, the cause of menstrual abnormalities in HIV infection is poorly understood. Stress, wasting, and other systemic illnesses can cause dysfunction in the hypothalamic axis and lead to amenorrhea. Some medications (*e.g.,* megestrol and opiates) can interfere with women's menstrual cycles. Platelet disorders associated with HIV, such as idiopathic thrombocytopenic purpura, may cause heavier menstrual cycles. Several clinical manuals suggest that work-up and management of menstrual problems should be pursued because many HIV-infected women react to menstrual disorders with discouragement and depression.

Tuberculosis (TB) affects men as well as women and deserves attention because, worldwide, it is the most important opportunistic infection associated with HIV. The devastating effect of high HIV prevalence on TB prevalence in poor nations, particularly those in Africa and Asia, has been well described [147,148]. In fact, these two continents together hold a majority of the world's population and bear most of the world's HIV infection and TB burden. Although TB is considered a curable disease, the treatment of HIV-infected persons is not always successful. Resistant strains have made TB control (even in some places with considerable resources, such as the United States) extremely difficult. The HIV epidemic poses a serious threat to TB control efforts in many parts of the world. For example, the HIV-related mortality due to TB in Chiang Rai Province in Thailand is estimated at 70%, compared with 10% among HIV-negative persons [148]. Because TB occurs earlier in the course of HIV infection than other opportunistic infections [146], effective treatment may greatly prolong productive life for coinfected persons. Isoniazid is cheap and effective for the preventive treatment of TB in HIV-infected people and may be the most effective HIV therapy in settings where antiretroviral agents or even prophylaxis for other opportunistic infections is unavailable [149].

VII. Prevention Issues—Public Health Impact

Since the onset of AIDS, one of the most remarkable success stories has been the efficacy of antiretroviral agents in reducing maternal-to-child transmission and HIV-related mortality and the progression of HIV to AIDS in adults (Fig. 26.1). The utility of antiretroviral therapy to prevent new infections in adults is less certain. However, a case-control study of zidovudine therapy for people occupationally exposed to HIV suggested that early antiretroviral therapy reduced new infections by 60% [150]. On the basis of this finding, postexposure prophylaxis after unsafe sex exposures has been promoted, although its effectiveness is unclear and difficult to assess [151]. It has also been suggested that antiretroviral therapies may prevent new cases of HIV infection through reducing viral shedding from infected to uninfected persons. Although an exciting premise, the extent to which these therapies will play a role in prevention is as yet unclear. Unfortunately, in the world's poorer nations, antiretroviral agents are still too costly to use in most settings, even for the limited regimens recommended to reduce maternal-to-child HIV transmission [30]. Studies evaluating shorter, less costly regimens have been undertaken and several have shown substantial reductions in maternal-to-child transmission [152]. However, even these are not affordable in some nations with high HIV prevalence.

Developing an effective, safe, and affordable vaccine is the ultimate goal of AIDS prevention research, but the challenge has proved formidable. Vaccine development will require a sustained scientific effort for the next several years. Nonetheless, several effective or possibly effective interventions are applicable to many women and should be applied whenever possible.

A. Community-based Programs for Injection Drug Users

Randomized controlled studies evaluating the impact of various community-based treatment programs for injection drug users have usually not been conducted. However, numerous observational studies have suggested that these strategies may prevent new HIV infections [19,20,153]. Community-based outreach programs have been associated with reductions in risky drug use and sex behaviors and with increased drug treatment referrals. Needle exchange programs and changes in prescription and paraphernalia laws have been significantly associated with reduced injection risk behaviors, increased drug treatment referrals, and reduced HIV, hepatitis B, and hepatitis C incidence [19,153]. Despite concerns that such programs might encourage drug use, this has not been observed to be the case.

B. Behavior Change Counseling

Growing evidence indicates that personalized risk reduction counseling can prevent sexually transmitted infections. A multicenter randomized trial done in the United States among HIV-negative STD clinic patients found that more men and women assigned to brief, interactive counseling interventions used condoms and significantly fewer developed new STDs over a year compared with patients assigned to information typical of current practice [154]. A randomized trial evaluating a similar HIV prevention counseling strategy was conducted in Kenya, Tanzania, and Trinidad [155]. Early results suggested that the personalized counseling approach was acceptable and feasible in these settings and that condom use by individuals and couples assigned to counseling increased significantly compared with those assigned to information alone. The effects of counseling on disease incidence have not been reported in that study. A randomized trial conducted in high risk women in San Antonio, Texas found a 40% reduction in STDs over a year for women who had received a four-session interactive counseling intervention aimed at reducing sexual risk behaviors compared with women who received typical counseling [156]. Also, a multisite randomized trial comparing a seven-session risk reduction counseling intervention to a one-hour AIDS education session found that the counseling intervention led to significant reductions in sexual risk behavior among low-income men and women attending public clinics. Although a subset analysis found significant reductions in new gonorrhea infections only in men, this was based on review of the medical chart and may have missed new infections in women, as women are more likely to be asymptomatic [157].

C. HIV Testing

Knowledge of one's serostatus is important. Uninfected women can avoid behaviors that place them at risk of acquiring HIV from their HIV-infected partners. Infected women can protect their uninfected sex and needle-sharing partners and, if they are pregnant, their babies. A large proportion of HIV-infected men and women do not realize they are infected until AIDS is diagnosed [158]. In the United States, for men, and particularly for women, with HIV, the most common reason for delaying testing was lack of recognition of their risks [159]. Although studies have been inconclusive about whether simply knowing one's serostatus leads to change in risk behavior, this is clearly a potentially powerful prevention strategy. The availability of new rapid tests which allow people to receive HIV test results within a few hours has the potential to greatly increase the number and proportion of people who learn their test result.

D. Barrier Methods

Well-focused and explicit condom promotion campaigns seem to increase condom use [160–162]. This result is perhaps best evidenced by the Thai national condom campaign aimed at commercial sex workers and their clients which is believed to have slowed the HIV epidemic in Thailand [162]. A potential problem for many women is discomfort in or inability to negotiate condom use with their male sex partners. In one study a substantial proportion of women (i.e., more than 40% of Hispanics and Haitians, 15% of whites, and 20% of African-Americans) reported they would have sex without a condom even with a partner who was HIV-infected [105]. Interventions that also focus on changing attitudes and skills related to communication about sex, partner negotiation, and use of condoms may be necessary to increase safer sex practices for women [104]. However, for those who are able to talk with their partners about condoms, an effective strategy may be simple, very explicit demonstrations of the correct use of condoms. This approach has reduced STDs in men [163], but more research is needed to determine its effectiveness for women.

Given concerns that underlying inequalities of power between the sexes may limit the ability of many women to protect themselves from HIV, one of the more intriguing research efforts is one directed toward promoting and developing HIV prevention technology that women can control themselves. The recently developed female condom, a lubricated polyurethane sheath that is inserted into the vagina, if used correctly should be as effective in preventing bacteria and viruses, including HIV, as the male latex condom [164]. The bulkiness of the condom has led some to worry that it will not be used; however, studies in various settings have suggested that when women are given enough encouragement and time to try the condom, they accept it fairly well. Serious problems with the device are that it requires at least passive compliance of the male partner and it is currently quite expensive (90¢ per condom). Although the condom theoretically is reusable, the manufacturer does not recommend reuse.

Vaginal microbicides have been an area of intensive but, thus far, unrewarding research. One microbicide, nonoxynol-9 (N-9), has been found to have a modest protective effect against some bacterial STDs [66,115], but so far no evidence indicates that N-9 or any other chemical barriers protect against HIV. A randomized controlled trial of N-9 in commercial sex workers in Cameroon compared a vaginal contraceptive film containing N-9 and a placebo film [165]. The N-9 film provided no additional protection against HIV and other STDs when it was provided as part of an overall HIV/STD prevention program that included behavioral change counseling, promotion and distribution of male latex condoms, and monthly screening and treatment for STDs. Condom use increased substantially in both groups of sex workers. Although this is good news about women's ability to negotiate condom use even in difficult settings, the trial did not support the use of microbicides for HIV prevention. Until the association between microbicide use and HIV risk is better understood, the use of microbicides alone to prevent HIV is not recommended. Furthermore, the frequent use of certain microbicides can cause vulvovaginal epithelial disruption, which might increase susceptibility to HIV [97,98]. Experts advise caution about frequent use of microbicides when HIV exposure is a concern [98]. A concern that has been raised about female-initiated protective barriers, whether mechanical or chemical, is that they assume that women with HIV risks understand their risks but may not be empowered to protect themselves [104,105]. It is also possible that women with real HIV risks may not perceive themselves to be at risk and, therefore, may not consider using any protective method, even if they were able.

E. Partner Notification and Referral

Partner notification, the practice of informing sex or needle-sharing partners that they have been exposed to HIV, is recommended in many settings, whether done by HIV-infected persons themselves or by health care providers. Part of the rationale for the practice is that it gives uninfected sex partners an opportunity to reduce their HIV exposure. Notifying women partners may be of particular benefit because many women who are heterosexually exposed to HIV (or other STDs) may not realize their exposure. Although partner notification has been found to be effective in preventing syphilis (which is curable), its efficacy in preventing new HIV infection is unknown [166]. Anecdotal reports have linked partner notification to partner violence against women [167,168], but controlled studies weighing the potential risks of partner notification against the possible benefits (particularly for high risk women) are needed to assess this relationship. In inner-city STD clinics in Newark and Miami, 13% of women reported partner violence during the past 12 months [169], but this history did not dissuade women from accepting an HIV test. Nonetheless, health care providers must be aware of the high rates of violence against many women patients and be ready to refer women who are in risky situations.

F. STD Control

The evidence is fairly persuasive that STD control can substantially reduce HIV incidence, at least in settings with high STD prevalence and where the persons at risk for HIV are the persons with STDs (see Section V. C.). The type of control program delivered and the populations targeted for STD control

are key in determining the efficacy of this strategy for HIV prevention [87,88]. STD control is likely less important for HIV prevention in communities where STD prevalence is low or the population links between STD and HIV are not strong. This strategy may be particularly important for HIV prevention in women living in societies that do not support women controlling the conditions under which they have sex, because it does not assume women choose less risky sex partners or protect themselves through barrier methods. STD control programs also have the benefit of reducing other adverse effects of sexually transmitted infections such as ectopic pregnancy, PID, and some congenital infections.

G. Structural Changes

Perhaps the interventions with the greatest potential for HIV prevention are those aimed at changing the social and environmental processes that facilitate HIV transmission or impede high risk individuals from taking advantage of prevention opportunities. These root determinants have not been particularly well studied in the context of AIDS, nor have they been well addressed in our public health responses to AIDS [37,107]. However, compelling precedents exist, for example, in the field of reproductive health, in which programs have historically focused on providing women with information and education about contraception and ensuring access to safe and effective contraceptive services. The inadequacy of this approach was recognized and addressed at the 1994 United Nations Conference on Population and Development, held in Cairo. The conference redefined its goal as ensuring that women are able to make free and informed decisions about their reproduction and are able to carry out their reproductive decisions. Recognizing the considerable societal, cultural, and environmental barriers that curtail women's ability to reach this goal, and that only as women's human rights are realized can the benefits of reproductive information and access to clinical services be fully realized, the conference described these impediments in the language of human rights. Population programs were called upon to help ensure that women's human rights, including equal rights in marriage, divorce, association, and political participation, were promoted. Modern human rights have been proposed as a backdrop to HIV/AIDS prevention as well, and some experts in this field have pointed out that public health programs should identify the human rights violations that enhance vulnerability to HIV and other sexually transmitted infections [107]. Other legal and medical experts on AIDS have written about the importance of governments' pursuit of noncoercive policies to prevent men and women from becoming infected, noting that if safer sex and safer drug use policies are to be both fair and effective, they must take into account the relative powerlessness of women in sexual and social contexts [170]. Additional research may be useful, but for many issues much is already known. Creative policies and strategies need to be tried out [36,37].

VIII. Summary and Conclusions/Future Directions

In 1981, the idea that AIDS would become an epidemic in women and children as well as in men would have been met with skepticism. In fact, throughout the world, regardless of a nation's resources, the mode of HIV acquisition is shifting to heterosexual transmission and the distribution of AIDS is moving to underserved populations. Worldwide, 40% of AIDS cases are in women, and it is expected that AIDS cases in women will soon exceed those in men [7,171]. For industrialized nations, the advent of effective antiretroviral therapies has led to greatly reduced progression of HIV to AIDS, and for the first time since the emergence of AIDS, deaths, both in men and in women, are declining. Despite these remarkable advances, large disparities in access to therapies and prevention services exist, even in wealthy nations. Of most concern, even as AIDS diagnoses and deaths decline in industrialized nations such as the United States, is that new infections, especially via heterosexual transmission, continue to occur. This suggests that our prevention interventions have seriously lagged behind our medical advances.

At this time of extremes in susceptibility to infection and disease and availability of treatment, the world faces serious challenges. For wealthy nations, today's challenges lie in encouraging HIV testing among persons at risk for HIV, ensuring appropriate and early antiretroviral therapy for all infected persons, and continuing to apply effective primary prevention strategies among people at greatest risk for HIV. Particular challenges lie in educating women about how their risk for HIV depends in part on their partners' risks, and about the effectiveness of various prevention strategies, in ensuring women access to health care, and in encouraging them to seek it. For most of the world, the cost of antiretroviral regimens is, and will remain, out of reach. The development of a safe, effective, and affordable vaccine must be a high priority for all nations of the world. Even after an effective vaccine exists, primary prevention efforts, such as information and education, condom promotion, and STD control programs, must continue to be advanced. Developing safe and affordable microbicides that prevent HIV should also remain a high priority.

Even with these measures, history suggests that unless women are supported in gaining sufficient power to safeguard their own health, the full potential of any educational or technological advances against HIV will not be realized. Interventions aimed at changing policies, cultural and societal norms, and other environmental factors that place women and men at risk for HIV infection or that impede them from fully using available prevention strategies will be crucial if we are to halt the HIV pandemic.

References

1. Centers for Disease Control and Prevention (1981). Pneumocystis pneumonia—Los Angeles. *Morbid. Mortal. Wkly. Rep.* **30,** 250–252.

2. Centers for Disease Control and Prevention (1981). Follow-up on Kaposi's sarcoma and pneumocytis pneumonia. *Morbid. Mortal. Wkly. Rep.* **30,** 409–410.

3. Holmes, K. K., Karon, J. M., and Kreiss, J. (1990). The increasing frequency of heterosexually acquired AIDS in the United States, 1983–1988. *Am. J. Public Health* **80,** 858–862.

4. Centers for Disease Control and Prevention (1996). Update: Mortality attributable to HIV infection among persons aged 25–44 years—United States, 1994. *Morbid. Mortal. Wkly. Rep.* **45,** 121–125.

5. Centers for Disease Control and Prevention (1998). HIV/AIDS Surveillance Midyear Report, 1998. **10(1),** 1–40.

6. Centers for Disease Control and Prevention (1995). Update: AIDS among women—United States, 1994. *Morbid. Mortal. Wkly. Rep.* **44,** 81–84.

7. Mann, J. M., and Tarantola, D. (1996). Global overview: A powerful HIV/AIDS pandemic. *In* "AIDS in the World II" (J. M. Mann and D. Tarantola, eds.), p. 11. Oxford University Press, New York.

8. Vuylsteke, B., Sunkutu, R., and Laga, M. (1996). Epidemiology of HIV and sexually transmitted infection in women. *In* "AIDS in the World II" (J. M. Mann and D. Tarantola, eds.), pp. 97–109. Oxford University Press, New York.

9. McNaughton, A. D., Hanson, D. L., Jones, J. L. *et al.* (1999). Effects of antiretroviral therapy and opportunistic illness primary chemoprophylaxis on survival after AIDS diagnosis. *AIDS,* **13.**

10. Connor E. M., Sperling R. S., Gelber R., *et al.* (1994). Reduction of maternal-infant transmission of human immunodeficiency virus type 1 with zidovudine treatment. *N. Engl. J. Med.* **331,** 1173–1180.

11. Wortley, P. M., Chu, S. Y., Diaz, T. *et al.* (1995). HIV testing patterns: Where, why, and when were persons with AIDS tested for HIV? *AIDS* **9,** 487–492.

12. Centers for Disease Control and Prevention (1997). Update: Trends in AIDS incidence, deaths, and prevalence—United States, 1996. *Morbid. Mortal. Wkly. Rep.* **46,** 165–173.

13. Wortley, P. M., and Flemming, P. L. (1997). AIDS in women in the United States. *JAMA, J. Am. Med. Assoc.* **278,** 911–916.

14. Davis, S. F., Rosen, D. H., Steinberg, S. *et al.* (1998). Trends in HIV prevalence among childbearing women, United States, 1989–1994. *J. Acquired Immune Defic. Syndr. Hum. Retrovirol.* **19,** 158–164.

15. Ward, J. W., Holmberg, S. D., Allen, J. R. *et al.* (1988). Transmission of human immunodeficiency virus (HIV) by blood transfusions screened as negative for HIV antibody. *N. Engl. J. Med.* **318,** 473–478.

16. Miller, H. G., Turner, C. F., and Moses, L. E., eds. (1990). "AIDS—the 2nd Decade." National Academy Press, Washington, DC.

17. Chiasson, R. E., Bacchetti, P., Osmond, D. *et al.* (1989). Cocaine use and HIV infection in intravenous drug users in San Francisco. *JAMA, J. Am. Med. Assoc.* **261,** 561–565.

18. Des Jarlais, D. C., Samuel, R. F., Perlis, T., *et al.* (1998). Declining seroprevalence in a very large HIV epidemic: Injecting drug users in New York City, 1991–1996. *Am. J. Public Health* **88,** 1801–1806.

19. Needle, R. H., Coyle, S. L., Normand, J., Lambert, E., and Cesari, H. (1998). HIV prevention with drug-using populations—current status and future prospects. *Public Health Rep.* **113,** 4–15.

20. Diaz, T., Chu, S. Y., Weinstein, B. *et al.* (1998). Injection and syringe sharing among HIV-infected injection drug users. *J. Acquired Immun. Defic. Syndr. Hum. Retrovirol.* **18,** S76–S81.

21. Chu, S. Y., Peterman, T. A., Doll, L. S. *et al.* (1992). AIDS in bisexual men in the United States: Epidemiology and transmission to women. *Am. J. Public Health* **82,** 220–224.

22. Chu, S. Y., Hammet, T. A., and Buehler, J. W. (1992). Update: Epidemiology or reported cases of AIDS in women who report sex only with other women, United States, 1980–1991 (letter). *AIDS* **6,** 518–519.

23. Chu, S. Y., Conti, L., Schable, B. A. *et al.* (1994). Female-to-female sexual contact and HIV transmission (letter). *JAMA, J. Am. Med. Assoc.* **272,** 433.

24. Wortley, P. M., Hammett, T. A., and Fleming, P. L. (1998). Donor insemination and human immunodeficiency virus. *Obstet. Gynecol.* **91**(4), 515–518.

25. Centers for Disease Control and Prevention (1990). HIV-infection and artificial insemination with processed semen. *Morbid. Mortal. Wkly. Rep.* **39,** 249–256.

26. Centers for Disease Control and Prevention (1988). Semen banking, organ and tissue transplantation, and HIV antibody testing. *Morbid. Mortal. Wkly. Rep.* **37,** 57, 58, 63.

27. Centers for Disease Control and Prevention (1994). Guidelines for preventing transmission of human immunodeficiency virus through transplantation of human tissue and organs. *Morbid. Mortal. Wkly. Rep.* **43**(RR-8), 1–17.

28. Boylan, L., and Stein, Z. (1996). The epidemiology of HIV infection in children and their mothers—Vertical transmission. *Epidemiol. Rev.* **13,** 143–177.

29. Schaefer, A., Friese, K., Lauper, U. *et al.* (1998). Influence of cesarean section before parturition and antiretroviral prophylaxis on the materno-fetal transmission of HIV (# 12466, Late Breaker). *12th World AIDS Conf.,* Geneva, *1998.*

30. Pazè, E. (1998). 12th World AIDS Conference. *Lancet* **352,** 1072.

31. Hamers, F. F., Downs, A. M., Infuso, A., and Brunet, J. B. (1998). The HIV/AIDS epidemic in Europe: Changing trends and new challenges (#522/13179), p. 121. *12th World AIDS Conf.,* Geneva, *1998.*

32. Mertens, T., Belsey, E., Stoneburner, R. L. *et al.* (1995). Global estimates and epidemiology of HIV-1 infections and AIDS: Further heterogeneity in spread and impact. *AIDS* **9**(Suppl. A), S259–S272.

33. Quinn, T. C. (1996). Global burden of the HIV pandemic. *Lancet* **348,** 99–106.

34. D'Cruz-Grote, D. (1996). Prevention of HIV infection in developing countries. *Lancet* **348,** 1071–1074.

35. Siriwasin, W., Shaffer, N., Roongpisuthiong, A., *et al.* (1998). HIV prevalence, risk and partner serodiscordance among pregnant women in Bangkok. *JAMA, J. Am. Med. Assoc.* **280,** 49–54.

36. Wawer, M. J. (1996). Urban-rural movement and HIV dynamics. *In* "AIDS in the World II" (J. M. Mann and D. Tarantola, eds.), p. 48. Oxford University Press, New York.

37. Stein, Z. A., and Kuhn, L. (1996). HIV in women: What are the gaps in knowledge? *In* "AIDS in the World II" (J. M. Mann and D. Tarantola, eds.), pp. 229–235. Oxford University Press, New York.

38. Gwinn, M., and Wortley, P. (1996). Epidemiology of HIV infection in women and newborns. *Clin. Obstet. Gynecol.* **39,** 292–304.

39. Valleroy, L. A., MacKellar, D. A., Karon, J. M., Janssen, S., and Hayman, C. R. (1998). HIV infection in disadvantaged out-of-school youth: U.S. Job Corps entrants, 1990 through 1996. *J. Acquired Immune Defic. Syndr.* **19,** 67–73.

40. Diaz, T., Buehler, J. W., Castro, K. G., and Ward, J. W. (1993). AIDS trends among Hispanics in the United States. *Am. J. Public Health* **83,** 504–509.

41. Diaz, T., Chu, S. Y., Conti, L. *et al.* (1994). Risk behaviors of persons with heterosexually acquired HIV infection in the United States. *J. Acquired Immune Defic. Syndr.* **7,** 958–963.

42. Edlin, B. R., Irwin, K. L., Faruque, S. *et al.* (1994). Intersecting epidemics: Crack cocaine use and HIV infection among inner-city young adults. *N. Engl. J. Med.* **331,** 1422–1427.

43. Gamble, V. N. (1997). Under the shadow of Tuskegee: African Americans and health care. *Am. J. Public Health* **87,** 1773–1778.

44. Hatch, J., Moss, N., Saran, A. *et al.* (1993). Community research: Partnership in black communities. *Commun. Res.* **9,** 27–31.

45. Thomas, J. C., Clark, M., Robinson, J. *et al.* (1999). The social ecology of syphilis. *Social Sci. Med.* **48,** 1081–1094.

46. Peterman, T. A., Wassherheit, J. N., and Cates, W. (1992). Prevention of the sexual transmission of HIV. *In* "AIDS Etiology, Diagnosis, Treatment and Prevention" (V. T. DeVita, S. Hellman, and S. A. Rosenberg, eds.), pp. 443–450. Lippincott, Philadelphia.

47. National AIDS Control Programme, Malawi (1996). "HIV/Syphilis Seroprevalence in Antenatal Clinic Attenders," Sentinel Surveill. Rep. Lilongwe, Malawi.

48. Donegan, E., Stuart, M., Nilan, J. C. *et al.* (1990). Infection with human immunodeficiency virus type 1 (HIV-1) among recipients of antibody-positive blood donations. *Ann. Intern. Med.* **113,** 733.

49. Padian, N., Marquis, L., Francis, D. P. *et al.* (1987). Male-to-female transmission of human immunodeficiency virus. *JAMA, J. Am. Med. Assoc.* **258,** 788–790.

50. Peterman, T. A., Stoneburner, R. L., Allern, J. R. *et al.* (1988). Risk of human immunodeficiency virus transmission from heterosexual adults with transfusion associated infections. *JAMA, J. Am. Med. Assoc.* **259,** 55–58.

51. DeGruttola, V., Seage, G. R., Mayer, K. H., and Horsburgh, C. R. (1989). Infectiousness of HIV between male homosexual partners. *J. Clin. Epidemiol.* **42,** 849–856.

52. deVincenzi, I. for The European Study Group (1994). A longitudinal study of human immunodeficiency virus transmission by heterosexual partners. *N. Engl. J. Med.* **331,** 341–346.

53. Padian, N. S., Shiboski, S. C., and Jewell, N. P. (1991). Female-to-male transmission of human immunodeficiency virus. *JAMA, J. Am. Med. Assoc.* **261,** 1664–1667.

54. Wiley, J. A., Herschkorn, S. J., and Padian, N. (1988). Heterogeneity in the probability of HIV transmission per sexual contact: The case of male-to-female transmission in penile-vaginal intercourse. *Stat. Med.* **8,** 93–102.

55. European Study Group on Heterosexual Transmission of HIV (1992). Comparison of female-to-male and male-to-female transmission of HIV in 563 stable couples. *Br. Med. J.* **304,** 809–813.

56. Downs, A. M., and deVincenzi, I. (1996). Probability of heterosexual transmission of HIV. *J. Acquired Immune Defic. Syndr. Hum. Retrovirol.* **11,** 388–395.

57. Mastro, T. D., Satten, G. A., Nopkesorn, T. *et al.* (1994). Probability of female-to-male transmission of HIV in Thailand. *Lancet* **343,** 204–207.

58. Detels, R., English, P., Visscher, B. R. *et al.* (1989). Seroconversion, sexual activity and condom use among 2915 HIV seronegative men followed for up to 2 years. *J. Aquired Immune Defic. Syndr.* **2,** 77.

59. Leynaert, B., Downs, A. M., deVincenzi, I. *et al.* (1998). Heterosexual transmission of human immunodeficiency virus. *Am. J. Epidemiol.* **148,** 88–96.

60. Soto-Ramirez, L. E., Renjifo, B., McLane, M. F. *et al.* (1996). HIV-1 Langerhans' cell tropism associated with heterosexual transmission of HIV. *Science* **271,** 1291–1293.

61. Levy, J. A. (1996). HIV heterogeneity in transmission and pathogenesis. *In* "AIDS in the World II" (J. M. Mann and D. Tarantola, eds.), pp. 177–185. Oxford University Press, New York.

62. Mayer, K. H., and Anderson, D. J. (1995). Heterosexual transmission of HIV. *In* "HIV Infection in Women" (H. Minkoff, J. A. DeHovitz, and A. Duerr, eds.), pp. 73–86. Raven Press, New York.

63. Schaffer, N., Roongpisuthipong, A., Siriwasin, W. *et al.* (1999). Maternal viral load and perinatal HIV-1 subtype E transmission, Thailand. *J. Infect. Dis.* **179**(3), 590–599.

64. Jacquez, J. A., Koopman, J. S., Simon, C. P., and Longini, I. M. (1994). Role of the primary infection in epidemics of HIV infection in gay cohorts. *J. Acquired Immune Defic. Syndr.* **7,** 1169–1184.

65. O'Brien, T. R., Shaffer, N., and Jaffe, H. (1992). Acquisition and transmission of HIV. *In* "The Medical Management of AIDS," pp. 3–17. Saunders, Philadelphia.

66. Elias, C. J., and Heise, L. L. (1994). Challenges for the development of female-controlled vaginal microbicides. *AIDS* **8,** 1–9.

67. Cohen, J. (1995). Women: Absent term in the AIDS research equation. *Science* **269,** 777–780.

68. Dezutti, C. D., and Lal, R. B. (1998). Mechanisms of HIV transmission through epithelial cell barriers (#32124). *12th World AIDS Conf.* Geneva, *1998,* p. 545.

69. Bourinbaiar, A. S., and Phillips, D. M. (1991). Transmission of human immunodeficiency virus from monocytes to epithelia. *J. Acquired Immune Defic. Syndr.* **4,** 56–63.

70. Miller, C. J., Alexander, N. J., Vogel, P. *et al.* (1992). Mechanism of genital transmission of SIV. *J. Med. Primatol.* **21,** 64–68.

71. Plummer, F. A., Simonsen, J. N., Cameron, D. W. *et al.* (1991). Co-factors in male-female transmission of HIV. *J. Infect. Dis.* **163,** 233–239.

72. Simonson, J. N., Plummer, F. A., Ngugi, E. N. *et al.* (1990). HIV infection among lower socioeconomic strata prostitutes in Nairobi. *AIDS* **4,** 139–144.

73. Lazzarin, A., Saracco, A., Musicco, M. *et al.* (1991). Man-to-woman sexual transmission of the human immunodeficiency virus. *Arch. Intern. Med.* **151,** 2411–2416.

74. Voeller, B., and Anderson, D. J. (1992). Vaginal pH and HIV transmission. *JAMA, J. Am. Med. Assoc.* **267**(14), 1917–1919.

75. Reid, E. (1992). Gender, knowledge, and responsibility. *In* "AIDS in the World" (J. M. Mann, D. Tarantola, and T. W. Netter, eds.), pp. 657–667. Harvard University Press, Cambridge, MA.

76. Moses, S., Plummer, F. A., Bradley, J. E. *et al.* (1994). The association between lack of male circumcision and risk for HIV infection: A review of the epidemiological data. *Sex. Transm. Dis.* **4,** 201–210.

77. Cameron, D. W., Simonsen, J. N., D'Costa, L. J. *et al.* (1989). Female to male transmission of human immunodeficiency virus type-1: Risk factors for seroconversion in men. *Lancet* **2,** 403–407.

78. Tyndall, M., Agoki, E., Malisa, W. *et al.* (1992). HIV-1 prevalence and risk of seroconversion among uncircumcised men in Kenya (#PoC4308). *8th World AIDS Conf.,* Amsterdam, The Netherlands, *1992,* p. C296.

79. Quinn, T. C., Glasser, D., Cannon, R. O. *et al.* (1988). Human immunodeficiency virus infection among patients attending clinics for sexually transmitted diseases. *N. Engl. J. Med.* **318,** 197–204.

80. Telzac, E. E., Chiasson, M. A., Bevier, P. J. *et al.* (1993). HIV-1 seroconversion in patients with and without genital ulcer disease. A prospective study. *Ann. Intern. Med.* **119,** 1181–1186.

81. Wasserheit, J. N. (1992). Epidemiologic synergy. Interrelationships between human immunodeficiency virus infection and other sexually transmitted diseases. *Sex. Transm. Dis.* **19,** 61–77.

82. Laga, M., Manoka, A., Kivuvu, M. T. *et al.* (1993). Non-ulcerative sexually transmitted diseases as risk factors for HIV-1 transmission in women: Results from a cohort study. *AIDS* **7,** 95–102.

83. Kassler, W. J., Zenilman, J. M., Erickson, B. *et al.* (1994). Seroconversion in patients attending sexually transmitted disease clinics. *AIDS* **8,** 351–355.

84. Kreiss, J. K., Coombs, R., Plummer, F. *et al.* (1989). Isolation of human immunodeficiency virus from genital ulcers in Nairobi prostitutes. *J. Infect. Dis.* **160,** 380–384.

85. Dyer, J. R., Eron, J. J., Hoffman, I. F. *et al.* (1998). Association of CD4 cell depletion and elevated blood and seminal plasma human immunodeficiency virus type 1 (HIV-1) RNA concentrations with genital ulcer disease in HIV-1-infected men in Malawi. *J. Infect. Dis.* **177,** 224–227.

86. Ghys, P. D., Fransen, K., Diallo, M. O. *et al.* (1997). The associations between cervicovaginal HIV shedding, sexually transmitted diseases and immunosuppression in female sex workers in Abidjan, Côte d'Ivoire. *AIDS* **11,** F85–F93.

87. Grosskurth, H., Mosha, F., Todd, J. *et al.* (1995). Impact of improved treatment of sexually transmitted diseases on HIV infection in rural Tanzania: Randomised trial. *Lancet* **346,** 530–536.

88. Wawer, M. J., Sewan Kambo, N. K., Serwadda, D., *et al.* (1999). Control of sexually transmitted diseases for AIDS prevention in Uganda: a randomized community trial. Rakai Project Study Group. *Lancet* **353,** 525–535.

89. Cates, W., and Stone, K. M. (1992). Family planning, sexually transmitted diseases, and contraceptive choice. *Fam. Plann. Perspect.* **24,** 75–84.

90. Saracco, A., Musicco, M., Nicolosi, A. *et al.* (1993). Man-to-woman sexual transmission of HIV: Longitudinal study of 343 steady partners of infected men. *J. Acquired Immune Defic. Syndr.* **6,** 497–502.

91. Warner, D. L., and Hatcher, R. A. (1998). Male condoms. *In* "Contraceptive Technology" (R. A. Hatcher, J. Trussell, F. H. Stewart, W. Cates, Jr., G. K. Stewart, F. Guest, and D. Kowal, eds.), pp. 325–356. Ardent Media, New York.

92. Zenilman, J. M., Weisman, C. S., Rompaol, A. M. *et al.* (1995). Condom use to prevent incident STDs: The validity of self-reported condom use. *Sex. Transm. Dis.* **22,** 15–21.

93. Peterman, T. A., Lin, L. S., Newman, D. *et al.* (1998). Does measured behavior reflect STD/HIV risk? (#257/14252). *12th World AIDS Conf.,* Geneva, *1998,* p. 232.

94. Cates, W., Stewart, F. H., and Trussell, J. (1992). The quest for women's prophylactic methods—hopes vs. science. *Am. J. Public Health* **82,** 1479–1482.

95. Stein, Z. A. (1995). More on women and the prevention of HIV infection. *Am. J. Public Health* **85,** 1485–1488.

96. Louv, W. C., Austin, H., Alexander, W. J. *et al.* (1988). A clinical trial of nonoxynol-9 for preventing gonococcal and chlamydial infections. *J. Infect. Dis.* **158,** 518–523.

97. Niruthisard, S., Roddy, R. E., and Chutivongse, S. (1992). Use of nonoxynol-9 and reduction in rate of gonococcal and chlamydial cervical infections. *Lancet* **339,** 1371–1375.

98. Cates, W., and Raymond, E. G. (1998). Vaginal spermicides. *In* "Contraceptive Technology" (R. A. Hatcher, J. Trussell, F. H. Stewart, W. Cates, Jr., G. K. Stewart, F. Guest, and D. Kowal, eds.), pp. 357–369. Ardent Media, New York.

99. Calsyn, D., Meinecke, C., Saxon, A., and Stanton, V. (1992). Risk reduction in sexual behavior: A condom giveaway program in a drug abuse treatment clinic. *Am. J. Public Health* **82,** 1536–1538.

100. MacGowan, R. J., Brackbill, R. M., Rugg, K. L. *et al.* (1997). Sex drugs, and HIV counseling and testing: A prospective study of behavior-change among methadone-maintentance clients in New England. *AIDS* **11,** 229–235.

101. Chaisson, M. A., Stoneburner, R. L., Hildebrandt, D. S. *et al.* (1991). Heterosexual transmission of HIV-1 associated with the use of smokable freebase cocaine (crack). *AIDS* **5,** 1121–1126.

102. Ellerbrock, T. V., Lieb, S., Harrington, P. E. *et al.* (1992). Heterosexually transmitted human immunodeficiency virus infection among pregnant women in a rural Florida community. *N. Engl. J. Med.* **327,** 1704–1709.

103. Fullilove, R. E., Fullilove, M. T., Bowser, B. P. *et al.* (1990). Risk of sexually transmitted disease among black adolescent crack users in Oakland and San Francisco, California. *JAMA, J. Am. Med. Assoc.* **263,** 851–855.

104. Moore, J. S., Harrison, J. S., and Doll, L. S. (1994). Interventions for sexually active, heterosexual women in the United States. *In* "Preventing AIDS: Theories and Methods of Behavioral Interventions" (R. J. DiClemente and J. L. Peterson, eds.), pp. 243–265. Plenum, New York.

105. Harrison, D. F., Wambach, K. G., Byers, J. B. *et al.* (1991). AIDS knowledge and risk behaviors among culturally diverse women. *AIDS Educ. Prev.* **3,** 79–89.

106. Mhloyi, M. M. (1992). Women, AIDS, and reproductive issues. *In* "AIDS in the World" (J. Mann, D. Tarantola, and T. W. Netter, eds.), pp. 373–375. Harvard University Press, Cambridge, MA.

107. Aral, S. O., and Mann, J. M. (1998). Commercial sex work and STD: The need for policy interventions to change societal patterns. *Sex. Transm. Dis.* **25**(9), 455–456.

108. Rothenberg, R., Woelfel, M., Stoneburner, R. *et al.* (1987). Survival with the acquired immunodeficiency syndrome. *N. Engl. J. Med.* **317,** 1297–1302.

109. Cotton, D. J., Finkelstein, D. M., He, W., and Feinberg, J. (1993). Determinants of accrual of women to a large, multi-center clinical trials program of human immunodeficiency virus infection. *J. Acquired Immune Defic. Syndr.* **6,** 1322–1328.

110. Creagh, T., Thompson, M., Morris, A. *et al.* (1993). Gender differences in the spectrum of HIV disease in Georgia (# PO-CO4-2657). *9th World AIDS Conf.* Berlin, *1993,* p. 660.

111. Lemp, G. F., Hirozawa, A. M., Cohen, J. B. *et al.* (1992). Survival for women and men with AIDS. *J. Infect. Dis.* **166,** 74–79.

112. Melnick, S. L., Renslow, S., Louis, T. A. *et al.* (1994). Survival and disease progression according to gender for patients with HIV infection. *JAMA, J. Am. Med. Assoc.* **272,** 1915–1921.

113. Chaisson, R. E., Keruly, J. C., and Moore, R. D. (1995). Race, sex, drug use, and progression of human immunodeficiency virus disease. *N. Engl. J. Med.* **333,** 751–756.

114. Guinan, M. E., and Leviton, L. (1995). Prevention of HIV infection in women: Overcoming barriers. *J. Am. Med. Women's Assoc.* **50,** 74–77.

115. Stein, Z. A. (1990). HIV prevention: The need for methods women can use. *Am. J. Public Health* **80,** 460–462.

116. Moore, R. D., Hidalgo, J., Sugland, B. W. *et al.* (1991). Zidovudine and the natural history of the acquired immunodeficiency syndrome. *N. Engl. J. Med.* **324,** 941–949.

117. Stein, M. D., Piette, J., Mor, V. *et al.* (1991). Difference in access to zidovudine (AZT) among symptomatic HIV-infected persons. *J. Gen. Intern. Med.* **6,** 635–640.

118. Moore, R. D., Stanton, D., Gopalan, R., and Chaisson, R. E. (1994). Racial differences in the use of drug therapy for HIV disease in a urban community. *N. Engl. J. Med.* **330,** 763–768.

119. Bastian, L., Bennett, C. L., Adams, J. *et al.* (1993). Differences between men and women with HIV-related Pneumocystis carinii pneumonia. *J. Acquired Immune Defic. Syndr.* **6,** 617–623.

120. Hellinger, F. J. (1993). The use of health services by women with HIV infection. *Health Serv. Res.* **28,** 544–561.

121. Bertolli, J., Simonds, R. J., Thomas, P. *et al.* (1998). Implementation of recommendations for the medical care of HIV-exposed infants in the first year of life, USA (#171/23269). *12th World AIDS Conf.,* Geneva, *1998,* p. 394.

122. Wortley, P. M., Fleming, P. L., Lindegren, M. L. *et al.* (1998). Prevention of perinatal transmission in the US: A population-based evaluation of prevention efforts in 4 states (#23289). *12th World AIDS Conf.,* Geneva, *1998,* p. 399.

123. Barbaccii, M. B., Dalabetta, G. A., Repke, J. T. *et al.* (1990). Human immunodeficiency virus infection in women attending an inner-city prenatal clinic: Ineffectiveness of targeted screening. *Sex. Transm. Dis.* **17,** 122–126.

124. Rogers, M. F., Mofenson, L. M., and Moseley, R. R. (1995). Reducing the risk of perinatal HIV transmission through zidovudine therapy: Treatment recommendations and implications for perinatal HIV counseling and testing. *J. Am. Med. Women's Assoc.* **50,** 78–82.

125. Centers for Disease Control and Prevention (1995). U.S. Public Health Service recommendations for human immunodeficiency virus counseling and voluntary testing for pregnant women. *Morbid. Mortal. Wkly. Rep.* **44**(RR-7), 1–12.

126. Chu, S. Y., Hanson, D. L., Jones, J. L. *et al.* (1996). Pregnancy rates among women infected with human immunodeficiency virus. *Obstet. Gynecol.* **87,** 195–198.

127. Selwyn, P. A., Carter, R. M., Schoenbaum, E. E. *et al.* (1989). Knowledge of HIV antibody status and decisions to continue or terminate pregnancy among intravenous drug users. *JAMA, J. Am. Med. Assoc.* **261,** 3567–3571.

128. Nieburg, P., and Stanecki, K. A. (1998). The global burden of mother-to-child transmission of HIV-1 (# 13591, Late Breaker). *12th World AIDS Conf.* Geneva, *1998,* p. 9.

129. DeHovitz, J. A. (1995). Natural history of HIV infection in women. *In* "HIV Infection in Women" (H. Minkoff, J. A. DeHovitz, and A. Duerr, eds.), pp. 57–71. Raven Press, New York.

130. Koonin, L. M., Ellerbrock, T. V., Atrash, H. K. *et al.* (1989). Pregnancy associated deaths due to AIDS in the United States. *JAMA, J. Am. Med. Assoc.* **261,** 1306–1309.

131. Fleming, P. L., Ciesielski, C. A., Byers, R. H. *et al.* (1993). Gender differences in reported AIDS-indicative diagnoses. *J. Infect. Dis.* **168,** 61–67.

132. Rhoads, J. L., Wright, C., Redfield, R. R. *et al.* (1987). Chronic vaginal candidiasis in women with human immunodeficiency virus infection. *JAMA, J. Am. Med. Assoc.* **257,** 3105–3107.

133. Imam, N., Carpenter, C. C., Mayer, K. H. *et al.* (1990). Hierarchical pattern of mucosal candida infections in HIV-seropositive women. *Am. J. Med.* **89,** 142–146.

134. Carpenter, C., Mayer, K., Stein, M. *et al.* (1991). HIV infection in North American women: Experience with 200 cases and a review of the literature. *Medicine (Baltimore)* **70,** 307–325.

135. Centers for Disease Control and Prevention (1998). Guidelines for treatment of sexually transmitted disease. *Morbid. Mortal. Wkly. Rep.* **47**(RR-1), 75.

136. Centers for Disease Control and Prevention (1997). USPHS/IDSA guidelines for the prevention of opportunistic infections in persons infected with human immunodeficiency virus. *Morbid. Mortal. Wkly. Rep.* **46**(RR-12).

137. Agency for Health Care Policy and Research (1994). "Evaluation and Management of Early HIV Infection," AHCPR Publ. No. 94-0572, Clin. Pract. Guidelines, No. 7. US DHHS, Public Health Service, Rockville, MD.

138. Korn, A. P., Landers, D. V., Green, J. R., and Sweet, R. L. (1993). Pelvic inflammatory disease in human immunodeficiency virus-infected women. *Obstet. Gynecol.* **82,** 765–768.

139. Irwin, K., O'Sullivan, M., Sperling, R. *et al.* (1995). The influence of HIV on initial presentation and course of pelvic inflammatory disease: Final results of a multicenter study (97-K94). *35th Intersci. Conf. Antimicrob. Agents Chemother,* San Francisco, *1995.*

140. Kamenga, M. C., DeCock, K. M., St. Louis, M. E. *et al.* (1995). The impact of human immunodeficiency virus on pelvic inflammatory disease. *Am. J. Obstet. Gynecol.* **172,** 919–925.

141. Capps, L., Peng, G., Doyle, M. *et al.* (1998). Sexually transmitted infections in women with the human immunodeficiency virus. *Sex. Transm. Dis.* **25,** 443–447.

142. Schiffman, M. H., Bauer, H. M., Hoover, R. N. *et al.* (1993). Epidemiologic evidence showing that human papilloma virus infection causes most cervical intraepithelial neoplasia. *J. Natl. Cancer Inst.* **85,** 958–964.

143. Vermund, S. T., Kelly, K. F., Klein, R. S. *et al.* (1992). High risk of human papillomavirus infection and cervical squamous intraepithelial lesions among women with syptomatic human immunoceficiency virus infection. *Am. J. Obstet. Gynecol.* **166,** 1232–1237.

144. Schafer, A., Friedmann, W., Mielke, M. *et al.* (1991). The increased frequency of cervical dysplasia-neoplasia in women infected with the human immunodeficiency virus is related to the degree of immunosuppression. *Am. J. Obstet. Gynecol.* **164,** 593–599.

145. Wright, T. C., Ellerbrock, T. V., Chiasson, M. A. *et al.* (1994). Cervical intraepithelial neoplasia in women infected with human immunodeficiency virus: Prevalence, risk factors, and validity of Papanicolaou smears. *Obstet. Gynecol.* **84,** 591–597.

146. Shah, P. N., Smith, J. R., Wells, C. *et al.* (1994). Menstrual symptoms in women infected the human immunodeficiency virus. *Obstet. Gynecol.* **83,** 397–400.

147. DeCock, K. M., Soro, B., Coulibaly, M., and Lucas, S. B. (1992). Tuberculosis and HIV infection in sub-Saharan Africa. *JAMA, J. Am. Med. Assoc.* **268,** 1581–1587.

148. Vanai, H., Uthaivoravit, W., Danich, V. *et al.* (1996). Rapid increase in HIV related tuberculosis, Chiang Rai, Thailand, 19990–1994. *AIDS* **10,** 527–531.

149. Pope, J. W., Jean, S. S., Ho, J. L. *et al.* (1993). Effect of isoniazid prophylaxis on incidence of active tuberculosis and progression of HIV infection. *Lancet* **342,** 268–272.

150. Cardo, D. M., Culver, D. H., Ciesielski, C. A. *et al.* (1997). A case-control study of HIV seroconversion in health care workers after percutanious exposure. *N. Engl. J. Med.* **337**(21), 1485–1490.

151. Roland, M. E. (1998). Post-exposure prophylaxis following sexual or injection drug use exposure to HIV: Current knowledge and future research strategies. *J. HIV Ther.* **3,** 17–20.

152. Centers for Disease Control and Prevention (1998). Administration of zidovudine during late pregnancy and delivery to prevent perinatal HIV transmission—Thailand, 1996–1998. *Morbid. Mortal. Wkly. Rep.* **47,** 151–154.

153. Hagan, H., Des Jarlais, D. C., Friedman, S. R. *et al.* (1995). Reduced risk of hapatitis B and hepatitis C among injection drug users in the Tacoma syringe exchange program. *Am. J. Public Health* **85,** 1531–1537.

154. Kamb, M. L., Fishbein, M., Douglas, J. M. *et al.* (1998) Efficacy of risk-reduction counseling to prevent human immunodeficiency virus and sexually transmitted diseases. *JAMA, J. Am. Med. Assoc.* **280,** 1161–1167.

155. Sangiwa, G., Balmer, D., Furlonge, C. *et al.* (1998). Voluntary HIV counseling and testing (VCT) reduces risk behavior in developing countries (# 133/ 33269). *12th World AIDS Conf.,* Geneva, *1998,* p. 646.

156. Shain, R. N., Piper, J. M., Newton, E. R., *et al.* (1999). A randomized, controlled trial of a behavioral intervention to prevent sexually transmitted disease among minority women. *N. Engl. J. Med.* **340,** 93–100.

157. NIMH Multisite HIV Prevention Trial Group (1998). The NIMH multisite HIV prevention trial: Reducing HIV sexual risk behavior. *Science* **280,** 1889–1894.

158. Hamers, F. F., Delmas, M. C., Alix, J. *et al.* (1998). Unawareness of HIV seropositivity before AIDS diagnosis in Europe (#137/ 43105) *12th World AIDS Conf.,* Geneva, *1998.*

159. Lehman, J. S., Hecht, F. M., Fleming, P. L. *et al.* (1998). HIV testing behavior among at-risk populations(#136/ 43103). *12th World AIDS Conf.,* Geneva, *1998.*

160. Mastro, T. D., and Limpakarnjanarat, K. (1995). Condom use in Thailand: How much is it slowing the HIV/AIDS epidemic? *AIDS* **9,** 523–525.

161. Ungphakorn, J., and Sittitrai, W. (1994). The Thai response to the HIV/AIDS epidemic. *AIDS* **8** (Suppl. 2), 5155–5163.

162. Cohen, D. A., Farley, T. A., Bedimo-Etame, J. R., Scribner, R., Ward, W., Kendall, C., and Rice, J. Implementation of condom social marketing in Louisiana, 1993–1996. *Am. J. Public Health,* **89,** 204–208.

163. Cohen, D. A., Dent, C., and MacKinnon, D. (1991). Condom skills education and sexually transmitted disease reinfection. *J. Sex Res.* **28**(1), 139–144.

164. Cavalieri d'Oro, L., Parazzini, F., Naldi, L., and La Vecchia, C. (1994). Barrier methods of contraception, spermicides, and sexually transmitted diseases: A review. *Genitourin. Med.* **70,** 410–417.

165. Roddy, R. E., Zekeng, L., Ryan, K. A. *et al.* (1998). A controlled trial of nonoxynol 9 film to reduce male-to-female transmission of sexually transmitted diseases. *N. Engl. J. Med.* **339,** 504–10.

166. Peterman, T. A., Toomey, K. E., Dicker, L. W. *et al.* (1997). Partner notification for syphilis. *Sex. Transm. Dis.* **24,** 511–518.

167. North, R. L., and Rothenberg, K. H. (1993). Partner notification and the threat of domestic violence against women with HIV infection. *N. Engl. J. Med.* **329**(16), 1194–1196.

168. Rothenberg, K., and Paskey, S. J. (1995). The risk of domestic violence and women with HIV infection: Implications for partner notification, public policy, and the law. *Am. J. Public Health* **85,** 1569–1576.

169. Maher, J., Seeman, G. M., Peterson, J. *et al.* (1998). Partner violence and women's decision to have an HIV test (# 43110). *12th World AIDS Conf.,* Geneva, *1998,* p. 869.

170. Faden, R. R., Kass, N. E., Acuff, K. *et al.* (1996). HIV infection and childbearing: A proposal for public policy and clinical practice. *In* "HIV, AIDS, and Childbearing: Public Policy, Private Lives" (R. R. Faden and N. E. Kass, eds.), pp. 447–461. Oxford University Press, New York.

171. U.S. Agency for International Development (1995). "USAID Responds to HIV/AIDS. A Report on the Fiscal Year 1994 HIV/ AIDS Prevention Programs of the United States Agency for International Development," pp. 1–8. USAID, Washington, DC.

27
Vaginal Infections

JANE R. SCHWEBKE
University of Alabama at Birmingham
Birmingham, Alabama

I. Introduction

Vaginal infections are the source of considerable morbidity among women. The three main causes of infectious vaginitis are vulvovaginal candidiasis (VVC), trichomoniasis, and bacterial vaginosis (BV). Each one may cause distressing vaginal symptomatology and the latter two may cause upper genital tract complications. These infections may also contribute to the transmission and acquisition of the human immunodeficiency virus (HIV).

It is estimated that there are more than 10 million office visits per year due to vaginal infections [1]. As none of these infections are reportable diseases, accurate numbers are unavailable. Despite the fact that affordable diagnostic tests exist, many women are misdiagnosed due to the failure of physicians to use these tests. Further, because of the common misconception perpetuated by the media that the most common cause of vaginitis is candidiasis (yeast), many women with vaginal discharge self-treat with over-the-counter (OTC) antifungal preparations, delaying appropriate therapy [2,3]. It is important to realize that the above comments apply to only symptomatic infections and that trichomoniasis and BV may be asymptomatic. Health care providers rarely screen asymptomatic women for vaginal infections, thus in asymptomatic women the diagnosis is often completely overlooked.

II. The Normal Vaginal Ecosystem

The ecosystem of the human vagina is complex and only beginning to be understood. Most of our current knowledge is about the microbiological inhabitants and only limited information is available about the factors that affect this flora. Information is also beginning to accumulate about the immunological and cellular defense mechanisms of the vagina.

Lactobacilli become the predominant inhabitants of the vagina at the time of puberty, presumably due to the effect of estrogens on the glycogen content of vaginal epithelial cells [4]. When estrogen levels fall, such as in postpartum or menopausal females, the prevalence of lactobacilli declines [5]. *In vitro,* lactobacilli have been shown to produce various potential microbial toxins, including hydrogen peroxide and more poorly defined bacteriocins [6,7]. Lactobacilli that produce H_2O_2 have been shown *in vitro* to inhibit various microorganisms, including *Gardnerella vaginalis,* anaerobes, *Neisseria gonorrhoeae,* and HIV [8–12] Epidemiologically, women with lactobacilli that produce H_2O_2 *in vitro* have been shown to be less likely to have BV and sexually transmitted disease (STD) pathogens such as *N. gonorrhoeae, Chlamydia trachomatis,* and

Trichomonas vaginalis [13–15]. Thus, it has been postulated that microbial production of hydrogen peroxide may play an important stabilizing role for vaginal ecology. Factors that modulate the production of H_2O_2 by lactobacilli are unknown. Other means by which lactobacilli may play a protective role within the vaginal ecosystem include competition for epithelial cell attachment sites and stimulation of the local immune system [5,16].

In addition to the mechanisms specific to lactobacilli, the vagina is equipped with other means of defense. These include the vaginal epithelial mucosa, the pH of the vaginal secretions, local antimicrobial secretions, and cell-mediated and humoral immune factors. The vaginal mucosa serves as a physical barrier to pathogens. The mucus stream, which is produced mainly by the cervix, carries antimicrobial factors and also physically washes microorganisms from the vagina [17]. New data are emerging concerning the active production of defensins and cytokines by the squamous epithelial cells in response to pathogens and other proinflammatory stimuli. These entities may help to initiate the immune response [18]. The normal pH of the vagina is less than 4.5, which is inhospitable to many bacteria. This acidity is maintained through the production of lactic acid as a metabolic by-product of the utilization of glycogen by the lactobacilli. The pH may be affected by changes in the vaginal flora as well as by menses, and by external factors such as douching and semen [19]. Nonspecific antimicrobial factors that are present in the vaginal secretions include lysosymes and lactoferrin, which competes with microbes for iron, zinc, and peroxidases [17]. Specific immune mechanisms include cell-mediated and humoral immunity. The reproductive tract is fully immunocompetent in a cyclic fashion subject to hormonal regulation [20].

Although lactobacilli are the predominant organisms in the healthy adult vagina, there is a diverse array of other bacteria present [15]. For example, *G. vaginalis,* which was previously thought to be diagnostic of BV, actually is present in lower concentrations in up to 55% of women without BV [21]. Many other anaerobic and facultatively anaerobic organisms, as well as several different species of lactobacilli, may inhabit the ecosystem [21,22]. The complex interactions of the various members of the vaginal flora are not well understood, but it is appreciated that there can be day-to-day variability in the composition of the flora even in women without evidence of infection. Studies performed using sequential vaginal cultures or vaginal smears have shown that in some women there are significant transient changes in the flora. The most unstable time appears to be around the time of menses but the exact factors responsible for these changes are unknown [23–28].

Table 27.1
Laboratory Features of Vaginitis

	Candida	*Trichomonas*	Bacterial Vaginosis
pH of discharge[a]	<4.5	>4.5[a]	>4.5
Amine odor with 10% KOH	Negative	Often positive	Positive
Wet prep, saline, or 10% KOH[d]	Leukocytes; budding yeast and pseudohyphae	Leukocytes; motile trichomonads	Clue cells[b,c]
Gram stain	Leukocytes; yeast and pseudohyphae	Leukocytes; may see fixed trichomonads	Clue cells[b], decreased numbers of lactobacilli; increased gram-variable rods and gram-variable curved rods (*Mobiluncus*)

[a]Not reliable during menses or after recent intercourse or douching; usually >4.5 but may be <4.5 for trichomonas.

[b]Clue cells are squamous epithelial cells with large numbers of adherent bacteria.

[c]Leukocytes are not present in bacterial vaginosis itself, but they may be present if a concurrent cervical infection such as *Neisseria gonorrhoeae* or *Chlamydia* is present.

[d]KOH is useful in this setting only for detection of yeast.

III. Diagnostic Work-up of Vaginitis

The appropriate management of vaginal infections mandates a specific, correct diagnosis. The most common forms of mismanagement include empiric diagnoses based on patient history or simple visualization of the discharge and failure to screen asymptomatic sexually active women for infection, particularly for trichomoniasis. Despite the fact that BV is the most prevalent infection, the most common empiric diagnosis by practitioners is yeast [29,30]. Although generalizations can be made about the appearance of the vaginal discharge associated with a particular infection, specific testing must be done to confirm the diagnosis. Mixed infections may occur [31]. Many women with BV or trichomoniasis are asymptomatic and will only be diagnosed by the use of routine screening. Although the treatment of asymptomatic BV is controversial, treatment of asymptomatic trichomoniasis is not, thus women diagnosed with trichomoniasis deserve counseling and treatment.

The bedside work-up of vaginitis is relatively quick and inexpensive. It includes description of the discharge, measurement of the vaginal pH, the "whiff" test for amines, and, most importantly, examination of the vaginal fluid under the microscope [32,33]. It is important that the sample for pH measurement be obtained from the vagina and not be contaminated with cervical secretions, which are normally more alkaline. Blood, semen, and recent douching may also interfere with this test. The sample is rubbed onto pH paper and the color matched to the color chart. A second vaginal swab specimen may be used for microscopy and the whiff test. This specimen is diluted in normal saline and 10% potassium hydroxide (KOH). The latter is then sniffed for the presence of an amine, fishy odor, which if present may indicate BV or trichomoniasis. The saline wet prep is examined under 400X magnification for the presence of budding yeast and/or pseudohyphae, motile trichomonads, white blood cells, and "clue" cells (squamous epithelial cells covered with bacteria to the extent that the edges of the cell are ob-

scured). The careful observer will also note the amount and morphotypes of the vaginal bacteria. White blood cells may occur with vulvovaginal candidiasis or trichomoniasis. They may also be emanating from the cervical secretions, suggesting the presence of cervicitis. In preparing and examining the sample for microscopy, it is important to obtain a sufficient amount of the specimen so that it will not be too dilute when mixed with the saline. Also, it may be necessary to examine more than one sample preparation, especially when looking for the pseudohyphae of yeast. By utilizing the information gained from these simple bedside tests, one can arrive at a specific diagnosis in the majority of cases (Table 27.1). Additional information regarding diagnosis may be found in the sections on specific infections.

IV. Vulvovaginal Candidiasis

Vulvovaginal candidiasis is the best known of the vaginal infections, especially among the general population. This notoriety is likely contributed to by the availability of over-the-counter (OTC) antifungal medications and the resultant advertising in the lay press. Many women assume in error that their vaginal symptoms are due to a yeast infection or are misdiagnosed by their physician.

It is estimated that yeast infections are responsible for 20–30% of cases of vaginitis [34]. The highest incidence of VVC occurs during the third and fourth decades of life [35].

Yeast may be colonizers of the vagina in about 15% of women and may also colonize the skin and alimentary tract [34]. Symptomatic VVC is thought to occur as the result of an overgrowth phenomenon in response to various triggers. Conditions that predispose a woman to VVC include diabetes mellitus, pregnancy, exogenous estrogens, use of broad-spectrum antimicrobials, immunosuppression, and possibly sexual intercourse. In most instances a trigger cannot be identified [31,35]. A small subset of women are subject to recurrent VVC [35].

Candida albicans is the species most likely to cause VVC, although it may be caused by *C. glabrata* or *C. tropicalis* and rarely by other species of *Candida* [34]. Several different strains of *C. albicans* appear capable of causing infection. In women with recurrent VVC, it is usually the same strain that persists [36]. Women with VVC do not have demonstrable alterations of their vaginal flora and maintain a predominance of lactobacilli [37].

Symptoms of VVC consist of vaginal discharge and pruritus. The discharge may have a thick, white, "cottage cheese" appearance and may be adherent to the mucosa, but signs such as erythema, edema, and excoriations may predominate [34]. An attempt should be made to confirm the diagnosis by direct microscopy of the vaginal secretions. The presence of the pseudohyphae of *Candida* is diagnostic. These may be better visualized by adding 10% KOH to the fluid which lyses the other cellular elements and bacteria. It may be necessary to examine multiple specimens in order to detect the pseudohyphae, as the sensitivity of these tests is suboptimal [35]. Occasionally, only budding yeast will be seen, especially in the patient with *C. glabrata* [31]. The vaginal pH is <4.5 and the whiff test is negative in patients with VVC unless there is a coinfection with BV or *T. vaginalis*. As this is an inflammatory condition, white blood cells are frequently present. Observation of a few budding yeast in the wet mount of the asymptomatic patient should be considered normal. The use of culture should be reserved for special circumstances such as confirming a clinical diagnosis in the patient with negative microscopy or in the patient with refractory infections where identification and susceptibility testing may be helpful [38]. Use of culture on a routine basis will lead to overdiagnosis, as yeast may colonize the vagina without causing infection.

Treatment options for VVC include oral and topical preparations, all with efficacy rates exceeding 80% [31]. The azole intravaginal creams and suppositories such as clotrimazole, terconazole, and butoconazole remain the most commonly prescribed and are used for 3–7 days. OTC formulations are available. Among the oral agents, ketoconazole has been largely replaced by fluconazole due to the more favorable side effect profile of the latter. Single-dose fluconazole has been shown to be as efficacious as the intravaginal products and has a similar cost, however, it should not be used during pregnancy [39]. Women with recurrent VVC require an induction treatment of 14 days duration and then benefit from weekly maintenance doses of medication for at least six months [35]. This regimen has been shown to prevent recurrences in 90% of women during the maintenance phase [35]. Although *Candida* species other than *albicans* tend to have higher minimum inhibitory concentrations to the azoles, the level of active drug present in the commonly used preparations usually is sufficient for cure [35].

V. Trichomoniasis

Of the three major causes of infectious vaginitis, trichomoniasis is the only one proven to be sexually transmitted. Trichomoniasis is not a reportable disease, thus the true number of cases is unknown, but the annual incidence in the U.S. is estimated at 3 million cases [34]. Although data derived from visits to physicians offices would suggest that the number of cases is declining, certain populations remain heavily affected. For example, the prevalence in sexually transmitted disease (STD) clinic populations may be over 25% [40]. Based on data from the National Disease and Therapeutic Index Survey, cases of vaginal trichomoniasis are geographically widespread but higher rates occur in the South. There appears to be a higher proportion of cases among African-Americans, however, nearly two-thirds of all physician visits for this problem were by white women. Infections are most prevalent among young, sexually active women [41].

The infection is caused by a parasite, *Trichomonas vaginalis*. Although survival on fomites is documented, the organism is believed to be almost exclusively transmitted by sexual activity [42]. The incubation period of this infection is unknown, but *in vitro* studies suggest an incubation period of 4–28 days [43]. *Trichomonas* frequently occurs as a coinfection with *Neisseria gonorrhoeae* and BV [42–44].

Women who are symptomatic from trichomoniasis complain of vaginal discharge, pruritus, and irritation. Signs of infection include vaginal discharge (42%), odor (50%), and edema or erythema (22–37%). The discharge is classically described as frothy in appearance but is actually frothy in only about 10% of patients. The color of the discharge may vary. Colpitis macularis ("strawberry cervix") is a specific clinical sign for this infection but is only detected by colposcopy and not during routine examination [44]. Other complaints may include dysuria and lower abdominal pain, the etiology of the latter being unclear [45]. Nearly half of all women with *T. vaginalis* are asymptomatic [46]. Thus, if these women are not screened, the diagnosis will be missed. The extent of the inflammatory response to the parasite determines the severity of the symptoms. Factors that influence the host's inflammatory response are not well understood but may include the coexisting vaginal flora and the strain and relative concentration of the organisms present in the vagina. Asymptomatic infections may ultimately become symptomatic [47]. The majority of infected men are asymptomatic but trichomoniasis may cause nongonococcal urethritis.

Trichomoniasis has been linked to serious sequelae such as preterm birth and HIV transmission/acquisition. In a multicenter study of vaginal infections during pregnancy, trichomoniasis was significantly associated with preterm delivery [48]. Interestingly, bovine venereal trichomoniasis is a similar infection in cattle and it causes infertility [49]. HIV infection has been associated with trichomoniasis in various studies [50,51]. Local inflammation and microabrasions caused by the parasite may facilitate acquisition of HIV through the mucosal barrier. Transmission of HIV may be enhanced by increased genital shedding of HIV in the setting of inflammation.

The most common means of diagnosis is the visualization of the motile trichomonads in a saline preparation of the vaginal fluid. The organisms are about the size of a white blood cell and may be actively motile or may be seen at rest beating their flagella. Although quick and inexpensive, the test has limited sensitivity, ranging from 60–70% [40,52,53]. There are often white blood cells in the vaginal fluid. The vaginal pH is elevated (greater than 4.5) in the majority of cases but may be normal. The whiff test is variable.

Currently, the gold standard for the diagnosis of trichomoniasis is culture. Traditionally, this has been accomplished though cultivation in Diamond's medium which is not widely available and thus was used mainly for research purposes. However, a new commercially available culture method comprised of liquid media in a clear pouch has been shown to be as good as the traditional research method. The media allows for the multiplication of the trichomonads to sufficient numbers so that they can be detected by microscopy. This method has been used successfully with both clinician-obtained and self-obtained specimens, the latter becoming quite useful in situations where pelvic examination is not possible or desirable (for example screening in adolescents, developing countries). Results from culture are available in 2–5 days [40,54,55]. Polymerase chain reaction (PCR) tests for *T. vaginalis* are currently under development. An office-based oligonucleotide probe test is also available which has a sensitivity of 80–90% and a specificity of 95% [56,57]. Trichomonads may be seen on Pap smears, with a sensitivity of about 60% and a specificity of 95% [52].

Metronidazole is the only efficacious antibiotic available in the U.S. for the treatment of trichomoniasis. The recommended dose is 2 g orally in a single dose. Sexual partners should also be treated. Metronidazole intravaginal gel has limited efficacy for trichomoniasis [58]. Although there continues to be some controversy about the safety of metronidazole during pregnancy, there has never been a documented case of fetal malformation attributed to its use, even during the first trimester [59,60]. Occasionally patients are allergic to metronidazole. Because there is no effective alternative, desensitization is the only option [58]. Another therapeutic dilemma is metronidazole resistance in *T. vaginalis*. This resistance is relative and can usually be overcome with higher doses of oral metronidazole [58,61]. Intravenous formulations offer no advantage over the oral drug. Women with asymptomatic infection should be treated. If left untreated, they may later become symptomatic and, without treatment, they continue to transmit the infection.

VI. Bacterial Vaginosis

BV is the most frequent cause of vaginal discharge in the U.S. [45]. Although not a reportable disease, it is estimated that over 3 million symptomatic cases occur annually [62]. Because approximately one-half of women with BV are asymptomatic [32], the actual annual number of cases may exceed 6 million. The prevalence of BV ranges from 25% in the general population to 50–60% among women attending STD clinics. It is nearly exclusively seen in sexually active women. Previously given little attention and called nonspecific vaginitis or *Gardnerella* vaginitis, BV is now known to be significantly associated with complications of pregnancy, including preterm rupture of membranes, preterm delivery, and low birth weight [63–65]. Additionally, it has been associated with gynecological complications such as postabortal endometritis, posthysterectomy vaginal cuff cellulitis, and pelvic inflammatory disease (PID) [65,66].

Although the pathogenesis of BV remains obscure, the microbiological correlates of BV are well described. Briefly, an unknown inciting event leads to dramatic changes in the vaginal flora. The normal vagina is inhabited primarily by lactobacilli within an acidic environment. In women with BV, lactobacilli, especially H_2O_2-producing strains, are greatly diminished and the vagina becomes overgrown with large numbers of anaerobic

and facultatively anaerobic bacteria [22,67]. These organisms include greatly increased numbers of *G. vaginalis, Mycoplasma hominis, Bacteroides spp., Prevotella spp, Peptostreptococci,* and *Mobiluncus spp.* [67–69]. Although most of these organisms may be present in small numbers within the normal vagina, *Mobiluncus spp,* are seldom seen in normals [70]. This tremendous increase in the number of anaerobic organisms within the vagina of women with BV has been described as comparable to an anaerobic abscess [71]. Several environmental consequences occur along with this anaerobic overgrowth, including reduced oxidation-reduction potential and accumulation of polyamines [71]. Polyamines are responsible for the odor of BV and may contribute to the vaginal epithelial cell exfoliation that characterizes BV [71,72]. Interestingly, there appears to be an absence of an inflammatory response with this condition, hence the term "vaginosis" rather than "vaginitis."

The precise etiology of BV is unknown. Additionally, the sequence of the microbiological events that occurs is poorly defined. Epidemiologic correlates of BV include use of an IUD, history of prior STDs (especially trichomoniasis), increased numbers of sexual partners, a new sexual partner within the month preceding the onset of symptoms, and douching [32,73–76]. The cause and effect relationship between BV and douching is unclear. It is not known if douching causes changes in the vaginal flora or if women douche in response to the odor caused by BV.

Sexual transmission of BV has been suggested frequently but has not been proven. BV is rare in sexually inexperienced females [77]. In their original report on *Haemophilus vaginalis* (now *G. vaginalis*), Gardner and Dukes stated unequivocally that the disease was sexually transmitted based on high recurrence rates and the isolation of the organism (*H. vaginalis*) from over 90% of the male partners [78]. *G. vaginalis* and anaerobic bacteria similar to those isolated from BV patients have been associated with balanoposthitis in males, an infection which is in some ways clinically and microbiologically similar to the syndrome of BV in females [79,80]. Further, *G. vaginalis* and anaerobes have been frequently isolated from the semen of asymptomatic males [81]. Finally, among lesbians in monogamous relationships, a high concordance rate of BV has been found [82]. This, as well as the presence of BV related microorganisms in the male genital tract, is supportive of sexual transmission.

In an opposing point of view, the presence of BV-associated bacteria within the rectum has been documented and could represent a potential reservoir of infection. Rectal carriage of *Mobiluncus spp, M. hominis,* or *G. vaginalis* was found in 45–62% of women with BV versus 10–14% of women without BV, suggesting that bacteria associated with BV emanated from a rectal reservoir rather than from sexual transmission [83]. Additionally, treatment of the male partner has not been reliably associated with decreased recurrence rates of BV [84–86].

Symptomatic BV causes vaginal discharge and/or odor. The odor usually is described as "fishy" and may be more noticeable after unprotected intercourse or during menses. Occasionally the patient may complain of vulvar irritation. Because a single etiologic agent has not been identified, clinicians rely on the so-called Amsel criteria [32]. These criteria state that BV is present if three of the four following conditions are met: (1) an elevated vaginal pH (>4.5), (2) positive amine odor with 10% KOH (whiff test), (3) presence of "clue cells" (squamous epithelial cells covered with adherent bacteria) in a saline preparation of the vaginal fluid, and (4) a homogenous vaginal discharge [32,33]. Although the original Amsel criteria did not specify the quantity of clue cells that must be present, it has been suggested by some that at least 20% of the epithelial cells present should be clue cells [33]. This modification does slightly increase the specificity of the diagnosis, but significantly decreases the sensitivity [87]. When examining the vaginal fluid under the microscope, the morphotypes of the bacteria should also be noted. For example, if only lactobacillus morphotypes are present (moderately long rods) it is unlikely that the patient has BV. On the other hand, motile curved rods which represent *Mobiluncus* are highly suggestive of the diagnosis [69,88]. White blood cells will not be present in the patient with BV alone, but their presence in the vaginal fluid should alert the clinician to the possibility of a coinfection either in the vagina or in the cervix. Although the Amsel criteria are the most commonly used diagnostic methods for BV, there are difficulties with the sensitivity and specificity of the individual parameters. For example, an elevated vaginal pH lacks specificity and the whiff test is subjective and lacks sensitivity [33]. Although the presence of clue cells is the most sensitive and specific of the individual parameters, it is subject to the skill and interpretation of the microscopist. Gram stains are a reliable means of diagnosing BV and have the advantage of being a permanent record which can be reviewed. Bacterial morphotypes consistent with lactobacilli, *Gardnerella*, and *Mobiluncus* are quantitated using this method. Standardized criteria have been developed that have facilitated interpretation and have resulted in good intra-and interobserver reproducibility [89]. In a multicenter study comparing the vaginal Gram stain to the Amsel criteria, the sensitivity and specificity of the Gram stain were found to be 89% and 83%, respectively. If the Gram stain was considered as the gold standard, the sensitivity and specificity of the Amsel criteria were 70 and 94%, respectively, suggesting that the use of the Amsel criteria may lead to underdiagnosis [87]. Other diagnostic tests for BV include detection of proline aminopeptidase in the vaginal fluid, an oligonucleotide probe technique based on high concentrations of *G. vaginalis* [57], and a rapid card test for detection pH and amines that may be useful for initial screening. Vaginal cultures for *Gardnerella* are of no value in diagnosing BV because *Gardnerella* may colonize up to 60% of healthy women [21].

Half of all women with BV are asymptomatic [32]. Previous studies have not systematically investigated whether women with asymptomatic BV are truly without symptomatology or whether poorly recognized or underreported symptoms may be present. One-fourth of women with BV admit to vaginal symptoms only after direct questioning [32]. Women may be reluctant to admit to the presence of unpleasant vaginal odor, or some women may have come to accept this odor as normal. Physicians may reinforce the latter by failing to diagnose and treat BV properly. The treatment of asymptomatic BV in the non-pregnant patient is controversial and is currently not the standard of practice [58]. Possible exceptions include women undergoing elective abortions or hysterectomies, as BV has been associated with infectious complications in these patients. Guidelines are beginning to be established for the screening and treatment of BV in the pregnant patient due to the association of BV with preterm labor and delivery. Currently, some experts recommend screening for and treating asymptomatic BV during the second trimester for women who are at high risk for preterm delivery (women who have had a previous preterm delivery) [58]. Treatment of these women has been shown to significantly decrease the incidence of preterm birth [90]. Studies addressing the same issues in women not at high risk for preterm delivery are ongoing.

Treatment of BV is aimed at eradicating the anaerobic and facultatively anaerobic bacteria that are present in this polymicrobial infection. Cure rates with agents such as metronidazole or clindamycin approach 90% [91,92]. Cures, however, often are not long-lasting, and recurrence rates of 30–40% within 3 months have been reported [93]. Recurrent disease may represent relapse or reinfection. In support of the former is the fact that even after successful eradication of the majority of abnormal bacteria, persistent abnormalities such as elevated vaginal pH and low levels of BV-associated bacteria may persist [94]. The issue of reinfection has been poorly studied, but in a longitudinal study of *G. vaginalis* biotypes, 90% of women who developed BV acquired a new biotype, suggesting a new infection rather than relapse [95]. Options for treatment include both oral and topical metronidazole and clindamycin. Although some clinicians favor using a 2 g one time oral dose of metronidazole to enhance compliance, 500 mg orally twice a day for 7 days has higher efficacy [84]. Topical preparations of metronidazole or clindamycin are as efficacious for treating BV as oral agents and avoid systemic side effects, although local side effects, such as vaginal yeast infections, may occur [58,96–98]. There is concern that topical agents may not be adequate therapy for the pregnant patient in whom upper tract colonization with BV-associated bacteria may have occurred, but no studies have addressed this issue specifically [99]. The use of clindamycin cream has been recommended against in the pregnant patient because it has been associated with an increase in preterm births [58]. Many clinicians are reluctant to use metronidazole during pregnancy, but there are no data to support these concerns [59,60]. Reconstitution of the vaginal flora with exogenous lactobacilli has been suggested as an adjunct to antibiotic therapy, but this requires use of a human-derived strain for effective colonization [31,100].

Bacterial vaginosis is frequently present as a coinfection with other cervical and vaginal infections. An association between BV and cervicitis has been well documented [74,101,102]. Up to 50% of women with cervicitis may have concomitant BV, a finding that has led to the hypothesis that the physiological changes associated with cervicitis create conditions favorable for the development of BV [101,103]. Parenthetically, this high coinfection rate creates a therapeutic dilemma because prior to the development of topical agents for BV treatment, simultaneous treatment of both infections would require administration of two oral agents each having gastrointestinal side effects. This problem, combined with the belief that BV might spontaneously resolve with treatment of cervicitis, led to recommendations to treat only the cervical infection [101]. New data show that such an approach does not lead to the resolution of BV and instead leads to an unnecessary and potentially undesirable delay in the reestablishment of normal vaginal flora [104].

Other associations between BV and STDs have been studied. In a retrospective follow-up study of female sexual contacts with men with gonorrhea, women without lactobacilli were more likely to be infected with *N. gonorrhoeae* than those with lactobacilli [9]. BV is strongly associated with the only proven cause of sexually transmitted vaginitis, *T. vaginalis*. Women with trichomonas are frequently coinfected with BV or BV-associated bacteria, with coinfection rates as high as 86% reported [44,105]. Information is also accumulating on the possible role of BV as a risk factor for HIV infection. Among female commercial sex workers in Thailand, HIV seropositivity was significantly correlated with BV [106]. Similar findings were documented from a study of women in Uganda [107]. Thus, an alternative explanation for the coexistence of BV and STDs is that the absence of protective lactobacilli found in patients with BV precedes and facilitates the acquisition of an STD.

Upper genital tract infections also are associated with abnormal vaginal flora. Bacterial vaginosis has been epidemiologically linked to pelvic inflammatory disease (PID) and BV-associated microorganisms often are isolated from tubal cultures of women with PID both in the presence and in the absence of *C. trachomatis* or *N. gonorrhoeae* [33,66,108,109]. In one study, anaerobic bacteria were the predominant isolates, occurring in 92% of patients with nongonococcal, nonchlamydial PID [110]. Endometritis represents the initial event in the pathogenesis of salpingitis (inflammation of the fallopian tube). In a study comparing different regimens for the treatment of PID, BV-related bacteria were isolated significantly more often from the endometrium of women who failed initial therapy for PID [111]. BV-related organisms appear to be able to colonize the endometrium. Among women without clinical symptoms of upper genital tract infection, women with BV were found to be more likely to have endometrial colonization with BV-associated organisms as well as histological evidence of endometritis than those women without BV [112]. Other upper genital tract infections that have been associated with BV include posthysterectomy endometritis, vaginal cuff cellulitis, and PID after first trimester abortion [65,66,113,114].

Colonization of the upper genital tract with BV-associated bacteria during pregnancy has also been demonstrated. In a study of chorioamnionitis in prematurity, microbial colonization of the chorioamnion was associated with preterm labor [64]. BV has been associated with low birth weight, premature rupture of the membranes, and prematurity in numerous epidemiologic studies [64,65,115–117]. In a randomized, controlled trial of the use of antibiotics (metronidazole and erythromycin) versus placebo in the second trimester for prevention of preterm delivery in high risk women, the use of antibiotics resulted in a decrease in preterm delivery, especially among those women who had BV [90]. Another study using metronidazole alone versus placebo in a group of high risk women with BV also found a decrease in preterm birth in the treated group [118]. Postpartum endometritis is also more common among women with BV [65].

VII. Conclusion

Increased awareness of the role of the normal vaginal flora and the adverse consequences associated with vaginitis have led to a new appreciation of the importance of timely, accurate diagnosis and treatment. Quick and inexpensive diagnostic tests are available and screening in high risk women should be undertaken. Additional information is needed to determine the benefit of screening and treatment in asymptomatic low risk populations.

References

1. Kent, H. Z. (1991). Epidemiology of vaginitis. *Am. J. Obstet. Gynecol.* **165,** 1168–1176.
2. Ferris, D. G., Dekle, C., and Litaker, M. S. (1996). Women's use of over-the-counter antifungal medications for gynecologic symptoms. *J. Fam. Pract.* **42,** 595–600.
3. Nyirjesy, P., Weitz, M. V., and Lorber, B. (1997). Over-the-counter and alternative medicines in the treatment of chronic vaginal symptoms. *Obstet. Gynecol.* **90,** 50–53.
4. Paavonen, J. (1983). Physiology and ecology of the vagina. *Scand. J. Infect. Dis.* **S40,** 31–35.
5. Redondo-Lopez, V., Cook, R. L., and Sobel, J. D. (1990). Emerging role of lactobacilli in the control and maintenance of the vaginal bacterial microflora. *Rev. Infect. Dis.* **12,** 856–872.
6. Whittenbury, R. (1964). Hydrogen peroxide formation and catalase activity in the lactic acid bacteria. *J. Gen. Microbiol.* **35,** 13–26.
7. Barefoot, S. F., and Klaenhammer, T. R. (1983). Detection and activity of lactacin B. A bacteriocin produced by *Lactobacillus acidophilus. Appl. Environ. Microbiol.* **45,** 1808–1815.
8. Skarin, A., and Sylwan, J. (1986). Vaginal lactobacilli inhibiting growth of *Gardnerella vaginalis, Mobiluncus* and other bacterial species cultured from vaginal content of women with bacterial vaginosis. *Acta Pathol. Microbiol. Immunol. Scand., Sect. B* **94B,** 399–403.
9. Saigh, J. H., Sanders, C. C., and Sanders, W. E. (1978). Inhibition of *Neisseria gonorrhoeae* by aerobic and facultatively anaerobic components of the endocervical flora: Evidence for a protective effect against infection. *Infect. Immun.* **19,** 704–710.
10. Klebanoff, S. J., and Coombs, R. W. (1991). Viricidal effect of *Lactobacillus acidophilus* on human immunodeficiency virus type I: Possible role in heterosexual transmission. *J. Exp. Med.* **174,** 289–292.
11. Klebanoff, S. J., and Hillier, S. L. (1991). Control of the microbial flora of the vagina by H$_2$O$_2$-generating lactobacilli. *J. Infect. Dis.* **164,** 94–100.
12. Zheng, H., Alcorn, T. M., and Cohen, M. S. (1994). Effects of H$_2$O$_2$-producing lactobacilli on *Neisseria gonorrhoeae* and catalase activity. *J. Infect. Dis.* **170,** 1209–1215.
13. Hillier, S. L., Krohn, M. A., Klebanoff, S. F., and Eschenbach, D. A. (1992). The relationship of hydrogen peroxide-producing lactobacilli to bacterial vaginosis and genital microflora in pregnant women. *Obstet. Gynecol.* **79,** 369–373.
14. Hillier, S. L., Krohn, M. A., Nugent, R. P., and Gibbs, R. S. (1992). Characteristics of three vaginal flora patterns assessed by Gram stain among pregnant women. *Am. J. Obstet. Gynecol.* **166,** 938–944.
15. Hillier, S. L., Krohn, M. A., Nugent, R. P., and Gibbs, R. S. (1993). The normal vaginal flora, H$_2$O$_2$-producing lactobacilli and bacterial vaginosis in pregnant women. *Clin. Infect. Dis.* **16**(S4), S273–S281.
16. Boris, S., Suarez, J. E., Vazquez, F., and Barber, C. (1998). Adherence of human vaginal lactobacilli to vaginal epithelial cells and interaction with uropathogens. *Infect. Immun.* **66,** 1985–1989.
17. Cohen, M. S., Weber, R. D., and Mårdh, P. A. (1990). Genitourinary mucosal defenses. *In* "Sexually Transmitted Diseases" (K. K. Holmes, P. A. Mårdh, and P. G. Sparling, eds., pp. 117–127. McGraw-Hill, New York.

18. Rasmussen, S. J., Echmann, L., Quayle, A. J., Shen, L., Zhang, Y., Anderson, D. J., Fierer, J. S., Stephens, J., Kagnoff, R. S., Stephens, R. S., and Kagnoff, M. F. (1997). Secretion of pro-inflammatory cytokines by epithelial cells in response to chlamydia infections suggests a central role for epithelial cells in chlamydial pathogenesis. *J. Clin. Invest.* **99,** 77–87.

19. Stevens-Simon, C., Jamison, J., McGregor, J. A., and Douglas, J. M. (1994). Racial variation in vaginal pH among healthy sexually active adolescents. *Sex. Transm. Dis.* **21,** 168–172.

20. White, H. D., Yeaman, G. R., Givan, A. L., and Wira, C. R. (1997). Mucosal immunity in the human female reproductive tract: Cytotoxic T lymphocyte function in the cervix and vagina of premenopausal and postmenopausal women. *Am. J. Reprod. Immunol.* **37,** 30–38.

21. Hillier, S. L. (1993). Diagnostic microbiology of bacterial vaginosis. *Am. J. Obstet. Gynecol.* **169,** 455–459.

22. Eschenbach, D. A., Davick, P. R., Williams, B. L., Klebanoff, S. J., Young-Smith, K., Critchlow, C. M., and Holmes, K. K. (1989). Prevalence of hydrogen peroxide-producing *Lactobacillus* species in normal women and women with bacterial vaginosis. *J. Clin. Microbiol.* **27,** 251–256.

23. Bartlett, J. G., Onderdonk, A. B., Drude, E., Goldstein, C., Anderka, M., Alpert, S., and McCormick, W. M. (1977). Quantitative bacteriology of the vaginal flora. *J. Infect. Dis.* **136,** 271–277.

24. Sautter, R. L., and Brown, W. J. (1980). Sequential vaginal cultures from normal young women. *J. Clin. Microbiol.* **11,** 479–484.

25. Brown, W. J. (1982). Variations in the vaginal bacterial flora. *Ann. Intern. Med.* **96,** 931–934.

26. Johnson, S. R., Petzold, C. R., and Galask, R. P. (1985). Qualitative and quantitative changes of the vaginal microbial flora during the menstrual cycle. *Am. J. Reprod. Immunol. Microbiol.* **9,** 1–5.

27. Priestley, C. J. F., Jones, B. M., Dhar, J., and Goodwin, L. (1997). What is normal vaginal flora? *Genitourin Med.* **73,** 23–28.

28. Schwebke, J. R., Morgan, S. C., and Weiss, H. L. (1997). The use of sequential self-obtained vaginal smears for detecting changes in the vaginal flora. *Sex. Transm. Dis.* **24,** 236–239.

29. Berg, A. O., Heidrich, F. E., Fihn S. D., Bergmann, J. J., Wood, R. W., Stamm W. E., and Holmes, K. K. (1984). Establishing the cause of genitourinary symptoms in women in a family practice: Comparison of clinical examination and comprehensive microbiology. *JAMA, J. Am. Med. Assoc.* **251,** 620–625.

30. Thomason, J. L., Gelbart, S. M., and Scaglione, J. J. (1991). Bacterial vaginosis: Current review with indications for asymptomatic therapy. *Am. J. Obstet. Gynecol.* **165,** 1210–1217.

31. Sobel, J. D. (1997). Vaginitis. *N. Engl. J. Med.* **337,** 1896–1903.

32. Amsel, R., Totten, P. A., Spiegel, C. A., Chen, K. C. S., Eschenbach, D., and Holmes, K. K. (1983). Non-specific vaginitis: Diagnostic and microbial and epidemiological associations. *Am. J. Med.* **74,** 14–22.

33. Eschenbach, D. A., Hillier, S., Critchlow, C., Stevens, C., DeRouen, T., and Holmes, K. K. (1988). Diagnosis and clinical manifestations of bacterial vaginosis. *Am. J. Obstet. Gynecol.* **158,** 819–828.

34. Sparks, J. M. (1991). Vaginitis. *J. Reprod. Med.* **36,** 745–752.

35. Sobel, J. D., Faro, S., Force, R. W., Forman, B., Ledger, W. J., Nyirjesy, P. R., and Reed, B. D. (1998). Vulvovaginal candidiasis: Epidemiologic, diagnostic and therapeutic considerations. *Am. J. Obstet. Gynecol.* **178,** 203–211.

36. Vazquez, J. A., Sobel, J. D., Demitriou, R., Vaishampayan, J., Lynch, M., and Zervos, M. J. (1994). Karyotyping of *Candida albicans* isolates obtained longitudinally in women with recurrent vulvovaginal candidiasis. *J. Infect. Dis.* **170,** 1566–1569.

37. Sobel, J. D., and Chaim, W. (1996). Vaginal microbiology of women with acute recurrent vulvovaginal candidiasis. *J. Clin. Microbiol.* **34,** 2497–2499.

38. Nyirjesy, P., Seeney, S. M., Grody, M. H. T., Jordan, C. A., and Brickley, H. R. (1995). Chronic fungal vaginitis: The value of cultures. *Am. J. Obstet. Gynecol.* **173,** 820–823.

39. Sobel, J. D., Brooker, D., Stein, G. E., Thomason, J. L., Wermeling, D. P., Bradley, B., Weinstein, L., and Study Group FV (1995). Single oral dose fluconazole compared with conventional clotrimazole topical therapy of *Candida* vaginitis. *Am. J. Obstet. Gynecol.* **172,** 1263–1264.

40. Schwebke, J. R., Morgan, S. C., and Pinson, G. B. (1997). Validity of self obtained vaginal specimens for diagnosis of trichomoniasis. *J. Clin. Microbiol.* **35,** 1618–1619.

41. Lossick, J. G. (1991). The descriptive epidemiology of vaginal trichomoniasis and bacterial vaginosis. *In* "Vaginitis and Vaginosis" B. J. Horowitz, and P. A. Mårdh, eds.), pp. 77–84. Wiley-Liss, New York.

42. Lossick, J. S. (1989). Epidemiology of urogenital trichomoniasis. *In* "Trichomonads Parasitic in Humans" (B. M. Honigberg, ed.), p. 313. Springer, New York.

43. Jirovec, O., and Petri, M. (1968). *Trichomonas vaginalis* and trichomoniasis. *Adv. Parasitol.* **6,** 117.

44. Wølner-Hanssen, P., Krieger, J. N., Stevens, C. E., Kiviat, N. B., Koutsky, L., Critchlow, C., DeRouen, T., Hillier, S., and Holmes, K. K. (1989). Clinical manifestations of vaginal trichomoniasis. *JAMA, J. Am. Med. Assoc.* **261,** 571–576.

45. Rein, M. F., and Holmes, K. K. (1983). "Non-specific vaginitis," vulvovaginal candidiasis, and trichomoniasis clinical features, diagnosis and management. *Curr. Clin. Top. Infect. Dis.* **4,** 281–315.

46. Fouts, A. C., and Kraus, S. J. (1980). *Trichomonas vaginalis:* Reevaluation of its clinical presentation and laboratory diagnosis. *J. Infect. Dis.* **141,** 137–143.

47. Rein, M. F., and Müller, M. (1990). *Trichomonas vaginalis* and *trichomoniasis. In* "Sexually Transmitted Diseases" (K. K. Holmes, P. A. Mårdh, and P. F. Sparling, eds.), pp. 481–490. McGraw-Hill, New York.

48. Cotch, M. F., Pastorek, J. G., Nugent, R. P., Hillier, S. L., Gibbs, R. S., Martin, D. H., Eschenbach, D. A., Edelman, R., Carey, J. C., Reegan, J. A., Krohn, M. A., Klebanoff, M. A., Rao, A. V., and Rhoads, G. G. (1997). *Trichomonas vaginalis* associated with low birth weight and preterm delivery. *Sex. Transm. Dis.* **24,** 361–362.

49. Skirrow, S. Z., and BonDurant, R. H. (1988). Bovine trichomoniasis. *Vet. Bull.* **58,** 591–603.

50. Klouman, E., and Massenga, E. J. (1997). HIV and reproductive tract infections in a total village population in rural Kilimanjaro, Tanzania: Women at increased risk. *J. AIDS Hum. Retrovirol.* **14,** 163–168.

51. Laga, M., Manoka, A., Kivuvu, M., Malele, B., Tuliza, M., Nzila, N., Goeman, J., Behets, F., Batter, V., and Alary, M. (1993). Non-ulcerative sexually transmitted diseases as risk factors for HIV-1 transmission in women: Results from a cohort study. *AIDS* **7,** 95–102.

52. Krieger, J. N., Tam, M. R., Stevens, C. E., Nielsen, I. O., Hale, J., Kiviat, N. B., and Holmes, K. K. (1988). Diagnosis of trichomoniasis. *JAMA, J. Am. Med. Assoc.* **259,** 1223–1227.

53. Pastorek, J. G., Cotch, M. F., Martin, D. H., and Eschenbach, D. A. (1996). Clinical and microbiological correlates of vaginal trichomoniasis during pregnancy. The Vaginal Infections and Prematurity Study Group. *Clin. Infect. Dis.* **23,** 1075–1080.

54. Draper, D., Parker, R., Patterson, E., Jones, W., Beutz, M., French, J., Borchardt, K., and McGregor, J. (1993). Detection of *Trichomonas vaginalis* in pregnant women with the InPouch TV system. *J. Clin. Microbiol.* **31,** 1016–1018.

55. Wawer, M. J., McNairn, D., Wabwire-Mangen, F., Paxton, L., Gray, R. H., and Kiwanuka, N. (1995). Self administered vaginal swabs for population-based assessment of *Trichomonas vaginalis* prevalence. *Lancet* **345,** 131–132.

56. DeMeo, M. R., Draper, D. L., McGregor, J. A., Moore, D. F., Petey, C. R., Kapernick, P. S., and McCormack, W. M. (1996). Evaluation of a deoxyribonucleic acid probe for the detection of *Trichomonas vaginalis* in vaginal secretions. *Am. J. Obstet. Gynecol.* **174**, 1339–1342.

57. Briselden, A. M., and Hillier, S. L. (1994). Evaluation of Affirm VP microbial identification test for *Gardnerella vaginalis* and *Trichomonas vaginalis*. *J. Clin. Microbiol.* **32**, 148–152.

58. Centers for Disease Control (1998). 1998 Guidelines for treatment of sexually transmitted diseases. *Morbid. Mortal. Wkly. Rep.* **47**(RR-1), 70–79.

59. Burtin, P., Taddio, A., Ariburnu, O., Einarson, T. R., and Koren, G. (1995). Safety of metronidazole in pregnancy: A meta-analysis. *Am. J. Obstet. Gynecol.* **172**, 525–529.

60. Schwebke, J. R. (1995). Metronidazole: Utilization in the obstetric and gynecologic patient. *Sex. Transm. Dis.* **22**, 370–376.

61. Lossick, J. G., Muller, M., and Gorrell, T. E. (1986). *In vitro* drug susceptibility and doses of metronidazole required for cure in cases of refractory vaginal trichomoniasis. *J. Infect. Dis.* **153**, 948–955.

62. Fleury, F. J. (1981). Adult vaginitis. *Clin. Obstet. Gynecol.* **24**, 407–438.

63. Martius, J., and Eschenbach, D. A. (1990). The role of bacterial vaginosis as a cause of amniotic fluid infection, chorioamnionitis and prematurity—a review. *Arch. Gynecol. Obstet.* **247**, 1–13.

64. Hillier, S. L., Martius, J., Krohn, M., Kiviat, N., Holmes, K. K., and Eschenbach, D. A. (1988). A case-control study of chorioamnionic infection and histologic chorioamnionitis in prematurity. *N. Engl. J. Med.* **319**, 972–978.

65. Eschenbach, D. A. (1993). Bacterial vaginosis and anaerobes in obstetric-gynecologic infection. *Clin. Infect. Dis.* **16**, S282–S287.

66. Paavonen, J., Teisala, K., Heinonen, P. K., Aine, R., Laine, S., Lehtinen, M., Miettinen, A., Punnonen, R., and Gronroos, P. (1987). Microbiological and histopathological findings in acute pelvic inflammatory disease. *Br. J. Obstet. Gynaecol.* **94**, 454–460.

67. Spiegel, C. A., Amsel, R., Eschenbach, D., Schoenknecht, F., and Holmes, K. K. (1980). Anaerobic bacteria in non-specific vaginitis. *N. Engl. J. Med.* **303**, 601–607.

68. Spiegel, C. A. (1991). Bacterial vaginosis. *Clin. Microbiol. Rev.* **4**, 485–502.

69. Thomason, J. L., Schreckenberger, P. C., Spellacy, W. N., Riff, L. J., and Le Beau, L. J. (1984). Clinical and microbiological characterization of patients with non-specific vaginosis associated with motile, curved anaerobic rods. *J. Infect. Dis.* **149**, 801–809.

70. Hillier, S. L., Critchlow, C. W., Stevens, C. E., Roberts, M. C., Wølner-Hanssen, P., Eschenbach, D. A., and Holmes, K. K. (1991). Microbiological, epidemiological and clinical correlates of vaginal colonization by *Mobiluncus* species. *Genitourin. Med.* **67**, 26–31.

71. Sobel, J. D. (1989). Bacterial vaginosis—an ecologic mystery. *Ann. Intern. Med.* **111**, 551–553.

72. Chen, K. C. S., Forsyth, P. S., Buchanan, T. M., and Holmes, K. K. (1979). Amine content of vaginal fluid from untreated and treated patients with nonspecific vaginitis. *J. Clin. Invest.* **63**, 828–835.

73. Barbone, F., Austin, H., Louv, W. C., and Alexander, W. J. (1990). A follow-up study of methods of contraception, sexual activity and rates of trichomoniasis, candidiasis and bacterial vaginosis. *Am. J. Obstet. Gynecol.* **163**, 510–514.

74. Moi, H. (1990). Prevalence of bacterial vaginosis and its association with genital infections, inflammation and contraceptive methods in women attending sexually transmitted disease and primary health clinics. *Int. J. STD AIDS* **1**, 86–94.

75. Paavonen, J., Miettinen, A., Stevens, C. E., Chen, K. C. S., and Holmes, K. K. (1983). *Mycoplasma hominis* in non-specific vaginitis. *Sex. Transm. Dis.* **45**, 271–275.

76. Wølner-Hanssen, P., Eschenbach, D. A., Paavonen, J., Stevens, C. E., Kiviat, N. B., Critchlow, C., DeRouen, T., Koutsky, L., and Holmes, K. K. (1990). Association between vaginal douching and acute pelvic inflammatory disease. *JAMA, J. Am. Med. Assoc.* **263**, 1936–1941.

77. Bump, R. C., and Buesching, W. J. (1988). Bacterial vaginosis in virginal and sexually active adolescent females: Evidence against exclusive sexual transmission. *Am. J. Obstet. Gynecol.* **158**, 935–939.

78. Gardner, H. L., and Dukes, C. D. (1955). *Haemophilus vaginalis* vaginitis. *Am. J. Obstet. Gynecol.* **69**, 962–976.

79. Kinghorn, G. R., Jones, B. M., Chowdhury, F. H., and Geary, I. (1982). Balanoposthitis associated with *Gardnerella vaginalis* infection in men. *Br. J. Vener. Dis.* **58**, 127–129.

80. Masfari, A. N., Kinghorn, G. R., and Duerden, B. I. (1983). Anaerobes in genitourinary infections in men. *Br. J. Vener. Dis.* **59**, 255–259.

81. Hillier, S. L., Rabe, L. K., Muller, C. H., Zarutskie, P., Kuzan, F. B., and Stenchever, M. A. (1990). Relationship of bacteriologic characteristics to semen indices in men attending an infertility clinic. *Obstet. Gynecol.* **75**, 800–804.

82. Berger, B. J., Kolton, S., Zenilman, J. M., Cummings, M. C., Feldman, J., and McCormick, W. M. (1995). Bacterial vaginosis in lesbians: A sexually transmitted disease. *Clin. Infect. Dis.* **21**, 1402–1405.

83. Holst, E. (1990). Reservoir of four organisms associated with bacterial vaginosis suggests lack of sexual transmission. *J. Clin. Microbiol.* **28**, 2035–2039.

84. Swedberg, J., Steiner, J. F., Deiss, F., Steiner, S., and Driggers, D. A. (1985). Comparison of single-dose vs. one-week course of metronidazole for symptomatic bacterial vaginosis. *JAMA, J. Am. Med. Assoc.* **254**, 1046–1049.

85. Vejtorp, M., Bollerup, A. C., Vejtorp, L., Fanoe, E., Nathan, E., Reiter, A., Anderson, M. E., Stromsholt, B., and Schroder, S. S. (1988). Bacterial vaginosis: A double-blind randomized trial of the effect of treatment of the sexual partner. *Br. J. Obstet. Gynaecol.* **95**, 920–926.

86. Mengel, M. B., Berg, A. O., Weaver, C., Herman, D. J., Herman, S. J., Hughes, V. L., Koepsell, T. K. (1989). The effectiveness of single-dose metronidazole therapy for patients and their partners with bacterial vaginosis. *J. Fam. Pract.* **28**, 163–171.

87. Schwebke, J. R., Hillier, S. L., Sobel, J. D., McGregor, J. A., and Sweet, R. L. (1996). Validity of the vaginal Gram stain for the diagnosis of bacterial vaginosis. *Obstet. Gynecol.* **88**, 573–576.

88. Livengood, C. H., Thomason, J. L., and Hill, G. B. (1990). Bacterial vaginosis: Diagnostic and pathogenetic findings during topical clindamycin therapy. *Am. J. Obstet. Gynecol.* **163**, 515–520.

89. Nugent, R. P., Krohn, M. A., and Hillier, S. L. (1991). Reliability of diagnosing bacterial vaginosis is improved by a standardized method of Gram stain interpretation. *J. Clin. Microbiol.* **29**, 297–301.

90. Hauth, J. C., Goldenberg, R. L., Andrews, W. W., DuBard, M. B., and Copper, R. L. (1995). Reduced incidence of preterm delivery with metronidazole and erythromycin in women with bacterial vaginosis. *N. Engl. J. Med.* **333**, 1732–1736.

91. Lossick, J. G. (1990). Treatment of sexually transmitted vaginosis/vaginitis. *Rev. Infect. Dis.* **12**(S6), S665–S681.

92. Pheiffer, T. A., Forsyth, P. S., Durfee, M. A., Pollock, H. M., and Holmes, K. K. (1978). Non-specific vaginitis. *N. Engl. J. Med.* **298**, 1429–1434.

93. Blackwell, A. L., Fox, A. R., Phillips, I., and Barlow, D. (1982). Anaerobic vaginosis (non-specific vaginitis): Clinical, microbiological and therapeutic findings. *Lancet* **2**, 1379–1382.

94. Cook, R. L., Redondo-Lopez, V., Schmitt, C., Meriweather, C., and Sobel, J. D. (1992). Clinical, microbiological, and biochemical

factors in recurrent bacterial vaginosis. *J. Clin. Microbiol.* **30,** 870–877.

95. Briselden, A. M., and Hillier, S. H. (1990). Longitudinal study of the biotypes of *Gardnerella vaginalis. J. Clin. Microbiol.* **28,** 2761–2764.

96. Hillier, S. L., Krohn, M. A., Watts, H., Wølner-Hanssen, P., and Eschenbach, D. (1990). Microbiologic efficacy of intravaginal clindamycin cream for the treatment of bacterial vaginosis. *Obstet. Gynecol.* **76,** 407–412.

97. Hillier, S. L., Lipinski, C., Brieselden, A. M., and Eschanbach, D. A. (1993). Efficacy of intravaginal 0.75% metronidazole gel for the treatment of bacterial vaginosis. *Obstet. Gynecol.* **81,** 963–967.

98. Ferris, D. G., Litaker, M. S., Woodward, L., Mathis, D., and Hendrich, J. (1995). Treatment of bacterial vaginosis: A comparison of oral metronidazole vaginal gel and clindamycin vaginal cream. *J. Fam. Pract.* **41,** 443–449.

99. McGregor, J. A., French, J. I., Jones, W., Milligan, K., McKinney, P. G., Patterson, E., and Parker, R. (1994). Bacterial vaginosis is associated with prematurity and vaginal fluid mucinase and sialidase: Results of a controlled trial of topical clindamycin cream. *Am. J. Obstet. Gynecol.* **170,** 1048–1060.

100. Hughes, V. L., and Hillier, S. L. (1990). Microbiologic characteristics of *Lactobacillus* products used for colonization of the vagina. *Obstet. Gynecol.* **178,** 203–211.

101. Paavonen, J., Critchlow, C. W., CeRouen, T., Stevens, C. E., Kiviat, N., Brunham, R. C., Stamm, W. E., Kuo, C., Hyde, K. E., Corey, L., Eschenbach, D. A., and Holmes, K. K. (1986). Etiology of cervical inflammation. *Am. J. Obstet. Gynecol.* **154,** 556–564.

102. Paavonen, J., Roberts, P. L., Stevens, C. E., Wølner-Hanssen, P., Brunham, R. C., Hillier, S., Stamm, W. E., Kuo, C., DeRouen, T., and Holmes, K. K. (1989). Randomized treatment of mucopurulent cervicitis with doxycycline or amoxicillin. *Am. J. Obstet. Gynecol.* **161,** 128–135.

103. Holmes, K. K., Chen, K. C. S., Lipinski, C. M., and Eschenbach, D. A. (1985). Vaginal redox potential in bacterial vaginosis (nonspecific vaginitis). *J. Infect. Dis.* **152,** 379–382.

104. Schwebke, J. R., Schulien, M. B., and Zajackowski, M. (1995). Pilot study to evaluate the appropriate management of patients with coexistent bacterial vaginosis and cervicitis. *Infect. Dis. Obstet. Gynecol.* **3,** 199–122.

105. Thomason, J. L., Gelbart, S. M., Sobun, J. F., Schulien, M. B., and Hamilton, P. R. (1988). Comparison of four methods to detect *Trichomonas vaginalis. J. Clin. Microbiol.* **26,** 1869–1870.

106. Cohen, C. R., Duerr, A., Pruithithada, N., Rugpao, S., Hillier, S., Garcia, P., and Nelson, K. (1995). Bacterial vaginosis and HIV seroprevalence among female commercial sex workers in Chiang Mai, Thailand. *AIDS* **9,** 1093–1097.

107. Sewankambo, N., Gray, R., Waiver, M. J., Paxton, L., McNairn, D., Wabruire-Mangen, F., Sewankambo, D., Li, C., Kievanuker, N., Hillier, S. L., Rabe, L., Gydos, C. A., Quinn, T. C., and Kondi-Lute, J. (1997). HIV-1 infection associated with abnormal vaginal flora morphology and bacterial vaginosis. *Lancet* **350,** 546–550.

108. Wasserheit, J. N., Bell, T. A., Kiviat, N. B., Wølner-Hanssen, P., Zabriskie, V., Kirby, B. D., Prince, E. C., Holmes, K. K., Stamm, W. E., and Eschenbach, D. A. (1986). Microbial causes of proven pelvic inflammatory disease and efficacy of clindamycin and tobramycin. *Ann. Intern. Med.* **104,** 187–193.

109. Eschenbach, D. A., Buchanan, T. M., Pollock, H. M., Forsyth, P. S., Alexander, E. R., Lin, J., Wang, S., Wentworth, B. B., McCormack, W. M., and Holmes, K. K. (1975). Polymicrobial etiology of acute pelvic inflammatory disease. *N. Engl. J. Med.* **293,** 166–171.

110. Sweet, R. L., Schachter, J., and Robie, M. O. (1983). Failure of beta-lactam antibiotics to eradicate *Chlamydia trachomatis* in the endometrium despite apparent clinical cure of acute salpingitis. *JAMA, J. Am. Med. Assoc.* **293,** 166–171.

111. Hemsell, D. L., Little, B. B., Faro, S., Sweet, R. L., Ledger, W., Berkeley, A. S., Eschenbach, D. A., Wølner-Hanssen, P., and Pastorek, J. G. (1994). Comparison of three regimens recommended by the Centers for Disease Control and Prevention for the treatment of women hospitalized with acute pelvic inflammatory disease. *Clin. Infect. Dis.* **19,** 720–727.

112. Korn, A. P., Bolan, G., Padian, N., Ohm-Smith, M., Schachter, J., and Landers, D. V. (1995). Plasma cell endometritis in women with symptomatic bacterial vaginosis. *Obstet. Gynecol.* **85,** 387–390.

113. Soper, D. E., Bump, R. C., and Hurt, W. G. (1990). Bacterial vaginosis and trichomoniasis vaginitis are risk factors for cuff cellulitis after abdominal hysterectomy. *Am. J. Obstet. Gynecol.* **163,** 1016–1023.

114. Larsson, P. G., Platz-Christensen, J. J., Thejls, H., Forsum, U., and Pahlson, C. (1992). Incidence of pelvic inflammatory disease after first trimester legal abortion in women with bacterial vaginosis after treatment with metronidazole: A double-blind, randomized study. *Am. J. Obstet. Gynecol.* **166,** 100–103.

115. Hillier, S. L., Nugent, R. P., Eschenbach, D. A., Krohn, M. A., Gibbs, R. S., Martin, D. H., Cotch, M. F., Edelman, R., Pastorek, J. G., Rao, A. V., McNellis, D., Regan, J. A., Carey, J. C., and Klebanoff, M. A. (1995). Association between bacterial vaginosis and preterm delivery of a low-birth-weight infant. *N. Engl. J. Med.* **333,** 1737–1742.

116. Holst, E., Goffeng, A. R., and Andersch, B. (1994). Bacterial vaginosis and vaginal microorganisms in idiopathic premature labor and association with pregnancy outcome. *J. Clin. Microbiol.* **32,** 176–186.

117. Martius, J. M., Krohn, A., Hillier, S. L., Stamm, W. E., Holmes, K. K., and Eschenbach, D. A. (1988). Relationships of vaginal *Lactobacillus* species, cervical *Chlamydia trachomatis* and bacterial vaginosis to preterm birth. *Obstet. Gynecol.* **71,** 89–95.

118. Morales, W. J., Schorr, S., and Albritton, J. (1994). Effect of metronidazole in patients with preterm birth in preceding pregnancy and bacterial vaginosis: A placebo-controlled double-blind study. *Am. J. Obstet. Gynecol.* **171,** 345–349.

28

Urinary Tract Infection

BETSY FOXMAN
Department of Epidemiology
University of Michigan
School of Public Health
Ann Arbor, Michigan

I. Overview

Urinary tract infection (UTI) is a multiagent syndrome that causes millions of women annually to suffer frequent, painful urination, suprapubic pressure, and an urgent need to urinate. Although men also are at risk, the frequency of UTI in women is as much as an order of magnitude higher than that in men [1]. Even among individuals at high risk of infection, such as catheterized patients, women have a higher risk of UTI than men because women have moist periurethral areas and a shorter distance from the urethral opening to the bladder. The moist areas of the vagina and the area around the urethral opening provide a niche for growth and multiplication of potential uropathogens. This, combined with the shorter distance to the bladder, increases the chance that a potential uropathogen can ascend to the bladder, multiply in the urine, and invade bladder walls or ascend further to the kidneys.

The vast majority of UTIs among otherwise healthy women are relatively mild and self-limiting. UTIs are easily treated with antibiotics. Unfortunately, they also have a propensity to recur. Following an apparently cured first UTI, about 20% of women will have another within six months [2]. Among sexually active women with a history of one or more UTIs, one-half to almost three-quarters will have another infection within a year [3]. Based on a random digit dialing survey of 2000 women representative of the United States population with telephones, one-third of all women have a UTI by age 24; the cumulative lifetime risk is 56.7% [4]. Thus, despite the relatively benign nature of UTIs, the sheer numbers impose a large burden on society.

The high annual costs of UTIs are another measure of the burden they imposes on society. In the United States, the annual total direct and indirect costs of UTIs in 1995 were estimated to be $1.6 billion [4]. Included in this calculation are cost of diagnosis and treatment, and of symptom, disability and bed days, time lost from work, childcare, and travel expenses to a health care facility. If a vaccine was developed to prevent either initial or recurrent UTI, the benefits to society would be substantial, even at a developmental cost of one billion dollars.

UTI has long been associated with sexual behavior (honeymoon cystitis), although whether the association is real or merely medical folklore has been the subject of some debate [5]. An association with vaginal intercourse has been clearly demonstrated in younger women, but in older women without chronic recurring UTI there is little evidence for the association. As will be described in detail later in this chapter, the lay and medical literature describe a wide range of potential risk factors ranging from sexual behavior, to diet, to exposure to cold, to hygiene habits.

This review will begin by describing the range of clinical entities that constitute UTI and the associated disease impact. The remainder of the discussion will focus on UTI among otherwise healthy women. In particular, it will describe the epidemiology and critically examine the literature regarding sexual behavior and contraceptive method on UTI risk. It closes with a brief overview of research issues and future directions.

II. Range of Clinical Entities and Associated Morbidity

UTI is an infection occurring anywhere in the urinary tract. While usually caused by bacteria, viral and fungal infections may occur, particularly in the kidney. In this review, only bacterial UTI, which account for the vast majority (>95%) of UTI among otherwise healthy women will be discussed [5]. The Enterobacteriaceae, a family of gram-negative facultative anaerobic bacilli commonly found in the large intestine, most frequently causes UTIs. The most common of the Enterobacteriaceae is *Escherichia coli,* which accounts for about 90% of all UTIs seen among ambulatory patients [5]. Other Enterobacteriaceae, including *Klebsiella* and *Proteus,* as well as members of the *Pseudomonas* family, also cause UTIs, especially among women with repeated infections [5]. *Staphylococcus* may cause 5–10% of UTIs in many populations [5].

The presence of bacteria in the urine is termed bacteriuria. When bacteriuria is accompanied by symptoms and confined to the bladder it is called cystitis or bladder infection. Bacteriuria accompanied by symptoms and kidney involvement is termed pyelonephritis or kidney infection. A UTI is considered recurring if a woman has three or more acute episodes during a 12 month period [6]. For research purposes, each of these conditions is defined using results of urine culture. Bacteriuria is defined as >100,000 colony forming units of a uropathogen per milliliter urine (cfu/ml); this is the definition used in population surveys. For clinical studies of UTI, cystitis is defined as the presence of one or more urinary symptoms in the presence of ≥1000 cfu of a uropathogen/ml. This definition has a sensitivity of ~80% and a specificity of ~90% [6].

A. Bacteriuria

Except during pregnancy, bacteriuria in the absence of symptoms is not considered a treatable condition. During pregnancy bacteriuria can increase the risk of symptomatic disease with kidney involvement and has been associated with prematurity and other adverse fetal outcomes [7].

B. Cystitis

The most common cystitis symptoms among the vast majority of otherwise healthy young women are frequent, painful urination, an urgent need to urinate, and suprapubic pressure. Forty percent may have blood in the urine [8]. Among older women, the presentation may be less specific to the urinary tract, with malaise and suprapubic pressure becoming more common presenting symptoms. Cystitis is generally self-limiting; in a placebo-controlled antibiotic trial of nonpregnant women, nearly all showed clinical improvement despite persistence of bacteriuria [9]. However, even with treatment, most women experience an average of 6.1 symptom days and 2.4 restricted activity days [10].

C. Pyelonephritis

Acute bacterial pyelonephritis, like cystitis, is more common in women than in men. Symptoms include fever, chills, and flank pain in addition to urinary symptoms. In general, patients recover without incident, although they may require hospitalization for administration of intravenous antimicrobials. In some cases bacterial pyelonephritis may be chronic, with the potential for life threatening complications [5].

D. UTI during Pregnancy

Pregnancy is a complicating factor for UTI; UTI treatment during pregnancy is considered separately in the medical literature. Asymptomatic bacteriuria during pregnancy is a risk factor for developing pyelonephritis and for premature birth, low birthweight infants, and fetal mortality [5,7]. Pyelonephritis during pregnancy is associated with bacteremia (bacteria in the blood) and severe illness. With appropriate treatment other complications and mortality are low. Thus, pregnant women are routinely screened for bacteriuria [5].

E. Nosocomial UTI

UTI accounts for 40% of all hospital-acquired infections. The vast majority are associated with catheterization and other manipulations of the urinary tract that occur during hospitalization. The patients' normal flora or exogenous bacteria from hospital personnel may be introduced into the bladder during catheterization. Bacteriuria associated with catheterization is important because 30–40% of all gram-negative bacteremias originate in the urinary tract. Among surgical patients UTIs increase length of stay by 2.4 days at a cost of $558 per patient [11].

III. UTI Distribution among Otherwise Healthy Women

Studies of UTI distribution among otherwise healthy women have been almost entirely based on the prevalence of significant bacteriuria (>100,000 cfu/ml) in a mid-stream urine culture. Most population-based prevalence surveys were conducted in the 1970s. Incidence studies are even rarer. I found only one study estimating incidence in the general population based on urine culture and physician diagnosis [12].

A. Prevalence of Bacteriuria by Age, Gender, Ethnic Group, and Geographic Area

Bacteriuria increases with age [1,13,14], is higher among women than men [1], and is found more commonly in working women than in noncloistered Roman Catholic nuns [14]. Estimates of the frequency of bacteriuria vary by region and ethnic group. Population prevalence estimates range from 3.5% among 8352 white, Italian-American working-class women in Boston [13] to 6.5% among 1806 women participating in the Rand Health Insurance Experiment (a sample chosen to represent the four census regions of the United States and an urban-rural mix) [15]. In Charlottesville-Albemarle County, Virginia the prevalence was almost twice as high among 396 African-American women (7.3%) than among 2302 white women (4.4%) [14]. In New Zealand, Maori women had twice the prevalence of bacteriuria (18.0% out of 124) than either Pacific Island (9.8% out of 256) or European (8.7% out of 1184) women [1].

B. Incidence of Symptomatic Infection

Symptomatic UTI is primarily an infection of young, sexually active women. In a random digit dialing survey of 2000 women in the United States, 18.3% (95% Confidence Interval (CI) 12.7–23.9%) of women aged 18–24 reported a UTI in the past 12 months [4]. Although the incidence decreased with age, it was still 9.8% (95% CI 6.4–13.2%) among women aged 65 and older. The lifetime cumulative probability of at least one UTI was 56.7% (95% CI 56.5–58.9%), with one-third of all women reporting at least one physician-diagnosed UTI by age 24.

The incidence is higher among women engaging in sexual activity. In a prospective cohort study of sexually active young women aged 18–40 years starting a new method of contraception [3], the incidence was 0.7 per person-year (or 7000/10,000 person-years) among women recruited from a university health clinic and 0.5 per person-year among women recruited from a health maintenance organization. Even in this group, incidence decreased with age, being lowest in those 31–40 years old (0.2 per person-year among university students and 0.4 per person-year among women members of a health maintenance organization). By contrast, the incidence among healthy university men in Seattle was estimated as 5 per 10,000 men per year, or 0.0005 per person-year [16].

In six municipalities of Northern Norway, all practicing physicians were asked to send urine samples from all patients with UTI symptoms. Because the study was population based, the population rate of UTI could be calculated. Overall, the one year incidence of symptomatic culture-confirmed UTI was 6.4% for women and 0.6% for men. In this study, the incidence increased with age among both women and men [12].

C. Seasonal Variation

UTI has been reported to occur more in the summer in Denmark [17] and Canada [18] and to be more common in the winter in Norway [12], England [19], and Finland [20]. In northern Sweden, the highest rate of UTI occurs in August. This peak is attributed to *Staphylococcus saprophyticus* which caused 28% of the reported episodes in August; all the excess episodes occurred in women [21].

IV. Risk Factors

UTI in otherwise healthy, nonpregnant women is thought to occur primarily from the ascendance of bacteria from the urethral meatus up the urethra to the bladder and kidneys [5]. Because most uropathogens are normal bowel flora, it has been presumed that the bowel is the reservoir for infection, although *E. coli* also can be cultured from the vagina and periurethral areas [22,23]. Based on this assumption, studies have focused on factors that would enhance bacterial movement from the bowel to the periurethral area and facilitate ascendance to the bladder. Host behaviors, host characteristics, and bacterial factors have been implicated.

A. Sexual Behavior

Sexual behavior, including contraceptive method, is the best described host behavior associated with UTI. Experimental studies suggest that virtually all women become bacteriuric following sexual intercourse [24,25], yet most do not develop urinary symptoms. Frequent sexual intercourse has been associated with UTI [3,8,10,26–28]. The hypothesized mechanism of infection is trauma and movement of bacteria from the vagina and/or bowel to the urethral opening. Sexual activity might also increase exposure to potential uropathogens.

Age at which a woman first engages in sexual intercourse has been associated with risk of first UTI. In a cross-sectional survey of history of physician-diagnosed UTI, there was a marked increase in UTI diagnosis around the average age at first sexual intercourse among both men and women [15]. In a random digit dialing survey among 2000 women in the United States, the incidence of UTI rose sharply around the age of first intercourse [2]. Among college students, initiating sexual activity was associated with a 3.5-fold (95% CI 2.5–5.1) increase in UTI risk [29]. A dual interpretation is possible: sexual activity may enhance bladder colonization by bacteria already present and/or uropathogens may be sexually transmitted. In a prospective study among sexually active college students, engaging in vaginal intercourse on three days during the previous week increased UTI risk by 4.8 times [3]. Engaging in a single sex act during the past two weeks increased risk of first UTI by 1.9 times (95% CI 1.4–2.5) [8].

Sexual practices are highly correlated. Although persons tend to engage in a set repertoire, few engage in only one practice, making it difficult to separate out the individual risks of vaginal, oral, and anal intercourse [30]. Few studies have examined the independent effects of each sexual practice. Vaginal intercourse following anal intercourse might introduce potential uropathogens and cause UTI, but there is little evidence to support or refute this hypothesis. After adjusting for frequency of vaginal intercourse, oral intercourse has no increased risk of first UTI over no intercourse [31]. Thus, lesbians (and others) who engage only in oral intercourse and digital stimulation and use no birth control method may be at lower risk of UTI. However, the literature on UTI risk among lesbians is virtually nonexistent.

After adjusting for frequency of vaginal intercourse, duration of the most recent sexual partnership is associated with UTI risk. Women who have engaged in sexual activity with the same sex partner for a year or more compared to women who have engaged in sexual activity with the same partner for less than one year have half the risk of first UTI [8]. This highlights that vaginal intercourse causes UTI only in the presence of a potential uropathogen and that the uropathogen may be acquired from a sex partner. If colonization by a particular bacteria protects against infection by a virulent strain either through development of cross-immunity and/or by prevention of colonization, more lifetime sex partners may be protective. Two case-control studies in college women observed no clear evidence for an association with number of sexual partners but showed a slightly protective effect with increasing numbers of partners [8,27].

Virtually all studies of the associations of sexual behavior and UTI have been conducted among premenopausal women, usually of college age. Bacteriuria increases with age but the incidence of symptomatic infection does not. However, the role sexual behavior plays in UTI risk in older age groups is unknown.

B. Contraceptive Use

Oral contraceptives double UTI risk relative to using no birth control method [31], even after adjusting for frequency of vaginal intercourse. Diaphragms and cervical caps, which are generally used in conjunction with spermicides, double UTI risk relative to using oral contraceptives. There is *in vitro* [32,33] and *in vivo* [3,34] evidence that spermicides alone also double UTI risk. Spermicides and/or diaphragm use alter the vaginal flora, enhancing growth by *E. coli* and other Enterobacteriaceae [10,26–28]. Women also may have allergic reactions to spermicides that facilitate bacterial invasion. Diaphragms decrease the speed of urine flow, thereby diminishing the body's ability to clear uropathogens from the bladder and/or increasing the amount of residual urine in the bladder [35].

The traumatic effects of vaginal intercourse on UTI risk may be increased by condom use [8,31,36]. A single sex act with an unlubricated condom compared to a sex act when using no birth control method was associated with four times the risk of acquiring first UTI among college women [31]. In this study, the UTI risk associated with condom use was markedly decreased in the presence of a lubricating agent, including spermicides. However, a sex act using spermicidal condoms or spermicides compared to a sex act using no birth control method did increase UTI risk. By contrast, a study among women in a large health maintenance organization found an increased risk of UTI with spermicidal condoms compared to other condom use [36]. Further studies are needed to clarify whether the condom use effect is primarily due to trauma or to use of spermicidal condoms.

C. Evidence of Sexual Transmission

Behavioral and microbiological evidence suggests that UTI may be sexually transmitted. Three case reports suggest sexual transmission of UTI from female to male [37–39]. In a study among servicemen in the British Army, 13 cases of UTI occurred among married men accompanied by their wives compared to one among unaccompanied men, for a relative risk of 11.9 (13/4530 versus 1/4150). Five of the wives of married men with UTI had UTI 2 to 8 weeks before their husbands' illnesses [40]. When midstream urine specimens were cultured, three of the wives exhibited significant growth of *E. coli* that were sensitive to the same antibiotics as the bacteria infecting their husbands. Unfortunately, serotyping was not performed. Stamey cultured the first voided urine of male consorts of females with recurring UTI. Four out of seven specimens cultured were positive and were colonized with bacteria of the same serotype as their sex partner [41]. A random initial void of the most recent male sex partner of 19 women with *E. coli* UTI found that four of the males carried the identical UTI pathogen in their urethral flora [29]. However, the infecting *E. coli* could not be cultured from the vagina of two of nineteen women or from rectal specimens of four of eighteen women from whom adequate specimens were obtained [29].

If the reservoir for uropathogens is the bowel flora, we would expect all urinary isolates also to be found in fecal samples (taking into account sampling error). A study conducted in 1969 [22] used O serotyping to compare fecal, urinary, vaginal, and periurethral *E. coli* isolates from 29 female outpatients with UTI. Demographic characteristics and sexual behavior variables were not presented. Of the 29 women, 26 had UTI caused by typeable *E. coli* strains. Among these 26, there were 5 (19%) urinary isolates not found in fecal samples; in four of these five cases urinary, vaginal, and periurethral isolates were identical by serotype.

D. Antibiotic Use

Evidence from two prospective studies suggests that antibiotic use may increase risk of UTI. These data must be viewed with caution because the risk of UTI is high among women with a history of one or more UTIs and antibiotics are used to treat UTI, thus antibiotic use may be confounded with the natural risk of UTI recurrence. Antibiotic use in the previous 15–28 days increased risk of UTI among women who took antibiotics either for treatment of a previous UTI or for some other illness [42]. In a cohort of women with first UTI, antibiotic use in the past 14 days increased risk of second UTI [43]. A possible mechanism is disruption of the vaginal flora, allowing overgrowth by potential uropathogens.

E. Cranberry Juice and Vitamin C

Folk wisdom suggests that drinking cranberry juice and/or taking vitamin C prevents UTI. This notion is supported by observational studies, where regularly drinking cranberry juice or taking vitamin C decreased risk of acquiring UTI by roughly one-half [8,28]. In a randomized, double-blind, placebo-controlled trial, cranberry juice decreased risk of bacteriuria among elderly women [44]. Cranberry juice may inhibit bacterial adherence [45,46] in addition to lowering pH. At present, no randomized controlled trial of the effects of regularly drinking cranberry juice or taking vitamin C on UTI risk has been conducted. However, drinking cranberry juice and taking vitamin C will certainly not hurt and may help.

F. Other Potential Risk Factors

Other potential risk factors hypothesized to follow from self-infection have not been supported by controlled studies. Using napkins for menstrual protection, taking tub baths, wearing tight clothing or pantyhose, wearing nonabsorbent underwear, and wiping the urogenital region from back to front have been hypothesized to increase risk of UTI by facilitating migration of bacteria from the bowel to the periurethral area. However, none of these factors has been shown to increase risk of UTI in controlled studies [8,27,28,47]. At best, the results have been suggestive and the effects have been small and have failed to reach statistical significance even in large studies.

In women with a history of recurrent UTI, exposure to cold appears to increase risk of symptomatic infection. In a case-control study, a higher relative frequency of women with UTI reported having cold hands, feet, or buttocks in the time prior to UTI than controls [48]. The feet of 29 healthy women with recurring UTI were cooled in a nonrandomized cross-over experiment; five developed bacteriologically verified symptomatic UTI following cooling as compared to none in the control period [49].

There is an increase in asymptomatic bacteriuria with age [12–14,50], and at least some of this increase has been attributed to normal hormonal changes associated with aging. During menopause the vaginal pH increases, thus facilitating growth of many bacterial species including known uropathogens [51]. Vaginal walls also become thinner and lose their ability to lubricate quickly. As a result, the vagina is less moist [52]. Decreased lubrication can increase the mechanical irritation associated with sexual intercourse and thereby facilitate invasion by uropathogens. It also may decrease urethral closure pressure making it easier for perineal pathogens to gain entry into the bladder. During menopause, the urethral epithelium as well as the vaginal epithelium atrophies [53]; atrophy itself may lead to urinary symptoms [54]. Estrogen replacement therapy minimizes these changes and is hypothesized to modify risk of UTI, both positively and negatively [51,53,55]. In a randomized, double-blind, placebo-controlled trial of a topically applied intravaginal estriol cream among postmenopausal women with a history of recurrent UTI, women using the cream had lower rates of UTI [56]. However, 20% of those treated, compared to 9% of the placebo group, discontinued treatment because of side effects.

G. Markers of Susceptibility

Women with chronic recurring UTI, defined as three or more episodes in a 12 month period [6], cannot be distinguished from women with intermittent or only one UTI on the basis of known behavioral risk factors. Thus, investigators have turned their attention to host factors, such as host response and genetic susceptibility, and bacterial characteristics, such as factors leading to increased virulence or duration of colonization.

Host characteristics such as blood phenotype may enhance the ability of bacteria with certain virulence characteristics to attach to cell surfaces [57,58], while secretion of cytokines or presence of immunoglobulins may limit the bacterial colonization or invasion into the mucosa [59]. In one large and two smaller studies, women with three or more UTIs in the past year were two the three times more likely than controls to be nonsecretors of blood group antigens [34,60,61]. Based on the relative risk from the largest study, approximately 68% of women with B or AB blood type who are nonsecretors and have recurring UTI can attribute their UTI to their blood type [61]. However, an association with blood type and secretor status was not observed among 99 Wisconsin women aged 18–82 with a history of recurring UTI [62]. The Wisconsin women had similar frequencies of ABO and secretor phenotypes as published reference populations. They also had similar frequencies of the following genetic phenotypes to those found in the University of Wisconsin Tissue Typing control data set: HLA-A, HLB-B, DR antigen, and -DQ. Thus, although there may be some genetic predisposition to recurring UTI, we do not currently have a marker or set of markers that clearly defines the susceptible group.

V. Bacterial Virulence Factors

The most common uropathogen, *E. coli,* is ubiquitous, and can be acquired indirectly or by intimate contact with others. Most *E. coli* are not uropathogens. In order to identify bacterial characteristics more common in uropathogens, investigators have compared urinary *E. coli* to fecal *E. coli. E. coli* associated with UTI belong to several serotypes that occur more frequently in urinary than in fecal isolates [63]. Three classes of bacterial virulence factors, adhesins, siderophores, and toxins, are found more frequently among urinary *E. coli* isolates [63], but no one virulence factor occurs in more than half of UTI isolates. Certain adhesins are associated with an increased risk of second UTI whereas toxins and other adhesins are associated with a decreased risk of second UTI [64].

Several different UTI types may exist, which may be clinically similar yet have different modes of pathogenesis. Genetically UTI isolates are quite diverse. There is no known virulence factor or set of virulence factors that uniquely identifies a uropathogen. Among 216 *E. coli* causing first UTI, grouping by the presence of nine known uropathogenic factors led to 36 distinct groups; using southern analysis, at least 125 different strains could be identified [65]. While this diversity may suggest that any *E. coli* can cause UTI, this is not the case. As noted earlier, uropathogenic *E. coli* are different from fecal *E. coli* on a number of parameters. Further, human experiments [66] have shown bladder colonization with known uropathogens to be difficult to establish.

Many bacterial strains may asymptomatically colonize the urine, periurethral area, and vaginal flora, and only cause symptoms in the presence of additional risk factors such as sexual intercourse. One-third to half of all second UTIs are caused by the same organism that caused the first infection [64,67]; among women with chronic recurring UTI, half of all recurrences are caused by the same organism that caused the previous infection [68]. Therefore, recurring UTI among women may either be caused by reinfection with the same organism from a sex partner or from the woman's own genitourinary or fecal flora.

VI. Clinical Issues: UTI among Otherwise Healthy Women

A. Diagnosis and Treatment

UTI is diagnosed on the basis of the combination of clinical symptoms and laboratory tests [69]. Among otherwise healthy women, UTI is generally easy to treat with any of a variety of antibiotics [5]. In the absence of complicating factors, empiric short-course therapy (single dose or 3-day therapy) with one of several antibiotics will cure most patients with a UTI that is confined to the bladder [69]. However, for the 20% with frequent recurring UTIs, each successive UTI becomes more difficult to treat as the bacteria may become resistant to antibiotic treatment [5]. Continuous antimicrobial prophylaxis, postcoital prophylaxis, and patient administered self-treatment are effective strategies for treating women with frequent or chronic recurrent infections [70].

B. Prevention

There is an extensive self-help literature on how to prevent UTIs [see for example, 71,72], but controlled studies supporting self-help advice such as increased fluid intake, frequent voiding, type of menstrual protection, and direction wiping following bowel movements are few and the evidence weak [70]. Regular consumption of cranberry juice and/or vitamin C may be helpful (see earlier discussion). As described earlier, many studies do support an association of diaphragm and spermicide use, and some support associations with oral contraceptives, condom use, and spermicide use alone. Therefore, it seems prudent for a woman with recurrent UTI to test her sensitivity to these risk factors. Decreasing frequency of sexual intercourse or increasing lubrication during intercourse might also be effective prevention strategies—although there are no studies testing these strategies.

Vaginal mucosal immunization [73], modifications of the vaginal flora with lactobacillus vaginal suppositories [74], or, among postmenopausal women, estrogen replacement therapy are possible avenues for preventing recurring infection. Vaginal mucosal immunization presumes that recurring infections are due to a decrease in local host response to colonization. Modifications of the vaginal flora are aimed at prohibiting colonization by uropathogens. The efficacy of these strategies remains to be shown.

VII. Epidemiologic Issues

UTI is a syndrome. Like other syndromes caused by multiple agents, this grouping probably masks several different etiologies that have very similar clinical presentations. Better classification enhances our ability to identify risk factors for chronic, recurring infection. For example, grouping by bacterial genus is informative: *E. coli* first UTIs are more likely than UTIs caused by other bacteria to be followed by a second UTI in the next six months (53/224 vs 3/39, p value = 0.02) [43].

E. coli can be differentiated by phenotypic characteristics, such as serotype and the expression of known virulence factors, and genotype, such as presence of genes for known virulence factors. The vast number of strains causing a similar syndrome combined with ease of treatment is at least partly responsible for the failure to identify different etiologies and more effective prevention strategies. Epidemiologic information, such as cocolonization of sex partners, combined with molecular genetic techniques provide a powerful strategy for differentiating between etiologies with similar clinical presentations and for discovering new genes of uropathogenic *E. coli.*

A. Measurement Issues

UTIs are acute infections with a propensity to recur. Once a woman has an initial UTI her risk of a second increases by an order of magnitude. Because the vast majority of clinical visits for UTI are from women with a history of more than one infection, it seems logical to focus on women with recurring UTI. Most studies have done so, comparing women with multiple UTIs to those with no history of multiple UTIs, adjusting for number of prior UTIs using a multivariate statistical model. However, this strategy may introduce inadvertent biases. Women with recurring UTIs often change their behavior in response to infection, so that a behavior that actually prevents UTIs appears to be a risk factor; this problem is most likely to occur using a case-control design [28]. Further, care is required when modeling UTI risk and controlling for number of prior UTIs. If the number of prior UTIs is included in the model as a single variable, there is an inherent assumption that the risk ratio between a first and second UTI is equal to that between the twentieth and twenty-first UTI. This may be a very poor assumption. Women with chronic recurring UTI are much more likely to have another than women with only a few UTI. Therefore, risk estimates will be distorted. The form of the relationship can be explored in a modeling setting, for example, by adding the square of the number of prior UTIs to the model.

B. Relevant Time Frame for Risk Factors

The time frame for development of UTI following colonization is not known and is probably quite variable. Separating risk factors for colonization from risk factors for development of symptoms is difficult because as yet, uropathogenicity is not easily defined and therefore is not easily measured. However, given that a uropathogen is present in the genitourinary tract, recent exposures to cofactors are probably most relevant for developing symptomatic infection.

VIII. Future Directions

For a common condition that imposes substantial burdens on society, our understanding of UTI pathogenesis and epidemiology is remarkably limited, encompassing primarily treatment and a description of risk factors. Molecular genetic techniques have shown that not all *E. coli* are equally capable of causing UTI. Therefore, it is difficult to determine which factors are associated with acquisition of uropathogens in the vaginal flora and urinary tract and which factors enhance colonization. A better understanding of vaginal ecology will assist in understanding UTIs, as it seems likely that disruptions in vaginal flora, such as those documented with spermicide use and suspected for antibiotic use, are important for vaginal colonization with *E. coli,* a critical first step in UTI pathogenesis.

The great mystery is why some women have chronic recurring UTI. Since asymptomatic bacteriuria is so common, is this a host or a bacterial effect (or a combination of the two)? One might hypothesize that women with chronic recurring UTI might be hypersensitive to bacteria in the bladder, so they ex-perience symptoms when other women would not. The bladder is an immunogenic organ generating local host response. Our knowledge of this local response is minimal. Some women may have more receptors to certain uropathogens or some other genetic susceptibility that makes them prone to colonization. Alternatively, or in addition, the recurrences may be due to bacterial characteristics. Some bacteria may be more effective at colonizing the genitourinary tract, making them harder to remove. Following infection with one organism, approximately half of the next infections are due to the same organism, even after an interim negative urine culture. Is this reinfection from self or from some other source? Careful, prospective studies that pay close attention to measurement issues and integrate information on bacterial characteristics with that of host behavior and host characteristics are required to answer these questions.

Acknowledgments

I thank my colleagues Karen Palin, Brenda Gillespie, and Carl Marrs, and members of my research staff, Bonnie Andree, Shannon Manning, Patricia Tallman, and Lixin Zhang for their comments and suggestions during the preparation of this chapter.

References

1. Metcalf, P. A., Baker, J. R., Scragg, R. K. R., Scott, A., Wild, C. J., and Dryson, E. (1993). Asymptomatic bacteriuria in a multiracial workforce. *Ethnicity Dis.* **3,** 270–277.
2. Foxman, B. (1990). Recurring lower urinary tract infections: Incidence and risk factors. *Am. J. Public Health* **80,** 331–333.
3. Hooton, T. M., Scholes, D., Hughes, J. P., Winter, C., Roberts, P. L., Stapleton, A. E., Stergachis, A., and Stamm, W. E. (1996). A prospective study of risk factors for symptomatic urinary tract infection in young adults. *N. Engl. J. Med.* **335,** 468–74.
4. Foxman, B., Barlow, R., d'Arcy, H., Gillespie, B., and Sobel, J. D. (1999). Urinary tract infection: Estimated incidence and associated costs. Submitted for publication.
5. Kunin, C. M. (1997). "Urinary Tract Infections: Detection, Prevention, and Management," 5th ed. Williams & Wilkins, Baltimore, MD.
6. Rubin, R. H., Shapiro, E. D., Andriole, V. T., Davis, R. J., and Stamm, W. E. (1992). General guidelines for the evaluation of new anti-infective drugs for the treatment of urinary tract infection. *Clin. Infect. Dis.* **15**(Suppl. 1), S216–S217.
7. Millar, L. K., and Cox, S. M. (1997). Urinary tract infections complicating pregnancy. *Infect. Dis. Clin. North Am.* **11,** 12–26.
8. Foxman, B., Geiger, A. M., Palin, K., Gillespie, B., and Koopman, J. S. (1995). First-time urinary tract infection and sexual behavior. *Epidemiology* **6,** 162–168.
9. Mabeck, C. E. (1972). Treatment of uncomplicated urinary tract infection in non-pregnant women. *Postgrad. Med. J.* **48,** 69–75.
10. Foxman, B., and Frerichs, R. R. (1985). Epidemiology of urinary tract infection I: Diaphragm use and sexual intercourse. *Am. J. Public Health* **75,** 1308–1313.
11. Givens, C. D., and Wenzel, R. P. (1980). Catheter-associated urinary tract infections in surgical patients: A controlled study on the excess morbidity and costs. *J. Urol.* **124,** 646.
12. Vorland, L. H., Carlson, K., and Aalen, O. (1985). An epidemiological survey of urinary tract infections among outpatients in northern Norway. *Scand. J. Infect. Dis.* **17,** 277–283.
13. Evans, D. A., Williams, D. N., Laughlin, L. W., Miao, L., Warren, J. W., Hennekens, C. H., Shimada, J., Chapman, W. G., Rosner, B., and Taylor, J. O. (1978). Bacteriuria in a population-based cohort of women. *J. Infect. Dis.* **138,** 768–773.
14. Kunin, C. M., and McCormack, R. C. (1968). An epidemiologic study of bacteriuria and blood pressure among nuns and working women. *N. Engl. J. Med.* **278**(12), 635–642.

15. Zielske, J. V., Lohr, K. N., Brook, R. H., and Goldberg, G. A. (1981). "Conceptualization and Measurement of Physiologic Health for Adults: Urinary Tract Infection," R-2262/16-HHS. Rand Corporation, Santa Monica, CA.

16. Kreiger, J. N., Ross, S. O., and Simonsen, J. M. (1993). Urinary tract infections in healthy university men. *J. Urol.* **149,** 1046–1048.

17. Steensberg, J., Bartels, E. D., Bay-Nielsen, H., Fanoe, E., and Hede, T. (1969). Epidemiology of urinary tract diseases in general practice. *Br. Med. J.* **4,** 390–394.

18. Anderson, J. E. (1983). Seasonality of symptomatic bacterial urinary tract infections in women. *J. Epidemiol. Commun. Health* **37,** 286–290.

19. Stansfield, J. M. (1966). Clinical observation relating to incidence and aetiology of urinary tract infections in children. *Br. Med. J.* **1,** 631–635.

20. Elo, J., Sarna, S., and Tallgren, L. G. (1979). Seasonal variations in the occurrence of urinary tract infections among children in an urban area in Finland. *Ann. Clin. Res.* **11,** 101–106.

21. Ferry, S., Burman, L. G., and Mattsson, B. (1987). Urinary tract infection in primary health care in northern Sweden. I. Epidemiology. *Scand. J. Primary Health Care* **5,** 123–128.

22. Gruneberg, R. N. (1969). Relationship of infecting urinary organism to the faecal flora in patients with symptomatic urinary infection. *Lancet* **2,** 766–768.

23. Stamey, T. A. (1987). Recurrent urinary tract infections in female patients: An overview of management and treatment. *Rev. Infect. Dis.* **9,** S195–S208.

24. Buckley, R. M., McGuckin, M., and MacGregor, R. R. (1978). Urine bacterial counts after sexual intercourse. *N. Engl. J. Med.* **298,** 321–324.

25. Nicolle, L. E., Harding, G. K. M., Preiksaitis, J., and Ronald, A. R. (1982). The association of urinary tract infection with sexual intercourse. *J. Infect. Dis.* **146,** 579–583.

26. Fihn, S. D., Latham, R. H., Roberts, P., Running, K., and Stamm, W. E. (1985). Association between diaphragm use and urinary tract infection. *JAMA, J. Am. Med. Assoc.* **254,** 240–245.

27. Strom, B. L., Collins, M., West, S. L., Kreisberg, J., and Weller, S. (1987). Sexual activity, contraceptive use, and other risk factors for symptomatic and asymptomatic bacteriuria. *Ann. Intern. Med.* **107,** 816–823.

28. Foxman, B., and Chi, J. W. (1990). Health behavior and urinary tract infection. *J. Clin. Epidemiol.* **43,** 329–337.

29. Foxman, B., Zhang, L., Tallman, P., Andree, B. C., Geiger, A. M., Koopman, J. S., Gillespie, B. W., Palin, K. A., Sobel, J. D., Rode, C. K., Bloch, C. A., and Marrs, C. F. (1997). Transmission of uropathogens between sex partners. *J. Infect. Dis.* **175,** 989–992.

30. Foxman, B., Aral, S. O., and Holmes, K. K. (1998). Heterosexual repertoire is associated with same-sex experience. *J. Sex. Transm. Dis.* **25,** 232–236.

31. Foxman, B., Marsh, J., Gillespie, B., Rubin, N., Koopman, J. S., and Spear, S. (1997). Condom use and first-time urinary tract infection. *Epidemiology* **8,** 637–641.

32. McGroarty, J. A., Chong, S., Reid, G., and Bruce, A. W. (1990). Influence of the spermicidal compound nonxynol-9 on the growth and adhesion of urogenital bacteria in vitro. *Curr. Microbiol.* **21,** 219–223.

33. Hooton, T. M., Fennell, C. L., Clark, A. M., and Stamm, W. E. (1991). Nonoxynol-9: Differential antibacterial activity and enhancement of bacterial adherence to vaginal epithelial cells. *J. Infect. Dis.* **164,** 1216–1219.

34. Hooton, T. M., Roberts, P. L., and Stamm, W. E. (1994). Effects of recent sexual activity and use of a diaphragm on the vaginal microflora. *Clin. Infect. Dis.* **19,** 274–278.

35. Fihn, S. D., Johnson, C., Pinkstaff, C., and Stamm, W. E. (1986). Diaphragm use and urinary tract infections: Analysis of urodynamic and microbiological factors. *J. Urol.* **136,** 853–856.

36. Fihn, S. D., Boyko, E. J., Normand, E. H., Chen, C. L., Grafton, J. R., Hunt, M., Yarbro, P., Scholes, D., and Stergachis, A. (1996). Association between use of spermicide-coated condoms and *Escherichia coli* urinary tract infection in young women. *Am. J. Epidemiol.* **144,** 512–520.

37. Wong, E. S., and Stamm, W. E. (1983). Sexual acquisition of urinary tract infection in a man. *JAMA, J. Am. Med. Assoc.* **250,** 3087–3088.

38. Bailey, R. R., Peddie, B. A., Swainson, C. P., and Kirkpatrick, D. (1986). Sexual acquisition of urinary tract infection in a man. *Nephron* **44,** 217–218.

39. Hebelka, M., Lincoln, K., and Sanberg, T. (1993). Sexual acquisition of acute pyelonephritis in a man. *Scand. J. Infect. Dis.* **25,** 141–143.

40. Simpson, B. (1981). Marital bacterial urinary infection in men. *J. R. Army Med. Corps* **127,** 139–140.

41. Stamey, T. A. (1972). "Urinary Infections." Williams & Wilkins, Baltimore, MD.

42. Smith, H. S., Hughes, J. P., Hooton, T. M., Roberts, P., Scholes, D., Stergachis, A., Stapleton, A., and Stamm, W. E. (1997). Antecedent antimicrobial use increases the risk of uncomplicated cystitis in young women. *Clin. Infect. Dis.* **25,** 63–68.

43. Foxman, B., Gillespie, B., Koopman, J., Zhang, L., Palin, K., Tallman, P., Marsh, J. V., Spear, S., Sobel, J. D., Marty, M. J., and Marrs, C. F. (1999). Epidemiology of urinary tract infection: Risk factors for second infection. Submitted for publication.

44. Avorn, J., Monane, M., Gurwitz, J. H., Glynn, R. J., Choodnovskiy, I., and Lipsitz, L. A. (1994). Reduction of bacteriuria and pyuria after ingestion of cranberry juice. *JAMA, J. Am. Med. Assoc.* **271**(10), 751–754.

45. Zafriri, D., Ofek, I., Adar, R., Pocino, M., and Sharon, N. (1989). Inhibitory activity of cranberry juice on adherence of type 1 and type P fimbriated *Escherichia coli* to eucaryotic cells. *Antimicrob. Agents Chemother.* **33,** 92–98.

46. Sobota, A. E. (1984). Inhibition of bacterial adherence by cranberry juice: Potential use for the treatment of urinary tract infections. *J. Urol.* **141,** 1013–1016.

47. Foxman, B., and Frerichs, R. R. (1985). Epidemiology of urinary tract infection II: Diet, clothing, and urination habits. *Am. J. Public Health* **75,** 1314–1317.

48. Baerheim, A., Laerum, E., and Sulheim, O. (1992). Factors provoking lower urinary tract infection in women. *Scand. J. Primary Health Care* **10,** 72–75.

49. Baerheim, A., and Laerum, E. (1992). Symptomatic lower urinary tract infection induced by cooling of the feet. A controlled experimental trial. *Scand. J. Primary Health Care* **10**(2), 157–160.

50. Bengtsson, C., Bengtsson, U., and Lincoln, K. (1980). Bacteriuria in a population sample of women. *Acta Med. Scand.* **208,** 417–423.

51. Parsons, C. L. (1987). Lower urinary tract infections in women. *Urol. Clin. North Am.* **14**(2), 247–250.

52. Golub, S. (1992). "Periods: From Menarche to Menopause." Sage, Newbury Park, CA.

53. Orlander, J. D., Jick, S. S., Dean, A. D., and Jick, H. (1992). Urinary tract infections and estrogen use in older women. *J. Am. Geriatr. Soc.* **40,** 817–820.

54. Iosif, C. S., and Bekassy, Z. (1984). Prevalence of genito-urinary symptoms in the late menopause. *Acta Obstet. Gynecol. Scand.* **63,** 257–260.

55. Brandberg, A., Mellström, D., and Samsioe, G. (1987). Low dose oral estriol treatment in elderly women with urogenital infections. *Acta Obstet. Gynecol. Scand., Suppl.* **140,** 33–38.

56. Raz, R., and Stamm, W. E. (1993). A controlled trial of intravaginal estriol in postmenopausal women with recurrent urinary tract infection. *N. Engl. J. Med.* **329,** 753–756.

57. Stapleton, A., Hooton, T. M., Fennell, C., Roberts, P. L., and Stamm, W. E. (1995). Effect of secretor status on vaginal and rectal

colonization with fimbriated *Escherichia coli* in women with and without recurrent urinary tract infection. *J. Infect. Dis.* **171,** 717–720.

58. Lomberg, H., Cedergren, B., Leffler, H., Nilsson, B., Carlstrom, A. S., and Svanborh-Eden, C. (1986). Influence of blood group on the availability of receptors for attachment of uropathogenic *Escherichia coli. Infect. Immun.* **51,** 919–926.

59. Agace, W., Hedges, S., Ceska, M., and Svanborg, C. (1993). IL-8 and the neutrophil response to mucosal Gram negative infection. *J. Clin. Invest.* **92,** 780–785.

60. Sheinfeld, J., Schaeffer, A. J., Cordon-Cardo, C., Rogatko, A., and Fair, W. R. (1989). Association of the Lewis blood-group phenotype with recurrent urinary tract infections in women. *N. Engl. J. Med.* **320,** 773–777.

61. Kinane, D. F., Blackwell, C. C., Brettle, R. P., Weir, D. M., Winstanley, F. P., and Elton, R. A. (1982). ABO blood group, secretor state, and susceptibility to recurrent urinary tract infection in women. *Br. Med. J.* **285,** 7–9.

62. Hopkins, W. J., Heisey, D. M., Lorentzen, K. F., and Uehling, D. T. (1998). A comparative study of major histocompatibility complex and red blood cell antigen phenotypes as risk factors for recurrent urinary tract infections in women. *J. Infect. Dis.* **177,** 1296–1301.

63. Johnson, J. R. (1991). Virulence factors in *Escherichia coli* urinary tract infection. *Clin. Microbiol. Rev.* **4,** 80–128.

64. Foxman, B., Zhang, L., Tallman, P., Palin, K., Rode, C., Bloch, C., Gillespie, B., and Marrs, C. F. (1995). Virulence characteristics causing first time urinary tract infection predict risk of second infection. *J. Infect. Dis.* **172,** 1536–1541.

65. Foxman, B., Zhang, L., Palin, K., Tallman, P., and Marrs, C. F. (1995). Bacterial virulence characteristics of *Escherichia coli* isolates from first-time urinary tract infection. *J. Infect. Dis.* **171,** 1514–1521.

66. Anderson, P., Engberg, I., Lidin-Janson, G., Lincoln, K., Hull, R., Hull, S., and Svanborg-Eden, C. (1991). Persistence of *Escherichia coli* bacteriuria is not determined by bacterial adherence. *Infect. Immun.* **59,** 2915–2921.

67. Ikaheimo, R., Siitonen, A., Karkkainen, U., Kuosmanen, P., and Mäkelä, P. H. (1993). Characteristics of *Escherichia coli* in acute community-acquired cystitis of adult women. *Scand. J. Infect. Dis.* **25,** 705–712.

68. Russo, T. A., Stapleton, A., Wenderoth, S., and Stamm, W. E. (1994). "An Evaluation of the Mechanism of Recurrent Urinary Tract Infections in Pre-menopausal Women," Abstr. 1994 Symp. Women's Urol. Health Res., 1994. Women's Urological Health Program, NIDDK and the Office of Research on Women's Health, NIH, Bethesda, MD.

69. Hooton, T. M., and Stamm, W. E. (1997). Diagnosis and treatment of uncomplicated urinary tract infection. *Infect. Dis. Clin. North Am.* **11,** 551–581.

70. Stapleton, A., and Stamm, W. E. (1997). Prevention of urinary tract infection. *Infect. Dis. Clin. North Am.* **11,** 719–731.

71. Gillespie, L. (1988). "You Don't Have to Live with Cystitis!" Avon, New York.

72. Kilmartin, A. (1980). "Cystitis, the Complete Self-help Guide." Warner Books, New York.

73. Uehling, D. T., Hopkins, W. J., Balish, E., Xing, Y., and Heisey, D. M. (1997). Vaginal mucosal immunization for recurrent urinary tract infection: Phase II clinical trial. *J. Urol.* **157,** 2049–2052.

74. Reid, G., Bruce, A. W., and Taylor, M. (1992). Influence of three-day antimicrobial therapy and lactobacillus vaginal suppositories on recurrence of urinary tract infections. *Clin. Ther.* **14,** 11–16.

29

Pelvic Inflammatory Disease

ROBERTA NESS* AND DEBORAH BROOKS-NELSON†

*Department of Epidemiology, University of Pittsburgh, Pittsburgh, Pennsylvania; †Center for Clinical Epidemiology and Biostatistics, University of Pennsylvania, Philadelphia, Pennsylvania

I. Introduction

Pelvic inflammatory disease (PID) is a general term that describes clinically suspected infection with resultant inflammation of the female upper reproductive tract including the tubes (salpingitis), ovaries (oophoritis), and uterine lining (endometritis). In the majority of cases, PID is the result of an ascension of microorganisms from the lower genital tract, with *Chlamydia trachomatis* and *Neisseria gonorrhoeae* being the most common. Inaccurate diagnosis of PID has resulted both in an underestimate of the magnitude of this condition and in untreated cases. Particularly if not rapidly treated, PID exacts a major toll in the form of infertility from obstructed fallopian tubes, ectopic pregnancy, chronic pelvic pain, and recurrent PID. This chapter will outline the frequency, microbial etiology, health consequences and medical costs, risk factors, diagnosis, treatment options, and prevention of PID.

II. Frequency

PID affects women worldwide in epidemic proportions. Rates have declined in some developed countries, in particular in Scandinavia, but there is no evidence of a decline in PID in developing nations [1]. In the United States there has also been a decline in the 1990s, but PID remains a common disease in this country. A national survey conducted in 1995 indicated that 8% of all women and 11% of African-American women reported that they have received treatment for PID at some time in their reproductive lives [2]. This compares with a 1988 national survey in which almost 11% of all women and 17% of African-American women of reproductive age reported that they had received treatment for PID [3].

It has been estimated that over one million American women seek treatment for PID annually. Hospitalizations occur in an estimated 200,000–300,000 women. One-third of the women hospitalized for acute PID will undergo surgery for their condition, frequently as a result of the formation of tuboovarian abscesses [4–6]. The remaining women diagnosed with PID are treated in private physicians' offices (over 400,000 annually), in public health clinics, or in hospital emergency rooms (approximately 140,000 annually) [4–7] (Fig. 29.1). However, hospitalization rates for PID have been falling. Data from a population-based review of medical records from Seattle, Washington showed that 13% of all women with diagnosed PID were hospitalized and that the strongest factors predicting hospitalization were seeking medical care at a hospital-affiliated facility, fevers/chills/sweats, infection with *Neisseria gonorrhoeae,* and lack of birth control use. Racial differences in rates of hospital-

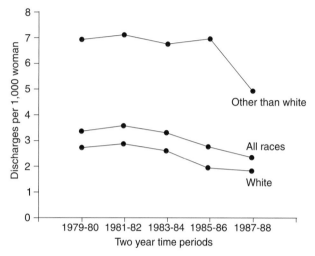

Fig. 29.1 Trends in hospitalizations for acute pelvic inflammatory disease among white and other-than-white women in the United States from 1979 to 1988. Acute pelvic inflammatory disease refers to ICD-9 codes designating acute or unspecified duration disease. Rates are presented as 2-year means. Source: [6].

ization continue to exist, with nonwhite women being 2.5 times more likely than white women to be hospitalized [6,8].

The occurrence of PID may be even more common than what is reflected in surveys of women seeking medical care. First, PID is not a reportable condition and its detection relies on passive surveillance and recall. Previous PID may not be consistently recalled. Furthermore, unrecognized or silent PID is now thought to account for half or more of all cases [7,9] (Figure 29.2).

III. Diagnosis

Laparoscopically confirmed visual findings of erythema and/or edema of the fallopian tubes, clubbing of the fimbriae, or presence of tuboovarian abscess are considered the gold standard for diagnosing PID [10,11]. However, because laparoscopy is an invasive procedure and not routinely performed on women with suspected PID, the great majority of cases of PID are diagnosed using clinical signs and symptoms. The Centers for Disease Control's minimal recommendation for the diagnosis of PID includes complaint of abdominal pain and clinical findings of lower abdominal, cervical motion, and adnexal tenderness [12]. Additional criteria, first proposed by Hager *et al.* [13],

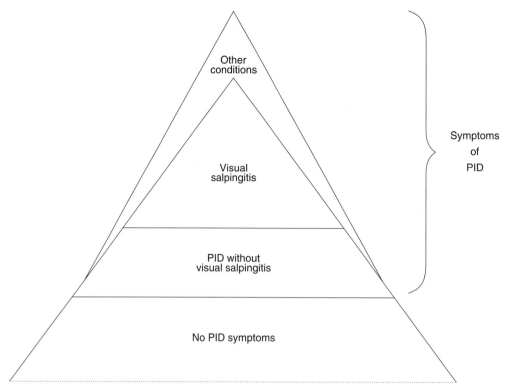

Fig. 29.2   Schematic representation of symptomatic and asymptomatic pelvic inflammatory disease (PID). Source: [7].

increase the specificity of the diagnosis of PID. These criteria include fever, elevated C-reactive protein, elevated erythrocyte sedimentation rate, laboratory documentation of cervical infection with *N. gonorrhoeae* or *C. trachomatis,* and/or abnormal cervical or vaginal discharge. The CDC's minimal criteria (abdominal pain plus abdominal/cervical/adnexal tenderness) correlate with laparoscopically demonstrable salpingitis only about one-half to two-thirds of the time [12,14–16]. The presence of the additional laboratory markers can substantially improve diagnostic accuracy [17]. For example, vaginal and cervical white blood cells (leukorrhea or cervical mucopus) have been shown to be quite specific for PID and moderately sensitive. Women without these findings, even if they have pelvic pain, are unlikely to have PID [18].

More expensive, yet still relatively noninvasive tests commonly used to diagnosis PID include ultrasonography and endometrial biopsy. Pelvic ultrasound detects enlargement of the ovaries and fallopian tubes [9]. However, ultrasound has been found to be neither sensitive nor specific and is more useful in detecting the presence of tuboovarian abscesses [19–21]. The best laboratory marker of PID is endometritis found on endometrial biopsy. A biopsy can be performed with a flexible openended cannula (termed a pipelle) with or without local anesthesia. It is relatively noninvasive, can be performed within a few minutes in an outpatient setting, and causes a minimal to modest amount of pain. Demonstration of inflammatory infiltrate in the endometrium generally predicts laparoscopically detected salpingitis with 70–93% sensitivity and with 67–89% specificity

[16,22,23]. One study reported a much lower sensitivity for endometrial biopsy (39%) [16]. In that study, the agreement of three pathologists was required to make the diagnosis of PID. Ness *et al.* [24] showed that one difficulty with using endometrial biopsy as a diagnostic tool is the lack of interpathologist agreement on the diagnosis of endometritis. However, a classification scheme whereby a diagnosis of PID is made on the basis of any plasma cells OR ≥5 neutrophils (rather than an alternative diagnostic scheme proposed by Kiviat *et al.* [23] that requires finding any plasma cells AND ≥5 neutrophils) improves interrater agreement. Another difficulty in using endometritis to diagnose PID is that the turn-around time for test results is typically at least 24 hours. This delay presents two problems. First, the longer the period between the onset of symptoms and the treatment for PID, the greater the risk of infertility [25]. Second, some patients may not be able to be reached and others may not return for treatment upon being notified of the results of a positive endometrial biopsy. Therefore, although endometrial biopsy represents an important research tool, the more realistic clinical strategy has been to treat all individuals who have clinically suspected disease knowing that some of them do not have PID.

IV. Microbial Etiology

PID is caused by the ascension of lower genital tract infections or bacteria into the upper genital tract, resulting in inflammation and infection of the uterus, tubes, and ovaries.

Numerous studies have reported a wide variety of microorganisms recovered from the upper genital tract of women with acute PID [26–32]. In the majority of cases, PID results from the ascension of untreated lower genital tract infection with *Chlamydia trachomatis* and *Neisseria gonorrhoeae;* however, mixed anaerobic and facultative bacteria have been identified, including *Prevotella sp., Prevotella bivius, Prevotella disiens, Peptostreptococii sp., Gardnerella vaginalis, Eschereschia coli, Haemophilus influenzae,* and aerobic streptococci and genital mycoplasmas (*M. hominis* and *U. urealyticum*).

Rice and Schachter noted that the majority of proven cases of PID (*i.e.,* laparoscopy or endometrial biopsy confirmed) are associated with *N. gonorrhoeae* and/or *C. trachomatis* [33]. However, in some reports only anaerobic and/or facultative bacteria were recovered in up to 30% of women [34]. Thus, while *N. gonorrhoeae* and *C. trachomatis* are the most common microorganisms associated with PID, the anaerobic and facultative bacteria from the endogenous vaginal flora may also play an important role. In fact, the recovery of anaerobic and facultative bacteria from the upper genital tract in combination with *N. gonorrhoeae* and/or *C. trachomatis* infection has occurred in up to two-thirds of hospitalized cases of PID. Research suggests that the ascension of *N. gonorrhoeae* and *C. trachomatis* into the upper genital tract results in inflammation which allows the normally nonpathogenic anaerobic and aerobic bacteria to further infect the upper genital tract [35,36]. On the other hand, alteration of the normal vaginal flora, as is seen in nonspecific vaginitis, may facilitate ascension of *N. gonorrhoeae* and *C. trachomatis* into the upper genital tract.

The role of nonspecific vaginitis, independent of infection with gonorrhea or chlamydia, in causing PID has been a major area of debate. The same bacteria found in the upper genital tract of women with PID have been associated with bacterial vaginosis (BV), the most common infectious cause of nonspecific vaginitis [37]. There is evidence that BV detected clinically and BV-associated bacteria in the upper genital tract are associated with salpingitis and endometritis, independent of *N. gonorrhoeae* or *C. trachomatis* infections [31,38,39]. Bacterial vaginosis is characterized by a 10^4-fold increase in the number of vaginal bacteria and by a disequilibrium in normally occurring vaginal microflora such that the normally predominant hydrogen peroxide-producing strains of lactobacilli are reduced and facultative and anaerobic organisms are increased. The production of lactic acid by lactobacilli has long been considered the basis of a protective role of these organisms against vaginal infection and anaerobic overgrowth [39]. Strains of hydrogen peroxide-producing lactobacilli may inhibit the growth of genital microorganisms [38]. Hydrogen peroxide-producing lactobacilli have been shown *in vitro* to kill bacteria, including *Gardnerella vaginalis, Prevotella bivia,* and *N. gonorrhoeae* [40,41]. In a longitudinal study, Hawes *et al.* showed that lactobacilli, and particularly those that produced hydrogen peroxide, protected against the development of BV [42]. Peipert *et al.* linked BV with PID [43], suggesting that a predominance of nonlactic-acid producing lactobacilli or a lack of lactobacilli altogether in the vaginal tract may predispose a woman to BV and, perhaps, to PID. On the other hand, Soper *et al.* [44], in a laparoscopic study entailing microbial examination of upper genital tract isolates, found that BV-related organisms were only found in

conjunction with *N. gonorrhoeae* and *C. trachomatis.* Thus, whether BV-organisms can initiate PID in the absence of sexually transmitted infections remains unknown.

Up to 40% of untreated lower genital tract infections may progress to PID, but the reason why some women with lower genital tract infections experience ascension to the upper genital tract and some women do not is unclear [45,46]. The cervix, which provides a functional barrier to prevent the ascension of microorganisms, has been shown to be less effective during ovulation and menstruation when the mucus plug is absent, and during periods of infection when the endocervical canal is inflamed [47,48]. Ascension of lower genital tract infections may also be promoted during menstruation by retrograde menstrual blood flow [49,50]. Studies have shown that symptoms of PID due to *N. gonorrhoeae* and *C. trachomatis* infection primarily begin during or immediately after menstruation [51]. The degree of epithelium exposed may also influence the rate of ascension. Gonococcal and chlamydial microorganisms must first attach to columnar epithelium in order to invade the upper genital tract. Such epithelium is not found in the vagina. It is only found in the area surrounding the cervical os. Teenage women generally have a larger zone of columnar epithelium around the cervix, a condition that is termed cervical ectopy. In comparison to older women, the larger zone of cervical ectopy in adolescents may place them at higher risk for ascension of vaginal infections and may explain their increased rates of PID [47]. In addition, putative risk factors for PID, including contraceptive methods and pregnancy, via elevations in blood estrogen levels, may promote ascension by enlarging the zone of cervical ectopy or by altering the normal vaginal microenvironment.

Infection in the upper genital tract induces inflammation and subsequent tissue damage. The best described model of this phenomenon is a primate model of *C. trachomatis* genital infection. In that model, a single infection with *C. trachomatis* is not pathogenic. However, repeated episodes of *C. trachomatis* infection trigger a chronic inflammatory response. The tubal cilia, which are responsible for gently moving the ovum down the tube toward the uterus before and after fertilization, become denuded and the area both in and around the fallopian tubes becomes filled with inflammatory exudate [52]. The eventual result is impaired tubal function and/or tubal obstruction which then manifests as tubal infertility, ectopic pregnancy, and/or chronic pelvic pain. At least some of the recurrent PID found among women with the disease may also be a function of autoimmunity due to cross-antigenicity between the chlamydial heat shock protein and the human 60 kilodalton (KDa) heat shock protein [53].

V. Health Consequences and Costs

Major medical sequelae result from PID. Short-term consequences include acute pelvic pain, tuboovarian abscess, tubal scarring, and adhesions [48]. Long-term consequences of PID include an increased risk of infertility, ectopic pregnancy, chronic pelvic pain, and recurrent episodes. In fact, PID is thought to be the leading cause of infertility worldwide and may be a major reason for the dramatic increase in ectopic pregnancies in the United States [54–56]. Only one long-term natural history study of PID has been conducted. Westrom and colleagues enrolled

Table 29.1

Retrospective Studies Linking Past Infection with Chlamydia and Infertility

Source	Number infertile/controls	Percent of women with tubal infertility with chlamydial antibodies	Percent of women without tubal infertility or controls with chlamydial antibodies
[59]	186 infertile patients	79% with abnormal HSG ab > 1:32. 73% with abnormal distal occlusion on laparoscope ab > 1:32	11% with normal HSG, ab > 1:32
[60]	123 infertile patients; 40 sterilized controls; 63 termination controls; 72 barrier contraception controls	75% with damaged tubes on laparoscope or HSG, ab > 1:32	24–48% ab > 1:32
[61]	245 infertile patients	83% with "residues of PID" by laparoscope, ab > 1:8	16% without residues, ab > 1:8
[62]	204 infertile patients	64% with old PID on laparoscope or HSG, ab > 1:16	28% without PID, ab > 1:16
[63]	105 infertile patients 90 pregnant controls	91% with tubal occlusion IgG > 1:8; 84% with tubal adhesions IgG > 1:8; 16 ≤ 24% with occlusion adhesions IgM > 1:8	47–55% IgG > 1:8 2–3% IgM > 1:8
[64]	164 infertile patients 162 family planning controls 38 sterilization controls	36% with tubal disease, ab ⊕ titer	11% control ab ⊕ 12% without tubal disease ab ⊕
[65]	172 infertile patients	60% with tubal disease, ab > 1:8	17% without tubal disease, ab > 1:8
[66]	128 infertile patients female contacts of males with nongonococcal urethritis	57% infertile ab > 1:16 86% female contacts ab > 1:16	29% pregnant females, ab > 1:16
[67]	104 infertile patients	46% damaged tubes ab ⊕ titer	7% normal tubes, ab ⊕
[68]	52 tubal factor infertility patients 114 sterilized controls 99 hysterectomy controls	79% with tubal infertility IgG ab > 1:400 or IGM ab > 1:800	38% sterilized ab ⊕, 38% hysterectomy ⊕
[69]	78 tubal factor infertility patients 155 other infertile controls 466 fertile controls	80% with tubal occlusion ab ⊕ for chlamydia or gonorrhea	60% fertile ab ⊕

Note. HSG, hysterosalpingogram; ab, antibody titer.

Swedish women who had clinically suspected disease from 1960–1984. All women were laparoscoped and treated with the variety of intravenous antibiotics that were considered most effective at that time. Over the course of the study, 2501 women were enrolled, 1844 of whom had abnormal laparoscopic findings (PID patients) and 657 of whom had normal findings (controls). Early reports from this study indicated that one-fifth of women who had at least one episode of PID experienced infertility and one-quarter of patients lived with chronic pelvic pain [57]. Of those women with three or more episodes of PID, over one-half subsequently became infertile [48,58]. Additionally, over 20% of patients had recurrent episodes of PID and almost one in ten women experienced an ectopic pregnancy in their first pregnancy after being diagnosed with PID compared to a rate of 1/66 for patients without a history of PID [48]. In the most recent report, after observing 13,400 women-years for PID patients and 3958 woman-years for controls, 16% of patients and 2.7% of controls were infertile; 10.8% of patients and 0.0% of controls had diagnosed tubal infertility [57].

Findings from case-control studies support the findings from the Scandinavian prospective cohort (Table 29.1). Serologic evidence of prior *C. trachomatis* infection has been found in 36–91% (median 75%) of women with tubal infertility as compared to 7–60% of control women [59–69]. A large proportion of women with ectopic pregnancy also have serum antibodies to *C. trachomatis,* suggesting past infection [70].

The financial costs associated with PID and its sequelae are enormous. Annual U.S. expenditures for the acute treatment of PID have been estimated at between $700 million and $2 billion. The indirect costs from PID may account for an additional $1 billion per year [71–73]. In 1990, the total estimated cost of PID was approximately $4.2 billion, which included the costs of acute PID care plus the direct and indirect costs of infertility and ectopic pregnancy attributed to PID. Functional impairment from chronic pelvic pain and recurrent episodes may increase the economic burden even further. Assuming a constant rate of inflation and a constant PID incidence rate, the total cost of PID, including both direct and indirect costs, is projected to be more than $9 billion in the year 2000 [71].

VI. Unrecognized PID

Cates suggested that women with clinical signs and symptoms suggestive of PID represent less than one-half of all incident cases (Fig. 29.2) [7]. Several lines of evidence suggest that a substantial proportion of PID is clinically unrecognized. First, tubal infertility is thought to be a direct consequence of PID, yet more than half of the women with tubal infertility and serologic evidence of prior *C. trachomatis* infection do not recall having been diagnosed or treated for PID [74]. Similarly, women with ectopic pregnancy and serum antibodies to *C. trachomatis* often have no recalled medical history of PID [75]. This evidence, although suggestive, is not compelling because medical events, particularly those that do not require hospitalization, are frequently forgotten. Therefore, it is not clear whether the lack of recalled PID among women with adverse reproductive sequelae is due to a lack of symptoms during the acute episode, atypical acute symptoms, or inaccurate recall.

Second, Henry-Suchet *et al.* reported isolation of *C. trachomatis* from the fallopian tubes of women with tubal infertility but without any recalled symptoms of PID [76]. Patton *et al.* demonstrated *C. trachomatis* DNA in 79% of women with postinfectious tubal infertility [77]. Patton also showed that the histologic pattern of damage in the tubes was similar for women with overt and unrecognized PID [78]. Chlamydia is not normally cultured from healthy upper pelvic structures. Third, many women with unsuspected upper genital tract infection have been found to have endometritis. Sweet *et al.* reported the persistence of *C. trachomatis* in endometrial specimens of women who had become pain-free after cephalosporin treatment for PID [28]. This finding demonstrates that *C. trachomatis* can cause chronic membrane inflammation without pelvic pain. Paavonen *et al.* found endometritis on histopathology in 47% of women with mucopurulent cervicitis without moderate pelvic pain [79]. In these same women, of those with positive cervical cultures for *C. trachomatis,* 65% were found to have endometritis. Ness *et al.* demonstrated that in women with untreated cervicitis or with infected partners, 44% had evidence of endometritis [24]. These findings suggest that large numbers of women who are thought to have uncomplicated lower genital tract infections actually have concomitant unrecognized PID.

VII. Risk Factors

Numerous demographic and behavioral factors have been associated with PID. These factors may act to increase the probability of infection with *N. gonorrhoeae* and *C. trachomatis,* or to increase the probability that infection will ascend from the lower genital tract into the upper genital tract or both. Alternatively, some factors may not, themselves, alter risk, but may be markers for increased diagnosis. A variety of factors will be discussed below with an explanation of the potential mechanism(s) by which each is associated with risk. Women of lower socioeconomic status, nonwhite women, and single or divorced women have the highest rate of both sexually transmitted diseases (STDs) and PID. Data from hospital admissions, physicians' offices, and self-reported questionnaires have consistently documented a higher rate of PID among nonwhite women. Black women were almost three times more likely to be hospitalized for PID in 1988 compared to white women, and black women reported a history of PID twice as frequently as white women [3,6,8]. One potential explanation for this racial disparity may be a higher degree of PID diagnosis among nonwhite women given the higher prevalence of STDs in this group [80].

Women of lower socioeconomic status are at higher risk for STDs and PID. Although low socioeconomic status is not causally related to PID, characteristics of individuals in lower socioeconomic groups increase the risk of developing PID once a cervical infection has been acquired. These characteristics include a reduced recognition of the signs and symptoms of cervical infections and PID, a lack of affordable accessible health care services, inadequate adherence to antibiotic therapy and follow-up appointments, and failure to seek antibiotic treatment for an infected sexual partner [35]. Prompt recognition of the signs and symptoms of PID is essential to reduce the sequela of this disease, and early medical intervention may prevent ascension of *N. gonorrhoeae* and *C. trachomatis.*

A. Age

Approximately 70% of women diagnosed with PID are 25 years of age or younger. Sexually active teenage women have consistently been found to have the highest rate of PID [81]. Sexually active teenagers are three times more likely to be diagnosed with PID compared to women aged 25–29 years [82]. One study reported a one in eight risk of PID among sexually active 15 year old women, a one in ten risk of PID among sexually active 16 year old women, and a one in eighty risk of PID among sexually active 24 year old women [81].

The high rate of PID among teenagers is mainly a function of their relatively large number of sexual partners and the increased prevalence of STDs in this age group. The rate of gonococcal infection has been reported to be three times higher among women 15–19 years of age compared to older women [83]. Although the rate of gonorrhea has declined substantially among adult women during the 1990s, teenage women experienced the least decline among any age group [83]. Teenage women also have a high prevalence of *C. trachomatis* infection. Genital chlamydial infection has surpassed gonorrhea as the number one sexually transmitted disease in the United States, with approximately 41 million new cases per year and teenage women experiencing the majority of the burden of disease [84].

Biologic factors may also help to explain the high rate of PID among teenage women. Younger women have a lower prevalence of protective chlamydial antibodies and a larger zone of cervical ectopy, placing them at increased risk for infection and, perhaps, for ascension of lower genital tract infections [7]. Teenage women have fewer ovulatory menstrual cycles, resulting in a high estrogen/ low progesterone state. Endogenous and exogenous progesterone production promote the development of a complex cervical mucus that restricts the ascension of sperm and bacteria to the upper genital tract; the presence of estrogen without progesterone in nonovulating teenage women results in a greater penetrability of cervical mucus by infectious organisms [85].

B. Sexual Behavior

Many sexual behaviors have been related to the risk of STDs and, therefore, PID, including a high number of current and lifetime sexual partners, a young age at first intercourse, and a high frequency of intercourse [86]. One study found that women who reported four or more sexual partners in the prior 6 months were 3.4 times more likely to be diagnosed with PID compared to women who reported only one sexual partner [87]. Other epidemiologic studies have confirmed the positive relationship between multiple sexual partners and PID [51,88]. Lee *et al.* found that women reporting a high frequency of intercourse, defined as six or more times per week, had an elevated risk of PID compared to women reporting intercourse less than once per week (OR = 1.6, 95% CI 1.1–2.3) [87]. Both multiple sexual partners and frequent intercourse increase the risk of cervical infections. Studies have found that the number of current sexual partners is related to infections with *N. gonorrhoeae* and *C. trachomatis* [89,90].

Male sexual partners with untreated *N. gonorrhoeae* or *C. trachomatis* infection provide a source for initial infection and reinfection with sexually transmitted diseases. One study screening the male sexual partners of women admitted to a hospital for acute PID found that nearly 60% of the men were infected with *C. trachomatis* (19%), *N. gonorrhoeae* (4%), or nonspecific urethritis (36%) [91]. Studies have reported that despite the fact that a large number of male sexual partners of women with PID report symptoms of urethritis (27–50%), only about 20% of men seek treatment for urethritis prior to the diagnosis of PID in their female partner [81,92]. The remaining one-half of men with urethritis do not experience any signs or symptoms of infection. This lack of detection and medical intervention among male sexual partners infected with *N. gonorrhoeae* or *C. trachomatis* may explain the high number of recurrent episodes of PID among women with initial episodes and the high rate of repeat infections with *N. gonorrhoeae* or *C. trachomatis* among these women.

C. Prior Infections

Women with a prior history of PID have an increased risk for a subsequent episode. One study reported that women diagnosed with an index episode of PID are twice as likely to have a prior history of PID as compared to women without index PID, after controlling for race, number of sexual partners, and IUD use [88]. In addition, nearly one-quarter of women with acute PID developed a second episode of PID due to reinfection by an untreated sexual partner or to inadequate adherence to antibiotic therapy. Repeat episodes of PID are important to prevent because the number of episodes is directly related to the risk of infertility. Westrom reported that the infertility rate among women increased from 13% given one episode of salpingitis to 35% following two episodes of salpingitis to 75% given three or more episodes of salpingitis [93].

Women with a previous infection with *N. gonorrhoeae* or *C. trachomatis* also have an increased risk for PID. Studies report a relative risk of PID given a prior STD ranging from 1.7 to 2.5 [3,51,87]. In the Women's Health Study, a hospital-based case-control study, women reporting a previous episode of gonorrhea had a significantly higher risk of PID compared to women without a history of gonorrhea (adjusted OR = 1.8, 95% CI 1.4–2.3) [87]. The increased risk for PID after a gonococcal infection may be related to repeat infection or to preexisting subclinical tubal infection. A defined population of individuals have a high rate of recidivism for gonorrhea as a result of reinfection [94]. This group, termed the "core group," is thought to act as a

reservoir for sexually transmitted pathogens, and the more frequent the sexual interrelations between the core group and other individuals, the more rapid the spread of infection. On the other hand, studies have suggested that a new gonococcal infection may reactivate latent chlamydial infection [95].

D. Contraception

Consistent use of spermicidal barrier methods of contraception, including condoms, the diaphragm, and the cervical cap, appear to prevent both infection with *N. gonorrhoeae* and *C. trachomatis* and the development of PID [96,97]. Data from the Women's Health Study found that women reporting consistent use of barrier methods of contraception had a reduced risk of an initial episode of PID compared to women using other forms of contraception or no contraception (RR = 0.6, 95% CI 0.5–0.9) [96]. In contrast, a prospective cohort study conducted in an STD clinic population designed to assess the validity of self-reported 30 day condom use found that 23.5% of women who reported always using condoms had a newly diagnosed sexually transmitted disease compared to almost the same percentage (26.8%) of women who reported never using condoms [98]. This suggests either that condoms were not used consistently and correctly or that women at higher risk of STDs were reporting condom use but were not actually using them [98,99].

Use of the intrauterine device (IUD) has been associated with an increased risk of PID, particularly in the first four months following insertion, and maybe related to the introduction of lower genital tract infections into the upper genital tract. A large prospective cohort study reported a 3.5-fold increased risk of PID among IUD users compared to non-IUD users [100]. Another hospital-based case-control study found that IUD use was an independent predictor of salpingitis and that IUD use doubled the probability of disease [88]. However, some studies have found that the increased risk of PID due to IUD use only occurs among women at risk for an STD because women determined to be at low risk for an STD did not have a substantial increase in IUD-associated PID [101]. Currently, experts suggest that IUDs can be used safely among women involved in monogamous, long-term relationships, particularly when antibiotics are given at the time of insertion.

Not surprisingly, IUD use has been related to an increase in tubal infertility via its association with PID. One case-control study compared prior IUD use among women seeking treatment for tubal infertility and fertile women [102]. This study found that women reporting prior IUD use had a 2.6 times greater risk of tubal infertility than women who reported never using an IUD (95% CI 1.3–5.2). The highest risk of tubal infertility was associated with the use of the Dalkon Shield (OR = 6.8, 95% CI 1.8–25.2). The copper-containing IUD, the most frequently used IUD, was not associated with an increase in tubal infertility (OR = 1.9, 95% CI 0.9–4.0).

Studies evaluating the role of oral contraceptives (OC) and PID due to chlamydia have been mixed. In retrospective cohort studies, women reporting oral contraceptive use had an increased prevalence of chlamydial infection compared to women reporting other methods of contraception. Studies have reported a two- to threefold increased risk of chlamydial infection among oral contraceptive users compared to women using other types of contraception [103–105]. Results from animal studies suggest that exposure to exogenous estrogen and progesterone stimulate the growth of *C. trachomatis* [106,107]. It is currently unclear whether the increase in *C. trachomatis* infection is due to an increase in detection among OC users or an actual increase in infection rates.

Paradoxically, numerous epidemiologic studies have reported a decreased risk of symptomatic PID among OC users [86,108]. One study found that the risk of symptomatic chlamydial PID was significantly lower among women reporting previous or current OC use compared to women reporting no method of contraception (OR = 0.30, 95% CI 0.10–0.89) [109]. This protective effect was significant after adjusting for age, number of pregnancies, and number of lifetime sexual partners. The protection afforded by oral contraceptive use may be due to hormonal influences. The estrogen and progesterone balance generated by oral contraceptive use produces a cervical mucus which is less permeable to bacterial penetration. In addition, the short time period of menstruation among OC users may reduce the amount of retrograde blood flow, further decreasing the risk of PID [81].

The inconsistency between an increase in lower genital tract chlamydial infection due to OC use and the reported decrease in symptomatic PID among OC users has led researchers to speculate that OC use may result in unrecognized, asymptomatic upper genital tract disease [33]. Ness *et al.* [24] studied whether current OC use was more common among women with unrecognized endometritis compared to women with recognized endometritis. The authors found that women with unrecognized endometritis were 4.3 times more likely (95% CI 1.6–11.7) than women with recognized endometritis to report current OC use [24], after adjusting for history of sexually transmitted diseases and the presence of *N. gonorrhoeae*. The findings from this research suggest that current OC use may be associated with unrecognized, asymptomatic cases of PID, perhaps due to a reduction in the inflammatory response and subsequent symptomatology among OC users.

E. Vaginal Douching

In the United States, vaginal douching is routinely practiced in certain subgroups of women. A nationwide survey of women over 18 years of age found that 32% of women self-reported vaginal douching within the previous week [2]. In this survey, the highest frequency and prevalence of douching was among nonwhite women and women of low socioeconomic status. A meta-analysis conducted to determine the association between vaginal douching and PID found that douching increased the risk of PID by 73% and the risk of ectopic pregnancy, a complication of PID, by 76% [110]. A case-control study comparing clinically-confirmed cases of PID and random controls found an increased risk of PID among women reporting vaginal douching in the previous two months (OR = 1.7, 95% CI 1.0–2.8) [111]. This relationship was significant after controlling for confounding factors including age, race, cigarette smoking, history of PID or STDs, frequency of intercourse, and contraceptive method. In addition, this study found a dose-response relationship between douching and PID. Women reporting douching once or twice a month had an odds ratio of 1.6 (95% CI 0.6–

4.0) and women douching three or more times a month had an odds ratio of 3.4 (95% CI 1.1–10.4) compared to women reporting douching less than once a month.

Another study examining the relationship between douching and PID found a twofold increase in PID among women reporting current douching (adjusted RR = 2.1, 95% CI 1.2–3.9) [112]. However, in this study the magnitude of the association between vaginal douching and PID was reduced when the reason for initiating douching was assessed. Women who reported initiating vaginal douching due to lower genital tract irritation or infection had a substantially increased risk of PID (RR = 7.9, 95% CI 2.6–24.2), but women who reported douching for other reasons had a nonsignificant increase in PID (RR = 1.6, 95% CI 0.8–3.0).

In theory, vaginal douching may increase the risk of PID mechanically by promoting the ascension of lower genital tract infections to the upper genital tract, by altering the vaginal environment to diminish protection against STDs, or by introducing nonpathogenic aerobes and anaerobes from the lower genital tract into the sterile, upper genital tract. Further studies are needed to determine if vaginal douching is causally related to PID or if douching is a marker for vaginal infection (that is, women may douche to alleviate the vaginal discharge associated with *N. gonorrhoeae* or *C. trachomatis* that would have led to PID even in the absence of douching).

F. Cigarette Smoking

Cigarette smoking has been identified as a risk factor for *N. gonorrhoeae* and *C. trachomatis* infection. A few studies have reported a positive association between current cigarette smoking and PID [113,114]. Data from the Women's Health Study found a twofold higher risk of PID both among current smokers (OR = 1.7, 95% CI 1.1–2.5) and among former smokers (OR = 2.3, 95% CI 1.3–4.2) [113].

Other studies have reported an increase in PID-related complications, such as tubal infertility and ectopic pregnancy, among women who smoke cigarettes. Chow *et al.* reported that women who reported smoking at the time of conception had a twofold increased risk of ectopic pregnancy compared to never smokers, after adjusting for gravity, race, IUD use, and vaginal douching (OR = 2.2, 95% CI 1.4–3.4) [115]. This study did not find a dose-response relationship and former smokers who reported quitting prior to conception did not have a significant increase in ectopic pregnancy (OR = 1.6, 95% CI 1.0–2.8). However, this study did not adjust for history of sexually transmitted diseases and relied on self-report. Another case-control study comparing current and never smokers related current cigarette use to a 60% increased risk of tubal infertility (RR = 1.6, 95% CI 1.1–2.1) [116]. This study also found an increase in cervical related infertility, defined as abnormal cervical mucus or nonprogressive sperm, among smokers but no increase in ovulatory or endometriosis related infertility.

The metabolites in cigarettes have been found in the cervical mucus of current smokers, which may explain the biologic mechanisms promoting the relationship between cigarette smoking and PID. A small study collected cervical mucus samples from smokers and nonsmokers and reported that nicotine and cotinine traces were detected in the cervical mucus of smokers [117].

The levels were equal or higher than the levels found in serum samples. These metabolites potentially may decrease vaginal immunity and/or limit ciliary motion and tubal mobility which may diminish effective clearing of infectious organisms in the cervix [86]. Animal studies also have found a delay in ovum transport to the uterus, blastocyst formation, and implantation among nicotine-treated rats [118].

VIII. Treatment

The goal of treatment for PID is to reduce the sequelae of reproductive morbidity. Despite the formulation and promotion of guidelines by the CDC, there is no clear consensus regarding an optimal approach to the treatment of PID [12,119]. The current CDC recommendations advise hospitalization for women whose acute illness appears to be more severe or in whom surgical emergencies such as appendicitis or tuboovarian abscess cannot be excluded [120]. Outpatient treatment is suggested for women with mild to moderate symptomatology in whom good compliance with the antibiotic therapy is assured [48]. These broad guidelines are based on consensus rather than the results of clinical trials. In addition, these guidelines are inconsistently interpreted, as illustrated by the widely divergent rates of hospitalization for PID among various geographic locations and healthcare settings [121].

Another problem with the CDC guidelines is that there is no established accurate test that predicts the development of reproductive complications among women with PID. Although the severity of inflammation upon direct inspection of the pelvic organs at laparoscopy is predictive of the development of infertility, laparoscopic findings do not directly correlate with clinical presentation. PID associated with *C. trachomatis* infection has a less dramatic presentation than that associated with *Neisseria gonorrhea* infection. However, *C. trachomatis* paradoxically results in a higher rate of infertility [26,122]. Therefore, the recommendation to hospitalize women with more severe disease may not actually reduce the burden of reproductive morbidity resulting from PID. Because the development of sequelae is unpredictable, some authorities have suggested that all women who wish to maintain their fertility potential be hospitalized [119]. One ongoing randomized clinical trial is the first to test the effectiveness of outpatient antimicrobial therapy as compared to inpatient therapy [123].

Whether medical treatment is administered in the inpatient or outpatient setting, antimicrobial therapy for PID requires a combination of medications that cover *N. gonorrhoeae*, *C. trachomatis,* and potentially vaginal anaerobes and facultative aerobes. Such regimens include, for example, one of the cephalosporins plus doxycycline, or clindamycin plus gentamycin, for inpatients and ofloxacin plus metronidazole, or one of the cephalosporins plus doxycycline, for outpatients [12]. The degree to which optimal treatment should include anaerobic coverage remains controversial. Treatment of sexual partners is a requisite part of good clinical care. This is important because reinfection is common when asymptomatic sexual partners remain untreated.

A final point regarding treatment of PID is the importance of its timeliness. Safrin *et al.* showed that delay in treatment of symptoms of PID was associated with an increased risk of subsequent infertility [124]. This finding has been confirmed in the Scandinavian cohort study. Women who waited more than three days after the onset of symptoms were threefold more likely to become infertile after PID than were women whose symptoms were more rapidly treated [25]. This may reflect the additional inflammatory response associated with an untreated, closed-space pelvic infection.

IX. Prevention

Timely medical treatment for PID is considered tertiary prevention with the goal of reducing the adverse reproductive consequences associated with PID. However, even in the Scandinavian cohort where most women were treated quickly and all women were treated with intravenous antibiotics, the frequency of infertility was high. Therefore, the development of effective primary and secondary prevention programs is critical. Primary prevention consists of reducing the risk of infection with a sexually transmitted disease in the first place. Secondary prevention consists of preventing cervicitis from ascending to the upper genital tract, resulting in PID. Approaches to primary prevention include promoting sexual abstinence or use of barrier methods of contraception. The secondary prevention approach is to improve screening for asymptomatic lower genital tract infections and to ensure rapid treatment of symptomatic lower genital tract infections. Current primary and secondary prevention activities have been shown to be effective in reducing rates of PID. In Scandinavia, reductions in gonococcal and chlamydial disease have resulted in a subsequent reduction in PID [1]. In Washington State, a randomized clinical trial conducted within a health maintenance organization found that active screening for *C. trachomatis* among asymptomatic young women, in addition to providing usual health care, reduced the incidence of PID by 60% over the course of one year [125]. This was the first experimental study to verify the effectiveness of secondary prevention for PID and it demonstrates the importance of the recognition, diagnosis, and treatment of lower genital tract infections.

X. Summary

Pelvic inflammatory disease is a remarkably common condition that affects women of reproductive age in the U.S. and worldwide. Although it is clearly caused by *N. gonorrhoeae* and *C. trachomatis,* the role of vaginal anaerobes and facultative aerobes is less clear, leading to some disagreement regarding appropriate treatment. Upper genital tract infection leads to inflammation and inflammation leads to tissue damage, the clinical result of which is infertility, ectopic pregnancy, chronic pelvic pain, and recurrent PID. These reproductive morbidities are both debilitating and costly. Despite the devastating reproductive consequences of PID and the existence of several known, modifiable risk factors, PID remains a common disease in the United States. Public health strategies such as those used in Scandinavia, which include aggressive primary and secondary prevention activities, could be applied in this country to effectively reduce the burden of morbidity from this preventable disease.

References

1. Kani, J., and Adler, M. W. (1992). Epidemiology of pelvic inflammatory disease. *In* "Pelvic Inflammatory Disease" G. S. Berger and L. V. Westrom, (eds.), Raven Press, New York.

2. National Survey of Family Growth (1995). "From Vital and Health Statistics. Data from the National Survey of Family Growth." U.S. Department of Health and Human Services, Public Health Service, National Center for Health Statistics, Hyattsville, MD.

3. Aral, S. O., Mosher, W. D., and Cates, W. (1991). Self-reported pelvic inflammatory disease in the U.S.: A common occurrence. *JAMA, J. Am. Med. Assoc.* **266,** 2570–2573.

4. Sweet, R. L. (1987). Pelvic inflammatory disease and infertility in women. *Infect. Dis. Clin. North Am.* **1,** 199.

5. Blount, J. H., Reynolds, G. H., and Rice, R. J. (1983). Pelvic inflammatory disease: Incidence and trends in private practice. *Morbid. Mortal. Wkly. Rep.* **32** (4SS), 27SS.

6. Rolfs, R. T., Galaid, E. I., and Zaidi, A. A. (1992). Pelvic inflammatory disease: Trends in hospitalizations and office visits in the U.S., 1979–88. *Am. J. Obstet. Gynecol.* **166**(3), 983–990.

7. Cates, W., Rolfs, R. T., and Aral, S. O. (1990). Sexually transmitted diseases, pelvic inflammatory disease, and infertility: An epidemiologic update. *Epidemiol. Rev.* **12,** 199–220.

8. Washington, A. E., Cates, W., and Zaidi, A. A. (1984). Hospitalizations for pelvic inflammatory disease. Epidemiology and trends in the United States, 1975 to 1981. *JAMA, J. Am. Med. Assoc.* **251,** 2529–2533.

9. Quan, M., Roadney, W. M., and Johnson, R. A. (1983). Pelvic inflammatory disease. *J. Fam. Pract.* **16**(1), 131–140.

10. Eschenbach, D. A., Wolner-Hanssen, P., Hawes, S. E., Pavletic, A., Paavonen, J., and Holmes, K. K. (1997). Acute pelvic inflammatory disease: Association of clinical and laboratory findings with laparoscopic findings. *Obstet. Gynecol.* **89,** 184–192.

11. Morcos, R., Frost, N., Hnat, M., Petrunak, A., and Caldito, G. (1993). Laparoscopic versus clinical diagnosis of acute pelvic inflammatory disease. *J. Reprod. Med.* **38**(1), 53–56.

12. Centers for Disease Control and Prevention (1998). Guidelines for treatment of sexually transmitted diseases. *Morbid. Mortal. Wkly. Rep.* **47**(RR-1), 79–86.

13. Hager, W. D., Eschenbach, D. A., Spence, M. R., and Sweet, R. L. (1983). Criteria for diagnosis and grading of salpingitis. *Obstet. Gynecol.* **61,** 113.

14. Kahn, J. G., Walker, C. K., Washington, E., Landers, D. V., and Sweet, R. L. (1991). Diagnosing pelvic inflammatory disease. A comprehensive analysis and considerations for developing a new model. *JAMA, J. Am. Med. Assoc.* **266,** 3594.

15. Paavonen, J., Aine, R., Teisala, K., Heinonen, P., and Punnonen, R. (1985). Comparison of endometrial biopsy and peritoneal fluid cytologic testing with laparoscopy in the diagnosis of acute pelvic inflammatory disease. *Am. J. Obstet. Gynecol.* **151**(5), 645.

16. Sellors, J., Mahony, J., Goldsmith, C., Rath, D., Mander, R., Hunter, B., Taylor, C., Groves, D., Richardson, H., and Chernesky, M. (1991). The accuracy of clinical findings and laparoscopy in pelvic inflammatory disease. *Am. J. Obstet. Gynecol.* **164,** 113–120.

17. Peipert, J. F., Boardman, L., Hogan, J. W., Sung, J., and Mayer, K. H. (1996). Laboratory evaluation of acute upper genital tract infection. *Obstet. Gynecol.* **87,** 730–736.

18. Ness, R. B., Soper, D. E., Peipert, J., Bass, D., Hemsell, D., and Shepherd, S. for the PEACH Study Group (1997). Mucopurulent cervicitis predicts the diagnosis of pelvic inflammatory disease (PID) in the absence of vaginal white blood cells excludes diagnosis. *Abstr. Int. Congr. Sex. Transm. Dis.* **P457,** 138. Seville, Spain.

19. Pattern, R. M., Vincent, L. M., Wolner-Hanssen, P., and Thorpe, E. (1990). Pelvic inflammatory disease endovaginal sonography and laparoscopic correlation. *J. Ultrasound Med.* **9,** 681.

20. Bulas, D. I., Ahlstrom, A., Sivit, C. J., Blask, A. R. N., and O'Donnell, R. M. (1992). Pelvic inflammatory disease in the adolescent: Comparison of transabdominal and transvaginal sonographic evaluation. *Radiology* **183,** 435–439.

21. Mikkelsen, A. L., and Felding, C. (1990). Laparoscopy and ultrasound examination of women with acute pelvic pain. *Gynecol. Obstet. Invest.* **30,** 162.

22. Wasserheit, J. N., Bell, T. A., Kiviat, N. B., Wolner-Hanssen, P., Zabriskie, V., Kirby, B., Prince, E. C., Holmes, K. K., Stamm, W. E., and Eschenbach, D. A. (1986). Microbial causes of proven pelvic inflammatory disease and efficacy of Clindamycin and Tobramycin. *Ann. Intern. Med.* **104,** 187.

23. Kiviat, N. B., Wolner-Hanssen, P., Eschenbach, D. A., Wasserheit, J. N., Paavonen, J. A., Bell, T. A., Critchlow, C. W., Stamm, W. E., Moore, D. E., and Holmes, K. K. (1990). Endometrial histology in patients with culture-proved upper genital tract infection laparoscopically diagnosed acute salpingitis. *Am. J. Surg. Pathol.* **14,** 167.

24. Ness, R. B., Keder, L. M., Soper, D. E., Amortegui, A. J., Gluck, J., Weisenfeld, H., Sweet, R. L., Rice, P. A., Peipert, J. F., Donegan, S. P., and Shakir, A. K. (1997). Oral contraception and the recognition of endometritis. *Am. J. Obstet. Gynecol.* **176,** 580–585.

25. Hillis, S. D., Joesoef, M. R., Marchbanks, P. A., Wasserheit, J. N., Cates, W., and Westrom, L. (1993). Delayed care of pelvic inflammatory disease as a risk factor for impaired fertility. *Am. J. Obstet. Gynecol.* **168,** 1503–1509.

26. Brunham, R. C., Binns, B., Guijon, F., and Danforth, D. (1988). Etiology and outcome of acute pelvic inflammatory disease. *J. Infect. Dis.* **158,** 510.

27. Sweet, R. L., Draper, D. L., Schachter, J., James, J., Hadley, W. K., and Brooks, G. F. (1980). Microbiology and pathogenesis of acute salpingitis as determine by laparoscopy. What is the appropriate site to sample? *Am. J. Obstet. Gynecol.* **138,** 985.

28. Sweet, R. L., Schachter, J., and Robbie, M. O. (1983). Failure of beta-lactam antibodies to eradicate *Chlamydia trachomatis* in the endometrium despite apparent clinical cure of acute salpingitis. *JAMA, J. Am. Med. Assoc.* **250,** 2641–2645.

29. Wasserhiet, J. N., Bell, T. A., Kiviat, N. B., Wolner-Hanssen, P., Zambriskie, V., Kisby, B. D., Prince, E. C., Holmes, K. K., Stamm, W. E., and Eschenbach, D. A. (1986). Microbial causes of proven pelvic inflammatory disease and efficacy of Clindamycin and Tobramycin. *Ann. Intern. Med.* **104,** 187.

30. Heinonnen, P. K., Teisala, K., Punnonen, R., Miettinen, A., Lehtinen, M., and Paavonen, J. (1985). Anatomic sites of upper genital tract infection. *Obstet. Gynecol.* **66,** 384–390.

31. Paavonen, J., Teisala, K., Heinonen, P. K., Aine, R., Laine, S., Lehtinen, M., Miettinen, A., Punnon, R., and Gronroos, P. (1987). Microbiological and histopathological findings in acute pelvic inflammatory disease. *Br. J. Obstet. Gynecol.* **94,** 454–460.

32. Soper, D. E., Brockwell, N. J., and Dalton, H. P. (1992). Microbial etiology of urban emergency room acute salpingitis: Treatment with ofloxacin. *Am. J. Obstet. Gynecol.* **167,** 653.

33. Rice, P. A., and Schachter, J. (1991). Pathogenesis of pelvic inflammatory disease: What are the questions? *JAMA, J. Am. Med. Assoc.* **266,** 2587–2593.

34. Jossens, M. O. R., Schachter, J., and Sweet, R. L. (1994). Risk factors associated with pelvic inflammatory disease of different microbial etiologies. *Obstet. Gynecol.* **83,** 989–997.

35. Eschenbach, D. A. (1980). Epidemiology and diagnosis of acute pelvic inflammatory disease. *Obstet. Gynecol.* **55**(Suppl.), 142S–151S.

36. Wasserheit, J. N. (1987). Pelvic inflammatory disease and infertility. *Md. State Med. J.* **36,** 58–63.

37. Thomason, J. L., Gelbart, S. M., and Scaglione, N. J. (1991). Bacterial vaginosis: Current review with indications for asymptomatic therapy. *Am. J. Obstet. Gynecol.* **165,** 1212–1217.

38. Eschenbach, D. A., Hillier, S. L., Critchlow, C. W., Stevens, C., De-Rouen, T., and Holmes, K. K. (1988). Diagnosis and clinical manifestations of bacterial vaginosis. *Am. J. Obstet. Gynecol.* **158,** 19–28.

39. Hillier, S. L., Krohn, M. A., Rabe, L. K., Klebanoff, S. J., and Eschenbach, D. A. (1993). The normal vaginal flora H_2O_2 producing lactobacilli, and bacterial vaginosis in pregnant women. *Clin. Infect. Dis.* **16**, S272–S281.

40. Klebanoff, S. J., Hillier, S. L., Eschenbach, D. A., and Waltersdorph, A. M. (1991). Control of the microbial flora of the vagina by H_2O_2—generating lactobacilli. *J. Infect. Dis.* **164**, 94–100.

41. Zheng, H. Y., Alcorn, T. M., and Cohen, M. S. (1994). Effects of H_2O_2—producing lactobacilli on Neisseria gonorrhoeae growth and catalase activity. *J. Infect. Dis.* **170**, 94–100.

42. Hawes, S. E., Hillier, S. L., Benedetti, J., Steven, C. E., Koutsky, L. A., Wolner-Hanssen, P., and Holmes, K. K. (1996). Hydrogen peroxide-producing lactobacilli and acquisition of vaginal infections. *J. Infect. Dis.* **174**, 1058–1063.

43. Peipert, J. F., Montagno, A. B., Cooper, A. S., and Sung, C. J. (1997). Bacterial vaginosis as a risk factor for upper genital tract infection. *Am. J. Obstet. Gynecol.* **177**, 1184–1187.

44. Soper, D. E., Brockwell, N. J., and Dalton, H. P. (1992). Microbial etiology of urban emergency department acute salpingitis: Treatment with ofloxacillin. *Am. J. Obstet. Gynecol.* **167**, 653–660.

45. Platt, R., Rice, P. A., and McCormack, W. M. (1983). Risk of acquiring gonorrhea and prevalence of abnormal adnexal findings among women recently exposed to gonorrhea. *JAMA, J. Am. Med. Assoc.* **250**(23), 3205–3209.

46. Stamm, W. E., Guinan, M. E., Johnson, C., Starcher, T., Holmes, K. K., and McCormack, W. M. (1984). Effect of treatment regimens for *Neisseria gonorrhoeae* on simultaneous infection with *Chlamydia trachomatis. N. Engl. J. Med.* **313**, 545–549.

47. Expert Committee on Pelvic Inflammatory Disease (1991). Pelvic inflammatory disease. Research directions in the 1990s. *Sex. Transm. Dis.* **18**, 46–64.

48. Westrom, L., and Mardh, P.-A. (1990). Acute pelvic inflammatory disease (PID). *In* "Sexually Transmitted Diseases" K. K. Holmes, P.-A. Mardh, and P. F. Sparling, (eds.), 2nd ed., p. 593. McGraw-Hill, New York.

49. Halme, J., Hammond, M. G., Hulka, J. F., Shailaja, G. R., and Talbert, L. M. (1984). Retrograde menstruation in healthy women and in patients with endometriosis. *Obstet. Gynecol.* **64**, 151.

50. Lee, N. C., Rubin, G. L., Ory, H. W., and Burkman, R. T. (1983). Type of IUD and the risk of PID. *Obstet. Gynecol.* **62**, 1.

51. Eschenbach, D. A., Harnish, J. P., and Holmes, K. K. (1977). Pathogenesis of acute pelvic inflammatory disease: Role of contraception and other risk factors. *Am. J. Obstet. Gynecol.* **128**, 838.

52. Patton, D., Wolner-Hanssen, P., Cosgrove, S. J., and Holmes, K. K. (1990). The effects of *Chlamydia trachomatis* on the female reproductive tract of the *Macaca nemestrina* after a single tubal challenge following repeated cervical inoculations. *Obstet. Gynecol.* **76**, 643–650.

53. Soper, D. E. (1994). Pelvic inflammatory disease. *Infect. Dis. Clin. North Am.* **8**, 821–840.

54. Cates, W., Jr., Farley, T. M. M., and Rowe, P. J. (1985). Worldwide patterns of infertility: Is Africa different? *Lancet* **2**, 596–598.

55. Spence, M. R., Genadry, R., and Rafael, L. (1981). Randomized prospective comparison of ampicillin and doxycycline in the treatment of acute pelvic inflammatory disease in hospitalized patients. *Sex. Transm. Dis.* **8**, 164.

56. Ory, H. W. (1981). The Women's Health Study: Ectopic pregnancy and intrauterine contraceptive devices: New perspectives. *Am. J. Obstet. Gynecol.* **57**, 137.

57. Westrom, L. V., Joesoef, R., Reynolds, G., Hogdu, A., and Thompson, S. E. (1992). Pelvic inflammatory disease and fertility. A cohort study of 1,844 women with laparoscopically verified disease and 657 control women with normal laparoscopic results. *Sex. Transm. Dis.* **19**, 185–192.

58. Westrom, L. (1980). Incidence, prevalence and trends of acute pelvic inflammatory disease and its consequences in industrialized countries. *Am. J. Obstet. Gynecol.* **138**, 880.

59. Moore, D. E., Foy, H. M., Daling, J. R., Spadoni, L. R., Wang, S. P., Kuo, C. C., Grayston, J. T., and Eshenbach, D. A. (1982). Increased frequency of serum antibodies to *Chlamydia trachomatis* in infertility due to distal tubal disease. *Lancet* **2**, 574–577.

60. Conway, D., Caul, E. O., Hull, M. G. R., Glazenen, C. M. A., Hodgson, J., Clarke, S. K. R., and Stirrat, G. M. (1984). Chlamydial serology in fertile and infertile women. *Lancet* **1**, 191–193.

61. Guderian, A. M., and Trobough, G. E. (1986). Residues of pelvic inflammatory disease in intrauterine device users: A result of the intrauterine device or *Chlamydia trachomatis* infection? *Am. J. Obstet. Gynecol.* **154**, 497–503.

62. Gump, D. W., Gibson, M., and Ashikaga, T. (1983). Evidence of prior pelvic inflammatory disease and its relationship to *Chlamydia trachomatis* antibody and intrauterine contraceptive device use in infertile women. *Am. J. Obstet. Gynecol.* **146**, 153–156.

63. Anestad, G., Lunde, O., Moen, M., and Dalaker, K. (1987). Infertility and chlamydial infection. *Fertil. Steril.* **48**, 787–790.

64. Kane, J. L., Woodland, R. M., Forsey, T., Darougar, S., and Elder, M. G. (1984). Evidence of chlamydial infection in infertile women with and without fallopian tube obstruction. *Fertil. Steril.* **42**, 843–848.

65. Jones, R. B., Ardery, B. R., Hui, S. L., and Clery, R. E. (1982). Correlation between serum antichlamydial antibodies and tubal factor as a cause of infertility. *Fertil. Steril.* **38**, 553–558.

66. Punnonen, R., Terho, P., Nikkanen, V., and Meurman, O. (1979). Chlamydial serology in infertile women by immunofluorescence. *Fertil. Steril.* **31**, 656–659.

67. Miettinen, A., Heinonen, P. K., Teisala, K., Hakkarainen, K., and Punnonen, R. (1990). Serologic evidence for the role of *Chlamydia trachomatis, Neisseria gonorrhoeae,* and *Mycoplasma hominis* in the etiology of tubal factor infertility and ectopic pregnancy. *Sex. Transm. Dis.* **17**, 10–14.

68. Sellors, J. W., Mahony, J. B., Chesnesky, M. A., and Rath, D. J. (1988). Tubal factor infertility: An association with prior chlamydial infection and asymptomatic salpingitis. *Fertil. Steril.* **49**, 451–457.

69. World Health Organization Task Force on the Prevention and Management of Infertility (1995). Tubal infertility: Serologic relationship to past chlamydial and gonococcal infection. *Sex. Transm. Dis.* **22**, 71–77.

70. Chow, J. M., Yonekura, L., Richard, G. A., Greenland, S., Sweet, R. L., and Schachter, J. (1990). The association between *Chlamydia trachomatis* and ectopic pregnancy. *JAMA, J. Am. Med. Assoc.* **263**, 3164–3167.

71. Washington, A. E., and Katz, P. (1991). Cost of and payment for pelvic inflammatory disease. Trends and projections, 1983 through 2000. *JAMA, J. Am. Med. Assoc.* **266**, 2565.

72. Curran, J. W. (1980). Economic consequences of pelvic inflammatory disease in the United States. *Am. J. Obstet. Gynecol.* **138**, 848.

73. Rendtorff, R. C., Curran, J. C., and Chandler, R. W. (1974). Economic consequences of gonorrhea in women. *J. Am. Vener. Dis. Assoc.* **1**, 40.

74. Cates, W., Joesoef, M. R., and Goldman, M. B. (1993). Atypical pelvic inflammatory disease: Can we identify clinical predictors? *Am. J. Obstet. Gynecol.* **169**, 341–346.

75. Svensson, L., Mårdh, P. A., °hlgren, M., and Nordenskjuld, F. (1985). Ectopic pregnancy and antibodies to *Chlamydia trachomatis. Fertil. Steril.* **44**, 313.

76. Henry-Suchet, J., Catalan, F., Loffredo, V., Sanson, M. J., Debache, C., Pigeau, F., and Coppin, R. (1981). *Chlamydia trachomatis* associated with chronic inflammation in abdominal specimens from women selected for tuboplasty. *Fertil. Steril.* **36**, 599.

77. Patton, D. L., Askienazy-Elbhar, M., Henry-Suchet, J., Campbell, L., Cappuccio, A., Tannais, W., Wang, S., and Kuo, L. (1994). Detection of *Chlamydia trachomatis* in fallopian tube tissue in women with postinfectious tubal infertility. *Am. J. Obstet. Gynecol.* **171,** 95–101.

78. Patton, D. L., Moore, D. E., Spadoni, L. R., Soules, M. R., Halbert, S. A., and Wang, S.-P. (1989). A comparison of the fallopian tube's response to overt and silent salpingitis. *Obstet. Gynecol.* **73,** 622–630.

79. Paavonen, J., Kiviat, N., Brunham, R. C., Stevens, C. E., Kuo, C., Stamm, W. E., Miettinen, A., Soules, M., Eschenbach, D. A., and Holmes, K. K. (1985). Prevalence and manifestations of endometritis among women with cervicitis. *Am. J. Obstet. Gynecol.* **152,** 280–286.

80. Aral, S. O., and Holmes, K. K. (1990). Descriptive epidemiology of sexual behavior and sexually transmitted diseases. *In* "Sexually Transmitted Diseases" K. K. Holmes, P.-A. Mårdh, and P. F. Sparling, (eds.), 2nd ed., pp. 19–36. McGraw-Hill, New York.

81. Arenas, J. M. B., and Lowe, C. R. (1990). Epidemiologic overview. *In* "Pelvic Inflammatory Disease: Epidemiology, Etiology, Management, Complications" H. H. Handsfield, (ed.) pp. 4–7. HP Publishing Company, New York.

82. Bell, T. A., and Holmes, K. K. (1984). Age-specific risks of syphilis, gonorrhea and hospitalized PID in sexually experienced women. *Sex. Transm. Dis.* **11,** 291–295.

83. Igra, V., Ellen, J., and Shafer, M.-A. (1997). Pelvic inflammatory disease in the adolescent female. *In* "Pelvic Inflammatory Disease" D. V. Landers and R. L. Sweet (eds.) pp. 116–138. Springer-Verlag, New York.

84. Cates, W., and Wasserheit, J. N. (1991). Genital chlamydial infections: Epidemiology and reproductive sequela. *Am. J. Obstet. Gynecol.* **164,** 1771–1781.

85. Enhorning, G., Huldt, L., and Melen, B. (1970). Ability of cervical mucus to act as a barrier against bacteria. *Am. J. Obstet. Gynecol.* **108,** 532–537.

86. Eschenbach, D. A. (1997). Epidemiology of pelvic inflammatory disease. *In* "Pelvic Inflammatory Disease" D. V. Landers and R. L. Sweet, (eds.) pp. 1–20. Springer-Verlag, New York.

87. Lee, N. C., Rubin, G. L., and Grimes, D. A. (1991). Measures of sexual behavior and the risk of pelvic inflammatory disease. *Am. J. Obstet. Gynecol.* **77**(3), 425–430.

88. Flesh, G., Weiner, J. M., Corlett, R. C., Boice, C., Mishell, D. R., and Wolf, R. M. (1979). The intrauterine contraceptive device and acute salpingitis. *Am. J. Obstet. Gynecol.* **135**(3), 402–408.

89. Handsfield, H. H. (1986). Criteria for selective screening for *Chlamydia trachomatis* infection in women attending family planning clinics. *JAMA, J. Am. Med. Assoc.* **255,** 1730–1735.

90. Schachter, J., Stoner, E., and Moncada, J. (1983). Screening for chlamydial infections in women attending family planning clinics. *West. J. Med.* **138,** 375–379.

91. Kamwendo, F., Johansson, E., Moi, H., Forslin, L., and Danielsson, D. (1993). Gonorrhea, genital chlamydial infection, and nonspecific urethritis in male partners of women hospitalized and treated for acute PID. *Sex. Transm. Dis.* **20**(3), 143–146.

92. Eschenbach, D. A., Buchanan, T. M., and Pollock, H. M. (1975). Polymicrobial etiology of acute pelvic inflammatory disease. *N. Engl. J. Med.* **293,** 166–171.

93. Westrom, L. (1975). Effect of acute pelvic inflammatory disease on fertility. *Am. J. Obstet. Gynecol.* **121,** 707.

94. Brooks, G. F., Darrow, W. W., and Day, J. A. (1978). Repeated gonorrhea: An analysis of importance and risk factors. *J. Infect. Dis.* **137,** 161–169.

95. Betteiger, B. E., Fraiz, J., Newhall, W. J., Katz, B. P., and Jones, R. B. (1989). Association of recurrent chlamydial infection with gonorrhea. *J. Infect. Dis.* **159,** 661–669.

96. Lee, N. C., Rubin, G. L., and Grimes, D. A. (1991). Measures of sexual behavior and the risk of pelvic inflammatory disease. *Am. J. Obstet. Gynecol.* **77,** 425–430.

97. Kelaghan, J., Rubin, G. L., Ory, H. W., and Layde, P. M. (1982). Barrier-method contraceptives and PID. *JAMA, J. Am. Med. Assoc.* **248,** 184–187.

98. Zenilman, J. M., Weisman, C. S., Rompalo, A. M., Ellish, N., Upchurch, D. M., Hook, E. W., and Celentano, D. (1995). Condom use to prevent incident of STD's: The validity of self-reported condom use. *Sex. Transm. Dis.* **22**(1), 15–21.

99. Nelson, D. B., Ness, R. B., Peipert, J. F., Soper, D. E., Amortegui, A. J., Gluck, J., Weisenfeld, H., and Rice, P. A. (1998). Factor predicting upper genital tract inflammation among women with lower genital tract infection. *J. Women's Health* **7,** 1033–1040.

100. Vessey, M. P., Doll, R., and Peto, R. (1976). A long term follow-up study of women using different methods of contraception—an interim report. *J. Biosoc. Sci.* **8,** 373–426.

101. Lee, N. C., Rebin, G. L., and Bonick, R. (1988). The intrauterine device and pelvic inflammatory disease revisited: New results from the women's health study. *Am. J. Obstet. Gynecol.* **72,** 1–6.

102. Daling, J. R., Weiss, N. S., Voigt, L. F., McKnight, B., and Moore, D. E. (1992). The untrauterine device and primary tubal infertility. *N. Engl. J. Med.* **326**(3), 203–204.

103. Cromer, A., and Heald, F. P. (1987). Pelvic inflammatory disease associated with *Neisseria gonorrhoeae* and *Chlamydia trachomatis:* Clinical correlates. *Sex. Transm. Dis.* **14,** 125–129.

104. Louv, W. C., Austin, H., Perlman, J., and Alexander, W. J. (1989). Oral contraceptive use and the risk of chlamydial and gonococcal infection. *Am. J. Obstet. Gynecol.* **160,** 396–402.

105. Avonts, D., Sercu, M., Heyerick, P., Vandermeeren, I., Meheus, A., and Piot, P. (1990). Incidence of uncomplicated genital infections in women using oral contraception or an IUD: A prospective study. *Sex. Transm. Dis.* **17,** 23–29.

106. Mårdh, P. A., Paavonen, J., and Puolakkainen, M. (1989). "Chlamydia." Plenum, New York.

107. Rank, R. G., White, H. J., Hough, A. J., Jr., Pasley, J. N., and Barron, A. L. (1982). Effect of estradiol on chlamydial genital infection of female guinea pigs. *Infect. Immun.* **38,** 699–705.

108. Svensson, L., Westrom, L., and Mårdh, P. A. (1984). Contraceptives and acute salpingitis. *JAMA, J. Am. Med. Assoc.* **251,** 2553–2555.

109. Spinillo, A., Gorini, G., Piazzi, G., Baltaro, F., Monaco, A., and Zara, F. (1996). The impact of oral contraception on chlamydial infection among patients with pelvic inflammatory disease. *Contraception* **54,** 163–168.

110. Zhang, J., Thomas, G., and Leybovich, E. (1997). Vaginal douching and adverse health effects: A meta-analysis. *Am. J. Public Health* **87**(7), 1207–1211.

111. Wolner-Hanssen, P., Eschenbach, D. A., Paavonen, J., Stevens, C. E., Kiviat, N. B., Critchlow, C., DeRouen, R., Koutsky, L., and Holmes, K. K. (1990). Association between vaginal douching and acute pelvic inflammatory disease. *JAMA, J. Am. Med. Assoc.* **263,** 1936–1941.

112. Scholes, D., Daling, J. R., Stergachis, A., Weiss, N. J., Wang, S. P., and Grayston, J. T. (1993). Vaginal douching as a risk factor for acute pelvic inflammatory disease. *Obstet. Gynecol.* **81,** 601–606.

113. Marchbanks, P. A., Lee, N. C., and Peterson, H. B. (1990). Cigarette smoking as a risk factor for pelvic inflammatory disease. *Am. J. Obstet. Gynecol.* **162,** 639–644.

114. Buchan, H., Villard-Mackintosh, L., Vessey, M., Yeates, D., and McPherson, K. (1990). Epidemiology of pelvic inflammatory disease in parous women with special reference to intrauterine device use. *Br. J. Obstet. Gynecol.* **97,** 780–788.

115. Chow, W.-H., Daling, J. R., Weiss, N. S., and Voigt, L. F. (1988). Maternal cigarette smoking and tubal pregnancy. *Obstet. Gynecol.* **71**(2), 167–170.

116. Phipps, W. R., Albrecht, B., Cramer, D. W., Gibson, M., Schiff, I., Berger, M. J., Belisle, S., Wilson, E., and Stillman, R. (1987). The association between smoking and female infertility as influence by cause of the infertility. *Fertil. Steril.* **48**(3), 377–382.

117. Sasson, J. M., Haley, N. J., Hoffman, D., Wynder, E. L., Hellborn, D., and Nilsson, S. (1985). Cigarette smoking and neoplasia of the uterine cervix: Smoke constituents in cervical mucus. *N. Engl. J. Med.* **312,** 315–319.

118. Yoshinaga, K., Rice, C., Krenn, J., and Pilot, R. L. (1979). Effects of nicotine on early pregnancy in rat. *Biol. Reprod.* **20,** 294.

119. Peterson, H. B., Walker, C. K., Kahn, J. G., Washington, A. E., Eschenbach, D., and Faro, S. (1991). Pelvic inflammatory disease: Key treatment issues and options. *JAMA, J. Am. Med. Assoc.* **266,** 2605.

120. Ness, R. B., Delaney, K., Rolfs, R. T., and Gale, J. L. (1995). Practice variability in the inpatient treatment of pelvic inflammatory disease. *J. Women's Health* **4,** 51–59.

121. Freeman, W. L., Green, L. A., and Becker, L. A. (1988). Pelvic inflammatory disease in primary care. *Fam. Med.* **20,** 192.

122. Svensson, L., Mårdh, P.-A., and Westrom, L. (1985). Infertility after acute salpingitis with special reference to *Chlamydia trachomatis. Fertil. Steril.* **40,** 322.

123. Ness, R. B., Soper, D. E., Peipert, J., Sondheimer, S. J., Holley, R. L., Sweet, R. L., Hemsell, D. L., Randall, H., Hendrix, S. L., Bass, D. C., Kelsey, S. F., Songer, T. J., and Lave, J. R. for the PID Evaluation and Clinical Health (PEACH) Study Investigators (1998). Design of the PID Evaluation and Clinical Health (PEACH) Study. *Controlled Clin. Trials* **19,** 499–514.

124. Safrin, S., Schactiter, J., Dahrouge, D., and Sweet, R. L. (1992). Long-term sequelae of acute pelvic inflammatory disease. *Am. J. Obstet. Gynecol.* **166,** 1300.

125. Scholes, D., Stergachis, A., Heidrich, F. E., Andrilla, H., Holmes, K. K., and Stamm, W. E. (1996). Prevention of pelvic inflammatory disease by screening for cervical chlamydial infection. *N. Engl. J. Med.* **334,** 1362–1366.

30

The Interrelationship of Reproductive Health and Sexually Transmitted Diseases

WILLARD CATES, JR.* AND NANCY S. PADIAN†

*Family Health International, Research Triangle Park, North Carolina; †University of California San Francisco,
San Francisco, California

I. Introduction

The term reproductive health has taken on expanded meaning [1]. As global concern about HIV and other STDs has grown, decisions about contraceptive use have increasingly involved the need to prevent STDs. In response, The United Nations 1994 International Conference on Population and Development in Cairo helped define a reproductive health agenda that encouraged family planning programs to add STD/HIV prevention services [2]. These services could include such clinical activities as presumptive treatment for clients with symptoms or signs of infection, laboratory screening and treatment for asymptomatic clients, and STD risk assessment to guide contraceptive counseling to reduce future risks of both unintended pregnancy and genital infections [3].

Using contraception for reproductive health has two main benefits: prevention of unplanned pregnancy and protection against sexually transmitted infections [4]. Abstinence from sexual intercourse provides nearly absolute protection against both outcomes. For those choosing to be sexually active, contraception reduces, but does not eliminate, the risk of unintended pregnancy. Unfortunately the contraceptives with the best record for pregnancy prevention provide minimal sexually transmitted disease (STD—including HIV) protection. Some contraceptives may even raise the risk of acquiring or transmitting certain infections, whereas the effect of other contraceptives on STDs remains unknown.

The choice of contraception is further complicated when considering its longer range reproductive implications. Contraceptive use has an influence not only on the acute risks of STD and unplanned pregnancy, but also on the eventual reproductive capacity of those making contraceptive decisions. For example, using condoms to prevent infections in the lower genital tract will reduce the likelihood of upper genital tract infection, ectopic pregnancy, and tubal infertility in the future [4,5]. Therefore, personal choices, community programs, and/or policy decisions made in the short run to prevent STD and unplanned pregnancy can simultaneously improve (or harm) chances of planned procreation in the long run [4,5]. This interrelationship of reproductive health choices for pregnancy prevention versus STD prevention (including HIV) forms the basis for this chapter.

II. The Effect of Contraceptive Use on Pregnancy

Our modern contraceptives have been portrayed as having both an ideal and an actual use effectiveness [6]. The *ideal* (or method) effectiveness describes the rate of pregnancy that oc-

curs when a contraceptive is used correctly and consistently according to standard directions. The *actual* effectiveness describes the pregnancy rates that occur when contraceptives are used under "typical" conditions. As such, the actual effectiveness depends on a variety of behavioral factors affecting adherence that produce a wide range of unintended pregnancy rates (Fig. 30.1). For example, based on data from the National Survey of Family Growth, young, unmarried persons who use barrier methods generally have higher unintended pregnancy rates than older, married individuals using comparable barrier methods [7]. Many of the same behavioral factors that produce the varying pregnancy rates for the compliance-dependent, coitally-related barrier methods of contraception are the same ones that affect the degree of STD/HIV protection these methods provide under typical use (see sections below).

III. The Effect of Contraceptive Use on STD/HIV

Unraveling the reproductive health interrelationships among contraception, sexually transmitted diseases (STDs), upper genital tract infection such as pelvic inflammatory disease (PID), and other sequelae is a complex exercise in causal reasoning (Fig. 30.2). We must examine correlations among variables with varying definitions, imprecise diagnoses, and different microbial organisms [8]. The causal pathway includes four distinct links in the chain (contraceptive choice as the antecedent condition, lower genital tract STDs and upper genital tract PID as intermediate conditions, and more distal sequelae as the outcome). The conditions each have a temporal lag in clinical expression (contraceptive use affecting STDs, STDs leading to PID, PID leading to sequelae). These relationships in turn are influenced by the overlapping effects of three environments: (1) the genital microbial environment, (2) the individual behavioral environment (*e.g.,* sexual behavior, contraceptive practice, and health care utilization patterns), and (3) the sociogeographic environment [9].

The scientific literature has been replete with reviews of the effects of different contraceptives on the risk of STD/HIV [8,10–14]. In general, they all come to the same conclusion (Table 30.1). However, study design, population sampled, and study methodologies have varied from study to study, making overall generalizations for each method uncertain. For example, no randomized controlled trials have tested the efficacy of male condoms to protect against STDs and HIV; instead other causal criteria, such as results from observational studies, or the biological plausibility (*in vitro* data) that condoms should be effective, have supplemented existing information.

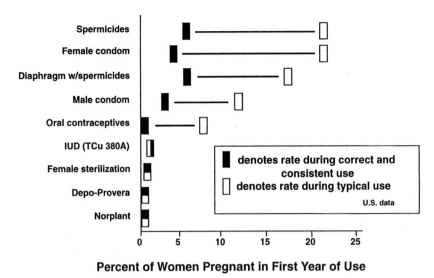

Percent of Women Pregnant in First Year of Use

Fig. 30.1 Contraceptive pregnancy rates. Source: Adapted by Family Health International from 1995 NSFG.

Male condoms used correctly and consistently provide good protection against most STDs, both bacterial and viral. The same is probably true for the female condom, which provides physical coverage of the vagina and cervix. Likewise, the diaphragm—whose mechanism of action involves coverage of the cervix only—shows some protection from bacterial STDs, but little is known about its effect with regard to HIV acquisition. Spermicides alone and in combination with mechanical/chemical methods can provide modest protection against bacterial STDs but little protection against HIV. Hormonal contraception may enhance some cervical infections, but its impact on *upper* genital tract infection (*e.g.,* PID) and HIV is still unsettled. The IUD is associated with acute PID, albeit primarily during the interval after its insertion.

A. Male Condoms

If men are willing and educated to use them properly, male condoms protect against transmission of STDs by preventing direct contact with semen, genital discharge, some genital lesions, or infectious secretions [15–17]. To be effective, con-

doms must be applied prior to genital contact, must remain intact, and most importantly, must be used *consistently and correctly with each act of intercourse* [15].

Laboratory studies confirm that latex and polyurethane condoms provide an impervious barrier to most STD pathogens. "Natural membrane" condoms, made of sheep intestinal membrane, may not be as effective as synthetic condoms in preventing STDs because HIV, hepatitis B virus (HBV), and herpes simplex virus (HSV) can pass through natural membrane condoms. This permeability may relate to the size of the pores in the intestinal membranes [18].

Table 30.1
Effects of Contraceptives on Bacterial and Viral STI[a]

Contraceptive methods	Bacterial STI	Viral STI
Condoms	Protective	Protective
Spermicides	Modestly protective against cervical gonorrhea and chlamydia	Not protective
Diaphragms	Protective against cervical infection; associated with vaginal anaerobic overgrowth	Protective against cervical infection
Hormonal	Associated with increased cervical chlamydia; protective against symptomatic PID, but higher risks of unrecognized endometritis	Not protective
IUD	Associated with PID in first month after insertion	Not protective
Natural Family Planning	Not protective	Not protective

[a]Source: [8].

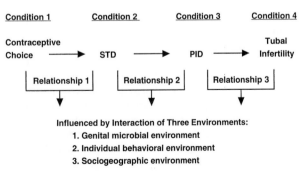

Influenced by Interaction of Three Environments:
1. Genital microbial environment
2. Individual behavioral environment
3. Sociogeographic environment

Fig. 30.2 Etiologic relationships among contraceptive choice, STD, PID, and tubal infertility.

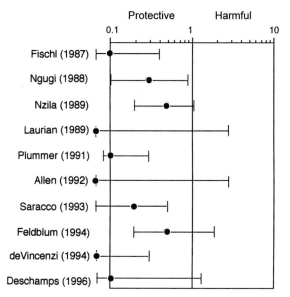

Fig. 30.3 Condom use and HIV infection in heterosexuals. Relative risk (log scale) and 95% confidence interval.

Epidemiologic studies show women are protected by male condoms against some but not all STDs [17]. In part, this may be due to the wider array of organisms studied, the varying consistency of the partners' condom use, and the incomplete protection by condoms against organisms shed from noncovered external genitalia. For example, in Colorado, women attending STD clinics whose partners had used condoms during the previous month were less likely to have gonorrhea or trichomoniasis but were just as likely to have chlamydia or bacterial vaginosis [19]. In addition, condoms probably do not confer protection against acquisition of organisms that are transmitted via "skin-to-skin" contact with external genitalia, such as HPV and HSV.

Much work has evaluated the male condom's influence in protecting against HIV (Fig. 30.3). Although low incidence of infection, even among those who do not use condoms or who use them intermittently, has reduced the power of many studies to demonstrate "statistically significant" associations, the data are remarkably supportive. Regular use of male condoms reduces the risk of acquiring or transmitting HIV [14]. For example, in Kenya, consistent use of condoms by the clients of commercial sex workers led to lower HIV seroconversion rates in women [20]. In France, none of the female partners of HIV seropositive men with hemophilia who always used condoms became infected with the virus [21]. In Rwanda, the rate of HIV seroconversion declined in a population of urban women who increasingly used condoms [22]. Finally, in the most definitive examples, the European and Haitian studies of HIV discordant couples found those who reported consistent condom use had minimal or no HIV transmission after multiple years of observation [23,24].

Taken together, these clinical studies strongly support the protective effect of consistent male condom use against some STDs. However, the variation in the data imply other factors also affect their impact. For example, because the man ulti-

mately decides on use of the male condom, the quality of partner communication is apparently the strongest predictor of consistent and correct use [25]. Couples who discussed the use of condoms prior to having initiated intercourse, and who practiced using different types until they found a preferred brand, tended to use condoms more consistently [18]. However, in spite of the best intentions, due to a variety of cultural and individual factors, not all women are able to successfully negotiate male condom use with their partner. Future investigations will help clarify the behavioral determinants of more effective condom use at the population level.

B. Female Condoms

Several female condoms ("intravaginal pouches," "vaginal condoms") have been licensed in the U.S. and elsewhere [26,27]. Basically, these products are pouches made of polyurethane which line the vagina. As with male condoms, protection against STD is likely only if these products are used consistently and stay properly positioned during intercourse. *In vitro* testing of the Reality female condom showed it to be an effective barrier to HIV and cytomegalovirus (CMV) [28].

Moreover, female condoms have also shown *in vivo* protection against STDs. In Virginia, female condoms significantly protected against recurrent trichomonal infection [29]. Women treated for trichomoniasis were assigned to either a female condom or control group. No reinfection occurred among those women who chose to use the female condom during each act of intercourse, compared to a 14% reinfection rate both in the control group and among those who used them intermittently. In Thailand, sex workers randomly assigned to a group receiving female condoms as a backup to male condoms had fewer unprotected coital acts and lower STD rates than those receiving male condoms only [30]. However, while these data, combined with the physical protection, appear promising, the higher cost of female condoms compared to male condoms is likely to limit their usefulness, especially in developing countries [31]. Future products involving female physical barriers need to be less expensive and more widely available to help reduce unintended pregnancy and STDs on a population level.

C. Diaphragms and Other Combined Physical/Chemical Barriers

Both chlamydia and gonorrhea preferentially infect the columnar epithelium of the endocervix [32,33]. Thus, coverage of the cervix may protect against acquisition of these pathogens. Although no experimental studies have tested this hypothesis, several observational studies have found that diaphragms used with spermicide decrease susceptibility to gonorrhea and its associated long-term reproductive sequelae of PID and infertility. Protective effects against gonorrhea have ranged from 0.1 to 0.7 depending on the population studied [34–37]. Data are more limited with regard to the effect of diaphragm use against other STD pathogens; using diaphragms has been associated with reduced levels of chlamydia, trichomoniasis, PID, and cervical cancer, but not of HPV [38–41]. Little is known about the primary anatomical site(s) for sexual transmission of HIV. Data support the relative importance of covering for the cervical, compared to the vaginal, epithelium.

Combined mechanical/chemical barrier methods may have harmful effects on normal genital flora. Both foam-condom users and diaphragm-spermicide users had higher levels of vaginal colonization and bacteriuria with *Escherichia coli* [42]. Some studies have found that women using diaphragm-spermicides or spermicide-coated male condoms have higher risks of acute urinary tract infections [43–45]. Potential mechanisms for the *E. coli* colonization include alteration of the vaginal ecosystem by the spermicide and/or a mechanical effect of the diaphragm; data that separate the independent effects of these combined methods have not been available. Because domination of vaginal flora by facultative anaerobic microbes (*e.g.,* bacterial vaginosis) has been associated both with urinary and with upper genital tract infection [44], the implications of this finding are of concern.

D. Spermicides

In vitro studies have shown that contraceptive spermicides kill or inactivate most STD pathogens. The main spermicidal agent that has been evaluated *in vivo* is nonoxynol-9 (N-9), a nonionic surfactant that damages the cell walls of sperm and STD pathogens. Laboratory tests have documented activity against *Neisseria gonorrhoeae, Trichomonas vaginalis,* herpes simplex virus, HIV, and *Treponema pallidum*. Reports on the effect of spermicides on *Chlamydia trachomatis* are conflicting [14].

In vitro microbicide activity, however, does not mean that spermicides can provide reliable *in vivo* protection. Data from epidemiologic studies of humans have been inconsistent regarding both the safety and effectiveness of N-9 against sexually transmitted diseases and human immunodeficiency virus. These studies have used different formulations and concentrations of spermicide and have been conducted in disparate populations. Thus, drawing clinically meaningful conclusions has been difficult [46].

The best studies of N-9 have been well done randomized controlled trials (RCTs). Through 1998, the three well-done RCTs have compared three different formulations containing N-9: a gel [47,48], a sponge [49], and a film [50]. Taken together, these studies have shown N-9 spermicide used alone reduces the risks both of gonorrhea and of chlamydia infection, albeit at modest levels of protection (Fig. 30.4). In Alabama, regular use of N-9 gel by women attending an STD clinic reduced cervical gonorrhea by 24%, cervical chlamydial infection by 22%, trichomoniasis by 17%, and bacterial vaginosis by 14% [47,48]. In Kenya, a sponge containing a high dose of N-9 apparently protected against gonorrhea but did not lower the risk of HIV [49]; the lack of effect on HIV risk may be due to increased ulceration assumed to enhance HIV susceptibility. In Cameroon, the N-9 film provided no better protection for any of the STDs, including HIV, than that provided by a placebo film [50]. As stated earlier, issues such as the varying doses and formulations of N-9 in each study make generalizations across studies difficult.

Because current ethical standards require that randomized, controlled trials of microbicides (topical products, such as spermicides, applied to the vagina or cervix to prevent acquisition

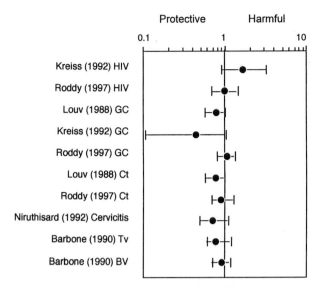

Fig. 30.4 Spermicides and STDs among women. Relative risk (log scale) and 95% confidence interval. Ct, *Chlamydia trachomatis;* Tv, *Trichomonas vaginalis;* BV, bacterial vaginosis.

of a variety of pathogens, sometimes including sperm) encourage study populations to use condoms as their primary method of HIV prevention [51], the resulting study design is only able to assess the marginal impact of spermicides in addition to condoms as a means of protecting against HIV acquisition. Moreover, methodologic issues such as the type of delivery system (*e.g.,* film versus gel) and the population studied (*e.g.,* sex workers versus more representative populations) also play a crucial role in allowing correct interpretation of studies involving spermicides and barriers. Finally, the large sample size required of trials to insure sufficient statistical power, and the resulting high costs to assess spermicide effectiveness against HIV, have also been a difficult hurdle.

A potential risk for spermicide users is chemical irritation of the vaginal epithelium, caused by N-9's membrane-disrupting properties [14]. However, inconsistencies also exist with regard to the population sampled and how the outcome is measured (*e.g.,* visual inspection, inspection enhanced via colposcopy, or biopsy). In addition, no gold standard exists as to what is considered abnormal, much less what is unsafe. Data regarding the clinical meaning of "abnormalities" determined through the invasive diagnostic regimes instituted as part of the study design are lacking. We have no population-based data to characterize a visually "normal" cervix, thus, the importance of redness or inflammation remains uncertain.

In Thailand, nearly half of sexually inactive women randomized to receive 150 mg of N-9 suppositories four times a day (albeit an artificially high level of exposure) suffered epithelial disruption of their vaginal or cervical mucosa [52]; none of those receiving a placebo did. A second randomized clinical trial in Thailand showed use of N-9 vaginal film was associated with a small increase in genital irritation, although this did not vary with number of inserts used, nor was it related to clinical signs [53].

E. Hormonal Contraceptives

Combined oral contraceptives (OCs) have an array of non-contraceptive health benefits; however, their influence on STD, HIV, PID, and eventual reproductive sequelae remains unsettled.

1. Hormones and Bacterial STD

The majority of studies, though not all, have found an increased risk of cervical infections with *C. trachomatis* among users of OCs as compared with nonusers [54–56]. In Alabama, OC users had higher rates of both *C. trachomatis* and *N. gonorrhoeae* detected in the cervix compared to women using no method of contraception [57]. The association between OC use and cervical infection may be mediated through the cervical ectopy commonly induced by OCs [58]. *Chlamydia trachomatis* has been isolated more frequently among women with ectopy than among women without ectopy, regardless of the method of birth control used [55].

The influence of OCs on the upper genital tract may be different than the influence on the lower genital tract. Studies from Europe and the United States [59] have revealed that women using OCs are half as likely to be hospitalized for PID compared with sexually active women who do not use contraception. Whether these findings are real, or if they represent an artifact of clinical detection of PID (*i.e.*, OC use may reduce symptomatic PID but may have no effect on asymptomatic upper genital tract infection), is unclear. OC users tend to have milder upper genital tract infection with *C. trachomatis,* as manifested by antibody response and laparoscopy [60]. In addition, OC users have increased risks of unrecognized endometritis [61].

Data on the impact of previous OC use on primary tubal infertility are conflicting. In the U.S. [62], among former users of most current low-estrogen OCs, tubal infertility was not reduced as might be expected if their use led to a decreased risk of PID. In fact, users of those OCs with estrogen levels higher than 50 mcg had an increased risk of tubal infertility. In the Lund, Sweden cohort, women who had been using OCs at the time their salpingitis was diagnosed had 70% lower risk of tubal infertility than women using other methods [63]. The reason for this discrepancy is unclear.

Possible mechanisms for the consistent protective effect of OCs on symptomatic PID remain speculative. The progestin component of combination OCs thickens the cervical mucus. Changes in either mucus composition or its immunologic properties might account for this protection. Likewise, the thinner endometrium and/or the decreased menstrual flow associated with pill use may play a role. Alternatively, if OC use tends to mask the symptoms of PID, OCs will appear protective although unrecognized inflammation may be occurring.

Use of hormonal contraception also seems to modify the acute clinical course of PID favorably. As judged by laparoscopic examination, women with PID who were using OCs had milder inflammation than women not using OCs [64]. Among women with chlamydial salpingitis, use of OCs protected against Fitz-Hugh-Curtis syndrome. Moreover, an IUD with progesterone protected against histological salpingitis compared to nonhormonal IUDs [65]. In monkey models, however, OCs containing both estrogen and progesterone did not alter the course of experimentally-induced chlamydial salpingitis [66]. Sex steroids can modify immunologic function [67], so their impact on infectious inflammation is plausible, though still unresolved.

2. Hormones and HIV

The effect of hormonal contraception on HIV transmission, acquisition, or disease progression remains unsettled [68]. Data from animal and epidemiologic studies have produced equivocal results. Most observational investigations have examined use of combined oral contraceptives (COC) and HIV, while attempting to control for such intervening variables as cervical ectopy or infectious cervicitis and behavioral confounders such as condom use. Of the multiple cross-sectional and prospective studies of COC, few were of high quality [68]. While no consistent pattern of COC association with HIV acquisition has emerged [68–72], several of the investigations that identified hormonal contraceptive exposures in a clearer temporal sequence have found elevated risks.

Animal models, using simian immunodeficiency virus (SIV) as the HIV surrogate, have raised concerns about progestin-only contraceptives [73]. In rhesus macaques, an increase in SIV acquisition occurred in monkeys with progesterone implants compared to monkeys exposed to SIV in the follicular phase of a normal menstrual cycle. However, the risk of SIV acquisition was lower if monkeys exposed to SIV during the full menstrual cycle were used as the comparison group [74]. This latter control group more closely approximates real-world HIV exposure occurring randomly throughout the menstrual cycle. Moreover, prospective human studies have not found consistent evidence supporting progestin-only risks [68].

In these observational studies, many methodologic problems exist with the comparability either between cases (women infected with HIV) and controls (uninfected women) or between the exposed (users of hormonal contraception) and the unexposed (nonusers or users of other methods). Accurate measurement of the interval of hormonal use has been difficult. Moreover, in most studies, a relatively small number of women used hormonal contraception, limiting the studies' abilities to detect any associations that might exist. Thus, because of power considerations, the point estimates are imprecise, with wide 95% confidence intervals. Finally, the myriad of existing observational studies applied crude definitions of hormonal contraceptive use, had different hormonal mixes, and were unable to control for crucial confounders.

Potential biologic mechanisms by which hormonal contraception might facilitate HIV transmission include: (1) increased cervical ectopy (and associated friability) caused by OC use; (2) increased cervical chlamydial infection (and associated purulence), possibly associated with ectopy; (3) systemic immunologic changes associated with some exogenous steroids, such as decreased cell-mediated immunity [67]; and (4) if long-acting hormonal contraception is used, irregular uterine bleeding and/or thinning of the vaginal epithelium. In addition, the particular types of estrogens and progestins contained in the OCs or other long-term hormonal contraceptives may be important factors relating to any impact on HIV acquisition or transmission [68].

Because of the crucial role that OC and other hormonal contraceptives play in international family planning programs, and because the HIV pandemic is one of the world's most important health problems, resolving this issue deserves a high research priority.

F. Intrauterine Devices

The precise risk of IUD use on STDs is unclear. Few investigations have examined the effect of the IUD on *lower* genital tract infections. One review article concluded that the IUD was unrelated to cervical chlamydial infection [75]. However, an analysis of cross-sectional data from a population-based sample of Seattle women found that those who had used an IUD had a significantly higher percentage of chlamydial antibodies than those who never used an IUD, even after stratifying for number of sex partners [76]. In contrast, a prospective study in Antwerp found IUD users had ninefold lower rates of chlamydia than OC users [77]. The same study also showed women using IUDs had nearly an eight times higher rate of bacterial vaginosis.

The possible association between IUD use and the development of *upper* genital tract infection is becoming clearer [78]. Current evidence suggests that the small, but still measurable, increased risk of PID associated with IUD use occurs around the time of insertion [79]. Thus, contamination of the endometrial cavity at insertion may be more responsible for IUD-related PID than the device itself. Because of the association of PID with the timing of IUD insertion, short-term antibiotics were evaluated to determine if they would reduce the risks. In Kenya [80], Nigeria [81], and the U.S. [82], randomized controlled trials of prophylactic antibiotics at the time of IUD insertion found much lower than expected rates of IUD-associated PID, even among the placebo group. Because PID occurred so infrequently in these trials, use of antibiotics at the time of insertion was not associated with lower PID incidence.

Based on these encouraging data, many perceive the IUD's safety today as vastly different from that of a decade ago. Evidence indicates the IUD poses little, if any, *added* STD or PID risk than if no contraceptive was being used. For example, among women who used copper IUDs and who had only one sex partner, no increase occurred in the risk either of PID or of tubal infertility [83]. Thus, the challenge for future researchers will be to establish criteria, based on a combination of demographic, behavioral, and clinical factors, that would expand the proportion of women for whom the IUD would be a good contraceptive choice. Moreover, because newer IUDs containing levonorgestrel apparently are associated with even lower rates of PID [84], they may be more appropriate contraceptives in populations where the risk of STDs is unknown.

G. Tubal Sterilization

Tubal sterilization protects against PID, but this protection is not absolute. Even though endometritis and proximal salpingitis are potentially possible, PID rarely is observed [85] among women after tubal sterilization. A likely mechanism for those few cases of poststerilization PID is iatrogenic contamination of the tubes during the operative procedure.

H. Abortion

Women who have cervical infection with either *N. gonorrhoeae* or *C. trachomatis* have an increased risk of endometritis following induced abortion performed under proper hygienic conditions. The risk of endometritis appears to be at least tripled in the presence of either organism [86]. A number of studies suggest that use of prophylactic antibiotics at the time of the abortion procedure reduces the risk of infection by one-half to two-thirds [87]. While preoperative screening for infection with these organisms is desirable, a brief perioperative course of an antibiotic such as azithromycin for all women seems both safe and cost-effective. Women later found to be infected by *N. gonorrhoeae* and/or *C. trachomatis* can be followed up with a full course of recommended antibiotics.

The greatest risk of upper genital tract infection associated with induced abortion occurs in circumstances where sterile conditions are not maintained. In countries where abortion services are restricted by law or practice, and especially in resource-poor regions where, even if legal, access to sanitary procedures is limited, postabortion infection poses risks not only to future fertility, but also to the woman's life [88]. More than half the estimated 100,000 abortion-related deaths worldwide occur in Southeast Asia, followed by sub-Saharan Africa, and then Latin America and the Caribbean [89].

Both in the developed and in the less-developed world, carrying a pregnancy to term leads to greater risks of infection and death than does terminating it through induced abortion [90,91]. Under sterile conditions, abortion is five-to-ten times safer than childbearing. Under less hygienic conditions, the risks of *both* pregnancy outcomes increase, and the gap between the infection risks from abortion and childbirth probably narrows. In these circumstances, use of any method of contraception to reduce pregnancy has simultaneous effects on reducing pregnancy-associated infection (see next section).

IV. Clinical and Policy Implications—Reproductive Health Trade-Offs

In ideal circumstances, consistent and correct use of male (and probably female) condoms can prevent *both* pregnancy and STD. However, in typical situations of inconsistent use, they provide lower rates of protection against these conditions. Moreover, based on currently available data, if couples choose to use contraceptive methods other than condoms, those with the best record for pregnancy prevention provide little STD protection [4]. Thus, trade-off choices are necessary.

For those whose families are not complete yet who do not wish to currently become pregnant, hormonal contraceptives and the IUD remain the most effective reversible methods available to prevent unintended pregnancy. However, they provide no protection against vaginal or cervical infection, and inserting the IUD might carry temporary risks to the upper genital tract. For persons who are not mutually monogamous, addition of a barrier method such as a condom will help reduce the risk of STD as well as unplanned pregnancy. Under typical conditions, however, barrier methods are substantially less effective in preventing conception than are hormonal methods, yet they offer important protection against STD. To maximize protection against both unintended pregnancy and STD, a barrier method should be used in conjunction with a hormonal method or IUD.

Because both mechanical and chemical barrier prophylaxis are coitally dependent, their efficacy in preventing either infection or unplanned pregnancy depends entirely on adherence by the couple. Some populations have demonstrated high levels of barrier method use. For example, in Thailand, a "100% condom policy" in commercial sex facilities has led to widespread use among the workers and has been associated with marked reductions in HIV and other STDs [92]. In the United States, women working in Nevada brothels have had similar success [93]. Unfortunately, to date, most heterosexual populations worldwide have not reported the same high level of condom use and have not experienced decreases in the traditional STDs [94].

Another important aspect to assessing trade-off concerns is whether conditions exist for safe, sterile childbirth and abortion. If pregnancy itself, regardless of whether it is terminated or continued, carries markedly high "iatrogenic" risks of genital infection, then the pregnancy prevention efficacy of the contraceptive choice takes on greater weight. By preventing undesired pregnancy, the contraceptive method(s) simultaneously protects against pregnancy-associated infections of the reproductive tract. In the developing world, these pregnancy-related etiologies of postabortal and puerperal infections are important causes of tubal infertility. However, different patterns of pregnancy-associated tubal infertility characterize different regions. In Africa, where the infectious etiology of infertility was most evident [95], genital infections occurring both before and after the first pregnancy were associated with tubal occlusion [96]. In Asia, abortion appeared to play a larger role than childbirth in contributing to infectious infertility, whereas in Latin America the reverse was found [96].

V. Conclusion

Because contraception affects not only the risk of unplanned pregnancy, but also that of sexually transmitted infections, the choice of particular methods is important both to current and to future reproductive health. However, important trade-offs exist. Moreover, epidemiologic studies are equivocal regarding the value of either recommending dual methods of contraception or relying on the condom alone to prevent both unplanned pregnancy and STDs [4]. Use of emergency (post-coital) contraception as a back-up to barrier methods is another approach that warrants evaluation.

Individual choices about reproduction may conflict with public health needs. For example, a woman who perceives an unintended pregnancy to be a greater risk than STD acquisition may choose her contraceptive method without regard to STD prophylaxis. However, the community in which she lives may be marketing condoms as the preferred contraceptive for their dual prophylactic effects both against pregnancy and against infection. The ultimate impact of this dynamic benefit-to-risk interrelationship among contraceptives for pregnancy protection versus STD prophylaxis is to create trade-off choices both for individual women and for public health policymakers. Continued biologic and behavioral research will be necessary to disentangle these complex reproductive health interrelationships.

References

1. Tsui, A. O., Wasserheit, J. N., and Haaga, J. G., eds. (1997). "Reproductive Health in Developing Countries: Expanding Dimensions, Building Solutions." National Academy Press, Washington, DC.

2. United Nations (1994). Programme of action of the 1994 International Conference on Population and Development (A/CONF.171/13). *Popul. Dev. Rev.* **21**, 187–213, 437–461.

3. Mayhew, S. (1996). Integrating MCH/FP and STD/HIV services: Current debates in future directions. *Health Policy Plann.* **11**, 339–353.

4. Cates, W., Jr. (1996). Contraception, unintended pregnancies, and sexually transmitted diseases: Why isn't a simple solution possible? *Am. J. Epidemiol.* **143**, 311–318.

5. Forrest, J. D. (1993). Timing of reproductive life stages. *Obstet. Gynecol.* **82**, 105–111.

6. Trussell, J. (1998). Contraceptive efficacy. *In* "Contraceptive Technology" R. A. Hatcher, J. Trussell, F. H. Stewart, W. Cates, Jr., G. K. Stewart, F. Guest, and D. Kowal, (eds.), 17th ed., pp. 779–844, Ardent Media, New York.

7. Henshaw, S. K. (1998). Unintended pregnancy in the United States. *Fam. Plann. Perspect.* **30**, 24–29, 46.

8. Cates, W., Jr. (1999). Contraceptive choices and sexually transmitted infections among women. *In* "Health and Disease Among Women: Biological and Environmental Influences" R. B. Ness and L. H. Kuller, (eds.), Chapter 15, pp. 401–419. Oxford University Press, New York.

9. Wasserheit, J. N. (1994). Effect of changes in human ecology and behavior on patterns of sexually transmitted diseases, including human immunodeficiency virus infection. *Proc. Natl. Acad. Sci. U.S.A.* **91**, 2430–2435.

10. Cates, W., Jr., and Stone, K. M. (1992). Family planning, sexually transmitted diseases and contraceptive choice: A literature update. Part I and Part II. *Fam. Plann. Perspect.* **24**(2), 75–84; **24**(3), 122–128.

11. Elias, C. J., and Leonard, A. (1995). Family planning and sexually transmitted diseases: The need to enhance contraceptive choice. *Curr. Issues Public Health* **1**, 191–199.

12. Carlin, E. M. and Boag, F. C. (1995). Women, contraceptives and STDs including HIV. *Int. J. STD AIDS* **6**, 373–386.

13. Daly, C. C., Helling-Giese, G. E., Mati, J. K., and Hunter, D. J. (1994). Contraceptive methods and the transmission of HIV: Implications for family planning. *Genitourin. Med.* **70**, 110–117.

14. Feldblum, P. J., Morrison, C. S., Roddy, R. E., and Cates, W., Jr. (1995). The effectiveness of barrier methods of contraception in preventing the spread of HIV. *AIDS* **9** (Suppl. A), 585–593.

15. Centers for Disease Control and Prevention (1993). Update: Barrier protection against HIV infection and other sexually transmitted diseases. *Morbid. Mortal. Wkly. Rep.* **42**, 589–591.

16. Consumer Reports (1995). How reliable are condoms? *Consum. Rep.,* May, pp. 320–325.

17. McNeill, E. T., Gilmore, C. E., Finger, W. R., Lewis, J. H., and Schellstede, W. P., eds. (1998). "The Latex Condom: Recent Advances, Future Directions." Family Health International, Research Triangle Park, NC.

18. Warner, D. L., and Hatcher, R. A. (1998). Male condoms. *In* "Contraceptive Technology" R. A. Hatcher, J. Trussell, F. H. Stewart, W. Cates, Jr., G. K. Stewart, F. Guest, and D. Kowal, eds., 17th ed., pp. 325–355. Ardent Media, New York.

19. Rosenberg, M. J., Davidson, A. J., Chen, J.-H., Judson, F. N., and Douglas, J. M. (1992). Barrier contraceptives and sexually transmitted diseases in women: A comparison of female-dependent methods and condoms. *Am. J. Public Health* **82**, 669–674.

20. Ngugi, E. N., Plummer, F. A., Simonsen, J. N. *et al.* (1988). Prevention of transmission of human immunodeficiency virus in Africa:

Effectiveness of condom promotion and health education among prostitutes. *Lancet* **2,** 887–890.

21. Laurian, Y., Peynet, J., and Verroust, F. (1989). HIV infection in sexual partners of HIV seropositive patients with hemophilia. *N. Engl. J. Med.* **320,** 183.

22. Allen, S., Tice, J., Van de Perre, P. *et al.* (1992). Effect of serotesting with counseling on condom use and seroconversion among HIV discordant couples in Africa. *Br. Med. J.* **304,** 1605–1609.

23. de Vincenzi, I. (1994). A longitudinal study of human immunodeficiency virus transmission by heterosexual partners. *N. Engl. J. Med.* **334,** 341–346.

24. Deschamps, M.-M., Pape, J. W., Hafner, A., and Johnson, W. D., Jr. (1996). Heterosexual transmission of HIV in Haiti. *Ann. Intern. Med.* **125,** 324–330.

25. Oakley, D., and Bogue, E. L. (1995). Quality of condom use as reported by female clients of a family planning clinic. *Am. J. Public Health* **85,** 1526–1530.

26. Gollub, E., and Stein, Z. (1993). The new female condom—item 1 on a woman's AIDS prevention agenda. *Am. J. Public Health* **83,** 498–500.

27. World Health Organization (1997). "The Female Condom: A Review." World Health Organization, Geneva.

28. Drew, W. L., Blair, M., Miner, R. C., and Conant, M. (1990). Evaluation of the virus permeability of a new condom for women. *Sex. Transm. Dis.* **17,** 110–112.

29. Soper, D. E., Shoupe, D., Shangold, G. A., Shangold, M. M., Gutmann, J., and Mercer, L. (1993). Prevention of vaginal trichomoniasis by compliant use of the female condom. *Sex. Transm. Dis.* **20,** 137–139.

30. Fontanet, A. L., Saba, J., Verapol, C., Chuanchom, S., Praphas, B., Rugpao, S. *et al.* (1998). Protection against sexually transmitted diseases by granting sex workers in Thailand the choice of using the male or female condom: Results from a randomized controlled trial. *AIDS* **12,** 1851–1859.

31. Farr, G., Gabelnick, H., Sturgen, K., and Dorflinger, L. (1994). Contraceptive efficacy and acceptability of the female condom. *Am. J. Public Health* **84,** 1960–1964.

32. Swanson, J., Eschenback, D. A., and Alexander, E. R. (1975). Light and electron microscopic study of *Chlamydia trachomatis* infection of the uterine cervix. *J. Infect. Dis.* **131,** 678–686.

33. Stratton, P., and Alexander, N. J. (1992). Prevention of sexually transmitted infections. Physical and chemical barrier methods. *Infect. Dis. Clin. North Am.* **4,** 8412–8857.

34. Berger, G. S., Keith, L., and Moss, W. (1975). Prevalence of gonorrhoea among women using various methods of contraception. *Br. J. Vener. Dis.* **51,** 307–309.

35. Quinn, R. W., and O'Reilly, K. R. (1985). Contraceptive practices of women attending the sexually transmitted disease clinic in Nashville, Tennessee. *Sex. Transm. Dis.* **12,** 99–102.

36. Rosenberg, M. J., Davison, A. J., Chec, J.-H., Judson, F. N., and Douglas, J. M. (1992). Barrier contraceptives and sexually transmitted diseases in women: A comparison of female-dependent methods and condoms. *Am. J. Public Health* **82,** 669–674.

37. Austin, H., Louv, V. W., and Alexander, W. J. (1984). A case-control study of spermicides and gonorrhea. *JAMA* **251,** 2822–2824.

38. Madger, L. S., Harrison, H. R., Ehret, J. M., Anderson, T. S., and Judson, F. N. (1988). Factors related to genital Chlamydia trachomatis and its diagnosis by culture in a sexually transmitted disease clinic. *Am. J. Epidemiol.* **28,** 298–308.

39. Becker, T. N., Wheller, C. N., McGough, N. S. *et al.* (1994). Contraceptive and reproductive risks for cervical dysplasia in southwestern Hispanic and non-Hispanic white women. *Int. J. Epidemiol.* **23,** 913–922.

40. Kjaer, S. K., Engholm, G., Teisen, C. *et al.* (1990). Risk factors for cervical human papillomavirus and herpes simplex virus infections in Greenland and Denmark: A population based study. *Am. J. Epidemiol.* **131,** 669–682.

41. Kelaghan, J., Rubin, G. L., Ory, H. W., and Layde, P. M. (1982). Barrier-method contraceptives and pelvic inflammatory disease. *JAMA* **248,** 184–187.

42. Hooton, T. M., Hillier, S., Johnson, C., Roberts, P. L., and Stamm, W. E. (1991). *Escherichia coli* bacteriuria and contraceptive method. *JAMA* **265,** 64–69.

43. Foxman, B., Geiger, A. M., Palin, K., Gillespie, B., and Koopman, J. S. (1995). First-time urinary tract infection and sexual behavior. *Epidemiology* **6,** 162–168.

44. Hooton, T. M., Scholes, D., Hughes, J. P. *et al.* (1996). A prospective study of risk factors for symptomatic urinary tract infection in young women. *N. Engl. J. Med.* **335,** 468–474.

45. Fihn, S. D., Boyko, E. J., Normand, E. H. *et al.* (1996). Association between use of spermicide-coated condoms and *Escherichia coli* urinary tract infection in young women. *Am. J. Epidemiol.* **144,** 512–520.

46. Roddy, R. E., Schulz, K. F., and Cates, W., Jr. (1998). Microbicides, meta-analysis, and the N-9?: Where's the research? *Sex. Transm. Dis.* **25,** 151–153.

47. Louv, W. C., Austin, H., Alexander, W. J., Stagno, S., and Cheeks, J. (1988). A clinical trial of nonoxynol-9 as a prophylaxis for cervical *Neisseria gonorrhoeae* and *Chlamydia trachomatis* infections. *J. Infect. Dis.* **158,** 518–523.

48. Barbone, F., Austin, H., Louv, W. C., and Alexander, W. J. (1990). A follow-up study of methods of contraception, sexual activity, and rates of trichomoniasis, candidiasis, and bacterial vaginosis. *Am. J. Obstet. Gynecol.* **163,** 510–514.

49. Kreiss, J., Ngugi, E., Holmes, K. K. *et al.* (1992). Efficacy of nonoxynol-9 contraceptive sponge use in preventing heterosexual acquisition of HIV in Nairobi prostitutes. *JAMA* **268,** 477–482.

50. Roddy, R. E., Zekeng, L., Ryan, K. A. *et al.* (1998). A randomized controlled trial of the effect of nonoxynol-9 film use on male-to-female transmission of HIV-1. *N. Engl. J. Med.* **339,** 504–510.

51. De Zoysa, I., Elias, C. J., and Bentley, M. E. (1998). Ethical challenges in efficacy trials of vaginal microbicides for HIV prevention. *Am. J. Public Health* **88,** 571–575.

52. Niruthisard, S., Roddy, R. E., and Chutivongse, S. (1991). The effects of frequent nonoxynol-9 use on the vaginal and cervical mucosa. *Sex. Transm. Dis.* **18,** 176–179.

53. Roddy, R. E., Cordero, M., Cordero, C., and Fortney, J. A. (1993). A dosing study of nonoxynol-9 and genital irritation. *Int. J. STD AIDS* **4,** 165–170.

54. Washington, A. E., Gove, S., Schachter, J., and Sweet, R. L. (1985). Oral contraceptives, *Chlamydia trachomatis* infection, and pelvic inflammatory disease. *JAMA* **253,** 2246–2250.

55. Cottingham, J., and Hunter, D. (1992). *Chlamydia trachomatis* and oral contraceptive use: A quantitative review. *Genitourin. Med.* **68,** 209–216.

56. Park, B. J., Stergachis, A., Scholes, D., Heidrich, F. E., Holmes, K. K., and Stamm, W. E. (1995). Contraceptive methods and the risk of *Chlamydia trachomatis* infection in young women. *Am. J. Epidemiol.* **142,** 771–778.

57. Louv, W. C., Austin, H., Perlman, J., and Alexander, W. J., Jr. (1989). Oral contraceptive use and risk of chlamydial and gonococcal infections. *Am. J. Obstet. Gynecol.* **160,** 396–400.

58. Critchlow, C. W., Wölner-Hanssen, P., Eschenbach, D. A., Kiviat, N. V., Koutsky, L. A., Stevens, C. E., and Holmes, K. K. (1995). Determinants of cervical ectopia and of cervicitis: Age, oral contraception, specific cervical infection, smoking, and douching. *Am. J. Obstet. Gynecol.* **173,** 534–543.

59. Westrom, L. (1980). Incidence, prevalence, and trends of acute pelvic inflammatory disease and its consequences in industrialized countries. *Am. J. Obstet. Gynecol.* **138,** 880–892.

60. Wölner-Hanssen, P. (1986). Oral contraceptive use modifies the manifestations of pelvic inflammatory disease. *Br. J. Obstet. Gynaecol.* **93**, 619–624.

61. Ness, R. B., Kader, L. M., Soper, D. E. *et al.* (1997). Oral contraception and the recognition of endometritis. *Am. J. Obstet. Gynecol.* **176**, 580–585.

62. Cramer, D. W., Goldman, M. B., Schiff, I. *et al.* (1987). The relationship of tubal infertility to barrier method and oral contraceptive use. *JAMA* **257**, 2446–2450.

63. Westrom, L. (1996). Chlamydia and its effect on reproduction. *J. Br. Fertil. Soc.* **1**, 23–28.

64. Svensson, L., Westrom, L., and Mårdh, P.-A. (1984). Contraceptives and acute salpingitis. *JAMA* **251**, 2553–2555.

65. Soderstrom, R. M. (1983). Will progesterone save the IUD? *J. Reprod. Med.* **28**, 305–308.

66. Patton, D. L., Sweeney, Y. T. C., and Kuo, C.-C. (1994). Oral contraceptives do not alter the course of experimentally-induced chlamydial salpingitis in monkeys. *Sex. Transm. Dis.* **21**, 89–92.

67. Sonnex, C. (1998). Influence of ovarian hormones on urogenital infection. *Sex. Transm. Infect.* **74**, 11–19.

68. Stephenson, J. M. (1998). Systematic review of hormonal contraception and risk of HIV transmission: When to resist meta-analysis. *AIDS* **12**, 545–553.

69. Plummer, F. A., Simonsen, J. N., Cameron, D. W. *et al.* (1991). Cofactors in male-female sexual transmission of human immunodeficiency virus type 1. *J. Infect. Dis.* **163**, 233–239.

70. Plourde, P. J., Pepin, J., Agoki, E., Ronald, A. R., Obette, J., Tyndall, M., Cheang, M., Ndinya-Achola, J. O., D'Costa, J., and Plummer, F. A. (1994). Human immunodeficiency virus type 1 seroconversion in women with genital ulcers. *J. Infect. Dis.* **170**, 313–317.

71. Sinei, S. K. A., Fortney, J. A., Kigondu, C. S., Feldblum, P. J., Kuyoh, M., Allen, M. Y., and Glover, L. H. (1996). Contraceptive use and HIV infection in Kenyan family planning clinic attenders. *Int. J. STD AIDS* **7**, 65–70.

72. Martin, H. L., Nyange, P., Richardson, B. A., Lavreys, L., Mandaliya, K., Jackson, D. J. *et al.* (1998). Hormonal contraception, sexually transmitted diseases and risk of heterosexual transmission of human immunodeficiency virus, type 1. *J. Infect. Dis.* **178**, 1053–1059.

73. Marx, P. A., Gettie, A., Dailey, P. *et al.* (1996). Progesterone implants enhance SIV vaginal transmission and early virus load. *Nat. Med.* **2**, 1084–1089.

74. Duerr, A., Warren, D., Smith, D. *et al.* (1997). Contraceptives and HIV transmission (letter). *Nat. Med.* **3**, 124.

75. Edelman, D. A. (1988). The use of intrauterine contraceptive devices, pelvic inflammatory disease, and *Chlamydia trachomatis* infection. *Am. J. Obstet. Gynecol.* **158**, 956–959.

76. Rossing, M. A., Daling, J. R., Weiss, N. W. *et al.* (1993). Past use of an intrauterine device and risk of tubal pregnancy. *Epidemiology* **4**, 245–251.

77. Avonts, D., Sercu, M., Heyerick, P., Vandermeeren, I., Meheus, A., and Piot, P. (1990). Incidence of uncomplicated genital infections in women using oral contraception or an intrauterine device: A prospective study. *Sex. Transm. Dis.* **17**, 23–29.

78. Grimes, D. A. (1987). Intrauterine devices and pelvic inflammatory disease: Recent developments. *Contraception* **36**, 97–109.

79. Farley, T. M. M., Rosenberg, M. J., Rowe, P. J., Chen, J.-H., and Meirik, O. (1992). Intrauterine devices and pelvic inflammatory disease: An international perspective. *Lancet* **339**, 785–788.

80. Sinei, S. K. A., Schulz, K. F., Lamptey, P. R. *et al.* (1990). Preventing IUD-related pelvic infection: The efficacy of prophylactic doxycycline at insertion. *Br. J. Obstet. Gynaecol.* **97**, 412–419.

81. Ladipo, O. A., Farr, G., Otolorin, E. *et al.* (1991). Prevention of IUD-related pelvic infection: The efficacy of prophylactic doxycycline at IUD insertion. *Adv. Contraception* **7**, 43–54.

82. Walsh, T. L., Grimes, D. A., Frezieres, R., Nelson, A., Bernstein, L., Coulson, A., Bernstein, G. *et al.* (1998). Randomized controlled trial of prophylactic antibiotics before insertion of intrauterine devices. *Lancet* **351**, 1005–1008.

83. Lee, N. C., Rubin, G. L., Ory, H. W. *et al.* (1983). Type of intrauterine device and the risk of pelvic inflammatory disease. *Obstet. Gynecol.* **61**, 1–6.

84. Luukkainen, T., and Toivonen, J. (1995). Levonorgestrel-releasing IUD as a method of contraception with therapeutic properties. *Contraception* **52**, 269–276.

85. Vessey, M., Huggins, G., Lawless, M., Yeates, D., and McPherson, K. (1983). Tubal sterilization: Findings in a large prospective study. *Br. J. Obstet. Gynaecol.* **90**, 203–209.

86. Burkman, R. T., Tonascia, J. A., Atienza, M., and King, T. M. (1976). Untreated endocervical gonorrhea and endometritis following elective abortion. *Am. J. Obstet. Gynecol.* **126**, 648–651.

87. Sawaya, G. F., Grady, D., Kerlikowske, K., and Grimes, D. A. (1996). Antibiotics at the time of induced abortion: The case for universal prophylaxis based on a meta-analysis. *Obstet. Gynecol.* **87**, 884–890.

88. Alan Guttmacher Institute (1994). "Clandestine Abortion: A Latin American Reality." Alan Guttmacher Institute, New York.

89. Henshaw, S. K. (1990). Induced abortion: A world review, 1990. *Fam. Plann. Perspect.* **22**, 76–89.

90. Cates, W., Jr. (1982). Legal abortion: The public health record. *Science* **215**, 1586–1590.

91. Tinker, A., and Koblinsky, M. A. (1993). "Making Motherhood Safe," World Bank Discuss. Pap., No. 202. World Bank, Washington, DC.

92. Hanenberg, R. S., Rojanapithayakorn, W., Kunasol, P. *et al.* (1994). Impact of Thailand's HIV-control programme as indicated by the decline of sexually transmitted diseases. *Lancet* **344**, 243–245.

93. Albert, A. E., Warner, D. L., Hatcher, R. A., Trussell, J., and Bennett, C. (1995). Condom use among female commercial sex workers in Nevada's legal brothels. *Am. J. Public Health* **85**, 1514–1520.

94. Gerbase, A. C., Rowley, J. T., Heymann, D. H. L., Berkley, S. F. B., and Piot, P. (1998). Global prevalence and incidence estimates of selected curable STDs. *Sex. Transm. Infect.* **74** (Suppl. 1), 512–616.

95. Cates, W., Jr., Farley, T. M. M., Rowe, P. J., and the WHO Task Force on Infertility (1985). Worldwide patterns of infertility: Is Africa different? *Lancet* **2**, 596–598.

96. WHO Task Force on Infertility (1987). Infections, pregnancies and infertility: Perspectives on prevention. *Fertil. Steril.* **47**, 964–968.

Section 5

INTERNATIONAL WOMEN'S HEALTH

Zena Stein
Columbia University
New York, New York

Throughout this comprehensive volume choices had to be made, overlaps avoided, and gaps filled. In this section, three criteria influenced selection of each disorder: frequencies in the less developed world substantially greater than in the developed world; manifestations and repercussions qualitatively different in different parts of the world; and distinctive features that can or could be ascribed to variations in environmental context.

One consequence of applying these criteria is a focus predominantly on the less developed world. The literature on the women of the developed world is abundant; that on those of the less developed world is sparse. Readers should find in the assembled chapters a realistic introduction to some present-day needs of women's health, especially in Africa. They might also become aware of linkages between these needs, a theme that will be emphasized in this introduction.

Maternal mortality is a continuing tragedy of the less developed world, one that has all but disappeared in developed countries. Maine and McGinn give an epidemiologic account of the issue [Chapter 31]. In countries with adequate resources, death connected with maternity is so uncommon as to be incommensurable; in countries with seriously deficient resources, such deaths are far from rare. The authors show that the distribution of appropriate resources and not the biological determinants of maternal deaths accounts for these huge differences in frequency across the world. Some decades ago, one said of the high infant mortality rates in poor urban and rural settings that medically we had their measure and we lacked only the resources to prevent them. Since then, the steady and considerable declines in rates of mortality in infants point to substantial present-day successes worldwide. Maternal mortality has not followed so gratifying a course. In hindsight, it is clear that maternal and infant mortality respond neither to the same biological and environmental factors nor to the same measures for prevention. Thus, infant deaths fall dramatically, given basic primary care (even when administered entirely by nurses, health educators, and village workers), available and accessible clean water, and appropriate feeding. On the maternal side, too, such measures may help to reduce the historic scourge of puerperal sepsis. What they do not do is deal effectively with the unpredictable crises of obstructed labor and of uterine hemorrhage. Even skilled antenatal care can play only a partial role in preventing these prominent causes of maternal death. To deal with such emergencies, the essentials are emergency services and the means for women to gain speedy access to them. The obvious and the simple only become so when pointed to, as Maine and McGinn do in their chapter.

The contrasting needs for resources required for reducing infant mortality rates on the one hand and maternal mortality rates on the other have tended to keep advocacy and planning for each separate and even competitive. The advent of the human immunodeficiency virus (HIV) epidemic, however, has stimulated a new debate among health planners. This debate revolves around the proposition that policies that address the current raging epidemic as it affects women and children in the less developed world might connect more closely with the critical existing health problem of maternal mortality, with advantage to both [1]. Although in theory everyone is in favor of bridging services, such propositions call for detailed examination of possible disadvantages and unanticipated consequences as well as benefits for each service arm. Chapters 31 and 33 prompt such examination.

Consider first what is needed to meet maternal–infant transmission of HIV, a critical issue for women in the less developed world. A range of possibilities begins in the antenatal period with voluntary testing and counseling, moves on in mid-pregnancy to administration of metronidazole (to counter chorioamnionitis) as well as supplementary vitamins, and, at delivery, vaginal lavage. Depending on policy and resources, these steps would be supplemented with antiviral medication, whether the full "076" regime or less expensive protocols. Other tasks may be even more demanding: support in the decision of whether or not to breastfeed in a traditional society where breastfeeding is the rule, counseling on early weaning when breastfeeding is chosen, and assisting women who choose not to breastfeed to provide alternate nourishment. For the HIV-infected woman herself, a regime of INH to reduce the risk of tuberculosis and of TMP-SMX (trimethoprim-sulfamethoxazole) to reduce some of the common opportunistic infections will be prescribed. Perhaps hardest of all is to counsel and support women in the face of adverse family and community responses.

If there is to be an emphasis on bridging, then it may be useful to add to the above tasks some information regarding elective caesarean section, which has been shown to reduce transmission in developed countries. This may, on the other hand, be artificial in less developed countries, faced as they will be with the list of responsibilities already mentioned, and the unlikely benefits HIV-positive women will receive with this intervention. The care of orphans, on the other hand, would be a common agenda item.

This set of tasks for primary health care facilities and their personnel is demanding in the extreme, requiring a large corps of well-trained staff to accomplish them. But as we have grown to expect, many of the poverty-stricken countries, and the same poor classes within them that are the most beset by maternal mortality and morbidity, are equally or even more beset by the HIV epidemic. They are also the same classes with the fewest resources for health. The correspondence in the distribution of these threats to the survival of women is far from complete. For instance, it is women in the far-flung rural areas who always are at highest risk of death from maternity but who also, usually, are at lower risk of HIV infection. Still, it is on the whole the same vulnerable groups for which facilities are most limited.

However, Maine and McGinn argue from their studies across the less developed world that the direct path to reducing the extraordinary disparities in frequency of maternal deaths requires that the structure and deployment of health services for deliveries converge to some degree on those in the developed world. There have to be adequate comprehensive services that include surgical facilities and access to them. The HIV epidemic, by contrast, poses different problems from those of maternal mortality. The contrasts emerge especially for infant transmission and the primary prevention of infection in women. These depend much less on straightforward structural change in health services that can render the necessary special care accessible. They depend instead on complex processes such as mobilizing social and political forces as well as intensifying and extending existing services; altering gender relations and norms of behavior; and advancing technology like vaginal microbicides (See Chapter 33). Ultimately, one can expect vaccines to eventuate, and thereafter, as with other immunizations, that distribution will follow. All this is likely to depend on social movements and political change, with their instruments being public

health programs and primary care. With HIV infection, of course, the numbers of women whose lives will be burdened and curtailed will be much larger than with maternal death.

These divergent requirements call for imagination from those health planners intent on creating services that bridge the differing nature of health and medical problems of poor people in poor countries. They need to ask whether and how the new requirements of the maternity services imposed by the HIV epidemic might also be used to inform women and health care professionals of the hazards of delivery. They would need to alert health and financial authorities of the necessity for higher level services to prevent this grave, if rarer, condition at the same time.

One genuine bridge between these two health problems is to place them both within a framework of human rights; that is, equity would forbid the neglect of either. Freedman, in Chapter 34, allows for such a strategy. She writes: "the fact is that women's movements around the world have also captured human rights discourse and methods—legal and otherwise—for their own purposes." A right of this kind must fall into Isaiah Berlin's category of "positive" rather than "negative" freedoms or rights, as does health generally [2]. Negative rights are readily protected by law; for example, they prevent infringements of the freedom of speech, assembly, and privacy. Positive rights are quite different. They depend on achieving social justice and equity. Positive rights may be promoted by law but are not readily subject to enforcement. They depend on social and political action to achieve the social justice that such rights require, as with universal education or access to health services. Even though access to such goods can be and is legislated, however, true social equity in such matters requires more than law alone.

The argument for a human rights agenda in the two spheres of maternal mortality and HIV infection has already begun, but independently for each. With regard to maternal mortality, Yamin and Maine [3] and others have presented a detailed argument, namely that societies that do not acknowledge the specific needs of women in labor neglect a fundamental human right of women. Moreover, they point out that because the necessary services can be described and enumerated, then simply to observe their lack is sufficient grounds for documenting neglect.

With regard to HIV, the late Jonathan Mann spent a decade illustrating the many facets of prevention and amelioration, which he insisted could and should be seen within a human rights perspective. In the same vein, the movement for prevention of HIV in women has been able to subsume health issues into a human rights agenda [Chapter 33; 4]. Commercial sex workers form a core group in the transmission of HIV. Many are poor women with families to support who must resort to selling sex for money. In a human rights perspective, to class such activities as causal in the epidemic is to stigmatize the women involved for what is essentially the compulsion of circumstance. Again, in many societies, women infected with HIV who are married and often monogamous bear the whole blame and the stigma from their own families and even from the husbands who infected them. The counseling message, "Use a condom," is by now universal. The countries and the classes most affected by the HIV epidemic are those in which women are most dependent, and not least in sexual encounters. For these women to insist on condom use can be to court violence or eviction from the home. Beyond personal and family issues, laws that mandate compulsory testing for HIV solely for pregnant women clearly raise an issue of discrimination against a group.

All such issues can be readily absorbed into a framework of advocacy for positive and negative human rights. What is called for is to remedy the plight of women exposed to inequities in health through an accumulation of social, political and legal action.

To inquire into maternal mortality and the HIV epidemic from this perspective is quickly to recognize injustice and inequity. Readers will appreciate Friedman's chapter, on Women's Health and Human Rights, which bears on the lives of women everywhere. She emphasizes that women's health concerns cannot be contained within pathological entities or physiological systems. The successive World Conferences on Women (and especially the Fourth, held in Beijing in 1995) did not only promote the right of women to regulate their reproduction, they also enabled women to join voices to claim those rights and, once won, to protect them and agitate for them when they are denied.

Some writers argue, further, that among the human rights of a woman, and one that is relevant to learning protective strategies against HIV, is the right to know her own body [E. Gollub, personal communication]. The enormous international enthusiasm for the Boston Collaborative publications [5] is a recognition of a felt need. The relevant anatomical and physiological facts are well known. Their translation into a right to knowledge is less so.

On grounds of equity, it is argued that a woman is entitled to the knowledge to appreciate her body, protect it, and enjoy it. Thus, women alone neither see nor touch anatomical structures engaged in their sexual encounters. Women alone experience hormonal cycles and menstrual bleeding. Women alone experience, for reasons not always clear to them, another organism growing within their own bodies. Indeed, without the requisite knowledge of her own anatomy and physiology, a woman is hardpressed to prevent unwanted pregnancies and infections. In the less developed world, even as general education in schools becomes more widespread, these facets of knowledge remain neglected.

Knowledge like this might be one of the supports a woman could bring to protect her against the antisocial aspects of a traditional culture. Chapter 32 discusses female "circumcision," a euphemism for a destructive assault on a woman's body, better described as female mutilation. This practice persists in many cultures, especially in Africa, and not without the support of prior initiates, now women of authority and bound by the tradition. Even young women raised in these cultures can find it harder to break away from family and peer pressure than to endure the mutilation. Thus, the practice has proved solidly rooted and difficult to eradicate, whether by law or by persuasion. Women outside the system, and many men, are aghast when faced with the more extreme consequences of these practices.

Physical assaults on women, often by male members of the family, are also sanctioned by some cultures. Admittedly, rape is rarely endorsed, although the level of tolerance varies by time and place, as do the punishments meted out to the rapist. Not rarely, relaxation of the usual social restraints seems to occur in group gatherings. These restraints may be ignored during wars especially, both in more and in less developed countries: witness the interethnic slaughters in Rwanda and in Yugoslavia. In both countries sexual atrocities were committed (and indeed continue at the time of writing) by the men of the aggressor group against the women of the group under attack, reaching epidemic proportions. Similar examples are on record from classical times. Why such mass rapes occur and what circumstances precipitate them is little studied. These enormities evoke speculation, but still daunt social analysts.

In peacetime and in the developed world, too, rape and male aggression are threats to women's health, and more often than is usually recognized. The danger is greater where women have little power. Such situations are all too common in less developed societies, both in women's own homes and in the wider social milieu.

Poverty is a particular social context that endangers women's health everywhere, but again more so for women in less developed countries. Among these women, poverty is often accompanied by continuous childbearing, infant losses, and an endless search for food for near-starving families on minimal resources. The women themselves often suffer emaciation and exhaustion, nutritional anemia, iodine deficiency disorders, and chronic infections.

Abdool Karim and Stein (Chapter 33), writing on HIV infection, describe the essential roles of women during the epidemic: the mother (and if not her, the grandmother) must nurse the sick through disabling illnesses of long duration, take into the home the orphaned children of friends and kin, and care for disabled and dying wage-earners of the family until death. Not infrequently, it is only through commercial sex work that such a woman can come by even that minimum amount of money needed to feed, clothe, and shelter her dependants. In the many developing countries beset by epidemic HIV and tuberculosis, sickness tests women to their limits.

The health problems discussed here are not easily resolved; that is the nature of all issues in which context must be taken into account. For the health scientist trained in the developed world, the problems in less developed countries are compounded by unfamiliarity with language and customs, and also with political and institutional intricacies. Still, health and medical workers in the international sphere can anticipate rewards hard to come by elsewhere. A large yield of knowledge often follows when modern tools and methods are employed in uncharted territory. Disorders little known and poorly described and understood enable epidemiologists to discover distributions, associations, causes and treatments, sometimes with applications worldwide. Moreover, to meet with and work with women and families of unfamiliar cultures is in itself an intense learning experience inaccessible at secondhand. Achieving even small gains in the quality of life in such communities rewards. Not uncommonly, though, gains and gratification rise far above expectation.

Acknowledgments

I acknowledge with pleasure the role of Mervyn Susser in clarifying for me some of the ideas developed here.

References

1. Graham, W. J., and Newell, M.-L. (1999). Seizing the opportunity: Collaborative initiatives to reduce HIV and maternal mortality. *Lancet* **353,** 836–839.
2. Susser, M. (1993). Health as a human rights Issue: An epidemiologist's perspective on the public health *Am. J. Public Health* **83,** 418–426.
3. Yamin, A. E., and Maine, D. P. (1999). Maternal mortality as a human rights issue: Measuring compliance with international treaty obligations. Human Rights Quarterly (in press).
4. Abdool Karim Q, Abdool Karim SS (1999). Epidemiology of HIV infection in South Africa. *International AIDS Society Newsletter* **12,** 4–7.
5. Boston Women's Health Book Collective (1984). Touchstone, New York.

31

Maternal Mortality and Morbidity

DEBORAH MAINE AND THERESE MCGINN

Center for Population and Family Health
Columbia University School of Public Health
New York, New York

I. Introduction and Background

For women giving birth in developed countries today, concerns tend to center on the health of the infant and the quality of the experience. However, in developing countries, the death of the woman is still a very real possibility. While maternal mortality ratios in many developed countries are now below 10 deaths per 100,000 live births, in parts of Africa and Asia they are more than 100 times higher [1]. Furthermore, because fertility also is higher in developing countries, women in developing countries are exposed to the risks of pregnancy and childbirth much more often. Consequently, their lifetime risk of maternal death is shockingly high. The United Nations estimates that in developing countries, one woman in 48 dies of maternal causes. In Africa, the estimate rises to one woman in 16. In contrast, in Northern Europe only one woman in 4000 dies of these causes.

As is true of maternal death, maternal morbidity is also a much more serious problem in developing than in developed countries. The study of maternal morbidity is still a relatively new field and there are methodological and definitional issues to be resolved, but it is clear that the problem is large. The World Bank, uses "disability-adjusted life years" (DALYs) to take into account the effects of both mortality and morbidity. It is estimated that among women of reproductive age, maternal conditions are the leading cause of DALYs lost in developing countries—nearly 28 million annually [2]. The largest cause of DALYs lost among men the same age in developing countries—HIV infection—accounts for only about 15 million annually.

Maternal mortality and morbidity will be discussed separately in this chapter because the literatures are (for the most part) separate and the methodological issues are different.

II. Definitions

A. Maternal Mortality

A maternal death is defined by the World Health Organization (WHO) as "the death of a woman while pregnant or within 42 days of termination of pregnancy, irrespective of the duration or site of the pregnancy, from any cause related to or aggravated by the pregnancy or its management, but not from accidental or incidental causes" [3].

A maternal death may be classified as direct or indirect. Direct obstetric deaths are due to "complications of the pregnant state [pregnancy, labor and puerperium] from interventions, omissions, incorrect treatment, or from a chain of events resulting from any of the above." Indirect obstetric deaths result from "previous existing disease, or disease that developed during pregnancy and which was not due to direct obstetric causes, but which was aggravated by the physiologic effect of pregnancy" [3].

Other organizations use different definitions of these and other terms. For example, WHO defines a "pregnancy-related death" as "the death of a woman while pregnant or within 42 days of termination of pregnancy, irrespective of the cause of death" [3]. The U.S. Centers for Disease Control and Prevention (CDC) use the same term, "pregnancy-related death," to refer to the death of a woman up to one calendar year after termination of pregnancy from complications of the pregnancy, the chain of events initiated by the pregnancy, or aggravation of an unrelated condition by the physiologic or pharmacologic effects of pregnancy [4]. Thus, WHO's and CDC's definitions of the term "pregnancy-related death" differ both in time frame (42 days vs one year) and in cause (any cause vs narrower range of causes).

Data on maternal deaths are used to construct several statistics, including the maternal mortality ratio and rate, and lifetime risk. The maternal mortality ratio is maternal deaths per 100,000 live births. The rate is maternal deaths per 100,000 women of reproductive age (usually 15–44) per year. In older articles, the ratio is sometimes called the "rate" [5].

B. Maternal Morbidity

This is a field of study that is still in its early stages. For example, there is no standard definition of maternal morbidity. There also is not a clear delineation among the related categories of obstetric, gynecologic, and contraceptive morbidities [6]. In this chapter, the definition of "maternal morbidity" reflects WHO's "obstetric morbidity," which parallels its definition of maternal mortality. Thus, maternal morbidity is defined as "morbidity in a woman who has been pregnant (regardless of the site or duration of the pregnancy) from any cause related to or aggravated by the pregnancy or its management, but not from accidental or incidental causes" [7].

There is, as yet, no standard list of the specific morbidities covered by the definition. Studies of maternal (or obstetric) morbidity have measured life-threatening conditions (*e.g.,* sepsis, obstructed labor, hemorrhage) as well as a range of less severe problems (*e.g.,* edema, anemia) [6].

Maternal morbidity may have direct, indirect, or psychological causes. Direct maternal morbidity results from obstetric complications during pregnancy, labor, or the postpartum period from interventions, omissions, or incorrect treatment. This includes temporary conditions (whether mild or severe) that occur during pregnancy or within 42 days of delivery, and chronic

Table 31.1
Numbers of Maternal Deaths and Maternal Mortality Ratios, by Region

Region	Annual number of maternal deaths	Maternal mortality ratio (maternal deaths per 100,000 live births)	Proportion of world's population	Proportion of world's maternal deaths
Africa	235,000	870	12%	40%
Asia	323,000	390	61%	55%
Latin America and Caribbean	23,000	190	8%	4%
Oceania	1,400	680	<1%	<1%
Europe	3,200	36	13%	<1%
North America	500	11	5%	<1%
Total (rounded)	585,000	430	100%	100%

Sources: [1,8].

conditions that result from pregnancy, abortion, or childbirth. Indirect maternal morbidity results from previously existing conditions or disease aggravated by the pregnancy. Such morbidity may occur at any time and continue beyond the reproductive years. Psychological maternal morbidity can result from obstetric complications, interventions, cultural practices, or coercion [6].

III. Distribution

A. *Maternal Mortality*

The United Nations estimates that 585,000 maternal deaths occur every year worldwide [1]. As noted earlier, the deaths are not evenly distributed throughout the world. Nearly all maternal deaths (99%) occur in the developing world, although these regions contain only about four-fifths of the world's population. African countries, and particularly those in sub-Saharan Africa, have the highest maternal mortality ratios, while Asia has the greatest absolute number of deaths (see Table 31.1) [1,8].

Of the commonly used health indicators, maternal mortality shows the greatest disparity between the developed and developing world. For example, the maternal mortality ratio in the United States is 12 deaths per 100,000 live births, whereas in Nigeria (the most populous country in Africa), it is 1000—more than 80 times higher [1]. In contrast, the infant mortality rate in the U.S. and Nigeria are 9 and 86 deaths per 1000 live births, respectively, a tenfold difference [9].

Causes of maternal death are remarkably consistent throughout the world. In most countries, 75–80% of maternal deaths are due to direct causes, principally five major direct causes: hemorrhage, sepsis, unsafe abortion, hypertensive disorders of pregnancy, and obstructed labor [10]. The consistency in causes is striking, even in settings where the levels of maternal mortality are vastly different. As Table 31.2 [1,10,11] illustrates, for the world as a whole the estimated maternal mortality ratio (MMR) is 430 deaths per 100,000 live births, of which hemorrhage accounts for 25%, sepsis 15%, and hypertensive disorders of pregnancy 12%. In the United States, with a MMR of 12 the same three causes account for 29%, 13%, and 18% of deaths, respectively [1,10,11].

It has long been known that maternal mortality ratios are highest among women at the beginning and end of their reproductive years (*e.g.,* younger than 20, older than 35) [12]. However, most births occur among women in their twenties. Consequently, women in the lowest risk group generally have the greatest absolute number of deaths. The same pattern holds for parity; while the ratios are higher among primiparous women and those having a fourth or later child, the largest numbers of deaths occur among women having their second or third child [13].

As there are substantial disparities in maternal mortality between rich and poor countries, so there are disparities within societies as well. Maternal mortality is generally highest among the least prosperous classes or racial groups. For example, in the

Table 31.2
Causes of Maternal Death, Worldwide and U.S.

Cause of maternal death	World estimates, 1993 percent of deaths	United States, 1987–1990 percent of deaths
Hemorrhage	25	29
Sepsis	15	13
Unsafe abortion	13	*
Hypertensive disorders of pregnancy and	12	18
Obstructed labor	8	*
Other direct causes	8	*
Embolism	*	20
Cardiomyopathy	*	6
Anesthesia	*	3
Indirect causes; Other unspecified	20	13
Total[a]	100	100
Maternal mortality ratio (Maternal deaths per 100,000 live births)	430	12

Note. *Negligible. Sources: [1,10,11].
[a]Figures do not add to totals due to rounding.

United States, the maternal mortality ratio among black women is more than four times higher than that among white women, and this disparity has more than doubled since the early 1900s [11,14]. This difference is largely due to differences in access to health services and information rather than to variation in the incidence of life-threatening complications.

B. Maternal Morbidity

Accurate information on the incidence and prevalence of maternal morbidity is difficult to obtain. A figure of 15 maternal morbidities per maternal death is often quoted, based on a small study in India [15]. Other studies suggest that this ratio substantially underestimates the true level of morbidity [6].

IV. Measurement Issues

Levels of maternal mortality and morbidity are difficult to ascertain for reasons that are described in the following sections. Consequently, efforts have been made to develop techniques of estimation, which are described as well.

A. Maternal Mortality

1. Problems of Measurement

There are a number of technical issues that make the study of maternal mortality in developing countries more difficult than the study of some other types of mortality, such as infant mortality. These issues are discussed under three headings: low incidence, underreporting, and misclassification.

a. LOW INCIDENCE.   Even though complications of pregnancy and childbirth are the leading cause of death among women of reproductive age in many developing countries, maternal deaths are rare events. For example, in 1982–1983, the Medical Research Council of Britain conducted a prospective study of 672 pregnant women in the Gambia. The maternal mortality ratio was extremely high: 2200 maternal deaths per 100,000 live births. There were, however, only 15 deaths [16].

As always, small numbers mean that confidence intervals will be wide, and this makes it difficult to measure trends.

b. UNDERREPORTING.   Registration of births and deaths (i.e., vital registration) is taken for granted in industrialized countries. In these countries, and in some developing countries, nearly all deaths are reported to the government. This is not the case, however, in most developing countries.

There are several reasons for this. One reason is that, in developing countries, most deaths do not take place in health facilities where personnel would be required to report them. Many people die at home or on their way to the hospital. Their deaths tend not to be recorded. Various population-based studies in Asia and the Middle East report 2–74% of maternal deaths occurring in hospitals in rural areas, and 54–82% in urban areas [17].

Furthermore, not all maternal deaths that occur in facilities are reported. Poor record-keeping, inefficient administrative practices, a desire not to be blamed for the death, and a desire

to avoid paperwork may provoke health personnel to return the woman's body to her family without formally reporting the death.

Finally, if a fee is required or if the bureaucratic requirements to register a death are onerous, the family may be deterred from doing so.

c. MISCLASSIFICATION.   In this context, misclassification means that the death is reported, but the fact that it was a *maternal* death is not. There are a number of factors that contribute to the frequent misclassification of maternal deaths in both developed and developing countries, including the timing of the death and the cause of death.

Timing of Death   As noted above, WHO specifies that a maternal death occur while the woman is "pregnant or within 42 days of termination of pregnancy . . ." Other definitions use different cutoff points [4]. Whatever the time limit used, it is necessary to know not only that the woman died, but that the death occurred within the pertinent time period. This may be difficult to ascertain, especially in situations where deaths are reported after substantial time has elapsed, if at all.

Some women die before they (or their relatives, who would report the death) know that they are pregnant, and autopsies are rare in developing countries. In addition, some women who eventually die of obstetric complications survive the 42-day period. Although such deaths are due to obstetric causes, they do not fit the standard definition and so are not counted as maternal deaths. This source of error is of greater importance where advanced medical technology is available and patients are kept alive longer [18].

Cause of Death   Cause of death is routinely reported in only 78 countries, covering approximately 35% of the world's population [1]. Even in those countries, however, the fact that a death is, in fact, a maternal death often is not noted on the death certificate, for both intentional and unintentional reasons. Intentional misreporting is common when the death is due to complications of illicit induced abortion. In many societies, abortion-related deaths are concealed to protect the reputation of the woman or her family [19]. In some countries, legal action is taken against people who perform abortions and/or against women who obtain them (if they live). Thus, fear of legal prosecution is also a cause of misreporting of maternal deaths [20].

Unintentional misreporting of maternal deaths is very common. Women often die of obstetric complications in emergency wards or medical wards, as opposed to maternity wards. Consequently, the obstetric origin of the bleeding or infection may not be recognized. Even when the health professionals are aware that the death was related to pregnancy, it may not be reported as a maternal death if the forms do not provide relevant categories. Such mistakes lead to massive underestimation of maternal deaths, even in developed countries [21]. In an analysis of pregnancy-related deaths in the United States during 1987–1990, researchers at the Centers for Disease Control and Prevention concluded that "more than half of such deaths . . . are probably still unreported" [11].

2. Methods of Estimation

A method that has been used to derive estimates of maternal mortality in developing countries is the "sisterhood method" [22]. This is a survey technique in which interviewers ask all

Table 31.3
Indicators of the Availability and use of Obstetric Services

Indicator	Minimum acceptable level

Amount of essential obstetric care (EOC):
  Basic EOC facilities
  Comprehensive EOC facilities
	For every 500,000 population, there should be:
	  At least 4 basic EOC facilities
	  At least 1 comprehensive EOC facility

Geographic distribution of EOC facilities	Minimum level for amount of EOC facilities is met in subnational areas
% of all births in basic or comprehensive EOC facilities	At least 15% of all births in the population take place in either basic or comprehensive facilities
Met need for EOC: women treated in EOC facilities as a % of those estimated to have obstetric complications	At least 100% of women estimated to have obstetric complications are treated in EOC facilities
Cesarean sections as a % of all births	Cesarean sections account for no less than 5% nor more than 15% of all births in the population
Case fatality rate: deaths among women with complications in EOC facilities	The case fatality rate among women with obstetric complications in EOC facilities is less than 1%

Source: [26].

adult members of sampled households how many sisters they had who survived to adulthood, and of these, how many died of pregnancy-related causes [22,23]. These data can then be used to construct the maternal mortality rate, ratio, and lifetime risk.

While the sisterhood method is the most efficient survey method available, it has important shortcomings. Because sisters who died did not necessarily live nearby, the results may not reflect maternal mortality in the study area. Moreover, the method usually yields an estimate of the level of maternal mortality for a period which has its midpoint 10–12 years before the study. This is because respondents are asked about their sisters' deaths, regardless of how long ago these occurred [22,24]. Thus, estimates derived using this technique must be interpreted carefully and have limited usefulness for monitoring trends. Efforts are underway to refine the sisterhood method so that mortality estimates for shorter time periods can be derived and/or the method simplified [25]. This may well mean larger sample sizes or larger confidence intervals, either of which would reduce the practical value of the method.

The most widely used figures for maternal mortality incidence are estimates derived by the World Health Organization and UNICEF in 1996. For countries with credible vital registration systems or reliable national studies, existing data were used "as is" or adjusted to account for likely underreporting and misclassification [1]. For countries with poor or no data, a model using total fertility rates and the proportion of births assisted by a trained person was developed to predict maternal mortality.

Using these techniques, WHO and UNICEF derived estimates of maternal mortality for the world's countries and regions. Although these estimates are more systematic than those previously available for many countries, the estimates are best interpreted as indicating orders of magnitude. They can not be used to monitor trends [1].

These estimates, of course, do not meet the need for information to be used in designing and monitoring programs. Consequently, the field is moving away from using "impact" indicators (such as the maternal mortality ratio and lifetime risk)

and toward "process" and "outcome" indicators [26]. Impact indicators are measures of maternal mortality, which (as noted earlier) is difficult and expensive to ascertain. Furthermore, impact indicators do not provide information about the health services that are crucial for reducing deaths. Process indicators (which measure activities) and outcome indicators (which measure behavior) can provide this kind of practical information.

In 1997, three United Nations agencies jointly issued a set of indicators for monitoring the availability and utilization of obstetric services [27]. The focus of these guidelines is treatment of life-threatening complications. As Table 31.3 shows, the guidelines set minimum and maximum acceptable levels for a series of process and outcome indicators. For example, the minimum acceptable standard for the availability of hospitals providing comprehensive obstetric care (i.e., including blood transfusion and cesarean section) is one hospital per 500,000 population. Early tests of these guidelines in Bangladesh showed that many areas of the country do not meet this standard because many district hospitals do not perform cesarean sections even though they have physicians on staff [28].

Another of the indicators developed for these guidelines is an estimate of "met need" for emergency obstetric care. The denominator of this statistic is the proportion of pregnant women who will develop complications serious enough to require medical treatment, which is estimated as 15% of live births (live births being one of the most widely available statistics). The numerator is the number of women who were admitted to a health facility for treatment of one of the major obstetric complications. One study in Egypt found low levels of met need. For example, in Akhmeim District in Upper Egypt, met need was only 17% [29].

While the "met need" indicator may be the most eloquent of this set of indicators, it is not the easiest to measure. This is because maternity ward registers in many developing countries do not have a column for maternal complications, although there are usually columns for the weight, number, and sex of infants. Thus, information on the number of women with obstetric complications admitted to the hospital (the numerator of unmet need) is often difficult or impossible to collect. Adding a column on obstetric complications (or reason for admission) to maternity registers seems like a modest and reasonable change to make if we are serious about improving the survival of women in developing countries.

These indicators can be extremely helpful to program planners and managers in developing countries as they seek to reduce maternal deaths. Moreover, they have a number of practical advantages over maternal mortality rates and ratios. These process indicators are easier and less expensive to measure, they can change quickly (*e.g.*, in 1–2 years), trends can be measured easily, and they can be used at district and national levels [30]. It is hoped that during the next decade, as these guidelines are applied, a substantial body of information will be accumulated. This information will help governments in developing countries make their efforts to reduce maternal deaths more effective.

B. Maternal Morbidity

1. Problems of Measurement

We face some of the same ascertainment problems in studying maternal morbidity as we do with maternal mortality. While low incidence is not an issue, underreporting and misclassification are. Moreover, as noted, there is no standard and specific definition of what constitutes maternal (or obstetric) morbidity, and there are plausible nonobstetric origins for many conditions that may be first reported during pregnancy.

There is an additional difficulty in assessing morbidity, not applicable to the study of mortality: death is an unmistakable event, but whether or not morbidity has occurred is not necessarily clear to women or to health providers. Nonrecognition or uncertainty is influenced by social and cultural factors, the nature of the conditions, and the resources available to assess them.

a. SOCIAL AND CULTURAL FACTORS.   Social and cultural norms affect perceptions of and response to illness. Potentially dangerous conditions (such as edema during pregnancy, light antepartum bleeding, and labor of long duration) often are viewed by women and their families as unremarkable, natural parts of pregnancy [31,32]. Problems that arise may not be perceived as related to the pregnancy but may instead be blamed on the woman's behavior (*e.g.*, infidelity or insubordination). In societies that value stoicism, women may not recognize or admit to pain and may be ashamed for having experienced obstetric complications [31,33]. Sequelae of abortion may go unreported if legal or social censure is likely to ensue. In situations such as these, treatment might not be sought even if it were available, and people might well not report obstetric morbidity when asked.

b. NATURE OF THE CONDITION.   Women are unlikely to recognize some morbid conditions. This reporting problem has long been recognized for gynecological morbidities, such as some reproductive tract infections. Obstetric morbidities, such as anemia, uterine prolapse, and hypertension may also go unrecognized by women experiencing them [6,34].

Some conditions may also be difficult for providers to recognize or to identify as obstetric in origin. For example, hypertension has several possible etiologies. Degrees of pain are hard for women to describe and for providers to assess. If antepartum bleeding or convulsions are not present at the time of an antenatal care visit, they can be missed [6].

c. RESOURCE AVAILABILITY.   Some conditions are difficult to assess accurately in low-resource settings. Blood tests for anemia may not be available; providers must rely on subjective measures such as a woman's report of fatigue or the color of the conjunctiva of her eye. Sphygmomanometers are not available or not functioning in many primary care facilities. Dipsticks for detecting albumin in urine are not available in many areas.

2. Methods of Estimation

Because facility-based records are known to be inadequate for ascertaining the incidence and prevalence of maternal morbidities, population-based measures have been sought. Unlike the study of mortality, the study of morbidity permits direct questioning of women who may have experienced it. It would therefore seem feasible to measure morbidity using standard population-based quantitative and qualitative research methods. Such studies have been undertaken, however, and investigators have concluded that the data are not reliable.

Retrospective population-based studies in India and the Philippines were done to obtain prevalence data on reproductive (not solely obstetric) morbidities. Women were asked about signs, symptoms, and diagnosed diseases related to reproduction. In India, about one-third of the women reported current symptoms suggestive of reproductive morbidity, and 41% reported disorders or problems associated with their last live birth [34]. In the Philippines, 11% of respondents had life-threatening obstetric complications and an additional 52% reported less serious obstetric complications [35].

A major question about estimates based on women's self-reports is the validity of the information provided. The Philippines study included a validation component in which diagnostic data based on hospital chart review were compared with women's recall of major complications during pregnancy. The sensitivity of the self-reports for four life-threatening complication ranged from 44% (for eclampsia) to 89% (for sepsis). Specificity was higher (78–97%) [36]. The accuracy of self-report was strongly questioned in a study in Egypt in which women's self-reports, diagnoses from clinical examinations, and reports from laboratory examination were compared for reproductive tract infections and prolapse. Overall, agreement was poor [33].

A task force evaluating the validity of women's self-reports for estimating prevalence and for program planning concluded that "women's retrospective self-report of complications is not an accurate means of estimating the proportion of women who needed medical treatment for obstetric complications" [37].

Another way that researchers have tried to quantify maternal morbidity is by estimating the death-to-case ratio. There are, however, very few studies on which to base such estimates. One such study on obstetric morbidity was conducted in India from 1974 to 1979 [15]. Studying 281 pregnancies, the researchers found 16.5 obstetric illnesses per maternal death (often cited and rounded to 15). Other population-based, retrospective studies in Egypt, Bangladesh, and India found 967, 643, and 541 morbidities per death, respectively. The figures declined to 591, 259, and 300 when only serious conditions were included, and declined even further (to 67, 114, and 24) when only life-threatening complications were counted [6].

A possible problem with this technique is that the proportion of women who will die or develop chronic problems will vary greatly depending on the level of obstetric care they receive. For example, a woman in the U.S. who has trouble delivering because of the position of the fetus would have a cesarean section, whereas a woman in the Sudan might well develop vesico-vaginal fistula or die. Surprisingly, data from the United States show a death-to-case ratio within the range of the studies cited above. During 1987–1990, there were 1453 pregnancy-related deaths in the United States and an estimated 31,000 "pregnancy-related" hospitalizations [11]. This is a ratio of 21 hospitalizations per death.

Thus, levels of incidence and prevalence of maternal morbidity are not well understood, for a variety of reasons: facility-based data have inherent biases; clinical monitoring of large populations of pregnant and postpartum women in developing countries is impractical (and, unless treatment is supplied, unethical); self-reports do not provide reliable information on prevalence (though they are useful for understanding women's perceptions of illness and health-seeking behavior); and sound methods of estimation are lacking.

The serious problems related to ascertainment of maternal mortality and morbidity are often ignored when health planners are faced with the urgent need for information. In order to help countries calculate the volume of services needed, a number of United Nations agencies use the following estimate: at least 15% of pregnant women will experience complications of pregnancy or delivery serious enough to require medical care. Because data on the number of pregnancies in a population are difficult to obtain, live births are used instead [27].

V. Trends

A. Maternal Mortality

In the seventeenth and eighteenth centuries in Europe, maternal mortality was apparently very high (*e.g.,* 900 maternal deaths per 100,000 live births), as studies using parish records have shown [38,39]. By the mid-nineteenth century, the maternal mortality ratio was between 400 and 700 deaths per 100,000 live births, a range which includes the United Nations estimate for the developing world today (480 deaths per 100,000 live births) [1,40]. The levels of maternal mortality in Europe remained largely unchanged between 1850 and about 1935. In the United States, the level also remained fairly stable between the early 1900s, when records first became available, and 1935. Then, beginning in the early 1930s, maternal mortality declined rapidly in Europe and the United States, and by 1950 it was no longer a major public health problem [40]. For example, in England and Wales, there were 441 maternal deaths per 100,000 live births in 1934. By 1950, the ratio had declined to 87, and by 1960 to 39.

The historical evolution of maternal mortality in now-developed countries—relatively stable, high rates until the 1930s followed by a steep decline—is unusual among major causes of death. Overall death rates in England and Wales, for example, began to fall in the early eighteenth century and continued to do so through the twentieth century [40]. Virtually all of the decline through the mid-1900's was due to a reduction in infectious diseases, largely attributable to a better standard of living, including better nutrition and sanitation [41]. In the United States, records show a steady decline in total mortality from 1900. This decline has been attributed to socioeconomic, rather than to medical, measures [42].

Maternal mortality did not respond to the same improvements in public health, it remained high while overall mortality rates fell. It is interesting to note that infant mortality, often assumed to be closely linked to maternal death, followed the more usual pattern, falling steadily from the mid-1800s to 1960 in England and Wales [43]. This indicates that maternal and infant mortality do not, in fact, respond to the same biological and/or external factors [44].

The post-1930 decline in maternal mortality appears to have been largely due to improvements in treatment of major obstetric complications, such as the widespread availability and use of antibiotics (sulphonamides and, later, penicillin) to treat infection, safer methods of anesthesia, better antiseptic technique, safer cesarean section, safer blood transfusion, and fewer unnecessary interventions by attendants during delivery [45]. Thus, maternal mortality seems to have declined in response to medical, rather to than socioeconomic, improvements.

B. Maternal Morbidity

As there are few data on maternal morbidity, there are fewer still on trends. It is clear, however, that morbidity, if it has declined at all, has not fallen to the same degree as mortality. Maternal morbidity is thought to be relatively high today, even in the developed world. That women rarely die or experience permanent disability from pregnancy-related problems in developed countries is due to the prompt and effective treatment they receive, rather than to a lower incidence of complications.

This can be illustrated using one serious and chronic pregnancy-related condition—vesico-vaginal fistulae (VVF). These are tears in the wall separating the bladder and vagina and are usually due to obstructed labor. Women with VVF constantly leak urine and are often ostracized [46]. Today, this condition still occurs in developing countries but is virtually unknown in the developed world [47].

It was not always so. VVF was a recognized problem in the United States and Europe through the mid- to late eighteenth century [48]. New York City had a hospital of international renown specifically for fistula repair [49]. The decline of VVF as a maternal health problem was due to two factors. First, a surgical technique to repair fistulae was developed in 1849 and was used thereafter [48]. Second, better techniques of managing obstructed labor (including, ultimately, cesarean sections) prevented fistulae from occurring. These treatments are still not available to most women in developing countries. For them, VVF, like other morbidities that can be successfully treated and sometimes prevented with adequate obstetric care, are an integral part of life.

VI. Influence of Women's Social Roles

Women's status in society can affect maternal mortality in a variety of ways. It can affect women's exposure to risk, access to care, and, possibly, the priority accorded to women's health services within health care programs.

In every society, most women are expected to bear children. In some societies, however, this is the primary way in which women can raise their standing in the community, and their social status rises with the number of children they bear (and, sometimes, with the number of sons). In these settings, women are exposed many times during their lives to the risk of maternal death.

For women who develop obstetric complications, social status can influence their access to health care in a variety of ways. One way is through their access to information about complications and about where to obtain medical care. In societies where many women are illiterate, their access to information may be severely limited.

Women also are limited (to varying degrees) in their access to money, transport, and decision-making authority. For example, in northern Nigeria (where many women are in "purdah"—*i.e.,* confined to the family house or compound), women must have their husbands' permission to travel, even in an emergency. Physicians in that area tell of numerous women who died or were terribly maimed because they developed obstetric complications (such as obstructed labor) when their husbands were away from home.

In summary, the myriad manifestations of women's low status limit their access to medical care when problems arise. Women's status may also influence the priority given to women's health services, though this is difficult to document.

VII. Future Directions

It is only since the late 1980s that complications of pregnancy and childbirth have received concerted attention in the international health field, even though they are the single largest cause of premature death and disability among women of reproductive age. The "Safe Motherhood Initiative," which focuses on maternal deaths, was launched in 1987. Morbidity began to receive more attention as part of a reproductive health campaign that came out of the United Nations International Conference on Population and Development, which was held in Cairo in 1995 [50].

Before these initiatives, it was assumed that "Maternal and Child Health" (MCH) programs would meet women's health needs. However, such programs were largely composed of activities designed to reduce infant and child mortality [51]. Now there is wider recognition of the fact that women and infants not only have different health problems, they often need quite different programs. For example, there is much that can be done outside the medical system to reduce infant deaths: village health workers can treat diarrhea and immunization campaigns can reduce infectious illnesses (such as measles) thatare often fatal to children in developing countries. Without prompt adequate treatment of obstetric complications, however, there is little that can be done to reduce deaths and disability among pregnant women. This means that women must have access to antibiotics for infection, anticonvulsants for eclampsia, and surgery for obstructed labor. It is not feasible to provide all of these services in small health centers staffed by paramedical personnel. While midwives and nurses in small facilities can play an important role, physicians and hospitals also are required. In other words, there must be a functioning health system.

The one exception to this statement concerns induced abortion. Once abortion is legalized and safe abortion services are allowed to flourish, abortion deaths quickly become extremely rare. For example, in the United States in 1972 (the year before the Supreme Court loosened the legal restrictions on abortion) there were 63 deaths from induced abortion [52]. By 1978, this number had fallen to 18, despite the doubling of the number of legal procedures performed.

To focus attention and resources on improving care in hospitals in developing countries goes against the philosophy that has dominated international public health for decades. The emphasis on primary health care (launched at the Alma-Ata conference in 1978) was an important advance [53]. It helped countries emerging from colonialism to develop models of health care that reached many more people than did the tertiary hospitals built by the colonial powers. However, this model, at least as it was commonly understood and applied, did not meet the needs of pregnant women, and neither did widespread prenatal care programs [16].

To reduce maternal deaths in developing countries it is not necessary to go back to the old "disease palace" model. However good the services in teaching hospitals may be, they are not accessible to most women in developing countries. What is needed is to upgrade small hospitals, especially in rural areas, and to make sure that the staff there (especially nurses and general practice physicians) are able and willing to treat obstetric complications. This does not usually require building new facilities or training new cadres of workers. Even in the poorest countries, there is often progress that can be made utilizing existing facilities and personnel [54]. Once the health facilities are able to help women with obstetric complications, then a variety of other activities may be appropriate, depending on the setting. These include: educational activities aimed at influential groups, such as religious leaders, husbands, and mothers-in-law; revolving loan funds in communities to help people overcome financial barriers to care; and the education and mobilization of transport workers, to enlist their help in carrying women to treatment facilities.

If, during its second decade, the Safe Motherhood Initiative is to fulfill its original purpose—reducing deaths among pregnant women—then efforts must focus on the treatment of obstetric complications.

References

1. World Health Organization and United Nations Children's Fund (1996). "Revised 1990 Estimates of Maternal Mortality: A New Approach by WHO and UNICEF," WHO/FRH/MSM/96.11; UNICEF/PLN/96.1. WHO, Geneva.
2. World Bank (1993). "World Development Report 1993: Investing in Health." World Bank, Oxford and New York.
3. World Health Organization (1992). "International Statistical Classification of Disease and Related Health Problems," 10th rev. WHO, Geneva.
4. Ellerbrock, T. V., Atrash, H. K., Hogue, C. J. R., and Smith, J. C. (1988). Pregnancy mortality surveillance: A new initiative. *Contemp. Obstet. Gynecol.,* **31**(6), 23–31.
5. Fortney, J. A. (1987). The importance of family planning in reducing maternal mortality. *Stud. Fam. Plann.* **18**(2), 109–113.

6. Fortney, J. A., and Smith, J. B., eds. (1996). "The Base of the Iceberg: Prevalence and Perceptions of Maternal Morbidity in Four Developing Countries." Family Health International, Research Triangle Park, NC.

7. World Health Organization (1990). "Measuring Reproductive Morbidity," Report of a Technical Working Group. WHO, Geneva.

8. United Nations Population Division (1992). "World Population, 1992." United Nations, New York.

9. UNICEF (1993). "The State of the World's Children 1993." UNICEF, Oxford University Press, New York.

10. World Health Organization (1994). Mother-Baby Package: Implementing Safe Motherhood in Countries," WHO/FHE/MSM/94.11. Maternal Health and Safe Motherhood Programme, Division of Family Health, WHO, Geneva.

11. Berg, C. J., Atrash, H. K., Koonin, L. M., and Tucker, M. Pregnancy-related mortality in the United States, 1987–1990 *Obstet. Gynecol.* **88**(2), 161–167.

12. Nortman, D. (1974). "Parental Age as a Factor in Pregnancy Outcomes," *Reports on Population/Family Planning.* Population Council, New York.

13. Chen, L. C., Melita, C., Gesche, S., Ahmed, A. I., Chowdhury, and Mosley, W. H. (1974). Maternal mortality in rural Bangladesh. *Stud. Fam. Plann.* **5**(11), 334–341.

14. U.S. Department of Health, Education and Welfare (DHEW) (1954). "Vital Statistics of the United States 1950. Vol. I. Analysis and Summary Tables with Supplemental Tables for Alaska, Hawaii, Puerto Rico, and Virgin Islands." U.S. Gov. Printing Office, Washington, DC.

15. Datta, K. K., Sharma, R. S., Razack, P. M. A., Ghosh, T. K., and Arora, R. R. (1980). Morbidity amongst women in Alwar, Rajasthan—a cohort study. *Health Popul.—Perspect. Issues* **3**, 282–292.

16. Greenwood, A. M. *et al.* (1987). A prospective survey of the outcome of pregnancy in rural area of the Gambia. *Bull. W.H.O.* **65**(5), 635–643.

17. Maine, D. (1991). "Safe Motherhood Programs: Options and Issues." Center for Population and Family Health, New York.

18. Walker, G. (1986). Maternal mortality and the postpartum interval. *Br. Med. J. Clin. Res.* **292**, 1524.

19. Rochat, R. W., Kramer, D., Senanayake, P., and Howell, C. (1980). Induced abortion and health programs in developing countries. *Lancet* **2**, 484.

20. Casas-Becerra, L. (1997). Women prosecuted and imprisoned for abortion in Chile. *Reprod. Health Matters* **9**, 29–36.

21. Campbell, O. M. R., and Graham, W. J. (1990). "Measuring Maternal Mortality and Morbidity: Levels and Trends." Maternal and Child Health Epidemiology Unit, London School of Hygiene and Tropical Medicine, London.

22. Graham, W., Brass, W., and Snow, R. Estimating maternal mortality in developing countries. *Lancet* 416.

23. Graham, W., Brass, W., and Snow, R. Estimating maternal mortality: The sisterhood method. *Stud. Fam. Plann.* **20**(3), 125–135.

24. Hanley, J. A., Hagen, C. A., and Shiferaw, T. (1996). Confidence intervals and sample-size calculations for the sisterhood method of estimating maternal mortality. *Stud. Fam. Plann.* **27**(4), 220–227.

25. Danel, I., Graham, W., Stupp, P., and Castillo, P. (1996). Applying the sisterhood method for estimating maternal mortality to a health facility-based sample: A comparison with results from a household-based sample. *Int. J. Epidemiol.* **25**(5), 1017–1022.

26. Bertrand, J., and Tsui, A. (1995). "Indicators for Reproductive Health Program Evaluation: Introduction." Carolina Population Center, Chapel Hill, NC.

27. Maine, D., Wardlaw, T. W., Ward, V. M., McCarthy, J., Birnbaum, A., Akalin, M. Z., and Brown, J. E. (1997). "Guidelines for Monitoring the Availability and Use of Obstetric Services." UNICEF/WHO/UNFPA, New York.

28. Haque, Y. A., and Mostafa, G. (1993). "A Review of Emergency Obstetric Care Functions of Selected Facilities in Bangladesh." UNICEF, Dhaka.

29. Wardlaw, T. (1997). "Preventing Maternal Deaths: Using Process Indicators for Obstetric Services," *Previews,* December. 1997. UNICEF, New York.

30. McGinn, T. (1997). Monitoring and evaluation of PMM efforts; what have we learned? *Int. J. Gynecol. Obstet.* **59**(Suppl. 2), S245–S251.

31. Prevention of Maternal Mortality Network (1992). Barriers to treatment of obstetric emergencies in rural communities of West Africa. *Stud. Fam. Plann.* **23**(5), 279–291.

32. Howson, C. P., Harrison, P. F., Hotra, D., and Law, M., eds. (1996). "In Her Lifetime: Female Morbidity and Mortality in Sub-Saharan Africa." National Academy Press, Washington, DC.

33. Zurayk, H., Khattab, H., Younis, N., Kamal, O., and El-Helw, M. (1995). Comparing women's reports with medical diagnoses of reproductive morbidity conditions in rural Egypt. *Stud. Fam. Plann.* **26**(1), 14–21.

34. Bhatia, J. C., and Cleland, J. (1995). Self-reported symptoms of gynecological morbidity and their treatment in south India. *Stud. Fam. Plann.* **26**(4), 203–216.

35. Stewart, M. K., and Festin, M. (1995). Validation study of women's reporting and recall of major obstetric complications treated at the Philippine General Hospital. *Int. J. Gynecol. Obstet.* **48**, S53–S66.

36. Stewart, M. K., Stanton, C. K., Festin, M. and Jacobson, N. (1996). Issues in measuring morbidity: Lessons from the Philippines Safe Motherhood Survey Project. *Stud. Fam. Plann.* **27**(1), 29–35.

37. Koblinsky, M., and Stewart, K. (1997). Statement from a Task Force Meeting on Validation of Women's Reporting of Obstetric Complications in National Surveys." Letter, May 1, 1997.

38. Hogberg, U. (1985). "Maternal Mortality in Sweden," Umea Univ. Med. Diss., No. 156 (ISSN 0346-6612). Umea University, Umea, Sweden.

39. Dobbie, B. M., and Willmott, (1982). An attempt to estimate the true rate of maternal mortality, Sixteenth to Eighteenth Centuries. *Med. Hist.* **26**, 79–90.

40. Loudon, I. (1992). "Death in Childbirth: An International Study of Maternal Care and Maternal Mortality 1800–1950." Clarendon Press, Oxford.

41. McKeown, T. (1976). "The Role of Medicine: Dream, Mirage or Nemesis." Nuffield Provincial Hospitals Trust, London.

42. McKinlay, J. B., and McKinlay, S. M. (1997). Medical measures and the decline of mortality. In "The Sociology of Health and Illness" (P. Conrad, ed.), 5th ed., pp. 10–23. St. Martin's Press, New York.

43. Loudon, I. (1991). On maternal and infant mortality 1900–1960. *Soc. Hist. Med.* **4**(1), 29–73.

44. Akalin, M. Z., Maine, D., de Francisco, A., and Vaughan, R. (1997). Why perinatal mortality cannot be a proxy for maternal mortality. *Stud. Fam. Plann.* **28**(4), 330–335.

45. Loudon, I. (1988). Maternal mortality: 1880–1950. Some regional and international comparisons. *Soc. Hist. Med.* **7**(2), 183–228.

46. Murphy, M. (1981). Social consequences of vesico-vaginal fistula in northern Nigeria. *J. Biosoc. Sci.* **13**, 139–150.

47. Lawson, J. (1989). Tropical obstetrics and gynaecology.3. Vesico-vaginal fistula—a tropical disease. *Trans. R. Soc. Trop. Med. Hyg.* **83**, 454–456.

48. Ojanuga, D. (1993). The medical ethics of the "Father of Gynaecology,' Dr J Marion Sims. *J. Med. Ethics* **19**, 28–31.

49. Tahzib, F. (1989). An initiative on vesicovaginal fistula. *Lancet,* June 10, pp. 1316–1317.

50. United Nations (1995). "Population and Development: Programme of Action Adopted at the International Conference on Population and Development, Cairo, 5–13 September 1994. Vol. 1, ST/ESA/SER.A/149. United Nations, Department for Economic and Social Information and Policy Analysis, Geneva.

51. Rosenfield, A., and Maine, D. (1985). Maternal mortality—A neglected tragedy: Where is the M In MCH? *Lancet,* July 13, pp. 83–85.

52. Cates, W. (1982). Legal abortion: The public health record. *Science* 1586–1590.

53. World Health Organization and United Nations Children's Fund (1978). "Alma-Ata 1978 Primary Health Care: Report of the International Conference on Primary Health Care, Alma-Ata, USSR, 6–12 September 1978." WHO, Geneva.

54. Maine, D. ed. (1997). Prevention of maternal mortality network. *Int. J. Gynecol. Obstet.* **59**(Suppl. 2).

32

Female Circumcision/Female Genital Mutilation

SUSAN IZETT AND NAHID TOUBIA

Research, Action and Information Network for the Bodily Integrity of Women (RAINBO)
New York, New York

I. Introduction

Female circumcision (FC), also known as female genital mutilation (FGM)[1], refers to several traditional practices that involve the cutting of female genitals. While FC/FGM is usually performed on girls anywhere between four and twelve years of age, in some cultures it is practiced as early as a few days after birth or later in a woman's life, such as just prior to marriage or after her first pregnancy [1]. The practice is considered part of a rite of passage to womanhood and serves to define a girl or woman within the social norms of her tribe or ethnic group.

Worldwide, an estimated 130 million women and girls have undergone FC/FGM, and at least two million girls a year are at risk [2]. While FC/FGM is primarily practiced in Africa across portions of the sub-Saharan region and the Nile valley, it has also been documented among a few ethnic groups in south Asia and has affected women who now live as refugees and immigrants in parts of Australia, Europe, and North America. The practice is present among several religious groups: Christians, Muslim, Ethiopian Jews (the Falasha), and indigenous African religions, although it is not the majority practice of any one religion.

Initially, attention to FC/FGM was focused mainly on the physical and psychological damage that it causes (as discussed in Section V). These consequences can be serious and debilitating, and the public health impact has been widely acknowledged by the international community, including the World Health Organization. However, arguments against the practice based solely on the health effects are problematic for two reasons. First, it could be inferred that in the majority of cases where no complications occur, the ritual is acceptable. Second, these arguments leave room for the false notion that the situation would improve if FC/FGM were performed by trained physicians or nurses under hygienic conditions.

Increasingly, debate about FC/FGM has revolved around discussion of human rights. The act itself—the cutting of healthy functional genital organs for nonmedical reasons—is at its essence a basic violation of girls' and women's rights to bodily integrity. This is true regardless of the degree of cutting or of the extent of the complications that may, or may not, ensue. In addition, the practice is performed mainly on children who are below the age of consent and often have no say in the matter. While parents that have their daughters circumcised do not intend to hurt them, the practice goes against basic human rights principles to protect children from harm.

As a result of these debates, FC/FGM has been widely recognized as a human rights violation in several United Nations conference documents [3–5], and organizations working to stop the practice are beginning to address it within the overall context of girls' and women's rights to their bodily integrity, rather than strictly as a health or medical issue.

II. Definition and Classification

Throughout history, forms of genital cutting procedures and surgeries have been employed among various societies for a number of medical, cosmetic, psychological, or social reasons. According to the World Health Organization (WHO) those procedures defined as female genital mutilation "comprise all procedures involving partial or total removal of the external female genitalia or other injury to the female genital organs for cultural or other non-therapeutic reasons" [6,7]. These procedures can vary greatly depending upon region, country, and ethnic group, and there has been much confusion and many inaccuracies in reporting on types of FC/FGM in past research and documentation of the practice. To address this issue, the WHO also recommended the following four type classification system to clarify and standardize typology [6,7]:

Type I: Excision of the prepuce, with or without excision of part or all of the clitoris (Fig. 32.1).

Type II: Excision of the prepuce and clitoris with partial or total excision of the labia minora (Fig. 32.2).

Type III: Excision of part or all of the external genitalia and stitching/narrowing of the vaginal opening (infibulation) (Fig. 32.3).

Type IV: Unclassified: includes pricking, piercing, or incising of the clitoris and/or labia; stretching of the clitoris and/or labia; cauterization by burning of the clitoris and surrounding tissues; scraping of tissue surrounding the vaginal orifice (angurya cuts) or cutting of the vagina (gishiri cuts); introduction of corrosive substances or herbs into the vagina to cause bleeding or for the purpose of tightening or narrowing it; and any other procedure that falls under the definition of female genital mutilation.

No single classification system is perfect, particularly when dealing with a range of cutting procedures performed by lay individuals. However, this definition and classification have been accepted and adopted by a wide range of international and national organizations and activists, health professionals, and researchers are encouraged to use it in future work.

---

[1]In this publication, the terms female cirumcision and female genital mutilation will be used jointly. The former is used to emphasize the cultural nature of the practice, which does not intend a deliberate violence, and the latter is used to emphasize the serious outcome of the act. For a further discussion of terminology, please see Section II.A.

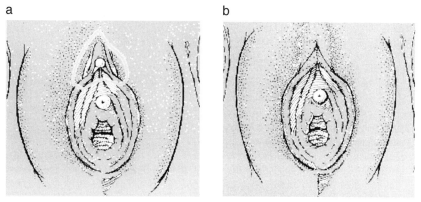

Fig. 32.1 Type I area cut (A); Type I healed (B). Illustrations by Stephen Gilbert for RAINBO.

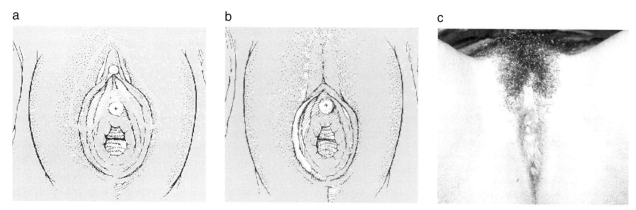

Fig. 32.2 (A) Type II area cut; (B) Type II healed; (C) Type II healed (from the National Committee Against Exision, Burkina Faso). Illustrations by Stephen Gilbert for RAINBO.

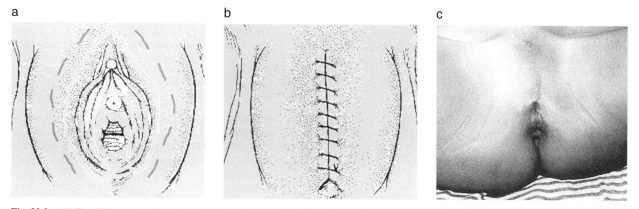

Fig. 32.3 (A) Type III area cut; (B) Type III area stitched; (C) Type III healed (photo: Nahid Toubia). Illustrations by Stephen Gilbert for RAINBO.

A. Terminology

In the literature from Africa, the term "circumcision" is predominant among the English-speaking countries, while in Francophone countries, FC/FGM is most commonly referred to as "excision." In addition, in some Arab-speaking countries, the term *sunna* is sometimes used to refer to clitoridectomy, while *pharaonic* usually denotes infibulation. This colloquial usage is imprecise and may not always correlate to the exact typology as outlined by the WHO. Among the general population, a great variety of terms in local dialects are used to describe the practice. These are often synonymous with purification or cleansing, such as the terms *tahara* in Egypt, *tahur* in Sudan, and *bolokoli* in Mali. (For a list of local terms used to describe FC/FGM, please see [8].)

Among the international community, the term "female genital mutilation" has been adopted by a wide range of women's health and human rights activists because it clearly indicates the damage caused by the practice. The term FGM was officially adopted by the Inter-African Committee on Harmful Traditional Practices[2] in 1990 [9] and has since been used in several United Nations conference documents [3–5].

Although use of the term FGM has been a very effective policy and advocacy tool, it can be offensive to populations that practice it, because it is not done with the intent to mutilate or harm girls. Out of respect for people's beliefs regarding the practice, those working in communities where FC/FGM is practiced have found that "female circumcision," "excision," or other local terminology from Africa is preferable for discussion of the subject. More recently, some agencies, including the United States Agency for International Development (USAID), have opted to use what they perceive as a more neutral phrase, "female genital cutting" (FGC).

B. Description of Procedures

In Type I—which is commonly referred to as clitoridectomy—the clitoris is held between the thumb and index finger, pulled out, and amputated with one stroke of a sharp object. Bleeding is usually stopped by packing and applying pressure bandage.

In Type II—also known as excision—the degree of cutting can vary considerably. Commonly the clitoris is amputated as described above and the labia minora are partially or totally shaved, often with the same stroke. Bleeding is stopped with packing and bandages or by a few circular stitches that may or may not cover the urethra and part of the vaginal opening. There are reported cases of extensive excisions that heal with fusion of the raw surfaces that result in pseudo-infibulation even though no stitching was applied [10–12]. Types I and II constitute 80–85% of all FC/FGM worldwide.

In Type III (infibulation), the clitoris is amputated, the labia minora are shaved off, and incisions are made in the labia majora to create raw surfaces. These edges of the labia majora are brought together and made to fuse using thorns, poultices, or stitching, and the girl's legs are tied together for two to six weeks [13]. The healed scar creates a "hood of skin" that covers the urethra and part or most of the vagina and acts as a physical barrier to intercourse [14]. A small opening is left at the back to allow for the flow of urine and menstrual blood; the opening is surrounded by skin and scar tissue and is usually two to three centimeters in diameter but may be as small as the head of a matchstick [15,16].

If, after infibulation, the posterior opening is large enough, sexual intercourse can take place after gradual dilation, which may take weeks, months or, in some recorded cases, as long as two years [17]. If the opening is too small to start the dilation, re-cutting (or *defibulation*) before intercourse is traditionally done by the husband or one of his female relatives using a sharp knife or a piece of glass. In almost all cases of infibulation [14,16,18] and many cases of severe Type II [12], defibulation

must also be performed during childbirth to allow exit of the fetal head without tearing the surrounding scar tissue. Traditionally, *re-infibulation* is performed after childbirth. The raw edges are re-stitched to create a small posterior opening, often the same size as that which existed before marriage. The severe Type III accounts for 15–20% of all FC/FGM and is most prevalent in Sudan, Somalia, and Djibouti and among their neighboring tribes in Kenya and Ethiopia.

C. Practitioners and Setting

For all types of FC/FGM, the procedures are generally performed by lay people, using a razor, knife, or sharpened stone or glass. These "traditional" practitioners—circumcisers, traditional birth attendants, barbers—are usually elder women or men whose vocation has been passed down within their families over generations. More recently, in some countries FC/FGM is increasingly being performed by doctors or nurses [19]. This trend is cause for concern among activists working to stop the practice. In Egypt, for example, it is an indication that some families see FC/FGM as a legitimate medical practice [20]. In addition, parents believe that going to a physician will reduce risk of complications, although cases have been reported of death due to uncontrolled bleeding after procedures performed by a physician [21].

FC/FGM is sometimes performed at the girl's home, or in a clinical setting, in cases where a doctor is the practitioner, but very often girls are circumcised with a group of peers in a special setting accompanied by a ceremony. In Sierra Leone, for example, FC/FGM is part of an initiation into a secret society, and girls are secluded for several weeks prior to the cutting [22]. In Kenya, circumcisions are celebrated by feasts with singing, dancing, and gift giving [23]. The significance of these ceremonies makes it difficult for families to consider giving up the practice. A full understanding of the social and cultural context is needed to understand how to effectively introduce change.

III. Social and Cultural Context

Because FC/FGM is practiced across a wide range of countries and ethnic groups, it is difficult to generalize about the social and cultural context. However, in reviewing studies on attitudes toward FC/FGM, the most frequently cited reason for support of the practice is "tradition." The practice is accepted among countless families because it has been performed by their ancestors for hundreds or thousands of years. In looking at the underlying norms and beliefs that perpetuate adherence to this tradition, some commonalities emerge.

A. Constructs of Femininity

Numerous findings on beliefs about FC/FGM suggest that they are linked to cultural norms: specifically, ideas about gender, sexuality, and power. One of the most common themes in stated beliefs about the practice is the overwhelmingly negative view of the female genitals. The clitoris is seen as dirty and as a male organ. Women must have it trimmed and "beautified" in order to be truly female [24]. These ideas of beauty and femininity are not so far removed from Western culture, where women consider

---

[2]The Inter-African Committee on Harmful Traditional Practices Affecting the Health of Women and Children (IAC) was established in 1984 to educate national governments as well as the general public about the harmful effects of FC/FGM. Headquarters are based in Geneva, and over the past 15 years IAC affiliates have been founded in over 22 African countries.

plastic surgery in an effort to conform to feminine ideals. Other views serve to demonize women's sexuality. Some ethnic groups in Mali believe the clitoris can prick a man and harm him during intercourse, and in Nigeria it is thought that the clitoris can kill a child if it touches the head of a baby during delivery [25].

These views also serve to justify the need to "control" women's sexuality. FC/FGM is meant to reduce sexual desire and curtail promiscuity, thus ensuring the fidelity of the wife or the virginity of the young woman and protecting the honor of a woman and her husband [26]. These notions ultimately perpetuate gender roles and stereotypes, where a woman is viewed as a vessel for reproduction while her role as a sexual being is denied. Overall, these views reflect patriarchal underpinnings that reinforce the secondary status of women. A recent study in Cairo noted a direct link between patriachal views and favorable attitudes about FC/FGM [27].

B. Kinship Communities

However, perpetuation of the practice goes beyond individual beliefs and attitudes. FC/FGM is strongly related to community dynamics, characterized by a hierarchy in which elders, traditional leaders, and other authority figures have a strong say over individual decisions. On a societal level, FC/FGM is meant to promote social cohesion [22]. It is considered important for the marriageability of young women and among some ethnic groups has been used to gain higher social status [28]. Consequently, individual rights are often subsumed by community priorities, and several people are usually involved in the decision to circumcise a daughter. For example, the extended family, particularly the husband's relatives, may hold more power than a mother over the fate of her daughter. This was true in the case of Fauziya Kassindja, a young woman from Togo who sought and was granted asylum in the United States. Her parents had opted not to circumcise their daughter. However, upon her father's death, decision making reverted to his brother and other members of the paternal family. They decided that Fauziya was to be circumcised and forced into an arranged marriage. It was only through the intervention of her mother and sister—who secretly helped her to flee the country—that she avoided circumcision [29].

These kinship communities also reflect a social hierarchy among women in which elders hold great sway over the lives of the younger women. FC/FGM is a way for younger women to gain peer respect and recognition and enhance their status and authority in their household and the larger community. This power can extend to their role with their spouse. For example, in some communities if a husband has difficulty penetrating his wife due to her "narrowed" opening, she can hold it over him as a sign of weakness or failure and thereby increase her leverage in their relationship [30]. However, this source of power is still based on a system in which women have to sacrifice their sexuality in order to advance their status.

C. Religion

Religious beliefs also figure largely in people's views of the practice. While FC/FGM is not an explicit requirement of any religion—it predates the arrival of Christianity and Islam in Africa—it is nevertheless strongly identified with Islam in several African countries. The Qur'an, the primary source for Islamic law, and the "hadith," collections of the sayings of the Prophet Mohammed, do not include a direct call for FC/FGM. However, debate over interpretations of statements from these sources continues, and many members of the Muslim community are advocates of the practice. Consequently, many practicing Muslims are convinced that FC/FGM is a necessary part of their religious life.

There are signs that these beliefs may change over time. In 1998, the International Conference on Population and Reproductive Health in the Muslim World was held at Al Azhar University, which is the oldest Islamic academic institution and acts as an authority in Quaranic interpretation and policy making. At the conference, a consensus was reached that certain harmful practices, including FC/FGM, were the result of misunderstandings of the provisions of Islam and were not scantioned by the religion [31].

IV. Distribution

A. Prevalence

1. Africa

Until the late 1970s, information on the prevalence of FC/FGM in Africa was primarily based on small studies undertaken by physicians. In 1979 Fran Hosken published the first report to present country-by-country estimates of prevalence [32]. Although the methodology by which the data were collected was not specified, these figures became a basis for global estimates. The first national survey on FC/FGM was conducted in Sudan by faculty at the University of Khartoum in 1979 [33]. The Sudan Fertility Survey of 1979 [34] and the Demographic and Health Survey of Sudan, 1990 [35] also included questions on FC/FGM. To date, Sudan is still the only country with reliable data spanning a ten-year interval. However, in 1993, the Demographic and Health Surveys (DHS) program[3] developed a module of questions related to FC/FGM. This module has been adapted by several African countries in their DHS, in many cases establishing the first reliable national data.

FC/FGM has been documented in 28 African countries in the sub-Saharan and north-eastern regions (Table 32.1 and Fig. 32.4). While most prevalence figures come from limited studies or anecdotal information, seven countries—Central African Republic, Egypt, Eritrea, Ivory Coast, Mali, Sudan, and Tanzania—have national data based on their DHS [36]. While 18 countries have a prevalence of 50% or higher, prevalence varies widely from country to country. There are also great disparities in prevalence within countries. In Mali, for example, prevalence in the remote northern regions of Timbuctu and Gao is less than 10%, while in the capital of Bamako and the city of Koulikoro, the rates are 95 and 99%, respectively [37]. As more countries adapt

---

[3]The Demographic and Health Surveys (DHS) Program, which is administered by Macro International, assists host country organizations in conducting nationwide surveys on population and health. The standard survey includes a women's questionnaire, administered to a nationally representative sample of women aged 15–49, which includes questions on a broad range of topics related to maternal and child health. Several supplemental modules have been introduced on issues such as maternal mortality, AIDS, women's status, and violence against women.

Table 32.1
Prevalence and Types of FC/FGM by Country

Prevalence	Country	Type(s) most commonly practiced
50%	Benin	Type II
70%	Burkina Faso	Type II
20%	Cameroon	Types I and II
43%	Central African Republic	Types I and II
60%	Chad	Type II (Type III only in eastern parts of the country bordering Sudan)
5%	Democratic Republic of Congo (formerly Zaire)	Type II
43%	Cote d'Ivoire (Ivory Coast)	Type II
98%	Djibouti	Types II and III
97%	Egypt	Types II (72%), I (17%), and III (9%)
95%	Eritrea	Types I (64%), III (34%), and II (4%)
90%	Ethiopia	Types I and II (Type III is practiced only in regions bordering Sudan and Somalia)
80%	Gambia	Type II (Type I practiced only in some parts)
30%	Ghana	Type II
50%	Guinea	Type II
50%	Guinea Bissau	Types I and II
50%	Kenya	Types I and II (Type III practiced in eastern regions bordering Somalia)
60%	Liberia	Type II
94%	Mali	Types I (52%) and II (47%) [Type III practiced in the southern part of the country (1%)]
25%	Mauritania	Types I and II
20%	Niger	Type II
60%	Nigeria	Types I and II (Type II is predominant in the South and Type III practiced only in the North)
20%	Senegal	Type II
90%	Sierra Leone	Type II
98%	Somalia	Type III
89%	Sudan-North	Types III (82%), I (15%), and II (3%)
18%	Tanzania	Types II and III
50%	Togo	Type II
5%	Uganda	Types I and II

Sources:

Carr, D. (1997). "Female Genital Cutting: Findings from Demographic and Health Surveys." Macro International Inc.

Office of Asylum Affairs. Bureau of Democracy, Human Rights & Labor. United States Department of State. Washington, DC January 1997. "Fact Sheet on Female Genital Mutilation in 16 African Countries: Benin, Burkina Faso, Cote d'Ivoire, Chad, Djibouti, Egypt, Ethiopia, The Gambia, Ghana, Liberia, Mali, Nigeria, Sierra Leone, Somalia, Sudan, Togo."

Toubia, N. (1995). "Female Genital Mutilation: A Call for Global Action." Women, Ink., New York.

the FC/FGM module in their DHS, comprehensive national data will become available. Plans are now under way to include questions on FC/FGM in the DHS in Ethiopia, Ghana, and Niger.

2. Middle East and South Asia

While there have been anecdotal reports of the practice of FC/FGM in countries of the Middle East including Bahrain, Saudi Arabia, and United Arab Emirates, the practice has only been documented in Yemen and among Yemini related tribes in southern Saudi Arabia, (S. Thadeus, unpublished data, 1992).

Reports from India indicate that FC/FGM is practiced by the small ethno-religious minority, the Daudi Bohra of the Ismaili Shia sect of Islam [38,39]. The total population is around half a

Fig. 32.4   Countries in Africa with communities practicing FC/FGM.

million, who live around the area of Bombay as well as in small immigrant communities in Africa and North America.

In Indonesia and Malaysia, genital cutting operations existed in the past but are no longer performed [40]. However, various non-cutting rituals—cleaning with herbal juice, symbolic cutting, and light puncture of the clitoris (Type IV)—still persist [7].

3. Immigrant Populations

The presence of increasing numbers of refugees and immigrants from countries where FC/FGM is practiced in Australia, Europe, and North America has brought much attention to this issue in the host countries. Unfortunately, there is scant hard data on the numbers and characteristics of African communities in these countries. Available statistics give only an estimate of immigrants from African countries where FC/FGM is practiced. It is difficult to derive prevalence figures from these data because (1) figures often do not include a breakdown of male or female immigrants and (2) given the wide disparity among regional and ethnic prevalence of FC/FGM within Africa, not all

women or girls coming from a country where it is practiced would necessarily have undergone the procedure.

a. AUSTRALIA.   Recent government estimates from the Department of Human Services indicate there are upwards of 87,000 women from African countries where FC/FGM is practiced living in Australia (Department of Human Services, unpublished data, 1996).

b. EUROPE.   The International Center for Reproductive Health of the University of Gent, Belgium recently completed a study on FC/FGM that included an assessment of the number of migrants from "FGM-risk" countries in European member states[4] [41]. These statistics are not completely representative. Only

[4]This study is part of the Daphne Project on Female Genital Mutilation in Europe and was carried out in 1998 under the Daphne Programme of the European Commission in collaboration with the Royal Tropical Institute of Amsterdam and Defence for Children International, the Netherlands.

official immigrants and registered refugees were counted, not the considerable number of illegal immigrants. Findings revealed that the number of African migrants is the highest in the United Kingdom (303,545), France (180,997), Italy (133,847) and Germany (77,795). Other countries with significant numbers of African immigrants include: the Netherlands (56,634), Sweden (31,798), Belgium (14,797), Portugal (12,785), and Denmark (11,105). Austria, Greece, and Spain have smaller numbers of immigrants, ranging from 3,000 to 8,000 in total.

c. NORTH AMERICA. The African Resource Centre in Ottawa, Canada reports 12,000 African immigrants in the city but gives no indication whether these immigrants come from countries where FC/FGM is practiced (African Resource Centre, unpublished data, 1993). No other data from Canada was available.

The United States receives immigrants and refugees from all African countries. The Centers for Disease Control (CDC) in Atlanta used the 1990 Census data to estimate that approximately 168,000 women and girls reported ancestry or birth in FC/FGM-practicing countries (CDC, unpublished data, 1997). One-fourth of them were under 18 years of age.

B. Socio-demographic Factors

In countries where the practice is highly prevalent, there are few variations in prevalence based on socio-demographic factors, as evidenced in the recent comparative report on the DHS findings [36]. The following general trends were reported across seven countries: prevalence varies little between urban and rural residences; there is no major decrease in prevalence in younger generations; little variation in prevalence was found based on education; however, support for the practice is lower among women with secondary education than among women with no education or only primary education. The only significant difference was regarding religion: Muslims were more likely than Christians to undergo the procedure.

However, findings from a few studies suggest that urban and rural residence may sometimes be associated with circumcision status. In Sudan, an assessment of interventions revealed that some ethnic sub-cultures in urban areas (e.g., Southern Sudanese, Nuba, and groups of West African origin) had started a new drive toward FC/FGM [42]. While in Burkina Faso, a trend was seen toward fewer circumcisions among urban dwellers [43]. A 1999 study from one village in Egypt found that one factor associated with families whose daughters were not circumcised was the involvement of one or more members of the household in development activities [44].

V. Health Consequences

All types of FC/FGM involve damage to the normal functioning of the external female genitalia, whether mild or severe. While the full range of possible physical complications has been well documented, primarily through clinical studies, far less is known about the psychological after-effects of the practice or about sexual problems associated with FC/FGM. (For a comprehensive review of the published literature on health consequences, please see [7].) However, prevalence of specific complications has been difficult to establish. In a recent review

of research on the topic, little evidence was found to substantiate claims often made about the frequency of cases of death or ill health due to FC/FGM [45]. But the review also noted that this finding was most probably due to the limitations of past research to investigate effects.

A. Physical Complications

The occurrence of physical complications depends on many factors, including the extent of cutting, the instruments used, the skill of the practitioner, and the physical condition of the girl. While serious consequences are possible with all types, long-term problems resulting from Type III (infibulation) are more severe and more frequent and are therefore discussed separately.

1. Immediate Complications, All Types

The majority of procedures are performed without an anesthetic, resulting in severe pain and often a state of shock. Even when local anesthesia is used, pain in the highly sensitive area of the clitoris returns within two to three hours of the operation [7]. Applying the local anesthesia itself is also extremely painful because the area of the clitoris and labia minora has a dense concentration of nerves and is highly sensitive.

Hemorrhage (severe bleeding) is the most common immediate complication, and evidence of its high incidence is abundant in the literature [16,17,46–48]. To stop bleeding of the clitoral artery, it must be packed tightly or tied with a running stitch, either of which may slip and lead to hemorrhage [49]. Secondary hemorrhage may occur after the first week as a result of sloughing of the clot over the artery due to infection. An acute episode of hemorrhage or protracted bleeding may lead to anemia [50].

Infection is very common and usually occurs within a few days as the cut area becomes soaked in urine or contaminated by feces [17]. The degree of infection varies widely from a superficial wound infection to a generalized blood infection.

Death may result from the pain and trauma (neurogenic shock), hemorrhaging, or from severe and overwhelming infection (septicemia). Although no large-scale studies have been undertaken on child mortality, cases of death have been reported in the literature [16,17,51].

Injury to neighboring organs, including the urethra, vagina, perineum, and rectum, can occur, particularly if the child moves or the circumciser is inexperienced [16]. These injuries may lead to the formation of fistulae through which urine or feces will leak continuously [52]. Urine retention can also develop due to pain, inflammation around the wound, or subsequent infection.

2. Long-Term Complications of Types I and II

Infection, separation by the urine flow, and movement during walking may prevent the wound edges from healing. A weeping wound oozing pus or a chronic infected ulcer may result. In cases where the infection is buried under the wound edges or an imbedded stitch fails to absorb, an abscess will form that will usually require surgical incision and repeat dressing over a period of time [7].

Dermoid cysts are among the most common long-term complications and result from embedded skin tissue in the scar. The gland that normally lubricates the skin will continue to se-crete under the scar and form a cyst or sac ranging from the size of a pea to that of a grapefruit, causing great distress for the woman [53].

Keloidal scars, or overgrowth of scar tissue, in the vulva (Fig. 32.5) are disfiguring and psychologically distressful. Treatment is often unsuccessful since surgical removal often provokes more keloidal growth.

Trapping of the clitoral nerve in a stitch or in the scar tissue of the healed wound can result in a neuroma (tumour consisting of neural tissue). This can make touching the vulva during in-tercourse or during washing very painful. Painful intercourse can also occur as a result of any of the above mentioned complications.

3. Long-Term Complications of Type III

In addition to the complications noted above, an added set of problems may occur in Type III as a result of the obstruction caused by the scarring covering the urethra and vagina and the extra damage caused by defibulation and re-infibulation.

Ascending infections from the vulva or due to retained dis-charge and blood can lead to pelvic inflammatory disease (PID) [54]. PID is not only painful but it also can lead to infertility as a result of scarring of the fallopian tubes.

A tight infibulation or urethral stricture resulting from acci-dental injury can cause obstruction of urethral flow, repeat infec-tions, and bladder stones. Dribbling of urine is common in infibulated women when the bladder is not completely emptied, and chronic infection under the skin hood makes control of urine difficult [17,55].

With infibulation, the artificial opening of skin covering the vagina may be so small as to close almost completely over time. This may cause incomplete voiding of urine, hematocolpus (re-tained menstrual blood), and impossibility of intercourse [17]. Products of miscarriage can also be retained in the vaginal canal leading to severe infection.

During childbirth, the infibulated woman must be defibulated to allow the fetal head to crown. If an experienced attendant is not available to perform defibulation, labor may become ob-structed (Fig. 32.6) [56]. Prolonged obstructed labor can cause moderate to severe complications for the mother and the child. Cases of ruptured vulval scar, perineal tears, fetal distress, and vesico-vaginal and vesico-rectal fistulae have been reported [52]. There have also been reports of severe lacerations involving the anal musculature and injuries to the urinary tract [57]. One case control study of mostly infibulated Sudanese women living in Saudi Arabia and delivering in a well-equipped hospital found significant delay in the second stage of labor, increased hemor-rhage, and increased occurrence of severe fetal asphyxia [56].

B. Psychological and Sexual Effects

1. Effect on the Psychological Health of Girls and Women

While women's memories of their circumcision often com-bine their excitement about the ceremony with their terror and physical suffering from the procedure, cumulative evidence suggests that the event is mostly remembered as extremely trau-matic and leaves a life-long emotional scar [58–61]. Research

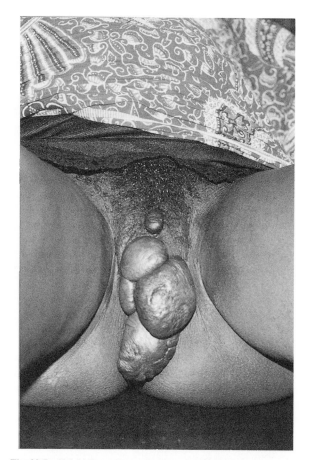

Fig. 32.5 Keloidal scar tissue. Courtesy of CNLPE, Burkina Faso.

findings also indicate a strong effect on self-identity. For exam-ple, a study in Somalia analyzed drawings by young girls about their experience at the moment of their circumcision and the period of convalescence afterwards. In analysis, the level of self-esteem and self-identity of the subjects reproduced ap-peared to be disturbed on the physical and psychological levels. The researchers also found that circumcision is not experienced as an event limited in time and then forgotten but as a factor that remains, even latently, in a girl's thoughts [62]. Observation of the lives of women in Sudan revealed a similar impact on the self perception of young girls who have undergone FC/FGM [63].

While not well-documented, there have also been a few re-ported cases of clinical psychological disease arising from FC/FGM. Diagnosis of a young girl's anxiety state was linked to her fears about circumcision. In adult women, cases of reactive depression and psychotic excitement were attributed to distress over complications due to FC/FGM [64].

2. Effect on Women's Sexuality

There are limited and mixed findings on the sexual effects of FC/FGM. Overall, research indicates that FC/FGM interferes to some degree with women's sexual response but that it does not necessarily abolish the possibility of sexual pleasure and cli-max for women. In a study of women from a clinic in northern Sudan, over 80% of those with Type III did not know of or

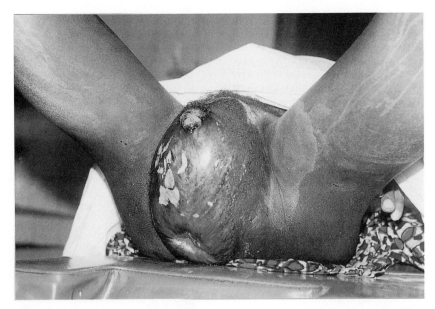

Fig. 32.6   Case of obstructed labor resulting in fetal death. Courtesy of Dr. Moustapha Toure, Mali.

experience orgasm, compared to 10% of those with Type I or uncircumcised [18]. A national survey in northern Sudan (where an estimated 90% of women are infibulated) found that that half of the women reported no sexual pleasure, a quarter were indifferent to sexual intercourse, and a quarter reported experiencing pleasure some or all of the time [47].

In research from Egypt, a study of circumcised clinic patients found that 30% did not experience any sexual satisfaction with intercourse, 30% experienced some satisfaction but did not reach orgasm, and 41% experienced satisfaction and orgasm frequently. The sample contained women with Types I, II, and III, but the difference in sexual experiences between women with different types was not assessed [65]. Another study of Egyptian women reported that 50% of uncircumcised women had sexual excitement in response to stimulation of the genitals compared with 25% of circumcised women (Types I and II) [66].

A study from Nigeria looking at the sexual experiences of young women found little difference between orgasm among those circumcised (Type II) and those uncircumcised (60% versus 70%). Findings also indicated that for some women, when the clitoris was removed, the labia minora and the breast became more sensitive and enhanced sexual stimulation [67].

In Sierra Leone, one study looked at women who had sexual experience before their circumcision and women who experienced sex only after circumcision (Types I and II) [68]. Those who experienced sex before had positive reactions to male sexual advances, and those who never had a sexual experience had a neutral response. No women in either group experienced intense arousal or orgasm, but those with previous experience were better able to detect a mild stimulation and arousal.

To date, no studies have attempted to differentiate between the physical and psychological causes of altered sexual response in circumcised women, and further research is needed to understand why some women with a particular type of FC/FGM may experience sexual pleasure while others do not.

VI. Health Care for Circumcised Women

An emphasis on making a health argument in favor of stopping the practice has resulted in a major gap in the knowledge of health care professionals (HCPs) on how to appropriately care for circumcised girls and women and avoid repeat suffering. The majority of published clinical and nonclinical studies enumerate the short- and long-term physical complications of FC/FGM with varying degrees of accuracy as a means of demonstrating the damaging effect of the practices. Little attention has been paid to documenting the experience of circumcised women physically, socially, and psychologically to assess their real needs in order that appropriate clinical protocols are designed to best suit their individual and social needs. In the context of immigrants living in Europe and North America, there is very little in research and documentation on how best to develop culturally competent care that will help the woman heal herself and empower her to protect her daughter from circumcision [69].

A small needs assessment study in the greater metropolitan New York area [70] among 49 HCPs and their African women clients found that the former were never educated on the practice or trained on how best to treat a women living with circumcision whether she came with a related or unrelated condition. The HCPs also expressed bewilderment and confusion about how to talk to women about their circumcision and how to offer help. The women reported that they went to the hospitals reluctantly, mostly for pre-natal care and delivery, and had very little expectation or trust that they would be understood or helped in ways that they needed. They also reported occasional incidents of humiliation when put on public display and treated as mutilated organs, not as human beings. This sense of loss, confusion, and hurt in a clinical context can be remedied with attention to better education and training. There is a need for investment in research and documentation of actual needs and of testing both standard and innovative approaches to resolving the knowledge, perception, and cultural gaps that exist between providers and their clients. It is important that members of the communities to be served are involved in all aspects of the research and developing of service models if they are to bring about better services for the clients. Many immigrant communities have nurses, social workers, physicians, and university and high school students who can be valuable resources for health care services. Creating partnerships with community-based organizations will build confidence and add social and cultural understanding and tangible services such as interpretation, patient escorts, and community outreach workers.

Physicians, physician assistants, nurses, nurse midwives, social workers, and students will need education on what FC/FGM is and how to provide the most appropriate physical treatment, counseling needs, education, and support to the woman and her family. In short, health care providers need to expand their cultural competence skills to include circumcised women, their families, and their communities. A technical manual for health care providers titled "Caring for Women with Circumcision" was developed by Toubia [71] with support from the Department of Health and Human Services of the Federal Government in the United States. It was made available in the spring of 1999 to the libraries of all medical, osteopathy, nursing, midwifery schools, and other relevant educational institutions. For more guidance on how to handle specific clinical situations, please refer to this manual.

In general, treating the presenting physical symptoms of a circumcised woman follows much of the common clinical sense used for managing all patients. The symptoms may or may not be related to her circumcision. The physician or nurse needs to know enough about the types of circumcision and their possible complications to feel comfortable making that distinction. It is also important that he or she treats the women with sensitivity, care, and reassurance that the HCP is not shocked or uncomfortable with her circumcision. Once the diagnosis is established, the patient must be fully informed of her condition, her options, and what may happen next. This should be attempted despite the possibility of linguistic barriers or cultural and family dynamics that may be unfamiliar to the provider, such as the presence of husband, mother-in-law, or others in the treatment room. When a decision is to be made on a course of treatment such as surgery for removal of a cyst or opening an infibulation scar through defibulation, the guiding principle is that the HCP is the woman's agent and should listen carefully and sensitively to her needs. If she wants to make an independent decision, she should be supported; if she wishes to consult with her husband or family members, the HCPs should not impose their point of view regardless of whether her decision is made out of love, respect, confusion, or fear. Regarding how much attention to pay to psychological or sexual effects that may be related to her circumcision, the balance to keep is to make sure that the woman knows that counseling and other forms of help are available if she needs them; to prepare accessible, caring, and competent help; and to give her the respect as a mature adult to accept or reject that help regardless of the HCP's own personal view. Patronizing women and imposing perceived problems that they have not identified, particularly those of a psychological and sexual nature, is counter productive. Failing to mention the availability of services that she may want or need, but does not

know about or is afraid to ask for, is negligence. It is a fine and difficult line to walk, but many clinical situations provide such challenges; caring for circumcised women from other cultures is one of them.

When any surgery to correct a complication (such as inclusion dermoid cyst or keloids) or to relieve a symptom (such as a painful neuroma) is contemplated, it is important to consider what further damage to the remaining sensitive tissue may result that may leave the patient even less satisfied with her physician and with her sexual life. For women with Type III FC/FGM or infibulation, defibulation is necessary for vaginal delivery. In many cases an unexpected intact clitoris has been found buried under the clitoral hood on defibulation and could easily have been damaged by a careless surgeon. Another surgical caution is that infibulated women should never be subjected to cesarean section solely because of their genital scar, because defibulation is a simple, reasonably safe procedure that can be done under local (or other form of anesthesia depending on patient and physician preferences) with few side effects. Defibulation is sometimes also performed to facilitate painless intercourse, to treat complications such as infections, ulcers, and stones, or simply because the patient requests it.

Figure 32.7 illustrates the six basic steps of defibulation technique under local anesthesia. A clinician desk "quick reference chart" in full color showing the three most common types and the technique of defibulation is also available [72].

Much has been said about the expectation of the women to be re-infibulated and the physician's responsibility to either respond to their request or deny it on medical principles. All HCPs would agree that a deliberate procedure by a physician that creates an obstruction to urinary and menstrual flow is unadvisable and probably unethical. The first course of action is counseling for the women. Experience in many countries and reports from health services in Denmark and Sweden serving Somali women have shown that explaining the dangers of obstruction with no added psychological or cultural counseling resulted in the majority of women, even those with minimal education, agreeing not to be re-infibulated. Some needed the information to be extended to their husbands by the physician; others were able to negotiate their own family discussion and reach the decision.

Development of pre-defibulation counseling protocols is an urgent need that health services must provide. This can be done by assigning counselors to listen to the women, understand their cultural and social background, identify and address their informational needs, alleviate their fears, and resolve their problems. Otherwise, many women may insist on refibulation at all costs and the system will be faced with the challenge of a decision. Post-defibulation counseling is also extremely important in coping with the outcome of defibulation. Many women experience severe pain or irritation in the defibulated area. This area had been devoid of sensation for years and may have become chronically inflamed. The brain is not used to sensations from this tissue, which may be intense and uncomfortable. This is especially the case when infibulated women have an intact clitoris under their hood. Post-defibulation care must include talking to the woman about the possibility of experiencing unusual and disturbing physical sensations as well as emotional feelings. The woman may feel guilt and shame for having reversed her parents' decision as well as many other issues. Counseling and

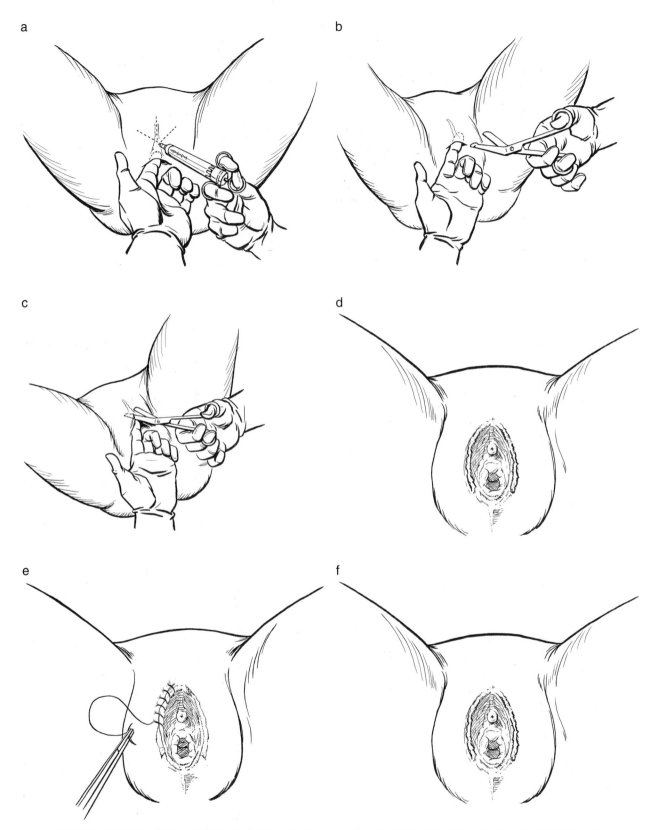

Fig. 32.7 Six steps of the defibulation technique. (a) Apply anaesthesia; (b) insert 1−2 fingers; (c) cut with bandage-like scissors avoiding injury to a buried clitoris; (d) inspect cut edges for bleeding points; (e) apply hemostatic running absorbable sutures; (f) defibulation complete. Illustrations by Jim Thorpe for RAINBO.

provision of simple palliative remedies such as warm salt and water bath and anesthetic or anti-inflammatory cream for a few weeks could help many women. Creating a pool of satisfied clients will also start a nucleus of peer educators who will influence the decision of other women. For the few who will insist on re-infibulation despite the best efforts of the HCP, a physician can abstain on moral principles, refer her to another provider, or leave her to fend for herself. All are decisions that must be left to the individual HCP. Unfortunately some countries have made re-infibulation illegal. While this takes away the dilemma of the HCP, it results in very little attention given to the women's pre- and post-defibulation counseling needs.

The medical–legal issues related to caring for families with children who are circumcised or who may be at risk of circumcision cannot be fully explored here. The special needs of adolescents who were circumcised as children and who will grow up struggling between two cultural identities and social norms are also not possible to cover in this brief overview. More information on these and many other topics can be found in [71].

VII. Methodology for Study

The majority of FC/FGM research to date has focused on the epidemiology and the physical complications of the practice. While this has provided much needed information at a basic level to alert national governments and the international community about the importance of the problem, a key consideration was overlooked. FC/FGM is not a disease. As discussed in Section III, it is a practice based on deeply rooted values, beliefs, and norms. At its essence it reflects human social behavior, and the methodology for study needs to focus on these behavioral aspects. Specifically, what are the processes a person goes through in making decisions about circumcision? What can help promote new behavioral choices? What are the barriers that impede change? In addition, it is necessary to look at processes of broader societal change, specifically, the elements that foster an environment conducive to promoting gender equity and women's empowerment. The following section provides an overview of the basic methods used in FC/FGM research, highlighting gaps in past studies and suggesting a new agenda for future investigations based on these concepts. For a comprehensive discussion of research issues, please see [73].

A. Surveys

Surveys done at a national or local level are necessary to establish prevalence and to collect data on age at circumcision, typology, and category of practitioner. Whenever possible, information should be gathered on daughters as well as on the women being interviewed to allow for comparisons between two generations. Careful attention must be paid to sample size and selection. Many past surveys did not utilize appropriate sampling techniques, and the resulting data were unreliable, in particular the estimates of prevalence. Including questions on FC/FGM as part of an existing national survey, such as the DHS, is the most effective way to gather information and also provides rich data on other aspects of maternal and child health for use in secondary analysis.

Data on prevalence is generated by asking women about their circumcision status, and recent research has confirmed that women's responses are generally reliable [74]. When asking about age of circumcision, approximate age ranges of five years can be utilized if exact age is not known. In the past there were problems collecting accurate information on typology, since no standard classification system was applied. For future studies, researchers are advised to use the four-type WHO classification (see Section II) when investigating the types of procedures in the community or region to be studied and to make approximate correlations with the standard classification. Regarding category of practitioner, attempts should be made to determine whether "traditional" or "medical" personnel performed the procedure.

Many previous national-level surveys included basic questions on attitudes and behavior, such as "Do you approve of the practice?" and "Do you intend to circumcise your daughter?" These provided crude indicators, but overall they were not meaningful measures. In societies where the practice is prevalent, the majority of interviewees answer "yes" to these questions. Likewise, asking why a person supports the practice has yielded vague, general responses that could not be used to guide programmatic decisions.

Using surveys to assess complications has also proved highly problematic because they rely on self-reporting by women. Most women do not have the medical knowledge to adequately assess a complication; they may have problems with recall; they also may not associate problems that develop later in their life with a circumcision that occurred when they were young.

In the future, surveys at a local level can serve an important function in the monitoring and evaluation of programs—a much needed part of the FC/FGM research agenda. Surveys can be used to establish baseline data and track trends in behavioral decisions regarding age, type, and practitioner. Is there a move toward "modern" practitioners? Is there shift in the type of FC/FGM being performed? Is there a trend toward younger age at circumcision? Answering such questions is crucial for the design of interventions. Additional data can be gathered to assess other processes of change within a community such as levels of attendance at education events.

Local surveys, informed by and supplemented with qualitative studies, can also be used to glean information on knowledge and beliefs to establish a baseline for comparison when implementing interventions. Two such studies in Egypt helped document the views of physicians and university teachers, respectively, regarding FC/FGM [27,75]. As organizations attempt to engage new sectors of the population in campaigns on FC/FGM, such information will reveal entry points for approaching these specific groups.

B. Clinical Studies

Clinical studies provide accurate and in-depth information on physical complications of FC/FGM. The majority of past studies have been in the form of case reports, using medical records and clinical examinations to document a range of complications from mild to severe. Case-control studies have also been undertaken to test the relationship between circumcision and a particular health problem. In countries where prevalence is high, there are often few women who are not circumcised. In this case, groups can be divided by circumcision type rather than status.

Clinical studies have also been used as a check on prevalence to help validate the results of larger studies [74]. This has become increasingly important as policy makers refuse to accept prevalence figures established in national surveys, as happened with the Egyptian DHS results. If physicians received adequate training on discerning the different types of FC/FGM (using the WHO classification), they could also confirm typology in a given area.

For future research, clinical studies could be employed to investigate the sexual and psychological aftereffects of the practice. The goal of such research would not be simply to document the problems but rather to initiate dialogues with women to learn more about how FC/FGM affects their lives as well as their needs for counseling or other support services. Such studies are possible, but they require a setting conducive to personal and private discussion as well as the involvement of trained counselors. In addition, past experience has shown that asking direct questions about sexual or psychological difficulties is less effective than indirect methods, such as asking a woman to relate her memories of important events related to her circumcision.

To date, most organizations addressing the practice have focused primarily on how to stop FC/FGM. There has been little emphasis on developing programs to assist those women who are already circumcised. New findings from such action-oriented clinical studies could redress this imbalance.

C. Qualitative Methods

In-depth interviews, focus group discussions, case studies, and other qualitative methods have all been used for assessing beliefs, attitudes, and behaviors regarding the practice of FC/FGM. However, the majority of studies have looked at whether or not a person supports the practice and the reasons why. Findings were often quantified and provided little meaningful information for program planners. To generate more useful data in the future, the focus needs to shift to investigating changes in the practice: where it is happening, why it is happening, and what the process entails. A key approach is to study those individuals, families, or communities who have abandoned the practice. This is a relatively new area of study, but there are a few examples to draw on.

In pioneering work, Dr. Amal Abd El-Hadi studied a village in Egypt, Dir El Barsha, where prevalence of FC/FGM had decreased markedly over one generation (from 95% among women surveyed to 50% among their daughters)[5] [44]. In addition to conducting household surveys, researchers conducted in-depth interviews with key people in the community, including women leaders and the first young woman to be married who was not circumcised. Questions centered on what influenced decisions to stop circumcision; information was also gathered about women's roles in the household and in the larger com-

munity. The findings revealed three main factors associated with the decrease in circumcision: (1) one or more persons from the household were involved in development activities; (2) the husband was working abroad and the wife was head of the household; and (3) the various Christian churches in the community had all spoken out against the practice.

Researchers in Kenya conducted case studies of women in one community who had stopped circumcising their daughters [76]. Similarities and differences across cases were assessed. There were two underlying commonalities: (1) the women held prominent positions in their community, and (2) they were members of a Christian church that took a stance against the practice.

A similar approach—known as the "Positive Deviance Inquiry"—is being employed by the Centre for Development and Population Activities in investigating factors in Egypt that enable some families not to circumcise their daughters. Community members are conducting the research, interviewing those who have renounced the practice, defined as "positive deviants," and using the information to develop interventions, focusing on what is already inherently working in the community (J. Masterson, CEDPA, Washington, DC, 1999, anecdotal information).

VIII. Future Directions

Within Africa, organizations have been raising awareness about the problem of FC/FGM for over 20 years, largely in the form of information and education programs. There have been very few evaluations of these activities. However, a recent review of interventions by Program for Appropriate Technology in Health indicates that the impact of such programs has been less than expected in terms of changing behaviors [77]. Findings revealed that the majority of the campaigns were too broad and general and centered on messages that FC/FGM was "bad" or "harmful." In the future, new educational strategies need to be developed—based on in-depth research in a community—to address specific beliefs, concerns, and misconceptions about FC/FGM and utilizing more positive messages that focus on families' love of their children and their desire to ensure healthy and productive lives for them. Such strategies also need to be tailored to specific audiences and communities, using research-based written and visual materials that are culturally appropriate to the particular setting.

Beyond education initiatives, new approaches that involve women's empowerment are called for in order to effectively foster change. In the example from Egypt cited earlier, a comprehensive program by the Coptic Evangelical Organization for Social Services (CEOSS) that involved literacy classes and income-generating activities helped to shape an environment in which women were able to exercise greater power over decisions. Likewise in Senegal, the organization Tostan's training on health and human rights and problem-solving gave rural women tools with which to negotiate new behavioral choices, including the decision to abandon FC/FGM [78]. Other strategies, such as creating alliances and building networks among likeminded families who choose alternatives to FC/FGM, can be explored.

Such approaches necessitate the involvement of the entire community, including the professional sector—those in the

---

[5]The majority of women surveyed were between the ages of 30 and 50. The circumcision status of their daughters was only recorded if the daughter was above 13 years of age, because in this village circumcision is customarily performed between the ages of 7 and 13.

fields of education, human rights, law, and medicine—as well as traditional, religious, and other community leaders. Organizations need to test actions that will encourage community dialogue about health, ethics, tradition, and individual rights and responsibilities versus community priorities. In this way, interventions can help more broadly to foster civil societies that are supportive of women's rights and that reject cultural practices such as FC/FGM that reinforce a secondary status for women.

The effectiveness of the law as a tool for social change regarding FC/FGM in Africa is still to be determined. The impact of recent legislation criminalizing the practice in Burkina Faso (1996), Ghana (1994), Cote d'Ivoire (1998), Senegal (1999), Tanzania (1998), and Togo (1998) has yet to be evaluated. Information from Ghana suggests that such laws may serve to stifle public discussion of the subject yet not curtail actual practice (A. Abdel Halim, Ghana, 1998, unpublished anecdotal information). As more countries consider legal measures, the means by which they are enforced and their effectiveness in conjunction with ongoing interventions need to be carefully assessed.

Within the immigrant context, there is little evidence that the practice of FC/FGM is actually occurring in the receiving countries on a wide scale. Nevertheless, laws are in place in seven countries: Australia, Canada, New Zealand, Norway, Sweden, the United Kingdom, and the United States [9]. A key concern is that the majority of these legal instruments were developed and implemented without consulting members of the immigrant communities. There is a danger that biased and discriminatory actions based on such legislation could hurt rather than protect women and their families.

In the future, there needs to be increased dialogue among policymakers, health and social service agencies, and the African immigrant communities in these countries. In the United States, Research, Action and Information Network for the Bodily Integrity of Women (RAINBO) has initiated the African Immigrant Program, based in New York City, to generate such cross-cultural dialogue and improve reproductive health services for circumcised women and girls. Similar initiatives are being undertaken in Australia, Canada, and parts of Europe. These interventions need to include education about the practice and its legal ramifications, as well as support services for women and girls, comprising not only health care but also counseling and other means of assisting immigrants in adapting to a new culture.

At the international level, United Nations agencies, bilaterals, and private foundations have all made commitments to address the practice of FC/FGM, and vehicles are in place through the platforms of action outlined in Cairo, Copenhagen, and Beijing. The key challenge is to find ways to tackle the issue as part of a broader agenda to promote the acknowledgment and protection of women's rights. Members of the international community can use the concepts of behavioral and societal change to generate new, cross-cutting strategies for policies and programs to empower women and address health and human rights violations, including FC/FGM. Such leadership will not only strengthen existing prevention efforts but also help to foster civil societies that truly reflect equity and opportunity for all women and girls.

References

1. Toubia, N. (1995). "A Call for Global Action" 2nd ed. RAINBO, New York.
2. UNICEF (1997). "The Progress of Nations." UNICEF, New York.
3. United Nations (1994). "Programme of Action." International Conference on Population and Development, Cairo.
4. United Nations (1996). "Report of the World Summit for Social Development." United Nations, New York.
5. United Nations (1995). "Declaration and Platform for Action." Fourth World Conference on Women, Beijing.
6. WHO (1996). "Female Genital Mutilation: Report of a WHO Technical Working Group, Geneva, July 17–19, 1995." World Health Organization, Geneva. (Unpublished document, available upon request from WHO, Family and Reproductive Health.)
7. WHO (1998). "Female Genital Mutilation: An Overview." World Health Organization, Geneva.
8. Toubia, N. (1999). "Caring for Women with Circumcision. A Technical Manual for Health Care Providers." RAINBO, New York.
9. IAC (1991). "Report of the IAC Regional Conference, Tanzania, 1990." Inter-African Committee on Traditional Practice Affecting the Health of Women and Children, Geneva.
10. Iregbulem, L. M. (1980). Post-circumcision vulval adhesions in Nigerians. *Br. J. Plastic Surgery* **33**, 83–86.
11. Diejomaoh, F. M. E., and Faal, M. K. B. (1981). Adhesions of labia minora complicating circumcisions in the neonatal period in a Nigerian community. *Tropical Geographical Medicine* **33**, 135–138.
12. Kere, L. A., and Tapsoba, I. (1994). Charity will not liberate women. *In* "Private Decisions, Public Debate." Panos Press, London.
13. Mustafa, A. Z. (1966). Female circumcision and infibulation in the Sudan. *J. Obstetrics Gynaecol. Br. Commonwealth* **73**, 302–306.
14. Daw, E. (1970). Female circumcision and infibulation complicating delivery. *The Practitioner* **204**, 559–563.
15. Worsley, A. (1938). Infibulation and female circumcision: A study of a little-known custom. *J. Obstetrics Gynaecol. Br. Empire* **45**, 686–691.
16. Aziz, F. A. (1980). Gynecologic and obstetric complications of female circumcision. *Int. J. Gynaecol. Obstetrics* **17**, 560–563.
17. Dirie, M. A., and Lindmark, G. (1992). The risk of medical complications after female circumcision. *E. African Med. J.* **69**, 479–482.
18. Shandall, A. A. (1967). Circumcision and infibulation of females, a general consideration of the problem and a clinical study of the complications in Sudanese women. *Sudan Med. J.* **5**, 178–212.
19. National Population Council (1996). "Egypt Demographic and Health Survey 1995." Macro International, Calverton, MD.
20. Ibrahim, S., Katsha, S., and Sedky, N. (1997). "Experiences of Nongovernmental Organizations Working toward the Elimination of Female Genital Mutilation in Egypt." The Center for Development and Population Activities and The Egyptian Society for Population and Development, Cairo.
21. United Press International (1996). Press notice, 25 August 1996.
22. Greene, P. (1996). "Synopsis of a Report on Knowledge, Attitude, and Practice Related to FGM in the Northern Province of Sierra Leone." Wesleyan Church of Sierra Leone, Freetown.
23. Njeru, E., and Program for Appropriate Technology in Health (PATH)/Kenya Staff (1996). "Female Circumcision in Nyeri, Embu, and Machakos Districts of Kenya, Report on Key Informant Interviews." PATH / Kenya, Nairobi.
24. Boddy, J. (1982). Womb as oasis: Pharaonic circumcision in rural northern Sudan. *Am. Ethnolog.* **9**, 682–98.
25. Lightfoot-Klein, H. (1989). "Prisoners of Ritual: An Odyssey into Female Genital Circumcision." Haworth Press, Binghamton, NY.
26. Hayes, R. O. (1975). Female genital mutilation, women's roles and the patrilineage in modern Sudan: A functional analysis. *Am. Ethnolog.* **2**, 617–633.

27. Abd El-Hadi, A. and Abd El-Salam, S. (1998). "Physicians' Perspectives Regarding Female Circumcision." Cairo Institute of Human Rights Studies, Cairo.

28. Grassivaro-Gallo, P., and Viviani, F. (1985). Female circumcision in Somalia: Anthropological traits. *Anthropologischer Anzeiger* **43**, 311–326.

29. Kassindja F. (1998). "Do They Hear You When You Cry?" Delacorte, New York.

30. Zimmerman, M., Radeny, S., and Abwao, S. (1996). "Female Circumcision: A Report on Focus Group Discussions from the Embu, Nyeri, and Machakos Districts of Kenya." Program for Appropriate Technology in Health/Kenya, Nairobi.

31. Al Azhar University, International Islamic Center for Population Studies and Research, and UNFPA (1998). "Report on the International Conference on Population and Reproductive Health in the Muslim World, Cairo, Egypt February 21–24 1998." Al Azhar University, Cairo.

32. Hosken, F. (1979). "The Hosken Report," 1st ed. Women's International Network News, Lexington, MA.

33. Bushner, N., and Rahman, A. (1979). "Epidemiological Study of Female Circumcision in the Sudan." Khartoum University Press, Khartoum.

34. Department of Statistics, Ministry of Economic and National Planning (1979). "Sudan Fertility Survey." Government of Sudan, Khartoum.

35. Departments of Statistics, Ministry of Economic and National Planning/Macro International (1991). "Sudan Demographic and Health Survey 1989/1990." Macro International, Calverton, MD.

36. Carr, D. (1998). "Female Genital Cutting: Findings from the Demographic and Health Surveys Program." Macro International, Calverton, MD.

37. Republic of Mali, Ministry of Health, Solidarity and Aged Persons/Macro International (1996). "Enquete Demographique et de Sante, Mali. 1995–96." Macro International, Calverton, MD.

38. Ghadially, R. (1992). Update on female genital mutilation in India. *Women's Global Network for Reproductive Rights Newsletter,* January–March 1992. Women's Global Network for Reproductive Rights, Amsterdam.

39. Srinivasan, S. (1991). Behind the Veil, the Mutilation. *The Independent (Times of India),* Sunday, April 14, 1991 (magazine section, "Vantage").

40. Pratiknya, A. W. (1989). Female circumcision in Indonesia: A synthesis profile for cultural, religious and health values. *In* "Female Circumcision: Strategies to Bring about Change. Proceedings of the International Seminar on Female Circumcision, Mogadishu, Somalia, 13–16 June 1988." Somali Women's Democratic Organization/Italian Association for Women in Development, Rome.

41. Leye, E., de Bruyn, M., and Meuwese, S. (1998). "Proceedings of the Expert Meeting on Female Genital Mutilation, Ghent, Belgium November 5–7 1998." International Centre for Reproductive Health, Ghent.

42. El-Beshir, H. (1995). "Women and the Agony of Culture: SNCTP and the Eradication of Harmful Traditional Practices in Sudan. An Impact Assessment for SNCTP Field Interventions." University of Khartoum, Khartoum.

43. Comite National de Lutte Contre le Pratique de l'Excision (1997). "Enquete National sur l'Excision au Burkina Faso. Rapport d'Analyse." CNLPE, Ouagadougou.

44. Abd El-Hadi, A. (1999). "We Are Decided: The Experience of an Egyptian Village to Stop FGM." Cairo Institute for Human Rights Studies, Cairo.

45. Obermeyer, C. (1999). Female genital surgeries: The known, the unknown, and the unknowable. *Med. Anthropol. Quar.* **13,** 79–106.

46. El Dareer, A. (1982). "Women Why Do You Weep?" Zed Press, London.

47. El Dareer, A. (1983). Epidemiology of female circumcision in the Sudan. *Tropical Doctor* **13,** 41–45.

48. Warsame, A. (1988). Social and cultural implications of infibulation in Somalia. *In* "Female Circumcision, Strategies to Bring About Change. Proceedings of the International Seminar on Female Circumcision, Mogadishu, Somalia, 13–16 June 1988." Somali Women's Democratic Organization/Italian Association for Women in Development, Rome.

49. Silberstein, A. J. (1977). Circoncision feminine en Cote d'Ivoire. *Annales de la Societe Belges de Medicine Tropicale* **57,** 129–135.

50. Fleischer, N. K. F. (1975). A study of traditional practices and early childhood anaemia in northern Nigeria. *Trans. Royal Soc. Tropical Med. Hygiene,* **69,** 198–200.

51. Asuen, M. I. (1977). Maternal septicaemia and death after circumcision. *Tropical Doctor* **7,** 177–178.

52. Sami, I. (1986). Female circumcision with special reference to the Sudan. *Ann. Tropical Pediatrics* **6,** 99–115.

53. Hathout, H. M. (1963). Some aspects of female circumcision with case report of a rare complication. *J. Obstetrics Gynaecol. Br. Empire* **70,** 505–507.

54. Rushwan, H. (1980). Etiologic factors in pelvic inflammatory disease in Sudanese women. *Am. J. Obstetrics Gynecol.* (December 1, 1980) 877–879.

55. Agugua, N. E. N., and Egwuatu, V. E. (1982). Female circumcision: Management of urinary complications. *J. Tropical Pediatrics* **28,** 248–252.

56. DeSilva, S. (1989). Obstetric sequelae of female circumcision. *Eur. J. Obstetrics, Gynaecol. Repr. Biol.* **32,** 233–240.

57. McCaffrey, M. (1995). "Female Genital Mutilation: Consequences for Reproductive and Sexual Health." British Association for Sexual and Marital Therapy, London.

58. Bijleved, C. (1985). "The Effect of Education on Sudanese Women's Attitudes towards Female Circumcision." University of Leiden, Subfaculty of Psychology, Leiden, The Netherlands.

59. Singhateh, S. K. (1985). "Female Circumcision, the Gambian Experience: A Study on the Social, Economic and Health Implications." The Gambia Women's Bureau, Banjul.

60. Sanderson, L. P. (1981). "Against the Mutilation of Women: The Struggle to End Unnecessary Suffering." Ithaca Press, London.

61. Lightfoot-Klein, H. (1989). The sexual experience and marital adjustment of genitally circumcised and infibulated females in the Sudan. *J. Sex Res.* **26,** 375–392.

62. Grassivaro Gallo, P., and Moro Boscolo, E. (1985). Female circumcision in the graphic reproduction of a group of Somali girls: Cultural aspects and psychological experiences. *Psychopathologie Africaine* **20,** 165–190.

63. Boddy, J. (1989). "Wombs and Alien Spirits. Women, Men, and the Zar Cult in Northern Sudan." University of Wisconsin Press, Madison.

64. Baasher, T. A. (1977). Psychological aspects of female circumcision. *In* "Fifth Congress of Obstetrical and Gynaecological Society of Sudan. Khartoum, Sudan, February 1977." World Health Organization, Eastern Mediterranean Regional Office, Alexandria.

65. Karim, M., and Ammar, R. (1966). Female circumcision and sexual desire. *Ain Shams Med. J.* **17,** 2–39.

66. Badawi, M. (1989). Epidemiology of female sexual castration in Cairo, Egypt. *The Truth Seeker,* July/August 1989, 31–34.

67. Megafu, U. (1983). Female ritual circumcision of Africa: An investigation of the presumed benefits among Ibos of Nigeria. *E. African Med. J.* **60,** 793–800.

68. Koso-Thomas, O. (1987). "The Circumcision of Women: A Strategy for Eradication." Zed Press, London.

69. RAINBO (1997). "Female Circumcision in the United States: Declaration of Values." RAINBO, New York.

70. Eyega, Z., and Conneely, E. (1997). Facts and fiction regarding female circumcision/female genital mutilation: A pilot study in New York City. *JAMWA* **52,** 174–187.

71. Toubia, N. (1999). "Caring For Women With Circumcision: A Technical Manual For Health Care Providers." RAINBO, New York.

72. RAINBO (1999). "Female Circumcision/Female Genital Mutilation: A Quick Reference Chart of the 3 Most Common Types of FC/FGM and the 6 Steps of Defibulation Technique." RAINBO, New York.

73. Izett, S., and Toubia, N. (1999). "Learning About Social Change: A Research and Evaluation Guidebook Using Female Circumcision as a Case Study." RAINBO, New York.

74. Hassan, E., and El-Nahal, N. (1996). "Clinic-Based Investigation of Self-Reporting and Typology of Female Genital Excision." Egyptian Fertility Care Society, Cairo.

75. Gadallah, A., Zarzour, A. H., El-Gibaly, O. M., Abdel Aty, M. A., and Monazea, I. M. (1996). "Knowledge, Attitude, and Practice of Women Teachers on Female Circumcision in Assiut Governorate: Final Report." Assiut University, Assiut, Egypt.

76. Njeru, E., Radeny, S., and Zimmerman, M. (1996). "Female Circumcision: Case Studies from Embu, Nyeri, and Machakos Districts of Kenya: Including Interviews with Several Religious Leaders." Program for Appropriate Technology in Health (PATH)/Kenya, Nairobi.

77. Program for Appropriate Technology in Health (1999). "Improving Women's Sexual and Reproductive Health. Review of Female Genital Mutilation Eradication Programs in Africa." World Health Organization, Geneva. In press.

78. Tostan (1999). "Breakthrough in Senegal: Ending Female Genital Cutting." Population Council, New York.

33

Women and HIV/AIDS: A Global Perspective
Epidemiology, Risk Factors and Challenges

Q. ABDOOL KARIM* AND Z. A. STEIN†

*Southern African Fogarty HIV/AIDS International Research and Training Programme and the South African Medical Research Council, Durban, South Africa; †HIV Center and New York State Psychiatric Institute, Columbia University School of Public Health, New York, New York

I. Introduction

Internationally recognition is growing, albeit slowly, that women are more vulnerable than men to HIV infection. Marginalization, discrimination, alienation, and impediments to the development of one's full potential are well recognized as factors contributing to increased vulnerability among certain individuals, groups, and regions of the world. The unfolding and expanding HIV epidemic among women starkly demonstrates this. The subordinate role of women in a predominantly patriarchal society underpins the excess burden of HIV infection in women. What has emerged in the past five years is a clearer definition of issues contributing to women's vulnerability to HIV infection. This chapter will describe the epidemiologic trends of HIV infection in women, biological and social factors contributing to enhanced transmission of HIV in women, and the implications of these issues in the context of current prevention initiatives. We conclude with the challenges posed by these factors and possible strategies to address them.

II. Epidemiology of HIV Infection

A. *Global Overview*

By the end of 1997, the Joint United Nations Programme on HIV/AIDS (UNAIDS) estimated that about 30.6 million people worldwide had been infected with HIV and about 11.7 million had died of AIDS [1]. The global distribution of HIV is presented in Table 33.1. The uneven regional distribution of HIV infection is influenced by numerous and varied factors such as

Table 33.1
Global Distribution of HIV Infection at the End of 1997

North America	860,000
Western Europe	480,000
Eastern Europe and Central Asia	190,000
South and South East Asia	5,800,000
East Asia and the Pacific	420,000
Australia and New Zealand	12,000
Sub-Saharan Africa	21,000,000
North Africa and the Middle East	210,000
Caribbean	310,000
Latin America	1,300,000
Total	30,600,000

Source: UNAIDS [1].

time of introduction of the virus, size of population at risk, and magnitude of biological, social, political, and economic factors promoting infection and transmission. Developing countries account for about 90% of new infections, with sub-Saharan Africa bearing a disproportionate burden of the HIV pandemic. Sexual transmission remains the predominant mode of transmission and accounts for about 80% of HIV transmission. Mother to child transmission accounts for about 5–10% and intravenous drug use (IDU) for about 5%. Transmission through blood and blood products is virtually eliminated in most parts of the world through stringent screening programs.

B. *HIV Infection in Women*

UNAIDS estimated that at the end of 1997 there were 12.2 million women living with HIV; this represents 41.5% of the global adult burden of infection [1]. The majority of women have been infected through unprotected sex. To date about 800,000 women have died of AIDS. Of the estimated five million new infections among adults contracted during 1997, two million (40.1%) were among women.

This global estimate masks the excess risk observed among women compared to men in countries where HIV has been predominantly heterosexually transmitted since the early days of the pandemic. Surveys among prenatal clinic attenders serve as a proxy measure for monitoring trends in epidemic progression in the heterosexual population and particularly among women. In several countries around the world, both developing and developed, true stabilization of the epidemic has been observed. However, in other parts of the world the spread of HIV continues, with the numbers of women ever increasing, especially in countries where previously the spread of HIV was limited to men who have sex with men and intravenous drug users. A closer examination of temporal trends in HIV transmission additionally highlights the excess HIV infection among women and the challenges these scenarios pose.

C. *New and Emerging Epidemics*

The HIV pandemic represents a complex mosaic of dynamic epidemics. There are countries and regions of the world that have until recently remained insulated from HIV infection but are currently experiencing phenomenally high rates of infection. In other countries a marked reduction or stabilization of HIV through some modes of transmission or communities may obscure newer modes of transmission or new subgroups who are

Table 33.2

HIV Seroprevalence among Prenatal Clinic Attenders in South Africa: 1990–1997

Year	HIV prevalence (%)
1990	0.76
1991	1.49
1992	2.69
1993	4.69
1994	7.57
1995	10.44
1996	14.07
1997	16.01

Source: [3].

Table 33.3

Annual HIV Incidence Rate Estimates among Seronegative Women (15–30 years) Attending Antenatal Clinics in Hlabisa

Year	Incidence rate
1993	3.8% per year
1995	7.4% per year
1997	11.9% per year

Source: [5].

acquiring HIV infection. Finally there are countries where the HIV prevalence is low but the impact in absolute numbers is disproportionate relative to the global total or where the emerging social milieu enhances HIV transmission. In order to unravel some of this complexity, the terms "explosive," "masked," and "emerging" have been coined [2] as a useful typology to capture the key features of the new and emerging HIV epidemics.

1. "Explosive" Epidemics

These include countries that have experienced or are experiencing an explosive spread of HIV, characterized by a relatively late introduction of HIV into the population, a rapid growth of the epidemic, and a high prevalence at plateau. This scenario is best epitomized by southern Africa and Cambodia.

a. SOUTHERN AFRICA. Southern Africa is experiencing one of the most rapidly growing epidemics. HIV is spreading predominantly through heterosexual contact. Compared to eastern and central African countries, southern Africa experienced a relatively late introduction to HIV. The HIV seroprevalence among prenatal clinic attenders in the majority of the countries in southern Africa is currently more than 20%. In 1996, the reported HIV prevalence rates in select southern African countries among prenatal clinic attenders were as follows: Zimbabwe (40%), Malawi (30%), and Botswana (30%) [P. Doyle, Metropolitan Life, personal communication, 1998].

b. HIV INFECTION IN SOUTH AFRICA. Data from South Africa demonstrate the spread of HIV, particularly among women, and the spread is likely to mirror the situation in southern Africa. In South Africa, annual, anonymous, national prenatal HIV seroprevalence surveys have been conducted since 1990. These surveys demonstrate a 21-fold rise in HIV infection within eight years, from 0.76% in 1990 to 16.01% in 1997 (Table 33.2) [3].

To date, in South Africa, there have been no large studies that have followed uninfected individuals over time to enable measurement of incidence rates. However, a variety of statistical techniques have been applied to estimate incidence rates from repeat cross-sectional surveys [4,5]. Incidence rates for women between the ages of 15 and 30 derived from repeat prenatal seroprevalence surveys conducted in Hlabisa, a rural community in the East coast of South Africa, from 1992–1997 (Table 33.3)

demonstrate rapidly rising incidence rates among young women [5]. In 1997, more than one in ten HIV seronegative women became infected with HIV. Even in the relatively low incidence rate year of 1993, one in 25 HIV negative women became infected with this virus.

Young women in the general population in South Africa are experiencing rates previously seen only in high-risk sex worker populations. In a cohort of sex workers from truck-stops in South Africa, the directly observed incidence of HIV was 11% during 1997 [G. Ramjee, personal communication, 1998].

The age specific HIV prevalence data from Hlabisa prenatal clinic surveys as well as the national prenatal surveys demonstrate higher rates of infection in younger women (Table 33.4) [6]. The large increases (in excess of 300%) in HIV prevalence from 1992 to 1995 in the 20–35 year old age group highlights the importance of youth in the HIV epidemic.

Other repeat, cross-sectional, community-based, anonymous HIV seroprevalence surveys conducted in rural areas of the East Coast of South Africa in 1990–1992 demonstrated that HIV infection was more prevalent among women (1.6%) compared to men (0.4%) (age-adjusted relative risk (RR) = 3.8; 95% confidence interval (CI) = 1.4–7.1) [7]. Figure 33.1, based on data from these surveys, demonstrates an early rise of infection in women compared to men. Note also that the peak HIV prevalence for women is in the 20–24 year age group, whereas that for men is in the 25–29 year age group. Women are not only experiencing high HIV prevalence, but they are also becoming infected at an earlier age than men [8].

The relative difference in HIV prevalence between men and women decreased as the epidemic progressed. A community-based, cross-sectional survey conducted in the same area in 1994 demonstrated a 2.3-fold difference in HIV prevalence between men and women [9].

Table 33.4

Progression in HIV Infection by Age in Hlabisa: 1992–1995

Age (years)	1992 Prevalence (%)	1995 Prevalence (%)
20–24	6.9	21.1
25–29	2.7	18.8
30–34	1.4	15.0
35–39	0.0	3.4

Source: [6].

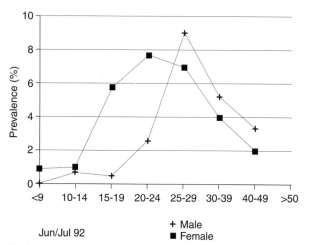

Fig. 33.1 HIV seroprevalence in rural KwaZulu-Natal (1992) by age—men and women compared.

c. MOBILITY AND HIV. Millions of southern African men (and women, to a lesser extent) are migrant laborers. The apartheid migrant labor system prevented workers from settling permanently and establishing families in the places where they worked. The result was the establishment of oscillatory migration patterns, where workers in urban areas maintained links with family in rural homesteads and moved between urban workplaces and rural homes on a weekly, monthly, or annual basis, depending on the distance. Oscillatory migration is a major factor in the spread of HIV and other sexually transmitted diseases in Southern Africa. Despite the demise of the apartheid government, oscillatory migration based now on economic and traditional factors rather than political factors, is still part of the reality of many Southern Africans' lives.

A community-based study conducted in Hlabisa in 1995 demonstrated that a woman's risk of HIV is substantially increased if her partner is a migrant worker. Women whose partners spent ten or fewer nights per month at home had an HIV prevalence of 13.7% compared to 0% infection in women who saw their partners more than ten times a month [9].

Using crude measurements of mobility/migration, population-based surveys conducted in rural KwaZulu-Natal found about 2.5 times more infections among "mobile" adults compared to adults resident in the area continuously for more than one year (Table 33.5) [7].

d. SEXUALLY TRANSMITTED DISEASES AND HIV. It is now well documented that coexisting sexually transmitted diseases (STDs) increase the risk of HIV transmission and, further, that the treatment of symptomatic STDs can result in a substantial impact on the HIV epidemic [10].

Data on STDs in Hlabisa provide an indication of the burden of disease in approximately 60,000 women aged 15–49 in this district [11]. On any given day about one in every four women in Hlabisa is infected with at least one STD. Of these, on any given day, 48% are asymptomatic, 50% are symptomatic but not seeking care, and only 2% seek care during an illness episode. Of the handful of women who seek care, only two out of three women is adequately treated for her STD.

Table 33.5

Mobility and HIV Risk in 15- 44-Year-Old Rural People in KwaZulu-Natal, South Africa (1990)

	N	Prevalence (%)
Stable	2365	1.5
Mobile	313	3.8
		RR = 2.5

Source: [7].

The key issues in this explosive typology are the intensely rapid growth in the HIV epidemic, the high HIV incidence rate in young people, the excess burden of HIV infection among women (and young women in particular), and mobility and high levels of other STDs.

2. "Masked" Epidemics

In this scenario, stable HIV prevalence is a key feature creating the impression that the HIV epidemic has stopped growing. In reality, one or more factors may be masking changes in the epidemic. These include: (1) high incidence rates in subgroups that are not being monitored; (2) an associated high mortality rate may be masking a high incidence rate; or (3) prevalence may be increasing in one subgroup while simultaneously declining in another subgroup—this scenario is epitomized by countries such as Rwanda and the U.S.

a. RWANDA. Fanned by a decade of ethnic war and social upheaval including massacres, rape, and refugees of monumental proportions, HIV is spreading rapidly in rural areas of Rwanda. A survey conducted in this central African country in 1986 demonstrated a seroprevalence of 9.4%; a 1997 government survey, ten years later, found that 15.2% of Rwandans were infected with HIV. About half of the women lost husbands and/or children during the tragic upheaval [Department of Health, Rwanda, unpublished report, 1998]. Data from surveys undertaken among prenatal clinic attenders from 1990 to 1996 demonstrate that HIV prevalence in urban areas reached a plateau and remained at between 15 and 20% [P. Doyle, Metropolitan Life, personal communication, 1998]. Currently most of the new infections among women are occurring in rural areas as demonstrated in Table 33.6.

Table 33.6

Prevalence of HIV Infection in Rwanda by Age and Region: 1986 and 1997

Age group	Rural 1986 (%)	Rural 1997 (%)	Urban 1986 (%)	Urban 1997 (%)
12–15	1.7	4.6	4.2	11.2
16–25	1.8	9.9	17	24.2
26–40	2.8	14.1	30	17.2
>40	0.6	12.7	14.9	6.4
Overall	1.7	10.3	16.5	14.7

Source: [Department of Health, Rwanda, unpublished report, 1998].

Table 33.7
Trends in HIV Prevalence in the United States in the 18–27 Year Age Group: 1988–1993

	No. of men		No. of women	
	1988	1993	1988	1993
Race/Ethnic category				
White	74,400	40,700	7,000	8,700
Black	42,100	42,300	13,700	22,400
Hispanic	27,800	25,000	6,700	7,100
Mode of transmission				
Homosexual contact	97,900	70,200		
Intraveous Drug Use (IDU)	24,400	19,100	13,100	11,000
Homosexual contact + IDU	27,000	9,200		
Heterosexual contact	3,700	9,200	12,900	25,000
Total	146,500	110,600	27,100	38,500

Source: Adapted from [13].

New infections among women in urban areas are in the younger age group and are declining in the older age groups. In contrast, in rural areas the increases are mainly among older women. There has been a more than five fold increase in HIV prevalence among rural Rwandan women in the decade from 1986 to 1997.

b. UNITED STATES.   The HIV epidemic in the U.S. has evolved from an initial outbreak among men who have sex with men in a few cities to an important killer of young Americans. Currently, of all incident HIV infections in the U.S.:

- two-thirds are in ethnic and racial minorities
- one-half are in heterosexuals
- almost one-third are in women
- Men having sex with men is still a powerful risk factor for HIV infection, but it accounts for fewer infections, and in most instances, it is occurring among younger men

The epidemic in the U.S. epitomizes two features of masked epidemics: high mortality masking increasing incidence rates in new subgroups and changes within subgroups at risk.

Several studies demonstrate the substantial fall in mortality rates due to AIDS and the dramatic impact of antiretroviral therapy on increasing survival among AIDS patients [12]. Data from Rosenberg and Biggar [13] demonstrate the overall decline in HIV seroprevalence in the 18–27 year age category (Table 33.7) from 1988 to 1993. However, a gender analysis of these data demonstrates a decline among men but an increase among women. Note also how declines in HIV prevalence among men who have sex with men masks the increase among black men and women through heterosexual transmission.

This situation highlights the importance of monitoring different subgroups, because declines in groups being monitored may mask new subgroups at increased risk: in this instance, young black women.

3. "Emerging" Epidemics

This scenario is characterized by low overall prevalence. However, with increasing incidence rates in some subgroups,

Table 33.8
STDs in Tamil Nadu: 1992

Population screened	Prevalence (%)
Female prisoners	36
Antenatal clinic attenders	10 (syphilis—2%)
Rural areas: Men	6 (syphilis—4%)
: Women	4

Source: [15].

there is potential for rapid spread or bridging into other groups, and the risk data suggest a milieu for a major epidemic. Despite a low prevalence of infection there is a disproportionate contribution to the global burden of infection; the future course of the epidemic is uncertain.

a. UKRAINE AND RUSSIA.   In the Ukraine and Russia there is a paucity of data on the spread of HIV in the general population. Rapid bridging between the IDU population and the nondrug-using general population of young adults is anticipated. Despite the lack of data on HIV there are indications that there may be rapid spread because of the major social, political, and economic changes that have led to a sharp rise in the IDU population.

In addition, sharp rises in STD rates suggest imminent risk of HIV spread. In Russia, the declines in syphilis rates through the 1980s have been reversed and they have increased rapidly since 1991, from 7900 cases in 1990 to close to 450,000 cases in 1997 alone [Ministry of Health, Republic of Russia, unpublished report, 1998].

b. INDIA.   In terms of absolute numbers of HIV infection in South and South-East Asia, India has the most identified infections. Relative to its total population of about 970 million people, the estimated 2.5 million HIV infections translates to a prevalence of <0.5%. Sexual transmission is the main driving force. Of the estimated one million female sex workers, 10% are in Mumbai (previously Bombay) alone. The HIV prevalence among female sex workers in Mumbai is 30% [14].

STDs constitute a major health problem in India. Surveys conducted in Tamil Nadu in 1992 demonstrate a high prevalence of STDs that could fuel the spread of HIV in this country. Already this state contributes about 20% of reported HIV infections [15].

The need for careful monitoring of various subgroups and a need to understand the risks in these subgroups, especially for targeting prevention interventions, are critical. The high numbers of sex workers, IDU, and STDs are important factors that can enhance transmission of HIV into the general population.

4. Increasing Feminization of the HIV Pandemic

The three scenarios depicting the new and emerging epidemics highlight a strong association between HIV infection and youth, gender, mobility, and poverty. With each of these factors a common feature is the accompaniment of an epidemic in women.

D. Tuberculosis

Tuberculosis (TB) is the most common presenting opportunistic infection among people with HIV disease and AIDS. In many countries, but especially in developing countries, the HIV epidemic has fuelled and expanded already existing TB epidemics. Data released by the World Health Organization (WHO) in 1998 indicate that TB is now globally the single greatest killer of women in the reproductive age group. It is estimated that in 1998 one million women will die from TB and a further 2.5 million will get sick. Further, differences in age groups most affected have been noted, depending on the status of the nation (*e.g.,* 25% of TB infections in industrialized countries are in people over 65; in contrast, in developing countries this group only constitutes 10% of infections [16]).

III. Factors Enhancing HIV Transmission among Women

A. *Biological Factors and Sexual Practices*

The biological factors that enhance transmission of HIV have been previously well described [17,18]. The presence of other STDs enhances HIV transmission, women are more exposed to HIV compared to men because of differences in their genital anatomy, and histologically the vaginal mucosal surfaces more efficiently transmit HIV than does penile skin.

Other factors that might enhance HIV transmission in women have been raised but are still not fully researched. These include hormonal contraceptive use, cervical ectopy, early sexual debut, implications of an immature reproductive tract, increased susceptibility among postmenopausal women, the role of female genital mutilation, dry sex practices, frequent douching, and intravaginal substance use.

The sexual practice of anal sex among heterosexual couples is much more common than hitherto assumed. Importantly, data suggest a strong association between anal sex as an independent risk factor for HIV transmission [19]. This has implications for targeting prevention interventions to prevent heterosexual transmission of HIV.

B. *Social Factors*

1. *Economic Vulnerability*

Inequality in the job market, including in some instances exclusion of women from the formal economy, has forced women to explore alternative survival strategies. Sex work is one such option taken up by women. Sex work in several countries is an illegal activity, which restricts women's recourse to protection or support when faced with adversity such as harassment, refusal to pay, or violence from clients. The definition of what constitutes sex work is complex; it includes clearly identifiable sex work in a variety of settings from escort agencies to truckstops. Women at the lower rungs of this hierarchy are at the highest risk of acquiring HIV. Less easily identifiable is sex work that takes place in informal settings in which sex is exchanged for gifts and favors. Informal sex work or sex for favors, including sex with sugar daddies, is fairly common.

2. *Age of Sexual Debut*

Sexual debuts in women as young as 14 or 15 who engage in sexual relations with men several years older is fairly common. The high rates of STDs and teenage pregnancy resulting from these sexual relations confirm that unprotected penetrative sex

takes place. Of equal concern are increasing reports of women's first sexual experience being violent. By the time many women reach age 20 they have had 4–5 sexual partners. Nonpenetrative sex appears to be rare among women who report being sexually active. High teenage pregnancy rates limit the teenage mother's ability to complete her education, prejudicing her future economic opportunities, thereby further entrenching her subordinate status.

3. *Poverty*

Women's perception of risk is influenced by their economic status. Women who have always known poverty, who have been marginalized or discriminated against, or who have little access to resources have always lived with risk of some kind. In light of HIV/AIDS, they do not consider their own personal risk and HIV becomes just another risk. HIV may be of relatively low concern for women who feel powerless to change the realities of their lives and where immediate survival needs take precedence over the spectre of AIDS looming in another 3–5 years.

4. *HIV Prevention Initiatives*

Current HIV prevention initiatives have components that include combinations of some form of behavior modification, barrier methods promotion, and treatment of other STDs. However, even in the few countries that do include a component on "women and AIDS," the emphasis continues to be the standard information of education, counselling, partner reduction, male condom promotion, and monogamy. These messages aimed at women disregard the power imbalances related to gender that preclude women from implementing most of these options. Further, sex workers continue to be the primary target of government efforts even though epidemiologic data demonstrate that young women in general are at highest risk of infection.

Data from a multitude of knowledge, attitude, practice, behaviour (KAPB) studies demonstrate that most women have at least a superficial knowledge of HIV/AIDS; they have heard of AIDs and know how it is spread. More importantly, these studies demonstrate two important findings: the poor correlation between knowledge and practice of risk-reducing behavior and the widespread denial of individual susceptibility to HIV infection.

In many societies, because mutually faithful monogamy is in practice uncommon it is too often an insufficient prop on which women may rely for protection. Male condom use, while increasing, is rarely consistent. Condom use is equated with a lack of trust and love in a partner; hence condoms may be used with casual partners but are seldom used with regular partners. Even among sex workers where condom use is reportedly higher, limited use in intimate relationships and regular clients poses a major risk for acquiring HIV.

While the role of prompt and early treatment of other STDs is well recognized, treatment seeking behaviors of women are influenced by a number of factors. Symptomatology in women, namely abdominal pain and vaginal discharge, may often be nonspecific and insidious at onset, thus delaying seeking treatment, or the infection may be asymptomatic, or symptoms may not be perceived as serious enough to seek treatment. Stigma in seeking treatment for STDs may also influence utilization of health services. Where health services are sought the episode of infection may be inadequately or inappropriately treated because of the poor quality of available health care services.

What we have learned to date is that knowledge and perceptions of HIV/AIDS are important prerequisites for behavior change but are in themselves insufficient. Intermediate factors such as having seen a person with AIDS, perception of self-risk, perception of partner's risk, and confidence in preventive measures are as critically important as access to means of prevention, such as access to condoms and STD treatment, societal norms relating to women seeking condoms or treatment for STDs, ability to communicate, and control over decision making to protect oneself. With regard to the latter, issues relating to individual power, such as experience of violence, perception of sexual roles, confidence in one's own ability, economic independence, influence of children, and existing group norms and support, all influence abilities and strategies for reducing risk. While many of these issues are recognized factors influencing women's vulnerability to HIV infection, there exist very little empiric data to guide decision making and policy formulation.

The overarching need to generate these empiric data was recognized by the International Center for Research on Women in 1990. Through support of 17 projects internationally, the first critical steps were taken in systematically documenting how societal constructions of gender increase women's vulnerability to HIV and in laying the foundations for strategic approaches for intervention. Despite the geographic, cultural, and social diversity among the 17 sites, a consistent picture emerged across sites—"good" women know and say little about sexual matters, risky behavior in men is tacitly condoned, women have little knowledge of their bodies (and of reproductive health in particular), and, in keeping with traditional gender roles, they remain subservient to their partners in sexual and other interactions [17].

It is clear that the vulnerability of women to HIV and other STDs as well as their reproductive health status is centrally related to the context of their lives within a patriarchal society. This male dominance pervades every aspect of their lives, such as family, religion, social, legal, and institutional, and influences their ability to protect themselves. Acquiring or avoiding infection with HIV and other STDs is more complicated for women compared to men. The circumstances under which women have sexual encounters with partners vary both with their own demographic background, such as age, religion, culture, socioeconomic status, sexual preference, geographical location (urban vs rural, developed vs developing), and also with the particular partnership. Characteristics such as male or female partner(s); casual or permanent; exchange of sex for money, drugs, or shelter; dominant or submissive role in the relationship; factors that influence the need for a relationship (*e.g.,* shelter, monetary, emotional); desire to have children; contraceptive choice; and the nature of the specific coital act (whether it was planned or unplanned and consensual or nonconsensual) all affect decisions and abilities to protect against HIV and other STDs.

IV. Challenges in Reducing Women's Vulnerability to HIV Infection

Women have multiple, largely unrecognized, roles in society such as being educators and caregivers in both formal and infor-

mal settings, being vanguards of societal values and norms, and ensuring continuity of society through procreation. These contributions are difficult to measure and their impact will be felt after they are lost. It will take many generations to recover from this loss. In the face of the AIDS pandemic, particularly where health care systems are weak, the burden of caring falls on women in many communities, yet knowledge of a woman's HIV status often leads to ostracization, violence, and loss of security [21,22].

The issue of addressing women's vulnerability has to be a central part of any country's response and: "the way in which we define a problem determines how we approach its solution. It is time to replace the notion that HIV is just a virus whose spread is affected by risk factors and biological cofactors with a broad concept of vulnerability to heterosexual HIV transmission. This shift will entail augmenting our existing strategies for prevention with programmes and policies that address the central inequities underlying vulnerability to HIV/AIDS" [J. Mann, in 20].

Short-, medium-, and long-term plans have to move beyond targeting the individual to addressing structural issues that redress the balance of power in society between men and women. Recognition of the need for specific programs to reduce women's vulnerability has been slow to evolve. In the past few years a clear definition of the problem has evolved. The challenge now lies in moving rapidly beyond rhetoric to action. This action is required at individual, community, and government levels and across several dimensions such as politics, science, culture, and service provision. Gender and power remain to a large extent abstract concepts with little meaning and the recognition of their role is insufficient to formulate policy or design interventions. Guided by a number of empiric studies, mechanisms to concretely act to reduce women's vulnerability to HIV infection and increase their level of autonomy have started to emerge.

A. Increasing Access to Information

There is a need to increase women's access to information as one form of mobilization and awareness-raising. Women need to be targets of messages and campaigns that include issues such as: you have rights, you can say no to unsafe sex and/or sex, increase awareness of risk of HIV infection and the importance of good sexual and reproductive health, and provide resources where information and or support can be accessed.

B. Increasing Access to Health Services

Health sector reform that prioritizes improving women's general health, including reproductive health, through integrated women-friendly services is an important point for reducing HIV risk. This type of transformation is starting to be seen in many developing countries through the integration of services into a minimum package to be offered at all primary health care centers that includes maternal and child health services, fertility control, STDs, HIV/AIDS support and counselling, and mental health services. This approach will maximize missed opportunities to reduce women's risk to HIV infection. The challenge lies in ensuring the implementation of these policies.

C. Increasing Access to Barrier Methods

While male condom use has been increasing over the years, wider access to condoms and consistent and correct use of condoms remain a challenge.

1. The Female Condom

Governments have been slow to respond to the urgent need for access to known technologies that women are more successfully able to implement, such as the female condom, and the development of newer technologies such as contraceptive and noncontraceptive microbicides. While the Zimbabwean government purchased 40,000 female condoms for distribution in the public sector in 1997[1] [23], few others have a policy on female condoms or are considering making them accessible within the public sector.

D. New Partnerships

Existing networks and alliances addressing women's rights, reproductive health issues, and gender need to start to incorporate HIV/AIDS into their agendas. Partnerships across sectors—government, nongovernment, and private—will synergize and enhance activities and the impact of these activities.

E. Young Women

It is important to ensure that lifeskills programs, specifically for young women become an integral part of socialization and address abstinence and postponement of sexual debut. Education should begin prepuberty. Of equal concern are the increasing reports of young girls' first sexual experiences being coerced and violent, and this issue needs to be addressed.

F. The Human Tragedy

Behind the numbers presented are the faces of young men and women. While the numbers highlight opportunities to prevent new infections, even with miraculous breakthroughs large numbers of men and women will continue to become ill and die. The provision of good quality care, including access to antiretrovirals, has to be planned for. A social environment that encourages and supports disclosure of HIV status as opposed to stigmatization and discrimination is necessary.

G. Biomedical and Technical Interventions

In the short term, the development of biomedical interventions that can be initiated by women needs to supported. Microbicides fall into this category. To date only two efficacy trials have been completed [24,25], but neither have demonstrated effectiveness. While the number of products entering Phase 1 trials have substantially increased since 1996, the need to move to Phase 2 and Phase 3 testing remains a shorter-term imperative. Lobbying and advocacy for more products to be developed and to proceed with human trials are a critical component to ensure that sufficient numbers of women survive to take on the

[1]Within two months all 40,000 units had been distributed.

challenges posed in redressing the gender imbalances that exist in society. Given the widespread use of hormonal contraceptives, and particularly long acting progesterone-only based contraceptives, there is an urgent need for research to be conducted to determine its impact on HIV transmission as a precursor to the development of safe guidelines for contraceptive use.

H. Helpless and Hopeless?

Despite the odds being stacked against them, women have displayed a resilience over the years in dealing with a number of social issues. Part of the solution in reducing women's vulnerability is to try to systematically understand sources of power women have at the individual and societal level that can be used to negotiate protection against HIV. How can traditional roles of nurturing, caring, and educating, for example, be used as a strategy to reduce risk of HIV infection? Unity in action has been proven to be useful in a number of instances in the past—how do we harness this to the struggle against HIV/AIDS?

I. Creating Safe Spaces for Dialogue, Opportunities to Learn Communication and Negotiation Skills, and Build Self-Esteem

Women need to know that they are not alone in the challenges that face them in trying to avoid infection with HIV. Opportunities that enable dialogue, debate and sharing of experiences, both positive and negative, will allow for group cohesion and unity to develop that can be supportive for the discussions that should take place with sexual partners. In addition, role-playing in safe environments on communication and negotiation will contribute to building self-esteem and enhancing women's ability to protect themselves and their health.

J. Supportive Legislative and Policy Frameworks

Protection, promotion, and enforcement of the rights of women are key to reducing women's vulnerability to HIV infection. The need for programs to take cognizance of the vulnerability of women and for governments especially to respond through the creation of supportive legislative and policy frameworks has already been underscored at the Cairo and Beijing conferences. The Platform for Action has yet to be implemented by most governments.

K. Men—The Missing Link in Addressing Women's Vulnerability

Implicit in redressing the balance of power is that men will have to give up some power for women to gain more power— what are the implications of this and what are the best strategic approaches to addressing this issue? Given the powerful role of men in society, HIV/AIDS interventions and strategies targeting men will have a substantial impact on reducing the vulnerability of women to HIV. The need to shift the balance of power, within relationships and society, between men and women remains paramount. This shift can be expedited by targeted interventions at men that clarify misconceptions about HIV, address sexual and women's reproductive health issues, and emphasize male responsibility within relationships, particularly with respect to disease prevention and rights of women.

V. Role of Research and Women as Subjects of Research

The issue of women as subjects of research needs to be examined. Currently, women are subjects of a large amount of research that may sometimes involve men, but what have we learned about HIV infection in women? The numerous vertical transmission studies most commonly regard women as incubators, issues related to them are ignored, and, worse, implications of interventions for women are given minimal attention. In countries where the new drug regimens are widely available, women still have limited access to these drugs or information on use of these drugs. Arguments used to justify the exclusion of women from treatment trials is that the effect on the fetus is unknown. Women are quite capable of using reliable fertility control methods to ensure that they do not become pregnant while on a trial. It is crucial that studies not only include women but address issues that will reduce their vulnerability. This inclusion is particularly important in the design of vaccine and drug trials so that appropriate extrapolations can be made expeditiously.

VI. Conclusion

Clearly, there are no quick fixes in reducing the vulnerability of women to HIV infection. There needs to be a comprehensive sustainable approach that addresses the three intertwined epidemics: HIV, AIDS, and impact of the premature loss of lives. Any attempt to reduce women's vulnerability and risk has to focus on structural changes/issues relating to development that could redress the power imbalance in society. There needs to be humility to recognize strengths and weaknesses of different modes of inquiry from the basic to the social sciences, as well as the community experiences garnered to date, and an openness and respect for sexual diversity and differences. A holistic and committed endeavour of politicians, leaders, civil society, scientists, and policymakers is needed to have a fully developed and sustained response that can address the complexities of women's vulnerability to HIV and AIDS.

References

1. UNAIDS/WHO (1998). "Report on the Global HIV/AIDS Epidemic," Joint United Nations Programme on HIV/AIDS (UNAIDS). UNAIDS/WHO, Geneva.
2. Abdool Karim, Q. (1998). The HIV pandemic: New and emerging epidemics. *Plenary Presentation, 12th World AIDS Conf.,* Geneva.
3. Department of Health, RSA (1998). "Eighth National Survey of Women Attending Antenatal Clinics in the Public Sector" (unpublished report). Dept. of Health, Pretoria, South Africa.
4. Gouws, E., Williams, B., Wilkinson, D., and Abdool Karim, S. S. "Estimating the Incidence Rate of HIV Infection from Repeat Cross-sectional Studies (unpublished report). Medical Research Council, Cape Town, South Africa.
5. Abdool Karim, S. S. (1997). "Establishing a South African Vaccine Trial Site through Intervention Trials for HIV Prevention among High Risk Rural Female Sexual Partners of Migrant Workers in Hlabisa."

HIVNET proposal submitted to Family Health International, Durham, North Carolina.
6. Coleman, R. L., and Wilkinson, D. (1997). Increasing HIV prevalence in a rural district of South Africa. *J. Acquired Immune Defic. Syndr. Hum. Retrovirol.* **16,** 50–53.
7. Abdool Karim, Q., Abdool Karim, S. S., Singh, B., Short, R., and Ngxongo, S. (1992). HIV Infection in rural South Africa. *AIDS* **6,** 1535–1539.
8. Abdool Karim, Q., and Abdool Karim, S. S. (1993). Epidemiology of HIV infection in Natal/KwaZulu. *In* "AIDS and Your Response," pp. 43–51. Institute of World Concerns, Coventry, UK.
9. Colvin, M., Abdool Karim, S. S., Connolly, C., Hoosen, A. A., and Ntuli, N. (1998). HIV infection and asymptomatic sexually transmitted infections in a rural South African community. *Int J. STD AIDS* **9,** 548–550.
10. Grosskurth, H., Mosha, F., Todd, J., Mwijarubi, E., Klokke, A., Senkoro, K., Mayaud, P., Changalucha, J., Nicoll, A., ka-Gina, G. (1995). Impact of improved treatment of sexually transmitted diseases on HIV infection in rural Tanzania: Randomised controlled trial. *Lancet* **346,** 530–536.
11. Wilkinson, D., Abdool Karim, S. S., Harrison, A., Lurie, M., Colvin, M., Connolly, C., and Sturm, A. W. (1999). Unrecognised sexually transmitted infections in rural South African Women—the hidden epidemic. *Bull. W.H.O.* **77,** 22–28.
12. Centers for Disease Control and Prevention (1997). Update: Trends in AIDS incidence, deaths, and prevalence—United States, 1996. *Morbid. Mortal. Wkly. Rep.* **46**(8), 165–172.
13. Rosenberg, P. S., and Biggar, R. J. (1998). Trends in HIV incidence among young adults in the United States. *JAMA,* 279(23), 1894–1899.
14. Asthana, S. (1996). AIDS-related policies, legislation and programme implementation in India. *Health Policy Plann.* **11,** 184–197.
15. National AIDS Control Programme India (1993). "Country Scenario: An Update." National AIDS Control Organization, Ministry of Health and Family Welfare, Government of India, Mumbai.
16. World Health Organization (1998). "TB is the Single Biggest Killer of Young Women" (unpublished report). WHO, Geneva.
17. Gupta, G. R., and Weiss, E. (1993). "Women and AIDS: Developing a New Health Strategy," Policy Ser. No. 1. International Center for Research on Women, Washington.
18. Elias, C. J., and Heise, L. (1993). "The Development of Microbicides: A New Method of HIV Prevention for Women," Working Papers No. 6. The Population Council, New York.
19. Abdool Karim, S. S., and Ramjee, G. (1998). Anal sex and HIV transmission in women. *Am. J. Public Health* **88,** 1265–1266.
20. Mann, J. (1993). "Women and AIDS: Developing a New Health Strategy" (G. R. Gupta and E. Weiss, eds.), Policy Ser. No. 1. International Center for Research on Women, Washington.
21. Temmerman, M., Ndinya-Achola, J., Ambani, J., and Piot, P. (1995). The right to know HIV-test results. *Lancet* **345,** 969–970.
22. Seidel, G., and Ntuli, N. (1996). HIV, confidentiality, gender and support in rural South Africa. *Lancet* **347,** 469.
23. Abdool Karim, Q., Tarantola, D., As Sy, E., and Moodie, R. (1997). Government responses to HIV/AIDS in Africa: What have we learnt? *AIDS* **11**(Suppl. B), S143–S149.
24. Kreiss, J., Ngugi, E., and Holmes, K. K. (1992). Efficacy of nonoxynol 9 contraceptive sponge use in preventing heterosexual acquisition of HIV in Nairobi prostitutes. *JAMA,* **268,** 477–482.
25. Roddy, R. E., Zekeng, L., Ryan, K. A., Tamoufe, U., Weir, S., and Wong, E. (1998). A controlled trial of Nonoxynol 9 film to reduce male to female transmission of sexually transmitted diseases. *N. Engl. J. Med.* **339,** 504–510.

34

Human Rights and Women's Health

LYNN P. FREEDMAN

Center for Population and Family Health
Joseph L. Mailman School of Public Health at Columbia University
New York, New York

I. Introduction

This chapter examines the growing and increasingly complex role that human rights is playing in the field of women's health, particularly at the international level. During the 1990s, human rights has become a primary organizing principle and/or rhetorical strategy employed by a range of different actors who engage with women's health issues from strikingly varied positions. For example, more than 180 governments attending the International Conference on Population and Development (ICPD) held in Cairo in 1994 and the Fourth World Conference on Women held in Beijing in 1995 endorsed final declarations that place human rights front and center: the very aim of women's health programs is to enable women to enjoy their human rights [1,2]. The United Nation's (UN) own health-related agencies—WHO, UNFPA, UNICEF—now describe the protection and fulfillment of human rights as basic to their work.[1] Mainstream nongovernmental organizations (NGOs) have tapped human rights language and concepts as well. The International Planned Parenthood Federation, for example, has produced a "Bill of Rights" explicitly grounded in human rights law, which it promotes widely as the guiding light of its program and a centerpiece of its mission [3]. Grassroots women's health movements also have made strategic use of human rights both in activist campaigns and in education and consciousness-raising projects.[2] Major universities, including public health and medical schools, now include "health and human rights" courses in their curricula, and a new journal focuses exclusively on health and human rights issues.[3]

---

[1]WHO now has a Women's Health and Development (WHD) Program that is committed to the promotion of women's health and human rights (http://www.who.int/frh-whd/index.html). UNICEF's and UNFPA's mission statements can be found on the World Wide Web at http://www.unicef.org/ about/ and http://www.unfpa.org/ABOUT/ MISSION.HTM.

[2]See, for example, the many positions and initiatives of grassroots groups described in the journals produced by the Latin American and Caribbean Women's Health Network (*Women's Health Journal,* Santiago Chile), the Women's Global Network for Reproductive Rights (*Newsletter,* Amsterdam, The Netherlands), and in the Final Report of the 8th International Women and Health Meeting [4].

[3]Public Health Schools at Boston University, Harvard University, University of California at Berkeley and San Francisco, Johns Hopkins University, and Columbia University all offer a course on health and human rights. In addition, medical schools at Harvard University and Yale University, as well as Georgetown University Law Center and Princeton University (undergraduate), have courses focusing on health and human rights. The journal *Health and Human Rights* is published by the Harvard School of Public Health and the François-Xavier Bagnoud Center for Health and Human Rights.

However, turning to the public health and medical journals where the lion's share of research on women's health is recorded and disseminated, there is an unmistakable sense of disconnect. Human rights language and concepts appear hardly to have touched the discipline of epidemiology. We ran multiple computer searches and found that, in the professional medical and public health literature, the terms "human rights" and "women's rights" barely surface at all.[4] When they do appear, it is most often in the context of articles about armed conflict or torture, the classic areas of human rights and humanitarian work [5].

In part the disconnect stems from the very real difficulties of doing truly interdisciplinary work [6]. Epidemiologists and public health researchers often go about their work focused on the technical issues of hypothesis testing, study design, and data analysis, perhaps recognizing some potential policy implications of their findings but only vaguely aware of how intimately their work connects to the politics of the times. Similarly, human rights activists may plunge into situations fraught with implications for women's health, barely conscious of the enormous body of health research that elucidates important dimensions of the violations they seek to address.

But there are some more basic reasons why segments of the women's health and human rights communities have kept their distance from each other. From the side of epidemiology and the perspective of many health professionals there are the perpetually thorny questions about the relationship between science and politics. The prestige and (in American culture, at least) the very persuasiveness of scientific discourse stems from its cultivated aura of objectivity, indeed from its deliberate distancing from politics. Although it is increasingly understood and accepted that the dichotomy between science and politics is a false one and that the supposed value-free neutrality of science is illusory [7,8], there clearly can be a price to pay for an open embrace between public health and human rights. See, for example, the bitter attacks that commentators affiliated with conservative think tanks have launched against progressive public health researchers who seek to link poor health with structural inequality and social injustice [9,10]. The sorry fates of officials

---

[4]Searches were conducted on materials appearing from 1965 to 1997 using the following 11 major epidemiology and medical journals: *The American Journal of Epidemiology, The International Journal of Epidemiology, Epidemiology, The Journal of Epidemiology and Community Health, The Journal of Clinical Epidemiology, Morbidity and Mortality Weekly, The American Journal of Public Health, Lancet, The New England Journal of Medicine, The Journal of the American Medical Association (JAMA).* One exception is the *Journal of the American Medical Women's Association,* which devoted an entire issue (Vol. 52, no. 4, Fall 1997) to "Human Rights and Women's Health."

such as (dismissed) Surgeon General Joycelyn Elders or research projects such as the (cancelled) government-funded sexuality study [11], which were deemed to have entered political space from the wrong angle, surely have the intended chilling effect on health researchers seeking funding, jobs, or influence.

On the other hand, many political activists, especially within the women's movement, are reluctant to tie their issues to the health professions or to health discourse at all. The history of medicalization of women's lives has been marked by the pathologizing of life events such as childbirth and menopause [12], by the relinquishing of control to male professionals and "experts" [13], and by the conversion of women's resistance into individualized problems of deviance construed as psychological or emotional disturbances [14,15]. For a women's movement committed to exposing the workings of patriarchy in women's lives, medical discourse and the medical establishment have often been seen as the "enemy," disguising the political nature of inequality and oppression behind the legitimizing mantle of biology.

While acknowledging the grounds for such reluctance, this chapter contends that the embrace between public health and human rights creates what is potentially one of the most powerful sets of theoretical, practical, and organizational tools for addressing the issues that loom largest in the international women's health arena at the dawn of the twenty-first century. That contention is based on an expansive view of both human rights and public health as two complementary discourses that have particular resonance and power across a wide range of different social settings and policy arenas at this moment in time. As such, they are subject to a good deal of debate and negotiation, making their content and significance open and fluid. The sections that follow give an overview of the nature of that debate and the range of different initiatives and perspectives that are beginning to flesh out the content of human rights in women's health. They describe a field in flux that reflects a world in flux, and, in so doing, they attempt to capture the dynamic and potentially transformative nature of human rights today.

II. Basic Premises and Critical Questions

This chapter builds from the basic premise, elaborated in many parts of this book, that women's lives are inextricably connected to and shaped by the social, economic, and cultural conditions in which they live, and that their health is therefore unavoidably linked to those conditions as well. The precise biological mechanisms by which such conditions actually influence health status—the process by which they are "internalized" by the body, to use Moss's phrase (Chapter 43)—remain the subject of lively debate and exciting new research. However, even a deeper understanding of the biological processes that yield specific disease states at the individual level will not obviate the need to confront fundamental social conditions that, if left unaddressed, will continue to result in poor health, unequally and unjustly distributed at the societal level [16].

Yet, simply documenting the fact of an association between social conditions and poor health does not yield a predetermined "correct" policy answer for addressing health issues. For example, witness the current disputes about whether the correlations between low socioeconomic status, poor "personal health

behaviors," and poor health should be approached as an issue of individual control and "personal responsibility" or as an issue of structural constraints and social injustice [17]. These and similar debates begin to reveal that both the framing of health issues and the conceptualization of related policies are inherently political processes. To put the point somewhat differently, while the experience of morbidity and mortality is verifiably "real" with biological and clinical markers that certainly can help establish effective treatments, the definition and framing of health problems is also a matter of social and cultural construction. As such, the science of women's health is unavoidably linked to the politics of women's health; ideological stances that determine political orientation can and do determine "health orientation" at the same time [18,19].

Human rights—as a body of formal law, a discursive system,[5] a set of moral and ethical imperatives, and an organizing strategy—can be understood as one such ideological stance that helps shape an approach to health and, simultaneously, an approach to policy and politics. Human rights places primary value on the inherent dignity of each individual human being (seeing "the highest attainable standard of physical and mental health" as one attribute of that dignity) and brings to the forefront—gives urgency and priority to—questions about how social structures can enable and protect the expression of that dignity. Taking a human rights approach to health helps us push beyond the epistemological debates that have occupied so many critiques of science and medicine. As Robert Proctor, writing the history of the idea of value-free science, put it, we need to develop a "political philosophy of science" that goes beyond the "purely epistemological question—How do we know?"—and asks the more fundamental and ultimately political questions:

"Why do we know what we know—and why don't we know what we don't know?"

"Why are our interests here and not there?"

"Who benefits from knowledge (or ignorance!) of a particular sort and who suffers?"

In short, we need to ask "How is science involved in patterns of dominance and/or exclusion?" [7, pp. 11–13].

Proctor's questions help orient us to one critical aspect of human rights work today. Everywhere in the world women's lives are constrained by deeply rooted inequalities based on gender, race, class, and, in some settings, by other social characteristics such as language and religion. Such inequalities dramatically influence women's health, functioning through a variety of mechanisms over different stages of the life span [20]. The exact contours of injustice and how it is mediated by social and cultural factors (*i.e.,* the precise constellation of mechanisms

---

[5]The terms "discourse" and "discursive system" refer to the ways in which language and the conventions surrounding its use help shape our perception, our understanding, and ultimately our experience of the world. Thus, the "discourse of public health" or the "discourse of human rights" highlights the fact that within each of these disciplines there is not only a shared body of knowledge, but also shared conventions for expressing that knowledge including, for example, metaphors, symbols, what counts as evidence, rules of narration, and so on. I am indebted to my colleague, Carole Vance, not only for this attempt at a concise explanation of discourse, but also for ongoing discussions about how these issues affect work in the health and human rights field.

through which it influences women's health) also vary between and within different societies. At the same time, those local structures link in complicated ways with processes of globalization that have plunged much of the world into deepening poverty and widened the gap between rich and poor—even as the global "communications revolution" enables virtually everyone to see and apprehend such consequences more fully.

We need to understand how and why these connections between the intricate workings of inequality and women's health remain hidden from the gaze of vast portions of the health research, practice, and policy communities. What is it about the way work is done in the health field that continually shields such matters from view or makes them impervious to standard health policy? Put differently, if women's health is indeed shaped by such fundamental social forces, and if we are serious about social change, then we need to ask—and continue over and over again to ask—what keeps such structures of inequality so firmly in place? Clearly, they are maintained through the obvious forms of coercive power lodged in the institutions of government and other socially significant authorities, such as religious institutions. Much of that coercive power is articulated and maintained through formal law and its related institutions. Thus, one essential piece of human rights work is to use the doctrines of international human rights law to effect changes in national and local laws.

However, to focus exclusively on formal law and state actors is to miss the most deeply rooted sources of women's oppression. The inequalities that shape so much of women's health are enforced not only through laws or even "cultural" and/or "traditional" practices, but also through the workings of power in the discourses that structure everyday life and are therefore reflected in health research, policy, and practice as well. Indeed, it is precisely because of their ability to appear so obvious and common sensical, to make socially constructed phenomena seem so self-evidently natural and inevitable, that discursive structures carry such enormous weight in shaping our worlds and the nature of our experience in them. Here the concept of hegemony is helpful. Hegemonic power is "that order of signs and practices, relations and distinctions, images and epistemologies—drawn from a historically situated cultural field—that come to be taken-for-granted as the natural and received shape of the world and everything that inhabits it." [21, p. 5, quoting Ref. 22, p. 23].

The question, then, is whether the discourse of public health/human rights can operate effectively to challenge this kind of hegemonic power. In other words, can public health/human rights discourse provide an effective "counter-hegemonic" strategy for advancing women's health and well-being internationally? Can it offer new ways to see and understand the world, to expose as socially constructed—and therefore as challengeable and changeable—precisely those workings of power that underlie conditions of poor health?

A full analysis of the potential of public health discourse in this respect is beyond the scope of this chapter. However, it is worth noting that a related debate has been going on within the professional epidemiology and public health literature. For example, assessments of the state of epidemiology have warned of its growing detachment from its roots in public health [23,24], and assessments of public health practice have warned of its detachment from its roots in social justice movements [25]. If

nothing else, such writing makes clear that public health is in ferment and its methodological tools, when wielded from within a pragmatic public health and social justice perspective, can function as a powerful lens through which to see and understand the complex determinants of health. As such, public health automatically becomes an indispensable partner for human rights in building a counter-hegemonic practice.

Thus public health and human rights, working together, reverse and move beyond Proctor's question ("How is science [or, for our purposes, conventional health practice] involved in patterns of domination and/or exclusion?") to address two further questions: How do patterns of dominance and exclusion actually influence women's health? What approach to women's health can both challenge and change those underlying patterns? These are the questions that have inspired the majority of initiatives described in the following sections.

III. Approaches to the Practice of Health and Human Rights

Painting a picture of human rights in the international women's health field today is a messy business. On the one hand, human rights is a body of formal law with rules that govern its codification, interpretation, and implementation. Like all bodies of law it can be used to uphold, as well as to challenge, the status quo and is therefore invoked in different times and places by groups with radically different stances toward the established order. That fact should counsel us to look critically at how human rights is used in any particular situation: who uses it to support what set of interests? [26,27]. However, human rights is also a discursive system that can and does help structure the ways in which we understand and experience the world. It thus has become subject to an entirely different level of contestation beyond its role as formal law. To ground the discussion of these different uses of human rights, I first give a very brief historical orientation to modern human rights. This is followed by a section surveying the ways in which those working from the perspective of formal human rights law have approached the issues of women's health. I then shift gears and attempt to show how, through human rights discourse, issues of women's health have been inserted into the far broader debates about globalization and diversity that loom so large today.

A. Evolution of the Modern Human Rights System

The modern human rights movement took shape with the formation of the UN in the years immediately following World War II, when the horrors of Nazi Germany and the atrocities committed by the Japanese in Asia were still fresh in mind. In its Charter, a binding treaty under international law, the UN proclaimed that one of its specific purposes would be "to promote . . . universal respect for, and observance of, human rights and fundamental freedoms for all, without distinction as to race, sex, language or religion" [28]. The first elaboration of those rights and freedoms came with the Universal Declaration of Human Rights, signed in 1948, and promulgated "as a common standard of achievement for all peoples and all nations."

The Universal Declaration set out a new concept in international law: the protection of the individual human person. Until then, international law had been the law that governed relationships between states, premised on respect for sovereignty and noninterference in domestic affairs. Governments could negotiate with each other about the laws of the sea, about the conduct of war, about the rules of trade or mail services; however, what any government did to its own citizens within its own borders was its own business. With the Universal Declaration, the individual person became, for the first time, the subject of international law. Its rationale was based on the fundamental notion that every human being, simply by virtue of being human, is entitled to certain basic protection. In the words of the Universal Declaration, signed by countries from every continent and from many different religious, cultural, and political traditions:

"All human beings are born free and equal in dignity and rights." (Art 1)
"Everyone is entitled to all the rights and freedoms set forth in this Declaration, without distinction of any kind, such as race, color, sex, language, religion, political or other opinion, national or social origin, property, birth or other status." (Art. 2)

It was not long before efforts to implement the lofty principles of the Universal Declaration became embroiled in the politics of the Cold War. As the UN undertook to codify those principles (*i.e.,* statements of human entitlement) into legally binding treaties (*i.e.,* statements of states' obligations), a major split emerged. Some governments, led by the United States, demanded priority for "civil and political" rights, for those rights aimed at limiting the state's ability to encroach on individual freedom. This included rights, similar to those found in the Bill of Rights to the U.S. constitution, such as freedom of speech and association; freedom from torture and from cruel, inhuman, or degrading treatment; the right to a fair trial; and to life, liberty, and security of person. Other governments, led by the Soviet Union, demanded priority for "economic, social, and cultural rights," matters related to the fulfillment of basic needs such as the rights to health, education, employment, and an adequate standard of living including food, clothing, housing, and health care.

As a result, in 1966 the UN ultimately promulgated two separate covenants: The International Covenant on Civil and Political Rights and the International Covenant on Economic, Social, and Cultural Rights. These two primary covenants have been followed by a series of additional conventions aimed at elaborating different aspects of the human rights regime. These include, for example, the Convention on the Elimination of All Forms of Discrimination Against Women (1979), the Convention on the Rights of the Child (1990), and the International Convention on the Elimination of All Forms of Racial Discrimination (1969), as well as regional conventions for Europe, the Americas, and Africa (although none has been developed within Asia) [28a]. These conventions and the increasingly elaborate set of institutions charged with implementing them are the backbone of the system of formal human rights law.

B. Working From the Perspective of the Formal Human Rights Law

Virtually every human right has health consequences when it is violated. A mere listing of a few rights codified in international law makes this immediately apparent: the right to food, the right to housing, the right to health care, the right to be free from torture. It takes little imagination to spin out a range of health consequences that might attend to the violation of each of these rights. Sometimes the connection requires a bit more elaboration. For example, there are studies showing how violations of freedom of speech and information are implicated in famines—with obvious ghastly implications for a population's health [29].

Thus, at the simplest level, human rights (or, rather, the violation of each of them) can be seen as a typology of risk factors for different health conditions. Taking the analysis one step further, we can think of violations of such rights as risks to which there is sometimes differential exposure based on characteristics such as gender, race, or class that are the grounds for discrimination claims. Importantly, we can also think of the fulfillment of different human rights as factors that contribute to good health in specific identifiable ways. For the most part, these kinds of analysis require us only to take the studies that already exist in the epidemiologic literature that document social, economic, and cultural determinants of health and to reconceptualize those social, economic, and cultural variables as human rights issues. While this approach to health and human rights is analytically fairly simple, there is still plenty of room for extending its scope beyond the existing literature. For example, Mann *et al.* point out that a "coherent vocabulary and framework to characterize dignity and different forms of dignity violations are lacking," and suggest that "a taxonomy and epidemiology of violations of dignity may uncover an enormous field of previously suspected, yet thus far unnamed and therefore undocumented, damage to physical, mental and social well-being." [30, p. 8].

Much of the earliest writing on women's health and human rights used this kind of analysis, setting forth various provisions of human rights treaties and then elucidating the relationship between the violation of the right and some condition of poor health for women [31,32]. Epidemiologists, demographers, and public health professionals familiar with the basic literature on the social determinants of women's health internationally will immediately see these connections. A nonexhaustive listing might include such rights as these:

- The right to decide freely and responsibly on the number and spacing of children
- The right to the benefits of scientific progress
- The right to be free from torture and cruel, inhuman, and degrading treatment
- The right to marry and found a family
- The right to an adequate standard of living
- The right to equal access to health care
- The right to life, liberty, and security of person

A dramatic example of actively generating such analysis is a report on Afghanistan by Physicians for Human Rights (PHR), which documents the disastrous health consequences that Taliban edicts banning education, limiting access to health care, and restricting freedom of movement—all conceptualized as human rights violations—have had on women [33,34]. The PHR report on Afghanistan is one of the few instances in the women's health field in which epidemiologic surveys have been designed and used to document systematically the pervasiveness of human rights violations and the prevalence of health consequences associated with such violations (such as the mental and physical symptoms associated with posttraumatic stress disorder). Another example is the work done by Swiss and colleagues on the prevalence and health consequences of rape during the civil war in Liberia [35].

The focus in this chapter is not so much on the dramatic emergency situations found in war and armed conflict—although even in these settings, where human rights violations might seem so clear and present, the particular violations committed against women *because* they are women (rape being the most obvious) have only very recently, with the horrific campaigns of "ethnic cleansing" in former-Yugoslavia and Rwanda, been exposed and analyzed as such [36]. Instead, the focus in this chapter is on the everyday "endemic" forms of injustice that shape women's health. Indeed, it is precisely because such injustices and their pervasive consequences for women's health are accepted as routine and unremarkable, as part of the inevitable nature of things, that a human rights perspective can be so important.

Building on the existing epidemiologic literature, a human rights perspective serves the extremely important function of converting scientific insights about the social determinants of poor health into claims of political entitlement and into demands for changes in social policy. Of course, the belief that social policy (and not just advances in medicine and medical care) can influence public health is not a new or particularly innovative one.[6] But the recent favoring of a *human rights* framework for policy does add important dimensions to the debate. Theoretically, health policy becomes not simply an action undertaken by the state to increase its own wealth and power for its own purposes; rather, taken from a human rights perspective, appropriate health policy becomes the *entitlement* of the people and the *obligation* of the state to its citizens [40].

The restatement of an epidemiologic "fact" into a demand for political change is a strategic move. For example, international agencies such as UNFPA and UNICEF, by invoking the language of human rights treaties when they explain the social and economic determinants of poor health, can tap the human rights discourse to raise the public profile of their agendas, to push their relevance beyond the narrow confines of Ministries of Health or Family Welfare, and to bolster their claims on government resources. For the reproductive health and reproductive rights movements, UNFPA, taking up the mandate of the 1994 Cairo conference, has done this most effectively [41].

Health and human rights advocates also use the formal human rights system and traditional human rights analysis to press states to take the necessary steps to improve the conditions that influence women's health. When a state ratifies a human rights treaty, it agrees to be legally bound by its provisions. However,

human rights law is not like U.S. domestic law, for example, where legislative provisions are usually quite specific to start with, and then, pursuant to such legislation, an executive agency such as the Department of Health and Human Services issues regulations elaborating procedures for implementation. The courts, by deciding specific cases, further flesh out the law through an accumulation of judicial decisions that function as precedents for analogous situations. By contrast, human rights provisions are generally stated in broad and essentially aspirational terms. For example, the primary statement of the right to health (which actually contains more detail than many human rights provisions) reads in full as follows (Article 12, Int'l Covenant on Economic, Social, and Cultural Rights) [28a]:

1. The States Parties to the present Covenant recognize the right of everyone to the enjoyment of the highest attainable standard of physical and mental health.

2. The steps to be taken by the States Parties to the present Covenant to achieve the full realization of this right shall include those necessary for:

(a) The provision for the reduction of the still-birth rate and of infant mortality and for the health development of the child;

(b) The improvement of all aspects of environmental and industrial hygiene;

(c) The prevention, treatment and control of epidemic, endemic, occupational and other diseases;

(d) The creation of conditions which would assure to all medical service and medical attention in the event of sickness.

Thus a critical area of work in the women's health and human rights field seeks to elaborate exactly what a state must do to meet the obligations set forth in human rights treaties in this general way. This requires legal analysis to elaborate the nature and scope of particular rights within the law, which typically means analyzing how aspects of the same general right are articulated in multiple treaties. The bulk of this work has been done by human rights scholars writing mostly in academic journals [32,42–45]. However, work in this vein is also being done within the treaty oversight committees themselves—often with input from nongovernmental organizations [46]—as they develop General Recommendations on health that are designed to flesh out the meaning of the general provisions found in their respective conventions (*e.g.,* CEDAW's General Recommendation on Article 12: Women and Health, CEDAW/C/1999/1/WG.11/WP.2).

There is an important role here for interdisciplinary collaboration between human rights scholars and health researchers. For example, in an article [47] on human rights and maternal mortality, Yamin and Maine show how process indicators, developed via epidemiologic research and analysis and issued jointly by WHO, UNICEF, and UNFPA as "Guidelines for Monitoring the Availability and Use of Obstetric Services," can be used as human rights indicators to monitor a state's fulfillment of its legal obligation under Article 2 of the International Covenant on Economic, Social, and Cultural Rights. That provision requires states to take steps "to the maximum of its available resources" toward the progressive realization of the rights recognized in the Covenant. The WHO/UNICEF/UNFPA Guide-

---

[6]Nevertheless, the question of whether *public health* policy can influence health trends significantly actually remains controversial. Much of that controversy has swirled around the highly influential work of Thomas McKeown, analyzing nineteenth century mortality patterns in Britain in his book, *The Modern Rise of Population.* Attacks on McKeown's work charge that, in analyzing the causes of dramatic declines in mortality, he failed—perhaps because of his own political predilections—to credit the impact of public health measures and the role of public health/social reform movements [37,38]. Others have raised the same questions in studying the history of tuberculosis control [39].

lines can help establish exactly *which* kinds of steps will actually help to realize a woman's right to conditions that enable her to give birth safely; they can also provide a method for assessing progress toward that goal [47]. If such indicators become not simply the mark of "good" health policy but also the measure of fulfillment of human rights obligations, the campaign to have them followed by states—and thus reduce maternal mortality—gains a new form of leverage.

Still working from the perspective of the formal law, more critical, sometimes feminist, work examines particular problems within human rights law and practice that have limited its usefulness for addressing the full range of women's health issues. For example, a number of articles consider the central problem of the distinction between the public and the private in international law and the ways in which it operates to obscure the massive violations against women that take place in the domestic sphere [43,48]. Other work analyzes how human rights law by its overwhelming focus on some issues in women's health—especially reproductive health and rights—misses other key health and human rights problems such as lesbian health issues [49].

In all of these areas, it is important to recognize that human rights law and practice is in its infancy. Moreover, the enforcement mechanisms now available within the formal human rights system for health-related rights are relatively weak because they rely almost exclusively on an "expose-and-embarrass" strategy. Thus, only as women's health and rights issues rise in importance on the public agenda generally does the deterrence value of exposing violations rise as well. However, despite the current weakness in enforcement and the highly theoretical or abstract nature of much of this writing, such work that basically seeks to put meat on the bones of the still very skeletal human rights system will ultimately be crucial for the usefulness of human rights law in the women's health field.

C. Mobilizing Health and Human Rights Discourse for Social Action

As important as these developments are, it is by now almost a truism that changes in formal law and the effective enforcement of formal law are necessary *but not sufficient* steps in effecting true social change. For those working within the women's health and human rights movements, the need for fundamental social change is abundantly clear. The inequalities associated with poor health for women are not just legislative aberrations; they are the result of distributions of power so deeply embedded in the social fabric—in the cultural and religious beliefs, economic relationships, and social hierarchies that shape everyday life—as to appear to be simply the natural and inevitable order of things. Going beyond formal law, then, can human rights discourse provide a useful set of tools for (1) exposing the patterns of dominance and exclusion that underlie poor health; and (2) building an affirmative approach to health and health services that will support and advance women's right to the highest attainable standard of physical and mental health?

Here I will use the issue of violence against women to illustrate, first, how a problem with truly massive and pervasive consequences both for women's physical and mental health and for their human rights can simply be *invisible*—hidden from public scrutiny and immune to social policy—and how a human

rights approach exposes that phenomenon. Second, by tracing the trajectory of the international movement opposing gender violence, I illustrate how human rights and public health discourses have been used together in the struggle to ensure that violence against women is not seen simply as a problem of individual deviant men connected to individual victimized women, but rather as a structural problem requiring structural solutions. This kind of effort to change the perception of the nature of a problem—to change the categories through which it is recognized, understood, and addressed—is a primary aim of human rights work that functions in tandem with the initiatives for formal legal change described in the previous section.

1. Exposing the Problem of Violence against Women

Acknowledging that there has indeed been real progress during the 1990s in the way some health providers deal with domestic violence issues, I nevertheless want to use here a study conducted in 1987 to illustrate my larger point about how hegemonic discourse functions and about the potential role of human rights in exposing and challenging its power.[7] Until recently the health care setting was one of the primary places in which domestic violence, though massively present, went routinely unacknowledged. Often the first point of contact for a battered woman, many health providers readily treated the physical symptoms—a shattered bone, a bruised limb, a burn, a knife or bullet wound—but failed to treat the person or even to recognize the violent event (much less the power dynamics) that lay behind the symptoms. The health system, reinforcing the powerlessness of the victim, became part of the problem.

What was it about the way that health professionals functioned, about the models through which they were trained to analyze and address their patients' problems, that made the medical system fail so miserably to meet the true needs of women who were experiencing domestic violence? In a study conducted at a large public hospital in Chicago, Warshaw examined medical records of all women seen in the emergency room for a two-week period in 1987. She selected 52 cases in which the women presented with injuries that were clearly caused by the deliberate actions of another person. Her review of the hospital charts focused on how information was recorded by the triage nurse and the treating physician, on the characteristics of the patient and the physician, and also on the ultimate disposition of the case. Statements on the charts were rated according to "degree of disembodiment (hit by thing versus by a body part versus by a person) and instrumentality (something happened to her versus someone did something to her)" [50, p. 508] Warshaw hypothesized that "the way statements were recorded on the chart would reflect how the triage nurse and the physician organized their perceptions about the patient," taking the most decontextualized or disembodied statements as indicative of a "biomedical" model. She further hypothesized that the charts with the most decontextualized statements would also be the ones with the least appropriate outcomes (*i.e.*, with failure to acknowledge abuse and to refer accordingly).

Despite the fact that this emergency room had a formal protocol for handling cases of women at risk for domestic abuse,

---

[7]Thanks to Lynne Stevens for drawing to my attention to this body of literature about patient–provider interactions and gender-based violence.

Warshaw found that "in all but one of these cases, the most serious and potentially life-threatening risk for the patient, that of ongoing domestic violence, was not reflected in the discharge diagnosis or in the disposition, despite its blatant presence on the chart and in the patient's life." [50, p. 510]. Analyzing how the biomedical model and its discursive forms structure a doctor–patient encounter, Warshaw highlights the disembodied language that focuses on the mechanism of injury—the collision of two body parts, fist and mouth—detached from the persons to whom they belong and from the context of the lives those persons live. She reproduces the following verbatim from one typical chart:

NURSE TRIAGE NOTE: Was hit on upper lip. Teeth loose and c̄ dislocation. Happened last night.

MD RECORD: Patient 25y/o BF c/o [complained of] swelling and pain on the mouth after was hit by a fist about 5 hours ago. No LOC [loss of consciousness], no visual symptoms, no vomiting, no nausea.

PHYSICAL EXAM: Afebrile, hydrated, conscious, oriented $\times$ 3 [time, place, person]. HEENT [head, eyes, ears, nose, throat]: has swelling in the upper lip and loose teeth. No evidence of Fx [fracture].

XRAY: No Fracture

DISCHARGE DIAGNOSIS; Blunt Trauma Face

DISPOSITION: Ice Packs, Oral Surgery Clinic appt, Motrin

It is what the chart *fails* to record, rather than the information it includes, that tells us most:

What is missing? The nurse's note does not mention who hit her, what her relationship to this person was, what the circumstances of the attack were, or why she waited five hours to seek medical help. We also see that there is no subject in this statement and that the woman is already out of the picture—only her lip and teeth are there. [50, p. 512]

Warshaw draws on the literature about doctor–patient interactions to show how the choice of language and category reflects and reinforces the health provider's inability or unwillingness to recognize and address the underlying pattern of violence. Pointing to the way in which "medicalizing" violence enables the physician to maintain control over the interaction with the woman and over the truncated handling of her "pathology" within the medical system, Warshaw draws the following conclusion—devastating from the perspective of the fundamental human rights of the battered woman:

Together, the doctor, nurse, and patient construct a "medical history" that is, in fact, ahistorical—one that extracts an event from its context, even from its status as an event, something that has happened in time and space. The dynamics of an abusive relationship are recreated in an encounter in which *the subjectivity and needs of the woman are reduced to categories that meet the needs of another, not her own, a relationship in which she as a person is neither seen nor heard.* [50, p. 513] (emphasis added).

Human rights begins right here. At its most fundamental level, human rights is about the entitlement of every person to be treated with dignity, as a subject, a full human being with a physical body, an emotional/psychological existence, a life complexly connected to others. It is about the right to be an agent in one's own life and not simply a tool or instrument to meet the needs or desires of others. It is the right to be seen, to be heard, to be listened to.

There are, of course, many reasons why a woman might not receive full, empathetic, holistic treatment in the emergency room of a large public inner-city hospital. But imagine the same encounters described by Warshaw, except cast them with women confident of their entitlement within the health care system and within their intimate relationships, and with health providers who see the clinical encounter as one in which the patient's interests, needs, and rights should be controlling [51]. The limitations in resources, the overcrowding in emergency rooms, the poor linkages with social service agencies, and the justice system—all the factors that conspire with the biomedical model of provider–patient interactions to muffle the violence and disempower the woman—would all be handled rather differently if the encounter were premised, first and foremost, on the need to protect and ensure a woman's assertion of her human rights.

Of course, that would require a fundamental reorientation of the operative principles that animate the health care delivery system—perhaps not something altogether impossible. For example, within the admittedly specialized area of family planning service delivery, this is exactly what The International Planned Parenthood Federation's (IPPF) "Bill of Rights" is intended to do. In many countries, provision of contraceptive services has, historically, been profoundly skewed by anti- or pronatalist state population policies or by dynamics within the health care setting that are blatantly dismissive of the individual woman client's needs and desires. The "Bill of Rights" is meant to be a set of principles to inspire and guide a fundamental reorientation of the dynamics of service provision in order to put fulfillment of the client's human rights first, as the controlling principle. The rhetorical commitment of IPPF from the headquarters is impressive; the actual impact in the private spaces of provider–client interactions in IPPF-affiliated family planning clinics around the world remains an open, and critically important, question.

2. Linking the Individual to Broader Structures of Inequality

Certainly a health care system will never respond well to violence against women unless it is reoriented to retrieve and preserve the individual's basic dignity. But that is not enough. A strictly medical approach (or even a multidisciplinary clinical approach), no matter how sensitive and caring, is still an approach primarily focused on the individual woman and the specific factors that contribute to violence in her particular case. If violence against women has its roots in the deeply-ingrained systems of gender inequality, then, in the long run, individual human well-being can not be detached from broader questions of social justice.

Surely the existence of domestic violence, as well as other forms of gender-based violence such as rape and sexual abuse, is well-known to women everywhere, for they experience it in virtually every corner of the world [52,53]. However, beginning in earnest in the mid-1970s and spurred in part by the UN Decade for Women (1975–1985), women's movements throughout the world took up the issue of domestic violence within their own countries, as both a personal and a political issue. A wide range of initiatives were developed in different countries designed to address particular aspects of the problem most relevant in their own contexts [54]. As these initiatives were beginning to make inroads at the local level, it was becoming increasingly problematic that violence against women had barely surfaced on any international agenda, even as international policies were coming to have greater and greater influence over the actual conditions that prevailed locally and nationally. Seeing this gap, women began to organize across borders to reframe violence against women simultaneously as a public health issue and as a human rights issue.

If medicalizing gender violence was the first step toward "invisibilizing" it, as Warshaw's work suggests, then converting gender violence into a public health issue was meant to do precisely the opposite. By showing the massive effect that violence has on a population's health, and consequently on productivity and overall development of the society too [53,55], gender violence—already pegged by women surveyed internationally as the number one issue of concern to them [53]—could be made part of the mainstream health, development, and human rights agendas.

These efforts to position gender violence as a public health problem, building on extensive organizing by women in the South and the North together, have shown real success in changing the international agenda. For example, in the international health field, the standard definition of reproductive rights used in UN documents beginning in the mid-1970s got a new formulation when women organized to influence the drafting process in the Cairo and Beijing conferences. "The right to decide freely and responsibly on the number and spacing of children, and to have the information, education, and means to do so," language present in the final declarations of the two previous UN Population Conferences (Bucharest 1974 and Mexico City 1984), articulated what was understood essentially as a fairly narrow right to family planning (see Freedman and Isaacs 1993). But in Cairo in 1994 that language became:

> . . . the right to decide freely and responsibly on the number, spacing and timing of their children and to have the information and means to do so, and the right to attain the highest standard of sexual and reproductive health. It also includes the right to make decisions concerning reproduction free of discrimination, coercion and violence, as expressed in human rights documents. [1, ¶7.3]

The following year at the Women's Conference in Beijing, the Platform for Action repeated essentially this formulation and also included an entire section on violence against women [2, Chapter IV].

Still, a major obstacle in the struggle to raise gender violence as a public health and human rights issue is the lack of solid, population-based data. Riding the momentum initiated by women's movements in the Cairo and Beijing conferences, WHO has now taken the lead in developing this information through the sponsorship of a multicountry study and the development of a Violence Against Women database (see *http://www.who.int/frh-whd*).

As women in the health field were mobilizing to extract violence against women from the medical model and to recast it as a public health issue, the women's rights movement was confronting similar issues in a different set of spaces. The virtually total absence of violence against women from the world's human rights agenda was the clearest evidence that the reality of women's lives still had not penetrated international policymaking processes. Significantly, even the Convention on the Elimination of All Forms of Discrimination Against Women (1979) [28a], the central legal document to emerge from the UN Decade for Women, nowhere even mentioned violence.[8] Not only governments, but also mainstream human rights NGOs, had neglected it altogether [56–58]. Even within the broader rights movement, progressive organizations committed to social change glossed over gender violence as a "private" issue, unworthy of concerted public attention, shunned as a problem too intertwined with the prickly issue of culture to risk approaching.

Yet here was a practice (violence against women) which, though it functioned in many different forms and with many different justifications across different cultures, was everywhere committed to support one basic goal: to enforce the control of men over women. With violence came terror, and the purpose of this terror, like the purpose of torture more generally, was to destroy the voice and the subjectivity of its victims [59–61]. From the perspective of its targets, here was a quintessential violation of human rights—yet virtually nowhere in the human rights laws, institutions, or movements was it recognized as such.

Women who had been organizing around gender violence within their own countries for years decided to come together to create an international initiative, designed to get gender violence on the international policy agenda. To do this, they used human rights discourse as an organizing tool and the formal human rights system—thus far deaf to the issue—as their target. Mobilizing around a simple rallying cry—"Women's Rights are Human Rights"—women's rights advocates used the human rights movement's own basic values and principles to expose how that movement had itself buried the issue. Their success was impressive. At the World Conference on Human Rights held in Vienna in 1993, women succeeded in having violence against women, including violence in the private sphere, recognized as a human rights violation [62]. This was followed by the adoption by the UN General Assembly in December 1993 of the Declaration on the Elimination of Violence Against Women and the appointment of a Special Rapporteur on the Causes and Consequences of Violence Against Women.

IV. Controversies

Violence against women was the issue that catalyzed an international movement. Significantly, this was not a process in which the direction of analysis or energy went North to South. Although not without its moments of controversy and tension, this was truly a movement built with consciousness of the legacy of ethnocentrism and colonialism (including within the women's movement) and with serious attempts to wrestle with the dangers that come from essentializing women, from stereotyping them

---

[8]This omission was rectified when the treaty oversight committee passed General Recommendation 19 which interprets many of the articles of the convention to cover violence against women (UN DOC HRI/GEN/1/REV.1 at 78 (1994)).

by race, ethnicity, nationality, religion, or class. The struggle for women activists was to affirm the universality of human rights as the fundamental entitlement of every human person, thereby affirming their commonalities, but also to recognize the specificities that shaped the actual experience of different women. Not least of those tensions between commonalities and specificities lay in the myriad, often contradictory, ways in which women have been affected by globalization and the ascendancy of free-market capitalism after the end of the Cold War.

Even domestic violence can not be delinked from the growing impoverishment experienced in vast portions of the world since the late 1980s. For example, studies conducted as early as 1988 documented an explicit connection between the implementation of International Monetary Fund and World Bank structural adjustment programs, the upheaval that the resulting impoverishment caused, and an upsurge in domestic violence [63]. There is another deeper level at which the connection between globalization and violence against women needs to be understood. The existence of globalization has meant that people are aware as never before of the world outside the boundaries of their communities. Such awareness clearly can cut two ways, either reinforcing the grip of (often imagined) tradition in the face of perceived threats from outside [64] or expanding the awareness of choice and of the possibilities of different ways to be in the world.[9] Thus, as Lock and Kaufert [21] point out:

> As a result of globalization, hegemonic power . . . is a shrinking domain. In other words, common sense—the unspoken authority of everyday life—becomes increasingly subject to disputation. Orthodoxy—that which is "naturalized," hegemonic, and taken as self-evident—is brought into consciousness and made recognizable as ideology, and is therefore laid bare for criticism.

There is some small and painful irony here. If "orthodoxy" includes the view that men's control over women is part of the natural common sense order of things, and that violence against women is a legitimate part of male prerogatives in maintaining that order, then the exposure of gender violence not as divinely-ordained and inevitable, but only as man-made ideology, can be a profound challenge. In a vicious and violent cycle, such challenges sometimes provoke more violence: "Individuals who dispute either physical violence or "symbolic violence"—the institutionalized violence of everyday life . . . —are considered dangerous to a conservative moral order, which is itself undergoing a renewed vitalization with the resurgence of various forms of global fundamentalism and the elaboration in North America of the New Right." [21, pp. 5–6].

The women's health and human rights movement, at its most transformative moments, challenges some of the most firmly entrenched inequalities that exist in any society. That primary fact should be the backdrop against which far more "academic" arguments about the appropriateness of an international human rights movement are considered. Rarely do opponents of human rights openly defend inequality or violence (although such inhibition often disappears in the case of women's human rights).

[9] See Shaheed [65] for an elaboration of how this dynamic works, for example, in activist organizing for women in Muslim countries and communities around the world.

Instead, the attacks against the human rights movement have centered on its focus on the individual and on the supposed incompatibility between the "individualism" of human rights and the values of non-Western societies.

Although any full discussion of these issues is obviously beyond the scope of the chapter, no account of human rights initiatives in the women's health field can afford to ignore them altogether. On the one hand, we need to acknowledge that the development of the human rights movement in tandem with the politics of the Cold War did create a mainstream human rights movement whose most visible and vocal proponents—though not its only proponents—were based in the West. For a range of different reasons, these groups and the human rights system as a whole focused on civil and political rights, employing a philosophical and political model associated with (Western) Enlightenment liberalism. The unfortunate legacy of this period is the common misperception of human rights as *necessarily* linked with a fierce individualism.

Indeed, to the extent its focus on the individual yields a narrow, conservative framework that aligns human rights with neoliberal economic principles, and with a view of human nature that conceives of each person as an isolated, autonomous individual, fighting to pursue his own self-interest, free from government interference, then its usefulness for women's health advocates will be severely limited. To the extent that human rights talk is perceived as a strategy being used cynically and selectively to extend the global reach of U.S. and European economic and political power—as a Trojan horse disguising what is merely the latest form of "cultural imperialism" [66]—it will be forcefully rejected. Nor will a human rights regime based on such radical individualism help to expose the workings of power that underlie poor health. As Scott Burris quite elegantly put it, "to accept the rhetorical structure of market individualism is to accept a political language that has no words for public health." [67, p. 1608].

Moreover, there is certainly a legitimate, necessary, and good-faith discussion that must be had about the "foreignness" of certain norms codified in international law and about the need to build—not simply assume—universality in a dynamic system [68]. There is also a serious argument that, however Western in origin human rights might be, rights-based strategies are ultimately not supportive of progressive social change in Western societies either [69]. Finally, there is further debate about whether, from a "postmodernist" perspective, it is possible to identify *any* propositions about the human condition as universal.

On the other hand, we need to be clear: the attacks on human rights by some authoritarian governments are often thinly-disguised manipulations to shield their own violations [70]. Attacks by other forces such as many religious fundamentalist groups are often equally cynical. When threatened by the governments they oppose, such groups readily invoke their own human rights and seek support and shelter from human rights advocates. But when women raise the same human rights provisions in defense of their own self-assertion, they are viciously attacked as "Western" and as traitors to their religion, nation, or ethnic group [27].

Perhaps the most useful way to approach this debate and to clarify the nature of human rights in the women's health field is from the perspective of the actual experience of using human rights within the women's health movement. Most would readily acknowledge that human rights is susceptible to cooptation in service of the most conventional of interests and the most hegemonic of powers. But the fact is that women's movements around the world have also captured human rights discourse and methods—legal and otherwise—for their own purposes. In doing so, they have forcefully used it in aid of the most basic goal of women's health movements everywhere: to expose and challenge the configurations of power that create the conditions that spawn ill health and that work their power by appearing natural, inevitable, and immutable.

Where the conventional discussion of universalism vs cultural relativism is carried on with the unspoken assumption that cultures (or communities) are homogeneous, bounded, separable, and isolated units, women's health and human rights movements have assumed precisely the opposite position. First, all cultures are dynamic, with internal struggle over meanings and definitions as a key feature [68]. This is particularly important for women who are so often marginalized in the process of "defining" their culture—even as culture is regularly used to keep them in their place and to quash any resistance. Yet resist they do. Studies show that across cultures and classes, women have a sense of their own entitlement as individual women, born out of the circumstance of their lives—and often they are willing to assert that entitlement, even if sometimes only in secretive or tentative ways [71].

Second, one of the implications of globalization is that virtually no culture is untouched by others. Thus, discussion about cultural differences, identity, and respect for others must proceed carefully. When diversity becomes an essentialized "otherness" it undermines the very possibility of seeing shared interests and mobilizing across community and cultural borders, trapping women in the "naturalized" role of bearers of tradition and markers of cultural "authenticity" [72]. The human rights framework, particularly when used as it has been by women's groups, helps guard against this trap.

The development of an international women's health and human rights movement has just begun to negotiate through these rough waters. But the fact that it finds itself faced with such dilemmas should be heartening. Irrelevance is comfortable and quiet. True change in the patterns of dominance and exclusion that underlie women's health is likely to be a noisy affair.

References

1. United Nations (1995). "Population and Development. Vol. 1: Programme of Action Adopted at the International Conference on Population and Development: Cairo: 5–13 September 1994" (Sales No. E.95.XIII.7). United Nations, New York.
2. United Nations (1996). "The Beijing Declaration and the Platform for Action: Fourth World Conference on Women: Beijing, China: 4–15 September 1995" (DPI/1766/Wom). Department of Public Information, New York.
3. International Planned Parenthood Federation (1996). "IPPF Charter on Sexual and Reproductive Rights." IPPF, London.
4. Secretaria Executiva da Rede Nacional Feminista de Saúde e Direitos Reproductivos (1997). "Final Document, 8th International Women and Health Meeting" (redesaude@uol.com.br). Rio de Janeiro, Brazil.
5. Swiss, S., and Giller, J. E. (1993). Rape as a crime of war. A medical perspective *JAMA, J. Am. Med. Assoc.* **270**(5), 612–615.
6. Freedman, L. P., and Maine, D. (1995). Facing facts: The role of epidemiology in reproductive rights advocacy. *Am. Univ. Law Rev.* **44**, 1085–1092.
7. Proctor, R. N. (1991). "Value-Free Science? Purity and Power in Modern Knowledge." Harvard University Press, Cambridge, MA.
8. Proctor, R. N. (1995). "Cancer Wars: How Politics Shapes What We Know and Don't Know About Cancer." Basic Books, New York.
9. Satel, S. (1996). The politicization of public health. *Wall Street J.,* December 12, p. A12.
10. MacDonald, H. (1998). Public health quackery. *City J.,* Autumn, pp. 40–52 (published by The Manhattan Institute).
11. Laumann, E. O., Michael, R. T., and Gagnon, J. H. (1994). A political history of the national sex survey of adults. *Fam. Plann. Perspect.* **26**(1), 34–38.
12. Oakley, A. (1993). "Essays on Women, Medicine and Health." Edinburgh University Press, Edinburgh.
13. Ehrenreich, B., and English, D. (1978). "For Her Own Good: 150 Years of Experts' Advice to Women." Doubleday, New York.
14. Smart, C. (1992). Disruptive bodies and unruly sex: The regulation of reproduction and sexuality in the nineteenth century. *In* "Regulating Womanhood: Historical Essays on Marriage, Motherhood and Sexuality" (C. Smart, ed.), pp. 7–32. Routledge, New York.
15. Ehrenreich, B., and English, D. (1990). The sexual politics of sickness. *In* "The Sociology of Health and Illness: Critical Perspectives" (P. Conrad and R. Kern, eds.), 3rd ed., pp. 270–284. St. Martin's Press, New York.
16. Link, B., and Phelan, J. (1995). Social conditions as fundamental causes of disease. *J. Health Soc. Behav., Extra Issue,* pp. 80–94.
17. Freedman, L. P. (1997). Human rights and the politics of risk and blame: Lessons from the international reproductive health movement. *J. Am. Med. Women's Assoc.* **52**(4), 165–168.
18. Tesh, S. N. (1988). "Hidden Arguments: Political Ideology and Disease Prevention Policy." Rutgers University Press, New Brunswick, NJ.
19. Kunitz, S. (1987). Explanations and ideologies of mortality patterns. *Popul. Dev. Rev.* **13**(3), 379–408.
20. Freedman, L. P., and Maine, D. (1992). Women's mortality: A legacy of neglect. *In* "The Health of Women: A Global Perspective" (M. Koblinsky *et al.,* eds.), pp. 147–170. Westview Press, Boulder, CO.
21. Lock, M., and Kaufert, P., eds. (1998). "Pragmatic Women and Body Politics." Cambridge University Press, Cambridge, UK.
22. Comaroff, J., and Comaroff, J. L. (1991). "Of Revelation and Revolution: Christianity, Colonialism and Consciousness in South Africa," Vol. 1. University of Chicago Press, Chicago.
23. Susser, M., and Susser, E. (1996). Choosing a future for epidemiology: II. From black box to Chinese boxes and eco-epidemiology. *Am. J. Public Health* **86**(5), 674–677.
24. Pearce, N. (1996). Traditional epidemiology, modern epidemiology, and public health. *Am. J. Public Health* **86**(5), 678–683.
25. Krieger, N., and Birn, A. (1998). A vision of social justice as the foundation of public health: Commemorating 150 years of the spirit of 1848. *Am. J. Public Health* **88**(11), 1603–1606.
26. Freedman, L. P., and Isaacs, S. (1993). Human rights and reproductive choice. *Stud. Fam. Plann.* **24**(1), 18–30.
27. Freedman, L. P. (1995). Reflections on emerging frameworks of health and human rights. *Health Hum. Rights* **1**(4), 314–348.
28. United Nations (1945). "Charter of the United Nations," 59 Stat. 1031. United Nations, New York.
28a. Center for the Study of Human Rights, Columbia University. (1994). "Twenty-five Human Rights Documents."
29. Article 19 (1990). "Starving in Silence: A Report on Famine and Censorship," Article 19. International Centre on Censorship, London.
30. Mann, J. *et al.* (1994). Health and human rights. *Health Hum. Rights* **1**(1), 6–23.

31. Cook, R. J. (1993). "Human Rights in Relation to Women's Health: The Promotion and Protection of Women's Health through International Human Rights Law," (WHO/DGH/93.1) Prepared for the World Health Organization, Geneva.

32. Cook, R. J. (1995). Human rights and reproductive self-determination. *Am. Univ. Law Rev.* **44**(4), 975–1016.

33. Physicians for Human Rights (1998). "The Taliban's War on Women: A Health and Human Rights Crisis in Afghanistan." PHR, Boston.

34. Rasekh, Z. *et al.* (1998). Women's health and human rights in Afghanistan. *JAMA, J. Am. Med. Assoc.* **280**(5), 449–455.

35. Swiss, S. *et al.* (1998). Violence against women during the Liberian civil conflict. *JAMA, J. Am. Med. Assoc.* **279**(8), 625–629.

36. Women Law & Development International (1998). "Gender Violence: The Hidden War Crime." Women Law & Development Intl., Washington, DC.

37. Szreter, S. (1997). Economic growth, disruption, deprivation, disease and death: On the importance of the politics of public health for development. *Popul. Dev. Rev.* **23**(4), 693–728.

38. Johansson, S. R. (1994). Food for thought: Rhetoric and reality in modern mortality history. *Hist. Methods* **27**(3), 101–125.

39. Fairchild, A., and Oppenheimer, G. (1998). Public health nihilism vs pragmatism: History, politics, and the control of tuberculosis. *Am. J. Public Health* **88**(7), 1105–1117.

40. Osborne, T. (1997). Of health and statecraft. *In* "Foucault, Health and Medicine" (A. Petersen and R. Bunton, eds.), pp. 173–188. Routledge, London.

41. United Nations Population Fund (UNFPA) (1997). "The State of the World's Population 1997: The Right to Choose: Reproductive Rights and Reproductive Health." UNFPA, New York.

42. Leary, V. (1994). The right to health. *Health Hum. Rights* **1**(1).

42a. Sullivan, D. (1995). The nature and scope of human rights obligations concerning women's rights to health. *Health Hum. Rights* **1**(4), 368–398.

43. Sullivan, D. (1995). The public/private distinction in international law. *In* "Women's Rights, Human Rights" (V. Peters and A. Wolper, eds.), pp. 126–134. Routledge, New York.

44. Rahman, A., and Pine, R. (1995). An international human right to reproductive health care: Toward definition and accountability. *Health Hum. Rights* **1**(4), 400–427.

45. Packer, C. (1996). "The Right to Reproductive Choice: A Study in International Law." Institute for Human Rights, Abo Akademi University, Turku/Abo, Finland.

46. Gruskin, S., and Sullivan, D. (1996). Proposal for draft general comment on the right to health under the ICESCRI (on file with the author).

47. Yamin, A. E., and Maine, D. (1999). Maternal mortality as a human rights issue: Measuring compliance with international treaty obligations. *Hum. Rights Q.* **21**(3), 563–607.

48. Yamin, A. E. (1997). Transformative combinations: Women's health and human rights. *J. Am. Med. Women's Assoc.* **52**, 169–173.

49. Miller, A., Rosga, A., and Satterthwaite, M. (1995). Health, human rights and lesbian existence. *Health Hum. Rights* **1**(4), 428–448.

50. Warshaw, C. (1989). Limitations of the medical model in the care of battered women. *Gender Soc.* **3**(4), 506–517.

51. Maine, D., Freedman, L., Shaheed, F., and Frautschi, S. (1994). Risk, reproduction and rights: The uses of reproductive health data. *In* "Population and Development: Old Debates, New Conclusions" (R. Cassen, ed.), pp. 203–227. Overseas Development Corporation, Washington, DC.

52. United Nations, Centre for social Development and Humanitarian Affairs (1989). "Violence Against Women in the Family" (ST/CSDHA/2) (Sales No. E.89.IV.5). United Nations, New York.

53. Heise, L., Pitanguy, J., and Germain, A. (1994). "Violence Against Women: The Hidden Health Burden," World Bank Discuss. Pap. No. 255, pp. 171–195. World Bank, Washington, DC.

54. Heise, L. (1993). Violence against women: The missing agenda. *In* "The Health of Women: A Global Perspective" (Koblinsky, Timyan, and Gay, eds.). Westview Press, Boulder, CO.

55. Heise, L. (1993). Reproductive freedom and violence against women: Where are the intersections? *J. Law, Med. Ethics* **21**(2), 206–216.

56. Ashworth, G. (1986). "Of Violence and Violation: Women and Human Rights. CHANGE, London.

57. Bunch, C. (1990). Women's rights as human rights: Toward a re-vision of human rights. *Hum. Rights Q.* **12**, 486–498.

58. Roth, K. (1994). Domestic violence as an international human rights issue. *In* "Human Rights of Women: National and International Perspectives" (R. Cook, ed.), pp. 326–339. University of Pennsylvania Press, Philadelphia.

59. Copelon, R. (1994). Intimate terror: Understanding domestic violence as torture. *In* "Human Rights of Women: National and International Perspectives" (R. Cook, ed.), pp. 116–152. University of Pennsylvania Press, Philadelphia.

60. Slaughter, J. (1997). A question of narration: The voice in international human rights law. *Hum. Rights Q.* **19**(2), 406–430.

61. Scarry, E. (1985). "The Body in Pain: The Making and Unmaking of the World." Oxford University Press, New York.

62. United Nations (1993). "Report of the Drafting Committee: Final Outcome of the World Conference on Human Rights," A/CONF.157/DC/1/Add.1. United Nations, New York.

63. Moser, C. (1989). The impact of recession and adjustment policies at the micro-level: Low income women and their households in Guayaquil, Ecuador. *In* "The Invisible Adjustment: Poor Women and the Economic Crisis," pp. 137–166. UNICEF, The Americas and The Caribbean Regional Office, Chile.

64. Freedman, L. P. (1996). The challenge of fundamentalisms. *Reprod. Health Matters* **8**, 55–69.

65. Shaheed, F. (1994). Controlled or autonomous: Identity and the experience of the network Women Living Under Muslim Laws. *Signs* **19**, 997–1019.

66. Esteva, G., and Prakash, M. S. (1998). "Grassroots Post-Modernism: Remaking the Soil of Cultures." Zed Books, New York.

67. Burris, S. (1997). The invisibility of public health: Population-level measures in a politics of market individualism. *Am. J. Public Health* **87**(10), 1607–1610.

68. An-Na'im, A. (1990). Toward a cross-cultural approach to defining international standards of human rights. *In* "Human Rights in Cross-Cultural Perspectives" (A. An-Na'im, ed.), pp. 19–43. University of Pennsylvania Press, Philadelphia.

69. Hunt, A. (1990). Rights and social movements: Counter-Hegemonic strategies. *J. Law Soc.* **17**(3), 309–328.

70. Wilson, R. (1997). Human rights, culture and context: An introduction. *In* "Human Rights, Culture and Context: Anthropological Perspectives" (R. Wilson, ed.), pp. 1–27. Pluto Press, Chicago.

71. Petchesky, R., and Judd, K. (1998). "Negotiating Reproductive Rights: Women's Perspectives Across Countries and Cultures." Zed Books, New York.

72. Hélie-Lucas, M. A. (1999). What is your tribe? Women's struggles and the construction of Muslimness. *In* "Religious Fundamentalisms and the Human Rights of Women" (C. Howland, ed.). St. Martin's Press, New York (in press).

Part III

OCCUPATIONAL, ENVIRONMENTAL, AND SOCIAL DETERMINANTS OF HEALTH

Section 6

WOMEN AT WORK

Shelia Hoar Zahm
Division of Cancer Epidemiology and Genetics
National Cancer Institute
Rockville, Maryland

Employment patterns among women have undergone dramatic changes since the 1950s. The participation of U.S. women in the work force has risen from 34% in 1950 to 60% in 1997 [1]. The proportions of women with young children and of older women in the work force have also increased. In 1950, 12% of women with children under age 6 and 37% of women aged 45–54 years worked outside the home, whereas 60% of women with children under age 6 and 71% of women aged 45–54 years were in the work force in 1990 [1,2]. Women are spending a greater proportion of their lives working outside the home. The potential health implications are profound. This section presents the changing patterns of employment in the U.S. and throughout the world and evaluates the impact of these changes on women's roles and responsibilities, reproductive health, musculoskeletal disorders, cancer risk, and diseases linked to indoor air quality in the workplace.

With the increasing prevalence of employment among women, there also have been changes in the occupations held by women, as presented by Walstedt (Chapter 35). Women continue to work primarily in sales or service occupations [2], with secretary still the leading occupation, but more women now hold jobs that are "gender neutral," neither traditional for women nor men. The percentage of women holding multiple jobs has tripled from 2% in 1970 to 6% in 1996 [2,3]. Labor force participation is expected to continue to increase among women, albeit at a slower rate than in the past [4,5].

On average, women are paid less and have fewer employee benefits than men. In 1996, among full-time wage and salary workers, women's median weekly earnings were 75% of men's [2]. Most women are employed in small businesses and service industries, which typically provide fewer benefits than larger businesses and manufacturing industries. In addition, part-time employment, common among women, has fewer benefits than full-time employment.

Women account for approximately one-third of the two million nonfatal occupational injuries and illnesses involving days away from work [6]. They are more likely than men to lose work time for repetitive motion disorders as well as for assaults and violent acts [7]. The five occupations with the greatest number of nonfatal injuries and illnesses among women are nursing aides and orderlies, cashiers, assemblers, nonconstruction laborers, and cooks. Approximately 500 women per year are killed on the job, with about one-third due to homicide [8]. The homicides are generally related to robberies, but about one-fifth of the homicides are committed by a husband, ex-husband, boyfriend, or ex-boyfriend. Highway traffic accidents are the second leading cause of work-related fatalities [8].

Many working women have substantial domestic work and childcare responsibilities, as described in the chapter by Messing (Chapter 36). These responsibilities are typically borne to a greater degree by women than men [9,10]. Multiple roles may give women more choices and result in better mental and physical health [11] but may be stressful [12] and result in more fatigue [13], accidents [14], and illnesses [15], both psychological and physical. Variable and unpredictable work schedules, common in the sales and service industries where most women are employed, compound the difficulties in balancing work and home responsibilities. Women may be reluctant to advocate changes in the workplace that would accommodate their family responsibilities for fear of negative repercussions. With the significant increase in the proportion of working mothers with young children, more widespread adoption of programs such as family leave, flexible hours, and alternative work locations (*e.g.,* at home) is needed to help women balance work and family obligations.

Messing also identifies elements of women's occupations that are often unnoticed but that have adverse health effects. An example is prolonged standing, which certainly does not appear to be dangerous, but which causes back, leg, and foot pain [16]. Prolonged standing among bank tellers, grocery cashiers, restaurant workers, and sales clerks is not a dramatic occupational health hazard, and its effects are largely unstudied, but it can be debilitating. Other occupational settings have multiple stressors that, considered individually, do not seem serious, but when considered together contribute to high levels of stress and illness. For example, teachers may experience prolonged standing, frequent bending, need for constant concentration, interruptions, simultaneous demands for attention, chalk dust exposure, noise, uncomfortable temperatures and humidity, hours of unpaid work at home, the distress of dealing with children with emotional or learning problems, the bureaucratic challenges of arranging special help for these children, and public criticism [17,18]. The health effects of combinations of factors, even those with little apparent risk, need to be examined further.

The relationship between reproductive health and chemical, physical, biological, and psychosocial exposures is reviewed by Lindbohm and Taskinen in Chapter 37. Reproductive health research has expanded and now encompasses pregnancy outcome, fertility, menstrual function, semen characteristics, and hormone levels. Postnatal development may also be affected by exposures prior to conception or during pregnancy.

Strong evidence exists linking organic solvents, antineoplastic agents, and some physical work load factors to adverse reproductive outcomes. Organic solvents pass through the placenta and into breast milk. Solvents have been associated with spontaneous abortion, congenital malformations, reduced fertility, menstrual disorders, stillbirth, and perinatal death, as well as leukemia and brain tumors in children of exposed workers [19]. The evidence is strongest for spontaneous abortions, particularly for ethylene glycol ethers [20–22], tetrachloroethylene [19,23], and toluene [19]. Antineoplastic agents, many of which are carcinogens, have adverse effects on menstrual and ovarian function and have been associated with spontaneous abortion, congenital malformations, and slightly lowered birth weight [24–27]. Shift work, which can interfere with the circadian regulation of human metabolism and hormone concentrations, has been reported to increase risk of menstrual disorders, subfecundity, preterm birth, low birth weight, and, in some studies, spontaneous abortion [28–31]. Strong physical effort and prolonged standing have been related to prematurity in several studies [32].

There is also evidence in some, but not all, studies of adverse reproduction outcomes and developmental problems in connection with exposure to metals, anesthetic gases, pesticides, the sterilizing agent ethylene oxide, the antiviral drug ribavirin, carbon monoxide, polychlorinated biphenyls, radiation, noise, viruses, and mycotoxins. For example, low level lead exposure has been linked to preterm delivery and low birth weight, although not consistently [33]. Reduced fertility and increased spontane-ous abortion, stillbirth, congenital malformations, and cancer in offspring have been linked to pesticide exposure [34–38].

Musculoskeletal disorders, reviewed by Punnett and Herbert (Chapter 38), are the single largest category of work-related illness in the United States, Scandinavia, and Japan. They most often affect the back and upper extremities and occur excessively in occupations involving rapid and repetitive motions, forceful exertions, nonneutral body positions, and vibration. High rates among working women have been observed in the garment industry, packing jobs, textile manufacturing, electronics manufacturing, poultry processing, health care, food preparation and laundry workers, grocery cashiers, cleaners, childcare workers, and clerical workers operating video display units. Women are often reported to have higher rates of occupational musculoskeletal disorders than men, but a large proportion of the gender difference appears to be attributable to differences in ergonomic exposures. Some, but not all, studies showed higher rates among women after adjustment for exposures. Other studies suggest that men may have a higher risk than women with increasing exposure, but women have a higher background risk. Gender differences might be related to psychosocial stress at work [39], domestic work [40,41], pain reporting and health care seeking behavior [42,43], body size and strength [44], tendon and muscle composition [45], and effects of circulating endocrine hormones.

Studies of occupational cancer among women have not been extensive in comparison to research on men [46; Zahm *et al.,* Chapter 39]. In a survey of 1233 occupational cancer epidemiology reports, only 14% presented any analyses of white women and only 2% presented data on nonwhite women [46]. Even fewer reports presented in-depth analyses (more than five risk estimates) for women (white women: 7%, nonwhite women: 1%). The existing research has identified increased cancer risk associated with employment in agriculture, selected service occupations, and manufacturing [46], but detailed data are limited. Potential occupational carcinogens of special importance to women include antineoplastic drugs, anesthetic gases, ionizing radiation, and electromagnetic fields among health care personnel; electromagnetic fields among office workers; hair dyes, hair sprays, and formaldehyde among cosmetologists; tetrachloroethylene and trichloroethylene among dry cleaners; building and household cleaners among janitorial and housekeeping workers; pharmaceutical exposures among drug manufacturers; passive tobacco smoke among waitresses and bartenders; pesticides among farmworkers; solvent exposure among aircraft repair and other industrial workers; asbestos, dyes, and other exposures in the textile industry; and paints, metal fumes, machinery fluids, and other agents in motor vehicle manufacturing [Zahm *et al.,* Chapter 39].

Although virtually every cancer has been linked to occupational exposures to some degree, certain cancers, such as leukemia and cancers of the lung and bladder, appear to have a stronger occupational component. Increased risk of leukemia has been observed among women exposed to benzene, other solvents, vinyl chloride, antineoplastic drugs, other chemicals, radiation, and pesticides or employed in food processing or the textile and garment industries [Zahm *et al.,* Chapter 39]. Lung cancer rates have been elevated among women exposed to asbestos and metals, including arsenic, chromium, nickel, and mercury, as well as women employed in motor vehicle manufacturing or as waitresses, bartenders, cooks, or cosmetologists [Zahm *et al.,* Chapter 39]. Occupations associated with elevated risk for bladder cancer among women include dyestuff workers and dye users, textile workers, rubber and plastic workers, leather workers, painters, dry cleaners, health care workers, and miscellaneous manufacturing jobs [Zahm *et al.,* Chapter 39]. Religious workers, teachers, chemists, physicians, nurses, other health care workers, and women in professional and technical occupations are at high risk for breast cancer, probably due in part to decreased or delayed childbearing [Zahm *et al.,* Chapter 39]. Some of these occupations, however, involve exposure to radiation, solvents, and some other chemicals that have been associated with increased breast cancer risk, suggesting that reproductive patterns may not explain all of the excess risk for these jobs. Ovarian cancer has been strongly associated with asbestos exposure based on studies of women employed as gas mask assemblers prior to and during World War II [47–50].

Most employed women work in indoor environments, which have been linked to interstitial lung diseases, including organic dust toxic syndrome and asthma; allergic rhinitis; mucosal irritation; infections from people such as tuberculosis, pneumococcal pneumonia, and viruses; infections from building systems such as Legionnaires' disease; infections from animals; dermatitis; pesticide poisonings, and other conditions [Hodgson and Storey, Chapter 40]. Hodgson and Storey (Chapter 40) review these conditions as well as the "sick-building syndrome," an ill-defined constellation of symptoms attributed to the air quality, temperature, noise, or other characteristics of the workplace. The syndrome generally includes eye irritation, unusual tiredness, headache, and nasal symptoms, with improvement away from work [51].

Health problems related to indoor air quality have been reported more commonly among women than men in most studies [51,52]. Two studies reported that women were exposed to levels of volatile organic compounds and particulates that were almost twice as high as those encountered by men [53,54], suggesting the gender patterns are related to exposure differences. Other studies also support the hypothesis that gender per se does not explain the gender differences in symptoms related to indoor air [55,56].

The Women at Work section ends with an international perspective on women's occupational health by Stellman and Lucas (Chapter 41). Despite economic differences, many factors affecting women are similar throughout the world, such as the clustering of women in a relatively small number of occupations, barriers to advancement, underemployment, low or no remuneration, lack of job security, the added burden of domestic and childcare responsibilities, and sexual harassment and violence.

In developing countries, women generally have no political and economic power and may be effectively barred from many occupations by custom or by law. Much of the work done by women is viewed as "housework," subsistence activities, such as collection of water and fuel, or processing of food, and is unpaid. Women participate widely in agriculture in developing countries. This work and many other types of labor often generate income for the family, paid to the male head of the household, not to the woman herself. The home-based arts and crafts cottage industries engaged in by many women in developing

countries generally do not have any health and safety oversight and can cause serious occupational injuries and illness among women and other families members as well. As traditionally female jobs are mechanized and upgraded, they are often transformed into men's work, with a net loss for women.

The regulation of occupational health among women varies internationally. While some countries have egalitarian policies, many countries have nonexistent, minimal, or rarely enforced standards, particularly in developing countries, where much of women's work is "invisible" and unpaid. Discriminatory and protectionist practices are also found in some countries, barring women from jobs that often are more lucrative.

Research on employment patterns and occupational health risks among women is limited by several factors. In developing countries, population censuses with occupation and industry are not as common as in developed countries, and they generally inadequately report women's unpaid work. Even in developed countries, the quality of occupational data on routine administrative databases may be less for women than for men. For example, in the U.S., the usual occupation is supposed to be entered on the death certificate regardless of the employment status at the time of death. However, "housewife" is often entered even if women held long-term full or part-time paid employment outside the home [57–59], especially for older women who are retired at the time of death [57]. Even when housewives are excluded, the accuracy of information on women's certificates is less than that on men's certificates [59].

Most occupational disease research is based on job title or industry, but information on specific exposures is needed to understand the etiology of disease and take preventive action. Chemical, physical, and biologic factors must be identified and exposure-response gradients evaluated. Biological markers of exposure and of genetic susceptibility can be invaluable but are available infrequently. Self-reported exposure data are notoriously inaccurate and result in underreporting of exposures [60]. In case-control studies, self-reporting may also be affected by case response bias. As an alternative to self-reported data, exposures may be inferred from the job title or industry by experts or through use of job-exposure matrices. Gender differences in tasks and exposures within the same job title must be considered, however. Disease differences between men and women in the same job may be due to a different distribution of tasks, not to gender-specific host responses. Future research also needs to pay greater attention to multiple exposures, both physical and psychological. The complex mixture of factors affecting health needs to be taken into account and disentangled, if possible. More rigorous and validated methods to assess domestic chores, childcare responsibilities, and other nontraditional "exposures" are needed.

Limited disease outcome data can hinder occupational health research. In many countries, but not all, registries exist for mortality, cancer incidence, and some reproductive outcomes, such as congenital malformations. Ironically, the best registries and disease surveillance systems tend to be in countries with the best industrial hygiene, while countries with poorer industrial hygiene and more occupational disease have few data resources with which to monitor and study problems. Many outcomes, such as reduced fertility or illnesses related to indoor air quality, are generally not recorded in registries and must be ascertained through special surveys and ad hoc studies. For many conditions, particularly ill-defined syndromes such as the sick building syndrome, population background rates are not readily available. Improvements in surveillance systems would benefit occupational health research.

Other methodologic issues that affect occupational health research on women are inadequate statistical power due to small numbers of exposed affected women; ill-defined populations at risk, such as the number of women at risk for an adverse reproductive outcome; difficulties in tracing women over long periods of time; lack of appropriate comparison population disease rates for working women; absence of data on potential confounders; and challenges in attributing illness to occupational exposures if symptoms arise at home or many years after the exposure. Better study designs and research tools that address these methodologic issues are needed.

Overcoming the obstacles to research on occupational health among women will lead to increased knowledge of the etiology of disease and to safer workplaces. This research is essential so that the economic and social gains women have made in the workplace are not lost to occupational illness and injury.

References

1. Wagener, D. K., Walstedt, J., Jenkins, L., Burnett, C., Lalich, N., and Fingerhut, M. (1997). Women: Work and Health (DHHS Publ. No. PHS 97-1415). *Vital Health Stat.* **3** (31), 1–91.

2. U.S. Department of Labor, Bureau of Labor Statistics (1998). "Employment and Earnings Report, 1997." U.S. Govt. Printing Office, Washington, DC.

3. U.S. Department of Labor, Bureau of Labor Statistics (1997). New data on multiple job holding available from the CPS. *Mon. Labor Rev.,* March, pp. 3–8.

4. U.S. Department of Labor, Bureau of Labor Statistics, (1997). December 3. "BLS Releases New 1996–2006 Employment Projections," News Release, U.S. Department of Labor, Washington, DC.

5. U.S. Department of Labor, Bureau of Labor Statistics (1997). Labor force 2006: Slowing down and changing composition. *Mon. Labor Rev.,* November, pp. 23–28.

6. U.S. Department of Labor, Bureau of Labor Statistics (1997). "Lost-worktime Injuries: Characteristics and Resulting Time Away from Work, 1995," News Release June 12, No. 97–188. U.S. Department of Labor, Washington, DC.

7. U.S. Department of Labor, Bureau of Labor Statistics (1998). "Occupational Injuries and Illnesses: Counts, Rates, and Characteristics, 1995," Bull. 2493. U.S. Govt. Printing Office, Washington, DC.

8. U.S. Department of Labor, Bureau of Labor Statistics (in press). "Table P-6, Job-related Fatalities of Women by Selected Characteristics, 1992–1996." U.S. Govt. Census of Fatal Occupational Injuries, Washington, DC.

9. Statistics Canada (1995). "Women in Canada," 3rd ed., Catalogue No. 89-503E, Table 6.24. Statistics Canada. Ottawa, Ontario, Canada.

10. Canadian Advisory Council on the Status of Women (1994). "110 Canadian Statistics on Work and Family." Canadian Advisory Council on the Status of Women, Ottawa.

11. Barnett, R. C., and Baruch, G. K. (1987). Social roles, gender and psychological distress. *In* "Gender and Stress" (R. C. Barnett, L. Biener, and G. K. Baruch, eds.), pp. 122–143. Free Press, New York.

12. Ilfeld, F. W. (1976). Methodological issues in relating psychiatric symptoms to social stressors. *Psychol. Rep.* **39**, 1251–1258.

13. Tierney, D., Romito, P., and Messing, K. (1990). She ate not the bread of idleness: Exhaustion is related to domestic and salaried work of hospital workers in Quebec. *Women Health* **16**, 21–42.

14. Hatch, M., and Moline, J. (1997). Women, work and health. *Am. J. Ind. Med.* **3**, 303–308.

15. U.S. Department of Health and Human Services (1997). "Women: Work and Health," DHHS Publ. No. (PHS) 98-1791. USDHHS, Washington, DC.

16. Seifert, A. M., Messing, K., and Dumais, L. (1997). Star wars and strategic defense initiatives: Work activity and health symptoms of unionized bank tellers during work reorganization. *Int. J. Health Serv.* **27**, 455–477.

17. Messing, K., Seifert, A. M., and Escalona, E. (1997). The 120-second minute: Using analysis of work activity to prevent psychological distress among elementary school teachers. *J. Occup. Health Psychol.* **2**, 45–62.

18. Escalona, E. (1997). Activité de travail des enseignantes de niveau primaire. Thesis submitted in partial fulfillment of the requirements for the M.Sc. in Biology, University of Quebec, Montreal.

19. Lindbohm, M.-L. (1995). Effects of parental exposure to solvents on pregnancy outcome. *J. Occup. Environ. Med.* **37**, 908–914.

20. Swan, S. H., Beaumont, J. J., Hammond, S. K., von Behren, J., Green, R. S., Hallock, M. F., Woskie, S. R., Hines, C. J., and Schenker, M. B. (1995). Historical cohort study of spontaneous abortion among fabrication workers in the semiconductor health study: Agent-level analysis. *Am. J. Ind. Med.* **28**, 751–769.

21. Correa, A., Gray, R. H., Cohen, R., Rothman, N., Shah, F., Seacat, H., and Corn, M. (1996). Ethylene glycol ethers and risk of spontaneous abortion and subfertility. *Am. J. Epidemiol.* **143**, 707–717.

22. Cordier, S., Bergeret, A., Goujard, J., Ha, M.-C., Ayme, S., Bianchi, F., Calzolari, E., De Walle, H., Knill-Jones, R., Candela, S., Dale, I., Cananche, B., de Vigan, C., Fevotte, J., Kiel, G., and Mandereau, L. (1997). Congenital malformations and maternal occupational exposure to glycol ethers. *Epidemiology* **8**, 355–363.

23. Doyle, P., Roman, E., Beral, V., and Brookes, M. (1997). Spontaneous abortion in dry cleaning workers potentially exposed to perchloroethylene. *Occup. Environ. Med.* **54**, 848–853.

24. Shortridge, L. A., Lemasters, G. K., Valanis, B., and Hertzberg, V. (1995). Menstrual cycles in nurses handling antineoplastic drugs. *Cancer Nurs.* **18**, 439–444.

25. Hoffman, D. M. (1986). Reproductive risks associated with exposure to antineoplastic agents: A review of the literature. *Hosp. Pharm.* **110**, 930–940.

26. Ahlborg, G., Jr., and Hemminki, K. (1995). Reproductive effects of chemical exposures in health professions. *J. Occup. Environ. Med.* **37**, 957–961.

27. Valanis, B., Vollmer, W., Labuhn, K., and Glass, A. (1997). Occupational exposure to antineoplastic agents and self-reported infertility among nurses and pharmacists. *J. Occup. Environ. Med.* **39**, 574–580.

28. Bisanti, L., Olsen, J., Basso, O., Thonneau, P., and Karmaus, W. (1996). Shift work and subfecundity: A European Multicenter Study. *J. Occup. Environ. Med.* **38**, 352–358.

29. Ahlborg, G., Jr., Axelsson, G., and Bodin, L. (1996). Shift work, nitrous oxide exposure and subfertility among Swedish midwives. *Int. J. Epidemiol.* **25**, 783–790.

30. Nurminen, T. (1995). Female noise exposure, shift work, and reproduction. *J. Occup. Environ. Med.* **37**, 945–950.

31. Axelsson, G., Ahlborg, G., Jr., and Bodin, L. (1996). Shift work, nitrous oxide exposure and spontaneous abortion among Swedish midwives. *Occup. Environ. Med.* **53**, 374–378.

32. Ahlborg, G., Jr. (1995). Physical work load and pregnancy outcome. *J. Occup. Environ. Med.* **37**, 941–944.

33. Bellinger, D. (1994). Teratogen update: Lead. *Teratology* **50**, 367–373.

34. Fuortes, L., Clark, M. K., Kirchner, H. L., and Smith, E. M. (1997). Association between female infertility and agricultural work history. *Am. J. Ind. Med.* **31**, 445–451.

35. Taskinen, J. K., Kyyronen, P. Liesivuori, J., and Sallmen, M. (1995). Greenhouse work, pesticides and pregnancy outcome. *Epidemiology* **6** (Suppl.), 109 (abstr.).

36. Abell, A., Juul, S., and Bonde, J. P. (1997). Time to pregnancy in female greenhouse workers. *In* "Book of Abstracts of International Symposium on Environment, Lifestyle, and Fertility, December 7–10, 1997, Aarhus, Denmark" (J. P. Bonde and P. Pors, eds.), p. 142.

37. Nurminen, T. (1995). Maternal pesticide exposure and pregnancy outcome. *J. Occup. Environ. Med.* **37,** 935–940.

38. Zahm, S. H., and Ward, M. H. (1998). Pesticides and childhood cancer. *Environ. Health Perspect.* **106**(Suppl. 3), 893–908.

39. Leino, P. I., and Hanninen, V. (1995). Psychosocial factors at work in relation to back and limb disorders. *Scand. J. Work Environ. Health* **21,** 134–142.

40. Messing, K., Tissot, F., Saurel-Cubizolles, M., Kaminski, M., and Bourgine, M. (1998). Sex as a variable can be a surrogate for some working conditions. *J. Occup. Environ. Med.* **40,** 913–917.

41. Bergqvist, U., Wolgast, E., Nilsson, B., and Voss, M. (1995). Musculoskeletal disorders among visual display terminal workers; individual, ergonomic and work organizational factors. *Ergonomics* **38,** 763–776.

42. Cunningham, L. S., and Kelsey, J. L. (1984). Epidemiology of musculoskeletal impairments and associated disability. *Am. J. Public Health* **74,** 574–579.

43. Bush, F. M., Harkins, S. W., Harrington, W. G., and Price, D. D. (1993). Analysis of gender effects on pain perception and symptom presentation in temporomandibular pain. *Pain* **53,** 73–80.

44. Kilbom, A. (1988). Isometric strength and occupational muscle disorders. *Eur. J. Appl. Physiol.* **57,** 322–326.

45. Pacifici, R., Rifas, L., McCracken, R., Vered, I., McMurtry, C., Avioli, L. V., and Peck, W. A. (1989). Ovarian steroid treatment blocks a postmenopausal increase in blood monocyte interleukin 1 release. *Proc. Natl. Acad. Sci. U.S.A.* **86,** 2398–2402.

46. Zahm, S. H., Pottern, L. M., Lewis, D. R., Ward, M. H., and White, D. W. (1994). Inclusion of women and minorities in occupational cancer epidemiologic research. *J. Occup. Med.* **36,** 842–847.

47. Wignall, B. K., and Fox, A. J. (1982). Mortality of female gas mask assemblers. *Br. J. Ind. Med.* **39,** 34–38.

48. Acheson, E., Gardner, M. J., Pippard, E. C., and Grime, L. P. (1982). Mortality of two groups of women who manufactured gas masks from chrysotile and crocidolite asbestos: A 40-year Follow-up. *Br. J. Ind. Med.* **39,** 344–348.

49. Newhouse, M. L., Berry, G., and Wagner, J. C. (1985). Mortality of factory workers in East London 1933–80. *Br. J. Ind. Med.* **42,** 4–11.

50. Edelman, D. A. (1992). Does asbestos exposure increase the risk of urogenital cancer? *Int. Arch. Occup. Environ. Health* **63,** 469–475.

51. Malkin, R., Wilcox, T., and Siebr, W. K. (1996). The NIOSH Indoor Environmental Evaluation Experience: Symptom prevalence. *Appl. Occup. Environ. Hyg.* **11,** 540–545.

52. Mendell, M. (1993). Nonspecific symptoms in office workers: A review and summary of the epidemiologic literature. *Indoor Air* **3,** 227–236.

53. Tenjoinsalo, J., Jaakola, J. J., and Seppanen, O. (1996). The Helsinki Office Study: Air change in mechanically ventilated buildings. *Indoor Air* **6,** 111–117.

54. Hodgson, M. J., Muldoon, S., Collopy, P., and Olesen, B. (1992). Work stress, symptoms, and microenvironmental measures. *In* "Indoor Air Quality 92: Environments for People," pp. 47–58. AHSRAE, Atlanta, GA.

55. Eriksson, N., Hoog, J., Mild, K. H., Sandstrom, M., and Stenberg, B. (1997). The psychosocial work environment and skin symptoms among visual display terminal workers: A case referent study. *Int. J. Epidemiol.* **26,** 1250–1257.

56. Broersen, J. P., de Zwart, B. C., van Dijk, F. J., Meijman, T. F., and van Veldhoven, M. (1996). Health complaints and working conditions experienced in relation to work and age. *Occup. Environ. Med.* **53,** 51–57.

57. Steenland, K., and Beaumont, J. J. (1984). The accuracy of occupational and industrial data on death certificates. *J. Occup. Med.* **26,** 288–296.

58. Gute, D. M., and Fulton, J. P. (1985). Agreement of occupational and industrial data on Rhode Island death certificates with two alternate sources of information. *Public Health Rep.* **100,** 65–72.

59. Schade, W. J., and Swanson, G. M. (1988). Comparison of death certificate occupational and industrial data with lifetime occupational histories obtained by interview: Variations in the accuracy of death certificate entries. *Am. J. Ind. Med.* **14,** 121–136.

60. Hemminki, K., Lindbohm, M.-L., and Kyyronen, P. (1995). Validity aspects of exposure and outcome data in reproductive studies. *J. Occup. Environ. Med.* **37,** 903–907.

35

Employment Patterns and Health among U.S. Working Women

JANE WALSTEDT
Women's Bureau
Department of Labor
Washington, DC

I. Introduction

Today over half of U.S. women are in the labor force, as are over half of mothers with children under one year of age [1,2]. Well over two-thirds of mothers of children under 18 years of age are in the labor force [2]. Most women work full-time, and they are increasingly holding multiple jobs [1,4]. They comprise the majority of workers in two major industry divisions: finance, insurance, and real estate; and services [1]. Although the majority of working women now hold jobs that are "gender neutral," 41% still hold traditionally female jobs. Secretary is still the leading occupation for women (i.e., the occupation with the greatest number of women).

Women sustain about a third of the most serious injuries and illnesses (those involving at least one day away from work), although they account for a larger share of certain categories of such injuries and illnesses [6,7]. In 1995, women comprised 86% of injured workers with lost worktime in nursing and personal care facilities, one of the industries with the highest rate of lost worktime injuries and illnesses [6]. Carpal tunnel syndrome was one of the four types of injuries or illnesses resulting in the greatest time away from work, and 71% of those losing time away from work because of this disabling condition were women [6].

II. Labor Force Participation of Women

In 1996, 62 million, or 59.3%, of the 104 million women aged 16 and over in the United States were in the civilian labor force, (i.e., working or looking for work), up from 33.9% in 1950, 37.7% in 1960, 43.3% in 1970, and 51.5% in 1980 [1,8]. 59 million, or 56%, were employed. Women comprised 46% of persons in the civilian labor force and 46% of employed persons in 1996 [1].

A. Full-Time/Part-Time

About three-fourths (73.1%) of employed women usually worked full-time, at least 35 hours per week, in 1996 [1]. In 1981, 72.9% of employed women worked full-time [3].

B. Multiple Job Holding

Multiple job holding by women grew between 1970 and 1980 and accelerated sharply in the 1980s, while multiple job holding by men fell [1,4]. The number of women holding more than one job simultaneously grew from 600,000 in 1970 to 3.6 million in 1996. The percentage of women holding multiple jobs during that time period rose from 2.2 to 6.2%, while the percentage of men holding such jobs declined from 7.0 to 6.1%. By 1996, the incidence of multiple job holding was virtually the same for both women and men (6.2% vs 6.1%).

The highest rate of multiple job holding among women is among women 20–24 years of age (7.4%). Women 65 years of age and over have the lowest rate of multiple job holding (2.7%).

There are significant differences in the types of multiple jobs women and men hold and the hours spent on the jobs. One-third of women holding more than one job work at multiple part-time jobs, compared to only 13% of men. Women who work more than one job average 43 hours per week at all of their jobs. Two-thirds of women who moonlight hold jobs in services and retail trade [4].

C. Hours at Work

In 1995, the average workweek for female nonagricultural workers was 35.8 hours, compared with 42.1 hours for men [9]. After removing the effect of the shifting age distribution of the U.S. working population, during the period 1976 to 1993 the average workweek for women increased by only one hour, while average weekly hours for men showed virtually no change. Persons aged 25–54 typically work more hours during the week than do younger and older workers. Between 1976 and 1993, although the average workweek for 25- to 54-year-old men and women were longer than in previous years, the net increase was much greater for women. Women's average workweek rose by nearly 2.5 hours.

Between 1985 and 1993, an increasing share of workers reported that they were at work 49 hours or more per week [9]. Although the shift into occupations in which long workweeks are the most prevalent accounted for only about 8.1% of the gain for men and women combined, it accounted for 12.7% of the gain among women versus 5.1% of the gain among men.

Men also worked more hours than women during the year, although since the mid-1970s men's annual hours have risen much less than those of women [9]. Employed women worked an average of nearly 20% more in 1993 than in 1976. After adjusting for the age shift, women worked an average of 15% more and men worked an average of 3% more.

D. Employment by Age

In 1996, women aged 25–54 had the highest employment rate of any age group of women (72.8%) [1]. Young women and young men aged 16–19 years had similar employment rates, 43.5% and 43.6%, respectively. Older women aged 65 years and over were about half as likely to be employed (8.2%) as men of the same age (16.3%).

E. Employment by Education

The more education a woman has, the greater the likelihood she will seek employment. In 1996, among women 25–64 years of age with four or more years of college, 85% were in the labor force. Among women of the same age group with less than four years of high school, only 43% were in the labor force.

F. Family Status

One of the most significant trends in the U.S. labor force has been the growth of working mothers. In 1996, well over two-thirds (70.8%) of the approximately 35 million women in the civilian noninstitutional population who had children under age 18 were in the labor force [2]. Fifty-four percent of mothers of children under one year of age were in the labor force [2].

The proportion of married mothers with children under 18 who worked during the year increased from 51.3% in 1970 to 72.9% in 1992 [10]. During the same time period, the proportion of married mothers with children under 18 working year-round full-time increased from 16.4 to 36.8%. The proportion of married mothers with children under six years of age who worked during the year increased from 44.4% in 1970 to 67.1% in 1992. The increase was particularly pronounced for those working year-round full-time, from 9.6% in 1970 to 30.6% in 1992. Black married mothers, both those with children under 18 and those with children under six, are more likely than white married mothers to work during the year and to do so year-round full-time.

G. Employment by Occupation

In 1996, 24 million, or 41%, of employed women 16 years and older worked in technical, sales, and administrative support occupations (Table 35.1). The occupational groups with the next largest proportions of women were management and professional specialty occupations (18 million, or 30%, of employed women) and service occupations (10.2 million, or 17%, of employed women) [1].

The majority of women workers now hold jobs that are neither traditional for women nor traditional for men. If we define "traditional women's jobs" as those in which women make up 75% or more of the employment and "traditional men's jobs" as those in which men make up 75% or more of the employment, then we find that most jobs no longer fall into either of these categories, but rather into the category of "gender neutral." In 1996, 54% of women workers held gender neutral jobs, up from 44% in 1991; 41% held traditionally female jobs, down from 48% in 1991; and 5% held traditionally male jobs, down from 8% in 1991.

In 1996, about 40% of all employed women were in the 20 leading occupations for women, down from over half (50.5%) in 1980. In 1996, the 20 leading occupations of employed women, (i.e., those with the greatest numbers of employed women) were, in descending order: secretary; cashier; managers

Table 35.1

Number of Employed Persons 16 Years and Older by Sex and Occupation, 1996 (in Thousands)

Occupation	Women	Men
Managers and professionals	17,754	18,744
Technical, sales, and administrative support	24,194	13,489
Technicians and related support	2,061	1,865
Sales	7,622	7,782
Administrative support, including clerical	14,511	3,842
Service occupations	10,210	6,967
Private household	764	41
Protective service	375	1,811
Food service	3,343	2,563
Health service	2,115	284
Cleaning and building service	1,388	1,737
Personal service	2,225	531
Precision production, craft, and repair	1,219	12,368
Operators, fabricators, and laborers	4,447	13,750
Farming, forestry, and fishing	677	2,889
Total	58,501	68,207

Source: Bureau of Labor Statistics [1].

and administrators, nec; registered nurse; sales supervisor and proprietor; nursing aide, orderly, and attendant; bookkeeper, accounting and auditing clerk; elementary school teacher; waiter and waitress; sales worker, other commodities; handler, equipment cleaner, helper, and laborer; receptionist; machine operator, assorted materials; cook; accountant and auditor; textile, apparel, and furnishing machine operator; janitor and cleaner; administrative support occupations; investigator and adjuster, excluding insurance; and secondary school teacher.

Table 35.2 compares the ten leading occupations for women in 1981 and 1996 in descending order. Secretary was the leading occupation of employed women in both years. The leading occupations among women are mainly concentrated in the technical, sales, and administrative support occupations. The leading occupations among men men differ from those of women and are uniformly distributed among six occupational groups: managerial

Table 35.2

Ten Leading Occupations among Women, 1981 and 1996

1981	1996
Secretary	Secretary
Bookkeeper	Cashier
Sales clerk, retail trade	Manager and administrator, nec[a]
Cashier	Registered nurse
Waitress	Sales supervisor and proprietor
Registered nurse	Nursing aide, orderly
Elementary school teacher	Bookkeeper, accounting and
Private household worker	auditing clerk
Typist	Elementary school teacher
Nursing aide	Waitress
	Sales worker, other commodities

[a]Not elsewhere classified.
Source: Bureau of Labor Statistics [1].

Table 35.3

Number of Employed Persons by Sex and Industry, 1996
(in Thousands)

Industry	Women	Men
Mining	75	494
Construction	796	7,147
Manufacturing	6,588	13,950
Durable goods	3,276	8,926
Nondurable goods	3,292	5,023
Transportation, communications and		
public utilities	2,525	6,293
Wholesale and retail trade	12,499	13,998
Finance, insurance, and real estate	4,717	3,359
Services	27,877	17,166
Public administration	2,754	3,228

Source: Bureau of Labor Statistics [1].

Table 35.4

Median Annual Earnings for Year-Round, Full-Time Workers,
by Sex and Ethnicity, 1996

All women	$23,710	All men	$32,144
White women	24,160	White men	32,966
Black women	21,473	Black men	26,404
Hispanic women	18,665	Hispanic men	21,056

Source: Bureau of Labor Statistics [1].

and professional specialty; technical, sales, and administrative support; service occupations; precision production, craft, and repair; operators, fabricators, and laborers; and farming, forestry, and fishing.

The detailed occupations with the greatest number of men, in descending order, are: managers and administrators; sales supervisors and proprietors; heavy truck drivers; engineers; janitors and cleaners; carpenters; sales representatives; laborers, except construction; farmers; and cooks, except short order.

H. Employment by Industry

Table 35.3 presents the number of persons employed in 1996 by industry division and sex. Women comprised the majority of employed persons in only two industries: finance, insurance, and real estate; and services. These two industries have historically employed large numbers of women because many occupations within them offer such features as flexible hours, ease of entry, and an abundance of job opportunities.

I. Projections for the Future: Women's Work Between the Years 1996 and 2006

Between the years 1996 and 2006, the labor force is projected to increase by 15 million (11%), from 134 million to 149 million [5,11]. The labor force participation rates of women in nearly all age groups are projected to increase, albeit at a slower rate than in the past, while those of men are projected to continue to decline. In 2006 women will constitute 47% of the civilian labor force, up from 46% in 1996.

The labor force participation rate of white women, which was lower than that of black women in 1996, is projected to exceed that of black women by the year 2006 [11]. Women of Hispanic origin have the lowest labor force participation rate of any race/ethnic group of women and this is expected to continue to be the case in 2006. Between 1996 and 2006, women of Hispanic origin are projected to have the greatest percentage increase in their labor force participation rate (3.8%) of any race/ethnic group of women, followed by white women (2.9%), Asian women (1.3%), and black women (0.9%).

Increases in the labor force will be concentrated in the older age groups and the labor force will continue to age, with the median age of persons in the labor force approaching 41 years by 2006.

The service-producing sector will account for virtually all of the job growth [5]. The services division will account for almost two-thirds of all newly created jobs. In 1996, women held 61.9% of all service jobs. In the goods-producing sector, only construction will add jobs, while mining and manufacturing will decline. Manufacturing is expected to account for only 12% of employment in 2006, down from 14% in 1996.

The retail trade division has replaced manufacturing as the second largest source of total employment. Women held 51.2% of retail trade industry jobs in 1996.

Between 1996 and 2006, professional specialty occupations, 53% of which were held by women aged 16 and over in 1996, are projected to increase the fastest and to add the most jobs (4.8 million). Service occupations, 59% of which were held by women aged 16 and over in 1996, are expected to grow by 3.9 million. Professional specialty occupations and service occupations are expected to account for 46% of total projected job growth.

Administrative support occupations, including clerical, are projected to grow by 1.8 million, much slower than the average and slower than they have grown in the past. In 1996, women held 79% of these jobs.

Of the 30 occupations expected to grow the fastest between 1996 and 2006, occupations where the majority of employees were women in 1996 include dental assistants (99.1% women); physical therapists (61.9% women); respiratory therapists (58.4% women); teachers, special education (84.8% women); and bill and account collectors (68.7% women) [12].

III. Earnings and Benefits

A. Earnings

In 1996, for full-time wage and salary workers, women's median weekly earnings were 75% of men's ($418 compared to $557 for men) [1]. The median annual earnings for women working full-time year-round (50 weeks or more) were 73.8% of men's ($23,710 compared to $32,144 for men) (Table 35.4).

Working wives contributed on average 36.5% of family earnings (unpublished data from the Current Population Survey, March, 1996). White wives contributed 35.8%, black wives contributed 43.7%, and Hispanic wives contributed 37.5%.

According to government survey data, in 1996 working wives in married couple families contributed on average 32.6% of total family income (earnings plus other income, such as unemployment compensation, Social Security and pensions, public assistance, interest and dividends, and child support) (unpublished data from the Current Population Survey, March, 1996). Working white wives contributed 31.9%, working black wives contributed 39.5%, and working Hispanic wives contributed 34.6%.

Health status and family income are strongly associated. In 1994 the age-adjusted percentage of persons with low family income (less than $14,000) who reported fair or poor health was five times that for persons with a high income of $50,000 or more (20% and 4%, respectively) [13].

B. Benefits

Since 1979, the Bureau of Labor Statistics has conducted an annual study of the incidence and detailed characteristics of employee benefits known as the Employee Benefits Survey. It measures benefits provided to full-time and part-time employees in small (less than 100 employees), medium (100 or more employees), and large (250 or more employees) private sector firms, and state and local governments. Smaller establishments and service-producing industries, where most women are employed, typically provide fewer benefits than larger establishments and goods-producing industries. For almost all benefits, part-time employees are less frequent recipients than full-time employees [14]. For example, employees in medium and large private establishments are more likely than employees in small private establishments to be eligible for wellness programs (37% of employees in medium and large establishments in 1993 compared to 6% of employees in small private establishments in 1994).

Medical care is one of the most widespread benefits for full-time employees in small private establishments. In 1994, 66% of workers in those establishments participated in such plans. It is even more prevalent in medium and large firms. Eighty-two percent of full-time employees in those firms participated in such plans in 1993.

With regard to family-related benefits, in 1994 only 2% of full-time employees in small private establishments had formal flexible work arrangements [15]. Only 1% of full-time employees were eligible for childcare assistance and for long-term care insurance; however, childcare resource and referral services were not included. Thirty-three percent of full-time employees were eligible for eldercare, including paid leave, leave of absence policies, employer-sponsored adult day care centers, and employer subsidized day care for the elderly. Eldercare resource and referral services were not included.

In contrast, in medium and large firms in 1993, 4% of full-time employees had formal flexible work schedules, 7% were eligible for childcare assistance, 6% were eligible for long-term care insurance, and 31% were eligible for eldercare [16]. Professional, technical, and related employees were most likely and blue collar and service employees were least likely to be eligible for childcare assistance (12% vs 4%).

A survey of 1050 major U.S. employers by Hewitt Associates found that in 1995, 85% offered some kind of childcare assistance, 26% offered eldercare programs, and 67% offered flexible scheduling arrangements [17]. Dependent care spending accounts and resource and referral services were the most prevalent forms of childcare assistance, offered by 96% and 40%, respectively. Resource and referral programs, offered by 77% of employers, was the most common form of eldercare assistance.

Starting in 1994, the Employee Benefits Survey ascertained the prevalence of formal family leave benefits [15]. In 1994, 47% of full-time employees in small private industry establishments had unpaid maternity leave. This represented an increase from the 18% of full-time employees in such firms who had unpaid maternity leave benefits and the 8% of full-time employees in such firms who had unpaid paternity leave benefits in 1992, prior to passage of the Family and Medical Leave Act. Full-time employees in medium and large private firms are more likely than full-time employees in small private firms to have unpaid family leave. In 1995 unpaid family leave was available to 84% of all full-time workers in medium and large private firms.

Women-owned firms with more than ten employees are more likely to offer flextime/job sharing (54% vs 33%) and child/elder care (9% vs 6%) than men-owned firms, according to a survey of 610 women- and men-owned businesses [18].

IV. Occupational Injuries and Illnesses

A. General Health Status

The 1989 National Health Interview Study found that of the total labor force, a greater proportion of women (7.1%) than men (6.4%) assessed their health as fair or poor [19]. There was a large variation among women in different occupational groups. Women employed in private household work had the highest percentage of women assessing their health as fair or poor (19%). They reported a high prevalence of high blood pressure, arthritis, and deformity or orthopedic impairment of the back. Note that private household worker is no longer one of the 20 leading occupations for women. In 1991, there were approximately 755,000 female private household workers (cleaners, servants, and childcare workers in the home) in the U.S., which represented 1.4% of U.S. working women, down from 2.1% in 1983.

The occupational group with the highest percentage of men assessing their health as fair or poor (10.8%) was farming, forestry, and fishing occupations.

More than one in ten women working as machine operators, assemblers, and inspectors, as well as those working as handlers, equipment cleaners, helpers, and laborers, also assessed their health as fair or poor.

For women, professional specialty occupations was the occupational group with the lowest percentage of women (3.6%) assessing their health as fair or poor. For men, professional specialty occupations was one of two occupational groups with the lowest percentage of men assessing their health as fair or poor, the other being technicians and related support occupations, both with 2.8%.

For the total labor force, 8.4% of the women were limited in activity as compared with 9.5% of the men. Limitation of activity refers to a person's inability to perform his or her major activity, a limitation in kind or amount of major activity, or a limitation in the kind and amount of other activities. Among both women and men, private household workers had the highest percentage with activity limitation (women: 17.3%; men:

12.0%); however, women accounted for 94% of the labor force in private household occupations. Among women, the occupational group with the lowest rate of activity limitation was administrative support (6.9%). Among men, the lowest rate was observed among precision production, craft, and repair occupations (8.5%), although the rates of activity limitation were more stable across occupations for men than for women.

Women in the labor force reported more restricted activity, bed disability, and work loss days per person than their male counterparts. Women had a higher incidence of acute conditions than men. They also had a higher prevalence than men of the following chronic conditions: hay fever, chronic sinusitis, arthritis, and deformity or orthopedic impairment of the back.

In 1994, for every 100 working women, 2.3 were injured at work. Men had more bed days due to work-related injuries than women—28.3 days per 100 men vs 19.5 days per 100 women [20]. Between 1982 and 1994, the number of bed days per 100 working women increased 24%, while for men the number decreased by 27%.

B. Nonfatal Occupational Injuries and Illnesses

There were approximately 6.2 million nonfatal injuries and illnesses reported in private industry in 1996, according to survey data compiled by the Bureau of Labor Statistics (Survey of Occupational Injuries and Illnesses) [21]. An occupational *injury* is one that results from a work-related accident or from exposure to a sudden event in the work environment. An occupational *illness* is any abnormal condition or disorder, other than one resulting from an occupational injury, caused by exposure to factors associated with employment. Included are acute and chronic illnesses or diseases that may be caused by inhalation, absorption, ingestion, or direct contact [22]. Occupational illnesses are thought to be underdiagnosed and understated [23,24].

Services, one of two industry groups in which women comprise the majority of employees, accounted for nearly one-fourth of the nonfatal injuries and illnesses in 1996 (Table 35.5) [21]. Manufacturing accounted for about one-third and wholesale and retail trade accounted for one-fourth.

Although three out of every ten injuries in 1996 occurred in manufacturing, the nine industries reporting the largest numbers of workplace injuries (at least 100,000 cases) included eight industries outside manufacturing: eating and drinking places; hospitals; trucking and courier services, except air; grocery stores; nursing and personal care facilities; department stores; air transportation, scheduled; and hotels and motels (Janice DeVine, Office of Safety, Health, and Working Conditions, Bureau of Labor Statistics, Department of Labor, unpublished data). Women constituted the majority of employees in all but trucking and air transportation [1]. They comprised 52.5% of employees in eating and drinking places; 76.1% of employees in hospitals; 52.2% of employees in grocery stores; 86.3% of employees in nursing and personal care facilities; 67.5% of employees in department stores; and 55.2% of employees in hotels and motels [1].

According to the Bureau of Labor Statistics Survey on Occupational Injuries and Illnesses, in 1995 women sustained about a third of the two million injuries and illnesses with days away from work [6]. Although women account for fewer cases

Table 35.5

Number and Percentage of Nonfatal Occupational Injuries and Illnesses Involving Days Away from Work, by Sex and Industry, 1996

Industry	Number of nonfatal injuries and illnesses (in thousands)	Women (percent)	Men (percent)
Total	1880.5	33.0	65.9
Agriculture, forestry, and fishing	38.3	17.2	65.9
Mining	15.1	2.1	97.6
Construction	182.3	2.4	97.2
Manufacturing	462.2	25.0	74.3
Transportation and public utilities	224.0	17.6	79.5
Wholesale trade	144.7	12.0	87.6
Retail trade	322.0	43.5	54.8
Finance, insurance, and real estate	42.8	50.4	48.8
Services	449.0	61.2	38.1

Source: [8].

of serious injuries and illnesses (defined as having at least one day away from work not counting the day the incident occured), they account for a larger share of certain categories of serious injuries and illnesses, such as assaults and violent acts and repetitive motion disorders, than do men [7].

Four types of injuries result in the greatest time away from work: carpal tunnel syndrome, amputation, hernia, and fracture. Of these, carpal tunnel syndrome, a disorder of the peripheral nervous system, is the only one in which a greater percentage of women than men lose work time (Fig. 35.1). Seventy-one percent of the injured losing time from work because of carpal

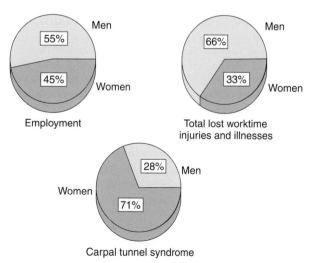

Fig. 35.1 Gender shares of employment, work injuries and illnesses, and carpal tunnel syndrome, 1995. Total may include cases for which sex of worker was not reported. Source: [7].

Table 35.6

Number and Percent Distribution of Nonfatal Occupational Injuries and Illnesses Involving Days Away from Work, by Sex for Selected Occupations, 1995

Occupation	Number of nonfatal injuries and illnesses (in thousands)	Percent women	Percent men
Total	2,040.9	32.7	66.4
Truck drivers	151.3	6.4	93.2
Non-construction laborers	115.5	14.3	84.6
Nursing aides, orderlies	100.6	89.2	10.1
Assemblers	55.5	36.2	63.3
Janitors and cleaners	52.6	28.5	70.9
Construction laborers	43.5	1.8	97.2
Cooks	35.4	46.8	52.5
Carpenters	35.0	1.1	98.3
Stock handlers and baggers	34.7	32.1	67.5
Cashiers	30.2	82.5	17.0

Source: [7].

tunnel syndrome are women. More women sustained carpal tunnel syndrome by operating machinery, on assembly lines, and tending retail stores than they did typing, keying, and performing other duties associated with office workers [6]. There were 31,457 cases of carpal tunnel syndrome involving days away from work, and workers with carpal tunnel syndrome took a median of 30 days to return to work [7].

Tendonitis, tenosynovitis, and ganglion/cystic tumor are some of the other specific disabling conditions for which there were at least 1500 cases in 1995 and which had a median of ten days or more away from work [7]. These are classified as musculoskeletal system and connective tissue diseases and disorders. There were 22,072 cases of tendonitis involving days away from work, and the median days away from work was 12. Although there were fewer cases of tenosynovitis (1564) and ganglion/cystic tumor (2294) involving days away from work, the median days away from work for each was 16 days and 13 days, respectively.

Although men account for twice as many injuries and illnesses with days away from work as women, this pattern differs in many specific industries [7]. These industry variations primarily reflect differences in where men and women typically work and differences in their jobs.

Nursing and personal care facilities is one of the industries with the highest rates of lost worktime injuries and illnesses, and 86% of the injured workers with lost worktime in this industry are women, primarily nurses and aides in nursing homes [6]. Patients are the primary source of injury or illness to nursing home staff [6].

In 1995, the ten occupations with the largest number of injuries and illnesses involving days away from work were truck drivers; laborers, nonconstruction; nursing aides and orderlies; assemblers; janitors and cleaners; construction laborers; cooks; carpenters; stock handlers and baggers; and cashiers (Table 35.6). These ten occupations accounted for a third of the case total, double their share of the employment total [7].

In 1995, only three individual occupations had at least 100,000 cases each of serious injuries and illnesses: truck drivers; nonconstruction laborers; and nursing aides and orderlies [7]. In 1995, women constituted the majority of one of these occupations—nursing aides, orderlies, and attendants [26].

C. Fatal Occupational Injuries and Illnesses

In 1996, there were 6112 workplace fatalities according to the Bureau of Labor Statistics Census of Fatal Occupational Injuries [27]. There were 507 workplace fatalities to women, up from 443 in 1992 but down from a high of 539 in 1995.

While highway traffic accidents are the leading cause of fatal work injuries to men, homicide is the leading cause of fatal work injuries to women, accounting for 170, or 34%, of women's workplace fatalities in 1996. This was a decrease from a high of 246 in 1995 and from 182 in 1992. Highway traffic accidents are the second leading cause of fatal work injuries to women.

Although robbery is the primary motive for job-related homicides, in 1996 31, or 18%, of the 170 workplace homicides occurring to women were perpetrated by a husband, ex-husband, boyfriend, or ex-boyfriend [27]. During the period 1992–1994, 17% of alleged attackers were current or former husbands or boyfriends [28]. For black women, the figure was 28%, and for Hispanic women, 20%.

The National Traumatic Occupational Fatality (NTOF) surveillance system of the National Institute for Occupational Safety and Health collects information from death certificates on traumatic occupational fatalities. During 1980–1994, there was a decrease in the rate of fatalities among men from 12.46 to 7.48 per 100,000 male workers. Among women, the rate decreased from 1.04 to 0.76 per 100,000 female workers (Suzanne Kisner, National Institute for Occupational Safety and Health, unpublished data). The industry divisions with the highest frequency of deaths among women workers were retail trade and services, followed by manufacturing (Table 35.7). The occupation divisions with the highest frequency of deaths were sales; service; clerical; and executive, administrative, and managerial occupations (Table 35.8). As with the Census of Fatal Occupational Injuries, homicide and motor vehicles were the leading causes of occupational fatalities among women during the period 1980–1994, according to NTOF data (Table 35.9).

V. Conclusions

In the U.S., women overall and mothers have been increasing their labor force participation and will continue to do so into the twenty-first century. Women are working in a wider variety of occupations, and more of them are holding multiple jobs and working longer hours. The shift from a goods-producing economy to a service-producing economy continues. All of these factors have implications for women's occupational safety and health.

Data on serious occupational injuries and illnesses among women are improving, and some better mechanisms to address women's health have been created. For example, in 1990 the Office of Research on Women's Health was established within the Office of the Director of the National Institutes of Health (NIH). Its mandate is to strengthen and enhance research related

Table 35.7

Number and Percentage of Occupational Traumatic Fatalities among Women, by Industry, 1980–1994

Industry	Frequency	Percent
Retail trade	1412	24.9
Services	1393	24.6
Manufacturing	546	9.6
Transportation, communications, and other public utilities	508	9.0
Finance, insurance, and real estate	265	4.7
Public administration	248	4.4
Agriculture	186	3.3
Construction	163	2.9
Wholesale trade	92	1.6
Mining	32	0.6
Not classified	817	14.4
Total	5662	100.0

Source: National Institute for Occupational Safety and Health, National Traumatic Occupational Fatalities Surveillance System.

Table 35.9

Number and Percentage of Occupational Traumatic Fatalities among Women, by Cause of Death, 1980–1994

Cause of death	Frequency	Percent
Homicide	2358	41.6
Motor vehicle	1435	25.3
Falls	288	5.1
Machine	265	4.7
Suicide	237	4.2
Air transport	217	3.8
Fires	139	2.5
Other	139	2.5
Nature/environment	104	1.8
Explosion	86	1.5
Struck by falling object	64	1.1
Drowning	63	1.1
Poisoning	60	1.1
Unknown	45	0.8
Electrocution	42	0.7
Suffocation	40	0.7
Flying object/caught	38	0.7
Water transport	32	0.6
Rail transport	10	0.2
Total	5662	100.0

Source: National Institute for Occupational Safety and Health, National Traumatic Occupational Fatalities Surveillance System.

to diseases and conditions that affect women and to ensure that research conducted and supported by the NIH adequately addresses issues regarding women's health and that women are appropriately represented in research studies. In 1991 the Office on Women's Health was created within the U.S. Public Health Service to advise the Assistant Secretary for Health on women's health issues and to coordinate women's health policies and programs across the Public Health Service [25]. In 1996, the National Institute for Occupational Safety and Health released a national agenda for workplace safety and health research (NORA). The agenda lists 21 research priorities under three major categories: disease and injury; work environment and workforce (including special populations); and research tools and approaches. Notwithstanding these advances, vigilance is needed to ensure that the occupational safety and health hazards

of women are adequately addressed. Information gaps still exist, and more attention needs to be focused on the safety and health hazards women face in the workplace.

Acknowledgments

The author gratefully acknowledges the assistance of the following persons for providing tables or reviewing the chapter: Jacqueline Bhola, Arline Easley, and Michael Williams from the Women's Bureau, U.S. Department of Labor; William L. Weber, Howard V. Hayghe, and Randy Ilg, from the Bureau of Labor Statistics, U.S. Department of Labor; and Suzanne Kisner, Division of Safety Research, National Institute for Occupational Safety and Health, Centers for Disease Control and Prevention, U.S. Department of Health and Human Services.

Table 35.8

Number and Percentage of Occupational Traumatic Fatalities among Women, by Occupation Division, 1980–1994

Occupation division	Frequency	Percent
Sales	923	16.3
Service	822	14.5
Clerical	606	10.7
Exec/adm/mgr	605	10.7
Prof/spec	527	9.3
Transport	391	6.9
Laborers	379	6.7
Operatives	198	3.5
Crafts	154	2.7
Farmers	145	2.6
Tech/support	124	2.2
Unknown	788	13.9
Total	5662	100.0

Source: National Institute for Occupational Safety and Health, National Traumatic Occupational Fatalities Surveillance System.

References

1. U.S. Department of Labor, Bureau of Labor Statistics, (1997). "Employment and Earnings." U.S. Govt. Printing Office, Washington, DC.
2. U.S. Department of Labor, Bureau of Labor Statistics (1997). "Employment Characteristics of Families: 1996," News Release. U.S. Govt. Printing Office, Washington, DC.
3. U.S. Department of Labor, Bureau of Labor Statistics (1988). "Labor Force Statistics Derived from the Current Population Survey, 1948–87," Bull. 2307, pp. 195, 300. U.S. Govt. Printing Office, Washington, DC.
4. Stinson, J. F. Jr. (1997). New data on multiple job holding available from the CPS. *Mon. Labor Rev.,* March, p. 5.
5. U.S. Department of Labor, Bureau of Labor Statistics (1997). "BLS Releases New 1996–2006 Employment Projections," News Release. U.S. Govt. Printing Office, Washington, DC.
6. U.S. Department of Labor, Bureau of Labor Statistics (1997). "Lost-worktime Injuries: Characteristics and Resulting Time Away

from Work, 1995," News Release 97–188. U.S. Govt. Printing Office, Washington, DC.

7. U.S. Department of Labor, Bureau of Labor Satistics (1998). "Occupational Injuries and Illnesses: Counts, Rates, and Characteristics, 1995," Bull. 2493. U.S. Govt. Printing Office, Washington, DC.

8. U.S. Department of Labor, Women's Bureau (1983). "Time of Change: 1983 Handbook on Women Workers," p. 11. U.S. Govt. Printing Office, Washington, DC.

9. Rones, P. L., Ilg, R. E., and Gardner, J. M. (1997). Trends in hours of work since the mid-1970s. *Mon. Labor Rev.,* April, pp. 3–14.

10. Hayghe, H. V., and Branchi, S. M. (1994). Married mothers' work patterns: The job-family compromise. *Mon. Labor Rev.,* June, p. 28.

11. Fullerton, H. N. (1997). Labor force 2006: Slowing down and changing composition. *Mon. Labor Rev.,* November, pp. 23–28.

12. Silvestri, G. T. (1997). Occupational employment projections to 2006. *Mon. Labor Rev.,* November, p. 77.

13. U.S. Department of Health and Human Services (1997). "Health United States, 1996–97 and Injury Chartbook," DHHS Publ. No. (PHS) 97-1232, p. 7. USDHHS, Washington, DC.

14. U.S. Department of Labor, Bureau of Labor Statistics (1995). "Employee Benefits Survey: A BLS Reader," Bull. 2 459. U.S. Govt. Printing Office, Washington, DC.

15. U.S. Department of Labor, Bureau of Labor Statistics (1996). "Employee Benefits in Small Private Establishments, 1994," Bull. 2475. U.S. Govt. Printing Office, Washington, DC.

16. U.S. Department of Labor, Bureau of Labor Statistics (1994). "Employee Benefits in Medium and Large Private Establishments, 1993," Bull. 2456. U.S. Govt. Printing Office, Washington, DC.

17. Hewitt Associates (1996). "Work and Family Benefits Provided by Major U.S. Employers in 1995."

18. The National Foundation for Women Business Owners (1997). "Retirement Plan Trends in the Small Business Market: A Survey of Women- and Men-owned Firms."

19. U.S. Department of Health and Human Services, Public Health Service, Centers for Disease Control, National Center for Health Statistics (1989). "Health Characteristics of Workers by Occupation and Sex: United States, 1983–85," Adv. Data, No. 168. USDHS, PHS, CDC, Washington, DC.

20. U.S. Department of Health and Human Services, Public Health Services, Centers for Disease Control and Prevention, National Center for Health Statistics (1996). "Health, United States, 1995," DHHS Publ. No. (PHS) 96-1232. USDHHS, PHS, CDC, Washington, DC.

21. U.S. Department of Labor, Bureau of Labor Statistics (1997). "Survey of Occupational Injuries and Illnesses, 1995," Summ. 97-7. U.S. Govt. Printing Office, Washington, DC.

22. U.S. Department of Labor, Bureau of Labor Statistics (1991). "Occupational Injuries and Illnesses in the United States by Industry, 1989," Bull. 2379. U.S. Govt. Printing Office, Washington, DC.

23. Landrigan, P. J., and Baker, D. B. (1991). The recognition and control of occupational disease. *JAMA, J. Am. Med. Assoc.* **266,** 676–680.

24. U.S. Department of Labor, Bureau of Labor Statistics (1997). "Occupational Injuries and Illnesses: Counts, Rates, and Characteristics, 1994," Bull. 2485. U.S. Govt. Printing Office, Washington, DC.

25. U.S. Public Health Service (1991/1992). "Women's Health is a National Priority," Prev. Rep. U.S. Govt. Printing Office, Washington, DC.

26. U.S. Department of Labor, Bureau of Labor Statistics (1996). "Employment and Earnings." U.S. Govt. Printing Office, Washington, DC.

27. U.S. Department of Labor, Bureau of Labor Statistics, (1992–1996). "Job-related Fatalities of Women by Selected Characteristics," Table P-6. U.S. Govt. Printing Office, Washington, DC.

28. U.S. Department of Labor, Women's Bureau (1996). "Domestic Violence: A Workplace Issue," Fact Sheet No. 96-3. U.S. Govt. Printing Office, Washington, DC.

36

Multiple Roles and Complex Exposures: Hard-To-Pin-Down Risks for Working Women

KAREN MESSING
CINBIOSE
Department of Biological Sciences
Université du Québec à Montréal
Montréal, Québec, Canada

I. Introduction

In treating occupational health problems, researchers and practitioners are asked to decide whether the cause of a problem lies in the workplace. For women, this attribution poses specific problems for two reasons. First, because of women's family responsibilities, women's exposures often belong, or appear to belong, to the "private" sphere, unrelated to the workplace. Second, women are rarely found in the occupations, such as mining and construction, where occupational health intervention was constructed so as to combat dramatic, visible risk factors. Women's exposures may seem unimpressive and trivial, therefore unlikely to cause health problems.

Because there is confusion about the "reality" of women's workplace exposures, it has been tempting for some health professionals to attribute women's reports of occupational health problems to "hysteria" [1], "neurosis" [2], lack of testosterone [3], or female reproductive physiology [4]. In speaking of mass hysteria, defined as "physical symptoms suggesting an organic illness but resulting from a psychological cause" [5], neurologists affirm that "women display the majority of symptoms." [1] In other words, a physical cause of symptoms may be harder to find in women's jobs. Women do, however, undergo exposures that, despite their subtlety, have real effects on their health. In view of these complexities, it is perhaps not surprising that those illnesses whose relation to the workplace are most hotly contested—repetitive strain injury [6], sick building syndrome [7], fibromyalgia [8], and multiple chemical sensitivity [9]—are found at least twice as often among women as among men.

This chapter will provide some data on the interactions between women's family roles and their occupational health, drawn from the available sources in the U.S., Canada, and Québec. It will present examples of types of complex exposures often found in jobs occupied by women and will conclude with some reports of approaches and methodologies that can be used to make women's exposures more visible. In particular, it will discuss the confusion between gender and workplace gender-specific exposures as risk factors for women's occupational health problems.

II. Public and Private Roles

A. Interactions between Professional and Family Responsibilities

In the United States, about 40% of women in the labor force have children under 18. To look at the numbers in another way,

60% of married women with children under 6 and 75% of those with children ages 6–17 are in the labor force [10]. In addition, some working people, mostly women, are responsible for the care of disabled family members, such as the elderly and mentally, physically or emotionally handicapped adults or children. The number of working women who care for infirm children or adults is not known, but we do know that 8% of all Canadian families with children under 13 in which one or both parents work have a child with a long-term disability, 16% of families care for elderly, handicapped, or disabled relatives, and 31% of those who care for elderly parents work [11]. It is therefore probable that at least an additional 5% of working women care for handicapped or ill adults.

The length of the domestic workday is difficult to measure, because the work usually takes place in private. Questionnaires that ask women and men to estimate the time they and their partners spend at certain tasks or the proportion of tasks done by each partner may be subject to over- or underestimation or just bad guessing. Asking spouses to report their own participation at different tasks yields slightly different results from those obtained when each reports on the partner's input [12,13]. Time budgets that require subjects to report hour by hour how they spend their time may take so much time and effort that they may be abandoned or invalid [14]. Another approach is to ask women and men who did a task the last time it was done, but this does not yield time estimates [13]. None of these methods enables the researcher to explore the demands of overlapping and simultaneous tasks, which are very common in households. Women also suggest that simple time budgeting is not sufficient to take into account the differences between women and men in overall level of responsibility for the smooth running of the household.

It is difficult to define what is meant by domestic work [15]. Is playing with children to be considered as leisure or work? Sewing or gardening may be leisure for some and work for others. Some studies distinguish between domestic work and childcare, while others do not. Thus, time estimates vary greatly. Common estimates of the amount of domestic work done by employed people per week range widely, depending on the number and ages of children and the degree of task sharing. Using interviews, Statistics Canada reports that women spend about twice the amount of time as men on domestic work and child care [16]. Employed women spend about 50% more time than men on domestic tasks [11]. Barnett and Shen [76] note that a number of studies, including their own, find that the total amount of time spent on paid and unpaid work differs little between men and women. However, women do more of the low-schedule-control unpaid work that is associated with psychological distress [17].

Access to multiple social roles may give women more choices and result in better mental and physical health [17]. It may, however, also occasion psychological distress, depending on paid and unpaid working conditions. In the Québec Health Survey, employed women experienced elevated psychological distress [18] significantly more often than employed men, and somewhat more often than housewives [19]. Married women with children were more likely than single women to be distressed if they worked [19, p. 145]. A survey of government workers showed that the younger a women's children, the more stress she reported [20].

Health may be affected by the combination of paid and unpaid work; sickness absence is higher among women with young children (although some of this may be disguised leave for care of sick children) [10, p.54; 21]. Work accidents may be more common among women with young children, possibly due to greater fatigue [22]. Nurses who work full time and have young children report more fatigue than those without children, especially where domestic tasks are not shared [23]. In addition to a long and tiring day, a "double workday" may imply longer exposure times. Thus exposure to cleaning substances, repetitive movements, or emotional demands may begin during the paid work day and be prolonged into the private sphere, resulting in higher exposures and shorter time for recuperation.

Psychological and physical health problems may also result from the actual effort taken to balance work and family. Psychologists emphasize the issues of role conflict experienced by working mothers [24]. Hessing [25] and also Hochschild [26] have used qualitative research techniques to describe the negotiating of housework with partners. Hochschild concludes that the struggles over housework in two-earner families are so painful that both partners are tempted to spend more time at work [27].

Specific efforts required to balance the time demands of work and family have been studied qualitatively by Hessing [25]. She describes various strategies used to "gain" time: extending the work day, anticipating and planning, overlapping tasks and lowering standards so as to diminish the time required for certain tasks. Guberman and colleagues have described how working women cope with eldercare—by arranging schedules, job flexibility, and leave [28]. Lee and colleagues found that women used organizational, planning, and negotiating skills in order to cope with simultaneous demands [20, p. 14].

Individual skills are not enough to cope with some working conditions. Women in several studies have reported that family leave, flexible hours, the ability to work at home, and supervisors' understanding were factors that helped them to balance work and family [20, p. 19; 29]. Surprisingly, Barnett and Brennan conclude that low control over work schedules was not associated with psychological distress among employed men and women; however, the same analysis retained as significant a measure of time pressure and conflicting tasks [30]. Thus, time pressure resulting from low control over work schedules was probably included in the schedule control measure.

A questionnaire study of six groups of male and female working parents provided detail on the effects of workplace policies

Table 36.1

Associations with Reported Difficulty Balancing Work and Family among Women with Children under 12

Variable	Odds ratio	95% Confidence interval	p Value
Works weekends	5.3	(1.2–24.0)	.02
Access to telephone limited at work	2.4	(1.0–5.8)	.04
Youngest child under 5	2.1	(1.0–4.3)	.05
Work schedule rigid	3.5	(1.2–10.0)	.02
Emotional support from partner			
little support[a]	1.9	(0.9–4.3)	ns
single parent	1.8	(0.6–5.6)	ns
Family income < $70,000/y	1.7	(0.7–3.3)	ns

Note. N = 187. From [31].
[a]Lowest tercile compared with others.

and conditions that can facilitate or make more difficult the ability to balance work and family [31]. In this study, women reported much more participation in household tasks than men did, and much more difficulty in balancing work and family. Table 36.1 shows several strong associations between various workplace conditions and reported difficulty in balancing work and home. In this study, telephone operators, who receive their very irregular and variable work schedules only four days before they become effective, report high levels of psychological distress. Seventy-one percent scored over the Québec Health Survey's limit for prepathological distress, a level chosen with reference to the most stressed 20% of the Québec population in 1987 [31]. The various techniques used by these operators to deal with an extremely irregular and unpredictable work schedule were recorded using a journal followed by phone interviews [32]. During a two-week period, thirty workers undertook an average of three actions per week to change (exchange or reschedule) their working hours and made an average of four attempts per week to rearrange their day care providers' work schedules. Over the short two-week period, they used an average of four and up to eight providers (including the spouse). Because it took an average of 5.3 attempts to succeed in changing a work schedule, this activity was particularly depressing and frustrating. Thus, they undertook one action a day, usually unsuccessful, just to be able to get to work, implying an ever-present worry. The operators also worried about the impact on the children forced to adapt to such a variety of providers.

B. Confusion between Public and Private Spheres of Women's Lives

Perhaps because of the real interpenetration between public and private spheres, there often seems to be confusion in the scientific literature about whether to treat women's exposures as occupational, wherever they occur. Some women's paid jobs (housekeeping, sex work) may appear to belong to the private sphere. For example, an article on De Quervain's tenosynovitis started its discussion entitled "Non-Occupational Factors" as follows [33]: "Historically, overexertion related to household duties was the most frequently reported exciting cause. Housewives and maids represented 79% of the cases [of the disease] summarized by Schneider.'' Thus, the exposures of not only housewives but also maids were considered to be nonoccupational.

Another example of this type of attitude is found in a suggestion that the male-female ratio in heart disease be used to identify occupational hazards [34]. This ratio was high (50% more men than women with heart disease) in higher social classes, leading the authors to suggest: "That wives of men in social classes I and II are not at as high a risk as their husbands suggests that it is *what happens at work* that produces a risk." In lower social classes, the ratio was low (three men with heart disease for every four women). Surprisingly, the authors did not conclude that the work of women in lower social classes was more dangerous. Their interpretation then became: "There may be particular risks for women married to men in unskilled jobs—*perhaps related to lifestyle*" (my italics). This article does not consider the possibility that women of lower social class may have occupational exposures that make it likely for them to get heart disease.[1] Other research, however, shows coronary heart disease to be particularly common among clerical workers who are mothers and are married to blue-collar men [35].

In women's work, some risks may be perceived as a contamination of the public work sphere by private concerns. Stress reactions to sexual harassment are an example, hence the ongoing debate about whether sexual harassment should be covered by occupational health legislation [36]. Similarly, occupational health intervention is just starting to respond to attacks on women in their workplaces by men with whom they have had a relationship in the private sphere [37].

Women's exposures may also appear to be private, personal, and irrelevant to occupational health studies because they relate to the nurturing and supportive roles [38,39] required of women in the workplace, rather than to the physical, ergonomic, chemical, and biological exposures usually considered to be part of occupational health. Studies show that service workers may find it particularly important to give good service to the public and become stressed when this is not possible [40]. They may even become impelled to work harder than they should in order to fill the gap between the functioning of institutions after personnel cuts and clients' needs [41].

It is possible that some exposures may appear to be private *because* they affect women specifically. For example, in Québec, the health and safety law provides that dangers should be eliminated at the source. One of only two exceptions is for dangers affecting pregnant women, where the endangered woman, rather than the chemical or physical hazard, is removed from the workplace.[2] Also, menstrual cycle anomalies and age at menopause may be seen as private health problems, rather embarrassingly associated with female gender. These reproductive anomalies may never be linked to the working conditions, such as irregular hours and cold temperatures, that produce them [42].

In her historical treatment of women's occupational health, Harrison shows that women's struggle to be accepted in the workplace places them between two unacceptable options. If they try to deny the specific problems they experience in reconciling work and family, they risk illness. If they emphasize their family responsibilities in the workplace and request that their needs be respected, they may ease their lives and those of their families in the short run, but they run a risk of being marginalized and kept from earning money that will contribute to creating a healthy family situation. They will be put at an economic and professional disadvantage and risk greater imbalance in domestic responsibilities [43].

De Koninck has suggested that women would do well not to deny the interrelationship between public and private spheres in their lives, but rather to try to allow and facilitate it for both sexes [44]. With growing exhaustion in the workplace, this option may become increasingly attractive to men. Several studies suggest that men's situation is becoming more like women's as their working conditions approach women's. Marshall *et al.* found that men in service jobs, whether or not the jobs were dominated by women, experienced the same stress as women in response to job demands and lack of rewards from service [40]. As hours of work increase, irregular and unpredictable work schedules are becoming more common for all [14,45]. Men's and women's time spent on domestic work is converging, especially with respect to childcare. This is true primarily because women are doing less, but also because men are doing more [14, p. 36], while families are delegating more domestic work to paid workers in restaurants, cleaning establishments, day care centers and the like [14, p. 85].

III. Exposures to "Invisible" Workplace Health Determinants

Women are still found concentrated in few occupations in the United States [46] and Canada [47]. (See Chapter 35). These occupations often are characterized by multiple, low-level exposures that are hard to identify as posing a health risk. This section will discuss two types of common exposures in women's work—prolonged standing and multiple low-level stressors—in order to show why and how they are so difficult to identify, describe, and study.

A. Prolonged Standing

In North America, service workers such as checkout clerks, bank tellers, and retail sales clerks, as well as some factory workers, quite often work standing without walking very much. The explanation usually given is that employees look more available, clients are not afraid to disturb them, and service is better if the employee stands up. In other countries such as France, Sweden, Morocco, and Brazil, chairs are provided in banks and often in supermarkets. However, most sales people and some factory workers in Europe also work standing.

Prolonged standing, like other static effort, is not a dramatic physical requirement. The observer is less impressed by the plight of a standing fast food server than by a worker washing the windows of a 17-story building. Yet the window washer has only a theoretical and quite small probability of accident,

---

[1]There is also an implicit assumption that working class men have jobs that are less stressful for the heart than upper class men. This is probably wrong. Other studies show that men of the upper social classes are less likely than working class men to have heart disease related to their jobs (see, for example, [34a]).

[2]See Article 64 of the Law on Occupational Health and Safety of Québec.

whereas both workers and experimental subjects exposed to constrained standing report high levels of discomfort [48,49].

Bank tellers studied by our research group passed 72% of their time standing (random sampling throughout the workday) and reported back, leg, and foot pain [50] significantly associated with the response "standing" given to the question "Do you spend most of your time sitting/standing/about half and half?"

Clinical and experimental studies [51–55] suggest that the discomfort and health damage from prolonged constrained standing arise from four main sources: (1) interference with the venous return from the lower limbs to the heart, resulting in tissue hypoxia; (2) interference with the venous return from the lower limbs to the heart, resulting in edema; (3) intervertebral disc stress from excessive lordosis; and (4) stretching of ligaments. In the occupational health system, however, simple discomfort does not often justify intervention or compensation; diseases must be diagnosed. Unfortunately, epidemiologic investigations of disease associated with prolonged standing are almost completely absent from the literature. No controlled study has sought a relationship between prolonged standing at work and possibly-related problems, common among women, such as fibromyalgia, fallen arches, cardiovascular disease, or stroke.

Prolonged standing *is* hard to study because workers rarely stand absolutely still [56]. Despite their prolonged standing at the wicket, for example, bank tellers spend only an average of 20 seconds without taking a step [50]. Similar patterns are found among restaurant workers and sales clerks [57]. Not only is the exposure undramatic and unimpressive, it is extremely difficult to characterize precisely, a far cry from the more obvious risks to the backs, life, and limbs of stevedores, construction workers, and miners. Thus, in order to demonstrate scientifically that there is a problem, careful observation of work activity has to be done for long periods of time. Otherwise, scientists are tempted to dismiss the clerks' complaints as not serious (no disease) or not linked to the workplace (not related to working posture).

B. "Stress"

In a 1996 international conference titled "Women, Health and Work," about one presentation in five was on stress [58]. The title of a 1991, *Women, Work and Health: Stress and Opportunities,* also expresses the extent to which stress is identified as *the* women's occupational health problem [59]. However, because stress reactions may arise from a number of factors, attribution of stress reactions to working conditions is more complicated than, for example, attributing mesothelioma to asbestos exposure. It is hard to prove the involvement of working conditions in determining stress, especially when there has been no trauma such as being attacked or robbed. As Lippel and colleagues have shown, Québec tribunals have tended to see stress in men's jobs as more dramatic than that in women's. In their analysis of 53 appeals decisions on compensation for chronic stress, they found that women were refused more than twice as often as men [60], the same result as was found in an analysis of American claims for compensation for stress-induced physical (*e.g.,* cardiovascular) problems [61]. In Québec, the claims of women with no psychiatric history prior to employment were refused more often than those of men with such a history. Reasons given for refusal often referred to the fact that the alleged stressful condi-

tion was a normal part of the women's work. For example, a man's claim for compensation because of the stress induced by having his vacation time changed without notice was accepted, while a woman's claim for compensation for stress induced by a change in work organization without sufficient training was refused. Dembe has described the history of these attitudes, dating from the beginning of the Industrial Revolution when women and minorities were considered to be more susceptible to nervous exhaustion. Any problems they had could be attributed to their unfitness for work rather than to problematic working conditions [62]. Thus, it seems that women's working conditions are perceived as easy, and any problems are attributed to the fragility of the worker.

In this context, it is important to have good studies establishing links between working conditions and measures of psychological distress. Unfortunately, women have not always been well served by existing studies. The classic research on the relationships between psychological distress and job conditions come from Karasek and Theorell, who initially derived their "Job Content Questionnaire (JCQ)" from studies of Swedish males [63]. They found that a combination of high job demands and low decision latitude (such as is found on a high-speed factory assembly line) are associated with psychological distress and cardiovascular disease. However, although women are overrepresented in such jobs [64], there was trouble later in applying these relationships to women [65]. Some of the difficulties may come from social roles or the specifics of jobs assigned to women; for example, the JCQ includes no questions on workplace policies concerning family. Other difficulties in applying the model to women may come from the role played by social support in women's lives as opposed to men's [66–68]. In addition, in a study of white, dual-earner couples, Marshall *et al.* [40] found that while the relationships apply to women and men in factory jobs, they did not explain psychological distress of either women or men in the service industries (where women are concentrated). Instead, the authors found the combination of high demand and less rewarding service to others to be an important determinant of distress for service workers.

Linking stress reactions to women's working conditions may be more complicated than just adding a gender-specific factor or two to a model. Stress is produced by multiple factors that vary with the situation and the worker. For example, school teachers in Québec (and elsewhere), members of one of women's ten most common professions, have very high levels of stress and burnout. These teachers, who have the lowest levels of psychological distress of any white-collar workers when they are young, have the highest levels of any when they reach the age of 45–54 [69]. For the population as a whole, 12.1% of invalidity is related to mental problems; for teachers, the figure rises to 33.5%.[3] However, in Québec (where armed pupil violence is rarer than in the U.S.), teachers are not often exposed to any obvious psychological risks such as robbery, trauma, or life-threatening events.

When researchers met with primary school teachers to discuss the reasons they felt exhausted and discouraged, they were unable to identify single risk factors. Instead, they referred to

---

[3]Calculated by Jacqueline Dionne-Proulx of the Université du Québec à Trois-Rivières from data supplied by the Québec Retirement Plan.

their "global" workload. Factors included were: environmental problems such as noise from the street or other classes, uncomfortable temperatures and too-low humidity levels that distracted the pupils, hours of work outside the paid work day, the numbers of procedures needed to arrange special help for children with emotional or learning problems, interruptions, demands for several simultaneous activities, public criticism of teachers and objections to their pedagogical conference days, time spent standing, awkward postures, and chalk dust exposure [70,71].

None of these stressors, taken individually, exceeded any threshold limit value or suggested exposure. For example, one first-grade teacher was exposed to the following conditions: (1) a class of 23 (the recommended size) of whom only six had no major health or learning problems, several were so poor that winter clothing had to be found for them, most came from first-generation immigrant families, others had parents who threatened her with violence (no recommended exposure level for any of the conditions); (2) one little boy was an incest victim whom she made a number of unsuccessful attempts to place, finally succeeding after 4 months (no recommended exposure level); (3) the classroom temperature often exceeded 28° C (82°F), making her and the children sleepy and uncomfortable (28° is well below the maximum allowable level for light work); (4) 84% of class time was spent talking and the average length of an explanation was 8.7 seconds, implying a very fast and unrelenting work speed (no suggested limit); (5) she had her eyes fixed on a pupil or on the class during more than 95% of class time, implying a high degree of concentration (no limit); (6) she spent 19 hours a week on average in preparation in addition to the 24 hours of presence in class (43 h/week is under the statutory work week), but was constantly preoccupied with the problems of her pupils (no recommended time limit for worrying); (7) due to discipline problems, poor insulation, and the proximity of a gymnasium, the noise level in class ranged from 45–69 dbA and was usually at 50–59 (well below the exposure limit for noise), making it difficult for her and the children to concentrate; (8) she always worked standing (no limit) and her back was flexed as she bent over the small children almost 20% of the time, putting pressure on the intervertebral discs (no limit for bending without a load); (9) she had two young children and found that when she got home she had to continue with the same activities she had done in school (not taken into account in the occupational health context). "I came home after helping 23 little children cut out 23 little pumpkins for Halloween, and my daughter greeted me with: 'Will you help me cut out a pumpkin, Mama?'—I couldn't answer."

No single risk factor could be viewed as anything more than the normal requirements of a teaching job, nor was the teacher exposed to any conditions that could be reported to a health and safety inspector. After the session spent dealing with the incest case, this teacher caught chicken pox at the school and was home for a week. At the end of the week, she found she felt unable to return to school and went on sick leave, not workers' compensation. Soon she had exhausted her sick leave and had to go back to school but she felt she was unable to do her job. She transferred out of regular elementary school teaching. This example points out the necessity of considering multiple risk factors in context. Because most jobs held by women do not

involve single, dramatic aggressors, more diffuse effects should be examined and remedies should be found.

IV. Changing Methods

In order that women's health problems may be considered in relation to the workplace environment, we need to develop new ways of identifying, describing, and analyzing their exposures. Several changes are needed. The first is to take appropriate account of the gendered division of labor in epidemiologic studies. For example, as mentioned earlier, female gender has been associated with an increased incidence of certain occupational health problems such as mass psychogenic illness, fibromyalgia, and multiple chemical sensitivity. It is therefore common, even recommended, to treat gender as a confounding variable and to stratify for it or to include it as such in logistic regression models [6, pp. 2–22, 5a–27, 6–38]. However, gender often is a proxy for exposure variables [21]. Results are beginning to accumulate showing that when gender-typed occupational exposures are taken into account, male-female differences in such outcomes as musculoskeletal problems [72,73], carpal tunnel syndrome [74], psychological distress [75], sick building syndrome [7], and even distress associated with domestic overload [76] disappear or are greatly attenuated. It is therefore important to examine the male-female distribution of tasks before treating gender as a confounder.

The second change that is needed is to find appropriate ways to consider the work-family interface in occupational health studies. Workplaces have gradually been obliged to take into account a certain number of workers' personal needs, inseparable from their characteristics as human beings. It is required that employers provide bathrooms, breaks, drinking water, and mealtimes as a recognition of the "personal need" for physical survival. Family needs have been included to some extent, and the fact that effects on the family differ by gender has not prevented governments from acting on some family-related problems. The pesticide dibromodichloropropane (DBCP) was outlawed in the U.S. shortly after its effects on lowering fertility of male workers became known [77].[4] In the United States, the Worker Family Protection Act provides for some protection of the families of (usually male) workers who may carry toxic dusts, such as asbestos, on their clothing.

Responsibility for childcare (usually women's) has not yet been protected by labor legislation. Employers are not required to provide access to a telephone, schedules compatible with childcare, or physical exposure limits that include attention to the combined paid and unpaid workday. Because the expanding demands of the workplace come increasingly into conflict with the domestic and family responsibilities of male and female workers, analyses of the determination of psychological and physical effects of work should not treat family situation as a confounder or an independent variable. Difficult though this task may be, occupational health researchers must identify elements of the work situation (schedules, telephone access,

---

[4]When lead was identified as a danger for pregnant women working with lead in batteries, the initial reaction was to exclude the women. A U.S. Supreme Court decision was necessary to give lead-exposed women the right to their jobs. No ban on lead in batteries has been proposed.

requirements for nurturing, exposure to cleaning chemicals) that may be synergistic with domestic exposures or problematic for workers with family and domestic responsibilities. These should then be analyzed as occupational health determinants.

Third, because women's occupational health is at the confluence of the sociological and ergonomic studies of women and work and the sociological and biomedical studies of women and health, quantitative, single-discipline studies must be supplemented by other approaches. For example, Swedish researchers are using biological indicators to follow small groups of male and female workers' reactions to stress (Töres Theorell, personal communication). Qualitative studies can permit identification of many components of complex environments. Such studies increasingly involve the workers under study as active participants in defining the research directions and interpreting the results [78,79].

Biomedical researchers and industrial hygienists accustomed to the canons of "evidence-based medicine" used to establish threshold limit values will be ill at ease with the necessarily less precise information that can be gained from multidisciplinary qualitative studies or those using biological prepathological indicators. The information gained through alternative methods may not be easily translated into numerical standards, used to support workers' compensation claims, or employed in employer-union confrontations, but it can be used as a base for more consensual prevention strategies. It also may be useful in mobilizing women workers [50] who may recognize their situation better when portrayed in a way that mirrors the complexities of their roles and problems.

Acknowledgments

I am grateful for a partnership grant from the Conseil québécois de la recherche sociale and for the active contributions of our partners, the women's committees of three unions, the Centrale de l'enseignement du Québec, the Confédération des syndicats nationaux, and the Fédération des travailleurs et travailleuses du Québec. I thank Louise Vandelac for suggesting an approach to work-family issues and Ana Maria Seifert and Åsa Kilbom for suggesting approaches to prolonged standing work.

References

1. Rothman, A. L., and Weintraub, M. I. (1995). The sick-building syndrome and mass hysteria. *Neurol. Clin.* **13**, 405–412.
2. Lucire, Y. (1986). Neurosis in the workplace. *Med. J. Aust.* **145**, 323–327.
3. Wiholm, C., and Arnetz, B. B. (1997). Musculoskeletal symptoms and headaches in VDU users—a psychophysiological study. *Work Stress* **11**, 239–250.
4. Voitk, A. J., Mueller, J. C., Farlinger, D. E., and Johnston, R. U. (1983). Carpal tunnel syndrome in pregnancy. *Can. Med. Assoc. J.* **128**, 277–281.
5. Small, G. W., and Nicholi, A. M. (1982). Mass hysteria among schoolchildren. *Arch. Gen. Psychiatry* **39**, 721–724 (cited in Ref. 1).
6. Bernard, B. P., ed. (1997). "Musculoskeletal Disorders and Workplace Factors," pp. B-3, B-4. National Institute of Occupational Safety and Health, Cincinnati, OH.
7. Stenberg, B., and Wall, S. (1995). Why do women report "sick building symptoms" more often than men? *Soc. Sci. Med.* **40**, 491–502.
8. Occupational Medicine Forum Committee (1992). Does fibromyalgia qualify as a work-related illness or injury? *J. Occup. Med.* **34**, 968.
9. Selvaggio, J. E. (1994). Psychological aspects of "environmental illness," "multiple chemical sensitivity," and building-related illness. *J. Allergy Clin. Immunol.* **94**, 366–370.
10. U.S. Department of Health and Human Services (1997). "Women: Work and Health," DHHS Pub. No. (PHS) 98–1791. USDHHS, Washington, DC.
11. Canadian Advisory Council on the Status of Women (1994). "110 Canadian Statistics on Work and Family." Canadian Advisory Council on the Status of Women, Ottawa.
12. Bernier, C., Laflamme, S., and Zhou, R. M. (1996). Le travail domestique: Tendances à la désexisation et à la complexification. *Can. Rev. Sociol. Anthropol.* **33**, 1–21.
13. Warde, A., and Hetherington, K. (1993). A changing domestic division of labour? Issues of measurement and interpretation. *Work, Employment Soc.* **7**, 23–45.
14. Schor, J. B. (1992). "The Overworked American," Chapter 4 and notes. Basic Books, New York.
15. Chadeau, A., and Fouquet, A. (1981. Peut-on mesurer le travail domestique? *Econ. Stat.* **136**, 29–42.
16. Statistics Canada (1995). "Women in Canada," 3rd ed., Catalogue No. 89–503E, Table 6.24. Statistics Canada, Ottawa.
17. Barnett, R. C., and Baruch, G. K. (1987). Social roles, gender and psychological distress. *In* "Gender and Stress" (R. C. Barnett, L. Biener, and G. K. Baruch, eds.), pp. 122–143. Free Press, New York.
18. Ilfeld, F. W. (1976). Methodological issues in relating psychiatric symptoms to social stressors. *Psychol. Rep.* **39**, 1251–1258.
19. Guyon, L. (1997.) "Derrière les apparences: Santé et conditions de vie des femmes." Ministry of Health and Social Services, Québec.
20. Lee, C., Duxbury, L., and Higgins, C. (1994). "Employed Mothers: Balancing Work and Family Life." Canadian Centre for Management Development, Ottawa.
21. Messing, K., Tissot, F., Saurel-Cubizolles, M. J., Kaminski, M., and Bourgine, M. (1998). Sex as a variable can be a surrogate for some working conditions: Factors associated with sickness absence. *J. Occup. Environ. Med.* **40**, 250–260.
22. Hatch, M., and Moline, J. (1997). Women, work and health. *Am. J. Ind. Med.* **3**, 303–308.
23. Tierney, D., Romito, P., and Messing, K. (1990). She ate not the bread of idleness: Exhaustion is related to domestic and salaried work of hospital workers in Québec. *Women Health* **16**, 21–42.
24. Nyquist, L., Slivken, K., Spence, J. T., and Helmreich, R. L. (1985). Household responsibilities in middle-class couples: The contribution of demographic and personality variables. *Sex Roles* **12**, 15–34.
25. Hessing, M. (1993). Mothers' management of their combined workloads: Clerical work and household needs. *Can. Rev. Soc. Anthropol.* **30**, 37–63.
26. Hochschild, A. (1989). "The Second Shift." Avon Books, New York.
27. Hochschild, A. (1997). "The Time Bind: When Work Becomes Home and Home Becomes Work." Henry Holt, New York.
28. Guberman, N., Maheu, P., and Maillé, C. (1993). "Travail et soins aux proches dépendants." Les éditions du remue-ménage, Montréal.
29. Fédération des travailleurs et des travailleuses du Québec (FTQ), (1996.) "Concilier l'inconciliable." FTQ, Montréal.
30. Barnett, R. C., and Brennan, R. T. (1995). The relationship between job experiences and psychological distress: A structural equation approach. *J. Organ. Behav.* **16**, 259–276.
31. Tissot, F., Messing, K., Vandelac, L., Garon, S., Prévost, J., Méthot, A. L., Pinard, R., Gingras, C., and de Grosbois, S. (1997). "Volet détresse psychologique des recherches *Concilier l'inconciliable* et *L'art de concilier l'inconciliable*." CINBIOSE, Montréal.
32. Prévost, J., and Messing, K. (1997). Quel horaire, What schedule? L'horaire de travail irrégulier des téléphonistes. *In* "Stratégies de résistance et travail des femmes" (A. Soares, ed.), pp. 251–270. Harmattan, Montréal.

33. Moore, J. S. (1997). De Quervain's tenosynovitis: Stenosing tenosynovitis of the first dorsal compartment. *J. Occup. Environ. Med.* **39,** 990–1002.

34. Heller, R. F., Williams, H., and Sittampalam, Y. (1984). Social class and ischaemic heart disease: Use of the male:female ratio to identify possible occupational hazards. *J. Epidemiol. Commun. Health* **38,** 198–202.

34a. Cassou, B., Derrienec, F., Lecuyer, G., and Amphoux, M. (1986). Déficience, incapacité et handicap dans un groupe de retraités de la Région Parisienne en relation avec la catégorie socio-professionnelle. *Rev. Epidemiol. Santé Publ.* **34,** 332–340.

35. Haynes, S. G., and Feinlib, M. (1980). Women, work and coronary heart disease: Prospective findings from the Framingham Heart Study. *Am. J. Public Health* **70,** 133–141.

36. Lippel, K. (1997). Le harcèlement sexuel au travail: Quel rôle attribuer à la C.S.S.T. et au Tribunal des droits de la personne suite à l'affaire Béliveau St-Jacques? *In* "Développements récents en responsabilité civile." Les éditions Yvon Blais, Cowansville, Quebec, Canada, pp. 99–139.

37. Duda, R. A. (1997). Workplace domestic violence. *AAOHN J.* **45,** 619–624.

38. Hochschild, A. (1983). "The Managed Heart." University of California Press, Berkeley.

39. Gutek, B. (1995). "The Dynamics of Service Work." Jossey-Bass, San Francisco.

40. Marshall, N. L., Barnett, R., and Sayer, A. (1997). The changing workforce, job stress and psychological distress. *J. Occup. Health Psychol.* **2,** 99–107.

41. Armstrong, P. (1998). Changes in work organization that affect the health of women workers. *Commun. Colloq. Improv. Women's Occup. Health,* Université du Québec à Montréal, *1998.* In press.

42. Messing, K., Saurel-Cubizolles, M. J., Kaminski, M., and Bourgine, M. (1992) Menstrual cycle characteristics and working conditions in poultry slaughterhouses and canneries. *Scand. Work, Environ. Health* **18,** 302–309.

43. Harrison, B. (1997). "Not only the 'Dangerous Trades': Women's Work and Health in Britain, 1880–1914." Taylor & Francis, London.

44. De Koninck, M. (1997). Le travail des femmes à travers le prisme de la continuité. *In* "Stratégies de résistance et travail des femmes" (A. Soares, ed.), pp. 251–270. Harmattan, Montréal.

45. Advisory Group on Working Time and the Distribution of Work (1994). "Report of the Advisory Group on Working Time and the Distribution of Work." Human Resources Development Canada, Ottawa.

46. Stellman, J. M. (1994). Where women work and the hazards they may face on the job. *J. Occup. Med.* **36,** 814–825.

47. Armstrong, P., and Armstrong, H. (1991). "Theorizing Women's Work." Garamond Press, Toronto.

48. Hansen, L., Winkel, J., and Jorgenson, K. (1997). Significance of mat and shoe softness during prolonged work in upright position. *Appl. Ergonomics* **29,** 217–224.

49. Ryan, G. A. (1989). The prevalence of musculo-skeletal symptoms in supermarket workers. *Ergonomics* **32,** 359–371.

50. Seifert, A. M., Messing, K., and Dumais, L. (1997). Star wars and strategic defense initiatives: Work activity and health symptoms of unionized bank tellers during work reorganization. *Int. J. Health Serv.* **27,** 455–477.

51. Buckle, P. W., Stubbs, D. A., and Baty, D. (1986). Musculo-skeletal disorders (and discomfort) and associated work factors. In N. Corlett, J. Wilson, and I. Manenica, eds.) "The Ergonomics of Working Postures: Models, Methods and Cases" pp. 19–30. Taylor & Francis, London.

52. Whistance, R. S., Adams, L. P., van Geems, B. A., and Bridger, R. S. (1995). Postural adaptations to workbench modifications in standing workers. *Ergonomics* **38,** 2485–2503.

53. Sejersted, O. M., and Westgaard, R. H. (1988). Occupational muscle pain and injury: Scientific challenge. *Eur. J. Appl. Physiol.* **57,** 271–274.

54. Edwards, R. H. T. (1988). Hypotheses of peripheral and central mechanisms underlying occupational muscle pain and injury. *Eur. J. Appl. Physiol.* **5,** 275–281.

55. Hansen, L., Winkel, J., and Jorgenson, K. (1998). Significance of mat and shoe softness during prolonged work in an upright position. *Appl. Ergonomics* **29,** 217–224.

56. Buckle, P. W., Stubbs, D. A., and Baty, D. (1986). Musculo-skeletal disorders (and discomfort) and associated work factors. *In* N. Corlett, J. Wilson, and I. Manenica, eds.), "The Ergonomics of Working Postures: Models, Methods and Cases" pp. 19–30. Taylor & Francis, London.

57. Messing, K., and Kilbom, °. (1999). Effects on feet of standing and very slow walking: A pilot study of pain-pressure threshold changes, subjective pain experience and work activity among sales and kitchen workers. Submitted for publication.

58. Conference on Women, Health and Work (1996). *Programme Abstr., Conf. Women, Health Work,* Barcelona, Spain, *1996.*

59. Frankenhaeuser, M., Lundbergh, U., and Chesney, M. (1991). "Women, Work and Health: Stress and Opportunities." Plenum, New York.

60. Lippel, K. (1999). Workers' compensation and stress: Gender and access to compensation. *Int. J. Law Psychiatry* **22,** 79–89.

61. Lippel, K. (1992). "Le stress au travail: L'indemnisation des atteintes à la santé en droit québécois, canadien et américain." Éditions Yvon Blais, Cowansville, Quebec, Canada.

62. Dembe, A. E. (1996). "Occupation and Disease: How Social Factors Affect the Conception of Work-Related Disorders," pp. 49–52. Yale University Press, New Haven, CT.

63. Karasek, R. A., Baker, D., Marxer, F., Ahlbom, A., and Theorell, T. (1981). Job decision latitude, job demands and cardiovascular disease: A prospective study of Swedish men. *Am. J. Public Health* **71,** 694–705.

64. Hall, E. M. (1989). Gender, work control and stress: A theoretical discussion and an empirical test. *Int. J. Health Ser.* **19,** 725–745.

65. Johnson, J. V., and Hall, E. M. (1988). Job strain, workplace social support and cardiovascular disease. *Am. J. Public Health* **78,** 1336–1342.

66. Hall, E. M. (1992). Double exposure: The combined impact of the home and work environments on psychosomatic strain in Swedish women and men. *Int. J. Health Serv.* **22,** 239–260.

67. Karasek, R. A., and Theorell, T. (1990). "Healthy Work: Stress, Productivity, and the Reconstruction of Working Life." Basic Books, New York.

68. Daniels, K., and Guppy, A. (1997). Stressors, locus of control, and social support as consequences of affective psychological well-being. *J. Occup. Health Psychol.* **2,** 156–174.

69. Gervais, M. (1993). "Bilan de santé des travailleurs québécois," p. 25. Institut de recherche en santé et en sécurité du travail, Montréal.

70. Messing, K., Seifert, A. M., and Escalona, E. (1997). The 120-second minute: Using analysis of work activity to prevent psychological distress among elementary school teachers. *J. Occup. Health Psychol.* **2,** 45–62.

71. Escalona, E. (1997). Activité de travail des enseignantes de niveau primaire. Thesis submitted in partial fulfillment of the requirements for the M.Sc. in Biology, Université du Québec à Montréal.

72. Messing, K., and Reveret, J. P. (1983). Are women in female jobs for their health? Working conditions and health symptoms in the fish processing industry in Québec. *Int. J. Health Serv.* **13,** 635–647.

73. Mergler, D., Brabant, C., Vézina, N., and Messing, K. (1987). The weaker sex? Men in women's working conditions report similar health symptoms. *J. Occup. Med.* **29,** 417–421.

74. Stetson, D. S., Albers, J. W., Silverstein, B. A., and Wolfe, R. A. (1992). Effects of age, sex, and anthropometric factors on nerve conduction measures. *Muscle Nerve* **15,** 1095–1104.

75. Mergler, D. (1995). Adjusting for gender differences in occupational health studies. *In* K. Messing, B. Neis, and L. Dumais, eds.), "Invisible: Issues in Women's Occupational Health and Safety/Invisible: La santé des travailleuses" pp. 236–251. Gynergy Books, Charlottetown, Quebec, Canada.

76. Barnett, R. C., and Shen, Y.-C. (1997). Gender, high- and low-schedule-control housework tasks and psychological distress. *J. Fam. Issues* **18,** 403–428.

77. Kenen, R. (1993). "Reproductive Hazards in the Workplace," pp. 18–19, cf. pp. 208–209. Haworth Press, New York.

78. Loewenson, R. H., Laurell, A. C., Hogstedt, L. C., and Wegman, D. H. (1995). Participatory approaches and epidemiology in occupational health research. *Int. J. Occup. Environ. Health* **1,** 121–130.

79. Messing, K. (1998). "One-eyed Science: Occupational Health and Women Workers." Temple University Press, Philadelphia.

37

Reproductive Hazards in the Workplace

MARJA-LIISA LINDBOHM* AND HELENA TASKINEN*,†

*Finnish Institute of Occupational Health, †Tampere School of Public Health, Tampere University, Helsinki, Finland

I. Introduction

The effect of occupational hazards on reproductive health is a rapidly evolving field of research. Reproductive health is a major aspect of human life, and different types of reproductive failures are common. One reason for the expansion of research is the high (and increasing) participation of women in the labor force. The average percentage of participation of women aged 15–64 years was 57% in the 15 countries of the European Union; it varied from 78% in Sweden to 43% in Italy in 1995 [1]. The percentage is also high among pregnant women. For example, in the United States, the proportion of women employed during their first pregnancy was 64.5% [2]. In the developing regions of the world, nearly half of all women are in the paid labor force.

The magnitude of workplace reproductive health problems is difficult to assess. Only a small proportion of the more than 60,000 chemicals on the market have been adequately tested for reproductive toxicity [3]. In addition, physical, biologic, and psychosocial stress factors may adversely affect reproductive health. There are no estimates on the number of women or men who are at risk of exposure to potential reproductive toxicants. In general, data on the prevalence of occupational exposures by sex are scarce, although the workforce is generally segregated into typical male and female jobs and, thus, women tend to be exposed to different occupational hazards than men. Exposure to potential reproductive hazards probably is most common among women in manufacturing and health care. Organic solvents and physical strain are likely to be the most prevalent harmful agents. In developing countries, the main problems are related to pesticides and industrial chemicals.

II. Effects of Occupational Reproductive Hazards

Human reproduction is a complex process that is a chain of interdependent events. Any of its phases can be disturbed by exogenous agents. Environmental exposure to a reproductive toxicant can affect the reproductive system of both men and women. Toxic exposures can cause direct cell damage to developing sperm and eggs, which are particularly susceptible to injury during meiotic division. Some exposures may be toxic to the organs that produce hormones and some may affect hormone levels by increasing amounts of the enzymes that regulate hormone excretion from the body [4]. Reproductive toxicity may manifest itself as alterations in sex hormone levels, diminished libido and potency, menstrual disorders, impairment of semen quality, ovarian dysfunction, infertility, or adverse pregnancy outcomes [5].

Environmental exposure to a developmental toxicant can induce adverse effects on the developing organism. Exposure can affect the germ cells of both men and women before conception and induce damage to the genetic material of the cells. Maternal exposure to hazardous agents during pregnancy may disturb the development of the fetus, thereby leading to an adverse pregnancy outcome. The exposure may affect the fetus directly or indirectly by interfering with maternal, placental, or fetal membrane function. Some direct-acting toxic agents are chemically reactive and affect the cellular components of tissues involved in reproduction, whereas indirect-acting agents require metabolic activation after exerting toxicity [6]. Estrogenic activity of some toxicants is also one mechanism that can produce an adverse reproductive effect. Developmental toxicity may appear as infertility, spontaneous abortion, stillbirth, intrauterine growth retardation, preterm birth, congenital malformation, postnatal death, functional disturbances, or childhood cancer [5].

Effects on reproduction and development are often the result of short-term exposure during the vulnerable periods of ovulation, spermatogenesis, or fetal organogenesis. The result of a toxic exposure may not become evident for years [3]. Some toxicants (*e.g.,* lead and polychlorinated biphenyls) accumulate in parental tissues and may be released many years later during pregnancy or lactation. Other toxicants (*e.g.,* antineoplastic agents) may deplete the number of female germ cells resulting in a shortened reproductive life span which is not evident until senescence. Adverse effects of preconceptional exposures (*e.g.,* ionizing radiation) may not become apparent until postnatal development, for example as childhood cancer.

Exposure may be associated with one or many reproductive outcomes, depending on the timing and duration of the exposure and the dose of the agent received. Spontaneous abortion and congenital malformation are the principal outcomes if exposure has occurred during the first trimester of pregnancy. Exposure occurring later in pregnancy is more likely to be associated with a shortened duration of gestation or low birth weight. With increased dose, the adverse effects may be shifted from malformations to spontaneous abortions or further to infertility, resulting in a nontraditional dose-response relationship. That is, a decreasing rate of adverse outcome of pregnancy with increasing dose may be observed in a study of only one reproductive outcome [7]. In order to identify and evaluate the risks of environmental exposure to reproduction, it is important to investigate a range of outcomes and exposure levels.

III. Employment and Reproductive Health

For the majority of women, ordinary work activity does not have a great effect on pregnancy. In general, women working during pregnancy seem to have, at least in some countries, fewer sociodemographic and behavioral risk factors for adverse pregnancy outcome than women who do not work, although there are

Table 37.1
Chemical Occupational Hazards Associated with Adverse Reproductive Effects

Hazard	Reported effects	Industry or occupational group
Anesthetic gases	Reduced fertility, fetal loss, birth defects[a]	Hospital, dental, and veterinary personnel
Nitrous oxide	Reduced fertility, fetal loss[a]	Hospital, dental, and veterinary personnel
Antineoplastic drugs	Menstrual disorders[a], reduced fertility[a], fetal loss, birth defects[a]	Hospital personnel, pharmaceutical industry, laboratory personnel
Ethylene oxide	Fetal loss[a]	Hospital personnel, chemical industry
Metals		
Inorganic mercury	Reduced fertility,[a] fetal loss[a]	Lamp industry, chloralkali industry, dental personnel
Lead	Reduced fertility[a], fetal loss[a], preterm birth[a], low birth weight[a], impaired cognitive development[a]	Battery industry, lead smelting, foundries, pottery industry, ammunition industry, and some other metal industries.
Solvents	Reduced fertility, fetal loss, birth defects[a]	Painting, several other fields of industry
Carbon disulfide	Menstrual disorders[a]	Viscose rayon industry
Ethylene glycol ethers	Reduced fertility, fetal loss	Electronics industry, silk screen printing, photography and dyeing, other industries
Formaldehyde	Reduced fertility[a], fetal loss[a]	Mechanical wood industry, pathology laboratories
Tetrachloroethylene	Reduced fertility[a], fetal loss	Dry cleaning, degreasing
Toluene	Reduced fertility[a], fetal loss	Shoe industry, painting, laboratory work
Aliphatic hydrocarbons	Fetal loss[a]	Graphic and shoe industry, painting
Estrogens	Menstrual disorders[a]. fetal loss[a]	Pharmaceutical industry
Pesticides	Reduced fertility[a], fetal loss[a], birth defects, preterm birth[a]	Agriculture, gardening, greenhouse work

[a]Inconclusive evidence.

clear differences by occupational sector [2]. In contrast to the favorable correlates of employment, there is some evidence that the past reproductive experience of working women is worse than that of nonworking women; this experience may affect the likelihood of entering or remaining in the work force, resulting in an overrepresentation of women with adverse pregnancy outcomes among workers [8]. It is also possible that the combination of the responsibilities related to paid employment and family needs may result in stress, fatigue, and health deficits among working women. All in all, these differences between employed and unemployed women have important methodologic consequences for occupational reproductive studies because they encourage the restriction of comparison groups to other employed women.

There is substantial heterogeneity in the work and work environments of employed women. Although participation in paid work itself should not affect the reproductive health of women, there are work-related factors—chemical, physical, biologic, and psychosocial—that may have adverse reproductive consequences. These potentially harmful agents need to be taken into account in order to protect the reproductive health of workers.

IV. Chemical Agents

A. Organic Solvents

Organic solvents are an important occupational reproductive hazard for women (Table 37.1). They are used widely in various industries. Solvents have been related to menstrual disorders, reduced fertility, spontaneous abortion, stillbirth, congenital malformations, and perinatal death, as well as to leukemia and brain tumors in children of exposed workers [9]. Several solvents pass through the placenta.

Data on the effects of solvent exposure on menstruation are inconclusive. The few existing studies suggest that exposure to toluene, perchlorethylene, carbon disulfide, dibromochloropropane, or glycol ethers is associated with various types of menstrual disorders [10]. Reduced fertility also has been indicated for women with high exposure to solvents [11–13].

The effects of solvent exposure on pregnancy outcome have been examined in several occupational groups exposed to different levels, types, or mixtures of solvents. An increased risk of spontaneous abortion has been observed in industrial populations exposed to high levels of solvents. These occupational groups include employees in the manufacturing, dry cleaning, painting, shoe, pharmaceutical, audio speaker, semiconductor, and laboratory industries [9]. No relation between solvent exposure and spontaneous abortion has been observed in workers whose level of exposure is usually low.

Occupational solvent exposure also has been related to an excess of congenital malformations among the children of exposed women. The excess of specific malformations has not been systematic, with different types of malformations reported in different studies. Exposure to various solvents has been linked to central nervous system defects, oral clefts, sacral agenesis, omphalocele and gastroschisis, renal-urinary and gastrointestinal defects, and ventricular septal defects [9].

Organic solvents are fat-soluble chemicals that can pass into breast milk. Contamination of breast milk with various solvents from occupational sources has been reported [14]. Organic solvents such as petroleum solvents, chloroprene, and styrene have also been suspected to inhibit the excretion of milk. There is, however, little information available at present on how the health and development of children who receive environmental contaminants in their milk might be affected.

Exposure to some specific types of solvents (*e.g.*, toluene, ethylene glycol ethers, and tetrachloroethylene) also has been associated with adverse reproductive effects. Occupational exposure to high levels of toluene has been related to spontaneous abortion among shoe workers, audio speaker factory workers, and laboratory workers [9]. In addition, an excess of urinary tract defects and reduced fertility has been reported for toluene exposure [11,15].

Ethylene glycol ethers, which have been used in a variety of industries and products, have caused adverse reproductive and developmental effects in several animal species exposed by different routes of administration. Among semiconductor manufacturing workers, exposure to ethylene glycol ethers was associated with an increased risk of spontaneous abortion and subfertility [12,13]. A European multicenter study indicated an association between glycol ether exposure and the risk of congenital malformations [16].

Several epidemiologic studies have pointed to an increased risk of spontaneous abortion from high exposure to tetrachloroethylene in dry cleaning work [9,17]. A time-to-pregnancy study also suggested decreased fertility in exposed women [11]. Despite the small sample size of these studies, the findings are consistent, indicating an association between a high level of exposure to tetrachloroethylene and adverse reproductive outcomes. An increased risk of spontaneous abortion also has been reported for exposure to aliphatic hydrocarbons in graphic, shoe, and pharmaceutical factory work, to petroleum ether in laboratory work, to methylene chloride in pharmaceutical factory work, to xylene and formalin in pathology and histology laboratories, and to paint thinners [9].

The most serious weakness in epidemiologic reproductive studies on the effects of organic solvents is inaccurate data on exposure, which is usually based on the workers' own reports. A few investigations used more objective data sources or methods in the exposure assessment. In these studies, exposure was assessed by industrial hygienists [12,15,18] or biologic monitoring data were used to support the reports of the workers [11,19]. Most of these studies showed an association between solvent exposure and adverse reproductive outcome.

Overall, the available evidence suggests that maternal exposure to high levels of solvents increases the risk of spontaneous abortion. The results for congenital malformations and fertility, although less conclusive, also suggest that solvent exposure may represent a hazard for the developing fetus and may impair fertility. The findings on individual solvents must be interpreted with caution because coincident exposure to multiple solvents is common among workers, making it difficult to ascribe an adverse effect to an individual agent. Nevertheless, the study results are supportive of adverse effects of ethylene glycol ethers, tetrachloroethylene, and toluene on female reproduction. It would be prudent to minimize exposure to organic solvents.

B. Metals

Historical descriptions of the reproductive outcome in women exposed to high levels of lead include an excess of miscarriages, neonatal deaths, prematurity, and low birth weight babies. Lead is transferred across the placenta during the 12th to 14th weeks of pregnancy. At birth the blood lead concentration in the umbilical cord of the child is close to that of the mother. Blood lead levels have also been observed to increase during pregnancy despite unchanged or decreasing environmental lead levels. The mobilization of lead from bone during pregnancy may explain the increase [20].

Studies on women occupationally exposed to low levels of lead have indicated no decrease in fecundability nor an increased risk of spontaneous abortion [21,22]. Low lead exposure has, however, been related to preterm delivery and low birth weight, although not consistently [23]. Impaired cognitive development in children exposed to lead during gestation also has been observed at blood lead levels below current workplace exposure standards. All in all, the reproductive impact of the current low exposure levels of lead is not clear, but because lead is known to be a strong toxicant, exposure to lead during pregnancy should be minimized.

Human data on the reproductive effects of other potentially harmful metals, such as inorganic mercury, cadmium, and nickel, are scanty. There is limited evidence based on experimental data that inorganic mercury is teratogenic in some animals. Human studies on the effects of exposure to metallic mercury in dentistry work or lamp work have not shown an increased risk of spontaneous abortion [22]; however, reduced fertility was observed in a study of dental assistants occupationally exposed to mercury [24]. Although the results in humans are inconsistent, mercury is a known potent toxicant and, therefore, occupational exposure should be restricted to a safe level.

Low birth weight in children of smokers has been connected with possible toxic effects of cadmium on the placenta. Smokers have greater amounts of cadmium in their ovaries than nonsmokers, and the placenta of smokers contains more cadmium than that of nonsmokers. In animal tests cadmium has been shown to be embryotoxic, fetotoxic, and teratogenic, and to affect postnatal development [25].

Adverse pregnancy outcomes have been reported in a preliminary study on nickel-exposed women working at a nickel hydrometallurgy refining plant [26]. Various adverse effects on reproduction also have been found in several animal studies, although not in all. Cadmium, nickel, and nickel compounds are classified as carcinogens. Thus, while there is no firm evidence of their reproductive toxicity in humans, there are grounds to restrict the exposure of pregnant women to these metals.

C. Pesticides

Pesticides are an important potential occupational and environmental reproductive hazard. The global use of pesticides in agriculture, greenhouses, and forestry amounts to 2 million tons a year. The U.S. Environmental Protection Agency estimated that there are about 600 active ingredients and 50,000 various formulations [27]. The great number of chemicals, multiexposure, and the possibility of considerable background exposure have made it difficult to investigate the effects of various compounds in pesticides, or occupational exposure to pesticides in general.

Exposure to pesticides from environmental sources or at work has been associated with several adverse effects related to fertility, pregnancy, and offspring. Excretion of organochlorine pesticides via human milk exposes developing infants to pesticides through breastfeeding, but potential adverse health effects of low-level exposure in infants are unknown. In a study of women medically diagnosed as infertile, employment in industries or occupations associated with agriculture seemed to be related to an increased risk of infertility [28], whereas in another study, exposure to pesticides in greenhouse or gardening work was not related to prolonged time to pregnancy [29]. However, one study reported reduced fecundability in women with potential exposure to pesticides in flower greenhouses [30]. There is also some evidence that agricultural work and pesticide exposure may be associated with elevated risk of spontaneous abortion and stillbirth [31]. Increased risks have been found among women in agricultural occupations and among gardeners who sprayed pesticides during pregnancy. In one study the risk was not increased, however, when proper respirators or respirators and protective clothing were used [29]. Another study indicated that occupational exposure to pesticides was associated with stillbirths due to all causes of death, stillbirths due to congenital malformations, or stillbirths due to abnormalities of the placenta, cord, or membranes [32].

Congenital malformations have been studied more often than other outcomes in relation to pesticide exposure [31]. An increase in birth defects has been associated with individual pesticides and with work in which pesticides are used. In Norway, a significant association between pesticides purchased for farm use and brain tumors in offspring was found [33]. Parental exposure to pesticides also has been related to childhood cancers, especially leukemia, lymphomas, and cancers of the brain and nervous system, in most, but not in all, studies [34].

The varying results of epidemiologic studies on pesticides may be due to methodologic problems (*e.g.*, difficulties in exposure assessment or lack of data on confounding factors) [31]. Despite these uncertainties, the results suggest that exposure to pesticides, at least during pregnancy, may be harmful. In animal experiments some pesticides also have been shown to have adverse reproductive effects (*e.g.*, carbaryl, benomyl, ethylenthiourea, maneb, zineb, thiram). Exposure to pesticides at work during pregnancy should be minimized or avoided to prevent potential adverse effects.

D. Anesthetic Gases

Numerous epidemiologic studies have examined the reproductive effects of exposure to trace concentrations of anesthetic gases among hospital personnel in operating rooms and delivery wards, and among dental and veterinary personnel. These studies have yielded conflicting results. Several studies have shown an increased risk of spontaneous abortion and congenital malformations in the offspring of women occupationally exposed to anesthetic gases [35]. Other studies, however, reported relative risks close to, or only slightly above, unity. One possible explanation for the contradictory findings is variation in the type and level of exposure due to differences in the substances used in operating rooms, the methods of administration, and in the scavenging equipment employed. Measurements of anesthetic gases have shown that the concentrations are high if no scavenging system is used or if the ventilation is inadequate.

It also has been suggested that methodologic shortcomings have contributed to the reported effects of the positive studies. The main weakness of these studies, particularly the early ones, has been the lack of reliable data on exposure and outcome. The author of one meta-analysis on spontaneous abortion and exposure to anesthetic gases concluded, however, that a real risk may be present [36]. The conclusion was based on the concordance of findings between animal and human data and on the most rigorous epidemiologic studies.

With the exception of nitrous oxide, the effects of individual anesthetic gases have not been examined in epidemiologic studies. Exposure to nitrous oxide has been related to an increased risk of spontaneous abortion in dental personnel [37] but not in midwives [38], and to reduced fertility [35]. Dental assistants exposed to high levels of unscavenged nitrous oxide and midwives assisting numerous nitrous oxide deliveries per month had lower fecundability as compared with unexposed women.

Although the epidemiologic evidence on reproductive effects of anesthetic gases is equivocal, measures to reduce exposure to these gases should be pursued. Efficient scavenging equipment and good ventilation and equipment for the administration of anesthetics are needed to keep the exposure levels low in operating rooms, delivery wards, dental offices, and veterinary surgeries.

E. Antineoplastic Agents

The treatment of female patients with antineoplastic drugs has induced gonadal damage and adverse effects on menstrual and ovarian function, postmenopausal levels of gonadotropins and estrogens, and symptoms of menopause [39]. Some case reports also have described malformations in the offspring of patients when antineoplastic therapy has been given during the first trimester of pregnancy [40]. Workers potentially exposed to these agents include hospital personnel involved in the preparation and administration of antineoplastic drugs, pharmaceutical factory workers involved in the processing and manufacturing of the drugs, and laboratory assistants and animal caretakers in animal laboratories.

An increased risk of spontaneous abortion has been observed in nurses who prepare injectable antineoplastic drug solutions for patients. In studies using a cruder definition of exposure, however, no association has been found. An excess of spontaneous abortion also has been reported among pharmaceutical factory workers, pharmacy assistants, and laboratory workers exposed to these agents. In addition, the handling of antineoplastic agents in hospitals has been associated with menstrual dysfunction, infertility, congenital malformations of the offspring, and slightly lowered birth weight [35,39,41]. A Danish study [42], however, found no increase in the risk for various adverse pregnancy outcomes among nurses handling antineoplastic drugs during pregnancy. This study was conducted in a setting where safety measures had been taken to protect the health of personnel against antineoplastic drugs and the level of exposure probably was low.

The findings of epidemiologic studies support the view that occupational exposure to antineoplastic drugs may be associated with adverse pregnancy outcome. Because many of these drugs also are carcinogens, exposure to antineoplastic drugs should be minimized by the use of protective garments and equipment and by good work practices. Current prevention measures do not, however, completely eliminate opportunities for exposure, and some countries have adopted a policy of transferring pregnant workers who dilute and prepare antineoplastic drug solutions to other jobs [43].

F. Other Chemical Agents

Ethylene oxide is widely used as a sterilizing agent and in the manufacture of chemicals. In animal experiments, ethylene oxide appears to have reproductive toxic effects at high concentrations. Epidemiologic observations among hospital staff engaged in sterilizing instruments suggest an association between exposure to ethylene oxide and an increased risk of spontaneous abortion [35]. Ethylene oxide is also classified as a carcinogen and, thus, exposure should be kept to a minimum.

Health care workers may be exposed to ribavirin, an antiviral drug, during its administration as an aerosol mist. Ribavirin has been found to be teratogenic and embryolethal in several animal species, but its adverse reproductive effects in humans are unknown. Because of its potential reproductive effects, avoidance of ribavirin prior to and during pregnancy and while breastfeeding has been recommended [44]. Other chemical agents reported to adversely affect female reproductive health include carbon monoxide and polychlorinated biphenyls.

V. Physical Agents

A. Ionizing Radiation

The adverse effects of exposure to high doses of ionizing radiation on reproductive health are well known. Electromagnetic ionizing radiation, X-rays, and gamma rays penetrate tissues and reach the sexual organs and fetus easily. Particulate radiation, that is from alpha (helium nuclei) and beta particles (electrons), does not penetrate tissues deeply but the particles generate ions along their short path in the tissue. Radionuclides also emit ionizing radiation.

Evidence on the adverse effects of ionizing radiation in humans has been gathered primarily from environmental exposure, medical diagnostics, and curative procedures rather than from occupational exposure (Table 37.2). In these studies, high exposure has been related to amenorrhoea and early menopause [45], growth retardation, spontaneous abortions, congenital malformations, mental retardation, childhood cancer, and leukemia [46]. The results of the studies on occupational exposure to diagnostic X-rays have been inconsistent. An increased risk of spontaneous abortion has been reported among veterinarians and veterinary assistants using diagnostic X-rays and among radiology technicians [44]. One study indicated no excess of congenital malformations or cancer in children of female radiographers, but a borderline excess of chromosomal anomalies other than Down's syndrome was noted [47].

Overall, evidence on the hazardous effects of radiation shows that the fetus is most sensitive to its effects. The exposure limits have been lowered in some countries following the recommendations of the International Commission on Radiological Protection [48].

B. Nonionizing Radiation

Nonionizing radiation includes optical radiation, radiofrequency radiation, microwaves, and low-frequency electromagnetic fields. The potential harmful effects of nonionizing radiation have been investigated among physiotherapists exposed to shortwaves, microwaves, and ultrasound, and among workers exposed to strong static magnetic fields while using magnetic resonance imaging (MRI) in diagnostic medicine. The results of studies associating the exposure of physiotherapists with adverse reproductive effects are conflicting. An increase in the risk of spontaneous abortion has been related to the use of shortwave and ultrasound equipment and to microwave use, whereas a low ratio of boys to girls and a low birth weight for male newborns

Table 37.2

Physical and Biological Occupational Hazards Associated with Adverse Reproductive Effects

Hazard	Reported effects	Industry or occupation
Ionizing radiation	Fetal loss[a]	Health care personnel
Microwaves, shortwaves, high-frequency EMF	Fetal loss[a], low birth weight[a], altered sex ratio[a]	Physiotherapists
Low-frequency EMF	Fetal loss[a]	Industrial or office work
Noise	Menstrual disorders[a], reduced fertility[a], preterm birth[a], growth retardation, low birth weight	Industrial manufacturing
Cosmic radiation	Menstrual disorders[a], fetal loss[a]	Flight attendants
Contagious diseases[b]	Fetal loss, growth retardation, birth defects	Health care personnel, childcare workers
Mycotoxins in grain	Midpregnancy delivery[a]	Grain farming

[a]Inconclusive evidence.
[b]Evidence based on studies in the general population.

have been reported for physiotherapists exposed to high-frequency electromagnetic radiation [44].

A survey of MRI technologists did not show any major reproductive hazards associated with MRI work, although a slight but nonsignificant excess of spontaneous abortions was observed [49]. Case reports on MRI examinations during pregnancy reported no increase in adverse pregnancy outcomes. All in all, evidence on the adverse reproductive effects of nonionizing radiation among physiotherapists and MRI technologists is inconclusive.

Exposure to the low frequency electromagnetic fields of video display terminals (VDTs) has been suggested as a potential reproductive hazard. Usually, the magnetic fields produced by modern VDTs are low and often are lower than the fields from other sources in the office environment. Most epidemiologic studies of VDT workers suggest that VDT work is not associated with spontaneous abortion, congenital malformations, fetal growth retardation, or other pregnancy complications [50]. A few studies have shown an excess of some reproductive outcomes, but recall bias may have contributed to the increased risks. In one study, an increased risk was found for a small group of workers who had used a VDT with a high level (>0.3 μT) of extremely low frequency magnetic fields [51]. In another study among semiconductor workers, no association was seen between exposure to extremely low frequency magnetic fields of 0.2–0.5 or >0.9 μT and spontaneous abortion [12]. The epidemiologic evidence does not, taken as a whole, suggest a strong association between exposure to low frequency magnetic fields and adverse reproductive outcome, although an effect at high levels of exposure cannot be excluded.

C. Noise

Most epidemiologic studies on the effects of occupational noise exposure concern preterm birth and low birth weight. Elevated risks of these outcomes have been observed among noise-exposed workers, but results showing no adverse effects also have been presented [52]. Studies that based exposure assessment on noise measurements at the workplace have revealed that exposure to high noise levels (8 hour time-weighted average of approximately 85 dB(A) or higher) is associated with fetal growth retardation. Excesses of hormonal disturbances, delayed conception, infertility, and spontaneous abortion also have been reported for occupational noise exposure. A possible mechanism for the adverse reproductive effects is a stress reaction induced by the increase of maternal catecholamine secretion, which may further stimulate or retard uterine contractions and affect uteroplacental blood flow.

Noise also has been suggested to have direct effects on the fetus by passing through the maternal abdomen, although the noise energy may be attenuated by the abdomen. Limited evidence suggests that occupational noise exposure during pregnancy may increase the risk of hearing deterioration in the children [53]. All in all, the current evidence suggests that high noise exposure should be considered as a potential reproductive hazard.

VI. Biological Agents

Health care workers and childcare workers may be exposed to infectious agents causing contagious diseases such as toxo-

Table 37.3
Physical and Psychosocial Work Load Factors Associated with Adverse Reproductive Effects

Hazard	Reported effects
Shift work	Fetal loss[a], preterm birth, low birth weight
Long working hours	Fetal loss[a], low birth weight
Physical strain	Reduced fertility[a], fetal loss[a], preterm birth, intrauterine growth retardation, low birth weight[a], birth defects[a]
Psychosocial strain	Menstrual disorders[a], preeclampsia[a], preterm births[a], fetal loss[a]

[a]Inconclusive evidence.

plasmosis, listeriosis, German measles (rubella), herpes, chickenpox (varicella), hepatitis B and C, cytomegalovirus infection, parvovirus infection, and HIV infection. Such agents can cause loss of pregnancy, fetal growth retardation, congenital malformations, mental retardation, or systemic disease [54]. Information on the effects of infectious agents on pregnancy outcome is based on studies in the general population, and there are few studies on the reproductive effects of occupational exposure to these agents. However, many workers have immunity against the usual viral diseases or have been vaccinated against rubella and hepatitis B. Other ways of prevention include good working practices and, in some cases, an alternative job for a nonimmune pregnant worker or special maternity leave. In some European countries, exposure to biologic agents has been taken into account in the legislation protecting reproductive health [43].

Another potential biologic hazard is exposure to mycotoxins from fungi. One study observed an excess of midpregnancy deliveries among grain farmers. The effect was strongest after the harvest and in seasons with poor harvest quality, when the probability of mycotoxin formation in grain was considered high [55]. Because several mycotoxins are reproductive toxicants, this new finding should be confirmed or refuted in future studies.

VII. Physical and Psychosocial Strain

A. Shift Work

Shift work has been considered as potentially harmful for reproduction because it may interfere with the circadian regulation of human metabolism (Table 37.3). This may lead to changes in hormonal concentrations that affect reproductive capability. Menstrual disorders have been reported in some studies, and two studies showed an increased risk of subfecundity for women with shift work [56,57]. An elevated risk of spontaneous abortion also has been indicated for shift work, although not consistently in all investigations. Shift work has been associated with preterm birth and low birth weight in several studies [52]. It has, however, remained unclear which forms of shift work—rotating or changing schedules, night work, irregular working hours, or shift work in general—may be harmful for reproductive health. The results of a study among midwives suggest that two-shift work is not harmful, whereas night work and three-shift work may increase the risk of spontaneous abortion [38]. Usually, shift workers also are exposed to other potential reproductive hazards, and in many studies it has been difficult to separate the effects of these other occupational exposures. Nevertheless, shift work should be considered as a potential risk to reproduction.

B. Physical Strain

Work entailing pronounced physical exertion may affect the reproductive health of women, although moderate physical activity is safe both for mother and for fetus. Strong physical effort may alter the hormonal balance and increase intrauterine pressure, thereby decreasing the nutritional blood flow to the fetus. In later pregnancy it also may promote uterine contractions and consequently increase the risk for early delivery.

The measures used in studies on the effects of physical work load include single risk factors (e.g., standing, walking, lifting, bending, and long working hours) or composite scores representing intensity of effort or fatigue. Most studies have relied on self-reported exposure data or surrogate measures of exposure. The results of these studies have been inconsistent [58]. The evidence is strongest for an effect on prematurity. Prolonged standing or walking has been found to increase the risk for preterm delivery in several studies. The evidence is less convincing for the effects of these factors on birthweight and spontaneous abortion. One study showed that physically demanding work combined with long work hours may reduce fetal weight [59]. Several other studies also have associated long working hours with adverse pregnancy outcomes, mainly prematurity and reduced fetal growth. Studies on the effects of work load on other end points are rare. It is possible that there is an association between physical activity and high blood pressure and preeclampsia. Physical work load also has been related, though not consistently, to certain types of congenital malformations [58].

Current evidence suggests that strenuous physical exertion should be considered in the assessment and prevention of occupational reproductive hazards. In early pregnancy, extremely heavy exertion should be avoided, and during the second and third trimesters the physical work load may need to be decreased, opportunities for rest organized, and continuous standing and walking avoided [58].

C. Psychosocial Job Stress

In studies of psychosocial job stress and pregnancy outcome, stress has been assessed either by questions on perceived work-related stress or by defining job stress as a combination of job demand and decision-making ability. Job stress has been associated with pregnancy-induced hypertension and preeclampsia [60,61]. Self-reported stress at work also has been related to spontaneous abortion [12] and menstrual dysfunction [39], and high-strain job to preterm delivery in some, but not in all, studies [62]. Further research is needed to assess the effects of job-related stress on reproductive health.

VIII. Methodological Issues

Epidemiologic reproductive studies present some specific methodologic problems related to reproductive failure [63], as well as general problems, such as misclassification of outcome and exposure, selection bias, information bias, confounding, and small sample sizes. These methodologic issues must be carefully considered when reviewing study results and evaluating the evidence for reproductive effects of occupational agents.

One of the most serious weaknesses in studies on occupational reproductive hazards has been inaccurate data on exposure. The studies have usually been retrospective, and in this type of study the quality of exposure data may be poor, especially if no records or registered data are available. A prospective study design would make it possible to collect measurement data on exposure during pregnancy. The use of biologic exposure markers also would be possible in prospective studies, although high expenses often limit the feasibility of this study design. In some studies on the effects of physical work load and psychosocial job strain on pregnancy outcome, exposure data have been collected prospectively during the pregnancy [59,62]. These kinds of studies, however, seldom yield sufficient numbers of pregnancies with exposure to a specific chemical or physical agent.

In some reproductive studies, exposure has been inferred from occupation and/or industry. These data are often too crude for exposure estimation and their use may result in nondifferential misclassification of exposure and dilution of the potential effects. They can, however, be used to generate hypotheses for further studies on specific exposures. For example, employment in the textile or leather industry or as a flight attendant has been related to adverse reproductive outcomes, although detailed information on specific hazardous agents is lacking [64,65].

Interviews have the advantage of providing more detailed information on work tasks and exposures than registers provide. Usually the workers give valid information on their jobs, but they may be unaware of specific exposures. Comparisons of different data sources have indicated that underreporting of exposure is a common problem in studies using self-reported data [66]. It also has been observed that underreporting is a greater problem than overreporting in occupational studies on reproductive health. In retrospective exposure assessment, the level of data quality may be increased by applying expert (e.g., industrial hygienist or chemist) judgment based on detailed job descriptions obtained by interview or questionnaire [67] and on available industrial hygiene measurements. This kind of approach has been applied in studies on the reproductive effects of exposure to chemical agents and noise [15,18,68]. A further improvement in data quality is the use of biologic exposure markers which provide quantitative data on internal exposure. Available biologic monitoring data were used in exposure assessment in studies on lead and solvent exposure [11,19,21].

Good and exact data on reproductive end points are essential prerequisites for valid study results. The few existing studies on the effects of occupational exposures on menstrual function have mainly been based on self-report. Data on menstruation may have involved recall of long-past menstrual cycles, possibly resulting in differential recall and response. One exception is a prospective study among semiconductor workers based on daily diaries and on urinary assays to exclude conceptions [69]. Results from this study indicated increased menstrual cycle variability and an elevated risk of short cycles in women working in photolithography, which involves potential exposure to organic solvents. Menstrual cycle data have, however, been considered to be insensitive indicators of change or disturbances in hormonal function resulting from occupational or environmental exposure because abnormalities can exist even with regular vaginal bleeding. Methods have been developed to measure female hormones in urine for use in studies of occupational hazards for women [70].

Time to pregnancy, which measures the degree of delay in conceiving, has been used to a growing extent in studies on occupational hazards and fertility. It is considered to be a sensitive method for detecting differences in fertility between the exposed and unexposed. Usually the studies have been retrospective. The main problems of these studies relate to difficulties in obtaining valid exposure data and to the fact that they do not include subjects who never become pregnant [71]. There are also several sources of bias that need to be considered in the interpretation of the study findings [72].

Medical records on pregnancy outcome are reliable data sources because the information usually is based on medical diagnoses and is not biased by selective recall or reporting of pregnancy outcome. In studies using medical records for the identification of spontaneous abortions, there is, however, potential for selection bias due to differing patterns in the use of medical services. The probability for seeking medical care varies with gestational age, and it may also depend on level of education, socioeconomic factors, and availability of medical services. Nationwide registries may be useful in the initial identification and restriction of the study population, as the data are readily accessible and cover large populations. A nationwide database on pregnancies has been used in several studies on spontaneous abortion and fertility [11,18,19].

Studies on congenital malformations, preterm birth, low birth weight, or fetal growth retardation are based mainly on information from registries or medical records. The problems associated with registered pregnancy data relate to the coverage and accuracy of the records. Ascertainment of malformations in the registries is incomplete, with minor malformations less likely to be reported. Small sample size has also been a frequent shortcoming in studies of congenital malformations. International collaboration may permit research on rare outcomes and infrequent exposures. An example of this is a multicenter study on glycol ethers and congenital malformations [16].

Interviews and questionnaires often have been used as sources of data on reproductive outcomes, especially in studies of fertility, menstruation, and spontaneous abortion. The weaknesses of these studies pertain to possible selection in recognition, recall, and reporting of reproductive outcomes, as well as in selective participation in the study. Recognition and recall may be influenced by education, medical experience, reproductive history, calendar time, and concern about possible harmful exposures [66]. However, carefully conducted questionnaire-based surveys have an important place in the collection of data on reproductive outcomes.

Biologic monitoring of early pregnancy loss is increasingly used in epidemiologic reproductive research, although it is seldomly used in occupational studies. The difficulties with these studies relate to their high cost and to problems in recruiting unselected subjects. In one study, pregnancy information was collected prospectively by measuring human chorionic gonadotropin in urine samples from a cohort of women employed in the semiconductor industry [73]. Despite a large base population, the final number of exposed pregnancies was small, reflect-

ing the difficulties of recruiting large numbers of subjects into this kind of study.

The human population as a whole is likely to be heterogeneous in response to reproductive hazards, but very little is known about individual variations in susceptibility. Susceptibility to the effects of reproductive and developmental toxicants may vary with demographic, nutritional, or genetic factors, as well as by health status and concomitant exposures [3]. Heritable differences in genes responsible for the metabolism of xenobiotics is one potential explanation for variations in susceptibility. The results of a study on the relation between spontaneous abortion and polymorphisms in two genes involved in the metabolism of xenobiotics provided little evidence of an association between these genes and risk of spontaneous abortion [74]. The number of genes involved in the organism's response to environmental hazards may, however, be very large, and further research is needed on the importance of metabolic polymorphisms to reproductive toxicity. Some studies have found evidence suggesting variation in susceptibility linked to reproductive history and environmental exposures. A history of preterm birth was used as an indicator of exposure susceptibility in a study among farmers. Heterogeneity was observed in the effect of grain farming on midpregnancy delivery; the effect was stronger for mothers with a history of preterm delivery than for those without [75]. In another study, excessive standing was associated with an elevated risk of spontaneous abortion among women with a history of two or more previous spontaneous abortions, but not among other women [76]. It was suggested that women with a history of losses may have a heightened physiologic response to standing.

IX. Risk Assessment and Risk Communication

Legislation in many countries obligates employers to ensure that the work environment is safe for the reproductive health of workers, for pregnant workers, and for offspring during intrauterine life. Thus, the assessment and management of reproductive risks at work is the employer's responsibility. To fulfill this demand the employer needs guidelines (*e.g.,* occupational exposure limits (OEL) based on reproductive risk assessment) and the expertise of an occupational physician and/or an industrial hygienist, or, in some special cases, a toxicologist.

In practice, the occupational health physicians (OHP) often are in a key position in risk assessment as experts of toxicologic and biomedical aspects of the work. This includes the identification of occupational exposures or agents that are potentially harmful to reproductive health and to the health of the offspring and the search for information on which to base the evaluation. Since the OELs presently in use do not always take reproductive effects into account, the OHPs and toxicologists need to estimate possible risks on the basis of experimental research. Although there are criteria for such an extrapolation from animals to humans [77], better estimation of safe exposure levels regarding reproduction need to be developed.

The employer, employees, and safety committees should be informed of the results of risk assessment by the assessors. Because employees have a right to know the possible health effects of their work exposure, the employer should inform new workers as well as present workers of the potential reproductive health risks. So far, very few employers provide such information, although they may ask the OHP to transfer the information on the risks and the ways to avoid them to the employees and safety committees. The OHP advises how to avoid the risks (*e.g.,* by substituting safe chemicals for hazardous ones, by improving the hygienic conditions of the work place, or by changing work tasks and work habits to safe ones, or, if needed, by transferring the worker to a safe environment for the time of pregnancy). If antenatal care services are given at public health clinics or elsewhere outside the occupational health services, information on occupational reproductive risks should be distributed to those clinics also.

The time dimension is important in the management of reproductive risks. In order to achieve the preventive goal regarding reproduction, the work environment should be safe for all workers, both men and women. This can be achieved through policymaking [77]. In counseling the pregnant worker, it is important to suggest preventive actions for the remainder of the pregnancy. When a worker asks for advice before pregnancy, this is a good opportunity to organize a safe work environment for her. After an adverse pregnancy outcome, counseling has to be done carefully, in a supportive way, so that it does not arouse unnecessary guilt or anxiety. The counseling must always be honest and truthful; also, the lack of accurate information can be discussed [78].

X. Prevention

Existing legislation on work safety and safety inspections, on occupational health services, and on special maternity leave (*i.e.,* maternity benefits if the pregnant woman has to stay at home during her pregnancy due to occupational risks) is the cornerstone of successful realization of preventive work. It is, for example, important that the safety inspectors and OHP have free access to check working conditions. The availability and obligation to follow relevant occupational exposure limits based on data on reproductive hazards are the common bases for safety goals in the work environment and for health policymaking. Risk management may include the formulation of a health policy to protect the reproductive health of workers and pregnant workers, the search for and distribution of information on the risks, and the participation of OHPs in the planning of the work environment. Guidelines for managing occupational reproductive health hazards and for developing a reproductive health policy that protects the reproductive health of female (and male) employees through nondiscriminatory risk-reduction strategies have been elaborated [6,79].

The duty to distribute information on known risks was mentioned earlier; in addition to the employer, the OHP is obliged to inform all the working parties continuously. The OHP should be active in risk management. The risks cannot always be eliminated (*e.g.,* infectious agents in the health care sector) because possible exposures cannot always be known beforehand. Therefore, safe working habits have to be taught to all workers because most biologic risks can be avoided through them. The OHP should assist the pregnant worker in negotiations with her superior on needed actions at the workplace.

In some countries (*e.g.,* Finland and Denmark), there are specific laws for the protection of reproductive health of workers and pregnant women. If the work exposure of a pregnant woman is seen as possibly harmful to the pregnancy or to the offspring, the woman is entitled to a transfer to a safe job or to changes in her job, or if that is not possible, to a special maternity leave (with the same benefits as normal maternity leave) from the beginning of the pregnancy. Experiences on use of the special maternity leave in the Nordic countries indicate that the use has been moderate, and it has not led to marked discriminative actions against female workers [43]. The work places where need for such a leave has occurred should be targets for hygienic improvements or changes in work organization.

Another challenge is to implement the protection of reproductive health without creating discriminative policies in work life. The increasing volume of information on the vulnerability of the male reproductive system, and on possible harmful effects to the offspring through the father, should help to balance the situation and direct prevention to all workers.

XI. Conclusions

Numerous epidemiologic studies have been conducted on occupational reproductive health hazards since the 1970s. Research has mainly focused on the effects of maternal exposure on pregnancy outcome. However, research has been expanded to include male-mediated reproductive effects and a wider scope of adverse outcomes. These previously neglected areas include fertility and biologic processes that may affect fertility such as menstrual function, semen characteristics, and hormone levels. The field of occupational exposures has also broadened to encompass biologic and psychosocial exposures, in addition to chemical and physical agents. Further research is also needed on the psychosocial and behavioral effects of occupational exposures on the offspring and on variations in individual susceptibility to reproductive toxicants. Epidemiologic research into the reproductive effects of occupational exposure is of special significance because of the potential for preventing adverse effects if the risks associated with exposure can be identified and eliminated.

Research on occupational reproductive hazards has mainly been carried out in developed countries. Information on these hazards in developing countries is almost nonexistent. The harmful effects may, however, have a greater impact on the female workers of the developing countries because of other concurrent problems, such as malnutrition, chronic and infectious diseases, sequelae of traumatic deliveries, genital mutilation, and lack of health services. Research on occupational reproduction hazards that takes these problems into account is needed in developing countries. The goal is to provide a safe work environment for all workers so that women and men of fertile age, as well as pregnant workers, can continue working.

References

1. European Commission (1996). "Employment in Europe 1996," COM(96) 485. Office for Official Publications of the European Communities, Luxembourg.
2. Moss, N., and Carver, K. (1993). Pregnant women at work: Sociodemographic perspectives. *Am. J. Ind. Med.* **23,** 541–557.
3. Marcus, M., Silbergeld, E., Mattison, D., and the Research Needs Working Group (1993). Reproductive hazards research agenda for the 1990s. *Environ. Health Perspect., Suppl.* **101,** 175–180.
4. Baird, D. D., and Wilcox, A. J. (1986). Effects of occupational exposures on the fertility of couples. *Occup. Med.: State of the Art Rev.* **1,** 361–374.

5. Kimmel, C. A. (1993). Approaches to evaluating reproductive hazards and risks. *Environ. Health Perspect., Suppl.* **101,** 137–143.

6. Committee Report (1996). ACOEM reproductive hazard management guidelines. *J. Occup. Environ. Med.* **38,** 83–90.

7. Selevan, S. G., and Lemasters, G. K. (1987). The dose-response fallacy in human reproductive studies of toxic exposures. *J. Occup. Med.* **29,** 451–454.

8. Savitz, D. A., Whelan, E. A., Rowland, A. S., and Kleckner, R. C. (1990). Maternal employment and reproductive risk factors. *Am. J. Epidemiol.* **132,** 933–945.

9. Lindbohm, M.-L. (1995). Effects of parental exposure to solvents on pregnancy outcome. *J. Occup. Environ. Med.* **37,** 908–914.

10. Gold, E. B., and Tomich, E. (1994). Occupational hazards to fertility and pregnancy outcome. *Occup. Med.: State of the Art Rev.* **9,** 435–469.

11. Sallmén, M., Lindbohm, M.-L., Kyyrönen, P., Nykyri, E., Anttila, A., Taskinen, H., and Hemminki, K. (1995). Reduced fertility among women exposed to organic solvents. *Am. J. Ind. Med.* **27,** 699–713.

12. Swan, S. H., Beaumont, J. J., Hammond, S. K., VonBehren, J., Green, R. S., Hallock, M. F., Woskie, S. R., Hines, C. J., and Schenker, M. B. (1995). Historical cohort study of spontaneous abortion among fabrication workers in the semiconductor health study: Agent-level analysis. *Am. J. Ind. Med.* **28,** 751–769.

13. Correa, A., Gray, R. H., Cohen, R., Rothman, N., Shah, F., Seacat, H., and Corn, M. (1996). Ethylene glycol ethers and risks of spontaneous abortion and subfertility. *Am. J. Epidemiol.* **143,** 707–717.

14. Jensen, A. A., and Slorach, S. A. (1990). "Chemical Contaminants in Human Milk." CRC Press, Boca Raton, FL.

15. McDonald, J. C., Lavoie, J., Coté, R., and McDonald, A. D. (1987). Chemical exposures at work in early pregnancy and congenital defect: a case-referent study. *Br. J. Ind. Med.* **44,** 527–533.

16. Cordier, S., Bergeret, A., Goujard, J., Ha, M.-C., Aymé, S., Bianchi, F., Calzolari, E., De Walle, H., Knill-Jones, R., Candela, S., Dale, I., Dananché, B., de Vigan, C., Fevotte, J., Kiel, G., and Mandereau, L. (1997). Congenital malformations and maternal occupational exposure to glycol ethers. *Epidemiology* **8,** 355–363.

17. Doyle, P., Roman, E., Beral V., and Brookes, M. (1997). Spontaneous abortion in dry cleaning workers potentially exposed to perchloroethylene. *Occup. Environ. Med.* **54,** 848–853.

18. Taskinen, H., Kyyrönen, P., Hemminki, K., Hoikkala, M., Lajunen, K., and Lindbohm, M.-L. (1994). Laboratory work and pregnancy outcome. *J. Occup. Med.* **36,** 311–319.

19. Lindbohm, M.-L., Taskinen, H., Sallmén, M., and Hemminki, K. (1990). Spontaneous abortions among women exposed to organic solvents. *Am. J. Ind. Med.* **17,** 449–463.

20. Lagerkvist, B. J., Ekesrydh, S., Englyst V., Nordberg, G. F., Söderberg, H.-Å, and Wiklund, D.-E. (1996). Increased blood lead and decreased calcium levels during pregnancy: A prospective study of Swedish women living near a smelter. *Am. J. Public Health* **86,** 1247–1252.

21. Sallmén, M., Anttila, A., Lindbohm, M.-L., Kyyrönen, P., Taskinen, H., and Hemminki, K. (1995). Time to pregnancy among women occupationally exposed to lead. *J. Occup. Environ. Med.* **37,** 931–934.

22. Anttila, A., and Sallmén, M. (1995). Effects of parental occupational exposure to lead and other metals on spontaneous abortion. *J. Occup. Environ. Med.* **37,** 915–921.

23. Bellinger, D. (1994). Teratogen update: Lead. *Teratology* **50,** 367–373.

24. Rowland, A. S., Baird, D. D., Weinberg, C. R., Shore, D. L., Shy, C. M., and Wilcox, A. J. (1994). The effect of occupational exposure to mercury vapour on the fertility of female dental assistants. *Occup. Environ. Med.* **51,** 28–34.

25. Barlow, S. M., and Sullivan, F. M. (1984). "Reproductive Hazards of Industrial Chemicals." Academic Press, London.

26. Chashschin, V. P., Artunina, G. P., and Norseth, T. (1994). Congenital defects, abortion and other health effects in nickel refinery workers. *Sci. Total Environ.* **148,** 287–291.

27. Baker, S. R., and Wilkinson, C. F. (1990). The effect of pesticides on human health. *In* "Advances in Modern Environmental Toxicology" (S. R. Baker and C. F. Wilkinson, eds.), Vol. 18. Princeton Scientific Publishing Co., Princeton, NJ.

28. Fuortes, L., Clark, M. K., Kirchner, H. L., and Smith, E. M. (1997). Association between female infertility and agricultural work history. *Am. J. Ind. Med.* **31,** 445–451.

29. Taskinen, H. K., Kyyrönen, P., Liesivuori, J., and Sallmén, M. (1995). Greenhouse work, pesticides and pregnancy outcome. *Epidemiology* **6,** Suppl. 109 (abstr.)

30. Abell, A., Juul, S., and Bonde J. P. (1997). Time to pregnancy in female greenhouse workers. In *Book Abstr., Int. Symp. Environ., Lifestyle Fertil.,* Aarhus, Denmark, *1997,* p. 142.

31. Nurminen, T. (1995). Maternal pesticide exposure and pregnancy outcome. *J. Occup. Environ. Med.* **37,** 935–940.

32. Pastore, L. M., Hertz-Picciotto, I., and Beaumont, J. J. (1997). Risk of stillbirth from occupational and residential exposures. *Occup. Environ. Med.* **54,** 511–518.

33. Kristensen, P., Andersen, A., Irgens, L. M., Bye, A. S., and Sundheim, L. (1996). Cancer in offspring of parents engaged in agricultural activities in Norway: Incidence and risk factors in the farm environment. *Int. J. Cancer* **65,** 39–50.

34. Zahm, S. H., and Ward, M. H. (1998). Pesticides and childhood cancer. *Environ. Health Perspect.* **106** (Suppl. 3), 893–908.

35. Ahlborg, G., Jr., and Hemminki, K. (1995). Reproductive effects of chemical exposures in health professions. *J. Occup. Environ. Med.* **37,** 957–961.

36. Boivin, J.-F. (1997). Risk of spontaneous abortion in women occupationally exposed to anaesthetic gases: A meta-analysis. *Occup. Environ. Med.* **54,** 541–548.

37. Rowland, A. S., Baird, D. D., Shore, D. L., Weinberg, C. R., Savitz, D. A., and Wilcox, A. J. (1995). Nitrous oxide and spontaneous abortion in female dental assistants. *Am. J. Epidemiol.* **141,** 531–538.

38. Axelsson, G., Ahlborg, G., Jr., and Bodin, L. (1996). Shift work, nitrous oxide exposure, and spontaneous abortion among Swedish midwives. *Occup. Environ. Med.* **53,** 374–378.

39. Shortridge, L. A., Lemasters, G. K., Valanis, B., and Hertzberg, V. (1995). Menstrual cycles in nurses handling antineoplastic drugs. *Cancer Nurs.* **18,** 439–444.

40. Hoffman, D. M. (1986). Reproductive risks associated with exposure to antineoplastic agents: A review of the literature. *Hosp. Pharm.* **110,** 930–940.

41. Valanis, B., Vollmer, W., Labuhn, K., and Glass, A. (1997). Occupational exposure to antineoplastic agents and self-reported infertility among nurses and pharmacists. *J. Occup. Environ. Med.* **39,** 574–580.

42. Skov, T., Maarup, B., Olsen, J., Rorth, M., Winthereik, H., and Lynge, E. (1992). Leukaemia and reproductive outcome among nurses handling antineoplastic drugs. *Br. J. Ind. Med.* **49,** 855–861.

43. Taskinen, H., Olsen, J., and Bach, B. (1995). Experiences in developing legislation protecting reproductive health. *J. Occup. Environ. Med.* **37,** 974–979.

44. Taskinen, H., and Lindbohm, M.-L. (1997). Pregnancy and work. *In* "Occupational Health Practice" (H. A. Waldron and C. Edling, eds.), 4th ed., pp. 183–199. Butterworth-Heinemann, Cornwall, England.

45. Lione, A. (1987). Ionizing radiation and human reproduction. *Reprod. Toxicol.* **1,** 3–16.

46. Bengtsson, G. (1991). Introduction: Present knowledge on the effects of radioactive contamination on pregnancy outcome. *Biomed. Pharmacother.* **45,** 221–223.

47. Roman, E., Doyle, P., Ansell, P., Bull, D., and Beral, V. (1996). Health of children born to medical radiographers. *Occup. Environ. Med.* **53,** 73–79.

48. International Commission on Radiological Protection (1991). 1990 Recommendations of the International Commission on Radiological Protection, ICRP Pub. 60. Pergamon, New York.

49. Evans, J. A., Savitz, D. A., Kanal, E., and Gillen, J. (1993). Infertility and pregnancy outcome among magnetic resonance imaging workers. *J. Occup. Med.* **35,** 1191–1195.

50. Delpizzo, V. (1994). Epidemiological studies of work with video display terminals and adverse pregnancy outcomes (1984–1992). *Am. J. Ind. Med.* **26,** 465–480.

51. Lindbohm, M.-L., and Hietanen, M. (1995). Magnetic fields of video display terminals and pregnancy outcome. *J. Occup. Environ. Med.* **37,** 952–956.

52. Nurminen, T. (1995). Female noise exposure, shift work, and reproduction. *J. Occup. Environ. Med.* **37,** 945–950.

53. Lalande, N. M., Hétu, R., and Lambert, J. (1986). Is occupational noise exposure during pregnancy a risk factor of damage to the auditory system of the fetus? *Am. J. Ind. Med.* **10,** 427–435.

54. Ekblad, U. (1995). Biological agents and pregnancy. *J. Occup. Environ. Med.* **37,** 962–965.

55. Kristensen, P., Irgens, L. M., Andersen, A., Bye, A. S., and Sundheim, L. (1997). Gestational age, birth weight, and perinatal death among births to Norwegian farmers, 1967–1991. *Am. J. Epidemiol.* **146,** 329–338.

56. Bisanti, L., Olsen, J., Basso, O., Thonneau, P., and Karmaus, W. (1996). Shift work and subfecundity: A European multicenter study. *J. Occup. Environ. Med.* **38,** 352–358.

57. Ahlborg, G., Jr., Axelsson, G., and Bodin, L. (1996). Shift work, nitrous oxide exposure and subfertility among Swedish midwives. *Int. J. Epidemiol.* **25,** 783–790.

58. Ahlborg, G., Jr. (1995). Physical work load and pregnancy outcome. *J. Occup. Environ. Med.* **37,** 941–944.

59. Hatch, M., Ji, B-T., Shu, X. O., and Susser, M. (1997). Do standing, lifting, climbing, or long hours of work during pregnancy have an effect on fetal growth? *Epidemiology* **8,** 530–536.

60. Klonoff-Cohen, H. S., Cross, J. L., and Pieper, C. F. (1996). Job stress and preeclampsia. *Epidemiology* **7,** 245–249.

61. Landsbergis, P. A., and Hatch, M. C. (1996). Psychosocial work stress and pregnancy-induced hypertension. *Epidemiology* **7,** 346–351.

62. Henriksen, T. B., Hedegaard, M., and Secher, N. J. (1994). The relation between psychosocial job strain, and preterm delivery and low birthweight for gestational age. *Int. J. Epidemiol.* **23,** 764–774.

63. Olsen, J., and Skov, T. (1993). Design options and methodological fallacies in the studies of reproductive failures. *Environ. Health Perspect., Suppl.* **101,** 145–152.

64. McDonald, A. D., McDonald, J. C., Armstrong, B., Cherry, N., Delorme, C., Nolin, A. D., and Robert, D. (1987). Occupation and pregnancy outcome. *Br. J. Ind. Med.* **44,** 521–526.

65. Bianchi, F., Cianciulli, D., Pierini, A., and Costantini, A. S. (1997). Congenital malformations and maternal occupation: a registry based case-control study. *Occup. Environ. Med.* **54,** 223–228.

66. Hemminki, K., Lindbohm, M.-L., and Kyyrönen, P. (1995). Validity aspects of exposure and outcome data in reproductive studies. *J. Occup. Environ. Med.* **37,** 903–907.

67. Siemiatycki, J., Fritschi, L., Nadon, L., and Gerin, M. (1997). Reliability of an expert rating procedure for retrospective assessment of occupational exposures in community-based case-control studies. *Am. J. Ind. Med.* **31,** 280–286.

68. Nurminen, T., and Kurppa, K. (1989). Occupational noise exposure and course of pregnancy. *Scand. J. Work Environ. Health* **15,** 117–124.

69. Gold, E. B., Eskenazi, B., Hammond, S. K., Lasley, B. L., Samuels, S. J., Rasor, M. O., Hines, C. J. Overstreet, J. W., and Schenker, M. B. (1995). Prospectively assessed menstrual cycle charasteritics in female wafer-fabrication and nonfabrication semiconductor employees. *Am. J. Ind. Med.* **28,** 799–815.

70. Kesner, J. S., Wright, D. M., Schrader, S. M., Chin, N. W., and Krieg, E. F., Jr. (1992). Methods of monitoring menstrual function in field studies: Efficacy of methods. *Reprod. Toxicol.* **6,** 385–400.

71. Joffe, M., and Asclepios Project (1997). Time to pregnancy: A measure of reproductive function in either sex. *Occup. Environ. Med.* **54,** 289–295.

72. Weinberg, C. R., Baird, D. D., and Wilcox, A. J. (1994). Sources of bias in studies of time to pregnancy. *Stat. Med.* **13,** 671–681.

73. Eskenazi, B., Gold, E. B., Lasley, B. L., Samuels, S. J., Hammond, S. K., Wight, S., Rasor, M. O., Hines, C. J., and Schenker, M. B. (1995). Prospective monitoring of early fetal loss and clinical spontaneous abortion among female semiconductor workers. *Am. J. Ind. Med.* **28,** 833–846.

74. Hirvonen, A., Taylor, J. A., Wilcox, A., Berkowitz, G., Schachter, B., Chaparro, C., and Bell, D. A. (1996). Xenobiotic metabolism genes and the risk of recurrent spontaneous abortion. *Epidemiology* **7,** 206–208.

75. Kristensen, P., Irgens, L. M., and Bjerkedal T. (1997). Environmental factors, reproductive history, and selective fertility in farmers' sibships. *Am. J. Epidemiol.* **145,** 817–825.

76. Fenster, L., Hubbard, A. E., Windham, G. C., Waller, K. O., and Swan, S. H. (1997). A prospective study of work-related physical exertion and spontaneous abortion. *Epidemiology* **8,** 66–74.

77. Taskinen, H., and Ahlborg, G., Jr. (1996). Assessment of reproductive risk at work. *Int. J. Occup. Environ. Health* **2,** 59–63.

78. Ahlborg, G., Jr., Bonde, J. P., Hemminki, K., Kristensen, P., Lindbohm, M.-L., Olsen, J., Schaumburg, I., Taskinen, H., and Viskum, S. (1996). Communication concerning the risks of occupational exposures in pregnancy. *Int. J. Occup. Environ. Health* **2,** 64–69.

79. Kaczmarczyk, J. M., and Paul, M. E. (1996). Reproductive health hazards in the workplace: Guidelines for policy development and implementation. *Int. J. Occup. Environ. Health* **2,** 48–58.

38

Work-Related Musculoskeletal Disorders: Is There a Gender Differential, and if So, What Does It Mean?

LAURA PUNNETT* AND ROBIN HERBERT†

*Department of Work Environment, University of Massachusetts Lowell, Lowell, Massachusetts; †Department of Community and Preventive Medicine, Mount Sinai School of Medicine, New York, New York

I. Background

Musculoskeletal disorders (MSDs) include the many conditions that affect the muscles, tendons, ligaments, and joint cartilage, as well as the anatomically associated peripheral motor and sensory nerves [1–5]. There is a strong and growing body of evidence associating a number of musculoskeletal disorders with workplace exposure to biomechanical and psychosocial factors [6,7]. Work-related musculoskeletal disorders (WMSDs) comprise a heterogeneous group of diagnoses that are often referred to collectively as cumulative trauma disorder, repetitive strain injury, occupational cervicobrachial disorder, or overuse syndrome. Specific clinical syndromes that have been associated with occupational exposure include nerve compression disorders (e.g., carpal tunnel syndrome), tendon inflammations and related conditions (e.g., tenosynovitis, epicondylitis, bursitis of the shoulder), disk disorders, and degenerative joint disease (Table 38.1). There are also a variety of less well-defined conditions such as myositis, fibromyalgia, focal dystonia, low back pain (LBP), and other regional pain syndromes. Musculoskeletal disorders may affect any part of the body: the trunk, the upper extremities, and the lower extremities. WMSDs of the back and upper extremity appear to be more common than those affecting the lower extremity.

Some disorders, such as carpal tunnel syndrome (CTS) and lumbosacral disc herniation, have well-described clinical features, "gold standard" diagnostic tests, and relatively good consensus on diagnostic criteria [8,9]. However, many others, such as the specific tendonitides of the distal forearm, are not easily distinguished from one another in the clinical setting and do not have well agreed upon diagnostic criteria for use in research. Additionally, when grouped anatomically, many of these disorders have common epidemiologic features, particularly with respect to symptoms and, in working populations, the nature of the ergonomic exposure(s). For most MSDs, both onset and duration are more often chronic or subchronic than acute. Typically, they tend to develop after months or years of overuse of the soft tissues, although muscle strains and tendon sprains may also occur as point-in-time injuries (meaning very soon after a sudden change in activity patterns). It has been observed that within a population occupationally exposed to particular ergonomic stressors, several different types of clinical disorders may arise in different individuals. Thus, it has often been statistically efficient to study MSDs as a group.

Among the known or suspected nonoccupational risk factors for MSDs are age, gender, socioeconomic status, race or ethnicity (any of which might be confounded by occupational de-

mands), history of acute trauma, various systemic diseases (e.g., endocrine, neurologic, autoimmune, and collagen disorders), smoking, alcohol use, obesity, and recreational sports [5]. For CTS in particular, studies also have implicated factors affecting female hormone levels, such as use of oral contraceptives and surgical removal of both ovaries.

Accurate data on the incidence and prevalence of MSDs are sparse. There are no population registries of MSDs and not all people affected seek medical care, so most available data probably underestimate the true magnitude of the disorders. Difficulties and inconsistencies in diagnosis compound this problem. However, it is clear that MSDs are widespread in many societies around the world and they account for much disability, pain, and suffering. In the United States, Canada, and Finland, for example, more people are disabled from working as a result of musculoskeletal disorders than from any other group of diseases [3,5,10,11]. It has been estimated that the direct and indirect economic burden of all musculoskeletal disease in the United States, including arthritis and repetitive trauma disorders, totaled $149 billion in 1992 [12].

A. Work-Related Musculoskeletal Disorders

Work-related musculoskeletal disorders (WMSDs) are a significant world-wide public health problem, although official statistics are difficult to compare across countries because of differences in record-keeping and case definitions. They are the single largest category of work-related illness in the United States, the Nordic countries, and Japan [3,6,13]. In the United States, workplace injuries and illnesses in the private sector, excluding firms with fewer than 11 employees, are reported to the U.S. Bureau of Labor Statistics (BLS). Musculoskeletal disorders associated with "repeated trauma" in the workplace have accounted for over 50% of occupational illnesses since 1992, annually totaling over 300,000 cases [6,14]. Among the Nordic countries, reported WMSDs account for 71% of all occupational illness in Sweden but only 1% in Iceland [15]. These statistics do not include back disorders, which are defined as injuries. In 1994, about one-third of all injuries and illnesses resulting in lost work days reported to the U.S. BLS were associated with overexertion or repetitive motion, and nearly one-half affected the back [6]. Two studies of carpal tunnel syndrome suggested that about one-half of the cases occurring in the general population were work-related [16,17]. The US National Institute of Occupational Safety and Health (NIOSH) has identified musculoskeletal disorders of the upper extremities as a national research priority.

Table 38.1

Common Musculoskeletal Disorders Associated with Occupational Exposure to Ergonomic Stressors

Neck and Upper Back Disorders

Cervical Disc Disease: cervical disc degeneration or herniation that may or may not be accompanied by nerve root compromise

Tension-Neck Syndrome (Trapezius Spasm): localized muscle fatigue due to static, sustained contraction with accumulation of metabolic end products in muscles combined with insufficient oxygen supply

Biomechanical risk factors: highly repetitive work, forceful exertion, static contraction

Shoulder Disorders

Rotator Cuff Disorder: degeneration of any of the four tendons of the rotator cuff, particularly the supraspinatous tendon

Bicipital Tendinitis: tendinitis of the biceps tendon, particularly the long head of the biceps

Biomechanical risk factors: repetitive work, sustained shoulder postures with greater than 60 degrees of flexion or abduction

Elbow Disorders

Lateral Epicondylitis ("tennis elbow"): inflammation of the insertion of the extensor carpi radialis brevis (muscle at the lateral epicondyle)

Medial Epicondylitis ("golfer's elbow"): inflammation of the insertion of flexor muscles at the medial epicondyle

Biomechanical risk factors: forceful exertion, especially in combination with other risk factors (*e.g.,* force and repetition, force and posture)

Hand and Wrist Disorders

Carpal Tunnel Syndrome: entrapment of the median nerve within the carpal canal of the wrist

Biomechanical risk factors: repetitive work, manual force, vibration transmitted to the hand, combination of risk factors (*e.g.,* force and repetition, force and wrist bending)

Trigger Finger: low grade inflammation of the flexor tendon sheath that results in tendon sheath thickening and local nodular swelling, with characteristic "sticking" of the digits

DeQuervain's Disease: thickening and narrowing of the tendon sheath of the abductor pollicis longus and extensor pollicis brevis tendons in the region of the anatomic "snuff box" of the thumb

Flexor Tenosynovitis: inflammation of flexor tendons and tendon sheaths of the forearm, particularly flexor carpi ulnaris

Extensor Tenosynovitis: inflammation of the extensor tendons and tendon sheaths of the forearm, particularly carpi radialis longus and brevis tendons

Biomechanical risk factors: repetitive work, forceful work, posture, combination of factors (*e.g.,* force and repetition, force and posture)

Low-Back Disorders

Low Back Syndrome: soft tissue injury characterized by back pain with localized symtpoms and no neurological deficits

Lumbosacral Disc Herniation Disease: lumbosacral disk degeneration or herniation, typically with nerve root compromise and associated neurological deficits

Biomechanical risk factors: heavy physical work, awkward posture, whole-body vibration

Lower Extremity

Hip Osteoarthritis: degeneration of the hip joint characterized by pain with associated X-ray changes

Biomechanical risk factors: manual material handling and other heavy work

Knee Osteoarthritis: degeneration of the knee joint characterized by pain with associated X-ray changes

Biomechanical risk factors: manual material handling, kneeling, stairclimbing

Source: [6,7].

The rates of musculoskeletal disorders reported to the BLS increased steadily throughout the 1980s and early 1990s [11,14,18], and similar trends have been reported in other countries. However, it is not clear to what extent this reflects a true increase in incidence, as compared with improved diagnosis, recognition of work-relatedness, or reporting. In the state of Washington, the annual incidence of paid workers' compensation claims for CTS increased significantly among women, but not men, from 1984 to 1988 [19]; it could not be determined whether this reflected a true differential incidence or different claim-filing behavior between men and women because of social or occupational factors.

Many industries and occupations have rates that are as much as three to four times higher than the overall frequencies. High-risk industries are as varied as nursing facilities, air transportation, mining, food processing, leather tanning, heavy manufacturing of vehicles, furniture, appliances and electrical equipment, and light manufacturing, including electronic products, textiles, shoes, apparel, and upholstery [6,14]. Upper extremity musculoskeletal disorders have been found to be highly prevalent in manual-intensive occupations such as clerical work, postal service work, janitorial work, industrial inspection, and packaging. Back and lower limb disorders often occur in occupations that require heavy work such as manual material handling, nonneutral trunk postures, or exposure to whole-body vibration, especially while seated (see section I.B.). At highest risk are truck drivers, operators of cranes and other large vehicles, warehouse workers, airplane baggage handlers, various construction trades, nurses, nursing aides, and other patient-care workers [3,11,13].

In 1989, U.S. workers' compensation costs were $11.4 billion for low back and $563 million for upper extremity disorders [20,21]. However, the frequency of WMSDs is likely to be underestimated by as much as 60% when relying on traditional administrative data sources such as workers' compensation records and OSHA logs of recordable injury and illness [22,23]. The reasons for these discrepancies probably include factors as varied as workers' access to medical care, information about work-related morbidity, perceived job security, expected success in filing compensation claims, and employer medical management and return-to-work programs. It has been estimated that the indirect or hidden costs of workplace injuries and MSDs can range from 2 to 3.5 times the workers' compensation costs paid by an employer [12,24,25].

Disability and lost work time account for a large proportion of workers' compensation costs and impact negatively on workers' household income and feelings of self-worth. Correlates of poor MSD outcomes have included physical demands of the job, lack of worker autonomy, job dissatisfaction, receiving workers' compensation payments, pending litigation, and psychological disturbance [26–35]. However, interpretation of these findings often is ambiguous because of interrelationships among risk factors and lack of information about the temporal sequence of events. For example, the apparent effect of compensation may be confounded by physical or psychosocial job features, education and availability of other work, disease severity, access to medical treatment, and other factors.

Because work disability is influenced by myriad psychosocial and workplace factors, it provides a limited view of the outcomes experienced by the injured worker. Furthermore, return

Fig. 38.1 Sewing machine operator positions garment for stitching. Copyright © Earl Dotter.

to work is not an irreversible or unidimensional event; workers may leave and return on multiple occasions, depending on changes in job assignment and working conditions, availability and quality of medical care, information about work-related morbidity, employers' medical management, and return-to-work programs [36]. Long-term effects on job security and career advancement have been described anecdotally but have not been studied quantitatively. In addition, injured persons experience social and psychological costs, including loss of self-esteem, ability to participate in family or community activities, and to care for self and family members [37,38]. Family members as well may incur costs as they take on new tasks to compensate for the disabled individual. Data on these burdens are also extremely sparse [39,40].

B. Occupational Exposures and MSD

The generic ergonomic exposures relevant to the occurrence of MSDs include both physical workload and the organization of work in general. Examples of the former are stereotyped repetition of motion patterns, rapid work pace, exertion of high and/or prolonged muscular forces, anatomically nonneutral body postures (either dynamic or static), mechanical stress concentra-

tions (direct pressure of hard surfaces or sharp edges on soft tissues), insufficient rest or recovery time during work, and contact with vibrating objects. Extensive laboratory evidence shows that there are multiple plausible mechanisms by which physical stressors injure the soft tissues of the musculoskeletal system [41,42]. For example, tendon strain and cell damage accumulate as a function of work pace (the frequency and duration of mechanical loading), the level of muscular effort, and recovery time between exertions [43,44].

The epidemiologic literature linking the physical features of work to the risk of MSDs is voluminous. In one review, NIOSH concluded that there is scientific evidence of cause-and-effect relationships between repetitive motion, forceful work, and postural stress for disorders of the back and upper extremities, including tendinitis and carpal tunnel syndrome [6]. The risk is particularly high when two or more of these physical job features are simultaneous and exert synergistic effects. Thus, primary prevention of MSDs in the workplace should include multifaceted ergonomics programs that emphasize engineering controls, especially the ergonomic design of workstations, equipment, tools, and work organization to fit the physical capabilities and limitations of the human body [15,45].

Work organization refers to the way that production or service activities are organized, allocated, and supervised so as to determine task structure, the division of labor, and skill utilization by individual workers. Work organization thus determines physical job features (*e.g.,* work pace, repetitiveness, duration of exposures, and recovery time) as well as psychosocial dimensions of the work environment such as decision latitude, psychological job demands, and social support from supervisors and among coworkers. Decision latitude reflects the worker's ability to make decisions about how the work is done, to control the work process, and to decide which skills to use to accomplish the job. Increased pace of work on an assembly line, time pressure in processing or responding to information, or a more complex maintenance problem increase psychological job demands. It is often difficult to distinguish between "physical" and "psychosocial" ergonomic risk factors. High psychological work demands typically involve both rapid physical work pace and feelings of time pressure, and highly stereotyped finger motion patterns occur when a manual job both is monotonous and offers little decision autonomy.

High psychological job demands in combination with low decision latitude result, over time, in chronic adverse health effects such as cardiovascular disease [46]. Because psychosocial job characteristics influence physiological parameters such as circulatory function, sympathetic nervous symptom arousal, and muscle tension, there are several plausible mechanisms by which they may increase musculoskeletal susceptibility to mechanical insult [47–49]. In a ten-year prospective study of workers in the metal industry, both social support and work content were predictive of musculoskeletal status [50]. However, most other studies on this question have been cross-sectional; because psychosocial factors may be perceived and reported differently *after* the development of MSDs, the reported associations are difficult to interpret with respect to etiology. Thus, there is still only limited epidemiologic evidence concerning the independent effect of occupational psychosocial factors on MSD incidence [51].

II. Gender and Musculoskeletal Disorders

Being female is often described as a "risk factor" for many musculoskeletal disorders because prevalences in the general population and in large groups of employees have been reported to be twice as high among women compared to men [10,16,17,52–64]. However, comparisons of other endpoints have shown no, or no statistically significant, gender differences [17,61,62,64–66], and occasionally, especially for low back pain (LBP), the risk has been lower for women than for men [10,62,67,68].

To the extent that there are gender differences in MSD occurrence, their interpretation is not straightforward. Different occupational exposures might be responsible because women are concentrated in jobs with certain physical features because of the failure of many work sites to accommodate female anthropometry, or because women either sustain increased exposure to, or have higher perceived exposure to, psychosocial job strain (see section III.A.). In light of widespread sexual divisions of labor, it is essential to distinguish between gender differences in *crude prevalence or risk* and differences in *the effects of occupational exposures* on MSD occurrence or outcome.

Factors that might affect the gender-specific relationships between occupational exposures and MSDs include differences in performance of household work, roles in the home and recreational activities, or gender differences in pain reporting and health care seeking behavior. The possible role of biological factors such as gender-related differences in strength, muscle mass, tendon and muscle composition, and hormonal fluctuation on soft tissue should also be considered.

The following sections attempt to summarize the literature pertinent to these related but different questions:

Do MSDs (overall, or those that can be attributed to working conditions) occur more often among women workers than among men workers?

Are women at higher risk of MSDs than men in jobs with the same or similar levels of occupational exposure to ergonomic stressors?

Do women with work-related MSDs have different outcomes (*e.g.,* pain severity and functional loss, disability, interference with social and family life, financial losses) than men?

Are there innate biological differences that might affect the risk of WMSDs in women compared with men?

A. Occurrence of MSDs among Women Workers

The evidence shows that working conditions explain a substantial proportion of MSDs among women workers. The same associations between MSDs and ergonomic stressors that have been reported in male and mixed occupational groups have also been observed in analyses exclusively of women. For example, women employed in garment manufacturing have a higher prevalence of upper extremity symptoms and long-term disability due to musculoskeletal disorders when compared to women in other occupations [69–73]; such MSDs exhibit an exposure-response relationship with length of employment but also appear to induce garment workers to leave employment [73–77].

Upper extremity muscle, tendon, and joint disorders were elevated among women workers in packing jobs [71,78], textile manufacturing [79,80], and electronics manufacturing [81,82]. Female assembly-line manufacturing workers had higher a prevalence of shoulder and hand pain compared to a random sample of women from the general population; both work pace and duration of employment showed exposure-response trends [83].

Women working in poultry processing were more than twice as likely to have impaired median nerve sensory latencies (age-adjusted) as women applying to work in the same facility [84], and those employed in jobs with high versus low exposure to repetitive and forceful motions had more than twice the prevalence of upper extremity disorders, especially those affecting the hand and wrist [85]. Similar effects have been reported among female fish processing employees [86–89].

In the service sector, work-related musculoskeletal disorders have been documented extensively among health care employees such as female nurses and home health aides, as well as among food preparation and laundry workers [90–95]. The prevalence of CTS in female cashiers is higher than that in other grocery store workers [96] and is associated with poor work station design [97] as well as with hours worked per week and years worked as a cashier [98]. Work-related MSDs have also been reported among women cleaners [99,100], childcare workers [101], and nursery school cooks [102]. Multiple studies of female clerical workers operating video display units (VDUs) have shown an elevated risk of neck/shoulder and hand/arm disorders compared with women in other occupations [103–106]. Among VDU operators, there is an increased risk with greater postural strain [107], more hours per day at the VDU [108], and faster call handling by telephone operators [109].

In the extensive literature on musculoskeletal disorders among working people, many investigations have examined whether there is an overall gender difference in the frequency of MSDs in employed populations. In a few of these studies it is possible to compare the risk in men and women after adjustment for differences in occupational exposures. A number of these studies are summarized here (Tables 38.2–38.4). This review is nonexhaustive; the primary selection criterion was the ability to control, at least partially, for gender differences in frequency or intensity of occupational ergonomic exposures in the analysis of their data.

It appears that the gender difference is least pronounced, or even nonexistent, for back pain (Table 38.2). The only study that showed a statistically significant increase in women was a large sample of the working population in the Netherlands [59]. However, another Dutch study showed a higher prevalence of back and lower limb disorders among men than among women, after stratification based on mental and physical job demands [56].

With regard to upper extremity disorders there is no obvious pattern. Some studies showed a much higher risk in women than in men, and others found no difference at all. Being female was associated with at least a doubling of MSD risk in about one-third of the analyses of the proximal (Table 38.3) and distal (Table 38.4) regions of the arm.

Some studies [86,110,111] found larger gender differences for distal endpoints (*e.g.,* wrist pain, CTS) than for more proximal endpoints (*e.g.,* shoulder disorders). However, others had similar results for both regions of the upper extremity [112–116].

Table 38.2

Whole Body, Back, and Lower Limb Musculoskeletal Disorders: Relative Risk (Prevalence Ratio or Odds Ratio) in Women Compared to Men, with 95% Confidence Interval[a]

Reference	Population	Gender effect (95% CI[b] or p-value)
[112]	VDU workers	
	low back symptoms	NS[c] (p>0.05)
[27]	Aircraft manufacturing workers	
	back disorders reported to employer	NS (p>0.05)
[59]	General working population, the Netherlands	
	back disorders	**1.6 (1.4–1.8)**
	back disorders (chronic)	1.2 (0.9–1.6)
	muscle or joint disorders (not specified)	**1.6 (1.4–1.9)**
[139]	Poultry slaughterhouse workers (cold exposed)	
	upper backache	1.2 (0.9–1.4)[d]
	leg pain	1.1 (0.9–1.3)[d]
	Poultry slaughterhouse workers (assembly line)	
	upper backache	1.0 (0.8–1.2)[d]
	leg pain	1.0 (0.8–1.2)[d]
[122]	Automobile manufacturing workers	
	cumulative trauma of lower back (medically treated)	1.1 (NS)
[111]	Workers in four manufacturing facilities	
	low back and upper limb pain combined	1.3 (p>0.05)
[116]	Slaughterhouse workers	
	back pain	1.2 (0.7–1.9)[d]

Note. Increase in women statistically significant at p≤0.05 (boldface).

[a]Adjusted for job title or ergonomic exposure, at minimum, and for age and other factors where possible, by stratified or multivariate analysis.

[b]CI = confidence interval.

[c]NS = not stated: point estimate or confidence interval not reported.

[d]Prevalence ratio (cross-sectional study), odds ratio (case-control study), and/or confidence interval calculated by authors from data in the paper.

Because of possible gender differences in pain reporting (see section V.B.), one could also hypothesize that the effect in women might be higher for symptoms than for endpoints requiring physical examination or more objective confirmation. For example, in a cross-sectional study at seven manufacturing facilities, Silverstein *et al.* found that hand/wrist disorders were strongly associated with exposure to both high manual forces and repetition rates [117]. Women had an increased risk compared to men, even adjusting for job demands (Table 38.4); most of the gender difference appeared to occur in jobs requiring high manual force but low repetition, where more male than female subjects were employed. When the outcome was restricted to carpal tunnel syndrome by symptoms and physical examination, there was no longer an effect of gender after adjusting for job demands [118]. Three other studies of CTS also reported no or

Table 38.3

Neck, Shoulder, and Combined Upper Extremity Musculoskeletal Disorders: Relative Risk (Prevalence Ratio or Odds Ratio) in Women Compared to Men, with 95% Confidence Interval[a]

Reference	Population and body region	Gender effect (95% CI[b] or p-value)
[222]	Workers in routinized visual display unit (VDU) work	
	neck symptoms	**1.5 (1.0–2.2)**
	shoulder symptoms	**2.2 (1.4–3.4)**
	Workers in CAD or programming VDU work	
	neck symptoms	**2.9 (1.6–5.5)**
	shoulder symptoms	**3.1 (1.6–5.8)**
[112]	VDU workers	
	tension neck syndrome (symptoms and signs)	2.0 (0.7–5.6)[e]
	any shoulder diagnosis (symptoms and signs)	**6.4 (1.9–21.5)[f]** **7.1 (1.6–32.2)**
[113]	Workers in VDU jobs with similar gender distribution	
	neck disorders (symptoms)	1.9 (0.8–4.5)
	shoulder disorders (symptoms)	1.5 (0.5–4.8)
[223]	Newspaper employees	
	neck disorders (symptoms)	**2.3 (1.5–3.5)**
[86]	Fish processing workers	
	shoulder girdle pain (various diagnoses)	1.1 (0.7–1.7)[d]
[120]	General population, Sweden	
	neck/shoulder disorders (recent medical care and signs)	**11.4 (4.7–28)**
[58]	General population, England (orthopedic clinic patients)	
	shoulder disorders	NS[c] (p>0.05)
[114]	Telecommunication workers	
	neck disorders (symptoms)	NS (p>0.05)
	shoulder disorders (symptoms)	NS (p>0.05)
[138]	Workers in two poultry processing facilities	
	neck and UE[g] disorders (symptoms)	1.2 (0.7–1.9)[d]
	neck and UE disorders (symptoms and signs)	1.1 (0.5–2.4)[d]
[60]	General population, Finland[h]	
	chronic neck pain (age 30–64)	**1.6 (1.4–1.9)[d]**
	chronic neck pain (age 65+)	**1.3 (1.0–1.7)[d]**
[115]	Textile manufacturing workers	
	epicondylitis	**NS (p=0.03)**
	shoulder conditions	NS (p>0.05)
	neck conditions	NS (p>0.05)
	myalgia	**NS (p=0.001)**
	arthralgia	NS (p>0.05)
[165]	Public sector office workers	
	neck/shoulder/back symptoms	**1.9 (p=0.005)**
[122]	Automobile manufacturing workers	
	cumulative trauma of the neck (medically treated)	1.1 (NS)
	rotator cuff syndrome (medically treated)	1.1 (NS)
	other UE cumulative trauma (medically treated)	1.0 (NS)

Table 38.3
(continued)

Reference	Population and body region	Gender effect (95% CI[b] or p-value)
[224]	Newspaper employees upper extremity disorders (symptoms)	**2.2 (1.5–3.3)**
[110,168]	Automobile manufacturing workers, cross-sectional shoulder/upper arm disorders (symptoms) shoulder/upper arm disorders (symptoms and signs)	**1.6 (1.1–2.2)** **1.7 (1.2–2.4)**
	Automobile manufacturing workers, 1-year follow-up upper extremity disorders (symptoms and signs)	1.4 (0.6–3.5)
[111]	Workers in four manufacturing facilities trapezius pain	10.1 (p>0.05)
[116]	Slaughterhouse workers neck and shoulder pain	1.3 (0.9–1.9)[d]

Note. Increase in women statistically significant at p ≤ 0.05 (boldface).
[a]Adjusted for job title or ergonomic exposure, at minimum, and for age and other factors where possible, by stratified or multivariate analysis.
[b]CI = confidence interval.
[c]NS = not stated: point estimate or confidence interval not reported.
[d]Prevalence ratio (cross-sectional study), odds ratio (case-control study), and/or confidence interval calculated by authors from data in the paper.
[e]Women without children under age 16 at home.
[f]Women with children under age 16 at home.
[g]UE = upper extremity.
[h]Ages 30–64: adjusted for physical demands in current job. Age 65+: adjusted for physical demands in longest held previous job.

only very weak associations with gender [71,96,119]. On the other hand, the occurrence of specific hand/wrist tendinitis conditions in the seven manufacturing plants showed a gender effect almost identical to that for all disorders [43]. In fact, several of the strongest associations with gender [112,120,121] were observed for more restrictive case definitions.

Thus, although women often are reported to have higher rates of (work-related) musculoskeletal disorders, after adjusting for occupational exposure there is no consistent pattern in the literature. Possible reasons for the discrepancies among studies include differences in study populations (*i.e.,* types and extent of occupational and non-occupational exposures) and how well ergonomic exposures were characterized. Adjustment only for plant type or department [115,116,121,122] probably results in greater nondifferential misclassification of exposure and resultant residual confounding of the gender comparison (see III.A.). Although some of these studies are consistent with the hypothesis that women have higher rates than men of MSDs, further research is required in studies designed to adjust adequately for exposure differences before general conclusions can be drawn.

Table 38.4
Wrist and Hand Musculoskeletal and Nerve Compression Disorders: Relative Risks (Prevalence Ratio or Odds Ratio) in Women Compared to Men, with 95% Confidence Interval[a]

Reference	Population and body region	Gender effect (95% CI[b] or p-value)
[43]	Workers in seven manufacturing facilities hand or wrist tendinitis (symptoms and signs)	**4.3 (p<0.05)**
[112]	Workers in routinized visual display unit (VDU) work any arm or hand diagnosis (symptoms and signs)	**5.2 (1.2–22.8)[e]**
[113]	Workers in VDU jobs with similar gender distribution wrist and hand symptoms	1.7 (0.8–3.6)
[158]	Automobile assembly line workers forearm/hand symptoms	**1.6 (1.2–2.1)[d]**
[86]	Fish processing workers carpal tunnel syndrome	**2.6 (1.3–5.2)[d]**
[58]	General population, England (orthopedic clinic patients) finger disorders wrist and forearm disorders	NS[c] (p>0.05) **1.8 (1.1–2.9)**
[137]	Automobile manufacturing workers forearm/hand symptoms[f]	**1.2 (1.0–1.4)**
[114]	Telecommunication workers elbow, wrist, and hand disorders (symptoms)	NS (p>0.05)
[71]	Workers in three industrial sectors carpal tunnel syndrome	1.5 (0.7–3.0)
[115]	Textile manufacturing workers carpal tunnel syndrome tendinitis ganglions	**NS (p=0.02)** **NS (p=0.01)** NS (p>0.05)
[119]	Workers from four industries, at 5-year follow-up slowed median NCV[g]	NS (p>0.05)
[165]	Public sector office workers hand/arm symptoms	NS (p>0.05)
[122]	Automobile manufacturing workers carpal tunnel syndrome (medically treated)	**2.3 (1.6–3.3)**
[110]	Automobile manufacturing workers, cross-sectional wrist/hand disorders (symptoms) wrist/hand disorders (symptoms and signs)	**2.2 (1.6–3.0)** **2.7 (2.0–3.7)**
[117]	Workers in six manufacturing facilities hand/wrist disorders (symptoms and signs)	**4.8 (p<0.05)**
[118]	Workers in seven manufacturing facilities CTS (symptoms and signs)	1.2 (0.3–4.7)[d]
[225]	Industrial automotive workers CTS symptoms (also had longer NCV latencies)	1.3 (0.9–1.7)[d]

(continued)

Table 38.4
(continued)

Reference	Population and body region	Gender effect (95% CI[b] or p-value)
[17]	General population employed during past year, U.S.	
	self-reported CTS	1.9 (1.6–2.4)
	self-reported medical history of CTS	2.2 (1.5–3.3)
[116]	Slaughterhouse workers	
	arm and hand pain	1.0 (0.7–1.4)[d]

Note. Increase in women statistically significant at p≤0.05 (boldface).

[a]Adjusted for job title or ergonomic exposure, at minimum, and for age and other factors where possible, by stratified or multivariate analysis.

[b]CI = confidence interval.

[c]NS = not stated: point estimate or confidence interval not reported.

[d]Prevalence ratio (cross-sectional study), odds ratio (case-control study), and/or confidence interval calculated by authors from data in the paper.

[e]Women with children under age 16 at home (only).

[f]Adjusted for repetitive hand movements > 4 hours daily. Adjustment for precision movements or repetitive finger movements, either > 45 minutes or > 4 hours per day, gave similar results, as did adjustment for manual handling activities.

[g]NCV = nerve conduction velocity.

B. Outcomes of Musculoskeletal Disorders among Women Workers

Women are usually reported as experiencing more work disability and losing more days from work due to CTS and other MSDs than men [30,52,62,123–127]. However, other studies have found no gender differences in clinical or occupational outcomes of MSDs [10,26,54]. A large study of U.S. automobile manufacturing workers found that being female was associated with a larger excess of upper extremity disability than back disorder disability, after adjusting for production department [121]. In Norway, absenteeism from work for more than 14 days in 1994 because of neck and shoulder MSD was more frequent among women than among men, but the gender difference was greatest among people over 45 years of age and in the lowest income group [128]. After adjustment for these two variables, men had a higher cumulative incidence of long-term absenteeism and little difference in mean duration. On the other hand, English men had more short-term MSD absenteeism (≤7 days) than women after adjustment for age and socioeconomic status, but there was no gender difference in longer absences [124].

The predictors of MSD disability and other outcomes have rarely been examined separately by gender. One study found that return to work after CTS treatment was markedly lower for women than for men, even taking into account prior symptom severity and functional impairment [127]. Another study showed that a multifaceted rehabilitation program for chronic back pain benefitted only the female participants [129]. Women workers differ from their male counterparts in several important respects that might influence outcomes. First, as noted earlier, women's jobs are more repetitive and offer less decision autonomy than men's jobs; thus, women may be at greater risk than men for

Fig. 38.2 Laundry worker stocks poorly designed hospital bed linen storage racks, forcing her to lift heavy bedding above shoulder height. Chicago, Illinois (1993). Copyright © Earl Dotter.

work-associated musculoskeletal disability because of difficulty with job performance or reinjury [130]. Second, women generally have more responsibility for household work and care of family members, so the family burden may have more influence on return-to-work decisions by women than by men; further, the cost of lost household services greatly outweighs wage losses for women [131]. Physicians' perceptions of the work demands in men's and women's jobs may also differentially affect medical treatment or the success of a compensation claim and thus the financial option to stay out of work [37,132].

III. Interpretation of Gender Differences: Occupational Physical Stressors

A. Exposure Differences

Because of workplace sex segregation, both formal and informal, women and men typically experience qualitatively and quantitatively different working conditions. In many epidemiologic studies of WMSDs, comparisons of the risk for women and for men may be confounded by differences in occupational exposures. In particular, job title or occupation has often been used to categorize ergonomic exposure. Use of such imprecise measures of exposure may create the appearance that gender is an independent risk factor for development of WMSDs when in fact occupational exposures may not have been adequately accounted for.

In a sample of the general population in Sweden, followed from 1970 to 1993, men had a higher index of physical work load overall than did women [133]. The difference was most pronounced among younger men during the period from 1970 to 1980 [134]. Physical work demands decreased over the 24 years among men, especially the younger subjects, whereas a small increase in physical demands was observed in younger women in the early 1980s. Similarly, in a large cross-sectional sample of the Finnish population, men aged 30 to 64 years were more likely to be in jobs with high exposure to physical stress at work, and women were more likely to be in jobs with intermediate exposure [60]. The same pattern was found among subjects aged 65 or more with respect to physical work load in the previous job of longest duration.

Such summary exposure indices may be insensitive to differences in types of exposures and often place more emphasis on whole-body exertions and energy expenditure than on localized, repetitive stresses to the upper extremities. In the industrial sector, including food processing, women are overrepresented in packing and assembly or other "light" jobs that require rapid, precise, repetitive hand motions [86,111,117,135–139]. Similar gender differences in ergonomic exposures have been reported in the service sector, including cleaning [99], grocery store work [96], and operation of VDUs [140].

Even within the same occupation, physical exposures may differ by gender. For example, women and men working in the same factories, with the same job titles, did not always perform tasks with the same physical requirements or work organization [141,142]. Women performed more repetitive work on average, whereas men were less likely to sit for prolonged periods. Men were significantly more likely than women to work irregular and unpredictable hours and more than 40 hours per week.

As a group, women also hold jobs with less control over the work process, less latitude for making autonomous decisions, and fewer opportunities for job modification [46,50,143–145]. These gender differences in decision autonomy and monotony have even been observed within female-dominated jobs [143].

Jobs requiring high static loading of the neck and shoulder, with repetitive use of small muscle groups, involve a high risk of upper extremity disorders such as carpal tunnel syndrome and hand-wrist tendinitis. During dynamic low-load manual work, both the speed of motion and high precision demands substantially increase measured muscle forces relative to their capacity [146,147]. Ironically, the physical demands of these female-intensive jobs are often perceived to be less strenuous than the jobs typically performed by men because they do not require high-force whole body exertions such as lifting and carrying very heavy objects [99,141,142]. In contrast, in the health care sector many women hold jobs such as nurse or nursing assistant, with high patient lifting demands and high rates of low back injury.

Women may sustain greater exposure to biomechanical stressors even when performing the same tasks as men because many workplaces have been designed on the basis of anthropometric data for men and are ergonomically inappropriate for women. Men and women, on average, differ in many aspects of physical body size and functional capacity, such as stature, body segment lengths, flexibility, and muscle strength [148]. These differences lead to a poorer fit, on average, of tools, equipment, workstations, gloves, and other personal protective equipment for women workers, with resulting increases in biomechanical disadvantage and postural strain [149–153]. As a result, gender differences in work technique or posture for performing the same tasks have been observed [43,154,155].

A strong association has been found between gender segregation at the occupational level and absence from work because of MSDs in both genders [156,157]. In the few gender-integrated occupations (40–60% men), both women and men had the lowest rates of sickness absence. Absenteeism for MSDs was higher when there was gender segregation in either direction and very high, especially for women, in the group of extremely male-dominated occupations (>90% men). Whether sex segregation per se is a risk factor (possibly as a source of psychosocial strain) or if it is confounded by (or acts as a proxy for) differences in physical working conditions has not been clarified.

B. Interaction between Exposure and Gender

Only a small number of MSD studies have compared the effects of similar job features on men and women to examine whether gender modifies the effect of ergonomic exposure on MSD risk. Among those reviewed here, the findings are markedly inconsistent. Some investigations found stronger associations for MSDs and physical stressors among women [122,158], whereas others reported no or few differences between men and women in the associations between MSDs and ergonomic exposures [86,117, 122,138] or that women are at higher risk from some exposures and men from others [57,159]. It is difficult to find any pattern among these results, even when examining specific exposure-endpoint relationships. For example, a group of French authors reported that carrying heavy loads was a risk factor for joint pain with restricted movement only in men [57], whereas heavy lifting was a risk factor for LBP only among women in an English study [160]. In the U.S. general population, osteoarthritis of the knee was associated with occupational demands for knee bending and manual material handling in men only; few women had jobs with similar physical demands [161]. However, two other studies showed a stronger association with occupational physical activity among women than among men [162,163].

In a few studies, risk ratios for the effect of exposure on MSDs have been higher among women than among men, although the findings varied by specific exposures and specific MSD endpoints [137,160]. Three prospective studies showed that MSD symptoms in women were more responsive to changes in occupational exposures. Two Nordic groups reported that women, but not men, experienced increased frequency of distal upper extremity symptoms following increased repetitiveness of manual motion patterns [89,164]. Among U.S. public sector office workers, neck, shoulder, and back symptoms were twice as likely to improve among female compared with male employees after office relocation [165].

Eight different studies have shown more MSDs among women than among men in the lowest stratum of ergonomic exposure,

a

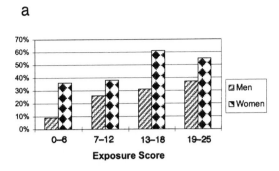

b

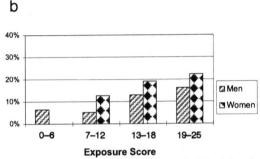

Fig. 38.3 Frequency of upper extremity musculoskeletal disorders on examination, by baseline exposure, among 1198 workers in automotive engine and stamping operations. (a) Baseline prevalence; (b) one-year cumulative incidence (data from references [110,167,168]).

with the difference decreasing in the higher exposure groups so that the effect of ergonomic stressors on MSD risk was greater among men [17,60,86,158,166,167]. For example, among automobile manufacturing employees, there was substantial effect modification of physical exposure by gender, in which women had a higher baseline risk than men in the lower exposure groups but about the same risk in the highest exposure group (Fig. 38.3a) [167]. However, at the one-year follow-up, the risk of new upper extremity disorders, adjusted for physical exposure level, varied much less between men and women (Fig. 38.3b) [168]. Possible explanations include that a healthy worker selection effect (HWE) might be operating predominantly among women employed in the most physically demanding jobs, or that gender differences in symptom reporting are modified by the intensity of ergonomic exposures.

Other findings are consistent with the possibility that women workers are disproportionately affected by HWE selection processes. Among automobile assembly line workers, women, but not men, with higher seniority had higher pressure pain thresholds in the forearm and hand [158]. Similarly, the proportion of female, but not male, package sorters decreased with increasing seniority [151]. Because the physical demands of the job were disproportionate to women's body size and strength, the authors hypothesized that selection out of the workplace had been more pronounced among women over time because of excessively heavy work. On the other hand, in two other studies, the effect of MSDs on the selection process appeared to be stronger in men than in women [160,169]. HWE patterns are obviously

influenced by opportunities to move into less demanding jobs, which may vary not only by gender but also by region and economic circumstances such as unemployment rates.

IV. Interpretation of Gender Differences: Occupational Psychosocial Exposures

As with physical ergonomic exposures, the effects of psychosocial job conditions on MSDs have rarely been examined in men and women separately. Leino and Hänninen [50] found that low work control at baseline predicted 10-year musculoskeletal morbidity of the neck/shoulder in white-collar women and of the low back in blue-collar women, whereas it had no predictive value among men. Monotonous work content was associated with signs of MSDs of the low back and lower limbs among blue-collar workers of both genders, and low social support at work was likewise a risk factor both for men and for women. All analyses were adjusted for a rough measure of physical workload that emphasized whole body effort rather than repetitive manual work, so intrastratum confounding by physical job demands could not be ruled out.

In two Swedish cross-sectional studies, multiple interactions between gender, psychological load, decision latitude, and social support were found. In one, musculoskeletal symptoms were generally elevated when high job demands were combined either with low job control or with low social support [170]. The prevalence ratios for psychosocial job features and neck symptoms were slightly higher among women than among men, whereas the effects on shoulder and back symptoms were stronger for men. In the other study, high decision latitude was also associated with an increased risk of neck and shoulder symptoms among women only [171]. In an examination of possible mechanisms of effect, psychological work demands were associated with physiological indicators of strain (self-reported muscle strain and cortisone levels); muscle tension was, in turn, associated with symptoms from the back, neck, and shoulders. There was also an unexpectedly high prevalence of MSD symptoms in the highest socioeconomic class among working women, which contradicted the traditional (albeit male-based) model that high decision latitude alleviates psychosocial stress. The authors noted that well-educated women in jobs with high decision latitude were found to perceive high stress related to responsibility at work, but this was confounded by responsibilities for the family. In addition, a Canadian study similarly found that women were more substantially affected by high job demand and more adversely affected by high job routinization than men [144]. These authors also hypothesized that women's greater vulnerability to job demands could be a function of demand at home (see section V.A.).

There are other types of psychosocial strain in the workplace that may be more common or more important for women than for men. These include responsibility for the well-being of others (health care, childcare, teaching); responsibility to multiple supervisors (clerical work); unpredictable, "flexible" scheduling that interferes with family responsibilities, especially caring for children; lack of social support, especially in traditionally male-dominated occupations such as construction and heavy manufacturing [152]; and sex discrimination and overt sexual harassment. These sources of strain typically are not measured by standardized instruments, such as the Job Content Questionnaire [172], and thus may represent sources of unmeasured confounding or effect modification in epidemiologic studies. The possible contribution of racial or ethnic discrimination as a source of stress has not been examined.

Women workers in traditionally male-dominated jobs often feel that they need to work harder or take more risks than their male colleagues take in order to prove themselves to their supervisors and coworkers [152]. On this issue the psychosocial aspects of the work environment, especially a lack of social support, have potential consequences for the physical exposures if overcompensation leads to workers injuring themselves, for example, by trying to lift loads that are too heavy.

V. Interpretation of Gender Differences: Extrinsic Factors

A. Double Exposure: Work and Home

Household responsibilities result in greater overall exposure to physically demanding activities and psychosocial strain, as well as a reduction in opportunities for recovery after the working day. The extra responsibilities of household chores and childcare have traditionally been borne more heavily by women, and there are significant gender differences in time spent on domestic work. For example, among French workers with children, women spent more time than men on domestic chores (21.8 hours per week compared with 4.9 for men), and women without children also performed significantly more housework than men (17.3 hours per week compared to only 5.3 for men) [141].

A Swedish study investigated gender differences among 12,777 workers in the distribution of home responsibilities and psychosocial work environment characteristics, as well as resources to deal with these potential stressors [173]. Exposure to nonoccupational stress was assessed through measurements of household work, number and age of children, and childcare resources. The study found that after adjusting for work and home characteristics, gender was no longer associated with psychosomatic strain. The authors noted that there were major differences in the life experiences of Swedish women and men that went far beyond their occupational exposures, and that these daily life circumstances might account for some of the observed disparity in the prevalence of musculoskeletal disease.

There are other data that lend support to the hypothesis that housework may at least partially explain the discrepancy in musculoskeletal disorders between male and female workers. Bergqvist *et al.* found that, among VDU users, women with young children at home had higher prevalences of MSDs than men or other women, after adjusting for work-related and other demographic factors [112]. In a large group of female sewing machine operators, having children was associated with about double the risk of chronic shoulder pain; the authors suggested that the reason might be the physical load of lifting and carrying small children [69]. Interestingly, in a prospective study, extra domestic work and few social contacts were predictors of LBP among men only, although unsatisfactory leisure time was a risk factor both in men and in women [159]. There has been little

study of the magnitude or variability of ergonomic stressors during home activities [174] or their effects on MSD risk.

B. Pain Reporting and Seeking Medical Care

Another possible explanation for the apparent higher rates of MSDs in women may be a greater willingness to report symptoms on interview, to report problems to workplace supervisors, or to seek medical care for conditions of the same pain severity. As already noted, the literature is inconclusive as to whether women are more often absent from work for MSDs, after adjusting for age, socioeconomic status, and job demands. Some investigators have explicitly suggested that women are prone to overreporting symptoms and malingering [175]. However, there have been few published comparisons of the validity of MSD symptoms reported by male and female workers. In a large sample of the U.S. general population, women reported more MSD symptoms than men but did not have more findings on physical examination [54]. In contrast, a Dutch study reported that gender differences in seeking medical care and use of medications were roughly similar to the differences in musculoskeletal symptom prevalence [59]. Bergqvist *et al.* [112] found that, as a group, men were actually more likely than women to report upper extremity and neck symptoms without physical examination findings in the same body region [176]. Among CTS patients, correlations between subjective assessment of hand weakness and measured grip strength were comparable between men and women [177].

Health care-seeking behavior often varies between men and women, although study findings disagree as to whether or not women and men with MSD symptoms seek care at different rates [38,66,178–180]. In a group of patients with tempromandibular pain, women were overrepresented but had similar ratings of chronic and experimental pain as men; the authors suggested that women might be more responsive to symptoms and more willing to admit health problems [181]. On the other hand, women often have surgery for degenerative joint disease at a later stage than men, despite significant pain and functional impairment, in part because of their caregiving roles in the family [182].

Several authors have examined pressure pain thresholds (PPT), which can be interpreted either as a measure of pain tolerance or of prior damage to the muscle tissue. PPT values are generally higher in men than women [183–185]. One study showed that pain thresholds in the neck and shoulder region were associated with the psychosocial work environment; the interpretation of the findings may involve plasma cortisol or endorphin levels or may be partially mediated by sleep disturbance [185].

Byström *et al.* reported that PPTs of the right hand and forearm were lower among auto assembly line workers with forearm or hand symptoms in the previous 7 days [158]. Men had about 50–60% higher thresholds than women, both overall and among asymptomatic workers. The difference in PPT between men with and without symptoms was much smaller than the corresponding difference in women, suggesting either that reported symptoms were more misclassified among men (men denied symptoms more often?) or that PPT did not reflect the same pain phenomenon in the two genders. In addition, the effect of assembly line work on physical examination findings of de Quervain's tendinitis was greater in men, but the reverse was true for clinical diagnosis (which also required symptoms), suggesting that men might have more preclinical conditions or be less likely to acknowledge symptoms than women.

Among 184 active workers with abnormal sensory nerve conduction studies, women were almost three times more likely than men to acknowledge symptoms consistent with CTS; subjects in more repetitive jobs were also more likely to report symptoms [186].

These results present a mixed picture. Some suggest that women are more likely to report symptoms or seek medical care for MSDs, whereas others disagree. However, the question is difficult to evaluate, because there is no objective gold standard for musculoskeletal pain. Clinical examination methods do not necessarily measure the same domains. Pain is inherently subjective, yet it is a legitimate entity for study and prevention. Furthermore, pain is the experience motivating the behaviors that result in MSD costs, such as seeking medical care, work absenteeism, and filing claims for compensation. If men deny pain and delay seeking medical attention, their disorders may progress further and be less amenable to medical treatment and workplace intervention. Whether pain perceptions and pain-related behavior actually vary between men and women, and in what ways, deserves further and careful study.

VI. Interpretation of Gender Differences: Physiological Factors

Physiological factors affect work performance and can possibly interact with ergonomic exposures to increase the likelihood of developing musculoskeletal disorders. Physiological factors with gender-related differences that may predispose a person to the development of WMSDs include muscle strength and muscle fiber type distribution, anthropometric measures, hormonal differences, and biological changes associated with pregnancy.

A. Muscle Strength and Physiology

Gender-related biological differences may result in differential vulnerability of women to physical workplace factors. Prominent among these differences is muscle strength and distribution. Women's total body strength is, on average, about two-thirds that of men's; however, the ratio in static strength (ability to move a stationary weight) ranges from 35 to 85%, depending on the tasks and muscles involved [148]. Women's average strength is relatively lower in the upper extremities [187] and closer to men's for leg exertions, dynamic lifting, pushing, and pulling activities, and manual handling of smaller containers. The difference is also smaller when men and women have similar industrial experience or athletic training [188]. Despite the average differences, there is also substantial overlap in the strength distribution between men and women, as much as 50% or more for certain muscle groups, especially because many muscle strength values are not normally distributed. In all, the factors of gender, age, weight, and height only explain about one-third of the variability in human strength data.

With respect to musculoskeletal disorders, the health implications of the average strength differential are not clear.

Kilbom reported a relationship between decreased neck disorders and isometric strength in women with heavy, varied jobs, but not in women in so-called light repetition jobs [189]. However, in another study there was no evidence that low muscle strength predicted upper extremity MSDs [82]. In a group of Swedish students followed from age 16 to age 34, physical capacity in adolescents was found in both genders to influence the development of neck, shoulder, and low back symptoms in adulthood [190]. Several measures of flexibility and grip strength were associated with decreased risk of neck, shoulder, and low back problems in adulthood, although bench press strength predicted an increased risk of LBP. Other studies have been similarly inconclusive as to whether or not muscle strength is protective against LBP [67,191,192], and at least one other study has shown high muscle force capacity to be a risk factor [193]. It should be noted that although stronger muscles are capable of generating higher internal forces, they do not imply greater strength in other soft tissues, such as nerves and spinal discs.

Another possible explanation for the low predictive value of muscle force capacity, especially in low-effort work, relates to the physiologic process of muscle fiber recruitment during contractions. The near-continuous firing of low-threshold motor units during static work has been proposed as a mechanism for the selective injury of individual muscle fibers, even when the muscle as a whole is not fatigued [194,195]. Women, on average, have a higher relative volume of slow-twitch (aerobic) to fast-twitch (anaerobic) muscle fibers than men [196]. It has been hypothesized, although not confirmed, that individuals with more slow-twitch fibers may be less likely to alternate among muscle motor units during low-force contractions [147,197]. This theory deserves further investigation as a possible explanation of gender differences in the occurrence of neck/shoulder disorders in jobs with high static muscle loading.

Strength testing has been proposed as a gender-neutral hiring criterion that might protect both men and women from WMSDs [193]. Because there is a large overlap between population distributions of the static strength of men and women, selection of individuals able to lift a given weight will not exclude all prospective female employees from heavy jobs. However, evidence is inconclusive concerning the effectiveness of static strength testing to predict an individual's likelihood of injury. Using such criteria might unnecessarily exclude persons who would not suffer injury and include those who would.

Furthermore, Stevenson et al. have argued that there is a gender bias inherent in standard strength testing procedures because motion patterns observed in the workplace are typically used to determine the relevant testing maneuvers, but women often use different strategies to perform physically demanding tasks [155,198]. In addition, the gender difference in static strength is greatly diminished after correction for body weight and fat composition, an appropriate procedure if women select work strategies to optimize their biomechanical and physiological advantage in ways related to these physical characteristics [149,188]. Furthermore, strength tests do not measure other important physiological capacities such as dynamic (aerobic) endurance or range of motion, all of which differ little between the genders or even favor women [196,198–200], and which also may be relevant to injury occurrence or susceptibility.

B. Tendons, Ligaments, and Connective Tissue

Tendons in women appear to have different responses to repetitive motion exposures than those in men. In particular, one study demonstrated that flexor digitorum profundus tendons from women were significantly stiffer than those from males when tension was applied [44]. Additionally, tissue creep (time-dependent elongation) was less pronounced in female than in male tendons.

It is possible that gender differences between the regulation of cell activity in women compared with men may contribute to differences in development of WMSDs. Overall, the literature indicates that tendon and ligament cells from female subjects are influenced by sex hormones as well as by pregnancy-related factors. Ligaments have receptors for sex hormones and are responsive to hormones and changes in hormone levels [201]. Both human and rabbit tendon tissues have been found to contain receptors for sex hormones and studies have revealed that cell activity in these tissues is influenced by pregnancy. Studies of sex hormone receptors in connective tissue indicate that hormonal influences that fluctuate during the menstrual cycle or during pregnancy may contribute to differences in regulation of ligaments and tendons and subsequent development of musculoskeletal disorders [202].

Sex hormones may also play a role in the degradation of cartilage through regulation of the inflammatory response [203]. Overuse injuries to tendons are thought to result primarily from fatigue of tissues followed by an inflammatory response [204]. There is a growing body of evidence to suggest that neurogenic inflammation may play a role in the induction of peritendinitis as well as the response to tendinitis [44]. Such responses can be generated by exposure to the neurotransmitters Substance P and calcitonin. Conceivably, gender-based differences in induction and elaboration of inflammatory responses might result in differential susceptibility through these neurogenic mechanisms.

C. Endocrine Factors

The role of female hormones in development of MSDs is very poorly understood. Although tendons and ligaments have receptors for female sex hormones, the implications of these findings are not known. The literature is inconsistent as to the relevance of endocrine influences, such as oral contraceptive use or bilateral oophorectomy, on the occurrence of carpal tunnel syndrome among employed persons [55,72,86,96,118,135,205]. LBP has been associated with menstruation, oral contraceptive use, induced abortion, number of live births, menopausal symptoms, and lower age at menopause, although the evidence is very sparse and the mechanisms of these associations are largely unknown [206–208]. One study has suggested a possible hormonal role in biochemical changes related to joint cartilage degeneration [163]. Data are lacking as to whether or not these indicators of hormonal status are important modifiers of the effect of ergonomic exposures at work on MSDs.

D. Pregnancy

A number of physical changes occur during pregnancy that may at least transiently place women at increased risk for musculoskeletal disorders. These include increases in body weight and changes in body weight distribution as well as an insufficiency of abdominal muscles. Additionally, pregnant women have altered connective tissue function, including increased peripheral joint laxity, possibly due to release of relaxin or other hormones.

Pregnancy-associated anthropometric changes may increase susceptibility of women to physical ergonomic stresses. Paul and Frings-Dresen described work postures in pregnant compared with nonpregnant women [209]. Pregnancy resulted not only in greater body weight and changed body mass distribution, but also in a difference in the fit between body and workplace dimensions. For example, pregnant women sat further from work surfaces compared to nonpregnant women. Their hips were typically positioned further backwards, with increased compensatory trunk flexion, arm anteflexion, and extension. Workplace layouts in which the work surface height was self-selected resulted in marked diminution of postural strain, highlighting the importance of workplace accommodation.

Although upper extremity strength is not substantially affected by pregnancy [210], whole-body lifting capacity is altered in later pregnancy as the center of gravity moves and as increased body size prevents objects from being lifted close to the body. In addition, the ligaments and muscles of the back and abdomen are stretched, making lifting potentially more hazardous. There is evidence of increased risk of low back injury for pregnant women performing specific tasks such as heavy lifting, standing, and frequent climbing of stairs [211], although LBP in pregnancy has not been associated with measures of physical fitness or strength [212,213]. There has also been little effort to distinguish LBP associated with pregnancy from an increase in risk of work-related low back pain during pregnancy.

Lastly, carpal tunnel syndrome associated with pregnancy is a well-described phenomenon [214–216], although the cause has not been completely elucidated. However, it has been hypothesized that fluid retention in pregnancy may be sufficient to lead to nerve compression at the wrist and subsequent carpal tunnel syndrome. Generally, carpal tunnel syndrome cases remit spontaneously after delivery, lending further support to the suggestion that biological changes during pregnancy are responsible. However, the effect of ergonomic exposures in the workplace on carpal tunnel syndrome during pregnancy has not been adequately investigated. It is not known if pregnant women working in jobs associated independently with increased risk of CTS are at greater risk of developing CTS than pregnant women who do not sustain such exposures. Additionally, the possible role of postpartum occupational exposures in preventing the usual postpartum spontaneous recovery has not been investigated. Thus, although pregnancy only accounts for a small portion of women's work lives, the relationship between the potential increased risk of MSD development in pregnancy and subsequent inability to heal due to sustained biomechanical insult in the workplace, even after pregnancy has ended, is troublesome, particularly in nations such as the United States that do not provide for prolonged maternity leave from work.

VII. Conclusions and Research Needs

Musculoskeletal disorders occur in relation to ergonomic exposures both in men and in women. There is substantial clinical, biomechanical, and epidemiologic evidence supporting the relationship between MSDs and ergonomic factors in the workplace, including high repetition, high manual forces, and vibration [6]. For example, the relative risk of CTS among individuals highly exposed to repetitive hand motion, forceful exertions, wrist bending, and segmental vibration has been estimated to range from 2 to 7 [55,118,217,218], implying an etiologic (*i.e.,* preventive) fraction ranging from 50 to 87% in worker populations with high prevalences of ergonomic exposures [219].

Thus, there is adequate scientific knowledge regarding specific occupational ergonomic stressors to prevent a large proportion of MSDs among working people. Preventive measures should not and need not be discriminatory with respect to gender. The best approach to eliminating musculoskeletal injuries from the workplace is the implementation of engineering controls, such as changes in equipment and workstation and job design, in the context of a comprehensive ergonomic program with participation from all levels of the enterprise [15,220,221].

More research is needed to elucidate whether MSD risk varies between women and men in jobs with the same occupational exposures, and whether work-related MSDs have the same outcomes in women and in men. Women often report MSDs more frequently than men; however, this difference appears to be less marked for low back disorders and when men and women are compared within homogenous job groups. Exposure-response relationships have been examined by gender in only a few studies, and the results have not been consistent. However, some studies suggest that men may have higher risk than women with increasing exposure to physical stressors, although women have a higher background risk. This may mean that other factors have a greater effect on women in low-exposure jobs and are less important when there is high physical loading, or because women with higher occupational exposures (physical and/or psychosocial) are more likely than men to leave employment or change jobs due to work-related MSDs. In contrast, the sparse literature suggests that women may both experience different levels of job strain and have increased vulnerability to similar levels of job strain compared with men, possibly because of the added demands of household responsibilities. All of these explanations remain tentative at present and require further study.

Despite the widespread belief that MSDs disproportionately affect women, the outcomes of these conditions have been examined primarily among men. In order to tailor rational interventions for primary or secondary prevention of disability, further study of the prognostic factors among women is an important research priority.

Analysis of gender as a "risk factor" for MSDs, or adjusting for gender differences, does not elucidate these issues. Future epidemiologic studies of MSDs should include subjects of both genders to avoid unnecessary constraints on the available exposure contrasts. The associations of musculoskeletal disorders with gender and occupational ergonomic exposures should be assessed separately in order to determine whether women are at increased risk when exposed to the same ergonomic stressors as men. Gender-stratified presentation of data is valuable because it permits examination, rather than smoothing over, of differences in the exposure-response relationships.

Acknowledgments

Rachel Kidman, for her assistance in compiling and managing the literature search; Karen Messing, for her insightful reading and synthesis of the literature; Ulf Bergqvist, Jeff Katz, Åsa Kilbom, Letitita Davis and Helen Wellman for discussions with LP about these issues.

References

1. Gerr, F., Letz, R., and Landrigan, P. J. (1991). Upper-extremity musculoskeletal disorders of occupational origin. *Ann. Rev. Public Health* **12,** 543–566.
2. Hagberg, M. (1994). Neck and shoulder disorders. *In* "Textbook of Occupational and Environmental Medicine" (L. Rosenstock and M. R. Cullen, eds.), pp. 356–364. Saunders, Philadelphia.
3. Pope, M. H., Andersson, G. B. J., Frymoyer, J. W., and Chaffin, D. B., eds. (1991). "Occupational Low Back Pain: Assessment, Treatment and Prevention," Mosby, St. Louis, MO.
4. Rempel, D. M., Harrison, R. J., and Barnhart, S. (1992). Work-related cumulative trauma disorders of the upper extremity. *JAMA, J. Am. Med. Assoc.* **267,** 838–842.
5. Riihimäki, H. (1995). Back and limb disorders. *In* "Epidemiology of Work Related Diseases" (C. McDonald, ed.), pp. 207–238. BMJ Publishing Group, London.
6. Bernard, B. P., ed. (1997). "Musculoskeletal Disorders and Workplace Factors: A Critical Review of Epidemiologic Evidence for Work-related Musculoskeletal Disorders of the Neck, Upper Extremity, and Low Back." U.S. Department of Health and Human Services, National Institute for Occupational Safety and Health, Cincinnati, OH.
7. Maetzel, A., Mäkelä, M., Hawker, G., and Bombardier, C. (1997). Osteoarthritis of the hip and knee and mechanical occupational exposure—a systematic overview of the evidence. *J. Rheum.* **24,** 1599–1607.
8. Agency for Health Care Policy and Research (1994). "Clinical Practice Guideline: Acute Low Back Problems in Adults." U.S. Department of Health and Human Services, Public Health Service, Agency for Health Care Policy and Research, Rockville, MD.
9. Rempel, D., Evanoff, B., Amadio, P. C. *et al.* (1998). Consensus criteria for the classification of carpal tunnel syndrome in epidemiologic studies. *Am. J. Public Health* **88,** 1447–1451.
10. Badley, E. M., and Ibañez, D. (1994). Socioeconomic risk factors and musculoskeletal disability. *J. Rheum.* **21,** 515–522.
11. Rempel, D. M., and Punnett, L. (1997). Epidemiology of wrist and hand disorders. *In* "Musculoskeletal Disorders in the Workplace: Principles and Practice" (M. Nordin, G. B. Andersson, and M. H. Pope, eds.), pp. 421–430. Mosby, Philadelphia.
12. Yelin, E., Callahan, L. F., and National Arthritis Data Work Group (1995). The economic cost and social and psychological impact of musculoskeletal conditions. *Arthritis Rheum.* **38,** 1351–1367.
13. Sjøgaard, G., Sejersted, O. M., Winkel, J., Smolander, J., Jørgensen, K. and Westgaard, R. H. (1993). Exposure assessment and mechanisms of pathogenesis in work-related musculoskeletal disorders: Significant aspects in the documentation of risk factors. *In* "Work and Health: Scientific Basis of Progress in the Working Environment" (O. Svane and C. Johansen, eds.), pp. 75–87. European Commission, Directorate-General V, Employment, Industrial Relations and Social Affairs, Copenhagen, Denmark.
14. Bureau of Labor Statistics (1996). "Occupational Injuries and Illnesses: Counts, Rates, and Characteristics, 1993." U. S. Department of Labor, Bureau of Labor Statistics, Washington, DC.
15. Westgaard, R. H., and Winkel, J. (1997). Ergonomic intervention research for improved musculoskeletal health: A critical review. *Int. J. Ind. Ergonomics* **20,** 463–500.

16. Cummings, K., Maizlish, N., Rudolph, L., Dervin, K., and Ervin, A. (1989). Occupational disease surveillance: Carpal tunnel syndrome. *Morbid. Mortal. Wkly. Rep.* **38,** 485–489.

17. Tanaka, S., Wild, D. K., Seligman, P. J., Halperin, W. E., Behrens, V. J., and Putz-Anderson, V. (1995). Prevalence and work-relatedness of self-reported carpal tunnel syndrome among U.S. workers: Analysis of the Occupational Health Supplement data of 1988 National Health Interview Survey. *Am. J. Ind. Med.* **27,** 451–470.

18. Brogmus, G. E., Sorock, G. S., and Webster, B. S. (1996). Recent trends in work-related cumulative trauma disorders of the upper extremities in the United States: An evaluation of possible reasons. *J. Occup. Environ. Med.* **38,** 401.

19. Franklin, G. M., Haug, J., Heyer, N., Checkoway, H., and Peck, N. (1991). Occupational carpal tunnel syndrome in Washington State, 1984–1988. *Am. J. Public Health* **81,** 741–746.

20. Webster, B. S., and Snook, S. H. (1994). The cost of compensable upper extremity cumulative trauma disorders. *J. Occup. Med.* **36,** 713–717.

21. Webster, B. S., and Snook, S. H. (1994). The cost of 1989 workers' compensation low back pain claims. *Spine* **19,** 1111–1116.

22. Fine, L. J., Silverstein, B. A., Armstrong, T. J., Anderson, C. A., and Sugano, D. S. (1986). Detection of cumulative trauma disorders of upper extremities in the workplace. *J. Occup. Med.* **28,** 674–678.

23. Nelson, N. A., Park, R. M., Silverstein, M. A., and Mirer, F. E. (1992). Cumulative trauma disorders of the hand and wrist in the auto industry. *Am. J. Public Health* **82,** 1550–1552.

24. Leigh, J. P., Markowitz, S. B., Fahs, M., Shin, C., and Landrigan, P. J. (1997). Occupational injury and illness in the United States: Estimates of costs, morbidity and mortality. *Arch. Intern. Med.* **157,** 1557–1568.

25. Oxenburgh, M. S. (1991). "Increasing Productivity and Profit through Health and Safety." Commerce Clearing House, Chicago.

26. Adams, M. L., Franklin, G. M., and Barnhart, S. (1994). Outcome of carpal tunnel surgery in Washington State workers' compensation. *Am. J. Ind. Med.* **25,** 527–536.

27. Bigos, S. J., Battié, M. C., Spengler, D. M., Fisher, L. D., Fordyce, W. E., Hansson, T. H., Nachemson, A. L., and Wortley, M. D. (1991). A prospective study of work perceptions and psychosocial factors affecting the report of back injury. *Spine* **16,** 1–6.

28. Burdorf, A., Naaktgeboren, B., and Post, W. (1998). Prognostic factors for musculoskeletal sickness absence and return to work among welders and metal workers. *Occup. Environ. Med.* **55,** 490–495.

29. Cotton, P. (1991). Symptoms may return after carpal tunnel surgery. *JAMA, J. Am. Med. Assoc.* **265,** 1922–1925.

30. Ekberg, K., and Wildhagen, I. (1996). Long-term sickness absence due to musculoskeletal disorders: The necessary intervention of work conditions. *Scand. J. Rehabil. Med.* **28,** 39–47.

31. Frymoyer, J. W. (1992). Predicting disability from low back pain. *Clin. Orthop. Relat. Res.* **279,** 101–109.

32. Kulick, M. I., Gordillo, G., Javidi, T., Kilgore, E. S., and Newmeyer, W. L. (1986). Longterm analysis of patients having surgical treatment for carpal tunnel syndrome. *J. Hand Surg.* **11A,** 59–66.

33. Leavitt, F. (1992). The physical exertion factor in compensable work injuries: A hidden flaw in previous research. *Spine* **17,** 307–310.

34. Nathan, P. A., Meadows, K. D., and Keniston, R. C. (1993). Rehabilitation of carpal tunnel surgery patients using a short surgical incision and an early program of physical therapy. *J. Hand Surg.* **18A,** 1044–1050.

35. Walsh, N. E., and Dumitru, D. (1988). The influence of compensation on recovery from low back pain. *Occup. Med.: State of the Art Revs.* **3,** 109–121.

36. Baldwin, M. L., Johnson, W. G., and Butler, R. J. (1996). The error of using return to work to measure the outcomes of health care. *Am. J. Ind. Med.* **29,** 632–641.

37. Reid, J., Ewan, C., and Lowy, E. (1991). Pilgrimage of pain: The illness experiences of women with repetition strain injury and the search for credibility. *Soc. Sci. Med.* **32,** 601–612.

38. Verbrugge, L. M., and Ascione, F. J. (1987). Exploring the iceberg: Common symptoms and how people care for them. *Med. Care* **25,** 539–569.

39. Pransky, G., and Himmelstein, J. (1996). Outcomes research: Implications for occupational health. *Am. J. Ind. Med.* **29,** 573–583.

40. Strunin, L. (1997). The human costs of occupational injuries. *In* "National Occupational Injury Research Symposium". NIOSH, Morgantown, WV.

41. Armstrong, T. J., Buckle, P., Fine, L. J. *et al.* (1993). A conceptual model for work-related neck and upper-limb musculoskeletal disorders. *Scand. J. Work Environ. Health* **19,** 73–84.

42. Gordon, S. L., Fine, L. J. and Blair, S., eds. (1995). "Repetitive Motion Disorders of the Upper Extremity." American Academy of Orthopedic Surgeons, Rosemont, IL.

43. Armstrong, T. J., Fine, L. J., Goldstein, S. A., and Silverstein, B. A. (1987). Ergonomics considerations in hand and wrist tendinitis. *J. Hand Surg.* **12A** (2, Pt. 2), 830–837.

44. Goldstein, S. A., Armstrong, T. J., Chaffin, D. B. *et al.* (1987). Analysis of cumulative strain in tendons and tendon sheaths. *J. Biomech.* **20,** 1–6.

45. Hagberg, M., Silverstein, B., Wells, R., Smith, R., Carayon, P., Hendrick, H., Perusse, M., Kuorinka, I., and Forcier, L., eds. (1995). "Work-related Musculoskeletal Disorders (WMSD): A Handbook for Prevention." Taylor & Francis, London.

46. Karasek, R. A., and Theorell, T. (1990). "Healthy Work. Stress, Productivity and the Reconstruction of Working Life." Basic Books, New York.

47. Schwartz, J. E., Pickering, T. G., and Landsbergis, P. A. (1996). Work-related stress and blood pressure: Current theoretical models and considerations from a behavioral medicine perspective. *J. Occup. Health Psychol.* **1,** 287–310.

48. Wærsted, M., Björklund, R. A., and Westgaard, R. H. (1991). Shoulder muscle tension induced by two VDU-based tasks of different complexity. *Ergonomics* **34,** 137–150.

49. Westgaard, R. H., and Björklund, R. (1987). Generation of muscle tension additional to postural muscle load. *Ergonomics* **30,** 911–923.

50. Leino, P. I., and Hänninen, V. (1995). Psychosocial factors at work in relation to back and limb disorders. *Scand. J. Work Environ. Health* **21,** 134–142.

51. Bongers, P. M., de Winter, C. R., Kompier, M. A. J., and Hildebrandt, V. H. (1993). Psychosocial factors at work and musculoskeletal disease. *Scand. J. Work Environ. Health* **19,** 297–312.

52. Blanc, P. D., Faucett, J., Kennedy, J. J., Cisternas, M., and Yelin, E. (1996). Self-reported carpal tunnel syndrome: Predictors of work disability from the National Health Interview Survey Occupational Health Supplement. *Am. J. Ind. Med.* **30,** 362–368.

53. Chrubasik, S., Junck, H., Zappe, H. A., and Stutzke, O. (1998). A survey on pain complaints and health care utilization in a German population sample. *Eur. J. Anaesthesiol.* **15,** 397–408.

54. Cunningham, L. S., and Kelsey, J. L. (1984). Epidemiology of musculoskeletal impairments and associated disability. *Am. J. Public Health* **74,** 574–579.

55. de Krom, M. C. T. F. M., Kester, A. D. M., Knipschild, P. G., and Spaans, F. (1990). Risk factors for carpal tunnel syndrome. *Am. J. Epidemiol.* **132,** 1102–1110.

56. de Zwart, B. C. H., Broersen, J. P. J., Frings-Dresen, M. H. W., and van Dijk, F. J. H. (1997). Musculoskeletal complaints in the Netherlands in relation to age, gender and physically demanding work. *Int. Arch. Occup. Environ. Health* **70,** 352–360.

57. Derriennic, F., Iwatsubo, Y., Monfort, C., and Cassou, B. (1993). Evolution of osteoarticular disorders as a function of past heavy physical work factors: Longitudinal analysis of 627 retired subjects living in the Paris area. *Br. J. Ind. Med.* **50**, 851–860.

58. English, C. J., Maclaren, W. M., Court-Brown, C., Hughes, S. P. F., Porter, R. W., Wallace, W. A., Graves, R. J., Pethick, A. J., and Soutar, C. A. (1995). Relations between upper limb soft tissue disorders and repetitive movements at work. *Am. J. Ind. Med.* **27**, 75–90.

59. Houtman, I. L. D., Bongers, P. M., Smulders, P. G. W., and Kompier, M. A. J. (1994). Psychosocial stressors at work and musculoskeletal problems. *Scand. J. Work Environ. Health* **20**, 139–145.

60. Mäkelä, M., Heliövaara, M., Sievers, K., Impivaara, O., Knekt, P., and Aromaa, A. (1991). Prevalence, determinants, and consequences of chronic neck pain in Finland. *Am. J. Epidemiol.* **134**, 1356–1367.

61. Petersson, C. J. (1983). Degeneration of the gleno-humeral joint. *Acta Orthop. Scand.* **54**, 277–283.

62. Praemer, A., Furner, S., and Rice, D. P. (1992). "Musculoskeletal Conditions in the United States." Academy of Orthopaedic Surgeons, Park Ridge, IL.

63. Stevens, J. C., Sun, S., Beard, C. M., O'Fallon, W. M., and Kurland, L. T. (1988). Carpal tunnel syndrome in Rochester, Minnesota, 1961 to 1980. *Neurology* **38**, 134–138.

64. Viikari-Juntura, E., Vuori, J., Silverstein, B. A., Kalimo, R., Kuosma, E., and Videman, T. (1991). A life-long prospective study on the role of psychosocial factors in neck-shoulder and low-back pain. *Spine* **16**, 1056–1061.

65. Burchfiel, C. M., Boice, J. A., Stafford, B. A., and Bond, G. G. (1992). Prevalence of back pain and joint problems in a manufacturing company. *J. Occup. Med.* **34**, 129–134.

66. Carey, T., Evans, A., Hadler, N., Lieberman, G., Kalsbeek, W., Jackman, A., Fryer, J., and McNutt, R. (1996). Acute severe low back pain: A population-based study of prevalence and care-seeking. *Spine* **21**, 339–344.

67. Kujala, U. M., Taimela, S., Viljanen, T., Jutila, H., Viitasalo, J. T., Videman, T., and Battié, M. C. (1996). Physical loading and performance as predictors of back pain in healthy adults: A 5-year prospective study. *Eur. J. Appl. Physiol.* **73**, 452–458.

68. Manninen, P., Riihimäki, H., and Heliövaara, M. (1996). Has musculoskeletal pain become less prevalent? *Scand. J. Rheumatol.* **25**, 37–41.

69. Andersen, J. H., and Gaardboe, O. (1993). Prevalence of persistent neck and upper limb pain in a historical cohort of sewing machine workers. *Am. J. Ind. Med.* **24**, 677–687.

70. Brisson, C., Vinet, A., and Vezina, M. (1989). Disability among female garment workers: A comparison with a national sample. *Scand. J. Work Environ. Health* **15**, 323–328.

71. Leclerc, A., Franchi, P., Cristofari, M. F., Delemotte, B., Mereau, P., Teyssier-Cotte, C., Touranchet, A., and Study Group on Repetitive Work (1998). Carpal tunnel syndrome and work organisation in repetitive work: A cross sectional study in France. *Occup. Environ. Med.* **55**, 180–187.

72. Punnett, L., Robins, J. M., Wegman, D. H., and Keyserling, W. M. (1985). Soft tissue disorders in the upper limbs of female garment workers. *Scand. J. Work Environ. Health* **11**, 417–425.

73. Schibye, B., Skov, T., Ekner, D., Christiansen, J. U., and Sjøgaard, G. (1995). Musculoskeletal symptoms among sewing machine operators. *Scand. J. Work Environ. Health* **21**, 427–434.

74. Andersen, J. H., and Gaardboe, O. (1993). Musculoskeletal disorders of the neck and upper limb pain among sewing machine operators: A clinical investigation. *Am. J. Ind. Med.* **24**, 689–700.

75. Blåder, S., Barck-Holst, U., Danielsson, S., Ferhm, E., Kalpamaa, M., Leijon, M., Lindh, M., and Markhede, G. (1991). Neck and shoulder complaints among sewing-machine operators: A study concerning frequency, symptomatology and dysfunction. *Appl. Ergonomics* **22**, 251–257.

76. Brisson, C., Vinet, A., Vezina, M., and Gingras, S. (1989). Effect of duration of employment in piecework on severe disability among female garment workers. *Scand. J. Work Environ. Health* **15**, 329–334.

77. Punnett, L. (1996). Adjusting for the healthy worker selection effect in cross-sectional studies. *Int. J. Epidemiol.* **25**, 1068–1075; **26**, 914.

78. Luopajärvi, T., Kuorinka, I., Virolainen, M., and Holmberg, M. (1979). Prevalence of tenosynovitis and other injuries of the upper extremities in repetitive work. *Scand. J. Work Environ. Health* **5**(Suppl. 3), 48–55.

79. Hadler, N. M., Gillings, D. B., Imbus, H. R., Levitin, P. M., Makuc, D., Utsinger, P. D., Yount, W. J., Slusser, D., and Moskovitz, N. (1978). Hand structure and function in an industrial setting. Influence of three patterns of stereotyped repetitive usage. *Arthritis Rheum.* **21**, 210–220.

80. Koskela, R.-S., Klockars, M., and Järvinen, E. (1990). Mortality and disability among cotton mill workers. *Br. J. Ind. Med.* **47**, 384–391.

81. Kilbom, Å., Persson, J., and Jonsson, B. G. (1986). Disorders of the cervicobrachial region among female workers in the electronics industry. *Int. J. Ind. Ergonomics* **1**, 37–47.

82. Jonsson, B. G., Persson, J., and Kilbom, Å. (1988). Disorders of the cervicobrachial region among female workers in the electronics industry: A two-year follow up. *Int. J. Ind. Ergonomics* **3**, 1–12.

83. Ohlsson, K., Attewell, R., and Skerfving, S. (1989). Self-reported symptoms in the neck and upper limbs of female assembly workers: Impact of length of employment, work pace, and selection. *Scand. J. Work Environ. Health* **15**, 75–80.

84. Schottland, J. R., Kirschberg, G. J., Fillingim, R., Davis, V. P., and Hogg, F. (1991). Median nerve latencies in poultry processing workers: An approach to resolving the role of industrial "cumulative trauma" in the development of carpal tunnel syndrome. *J. Occup. Med.* **33**, 627–631.

85. Hales, T. R., and Fine, L. J. (1989). "Health Hazard Evaluation Report: Cargill Poultry Division, Buena Vista GA." National Institute for Occupational Safety and Health, Cincinnati, OH.

86. Chiang, H.-C., Ko, Y.-C., Chen, S.-S., Yu, H.-S., Wu, T. N., and Chang, P.-Y. (1993). Prevalence of shoulder and upper-limb disorders among workers in the fish-processing industry. *Scand. J. Work Environ. Health* **19**, 126–131.

87. Messing, K., and Reveret, J. P. (1983). Are women in female jobs for their health? A study of working conditions and health effects in the fish-processing industry in Quebec. *Int. J. Health Serv.* **13**, 635–647.

88. Ohlsson, K., Hansson, G., Balogh, I., Strömberg, U., Pålsson, B., Nordander, C. *et al.* (1994). Disorders of the neck and upper limbs in women in the fish processing industry. *Occup. Environ. Med.* **51**, 826–832.

89. Olafsdóttir, H., and Rafnsson, V. (1995). New technique increases the prevalence of musculoskeletal symptoms from upper limbs among women in fish fillet plants. "2nd International Scientific Conference on Prevention of Work-Related Musculoskeletal Disorders (PREMUS 95)." Montréal, Québec.

90. Ahlberg-Hultén, G. K., Theorell, T., and Sigala, F. (1995). Social support, job strain and musculoskeletal pain among female health care personnel. *Scand. J. Work Environ. Health* **21**, 435–439.

91. Engkvist, I.-L., Hagberg, M., Lindén, A., and Malker, B. (1992). Over-exertion back accidents among nurses' aides in Sweden. *Saf. Sci* **15**, 97–108.

92. Estryn Behar, M., Kaminski, M., Peigne, E., Maillard, M. F., Pelletier, A., Berthier, C. *et al.* (1990). Strenuous working condi-

tions and musculo-skeletal disorders among female hospital workers. *Int. Arch. Occup. Environ. Health* **62,** 47–57.

93. Jensen, R. C. (1990). Back injuries among nursing personnel related to exposure. *Appl. Occup. Environ. Hyg.* **5,** 38–45.

94. Ono, Y., Lagerström, M., Hagberg, M., Lindén, A., and Malker, B. (1995). Reports of work related musculoskeletal injury among home care service workers compared with nursery school workers and the general population of employed women in Sweden. *Occup. Environ. Med.* **52,** 686–693.

95. Punnett, L. (1987). Upper extremity musculoskeletal disorders in hospital workers. *J. Hand Surg.* **12A**(2. Pt. 2), 858–862.

96. Osorio, A. M., Ames, R. G., Jones, J., Castorina, J., Rempel, D., Estrin, W. *et al.* (1994). Carpal tunnel syndrome among grocery store workers. *Am. J. Ind. Med.* **25,** 229–245.

97. Vézina, N., Geoffrion, L., Chatigny, C., and Messing, K. (1994). A manual materials handling job: Symptoms and working conditions among supermarket cashiers. *Chron. Dis. Can.* **15,** 17–22.

98. Morgenstern, H., Kelsh, M., Kraus, J., and Margolis, W. (1991). A cross-sectional study of hand/wrist symptoms in female grocery checkers. *Am. J. Ind. Med.* **20,** 209–218.

99. Messing, K., Chatigny, C., and Courville, J. (1998). 'Light' and 'heavy' work in the housekeeping service of a hospital. *Appl. Ergonomics* **29,** 451–459.

100. Pierre-Jerome, C., Bekkelund, S. I., Mellgren, S. I., and Torbergsen, T. (1996). Quantitative magnetic resonance imaging and the electrophysiology of the carpal tunnel region in floor cleaners. *Scand. J. Work Environ. Health* **22,** 119–123.

101. Stock, S. R. (1995). A study of musculoskeletal symptoms in daycare workers. *In* "Invisible: Issues in Women's Occupational Health/Invisible: La santé des travailleuses" (K. Messing, B. Neis, and L. Dumais, eds.), pp. 62–74. Gynergy Books, Charlottetown, Quebec Canada.

102. Ono, Y., Nakamura, R., Shimaoka, M., Hiruta, S., Hattori, Y., Ichihara, G., Kamijima, M., and Takeuchi, Y. (1998). Epicondylitis among cooks in nursery schools. *Occ. Environ. Med.* **55,** 172–179.

103. Camerino, D., Lavano, P., Ferrario, M., Ferretti, G., and Molteni, G. (1995). Musculoskeletal disorders, working posture, psychosocial environment in female VDU operators and conventional office workers. *In* "Work with Display Units 94. Selected Papers of the Fourth International Scientific Conference on Work with Display Units, Milan, Italy, 2–5 October, 1994" (A. Grieco, G. Molteni, B. Piccoli, and E. Occhipinti, eds.), pp. 67–72. Elsevier, Amsterdam.

104. Marcus, M., and Gerr, F. (1996). Upper extremity musculoskeletal symptoms among female office workers: Associations with video display terminal use and occupational psychosocial stressors. *Am. J. Ind. Med.* **29,** 161–170.

105. Murata, K., Araki, S., Okajima, F., and Saito, Y. (1996). Subclinical impairment in the median nerve across the carpal tunnel among female VDT operators. *Int. Arch. Occup. Environ. Health* **68,** 75–79.

106. Stellman, J., Klitzman, S., Gordon, G. C., and Snow, B. R. (1987). Work environment and the well-being of clerical and VDT workers. *J. Occup. Behav.* **8,** 95–114.

107. Sauter, S. L., Schleiffer, L. M., and Knutson, S. J. (1991). Work posture, workstation design, and musculoskeletal discomfort in a VDT data entry task. *Hum. Factors* **33,** 151–167.

108. Oxenburgh, M. S. (1987). Musculoskeletal injuries occurring in word processor operators. *In* "Readings in RSI: The Ergonomic Approach to Repetition Strain Injuries" (M. Stevenson, ed.), pp. 91–95. New South Wales University Press, Kensington, NSW, Australia.

109. Ferreira, J. M., Conceicao, G. M., and Saldiva, P. H. (1997). Work organization is significantly associated with upper extremities

110. musculoskeletal disorders among employees engaged in interactive computer-telephone tasks of an international bank subsidiary in Sao Paulo, Brazil. *Am. J. Ind. Med.* **31,** 468–473.

110. Punnett, L. (1998). Ergonomic stressors and upper extremity disorders in vehicle manufacturing: Cross-sectional exposure-response trends. *Occup. Environ. Med.* **55,** 414–420.

111. Schierhout, G. H., Myers, J. E., and Bridger, R. S. (1993). Musculoskeletal pain and workplace ergonomic stressors in manufacturing industry in South Africa. *Int. J. Ind. Ergonomics* **12,** 3–11.

112. Bergqvist, U., Wolgast, E., Nilsson, B., and Voss, M. (1995). Musculoskeletal disorders among visual display terminal workers; individual, ergonomic and work organizational factors. *Ergonomics* **38,** 763–776.

113. Bernard, B. P., Sauter, S., Fine, L. J., Petersen, M., and Hales, T. (1994). Job task and psychosocial risk factors for work-related musculoskeletal disorders among newspaper employees. *Scand. J. Work Environ. Health* **20,** 417–426.

114. Hales, T. R., Sauter, S. L., Peterson, M. R., Putz-Anderson, V., Fine, L. J., Ochs, T. T., Schleifer, L. R., and Bernard, B. P. (1992). "Health Hazard Evaluation Report: U.S. West Communications, Phoenix AZ, Minneapolis/St Paul MN, Denver CO." National Institute for Occupational Safety and Health, Cincinnati, OH.

115. McCormack, R. R., Inman, R. D., Wells, A., Berntsen, C., and Imbus, H. R. (1990). Prevalence of tendinitis and related disorders of the upper extremity in a manufacturing workforce. *J. Rheum.* **17,** 958–964.

116. Viikari-Juntura, E. (1983). Neck and upper limb disorders among slaughterhouse workers. *Scand. J. Work Environ. Health* **9,** 283–290.

117. Silverstein, B. A., Fine, L. J., and Armstrong, T. J. (1986). Hand wrist cumulative trauma disorders in industry. *Br. J. Ind. Med.* **43,** 779–784.

118. Silverstein, B. A., Fine, L. J., and Armstrong, T. J. (1987). Occupational factors and carpal tunnel syndrome. *Am. J. Ind. Med.* **11,** 343–358.

119. Nathan, P. A., Keniston, R. C., Myers, L. D., and Meadows, K. D. (1992). Longitudinal study of median nerve sensory conduction in industry: Relationship to age, gender, hand dominance, occupational hand use, and clinical diagnosis. *J. Hand Surg.* **17A,** 850–857.

120. Ekberg, K., Björkqvist, B., Malm, P. *et al.* (1994). Case-control study of risk factors for disease in the neck and shoulder area. *Occup. Environ. Med.* **51,** 262–266.

121. Park, R. M., Krebs, J. M. and Mirer, F. E. (1996). Occupational disease surveillance using disability insurance at an automotive stamping and assembly complex. *J. Occup. Environ. Med.* **38,** 1111–23.

122. Park, R. M., Nelson, N. A., Silverstein, M. A., and Mirer, F. E. (1992). Use of medical insurance claims for surveillance of occupational disease: An analysis of cumulative trauma in the auto industry. *J. Occup. Med.* **34,** 731–737.

123. Cheadle, A., Franklin, G., Wolfhagen, C., Savarino, J., Liu, P. Y., Salley, C., and Weaver, M. (1994). Factors influencing the duration of work-related disability: A population-based study of Washington State workers' compensation. *Am. J. Public Health* **84.**

124. Feeney, A., North, F., Head, J., Canner, R., and Marmot, M. (1998). Socioeconomic and sex differentials in reason for sickness absence from the Whitehall II study. *Occup. Environ. Med.* **55,** 91–98.

125. Feuerstein, M., Berkowitz, S. M., and Peck, C. A. (1997). Musculoskeletal-related disability in U.S. army personnel: Prevalence, gender, and military occupational specialties. *J. Occup. Environ. Med.* **39,** 68–78.

126. Hashemi, L., Webster, B. S., Clancy, E. A., and Courtney, T. K. (1998). Length of disability and cost of work-related musculoskeletal disorders of the upper extremity. *J. Occup. Environ. Med.* **40,** 261–269.

127. Katz, J. N., Keller, R. B., Fossel, A. H., Punnett, L., Bessette, L., Simmons, B. P., and Mooney, N. (1997). Predictors of return to work following carpal tunnel release. *Am. J. Ind. Med.* **31,** 85–91.

128. Brage, S., Nygård, J. F., and Tellnes, G. (1998). The gender gap in musculoskeletal-related long term sickness absence in Norway. *Scand. J. Soc. Med.* **26,** 34–43.

129. Jensen, I. B., Nygren, A., and Lundin, A. (1994). Cognitive-behavioural treatment for workers with chronic spinal pain: a matched and controlled cohort study in Sweden. *Occup. Environ. Med.* **51,** 145–151.

130. Yelin, E., Nevitt, M., and Epstein, W. (1980). Toward an epidemiology of work disability. *Epidemiol. Work Disabil.* **58,** 386–415.

131. Levenstein, C. (1999). Economic losses from repetitive strain injuries. *In* "Occupational Medicine: State of the Art Reviews" (M. Cherniack, ed.), pp. 149–161. Hanley & Belfus, Philadelphia.

132. Dembe, A. E. (1997). The history of carpal tunnel syndrome. *New Solutions* **7,** 15–22.

133. Torgén, M., and Kilbom, Å. (1997). The REBUS study: Physical work load between 18 and 58 years of age—does it change? *In* "Proceedings of the 13th Triennial Congress of the International Ergonomics Association," pp. 544–546. Finnish Institute of Occupational Health, Tampere, Finland.

134. Torgén, M., Fredriksson, K., Bildt Thorbjörnsson, C., Alfredsson, L., Vingård, E., and Kilbom, Å. (1998). Age in relation to physical and psychosocial exposures, physical capacity and health in 484 Swedish middle-aged men and women: The REBUS study. *In* "Second International ICOH Conference on Aging and Work—Work Ability of Elderly Workers: A Challenge for Occupational Health," p. 17. The Danish Working Environment Fund, Elsinore, Denmark.

135. Barnhart, S., Demers, P. A., Miller, M., Longstreth, W. T. J., and Rosenstock, L. (1991). Carpal tunnel syndrome among ski manufacturing workers. *Scand. J. Work Environ. Health* **17,** 46–52.

136. Courville, J., Dumais, L., and Vézina, N. (1994). Conditions de travail de femmes et d'hommes sur une chaîne de découpe de volaille et développement d'atteintes musculo-squelettiques (Working conditions and development of musculoskeletal disorders among women and men on a poultry cutting line). *Trav. Santé* **10,** S17–S23.

137. Fransson Hall, C., Byström, S., and Kilbom, Å. (1995). Self-reported physical exposure and musculoskeletal symptoms of the forearm-hand among automobile assembly line workers. *J. Occup. Environ. Med.* **37,** 1136–1144.

138. Kiken, S., Stringer, W., Fine, L., Sinks, T., and Tanaka, S. (1990). "Health Hazard Evaluation Report: Perdue Farms, Inc., Lewiston NC, Roberson NC." National Institute for Occupational Safety and Health, Cincinnati, OH.

139. Mergler, D., Brabant, C., Vézina, N., and Messing, K. (1987). The weaker sex? Men in women's working conditions report similar health symptoms. *J. Occup. Med.* **29,** 417–421.

140. Evans, J. (1987). Women, men, VDU work and health: a questionnaire survey of British VDU operators. *Work Stress* **1,** 271–283.

141. Messing, K., Dumais, L., Courville, J., Seifert, A. M., and Boucher, M. (1994). Evaluation of exposure data from men and women with the same job title. *J. Occup. Med.* **36,** 913–917.

142. Messing, K., Tissot, F., Saurel-Cubizolles, M., Kaminski, M., and Bourgine, M. (1998). Sex as a variable can be a surrogate for some working conditions. *J. Occup. Environ. Med.* **40,** 250–260.

143. Hall, E. M. (1990). "Women's Work: An Inquiry into the Health Effects of Invisible and Visible Labor." Ph.D. thesis, Division of Behavioral Sciences and Health Education, Health Policy and Management Department, John Hopkins University School of Hygiene and Public Health, Baltimore, Maryland; and Karolinska Institute, Stockholm.

144. Roxburg, S. (1996). Gender differences in work and well-being: Effects of exposure and vulnerability. *J. Health Soc. Behav.* **37,** 265–277.

145. Zetterberg, C., Forsberg, A., Hansson, E., Johansson, H., Nielsen, P., Danielsson, B., Inge, G., and Olsson, B.-M. (1997). Neck and upper extremity problems in car assembly workers: A comparison of subjective complaints, work satisfaction, physical examination and gender. *Int. J. Ind. Ergonomics* **19,** 277–289.

146. Martin, B. J., Armstrong, T. J., Foulke, J. A., Natarajan, S., Klinenberg, E., Serina, E., and Rempel, D. (1996). Keyboard reaction force and finger flexor electromyograms during computer keyboard work. *Hum. Factors* **38,** 654–664.

147. Sjøgaard, G., and Søgaard, K. (1998). Muscle injury in repetitive motion disorders. *Clin. Orthop. Relat. Res.* **351,** 21–31.

148. Chaffin, D. B., and Andersson, G. B. J. (1991). "Occupational Biomechanics." Wiley, New York.

149. Chatigny, C., Seifert, A. M., and Messing, K. (1995). Repetitive strain in nonrepetitive work: A case study. *Int. J. Occup. Saf. Ergonomics* **1,** 42–51.

150. Courville, J., Vézina, N., and Messing, K. (1991). Comparison of the work activity of two mechanics: A woman and a man. *Int. J. Ind. Ergonomics* **7,** 163–174.

151. Courville, J., Vézina, N., and Messing, K. (1992). Analyse des facteurs pouvant entraîner l'exclusion des femmes du tri des colis postaux (Ergonomic factors which may exclude women from package sorting jobs). *Trav. Hum.* **55,** 119–134.

152. Goldenhar, L. M., and Sweeney, M. H. (1996). Tradeswomen's perspectives on occupational health and safety: A qualitative investigation. *Am. J. Ind. Med.* **29,** 516–520.

153. Green, R. A., and Briggs, C. A. (1989). Anthropometric dimensions and overuse injury among Australian keyboard operators. *J. Occup. Med.* **31,** 747–750.

154. Messing, K., Courville, J., Boucher, M., Dumais, L., and Seifert, A. M. (1994). Can safety risks of blue-collar jobs be compared by gender? *Saf. Sci.* **18,** 95–112.

155. Stevenson, J. M., Greenhorn, D. R., Bryant, J. T., Deakin, J. M., and Smith, J. T. (1996). Gender differences in performance of a selection test using the incremental lifting machine. *Appl. Ergonomics* **27,** 45–52.

156. Alexanderson, K., Leijon, M., Åkerlind, I., Rydh, H., and Bjurulf, P. (1994). Epidemiology of sickness absence in a Swedish county in 1985, 1986 and 1987. *Scand. J. Soc. Med.* **22,** 27–34.

157. Leijon, M., Hensing, G., and Alexanderson, K. (1999). Sickness absence due to musculoskeletal disorders in different occupational groups—the role of gender segregation. *Eur. J. Public Health* (to be published).

158. Byström, S., Fransson Hall, C., Welander, T., and Kilbom, Å. (1995). Clinical disorders and pressure-pain threshold of the forearm and hand among automobile assembly line workers. *J. Hand Surg. (Br. Eur. Vol.)* **20B,** 782–90.

159. Bildt Thorbjörnsson, C. O., Alfredsson, L., Fredriksson, K., Köster, M., Michélsen, H., Vingård, E., Torgén, M., and Kilbom, Å. (1998). Psychosocial and physical risk factors associated with low back pain: A 24 year follow up among women and men in a broad range of occupations. *Occup. Environ. Med.* **55,** 84–90.

160. Macfarlane, G. J., Thomas, E., Papageorgiou, A. C., Croft, P. R., Jayson, M. I. V., and Silman, A. J. (1997). Employment and physical work activities as predictors of future low back pain. *Spine* **22,** 1143–1149.

161. Felson, D. T., Hannan, M. T., Naimark, A., Berkeley, J., Gordon, G., Wilson, P. W. F., and Anderson, J. (1991). Occupational physical demands, knee bending, and knee osteoarthritis: Results from the Framingham Study. *J. Rheum.* **18,** 1587–1592.

162. Anderson, J. J., and Felson, D. (1988). Factors associated with OA of the knee in the first HANES. *Am. J. Epidemiol.* **128,** 179–189.

163. Imeokparia, R. L., Barrett, J. P., Arrietta, M. I., Leaverton, P. E., Wilson, A. A., Hall, B. J., and Marlowe, S. M. (1994). Physical activity as a risk factor for osteoarthritis of the knee. *Ann. Epidemiol.* **4**, 221–230.

164. Bergqvist, U. (1995). Visual display terminal work—a perspective on long-term changes and discomforts. *Int. J. Ind. Ergonomics* **16**, 201–209.

165. Nelson, N. A., and Silverstein, B. A. (1998). Workplace changes associated with a reduction in musculoskeletal symptoms in office workers. *Hum. Factors* **40**, 337–350.

166. Leino, P. I., Hasan, J., and Karppi, S. L. (1988). Occupational class, physical workload, and musculoskeletal morbidity in the engineering industry. *Br. J. Ind. Med.* **45**, 672–681.

167. Punnett, L. (1997). Upper extremity disorders among male and female automotive manufacturing workers. *In* "Proceedings of the 13th Triennial Congress of the International Ergonomics Association," pp. 512–514. Finnish Institute of Occupational Health, Tampere, Finland.

168. Punnett, L. (1998). Comparison of analytic methods for a follow-up study of upper extremity disorders. *In* "Third International Scientific Conference on Prevention of Work-Related Musculoskeletal Disorders/International Symposium on Epidemiology in Occupational Health (PREMUS-ISEOH 98)." Finnish Institute of Occupational Health, Helsinki, Finland.

169. Östlin, P. (1988). Negative health selection into physically light occupations. *J. Epidemiol. Commun. Health* **42**, 152–156.

170. Johansson, J. Å. (1995). The impact of decision latitude, psychological load and social support at work on musculoskeletal symptoms. *Eur. J. Public Health* **5**, 169–174.

171. Theorell, T., Harms-Ringdahl, K., Ahlberg-Hultén, G., and Westin, B. (1991). Psychosocial job factors and symptoms from the locomotor system—A multicausal analysis. *Scand. J. Rehabil. Med.* **23**, 165–173.

172. Karasek, R. A. (1985). Job Content Questionnaire and User's Guide, Department of Work Environment, University of Massachusetts Lowell, Lowell, MA.

173. Hall, E. M. (1992). Double exposure: The combined impact of the home and work environments on psychosomatic strain in Swedish women and men. *Int. J. Health Serv.* **22**, 239–260.

174. Romito, P., and Hovelacque, F. (1987). Changing approaches in women's health: New insights and new pitfalls. *Int. J. Health Serv.* **17**, 241–258.

175. Cleland, L. (1987). "RSI:" A model of social iatrogenesis. *Med. J. Aust.* **147**, 236–239.

176. Punnett, L., and Bergqvist, U. (1997). "Visual Display Unit Work and Upper Extremity Musculoskeletal Disorders. A Review of Epidemiological Findings." National Institute of Working Life, Solna, Sweden.

177. Katz, J. N., Punnett, L., Simmons, B. P., Fossell, A. H., Mooney, N., and Keller, R. B. (1996). Workers' compensation recipients with carpal tunnel syndrome: The validity of self-reported health measures. *Am. J. Public Health* **86**, 52–56.

178. Harreby, M., Kjer, J., Hesselsoe, G., and Neergaard, K. (1996). Epidemiological aspects and risk factors for low back pain in 38-year-old men and women: A 25-year prospective cohort study of 640 school children. *Eur. Spine J.* **5**, 312–318.

179. Hurwitz, E. L., and Morgenstern, H. (1997). The effects of comorbidity and other factors on medical versus chiropractic care for back problems. *Spine* **22**, 2254–2263.

180. Von Korff, M., Wagner, E. H., Dworkin, S. F., and Saunders, K. W. (1991). Chronic pain and use of ambulatory health care. *Psychosom. Med.* **53**, 61–79.

181. Bush, F. M., Harkins, S. W., Harrington, W. G., and Price, D. D. (1993). Analysis of gender effects on pain perception and symptom presentation in temporomandibular pain. *Pain* **53**, 73–80.

182. Karlson, E. W., Daltroy, L. H., Liang, M. H., Eaton, H. E., and Katz, J. N. (1997). Gender differences in patient preferences may underlie differential utilization of elective surgery. *Am. J. Med.* **102**, 524–530.

183. Hogeweg, J. A., Langereis, M. J., Bernards, A. T. M., Faber, J. A. J., and Helders, P. J. M. (1992). Algometry: Measuring pain threshold, method and characteristics of healthy subjects. *Scand. J. Rehabil. Med.* **24**, 99–103.

184. Takala, E.-P. (1990). Pressure pain threshold on upper trapezius and levator scapulae muscles. *Scand. J. Rehabil. Med.* **22**, 63–68.

185. Theorell, T., Nordemar, R., Michelsén, H., and Stockholm MUSIC I Study Group (1993). Pain thresholds during standardized psychological stress in relation to perceived psychosocial work situation. *J. Psychosom. Res.* **37**, 299–305.

186. Werner, R. A., Franzblau, A., Albers, J. W., and Armstrong, T. J. (1998). Median mononeuropathy among active workers: Are there differences between symptomatic and asymptomatic workers? *Am. J. Ind. Med.* **33**, 374–378.

187. Bohannon, R. W. (1997). Reference values for extremity muscle strength obtained by hand-held dynamometry from adults aged 20–79 years. *Arch. Phys. Med. Rehabil.* **78**, 26–32.

188. Messing, K. and Stevenson, J. (1996). Women in Procrustean beds: Strength testing and the workplace. *Gender, Work Organ.* **3**, 156–167.

189. Kilbom, Å. (1988). Isometric strength and occupational muscle disorders. *Eur. J. Appl. Physiol.* **57**, 322–326.

190. Barnekow-Bergkvist, M., Hedberg, G. E., Janlert, U., and Jansson, E. (1998). Determinants of self-reported neck-shoulder and low back symptoms in a general population. *Spine* **23**, 235–243.

191. Biering-Sørensen, F. (1984). Physical measurements as risk indicators for low-back trouble over a one-year period. *Spine* **9**, 106–118.

192. Leino, P., Aro, S., and Hasan, J. (1987). Trunk muscle function and low back disorders: A ten-year follow-up study. *J. Chron. Dis.* **40**, 289–296.

193. Keyserling, W. M., Herrin, G. D., and Chaffin, D. B. (1980). Isometric strength testing as a means of controlling medical incidents on strenuous jobs. *J. Occup. Med.* **22**, 332–336.

194. Hägg, G. (1991). Static work loads and occupational myalgia—a new explanation model. *In* "Electromyographical Kinesiology" (P. A. Anderson, D. J. Hobart, and J. V. Danoff, eds.), pp. 141–144. Elsevier, Amsterdam.

195. Hägg, G. M., and Åström, A. (1997). Load pattern and pressure pain threshold in the upper trapezius muscle and psychosocial factors in medical secretaries with and without shoulder/neck disorders. *Int. Arch. Occup. Environ. Health* **69**, 423–432.

196. Nygaard, E., and Hede, K. (1987). Physiological profiles of the male and the female. *In* "Exercise: Benefits, Limits and Adaptations" (D. Macleod *et al.*, eds.), pp. 289–307. Spon, London.

197. Søgaard, K., Christensen, H., Fallentin, N., Mizuno, M., Quistorff, B., and Sjøgaard, G. (1998). Motor unit activation patterns during concentric wrist flexion in humans with different muscle fibre composition. *Eur. J. Appl. Physiol.* **78**, 411–416.

198. Stevenson, J. M. (1995). Gender-fair employment practices: Developing employee selection tests. *In* "Invisible: Issues in Women's Occupational Health/Invisible: La santé des travailleuses" (K. Messing, B. Neis, and L. Dumais, eds.), pp. 306–320. Gynergy Books, Charlottetown, P.E.I., Canada.

199. Åstrand, P.-O., and Rodahl, K. (1977). "Textbook of Work Physiology." McGraw-Hill, New York.

200. Dvorak, J., Antinnes, J. A., Panjabi, M., Loustalot, D., and Bonomo, M. (1992). Age and gender related normal motion of the cervical spine. *Spine* **10**, S393–S398.

201. Hart, D. A., Archambault, J. M., Kydd, A., Reno, C., Frank, C. B., and Herzog, W. (1998). Gender and neurogenic variables in ten-

don biology and repetitive motion disorders. *Clin. Orthop. Relat. Res.* **351**, 44–56.

202. Pacifici, R., Rifas, L., McCracken, R., Vered, I., McMurtry, C., Avioli, L. V., and Peck, W. A. (1989). Ovarian steroid treatment blocks a postmenopausal increase in blood monocyte interleukin 1 release. *Proc. Nat. Acad. Sci. U.S.A.* **86**, 2398–2402.

203. Da Silva, J. A. P., Larbre, J., Seed, M. P., Cutolo, M., Villaggio, B., Ascott, D. L., and Willoughby, D. A. (1994). Sex differences in inflammation induces cartilage damage in rodents: the influence of sex steroids. *J. Rheum.* **21**, 330–337.

204. Hart, D. A., Frank, C. B., and Bray, R. C. (1995). Inflammatory processes in repetitive motion and overuse syndrome: potential role of neurogenic mechanisms in tendons and ligaments. *In* "Repetitive Motion Disorders of the Upper Extremity" (S. L. Gordon, L. J. Fine, and S. Blair, eds.), pp. 247–262. American Academy of Orthopedic Surgeons, Rosemont, IL.

205. Cannon, L. J., Bernacki, E. J., and Walter, S. D. (1981). Personal and occupational factors associated with carpal tunnel syndrome. *J. Occup. Med.* **23**, 255–258.

206. Adera, T., Deyo, R. A., and Donatelle, R. J. (1994). Premature menopause and low back pain: A population-based study. *Ann. Epidemiol.* **4**, 416–422.

207. Svensson, H.-O., Andersson, G. B. J., Hagstad, A., and Jansson, P.-O. (1990). The relationship of low-back pain to pregnancy and gynecologic factors. *Spine* **15**, 371–375.

208. Wreje, U., Isacsson, D., and Aberg, H. (1997). Oral contraceptives and back pain in women in a Swedish community. *Int. J. Epidemiol.* **26**, 71–74.

209. Paul, J. A., and Frings-Dresen, M. H. W. (1994). Standing working posture compared in pregnant and non-pregnant conditions. *Ergonomics* **37**, 1563–1575.

210. Masten, W. Y., and Smith, J. L. (1988). Reaction time and strength in pregnant and nonpregnant employed women. *J. Occup. Med.* **30**, 451–456.

211. Kelsey, J. L., Greenberg, R. A., Hardy, R. J., and Johnson, M. F. (1975). Pregnancy and the syndrome of herniated lumbar intervertebral disc; an epidemiological study. *Yale J. Biol. Med.* **48**, 361–368.

212. Dumas, G. A., Reid, J. G., Wolfe, L. A., Griffin, M. P., and McGrath, M. J. (1995). Exercise, posture, and back pain during pregnancy: Part 1. Exercise and posture. *Clin. Biomech.* **10**, 98–103.

213. Fast, A., Weiss, L., Ducommun, E. J., Medina, E., and Butler, J. G. (1989). Low-back pain in pregnancy: Abdominal muscles, sit-up performance, and back pain. *Spine* **15**, 28–30.

214. Ekman-Ordeberg, G., Salgeback, S., and Ordeberg, G. (1987). Carpal tunnel syndrome in pregnancy: a prospective study. *Acta Obstet. Gynecol. Scand.* **66**, 233–235.

215. Wand, J. S. (1990). Carpal tunnel syndrome in pregnancy and lactation. *J. Hand Surg.* **15B**, 93–95.

216. Voitk, A. J., Mueller, J. C., Farlinger, D. E., and Johnston, R. U. (1983). Carpal tunnel syndrome in pregnancy. *Can. Med. Assoc. J.* **128**, 277–279.

217. Nilsson, T., Hagberg, M., Burström, L., and Kihlberg, S. (1994). Impaired nerve conduction in the carpal tunnel of platers and truck assemblers exposed to hand-arm vibration. *Scand. J. Work Environ. Health* **20**, 189–99.

218. Wieslander, G., Norbäck, D., Göthe, C.-J., and Juhlin, L. (1989). Carpal tunnel syndrome (CTS) and exposure to vibration, repetitive wrist movements, and heavy manual work: A case-referent study. *Br. J. Ind. Med.* **46**, 43–47.

219. Hagberg, M., Morgenstern, H., and Kelsh, M. (1992). Impact of occupations and job tasks on the prevalence of carpal tunnel syndrome. *Scand. J. Work Environ. Health* **18**, 337–345.

220. Cohen, A. L., Gjessing, C. C., Fine, L. J., Bernard, B. P., and McGlothin, J. D. (1997). "Elements of Ergonomics Programs: A Primer Based on Workplace Evaluations of Musculoskeletal Disorders." National Institute for Occupational Safety and Health, Cincinnati, OH.

221. Grant, K., and Habes, D. (1995). Summary of studies on the effectiveness of ergonomic interventions. *Appl. Occup. Environ. Hyg.* **10**, 523–530.

222. Aronsson, G., Bergqvist, U., and Almers, S. (1992). "Arbetsorganisation och muskuloskeletala besvär vid bildskärmsarbete (Work organization and musculoskeletal discomforts in VDT work, in Swedish)." National Institute of Working Life, Solna Sweden.

223. Burt, S., Hornung, R., and Fine, L. J. (1990). "Health Hazard Evaluation Report: Newsday Inc., Melville, NY." National Institute for Occupational Safety and Health, Cincinnati, OH.

224. Polanyi, M. F. D., Cole, D. C., Beaton, D. E., *et al.* (1997). Upper limb work-related musculoskeletal disorders among newspaper employees: Cross-sectional survey results. *Am. J. Ind. Med.* **32**, 620–628.

225. Stetson, D. S., Silverstein, B. A., Keyserling, W. M., Wolfe, R. A., and Albers, J. W. (1993). Median sensory distal amplitude and latency: Comparisons between nonexposed managerial/professional employees and industrial workers. *Am. J. Ind. Med.* **24**, 175–189.

39

Occupational Cancer

SHELIA HOAR ZAHM, MARY H. WARD, AND DEBRA T. SILVERMAN
Division of Cancer Epidemiology and Genetics
National Cancer Institute
Rockville, Maryland

I. Introduction

Occupational exposures are thought to account for less than 5% of human cancers [1]; however, this estimate was derived from research that focused almost entirely on men and on exposures sustained in the 1950s and 1960s [2]. Since the 1950s there have been dramatic changes in the employment patterns of women which suggest that the role of occupational cancers among women should be reconsidered. Many more women are in the workforce now than in the early years after World War II. In 1950, only 33.9% of U.S. women were in the civilian labor force (Table 39.1) [3]. By 1997, 59.8% of U.S. women were in the labor force [4]. The number of years women spend working outside the home also has increased. The once-traditional pattern of leaving the labor force when children are born is no longer the norm in the U.S. [3]. In 1950, 12% of women with children under age 6 and 29% of women with children aged 6–17 worked outside the home. By 1990, 60% of women with children under age 6 and 75% of women with children aged 6–17 were in the labor force. Participation in the labor force has also increased among older women [3]. In addition, more women are employed in some nontraditional jobs that may involve hazardous chemical exposures. The number of women employed in skilled trades in 1980 was almost four times the number in 1960 [5]. The proportion of female mechanics and repairers increased almost threefold between 1960 and 1990 [6,7]. More women are working, working longer, and working in nontraditional jobs, which means that women may have increased potential for long-term exposure to occupational carcinogens compared to women in the 1950s and 1960s. This chapter will review the known and suspected occupational carcinogens women may face on the job, discuss the limitations of the existing research, particularly given the changing employment patterns of women, and identify opportunities for future research.

II. Employment of Women by Industry

In 1997, approximately two-thirds of women in the U.S. labor force were employed in sales or service industries (Table 39.2) [4]. Many of these industries are unlikely to involve exposure to potential carcinogens. However, women employed in particular service industries, such as cosmetology, dry cleaning, food service, housekeeping and janitorial services, and health care

Table 39.2

Distribution of Employed Women by Industry, 1997[a]

Industry	Number of women[b]	% of employed women[c]	Women % of industry[d]
Agriculture	847	1.4	24.9
Mining	92	0.2	14.4
Construction	784	1.3	9.4
Manufacturing	6,683	11.2	32.1
Transportation, communications, and other public utilities	2,643	4.4	28.8
Sales	12,659	21.1	47.3
Wholesale	1,454	2.4	29.6
Retail	11,205	18.7	51.2
Finance, insurance, real estate	4,848	8.1	58.4
Services	28,766	48.0	62.0
Private household services	837	1.4	90.9
Business and repair services	3,145	5.3	37.2
Personal services	2,198	3.7	63.1
Entertainment and recreational services	1,098	1.8	44.5
Professional and related services	21,458	35.8	69.4
Public Administration	2,552	4.3	44.5
Total	**59,873**	**100.0**	**46.2**

[a]Source: [4].
[b]Number in thousands.
[c]% of employed women = number of women in specific industry/ number of employed women in all industries.
[d]Women % of industry = number of women in specific industry/ number of women and men in specific industry.

Table 39.1

Percentage of U.S. Women in the Civilian Labor Force, 1950–1997[a]

Year	Total women	Women with children aged < 6 years	Women with children aged 6–17 years
1950	34	12	29
1960	38	17	38
1970	43	25	49
1980	52	44	63
1990	57	60	75
1997	60	65	78

[a]Sources: [3,4].

Table 39.3

Selected Industries, Known or Potential Carcinogens, and Reported Cancer Excesses

Industry	Known/potential carcinogens	Cancers
Agriculture	Pesticides, sunlight, fuels	Brain, cervix, gallbladder, leukemia, liver, lymphoma, multiple myeloma, ovary, stomach
Service industries		
Cosmetology	Hair dyes, hair sprays, formaldehyde	Bladder, brain, leukemia, lung, lymphoma, ovary
Dry cleaning	Carbon tetrachloride, trichloroethylene, tetrachloroethylene, other solvents	Bladder, cervix, esophagus, kidney, liver, lung, ovary, pancreas
Food service	Tobacco smoke, cooking fumes	Bladder, cervix, esophagus, lung
Health care	Antineoplastic drugs, anesthetic gases, ionizing radiation, viruses	Bladder, brain, breast, leukemia, lung, lymphoma
Manufacturing		
Computers, electronics	Solvents, metal fumes	Brain
Furniture	Wood dust, solvents, glues, formaldehyde	Lung, pancreas, sinonasal
Motor vehicle manufacturing	Paints, metal fumes, solvents, machining fluids	Colorectum, lung, stomach
Chemical/Plastics/Rubber	Vinyl chloride, 1,3-butadiene, benzene, other solvents, nitrosamines	Bladder, brain, breast (?), leukemia, lung, lymphoma, ovary
Textile, apparel	Asbestos, dyes, lubricating oils	Biliary tract, bladder, leukemia, lung, lymphoma, mesothelioma

may have exposures to chemicals that have been linked to cancer in epidemiologic or animal studies.

Manufacturing accounted for 11.2% of employed women in 1997 [4]. The proportion of women in this sector has decreased slightly over the past few decades [3] as more manufacturing has shifted to foreign locations. Within U.S. manufacturing, however, there is a greater proportion of women employed currently than in the 1950s [7,8]. Manufacturing industries that typically employ many women include the textile and apparel industries and the electronics industry, particularly the semiconductor industry [5].

III. Carcinogens in the Workplace

Table 39.3 presents selected industries, exposures, and reported cancer excesses among women. Women in agricultural work may have exposure to pesticides, fertilizers, fuels and exhausts, solvents, and other potential carcinogens and experience increased risks of lymphatic and hematopoietic tumors, similar to those of men employed in agriculture [9–15]. Brain cancer was excessive in studies of female farmers in the U.S. [16] and in China [17]. Hispanic women who were farmworkers had excess cancers of the stomach, liver, gallbladder, and cervix [18]. Occupational exposure to herbicides was linked to ovarian cancer in a case-control study in Italy [19].

Jobs in the service industries are not usually thought of as hazardous, but many of these jobs have exposures to potential carcinogens. Cosmetologists and beauticians may be exposed to hair dyes, hair sprays, formaldehyde, and some solvents. The International Agency for Research on Cancer (IARC) has classified the occupation of hairdresser as posing a probable carcinogenic risk based primarily on the evidence of excess bladder cancer [20]. Other cancers found to be excessive among cosmetologists include leukemia, lymphoma, multiple myeloma,

and cancers of the lung, pancreas, brain, ovary, and salivary gland [16,20–25].

Dry cleaners may be exposed to tetrachloroethylene (perchloroethylene) and, historically, trichloroethylene, which are considered probable human carcinogens by IARC [26]. A cohort study of dry-cleaning workers [27] found excess mortality from esophagus, bladder, pancreas, and cervical cancer among women. Excesses at these tumor sites and others, including the kidney, liver, lymphohematopoietic system, and ovary, have also been reported [25,28–32].

More than 80% of health care workers are women [5,8]. They are potentially exposed to antineoplastic drugs, anesthetic gases, ethylene oxide, mercury, viruses, ionizing radiation, and electromagnetic fields [33,34]. Excess hematopoietic and lymphoproliferative cancers have been reported for female dentists, nurses, and pharmacy workers [11,35–38], for nurses handling antineoplastic drugs [39], and for health workers using diagnostic X-rays [40]. A registry-based study of brain tumors in China [17] found an elevated risk for female nurses which had not been reported previously. Breast cancer was elevated in several studies of nurses [35,36,38,41–48] including a cohort of Icelandic nurses in which the excess was not explained by maternal age at first birth or number of children [36]. The risk was greatest among women who handled cytotoxic drugs [48]. Breast cancer was also elevated among radiologic technologists in a small incidence study in China [40,49] and in a large mortality study in the U.S. [50], but not in the nested case-control study from the latter cohort for which detailed radiation exposure and reproductive information was available [51,52]. Increased risk of bladder cancer has been observed among female health care workers, including nurses and pharmacists [25,53–55].

Restaurant and bar personnel have exposure to environmental tobacco smoke and have excesses of smoking-related cancers including lung, esophagus, oral, and bladder [44,56–59]. Some of these reports of increased risks of smoking-related cancers, however, did not adjust for active smoking. Excess cervical cancer has also been observed among waitresses [25].

Jobs in the manufacturing industries have potential for exposure to a wide variety of carcinogens. Women work in large numbers in the furniture industry where they are exposed to solvents, glues, and wood dust (a recognized human carcinogen) [60]. Woodworking occupations and wood dust exposure have been linked to an increased risk of nasosinal cancer among men and women, although the findings for women were based on very small numbers [61–63]. Elevated rates of pancreas and lung cancer were reported for women working in wood furniture plants in the U.S. [64], but this finding was not confirmed by a combined analysis of this study with four other cohorts of wood workers [62].

The manufacture of computer and electronic equipment includes exposures to solvents, metal dusts and fumes, and electromagnetic radiation. More than 75% of the semiconductor production workers in Santa Clara County, California ("Silicon Valley") are women [5]. Most studies of electronics workers have not evaluated cancer risks separately for women, but a study by Fear et al. [65] of electrical workers in England observed a significant doubling of brain cancer risk among the women employees, which is consistent with reports of increased brain cancer among male electrical and electronics workers [66].

Substantial numbers of women work in motor vehicle manufacturing in which paints, metal fumes, solvents, and machining fluids are used. Most studies of workers have focused on men and have shown increased risks of lung, stomach, and colorectal cancer in foundry, metal processing, and wood pattern-making operations, respectively. A study that evaluated risks for women [67] found small excesses of lung cancer in assembly jobs and colorectal cancer in nonproduction jobs. Women in a closely related industry, aircraft maintenance and repair, had excess risks of multiple myeloma and non-Hodgkin's lymphoma related to solvent exposure [68].

In the U.S. National Bladder Cancer Study, occupational exposures accounted for 11% of bladder cancers occurring among women, compared to 20–25% of bladder cancer among men [69]. Although the attributable risks were lower for women, the occupations linked to bladder cancer among women were similar to those observed among men [69]. Increased bladder cancer risk was found among women employed in occupations involved with metal working and fabrication, chemical processing, and rubber processing [69].

Women are employed in many types of jobs in the plastics and rubber industries where exposures can include aromatic amines, acrylonitrile, benzene, 1,3-butadiene, vinyl chloride, styrene, polyethylene, formaldehyde, phenol, and urea. Work in the rubber industry has been associated with increased risks of brain cancer [17], hematopoietic and lymphoproliferative malignancies [70], and bladder cancer [25] among women. The aromatic amine, 4-aminobiphenyl, was used in rubber manufacturing in the U.S. and is an established human bladder carcinogen [71]. Benzene is also an established human carcinogen [71] and 1,3-butadiene is a probable human carcinogen [72]. Both are used extensively in the rubber industry and have been linked to increased risks of leukemia and lymphoma [70,73]. Benzene and 1,3-butadiene cause breast and ovarian cancer in animal

studies [74] and evaluation of possible links to these tumors in occupationally exposed women is warranted.

Vinyl chloride, which is used in the plastics industry, is a human carcinogen [71] and exposure has been linked to an increased risk of brain, liver, and lung cancer among men [75,76]. Low levels of vinyl chloride cause mammary tumors in rats and a moderate excess of breast cancer mortality was observed in one study among women with low exposures [77].

Women comprise a majority of the labor force in textile and apparel manufacturing and can have exposures to cotton dust, dyes, lubrication oils, formaldehyde, and asbestos, although the latter is now rare. Women employed as textile workers had increased risks of nasosinal cancer in studies in France [78], nasopharyngeal [79] and biliary tract [80] cancer in China, and bladder cancer in Europe [25]. In the U.S., female and male textile workers who performed operations unrelated to dyeing had decreased risk of bladder cancer [69]. Lung cancer mortality was increased among men and women in textile operations in a U.S. study of furniture workers [64], although cotton textile workers in Shanghai experienced decreased risk of lung cancer [81]. Among women manufacturing asbestos textile products, risks of lung cancer and mesothelioma were elevated [82,83]. Elevated risks of hematopoietic and lymphoproliferative tumors have also been observed among female textile workers [11,84].

IV. Selected Cancer Sites

Although virtually every cancer has been linked to occupational exposures to some degree, certain cancers, such as leukemia and cancers of the lung, bladder, and brain, appear to have a stronger occupational component. The occupational associations that have been observed among women for these cancers will be reviewed. In addition, the possible role of occupational exposures in the etiology of breast and ovarian cancer will be reviewed.

A. Leukemia

The most well established occupational chemical leukemogen is benzene [85]. Although studied less frequently among women than among men, women exposed to benzene have been observed to have seven- to eightfold increased risks of leukemia [70,86]. Other women occupationally exposed to solvents [28,87], including chemists, engineers, and laboratory personnel [44,88–90], dry cleaners [90], and beauticians [91], as well as women exposed to vinyl chloride [92] or employed near capacitors [93] or in food processing [44,90], have been reported to be at an increased risk of leukemia. Excess leukemia has been reported in many studies of health care workers [11,39,44,89,90,94], including oncology nurses handling antineoplastic drugs [39,94], clinical laboratory technicians [89], and diagnostic X-ray workers [40]. Women workers in agriculture [11,12,14,91] and in the textile and garment industries [11,95,96] also have reported excesses of leukemia, which is consistent with similar excesses observed among men.

B. Lung

Women exposed to asbestos, including gas mask assemblers and textile workers, had an increased risk of lung cancer

[82,83,97–103]. Lung cancer was also elevated among women in construction [25,56,104–108], where exposures include asbestos and other hazardous exposures. Several metals, including arsenic, chromium, and nickel, as well as polycyclic aromatic hydrocarbons encountered in the processing and machining of metals, are lung carcinogens [109]. Women employed in metal working, grinding, or assembly [25,56,107,110–115], such as in motor vehicle manufacturing [67,96,107,116,117], had greater than expected rates of lung cancer. Exposure to mercury in the fur hat industry, which occurred up to the 1950s, was associated with lung cancer among women [118]. Women in the textile industry without exposure to asbestos have been reported to have excess lung cancer mortality in some studies [64,107,119,120], but decreased risk in another study [81]. Waitresses, bartenders, and cooks have been observed consistently to have increased rates of lung cancer, which may be due to exposure to environmental tobacco smoke and polycyclic aromatic hydrocarbons in cooking fumes [25,44,56,58,59,89,110,115], but better adjustment for active smoking is needed to properly evaluate these excesses. The excess lung cancer among cosmetologists [56,110,121–123] may also be due in part to environmental tobacco smoke or inhalation of solvents in hair sprays and other beauty products.

C. Bladder

The major occupational risk factors for bladder cancer observed among men have also been observed among women [69,124]. These occupations include dyestuff workers and dye users, textile workers [25,69,84,125–130], rubber and plastic workers [25,69,125,131,132], leather workers [133,134], and painters [54]. Excess bladder cancer has also been observed among women employed in dry cleaning [30,32], health care [25,53–55], some office jobs [128,135–137], miscellaneous manufacturing jobs [25,69,115,125,131], gardeners [54], teachers [138], waitresses [54], maids [139], seafarers [140], and telephone operators [141], although many reports could not adjust for smoking.

D. Brain

Few studies of brain cancer and occupational risk factors among women have been conducted, with any given association observed in only one or, at most, two studies. Increased risk of brain cancer has been reported for women employed in the electronics industry [65,142], the textile industry [17,143], candy manufacturing [143], other manufacturing [17,144,145], occupations with electromagnetic field exposure [17,146], construction [143], teaching [43,138], the telephone industry [141], and grain farming [17].

E. Breast

One of the earliest clues to the etiology of breast cancer was the observation by Ramazzini in 1700 that breast cancer risk was increased among nuns [147]. The increased risk was thought to be due to reproductive patterns and other lifestyle differences. In modern times, religious workers still have an increased risk of breast cancer [43,46]. Other occupational groups at high risk, probably due in part to delayed or decreased childbearing, include teachers [25,42,43,45–47,138,148,149], chemists [88,150], physicians [42,46], nurses [35,36,41–47], other health care workers [40,115,151,152], and professional and technical occupations [43–46,96,114,141,148,149,153–155].

Some of these groups, however, such as chemists and health care workers, may have chemical and physical occupational exposures that affect breast cancer risk. For example, Labreche and Goldberg [156] have hypothesized that solvents, several of which are mammary carcinogens in animals, may play a role in human breast cancer, particularly due to their toxicokinetic distribution favoring accumulation in and around the breast. Excess breast cancer has been observed among women employed in chemical manufacturing [43,157–159], drug manufacturing [158,160,161], and cosmetologists [43,114,123,162]. Exposures that affect hormone levels or metabolism may affect breast cancer risk. Women manufacturing hormones and other drugs had excess breast cancer [161].

Radiation, a well-established risk factor for breast cancer [163], was responsible for the increased breast cancer observed among radium dial painters employed in the 1920s [164,165] diagnostic X-ray workers [40,49,50], and possibly airline cabin attendants [166]. Elevated rates of breast cancer have also been observed among women in electrical, electronics, and other occupations in which there was potential exposure to electromagnetic fields, solvents, and other agents [43–45,96,141,158, 159,167–169].

F. Ovary

Women employed as gas mask assemblers prior to and during World War II were exposed to asbestos and experienced excess ovarian cancer that increased with intensity and duration of exposure [97–99,170]. No other occupational association is as well established for ovarian cancer. Increased risk of this malignancy has been observed in several studies among women employed as chemists, engineers, or in laboratory occupations [45,88,141,155], and as cosmetologists [11,123,162,171], teachers [42,45,47], carpenters, and woodworkers, and in other construction trades [45,48,105,142,149]. There are single reports linking ovarian cancer to employment in textile manufacturing [149], stone, clay, and glass work [158], dry cleaning [25], telephone operators [141], nursing [42], and to herbicide exposure [19].

V. Limitations of Current Epidemiologic Knowledge

Our current knowledge of occupational cancer is limited by the general lack of research on women. In a survey of 1233 occupational cancer epidemiology reports, only 14% presented any analyses of white women and only 2% presented data on nonwhite women [2]. Even fewer reports presented in-depth analyses (more than five risk estimates) for women (white women: 7%, nonwhite women: 1%). While we assume risks identified in men also pertain to women, there are important reasons to study women specifically. Women may respond differently than men to the same occupational exposures due to metabolic, genetic, or other differences. Also, the risk of breast cancer and tumors of the female reproductive system cannot be evaluated in studies of men. In addition, women may also be experiencing hazardous occupational exposures in recently developed industries that have not yet been evaluated fully for cancer risks among men (*e.g.*, the semiconductor industry).

There are, however, several obstacles to epidemiologic research on occupational cancer among women. The most important obstacle is that, in general, only a small number of women are employed in jobs of interest in any given study. National employment statistics [8] suggest that typically five-sixths of the women in case-control studies will be either housewives or employed in the service or retail industry, leaving few women in jobs with potential carcinogenic exposures. Similarly, most cohort studies of industrial workforces are composed primarily of men. The few women in these workforces are likely to be in lower exposed jobs or be more recent hires than the men, often with inadequate latency for studies of most solid tumors. Women miners, for example, are a very small proportion of the total miner workforce and are relatively recent hires.

Death certificates and other routinely collected administrative data on occupation may be of poorer quality for women than for men. For example, the usual occupation is supposed to be entered on the death certificate regardless of the employment status at the time of death. However, "housewife" is often entered even if women held long-term full or part-time paid employment outside the home [172–174], especially among older women who are retired at the time of death [172]. When housewives are excluded, the accuracy of information on women's certificates is less than that on men's certificates [174].

Retrospective exposure assessment, which typically is based on job titles, historical monitoring data, and information on use of protective equipment and engineering controls, poses additional challenges in studies including women workers. In some instances, men and women with the same job title do not have the same duties and exposures [175–178]. Most available historical monitoring data have been gathered on men and there may be work practice and personal hygiene differences that make extrapolation of these data from men to women inappropriate [179]. Protective equipment and engineering controls, designed and tested for men, may not fit or be as efficacious for women [180,181]. Exposure assessment algorithms that estimate the impact of these control measures may be less accurate for women than for men. The same external exposure may result in different internal dose because of gender-specific absorption, distribution, kinetic, and metabolic differences [179]. The lack of gender-specific exposure assessment methods may have limited past research.

Women may be more difficult to trace due to name changes. In a study of dry cleaners, the percentage of people lost to follow-up was 8% for white men, 9% for nonwhite men, 12% for white women, and 15% for nonwhite women [30].

The best comparison population for occupational studies of women is not always clear. In occupational studies of men, the reference population is often the total general population of the United States or of a restricted study area. Although the healthy worker effect may be a consideration, the general population usually serves well as the reference. In occupational studies of women, however, the healthy worker effect is greater than that in studies of men [182–184]. To a greater extent than among men, employed women differ from unemployed women in

terms of age, marital status, family socioeconomic status, fertility, drug use, alcohol consumption, and other factors [45,182–185]. For studies of cancers related to these factors, other employed women might be a more comparable and desirable reference population than the general population. In the U.S., however, mortality data by employment status are difficult to obtain.

Cohort studies usually lack information on factors that might confound the observed occupational associations. For example, a study based on occupational records is unlikely to have information on reproductive history, which would be needed for a thorough evaluation of cancer of the breast or female reproductive organs. In addition, the nonoccupational exposures to household cleaning products and other exposures that might be more common in women may go unstudied.

VI. Future Research

Research on occupational cancer among women provides the opportunity to identify avoidable carcinogens and to shed light on carcinogenic mechanisms. There are several changes in approach, however, that could optimize the usefulness of future research. First, every effort should be made to fully utilize the data on women in all study populations. A survey of occupational cancer studies reported that some studies that presented no analyses for women had populations that appeared large enough to evaluate them [2]. Some reports seemed to dismiss the women because the numbers were small relative to the white men in that study, but the numbers were as large or larger than the numbers of men in many other studies. We may be able to learn more about occupational cancer among women simply by looking at existing data.

Second, studies of occupational cancer among women must move beyond analyses based merely on job title or industry to analyses based on specific chemical and physical exposures. Job or industry categories that group workers with heterogeneous exposures introduce misclassification of exposure and dilute risk estimates. Analyses based on specific exposures have greater sensitivity to detect true associations [186]. Job-exposure matrices, exposure assessment by expert industrial hygienists, and exposure registries have been used to enhance the precision and sensitivity of occupational studies [28,187–189], although these methods have been used primarily in studies of men. These methods need to be applied to studies of women as well.

Much future research on occupational cancer among women will probably need to be conducted outside the U.S. in locations with high employment rates among women, particularly in manufacturing occupations with potentially harmful exposures. For example, over 90% of women in Shanghai work outside the home [W. H. Chow, personal communication, 1997] as compared to 60% of U.S. women [4]. In Poland, about 60% of women in Warsaw and over 90% of women in Lodz hold jobs in manufacturing [W. H. Chow, personal communication, 1998], which are much higher percentages than the 11% of U.S. women in manufacturing [4]. Also, exposures may be greater in other countries with less stringent regulations and enforcement, for example, in manufacturing sites in Mexico near the U.S. border [190].

Finally, future research needs to incorporate biologic measures of susceptibility to assess genetic-environment interactions.

Genetically determined metabolic polymorphisms may be key to identifying occupational carcinogens and susceptible subgroups of the population. Polymorphisms of hormone metabolizing genes may also be critical for understanding breast and reproductive cancers. Collection of genomic DNA, obtained through blood, buccal cells, or other appropriate biological samples, should be incorporated into future research whenever possible.

VII. Conclusions

Studies of occupational cancer among women, although not extensive, have identified increased cancer risk associated with employment in agriculture, selected service occupations, and manufacturing. Potential occupational carcinogens of special importance to women include antineoplastic drugs, anesthetic gases, ionizing radiation, and electromagnetic fields among health care personnel; electromagnetic fields among office workers; hair dyes, hair sprays, and formaldehyde among cosmetologists; tetrachloroethylene and trichloroethylene among dry cleaners; building and household cleaners among janitorial and housekeeping workers; pharmaceutical exposures among drug manufacturers; passive tobacco smoke among waitresses and bartenders; pesticides among farmworkers; solvent exposure among aircraft repair and other industrial workers; asbestos, dyes, and other exposures in the textile industry; and paints, metal fumes, machinery fluids, and other agents in motor vehicle manufacturing. More research is needed to identify risks among women, particularly in countries where many women are employed in manufacturing, and to improve the assessment of workplace exposures among women. Occupational cancers among women, as well as among men, are completely preventable and deserve our attention.

References

1. Doll, R., and Peto, R. (1981). "The Causes of Cancer: Quantitative Estimates of Avoidable Risks of Cancer." Oxford University Press, New York.
2. Zahm, S. H., Pottern, L. M., Lewis, D. R., Ward, M. H., and White, D. W. (1994). Inclusion of women and minorities in occupational cancer epidemiologic research. *J. Occup. Med.* **36,** 842–847.
3. Wagener, D. K., Walstedt, J., Jenkins, L., Burnett, C., Lalich, N., and Fingerhut, M. (1997). Women: Work and health (DHHS Publication no. PHS 97-1415). *Vital Health Stat.* **3**(31), 1–91.
4. Bureau of Labor Statistics (1998). "Employment and Earnings Report, 1997, January 1998." U.S. Govt Printing Office, Washington, DC.
5. Quinn, M. M., Woskie, S. R., and Rosenberg, B. J. (1995). Women and work. *In* "Occupational Health: Recognizing and Preventing Work-related Diseases" (B. S. Levy and D. H. Wegman, eds.), 3rd ed., pp. 619–638. Little, Brown, Boston.
6. Bureau of the Census (1963). "U.S. Census of the Population: 1960. Detailed Characteristics. United States Summary. Final Report PC (1)-1D." U.S. Govt. Printing Office, Washington, DC.
7. Bureau of the Census (1993). "1990 Census of the Population. Social and Economic Characteristics. United States." U.S. Govt. Printing Office, Washington, DC.
8. Bureau of Labor Statistics (1995). "Employment and Earnings Report, 1994," Vol. 42, No. 1. U.S. Govt. Printing Office, Washington, DC.
9. McDuffie, H. H. (1994). Women at work: Agriculture and pesticides. *J. Occup. Med.* **36,** 1240–1246.
10. Blair, A., and Zahm, S. H. (1991). Cancer among farmers. *Occup. Med.: State of the Art Rev.* **6,** 335–354.
11. Linet, M. S., McLaughlin, J. K., Malker, H. S., Chow, W. H., Weiner, J. A., Stone, B. J., Ericsson, J. L. E., and Fraumeni, J. F. (1994). Occupation and hematopoietic and lymphoproliferative malignancies among women: A linked registry study. *J. Occup. Med.* **36,** 1187–1198.
12. Blair, A., Dosemeci, M., and Heineman, E. F. (1993). Cancer and other causes of death among male and female farmers from twenty-three states. *Am. J. Ind. Med.* **23,** 729–742.
13. Zahm, S. H., Weisenburger, D. D., Saal, R. C., Vaught, J. B., Babbitt, P. A., and Blair, A. (1993). The role of agricultural pesticide use in the development of non-Hodgkin's lymphoma in women. *Arch. Environ. Health* **48,** 353–358.
14. Ronco, G., Costa, G., and Lynge E. (1992). Cancer risk among Danish and Italian farmers. *Br. J. Ind. Med.* **49,** 220–225.
15. Zahm, S. H., Blair, A., and Weisenburger, D. D. (1992). Sex differences in the risk of multiple myeloma associated with agriculture. *Br. J. Ind. Med.* **49,** 815–816.
16. Cocco, P., Dosemeci, M., and Heineman, E. F. (1998). Occupational risk factors for cancer of the central nervous system: A case-control study on death certificates from 24 U.S. states. *Am. J. Ind. Med.* **33,** 247–255.
17. Heineman, E. F., Gao, Y.-T., Dosemeci, M., and McLaughlin, J. K. (1995). Occupational risk factors for brain tumors among women in Shanghai, China. *J. Occup. Environ. Med.* **37,** 288–293.
18. Zahm, S. H., and Blair, A. (1993). Cancer among migrant and seasonal farmworkers: An epidemiologic review and research agenda. *Am. J. Ind. Med.* **24,** 753–766.
19. Donna, A., Betta, P. B., Robutti, F., Crosignani, P., Berrino, F., and Bellingeri, D. (1984). Ovarian mesothelial tumours and herbicides: A case-control study. *Carcinogenesis* (*London*) **5,** 941–942.
20. International Agency for Research on Cancer (1993). "IARC Monograph on the Evaluation of the Carcinogenic Risk to Humans. Occupational Exposures of Hairdressers and Barbers and Personal Use of Hair Colourants; Some Hair Dyes, Cosmetic Colourants, Industrial Dyestuffs and Aromatic Amines," Vol. 57. IARC, Lyon.
21. Boffetta, P., Andersen, A., Lynge, E., Barlow, L., and Pukkala, E. (1994). Employment as hairdresser and risk of ovarian cancer and non-Hodgkin's lymphomas among women. *J. Occup. Med.* **36,** 61–65.
22. Skov, T., and Lynge, E. (1994). Cancer risk and exposures to carcinogens among hairdressers. *Skin Pharmacol.* **7,** 94–100.
23. Mele, A., Szklo, M., Visani, G., Stazi, M. A., Castelli, G., Pasquini, P., and Mondelli, F. (1994). Hair dye use and other risk factors for leukemia and pre-leukemia: A case-control study, Italian Leukemia Study Group. *Am. J. Epidemiol.* **139,** 609–619.
24. Swanson, G. M., and Burns, P. B. (1997). Cancers of the salivary gland: Workplace risks among women and men. *Ann. Epidemiol.* **7,** 369–374.
25. Carpenter, L., and Roman, E. (1995). Identifying associations between cancer and occupation in women in Europe. The role of routinely collected national data. *Med. Lav.* **86,** 252–255.
26. International Agency for Research on Cancer (1995) "IARC Monograph on the Evaluation of the Carcinogenic Risk to Humans. Dry Cleaning, Some Chlorinated Solvents and Other Industrial Chemicals," Vol. 63. IARC, Lyon.
27. Ruder, A. M., Ward, E. M., and Brown, D. P. (1994). Cancer mortality in female and male dry-cleaning workers. *J. Occup. Med.* **36,** 867–874.
28. Anttila, A., Pukkala, E., Sallmén, M., Hernberg, S., and Hemminki, K. (1995). Cancer incidence among Finnish workers exposed to halogenated hydrocarbons. *J. Occup. Environ. Med.* **37,** 797–806.

29. Lynge, E., and Thygesen, L. (1990). Primary liver cancer among women in laundry and dry-cleaning work in Denmark. *Scand. J. Work Environ. Health* **16,** 108–112.

30. Blair, A., Stewart, P. A., Tolbert, P. E., Grauman, D., Moran, F. X., Vaught, J., and Rayner, J. (1990). Cancer and other causes of death among a cohort of dry cleaners. *Br. J. Ind. Med.* **47,** 162–168.

31. Duh, R., and Asal, N. R. (1984). Mortality among laundry and dry cleaning workers in Oklahoma. *Am. J. Public Health* **74,** 1278–1280.

32. Katz, R. M., and Jowett, D. (1981). Female laundry and dry cleaning workers in Wisconsin: A mortality analysis. *Am. J. Public Health* **71,** 305–307.

33. Hewitt, J. B., Misner, S. T., and Levin, P. F. (1993). Health hazards of nursing: Identifying workplace hazards and reducing risks. *Association of Women's Health, Obstetric, and Neonatal Nurses Clin. Issues Perinat. Women's Health Nurs.* **4,** 320–327.

34. Philips, K. L., Morandi, M. T., Oehme, D., and Cloutier, P. A. (1995). Occupational exposure to low frequency magnetic fields in health care facilities. *Am. Ind. Hyg. Assoc. J.* **56,** 677–685.

35. Sankila, R., Karjalainen, S., Laara, E., Pukkala, E., and Teppo, L. (1990). Cancer risk among health care personnel in Finland, 1971–1980. *Scand. J. Work Environ. Health* **16,** 252–257.

36. Gunnarsdottir, H., and Rafnsson, V. (1995). Cancer incidence among Icelandic nurses. *J. Occup. Environ. Med.* **37,** 307–312.

37. Hansen, J., and Olsen, J. H. (1994). Cancer morbidity among Danish female pharmacy technicians. *Scand. J. Work Environ. Health* **20,** 22–26.

38. Peipins, L. A., Burnett, C., Alterman, T., and Lalich, N. (1997). Mortality patterns among female nurses: A 27-state study, 1984 through 1990. *Am. J. Public Health* **87,** 1539–1543.

39. Skov, T., Maarup, B., Olsen, J., Rorth, M., Winthereik, H., and Lynge, E. (1992). Leukaemia and reproductive outcome among nurse handling antineoplastic drugs. *Br. J. Ind. Med.* **49,** 855–861.

40. Wang, J.-X., Boice, J. D., Jr., Li, B.-X., Zhang, J. Y., and Fraumeni, J. F., Jr. (1988). Cancer among medical diagnostic x-ray workers in China. *J. Natl. Cancer Inst.* **80,** 344–350.

41. Katz, R. M. (1983). Causes of death among registered nurses. *J. Occup. Med.* **25,** 760–762.

42. King, A. S., Threlfall, W. J., Band, P. R., and Gallagher, R. P. (1994). Mortality among female registered nurses and school teachers in British Columbia. *Am. J. Ind. Med.* **26,** 125–132.

43. Morton, W. E. (1995). Major differences in breast cancer risks among occupations. *J. Occup. Environ. Med.* **37,** 328–235.

44. Bulbulyan, M., Zahm, S. H., and Zaridze, D. G. (1992). Occupational cancer mortality among urban women in the former USSR. *Cancer Causes Control* **3,** 299–307.

45. Roman, E., Beral, V., and Inskip, H. (1985). Occupational mortality among women in England and Wales. *Br. Med. J.* **291,** 194–196.

46. Rubin, C. H., Burnett, C. A., Halperin, W. E., and Seligman, P. J. (1993). Occupation as a risk identifier for breast cancer. *Am. J. Public Health* **83,** 1311–1315.

47. Threlfall, W. J., Gallagher, R. P., Spinelli, J. J., and Pierre, R. B. (1985). Reproductive variables as possible confounders in occupational studies of breast and ovarian cancer in females. *J. Occup. Med.* **27,** 448–450.

48. Gunnarsdottir, H. K., Aspelund, T., Karlsson, T., and Rafnsson, V. (1997). Occupational risk factors for breast cancer among nurses. *Int. J. Occup. Environ. Health* **3,** 254–258.

49. Wang, J.-X., Inskip, P. D., Boice, J. D., Jr., Li, B.-X., Zhang, J. Y., and Fraumeni, J. F., Jr. (1990). Cancer incidence among medical diagnostic x-ray workers in China, 1950 to 1985. *Int. J. Cancer* **45,** 889–895.

50. Doody, M. M., Mandel, J. S., Lubin, J. H., and Boice, J. D., Jr. (1998). Mortality among U.S. radiologic technologists, 1926–90. *Cancer Causes Control* **9,** 67–75.

51. Doody, M. M., Mandel, J. S., and Boice, J. D., Jr. (1995). Employment practices and breast cancer among radiologic technologists. *J. Occup. Environ. Med.* **37,** 321–327.

52. Boice, J. D., Jr., Mandel, J. S., and Doody, M. M. (1995). Breast cancer among radiologic technologists. *J. Am. Med. Assoc.* **274,** 394–401.

53. Anthony, H. M., Thomas, G. M., Cole, P., and Hoover, R. (1971). Tumors of the urinary bladder: An analysis of the occupations of 1,030 patients in Leeds, England. *J. Natl. Cancer Inst. (U.S.)* **46,** 1111–1113.

54. Silverman, D. T., McLaughlin, J. K., Malker, H. S., and Ericsson, J. L. (1989). Bladder cancer and occupation among Swedish women. *Am. J. Ind. Med.* **16,** 239–240.

55. Levin, L. I., Holly, E. A., and Seward, J. P. (1993). Bladder cancer in a 39-year old female pharmacist. *J. Natl. Cancer Inst.* **85,** 1089–1090.

56. Rubin, C. H., Burnett, C. A., Halperin, W. E., Seligman, P. J. (1994). Occupation and lung cancer mortality among women: Using occupation to target smoking cessation programs for women. *J. Occup. Med.* **36,** 1234–1238.

57. Siegel, M. (1993). Involuntary smoking in the restaurant workplace. A review of employee exposure and health effects. *J. Am. Med. Assoc.* **279,** 490–493.

58. Dimich-Ward, H., Gallagher, R. P., Spinelli, J. J., Threlfall, W. J., and Band, P. R. (1988). Occupational mortality among bartenders and waiters. *Can. J. Public Health* **79,** 194–197.

59. Kjaerheim, K., and Andersen, A. (1994). Cancer incidence among waitresses in Norway. *Cancer Causes Control* **5,** 31–37.

60. International Agency for Research on Cancer (1981). "IARC Monograph on the Evaluation of the Carcinogenic Risk to Humans. Wood, Leather and Some Associated Industries," Vol. 25. IARC, Lyon.

61. Hayes, R. B., Gerin, M., Raatgever, J. W., and de Bruyn, A. (1986). Wood-related occupations, wood dust exposure, and sinonasal cancer. *Am. J. Epidemiol.* **124,** 569–577.

62. Demers, P. A., Boffetta, P., Kogevinas, M., Blair, A., Miller, B. A., Robinson, C. F., Roscoe, R. J., Winter, P. D., Colin, D., Matos, E., and Vainio, H. (1995). A pooled re-analysis of cancer mortality among five cohorts of workers in wood-related industries. *Scand. J. Work Environ. Health* **21,** 179–190.

63. Demers, P. A., Kogevinas, M., Boffetta, P., Leclerc, A., Luce, D., Gerin, M. Battista, G., Belli, S., Bolm-Audorf, U., Brinton, L. A., Colin, D., Comba, P., Handell, L., Hayes, R. B., Magnani, C., Merler, E., Morcet, J. F., Preston-Martin, S., Matos, E., Rodella, S., Vaughar, T. L., Zheng, W., and Vainio, H. (1995). Wood dust and sino-nasal cancer: a pooled re-analysis of twelve case-control studies. *Am. J. Ind. Med.* **28,** 151–166.

64. Miller, B. A., Blair, A., and Reed, E. J. (1994). Extended mortality follow-up among men and women in a U.S. furniture workers union. *Am. J. Ind. Med.* **25,** 537–549.

65. Fear, N. T., Roman, E., Carpenter, L. M., Newton, R., and Bull, D. (1996). Cancer in electrical workers: An analysis of cancer registrations in England, 1981–87. *Br. J. Cancer* **73,** 935–939.

66. Thomas, T. L., Stolley, P. D., Stemhagen, A., Fontham, E. T., Bleecker, M. L., Stewart, P. A., and Hoover, R. N. (1987). Brain tumor mortality risk among men with electrical and electronic jobs: A case-control study. *J. Natl. Cancer Inst.* **79,** 233–238.

67. Delzell, E., Beall, C., and Macaluso, M. (1994). Cancer mortality among women employed in motor vehicle manufacturing. *J. Occup. Med.* **36,** 1251–1259.

68. Blair, A., Hartge, P., Stewart, P., McAdams, M., and Lubin, J. (1998). Mortality and cancer incidence of aircraft maintenance workers exposed to trichloroethylene and other organic solvents and chemicals: Extended follow-up. *Occup. Environ. Med.* **55,** 161–171.

69. Silverman, D. T., Levin, L. I., and Hoover, R. N. (1990). Occupational risks of bladder cancer among white women in the U.S. *Am. J. Epidemiol.* **132,** 453–461.

70. Li, G.-L., Linet, M. S., Hayes, R. B., Yin, S. N., Dosemeci, M., Wang, Z. Y., Chow, W. H., Jiang, Z. L., Wacholder, S., Zhang, W. U., Dai, T. R., Choa, X. J., Zhang, X. C., Ye, P. Z., Kou, Q. R., Meng, J. F., Zho, J. S., Lin, X. F., Ding, C. Y., Wu, C., and Blot, W. J. (1994). Gender differences in hematopoietic and lymphoproliferative disorders and other cancer risk by major occupational group among workers exposed to benzene in China. *J. Occup. Med.* **36,** 875–881.

71. International Agency for Research on Cancer (1987). "IARC Monograph on the Evaluation of the Carcinogenic Risk to Humans. Overall Evaluation of Carcinogenicity: An Updating of IARC Monographs," Vols 1–42, Suppl. 7. IAEA, Lyon.

72. International Agency for Research on Cancer (1992). "IARC Monograph on the Evaluation of the Carcinogenic Risk to Humans. Occupational Exposures to Mists and Vapours from Strong Inorganic Acids; and Other Industrial Chemicals," Vol. 54. IAEA, Lyon.

73. Boffetta, P., Kogevinas, M., Simonato, L., Wilbourn, J., and Saracci, R. (1995). Current perspectives on occupational cancer risks. *Int. J. Occup. Environ. Health* **1,** 315–325.

74. Griesemer, R. A., and Eustic, S. L. (1994). Gender differences in animal bioassays for carcinogenicity. *J. Occup. Med.* **36,** 855–859.

75. Wu, W., Steenland, K., Brown, D., Wells, V., Jones, J., Schulte, P., and Halperin, W. (1989). Cohort and case-control analyses of workers exposed to vinyl chloride: An update. *J. Occup. Med.* **31,** 518–523.

76. Wong, O., Whorton, M. D., Foliart, D. E., and Ragland, D. (1991). An industry-wide epidemiologic study of vinyl chloride workers, 1942–1982. *Am. J. Ind. Med.* **20,** 317–334.

77. Chiazze, L., Jr., Wong, O., Nichols, W. E., and Ference, L. D. (1980). Breast cancer mortality among PVC fabricators. *J. Occup. Med.* **22,** 677–679.

78. Luce, D., Leclerc, A., Morcet, J.-F., Casal-Lares, A., Gerin, M., Brugere, J., Haguenoer, J. M., and Goldberg, M. (1992). Occupational risk factors for sinonasal cancer: A case-control study in France. *Am. J. Ind. Med.* **21,** 163–175.

79. Zheng, W., McLaughlin, J. K., Gao, R. G., and Blot, W. J. (1992). Occupational risks for nasopharyngeal cancer in Shanghai. *J. Occup. Med.* **34,** 1004–1007.

80. Chow, W.-H., Ji, B.-T., Dosemeci, M., McLaughlin, J. K., Gao, Y.-T., and Fraumeni, J. F., Jr. (1996). Biliary tract cancers among textile and other workers in Shanghai, China. *Am. J. Ind. Med.* **30,** 36–40.

81. Levin, L. I., Gao, Y. T., Blot, W. J., Zheng, W., and Fraumeni, J. F., Jr. (1987). Decreased risk of lung cancer in the cotton textile industry of Shanghai. *Cancer Res.* **47,** 5777–5781.

82. Brown, D. P., Dement, J. M., and Okun, A. (1994). Mortality patterns among female and male chrysotile asbestos textile workers. *J. Occup. Med.* **36,** 882–888.

83. Rösler, J. A., Woitowitz, H.-J., Lange, H.-J., Woitowitz, R. H., Ulm, K., and Rödelsperger, K. (1994). Mortality rates in a female cohort following asbestos exposure in Germany. *J. Occup. Med.* **36,** 889–893.

84. Delzell, E., and Grufferman, S. (1983). Cancer and other causes of death among female textile workers, 1976–1978. *J. Natl. Cancer Inst.* **71,** 735–740.

85. Linet, M. S., and Cartwright, R. A. (1996). The leukemias. *In* "Cancer Epidemiology and Prevention" (D. Schottenfeld and J. F. Fraumeni, Jr., eds.), 2nd ed., pp. 841–892. Oxford University Press, New York.

86. Yin, S. N., Li, G. L., Tain, F. D., Fu, Z. I., Jin, C., Chen, Y. J., Luo, S. J., Ye, P. Z., Zhang J. Z., Wang, G. C., Zang, X. C., Wu, H. N., and Zhong, Q. C. (1987). Leukemia in benzene workers: A retrospective cohort study. *Br. J. Ind. Med.* **44,** 124–128.

87. Oleske, D., Golomb, H. M., Farber, M. D., and Levy, P. S. (1985). A case-control inquiry into the etiology of hairy cell leukemia. *Am. J. Epidemiol.* **121,** 675–683.

88. Walrath, J., Li, F. P., Hoar, S. K., Mead, M. W., and Fraumeni, J. F., Jr. (1985). Causes of death among female chemists. *Am. J. Public Health* **75,** 883–885.

89. Burnett, C. A., and Dosemeci, M. (1994). Using occupational mortality data for surveillance of work-related diseases of women. *J. Occup. Med.* **36,** 1199–1203.

90. Morton, W., and Marjanovic, D. (1984). Leukemia incidence by occupation in the Portland-Vancouver metropolitan area. *Am. J. Ind. Med.* **6,** 185–205.

91. Giles, G. G., Liciss, J. N., Baikie, M. J., Lowenthal, R. M., and Panton, J. (1984). Myeloproliferative and lymphoproliferative disorders in Tasmania, 1972–80: Occupational and familial aspects. *J. Natl. Cancer Inst.* **72,** 1233–1240.

92. Smulevich, V. B., Fedotova, I. V., and Fikativa, V. S. (1988). Increasing evidence of the rise of cancer in workers exposed to vinyl chloride. *Br. J. Ind. Med.* **45,** 93–97.

93. Bertazzi, P. A., Riboldi, L., Pestori, A., Radice, L., and Zocchetti, C. (1987). Cancer mortality of capacitor manufacturing workers. *Am. J. Ind. Med.* **11,** 165–176.

94. Lynge, E. (1994). Danish Cancer Registry as a resource for occupational research. *J. Occup. Med.* **36,** 1169–1173.

95. Stayner, L., Smith, A. B., Reeve, G., Blade, L., Elliott, L., Keenlyside, R., and Malprin, W. (1985). Proportionate mortality study of workers in the garment industry exposed to formaldehyde. *Am. J. Ind. Med.* **7,** 229–240.

96. Aronson, K. J., and Howe, G. R. (1994). Utility of a surveillance system to detect associations between work and cancer among women in Canada, 1965–1991. *J. Occup. Med.* **36,** 1174–1179.

97. Wignall, B. K., and Fox, A. J. (1982). Mortality of female gas mask assemblers. *Br. J. Ind. Med.* **39,** 34–38.

98. Acheson, E., Gardner, M. J., Pippard, E. C., and Grime, L. P. (1982). Mortality of two groups of women who manufactured gas masks from chrysotile and crocidolite asbestos: A 40-year follow-up. *Br. J. Ind. Med.* **39,** 344–348.

99. Newhouse, M. L., Berry, G., and Wagner, J. C. (1985). Mortality of factory workers in East London 1933–80. *Br. J. Ind. Med.* **42,** 4–11.

100. Dement, J. M., Brown, D. P., and Okun, A. (1994). Follow-up study of chrysotile asbestos textile workers: Cohort mortality and case-control analyses. *Am. J. Ind. Med.* **26,** 431–447.

101. Newhouse, M. L., Berry, B., Wagner, J. C., and Turok, M. E. (1972). A study of the mortality of female asbestos workers. *Br. J. Ind. Med.* **29,** 134–141.

102. Botta, M., Magnani, C., Terracini, B., Bertolone, G. P., Castagneto, B., Cocito, V., DeGiovanni, D., and Paglieri, P. (1991). Mortality from respiratory and digestive cancers among asbestos cement workers in Italy. *Cancer Detect. Prev.* **15,** 445–447.

103. Brownson, R. C., Alavanja, M. C. R., and Chang, J. C. (1993). Occupational risk factors for lung cancer among nonsmoking women: A case-control study in Missouri (United States). *Cancer Causes Control* **4,** 449–454.

104. Gunnarsdottir, H., and Rafnsson, V. (1992). Mortality among female manual workers. *J. Epidemiol. Commun. Health* **46,** 601–604.

105. Gunnarsdottir, H., Aspelund, T., and Rafnsson, V. (1995). Cancer incidence among female manual workers. *Epidemiology* **6,** 439–441.

106. Robinson, C. F., Stern, F., Halperin, W. E., Venable, H., Petersen, M., Frazier, T., Burnett, C. A., Lalich, N., Salg, J., and Sestito, J. (1995). Assessment of mortality in the construction industry in the United States, 1984–1986. *Am. J. Ind. Med.* **28,** 49–70.

107. Wu-Williams, A. H., Xu, Z. Y., Blot, W. J., Dia, X. D., Louie, R., Xiao, H. P., Stone, B. J., Sun, X. W., Yu, S. F., and Feng, Y. P. (1993). Occupation and lung cancer risk among women in Northern China. *Am. J. Ind. Med.* **24,** 67–79.

108. Robinson, C. F., and Burnett, C. A. (1994). Mortality patterns of U.S. female construction workers by race, 1979–1990. *J. Occup. Med.* **36,** 1228–1233.

109. Blot, W. J., and Fraumeni, J. F., Jr. (1996). Cancers of the lung and pleura. *In* "Cancer Epidemiology and Prevention" (D. Schottenfeld and J. F. Fraumeni, Jr., eds.), 2nd ed., pp. 637–665. Oxford University Press, New York.

110. Menck, H. R., Pike, M. C., Henderson, B. E., and Jing, J. S. (1977). Lung cancer risk among beauticians and other female workers: Brief communication. *J. Natl. Cancer Inst.* **59,** 1423–1425.

111. Lubin, J. H., and Blot, W. J. (1984). Assessment of lung cancer risk factors by histologic category. *J. Natl. Cancer Inst.* **73,** 383–389.

112. Park, R. M., Wegman, D. H., Silverstein, M. A., Maizlish, N. A., and Mirer, F. E. (1988). Causes of death among workers in a bearing manufacturing plant. *Am. J. Ind. Med.* **13,** 569–580.

113. Wu-Williams, A. H., Dai, X. D., Blot, W., Xu, Z. Y., Sun, X. W., Xiao, H. P., Stone, B. J., Yu, S. F., Feng, Y. P., and Ershow, A. G. (1990). Lung cancer among women in north-east China. *Br. J. Cancer* **62,** 982–987.

114. Kato, I., Tominaga, S., and Ikari, A. (1990). An epidemiological study on occupation and cancer risk. *Jpn. J. Clin. Oncol.* **20,** 121–127.

115. Olsen, J. H., and Jensen, O. M. (1987). Occupation and risk of cancer in Denmark. An analysis of 93,810 cancer cases, 1970–1979. *Scand. J. Work Environ. Health* **13**(Suppl. 1), 1–91.

116. Beall, C., Delzell, E., and Macaluso, M. (1995). Mortality patterns among women in the motor vehicle manufacturing industry. *Am. J. Ind. Med.* **28,** 325–337.

117. Wang, Q. S., Boffetta, P., Parkin, D. M., and Kogevinas, M. (1995). Occupational risk factors for lung cancer in Tianjin, China. *Am. J. Ind. Med.* **28,** 353–362.

118. Merler, E., Boffetta, P., Masala, G., Monchi, V., and Bani, F. (1994). A cohort study of workers compensated for mercury intoxication following employment in the fur hat industry. *J. Occup. Med.* **36,** 1260–1264.

119. Buiatti, E., Kreibel, D., Geddes, M., Santucci, M., and Pucci, N. (1985). A case control study of lung cancer in Florence, Italy. I. Occupational risk factors. *J. Epidemiol. Commun. Health* **39,** 244–250.

120. Kabat, G. C., and Wynder, E. L. (1984). Lung cancer in nonsmokers. *Cancer (Philadelphia)* **53,** 1214–1221.

121. Milne, K. L., Sandler, D. P., Everson, R. B., and Brown, S. M. (1983). Lung cancer and occupation in Alameda County: A death certificate case-control study. *Am. J. Ind. Med.* **4,** 565–575.

122. Skov, T., Anderson, A., Malker, H., Pukkala, E., Weiner, J., and Lynge, E. (1990). Risk for cancer of the urinary bladder among hairdressers in the Nordic Countries. *Am. J. Ind. Med.* **17,** 217–223.

123. Pukkala, E., Noksol-Koivisto, P., and Roponen, P. (1992). Changing cancer risk pattern among Finnish hairdressers. *Int. Arch. Occup. Environ. Health* **64,** 39–42.

124. Silverman, D. T., Morrison, A. S., and Devesa, S. S. (1996). Bladder cancer. *In* "Cancer Epidemiology and Prevention" (D. Schottenfeld and J. F. Fraumeni, Jr., eds.), 2nd ed., pp. 1156–1179. Oxford University Press, New York.

125. Zheng, W., McLaughlin, J. K., Gao, Y. T., Silverman, D. T., Gao, R. N., and Blot, W. J. (1992). Bladder cancer and occupation in Shanghai, 1980–1984. *Am. J. Ind. Med.* **21,** 877–885.

126. You, X.-Y., Chen, J.-G., and Hu, Y.-N. (1990). Studies on the relation between bladder cancer and benzidine or its derived dyes in Shanghai. *Br. J. Ind. Med.* **47,** 544–552.

127. Bulbulyan, M. A., Figgs, L. W., Zahm, S. H., Savitskaya, T., Goldfarb, A., Astashevsky, S., and Zaridze, D. (1995). Cancer incidence and mortality among beta-naphthylamine and benzidine dye workers in Moscow. *Int. J. Epidemiol.* **24,** 266–275.

128. Gonzalez, C. A., Lopez-Abente, G., Errezola, M., Escolar, A., Riboli, E., Izaraugaza, I., and Nebot, M. (1989). Occupation and bladder cancer in Spain: A multi-centre case-control study. *Int. J. Epidemiol.* **18,** 569–577.

129. Cordier, S., Clavel, J., Limasset, J. C., Boccon-Gibod, L., Le Moual, N., Mandereau, L., and Hemon, D. (1993). Occupational risks of bladder cancer in France: A multi-centre case-control study. *Int. J. Epidemiol.* **22,** 403–411.

130. Maffi, L., and Vineis, P. (1986). Occupation and bladder cancer in females. *Med. Lav.* **77,** 511–514.

131. Swanson, G. M., and Burns, P. B. (1995). Cancer incidence among women in the workplace: A study of the association between occupation and industry and eleven cancer sites. *J. Occup. Environ. Med.* **37,** 282–287.

132. Solionova, L. G., and Smulevich, V. B. (1993). Mortality and cancer incidence in a cohort of rubber workers in Moscow. *Scand. J. Work Environ. Health.* **19,** 96–101.

133. Garabrant, D. H., and Wegman, D. H. (1984). Cancer mortality among shoe and leather workers in Massachusetts. *Am. J. Ind. Med.* **5,** 303–314.

134. Decoufle, P. (1979). Cancer risks associated with employment in the leather and leather products industry. *Arch. Environ. Health* **34,** 33–37.

135. Bond, G. G., McLaren, E. A., Cartmill, J. B., Wymer, K. T., Lipps, T. E., and Cook, R. R. (1987). Mortality among female employees of a chemical company. *Am. J. Ind. Med.* **12,** 563–578.

136. Hours, M., Dananche, B., Fevotte, J., Bergeret, A., Ayzac, L., Cardis, E., Etard, J. F., Pallen, C., Roy, P., and Fabry, J. (1994). Bladder cancer and occupational exposures. *Scand. J. Work Environ. Health* **20,** 322–330.

137. Cole, P., Hoover, R., and Friedell, G. H. (1972). Occupation and cancer of the lower urinary tract. *Cancer (Philadelphia)* **29,** 1250–1260.

138. Rosenman, K. D. (1994). Causes of mortality in primary and secondary school teachers. *Am. J. Ind. Med.* **25,** 749–758.

139. Davis, L. K., and Martin, T. R. (1988). Associations between cancer and employment [letter]. *Am. J. Public Health* **78,** 594.

140. Pukkala, E., and Saarni, H. (1996). Cancer incidence among Finnish seafarers, 1967–92. *Cancer Causes Control* **7,** 231–239.

141. Dosemeci, M., and Blair, A. (1994). Occupational cancer mortality among women employed in the telephone industry. *J. Occup. Med.* **36,** 1204–1209.

142. Park, R. M., Silverstein, M. A., Green, M. A., and Mirer, F. E. (1990). Brain cancer mortality at a manufacturer of aerospace electromechanical systems. *Am. J. Ind. Med.* **17,** 537–552.

143. McLaughlin, J. K., Malker, H. S. R., Blot, W. J., Malker, B. K., Stone, B. J., Weiner, J. A., and Ericsson, J. L. (1987). Occupational risks for intracranial gliomas in Sweden. *J. Natl. Cancer Inst.* **78,** 253–257.

144. Moser, K. A., and Goldblatt, P. O. (1991). Occupational mortality of women aged 15–59 years at death in England and Wales. *J. Epidemiol. Commun. Health* **45,** 117–124.

145. McLaughlin, J. K., Thomas, T. L., Stone, B. J., Blot, W. J., Malker, H. S. R., Weiner, J. A., Ericsson, J. L. E., and Malker, B. K. (1987). Occupational risks for meningiomas of the CNS in Sweden. *J. Occup. Med.* **29,** 66–68.

146. Ryan, P., Lee, M. W., North, J. B., and McMichael, A. J. (1992). Risk factors for tumors of the brain and meninges: Results from the Adelaide Adult Brain Tumor Study. *Int. J. Cancer* **51,** 20–27.

147. Ramazzini, B. (1700). "DeMorbis Artificum Diatriba."

148. Habel, L. A., Stanford, J. L., Vaughn, T. L., Rossing, M. A., Voigt, L. F., Weiss, N. S., and Daling, J. A. (1995). Occupation and breast cancer risk in middle-aged women. *J. Occup. Environ. Med.* **37,** 349–356.

149. Constantini, A. S., Pirastu, R., Lagorio, S., Miligi, L., and Costa, G. (1994). Studying cancer among female workers: Methods and preliminary results from a record-linkage system in Italy. *J. Occup. Med.* **36,** 1180–1186.

150. Dosemeci, M., Alavanja, M. C. R., Vetter, R., Eaton, B., and Blair, A. (1992). Mortality among laboratory workers employed at the U.S. Department of Agriculture. *Epidemiology* **3,** 258–262.

151. Norman, S. A., Berlin, J. A., and Soper, K. A., Middendorf, B. F., and Stolley, P. D. (1995). Cancer incidence in a group of workers potentially exposed to ethylene oxide. *Int. J. Epidemiol.* **24,** 276–284.

152. Belli, S., Comba, P., De Santis, M., Grignoli, M., and Sasco, A. J. (1992). Mortality study of workers employed by the Italian National Institute of Health, 1960–1989. *Scand. J. Work Environ. Health* **18,** 64–67.

153. Fleming, N. T., Armstrong, B. K., and Sheiner, H. J. (1982). The comparative epidemiology of benign breast lumps and breast cancer in Western Australia. *Int. J. Cancer* **30,** 147–152.

154. Talamini, R., La Vecchia, C., Decarli, A., Franceschi, S., Grattoni, E., Griogoletto, E., Liberati, A., and Tognoni, G. (1984). Social factors, diet and breast cancer in a northern Italian population. *Br. J. Cancer* **49,** 723–729.

155. Zheng, W., Shu, X. O., McLaughlin, J. K., Chow, W. H., Gao, Y. T., and Blot, W. J. (1993). Occupational physical activity and the incidence of cancer of the breast, corpus uteri, and ovary in Shanghai. *Cancer* (*Philadelphia*) **71,** 3620–3624.

156. Labreche, F. P., and Goldberg, M. S. (1997). Exposure to organic solvents and breast cancer in women: A hypothesis. *Am. J. Ind. Med.* **32,** 1–14.

157. O'Berg, M. T., Burke, C. A., Chen, J. L., Walrath, J., Pell, S., and Gallie, C. R. (1987). Cancer incidence and mortality in the DuPont Company: An update. *J. Occup. Med.* **29,** 245–252.

158. Hall, N. E., and Rosenman, K. D. (1991). Cancer by industry: Analysis of a population-based cancer registry with an emphasis on blue-collar workers. *Am. J. Ind. Med.* **19,** 145–159.

159. Cantor, K. P., Stewart, P. A., Brinton, L. A., and Dosemeci, M. (1995). Occupational exposures and female breast cancer mortality in the United States. *J. Occup. Environ. Med.* **37,** 336–348.

160. Thomas, T. L., and Decoufle, P. (1979). Mortality among workers employed in the pharmaceutical industry: A preliminary investigation. *J. Occup. Med.* **21,** 619–623.

161. Hansen, J., Olsen, J. H., and Larsen, A. I. (1994). Cancer morbidity among employees in a Danish pharmaceutical plant. *Int. J. Epidemiol.* **23,** 891–898.

162. Teta, M. J., Walrath, J., Meigs, J. W., and Flannery, J. (1984). Cancer incidence among cosmetologists. *J. Natl. Cancer Inst.* **72,** 1051–1057.

163. Boice, J. D., Jr., Land, C. E., and Preston, D. L. (1996). Ionizing radiation. *In* "Cancer Epidemiology and Prevention" (D. Schottenfeld and J. F. Fraumeni, Jr., eds.), 2nd ed., pp. 319–354. Oxford University Press, New York.

164. Stebbings, J. H., Lucas, H. F., and Stehney, A. F. (1984). Mortality from cancers of major sites in female radium dial workers. *Am. J. Ind. Med.* **5,** 435–459.

165. Adams, E. E., and Brues, A. M. (1980). Breast cancer in female radium dial workers first employed before 1930. *J. Occup. Med.* **22,** 583–587.

166. Pukkala, E., Auvinen, A., and Wahlberg, G. (1995). Incidence of cancer among Finnish airline cabin attendants, 1967–92. *Br. Med. J.* **311,** 649–652.

167. Cantor, K. P., Dosemeci, M., Brinton, L. A., and Stewart, P. A. (1995). Re: Breast cancer mortality among female electrical workers in the United States. *J. Natl. Cancer Inst.* **87,** 227–228.

168. Loomis, D. P., Savitz, D. A., and Anath, C. V. (1994). Breast cancer mortality among female electrical workers in the United States. *J. Natl. Cancer Inst.* **86,** 921–925.

169. Chow, W. H., Ji, B.-T., Dosemeci, M., McLaughlin, J. K., Gao, Y. T., and Fraumeni, J. F., Jr. (1996). Biliary tract cancers among textile and other workers in Shanghai, China. *Am. J. Ind. Med.* **30,** 36–40.

170. Edelman, D. A. (1992). Does asbestos exposure increase the risk of urogenital cancer? *Int. Arch. Occup. Environ. Health* **63,** 469–457.

171. Spinelli, J. J., Gallagher, R. P., Band, P. R., and Threlfall, W. J. (1984). Multiple myeloma, leukemia, and cancer of the ovary in cosmetologists and hairdressers. *Am. J. Ind. Med.* **6,** 97–102.

172. Steenland, K., and Beaumont, J. J. (1984). The accuracy of occupational and industrial data on death certificates. *J. Occup. Med.* **26,** 288–296.

173. Gute, D. M., and Fulton, J. P. (1985). Agreement of occupational and industrial data on Rhode Island death certificates with two alternate sources of information. *Public Health Rep.* **100,** 65–72.

174. Schade, W. J., and Swanson, G. M. (1988). Comparison of death certificate occupational and industrial data with lifetime occupational histories obtained by interview: Variations in the accuracy of death certificate entries. *Am. J. Ind. Med.* **14,** 121–136.

175. Messing, K., Dumais, L., Courville, J., Seifert, A. M., and Boucher, M. (1994). Evaluation of exposure data from men and women with the same job title. *J. Occup. Med.* **36,** 913–917.

176. Messing, K., Doniol-Shaw, G., and Haentjens, C. (1993). Sugar and spice: Health effects of the sexual division of labour among train cleaners. *Int. J. Health Serv.* **23,** 133–46.

177. Mergler, D., Brabant, C., Vezina, N., and Messing, K. (1987). The weaker sex? Men in women's working conditions report similar health symptoms. *J. Occup. Med.* **29,** 417–421.

178. Dumais, L., Messing, K., Seifert, A. M., Courville, J., and Vezina, N. (1993). Make me a cake as fast as you can: Determinants of inertia and change in the sexual division of labour of an industrial bakery. *Work, Employment Soc.* **7,** 363–382.

179. Greenberg, G. N., and Dement, J. M. (1994). Exposure assessment and gender differences. *J. Occup. Med.* **36,** 907–912.

180. Goldenhar, L. M., and Sweeney, M. H. (1999). Tradeswomen's perspectives on occupational health and safety: A qualitative investigation. *Am. J. Ind. Med.* (in press).

181. U.S. Department of Labor (1993). "Women Workers: Trends and Issues," p. 183. U.S. Govt. Printing Office, Washington, DC.

182. Herold, J., and Waldron, I. (1985). Part-time employment and women's health. *J. Occup. Med.* **27,** 405–412.

183. McMichael, A. J. (1976). Standardized mortality ratios and the "healthy worker effect": Scratching beneath the surface. *J. Occup. Med.* **18,** 165–168.

184. Sorlie, P. D., and Rogot, E. (1990). Mortality by employment status in the National Longitudinal Mortality Study. *Am. J. Epidemiol.* **132,** 983–993.

185. Kryston, K., Higgins, M., and Keller, J. (1983). Characteristics of working and non-working women in Tecumseh. *Am. J. Epidemiol.* **188,** 419 (abstr.).

186. Hoar, S. K., Morrison, A. S., Cole, P., and Silverman, D. T. (1980). An occupation and exposure linkage system for the study of occupational carcinogens. *J. Occup. Med.* **22,** 722–726.

187. Stewart, P. (1993). Assessing occupational exposures in epidemiology studies. *Proc. 24th Meet. Int. Congr. Occup. Health,* Nice, France, pp. 137–149.

188. Herrick, R. F., Stewart, P. A. (1991). International workshop on retrospective exposure assessment for occupational epidemiologic studies: preface. *Appl. Occup. Environ. Hyg.* **6,** 417–420.

189. Gerin, M. (1990). Recent approaches to retrospective exposure assessment in occupational cancer epidemiology. *Recent Results Cancer Res.* **120,** 39–49.

190. Cedillo Becerril, L. A., Harlow, S. D., Sanchez, R. A., and Sanchez Monroy, D. (1997). Establishing priorities for occupational health research among women working in the maquiladora industry. *Int. J. Occup. Environ. Health* **3,** 221–230.

40

Indoor Air Quality

MICHAEL HODGSON AND EILEEN STOREY
Division of Occupational and Environmental Medicine
University of Connecticut Health Center
Farmington, Connecticut

I. Introduction

This chapter will address what we know about measurement and treatment of symptoms and health effects in office workers. Do symptoms or perceived health effects represent important outcomes? Are measurement techniques valid? What are their causes? What do we need to know to intervene and prevent them? Are the interventions that are commonly undertaken effective or healthful?

Office work has, stereotypically, been considered women's work. Over 50% of women work in indoor environments (see Chapter 36). The single largest occupational category remains, as it was in 1981, "secretary." Because the proportion of women in the workforce has been rising and is projected to continue, adverse health risks in this occupation remain of increasing concern. This chapter summarizes current thinking on indoor environmental health. A first section outlines building-related diseases, diseases with well-defined criteria for recognition, readily available attribution strategies, and remediation strategies. A second section summarizes our current knowledge on the "sick-building syndrome," a term that remains without operational definition but is used for symptom conglomerations that individually often have equally well-defined mechanisms but whose "illness" potential remains poorly described. A third section summarizes briefly our knowledge of engineering considerations for these complaints. A final section outlines the current etiologic theories.

II. Diseases and Mechanisms

Diseases related to indoor environments do not differ from those same diseases related to other factors. Documentation of disease, per se, is therefore no different. Diseases related to indoor environments do differ from nonoccupational and environmental forms in that their etiology is related to an exposure that is remediable. Identification of the cause, based on a specific linkage strategy, is therefore of paramount importance. Such linkages are far more persuasive to both the professional and regulatory community and more clearly define economic implications than the complex of symptoms commonly labeled "sick building syndrome". Table 40.1 presents a summary of those diseases, diagnostic criteria, and linkage strategies. In practice, patients present with some, but rarely with even most, of the pertinent symptoms. In group presentations, outbreaks, or epidemics, recognition is far more straightforward than in sporadic isolated cases where identification of the specific source (home versus work) may prove more difficult.

Many physicians practicing occupational and environmental health adhere to the sentinel health events mode of practice. This implies that all work-related diseases lead investigators to an exposure that is remediable and to an exposed cohort that requires some screening for further disease. The failure to identify additional cases of disease, and to look for them, lets us avoid recognizing the true frequency of disease.

A. Interstitial Lung Disease

1. Epidemiology

The interstitial lung diseases (ILD), such as interstitial pneumonitis (UIP), hypersensitivity pneumonitis (HP), sarcoidosis, and others, occur at a prevalence of about 70 per 1,000,000 persons and at an incidence of 30 per 1,000,000 person-years [1]. Less than 3% of these cases represent hypersensitivity pneumonitis, the disease most clearly associated with moisture and mold; over 80% are considered idiopathic interstitial pneumonitis. Nevertheless, the single most frequently discussed building-associated lung disease remains hypersensitivity pneumonitis (HP).

Since the report of Pestalozzi [2] that attributed an inhalation fever to a humidifier in 1959, immunologic lung disease has been attributed to indoor environments. A first review [3] identified outbreaks and case reports with a common thread of unwanted moisture, generally associated with mechanical ventilation of air conditioning. The National Institute for Occupational Safety and Health (NIOSH) investigated several additional outbreaks [4–6] and suggested that moisture, not just in ventilation systems but also in construction, was associated with HP, and it outlined seven steps to preventing disease [7]. More recent outbreaks continue to support this view [Jarvis, personal communication; 8]. There is no question about this relationship.

Some evidence suggests that other forms of interstitial lung disease (ILD) might be related to buildings. In a case-control study of UIP (excluding the pneumoconioses), individuals with mold exposure in their workplace, including buildings, were approximately 15 times more likely to develop such diseases [9]. A case report of nonspecific interstitial pneumonitis without granulomata, and therefore distinct from HP, has been associated with fungal exposure [10]. Sarcoidosis is meanwhile considered likely to represent an inhalation disorder with a different immunology from hypersensitivity pneumonitis [11]. Reports of disease at least initially appearing to be sarcoidosis have been related to a wet building [12,13]. In addition, in a replication of the case-control study of UIP that used sarcoidosis patients as

Table 40.1

Diseases Associated with Buildings

Disease	Symptoms	Mechanisms	Diagnostic criteria	Diagnostic tests	Linkage strategies	Causes
Rhinitis, sinusitis	Stuffy or runny nose; postnasal drip; "sinus congestion"; nosebleeds	Allergy (IgE)	Eosinophilic granulocytes in nasal secretions; specific changes on nasal provocation; symptoms	Nasal secretions; eosinophiles; CT scan for chronic inflammatory changes; acoustic rhinometry; rhinomanometry (anterior and posterior)	IgE-based: RAST or skin prick tests; nasal challenge symptom patterns	Sensitizers in the workplace (allergens) including molds, carbonless copy paper, photoactive processes (toners), and secondary exposures, e.g., cat danders brought to work on clothing; pesticides (OPs, pyrethrins)
		Irritation	None described	None described	Symptom patterns	Irritant exposures, including cleaning agents, volatile organic compounds, dust, molds and bacteria; low relative humidity
Allergic fungal sinusitis	Stuffy nose; postnasal drip; headache	Allergy (IgE); other?	CT-scan evidence of inflammation; eosinophilic mucus; fungal hyphae visible on staining	Surgical tissue for eosinophilic; staining in mucous; fungal bodies	Recurrences at work; same organisms in workplace	Bioaerosols at work?
Asthma (airways disease)	Coughing; wheezing; shortness of breath; chest tightness	Allergy (IgE, IgG)	Airways hyper-reactivity (more than 15% change in FEV1); reversibility of obstruction with bronchodilators	Physical examination; spirometry with bronchodilators; methacholine challenge; substance specific challenge	Temporal relationship of lung function decrements at work (PEFR, spirometry); immunologic tests (skin prick tests, RAST tests);	Allergic rhinitis
		Pharmacologic irritant	See above; history of exposure during onset of asthma	See above	See above; clinical history	OPs; irritant exposures (cleaning agents, accidental spills); fungi and bacteria (cell wall components, secondary metabolites)
Hypersensitivity Pneumonitis	Cough; shortness of breath; muscle aches; feverishness	Type IV allergy (cell-mediated immunity)	Granulomas on biopsy; reversible restrictive changes (DLCO, TLC, CXR); CT scan: thin-section CT scan	Granulomas; restrictive changes in lung function in a convincing clinical setting	Physiologic linkage (acute disease: reversible patterns); immunologic linkage (IgG-exposure only); lymphocyte transformation to specific antigens	Molds and thermotolerant bacteria related to moisture
Organic dust toxic syndrome (inhalation fevers)	Cough; shortness of breath; muscle aches; feverishness	Endotoxin response (macrophage receptor based effect)	DLCO; TLC; spirometry; white blood cell count elevations	Bronchial lavage; DLCO: TLC; spirometry; white blood cell count elevations related to work	Temporal pattern and exposure documentation	Gram negative bacteria, molds, and polymers in thermal degradation
Contact dermatitis (allergic)	Dry, itching, scaling, red skin	Type IV skin allergy	Inspection; skin biopsy; patch testing	Inspection, skin biopsy; patch testing	Patch testing; temporal pattern	Formaldehyde; molds, laser toners; behenic acid (photoactive process)
Irritant contact dermatitis	Dry, itching, or weeping skin	Irritation	Inspection; skin biopsy	Clinical history; "lactic acid application" ("stinger test")	Temporal pattern	Office products, VOC based
Contact urticaria	Red, irritated skin; hives	Type I allergy	Inspection; RAST or skin prick testing	Clinical history; RAST test	Temporal pattern; RAST or skin prick testing	Office products
Conjunctivitis (allergic)	Eye irritation, dryness, tearing	Type I allergy	Inspection; RAST or skin prick testing; tear-film break-up time; conjunctival staining	Punctate conjunctivitis; shortened tear-film break-up time; RAST or skin prick testing; cobblestoning on physical examination	Clinical impression	Molds; sensitizers
Conjunctivitis (irritant)	Eye irritation, dryness, tearing	Irritation	Inspection; tear-film break-up time	Inspection; tear film break-up time	Clinical impression	Irritants (VOCs, dust, low relative humidity); failure to blink at VDUs
CNS toxicity	Headaches; cognitive impairment	Carbon monoxide	Elevated carboxyhemoglobin (COhgb)	Neuropsychological tests; COhgb levels	COhgb > 3% in non-smokers; > 8% in smokers	Fossil fuel sources. Home: attached garages; backdrafting appliances; barbecues. Commercial/public buildings: entrained exhaust, indoor CO sources
		VOCs	Abnormal neuropsychological tests	Abnormal neuropsychological tests	Abnormal neuropsychological tests	Imbalance between local VOC sources and exhaust ventilation
		Heat: noise	Calculated heat indices outside range; noise levels above comfort range	None	Clinical impression	Inadequate control strategies
Legionnaire's disease	Coughing; phlegm production; fevers	Legionella exposure; susceptible host?	Four-fold rise in antibody titers; culture of same Legionella strain from source as from tissue; epidemiologic clustering	Four-fold rise in antibody titers; culture from human samples; epidemiology	Same organisms from putative source and from patient; clinical pattern	Aerosol dispersion; aspiration
Tuberculosis	Coughing; phlegm production; fevers; weight loss	Infection with *Mycobacterium tuberculosis*	Isolation of organisms; positive skin test with typical chest x-ray pattern	PCR similarity between organism from patient and from source	PCR similarity between organism from patient and from source	Inadequate control strategies

Note. FEV1, forced expiratory volume in the first second; PCR, polymerase chain reaction-based lab test; RAST, radioallergosorbent test; DLCO, single breath carbon monoxide diffusing capacity; TLC, total lung capacity; CXR, chest x-ray.

cases, subjects were eleven times more likely to have mold exposures in their work places, including buildings, than controls [14]. Both of these case-control studies identified elevated risks associated with wet basements [9,14].

The data are not yet robust enough to allow calculation of attributable risks because of the small size of the studies, but standard calculations suggest population-attributable risks in the range of 8–10%.

2. Clinical Aspects

HP has been reviewed by Daroowalla and Raghu [15] and Salvaggio [16]. In the past, almost all patients with evidence of interstitial lung disease underwent biopsy to identify the nature of disease. This occurs less frequently now despite increasing evidence that radiology is quite insensitive [17]. In the one recent population-based investigation, less than 10% of chest X-rays were abnormal and less than 50% of thin-section computer tomography (CT) scans were abnormal [18]. These data suggest that few patients had abnormal lung function tests [19]. We are left with a relatively invasive technique, lung biopsy, as the only specific diagnostic technique to characterize disease that has been identified through sensitive screening instruments such as questionnaires.

Linkage of any of these lung diseases to specific exposures is difficult because there are only a few pathognomonic findings on lung biopsy (e.g., silica nodules and silica, Giant cells and cobalt). Although some authors have attempted to use antibodies, these have long been considered markers of exposure rather than of effect [20]. The linkage to an exposure generally relies on a history of exposure to an agent that has been implicated epidemiologically. This is tedious and makes the recognition of new causes problematic. Nevertheless, for ILD in most of its forms this is the only available technique.

3. Organic Dust Disorders

Some evidence exists that office workers may develop yet another form of interstitial pulmonary response, possibly more frequently than commonly assumed, called an "organic dust toxic syndrome," (the same disease as humidifier fever). This evidence was reviewed by Milton [21]. Beginning with Pickering's search for humidifiers as a cause of symptoms among office workers [22], excess rates of chest tightness and flu-like illness have been seen. A reanalysis of several older data sets [23] suggested that a symptom cluster of chest tightness, difficulty breathing, and flu-like illness was distinct from that of mucosal irritation, central nervous system symptoms, and skin irritation. Although this syndrome is considered benign and self-limited, there is some evidence for impairment of lung function [21].

4. Asthma

Between 5 and 8% of the U.S. population have asthma. In children and in adults, asthma is associated with moisture in the home [24]. Cooper [25] estimated that the etiologic fraction might be approximately 25%. In adults, about 20 per 100,000 persons will develop new onset asthma from their workplace [26], and almost 1% of the U.S. population is thought to have occupational asthma [27]. According to reports for state health departments where asthma is a reportable disease, approxi-

mately 12% of work-related asthma [27a] was attributed to buildings between 1993 and 1995 using SENSOR criteria [28]. Only one outbreak of building-related asthma has been described [29]. Nevertheless, in a series of NIOSH building investigations, chest symptoms appeared to be related to specific aspects of building operations including moisture, dirt, and debris in the ventilation system [30]. This suggests that building-related asthma may in fact be much more common than is now recognized.

Very few cases of occupational asthma related to buildings are recognized as such. Nevertheless, a Swedish [31] study of office workers suggested that they die of asthma more frequently than the general population. Premature mortality costs also are not covered in direct medical benefits. Nevertheless, a follow-up study did not confirm this elevation. That means that of the $4–6 billion in annual medical costs, about $0.52–1.56 billion of direct medical costs may be related to buildings. If 25% of adult asthmatics (⅓ of medical costs) have building related asthma, direct medical costs attributable to buildings range from $42.9 million to $129 million.

5. Clinical Aspects

Linkage strategies for asthma are well developed (Asthma chapter this volume). In addition to a careful history, the primary linkage rests on documentation of temporal relationships. Initial onset or reoccurrence of asthma after beginning work in a building, or exacerbation of previously stable asthma, may lead to suspicion of work-relatedness. Examination of spirometry or peak expiratory flow rate patterns at and away from work are the usual linkage strategy.

The standard calculations for population attributable risk (or "etiologic fraction") suggest that the population attributable risk of asthma ranges from 13 to 26% for moisture in buildings, including homes and work places [25].

B. Allergic Rhinitis

Twenty to thirty percent of the U.S. population have complaints of a stuffy or runny nose [32]; approximately half of these cases are thought to be allergic rhinitis and some proportion of these are chronic irritation. Although the disease is clearly recognizable, no good linkage strategies have been developed for clinical use. Such symptoms among office workers are meanwhile thought to reflect both allergy and simple mucosal irritation. Studies of the upper respiratory tract suggest that at least some measures of effect, including cells in nasal lavage fluid, physiologic changes in the nose, or biopsy evidence [33–38] are present when sought with sophisticated techniques. Similarly, challenges with specific substances are well known and widely accepted as accurate, although they are rarely performed and are not widely available clinically.

The direct medical costs for allergic rhinitis are estimated at $1.8 billion. People with allergic rhinitis may lose several days of work per year or, for the period when they are symptomatic, have decreased productivity. Cost estimates separating work from home are unavailable. Under the assumption that half of the allergic rhinitis cases are associated with work (based on questionnaire surveys) and under the assumption that these are associated with moisture indicators, population attributable risks suggest an etiologic fraction of less than 6% for the workplace. Menzies [39] suggested that only approximately 5% (total: 1%) of the 20% of office workers with work-related nasal symptoms developed their symptoms on the basis of IgE-based mechanisms, although other researchers [40] identify a greater proportion. Organic dusts, specifically endotoxin, are associated with mucosal irritation [41] and may represent a second bioaerosol-based mechanism by which these symptoms occur. Mucosal irritation, as described in the next section, is likely to represent a third possible mechanism.

At present, no specific diagnostic techniques have been described to link such disease to the work place in a clinical setting. It is possible to use the techniques outlined earlier to document changes at work, although the implementation of such studies is likely to be difficult and expensive.

Allergic fungal sinusitis is a third form of nasal disease likely related to indoor environmental exposures. The frequency is at present unclear. Diagnosis is made through documentation of mucosal thickening on CT scan or other primary diagnostic procedures, the documentation of eosinophilic mucin in sinus drainage or biopsy material, and the presence of fungal organisms in those same materials [42].

C. Mucosal Irritation

Eye irritation is a common complaint among office workers and is widely thought by some to be the primary driver for the widespread interest in indoor environmental health effects. A wide range of frequencies of work-related mucosal irritation has been described, ranging from 8% to well over 30% in randomly selected buildings. In buildings with known problems, this frequency is even higher [40]. Additional individuals have persistent eye irritation without relief on weekends. These symptoms may be due to factors other than external exposures, or these individuals may have similar exposure problems at home. Some data suggest that residential mucosal irritation is in fact far more frequent than assumed [43]. Epidemiologic data have long suggested that atopic individuals describe more mucosal irritation than nonatopic individuals. Chamber studies have confirmed that atopic individuals describe higher degrees of mucosal irritation at any given level of irritant exposures than do nonatopic individuals [44].

Studies on the physiology of mucosal irritation are well known. Work by Franck, Kjaergard, Skov, and Molhave [45] suggests that complaints of eye irritation are measured reliably by different questionnaires, that there are physiologic indicators of eye irritation, and that at the very least subjects with those markers are at greater risk of eye irritation than subjects without. Whether tear film break-up time, fat foam thickness, or canthal foam represent markers of susceptibility, markers of chronic effects, or mechanisms of disease remains unclear. Punctate conjunctivitis, documented with lissamine green or fluorescein staining, represents a commonly recognized acute effect of irritants.

The work by Cain and Cometto-Muniz [48] has shed more light on the consideration that mucosal symptoms may represent simple irritation by commonly encountered volatile organic compounds. Quantitative structure activity relationships suggest a predictable, dose-dependent effect [47], at least for "nonreactive", or relatively inert, compounds. Individual agents are

therefore clearly shown to have an irritant threshold, usually two to four orders of magnitude below established criteria such as Permissible Exposure Levels (PELs) set by the Occupational Safety and Health Administration, the American Conference of Governmental Industrial Hygienists, or other standard setting bodies. In addition, Cometto-Muniz and Cain [48] have suggested that combinations of subthreshold concentrations are more than additive, so that complex mixtures of mild irritants may be far more irritating in offices than predicted on the basis of their individual toxicities.

D. Infections

Much interest has developed about infections indoors because of concerns that these might be transmitted through ventilation systems. These infections include human source infections, such as tuberculosis, pneumococcal pneumonias, and viruses; infections emanating from building systems, such as Legionnaire's disease from cooling towers or potable water systems; and infections from other sources such as animal research facilities.

Outbreaks of tuberculosis in hospital settings are well known. Such ventilation-system associated clusters also have been reported for office workers [49] and attributed to dysfunctional ventilation systems in jails [50]. Outbreaks of animal-source diseases, such as Q-fever or histoplasmosis from pigeon droppings, have been attributed to entrainment into ventilation systems [3]. At least one outbreak of pneumococcal pneumonia has been attributed to crowding in jails [51], although this was thought to be due to person to person transmission. Similarly, viral epidemics of highly contagious agents such as chicken pox and measles have occurred via dissemination through ventilation systems.

More interest has arisen about whether common respiratory tract viruses, primarily rhino viruses, are likely to cause excess preventable respiratory tract disease among office workers that may be reduced by changes in ventilation rates [52] (although this manuscript does not distinguish between the viruses that may and those that may not be transmitted through droplet aerosols). Older literature supported hand-to-hand transmission through secretions and suggested, based on experimental and field studies, that this was a more effective form of transmission than the airborne route [53,54]. At the same time, experimental evidence suggested that droplet transmission might in fact contribute to the burden of disease [55]. Because of the known associations with emotional states and fomite transmission, the importance of this model A field study in barracks appeared to support this hypothesis, although no data on ventilation or air exchange rates were available [56], and a follow-up study failed to find supporting evidence [57]. Even if disease is transmitted through droplets, reduction in frequency through general dilution ventilation may not come cheaply because of the inefficiency of general dilution ventilation in reducing local exposures [49] and because it may not intervene on the disease burden associated with emotional states.

E. Dermatitis

Contact dermatitis is recognized as a common problem among office workers. This problem ranges from allergic contact dermatitis and urticaria [58] to skin irritation from dryness. Rare cases of vasculitis ("palpable purpura") have been associated with photoactive copy paper [59].

Interest has focused on whether low relative humidity is associated with irritation alone or imitation in combination with irritant exposures, such as volatile organic compounds. Close scrutiny of chamber studies clearly documents mucosal irritation at levels of relative humidity below 20% [60]. Field studies have suggested that mucosal irritation increases as relative humidity drops below 35% [61].

F. Miscellaneous Disorders

The building environment is put to many uses so that it is impossible to predict all of the potential exposures and diseases that may occur. "Mixed-use" buildings are those with work processes that are not just restricted to traditional office work. Printing shops, auto body shops, and dry-cleaning establishments may contribute to the levels of volatile organic compounds in offices. Garages, loading docks, and fossil-fuel powered floor buffers may contribute to carbon monoxide levels.

Other processes indoors may lead to problems. Organophosphate poisoning has been attributed to entrained organophosphates [62,63]; such events may be more frequent than is commonly assumed [64]. Emissions from architectural blue print machines have been associated with "palpable purpura," a form of blood vessel allergy [59]. Carbonless copy paper has been associated with contact urticaria and asthma [65], although its primary effect may be simple mucosal irritation [66].

Equally importantly, physical conditions indoors may lead to comfort problems. The combination of high indoor temperatures and low water consumption among teachers has been associated with mild heat illness, which presents as headaches and fatigue. Transmitted vibration from mechanical building systems has been associated with headaches, dizziness, and irritability [67].

When strong suspicion arises about potential building-relatedness of complaints, very thoughtful approaches may be required to identify true causal connections.

III. The Sick Building Syndrome

The term Sick Building Syndrome (SBS) has no operational definition. It remains unclear whether a single individual or a group of building occupants develop the problem, how frequently, on average how many individuals must suffer symptoms, and what proportion of individuals must describe symptoms. Although the symptoms are often labeled as "nonspecific," well-documented specific mechanisms are clearly adequate to explain them. In addition, potential causes for such symptoms are widely recognized as being present indoors, but causal linkage strategies often have not been developed on an individual level so that attribution in specific cases remains difficult. Jaakola [68] suggests that the term "SBS" represents a construct for discussion purposes only.

Nevertheless, some authors have begun using numerical criteria that include: a percentage of affected, such as 20%; temporal characteristics of symptoms, such as at least 3 days per week and improvement on weekends; and number of symptoms

or symptom categories. These definitions have no external scientific validity. Scrutiny of studies suggests that in all buildings, at least 6% of subjects will describe symptoms occurring most days of the week. Often 50–60% of subjects describe symptoms occurring more than monthly that they attribute to work. In general, subjects in buildings with recognized outbreaks of disease have higher rates of symptoms, and frequencies of 70–80% are recognized in buildings with clearly identified disease.

Few publications provide data to scrutinize the range of symptoms meaningfully. In the U.S., the only published data set [69] of a large number of buildings examined with the same questionnaire suggested that 30% of workers overall experienced eye irritation, 25% experienced unusual tiredness, 25% had headache, and 21% had nasal symptoms on almost every day, with improvement away from work. In general, women were 50% more likely to describe all of these symptoms. Comparison of buildings from a national "problem-building" data set [30] with a set without known complaints that used the same questionnaire [Brightman, unpublished data] suggests that workers from the former report, on average, 25% higher rates of symptoms.

Measurement of Symptoms among Office Workers: Questionnaires and Validity

Symptoms often are assumed to represent an important outcome in and of themselves, independent of the underlying mechanisms. Questionnaire responses represent the symptoms being measured, and are influenced by factors such as affect, work stress experience, perceptions of environmental factors (temperature, noise), or disease. Although questionnaires are used as the standard measurement instrument, their interpretation is not straightforward.

Several different questionnaires have been used widely, including one developed and modified by the EPA and NIOSH [30,70], one used in the UK [71], and several used in Sweden and Denmark that also exist in English language versions. At least one study used direct interviews [22]. These all inquire about the frequency of symptoms. Separately, several investigators have used linear or visual analog scales [72,73] in an attempt to identify specific factors. Nevertheless, none of the instruments distinguish the postulated causes of symptoms. Physicians frequently distinguish among "tension-type" headaches, migraines, cluster headaches, and others such as CO or solvent-induced headaches; some formal tools do exist for these [74], including in field studies [75,76], but they have not been used in the study of office workers. Similar considerations exist concerning eye and nose irritation, chest symptoms, and skin symptoms. Where validation has been mentioned [77,78], usually no numerical or statistical indicators of reliability and precision were presented. Although a reasonable correlation does exist between the general symptom categories on the most widely used questionnaires [23], temporal characteristics such as weekend improvement differ considerably, with kappa scores of less than 0.15.

Several authors have compared the range of questionnaires and data analysis methods. Although somewhat different results are obtained with the various instruments, in general they do perform similarly for symptom complexes that we understand. Hodgson and Turner (Hodgson 1990) examined whether use of the Work Environment Scale (WES) and the EPA BASE questionnaires would lead to identification and classification of office areas as higher and lower complaint areas. There was a general tendency for the questionnaires to rank areas similarly, although the rank order often was different. Apter [23] examined the relationship between symptoms defined with the EPA BASE and the WES on an individual level. Symptoms of mucosal irritation scores correlated reasonably well, with kappa values above 0.35, despite minor differences in wording. On the other hand, almost no relationship was seen between "general symptoms" (*i.e.,* headaches and fatigue), with similar minor difference in wording. There appears to be "convergence validity" for mucosal irritation symptoms, as similar results are achieved with different instruments, but this is not the case for headaches. This obviously does not mean that headaches or fatigue are meaningless or less important among office workers, but simply that the results of studies are more difficult to interpret. Headaches remain the most common symptom that is attributed to the indoor work environment.

A larger problem is that of "work-relatedness." Susceptible subjects may develop their symptoms at work and at home, so that excluding them from analysis may not be necessary. On the other hand, questions concerning work-relatedness may not give reliable answers, in part for the reasons listed above. Some empirical evidence was collected and suggests that the EPA BASE instrument and the WES perform differently in assessing work-relatedness. Kappa statistics were below 0.15, indicating poor agreement [23]. This is of some concern to investigators relying on questionnaires alone for such determinations.

These considerations may lead us to think more carefully about how we study these symptoms, how we interpret the results of questionnaire surveys, and how we seek their explanations. They certainly support the idea that a search for physiologic underpinnings for these diagnoses is important.

IV. Engineering and Health

Ventilation has been mandated for some building environments since at least 1895 [79] for health considerations. From the 1930s to the 1950s, Yaglou, in the American Society of Heating, Refrigerating, and Air Conditioning Engineers (ASHRAE) Laboratory at the Harvard School of Public Health Ventilation documented relationships between comfort and ventilation. The study included odor control requirements and even addressed tobacco odor. In 1973, ASHRAE published its first ventilation standard. At present, no federal law addresses ventilation requirements, although OSHA proposed a standard on indoor air in 1995. State laws exist in several states, including California, New Jersey, and Washington.

The current ASHRAE standard [80] suggests two possible approaches, one based on prescriptive rates per occupant ("ventilation rate procedure") and one based on a mass-balance approach. The latter approach requires selection of a target concentration, identification of emission rates of pertinent pollutants, and calculation of the required dilution rates.

Despite the presence of such a standard, many buildings do not meet the recommended guidelines. This number has varied from about 60% of problem buildings in the 1980s to over 30% in the late 1990s (see Table 40.2) [80a]. Importantly, other prob-

Table 40.2

Frequency of Engineering (Design, Construction, Maintenance) Problems in the Case Series

	Woods	Robertson	NIOSH (1996)
Outside air	75	64	40
Distribution systems	75	46	24
Drain pan	60	63	17
Filters	65	57	32
Contamination	45	38	38

Note. Each cell represents the percentage of buildings with problems in that known category. See [30] and [80a].

lems besides inadequate ventilation alone may contribute to perceived inadequate air quality. Although the view that odors are not a solvable problem is widespread, one group documented that specific likely causes were identified for odor problems when an aggressive, standardized approach was implemented [81].

In the home, some evidence suggests that commercial building approaches have changed ventilation rates over time [82], with the implication of inadequate air exchange for comfort. This reduced air exchange may also put individuals at risk for back-drafting-associated carbon monoxide poisoning [83]. Finally, in addition to the work on moisture and lung disease, work has begun to address the relationships between engineering design and construction defects and bioaerosol exposures [84,85].

V. Associations of Health Outcomes and Exposure

The SBS must be dealt with on three levels: office workers who are sick, the building systems, and the work process itself (a fundamental belief in occupational health where work, the worker, and the workplace are considered three distinct but key ingredients) [86]. The latter represents not just the work process in the physical sense—with engineering generation and control strategies—but also the social and organizational structure in the workplace. Unless problems are addressed at all three levels, it is hard to come to understand the processes or to intervene effectively. In a topic such as this, the true etiology is of interest because it may lead to intervention strategies in specific buildings on an individual level and, in the long run, guide engineering strategies. In addition, cost considerations are of interest because they may generate the attention of engineers and managers in a way that may justify intervention.

Through cross-sectional and quasi-experimental field studies, a number of environmental factors appear to be consistently associated with office worker symptoms. There are approximately 30 cross-sectional surveys using between 12 and 70 office buildings as the sampling unit defining worker complaint frequencies [87]. The most striking result of these studies is that symptoms were consistently associated with mechanical ventilation systems, a finding seen in subsequent cross-sectional studies [88]. A range of additional associations is widely recognized. Table 40.3 summarizes these additional "risk factors" for symptoms. Almost universally, these investigations have re-

Table 40.3
Factors Associated with Symptoms

Building factors
    Mechanical ventilation
    Dirty ducts
    Inadequate maintenance
    Recent carpeting
    Fuzzy surfaces
Work process
    Work stress
    Carbonless copy paper
    Photocopying
Individual susceptibility
    Atopy
    Gender
    Seborrheic dermatitis

lied on self-administered questionnaires to identify both outcome (symptoms) and causes (environmental factors). In epidemiologic investigations, such approaches to exposure assessment generally are viewed with skepticism because they are subject to a number of biases that are labeled the common instrument fallacy. Because office workers are aware of the reason for the investigations, their self-awareness may lead symptomatic individuals to inflate symptom magnitude. Because of this same awareness, respondents may overemphasize specific exposures when presented with semiquantitative choices such as "hours of work with . . ." or "discomfort from . . .". Finally, it is not unreasonable to assume that self-estimated measures of productivity are subject to the same problem.

Investigations of problem buildings have consistently demonstrated problems with building systems. Table 28.2 presents a comparison of the frequencies identified in three well known series of building investigations. All buildings had more than one problem identified. This suggests that more than one problem may need to be addressed before building occupants can be expected to describe substantially fewer complaints.

Although there is a widespread perception that symptoms are not associated with specific exposures, several field studies confirm the laboratory-based hypothesis that symptoms may be associated with complex mixtures of volatile organic compounds (VOCs) [89,90]. In addition, quasi-experimental field studies, that is, manipulation of real buildings in a quasi-experimental way with repeated measures of symptoms, have demonstrated effects of relative humidity, particulates, and thermal comfort on symptoms. Nevertheless, of these pollutants, VOCs have generated the largest body of interest and literature.

A. The Ventilation/VOC Hypothesis

Support for the ventilation/VOC hypothesis exists in field studies, chamber studies, and theoretical derivation [91]. Mendell [87], after reviewing published studies, suggested that symptoms began to increase as quantities of outside air per occupant decreased to below 10 cfm. Similarly, Sundell [92], in a careful population-based study, showed a dose-response relationship between symptoms and outside air measured with tracer gas dilution. Quasi-experimental field studies have demonstrated decreases in symptoms with increases in delivered ventilation. Equally importantly, Teijonsalo [93] suggested that although the overall ventilation rate into buildings might be adequate, there was strong evidence of inhomogeneity across different rooms in the relationship of delivered quantities of air and occupant needs within any given building.

"Inadequate ventilation" cannot by itself cause symptoms but requires some individual or combination of pollutants to trigger symptoms. Two main candidates have been identified. The complex mixture of VOCs encountered in indoor air is clearly irritating and has been well documented through the work of Molhave et al. [37] and Cain [46]. Estimation of the irritant potency of such mixtures on the basis of quantitative structure activity relationships is possible but has not been published in the peer-reviewed literature [47]. Field data do support the presence of compounds with a wide range of irritation potential, though their assessment remains quite controversial. Beginning in 1992, Wechsler and Shields [94,95] documented the interaction of VOC emissions from various sources indoors, including carpets, paints, caulks, sealants, and office processes that lead to the production of aldehydes in a ratio of 2:1 for each molecule of ozone. Aldehydes are not only far more irritating but the numeric multiplier effect is quite dramatic. The topic has been reviewed by Wakoff et al. [96]. This reformulation of the VOC hypothesis explains the problem of Sundell's "lost VOCs." Chamber studies clearly document the ability of complex mixtures to cause such effects and at least two field studies [89,90] did identify dose-response relationships.

B. Microbial Hypothesis

Before Fanger introduced the "olf" and "decipol" units, ventilation systems were not included in the list of potential culprits of work-related illness. Although the units may have fallen out of favor, his work suggested for the first time that building components such as mechanical ventilation systems were actually very important sources of occupant discomfort. He developed a rating method using visitor panels to rate air quality in buildings under three different conditions: unoccupied/unventilated, occupied/unventilated, and occupied/ventilated. Almost one-half of the perceived pollution load resulted from ventilation systems [97]. Although this method is no longer used, the documentation of the potential for systematic contribution of ventilation systems to the problem was startling to most professionals. A subsequent review [98] provided ample evidence of this possibility. Interestingly, more detailed examination of studies that presented symptom rates by type of ventilation system suggests that symptoms also are associated with humidification systems [22,71]. In addition [99], data suggest that enthalpy (temperature and relative humidity) is more strongly associated with symptoms than either temperature or relative humidity alone. The second major hypothesis implicating ventilation systems postulates that there are microbial products that make people complain of illness. There is some evidence to support this hypothesis, although there is substantially less evidence than for the VOC hypothesis. Work that supports the second theory includes both bacterial and fungal hypotheses.

The evidence linking bacterial exposure to office worker symptoms exists both on a mechanistic and on epidemiologic levels. The cell wall components of bacteria generally consist of endotoxins, a class of lipopolysaccharides that react with cell wall receptors, especially on alveolar macrophages. These cells release tumor necrosis factor, a variety of other cytokines, and platelet activating factor [21]. Epidemiologic evidence suggests some association of symptoms with measures of organic residues. Teeuw [100] found surprisingly high levels of endotoxins in a study of 19 buildings and demonstrated dose-response relationships between increasing concentrations of measured endotoxin and symptoms. Rylander [101] found a similar result. Harrison [102] suggested a weak relationship between airborne bacterial counts and symptoms. A reanalysis of the Danish Town Hall study demonstrated higher rates of symptoms related to higher gram negative bacteria, though not to endotoxin levels in dust [103]. These data suggest that symptoms are related to bacterial products.

Fungi can cause similar effects *in vitro* and some evidence supports an effect *in vivo*. The cell wall of fungi consists primarily of glucans, which, by itself, may induce inflammatory responses [104,105]. In addition, secondary metabolites of fungi may cause a broad range of adverse health effects. Possible mechanisms include low-molecular weight compounds that may cause irritation, traditional allergic pathways, and less volatile mycotoxins that may induce mucosal inflammation and fibrosis. Little field evidence supports this hypothesis, although several outbreaks in office workers suggest that in specific outbreaks quantifiable adverse health effects are associated with disease. Since 1984, studies of pulmonary disease in large office buildings have documented not just disease but also higher "nonspecific" symptoms in buildings with known disease and have led to warnings concerning moisture control. Ruotsalainen [106], in a study of Finnish daycare centers, demonstrated an increase in eye and chest symptoms among workers in centers with moisture problems; 70% of daycare centers had such problems. A series of field investigations of problem buildings by NIOSH suggested that moisture problems were associated with lower respiratory symptoms [30].

C. Work Stress

Consistently, in cross-sectional studies, measures of work stress are associated with indoor environmental symptoms [87]. This is true whether questionnaires simply asked ratings of work stressfulness or used formal validated work stress questionnaires. The scientific community generally agrees that work stress must be viewed as an effect modifier. That is, it may make the perception of irritation worse but it will not lead to the perception of irritation without the presence of irritants. This stands in contrast to some work in experimental psychology that suggests symptoms may arise without underlying causes but simply from expectation. In a case-control study, Eriksson [107] suggested that control over work was the single most important factor. In any case, "work stress" is amenable to intervention.

The use of two- or three-component models (i.e., investigator's beliefs about the nature of stress) determines the components that may be treated. Two-component models argue for the presence of two factors: job constraints versus decision latitude. Three-component models generally include measures of person-

ality and individual susceptibility. The first implies an organizational analysis and identification of possible intervention strategies focused solely on workplace factors. This would imply the need for management training in better supervisory practices. The second argues for the importance of scrutiny of the individual person, the environment, and the fit of the two. Solutions would then focus on the individual, including reeducation, personal stress management techniques, and coping skills. A more sophisticated view of "work stress" argues that educating both workers and management about building needs and appropriate behaviors is appropriate for risk communication and stress management and is a good worker education practice. Such education could then focus on management and worker styles, address the recognized limitations of the current complex systems in the office, and lead to changed expectations.

A lesson may also be learned from the world of hazardous waste sites, community outrage, and public health. The needs of communities, their perception of pollutedness, and their intellectual dependency on regulatory agencies, outside experts, the media, and other "forces beyond their control" lead to major distress on their part. A set of guidelines has evolved to help such communities recover. They rely on "empowerment," a no longer trendy but still important concept. In health psychology, the locus of control is an important concept and one that allows professionals to work with individuals in groups who fell ill. Education, knowledge, skills, and understanding are the only weapons we have against media, concern, information mismanagement, and failure to educate.

D. Gender

Consistently in cross-sectional studies of office workers, women have described higher rates of symptoms. This has been demonstrated across studies [87] and is seen in individual studies with large samples of buildings [69]. A detailed analysis of one of these studies suggests that chest symptoms occur in excess and can not be attributed simply to gender [108]. Two of the few studies to examine personal exposures in the office suggested that women were exposed to VOCs and particulates at concentrations almost twice as high as those to which men were exposed [93,109]. Broerson [110] suggests that gender per se is an inadequate explanation for gender discrepancies elsewhere among working women.

VI. Prevention

A number of professional standards guide work on indoor environments. Standards published by ASHRAE, some with the American National Standards Institute and the International Illuminating Engineering Society, or the International Standards Organization (concerning ventilation, thermal comfort, lighting, and acoustics) guide indoor environmental design work but are not enforceable. Equally important, they are often simply not implemented in offices in the U.S. The OSHA Notice of Proposed Rulemaking (NPR) on indoor air has long stalled. Several "voluntary consensus standards" are underway, with very different strategies; one is likely to focus on VOC emissions without addressing broader problems. Until better linkage strategies are formalized and used that identify and discriminate between "sick" and "well" people, that allow attribution of disease to specific causes, and that allow specific estimation of economic consequences, it may be difficult to persuade regulators and public health agencies that disease is real and worth addressing.

References

1. Coultas, D. B., Zumwalt, R. E., Black, W. C., and Sobonya, R. E. (1994). The epidemiology of interstitial lung disease. *Am. J. Respir. Crit. Care Med.* **150,** 967–972.

2. Pestalozzi, C. (1959). Febrile Gruppenerkranking in einer Modellschreinerei durch Inhalation von mit Schillmelpilzen kontaminiertem Befeuchterwasser ("Befeuchterfieber"). *Schweiz. Med. Wochensch.* **27,** 710–714.

3. Kreiss, K., and Hodgson, M. J. (1984). Building-associated epidemics. *In* "Indoor Air Quality" (C. S. Walsh, P. J. Dudney, and E. Copenhaever, eds.), pp. 87–106. CRC Press, Boca Raton, FL.

4. Bernstein, R. S., Sorenson, W. G., Garabrant, D., Reaux, C., and Treitman, R. D. (1983). Exposures to respirable, airborne Penicillium from a contaminated ventilation system: Clinical, environmental and epidemiological aspects. *Am. Ind. Hyg. Assoc. J.* **44,** 161–169.

5. Hodgson, M. J., Morey, P. R., Simon, J., Waters, T., and Fink, J. N. (1987). Acute and chronic hypersensitivity pneumonitis from the same source. *Am. J. Epidemiol.* **125,** 631–638.

6. Hodgson, M. J., Morey, P., Attfield, M., Fink, J. N., Sorensen, W., and Visvesvara, G. (1985). Pulmonary disease in an office building: Single-breath carbon-monoxide diffusing capacity as a cross-sectional field tool. *Arch. Environ. Health* **40,** 96–101.

7. Division of Respiratory Disease Studies (1984). Outbreaks of respiratory illness among employees in large office buildings. *Morbid. Mortal. Wkly. Rep.* **33,** 506–513.

8. Welterman, B., Hodgson, M. J., Storey, E., Cartter, M., Bracker, A., Groseclose, S., and Philips, D. (1998). Hypersensitivity pneumonitis: Sentinel health events in clinical practice. *Am. J. Ind. Med.* **34,** 718–723.

9. Mullen, J., Hodgson, M., DeGraff, C. A., and Godar, T. (1998). A case-control study of diffuse interstitial pneumonitis. *J. Occup. Environ. Med.* **40**(4), 1–5.

10. Lonneux, M., Nolard, N., Philippart, I., Henkinbrant, A., Hamels, J., Fastrez, J., and d'Odemont, J. P. (1995). A case of lymphocytic pneumonitis, myositis, and arthritis associated with exposure to Aspergillus niger. *J. Allergy Clin. Immunol.* **95,** 1047–1049.

11. Newman, L. S., Maier, L. A., and Rose, C. S. (1997). Sarcoidosis. *N. Engl. J. Med.* **336,** 1224–1234.

12. Thorn, A., Lewne, M., and Belin, L. (1996). Allergic alveolitis in a school environment. *Scand. J. Work Environ. Health* **22,** 311–314.

13. Forst, L. S., and Abraham, J. (1993). Hypersensitivity pneumonitis presenting as sarcoidosis. *Brit. J. Ind. Med.* **50,** 497–500.

14. Ortiz, C., Hodgson, M. J., McNally, D., and Storey, E. (1999). A case-control study of sarcoidosis. In press.

15. Daroowalla, F., and Raghu, G. (1997). Hypersensitivity pneumonitis. *Compr. Ther.* **23,** 244–248.

16. Salvaggio, J. E. (1997). Extrinsic allergic alveolitis (hypersensitivity pneumonitis): Past, present and future. *Clin. Exp. Allergy* **27** (Suppl. 1), 18–25.

17. Hodgson, M. J., Parkinson, D. K., and Karpf, M. (1989). Chest x-rays and hypersensitivity pneumonitis: A secular trend in sensitivity. *Am. J. Ind. Med.* **16,** 45–63.

18. Lynch, D. A., Way, D., Rose, C. S., and King, T. E., Jr. (1992). Hypersensitivity pneumonitis: Sensitivity of high-resolution CT in a population-based study. *AJR, Am. J. Roentgenol.* **159**(3), 469–472.

19. Rose, C. S., Martyny, J. W., Newman, L. S., Milton, D. K., King, T. E., Jr., Beebe, J. L., McCammon, J. B., Hoffman, R. E., and Kreiss, K. (1998). "Lifeguard lung": Endemic granulomatous

pneumonitis in an indoor swimming pool. *Am. J. Public Health* **88,** 1795–1800.

20. Burrell, R., and Rylander, R. (1981). A critical review of the role of precipitins in hypersensitivity pneumonitis. *Eur. J. Respir. Dis.* **62,** 332–343.

21. Milton, D. (1996). Endotoxins. *In* "Indoor Air Quality II (Oak Ridge Symposium)" (R. Gammage, ed.). Lewis Publishers/CRC Press, Boca Raton, FL.

22. Finnegan, M., Pickering, C. A. C., and Burge, P. S. (1984). The sick-building syndrome: Prevalence studies. *Br. Med. J.* **289,** 1573–1575.

23. Apter, A., Hodgson, M., Lueng, W.-Y., and Pichnarcik, L. (1997). Nasal symptoms in the "Sick Building Syndrome." *Ann. Allergy, Asthma, Immunol.* **78,** 152 (abstr.).

24. Spengler, J. D., Burge, H., and Su, H. J. (1992). Biological agents and the home environment. *In* "Bugs, Mold & Rot. Proceedings of the Moisture Control Workshop" (E. Bales and W. B. Rose, eds.), pp. 11–18. National Institute of Building Sciences, Washington, DC.

25. Cooper, K., Hodgson, M., and Demby, S. (1997). Moisture and lung disease: Population-attributable risk calculations. *Healthy Build./Indoor Air Quality.* **1,** 213–218.

26. Milton, D. K., Solomon, G. M., Rosiello, R. A., and Herrick, R. F. (1998). Risk and incidence of asthma attributable to occupational exposure among HMO members. *Am. J. Ind. Med.* **33,** 1–10.

27. Venables, K. M., and Chan-Yeung, M. (1997). Occupational asthma. *Lancet* **349,** 1465–1469.

27a. Jajosky, R. A., Harrison, R., Reinish, F., Flattery, J., Chan, J., Tumpowsky, C., Davis, L., Reilly, M. J., Rosenman, K., Kalinoski, D., Stanbury, M., Schill, D. P., and Wood, J. (1999). Surveillance of work-related asthma in the U.S. *MMWR* **SS3,** 1–20.

28. Matte, T. D., Hoffman, R. E., Rosenman, K. D., and Stanbury, M. (1990). Surveillance of occupational asthma under the SENSOR model. *Chest* **98**(5 Suppl.), 173S–178S.

29. Hoffmann, R. E., Wood, R. C., and Kreiss, K. (1993). Building-related asthma in Denver office workers. *Am. J. Public Health* **83,** 89–93.

30. Sieber, W. K., Stayner, L. T., Malkin, R. *et al.* (1996). The NIOSH Indoor Evaluation experience: Associations between environmental factors and self-reported health conditions. *Appl. Occup. Environ. Hyg.* **11,** 1387–1392.

31. Toren, K., Horte, L. G., and Jarvholm, B. (1991). Occupational and smoking adjusted mortality due to asthma among Swedish men. *Br. J. Ind. Med.* **48,** 323–326.

32. Gergen, P. J., and Turkeltaub, P. C. (1992). The association of individual allergen reactivity with respiratory disease in a national sample: Data from the second National Health and Nutrition Examination Survey, 1976–80 (NHANES II). *J. Allergy Clin. Immunol.* **90,** 579–588.

33. Koren, H. S., and Devlin, R. B. (1992). Human upper respiratory tract responses to inhaled pollutants with emphasis on nasal lavage. *Ann. N. Y. Acad. Sci.* **641,** 215–224.

34. Ohm, M., Juto, J. E., Andersson, K., and Bodin, L. (1997). Nasal histamine provocation of tenants in a sick-building residential area. *Am. J. Rhinol.* **11,** 67–75.

35. Ohm, M., Juto, J. E., and Andersson, K. (1993). Nasal hyperreactivity and sick building syndrome. *In* "Indoor Air Quality: Environments for People." ASHRAE, Atlanta, GA.

36. Willes, S. R., Bascom, R., and Fitzgerald, T. K. (1992). Nasal inhalation challenge studies with sidestream tobacco smoke. *Arch. Environ. Health* **47,** 223–230.

37. Molhave, L., Liu, Z., Jorgensen, A. H., Pederson, O. F., and Kjaergard, S. (1993). Sensory and physiologic effects on humans of combined exposures to air temperatures and volatile organic compounds. *Indoor Air* **3,** 155–169.

38. Meggs, W. J., Albernaz, M., Elsheik, T., Bloch, R. M., and Metzger, W. J. (1996). Nasal pathology and ultrastructure in patients with chronic airway inflammation (RADS and RUDS) following an irritant exposure. *J. Toxicol. Clin. Toxicol.* **34**, 383–396.

39. Menzies, D., Comtois, P., Pasztor, J., Nunes, F., and Hanley, J. A. (1998). Aeroallergens and work-related respiratory symptoms among office workers. *J. Allergy Clin. Immunol.* **101**, 38–44.

40. Malkin, R., Martinez, K., Marinkovich, V., Wilcox, T., Wall, D., and Biagini, R. (1998). The relationship between symptoms and IgG and IgE antibodies in an office environment. *Environ. Res.* **76**, 85–93.

41. Pirhonen, I., Nevalainen, A., Husman, T., and Pekkanen, J. (1996). Home dampness, moulds and their influence on respiratory infections and symptoms in adults in Finland. *Eur. Respir. J.* **26**, 18–22.

42. deShazo, R. D., Chapin, K., and Swain, R. E. (1997). Fungal sinusitis. *N. Engl. J. Med.* **337**, 254–259.

43. Kodama, A. M., and McGee, R. I. (1986). Airborne microbial contaminants in indoor environments. Naturally ventilated and air-conditioned homes. *Arch. Environ. Health* **41**, 306–311.

44. Kjaergard, S. (1992). Assessment methods and causes of eye irritation in humans in indoor environments. *In* "Chemical, Microbiological, Health, and Comfort Aspects of Indoor Air Quality" (H. Knoeppel and P. Wolkoff, eds.), pp. 115–127. ECSC, EEC, EAEC, Brussels.

45. Kjaergard, S., Rasmussen, T. R., Molhave, L., and Pedersen, O. F. (1995). An experimental comparison of indoor air VOC effects on hayfever- and healthy subjects. *Proc. Healthy Build.* **1**, 564–569.

46. Cain, W. S. (1996). Odors and irritation in indoor air pollution. *In* "Indoor Air and Human Health" (R. B. Gammage, ed.). Lewis Publishers/CRC Press, Boca Raton, FL.

47. Abraham, M. (1996). Potency of gases and vapors: QSARs. RB Gammage, ed. *In* Indoor Air and Human Health" (R. B. Gammage, ed.). Lewis Publishers/CRC Press, Boca Raton, FL.

48. Cometto-Muniz, J. E., Cain, W. S., and Hudnell, H. K. (1997). Agonistic sensory effects of airborne chemicals in mixtures: odor, nasal pungency, and eye irritation. *Percepts. Psychophys.* **59**, 665–674.

49. Nardell, E. A., Keegan, J., Cheney, S. A., and Etkind, S. C. (1991). Airborne infection. Theoretical limits of protection achievable by building ventilation. *Am. Rev. Respir. Dis.* **144**, 302–306.

50. Steenland, K., Levine, A. J., Sieber, K., Schulte, P., and Aziz, D. (1997). Incidence of tuberculosis infection among New York State prison employees. *Am. J. Public Health* **87**, 2012–2014.

51. Hoge, C. W., Reichler, M. R., Dominguez, E. A., Bremer, J. C., Mastro, T. D., Hendricks, K. A., Musher, D. M., Elliott, J. A., Facklam, R. R., and Breiman, R. F. (1994). An epidemic of pneumococcal disease in an overcrowded, inadequately ventilated jail. *N. Engl. J. Med.* **331**, 643–648.

52. Fisk, W., and Rosenfeld, A. H. (1997). Estimates of improved productivity and health from better indoor environments. *Indoor Air* **7**, 158–172.

53. Hendley, J. O., and Gwaltney, J. M., Jr. (1988). Mechanisms of transmission of rhinovirus infections. *Epidemiol. Rev.* **10**, 243–258.

54. Gwaltney, J. M., Jr., and Hendley, J. O. (1978). Rhinovirus transmission: One if by air, two if by hand. *Am. J. Epidemiol.* **107**, 357–361.

55. Dick, E. C., Jennings, L. C., Mink, K. A., Wartgow, C. D., and Inhorn, S. L. (1987). Aerosol transmission of rhinovirus colds. *J. Infect. Dis.* **156**, 442–448.

56. Brundage, J. F., Scott, R. M., Lednar, W. M., Smith, D. W., and Miller, R. N. (1988). Building-associated risk of febrile acute respiratory diseases in Army trainees. *JAMA, J. Am. Med. Assoc.* **259**, 2108–2112.

57. Rose, D. M., Airah, M., and Wilke, S. E. (1992). An investigation of sick absences from work in office buildings. *Aust. Refrig.*, November, pp. 27–33.

58. Marks, J. G., Jr., Trautlein, J. J., Zwillich, C. W., and Demers, L. M. (1984). Contact urticaria and airway obstruction from carbonless copy paper. *JAMA, J. Am. Med. Assoc.* **252**, 1038–1040.

59. Tencati, J. R., and Novey, H. S. (1983). Hypersensitivity angiitis caused by fumes from heat-activated photocopy paper. *Ann. Intern. Med.* **98**, 320–322.

60. Andersen, I., Lundqvist, G. R., Jensen, P. L., and Proctor, D. F. (1974). Human response to 78-hour exposure to dry air. *Arch. Environ. Health* **29**, 319–324.

61. Nordstrom, K., Norback, D., and Akselsson, R. (1994). Effect of air humidification on the sick building syndrome and perceived indoor air quality in hospitals: A four month longitudinal study. *Occup. Environ. Med.* **51**, 683–688.

62. Hodgson, M. J., Block, G., and Parkinson, D. K. (1986). An outbreak of organophosphate pesticide poisoning in office workers. *J. Occup. Med.* **28**, 435–437.

63. Cone, J. E., and Sult, T. A. (1992). Acquired intolerance to solvents following pesticide/solvent exposure in a building: a new group of workers at risk for multiple chemical sensitivities? *Toxicol. Ind. Health* **8**, 29–39.

64. Muldoon, S., and Hodgson, M. J. (1992). Non-occupational organophosphate poisoning. *J. Occup. Med.* **34**, 38–44.

65. Marks, J. G., Jr. (1981). Allergic contact dermatitis from carbonless copy paper. *JAMA, J. Am. Med. Assoc.* **245**(22), 2331–2332.

66. NIOSH Hazard Review (1999). "Adverse Health Effect from Carbonless Copy Paper," HHS Doc. (to be published).

67. Hodgson, M. J., Permar, E., Squires-Bertocci, G., Allen, A., and Parkinson, D. K. (1987). Vibration as a contributing factor to the sick-building syndrome. *Inst. Water, Soil, Air Hyg.* **2**, 449–452.

68. Jaakola, J. J. (1998). The office environment model: A conceptual analysis of the sick building syndrome. *Indoor Air, Suppl.* **4**, 7–16.

69. Malkin, R., Wilcox, T., and Sieber, W. K. (1996). The NIOSH Indoor Environmental Evaluation Experience: Symptom prevalence. *Appl. Occup. Environ. Hyg.* **11**, 540–545.

70. Nelson, C., Wallace, L. A., Clayton, C. A., Highsmith, R., Kollande, M., Bascom, R., and Leaderer, B. P. (1991). Indoor Air Quality and Work Environment Survey: Relationships of employee's self-reported symptoms and direct indoor air quality measurements. IAQ 91, *In* "Indoor Air Quality" pp. 22–32. ASHRAE, Atlanta, GA.

71. Burge, P. S., Hedge, A., Wilson, S., Bass, J. H., and Robertson, A. (1987). Sick-building syndrome: A study of 4373 office workers. *Ann. Occup. Hyg.* **31**, 493–504.

72. Wyon, D. (1992). Sick buildings and the experimental approach. *Environ. Technol.* **13**, 313–322.

73. Hodgson, M. J. (1992). A series of field studies on the sick-building syndrome. *Ann. N. Y. Acad. Sci.* **641**, 21–36.

74. International Society for the Study of Headache (1993). "Classification Criteria."

75. Schwartz, B. S., Stewart, W. F., and Lipton, R. B. (1997). Lost workdays and decreased work effectiveness associated with headache in the workplace. *J. Occup. Environ. Med.* **39**, 320–327.

76. Schwartz, B. S., Stewart, W. F., Simon, D., and Lipton, R. B. (1998). Epidemiology of tension-type headache. *JAMA, J. Am. Med. Assoc.* **279**, 381–383.

77. Burge, P. S., Robertson, A. S., and Hedge, A. (1991). Comparison of a self-administered questionnaire with physician diagnosis in the diagnosis of the sick-building syndrome. *Indoor Air* **1**, 422–427.

78. Burge, P. S., Robertson, A., and Hedge, A. (1990). The development of a questionnaire suitable for the surveillance of office buildings to assess the building symptom index a measure of the sick building syndrome. *Proc. Indoor Air* **1**, 731–736.

79. Jansen, J. (1994). The "V" in AHSVE: A historical perspective. *ASHRAE J.* 126–132.

80. American Society of Heating, Refrigerating, and Airconditioning Engineers (1989). "Ventilation for Aceptable Air Quality." ASHRAE, Atlanta, GA.

80a. Woods, J. (1989). Cost avoidance and productivity. *STAR-Occup. Med.* **10,** 4.

81. Boswell, T., DiBerardinis, and Ducatman, A. (1994). Descriptive epidemiology of indoor odor complaints at a large teaching institution. *Appl. Occup. Environ. Hyg.* **9,** 281–286.

82. Pandian, M. D., Ott, W. R., and Behar, J. V. (1993). Residential air exchange rates for use in indoor air and exposure modeling. *J. Exposure Anal. Environ. Epidemiol.* **3,** 407–416.

83. Conibear, S., Geneser, S., and Carnow, B. W. (1996). Carbon monoxide levels and sources found in a random sample of households in Chicago during the 1994–1995 heating season. *In* "Proceedings of Indoor Air Quality: Practical Engineering for Indoor Air Quality," pp. 111–118. ASHRAE, Atlanta, GA.

84. Lawton, M. D., Dales, R. E., and White, J. (1998). The influence of house characteristics in a Canadian community on microbiological contamination. **8,** 2–11.

85. Nevalainen, A., Partanen, P., Jaaksainen, E., Hyvarinen, A., Koskinen, O., Meklin, T., Vahteristo, J., Koivisto, J., and Husman, T. (1998). The prevalence of moisture problems in Finnish houses. *Indoor Air, Suppl.* **8,** 40–44.

86. Stenberg, B. (1994). "Office Illness: The Worker, the Work, and the Workplace." NIOH, Sweden.

87. Mendell, M. (1993). Nonspecific symptoms in office workers: A review and summary of the epidemiologic literature. *Indoor Air* **3,** 227–236.

88. Mendell, M., Fisk, W., Deddens, J. A., *et al.* (1996). Elevated symptom prevalence associated with ventilation type in office buildings. The California Health Building Study. *Epidemiology* **7,** 583–589.

89. Ten Brinke, J. (1995). Development of new VOC exposure matrices related to "Sick Building Syndrome" symptoms in office workers (Doctoral dissertation). *Lawrence Berkeley Lab. [Rep.] LBL* **LBL-37652.**

90. Hodgson, M. J., Frohliger, J., Permar, E., Tidwell, C., Traven, C., Olenchuk, S., and Karpf, M. (1991). Symptoms and microenvironmental measures in non-problem buildings. *J. Occup. Med.* **33,** 527–533.

91. Anderson, K., Bakke, J., Bjorseth, O., Bornehag, C. G., Clausen, G., Hongslo, J. K., Kjellman, M., Kjaergard, S., Levy, F., Molhave, L., Skerfving, S., and Sundell, J. (1997). TVOC and health in non-industrial indoor environments. Report from a Nordic Scientific Consensus Meeting in Stockholm 1996. *Indoor Air* **7,** 78–91.

92. Sundell, J. (1994). On the association between building ventilation characteristics, some indoor environmental exposures, some allergic manifestations, and subjective symptom reports. *Indoor Air, Suppl.* **2,** 9–148.

93. Teinjoinsalo, J., Jaakola, J. J., and Seppanen, O. (1996). The Helsinki Office Study: Air change in mechanically ventilated buildings. *Indoor Air* **6,** 111–117.

94. Wechsler, C. J., and Shields, H. C. (1996). Production of the hydroxyl radical in indoor air. *Environ. Sci. Technol.* **30,** 3250–3258.

95. Wechsler, C. J., and Shields, H. C. (1998). Indoor ozone/terpene reactions as a source of indoor particles. *AWWMA Meet.,* p. 98–A949.

96. Wolkoff, P., Clausen, G., and Fanger, P. O. (1997). Are we measuring the right pollutants *Indoor Air* **7,** 92–106.

97. Fanger, P. O., Lauridsen, J., Bluyssen, P., and Clausen, G. (1988). Air pollution sources in offices and assembly halls, quantified by the olf unit. *Energy Build.* **12,** 7–12.

98. Batterman, S., and Burge, H. A. (1995). HVAC systems as emission sources affecting indoor air quality: A critical review. *Int. J. HVAC R Res.* **1,** 61–81.

99. Fang, L., Clausen, G., and Fanger, P. O. (1996). The impact of temperature and humidity on perseption and emmission of indoor pollutants. *Proc. Indoor Air* **4,** 349–354.

100. Teeuw, K., and Vandenbroucke. (1994). Airborn gram negative bacteria and endotoxin in SBS: A study in Dutch office buildings. *Arch. Intern. Med.* **154,** 2339–2345.

101. Rylander, R., Persson, K., Goto, H., Yuasa, K., and Tanaka, S. (1992). Airborne beta-1,3-glucan may be related to symptoms in sick buildings. *Indoor Environ.* **1,** 263–267.

102. Gyntelberg, F., Suadicani, P., Wolkoff, P., Nielsen, J. W., Skov, P., Valbjrn, P. A., Nielsen, T., Schneider, T., Jorgenson, O., Wilkins, C. A., Gravesen, S., and Norn, S. (1994). Dust and the Sick Building Syndrome. *Indoor Air* **4,** 223–228.

103. Harrison, J., Pickering, C. A., Faragher, E. B., Austwick, P. K., Little, S. A., and Lawton, L. (1992). An investigation of the relationship between microbial and particulate indoor air pollution and the sick building syndrome. *Respir. Med.* **86,** 225–235.

104. Thorn, J., and Rylander, R. (1998). Airways inflammation and glucan in a rowhouse area. *Am. J. Respir. Crit. Care Med.* **157,** 1179–1803.

105. Fogelmark, B., Sjöstrand, O., Williams, D., and Rylander, R. Inhalation toxicity of beta(1->3)-D- glucan: Recent advances.

106. Ruotsalainen, R., Jaakola, N., and Jaakola, J. J. (1995). Dampness and molds in day-care centers as an occupational health problem. *Int. Arch. Occup. Environ. Health* **66,** 369–374.

107. Stenberg, B., and Wall, S. (1995). Why do women report 'sick building symptoms' more often than men? *Soc. Sci. Med.* **40,** 491–502.

108. Eriksson, N., Hoog, J., Mild, K. H., Sandstrom, M., and Stenberg, B. (1997). The psychosocial work environment and skin symptoms among visual display terminal workers: A case referent study. *Int. J. Epidemiol.* **26,** 1250–1257.

109. Hodgson, M. J., Muldoon, S., Collopy, P., and Olesen, B. (1992). Work stress, symptoms, and microenvironmental measures. *In* Indoor Air Quality: Environments for People," pp. 47–58. ASHRAE, Atlanta, GA.

110. Broersen, J. P., de Zwart, B. C., van Dijk, F. J., Meijman, T. F., and van Veldhoven, M. (1996). Health complaints and working conditions experienced in relation to work and age. *Occup. Environ. Med.* **53,** 51–57.

41

Women's Occupational Health: International Perspectives

JEANNE MAGER STELLMAN AND ANDREA LUCAS
Joseph L. Mailman School of Public Health
Columbia University
New York, New York

I. Introduction

The conditions under which women work and the hazards they may face on the job are, to a great extent, a function of the real and perceived roles of women in society. Women's lives differ dramatically around the world and are shaped largely by their country's economic status and sociopolitical culture, yet many of the qualities of the work women do are remarkably consistent throughout human society. Despite the vast economic differences that can be found among nations, the basic socioeconomic factors affecting women as workers are startlingly similar and include the following:

1. *Occupational segregation:* Women tend to be heavily clustered in a relatively small number of occupations, and to be underrepresented in the majority of others.

2. *Barriers to advancement/attainment:* These can range from the "glass ceiling," an invisible, unspoken barrier to advancement in job status typical of the most industrialized countries, to a "brick wall," or clearly defined limit, often preventing entry to an occupation or industry.

3. *Underemployment:* Many women work part time, in seasonal or otherwise "temporary" jobs, or in fields or positions far below their educational level. Others are not permitted to work at all, often after marriage or through the widespread custom of *purdah*, which forces women, married or single, to remain enclosed in their homes.

4. *Low or no remuneration:* Whether salaried jobs or unpaid traditional "women's work," economic and subsistence activities performed by women are undervalued.

5. *Instability/lack of tenure:* Women are particularly prone to job loss through a "last in/first out" syndrome—the most recently hired employees being terminated first during layoffs. Relatively few women are members of trade unions with access to grievance procedures and contractual protections.

6. *"Multiple burdens" of simultaneous family and household obligations:* Employed women work many more hours on household tasks—childcare, elder care, subsistence activities, and housework—than do employed men.

7. *Sexual harassment and violence:* These can both be "incidental" to work or used as a specific means of control over work-related demands.

The sociopolitical patterns of women's employment described here have implications for the occupational safety and health of women workers. The employment patterns of women workers in female job ghettos and women's comparative lack of power and representation have led to a consistent pattern of neglect of research and regulation of the occupational health and safety problems associated with female-dominated jobs. Or, perhaps even more ominously for women's health, the marginalization of women as workers has fostered the unwarranted belief that women's work is "safe."

Social attitudes impinge on women's work lives in other ways, as well. Disease and injury, for example, have significantly greater effects on women's ability to earn a living than on men's. In India, women with leprosy are thrown out of their homes far more often than men, have much greater difficulty finding work, and, when they do hold jobs, earn a quarter of what other women, working fewer days, do—and they earn only 5% of what men with leprosy earn [1]. In the United States, as another example, a study of disfiguring disability and reemployment found women far less likely to be rehired at their same or equivalent jobs than men who suffered the same occupational injuries [2].

Overall, the average proportion of women in the worldwide labor force increased significantly from 1970 to 1990. Two exceptions, however, are East Asia, where there was only a 1% increase in the number of women workers, and sub-Saharan Africa, where a slight decrease apparently paralleled a general economic decline. Although exact statistics are sparse, women's share of the labor force averaged, as shown in Table 41.1, some 40% or more in developed countries, the Caribbean, and parts of Asia. North Africa and West Asia had the lowest rates, 21% and 25%, respectively. During the same period, the average percentage of economically active men dropped slightly, so that the relative proportion of women in the labor force increased.

II. The "Developed/Developing" Split

Internationally, the demographics of women at work generally fall into one of two patterns, based on the division of nations' economies into "developed" (industrial or, increasingly, postindustrial) and "developing" (whether rural, urban, or mixed). There is an irony in the imagery surrounding women at work in these two types of economies. In developing countries, including those where there is not even the pretense of gender equality, it is often the brawn and endurance of women that is prized. In industrialized countries, where women have long been banished from mining and have only recently made inroads into the construction industry, and in the multitudinous small factories of newly industrializing nations' light industries, it is women's "fine finger dexterity" and patience with repetitive work that is extolled and capitalized on. In heavy industry

Table 41.1
Women in the Paid and Unpaid Total Workforce[a]

Country	GDP (1992) per capita ($)	Total active workforce	Women as % workforce	% of total workforce unpaid	Women as % unpaid workforce	Ratio (%) women unpaid/ total workforce unpaid
Ethiopia	52	18,235,708	39.08	32.38	66.77	2.06
Malawi	184	3,300,198	51.89	12.60	58.06	4.61
Haiti	235	2,679,140	39.59	11.14	37.28	3.35
Nigeria	256	30,765,500	33.39	10.84	46.24	4.27
Cen. Afr. Rep.	514	1,186,972	48.68	8.90	54.70	6.15
Indonesia	671	75,508,082	39.85	29.67	65.99	2.22
Egypt	746	16,033,600	28.67	21.12	62.34	2.95
Philippines	807	24,525,000	36.89	14.29	52.52	3.67
Bolivia	839	924,962	41.82	10.91	79.32	7.27
El Salvador	1,109	982,802	45.28	6.57	57.88	8.81
Poland	2,356	18,452,230	45.42	10.69	76.25	7.13
Turkey	2,647	21,146,004	30.64	30.85	69.01	2.24
Botswana	3,003	443,455	53.67	21.22	34.76	1.64
Mexico	3,736	24,063,283	23.46	2.49	11.21	4.50
Korea, Rep. of	6,721	18,487,000	40.29	10.89	87.06	7.99
Argentina	6,912	10,033,798	27.83	2.70	23.60	8.74
Greece	7,686	3,999,811	36.86	12.14	75.68	6.23
Australia	16,715	8,412,500	41.60	0.78	58.80	75.46
United Kingdom[b]	18,182	28,893,000	43.34			
Canada	20,600	13,681,000	44.97	0.48	80.30	167.40
France	23,149	23,972,302	43.45	3.18	81.94	25.77
United States	23,332	126,424,000	45.40	0.29	76.24	263.91
Denmark	27,626	2,912,428	46.06	1.83	96.73	52.88
Japan	29,387	63,840,000	40.77	7.95	82.01	10.32

The 24 countries listed represent all regions of the world and the full spectrum of development, from strictly agricultural to postindustrial.

[a]Source: [1].

[b]United Kingdom asserts that there are no unpaid workers.

in many of the most developed countries, women have been "protected" by exclusion from these more highly paid jobs, a policy discussed more fully below.

III. Traditional Tasks

In developing regions, women generally have almost no political and economic power (*e.g.,* they cannot own land in many countries), and they may be effectively barred from many occupations and sectors by custom or by law. This lack of power is also reflected in the *invisibility* of much of their work, which as often as not consists of subsistence activities hidden under the rubric of "housework." These laborious and often dangerous tasks—such as collection and transport of water and wood, gathering and processing of cow-patties for fuel, processing and storing foodstuffs—are critical for family survival, yet are unpaid and often not considered in policy development and systems of national accounts, much less health and safety programs. Similarly, women's participation in agricultural work (and they are usually responsible for the most arduous "stoop labor"), whether for subsistence or cash crops, is also most often unpaid, usually because they are working in the family's holding but

sometimes, where they work as day laborers, because a "family wage" is paid to the male head of household. In *all* categories with a sizable proportion of unpaid workers, women are overwhelmingly the source of that free labor. Much of the world's labor is unpaid and women are greatly overrepresented in the unpaid workforce. Some statistics are given in Table 41.2.

It is important to note that in developing agrarian countries in particular, environmental health and occupational health are closely related. Sims has systematically explored the relationship between women's health and work and environment, with a focus on international trends and women's traditional work [3]. She was unable to locate any studies entirely devoted to this theme. Rather, her annotated bibliography developed into a carefully constructed potpourri of articles describing various aspects of women's paid work, of environmental hazards associated with "traditional" women's work, and of the health of women workers, both paid and unpaid. The scant literature that exists demonstrates the arduous nature of many of these tasks, the presence of associated hazards (*e.g.,* fetching water and water contamination), as well as the simultaneous discrimination against women. For example, women are routinely denied food and shelter in order that it be preferentially available to the

Table 41.2

Women Paid and Unpaid Family Workers, by Industrial Sector, in Selected Countries, 1992[a]

Country	Agriculture, animal husbandry, forestry, hunting, and fishing			Manufacturing			Trade, restaurants, and hotels			Community, social, and personal services		
	Total workforce in sector	Women as % of sector workforce	Women as % of unpaid in sector	Total workforce in sector	Women as % of sector workforce	Women as % of unpaid in sector	Total workforce in sector	Women as % of sector workforce	Women as % of unpaid in sector	Total workforce in sector	Women as % of sector workforce	Women as % of unpaid in sector
Ethiopia	16,101,011	41.08		286,898	41.56		696,018	63.96		933,496	37.66	
Malawi	2,699,900	55.35	56.92	95,594	20.40	77.99	103,203	37.02	71.68	137,142	29.24	66.40
Haiti	1,535,444	29.84	34.48	151,387	45.05	38.08	352,970	76.87	81.15	155,347	47.28	44.22
Nigeria	13,259,000	26.08	46.52	1,263,700	36.19	50.72	7,417,400	63.91	50.06	4,902,100	19.64	0.00
India	172,713,291	30.92	17.26	26,554,517	18.13	26.06	12,638,204	8.06	11.01	18,514,810	18.04	33.72
Cen. Afr. Rep.	880,637	52.58	54.47	17,346	7.21	16.37	91,998	59.31	65.24	70,301	12.14	26.62
Indonesia	41,284,232	40.50	64.91	7,334,874	46.48	73.56	10,890,729	52.98	73.10	8,869,082	35.20	59.15
Egypt	6,335,200	41.05	63.64	1,958,700	17.59	72.39	1,340,000	17.72	37.54	3,115,500	26.10	10.89
Philippines	10,185,000	25.18	49.30	2,188,000	45.75	67.29	3,145,000	63.69	74.15	4,220,000	55.59	60.98
Bolivia	10,365	6.91	2.72	133,756	28.44	47.59	220,697	68.91	72.44	319,753	46.99	88.94
El Salvador	80,212	16.08	17.99	213,145	43.68	67.94	253,637	63.18	71.59	265,068	53.32	14.03
Poland	5,133,826	46.00	76.49	4,544,124	39.94	63.75	1,493,045	72.27	66.34	3,557,951	63.58	72.44
Turkey	9,221,229	52.05	73.97	2,867,172	19.57	29.63	2,220,639	6.96	13.08	2,865,026	19.17	12.76
Botswana	100,446	27.54	33.20	26,635	50.07	79.17	34,322	60.41	70.62	103,045	55.55	39.53
Mexico	5,300,114	3.57	2.30	4,493,279	23.53	27.02	3,875,100	33.97	47.97	5,083,779	43.12	34.87
Korea, Rep. of	3,292,000	45.50	85.65	4,847,000	42.29	87.57	3,920,000	52.93	89.49	2,630,000	42.97	87.64
Argentina	1,200,992	6.48	15.89	1,985,995	21.65	27.05	1,702,080	28.26	24.22	2,399,039	56.47	79.14
Greece	892,694	44.54	78.56	746,834	31.13	65.84	672,108	38.61	73.39	720,566	47.30	90.33
Australia	447,700	28.26	40.93	1,281,000	26.53	78.57	1,695,000	45.00	51.98	2,420,100	59.54	72.31
Canada	451,000	32.15	76.92	2,198,000	29.62		2,428,000	45.30	90.91	5,496,000	60.08	85.71
France	1,594,689	35.11	76.40	4,647,415	30.07	93.65	3,448,670	46.84	86.17	6,685,894	59.68	95.30
United States	3,566,000	21.31	63.89	22,464,000	32.91	88.89	25,811,000	47.62	78.79	42,215,000	59.17	85.48
Denmark	155,856	23.71	98.52	579,065	33.58	95.82	428,685	48.69	92.48	1,018,504	66.34	100.00
Japan	4,510,000	47.67	83.42	15,050,000	39.53	80.70	14,150,000	48.69	83.23	13,320,000	47.90	83.33

[a]Source: [4].

males in the family, while at the same time they are required to work at arduous tasks. No data are available on the prevalence of health outcomes, which are undoubtedly related to such conditions, nor are there data on any effective preventive strategies or remedial practices which could avoid such effects. Even in developed countries, Sims found a dearth of data on the integrated issue of the effects of women's work (both paid and unpaid) and of environmental hazards on women's health [4].

IV. The Informal Sector

For women in developing countries who *are* paid for their labor, the informal sector is a very significant source of work. The distinction between "informal" and "unpaid" is important. In developed countries, "informal" often connotes unpaid work, such as volunteerism. In the developing world, the informal sector comprises "own-account" household-based enterprises that occasionally provide employment for family members or other workers, but most often generate work and income only for the proprietor. Although market oriented, these businesses are generally small-scale, low-tech, and at "a low level of organization" [1, p. 116]. They may be in the service sector, manufacturing (often piece-work for small textile and other factories), or in arts and crafts cottage industries such as pottery, weaving, jewelry-making, or woodworking.

Such home-based arts and crafts cottage industries generally escape any health and safety oversight, and the artisans themselves may not recognize the hazards of their work. Occupational illness associated with traditional crafts includes lead poisoning among potters and respiratory and ergonomic problems among carpet weavers. The adoption of modern chemicals and processes for use with other crafts has led to peripheral neuropathy, paralysis, and other illness and injury. Because production takes place in the home, these occupational hazards may place entire families at risk, not just those actively working [5].

Further, developing countries, in contrast to virtually all developed countries, often have fewer or lower occupational health standards or, where adequate regulations *are* on the books, little ability—organizationally and economically—to enforce standards. The problems are especially severe with regard to small-scale manufacturing and other industries such as the numerous factories with fewer than 100 workers, which in parts of Asia, for example, employ one-quarter to one-third of the work force and account for a majority of production [6].

The informal sector is important to women workers largely because salaried work is so scarce. Throughout Africa (with the exception of Egypt), at least one-third of women working in an area other than agriculture run "own-account" businesses. In Asia there is more variation, with women's informal-sector participation ranging from less than 10% to more than 65%. In a number of countries, although there are far fewer women workers in general, women outnumber men in the informal sector. In general, the more important the informal sector is to the overall economy of a country, the greater the participation in it by women. For example, in Indonesia about 65% of economically active women are in the informal sector, which is responsible for nearly half of all production (Table 41.2) [1, pp. 115–116].

Whether unpaid or in the informal sector, much of the work women do is the physically demanding labor traditionally seen as "men's work." In some developing countries, such as India and China, women make up a substantial portion of the ranks of construction and natural-resource workers, laboring in road crews and mines. And the "women's work" noted above—the subsistence activities of gathering water and fuel, or processing foodstuffs such as cassava—is equally strenuous.

There are no credible data on the extent to which women suffer injury and illness on these jobs. However, industries such as construction and mining have the highest accident and injury rates and women working in these sectors are provided with little or no appropriate work clothing or protective gear. The routine use of standard safety procedures, such as safe scaffolding and reinforced trenching operations, is usually nonexistent. Descriptions of hazards in mining and construction provide more insight into the situation [7].

V. Transformation of Women's Work

In newly developing countries, just as in industrialized nations, many of the traditional jobs that have always been in the domain of women's work are transformed into more mechanized, marketplace activities and, also as in the industrialized nations, they are also transformed into men's work. In Thailand, for example, women had been responsible for turning over cassava chips as they dried, until the flatbeds of pickup trucks were modified so that they could handle this step. Now men are hired to operate the trucks that have replaced women's manual labor [8]. To some degree, this pattern holds throughout both developed and developing countries. Work of higher status and higher pay is almost invariably men's work; work that is being upgraded *becomes* men's work. Conversely, when women enter previously male-dominated occupations or achieve high-level positions in those occupations, both the occupations' statuses and their average pay decline [1, pp. 126–127].

Most tellingly, where traditional women's activities have been transformed into industrialized large-market "jobs," recognition of the economic value of women's nonmarket labor at these same tasks has, if anything, been reduced. As food growing, processing, preparation, and serving have entered the market economy in industrialized countries, for example, women's unpaid cooking and serving is less readily seen as a valued contribution because the "real work" is done elsewhere, in the agribusiness, manufacturing, and service sectors. Two other important domains of women's work—caring for the sick and the weaving of cloth and sewing of clothing—have seen similar transformations.

Unfortunately, the transformation of these traditional tasks has been accompanied by the introduction or exacerbation of many occupational health hazards. For example, the industrialization of textile trades has led to high rates of production and accompanying high levels of dust, noise, and repetitive, stressful use of hands and feet. These hazards often result in lung disease, hearing loss, and repetitive strain injuries, respectively. Similarly, working in the health care industry may involve heavy and awkward lifting in the handling of patients, exposure to infectious agents, occupational stress, and other hazards. Food manufacturing is associated with many serious repetitive strain injuries, a high accident rate, and exposure to biological hazards.

Table 41.3
Percentage of Male/Female Workforce in Occupational Groups by Region, 1990[a]

	Professional and technical; administrative and managerial		Clerical, sales, and service		Agriculture and related		Production and transport workers and laborers	
	M	W	M	W	M	W	M	W
Developed regions	20	23	22	48	9	8	43	15
Northern Africa and Western Asia	11	21	29	35	18	27	35	9
Sub-Saharan Africa	5	6	14	23	50	53	21	9
Latin America and Caribbean	11	15	25	55	21	5	36	14
Eastern and Southeastern Asia	9	9	23	38	37	35	29	14
Southern Asia	5	11	20	12	39	44	26	19
Oceania	13	17	15	37	32	21	28	13

[a]Adapted from [1], p. 124.

Developed countries offer a wide range of occupations within their economies with relatively few overt, hard-and-fast barriers to women. However, although women workers have gained entry to most occupations and sectors, the majority of women remain concentrated in a comparatively narrow set of jobs. The Organization for Economic Co-operation and Development (OECD) measures occupational segregation by dividing women's share of a particular occupational category (i.e., % women workers in Table 41.2 and Table 41.3) by their share of total employment (i.e., % women in total workforce in Table 41.1). A ratio greater than one indicates overrepresentation of women in a field, whereas a ratio less than one represents employment of fewer women than would be expected [9]. Looking at the data in Tables 41.1 and 41.2 for the United States, we can see that women represented 45% of the total workforce in 1992 but only 33% of the manufacturing sector (i.e., they were underrepresented). By contrast, women were 59% of the community/service sector (ratio is greater than one) and were overrepresented. In general, we can cross the industrialized/developing divide for women. Throughout virtually all regions, women are underrepresented in managerial and administrative positions and are overrepresented in clerical, service, and professional/technical work. Table 41.3 provides some summary statistics.

VI. Developed Nations

Virtually all long-developed nations have established some job-related safety and health standards and administrative and regulatory structures to assure at least a minimum level of compliance with those standards. Most of these countries also have requirements for worker training and education and a reporting system for registering industrial accidents, injuries, and diseases. Industrial processes themselves are more modern and well controlled in richer nations. Perhaps most important, a major transformation of industrial economies away from basic manufacturing and toward service industries, with the older, often more hazardous, jobs being exported to developing countries is occurring.

In general, there is little reason to believe that occupational exposures affect women differently than they affect men. However, the risks that women and men face, even when they are working in the same industry, and perhaps even the same workplace with the same job title, may differ markedly because women and men have different tasks and consequently can suffer different exposures [9, pp. 42–44]. It is important to emphasize that, although employment statistics indicate the occupational segregation of women workers, they generally underestimate the extent to which such segregation reaches the very tasks that women perform. One study of poultry industry workers, for example, found that women's jobs involved more repetitive motions, less mobility around the plant, and the use of equipment that was less ergonomically adapted to their anatomy and physiology than did men's jobs in the same factory [10]. Thus, employment statistics might show that women and men are employed in the same food processing operation, yet their work environments may not be comparable. This phenomenon makes the understanding and analysis of women's occupational health hazards very difficult.

Social factors also play an important role in women's occupational health. Women workers tend to have less on-the-job experience, for example, because they are often more recently hired or hold only part-time jobs. They also are mentored less frequently than men. It also is younger and/or inexperienced workers who suffer the greatest accident rates. Women also bear greater social stresses with less support; fewer women than men are covered by health benefits, pension plans, and trade union agreements [11].

Tools, equipment, work stations, and protective devices are, by and large, not designed for the size and physiological needs of women workers. These problems were documented in 1946 by Baetjer [12], yet they continue to persist today. The irony of such poor workplace design is that few persons of either sex actually fit the "average" very well, especially with ethnic and racial variations in anthropomorphic characteristics. Certainly the size and weight of the average white male American or northern European is greater than that of his southeast Asian counterpart. Similarly, the northern European female is also larger than many "average" males in other parts of the world [13]. Equipment and processes that are inadequately designed for the western female will, therefore, assuredly pose problems

Table 41.4

Distribution of Hours of Paid Work and Unpaid Housework, by Gender, in Selected Countries[a]

		Work (hours per week)				Unpaid housework (% share of women and men)							
		Paid		Unpaid		Preparing meals		Child care		Shopping		Other housework	
		W	M	W	M	W	M	W	M	W	M	W	M
Austria	1981	15.2	35.8	36.5	10.6	95	5	76	24	73	27	77	23
Canada	1986	17.5	32.9	28.9	13.5	81	19	76	24	58	42	67	33
	1992	18.7	31.5	28.9	15.6	76	24	71	29	59	41	59	41
Denmark	1987	21.8	35.0	22.5	11.2	73	27	95	36	60	40	65	35
Finland	1979	21.8	30.0	25.6	11.7	82	18	77	23	57	43	54	46
	1987	23.1	31.7	24.4	12.6	78	22	75	25	57	44	58	42
Former USSR	1965	43.0	53.2	35.9	15.4	87	13	72	28	50	50	67	33
	1986	38.5	49.0	30.1	16.1	75	25	75	25	62	38	59	41
Germany, Federal Rep. of	1965	13.3	42.4	44.2	11.1	94	6	84	16	75	25	74	26
	1991/92	14.7	29.5	30.0	12.3	77	23	71	29	61	39	69	31
Israel	1991/92	12.8	32.7	30.0	10.0	90	10	75	25	52	48	79	21
Japan	1976	23.5	42.4	23.1	0.9	—	—	—	—	90	10	96	4
	1981	22.3	42.5	23.7	0.9	—	—	—	—	86	14	96	4
	1986	21.2	41.8	24.3	1.3	—	—	93	7	82	18	90	10
	1991	19.5	40.8	27.1	2.8	—	—	87	12	79	21	94	6
Korea, Republic of	1987	22.5	34.8	19.0	2.3	98	1	90	10	89	11	82	18
	1990	21.4	35.4	17.6	2.1	98	2	79	20	90	11	83	17
Netherlands	1980	7.1	23.9	33.4	8.8	80	20	79	21	63	37	86	14
	1985	14.6	39.4	33.2	10.3	77	23	76	24	66	34	80	20
	1987	10.5	25.4	34.9	17.5	75	25	73	27	62	38	80	20
	1988	10.4	26.6	34.2	17.9	75	25	72	28	61	39	78	22
Norway	1980/81	17.1	34.2	29.8	9.2	81	19	70	30	57	43	82	18
	1990	19.3	30.8	30.6	18.3	75	26	71	29	58	42	58	42
Spain	1991	11.4	29.4	52.4	11.2	89	11	86	14	73	27	79	21
Sweden	1990/91	27.3	41.1	33.2	20.2	70	30	72	28	60	40	60	40
United Kingdom	1984	14.1	26.8	30.0	11.4	74	26	76	24	60	40	76	24
United States	1965	18.7	48.3	37.8	10.0	90	10	82	18	66	34	78	22
	1986	24.5	41.3	31.9	18.1	78	22	73	28	60	40	61	39

[a]Adapted from [1], p. 132.

for those males who are also smaller than the white male norm. We can still draw the same conclusion as Baetjer's World War II study; improving working conditions and equipment so that women can do their jobs safely and efficiently will also improve the safety and efficiency of working men.

Women also bear the greater burden for homemaking and care of others, which is sometimes called multiple role occupancy. These multiple burdens may not be detrimental in and of themselves when the roles are wanted ones that yield satisfaction, and when there is adequate support from community, family, and workplace. However, when some or all of the multiple roles are unwanted or simply overwhelming, adverse effects can be expected [14]. Employed women who are mothers of young children have been found to suffer the greatest stress effects. With the continuing increase in the number of older adults requiring care, which is usually provided by their employed daughters, there is a growing concern that such "elder care" will cause similar stress [15]. Multiple role occupancy must be factored into any consideration of women's occupational health. Some comparative data on women's share of unpaid housework tasks are given in Table 41.4.

VII. Social Responses and Future Needs

The role of women workers has been the focus of intense social scrutiny since the earliest days of the industrial revolution. As the numbers of girls and women employed in newly developed industries swelled, so, too, did the cadres of well-meaning social activists who sought relief and regulation for these workers.

Much attention has been paid, both in the press and in the courts, to the particular vulnerability (and responsibility) of women workers with respect to maternity and occupational health issues. It is clear that men, too, can suffer adverse effects from exposure to chemical substances and physical hazards, such as excess heat or ionizing radiation. However, on a global basis, protections related to childbearing have focused primarily on the female worker (although, as in the case of ionizing radiation, there is evidence of greater male vulnerability to some hazards) [16].

Pregnancy, childbirth, breastfeeding, and other parenting tasks have, in general, remained the exclusive domain of women, even where biology has permitted the participation of men (*e.g.,* postchildbirth care and bottle-feeding of pumped breast milk). The assurance of prenatal health, while also a concern for males (males are not immune to prenatal reproductive hazards), has generally been viewed as solely a female risk and obligation. Some countries have instituted so-called "fetal protection policies" wherein all women physiologically capable of bearing children are excluded from employment regardless of their marital status and childbearing intent. Similar all-inclusive safeguards for men do not exist. The United States is a notable exception to fetal protection policy programs. The U.S. Supreme Court decision *UAW vs. Johnston Controls* unanimously barred fetal protection policies in the workplace. Part of the Court's reasoning was that such policies do not take possible male reproductive hazards into consideration [17].

The International Labor Office has summarized in some detail the legislated maternity protection that exists around the world [18]. Most countries provide periods of paid leave and monetary maternity allowances. Often this leave is mandatory—that is, pregnant women are automatically forced to leave the job. Many countries have breastfeeding provisions wherein women's work times are adjustable during the breastfeeding period, and some countries provide paid paternal leave as an option (although the percentage of men utilizing the parental leave is much lower than that of women).

In addition to maternity provisions, regulations governing the employment of women have generally covered the permitted hours of work, particularly night work and overtime, exposure to specific substances or conditions, maximum weight to be lifted or carried, and maternity/parenting leaves and benefits. Regulation of and standards for women's work lives and work-related health include a broad range of approaches:

1. *Nonexistent, minimal, or rarely enforced, as in many developing countries*—especially where women's work roles are largely ignored.
2. *Egalitarian,* as in Canada's rules on ionizing radiation exposure, which show it to be the most even-handed of all countries [19].
3. *Discriminatory,* as in Italy's prohibition of any night work for women (except where trade unions have specifically negotiated the right for women to work at night), thus barring women from many economic opportunities.
4. *"Protectionist,"* often masking economic/political decisions and expediency, and ironically failing to protect women working in traditional roles, where their labor is needed, while keeping them from entering lucrative jobs (the exemption of nurses, who face hazards of heavy lifting and exposure to toxic agents, is a classic example).

VIII. Dearth of Data

Figuring out where women work and under what conditions requires an ingenious assault on a multitude of data banks—a sort of sewing circle to assemble the "patchwork quilt" of the bits and pieces known about women's health on the job. The systematic and insightful gathering of data is a key research issue for the future. The following are some factors which must be addressed:

No comprehensive, thoroughly trustworthy data exist on *where* women work.

Measurement tools, such as survey questionnaires and job analyses, are either nonexistent or inadequate to explore *how* and *how much* women work.

Data are also lacking on *morbidity* and *mortality* within each sphere of women's work, and on gender differences in these vital statistics within integrated work areas.

Study designs, to avoid the arbitrary exclusion of women from industry-wide or other occupational studies [19].

In addition to the lack of data and research on women's paid employment and its relationship to women's health is the fundamental methodological problem of the classification of women's work as "work." That is, economists have imposed an artificial dichotomy that differentiates "real work" from "women's work," the extent of which is given in Table 41.1. This dichotomy is particularly inappropriate in those less developed countries where traditional "women's work" does provide the means by which the society more or less survives. Even in advanced industrial economies, the unpaid yet absolutely essential labor of women only takes on economic "value" when one has to replace it by the hiring of paid personnel (who very often are also "off the books" and are female). A more equitable way to look at the work that women do—both paid and unpaid, formal and informal—would be to use a modification of the classification scheme developed by Leewenhak, wherein work is organized into spheres of activity analogous to women's traditional roles. We have added some commercial activities to her scheme:

Motherhood and childcare

Water, sanitation, and cleanliness

Health care

Food production

Food processing, catering, and service

Textiles and clothing

Fuel and shelter

Commerce and distribution

Education

Personal services other than child- or health-related care—anything from hairdressing to sex work

Communications, mass media, and power production

Light manufacturing (*e.g.,* electronics, toys, other consumer goods)

Such a scheme fits both developed and developing countries and would provide a framework for "counting up" the paid and unpaid contributions made to economic functioning of a society. It broadens the definition of industrial sectors such as "agriculture," "trade," and "community, social, and personal services" by capturing the ignored unpaid and the informal-sector work of women. Currently "official" statistics do *not* reflect women's primarily "unofficial" work experiences.

Table 41.5 summarizes the major hazards that women could expect to encounter in these domains of work. The hazards cover the full gamut of exposures to chemical and physical agents, infectious agents, ergonomic injury, and traumatic injury. In addition, it should not be forgotten that the vast majority

Table 41.5
Examples of Hazards That Can Be Found in Areas of Women's Work

Area of work[a]	Biological hazards	Physical hazards	Chemical hazards	Stress
Motherhood and childcare	Infectious diseases (particularly respiratory)	Injuries associated with lifting and carrying	Household cleaning agents	Stress associated with caring occupations; burnout
Water, sanitation, and cleanliness	Infectious diseases (particularly water-borne)	Injuries associated with lifting and carrying	Household cleaning agents	
Health care	Infectious diseases (especially airborne and blood-borne)	Injuries associated with lifting and carrying; ionizing radiation	Cleaning, sterilizing, and disinfecting agents; laboratory agents and drugs	Stress associated with caring occupations; burnout
Food production	Infectious diseases (especially animal-borne and those associated with molds, spores and other organic dusts)	Repetitive motions (e.g., slaughterhouse and meatpacking); knife wounds; cold temperatures; noise; microwaves	Pesticide residues; sterilizing agents; sensitizing spices and additives	Stress associated with repetitive assembly line work
Food processing, catering, and service	Infectious diseases (from contact with public); Dermatitis	Injuries associated with lifting and carrying; wet hands; slipping and tripping; microwaves and heat	Secondhand cigarette smoke	Stress associated with dealing with the public; sexual harassment
Textiles and clothing	Organic dusts	Noise; repetitive motions	Formaldehyde in permanent press; dyes	Stress associated with assembly line work
Fuel and shelter		Injuries associated with lifting and carrying; exposure to elements; cuts and bruises from collecting and transporting materials	Polycyclic aromatic hydrocarbons from incomplete combustion of fuels	Stress associated with arduous labor
Commerce and distribution		Repetitive stress injuries and eyestrain (VDTs)	Poor indoor air quality	Stress associated with dealing with public
Education	Infectious diseases (particularly respiratory; measles)	Violence	Poor indoor air quality	Stress associated with caring occupations; burnout
Personal services other than child- or health-related care—anything from hairdressing to sex work	Infectious diseases (e.g., skin infections; AIDS and other sexually transmitted diseases)	Standing; lifting and carrying; violence	Chemical cleaning agents; hairdressing chemicals	Stress associated with caring occupations; burnout
Communications		Violence (journalism); repetitive motions (data entry); excessive sitting or standing	Poor indoor air quality	Electronic performance monitoring; fear of redundancy/unemployment
Light manufacturing		Repetitive motions (e.g., assembly line work); standing	Exotic chemicals in microelectronics	Stress associated with repetitive assembly line work

[a]These categories are an adaptation of the classification scheme of Leewenhak [8].

of women laborers bear the burden of multiple roles and are thus perhaps exposed to the hazards of more than one sector.

The absence of definitive data on where women work, the hazards they face, and the extent to which they may develop disease or disability reflects the discrimination against women workers that exists today. Furthermore, continuing to use schema that shortchange the actual physical, social, and economic contribution that women make perpetuates a major impediment to the creation of occupational safety and health programs for tomorrow.

References

1. United Nations (1995). "The World's Women 1995: Trends and Statistics, Social Statistics and Indicators," Ser. K, No. 12, p. 73. United Nations, New York.

2. Barry, J. (1983). Compensating pay differentials in hazardous work situations: A labor market segmental analysis. Ph.D. Dissertation, New School for Social Research, New York.

3. Sims, J. (1994). "Women, Health and Environment" (WHO/EHG/94.11). World Health Organization, Geneva.

4. United Nations (1995). "Women's Indicators and Statistics Database." Version 3 (Wistat-CD). United Nations, New York.

5. McCann, M. (1996). Hazards in cottage industries in developing countries. *Am. J. Ind. Med.* **30**(2), 125–129.

6. Reverente, B. R. (1992). Occupational health services for small-scale industries. In "Occupational Health in Developing Countries" (J. Jeyaratnam, ed.). Oxford University Press, New York.

7. Stellman, J. M. (1998). "Encyclopedia of Occupational Health and Safety," 4th ed. International Labor Office, Geneva.

8. Lewenhak, S. (1992). "The Revaluation of Women's Work," rev. ed., p. 119. Earthscan Publications Ld., London.

9. Organisation for Economic Co-operation and Development (1980). "Women and Employment: Policies for Equal Opportunities," pp. 39–40. OECD, Paris.

10. Mergler, D., Brabant, C., Vezina, N., and Messing, K. (1987) The weaker sex? Men in women's working conditions report similar health symptoms. *J. Occup. Med.* **29,** 417–421.

11. Stellman, J. M. (1995). Social factors: Women and cancer. *Semin. Oncol. Nurs.* **11**(2); 103–108.

12. Baetjer, A. (1946). "Women in Industry: Their Health and Efficiency." Saunders, New York.

13. Jürgens, H., Aune, I., and Pieper, U. (1990). "International Data on Anthropometry." International Labor Organization, Geneva.

14. Messing, K. (1997). Women's occupational health: A critical review and discussion of current issues. *Women Health* **25**(4), 39–68.

15. Messing, K. (1997). *Women Health* **25**(4), 55.

16. Stellman, J. M. (1987). Protective legislation, ionizing radiation and health: A new appraisal and international survey. *Women Health* **12**(1), 105–125.

17. International Union (1991). *UAW v Johnson Controls.* U.S. Supreme Court, Washington, DC. 1196.

18. International Labor Office (1994). "Maternity and Work," Conditions of Work Digest, Vol. 13 (the volume presents a compendium of worldwide regulation). International Labor Office, Geneva.

19. Zahm, S. H., Pottern L. M., Lewis, D. R., Ward, M. H., and White, D. W. (1994). Inclusion of women and minorities in occupational cancer epidemiological research. *J. Occup. Med.* **36**(8), 842–847.

Section 7

SOCIAL DETERMINANTS OF HEALTH

Glorian Sorensen

Harvard School of Public Health and Center for Community-Based Research
Dana-Farber Cancer Institute
Boston, Massachusetts

Social determinants of health operate through a range of social pathways, including social integration, social structure, neighborhood characteristics, and the division of labor [1–6]. Gender itself is one of the most profound social determinants of health. Gender includes both a basic biological distinction, labeled *sex,* and a fundamentally social one, labeled *gender.* Studies of gender illuminate the potency of social structure, focusing on economic, political, and social power, and raise questions about how social structure may influence personal choices [7]. As noted by Hall, studies of health can ignore gender, control for gender, or analyze for gender [8]. It is necessary to go beyond simply controlling for the effects of gender in order to understand the role of gender as an organizing social force, thereby uncovering unstated assumptions and explicating the social processes shaping women's lives.

The study of social stratification by gender has a long history [7]. Borrowing from prior research, R. W. Connell synthesized this literature into three fundamental structures that influence the relationships between men and women [9]. First, the division of labor structures the organization of housework, the sex segregation of the labor force, the distinction between paid and unpaid work, and inequalities in pay between men and women. Second, power structures determine the nature of authority and control within society and influence the dynamics of power within domestic relationships, institutional and interpersonal violence, and business hierarchies. Third, a structure that Connell labeled cathexis defines the social patterning of relationships that are based on sexual or emotional attachments, including laws, taboos, and prohibitions that define sexuality, attractiveness, and attachments.

The chapters in this section provide illustrations of these social structures at work in the day-to-day lives of women and demonstrate the impact of social structures on health. For instance, the structures of power and the division of labor influence the types of jobs women hold, the amount they are paid, and the patterning of household responsibilities, factors contributing to social inequalities in women's health and the stressors in women's lives. These chapters also provide examples of the ways in which distal social forces may influence personal health behaviors. These socially structured pathways lead to behaviors such as eating patterns, smoking, and alcohol use that further influence women's health.

The structure of authority and power is perhaps most starkly observed in patterns of violence against women, described in Chapter 42. Violence against women is prevalent in many different forms across virtually all cultures [10]. The 1995 National Crime Victimization Survey found that 4% of women in the U.S. reported experiences of violent crime during the past year [11]. In the National Women's Study conducted in 1992, 13% of the women reported at least one incident of a completed rape during their lifetime [12]. Other studies have focused specifically on rates of partner violence, the second leading cause of injuries to women of all ages and the leading cause of injuries to women between the ages of 15 and 44 [13]. Also common to many women's lives is a pattern of psychological and emotional abuse, ranging from derogatory and demeaning epithets and harassment to more physically threatening psychological abuse such as stalking. Different forms of physical and emotional abuse may be present with or without physical abuse, although when physical abuse is present, it is virtually always accompanied by psychological or emotional abuse [14,15]. The health outcomes associated with violence against women extend beyond the immediate injuries that may accompany a violent encounter. Violent victimization may lead to increased stress, increased risk of mental health problems, impaired immune system functioning, increased health risk behavior, and inappropriate health care utilization, overall placing the individual at increased health risk [16].

Although violence against women is prevalent in all social strata, there is evidence that the rates are particularly high among lower social strata [17–20]. An increasing body of research documents the broad effects of socioeconomic factors on health, as described in Chapter 43. In addition to shaping exposure to violence and abuse, socioeconomic status influences health habits, the prevalence of depression, the burden of caretaking and number of working hours, access to health insurance and medical care, and security in old age [Moss, Chapter 43]. Socioeconomic position has further been shown to affect nutrition, growth, and development; determine life chances including later education and income; and contribute at a population level to striking differences between rich and poor geographic areas and between developed and developing countries. On a social structural level, men earn more than women, and women continue to perform most domestic work [21–23]. Among women, those with more education and higher incomes live longer and are healthier than women with fewer years of schooling and less income [24]. In addition to the adverse effects of poverty, evidence indicates that society-level income inequality and the presence of socioeconomic gradients negatively influences morbidity and life expectancy [25,26].

Socioeconomic effects may differ over the life course and may depend on household configuration and marital status. To understand the relationship between socioeconomic factors and women's health requires viewing patterns of health and disease among persons in different groups as the product of social relationships between the groups. Chapter 43 provides alternative theoretical models to guide the formulation of research questions about this relationship and the selection of the appropriate socioeconomic measures, whether at the individual or community level.

Living in poverty exposes individuals to a number of stressors, including chronic stressors such as economic strain and living in areas with high rates of crime and residential mobility [27,28]. Because the responsibilities of caring for dependent children may interfere with full-time employment, about half of all single mothers receive welfare benefits to help meet the financial needs of their families [29]. In addition, poor women may be more vulnerable to the negative effects of stress exposure, perhaps in part due to fewer personal and social resources [30].

These and other stressors are described in Chapter 44. Women have higher rates of psychological distress than men and report higher levels of exposure to chronic stress than do men [31]. This chapter examines the specific social–environmental stressors that women commonly encounter and the pathways through which these stressors affect women's health and well-being. Differences in the experience of stress result both from differing exposures to stress and from differences in vulnerability to stress. Research on stress distinguishes life events or acute changes requiring major behavioral adjustments in a relatively short period of time from chronic strains, which require readjustments over prolonged periods of time [32]. Vulnerability to stress, or the extent to which stress exposure results in psychological and biological changes that undermine health and wellbeing, may be influenced by a range of personal and contextual factors, including social support and personal control.

Stress in women's lives springs from a range of sources. Personal relationships may be a source of both strain and social support. Compared to men, women have more emotional support provided in close personal relationships. Social support provides a buffer that protects individuals from the negative health effects of stress exposure, largely by reducing the likelihood that undesirable events and life conditions will be appraised as stressful [33]. Negative dimensions of personal relationships may have greater effects on physical and mental health than positive components, however, and include marital and intimate relationships, relationships with dependent children, and relationships with parents that involve the provision of care. Studies show that marriage is beneficial to mental and physical health, largely because it provides economic resources and social support [34]. However, the benefits of marriage depend on the quality of the marital relationship [35]. Marital strain for women arises in part from the inequality and power differential inherent in traditional gender roles, which includes strain from unequal distribution of household labor and from domestic violence [36,37].

Additional sources of stress for women are related to women's care giving roles. Parenting of minor children is a source of stress for both married and single women. Strains associated with parenting arise from both financial strains and from providing and arranging for childcare [38–40]. The salience of the parental role may also shape a woman's vulnerability to parenting-related stressors; most evidence suggests that the parenting role is more salient to women than to men [41]. Providing care to parents may be an additional source of strain. Women shoulder the bulk of the responsibility for providing care to elderly family members. About 55% of women between the ages of 45 and 59 with one living parent can expect to provide some level of care to a parent in the next 25 years [42].

Given that employed women may be balancing job roles with family care giving, it is not surprising that most studies find that employed women have higher levels of stress and distress than employed men [43]. Paid work is generally beneficial to the health of both men and women; nonetheless, women accrue fewer benefits than men from paid employment [44]. More relevant than women's employment status per se are the specific characteristics of the jobs women tend to occupy [45]. Many of the stressors to which women are exposed are rooted in the sex-segregation of the labor force. Women are overwhelmingly concentrated in the service sector, in jobs that include the provision of care to others (*e.g.*, nurses, social workers, teachers) [46,47]. On average, men and women are exposed to different workplace environments and different types of demands and strains. Female-dominated jobs are characterized by low pay, low levels of autonomy, low levels of authority and power, low levels of complexity and high levels of routinization, and responsibility for providing care and support for others [48–51].

The strategies that women use to cope with life's stressors vary to some extent by socioeconomic status. Women in lower socioeconomic strata may have fewer personal and social resources than women of higher social class [52]. The work of Graham suggests that low income women may use smoking as a means of coping with their economic pressures and the resulting demands placed on them to care for others [53]. Indeed, she observed that spending on cigarettes appears to be protected because it is viewed as a necessary luxury. Compared to their nonsmoking counterparts, working class mothers who smoke generally are caring for more children and for children in poorer health and are more likely to be providing that care alone; a larger proportion of smokers had insufficient resources to meet the basic needs of their families and lived in less desirable neighborhoods. Graham concluded that smoking among working class women was linked to the caring responsibilities and material circumstances that shape the everyday lives of these women.

Chapter 45 documents the importance of smoking habits in women's health outcomes. Cigarette smoking is the leading preventable cause of death among women in America, killing more than 152,000 women annually, primarily from cardiovascular disease, cancer, and respiratory diseases [54]. In 1995, 22.6% of women in the U.S. were smokers [55]. Smoking prevalence is highest among American Indian and Alaska Native women and among women with 9–11 years of education. Women are now attempting cessation and maintaining abstinence at the same rate as men [56]. Smoking increased dramatically among girls during the 1990s. In 1997, 35% of twelfth grade girls were current smokers, defined as smoking in the last 30 days [57]. Persons who start smoking at a young age are more likely to smoke heavily, more likely to become dependent on nicotine, and less likely to quit smoking [58–60]. Most girls say they want to quit smoking, and more than half of all adolescent smokers try to quit each year [61,62]. Adolescents and children report the same problems with quitting as do adults, indicating that quitting is no easier for youth than it is for adults [63,64].

Weight concerns pose a barrier to smoking cessation for many women and may increase relapse rates among women who quit [65,66]. The perception that smoking may help with weight control may also contribute to the decision to smoke for many adolescent girls [61]. The "slimming" effects of smoking are marketed heavily by the tobacco industry. Such promotion dates back to as early as 1928, when Lucky Strike promoted smoking for women as a weight control measure with the slogan "Reach for a Lucky instead of a sweet" and continues to be a focus of today's "women-only" cigarette brands [7].

Women's preoccupation with weight reflects gender norms about attractiveness, as outlined in Chapter 46. Many American women exhibit varying degrees of body image distortion and dissatisfaction with their bodies. A poor body image may translate into reductions in self-esteem and may lead to unhealthy eating habits and other negative behaviors designed to reduce body weight and body fat [67,68]. While obesity and overweight are risk factors for a variety of diseases, including cardiovascular disease, diabetes, and osteoarthritis [69–71], it is also apparent that underweight and thinness carry some important morbidity and mortality consequences [72]. This chapter discusses disordered eating patterns, including anorexia nervosa, bulimia nervosa, and binge eating disorder, disorders that are clear indicators of body image disturbances and nutritional problems.

The final chapter in this section addresses alcohol use and abuse. As described in Chapter 47, traditionally relatively little attention has been given to alcohol use and abuse in women. This field has been faced with numerous challenges, including inconsistencies in the measurement of alcohol use and in definitions of abuse, given that use is culturally normed. Patterns of higher alcohol use or alcohol abuse have been associated with psychopathology [73] and the experience of violent victimization [74]. Having alcoholic parents is a major risk factor for alcohol abuse, with evidence indicating both the heritability of risk for alcohol abuse and the negative impact of growing up in an alcoholic family environment [75–78]. Once problem drinking is established, sexual dysfunction may be an important predictor of continued alcohol abuse [79]. Alcohol problems may be less likely to be identified in women compared to men. Further barriers to treatment include lack of childcare, fear of losing child custody, and lack of insurance coverage. When these barriers are overcome, evidence indicates that women respond well to treatment [80].

These chapters point to important areas of future research in the social determinants of women's health. A major theme across all chapters is the need for improved measures of social determinants of women's health—measures that are guided by theory and provide models for consistent assessment in order to allow comparisons across studies. There is further need for understanding moderating variables and mediating mechanisms influencing these relationships. Such research, for example, might illuminate the mechanisms by which stress might influence physical and mental health, the ways in which socioeconomic status may be internalized to produce the observed differences in health, or the processes by which adolescent girls take up smoking. Study is also needed to assess the extent to which the social determinants discussed here may have different effects on men's and women's health outcomes and health behaviors. Biological differences between men and women may shape some of the observed gender differences and deserve further examination. Evidence indicates, for example, that the bioavailability of alcohol is higher in women than in men, resulting in women's greater sensitivity to alcohol's effects. This greater sensitivity may mean that fewer women than men have alcohol-related health problems and may also place women who do drink heavily at increased risk of alcohol-related disease.

Research on gender issues in health has focused chiefly on explaining women's mortality advantage over men, men's morbidity advantage, and trends in these relationships. These comparisons mask the very real possibility that both men and women might be healthier in a society better structured for economic and social equity, shared power and responsibility, and greater tolerance of diversity in sexuality and other areas of personal expression [7]. Ultimately, influencing women's health outcomes may rely heavily on social structural changes that may likewise prove beneficial for men's health outcomes.

References

1. Amick, B. C., Levine, S., Tarlov, A. R., and Walsh, D. C. (1995). Introduction. *In* "Society and Health" (D. L. Patrick and T. M. Wickizer, eds.), pp. 46–73. Oxford University Press, Oxford.

2. Anderson, N., and Armstead, C. (1995). Toward understanding the association of socioeconomic status and health: A new challenge for the biopsychosocial approach. *Psychosom. Med.* **57,** 213–225.

3. Ashton, J., and Seymour, H. (1988). "The New Public Health." Open University Press, Buckingham, UK.

4. Kaplan, G. (1995). Where do shared pathways lead? Some reflections on a research agenda. *Psychosom. Med.* **57,** 208–212.

5. McKinlay, J. (1995). The new public health approach to improving physical activity and autonomy in older populations. *In* "Preparation for Aging" (E. Heikkinon, ed.) pp. 87–103. Plenum, New York.

6. Robertson, A., and Minkler, M. (1994). New health promotion movement: A critical examination. *Health Educ. Q.* **21,** 295–312.

7. Walsh, D., Sorensen, G., and Leonard, L. (1995). Gender, health, and cigarette smoking. *In* "Society and Health" (B. I. Amick *et al.,* eds.), pp. 131–171. Oxford University Press, New York.

8. Hall, S. M., *et al.* (1992). Weight gain prevention and smoking cessation: Cautionary findings. *Am. J. Public Health* **82**(6), 799–803.

9. Connell, R. W. (1987). "Gender and Power." Stanford University Press, Stanford, CA.

10. Fishbach, R. L., and Herbert, B. (1997). Domestic violence and mental health: Correlates and conundrums within and across cultures. *Soc. Sci. Med.* **45**(8); 1161–1176.

11. Bachman, R., and Saltzman, L. E. (1995). Violence against women; Estimates from the redesigned survey. *In* "National Crime Victimization Survey." U.S. Dept. of Justice; Office of Justice Programs, Bureau of Justice Statistics; Washington, DC.

12. Center, N. V. (1992). "Rape in America: A Report to the Nation." Crime Victims Research Center; Arlington, VA.

13. Barrier, P. A. (1998). Domestic violence. *Mayo Clin. Proc.* **73**(3), 271–274.

14. Bograd, M. (1988). Feminist perspectives on wife abuse: An introduction. *In* "Feminist Perpectives on Wife Abuse" (M. Bograd and K. Yllo, eds.) Sage; Newbury Park, CA.

15. Adams, D. (1988). Treatment models of men who batter: A profeminist analysis. *In* "Feminist Perspectives on Wife Abuse" (M. Bograd and K. Yllo, eds.). Sage; Newbury Park, CA.

16. Acierno, R., Resnick, H. S., and Kilpatrick, D. G. (1997). Health impact of interpersonal violence 1: Prevalence rates, case identification, and risk factors for sexual assault, physical assault, and domestic violence in men and women. *Behav. Med.* **23**(2), 53–64.

17. Council on Scientific Affairs, A.M.A. (1992). Violence against women. *JAMA, Am. Med. Assoc.* **267,** 3184–3189.

18. Sugg, N. K., and Inui, T. (1992). Primary care physicians' response to domestic violence: Opening Pandora's box. *JAMA, J. Am. Med. Assoc.* **267,** 3157–3160.

19. Ernst, A. A., *et al.* (1997). Domestic violence in an inner-city ED. *Ann. Emerg. Med.* **30**(2); 190–197.

20. Abbott, J., *et al.* (1995). Domestic violence against women: Incidence and prevalence in an emergency department population. *JAMA, J. Am. Med. Assoc.* **273**(22), 1763–1767.

21. Marini, M. M. (1980). Sex differences in the process of educational attainment: A closer look. *Soc. Sci. Res.* **9,** 307–361.

22. McGuire, G. M., and Reskin, B. (1993). Authority hierarcies at work: The impacts of race and sex. *Gender Soc.* **7,** 487–506.

23. Browner, C. H. (1989). Women, household and health in Latin America. *Soc. Sci. Med.* **28,** 461–473.

24. U.S. Department of Health and Human Services and National Center for Health Statistics (1998). "Health United States 1998 with Socioeconomic Status and Health Chartbook." National Center for Health Statistics, Hyattsville, MD.

25. Kennedy, B. P., Kawachi, I., and Prothrow-Stith, D. (1996). Income distribution and mortality: Test of the Robin Hood Index in the United States. *Br. Med. J.* **312,** 1004–1007.

26. Kaplan, G. A., *et al.* (1996). Inequality in income and mortality in the United States: analysis of mortality and potential pathways. *Br. Med. J.* **312,** 999–1005.

27. Garfinkel, I., and McLanahan, S. S. (1986). "Single Mothers and Their Children: A New American Dilemma." Urban Institute Press, Washington, DC.

28. Williams, D. R. (1990). Socioeconomic differential in health: A review and redirection. *So. Psychol. Q.* **53,** 81–99.

29. Rodgers, H. R. (1996). "Poor Women, Poor Children: American Poverty in the 1990s." M. E. Sharpe, Armonk, NY.

30. Turner, R. J., and Noh, S. (1983). Class and psychological vulnerability among women: The significance of social support and personal control. *J. Health Soc. Behav.* **24,** 2–15.

31. Turner, R. J., Blair, W., and Lloyd, D. A. (1995). The epidemiology of social stress. *Am. Sociol. Rev.* **60,** 104–125.

32. Thoits, P. (1995). Stress, coping, and social support processes: Where are we? What next? *J. Health Soc. Behav. Extra Issue,* pp. 53–79.

33. Cohen, S. (1988). Psychosocial models of the role of social suport in the etiology of physical disease. *Health Psychol.* **7,** 269–297.

34. Umberson, D., and Williams, K. (1999). Family status and mental health. *In* "Handbook on the Sociology of Mental Health" (C. S. Aneshensel, ed.). Plenum, New York (to be published).

35. Gove, W. R., Hughes, M., and Style, C. B. (1983). Does marriage have positive effects on the psychological well-being of the individual? *J. Health Soc. Behav.* **24,** 122–131.

36. Suitor, J. J. (1991). Marital quality and satisfaction with the division of household labor across the family life cycle. *J. Marriage Fam.* **53,** 221–230.

37. Stark, E., and Flitcraft, A. H. (1991). Spouse abuse. *In* "Violence in America: A Public Health Approach" (M. L. Rosenberg and M. A. Fenley, eds.), pp. 123–157. Oxford University Press, New York.

38. Simon, R. (1998). Assessing sex differences in vulnerability among employed parents: The importance of marital status. *J. Health Soc. Behav.* **39,** 38–54.

39. Bird, C. E. (1997). Gender differences in the social and economic burdens of parenting and psychological distress. *J. Marriage Fam.* **59,** 809–823.

40. Ross, C. E., and Van Willigen, M. (1996). Gender, parenthood, and anger. *J. Marriage Fam.* **58,** 572–584.

41. Simon, R. (1992). Parental role strains, salience of parental identity and gender differences in psychological distress. *J. Health Soc. Behav.* **33,** 25–35.

42. Himes, C. (1994). Parental caregiving by adult women: A demographic perspective. *Res. Aging* **16,** 191–211.

43. Roxburgh, S. (1996). Gender differences in work and well-being: Effects of exposure and vulnerability. *J. Health Soc. Behav.* **37,** 265–277.

44. Aneshensel, C. S., Frerichs, R. R., and Clark, V. A. (1981). Family roles and sex differences in depression. *J. Health Soc. Behav.* **22,** 379–393.

45. Aneshensel, C. S. (1986). Marital and employment role streain, social support, and depression among adult women. *In* "Stress, Social Support, and Women" (S. E. Hobfoll, ed.), pp. 99–114. Hemisphere Publishing, New York.

46. Lennon, M. C. (1987). Sex differences in distress: The impact of gender and work roles. *J. Health Soc. Behav.* **28,** 290–305.

47. Marshall, N. L., *et al.* (1991). More than a job: Women and stress in caregiving occupations. *Curr. Res. Occup. Prof.* **6,** 61–81.

48. Starrels, M. E., Bould, S., and Nichols, L. J. (1994). The feminization of poverty in the United States: Gender, race, ethnicity, and family factors. *J. Fam. Issues* **15,** 590–607.

49. Pugliesi, K. (1995). Work and well-being: Gender differences in the psychological consequences of employment. *J. Health Soc. Behav.* **36,** 57–71.

50. Wright, O. E., Baxter, J., and Birkelund, G. E. (1995). The gender gap in workplace authority: A Cross-National Study. *Am. Sociol. Rev.* **60,** 407–435.

51. Bulan, H. F., Erickson, R. J., and Wharton, A. S. (1997). Doing for others on the job: The affective requirements of service work, gender, and emotional well-being. *Soc. Probl.* **44,** 235–256.

52. House, J. S., and Williams, D. R. (1995). Psychosocial pathways linking SES and CVD. *In* "Report of the Conference on Socioeconomic Status and Cardiovascular Health and Disease," pp. 119–124. National Institutes of Health, National Heart, Lung, and Blood Institute; Bethesda, MD.

53. Graham, H. (1994). Gender and class as dimensions of smoking behavior in Britain: Insights from a survey of mothers. *Soc. Sci. Med.* **38,** 691–698.

54. Centers for Disease Control and Prevention (1997). Smoking-attributable mortality and years of potential life lost—United States, 1984. *Morbid. Mortal. Wkly. Rep.* **46,** 444–451.

55. Centers for Disease Control and Prevention (1997). Cigarette smoking among adults—United States, 1995. *Morbid. Mortal. Wkly. Rep.* **46,** 1217–1220.

56. Centers for Disease Control and Prevention (1993). Smoking cessation during previous year among adults—United States, 1990 and 1991. *Morbid. Mortal. Wkly. Rep.* **42,** 504–507.

57. University of Michigan (1997). Cigarette smoking rates may have peaked among younger teens (press release). University of Michigan News and Information Services, Ann Arbor.

58. Taioli, E., and Wynder, E. L. (1991). Effect of the age at which smoking begins on frequency of smoking in adulthood. *N. Engl. J. Med.* **325,** 968–969.

59. Breslau, N., Kilbey, M. M., and Adreski, P. (1993). Vulnerability to psychopathology in nicotine-dependent smokers: An epidemiologic study of young adults. *Am. J. Psychiatry,* **150,** 941–946.

60. Breslau, N., and Peterson, E. L. (1996). Smoking cessation in young adults: Age at initiation of cigarette smoking and other suspected influences. *Am. J. Public Health* **86,** 214–220.

61. Moss, A. J., *et al.* (1992). "Recent Trends in Adolescent Smoking, Smoking-Uptake Correlates, and Expectation about the Future," Adv. Data No. 221. U.S. Department of Health and Human Services, Centers for Disease Control and Prevention, National Center for Health Statistics, Hyattsville, MD.

62. Pirie, P. L., Murray, D. M., and Leupker, R. V. (1991). Gender differences in cigarette smoking and quitting in a cohort of young adults. *Am. J. Public Health* **81,** 324–327.

63. McNeill, A. D., *et al.* (1986). Cigarette withdrawal symptoms in adolescent smokers. *Psychopharmacology* **90,** 533–536.

64. Ershler, J., *et al.* (1989). The quitting experience for smokers in sixth through twelfth grades. *Addict. Behav.* **14,** 365–378.

65. Pomerleau, C. S., and Pomerleau, O. F. (1994). Gender differences in frequency of smoking withdrawal symptoms. *Ann. Behav. Med.* **16,** S118.

66. Gritz, E. R., Nielsen, I. R., and Brooks, L. A. (1996). Smoking cessation and gender: The influence of physiiological, psychological, and behavioral factors. *J. Am. Med. Women's Assoc.* **51,** p. 35–42.

67. Freedman, R. (1984). Reflections on beauty as it relates to health in adolescent females. *Women's Health* **9,** 29–45.

68. Lewis, V. J., and Blair, A. J. (1993). Women, food, and body image. *In* "The Health Psychology of Women" (C. Niven and D. Carroll, eds.), pp. 107–120. Harwood Academic Publishers; Chur, Switzerland.

69. Manson, J. E., Willet, W. C., Stampfer, M. J., Colditz, G. A., Hunter, D. J., Hankinson, S. E., Hennekens, C. H., and Speizer, F. E. (1995). Body weight and mortality among women. *N. Engl. J. Med.* **333,** 677–685.

70. Bender, R., Trautner, C., Spraul, M., and Berger, M. (1998). Assessment of excess mortality in obesity. *Am. J. Epidemiol.* **147**(1), 42–48.

71. Diehr, P., Bild, D., Harris, T., Duxbury, A., Siscovick, D., and Rossi, M. (1998). Body mass and mortality in nonsmoking older adults: The Cardiovascular Health Study. *Am. J. Public Health* **88,** 623–629.

72. Heaney, R. P. (1996). Osteoporosis. *In* "Nutrition in Women's Health" (D. Krummel and P. Kris-Etherton, eds.). Aspen Publications, Gaithersburg, MD.

73. Wilsnack, S. C. (1995). Alcohol use and alcohol problems in women. *In* "The Psychology of Women's Health: Progress and Challenges in Research and Application" (A. L. Stanton and S. J. Gallant, eds.), pp. 381–443. American Psychological Association; Washington, DC.

74. Miller, B. A. (1996). Women's alcohol use and their violent victimization. *In* NIAAA "Women and Alcohol: Issues for Prevention Research" (J. M. Howard *et al.,* eds.), NIAA Res. Monogr. 32, pp. 239–260. U.S. Department of Health and Human Services, Bethesda, MD.

75. McGue, M., and Slutske, W. (1996). The inheritance of alcoholism in women. *In* "Women and Alcohol: Issues for Prevention Research" (J. M. Howard *et al.,* eds.), NIAAA Res. Monogr. 32, pp. 65–92. U.S. Department of Health and Human Services, Bethesda, MD.

76. Heath, A. C., Slutske, W. S., and Madden, P. A. F. (1997). Gender differences in the genetic contribution to alcoholism risk and to alcohol consumption patterns. *In* "Gender and Alcohol: Individual and Social Perspectives" (R. W. Wilsnack and S. C. Wilsnack, eds.), pp. 114–149. Rutgers Center of Alcohol Studies; New Brunswick, NJ.

77. Seilhamer, R. A., and Jacob, T. (1990). Family factors and adjustment of children of alcoholics. *In* "Children of Alcoholics: Critical Perspectives" (M. Windle and J. S. Searles, eds.), pp. 168–186. Guilford Press, New York.

78. U.S. Department of Health and Human Services (1997). "Ninth Special Report to the U.S. Congress on Alcohol and Health." U.S. Govt. Printing Office; Washington, DC.

79. Wilsnack, S. C., Klassen, A. D., Schur, B. E., and Wilsnack, R. W. (1991). Predicting onset and chronicity of women's problem drinking: A five-year longitudinal analysis. *Am. J. Public Health* **81,** 305–318.

80. Walitzer, K. S., and Connors, G. J. (1997). Gender and treatment of alcohol-related problems. *In* "Gender and Alcohol: Individual and Social Perspectives" (R. W. Wilsnack and S. C. Wilsnack, eds.), pp. 445–461. Rutgers Center of Alcohol Studies, New Brunswick, NJ.

42

Violence against Women

BRENDA A. MILLER* AND WILLIAM R. DOWNS†

*Center for Research on Urban Social Work Practice, University at Buffalo, Buffalo, New York; †Center for the Study of
Adolescence, University of Northern Iowa, Cedar Falls, Iowa

I. Introduction

Across virtually all cultures, violence against women is prevalent in many different forms, including physical, psychological, and emotional abuse. A review of diverse cultures suggests that gender-based violence is a pervasive, global issue that contributes to preventable morbidity and mortality of women everywhere [1]. The types of violence that affect women around the world include war-related rape and sexual assault [2], nonmedically induced abortion [3], sex-selective abortion of female fetuses [4], female genital mutilation [5], partner violence [6,7], and rape [8]. Although women experience criminal violence from outside the family (*e.g.,* rape from strangers), most violence against women is relationship-based and has been culturally institutionalized. For centuries women's experiences of sexual, physical, and psychological violence from male partners have been fostered by cultural norms and expectations [9]. However, a variety of strategies has emerged worldwide to promote social change, beginning with a recognition that the status of women needs to be improved and women need more control over their bodies [1].

The prevalence of violence against women has been examined in population-based studies as well as in various clinic and special populations. Violence appears to be particularly common among women seeking services in health care settings.

Links between violence against women and negative mental health consequences have been reported in numerous clinical and epidemiologic studies. Experiences of violence have been found to be related to elevated depressive symptoms, lower self-esteem, lower perceived health status and lower health care quality [10], elevated psychiatric symptoms (*i.e.,* depression, anxiety, somatization, and interpersonal sensitivity), alcohol or drug abuse [11–15], suicide attempts [16], and higher levels of psychological disturbance [17], including the development of posttraumatic stress disorder (PTSD) [11,18,19].

There is evidence that physical violence increases the likelihood of subsequent physical problems as well. Some are easily linked to the violent episode, such as the emergence of HIV infection as a result of rape or other sexual assaults. Other conditions are more distally linked to episodes of violence. There is evidence that chronic headaches, chronic pelvic pain, and gastrointestinal disorders emerge more frequently among women with trauma histories [20–26]. A variety of somatic problems, chronic pain, and other chronic diseases have been linked to experiences of violent victimization [27]. Women who have experienced some form of sexual assault or multiple assaults including sexual trauma were found to have poorer health outcomes involving all body systems except eyes and skin, as compared to women without these forms of victimization [28]. Risk for gynecological symptoms was especially pronounced for women with histories of sexual assaults. Rates of visits to physicians approximately doubled for victims of sexual assault and costs more than doubled. Particularly of interest is the fact that the greatest utilization of services occurred in the second year following the victimization, suggesting that physical consequences are, indeed, long term [28]. Another study, using structural equation modeling and longitudinal data, found that prior experiences of partner physical abuse predicted higher levels of future physical health symptoms and that the effects of abuse on physical health were mediated through changes in anxiety and depression over time [29].

In this chapter, we present an overview of the range and extent of violence against adult women. Difficulties in ascertaining rates of violence against women are discussed and barriers to ascertaining violence are identified, both for the health care system that responds to women and for individuals self-reporting their experiences of violence. These barriers are better understood within the cultural and historical responses to violence and a short summary of this history is presented. Opportunities for identifying and treating women who have been victimized are discussed, with a particular focus on the health care system as an important point of intervention for women. Women's exposure to violence occurs throughout the life span, yet there are numerous opportunities, particularly within physical and mental health care settings, to identify victims of violence and offer appropriate interventions.

II. Difficulties in Assessing Rates of Violence against Women

There are a number of different reasons why rates of violence against women may vary across different studies. Violence against women includes many different types of incidents, including assaults, sexual assaults, and psychological and emotional abuse. Studies vary in the types of research questions addressed, samples that are included, and how the research is conducted. Some variation also may occur depending on when the study was conducted, either due to real differences across time or due to differences in respondents' willingness to discuss issues of victimization. It is possible that as victimization issues become more widely discussed in the general population, the stigmatization that occurs with victimization may be reduced and more victims will be willing to self-report. As detailed below, variations in rates across different studies need to be understood in the context of these methodological difficulties.

Differences across studies are apparent in the actual construction of the research questions and the instruments designed and used to answer these questions. For example, asking respondents if they have ever been a victim of rape or sexual assault necessitates that women define themselves as "victims" and identify themselves as having experienced "rape or sexual assault." Thus, how the questions are asked regarding victimization can influence the results. Variations in timeframes employed for assessing victimization also occur, with some studies focusing on lifetime rates and other studies focusing on a more recent time period (e.g., last year). Further, there may be differences in self-reporting victimization depending on the age of the woman at the time the questions are asked.

Studies vary according to type of victimization assessed. Some studies examine only one type of victimization and other studies include multiple forms of victimization. Definitional problems also exist. For example, partner violence can be narrowly defined as marital violence or as violence between couples living together. Given the prevalence of violence among couples not married and not living together, the extent of violence within women's lives can be greatly underestimated if such narrow definitions are employed. Adding to the complexity is recognition that violence also can occur within dating relationships.

Still other differences emerge due to characteristics of samples included in different studies. For instance, rates may vary across clinical and general population studies because of diversity in ethnic backgrounds, family structures, socioeconomic status, age ranges, geographic regions, and other social setting variables. While violence against women occurs across all social strata, ethnic groups, and age ranges, our social contexts and social beliefs influence what acts are defined and reported as violence, as well as women's reactions to their experiences. Some large-scale epidemiologic studies have focused on asking women about specific experiences they have encountered without asking them to define themselves as victims. While this is an important step in systematically inquiring about violence in women's lives, it does not address the problem that women's interpretations of their experiences vary across different cultural backgrounds. Women's interpretations, meanings, and understandings of their experiences are important, not only for long- and short-term consequences that women experience, but also, in guiding our societal responses to violence.

Rates of violence reported for women vary, in part, due to the interaction between respondent and the research process. For instance, different social contexts or settings where victimization experiences are assessed may influence the reporting of the violent victimization. Telephone surveys may be influenced by the presence of a male who is a perpetrator of violence. How victimization questions are asked in the context of other questions also can influence the responses. Asking questions about rape and sexual assault in the context of other criminal victimization questions may influence women's perceptions that only "crimes" are of interest, requiring that the experience be labeled as a crime. With sensitive issues such as victimization, respondents' perceptions of reasons for asking the questions are important. The respondents' ability to trust the research process and their perceptions of the confidentiality of this process can all affect the responses. More attention needs to be given to these important conceptual, definitional, and process issues.

III. Violence against Women: Prevalence, Range, and Scope of the Problem

A. Physical Victimization from All Types of Perpetrators

The 1995 National Crime Victimization Survey (NCVS) was conducted by the U.S. Department of Justice, Bureau of Justice Statistics and is a nationally representative survey of all adults ages twelve and older. It was conducted to estimate the prevalence of victimization in the United States. This survey assessed violent victimization in the context of other questions regarding victimization from different types of crime in 1994. In this survey, approximately 4% of women in the U.S. reported experiences of violent crime (e.g., homicides, sexual assaults, rape, robbery, assaults) during the past year [30]. The estimated number of incidents perpetrated against women is over 4.7 million. The number of rapes in 1994 was estimated at 500,200, or a rate of 4.6 per 1000 women.

In 1992, the National Women's Study surveyed 4008 adult women, approximately half of whom were between ages 18 and 34 and half of whom were a cross-section of women 18 and older [31]. Thirteen percent of respondents reported at least one incident of a completed rape during their lifetime, resulting in an estimated number of 12.1 million women having had an experience of forcible rape. During 1991, the number of rapes was estimated at 683,000, or a rate of 7 per 1000 women. The higher reported rate of rapes during the preceding year for the National Women's Study, in comparison to other national surveys (e.g., NCVS), may be due to the more extensive questions and focus on rape in the National Women's Study and the different age ranges included in the surveys.

In the National Co-morbidity Study, questions about violence are embedded within a series of questions regarding different mental health symptoms. In this national probability sample, 9% of female respondents reported experiences of rape and 12% reported experiencing molestation during their lifetime [32].

Rates of violent victimization generally are higher when women from younger age ranges are sampled. For example, a study of women in colleges across the United States indicates that 15% had been raped and 12% had been victims of attempted rape during their lifetime [33]. In a review of cross-national studies of rape for college-age women, Koss et al. [8] noted that the lifetime rate of completed rape ranged from a low of 7.7% in South Korea to a high of 15.4% in the United States. When rates of attempted rape were added to these figures, the rate of completed and attempted rape ranged from a low of 19.4% in the United Kingdom to a high of 27.5% in the United States.

Comparisons of clinical samples with the general population suggest that rates of violent victimization are particularly high among women from various mental health settings and among women with mental health problems. Alcohol and/or other drug problems are highly associated with violent victimization, based on data from large epidemiologic studies [13,34]. Women in residential treatment settings for chemical dependency report high rates of prior victimization. Teets [35] found that 73% of women in residential treatment for chemical dependency had been raped (including childhood rapes) and 45% had been raped more than once.

Several studies have focused on women seeking emergency care to determine rates of violence against women. Estimates range between a low of 7% to a high of a third of all emergency room visits by women being due to violent victimization [36]. In general, urban settings that include women from poor inner city settings have some of the highest rates of women seeking medical treatment for violent victimization.

B. Physical Victimization from Partners

While the preceding rates of violent victimization include all types of perpetrators (*e.g.,* strangers, acquaintances, and partners), other studies have focused specifically on the rates of partner violence, a particularly common experience for women. Partner violence is the second leading cause of injuries to women of all ages and the leading cause of injuries to women between the ages of 15 and 44 [37]. Women are at greater risk for being repeatedly attacked, raped, injured, or killed by male partners as compared to other types of perpetrators [38,39].

Using the Conflict Tactics Scale (CTS), in a nationally representative sample of American families recruited through telephone interviews in 1985, the rate of overall partner-to-woman violence was approximately 12% in the year before the study [40]. The CTS is one of the most commonly used measures of partner violence and assesses specific types of violent behaviors that have occurred within families. Severe partner violence from this scale includes experiences such as: being kicked, hit, or hit with a fist; being hit with an object; being beaten up; having life threatened with a knife or gun; or having a knife or gun used on you. Moderate violence behaviors include being slapped, having an object thrown at you, and being pushed, grabbed, or shoved. Severe violence was reported by 4% of the women in the year before the survey [40]. The specificity of the questions undoubtedly improved the reporting of violent crime by partners.

Despite the concerns raised by the estimated number of women impacted by either moderate or severe violence during a one-year window, this survey [40] may actually underrepresent the level of violence in the United States. First, questions asked during telephone surveys may not reveal violence in the most violent homes where partners are monitoring telephone calls and exerting control over the interactions of women. Second, some types of severe violence, such as partner rape, were not included in the list of violent experiences encountered. Third, telephone surveys do not include households without telephones, thus excluding the lowest socioeconomic strata of the population.

Although violence is prevalent in all social strata, there is evidence that rates are particularly high among lower social classes. For example, data from an inner city emergency department revealed that one-third of the women (32%) reported lifetime physical abuse from a partner and nearly one-fifth (19%) reported current physical abuse [41]. Similar rates were found in an Australian public hospital's emergency room department, with lifetime rates of physical abuse from a partner at 24% [42]. In another study of emergency room patients, over half of the

women (54%) experienced violence from a male partner during their lifetime and 12% reported current partner violence as the reason for their visit [43].

Women in other types of health care settings also report high levels of violent victimization. For example, one-third (34%) of women in university-affiliated ambulatory care and internal medicine clinics reported histories of partner violence and nearly one-fifth (17%) reported current partner violence [44]. In a sample of general practice clinics designed to represent all social classes and geographic regions in a large metropolitan area in Australia, 22% of women experienced partner physical violence in the past year [45]. Ten percent had experienced severe partner violence and 13% experienced rape or attempted rape since age 16 [45]. One-third (32%) of women in a suburban family practice clinic reported an annual prevalence of any physical aggression and 15% reported having sustained an injury or being afraid of partners due to violence in the past year [46]. Among a sample of 280 women in a rural primary care clinic, one-third (34%) of the patients reported a past history of emotional or physical abuse [47]. Eight percent reported physical abuse in the past year and 5% reported being currently afraid of their partners. Thus, evidence suggests that women in all social strata and social contexts are vulnerable to violence.

Pregnant women have been identified as being particularly vulnerable to partner violence. Nearly one-fifth (17%) of pregnant women from public prenatal clinics (95% with incomes below the poverty line), which included an approximately equal proportion of African Americans, Hispanics, and whites, experienced either physical or sexual abuse [48]. Women in an urgent care obstetrics and gynecology clinic reported lifetime and current rates of 46% and 10%, respectively, for physical or sexual abuse [49].

Rates of severe partner violence vary widely across different types of clinic populations, but generally, women in alcoholism and drug treatment settings have higher rates than those found in other clinic samples. Among predominantly inner city African-American and Hispanic women who were methadone maintenance patients, over half of the women (60%) reported physical or sexual abuse by a spouse or boyfriend [50]. Women in alcoholism treatment were found to report higher levels of partner violence than women in comparison groups. Forty-one percent of the women in alcoholism treatment programs, compared to 23% of women in mental health outpatient treatment, 12% of women in drinking and driving classes, and 9% of women in a general household population reported severe partner violence [51]. In another study comparing women from alcoholism treatment settings (N = 45) with a random household sample of women (N = 40), partner violence was reported by alcoholic women at significantly higher rates for virtually every category of severe violence assessed by the CTS; alcoholic women were four to five times more likely to report experiences of severe violence as compared to women in a randomly selected community group [15].

High rates of partner violence are also found in other specialized populations. Among homeless women (N = 436), 61% reported severe violence by a male partner during their lifetime [52]. In a study of 82 male parolees and their partners, high rates of parolee to partner violence were reported [53]. During a

3-month period preceding the study, three-fourths of the parolees perpetrated moderate violence and one-third perpetrated severe violence toward their partner.

C. Psychological and Emotional Abuse

Various forms of psychological and emotional abuse also are common in women's lives. These experiences range from derogatory and demeaning epitaphs and harassment to more physically threatening psychological terror such as stalking. Psychological and emotional abuse may potentially cause physical harm by affecting women's psychological and physical health, and they can also be a precursor to physical violence. Psychological and emotional abuse have been commonly identified as pervasive and important elements in their own right and also as part of a pattern of escalating violence. These different forms of psychological and emotional abuse may be evident with or without physical abuse. However, when there is physical abuse, some form of psychological or emotional abuse is virtually always present [54–56]. As may be expected, the effects of psychological abuse are enhanced in the presence of physical violence. In particular, physical violence may be used with the intention of reinforcing the effects of psychological abuse [57]. For example, a glare from a partner with no prior history of physical violence may communicate only his anger or frustration. However, the same glare from a partner with a history of physical violence may be much more intimidating, communicating the possibility of renewed violence if the female partner fails to alter her behavior to conform to the desires of the male partner. Even more passive forms of coercion, such as standing in the door to prevent a female partner from leaving a room, may be more intimidating if there is a history of violence as compared to no previous violence.

When men use psychological and emotional abuse, they also rely on power and control over women's lives. This form of abuse has been conceptualized as a power/control wheel, consisting of eight quadrants [57]. (1) intimidation (e.g., glaring or staring at the woman); (2) emotional abuse (e.g., telling her that she is sexually undesirable); (3) isolation (e.g., controlling who she can have for friends and when she can see family); (4) minimizing, denying, or blaming (e.g., indicating that it is her fault that he got angry because she did not do something or did not do it the right way); (5) using children (e.g., threatening that if she leaves him, he will not let her have the children); (6) male privilege (e.g., asserting that it is a man's privilege to have a night out, drink excessively); (7) economic abuse (e.g., control money and demand to know how it is spent); and (8) using coercion or threats (e.g., threatening to harm children, family members, or pets).

One study of psychological abuse among women who identified themselves as being in a "bad" or "stressful" heterosexual relationship revealed that there were different clusters of psychological abuse patterns [58]. Two types of psychological abuse that were identified with low levels of violence but high levels of psychological abuse included a subtle controlling pattern (e.g., isolation, enforced secrecy) and a more overt controlling pattern (e.g., controlling money). Psychological abuse and threats of violence were found to be related to poorer health and more help seeking [58].

Emotional and psychological abuse among women presenting to health care settings is less well documented than violence. However, women patients in an inner city emergency department reported lifetime rates of 22% and current rates of 15% for emotional and psychological abuse by a partner [41]. Among women patients in general practice clinics, 20% reported partner emotional abuse in the past year [45].

D. Fear of Violence

There also has been interest in stalking, a form of psychological violence against women. Stalking is comprised of the following types of behaviors: repeat occurrences of following and harassing, surveillance of another person's actions or whereabouts, nonconsensual communication (including telephone harassment), and property destruction [59,60]. Many states require evidence of a credible threat as part of the definition of stalking, although victims typically experience high levels of fear in the absence of direct threats and the recommendation has been made to eliminate this criterion [60]. The National Violence Against Women survey was conducted by telephone interviews of a sample of randomly selected households, stratified by U.S. census region [60]. In this survey, stalking was defined as "a course of conduct directed at a specific person that involves repeated (at least twice) visual or physical proximity, nonconsensual communication, or verbal, written, or implied threats, or a combination thereof, that would cause a reasonable person fear" [60]. Based on these definitions and criteria, 8.1% of women in the U.S. have been stalked in their lifetime and 1% have been stalked during the past year [60].

Even among women who have never experienced physical violence or stalking, the fear of violence is recognized as a salient and real force in women's lives. Madriz [61] reports on the prevalence of fear of crime in women's lives. Women engage in a variety of actions to cope with these fears. Strategies identified by women to cope with their fears and to protect themselves include: self-isolation, hardening the target (e.g., crime-proofing the home, participating in neighborhood watch programs), strategies of disguise (e.g., dressing in nondescript manner), looking for guardians, ignoring or denying fears, carrying protection, and fighting. While these strategies might serve to provide some protection from victimization by strangers, they are less effective for family violence.

IV. Barriers to Identifying Violence within Women's Lives

A. Barriers in the Health Care System

Despite high rates of violence found among women of all socioeconomic and cultural backgrounds, there is reluctance both to ask about violence and for women to share their experiences about violence unless prompted. Within health care settings, actual identification of cases of violence is quite low. One study estimated that only 5% of partner violence victims are actually identified within emergency departments [62]. A major reason for this lack of case-finding is that only 13% of women with current partner violence experiences either told medical staff or were asked by medical staff about these experiences [43]. Another study found that fewer than 10% of patients were ever asked about physical or sexual abuse during routine physician inquiries [63]. Physicians are reluctant to ask about these problems in women's lives. In another study [64], fewer than 2% of all women seeking services in a Midwestern community clinic setting had been asked about partner violence from their physicians, despite a lifetime prevalence male partner assault rate of 39% percent and a past year rate of 25% among women potentially at risk from partner violence. Another problem is that women who experience partner violence typically present in an emergency department during evening hours when social work services are not available [42].

Physicians' reasons for not asking include concerns about time constraints, lack of comfort in asking, fear of offending patients, and perceived loss of control over the treatment plan (largely due to lack of knowledge concerning the dynamics of partner violence) [65]. Health care providers may be overwhelmed by the demands being placed on them to give quality care in prescribed units of time set by managed care insurance systems. Additionally, increased paperwork and concerns about being involved in the legal system may further impede physicians' attempts to identify partner violence. Sugg and Inui [65] report that most physicians fear being forced to respond to difficult issues that are not easily addressed in the medical model. Burge [66] suggests that physicians may believe that working with victims is a "hopeless cause," violence is not a health issue, or that violence is a private issue not within the domain of the health care system.

The failure to identify violence against women in our health care systems may also be due to patriarchal perspectives and attitudes that permeate the medical system. Stark and Flitcraft [67] identify three stages of the medical system's response to battered women that limit our abilities to provide appropriate and sensitive treatment to women's needs. The first stage occurs when the injury appears in the medical context. During this stage when the injury first appears, there is little attempt to identify, let alone address, the issues of interpersonal violence that cause injuries. In part, this can be understood by the fact that physicians are trained to treat medical problems and not to comment on the complex psychosocial factors that might surround the injury. The second stage is identified by isolation of the women. During this stage, women change from people with discrete injuries to individuals with a complex assortment of medical and psychological problems. Women are viewed as individuals engaged in a series of self-destructive behaviors, possibly including such problems as alcohol and/or drugs. Movement from isolation to entrapment then defines the third stage. At this point, women, like their physicians, may perceive their victimization as emerging from their own pathology, dependency, or helplessness. This pejorative and negative labeling of women's behaviors may continue as they are sent to other community resources.

Reliance on injuries alone is likely to miss many cases of violent victimization. Routine screenings of women in internal medicine, gynecology/obstetrics, or family practice clinics, and in emergency departments or substance abuse treatment centers are more likely to identify cases of partner violence or abuse and are likely to increase the positive identification rate of partner abuse victims. At least one study suggests that women patients would welcome such inquiries about their lives; approximately three-fourths of women patients favored physicians inquiring about physical abuse and approximately two-thirds favored inquiries regarding sexual abuse [63].

B. Women's Reluctance to Self-Identify Violent Victimization

Even though women may want to be asked, there may be reluctance to disclose experiences of violent victimization. Fear of retribution, fear of stigmatization and labeling, fear of being blamed, past negative outcomes from disclosure (*e.g.,* lack of follow-up care), lack of encouragement to disclose, and fear of revisiting a traumatic event all are reasons for nondisclosure [68]. Additional concerns may include feeling protective of their partners, thinking that they deserved the abuse, and believing that their injuries are not severe enough to warrant attention [69]. Other concerns may be related to delivery of services. For example, discomfort in disclosing abuse to male health care or service providers, different expectations that the medical professional, social service provider, and women may have regarding the outcome of treatment, and concern that their health care insurance may be canceled or their rates raised because they have been identified as victims of partner violence are all concerns related to the delivery of services [27]. Also, women fear losing custody of their children if they reveal violence in the home.

A variety of steps must be successfully negotiated before identification of victimization experiences occurs. In a model proposed by Acierno *et al.* [68], seven steps must be negotiated prior to a successful case identification of violence: (1) victim must perceive event and label as assault; (2) victim must encode the assault into her memory; (3) clinician must inquire about assault using label similar to victim's; (4) clinician inquiry must cue victim memory about event; (5) victim must be willing/able to disclose that assault has occurred; (6) victim must be able to safely disclose that assault has occurred; (7) clinician must define disclosed event as assault. Thus, there are many points in the process where women may not be identified or self-disclose as victims of violence.

V. Multicultural Heritage of Violence against Women

Not only are there barriers for individual women to reveal their histories of violence and victimization, but also cultural contexts in which women reside provide powerful and complex forces that influence the reporting of violent victimization. Cultural contexts also shape and influence policies that determine whether women are asked about violent victimization and social responses that emerge when violence is reported. It has only been since the 1970's that widespread legal and social efforts have been made to address relationship violence. Even recently, relationship violence has been either actively supported, or at the very least, tacitly accepted, across multiple cultural settings. A brief review of some of the highlights of the recent historical events that influenced and changed social norms is presented here, providing some insight into the power of social and cultural environments in implementing change.

A. Women's Rights and Men's Violence—The Early History

Women historically have been viewed as the property of men; daughters belonged to their fathers until marriage and subsequently to their husbands [70–72]. Rape laws were initially enacted to protect men's property (daughters and wives) from other men rather than to protect the rights of women [70]. Rape laws did not protect wives from marital rape because husbands retained the right to control their property [54,72].

During the nineteenth century, a strong women's rights movement was organized to provide legal protection to women who experienced physical and sexual abuse in their relationships with men [73,74]. This movement was contending with historically entrenched European cultural values that legitimized men's legal right to control women through force. Chastisement, or the right of men to use violence to control or discipline their wives, traces its history to early Roman law, was accepted in English common law, and was adopted within the United States. Women were never provided the right to chastise husbands, resulting in asymmetrical power relationships [54]. Women's rights advocates fought against chastisement, finally gaining some measures of success toward the end of the nineteenth century.

Despite this movement, wife beating remained common throughout the twentieth century both in England and in the United States and continues to be prevalent in many cultures throughout the world today. Laws against the physical and sexual abuse of wives by husbands have been largely ignored and not enforced. After 1920, the women's movement was deemphasized as society tired of reform movements, and experienced prosperity, the economic depression, and war [75]. Despite high expectations, the right to vote did not provide for equal rights for women; in particular, equality with men on various legal, occupational, and economic issues (*e.g.,* equal pay for equal work) was not provided.

During the late 1960s another wave of feminism began [76,77]. This movement was concerned with issues such as the Equal Rights Amendment, equal pay for women, and sexual autonomy [75]. This second wave of feminism set the stage and provided the context for a renewed public debate of equal treatment for women within the family, in particular addressing the different forms of violence and abuse that women experience in the family [54,76]. Warrior [78] and Pizzey [79] began the public debate on the issues of wife-beating and, more generally, battered women during this period. Based on their work, the content of this public debate focused on the labeling, public recognition, discussion, and analysis of problems women encountered on a daily basis but yet had been largely ignored by researchers, the criminal justice system, and the medical profession.

B. Social Action during the 1970s and 1980s

The number of shelters and services for battered women increased during the 1970s as grass-roots groups and coalitions were initiated and developed. These groups and coalitions represented a political movement to promote gender equality and to reform society, in particular roles for both males and females, as well as to provide for the safety of individual women who had experienced violence from their partner [76]. The very process of providing services was considered as important as the end product of the services. There was a commitment to diversity (*i.e.,* the participation of all women, regardless of age, culture, social class, or sexual orientation), and this commitment brought forward an understanding of how violent victimization

manifests itself in the different cultural backgrounds. The inclusion of battered women in the planning and development of programs helped to maintain sensitivity towards empowering women with their own voice and provided a further balancing of actions across age, ethnic groups, and social strata. Particularly important to the women's movements' efforts to provide services to women was an emphasis on collective decision making and nonhierarchical organizational models for governing operations. These frameworks were seen as counteracting traditional social structures that are stratified along gender and cultural lines and help to support patriarchal attitudes. Counseling the battered woman was seen as giving women permission to take back control of their own lives and the lives of their children [76].

During the 1970s and 1980s, the predominant view was that women's experiences of violence were caused primarily by men's needs to exert power and control over women's lives. It was as if the legal right to chastisement was still operative. Although men were no longer legally entitled to use violence as a mechanism of control over women, there remained a reluctance on the part of criminal justice systems to intervene in cases of partner violence or rape. This reluctance was viewed as granting tacit support to male violence against women. In addition to chastisement, there have been other cultural influences that have supported the right for men to control women's lives. Causes of violence against women were thus viewed as primarily cultural and sociopolitical and were supported by a criminal justice system that was reluctant to intervene. Gender stratification further contributed to an inequitable power balance, with men having more economic, social, political, and interpersonal power than women [54,57,80,81]. Rather than trying to explain why individual, specific men are violent to specific women, feminist views of male violence sought to explain why men as a class are violent to women as a class [55,82,83]. Consequently the need for social change was emphasized, which was an extension of the second wave of feminism.

During the 1970s and 1980s, a major intervention strategy for women victims was to provide advocates at various stages of the process and in different settings throughout the legal and social service sector [57]. Advocates provided guidance through the social service, criminal justice, and health care systems that were viewed as reluctant to intervene or, at best, as intervening in ways that failed to take into account the gender-based power imbalances. At the societal level, advocacy included a diverse array of attempts to foster social change, including the promotion of laws criminalizing partner violence and of policies to improve the criminal justice system responses. Prosecution of sexual assault cases and partner violence cases increased. Advocacy services included insuring appropriate medical care for victims of violence. The health care providers and the general public were a focus for educational campaigns and media advertisements to increase awareness.

Although chastisement was not officially condoned during the 1900s, in fact social policies and responses to partner violence appeared to "permit" such action because of a failure to take any legal or official action against partner violence. In part as a reaction to this failure to act, a social movement emerged in the 1980s that was also based on some initial empirical evidence in support of such action. In a landmark study, Sherman and Berk [84] found that arrest had a greater deterrent effect for men's violence against partners than did other police responses (*i.e.,* sending the suspect away for 8 hours, advising the couple to get help). These researchers urged caution, arguing that this was only one study in a specific geographic area and that replication was needed prior to widespread application of the results [84]. However, social change proponents used the results of this study to advocate for arrest of men found to perpetrate partner violence. In a criminal justice system that had become conducive to procedural changes, mandatory arrest or pro-arrest policies were adopted in many jurisdictions in the case of partner violence. Ironically, later research found generally that arrests reduced violence in the short run but not necessarily the long run, and only for some groups of perpetrators (*e.g.,* employed men, and those whose victims were white or Hispanic) [85,86]. Despite these later results, mandatory arrest or pro-arrest policies were retained for a number of reasons. Primarily it was believed that these arrest policies may reduce overall domestic violence by "sending a clear message that battering is unacceptable" [87].

It has only been in the 1990s that rape within marriage has been recognized as a social problem that women experience [71,72]. Historically, men have been exempted from charges of rape based on legal concepts that wives were the property of their husbands and therefore husbands cannot rape their own property. Thus, women consented to sexual intercourse with husbands by virtue of the marriage contract [71,72]. The women's movement challenged this concept during the 1970s and 1980s, and by 1993, wife rape had become a crime in every state [72]. Legal definitions of rape in the late 1970s typically included criteria of forced intercourse, intercourse obtained by threat of force, or intercourse when consent is not possible [87]. However, these definitions have been expanded to include a range of sexual acts in addition to intercourse such as oral sex, anal sex, and digital penetration [71].

There was also a growing recognition that battered women needed a broad range of services in addition to the empowerment counseling and advocacy that were the hallmarks of intervention during the 1970s. A high percentage of women in treatment for problems such as substance abuse and mental health needs were found to have experienced recent severe violence from partners [15,51,88,89]. There was a growing recognition of the need to integrate services for battered women with other service programs. Findings such as these implied the need to develop screening programs and assessments to identify battered women as they present in other service arenas, such as emergency rooms or other medical settings. Thus, compared with the initial intervention model that emphasized advocacy and nonhierarchical organization of services, there have been considerable changes in the movement to provide services for battered women. The changes most in focus at the present time are the concerns with integrating health care services with other types of services (employment, criminal justice system, and social services).

VI. New Directions for Intervening on Violence against Women

Since the 1980s, grass root efforts to support women victims of violence emerged in the contexts of the criminal justice system and of community based services. Local communities established networks of women's shelters to provide immediate crisis services to women and their children. These systems of response have grown and become more institutionalized over the years. What has not emerged simultaneously is a comprehensive and well-developed system of response within the health care system. Opportunities for responding to women's crises and to averting more serious consequences are numerous within health care settings. A particular advantage of health care settings is that women can be identified while seeking medical care either for themselves or for family members.

The emergency room is a particularly obvious setting for identifying victims and for offering services to women. Many emergency settings do not have a protocol for asking about violence or keep minimal information about causes of injury. Studies based on record data reveal violence, although an important contributor to women's injuries, is not well documented in medical records. For example, in a sample of 481 women who sought treatment in an urban emergency room and for whom full medical records were available for review, only 2.8% were positively identified in the medical records as victims of partner violence [67]. However, on closer examination, another 5.2% were deemed to be probable victims of partner violence (assault by another person but relationship was not disclosed in medical records), and still another 9.8% were suggestive of partner violence (at least one injury was inadequately explained by the recorded medical history) [67]. These same records were reviewed for lifetime history of partner violence and only 9.6% were positively identified as having experienced partner violence based on the medical records. Another 4.8% were probable and 10.6% were positive.

A. Screening for Victimization

Patient histories do not routinely include questions about violent victimization. However, this opportunity could easily provide a more comprehensive assessment of the range and extent of violence in women's lives. Care given to handling the issues in a sensitive manner can reduce concerns about offending patients. Acierno *et al.* [68] suggest that "contextually orienting preface statements" or empathic statements prior to questioning are critical. An example of such a preface statement is "Because abuse and violence are so common in women's lives, I've begun to ask about it routinely" [69]. Further, it is important that the medical profession convey an awareness that interpersonal violence is a common occurrence and that it is appropriate and relevant to ask such questions in a medical setting [68]. Following an empathetic opening, behaviorally specific questions that can be answered with simple closed-ended responses are needed to elicit accurate information. Care should be taken to avoid requiring the patient to provide a detailed, explicit account of the event. To ensure women's safety, women must be questioned alone and partners must not be present or nearby. Also, partners must not be used as interpreters when there are language barriers. It is equally important to avoid victim-blaming questions such as "Why did you provoke him when he was drinking?" and "Why do you stay with him?" Validation of patient concerns and feelings regarding partner violence is also an important part of the interview process [67,90]. Women who experience violence often believe they are alone; knowledge that others share their experiences can be empowering. Some women may think that they deserve what happened to them and must be assured that no woman should experience violence. Finally, given the potential concerns of women disclosing partner abuse to male providers [27], assessments in health care settings should be considered with female providers providing routine interviewing and screening for cases of partner abuse. Nurse practitioners or registered nurses might provide appropriate professional staff for detecting the problems of violent victimization.

More extensive education about the problems of violence will also provide a greater comfort level for health care providers in handling the issues. Given that health care providers already handle sensitive and potentially embarrassing questions about health problems, there is expertise and experience for addressing sensitive topics. What has been recognized as critical is the use of direct, specific questions for ascertaining the problems [68].

The American Medical Association (AMA) [91] has provided a series of ten questions that may be used to screen for women's experiences of physical or psychological abuse from partners [69].[1]

1. Are you in a relationship in which you have been physically hurt or threatened by your partner? Have you ever been in such a relationship?

2. Are you (have you ever been) in a relationship in which you were treated badly? In what ways?

3. Has your partner ever destroyed things that you cared about?

4. Has your partner ever threatened or abused your children?

5. Has your partner ever forced you to have sex when you didn't want to? Does he ever force you to engage in sex that makes you feel uncomfortable?

6. We all fight at home. What happens when you and your partner fight or disagree?

7. Do you ever feel afraid of your partner?

8. Has your partner ever prevented you from leaving the house, seeking friends, getting a job, or continuing your education?

9. You mentioned that your partner uses drugs/alcohol. How does he act when he is drinking or on drugs? Is he ever verbally or physically abusive?

10. Do you have guns in your home? Has your partner ever threatened to use them when he was angry?

In addition to the AMA-recommended questions, several brief screening instruments have been developed to assess women's experiences of partner violence. Pan et al. [46] developed an 11-item Partner Abuse Interview (PAI) form which takes about 3 minutes to complete. The Confidential Abuse Assessment Screen (CAAS) consists of five items and has a body map on which the female patient may indicate the location of injuries [47].

These screening instruments are not as detailed and comprehensive as research-based scales such as the Conflict Tactics Scale [40,92] or the Abusive Behavior Inventory [93]. There is some debate regarding whether screening with self-administered instruments results in lower rates of identification than screening with interviewer administered instruments. For example,

[1]Reprinted from *Archives of Family Medicine* **1**, 39–47, Copyright 1992, American Medical Association, with permission.

one study found substantially higher rates of reported partner physical abuse when the four screening questions were asked in an interviewer format as compared to a self-administered format (29.3 vs 7.3%, respectively) [94]. However, until further research is done to assess comparatively the different screening instruments and the different methods for administering, assessments should be done with some consideration of the population of focus, the time available, and the approach that is most coherent within the particular setting. Once asked, however, there should be a plan for assisting women who want help.

B. Physical Signs of Abuse

Despite increases in detection that simple screening instruments might provide, there are women who will not respond to screening instruments in a manner that allows detection of their victimization. Thus, there must be sensitivity to recognizing other potential signs of hidden violence. Patterns of old injuries, multiple injuries, and injuries to the periphery of the body may be signs of partner abuse [27,91]. In a study of emergency department patients, women in physically abusive relationships also were more likely to be diagnosed with urinary tract infections, neck pain, foot wounds, vaginitis, finger fractures, and suicide attempts as compared to women not in physically abusive relationships [16]. Injuries to the breasts, chests, or abdomens are 13 times more evident among female victims of domestic violence as compared to accident victims [95]. Studies of women in emergency departments across various geographic settings (e.g., inner-city, urban, and suburban) suggested that battered women were significantly more likely to present with injuries to the head, face, neck, thorax, and abdomen than were women not injured by violence [16]. Multiple injuries, as well as evidence of old injuries in multiple stages of healing, are also more evident among women victims of violence as compared to accident victims [66]. Women victimized by partner violence may seek services from the health care delivery system for somatic problems or chronic pain and for treatment for ancillary problems such as substance abuse [27,51,96]. Violence may also lower the immune system response and further place women at risk for other opportunistic diseases [97].

According to Acierno and colleagues [68], a hypothetical model explaining the development of violence-related health problems can be conceptualized as follows. Violent victimization leads to acute physical injury, increased stress, and increased risk of mental health problems. Together these consequences lead to chronic physical injury, impaired immune system functioning, increased health risk behavior, and inappropriate health care utilization. The final outcome is an increased risk of health problems for the given individual.

C. Intervening with Victims of Violence

Private medical practices, general health care settings, mental health treatment settings, and alcohol and other drug treatment settings all provide important contexts for identifying victims of violence [98]. Women may have the goal of ending the violence whereas these systems may have different goals (e.g., treating the injury) [27]. It is important that a range of services are offered to women, and a treatment team consisting of a doctor, nurse, and social worker may be appropriate to accomplish this goal [27]. A primary guideline in offering services to women is that women remain in control of which services they accept. This guideline may be contrary to the usual medical model in which the doctor, as the expert, prescribes the medical services needed. However, women are likely to know their partner and his likelihood of perpetrating further abuse on her. Women know which actions are safe and which are unsafe. In addition, allowing women to control their service plan regarding violent victimization is self-empowering. Within this guideline, many services can be offered.

First and foremost is the need for women to have a safety plan. El-Bayoumi *et al.* [90] indicated that risk assessment and safety planning are an important part of services to be offered to women. The AMA guidelines also address the issue of whether women have safe havens and plans for how to arrive at these safe havens in case of impending violence [69]. Barrier [37] has recommended asking whether women have an emergency escape plan (Are you in danger now, and would you like to go to a shelter or talk with someone? Do you have a place where you and your children could go in an emergency?).

Second, women may need to be linked with local resources or agencies that provide services to women victimized by violence. In many areas the local shelter for battered women or rape crisis services provide advocates on a 24-hour per day basis. Because hospital-based social work services are unlikely to be available when women access the emergency department, victim advocates may be the most available source for help [42]. Also, victim advocates are likely to be experts on the dynamics of partner abuse as well as the range of issues. They are likely to be able to assist women in safety planning as well as facilitate plans for action.

Third, education can be provided about community resources as well as state or national hotlines that women can access and obtain information about rape or partner violence. This information should be provided only if retaining such information does not lead to further danger. Abusive partners invade the boundaries of women and even the presence of such materials could endanger some women. In addition, hospitals and other health care agencies can maintain current information on partner violence and on local resources by joining and participating in local coalitions against family violence [27].

Fourth, health care systems can develop their own in-house programs as well. Hotch *et al.* [99] describe a hospital-based domestic violence program that includes several steps. First, a nurse screens female patients for experiences of partner violence. If there are indications of partner violence, patients are moved to a safe area within the emergency department. In addition, there is coordination of services by physicians, nurses, and social workers, as well as by contacts with the police and criminal justice system upon the woman's request [99]. About half of the women offered their hospital-based program declined the services offered [99], suggesting that multiple points of accessing services are needed. Norton and Schauer [100] describe a hospital-based drop-in group for female victims of partner violence. Women in the group are more likely to report current needs of counseling (53%) and support from other women (46%) than material needs (11%), legal assistance (26%), or financial assistance (28%).

Sensitivity in responding to violent victimization must also include recognition that treating reactions to trauma with psychoactive medications may remove from women their ability to reason and think, thus endangering them further if they return to a violent home. According to Stark and Flitcraft's [67] study conducted on battered women who sought services in an emergency room, 24% of the battered women received minor tranquilizers or pain medications, as compared to 9% of the nonbattered women. Battered women were much more likely to be described with "pseudo-psychiatric labels," such having multiple vague complaints.

Kilpatrick and colleagues [98] suggest that official policies be employed to guide the coordination of services, such as cooperative agreements. Other mechanisms can be put in place that allow for the opening of boundaries between systems and the "hosting" of other agencies within the health care setting to promote the availability of community resources. One innovative model for providing services to victims of childhood sexual abuse ensures that all outside agencies meet with the child and his/her parents within the treatment setting for addressing the problem. Thus, one physical setting provides the services to the family rather than demanding that the family physically appear in separate locations for interviews with police, the medical profession, and other social service agencies as required. By physically locating services in the medical settings, we might provide not only better services to victims of violence but also increase the chances of healing more quickly and completely. This model might serve as an example for providing innovative services to women.

VII. Summary

Research indicates that many forms of violence are prevalent in women's lives. Both physical and psychological violence can occur from perpetrators who are strangers, acquaintances, family members, or partners. These experiences exist throughout the life span. Women never "outgrow" their concerns, fears, and need to cope with violence. Thus, fear of violence permeates the lives of all women, influencing their decisions, impacting their psychological and physical health, and limiting their options. Violence exists across cultures and women's perspectives of violence have been shaped not only by this widespread awareness of violence but also by historical events that continue to influence women's lives today. No woman is immune from the impact of violence on her life.

Prior efforts to change the culture of violence towards women have focused on legal changes and community-based grass roots efforts to provide safe alternatives and support to women. Health care settings have largely been relegated the role of attending to the acute physical consequences with little involvement in prevention or intervention efforts. However, the health care system offers, potentially, one of the most powerful systems for implementing change. Violence remains one of the most important public health crises in our communities today. Intervention and prevention efforts can easily be justified on the basis of costs. Of greater importance is the need to end the cycles of violence and victimization that occur within our communities and impact not only the lives of individuals in these communities, but also the lives of future generations.

References

1. Fishbach, R. L., and Herbert, B. (1997). Domestic violence and mental health: Correlates and conundrums within and across cultures. *Soc. Sci. Med.* **45**(8), 1161–1176.

2. Swiss, S., Jennings, P. J., Aryee, G. V., Brown, G. H., Jappah-Samukai, R. M., Kamara, M. S., Schakk, D. H. R., and Turay-Kanneh, R. S. (1998). Violence against women during the Liberian civil conflict. *JAMA, Am. Med. Assoc.* **279**, 625–629.

3. Fauveau, V., and Blanchet, T. (1989). Deaths from injuries and induced abortion among rural Bangladeshi women. *Soc. Sci. Med.* **29**(9), 1121–1127.

4. Coale, A. J., and Banister, J. (1994). Five decades of missing females in China. *Demography* **31**(3), 459–479.

5. Council on Scientific Affairs, American Medical Association (1995). Female genital mutilation. *JAMA, J. Am. Med. Assoc.* **274**(21), 1714–1716.

6. Wessel, L., and Campbell, J. C. (1997). Providing sanctuary for battered women: Nicaragua's Casas de la Mujer. *Issues Ment. Health Nurs.* **18**, 455–476.

7. Finkler, K. (1997). Gender, domestic violence and sickness in Mexico. *Soc. Sci. Med.* **45**, 1147–1160.

8. Koss, M. P., Heise, L., and Russo, N. F. (1994). The global health burden of rape. *Psychol. Women Q.* **18**, 509–537.

9. Koss, M. P., Goodman, L. A., Browne, A., Fitzgerald, L. R., Keita, G. P., and Russo, N. F. (1994). "No Safe Haven: Male Violence against Women at Home, at Work, and in the Community." American Psychological Association, Washington, DC.

10. Russo, N. F., Denious, J. E., Keita, G. P., and Koss, M. P. (1997). Intimate violence and black women's health. *Women's Health: Res. Gender, Behav. Policy* **3** (3 and 4), 315–348.

11. Kilpatrick, D. G., Best, C. L., Veronen, L. J., Amick, A. E., Ville-ponteaux, L. A., and Ruff, G. A. (1985) Mental health correlates of criminal victimization: A random community survey. *J. Consult. Clin. Psychol.* **53**, 866–873.

12. Kantor, G. K., and Straus, M. A. (1989). Substance abuse as a percipient of wife abuse victimizations. *Am. J. Drug Alcohol Abuse* **15**(2), 173–189.

13. Wilsnack, S. C., Vogeltanz, N. D., Klassen, A. D., and Harris, T. R. (1997). Childhood sexual abuse and women's substance abuse: National survey findings. *J. Stud. Alcohol* **58**(3), 264–271.

14. Miller, B. A., Downs, W. R., and Testa, M. (1993). Interrelationships between victimization experiences and women's alcohol use. *J. Stud. Alcohol, Suppl.* **11**, 109–117.

15. Miller, B. A., Downs, W. R., and Gondoli, D. M. (1989). Spousal violence among alcoholic women as compared to a random household sample of women. *J. Stud. Alcohol* **50**(6), 533–540.

16. Muellman, R. L., Lenaghan, P. A., and Pakieser, R. A. (1998). Nonbattering presentations to the ED of women in physically abusive relationships. *Am. J. Emerg. Med.* **16**(2), 128–131.

17. Rollstin, A. O., and Kern, J. M. (1998). Correlates of battered women's psychological distress; Severity of abuse and duration of the postabuse period. *Psychol. Rep.* **82**, 387–394.

18. van der Kolk, B. A. (1996) The complexity of adaptation to trauma: Self regulation, stimulus discrimination, and characterological development. In "Traumatic Stress: The Effects of Overwhelming Experience on Mind, Body, and Society" (B. A. van der Kolk, A. C. McFarlane, and L. Weisaeth, eds.), pp. 182–213. Guilford Press, New York.

19. Herman, J. L. (1992). "Trauma and Recovery." HarperCollins, New York.

20. Domino, J. V., and Haber, J. D. (1987). Prior physical and sexual abuse in women with chronic headache: Clinical correlates. *Headache* **2**(27), 310–314.

21. Harrop-Griffiths, J., Katon, W., Walker, E., Holm, L., Russo, J., and Hickok, L. (1988). The association between chronic pelvic pain, psychiatric diagnoses, and childhood sexual abuse. *Obstet. Gynecol.* **71**(4), 589–594.

22. Drossman, D. A., Zhiming, L., Toomey, T. C., Nachman, G., and Glogau, L. (1997). Sexual and physical abuse history in gastroenterology practice: How types of abuse impact health status. *Psychosom. Med.* **58**, 4–15.

23. Rapkin, A. J., Kames, L. D., Darke, L., Stampler, F. M., and Naliboff, B. D. (1990). History of physical and sexual abuse in women with chronic pelvic pain. *Obstet. Gynecol.* **76**(1), 92–96.

24. Haber, J. D., and Roos, C. (1985). Effects of spouse abuse and or sexual abuse in the development and maintenance of chronic pain in women. *Adv. Pain Res. Ther.* **9**, 889–895.

25. Walker, E. A., and Stenchever, M. A. (1993) Sexual victimization and chronic pelvic pain. *Obstet. Gynecol. Clin. North Am.* **20**, 795–807.

26. Leserman, J., Olden, K. W., and Barreiro, M. A. (1995). Sexual and physical abuse and gastrointestinal illness: Review and recommendations. *Adv. Intern. Med.* **123**(10), 782–794.

27. Schornstein, S. L. (1997). "Domestic Violence and Health Care: What Every Professional Needs to Know." Sage, Thousand Oaks, CA.

28. Koss, M. P., Koss, P. G., and Woodruff, W. J. (1991) Deleterious effects of criminal victimization on women's health and medical utilization. *Arch Intern. Med.* **151**, 342–347.

29. Sutherland, C., Bybee, D., and Sullivan, C. (1998). The long-term effects of battering on women's health. *Women's Health: Res. Gender, Behav., Policy* **4**(1), 41–70.

30. Bachman, R., and Saltzman, L. E. (1995). "Violence Against Women; Estimates from the Redesigned Survey," National Crime Victimization Survey. U.S. Dept. of Justice, Office of Justice Programs, Bureau of Justice Statistics, Washington, DC.

31. National Victim Center (1992). "Rape in America: A Report to the Nation." Crime Victims Research Center, Arlington, VA.

32. Kessler, R. C., Sonnega, A., Bromet E., Hughes, M., and Nelson, C. B. (1995). Posttraumatic stress disorder in the National Comorbidity Survey. *Arch. Gen. Psychiatry* **52**(12), 1048–1060.

33. Koss, M. P., Gidycz, C., and Wisniewski, N. (1987) The scope of rape: Incidence and prevalence of sexual aggression and victimization in a national sample of higher education students. *J. Consult. Clin. Psychol.* **55**(2), 162–170.

34. Kilpatrick, D. G. (1993). "Violence and Women's Substance Abuse." Crime Victims Research and Treatment Center, Charleston, SC.

35. Teets, J. M. (1997). The incidence and experience of rape among chemically dependent women. *J. Psychoact. Drugs* **29**(4), 331–336.

36. Plichta, S. (1992). The effects of women abuse on health care utilization and health status: A literature review. *Jacobs Inst. Women's Health* **2**(3), 154–163.

37. Barrier, P. A. (1998). Domestic violence. *Mayo Clinic Proc.* **73**(3), 271–274.

38. Browne, A., and Williams, K. R. (1989). Exploring the effect of resource availability and the likelihood of female-perpetrated homicides. *Law Soc. Rev.* **23**(1), 75–94.

39. Browne, A., and Williams, K. R. (1993). Gender, intimacy, and lethal violence: Trends from 1976 through 1987. *Gender Soc.* **7**(1), 78–98.

40. Straus, M. A., and Gelles, R. J. (1990). How violent are American families? Estimates from the National Family Violence Resurvey and other studies. *In* "Physical Violence in American Families: Risk Factors and Adaptations to Violence in 8,145 Families" (M. Straus and R. Gelles, eds.), pp. 95–112. Transaction, New Brunswick, NJ.

41. Ernst, A. A., Nick, T. G., Weiss, S. J., Houry, D., and Mills, T. (1997). Domestic violence in an inner-city ED. *Ann. Emerg. Med.* **30**(2), 190–197.

42. Roberts, G. L., O'Toole, B. I., Raphael, B., Lawrence, J. M., and Ashby, R. (1996). Prevalence study of domestic violence victims in an emergency department. *Ann. Emerg. Med.* **27**(6), 747–753.

43. Abbott, J., Johnson, R., Koziol-McLain, J., and Lowenstein, S. R. (1995). Domestic violence against women: Incidence and prevalence in an emergency department population. *JAMA, J. Am. Med. Assoc.* **273**(22), 1763–1767.

44. Gin, N. E., Rucker, L., Frayne, S., Cygan, R., and Hubbell, F. A. (1991). Prevalence of domestic violence among patients in three ambulatory care internal medicine clinics. *J. Gen. Intern. Med.* **6**, 317–322.

45. Mazza, D., Dennerstein, L., and Ryan, V. (1996). Physical, sexual and emotional violence against women: A general practice-based prevalence study. *Med. J. Aust.* **164**, 14–17.

46. Pan, H. S., Ehrensaft, M. K., Heyman, R. E., O'Leary, K. D., and Schwartz, R. (1997). Evaluating domestic partner abuse in a family practice clinic. *Fam. Med.* **29**(7), 492–495.

47. Johnson, D., and Elliott, B. (1997). Screening for domestic violence in a rural family practice. *Minn. Med.* **80**(10), 43–45.

48. McFarlane, J., Parker, B., Soeken, K., and Bullock, L. (1992). Assessing for abuse during pregnancy: Severity and frequency of injuries and associated entry into prenatal care. *JAMA, J. Am. Med. Assoc.* **267**(23), 3176–3178.

49. McGrath, M. E., Hogan, J. W., and Peipert, J. F. (1998). A prevalence survey of abuse and screening for abuse in urgent care patients. *Gynecol. Obstet.* **91**(4), 511–514.

50. Gilbert, L., El-Bassel, N., Schilling, R. F., and Friedman, E. (1997). Childhood abuse as a risk for partner abuse among women in a methadone maintenance. *Am. J. Drug Alcohol Abuse* **23**(4), 581–595.

51. Miller, B. A., and Downs, W. R. (1993). The impact of family violence on the use of alcohol by women. *Alcohol Health Res. World* **17**(2), 137–143.

52. Browne, A., and Bassuk, S. S. (1997). Intimate violence in the lives of homeless and poor housed women: Prevalence and patterns in an ethnically diverse sample. *Am. J. Orthopsychiatry* **67**(2), 261–278.

53. Miller, B. A., Nochajski, T. H., Leonard, K. E., Blane, H. T., Gondoli, E., and Bowers, P. M. (1990). Spousal violence and alcohol/drug problems among parolees and their spouses. *Women Criminal Justice* **1**(2), 55–72.

54. Dobash, R. E., and Dobash, R. (1979). "Violence against Wives: A Case against the Patriarchy." Free Press, New York.

55. Bograd, M. (1988). Feminist perspectives on wife abuse: An introduction. *In* "Feminist Perspectives on Wife Abuse" (K. Yllo and M. Bograd, eds.). Sage, Newbury Park, CA.

56. Adams, D. (1988). Treatment models of men who batter: A profeminist analysis. *In* "Feminist Perspectives on Wife Abuse" (K. Yllo and M. Bograd, eds.). Sage, Newbury Park, CA.

57. Pence, E., and Paymar, M. (1993). "Education Groups for Men who Batter: The Duluth Model." Springer, New York.

58. Marshall, L. L. (1996). Psychological abuse of women: Six distinct clusters. *J. Fam. Violence* **11**, 379–409.

59. National Institute of Justice (1996). "Domestic Violence, Stalking, and Antistalking Legislation: An Annual Report to Congress under the Violence Against Women Act." U.S. Department of Justice, National Institute of Justice, Washington, DC.

60. Tjaden, P., and Thoennes, N. (1998) "Stalking in America: Findings from the National Violence Against Women Survey" (under Grant No. 93-IJ-CX-0012). U.S. Department of Justice, National Institute of Justice and the Centers for Disease Control and Prevention, Washington, DC.

61. Madriz, E. (1997). "Nothing Bad Happens to Good Girls: Fear of Crime in Women's Lives." University of California Press, Berkeley.

62. Council on Scientific Affairs, American Medical Association (1992). Violence against women. *JAMA, J. Am. Med. Assoc.* **267**, 3184–3189, cf. Barrier [37].

63. Friedman, L. S., Samet, J. H., Roberts, M. S., Hudlin, M., and Hans, P. (1992). Inquiry about victimization experiences: A survey of patient preferences and physician practices. *Arch. Intern. Med.* **152**(6), 1186–1190.

64. Hamberger, L. K., Sauders, D. G., and Hovey, M. (1992). The prevalence of domestic violence in a community practice and rate of physician inquiry. *Fam. Med.* **24**(4), 283–287.

65. Sugg, N. K., and Inui, T. (1992). Primary care physicians' response to domestic violence: Opening Pandora's box. *JAMA, J. Am. Med. Assoc.* **267**, 3157–3160.

66. Burge, S. K. (1989). Violence against women as a health care issue. *Fam. Med.* **21**(5), 368–373.

67. Stark, E., and Flitcraft, A. (1996). "Women at Risk." Sage, Thousand Oaks, CA.

68. Acierno, R., Resnick, H. S., and Kilpatrick, D. G. (1997). Health impact of interpersonal violence 1: Prevalence rates, case identification, and risk factors for sexual assault, physical assault, and domestic violence in men and women. *Behav. Med.* **23**(2), 53–64.

69. Flitcraft, A. H., Hadley, S. M., Hendricks-Matthews, M. K., McLeer, S. V., and Warshaw, C. (1992). American Medical Association diagnostic and treatment guidelines on domestic violence. *Arch. Fam. Med.* **1**, 39–47.

70. Pagelow, M. (1984). "Family Violence." Praeger, New York.

71. Russell, D. E. H. (1990). "Rape in Marriage." Indiana University Press, Bloomington.

72. Bergen, R. K. (1996). "Wife Rape: Understanding the Response of Survivors and Service Providers." Sage, Thousand Oaks, CA.

73. Pleck, E. (1987). "The Making of American Social Policy against Family Violence from Colonial Times to the Present." Oxford University Press, Oxford.

74. Sanchez-Eppler, K. (1993). "Touching Liberty: Abolition, Feminism, and the Politics of the Body." University of California Press, Berkeley.

75. Banner, L. W. (1974). "Women in Modern America: A Brief History." Harcourt Brace Jovanovich, New York.

76. Whalen, M. (1996). "Counseling to End Violence against Women." Sage, Thousand Oaks, CA.

77. Deckard, B. (1979). "The Women's Movement: Political, Socioeconomic and Psychological Issue." Harper & Row, New York.

78. Warrior, B. (1976). "Wifebeating." New England Free Press, Somerville, MA.

79. Pizzey, E. (1974). "Scream Quietly or the Neighbors will Hear You." Penguin, London.

80. Dobash, R. P., Dobash, R. E., Wilson, M., and Daly, M. (1992). The myth of sexual symmetry in marital violence. *Soc. Probl.* **39**(1), 71–91.

81. Yllo, K. A. (1993). Through a feminist lens: Gender, power, and violence. *In* "Current Controversies on Family Violence" (R. J. Gelles and D. R. Loseke, eds.), pp. 47–62. Sage, Newbury Park, CA.

82. Gondolf, E. W. (1985). "Men who Batter: An Integrated Approach for Stopping Wife Abuse." Learning Publications, Holmes Beach, FL.

83. Bograd, M. (1984). Family systems approach to wife battering: A feminist critique. *Am. J. Orthopsychiatry* **54**, 558–568.

84. Sherman, L. W., and Berk, R. A. (1984). The specific deterrent effects of arrest for domestic assault. *Am. Sociol. Rev.* **49**, 261–272.

85. Sherman, L. W., Smith D. A., Schmidt, J. D., and Rogan, D. P. (1992). Crime, punishment and stake in conformity: Legal and extralegal control of domestic violence. *Am. Sociol. Rev.* **57,** 680–690.

86. Schmidt, J. D., and Sherman, L. W. (1996). Does arrest deter domestic violence? *In* "Do Arrests and Restraining Orders Work?" (E. S. Buzawa and C. G. Buzawa, eds.), pp. 43–53. Sage, Thousand Oaks, CA.

87. Stark, E. (1996). Mandatory arrest of batterers: A reply to its critics. *In* "Do Arrests and Restraining Orders Work?" (E. S. Buzawa and C. G. Buzawa, eds.), p. 128. Sage, Thousand Oaks, CA.

88. Bennett, L. W. (1995). Substance abuse and the domestic assault of women. *Soc. Work* **40,** 760–771.

89. Gleason, W. J. (1993). Mental disorders in battered women: An empirical study. *Violence Victims* **8**(1), 53–68.

90. El-Bayoumi, G., Borum, M. L., and Haywood, Y. (1998). Domestic violence in women. *Med. Clin. North Am.* **82**(2), 391–401.

91. American Medial Association (1992). "Diagnostic and Treatment Guidelines on Domestic Violence." AMA, New Haven, CT.

92. Straus, M. A. (1979). Measuring intrafamily conflict and violence: The Conflict Tactics (CT) Scales. *J. Marriage Fam.* **41,** 75–88.

93. Shepard, M. F., and Campbell, J. A. (1992). The abusive behavior inventory: A measure of psychological and physical abuse. *J. Interpersonal Violence* **7**(3), 291–305.

94. McFarlane, J., Christoffel, K., Bateman, L., Miller, V., and Bullock, L. (1991). Assessing for abuse: Self-report versus nurse interview. *Public Health Nurs.* **8**(4), 245–250.

95. Stark, E., Flitcraft, A., and Frazier, W. (1979) Medicine and patriarchal violence: The social construct of a private event. *Int. J. Health Serv.* **9**(3), 461–498.

96. Downs, W. R., Miller, B. A., and Panek, D. D. (1993). Differential patterns of partner-to-woman violence: A comparison of samples of community, alcohol-abusing, and battered women. *J. Fam. Violence* **8**(2), 113–135.

97. Maier, S. F., Watkins, L. R., and Fleshner, M. (1994) Psychoneuroimmunology: The interface between behavior, brain, and immunity. *Am. Psychol.* **49**(12), 1004–1017.

98. Kilpatrick, D. G., Resnick, H. S., and Acierno, R. (1997). Health impact of interpersonal violence. 3. Implications for clinical practice and public policy. *Behav. Med.* **23**(2), 65–78.

99. Hotch, D., Grunfeld, A. F., Mackay, K., and Cowan, L. (1996). An emergency department-based domestic violence intervention program: Findings after one year. *J. Emerg. Med.* **14**(1), 111–117.

100. Norton, I. M., and Schauer, J. (1997). A hospital-based domestic violence group. *Psychiatr. Serv.* **48**(9), 1186–1190.

43

Socioeconomic Inequalities in Women's Health

NANCY E. MOSS
Pacific Institute for Women's Health
Los Angeles, California

I. Introduction: Socioeconomic Inequalities and Women's Health

In 1991, a group of nationally prominent health scientists, most of them female, met at Hunt Valley, Maryland, under the auspices of the National Institutes of Health to plan the research agenda for the newly formed Office of Research on Women's Health (ORWH). Despite a cursory recognition of socioeconomic factors, the emphasis of the group—and the resulting research agenda—was largely biomedical and clinical in orientation. How socioeconomic factors influence women's health received only cursory attention [1]. By 1997, when a series of ORWH-sponsored meetings revisited the national research agenda for women's health, socioeconomic issues, and particularly poverty, received significantly more attention, with a special working group devoted to the topic. The period between the two meetings witnessed a dramatic increase in the United States in interest in how socioeconomic factors affect health. This interest extends beyond simply documenting the relationship to trying to understand why the relationship exists (how socioeconomic status (SES) gets into the body), and how it varies across disease and health endpoints, across the life course, and by age, ethnicity, and place of birth.

In fact, the relationship between socioeconomic position or social class and health is one of the most robust and well documented in epidemiology and public health and has been noted by observers since the time of the Romans. Social class, or SES, captures "living conditions and life chances, skill levels and material resources, relative power and privilege" [2]. It is this very capturing of so many aspects of life history and everyday experience that makes SES such a powerful predictor of health status.

Why is the relationship between SES and health critically important? Many studies show that SES (as measured by education, income, occupation, or a socioeconomic prestige measure) affects nutrition, growth, and development prenatally and through childhood and young adulthood; determines life chances including later education and income; and contributes at a population-level to striking differences between rich and poor geographic areas and between developed and developing countries. SES shapes decision making and health habits; exposure to violence, abuse, depression; burden of caretaking and number of working hours; health insurance and access to care. It also profoundly affects security in old age through differing lifelong patterns of savings between rich and poor, men and women, couples and singles.

This chapter reviews the ways in which women's SES, or social class, affects their health and life expectancy in the United States and other countries, with an emphasis on different conceptual and methodological approaches to this topic. It suggests some areas in which further research is needed and touches briefly on the policy implications of studying the links between socioeconomic factors and women's health. The chapter emphasizes the ways in which social class affects women in the context of their everyday lives.

II. Socioeconomic, Racial/Ethnic, and Gender Inequalities

Poverty and affluence are intrinsically entwined with gender and ethnicity and "race" as well as with marital status and household configuration [3]. In 1993, female-headed households had a median income of $18,545 compared to $43,129 for married-couple families, and 35.6% of female-headed households fell below the Federal poverty level [4]. In 1996, the poverty rate for whites was 11.6% but it was 29.4% for Hispanics and 28.4% for blacks. Among female-headed households, 51% of Hispanics and 44% of blacks are below the poverty line compared to 27% of whites [5]. One-third of black women are considered poor, with the brunt of poverty falling on single-parent, female-headed households—their median income was only about $15,600 in 1995. The picture for American Indians/Alaska Natives is even more dismal. Of total American Indian/Alaska Native households, 27% are below the federal poverty level, but 50% of those headed by women are considered poor by the same standard (1990 data) [6]. These data suggest that for many U.S. women, social class, gender, and "race" or ethnicity constitute a triple jeopardy shaping daily experience, including health and death [3].

On a social structural level, women face many socioeconomic obstacles. Men outpace women in earnings and in workplace authority, even though women have higher educational levels than men in similar occupations [7,8]. Despite progress made in the equitable allocation of household responsibilities between men and women, women continue to carry the lion's share of domestic work. In developing countries women often take full responsibility for domestic work, even when they also have substantial responsibility for economic maintenance of the household. The distribution of resources and power within the household is intimately tied to women's health and survival [9]. Not only are single-headed households headed by women far more likely than other households to be poor, being a single household head is a source of multiple stressors. While women, especially those heading households, may shift in and out of poverty [10,11], socioeconomic disadvantage and the irregular career trajectories experienced by women during their middle years shape the availability of pensions and savings in later life. Economic advantages accrue to older married couples in a manner disproportionate to those that accrue to older single women (an apparent "marriage

Table 43.1

Life Expectancy among Women 45 and 65 Years of Age by Family Income and Race: United States, Average Annual, 1979–1989

	White women	White women	Black women	Black women
Family income	45 years	65 years	45 years	65 years
<$10,000	35.8	19.7	32.7	18.6
$10,000–14,999	37.4	20.4	33.5	18.0
$15,000–24,999	37.8	20.5	36.3	19.4
$25,000 or more	38.5	20.7	36.5	19.9

Source: U.S. Bureau of the Census and National Institutes of Health, National Heart, Lung, and Blood Institute, National Longitudinal Mortality Study [15, data table for Figure 25, p. 152].

Table 43.2

Death Rates for Selected Causes among Women 25–64 Years of Age by Education: Selected States, 1995

Education	Chronic disease	Injuries	Communicable diseases		
			Total	HIV	Other
Less than 12 years	317.9	41.9	36.0	19.8	16.2
12 years	273.8	33.1	24.8	13.8	11.0
13 years or more	147.6	17.9	8.6	3.4	5.2

Source: Centers for Disease Control and Prevention, National Center for Health Statistics, Vital Statistics System [15, data table for Figure 26, p. 152].

effect" that promotes savings) [12,13]. These adverse conditions of women's lives are structural rather than behavioral—they are determined by the nature of our social organization, policies, and institutions. In the United States, being nonwhite compounds the structural disadvantages that accrue to women [3,14]. These socioeconomic structural relationships—often abbreviated to "years of education," "usual occupation," or "family income"—play a major role in shaping women's health.

III. Socioeconomic Inequalities in Women's Health: Life Expectancy and Chronic Disease

Women with more education and higher incomes live longer than women with fewer years of schooling and less income and they are healthier along the way. In the United States, life expectancy, death rates, perceived health, and functional status are directly related to SES [15]. Table 43.1 shows that for both white and black women, life expectancy at 45 and 65 years of age is directly related to family income (note that the life expectancy for black women is lower than that for whites at both ages.) The income effect is stronger during the middle years but is still apparent at age 65.

When we examine cause-specific death rates by education, we see the same general pattern (Table 43.2). Women with 13 or more years of education have strikingly lower rates of death from chronic disease, injuries, and HIV than women with a high school education or less.

Looking at individual chronic diseases, we see that there are strong income gradients for heart disease and diabetes and for lung cancer except among women age 65 and older (due to the social class characteristics of smokers in earlier cohorts (Table 43.3)). For instance, women with annual family incomes of $25,000 or more have one-third the likelihood of death by heart disease as women with incomes under $10,000.

Whether the socioeconomic measure is education or income—in the United States or in other countries—the story is the same. For almost every disease for which a socioeconomic gradient exists, with the exception of breast cancer, women at higher levels of education and income have longer lives and are freer from chronic and communicable disease than women with less socioeconomic attainment. Lower income is also associated

with worse perceived health and with activity limitation among U.S. working age adults, as shown in Tables 43.4 and 43.5.

Irrespective of race, women who are poor or near poor are far more likely to describe themselves as being in fair or poor health than are women with middle or higher incomes (Table 43.4). Among elderly women, irrespective of available help, there are striking differences in functioning by income category. Poor women are one-third more likely than higher income women to report activity limitations (Table 43.5). Bivariate data obtained at one point in time obscure the possibility that poor health is causing diminished income. However, there is limited evidence for this "drift hypothesis." The relationship of health to completed education is unlikely to be affected by downward socioeconomic drift, yet years of education are almost invariably associated with better health.

The simple descriptive data presented in Tables 43.1 through 43.5 are confirmed in many multivariate studies from the United States and other developed nations (inequalities in women's health in poorer nations are also profound). Using data from the National Longitudinal Mortality Study, Elo and Preston showed that after adjusting for race, region, urban residence, marital status, and income (logged), nonelderly women with less than 12 years of education had significantly higher mortality than women with at least a high school degree [16]. Similar relationships between educational attainment and mortality have been observed in Europe, as well [17]. Why are these relationships

Table 43.3

Heart Disease, Lung Cancer, and Diabetes Death Rates for Women by Family Income

Family income	Heart disease		Lung cancer		Diabetes
	25–64 years	65+ years	25–64 years	65+ years	45+ years
Less than $10,000	126.9	1346.0	29.1	122.1	42.6
$10,000–$14,999	74.1	1043.6	30.6	118.8	31.3
$15,000–$24,999	51.9	969.7	18.1	140.7	20.3
$25,000 or more	37.3	963.4	20.9	153.4	14.1

Source: U.S. Bureau of the Census and National Institutes of Health, National Heart, Lung, and Blood Institute, National Longitudinal Mortality Study [15, data tables for Figures 27, 28, 29, pp. 152–153].

Table 43.4

Fair or Poor Health among Women 18 Years of Age and Over by Family Income, Race, and Hispanic Origin

Family income	All races		White non-Hisp		Black non-Hisp		Hispanic	
	Percent	SE	Percent	SE	Percent	SE	Percent	SE
Poor	32.4	0.8	30.2	1.2	38.2	1.6	30.4	1.4
Near poor	19.8	0.6	17.9	0.7	26.1	1.6	24.3	1.4
Middle income	9.9	0.3	9.2	0.3	14.6	1.1	13.5	1.2
High income	6.0	0.3	5.8	0.3	9.2	1.7	7.0	1.2

SE = standard error.

Source: Centers for Disease Control and Prevention, National Center for Health Statistics, National Health Interview Survey [15, data table for Figure 32, p. 154].

relatively universal among "developed," Western industrialized nations? Education in higher and middle income countries captures a number of features of women's lives: their relative poverty or privilege growing up, their likely occupational standing and marriage prospects, their access to information, and their cognitive ability. Note, however, that the relationship between socioeconomic measures and health or mortality is almost always weaker for women than for men and for women over 65 years of age compared to the nonelderly. (An excellent summary of these issues is found in Williams and Collins [2].) The relationship of SES to women's health and mortality has been complicated by issues of measurement—whether a woman's own occupation or her husband's is used as the measure of choice may affect the strength of the relationship, especially for older cohorts [18,19]. A second explanation is that women have greater access than men to nonmaterial sources of well-being such as social ties, spirituality, and community activities. Third, even poor women may also have had greater access to and use of health services than men because past cohorts of women have been eligible for Medicaid due to pregnancy or participation in the Aid to Families with Dependent Children program. The availability of health insurance may be more likely to modify the SES–mortality relationship among women than among men [20], although evidence that SES effects are mediated by access to health care is weak.

IV. Socioeconomic Status in Epidemiologic Research

Because SES is unchallenged as a predictor of women's health, it is almost always included in clinical and population-based research. However, this often is done without attention to how socioeconomic factors shape the complex determinants of health at many levels. In much of epidemiology literature SES is "controlled" so that outcomes among women in different socioeconomic strata can be compared. This approach is helpful if all that is required are some basic descriptive data about differences among groups or if the goal is to assess the relative weight of various risk factors. Krieger [21] has written that many epidemiologists are more concerned with modeling complex relationships among risk factors than with understanding their causation. However, understanding the relationship of socioeconomic factors to women's health requires viewing patterns of health and disease among persons in different groups or categories as the product of

social relationships *between* the specified groups, with these relationships expressed through people's everyday living and working conditions, including daily interactions with others [21].*

In this view, economic and social relations are causal, explicitly shaping the production and distribution of individual and population health and disease at many different points across the life span [22]. This complex, theoretically motivated causal approach to epidemiologic research situates "risk factors" and especially behaviors in the context of an individual's social and economic position and the socioeconomic characteristics of a community [23].

V. Increasing Socioeconomic Disparities and Their Effect on Health

Part of the reason for the interest in socioeconomic determinants of women's health is the growing recognition that the two decades between 1973 and 1993 were a period of striking growth in income and wealth inequality in the United States and in other developed nations [24,25], which was paralleled by increasing socioeconomic disparities in health. A number of studies in the United States and in Europe have shown that

*Reprinted from *Social Science and Medicine* **39**, N. Krieger, Epidemiology and the web of causation, pp. 887–903, copyright 1994 with permission from Elsevier Science.

Table 43.5

Difficulty with One or More Activities of Daily Living among Women 70 Years of Age and Over by Family Income

Family income	Any difficulty		Receives help		No help	
	Percent	SE	Percent	SE	Percent	SE
Poor	42.8	1.9	19.9	1.5	22.9	1.7
Near poor	31.8	1.3	14.5	1.1	17.3	1.1
Middle/high	27.5	1.1	13.6	0.9	13.9	0.9

SE = standard error.

Source: Centers for Disease Control and Prevention, National Center for Health Statistics, National Health Interview Survey [15, data table for Figure 34, p. 154].

growing income and wealth inequality is associated with widening differentials in mortality (although estimates of the impact on women have varied). One study found that from 1960 to 1986 the death rates in the United States for blacks and whites, men and women showed an overall decline, but the difference in mortality rates between those in higher and lower income and education categories actually increased. By 1986 there was actually a greater disparity between mortality rates of women in the higher and lower educational categories than there had been in 1960 [26]. Using somewhat different data and measures, Preston and Elo found that inequality in mortality did, in fact, increase for men but was stationary among women ages 65–74 [20]. Among women aged 25–64, educational differences in mortality appear to have narrowed rather than widened in the time period. They suggest that women's recent economic gains, owing to greater labor market participation and opportunities, may have contributed to reduced inequality among this age group.

Significantly, the association of societal and state level income inequality with mortality also appears to be independent of the proportion of the population engaged in risk behaviors such as smoking and of access to health services. In other words, something in the nature of inequality itself appears to be driving socioeconomic differences in mortality patterns.

More specifically, a number of studies suggest that it is relative, rather than absolute, income and poverty that affect life expectancy and health. Using state-level data and controlling for smoking and poverty rates, Kennedy et al. found that the size of the gap between rich and poor has a greater effect on mortality rates than the absolute extent of poverty [27]. In addition to its influence on mortality, state-level income inequality is associated with higher rates of low birth weight, homicide, violent crime, medical care expenditures, smoking, and sedentary activity [28]. At the national level, there is a relationship between how income is distributed in the population (percentage of income going to a particular segment of the population) and life expectancy, such that countries where a larger share of income goes to the less well-off have higher life expectancy than countries where income distribution is skewed to the better-off. Countries with a more equitable income distribution (such as Japan) enjoy higher life expectancy. The fact that it appears to be relative (hierarchical or rank-related) rather than absolute aspects of income that affect people suggests a strong psychosocial component [29,30]. In other words, it may not be occupation and its rewards, per se, that determine health, but job characteristics (e.g., job strain, low control, high demands), limited psychological and social resources, perceived hostility and discrimination, lifestyle ''incongruity,'' and related frustration. The literature is well summarized in Krieger et al. [3].

VI. SES Does Not Only Affect the Poor: The Importance of the Gradient

There is compelling evidence that it is not only poor women who suffer health effects from socioeconomic inequality; the very existence of socioeconomic differences appears to pose a threat to health and well-being and women at the upper as well as the lower echelons of society experience worse outcomes than women who outrank them. The relationship between inequality and adverse health outcomes is observed at both the population and individual levels. The strength of the effects of socioeconomic gradients at all levels on a number of indicators of health—even for an employed, primarily white population with secure jobs—is evident in data from an ongoing study of civil servants in the UK [31]. In this study, many outcome measures demonstrate that the health of employees just below the top ranks, but still in high SES positions, is worse than those who outrank them.

The growing body of literature relating inequality to mortality risk suggests that relative position in a hierarchy substantially affects the risk of disease and death [32] and that this relationship occurs at least partly through the neuroendocrine system—through the twin mechanisms of stress and resilience. This exciting area of research has the potential for linking position in structural, socioeconomic (and other) hierarchies with biological response [33,34]. Some of the strongest evidence for the effect of SES on disease via psychosocial pathways comes from the literature on hypertension and cardiovascular disease. A model developed by Kaplan identifies the pathways by which SES is associated with hostility/anger, depression, job strain, and hopelessness, which are in turn correlated with more proximate determinants of cardiovascular disease such as smoking, stress, and distress [34,35]. House and Williams state:

> . . . relationships of these psychosocial risk factors to SES are remarkably consistent. . . . [I]t is almost always the case that lower-SES groups are exposed to more psychosocial risk. This holds for education and other SES indicators that are clearly antecedent to psychosocial risk factors, and for income and other SES variables that are more proximal to the psychosocial risk variables, yet probably also antecedent to them. [36, p. 121]

It is understandable why individuals lower in social hierarchies would experience greater risk of disease and death. However, what about those just below the top who are not at all deprived? Answers seem to lie at the level of cell biology, with evidence that even slight gradations of rank and control influence immune response and blood fibrinogen, elevated levels of which are associated with coronary heart disease and rheumatoid arthritis [37].

The effect of social ties and support networks on women's health depends on a variety of host and environmental factors such as the woman's age, her occupational position and role including discretion and control on the job, whether she lives in an urban or rural setting, her genetic resilience, and upon the particular endpoint. Qualitative research conducted by Graham in the UK suggests that women's choice among coping mechanisms (e.g., smoking vs physical exercise) is socioeconomically determined [38]. Powerful market forces shape the behaviors of different socioeconomic groups through the manipulation of cultural symbols. For example, billboards in African-American communities promote smoking as a form of relaxation and as a symbol of affluence for women and NIKE advertisements in middle class women's magazines promote exercise wear as symbols of liberation. Market forces reinforce existing socioeconomic patterns of psychosocial response for women; class-based strategies sell products that demonstrate rank, reduce stress, or assist in coping.

Table 43.6
Cigarette Smoking, Overweight, and Sedentary Lifestyle among Women by Family Income, Race, and Hispanic Origin

Family income	All races		White non-Hisp		Black non-Hisp		Hispanic	
	Percent	SE	Percent	SE	Percent	SE	Percent	SE
a. Cigarette smoking among women 18 years of age and over								
Poor	31.2	1.7	38.6	2.5	29.3	2.7	16.6	2.4
Near poor	28.0	1.3	31.6	1.7	24.9	3.0	14.7	1.9
Middle income	24.6	1.0	22.2	0.8[a]	15.7	2.0[a]	13.9	1.8[a]
High income	16.8	1.2						
b. Overweight among women 20 years of age and over								
Poor	46.3	1.9	42.0	3.2	55.0	2.3	54.9	2.4
Near poor	40.4	1.8	36.6	2.4	51.0	1.9	48.7	2.2
Middle income	33.7	1.3	30.0	1.2[a]	52.4	2.4[a]	45.3	2.2[a]
High income	28.2	1.7						
c. Sedentary lifestyle among women 18 years of age and over								
Poor	N/A		35.5	1.7	39.0	2.5	47.8	2.8
Near poor	N/A		30.6	1.2	36.4	2.6	45.3	4.5
Middle income	N/A		22.6	0.8	32.6	2.1	32.2	3.0
High income	N/A		17.1	1.0	24.7	4.0	21.1	4.3

SE = standard error.

[a]Middle and high income categories combined.

Sources: a and c: Centers for Disease Control and Prevention, National Center for Health Statistics, National Health Interview Survey; b: Centers for Disease Control and Prevention, National Center for Health Statistics, National Health and Nutrition Examination Survey [15, data tables for Figures 36,39,40, pp. 155–157].

VII. SES Is Strongly Related to Risks for Disease but Behavioral Risks Do Not Explain the SES–Health Relationship

Many studies suggest that behavioral risk factors such as cigarette smoking, physical activity, and body mass do not, in fact, account for much of the relationship between SES and health [27,28,39,40]. A portion, but only a portion, of the relationship between SES and health can be explained by SES-based differences in behavioral risk factors such as cigarette smoking, poor nutrition, lack of exercise, and excess alcohol use. However, these health habits show clear socioeconomic gradations in the U.S. and the UK. Table 43.6 shows cigarette smoking and sedentary lifestyle by family income for adult American women by race and ethnicity.

Among women ages 18 and over, cigarette smoking is strongly and inversely related to family income, with the exception of Hispanics, who have a lower overall rate of smoking. High income women are half as likely as poor women to smoke (Table 43.6a). A similar, though somewhat weaker pattern, emerges for overweight women (Table 43.6b). Among white women, 30% of those with middle and high income are overweight compared to 42% of poor women. A narrower spread is evident for blacks and Hispanics. Overweight is probably linked to sedentary lifestyle (Table 43.6c), where again a strong income gradient is visible for all ethnic groups.

Table 43.7 shows data on women's health behaviors from the British Whitehall II study where, again, a clear socioeconomic gradient is visible (note that the gradient is reversed for alcohol consumption) and all differences are statistically significant [31].

Those in the inferior civil service grades studied by Marmot were not only likely to report worse health behaviors, but they also reported more financial and housing difficulties, less social

Table 43.7
Health Behaviors among Women by UK Civil Service Employment Grade Category (Age-Adjusted)

Health behaviors	Employment Category						
	1	2	3	4	5	6	n
Current smokers (%)	18.3	11.6	15.2	20.3	22.7	27.5	3408
Units alcohol last 7 days	12.1	9.8	9.3	7.0	5.2	3.6	3375
Little/no exercise (%)	12.0	14.7	10.8	13.2	19.7	31.1	3221
Use low/non-fat milk (%)	39.5	48.3	49.8	46.2	40.5	34.4	3389
Mainly whole-grain bread (%)	57.2	52.9	58.2	55.4	43.8	35.5	3380
Fresh fruit or veg <1/day (%)	17.7	20.4	28.4	29.7	36.4	43.6	3400

Note: Category 1 represents the highest status jobs and category 6 the lowest; some categories have been combined.

From Marmot *et al.* (1991), Health inequalities among British civil servants, *The Lancet* **337**, 1387–1393 [31], © by The Lancet Ltd., with permission.

support, and work environments with less control and opportunity to learn new skills; all are factors that have been shown to correlate with worse functional status among women [41].

Behaviors do not occur in isolation from people's social lives. A perspective advocated by the British sociologist Peter Townsend and others [19,42,43] frames behaviors within the material conditions of people's lives. Different forms of social disadvantage, including gender and race, as well as poverty and affluence, shape the opportunities and choices available to people [43]. It is not always easy to disentangle what are "class effects" from what are "behavioral effects" because the two may be so strongly correlated. Behaviors can be associated with unmeasured aspects of class position, capturing some of the unexplained variance that is really due to social class differences [44, p. 205]. Graham's research shows that low-income mothers give budget priority to payments for housing and utilities and cigarettes, while saving on the food budget. Spending on cigarettes is protected because cigarettes, a coping strategy, are viewed as a necessary luxury [38]. Graham's work among women in the UK shows strong class gradients in smoking and nutrition as evidenced by the data from the UK Health and Lifestyle survey shown in Table 43.8. Smoking is nearly three times as common among women who are manual laborers compared with professionals. Conversely, women managers and professionals are far more likely to eat whole grain bread and fruit than women in lower-ranked occupations.

VIII. Socioeconomic Differences in Breast Cancer Incidence: The Inverted Gradient

SES is also positively associated with some health outcomes. The most widely known of these is breast cancer. Devesa and Diamond reported a strong positive association of breast cancer with income and education for white women, and with education but not income for black women [45]. Bacquet *et al.* also showed a linear relationship between years of education and breast cancer incidence for whites and a weak positive relationship for black women. Among whites, there is a also a strong positive income gradient, but the relationship for black women is more ambivalent [46]. Characteristics of higher SES women such as later and lower fertility [47] are believed to contribute to the positive relationship with incidence of breast cancer.

IX. Theoretical Approaches to Socioeconomic Effects on Women's Health

As with any problem in science, examining socioeconomic effects on health is more interpretable when the researcher begins with a theoretical and analytic framework [48]. Two analytic approaches developed to explain socioeconomic differences in women's health are presented below.

Arber has created an analytic framework that integrates women's roles in the household (and the consumption patterns that follow from these roles) with the structural measure of occupational class [19]. This approach is particularly useful for capturing socioeconomic factors in women's lives because paid employment is both a potential stressor (when added to child-care and marital roles) and also an indicator of potential material and social resources. A strength of Arber's approach is that it

Table 43.8

Women's Smoking and Consumption of "Brown" Bread and Fruit by Household Socioeconomic group: UK, 1985 (Percentage)

Household socioeconomic group	Current smokers	Consume regularly	
		Brown bread	Fruit in summer
Professional	14	72	74
Employers/managers	28	56	73
Other nonmanual	26	54	64
Skilled manual	35	40	59
Semiskilled manual	37	34	54
Unskilled manual	45	30	42

From Graham (1990), in "Women's Health Counts," [44], by permission of Routledge Publishers.

encompasses women who are not in paid employment such as housewives, the unemployed, and those unable to work due to physical or mental disabilities. The importance of Arber's work is that it takes into account the differing day-to-day realities of men's and women's lives within a socioeconomic framework. Arber argues that using consumption measures, such as housing tenure and car ownership, may be equally or more revealing of a woman's class position than occupation, perhaps because they are resources that make a difference in women's everyday lives. For example, using the British General Household Survey to look at limiting long-standing illness (LLI), Arber finds that the most disadvantaged (vulnerable) women are those aged 55 or older who are previously married, unemployed, living in public housing, and doing unskilled manual work; she also finds that they are worse off than men in similar positions. Over 50% of this most disadvantaged group report LLI, compared to 5% of the most "favored category," who are women aged 25 who are married, employed, childless professionals who own their own homes. Arber's work also suggests that the socioeconomic factors as well as the family and occupational roles that influence health in women are different from the factors that influence health in men. Women are more likely to experience role strain and overload that occurs when familial responsibilities are combined with occupation-related stress. These are compounded (or alleviated) by material circumstances. Arber finds that for men, occupational class and employment status explain more of the variation in health than do familial roles and responsibilities.

Another useful analytic framework for understanding the impact of social inequalities on women's health—more focused on racial and ethnic than on gender-based differences—is the "weathering" hypothesis proposed by Geronimus [14]. The weathering hypothesis, or analytic framework, proposes that women age in different ways depending on how different life circumstances undermine or promote health during the period, and that women's health and mortality experience reflects a cumulation or cascade of advantages and disadvantages. The theory helps to explain racial and ethnic as well as socioeconomic differences in the development of chronic conditions, because women of different ethnic backgrounds "weather" at different rates, and it encompasses age-based trajectories of behaviors such as smoking, as well as environmental exposures (*e.g.,* to lead-based paint) and access to health services [49,50]. The

Table 43.9

Mammography Within the Past 2 Years among Women 50 Years of Age and Over

Family income	All races		White non-Hisp		Black non-Hisp		Hispanic	
	Percent	SE	Percent	SE	Percent	SE	Percent	SE
Poor	44.7	2.2	37.5	2.5	60.0	3.7	49.3	6.8
Near poor	48.2	1.5	47.9	1.8	57.4	3.7	38.2	5.1
Middle income	67.2	1.0	70.4[a]	0.9	72.2[a]	3.4	64.7[a]	4.8
High income	76.3	1.8						

SE = standard error.

Source: Centers for Disease Control and Prevention, National Center for Health Statistics, National Health Interview Survey [15, data table for Figure 45, p. 159].

weathering framework also helps to explain the intergenerational transmission of the link between SES and health by connecting "physical manifestations" of weathering such as hypertension, cigarette smoking, and blood-lead levels to infant's and children's health. Most recently, Geronimus and her collaborators have examined the effects on excess mortality of residence in geographic areas that vary in population-level patterns of disadvantage and segregation. They show that environmental and psychosocial factors in different communities create sharp disparities in death rates among black and white men and women [51]. Their work argues for looking beyond individual to population-level indicators of hardship, in addition to the chronic burdens of everyday deprivation.

X. Socioeconomic Inequalities and Access to Health Services

Most epidemiologists and demographers have found, in fact, that the contribution of universal access to medical care to reduction of inequalities in health is minimal. The same disparities in outcomes for individuals in different socioeconomic categories hold for countries such as the UK and Sweden, which have universal health insurance and access to medical care, and in the U.S., with its patchwork of health insurers and providers and substantially large uninsured population. There is little or no evidence that universal access to health services is sufficient to reduce inequalities in disease and mortality. For example, evidence from the UK, where the National Health Service with universal access was begun in 1947, shows that mortality differentials among the five occupational classes actually widened in the fifty-year period. However, availability of health insurance and access to health services for women are very much affected by SES. In the U.S. in 1993, poor women 25–64 years of age were 3.2 times as likely to have no health care coverage as nonpoor women [52]. In particular, poor women have often lacked access to preventive services such as breast and cervical cancer screening and poverty results in lower use of preventive services such as mammograms (Table 43.9). In California, for instance, during the decade 1984–1994, women over 50 years of age without health insurance were 1.8 times more likely never to have had a mammogram (47.7% of the uninsured, compared with 28% of insured women) [53]. For the U.S. as a whole, uninsured women were 2.8 times more likely than the

insured never to have received a Pap test. Among women aged 50–64 years, the percentage reporting a recent mammogram was lowest among the uninsured (20%) and highest among those enrolled in an HMO (59%) [54]. There is a strong income gradient in use of mammography (Table 43.9) [15]. Among women aged 50 and over, 76% of those with high income had a mammogram within the past 2 years, compared to 45% of poor women and 48% of the near poor.

The advent of HMOs and Medicaid has placed greater emphasis on prevention. Nevertheless, the prevalence of sexually transmitted diseases (STDs) is particularly associated with poverty and problems in access to health services. Even poor women with health insurance coverage may not have adequate diagnosis and treatment of STDs because of copayments and deductibles [55]. Poverty and near poverty make it harder to pay for medication, and material circumstances in which poor women live may make it difficult to follow-through with treatment regimens. Thus, it is not surprising that there is a strong direct relationship between SES and the incidence of cervical cancer, and that cervical cancer and other STD rates, including HIV, are disproportionately high among poor women of color [15].

The result of socioeconomic barriers to health care is that there are unmet needs for health services in the U.S., particularly among the under-65 year old population (Table 43.10). Table 43.10 shows a sharp income gradient in the extent of the need for care. Only 8% of high income younger women have

Table 43.10

Unmet Need for Health Care during the Past Year among Women 18–64 Years of Age and Over

Family income	18–44 years		45–64 years	
	Percent	SE	Percent	SE
Poor	31.3	0.7	41.4	1.3
Near poor	30.5	0.6	31.7	0.9
Middle income	19.3	0.4	15.4	0.5
High income	7.9	0.3	6.4	0.3

SE = standard error.

Source: Centers for Disease Control and Prevention, National Center for Health Statistics, National Health Interview Survey [15, data table for Figure 46, p. 159].

unmet needs compared to nearly a third of the poor. Among middle-aged women, 41% of the poor are unable to obtain needed care compared to 6% of those with high incomes.

Reasons for unmet need are suggested by qualitative research among poor women. A study of low-income women in California found that lack of coverage, inadequate coverage of services, and copayments delayed or prevented initiation of care. Barriers such as time constraints, transportation problems, and appointment systems continue to impede access to health services. Referral to specialists is especially problematic. All of these issues are particularly acute for women in rural areas [56].

SES, particularly occupational position and employment status, influences a woman's health insurance and access to health care, which are in turn associated with important correlates of morbidity and mortality such as timely and effective disease screening. Despite observed SES differences in insurance coverage and access to health services, there is abundant evidence that, like risk behaviors, they explain only a small portion of the socioeconomic variation in morbidity and mortality [57].

XI. Socioeconomic Effects on Women's Health in Lower and Middle Income Nations

Among the developing and middle income nations of the world, the effect of poverty on women's health is even more acute than it is in Europe and the U.S. The 1980s and early 1990s, a period of increasing income inequality within developed nations, was a time of "structural adjustment" elsewhere. Under restrictions posed by international lending agencies in exchange for the restructuring of national debt, countries slashed their health and welfare subsidies. Many developing countries also have a large proportion of households headed by women. Migration and urbanization, together with decreases in fertility, have contributed to changes in support patterns for women of different ages. These societal-level changes, in combination with cultural practices and intense socioeconomic disparities, have placed women at high risk of poor diet, disease and injury, lack of access to health services, and occupational and environmental hazards [58]. Women's lack of education and poverty go hand-in-glove with iniquitous distribution of resources, including food, within the household, and, often, denial of or lack of access to Westernized medical care for women and female offspring [9]. Poverty, gender discrimination, and migration are associated with high rates of sexually transmitted disease, including HIV, among women.

Poverty and low SES are not inversely associated with poor health for all disease outcomes. In developing and middle income countries, the relationship between socioeconomic distribution and health is reversed for some diseases. For example, cardiovascular disease and lung cancer, which have an inverse relationship with income and education in the U.S., have a positive correlation with SES in countries such as China which have not passed through what is called the "health transition." The positive relationship of disease endpoints with SES is partly mediated by diet and smoking in societies where consumption of fats and tobacco are symbols of affluence. Although women have lagged behind men in adopting these behaviors, their behaviors are changing rapidly in some areas as the result of targeted marketing to women [58].

XII. Socioeconomic Measures in Women's Health Research—A Methodological Note

One of the practical problems that epidemiologists and public health researchers face is selection of an appropriate socioeconomic measure in studies of women's health [48]. This section briefly reviews different socioeconomic measures: income, education, and occupation; material and social resources and deprivation; effects of earlier on later SES; income transitions; and social class at the neighborhood and community levels.

A. Income, Education, and Occupation

Work that is theoretically motivated by a particular framework of women's health "production" may guide the selection of the appropriate socioeconomic measure. Often, however, it is more a matter of expedience because each item in a survey or clinic-based data set can be costly of time and scarce funds. Income, education, and occupation are the most common indicators in Western developed nations; while each contributes to explaining socioeconomic variation in health outcomes, none alone may be sufficient, and there are multiple ways of measuring each of these commonly used indicators. For example, using data from the Wisconsin Longitudinal Study, Marks found that among a relatively well-off sample of white American women, income, education, and occupational status all play distinct roles in shaping women's health behaviors, functioning, and self-reported health in midlife [59].

B. Material and Social Resources

The work of the British sociologist Peter Townsend and others in the UK has called attention to the importance of material and social deprivation and resources as key influences on health [42,48]. Arber's work has shown that consumption measures such as car and home ownership may be more important predictors of functional disability for women than for men [19], and for older women, income, car ownership, and housing tenure have a significant effect on self-assessed health and disability (functional status), after controlling for age and occupation [41]. The relevance of particular socioeconomic measures varies by outcome: occupational class is a more important determinant of older women's functional status but car and home ownership contribute to better self-assessed health and well-being [41, p. 44].

C. Lifecourse Considerations in Effect of SES on Adult Health

A number of studies suggest links between fetal and childhood health and nutrition and outcomes in later life [60–62], although Barker has found that social class at birth does not affect disease in later life, once other factors, such as maternal nutrition, are controlled [63]. Childhood socioeconomic circumstances may affect adult health and mortality through schooling and adult SES [61], but whether childhood SES has stronger effects for men or for women on later life outcomes is unclear. The effects of childhood social class on health inequalities measured at ages 23 and 33 differ by outcome and for men and women, suggesting that the pathways for social class effects are complex and confirming that lifecourse effects are likely to be gender-specific [64]. For women, the impact of childhood social class on self-rated health at age 33 is modified by childhood housing, adult income, and age at first child [65].

D. Mobility and Income Transitions

No individual woman's socioeconomic position is completely static throughout life. It is likely to change as she moves from childhood to independent adulthood, upon partnership formation, and in the context of her own career development. There is growing interest in modeling how these changes affect women's health and well-being [34]. Socioeconomic position in family of origin and in adult life should each be considered separately. A measure of economic mobility or income dynamics may also be important. Income may be the most important component of SES for health, and negative changes in income (often associated with other adverse events such as job loss and widowhood) contribute to adverse health outcomes; changes in income may hit middle-income individuals hardest [66]. Using data from the Panel Study on Income Dynamics, Duncan showed that the average level of women's income-to-needs ratio declines from middle to older ages [67]. Over a third of women aged 56–65 experience at least a 5% decline in the income-to-needs ratio. (The income-to-needs ratio is arrived at by dividing family income by the U.S. poverty threshold for a given year. A decline in the ratio suggests that the family is poorer.) In the same age group, 27% of women (compared to 17% of men) have been poor at least once; the comparable figure for women aged 66–75 is 35%. These studies have two methodological implications: income changes should be taken into account in estimating SES effects on health *and*, if possible, it is important to have an income measure that captures changes over time. The latter study suggests the likely increase in poverty as women age and their disadvantageous economic status relative to men.

E. Neighborhood/Community-Level Effects on Health

During the 1990s, a growing body of research has assessed the effects of different kinds of neighborhoods on health [39,51, 68,69]. A longitudinal study of Alameda County, CA residents showed that nonelderly women living in poverty areas had substantially higher mortality rates than equivalent women in better neighborhoods [39]. Some communities offer safe and uncrowded schools, parks and recreational areas that invite physical activity, clinics and well-stocked grocery markets that promote healthy behaviors, and libraries that serve as sources of information, whereas other neighborhoods have few or none of these resources. Streets free from violence, trash, and pollution create an environment that promotes social ties and mobility, which in turn affect health positively. Healthy neighborhood environments may be especially important to women because it is often women who are responsible for grocery shopping, obtaining health care, and contacts with schools and community organizations [70]. Neighborhood characteristics (*e.g.*, percentage of households in poverty, percentage of households headed by females) can be used to proxy individual level social class characteristics when the latter are not available, for example in cancer registry data [71,72]. The community context also affects

people above and beyond their individual characteristics [73].

Using census-based neighborhood (*e.g.*, block group) characteristics can also help epidemiologists to explore the contribution of social class to disease incidence and outcomes. This is done by linking census or other geographically-based data to individual-level records. Using this approach, Krieger showed that simply examining racial differences in breast cancer incidence without taking social class context into account obscured important insights into disease etiology [74].

There are a number of analytic considerations when taking neighborhood socioeconomic characteristics into account, including level of aggregation, statistical techniques, selection effects, and choice of appropriate variables [e.g., 48,75–77]. Individual-level socioeconomic characteristics may be as or more important than neighborhood indicators; results may be very sensitive to the specific measures used [78]. Improving our substantive and methodological approaches will help us to better understand community-level social class effects on women's health. The advent of geographical information systems has made them readily available to epidemiologists.

XIII. Unanswered Questions in the Epidemiology of Socioeconomic Effects on Women's Health

While there has been a long tradition in epidemiology and public health of assessing socioeconomic effects on health, it is only fairly recently that gender-specific frameworks and relationships have been singled out for study. This brief review suggests a sampling of research questions, the answers to which may make a significant difference to women's health:

1. What is it about inequality per se that is so injurious to health, apart from the absolute adverse effects of poverty?

2. Can we deepen our understanding of how relations of social class and social hierarchies "get into the body?" To what extent are the endocrine, behavioral, and social processes through which this occurs gender-specific? Do hierarchical relations have different consequences for men and for women and how do these patterns affect health? Are there other creative paradigms that can bring biology and "social" effects closer together?

3. Why do socioeconomically patterned differences in behavioral risk not account for the SES-health relationship at either individual or population levels?

4. Can we develop gender-specific models that bring together macrolevel influences, socioeconomic factors, culture, biology and behaviors, that are better able to predict health endpoints for women?

5. How and why do socioeconomic effects vary over the life course, and are there ways in which this is gender specific? How do these interact with other demographic influences, such as race/ethnicity and nativity?

6. What is the role of protective resources such as social networks and spirituality in countering adverse socioeconomic effects on women's health?

7. Are women particularly vulnerable to neighborhood or "contextual" effects on health, and if so, why?

A necessary step is to continue to develop and test socioeconomic measures that are most appropriate for women and to insure that survey, registry, and clinic-based data contain such measures so that socioeconomic effects on women's health can be appropriately monitored and studied. In order to understand the sources of differences in women's health at the individual and population levels, it is vitally important that epidemiologists include socioeconomic measures in their data and examine the relationship of socioeconomic gradients to the outcome of interest, as recommended in the 1993 National Institutes of Health Guidelines on the Inclusion of Women and Minorities in Research [79]. Guidelines for public health researchers on how to go about measuring social class and socioeconomic position are found in Krieger et al. [48].

XIV. Implications for Health Policy

All too often the focus of policy is on the health outcome or on behavioral mediators rather than on the underlying or fundamental social and economic causes of different states of health [23,43]. Insuring that all Americans, both men and women, have continuing access to equitable sources of health care, the maintenance of dignified income supports for the poor, elderly, disabled, and unemployed, and the use of taxation to redistribute income and reduce inequalities would do much to reduce the impact of SES on health. The rhetoric of equality in the U.S. competes with the ideology of individualism; our economic and social policies do little to moderate inequality at the individual and household level. Policy shifts, such as the devolution of welfare programs to the states [80] as well as technological changes that shape employment opportunities suggest that the impact of SES differences on women's health may intensify during coming years. The reduction of the state sector in developed and developing countries, with concomitant changes in welfare and social insurance policies, may exacerbate social class differences and make the health costs of poverty and social gradients even higher for women. It is important to keep in mind, too, that policy change aimed at reducing socioeconomic inequalities may not have an immediate impact on women's health. There are individual, social structural, political, and cohort-related obstacles (e.g., dependent care responsibilities, residential location, transportation limitation, labor market opportunities) to improving survival prospects [81].

XV. Summary and Conclusions

Socioeconomic disparities have increased sharply since the 1970s with a concomitant widening of health disparities both for men and for women. An ever-growing body of research substantiates the effect of socioeconomic factors on health. In addition to the adverse effects of poverty, the existence of societal level income-inequality and the presence of socioeconomic gradients has an adverse effect on morbidity and life expectancy, and the way in which socioeconomic differences are "internalized" by the body is a topic of great interest. Many risk behaviors follow a socioeconomic gradient but they are insufficient to explain socioeconomic differences in health. The effects of socioeconomic factors on women's health are interrelated with race/ethnicity and gender statuses within the household and in society as a whole. Socioeconomic effects may differ over the life course and they depend on household configuration and marital status.

Epidemiologic analyses of socioeconomic effects are assisted by theoretical approaches (e.g., a perspective that interrelates occupational roles with material and social resources) and a framework that uses racial/ethnic and age-based trajectories and contextual location to explain differences in risk factors, morbidity, and mortality. Theoretical and analytic clarity should guide selection of the appropriate socioeconomic measure, whether at the individual or community level. Socioeconomic disparities in nonindustrialized societies are extreme and women's educational attainment and status are often low; the results are numerous threats to health within the household and the community. These have been exacerbated by macrolevel policies that often have adverse effects on poor women. Achieving socioeconomic equity for women would assist in reducing the unnecessary financial and human costs of avoidable ill health. Policy shifts to reduce societal and individual level socioeconomic inequalities will make this goal more feasible. Epidemiologists and public health researchers can contribute to this effort with studies that address how structural inequalities contribute to an unfair burden of disability and disease.

References

1. Office of Research on Women's Health (1991). "Opportunities for Research on Women's Health." U.S. Department of Health and Human Services, Public Health Service, Hunt Valley, MD.
2. Williams, D. R., and Collins, C. (1995). US socioeconomic and racial differences in health: Patterns and explanations. Annu. Rev. Soc. 21, 349–386.
3. Krieger, N., Rowley, D. L., Herman, A. A., Avery, B., and Phillips, M. T. (1993). Racism, sexism and social class: Implications for studies of health, disease, and well-being. Am. J. Prev. Med. 9, 82–122.
4. U.S. Bureau of the Census (1995). "Income, Poverty and Valuation of Noncash Benefits: 1993," Curr. Popul. Rep., Ser. P-60,188. U.S. Govt. Printing Office, Washington, DC.
5. U.S. Bureau of the Census (1997). "Poverty in the United States: 1996," Curr. Popul. Rep., Ser. P-60, 198. U.S. Govt. Printing Office, Washington, DC.
6. Leigh, W. A., and Lindquist, M. A. (1998). "Women of Color Health Data Book," NIH Publ. No. 98-4247. National Institutes of Health, Office of Research on Women's Health, Bethesda, MD.
7. Marini, M. M. (1980). Sex differences in the process of educational attainment: A closer look. Soc. Sci. Res. 9, 307–361.
8. McGuire, G. M., and Reskin, B. (1993). Authority hierarchies at work: The impacts of race and sex. Gender Soc. 7, 487–506, cited in Warren, J. R., Sheridan, J. T., and Hauser, R. M. (1996). "How do Indexes of Occupational Status Affect Analyses of Gender Inequality in Occupational Attainment?" (Working Paper Ser. 96-10). Center for Demography and Ecology, University of Wisconsin, Madison.
9. Browner, C. H. (1989). Women, household and health in Latin America. Soc. Sci. Med. 28, 461–473.
10. Garfinkel, I., and McLanahan, S. S. (1986). "Single Mothers and Their Children: A New American Dilemma." Urban Institute Press, Washington, DC.
11. Burkhauser, R., and Duncan, G. J. (1988). Economic risks of gender roles: Income loss and life events over the life course. Soc. Sci. Q. 70, 3–23.
12. Smith, J. P. (1995). "Marriage, Assets, and Savings" (Work. Pap. Ser. 95-08). Labor and Population Program, RAND Corporation, Los Angeles.
13. Waite, L. J. (1995). Does marriage matter? Demography 32, 483–507.

14. Geronimus, A. T. (1992). The weathering hypothesis and the health of African-American women and infants: Evidence and speculations. *Ethnicity Dis.* **2,** 207–221.

15. U.S. Department of Health and Human Services (1998). "Health United States 1998 with Socioeconomic Status and Health Chartbook." National Center for Health Statistics, Hyattsville, MD.

16. Elo, I. T., and Preston, S. H. (1996). Educational differences in mortality: United States, 1979–85. *Soc. Sci. Med.* **42,** 47–57.

17. Valkonen T. (1989). Adult mortality and level of education: A comparison of six countries. *In* "Health Inequalities in European Countries" (J. Fox, ed.), pp. 142–162. Gower, Aldershot, UK.

18. Arber, S. (1987). Social class, non-employment, and chronic illness: Continuing the inequalities in health debate. *Br. Med. J.* **294,** 1069–1073.

19. Arber, S. (1991). Class, paid employment and family roles: Making sense of structural disadvantage, gender, and health status. *Soc. Sci. Med.* **32,** 425–436.

20. Preston, S. H., and Elo, I. T. (1995). Are educational differentials in adult mortality increasing in the United States? *J. Aging Health* **7,** 476–496.

21. Krieger, N. (1994). Epidemiology and the web of causation: Has anyone seen the spider? *Soc. Sci. Med.* **39,** 887–903.

22. Evans, R. G. (1994). Introduction. *In* "Why Are Some People Healthy and Others Not? The Determinants of Health of Populations" (R. G. Evans, M. L. Barer, and T. R. Marmor, eds.), Chapter 1. de Gruyter, New York.

23. Link, B. G., and Phelan, J. (1995). Social conditions as fundamental causes of disease. *J. Health Soc. Behav.,* Extra Issue, pp. 80–94.

24. Karoly, L. A. (1996). Anatomy of the U.S. income distribution. *Oxford Rev. Econ. Policy* **12,** 77–96.

25. Wolff, E. (1995). "Top Heavy. A Study of the Increasing Inequality of Wealth in America." Twentieth Century Fund Press, New York.

26. Pappas, G., Queen, S., Hadden, W., and Fisher, G. (1993). The increasing disparity between socioeconomic groups in the United States, 1960 and 1986. *N. Engl. J. Med.* **329,** 103–109.

27. Kennedy, B. P., Kawachi, I., and Prothrow-Stith, D. (1996). Income distribution and mortality: Test of the Robin Hood Index in the United States. *Br. Med. J.* **312,** 1004–1007.

28. Kaplan, G. A., Pamuk, E. R., Lynch, J. W., Cohen, R. D., and Balfour, J. L. (1996). Inequality in income and mortality in the United States: Analysis of mortality and potential pathways. *Br. Med. J.* **312,** 999–1005.

29. Wilkinson, R. G. (1992). Income distribution and life expectancy. *Br. Med. J.* **304,** 165–168.

30. Haan, M. N., Kaplan, G. A., and Syme, S. L. (1989). Socioeconomic status and health: Old observations and new thoughts. *In* "Pathways to Health: The Role of Social Factors" (J. P. Bunker, D. S. Gomby, and B. H. Kehrer, eds.), pp. 76–135. Henry J. Kaiser Family Foundation, Menlo Park, CA.

31. Marmot, M. G., Davey Smith, G., Stansfeld, S., Patel, C., North, F., Head, J., White, I., Brunner, E., and Feeney, A. (1991). Health inequalities among British civil servants: The Whitehall II study. *Lancet* **337,** 1387–1393.

32. Adler, N., Boyce, T., Chesney, M. A., Cohen, S., Folkman, S., Kahn, R. L., and Syme, S. L. (1994). Socioeconomic status and health: The challenge of the gradient. *Am. J. Psychol.* **49,** 15–24.

33. Sapolsky, R. (1990). Stress in the wild. *Sci. Am.* **262,** 116–123.

34. Singer, B., and Ryff, C. D. (1995). Social hierarchies and health: Pathways and outcomes. Paper presented at the National Bureau of Economic Research Seminar on Economics of Aging, Carefree, AZ.

35. Kaplan, G. A. (1995). Biologic and methodologic approaches to the association between socioeconomic factors and cardiovascular disease. *In* "Report of the Conference on Socioeconomic Status and Cardiovascular Health and Disease," pp. 15–22. National Institutes

36. House, J. S., and Williams, D. R. (1995). Psychosocial pathways linking SES and CVD. *In* "Report of the Conference on Socioeconomic Status and Cardiovascular Health and Disease," pp. 119–124. National Institutes of Health, National Heart, Lung, and Blood Institute, Bethesda, MD.

37. Marmot, M. G., and Mustard, J. F. (1994). Coronary heart disease from a population perspective. *In* "Why Are Some People Healthy and Others Not? The Determinants of Health of Populations" (R. G. Evans, M. L. Barer, and T. R. Marmor, eds.), pp. 189–214. de Gruyter, New York.

38. Graham, H. (1994). Gender and class as dimensions of smoking behavior in Britain: Insights from a survey of mothers. *Soc. Sci. Med.* **38,** 691–698.

39. Haan, M., Kaplan, G. A., and Camacho, T. (1987). Poverty and health: Prospective evidence from the Alameda County study. *Am. J. Epidemiol.* **125,** 989–998.

40. Rose, G., and Marmot, M. G. (1981). Social class and coronary heart disease. *Br. Heart J.* **45,** 13–19.

41. Arber, S., and Ginn, J. (1993). Gender and inequalities in health in later life. *Soc. Sci. Med.* **36,** 33–46.

42. Townsend, P., Davidson, N., and Whitehead, M. (1990). "Inequalities in Health: The Black Report and the Health Divide." Penguin Books, London.

43. Hughes, D., and Simpson, L. (1995). The role of social change in preventing low birth weight. *Future Children* **5,** 87–102.

44. Graham, H. (1990). Behaving well: Women's health behavior in context. *In* "Women's Health Counts" (H. Roberts, ed.), Chapter 8, pp. 195–221. Routledge, London.

45. Devesa, S. S., and Diamond, E. L. (1980). Association of breast cancer and cervical cancer incidences with income and education among whites and blacks. *JNCI, J. Natl. Cancer Inst.* **65,** 515–528.

46. Bacquet, C. R., Horm, J. W., Gibbs, T., and Greenwald, P. (1991). Socioeconomic factors and cancer incidence among blacks and whites. *J. Natl. Cancer Inst.* **83,** 551–557.

47. Robbins, A. S., Brescianini, S., and Kelsey, J. L. (1997). Regional differences in known risk factors and the higher incidence of breast cancer in San Francisco. *J. Natl. Cancer Inst.* **89,** 960–965.

48. Krieger, N., Williams, D. R., and Moss, N. E. (1997). Measuring social class in U.S. public health research: concepts, methodologies, and guidelines. *Annu. Rev. Public Health* **18,** 341–378.

49. Geronimus, A. T., Anderson, H. F., and Bound, J. (1991). Differences in hypertension prevalence among U.S. black and white women of childbearing age. *Public Health Rep.* **106,** 393–399.

50. Geronimus, A. T., and Hillemeier, M. M. (1992). Patterns of blood lead levels in U.S. black and white women of childbearing age. *Ethnicity Dis.* **2,** 222–231.

51. Geronimus, A. T., Bound, J., Whitman, T. A., Hillemeier, M. M., and Burns, P. B. (1996). Excess mortality among blacks and whites in the United States. *N. Engl. J. Med.* **335,** 1552–1558.

52. National Center for Health Statistics (1996). "Health, United States, 1995." Public Health Service, Hyattsville, MD.

53. California Department of Health Services (1997). "Profile of Women's Health Status in California, 1984–94." Office of Women's Health, Sacramento, CA.

54. Makuc, D., Fried, V. M., and Parsons, P. E. (1994). Health Insurance and Cancer among Women. Advance Data from Vital and Health Statistics, No 254. National Center for Health Statistics, Hyattsville, MD.

55. Eng, T. R., and Butler, W. T. (1997). "The Hidden Epidemic. Confronting Sexually Transmitted Diseases." Institute of Medicine, Division of Health Promotion and Disease Prevention, Committee on Prevention and Control of Sexually Transmitted Diseases, National Academy Press, Washington, DC.

56. Wyn, R., Leslie, J., Glik, D., and Solis, B. (1997). "Low-income Women and Managed Care in California." UCLA Center for Health Policy Research and Pacific Institute for Women's Health, Los Angeles.

57. Adler, N. E., Boyce, T., Chesney, M. A., Folkman, S., and Syme, S. L. (1993). Socioeconomic disparities in health: No easy solution. *JAMA, J. Am. Med. Assoc.* **269,** 3140–3145.

58. The World Bank (1994). "A New Agenda for Women's Health and Nutrition." The World Bank, Washington, DC.

59. Marks, N. F. (1996). Socioeconomic status, gender, and health at midlife: Evidence from the Wisconsin Longitudinal Study. *Res. Sociol. Health Care* **13,** 133–150.

60. Kaplan, G. A., and Salonen, J. T. (1990). Socioeconomic conditions in childhood and ischemic heart disease during middle age. *Br. Med. J.* **301,** 121–123.

61. Mare, R. D. (1990). Socioeconomic careers and differential mortality among older men in the United States. *In* "Measurement and Analysis of Mortality: New Approaches" (J. Vallin, S. D'Souza, and A. Palloni, eds.), pp. 362–387. Clarendon Press, Oxford.

62. Elo, I. T., and Preston, S. H. (1992). Effects of early-life conditions on adult mortality: A review. *Popul. Index* **58,** 186–212.

63. Barker, D. J. P. (1998). "Mothers, Babies, and Health in Later Life." Churchill-Livingstone, Edinburgh.

64. Power, C., Hertzman, C., Mathhews, S., and Manor, O. (1997). Social differences in health: Life-cycle effects between ages 23 and 33 in the 1958 British birth cohort. *Am. J. Public Health* **87,** 1499–1503.

65. Power, C., Mathhews, S., and Manor, O. (1998). Inequalities in self-rated health: Explanations from different stages of life. *Lancet* **351,** 1009–1014.

66. McDonough, P., Duncan, G. J., Williams, D., and House, J. (1997). Income dynamics and adult mortality in the United States, 1972 through 1989. *Am. J. Public Health* **87,** 1476–1483.

67. Duncan, G. J. (1996). Income dynamics and health. *Int. J. Health Serv.* **26,** 419–444.

68. MacIntyre, S., MacIver, S., and Sooman, A. (1993). Area, class and health: Should we be focusing on places or people? *J. Soc. Policy* **22,** 213–234.

69. Kaplan, G. (1996). People and places: Contrasting perspectives on the association between social class and health. *Int. J. Health Serv.* **26,** 507–519.

70. Hondagneu-Sotelo, P. (1994). Women consolidating settlement. *In* "Gendered Transitions: Experiences of Mexican Immigration," Chapter 6, pp. 148–185. University of California Press, Berkeley.

71. Krieger, N. (1991). Women and social class: A methodological study comparing individual, household, and census measures as predictors of black/white differences in reproductive history. *J. Epidemiol. Commun. Health* **45,** 35–42.

72. Geronimus, A. T., Bound, J., and Neidert, L. J. (1996). On the validity of using census geocode data to proxy individual socioeconomic characteristics. *J. Am. Stat. Assoc.* **91,** 529–537.

73. Wilson, W. J. (1987). "The Truly Disadvantaged: The Inner City, the Underclass and Public Policy." University of Chicago Press, Chicago.

74. Krieger, N. (1990). Social class and the black/white crossover in the age specific incidence of breast cancer: A study linking census-derived data to population-based registry records. *Am. J. Epidemiol.* **131,** 804–819.

75. Tienda, M. (1991). Poor people and poor places: Deciphering neighborhood effects on poverty outcomes. *In* "Macro-Micro Linkages in Sociology" (J. Huber, ed.), pp. 244–262. Sage, Newbury Park, CA.

76. Bryk, A. S., and Raudenbush, S. W. (1992). "Hierarchical Linear Models: Applications and Data Analysis Methods." Sage, Newbury Park, CA.

77. Mayer, S. E., and Jencks, C. (1989). Growing up in poor neighborhoods: How much does it matter? *Science* **243,** 1441–1445.

78. Sloggett, A., and Joshi, H. (1994). Higher mortality in depressed areas: Community or personal disadvantage? *Br. Med. J.* **309,** 1470–1474.

79. National Institutes of Health (1995). The NIH Guidelines on the Inclusion of Women and Minorities as Subjects in Clinical Research. *Fed. Regist.* **59,** 14508–14513.

80. Children's Defense Fund (1996). "Summary of the New Welfare Legislation (Public Law 104-193). Children's Defense Fund, Washington, DC.

81. Rogers, R. (1992). Living and dying in the USA. *Demography* **29,** 287–303.

44

Women, Stress, and Health

KRISTI WILLIAMS AND DEBRA UMBERSON
Department of Sociology and Population Research Center
The University of Texas
Austin, Texas

I. Introduction

Thirty years of epidemiologic research reveal consistent gender differences in mental and physical health. Compared to men, women exhibit greater physical morbidity and higher rates of psychological distress. A vast body of theoretical and empirical work in psychosocial epidemiology addresses the observed gender gap in physical and mental health. The predominant framework guiding this research is the social stress paradigm. According to this model, gender differences in health and well-being reflect socially patterned differences in exposure and vulnerability to stress. In this chapter we examine the specific social-environmental stressors that women commonly encounter and the pathways through which these stressors affect women's health and well-being.

We begin with an overview of the social stress paradigm. We then turn to an examination of the key sources of stress in women's lives and the impact of these stressors on women's mental and physical health. Finally, we describe gaps in current knowledge about gender, stress, and health and suggest ways in which interdisciplinary research can help to identify the full range of psychosocial variables that are relevant to women's health and well-being.

In order to explore each of these issues in greater depth, we narrow the focus of this chapter in several ways. First, we concentrate almost exclusively on women. This approach should not suggest that men live stress-free lives. In fact, it is clear that certain types of stress and certain expressions of distress (e.g., excessive alcohol consumption) are more common among men than among women. However, given the greater psychological distress and physical morbidity of women, it is important to identify the specific types of stress that women face, the domains of women's lives in which they occur, and the pathways through which these sources of stress affect health. Second, we focus on sociological research on stress. The sociological model is unique in its emphasis on the power of the social environment to affect physical and mental health. For example, psychologists consider how characteristics of the individual (e.g., personality) affect health behavior and health whereas sociologists consider how characteristics of the social environment (e.g., poverty) affect health behavior and health. Sociological research emphasizes *group* differences in exposure and vulnerability to stress. For example, sociologists argue that men and women are exposed to different opportunities, constraints, and demands throughout their lives in ways that ultimately influence their health. The social environment of men and women differs in certain characteristic ways that we review in this chapter. Third, the sociological literature—and consequently, our review—is dominated

by a focus on the mental health consequences of stress, typically depression and psychological distress. Mental health outcomes are relevant to physical health because mood is a primary pathway through which stress affects physical health [1]. Where possible, however, we include research that directly examines the relationship between the specific sources of stress that women encounter and physical health outcomes.

II. The Social Stress Model

Stress is defined as a "state of arousal resulting either from the presence of socioenvironmental demands that tax the ordinary adaptive capacity of the individual or from the absence of the means to attain sought after ends [2, p. 16]." Epidemiologic research indicates that social groups differ in the extent to which they experience stress. These group differences in stress are a function of two key processes: (1) differences in exposure to stress and (2) differences in vulnerability to stress. Exposure to stress occurs when an individual encounters an objectively defined socioenvironmental demand—also referred to as a *stressor.* Most research on stress distinguishes between two categories of stressors that differ primarily in their intensity and duration: life events and chronic strains:

> Life events are acute changes which require major behavioral adjustments within a relatively short period of time (*e.g.,* birth of first child, divorce). Chronic strains are persistent or recurrent demands which require readjustments over prolonged periods of time (*e.g.,* disabling injury, poverty, marital problems) [3, p. 54].

Similar levels of stress exposure do not always have similar effects on health. Vulnerability to stress refers to the extent to which stress exposure results in the psychological and biological changes that undermine health and well-being. A range of personal and contextual factors are thought to influence individual and group differences in vulnerability to stress, primarily through their effect on stress appraisal. Appraisal refers to the subjective processes through which an individual judges an environmental demand as threatening or benign (*i.e.,* primary appraisal) and assesses his or her ability to effectively manage or cope with these demands (*i.e.,* secondary appraisal). Objectively defined stressors (*e.g.,* divorce, poverty, etc.) that are subjectively appraised as benign or that are perceived by the individual as manageable generally have little effect on health and well-being.

The appraisal process and, more generally, vulnerability to stress are influenced by personal and social resources that moderate the impact of stress on health. Thus, the magnitude of the effect of stress on health differs across individuals and social groups depending on the degree to which these psychosocial resources are present. The moderators of the stress-health relationship that have received the most attention are social support and personal control. Social support, defined as the "commitment, caring, advice, and aid provided in personal relationships [4, p. 1062]," may help individuals to perceive a stressor as manageable and reduce the probability that undesirable events and conditions will be appraised as stressful:

> Support may intervene between the potentially stressful event (or expectation of that event) and a stress reaction by attenuating or preventing a stress appraisal response. That is, the perception that others can and will provide necessary resources may redefine the potential for harm posed by a situation and/or bolster one's perceived ability to cope with imposed demands and hence may prevent a particular situation from being appraised as highly stressful. [5, p. 278]

Personal control, the belief that one's own intentions and behaviors can impose control over one's environment, is thought to have a similar effect on coping and appraisal processes by encouraging individuals to actively solve (or to perceive that they can solve) the problems that lead to and result from stress exposure [6]. The meaning or salience of the stressor to the individual also influences individual reactivity to stress. Stressors that occur in domains that are highly important have greater effects on well-being than those that are perceived as relatively trivial [7].

An individual's social environment influences both the magnitude and frequency of stressors that are encountered as well as the availability of personal and social resources that moderate the effect of stress on health. As a result, substantial variation exists across social groups in exposure and vulnerability to stress. For example, a study of the social distribution of stress indicates that relative to men, women report higher levels of exposure to chronic stressors and to some types of stressful life events [8]. Turner *et al.* [8] also find evidence of a sex difference in vulnerability to stress, as similar levels of stress exposure are shown to have a greater negative effect on women's mental health than on men's. Turner *et al.* [8] estimate that these differences in exposure and vulnerability account for 23% and 50% of the gender gap in depression, respectively. Given that women appear to have higher overall levels of exposure and vulnerability to stress, it is important to identify the specific dimensions of women's lives in which these stressors are likely to occur and to examine the specific psychosocial resources (or lack thereof) that might account for their greater vulnerability to stress.

III. Key Sources of Stress in Women's Lives: Exposure and Vulnerability

A. Personal Relationships

Personal relationships may be a source both of strain and of social support. Numerous studies identify the beneficial aspects of personal relationships for health and well-being, but a growing body of evidence indicates that the negative dimensions of personal relationships—those that involve demands, strains, and conflict—have greater effects on physical and mental health than do the positive components [9]. Moreover, some evidence

indicates that women may be particularly reactive to the stressors that they encounter in their personal relationships [9, but see 10]. We focus on three types of relationships that tend to expose women to stress: (1) marital and intimate relationships; (2) relationships with dependent children; and (3) relationships with parents that involve the provision of care.

1. Marital and Intimate Relationships

Research consistently shows that marriage is beneficial to mental and physical health, largely because it provides individuals with economic resources and social support, both of which can reduce exposure and vulnerability to stressors (see a review by Umberson and Williams [11]). However, the benefits of marriage are highly dependent on the quality of the marital relationship. Married individuals with poor marital quality have higher rates of psychological distress than do the unmarried. Moreover, marital quality is more important to the psychological well-being of women than it is to men [12].

Although the precise mechanisms through which marital quality affects psychological well-being have not been clearly established, evidence suggests that marital conflict and the relative lack of social support exchanged in problematic relationships partially accounts for the adverse effects of poor marital quality on well-being [9]. A wide range of factors contribute to marital conflict, undermine marital quality, and produce stress in marriage. Although a review of these factors is beyond the scope of this chapter, a few marital stressors stand out as particularly relevant to women's lives.

Most sociological research on the family suggests that marital strain experienced by women arises, in part, from the inequality and power differential inherent in traditional gender roles. One particular source of stress for women in marriage is the unequal distribution of household labor. Data indicate that, excluding childcare, married and cohabiting women spend almost twenty hours more per week on housework than their male partners [13]. Perceived inequity, however, is probably more important to the stress process than absolute differences in hours of housework performed. Among married couples, women are more likely than men to perceive the household distribution of labor as unfair [14]. Numerous studies indicate that the level of housework inequity perceived by the wife is positively related to marital strain and conflict [14,15]—variables that, in turn, predict depression and psychological well-being among women [16].

Domestic violence is a significant source of stress for women in marital and intimate relationships. Estimates indicate that between 20 and 30% of adult women will be physically abused by a male partner on at least one occasion [17]. Domestic violence poses a direct threat to the health and safety of female victims because it can result in serious physical injury or death (see Chapter 42). One study indicates that between 25 and 30% of female emergency room trauma cases are the result of domestic violence [18]. Perhaps more pervasive, however, are the more indirect effects of partner abuse on women's physical and mental health—outcomes that result from the high levels of stress that domestic violence creates in abused women's lives. Female victims of domestic violence have higher rates of stress-related mental health disorders including depression, substance abuse, and suicide [17].

The chronic strains associated with domestic violence occur in many domains. Acts of violence are themselves a source of stress. The fear of future violence is a chronic strain for many victims of partner abuse. Domestic violence may also deprive women of the social and psychological resources needed to effectively cope with the strains that abuse produces. For example, one study suggests that domestic violence undermines women's personal control [10a]. The loss of personal control is, in turn, associated with psychological distress and depression [4]. The feelings of helplessness associated with a reduced sense of personal control may, in turn, make it more difficult for women to leave an abusive relationship [19]. Social isolation also tends to result from repeated abuse, either because women withdraw to hide their abuse or because male perpetrators of domestic violence exercise control over other aspects of women's lives, often by limiting their partner's contact with supportive networks [20]. The resulting lack of social support may further erode women's ability to cope with the strains of domestic violence and prohibit their escape from the relationship.

Although most evidence suggests that many of the specific stressors encountered in marriage differ for men and women, it is unclear whether women are exposed to more numerous and more intense marital strains or if they are more vulnerable to the stressors to which marriage exposes them. One of the most consistent findings in research on marriage and health is that, on average, women benefit less from marriage than men in terms of both physical and mental health [4]. Theoretical work on gender differentials in health emphasize the greater demands and obligations that are inherent in women's social roles and statuses, particularly in marriage [21]. Some support for this theory is provided by empirical evidence showing that women report more problems in marriage than men [22]. Other research, however, indicates that men and women report similar levels of spousal strain, at least when strain is measured as the degree to which the spouse is critical or demanding [10].

The second explanation for married women's lower levels of well-being relative to married men is that women are more vulnerable or reactive to strains encountered in marriage, and this hypothesis has received mixed support. Theory suggests that, in general, women should be more reactive than men to strains encountered in personal relationships because women's socialization experiences lead them to place greater importance on relationships [23]. Supporting this possibility, the occurrence of negative life events among social network members has a greater negative effect on women's psychological well-being than on men's [24]. With respect to marital strains, per se, the evidence for gender differences in vulnerability is somewhat mixed. Several studies indicate that marital strain has a greater negative effect on psychological well-being for women than for men. This association has been demonstrated among subsamples of older adults [25], younger couples [9], and married parents [26]. In contrast to these findings, Umberson et al. [10] conclude from an analysis of national survey data that there are no gender differences in vulnerability to strain in the spousal relationship. Some of these discrepancies may be due to the way that strain is measured. All of the studies that found greater vulnerability among women examined marital strain (e.g., marital conflict and a lack of marital happiness), whereas Umberson et al. [10] focused on spousal strain (e.g., having a critical, demanding spouse).

The degree of emotional support received in marriage may also contribute to a gender difference in vulnerability to marital strain. Women receive less social support in marriage than men [10]. Because social support is generally associated with better physical and mental health, this lack of spousal support among women may partially contribute to an explanation of married women's higher rates of physical and psychological morbidity relative to men. Further, Horwitz et al. [9] find that the balance of positive and negative aspects of marriage is a stronger predictor of mental health than the negative dimensions alone and that an imbalance in positive and negative aspects is more strongly associated with well-being for women than for men.

2. Dependent Children

A large body of evidence indicates that the parenting of minor children is a source of stress for both married and single women. Most research focuses on the relationship between parental status and psychological well-being and assumes that the observed association is indicative of the presence of stress. Some studies, however, identify specific stressors that parents of dependent children tend to encounter. For example, the parenting-related strains to which women are particularly likely to be exposed occur primarily in two domains: (1) strains related to providing and arranging childcare and (2) financial strains. In each of these domains, women face disproportionate stress compared to men [26–29], largely because they assume the majority of childcare responsibilities and, thus, are exposed to more numerous and more intense parenting demands [29].

Perhaps the greatest parenting stressors that women face are in the area of childcare provision. Despite greater equality in gender roles, women still provide an overwhelming majority of childcare. Studies indicate that women perform between 76 and 81% of all childcare tasks [28,29].

The number and intensity of the stressors that women experience in parenting minor children are influenced by employment and marital status. One of the potentially stress-provoking responsibilities faced by employed mothers, both married and single, is that of making arrangements for childcare [28,29]. Arranging care for children can be a source of stress, not only because it requires substantial levels of time and effort, but also because it is often accompanied by worries about the cost and quality of care. Moreover, the stress that women experience in making childcare decisions is often compounded by the lack of assistance received from spouses and others. One estimate suggests that mothers assume approximately 90% of the responsibility for making childcare arrangements [30].

Employed women with dependent children may experience added strains from occupying both the worker and the parent role. Married women who are employed in the paid labor force shoulder most of the responsibility for childcare [31]. Most single women assume sole responsibility for both the care and financial support of their children [32]. In general, however, there is no clear consensus about how employment affects the stress process among mothers. Some evidence suggests that employment protects women from the negative effects of parenting stress [33]. Other studies indicate that combining the work and parent role results in role strain, which refers to stress associated with balancing demands from the often incompatible roles of employee and mother [27]. Overall, most evidence suggests that the benefits of employment for psychological well-being (e.g., social integration, personal control, fewer financial strains) outweigh the costs [16]. In Chapter 36 of this volume, Messing

provides a more comprehensive discussion of the effects of occupying multiple roles on women's health.

Financial and economic strains are a second source of stress
for many women with dependent children. Most mothers, regardless of employment or marital status, experience some level
of economic strain because, in general, children increase household expenditures. Single custodial mothers, however, are especially burdened. Estimates indicate that approximately 50%
of single mothers and their children live in poverty [32].

Although married mothers, regardless of their own employment status, typically have the security of a husband's income,
they may nonetheless be exposed to economic and financial
strain [29]. Even when women do not earn the majority of the
household income, they tend to be responsible for budgeting
and allocating this money in order to meet the increased financial demands associated with childrearing [29]. Moreover, the
financial demands that accompany parenting may undermine
marital satisfaction and create additional strain associated with
marital conflict [9], particularly for women [28].

The meaning or salience of the parental role also shapes vulnerability to parenting-related stressors [27]. The best evidence
indicates that the parenting role is more salient to women than
to men [27]. However, the greater importance of the parental
role to women does not appear to result in greater vulnerability
to parenting strains relative to men [27]. This does not mean
that role salience is unimportant to the stress process among
mothers of dependent children, only that it does not appear to
explain gender differences in outcomes. Overall, this research
suggests that women (and men) who view the parental role as
highly important to their identity are more negatively affected
by parental role strain [27].

Although we know a great deal about the parenting strains to
which women are exposed, we know very little about the moderating variables that might affect women's response to parenting stress. For example, we are aware of no studies that have
examined how the parenting of dependent children affects other
personal resources such as personal control. Moreover, while
numerous studies establish that the parenting of minor children
undermines women's psychological well-being, very little is
known about the effect of parenting strain on physical health.
Knowledge about the stress process among mothers of dependent children would be greatly improved by more research on
the mechanisms through which parenting stress affects both
physical and mental health and the psychosocial moderators of
this relationship. Moreover, as research on parental role salience
progresses, it would be useful to determine if certain subpopulations of women (perhaps differentiated by race, age, marital
status, or employment status) differ in the meaning that they assign to the parental role and their vulnerability to the parenting-
related stressors that they encounter.

3. Relationships with Parents and Caregiving

After spending many years providing care to minor children,
many women find themselves in yet another caregiving role:
providing assistance to elderly family members. About 55% of
women currently between the ages of 45 and 59 with one living
parent can expect to provide some level of care to a parent in
the next 25 years [34]. If mortality rates continue to decline,
that percentage may increase to as much as 74% [34]. Nu-

merous studies indicate that women, particularly daughters and
daughters-in-law, shoulder the bulk of the responsibility for providing care to elderly family members [35,36]. Caregiving is
associated with higher rates of depression and lower levels of
self-rated health [37]. Moreover, the negative effects of caregiving on mental health appear to be greater for women than for
men [38]. These effects are probably largely due to the stress
associated with providing care.

Caregivers are exposed to both primary and secondary stressors. Primary stressors are the frequent strains and daily hassles
associated with performing caregiving tasks. Secondary stressors are the chronic strains that arise in other areas of women's
lives as a result of caregiving demands [39,40].

The intensity of the primary stressors experienced by caregivers varies depending on the level of disability of the care recipient and on the type and frequency of care provided. Female
caregivers provide a wide range of care. However, the specific
tasks that are most often performed by women as compared to
men tend to be the most time consuming and the most stressful
[36]. These include the frequent, routine provision of daily care
and the regular performance of household tasks. The assistance
provided by male caregivers, in contrast, tends to be more sporadic and more centered on financial matters [36]. Most studies
find that the frequency and intensity of care is positively related
to caregiver strain [36]. Not surprisingly, women who provide
care to elderly family members report more strain than their
male counterparts [10].

The secondary stressors associated with providing care to an
elderly family member may pose the greatest threat to health
and well-being because they impinge on so many dimensions of
the caregiver's life. Particularly relevant to employed women
are the strains that caregiving creates at work. Among caregivers, women are more likely than men to miss work because of
their responsibilities to an elderly family member [41]. This is
not surprising given the evidence that, among female caregivers,
those employed in the paid labor force provide approximately
the same amount of care as their unemployed counterparts [36].
Furthermore, some women quit their jobs or reduce their work
hours in response to the competing demands of paid employment and caregiving [42]. These adjustments to work may, in
turn, create financial stressors and other strains caused by isolation from potentially supportive social networks.

The negative effect of the primary and secondary stressors
that women face are often compounded by a lack of support
from spouses and care recipients. Women who provide care to a
parent receive less spousal support than men in similar roles receive [43]. Moreover, daughters who provide care to a dependent
parent are less likely than sons to receive socioemotional support from the care recipient [35]. Starrels and colleagues [35]
surmise that this is due to normative expectations about who
should provide care. Because women are more frequently care
providers, those whom they help may be more likely to take the
assistance for granted.

B. Work and Financial Strain

1. Women's Paid Work

Most studies find that employed women have higher levels of
stress and distress than do employed men [44]. Although paid

work is generally beneficial to the health of both women and men, women appear to receive fewer benefits from paid employment [45]. Most attempts to explain the gender difference in well-being among the employed has traditionally examined the interaction of women's family roles with their work roles in producing stress and distress. This research, however, has not produced a consistent pattern of findings (see Chapter 36). Recent research indicates that it is not employment, per se, that is most important to women's health and well-being [16]. More relevant are the specific characteristics of the jobs that women tend to occupy. As a result, research has begun to move beyond the traditional focus on employment status and well-being to examine the nature and structure of women's paid work and the stressors that the work environment creates in women's lives.

Many of the stressors to which women are particularly likely to be exposed are rooted in the sex-segregation of the labor force. Women are overwhelmingly concentrated in service sector jobs [46], many of which include the provision of care to others (e.g., nurses, social workers, teachers, etc.) [47]. These jobs differ in a number of important ways from those most commonly occupied by men. Thus, on average, men and women are exposed to different workplace environments and different types of demands and strains [46]. Female-dominated jobs are characterized by low rates of pay [48], low levels of autonomy and control over work conditions [44,49,50], low levels of authority and power [51], low levels of complexity and high levels of routinization [46,50], and responsibility for providing care and support to others [52].

Many of these characteristics affect women's exposure and vulnerability to stress. For example, several studies indicate that women are disadvantaged relative to men in terms of job autonomy [44,49], primarily because their jobs are concentrated in lower levels of the organizational structure. Lack of autonomy is in turn positively associated with job strain and negatively associated with personal control and psychological well-being [49,53]. Very little is known about the effects of stress associated with low job autonomy on physical health.

High levels of routinization and lack of substantive complexity of work, both of which are common characteristics of women's jobs, are negatively associated with psychological well-being [50]. Further, low rates of pay can result in economic strain. Because the yearly earnings of women employed full-time are, on average, only about 71% of men's earnings [54], all other things being equal, employed women are more likely than their male counterparts to be exposed to financial and economic strain.

Despite the predominance of women in jobs involving the provision of care and support to others (e.g., nursing, social work, teaching), very little research has examined how various dimensions of these jobs affect the stress process for women. Bulan et al. [52] observed that service sector jobs in general often require women to sublimate their own needs to the emotional needs of others. Research on nursing and social work, in particular, indicates that these jobs expose women to many emotional demands and strains—factors that are, in turn, negatively associated with mental and physical health [47]. However, the costs to women in caregiving occupations, in terms of stress exposure, may in some circumstances be partially offset by rewards associated with helping others [47]. More research is needed to identify the social and psychological processes through which supporting and caring for others on the job affect women's exposure and vulnerability to stress and, ultimately, their physical and mental well-being.

2. Poverty and Single Motherhood

Among the stressors that disproportionately affect women, poverty is perhaps the most pervasive. Approximately 80% of all families living in poverty consist of women and their children [54]. The "feminization of poverty" results from sex segregation of the workforce, wage inequity, discrimination, divorce, nonmarital childbearing, and widowhood [48]. Of these, single parenthood, which typically occurs through divorce or nonmarital childbirth, poses the greatest threat to the economic well-being of women. Data indicate that approximately 46% of single custodial mothers live in poverty [55].

Living in poverty exposes all individuals, regardless of their sex or parental status, to a number of stressors. Studies examining the link between socioeconomic status and stress exposure indicate that individuals living in poverty are more likely to be exposed to chronic stressors such as economic strain, living in areas with high rates of crime, and residential mobility [32,56]. Despite the increasing feminization of poverty, we know very little about the specific stressors that are encountered by economically disadvantaged women or how women may be uniquely affected by the economic stressors that they encounter. We do know that women who live in poverty differ in a number of important ways from their male counterparts, particularly in the likelihood of being single custodial parents. Structural contingencies such as this affect both the total number and the intensity of stressors with which poor women must cope as well as the meaning of poverty. For example, women with dependent children who live in poverty face the additional stressors of balancing the demands of their roles as caregivers with their obligations as providers. Many women find that the high cost of childcare combined with a lack of flexibility and health care in available jobs make working and caring for young children incompatible [57,58]. As a result, only 40% of single mothers maintain full-time employment throughout the year [55]. Unemployment, in turn, can lead to strains associated with financial worries and social isolation.

Because the responsibilities of caring for dependent children often interfere with full-time employment, about half of all single mothers in the U.S. receive welfare benefits to help meet the financial needs of their families [59]. Relying on public assistance, however, is associated with an additional set of stressors. The income that single mothers receive from welfare often is not sufficient to cover their monthly expenses [58]. Of all single mothers receiving public assistance, welfare removes only about one-third from poverty [60]. This means that these women must often rely on supplemental forms of income from community groups, network members, and from nonreported part-time jobs [58]. Although very little research has directly examined the strains associated with an ongoing reliance on supplementary income, it is likely that the economic uncertainty inherent in this situation is a significant source of stress. Moreover, relying on family and other network members for financial support may produce strain in these close relationships, ultimately undermining social support and heightening vulnerability to the numerous stressors that disadvantaged women encounter. Despite the stressors to which single mothers who receive welfare benefits are exposed, it is unlikely that recent changes in the welfare system will improve their situation. Welfare reform legislation limits the amount of time most individuals can receive benefits to two years. As Edin and Lein [58] point out, after the two year limit, single mothers who lack job skills will have to return to work in low paying jobs—jobs that do not provide the financial resources needed to pay for childcare. Furthermore, because of the time spent working, women will have less time to seek out the supplemental forms of income that kept them afloat under the old system. Thus, under the new welfare system, single mothers may face more intense and more numerous economic strains.

Some evidence suggests that poor women may be particularly vulnerable to the negative effects of stress exposure. Studies of socioeconomic differentials in psychological well-being find that even when the number of stressors that women encounter is controlled, those of low socioeconomic status report significantly higher levels of psychological distress than their middle or upper class counterparts [61]. In other words, exposure to stress does not fully account for the class difference in psychological well-being, suggesting that women react differently to stressors depending on their social class.

Personal and social resources, such as social support and personal control, may explain part of the class difference in vulnerability to stress among women. For example, poor women appear to have smaller social networks and to receive less social support than the more economically advantaged [62]. Moreover, the social networks of those who live in poverty tend to consist of more demands and strains [62]. The poor are also disadvantaged in terms of personal control, but the effect of this disadvantage on vulnerability to stress is not clear. In a study of married women with children, Turner and Noh [61] found that personal control moderated the relationship between exposure to stressors and psychological well-being among middle class women, but not among lower class women. It is possible, however, that the relationship between socioeconomic status, stress, personal control, and well-being differ depending on marital and parental status. Future research should more fully examine how personal control and other psychosocial variables affect the stress reactivity of a wider range of economically disadvantaged women.

3. Divorce and Widowhood

Studies that examine the association of marital status with physical and mental health consistently find that, compared to the married, the divorced and widowed have higher levels of morbidity [63], a greater mortality risk [64], and higher levels of psychological distress and disorder [65,66]. Moreover, some evidence suggests that, compared to men, women are more negatively affected by divorce [67] (for an exception, see Booth and Amato [68]) and less negatively affected by widowhood [65], at least in terms of psychological well-being. Explanations of these associations generally point to the stressful nature of divorce and widowhood. These transitions are typically characterized as stressful life events—abrupt changes that entail substantial adjustments in numerous dimensions of individuals' lives. Although life events have received much attention in research on stress, theoretical work suggests that the chronic strains that life events provoke are more strongly related to well-being than are the events themselves [69]. Thus, attempts to explain health and well-being differentials between the married, divorced, and widowed often focus on the more enduring strains that these life transitions engender.

With respect to both divorce and widowhood, the primary chronic strains that women face arise from decreases in economic and financial resources [65,70]. For women, the decline in income associated with divorce is estimated to be approximately 27% on average [71], and is as high as 37% for women with children [72]. Studies of the effects of widowhood indicate that it is associated with approximately an 18% reduction in women's living standards [73]. The role of perceived and objective economic disadvantage in the stress process is well-documented, with many studies showing that socioeconomic disadvantage is associated with greater stress exposure, lower personal control, and more negative physical and mental health outcomes [6,8,56].

It is important to note that longitudinal evidence indicates that the negative impact of divorce on women's psychological well-being only persists for two to three years following divorce [68]. After this time, levels of psychological distress appear to return to predivorce levels. The best evidence indicates that the well-being of women who remarry following divorce does not improve significantly more than the well-being of those who remain single [63]. Both groups experience improvements in well-being over time. Similarly, the negative effects of widowhood on psychological well-being diminish over time, although not as quickly for women as for men [65]. The implications of these findings for the duration of financial strain after divorce and widowhood are unclear. It may be that many women recover from the initial drop in financial resources after divorce and widowhood by making adjustments to their employment and living circumstances. This is unlikely, however, at least for most divorced women with children, because the responsibilities of childcare often interfere with full-time work.

It is more likely that both divorced and widowed women experience a recovery in other resources that enable them to cope more effectively with the more enduring financial strain that accompanies the loss of a spouse. If the effects of divorce and widowhood on social support and personal control are relatively short-lived, then as time progresses these resources may enable women to effectively cope with any persistent financial strain. In fact, some evidence suggests that successfully coping with the strains of divorce and widowhood provides some women with a greater sense of self-confidence and control over their lives [65,74]. Research on the effects of divorce and widowhood on health over time is an important endeavor, but one that is still developing. Future studies should examine the duration of divorce- and widowhood-related stress reactions as well as the time-contingent effects of divorce and widowhood on potentially protective social and psychological resources.

C. Women and Aging

Women have a greater life expectancy than men and, thus, spend more time in old age [75]. Stressors that are generally associated with old age then accrue disproportionately to women. For example, women are more likely than men to be widowed, to live alone after the age of 65, to experience substantial declines in financial resources, and to suffer from chronic health problems and functional disabilities in later life [76]. Older women also encounter a number of stressful life events, including the illness and death of a spouse and of other close network members. Evidence that women are exposed to an increasing number of stressors at older ages is consistent with the findings of studies that examine gender differences in psychological well-being across the life course. For example, Mirowsky [76] finds that the gender gap in depression increases with age. This likely reflects, in part, the more stressful characteristics of older women's lives.

Financial strain is a particularly pervasive stressor among older women. Women comprise about 58% of the population over the age of 65, but almost 75% of the elderly poor [77]. Moreover, the median annual income for women over 65 was only 28% above the poverty line in 1990, resulting in almost one-third of elderly women being categorized as "poor" or "near poor" [78]. The risk of exposure to financial and economic stress is particularly high for the widowed and for other women who live alone. For example, Umberson and colleagues [65] report that financial strain is the primary variable accounting for widowed women's higher level of depression relative to that of married women. Although few studies examine how various combinations of stressors affect elderly women, it is likely that the social isolation associated with living alone makes widows particularly vulnerable to the financial strains that they encounter.

Health problems are another source of stress for elderly women. Although men and women at later stages of the life cycle develop similar chronic conditions, women's health problems tend to be less immediately life threatening. As a result, women spend more time coping with the stressors that accompany chronic illness [75] Moreover, most evidence suggests that older women have more functional limitations than men in a wide range of daily activities [79]. Indirect evidence that health problems provoke stress is provided by studies showing that functional disabilities and chronic illness are negatively associated with perceived health status and psychological well-being [79]. Although few studies directly examine the specific stressors associated with declining health, some evidence suggests that the loss of autonomy that accompanies illness and disability is responsible for the negative association between health problems and well-being [79].

Health problems and the lack of autonomy that they may produce also make older individuals more vulnerable to other stressors to which they are exposed. In a study examining the moderators of stress among those over 84 years of age, Roberts and colleagues [80] found that those with high levels of functional disability appeared to be more negatively affected by exposure to stressors than their more independent counterparts. Although more evidence is needed, this research highlights the importance of examining the interrelationships between various characteristics of elderly women's lives and how these characteristics combine to affect their risks of exposure and their vulnerability to stress.

In sum, elderly women are at an especially high risk of exposure to strains associated with poverty and economic disadvantage, social isolation, health problems, and functional disabilities. The negative consequences of these stressors on the health and well-being of the general population is well documented, but we know much less about their possibly differential impact across the life course. Furthermore, the consequences of the accumulation of numerous stressors in later life remain largely unexamined. It is possible that as stressors pile up, they have a multiplicative, rather than a simple additive, effect on women's health and well-being. Knowledge about the stress process among aging women in particular, and the elderly in general, would be greatly improved by more attention to these issues. We also need to know more about how the changes that women experience in later life affect various social and psychological resources such as personal control and social support, how the meaning or salience of various stressors changes across the life course, and whether these moderators have stronger or weaker effects on the health of older, compared to younger, women.

IV. Positive Features of the Social Environment

Although women encounter stress in many dimensions of their lives, some aspects of women's social environments have beneficial effects on their health and well-being. Women are particularly advantaged relative to men in terms of social support, especially with respect to the emotional support provided in close personal relationships. A great deal of evidence indicates that, relative to men, women have more intimate relationships, receive more support from these relationships, and are more likely to have a confidant [10,81]. Thus, although women may be exposed to more stressors in their everyday lives, they are also exposed to more social support.

What are the implications of social support for women's health and well-being? Social support provides a buffer that protects individuals from the negative health effects of stress exposure, largely by reducing the likelihood that undesirable events and life conditions will be appraised as stressful [5]. In general, there is little evidence that women are more strongly affected by social support than men. That is, when the quality and quantity of the social support received by men and women are similar, they appear to receive similar psychological benefits [10]. However, most evidence indicates that men and women do, in fact, differ in the quality and the quantity of social support they receive. If women did not have this advantage in social support, the gender gap in psychological well-being would be larger than it otherwise is:

> If women did not have the higher levels of support and integration that they have relative to men, they would exhibit even higher levels of depression relative to men than they currently do. Thus, women's greater involvement in certain types of relationships—especially the positive aspects of those relationships—protects them psychologically. [10]

Research on the physical health consequences of social support suggests that, although both men and women benefit from social involvement, men actually benefit more than women from similar levels of emotional support and social involvement [82].

In sum, it is important to recognize that the social environment can have both negative and positive effects on women's well-being. Social support may affect physical health in much the same way it affects mental health—via the appraisal process.

Social support also affects immune competence, which may benefit physical health, although this benefit may be greater for men than for women [82]. Social support may have positive effects on mood which, in turn, may minimize the negative immunological consequences of stress exposure (see Cohen [5] for a review of this research). Researchers should continue to examine the pathways that link social support to health and determine whether these effects operate similarly for men and for women.

V. The Measurement of Stress

The way that stress is measured and the specific outcomes that are examined have implications for observed gender differences in vulnerability to stress. Evidence indicates that men and women express emotional upset in different ways. For example, for women psychological distress appears to be expressed as depression and other affective disorders, whereas for men it appears to expressed as alcohol and substance use [83]. Because most sociological stress research only examines depression as an outcome, it is likely that observed gender differences in vulnerability to stress have been at least somewhat overstated. In a test of this hypothesis, Aneshensel and colleagues [83] demonstrated that women appeared to be more vulnerable than men to financial strain when affective and anxiety disorders were examined, whereas men were more negatively affected by stressful life events than women when alcohol and substance-use were examined. They concluded that "gender differences in the impact of stress are disorder-specific and do not indicate general differences between men and women in stress-reactivity [83, p. 176]." In sum, we cannot conclude that women are more vulnerable than men to certain stressors unless we assess the key ways in which both groups express distress. Moreover, gender differences in the impact of stress on physical health may also be outcome-specific. Researchers should continue to identify the specific physical and mental health consequences of stress both for men and for women and include these multiple measures in any comparative analyses.

VI. Directions for Research

Sociological research on women, stress, and health provides strong evidence that the demands, constraints, and opportunities that women encounter in their social environments have important implications for their exposure and vulnerability to stress and, ultimately, for their health. Despite the advances that sociological research has made in identifying the sources of stress in women's lives—including personal relationships, work environments, financial difficulties, and health problems in later life—several questions about the relationship between stress and health among women remain unanswered.

In general, we know very little about the impact that the specific stressors that women encounter have on their physical health or about the mechanisms through which exposure to these stressors leads to morbidity. Although sociological research indicates that differences in exposure and vulnerability to stress partially account for the gender gap in psychological well-being, it is unclear whether similar differences explain the gender gap in morbidity. According to the psychological stress model, stress can affect health by resulting in a negative emo-

tional state—such as depression or anxiety—which, in turn, causes hormonal changes that undermine immune functioning and increase susceptibility to disease [1]. A particularly promising area of research in health psychology examines how specific stressors affect the immune system and health outcomes of men and women. Research on marital conflict, for example, indicates that marital discord undermines the immune system of women more than men, perhaps because women are more aware of and, thus, are more greatly affected by negative spousal behavior or relationships [84]. Thus, it is clear that the stress process involves a complex interplay of social, psychological, and biological factors. More research is needed to determine how specific aspects of women's social environments affect their immunity and susceptibility to disease. Sociological research of the kind outlined in this chapter can inform psychological research by identifying the particular sources of stress in the social environment that might compromise women's immune systems. Psychologists, in turn, can inform sociological work by specifying the particular diseases that might result from stress exposure for women and the mechanisms through which these physical health outcomes occur.

According to the psychological stress model, health behavior is another mechanism through which stress affects physical health [1]. Health behaviors refer to specific actions taken by individuals that directly or indirectly promote health (e.g., getting adequate sleep and nutrition) or undermine health (e.g., cigarette smoking, taking risks). Exposure to stress may lead individuals to practice fewer health protective behaviors and more health compromising behaviors [1]. What remains to be studied, however, is whether stress exposure has different effects on men's and women's health behaviors, or on the health behaviors of women in differing social circumstances. Research suggests that the probability of engaging in positive health behaviors during times of stress is associated with the availability of social support [5] but, again, it is unclear whether this association varies across social groups in general or for men and women in particular. Those who are disadvantaged in terms of social support (e.g., men, socially isolated women, etc.) may be more vulnerable to the physical health consequences of stress exposure because stress may have a greater negative effect on their health behaviors. Interdisciplinary research should establish whether and how social support and health behaviors interact in times of stress to produce differential physical health outcomes for men and women.

Psychological research on cognitive appraisal of stress can also inform the sociological study of gender differences in the effect of stress on physical and mental health. Although sociological research indicates that women may be more vulnerable than men to the negative effects of some stressors, we know very little about how and why these differences arise. According to the psychological model, individual differences in response to a stressor are largely explained by differences in the way that stress is appraised. Gender differences in vulnerability to stress may be strongly linked to gender differences in the way that stress is subjectively appraised by men and women. Sociological research indicates that variables that affect the way that stress is appraised—personal control, social support, and the symbolic meaning of particular stressors—may be differentially distributed among men and women. Sociological research should examine more directly how these differences are produced. It is likely that the processes of stress appraisal are linked to the differing social environments in which men and women live.

In sum, we need more information on how the constellation of factors that affect the exposure and vulnerability of women to stress combine to affect their physical and psychological well-being. More information is also needed on how these factors come to arrive differentially in the lives of women. Interdisciplinary work offers great promise for future research on women, stress, and health.

References

1. Cohen, S., Kessler, R. C., and Gordon, L. U. (1995). Strategies for measuring stress in studies of psychiatric and physical disorders. *In* "Measuring Stress: A Guide for Health and Social Scientists", (S. Cohen, R. C. Kessler, and L. U. Gordon, eds.), pp. 3–26. Oxford University Press, New York.
2. Aneshensel, C. S. (1992). Social stress: Theory and research. *Annu. Rev. Sociol.* **18,** 15–38.
3. Thoits, P. (1995). Stress, coping, and social support processes: Where are we? What next? *J. Health Soc. Behav.,* Extra Issue); pp. 53–79.
4. Ross, C. E., Mirowsky, J., and Goldsteen, K. (1990). The impact of the family on health: The decade in review. *J. Marriage Fam.* **52,** 1059–1078.
5. Cohen, S. (1988). Psychosocial models of the role of social support in the etiology of physical disease. *Health Psychol.* **7,** 269–297.
6. Mirowsky, J., and Ross, C. E. (1986). Social patterns of distress. *Annu. Rev. Sociol.* **12,** 23–45.
7. Thoits, P. (1994). Identity-relevant events and psychological symptoms: A cautionary tale. *J. Health Soc. Behav.* **36,** 72–82.
8. Turner, R. J., Wheaton, B., and Lloyd, D. A. (1995). The epidemiology of social stress. *Am. Sociol. Rev.* **60,** 104–125.
9. Horwitz, A. V., McLaughlin, J., and White, H. R. (1998). How the negative and positive aspects of partner relationships affect the mental health of young married people. *J. Health Soc. Behav.* **39,** 124–136.
10. Umberson, D., Chen, M. D., House, J. S., Hopkins, K., and Slaten, E. (1996). The effect of social relationships on psychological well-being: Are men and women really so different? *Am. Sociol. Rev.* **61,** 836–856.
10a. Umberson, D., Anderson, K., Glick, J., and Shapiro, A. (1998). Domestic violence, personal control, and gender. *J. Marriage Fam.* **60,** 442–452.
11. Umberson, D., and Williams, K. (1999). Family status and mental health. *In* "Handbook on the Sociology of Mental Health" (C. S. Aneshensel and J. Phelan, eds.). Plenum, New York.
12. Gove, W. R., Hughes, M., and Style, C. B. (1983). Does marriage have positive effects on the psychological well-being of the individual? *J. Health Soc. Behav.* **24,** 122–131.
13. Blair, S. L., and Lichter, D. T. (1991). Measuring the division of household labor: Gender segregation of housework among American couples. *J. Fam. Issues* **12,** 91–113.
14. Suitor, J. J. (1991). Marital quality and satisfaction with the division of household labor across the family life cycle. *J. Marriage Fam.* **53,** 221–230.
15. Perry-Jenkins, M., and Folk, K. F. (1994). Class, couples, and conflict: Effects of the division of labor on assessments of marriage in dual-earner couples. *J. Marriage Fam.* **56,** 165–180.
16. Aneshensel, C. S. (1986). Marital and employment role-strain, social support, and depression among adult women. In "Stress, Social Support, and Women" (S. E. Hobfoll, ed.), pp. 99–114. Hemisphere Publishing, New York.

17. Stark, E., and Flitcraft, A. H. (1991). Spouse abuse. *In* "Violence in America: A Public Health Approach" (M. L. Rosenberg and M. A. Fenley, eds., pp. 123–157. Oxford University Press, New York.
18. McCleer, S. V., and Anwar, R. (1989). A study of women presenting in an emergency department. *Am. J. Public Health* **79,** 65–67.
19. Walker, L. (1984). "The Battered Woman's Syndrome" Springer, New York.
20. Stark, E. (1992). Framing and reframing battered women. *In* "Domestic Violence: The Criminal Justice Response" (E. Buzawa, ed.), pp. 271–292. Auburn House, Westport, CT.
21. Gove, W. R., and Tudor, J. F. (1973). Adult sex roles and mental illness. *Am. J. Sociol.* **78,** 812–835.
22. Amato, P. R., and Rogers, S. J. (1997). A longitudinal study of marital problems and subsequent divorce. *J. Marriage Fam.* **59,** 612–624.
23. Gilligan, C. (1982). "In a Different Voice: Psychological Theory and Women's Development." Harvard University Press, Cambridge, MA.
24. Kessler, R. C., and McLeod, J. D. (1984). Sex differences in vulnerability to undesirable life events. *Am. Sociol. Rev.* **49,** 620–631.
25. Turner, H. (1994). Gender and social support: Taking the bad with the good. *Sex Roles* **30,** 521–541.
26. Simon, R. (1998). Assessing sex differences in vulnerability among employed parents: The importance of marital status. *J. Health Soc. Behav.* **39,** 38–54.
27. Simon, R. (1992). Parental role strains, salience of parental identity and gender differences in psychological distress. *J. Health Soc. Behav.* **33,** 25–35.
28. Bird, C. E. (1997). Gender differences in the social and economic burdens of parenting and psychological distress. *J. Marriage Fam.* **59,** 809–823.
29. Ross, C. E., and Van Willigen, M. (1996). Gender, parenthood, and anger. *J. Marriage Fam.* **58,** 572–584.
30. Lamb, M. E. (1987). "The Father's Role: Cross-Cultural Perspectives." Erlbaum, Hillsdale, NJ.
31. Nock, S. L., and Kingston, P. W. (1988). Time with children: The impact of couple's work-time commitments. *Soc. Forces* **67,** 59–85.
32. Garfinkel, I., and McLanahan, S. S. (1986). "Single Mothers and Their Children: A New American Dilemma." Urban Institute Press, Washington, D C.
33. Wethington, E., and Kessler, R. C. (1989). Employment, parental responsibility, and psychological distress. *J. Fam. Issues* **10,** 527–546.
34. Himes, C. (1994). Parental caregiving by adult women: A demographic perspective. *Res. Aging* **16,** 191–211.
35. Starrels, M. E., Ingersoll-Dayton, B., Dowler, D. W., and Neal, M. B. (1997). The stress of caring for a parent: Effects of the elder's impairment on an employed adult child. *J. Marriage Fam.* **59,** 860–872.
36. Walker, A. J., Pratt, C. C., and Eddy, L. (1995). Informal caregiving to aging family members. *Fam. Relat.* **44,** 402–411.
37. Schultz, R., Visintainer, P., and Williamson, G. M. (1990). Psychiatric and physical morbidity effects of caregiving. *J. Gerontol.: Psychol. Sci.* **45,** 181–91.
38. Tennstedt, S., Cafferata, G. L., and Sullivan, L. (1992). Depression among caregivers of impaired elders. *J. Aging Health* **4,** 58–76.
39. Pearlin, L. I., Mullan, J. T., Semple, S. J., and Skaff, M. M. (1990). Caregiving and the stress process. *In* "Stress and Health: Issues in Research Methodology" (S. V. Kasl and C. L. Cooper, eds.), pp. 143–165. Wiley, New York.
40. Aneshensel, C. S., Pearlin, L. I., and Schuler, R. H. (1993). Stress, role captivity, and the cessation of caregiving. *J. Health Soc. Behav.* **34,** 54–70.
41. Anastas, J. W., Gibeau, J. L., and Larson, P. J. (1990). Working families and eldercare: A national perspective in an aging America. *Soc. Work* **35,** 405–411.

42. Stone, R. I., and Short, P. F. (1990). The competing demands of employment and informal caregiving to disabled elders. *Med. Care* **28,** 513–526.

43. Horowitz, A. (1985). Sons and daughters as caregivers to older parents: Differences in role performance and consequences. *Gerontologist* **25,** 612–617.

44. Roxburgh, S. (1996). Gender differences in work and well-being: Effects of exposure and vulnerability. *J. Health Soc. Behav.* **37,** 265–277.

45. Aneshensel, C. S., Frerichs, R. R., and Clark, V. A. (1981). Family roles and sex differences in depression. *J. Health Soc. Behav.* **22,** 379–393.

46. Lennon, M. C. (1987). Sex differences in distress: The impact of gender and work roles. *J. Health Soc. Behav.* **28,** 290–305.

47. Marshall, N. L., Barnett, R. C., Baruch, G. K., and Pleck, J. H. (1991). More than a job: Women and stress in caregiving occupations. *Curr. Res. Occup. Prof.* **6,** 61–81.

48. Starrels, M. E., Bould, S., and Nichols L. J. (1994). The feminization of poverty in the United States: Gender, race, ethnicity, and family factors. *J. Fam. Issues* **15,** 590–607.

49. Ross, C. E., and Mirowsky, J. (1992). Households, employment, and the sense of control. *Soc. Psychol. Q.* **55,** 217–235.

50. Pugliesi, K. (1995). Work and well-being: Gender differences in the psychological consequences of employment. *J. Health Soc. Behav.* **36,** 57–71.

51. Wright, E. O., Baxter, J., and Birkelund, G. E. (1995). The gender gap in workplace authority: A cross-national study. *Am. Sociol. Rev.* **60,** 407–435.

52. Bulan, H. F., Erickson, R. J., and Wharton, A. S. (1997). Doing for others on the job: The affective requirements of service work, gender, and emotional well-being. *Soc. Prob.* **44,** 235–256.

53. Lennon, M. C., and Rosenfield, S. (1992). Women and mental health: The interaction of job and family conditions. *J. Health Soc. Behav.* **33,** 316–327.

54. U.S. Bureau of the Census (1993). "Money Income of Households, Families, and Persons in the United States: 1992," Curr. Popul. Rep. Ser. P60, No. 184. U.S. Department of Commerce, U.S. Gov. Printing Office, Washington, DC.

55. U.S. Bureau of the Census (1995). "Income, Poverty, and Valuation of Noncash Benefits: 1993," Curr. Popula. Rep., Ser. P60, No. 188. U.S. Govt. Printing Office, Washington, DC.

56. Williams, D. R. (1990). Socioeconomic differential in health: A review and redirection. *Soc. Psychol. Q.* **53,** 81–99.

57. Harris, K. M. (1996). Life after welfare: Women, work, and repeat dependency. *Am. Sociol. Rev.* **61,** 407–426.

58. Edin, K., and Lein, L. (1997). Work, welfare, and single mothers' economic survival strategies. *Am. Sociol. Rev.* **62,** 253–266.

59. Rodgers, H. R., Jr. (1996). "Poor Women, Poor Children: American Poverty in the 1990s." M. E. Sharpe, Armonk, NY.

60. Danziger, S. H., Sandefur, G. D., and Weinberg, D. H. (1994). "Confronting Poverty: Prescriptions for Change." Harvard University Press, Cambridge, MA.

61. Turner, R. J., and Noh, S. (1983). Class and psychological vulnerability among women: The significance of social support and personal control. *J. Health Soc. Behav.* **24,** 2–15.

62. Belle, D. (1982). "Lives in Stress: Women and Depression." Sage, Beverly Hills, CA.

63. Wyke, S., and Ford, G. (1992). Competing explanations for associations between marital status and health. *Soc. Sci. Med.* **34,** 523–532.

64. Zick, C. D., and Smith, K. R. (1991). Marital transitions, poverty, and gender differences in mortality. *J. Marriage Fam.* **53,** 327–336.

65. Umberson, D., Wortman, C. B., and Kessler, R. C. (1992). Widowhood and depression: Explaining long-term gender differences in vulnerability. *J. Health Soc. Behav.* **33,** 10–24.

66. Williams, D., Takeuchi, D. T., and Adair, R. K. (1992). Marital status and psychiatric disorders among blacks and whites. *J. Health Soc. Behav.* **33,** 140–157.

67. Aseltine, R. H., Jr., and Kessler, R. C. (1993). Marital disruption and depression in a community sample. *J. Health Soc. Behav.* **34,** 237–251.

68. Booth, A., and Amato, P. R. (1991). Divorce and psychological stress. *J. Health Soc. Behav.* **32,** 396–407.

69. Pearlin, L. (1989). The sociological study of stress. *J. Health Soc. Behav.* **30,** 241–256.

70. Shapiro, A. D. (1996). Explaining psychological distress in a sample of remarried and divorced persons: The influence of economic distress. *J. Fam. Issues* **17,** 186–203.

71. Peterson, R. R. (1996). A re-evaluation of the economic consequences of divorce. *Am. Sociol. Rev.* **61,** 528–536.

72. U.S. Bureau of the Census (1991). "Survey of Income and Program Participation," Curr. Popul. Rep., Ser. P70, No. 23. U.S. Gov. Printing Office, Washington, DC.

73. Bound, J., Duncan, G. J., Laren, D. S., and Oleinick, L. (1991). Poverty dynamics in widowhood. *J. Gerontol.: Soc. Sci.* **46,** S115–S124.

74. Riessman, C. K. (1990). "Divorce Talk: Men and Women Make Sense of Personal Relationships." Rutgers University Press, New Brunswick, NJ.

75. Verbrugge, L. M. (1989). Gender and health: An update on hypotheses and evidence. *J. Health Soc. Behav.* **26,** 156–182.

76. Mirowsky, J. (1996). Age and the gender gap in depression. *J. Health Soc. Behav.* **37,** 362–380.

77. U.S. Bureau of the Census (1991). "Poverty in the United States: 1990," Curr. Popul. Rep., Ser. P60, No. 175. U.S. Gov. Printing Office, Washington DC.

78. Malveaux, J. (1993). Race, poverty, and women's aging. *In* "Women on the Front Lines: Meeting the Challenges of an Aging America" (J. Allen and A. Pifer, eds.), pp. 167–190. Urban Institute Press, Washington, DC.

79. Penning, M. J., and Strain, L. A. (1994). Gender differences in disability, assistance, and subjective well-being in later life. *J. Gerontol.: Soc. Sci.* **49,** S202–S208.

80. Roberts, B., Dunkle, R., and Haug, M. (1994). Physical, psychological, and social resources as moderators of the relationship of stress to mental health of the very old. *J. Gerontol.: Soc. Sci.* **49,** S35–S43.

81. Turner, R. J., and Marino, F. (1994). Social support and social structure: A descriptive epidemiology. *J. Health Soc. Behav.* **35,** 193–212.

82. Seeman, T. E., and McEwan, B. S. (1996). Impact of social environment characteristics on neuroendocrine regulation. *Psychosom. Med.* **58,** 459–471.

83. Aneshensel, C. S., Rutter, C. M., and Lachenbruch, P. A. (1991). Social structure, stress, and mental health: Competing conceptual and analytic models. *Am. Sociol. Rev.* **56,** 166–178.

84. Kiecolt-Glaser, J. K., Malarkey, W. B., Chee, M. A., Newton, T., Capioppo, J. T., Mao, H.-Y., and Glaser, R. (1993). Negative behavior during marital conflict is associated with immunological down-regulation. *Psychosom. Med.* **55,** 395–409.

45

Cigarette Smoking: Trends, Determinants, and Health Effects

CORINNE G. HUSTEN AND ANN MARIE MALARCHER
Office on Smoking and Health
Centers for Disease Control and Prevention
Atlanta, Georgia

I. Introduction

Information on women using tobacco in the United States before 1935 is anecdotal. Women and girls in colonial times were known to smoke pipes, and some women in the 1800s used snuff [1]. Cigarette smoking among women began to increase in the 1920s, when social and cultural changes reflecting the liberalization of norms about women's roles and behavior led to increasing acceptance of tobacco use by women [2,3]. The prevalence of cigarette smoking among women (females aged ≥ 18 years) has been estimated at 6% in 1924, 16% in 1929 [3], 20% in 1935, 26% in 1940 [4], and 33% in 1949 [5]. Cigarette smoking among women peaked in the 1960s [6] and subsequently declined; smoking among girls (females aged <18 years), however, increased dramatically in the 1990s [7]. This chapter focuses on the trends in tobacco use among women and girls, the determinants of cigarette smoking among women and girls, and the health implications of tobacco use for women.

II. Trends in Tobacco Use among Women and Girls

A. *Trends in Cigarette Smoking*

1. Trends in Cigarette Smoking among Women

Cigarette smoking is now the leading preventable cause of death among women in America, killing more than 152,000 women each year, primarily from cardiovascular disease, cancer, and respiratory diseases [8]. Lung cancer now exceeds breast cancer as the leading cause of cancer deaths among women [9]. In 1965, 42% of women in the United States had ever smoked at least 100 cigarettes in their lifetime; this percentage increased to 46% in 1985 [10], then declined to 42% in 1995 [11]. The prevalence of current smoking (defined from 1965 to 1992 as ever smokers who reported that they "smoke now" and defined from 1992 to 1995 as ever smokers who reported that they "now smoke every day or some days") among women was 34% in 1965 [10], 30% in 1983, and 23% in 1995 [12]; most of this decline occurred from 1983 to 1990 [10] (Fig. 45.1). In 1995, 22 million women smoked [12], including 14 million women of reproductive age [11]. Smoking prevalence is highest among women 25–44 years of age [12]. Among racial and ethnic groups, American Indian and Alaska Native women had the highest smoking prevalence (35%) and Hispanic (15%) and Asian and Pacific Islander women (4%) had the lowest smoking prevalence in 1995. In 1995, women with 9–11 years of education were the most likely to smoke (34%) and women with 16 or more years of education were the least likely to smoke (14%) [12]. For women, trends in smoking are

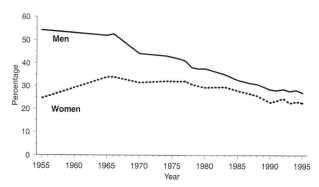

Fig. 45.1 Percentage of adults aged ≥ 18 years who are current cigarette smokers, by Sex, United States, 1955–1995. Sources: Bureau of the Census, Current Population Survey, 1955, unpublished data; National Center of Health Statistics, National Health Interview Surveys, 1965–1995, unpublished data. Estimates since 1992 incorporate someday smoking.

most strongly correlated with education; income and occupation are more weakly correlated [13].

Smoking during pregnancy has declined from 20% in 1989 to 14% in 1995 [14,15]. However, these data are based on information on the birth certificate, and pregnant women may underreport smoking to their clinicians during pregnancy [16]. Other data from surveys suggest that 22% of pregnant women smoke [17]. Smoking during pregnancy declined for all racial and ethnic groups from 1989 to 1995, and the number of cigarettes smoked per day by pregnant women has declined over the same period [14,15].

The prevalence of current smoking has decreased more dramatically from 1965 to 1995 among men than among women, which has sometimes been interpreted to mean that women have been less responsive than men to societal changes that have occurred since the 1964 Surgeon General's Report outlining the dangers of smoking. However, women, as a group, were at an earlier stage than men of smoking uptake when the antismoking campaign began after the 1964 Surgeon General's Report. In the absence of the campaign, smoking prevalence among women would have continued to rise dramatically instead of plateauing and decreasing [18]. Since 1983, smoking prevalence has decreased at a comparable rate for women and men (Fig. 45.1) [19,20].

2. Trends in Cigarette Smoking among Girls

The prevalence of ever smoking (defined as lifetime use) among twelfth grade girls declined from 75% in 1977 [21] to

62% in 1992 [22]; it then increased to 65% in 1997 [7]. Older girls are more likely than younger girls to have ever smoked [7,23]. Gender differences in ever smoking are small [23].

Current smoking (defined as smoking in the last 30 days) among twelfth grade girls declined from 40% in 1977 to 26% in 1992 but then increased to 35% in 1997. Smoking among eighth grade girls increased from 14% in 1991 to 19% in 1997, and smoking among tenth grade girls increased from 21% in 1991 to 31% in 1997 (Fig. 45.2) [7]. Among African-American girls, smoking prevalence declined dramatically from 1976 to 1992 then increased slightly by 1997 [19,20,24]. Smoking prevalence was higher for girls than boys until the late 1980s; since then gender differences in smoking have been small [19,20].

B. Trends in Frequent and Heavy Cigarette Smoking

1. Heavy Smoking among Women

Number of cigarettes smoked per day is directly associated with addiction and adverse health outcomes; it is inversely associated with quitting [25–28]. It is estimated that 6–7% of adult women smoked 21 or more cigarettes per day in 1955 and that 10–13% smoked this heavily in 1966 [29]. The proportion of women who smoked 25 or more cigarettes per day was 19% in 1974, 23% in 1980, and 16% in 1991 [10]. White women are more likely to be heavy smokers than black or Hispanic women, and highly educated women are the least likely to be heavy smokers [17]. Women are less likely than men to be heavy smokers [10].

2. Frequent or Heavy Smoking among Girls

Frequent cigarette use (use on 20 or more days in the past 30 days) was reported by 16% of high school girls [30]. White girls (20%) were more likely than African-American (4%) or Hispanic (8%) girls to be frequent smokers. The only gender difference noted was among African-Americans: African-American girls were less likely than African-American boys to be frequent smokers. The prevalence of frequent smoking increased with age. The prevalence of heavy smoking (a half pack per day or more) among twelfth grade girls for the combined years 1985–1989 was an estimated 23% for Native Americans, 13% for whites, 4% for Asians, 2–4% for Hispanics, and 2% for African-Americans [31].

C. Cigarette Smoking Initiation

People who start smoking at a young age are more likely to smoke heavily [32], more likely to become dependent on nicotine [33], and less likely to quit smoking [34]; thus they are at increased risk for smoking-attributable illness or death [35]. The mean age of initiation of regular smoking has decreased over time; it was 35 years for women born between 1900 and 1910, 20 years for women born between 1931 and 1940, and less than 18 years for women born between 1951 and 1960. Women initiated regular smoking at an older age than men for all of these cohorts, but the gender gap has narrowed considerably over time [3]. In 1995, among people who had ever tried a cigarette, 82% of both men and women tried their first cigarette before age 18 [36].

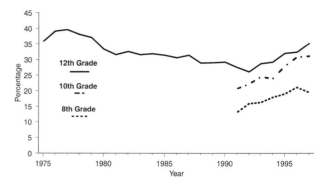

Fig. 45.2 Current cigarette smoking (smoking one or more cigarettes per day during the previous 30 days) among girls by grade in school, United States, 1975–1997. Source: [7].

D. Cigarette Smoking Cessation

1. Quitting Behavior among Women

A commonly used measure of quitting is the percentage of ever smokers who have quit. In 1955, 11% of women smokers had quit [37]; in 1965, 19% had quit [10]; and in 1994, 46% had quit [20]. The proportion of ever smokers who have quit increases with increasing age [10,26] because as ever smokers age, a greater percentage quit smoking [28,38,39], and because continuing smokers are more likely than former smokers to die [35,39]. The percentage of ever smokers who have quit is higher for white women than for African-American women. Women with 16 or more years of education are the most likely to have quit smoking; women with 12 or fewer years of education are the least likely to have quit [12,20].

In 1955, the percentage of ever smokers who had quit was 11% for both women and men [37]. However, over the next decade, a large gender difference developed: by 1965, the percentage of ever smokers who had quit was 19% for women and 28% for men [10]. The gender gap was greatest from 1974 to 1978 (11 percentage points) and has subsequently narrowed (5 percentage points in 1994) (Fig. 45.3) [10,20].

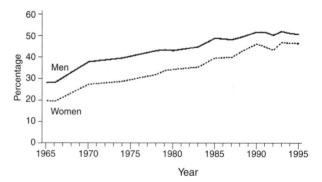

Fig. 45.3 Percentage of ever smokers (ever smoked ≥ 100 cigarettes) who have quit, adults aged ≥ 18 years, by sex, United States, 1965–1995. This is also known as the quit ratio; estimates since 1992 incorporate some-day smoking. Source: National Center for Health Statistics, National Health Interview Surveys, 1965–1995, unpublished data.

This gender difference has sometimes been interpreted to mean that women are less likely to quit smoking than men, but it is actually a product of several factors. First, this measure of quitting does not take into account other tobacco use. Men are more likely than women to switch to, or to continue to use, other tobacco products after stopping smoking cigarettes. If adjustment is made for other tobacco use, the gender gap is reduced by 75% [35,40]. Second, the gender difference in quitting also reflects gender differences in the historical pattern of smoking; just as men began smoking in greater numbers before women did, men started quitting in greater numbers before women did. Thus, smoking prevalence peaked in the 1950s for men but not until the 1960s for women [19]. Because the percentage of ever smokers who have quit is a cumulative measure across cohorts of smokers, the percentage has been greater for men than for women because the early cohorts of quitters were primarily men [19,20,39]. The change in the percentage of ever smokers who have quit since 1985 is at least as great among women as men [10].

Other data support the premise that overall gender differences in quitting reflect gender differences in quitting behavior from 1950 to 1980. Women were less likely than men to quit in the 1950s and 1960s [25,27]. By the early 1980s, however, the probability of attempting to quit and succeeding were comparable for women and men [3,39,41,42]. Recent data on quitting also show comparable attempts to quit and quitting success for women and men [38].

2. Quitting Behavior among Pregnant Women

In 1990, 23% of pregnant women quit after learning of their pregnancy; the percentage of women who quit during pregnancy increased as level of education increased [43]. Pregnant women generally quit because of concerns about adverse fetal outcomes [44]. Unfortunately, many women appear to consider cessation during pregnancy as a temporary abstinence. Up to 75% of mothers resume smoking within six months after delivery [43,45]. Age, race, marital status, and education do not appear to predict postpartum relapse [44,45].

3. Quitting Behavior among Girls

Most girls say they want to quit smoking [46,47]. More than half of all adolescent smokers try to quit each year [46–50]. Quitting does not appear to be any easier for children or adolescents than for adults; they report the same problems quitting that adults do [49,50]. Among adolescents who try to quit smoking, 65% relapse within one month and 82% within six months [50,51]. Only 14% of high school girls who ever smoked regularly no longer smoke [52]. Girls and boys are equally likely to fail in their attempts to quit [47,50,53].

E. Use of Other Tobacco Products

1. Cigars

Cigar smoking increases the risk for laryngeal, oral and pharyngeal, esophageal, pancreatic, and lung cancer; coronary heart disease; aortic aneurysm; and chronic obstructive pulmonary disease (COPD) [54,55]. In 1964, 0.2% of women smoked cigars [56]. Until recently, cigar use among women remained low

[10], but reports suggest that cigar smoking has increased among women [54].

More information is available regarding cigar use among adolescents. In 1996, 16% (1.7 million) of adolescent girls had smoked a cigar in the previous year [57] and in 1997, 11% of high school girls had smoked a cigar in the past month [30]. Girls are less likely than boys to smoke cigars [30,57].

2. Smokeless Tobacco

Smokeless tobacco use is causally associated with oral leukoplakia and oral cancer [58]. It also increases the risk of tooth loss, nicotinic stomatitis, periodontitis, and gingival recession [59]. In 1964, 2% of women used snuff and 0.4% used chewing tobacco [56]. In 1985, 3% of adult women reported ever using smokeless tobacco, and 1% reported using it the past year [60]. Smokeless tobacco use among women has remained low (<1%) [10]. Smokeless tobacco use is more prevalent among older than younger women [60,61]. African-American women are more likely than white women to use smokeless tobacco [10,60,61], particularly older African-American women [60]. Smokeless tobacco use is also higher among American Indians and Alaska Native women [10] and among women who are older, have less than 12 years of education, and have a low income [62]. In all racial and ethnic groups except African-Americans, women are less likely than men to use smokeless tobacco [10,63].

In 1995, 3% of eighth grade girls, 2% of tenth grade girls, and 2% of twelfth grade girls had used smokeless tobacco in the previous month [64]. In 1997, 2% of girls in grades 9–12 had used smokeless tobacco in the previous month; white, African-American, and Hispanic girls were equally likely to have used smokeless tobacco [30]. Smokeless tobacco use is higher among girls in the Southeast and in rural Alaska [65,66] and among American Indian or Alaska Native girls [66,67]. Girls are less likely than boys to use smokeless tobacco [23,30].

3. Pipes

Pipe use increases the risk for laryngeal, oral, esophageal, and lung cancer [55]. In 1964, 0.3% of adult women smoked pipes [56], and in 1991, 0.03% of women in the United States smoked pipes [68]. Women are less likely than men to smoke pipes.

III. Determinants of Cigarette Smoking among Women and Girls

A. Introduction

The discussion in this section is limited to cigarette smoking because there are few studies of the determinants of other types of tobacco use among women and girls [23,54,62,65–67]. The determinants of smoking among women are complex and include risk factors for initiation among adolescent girls, the physiologic effects of nicotine, and the influence of behavioral and psychological factors on cessation. Initiation of tobacco use almost always occurs during adolescence, and the prevalence of tobacco use among girls has increased [23,30]. Cigarette smoking clearly provides functional value for many girls undergoing developmental transition to adulthood. Reviews of gender differences in risk factors of adolescent smoking have concluded that some determinants have a similar impact on smoking among both girls and boys, whereas other factors seem especially predictive among girls [69,70]. Although more prospective research is needed to identify risk factors for smoking initiation among girls, several sociodemographic, personal, behavioral, and environmental determinants of cigarette smoking among girls have been identified.

B. Race, Ethnicity, and Socioeconomic Status

Large racial, ethnic, and socioeconomic differences in smoking prevalence exist among women. Currently, American Indians and Alaska Natives have the highest prevalence of cigarette smoking among women in the United States, followed by white, African-American, Hispanic, and Asian and Pacific Islander women (see section II.A.1) [12,63]. However, among girls in grades 9–12, the prevalence of current smoking is higher for whites than for Hispanics and is lowest for blacks [30]. Among high school seniors, from 1990 through 1994, smoking in the past month was highest for American Indian and Alaska Native girls (39.4%) and white girls (33.1%), intermediate for Hispanic girls (19.2%), and lowest for Asian American and Pacific Islander girls (13.8%) and African-American girls (8.6%) [63]. Few studies have investigated the racial and ethnic differences in the value and determinants of smoking among girls. Risk factors may be more prevalent (*e.g.,* weight concerns among white girls) or more predictive in some racial and ethnic groups than in others [63].

Current smoking prevalence is highest among women with 9–11 years of education and lowest among women with at least 16 years of education and is higher among women living below the poverty level than among those living at or above the poverty level [12]. The inverse relationship between education and smoking reflects, in part, a more rapid decline in smoking prevalence among higher educated groups from the mid-1970s through the 1980s [71]. Several theories have been suggested to explain this gradient, including that persons with more education (and hence greater longevity) may be more aware and accepting of information about behaviors that affect mortality later in life [71]. Women with higher education may also have more access to and understanding of prevention messages [71]. A inverse relationship also exists within socioeconomic groups; this relationship was observed by Graham [72] among women with young children in households where the head of household was in a manual occupation or unemployed. Among this population, smokers were concentrated among those with heavier caring responsibilities (*i.e.,* caring for more children, for children with poorer health, and for other family members) and among the most materially disadvantaged households (*i.e.,* receiving government assistance and not being financially able to meet basic necessities) [72].

Socioeconomic status and household structure are also determinants of adolescent smoking [23]. After controlling for age, sex, race and ethnicity, and school enrollment status, smoking among 12–17 year olds was observed to decrease as years of education completed by the responsible adult and family income increased [73]. Data from the combined 1991–1995 Monitoring the Future Studies indicate that high school senior girls who lived alone had a high rate of current smoking (47.1%)

compared with girls who lived with both parents (28.2%) [64]. Data from another national survey indicate that girls and boys aged 12–17 years who lived in a family structure other than a two-biologic-parent family were more likely than those who lived with both biological parents to have smoked in the past year [74]. Youth from low-socioeconomic-status families may experience more stressful events, lack the resources to deal effectively with stressors, and have more role models and friends who smoke [23,75].

C. Determinants of Smoking among Women

1. Nicotine Dependence

Time to first cigarette of the day is an established measure of nicotine dependence [76–78]. In 1987, 62% of women smokers 25 years of age and older reported smoking within 30 minutes of awakening, and 38% reported smoking within 10 minutes of awakening. The proportion of women who smoked within 30 minutes of awakening was directly related to the number of cigarettes smoked per day. No gender difference in time to first cigarette was noted [19].

In 1991/1992 about 75% of women smokers 25 years of age and older reported feeling dependent on cigarettes, 78% reported being unable to cut down on their smoking, and 35% reported feeling sick when they tried to cut down on their smoking [19,79]. Even light smokers reported these indicators of nicotine dependence; among women smoking five or fewer cigarettes per day, 43% felt dependent on cigarettes, 54% felt unable to cut down on their smoking, and 22% felt sick when they tried to cut down on their smoking. African-American and Hispanic women who smoked were less likely than white women to report one of more indicators of nicotine dependence [80] but these findings may reflect racial and ethnic differences in number of cigarettes smoked per day. Women were more likely than men to report feeling dependent on cigarettes [19], but no other gender differences in measures of nicotine dependence have been noted. Some studies have found no gender differences in withdrawal symptoms [47,81,82]; others have found that women reported more symptoms of nicotine dependence, more severe withdrawal symptoms, or longer duration of withdrawal symptoms [80,83–85].

Gender differences may exist in the pharmacokinetic properties of nicotine, which may result in higher sensitivity to the subjective euphoric and cognitive effects of nicotine among women [85,86]. After controlling for body size, the clearance of nicotine appears to be faster in men than in women [86]. With an equivalent daily intake of nicotine, women may therefore be exposed to higher levels of plasma nicotine than men are because of women's slower metabolic clearance [87]. This difference may partially explain why women smoke fewer cigarettes per day on average than men do [10].

The intensity and the types of symptoms experienced during withdrawal may also be different for women and men. Women may be more likely to experience depression and cigarette craving during withdrawal [81,86,88]. Withdrawal symptoms and feelings of anxiety and craving may be stronger predictors of relapse among women than among men [85]. The timing of quit attempts during the menstrual cycle may also be an important predictor of successful quitting among women (with higher rates of cessation and lower reported withdrawal symptoms in the follicular than the luteal phase) [85,89].

2. Other Factors

The prevalence of smoking is higher among persons suffering from depression [33,90,91] and current smokers are more likely than never smokers to report feelings of depression [92]. However, the causal direction of this association is unclear. Depressed smokers are less likely to quit [90,91], and smokers with a history of depression are more likely to relapse after a cessation attempt [93]. Studies have shown an association between heavy smoking and depression among adolescents [94], and depressed adolescents are more likely than nondepressed adolescents to report daily smoking nine years later [80]. Because the prevalence of depression is two times higher among women than among men [95,96], these associations may be particularly important for women.

The prevalence of cigarette smoking is also higher for persons having other psychiatric disorders, such as schizophrenia, mania, personality disorders, anxiety, and panic disorders [33,97,98]. Again, the causal direction of these associations is unclear. Thirty percent of smokers using smoking cessation services may have a history of alcohol abuse or dependence [93]. Smoking cessation may exacerbate existing psychiatric disorders, and smokers with psychiatric comorbidity are more likely to relapse after a cessation attempt.

Smokers report that cigarettes help them to cope with feelings of anxiety, tension, anger, and aggression [85]. Women are more likely than men to report smoking to decrease negative emotions [99]. Women may also be particularly susceptible to relapse during stressful situations [85].

Women's concerns about weight may be a barrier to smoking cessation and may increase relapse rates among women who quit [84,85]. However, a population-based study observed that history of dieting, desired weight loss, and personal relative weight preference (heavier, moderate, leaner) were not related to smoking cessation or relapse among over 1200 employed women after two years of follow-up [100]. The authors concluded that further population studies should examine the effect of weight concerns that are specific to smoking cessation and that are closer temporally to when the changes in smoking behavior are occurring [100]. Although smokers lose weight when they start smoking, weigh less than nonsmokers, and gain weight after quitting smoking, changes in body weight with changes in smoking status are generally small, and the health benefits of quitting greatly outweigh any risks associated with weight gain [101].

D. Determinants of Smoking among Girls

1. Personal Factors

Self-esteem, self-image, psychological health, and attitudes and beliefs about the utility of smoking are important determinants of adolescent smoking [23]. Although some studies indicate that girls who smoke are more self-confident, rebellious, and socially skilled than nonsmokers [69,70,102–104], other

studies have reported associations among girls between smoking and low self-esteem, high susceptibility to peer pressure to smoke, and depression [70,105–107]. Among adolescents in general, confidence or efficacy in resisting offers to smoke is inversely related to cigarette smoking [23,108]. Adolescent smokers are more likely than nonsmokers to believe that smoking has social, psychological, and physiologic benefits and that the consequences of smoking are generally positive [23,109]. Adolescents who view smokers or smoking as having desirable attributes or benefits may be more likely to use smoking to improve their self-image [23,110].

Beliefs and attitudes about the benefits of smoking regarding weight control are a particular issue for initiation of smoking among girls. Several studies of adolescents have found relationships between smoking and body image, body weight, and dieting behavior [46]. Nationally, girls and boys aged 12–18 years who smoked were more likely than nonsmokers to believe that smoking helps keep weight down. Also, girls in grades 9–12 who currently smoke (69.3%) were more likely to be attempting weight loss than were former smokers (56.8%) or never smokers (54.7%) [111]. Similar relationships between smoking and weight concerns were found in several school-based studies. For example, smoking prevalence increased with frequency of dieting among the 17,135 seventh to twelfth grade girls who participated in a health behavior survey in 1987 while attending Minnesota public schools [112]. A longitudinal study of 1705 seventh to tenth grade students indicated that weight concerns and dieting behaviors (*i.e.,* constant thoughts about weight and trying to lose weight) were positively related to smoking initiation for girls only [113]. Fear of weight gain, the desire to be thin, and trying to lose weight were also positively related to current smoking at baseline among girls [113]. In another study among 1915 students in grades 10–12 in one school district in Mississippi, girls but not boys who smoked were more likely than nonsmokers to perceive themselves as fat [114]. Both girls and boys who smoked were less satisfied with their weight than were nonsmokers [114].

2. Behavioral Factors

Behavioral factors related to smoking initiation among adolescents include academic performance, risk taking, rebelliousness, deviant behaviors, and religious and athletic participation [23,115]. Smoking is associated with a host of problem behaviors among adolescent girls, including alcohol and drug abuse, early sexual relations, poor school achievement, and dropping out of school [23,116]. Cigarette smoking may be viewed as a method of coping with anxiety, frustration, or other psychological distress induced by lack of academic success or other environmental stressors [23,115]. Participation in organized sports and religion are protective for cigarette smoking because they may serve as alternatives for dealing with stress [23,70,104,117].

3. Social and Environmental Influences

Smoking behavior is also influenced by external factors including adult, sibling, and peer smoking; the adolescent's perceptions of norms and expectations for smoking; the societal acceptability of smoking; and the availability of cigarettes in the environment. Both parental and peer smoking are important predictors of adolescent smoking [23]. Peer smoking may be especially important among younger girls, and maternal smoking may be a more important predictor among girls than boys [70,118]. Gender differences may also exist in the influence of sibling smoking [23,115,119]. For example, a sister's smoking might be more important for white females than white males [119]. However, further research is needed to confirm these findings. A positive, warm, parent–child relationship, parental monitoring, and parental rules against smoking are protective for youth smoking [120]. In contrast, a high level of peer-centered bonding (particularly with peers who smoke) and participation in antisocial behaviors (*i.e.,* fighting and destroying property) is directly associated with smoking onset among adolescents [23,120].

Adolescents tend to overestimate the percentage of adults and youth who smoke. The perception that a large percentage of the population smokes is related to future smoking in longitudinal studies of adolescents [23]. Perceived approval or disapproval of adolescent smoking by parents, siblings, adult role models, and peers has also been associated with future smoking. For example, in a longitudinal study of girls and boys 11–13 years of age in England, those who believed that their friends and teachers would not be concerned about their smoking were more likely than those who thought their friends and teachers were critical of smoking to become regular smokers 30 months later [23,121].

a. ADVERTISING AND PROMOTION. The tobacco industry, through advertising (*e.g.,* broadcasts, magazines, newspapers, outdoor advertising, and transit advertising) and promotions (*e.g.,* direct-mail promotions, allowances, coupons, premiums, point-of-purchase displays, and entertainment sponsorships), has had a major influence on the acceptability of cigarettes for girls and women [23]. In 1996, cigarette advertising expenditures were $578 million and promotional expenditures were $4.5 billion [122], making cigarettes one of the most heavily marketed products in the United States. In 1988, cigarettes ranked first among products advertised in outdoor media, second in magazines, and sixth in newspapers [123]. Aside from women's and girls' direct exposure to advertising in women's magazines, cigarette advertising has an indirect effect on exposure to information on the health consequences of smoking. Among women's magazines, those that did not carry cigarette advertisements were more than 230% more likely to report the health consequences of smoking than were magazines that carried such advertising [124].

In the early 1900s, cigarette smokers were mostly male, and females represented an untapped market for the tobacco industry. Cigarette makers were able to capitalize on the changing roles of women and the rising consumer culture at this time to position cigarettes as a symbol for both feminists and flappers [125]. One of the first major marketing efforts directed at women began in the 1920s: the American Tobacco Company's Lucky Strike campaign featured the slogan "Reach for a Lucky Instead of a Sweet" [126]. Several major public relations efforts were associated with this campaign, including organizing women to smoke in public during the 1929 New York Easter Parade where women carried placards identifying their cigarettes as

"torches of liberty" [127]. Around this time other tobacco companies also began marketing to women; for example, an advertisement for Liggett & Myers' Chesterfield brand featured a young women saying "Blow Some My Way" [6]. Such advertising helped to speed up the social acceptance of women's smoking, and the widespread campaigns made cigarettes respectable through society's "friendly familiarity" with the product [23,128].

Although advertising strategies have changed over the years by relying less on words, information, and health claims and more on pictures, images, and lifestyle portrayals, the themes of smoking and slenderness and smoking and liberation have been continually exploited by the tobacco industry in its marketing to women [23]. The transition to image advertising led the industry to give cigarette brands distinctly male or female identities. For example, the Virginia Slims brand was launched with a 100 mm "slimmer than usual" cigarette and the "You've Come a Long Way, Baby!" slogan in 1968 [129]. In this campaign, as well as in most modern advertising strategies, smoking is made to appear universally acceptable, attractive, and desirable by emphasizing themes of popularity, youthful vigor, and social and professional success [23,69].

Several studies have observed a high level of recognition of cigarette advertisements among children and adolescents, and the most heavily used brands among youth correspond to the most widely advertised brands [130]. A study of smoking trends among girls and women identified two major periods of increased initiation among girls and young women: one in the mid-1920s coincident with the Lucky Strike and Chesterfield women's marketing campaigns, and one in the late 1960s coincident with large-scale marketing of women's brands [131]. Male smoking initiation did not increase during these periods. Recently released corporate documents confirm the tobacco industry's marketing of cigarettes to youth [23], as does the rapid growth in cigarette promotional expenditure, which may have particular influence on youth.

b. OTHER SOCIETAL INFLUENCES. The appeal of smoking is further influenced by the widespread use of cigarettes in movies, music videos, and television and is reinforced by community norms and policies that make tobacco products relatively accessible for adolescents. Several community and school-based tobacco use prevention strategies have been identified, including increasing tobacco prices, reducing access by implementing and adequately enforcing restrictions on minors' access, reducing the appeal of tobacco products by restricting advertising and promotion, and conducting youth-oriented mass media campaigns and school-based tobacco use prevention programs [23,132]. Establishing health-oriented social norms through smoke-free indoor air provisions and decreased modeling of tobacco use by parents, teachers, and celebrities also will contribute to prevention among girls [23].

4. Nicotine Dependence

Many girls who currently smoke are addicted to nicotine. The pharmacologic and behavioral processes that determine nicotine addiction are similar to those that determine addiction to other drugs, such as heroin and cocaine [133]. Nicotine has psychoactive properties, including providing pleasurable effects, and it can serve to reinforce tobacco-seeking and regular and compulsive patterns of use. Nicotine tolerance develops such that repeated use results in diminished effects and can be accompanied by increased intake. Physical dependence on nicotine is characterized by a withdrawal syndrome on tobacco abstinence [133]. Withdrawal symptoms include increased appetite; depressed mood; insomnia; inability to concentrate; irritability, frustration, or anger; anxiety; restlessness; and craving for tobacco [81,134]. Some data suggest that smokers who are nicotine dependent are less likely to quit smoking [33,34,47,76,133].

Time to first cigarette of the day is a measure of nicotine dependence and has been correlated with quitting success [77,78]. However, time to first cigarette may be a more valid measure of nicotine dependence for adults than for adolescents, who may have to wait until they can leave the house to smoke their first cigarette. In 1992, 33% of adolescents reported smoking within 30 minutes of awakening (see section III.C.1 for measures among women) [135].

In 1991/1992, girls and young women who smoked often were as likely as older female smokers to report feeling dependent on cigarettes (75%), feeling unable to cut down on their smoking (82%), and feeling sick when they cut down on their smoking (35%) [79]. Girls and young women who smoked were more likely than older female smokers to report needing more cigarettes for the same effect (18% vs 13%, respectively). Even light smokers reported these indicators of nicotine dependence; among girls and young women who smoked five or fewer cigarettes per day, 52% felt dependent on cigarettes, 67% felt unable to cut down on their smoking, and 22% felt sick when they tried to cut down on their smoking. Another study found that 54% of adolescent smokers felt worse upon quitting smoking and that heavy smokers were most likely to report adverse symptoms [50].

The reporting that smoking helps calm the smoker is associated with increased withdrawal symptoms when attempting to quit [136]. One study found that 55% of girls aged 11–13 and 75% of girls aged 14–16 who smoked one or more cigarettes per week reported that they smoked to calm their nerves [137]. The most common effect reported by children who smoke is feeling calmer [136].

IV. Health Effects of Cigarette Smoking among Women

A. Introduction

In general, the risks for the health effects of smoking are similar for women and men (*e.g.,* coronary heart disease, COPD, and cancers of the oral cavity, pharynx, larynx, esophagus, pancreas, kidney, and bladder). Early studies indicate that the relative risk for smoking and stroke many be slightly higher among women than among men [138]; however, further studies are needed to confirm this finding. For lung cancer, female smokers have historically had a lower risk of lung cancer than male smokers [139], a pattern that persists in nearly all recent cohort studies [140]; however, some recent case-control studies have found a comparable risk or even higher risk for female smokers than male smokers [141–144]. Possible explanations include a lower risk of lung cancer among nonsmoking women

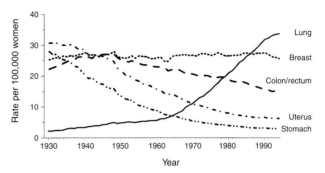

Fig. 45.4 Cancer death rates among women, by cancer site, United States, 1930–1994. Rates are adjusted to the 1970 census population. Source: Vital Statistics of the U.S., National Center for Health Statistics, CDC; and [146a].

(perhaps due to occupational exposures that increase the nonsmoker risk in males), increased female susceptibility to carcinogens in tobacco smoke, and increased exposure levels among women [141,145,146].

B. Lung Cancer

Beginning in the 1960s, an epidemic of lung cancer has occurred among women (Fig. 45.4) [146a]. From 1972 to 1992, lung cancer deaths among women increased by an average of 4.9% per year [147]. Lung cancer is now the leading cancer cause of death among both African-American and white women [9]; an estimated 67,000 women will die of this disease in 1998 [148]. The five-year survival rate for women with lung cancer has been at 15–17% since the mid 1970s. Survival is higher for women with localized disease, but only 16% of women with lung cancer present with localized disease [147].

Lung cancer mortality in women increases with the amount smoked, duration of smoking, and depth of inhalation [3,35]. Filtered cigarettes and low-tar cigarettes may reduce lung cancer risk among women slightly, but even smokers of the lowest tar cigarette have a significantly increased risk of lung cancer compared with never smokers [149]. Smoking induces all four major histologic types of lung cancer (squamous cell, small cell, large call, and adenocarcinoma). Adenocarcinoma is the most common type of lung cancer in women [150,151] and its incidence has been increasing in the United States. One analysis suggests that this increase parallels changes in cigarette design and smoking behavior. The smoke from nonfiltered, high-tar cigarettes may be too harsh to be inhaled deeply, resulting in particulate deposits in the central bronchi, where squamous cell cancers occur; however, smoke from filtered, low-tar cigarettes can be inhaled more deeply into the periphery of the lung, and particulate deposits here result in adenocarcinomas [152,153].

Smoking cessation reduces the risk of lung cancer. After ten years, the risk among former smokers is 30–50% that of continuing smokers. The risk never completely returns to that of never smokers, however [35].

C. Other Cancers

Smoking also causes oral, laryngeal, and esophageal cancer [35,139,154]. Most cases of these cancers, and more than 3500 deaths each year in women, are attributable to smoking [59]. Smoking cessation halves the risk of oral and esophageal cancer within five years of quitting; longer abstinence reduces the risk further. The risk of laryngeal cancer is reduced after 3–4 years of abstinence but remains higher than that of never smokers [35].

About 30% of bladder and renal cancers in women are attributable to smoking and these smoking attributable cancers cause more than 1400 deaths per year [59,139]. Smoking cessation halves the risk of bladder cancer within a few years [35]. Smoking also increases the risk of pancreatic cancer in women; these smoking-attributable cases kill more than 3600 women each year [59].

Another 1300 women each year die from smoking-attributable cervical cancer [59]. Although other factors such as sexually transmitted diseases are believed to be causally related to cervical cancer, components of tobacco smoke have been found in cervical mucus and have mutagenic activity. After one year of smoking cessation, former smokers are at lower risk for cervical cancer; this reduction in risk supports the designation of cigarette smoking as a contributing cause of cervical cancer [35].

D. Cardiovascular Disease

Of the various diseases associated with tobacco use among women, coronary heart disease (CHD) is the leading cause of death; smoking-attributable cardiovascular disease is estimated to kill more than 60,000 women each year [59]. Smoking is estimated to cause 85% of CHD death among women 45–49 years of age, 37% of ischemic heart disease death in women under 65 years of age, and 12% among women 65 years and older [8,155]. Risk of death from CHD increases with early smoking initiation, number of cigarettes smoked per day, and depth of smoke inhalation [35,139,156,157]. In studies of women using high-dose oral contraceptives, increased cardiovascular risk was reported among smokers; it is unclear if this pattern holds for users of the newer, low-dose pills as well [158,159]. Smoking cessation substantially reduces the risk of CHD among women of all ages. The excess risk is reduced by half after one year of abstinence, then declines more gradually. After 15 years of abstinence, the risk is comparable to that of never smokers [35].

The strongest risk factor for atherosclerotic peripheral arterial occlusive disease is cigarette smoking [35,160]. Smoking is strongly associated with both aortoiliac and femoropopliteal disease [161]. Former smokers have a lower risk of peripheral arterial occlusive disease and slower progression of existing disease than do continuing smokers [35,162]. In one study, however, progression of atherosclerosis was not reduced by abstinence after adjustment was made for amount smoked, which suggests that some effects of smoking may be cumulative and irreversible [162].

Smoking is causally associated with stroke, particularly cerebral infarction and subarachnoid hemorrhage [138,139,163]. Women smokers who use high-dose oral contraceptives are at even higher risk of stroke [164], but the risk with the newer low-dose formulations may be different [35,158]. Smoking cessation reduces the risk of both ischemic stroke and subarachnoid hemorrhage. In some studies, risk returned to that of never smokers within five years; in other studies, this reduction did not occur until up to 15 years of abstinence [35].

E. Chronic Lung Disease

More than 28,000 women die each year from chronic obstructive pulmonary disease (COPD) caused by smoking [59]. Mortality among women from COPD has increased steadily over the past 25 years [3,139]. The risk of COPD is nearly as strong as that for lung cancer, with 79% of COPD in women attributable to smoking [139,165]. Risk of death from COPD is associated with age of initiation, number of cigarettes smoked daily, and depth of inhalation [3,139,166]. Smokers also have more respiratory symptoms than nonsmokers do [35]. Smoking by children and adolescents is associated with reduced lung function, which may place these adolescents at future risk for COPD (see section V.B for ETS effects) [23].

Smoking cessation reduces respiratory symptoms and respiratory infections [3,166], and smokers who quit have better pulmonary function than do continuing smokers. COPD mortality is lower among former smokers than continuing smokers [35,167], but even after 20 years of abstinence, the risk is still greater than that of never smokers [166].

F. Reproductive Outcomes

Several studies have reported reduced fertility and increased menstrual problems in women who smoke. Alterations in sperm quality and quantity and reduced testosterone production have been noted among males as well [3,35].

Infants born to women who smoke during pregnancy weigh, on average, 200 grams less than infants born to nonsmokers [35]. The risk of having a low birth weight infant is doubled in smokers; this effect is independent of other factors known to cause low birth weight. An estimated 17–26% of low birth weight births could be prevented by eliminating smoking during pregnancy. Women who stop smoking before becoming pregnant or in the first 3–4 months of pregnancy have infants of the same birth weight as never smokers. Even stopping smoking before the thirtieth week of gestation results in a higher birth weight infant compared with continuing smoking [35].

Preterm delivery (<37 weeks gestation) is associated with maternal smoking [35]. An estimated 7–10% of these deliveries could be prevented by eliminating smoking during pregnancy. In addition, the risk of intrauterine growth retardation is four times higher among women who smoke during their pregnancy. The primary mechanism for the reduction in fetal growth is thought to be intrauterine hypoxia.

Maternal smoking is also associated with higher fetal, neonatal, and infant mortality. Smoking during pregnancy increases the risk of placenta previa and abruptio placentae [35]. These two outcomes alone appear to cause at least half of the excess fetal and neonatal deaths associated with smoking during pregnancy [168]. Maternal smoking during pregnancy is also a strong risk factor for sudden infant death syndrome (SIDS) [169–171].

G. Osteoporosis

Osteoporosis is an important medical problem for older women. Because smoking is associated with earlier menopause (smokers undergo menopause 1–2 years earlier than never smokers) and because lower weight is associated with an increased risk of osteoporosis (smokers weigh less than nonsmokers) [35], the relationship between smoking and osteoporosis has been of great interest. The studies to date examining bone mineral content in smokers and nonsmokers have produced mixed results. Some case-control studies of the relationship between smoking status and osteoporosis have found a higher risk of osteoporotic fractures among smokers, but many have not adjusted for important confounders such as estrogen use, weight, and age; these adjustments tend to reduce the magnitude of the effect found [172,173]. Cohort studies have also reported mixed results regarding the association between smoking and osteoporotic fractures [35].

V. Health Effects of Environmental Tobacco Smoke

Environmental tobacco smoke (ETS) is a combination of diluted mainstream smoke exhaled by smokers and sidestream smoke from the burning end of the cigarette. ETS is chemically similar to inhaled smoke. All of the five known human carcinogens, nine of the probable human carcinogens, and three of the animal carcinogens are emitted at higher levels in sidestream smoke than in mainstream smoke. Other toxic compounds (*e.g.,* ammonia and carbon monoxide) are also emitted at higher levels in sidestream smoke [174].

A large national study found that 88% of people who did not use tobacco had detectable levels of serum cotinine (a nicotine metabolite), although only 37% of the adults and 43% of the children were aware that they were exposed to ETS [175]. Women are more likely than men to report no home or workplace ETS exposure; women are more likely than men to report home ETS exposure whereas men are more likely than women to report workplace ETS exposure [175]. Home ETS exposure generally results in higher cotinine levels than workplace exposure [175]. In 1992/1993, one-third to one-half of adult cigarette smokers had children living in the home, and 70% of smokers with children allow smoking in some or all areas of the home [176].

A. ETS Effects in Adults

Most studies of the health effects of ETS involve nonsmoking women married to smokers. However, some studies have begun to examine the role of ETS exposure *in utero* and during childhood and the risk of adult cancers and respiratory disease [177]. The most common effect of ETS in adults is irritant effects of the eye and mucous membranes of the nose, throat, and lower respiratory tract. Exposure to ETS can precipitate and aggravate allergic attacks in persons with respiratory allergies [178,179]. ETS exposure may also result in more frequent respiratory symptoms in adults; respiratory symptoms are an estimated 30–60% higher in nonsmokers exposed to ETS than in nonsmokers not exposed to ETS [174]. Healthy adults exposed to ETS may have small, but likely insignificant, reductions in pulmonary

function from the irritant effects of ETS [178]; however, ETS may add to the burden of respiratory environmental insults that can cause chronic lung disease [180].

Most studies of ETS and lung cancer have shown a positive association, although this association was not always statistically significant due to small sample sizes. Because of (1) biologic plausibility, (2) supporting evidence from animal studies, (3) consistency of findings, (4) data from multiple countries, (5) positive dose-response relationships, (6) increased risk at all exposure levels, (7) remaining effects after adjustment for potential bias, and (8) strong associations found at high exposure levels, and because (9) confounding factors do not explain the findings, ETS is considered to be causally associated with lung cancer in nonsmoking adults [174,177,178,180,181]. The largest U.S. study found a statistically significant increase in lung cancer risk among women ever exposed to ETS and a dose-response relationship between exposure and lung cancer [182]. ETS has also been classified as a "Group A" (known human) carcinogen and is estimated to cause 3000 deaths each year in nonsmoking adults [174].

Evidence has been growing that ETS is associated with cardiovascular disease. A meta-analysis of 12 studies concluded that CHD mortality was 23% higher among never smokers exposed to ETS than for nonexposed never smokers [183]. Risk is comparable for male and female never smokers exposed to ETS [184]. In 1997, the California Environmental Protection Agency (EPA) reported that there was sufficient evidence to conclude that ETS is causally associated with cardiovascular disease [177].

B. ETS Effects in Children

ETS exposure is associated with respiratory conditions in children, particularly young children [178,180]. The U.S. EPA concluded that ETS is causally associated with an increased risk of lower respiratory infections, such as pneumonia and bronchitis, in children; an estimated 150,000–300,000 cases each year among infants and children up to 18 months of age are attributable to ETS exposure. ETS is also causally associated with fluid in the middle ear, symptoms of upper respiratory tract irritation, reduced lung function, and additional episodes and increased severity of asthma in children. An estimated 200,000–1,000,000 asthmatic children have their condition worsened by exposure to ETS, and ETS is also a risk factor for new onset of asthma among previously asymptomatic children [174,177].

Children exposed to ETS have, on average, 1.9 more days of restricted activity, 1.1 more days in bed, and 1.4 more days absent from school per year than children not exposed to ETS [185]. Nationally, this represents 18 million days of restricted activity, 10 million days of bed confinement, and 7 million days of school absence attributable to ETS exposure.

Several studies have shown an association between ETS exposure and SIDS that is independent of maternal smoking during pregnancy [169–171]. This association has been found for maternal smoking, paternal smoking, and smoking by others in the household. A dose-response relationship with increasing number of smokers has also been reported [169,171]. The California EPA reported that the evidence was sufficient to consider the relationship between ETS exposure and SIDS to be causal

[177]. Other studies have found an association between maternal ETS exposure during pregnancy and lower birth weight and intrauterine growth retardation [186,187].

VI. Conclusions

Currently, 22 million women in the United States (22.6%) smoke cigarettes and the prevalence of cigarette smoking among high school girls has increased since 1992. Several important determinants of smoking among women and girls have been identified, including the role smoking plays in meeting the developmental needs of the adolescent, weight concerns, and nicotine addiction. Effective community and school-based strategies to prevent tobacco use have been identified and need to be implemented at every level of society. The health effects of an adolescent's decision to smoke cigarettes are devastating: each year, over 152,000 women die from cigarette smoking. Effective smoking cessation treatments have been identified to break the cycle of nicotine addiction, and smoking cessation clinical trials indicate that the same treatments benefit both women and men. These treatment strategies are reviewed and summarized in the Agency for Health Care Policy's clinical practice guideline for smoking cessation [93] which concludes that smoking cessation interventions offer health care providers their greatest opportunity to improve the current and future health of all Americans. The combined efforts of the public, scientists, government officials, and voluntary and professional health organizations are needed to improve women's health through prevention of tobacco use among children and adolescents and through treating girls and women addicted to nicotine.

References

1. Robert, J. C. (1952). "The Story of Tobacco in America." Knopf, New York.
2. Waldron, I. (1991). Patterns and causes of gender differences in smoking. *Soc. Sci. Med.* **32,** 989–1005.
3. U.S. Department of Health and Human Services (1980). "The Health Consequences of Smoking for Women. A Report of the Surgeon General." U.S. Department of Health and Human Services, Public Health Service, Office on Smoking and Health, Washington, DC.
4. Milwaukee Journal (1979). Consumer analysis of the greater Milwaukee market, 1924–1979. *Milwaukee J.,* pp. 76–78.
5. American Institute of Public Opinion (Gallup) (1972). "The Gallup Poll: Public Opinion 1935–1971. Vol. Two. 1949–1958." Random House, New York.
6. Howe, H. (1984). An historical review of women, smoking and advertising. *Health Educ.* **15,** 3–9.
7. University of Michigan (1997). "Cigarette Smoking Rates may have Peaked among Younger Teens [press release]." University of Michigan News and Information Services, Ann Arbor.
8. Centers for Disease Control and Prevention (1997). Smoking-attributable mortality and years of potential life lost—United States, 1984. *Morbid. Mortal. Wkly. Rep.* **46,** 444–451.
9. Centers for Disease Control and Prevention (1993). Mortality trends for selected smoking-related cancers and breast cancer—United States, 1950–1990. *Morbid. Mortal. Wkly. Rep.* **42,** 857, 863–866.
10. Giovino, G. A., Schooley, M. W., Zhu, B. P., *et al.* (1994). Surveillance for selected tobacco-use behaviors—United States, 1900–1994. *Morbid. Mortal. Wkly. Rep.* **43,** 1–43.
11. National Center for Health Statistics (1995). "National Health Interview Survey" (unpublished data).
12. Centers for Disease Control and Prevention (1997). Cigarette smoking among adults—United States, 1995. *Morbid. Mortal. Wkly. Rep.* **46,** 1217–1220.
13. Wagenecht, L. E., Perkins, L. L., Cutter, G. R., *et al.* (1990). Cigarette smoking behavior is strongly related to educational status: The CARDIA Study. *Prev. Med.* **19,** 158–169.
14. National Center for Health Statistics (1992). Advance report of new data from the 1989 birth certificate. *In* "Monthly Vital Statistics Report 40 (Suppl. 12)." National Center for Health Statistics, Hyattsville, MD.
15. Ventura, S. J., Martin, J. A., Curtin, S. C., and Mathews, T. J. (1997). Report of final natality statistics, 1995. *In* "Monthly Vital Statistics Report 45 (Suppl. 11)." National Center for Health Statistics, Hyattsville, MD.
16. Kendrick, J. S., Zahniser, S. C., Miller, N., *et al.* (1995). Integrating smoking cessation into routine public prenatal care: The smoking cessation in pregnancy project. *Am. J. Public Health* **85,** 217–222.
17. U.S. Department of Health and Human Services (1997). "Substance Use among Women in the United States," DHHS Publ. No. (SMA) 97-3162. U.S. Department of Health and Human Services, Substance Abuse and Mental Health Services Administration, Rockville, MD.
18. Warner, K. E. (1989). Effects of the antismoking campaign: An update. *Am. J. Public Health* **79,** 144–151.
19. Husten, C. G., Chrisman, J. H., and Reddy, M. N. (1996). Trends and effects of cigarette smoking among girls and women in the United States, 1965–1993. *J. Am. Med. Women's Assoc.* **51** 11–18.
20. Husten, C. G. (1997) Cigarette smoking among girls and women. *J. Med. Assoc. Ga.* **86,** 213–216.
21. Johnston, L. D., Bachman, J. G., and O'Malley, P. M. (1980). "Monitoring the Future: Questionnaire Responses from the Nation's High School Seniors, 1977." Institute for Social Research, University of Michigan, Ann Arbor.
22. Bachman, J. G., Johnston, L. D., and O'Malley, P. M. (1993). "Monitoring the Future: Questionnaire Responses from the Nation's High School Seniors, 1992." Institute for Social Research, University of Michigan, Ann Arbor.
23. U.S. Department of Health and Human Services (1994). "Preventing Tobacco Use among Young People: A Report of the Surgeon General." U.S. Department of Health and Human Services, Centers for Disease Control and Prevention, Office on Smoking and Health, Atlanta, GA.
24. University of Michigan (1997). "Monitoring the Future Study" (unpublished data).
25. Gordon, T., Kannel, W. B., Dawber, T. R., and McGee, D. (1975). Changes associated with quitting cigarette smoking: The Framingham Study. *Am. Heart J.* **90,** 322–328.
26. Freund, K. M., D'Agostino, R. B., Belaner, A. J., Kannel, W. B., and Stokes, J., III (1992). Predictors of smoking cessation: The Framingham Study. *Am. J. Epidemiol.* **135,** 957–964.
27. Hammond, E. C., and Garfinkel, L. (1968). Changes in cigarette smoking, 1959–1965. *Am. J. Public Health* **58,** 30–45.
28. McWhorter, W. P., Boyd, G. M., and Mattson, M. E. (1990). Predictors of quitting smoking: The NHANES I Followup Experience. *J. Clin. Epidemiol.* **43,** 1399–1405.
29. Ahmad, P. I., and Gleeson, G. A. (1970). Changes in cigarette smoking habits between 1955 and 1966. *In* "Vital and Health Statistics, Series 10, No. 59," Publ. No. (PHS)1000. U.S. Department of Health, Education, and Welfare, Health Services and Mental Health Administration. National Center for Health Statistics, Rockville, MD.

30. Centers for Disease Control and Prevention (1998). Tobacco use among high school students—United States, 1997. *Morbid. Mortal. Wkly. Rep.* **47,** 229–233.

31. Bachman, J. G., Wallace, J. M., O'Malley, P. M., *et al.* (1991). Racial/ethnic differences in smoking, drinking, and illicit drug use among American high school seniors, 1976–1989. *Am. J. Public Health* **81,** 372–377.

32. Taioli, E., and Wynder, E. L. (1991). Effect of the age at which smoking begins on frequency of smoking in adulthood. *N. Engl. J. Med.* **325,** 968–969.

33. Breslau, N., Kilbey, M. M., and Andreski, P. (1993). Vulnerability to psychopathology in nicotine-dependent smokers: An epidemiologic study of young adults. *Am. J. Psychiatry* **150,** 941–946.

34. Breslau, N., and Peterson, E. L. (1996). Smoking cessation in young adults: Age at initiation of cigarette smoking and other suspected influences. *Am. J. Public Health* **86,** 214–220.

35. U.S. Department of Health and Human Services (1990). "The Health Benefits of Smoking Cessation: A Report of the Surgeon General, 1990," DHHS Publ. No. (CDC)90-8416. U.S. Department of Health and Human Services, Centers for Disease Control and Prevention, Office on Smoking and Health, Atlanta, GA.

36. Substance Abuse and Mental Health Services Administration (1995). "National Household Survey on Drug Abuse" (unpublished data).

37. Haenszel, W., Shimkin, M. B., and Miller, H. P. (1956). "Tobacco Smoking Patterns in the United States," Public Health Monogr. No. 45, Publ. No. (PHS)463. U.S. Department of Health, Education, and Welfare, Public Health Service, Washington, DC.

38. Centers for Disease Control and Prevention (1993). Smoking cessation during previous year among adults—United States, 1990 and 1991. *Morbid. Mortal. Wkly. Rep.* **42,** 504–507.

39. Pierce, J., Giovino, G., Hatziandreu, E., and Shopland, D. (1989). National age and sex differences in quitting smoking. *J. Psychoac. Drugs* **21,** 293–298.

40. Jarvis, M. (1984). Gender and smoking: Do women really find it harder to give up? *Br. J. Addict.* **79,** 383–387.

41. Orlandi, M. A. (1986). Gender differences in smoking cessation. *Women's Health* **11,** 237–251.

42. Cohen, S., Lichtenstein, E., Prochaska, J. O., *et al.* (1989). Debunking myths about self-quitting: Evidence from 10 prospective studies of persons who attempt to quit smoking by themselves. *Am. Psychol.* **44,** 1355–1365.

43. Floyd, R. L., Rimer, B. K., Giovino, G. A., Mullen, P. D., and Sullivan, S. E.. (1993). A review of smoking in pregnancy: Effects on pregnancy outcomes and cessation efforts. *Annu. Rev. Public Health* **14,** 379–411.

44. O'Campo, P., Faden, R. R., Brown, H., *et al.* (1992). The impact of pregnancy on women's prenatal and postpartum smoking behavior. *Am. J. Prev. Med.* **8,** 8–13.

45. Fingerhut, L. A., Kleinman, J. C., and Kendrick, J. S. (1990). Smoking before, during, and after pregnancy. *Am. J. Public Health* **80,** 541–544.

46. Moss, A. J., Allen, K. F., Giovino, G. A., and Mills, S. L. (1992). "Recent Trends in Adolescent Smoking, Smoking-Uptake Correlates, and Expectation about the Future," Adv. Data No. 221. U.S. Department of Health and Human Services, Centers for Disease Control and Prevention, National Center for Health Statistics, Hyattsville, MD.

47. Pirie, P. L., Murray, D. M., and Leupker, R. V. (1991). Gender differences in cigarette smoking and quitting in a cohort of young adults. *Am. J. Public Health* **81,** 324–327.

48. Adams, P. F., Schoenborn, C. A., Moss, A. J., Warren, C. W., and Kann, L. (1995). Health risk behaviors among our Nation's youth: United States, 1992. *In* "Vital and Health Statistics, Series 10, No. 192," DHHS Publ. No. (PHS)95-1520. U.S. Department of Health

and Human Services, Centers for Disease Control and Prevention, National Center for Health Statistics, Hyattsville, MD.

49. McNeill, A. D., West, R. J., Jarvis, M., Jackson, P., and Bryant, A. (1986). Cigarette withdrawal symptoms in adolescent smokers. *Psychopharmacology* **90,** 533–536.

50. Ershler, J., Leventhal, H., Fleming, R., and Glynn, R. (1989). The quitting experience for smokers in sixth through twelfth grades. *Addict. Behav.* **14,** 365–378.

51. Hansen, W. B. (1983). Behavioral predictors of abstinence: Early indicators of a dependence on tobacco among adolescents. *Int. J. Addict.* **18,** 913–920.

52. Centers for Disease Control and Prevention (1998). Selected cigarette smoking initiation and quitting behaviors among high school students. *Morbid. Mortal. Wkly. Rep.* **47,** 386–389.

53. Waldron, I., Lye, D., and Brandon, A. (1991). Gender differences in teenage smoking. *Women's Health* **17,** 65–90.

54. U.S. Department of Health and Human Services (1998). "Cigars: Health Effects and Trends," Natl. Cancer Inst. Monogr. No. 9. U.S. Department of Health and Human Services, National Cancer Institute, Bethesda, MD.

55. U.S. Department of Health and Human Services (1982). "The Health Consequences of Smoking: Cancer. A Report of the Surgeon General," DHHS Publ. No. (PHS) 82-50179. U.S. Department of Health and Human Services, Office on Smoking and Health, Rockville, MD.

56. U.S. Department of Health, Education, and Welfare (1969). "Use of Tobacco: Practices, Attitudes, Knowledge, and Beliefs, United States—Fall 1964 and Spring 1966." U.S. Department of Health, Education, and Welfare, Health Services and Mental Health Administration, National Clearinghouse for Smoking and Health, Washington, DC.

57. Centers for Disease Control and Prevention (1997). Cigar smoking among teenagers—United States, Massachusetts, and New York, 1996. *Morbid. Mortal. Wkly. Rep.* **46,** 433–440.

58. U.S. Department of Health and Human Services (1986). "The Health Consequences of Using Smokeless Tobacco. A Report of the advisory committee to the Surgeon General, 1986," NIH Publ. No. 86-2874. U.S. Department of Health and Human Services, National Institutes of Health, Bethesda, MD.

59. Novotny, T. E., and Giovino, G. A. (1998). Tobacco use. *In* "Chronic Disease Epidemiology and Control" (R. E. Brownson, P. L. Remington, and J. R. Davis, eds.), 2nd ed. pp. 117–148. American Public Health Association Press, Washington, DC.

60. U.S. Department of Health and Human Services (1989). "Smokeless Tobacco Use in the United States," Natl. Cancer Inst. Monogr. No. 8. U.S. Department of Health and Human Services, National Cancer Institute, Bethesda, MD.

61. Giovino, G. A., Henningfield, J. E., Tomar, S. L., Escobedo, L. G., and Slade, J. (1995). Epidemiology of tobacco use and dependence. *Epidemiol. Rev.* **17,** 48–65.

62. Spangler, J. G., Dignan, M. B., and Michielutte, R. (1997). Correlates of tobacco use among Native American women in western North Carolina. *Am. J. Public Health* **87,** 108–111.

63. U.S. Department of Health and Human Services (1998). "Tobacco Use Among U.S. Racial/Ethnic Minority Groups—African Americans, American Indians and Alaska Natives, Asian Americans and Pacific Islanders, and Hispanics: A Report of the Surgeon General." U.S. Department of Health and Human Services, Centers for Disease Control and Prevention, Office on Smoking and Health, Atlanta, GA.

64. Johnston, L. D., O'Malley, P. M., and Bachman, J. G. (1996). "National Survey Results on Drug Use from the Monitoring the Future Study, 1975–1995: Vol. I. Secondary School Students," NIH Publ. No. 96-4139. U.S. Department of Health and Human

Services, National Institutes of Health, National Institute on Drug Abuse, Rockville, MD.

65. Centers for Disease Control (1987). Smokeless tobacco use in rural Alaska. *Morbid. Mortal. Wkly. Rep.* **36,** 140–143.

66. Riley, W. T., Barenie, J. T., Mabe, A., and Myers, D. R. (1990). Smokeless tobacco use in adolescent females: Prevalence and psychosocial factors among racial/ethnic groups. *J. Behav. Med.* **13,** 207–220.

67. Backinger, C. L., Bruerd, B., Kinney, M. B., and Szpunar, S. M. (1993). Knowledge, intent to use, and use of smokeless tobacco among sixth grade schoolchildren in six selected U.S. sites. *Public Health Rep.* **108,** 637–642.

68. Nelson, D. E., Davis, R. M., Chrismon, J. H., and Giovino, G. A. (1996). Pipe smoking in the United States, 1965–1991: Prevalence and attributable mortality. *Prev. Med.* **25,** 91–99.

69. French, S. A., and Perry, C. L. (1996). Smoking among adolescent girls: Prevalence and etiology. *J. Am. Med. Women's Assoc.* **51,** 25–28.

70. Clayton, S. (1991). Gender differences in psychosocial determinants of adolescent smoking. *J. Sch. Health* **61,** 115–120.

71. Walsh, D. C., Sorensen, G., and Leonard, L. (1995). Gender, health, and cigarette smoking. *In* "Society and Health" (B. C. Amick, S. Levine, A. R. Tarlov, and D. C. Walsh, eds.), pp. 131–170. Oxford University Press, Oxford.

72. Graham, H. (1994). Gender and class as dimensions of smoking behavior in Britain: Insights from a survey of mothers. *Soc. Sci. Med.* **38,** 691–698.

73. Lowry, R., Kann, L., Collins, J., and Kolbe, L. J. (1996). The effect of socioeconomic status on chronic disease risk behaviors among U.S. adolescents. *JAMA, J. Am. Med. Assoc.* **276,** 792–797.

74. Substance Abuse and Mental Health Services Administration (1997). "Preliminary Estimates from the 1995 National Household Survey on Drug Abuse," Adv. Rep. No. 10. U.S. Department of Health and Human Services, Substance Abuse and Mental Health Services Administration, Office of Applied Studies, Rockville, MD.

75. Perry, C. L., Kelder, S. H., and Komro, K. A. (1993). The social world of adolescents: Family, peers, schools, and the community. *In* "Promoting the Health of Adolescents. New Directions for the Twenty-First Century" (S. G. Millstein, A. C. Petersen, and E. O. Nightingale, eds.), pp. 73–96. Oxford University Press, New York.

76. Hymowitz, N., Cummings, K. M., Hyland, A., Lynn, W. R., Pechacek, T. F., and Hartwell, T. D. (1997). Predictors of smoking cessation in a cohort of adult smokers followed for five years. *Tob. Control* **6,** S57–S62.

77. Fagerström, K. O. (1978). Measuring degree of physical dependence to tobacco smoking with reference to individualization of treatment. *Addict. Behav.* **3,** 235–241.

78. Kabat, G. C., and Wynder, E. L. (1987). Determinants of quitting smoking. *Am. J. Public Health* **77,** 1301–1305.

79. Centers for Disease Control and Prevention (1995). Indicators of nicotine addiction among women—United States, 1991–1992. *Morbid. Mortal. Wkly. Rep.* **44,** 102–105.

80. Kandel, D., Chen, K., Warner, L. A., Kessler, R. C., and Grant, B. (1997). Prevalence and demographic correlates of symptoms of last year dependence on alcohol, nicotine, marijuana and cocaine in the U.S. population. *Drug Alcohol Depend.* **44,** 11–29.

81. Svikis, D. S., Hatsukami, D. K., Hughes, J. R., Carroll, K. M., and Pickens, R. W. (1986). Sex differences in tobacco withdrawal syndrome. *Addict. Behav.* **11,** 459–462.

82. Gunn, R. C. (1986). Reactions to withdrawal symptoms and success in smoking cessation clinics. *Addict. Behav.* **11,** 49–53.

83. Shiffman, S. M. (1979). The tobacco withdrawal syndrome. *In* "Cigarette Smoking as a Dependence Process" (N. A. Krasnegor,

ed.), pp. 158–184. Res. Monogr. 23. U.S. Department of Health, Education and Welfare, National Institute on Drug Abuse, Rockville, MD.

84. Pomerleau, C. S., and Pomerleau, O. F. (1994). Gender differences in frequency of smoking withdrawal symptoms. *Ann. Behav. Med.* **16,** S118.

85. Gritz, E. R., Nielsen, I. R., and Brooks, L. A. (1996). Smoking cessation and gender: The influence of physiological, psychological, and behavioral factors. *J. Am. Med. Women's Assoc.* **51,** 35–42.

86. Fant, R. V., Everson, D., Dayton, G., Pickworth, W. B., and Henningfield, J. E. (1996). Nicotine dependence in women. *J. Am. Med. Women's Assoc.* **51,** 19–24.

87. Benowitz, N. L., and Jacob, P. (1984). Daily intake of nicotine during cigarette smoking. *Clin. Pharmacol. Ther.* **35,** 499–504.

88. Swan, G., Rosenman, R., Parker, S., and Denk, C. (1985). "Identification of Variables Associated with Maintenance of Nonsmoking in Ex-Smokers." SRI International, Menlo Park, CA.

89. O'Hara, P., Portser, S. A., and Anderson, B. P. (1989). The influences of menstrual cycle changes on the tobacco withdrawal syndrome in women. *Addict. Behav.* **14,** 595–600.

90. Anda, R. F., Williamson, D. F., Escobedo, L. G., Mast, E. E., Giovino, G. A., and Remington, P. L. (1990). Depression and the dynamics of smoking: A national perspective. *JAMA, J. Am. Med. Assoc.* **264,** 1541–1545.

91. Glassman, A. H., Helzer, J. E., Covey, L. S., *et al.* (1990). Smoking, smoking cessation, and major depression. *JAMA, J. Am. Med. Assoc.* **264,** 1546–1549.

92. Schoenborn, C. A., and Horm, J. (1993). "Negative Moods as Correlates of Smoking and Heavier Drinking: Implications for Health Promotion," Adv. Data, Vital Health Stat. No. 236. U.S. Department of Health and Human Services, Centers for Disease Control and Prevention, National Center for Health Statistics, Hyattsville, MD.

93. Agency for Health Care Policy and Research (1996). "Smoking Cessation: Clinical Practice Guideline," AHCPR Publ. No. 96-0692. U.S. Department of Health and Human Services, Agency of Health Care Policy and Research, Rockville, MD.

94. Covey, L. S., and Tam, D. (1990). Depressive mood, the single-parent home, and adolescent cigarette smoking. *Am. J. Public Health* **80,** 1330–1333.

95. American Psychiatric Association (1994). "Diagnostic and Statistical Manual of Mental Disorders: DSM-IV," 4th ed., pp. 320–345. American Psychiatric Association, Washington, DC.

96. Weissman, M. M., Bruce, M. L., Leaf, P. J., Florio, L. P., and Holzer, C. (1991). Affective disorders. *In* "Psychiatric Disorders in America: The Epidemiologic Catchment Area Study," (L. N. Robins and D. A. Regier, eds), pp. 53–80. Free Press, Toronto.

97. Hughes, J. R., Hatsukami, D. K., Mitchell, J. E., and Dahlgren, L. A. (1986). Prevalence of smoking among psychiatric outpatients. *Am. J. Psychiatry* **143,** 993–997.

98. Pohl, R., Yeragani, V. K., Balon, R., Lycaki, H., and McBride, R. (1992). Smoking in patients with panic disorder. *Psychiatry Res.* **43,** 253–262.

99. Livson, N., and Leino, E. V. (1988). Cigarette smoking motives: Factorial structure and gender differences in a longitudinal study. *Int. J. Addict.* **23,** 535–544.

100. French, S. A., Jeffrey, R. W., Klesges, L. M., and Forster, J. L. (1995). Weight concerns and change in smoking behavior over two years in a working population. *Am. J. Public Health* **85,** 720–721.

101. Flegal, K. M., Troiano, R. P., Pamuk, E. R., Kuczmarski, R. J., and Campbell, S. M. (1995). The influence of smoking cessation on the prevalence of overweight in the U.S. *N. Engl. J. Med.* **333,** 1165–1170.

102. Yankelovich, Skelly, and White Inc. (1977). "Cigarette Smoking Among Teenagers and Young Women: Summary of Findings," Publ. No. (NIH) 77-1203. U.S. Gov. Printing Office, Washington, DC.

103. Urberg, K., and Robbins, R. L. (1981). Adolescents' perceptions of the costs and benefits associated with cigarette smoking: Sex differences and peer influences. *J. Youth Adolesc.* **10**, 353–361.

104. Swan, A. V., Creeser, R., and Murray M. (1990). When and why children first start to smoke. *Int. J. Epidemiol.* **19**, 323–330.

105. Murphy, N. T., and Price, C. J. (1988). The influences of self-esteem, parental smoking, and living in a tobacco production region on adolescent smoking behaviors. *J. Sch. Health* **58**, 401–405.

106. Friedman, L. S., and Lichtenstein, E. (1985). Smoking onset among teens: An empirical analysis of initial situations. *Addict. Behav.* **10**, 1–13.

107. Kandel, D. B., and Davies, M. (1986). Adult sequelae of adolescent depressive symptoms. *Arch. Gen. Psychiatry* **43**, 255–262.

108. Conrad, K. M., Flay, G. R., and Hill, D. (1992). Why children start smoking cigarettes: Predictors of onset. *Br. J. Addict.* **21**, 897–913.

109. Bauman, K. E., Fisher, L. A., Bryan, E. S., and Chenoweth, R. L. (1984). Antecedents, subjective expected utility, and behavior: A panel study of adolescent cigarette smoking. *Addict. Behav.* **9**, 121–136.

110. Chassin, L., Presson, C. C., and Sherman, S. J. (1991). Social psychological contributions to the understanding and prevention of adolescent cigarette smoking. *Pers. Soc. Psychol. Bull.* **16**, 133–151.

111. Office on Smoking and Health (1995). "Youth Risk Behavior Survey" (unpublished data).

112. French, S. A., Story, M., Downes, B., Resnick, M. D., and Blum, R. W. (1995). Frequent dieting among adolescents: Psychosocial and health behavior correlates. *Am. J. Public Health* **85**, 695–701.

113. French, S. A., Perry, C. L., Leon, G. R., and Fulkerson, J. A. (1994). Weight concerns, dieting behavior, and smoking initiation among adolescents: A Prospective Study. *Am. J. Public Health* **84**, 1818–1820.

114. Page, R. M., Allen, O., Moore, L., and Hewitt, C. (1993). Weight-related concerns and practices of male and female adolescent cigarette smokers and nonsmokers. *J. Health Educ.* **24**, 339–346.

115. Brunswick, A. F., and Messeri, P. (1983). Causal factors in onset of adolescents' cigarette smoking: A prospective study of urban black youth. *Adv. Alcohol Subst. Abuse* **3**, 35–52.

116. Gritz, E. R. (1984). Cigarette smoking by adolescent females: Implications for health and behavior. *Women's Health* **9**, 103–115.

117. Krohn, M. D., Haugthton, M. J., Skinner, W. F., Becker, S. L., and Lauer, R. M. (1986). Social disaffection, friendship patterns and adolescent cigarette use: The Muscatine Study. *J. Sch. Health* **56**, 146–150.

118. Chassin, L., Presson, C. C., Sherman, S. J., Montello, D., and McGrew, J. (1986). Changes in peer and parental influence during adolescence: Longitudinal versus cross-sectional perspectives on smoking initiation. *Dev. Psychol.* **22**, 327–334.

119. Hunter, S. M., Croft, J. B., Burke, G. L., Parker, F. C., Webber, L. S., and Berenson, G. S. (1987). Longitudinal patterns of cigarette smoking and smokeless tobacco use in youth: The Bogalusa Heart Study. *Am. J. Public Health* **76**, 193–195.

120. Cohen, D. A., Richardson, J. LaBree, L. (1994). Parenting behaviors and the onset of smoking and alcohol use: A Longitudinal Study. *Pediatrics* **94**, 368–375.

121. McNeill, A. D., Jarvis, M. J., and Stapleton, J. A., *et al.* (1988). Prospective study of factors predicting uptake of smoking in adolescents. *J. Epidemiol. Commun. Health* **43**, 72–78.

122. Federal Trade Commission (1997). "Report to Congress: Pursuant to the Federal Cigarette Labeling and Advertising Act." Federal Trade Commission, Washington, DC.

123. Center for Disease Control and Prevention (1990). Cigarette advertising—United States, 1988. *Morbid. Mortal. Wkly. Rep.* **39**, 261–265.

124. Warner, K. E., Goldenhar, L. M., and McLaughlin, C. G. (1992). Cigarette advertising and magazine coverage of the hazards of smoking: A statistical analysis. *N. Engl. J. Med.* **32**, 305–309.

125. Brandt A. M. (1996). Recruiting women smokers: The engineering of consent. *J. Am. Med. Women's Assoc.* **51**, 63–69.

126. Genther, J. (1960). "Taken at the Flood: The Story of Albert D. Lasker." Harper, New York.

127. Bernays, E. L. (1965). "Biography of an Idea: Memoirs of Public Relations Counsel Edward L. Bernays." Simon & Schuster, New York.

128. Burnett, L. (1961). "Communications of an Advertising Man." Burnett, Chicago.

129. O'Keefe, A. M., and Pollay, R. W. (1996). Deadly targeting of women in promoting cigarettes. *J. Am. Med. Women's Assoc.* **51**, 67–69.

130. Pierce, J. P., Gilpin, E., Burns, D. M., *et al.* (1991). Does tobacco advertising target young people to start smoking? Evidence from California. *JAMA, J. Am. Med. Assoc.* **266**, 3154–3158.

131. Pierce, J. P., and Gilpin, E. (1995). A historical analysis of tobacco marketing and uptake of smoking by youth in the United States: 1890–1977. *Health Psychol.* **14**, 500–508.

132. Food and Drug Administration (1996). Regulations restricting the sale and distribution of cigarettes and smokeless tobacco to children and adolescents: Final rule. *Fed. Regist.* **61**, 44395–44618.

133. U.S. Department of Health and Human Services (1988). "The Health Consequences of Smoking: Nicotine Addiction. A Report of the Surgeon General." U.S. Department of Health and Human Services, Centers for Disease Control, Office on Smoking and Health, Rockville, MD.

134. Hatsukami, D., Hughes, J. R., and Pickens, R. (1985). Characterization of tobacco withdrawal: Physiological and subjective effects. *In* "Pharmacological Adjuncts in Smoking Cessation," (J. Grabowski and S. M. Hall, eds.), Monogr. No. 53, DHHS Publ. No. (ADM) 85-1333. U.S. Department of Health and Human Services, National Institute on Drug Abuse, Rockville, MD.

135. George H. Gallup International Institute (1992). "Teen-Age Attitudes and Behavior Concerning Tobacco: Report of the Findings." George H. Gallup International Institute, Princeton, NJ.

136. McNeill, A. D., Jarvis, M. J., and West, R. J. (1987). Subjective effects of cigarette smoking in adolescents. *Psychopharmacology* **92**, 115–117.

137. Charlton, A. (1984). Children's opinions about smoking. *J. R. Coll. Gen. Pract.* **34**, 483–487.

138. Shinton, R., and Beevers, G. (1989). Meta-analysis of relation between cigarette smoking and stroke. *Br. Med. J.* **298**, 789–794.

139. U.S. Department of Health and Human Services (1989). "Reducing the Health Consequences of Smoking: 25 Years of Progress. A Report of the Surgeon General," DHHS Publ. No. (CDC)89-8411. U.S. Department of Health and Human Services, Public Health Service, Centers for Disease Control and Prevention, Office on Smoking and Health, Rockville, MD.

140. Thun, M. J., Day-Lally, C. A., Calle, E. E., Flanders, W. D., and Heath, C. W. (1995). Excess mortality among cigarette smokers: Changes in a 20-year interval. *Am. J. Public Health* **85**, 1223–1230.

141. Risch, H. A., Howe, G. R., Jain, M., Burch, J. D., Holowaty, E. J., and Miller, A. B. (1993). Are female smokers at higher risk for lung cancer than male smokers? *Am. J. Epidemiol.* **138**, 281–293.

142. Zang, E. A., and Wynder, E. L. (1996). Differences in lung cancer risk between men and women: Examination of the evidence. *J. Natl. Cancer Inst.* **88**, 183–192.

143. Brownson, R. C., Chang, J. C., and Davis, J. R. (1992). Gender and histologic type variations in smoking-related risk of lung cancer. *Epidemiology* **3**, 61–64.

144. Osann, K. E., Anton-Culver, H., Kurosaki, T., and Taylor, T. (1993). Sex differences in lung-cancer risk associated with cigarette smoking. *Int. J. Cancer* **54**, 44–48.

145. Wilcox, A. J. (1994). Letter to the editor. *Am. J. Epidemiol.* **140**, 186.

146. Hoover, D. R. (1994). Letter to the editor. *Am. J. Epidemiol.* **140**, 186–187.

146a. Landis, S. H. *et al.* (1998). Cancer Statistics, 1998. *Ca: Cancer J. Clin.* **48**, 6–29.

147. Kosary, C. L., Ries, L. A. G., Miller, B. A., *et al.* (1995). "SEER Cancer Statistics Review, 1973–1992: Tables and Graphs, NIH Publ. No. 96-2789. National Cancer Institute, Bethesda, MD.

148. American Cancer Society (1998). "Cancer Facts and Figures—1998." American Cancer Society, Atlanta, GA.

149. Samet, J. M. (1996). The changing cigarette and disease risk: Current status of the evidence. *In* "The FTC Cigarette Test Method for Determining Tar, Nicotine, and Carbon Monoxide Yields of U.S. Cigarettes: Report of the NCI Expert Committee. Smoking and Tobacco Control, Monogr. No. 7, NIH Publ. No. 96-4028. U.S. Department of Health and Human Services, National Institutes of Health, Bethesda, MD.

150. Ernster, V. L. (1996). Female lung cancer. *Annu. Rev. Public Health* **17**, 97–114.

151. Travis, W. D., Travis, L. B., and Devesa, S. S. (1995). Lung cancer. *Cancer (Philadelphia), Suppl.* **75**, 191–201.

152. Levi, F., Franceschi, S., La Vecchia, C., *et al.* (1997). Lung carcinoma trends by histologic type in Vaud and Neuchatel, Switzerland, 1974–1994. *Cancer (Philadelphia)* **79**, 906–914.

153. Thun, M. J., Lally, C. A., Flannery, J. T., Calle, E. E., Flanders, W. D., and Heath, C. W. (1997). Cigarettes smoking and changes in the histopathology of lung cancer. *J. Natl. Cancer Inst.* **89**, 1580–1586.

154. Centers for Disease Control and Prevention (1993). Cigarette smoking-attributable mortality and years of potential life lost—United States, 1990. *Morbid. Mortal. Wkly. Rep.* **42**, 645–649.

155. Thun, M. J., Myers, D. G., Day-Lally, C., *et al.* (1997). Age and exposure-response relations between cigarette smoking and premature death in Cancer Prevention Study II. *In* "Changes in Cigarette Related Disease Risks and Their Implication for Prevention and Control. Smoking and Tobacco Control (D. M. Burns, L. Garfinkel, and J. Samet, eds.), Monogr. No. 8, NIH Publ. No. 97-1213. U.S. Department of Health and Human Services, National Cancer Institute, Rockville, MD.

156. Kawachi, I., Colditz, G. A., Stampfer, M. J., *et al.* (1994). Smoking cessation and time course of decreased risks of coronary heart disease in middle-aged women. *Arch. Intern. Med.* **154**, 169–175.

157. Willet, W. C., Green, A., Stampfer, M. J. *et al.* (1987). Relative and absolute excess risks of coronary heart disease among women who smoke cigarettes. *N. Engl. J. Med.* **317**, 1301–1309.

158. McBride, P. E.. (1992). The health consequences of smoking. *Med. Clin. North Am.* **76**, 333–353.

159. Rosenberg, L., Kaufman, D. W., Helmrich, S. P., *et al.* (1985). Myocardial infarction and cigarette smoking in women younger than 50 years of age. *JAMA, J. Am. Med. Assoc.* **253**, 2965–2969.

160. Stokes, J., Kannel, W. B., Wolf, P. A., *et al.* (1987). The relative importance of selected risk factors for various manifestations of cardiovascular disease among men and women from 35 to 64 years old: 30 years of follow-up in the Framingham Study. *Circulation* **75**, V-65–V-73.

161. Levy, L. A. (1992). Smoking and peripheral vascular disease. *Clin. Podiatr. Med. Surg.* **9**, 165–171.

162. Howard, G., Wagenknecht, L. E., Burke, G. L., *et al.* (1988). Cigarette smoking and progression of atherosclerosis. *JAMA, J. Am. Med. Assoc.* **279**, 119–124.

163. Wolf, P. A., D'Agostino, R. B., Kannel, W. B., *et al.* (1988). Cigarette smoking as a risk factor for stroke: The Framingham Study. *JAMA, J. Am. Med. Assoc.* **259**, 1025–1029.

164. Bonner, L. L., Kanter, D. S., and Manson, J. E. (1995). Primary prevention of stroke. *N. Engl. J. Med.* **333**, 1392–1400.

165. Davis, R. M., and Novotny, T. E. (1989). The epidemiology of cigarette smoking and its impact on chronic obstructive pulmonary disease. *Am. Rev. Respir. Dis.* **140**, S82–S84.

166. U.S. Department of Health and Human Services (1984). "The Health Consequences of Smoking: Chronic Obstructive Lung Disease. A Report of the Surgeon General," DHHS Publ. No. (PHS) 84-50205. U.S. Department of Health and Human Services, Public Health Service, Office on Smoking and Health, Washington, DC.

167. Doll, R., Gary, R., Hafner, B., and Peto, R. (1980). Mortality in relation to smoking: 22 years' observations on female British doctors. *Br. Med. J.* **280**, 967–971.

168. Naeye, R. L. (1980). Abruption placentae and placenta previa: Frequency, perinatal mortality, and cigarette smoking. *Obstet. Gynecol.* **55**, 701–704.

169. Klonoff-Cohen, H. S., Edelstein, S. L., Lefkowitz, E. S., *et al.* (1995). The effect of passive smoking and tobacco exposure through breast milk on Sudden Infant Death Syndrome. *JAMA, J. Am. Med. Assoc.* **273**, 795–798.

170. Schoendorf, K. C., and Kiely, J. L. (1992). Relationship of Sudden Infant Death Syndrome to maternal smoking during and after pregnancy. *Pediatrics* **90**, 905–908.

171. Blair, P. S., Fleming, P. J., Bensley, D., *et al.* (1996). Smoking and the sudden infant death syndrome: Results from 1993–5 case-control study for confidential inquiry into stillbirths and deaths in infancy. *Br. Med. J.* **313**, 195–198.

172. Cummings, S. R., Nevitt, M. C., Borwner, W. C. *et al.* (1995). Risk factors for hip fracture in white women. *N. Engl. J. Med.* **332**, 767–773.

173. Bauer, D. C., Browner, W. S. *et al.* (1993). Factors associated with appendicular bone mass in older women. *Ann. Intern. Med.* **118**, 657–665.

174. U.S. Environmental Protection Agency (1993). "Respiratory Health Effects of Passive Smoking: Lung Cancer and Other Disorders," Rep. No. EPA/600/6-90/006F. U.S. Environmental Protection Agency, Office of Research and Development, Office of Atmospheric and Indoor Air Programs, Indoor Air Division, Washington, DC.

175. Pirkle, J. L., Flegal, K. M., Bernert, J. T., *et al.* (1996). Exposure of the U.S. populations to environmental tobacco smoke. *JAMA, J. Am. Med. Assoc.* **275**, 1233–1240.

176. Centers for Disease Control and Prevention (1997). State-specific prevalence of cigarette smoking among adults, and children's and adolescent's exposure to environmental tobacco smoke—United States, 1996. *Morbid. Mortal. Wkly. Rep.* **46**, 1038–1043.

177. California Environmental Protection Agency (1997). "Health Effect of Exposure to Environmental Tobacco Smoke." California Environmental Protection Agency, Office of Environmental Health Hazard Assessment, Sacramento.

178. U.S. Department of Health and Human Services (1986). "The Health Consequences of Involuntary Smoking: A Report of the Surgeon General," DHHS Publ. No. (CDC)87-8398. U.S. Department of Health and Human Services, Centers for Disease Control, Rockville, MD.

179. Speer, F. (1968). Tobacco and the nonsmoker: A study of subjective symptoms. *Arch. Environ. Health* **16**, 443–446.

180. National Research Council, Committee on Passive Smoking (1986). "Environmental Tobacco Smoke: Measuring Exposures and Assessing Health Effects." National Academy Press, Washington, DC.

181. Interagency Task Force on Environmental Cancer, Heart and Lung Disease (1985). "Environmental Cancer and Heart and Lung Disease: Annual Report to Congress (8th)." Interagency Task Force on Environmental Cancer, Heart and Lung Disease, Washington, DC.

182. Fontham, E. T., Correa, P., Reynolds, P., *et al.* (1994). Environmental tobacco smoke and lung cancer in nonsmoking women: A Multicenter Study. *JAMA, J. Am. Med. Assoc.* **271,** 1752–1759.

183. Law, M. R., Morris, J. K., and Wald, N. J. (1997). Environmental tobacco smoke exposure and ischaemic heart disease: An evaluation of the evidence. *Br. Med. J.* **315,** 973–980.

184. Steenland, K., Thun, M., Lally, C., and Heath, C., Jr. (1996). Environmental tobacco smoke and coronary heart disease in the American Cancer Society CPS-II Cohort. *Circulation* **94,** 622–628.

185. Mannino, D. M., Siegel, M., Husten, C., *et al.* (1996). Environmental tobacco smoke exposure and health effects in children: Results from the 1991 National Health Interview Survey. *Tob. Control* **5,** 13–18.

186. Martinez, F. D., Wright, A. L., and Taussig, L. M. (1994). Effect of paternal smoking on the birthweight of newborns whose mother did not smoke. *Am. J. Public Health* **84,** 1489–1491.

187. Windham, G. C., and Waller, E. A. (1995). Is environmental tobacco smoke exposure related to low birthweight? *Epidemiology* **6,** S41 (abstr.).

46

Nutrition, Weight, and Body Image

SHEILA HILL PARKER

Arizona Prevention Center
University of Arizona
Tucson, Arizona

I. Introduction

The 1990s have brought an explosion of health information and a variety of products to the public, requiring greater awareness, intelligence, and vigilance among the general public and health care professionals to determine what is truly "healthy" and "nutritious." The health professions, manufacturers, food establishments, fitness centers, and the media are all promoting their often conflicting and confusing messages about what the terms "healthy" and "nutritious" mean. Along with these messages come the images of the right size and shape of the "healthy" American woman: the extremely thin woman whose food behavior and intake often counters the accepted scientifically based recommendations for a balanced dietary pattern [1]. Unfortunately, many American women are more accepting of the social and media-generated image of the "ideal" American woman rather than the ideal dietary pattern for health maintenance and the prevention of chronic diseases.

Optimal nutrition for women is basic to health and is influenced by a variety of factors. In this chapter, some primary factors that determine nutritional status and health are presented: the nutritional demands of the life cycle, body weight and its influence on morbidity and mortality, the major effects of body image in nutritional choices and health behaviors, and the social and cultural dictates that shape body image among women. Therefore, the chapter first provides an overview of the nutritional needs and concerns of adolescents, adult women, pregnant women, and elderly women. Secondly, there is discussion of body composition and weight, the relationship among body weight, body fat, and caloric intake, and the nutritional and health implications of body weight and disordered eating behavior. Finally, body image and how it has historically influenced women's behaviors that directly impact women's health are examined. Health professionals will recognize the relationship of these topics to diagnosing, treating, and generally serving female clients for improving their health status. As in some other areas of women's health the epidemiologic data have been sparse, but clearly there is urgent need for promoting greater research on women's health and nutrition. Additional relevant research in the areas of nutrition, weight, and body image has particular relevance in clinical and public health, as health professionals assist women in making life-enhancing changes in their health behaviors.

II. Nutrition

Central to the health of women is nutrition. Nutritional needs and concerns of women are based on a variety of determinants such as age, stages of growth and development, socioeconomic factors, and culture. Beyond these determinants, women have distinct needs related to menarche, reproduction, lactation, and menopause. The appropriate dietary pattern for women supplies adequate amounts of all nutrients essential for life and optimal functioning. The essential nutrients must be supplied by a balanced food intake that is easily digestible, affordable, in proper quantities, and meets the individual's need for satiety and cultural gastronomic expectations [2]. Proper nutrition is central to health promotion, chronic disease prevention, and disease management. The Food and Nutrition Board of the National Research Council has responsibility for establishing recommended dietary allowances (RDAs) representing estimated standards for essential nutrients of protein, 11 vitamins, and 7 minerals to meet the documented nutritional needs of most healthy individuals of both genders at different ages [2]. The challenge has been to interpret the RDAs to the general public in a recommended eating pattern that can be universally understood and applied.

Government and volunteer agencies have developed various methods of dietary guidelines and dietary recommendations. The Committee of Diet and Health of the Food and Nutrition Board published the current recommendations to reduce risk of diet-related chronic diseases, such as coronary heart disease, cancers, obesity, osteoporosis, and diabetes [3]. The recommendations can be used as an optimal dietary pattern to meet the nutritional needs of most women in the premenopausal period (19 to 50 years of age) for maintaining optimal health. The recommendations also offer guidelines for the evaluation of dietary patterns in chronic disease prevention [1] and in nutrition intervention programs.

- Reduce total fat intake to 30% or less of total calories. Reduce saturated fatty acid intake to less than 10% of calories and the intake of cholesterol to less than 300 mg daily.
- Eat five or more servings of vegetables and fruits daily, especially green and yellow vegetables and citrus fruits. Also increase intake of starches and other complex carbohydrates by eating six or more servings of a combination of breads, cereals, and legumes.
- Maintain protein intake at moderate levels.
- Balance food intake and physical activity to maintain appropriate body weight.
- Alcohol consumption is not recommended. For those who drink alcoholic beverages, consumption should be limited to the equivalent of less than 11 ounces of pure alcohol in a single day. Pregnant women should avoid alcoholic beverages.
- Limit daily intake of salt (sodium chloride) to 6 g or less.

- Maintain adequate calcium intake.
- Avoid taking dietary supplements in excess of the RDA in any one day.

Women in adolescence, pregnancy, and the senior years will have nutritional needs specific to developmental tasks and changes.

A. Adolescence

National surveys and studies have identified inadequate dietary intakes among adolescent girls in the United States. Adolescent girls are more likely than boys to have low dietary intakes of the essential vitamins of folate, vitamin A, vitamin E, and vitamin B6, and minerals of iron, calcium, zinc, phosphorus, and magnesium [4]. Most adolescent girls consume adequate amounts of protein. However, their dietary intakes of fiber are low. The dietary excesses found among all groups of adolescent girls, regardless of income or ethnicity, are total fat, saturated fat, cholesterol, sodium, and sugar. Nutritional deficiencies during adolescence can retard potential growth, delay sexual maturation, and affect reproductive health, influencing the risk of fetal malformations and spontaneous abortion. Nutritional excesses and deficiencies are risk factors for chronic diseases in adulthood. The most common predictors for nutritional deficiencies are race and region. African-American girls and girls from the southern United States are most likely to present deficiencies in the essential vitamins and minerals [1,4–7].

The energy needs of adolescent females are based on the basal metabolic rate, growth rates, body composition, and physical activity. U.S. adolescent girls have mean energy intakes that are substantially below the RDA. This is of great concern because energy intake is the greatest predictor of nutritional adequacy and diet quality. Low energy intakes are associated with inadequate nutrient intakes, and higher energy intakes indicate higher intakes of essential nutrients. The growth spurt during adolescence is especially sensitive to energy and nutrient deprivation.

It is during adolescence that food behavior and body image issues are most pronounced and critical. At the time of adolescence, girls form their personal identity and self-concept as well as go through major physical growth and development [8]. Instrumental in the development of self-concept is the girl's positive body image. Caucasian girls suffer greatly from dissatisfaction with their bodies and may be involved in the "epidemic of dieting" [8]. This behavior certainly has major impact on caloric and nutrient intake at this time. Berg estimates that as many as 60–80% of American adolescent girls are dieting at any given time [9]. This does not appear to be true for all ethnic groups. Parker et al. found that among African-American adolescent girls, 70% of the informants said that they were satisfied or very satisfied with their current weight and shape. Almost 90% of the white adolescent girls in the study expressed some degree of dissatisfaction with body weight and shape, even when their weight was normal or below normal [10].

It is important that health care professionals be cognizant of the growth, development, body image issues, and sociocultural features that are characteristic of early adolescence (10 to 14 years old), middle adolescence (15 to 17 years of age), and late adolescence (18 to 21 years of age). The health care professional will then have the context in which to understand and intervene in the health behavior issues of each age group. Health and nutrition interventions that do not address the social, cognitive, and psychological needs in a culturally appropriate context will probably be ineffective and possibly harmful. For relevance and sociocultural competence, it is essential that professionals include representatives of the target population in the decision-making processes throughout the planning, implementation, and evaluation of interventions, programs, and services.

B. Adult Women

Kris-Etherton and Krummel [11] identified several factors that impact women's eating patterns and nutritional status: greater employment of women; the increase in female-headed households; increased access to commercial food establishments, especially fast food restaurants; increased availability and use of convenience foods; and the increased use of tobacco among women. By the year 2000 approximately 50% of the work force will be women, and 60% of all women will work outside of the home. In 1987, 73% was the maternal rate of employment outside the home in two-parent households [12]. For many women these trends are expected to have significant impacts on their own nutritional status and that of their families.

As greater numbers of women are employed outside the home, the income available to eat away from home increases. The results on the USDA Nationwide Food Consumption Survey (NFCS) and the USDA Continuing Survey of Food Intakes by Individuals (CSFII) show that eating away from home does affect the quality of women's diets. As the number of meals eaten away from home increases, there is higher consumption of calories, total fat and saturated fat, cholesterol, and sodium and significantly lower intakes of iron, calcium, vitamin C, and fiber. The use of convenience foods also increases with family income, women's employment, family size, the presence of children, and the age of the women. The effect of prepared food use on the quality of the diet has not been determined. Women who work outside of the home devote less time to family work and meal preparation. The family consumption of convenience or prepared foods also increases when women are employed, but the nutritional effect of this on women is not clear. Johnston et al. [12] found that the mothers' employment does not negatively affect the diets of their children.

Between 1970 and 1992, female-headed households grew from 12% of all family groups with children under age 18 years to 26% [13]. In 1991, the U.S. Census's report on poverty found that 55% of female-headed households were below the poverty line compared to 14% of all other family groups. This group of women has less formal education and experiences higher unemployment. Because many of these women face the severity of poverty, their risks for nutritionally inadequate diets increase. Research finds that food insufficiency (inadequate amounts of quality food supplies) is not limited to very low-income persons, specific ethnic or racial groups, family types, or the unemployed. Alaimo et al. [14] examined data from the National Health and Nutrition Examination Surveys III (NHANES III) to determine the prevalence and characteristics of food insufficiency in the United States. Individuals were classified as "food insufficient" if a family respondent reported that the family often or sometimes did not have enough food to eat. Food insufficiency from 1988 through 1994 had a prevalence of 4.1% (approximately 9–12 million people). The highest food insufficiency prevalence was among Mexican-Americans (15.2%), followed by non-Hispanic black Americans (7.7%), and was lowest among non-Hispanic white Americans (2.5%). Overall, 53.5% of food insufficient individuals lived in families headed by employed individuals. Federally funded programs, such as the Food Stamp Program, the Special Supplemental Program for Women, Infants, and Children (WIC), and the Cooperative Extension Expanded Food and Nutrition Education Program (EFNEP), attempt to provide necessary support for improving food and nutrition behaviors and health status of women and their families.

Jeffery and French investigated the association between television viewing, fast food eating, and body mass index for one year in 1059 men and women [15]. They found that television viewing hours were positively associated with fast food eating, energy intake, percentage of energy intake from fat, and body mass index in women but not in men. Television viewing predicted weight gain in high-income women.

The prevalence rates for smoking among women 20 years of age or older has declined in the United States since 1974, but more slowly than among men. Women smokers have significantly poorer diets than do nonsmoking women, according to the findings of the CSFII 1985. Women smokers had lower intakes of protein, fruits, vegetables, whole grain products, fiber, vitamin C, thiamin, and carotenoids. Among women who smoke, the intakes of meat, alcohol, fat, soft drinks, and cholesterol are higher than intakes among nonsmokers. The dietary patterns of those who successfully stop smoking more closely resemble the diets of nonsmokers than current smokers [1]. The use of tobacco is itself a major high-risk behavior for chronic disease development and pregnancy complications. The high risk for health problems becomes even greater when smoking is associated with poor nutritional status in women.

The relationship between diet, health maintenance, and disease prevention is well established by the consistent findings of scientific studies. Epidemiologic findings consistently describe the positive relationship between total dietary fat consumption and cardiovascular disease. Populations with high intakes of plant foods (fruits, vegetables, and grain products) have lower risk of some cancers and coronary heart disease because the diets are also low in total fat, saturated fat, and cholesterol and high in dietary fiber, vitamin C, and the carotenoids.

Early results of the NHANES III demonstrate that females of all racial and ethnic groups consumed 34% of their total caloric intake from total fat and 12% from saturated fat, which far exceeds the current dietary recommendations. For calcium intake, the mean consumption was 744 mg for all females, well below the recommended 1200 mg for females 11 to 24 years old and pregnant and lactating women, and below the 800 mg for women 25 years of age and over. The mean intake of iron (12.37 mg) was below the recommendation of 15 mg for nonpregnant women under 50. The mean intakes of vitamin A (884 retinol equivalents) and vitamin C (95 mg) exceeded the recommendations for most nonpregnant and nonlactating adult women [6,7].

C. Pregnant Women

Pregnancy and lactation are critical times for women because of special nutritional demands. Proper nutrition during these times is critical for fetal growth and development as well as maternal health during the pregnancy. During pregnancy, nutrient needs may be so high that supplementation may be necessary. Appropriate nutrient intake is strongly associated with energy intake and weight gain during pregnancy.

A retrospective study compared infant birth outcomes (small for gestational age (SGA), large for gestational age (LGA), and cesarean delivery) to the Institute of Medicine (IOM) weight gain recommendations [16]. The IOM prenatal weight gain recommendations are: for women with low prepregnancy body mass index (BMI) (<19.8), the recommendation is 12.5–18 kg (28–40 lbs); for women of normal BMI (19.8–26.0), the recommendation is 11.5–16 kg (25–35 lbs); and for women of high BMI (26.0–29.0), the recommendation is 7–11.5 kg (15–25 lbs). The results revealed reduced risks of SGA, LGA, or cesarean delivery when women gained weight during pregnancy within the IOM guidelines. Both inadequate and excessive prenatal weight gains are associated with poor infant outcomes [16]. The IOM recommends that young adolescents and black women should gain weight at the upper end of the recommended range (12.5–18 kg; 28–40 lbs) based on reports that black infants tend to be smaller than white babies for the same weight gain of the mothers. For obese women, low maternal weight gain was significantly related to birth of SGA infants [16].

Weight retention after pregnancy has important implications for what is considered normal or ideal weight for women who have experienced childbirth. Weight charts and sometimes health professionals treat all women the same rather than recognizing differences that may exist among those who are childbearing and those who are not and among those of various racial groups. Normal and excessive prenatal weight gain may result in greater postpartum weight retention. Researchers surveyed 2000 women with normal prepregnancy BMI. The women who gained more than 35 lbs prenatally experienced significantly greater weight retention (20 lbs or more) for the 10–24 months postpartum than those who gained less weight. Interestingly, black women retained more weight than white women. Studies regarding postpartum weight loss indicate that black and white women are most likely to increase their body weight after the first pregnancy and not after later pregnancies [17]. Evidence has shown that prolonged lactation for greater than three months enhances postpartum weight loss. The study included no lactating women who were intentionally dieting to lose weight.

D. Elderly Women

As the American population continues to grow, the older adult population of 65 years of age and above also continues to grow. By the year 2020, there will be approximately 50 million people over 65 and more than 70% of these will be women [18]. Nutrition contributes greatly to the maintenance of optimal health in older adults. There is concern about the association of emotional distress, health, and nutrition. Depression is a significant factor in the loss of appetite among the elderly. Intervention to mediate depression is vital because continued loss of appetite leads to chronically decreased energy and nutrient intake causing subclinical malnutrition. If the individual suffers a traumatic event the consequences of malnutrition can complicate the outcome [18].

Aging brings a decrease in energy requirements, responding to the reduction in lean body mass and all the shrinking protein compartments in the body. Other reasons for the decrease in energy requirements are reduction in the time and intensity involved in physical activity and changes in hormone production, especially growth hormone and testosterone [19]. Studies by Kohrt *et al.* [20] and Owens *et al.* [21] have confirmed that exercise or physical training is related to higher levels of energy and nutrient intake in elderly women. Physical exercise can contribute to the retention of lean body mass, muscle, and bone density, as well as contribute to women's emotional well-being.

While the energy requirement decreases in aging, the requirements for nutrients do not decline. The inclusion of nutrient-dense foods in the dietary pattern and at the same time lower calories should be encouraged for the elderly. The research on nutritional status of elderly women has been inadequate and variable in results. The need to raise the dietary calcium recommendation has been debated, given the high incidence of osteoporosis among postmenopausal women. Recommendations are to increase the calcium recommendation from 800 mg/day to 1000 mg/day for postmenopausal women receiving hormone replacement therapy and to 1500 mg/day for women not receiving hormone replacement therapy. Other factors necessary for preserving bone density are recommended intakes of vitamin D and weight-bearing exercise [22]. Studies also show that weight and body mass index affect bone density. Felson *et al.* [23] and Tremollieres *et al.* [24] documented that postmenopausal bone loss is significantly reduced in overweight women. Research also suggests that differences in calcium requirements should be studied within ethnic and cultural groups and not across them. These studies identify ethnic differences that are currently poorly understood. For example, it appears that black Americans, while consuming lower calcium intakes than white Americans, use and conserve calcium more efficiently than whites. Blacks preserve more bone and have fewer fractures. Asians also have low calcium intakes and lose bone at the same rates as whites but experience only half the hip fracture risk of whites. Poor populations that have low consumptions of meat and salt have much lower requirements for calcium than populations that are affluent and have higher intakes of meat and salt [25].

III. Weight

A. Body Composition and Weight

Understanding body composition helps to explain the recommendations for body size and body fat. However, there is often misunderstanding among the general population and sometimes health professionals about how this information is translated and used in recommendations for optimal health. Body composition can be defined by the chemical composition of the body's stores of water, bone tissue, connective tissue, soft tissue, and adipose tissue. The study of body composition takes into ac-

count the functional, physiological, genetic, and nutritional factors that are significant determinants of body tissue structure. Body composition also reflects the individual's age and stage of development. Several methods exist for measuring body composition. Those methods reveal that approximately 60% of a healthy adult's weight is water, 6–22% is protein, and 3% is minerals. Body fat accounts for most of the remaining weight. The adult male consists of about 5–15% fat, while the adult woman consists of 18–25% fat. An overview of some of the most common methods for determining body composition are presented in this chapter.

The difficulties of studying the human body by direct chemical analysis and the desirability of studying *in vitro* measurements has led to the development of indirect methods of determining body composition, body density, and body fat. Anthropometric body measurements provide a good estimation of fat stores and gives indications of body composition. Traditionally, height and weight measurements have been utilized to relate physical size to an estimated set of standards established by the National Research Council. However, bone, muscle, and fat cannot be quantified accurately by these gross measurements. The use of skinfold calipers can aid in the estimation of body fat stores via measurement of the triceps, abdominal, and mid-thigh regions.

Hydrostatic weighing, or underwater weighing, is one of the most reliable methods used for estimating the size of body fat and fat-free compartments. Submerging the entire body in a tank full of water and then recording the air volume in the lungs and measuring the rise in water volume can provide statistics that yield the body density of the subject through mathematical equations. Body fat is less dense than lean tissue or water.

Bioelectrical impedance uses electrical currents to determine body composition. Water and certain minerals conduct electrical currents but fat does not. Bioelectrical impedance has been effective in estimating individual percentage of body fat.

Medical experts in 1959 developed tables known as the Metropolitan Life Insurance Height/Weight Tables that were used as the standards for acceptable weight for Americans, listing desirable weights for adults related to heights. These were based on the analysis of health data from several life insurance policyholders. The data demonstrated an association between body weight, an indirect measurement of body fat, and mortality rates. In 1983 the Metropolitan Life Insurance Company revised the tables reflecting "desirable" weights for men and women that were heavier than the 1959 standards. The changes were based on increased weights of Americans. Many health professionals concerned about the weight gain of the increasingly sedentary American population criticized the changes.

Metropolitan Life Insurance norms are often used to define underweight, normal weight, overweight, and obesity. However, it has become more common to compute the body mass index (BMI), which is a person's weight (in kilograms) divided by their height (in meters) squared. A BMI of 19–24 corresponds to "ideal" body weight, a BMI of 25–29 is moderately overweight (approximately 15–30% over ideal weight), and a BMI of more than 31 is considered severely obese (more than 40% over ideal body weight). There continues to be debate among health professionals about what "ideal weight" actually means.

B. Weight, Body Fat, and Caloric Intake

Deviations from the accepted weight norms depend on energy balance. Maintaining caloric or energy balance is a challenge throughout the life cycle. As long as caloric intake is equal to caloric expenditure there should be no change in body weight beyond a normal day-to-day fluctuation. As soon as caloric intake deviates from caloric expenditure, calories are either derived from body stores to meet a deficit or are added to body stores in cases of surplus. When caloric intake is less than caloric expenditures, the body loses weight as its cells burn the fat stored in fat cells. If caloric intake is more than caloric expenditures, the body conserves much of the excess calories as body fat, and weight gain occurs. The physics of weight control and energy balance are straightforward, but beyond the physics and the physiological factors related to body fat and body weight are psychological and sociocultural factors that relate to the weight or the amount of fat that a woman carries on her body. The health professions as well as society-at-large place great importance on weight as a way to define what is healthy, what is ideal, what is acceptable, and even what is beautiful. This in turn has generated behaviors and attitudes related to body fat, body size, and body shape among women that seriously impact their self-esteem and self-worth. Their physical and psychological health can be impaired with flawed understandings about body weight, body fat, and energy requirements for life. The words "ideal weight," "thinness," "overweight," and "obesity" have meanings far beyond describing an individual's body composition. These terms have various meanings for individuals given their sociocultural background and experiences. Some of these will be addressed in Section IV.

C. Body Weight, Body Fat, and Health Implications

For women throughout the life cycle, body weight is strongly associated with a variety of risk factors and health outcomes. Energy intake directly affects body weight, body fat, essential nutrient intakes, and nutrient stores. Most of the literature has shown that obesity and overweight are risk factors for a variety of diseases, such as cardiovascular disease, and indirectly contribute to diabetes, hypertension, dyslipidemia, and osteoarthritis [26]. The terms obesity and overweight are often used interchangeably in the literature, although they do not mean or describe the same conditions. Obesity can best be described as a bodily state characterized by the excessive accumulation of fatty tissue beneath the skin and within the tissues of body organs. Obesity becomes a disease when it interferes with the body and its parts. Overweight generally means exceeding an "ideal" weight related to gender, height, and frame size. Standards for "ideal" weight do not address differences in body composition, activity level, genetics, or culture [27]. Most importantly, the scientific community cannot precisely define the ideal weight related to disease prevention that is separate from other risk factors. The definition for an ideal, optimal, healthy, or normal weight continues to be debated in the scientific community.

Studies have presented results that conflict with long held beliefs about the relationship of obesity, mortality, and morbidity. For example, hypertension has been identified as a crucial determinant of ill health and cardiovascular disease among Americans, especially black Americans, and it accounts for a large part of the black–white differential in health. Cooper et al. sought to describe the pattern of hypertension prevalence determined in populations of West African descent in Africa, the Caribbean, and the United States [28]. They examined, through a standardized protocol, the cross-cultural association of hypertension with obesity and sodium and potassium intakes among populations sharing the same genetic ancestry but living under very different social and economic conditions. The prevalence of hypertension follows the migration pattern of people of African origin from Africa to the United States, rising from 16% in West Africa, to 26% in the Caribbean, to 33% in the United States. Measures of body size (mean weight, body mass index, and waist-to-hip ratio) and sodium and potassium intakes varied consistently with hypertension prevalence across the sociocultural gradient, with the lowest values in Africa and the highest in the United States. The study demonstrated the role of social conditions and accompanying stress in the evolution of hypertension risk in populations of African genetic ancestry.

Some research reminds us that weight gain is not always a negative health factor. Smoking impacts negatively on women's nutritional status. One of the factors that hinder women from smoking cessation is the fear of weight gain. Burnette and colleagues studied the relationship between smoking cessation and weight gain in women from premenopause to the first and second years postmenopause in a prospective study measuring coronary heart disease risk factor changes. Although exsmokers gained substantially more weight than nonsmokers, they did not experience a greater increase in cardiovascular risk factors. In fact, the weight gain appeared to be associated with positive changes in HDL cholesterol levels in exsmokers [29].

Hip fractures in the elderly are related to reduced bone density, nutritional status, and weight-bearing exercise. As presented earlier in this chapter, the mean calcium intake for all adult women is well below the recommended 1200 mg/day. Poor calcium intake and postmenopausal bone loss especially in underweight women have important implications for predisposition to hip fractures. A prospective study assessed the independent effect of hip fracture on mortality, hospitalization, and functional status among 7527 individuals over 70 years of age. Between 1984 and 1991, 368 persons had hip fractures. The study demonstrated that the health of older adults deteriorates after a hip fracture. Hip fracture significantly increased the likelihood of subsequent hospitalization and is significantly related to mortality. Hip fracture also increased the number of functional status dependencies [30].

Diehr et al. [31] investigated the relationship of body mass index to 5-year mortality in a cohort of 4317 nonsmoking men and women aged 65 to 100 years old. They found an inverse relationship between body mass index and mortality. Mortality rates were higher for those who had the lowest BMI and who had lost 10% or more of their body weight since age 50. The association between higher body mass index and mortality often found in middle-aged adults was not found in this large cohort of older adults. Women with a BMI of 20 or lower had higher mortality than others. There was not a significant relationship between BMI and mortality in men.

Bender and colleagues [32] followed 6193 men and women in various areas of Germany for approximately 14 years. Mortality data on these patients, with BMIs ranging from 25–74 kg/m², were compared to data on the general population. Researchers found that moderate obesity, or BMI between 25 and 32, did not raise the risk of mortality at all in women, and only slightly in men. Grossly obese women, with BMIs between 32 and 40, only showed a 20–27% increase in risk of mortality, while men had risk increases of about 30% for BMIs of 32 to less than 36. For men with BMIs of 36 to less than 40, the risk of mortality rose to about 90%. Men and women defined as morbidly obese, with BMIs above 40, had the risk of premature death increased substantially, but the risks still are considerably lower than previously assumed in the research literature [32].

Manson *et al.* studied the relationship between body weight and mortality of U.S. women participating in the Nurses Health Study [32a]. The study included 4726 deaths among the cohort of 115,195 women who began the study in 1976 when they were between 30 and 55 years of age. Initially all of the women were free of cardiovascular disease and cancer. Over the sixteen years of follow-up, Manson *et al.* found no increase in risk among the leaner women who had never smoked. However, as the BMI increased among these women so did the relative risk. Among women who never smoked with a BMI of 32 or higher, the relative risk of death from cardiovascular disease was 4.1 and that of death from cancer was 2.1 compared to women with a BMI below 19. The lowest mortality was among women who weighed at least 15% less than the U.S. average weight for women of similar age and whose weight had been stable since early adulthood.

Clearly, there are conflicting findings in the scientific community regarding the health implications of obesity and overweight. It is also apparent that underweight or thinness carry some important morbidity and mortality consequences. Women are placing themselves in jeopardy of poor health and increased risk of mortality to achieve an "ideal body." While excessive leanness is an ideal for many women, it is detrimental for the skeleton and for bone density. Semi-starvation diets result in inadequate intakes of essential nutrients. Reduction in body fat results in states of relative estrogen deficiency well before menopause. Thin women lose a greater fraction of bone at menopause than do heavy women. Finally, leanness is harmful because of insufficient soft tissue padding around bony prominences to protect them adequately in the event of injury [25]. The preoccupation with thinness and weight reduction certainly poses problems for women's nutritional status as well as for their psychological well being. Disordered eating and eating disorders have become problematic resolutions to significant issues intrinsic to growing up female in today's society. Anorexia nervosa (AN), bulimia nervosa (BN), and binge eating disorder (BED) are three major eating disorders recognized in the health field today [33]. These disorders are clear indicators of body image disturbances and nutritional problems.

The core feature of AN is the relentless pursuit of thinness, resulting in severe emaciation, nutritional deficiencies, and amenorrhea in women. Diagnostic criteria include refusal to maintain a weight above a minimally normal level for age and height (operationalized as 85% of ideal body weight), intense fears of gaining weight or becoming fat, body image disturbance such as denial of the seriousness of the current low body weight or undue influence of weight and shape on self-evaluation, and finally, in postmenarcheal women, the absence of at least three consecutive menstrual cycles. There are two clinical subtypes of AN. The restricting subtype of AN uses dieting, fasting, and excessive exercise as a means of losing weight. Individuals with the binge-eating/purging subtype of AN regularly engage in binge eating or purging behavior (self-induced vomiting, laxative abuse, and diuretics) [34]. Girls or women account for more than 90% of the reported cases of AN. Among female adolescents and young adults, the prevalence of AN is approximately 1% [35].

BN is characterized by recurrent episodes of binge eating followed by extreme behaviors aimed at controlling body weight and shape, such as self-induced vomiting; the abuse of laxatives, diuretics, or enemas; extreme dieting or fasting; or excessive exercise. Additionally, individuals with BN report a persistent preoccupation with thinness, judging their self-worth excessively in terms of body shape and weight. Individuals with the purging subtype of BN regularly self-induce vomiting or abuse laxatives, diuretics, or enemas. Individuals with the nonpurging subtype of BN regularly use other inappropriate compensatory behavior, such as restrictive dieting, fasting, or excessive exercise [33]. Young adult females make up the overwhelming majority of patients with BN, with males accounting for one-tenth of BN cases. Epidemiologic studies of BN report prevalence rates of 1–2% of adult females [36–38]. BN and AN are identified as disorders of white, affluent Western cultures, yet eating-related problems in general, and perhaps more specifically BN, also are present and possibly increasing in non-white groups, including African-Americans and women in developing non-Western cultures preoccupied with eating and weight concerns [39,40]. In the Teen Lifestyle Project, the very few participating Hispanic-American girls and Asian-American girls expressed the same "ideal" body as the white adolescent girls [10].

BED is one example of an eating disorder not otherwise specified (EDNOS) and is characterized by recurrent episodes of binge eating. The binge eating disorder includes eating an unusually large amount of food accompanied by feelings of loss of control, without regular attempts to engage in the extreme compensatory weight control practices found in BN, such as purging or fasting. In addition to recurrent binge eating, behavioral indicators also must be present, such as eating faster than normal, eating until uncomfortably full, eating a large amount of food when not physically hungry, eating alone because of social embarrassment about the amount of food eaten, and feeling disgust, depression, or guilt after overeating.

Approximately 2% of adult women meet criteria for BED. BED is found across all weight groups, although initial reports of community-based samples suggest that about half of those with BED are obese [41,42]. BED affects 5–10% of obese individuals in the community [43]. The relationship between binge eating and obesity is not well understood and future research is needed to clarify the ways in which binge eating may be linked causally to obesity and vice versa. Individuals with BED are more diverse, varying across categories of gender, age, and race, than is the case with the other eating disorders. Black women and white women seem to be at equal risk for BED [44]. Mean age of BED is 26 years [45]. Two distinct patterns develop. In approximately half the cases, dieting preceded the onset of BED. For the other half, onset occurred without dieting to lose weight.

The eating disturbances presented have direct impacts on women's health and nutritional status, including menstrual and reproductive problems, especially among those with AN and BN; low bone-mineral density, which is associated with an increased risk of developing stress fractures, crush fractures, and osteoporosis; and gastrointestinal problems associated with vomiting and laxative abuse. Cardiovascular abnormalities such as bradycardia and hypotension are present in many AN and BN patients [46–48]. Other difficulties that impact nutritional and health status are severe dental erosion, greater prevalence of dental decay, increased tooth sensitivity, and periodontal disease due to regular vomiting. Fluid and electrolyte imbalance appears most commonly among individuals with bulimic symptoms. About 50% of this group experience some sort of electrolyte abnormality. Because of depletion of body fluids lost through vomiting, metabolic alkalosis and hypokalemia are seen among individuals who vomit regularly. Those abusing laxatives frequently exhibit the opposite phenomenon of metabolic acidosis, or acidic blood, which also is due to a depletion of the body's fluids.

Not only will women with eating disturbances affect their own nutrition and health, but there is evidence that women with eating disturbances are at greater risk for providing inadequate nutrition to their children [49]. Compared to women without eating disturbances, women with eating disturbances tend to be more intrusive with their infants during play and meal times by being more critical and demonstrating more negative affect during meal times, often expressing anxiety and disgust regarding the mess that their children make while eating [49]. Moreover, eating disturbances among adolescent girls appear to be associated with increased criticisms about weight and appearance by their mothers [50].

IV. Body Image

A. Physical Appearance Stereotyping

The pursuit of youth, beauty, and thinness is the absorption of many in American society. The literature reveals that this pursuit can lead to increased risk for premature mortality and morbidity. Often the preoccupation is accompanied by a distorted body image. Body image is the mental picture that a person has of his or her body that directly affects one's self-concept, self-esteem, and body esteem [51]. Most people will make judgements about their own body image by comparing it to a standard that is determined to be the "ideal body." The ideal body is influenced by one's culture, society's standards of appearance, and by those who are significant in the individual's relationships. Many women in Western societies are increasingly concerned with their bodies. They view themselves as overweight or obese with the associated negative stereotypes and suffer from poor self-esteem, whether in reality they are or are not overweight or obese. They believe the many messages in our society from a variety of agents (parents, teachers, peers, employers, health professionals, media, etc.) that may promote

an unrealistic ideal of thinness. An unrealistic ideal of thinness is associated with social and personal acceptance, elevated status, sexual attractiveness, beauty, youthfulness, and personal power to control one's appearance and success based on appearance. The failure to attain this ideal causes some women to determine that they are unattractive and failures, lowering their self-esteem. One's physical appearance receives great attention, is assigned value, and leads to physical attractiveness stereotyping by the general population and among many health care providers as well.

Human beings indulge in a variety of stereotyping behaviors that relate to sex, age, ethnicity, color, body size, shape, etc. Since the 1980s the research literature indicates that an individual's physical attractiveness does affect the impressions that others may ascribe to him or her. Dion et al. [52] found physical attractiveness stereotyping among male and female university students to be a function of giving their favorable evaluations of peers in a variety of social, experiential, and personality domains, except for "expected parental competence." The researchers observed a pattern that they summarized as "What is beautiful is good." Others have documented similar stereotyping related to physical attractiveness, especially among white middle class females [10,53,54]. In these studies, slenderness is essential to attractiveness and is a key component for interpersonal success. Parker et al. [10] determined that among white adolescent girls in the Teen Lifestyle Project, images of beauty were strongly influenced by thinness and equated to being "perfect" in their appearance, relationships, and personal lives.

Possibly from birth, as a function of socialization or social learning, individuals learn to make differential judgements about people who differ in physical attractiveness [55]. A variety of social agents have important roles influencing children, youth, and adults in judging differences in attractiveness and assigning value, expectations, rewards, and penalties to those differences. Various studies document that adults make stereotypical judgements about children and their behavioral expectations of children. Langlois found that attractive children are thought to possess and exhibit more positive traits and behaviors than unattractive children [56]. Some stereotypes are learned and may be consciously reinforced for specific social identities and roles (i.e., traditional gender roles). Other stereotypes, such as physical attractiveness stereotypes, may develop relatively informally, but are no less profound in impact. Some people may differ with this opinion of informality, given the very conscious and high-powered efforts of marketing in the cosmetic, beauty, weight loss, and food industries in the media to reach its target audiences. Even in traditional stories for children the hero and heroine are usually portrayed as physically attractive with positive attributes. The villains are typically presented as unattractive with negative or evil attributes.

Obviously, stereotyping of physical attractiveness as it develops through social learning has important advantages for some and detrimental consequences for others. Physical attractiveness is most certainly influenced by sociocultural factors. Cultures that strongly value individualism may also place greater emphasis on this stereotype. Societies and cultures that place greater importance on collective or group values may place less emphasis on physical stereotyping [57,58]. Researchers also find that attractiveness seems to provide a consistent social advantage for girls but not boys [56]. Physical attractiveness may also carry negative consequences in terms of the unwanted attention that it brings to girls and women.

B. Historical Perspective and Trends

Attractiveness, beauty, and physical appearance standards are largely established by societal preferences and social pressures for conformity to those standards. Women throughout history have been the recipients of specific dictates, prescriptions, and proscriptions about their physical appearance, body size, and shape in order to attain social acceptability. Lewis and Blair suggest that female lines and contours, or the lack of them, have been marketed as commodities like the lines and contours of automobiles [59]. Often the social ideal and what is marketable in one decade is stigmatized in the next [59]. Currently, society's strongest agent of images and change is the media, which dictates the context and standards by which women become objects confounded by distressed eating and distorted body images. An overview of history, especially Western fashion history, illustrates the experiences of women trying to adhere to social and fashion decrees.

> As clothing forms and fashions of each age changed, figure emphasis areas and pose patterns developed. There has been a direct relationship between the part of the figure receiving the focus of attention and the resulting posturing, pose pattern, or movement [60].

To demonstrate women's current relationships to the physical attractiveness stereotyping, the resulting body images, and health-related behaviors, history can offer many examples. In sixteenth century Italy, women with waists measuring more than 13 inches were barred from the court of Catherine de'Medici. The seventeenth century brought the Rubenesque body type as the aesthetic ideal. The artists, Rubens and Rembrandt, depicted women possessing body weights that would be considerably overweight by current standards. These aesthetic ideals for women's body shapes continued until well into the nineteenth century.

By the nineteenth century, some dissatisfaction was expressed about the ways that women's roles and potential contributions to society were restricted. Katherine Frank writes that the author Emily Bronte was anorexic as a result of her rebellion against her powerlessness and lack of opportunities [61]. The poetess Elizabeth Browning (1809–1861) may also have suffered from anorexia nervosa. Anorexia nervosa may have been more prevalent historically than has been documented, but the motivations for such behavior may have been quite different in past centuries than currently. Social rebellion, political protests, or religious fasting seem to have been reasons for disordered eating, as opposed to the body image distortions that are now associated with such eating behaviors [59].

The twentieth century brought many changes for both men and women in Western societies. The movement for women's suffrage became active, resulting in the right to vote. The aesthetic ideal for women's body shapes was also changing. The Duchess of Windsor, Wallis Simpson, was quoted as saying that a woman could never be "too rich or too thin [59]." With the vote, women gradually began to take advantage of higher education in increasing numbers prior to World War II. With World War II, women were recruited into the work force for jobs previously held predominantly by men until they became more involved in military service. During these years, women experienced greater paid employment and relative independence for positions of responsibility and decision making outside of their homes. After the war, as men returned from military service, women were encouraged to give up their paid positions and return to their homes to their more traditional roles. Orbach notes that with this overt campaign to get women back into the home came a focus on the woman's traditional role in the household [62]. Also since World War II, the documented incidence of difficulties associated with food and body image continued to increase [63].

In the 1960s, a new role model emerged for women, especially for the young white Western middle-class female. Her name was Twiggy, a very thin British fashion model. Society, through fashion and media, sent conflicting messages to young middle class women, especially white women, to be thin, but nurturing to others; be educated, but sacrifice your training in order to care for others; be mature, but look like a prepubescent female. Expanding new opportunities made it possible for women to be all: mother, career person, and superwoman [59]. With the new social prescriptions for how women should look and how they should behave, there has been major growth in various industries of products and services to help women achieve the thin body shape that will supposedly help women to be all and to do it all. There are a myriad of diet, weight loss, exercise, and slimming products, regimens, and publications available to the public, generating billions of dollars. Women with distorted body images are prime targets for these industries because many are in a continuous state of dissatisfaction with their bodies and themselves.

Not all women participate in this pursuit of thinness. African-American women across all groups have an overweight prevalence of 48.6% as compared to 32.1% for white women [64]. It is generally known that African-American women tend to accept a larger body size than white American women, finding the usual social image to be inconsistent with their own body image [10,27,65]. Walcott-McQuigg studied weight and weight control behavior among middle income African-American women. When asked how they determined their ideal weight, more than 50% of the women did so by reflecting on a weight at which they "felt lighter," "felt healthier," and "felt comfortable." The remaining women determined their ideal weight based on varied resources: information from health professionals, health club personnel, height and weight charts, and weight management staff [66]. Most women were reluctant to accept or value the Metropolitan Life Insurance standards as relevant for them. They opted to use "an internal barometer" to achieve and maintain body weight [65]. They depended on the "self-diet" and their own weight management regimens to monitor and manage their weight. Accepting the social image of dieting and weight management would involve accepting opinions of others and external body size criteria, such as those promoted in the media and height-weight tables. The women were motivated by internal cues to reduce and avoid self-perceived weight problems and to engage in weight control behaviors. Women reported that time constraints associated with work and family responsibilities, occupational endeavors, and a variety of life events would impact or sabotage attempts at weight management. Powell and Kahn [66] found that African-American men are less likely than white American men to report sanctions for dating larger than "ideal-sized" women. Among African-American women, weight was not perceived as a barrier to dating, dancing, exercising, and sports. If women do not perceive overweight as a problem, sanctions for the failure to achieve successful weight loss may be minimal.

V. Conclusions

The health and scientific communities have become significant agents in health prescriptions and in influencing the social prescriptions to women, along with other social agents, about their body size and shape. Often the prescriptions for health are neither clear for the professional nor for the population-at-large. Sometimes scientific research results are used and misused by the media and other social agents who present confusing and conflicting messages about women's health, nutrition, physical activity, weight, fat, etc. This chapter has presented a review of the scientific literature that demonstrates what is known about nutrition, body weight, and body image and how these are interrelated in their impact on women's health.

A woman's nutritional status is central to her overall health, and it is affected by the demands of the life cycle, socioeconomic factors, culture, and her body image. While official health organizations have developed various guidelines for the public to use in making health-enhancing dietary choices, many American women exhibit nutritional excesses and deficiencies that are risk factors for chronic diseases. Adolescent women, regardless of income and ethnicity, show excesses in their intakes of total fat, saturated fat, cholesterol, sodium, and sugar. Their diets are usually deficient in fiber, folate, vitamin A, vitamin E, vitamin B6, iron, calcium, zinc, phosphorus, and magnesium. It is also during adolescence, while forming their personal identity and self-concept, that food behavior and body image issues are most pronounced. Many women continue to struggle with food behavior and body image issues throughout adulthood, as they also struggle with physiologic demands, personal identity, role issues, employment, and family responsibilities. Females of all races and ethnicities consume excessive amounts of total fat and saturated fat. The intakes of calcium and iron are below recommended levels for adult women. Among pregnant women, weight gain is the best indicator of consumption of essential nutrients. What would be considered above normal weight gain for white women is actually necessary for African-American women to give birth to infants who are not low birth weight. Observations such as this, and the differences in consumption and metabolism of calcium among racial groups, indicate that there are differences among women regarding nutritional requirements. There may even be differences in what is defined as ideal body weight among women and its relationship to disease prevention and health promotion. However, there is limited understanding about these observations in the scientific community.

The literature definitely documents the social and cultural differences among women related to food behavior, body size preferences, and body image. Researchers must yield even greater understanding of the physiological differences and the sociocultural differences affecting the nutritional needs among women so that health professionals can provide appropriate guidance to women in meeting their nutritional needs. It is evident that poverty and distorted body image have negatively impacted the nutritional and health status of women. The nutritional and health effects of increased numbers of women in the workforce, the stresses of fast-paced living, and the greater use of convenience foods is much less understood.

Surely there is need for greater research in the areas of nutrition, weight, and body image for women for more effective women's health care. Health professionals must provide very distinct messages to women about body weight, body fat, and eating behavior that are conducive to optimal health. Greater and more accurate communication is needed to teach women about the relationship of obesity to morbidity and mortality and that being too thin and constant regimens of weight loss can also be detrimental to one's health, leading to increased morbidity and mortality.

The current conflicts, controversies, and contradictions about food, weight, and weight controlling behavior are paralleled by women trying to define themselves and their roles in families, workplaces, and societies, as well as striving for equal opportunities with men. For many women these issues become intertwined with body image concerns. Significant numbers of women, especially adolescent and young adult females who enter this state of apparent role conflict and body image confusion, retreat into detrimental food behaviors and focus on their bodies. According to the research literature, such retreats are not without consequences for poorer women's health and lowered nutritional status.

References

1. Johnson, R. (1996). Normal nutrition in premenopausal women. *In* "Nutrition in Women's Health" (D. A. Krummel and P. M. Kris-Etherton, eds.). Aspen Publishers, Gaithersburg, MD.

2. Clifford, C. K (1996). Major diet-related risk factors in American women. *In* "Nutritional Concerns of Women" (I. Wolinsky and D. Klimis-Tavantzis, eds.), pp. 15–23. CRC Press, Boca Raton, FL.

3. Committee of Diet and Health, Food and Nutrition Board, Commission of Life Sciences, National Research Council (1989). "Diet and Health, Implications for Reducing Chronic Disease Risk." National Academy Press, Washington, DC.

4. Wright, H. S., Gutherie, H. A., Wang, M. Q., and Bernardo, V. (1991). The 1987–88 nationwide food consumption survey: An update on the nutrient intake of respondents. *Nutr. Today* **26,** 21–27.

5. Johnson, R. K., Johnson, D. G., Wang, M. Q., *et al.* (1994). Characterizing nutrient intakes of adolescents by sociodemographic factors. *J. Adolesc. Health* **15,** 149–154.

6. Alaimo, K., McDowell, M. A., Briefel, R. R., *et al.* (1994). "Dietary Intakes of Vitamins, Minerals and Fiber of Persons Ages 2 Months and Over in the United States: Third National Health and Nutrition Examination Survey, Phase I, 1988–1991." National Center for Health Statistics, Hyattsville, MD.

7. McDowell, M. A., Briefel, R. R., Alaimo, K. *et al.* (1994). "Energy and Macronutrient Intakes of Persons 2 Months and Over in the United States: Third National Health and Nutrition Examination Survey, Phase I, 1988–1991." National Center for Health Statistics, Hyattsville, MD.

8. Rosen, J. C., and Gross, J. (1987). Prevalence of weight reducing and weight gaining in adolescent girls and boys. *Health Psychol.* **6,** 131–147.

9. Berg, F. (1992). Harmful weight loss practices are widespread among adolescents. *Obes. Health,* July/August, pp. 69–72.

10. Parker, S., Nichter, M., Nichter, M., Vuckovic, N., Sims, C., and Ritenbaugh, C. (1995). Body image and weight concerns among African American and White adolescent females: Differences that make a difference. *Hum. Organ.* **54**(2), 103–114.

11. Kris-Etherton, P. M., and Krummel, D. (1993). Role of nutrition in the prevention and treatment of coronary heart disease in women. *J. Am. Diet. Assoc.* **93**, 987–993.

12. Johnston, R., Smiciklas-Wright, H., and Crouter, A. (1992). The effect of maternal employment on the quality of young children's diets—the CSFII experience. *J. Am. Diet. Assoc.* **92**, 213–214.

13. Rawlings, S. W. (1993). "Household and Family Characteristics: March 1992. Current Population Reports, Population Characteristics." U.S. Department of Commerce, Bureau of the Census, Washington, DC.

14. Alaimo, K., Briefel, R., Frongillo, E., and Olson, C. (1998). Food insufficiency exists in the United States: Results from the third National Health and Nutrition Examination Survey (NHANES III). *Am. J. Public Health* **88**, 419–426.

15. Jeffery, R. W., and French, S. A. (1998). Epidemic obesity in the United States: Are fast foods and television viewing contributing? *Am. J. Public Health* **88**, 277–280.

16. Parker, J. D., and Abrams, B. (1992). Prenatal weight gain advice: An examination of the recent prenatal weight gain recommendations of the Institute of Medicine. *Obstet. Gynecol.* **79**, 664–669.

17. Parker, J. D., and Abrams, B. (1993). Differences in postpartum weight retention between black and white mothers. *Obstet. Gynecol.* **81**, 768–774.

18. Chernoff, R. (1991). Demographics of aging. *In* "Geriatric Nutrition: The Health Professional's Handbook" (R. Chernoff, ed.), pp. 1–10. Aspen Publishers, Gaithersburg, MD.

19. Reilly, J. J., Lord, A., Bunker, W. W., *et al.* (1993). Energy balance in healthy elderly women. *Br. J. Nutr.* **69**, 21–27.

20. Kohrt, W. M., Obert, K. A., and Holloszy, J. O. (1992). Exercise training improves fat distribution patterns in 60-to 70-year-old men and women. *J. Gerontol. Med. Sci.* **47**(4), M99–M105.

21. Owens, J. F., Matthews, K. A., Wing, R. R., *et al.* (1992). Can physical activity mitigate the effects of aging in middle-aged women? *Circulation* **85**, 1265–1270.

22. Nelson, M. E., Fisher, E. C., Dilamnian, F. A. *et al.* (1991). A 1 year walking program and increased dietary calcium in postmenopausal women: Effects on bone. *Am. J. Clin. Nutr.* **53**, 1034–1311.

23. Felson, D. T., Zhang, Y., Hannan, M. T., *et al.* (1993). Effects of weight and body mass index on bone mineral density in men and women: The Framingham Study. *J. Bone Miner. Res.* **8**(5), 567–573.

24. Tremollieres, F. A., Pouilles, J.-M., and Ribot, C. (1993). Vertebral postmenopausal bone loss is reduced in overweight women: a longitudinal study in 155 early postmenopausal women. *J. Clin. Endocrinol. Metab.* **77**, 683–686.

25. Heaney, R. P. (1996). Osteoporosis. *In* "Nutrition in Women's Health" (D. Krummel and P. Kris-Etherton, eds.). Aspen Publications, Gaithersburg, MD.

26. Pi-Sunyer, F. X. (1995). Medical complications of obesity. *In* "Eating Disorders and Obesity" (K. D. Brownell and C. G. Fairburn, eds.), pp. 401–405. Guilford Press, New York.

27. Kumanyika, S., Wilson, J., and Guilford-Davenport, M. (1993). Weight-related attitudes and behaviors of black women. *J. Am. Diet. Assoc.* **93**(4), 416–422.

28. Cooper, R., Rotini, C., Ataman, S., McGee, Osotimehin, B., Kadiri, W. M., Kingue, S., Fraser, H., Forrester, T., Bennet, F., and Wilks, R. (1997). The prevalence of hypertension in seven populations of West African origin. *Am. J. Public Health* **87**, 160–168.

29. Burnette, M, Meilahn, E., Wing, R., and Kuller, L. (1998). Smoking cessation, weight gain and changes in cardiovascular risk factors during menopause: The Healthy Women Study. *Am. J. Public Health* **88**, 93–96.

30. Wolinsky, F. D., Fitzgerald, J. F., and Stump, T. E. (1997). The effect of hip fracture on mortality, hospitalization, and functional status: A Prospective Study. *Am. J. Public Health* **87**, 398–403.

31. Diehr, P., Bild, D., Harris, T., Duxbury, A., Siscovick, D., and Rossi, M. (1998). Body mass and mortality in nonsmoking older adults: The Cardiovascular Health Study. *Am. J. Public Health* **88**, 623–629.

32. Bender, R., Trautner, C., Spraul, M., and Berger, M. (1998). Assessment of excess mortality in obesity. *Am. J. Epidemiol.* **147**(1), 42–48.

32a. Manson, J. E., Willet, W. C., Stampfer, M. J., Colditz, G. A., Hunter, D. J., Hankinson, S. E., Hennekens, C. H., and Speizer, F. E. (1995). Body weight and mortality among women. *N. Engl. J. Med.* **333**(11), 677–685.

33. Pike, K. M., and Striegel-Moore, R. H. (1997). Disordered eating and eating disorders. *In* "Health Care for Women" (S. J. Gallant, G. P. Keita, and R. Royak-Schaler, eds.), pp. 97–114. American Psychological Association, Washington, DC.

34. Halmi, K. A., Eckert, E., Marchi, P., Sampugnaro, V., Apple, R., and Cohen, J. (1991). Comorbidity of psychiatric diagnoses in anorxia nervosa. *Arch. Gen. Psychiatry* **48**, 712–718.

35. Walters, E. E., and Kendler, K. S. (1995). Anorexia nervosa and anorexic-like syndromes in a population-based female twin sample. *Am. J. Psychiatry,* **152**, 64–71.

36. Fairburn, C. G., and Beglin, S. J. (1990). Studies of the epidemiology of bulimia nervosa. *Am. J. Psychiatry* **147**, 401–408.

37. Hoek, H. W. (1991). Review of the epidemiological studies of eating disorders. *Int. Rev. Psychiatry* **5**, 61–74.

38. Kendler, K. S., Maclean, C., Neale, M., Kessler, R., Heath, A., and Eaves, L. (1991). The genetic epidemiology of bulimia nervosa. *Am. J. Psychiatry* **148**, 1627–1637.

39. Striegel-Moore, R. H., and Smolak, L. (1996). The role of race in the development of eating disorders. *In* "The Developmental Psychopathology of Eating Disorders: Implications for Research, Treatment, and Prevention" (L. Smolak, M. Levine, and R. H. Striegel-Moore, eds.), pp. 259–284. Erlbaum, Hillsdale, NJ.

40. Pate, J., Pumariega, A., Hester, C., and Garner, D. (1992). Cross-cultural patterns in eating disorders: A review. *J. Am. Acad. Child and Adolesc. Psychiatry* **31**, 802–809.

41. Spitzer, R. L., Devlin, M. J., Walsh, B. T., Hasin, D., Wing, R. R., Marcus, M. D., Stunkard, A., Wadden, T. A., Yanovski, S., Agras, W. S., Mitchell, J., and Nonas, C. (1992). Binge eating disorder: A multisite field trial for the diagnostic criteria. *Int. J. Eat. Disord.* **11**, 191–203.

42. Spitzer. R. L., Yanovski, S., Wadden, T., *et al.* (1993). Binge eating disorder: Its further validation in a multisite trial. *Int. J. Eat. Disord.* **13**(2), 137–153.

43. Yanovski, S. Z. (1993). Binge eating disorder: Current knowledge and future directions. *Obes. Res.* **1**, 305–324.

44. Marcus, M. D., Wing, R. R., Ewing, L., Kern, E., Gooding, W., and McDermott, M. (1995). Psychiatric disorders among obese binge eaters. *Int. Eat. Disord.* **9**, 69–77.

45. Mussell, M. P., Mitchell, J. E., Weller, C. L., Raymond, N. C., Crow, S. J., and Crosby, R. D. (1995). Onset of binge eating, dieting, obesity, and mood disorders among subjects seeking treatment for binge eating disorder. *Int. Eat. Disord.* **17**, 395–401.

46. Stewart, D. (1992). Reproductive functions in eating disorders. *Ann. Med.* **24**, 287–291.

47. Seeman, E., Szmukler, G., Formica, C. Tsalamandris, C., and Mestrovic, R. (1992). Osteoporosis in anorexia nervosa: The influence of peak bone density, bone loss, oral contraceptive use and exercise. *J. Bone Miner. Res.* **7**, 1467–1474.

48. Brotman, A., Rigotti, N., and Herzog, D. (1985). Medical complications of eating disorders: Outpatient evaluation and management. *Compr. Psychiatry* **26**, 258–272.

49. Stein, A. (1995). Eating disorders and childrearing. *In* "Eating Disorders and Obesity" (K. D. Brownell and C. G. Fairburn, eds.), pp. 188–195. Guilford Press, New York.

50. Pike, K. M., and Rodin, J. (1991). Mothers, daughters, and disordered eating. *J. Abnorm. Psychol.* **100,** 198–204.

51. Edlin, G., Golanty, E., and McCormack Brown, K. (1998). "Health and Wellness," pp. 123–124. Jones & Barlett, Sudbury, MA.

52. Dion, K. K., Berscheid, E., and Walster, E. (1972). What is beautiful is good. *J. Pers. Soc. Psychol.* **24,** 285–290.

53. Freedman, R. (1984). Reflections on beauty as it relates to health in adolescent females. *Women's Health* **9,** 29–45.

54. Hawkins, R. C., and Clement, P. F. (1980). Development and construct validation of a self-report measure of binge eating tendencies. *Addict. Behav.* **5,** 219–226.

55. Dion, K. (1981). Stereotyping based on physical attractiveness: Issues and conceptual perspectives. *In* "Physical Appearance, Stigma, and Social Behavior" (C. P. Herman, M. P. Zanna, and E. T. Higgins, eds.), pp. 7–21. Erlbaum, Hillsdale, NJ.

56. Langlois, J. H. (1986). From the eye of the beholder to behavioral reality: Development of social behaviors and social relations as a function of physical attractiveness. *In* "Physical Appearance, Stigma, and Social Behavior" (C. P. Herman, M. P. Zanna, and E. T. Higgins, eds.), pp. 23–51. Erlbaum, Hillsdale, NJ.

57. Barnlund, D. C. (1975). "Public and Private Self in Japan and the United States," pp. 153–154. Simul Press, Tokyo.

58. Sampson, E. E. (1977). Psychology and the American ideal. *J. Pers. Soc. Psychol.* **35,** 767–782.

59. Lewis, V. J., and Blair, A. J. (1993). Women, food, and body image. *In* "The Health Psychology of Women" (C. Niven and D. Carroll, eds.), pp. 107–120. Harwood Academic Publishers, Chur, Switzerland.

60. Bigelow, M. S. (1970). "Fashion in History." Burgess, Minneapolis, MN.

61. Frank, K. (1990). "Emily Bronte: A Chainless Soul." Hamish Hamilton, London.

62. Orbach, S. (1986). "Hunger Strike." Faber & Faber, London.

63. Orbach, S. (1982). "Fat is a Feminist Issue 2." Hamlyn Paperbacks, London.

64. Kuczmarski, R. S., Flegal, K. M., Campbell, S. M., and Johnson, C. L. (1994). Increasing prevalence of overweight among U.S. adults: The national health and examination surveys, 1960 to 1991. *JAMA, J. Am. Med. Assoc.* **272,** 205–211.

65. Walcott-McQuigg, J. (1997). Methodological issues in triangulation: Measuring weight control behavior of African American women. *In* "Oral Narrative Research with Black Women" (K. M. Vaz, ed.), pp. 119–142. Sage Publications, Thousand Oaks, CA.

66. Powell, A. D., and Kahn, A. S. (1995). Racial differences in women's desires to be thin. *Int. J. Eat. Disord.* **17,** 191–195.

47

Alcohol Use and Abuse

MARCIA RUSSELL,* MARIA TESTA,* AND SHARON WILSNACK†
*Research Institute on Addictions, Buffalo, New York; †University of North Dakota School of Medicine, Grand Forks, North Dakota

I. Introduction and Background, including Historical Evolution/Trends

Alcohol use is widespread throughout the world and has been throughout history. In many developed western societies there have been cycles in which social approval of drinking alcohol led to gradually increasing use until alcohol-related problems became so pervasive and serious that there was a backlash [1]. The most dramatic U.S. backlash resulted in the passage of the Volstead Act, instituting Prohibition in 1919. Illegal activity to circumvent the law led to its being revoked in 1933. Alcohol use took a big jump after World War II and then gradually increased throughout the 1950s and 1960s, peaking in the 1970s. Current trends toward lower alcohol intakes have been influenced by societal reactions to high mortality and morbidity rates related to drunk driving, as evidenced by the formation of activist groups such as Mothers Against Drunk Drinking, and by increased recognition of the negative effects of excessive drinking on health and social functioning.

Until relatively recently, little attention was given to alcohol use and abuse in women. Only 28 studies on alcoholic women were published between 1929 and 1970 in English, and the National Institute on Alcohol Abuse and Alcoholism (NIAAA) funded only three studies on women and alcohol in its first five years (1971–1976). There are several reasons for this neglect. Much of the early research on alcohol abuse took place in treatment settings, and most alcoholism treatment took place in VA or state hospitals which housed few female patients. Therefore, women were less accessible as research subjects. There were higher rates of alcoholism in males, and their alcohol-related problems were more prevalent and more visible. Even today this may lead researchers to assume that male alcoholism is more socially relevant and more in need of study. Most alcohol researchers were male, and for this reason they may have been less interested in women's problems. Finally, there was a tradition of relying on male subjects, both human and nonhuman, to avoid variability influenced by physiological changes related to menstrual and reproductive cycles.

However, by the late 1970s several factors combined to stimulate an interest in research on women and alcohol. One was a perception that there was an ongoing "epidemic" of alcoholism in women. This perception may have been a delayed reaction to the precipitous increase in women's alcohol consumption that took place after the Second World War, even though it had leveled off in the 1970s. The impression of an epidemic may also have been heightened by increased visibility of women's alcohol problems as more women entered treatment and by social discomfort with changes in traditional sex roles. Finally, description of the Fetal Alcohol Syndrome in the English medical literature in 1973 stimulated public concern about alcohol drinking among women of child-bearing age. Progress in understanding the factors that influence women's alcohol use and abuse has been documented in three national research conferences held in 1978 [2], 1984 [3], and 1993 [4]. In addition, NIAAA funded a National Longitudinal Study of Women and Alcohol, which was initiated in 1981, with follow-ups in 1986, 1991, and 1996 [5]. There is now a rich body of research to draw upon in discussing the epidemiology of alcohol use and abuse in women.

Here we will provide a broad overview of the available literature and refer readers to more detailed reviews for information about individual research studies that have addressed specific questions as they relate to women and alcohol. To better equip readers to interpret this literature, we will discuss methodological issues related to the measurement of alcohol use and abuse and review a broad range of factors that influence women's drinking, including host and environmental determinants, social roles and context, and risk factors. Gender-related biological vulnerability to alcohol's effects, health consequences of drinking, and clinical issues related to diagnosis and treatment also will be discussed.

II. Definition of Alcohol Use and Abuse

A. Alcohol Use

Alcohol consumption is usually assessed by asking women how often they drink any beverage containing alcohol (or Frequency) and how many drinks they usually have on a day when they drink (or Quantity). Quantity-Frequency (QF) questions often are introduced by defining a standard drink as a 12-oz. can or bottle of beer, a 5 oz. glass of wine, or a drink containing 1.5 oz. of liquor. Within this general framework, there is considerable variability in the way QF questions are formulated. Sometimes they are repeated for specific alcoholic beverages, wine, beer, and liquor; or respondents may be given the option of defining their own drink size. Questions on usual QF may be supplemented by additional QF questions on times when respondents drink more than usual. However they are asked, the resulting data are routinely used to estimate volume of alcohol consumption (ounces of ethanol per day) by multiplying drinking days in a year by drinks per drinking day by ounces of ethanol in a drink and dividing by days in a year. Ethanol in a drink is estimated by applying conversion factors based on the amount of ethanol in an ounce of the average beer (0.045), wine (0.121), or liquor (0.409) to the ounces in a drink of each beverage. There is approximately 0.5 oz. of ethanol in a standard drink in the United States, equivalent to about 12 grams or 15 ml of ethanol, but this varies somewhat in other countries [6].

Women's alcohol drinking is strongly influenced by culture, historical trends, and medical factors, making it difficult to define norms. In some cultures alcohol use is proscribed for both men and women; in others men's drinking is condoned and women's is condemned. In countries such as France and Italy, wine is considered part of the diet and is given even to children. The definition of normal alcohol use varies, not only across cultures, but also within cultures. As already mentioned, reaction to the alcohol problems associated with a given alcohol intake can influence subsequent societal approval of drinking, causing consumption to fluctuate over time. Finally, the available data indicate that alcohol use is associated both with health benefits and with risks, making it necessary to tailor medical advice to a given woman's circumstances, such as whether she is pregnant or likely to become pregnant, whether she is vulnerable to developing an alcohol disorder or an alcohol-related disease, or if she might be expected to benefit from alcohol effects on her cardiovascular system (*e.g.*, because of age or cardiovascular risk factors). Such caveats complicate the formulation of guidelines for moderate drinking for men as well as for women. For example, moderate drinking may be contraindicated for individuals with a history of alcohol abuse, a genetic predisposition toward alcohol disorders, and certain preexisting physical conditions, such as hypertension or liver disease, as well as medication status and intention to drive or operate machinery. It has been estimated that contraindications for moderate drinking such as these may apply to as many as 31% of the population at any one time [7].

B. Alcohol Abuse and Dependence

Alcohol abuse and dependence are psychiatric disorders, and the American Psychiatric Association has developed criteria for their diagnosis that were most recently revised in 1994 [8]. Abuse is diagnosed if a person's maladaptive alcohol use causes clinically important distress or impairment, as shown in a single 12-month period by one or more of the following: failure to carry out major obligations at work, home, or school because of repeated alcohol use; repeated use of alcohol even when it is physically dangerous to do so; repeated experience of legal problems; or continued use of alcohol despite knowing that it has caused or worsened social or interpersonal problems. Dependence is diagnosed when a person's maladaptive pattern of alcohol use leads to clinically important distress or impairment, as shown in a single 12-month period by three or more of the following: tolerance; withdrawal; amount or duration of use often greater than intended; repeatedly trying without success to control or reduce alcohol use; spending much time using alcohol, recovering from its effects, or trying to obtain it; reducing or abandoning important work, social, or leisure activities because of alcohol use; or continuing to use alcohol, despite knowing that it has probably caused ongoing physical or psychological problems. Structured diagnostic interviews have been developed to permit lay interviewers to assess these and other criteria, such as those developed under the auspices of the World Health Organization, to support diagnoses of alcohol abuse and dependence in national and international epidemiologic studies [9].

C. Alcohol-Related Morbidity and Mortality

Repeated use of alcohol even when it is physically dangerous to do so and continued use of alcohol despite knowing that it has probably caused physical problems are criteria that contribute to a diagnosis of alcohol abuse or dependence, respectively. Accordingly, alcohol-related morbidity and mortality are recognized as consequences of an alcohol psychiatric disorder. However, one does not need to be dependent on alcohol to have an alcohol-related health problem. Some individuals tolerate alcohol well and do not experience withdrawal symptoms, social consequences, or psychological symptoms related to their drinking. Thus, even though they may drink enough to suffer alcohol-related organic damage, if they are unaware of the physical harm they are incurring and do not have social or legal consequences, they may not meet criteria for a diagnosis of dependence. In addition, even isolated episodes of intoxication can result in traffic crashes or falls that lead to serious injury or death. In summary, alcohol-related morbidity and mortality include acute and chronic health consequences of alcohol misuse that may or may not be associated with a psychiatric diagnosis of alcohol abuse or dependence.

III. Issues Related to Ascertainment

A. Validity

Validity is a major concern for epidemiologic studies of alcohol consumption. Alcohol consumption is difficult to remember and summarize accurately, and fear of stigmatization may inhibit disclosure of heavy drinking. Evidence that respondents underestimate their consumption in alcohol surveys has been provided by studies comparing the amount of alcohol sold in a given region with the consumption estimated from survey data; self-reported alcohol consumption typically accounts for only 40–60% of the alcohol sold, although reporting error is not the only reason for this discrepancy [10]. If all drinkers underestimated their intakes to the same extent, the amounts reported could be simply adjusted upward to compensate; however, investigation of this point has yielded inconclusive results. Some researchers have reported fairly good agreement between self-reports and independent assessments of alcohol intake, whereas others found that heavier drinkers and drinkers with alcohol-related problems underestimated their intakes more than light drinkers [10]. Despite these limitations, survey researchers have continued to rely on self-report of alcohol-related behavior. They have been forced to do so by the lack of practical alternatives, but they have also been encouraged by studies demonstrating that the validity of self-report data can be maximized by employing good assessment techniques [11]. These techniques include such strategies as putting respondents at ease, providing privacy, guaranteeing confidentiality, using clearly stated questions, and conducting the interview in a neutral, nonjudgmental manner.

B. Measurement of Alcohol Use

Despite the long tradition of employing QF measures of alcohol consumption in alcohol research, many of the best-known epidemiologic studies have used alcohol measures that fall short of this standard [12]. Much of our information on alcohol and women's health comes from the Nurses Health Study, in which alcohol consumption is assessed as part of a semiquantitative food frequency questionnaire (SQFF) in which respondents are asked a series of questions about how often they have one drink of white wine/red wine/beer/liquor [13]. Indeed, one metaanalysis of data from studies on alcohol and breast cancer is based on six studies, all of which used SQFF to assess alcohol intake [14]. Limiting SQFF questions to the frequency of consuming one drink complicates reporting of alcohol intake for women who have more than one drink on days when they drink, but who do not drink daily. For example, a woman who has two drinks a day on three days a week needs to double her frequency to adjust for the fact that she has two drinks, i.e., one drink six times a week. However, methodological studies indicate that subjects who have more than one drink a day often report the same frequency for both QF and SQFF, leading to underreporting. If the woman in our example reported that she had two drinks a day, it would be assumed that she did this daily, leading to overreporting. The potential for both under- and overestimates on the SQFF makes it difficult to predict how use of this measure might affect conclusions based on these studies. It has been argued that the SQFF is an adequate measure of alcohol use in the Nurses Health Study because nurses are health conscious professionals who are unlikely to have more than one drink. However, studies of substance use among predominantly female samples of nurses indicate that their rates of binge drinking are comparable to those reported by women in general population surveys [15].

C. Measurement of Alcohol-Related Morbidity and Mortality

Data on alcohol-related morbidity and mortality for diseases directly caused by alcohol, such as alcoholic cirrhosis of the liver, are readily available from the National Hospital Discharge Survey and from vital statistics, respectively [16]. However, diseases or injuries indirectly attributable to alcohol often are not taken into consideration, thereby underestimating the total effect of alcohol on morbidity and mortality in the U.S. In one effort to document such risks, the Department of Transportation set up the Fatal Accident Reporting System (FAR), which contains information on the extent to which alcohol is involved in traffic crashes in which at least one person dies within 30 days of the crash [16]. To further investigate the effect of alcohol on mortality, next of kin or knowledgeable informants were interviewed in the 1986 National Mortality Followback Study regarding alcohol use and abuse in a sample of individuals who had died in that year. Based on these and other studies, disease or injury diagnoses causally linked to alcohol use/misuse have been identified and estimates have been derived for the proportions of deaths from these diagnoses that could be attributed to alcohol (i.e., alcohol-attributable fractions, AAFs). AAFs are then used to correct for underestimates when estimating the overall impact of alcohol on mortality [17,18]. For example, if it were determined that a certain proportion of the mortality from breast cancer could be attributed to heavy drinking, this proportion could be applied to the number of deaths from breast cancer to improve estimates of the overall impact of alcohol on

mortality. To date, little attention has been given to the development of gender-specific AAFs to aid in understanding the overall impact of alcohol use on mortality in women.

D. Analysis

Although the practice of using volume as a measure of alcohol consumption in epidemiologic studies of health outcomes is widespread and convenient, it has been criticized because it obscures differences in drinking pattern. Thus, individuals having two drinks every day would have the same volume measure as a person having 14 drinks every Saturday night, but the health implications of these two patterns of drinking are likely to differ. For example, animal models of prenatal alcohol exposure demonstrate that binge drinking is more harmful to the fetus than the same amount of alcohol consumed in small amounts over a longer period of time [19]. Volume measures are particularly problematic when investigating health consequences associated with alcohol consumption in countries, such as the United States, in which daily drinking is relatively infrequent [12,20]. Under these circumstances, health outcomes are correlated with a statistical artifact that does not correspond well to actual drinking patterns of the study population, thus providing a poor basis for the development of guidelines for moderate drinking that are consistent with good health. The development of methods to assess and analyze drinking patterns in the United States is an active area of research in alcohol epidemiology. A measure to assess lifetime drinking patterns, the Cognitive Lifetime Drinking History, was recently introduced for use in case-control studies of alcohol and health [21]. As indicated by its name, it employs cognitive techniques for improving recall of alcohol consumption in the past.

IV. Host and Environmental Determinants

A. Gender and Age

The influence of gender and age on alcohol use and abuse is powerful and consistent [22–24]. In every society that has been studied, men drink more than women and men's drinking leads to more social problems than does women's drinking [25]. The consistency of this finding suggests a biological influence, as will be discussed later; however, variability in the ratio of men's drinking to women's across cultures indicates the potential of sociocultural factors to interact with biological factors [26]. In general, alcohol use and abuse decrease with age [22]. Thus, younger women are more likely than older women to drink, to drink heavily, to binge drink, and to report adverse consequences [23]. It has been postulated that older women in the U.S. may drink less because they were influenced by Prohibition, a cohort effect. However, the fact that age-related decreases in the amount women drink per drinking occasion have been observed in 15 countries suggests that biological factors play an important role [25]. Potential age-related reasons for decreasing alcohol intake include chronic health problems aggravated by drinking, the potential for alcohol interactions with medications being taken, an increased vulnerability to alcohol's effects, and the need to adjust for changes in body composition that result in higher blood alcohol levels when amounts drunk at younger ages are consumed. Social isolation and loss of income may also encourage abstinence in older women [24].

B. Race and Ethnicity

Study of race and ethnicity as they relate to alcohol use and abuse has lagged behind study of gender effects because most surveys included too few members of a given minority group for separate analysis [27]. The 1984 National Alcohol Survey oversampled blacks and hispanics to address this issue. It was found that black women are more likely to abstain than whites; however, among drinkers, the proportion drinking heavily and experiencing alcohol-related problems tends to be more similar, and there is evidence to suggest that alcohol-related health problems among blacks are more prevalent and more severe [28,29]. It has been suggested that alcohol effects on health may be potentiated in blacks by adverse conditions associated with poverty and a greater tendency for blacks to maintain heavier drinking patterns through middle age. Analysis of drinking in hispanics revealed differences related to country of origin and to acculturation [30]. Mexican-American and Cuban-American women were more likely to abstain than were Puerto Rican women (46% and 42%, respectively, compared to 33%); heavier drinking was more prevalent among Mexican-American women than among Cuban-American or Puerto Rican women (14% compared to 7% and 5%); and acculturation was associated with increased drinking [27]. Similar differences limit the extent to which alcohol use and abuse can be characterized for other racial and ethnic groups. Thus, the limited data on alcohol available for Native Americans tends to focus on high rates of alcoholism seen in some tribes, without appreciating the fact that the federal government recognizes over 300 different tribes, many of which drink moderately with few problems [31]. Asian women tend to drink moderately, if at all, but there are substantial differences in men's drinking patterns among Chinese, Japanese, Koreans, Filipinos, and other Asian groups [27]. In addition to the impact of culture and gender on alcohol use and abuse in different racial and ethnic groups, there are racial differences in the enzymes that metabolize ethanol that produce variable physiological responses that may influence both drinking patterns and the likelihood of adverse health consequences [32]. In summary, ethnic differences in women's and men's drinking are influenced by a number of factors that have often been ignored in these studies. These include: socioeconomic and regional differences in the distribution of racial or ethnic groups being compared; differences related to acculturation among immigrant groups; national or tribal differences between individuals having backgrounds often categorized simply as Asian, Hispanic, or Native American; and genetic differences in the enzymes that metabolize alcohol.

V. Distribution in Women of Alcohol Use and Alcohol Abuse or Dependence

Data on the prevalence of drinking patterns and a DSM-IV (Diagnostic and Statistical Manual of Mental Disorders, 4th Edition) diagnosis of alcohol abuse or dependence in the past 12 months are summarized according to ethnicity, sex, and age

in Table 47.1. These data are from the 1992 National Longitudinal Alcohol Epidemiologic Survey (NLAES) conducted by the Bureau of the Census under the auspices of the National Institute on Alcohol Abuse and Alcoholism in 42,862 respondents, 18 years or older, in the contiguous United States and the District of Columbia [33]. In NLAES, abstention is defined as having had fewer than 12 drinks in the past 12 months. Estimates of volume are based on beverage-specific QF questions on the usual and the largest number of drinks, letting respondents define drink size [34]. Volume is used to define lighter, moderate, and heavier drinking as <0.22, $\geqslant 0.22 < 1.00$, and $\geqslant 1.00$ oz. ethanol per day, respectively, where 1.00 oz. of ethanol is equal to about two standard drinks. Binge drinking is defined as having five or more drinks in a day at least weekly.

The NLAES data summarized in Table 47.1 illustrate some of the relations between gender, ethnicity, and age and alcohol use and abuse mentioned previously. Abstention is more prevalent among women than men and among blacks than nonblacks. Accordingly, almost 80% of black women abstain compared to 64% of nonblack women, and 54% of black men abstain compared to only 43% of nonblack men. Moderate drinking is correspondingly more prevalent among men than women and among nonblacks than blacks. Men are more likely than women to report heavier drinking, but ethnicity was not related to the prevalence of heavier drinking. Abstention increases with age both in males and in females, and the proportion of the population reporting lighter, moderate, or heavier drinking decreases correspondingly. In several community surveys of alcohol consumption, an interaction between age and ethnicity has been observed [27]. Nonblacks are more likely than blacks to drink and drink heavily at younger ages and then decrease their alcohol consumption. In contrast, younger blacks are less likely to use or abuse alcohol, but middle-aged and older blacks are more likely than nonblacks to initiate or maintain patterns of heavier drinking. This is illustrated in NLAES by data on the proportion of the population who reported having five or more drinks a day at least weekly in the past 12 months for both men and women and by data on the prevalence of alcohol abuse and/or dependence in women.

VI. Influence of Women's Social Roles or Context

A. Marital Status

Cross-sectional, epidemiologic surveys show heavier drinking among single and divorced and separated women relative to married or widowed women [22]. In addition, there is longitudinal evidence that marriage predicts subsequent declines in drinking both for women and for men, although some of the apparent effect of marital status may in fact be an effect of age [35]. Cohabitation is associated with heavy drinking and it predicts onset of subsequent problem drinking in women, perhaps because of stress related to instability and lack of institutional support for the relationship, or perhaps because women in cohabitating relationships may be more unconventional and have higher drinking norms than more conventional women. Within marital or cohabitating relationships, a woman's drinking may be influenced by the drinking level of her partner. Because men tend to drink more than women on average, husband-to-wife influence tends to result in increased drinking for the woman

Table 47.1

Alcohol Consumption[a] and Alcohol Abuse and/or Dependence[b] in the Past 12 Months, According to Age, Gender, and Ethnicity, United States, 1992

	Abstainers	Lighter drinkers	Moderate drinkers	Heavier drinkers	Drank 5+ drinks weekly	Alcohol abuse and/or dependence
Nonblack females	64.28 (0.53)	18.40 (0.35)	13.06 (0.32)	4.26 (0.17)	2.07 (0.13)	4.25 (0.20)
18–29	53.44 (0.99)	21.75 (0.70)	19.06 (0.72)	5.75 (0.43)	4.55 (0.40)	10.99 (0.64)
30–44	58.40 (0.76)	23.22 (0.61)	14.60 (0.49)	3.78 (0.26)	1.95 (0.20)	3.94 (0.27)
45–64	67.82 (0.89)	16.70 (0.65)	10.75 (0.54)	4.73 (0.35)	1.11 (0.17)	1.45 (0.19)
65+	82.83 (0.68)	8.42 (0.46)	6.15 (0.42)	2.60 (0.25)	0.52 (0.11)	0.29 (0.09)
Black females	79.06 (0.79)	10.63 (0.57)	6.95 (0.48)	3.36 (0.37)	2.36 (0.29)	2.88 (0.32)
18–29	79.31 (1.33)	11.78 (1.03)	5.82 (0.76)	3.09 (0.63)	1.71 (0.42)	3.32 (0.60)
30–44	71.37 (1.48)	12.86 (1.06)	11.40 (1.06)	4.38 (0.62)	3.56 (0.61)	4.18 (0.65)
45–64	81.40 (1.62)	10.32 (1.28)	4.87 (0.91)	3.41 (0.76)	2.30 (0.66)	1.92 (0.54)
65+	95.57 (0.83)	2.37 (0.64)	1.00 (0.32)	1.06 (0.42)	0.68 (0.37)	0.00 (0.00)
Total females	66.07 (0.48)	17.46 (0.32)	12.32 (0.28)	4.15 (0.16)	2.10 (0.12)	4.08 (0.18)
18–29	57.29 (0.88)	20.27 (0.61)	17.08 (0.63)	5.36 (0.38)	4.12 (0.35)	9.84 (0.56)
30–44	60.10 (0.70)	21.86 (0.56)	14.18 (0.45)	3.86 (0.24)	2.16 (0.19)	3.98 (0.25)
45–64	69.28 (0.81)	16.01 (0.59)	10.12 (0.49)	4.59 (0.33)	1.23 (0.17)	1.50 (0.18)
65+	83.91 (0.64)	7.91 (0.43)	5.72 (0.39)	2.47 (0.24)	0.53 (0.11)	0.27 (0.09)
Nonblack males	43.13 (0.56)	20.16 (0.40)	22.87 (0.40)	13.84 (0.35)	9.09 (0.29)	11.33 (0.34)
18–29	33.80 (0.95)	20.69 (0.73)	28.19 (0.83)	17.32 (0.75)	15.87 (0.70)	23.48 (0.84)
30–44	38.05 (0.84)	22.78 (0.65)	26.09 (0.68)	13.08 (0.55)	8.60 (0.45)	10.89 (0.47)
45–64	48.26 (1.02)	19.83 (0.80)	18.09 (0.70)	13.82 (0.67)	6.70 (0.49)	5.61 (0.44)
65+	62.34 (1.13)	13.63 (0.75)	14.49 (0.75)	9.54 (0.62)	2.58 (0.32)	1.21 (0.23)
Black males	54.09 (1.35)	14.29 (1.02)	18.15 (0.99)	13.47 (0.92)	9.53 (0.84)	8.25 (0.72)
18–29	49.36 (2.55)	15.44 (1.72)	18.80 (1.88)	16.39 (2.13)	11.67 (1.89)	12.33 (1.70)
30–44	49.48 (2.20)	14.80 (1.70)	21.33 (1.84)	14.38 (1.52)	10.05 (1.41)	8.75 (1.21)
45–64	57.47 (2.80)	14.51 (2.05)	16.13 (1.95)	11.89 (1.43)	8.46 (1.29)	5.19 (0.97)
65+	77.71 (3.48)	8.41 (1.89)	9.24 (2.27)	4.64 (1.30)	3.40 (1.16)	0.82 (0.51)
Total males	44.28 (0.51)	19.55 (0.37)	22.37 (0.38)	13.80 (0.33)	9.14 (0.27)	11.00 (0.32)
18–29	35.76 (0.91)	20.04 (0.67)	27.01 (0.79)	17.20 (0.72)	15.34 (0.67)	22.07 (0.77)
30–44	39.32 (0.78)	21.89 (0.62)	25.56 (0.64)	13.23 (0.52)	8.76 (0.43)	10.65 (0.45)
45–64	49.11 (0.97)	19.34 (0.75)	17.91 (0.66)	13.64 (0.63)	6.86 (0.46)	5.57 (0.41)
65+	63.51 (1.07)	13.23 (0.72)	14.09 (0.73)	9.17 (0.57)	2.65 (0.31)	1.18 (0.22)
Total	55.62 (0.40)	18.46 (0.24)	17.14 (0.26)	8.78 (0.19)	5.48 (0.16)	7.41 (0.20)
18–29	46.56 (0.70)	20.15 (0.44)	22.03 (0.55)	11.26 (0.42)	9.72 (0.41)	15.94 (0.53)
30–44	49.84 (0.54)	21.88 (0.41)	19.80 (0.40)	8.48 (0.29)	5.42 (0.24)	7.27 (0.26)
45–64	59.54 (0.67)	17.62 (0.48)	13.88 (0.43)	8.96 (0.35)	3.95 (0.24)	3.47 (0.22)
65+	75.46 (0.63)	10.11 (0.40)	9.19 (0.38)	5.24 (0.27)	1.41 (0.14)	0.64 (0.10)

Note. Percentages (SE).

[a]D.A. Dawson, personal communication, March 25, 1998. Abstainers drank less than 12 alcoholic drinks in the past 12 months; lighter drinkers consumed an average of ≤0.22 ounces of ethanol per day; moderate drinkers consumed an average of >0.22 and <1.00 ounces of ethanol per day; and heavier drinkers consumed an average of 1.00 or more ounces (14 or more drinks per week).

[b]Source: [33].

[35]. However, wives may also influence husbands' drinking, and drinking may increase in response to marital dysfunction both among women and among men [35]. Several studies show that discrepancies in drinking patterns between husband and wife are associated both with marital distress and with problematic drinking behavior [36]. Longitudinal studies of the effect of divorce on subsequent alcohol use are inconsistent, with some reporting increases in drinking by women after divorce [25] and others suggesting that heavy-drinking women may decrease al-

cohol use if their marriages were characterized by sexual dysfunction or if they divorced heavy-drinking spouses [36].

B. Employment

In general, employed women are less likely to abstain than are homemakers, but their rates of heavy drinking and drinking problems are similar [23]. However, there is evidence that women employed in nontraditional or male-dominated professions drink more than women in female-dominated professions. Possible explanations for these findings include stress related to women's minority status in these jobs, influence of heavier drinking norms, or drinking by women in predominately male jobs to assert power or equality. It has been hypothesized that combining employment with family responsibilities would be stressful and that women might drink more to relieve such stress. Although a number of studies have failed to demonstrate that simply having multiple roles is associated with heavier alcohol use, a limitation of many of these studies is that inter-role conflict has been assumed but has not been examined. Research in which conflict between work and family has been directly assessed shows that work-family conflict is positively related to heavy drinking both among women and among men [37].

C. Sexual Orientation

One review of alcohol use and abuse among lesbians concluded that widespread use of unrepresentative samples (*e.g.,* patrons of gay bars) made findings unreliable [38]. Nonetheless, the better studies available suggest that lesbians are more likely to drink and to report alcohol problems than are heterosexual women. Further, the usual declines in drinking with increasing age typically observed among heterosexual women are not apparent in studies of lesbians. Alcohol use and abuse among lesbian women appear to be influenced by many of the same risk factors that contribute to heavy drinking and drinking problems among heterosexual women, such as high rates of childhood sexual abuse, high rates of employment in male-dominated occupations, low rates of marriage and child-bearing, high cohabitation rates, and relationships with violent or heavy drinking partners. Further, the additional stressors associated with being lesbian (*e.g.,* discrimination, hiding one's status as lesbian) coupled with the prominence of gay bars as social settings may contribute to heavier drinking as a coping strategy.

VII. Risk Factors

A. Psychopathology

The literature is consistent in showing that alcohol abuse and dependence are likely to co-occur with other psychopathology and that these associations are likely to be stronger among women than among men [36]. For example, the National Comorbidity Study [39] revealed substantial co-occurrence of alcohol abuse and dependence with various other psychiatric disorders, such as drug disorders, conduct disorder and antisocial personality disorder, and anxiety and affective disorders, including depression, phobia, and posttraumatic stress disorder (PTSD). This study involved retrospective assessment of lifetime psychiatric problems among a sample of 8098 men and women between 15 and 54 years of age. In a multivariate model, social phobia, simple phobia, depression, and drug dependence all emerged as significant predictors of subsequent alcohol abuse among women. In general, evidence of comorbidity was stronger for women than for men. While this study is important in that it clearly demonstrates a link between substance use disorders and other psychiatric disorders, prospective studies are needed to

more firmly establish the causal relationships between psychopathology and substance abuse.

Recently Hartka *et al.* [40] examined the relationship between depressive symptoms and alcohol consumption in eight longitudinal general population studies conducted in the United States, Great Britain, and Canada. Although these studies examined symptoms rather than diagnosis of depression, they still provide useful information on the direction of causality. Hartka *et al.* [40] concluded that for women, but not for men, depressive symptoms predicted subsequent quantity of alcohol consumption per occasion. Second, symptoms of depression and quantity of alcohol consumed per occasion predicted subsequent depression for both sexes; however, the relationship was stronger for women. These findings suggest that women may be more likely than men to drink in response to depressive symptoms. Alternatively, it may be that heavy drinking in men is more reflective of social norms, but that in women drinking is less socially acceptable and, hence, more associated with social and psychological problems.

B. Violent Victimization

There is growing evidence that experiencing violent victimization increases the risk of developing subsequent alcohol problems [41]. Several studies suggest that childhood physical and sexual abuse are associated with higher rates of substance abuse in adulthood [42] and that alcoholic women in treatment exhibit higher rates of childhood physical and sexual abuse than do women in the general population [41]. Adult sexual victimization also appears to be linked to subsequent problem drinking. For example, in two large probability samples, sexual assault was found to increase the likelihood of subsequent alcohol abuse [43,44]. PTSD may play a role in linking sexual victimization with subsequent alcohol abuse. The prevalence of alcohol disorders is twice as high among women with PTSD as it is among women without PTSD, and for women, the most common precipitating event for PTSD is sexual victimization [45]. It appears that heavier or problem drinking increases vulnerability to sexual victimization, an outcome that may result from the effects of acute intoxication on the ability to perceive and resist sexual aggression (see Testa and Parks [46] for a review).

C. Heredity

Having alcoholic parents is a major risk factor for alcohol abuse and dependence both in women and in men [47,48]. Studies of twins, adoptees, and genetic markers have established a firm foundation for the heritability of risk for alcohol abuse, but the lower prevalence of alcohol disorders among women, together with other methodological limitations in studies to date, have resulted in inconsistent findings regarding its magnitude in women. A multicenter effort, the Collaborative Study on the Genetics of Alcoholism, to identify the genes involved in determining innate alcohol risk is ongoing [49].

In addition, growing up in an alcoholic family environment may expose children to marital conflict, physically abusive parenting, and harsh or inconsistent disciplining [50], which have a negative impact on their development [51] and contribute to future risk of alcohol abuse. Current developmental models of alcoholism attempt to integrate genetic, psychological, and sociocultural influences on alcohol use and abuse [52].

D. Sexual Experience

Sexual problems and sexual dissatisfaction are common in clinical samples of alcoholic women, but temporal sequences of alcohol abuse and sexual dysfunction are difficult to determine. Longitudinal data suggest that once problem drinking is established, sexual dysfunction may be an important predictor of continued alcohol abuse [53]. The majority of U.S. women believe that alcohol facilitates sexual experience, with this belief most prevalent among the heaviest drinkers [36]. Women who experience sexual problems and who believe that alcohol can enhance sexual functioning may drink alcohol to cope with or reduce sexual difficulties. Such a pattern might help to explain the finding that women drinkers who believed that alcohol reduces sexual inhibition were 2.5 times more likely than other women drinkers to develop symptoms of alcohol dependence over a ten-year period [54].

VIII. Biological Markers of Susceptibility or Exposure

With increasing attention being given to alcohol use and abuse in women, the realization has come that women seem to be more vulnerable to alcohol's effects than men [55]. Notably, this has been documented in terms of an increased susceptibility to liver disease, with women tending to develop cirrhosis after drinking smaller amounts of alcohol over shorter periods of time [56]. One explanation for this is the higher bioavailability of alcohol in women compared to men because: (1) alcohol is distributed in body water, and women tend to have smaller amounts of body water than men because they have higher proportions of body fat and tend to be smaller in size, and (2) men metabolize more alcohol in their stomach linings than women do, so women absorb more of the alcohol they drink. It is also hypothesized that other factors, such as altered hepatic metabolism, may play a role [55]. It seems likely that gender differences in ethanol bioavailability contribute to the consistently lower alcohol intakes reported by women and put women at higher risk of adverse alcohol effects on health; however, the observation that women reported higher scores on a sedation index than men did at the same blood alcohol levels suggests that additional mechanisms may be involved [55].

IX. Clinical Issues (Diagnosis, Treatment)

There are a number of factors that decrease the likelihood that signs and symptoms of alcohol disorders in women will lead to diagnosis of their problem and referral for treatment. For example, women are less likely than men to be arrested for driving while intoxicated. Although this may seem like an advantage, it also tends to exclude them from programs designed to get problem drinkers into treatment early in the development of alcoholism. Similarly, Employee Assistance Programs designed to detect and refer early problem drinkers tend to be male-oriented [57]. Family members may discourage women from seeking help for an alcohol problem, either because they are ashamed or because they want a drinking partner. Finally, even though

women are more likely than men to visit a physician, doctors often fail to detect alcohol abuse in their female patients. Relatively low rates of heavy drinking and alcohol abuse in women result in low levels of suspicion on the part of their caregivers. Brief questionnaires to screen for risk drinking during pregnancy and for alcohol-related problems have been developed for use with women [58], and brief interventions have been developed for use in primary care medical practices that have demonstrated the ability to reduce alcohol consumption by women who are not dependent on alcohol [59]. However, physicians and other caregivers may be reluctant to raise the issue of alcohol use because they may fear that women would be offended or feel threatened and terminate care, they may not feel it is worthwhile to bring it up because they do not think that treatment is effective, or they may not feel comfortable bringing it up because they lack experience in diagnosis and treatment or referral. In recognition of these barriers to diagnosis and referral, the American Medical Association has drafted a statement on alcoholism and alcohol abuse among women to try to increase early identification, treatment, and prevention [60].

Even if an alcohol disorder is recognized, many alcoholics do not get treatment. Special barriers for women include lack of childcare, fear of losing child custody, and lack of insurance coverage. When these barriers are overcome, available evidence indicates that women dependent on alcohol respond well to treatment [61]. In some treatment settings dominated by males, hostility may be expressed toward women, their problems may be belittled, and they may be sexually exploited, suggesting that it may be useful to have gender-specific services for women, especially in the early stages of recovery. Comprehensive services are needed because alcoholism impacts many aspects of women's lives.

X. Epidemiologic Issues, Including Methodology for Study and Public Health Impact

The influence of alcohol on women's health has not been well studied, partly because excessive drinking was until recently considered less of a problem for women than for men. In addition to including more women in future studies of alcohol and health, there is a need to employ state-of-the-art methods of assessing and analyzing alcohol consumption in these studies. There is also a need to give more attention to screening for risk drinking and alcohol problems among women. From a methodological point of view, the routinization of screening should help to break down barriers to ascertainment and provide more accessible and more valid study data. Such screening would improve our understanding of alcohol's impact on public health, particularly in areas of special concern for women, such as the Fetal Alcohol Syndrome, sexual victimization, and HIV/STD infection.

XI. Summary and Conclusions/Future Directions

The systematic study of alcohol use and abuse is a young field, and there is a need to integrate alcohol and epidemiology research methodologies, particularly in the area of measuring alcohol consumption. Relatively little attention has been given to gender differences in the etiology and health consequences of alcohol use and many areas are in need of further study. For example, available data indicate that the bioavailability of alcohol is higher in women than in men and suggest that women may be more sensitive to alcohol's effects than men. This greater responsivity to alcohol may result in women consuming less alcohol and having fewer alcohol-related problems; however, it may place those women who do drink heavily at higher risk of alcohol-related disease.

In terms of risk factors, women who abuse alcohol appear even more likely than men to suffer from comorbid psychopathology such as depression and posttraumatic stress disorder. Further, there is a high prevalence of heavy drinking and alcohol abuse among women who have experienced victimization, either in childhood or adulthood. The identification of risk factors for alcohol abuse and dependence in women provides a number of opportunities for the development and testing of intervention programs. Risk factors, such as childhood abuse and neglect, childhood sexual abuse, and sexual victimization warrant increased attention and prevention efforts in their own right.

Like men, women who drink may experience potential health benefits (*e.g.,* protection from heart disease) as well as potential risks (*e.g.,* liver damage, family disruption, accidents). However, drinking also poses special risks for women. For example, it has been linked to increased risk of breast cancer and to increased vulnerability to sexual victimization. Further, women of childbearing years who drink heavily run the risk of birth complications and delivering a child with alcohol-related birth defects. Developing a feasible policy to prevent alcohol-related birth defects or alcohol-related neurological deficits is complicated by the facts that 50% of pregnancies are unplanned, most women do not recognize that they are pregnant right away, and the fetus is vulnerable to adverse effects of prenatal alcohol exposure early in pregnancy.

A better understanding of the mechanisms underlying gender differences in consumption patterns is needed, together with prospective studies to better define factors influencing alcohol use and its effect on the health of women and their children. The roles of factors such as ethnicity, age, marital status, and sexual orientation also need to be considered, as they significantly influence women's drinking. Health services research is needed to develop effective prevention programs and to identify methods of breaking down barriers that keep women from seeking and obtaining alcohol treatment and to evaluate the potential benefits of gender-sensitive and gender-specific treatment. In this era of accountability for health spending, studies of health economics are needed to document costs associated with lack of programs to increase case-finding for women abusing alcohol and barriers that reduce the likelihood that they will receive treatment.

References

1. Room, R. (1991). Cultural changes in drinking and trends in alcohol problems indicators: Recent U.S. experience. *In* "Alcohol in America: Drinking Practices and Problems" (W. B. Clark and M. E. Hilton, eds.), pp. 149–162. State University of New York Press, Albany.
2. National Institute on Alcohol Abuse and Alcoholism (1980). "NIAAA Research Monograph 1. Alcoholism and Alcohol Abuse among Women: Research Issue" (DHEW Publ. No. ADM 80-835). U.S. Govt. Printing Office, Washington, DC.
3. National Institute on Alcohol Abuse and Alcoholism (1986). "NIAAA Research Monograph 16. Women and Alcohol: Health-related Issues," DHHS Publ. No. ADM 86-1139. U.S. Govt. Printing Office, Washington, DC.
4. National Institute on Alcohol Abuse and Alcoholism (1996). "NIAAA Research Monograph 32. Women and Alcohol: Issues for Prevention Research," NIH Publ. No. 96-3817. U.S. Department of Health and Human Services, Bethesda, MD.
5. Wilsnack, S. C., and Wilsnack, R. W. (1995). Drinking and problem drinking in U.S. women: Patterns and recent trends. *In* "Recent Developments in Alcoholism" (M. Galanter, ed.), vol. 12, pp. 29–60. Plenum, New York.
6. Turner, C. (1990). How much alcohol is in a "standard drink"?: An analysis of 125 studies. *Br. J. Addict.* **85,** 1171–1175.
7. Hawks, D. (1994). A review of current guidelines on moderate drinking for individual consumers. *Contemp. Drug Probl.* **21,** 223–238.
8. American Psychiatric Association (1994). "Diagnostic and Statistical Manual of Mental Disorders, 4th ed." American Psychiatric Association, Washington, DC.
9. Ustun, B., Compton, W., Mager, D., Babor, T., Baiyewu, O., Chatterji, S., Cottler, L., Gogus, A., Mavreas, V., Peters, L., Pull, C., Saunders, J., Smeets, R., Stipec, M. R., Hasin, D., Room, R., van den Brink, W., Regier, D., Blaine, J., Grant, B. F., and Sartorius, N. (1997). WHO Study on the reliability and validity of the alcohol and drug use disorder instruments: Overview of methods and results. *Drug Alcohol Depend.* **47,** 161–169.
10. Midanik, L. T. (1988). Validity of self-reported alcohol use: A literature review and assessment. *Br. J. Addict.* **83,** 1019–1030.
11. Room, R. (1991). Measuring alcohol consumption in the U.S.: Methods and rationales. *In* "Alcohol in America: Drinking Practices and Problems" (W. B. Clark and M. E. Hilton, eds.), pp. 26–50. State University of New York Press, Albany.
12. Knupfer, G. (1987). Drinking for health: The daily light drinker fiction. *Br. J. Addict.* **82,** 547–555.
13. Fuchs, C. S., Stampfer, M. J., Colditz, G. A., Giovannucci, E. L., Manson, J. E., Kawachi, I., Hunter, D. J., Hankinson, S. E., Hennekens, C. H., Rosner, B., Speizer, F. E., and Willett, W. C. (1995). Alcohol consumption and mortality among women. *N. Engl. J. Med.* **332,** 1245–1250.
14. Smith-Warner, S. A., Spiegelman, D., Yaun, S.-S., van den Brandt, P. A., Folsom, A. R., Goldbohm, A., Graham, S., Holmberg, L., Howe, G. R., Marshall, J. R., Miller, A. B., Potter, J. D., Speizer, F. E., Willett, W. C., Wolk, A., and Hunter, D. J. (1998). Alcohol and breast cancer in women: A pooled analysis of cohort studies. *JAMA, J. Am. Med. Assoc.* **279,** 535–540.
15. Trinkoff, A. M., and Storr, C. L. (1998). Substance use among nurses: Differences between specialties. *Am. J. Public Health* **88,** 581–585.
16. U.S. Department of Health and Human Services (1994). "Eighth Special Report to the U.S. Congress on Alcohol and Health," NIH Publ. No. ADM 94-3699. U.S. Govt. Printing Office, Washington, DC.
17. Centers for Disease Control and Prevention (1990). Alcohol-related mortality and years of potential life lost-United States, 1987. *Morbid. Mortal. Wkly. Rep.* **39,** 173–178.
18. Stinson, F. S., and DeBakey, S. F. (1992). Alcohol-related mortality in the United States, 1979–1988. *Br. J. Addict.* **87,** 777–783.
19. Bonthius, D. J., and West, J. R. (1988). Blood alcohol concentration and microencephaly: A dose-response study in the neonatal rat. *Teratology* **37,** 223–231.
20. Russell, M., Cooper, M. L., Frone, M. R., and Welte, J. W. (1991). Alcohol drinking patterns and blood pressure. *Am. J. Public Health* **81,** 452–457.

21. Russell, M., Marshall, J. R., Trevisan, M., Freudenheim, J., Chan, A. W. K., Markovic, N., Vana, J. E., and Priore, R. L. (1997). Test-retest reliability of the cognitive lifetime drinking history. *Am. J. Epidemiol.* **146,** 975–981.

22. Midanik, L. T. and Clark, W. B. (1994). The demographic distribution of U.S. drinking patterns in 1990: Description and trends from 1984. *Am. J. Public Health* **84,** 1218–1222.

23. Wilsnack, S. C. (1996). Patterns and trends in women's drinking: Recent findings and some implications for prevention. *In* "NIAAA Research Monograph 32. Women and Alcohol: Issues for Prevention Research" (J. M. Howard, S. E. Martin, P. D. Mail, M. E. Hilton, and E. D. Taylor, eds.), NIH Publ. No. 96-3817, pp. 19–63. U.S. Department of Health and Human Services, Bethesda, MD.

24. Wilsnack, S. C., Vogeltanz, N. D., Diers, L. E., and Wilsnack, R. W. (1995). Drinking and problem drinking in older women. *In* "Alcohol and Aging" (T. Beresford and E. Gomberg, eds.), pp. 263–292. Oxford University Press, New York.

25. Fillmore, K. M., Golding, J. M., Leino, E. V., Motoyoshi, M., Shoemaker, C., Terry, H., Ager, C. R., and Ferrer, H. P. (1997). Patterns and trends in women's and men's drinking. *In* "Gender and Alcohol: Individual and Social Perspectives" (R. W. Wilsnack and S. C. Wilsnack, eds.), pp. 21–48. Rutgers Center of Alcohol Studies, New Brunswick, NJ.

26. Wilsnack, R. W., Vogeltanz, N. D., Wilsnack, S. C., Ahlstrom, S., Andrews, F., Bondy, S., Csemy, L., Ferrence, R., Ferris, J., Fleming, J., Graham, K., Greenfield, T., Guyon, L., Haavio-Mannila, E., Knibbe, R., Kubicka, L., Lukomskaya, M., Mustonen, H., Nadeau, L., Narusk, A., Neve, R., Rahav, G., Spak, F., Teichman, M., Trocki, K., Webster, I., and Weiss, S. (1998). Gender differences in alcohol consumption and adverse drinking consequences: Cross-cultural patterns. *Addictions* (in press).

27. Gilbert, M. J. and Collins, R. L. (1997). Gender, stress, coping, and alcohol use. *In* "Ethnic Variation in Women's and Men's Drinking" (R. W. Wilsnack and S. C. Wilsnack, eds.), pp. 357–378. Rutgers Center of Alcohol Studies, New Brunswick, NJ.

28. Herd, D. (1988). Drinking by black and white women: Results from a national survey. *Soc. Probl.* **35,** 493–505.

29. Herd, D. (1993). An analysis of alcohol-related problems in black and white women drinkers. *Addict. Res.* **1,** 181–198.

30. Caetano, R. (1987). Acculturation and drinking patterns among U.S. Hispanics. *Br. J. Addict.* **82,** 789–799.

31. May, P. A. (1989). Alcohol abuse and alcoholism among American Indians: An overview. *In* "Alcoholism in Minority Populations" (T. D. Watts and R. Wright, eds.), pp. 95–119. Thomas, Springfield, IL.

32. Bosron, W. F., Ehrig, T., and Li, T.-K. (1993). Genetic factors in alcohol metabolism and alcoholism. *Semin. Liver Dis.* **13,** 126–132.

33. Grant, B. F., Harford, T. C., Dawson, D. A., Chou, P., Dufour, M., and Pickering, R. (1994). Epidemiologic Bulletin No. 35: Prevalence of DSM-IV alcohol abuse and dependence, United States, 1992. *Alcohol Health. Res. World* **18,** 243–248.

34. Dawson, D. A. (1998). Volume of ethanol consumption: Effects of different approaches to measurement. *J. Stud. Alcohol* **59,** 191–197.

35. Roberts, L. J., and Leonard, K. E. (1997). Gender, stress, coping, and alcohol use. *In* "Gender and Alcohol: Individual and Social Perspectives" (R. W. Wilsnack and S. C. Wilsnack, eds.), pp. 289–311. Rutgers Center of Alcohol Studies, New Brunswick, NJ.

36. Wilsnack, S. C. (1995). Alcohol use and alcohol problems in women. *In* "The Psychology of Women's Health: Progress and Challenges in Research and Application" (A. L. Stanton and S. J. Gallant, eds.), pp. 381–443. American Psychological Association, Washington, DC.

37. Frone, M. R., Russell, M., and Barnes, G. M. (1996). Work-family conflict, gender, and health-related outcomes: A study of employed parents in two community samples. *J. Occup. Health Psychol.* **1,** 57–69.

38. Hughes, T. L., and Wilsnack, S. C. (1997). Use of alcohol among lesbians: Research and clinical implications. *Am. J. Orthopsychiatry* **67,** 20–36.

39. Kessler, R. C., Crum, R. M., Warner, L. A., Nelson, C. B., Schulenberg, J., and Anthony, J. C. (1997). Lifetime co-occurrence of DSM-III-R alcohol abuse and dependence with other psychiatric disorders in the National Comorbidity Survey. *Arch. Gen. Psychiatry* **54,** 313–321.

40. Hartka, E., Johnstone, B., Leino, E. V., Motoyoshi, M., Temple, M., and Fillmore, K. M. (1991). A meta-analysis of depressive symptomatology and alcohol consumption over time. *Br. J. Addict.* **86,** 1283–1298.

41. Miller, B. A. (1996). Women's alcohol use and their violent victimization. *In* "NIAAA Research Monograph 32. Women and Alcohol: Issues for Prevention Research" (J. M. Howard, S. E. Martin, P. D. Mail, M. E. Hilton, and E. D. Taylor, eds.), NIH Publ. No. 96-3817, pp. 239–260. U.S. Department of Health and Human Services, Bethesda, MD.

42. Wilsnack, S. C., Vogeltanz, N. D., Klassen, A. D., and Harris, T. R. (1997). Childhood sexual abuse and women's substance abuse: National survey findings. *J. Stud. Alcohol* **58,** 264–271.

43. Burnam, M. A., Stein, J. A., Golding, J. M., Siegel, J. M., Sorenson, S. B., Forsythe, A. B., and Telles, C. A. (1988). Sexual assault and mental disorders in a community population. *J. Consult. Clin. Psychol.* **56,** 843–850.

44. Kilpatrick, D. G., Acierno, R., Resnick, H. S., Saunders, B. E., and Best, C. L. (1997). A 2-year longitudinal analysis of the relationships between violent assault and substance use in women. *J. Consult. Clin. Psychol.* **65,** 834–847.

45. Kessler, R. C., Sonnega, A., Bromet, E., Hughes, M., and Nelson, C. B. (1995). Posttraumatic stress disorder in the National Comorbidity Survey. *Arch. Gen. Psychiatry* **52,** 1048–1060.

46. Testa, M. and Parks, K. A. (1996). The role of women's alcohol consumption in sexual victimization. *Aggression Violent Behav.* **1,** 217–234.

47. McGue, M. and Slutske, W. (1996). The inheritance of alcoholism in women. *In* "NIAAA Research Monograph 32. Women and Alcohol: Issues for Prevention Research" (J. M. Howard, S. E. Martin, P. D. Mail, M. E. Hilton, and E. D. Taylor, eds.), NIH Publ. No. 96-3817, pp. 65–92. U.S. Department of Health and Human Services, Bethesda, MD.

48. Heath, A. C., Slutske, W. S., and Madden, P. A. F. (1997). Gender differences in the genetic contribution to alcoholism risk and to alcohol consumption patterns. *In* "Gender and Alcohol: Individual and Social Perspectives" (R. W. Wilsnack and S. C. Wilsnack, eds.), pp. 114–149. Rutgers Center of Alcohol Studies, New Brunswick, NJ.

49. Schuckit, M. A., Daeppen, J.-B., Tipp, J. E., Hesselbrock, M., and Bucholz, K. K. (1998). The clinical course of alcohol-related problems in alcohol dependent and nonalcohol dependent drinking women and men. *J. Stud. Alcohol* **59,** 581–590.

50. Johnson, J. L., Sher, K. J., and Rolf, J. E. (1991). Models of vulnerability to psychopathology in children of alcoholics: An overview. *Alcohol Health Res. World* **15,** 33–42.

51. Seilhamer, R. A. and Jacob, T. (1990). Family factors and adjustment of children of alcoholics. *In* "Children of Alcoholics: Critical Perspectives" (M. Windle and J. S. Searles, eds.), pp. 168–186. Guilford Press, New York.

52. U.S. Department of Health and Human Services (1997). "Ninth Special Report to the U.S. Congress on Alcohol and Health," NIH Publ. No. 97-4017. U.S. Govt. Printing Office, Washington, DC.

53. Wilsnack, S. C., Klassen, A. D., Schur, B. E., and Wilsnack, R. W. (1991). Predicting onset and chronicity of women's problem

drinking: A five-year longitudinal analysis. *Am. J. Public Health* **81,** 305–318.

54. Wilsnack, R. W., Wilsnack, S. C., Kristjanson, A. F., and Harris, T. R. (1998). Ten-year prediction of women's drinking behavior in a nationally representative sample. *Women's Health: Res. Gender, Behav. Policy* **4,** 199–230.

55. Schenker, S. (1997). Medical consequences of alcohol abuse: Is gender a factor? *Alcohol Clin. Exp. Res.* **21,** 179–181.

56. Gavaler, J. S., and Arria, A. M. (1995). Increased susceptibility of women to alcoholic liver disease: Artifactual or real? *In* "Alcoholic Liver Disease: Pathology and Pathogenesis" (P. Hall, ed.), pp. 123–133. Arnold, London.

57. Blume, S. B. (1997). Women and alcohol: Issues in social policy. *In* "Gender and Alcohol: Individual and Social Perspectives" (R. Wilsnack and S. Wilsnack, eds.), pp. 462–489. Rutgers Center of Alcohol Studies, New Brunswick, NJ.

58. Russell, M., Martier, S. S., Sokol, R. J., Mudar, P., Jacobson, S., and Jacobson, J. (1996). Detecting risk drinking during pregnancy: A comparison of four screening questionnaires. *Am. J. Public Health* **86,** 1435–1439.

59. Fleming, M. F., Barry, K. L., Manwell, L. B., Johnson, K., and London, R. (1997). Brief physician advice for problem alcohol drinkers. *JAMA, J. Am. Med. Assoc* **277,** 1039–1045.

60. American Medical Association (1997). "Council on Scientific Affairs Report. Alcoholism and Alcohol Abuse among Women" AMA, Chicago.

61. Walitzer, K. S., and Connors, G. J. (1997). Gender and treatment of alcohol-related problems. *In* "Gender and Alcohol: Individual and Social Perspectives" (R. W. Wilsnack and S. C. Wilsnack, eds.), pp. 445–461. Rutgers Center of Alcohol Studies, New Brunswick, NJ.

Section 8

ENVIRONMENTAL EXPOSURES

Ellen K. Silbergeld
University of Maryland
Baltimore, Maryland

Because women's lives take place in the context of many environments—personal and societal—the definition of women's environmental health is complex. In this section, we have decided to limit the scope of "environment" to those physical or chemical stressors that occur primarily in the nonoccupational environment. This limitation unavoidably obscures connections between these environmental exposures and other external influences on health, such as nutrition and psychosocial, cultural, andeconomic factors. This decision was made consciously because of the continued lack of scientific and clinical attention to women's environments as determinants of health. Even now, more often than not, epidemiologic studies (and clinical medicine) fail to ascertain information on women's employment. The large scale Women's Health Initiative, for instance, is collecting little information on women's occupations despite the fact that since the authors of this book were born, work has become an important aspect of women's lives. Women's work is considered hazard free, even though the "pink collar" world is filled with chemicals and poor ventilation; nursing and health care are notoriously risky occupations (drugs, pathogens, injury); home craftwork can be highly dangerous; and ergonomic stresses have greater impact on women, who often work with poorly designed equipment. Environmental exposures are even less well studied for women, despite information on the complexity and intensity of women's exposures to toxic waste, pesticides in food, or contaminants in indoor air. One aim of this section is to stimulate researchers, clinicians, and women to ask questions of each other, to design appropriate studies, and to examine environments rigorously as sources for identifying preventable causes of disease.

Women's environments, exposures, and responses vary, in part as a result of life stage, and the extent of this variation among women is at least as great as the variation between women and men. Similarly, on a population basis, the range of physiological responses and genetic susceptibilities of women as compared to men is no greater than that among women. Nevertheless, there are environmental risks that are more often, and sometimes more intensively, experienced by women. Yet there are aspects of women's biology and genetics that affect the distribution of women's responses to environmental risks differentially. Perhaps no issue highlights these complexities more clearly than the environmental endocrine disruptor hypothesis [1]. The effects of xenobiotic hormone-like agents on human health are influenced by the differences in the background of normal hypothalamic–pituitary–gonadal physiology and genetics in males and females. Moreover, the different trajectory of development in males and females, influenced by endocrinological ontogeny, affects responses to these agents [2]. Thus, for example, it has been hypothesized that in males fertility may be susceptible to endocrine disruptors, while in females the hormone-dependent cancers (such as ovarian and breast) may be influenced by these chemicals.

The chapters in this section highlight our knowledge and our ignorance. At the end of the twentieth century, despite enormous public concern over the environment and the creation of new institutions to respond to and protect against environmental risks, we know relatively little of the environmental determinants of women's health and disease. As noted before, epidemiologic studies infrequently attend to environmental factors in women's health outcomes. Toxicologists also fail to study effects of chemicals in female animals and usually include both sexes only in examining reproductive toxicity. For the general public, too often women's health is confused with the health of children. For instance, studies on the release of lead from bone have mostly focused on the potential impacts of this on the fetus or nursing infant, without consideration of the risks presented to women when their circulating levels of lead rise sharply during lactation or after menopause [3].

In this section, we discuss three aspects of women's environmental health: the context of women's health in the global environment; the particular interactions between the environment and women's reproductive biology; and the role of environment in cancer in women. But in an important way, the role of the environment in women's health permeates this book, beyond this section. In every chapter dealing with a major health endpoint the role of environmental factors must be considered unless genetic determinism is unequivocal. There is considerable confusion about the implications of the new genetics for understanding risk factors for disease, and no more so than in the case of "cancer genes," such as BRCA1 and BRCA2 (see Chapter 69). These genes impart a high risk of cancer to those women who carry mutations, but for many mutations the risks are no higher than those of nonfamilial cancer. Most "cancer genes" are susceptibility genes, that is, they are associated with increased risks of cancer in the context of specific exposures [4].

It is the *gene in the environment* that causes disease; as Dr Kenneth Olden, Director of the National Institute of Environmental Health Sciences, has stated: "Genetics loads the gun but the environment pulls the trigger." [5]. We cannot understand patterns of disease in women unless we understand the role of genetics in modulating individual and population response. But of equal importance, we cannot prevent disease by repairing genes if we do not clean up our environments.

References

1. Colborn, T., Myers, J. P., and Dumanoski, D. (1996). "Our Stolen Future." Little, Brown, Boston.
2. McLachlan, J. A. (1993). Functional toxicology: A new approach to detect biologically active xenobiotics. *Environ. Health Perspect.* **101,** 386–387.
3. Silbergeld, E. K., Sauk, J., Somerman, M., Todd, A., McNeill, F., Fowler, B., Fontaine, A., and van Buren, J. (1993). Lead in bone: Storage site, exposure source, and target organ. *Neurotoxiicology* **14,** 225–236.
4. Harris, C. C., Weston, A., Willey, J. C., Trivers, G. E., and Mann, D. L. (1987). Biochemical and molecular epidemiology of human cancer: Indicators of carcinogen exposure, DNA damage, and genetic predisposition. *Environ. Health Perspect.* **75,** 109–119.
5. Olden, K. (1999). Environmental risks and health disparities. *Speech Soc. Toxicol. Ann. Meet. 1999.*

48

The Environment and Women's Health: An Overview

ELLEN K. SILBERGELD
University of Maryland School of Medicine
Program in Human Health and the Environment
Baltimore, Maryland

I. Introduction

Concerns over women's health and concerns over environmental health risks are major issues for the American public. Despite this interest and the substantial investments in research and public policy in these areas, relatively little effort has been made to examine areas of potential overlap. By way of introduction to the specific topics covered in this section, this chapter discusses those aspects of environmental health that are of concern to women and those aspects of women's health that may relate to environmental factors. The chapter's goals are to provide definitions for the domain of overlap between women's health and environmental health and to identify some critical issues for clinical medicine, biomedical research, and public policy.

Environmental risks are an important set of risks because, if identified, they are preventable, often through concerted action, including regulation and voluntary changes by consumers and industry. Studying environmental factors specifically in the context of women's health is important because it cannot be assumed that the environment will affect women and men in the same way. First, systematic differences in the way men and women live and work can result in significant differences in environmental exposures. Second, women's lives involve hormonal and metabolic changes that create opportunities for differential effects of similar environmental exposures as compared to men. Many of these biologic events influence the uptake, distribution, and storage of chemicals.

Men and women do differ in readily apparent biological ways, although in many cases the differences *between* men and women are no greater than those *among* women or men. It is important to ask which of these differences, if any, matter for public health social policy, and under what circumstances. Decisions about the policy relevance of such differences should ideally be made prior to assessment of specific cases to avoid introducing bias in the event of controversy. When such differences are closely associated with gender, then studying gender-specific exposures and outcomes is an important strategy to identify and prevent risks of disease.

II. Definition of the Topic

Defining the "environment" is a difficult task. In a biological context, the "environment" may be considered everything outside the genome, that is, all events that are acquired by the cell through experience. More practically, for purposes of this chapter, the "environment" is defined as the external social, biological, physical, and chemical milieu in which individuals and

populations exist. The human environment includes economic and psychological forces that affect individuals and societies (see Rowland Hogue, Chapter 2). This section is limited to the health impacts of chemical exposures from contact with air, water, soils and dusts, from drinking water and food, and from using products containing biologically active chemicals. Genetics, the nutritional content of diet, and other exposures, some of which are voluntary (such as smoking), also affect health, and these may interact with traditionally defined environmental risks.

The distinction between inherent and acquired risks to health is blurred by recent research. Many inherent characteristics, associated with genotype, do not in themselves "cause" disease, but rather modulate the response of the individual to an acquired, or environmental, risk. Inheritance of a genetic mutation predisposing a person to disease risk, such as a "breast cancer gene," may only become important for an individual's health when an environmental factor elicits its expression or adds an additional mutation. Similarly, the fact that women's immunology differs from that of men's does not in itself "cause" disease [1]; however, it may result in differential responses to pathogen challenge between men and women (see Section 8 in this book). In addition, biological factors that affect susceptibility may have nongenetic origins. For example, women are reportedly more susceptible to lead-induced anemia (WHO 1995); however, the cause of this differential response is likely to be due in large part to the fact that women are more likely to be iron-deficient than are men.

When environmental risk factors are studied in the context of women's health, it is often assumed that women are intrinsically more susceptible than are men to the same exposures. For many reasons, this assumption must be critically examined, especially since it has been cited as a basis for discriminatory policies in the workplace [2]. Of course, if women are truly more susceptible than men to an occupational risk, this is not a reason in itself to exclude women from jobs where such exposures may occur, but rather it is a reason to promulgate and enforce standards that are protective of all workers. Nevertheless, despite U.S. court decisions upholding legal requirements for equal employment opportunities, this has not been the practice in occupational health policy in the U.S.

It is equally important to expand the definition of environmental risks beyond issues related to intrauterine and postpartum development of the fetus and child. The effect of the environment on the growth and development of children should be a concern both to men and to women, and should not be considered solely as a "women's health" issue. Environmental impacts on reproduction and development can be mediated through either or both biological parents [3]. Too often when environmental health research and policy address women's health, the discussion becomes restricted to reproduction and child development and fetal toxicity is confused with women's susceptibility. For example, the U.S. Environmental Protection Agency guidelines for the assessment of neurotoxins identifies pregnant women as "specially susceptible" on the basis of potential developmental neurotoxicity of the fetus. This confuses the fetus with the woman; for pregnant women to be "specially susceptible" to neurotoxins would mean that pregnancy in itself changes the response of *women* to these chemicals. Whether and under what conditions the fetus may be "specially susceptible" is important to determine as a separate matter of health policy.

III. Gender and the Environment

A. *Nonbiological Issues*

1. Socioeconomic Factors

Gender is both a biological condition and a social construct. The social dimensions of gender affect the interactions of the individual with her environment in many ways (see also Section 3 of this book). Socioeconomic status is a major determinant of environmental exposures. As has been documented in many studies during the 1990s, poverty has become feminized in the U.S. and women are overrepresented among the heads of poor families. This has important implications for exposures to toxic chemicals in the environment. The environmental justice movement has documented many instances where exposures are inversely related to income [4]. One of the clearest examples is lead. In the U.S. population, blood lead levels in women and men significantly differ by income, socioeconomic status, and place of residence [5]. The fact that socioeconomic status varies between men and women in the U.S. makes these kinds of analyses even more complex, even outside the workplace, where gender is a significant determinant in employment patterns, workplace assignment, and occupational exposures (see Chapter 36).

Gender also affects exposures by influencing lifestyle. Studies by ATSDR have found that women are more often exposed, usually in the home, to contaminants from hazardous waste sites such as benzene, trichloroethylene, and trichloroethane, because of different responsibilities at the home, interactions with children and responsibilities for family care giving, and other variables [6]. Activities in the home, such as cleaning, cooking, hobbies, and "do it yourself" repairs and renovations can involve exposures to toxic agents, although this is not often realized. Indeed, in some cases, these exposures may exceed at peak the exposures in a well-run workplace, and in many cases, persons exposed in the home are not informed as to appropriate conditions of ventilation or protective equipment to reduce exposures. Home exposures to solvents are an important example. As shown in Table 48.1, in a study of risk factors for cardiac birth defects, it was found that the women surveyed (mothers of both cases and controls) reported more "frequent" exposures to solvents and pesticides from home activities than from occupation [7]. Cooking with gas can involve substantial exposures to complex hydrocarbons, particularly in unvented small spaces such as kitchens. Other home based exposures include microwave radiation and radon. Compounds found in indoor air, re-

Table 48.1

Home Exposures Reported by Women in the Baltimore-Washington Infant Study, 1981–1989[a]

Exposure	% Reporting	
	Cases[b]	Controls[c]
Paints[d]	9.2	9.1
Paint Stripping[d]	2.0	2.2
Varnish[d]	2.3	2.5
Dry Cleaning Solvents[d]	0.3	0.3
Degreasers	0.4	0.4
Miscellaneous Solvents	1.2	0.6
Arts and Crafts Paint	3.5	3.2
Lead Solder	0.2	0.2
Jewelry Making[d]	0.1	0.1
Stained Glass[d]	0.3	0.2
Pesticides[d]	25.7	24.2

[a]Data from the Baltimore-Washington Infant Study [7].

[b]Cases consist of 3377 mothers of liveborn infants with congenital heart defects.

[c]Controls consist of 3572 mothers randomly selected from area hospitals and matched by year of birth.

[d]Home exposures reported more frequently than occupational exposures.

leased from off-gassing of products (such as formaldehyde from particle board) or other sources, may be more intensively experienced by those who spend greater periods of time at home indoors. This may often, but not always, include homemakers and those who care for preschool children, who are often but not always women.

2. Diet and Other Factors

Gender differences in diet may also modulate responses to environmental exposures. The diets of men and women tend to differ within sociocultural groups, although across the multicultural patterns of diet in this country there is considerable overlap. It is not possible to review here all the interactions between nutrients and toxic chemicals. For example, most of the toxic metals (lead, cadmium, arsenic, manganese, mercury) interact in terms of absorption and distribution, if not toxicity, with the essential trace elements calcium, iron, magnesium, and possibly phosphorus [8]. Moreover, food itself is often a medium of exposure to such contaminants as pesticides and metals.

Ethnicity and socioeconomic status are important determinants of food consumption that can cause large variabilities in both exposures, when food is contaminated, and response, because nutritional status can affect the absorption of toxicants (*e.g.,* trace elements and cadmium absorption in women, studied by Vahter *et al.* [9]; dietary iron and calcium and blood lead levels in women, studied by Van Buren [10]). Fish are particularly significant vectors of exposure to persistent organic pollutants [11]. Several studies have drawn attention to the potential for increased exposures to contaminants in freshwater fish consumed by Native Americans, such as the Akwasane in the Great Lakes region of New York, and others who depend on this protein source to a much greater extent than the "average" or "median" North American consumer [12,13]. Sport fish consumption from contaminated ecosystems in the Great Lakes can increase the dose of PCBs more than 4000 times above "background" for the general population [14]. For women, these exposures may be particularly dangerous because since PCBs and other persistent organochlorines are preferentially accumulated and stored in adipose tissue (this quality is known as lipophilicity). Women, many of whom have higher body fat composition ratios than men, may retain proportionally greater amounts of lipophilic chemicals, such as PCBs, DDT, and chlorinated dioxins.

A variety of other cultural practices expose women in certain subpopulations to increased risk. The use of traditional cosmetics and medicines made with lead by some Hispanic Americans (*azarcon* and *greza*) and by some groups of Muslims (*kohl*) has been associated with clinical lead toxicity in women and young children [15].

B. Biological Issues—Uptake, Absorption, Metabolism

Sex influences the transfer of compounds from the external environment and the fate of absorbed substances within the body. The first set of factors relate to uptake, or the rates of contact between the individual and substances in air, water, food, and soils. As is now recognized, the "medical model" of the 70 kg male does not adequately describe the distribution of such functions as ventilation (m^3 air breathed per hour or day) or water intake. In addition, women may differ in transfer functions such as ventilation efficiency, absorption across the alveoli, skin absorption, and blood flow among tissue compartments. There may also be important gender differences in excretion of toxins, as has been shown for mercury [15a]. The second set of factors relates to the distribution and metabolism of toxic chemicals. Some major metabolic pathways differ among men and women and are influenced by hormonal status among women. For toxicology, two major sets of metabolizing pathways are Phase I and Phase II metabolism, or oxidation and conjugation reactions. Of these, the CYP450 enzymes are of great importance in detoxifying certain compounds and activating others. Many of the CYP450 enzymes differ in activity between males and females; some of them are affected by circulating levels of estrogen [16,17]. The conjugating enzymes include glucuronidases, methyl transferases, and glutathione-S-transferases. Some of these enzymes are also affected by changes in circulating levels of hormones. As shown in Table 48.2, the effects of benzene and ethanol on hematopoiesis differ between male and female mice and between pregnant and nonpregnant females [18].

Distribution of absorbed toxicants differs among individuals. A major difference relates to compartmentation into adipose tissue, a characteristic of lipophilic compounds such as DDT, PCBs, and dioxins. On average, women have a higher body fat percentage than men, and thus may carry higher body burdens of these fat seeking compounds, given equal exposures. However, because men are often more exposed than women to chemical toxicants (owing to diet and occupation, as well as avocational activities such as fishing), these physiological factors may be less important. The retention of compounds in adipose compartments also differs between men and women, often in response to changes in physiological status. Some women

Table 48.2
Effects of Benzene or Ethanol on Hematopoieses (Erythroid Colony Forming Units) in Mice

Group	Benzene (10 ppm)	Ethanol (5%)
Adult female	No effect	No effect
Adult female	Reduced	Reduced
Adult female exposed *in utero*	No effect	No effect
Adult male exposed *in utero*	Reduced	Reduced
Pregnant female	No effect	Increased

Data from Corti and Snyder, 1996, cited in Setlow *et al.* [18].

experience dramatic changes in body fat because of pregnancy, diet, and socially determined patterns of weight gain and loss.

Another major compartment for the long term storage of toxicants is bone. Several toxic metals are accumulated in the mineral matrix of bone (strontium, mercury, lead, aluminum, and cadmium) [18a]. The skeleton is a sexually dimorphic organ in humans; its growth, patterns at maturation, and senescence differ between males and females. Bone is highly responsive to hormonal signals of estrogen and progesterone. The greater liability of bone in females, in response to hormonal changes, may influence the storage and release of bone-seeking metals. Aufderheide *et al.* [19] showed that men accumulate higher levels of lead in bone and that accumulation continues over the lifespan in men as compared to women.

Physiological stages in women's lives can influence both the bone and adipose tissue compartments. Pregnancy, lactation, and menopause are normal stages of many women's lives. These stages can be affected by exposure to toxic chemicals, as discussed by Flaws *et al.* (Chapter 50). At the same time, changes in women's physiological state throughout the lifespan can modulate risks to women themselves [20].

Pregnancy is a period of significant physiologic change for women. Blood volume increases substantially during pregnancy (about 40% overall); body composition changes and weight increases; diet and intermediary metabolism change; absorption and retention of specific nutrients and minerals change; the kinetics of hematopoiesis change; and hepatic metabolism changes. The fetus also imposes an energy demand on the pregnant woman, which changes her fat metabolism. During early pregnancy, fat stores are increased. Circulating plasma levels of triglycerides and cholesterol increase gradually over pregnancy and all three classes of lipoproteins are enriched in these fractions. Later in pregnancy increased lipolysis in adipose tissue releases fatty acids into plasma to support triglyceride synthesis in the liver.

Lactation is another period of substantial physiologic stress. The hormonal changes accompanying lactation affect ovarian function (suppressing ovulation) and other organs. Lactation appears to exert more of a stress on calcium and fat stores than does pregnancy. There can be measurable decreases in bone mineral content during lactation [21]. Fat, specifically triglycerides, is required for the production of breast milk. Most fat is drawn from increased hepatic synthesis. In addition, lipolysis in adipocytes (fat stores) increases and total adipocyte storage enzyme decreases. Even if these stores are replenished later, the chemicals stored earlier will have been mobilized back into the circulation.

Menopause, the cessation of reproductive fertility in women, occurs when the supply of oocytes for ovulation is depleted. At menopause, the physiology of bone and adipose tissue changes significantly. Fat metabolism changes during menopause and there is an almost universal weight gain at this stage, with a relative increase in adipose tissue [22]. Bone density decreases as the skeleton loses calcium. This change causes significant mobilization of metals stored in bone, as demonstrated in several studies of lead. As shown in Table 48.3, in three separate cross-sectional studies, postmenopausal women were found to have significantly increased levels of lead in blood [23].

These three life stages in women are associated with significant changes in calcium metabolism, fat storage and metabolism, and the absorption and retention of nutrients and minerals by the gut and kidney. This suggests that there are interactions between life stages and toxicants that are persistent and stable and that are stored in long term compartments such as bone and fat. The possibility that changes in internal distribution of such toxicants might contribute to the morbidity and mortality associated with pregnancy and menopause deserves consideration. One of the most obvious outcomes is the possibility that increases in circulating levels of lead may increase blood pressure. Hypertension is an important risk during pregnancy, and risks of hypertensive heart disease increase substantially in women after menopause [22]. The possibility that changes in the compartmentation of lead during these physiological stages might contribute to hypertension has not received much study.

Another possible area of interaction is the mobilization of lipophiles in adipose tissue during breastfeeding. This might increase the exposure of breast tissue to these chemicals. This raises interesting questions about the interactions between reproductive history, breastfeeding, and past chemical exposures as risk factors for such diseases as breast cancer [24].

IV. Environmental Health Concerns for Women

A. Endocrine Disruption

Endocrine disruption is a new concern in environmental health, articulated by Colborn and others [25], drawing upon a long history of research in ecology going back to the work of Rachel Carson in the 1960s. The stated thesis is that there are chemicals, many of which are persistent and synthetic, that can adversely affect human health by interrupting normal endocrine function through any of several mechanisms: binding to hormone receptors, blocking synthesis or catabolism of hormones, affecting hormone release or binding to transport proteins, or affecting hormone dependent signal transduction within responsive cells and tissues. While one of the first concerns raised by proponents of this thesis was breast cancer [24], more recent concerns have been focused on the male and include reduced sperm counts and increased rates of hypospadias in male infants. Some of the chemicals identified as endocrine disruptors— TCDD, DDT, PCBs, kepone, methoxychlor—have been found to induce alterations in the development of the reproductive

Table 48.3
Blood Lead Levels in Women by Age

Source	Age range	Blood lead (geometric mean)
Western Europe 1988–1990	30–39	6.3 μg/dL
	40–49	6.5
	50–59	8.9
Western Europe 1990–1991	28–37	6.0
	38–47	6.3
	48–57	6.9
United States 1976–1980	Premenopausal	11.6
	Postmenopausal	14.2
United States 1988–1991	20–49	1.66
	50–69	3.01

system both in male and in female rats [26]. No clear evidence of such effects has been reported for humans, although ecological studies are replete with descriptions of intersex fish, birds, and amphibians.

B. Autoimmune Disorders

Autoimmune disorders are more prevalent in women and their age-related incidence patterns differ in women as compared to men (see Section 8). The immune system in general is modulated by sex steroids [26a,27]. Autoimmune diseases are caused by failures in the regulation of the immune system [28] such that autologous proteins are not recognized and the biological responses to foreign proteins are inappropriately mobilized. Devastating organ-specific effects can ensue, such as rheumatoid arthritis, lupus, and glomerulonephritis. Most autoimmune diseases are acquired, that is they do not result from inborn errors in the immune system [27]. The best model for an environmentally induced autoimmune condition is mercury-induced glomerulonephritis, in which the immune system produces autoantibodies to glomerular basement membrane protein (GBMP). In this disease, GBMP autoantibodies are deposited within the glomerulus and renal dysfunction ensues. In animals, females are more sensitive to mercury-induced autoimmunity and immunotoxicity than are males [29]. However, human data are lacking on sex differences because most studies of mercury autoimmunity have been conducted on workers exposed to mercury and until recently, these cohorts were almost exclusively male.

C. Reproductive Toxicology

Other aspects of physiology specific to women that are capable of being affected by environmental risks are related to reproduction. As reviewed by Flaws *et al.* (Chapter 50), sexual maturation, menstrual cycling, ovulation, pregnancy, lactation, and menopause can all be affected by chemical exposures. Some of these chemicals are gonadal toxins and some act at the level of the hypothalamus or pituitary. Lactation is also sensitive to environmental chemical exposures. Gladen and Rogan [30] have reported that exposure to DDT and body burdens of its metabolites (DDD and DDE) are correlated with length of lactation among women in Mexico. Labor and delivery also can be affected by chemical exposure. Early delivery has been reported in women with elevated lead exposures [31], and more difficult deliveries have been reported in women exposed to PCBs [32]. Loch-Caruso [33] has investigated some of the physiological bases for effects of PCBs on labor and has proposed that PCBs can affect the onset of labor through actions on uterine muscle.

Menopause also is sensitive to environmental risks. Smoking and chemicals found in cigarette smoke have been associated with earlier age of menopause and also with a shortened menopausal transition [34; Flaws et al., (Chapter 50) this book]. In animal models, chemicals that kill follicles hasten the onset of reproductive failure; these chemicals include several solvents, polycyclic aromatic hydrocarbons, and other synthetic organic chemicals [35,36]. There are few data on humans, again because of the exclusion of women from occupations where such exposures have occurred.

Health status *after* menopause may also be affected by earlier chemical exposure. Osteoporosis, discussed in detail in Chapter 94, is a condition of bone density loss that can compromise the structural integrity of bone to an extent where the risk of disabling fractures is substantially increased. The frequency of lowered bone mineral density increases dramatically in women after menopause. Much is known of the pathophysiology of this and its relation to the loss of the gonadal hormones estrogen and progesterone; hormone replacement after menopause can slow the rate of bone density loss significantly, with substantial benefit in terms of reduced incidence of fracture.

Environmental exposures interact with bone density loss in several ways. First, certain environmental exposures may cause bone resorption or increase the risks of fracture associated with postmenopausal bone density loss by placing premenopausal women at lower bone density prior to menopause. Second, some toxic agents stored in bone may affect the way in which bone responds to the alterations in hormonal homeostasis at menopause. Third, toxic chemicals stored in the skeleton can be released when bone is resorbed. Cadmium and aluminum are two toxic metals associated with osteopenia under conditions of high level environmental exposures [18a]. Data from the study of Japanese women exposed to cadmium from contaminated fertilizers showed that exposure resulted in a clinical disease called "itai itai" [it hurts] disease, which is characterized by acute joint pain. Clinical measures indicated substantial bone resorption in these women, and at menopause they were more likely to experience osteoporosis than were women without excessive cadmium exposure [37]. Aluminum exposure has also been associated with osteodystrophy and bone mineral density change [18a]. A brief report suggests a greater risk of clinical osteoporosis in women who had been heavily exposed to lead as young adults [38].

V. Summary and Future Directions

Environmental risk factors can affect almost every aspect of women's health, more often because of patterns of exposure than because of gender-based differences in biological response. Fetal exposure mediated by women's exposures has at times resulted in unfortunate public policy ostensibly designed to promote health but effectively supporting discrimination in employment.

Most studies have focused on reproductive health and cancer, which is one reason for the lack of information on environment and women's health. Another reason is the relative lack of study of women in occupational settings, which is often the origin of information on chemical risks to health because of the relatively more intense exposures that usually occur in the workplace and the ease of defining cohorts for study. In addition, there is a continuing failure to appreciate that chemical exposures, in the workplace or the general environment, are important to women's health. Case-control studies of specific endpoints in women frequently pay little attention to occupational exposures. Even more importantly, there is continued neglect of the exposures that can occur to women outside the workplace. As shown in Table 48.1, household products and avocational activities can involve exposures to highly toxic agents, often without the precautions or information available to workers. In a study conducted among women in Baltimore to ascertain sources of lead, the pregnant woman with the highest blood lead was exposed via a home based hobby of jewelry making (J. Van Buren and E. K. Silbergeld, unpublished data).

The paucity of data on environment and women's health arises from two additional sources: the failure to collect surveillance information on women's health that can be correlated with environmental indicators such as air and water pollution or releases of toxic chemicals, and the continuing limits of toxicological studies. Too often, toxicologists only study male animals and thus the opportunity to gather data on possible differences in susceptibility or even types of toxic effects is not available. The important role of hormonal status usually is considered a complication rather than an opportunity to gain further understanding to help predict and prevent possible effects in humans.

The chapters in this section represent a state-of-the-art analysis of information on women's health and environmental risks. That the data are still so sparse challenges us to better understand the reality of women's lives in order to identify, through improved study designs and clinical practice, preventable environmental risks important for women's health.

Acknowledgments

Research described in this chapter has been supported by grants from the Heinz Family Foundation and the Winslow Foundation. The contributions of Jenna S Roberts, Devra Davis, Joan Bertin, Jude van Buren, Denis Nash, and Jodi Flaws to the development of concepts of women's health and the environment are deeply appreciated.

References

1. Garenne, M., and Lafon, M. (1998). Sexist diseases. *Perspect. Biol. Med.* **41**(2), 176–189.
1a. World Health Organization (WHO) (1995). "*Environmental Health Criteria Document: Lead.*" WHO, Geneva.
2. Bertin, J. E., and Werby, E. A. (1993). Legal and policy issues. *In* "Occupational and Environmental Reproductive Hazards" (M. Paul, ed.), pp. 15–162. Williams & Wilkins, Baltimore, MD.
3. Silbergeld, E. K., and Gandley, R. (1994). Male-mediated reproductive toxicity: Effects on the nervous system of offspring. *In* "Male-Mediated Developmental Toxicity" (A. Olshan and D. Mattison, eds.). pp. 141–151. Liss, New York.
4. Sexton, K., Gong, H., Bailar, J. C., Ford, J. G., Gold, D. R., Lambert, W. E., and Utell, M. J. (1994). Air pollution health risks: Do class and race matter? *Toxicol. Ind. Health* **9**(5), 843–878.

5. Brody, D. J., Pirkle, J. L., Kramer, R. A., Flegal, K. M., Matte, T. D., Gunter, E. W., and Paschal, D. C. (1994). Blood lead levels in the U.S. population. *JAMA, J. Am. Med. Assoc.* **272**(4), 277–283.

6. Burg, J. A., and Gist, G. L. (1997). The potential impact on women from environmental exposures. *J. Women's Health* **6**(2), 159–161.

7. Ferencz, C., Correa-Villasenor, A., Loffredo, C. A., and Wilson, P. D. (1993). "Genetic and Environmental Risk Factors of Major Cardiovascular Malformations." Futura Publ. Co., Armonk, NY.

8. Mahaffey, K. R. (1990). Environmental lead toxicity: Nutrition as a component of intervention. *Environ. Health Perspect.* **89**, 75–78.

9. Vahter, M., Berglund, M., Nermell, B., and Akesson, A. (1996). Bioavailability of cadmium from shellfish and mixed diet in women. *Toxicol. Appl. Pharmacol.* **136**(5), 332–334.

10. Van Buren, J. (1996). Interactions between lead and calcium in pregnant women. Thesis submitted for the Ph.D., Johns Hopkins University School of Hygiene and Public Health, Baltimore, MD.

11. Asplund, L., Svensson, B. G., Nilsson, A., Eriksson, U., Jansson, B., Jensen, S., Wideqvist, U., and Skerfving, S. (1994). Polychlorinated biphenyls, 1,1,1-trichloro-2, 2-bis(p-chlorophenyl)-ethane (p,p'-DDE) in human plasma related to fish consumption. *Arch. Environ. Health* **49**(6), 477–486.

12. Mendola, P., Buck, G. M., Vena, J. E., Zielezny, M., and Sever, L. E. (1995). Consumption of PCB- contaminated sport fish and risk of spontaneous fetal death. *Environ. Health Perspect.* **103**, 598–502.

13. Environment Canada (1991). "Toxic Chemicals in the Great Lakes and Associated Effects," Vol. 2. Environment Canada, Ottawa.

14. Schantz, S. L., Jacobson, J. L., Humphrey, H. E. B. *et al.* Determinants of plychlorinated biphenyls (PCBs) in the sera of mothers and children from michigan farms with PCB-contamined silos. *Arch. Environ. Health* **49**(6), 452–458.

15. Mojdehi, G. M., and Gurtner, J. (1996). Childhood lead poisoning through kohl. *Am. J. Public Health* **86**(4), 587–588.

15a. Thomas, D. J., Fisher, H. L., Sumler, M. R., Mushak, P., and Hall, L. L. (1987). Sexual differences in the excretion of organic and inorganic mercury by methyl-mercury treated rats. *Environ. Res.* **43**, 203–216.

16. Hunt, C. M., Westerkam, W. R., and Stave, G. M. (1991). Effect of age and gender on the activity of human hepatic microsomal mono oxygenases. *Biochem. Pharmacol.* **44**, 275–283.

17. Wilson, K. (1984). Sex-related differences in drug disposition in man. *Clin. Pharmacokinet* **9**, 189–202.

18. Setlow, V. P., Lawson, C. E., and Woods, N. F., eds. (1998). "Gender Differences in Susceptibility to Environmental Factors: A Priority Assessment." NAS Press, Washington, DC.

18a. Bhattacharyya, M., Wilson, A. K., Silbergeld, E. K., Watson, L., and Jeffery, E. (1995). Metal-induced osteotoxicities. *In* "Metal Toxicity" R. A. Goyer, M. P. Waalkes, and C. D. Klaassen, eds.), pp. 465–510. Academic Press, San Diego, CA.

19. Aufderheide, A. C., Neiman, F. D., Wittmers, L. E., and Rapp, G. (1981). Lead in bone. *Am. J. Phys. Anthropol.* **55**, 285–291.

20. Roberts, J. S., and Silbergeld, E. K. (1995). Pregnancy, lactation, and menopause: How physiology and gender affect the toxicity of chemicals. *Mt. Sinai J. Med.* **62**(5), 343–355.

21. Sowers, M. F. R., Corton, G., Shapiro, B., Jannavsch, M. L., Crutchfield, M., Smith, M. L., Randolph, J. F., and Hollis, B. (1993). Changes in bone density with lactation. *JAMA, J. Am. Med. Assoc.* **269**, 3130–3135.

22. Sowers, M. F. R., and La Pietra, M. T. (1995). Menopause: Its epidemiology and potential association with chronic diseases. *Epidemiol. Rev.* **17**(2), 287–302.

23. Watson, L., and Silbergeld, E. K. (1995). Exposure to lead during reproduction and menopause. *Funda. Appl. Toxicol.* **25**, 167–168.

24. Davis, D. L., Bradlow, H. L., Wolff, M. S., Woodruff, T., Hoel, D. G., and Anton-Culver, H. (1993). Medical hypothesis: Xenoestrogens as preventable causes of breast cancer. *Environ. Health Perspect.* **101**, 372–377.

25. Colborn, T., Myers, J. P., and Dumanoski, D. (1996). "Our Stolen Future." Little, Brown, Boston.

26. Flaws, J. A., Sommer, R. J., Silbergeld, E. K., Peterson, R. E., and Hirshfield, A. N. (1997). In utero and lactational exposure to 2,3,7.8-tetrachlorodibenzo-p-dioxin (TCDD) induces genital dysmorphogenesis in the female rat. *Toxicol. Appl. Pharmacol.* **147**(2), 351–362.

26a. Grossman, C. J. (1984). Regulation of the immune system by sex steroids. *Endocr. Rev.* **5**, 435–455.

27. Lahita, R. G. (1996). The connective tissue diseases and the overall influence of gender. *Int. J. Fertil.* **41**, 156–165.

28. Kilburn, K. H., and Warshaw, R. H. (1994). Chemical induced autoimmunity. *In* "Immunotoxicology and Immunopharmacology" J. H. Dean, M. I. Luster, A. E. Munson, and I. Kimter, eds.), 2nd ed., pp. 523–538.

29. Goter Robinson, C., Balazs, T., and Egorov, I. K. (1986). Mercuric chloride, gold, sodium thromalate, and D-penicillamine-induced antinuclear antibodies in mice. *Toxicol. Appl. Pharmacol.* **86**, 159–169.

30. Gladen, B. C., and Rogan, W. J. (1995). DDE and shortened duration of lactation in a northern Mexican town. *Am. J. Public Health* **85**, 404–408.

31. Wood, J. W. (1994). "Dynamics of Human Reproduction." de Gruyter, New York.

32. Taylor, P. R., Lawrence, G., Hwang, H. L., and Paulson, A. S. (1984). PCBs: Influence on birthweight and gestation. *Am. J. Public Health* **74**, 1153–1154.

33. Tsai, M. L., Webb, R. C., and Loch-Caruso, R. (1996). Congener-specific effects of PCBs on contractions of pregnant rat uteri. *Reprod. Toxicol.* **10**(1), 21–28.

34. Mattison, D. R., and Thorgeirsson, S. S. (1978). Smoking and industrial pollution, and their effects on menopause and ovarian cancer. *Lancet* **1**, 187–188.

35. Flaws, J. A., Doerr, J. K., Sipes, I. G., and Hoyer, P. B. (1994). Destruction of preantral follicles in adult rats by 4-vinyl-1-cyclohexene diepoxide. *Reprod. Toxicol.* **8**(6), 509–514.

36. Silbergeld, E. K., and Flaws, J. A. (1999). Chemicals and menopause: effects on age at menopause and on health status in the postmenopausal period. *J. Women's Health* **8**, 227–234.

37. Nogawa, K. (1981). *Itai-itai* disease and follow-up studies. *In* "Cadmium in the Environment" (J. Nriagu, ed.), pp. 1–37. Wiley, New York.

38. Agency for Toxic Substances and Disease Registry (1996). "Study of Female Former Workers at a Lead Smelter." U.S. Department of Health and Human Services, Atlanta, GA.

49

Evidence for the Role of Environment in Women's Health: Geographical and Temporal Trends in Health Indicators

RUTH H. ALLEN

U.S. Environmental Protection Agency
Washington, DC

I. Introduction/Background

Women's health indicators are measures that can be used to reflect the state of health of women in a given area [1]. Some women's health indicators (*e.g.,* breast cancer incidence) have been increasing in some developed and developing countries since the 1970s, coinciding with widespread changes in the global environment. Such geographic and temporal trends can present research and prevention opportunities.

Changes in women's health indicators by time, place, and person provide clues that can be exploited by epidemiologists. As shown in Figure 49.1, epidemiology is evolving rapidly, with a new era of integrated studies that will better address the role of the environment in influencing health indicators. These new integrated approaches include measurement of genetic susceptibility and biomarkers of dose, as well as direct environmental measures of exposure from toxic compounds and mixtures. In geographic areas with high rates of disease, spatial analysis and ecologic studies of exposure pathways may have a role.

Our knowledge of trends in women's health indicators depends on the quality and extent of data collected by population-based surveillance systems. While data from both health surveillance and environmental monitoring are improving in quality and timeliness, globally data are still uneven for exposure to important agents such as pesticides, air and water pollutants, metals, and solvents.

This chapter examines evidence for the role of the environment in women's health. First, geographical and temporal trends in women's health indicators are evaluated using published data from population-based surveillance systems. Breast cancer rates are the main example presented. Noncancer women's health in-

dicators (*e.g.,* rates of depression (bipolar disorder), multiple sclerosis, diabetes, pelvic inflammatory disease, infertility (from chlamydia and gonorrhea), and heart disease (first heart attack and stroke)) are also presented to provide a wider perspective. All women's health indicators are plotted to highlight comparisons among diseases, across time periods, and over large geographic regions. Two examples of studies in progress are then presented. Finally, measurement improvements are described for the subtle, complex interplay of known and suspected risk factors: race/ethnicity, sociodemographic and cultural factors, personal/lifestyle choices, medical and reproductive history, diet and nutrition, occupational exposures, and environmental exposures.

II. Breast Cancer as a Women's Health Indicator

The focus of this section is on breast cancer incidence rates, an example of a well-studied woman's health indicator. Figures 49.2 and 49.3 summarize age-adjusted global female breast cancer incidence rates and trends from 1970 to 1985 as compiled by the International Agency for Research on Cancer (IARC) of the World Health Organization (WHO). Figures 49.4 and 49.5 present the data for breast cancer mortality and the trends by country from 1965 to 1985 [2]. Mortality data are used as indicators because they are available for places of interest that may not yet have full incidence reporting.

Many countries follow international disease surveillance reporting standards. Some also publish separate environmental monitoring reports or rely on WHO/IARC environmental health reports on particular chemicals. Cancer statistics are generally of good quality. There is histological confirmation of tumors (except nonmelanoma skin cancer) in 70–99% of the Americas, Oceania, Asia, and European Community (EC) and non-EC nations. Confirmation is generally above 50% for Eastern Europe, China, India, and Cuba. Population estimates for denominators in disease rates are reported for 1980, and confirmation on mortality and incidence rates is 100% for national data but it is less if disease registries only cover a city or province within a country.

Reliable U.S. cancer incidence rates are available on the Internet from the National Cancer Institute (NCI) Surveillance, Epidemiology, and End Results (SEER) geographic sites from as early as the 1930s for Connecticut and from the 1960s and 1970s or later for other SEER sites. SEER data can be found on the NCI web page at *www.nih.nci.gov,* which hyperlinks to WHO/IARC and several state cancer registries through the North American Association of Central Cancer Registries. Environmental data with reliable geographical coding are available

---

1. **Population-based health indicators**	2. **Environmental risk factors**
• Breast cancer incidence	• Physical, chemical, biologic agents
• Breast cancer mortality	• Completed exposure pathways
• Gynecologic cancers	• Carcinogenic, genotoxic,
• Reproductive system infection rates	immunotoxic chemical substances
• Genetic diseases and other chronic	• Improved environmental monitoring
diseases and health endpoints	in high disease regions
3. Gene-environment interactions	**4. Disease prevention**
• Genetic susceptibility studies	• Environmental exposure reductions
• Population-based family registries	• Genetic heterozygosity retained
• DNA structure and function changes	• Measurement of risk improved
• Cell communication and architecture	• Public health education expanded
changes (e.g., endocrine disruption)	• Improved health indicators/statistics

Fig. 49.1 Overview of environment and ecogenetic epidemiology in women's health.

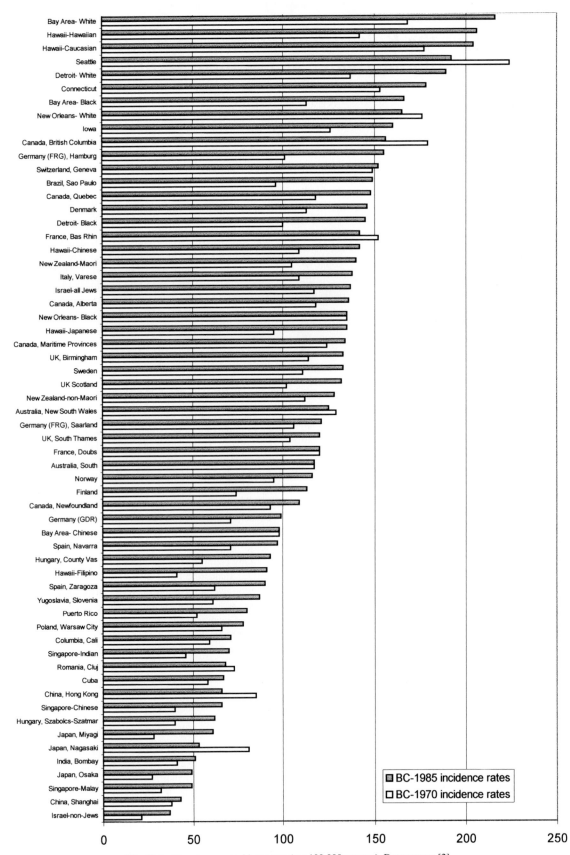

Fig. 49.2   Breast cancer incidence rate (per 100,000 women). Data source: [2].

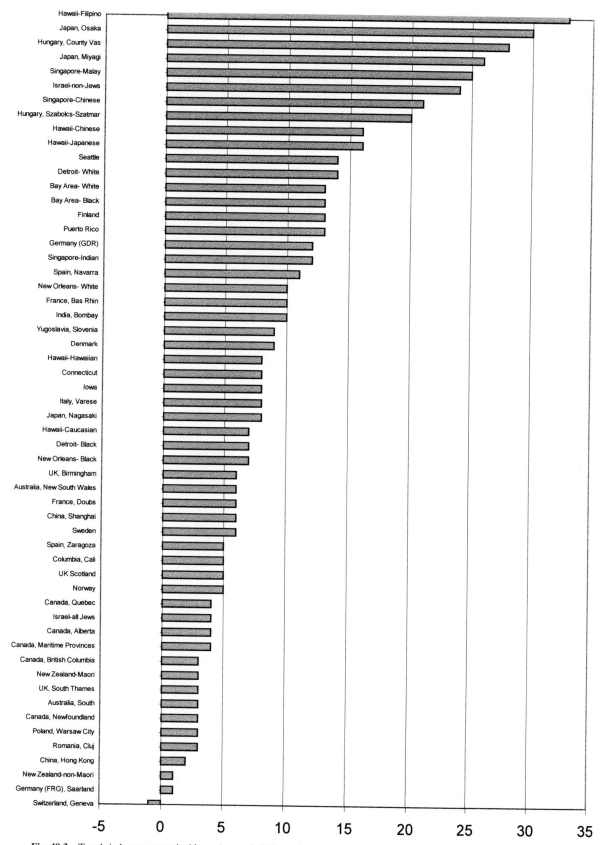

Fig. 49.3 Trends in breast cancer incidence (percent). Estimated mean percentage change per five-year period in the age-specific rates (30–74 years) over the period 1973–1984. Data source: [2].

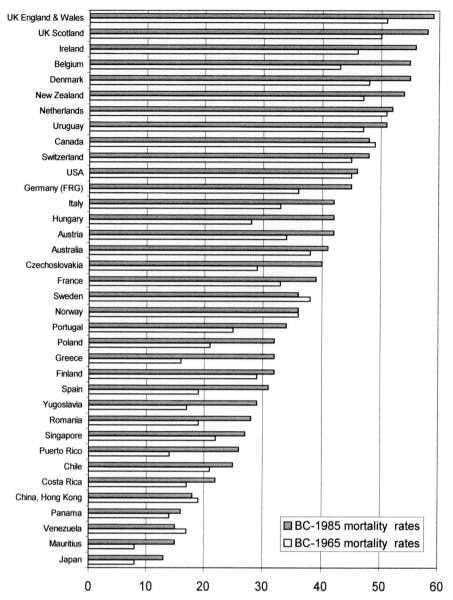

Fig. 49.4   Breast cancer mortality rate (per 100,000 women). Data source: [2].

from the Environmental Protection Agency (EPA) web page at *www.epa.gov;* larger volume emitters (more than 10,000 pounds per year) are found in the Toxic Release Inventory, and selected large industrial sectors are found in the Sector Facility Indexing Project. New epidemiology studies should exploit both data sets.

III. Global Incidence, Prevalence, and Mortality Statistics

The environment is an integral part of individual and population level risk factors, and cancer statistics reflect the sum total of environmental risk, mediated by lifestyle choices, genetic inheritance, and metabolic factors. There is an increasing awareness of the role of the environment (carcinogens, genotoxic agents, immune system toxins, and endocrine disruptors) as factors in the occurrence of disease, and this is one reason for many

legally enforced environmental control measures. As one consequence, widespread use of persistent organochlorine pesticides has decreased due to the introduction of less toxic or persistent alternatives following increased insect resistance and subsequent reduced effectiveness of dichlorodiphenyltrichloroethane (DDT). There are a few cautions in looking at international geographical and temporal trends. First, environmental risk factors may have a long latency between earlier childhood exposure and disease in mid-life and later. Second, because the U.S. population is older than much of the world's population, health indicator comparisons between the two could be misleading. The maps of U.S. disease rates from the Centers for Disease Control and Prevention (CDC) [3] are standardized to the 1970 U.S. population. The previously cited WHO/IARC volume has rates standardized to the younger world population. Such differ-

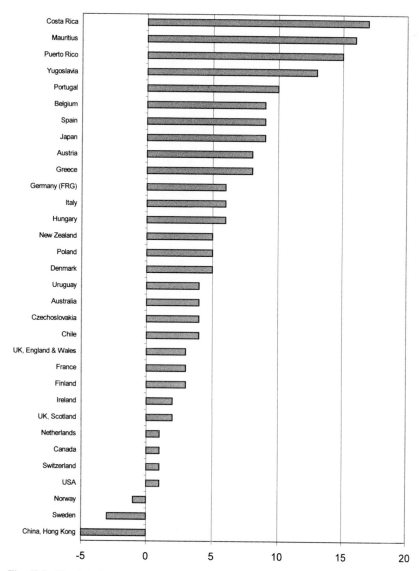

Fig. 49.5 Trends in breast cancer mortality (percent). Estimated mean percentage change per five-year period in the age-specific rates (30–74 years) over the period 1975–1988. Data source: [2].

ences mean comparisons are only valid within a data set. Third, it is important to understand the disease biology in order to make any meaningful interpretations of geographical and temporal differences. An exhaustive summary of biology and epidemiology for every cancer site is beyond the scope of this work. Readers are referred to Schottenfeld and Fraumeni [4] for comprehensive coverage of the epidemiology of cancer by tissue of origin and environmental exposures.

To compare countries and time more easily, breast cancer data that are age-standardized to the world population by WHO/IARC were plotted together. These data reflect the best model fits to age, cohort, and time periods. Statistical significance as applied in the following examples is based on WHO/IARC methodology, and best model fit is used to estimate missing earlier data points in cancer trend data. Three regions are discussed separately: the Americas, Asia and Oceania, and Europe.

A. Breast Cancer in the Americas

For a regional perspective, female breast cancer incidence rates in the Americas are described for parts of Canada, Latin America, and the U.S.. Selected results from 1970 and 1985 in Figure 49.2 are reviewed for incidence rates of more than 100 per 100,000 (the choice of 100 as a level of interest is arbitrary). Percent change in incidence trends are shown in Figure 49.3.

In Canada, Quebec had the highest number of incident cases (n = 1433) or an incidence rate of 117.6 per 100,000 in 1970 and 2419 incident cases or 148.4 per 100,000 women in 1985, as shown in Figure 49.2. British Columbia had 1150 incident cases in 1985 and rates of 179 per 100,000 in 1970, which decreased to 156 in 1985. Trends in Canada show increases from 1970 to 1985 of from 2.7% for Newfoundland (not significant at the 5% level) to 4.4% for Quebec. Breast cancer incidence rates in Newfoundland increased from 92.8 per 100,000 women in 1970 to 109.1 per 100,000 women in 1985. Is there a biological reason why incidence rates declined 23 per 100,000 only in British Columbia and went up in all other Canadian reporting areas? Seattle also exhibited this anomaly, with a decline of 31 per 100,000. Both areas have >100 breast cancer cases, so small numbers are an unlikely answer. Because this is the desired direction of change, follow-up might be needed. The universal availability of health care may be involved in the modest downturn in female breast cancer mortality rates in Canada in Figure 49.4 but would be expected to have less impact on incidence due to disease latency.

With a 13% increase, Puerto Rico has the highest trend increase in breast cancer incidence despite having a younger population and protective childbearing practices, including large families. Cali, Columbia has the next highest increase at 5%, which is not statistically significant. In São Paulo, Brazil, the female breast cancer incidence rate rose from 95.6 per 100,000 women in 1970 to 149.1 per 100,000 women in 1985 (not significantly different at the 5% level). These increases in breast cancer, along with increases in women's health indicators, represent a significant disease burden in developing countries.

In the U.S., female breast cancer incidence rates are reported in Figure 49.2 for six geographic areas and in certain areas by race/ethnicity groupings. The reported trends in incidence are in Figure 49.3 and highlights follow:

- San Francisco, Bay Area, blacks up between 1970 and 1985 from 112.8 to 165.2 per 100,000 women.
- San Francisco, Bay Area, Chinese have incomplete or missing trend data at 97.8 per 100,000 for the period of study due to a small number of cases (<100).
- San Francisco, Bay Area, whites up from 167.7 to 216.1 per 100,000 women.
- Connecticut up from 152.8 to 177.9 per 100,000 women.
- Detroit, blacks up from 99.9 to 144.7 per 100,000 women.
- Detroit, whites up from 136.9 to 188.5 per 100,000 women.
- Iowa up from 125.7 to 160.0 per 100,000 women.
- New Orleans, blacks the same for both time periods at 134.8 per 100,000 women, a value to be interpreted with caution due to fewer than 100 cases.
- New Orleans, whites down from 176.2 per 100,000, a value that should be interpreted with caution, to 164.5 per 100,000 women.
- Seattle down from 223.6 per 100,000 women in 1970, a number also to be interpreted with caution, to 192.0 per 100,000 women.

These data highlight ethnic group differences, the epidemic aspects of breast cancer in the U.S., and reasons for community and consumer group concern. The challenge ahead is to identify risk factors and implement prevention opportunities. Research may be needed in areas where rates are particularly high or low.

B. Breast Cancer in Asia and Oceania

For Asia and Oceania, female breast cancer incidence rates show interesting increases and ethnic variations. Overall, Asian rates are lower than in Europe or the Americas, and where they have begun climbing, gene–environment interaction studies may be especially productive. Incidence rates increased between 1970 and 1985 from a range of 5.9% in Shanghai, China to 29.9% for Osaka, Japan. For example, in Shanghai, China, female breast cancer incidence rates showed a modest increase from over 37 to more than 43 per 100,000 women between 1970 and 1985. For Chinese in Singapore, the increase was 21.2%, from 40.4 per 100,000 women in 1970 to 66.2 per 100,000 women in 1985. For Chinese in Hawaii, the increase for the same time period was more than 15.8%, from 109.4 per 100,000 in 1970 to 142.4 per 100,000 in 1985. By comparison, the female breast cancer incidence rates for Hawaiians in Hawaii also increased 8.1% from 142.0 per 100,000 in 1970 to more than 206.4 per 100,000 women in 1985. The value of 8.1% is not statistically significant at the 5% level, perhaps due to the small number of cases. However, a better understanding of tumor biology, genetic susceptibility, and environmental risk factors may be achievable for a finite island people.

Incidence rates of female breast cancer in Hong Kong, China showed a small decline of 2.2% (not significant at the 5% level) from a rate of 84.7 per 100,000 women in 1970 to 66.2 per 100,000 women in 1985. This rate difference may reflect migration or demographic shifts to a younger healthier population with increased wealth and access to better medical care.

In Bombay, India, female breast cancer incidence rates for the same time period were up 10.3%, from 41.1 per 100,000 to 51.3 per 100,000 women. For Indians in Singapore, the increase in female breast cancer rates (not significant at the 5% level) was 11.7% for incidence rates of 46.3 per 100,000 in 1970 and 69.8 per 100,000 in 1985. By comparison, for Malay in Singapore, the female breast cancer incidence rate increased 24%, from 31.8 per 100,000 in 1970 to 48.7 per 100,000 women in 1985.

In Oceania, the largest increase in incident cases from 1970 to 1985 was in New South Wales, Australia; rates increased 6%, from 1441 cases (an unstable estimate) to 1603 cases. There is a striking increase in Asia/Oceania breast cancer incidence rates of more than 100 per 100,000 from 1970 to 1985 in Hawaii for Caucasian, Japanese, Chinese, and Filipino; in New Zealand for non-Maori and Maori; and in South and New South Wales, Australia. Sometimes breast cancer incidence rates were below 100 per 100,000, but trends (rounded to the nearest percent) were above 10%, illustrating the effects of large increases in a small absolute number of cases: Japan with Osaka up 30% (407 to 1134 cases) and Miyagi up 26% (118 to 361 cases); Hawaii with Filipino up 33% (from 4 to 24 cases), Japanese up 16% (54 to 135 cases), and Chinese up 16% (not statistically significant (10 to 22 cases); Singapore with Chinese up 21% (96 to 248 cases), Malay up 25% (10 to 26 cases), and Indian up 12% (4 to 15 cases); India with Bombay up 10% (232 to 542 cases); New South Wales, Australia 129 (an unstable best model estimate) to 125 up 6 percent. These increased cases could represent a significant public health challenge and prevention opportunity.

In Japan, the A-bomb aftermath produced an anomaly in breast cancer rates for Nagasaki. Female breast cancer incidence rates dropped more than 8% from 81.1 per 100,000 in 1970 to 53.1 per 100,000 Japanese women in 1985 (not significant at the 5% level). For two other reporting districts in Japan, Miyagi and Osaka, female breast cancer incidence rates increased 26.4% and 29.9%, respectively, from 27.8 per 100,000 and 26.9 per 100,000 women in 1970 to 61.2 and 48.8 per 100,000 women respectively in 1985. By comparison, for Japanese in Hawaii the increase in female breast cancer incidence rates was 15.9% for all ages (30–74) and 48% for Japanese women between 63 and 74 years of age. Female breast cancer incidence rates in Japanese women in Hawaii increased from 94.9 per 100,000 women in 1970 to 134.7 per 100,000 in 1985. These data for Japanese women in Japan and Hawaii illustrate opportunities for gene–environment interaction studies in breast cancer etiology.

C. Breast Cancer in Europe (EC and non-EC)

Trends from 1970 to 1985 for non-EC nations' female breast cancer incidence rates were up 13.4% in Finland; 20.1% and 26.9% in two geographic sites in Hungary; and 23.7% for non-Jews in Israel. Female breast cancer incidence rates were up almost 10% for Norway, Poland, Romania, Sweden, and for all Jews in Israel. Only for Geneva, Switzerland, where overall rates are among the highest in the world, was there a small decline of 1.3%. In 1970, female breast cancer incidence rates in Europe (non-EC) ranged from 21.2 per 100,000 women for non-Jews in Israel to 148.9 per 100,000 women in Geneva, Switzerland, with a median value of 72.6 per 100,000 for Cluj, Romania. This latter value should be interpreted with caution due to incomplete or missing data. By 1985 in Europe, female breast cancer incidence rates ranged from 36.5 per 100,000 to 152.3 per 100,000, with a median value of 115.6 per 100,000 women. These data highlight the disparity in breast cancer incidence rates across non-EC countries with varying environmental monitoring and control standards.

In EC nations from 1970 to 1985, Denmark, France, Germany (FRG), Italy, Spain, and the United Kingdom reported an increase in female breast cancer ranging from 2.8% in South Thames, UK, to 12.9% in Germany (GDR). Differing environmental pollution and standards of health care may be reflected in these disease rate differences. For 1985, female breast cancer incidence rates were between 89.8 in Zaragoza, Spain and 154.6 per 100,000 women in Hamburg, Germany (FRG). This latter rate should be interpreted with caution due to incomplete or missing data. The 1985 Europe (EC) median value is in Scotland, at 132.1 per 100,000 women.

Between 1965 and 1985, in spite of many important technological advances, breast cancer mortality climbed in many parts of Europe, from 18.9 to 59.0 per 100,000, as shown in Figure 49.4. As previously indicated in Figure 49.3, between 1970 and 1985 the trends in female breast cancer incidence rates climbed from 2.8% to as much as 26.8% despite age standardization to the younger world population. Incidence rates range from 21.2 to 154.6 per 100,000. Overall, such rates indicate the need for studies in geographic areas where rates are above and below a cut point set by society or by responsible scientific groups.

IV. Host and Environmental Determinants of Breast Cancer

A. Race and Ethnicity

African-American women have a lower incidence rate of breast cancer than whites but a twofold excess in mortality [5].

The reasons for this may include differential exposure to toxic agents, social factors related to wealth or poverty, differences in mammography utilization, and increased adiposity that limits the effectiveness of early detection [6]. Similarly, the lower rates of breast cancer in Japanese women in Japan compared to women who migrate to Hawaii or the western U.S. are not yet fully explained, and the fact that disease rates change within a generation suggests the importance of environmental, as opposed to strictly genetic, factors in disease etiology.

B. Socioeconomic Status

Women of lesser means in developed countries and women in developing countries experience a poorer health outcome than women with better access to health services. In the developing world, women, many of them illiterate, are at risk for any number of reasons. They often do the farming barefoot and apply pesticides without protective equipment or proper knowledge of printed label precautions. To correct this situation, international aid agencies and lending banks have emphasized educating workers about proper disposal of pesticides and pesticide containers. This education is designed to prevent contamination of the environment and reuse of pesticide containers for carrying water or storing food.

C. Radiation

Disease processes arise in the body's microenvironment and, as the following breast cancer and radiation story illustrates, this microenvironment is directly influenced by changes in the macroenvironment. The lessons from radiation science about the sensitivity of breast tissue are instructive as a point of reference. Breast cancer was induced in X-radiated mice in 1936. In 1965, the link between X-ray exposure and excess breast cancer was confirmed in humans exposed to repeated fluoroscopic examination, and it was confirmed again in Japanese atomic bomb survivors in 1968. Data consistently showed that (1) the subsequent development of cancer was critically dependent on the hormonal status of target cells; (2) the age distribution and pathology of the radiation-induced breast cancers were similar to those of breast cancer from other or unknown causes; (3) women irradiated before 20 years of age showed a higher relative risk than those exposed to radiation later in life; and (4) the yield in tumors was similar whether the total radiation dose was experienced in a single brief dose or in multiple exposures [7].

D. Pesticides

Tons of persistent organochlorine pesticides came into widespread use starting in the late 1940s. Following the recognition of disturbing adverse effects on wildlife, the U.S. Public Health Service human tissue monitoring program began to track residues in human tissues in 1967. The program was transferred to the U.S. EPA in 1970, and concurrent with protracted pesticide cancellation proceedings, population levels were reported in the 1970s at 7–12 ppm for DDT and its persistent +metabolite DDE (dichlorodiphenyldichloroethylene) in human adipose tissue [8]. Measured levels of DDE in human tissue reflect variations in cumulative past exposure to DDT or chemicals that produce

Table 49.1

Inconsistencies in Smaller Human Studies of DDE and Breast Cancer Led to Larger Studies

Study	Results	Notes
[35a] Inverse association	OR = 0.35; CI = 0.18–0.70 1.53 ppm cases 4.32 ppm controls	n = 9/5; DDE in adipose, small older study, higher exposure in controls
[35b] Positive association	OR = 1.60; CI = 1.09–2.34 1877 ng/g cases 1174 ng/g controls	n= 20/20; DDE in sera small study, 60% excess risk, confidence intervals >1
[35c] Positive association	OR = 3.68; CI = 1.01–13.50 11.0 ng/ml cases 7.7 ng/ml controls	n = 58/171; DDE in sera, 268% excess risk, confidence intervals > 1, nested study design with careful control for other breast cancer risk factors, New York women
[35d] No association	OR 1.33; CI = 0.68–2.62 43.3 ± 25.9 ppb cases 43.1 ± 23.7 ppb controls	n = 150/150; DDE in stored sera, results differed in three racial groups individually, similar mean exposure levels in cases and controls, California women
[35e] Positive association	OR = 1.79; CI = 0.83–3.88 1370.6 μg/kg cases 765.3 μg/kg controls	n = 18/17; DDE in sera Canadian women, positive for some PCBs
[35f] No association	OR = 0.69; CI = 0.55–1.70 562.48 ppb cases 505.46 ppb controls	n = 141/141; DDE in sera Mexican women, high exposure in cases and controls, results unaffected by adjustment for body mass, breastfeeding, menopause, or other breast cancer risk factors
[35g] Inverse association	OR = 0.48; CI = 0.25–0.95 1.35 μg/g cases 1.51 μg/g controls	n = 265/331; DDE in buttocks adipose tissue, postmenopausal European women, DDE 9.2% lower in cases, may be secondary to disease
[35h] No association	OR = 0.72 L; CI = 0.37–1.40 OR 0.43 H; CI = 0.13–1.44 4.71 ppb cases 5.35 ppb controls	n = 236/236; DDE in sera American women volunteers in large Nurses Health Study, age and body mass were significant, results do not rule out other environmental risk factors
[35i] No association, DDE; Positive association, dieldrin	OR = 2.05, CI = 1.17–3.57 (dieldrin) 24.42 μg/g ± 42.40 μg/g[a]	No overall excess risk assocaition between DDE and breast cancer in women from Copenhagen, Denmark
[35j] Positive association	OR = 1.95, CI = 1.10–3.52 3.30 μg/ml cases 2.50 μg/ml controls	n = 153 incident breast cancer cases and 153 age matched controls in Santa Fe de Bogota, Colombia. The test for trend was not statistically significant (p = 0.09)
[35k] No association	OR = 1.34, CI = 0.71–2.55 13.16 μg/g cases[b] 10.82 μg/g controls[b]	n = 154 postmenopausal breast cancer cases and 192 postmenopausal community controls. Adjusted for age and serum lipids.

Note. OR = odds ratio; compares exposure measurement in case with disease and controls without disease (*e.g.*, 1.6 equals 60% excess). CI = confidence interval; 95% of values expected to be in this range; lower bound above 1.0 means results are considered statistically significant for a population studied that is distinct from biological significance and individual disease risk. ppm = parts per million; higher levels seen in human adipose tissue in 1970s before cancellation of DDT. ppb = parts per billion; also equivalent to ng/g or μg/kg or ~ng/ml. n = study number of cases/controls.

[a]Irrespective of breast cancer status.

[b]Never lactated; ever lactated values were similar, e.g., 10.82 μg/g cases and 10.44 μg/g controls.

DDE metabolites and other factors, such as weight gain or loss. As summarized in Table 49.1, subsequent epidemiology studies of varying statistical power found positive (two to fourfold excess), negative, or no association for breast cancer and levels of DDE (ppb) in blood or adipose tissue. Study inconsistencies are attributed to small sample size, measurement method issues, sampling different sources (adipose tissue vs sera), and incomplete characterization or misclassification of exposure and key variables such as lactation history or stage of disease [9,10]. In addition, differences in pesticide exposure are noteworthy, with levels reaching parts per million in the earlier studies. Moreover,

the significance of body burdens in highly exposed individuals may not be reflected in measures based on group means.

E. Smoking

Although smoking was not generally thought to be associated with breast cancer based on earlier studies, mutagens from cigarettes were later found in breast fluids of nonlactating women, and excess cancers were observed in women smokers at sites without contact with cigarette smoke, including the cervix, pancreas, and bladder [11]. Therefore, better exposure measure-

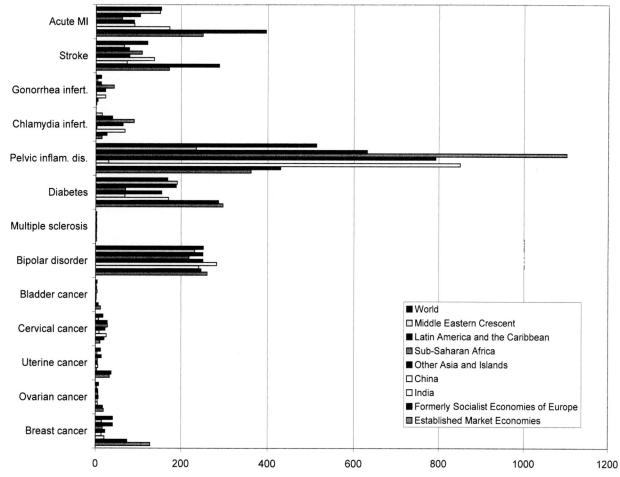

Fig. 49.6   Disease incidence rates per 100,000 women. Data source: [20a].

ments are required to establish the validity of smoking classifications and to better understand biologic mechanisms and hormonal interactions. Uncertainty about age of onset of smoking, passive smoking, and the need for reduction of bias, confounding, and nondifferential misclassification all argue the need for a more sophisticated approach. Therefore, smoking and breast cancer are now being examined as an example of gene–environment interaction. A number of molecular assays exist to determine enzyme detoxification phenotypes and genotypes, such as Phase I (cytochrome P450 or CYP) and Phase II [*e.g.,* glutathione-S-transferase (GST) and N-acetyl Transferase (NAT)] enzymes [12]. The hypothesized mechanisms that convey risk are high exposure, metabolic activation of more toxic biochemical intermediates, and genetically controlled absence of detoxifying enzymes. A lesson from lung cancer research on smoking is that evidence is accumulating to implicate polymorphic enzyme variants of GST and P450 [13]. The decline in male lung cancer following smoking cessation and the rise in lung cancer in women following increased female smoking illustrates the role of environmental factors in disease and the difficulty in changing individual and population level practices to reach public health goals.

F. Women's Social Roles

The nutritional stress of pregnancy and lactation can put women at increased health risk. The social stress of being a female head of household or stigmatization over lifestyle choices and sexual orientation may affect immune function in ways that are not yet fully understood. Environmental exposures to chemical toxins can potentially increase immune system mediated reproductive diseases such as endometriosis [14] and adversely affect fertility [15,16], lymphatic system cancers [17], and other common chronic diseases in women [18–20]. In Figure 49.6 [20a], the high rates of pelvic inflammatory disease in Sub-Saharan Africa of over 1000 per 100,000 highlight the potential of women who have early rates of infection to be at increased risk from other diseases later in life. The median age for pelvic inflammatory disease is 22 years. In Africa, cultural practices are changing, but female circumcision still exists, adding to the lifetime risk of recurring infection.

The choices that young women make about smoking, drugs, alcohol, and unprotected sexual activity are becoming important targets of prevention, and more research is needed to better understand these risky behaviors. Consequently, behavioral sciences are a major focus of increased research emphasis in prevention, population sciences, and cancer control.

V. Examples of Studies in Progress

Large population-based epidemiology studies are in progress to evaluate human exposure to organochlorine pesticides and other persistent toxic chemicals with endocrine disrupting properties. Examples of new era epidemiology studies include the Long Island Breast Cancer Study Project (LIBCSP) and the Agricultural Health Study (AHS). Each is a multiagency partnership involving multidisciplinary teams of investigators, and each was designed in response to scientific questions as well as citizen, local, regional, and State government and private sector questions. These research efforts may help to uncover complex reasons why disease rates may be higher in particular geographic areas.

The LIBCSP involves three concurrent approaches: (1) hypothesis testing on the role of persistent organochlorine pesticides and polycyclic hydrocarbons (PAHs) in breast cancer etiology using a common population in Nassau and Suffolk Counties, New York, and two comparison counties in New York and Connecticut. The epidemiology portion of the Congressionally mandated study uses a case-control study design with separate protocols for biomarkers of chemical exposure and electromagnetic fields; (2) comprehensive statistical and chemical characterization of current and cumulative environmental exposure, including complex mixtures across time and space; and (3) geographical information analysis to integrate findings, explore patterns, and generate new testable hypotheses.

Organochlorine pesticide residues provide one indicator of past human exposure to pesticides via food chain bioaccumulation and other routes of exposure. The Long Island study is a unique mix of scientific and citizen concerns about why there is excess breast cancer on Long Island. The Long Island population-based case-control study was mandated in 1993, and because of intense Congressional and consumer group interest, the study has necessitated careful management and full cooperation both on the data collection, which involves complex partnerships across over a dozen institutions and two dozen investiga-

tors, and on integration of scientific findings. In other studies, pesticide contaminants have been found in human milk [21], infant and adult foods [22–26], the home environments of childhood cancer cases [27], human urine [28], air and residential house dust tracked in from contaminated yard soil [29], on the surface of children's toys [30,31], and in schools [32,33]. It is no small task to characterize current and cumulative chemical exposure on Long Island or in any other urbanized area; therefore, new methods have had to be developed, and this work is ongoing to support the planned geographical information analyses.

In the LIBCSP, more than 220 toxic chemicals are being analyzed to evaluate exposure pathways for identified carcinogens and to characterize the extent of human exposure to pesticides in nearly sixteen hundred cases and sixteen hundred community controls. The complexity of the work and the range of expertise needed are reflected in the lists of chemicals: seventeen persistent organochlorine pesticides widely used on Long Island for several decades and subsequently canceled for health or environmental reasons including food chain bioaccumulation; seven hazardous air pollutants; twenty-seven inorganic chemicals in water including arsenic and chromium; thirteen industrial solvent chemicals; eight metals; various currently registered pesticides (ten fungicides, six herbicides, twenty-one insecticides); various PAHs and eight polychlorinated biphenyls (PCBs); five pesticide metabolites; twenty three superfund (SARA) chemicals; fourteen pesticides in special review for health and environmental concerns; four trihalomethanes; six volatile aromatics; eighteen volatile hydrocarbons; and twenty-eight volatile organic compounds (VOCs). There are twenty-six superfund sites on Long Island with metals and solvents as major potential risk factors. An environmental exposure working group is ranking the chemicals for priority. They are recommending follow-up measures, based on published findings on carcinogenicity, genetic toxicity, immunotoxicity (summarized in Table 49.2), and other relevant factors such as expected extent of exposure and use history on Long Island. Working closely with scientists and community leaders and being responsive to the concerns of each group has helped to build and maintain the necessary level of trust. This level of trust is needed on an ongoing basis to maximize integration of macroecological and geographical exposure patterns,

Table 49.2
Selected Pesticides and Toxics from LIBCSP Integrated Chemical List

Chemical (Trade name/R/TM)	IARC/EPA	Genetic activity profile	Immunotoxicity
2,4,5-Trichlorophenoxyacetic acid	2B	Yes (8/33) G,C,A,S	—
2,4-Dichlorophenoxyacetic acid (2,4-D)	2B	Yes (27/76) R,G,C,D,S	Yes SD, TU, LyD, AD, BU
Acephate (Orthene)	2B	Yes (18/49) G,R,C,S,D,M,A	—
Alachlor	2A/B2	Yes (3/28) R,M,C	—
Aldicarb (Temik)	2B	Yes (7/23) G,S,C	Yes TD, MD, LyDU, ADU
Aldrin	2B	Yes (7/16) G,D,C,A	—
Amitrole	2B	Yes (13/75) D,R,G,A,C,T,H	—
Arsenic +3	1,2A	Yes (30/58) G,R,S,T,M,C,D	—
Arsenic +5	1,2A	Yes (11/33) R,S,C,T	—
Atrazine	2B	Yes (18/88) G,A,C,H,D	Yes ThD, SD, TD, LeD, LyD, MPD
Benzene	1	Yes (44/144) G,R,D,C,A,T,S,M	—
Beryllium Compounds	1	Yes (20/75) D,G,S,C,T	—

(continued)

Table 49.2

(continued)

Chemical (Trade name/R/TM)	IARC/EPA	Genetic activity profile	Immunotoxicity
BHC, beta-	2B	—	—
BHC, gamma- [Lindane isomer]	-/B2	Yes (6–36) A.C,D,R	—
Bromodichloromethane	2B	Yes (10/32) G,R,C,S	—
Cadmium compounds	2A	Yes (76/155) D,G,R,A,M,C,T	—
Captan	2B,3	Yes (57/74) D,G,R,S,H,M	Yes ThD, BD, TD, LyD, AD
Carbon Tetrachloride	2B	Yes (10/44) R,G,D,A,T	—
Chloroform	2B	Yes (5/53) D,R,G,T	—
Chromium +6	1	Yes (106/131) D,G,R,A,S,T,C,M	—
DDT-p,p¹ (and metabolites DDD-p,p¹, DDE-p,p¹)	2B	Yes (23/115) C,A,T	Yes
Dibromo-3-chloropropane, 1,2-(DBCP)	2A,2B	Yes (15/32) C,D,S,A,G,R	—
Dibromoethane 1,2- +	2A	—	—
Dicamba [dimethylamine salt] (Banvel)	2B	—	—
Dichlorobenzene, 1,3-	2B	Yes (3/10) D,R,M	—
Dichloroethane, 1,2-	2B	Yes (18/44) D,G,T,F,M,A	—
Dichloromethane	2B	Yes (23/34) G,R,C,S,T	—
Dichloropropane, 1,3-	2B	—	—
Dichloropropene, 1,1-	2	Yes	—
Dieldrin	2B,3	Yes (5/40) G,D,A	Yes SD, TD, MDU, LyD, AD, MPD, VD
Diethylhexyphthalate	2B	Yes (16/100) G,S,T,C,R,A	—
Ethylbenzene	2B	Yes (53/145) G,R,D,S,C,M,H,A	—
Ethylene Dibromide	2A	Yes (55/73) D,G,M,S,C,A	—
Hexachlorobenzene (HCB)	2B	Yes (1/8) G	Yes TD, MD, ADU, HDU, BaFPD
Lead (Lead Powder)	2B	Yes (12/45) C,T,M,S,A	—
Lead Chromate	2B	Yes (16/32) G,R,S,C,T,M	—
Mancozeb (ETU metabolite)	2B	—	—
Maneb (ETU metabolite)	-/B2	—	Yes AD, BaFD
Mecoprop	2B	—	—
Mirex	2B	—	Yes LyD
Naphthol, 1- (metabolite of carbaryl)	2A	—	—
Nickel Acetate	1	Yes (5/12) D,C,T	—
Nickel Chloride	1	Yes (25/53) D,R,G,S,C,T,M	—
Nickel Nitrate	1	Yes (0/13)	—
Nickel Oxides	1	Yes (6/9) T,C	—
Nickel Subsulfides (Cryst.)	1	Yes (13/15) C,D,G,T,S	—
Nickel Sulfate	1	Yes (22/33) G,A,S,C,T	Yes AD, MPD
Nickel Sulfides (Amorph.)	1	Yes (1/6) G	—
Nickel Sulfides (Cryst.)	1	Yes (15/15) C,D,S,T	—
Nickel, Metallic (Nickel Powder)	1	Yes (2/3) T	—
Nitrofen	2B	—	—
Pentachlorophenol (PCP)	2B	Yes (9/34) D,R,G,S,C	Yes ThD, TD, TU, MD, LD, MPU
Polybrominated Biphenyls (PBBs)	2B	Yes (2/27) A	—
Polychlorinated Biphenyls (PCB's)	2A	Yes (3/37) D,A	—
Styrene (see ethylbenzene)	2B	Yes (53/145) G,R,S,C,M,H,D,A	—
Styrene Oxide	2A	Yes (77/112) D,G,R,S,M,C,H,A	—
TCDD, 2,3,7,8-(Dioxin)	2B	Yes (3/29) G,C	—
Tetrachloroethylene	2B	Yes (10/59) R,G,A,M,T,D	—
Toxaphene	2B	Yes (8/12) D,G,S,C,A	Yes AD, MPD
Trichloropropane, 1,2,3-	2A	Yes (3/27) T,D,A, & (15,27) G,S	—
Vinyl Chloride	1	Yes (35/81) G,D,T,H,M,S,CA	—

IARC/EPA cancer codes: 1, carcinogenic to humans; 2A, probably carcinogenic to humans; 2B, possibly carcinogenic to humans; 3, not classifiable as to carcinogenicity to humans.

Genetic activity profile codes: G = gene mutation; C = chromosome aberration; A = aneuploidy; S = sister chromatid exchange; R = recombination; D = DNA damage; T = cell transformation; F = body fluids; M = micronucleus; H = host-mediated assay; [8/33 = no. positive tests per total no. tests; different tests and different species]. Data source: [35l] (see http://www.epa.gov/mdwgapdb/index.htm).

Immunotoxicity codes: Tier 1 test: IP1 = immunopathology; Th = thymus weight; S = spleen weight; B = B-cell proliferation; T = T-cell proliferation; M = macrophage/neutrophil/natural killer cell act; U = up; D = down direction of change. Tier 2 test: IP2 = immunopathology; L = leucocyte count; Ly = lymphocyte count; A = secondary antibody activity; CMI2 = cell-mediated immunity; H = hypersensitivity; NSI2 = nonspecific immunity; MP = macrophage phagocytosis; V = virus; Ba = bacteria; F = fungi; P = parasites. Data source: [35m] (see http://www.wri.org and http://www.igc.org/wri/health/pis-high.html).

with microenvironmental, cellular, and subcellular biomarker health data. It is also a prerequisite for building the era of ecogenetic epidemiology.

Lessons on new ways to estimate cumulative exposure to pesticides are also being derived from ongoing environmental exposure characterization work in the interagency AHS, a prospective cohort study of 90,000 certified pesticide applicators and their spouses and families in North Carolina and Iowa. The AHS is in its second five years and it will include further characterization of environmental exposure to more than fifty pesticides. Work practices, personal protective measures, and occupational and residential exposures are being characterized with periodic population surveys, as well as environmental monitoring in subpopulations. The development of new tools to better characterize chemical exposure in epidemiology studies of human populations is an essential response to questions from citizens, scientists, and civil authorities about unexplained rising cancer rates.

VI. Better Measurement of Risk Factors

Breast cancer risk factors, especially environmental and occupational exposure to cancer-causing agents, illustrate the problem of synthesis and integration in current chronic disease epidemiology. The simultaneous failure of conventional risk factors to fully explain or guide changes in women's health indicators and the rise of consumer groups' involvement in setting public health priorities have changed the context of cancer etiology studies. To improve the capacity to systematically measure competing risk factors for breast cancer and other diseases, the President's National Action Plan on Breast Cancer (NAPBC), a public–private partnership with extensive consumer involvement, sponsored the Breast Cancer Comprehensive Questionnaire Project (BCCQ). It has a short self-administered breast cancer core questionnaire that can be used internationally and an in-depth interview-administered version, to measure and evaluate connections among different risk factors. The purpose of Internet publication of the Breast Cancer Comprehensive Questionnaire is to foster new collaborations and more integration by simultaneous careful measurement of known and suspected etiologic agents. This research can facilitate meta-analysis that is needed to sort out the complexity of disease origins and to evaluate prevention strategies and the long term effectiveness of environmental interventions and control measures. The web site for the BCCQ is *www.napbc.org.*

Six questionnaire modules were developed by teams of subject matter experts and consumers. The instrument was focus group tested in various geographic areas and ethnic groups and professionally prepared with an easy to use format that is comparable across modules [34].

A. Occupational Exposures

Certain female occupational groups (*e.g.,* teachers with required tuberculosis X-rays, nurses and dental hygienists using anesthetic gases, hair dressers using solvents and dyes) have a higher incidence of breast cancer. A classic example of occupational exposure and cancer in women is the radium watch dial painters in whom high exposure via ingestion of many micro-grams (microcuries) of radium led to development of oral (paranasal sinus) cancers [35].

B. Environmental Exposures

As indicated in Table 49.1 [35a–35k], certain studies are positive for excess breast cancer risk with cumulative chemical exposure to organochlorine pesticides, and other studies are negative or show no association. Therefore, research is being done in the LIBCSP, the AHS, and elsewhere to see if residential, occupational, and/or dietary pesticide exposure is on the pathway to cancer or merely a marker of another relevant exposure, such as residential exposure to solvents in drinking water [36,37].

Another example is from the 1950s, when breast cancer excesses of 1.5 (observed to expected ratio) were seen in a population of 1047 women following fluoroscopic examination of their chests as young girls in the 1930s and 1940s. High doses of radiation were given during repeated examinations of tuberculosis sufferers receiving air-collapse therapy [38]. The radiation example points out the long time spans necessary to do epidemiology and the need for access to appropriate records. The environmental module includes a lifetime residential history.

C. Reproductive and Medical

Reproductive characteristics that are associated with a twofold or less excess breast cancer risk include early menarche, late menopause, late childbearing, and nulliparity. Oophorectomy before 40 years of age confers a 50% reduction in risk. Many studies also show increased risk of breast cancer with long-term exposure to oral contraceptives, estrogen replacement therapy, diethylstilbestrol use during pregnancy, increased height, and alcohol consumption [39].

D. Diet and Nutrition

The role of diet and protective dietary factors in breast cancer etiology is under intensive study, especially the role of phytoestrogens, vitamin E, selenium, beta-carotene, and trans-fatty acids. Large international differences in breast cancer rates and migrant studies suggest that diet is a likely environmental factor. A high fat and low fiber diet in adults was not shown to be a risk factor, but the role of childhood diet is less clear [40]. Micronutrients such as vitamin A are being investigated for antiproliferative properties and for their relationship to sex hormone-binding globulin [41].

E. Personal and Lifestyle Choices

Strenuous exercise in college athletes and physical activity measured from usual job title are associated with a slight decrease in relative risk (50%) for breast cancer. The hypothesized mechanism is through estrogen or estrogen-related pathways [42,43]. However, based on inconsistent physical activity findings in the Nurses Health Study, it appears that greater precision and accuracy are needed to better quantify past physical activity. Also, in different countries and over time women are changing

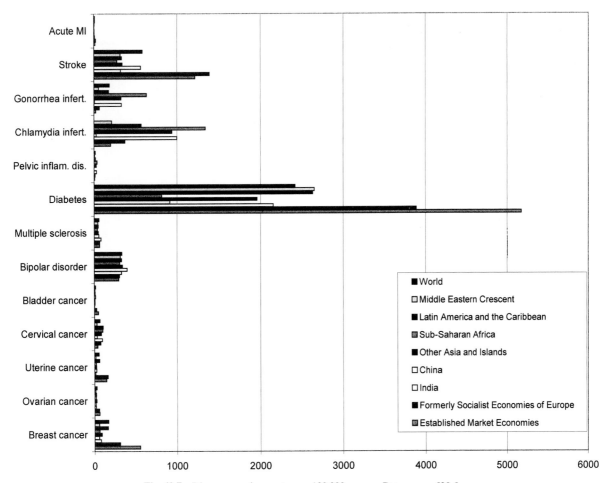

Fig. 49.7   Disease prevalence rates per 100,000 women. Data source: [20a].

their exercise practices for various reasons, including improved mental health.

As shown in Figures 49.6–8, the rates of bipolar disorder, or what used to be called manic depression, are comparable for women living under many different economic systems and in widely different geographic regions. Also, personal and lifestyle risk factors are gender specific; for example, depression is twice as common in women, while alcohol abuse, which may have an underlying depression component, is more common in men. Over the life span, 8% of the population will experience depression. The NAPBC questionnaire provides for improved measurement of exercise in the lifestyle module and mental health in the medical module.

F. Sociodemographic and Cultural

Income, education, ethnicity, place of residence, and access to wealth are all associated with differential disease incidence or survival [44]. Some evidence argues for conventional risk factors as the main reason behind geographic variation in mortality from breast cancer [45]. More research is needed to use this information to achieve prevention.

Cultural practices and literacy determine whether prevention messages are received. Founder effects can be a factor associated with increased incidence of breast cancer in Jewish ancestry women with BRCA1 and BRCA2 genes. Decreased risk of cancer is seen in Mormons and Seventh Day Adventists with favorable diet and healthy lifestyle choices. Asian women also have a lower breast cancer rate (but higher stomach cancer rate); nonetheless, high socioeconomic status with lifestyle and dietary changes in this population upon migration to higher disease rate areas are associated with increased risk of disease. Cancer incidence rates have increased in the United States and globally. Rates vary widely from site to site, as do sociodemographic and cultural risk factors, and the reasons for this variability are not always clear [46]. The rising rates of disease in Asia deserve an adequate explanation.

VII. Biologic Markers of Susceptibility and Human Exposure to Persistent Chemical Toxins

Biomarkers of susceptibility or exposure are only beginning to be examined following measurement advances in the 1990s. Biomarkers are generally sought in addition to case histories

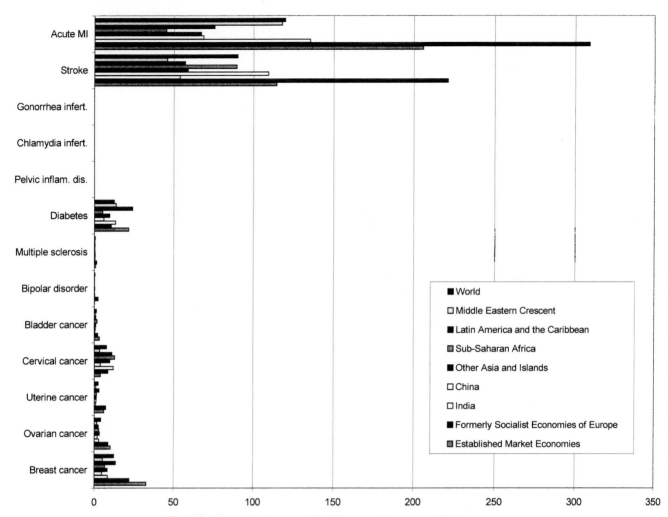

Fig. 49.8   Disease death rates per 100,000 women. Data source: [20a].

because they can indicate an individual's cumulative exposure and their ability (or lack of ability) to metabolize chemical substances in the environment (*e.g.,* pesticide residues in food, air, and water). Some promising approaches include human cytogenetic cancer markers in breast cancer [47], studies of loss of genetic heterozygosity in normal tissue adjacent to breast carcinomas [48], DNA repair and mutagen sensitivity in patients with triple primary cancers [49], Fourier-transform infrared spectroscopy of breast cell DNA changes [50], hormone ratios of 16α/2-hydroxyestrone [51], structure-activity and pesticide exposure [52], and mutational spectra technology [53].

Pesticides in national surveys of human breast milk provide a graphic example of human exposure to persistent toxic chemicals that bioaccumulate as they move through the food chain. Persistent organochlorine pesticides, used widely in the late 1940s, were phased out in the mid-1970s. In the early 1980s, levels of these pesticides from dietary sources were measured in breast milk at 2-100µg/L for PCBs and 1294 ppb and 1558 ppb for DDT [54]. Pesticide levels in children increased with months of breast feeding in the U.S. and Japan. They are

still elevated in Russia, Asia, and Latin America, where EPA-canceled persistent pesticides continue to be widely used.

Pesticides in human urine were measured in 1000 adults living in the United States and four compounds in wide use were seen in 64–98% of those adults tested. National health indicators could (but do not) separate male and female pesticide levels in urine but do show mean concentrations for selected population percentiles. Specific analytes or pesticide residues form a complex picture of the population distribution of carcinogenic, immunotoxic, and genotoxic chemicals in humans because multiple compounds of interest may have the same metabolite. For compounds with a high frequency of detection (in 98% of the 1000 urine samples), metabolites of 1,4-dichlorobenzene were seen in the 790 ppb to 8700 ppb range at the 95 and 100% upper end of the distribution. Many pesticides have multiple and sometimes chemically similar metabolites, and additional work is needed to characterize the subtle physiologic significance of the prevalent body burdens of pesticides and their metabolites. The upper end of the population distribution for exposure is where any health effects may be most pronounced, especially for children, the elderly, and other sensitive subpopulations. According to the previously referenced article by Hill *et al.* [28], this study of pesticides in urine was a convenience sample in the National Health and Nutrition Examination Survey (NHANES III), a nationally representative periodic study of the general U.S. population.

Pesticides in drinking water from agricultural chemical use are one source of contamination of surface and ground water supplies that is amenable to geographic analysis [55]. Pesticide residues in food are currently a special focus of regulatory attention. Because pesticides are intentionally used on food, elaborate commodity monitoring exists [56]. An estimated 2.6 billion pounds of pesticides were used in the U.S. in 1991, and it is nearly impossible to avoid low level exposure to multiple chemical compounds. In 1991, more than one billion pounds of pesticides were used on food, and in 1995 the estimate was more than 1.2 billion pounds on food. Because of differential physiologic, metabolic, and behavioral characteristics, children are a particular concern and the focus of new regulatory approaches by the U.S. EPA and Department of Agriculture (USDA). This includes mandatory tolerance reassessment of more than 9000 pesticide registrations for food-use chemicals.

Dietary factors including total calories, weight, and height are correlated with breast cancer risk [57]. Therefore, residual pesticide levels require biochemical assessment where diet is a significant risk factor for disease. Fish and shellfish are susceptible to food chain bioaccumulation of persistent organochlorine pesticides and heavy metals. Both agents can interact with micronutrients that are important in proper immune function and cell repair.

Pesticides in house dust come from previously treated soil that is tracked into the home and from spray drift. These sources of residential exposure to pesticides in house dust are a special concern for children and are now being measured in several ongoing studies [29]. The health significance of pesticide residues in house dust is not yet well understood, but the lessons from lead in house dust come to mind.

VIII. Epidemiologic Issues

A. Aggregate and Cumulative Risk Assessment

Two epidemiologic measures are emerging in an effort to improve exposure assessment: aggregate exposure and cumulative exposure. Both rely on increased use of innovative statistical analyses [58–61] and understanding differences in risk factors [62]. Aggregate exposure refers to estimation of exposure to the same chemical by multiple routes (*e.g.,* occupational, dietary, air, and drinking water exposure). Cumulative exposure examines exposure to multiple compounds with a common mechanism of toxicity (*e.g.,* organophosphate pesticides that work by cholinesterase inhibition).

B. Historical Reconstruction of Occupational and Residential Pesticide Exposure

The AHS and LIBCSP have common exposure design considerations, both for measurement of current exposures and for reconstruction and estimation of historical timing of exposures that are relevant for disease etiology. A specific geographic focus (North Carolina and Iowa for the former and Nassau and Suffolk Counties, New York for the latter) allows for specific use of historical agricultural production data by crop, pesticide use label rates. Label application rates, percent crop treated, various mixtures in actual practices, and historical changes by crop and locality can be reconstructed from focus groups of state and county agricultural experts and knowledgeable local farmers. While not as comprehensive as dose reconstruction in radiation studies, such as the examination of iodine 131 exposure available on the National Cancer Institute (NCI) web site, pesticide dose reconstruction provides a useful new tool to know where to look for likely gene–environment interactions and where to do intensive genetic follow-ups within high cancer rate areas.

New boundaries of analysis are made possible with the application of Geographical Information Systems (GIS) analyses in the service of public health and prevention [63], with appropriate cautions [64] and the benefit of international experience [65]. With the rapid spread of GIS technology, state, county, and local groups are taking advantage of local knowledge to accelerate the rate of change in understanding how to reduce exposure to harmful chemical contaminants.

IX. Summary and Conclusions/Future Directions

In summary, this chapter looked at the magnitude of the global breast cancer increases as one example of a women's health indicator, and it examined evidence for the environment and other risk factors in diseases common in women. It highlighted examples of such research. It provided geographic and temporal specifics on disease rates to focus public and scientific discussion on tracking progress on disease measures at the individual and population level. It points to the need to collect appropriate biologic samples, including genomic DNA, for the assessment of exposure, effects, and susceptibilities in order to understand the role of gene–environment interactions in disease prevention.

Several topics of ongoing and needed research can be derived from the analysis above:

• Better explanations are needed for observed differences in female breast cancer incidence rates among women of black, white, and Asian ancestry. Processes under genetic mechanisms are thought to explain less than one in ten breast cancers and many women with disease have none of the known risk factors. In addition, risk factors are not necessarily causes of disease or definite clues for prevention.

• In Hawaii and other locations with rapid increases in women's health indicators, better studies and risk communication tools are needed for simultaneous environmental exposures (*e.g.,* pesticides, radiation, and air pollution). Concurrent laboratory and population level gene–environment interaction studies are needed for prevalent chemical exposures and compounds that may act as direct carcinogens, cancer promoters, or via endocrine disruption.

• Better statistical methods are needed to explain disease incidence and predict individual risk and to address women's cumulative risk from exposure over the life span. Risk factors at the population level do a poor job when applied to individual women.

• Better statistical methods are needed to identify spatial clusters and to distinguish real geographic disease clusters from statistical artifacts due to unstable rates in small geographic areas.

• An assessment is needed of Pacific Rim areas with high breast cancer rates: British Columbia; San Francisco Bay Area; Seattle; Hawaii; Nagasaki, Japan; Hong Kong, China; and New South Wales, Australia. These anomalies may be due to artifacts, unstable rates, or possibly prior radiation exposure in Nagasaki.

• Diet and nutrition and the other five conventional risk factor groupings in the BCCQ need to be better characterized, including consideration of biomarkers as mediating factors in the onset, severity, and outcome of breast cancer with and without comorbidity.

• Inflammatory breast cancer would be an appropriate focus, since unexplained geographic and temporal patterns have been reported for this relatively rare but aggressive form of breast cancer. With a poor prognosis, it meets the criteria of tackling the worst first. Those at the unfortunate extreme end of the disease severity distribution may also represent the extreme end of the distribution for exposures, effects, or susceptibility.

• Gene–environment interaction studies should be conducted in places with high or low rates of breast cancer such as Hawaii with genetically diverse populations, good cancer surveillance from the 1960s, access to health care and screening, and unique patterns of exposure to etiologic agents (*e.g.,* persistent pesticides, air pollutants, and radiation exposure).

• Aging populations require additional study because any consequences of environmental contaminants on health indicators tends to be cumulative. Chemical residues are present in food, air, water, house dust, and yard and garden soil, and at higher levels may adversely impact the growing number of elderly and cancer survivors, many of whom are women.

• Girls and women in developing countries who may lack adequate nutrition due to cultural biases and interruptions in their food supply also need to be the focus of culturally sensitive research on the role of environmental factors on breast cancer etiology.

• Occupational epidemiology studies of chemical poisoning and residential epidemiology studies of pesticides are needed to better understand high-end exposure patterns that are often lognormal statistical curves, indicating that some people have much higher than average exposures.

• Studies are needed of seasonal variations in exposure to toxic chemicals, such as springtime surges of agricultural pesticide runoff into drinking water supplies. From both consumer and scientific views, significant "prevention opportunities" may emerge by lowering overall exposure [66].

Overall, the reason for collection and analysis of women's health indicators is to guide long-term prevention efforts globally. Success at breast cancer prevention will take a whole new set of large-scale studies using creative ecogenetic epidemiology tools to uncover gene–environmental interactions [67].

Future directions should also include public health measures aimed at the rapidly growing aged population. Statistics from the National Institute on Aging indicate the magnitude of the challenges ahead [68]. In the developing world the issue is only beginning to be addressed. A Pan American document titled "Health of the Elderly: A Concern for All" [69] notes that: "Aging is one of the most significant changes in the structure of the world population." The AHS [70–73], the LIBCSP and the BCCQ all provide differing opportunities to study the role of aging and environment in women's health.

Acknowledgments

Stimulus and constructive comments were provided by Drs. Michael Alavanja, Aaron Blair, Brenda Edwards, Nancy Evans, Capri-Mara Fillmore, Ray Kent, Barbara Mandula, G. Iris Obrams, Amy Rispin, and Janette Sherman. This research is based in part on work completed as Program Director for the Long Island Breast Cancer Study Project (LIBSCP), Epidemiology and Genetics Program, Division of Cancer Control and Population Sciences, National Cancer Institute. The views expressed are solely those of the author and do not necessarily reflect the policies of the U.S. Environmental Protection Agency or the National Cancer Institute.

References

1. Last, J. M., ed. (1988). "A Dictionary of Epidemiology," 2nd ed. Oxford University Press, New York.
2. World Health Organization (1993). "Trends in Cancer Incidence and Mortality," IARC Sci. Publ. No. 121. IARC, Lyon, France.
3. Pickle, L., Mungiole, M., Jones, G. K., and White, A. A. (1996). "Atlas of United States Mortality," DHHS Publ. No. (PHS) 97-1015. U.S. Department of Health and Human Services, Centers for Disease Control and Prevention, National Center for Health Statistics, Hyattsville, MD.
4. Schottenfeld, D., and Fraumeni, J. F., Jr. (1996). "Cancer Epidemiology and Prevention," 2nd ed. Oxford University Press, New York.
5. Miller, B. A., Kolonel, L. N., Bernstein, L., Young, J. L., Jr., Swanson, G. M., West, D., Key, C. R., Liff, J. M., Glover, C. S., Alexander, G. A., et al., eds. (1996). "Racial/Ethnic Patterns of Cancer in the United States 1988–1992," NIH Publ. No. 96-4104. National Cancer Institute, Bethesda, MD.
6. Trock, B. J. (1996). Breast cancer in African American women: Epidemiology and tumor biology. *Breast Cancer Res. Treat.* **40,** 11–24.
7. National Research Council (1990). "Health Effects of Exposure to Low Levels of Ionizing Radiation BEIR V." National Academy Press, Washington, DC.
8. National Research Council (1991). "Monitoring Human Tissues for Toxic Substances." National Academy Press, Washington, DC.
9. Allen, R. H., Gottlieb, M., Clute, E., Pongsiri, M. J., Sherman, J., and Obrams, G. I. (1997). Breast cancer and pesticides in Hawaii: The need for further study. *Environ. Health Perspect.* **105** (Suppl. 3), 679–683.
10. Allen, R. H. (1995). "Environmental Aspects of Breast Cancer: The Role of Statistics," Proc. Sect. Stat. Environ., pp. 11–20. American Statistical Association, Alexandria, VA.
11. Palmer, J. R., and Rosenberg, L. (1993). Cigarette smoking and risk of breast cancer. *Epidemiol. Rev.* **15,** 145–156.
12. Wolff, M. S., Collman, G. W., Barrett, J. C., and Huff, J. (1996). Breast cancer and environmental risk factors: Epidemiological and experimental findings. *Annu. Rev. Pharmacol. Toxicol.* **36,** 573–596.
13. Gonzalez, F. J., and Gelboin, H. V. (1993). Role of human cytochrome P450s in risk assessment and susceptibility to environmentally based disease. *J. Toxicol. Environ. Health* **40,** 289–308.
14. Rier, S. E., Martin, D. C., Bowman, R. E., and Becker, J. L. (1995). Immunoresponsiveness in endometriosis: Implications for estrogenic toxicants. *Environ. Health Perspect.* **103** (Suppl. 7), 151–156.

15. Carlsen, E., Giwercman, A., Keiding, N., and Skakkebaek, N. (1992). Evidence for the declining quality of semen over the last 50 years. *Br. Med. J.* **305,** 609–613.

16. Glass, R. H. (1991). Infertility. *In* "Reproductive Endocrinology: Physiology, Pathophysiology and Clinical Management" (S. S. C. Yen and R. B. Jaffe, eds.), pp. 689–709. Saunders, Philadelphia.

17. Blair, A., Linos, A., Stewart, P. A., Burmeister, L. F., Gibson, R., Everett, G., Schuman, L., and Cantor, K. P. (1992). Comments on occupational and environmental factors in the origin of non-Hodgkin's lymphoma. *Cancer Res., Suppl.* **52,** 5501s–5502s.

18. Kuller, L. H., Laporte, R. E., and Orchard, T. J. (1986). Diabetes. *In* "Maxcy-Rosenau Public Health and Preventive Medicine" (J. M. Last, ed.), pp. 1225–1240. Appleton-Century-Crofts, Norwalk, CT.

19. Kurtzke, J. F., and Page, W. F. (1997). Epidemiology of multiple sclerosis in U.S. veterans: VII. Risk factors for MS. *Neurology* **48,** 204–213.

20. Blumenthal, S. J. (1994). Women and depression. *J. Women's Health* **3,** 467–479.

20a. Lopez, A. D. and Murray, C. J. (Eds.). (1996). "Global Health Statistics: A Compendium of Incidence, Prevalence and Mortality Estimates for over 200 Conditions (Global Burden of Disease and Injury). Harvard University Press, Boston, MA.

21. Rogan, W. J. (1996). Pollutants in breast milk. *Arch. Pediatr. Med.* **150,** 981–990.

22. Jacobson, J. L., Humphrey, H. E. B., Jacobson, S. W., Schwartz, S. L., Mullin, M. D., and Welch, R. (1989). Determinants of polychlorinated biphenyls, polybrominated biphenyls, and dichlorodiphenyl tricholorethane in the sera of young children. *Am. J. Public Health* **79,** 1401–1404.

23. National Research Council (1993). "Pesticides in the Diets of Infants and Children." National Academy of Sciences Press, Washington, DC.

24. Wargo, J. W. (1996). "Our Children's Toxic Legacy." Yale University Press, New Haven, CT.

25. Wallinga, D. (1998). "Putting Children First: Making Pesticide Levels in Food Safer for Infants and Children." Natural Resources Defense Council, New York.

26. Wiles, W., Davies, K., and Campbell, C. (1998). "Overexposed: Organochlorine Insecticides in Children's Food." Environmental Working Group, Washington, DC.

27. Leiss, J. K., and Savitz, D. A. (1995). Home pesticide use and childhood cancer. *Am. J. Public Health* **85,** 249–252.

28. Hill, R. H., Head, S. L., Baker, S., Gregg, M., Shealy, D. B., Bailey, S. L., Williams, C. C., Sampson, E. J., and Needham, L. L. (1995). Pesticide residues in urine of adults living in the United States: Reference range concentrations. *Environ. Res.* **71,** 99–108.

29. Whitmore, R. W., Immerman, F. W., Camann, D. E., Bond, A. E., Lewis, R. G., and Schaum, J. L. (1994). Non-occupational exposures to pesticides for residents of two U.S. cities. *Arch. Environ. Contam. Toxicol.* **26,** 47–59.

30. Gurunathan, S., Robson, M., Freeman, N., Buckley, B., Roy, A., Meyer, R., Bukowski, J., and Lioy, P. J. (1998). Accumulation of chlorpyrifos on residential surfaces and toys accessible to children. *Environ. Health Perspect.* **106,** 1–6.

31. Lewis, R. G., Fortmann, R. C., and Camann, D. E. (1994). Evaluation of methods for monitoring the potential exposure of small children to pesticides in the residential environment. *Arch. Environ. Contam. Toxicol.* **26,** 37–46.

32. Vacco, D. C. (1996). "Pesticides in Schools: Reducing the Risks." Attorney General of New York State, New York Department of Law.

33. Sesline, D., Ames, R. G., and Howd, R. A. (1994). Irritative and systemic symptoms following exposure to Microban disinfectant through a school ventilation system. *Arch. Environ. Health* **49,** 439–444.

34. Allen, R. H., and Werner, E. (1997). "The Breast Cancer Core Questionnaire: Overview," Ann. Meet. Program Abstr. American Public Health Association, Washington, DC. P,100, P156, 122, 278.

35. Rowland, R. E., Stehney, A. F., and Lucas, H. F. (1978). Dose-response relationships for female radium dial workers. *Radiat. Res.* **76,** 368–383.

35a. Wassermann, M., Nogueira, D. P., Tomatis, L., Mirra, A. P., Shibata, H., Arie, G, Cucos, S., Wassermann, D. (1976). Organochlorine compounds in neoplastic and apparently normal breast tissue. *Bull. Environ. Contam. Toxicol.* **15,** 478–484.

35b. Falck, F., Jr., Ricci, A., Jr., Wolff, M. S., Godbold, J., and Deckers, P. (1992). Pesticides and polychlorinated biphenyl residues in human breast lipids and their relation to breast cancer. *Arch. Environ. Health* **4,** 143–146.

35c. Wolff, M. S., Toniolo, P. G., Lee, E. W., Rivera, M., and Dubin, N. (1993). Blood levels of organochlorine residues and risk of breast cancer. *J. Natl. Cancer Inst.* **85,** 648–652.

35d. Krieger, N., Wolff, M. S., Hiatt, R. A., Rivera, M., Vogelman, J., and Orentreich, N. (1994). Breast cancer and serum organochlorines: A prospective study among white, black and Asian women. *J. Natl. Cancer Inst.* **86,** 589–599.

35e. Dewailly, E., Dobin, S., Verreault, R., Ayotte, P., Sauve, L., Morin, J., and Brisson, J. (1994). High organochlorine body burden in women with estrogen-receptor positive breast cancer. *J. Natl. Cancer Inst.* **86,** 232–234.

35f. López-Carrillo, L., Blair, A., López-Cervantes, M., Cebrian, M., Rueda, C., Reyes, R., Mohar, A., and Bravo, J. (1997). Dichlorodiphenyltrichlorethane serum levels and breast cancer risk: A Case-control Study from Mexico. *Cancer Res.* **57,** 3728–3732.

35g. van't Veer, P., Lobbezoo, I. E., Martin-Moreno, J. M., Guallar, E., Gomez, A. R., Aracena, J., Kardinaal, A. F., Kohlmeier, L., Martin, B. C., Strain, J. J., Thamm, M., van Zoonen, P., Baumann, B. A., Huttunen, J. K., and Kok, F. J. (1997). DDT (dicophane) and postmenopausal breast cancer in Europe: Case-control Study. *Br. Med. J.* **315,** 81–85.

35h. Hunter, D. J., Hankinson, S. E., Laden, F., Colditz, G. A., Manson, J. E., Willett, W. C., Speizer, F. E., and Wolff, M. S. (1997). Plasma organochlorine levels and the risk of breast cancer. *N. Engl. J. Med.* **337,** 1253–1258.

35i. Hoyer, A. P., Grandjean, P., Jorgensen, T., Brock, J. W., and Hartvig, H. B. (1998). Organochlorine exposure and risk of breast cancer. *Lancet* **352,** 1816–1820.

35j. Olaya-Contreras, P., Rodriguez-Villamil, J., Posso-Valencia, H. J., and Cortez, J. E. (1998). Organochlorine exposure and breast cancer risk in Colombian women. *Cad Saude Publ.* **14** (Suppl. 3), 125–132.

35k. Moysich, K. B., Ambrosone, C. B., Vena, J. E., Shields, P. G., Mendola, P., Kostyniak, P., Greizerstein, H., Graham, S., Marshall, J. R., Schisterman, E. F., and Freudenheim, J. L. (1998). Environmental organochlorine exposure and postmenopausal breast cancer risk. *Cancer Epidemiol. Biomarkers Prev.* **7**(3), 181–188.

35l. Jackson, M. A., Stark, H. F., and Waters, M. D. (1996). "Genetic Activity Profiles for the Long Island Breast Cancer Study Project." U.S. EPA, Research Triangle Park, NC.

35m. Repetto, R., and Baliga, S. S. (1996). "Pesticides and the Immune System: The Public Health Risks." World Resources Institute, Washington, DC.

36. Aschengrau, A., Ozonoff, D., Paulu, C., Coogan, P., Vezina, R., Heeren, T., and Zhang, Y. (1993). Cancer risk and tetrachloroethylene-contaminated drinking water in Massachusetts. *Arch. Environ. Health* **48,** 284–292.

37. Newman, B., Moorman, P. G., Millikan, R., Qaqish, B. F., Geradts, J., Aldrich, T. E., and Liu, E. T. (1995). The Carolina Breast Cancer Study: Integrating population-based epidemiology and molecular biology. *Breast Cancer Res. Treat.* **35,** 51–60.

38. Monson, R. R. (1980). "Occupational Epidemiology," CRC Press, Boca Raton, FL.

39. Kelsey, J. L. (1993). Breast cancer epidemiology: Summary and future directions. *Epidemiol. Rev.* **15,** 256–263.

40. Hunter, D. J., and Willett, W. C. (1993). Diet, body size, and breast cancer. *Epidemiol. Rev.* **15,** 110–132.

41. Gates, J. R., Parpia, B., Campbell, T. C., and Junshi, C. (1996). Association of dietary factors and selected plasma variables with sex hormone-binding globulin in rural Chinese women. *Am. J. Clin. Nutr.* **63,** 22–31.

42. Frisch, R. E., Wyshak, G., Albright, N. L., Albright, T. E., Schiff, I., and Witschi, J. (1987). Lower lifetime occurrence of breast cancer and cancers of the reproductive system among former college athletes. *Am. J. Clin. Nutr.* **45,** 328–335.

43. Bernstein, L., Ross, R. K., Lobo, R. A., Hanisch, R., and Krailo, M. D. (1987). The effects of moderate physical exercise on menstrual cycle patterns in adolescence: Implications for breast cancer prevention. *Br. J. Cancer* **55,** 681–685.

44. Kelsey, J. L., and Horn-Ross, P. L. (1993). Breast cancer: Magnitude of the problem and descriptive epidemiology. *Epidemiol. Rev.* **15,** 7–16.

45. Sturgeon, S. R., Schairer, C., Gail, M., McAdams, M., Brinton, L. A., and Hoover, R. N. (1995). Geographic variation in mortality from breast cancer among white woman in the United States. *J. Natl. Cancer Inst.* **87,** 1846–1853.

46. Devesa, S. S., Blot, W. J., Stone, B. J., Miller, B. A., Tarone, R. E., and Fraumeni, J. F. (1995). Recent cancer trends in the United States. *J. Natl. Cancer Inst.* **87,** 175–182.

47. Slovak, M. L. (1997). Breast tumor cytogenetic markers. *In* "Human Cytogenetic Cancer Markers" (S. W. Wolman and S. Sell, eds.), pp. 111–149. Humana Press, Totowa, NJ.

48. Deng, G., Lu, Y., Zlotnikov, G., Thor, A. D., and Smith, H. S. (1996). Loss of heterozygosity in normal tissue adjacent to breast carcinomas. *Science* **274,** 2057–2059.

49. Miller, D. G., Tiwari, R., Pathak, S., Hopwood, V. L., Gilbert, F., and Hsu, T. C. (1998). DNA repair and mutagen sensitivity in patients with triple primary cancers. *Cancer Epidemiol. Biomarkers Prev.* **7,** 321–327.

50. Malins, D. C., Polissar, N. L., Nishikida, K., Holmes, E. H., Gardiner, H. S., and Gunselman, S. J. (1994). The etiology and prediction of breast cancer. *Cancer (Philadelphia)* **75,** 503–517.

51. Bradlow, H. L., Davis, D. L., Lin, G., Sepkovic, D., and Tiwari, R. (1995). Effects of pesticides on the ratio of 16α/2-hydroxyestrone: A biologic marker of breast cancer risk. *Environ. Health Perspect.* **103,** 147–150.

52. Sherman, J. D. (1994). Tamoxifen: Structure-activity and public health concerns. *J. Occup. Med. Toxicol.* **3,** 91–94.

53. Coller, H. A., and Thilly, W. G. (1994). Development and application of mutational spectra technology. *Environ. Sci. Technol.* **28,** 478A–487A.

54. Lewis, D. R., and Vandermer, H. (1997). "The Workshop on Human Milk Contamination and Endocrine Disruption." U.S. EPA and SRA Technologies, Falls Church, VA.

55. Kellogg, R. L., Maizel, M. S., and Goss, D. W. (1992). "Agricultural Chemical Use and Ground Water Quality: Here Are the Potential Problem Areas?" U.S. Department of Agriculture, Washington, DC.

56. Marketing and Regulatory Program (1998). "Pesticide Data Program—Annual Summary Calendar Year 1996." U.S. Department of Agriculture, Washington, DC.

57. Ziegler, R. G., Hoover, R. N., Nomura, A. M. Y., West, D. E., Wu, A. H., Pike, M. C., Lake, A. J., Horn-Ross, P. L., Kolonel, L. N., Siiteri, P. K., and Fraumeni, J. F., Jr. (1996). Relative weight, weight change, height, and breast cancer risk in Asian-American women. *J. Natl. Cancer Inst.* **88,** 650–659.

58. Kulldorff, M., and Nagarwalla, N. (1995). Spatial disease clusters: Detection and inference. *Stat. Med.* **14,** 799–810.

59. Munasinghe, R. L., and Morris, R. D. (1996). Localization of disease clusters using regional measures of spatial autocorrelation. *Stat. Med.* **15,** 893–905.

60. Pichlmeier, U., and Gefeller, O. (1997). Conceptual aspects of attributable risk with recurrent disease events. *Stat. Med.* **16,** 1120–1120.

61. Järvholm, B. (1992). Dose-response in epidemiology—Age and time aspects. *Am. J. Ind. Med.* **21,** 101–106.

62. Lele, C., and Whittemore, A. S. (1997). Different disease rates in two populations: How much is due to differences in risk factors? *Stat. Med.* **16,** 2543–2554.

63. Reichhardt, T. (1996). Environmental GIS: The world in a computer. *Environ. Sci. Technol.* **30,** 340A–343A.

64. Moulton, L. H., Foxman, B., Wolfe, R. A., and Port, F. K. (1994). Potential pitfalls in interpreting maps of stabilized rates. *Epidemiology* **5,** 297–301.

65. Worrall, L., and Bond, D. (1997). Geographical information systems, spatial analysis and public policy: The British experience. *Int. Stat. Rev.* **65,** 365–379.

66. Steingraber, S. (1997). "Living Downstream: An Ecologist Looks at Cancer and the Environment." Addison-Wesley, New York.

67. Rothman, N., and Hayes, R. B. (1994). Using biomarkers of genetic susceptibility to enhance the study of cancer etiology. *Environ. Health Perspect.* **103**(Suppl. 8), 291–295.

68. National Institute on Aging (1998). "Databases on Aging—Survey Summaries: A Selected Compilation of Archived Surveys Relevant to the Demography, Economics and Epidemiology of Aging." National Institutes on Aging, Bethesda, MD.

69. Pan American Health Organization (1992). "Health of the Elderly: A Concern for All," Commun. Health Ser. No. 3. World Health Organization, Washington, DC.

70. Alavanja, M. C. R., Sandler, D. P., McMaster, S. B., Zahm, S. H., McDonnell, C. J., Lynch, C. F., Pennybacker, M., Rothman, N., Dosemeci, M., Bond, A., and Blair, A. (1996). The Agricultural Health Study. *Environ. Health Perspect.* **104,** 362–369.

71. Tarone, R. E., Alavanja, M. C. R., Zahm, S. H., Lubin, J., Sandler, D. P., McMaster, S. B., Rothman, N., and Blair, A. (1997). The Agricultural Health Study: Factors affecting the completion and return of self-administered questionnaires in a large prospective cohort study of pesticide applicators. *Am. J. Ind. Med.* **31,** 233–242.

72. Alavanja, M. C. R., Sandler, D. P., McDonnell, C. J., Lyneh, C. F., Pennybacker, M., Zahm, S. H., Lubin, J., Mage, D., Steen, W. C., Wintersteen, W., and Blair, A. (1998). Factors associated with self-reported, pesticide-related visits to health care providers in the Agricultural Health Study. *Environ. Health Perspect.* **106,** 415–420.

73. Alavanja, M. C. R., Sandler, D. P., McDonnell, C. J., Mage, D. T., Kross, B. C., Rowland, A. S., and Blair, A. (1999). Characteristics of persons who self-reported a high pesticide exposure event in the Agricultural Health Study. *Environ. Res.* **80,** 180–186.

50

Environmental Exposures and Women's Reproductive Health

JODI A. FLAWS,* FADY I. SHARARA,† ELLEN K. SILBERGELD,* AND ANNE N. HIRSHFIELD‡

*Department of Epidemiology and Preventive Medicine, University of Maryland School of Medicine, Baltimore, Maryland;
†Reproductive Endocrinology and Infertility, University of Maryland, Baltimore, Maryland; ‡Department of Anatomy and
Neurobiology, University of Maryland School of Medicine, Baltimore, Maryland

I. Introduction

The reproductive system of women is important for the production of germ cells (oocytes), the transport of oocytes to the fallopian tubes for fertilization, growth and development of the fetus, and nourishment of young offspring [1–4]. This system consists of the following tissues: ovaries, vagina, cervix, uterus, breasts, and the hypothalamic-pituitary unit. These tissues must function properly to insure normal female reproduction and propagation of the human species.

Environmental chemicals have the potential to alter female reproductive tissues and thus affect the ability of women to conceive healthy offspring. During the twentieth century, industrial and technological advances have lead to the production and use of many new chemicals [5–8]. Women may be exposed to these chemicals in the workplace, and through soil, air, water, and food. The physiological consequences of such exposures are unclear because the effects of chemicals on women's reproductive health are often difficult to evaluate.

Relatively few studies have examined associations between chemical exposures and adverse reproductive outcomes in women. Observations in wildlife have suggested the specific possibility that human exposure to environmental compounds with endocrine-disrupting capability may have adverse effects on reproduction. Little research on this topic has been conducted in relation to female reproduction but some studies, reviewed in this chapter, have focused on exposures with estrogenic properties, such as diethylstilbesterol and dioxin.

II. Definition of the Topic

The purpose of this chapter is to review the data that exist on the associations between chemical exposures and reproductive outcomes in women. The chapter will focus on what is known about the impact of chemicals on reproductive structures, menarche, menstrual cyclicity, libido, pregnancy, and menopause. Because limited data are available, we will also highlight areas for future research in women's reproductive health. The chapter will not cover reproductive cancers, which are discussed either in other chapters or in review articles [9–13]. Occupational exposures are included when they are relevant to environmental exposure.

III. Methodologic Issues

Studies of the associations between chemical exposures and abnormal reproductive structures, menarche, menstrual cyclicity, libido, pregnancy, and menopause present a number of difficulties [14–17]. Some women may choose not to participate in reproductive studies because of an unwillingness to discuss personal topics such as menstrual cyclicity, sexual interest, and abortion. Other women may intentionally lie about sexual behaviors, abortions, and miscarriages for cultural reasons.

Further difficulties arise from the fact that current research methods lack the sensitivity required to measure some reproductive endpoints [14]. A spontaneous abortion may not be recognized if it occurs before pregnancy is detectable with clinical assays. Infertility may not be noted if a women intentionally prevents pregnancy by taking oral contraceptives or practicing abstinence. Investigators also may fail to examine a variety of reproductive endpoints and to use specific measures of exposure.

Exposure data also can be hard to collect [14,15]. In general, there is a lack of national databases on chemical exposures in women. In addition, many studies rely on questionnaire data to estimate the amount of exposure as well as the reproductive outcome. These studies may not accurately reflect the actual amount of chemical that enters the body and reaches a target organ. Studies may also overlook some adverse outcomes because they lack the ability to detect early pregnancy losses, subtle birth defects, or slight changes in cyclicity. Women may not remember their exact age at menarche or menopause or whether their cycles were abnormally long or short.

Another difficulty is that certain potential biases may arise in reproductive studies [14]. Exposed women with adverse reproductive outcomes may be more likely to participate in studies compared to exposed women with normal reproductive outcomes. Women with adverse outcomes may also be more likely to remember chemical exposures compared to women with normal outcomes. Investigators may be more likely to publish data that show an association between chemical exposures and adverse outcomes than to publish negative findings.

Problems in reproductive studies also occur because of confounding factors [16,17]. Women may be exposed to several chemicals simultaneously, making it difficult to determine which chemical is associated with a specific adverse outcome. They may also have other risk factors, such as family history, socioeconomic status, and age, that may confound the associations between chemical exposures and reproductive outcomes.

IV. Distribution in Women: Incidence and Mortality Statistics

The incidence of many reproductive disorders is unknown. However, data from Global Health Statistics show that the incidence of infertility for women of all ages varies by country (Table 50.1) [18]. The lowest incidence is reported in China

Table 50.1

Incidence and Mortality of Selected Adverse Reproductive Outcomes, 1990[a]

	North America, Western Europe, Australia	Eastern Europe, Russia	Latin America	Africa	China	India	World
Infertility							
Incidence	1.0	32	62	151	0	117	55
Mortality	0	0	0	0	0	0	0
Abortion							
Incidence	31	1082	2119	1405	0	1047	788
Mortality	0	0.6	2.0	9.8	0.5	4.2	2.3
Low birth weight							
Incidence	0.4	0.8	4.4	11	0.9	18	5.6
Mortality	1.1	1.2	5.0	35	3.5	35	16

[a]Data from [18]. Values represent rates per 100,000 women of all ages in 1990.

(0 cases per 100,000) and the highest incidence is found in Africa (151 cases per 100,000 women). The incidence of spontaneous abortion also varies by country. The rate is lowest in the United States (31 per 100,000 women) and highest in Latin American (2,119 per 100,000 women). Similarly, the incidence of low birth weight varies across the world. It is lowest in North America (0.4 cases per 100,000) and highest in India (18 cases per 100,000).

The mortality from most adverse reproductive outcomes is estimated to be negligible [18]. Low birth weight may contribute to mortality of infants, however. (Table 50.1) [18]. In 1990, 1.1 per 100,000 infants died from low birth weight in the United States and 35 per 100,000 infants died from low birth weight in Africa. Similarly, abortion contributes to mortality rates in women. The rates range from 0 per 100,000 women in North America to 9.8 per 100,000 women in Africa (Table 50.1) [18]. An early age at menopause also may contribute to mortality rates [19,20]. One report indicates that the risk of dying increases twofold in women with an early age at menopause compared to women with a normal age at menopause [20].

V. Environmental Determinants and Risk Factors

Chemical exposures can interfere with the structure and function of the ovaries, uterus, breasts, cervix, and the hypothalamic-pituitary unit (14,21–23]. Chemicals also may affect reproductive health by altering reproductive processes such as menarche, menstrual cyclicity, pregnancy, libido, and menopause [24–27]. A summary of the associations between chemical exposures and adverse reproductive outcomes is presented in the following sections and in Table 50.2.

A. Structure of the Reproductive System

Little information is available regarding whether environmental chemicals cause structural abnormalities in reproductive tissues. A few studies suggest that the structure of the vagina, uterus, and cervix may be altered by exposure to diethylstilbestrol (DES) [10]. Prior to 1970, DES was widely prescribed for use in high-risk pregnancies during the first trimester. After

1970, it was banned when studies suggested that *in utero* DES exposure was associated with increased risk of vaginal clear-cell carcinoma [28]. Investigators have since examined the effects of *in utero* DES exposure on reproductive organs in more detail [10,29]. The National Cooperative Diethylstilbestrol Adenosis (DESAD) study enrolled women exposed to DES during gestation. Each participant received annual exams for about 10 years and completed self-administered questionnaires for 6 years after the annual exams. The results indicated that DES exposure increased the risk of reproductive tract structural anomalies such as gross anatomical changes in the cervix, T-shaped uteri, and hypoplastic uteri [29]. Such abnormalities are thought to contribute to the infertility and subfecundity experienced by women exposed to DES during development [10,29].

Other chemicals have been associated with abnormalities in uterine structure. Two industrial by-products with estrogenic activity (2,3,7,8-tetrachlorodibenzo-*p*-dioxin and polychlorodiphenyl) have been found to increase the risk of endometriosis [30]. This condition is characterized by the growth of endometrial cells outside the uterus, leading to pain and often to infertility. Cadmium and solvents have been associated with an increased risk of uterine fibroids, a condition that can lead to pain, abnormal menstrual patterns, and hysterectomy [31,32].

A few studies suggest that chemicals in cigarette smoke alter ovarian structure [33–35]. Polyaromatic hydrocarbons are thought to deplete ovarian follicles, which contain the female germ cells and produce hormones such as estrogen [32,36]. This poses a health concern because loss of germ cells causes infertility and low estrogen increases the risk of vaginal dryness, incontinence, hot flashes, cardiovascular disease, and osteoporosis [19,37].

Although some reports indicate that environmental factors are associated with defects in the reproductive system, the data are limited and cannot be used to make firm conclusions. Published studies rely on ecological data, extremely small sample sizes, insensitive measures of outcomes, or indirect exposure assessments. These studies often present conflicting results; one study will report an association while another will indicate a lack of association between chemicals and structural abnormalities.

Table 50.2

Associations Noted between Adverse Reproductive Outcomes and Chemical Exposures

Adverse outcome	Chemicals	Source of exposure
Structural abnormalities		
Cervix	Diethylstilbestrol*	Prescribed for difficult pregnancies
Ovary	Polycyclicaromatic hydrocarbons	Cigarette smoke
Uterus	Diethylstilbestrol*	Prescribed for difficult pregnancies
	2,3,7,8-Tetrachlorodibenzo-p-dioxin*	Industrial waste
	Polychlorodiphenyl*	Industrial waste
	Cadmium	Workplace and environmental contamination
	Solvents	Workplace and environmental contamination
Menstrual disorders	Benzene, toluene, styrene, carbon disulfide, hair dyes, formaldehyde	Workplace
Infertility	Polycyclicaromatic hydrocarbons	Cigarette smoke
	Pesticides	Workplace
	Ionizing radiation	Chemotherapy and workplace
	Diethylstilbestrol*	
Birth defects	Polychlorinated biphenyls*	Capacitators, transformers, hydraulics, lubricants, sealants, pesticides, paints, plastics, dyes, fish.
	Ionizing radiation	Nuclear plant accident at Chernobyl
	Mercury	Contaminated fish, soil, water
	Glycol ethers	Semiconductors, workplace
Spontaneous abortion	Drinking water	Home
	Glycol ethers	Semiconductors, workplace
	Dibromochloropropane	Agriculture (banana crops)
Low sexual interest	Diethylstilbestrol*	Prescribed for difficult pregnancies
Menopausal disorders		
Abnormal age	Polycyclicaromatic hydrocarbons	Cigarette smoke
Abnormal symptoms	Polycyclicaromatic hydrocarbons	Cigarette smoke

Note. Chemicals with an asterisk are banned in the United States.

B. Menarche

Menarche symbolizes the onset of sexual maturity and is characterized by the onset of the first menstrual bleeding. The average age at menarche is 13.8 years; however, it ranges from 9 to 18 years and varies by race and ethnicity [1]. The median age at menarche for European girls ranges from 12.6 years (Italy and Greece) to 15.2 years (Russia). The median age for North American girls varies from 12.5 years (U.S. blacks) to 13.8 years (U.S. Eskimos), the median age for Asian girls ranges from 12.7 years (Singapore) to 18.1 years (Nepal), and the median age for African girls varies from 13.2 to 17.0 years (Rwanda) [1].

The physiological consequences of early or late menarche are equivocal. One study indicates that girls with late age at menarche require a longer time period before they reach ovulatory cycles than that required by girls with early age at menarche [38]. Other studies indicate that late age at menarche is associated with a decreased time to first conception [38a]. Although data from these two studies appear contradictory, they imply that either a late age or early age at menarche might affect fecundity and fertility.

Currently, it is unclear whether chemical exposures alter the age at menarche. Ecological data suggest that chemical exposure has increased over time and that this increase may be correlated with a decline in the age at menarche [1,39–41].

European studies indicate that age of menarche has declined from 17 years in 1840 to about 13 years in 1970 [1,40]. North American studies show that the age at menarche declined from about 15 years in 1890 to approximately 13 years in 1920 [1,41]. However, it should be noted that no data directly show that chemicals are responsible for the decline in age at menarche. Alternative hypotheses, endorsed by many, state that the decline in age at menarche stems from changes in nutrition, weight, stress, or the accuracy of recording age at menarche over time [1,41–45].

C. Menstrual Cyclicity

The reproductive system of most women undergoes a series of cyclical changes known as the menstrual cycle. This cycle involves structural and functional changes at the level of the hypothalamus, pituitary, ovary, and uterus [1,46]. These cycles are characterized by periodic menstrual bleeding for 3–5 days, with approximately 28 days between the onset of one menstrual bleeding and the next bleeding.

Cycle length is variable within individual women, particularly at certain ages [1]. A study of menstrual cycle length in 23,016 white women from the United States and Canada found that the mode of cycle length was 27 days, the mean was 29.1 days and the variance was 55.7 days. The large variance in this study was attributed to irregularity from cycle to cycle within individual women [1].

Cycle quality also varies between individual women [1]. In general, cycle quality is considered poor if the cycle results in anovulation or a defective luteal phase. Anovulation is characterized by degeneration of ovarian follicles before the final maturation of the egg. A defective luteal phase occurs because the corpus luteum is either short lived or incapable of producing adequate amounts of progesterone.

The associations between environmental chemicals and cycle length or quality are not clear [48,49]. Some studies report an increased risk of menstrual disorders in women who work with benzene, toluene, xylene, styrene, carbon disulfide, hair dyes, and formaldehyde [5,48,49]. In contrast, other studies find no association between chemical exposures and menstrual disorders [50,51]. There was no difference in menstrual cycle length, duration, or flow between workers exposed to 2-ethoxyethylacetate in the liquid crystal display manufacturing industry and unexposed workers [50]. There also was no association observed between styrene exposure and menstrual disorders in a study of 1500 blue-collar workers in the United States [51]. These conflicting findings may stem from differences in the types and amounts of chemical exposure, from increased susceptibility of some women to menstrual disorders, or from methodological problems (*e.g.,* inaccurate measurement of exposure and outcome, biases, and confounding factors).

D. Pregnancy

Pregnancy is a complex biologic process that involves fertilization, implantation, embryogenesis, and organogenesis [46]. Several studies indicate that chemical exposures can adversely affect pregnancy outcomes. The type of effect depends on the chemical itself, the timing of exposure, and whether the chemical affects the mother or the fetus [14]. Chemical exposures in the first few weeks after conception may lead to spontaneous abortions because this is a time prior to the development of organs and tissues that help to detoxify chemicals and prevent death [1,14,22,23]. Exposure in the first 3–12 weeks of gestation may result in birth defects because it is during this period that the major organs are developing [1,14,22,23]. For example, the central nervous system and heart develop between 3 and 6 weeks gestation. The limbs, ears, eyes, and teeth form during 4.5–8.5 weeks. The palate and external genitalia develop during 6–12 weeks of gestation. Exposure after 9 weeks of gestation may cause growth deficits and minor abnormalities because organogenesis is almost complete [1,14,22,23].

Environmental chemicals known as polychlorinated biphenyls (PCBs) have been associated with birth defects [52–58]. From 1930 to 1970, PCBs were used as dielectric fluids in capacitators and transformers. They also were used as heat transfer fluids, hydraulic fluids, lubricants, sealants, extenders in pesticides, and additives in paints, plastics, and dyes. In 1970, these chemicals were banned in the United States because they were associated with adverse pregnancy outcomes [59]. In 1968, 1000 Japanese people consumed rice oil contaminated with PCBs [56]. Women who were pregnant while they consumed the rice oil gave birth to babies with low birth weight, dark

brown pigmentation, orbital edema, gingival hyperplasia, natal teeth, abnormal calcification of the skull, and rocker bottom heel [56,60]. Some children also exhibited growth impairment, lack of endurance, hypotonia, jerky movements, clumsy movement, apathy, and low IQ [56]. In 1979, Taiwanese women were exposed to PCBs through contaminated rice oil [56]. Pregnant women exposed to the rice oil gave birth to babies with brown pigmentation, growth retardation, and low IQ [56,61]. Women in the U.S. have been exposed to PCBs through the consumption of contaminated fish. Some of these women gave birth to babies with low birth weight, small heads, weak reflexes, and poor memory [55–57].

Exposure to drinking water has been associated with spontaneous abortion and birth defects [25,26,62–66]. A case-control study in North Carolina reported that ingestion of water was associated with increased risk of miscarriage, preterm delivery, and low birth weight [66]. A cohort study in Northern California found that the risk of spontaneous abortion was 1.5 times greater for consumers of tap water compared to consumers of bottled water [65]. Another study observed a fourfold greater odds of spontaneous abortion among women who consume tap water during the first trimester compared to women without tap water consumption [63]. In addition, women exposed to water contaminated with trichlorethylene had an increased risk of spontaneous abortion and birth defects compared to unexposed women. The odds ratio for spontaneous abortion in exposed women was 2.3 (confidence interval (CI) 1.1–10.4) [25].

Several miscellaneous chemicals have been associated with spontaneous abortions and birth defects [67–78]. A survey of female members of the Oil, Chemical, and Atomic Workers International Union found that, among those responding, women exposed to halogenated hydrocarbons had an increased risk of infant mortality compared to unexposed women [69]. A study of female semiconductor manufacturers reported a 45% excess risk of abortion in workers exposed to glycol ethers and fluoride compounds compared to unexposed workers [68]. Women who were exposed both to glycol ethers and to fluoride had a relative risk of spontaneous abortion of 3.21 (CI = 1.29–5.96) compared to women who were not exposed to either chemical [68]. Glycol ether exposure has also been reported to increase the risk of neural tube defects (odds ratio (OR) = 1.94; CI = 1.16–3.24), multiple anomalies (OR = 2.00; CI 1.24–3.23), and cleft lip (OR = 2.03; CI 1.11–3.73) [79].

Women exposed to ionizing radiation in Chernobyl appear to have an increased risk of giving birth to children with anemia, congenital abnormalities, and early death compared to unexposed women [78]. Studies also show that children born to women exposed to mercury have an increased risk of congenital abnormalities [76,80]. Increases in the risk of infertility, miscarriage, premature membrane rupture, preeclampsia, pregnancy hypertension, and premature delivery have been seen among women exposed to lead [67]. A study of women living in three different regions of France found an increased risk of ectopic pregnancy with exposure to X-rays, benzene, or pesticides (relative risk (RR) = 6.6; CI 1.3–30.4) [81]. A historical cohort study of hairdressers in the Netherlands found an increased risk of spontaneous abortions compared to women who were not hairdressers (OR = 1.6; CI = 1.0–2.4) [72]. Women working with banana crops who were exposed to the pesticide dibromochloropropane had an increased risk of spontaneous abortion compared to unexposed women [75].

A few case studies also suggest that chemical exposures may be associated with adverse pregnancy outcomes [70,73]. One pregnant woman was accidentally exposed to N-methyl-2-pyrolidine, a solvent used in petroleum refining, microelectronics, veterinary medicine, and pesticide formulation. This exposure was associated with intrauterine growth retardation followed by fetal death at 31 weeks [73]. Another pregnant woman who was exposed to a cauliflower field that was contaminated with organophosphate and carbamate insecticides gave birth to a fetus with multiple cardiac defects, micropthalamia of the left eye, brain atrophy, and facial abnormalities [70].

Although a number of studies suggest that chemical exposures are associated with spontaneous abortions and birth defects, some others report no association between chemicals and adverse pregnancy outcomes. Radiowicki and Wierzba conducted a study of 146 women exposed to phenol in a polluted area in Poland [82] and saw no association between phenol exposure and adverse pregnancy outcomes. A study of female members of the Union of Rubber and Leather Workers found no significant association between exposure to chemicals in the rubber industry and adverse pregnancy outcomes such as malformations and spontaneous abortions [83]. A prospective study of 3901 Swedish women also found no association between chemical exposures and adverse pregnancy outcomes (OR = 1.28; CI = 0.91–1.80) [84]. A case-control study of dental assistants and gardening workers showed no association between occupational exposure to chemicals and spontaneous abortions [24]. Three natural experiments involving environmental exposure to chemicals (Love Canal, New York; Woburn, Massachusetts; and Seveso, Italy) have found limited evidence of adverse pregnancy outcomes [22,23]. The reasons for conflicting reports are not known but methodologic limitations may be part of the explanation.

E. Sexual Interest (Libido)

Studies on the association between environmental exposures and libido are particularly difficult to conduct because little is known about the normal factors that determine sexual interest [1]. This difficulty often precludes investigators from knowing which variables to measure during sexual interest studies. A few studies imply that age determines sexual interest and frequency [1]. These studies show that the frequency of sex declines with age and that this decline corresponds to diminishing levels of testosterone. Other studies indicate that the day of the menstrual cycle, drug use, hormone use, sympathetic activation, and emotional variables alter sexual interest [1,85–89].

Other problems in studying sexual interest include potential recall error, selection bias, and deception. Women may be unable to remember how many times they had sexual intercourse in the past few months or years. Women who participate in studies on libido may be more willing to discuss sexual interest than women who choose not to participate. However, some participants may be uncomfortable about their level of sexual activity or preference and thus may give misleading responses.

Because of difficulty in studying this endpoint, there are few reports on the associations between chemical exposures and libido. One study examined 30 women who were exposed to diethylstilbesterol during gestation [90]. These women reported fewer well-established sex-partner relationships, less sexual desire and enjoyment, decreased sexual excitability, and lower orgasmic coital function compared to unexposed women. Another study examined women working in the clothing industry [91]. Women who were chronically exposed to carbon disulfide reported a loss of libido.

F. Menopause

Menopause is a developmental event signifying the loss of reproductive capability and enhanced susceptibility to adverse health outcomes [19]. Postmenopausal women have higher rates of osteoporosis and coronary heart disease compared to premenopausal women [19,92–97]. Postmenopausal status also has been associated with many conditions that influence the quality of life, including depression, irritability, fatigue, insomnia, forgetfulness, decline in cognitive functioning, vaginal and skin dryness, urinary incontinence, urinary tract infections, and arthritis [98–104].

The age at menopause plays a critically important role in determining the general health of aging women [19]. The median age at menopause in the United States is approximately 50 years, with about 13% of the population experiencing "early menopause" (menopause between ages 40 and 47) [1,19,47,105]. Early menopause is an indicator of significant health risk because it is associated with an increased risk of developing chronic diseases and other morbidities. [9,19,95,106,107]. More importantly, the life span of women who experience an early menopause appears to be significantly reduced compared to women who undergo normal menopause [20].

The factors that determine the age at menopause are not well understood, but several studies indicate that the size of the primordial follicle pool is a key determinant [19,108,109]. Chemical exposures may increase the rate of loss of primordial follicles, thus accelerating the onset of menopause, but there is limited direct evidence that chemicals deplete primordial ovarian follicles. However, epidemiologic studies show that chemicals in cigarette smoke accelerate the onset of menopause [33,110–112]. In one study, the mean age at natural menopause was 49.7 years for smokers and 50.8 years for nonsmokers [33]. In addition, the mean age at menopause of nonsmokers whose spouses did not smoke was 51.9 years, whereas the mean age at menopause of nonsmokers whose spouses smoked was 49.8 years [33].

Menopausal symptoms also may alter the quality of life in women. One study indicates that cigarette smoking increases the odds of frequent hot flashes in menopausal and perimenopausal women (OR = 2.7; CI = 1.4–5.2) [113]. A study of women working in an industrial clothing factory demonstrates that women exposed to carbon disulfide experienced different menopausal symptoms compared to unexposed women [91]. The exposed women reported more headaches and weight gain than unexposed women. Exposed women also experienced decreased hot flash frequency compared to unexposed women.

Although a few studies have identified associations between chemical exposures and menopausal age and symptoms, these studies have not been widely repeated using a variety of study designs. Previous studies have used a retrospective approach, which may lead to errors in recall. Women who are postmenopausal may not remember their exact age at menopause or what their symptoms were like during the transition. Women also may not remember whether they were exposed to chemicals during menopause and the level of their exposures, or may remember differently depending on their level of symptoms.

VI. Biological Markers of Susceptibility or Exposure

Biomarkers may be useful for studies on the associations between chemical exposures and reproductive outcomes because they help determine the internal dose or biologically effective dose of a chemical. They also help to monitor early responses to chemicals, detect alterations in reproductive structure or function, and measure the onset of disease [114,115]. For example, serum levels of steroids and pituitary hormones may be used to monitor menstrual cyclicity, fertility, and menopause [114]. Serum levels of human chorionic gonadotropin are useful for detecting pregnancy during the first few weeks of gestation [114]. Serum and urinary levels of chemicals can be used to determine internal dose and metabolic capacity [115]. Tissue samples may be used to detect chromosomal abnormalities, DNA damage, metabolic capacity, and histological abnormalities. Ultrasound often is useful for measuring structural changes in ovarian, vaginal, cervical, or uterine tissues in a noninvasive manner.

Although biomarkers may be useful for reproductive studies, they also may lead to some methodological problems. Some of these problems stem from the fact that it is difficult to obtain accurate measures of steroid and pituitary hormones. The normal hormone levels vary within individuals, over time, between women, and between assays. Problems with biomarkers also occur because it may be difficult to obtain human tissues. Women may not participate in studies that require tissue biopsies because biopsies are invasive and often painful. Women may be reluctant to donate fetal tissues for research because of ethical and legal considerations. Lastly, biomarkers may lack the sensitivity required to monitor reproductive processes and predict impending disease. For example, no markers sensitively measure cycle quality, libido, menopausal symptoms, or ovarian follicle numbers in a noninvasive manner. Therefore, it is difficult to predict abnormal cyclicity, low sexual interest, abnormal menopausal symptoms, or impending early menopause.

VII. Clinical Issues

It is estimated that 8–13% of all couples are infertile and that 15–20% of all pregnancies result in spontaneous abortion [1,22,23]. Further, it is thought that 2–3% of infants are born with major abnormalities, 5% of infants are preterm births, and 7% of infants have low birth weight [1,22,23]. The reasons for these adverse outcomes are unknown, but even if a small percentage is due to environmental chemicals, the effect on the population may be huge. For these reasons, chemical exposures should be controlled or eliminated for females of reproductive age. Physicians should be informed about toxicants, learn to sort

through animal data and evaluate its relevance to human patients, and obtain an environmental history of infertile women or pregnant women who give birth to children with birth defects as well as from their healthy patients. Physicians also should be prepared to advise patients about chemical exposures that do not alter the risk of adverse outcomes and to suggest appropriate medical care for exposures that are definitively linked to adverse outcomes.

VIII. Epidemiologic Issues

Several epidemiologic issues must be considered during studies on chemicals and reproductive health. Reproductive outcomes are often difficult to measure in human populations. For example, it may be impossible to determine whether a woman experienced a spontaneous abortion if she did not know she was pregnant. In addition, accurate exposure data are difficult to obtain using indirect measurements (e.g., questionnaires and interviews). The actual internal dose or concentration of a chemical in a target organ cannot be assessed by indirect methods. Instead, blood, urine, or tissue samples must be collected and analyzed using laboratory techniques. Reproductive studies also must consider host factors, genetics, and metabolic capacity. These considerations require the use of biomarkers and interdisciplinary collaboration with clinicians and basic scientists. Lastly, there is a lack of national databases that contain information about the incidence, prevalence, and mortality of adverse reproductive outcomes. This makes it difficult to evaluate whether a particular chemical exposure alters natural incidence, prevalence, or mortality rates and poses a real human health hazard.

IX. Summary and Future Directions

The focus of this chapter is on the published associations between chemical exposures and reproductive outcomes. Current data suggest that some chemicals adversely affect the female reproductive system. However, it is important to remember that many of the published studies are flawed, with small or biased samples, uncontrolled confounding factors, and inaccurate measurements of exposure or outcome. Crucial confirmatory studies using different populations or study designs are lacking. For these reasons, it is difficult to assess the "real" impact of chemical exposures on the reproductive health of women. The current data, however, are strong enough to warrant caution and more thorough investigations of this issue.

Future studies should focus on the associations between chemical exposures and abnormalities in the structures of reproductive organs. Such studies might use noninvasive methods (ultrasound) to detect gross changes in uterine or ovarian structure or invasive methods (biopsy) to identify histological changes in tissues. Future studies also should determine whether chemical exposures affect the process of menarche or menopause. Perhaps studies could evaluate the age at menarche or menopause in women exposed to diethylstilbestrol or chemicals in cigarette smoke using available databases. Studies also could evaluate the age at menarche or menopause in women exposed to chemicals in the workplace (e.g., solvents, metals, or polychlorinated biphenyls) during gestation, puberty, or adulthood. Future studies also should evaluate the physiological consequences of menstrual disorders. It is not known whether short-term alterations in cycle length or quality significantly impact fertility. Such information would help determine whether women should be concerned about chemically induced alterations in their menstrual cycles. Future studies also should examine the ovaries of women exposed to chemicals that are known to destroy ovarian follicles in animal models. Perhaps follicle numbers could be assessed in exposed women using ovarian ultrasound or tissues obtained from oophorectomy. Such studies would provide information regarding the effects of chemicals on ovarian follicle numbers and thus on fertility and age at menopause. Lastly, future studies should focus on biological mechanisms of chemically induced reproductive outcomes using human tissues and animal models. Such studies will help identify early markers of impending disease and possibly help in the development of treatment strategies.

References

1. Wood, J. W. (1994). "Dynamics of Human Reproduction: Biology, Biometry, Demography." de Gruyter, New York.

2. Hirshfield, A. N. (1991). Development of follicles in the mammalian ovary. *Int. Rev. Cytol.* **124,** 43–101.

3. Erickson, G. F. (1978). Normal ovarian function. *Clin. Obstet. Gynecol.* **21,** 31–52.

4. Richards, J. S. (1980). Maturation of ovarian follicles: Actions and interactions of pituitary and ovarian hormones on follicular cell differentiation. *Physiol. Rev.* **68,** 51–85.

5. Gold, E. B., Lasley, B. L., and Schenker, M. B. (1994). Introduction: Rationale for an update. Reproductive hazards. *Occup. Med.* **9,** 363–372.

6. McLachlan, J., and Arnold, S. F. (1996). Environmental estrogens. *Am. Sci.* **84,** 452–461.

7. Cooper, R. L., and Kavlock, R. J. (1997). Endocrine disruptors and reproductive development: A weight-of-evidence overview. *J. Endocrinol.* **152,** 159–66.

8. Wells, V. E., Schnorr, T. M., and Halperin, W. E. (1988). NIOSH selection of chemicals and study publications: Self profiles for reproductive research. *Reprod. Toxicol.* **2,** 289–290.

9. Cramer, D. W. (1990). Epidemiologic aspects of early menopause and ovarian cancer. *Ann. N.Y. Acad. Sci.* **592,** 363–375.

10. Giusti, R. M., Iwamoto, K., and Hatch, E. E. (1995). Diethylstilbestrol revisited: A review of the long-term health effects. *Ann. Intern. Med.* **122,** 778–788.

11. Wolff, M., and Toniolo, P. G. (1995). Environmental organochlorine exposure as a potential etiologic factor in breast cancer. *Environ. Health Perspect.* **103,** 141–145.

12. Wolff, M., and Weston, A. (1997). Breast cancer risk and environmental exposures. *Environ. Health Perspect.* **105,** 891–896.

13. Schottenfeld, D., and Fraumeni, J. F. (1996). "Cancer Epidemiology and Prevention," 2nd ed. Oxford University Press, New York.

14. Whelan, E. A. (1997). Risk assessment studies: Epidemiology. *In* "Comprehensive Toxicology" (K. Boekelheide, R. E. Chapin, P. B. Hoyer, and C. Harris, eds.), Vol. 10. pp. 359–366. Cambridge University Press, Cambridge, UK.

15. Correa, A., Stewart, W. F., and Yeh, H. C. (1994). Exposure measurement in case-control studies: Reported methods and recommendations. *Epidemiol. Rev.* **16,** 18–32.

16. Joffe, M. (1985). Biases in research on reproductive women's work. *Int. J. Epidemiol.* **14,** 118–23.

17. Weinberg, C. R. (1993). Toward a clearer definition of confounding. *Am. J. Epidemiol.* **137,** 1–8.

18. Murray, C. L., and Lopez, A. D. (1996). "Global Health Statistics: A Compendium of Incidence, Prevalence and Mortality Estimates for over 200 Conditions." Harvard School of Public Health, Cambridge, MA.

19. Sowers, M., and LaPietra, M. T. (1996). Menopause: Its epidemiology and potential association with chronic diseases. *Epidemiol. Rev.* **17,** 287–302.

20. Snowden, D. A., Kane, R. L., Beeson, W. L., Burke, G. L., Sprafka, M., Potte, J., Iso, H., Jacobs, D. R., and Phillips, R. L. (1989). Is natural menopause a biologic marker of health and aging? *Am. J. Public Health* **79,** 706–714.

21. Lebel, G., Dodin, S., Ayotte, P., Marcoux, S., Ferron, L. A., and Dewailly, E. (1998). Organochlorine exposure and the risk of endometriosis. *Fertil. Steril.* **69,** 221–228.

22. Paul, M. (1993). "Occupational and Environmental Reproductive Hazards: A Guide for Clinicians." Williams & Wilkins, Baltimore, MD.

23. Pope, A. M., and Rall, D. P. (1995). "Environmental Medicine: Integrating a Missing Element into Medical Education." National Academy Press, Washington, DC.

24. Heidam, L. Z. (1984). Spontaneous abortions among dental assistants, factory workers, painters, and gardening workers: A follow-up study. *J. Epidemiol. Commun. Health* **38,** 149–55.

25. Deane, M., Swan, S. H., Harris, J. A., Epstein, D. M., and Neutra, R. R. (1989). Adverse pregnancy outcomes in relation to water contamination, Santa Clara County, California, 1980–1981. *Am. J. Epidemiol.* **129,** 894–904.

26. Wrensch, M., Swan, S. H., Lipscomb, J., Epstein, D. M., Neutra, R. R., and Fenster, L. (1992). Spontaneous abortions and birth defects to tap and bottled water use, San Jose, California, 1980–1985. *Epidemiology* **3,** 98–103.

27. Hornsby, P. P., Wilcox, A. J., and Herbst, A. L. (1995). Onset of menopause in women exposed to diethylstilbestrol in utero. *Am. J. Obstet. Gynecol.* **172,** 92–95.

28. Herbst, A., and Sculley, R. E. (1971). Adenocarcinoma of the vagina. Association of maternal stilbesterol therapy with tumor appearance in young women. *N. Engl. J. Med.* **284,** 878–881.

29. Labarthe, D., Adam, E., Noller, K. L., O'Brien, P. C., Robboy, S. J., Tilley, B. C. et al. (1978). Design and preliminary observations of the National Cooperative Diethylstilbestrol Adenosis Project. *Obstet. Gynecol.* **51,** 453–458.

30. Rier, S. E., Martin, D. C., Bowman, R. E., and Becker, J. L. (1995). Immunoresponsiveness in endometriosis: Implications of estrogenic toxicants. *Environ. Health Perspect.* **103,** 151–156.

31. Gerhard, I., and Runnebaum, B. (1992). The limits of hormone substitution in pollutant exposure and fertility disorders. *Zentralbl. Gynaekol.* **114,** 593–602.

32. Smith, E. M., Hammonds-Ehlers, M., Clark, M. K., Kirchner, H. L., and Fuortes, L. (1997). Occupational exposures and risk of female infertility. *J. Occup. Environ. Med.* **39,** 138–47.

33. Everson, R. B., Sandler, D. P., Wilcox, A. J., Schreinemachers, D., Shore, D. L., and Weinberg, C. (1986). Effect of passive exposure to smoking on age at natural menopause. *Br. Med. J.* **293,** 792.

34. Mattison, D. R. (1980). Morphology of oocyte and follicle destruction by polyaromatic hydrocarbons in mice. *Toxicol. Appl. Pharmacol.* **53,** 249–259.

35. Miller, M. M., Plowchalk, D. R., Weitzman, G. A., London, S. N., and Mattison, D. R. (1992). The effect of benzo(a)pyrene on murine ovarian and corpora lutea volumes. *Am. J. Obstet. Gynecol.* **166,** 1535–1541.

36. Flaws, J. A., and Hirshfield, A. N. (1997). Reproductive, developmental, and endocrine toxicology. *In* "Comprehensive Toxicology" (K. Boekelheide, R. E. Chapin, P. B. Hoyer, and C. Harris, eds.), Vol. 10, pp. 283–291. Cambridge University Press, Cambridge, UK.

37. Bolumar, F., Boldsen, J., and European Study Group on Infertility and Subfecundity (1996). Smoking reduces fecundity: A

European multicenter study on infertility and subfecundity. *Am. J. Epidemiol.* **143,** 578–587.

38. Vihko, R. K., and Apter, D. L. (1986). The epidemiology and endocrinology of the menarche in relation to breast cancer. *Cancer Surv.* **5**(3), 561–571.

38a. Johnson, P. L., Wood, J. W., and Weinstein, M. (1990). Female fecundity in highland Papua New Guinea. *Soc. Biol.* **37**(1–2), 26–43.

39. Sanchez-Andres, A. (1997). Genetic and environmental factors affecting menarcheal age in Spanish women. *Anthropol. Anz.* **55,** 69–78.

40. Boldsen, J. L., Jeune, B., Bach-Ramussen, K. L., Sevelsted, M., and Vinther, E. (1993). Age at menarche among schoolgirls in Odense. Is age at menarche still decreasing in Denmark? *Ugeskr. Laeg.* **155,** 482–484.

41. Campbell, B. C., and Udry, J. R. (1995). Stress and age at menarche of mothers and daughters. *J. Biosoc. Sci.* **27,** 127–34.

42. Graber, J. A., Brooks-Gunn, J., and Warren, M. P. (1995). The antecedents of menarcheal age: Heredity, family environment, and stressful life events. *Child Dev.* **66,** 346–359.

43. Stark, O., Peckham, C. S., and Moynihan, C. (1989). Weight and age at menarche. *Arch. Dis. Child.* **64,** 383–387.

44. Wierson, M., Long, P. J., and Forehand, R. L. (1993). Toward a new understanding of early menarche: the role of environmental stress in pubertal timing. *Adolescence* **28,** 913–924.

45. Rimpela, A. H., and Rimpela, M. K. (1993). Towards an equal distribution of health? Socioeconomic and regional differences of the secular trend of the age of menarche in Finland from 1979 to 1989. *Acta Paediatr.* **82,** 87–90.

46. La Barbera, A. R. (1997). Differentiation and function of the female reproductive system. *In* "Comprehensive Toxicology" (K. Boekelheide, R. E. Chapin, P. B. Hoyer, and C. Harris, eds.), Vol. 10, pp. 255–272. Cambridge University Press, Cambridge, UK.

47. Cramer, D. W., Xu, H., and Harlow, B. L. (1995). Family history as a predictor of early menopause. *Fertil. Steril.* **64,** 740–745.

48. Halbreich, U. (1995). Menstrually related disorders: What we do know, what we only believe that we know, and what we know that we do not know. *Crit. Rev. Neurobiol.* **9,** 163–175.

49. Keye, W. R., Jr. (1984). Environmental exposure and altered menstrual function. *Prog. Clin. Biol. Res.* **160,** 203–209.

50. Chia, S. E., Foo, S. C., Khoo, N. Y., and Jeyaratnam, J. (1997). Menstrual patterns of workers exposed to low levels of 2-ethoxyethylacetate (EGEEA). *Am. J. Ind. Med.* **31,** 148–152.

51. Baranski, B. (1993). Effects of the workplace on fertility and related reproductive outcomes. *Environ. Health Perspect.* **101,** 81–90.

52. Vena, J. E., Buck, G. M., Kostyniak, P., Mendola, P., Fitzgerald, E., Sever, L., Freudenheim, J., Greizerstein, H., Zielezny, M., McReynolds, J., and Olson, J. (1996). The New York Angler Cohort Study: Exposure characterization and reproductive and developmental health. *Toxicol. Ind. Health* **12,** 327–334.

53. Jacobson, J. L., Jacobson, S. W., and Humphrey, H. E. B. (1990). Effects of exposure to PCBs and related compounds on growth and activity in children. *Neurotoxicol. Teratol.* **12,** 319–326.

54. Seegal, R. F. (1996). Can epidemiological studies discern subtle neurological effects due to perinatal exposure to PCBs? *Neurotoxicol. Teratol.* **18,** 251–254.

55. Jacobson, J. L., and Jacobson, S. W. (1996). Sources and implications of interstudy and interindividual variability in the developmental neurotoxicity of PCBs. *Neurotoxicol. Teratol.* **18,** 257–264.

56. Schantz, S. L. (1996). Developmental neurotoxicity of PCBs in humans: What do we know and where do we go from here? *Neurotoxicol. Teratol.* **18,** 217–227.

57. Jacobson, J. L., Jacobson, S. W., and Humphrey, H. E. B. (1990). Effects of in utero exposure to polychlorinated biphenyls and re-

lated contaminants on cognitive functioning in young children. *J. Pediatr.* **116,** 38–45.

58. Rice, D. C. (1996). PCBs and behavioral impairment: Are there lessons we can learn from lead? *Neurotoxicol. Teratol.* **18,** 229–232.

59. Tanabe, S. (1988). PCB problems in the future: Foresight from current knowledge. *Environ. Pollut.* **50,** 5–28.

60. Urabe, H., Koda, H., and Asahi, M. (1979). Present state of Yusho patients. *Ann. N.Y. Acad. Sci.* **320,** 273–276.

61. Hsu, S.-T., Ma, C.-I., Hsu, S.-H., Wu, S. S., Hsu, N.-M., Yeh, C. C., and Wu, S. B. (1985). Discovery and epidemiology of PCB poisoning in Taiwan: A four year follow up. *Environ. Health Perspect.* **59,** 5–10.

62. Wrensch, M., Swan, S., Lipscomb, J., Epstein, D., Fenster, L., Claxton, K., Murphy, P. J., Shusterman, D., and Neutra, R. (1990). Pregnancy outcomes in women potentially exposed to solvent-contaminated drinking water in San Jose, California. *Am. J. Epidemiol.* **131,** 283–300.

63. Center for Disease Control, Ward, J. W. (ed.) (1996). Spontaneous abortions possibly related to ingestion of nitrate-contaminated well water -LaGrange County, Indiana, 1991–1994. *Morbid. Mortal. Wkly. Rep.* **45,** 569–572.

64. Hertz-Picciotto, I., Swan, S. H., Neutra, R. R., and Samuels, S. J. (1989). Spontaneous abortions in relation to consumption of tap water: An application of methods from survival analysis to a pregnancy follow-up study. *Am. J. Epidemiol.* **130,** 79–93.

65. Cohn, P. D., Fagliano, J. A., and Klotz, J. B. (1994). Assessing human health effects from chemical contaminants in drinking water. *N. J. Med.* **91,** 719–722.

66. Savitz, D. A., Andrews, K. W., and Pastore, L. M. (1995). Drinking water and pregnancy outcome in central North Carolina: Source, amount, and trihalomethane levels. *Environ. Health Perspect.* **103,** 592–596.

67. Winder, C. (1993). Lead, reproduction and development. *Neurotoxicology* **14,** 303–317.

68. Swan, S. H., Beaumont, J. J., Hammond, S. K., VonBehren, J., Green, R. S., Hallock, M. F., Woskie, S. R., Hines, C. J., and Schenker, M. B. (1995). Historical cohort study of spontaneous abortion among fabrication workers in the semiconductor health study: Agent-level analysis. *Am. J. Ind. Med.* **28,** 751–769.

69. Savitz, D. A., Harley, B., Krekel, S., Marshall, J., Bondy, J., and Orleans, M. (1984). Survey of reproductive hazards among oil, chemical, and atomic workers exposed to halogenated hydrocarbons. *Am. J. Ind. Med.* **6,** 253–264.

70. Romero, P., Barnett, P. G., and Midtling, J. E. (1989). Congenital anomalies associated with maternal exposure to oxydemetonmethyl. *Environ. Res.* **50,** 256–261.

71. Ahlborg, G., Jr., and Hemminki, K. (1995). Reproductive effects of chemical exposures in health professions. *J. Occup. Environ. Med.* **37,** 957–961.

72. Kersemaekers, W. M., Roeleveld, N., and Zielhuis, G. A. (1995). Reproductive disorders due to chemical exposure among hairdressers. *Scand. J. Work Environ. Health* **21,** 325–334.

73. Solomon, G. M., Morse, E. P., Garbo, M. J., and Milton, D. K. (1996). Stillbirth after occupational exposure to N-methyl-2-pyrrolidone. A case report and review of the literature. *J. Occup. Environ. Med.* **38,** 705–713.

74. Sever, L. E., Arbuckle, T. E., and Sweeney, A. (1997). Reproductive and developmental effects of occupational pesticide exposure: the epidemiologic evidence. *Occup. Med.* **12,** 305–325.

75. Goldsmith, J. R. (1997). Dibromochloropropane: Epidemiological findings and current questions. *Ann. N. Y. Acad. Sci.* **837,** 300–306.

76. Elghany, N. A., Stopford, W., Bunn, W. B., and Fleming, L. E. (1997). Occupational exposure to inorganic mercury vapour and reproductive outcomes. *Occup. Med. (Oxford)* **47,** 333–336.

77. Kersemaekers, W. M., Roeleveld, N., and Zielhuis, G. A. (1997). Reproductive disorders among hairdressers. *Epidemiology* **8,** 396–401.

78. Petrova, A., Gnedko, T., Maistrova, I., Zafranskaya, M., and Dainiak, N. (1997). Morbidity in a large cohort study of children born to mothers exposed to radiation from Chernobyl. *Stem Cells* **15,** 141–150.

79. Cordier, S., Bergeret, A., Goujard, J., Ha, M. C., Ayme, S., Bianchi, F., Calzolari, E., De Walle, H. E., Knill-Jones, R., Candela, S., Dale, I., Dananche, B., de Vigan, C., Fevotte, J., Kiel, G., and Mandereau, L. (1997). Congenital malformation and maternal occupational exposure to glycol ethers. Occupational exposure and congenital malformations working group. *Epidemiology* **8,** 355–363.

80. Cordier, S., Deplan, F., Mandereau, L., and Hemon, D. (1991). Paternal exposure to mercury and spontaneous abortions. *Br. J. Ind. Med.* **48,** 375–381.

81. Thonneau, P., Ducot, B., and Spira, A. (1993). Risk factors in men and women consulting for infertility. *Int. J. Fertil. Menopausal Stud.* **38,** 37–43.

82. Radowicki, S. and Wierzba, W. M. (1997). The duration of pregnancy in ecologically challenged area. *Ginekol. Pol.* **68,** 53–58.

83. Lindbohm, M. L., Hemminki, K., Kyyronen, P., Kilpikari, I., and Vainio, H. (1983). Spontaneous abortions among rubber workers and congenital malformations in their offspring. *Scand. J. Work Environ. Health* **9,** 85–90.

84. Ahlborg, G., Jr., Hogstedt, C., Bodin, L., and Barany, S. (1989). Pregnancy outcome among working women. *Scand. J. Work Environ. Health* **15,** 227–233.

85. Vandel, P., Bizouard, P., Vandel, S., David, M., Nezelof, S., Bonin, B., François, T., Bertschy, G., and Sechter, D. (1995). Undesirable effects of drugs. Epidemiologic study at a psychiatric service of a university hospital. *Therapie* **50,** 67–72.

86. Grimm, R. H., Jr., Grandits, G. A., Prineas, R. J., McDonald, R. H., Lewis, C. E., Flack, J. M., Yunis, C., Svendsen, K., Liebson, P. R., and Elmer, P. J. (1997). Long-term effects on sexual function of five antihypertensive drugs and nutritional hygienic treatment in hypertensive men and women. Treatment of Mild Hypertension Study (TOMHS). *Hypertension* **29,** 8–14.

87. Campbell, B. C., and Udry, J. R. (1994). Implications of hormonal influences on sexual behavior for demographic models of reproduction. *Ann. N.Y. Acad. Sci.* **709,** 117–27.

88. Meston, C. M., and Gorzalka, B. B. (1996). Differential effects of sympathetic activation on sexual arousal in sexually dysfunctional and functional women. *J. Abnorm. Psychol.* **105,** 582–591.

89. Koukounas, E., and McCabe, M. (1997). Sexual and emotional variables influencing sexual response to erotica. *Behav. Res. Ther.* **35,** 221–230.

90. Meyer-Bahlburg, H. F., Ehrhardt, A. A., Feldman, J. F., Rosen, L. R., Veridiano, N. P., and Zimmerman, I. (1985). Sexual activity level and sexual functioning in women prenatally exposed to diethylstilbestrol. *Psychosom. Med.* **47,** 497–511.

91. Pieleszek, A. (1997). The effect of carbon disulphide on menopause, concentration of monoamines, gonadotropins, estrogens and androgens in women. *Ann. Acad. Med. Stetin.* **43,** 255–267.

92. Hui, S. I., Slemenda, C. W., Johnston, C. C., and Appledorn, C. R. (1987). Effects of age and menopause on vertebral bone density. *Bone Miner.* **2,** 141–146.

93. Johnston, C. C., Hui, S. L., Witt, R. M., Appledorn, C. R., Baker, R. S., and Longcope, C. (1985). Early menopausal changes in bone mass and sex steroids. *J. Clin. Endocrinol. Metab.* **61,** 905–911.

94. Nilas, L., and Christiansen, C. (1989). The pathophysiology of peri- and postmenopausal bone loss. *Br. J. Obstet. Gynecol.* **96,** 580–587.

95. Bagur, A. C., and Mautalen, C. A. (1992). Risk for developing osteoporosis in untreated premature menopause. *Calcif. Tissue Int.* **51,** 4–7.

96. Matthews, K. A., Meilahn, E., Kuller, L. H., Kelsey, S. F., Caggiula, A. W., and Wing, R. R. (1989). Menopause and risk factors for coronary heart disease. *N. Engl. J. Med.* **321,** 641–646.

97. Kannel, W. B., Hjortland, M. C., McNamara, P. M., and Gordon, T. (1976). Menopause and risk of cardiovascular disease: The Framington Study. *Ann. Intern. Med.* **85,** 447–452.

98. Dennerstein, L. (1987). Depression in the menopause. *Clin. North Am.* **14,** 13–18.

99. Winokur, G. (1973). Depression in the menopause. *Am. J. Psychiatry* **130,** 92–93.

100. Brincat, M., Kabalan, S., and Studd, J. W. (1985). A study of the decrease of skin content, skin thickness, and bone mass in postmenopausal women. *Obstet. Gynecol.* **70,** 840–845.

101. Versi, E. (1990). Incontinence in the climatic. *Clin. Obstet. Gynecol.* **33,** 392–398.

102. Iosif, S., Henriksson, L., and Ulmsten, U. (1981). The frequency of disorders of the lower urinary tract, urinary incontinence in particular, as evaluated by questionnaire survey in a health control population. *Acta Obstet. Gynecol.* **60,** 71–76.

103. Linos, A., Worthington, J. W., O'Fallon, W. M., and Kurland, L. T. (1980). The epidemiology of rheumatoid arthritis in Rochester, Minnesota: A study of incidence, prevalence, and mortality. *Am. J. Epidemiol.* **111,** 87–98.

104. Hannan, M. T., Felson, D. T., Anderson, J. J., Naimark, A., and Kannel, W. B. (1990). Estrogen use and radiographic osteoarthritis of the knee in women: The Framington Arthritis Study. *Arthritis Rheum.* **33,** 525–532.

105. Coulam, C. B., Adamson, S. C., and Annegers, J. F. (1986). Incidence of premature ovarian failure. *Obstet. Gynecol.* **67,** 604–606.

106. Neugarten, B. L., and Kraines, R. J. (1965). Menopausal symptoms in women of various ages. *Psychosom. Med.* **27,** 266–273.

107. Gardsell, P., Johnell, O., and Nilsson, B. E. (1991). The impact of menopausal age on future fragility fracture risk. *J. Bone Miner. Res.* **6,** 429–433.

108. Richardson, S. J., and Nelson, J. F. (1990). Follicular depletion during the menopausal transition. *Ann. N.Y. Acad. Sci.* **592,** 13–20.

109. Richardson, S. J., Senikas, V., and Nelson, J. F. (1987). Follicular depletion during the menopausal transition: evidence for accelerated loss and ultimate exhaustion. *J. Clin. Endocrinol. Metab.* **65,** 1231–1237.

110. Cramer, D. W., Barbieri, R. L., Xu, H., and Reichardt, K. V. (1994). Determinants of basal follicle-stimulating hormone levels in premenopausal women. *J. Clin. Endocrinol. Metab.* **79,** 1105–1109.

111. Cooper, G. S., Baird, D. D., Hulka, B. S., Weinberg, C., Savitz, D. A., and Hughes, C. L. (1995). Follicle-stimulating hormone concentrations in relation to active and passive smoking. *Obstet. Gynecol.* **85,** 407–411.

112. Mattison, D. R., and Thorgeirsson, S. S. (1978). Smoking and industrial pollution, and their effects on menopause and ovarian cancer. *Lancet* **1,** 187–188.

113. Staropoli, C. A., Flaws, J. A., Bush, T. L., and Moulton, A. (1998). Predictors of menopausal hot flashes. *J. Women's Health* **7,** 1149–1155.

114. National Research Council (1989). "Biologic Markers in Reproductive Toxicology." National Academy Press, Washington, DC.

115. Mortimer, L., Mendesohn, J. P., and Normandy, M., eds. (1995). "Biomarkers and Occupational Health." Joseph Henry Press, Washington, DC.

51

Environmental Exposures and Cancer

EILEEN V. MOY AND DAVID C. CHRISTIANI
Harvard School of Public Health
Boston, Massachusetts

I. Introduction

In this chapter, we will present the environmental exposures and issues that are of greatest concern to women in terms of their potential to cause cancer. Most of these environmental exposures pertain to both women and men, but some may have special relevance to women because of the enhanced susceptibilities of female organs or because of gender-related social issues.

We will also explore some of the interesting questions being framed at the frontiers of research on environmental cancer. Since the 1950s, the scientific community has attempted to identify specific carcinogens in the environment and quantify the cancer risks from given levels of exposure to these agents. Future risk paradigms, however, should go beyond issues of exposure and begin to incorporate the concept that individuals carry different risks of developing cancer.

The information on environmental exposures and cancer is incomplete, and many misconceptions prevail in the public sphere. Historically, the medical community largely overlooked gender as a potential effect modifier of disease. New data are becoming available because of the emergence of women's health initiatives. However, the reconstruction of temporal trends has been difficult, and information gaps persist because environmental cancer is rapidly becoming a disease of developing countries.

II. Definition of the Topic: Environmental Exposures and Cancer

The attribution of cancer to environmental causes has been difficult. There are uncertainties because of the long latency period in the development of cancer, the difficulties in characterizing past exposures, the lack of scientific evidence about the cancer risk of most environmental toxicants, and the growing awareness of cancer-predisposing genes. It seems that cancer reflects the confluence of many processes. In the simplistic two-step model of cancer, the development of clinical cancer requires both initiation and promotion, but the latency period between initiation and promotion could be as long as thirty years.

Table 51.1 contains a list of cancers that have been linked to endogenous influences that can be classified as "environmental" factors. In this list, solar radiation, active smoking, environmental tobacco smoke, ionizing radiation, radon, and arsenic are established environmental carcinogens. The remaining agents are best known as occupational carcinogens because workers who were exposed to high levels of these agents experienced high rates of cancer. The general public, especially populations residing in the vicinity of industries that emit these agents, may sustain lower levels of exposures through air, food, and water and consumer products. Currently, investigators are unable to

Table 51.1

Some Cancers and Their Associated Environmental Exposures

Cancer	Environmental exposure
Skin	Solar radiation
	Arsenic
	Coal tar
	Soot
Lung	Environmental tobacco smoke
	Asbestos
	Ionizing radiation—radon
	Arsenic
	Nickel
Pleura (lining of the lung)	Asbestos
Breast	Ionizing radiation
Thyroid	Ionizing radiation
Nasal cavity and sinus	Chromium
	Nickel
	Wood dust
	Leather dust
Liver	Arsenic
	Vinyl chloride
	Aflatoxin
Bone marrow	Ionizing radiation
	Benzene
Bladder	Arsenic
	Aromatic amines

assign definite cancer risks from chronic, low-level exposures to these carcinogens.

Exposure to environmental carcinogens can also occur as a result of direct contact with agents at abandoned or unsecured sites, volatilization or conversion of agents to dust which is subsequently carried in air, seepage of agents into groundwater used by wells, contamination of the food chain through marine organisms living in polluted waters and plants and animals living on contaminated soil, and dissemination of agents by fire and explosion [1].

III. Epidemiologic Evidence

A. Sunlight Exposure and Cancer

Sunlight exposure ranks as the most important environmental carcinogen, as there is strong evidence that excessive sun exposure causes skin cancer. The component of sunlight thought to be responsible for skin cancer is UVB, ultraviolet radiation in the range of 280 to 315 nm. UVB presumably induces DNA damage, which may in turn lead to genetic mutations. These mutations can culminate in three types of skin cancer: malignant melanoma, basal cell carcinoma, and squamous cell carcinoma [2]. Some epidemiologic studies are suggesting that patients with basal cell and squamous cell skin cancers are more susceptible to developing other cancers [3,4], a possible consequence of UVB-induced generalized immune suppression.

There are six lines of evidence supporting the proposition that excessive sun exposure is highly correlated with skin cancer: (1) skin cancer occurs more frequently in residents of areas of high solar radiation; second, skin cancer occurs more frequently in sun-sensitive people; (3) skin cancer occurs more frequently in sun-exposed body sites; (4) skin cancer occurs more frequently in people with a history of sunburn; (5) skin cancer occurs more frequently in people who have a benign sun-related skin condition; and (6) skin cancer is reduced by protection of skin against the sun [2]. However, the relationship between sunlight exposure and skin cancer is complex. Paradoxically, occupations that involve extensive sun exposure do not necessarily correlate with increased incidence or mortality from melanoma [5] or basal cell carcinoma [6]. These and other inconsistencies may reflect the inherent difficulty in characterizing the exact patterns of sun exposure over a lifetime.

Malignant melanoma is the worst form of skin cancer, and its worldwide incidence has been increasing since the 1950s. There is disagreement about the ultraviolet waveband (action spectra) of malignant melanoma. It appears that sunscreens may not be effective in protecting against malignant melanoma. People who use sunscreens seem to be exposing themselves to sunlight for longer hours, thus dramatically increasing their exposure to melanoma-inducing wavelengths [7].

B. Environmental Tobacco Smoke, Active Smoking, and Cancer

Environmental tobacco smoke (ETS), also commonly known as second-hand smoke or passive smoke, originates from the exhalations of smokers and the sidestream emissions from cigarettes, cigars, or pipes between smokes. ETS contains all of the established and suspected carcinogens found in mainstream smoke generated by active smoking [8,9].

The medical community in the United States, by tradition, viewed ETS as a woman's health issue because women in the United States rarely smoked until the 1960s [10]. Conveniently, men who smoked were the focus of studies on active smoking and lung cancer, while never-smoking women married to men who smoked were the study population of choice for ETS and lung cancer. The pooled results from these studies to date are showing a strong association between active smoking and lung cancer in men (relative risk [RR] = 13) and a weak association between ETS and lung cancer in women (RR = 1.5) [11,12].

In 1993, officials at the United States Environmental Protection Agency (EPA) designated ETS as a Group A Carcinogen, a category strictly reserved for known human carcinogens [13]. Critics of the designation argue that the epidemiologic evidence does not support such a strong designation. Supporters, on the other hand, counter that even if the current epidemiologic evi-

Table 51.2

Major Established and Suspected Carcinogens in Mainstream Smoke

Benzene	2-naphthalamine
Formaldehyde	4-aminobiphenyl
Hydrazine	Benz[a]anthracene
Butadiene	Benz[a]pyrene
N-nitrosamines	NNK
Aniline	

From [13].

dence tends to be weak, the biological plausibility of an association between environmental tobacco smoke and lung cancer cannot be disregarded [11].

ETS appears to be more compelling as a cause of lung cancer than other environmental exposures. Some researchers estimate that nonsmokers have a 24% higher risk of developing lung cancer from exposure to ETS than from indoor radon and maintain that this risk is 57 times greater than the combined estimated lung cancer risk from all of the hazardous outdoor air pollutants currently regulated by the EPA: airborne radionuclides, asbestos, arsenic, benzene, coke oven emissions, and vinyl chloride [14].

There is also great interest in the possible association between ETS and cancers of other sites, as there is evidence that active cigarette smoking causes cancers of the head and neck, esophagus, stomach, pancreas, kidney, bladder, and possibly cancer of the uterine cervix and leukemia [12,15,16]. Some limited data are suggesting an association between ETS and head and neck cancers [17].

C. Air Pollution and Lung Cancer

Both researchers and the general public have long suspected air pollution to be a cause of cancer, in particular lung cancer. However, there are many difficulties in identifying airborne carcinogens and calculating their cancer risks. Outdoor air pollution arises mainly from environmental tobacco smoke and products of incomplete combustion of fossil fuels, including automobile exhaust and coal emissions [18,19]. Over 2800 chemicals have been identified in air or emission sources, but only about 10% of these chemicals have undergone evaluation for their potential to cause genetic or carcinogenic effects, with equivocal results [19]. Moreover, cumulative exposures to air contaminants occurring at low levels over several decades are highly difficult to characterize. Even if it were possible to assign unique cancer risks to specific toxicants, as the EPA did for the main outdoor air pollutants (excluding ETS and radon) in Table 51.2, these cancer risks may change when placed in the context of a mixed chemical environment.

The EPA has promoted the use of the five criteria pollutants listed in Table 51.3 [19a] as standardized measurements of outdoor air pollution, since the number of airborne chemicals is vast. A major weakness to limiting measurements to these five criteria pollutants is the omission of data on many known and suspected carcinogens, such as benzo(a)pyrene. Nevertheless, epidemiologic studies are demonstrating a specific association

Table 51.3

Criteria Pollutants

Carbon monoxide	Sulfur dioxide
Nitrogen dioxide	Suspended particulate matter
Ozone	Toxic metals

From [19a].

between fine respirable particles in polluted air and lung cancer, although the relative risk is low at 1.5 [20]. There may be an interaction between outdoor air pollution and tobacco smoking that significantly increases lung cancer risk in smokers in many of these studies [21]. More supportive evidence of a link between air pollution and lung cancer comes from animal studies in which smoking is not a problem, such as one study in which air pollution acts as a lung cancer promoter in urethane-exposed mice [22].

Some possible flaws in epidemiologic studies of air pollution are incomplete control of confounding factors and incomplete exposure data. Confounders may include smoking and exposure to indoor pollution as well as other unmeasured risk factors [20]. Exposure data can be inaccurate because pollutant levels are often assigned retrospectively [22]. The tendency to use proxy measurements, for example from fixed area monitoring stations [23], can also be problematic.

Indoor air pollution seems to be a significant risk factor for lung cancer in women in developing countries like China, where women commonly sustain exposures to "smoky" coal used for cooking and heating beds, cooking fumes, and volatile chemicals emitted from cooking oils [24–26]. Because Chinese men smoke heavily, these women also experience significant exposures to ETS [27].

The data on air pollution and lung cancer in women is highly incomplete. Historically, investigators have failed to include women as subjects for several of the major, ongoing studies on air pollution and health.

D. Drinking Water and Cancer

Many contaminants in drinking water are suspected carcinogens, including inorganic solutes such as arsenic and nitrates; organic chemicals such as chlorine byproducts and other substances from industrial, agricultural, and domestic sources as well as from hazardous waste sites; radionuclides; and solid particulates such as asbestos. Most of the epidemiologic data linking drinking water contaminants and cancer are equivocal, with a few notable exceptions [28,29]. At this point, there are no meaningful data on the effect of gender on drinking water and cancer.

The evidence is more convincing for waterborne arsenic as a carcinogen. Studies link cancers of the skin (nonmelanoma), bladder, and lung to waterborne arsenic exposure in countries where natural wells and underground aquifers are heavily contaminated with arsenic, such as in Taiwan [30] and Chile [31].

There is also increasing concern about nitrates in drinking water and cancer, in particular cancers of the gastrointestinal tract. Even though the existing epidemiologic data are inconsistent, investigators have observed in the laboratory setting that

nitrates are procarcinogens that can undergo transformation into the known carcinogens N-nitrosamines and N-nitrosamides. The concentration of nitrates in drinking water is increasing worldwide because of the growing use of nitrogen fertilizers and intensive animal husbandry [28,29].

Another concern is chlorination by-products in drinking water and cancer, especially bladder and colon cancers. Chlorination by-products are present in all chlorinated drinking water and arise from complex interactions of chlorine with chemicals from decomposing plant matter and other organic chemicals. The most common class of by-products is the trihalomethanes (THMs). Generally, associations of chlorination by-products with cancer are weak and inconsistent [32,33]. However, the inability to demonstrate any strong associations may be due to confounders and geographical differences in the composition of by-product mixtures.

Contamination of underground and surface waters with organic chemicals such as pesticides, solvents, and chemicals from hazardous waste sites is also on the rise worldwide. Exposure levels are difficult to characterize, as these agents often occur as an admixture [28].

E. Ionizing Radiation

The natural sources of ionizing radiation exposure are radon and cosmic rays. Together, these two sources account for only about 3% of all cancers. Radon alone contributes to over half of all radiation exposures experienced by the general population [34].

Radon is formed from rocks and occurs as a gas. Radon gas can accumulate to high levels in enclosed spaces, such as underground mines, basements, and first floors of buildings. Decay products of radon (radon progeny) can attach to dust particles. When the decay products are inhaled, further decay exposes the epithelia (lining) of the lung to alpha-type irradiation, thus predisposing the tissue to DNA mutations and cancer [34].

Residential radon exposures vary greatly in intensity and depend on geography, the adequacy of ground-level ventilation, and time spent in the exposed area. Residential radon can rise to extremely high levels and probably accounts for over half of all radiation exposures received by the general population. There is epidemiologic evidence that radon causes lung cancer in male underground uranium miners, but epidemiologic data are equivocal for lung cancer and residential radon exposures for both women and men. The data for radon exposure in uranium mines, when extrapolated to the residential setting, suggest that 10–14% of all lung cancer deaths in the United States could be due to indoor radon. In addition, the interaction of radon exposure with smoking greatly enhances lung cancer risk in smokers exposed to radon [35–37].

There are extensive data on the health effects of ionizing radiation from the 1945 Hiroshima and Nagasaki atomic bomb detonations in Japan because the Radiation Effects Research Foundation has been closely following the survivors. There is an excess of many cancers relative to the general population, including significantly higher rates of leukemia, thyroid nodules and cancer, breast cancer, and many other solid tumors. Women are especially affected because both thyroid and breast cancers occur predominantly in women. However, dose, age at exposure, and genetic susceptibility are factors that may alter risk and warrant further investigation. Women who were less than 10 at the time of bombing subsequently developed the highest rate of breast cancer [38,39].

Nuclear facilities have contributed to exposures to ionizing radiation in many parts of the world. High levels of radioactive fallout from nuclear weapons testing and nuclear accidents are linked to increased cancer rates, and the association between exposure and cancer occurrence appears to be dose-related. The accidental exposures to radioactive iodides and gamma radiation during the BRAVO test in the Marshall Islands in 1954 resulted in documented increases in incidence of thyroid cancer or benign thyroid nodules [34,40]. The Chernobyl nuclear reactor accident in the Ukraine in 1986 resulted in massive releases of radioactive iodides into the atmosphere, but the health impact of the accident to the surrounding region is still unclear in view of the short follow-up period [41]. Analysis of data related to other unfortunate nuclear accidents that have occurred in the former Soviet Union since the late 1940s is also taking place [42]. In the United States, the data for cancer and exposure to radioactive fallout resulting from the Nevada Test Site detonations during the 1950s remains equivocal [34,43].

F. Nonionizing Radiation

Low-frequency electromagnetic fields (EMF) can cause human injury through generation of excess heat or by shock from direct contact with electric current. There has been an additional widespread public perception that EMF exposures through power lines, at levels slightly above ambient background, may cause cancer. Researchers have hypothesized that exposure to EMF, pulsed electromagnetic fields, and light (visible electromagnetic fields) may reduce the pineal gland's nocturnal production of the hormone melatonin and increase susceptibility to sex hormone-related cancers such as breast cancer [44].

Despite nearly two decades of studies, there is no clear evidence linking higher-than-background residential or occupational EMF exposures to cancer in either adults or children. Even in studies with large sample sizes, the results have generally been null or inconclusive [45–49]. Exposure misclassification may have weakened the findings, as precise EMF exposure measurements are difficult to characterize.

There have also been ongoing concerns about the use of electric blankets and cancer, notably breast cancer in women, but again the results are inconclusive [50]. Currently, there is no scientific data to support the setting of occupational or environmental standards for exposures to electromagnetic fields.

In the future, there will be more questions about the cancer risks of high-frequency electromagnetic waves. The use of radiowave communication systems has been expanding rapidly in the modern world through the widening use of radio, television, radar, wireless communication, and satellite communication. Currently, one area of active study is the possible association between cellular phones and brain tumors.

IV. Historical Evolution/Trends

A. Lung Cancer and Smoking

The incidence and mortality of lung cancer, a relatively rare disease at the turn of the twentieth century in the United States,

have generally shadowed the trends in cigarette smoking. Initially, a larger percentage of men smoked. During the 1930s, 50% of men and 20% of women smoked. In the ensuing decades, more women began to take up the habit so that by the 1960s, 50% of men and 30% of women smoked. In 1990, smoking had declined in both men and women, but at a slower rate among women; 28% of men and 23% of women smoked [51,52]. There is some data suggesting that some young women, even those currently seeking a college education, may be more likely to smoke than their male counterparts [53].

Figure 51.1 summarizes the lung cancer trends for men and women. For men, there was a dramatic increase in lung cancer during the middle of the 1900s followed by a stabilization of rates toward the end of the century. For women, the increase in lung cancer over the past few decades has not yet peaked by 1999.

There are ethnic, cultural, social class, and geographical differences in smoking prevalence. Smoking prevalence is declining faster in black and Hispanic women compared to white women. However, smoking remains high or is on the rise in women of other ethnic groups, such as American Indians, Alaskan natives, and Cuban-Americans, all of whom have smoking prevalences close to or greater than 50% [54,55]. Furthermore, when comparing different areas of the United States, researchers have noted that both women and men in the southern states seem to smoke more than their counterparts in the western states, possibly because of a difference in level of education [56]. Other researchers have noted that in general, women who have less education [57] or are living in poverty [58,59] may be more likely to smoke and less likely to participate in smoking cessation programs.

The effect of other factors, notably sexual orientation, on cigarette smoking has not been adequately studied. One study that included lesbian women in a southern state showed alarmingly high rates of smoking, with over 40% in any age, educational, or socioeconomic category, although the reasons remain unclear [60].

B. Skin Cancer and Sunbathing

While lung cancer is currently increasing for women, skin cancer is on the rise for both women and men in the United States, northern Europe, and other countries with a significant European migrant population [61]. Stratospheric ozone depletion due to halocarbon emissions in conjunction with an upsurge in popularity of sunbathing since the 1920s seem to account for the increase [62]. However, sunbathing was not a traditional pastime in any culture. The tank style bathing suit and the suntanned look were supposedly invented on the French Riviera during the 1920s and may have changed sunbathing patterns around the world. During the 1960s and 1970s, the fashion media aggressively promoted the suntanned look as the contemporary standard of beauty for American and northern European women. Suntans were equated with high socioeconomic status, leisure, and good health. Later, these perceptions led to the widespread popularity of sunbathing for young women and men of all socioeconomic classes.

With growing awareness of the gradual depletion of the ozone layer and of the deleterious effects of excessive sun exposure, sunbathing seems to be waning in popularity. Since the 1970s, sunbathers have been avoiding midday sun and using

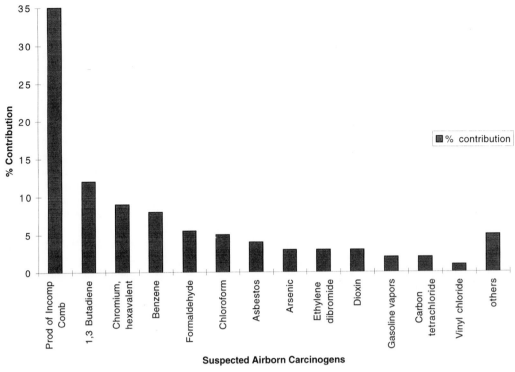

Suspected Airborn Carcinogens

Fig. 51.1

more sunscreens, hats, and protective clothing. Many sunbathers do not realize, however, that high protection factor sunscreens are protective for sunburn but not necessarily for skin cancer [2,63].

C. Comparison of Trends in Developed and Developing Countries

Developed and developing countries have marked differences in cancer rates. In general, for developed countries in North America, Western Europe, East Asia, Oceania, and Nordic countries that have undergone the demographic transition, cancer morbidity and mortality for women have increased, especially for lung and breast cancers, and decreased for gastric cancer during the past few decades [64].

High gastric cancer rates seem to be a marker of a lower state of development and industrialization, with the notable exceptions of Japan and Korea. Among developing countries, gastric cancer is the third most important cancer for women [61]. Gastric cancer has also been linked to poverty and unsanitary living conditions in developed countries, and it is possible that living closer to the natural environment may promote chronic bacterial infection with *Helicobacter pylori,* a significant risk factor for gastric cancer [65]. There are speculations that the transmission of *Helicobacter pylori* may be related to the presence of volatile N-nitrosamines in food [66]. Furthermore, it is possible that the replacement of traditional food storage practices by refrigeration may have resulted in decreased incidence of gastric cancer in many parts of the developing world, except for Japan and

Korea [61]. There may also be genetic susceptibility to gastric cancer, as familial aggregation has been found; however, no genes have been clearly implicated to date [67–69].

Liver cancer is another important cancer whose incidence has decreased markedly in developed countries. Aside from hepatitis B infection, chronic exposure to fungal aflatoxin may play a role in liver cancer. Foods such as peanuts and grains harbor aflatoxin. As heat and moisture enhance aflatoxin growth, underground storage and other traditional methods of storing food in tropical and subtropical countries may play important roles [70,71].

In the former Soviet Bloc, including Hungary and the Czech Republic, there are high rates of many forms of cancer in women, especially lung and breast cancers [64]. Cigarette smoking and aging cannot entirely account for these trends. Unfortunately, epidemiologic studies in Eastern Europe have been inadequate, despite the high level of disease.

As a whole, women in developing countries are more likely to experience and be affected by the environmental exposures we have described in this chapter. UVB radiation and skin cancer remain mainly a problem among European migrants to areas of high ambient solar radiation. Indoor and outdoor air pollution are problems of increasing magnitude. While coal burning and poor ventilation are age-old phenomena, active tobacco smoking and environmental tobacco smoke exposure are relatively new to developing countries. It is likely that the increases in lung cancer incidence and mortality in some developing countries are due in part to interactions between air pollution and tobacco smoke [22].

In countries such as Singapore, where rapid industrialization is taking place and whose population is adopting many aspects of a Western lifestyle, there has been a dramatic shift in the pattern of cancer incidence. Stomach and esophageal cancer rates have declined, but lung, colon, rectum, skin (excluding melanoma), breast, and ovary cancer rates are rising [72]. In particular, the incidence of breast cancer has risen sharply [73]. These cancer trends are likely to spread to other countries as they become industrialized. In countries where industrialization is occurring at an uneven pace, patterns of cancer occurrence typical of both developed and developing countries may coexist. In the most extreme case, uncontrolled environmental exposures coupled with increasing disparities and subsistence-level poverty in some developing countries may cause increases in all types of cancers.

V. Distribution of Environmental Cancer in Women, Including Incidence, Prevalence, and Mortality Statistics

The worldwide incidence and mortality from all cancers are higher in men than in women. Incidence and mortality from lung cancer, however, are rising in women in both developed and developing countries. Skin cancer and lung cancer rank as the most important environmental cancers, and their incidence and mortality in selected countries are shown in Figures 51.2 and 51.3. The highest incidence and mortality from skin cancer in women occur in the most genetically-susceptible populations: fair-skinned migrants who settle near the equator where UVB irradiation is most intense, such as in countries like New Zealand and Australia. In Figure 51.2, a comparison of Maori and non-Maori New Zealanders shows a dramatic difference based on ethnicity. On the other hand, Figure 51.3 demonstrates that the highest incidences of lung cancer occur generally in women living in industrialized countries who were smokers or were exposed to environmental tobacco smoke for several decades; North America, Western Europe, and some Nordic countries fall into this category [2,64,72,74].

VI. Host Determinants

Host factors of potential importance to environmentally induced cancer in women include constitutional factors, such as hormonal and biochemical factors, and genetic factors represented by race and ethnicity. It is likely that a multiplicity of factors serve to either predispose individuals to or protect individuals against environmentally induced cancers.

Constitutional factors related to female gender that are important to environmental cancer are ill defined at this time. In the United States and in many other developed countries, women live longer and experience less age-adjusted mortality from cancer [72,74,75]. However, it is possible that estrogens in women, either endogenous or exogenous, can affect metabolism of toxicants by the liver P450 enzyme systems. Differences in metabolism may be either deleterious or protective, as toxicants may be either direct carcinogens or procarcinogens that require conversion to their active forms. Normal cellular processes may also produce carcinogens or procarcinogens. Therefore, cancer can also be viewed as a consequence of aging and/or decreased capacity for handling exogenous and endogenous carcinogens.

At the present time, only skin cancer is widely accepted to have a strong genetic component. All three types of skin cancer occur predominantly in light-skinned populations of European origin. In the United States, the incidence of malignant melanoma is 20-fold higher in whites than in blacks. In Australia and New Zealand, northern European migrants experience exceptionally high rates of malignant melanoma, basal cell carcinoma, and squamous cell carcinoma [2,72].

As UVB radiation appears to cause DNA damage, effective DNA repair seems to be a major defense mechanism against skin cancer. Investigators have gained some insights into the importance of DNA repair by studying people with a genetically-determined disorder known as xeroderma pigmentosum (XP), which is the manifestation of a deficiency in the repair of photoproducts formed in DNA after exposure to ultraviolet radiation. In normal individuals, these photoproducts are excised by intact DNA repair systems. By contrast, these photoproducts persist in individuals with XP and cause mutations that subsequently result in multiple basal cell and squamous cell carcinomas of the skin at an early age [76].

Lung cancer is another disease in which host biological factors may be important. Even though active or passive smoking and air pollution do not seem to carry gender-specific significance for cancer in women, it is possible that women respond differently to environmental carcinogens because of hormonal influences on enzymes that either activate or detoxify these carcinogens. There is also some evidence of genetic susceptibility to lung cancer; family history is related to risk of lung cancer in nonsmoking women, and a threefold increase in risk was found in women with mothers or sisters diagnosed with lung cancer [77]. The specific genes that confer susceptibility to lung cancer are probably numerous and have not yet been clearly identified.

VII. Influence of Women's Social Roles or Context on Environmental Cancer

Gender inequalities of health originate in traditional society where health practices and concepts reflect the subordinate social status of women. In many cultures, women's illnesses are often attributed to behavioral lapses rather than to external causes [78].

Initially, the rising popularity of cigarette smoking and sunbathing in women may have been related to the relaxation of restrictions on women's behavior and the opening up of opportunities for women to explore activities once reserved for men. Another inducement for smoking may have stemmed from the perception that smoking is an effective method of weight control [10]. Tobacco companies have strongly reinforced these and other perceived social advantages of smoking through cigarette advertising, as these companies have found women, especially younger, less educated, and disadvantaged women, to be viable targets of advertising. The glamorization of smoking is especially flourishing under economic globalization. Women who smoke in cigarette advertisements tend to be young, sophisticated, extroverted, wealthy, athletic, independent, and physically alluring. Indeed, these characteristics can also describe women in bathing suit or suntan lotion advertisements.

This projected view of women who smoke is in harsh conflict with reality. Since the 1970s, there has been a feminization of poverty. In the United States, one-third of families headed by

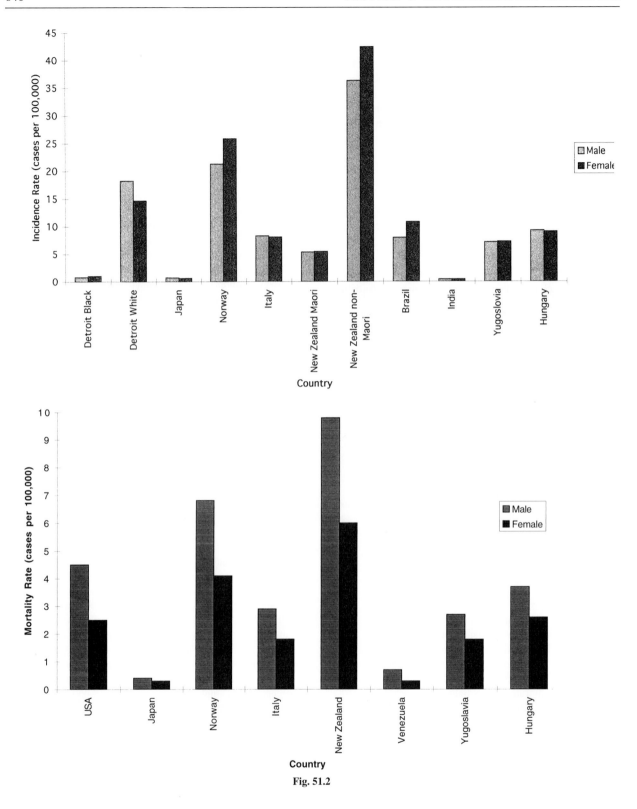

Fig. 51.2

women currently live in poverty, and the fraction is greater than one-half for African-American and Latina women. In general, women are more likely to have low-paying, nonunion jobs that do not provide health insurance, which may result in differences in the quality of health care delivery [79] as well as preventive health services. Lesbian women may experience additional barriers to both preventive health care and primary care services because of social discrimination. Furthermore, women in devel-

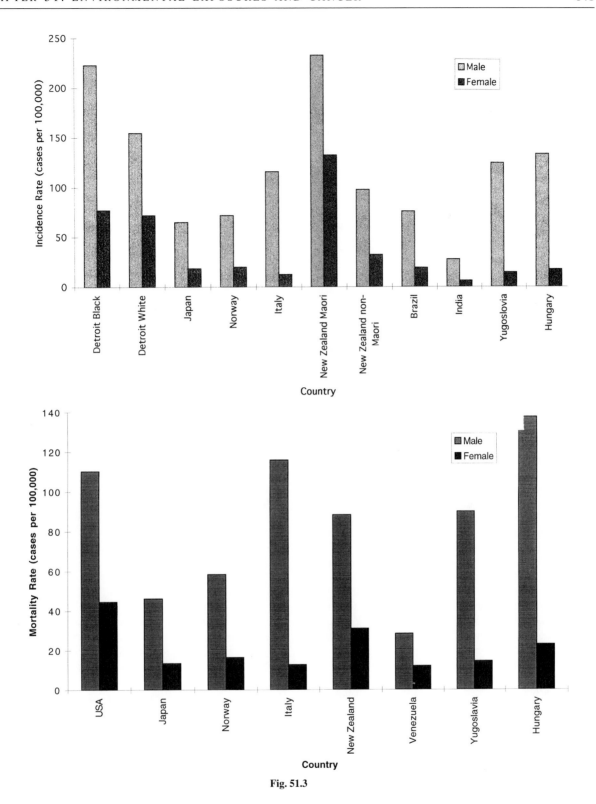

Fig. 51.3

oping countries may be facing unprecedented levels of contamination in air, water, and food in the future. Increasingly, women may be bearing an increasing burden of environment exposures because of poverty and lack of power, education, and mobility.

VIII. Risk Factors

The risk factors for environmentally induced cancer in women in developed and developing countries are summarized in Table

Table 51.4

Risk Factors for Environmentally Induced Cancer in Developed and Rapidly Developing Countries

Level of susceptibility	Associated risk factors
Genetic	Unidentifiable at this time
Hormonal	Uncertain at this time
Behavioral	Smoking
	Excessive sun exposure
	Excessive alcohol
	Excessive ingestion of saturated fats
Environmental	Exposure to environmental tobacco smoke
	Exposure to coal burning emissions
	Exposure to water contaminated with arsenic
	Exposure to severe air pollution

51.4. An expanded discussion of gene–environmental interaction follows in Section IX.

IX. Biological Markers of Susceptibility or Exposure

Several lines of research involving biomarkers of exposure and biologically effective dose relative to lung cancer are taking place simultaneously. One strategy focuses on the detection and validation of adducts in lung and surrogate tissues that are formed when metabolites of cigarette smoke or other environmental carcinogens react with DNA; important adducts include PAH adducts and 4-aminobiphenyl-hemoglobin adducts. Another approach is to find mutations in specific genes that can be used as "fingerprints" of specific exposures, as surrogate end points of cancer, or for further subtyping of cancer to clarify causal associations. Genes of interest include the liver P450 metabolizing genes, the tumor-suppressor genes (p53), and oncogene families *ras* and *myc* [80,81].

Liver P450 metabolic enzymes can either activate or detoxify carcinogens; metabolism of carcinogens may be important to lung cancer susceptibility. Currently, investigators are trying to find out if genetic polymorphisms within the genes that encode different metabolic enzymes, such as CYP1A1, CYP2D6, and GSTM1, can explain some of the variability in lung cancer susceptibility among different ethnic groups [80,82,83].

Gene–environment interactions may be important to cancer susceptibility. For example, mutations in the p53 gene may interact with both asbestos exposure and cigarette smoking in causing lung cancer [81].

Increasingly, researchers are looking for ways to elucidate factors that modestly increase the risk of complex diseases. Traditional monogenic susceptibility to cancer is related to rare diseases or small subgroups of common diseases. Polygenic susceptibility involves frequently occurring genetic polymorphisms, confers a low to medium elevation of risk for frequent diseases, and is identified through case-control studies. These factors are difficult to detect in epidemiologic studies because the relative risk for low-level exposed, versus unexposed, people is only slightly elevated. Complicating this problem is the generic aim of exposure-dose assessment in field studies; imprecise methodology to quantify exposures, especially exposures

that may have occurred many years ago, can lead to misclassification and bias. Moreover, the temporal pace of the traditional cohort study, which must await data from large populations exposed for long periods, has become an obstacle [80].

Since the concept of molecular epidemiology was popularized in the early 1980s to describe an evolving approach to human environmental health research, much work has been done on biomarker development. The relationship of biomarkers to exposure, susceptibility, and disease is shown in Figure 51.4. The theoretical foundation of this work relies on three biologic tenets: (1) Early biologic effects from a toxic exposure are more prevalent and detectable in the population at risk than clinical disease is; (2) With technological advances, many toxins can be either directly quantified in body fluids or tissues or indirectly measured by identification of some predictable, dose-related (early) biological response; and (3) either (or both) of the above may be influenced by susceptibility phenomena (heritable or nonheritable), which can also be accounted for with recent advances in molecular biology. Combining these three principles, molecular epidemiology represents a new approach to environmental health research that expands the traditional toxicologic or epidemiologic paradigm [80].

Biomarkers of exposure, effect, and individual susceptibility may encompass (1) DNA adducts; (2) mutations within cancer-related genes, including tumor suppressor genes and DNA repair genes; and (3) mutations or aberrant levels of cancer-related proteins [85]. There are several limitations to the use of biomarkers. Many of the proposed methods used for biomarker detection have not been extensively validated, and it is not known to what extent these biomarkers can be used for cancer risk assessment. In the future, biomarkers with both high sensitivity and high specificity will be important in the screening of mutations in a reliable and rapid fashion within at-risk groups and in the general population. At the present time, DNA adducts may be a promising biomarker, although it will be important to distinguish between DNA adducts formed by normal cellular processes and adducts resulting from environmental exposures. DNA adducts may prove to be the best biomarker for characterizing exposures to complex mixtures [86].

Ideally, biomarkers should be able to account for endogenous (metabolic/hormonal) as well as exogenous (environmental) exposures and individual responses to these exposures. We are also hoping that biomarkers will provide better tools for evaluating the contribution of gender to cancer susceptibility or protection.

X. Clinical Issues

Patients and health care practitioners often question whether there are associations between past environmental exposures and the manifestation of cancer. For most environmental exposures, the associations with cancers are unclear. However, the clinical setting has the potential to yield important new information on environmental exposures and cancer.

Early diagnosis is relevant to the outcome of some cancers, especially malignant melanoma. During the earlier stages, a surgical cure is possible. The challenge is to distinguish malignant melanoma from benign forms of skin cancer. Women more frequently have malignant melanoma on their legs, where self-recognition and prognosis are better, whereas men commonly

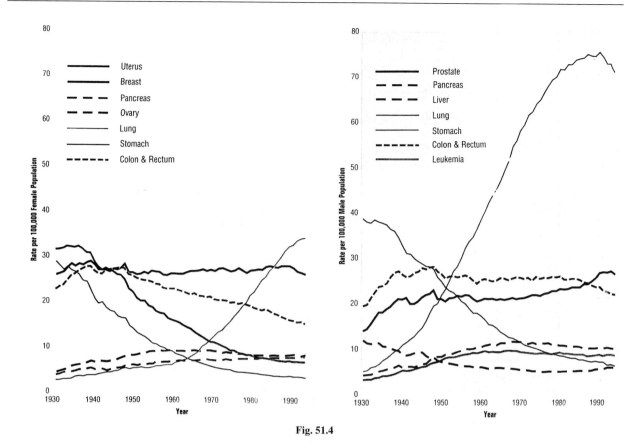

Fig. 51.4

have lesions on their back. Metastasis occurs when tumor cells have spread by the bloodstream or by the lymphatic system. Generally, malignant melanoma is lethal once metastases have occurred, so early detection is essential [87].

For lung cancer, early diagnosis is rarely possible. Lung cancer is a clinically silent disease. Most people present to clinicians at a late stage, when they have symptoms, and often when metastases have already occurred. Radiographic screening is being considered in high-risk men over 45 who smoke greater than 40 cigarettes per day. However, there may not be any difference in survival when screened and unscreened men are diagnosed [88]. Similarly, radiographic screening may be of little value to women who are at risk of lung cancer.

Given the lack of treatment for advanced melanomas and lung cancers, primary prevention of environmental cancer is an important clinical issue. Even though the etiologic factors for many cancers have not been clearly delineated, preventive health care for an individual patient should take into consideration the model of gene–environment interactions in disease formation. Information should be obtained on diet, exercise, smoking history, occupational and environmental exposures, and family history. Advice on prevention should emphasize the awareness and avoidance of known environmental carcinogens as much as possible.

On the other hand, the careful clinical follow-up of patients who have sustained exposures to known environmental carcinogens is important. For example, since the Chernobyl accident

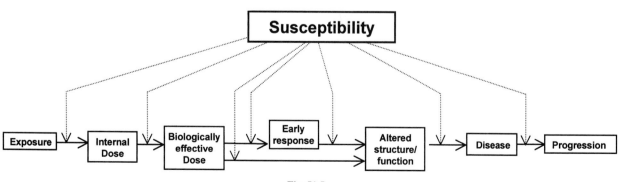

Fig. 51.5

in the former Soviet Union in 1986, clinicians worldwide have been encountering patients with extreme concerns about the health consequences of ionizing radiation. This phenomenon has become widespread because of the large geographical scale of the exposure and extensive emigration from the affected areas. The clinical setting is a good venue for educating patients about the risks of environmental exposures, including the dispelling of some common misconceptions. In addition, the meticulous follow-up of Chernobyl victims is essential for early disease detection and accurate disease reporting and surveillance. For female victims, particular importance should be placed on the breast and thyroid.

It is possible that there are differences in the early detection and clinical evaluation and treatment of men and women with cancer. Although women experience less mortality from many diseases, including cancer, they seem to have more morbidity or complications from disease. Women are also less likely to receive invasive procedures, yet they make more visits to physicians and utilize more pharmaceutical agents [79].

XI. Epidemiologic Issues

Epidemiologic studies have rarely been able to show a causal relationship between exposure to an environmental toxin and cancer. The data are most convincing for cigarette smoking and cancer and UV exposure and skin cancer. However, even the association between environmental tobacco smoke and lung cancer is controversial and is more convincing from the perspective of biological plausibility [11].

Even when epidemiologic studies are able to show a positive association, the results are not always repeatable; other studies may show negative or equivocal results. These inconsistencies may be due to selection bias through the use of study populations with different inclusion criteria or to confounding factors. Smoking is a classic confounder, especially when studying an airborne carcinogen and lung cancer.

There are two major limitations to exposure assessment. The first limitation is possible misclassification of exposure levels. It is thought that cancers generally arise after at least 30 years of exposure. Prospective measurements of exposure levels are not feasible, so retrospective studies are often undertaken. However, the proxy measurements used in retrospective protocols are rarely representative of the actual exposures experienced by the study population. Another problem is that environmental toxins often occur as a complex mixture. Currently, neither the statistical tools nor the biological models of carcinogenesis are adequate to tease out the contribution made by individual components of the mixture [89]. A new and promising approach to human environmental health research is molecular epidemiology and the use of biomarkers to characterize exposure as well as internal biological effects that are in the earlier part of the presumed pathway to cancer.

XII. Is Breast Cancer an Environmental Cancer?

Breast cancer is currently the leading cause of cancer death in women worldwide [61]. For women in the United States, breast cancer constitutes the second most important cause of cancer death after lung cancer [90,91].

Within the United States, there are regional, ethnic, and socioeconomic differences in breast cancer prevalence and mortality. For example, breast cancer mortality in white women is highest in the northeast, but this is not true for black women. This racial difference may argue against the importance of "environmental" risk factors. However, the "lifestyles" practiced by different groups of women may not necessarily exclude different exposures to the physical environment [92].

Time trends seem to be important to breast cancer. Statistics show that breast cancer incidence had been increasing for women born in successive cohorts after 1900 but now is decreasing slightly for women born after 1950 [92]. Therefore, there are likely to be distinct environmental and lifestyle risk factors for the different birth cohorts.

Compared to American women, breast cancer traditionally is a rare disease for Chinese and Japanese women. However, when women from these countries migrate to the United States, exposure to a Western environment/lifestyle has resulted in increasing breast cancer risk over successive generations [93]. Again, the exact changes in risk profile have not been identified.

The etiology of the majority of human breast cancers is multifactorial and poorly understood. Known risk factors such as childbearing practices account for only one-third of the new cases diagnosed in the United States. To date, ionizing radiation seems to be the only plausible environmental etiologic factor. Japanese women who survived the Hiroshima and Nagasaki atomic bomb blasts in 1945 are showing an increased risk of breast cancer when compared to nonexposed controls, but age at time of exposure, reproductive history, and genetic factors may act in concert to modify this risk [39]. The data for breast cancer and nonionizing radiation, as in electromagnetic fields, are inconsistent but warrant further investigation [94].

There have been speculations about endocrine disrupters as a cause of breast cancer. In particular, researchers have been focusing on organochlorine pesticides such as polychlorinated biphenyls (PCBs) and 1,1-dichloro-2,2-bis (p-chlorophenyl) ethylene (DDE) because they mimic the growth promoting effects of estrogen [95]. However, there is lack of conclusive evidence to support this theory, and epidemiologic studies from around the world are showing negative or equivocal results. Notably, in one study based on the Nurses Health Study cohort that controlled for reproductive factors, women who developed breast cancer showed no significant difference in plasma levels of DDE and PCB prior to diagnosis when compared to controls who did not develop breast cancer [96].

The uncertainty surrounding breast cancer etiology compels us to reconsider ETS as a possible contributing factor because of the abundance of carcinogens in cigarette smoke. As mentioned in the last section, the role of cigarette smoking, either passive or active, remains unclear. Many large-scale epidemiologic studies are finding some heightened risk among smokers, especially for heavy smokers who smoke 40 or more cigarettes per day [97] or for 30 or more years [98], but other well-designed studies are detecting no change in risk for any level of smoking [99–102]. The data on ETS exposure and breast cancer is sparser; one study suggests a significantly increased risk [odds ratio (OR) =

3.2] for passive smokers who were exposed for 2 hours per day for 25 years [103].

Genetic variability may also determine an individual's ability to detoxify carcinogens in cigarette smoke. For example, some investigators are focusing on the associations of cytochrome P450E1 genetic polymorphism and tobacco smoking to breast cancer susceptibility [104]. Others are studying the genetic polymorphism within the N-acetyltransferase gene and speculate that a slow acetylator polymorphism that is less favorable to detoxification may confer an increase in breast cancer risk [105].

Another approach to exploring the role of environmental risk factors in breast cancer is to characterize the actual DNA damage in affected breast tissue. It is thought that DNA is damaged through the direct combination of the carcinogen with DNA, forming DNA adducts. Identifying these adducts may allow researchers to deduce the chemical exposures. Because cigarette smoke contains a mixture of adduct-forming carcinogens, it will be difficult to rank the risks of different adducts [106]. Aromatic amine DNA adducts [107], benzo(a)pyrene adducts [108], as well as DNA adducts that are formed endogenously, seem to accumulate in breast tissues with cancer.

As mentioned previously, time of life when exposures take place seems to be important, thus prenatal, neonatal, and adolescent environmental exposures must be characterized. The data from the Japanese atomic bomb survivors and Marshall Islanders suggest that breast cancer risk is heightened among women who were exposed to ionizing radiation as children [109]. It is also possible that transplacental environmental exposures or exposures through the lipid fraction of human milk can affect breast cancer rates in female infants when they become adults [110].

Other research has focused on specific inherited genetic mutations because investigators have been able to establish that family history increases a woman's risk of breast cancer. Mutations in either the BRCA1 or BRCA2 genes may account for a significant percentage of familial breast cancer cases, and the BRCA1 gene may also confer increased susceptibility to combined breast and ovarian cancer [111–113]. Some studies suggest that the prevalence of BRCA1 and BRCA2 mutations may be higher in Ashkenazi Jews [111]. The CYP1A1 gene may also be important to breast cancer, as this gene is involved in estrogen metabolism, and there may be polymorphisms within CYP1A1 that encode for different levels of enzyme expression [114]. In addition to these genes, other genes are currently under investigation for possible links with breast cancer.

The effect of the environment in mediating breast cancer in those who are already genetically susceptible to the disease is unknown. Ultimately, scientists will be exploring new directions for research that will involve a synthesis of both genetic and environmental approaches.

XIII. Conclusions

Cancer remains a major cause of morbidity and mortality, despite several decades of intense research. Based on current knowledge, the most prudent way to prevent environmental cancers is for individuals to avoid active smoking, ETS, and intense sunlight. Even though education holds an important key to preventing new cases of environmentally induced cancer, public concern and resource allocation for reductions in chemical exposures should continue. In actuality, the environmental risk factors for cancer have not been fully characterized.

We believe that developing lung cancer prevention programs and enforcing tobacco legislation should be priorities for all countries, in addition to strengthening research on lung cancer in women. General awareness of active and passive smoking as risk factors for lung cancer in women is low. A survey in the United States, for example, revealed that few books on women's health provide information on lung cancer [115].

The trends toward glamorization of smoking and increased industrialization in the world will result in the increasing release of a multitude of chemicals into the environment. It will be impossible to avoid exposures to cigarette smoke or industrial chemicals, especially in the developing world. One of the health dilemmas of the twenty-first century will be how to attain some sense of environmental equity in the face of massive exportation to developing countries of tobacco as well as heavy industries that emit hazardous chemicals. Regulatory standards for carcinogens differ from country to country, and there is a tendency for more laxity in developing countries. However, none of us will ever be free from the burden of environmental cancer because environmental carcinogens are incorporated into imported products and are dispersed by air and water.

Acknowledgments

This work was supported by NIH grants ES00002 and CA 9863 and the American Cancer Society.

References

1. Hu, H., and Speizer, F. (1994). Physicians and hazards of the environment and workplace. *In* "Harrison's Principles of Internal Medicine" (K. Isselbacher, E. Braunwald, J. Wilson, J. Martin, A. Fauci, and D. Kasper, eds.), 13th ed., Chapter 394. McGraw-Hill, New York.
2. English, D. R., Armstrong, B. K., Kricker, A., and Fleming, C. (1997). Sunlight and cancer. *Cancer Causes Control* **8,** 271–283.
3. Karagas, M. R., Greenberg, E. R., Mott, L. A., Baron, J. A., and Ernster, V. L. (1998). Occurrence of other cancers among patients with prior basal cell and squamous cell skin cancer. *Cancer Epidemiol., Biomarkers Prev.* **7,** 157–161.
4. Frisch, M., and Melbye, M. (1995). New primary cancers after squamous cell skin cancer. *Am. J. Epidemiol.* **141,** 916–922.
5. Armstrong, B., Thériault, G., Guénel, P., Deadman, J., Goldberg, M., and Héroux, P. (1994). Association between exposure to pulsed electromagnetic fields and cancer in electric utility workers in Quebec, Canada, and France. *Am. J. Epidemiol.* **140,** 805–820.
6. Gallagher, R. P., Hill, G. B., Bajdak, C. D., *et al.* (1995). Sunlight exposure, pigmentary factors, and risk of nonmelanocytic skin cancer. I. Basal cell carcinoma. *Arch. Dermatol.* **131,** 157–163.
7. Setlow, R. B., and Woodhead A. D. (1994). Temporal changes in the incidence of malignant melanoma: Explanation from action spectra. *Mutat. Res.* **307,** 365–374.
8. Baker, R. R., and Proctor, C. J. (1990). The origins and properties of environmental tobacco smoke. *Environ. Int.* **16,** 231–245.
9. Guérin, M. R., Jenkins, R. A., and Tomkins, B. A. (1992). "The Chemistry of Environmental Tobacco Smoke: Composition and Measurement." Lewis Publishers, Chelsea, MI.
10. Waldron, I. (1991). Patterns and causes of gender differences in smoking. *Soc. Sci. Med.* **32,** 989–1005.

11. Dockery, D. W., and Trichopoulos, D. (1997). Risk of lung cancer from environmental exposures to tobacco smoke. *Cancer Causes Control* **3**, 333−345.

12. Ernster, V. L. (1996). Female lung cancer. *Annu. Rev. Public Health* **17**, 97−114.

13. U.S. Environmental Protection Agency (1990). "Respiratory Health Effects of Passive Smoking: Lung Cancer and Other Disorders," EPA/600/6-90/006F. USEPA, Washington, DC.

14. Repace, J. L., and Lowrey, A. H. (1990). Risk assessment methodologies for passive smoking-induced lung cancer. *Risk Anal.* **10**, 27−37.

15. Boyle, P. (1997). Cancer, cigarette smoking and premature death in Europe: A review including the recommendations of European cancer experts consensus meeting, Helsinki, October 1996. *Lung Cancer* **1**, 1−60.

16. Prokopczyk, B., Cox, J. E., Hoffmann, D., and Waggoner, S. E. (1997). Identification of tobacco-specific carcinogen in the cervical mucus of smokers and nonsmokers. *J. Natl. Cancer Inst.* **89**, 868−873.

17. Tan, E. H., Adelstein, D. J., Droughton, M. L., Van Kirk, M. A., and Lavertu, P. (1997). Squamous cellhead and neck cancer in nonsmokers. *Am. J. Clin. Oncol.* **2**, 146−150.

18. Cohen, A. J., and Pope, A. C., III (1995). Lung cancer and air pollution. *Environ. Health Perspect.* **103**(Suppl. 8), 219−224.

19. Lewtas, J. (1993). Airborne carcinogens. *Pharmacol. Toxicol.* **72**(Suppl. 1), 55−63.

19a. World Health Organization (1987). "Air Quality Guidelines for Europe," WHO Reg. Publ., Eur. Ser. No. 23. WHO, Copenhagen.

20. Dockery, D., Pope, A., III, Xu, X., Spengler, J. D., Ware, J. H., *et al.* (1993). An association between the air pollution and mortality in six U.S.. cities. *N. Engl. J. Med.* **329**, 1753−1759.

21. Katsouyanni, K., and Pershagen, G. (1997). Ambient air pollution exposure and cancer. *Cancer Causes Control* **8**, 284−291.

22. Reymao, M. S., Cury, P. M., Lichtenfels, A. J., Lemos, M., Battlehner, C. N., Conceicao, G. M., Capelozzi, V. L., Montes, G. S., Júnior, M. F., Martins, M. A., Böhm, G. M., and Saldiva, P. H. (1997). Urban air pollution enhances the formation of urethane-induced lung tumors in mice. *Environ. Res.* **74**, 150−158.

23. Barbone, F., Bovenzi, M., Cavallieri, F., and Stanta, G. (1995). Air pollution and lung cancer in Trieste, Italy. *Am. J. Epidemiol.* **141**, 1161−1169.

24. Gao, Y. T. (1996). Risk factors for lung cancer among nonsmokers with emphasis on lifestyle factors. *Lung Cancer, Suppl.* **1**, 39−45.

25. Lei, Y. X., Cai, W. C., Chen, Y. Z., and Du, Y. X. (1996). Some lifestyle factors in human lung cancer: A case-control study of 792 lung cancer cases. *Lung Cancer, Suppl.* **1**, 121−136.

26. Wu-Williams, A. H., Dai, X. D., Blot, W., Xu, Z. Y., Sun, X. W., Xiao, H. P., Stone, B. J., Yu, S. F., Feng, Y. P., Ershow, A. G., *et al.* (1990). Lung cancer among women in north-east China. *Br. J. Cancer* **62**, 982−987.

27. Liu, Q., Sasco, A. J., Riboli, E., and Hu, M. X. (1993). Indoor air pollution and lung cancer in Guangzhou, People's Republic of China. *Am. J. Epidemiol.* **137**, 145−154.

28. Cantor, K. P. (1996). Drinking water and cancer. *In* "Cancer Epidemiology and Prevention" (D. Schottenfeld and J. F. Fraumeni, eds.), 2nd ed. Oxford University Press, New York.

29. Cantor, K. (1997). Drinking water and cancer. *Cancer Causes Control* **8**, 292−308.

30. Chiou, H. Y., Hsueh, Y. M., Liaw, K. F., Horng, S. F., Chiang, M. H., Pu, Y. S., Lin, J. S., Huang, C. H., and Chen, C. J. (1995). Incidence of internal cancers and ingested inorganic arsenic: A seven-year follow-up study in Taiwan. *Cancer Res.* **55**, 1296−1300.

31. Smith, A. H., Goycolea, M., Haque, R., and Biggs, M. L. (1998). Marked increase in bladder and lung cancer mortality in a region of Northern Chile due to arsenic in drinking water. *Am. J. Epidemiol.* **147**, 660−669.

32. Cantor, K. P., Lynch, C. F., Hildesheim, M. E., Dosemeci, M., Lubin, J., Alavanja, M., and Craun, G. (1998). Drinking water source and chlorination byproducts. I. Risk of bladder cancer. *Epidemiology* **9**, 21−28.

33. Hildesheim, M. E., Cantor, K. P., Lynch, C. F., Dosemeci, M., Lubin, J., Alavanja, M., and Craun, G. (1998). Drinking water source and chlorination byproducts. II. Risk of colon and rectal cancers. *Epidemiology* **9** 29−35.

34. Boice, J. D., Jr., and Lubin, J. H. (1997). Occupational and environmental radiation and cancer. *Cancer Causes Control* **8**, 309−322.

35. Lubin, J. H., and Boice, J. D. (1997). Lung cancer risk from residential radon: Meta-analysis of eight epidemiologic studies. *J. Natl. Cancer Inst.* **89**, 49−57.

36. Lubin, J. H., Boice, J. D., Jr., Edling, C., Hornung, R. W., Howe, G. R., Kunz, E., Kusiak, R. A., Morrison, H. I., Radford, E. P., and Samet, J. M. (1995). Lung cancer in radon-exposed miners and estimation of risk from indoor exposure. *J. Natl. Cancer Inst.* **87**, 817−827.

37. Lubin, J. H., Liang, Z., Hrubec, Z., Pershagen, G., Schoenberg, J. B., Blot, W. J., Klotz, J. B., Xu, Z. Y., and Boice, J. D., Jr. (1994). Radon exposure in residences and lung cancer among women: Combined analysis of three studies. *Cancer Causes Control* **2**, 114−128.

38. Pierce, D. A., Shimizu, Y., Preston, D. L., Vaeth, M., and Mabuchi, K. (1996). Studies of the mortality of atomic bomb survivors. Report 12. Part I. Cancer: 1950−1990. *Radiat. Res.* **146**, 1−27.

39. Land, C. E. (1995). Studies of cancer and radiation dose among atomic bomb survivors: The example of breast cancer. *JAMA, J. Am. Med. Assoc.* **274**, 402−407.

40. Takahashi, T., Trott, K. R., Fujimori, K., Simon, S. L., Ohtomo, H., Nakashima, N., Takaya, K., Kimura, N., Satomi, S., and Schoemaker, M. J. (1997). An investigation into the prevalence of thyroid disease on Kwajalein Atoll, Marshall Islands. *Health Phys.* **73**, 199−213.

41. Bleuer, J. P., Averkin, Y. I., and Abelin, T. (1997). Chernobyl-related thyroid cancer: What evidence for role of short-lived iodines? *Environ. Health Perspect.* **105**(Suppl. 6), 1483−1486.

42. Goldman, M. (1997). The Russian radiation legacy: Its integrated impact and lessons. *Environ. Health Perspect.* **105**(Suppl. 6), 1385−1391.

43. Zeighami, E. A., and Morris, M. D. (1986). Thyroid cancer risk in the population around the Nevada Test Site. *Health Phys.* **50**, 19−32.

44. Reiter, R. J. (1994). Melatonin suppression by static and extremely low frequency electromagnetic fields: Relationship to the reported increased incidence of cancer. *Rev. Environ. Health* **10**, 171−186.

45. Loomis, D. P., Savitz, D. A., and Ananth, C. V. (1994). Breast cancer mortality among female electrical workers in the United States. *J. Natl. Cancer Inst.* **86**, 921−925.

46. Tynes, T., Reitan, J. B., and Anderson, A. (1994). Incidence of cancer among workers in Norweigan hydroelectric power companies. *Scand. J. Work Environ. Health* **20**, 339−344.

47. Harrington, J. M., McBride, D. I., Sorahan, T., Paddle, G. M., and van Tongeren, M. (1997). Occupational exposure to magnetic fields in relation to mortality from brain cancer among electricity generation and transmission workers. *Occup. Environ. Med.* **54**, 7−13.

48. Savitz, D. A., Dufort, V., Armstrong, B., and Thériault, G. (1997). Lung cancer in relation to employment in the electrical utility industry and exposure to magnetic fields. *Occup. Environ. Med.* **54**, 396−402.

49. Verkasalo, P. K., Pukkala, E., Kaprio, J., Heikkilä, K. V., and Koskenvuo, M. (1996). Magnetic fields of high voltage power lines and risk of cancer in Finnish adults: Nationwide Cohort Study. *Br. Med. J.* **7064,** 1047–1051.

50. Vena, J. E., Freudenheim, J. L., Marshall, J. R., Laughlin, R., Swanson, M., and Graham, S. (1994). Risk of premenopausal breast cancer and use of electric blankets. *Am. J. Epidemiol.* **140,** 974–979.

51. Giovino, G. A., Schooley, M. W., Zhu, B. P., Chrismon, J. H., Tomar, S. L., Peddicord, J. P., Merritt, R. K., Husten, C. G., and Eriksen, M. P. (1994). Surveillance for selected tobacco-use behaviors—United States, 1900–1994. *MMWR CDC Surveill. Summ.* **43,** 1–43.

52. Ernster, V. L. (1994). The epidemiology of lung cancer in women. *Ann. Epidemiol.* **4,** 102–110.

53. Emmons, K. M., Wechsler, H., Dowdall, G., and Abraham, M. (1998). Predictors of smoking among U.S. college students. *Am. J. Public Health* **88,** 104–107.

54. Husten, C. G., Chrismon, J. H., and Reddy, M. N. (1996). Trends and effects of cigarette smoking among girls and women in the United States, 1965–1993. *J. Am. Med. Women's Assoc.* **51,** 11–18.

55. Haynes, S. G., Harvey, C., Montes, H., Nickens, H., and Cohen, B. H. (1990). Patterns of cigarette smoking among Hispanics in the United States: Results from HHANES 1982–84. *Am. J. Public Health* **80** (Suppl.), 47–53.

56. Shopland, D. R., Hartman, A. M., Gibson, J. T., Mueller, M. D., Kessler, L. G., and Lynn, W. R. (1996). Cigarette smoking among U.S. adults by state and region: Estimates from the current population survey. *J. Natl. Cancer Inst.* **88,** 1748–1758.

57. Zhu, B. P., Giovino, G. A., Mowery, P. D., and Eriksen, M. P. (1996). The relationship between cigarette smoking and education revisited: Implications for categorizing persons' educational status. *Am. J. Public Health* **86,** 1582–1589.

58. Flint, A. J., and Novotny, T. E. (1997). Poverty status and cigarette smoking prevalence and cessation in the United States, 1983–1993: The independent risk of being poor. *Tob. Control* **6,** 14–18.

59. Stewart, M. J., Brosky, G., Gillis, A., Jackson, S., Johnston, G., Kirkland, S., Leigh, G., Pawliw-Fry, B. A., Persaud, V., and Rootman, I. (1996). Disadvantaged women and smoking. *Can. J. Public Health* **87,** 257–260.

60. Skinner, W. F. (1994). The prevalence and demographic predictors of illicit and licit drug use among lesbians and gay men. *Am. J. Public Health* **84,** 1307–1301.

61. Pisani, P., Parkin, D. M., and Feerlay, J. (1993). Estimates of the worldwide mortality from eighteen major cancers in 1985. Implications for prevention and projections of future burden. *Int. J. Cancer* **55,** 891–903.

62. Madronich, S., and de Gruijl, F. R. (1994). Stratospheric ozone depletion between 1979 and 1992: Implications for biologically active ultraviolet-B radiation and non-melanoma skin cancer incidence. *Photochem. Photobiol.* **59,** 541–546.

63. Ley, R. D., and Reeve, V. E. (1997). Chemoprevention of ultraviolet radiation-induced skin cancer. *Environ. Health Perspect.* **105**(Suppl. 4), 981–984.

64. Hoel, D. G., Davis, D. L., Miller, A. B., Sondik, E. J., and Swerdlow, A. J. (1991). Trends in cancer mortality in 15 industrialized countries, 1969–1986. *J. Natl. Cancer Inst.* **84,** 313–320.

65. Barker, D. J., Coggon, D., Osmond, C., and Wickham, C. (1990). Poor housing in childhood and high rates of stomach cancer in England and Wales. *Br. J. Cancer* **61,** 575–578.

66. You, W. C., Chang, Y. S., Yang, Z. T., Zhang, L., Xu, G. W., Blot, W. J., Kneller, R., Keefer, L. K., and Fraumeni, J. F., Jr. (1991). Ecological research on gastric cancer and its precursor lesions in Shandong, China. *IARC Sci. Publ.* **105,** 33–38.

67. Peddanna, N., Holt, S., and Verma, R. S. (1995). Genetics of gastric cancer. *Anticancer Res.* **15,** 2055–2064.

68. Nagase, H., Ogino, K., Yoshida, I., Matsuda, H., Yoshida, M., Nakamura, H., Dan, S., and Ishimaru, M. (1996). Family history-related risk of gastric cancer in Japan: A Hospital-based Case-control Study. *Jpn. J. Cancer Res.* **87,** 1025–1028.

69. Shinamura, K., Yin, W., Isogaki, J., Saitoh, K., Kanazawa, K., Koda, K., Yokota, J., Kino, J., Arai, T., and Sugimura, H. (1997). Stage-dependent evaluation of microsatellite instability in gastric carcinoma with familial clustering. *Cancer Epidemiol., Biomarkers Prev.* **6,** 693–697.

70. Saracco, G. L. (1995). Primary liver cancer is of multifactorial origin: Importance of hepatitis B virus infection and dietary aflatoxin. *J. Gastroenterol. Hepatol.* **10,** 604–608.

71. Peers, F., Bosch, X., Kaldor, J., Linsell, A., and Pluijmen, M. (1987). Aflatoxin exposure, hepatitis B infection, and liver cancer in Swaziland. *Intl. J. Cancer* **39,** 545–553.

72. Parkin, D. M., Muir, C. S., Whelan, S. L., Gao, Y. T., Ferlay, J., and Powell, J., eds. (1992). "Cancer Incidence in Five Continents," IARC Sci. Publ. No. 120. IARC, Lyon, France.

73. Seow, A., Duffy, S. W., Ng, T. P., McGee, M. A., and Lee, H. P. (1998). Lung cancer among Chinese females in Singapore 1968–1992: Time trends, dialect group differences and implications for aetiology. *Int. J. Epidemiol.* **27,** 167–172.

74. Coleman, M. P., Esteve, J., Damiecki, P., Arslan, A., and Renard, H. (1993). "Trends in Cancer Incidence and Mortality," *IARC Sci. Publ. No.* **121.** IARC, Lyon, France.

75. Landis, S. H., Murray, T., Bolden, S., and Wingo, P. A. (1998). Cancer statistics, 1998. *Ca–Cancer J. Clin.* **48,** 6–29.

76. Kraemer, K. H., Levy, D. D., Parris, C. N., Gozukara, E. M., Moriwaki, S., Adelberg, S., and Seidman, M. M. (1994). Xeroderma pigmentosum and related disorders: Examining the linkage between defective DNA repair and cancer. *J. Invest. Dermatol.* **103**(Suppl.), 96–101.

77. Wu, A. H., Fontham, E. T., Reynold, P., Greenberg, R. S., Buffler, P., Liff, J., Boyd, P., and Correa, P. (1996). Family history of cancer and risk of lung cancer among lifetime nonsmoking women in the United States. *Am. J. Epidemiol.* **143,** 535–542.

78. Okojie, C. E. (1994). Gender inequalities of health in the Third World. *Soc. Sci. Med.* **39,** 1237–1247.

79. Komaroff, A., Robb-Nicholson, C., and Woo, B. (1998). Women's Health. *In* "Harrison's Principles of Internal Medicine" (A. Fauci, E. Braunwald, K. Isselbacher, J. Wilson, J. Martin, D. Kasper, S. Hauser, and D. Longo, eds.). Chapter 6. McGraw-Hill, New York.

80. Christiani, D. C. (1995). Utilization of biomarker data for clinical and environmental intervention. *Environ. Health Perspect.* **104** (Suppl. 5), 921–925.

81. Wang, X., Christiani, D. C., Wiencke, J. K., Fischbein, M., Xu, X. P., Cheng, T. J., Mark, E., Wain, J. C., and Kelsey, K. T. (1995). Mutations in the p53 gene in lung cancer are associated with cigarette smoking and asbestos exposure. *Cancer Epidemiol., Biomarkers Prev.* **4,** 543–548.

82. Xu, X., Kelsey, K. T., Wiencke, J. K., Wain, J. C., and Christiani, D. C. (1996). Cytochrome P450 CYP1A1 MspI polymorphism and lung cancer susceptibility. *Cancer Epidemiol., Biomarkers Prev.* **5,** 687–692.

83. London, S. J., Daly, A. K., Leathart, J. B., Navidi, W. C., Carpenter, C. C., and Idle, J. R. (1997). Genetic polymorphism of CYP2D6 and lung cancer risk in African-Americans and Caucasians in Los Angeles County. *Carcinogenesis (London)* **18,** 1203–1214.

84. Kelsey, K. T., Spitz, M. R., Zuo, Z. F., and Wiencke, J. K. (1997). Polymorphisms in the glutathione S-transferase class mu and theta genes interact and increase susceptibility to lung cancer in minor-

ity populations (Texas, United States). *Cancer Causes Control* **8**, 554–559.

85. Hemminki, K., Kumar, R., Bykov, V. J., Louhelainen, J., and Vodicka, P. (1996). Future research directions in the use of biomarkers. *Environ. Health Perspect.* **105**(Suppl. 3), 459–464.

86. Nestmann, E. R., Bryant, D. W., and Carr, C. J. (1996). Toxicological significance of DNA adducts: Summary of discussions with an expert panel. *Regul. Toxicol. Pharmacol.* **24**, 9–18.

87. Sober, A. J., Koh, H., Tran, N. L. T., and Washington, C. (1998). Melanoma and other skin cancers. *In* ''Harrison's Principles of Internal Medicine'' (A. Fauci, E. Braunwald, K. Isselbacher, J. Wilson, J. Martin, D. Kasper, S. Hauser, and D. Longo, eds.), Chapter 88. McGraw-Hill, New York.

88. Minna, J. D. (1998). Neoplasms of the lung. *In* ''Harrison's Principles of Internal Medicine'' (A. Fauci, E. Braunwald, K. Isselbacher, J. Wilson, J. Martin, D. Kasper, S. Hauser, and D. Longo, eds.), Chapter 90. McGraw-Hill, New York.

89. Moolgavkar, S. H., and Luebeck, E. G. (1996). A critical review of the evidence on particulate air pollution and mortality. *Epidemiology* **7**, 420–428.

90. Centers for Disease Control and Prevention (1993). Mortality trends for selected smoking-related cancer and breast cancer—United States 1950–1992. *Morbid. Mortal. Wkly. Rep.* **43**, 789–791.

91. Devessa, S. S., Blot, W. J., Stone, B. J., Miller, B. A., Tarone, R. E., and Fumeni, J. F., Jr. (1995). Recent cancer trends in the United States. *J. Natl. Cancer Inst.* **87**, 175–182.

92. Tarone, R. E., Chu, K. C., and Gaudette, L. A. (1997). Birth cohort and calendar period trends in breast cancer mortality in the United States and Canada. *J. Natl. Cancer Inst.* **89**, 251–256.

93. Ziegler, R. G., Hoover, R. N., Pike, M. C., Hildesheim, A., Nomura, A. M., West, D. W., Wu-Williams, A. H., Kolonel, L. N., Horn-Ross, P. L., and Rosenthal, J. F. (1993). Migration patterns and breast cancer risk in Asian-American women. *J. Natl. Cancer Inst.* **85**, 1819–1827.

94. Wolf, M. S., Collman, G. W., Barrett, J. C., and Huff, J. (1996). Breast cancer and environmental risk factors: Epidemiological and experimental findings. *Annu. Rev. Pharmacol. Toxicol.* **36**, 573–596.

95. Allen, R. H., Gottlieb, M., Clute, E., Pongsiri, M. J., Sherman, J., and Obrams, G. I. (1997). Breast cancer and pesticides in Hawaii: The need for further study. *Environ. Health Perspect.* **106**(Suppl. 3), 679–683.

96. Hunter, D. J., Hankinson, S. E., Laden, F., Colditz, G. A., Manson, J. E., Willett, W. C., Speizer, F. E., and Wolff, M. S. (1997). Plasma organochlorine levels and the risk of breast cancer. *N. Engl. J. Med.* **337**, 1253–1258.

97. Calle, E. E., Miracle-McMahill, H. L., Thun, M. J., and Heath, C. W. (1994). Cigarette smoking and risk of fatal breast cancer. *Am. J. Epidemiol.* **139**, 1001–1007.

98. Bennicke, K., Conrad, C., Sabroae, S., and Sorensen, H. T. (1995). Cigarette smoking and breast cancer. *Br. Med. J.* **310**, 1431–1433.

99. Ewert, M. (1990). Smoking and breast cancer risk in Denmark. *Cancer Causes Control* **1**, 31–37.

100. Vatten, L. J., and Kvinnsland, S. (1990). Cigarette smoking and risk of breast cancer: A propective study of 24,329 Norweigan women. *Eur. J. Cancer* **26**, 830–833.

101. Field, N. A., Baptiste, M. S., Nasca, P. C., and Metzger, B. B. (1992). Cigarette smoking and breast cancer. *Int. J. Epidemiol.* **21**, 842–848.

102. Braga, C., Negri, E., La Vecchia, C., Filiberti, R., and Franceschi, S. (1996). Cigarette smoking and the risk of breast cancer. *Eur. J. Cancer Prev.* **5**, 159–164.

103. Morabia, A., Bertein, M., Heritier, S., and Khatchatrian, N. (1996). Relation of breast cancer with passive and active exposure to tobacco smoke. *Am. J. Epidemiol.* **143**, 918–928.

104. Shields, P. G., Ambrosone, C. B., Graham, S., Bowman, E. D., Harrington, A. M., Gillenwater, K. A., Marshall, J. R., Laughlin, R., Nemoto, T., and Freudenheim, J. L. (1996). A cytochrome P4502E1 genetic polymorphism and tobacco smoke in breast cancer. *Mol. Carcinog.* **17**, 144–150.

105. Ambrosone, C. B., Freudenheim, J. L., Graham, S., Marshall, J. R., Vena, J. E., Brasure, J. R., Michalek, A. M., Laughlin, R., Nemoto, T., Gillenwater, K. A., Harrington, A. M., and Shields, P. G. (1996). Cigarette smoking, N-acetyltransferase 2 genetic polymorphism, and breast cancer risk. *JAMA, J. Am. Med. Assoc.* **276**, 1494–1501.

106. Perera, F. P., Estabrook, A., Hewer, A., Channing, K., Rundle, A., Mooney, L. A., Whyatt, R., and Phillips, D. H. (1995). Carcinogen-DNA adducts in human breast tissue. *Cancer Epidemiol., Biomarkers Prev.* **4**, 233–238.

107. Li, D., Wang, M., Dhingra, K., and Hittelman, W. N. (1996). Aromatic DNA adducts in adjacent tissues of breast cancer patients: Clues to breast cancer etiology. *Cancer Res.* **56**, 287–293.

108. Wang, M., Dhingra, K., Hittelman, W. N., Liehr, J. G., de Andrade, M., and Li, D. (1996). Lipid peroxidation-induced putative malondialdehyde-DNA adducts in human breast tissues. *Cancer Epidemiol., Biomarkers Prev.* **5**, 705–710.

109. Miller, R. W. (1995). Special susceptibility of the child to certain radiation-induced cancers. *Environ. Health Perspect.* **103**(Suppl. 6) 41–44.

110. Perera, F. P. (1996). Molecular epidemiology: Insights into cancer susceptibility, risk assessment, and prevention. *J. Natl. Cancer Inst.* **88**, 496–509.

111. Roa, B. B., Boyd, A. A., Volcik, K., and Richards, C. S. (1996). Ashkenazi Jewish population frequencies for common mutations in BRCA1 and BRCA2. *Nat. Genet.* **14**, 185–187.

112. Muto, M. G., Cramer, D. W., Tangir, J., Berkowitz, R., and Mok, S. (1996). Frequency of the BRCA1 185delAG mutation among Jewish women with ovarian cancer and matched population controls. *Cancer Res.* **56**, 1250–1252.

113. Eby, N., Chang-Claude, J., and Bishop, D. T. (1994). Familial risk and genetic susceptibility for breast cancer. *Cancer Causes Control* **5**, 458–470.

114. Taioli, E., Trachman, J., Chen, X., Toniolo, P., and Garte, S. J. (1995). A CYP1A1 restriction fragment length polymorphism is associated with breast cancer in African-American women. *Cancer Res.* **55**, 3757–3758.

115. Sarna, L. (1995). Lung cancer: The overlooked women's health priority. *Cancer Pract.* **3**, 13–18.

Part IV

CHRONIC DISEASE

Section 9

AUTOIMMUNE DISORDERS

Rosalind Ramsey-Goldman

Department of Medicine, Division of Arthritis and Connective Tissue Diseases
Northwestern University
Chicago, Illinois

Diseases of the immune system are excellent examples of conditions in which gender differences are noted in incidence, age at presentation, natural history, or severity of disease. Distinct environmental differences between women and men contribute to variations observed in immune system behavior [1]. These dissimilarities are illustrated in the cytokines measured, the degree of immune responsiveness, and the presence of sex hormones. Consequently, not only are women's immunological reactions to infectious diseases different from those of men, but their greater immune responses also increase their susceptibility to autoimmune diseases [2].

In this overview, the immune system is briefly reviewed, the epidemiology of autoimmune diseases is summarized, and a discussion of why women preferentially develop autoimmune disease is presented. The following chapters cover several common rheumatic (rheumatoid arthritis, systemic lupus erythematosus, and Sjögren's syndrome) and organ-specific (thyroid diseases and multiple sclerosis) autoimmune diseases. One chapter also reviews the epidemiology of asthma and highlights the important differences between men and women in age of onset and risk factors for this common immunologically mediated disease.

I. What Is Autoimmunity?

Autoimmunity results from a breakdown or failure of the mechanisms that are normally responsible for maintaining self-tolerance or the inability of the immune system to recognize self-antigens [3–5]. The common theme connecting autoimmune disorders is the presence of an autoimmune response based on genetic risk factors that interacts with environmental triggers.

The external instigators might be exposures to infections, chemicals, physical stresses, or other unknown exposures [6].

II. Why Do Autoimmune Diseases Occur?

Because reactivity to self by T cells during childhood and adulthood is physiologic and important to the formation of normal immune responses [7,8], it becomes crucial to understand the mechanism whereby this normal physiologic process becomes pathogenic. Both major histocompatability complex (MHC) and non-MHC genes can influence the immune response, shape the immune repertoire, regulate the immune response, or regulate the vulnerability of the target organ [6]. There is likely to be a role for somatic mutations [9]. Normally quiescent T cells become activated and induce B cells to produce pathogenic autoantibodies in autoimmune diseases [10–12]. These antibodies are typically IgG and, unlike the low-affinity IgM antibodies produced during nonspecific activation, are high-affinity pathogenic antibodies directed at specific antigens. It is hypothesized that patients with systemic lupus erythematosus also lack the ability to regulate hyperactivated B and T cells. The overproduction of pathogenic autoantibodies, coupled with impairment of the normal downregulation of B- and T-cell hyperactivity, could lead to immune complex (antigen/antibody) formation, inflammation, and tissue injury [13].

In summary, an individual with a strong genetic load may need only minor external triggers to develop autoimmune disease compared with another person in whom a weak genetic burden requires multiple environmental exposures [6]. An example of the possible interaction between genetics and environment follows. Various complement deficiencies exhibit some of the most powerful relationships to disease expression; these include homozygous C_1, C_4, and C_2 deficiency. Both C_4 and C_2 deficiency are associated with a disease closely resembling lupus. Because these genes reside on chromosome 6 and are close to the MHC genes that strongly influence the production of autoantibodies, those genetic relationships may find their meaning in that chromosomal proximity. In addition, complement deficiency could increase the susceptibility to lupus by enhancing survival of viruses because the early components of complement are involved in viral neutralization [13a].

III. Why Are Autoimmune Diseases Important from the Public Health Perspective?

Autoimmune diseases are a challenge to study on a population level because they occur rarely, are difficult to define, and because ascertainment of cases can be difficult even if a large population is sampled. Although autoimmune diseases are usually studied as separate entities [14], from a public health, biologic, and health services standpoint, it is important to consider the prevalence and incidence of these disorders as a group.

In one study [14], review of available data on prevalence and incidence of autoimmune diseases was restricted to well-characterized disorders for which there is substantial evidence of an autoimmune etiology and well-documented epidemiologic data. The weighted mean prevalence in 1996 of 24 selected autoimmune diseases in the United States was estimated at 8.5 million persons or about 1 in 31 Americans. Women accounted for 75% of the affected individuals. Some of the most prevalent diseases included thyroid disease, rheumatoid arthritis, insulin-dependent diabetes mellitus (IDDM), multiple sclerosis, systemic lupus erythematosus, and Sjögren's syndrome. All of these diseases, except for IDDM, are more prevalent in women than in men. The incidence of autoimmune disease was predicted from studies with population-based data; using these studies, 237,203 Americans were estimated to have developed an autoimmune disease in 1996. Women were at greater risk, in that there would be 2.7 times more women than men affected with newly diagnosed autoimmune diseases. If these incidence rates remain constant, there will be 1,186,015 new cases of autoimmune disease by 2001 [14].

This overview and the chapters in this section illustrate the relative paucity of incidence and prevalence data for rheumatic and organ-specific autoimmune diseases. Questions about case definition, particularly in the rheumatic diseases, thyroid diseases, and multiple sclerosis, emphasize the need for more studies to refine these estimates. Specialists treating these disorders need to form consensus panels to develop standardized case definitions that can be applied to epidemiologic research. Investigators must be able to identify cases using commonly available clinical and laboratory data [14].

IV. Why Do Autoimmune Diseases Preferentially Develop in Women?

Animal data show that distinct immune environments are established when different cytokines are released by two types of helper cells, T_H1 and T_H2 cells [1,9]. T_H1 cell products promote

a proinflammatory environment. T_H2 cell products promote autoantibody production. In addition, both T_H1- and T_H2-secreted cytokines exert cross-regulatory influences on each other. Females are more likely to develop a T_H1 response after challenge with an infectious agent or antigen, except during pregnancy when a T_H2 environment prevails. These observations have not been documented in humans.

The degree of immune responsiveness differs between women and men. Sex hormones have been suggested as a causal factor of autoimmune diseases that predominate in women. However, in articles by Lockshin, this theory has been challenged as too simplistic because some rheumatic diseases, such as Wegener's granulomatosus and ankylosing spondylitis, are not demonstrably autoimmune yet have gender skew. In addition, other chronic inflammatory diseases, such as Crohn's disease and ulcerative colitis, have no gender skew. If gonadal hormones induce autoimmunity in states of chronic inflammation, the absence of sexual dimorphism of these illnesses needs to be explained [15,15a].

An alternative hypothesis suggested by this author and others is that sex hormones *modulate* susceptibility to autoimmune disease [1,15]. Other explanations for the biological basis of female predominance for most autoimmune diseases were summarized in a commentary on gender-based biology [16]. The hypotheses suggested include gender differences in metabolism of sex hormones [17], mosaicism of the X chromosome [15], graft versus host disease resulting from chimerism related to pregnancy [18], and neuroendocrine functions interacting with the immune system [19]. The candidate molecules that may be involved are the pituitary hormones, prolactin and growth hormone, and liver-derived insulin-like growth factor-1. These hormones may act on immune cells through interactions with cell-surface receptors or they may mediate their effects through modulation of the hypothalamic–pituitary–adrenal axis [1,15,19].

V. What Is the Relationship Between Sex Hormones and Autoimmune Diseases?

Classical teaching states that physiological levels of estrogens were thought to stimulate the immune response and male hormones to suppress it [2,20,21]. Estrogens account in part for the higher immune reactivity in females against a variety of antigens and the more rapid rejection of allografts [22].

Estrogens were demonstrated *in vitro* to stimulate B-cell response and decrease suppressor T-cell reactivity, leading to an increase in autoantibody production [23]. However, this model did not explain estrogen's apparent contradictory effects on other normal immune responses, such as inhibiting immune responses during pregnancy.

Reconciliation of these discrepant observations in estrogen effect may be particularly relevant to understanding the development of autoimmune diseases. Estrogens have biphasic dose effects, with lower levels enhancing and higher levels (in pregnancy) inhibiting specific immune activities. Current evidence suggests that sex hormones may modulate T-cell receptor signaling, expression of activation molecules on T lymphocytes and antigen-presenting cells, transcription or translation of cytokine genes, or lymphocyte homing [24–26]. Important questions remain about the differences between women and men in production and secretion of immunomodulatory mediators such as TGF-β, interferons, prostaglandins, and individual T_H1 and T_H2 cytokines [1]. Several observations have launched new areas of investigation including the identification of a second estrogen receptor, the activity of progesterone preferentially favoring T_H2 over T_H1 cells, and the anti-inflammatory properties noted in testosterone [1].

VI. Future Research Challenges

Autoimmune diseases and immunologic disorders (*e.g.*, asthma) are frequently chronic, can adversely affect quality of life, and incur a significant financial burden for the patient and society. The public health burden of these diseases is underestimated and unappreciated. The chapters in this section highlight the research challenges for the future. New therapeutic options are becoming available for many patients. Emerging insights into the functions and interactions of the various components of the immune system with other systems, such as neuroendocrine and genetics, herald an exciting future for patients and their physicians.

Acknowledgments

This work is supported by grants from the National Institutes of Health, National Institute of Arthritis and Musculoskeletal and Skin Diseases Branch (2P60 AR 30692-15); the Lupus Foundation, Illinois Chapter; an Arthritis Foundation Clinical Science Grant; and the Arthritis Foundation, Illinois Chapter.

References

1. Whitacre, C. C., Reingold, S. C., O'Looney, P. A., and the Task Force on Gender, Multiple Sclerosis and Autoimmunity (1999). A gender gap in autoimmunity. *Science* **283**, 1277–1278.
2. Cutolo, M., Sulli, A., Seriolo, B., Accardo, S., and Masi, A. T. (1995). Estrogens, the immune response and autoimmunity. *Clin. Exp. Rheumatol.* **13**, 217–226.
3. Theofilopoulos, A. N. (1995). The basis of autoimmunity: Part I. Mechanisms of aberrant self-recognition. *Immunol. Today* **16**, 90–98.
4. Theofilopoulos, A. N. (1995). The basis of autoimmunity: Part II. Genetic predisposition. *Immunol. Today* **16**, 150–159.
5. Hahn, B. H. (1997). An overview of the pathogenesis of systemic lupus erythematosus. *In* "Dubois' Lupus Erythematosus" (D. J. Wallace and B. H. Hahn, eds.), pp. 69–75. Williams & Wilkins, Baltimore, MD.
6. Rose, N. R., and Mackay, I. R. (1998). Prelude. *In* "The Autoimmune Diseases" (N. R. Rose and I. R. Mackay, eds.), pp. 1–4. Academic Press, San Diego, CA.
7. Burkly, L. C., Lo, D., and Flavell, R. A. (1990). Tolerance in transgenic mice expressing major histocompatibility molecules extrathymically on pancreatic cells. *Science* **248**, 1364–1368.
8. Horowitz, D., Stahl, W., and Gray, J. D. (1997). T lymphocytes, natural killer cells, cytokines and immune regulation. *In* "Dubois' Lupus Erythematosus" (D. J. Wallace and B. H. Hahn, eds.), pp. 155–194. Williams & Wilkins, Baltimore, MD.
9. Mackay, I. R., and Rose, N. R. (1998). Autoimmunity yesterday, today, and tomorrow. *In* "The Autoimmune Diseases" (N. R. Rose and I. R. Mackay, eds.), pp. 854–875. Academic Press, San Diego, CA.
10. Gharavi, A. E., Chu, J. L., and Elkon, K. B. (1988). Autoantibodies to intracellular proteins in human SLE are not due to random polyclonal B cell activation. *Arthritis Rheum.* **31**, 1337–1345.
11. Shivakumar, S., Tsokos, G. C., and Datta, S. K. (1989). T cell receptor alpha/beta expressing double negative (CD4−CD8−) and CD4+ T helper cells in humans augment the production of pathogenic anti-DNA autoantibodies associated with lupus nephritis. *J. Immunol.* **143**, 103–112.
12. Takeuchi, T., Abe, T., Koide, J., Hosono, O., Morimotos, C., and Homma, M. (1984). Cellular mechanism of DNA-specific antibody synthesis by lymphocytes from systemic lupus erythematosus patients. *Arthritis Rheum.* **27**, 766–773.
13. Manzi, S., and Ramsey-Goldman, R. (1999). Autoimmune diseases. *In* "Health and Diseases among Women. Biological and Environmental Influences" (R. B. Ness and L. H. Kuller, eds.), pp. 342–372. Oxford University Press, New York.
13a. Reichlin, M. (1998). Systemic lupus erythematosus. *In* "The Autoimmune Diseases" (N. R. Rose and I. R. Mackay, eds.), pp. 289–290. Academic Press, San Diego, CA.
14. Jacobson, D. L., Gange, S. J., Rose, N. R., and Graham, N. M. H. (1997). Epidemiology and estimated population burden of selected autoimmune diseases in the United States. *Clin. Immunol. Immunopathol.* **84**, 223–243.
15. Lockshin, M. D. (1997). Why women? *Lupus* **6**, 625–632.
15a. Lockshin, M. D. (1998). Why do women have rheumatic disease? *Scand. J. Rheumatol.* **27**, (Suppl. 107), 5–9.
16. Thompson, P. M., and Wolf, J. L. (1999). The sexual revolution in science: What gender-based research is telling us. *J. Invest. Med.* **47**, 106–113.
17. Lahita, R. G. (1997). Predisposing factors to autoimmune disease. *Int. J. Fertil.* **42**, 115–119.
18. Nelson, J. L. (1996). Viewpoint. Maternal-fetal immunology and autoimmune disease. Is some autoimmune disease auto-allo or allo-auto immune? *Arthritis Rheum.* **39**, 191–194.
19. Chrousos, G. P. (1995). The hypothalamic-pituitary-adrenal axis and immune-mediated inflammation. *N. Engl. J. Med.* **332**, 1351–1362.
20. Sthoeger, Z. N., Chiorazzi, R. G., and Lahita, R. G. (1988). Regulation of the immune response by sex steroids. *J. Immunol.* **141**, 91–98.
21. Grossman, C. J. (1985). Interaction between the gonadal steroids and the immune system. *Science* **227**, 257–261.
22. Graff, R. J. (1969). The influence of the gonads and adrenal glands on the immune response to skin grafts. *Transplantation* **7**, 105–111.
23. Carlsten, H., Nilsson, N., Jonsson, R., Backman, K., Holmdahl, R., and Tarkowski, A. (1992). Estrogen accelerates immune complex glomerulonephritis but ameliorates T-cell mediated vasculitis and sialoadenitis in autoimmune MRl LPR/LPR mice. *Cell. Immunol.* **144**, 190–202.
24. Yung, R., Williams, R., Johnson, K., Phillips, C., Stoolman, L., Chang, S., and Richardson, R. B. (1997). Mechanisms of drug-induced lupus. III. Sex-specific differences in T cell homing may explain increased disease severity in female mice. *Arthritis Rheum.* **40**, 1334–1343.
25. Theofilopoulos, A. N., and Dixon, F. J. (1981). Etiopathogenesis of murine systemic lupus erythematosus. *Immunol. Rev.* **55**, 179–216.
26. Cohen, J. H. M., Danel, L., Cordier, G., Saez, S., and Revillard, J. P. (1983). Sex steroid receptors in peripheral T cells: Absence of androgen receptors and restriction of estrogen receptors to OKT8-positive cells. *J. Immunol.* **131**, 2767–2771.

52

Thyroid Diseases

PETER KOPP

Division of Endocrinology, Metabolism, and Molecular Medicine

Northwestern University

Chicago, Illinois

I. Introduction

Thyroid disorders affect a substantial proportion of the general population, but most have a remarkably higher incidence in women. Worldwide, the most common cause of thyroid disorders is iodine deficiency. Although eradicated in many parts of the world, it is still a major health problem affecting almost a third of the world's population. About twenty million people are believed to be significantly mentally handicapped as a result of iodine deficiency. In iodine-sufficient regions, the autoimmune thyroid diseases, Hashimoto's thyroiditis and Graves' disease, are the most common clinical entities affecting the thyroid. The prevalence of thyroid disorders depends on the diagnostic criteria used and the surveyed population's age and sex. The prevalence of overt hypothyroidism is $\sim$ 3–20 per 1000 females and $\sim$ 1–7 per 1000 males. The most common cause of hypothyroidism is Hashimoto's thyroiditis. The prevalence of hyperthyroidism is $\sim$ 2–19 per 1000 females and $\sim$ 1–2 per 1000 males. Graves' disease is its most frequent cause. The differences in the prevalences of hypo- and hyperthyroidism in men and women are most commonly explained by influences of sex steroids on immunoregulatory mechanisms. Thyroid carcinomas are relatively rare neoplasms that account for 0.6–1.6% of all malignancies. The annual incidence of papillary, follicular, and differentiated thyroid carcinomas cancer is 1–10 per 100,000 people. They are about three times more common in women than men.

In addition to the striking gender differences in the frequency of autoimmune thyroid diseases and thyroid cancer, several other aspects of thyroid physiology and disease are distinct in women. Pregnancy is characterized by intricate physiological changes in iodine balance, thyroid activity, thyroid hormone transport, and peripheral metabolism. Pregnancy also has modulating influences on autoimmune thyroid disorders and can be complicated by a distinct form of hyperthyroidism, gestational hyperthyroidism, which is caused by the secretion of the placental hormone human chorionic gonadotropin (hCG). The hCG-induced rise in thyroid hormone levels has also been implicated in the pathogenesis of hyperemesis gravidarum (nausea and vomiting of pregnancy). After delivery, postpartum thyroiditis may result in either hyper- or hypothyroidism.

Among the multiple peripheral actions of thyroid hormones, the effects on the cardiovascular system, lipid metabolism, and bone are particularly prominent. It is well recognized that cardiovascular disease and osteoporosis have distinct characteristics in women, and thyroid hormone excess or deficiency may further modulate these important health problems.

Hence, thyroid diseases continue to present challenges to researchers and clinicians, and many of these disorders are particularly relevant to the health of women.

II. Historical Aspects

Goiter, enlargement of the thyroid gland, has been documented in sculptures as early as the second century [1]. In the sixteenth century, Leonardo da Vinci and Andreas Vesalius identified the thyroid gland, and the connection between endemic goiter and cretinism was recognized [1,2]. Although it was not yet known that a deficiency of iodine was their cause, iodine was first used in the treatment of goiters in the nineteenth century. Despite remarkable effects on shrinking the size of goiters, these therapeutic attempts were met with skepticism because some of the patients developed thyrotoxicosis. It was not until the second decade of the twentieth century that systematic iodine prophylaxis was instituted to prevent goiter. For example, Marine and Kimball conducted a trial of 0.2 g of sodium iodide for 10 days twice a year in 1000 girls aged 11 to 18 in Akron, Ohio [3]. This study proved to be highly successful, and in 1924 iodine prophylaxis was introduced in the state of Michigan. However, opposition to the use of iodine was harsh. The U.S. Department of Agriculture required that the "skull and crossbones" should figure on packs of iodized salt. Independent, simultaneous trials performed by several investigators in Switzerland from 1915 to 1929 resulted in successful treatment of goiters in school children, and goiter prophylaxis was first introduced in 1922 [4]. Iodine prophylaxis was introduced in most developed countries prior to 1960. Today, the International Council for the Control of Iodine Deficiency Disorders (ICCIDD) of the World Health Organization (WHO), UNICEF, and other professional organizations are leading the ongoing efforts aimed at the eradication of iodine deficiency in all countries of the world [5].

Descriptions of patients with the characteristic signs of thyroid hormone excess (thyrotoxicosis or hyperthyroidism) and thyroid hormone deficiency (hypothyroidism or myxedema) were published by several authors in the nineteenth century [2]. However, it was only recognized later that these disorders were caused by a dysfunction of the thyroid. Toward the end of the nineteenth century, much progress was made in surgical procedures to remove the thyroid gland. Reverdin and Kocher reported the dramatic consequences of total thyroidectomy leading to a severely hypothyroid state [6,7]. These observations ultimately resulted in the recognition that cretinism, myxedema, and postsurgical changes were the result of a deficiency in one or several substances secreted by the thyroid. In 1891, Murray demonstrated that the injection of thyroid extract could cure hypothyroidism [8]. The discovery of iodine in the thyroid by Baumann in 1895, the isolation of the thyroid hormone thyroxine by Kendall in 1914, and its synthesis by Harington in 1927 opened the door for a more rational understanding of thyroid physiology, thyroid hormone action, and thyroid hormone replacement [2]. In 1949, therapy with synthetic levothyroxine became available with the introduction of its large scale production. Moreover, thionamide drugs and radioiodine, developed in the mid-1940's, offered, for the first time, nonsurgical methods for the treatment of thyrotoxicosis. Radioiodine became a cornerstone in the diagnosis and treatment of differentiated thyroid carcinomas [2].

Systematic neonatal screening for congenital hypothyroidism was introduced in the 1970s and proved to be extremely successful in avoiding the mental retardation associated with untreated hypothyroidism [9,10]. Following the characterization of thyroid disorders at the clinical and biochemical level, many of the key genes responsible for thyroid hormone synthesis and action were cloned during the 1990s [11]. Gene identification and the rapidly growing knowledge of signal transduction pathways continue to broaden our understanding of the underlying molecular defects and the mechanisms governing thyroid cell growth and function under physiologic or pathologic thyroid conditions.

III. Thyroid Physiology

The thyroid is controlled by the hypothalamic-pituitary axis (Fig. 52.1). The hypothalamic tripeptide thyrotropin-releasing hormone (TRH) stimulates the production and secretion of the glycoprotein thyroid-stimulating hormone (TSH) in the pituitary. TSH in turn stimulates growth and function of thyroid follicular cells resulting in the production of the thyroid hormones, thyroxine (T4) and triiodothyronine (T3), which have multiple effects on differentiation, growth, and metabolism of peripheral tissues [12,13].

The thyroid consists of follicles of varying size that contain colloid produced by the surrounding monolayer of follicular cells. Thyroid hormone synthesis occurs at the apical membrane and requires sufficient amounts of the essential trace element, iodine, which is actively transported into the cells by the sodium-iodide symporter (NIS). Iodination of the intrafollicular colloidal protein thyroglobulin by the enzyme thyroperoxidase on the apical membrane results in the synthesis of T4 and T3. After micro- and macropinocytosis of thyroglobulin and digestion in lysosomes, T4 and T3 are released into the blood stream [14].

Thyroid hormones circulate bound to several plasma proteins [15,16]. Under physiological conditions, only 0.03% of T4 and 0.3% of T3 circulate as free hormone. T4 and T3 enter the peripheral cells in unbound form. T4 is metabolized into the more active compound T3 by intracellular 5'-monodeiodination or into the inactive metabolite reverse T3 (rT3) by 5-monodeiodination.

At the level of the cell, thyroid hormone action is primarily mediated by specific nuclear receptors, the thyroid hormone receptors α and β [17]. These transcription factors belong to the superfamily of nuclear hormone receptors for steroid and thy-

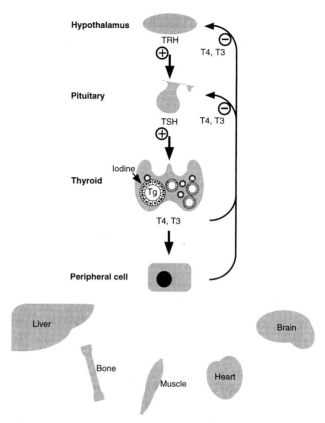

Fig. 52.1 The hypothalamic-pituitary axis. The hypothalamic hormone, TRH, enhances production and secretion of the pituitary hormone, TSH, which then stimulates thyroid hormone synthesis in thyroid cells that are organized in follicles. Iodine is concentrated in follicular cells and hormone synthesis occurs at the luminal membrane on thyroglobulin (Tg). Thyroid hormones act on multiple target genes in peripheral cells through nuclear receptors that regulate gene transcription. T4 and T3 regulate TSH and TRH through a negative feedback mechanism.

roid hormones, retinoids, vitamin D, and peroxisomal activators [18]. Numerous genes that respond to thyroid hormone, either by being activated or repressed, have been identified [17]. For example, T3 stimulates the transcription of genes involved in cholesterol metabolism (VLDL receptor, HMG-CoA reductase, malic enzyme). Hypercholesterinemia is commonly seen in hypothyroidism, and a reduction in cholesterol can be found in hyperthyroidism. The classical example of a negatively regulated gene is the TSH β gene (Fig. 52.1). If T3 levels are low (e.g., because of autoimmune destruction of the thyroid gland), TSH will increase. Conversely, TSH will be low or undetectable if T3 is elevated.

IV. Overview of Thyroid Disease

The following overview aims at providing a brief characterization of the most important public health issues and clinical entities (Table 52.1) but does not attempt to present all thyroid disorders comprehensively. All aspects of thyroid physiology and disease, including clinical presentation, diagnosis, and ther-

Table 52.1
Overview of Important Thyroid Disorders

Disorder/Cause	Significance and frequency	Gender specific problems
Iodine deficiency	~1570 million people (~30% of the world population) ~750 million with goiter ~20–40 million with brain damage	Women during pregnancy at higher risk for developing goiter and hypothyroidism Iodine deficiency possibly associated with fibrocystic breast disease (?)
Congenital hypothyroidism	1/3000 to 1/4000 newborns Screening of newborns essential to start early treatment and to avoid mental retardation	Female to male ratio ~2:1
Hypothyroidism	~3–20 per 1000 in females, ~1–7 per 1000 in males	Significantly more frequent in females Screening of women above age 40 recommended by some, but not all, specialists
Hyperthyroidism	~2–19 per 1000 in females, ~1–2 per 1000 in males	Significantly more frequent in females Pregnancy: aggravation during 1st trimester, amelioration during second and third trimester Frequent recurrence after delivery
Gestational thyrotoxicosis (caused by the placental hormone human chorionic gonadotropin)	Incidence not established More frequent in twin pregnancies Associated with hydatidiform mole and choriocarcinoma	Only occurs in pregnancy Usually mild hyperthyroidism Associated with hyperemesis gravidarum
Subacute thyroiditis	Incidence variable Associated with viral infections	
Postpartum thyroiditis	Incidence ~3–16% of all pregnancies 25% of affected patients develop primary thyroid failure	Associated with pregnancy
Thyroid nodules and multinodular goiter	Very common; up to 60% of the population Radiation major risk factor	

Thyroid carcinomas originating from thyroid follicular cells
    Papillary thyroid carcinoma
    Follicular thyroid carcinoma
    Anaplastic thyroid carcinoma
	~0.6–1.6% of all malignancies
	Incidence ~1 to 10 per 100,000 people
	More frequent in iodine-replete regions
	More frequent in iodine-deficient regions
	Rare
	Radiation major risk factor
		About three times more frequent in females

Thyroid carcinomas originating from calcitonin-producing C cells Medullary thyroid carcinoma	Autosomal dominant inheritance in Multiple Endocrine Neoplasia Syndrome type 2 (MEN-2) and Familial Medullary Thyroid Carcinoma (FMTC)	

apeutic management, are thoroughly discussed in several textbooks [19–22].

Thyroid disease can be categorized very broadly into disorders of function and growth (Fig. 52.2). The two mechanisms may occur independently or in combination. For example, in patients with Graves' disease, the thyroid is typically diffusely enlarged and secretes excessive amounts of thyroid hormone. Detailed algorithms for thyroid testing have been published by several groups of experts [23–26].

The measurement of serum TSH represents the best biochemical marker of thyroid function when measured with a sensitive assay (Fig. 52.3). TSH and free T4 (FT4) have a log-linear relationship and a small decrease or increase in FT4 is thus associated with an exponential change in TSH levels. The original radioimmunoassays had a lower detection limit of about 0.5 to 1 mU/l (first generation assays). The so-called "sensitive" second generation immunometric assays have a detection limit of 0.1 to 0.2 mU/l. With these assays, TSH levels of euthyroid and hyperthyroid patients can be reliably differentiated. "Supersensitive" third generation assays have even lower detection limits (<0.02 mU/l). Today, most clinical laboratories use assays where values less than 0.05 to 0.2 mU/l are reported to be undetectable (normal values are typically 0.5 to 4 mU/l) [26]. These differences in assay sensitivity in the low range may have underestimated the true prevalence of low TSH levels in older studies. With the introduction of sensitive TSH assays, dynamic

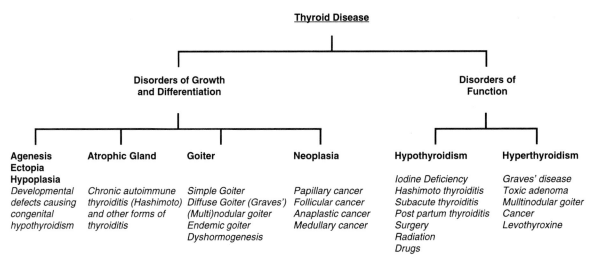

Fig. 52.2 Alterations of growth and function in thyroid disease.

testing of the pituitary-thyroid axis with TRH has assumed a limited role and is only required in special situations such as pituitary dysfunction.

Other biochemical markers include antibodies against thyroid peroxidase (anti-TPO Ab) which are typically elevated in autoimmune thyroiditis, antibodies against the TSH receptor in Graves' disease, and thyroglobulin as a tumor marker in differentiated thyroid carcinoma [27]. Cholesterol, sex hormone binding globulin, and creatine phosphokinase, among others, can be useful as markers of peripheral thyroid hormone action [27]. Radioisotope scanning ([123]iodine, [131]iodine, [99]technetium), ultrasound, and fine needle aspiration are used in the evaluation of thyroid nodules and cancer [24].

A. Iodine Deficiency Disorders (IDD): Endemic Goiter and Cretinism

Iodine is an essential component of thyroid hormones. Insufficient iodine supply causes endemic goiter, the compensatory growth of the thyroid gland. If it is more severe, it can result in hypothyroidism and severe developmental retardation (endemic cretinism) [28,29]. It has been reported that iodine-deficient communities have a lower intellectual level, decreased work output, and decreased per capita income [30]. Iodine Deficiency Disorders (IDD) are probably the most common endocrine disease. Roughly 1570 million people, or about 30% of the world population, are still at risk of IDD [29]. About 750 million peo-

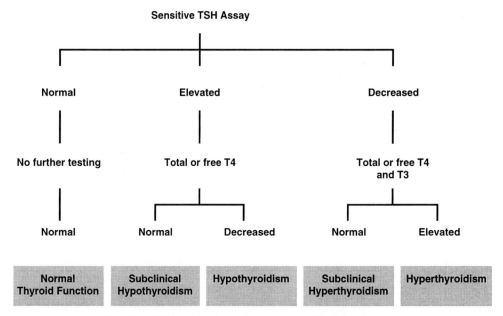

Fig. 52.3 TSH-based biochemical assessment of thyroid function.

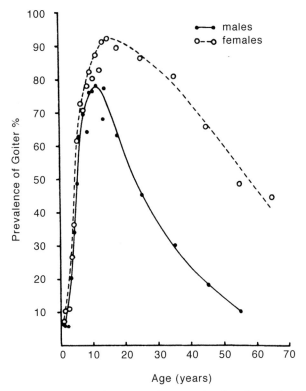

Fig. 52.4 Pattern of goiter prevalence as a function of age and sex in the inhabitants of the Idjwi Island, an endemic goiter area in Zaire. (From [29] with permission.)

Table 52.2
Recommended Daily Iodine Intake in Different Age Groups

Adolescents and adults	150 μg/day
Pregnant women	175 μg/day
Lactating women	200 μg/day
Children ages 1 to 10	70–120 μg/day
Infants ages 6 to 12 months	50 μg/day
Infants up to age 6 months	40 μg/day

ple are thought to have endemic goiter due to iodine deficiency. The Pan American Health Organization (PAHO) defines endemic goiter as areas where more than 10% of the children aged 6 to 12 have a goiter [29]. The WHO/UNICEF/ICCIDD proposed to decrease this arbitrary threshold from 10 to 5%. Goiter prevalence is strongly influenced by age and sex and is significantly higher in women (Fig. 52.4) [29]. It has to be emphasized that a goiter endemia should not only be assessed by the frequency of goiter, but it should also take the severity of iodine deficiency into account.

Particularly impressive are the estimates of the prevalence of IDD-related brain damage and mental retardation which range from 20 to 43 million people [29,31]. This defines iodine deficiency as the most prevalent cause of mental retardation. Moreover, iodine deficiency results in reproductive impairment and decreased child survival [29]. Interestingly, it has also been suggested that fibrocystic breast disease is associated with iodine deficiency [32,33].

Following the introduction of iodine prophylaxis between 1920 and 1930 in a few countries, most developed countries had established similar programs by 1960 [29]. Iodine supplementation is most commonly performed by the large-scale iodization of salt. Additional strategies may be required in regions where groups at risk of IDD have no access to iodized salt. These may include the free distribution of iodized salt to vulnerable population groups, particularly women of childbearing age, the use of oral, high-dose, iodized oil capsules, iodization

of drinking-water, or injectable iodized oils, are technologies that are well established [34].

The recommended daily allowance (RDA) for iodine is 150 μg/day for adults (Table 52.2). Taking into consideration that iodine clearance is higher during pregnancy, the RDA is increased to 175 to 200 μg/day in pregnant and lactating women [5,35]. Lower RDAs have been established for children.

Once iodine prophylaxis has been established, continuous monitoring of the iodine intake is absolutely key to assure and maintain an adequate supply [30]. Because ~90% of iodine is eventually excreted in urine, monitoring can be easily performed by measuring urinary iodine concentrations. The minimal daily urinary iodine excretion has been set at 100 μg/l, corresponding roughly to an intake of 150 μg/day. The importance of monitoring as a public health measure is emphasized by the following observations. In several countries (*e.g.,* Guatemala, Colombia, Mexico, and Thailand), iodine prophylaxis was successfully introduced. However, monitoring and ongoing support of these programs failed and led to the reappearance of IDD. A survey in Europe published in 1993 indicated that iodine deficiency was entirely under control in only five countries, marginal intake was found in many regions, and even severe iodine deficiency was still prevalent [36]. In the U.S., the survey published in 1998 on iodine nutrition obtained from the National Health and Nutrition Examination Survey NHANES III indicated a median urinary iodine excretion of 145 μg/l during the years 1988 to 1994 [37]. This is significantly lower than the iodine excretion of 321 μg/l determined in NHANES I between 1971 to 1974. The current numbers do not indicate iodine deficiency and are still close to the RDA (Table 52.2). They indicate, however, a marked decrease in iodine intake in comparison to the levels determined in NHANES I. Remarkably, in NHANES III low iodine concentrations (<50 μg/l) were found in 11.7% of the population, in 6.7% of pregnant women, and in 14.9% of women of childbearing age. These observations emphasize that further monitoring and ongoing public education is mandatory. In the plans for NHANES IV, unfortunately, assessment of iodine concentrations are currently not included [31].

B. Congenital Hypothyroidism

In iodine-sufficient regions, permanent sporadic congenital hypothyroidism affects about 1/3000 to 1/4000 newborns [9,38]. It appears to have a female to male ratio of 2:1. Congenital hypothyroidism can be caused by developmental abnormalities of the thyroid (thyroid dysgenesis) or, more rarely, by mutations in genes involved in hormonogenesis [11]. The majority of

Table 52.3

Hypothyroidism

Etiology
   Iodine deficiency
   Chronic autoimmune thyroiditis: Hashimoto's thyroiditis
   Therapy for hyperthyroidism (radioablation, surgery)
   Developmental defects of the thyroid (athyreosis, ectopy,
      hemiagenesis)
   Dyshormogenesis
   Drugs (lithium, iodine in high doses)
Clinical presentation
   Symptoms
      Weakness, lethargy, depression, cognitive dysfunction, cold
         intolerance, decreased appetite, weight gain, constipation,
         paresthesias, muscle cramps, irregular menses
   Signs
      Adults: slow appearance, bradycardia, pericardial effusion,
         hyporeflexia, dry yellow skin, hoarseness, myxedema
      Intrauterine/neonatal: mental retardation, neurological deficit,
         retarded bone age/growth
Biochemical Findings
   Primary: TSH ↑, T4 ↓, T3 ↓
   Secondary (pituitary, hypothalamic): TSH ↓, T4 ↓, T3 ↓
Therapy
   Newborns: Levothyroxine 10–15 μg/kg/day
   Adults: Levothyroxine 1.6 μg/kg body weight (~112 μg per day)

newborns with sporadic congenital hypothyroidism appear normal at birth, and fewer than 10% will be diagnosed as hypothyroid based solely on clinical signs. Because diagnosis may be difficult and early treatment is essential in order to avoid permanent mental retardation, neonatal screening programs have been established in most developed countries [10]. With early and well-monitored treatment, infants with congenital hypothyroidism have the same prospects as their unaffected siblings. Screening for congenital hypothyroidism is performed by measuring a total T4 or TSH on a capillary blood sample obtained shortly after birth. The screening for congenital hypothyroidism is performed together with the screening for other inborn errors of metabolism such as phenylketonuria, and, in certain countries, galactosemia, maple syrup disease, and congenital adrenal hyperplasia. Some aspects of screening strategies in newborns are discussed in further detail in the comments on screening for thyroid disease. Congenital hypothyroidism is treated with 10–15 μg/kg/day levothyroxine given at least 30 minutes before food intake.

C. Hypothyroidism

Hypothyroidism, the decreased production of thyroid hormone, is the most common disorder of thyroid function. Its main etiologies, clinical, and biochemical characteristics are summarized in Table 52.3. In iodine-sufficient regions, hypothyroidism is most frequently caused by chronic autoimmune inflammation of the thyroid gland, Hashimoto's thyroiditis. Other important etiologies include hypothyroidism after therapy for hyperthyroidism with radioiodine or surgery, and, as discussed above,

iodine deficiency and congenital hypothyroidism. Subacute, silent, and postpartum thyroiditis can result in transient hyperthyroidism, followed by transient hypothyroidism. Only rarely, hypothyroidism results from a pituitary or hypothalamic defect leading to a decrease in TSH secretion [39].

The pathogenic mechanisms leading to autoimmune thyroid disease are still poorly understood. Hashimoto's thyroiditis has been associated with certain HLA haplotypes (e.g., HLA-DR3, DR4, and DR5). The high propensity of women for autoimmune thyroid disease is thought to be caused by effects of sex steroids on the immune system [40]. It appears that estrogen and progesterone have effects on the differentiation and maturation of lymphocytes and modulate the induction of the autoimmune response. However, many autoimmune diseases improve during the second and third trimester of pregnancy (i.e., during a high estrogen state), and they exacerbate postpartum [41,42]. Therefore, it has been hypothesized that cytokines inhibit the immune response during gestation. Consequently, the decrease of such inhibitors could allow the exacerbation or the first manifestation of autoimmune thyroid disease postpartum.

1. Clinical Presentation

The classic symptoms of overt hypothyroidism include fatigue, cold intolerance, constipation, decreased appetite, weight gain, muscle cramps, hair loss, depression, cognitive dysfunction, menstrual irregularities, and infertility (Table 52.3) [39]. Not infrequently, however, the clinical presentation is nonspecific. Patients with central hypothyroidism will typically present with deficiencies in other hormonal axes (growth-hormone insufficiency, adrenal insufficiency, and hypogonadism). In infants and children, hypothyroidism will result in mental retardation, neurological deficits, and retarded growth and bone maturation. At all ages bradycardia, hyporeflexia, myxedema, and hoarseness may be present. Presence or absence of a goiter does not distinguish between primary and central hypothyroidism, and in Hashimoto's thyroiditis the thyroid is initially often enlarged and becomes atrophic later in the course of the disease (Fig. 52.2).

2. Diagnosis

TSH is elevated in primary hypothyroidism and the diagnosis is confirmed by a decreased (free) T4 level (Fig. 52.3). Subclinical hypothyroidism is defined as elevated TSH but normal T4 and T3 concentrations. In central hypothyroidism, both TSH and T4 will be decreased. Elevated anti-TPO antibodies can confirm the diagnosis of autoimmune hypothyroidism. Thyroid function tests undergo complex alterations in severely ill patients and result in constellations found in hypothyroid subjects [43].

3. Therapy

Therapy for hypothyroidism is generally straightforward. In adults the average replacement dose is 1.6 μg/kg body weight (~112 μg per day). The adequacy of the replacement dose is monitored by measuring TSH. After a hypothyroid patient is treated with thyroxine, it may take 4–6 weeks to normalize the TSH levels. In secondary (pituitary) hypothyroidism, one has to rely on the peripheral hormone levels. In case of suspected adrenal insufficiency, levothyroxine should not be administered before evaluating and treating hypocortisolism, and particular care should also be taken in patients with cardiac disease. In long-standing hypothyroidism and elderly patients, treatment is often started with 25 μg levothyroxine per day and increased every two weeks. Excessive levothyroxine (L-T4) treatment with suppression of TSH may have negative impact on the heart and bone [44,45], issues that are discussed in the context of subclinical hyperthyroidism.

4. Subclinical Hypothyroidism

Widespread use of improved measurements for serum thyroid hormones and TSH has resulted in characterization of the syndrome we now call subclinical hypothyroidism, characterized by normal free T4 concentrations and raised serum TSH. Most patients have chronic autoimmune thyroiditis, as defined by positive anti-TPO antibodies. Several studies have clearly demonstrated that patients with so-called subclinical hypothyroidism present with discrete changes. For example, women with a mean age of 50 years, elevated TSH, and normal peripheral hormones were found to have significant changes on a clinical scale (Billewicz scale). Biochemical markers such as apolipoprotein A-I, TRH-stimulated prolactin, serum myoglobin, and creatine phosphokinase were elevated proportionally to the TSH elevation in comparison to age-matched controls [46]. In addition, the frequency of elevated LDL-cholesterol levels was higher in patients with TSH levels above 12 mU/l in this study, while total cholesterol, triglycerides, apoprotein B, and systolic time interval were only elevated in overt hypothyroidism [46].

Levothyroxine treatment may reduce serum cholesterol levels in patients with subclinical hypothyroidism and thereby decrease the incidence of coronary artery disease, stroke, and peripheral vascular disease [47–49]. Other vascular risk factors, such as lipoprotein(a) levels, were found to be elevated in some [50,51], but not all, studies of patients with subclinical hypothyroidism [52]. Additional factors in the complex equation leading to cardiovascular disease are nutrition, genetic predisposition, estrogen status, and smoking. Smokers with subclinical hypothyroidism have significantly more abnormal lipid patterns than nonsmokers and may benefit more from therapy with levothyroxine [53].

In addition to these effects on lipid metabolism and cardiovascular risk, subclinical hypothyroidism may have a negative impact on several other organ systems and may negatively influence reproductive function [54]. Moreover, it has also been associated with increased intraocular pressure [55].

Patients with subclinical hypothyroidism may present with alterations in memory, mood, anxiety, and depression, and show significant improvement after therapy with levothyroxine [56]. Over time, patients with subclinical hypothyroidism may progress to overt hypothyroidism with decreased T4 levels (Fig. 52.3). Risk factors for progression are the presence of thyroid autoantibodies, a high TSH level (>10 mU/l) and age. Early treatment may prevent symptoms and signs, many of them extremely subtle, that may develop with the gradual progression to overt hypothyroidism.

Only a small number of randomized trials of treatment for subclinical hypothyroidism have been performed and all were of short duration [57–59]. The results do not allow definite conclusions. Large randomized clinical trials have not been performed. Many clinicians are in favor of treating subclinical hypothyroidism with levothyroxine, at least in the form of a

Table 52.4
Hyperthyroidism

Etiology
  Graves' disease
  Multinodular goiter
  Toxic adenoma
  hCG-Induced hyperthyroidism (pregnancy, hydatidiform mole, choriocarcinoma)
  Thyroiditis (subacute, postpartum)
  Intake of thyroid hormone (thyrotoxicosis factitia)
  Thyroid hormone secreting thyroid carcinoma (rare)
  TSH-secreting pituitary adenoma (rare)
Clinical Presentation
  Symptoms
    Nervousness, heat intolerance, palpitations, weakness, fatigue, hair loss, eye symptoms
  Signs
    Adults: hyperactivity, increased temperature, increased perspiration, tachycardia, arrhythmias, hyperreflexia, weight loss, muscle weakness, tremor, eyelid retraction, exophthalmos, diffuse or nodular goiter
    Intrauterine/neonatal: advanced bone age, mental retardation possible
  Biochemical findings
    Primary: TSH ↓, T4 (↑), T3 ↑
    Secondary (pituitary): TSH normal or ↑, T4 ↑, T3 ↑
  Therapy
    Antithyroid drugs: carbimazole or propylthiouracil
    Betablockade with propanolol
    Radioablation with [131]iodine
    Surgery

time-limited trial [60,61]. Therapy with levothyroxine is safe if well monitored. Adverse effects occur only rarely and include signs and symptoms commonly found in hyperthyroidism.

Controversy exists as to whether subclinical thyroid disease is of sufficient clinical importance to warrant screening or case finding and whether therapy is justified. However, it is more and more accepted that measurement of serum TSH levels in patients who come to physicians for other reasons, particularly in women above age 40 to 50, is justified (case finding) [60,62,63]. Helfand and Redfern performed an extensive analysis of the literature to determine whether screening for subclinical thyroid dysfunction should be recommended, and whether persons with mildly abnormal TSH levels can benefit from treatment with levothyroxine [63]. This study provided a valuable summary of the studies pertinent to this complex question, although some of the conclusions are controversial, particularly from a clinical perspective [61].

D. Hyperthyroidism

Hyperthyroidism, the secretion of excessive amounts of thyroid hormone, is most commonly caused by Graves' disease, an autoimmune disorder, followed by toxic adenomas and toxic multinodular goiters (Table 52.4) [64]. Hyperthyroidism is less common than hypothyroidism. As it is the case for hypothyroidism, thyrotoxicosis occurs significantly more often in women than men. Graves' disease is caused by autoantibodies that stimulate the TSH receptor on thyroid follicular cells. This results in increased growth and hormone production. The etiology of Graves' disease is still elusive, but, in analogy to other autoimmune diseases, it is thought that it has a polygenic inherited basis [65]. The hereditary predisposition to Graves' disease is strongly supported by twin studies that demonstrate a high concordance for monozygotic, but not dizygotic, twins [66]. As in other autoimmune diseases, associations have been established between the disease and the presence of certain constellations of human leukocyte antigens (HLA), and Graves' disease was one of the first autoimmune disorders with documented associations with certain HLA types [67]. In addition to certain HLA-associated predispositions, other genes have been implicated [68–70], while the genes encoding well known thyroid antigens like the TSH receptor, thyroid peroxidase, and thyroglobulin have been excluded [69,71].

1. Ophthalmopathy

One of the peculiarities of Graves' disease is the fact that it can be associated with endocrine ophthalmopathy [72]. Endocrine ophthalmopathy is defined as an inflammatory process of the eyes which leads to soft tissue involvement (periorbital edema, congestion, and swelling of the conjunctiva), proptosis (anterior deplacement of the eye), extraocular muscle involvement leading to double-vision (diplopia), corneal lesions, and compression of the optic nerve. The pathogenesis of endocrine ophthalmopathy is poorly understood [73].

2. Clinical Presentation

Clinically, thyrotoxicosis is frequently associated with nervousness, palpitations, heat intolerance, increased appetite with concomitant weight loss, hair loss, weakness, and, in the case of Graves' disease, eye symptoms [64]. Menstrual irregularity is a frequent complaint in thyrotoxic women, but fertility is probably not significantly impaired in mild hyperthyroidism [74]. Classical signs include tachycardia, atrial fibrillation, hyperthermia, and muscle weakness and atrophy, as well as increased reflexes. The thyroid is diffusely enlarged in Graves' disease. Many patients with hyperthyroidism have a stare because of the increased sympathetic activity; this should be separated from the findings of true endocrine ophthalmopathy [72]. One also has to bear in mind that some older hyperthyroid patients with relatively severe thyrotoxicosis have few symptoms ("apathetic thyrotoxicosis").

3. Diagnosis

The biochemical diagnosis of hyperthyroidism is usually straightforward and characterized by a suppressed TSH with elevated peripheral hormones. However, isolated elevations of T3 with a normal T4 occur in "T3-thyrotoxicosis." Hyperthyroidism caused by the pathologic secretion of TSH is a rarity. In order to determine the etiology in the absence of eye disease (Graves' disease) or a palpable nodular goiter, scanning with radioiodine can be diagnostic. Autoantibodies directed against the TSH receptor may also be useful in discriminating between the different etiologies.

4. Therapy

The goal in the treatment of hyperthyroidism is a permanent normalization of thyroid hormone levels. Treatment options include thionamide drugs (propylthiouracil, methimazole, carbimazole), radioiodine, and rarely surgical removal of the hyperactive thyroid tissue. Betablockers can be helpful as adjuvant therapy in patients with marked increase in sympathetic activity. Preference is usually given to propranolol which also inhibits T4 to T3 conversion to some degree. Thionamides inhibit thyroid hormone synthesis [75]. Radioiodine therapy is the most frequently chosen form of therapy for hyperthyroidism in the United States, but it is not available or is less popular in other parts of the world [76]. It destroys thyroid follicular cells by β-irradiation. The effects have a slow onset (2–6 months). Thionamide treatment is therefore generally required prior to radiotherapy. Acute complications include radiation-induced thyroiditis and sialoadenitis. An absolute contraindication for radiotherapy is pregnancy, and women of childbearing age are advised not to get pregnant for 6–12 months after radiotherapy. The use of radioactive iodine in the treatment of children with hyperthyroidism is controversial [77].

The long-term safety of radioactive iodine therapy has been evaluated in several epidemiologic investigations beginning with the Cooperative Thyrotoxicosis Therapy Study in 1961 [77a–77c]. An elevated rate ratio (RR = 1.8, 95% confidence interval (CI) = 1.1–3.2) was observed for cancers of the organs that are known to concentrate iodine (salivary glands, digestive tract, kidney, and bladder), among patients at the Mayo Clinic who were treated with [131]I [77b]. A study of hyperthyroid patients who were treated at the Massachusetts General Hospital observed an increased risk of breast cancer in patients treated with radioactive iodine compared with hyperthyroid patients who received other treatments (standardized rate ratio = 1.9, 95% CI = 0.9–4.1) [77c]. Another report found no significantly increased risk of total cancer mortality in adults with hyperthyroidism who were treated with ^{131}I, but the risk of thyroid cancer mortality was elevated (RR = 3.9, 95% CI = 2.5–5.9) [77d]. The role of the underlying thyroid disease cannot be assessed in these studies and may have contributed to the observed risks. A review of the relationship between thyroid diseases and breast cancer has been published [77e].

Surgery can be indicated in the case of allergy to thionamides, nonacceptance of radioiodine treatment, pregnancy requiring large doses of thionamides, large goiters, and the suspicion of a malignant process. Hypoparathyroidism and injury of the recurrent nerve are potential complications. Ablation of the thyroid gland by radioactive iodine or surgery may result in posttherapeutic hypothyroidism requiring therapy with levothyroxine. Besides correction of thyroid dysfunction, the therapies for endocrine ophthalmopathy include steroids, radiotherapy of the orbita, and various forms of surgical decompression.

5. Subclinical Hyperthyroidism

In subclinical hyperthyroidism, patients have a decreased serum TSH but a T4 and T3 level in the normal range. In terms of symptoms and signs in patients with subclinical hyperthyroidism, some studies suggested an increase in frequency of nervousness and heat intolerance [78] and others documented mild symptoms in elderly persons on clinical scales (Wayne index) [79]. Given the fact that overt hyperthyroidism is associated with tachycardia, atrial fibrillation, and systolic hypertension, the question of whether subclinical hyperthyroidism affects the cardiovascular system is of particular importance. Atrial fibrillation was more prevalent in patients with subclinical hyperthyroidism followed in the Framingham population [45]. A higher heart rate, an increase in premature atrial beats, and a greater left ventricular mass were found in patients with thyroxine-induced subclinical hyperthyroidism [80]. Another study suggested the presence of an increased left ventricular mass in patients with subclinical hyperthyroidism [81].

Of further concern are the effects of excessive amounts of thyroid hormone on bone [82]. In a meta-analysis of thirteen studies of women with suppressed serum TSH due to L-T4 treatment and a control group, bone loss appeared to be increased in postmenopausal women on thyroxine [83]. In a randomized prospective study of postmenopausal women with subclinical hyperthyroidism, distal forearm bone density increased slightly in the group treated with methimazole, whereas it declined by 5% in the control group [84]. A similar trend was observed in a nonrandomized prospective analysis of bone density in the spine and the hip [85].

Debate exists as to whether subclinical hyperthyroidism should be treated. This decision may depend on other clinical findings such as cardiac alterations, postmenopausal osteoporosis, or the finding of a toxic adenoma or a multinodular goiter [60,86]. Defined risk factors for progression to overt hyperthyroidism are multinodular goiters and hyperfunctioning solitary nodules [79].

E. Thyroid Nodules and Multinodular Goiter

Nodular thyroid disease is defined by the presence of one or multiple nodules in the gland. It is extremely common and, depending on the sensitivity of the detection method and the population studied, it has a prevalence of up to 60%. Risk factors for the development of thyroid nodules are iodine deficiency and radiation, and there is a progressive increase with age. Thyroid nodules may be hyperfunctioning and secrete excessive amounts of thyroid hormone (toxic adenoma). Many questions remain as to why these benign neoplasms develop [87]. Some studies demonstrated that many thyroid nodules are true monoclonal tumors, suggesting that somatic mutations in genes involved in growth regulation are at their onset. Toxic adenomas are most commonly caused by activating mutations in the TSH receptor gene; more rarely they harbor mutations in the α-subunit of the stimulatory GTP-binding protein (Gsα). Malignant alterations are rare but depend on age and previous exposure to radiation. Besides biochemical testing, the evaluation often includes fine needle aspiration biopsy, thyroid scanning, and ultrasound [88]. Toxic adenomas can be treated with radioiodine. Surgery is the therapy of choice for lesions that are suspicious for malignancy. The success of growth suppression with levothyroxine is controversial.

F. Thyroid Cancer

Thyroid cancers are relatively infrequent neoplasms and account for about 0.6–1.6% of all malignancies [89]. The annual incidence of thyroid cancer is 1–10 per 100,000 persons in most countries. Thyroid cancer is about three times more frequent in females than males. Sixty to ninety percent of all thyroid cancers are papillary thyroid carcinomas (PTC) or follicular thyroid carcinomas (FTC), which originate from thyroid follicular cells. Their relative distribution is variable and the incidence of FTC

increases in regions with iodine deficiency. Besides the clinically relevant forms of PTC, occult PTC (<1 cm) are found frequently at autopsy. PTC may also occur in the setting of some rare autosomal dominant syndromes with disseminated neoplasia (Gardner's syndrome, Cowden's syndrome) [90]. The highly malignant anaplastic carcinoma (ATC) and several unusual variants of thyroid cancers are much rarer.

A major risk factor for the development of benign and malignant thyroid neoplasms is exposure to radiation. This has been recognized after use of external radiation for medical treatment of benign and malignant conditions in the neck, and more dramatically after accidental releases of ionizing radiation from atomic explosions and accidents in nuclear facilities [91,92].

Many molecular mechanisms involved in the development of thyroid neoplasms have been elucidated during the 1990s [67]. Analogous to colon cancer, a multistep progression from polyclonal hyperplastic lesion to monoclonal adenoma and to carcinoma has been proposed for thyroid malignancies [93]. While some elements are indeed consistent with such a scenario, the picture is far from being complete and further efforts are needed to fully understand thyroid oncogenesis.

Another form of cancer is medullary thyroid cancer (MTC) which originates from the calcitonin-producing C-cells in the thyroid gland. MTC can occur together with other endocrine tumors in the hereditary cancer syndrome, multiple endocrine neoplasia type 2, (MEN 2) [94]. MEN 2A is an autosomal dominant disease characterized by the association of MTC, parathyroid adenomas, and phaeochromocytoma [94]. In MEN 2B, parathyroid disease is absent, but the patients develop mucocutaneous ganglioneuromas and some have a marfanoid habitus. MTC can be sporadic or occur within families (familial medullary thyroid cancer, FMTC) without other manifestations of the MEN complex. The elucidation of the underlying genetic defect now allows determination of the carrier status within affected families and thus early therapy (see Section VI) [95].

1. Diagnosis

Particularly helpful in the diagnosis of thyroid cancer is fine needle aspiration. Imaging techniques include scanning with radioiodine, ultrasound, computerized tomography, and magnetic resonance imaging.

2. Therapy

The first step in the therapy of thyroid cancer is usually a near-total thyroidectomy with regional lymph node exploration. In the case of differentiated thyroid cancer, this is followed by postoperative radioiodine therapy. Levothyroxine is then administered to suppress TSH which is a stimulator of thyroid cell growth (Fig. 52.1). After surgical procedures, patients with differentiated carcinomas of the thyroid are routinely evaluated for metastatic disease using radioiodine scans and measurements of thyroglobulin. To achieve high sensitivity, these tests are traditionally performed after withdrawal of the suppressive treatment with thyroxine for several weeks. This results in an increase in TSH and a high uptake of radioiodine in residual tissue. Withdrawal of thyroxine leads to hypothyroidism and, potentially, to all the clinical problems associated with it. Moreover, the increase in TSH may potentially lead to proliferation of residual tumor cells. To avoid these problems, exogenous TSH was administered during uninterrupted treatment with thyroxine prior to performing radioiodine scans and thyroglobulin measurements. These procedures were performed using bovine TSH or human TSH from autopsies, sources that are no longer acceptable because of the possibility of anaphylactic reactions and the transmission of infectious disease (Jakob-Creutzfeld). After the cloning of the α and β subunits of human TSH, the production of pure recombinant human TSH (rhTSH) in sufficient quantities is now a reality. Administration of rhTSH has been evaluated in a clinical trial of patients with differentiated thyroid cancer [96] and is being introduced for clinical use. In the future, other applications may involve TSH administration during therapy with radioiodine for thyroid cancer, as well as in selected cases of hyperthyroidism, in order to increase the uptake of the isotope.

G. Thyroid Dysfunction Associated with Pregnancy

The physiological alterations found in pregnancy result in several unique changes in thyroid metabolism [97,98]. Iodine clearance increases because of the increase in renal glomerular filtration. The pregnancy-induced increase in estrogens results in higher serum levels of thyroxine binding globulin (TBG), the major hormone binding protein [99]. Consequently, the total T4 and T3 levels increase, whereas the free fractions remain in the normal range.

The increase in iodine clearance usually has few consequences. In iodine-sufficient regions there appears to be no increased incidence of clinically detectable goiter during pregnancy [100]. Ultrasound studies reveal either a small increase in thyroid volume during pregnancy or no detectable change. If iodine is scarce, pregnancy may induce a relative iodine-deficient state that may lead to goiter development and maternal hypothyroidism [101]. If iodine-deficiency is severe, not only maternal hypothyroidism, but also fetal hypothyroidism and cretinism, may result [98,99].

1. Graves' Disease during Pregnancy

Hyperthyroidism is second to diabetes mellitus as the most common endocrinopathy in pregnancy. Its prevalence is about 0.05–0.20% [70,102]. Clinical assessment may be difficult because many of the symptoms of hyperthyroidism are also associated with normal pregnancy. Human choriogonadotropin (hCG) may cause mild hyperthyroidism during the first trimester of gestation [103]. hCG interacts with the TSH receptor on thyroid follicular cells and can result in mildly elevated free thyroid hormones and suppressed TSH. The natural history of Graves' disease in pregnancy is typically characterized by aggravation during the first trimester, amelioration in the second and third trimester, and recurrence a few months after pregnancy [103]. Antithyroid drug treatment, usually with propylthiouracil, is the treatment of choice. Severe maternal thyrotoxicosis during pregnancy is associated with a significant increase in low birth weight babies, contributing to the neonatal mortality rate [97]. Severe maternal hyperthyroidism has been reported to be associated with a higher risk for miscarriage, but there is little evidence that fetal mortality is increased in comparison with pregnancy outcome in euthyroid women [97].

In mothers with autoimmune thyroid disease, the fetus may rarely present with congenital hyperthyroidism caused by

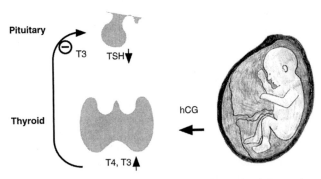

Fig. 52.5 Human choriogonadotropin-induced gestational thyrotoxicosis. The placental hormone hCG can cross-react with the TSH receptor on thyroid follicular cells. This leads to an increase in T4 and T3 secretion and suppression of TSH. This mechanism can lead to gestational thyrotoxicosis.

maternal-to-fetal transfer of thyroid-stimulating antibodies. This form of hyperthyroidism is transient because the stimulating antibodies are cleared from the fetal circulation. Very rarely, congenital hyperthyroidism may be persistent and due to activating mutations in the TSH receptor [104].

2. hCG-Induced Gestational Hyperthyroidism and Hyperemesis Gravidarum

Secretion of the placental hormone human chorionic gonadotropin (hCG) may result in additional changes in the pituitary-thyroid axis (Fig. 52.5). At high doses, hCG cross-reacts with the TSH-receptor on the thyroid cells. This may result in increased secretion of T4 and T3, and consequently in a decrease in TSH secretion [99,103]. The increased secretion of hCG may result in suppression of TSH or in overt hyperthyroidism. Because of its transient nature during pregnancy, it is called *gestational thyrotoxicosis* or *transient hyperthyroidism of hyperemesis gravidarum*. This form of hyperthyroidism is often associated with hyperemesis gravidarum [98,103]. The hyperemesis might be related to high levels of hCG-induced estradiol [105]. Elevations of hCG are particularly pronounced in twin pregnancies and in the presence of hydatidiform moles [106].

3. Postpartum Thyroiditis

Pregnancy is associated with a period of generalized immunosuppression. Autoimmune diseases often relapse after delivery. In women with a history of autoimmune thyroid disease, antithyroid antibodies reach maximal levels between three and seven months after delivery [107]. This can lead to the first manifestation or relapse of Graves' disease or postpartum thyroiditis. Postpartum thyroiditis is characterized by transient mild hyper- and/or hypothyroidism. Its prevalence is estimated to be between 3 and 16% [107]. The diagnosis can be confirmed by measuring anti-TPO antibodies, or, in the nonbreastfeeding mother by low or absent 24-hour radioiodine uptake. Treatment depends on severity. Follow-up is warranted because up to 25% of these patients will develop primary thyroid failure in the long term. Therapy with levothyroxine may be necessary during the hypothyroid phase of the illness.

4. Thyroid Cancer and Pregnancy

Pregnancy may modify the risk and management of thyroid cancer. Epidemiologic studies have suggested an association between thyroid cancer and parity [108–110]. The risk of cancer in thyroid nodules that develop during pregnancy appears to be increased [111–113]. There is controversy whether the natural course of thyroid cancer is altered during pregnancy, but the small number of studies does not allow definite conclusions [113–115]. Evaluation of a suspicious nodule can be performed with fine needle aspiration and ultrasound; radionuclide scanning is contraindicated during pregnancy. If a biopsy is suspicious or diagnostic for cancer, operation can be performed during pregnancy or immediately postpartum depending on the nature of the findings [116]. Radiotherapy has to be performed postpartum.

5. Placental Passage

TRH, iodine, thionamides, and thyroid antibodies, but not TSH, cross the placenta. Thionamides are used for the treatment of hyperthyroidism during pregnancy, while the administration of radioiodine for diagnostic or therapeutic reasons is contraindicated because of its transplacental passage. The extent of placental crossing of T4 and T3 has been a subject of controversy for many years [97]. T4 levels measured in neonates born with a complete organification defect (inability to synthesize T4) had T4 levels which that between 20 and 50% of normal, indicating that maternal-to-fetal transfer does occur [117]. The clinical observation that most infants with congenital hypothyroidism are asymptomatic at delivery further supports the concept that thyroid hormones do cross the placenta.

6. Lactation

Although thionamides are secreted in milk (methimazole, propylthiouracil), breastfeeding under thionamide therapy is considered safe if only small or moderate doses are required. Monitoring of the thyroid function of the infant is helpful in this instance. Lactation is an absolute contraindication to radioiodine therapy because iodine is secreted in the milk.

V. Epidemiologic Study Issues and Public Health Impact

The importance and prevalence of iodine deficiency disorders have already been mentioned (Section IV.A). The following discussion gives a short overview of the epidemiologic studies that address the incidence and prevalence of hypothyroidism, hyperthyroidism, thyroid nodules, and thyroid carcinomas. Comprehensive reviews on the epidemiology of thyroid disease are available [63,118,119].

The classic study in this field is the Whickham study performed in northeastern England [120]. It surveyed a representative sample of the adult population in 1972 which was then followed over a 20-year period [121]. This allowed the estimate of the incidence and prevalence of overt hyper- and hypothyroidism, subclinical hypothyroidism, and goiter in a community of approximately 21,000 adults. This sample population is thought to reflect the adult population of the United Kingdom and may be representative of many Western countries. While there is a paucity of longitudinal studies, the findings of the

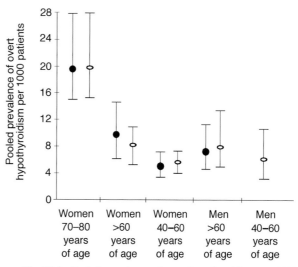

Fig. 52.6 Pooled prevalence of overt hypothyroidism per 1000 patients (means ±95% confidence intervals) in different age and sex categories. (From [63] with permission).

Whickham study have been confirmed by smaller studies from Europe, the United States, and Japan [63,119].

A. Hypothyroidism

The prevalence of hypothyroidism is age- and sex-dependent. It is particularly frequent in older women and lower in younger women and men. Figure 52.6 shows the pooled prevalence of overt hypothyroidism in women and men as a function of age [63]. When hypothyroidism was defined solely as an elevated TSH level, it was found in 7.5% of women and 2.8% of men in the Whickham population [120]. Overt hypothyroidism was found in 1.8% of women and 0.1% of men. In the Framingham study, 5.9% of women and 2.3% of men over age 60 had an elevated TSH [122]. Overall, overt thyroid dysfunction appears to be relatively rare in women younger than 40 years of age and men younger than 60 years of age [123]. Subclinical hypothyroidism is the most common condition found when screening with thyroid function tests; 5–10% of adult women have an elevated TSH level [63]. In the Whickham survey, 7.5% of women and 2.8% of men of all ages had an elevated TSH (6 mU/l or higher). The results of screening in a general office population are similar [124]. Hypothyroidism was found to be significantly more prevalent in a small study of geriatric inpatients, mostly males, in comparison to elderly ambulatory patients (males: 7.8% versus 0.7%; females: 17% versus 2.4%) [125].

The incidence of hypothyroidism was also assessed in the Whickham survey. Two-thirds of the women with an elevated TSH level had antithyroid antibodies in serum. During 20 years of follow-up, 55% of women with a TSH level of 6 mU/l or higher and a positive antibody test result developed overt hypothyroidism [121]. About 25% of those women with an initial TSH level of 10 mU/l or higher (~25%) had a risk for overt hypothyroidism over 20 years of ~90%, or about 0.11 per year.

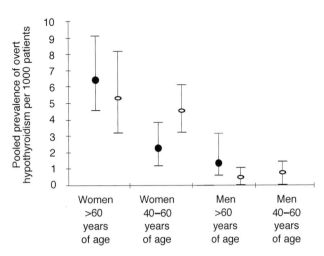

Fig. 52.7 Pooled prevalence of overt hyperthyroidism per 1000 patients (means ±95% confidence intervals) in different age and sex categories. (From [63] with permission).

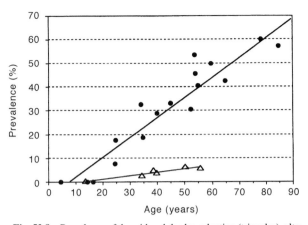

Fig. 52.8 Prevalence of thyroid nodules by palpation (triangles), ultrasound or autopsy (filled circles) in patients without radiation exposure or known thyroid disease. (From [126]. Copyright ©1993 Massachusetts Medical Society. All rights reserved.)

B. Hyperthyroidism

Analogous to the situation in hypothyroidism, the prevalence of hyperthyroidism is age- and sex-dependent, and it is lower in younger women and in men. Figure 52.7 shows the pooled prevalence of overt hyperthyroidism in women and men as a function of age [63]. In the Whickham study, overt hyperthyroidism was found in 2.8% of women and 0.2% of men [120]. In the Framingham study, TSH was found to be <0.1 mU/l in 1% of persons over age 60 and was associated with a threefold higher risk for the development of atrial fibrillation in the subsequent decade [45]. As illustrated in Fig. 52.7, the overall prevalence of overt hyperthyroidism was 2–4 per 1000 women between 40 and 60 years of age, and it showed a further increase above age 60. In contrast, the prevalence of hyperthyroidism is low in men.

The prevalence of subclinical hyperthyroidism is ~1.5% in women and ~1% in men older than 60 years of age [63]. These results may, in part, be influenced by the sensitivity of the TSH assays. Patients with clinical findings of thyroid disease, such as a functioning solitary nodule or a pronounced multinodular goiter, have a high risk for progression to overt hyperthyroidism, but it seems much lower in patients with subclinical hyperthyroidism on screening [63]. In screening studies, the yearly rate of progression to overt hyperthyroidism was found to be ~1.5% in women and ~0% in men [63].

C. Sporadic Goiter

In the Whickham survey, 16% of individuals were found to have a small but easily palpable diffuse or multinodular goiter [120]. Twenty-six percent of the women had a goiter and there was a marked decline with age (under age 45 years: 31%; over age 75 years: 12%). In men under age 25, goiter prevalence was 7%, between ages 64 and 75 it was 4 %, and no goiters were detected in men over age 75. This decline in frequency of goiters with increasing age was confirmed longitudinally in the Whickham study and was also observed in smaller studies in

the United States and Japan [118,121]. It contrasts with the increase in nodular lesions with increasing age [118].

D. Thyroid Nodules

The prevalence of nodules depends on the modality of screening and the population being evaluated. The prevalence of nodular thyroid disease has been assessed by palpation, with imaging techniques (ultrasound, scanning, computerized tomography, magnetic resonance imaging), and in autopsies [119]. In addition to the obvious differences in the sensitivity of the methodological approach, the prevalence depends on iodine intake and exposure to radiation. The prevalence of thyroid nodules in nonendemic goiter areas and unselected populations varies between 2.2 and 51 per 1000 people if investigated by palpation, between 190 and 347 per 1000 people by ultrasound, and between 82 and 646 per 1000 people in autopsy studies (Fig. 52.8) [126]. These studies indicate that thyroid nodules are very prevalent in the general population. The prevalence is higher in women and increases with age. The clinical significance of these findings are not clear, however, because most thyroid nodules are benign.

The incidence of thyroid nodules is poorly defined. For example, in the Framingham study, 4909 subjects were initially without evidence of thyroid nodules by palpation. During the 15 year follow-up, 67 of these 4909 individuals developed nodules as assessed by palpation (annual incidence ~0.09%). Women had a higher incidence (~0.11%) than men (0.06%). There are no reported studies on the incidence of thyroid nodules using more sensitive detection methods. Although difficult to assess in clinical practice, progression from a benign nodule to cancer seems to be a rare event, at least under conditions of normal iodine intake.

E. Cancer

Clinically relevant thyroid cancer is relatively rare. In contrast, occult carcinomas of the thyroid gland are common find-

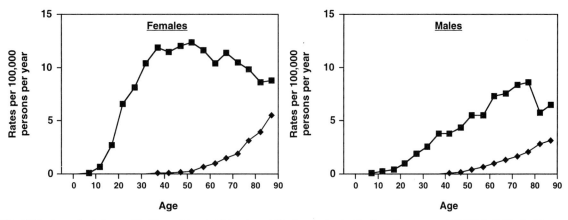

Fig. 52.9 Age-adjusted annual incidence (rhomboids) and mortality (squares) rates for thyroid cancer per 100,000 individuals. (Data from [128]).

ings in autopsy series. Occult thyroid cancer is defined as a cancer of less than 1 cm in diameter without clinical manifestations. The large majority of occult thyroid cancers are papillary carcinomas. Their prevalence ranges from 5 to 130 per 1000 people in autopsy series (mean-36 per 1000 people) [119]. Clinical thyroid cancer accounts for 0.6–1.6% of all malignancies with an annual incidence of ~0.5–10 per 100,000 people worldwide [127]. In 1996, the American Cancer Society estimated a total of 15,600 new cases or thyroid carcinomas in the United States, 11,600 in women and 4000 in men [128]. The age-adjusted annual incidence rate in the United States was 4.5 per 100,000 people between 1986 and 1990 [129]. For females the annual rate was 6.4 per 100,000 and for males it was 2.5 per 100,000. As illustrated in Fig. 52.9, there is an increase in the annual incidence rate with age. The 10-year survival rates are ~95% for papillary carcinomas and 90% for follicular carcinomas. The survival rates are higher in children and lower in older patients. In addition, mortality rates are influenced by the initial stage, the histological subtype, and the form of treatment. As discussed in Section IV.F, the relative distribution of papillary and follicular thyroid carcinomas depends on iodine intake.

VI. Screening for Thyroid Disease

A. Neonatal Screening

The majority of newborns with sporadic congenital hypothyroidism appear normal at birth, and fewer than 10% will be diagnosed as hypothyroid based solely on clinical signs. Because diagnosis may be difficult and early treatment is essential in order to avoid permanent mental retardation, neonatal screening programs have been established in most developed countries [10]. All neonates should be screened with a capillary blood sample collected on filter paper after birth for either a total T4 or TSH. Screening with T4, primarily used in North America, determines a T4 level with a high cut-off rate near the tenth percentile. A confirmatory test is performed by measuring TSH. In Europe and Japan, most screening programs rely on determination of TSH and use T4 for confirmation of the diagnosis. Both methodological approaches have been shown to yield small

numbers of false-negative and false-positive results, although concomitant measurement of TSH and T4 would be ideal if possible from a financial perspective. These programs are combined with tests for other congenital diseases and have proven to be successful and affordable.

B. Screening in Adults

Given the high prevalence and incidence of abnormal thyroid function tests, particularly in women above the age of 50, the question arises whether older women should be systematically screened using an ultrasensitive TSH assay. There is currently considerable debate about whether screening or case-finding is justified [61,63]. The prevalence and incidence of nonovert hypothyroidism and hyperthyroidism are low [121].

Screening and case-finding will thus mainly detect subclinical disease, particularly subclinical hypothyroidism. Is the treatment of subclinical hypothyroidism indicated, because it may be asymptomatic and treatment may have its own risks? There is, however, evidence that nonspecific symptoms may be improved by T4 treatment and that the disability only becomes apparent retrospectively [57]. It may also be of benefit to the cardiovascular system and to lipid metabolism [58,130,131]. In this context it is also of importance to emphasize that signs and symptoms of thyroid dysfunction may be discrete and misleading both in neonates and in adults. Therefore, many clinicians are in favor of therapy for subclinical hypothyroidism with levothyroxine, at least in the form of a time-limited trial [60,61]. Concerns have been raised as to the costs following the diagnosis of subclinical thyroid disease. It has been estimated that screening and treatment for all U.S. women older than 50 years of age with subclinical hypothyroidism would generate 4 million new lifetime prescriptions for L-thyroxine in the first year and an additional 600,000 to 1 million prescriptions every five years [63]. However, it has been suggested that screening for thyroid dysfunction in patients aged 35 years and older undergoing routine periodic health examinations may be cost-effective [62].

What should then be done with patients with subclinical hyperthyroidism identified by screening? Treatment should be considered in patients with atrial fibrillation, coronary heart disease, or osteoporosis. Other patients should be followed on a regular basis in order to detect progression to overt hyperthyroidism at an early stage.

While screening is not generally accepted, thyroid function should be assessed in all patients with dyslipidemias or atrial fibrillation and before starting treatment with amiodarone. Periodic measurement of TSH is indicated in patients on lithium therapy. Annual follow-up is also indicated in patients who have been treated for hyperthyroidism and who are at risk for recurrence of hyperthyroidism or posttherapeutic hypothyroidism. Patients with a history of irradiation of the neck or head are at higher risk for developing hypothyroidism and thyroid cancer and should be followed carefully as well. Assessment of thyroid function is also warranted in the context of other autoimmune diseases such as diabetes mellitus type 1, pernicious anemia, Addison's disease, or polyglandular autoimmune syndromes.

C. Genetic Testing

The availability of reliable genetic testing has had a major impact on the management of families with MEN-2 or FMTC [94,95]. Until recently, siblings of affected patients had to undergo annual clinical and biochemical testing. These tests were hampered by false-negative and false-positive results. Because it is now known that mutations in the ret proto-oncogene cause these cancer syndromes, the gene can be directly sequenced. Children with negative test results in the genetic analyses can be excluded from further screening studies. In the situation of a positive test result, total thyroidectomy can be performed early in childhood in order to avoid the development of a clinically significant MTC. Additional biochemical testing will allow detection of tumors of other glands at an early stage.

VII. Host and Environmental Determinants Influencing Thyroid Disorders

A. Gender and Ethnicity

Many thyroid diseases are caused or modulated by genetic, environmental, and nutritional factors. The basis for the pronounced gender difference in many thyroid diseases is only partially understood. Hormonal influences (e.g., estrogen and hCG) do play a role. Estrogen receptors have also been documented in normal and cancerous thyroid follicular cells. Linkage studies suggest that a locus on the X-chromosome is associated with autoimmune thyroid disease [132]. If confirmed, this may add another piece to the still incomplete puzzle. Ethnic background can also be relevant to the frequency of various thyroid diseases. For example, in the United States congenital hypothyroidism is much less prevalent in African-Americans but is more frequent in children of Hispanic origin. This observation supports the notion that congenital hypothyroidism has a genetic basis. Asians seem to be more susceptible to one complication of hyperthyroidism, thyrotoxic periodic paralysis, characterized by generalized muscular weakness or paralysis associated with decreased potassium levels. Roughly 50% of Australian Aborigines have an abnormal thyroxine binding globulin (TBG-A), and other variants are prevalent in populations from Africa and the Pacific Islands (TBG-S) [133].

B. Nutrition

Factors other than iodine deficiency have further impact on the development of endemic goiter [134]. They include environmental goitrogens, microbial pollution, and protein–calorie malnutrition. Naturally occurring agents with goitrogenic and/or antithyroid effects include sulfated organic compounds (thiocyanates, thyoglucosides), polyphenols, phenolic compounds, phthalate esters, and inorganic compounds such as lithium. Selenium deficiency, in combination with iodine deficiency, is thought to modulate the phenotypic expression of cretinism (myxoedematous versus neurologic cretinism).

C. Radiation

Exposure to radiation was first shown to cause thyroid cancer in 1950 [91]. It then became apparent that radiation exposure during childhood, which was widely used for the treatment of conditions like tonsil and thymic enlargement, acne, and fungal infections of the scalp (tinea capitis), is a significant risk factor for the development of thyroid nodules and thyroid cancer [92]. Besides radiation from exposure to X-rays or terrestrial sources, other more drastic exposures to radiation resulted from atomic bomb explosions (Hiroshima, Nagasaki, test areas in the Marshall islands and Nevada), nuclear facilities (Hanford, WA), and the accident in the nuclear power plant in Chernobyl in April 1996 [92,135]. Information about these exposures was often significantly delayed or withheld. For example, the releases of large amounts of radioiodine from the Hanford nuclear power plant in Washington during the 1940s and 1950s were only confirmed during the 1980s. In Chernobyl, millions of curies of radioactive material were released into the atmosphere, including ~40 million curies of ^{131}iodine, as well as short-lived iodines [135]. The delays in informing the public impeded attempts to administer potassium iodide prophylaxis. This led to significant exposures, particularly in children. In the years following the accident, a marked increase in the frequency of pediatric thyroid cancer has been observed in Belarus [136]. Between 1991 and 1992, the annual incidence of thyroid carcinomas was 62 times higher than in the decade preceding the nuclear accident. Histologically, more than 95% of these cancers were papillary thyroid carcinomas with a high prevalence of solid components. At the molecular level, post-Chernobyl carcinomas have a very high prevalence of particular DNA damage, the ret/PTC-3 rearrangement.

D. Drugs

Numerous drugs influence thyroid function tests [137]. This can be caused by elevations or decreases in the level of binding proteins, interference with protein binding, or modulation of pituitary secretion of TSH. Other mechanisms include alterations of thyroid cell function or monodeiodination of T4. For example, oral contraceptives and other forms of estrogen increase the levels of thyroxine binding globulin. This leads to an increase in total T4, a constellation that is still frequently mistaken to reflect a hyperthyroid state. TSH and free hormone levels are, however, normal. Two frequently used medications that interfere with thyroid function are lithium and amiodarone.

Lithium, commonly used for the treatment of depressive and manic states, inhibits thyroid hormone secretion and can result in hypothyroidism and goiter. The antiarrhythmic drug, amiodarone, contains 75 mg iodine per 200 mg tablet. It has multiple interactions with thyroid hormone formation and action. It inhibits TSH secretion and 5'-monodeiodination. In iodine deficient regions it induces hyperthyroidism in ~10% of the treated patients. In iodine replete areas, hypothyroidism occurs in ~20% of patients under amiodarone. Amiodarone seems also to interact with thyroid hormone receptors in the cell nucleus. Dopaminergic substances, frequently used in intensive care medicine, decrease TSH secretion.

E. Smoking

The effect of cigarette smoking on thyroid hormone action was compared in a study on women with various grades of hypothyroidism and a control group of euthyroid women [53]. Peripheral thyroid hormone action was assessed by a clinical score and measurements of ankle-reflex time and serum lipids and creatine kinase. Thyroid function was evaluated by measurement of serum thyrotropin, free T4, and T3. Among the women with subclinical hypothyroidism, smokers had a higher mean serum thyrotropin concentration and a higher T3/free T4 ratio than nonsmokers. Total cholesterol and low-density lipoprotein cholesterol were higher in smokers. Among the women with overt hypothyroidism, serum concentrations of thyrotropin, free thyroxine, and triiodothyronine were similar in smokers and nonsmokers. As compared with nonsmokers, smokers had a clinical score indicating a greater degree of hypothyroidism, higher serum concentrations of total and low-density lipoprotein cholesterol, longer ankle-reflex time, and higher serum concentrations of creatine kinase. This study suggests that smoking increases the metabolic effects of hypothyroidism [53].

VIII. Summary and Perspective

Disorders of the thyroid gland are common, particularly in women. The clinical spectrum is broad and these diseases have relevance to a wide variety of fields, not only in medicine but also in aspects of nutrition and radiation exposure.

Which problems and questions need further attention and research? The adequate iodine supplementation of the world population is of paramount importance for the eradication of endemic goiter and cretinism, the most prevalent cause of mental retardation. IDD is theoretically entirely preventable, but like other major public health issues, it is hampered by political, cultural, and socioeconomic hurdles. In countries with existing iodine supplementation, iodine intake needs to be monitored prospectively to assure its adequacy in the future. Neonatal screening for sporadic congenital hypothyroidism, well established in many countries, is not available in all regions of the world. Clinical research needs to answer the important question as to whether screening for thyroid disease should be performed routinely in women. This is intimately associated with the question of whether the detection and treatment of subclinical thyroid disease is beneficial and cost-effective. The role of systematic screening for thyroid dysfunction during pregnancy is also a question. Longitudinal studies on thyroid disease are scarce,

with the majority of studies performed in Caucasians. A better understanding of the pathogenesis of thyroid neoplasms and further improvements in their management remain challenging tasks. We know that radiation is a major risk factor for the development of thyroid nodules and malignancies, a risk factor which needs rigorous control. Basic research is providing many insights into the molecular basis of thyroid development and thyroid hormone synthesis and action. This rapidly moving field has growing impact on the diagnosis and treatment of thyroid disease, as illustrated by recombinant thyrotropin (rhTSH) which is currently being introduced for the follow-up management of patients with thyroid cancer. The hope remains that a better characterization of thyroid disorders at the epidemiologic, clinical, and molecular levels will ultimately result in further improvements in prophylaxis and therapy.

Acknowledgments

I thank Laird D. Madison, M.D., Ph.D., Northwestern University, Chicago, for critically reading the manuscript. This work was in part supported by a Northwestern University Medical School New Investigator Award from the Howard Hughes Medical Institute.

References

1. Merke, F. (1984). "History and Iconography of Endemic Goitre and Cretinism." Hans Huber, Bern.
2. Sawin, C. T. (1996). The heritage of the thyroid. In "The Thyroid" (L. E. Braverman and R. D. Utiger, eds.), 7th ed., pp. 2–5. Lippincott-Raven, Philadelphia.
3. Marine, D., and Kimball, O. P. (1920). Prevention of simple goiter in man. Arch. Intern. Med. **25**, 661–672.
4. Bürgi, H., Supersaxo, Z., and Selz, B. (1990). Iodine deficiency diseases in Switzerland one hundred years after Theodor Kocher's survey: A historical review with some new goitre prevalence data. Acta Endocrinol. (Copenhagen) **123**, 577–590.
5. WHO/UNICEF/ICCIDD (1996). "Recommended Iodine Levels in Salt and Guidelines for Monitoring their Adequacy and Effectiveness." Nutrition Unit, World Health Organization, Geneva.
6. Reverdin, J. L. (1882). Accidents consécutifs à l'ablation totale du goitre. Rev. Méd. Suisse Romande **2**, 539–540.
7. Kocher, T. (1883). Ueber Kropfexstirpation und ihre Folgen. Arch. Klin. Chir. **29**, 254–337.
8. Murray, G. R. (1891). Note on the treatment of myxoedema by hypodermic injection of an extract of the thyroid gland of a sheep. Br. Med. J. **2**, 796–797.
9. Fisher, D. A., Dussault, J. H., Foley, T. P. J., Klein, A. H., LaFranchi, S., Larsen, P. R., Mitchell, M. L., Murphey, W. H., and Walfish, P. G. (1979). Screening for congenital hypothyroidism: Results of screening of one million North American infants. J. Pediatr. **94**, 700–705.
10. LaFranchi, S. (1994). Congenital hypothyroidism: A newborn screening success story? Endocrinologist **4**, 477–486.
11. Kopp, P., and Jameson, J. L. (1998). Thyroid disorders. In "Principles of Molecular Medicine" (J. L. Jameson, ed.), pp. 459–473. Humana Press, Totowa, NJ.
12. Magner, J. A. (1990). Thyroid-stimulating hormone: Biosynthesis, cell biology, and bioactivity. Endocr. Rev. **11**, 354–385.
13. Grossmann, M., Weintraub, B. D., and Szkudlinski, M. W. (1997). Novel insights into the molecular mechanisms of human thyrotropin action: Structural, physiological, and therapeutic implications for the glycoprotein hormone family. Endocr. Rev. **18**, 476–501.
14. Taurog, A. (1996). Hormone synthesis. In "The Thyroid" (L. E. Braverman and R. D. Utiger, eds.), pp. 47–81. Lippincott-Raven, Philadelphia.

15. Bartalena, L. (1990). Recent achievements in studies on thyroid hormone-binding proteins. *Endocr. Rev.* **11**, 47–64.

16. Bartalena, L. (1994). Thyroid binding proteins: Update 1994. *In* "Endocrine Review Monographs" (L. Braverman and S. Refetoff, eds.), pp. 140–142. Endocrine Society, Rockville, MD.

17. Motomura, K., and Brent, G. A. (1998). Mechanisms of thyroid hormone action. *Clin. Endocrinol. Metab. North Am.* **27**, 1–19.

18. Mangelsdorf, D. J., Thummel, C., Beato, M., Herrlich, P., Schutz, G., Umesono, K., Blumberg, B., Kastner, P., Mark, M., Chambon, P., and Evans, R. M. (1995). The nuclear receptor superfamily: The second decade. *Cell (Cambridge, Mass.)* **83**, 835–839.

19. Braverman, L. E., and Utiger, R. D., eds. (1996). "The Thyroid." Lippincott-Raven, Philadelphia.

20. DeGroot, L. J., Larsen, P. R., and Hennemann, G. (1996). "The Thyroid and its Diseases." Churchill-Livingstone, New York.

21. Falk, S. A. (1997). "Thyroid Disease: Endocrinology, Surgery, Nuclear Medicine, and Radiotherapy." Lippincott-Raven, Philadelphia.

22. Braverman, L. E. (1997). "Diseases of the Thyroid." Humana Press, Totowa, NJ.

23. Singer, P. A., Cooper, D. S., Levy, E. G., Ladenson, P. W., Braverman, L. E., Daniels, G., Greenspan, F. S., McDougall, I. R., and Nikolai, T. F. (1995). Treatment guidelines for patients with hyperthyroidism and hypothyroidism. Standards of Care Committee, American Thyroid Association. *JAMA, J. Am. Med. Assoc.* **273**, 808–812.

24. Singer, P. A., Cooper, D. S., Daniels, G. H., Ladenson, P. W., Greenspan, F. S., Levy, E. G., Braverman, L. E., Clark, O. H., McDougall, I. R., Ain, K. V., and Dorfman, S. G. (1996). Treatment guidelines for patients with thyroid nodules and well-differentiated thyroid cancer. American Thyroid Association. *Arch. Intern. Med.* **156**, 165–172.

25. Vanderpump, M. P., Ahlquist, J. A., Franklyn, J. A., and Clayton, R. N. (1996). Consensus statement for good practice and audit measures in the management of hypothyroidism and hyperthyroidism. *Br. Med. J.* **313**, 539–544.

26. Klee, G. G., and Hay, I. D. (1997). Biochemical testing of thyroid function. *Endocrinol. Metab. Clin. North Am.* **26**, 763–775.

27. Sarne, D. H., and Refetoff, S. (1994). Thyroid function tests. *In* "Endocrinology" (L. J. De Groot, ed.), pp. 617–664. Saunders, Philadelphia.

28. Hetzel, B. S., and Dunn, J. T. (1989). The iodine deficiency disorders: Their nature and prevention. *Annu. Rev. Nutr.* **9**, 21–38.

29. Delange, F. (1994). The disorders induced by iodine deficiency. *Thyroid* **4**, 107–128.

30. Dunn, J. T. (1996). Seven deadly sins in confronting endemic iodine deficiency, and how to avoid them. *J. Clin. Endocrinol. Metab.* **81**, 1332–1335.

31. Dunn, J. T. (1998). What's happening to our iodine. *J. Clin. Endocrinol. Metab.* **83**, 3398–3400.

32. Ghent, W. R., Eskin, B. A., Low, D. A., and Hill, L. P. (1993). Iodine replacement in fibrocystic disease of the breast. *Can. J. Surg.* **36**, 453–460.

33. Eskin, B. A., Grotkowski, C. E., Connolly, C. P., and Ghent, W. R. (1995). Different tissue responses for iodine and iodide in rat thyroid and mammary glands. *Biol. Trace Elem. Res.* **49**, 9–19.

34. Dunn, J. T., Thilly, C. H., and Pretell, E. (1986). Iodized oil and other alternatives to iodized salt for the prophylaxis of endemic goiter and cretinism. *In* "Towards Eradication of Endemic Goiter, Cretinism, and Iodine Deficiency" (J. T. Dunn, E. A. Pretell, C. H. Daza, and F. E. Viteri, eds.), pp. 170–181. Pan American Health Organization (PAHO), Washington, DC.

35. Dunn, J. T., Semigran, M. J., and Delange, F. (1998). The prevention and management of iodine-induced hyperthyroidism and its cardiac features. *Thyroid* **8**, 101–106.

36. Delange, F., Dunn, J. T., and Glinoer, D. (1993). "Iodine Deficiency in Europe. A Continuing Concern." Plenum, New York.

37. Holowell, J. G., Staehling, N. W., Hannon, W. H., Flanders, D. W., Gunter, E. W., Maberly, G. F., Braverman, L. E., Pino, S., Miller, D. T., Garbe, P. L., DeLozier, D. M., and Jackson, R. J. (1998). Iodine nutrition in the United States. Trends and public health implications: iodine excretion data from National Health and Nutrition Examination Surveys I and III (1971–1974 and 1988–1994). *J. Clin. Endocrinol. Metab.* **83**, 3401–3408.

38. Kaplan, E. L., Shukla, M., Hara, H., and Ito, K. (1994). Developmental abnormalities of the thyroid. *In* "Endocrinology" (L. J. De Groot, ed.), pp. 893–899. Saunders, Philadelphia.

39. Utiger, R. D. (1994). Hypothyroidism. *In* "Endocrinology" (L. J. DeGroot, ed.), Vol. 1, pp. 752–768. Saunders, Philadelphia.

40. Ahmed, S. A., Penhale, W. J., and Talal, N. (1985). Sex hormones, immune responses, and autoimmune diseases. *Am. J. Pathol.* **121**, 531–551.

41. Chiovato, L., Lapi, P., Fiore, E., Tonacchera, M., and Pinchera, A. (1993). Thyroid autoimmunity and female gender. *J. Endocrinol. Invest.* **16**, 384–391.

42. Froelich, C. J., Goodwin, J. S., Bankhurst, A. D., and Williams, R. C. J. (1980). Pregnancy, a temporary fetal graft of suppressor cell in autoimmune disease. *Am. J. Med.* **69**, 329–331.

43. Nicoloff, J. T., and LoPresti, J. S. (1996). Nonthyroidal illness. *In* "The Thyroid" (L. E. Braverman and R. D. Utiger, eds.), pp. 286–296. Lippincott-Raven, Philadelphia.

44. Toft, A. D. (1994). Thyroxine therapy. *N. Engl. J. Med.* **331**, 174–180.

45. Sawin, C. T., Geller, A., Wolf, P. A., Bélanger, A. J., Baker, E., Bacharach, P., Wilson, P. W., Benjamin, E. J., and D'Agostino, R. B. (1994). Low serum thyrotropin concentrations as a risk factor for atrial fibrillation in older persons. *N. Engl. J. Med.* **331**, 1249–1252.

46. Staub, J. J., Althaus, B. U., Engler, H., Ryff, A. S., Trabucco, P., Marquardt, K., Burckhardt, D., Girard, J., and Weintraub, B. D. (1992). Spectrum of subclinical and overt hypothyroidism: Effect on thyrotropin, prolactin, and thyroid reserve, and metabolic impact on peripheral target tissues. *Am. J. Med.* **92**, 631–642.

47. Tanis, B. C., Westendorp, G. J., and Smelt, H. M. (1996). Effect of thyroid substitution on hypercholesterolaemia in patients with subclinical hypothyroidism: A reanalysis of intervention studies. *Clin. Endocrinol. (Oxford).* **44**, 643–649.

48. Diekman, T., Lansberg, P. J., Kastelein, J. J., and Wiersinga, W. M. (1995). Prevalence and correction of hypothyroidism in a large cohort of patients referred for dyslipidemia. *Arch. Intern. Med.* **155**, 1490–1495.

49. Althaus, B. U., Staub, J. J., Ryff-De Leche, A., Oberhansli, A., and Stahelin, H. B. (1988). LDL/HDL-changes in subclinical hypothyroidism: Possible risk factors for coronary heart disease. *Clin. Endocrinol. (Oxford)* **28**, 157–163.

50. Kung, A. W., Pang, R. W., and Janus, E. D. (1995). Elevated serum lipoprotein(a) in subclinical hypothyroidism. *Clin. Endocrinol. (Oxford)* **43**, 445–449.

51. Yildirimkaya, M., Ozata, M., Yilmaz, K., Kilinc, C., Gundogan, M. A., and Kutluay, T. (1996). Lipoprotein(a) concentration in subclinical hypothyroidism before and after levo-thyroxine therapy. *Endocr. J.* **43**, 731–736.

52. Arem, R., Escalante, D. A., Arem, N., Morrisett, J. D., and Patsch, W. (1995). Effect of l-thyroxine therapy on lipoprotein fractions in overt and subclinical hypothyroidism, with special reference to lipoprotein(a). *Metab., Clin. Exp.* **44**, 1559–1563.

53. Müller, B., Zulewski, H., Huber, P., Ratcliffe, J. G., and Staub, J. J. (1995). Impaired action of thyroid hormone associated with smoking in women with hypothyroidism. *N. Engl. J. Med.* **333**, 964–969.

54. Arem, R., and Escalante, D. (1996). Subclinical hypothyroidism: Epidemiology, diagnosis, and significance. *Adv. Intern. Med.* **41**, 213–250.

55. Centanni, M., Cesareo, R., Verallo, O., Brinelli, M., Canettieri, G., Viceconti, N., and Andreoli, M. (1997). Reversible increase of intraocular pressure in subclinical hypothyroid patients. *Eur. J. Endocrinol.* **136**, 595–598.

56. Monzani, F., Del Guerra, P., Caraccio, N., Pruneti, C. A., Pucci, E., Luisi, M., and Baschieri, L. (1993). Subclinical hypothyroidism: Neurobehavioral features and beneficial effect of l-thyroxine treatment. *Clin. Invest.* **71**, 367–371.

57. Cooper, D. S., Halpern, R., Wood, L. C., Levin, A. A., and Ridgway, E. C. (1984). L-thyroxine therapy in subclinical hypothyroidism. A double-blind, placebo-controlled trial. *Ann. Intern. Med.* **101**, 18–24.

58. Nyström, E., Caidahl, K., Fager, G., Wikkelso, C., Lundberg, P. A., and Lindstedt, G. (1988). A double-blind cross-over 12-month study of L-thyroxine treatment of women with "subclinical' hypothyroidism. *Clin. Endocrinol. (Oxford)* **29**, 63–75.

59. Jaeschke, R., Guyatt, G., Gerstein, H., Patterson, C., Molloy, W., Cook, D., Harper, S., Griffith, L., and Carbotte, R. (1996). Does treatment with L-thyroxine influence health status in middle-aged and older adults with subclinical hypothyroidism? *J. Gen. Intern. Med.* **11**, 744–749.

60. Surks, M. I., and Ocampo, E. (1996). Subclinical thyroid disease. *Am. J. Med.* **100**, 217–223.

61. Cooper, D. S. (1998). Subclinical thyroid disease: A clinician's perspective. *Ann. Intern. Med.* **129**, 135–138.

62. Danese, M. D., Powe, N. R., Sawin, C. T., and Ladenson, P. W. (1996). Screening for mild thyroid failure at the periodic health examination: A decision and cost-effectiveness analysis. *JAMA, J. Am. Med. Assoc.* **276**, 285–292.

63. Helfand, M., and Redfern, C. C. (1998). Clinical guideline. Part 2. *Ann. Intern. Med.* **129**, 144–158.

64. McKenzie, J. M., and Zakarija, M. (1994). Hyperthyroidism. *In* "Endocrinology" (L. J. DeGroot, ed.), Vol. 3, pp. 676–711. Saunders, Philadelphia.

65. Davis, T. F. (1998). Autoimmune thyroid disease genes come in many styles and colors. *J. Clin. Endocrinol. Metab.* **83**, 2291–3393.

66. Brix, T. H., Christensen, K., Holm, N. V., Harvald, B., and Hegedus, L. (1998). A population-based study of Graves' disease in Danish twins. *Clin. Endocrinol. (Oxford)* **48**, 397–400.

67. Farid, N., Shi, Y., and Zou, M. (1994). Molecular basis of thyroid cancer. *Endocr. Rev.* **15**, 202–232.

68. Yanagawa, T., Hidaka, Y., Guimaraes, V., Soliman, M., and DeGroot, L. J. (1996). CTLA-4 gene polymorphism associated with Graves' disease in a Caucasian population. *J. Clin. Endocrinol. Metab.* **80**, 41–45.

69. Tomer, Y., Barbesino, G., Greenberg, D. A., Concepcion, E. S., and Davis, T. F. (1998). Linkage analysis of candidate genes in autoimmune thyroid disease: 3. Detailed analysis of chromosome 14 localizes GD-1 close to MNG-1. *J. Clin. Endocrinol. Metab.* **83**, 4321–4327.

70. Davis, L. E., Lucas, M. J., and Hankins, G. D. V. (1989). Thyrotoxicosis complicating pregnancy. *Am. J. Obstet. Gynecol.* **160**, 63–70.

71. De Roux, N., Shields, D. C., Misrahi, M., Ratanachaiyavong, S., McGregor, A. M., and Milgrom, E. (1996). Analysis of the TSH receptor as candidate gene in familial Graves' disease. *J. Clin. Endocrinol. Metab.* **81**, 3483–3486.

72. Gladstone, G. J. (1998). Ophthalmologic aspects of thyroid-related orbitopathy. *Clin. Endocrinol. Metab. North Am.* **27**, 91–100.

73. Paschke, R., Vassart, G., and Ludgate, M. (1995). Current evidence for and against the TSH receptor being the common antigen

in Graves' disease and thyroid associated ophthalmopathy. *Clin. Endocrinol. (Oxford)* **42**, 565–569.

74. Thomas, R., and Reid, R. L. (1987). Thyroid disease and reproductive dysfunction. A review. *Obstet. Gynecol.* **70**, 789–798.

75. Cooper, D. S. (1998). Antithyroid drugs for the treatment of hyperthyroidism caused by Graves' disease. *Clin. Endocrinol. Metab. North Am.* **27**, 225–247.

76. Kaplan, M. M., Meier, D. A., and Dworkin, H. J. (1998). Treatment of hyperthyroidism with radioactive iodine. *Clin. Endocrinol. Metab. North Am.* **27**, 205–223.

77. Zimmerman, D., and Lteif, A. (1998). Thyrotoxicosis in children. *Clin. Endocrinol. Metab. North Am.* **27**, 109–126.

77a. Dobyns, B. M, Sheline, G. E., Workman, J. B., Tompkins, E. A., McConahey, W. M., and Becker, D. V. (1974). Malignant and benign neoplasms of the thyroid in patients treated for hyperthyroidism: A report of the Cooperative Thyrotoxicosis Therapy Follow-up Study. *J. Clin. Endocrinol. Metab.* **38**, 976–998.

77b. Hoffman, D. A., McConahey, W. M., Fraumeni, J. F., and Kurland, J. T. (1982). Cancer incidence following treatment of hyperthyroidism. *Int. J. Epidemiol.* **11**, 218–224.

77c. Goldman, M. B., Maloof, F., Monson, R. R., Aschengrau, A., Cooper, D. S., and Ridgway, E. C. (1988). Radioactive iodine therapy and breast cancer. A follow-up study of hyperthyroid women. *Am. J. Epidemiol.* **127**, 969–980.

77d. Ron, E., Doody, M. M., Becker, D. V., Brill, A. B., Curtis, R. E., Goldman, M. B., Harris, B. S. H., Hoffman, D. A., McConahey, W. M., Maxon, H. R., Preston-Martin, S., Warshauer, E., Wong, F. L., and Boice, J. D. (1998). Cancer mortality following treatment for adult hyperthyroidism. *JAMA, J. Am. Med. Assoc.* **280**, 347–355.

77e. Goldman, M. B. (1990). Thyroid diseases and breast cancer. *Epidemiol. Rev.* **12**, 16–28.

78. Schlote, B., Nowotny, B., Schaaf, L., Kleinbohl, D., Schmidt, R., Teuber, J., Paschke, R., Vardarli, I., Kaumeier, S., and Usadel, K. H. (1992). Subclinical hyperthyroidism: Physical and mental state of patients. *Eur. Arch. Psychiatry Clin. Neurosci.* **241**, 357–364.

79. Stott, D. J., McLellan, A. R., Finlayson, J., Chu, P., and Alexander, W. D. (1991). Elderly patients with suppressed serum TSH but normal free thyroid hormone levels usually have mild thyroid overactivity and are at increased risk of developing overt hyperthyroidism. *Q. J. Med.* **78**, 77–84.

80. Biondi, B., Fazio, S., Carella, C., Amato, G., Cittadini, A., Lupoli, G., Sacca, L., Bellastella, A., and Lombardi, G. (1993). Cardiac effects of long term thyrotropin-suppressive therapy with levothyroxine. *J. Clin. Endocrinol. Metab.* **77**, 334–338.

81. Shapiro, L. E., Sievert, R., Ong, L., Ocampo, E. L., Chance, R. A., Lee, M., Nanna, M., Ferrick, K., and Surks, M. I. (1997). Minimal cardiac effects in asymptomatic athyreotic patients chronically treated with thyrotropin-suppressive doses of l-thyroxine. *J. Clin. Endocrinol. Metab.* **82**, 2592–2595.

82. Ross, D. (1994). Hyperthyroidism, thyroid hormone therapy, and bone. *Thyroid* **4**, 319–326.

83. Faber, J., and Galloe, A. M. (1994). Changes in bone mass during prolonged subclinical hyperthyroidism due to L-thyroxine treatment: A meta-analysis. *Eur. J. Endocrinol.* **130**, 350–356.

84. Mudde, A. H., Houben, A. J., and Nieuwenhuijzen Kruseman, A. C. (1994). Bone metabolism during anti-thyroid drug treatment of endogenous subclinical hyperthyroidism. *Clin. Endocrinol. (Oxford)* **41**, 421–424.

85. Faber, J., Jensen, I. W., Petersen, L., Nygaard, B., Hegedus, L., and Siersbaek-Nielsen, K. (1998). Normalization of serum thyrotrophin by means of radioiodine treatment in subclinical hyperthyroidism: Effect on bone loss in postmenopausal women. *Clin. Endocrinol. (Oxford)* **48**, 285–290.

86. Marqusee, E., Haden, S. T., and Utiger, R. D. (1998). Subclinical thyrotoxicosis. *Endocrinol. Metab. Clin. North Am.* **27**, 37–49.

87. Studer, H., and Derwahl, M. (1995). Mechanisms of non-neoplastic endocrine hyperplasia and hyperfunction: A changing concept. *Endocr. Rev.* **16,** 411–426.

88. Gharib, H. (1997). Changing concepts in the diagnosis and management of thyroid nodules. *Endocrinol. Metab. Clin. North Am.* **26,** 777–800.

89. Schlumberger, M. J. (1998). Papillary and follicular thyroid carcinoma. *N. Engl. J. Med.* **338,** 297–306.

90. Ain, K. B. (1995). Papillary thyroid carcinoma. *Endocrinol. Metab. Clin. North Am.* **24,** 711–760.

91. Duffy, B. J., and Fitzgerald, P. J. (1950). Cancer of the thyroid in children: A report of 28 cases. *J. Clin. Endocrinol. Metab.* **10,** 1296–1308.

92. Sarne, D., and Schneider, A. B. (1996). External radiation and thyroid neoplasia. *Clin. Endocrinol. Metab. North Am.* **25,** 181–195.

93. Fagin, J. (1992). Molecular defects in thyroid neoplasia. *J. Clin. Endocrinol. Metab.* **75,** 1398–1400.

94. Raue, F., Frank-Raue, K., and Grauer, A. (1994). Multiple endocrine neoplasia type 2. *Endocrinol. Metab. Clin. North Am.* **23,** 137–156.

95. Wohllk, N., Cote, G., Evans, D. B., Goepfert, H., Ordonez, N. G., and Gagel, R. F. (1996). Application of genetic screening information to the management of medullary thyroid carcinoma and endocrine neoplasia type 2. *Endocrinol. Metab. Clin. North Am.* **25,** 1–25.

96. Ladenson, P. W., Braverman, L. E., Mazzaferri, E. L., Brucker-Davis, F., Cooper, D. S., Garber, J. R., Wondisford, F. E., Davies, T. F., DeGroot, L. J., Daniels, G. H., Ross, D. S., and Weintraub, B. D. (1997). Comparison of administration of recombinant human thyrotropin with withdrawal of thyroid hormone for radioactive iodine scanning in patients with thyroid carcinoma. *N. Engl. J. Med.* **337,** 888–896.

97. Burrow, G. (1993). Thyroid function and hyperfunction during gestation. *Endocr. Rev.* **14,** 194–202.

98. Glinoer, D. (1997). The regulation of thyroid function in pregnancy: Pathways of endocrine adaptation from physiology to pathology. *Endocr. Rev.* **18,** 404–433.

99. Burrow, G. N., Fisher, D. A., and Larsen, P. R. (1994). Maternal and fetal thyroid function. *N. Engl. J. Med.* **331,** 1072–1078.

100. Levy, R. P., Newman, D. M., Rejali, L. S., and Barford, D. A. (1980). The myth of goiter in pregnancy. *Am. J. Obstet. Gynecol.* **137,** 701–703.

101. Glinoer, D., De Nayer, P., Delange, F., Lemone, M., Toppet, V., Spehl, M., Grun, J. P., Kinthaert, J., and Lejeune, B. (1995). A randomized trial for the treatment of mild iodine deficiency during pregnancy: Maternal and neonatal effects. *J. Clin. Endocrinol. Metab.* **80,** 258–269.

102. Burrow, G. N. (1985). The management of thyrotoxicosis in pregnancy. *N. Engl. J. Med.* **313,** 562–565.

103. Glinoer, D. (1998). Thyroid hyperfunction during pregnancy. *Thyroid* **8,** 859–864.

104. Kopp, P., van Sande, J., Parma, J., Duprez, L., Gerber, H., Joss, E., Jameson, J. L., Dumont, J. E., and Vassart, G. (1995). Congenital hyperthyroidism caused by a mutation in the thyrotropin-receptor gene. *N. Engl. J. Med.* **332,** 150–154.

105. Yoshimura, M., and Hershman, J. M. (1995). Thyrotropic action of human chorionic gonadotropin. *Thyroid* **5,** 425–434.

106. Grun, J. P., Meuris, S., De Nayer, P., and Glinoer, D. (1997). The thyrotrophic role of human chorionic gonadotrophin (hCG) in the early stages of twin (versus single) pregnancies. *Clin. Endocrinol. (Oxford)* **46,** 719–725.

107. Roti, E., and Emerson, C. H. (1992). Clinical review 29: Postpartum thyroiditis. *J. Clin. Endocrinol. Metab.* **74,** 3–5.

108. Levi, F., Franceschi, S., Gulie, C., Negri, E., and La Vecchia, C. (1993). Female thyroid cancer: The role of reproductive and hormonal factors in Switzerland. *Oncology* **50,** 309–315.

109. Kravdal, O., Glattre, E., and Haldorsen, T. (1991). Positive correlation between parity and incidence of thyroid cancer: New evidence based on complete Norwegian birth cohorts. *Int. J. Cancer* **49,** 831–836.

110. Preston-Martin, S., Bernstein, L., Pike, M. C., Maldonado, A. A., and Henderson, B. E. (1987). Thyroid cancer among young women related to prior thyroid disease and pregnancy history. *Br. J. Cancer* **55,** 191–195.

111. Rosen, I. B., Walfish, P. G., and Nikore, V. (1985). Pregnancy and surgical thyroid disease. *Surgery* **98,** 1135–1140.

112. Rosen, I. B., and Walfish, P. G. (1986). Pregnancy as a predisposing factor in thyroid neoplasia. *Arch. Surg. (Chicago)* **121,** 1287–1290.

113. Kobayashi, K., Tanaka, Y., Ishiguro, S., and Mori, T. (1994). Rapidly growing thyroid carcinoma during pregnancy. *J. Surg. Oncol.* **55,** 61–64.

114. Herzon, F. S., Morris, D. M., Segal, M. N., Rauch, G., and Parnell, T. (1994). Coexistent thyroid cancer and pregnancy. *Arch. Otolaryngol. Head Neck Surg.* **120,** 1191–1193.

115. Mestman, J. H., Goodwin, T. M., and Montoro, M. M. (1995). Thyroid disorders of pregnancy. *Endocrinol. Metab. Clin. North Am.* **24,** 41–71.

116. McClellan, D. R., and Francis, G. L. (1996). Thyroid cancer in children, pregnant women, and patients with Graves' disease. *Endocrinol. Metab. Clin. North Am.* **25,** 27–48.

117. Vulsma, T., Gons, M. H., and De Vijlder, J. J. M. (1989). Maternal-fetal transfer of thyroxine in congenital hypothyroidism due to a total organification defect or thyroid agenesis. *N. Engl. J. Med.* **321,** 13–16.

118. Turnbridge, M. P. J., and Vanderpump, W. M. G. (1996). The epidemiology of thyroid disease. *In* "The Thyroid" (L. E. Braverman and R. D. Utiger, eds.), pp. 474–482. Lippincott-Raven, Philadelphia.

119. Wang, C., and Crapo, L. M. (1997). The epidemiology of thyroid disease and implications for screening. *Endocrinol. Metab. Clin. North Am.* **26,** 189–218.

120. Tunbridge, W. M., Evered, D. C., Hall, R., Appleton, D., Brewis, M., Clark, F., Evans, J. G., Young, E., Bird, T., and Smith, P. A. (1977). The spectrum of thyroid disease in a community: The Whickham Survey. *Clin. Endocrinol. (Oxford)* **7,** 481–493.

121. Vanderpump, M. P., Tunbridge, W. M., French, J. M., Appleton, D., Bates, D., Clark, F., Grimley Evans, J., Hasan, D. M., Rodgers, H., Turnbridge, F., and Young, E. T. (1995). The incidence of thyroid disorders in the community: A twenty-year follow-up of the Whickham survey. *Clin. Endocrinol. (Oxford)* **43,** 55–68.

122. Sawin, C. T., Chopra, D., Azizi, F., Mannix, J. E., and Bacharach, P. (1979). The aging thyroid. Increased prevalence of elevated serum thyrotropin levels in the elderly. *JAMA, J. Am. Med. Assoc.* **242,** 247–250.

123. Schaaf, L., Pohl, T., Schmidt, R., Vardali, I., Teuber, J., Schlote-Sauter, B., Nowotny, B., Schiebeler, H., Zober, A., and Usadel, K. H. (1993). Screening for thyroid disorders in a working population. *Clin. Invest.* **71,** 126–31.

124. Eggertsen, R., Petersen, K., Lundberg, P. A., Nystrom, E., and Lindstedt, G. (1988). Screening for thyroid disease in a primary care unit with a thyroid stimulating hormone assay with a low detection limit. *Br. Med. J.* **297,** 1586–1592.

125. Livingston, E. H., Hershman, J. M., Sawin, C. T., and Yoshikawa, T. T. (1987). Prevalence of thyroid disease and abnormal thyroid tests in older hospitalized and ambulatory persons. *J. Am. Geriatr. Soc.* **35,** 109–114.

126. Mazzaferri, E. L. (1993). Management of a solitary nodule. *N. Engl. J. Med.* **328,** 553–559.

126a. Vander, J. B., Gaston, E. A., and Dawber, T. R. (1968). The significance of nontoxic thyroid nodules: Final report of a 15 year study of the incidence of thyroid malignancy. *Ann. Intern. Med.* **69,** 537–540.

127. Franceschi, S., and La Vecchia, C. (1994). Thyroid cancer. *Cancer Surv.* **19,** 393–422.

128. Parker, S. L., Tong, T., Bolden, S., and Wingo, P. A. (1996). Cancer statistics, 1996. *Ca—Cancer J. Clin.* **46,** 5–27.

129. National Cancer Institute (1993). "SEER Cancer Statistics Review." NIH, Bethesda, MD.

130. Bell, G. M., Todd, W. T., Forfar, J. C., Martyn, C., Wathen, C. G., Gow, S., Riemersma, R., and Toft, A. D. (1985). End-organ responses to thyroxine therapy in subclinical hypothyroidism. *Clin. Endocrinol. (Oxford)* **22,** 83–89.

131. Franklyn, J. A., Daykin, J., Betteridge, J., Hughes, E. A., Holder, R., Jones, S. R., and Sheppard, M. C. (1993). Thyroxine replacement therapy and circulating lipid concentrations. *Clin. Endocrinol. (Oxford)* **38,** 453–459.

132. Barbesino, G., Tomer, Y., Concepcion, E. S., Davis, T. F., and Greenberg, D. A. (1998). Linkage analysis of candidate genes in autoimmune thyroid disease: 2. Selected gender-related genes and the X-chromosome. *J. Clin. Endocrinol. Metab.* (in press).

133. Refetoff, S. (1994). Inherited thyroxine-binding globulin abnormalities in man: Update 1994. *Endocr. Rev. Monogr.* **3,** 162–164.

134. Gaitan, E., Cooksey, E. C., and Lindsay, R. H. (1986). Factors other than iodine deficiency in endemic goiter: Goitrogenesis and protein-calorie malnutrition (PCM). *In* "Towards Eradication of Endemic Goiter, Cretinism, and Iodine Deficiency" (J. T. Dunn, E. A. Pretell, C. H. Daza, and F. E. Viteri, eds.), pp. 28–45. Pan American Health Organization (PAHO), Washington, DC.

135. Robbins, J. (1997). Lessons from Chernobyl; the event, the aftermath fallout: Radioactive, political, social. *Thyroid* **7,** 189–192.

136. Nikiforov, Y., Gnepp, D., and Fagin, J. (1996). Thyroid lesions in children and adolescents after the Chernobyl disaster: Implications for the study of radiation carcinogenesis. *J. Clin. Endocrinol. Metab.* **81,** 9–14.

137. Meier, C., and Burger, A. (1996). Effects of pharmacologic agents on thyroid hormone homeostasis. *In* "The Thyroid" (L. E. Braverman and R. D. Utiger, eds.), pp. 276–286. Lippincott-Raven, Philadelphia.

53

Rheumatoid Arthritis

CARIN E. DUGOWSON
Bone and Joint Center
University of Washington Medical School
Seattle, Washington

I. Introduction

Rheumatoid arthritis (RA) is an autoimmune disease that has been recognized as a clinical entity for over two centuries. The most common presentation of RA is a symmetrical inflammatory polyarthritis, particularly of the hands and feet, although any synovial joint is at risk. Rheumatoid arthritis is also a systemic illness, with extra-articular manifestations occurring commonly, including subcutaneous nodules, pulmonary disease, vasculitis, and neuropathy. RA is the most common of the inflammatory arthritides and, combined with its multiple systemic effects and complications of therapy, causes significant morbidity [1–3].

Some researchers believe that RA is a modern disease because there are few bony artifacts showing the characteristic bony changes of RA, compared to clear skeletal evidence of spondylitis. Skeletal remains in North America suggest that RA was present there in pre-Columbian times. Such discussions are potentially important in the continuing search for the cause of RA.

As with many autoimmune diseases, women are at greater risk than men for the development of RA. Beyond gender differences in incidence rates, there is also a complex relationship among rheumatoid arthritis, female sex hormones, and reproductive status in modulating both the risks for disease and its clinical course in women [4]. Genetic susceptibility also plays a role, with increased disease risk and severity in certain Human Leukocyte Antigen (HLA) haplotypes (see Section III, Immunogenetics). These factors make understanding the epidemiology of RA particularly relevant in discussion of women's health issues.

The initiating event in RA is still unknown. It is now understood, however, that the time course of joint damage occurs early in the disease for many patients [5]. Studies of morbidity and mortality in rheumatoid arthritis have demonstrated decreased life expectancy and lost wages that are comparable to other major illnesses such as stroke [6]. For this reason, early intervention with remittive therapies has become the standard of care for active, recent onset RA. Combinations of remittive drugs are often used for active disease. Improved surgical interventions, particularly in total joint replacement, offer pain relief and improved function to many patients.

Studies also demonstrate the importance of psychosocial issues for RA patients. These include the issues associated with child-rearing and job performance for women with active disease. Additionally, studies of fatigue, depression, and self-efficacy in people with rheumatoid arthritis have shown the need for interventions to control or minimize these factors. A multidisciplinary approach, including an assessment of the psychosocial and economic impact, is important to manage RA optimally.

II. Pathogenesis

Understanding the pathogenesis of RA has been difficult because of the complexity of immune and nonimmune responses, as well as the diversity of tissues affected. In the 1990s, however, consensus has emerged that the initial event in development of RA is antigen-dependent activation of T cells which then causes a host of secondary physiologic and morphologic effects. Primary among these effects is the production of autoantibodies with changes in joint synovium and the secretion of proinflammatory cytokines. For unknown reasons, this initial immune response is located in the synovium and is responsible for the joint swelling and rheumatoid factor production.

One of the most important of the proinflammatory cytokines released in these early synovial events is tumor necrosis factor (TNF)-alpha. TNF-alpha appears to be a principal mediator in the development of synovitis. This proliferative synovitis is known as pannus and is seen only in RA. Once pannus has formed, the next step in joint damage is the invasion and destruction of cartilage by this activated synovium. Synovial fluid containing destructive inflammatory molecules is also produced. Cartilage damage occurs from the edges of the cartilage centrally and, like the initial antigenic event, has a complex causal pathway, including inflammatory mediators and immune complex activity. It is through these mechanisms that the damage and destruction of cartilage and bone occurs.

Since RA was first described as a clinical entity, scientists have searched for an infectious cause without success. Several viruses have been considered as possible causative agents in a susceptible host. The infectious disease model proposes a ubiquitous agent, such as Epstein-Barr virus (EBV) or human parvovirus B19, to trigger an autoimmune reaction in the susceptible host. The mechanism is theorized to be an immune reaction caused by homology between some part of the virion and an antigenic component of the joint, such as cartilage or synovium. Efforts to link RA causally with EBV have failed to demonstrate a consistent pattern. Parvovirus B19 causes a clinical syndrome quite similar to RA in women [7], but it is not thought to be a cause of persistent disease. In addition, parvovirus infections are not associated with the increase in HLA-DR4 characteristically seen in RA [8]. Other putative agents, including mycobacteria and retroviruses, have also been studied with similar inconclusive results.

III. Immunogenetics

The association between genetic traits of an individual and their risk for certain autoimmune rheumatic diseases has been recognized since the 1980s. Genetic associations with clinical aspects of RA, including risk for development of disease, severity of disease, and remission during pregnancy, have provided us with insight into immunologic events in RA [9]. Chromosome 6 contains a region known as the Major Histocompatibility Complex (MHC), which in humans includes the two classes of transplantation antigens or HLA types. Class I HLAs include the classic transplantation antigens HLA-A, HLA-B, and HLA-C. The first association with rheumatic disease, including ankylosing spondylitis and Reiter's syndrome, was with HLA-B27, a Class I gene.

The Class II complex is organized into three major areas: HLA-DR, HLA-DQ, and HLA-DP. Susceptibility to RA is associated with DR4 and related DR4 alleles. Certain DRB1 genes encode for a shared amino acid sequence associated with increased risk of RA in Caucasians. Additionally, a DR14 allele associated with RA risk in Native Americans also shares this sequence. The association of the shared amino acid sequence and increased risk for RA is called the shared epitope theory. Studies suggest that when this shared epitope is present there is increased susceptibility for severe RA, although it is uncertain if the susceptibility is to severe disease or to the disease itself [9,10]. Possible mechanisms of increased risk include an effect of these HLA alleles on shaping the T-cell receptor repertoire or structural changes in an area of antigen binding. This HLA-DRB1 susceptibility association has been documented in multiple populations, including relative risks of 6 in Caucasians, 3.5 in Japanese, and 2.6 in Native Americans [11,12]. The amount, or "dosage," of the shared epitope also appears to be important, with homozygosity for DR4 conferring the greatest increased risk. The shared epitope is thought to be associated with severe disease rather than directly to rheumatoid factor status.

The clinical implications of this observation are unclear. Early identification of patients at risk for erosive or otherwise severe disease would allow earlier institution of more aggressive therapy and spare those patients not in need of additional intervention. Compared to earlier attempts to stratify risk through full HLA typing, testing for these more specific HLA alleles is more easily done. Additionally, HLA typing does not change with disease activity. Studies are currently investigating the value of DR4 typing alone or in conjunction with other clinical information, including rheumatoid factor, X-rays, and clinical presentation. Such algorithms hold significant promise for a more focused therapeutic approach. At this time, however, routine HLA typing is not recommended.

IV. Epidemiology

A. Disease Occurrence

This section includes geographic distribution, the frequency measures of incidence and prevalence, secular trends (changes in disease occurrence over time), and prognosis or outcome.

1. Geographic Distribution

RA has been identified worldwide [13,14]. It is commonly cited that RA has a prevalence of 1%, but published ranges actually vary widely [15–17]. The highest prevalence has been described in North American native tribes like the Yakima and Pima [18], with intermediate frequencies in Caucasian populations both in Europe and the United States. The lowest prevalence is seen in African populations [19].

Incidence studies in RA have been difficult due to its relative infrequency. Overall, an average annual incidence rate of 28.1/100,000 population for males and of 65.7/100,000 for females was reported for the period 1950 to 1974 [16]. Age-specific rates increased with age. In women, the incidence was two to three times higher than in men of the same age. [16,17]. In the decade from 1972 to 1981, rates for females declined while rates for males remained relatively stable [17].

2. Time Trends of Incidence

A study from Rochester, Minnesota first reported a decline in the incidence of RA in recent decades [17]. Studies in England and America have confirmed these findings [17,20,21] As the decline in incidence was seen only in women, one area of intense research has been the association of the observed decline in the disease to use of oral contraceptives and postmenopausal hormones. The magnitude of decline reported approximates 50% between the 1970s and 1990s. However, studies in the United States and Europe have produced conflicting results and researchers have questioned whether there was an actual decline in incidence or changes in methodology through diagnostic criteria revision.

Many rheumatologists also believe the pattern of severity has been changing in recent decades, with less frequent occurrence of extra-articular disease such as vasculitis. One study found a peak in seropositivity and disease severity during the 1960s, with less RA and less severe RA being diagnosed currently [22]. Whether this is true or a consequence of changes in diagnosis and/or treatment remains uncertain.

B. Age

In contrast to earlier teaching, it appears that the incidence of RA does not peak in women at middle age. Multiple studies now show that incidence continues to rise with age in women [16,17,20]. However, one study showed a plateau for women between ages 45 and 75, but incidence continued to rise with age in men [21].

C. Race

The study of geographic variation is closely related to variation by race or ethnicity, but multiple studies confirm the universality of the disease. Although higher than average prevalence has been found in the Yakima and Pima Native American communities [18], this is not seen in all Native American groups. This differential risk is at least partially explained by the presence of certain HLA types. Other populations have shown a particularly low prevalence of RA, including Chinese populations in several locations, Japanese, Indonesian groups, and rural African groups. The risk of developing RA in adult white populations of Europe and North America shows the least variation by geographic area.

D. Gender Differences in Disease Manifestations

A study of differences in clinical presentation between men and women found that men have more erosive disease with

Table 53.1
Reproductive Variables and Rheumatoid Arthritis

Parity
    Nulliparous women have increased risk of developing RA
    A single live birth appears to permanently decrease risk of
        developing RA

Established RA
    Remission: Symptoms improve in 75% of pregnancies
    Postpartum flare: Worsening in first six weeks postpartum
    Maternal–fetal HLA mismatch predictive of clinical remission

New-onset RA
    Decreased risk for developing RA during pregnancy
    Increased risk for developing RA in first year postpartum

Oral contraceptives
    May decrease risk for developing RA
    No known treatment effect

earlier age at onset [23] and overall a worse outcome. Although mortality in RA is increased compared to the general population, women have a better prognosis in this regard than men [24]. Extra-articular disease is more common in men, including rheumatoid nodules and lung disease, which are known predictors of increased mortality. Women more often have keratoconjunctivitis sicca, an autoimmune syndrome of dryness of the eyes and/or mouth. There was no difference in treatment patterns between men and women. Men with RA have more large joints affected, whereas women have more involvement of the small joints of the hand. Women have more joint surgeries, perhaps because of the greater hand involvement.

1. Reproductive Variables in RA

Since the first observation by Hench in 1938 of the remission of RA symptoms with pregnancy, it has been recognized that a woman's reproductive characteristics can affect the course of rheumatoid arthritis [25]. Research in the 1990s has provided remarkable discoveries in this area. Table 53.1 summarizes these findings.

a. PREGNANCY AND RA.   Women with established RA who become pregnant experience relief of symptoms about 75% of the time [26,27]. Relief usually begins in the first trimester with symptoms returning by about the third month after delivery. Improvement in one pregnancy often predicts remission in future pregnancies. The risk of developing RA is also affected by reproductive status. There is a decreased risk of onset of RA during pregnancy [28–30].

Nulliparity was found to be a risk for development of RA in several studies [29,31,32] but not in another [33]. In at least one study, the protective effect of pregnancy was found to last many years after pregnancy completion. The effect of parity as well as its biological significance are still controversial.

There is no evidence of adverse outcomes of pregnancy in women with RA [27]. The question of whether fertility or fecundity is affected remains somewhat unclear; smaller family sizes have been found in women with RA compared to controls, and time to conception has also been found to be longer. If there is an effect, however, it is not a large one [34].

b. LACTATION.   As with other reproductive variables, breast-feeding has been reported to have a variety of effects. One study found breastfeeding increased the risk of developing RA, although this is complicated by the fact that lactation occurs in the immediate postpartum period when RA is known to flare independent of breastfeeding. Additionally, it was found that lactation might have a later protective effect on the development of RA [35,36].

c. EXOGENOUS HORMONES.   As a modulator of hormonal status, exogenous hormone use has received a great deal of attention. Oral contraceptives are a combination of an estrogen and progesterone which are taken to prevent ovulation in women. In 1978, Wingrave proposed that the use of such medications had decreased the risk of developing RA [37]. Since then, studies in the United States and Europe have obtained conflicting results about the existence and magnitude of such an effect. A meta-analysis reviewing these studies concluded there was a protective effect, with a 30% decrease in risk of developing RA in oral contraceptive users [38]. The effect of postmenopausal hormone replacement therapy on RA risk in older women is modest, if it exists at all [39].

E. Lifestyle Factors

1. Socioeconomic Status

Socioeconomic status and formal education level are predictive markers for both morbidity and mortality in RA [40]. Lower socioeconomic status and level of education are associated with worse functional status. Neither socioeconomic status nor formal education level is a risk factor for development of the disease.

2. Diet

There are many beliefs about the influence of diet on arthritis in general and RA in particular. Dairy products, tomatoes, potatoes, and meat products all have been invoked as causative or exacerbating. High dietary intake of fish-oil fatty acids is associated with a significantly decreased risk for developing RA [41].

Dietary modification has been advocated for management of arthritis for decades, including but not limited to RA. Since that time, several possible mechanisms have been studied to relate oral intake to activity or prevention of RA. Food hypersensitivity appears to be a possibility for at least some patients. In 1986, a case report described a patient with possible milk allergy exacerbating RA. Exclusion of dietary antigens was found to decrease disease activity in some patients [42–44]. Fasting [45] and a vegan diet [46] were shown to be mildly beneficial in two controlled studies.

Dietary supplementation has had mixed results. A trial of supplementation with alpha-linoleic acid using flaxseed and safflower seed failed to show any improvement in three months [47]. Two reviews agree that marine and botanical lipids were of some benefit in active RA [48,49]. The most promising approach appears to be a diet enriched with fish-oil fatty acids. Some improvement in symptoms was found in 11 of 12 studies reviewed using fish-oil supplementation. An objective decrease in several measures of disease activity was also seen [50]. A

Table 53.2

Classification Criteria for Rheumatoid Arthritis*[a]*

1. Morning stiffness lasting at least one hour
2. Simultaneous arthritis of three or more joints
3. Arthritis of hand joints
4. Symmetrical arthritis
5. Rheumatoid nodules
6. Abnormal serum rheumatoid factor
7. Radiographic changes typical of RA on posteroanterior hand and wrist radiographs

[a] Four of seven criteria must be satisfied for a diagnosis of RA; criteria 1 through 4 must be continuous for at least six weeks.

Adapted with permission from [55].

proposed mechanism is modification of arachidonic-acid metabolites to less inflammatory products, thus decreasing the symptoms of RA or other rheumatic disease. Vitamin treatment has not been shown to improve disease. Of the trace metals, low serum zinc levels are found in RA patients and patients experienced some benefit with repletion [51]. There is no information on other trace metals. Dietary modification and supplementation remain of interest but of modest clinical impact at this time.

Smoking has been associated with increased risk for development of RA [52–54]. Increased alcohol use in women was associated with a decreased risk and obesity with an increased risk in the latter study [54].

V. Classification Criteria

Several criteria sets for definition of RA have been developed over the years and all have been hampered by the lack of a definitive laboratory or clinical test for the disease. In 1987, revised criteria developed by the American College of Rheumatology were published [55]. A primary use of these criteria is for reproducibility in clinical studies, particularly to distinguish RA from other autoimmune diseases and to remove the modifiers reflecting confidence in the diagnosis, such as classical, definite, or probable RA. They were also developed in patients having established disease. Studies of these revised criteria indicate that they are more specific but less sensitive than earlier criteria [56]. This lesser sensitivity means that they can be less useful in the diagnosis of rheumatoid arthritis in early or mild forms. In an important change from earlier criteria sets, patients with other clinical diagnoses such as psoriatic arthritis, systemic lupus erythematosus, or fibromyalgia are not excluded from the diagnosis of RA.

There are seven criteria for diagnosis; four must be fulfilled for classification as RA (see Table 53.2). Morning stiffness of greater than one hour, symmetrical arthritis, arthritis in at least three areas, and hand arthritis are the four commonly occurring clinical criteria. Presence of rheumatoid nodules is the fifth clinical feature but is less commonly seen; this finding is generally quite specific for RA but not often seen early in disease. Rheumatoid factor positivity and radiographic changes are the two laboratory findings, which complete the criteria set. These criteria are frequently not seen until a year or more of active dis-

ease has passed. Patients who are initially seronegative often subsequently develop a positive rheumatoid factor [57,58].

It is useful to refer to these criteria when evaluating a patient with recent onset disease but one should remember that at times no definitive diagnosis can be made or the initial diagnosis may need reevaluation. Conversely, patients can develop other symptoms or signs that definitively diagnose a different condition. Passage of time can be a diagnostic tool that allows reassessment and reclassification of the initial diagnosis.

VI. Evaluation and Management of RA

A. Clinical Presentation

The characteristic manifestation of RA is a symmetrical inflammatory polyarthritis [59]. Onset of symptoms is most commonly gradual, although acute onset is infrequently seen [60,61]. Multiple joints are usually affected and symmetry of involvement is the rule. The most commonly affected joints are the smaller peripheral joints, which in the hands are the metacarpophalangeal joints (MCPs), proximal interphalangeal joints (PIPs), and wrists. The metatarsophalangeal joints (MTPs) of the feet are often affected, although patients often do not recognize these symptoms as part of a systemic illness.

RA can affect virtually any synovial joint, including such locations as the jaw (temporomandibular joint) or ribs (costochondral joints). The back, however, is rarely involved. Systemic symptoms such as fatigue and morning stiffness are the rule; fever and weight loss are seen in a minority of patients. Early in the illness, joint swelling is common. Joint deformity and extra-articular complications, such as vasculitis, splenomegaly, or lung disease, are not usually seen early in presentation.

Symptoms are characteristically much worse in the morning and will often last for hours, in contrast to the symptoms of noninflammatory illnesses such as osteoarthritis. Palindromic rheumatism is a pattern of abrupt onset with a migratory pattern. Short-lived attacks on one or two joints resolve spontaneously, only to recur in a different location weeks or months later. In about one-third of patients, these episodes become more frequent and long lasting, ultimately to evolve into typical RA.

As variable as the onset, the clinical course can take one of several patterns. Three general courses of disease are often described. Polycyclic is the most common course. This pattern is seen in about 70% of cases and is characterized by continuing disease with flares of varying degrees of severity and remissions, which are often incomplete. Monocyclic describes a single cycle with remission lasting over one year; this is seen in about 20% of cases. A progressive course of unremitting disease is the worst, with additive joint involvement over years. This is the least common, seen in only about 10% of patients in published studies; these may be the cases seen less frequently with the decreased severity of disease reported in recent years.

Clinical manifestations are not only articular. The systemic symptoms of morning stiffness and fatigue have already been mentioned. Rheumatoid nodules are the most frequently occurring extra-articular problem. The nodules are usually located on extensor surfaces such as the elbow, dorsum of the fingers, or Achilles tendon. Neurologic involvement is fairly common, with carpal tunnel syndrome, which is caused by entrapment of the median nerve at the wrist, the most often seen. It can be a presenting symptom in patients where wrist involvement is prominent. Cervical spine disease can be confusing in the older patient with known osteoarthritis, but the characteristic morning stiffness helps distinguish this from noninflammatory disease. It is not generally an early finding, but rather one of later, erosive disease. Cervical spine symptoms should be investigated carefully; preoperative evaluation before elective surgery, particularly where general anesthesia will be used, is essential.

Hematologic abnormalities other than anemia include thrombocytosis, which is rarely a clinical problem, and Felty's syndrome. The latter, a triad of rheumatoid arthritis, splenomegaly, and leukopenia, is a complication of long-standing disease, seen usually in patients with erosions and high-titer rheumatoid factor. Some Felty's patients later develop lymphoproliferative malignancy, including large granular lymphocyte (LGL) leukemia.

Virtually any organ system can be affected. Pulmonary involvement is common but clinically significant respiratory disease is not. Small pleural effusions, usually transudates, can be found on X-ray. Rheumatoid nodules in the lung require evaluation to rule out malignancy or infection. Diffuse interstitial fibrosis and bronchiolitis are uncommon and worsen the prognosis.

Cutaneous vasculitis was a more common finding in earlier times and represents part of the spectrum of small vessel vasculitis, which can include periungual infarcts and leg ulcers. Some studies have suggested an association between the use of corticosteroids and vasculitis but whether this is causal or rather that sicker patients are more likely to have both problems is still not certain.

The most common eye problem is keratoconjunctivitis sicca, or dry eye syndrome, which affects 10–30% of patients. Other uncommon syndromes can occur. Systemic amyloidosis is another uncommon complication of long-standing disease.

Patients with RA are at high risk for development of osteoporosis and all RA patients should be regularly assessed as to their bone health. RA itself causes osteopenia; corticosteroid use and immobility worsen the risk. In addition, many postmenopausal women are not on hormone therapy or other agents to maintain bone density. There are now proven therapies for prevention and treatment of corticosteroid-induced osteoporosis, including hormone replacement and bisphosphonate therapy. Bone density should be measured by dual photon X-ray absorptiometry (DEXA) in women at increased risk for osteoporosis and, if abnormal, should be monitored for appropriate response to therapy.

B. Laboratory Testing

Laboratory testing can be helpful in diagnosis. An elevated sedimentation rate (ESR) or C-reactive protein level, nonspecific measures of inflammation, is often seen. Hematologic studies often show a mild anemia with indices suggesting inflammatory block. Iron deficiency is not usually seen except as associated with other problems such as antiinflammatory use.

Rheumatoid factor is an autoantibody directed against a portion of IgG and was first found in the serum of RA patients by Waaler in 1940 [62]. Blood tests show rheumatoid factor to be present in elevated amounts in half of the patients at initial diagnosis [63]. Whether these autoantibodies are pathologic or

epiphenomenal still remains unclear. Patients with early development of positive and high titers of rheumatoid factor tend to have more severe disease. However, this correlation is far from perfect; rheumatoid factor negative patients can have erosive disease and the true nature or function of this antibody is still unknown.

Patients who make high amounts of rheumatoid factors often have certain HLA haplotypes, which are themselves associated with increased risk for RA. Many patients with RA are rheumatoid factor negative when first tested; repeat testing in a year shows conversion to seropositivity in about 80% of patients [57]. From a clinical perspective, the absence of this antibody, particularly on initial or early testing, does not eliminate RA as the correct diagnosis. In addition, rheumatoid factors are found in other chronic diseases with immune responses.

Other laboratory tests, including routine chemistries and immunologic testing, such as complement levels, ANA (antinuclear antibody) testing, or anti-DNA binding, are usually normal. In RA of relatively recent onset, almost all tests can be normal or negative and diagnosis will depend on the clinical history and examination.

C. Radiologic Evaluation

Radiographic studies should include X-rays of affected joints, usually the hands, wrists, and feet. These may show osteopenia (loss of bone) in affected areas but are otherwise normal the majority of the time. Although hand involvement is readily observed and frequently the presenting symptomatic joint, the MTPs of the feet have been shown to be the most sensitive areas for early identification of joint erosions [65]. It is important to evaluate these areas for erosions, for this alters prognosis and appropriate therapy. A radionucleotide bone scan is uncommonly done for diagnosis but can show inflammation in affected areas when the physical examination is negative and further investigation is indicated based on the patient's symptoms.

D. Differential Diagnosis

The elderly can present with polyarthralgias and polymyalgias of the shoulder and pelvic girdle. When these occur in association with a high ESR, it may be difficult to distinguish RA from polymyalgia rheumatica. These patients are generally seronegative and X-rays are nondiagnostic. Treatment often begins with low doses of corticosteroids until more characteristic articular involvement appears.

Monoarticular presentation, especially in a knee or shoulder, can be confusing. If the diagnosis is RA, other joints become involved over time to clarify the diagnosis. If a patient has an acute monoarticular problem, however, gout or infection is possible. Evaluation for a septic joint, though uncommon, should include questioning for clues to an organism's point of entry, such as dental work or other procedures. If there is any question, the joint should be aspirated. The white blood cell count in RA joint fluid can occasionally be quite high but this is usually not the case. Peripheral blood and joint white cell count and differential joint fluid crystal and gram stain should be done. Culture of the fluid will ultimately make or exclude the diagnosis of infection.

Parvovirus is an acute infection which is almost identical in clinical manifestations to RA. Although cases are often sporadic, it is sometimes seen in an epidemic presentation, since it is responsible for Fifth's disease in children. Exposure in a school or daycare setting can be a useful clue. The diagnosis is confirmed by serology showing an acute parvovirus infection. Other infections considered in the differential diagnosis could include sexually transmitted diseases such as disseminated gonococcal infection.

Systemic lupus erythematosus (SLE) is an uncommon systemic autoimmune illness that should be considered in the differential diagnosis. Criteria for lupus include inflammatory arthritis, positive ANA, photosensitivity, and other organ involvement, particularly renal or hematologic. Rheumatologic consultation should be sought promptly if these findings are present.

E. Treatment of RA

Treatment of RA has changed substantially over the years. There are new remittive drugs, anti-inflammatory drugs, and new approaches to medication therapy. In addition, the pregnant or lactating woman has special concerns and risks. Familiarity with the use of these medications is critical for anyone contemplating treating women with RA. Review articles summarize what is known about medication use in pregnant and lactating women. [66,67] Nonetheless, for the safety of the woman and her child,

treatment should only be offered by or in conjunction with a practitioner experienced in the management of these women.

Once a patient has been diagnosed with RA, the physician should make every effort to use a standard assessment to monitor her status. Subjective information, including functional disability, should be obtained in as reproducible a way as possible. Two measurement instruments, the Health Assessment Questionnaire (HAQ) and the Arthritis Impact Measurement Scales (AIMS), have been validated for RA and are sensitive to clinical change [68]. Physicians can also use a specific activity to assess a patient's current status.

Treatment needs to be customized for each patient by careful evaluation of her physical examination, laboratory information, radiographs, and psychosocial situation. A key assessment is what the likely prognosis is given the patient's constellation of findings. The presence of joint erosions is the single most important negative finding as erosions lead to deformity and predict the greatest loss of function. The evidence documenting development of erosions as an early event in RA is compelling (See Figure 53.1) [66] and early treatment with remittive drugs is designed to arrest that process as early and effectively as possible in the at-risk patient. Patients can have synovitis without erosions and control of that synovitis is also important. Disease-modifying agents that prevent these complications will

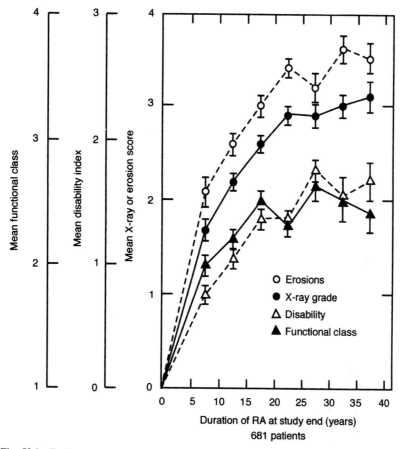

Fig. 53.1 Erosion score, radiologic grade, functional class, and disability as a function of disease duration.

also treat the constitutional symptoms of RA, notably the morning stiffness, swelling, and fatigue.

1. Medications for RA

a. DRUG THERAPY FOR RA.   Drug therapy for RA is divided into four categories: nonsteroidal anti-inflammatory drugs (NSAIDs), disease-modifying drugs, corticosteroids, and analgesics. NSAIDs are also called first-line drugs. As the name implies, they are used for initial therapy and are characterized by rapid onset of action for pain relief and swelling. Disease-modifying antirheumatic drugs (DMARDs) are also called remittive, slow-acting drugs or second-line drugs. These take a variable amount of time before benefit is seen, ranging from several weeks to many months. They are used in more active disease to control the underlying immunologic process, to improve function, and to prevent joint damage and deformity. Second-line drugs include such agents as methotrexate, sulfasalazine, hydroxychloroquine, and azathioprine.

Corticosteroids are often used in the treatment of RA and are in a category of their own because of their particular combination of risks and benefits. The use of corticosteroids is quite common regardless of disease duration because of the potency of their anti-inflammatory action. Several studies suggest that corticosteroid use can prevent progression of the bony disease but this is not universally accepted [67]. In addition, the multiple side-effects of corticosteroids make them an undesirable maintenance medication in most situations.

NSAIDs include such older medications as ibuprofen, indomethacin, and naproxen, as well as about two dozen others. Some of these medications are available in both over-the-counter and prescription doses, although it is difficult to get an anti-inflammatory effect in the doses available without prescription. NSAIDs are the most common therapy in RA, being used both initially and then in combination with other drugs.

The most common adverse effects of NSAIDs are on the gastrointestinal system, with dyspepsia occurring in about 10% of all patients on NSAIDs. The most serious gastrointestinal (GI) side effects are development of stomach or duodenal ulcers with or without bleeding. Estimates of the risk of GI problems show significant variability between patient groups and individual medications but overall demonstrate an increased risk of hospitalization for GI toxicity in RA users [69]. Risk factors for complications include age > 60 years, a past history of GI problems, anticoagulation therapy, and concurrent corticosteroid use [70]. These medications are often a key part of the management of a woman with RA and can be used quite successfully. It is essential, however, to monitor these drugs carefully even in the otherwise healthy woman and it is of critical importance when treating an older woman or one with multiple medical problems and/or medications.

The mechanism of anti-inflammatory action in NSAIDs is believed to be inhibition of prostaglandin synthesis, although the differential effect and toxicities of different NSAIDs has not been explained. Some drugs have been believed to be less likely to cause renal toxicity or GI problems because of their pharmacological properties. It was recognized that cyclo-oxygenase (COX), a proinflammatory compound, has two forms [71]. The Food and Drug Administration (FDA) has approved medications that are specific COX-2 inhibitors that have demonstrated much

less GI toxicity. With the availability of this category of medication, physicians should be able to provide anti-inflammatory medication to patients previously unable to tolerate such drugs.

In pregnancy, NSAIDs have demonstrated a variety of adverse effects, including oligohydramnios. The most significant concern is premature closure of the ductus arteriosus in the fetal heart, which can occur any time after 32 weeks of gestation. NSAIDs are contraindicated after that time. In general, corticosteroids are the anti-inflammatory of choice in the pregnant or lactating woman. These patients need comanagement with an experienced obstetrician and pediatrician to optimize care for mother and child.

DMARDs are medications that have been shown to alter favorably the clinical course of RA. Such medications take from six weeks to six months to become effective but decrease the likelihood and severity of radiologically defined joint damage and they provide relief of clinical symptoms. Analyses of therapy trials demonstrate fewer erosions and improved quality of life through use of these drugs [72,73]. Unfortunately, a fairly low percentage of patients are able to continue on any given remittive therapy over the course of their illness, for reasons of toxicity or decreasing efficacy [74].

Gold, given by intramuscular injection, is the first drug for which randomized studies demonstrated remittive efficacy and was the "gold standard" of remittive therapy for many years [75]. In the 1990s, however, gold has been almost totally eclipsed by better tolerated and more convenient medications. It is also available in an oral form, auranofin, with substantially less toxicity. Efficacy of the oral form is also low and it is generally used only in combination, if at all. Gold is generally avoided during pregnancy because of a case report of cleft palate and concern about postpartum hematologic or GI problems.

Methotrexate (MTX), a folic acid antagonist, was used to treat malignant disease for many years. Subsequently it was used as a weekly medication for psoriasis on the theory that it would selectively affect the more rapidly dividing cells in psoriatic disease. It was reported to be efficacious in RA and psoriasis as early as 1951 but little attention was given it until two reports from the Pacific Northwest were published in the early 1980s [76,77]. Since that time, MTX has become the most commonly used remittive drug for treatment of RA.

MTX is given orally in tablet or liquid form or by injection but is taken weekly rather than daily. The time to initial improvement is about six weeks, with continued improvement over time and as the dosage is increased. In long-term studies monitoring both efficacy and toxicity, MTX has been shown to effect sustained clinical improvement, although protection against radiologic progression is still uncertain [78,79]. MTX is excreted by the kidneys and renal impairment will cause increased serum levels and toxicity. The dose must be adjusted if MTX is used in patients with creatinine clearance under 60 ml/min [80]. Some physicians feel patients have less GI distress with divided doses over 24–36 hours, although patient convenience and compliance is aided by a single dose. It has been shown that there is increased toxicity if MTX is given daily and this is not an acceptable regimen.

Oral medication is more convenient than injections. The latter is reserved for patients who have responded to MTX but have GI symptoms that make continued therapy difficult. It is also commonly believed that parenteral administration makes liver abnormalities less likely, although confirmatory data are lacking.

The major adverse effect of MTX is on the liver and careful monitoring of liver function and clinical symptoms is mandatory. Other complications can include hematologic abnormalities and pulmonary disease. If monitored carefully and regularly, MTX can be given with a minimum of toxicity. Women should have their blood counts and renal and liver function measured at the start of therapy and be seen every two to four weeks at the initiation of therapy, both for monitoring of toxicity and dose adjustment. Alcohol intake is a risk factor for liver toxicity and most rheumatologists counsel their patients to avoid alcohol entirely while on MTX. Many rheumatologists also routinely supplement their patients' diet with folic acid to avoid hematologic or other complications of therapy.

MTX is contraindicated during pregnancy because of demonstrated toxicity to fetal tissues and cranial abnormalities in infants exposed *in utero*. Women with inadvertent exposure during pregnancy need to be counseled on possible fetal abnormalities. Women planning pregnancy should discontinue MTX at least two to three months prior to attempted conception [81]. MTX is found in breast milk and is, therefore, also contraindicated during breastfeeding [81].

Sulfasalazine (SSZ) is the next most frequently used remittive drug. Like MTX, it was first used several decades ago. Controlled studies in the 1970s and 1980s showed it to be an effective drug [82,83] with slowing of radiologic progression [84]. It has been used for inflammatory bowel disease for decades and has a well-defined safety profile. The mechanism of action is unclear. SSZ is metabolized in the GI tract. It is broken down after ingestion to sulfapyridine, which is a sulfonamide antibiotic, and a salicylate. Studies suggest that sulfapyridine is the active agent in treatment of RA [85].

SSZ is better tolerated when therapy is begun at a low dose and the dose is gradually increased. Most patients require 1.5–2 g daily to a maximum of 3 g. For the majority of women who do not have side effects, it is an easy medication to take, requiring little monitoring and useful over a long time period. Blood count and liver function should be measured at the start of therapy and at intervals of three to six months. Some rheumatologists also check a G6PD to minimize the chance of hemolytic anemia.

Side-effects are fairly common with SSZ and force discontinuation in 20–25% of RA patients. SSZ is sulfa-containing and cannot be used in women with a known sulfa allergy. The side-effects can be considered idiosyncratic or dose-related. The idiosyncratic ones are skin reactions, hematologic catastrophes of agranulocytosis or aplastic anemia, hepatitis, or pneumonitis. The drug should be stopped and the woman not rechallenged. More common and less severe reactions include GI discomfort, headaches, leukopenia, and skin rashes.

No teratogenicity or perinatal mortality has been reported [86]. However, many physicians will discontinue SSZ in the third trimester because of concerns about the fetus' ability to conjugate the drug and the potential for hyperbilirubinemia as well as gastric or hematologic problems in the perinatal period.

Hydroxychloroquine (HCQ) was initially used for treatment of malaria and was serendipitously found to treat inflammatory arthritis. It has been in use since the 1960s and is less potent than most other remittive agents. Because of the current emphasis on early treatment and the drug's favorable toxicity profile, it is used in treatment of early nonerosive disease and in combination therapy for more severe active disease.

The time to clinical improvement for HCQ is longer than that of any other DMARD and treatment for three to six months is necessary before efficacy can be assessed. Similarly, the drug has a very long half-life and time to steady-state tissue concentrations. The mechanism of action is not clear, although effects on lysosomes are well documented and are presumed to be central to the drug's efficacy.

Many studies have demonstrated the efficacy of HCQ, either alone or in combination. As a single agent, it is more effective than placebo [87]. However, it is not as powerful as SSZ in prevention of erosions or progression of erosions [84]. HCQ is now probably most often used in combination therapy. The literature recommends 6 mg/kg daily based on lean body weight, although in practice most patients get 400 mg daily. Side-effects of HCQ are generally mild and include dyspepsia and rash; however, earlier use of chloroquine caused significant retinal toxicity and has mandated careful follow-up of patients on this drug. Eye examinations should be done every six to twelve months. Uncommon toxicities are hair loss and, rarely, myopathy.

HCQ is one of the medications that some rheumatologists have considered safe for pregnant women. There are no prospective studies but a few case series have shown no significant abnormalities or complications [88]. The decision to use any medication during pregnancy must weigh the potential complications of untreated illness against the toxicity of continued medication.

Immunosuppressive drugs used to treat RA have included azathioprine, d-penicillamine, cyclophosphamide, and cyclosporin [89–91]. Other, more attractive, medication options are usually available, largely because of toxicities associated with these medications. Treatment with azathioprine appears to cause a small increased risk of lymphoma but it is sometimes used, particularly as a "steroid-sparing" agent. Cyclosporin has a narrow treatment/toxicity index because of renal effects. However, both azathiprine and cyclosporin have been used in pregnant women with systemic lupus and renal transplants [80].

Corticosteroids are the synthetic equivalent of cortisol, the potent anti-inflammatory produced by the adrenal gland. Hench found that pregnant women produced large amounts of glucocorticoids and associated it with the phenomenon of remission of RA during pregnancy [25]. That observation ushered in the modern era of treatment for RA. Corticosteroids can reliably decrease swelling, pain, and constitutional symptoms in RA. In addition, some studies demonstrate a decrease in radiologic progression [92].

Corticosteroids are usually given orally for chronic management of RA. To minimize GI symptoms they are taken with food; women at risk are sometimes given misoprostol in addition. To minimize adrenal suppression it should be given in a single dose in the morning. However, when disease is very active, divided doses are sometimes needed. Pulse, or intravenous, steroids are not often used. Probably the most common use of corticosteroids is intra-articular injection to decrease pain and swelling in a single or a few affected joints. If carefully done, patients can experience good relief with little complication.

Repeated frequent injections in the same joint are of concern because of the possibility of accelerated joint damage and other complications.

If the treating provider believes the patient has a favorable prognosis, meaning nonerosive nondeforming RA, nonsteroidal anti-inflammatory drugs may be used as the initial and, occasionally, only therapy. This is reasonable if radiographs of hands and other affected areas are normal, if synovitis and constitutional symptoms are mild, and if rheumatoid factor is negative. If there is uncertainty, often a mild remittive drug such as HCQ will be used to treat constitutional symptoms. Regular reevaluation is essential to be certain that the patient's course is not worsening on such a regimen.

If the patient is significantly affected by constitutional symptoms and swollen joints, even with negative X-rays, a more rapid-acting and stronger disease-modifying drug should be used. MTX is the most used remittive drug; it and SSZ are comparable to each other as single agents, each having certain attractions and problems. Both begin working in about six weeks and patients have an equal chance of responding. For patients with very active disease, including erosions and high-titer rheumatoid factor, more aggressive early intervention is considered essential for the prevention of later deformity.

b. NEW THERAPY FOR RA. With the understanding of the pathologic role of cytokines in RA, inhibition of specific molecules became the urgent agenda for new therapies. In 1998, a drug that inhibits TNF-alpha was approved by the FDA. Etanercept (Enbrel) is a humanized monoclonal antibody to TNF that blocks its interaction with cell surface TNF receptors. Etanercept has demonstrated efficacy in treating RA both as a single agent and in combination with MTX [93,94]. It is given by subcutaneous injection twice weekly. Other TNF-alpha antagonists in clinical trials also show promise. In developmental toxicity studies, etanercept has not been shown to cause damage in fetal rats or rabbits, but no studies in pregnant women have been done. Similarly, no data are available on its secretion into breast milk and therefore its use should be avoided in lactating women.

Leflunomide, an inhibitor of pyrimidine synthesis and an oral medication, was also released for treatment of RA in 1998. It has been shown in double-blind placebo controlled studies to decrease the activity of RA [95]. It is known to cause harm to fetal animals and is contraindicated in pregnancy. It is also contraindicated in lactation, as information is not available on its secretion into breast milk.

Because these medications are both so recently released, their role in therapy is still evolving in terms of short- and long-term side effects, efficacy, and cost. Nonetheless, they appear to represent a significant advance in our treatment of this disease.

c. COMBINATION THERAPY. Double or triple therapy represents the current state-of-the-art treatment of RA, with the incorporation of new medications into combinations. The most common combinations are SSZ and HCQ, MTX and HCQ, and the three-drug combination of SSZ, MTX, and HCQ [96]. This "triple therapy" has demonstrated a promising approach that may provide more efficacy without an equivalent increase in toxicity. One study showed that addition of etanercept to MTX significantly improved control of disease activity in a group of RA patients with persistently active disease on maximum MTX dose [94]. Although this was a short-term study, it appears that combination of TNF antagonist therapy and older disease-modifying drugs offers another avenue of combination treatment for better efficacy with decreased toxicity.

d. ANTIMICROBIAL THERAPY FOR RA. The thesis that an infectious agent is the cause of RA has been persistently popular but remains unproven. Various antibiotics, based on the putative agent involved, have been tried. Other possible mechanisms of antibiotic action on disease activity in RA have been postulated, including decreased collagenase activity. Two double-blind studies of minocycline in RA found a small but statistically significant benefit to treatment [97,98]. One group used minocycline 200 mg/day for 48 weeks and showed improvement in clinical and laboratory parameters with few side effects.

2. Chondroprotective Agents

Oral administration of cartilage-derived type II collagen is currently in clinical trials [99]. This approach is intended to induce antigen-specific peripheral immune tolerance by oral ingestion of such antigens. A short-term study of this therapy showed some positive effects with minimal side effects. Further studies are ongoing.

3. Physical Modalities

Occupational and physical therapy can play an important role in the care of the RA patient. Splints can provide pain relief, improve grip strength in the hand, and improve and protect joint range of motion. Careful attention to muscle strengthening and balance is particularly important for older women where the risk of falling is high. Strengthening leg muscles also decreases the impact of weight-bearing when walking. Aerobic conditioning at an appropriate level is to be encouraged in all women. This helps with muscle strength and can improve sleep and mood as well as provide the known cardiac benefits.

4. Surgery in RA

Appropriate surgery has much to offer selected patients. Carpal tunnel release is a common, safe procedure that can relieve the numbing and tingling of refractory carpal tunnel syndrome. For isolated joints unresponsive to medicine or injection, synovectomy can relieve pain and improve range of motion.

Total joint replacement has had a profound impact in RA care. Total knee and total hip replacement have given mobility back to many patients. In addition, this surgery provides excellent relief of pain from end-stage knee or hip disease. The life expectancy of the prosthesis is usually at least a decade and surgical techniques continue to improve both morbidity and long-term outcomes of this procedure. Finger and toe reconstruction can also provide improved function to severely impaired joints.

5. Complementary Therapies

A significant portion of patients with chronic illnesses use complementary or alternative therapies and patients with RA are no exception [100]. Physician reaction to patient requests for such therapy ranges from active assistance to skeptical refusal, but negative reaction has not stopped patients from requesting these therapies or using them without informing the practitioner.

Such therapies can be used as adjuncts or substitutes for allopathic therapy. The treatments offer a variety of modalities, such as acupuncture, dietary modification and supplements, antimicrobial agents, biological agents, and unapproved uses of existing pharmaceuticals. The cost of such therapies includes the direct costs, costs of delay of treatment with conventional therapy where necessary, and treatment of complications. Scientific studies of alternative therapies are now being done. Such studies are important to bridge the gap between patient concerns and traditional scientific methodology. Patients must feel comfortable with telling their providers all therapies they use to prevent complications from unintended interactions.

Interventions such as acupuncture and yoga appear to offer some benefit in rheumatic conditions, although there are no specific data on effects on RA [101,102]. Chiropractic, herbal medicine, homeopathy, and naturopathy are approaches to care that are often used but infrequently studied.

VII. Prognosis and Outcomes in RA

RA can be a fatal illness [1,103,104]. One study found mortality was increased twofold in people with RA, resulting in a decreased lifespan of seven to ten years [105]. Mortality directly due to RA represented only 10% of the deaths and was from such causes as respiratory failure or cervical spine instability. More often, death was due to cardiovascular or cerebrovascular disease, infection, or GI bleeding. Independent predictors of mortality in these studies include age, education, and male gender, as well as clinical and laboratory parameters associated with clinical severity and extra-articular disease. Lymphoproliferative disorders are increased by two- to eightfold in RA patients. Overall, the decreased survivorship in people with RA found in this study is comparable to that of chronic heart disease, although these studies include many patients with disease onset over two decades ago [106]. There is no information about the impact of changes in therapy on mortality for disease of more recent onset.

Morbidity and functional status are also affected in patients with RA. Functional status in RA is graded on a I-IV scale, from no interference with daily activities to bed-bound. Significant decrease in functional status has been documented in RA patients. In a longitudinal study of 1284 RA patients followed for up to 12 years, half showed a moderate loss of function in one to two years, severe in two to six years, and very severe in ten years [107]. Work loss and depression are also seen. Part of this is related to other medical comorbidity. In one study, 54% of patients identified had one or more other chronic illnesses. Twenty percent of respondents reported that at least one of these was severe.

VIII. Future Directions in Rheumatoid Arthritis

This is an exciting time in the care of women with RA. There is now better understanding that untreated RA is not a benign illness and that early intervention with combinations of new and existing therapies can improve the clinical outcome. New therapies can arrest the inflammatory process more effectively than

previously. The new generation of targeted therapies may provide more specific and less toxic therapy. In the meantime, meticulous attention to the full array of issues—medical, psychological, social—facing women with RA offers more hope to our patients than ever before.

References

1. Pincus, T., and Callahan, L. F. (1986). Taking mortality in rheumatoid arthritis seriously—predictive markers, socioeconomic status and co-morbidity. *J. Rheumatol.* **13,** 841–845.
2. Scott, D. L., Coulton, B. L. *et al.* (1987). Long-term outcome of treating rheumatoid arthritis: Results after 20 years. *Lancet* **I,** 1108–1111.
3. Spector, T. D. (1990). Rheumatoid arthritis. *Rheum. Dis. Clin. North Am.* **16,** 513–537.
4. Nelson, J. L., and Ostensen, M. (1997). Pregnancy and rheumatoid arthritis. *Rheum. Dis. Clin. North Am.* **23,** 195–212.
5. Sherrer, Y. S., Bloch, D. A., Mitchell, D. M., *et al.* (1986). The development of disability in rheumatoid arthritis. *Arthritis Rheum.* **29,** 494–500.
6. Wolfe, F., Mitchell, D. M., *et al.* (1994). The mortality of rheumatoid arthritis. *Arthritis Rheum.* **37,** 481–494.
7. Naides, S. J., Scharosch, L. L. *et al.* (1990). Rheumatologic manifestations of human parvovirus B19 infection in adults. Initial two-year clinical experience. *Arthritis Rheum.* **33,** 1297–1309.
8. Naides, S. J. (1992). Parvoviruses. *In* "Clinical Virology Manual" (S. Specter and G. Lancz, eds.), 2nd ed., pp. 547–569. Elsevier, New York.
9. Nepom, B. S., and Nepom, G. T. (1995). Polyglot and polymorphism: An HLA update. *Arthritis Rheum.* **38,** 1715–1721.
10. van Zeben, D., Hazes, J. M. W., Zwiderman, A. H., *et al.* (1991). Association of HLA-DR4 with a more progressive disease course in patients with rheumatoid arthritis. *Arthritis Rheum.* **34,** 822–830.
11. Nelson, J. L., Boyer, G., Templin, D., *et al.* (1992). HLA antigens in Tlingit Indians with rheumatoid arthritis. *Tissue Antigens* **40,** 57–63.
12. Williams, R. C., Jacobsson, L. T. H., Knowler, W. C., *et al.* (1995). Meta-analysis reveals association between most common class II haplotype in full-heritage Native Americans and rheumatoid arthritis. *Hum. Immunol.* **42,** 90–94.
13. Hochberg, M. C., and Spector, T. D. (1990). Epidemiology of rheumatoid arthritis. *Epidemiol. Rev.* **12,** 247–252.
14. Alarcon, G. S. (1995). Epidemiology of rheumatoid arthritis. *Rheum. Dis. Clin. North Am.* **21,** 589–599.
15. Kato, H., Duff, I. F., *et al.* (1971). Rheumatoid arthritis and gout in Hiroshima and Nagasaki, Japan: A prevalence and incidence study. *J. Chronic Dis.* **23,** 659–670.
16. Linos, A., Worthington, J. W. *et al.* (1980). The epidemiology of rheumatoid arthritis in Rochester Minnesota: A Study of Incidence, Prevalence and Mortality. *Am. J. Epidemiol.* **111,** 87–98.
17. Hochberg, M. C. (1990). Changes in the incidence and prevalence of rheumatoid arthritis in England and Wales, 1970–1982. *Semin. Arthritis Rheum.* **19,** 294–302.
18. Del Puente, A., Knowler, W. C., *et al.* (1989). High incidence and prevalence of rheumatoid arthritis in Pima Indians. *Am. J. Epidemiol.* **129,** 1170–1178.
19. Moolenburgh, J. D., Valkenburg, H. A., *et al.* (1986). A population study on rheumatoid arthritis in Lesotho, southern Africa. *Ann. Rheum. Dis.* **4,** 691–695.
20. Dugowson, C. E., Koepsell, T. D., *et al.* (1991). Rheumatoid arthritis in women: Incidence rates in Group Health Cooperative, Seattle, Washington, 1987–1989. *Arthritis Rheum.* **34,** 1502–1507.
21. Symmons, D. P. M., Barrett, E. M., Bankhead, C. R., *et al.* (1994). The occurrence of rheumatoid arthritis in the United Kingdom:

Results from the Norfolk Arthritis Register. *Br. J. Rheumatol.* **33**, 735–739.

22. Silman, A. J., Davies, P., Currey, H. L. F., *et al.* (1983). Is rheumatoid arthritis becoming less severe? *J. Chronic Dis.* **36**, 891–897.

23. Weyand, C. M., Schmidt, D., *et al.* (1998). The influence of sex on the phenotype of rheumatoid arthritis. *Arthritis Rheum.* **41**, 817–822.

24. Wolfe, F., Mitchell, D. M., *et al.* (1994). The mortality of rheumatoid arthritis. *Arthritis Rheum.* **37**, 481–494.

25. Hench, P. S. (1938). The ameliorating effect of pregnancy on chronic atrophic (infectious rheumatoid) arthritis, fibrosis and intermittent hydrarthrosis. *Mayo Clin. Proc.* **13**, 161–167.

26. Oka, M. (1953). Effect of pregnancy on the onset and course of rheumatoid arthritis. *Ann. Rheum. Dis.* **12**, 227–229.

27. Ostensen, M., and Husby, G. (1983). A prospective clinical study of the effect of pregnancy on rheumatoid arthritis and ankylosing spondylitis. *Arthritis Rheum.* **26**, 1155–1159.

28. Koepsell, T. D., Dugowson, C. E., Voigt, L. F., *et al.* (1990). Reduced incidence of rheumatoid arthritis during pregnancy. *Arthritis Rheum.* **32**, 29.

29. Hazes, J. M. W., Dijkmans, B. A. C., Vandenbroucke, J. P., *et al.* (1990). Pregnancy and the risk of developing rheumatoid arthritis. *Arthritis Rheum.* **33**, 1770–1775.

30. Silman, A. J., Kay, A., and Brennan, P. (1992). Timing of pregnancy in relation to the onset of rheumatoid arthritis. *Arthritis Rheum.* **35**, 152–155.

31. Spector, T. D., Roman, E., and Silman, A. J. (1990). The pill, parity, and rheumatoid arthritis. *Arthritis Rheum.* **33**, 782–789.

32. Dugowson, C. E., Nelson, J. L., Koepsell, T. D., *et al.* (1991). Nulliparity as a risk factor for rheumatoid arthritis. *Arthritis Rheum.* **34**, 48.

33. Heliovaara, M., Aho, K., *et al.* (1995). Parity and the risk of rheumatoid arthritis in Finnish women. *Br. J. Rheumatol.* **34**, 215–220.

34. Nelson, J. L., Koepsell, T. D., Dugowson, C. E., *et al.* (1993). Fecundity before disease onset in women with rheumatoid arthritis. *Arthritis Rheum.* **36**, 7–14.

35. Brennan, P., and Silman, A. J. (1994). Breast-feeding and the onset of rheumatoid arthritis. *Arthritis Rheum.* **37**, 808–813.

36. Brun, J. G., Nilssen, S., *et al.* (1995). Breast-feeding, other reproductive factors and rheumatoid arthritis: A Prospective Study. *Br. J. Rheumatol.* **34**, 542–546.

37. Wingrave, S. J., and Kay, C. R. (1978). Royal College of General Practitioners Study. Reduction in incidence of rheumatoid arthritis associated with oral contraceptives. *Lancet* **I**, 569–571.

38. Spector, T. D., and Hochberg, M. C. (1990). The protective effect of the oral contraceptive pill on rheumatoid arthritis. An overview of the analytic epidemiological studies using meta-analysis. *J. Clin. Epidemiol.* **43**, 1221–1230.

39. Koepsell, T. D., Dugowson, C. E., Nelson, J. L. *et al.* (1994). Noncontraceptive hormones and the risk of rheumatoid arthritis in menopausal women. *Int. J. Epidemiol.* **23**, 1248–1255.

40. Pincus, T., and Callahan, L. F. (1985). Formal education as a marker for increased mortality and morbidity in rheumatoid arthritis. *J. Chronic Dis.* **38**, 973–984.

41. Shapiro, J. A., Koepsell, T. D., *et al.* (1996). Diet and rheumatoid arthritis in women: A possible protective effect of fish consumption. *Epidemiology* **7**, 256–263.

42. Panush, R. S., Stroud, R. M., and Webster, E. (1986). Food-induced (allergic) arthritis). Inflammatory arthritis exacerbated by milk. *Arthritis Rheum.* **29**, 220–226.

43. Beri, D., and Malaviya, A. N. (1988). Effect of dietary restriction on disease activity in rheumatoid arthritis. *Ann. Rheum. Dis.* **47**, 69–72.

44. Panush, R. S., and Carter, R. L. (1983). Diet therapy for rheumatoid arthritis. *Arthritis Rheum.* **26**, 462–471.

45. Kjeldsen-Kragh, J., Haugen, M., *et al.* (1991). Controlled trial of fasting and a one-year vegetarian diet in rheumatoid arthritis. *Lancet* **338**, 899–902.

46. Peltonen, R., Nenonen, M., *et al.* (1997). Fecal microbial flora and disease activity in rheumatoid arthritis during a vegan diet. *Br. J. Rheumatol.* **36**, 64–68.

47. Nordstrom, D. C., Honkanen, V. E., *et al.* (1995). Alpha-linoleic acid in the treatment of rheumatoid arthritis. A double-blind, placebo-controlled and randomized study: Flaxseed *vs.* safflower seed. *Rheumatol. Int.* **14**, 231–234.

48. James, M. J., and Cleland, L. G. (1997). Dietary n-3 fatty acids and therapy for rheumatoid arthritis. *Semin. Arthritis Rheum.* **27**, 85–97.

49. DeLuca, P., Rothman, D., and Zurier, R. B. (1995). Marine and botanical lipids as immunomodulatory and therapeutic agents in the treatment of rheumatoid arthritis. *Rheum. Dis. Clin. North Am.* **21**, 759–777.

50. Kremer, J. M., Jubiz, W., *et al.* (1987). Fish-oil fatty acid supplementation in active rheumatoid arthritis. *Ann. Intern. Med.* **106**, 497–503.

51. Simkin, P. A. (1997). Zinc again. *J. Rheumatol.* **24**, 626–628.

52. Heliovaara, M., Aho, K., *et al.* (1993). Smoking and risk of rheumatoid arthritis. *J. Rheumatol.* **20**, 1830–1835.

53. Silman, A. (1993). Smoking and the risk of rheumatoid arthritis. *J. Rheumatol.* **20**, 1815–1816.

54. Voigt, L. F., Koepsell, T. D., Nelson, J. L. *et al.* (1994). Smoking, obesity, alcohol consumption and the risk of rheumatoid arthritis. *Epidemiology* **5**, 525–532.

55. Arnett, F. C., Edworthy, S. M., Bloch, D. A., *et al.* (1988). The American Rheumatism Association 1987 revised criteria for the classification of rheumatoid arthritis. *Arthritis Rheum.* **31**, 315–324.

56. Dugowson, C. E., Nelson, J. L., *et al.* (1990). Evaluation of the 1987 revised criteria for rheumatoid arthritis in a cohort of newly diagnosed female patients. *Arthritis Rheum.* **33**, 1042–1046.

57. Cats, A., and Hazevoet, H. M. (1970). Significance of positive tests for rheumatoid factor in the prognosis of rheumatoid arthritis: A Follow-up Study. *Ann. Rheum. Dis.* **29**, 254–259.

58. van der Heijde, D. M., van Riel, P. L., *et al.* (1988). Influence of prognostic features on the final outcome in rheumatoid arthritis: A review of the literature. *Semin. Arthritis Rheum.* **17**, 284.

59. Masi, A. T., Feigenbaum, S. L., and Kaplan, S. B. (1983). Articular patterns in the early course of rheumatoid arthritis. *Am. J. Med.* **75**, S6A–S26.

60. Fleming, A., Crown, J. M., and Corbett, M. (1976). Early rheumatoid disease. I. Onset. II. Patterns of joint involvement. *Ann. Rheum. Dis.* **35**, 357.

61. Jacoby, R. K., Jayson, M. I. V., and Cosh, J. A. (1973). Onset, early stages and prognosis of rheumatoid arthritis: A Clinical Study of 100 Patients with 11 year Follow-up. *Br. Med. J.* **2**, 96–100.

62. Waaler, E. (1940). On the occurrence of a factor in human serum activating the specific agglutination of sheep blood corpuscles. *Acta Pathol. Microbiol. Scand.* **17**, 172–188.

63. Masi, A. T., Maldonado-Cocco, J. A., Kaplan, S. B., *et al.* (1976). Prospective study of the early course of rheumatoid arthritis in young adults: Comparison of patients with and without rheumatoid factor positivity at entry and identification of variables correlating with outcome. *Semin. Arthritis Rheum.* **4**, 299–326.

64. Halla, J. T., Fallahi, S., and Hardin, J. G. (1986). Small joint involvement: A Systematic Roentgenographic Study in Rheumatoid Arthritis. *Ann. Rheum. Dis.* **45**, 327–330.

65. Sherrer, Y. S., Bloch, D. A., Mitchell, D. M., *et al.* (1986). The development of disability in rheumatoid arthritis. *Arthritis Rheum.* **29**, 494–500.

66. Ramsey-Goldman, R., and Schilling, E. (1997). Immunosuppressive drug use during pregnancy. *Rheum. Dis. Clin. North Am.* **23**, 149–167.

67. Bermas B. L., and Hill J. A. (1995). Effects of immunosuppressive drugs during pregnancy. *Arthritis Rheum.* **38**, 1722–1732.

68. Fries, J. F., and Ramey, D. R. (1997). "Arthritic specific" global health analog scales assess "generic" health related quality-of-life in patients with rheumatoid arthritis. *J. Rheumatol.* **24**(9), 1697–1702.

69. Kirwan, J. R. (1995). The effect of glucocorticoids on joint destruction in rheumatoid arthritis. The Arthritis and Rheumatism Council Low-Dose Glucocorticoid Study Group. *N. Engl. J. Med.* **333**, 142–146.

70. Fries, J. F., Williams, C. A., *et al.* (1993). The relative toxicity of alternative therapies for rheumatoid arthritis: Implications for the therapeutic progression. *Semin. Arthritis Rheum.* **23**(S2), 68–73.

71. Gabriel, S. E., Jaakkimainen, L., and Bombardier, C. (1991). Risk for serious gastrointestinal complications related to use of non-steroidal anti-inflammatory drugs. *Ann. Intern. Med.* **115**, 787–796.

72. Frolich, J. C. (1995). Prostaglandin endoperoxide synthetase isoenzymes: The clinical relevance of selective inhibition. *Ann. Rheum. Dis.* **54**, 942–943.

73. Felson, D., Anderson, J., and Meenan, R. (1994). The comparative efficacy and toxicity of second-line drugs in rheumatoid arthritis. *Arthritis Rheum.* **33**, 1449–1461.

74. van der Heide, A., Jacobs, J. W. G., Bijlsma, J. W. I., *et al.* (1996). The effectiveness of early treatment with "second-line" antirheumatic drugs. A randomized controlled trial. *Ann. Intern. Med.* **124**, 699–707.

75. Wolfe, F., Hawley, D., *et al.* (1990). Termination of slow-acting antirheumatic therapy in rheumatoid arthritis: A 14-year prospective evaluation of 1017 consecutive starts. *J. Rheumatol.* **17**, 994–1002.

76. Sigler, J. W., Bluhm, G. B., *et al.* (1974). Gold salts in the treatment of rheumatoid arthritis, a double-blind study. *Ann. Intern. Med.* **80**, 21–26.

77. Willkens, R. F., and Watson, M. A. (1982). Methotrexate: A perspective of its use in the treatment of rheumatic diseases. *J. Lab. Clin. Med.* **100**, 314–321.

78. Hoffmeister, R. T. (1983). Methotrexate therapy in rheumatoid arthritis: 15 years experience. *Am. J. Med.* **12**, 69–73.

79. Weinblatt, M., Kaplan, H., Germain, B. F., *et al.* (1994). Methotrexate in rheumatoid arthritis: A five year prospective multicenter study. *Arthritis Rheum.* **37**, 1492–1498.

80. Kremer, J. M. (1997). Safety, efficacy, and mortality in a long-term cohort of patients with rheumatoid arthritis taking methotrexate: Follow-up after a mean of 13.3 years. *Arthritis Rheum.* 984–985.

81. Kremer, J. (1998). Use of methotrexate in the treatment of rheumatoid arthritis. "*UptoDate*" **6**(2).

82. McConkey, B., Amos, R. S., Durham, S., *et al.* (1980). Sulphasalazine in rheumatoid arthritis. *Br. Med. J.* **280**, 442–444.

83. Pullar, T., Hunter, J. A., and Capell, H. A. (1983). Sulphasalazine in rheumatoid arthritis: A double blind comparison of sulphasalazine with placebo and sodium aurothiomalate. *Br. Med. J.* **287**, 1102–1104.

84. van der Heijde, D. M., van Riel, P. L., Nuver-Zvart, I. H., *et al.* (1989). Effects of hydroxychloroquine and sulphasalazine on progression of joint damage in rheumatoid arthritis. *Lancet* **I**, 1036–1038.

85. Bird, H. A. (1995). Sulphasalazine, sulphapyridine or 5-aminosalicylic acid—which is the active moiety in rheumatoid arthritis? *Br. J. Rheumatol.* **34**, 16–19.

86. Day, R. O. (1998) "SAARDS" *In* "Rheumatology" (J. H. Klippel, and P. A. Dieppe, eds.), 2nd ed. Mosby, London.

87. Clark, P., Casa, E. *et al.* (1993). Hydroxychloroquine compared with placebo in rheumatoid arthritis. A randomised controlled trial. *Ann. Intern. Med.* **119**, 1067–1071.

88. Parke, A. L. (1988). Antimalarial drugs, systemic lupus erythematosus and pregnancy. *J. Rheumatol.* **15**, 607–610.

89. Tugwell, P., Bombardier, C., *et al.* (1990). Low-dose cyclosporine versus placebo in patients with rheumatoid arthritis. *Lancet* **335**, 1051–1055.

90. Salaffi, F., Carotti, M., and Cervini, C. (1996). Combination therapy of cyclosporine A with methotrexate or hydroxychloroquine in refractory rheumatoid arthritis. *Scand. J. Rheumatol.* **25**, 16–23.

91. Clements, P. J. (1991). Alkylating agents. *In* "Second Line Agents in the Treatment of Rheumatic Diseases" (J. Dixon and D. E. Furst, eds.). Dekker, New York.

92. Kirwan, J. R., and the Arthritis and Rheumatism Council Low-Dose Glucocorticoid Study Group (1995). The effect of glucocorticoids on joint destruction in rheumatoid arthritis. *N. Engl. J. Med.* **333**, 142–146.

93. Moreland, L. W. (1998). Soluble tumor necrosis factor receptor (p75) fusion protein (ENBREL) as a therapy for rheumatoid arthritis. *Rheum. Dis. Clin. North Am.* **24**, 579–591.

94. Weinblatt, M. E. *et al.* (1999). A trial of etanercept, a recombinant tumor necrosis factor:Fc fusion protein, in patients with rheumatoid arthritis receiving methotrexate. *N. Engl. J. Med.* **340**, 253–259.

95. Rozman, B. (1998). Clinical experience with leflunomide in rheumatoid arthritis. Leflunomide Investigators' Group. *J. Rheumatol. Suppl.* **53**, 27–32.

96. O'Dell, J. R., Haire, C. E., Erikson, N., *et al.* (1996). Treatment of rheumatoid arthritis with methotrexate alone, sulfasalazine and hydroxychloroquine or a combination of all three medications. *N. Engl. J. Med.* **334**, 1287–1291.

97. Kloppenburg, M., Breedveld, F. C., Terwiel, J., *et al.* (1994). Minocycline in active rheumatoid arthritis. A double-blind controlled trial. *Arthritis Rheum.* **37**, 629–636.

98. Tilley, B. C., Alarcon, G. S., Heyse, S. P., *et al.* (1995). For the MIRA trial group: Minocycline in rheumatoid arthritis. A 48 week, double-blind, placebo-controlled trial. *Ann. Intern. Med.* **122**, 81–89.

99. Barnett, M. L., Kremer, J. M., *et al.* (1998). Treatment of rheumatoid arthritis with oral type II collagen. *Arthritis Rheum.* **41**, 290–297.

100. Eisenberg, D. M., Kessler, R. C., Foster, C., *et al.* (1993). Unconventional medicine in the United States. Prevalence, costs and patterns of use. *N. Engl. J. Med.* **328**, 246–252.

101. Berman, B. M., Lao, L., *et al.* (1995). Efficacy of traditional Chinese acupuncture in the treatment of symptomatic knee osteoarthritis: A 24-week Pilot Study. *Osteoarthritis Cartilage* **3**, 139–142.

102. Garfinkel, M. S., Schumacher, H. R., Husain, A., *et al.* (1989). Evaluation of a yoga based regimen for treatment of osteoarthritis of the hands. *J. Rheumatol.* **21**, 2341–2343.

103. Erhardt, C. C., Mumford, P. A., Venables, P. J. W., and Maini, R. N. (1984). Factors predicting a poor life prognosis in rheumatoid arthritis: An Eight Year Prospective Study. *Ann. Rheum. Dis.* **48**, 7–13.

104. Vandenbroucke, J. P., Hazevoet, H. M., *et al.* (1994). Survival and cause of death in rheumatoid arthritis: a 25-year prospective follow-up. *J. Rheumatol.* **11**, 158–161.

105. Wolfe, F., Mitchell, D. M., *et al.* (1994). The mortality of rheumatoid arthritis. *Arthritis Rheum.* **37**, 481–494.

106. Wolfe, F., and Cathey, M. A. (1991). The assessment and prediction of functional disability in rheumatoid arthritis. *J. Rheumatol.* **18**, 1298–1306.

107. Berkanovic, E., and Hurwicz, M. L. (1990). Rheumatoid arthritis and comorbidity. *J. Rheumatol.* **17**, 882–892.

54

Multiple Sclerosis

MICHAEL J. OLEK AND SAMIA J. KHOURY

Brigham and Women's Hospital
Harvard Medical School
Center for Neurologic Diseases
Multiple Sclerosis Unit
Boston, Massachusetts

I. Overview of Multiple Sclerosis

A. Definition

Multiple sclerosis (MS) is a process of chronic demyelination in the central nervous system (CNS) resulting in progressive neurological impairment. The destruction of myelin is thought to be orchestrated by autoreactive T lymphocytes. The loss of the myelin sheath results in decreased axonal transmission and eventually in axonal loss [1,2]. Clinical manifestations surface if the destruction occurs in functionally relevant areas of the brain. While early definitions of MS relied solely on clinical criteria, technological advances have enabled neurologists to accurately diagnose the disease at an earlier juncture where new therapeutic agents may be more beneficial.

B. Epidemiology

Autoimmune diseases in general and MS in particular affect more females than males. In a summary of 30 incidence/prevalence studies of MS there was a cumulative female-to-male ratio of 1.77 to 1.00 [3]. A total of 80 prevalence studies were conducted from 1965 to the present: 49 in Europe, 16 in the United States or Canada, and 15 in other countries. Based on the U.S. Census Bureau population statistic in 1996 of 264,775,000 people, the prevalence of MS in the United States was 58.3/100,000 people. The percentage of female patients was 64.2%. There were 28 incidence studies that met the strict criteria of Jacobson *et al.* [4], with 20 conducted in Europe, 5 in the United States or Canada, and 3 in other countries. Again, based on the 1996 population estimate, the incidence of MS in the United States in 1996 was 3.2/100,000 people.

The mortality statistics for MS are difficult to ascertain due to poor data reporting and collecting. The U.S. Department of Health and Human Services report on the 1992 mortality statistics documented that 1900 U.S. citizens died of MS that year for a U.S. mortality rate of 0.7/100,000 people [5]. A breakdown of the numbers shows that 1187 women (89% white/10% black) and 713 men (90% white/9% black) died of MS in the U.S. in 1992. From these data, the mean age at death of MS patients was 58.1 years compared to 70.5 years for all causes of death. The life expectancy for MS patients was 82.5% of normal [5].

In another study, mortality from MS in England and Wales for the years 1963–1990 was reported [6]. Analysis revealed a steady and consistent improvement in mortality compared to the general population, with a decrease from 556 to 360 deaths up to age 59. This was balanced by an increase from 275 to 440 deaths for MS patients aged 60 and older. There was also an increase in the MS mortality percentage from 33.1% in 1963 to 55.0% in 1990; however, the total number of deaths remained essentially stable (831 deaths in 1963 and 800 deaths in 1990) due to a decrease in the total number of MS patients [6].

Most studies report that the median age at onset of MS is 23.5 years [7–9]. The peak age of onset for women is the early twenties, whereas for men it is the late twenties. The mean age of onset overall is 30 years. The relapsing–remitting group has a mean age of onset of 25–29 years, the relapsing–remitting progressive group has a mean age of onset of 25–29 years (with a mean age of conversion to the progressive type of 40–44 years), and the primary progressive group has a mean age of onset 35–39 years [10–12].

C. Genetic Determinants

There are several candidate genes for MS including human leukocyte antigen (HLA), T-cell receptor, myelin basic protein (MBP), immunoglobulin, complement, tumor necrosis factor (TNF), and mitochondrial genes. The results of three entire genomic scans have been reported [13–15]. The HLA region was strongly positive in two of these studies. Only the HLA complex has been widely accepted as a susceptibility complex with the studies showing association and linkage with HLA-DR15,DQ6 haplotype. This haplotype, however, is neither sufficient nor necessary for the development of MS [16].

In terms of genetic heritage, MS is very rare in Japan, China, and Africa blacks [17]. It has never been reported in ethnically pure Eskimos, Inuits, North and South Amerindians, Australian aborigines, New Zealand Maoris, Pacific Islanders, or Lapps. Worldwide, the most common thread for the development of MS is in women of Scandinavian and northern European ancestry.

In one study, the highest risks for developing MS were found in daughters of either female or male patients with MS (Table 54.1) [18]. For a female patient, a daughter had an age-adjusted lifetime risk of 4.96% of developing MS. The combined cumulative incidence for first-, second-, and third-degree relatives of MS index cases in this cohort, as well as in another study [19], approached 20%. These risks must be understood in the context of the incidence of MS in the general population, which is estimated at 3.2/100,000 people.

Table 54.1
Familial Risks for the Development of Multiple Sclerosis

MS patient family member	Female MS patient		Male MS patient	
	Number of family members affected	Age-adjusted lifetime risk (%)	Number of family members affected	Age-adjusted lifetime risk (%)
Mother	14/383	3.71 ± 0.97%	7/184	3.84 ± 1.42%
Father	6/303	2.00 ± 0.81%	1/128	0.79 ± 0.79%
Daughter	5/386	4.96 ± 2.17%	2/223	5.15 ± 3.53%
Son	0/411	0.00 ± 0.90%	0/248	0.00 ± 1.49%
Sister	25/608	5.65 ± 1.10%	9/340	3.46 ± 1.14%
Brother	10/612	2.27 ± 0.71%	10/326	4.15 ± 1.28%
Aunt/Uncle	23/1491	1.59 ± 0.33%	15/560	2.68 ± 0.68%
Niece/Nephew	7/1789	1.83 ± 0.69%	3/1000	1.47 ± 0.84%
First Cousins	34/2347	2.37 ± 0.40%	7/795	1.53 ± 0.57%

Modified according to [18].

D. Environmental Determinants

The world geographic distribution of multiple sclerosis may provide clues to environmental determinants (Table 54.2) [17]. The incidence, prevalence, and mortality rates of MS vary with latitude. MS is rare in tropical and subtropical areas. Within temperate zones, disease rates increase with increasing latitude in both the Northern and Southern Hemispheres. A North–South prevalence gradient has been detected in Europe, the United States, Japan, Australia, and New Zealand. In both Europe and the United States, the prevalence of MS reflects the degree of Scandinavian and northern European heritage in resident populations [16]. The prevalence rises as one moves away from the equator. However, prevalence varies widely even in populations living at the same latitude.

Studies in the United States confirmed that the MS risk decreased with migration southward and that this change was most prominent when migration occurred before 10 years of age [20,21]. Although an increased MS risk has been found with northward migration in the United States, the relationship with age at migration is less clear [16]. Migration between countries of different prevalences does not seem to result in a change in MS risk. Japanese and other Asians retain their relatively low susceptibility after immigrating to the United States. However, U.S. Asians seem to have higher rates of MS than do Asians in Japan [16]. The issue of environmental

Table 54.2
World Geographic Distribution of Multiple Sclerosis

Location	Latitude	Prevalence (per 100,000)	Location	Latitude	Prevalence (per 100,000)
Iceland	65-North	99.4	Alcoy, Spain	39-North	17
Shetland Islands	61-North	129	Seoul, Korea	38-North	2
Winnipeg, Canada	50-North	35	Malta	36-North	4
Seattle, WA	47-North	69	Charlestown, SC	33-North	14
Switzerland	47-North	52	Israel (Natives)	32-North	9.5-Sephardi 35.6-Ashkenazi
Asahikawa, Japan	44-North	2.5	New Orleans, LA	30-North	6
Arles, France	44-North	9	Kuwait (Arabs)	30-North	9.5-Kuwaitis 24-Palestinians
Parma, Italy	44-North	11	Canary Islands	29-North	18.3
Copparo, Sardinia	44-North	31.1	Okinawa, Japan	26-North	1.9
Krk, Croatia	44-North	44	Hong Kong	23-North	0.8
Olmstead County, MN	44-North	122	Bombay (Parsis)	19-North	26
Hautes Pyrenees, France	43-North	39.6	Newcastle, Australia	33-South	32.5
Boston, MA	42-North	41	Capetown, S. Africa	36-South	10.9-Afrikaner 3-"Colored"/Orientals
Sassari, Sardinia	41-North	69	Hobart, Tasmania	43-South	68

Modified according to [17].

Table 54.3
Classification of Multiple Sclerosis

	Criteria	Attacks	Clinical lesions	Paraclinical lesions	Cerebrospinal fluid
Clinically definite MS	1.	Two	Two Separate		
	2.	Two	One	Another Separate	
Laboratory-supported definite MS	1.	Two	One		OCB or Increased IgG
	2.	Two		One	OCB or Increased IgG
	3.	One	Two Separate		OCB or Increased IgG
	4.	One	One	One	OCB or Increased IgG
Clinically probable MS	1.	Two	One		
	2.	One	Two Separate		
	3.	One	One	One	
Laboratory-supported probable MS	1.	Two			OCB or Increased IgG

Modified according to [51]. Clinical lesion, lesions detected on history and physical examination; paraclinical lesion, lesions detected by evoked-response testing or neuroimaging studies; OCB, oligoclonal banding; IgG, immunoglobulin G.

determinants as a factor for the development of MS remains unresolved.

Geographic clusters in MS have been identified and studied. The most common cluster studies are those attempting to interpret an apparent excess of cases within a small geographic area. This is referred to as post hoc cluster analysis [22]. Nine post hoc clusters studies spanning the years 1937 to 1990 and involving 155 patients were identified [23–38]. These studies do not provide enough evidence to establish any proposed agent as a specific risk factor for MS. There are also reports of clustering among groups of workers in the general population. Four MS cases among seven veterinarians working on swayback disease, a nervous system disease in sheep [39,40], and an excess number of MS cases among workers at a zinc plant in upstate New York, have been reported [41,42]. A second type of analysis is space-time cluster analysis, an epidemiologic model used for investigating an infectious etiology. There have been six published studies [43–48] on space-time analysis in MS and only two have found significant clustering. The study in Norway [48] showed clustering between 13 and 20 years of age with a peak at age 18. The study in the Orkney and Shetland Islands [46] showed clustering at 21–23 years of age.

II. Pathophysiology

Although the cause and pathogenesis of MS are unknown, the most commonly held view is that it is an autoimmune disease triggered by a viral infection. The inflammatory response in the CNS consists predominantly of activated T lymphocytes and macrophages resulting in the local secretion of cytokines and the synthesis of oligoclonal immunoglobulin within the

CNS. Immune abnormalities have been described in the peripheral blood of MS patients, and these include the loss of suppressor function, presence of activated T cells, alterations of T-cell populations, and increased secretion of inflammatory cytokines [49]. The two major hypotheses to explain CNS inflammation are either a cell-mediated autoimmune attack against myelin antigens or the presence of a persistent virus or infectious process within the nervous system against which the inflammatory response is directed. Despite many attempts, an infectious agent has not been identified in MS.

III. Diagnostic Evaluation

A. History and Physical Examination

The cornerstone of the diagnosis of MS is the neurologic history and examination. The first widely accepted criteria based on history and physical examination were described by Allison and Millar in 1954 [50] but were not well standardized. The original criteria from Schumacher et al. in 1965 [8] were devised for clinical trials and had a more restrictive definition. Their criteria for definite MS included at least two episodes of neurologic impairment separated by at least one month and lasting at least 24 hours, with objective abnormalities on neurologic examination. With the advent of new technologies, Poser et al. [51] devised new criteria to include ancillary testing as well as possible and probable categories (Table 54.3). Neurologic examinations are scored using the Kurtzke Expanded Disability Status Scale (EDSS) [52] (Table 54.4). This is a nonlinear scale ranging from 0 to 10, with zero representing a normal neurologic exam and ten representing death due to MS. Other useful scales

Table 54.4

Kurtzke EDSS (Expanded Disability Status Scale)

0	Normal neurological exam (all grade 0 in functional systems (FS); cerebral grade 1 acceptable).
1.0	No disability, minimal signs in one FS (*i.e.,* one grade 1 excluding cerebral grade 1).
1.5	No disability, minimal signs in more than one FS (more than one grade 1 excluding cerebral grade 1).
2.0	Minimal disabilith in one FS (one FS grade 2, others 0 or 1).
2.5	Minimal disability in two FS (two FS grade 2, others 0 or 1).
3.0	Moderate disability in one FS (one FS grade 3, others 0 or 1), or mild disability in three or four FS (three-four FS grade 2, others 0 or 1).
3.5	Fully ambulatory but with moderate disability in one FS (one grade 3 and one or two FS grade 2) or two FS grade 3, others 0 or 1, or five FS grade 2, others 0 or 1.
4.0	Fully ambulatory without aid, self-sufficient, up and about some 12 hours a day despite relatively severe disability consisting of one FS grade 4 (others 0 or 1), or combinations of lesser grades exceeding limits of previous steps. Able to walk without aid or rest some 500 meters (0.3 miles).
4.5	Fully ambulatory without aid, up and about much of the day, able to work a full day, may otherwise have some limitation of full activity or require minimal assistance; characterized by relatively severe disability, usually consisting of one FS grade 4 (others 0 or 1) or combinations of lesser grades exceeding limits of previous steps. Able to walk without aid or rest for some 300 meters (975 ft).
5.0	Ambulatory without aid or rest for about 200 meters (650 feet); disability severe enough to impair full daily activities (*e.g.,* to work full day without special provisions). (Usual FS equivalents are one grade 5 alone, others 0 or 1; or combinations of lesser grades usually exceeding specifications for step 4.0).
5.5	Ambulatory without aid or rest for about 100 meters (325 ft); disability severe enough to impair full daily activities. (Usual FS equivalents are one grade 5 alone, others 0 or 1; or combinations of lesser grades usually exceeding specifications for step 4.0).
6.0	Intermittent or constant unilateral assistance (cane, crutch, brace) required to walk about 100 meters (325 ft) with or without resting. (Usual FS equivalents are combinations with more than two FS grade 3+).
6.5	Constant bilateral assistance (canes, crutches, braces) required to walk about 2.0 meters (65 ft). (Usual FS equivalents are combinations with more than two FS grade 3+).
7.0	Unable to walk beyond about 5 meters (16 ft) even with aid, essentially restricted to wheelchair; wheels self in standard wheelchair a full day and transfers alone; up and about in wheelchair some 12 hours a day. Usual FS equivalents are combinations with more than one FS grade 4+; very rarely pyramidal grade 5 alone.
7.5	Unable to take more than a few steps; restricted to wheelchair; may need aid in transfers, wheels self but cannot carry on in standard wheelchair a full day; may require motorized wheelchair; usual FS equivalents are combinations with more than one FS grade 4+.
8.0	Essentially restricted to bed or chair or perambulated in wheelchair, but may be out of bed much of the day; retains many self-care functions; generally has effective use of arms. Usual FS equivalents are combinations, generally grade 4+ in several systems.
8.5	Essentially restricted to bed for much of the day; has some effective use of arm(s); retains some self-care functions. Usual FS equivalents are combinations, generally grade 4+ in several systems.
9.0	Helpless bed patient; can communicate and eat. Usual FS equivalents ae combinations, mostly grade 4.
9.5	Totally helpless bed patient; unable to communicate effectively or eat/swallow. Usual FS equivalents are combinations, almost all grade 4+.
10	Death due to MS.

Modified according to [52].

used in clinical trials include the Ambulation Index (AI) (Table 54.5) [52a] and the Disease Steps Scale (DS) (Table 54.6) [52b].

B. Ancillary Testing

Evoked potentials provide a noninvasive technique to evaluate nerve conduction velocities in the CNS. Visual, brainstem, and somatosensory evoked potentials have been useful in confirming the diagnosis of MS. Visual evoked potentials are abnormal in 85%, brainstem evoked potentials are abnormal in 67%, and somatosensory evoked potentials are abnormal in 77% of definite MS patients [53]. Cerebrospinal fluid (CSF) evaluation is another useful diagnostic tool. Typically in MS patients, the CSF shows normal glucose, a few white blood cells, elevated total protein, and oligoclonal banding (OCB).

The OCB can be seen in 95% of patients with definite MS [54]. However, OCB is not exclusive to MS and can be seen in infections (bacterial, mycobacterial, fungal, viral, borrelial, treponemal), Behcet's disease, HTLV-1 associated myelopathy, lymphoproliferative disorders involving the leptomeninges, lupus, and cerebrovascular accidents.

The imaging method of choice for the diagnosis of MS, a disease characterized by plaques in the white matter with edema, inflammation, and demyelination, is magnetic resonance imaging (MRI). Multiple rounded or oval plaques around the ventricles of the brain appear as areas of increased signal intensity on T_2-weighted MR images [55,56]. Lesions often have irregular outlines, a "lumpy-bumpy" appearance, and small size [56–58]. They may appear homogenous or possess a thin rim of altered signal intensity [59]. MS may also produce similar

Table 54.5

Ambulation Index (AI)

Instructions: Time the patient's walk of 25 feet, after telling the patient to walk as quickly as he/she safely can.

0	**Asymptomatic;** fully active.
1	**Walks normally but reports fatigue** which interferes with athletic or other demanding activities.
2	**Abnormal gait** or episodic imbalance; gait disorder is noticeable to family and friends. Able to walk 25 feet in **10 seconds or less.**
3	Walks **independently;** able to walk 25 feet in **between 11 and 20 seconds.**
4	Requires **unilateral support** (cane, single crutch) to walk. Walks 25 feet in **20 seconds or less.**
5	Requires **unilateral support** but walks 25 feet in **greater than 20 seconds** or requires **bilateral support** (canes, crutches, walker) and walks 25 feet in **20 seconds or less.**
6	Requires **bilateral support** and walks 25 feet in **greater than 20 seconds.** (May use wheelchair on occasion)[a]
7	Walking limited to **several steps with bilateral support;** unable to walk 25 feet. (May use wheelchair most of the time)
8	Restricted to wheelchair; able to **transfer independently.**
9	Restricted to wheelchair; **unable to transfer independently.**
10	**Bedridden.**

[a]The use of a wheelchair may be determined by a patient's lifestyle and motivation. It is expected that patients in grade 7 will use a wheelchair more frequently than patients in grade 5 or 6. Assignment of a grade, however, in the 5–7 range is determined by the ability of a patient to walk a given distance and not by the extent to which a patient uses a wheelchair.

Modified according to [52a].

lesions in the brain stem and spinal cord [60,61]. MRI scans are positive in 70–95% of patients with clinically definite MS and in approximately 50% of patients with an initial presentation of optic neuritis [62]. Patients who present with optic neuritis have lesions in similar areas to those patients who present with other symptoms and, in addition, may have lesions in the optic nerves that can only be detected by special MRI sectioning through the optic nerve region. Abnormalities displayed on MR

Table 54.6

Disease Steps Scale (DS)

0	**NORMAL functionally normal; no limitations on activity or lifestyle;** RR; EDSS = 01.5; AI = 0; may have minor abnormality on exam (*e.g.,* nystagmus, extensor plantar); attack frequency <1/year; return to baseline with/without treatment and generally not treated.
1	**MILD DISABILITY minor, but noticeable signs and/or symptoms** (*e.g.,* sensory symptoms, minor bladder problems, fatigue); RR or very early CP; EDSS = 2–2.5; AI = 0–1 (*i.e.,* no visible abnormality of gait); attack frequency = 1–2/year.
2	**MODERATE DISABILITY main feature—abnormality of gait;** AI = 2–3; EDSS = 2–4.5; able to work full day, but may find athletic activities difficult; RR-P or early CP; may need cane with attack but recover to independent ambulation.
3	**EARLY CANE EDSS = 6; RR-P or CP; AI = 2–3; may use a cane/unilateral support** (*e.g.,* **spouse's arm) outside** for greater distances, but can walk 25 feet in 20 seconds or less without a cane.
4	**LATE CANE EDSS = 6; RR-P or CP; AI = 4–5; unable to walk 25 ft. without a cane/unilateral support** (*e.g.,* may hang onto furniture inside home or touch wall in clinic; use cane outside; may need a scooter for greater distances (*e.g.,* in malls).
5	**BILATERAL SUPPORT EDSS = 6.5; AI = 5–6; require bilateral support to walk 25 feet** (*e.g.,* two canes or crutches or walker); may use a scooter for greater distances.
6	**WHEELCHAIR EDSS > 7; AI > 7; confined to wheelchair; may be able to take 1 or 2 steps,** but cannot ambulate 25 feet even with bilateral support; may have further progression including problems with use of hands, inability to transfer.
U	**UNCLASSIFIABLE used for patients who do not fit above classification;** *e.g.,* significant cognitive or visual impairment. **State reason that patient is unclassifiable.**

Modified according to [52b].

images may appear due to the increased water content resulting from inflammation and edema around the MS lesion; inflammation and edema are commonly associated with acute plaque formation in MS [63]. The T_2-weighted hyperintense signals of MS plaques are not specific for MS. Signal abnormalities reflect changes in water content and gliosis and can be seen in normal elderly individuals [64], patients with small vessel disease [65], sarcoidosis [66], and a variety of other inflammatory and infectious CNS diseases [67,68].

The advent of gadolinium-based contrast agents did not greatly improve the diagnosis of MS with MRI, but these paramagnetic substances do enhance the identification of new MS lesions [69,70]. Gadolinium contrast agents pass through the disrupted blood-brain barrier in new, active MS plaques, enhancing the relaxation of the hydrogen protons of water molecules in the edema that accompanies new MS plaques [69,70]. Acute disseminated encephalomyelitis (ADEM), a monophasic demyelinating illness, has an MRI appearance that is very similar to that of MS [71]. Follow-up examination and imaging studies are often needed to show resolution of some lesions in ADEM and appearance of new lesions in cases of MS.

In order to increase the specificity of MRI for the diagnosis of MS, different sets of criteria have been established. Paty and colleagues used the term "strongly suggestive of MS" when three lesions greater than 3 mm in diameter are present, one of them periventricular, or when a total of four white matter lesions are present [62]. Fazekas and colleagues proposed a different set of criteria that require the presence of two of the following three items: lesion size greater than 6 mm, lesions abutting the lateral ventricles, and lesions in the posterior fossa [72]. A study comparing the use of these criteria showed that the Fazekas criteria have a higher specificity (96%) and lower false-positive rate (22% of patients under 60 years) than the Paty criteria [73].

MRI has been used not only to diagnose MS, but also to predict its development in certain patients. Two long-term studies have demonstrated that the extent of MRI abnormalities detected at initial presentation of a clinically isolated syndrome suggestive of MS (such as optic neuritis) provides important prognostic information regarding the potential development of MS [74–77]. The appearance of MS lesions on MRI correlate with clinical measures such as exacerbations and EDSS scores in several studies. A large study of 281 patients compared change in EDSS with the appearance of new or enlarging lesions on T_2-weighted images on two MRI scans separated by an interval of 24 to 36 months [78]. This study showed a weak but significant correlation between the number of new or enlarging lesions on MRI and clinical disability over 2–3 years. A National Institute of Health (NIH) group headed by McFarland conducted another study examining the correlation between serial monthly gadopentetate dimeglumine (Gd-DTPA) enhanced MRIs and clinical disability in ten patients with relapsing–remitting MS and two patients with chronic–progressive MS examined for 12 to 55 months [79]. Although the frequency of enhancing lesions varied from patient to patient, "bursts" of enhancing lesions—both lesion number and area—were usually accompanied by an increase of 0.5 or more in EDSS score and clinical worsening. This study also supported the relationship between disease activity as detected by contrast-enhanced MRI and clinical disability as well as the use of contrast-enhanced MRI as an outcome measure in clinical trials assessing MS therapies [79].

In another study, the correlation between lesion burden on contrast-enhanced or nonenhanced MRI and clinical disability, as measured by the EDSS and AI scales, was investigated in 18 patients with MS followed for one year at the Brigham and Women's Hospital in Boston [80]. The study found a significant correlation between changes in EDSS and AI scores and changes in lesion burden as measured by the number of enhancing lesions or number of lesions on T_2-weighted images. In addition, a cumulative increase in EDSS or AI scores correlated with an increase in the cumulative number of lesions on MRI; similarly, an improvement (decrease) in EDSS and AI scores correlated with a decrease in lesion number. On average, an increase by one enhancing lesion on MRI correlated with a deterioration of 0.1 points in EDSS and AI scores, but the appearance of new lesions on MRI was more frequent than the detectable clinical activity of MS. Of a total of 68 scans that showed an increase in the number of lesions, only 16 were associated with clinical worsening (increase in EDSS or AI scores), confirming that MRI was more sensitive than clinical measures for detection of disease activity and that it may be useful as an outcome measure in clinical trials.

MRI is also useful in predicting progression to MS in patients with a clinically isolated syndrome of the optic nerves, brain stem, or spinal cord who are at increased risk for developing MS. The presence of MRI lesions at initial presentation in monosymptopmatic patients increases the risk of developing MS and will be discussed further in the immunology section. MRI has been used as an outcome measure in clinical trials assessing potential new MS therapies, where it confirmed the clinical findings. It can be also be useful in quickly screening potential new therapies without having to enroll large numbers of patients.

While conventional MRI is limited by variability of lesion appearance between patients and within the same patient over time, interobserver variability, patient repositioning errors, and an inability to distinguish among the various pathological appearances of MS (edema, inflammation, demyelination, and axonal loss), more advanced technologies may resolve some of these problems. Such newer applications include: (1) fluid-attenuated inversion recovery (FLAIR) scanning which can identify MS lesions in the spinal cord and brain stem that remain undetectable on conventional spin-echo MRI; (2) volumetric analysis, a three-dimensional method that can permit more accurate quantitation of disease activity in MS; and (3) magnetization transfer imaging and magnetic resonance spectroscopy (MRS) which permit the differentiation of demyelination from other MS pathologies and detect biochemical changes. Such new technologies may be used in conjunction with conventional MRI to more accurately stage lesions and gauge the progress of disease in patients with MS.

IV. Clinical Parameters

A. Disease Categories

Older clinical definitions are described here in order to interpret studies that used this classification system:

Relapsing–remitting disease (RR): Patients have discrete motor, sensory, cerebellar, or visual attacks that come on over a 1–2 week period and resolve over a 4–8 week period with or

Table 54.7
Presenting Symptoms in Multiple Sclerosis Patients

Symptom	Female percentage (68.9% of cohort)	Male percentage (31.1% of cohort)	Total percentage (cohort of 1702)	Mean age at onset (total)
Sensory in limbs	33.2	25.1	30.7	28.6
Visual loss	16.3	15.1	15.9	28.2
Motor (subacute)	8.3	10.4	8.9	36.4
Diplopia	6	8.5	6.8	26.4
Gait disturbance	3.2	8.3	4.8	34.9
Motor (acute)	4.4	4.2	4.3	28.4
Balance problems	2.5	4	2.9	33.2
Sensory in face	2.9	2.5	2.8	30
Lhermitte's	1.6	2.3	1.8	27.4
Vertigo	1.8	1.5	1.7	28.6
Bladder problems	0.9	1.1	1	29.4
Limb ataxia	0.9	1.3	1	30.5
Acute transverse myelopathy	0.8	0.6	0.7	27.6
Pain	0.3	0.8	0.5	39.8
Other	2.6	2.5	2.5	30.2
Polysymptomatic onset	14.5	11.9	13.7	30.8

Modified according to [7] and [9].

without corticosteroid treatment. Patients in this category return to their pre-attack baseline.

Relapsing–remitting progressive disease (RR-P): Patients have attacks but do not return to their baseline and accumulate stepwise disability.

Chronic progressive disease (CP): Patients have progressive worsening with no periods of stability. Secondarily progressive patients are those who began with an RR course of disease.

Primary progressive disease (PP): Patients who have never had a relapse but have gradual worsening of their neurological function over an extended period of time.

There has been a standardization of terminology to describe the pattern and course of MS [81]:

Relapsing–remitting (RR) MS: Clearly defined relapses with full recovery or with sequelae and residual deficit upon recovery; the periods between disease relapses are characterized by a lack of disease progression.

Primary–progressive (PP): Disease progression from onset with occasional plateaus and temporary minor improvements allowed.

Secondary–progressive (SP) MS: Initial RR disease course followed by progression with or without occasional relapses, minor remissions, and plateaus.

Progressive–relapsing (PR) MS: Progressive disease from onset, with clear acute relapses, with or without full recovery; the periods between relapses are characterized by continuing progression.

There are also two severity outcomes described:

Benign MS: Disease in which the patient remains fully functional in all neurologic systems 15 years after disease onset.

Malignant MS: Disease with a rapid progressive course, leading to significant disability in multiple neurologic systems or death in a relatively short time after disease onset.

B. Disease Course

There are several determinants early in the disease that influence the overall course in MS patients. In general, poor outcomes are seen in patients who develop symptoms after the age of 40. Also, overall, men have a more severe course of MS than women. Initial MS symptoms of optic neuritis and isolated sensory symptoms predict a more favorable outcome than initial cerebellar symptoms. The frequency of attacks was evaluated in a retrospective study of 1099 patients with MS. Patients with five or more attacks in the first two years took seven years to reach an EDSS of 6, while patients with two to four attacks took 13 years and patients with one attack took 18 years to reach an EDSS of 6. Patients with lower EDSS scores after 5 years had a more favorable outcome than those patients with higher scores [82].

The annual cost of MS, including personal services, alterations to home and vehicle, purchase of special equipment, and earnings loss was nearly $35,000, translating into a national annual cost of $9.7 billion in 1994 dollars [83]. The average total cost for individuals with chronic progressive MS was $50,000/year and for relapsing-remitting disease was $30,500/year. Mean lifetime cost per case was $2.5 million in 1994.

C. Symptomatology

Patients with MS present with a myriad of complaints. A list of the most common presenting symptoms can be found in Table 54.7. Paresthesias, optic neuritis, and internuclear ophthalmoplegia (INO) continue to be the most common presenting signs in MS patients. Unusual symptoms can occur including pain, paroxysmal symptoms, movement disorders, seizures, aphasia, and autonomic disturbances.

Many symptoms are difficult to disclose and clinic-based studies suggest that 45–74% of women with MS experience sexual dysfunction. These symptoms have been associated with depression, bowel dysfunction, fatigue, spasticity, and pelvic floor weakness [84–89]. There was no association between duration of disease, type of disease, recent exacerbations, or disability scores. Sexual dysfunction may respond to steroid therapy or abate spontaneously [85,88].

Prevalence of depression in MS patients ranges from 14 to 57% [90–92] as compared to 1.3 to 3.7% in the general population [93,94]. The lifetime prevalence of depression in a group of patients with chronic medical disorders was 12.9% [95]. The nature of a chronic debilitating neurologic disorder contributes to depressive symptoms and patients experience coping problems [96]. Studies designed to correlate disease burden with depression have had mixed results. Patients taking multiple medications are prone to depression, and the side effect profile of the newly approved beta interferon medications includes depression. A study evaluated the potential common etiology between MS and depression using DSM-III-R criteria in MS patients and their first-degree relatives. Of the 221 cases, 34.4% of the patients (20 men and 56 women) had a current or lifetime diagnosis of depression. The age-corrected cumulative risk for depression was 50.3% by age 59 years [97].

Symptomatic bladder dysfunction can be identified at some time during the course of MS in 50–80% of patients [98–101]. The severity of bladder symptoms is unrelated to the duration of disease but parallels the severity of other neurologic symptoms, particularly those due to pyramidal tract involvement. The lateral and posterior cervical spinal tracts are the most common sites for demyelination and are also the same pathways involved in normal voiding. Detrusor hyperreflexia due to sacral and/or suprapontine plaques is reported to occur in 50–90% of patients with MS and half of these patients also have sphincter dyssynergia [102]. Detrusor areflexia occurs in 20% of patients with a sacral plaque [100,101]. Common disorders such as urinary tract infections, prostate and bladder cancer, and benign prostatic hypertrophy may mimic symptoms of neurologic dysfunction and should be excluded. Initial management includes fluid management, timed voiding, and bedside commode. For detrusor hyperreflexia without outlet obstruction, medications such as anticholinergics, tricyclic antidepressants (imipramine), or desmopressin are helpful. Detrusor hyperreflexia with outlet obstruction may respond to bladder manipulation, antispasticity medications, or anticholinergics in combination with alpha sympathetic blocking agents. Detrusor areflexia may respond to bethanechol, bladder manipulation, or alpha sympathetic blocking agents. Appropriate patients may be candidates for surgical correction. Catheterization may be employed if the above measures are ineffective; however, the long-term effects of catheterization must be considered. Squamous metaplasia of the bladder was significantly greater in patients who had been catheterized for more than 10 years (80%) compared to those catheterized for less than 10 years (42%), or patients without catheters (20%) [103].

Motor system dysfunctions include weakness, spasticity, hyperactivity of deep tendon reflexes, and clonus. Weakness usually involves the lower extremities where symptoms appear earlier than in the upper extremities. When weakness occurs in the upper extremities it is distal in nature. A compound called 3,4-diaminopyridine (3,4-DAP) has been evaluated in MS patients. A randomized, double-blind, placebo-controlled crossover trial showed improvement in leg weakness [104], but the toxicity of 3,4-DAP limits current widespread usage. Spasticity

may respond to treatment with dantrolene, baclofen, or benzodiazepines. Tizanidine (Zanaflex), a central alpha-2 noradrenergic agonist, was approved for use in MS after a positive double-blind, placebo-controlled trial in the United States [105]. This medication can also be used in combination with baclofen because their mechanisms of action are different.

Somatosensory symptoms include both positive and negative sensations and do not correspond to any anatomical distribution. Vibration and proprioceptive loss predominate over light touch and pinprick abnormalities. A constrictive "banding" feeling in the trunk is a common complaint. Lhermitte's sign, a sudden electric-like sensation radiating down the spine for a brief period after flexion of the cervical spine, is a highly specific complaint of MS patients. Dysesthetic pain may also occur. Unusual somatosensory symptoms include headache, low-back pain, or radicular pain.

Brainstem signs include impairment of ocular motility, dysarthria, facial paresis, and auditory disturbances. Internuclear ophthalmoplegia, either unilateral or bilateral, is a heralding sign in MS. Nystagmus is common and is most often horizontal but is usually asymptomatic. Chronic patients develop dysarthria and some exhibit scanning speech in which a staccato emphasis is given to each syllable. Vertigo is a frequent complaint and is usually associated with an attack and other brainstem symptoms. Less frequent symptoms include pseudobulbar palsy, facial paresis, blepharospasm, and auditory disturbances.

Visual symptoms include optic neuritis and visual field defects. Optic neuritis is one of the most common presenting signs of MS. Vision is usually decreased unilaterally with accompanying photophobia and pain on eye movement. Total visual loss is rare and recovery is good [106]. Clinically, one may detect optic atrophy, decreased visual acuity, impairment of color vision, abnormal visual fields, or an afferent pupillary defect. Subtle abnormalities may surface with the use of visual evoked potentials.

Cerebellar symptoms include ataxia and tremors and are difficult to control. Gait ataxia is frequent and disabling. Appendicular tremor may limit the activities of daily living but may respond to isoniazid, ondancetron, anticonvulsants, or weighted wrist bracelets.

Fatigue is seen in up to 78% of patients [107] and interferes with daily activities. Fatigue must be separated from depression, medication side effects, or physical exhaustion from gait alterations. Treatment options include amantidine, selective serotonin reuptake inhibitors, or amphetamines.

Problems with cognition are increasingly being recognized as an important deficit affecting MS patients. Using neuropsychological testing procedures, Huber and coworkers evaluated 32 patients with MS for a correlation between dementia and MRI findings [108]. Neuropsychological evaluation of dementia included assessment of language, memory, cognition, visuospatial skills, and depression. Patients with dementia did not have a greater number of lesions on MRI, lesion distribution, or degree of cerebral atrophy than the patients without dementia. However, atrophy of the corpus callosum was significantly more common in patients with dementia, suggesting that the observation of such atrophy should alert physicians to the possible association of dementia with MS. A one-year study of 44 patients of mixed disease categories showed a correlation between worsening of the lesion burden on MRI with cognitive decline

[109]. In a study of 53 patients, Rao and coworkers found a strong correlation between severity of cerebral pathology on MRI and cognitive dysfunction [110]. Similar correlations were seen in another study that identified that various patterns of cognitive impairment depended on the location of cerebral lesions [111]. Of 60 patients with chronic–progressive MS who underwent MRI scans and a neuropsychological screening battery (NSB), those who were impaired according to the NSB had significantly more cerebral lesions than those who were judged unimpaired [112]. The investigators concluded that NSB is a valuable screening test for cognitive dysfunction in patients with chronic–progressive MS.

D. Prognosis

In addition to the factors that contribute to the disease course, data on progression is important for the overall prognosis of individual patients. A large epidemiologic study [113] of 308 patients with a 25-year follow-up period evaluated several clinical factors. In approximately 40% of the patients favorable prognostic factors included: (1) relapsing–remitting disease at onset; (2) young age; (3) optic neuritis or sensory symptoms at onset; (4) high degree of remission; and (5) few functional neurologic systems affected. In approximately 60% of patients, no single independent factor indicated a 25-year risk for the development of the progressive phase of MS. Even with a combination of favorable factors [82], the risk for progression was close to 50%. The data also showed that 14.1% of patients have a progressive course from the onset. One benchmark for prognosis is an EDSS score of 6.0 which indicates that the patient requires intermittent or constant assistance to walk about 100 meters with or without resting. The median time to reach an EDSS of 6.0 in the primary progressive group was 6 years and in the secondary progressive group was 5.2 years [82].

V. Reproductive Life: MS, Fertility, and Pregnancy

Irregular menstrual cycles have been reported in 14–15% of MS patients, a figure comparable to that of the general population [114,115]. In one study, despite normal menstrual patterns and fertility, MS patients had statistically significant increases in mean serum prolactin, follicle-stimulating hormone, luteinizing hormone, total and free testosterone, 5-alpha-dihydrotestosterone, and delta-4-androstenedione levels, and decreased levels of estrone sulfate [116]. The effect of the menstrual cycle on MS symptoms is variable [117,118]. Women who worsen during their menstrual cycles have shown improvement with oral contraceptives or estrogen treatment [119]. Oral contraceptive use does not have adverse effects on the incidence, overall prognosis, or the degree of disability of MS [120,121].

Nearly two-thirds of patients with MS are women and the onset of their disease overlaps the childbearing years. However, the onset of MS during pregnancy is uncommon, occurring in less than 10% of patients with MS. The frequency of childlessness is greater in the MS population [122].

Several studies have reported on the short-term effects of pregnancy on MS. Many of the studies are small, uncontrolled, retrospective analyses. Of the larger, prospective, more controlled trials [123–128], the annualized relapse rate during pregnancy was consistently lower than that of the nonpregnant control period. In contrast, the annualized relapse rate in the six month postpartum period was increased. To corroborate, a study involving 254 women followed prospectively found that the frequency of relapses decreased during pregnancy and increased in the first three months postpartum [129]. MS has no demonstrable effect on the course or outcome of pregnancy. There are no meaningful data on the severity of exacerbations during pregnancy. Subclinical disease activity as measured by MRI during pregnancy has not been studied except in case reports [122]. In terms of lactation, the results of a case-control study involving 93 cases and 93 controls reported that patients with MS were less likely than controls to have been breastfed for greater than seven months [130]. Prolonged breastfeeding may be associated with a decreased risk of MS for several reasons, including the fact that cow's milk contains lower amounts of unsaturated fatty acids thereby negatively affecting the formation of membranes and the protective influences of human milk on the immune system [131].

Studies show no adverse effects of pregnancy on long-term disability from MS [124,125,127,132,133]. In fact, one population-based prospective study [127] showed a decrease in disability. In addition, no association between disability and parity was seen. The timing of pregnancies relative to the onset of MS was not associated with disability as reported in one study [124]. Disability was also evaluated in patients with pregnancies before the onset of MS as compared to patients whose pregnancies occurred after the onset of MS [124,125,127]. No overall difference was found between these two patient groups. The current data fail to demonstrate that pregnancy negatively affects the long-term outcome of MS.

During delivery, general anesthesia does not increase the risk of a relapse of MS in the subsequent months [134]. Epidural anesthesia also appears to be as safe as general or local anesthesia during labor [135,136]. Spinal anesthesia was associated with a postoperative exacerbation in one of nine procedures in one series [134] and two of nineteen procedures in another series [137] and therefore has generally been avoided.

Overall, there is no apparent effect of MS on fertility, course of pregnancy, labor and delivery, spontaneous abortions, stillbirths, fetal malformations, complications of pregnancy and delivery, or mean birth dimensions of the baby [125,138].

VI. Disease Monitoring

The major problem confronting therapy of MS remains its unpredictable course and the length of time over which disability accumulates. Even for treatments that have proven efficacy, it may be difficult to determine in an individual patient whether the treatment is having a therapeutic effect unless the disease becomes truly quiescent clinically or the patient improves on therapy. In some instances it is clear that a patient is either a responder or a nonresponder to a particular treatment. We have occasionally seen such positive clinical effects in younger patients with rapidly progressive, steroid unresponsive disease treated with pulse cyclophosphamide. The hope is to find a surrogate marker that is linked to the underlying disease process that will allow a more rational approach to assessing therapy

Table 54.8
Multiple Sclerosis Treatment Strategies

Disease course/disease stage	Treatment options
Optic neuritis	Intravenous (IV methylprednisolone 1000 mg for 5 days with or without an oral taper
Relapsing–remitting with <2 attacks/year and low MRI activity	IV methylprednisolone for attacks
Relapsing–remitting with >2 attacks/year and/or high MRI activity	IV methylprednisolone for attacks plus either: 1. Avonex 30 mcg intramuscularly (IM) weekly 2. Betaseron 1 cc subcutaneously (SQ) every other day 3. Copaxone 20 mcg SQ daily
Relapsing–remitting (Avonex/Betaseron/Copaxone nonresponders)	Add IV methylprednisolone monthly boosters to Avonex/Betaseron/Copaxone therapy
Relapsing–remitting with accumulating disability (Avonex/Betaseron/Copaxone/Methylprednisolone nonresponders)	IV cyclophosphamide/methylprednisolone pulse therapy
Rapid progressing disability	IV cyclophosphamide/methylprednisolone induction followed by pulse therapy
Fulminating disability	Plasma exchange
Secondary progressive	1. IV methylprednisolone monthly boosters 2. IV cyclophosphamide/methylprednisolone pulses 3. Methotrexate (oral or SQ) weekly with or without IV methylprednisolone monthly boosters
Primary progressive	1. IV methylprednisolone monthly boosters 2. Methotrexate (oral or SQ) weekly with or without IV methylprednisolone monthly boosters 3. IVIg monthly 4. Cladribine (IV or SQ)

Modified according to [147a].

apart from clinical assessment. MRI has provided such a surrogate marker and was one of the major factors in approval of beta interferon for the treatment of relapsing–remitting MS [80,139–142]. Even though MRI is not a perfect correlate of disease activity, it may ultimately serve as the best objective measure of ongoing disease in the nervous system given that it is now known that MS is chronically active even when clinical activity is not present. However, apart from formal clinical trials, MRIs are costly and impractical to perform on a frequent basis in a general neurology practice. There is a large body of evidence showing abnormalities of immune function in MS including activated T cells, loss of suppressor function, the presence of myelin-reactive T cells in the peripheral blood, and abnormal cytokine patterns [143]. Studies have shown an increase in IL-12 expression in progressive patients [144]. IL-12 is a key cytokine that drives the secretion of interferon gamma and promotes inflammation. IL-12 production was decreased in progressive patients treated with cyclophosphamide, suggesting a role for measuring this cytokine in patient management [145]. One of the most intriguing observations is that the cytokine TNF alpha appears to correlate with disease activity and disability [146,147]. TNF alpha may play a role in the disease process as it is an important inflammatory cytokine that can result in damage to CNS myelin. Ultimately, however, immune measures such as TNF must be correlated with response to therapy as measured by clinical and MRI criteria.

VII. Treatment Strategies

Therapy is difficult because MS is a chronic, relapsing, or progressive disease with an unpredictable clinical course that generally spans 10–20 years during which time neurologic disability accumulates. Table 54.8 [147a] outlines the treatment paradigm utilized at Brigham and Women's Hospital. Other centers and individuals have various adaptations on the use of these medications and may use medications not mentioned here. Table 54.9 presents selected medications used in MS and their corresponding Food and Drug Administration (FDA) pregnancy categories.

A. Approved Therapies

The first medication approved by the FDA for use in MS was recombinant interferon-beta-1b (Betaseron), which has been shown to decrease the frequency of relapses from 1.27/year to 0.84/year after two years in relapsing–remitting patients receiving eight MIU [148]. It did not affect clinical disability in patients followed over the three-year period, but it did significantly affect accumulation of lesions on the MRI [74,139], suggesting that the underlying disease process was affected. Five-year follow-up data report that disease progression was less in the interferon-beta-1b group (35%) compared with the placebo group (46%) [149]. Also seen was a 30% decrease in

Table 54.9

FDA Pregnancy Classification for Drugs Used in the Treatment of Multiple Sclerosis

Category B	Category D
Copaxone (Copolymer)	Azathioprine
Pemoline	Cladribine
Oxybutynin	Cyclophosphamide
Fluoxetine	Category X
Desmopressin	Methotrexate
Category C	
Corticosteroids	
Interferon-beta-1-a	
Interferon-beta-1-b	
Baclofen	
Amantidine	
Tizanidine	
Carbamazepine	

Category B: Animal data show no harm to the fetus/No human data available. Category C: Animal data show harm to the fetus/No controlled human studies. Category D: Known to cause fetal harm when administered to pregnant women. Category X: Contraindicated for use during pregnancy.

Modified according to [122].

the annual exacerbation rate in the treated group. The MRI data showed no significant change in median MRI lesion burden (3.6%) in the interferon-beta-1b patients, whereas the placebo patients had a change in median MRI lesion burden of 30.2% over five years. The medication is administered every other day by self-injection. Side effects include flu-like symptoms and reactions at the injection site, but these tend to diminish with time. Presently it is not known how interferon-beta-1b will impact the progression of MS and accumulation of disability over time, although experience with the drug over the years will help clarify this issue. Not all patients respond to the drug, and with time all patients have had additional attacks. There are no published data yet on the effect of interferon-beta-1b in patients with progressive disease. The mechanism of action of interferon-beta-1-b is currently unknown [150]. Possible mechanisms are: (a) interferon-beta may be counteracting the deleterious effects of interferon gamma which has been shown to increase MS attacks [151,152], (b) interferon-beta-1b may help reverse a suppressor cell defect that has been described in MS [153], although the defect is more common in progressive MS patients [154], (c) viral infections have been associated with increased MS attacks [155] and beta interferon could be preventing viral infections [150,156], although there is no evidence from the trial that interferon-beta-1b worked by affecting viral infections, and (d) interferon beta may favor increased IL-4 secretion and decreased gamma interferon secretion [157].

A second major study in 301 relapsing–remitting patients investigated the efficacy of weekly intramuscular injections of 6 million units of interferon-beta-1a (Avonex), a glycosylated recombinant beta-interferon [158]. After two years the MRI data revealed a lesion volume of 122.4 (mean) in the placebo group compared with 74.1 (mean) in the Avonex group. The number of enhancing lesions on MRI over two years was 1.65 (mean) in the placebo group and 0.80 (mean) in the Avonex group. The proportion of patients progressing by the end of 104 weeks of

the trial was 34.9% in the placebo group and 21.9% in the Avonex group. Over two years the annual exacerbation rate was 0.90 in the placebo group and 0.61 in the Avonex group. Adverse events included mild flu-like symptoms and mild anemia. No skin reactions occurred because the injection is deep. Laboratory monitoring is suggested but not mandatory because no serious liver toxicities occurred.

Glatiramer acetate/Copolymer 1 (Copaxone), a daily subcutaneous injectable synthetic polymer, showed positive effects in a small double-blind trial of RR MS [159] but not in progressive disease [160]. A large double-blind trial in RR disease involving 251 randomized patients also has been completed [161]. The Copaxone patients had a two-year relapse rate of 1.19, compared with 1.68 for patients receiving placebo. There was a 29% reduction in the relapse rate over two years for those using Copaxone. Side effects included local injection site reactions and transient systemic postinjection reactions including chest pain, flushing, dyspnea, palpitations, and/or anxiety. No laboratory monitoring is necessary.

This molecule is a random polymer initially based on the amino acid composition of myelin basic protein. The mechanism by which Copolymer 1 may work in humans is unknown. It could function by generating regulatory cells, as this has been shown to happen in animals [162]. In addition, proliferation of myelin basic protein (MBP)-specific T-cell clones has also been reported to be suppressed by Copolymer 1 and it theoretically could interfere with binding of MBP or other peptides to the major histocompatibility (MHC) cleft [163,164].

B. Immunosuppressants

Corticosteroids given in short courses may be used to treat acute attacks. Indications for treatment of a relapse include functionally disabling symptoms with objective evidence of neurological impairment. Thus, mild sensory attacks are typically not treated. In the past, adrenocorticotropic hormone (ACTH) and oral prednisone were primarily used. More recently, treatment with short courses of intravenous (IV) methylprednisolone, 500–1000 mg daily for 3–7 days, with or without a short prednisone taper is commonly used [165]. Optic neuritis may occur anytime during the course of MS or may be one of the initial symptoms. A trial on optic neuritis demonstrated that patients treated with oral prednisone alone were more likely to suffer recurrent episodes of optic neuritis as compared to those treated with methylprednisolone followed by oral prednisone [166,167]. These results now make intravenous methylprednisolone the primary treatment used for optic neuritis and have further supported its use for major attacks. Furthermore, as part of the optic neuritis study, it was found that treatment with a three-day course of high-dose IV methylprednisolone reduced the rate of development of MS over a two-year period [74,167]. Five-year follow-up data have shown that the development of clinically definite MS following optic neuritis was 30% and did not differ by treatment group. MRI was a strong predictor of the development of MS over five years, ranging from 16% in the 202 patients with no MRI lesions to 51% in the 89 patients with three or more lesions [168]. These data support the use of high-dose IV methylprednisolone for acute MS attacks. High-dose IV methylprednisolone appears to be accompanied by relatively few side effects in most patients, although a number of side effects have been reported including mental changes, unmasking of infections, gastric disturbances, and an increased incidence of fractures. Anaphylactoid reactions and arrhythmias may also occur. The immunologic mechanisms of high-dose corticosteroids include reduction of CD4+ cells, decrease in cytokine release from lymphocytes, and decrease of cytokines including TNF, gamma interferon, and decreased class II expression [165]. Steroids have been shown to decrease IgG synthesis in the CNS and to reduce CSF antibodies to MBP and oligoclonal bands. Intravenous methylprednisolone may decrease the entry of cells into the brain and may also affect cytokine patterns.

Treatment directed at the progressive phase is the most difficult because the disease may be harder to affect once the progressive stage has been initiated. Cyclophosphamide, total lymphoid radiation (TLI), cyclosporine, methotrexate, and 2-chlorodeoxyadenosine (2-CdA) have shown some positive clinical effects in progressive disease. During the 1990s, we have investigated the use of cyclophosphamide to treat progressive MS based on initial reports of European investigators [169,170]. Our initial study demonstrated a positive effect in patients treated with a two-week course of cyclophosphamide/ ACTH as compared to patients who received ACTH alone [52a]. However, within one to three years, most patients began to reprogress [171]. These findings led to the study of the Northeast Cooperative Treatment Group, a randomized single-blind trial that tested the efficacy of outpatient cyclophosphamide pulses every two months in 236 patients who received an initial cyclophosphamide/ACTH induction [172]. The results demonstrated a modest positive clinical effect in those patients receiving pulses as opposed to those who did not. Most striking was the finding that younger patients (under 40 years of age) tended to respond to therapy whereas older patients did not. Other data suggest that the possible effects of cyclophosphamide are seen in the patients with a short duration of progression [173]. The results to date and the toxicities of cyclophosphamide, which are well known, make the use of the drug only appropriate for carefully selected patients with actively progressive disease that has not responded to other treatment regimens. In patients with progressive disease who have not had an adequate trial of steroids, we use methylprednisolone induction followed by monthly methylprednisolone pulses before initiating cyclophosphamide therapy.

TLI has potent immunosuppressive effects. A double-blind study of lymphoid irradiation reported benefit in patients with progressive MS [174]. The absolute lymphocyte count appeared to be a crude indication of therapeutic efficacy, with greater efficacy in patients with lower counts. Many patients began reprogressing after initial therapy. A major limitation of TLI is that it may preclude the use of other treatments that affect the immune system at a subsequent time for those who reenter the progressive phase despite radiation therapy. A large multicenter trial of cyclosporine in the United States [175] and a trial in London [176] indicate that cyclosporine has a beneficial, albeit modest, effect in ameliorating clinical disease progression, but it has not found clinical use because of the narrow benefit-to-risk ratio.

Weekly low dose oral methotrexate (7.5 mg) was studied in a randomized, double-blind, placebo-controlled trial in 60 patients with chronic progressive disease and has been reported to positively affect measures of upper extremity function in pro-

gressive MS. Lower extremity function as measured by ambulation and disability scales was not affected [177]. An advantage of methotrexate is that it is a well-known immunomodulator used extensively in other conditions, such as rheumatoid arthritis, with a known side-effect profile. Whether higher doses would be more effective in MS is unclear.

Cladribine (2-CdA, Leustatin), a potent immunosuppressive agent useful in the treatment of hairy cell leukemia, was reported to be of benefit in chronic progressive MS [178]. In a one-year double-blind study, 48 matched pairs received four monthly courses of 2-CdA through a central venous catheter over a seven-day period. Seven out of 23 evaluable patients on placebo experienced a one point or more worsening in their EDSS score at one year, whereas only one out of 24 cladribine patients had worsened. Treated patients had improvement on disability scores, no increase in brain MRI lesions, and decreased CSF oligoclonal bands. Side effects included two cases of herpes zoster, a fatal case of hepatitis (not clearly related to the treatment), and persistently lowered CD4+ counts. A multicenter trial in six centers in the United States is currently underway using subcutaneous cladribine.

C. Emerging Therapies

A great deal has been learned about the mechanism of oral tolerance. The use of oral tolerance has been successfully applied to several animal models of autoimmune diseases, including animal models of MS, arthritis, uveitis, diabetes, thyroiditis, myasthenia, and transplantation [179]. Oral tolerance is also being tested clinically in the human disease states of MS, rheumatoid arthritis, uveitis, and juvenile diabetes. Depending on the amount of antigen fed, orally administered antigens result in the generation of active suppression or anergy. Low doses favor active suppression whereas higher doses favor anergy [180]. The doses and strategy being used in clinical trials aim to generate regulatory T cells that suppress inflammation at the target organ. Oral tolerance also has the advantage of being an orally administered medication with no apparent side effects. An initial trial in relapsing–remitting MS demonstrated a decrease in MBP reactive cells in patients fed bovine myelin and a suggestion of a clinical effect on the number of attacks, although the sample size was too small to draw conclusions [181]. A phase II/III pivotal multicenter clinical trial in relapsing–remitting patients recently ended and Myloral and the placebo group showed a 50% reduction in attack rate over two years [182]. Further testing using a recombinant myelin compound is planned.

The use of high doses of systemic recombinant interferon alfa-2a (rIFNA) was evaluated in 20 patients with relapsing–remitting MS [183]. The rIFNA group experienced significantly fewer clinical exacerbations than the placebo group and had fewer new and enlarging lesions on MRI. It was concluded that rIFNA might reduce clinical and MRI signs of disease activity in relapsing–remitting MS and should be evaluated in a larger clinical trial.

Immune globulin may help a number of autoimmune diseases, including chronic relapsing polyneuropathy and dermatomyositis [184], and has been tried in initial MS trials [185]. Its mechanism of action is unclear but may relate to anti-idiotypic effects [186] and suppression of TNF alpha [187]. A randomized placebo-controlled trial of monthly intravenous immunoglobulin (IVIg) in RR MS was reported [188]. This placebo-controlled trial involved 150 patients over two years. In the placebo group there were 116 relapses compared to 62 in the IVIg group, and 36% of the placebo group were relapse-free compared to 53% of the IVIg group (p value = 0.03). Only three IVIg patients (4%) experienced side effects that included cutaneous reactions and depression, although these were not unequivocally related to the IVIg.

Other strategies have focused on a more aggressive approach of immunosuppression such as autologous bone marrow transplantation (BMT). Studies in experimental models of MS have shown promise for this therapy [189,190]. There has been one case report of an MS patient who developed chronic myelogenous leukemia and was treated with allogenic BMT [191]. Another MS patient was treated after she failed to respond to corticosteroids, beta-interferon, azathioprine, methotrexate, and/or cyclophosphamide [192]. After autologous BMT the patient showed mild improvement in her strength, but died seven months posttransplant after acute bronchopneumonia associated with cardiocirculatory insufficiency and elevated liver enzymes. In another controlled trial, two patients with rapidly progressive disease were given cyclophosphamide followed by total body irradiation and methylprednisolone. The patients were then given autologous stem cell transplantation. At the two-month follow-up period both patients improved clinically and were off all immunosuppressive medications [193]. We await larger trials to evaluate the benefit-to-risk ratio of this approach.

Future treatment directions in MS, such as specific immune system modulation or immunosuppression, will require a better understanding of the pathophysiology of the disease. These therapies will be studied in combination to assess their efficacy. Future therapies will also rely on more accurate clinical markers, such as improved MRI techniques and/or specific immune markers to follow the disease progression.

References

1. Trapp, B. D., Peterson, J., Ransohoff, R. M., Rudick, R., Mork, S., and Bo, L. (1998). Axonal transection in the lesions of multiple sclerosis. *N. Engl. J. Med.* **338**, 278–285.

2. Ferguson, B., Matyszak, M. K., Esiri, M. M., and Perry, V. H. (1997). Axonal damage in acute multiple sclerosis lesions. *Brain* **120**, 393–399.

3. Irizarry, M. C. (1997). Multiple sclerosis. *In* "Neurologic Disorders in Women" (M. E. Cudkowicz and M. C. Irizarry, eds.), pp. 85–99. Butterworth-Heinemann, Boston.

4. Jacobson, D. L., Gange, S. J., Rose, N. R., and Graham, N. M. H. (1997). Epidemiology and estimated population burden of selected autoimmune diseases in the United States. *Clin. Immunol. Immunopathol.* **84**, 223–243.

5. National Center for Health Statistics (1996). Vital statistics of the U.S. 1992—Mortality. *In* "Public Health Service," Vol. 2, Part A. U.S. Department of Health and Human Services, U.S. Govt. Printing Office, Washington, DC.

6. Neilson, S., Robinson, I., and Hunter, M. (1993). Static and dynamic models of interdisease competition: Past and projected mortality from amyotrophic lateral sclerosis and multiple sclerosis. *Mech. Ageing Dev.* **66**, 223–241.

7. Paty, D., Studney, D., Redekop, K., and Lublu, F. (1994). MS COSTAR: A computerized patient record adapted for clinical research purposes. *Ann. Neurol.* **36** (Suppl.), S134–135.

8. Schumacher, G. A., Beebe, G., Kibler, R. F., Kurland, L. T., Kurtzke, J. F., McDowell, F., Nagler, B., Sibley, W. A., Tourtellotte, W. W., and Willmon, T. L. (1965). Problems of experimental trials of therapy in multiple sclerosis: Report by the panel on the evaluation of experimental trials of therapy in multiple sclerosis. *Ann. N.Y. Acad. Sci.* **122**, 552–568.

9. Paty, D., Studney, D., Redekop, K., and Lublin, F. (1994). MS COSTAR: A computerized patient record adapted for clinical research purposes. *Ann. Neurol.* **36**, S134–S135.

10. Weinshenker, B. G., Bass, B., Rice, G. P., Noseworthy, J., Carriere, W., Baskerville, J., and Ebers, G. C. (1989). The natural history of multiple sclerosis: A geographically based study. I. Clinical course and disability. *Brain* **112**, 133–146.

11. Noseworthy, J. H., Paty, D. W., Wonnacott, T., Feasby, T. E., and Ebers, G. C. (1983). MS after age 50. *Neurology* **33**, 1537–1544.

12. Paty, D. W., and Ebers, G. C. (1998). Multiple sclerosis. *In* "Contemporary Neurology Series" (D. W. Paty and G. C. Ebers, eds.), Vol. 50, p. 150. Davis, Philadelphia.

13. Sawcer, S., Jones, H. B., Feakes, R., Gray, J., Smaldon, N., Chatway, J., Robertson, N., Clayton, D., Goodfellow, P. N., and Compston, A. (1996). A genome screen in multiple sclerosis reveals susceptibility loci on chromosome 6p21 and 17q22. *Nat. Genet.* **13**, 464–468.

14. The Multiple Sclerosis Genetics Group (1996). A complete genomic screen for multiple sclerosis underscores a role for the major histocompatibility complex. *Nat. Genet.* **13**, 469–471.

15. Ebers, G. C., Kukay, K., Bulman, D. E., Sadovnick, A. D., Rice, G., Anderson, C., Armstrong, H., Cousin, K., Bell, R. B., Hader, W., Paty, D. W., Hashimoto, S., Oger, J., Duquette, P., Warren, S., Gray, T., O'Connor, P., Nath, A., Auty, A., Metz, L., Francis, G., Paulseth, J. E., Murray, T. J., Pryse-Philips, W., Nelson, R., Freedman, M., Brunet, D., Bouchard, J.-P., Hinds, D., and Risch, N. (1996). A full genomic search in multiple sclerosis. *Nat. Genet.* **13**, 472–476.

16. Hogancamp, W. E., Rodriguez, M., and Weinshenker, B. G. (1997). Identification of multiple sclerosis-associated genes. *Mayo Clin. Proc.* **72**, 965–976.

17. Poser, C. M. (1994). The epidemiology of multiple sclerosis: A general overview. *Ann. Neurol.* **36**, S180–S193.

18. Sadovnick, A. D., Baird, P. A., and Ward, R. H. (1988). Multiple sclerosis: Updated risks for relatives. *Am. J. Med. Genet.* **29**, 533–541.

19. The Multiple Sclerosis Genetics Group (1998). Clinical demographics of multiplex families with multiple sclerosis. *Ann. Neurol.* **43**, 530–534.

20. Detels, R., Visscher, B. R., Haile, R. W., Malmgren, R. M., Dudley, J. P., and Coulson, A. H. (1978). Multiple sclerosis and age at migration. *Am. J. Epidemiol.* **108**, 386–393.

21. Kurtzke, J. F., Beebe, G. W., and Norman, J. E., Jr. (1985). Epidemiology of multiple sclerosis in U.S. veterans. III. Migration and the risk of MS. *Neurology* **35**, 672–678.

22. Riise, T. (1997). Cluster studies in multiple sclerosis. *Neurology* **49**(Suppl. 2), S27–S32.

23. Deacon, W. E., Alexander, L., Siedler, H. D., and Kurland, L. T. (1959). Multiple sclerosis in a small New England community. *N. Engl. J. Med.* **261**, 1059–1061.

24. Eastman, R., Sheridan, J., and Poskanzer, D. C. (1973). Multiple sclerosis clustering in a small Massachusetts community, with possible exposure 23 years before onset. *N. Engl. J. Med.* **289**, 793–794.

25. Kock, M. J., Reed, D., Stern, R., and Brody, J. A. (1974). Multiple sclerosis. A cluster in a small northwestern United States community. *JAMA, J. Am. Med. Assoc.* **228**, 1555–1557.

26. Cook, S. D., and Dowling, P. C. (1982). Distemper and multiple sclerosis in Sitka, Alaska. *Ann. Neurol.* **11**, 192–194.

27. Murray, T. J. (1976). An unusual occurrence of multiple sclerosis in a small rural community. *Can. J. Neurol. Sci.* **3**, 163–166.

28. Hader, W. J., Irvine, D. G., and Schiefer, H. B. (1990). A cluster-focus of multiple sclerosis at Henribourg, Saskatchewan. *Can. J. Neurol. Sci.* **17**, 391–394.

29. Irvine, D. G., and Schiefer, H. B. (1989). Geotoxicology of multiple sclerosis: The Henribourg, Saskatchewan, cluster focus. I. The water. *Sci. Total Environ.* **84**, 45–59.

30. Irvine, D. G., and Schiefer, H. B. (1988). Geotoxicology of multiple sclerosis: The Henribourg, Saskatchewan, cluster focus. II. The soil. *Sci. Total Environ.* **77**, 175–188.

31. Ingalls, T. H., Huguenin, I., and Ghent, T. (1989). Clustering of multiple sclerosis in Galion, Ohio 1982–1985. *Am. J. Forensic Med. Pathol.* **10**, 213–215.

32. Sheremata, W. A., Poskanzer, D. C., Withum, D. G., MacLeod, C. L., and Whiteside, M. E. (1985). Unusual occurrence on a tropical island of multiple sclerosis. *Lancet* **2**, 618.

33. Ingalls, T. H. (1986). Endemic clustering of multiple sclerosis in time and place, 1934–1984. *Am. J. Forensic Med. Pathol.* **7**, 3–8.

34. Roman, G. C., and Sheremata, W. A. (1987). Multiple sclerosis (not tropical spastic paraparesis) on Key West, Florida. *Lancet* **1**, 1199.

35. Cook, S. D., Blumberg, B., Dowling, P. C., Deans, W., and Cross, R. (1987). Multiple sclerosis and canine distemper virus on Key West, Florida. *Lancet* **1**, 1426–1427.

36. Helmick, C. G., Wrigley, M., Zack, M. M., Bigler, W. J., Lehman, J. L., Janssen, R. S., Hartwig, E. C., and Witte, J. J. (1989). Multiple sclerosis in Key West, Florida. *Am. J. Epidemiol.* **130**, 935–949.

37. MacGregor, H. S., and Latiwonk, Q. I. (1992). Search for the origin of multiple sclerosis by first identifying the vector. *Med. Hypotheses* **37**, 67–73.

38. Binzer, M., Forsgren, L., Holmgren, G., Drugge, U., and Fredrikson, S. (1994). Familial clustering of multiple sclerosis in a northern Swedish rural district. *J. Neurol., Neurosurg. Psychiatry* **57**, 497–499.

39. Campbell, A. M. G., Daniel, P., Porter, R. J., Russell, W. R., Smith, H. V., and Innes, J. R. M. (1947). Disease of the nervous system occurring among research workers on swayback in lambs. *Brain* **70**, 50–58.

40. Dean, G., McDougall, E. I., and Elian, M. (1985). Multiple sclerosis in research workers studying swayback disease in lambs: An updated report. *J. Neurol. Neurosurg. Psychiatry* **48**, 859–865.

41. Stein, E. C., Schiffer, R. B., Hall, W. J., and Young, N. (1987). Multiple sclerosis and the workplace: Report of an industry-based cluster. *Neurology* **37**, 1672–1677.

42. Schiffer, R. B., Weitkamp, L. R., Ford, C., and Hall, W. J. (1994). A genetic marker and family history of upstate New York multiple sclerosis cluster. *Neurology* **44**, 329–333.

43. Ashitey, G. A., and Mackenzie, G. (1970). "Clustering" of multiple sclerosis cases by date and place of birth. *Br. J. Prev. Soc. Med.* **24**, 163–168.

44. Hargreaves, E. R., and Merrington, M. (1973). A note on the absence of "clustering" of multiple sclerosis cases. *J. Chronic Dis.* **26**, 47–50.

45. Neutel, C. I., Walter, S. D., and Mousseau, G. (1977). Clustering during childhood of multiple sclerosis patients. *J. Chronic Dis.* **30**, 217–224.

46. Poskanzer, D. C., Walter, A. M., Prenney, L. B., and Sheridan, J. L. (1981). The aetiology of multiple sclerosis: Temporal-spatial clustering indicating two environmental exposures before onset. *Neurology* **31**, 708–713.

47. Larsen, J. P., Riise, T., Nyland, H., Kvale, G., and Aarli, J. A. (1985). Clustering of multiple sclerosis in the county of Hordaland, western Norway. *Acta Neurol. Scand.* **71**, 390–395.

48. Riise, T., Gronning, M., Klauber, M. R., Barrett-Conner, E., Nyland, H., and Albrektsen, G. (1991). Clustering of residence of multiple sclerosis patients at age 13 to 20 years in Hordaland, Norway. *Am. J. Epidemiol.* **133**, 932–939.

49. Hafler, D. A., and Weiner, H. L. (1995). Immunologic mechanisms and therapy in multiple sclerosis. *Immunol. Rev.* **144**, 75–107.

50. Allison, R. S., and Millar, J. H. D. (1954). Prevalence and familial incidence of disseminated sclerosis (a report to the Northern Ireland Hospitals Authority on the results of a three-year survey): Prevalence of disseminated sclerosis in Northern Ireland. *Ulster Med. J.* **23**, 5–27.

51. Poser, C. M., Paty, D. W., Scheinberg, L., McDonald, W. I., Davis, F. A., Ebers, G. C., Johnson, K. P., Sibley, W. A., Silberberg, D. H., and Tourtellotte, W. W. (1983). New diagnostic criteria for multiple sclerosis: Guidelines for research protocols. *Ann. Neurol.* **13**, 227–231.

52. Kurtzke, J. F. (1983). Rating neurologic impairment in multiple sclerosis: An expanded disability status scale (EDSS). *Neurology* **33**, 1444–1452.

52a. Hauser, S. L., Dawson, D. M., Lehrich, J. R., Beal, M. F., Kevy, S. V., Propper, R. D., Mills, J. A., and Weiner, H. L. (1983). Intensive immunosuppression in progressive multiple sclerosis: A randomized, three-arm study of high dose intravenous cyclophosphamide, plasma exchange and ACTH. *N. Engl. J. Med.* **308**, 173–180.

52b. Hohol, M., Khoury, S. J., Hafler, D. A., Dawson, D. A., Tourbah, A., Lubetzki, C., Lyon-Caen, O., and Weiner, H. L. (1993). Disease steps in multiple sclerosis: A simple approach to classify patients and evaluate disease progression. *Neurology* **43**, A203.

53. Nuwer, M. R., Packwood, J. W., Myers, L. W., and Ellison, G. W. (1987). Evoked potentials predict the clinical changes in multiple sclerosis drug study. *Neurology* **37**, 1754–1761.

54. McLean, B. N., Luxton, R. W., and Thompson, E. J. (1990). A study of immunoglobulin G in the cerebrospinal fluid of 1007 patients with suspected neurological disease using isoelectric focusing and the Log IgG-index: A comparison and diagnostic applications. *Brain* **113**, 1269–1289.

55. Horowitz, A. L., Kaplan, R. D., Grewe, G., White, R. T., and Salberg, L. M. (1989). The ovoid lesion: A new MR observation in patients with multiple sclerosis. *Am. J. Neuroradiol.* **10**, 303–305.

56. Runge, V. M., Price, A. C., Kirshner, H. S., Allen, J. H., Partain, C. L., and James, A. E., Jr. (1986). The evolution of multiple sclerosis by magnetic resonance imaging. *Radiographics* **6**, 203–212.

57. Ormerod, I. E., Miller, D. H., McDonald, W. I., du Boulay, E. P., Rudge, P., Kendall, B. E., Moseley, I. F., Johnson, G., Tofts, P. S., Halliday, A. M. *et al.* (1987). The role of NMR imaging in the assessment of multiple sclerosis and isolated neurological lesions. A quantitative study. *Brain* **110**, 6.

58. Runge, V. M. (1992). Magnetic resonance imaging contrast agents. *Curr. Opin. Radiol.* **4**, 3–12.

59. Nesbit, G. M., Forbes, G. S., Scheithauer, B. W., Okazaki, H., and Rodriguez, M. (1991). Multiple sclerosis: Histopathologic and MR and/or CT correlation in 37 cases at biopsy and three cases at autopsy. *Radiology* **180**, 467–474.

60. Adams, R. D., and Kubik, C. S. (1952). The morbid anatomy of demyelinative diseases. *Am. J. Med.* **12**, 510–546.

61. Oppenheimer, D. R. (1978). The cervical cord in multiple sclerosis. *Neuropathol. Appl. Neurobiol.* **4**, 151–162.

62. Paty, D. W., Oger, J. J. F., Kastrukoff, L. F., Hashimoto, S. A., Hooge, J. P., Eisen, A. A., Eisen, K. A., Purves, S. J., Low, M. D., Brandejs, V. *et al.* (1988). Magnetic resonance imaging in the diagnosis of multiple sclerosis (MS): A prospective study with comparison of clinical evaluation, evoked potentials, oligoclonal banding and CT. *Neurology* **38**, 180–185.

63. Koopmans, R. A., Li, D. K. B., Oger, J. J. F., Mayo, J., and Paty, D. W. (1989). The lesion of multiple sclerosis: Imaging of acute and chronic stages. *Neurology* **39**, 959–963.

64. Fazekas, F., Kleinert, R., Offenbacher, H., Schmidt, R., Kleinert, G., Payer, F., Radner, H., and Lechner, H. (1993). Pathologic correlates of incidental MRI white matter signal hyperintensities. *Neurology* **43**, 1683–1689.

65. Lechner, H., Schmidt, R., Bertha, G., Justich, E., Offenbacher, H., and Schneider, G. (1988). Nuclear magnetic resonance image white matter lesions and risk factors for stroke in normal individuals. *Stroke* **19**, 263–265.

66. Miller, D. H., Kendall, B. E., Barter, S., Johnson, G., MacManus, D. G., Logsdail, S. J., Ormerod, I. E., and McDonald, W. I. (1988). Magnetic resonance imaging in central nervous system sarcoidosis. *Neurology* **38**, 378–383.

67. Pachner, A. R., Duray, P., and Steere, A. C. (1989). Central nervous system manifestations of Lyme disease. *Arch. Neurol.* **46**, 790–795.

68. Miller, D. H., Ormerod, I. E., Gibson, A., du Boulay, E. P., Rudge, P., and McDonald, W. I. (1987). MR brain scanning in patients with vasculitis: Differentiation from multiple sclerosis. *Neuroradiology* **29**, 226–231.

69. Gonzalez-Scarano, F., Grossman, R. I., Galetta, S., Atlas, S. W., and Silberberg, D. H. (1987). Multiple sclerosis disease activity correlates with gadolinium-enhanced magnetic resonance imaging. *Ann. Neurol.* **21**, 300–306.

70. Grossman, R. I., Braffman, B. H., Bronson, J. R., Goldberg, H. I., Silberberg, D. H., and Gonzalez, S. F. (1988). Multiple sclerosis: Serial study of gadolinium-enhanced MR imaging. *Radiology* **169**, 117–122.

71. Kesselring, J., Miller, D. H., Robb, S. A., Kendall, B. E., Moseley, I. F., Kingsley, D., du Boulay, E. P., and McDonald, W. I. (1990). Acute disseminated encephalomyelitis. MRI findings and the distinction from multiple sclerosis. *Brain* **113**, 291–302.

72. Fazekas, F., Offenbacher, H., Fuchs, S., Schmidt, R., Niederkorn, K., Horner, S., and Lechner, H. (1988). Criteria for an increased specificity of MRI interpretation in elderly subjects with suspected multiple sclerosis. *Neurology* **38**, 1822–1825.

73. Offenbacher, H., Fazekas, F., Schmidt, R., Freidl, W., Flouh, E., Payer, F., and Lechner, H. (1993). Assessment of MRI criteria for diagnosis of MS. *Neurology* **43**, 905–909.

74. Beck, R. W., Cleary, P. A., Trobe, J. D., Kaufman, D. I., Kuppersmith, M. J., Paty, D. W., Brown, C. H., and the Optic Neuritis Study Group (1993). The effect of corticosteroids for acute optic neuritis on the subsequent development of multiple sclerosis. *N. Engl. J. Med.* **239**, 1764–1769.

75. Filippi, M., Horsefield, M., Morrissey, S., MacManus, D., Rudge, P., McDonald, W., and Miller, D. (1994). Quantitative brain MRI lesion load predicts the course of clinically isolated syndromes suggestive of multiple sclerosis. *Neurology* **44**, 635–641.

76. Morrissey, S. P., Miller, D. H., Kendall, B. E., Kingsley, D. P. E., Kelly, M. A., Francis, D. A., MacManus, D. G., and McDonald, W. I. (1993). The significance of brain magnetic resonance imaging abnormalities at presentation with clinically isolated syndromes suggestive of multiple sclerosis. *Brain* **116**, 135–46.

77. Miller, D. H., Ormerod, I. E., McDonald, W. I., MacManus, D. G., Kendall, B. E., Kingsley, D. P., and Moseley, I. F. (1988). The early risk of multiple sclerosis after optic neuritis. *J. Neurol., Neurosurg. Psychiatry* **51**, 1569–1571.

78. Filippi, M., Paty, D. W., Kappos, L., Barkhof, F., Compston, D. A., Thompson, A. J., Zhao, G. J., Wiles, C. M., McDonald, W. I., and Miller, D. H. (1995). Correlations between changes in disability and T2-weighted brain MRI activity in multiple sclerosis: A Follow-up Study. *Neurology* **45**, 255–260.

79. Frank, J. A., Stone, L. A., Smith, M. E., Albert, P. S., Maloni, H., and McFarland, H. F. (1994). Serial contrast-enhanced magnetic resonance imaging in patients with early relapsing-remitting multiple sclerosis: implications for treatment trials. *Ann. Neurol.* **36**(Suppl.), S86–S90.

80. Khoury, S. J., Guttmann, C. R. G., Orav, E. J., Hohol, M. J., Ahn, S. S., Hsu, L., Kikinis, R., Jolesz, F. A., and Weiner, H. L. (1994). Longitudinal MRI imaging in multiple sclerosis: correlation between disability and lesion burden. *Neurology* **44**, 2120–2124.

81. Lublin, F. D., Reingold, S. C., and the National Multiple Sclerosis Society, USA Advisory Committee on Clinical Trials of New Agents in Multiple Sclerosis (1996). Defining the clinical course of multiple sclerosis: Results of an international survey. *Neurology* **46**, 907–911.

82. Weinshenker, B. G. (1994). Natural history of multiple sclerosis. *Ann. Neurol.* **36**(Suppl.), S6–S11.

83. Whetten-Goldstein, K., Sloan, F., Conover, C., Viscusi, K., Kulas, B., and Chessen, H. (1996). The economic burden of multiple sclerosis. *MS Manag.* **3**, 33–37.

84. Lilius, H. G., Valtonen, E. J., and Wikström, J. (1976). Sexual problems in patients suffering from multiple sclerosis. *J. Chronic Dis.* **29**, 643–647.

85. Lundberg, P. O. (1981). Sexual dysfunction in female patients with multiple sclerosis. *Int. Rehabil. Med.* **3**, 32–34.

86. Minderhoud, J. M., Leemhuis, J. G., Kremer, J., Laban, E., and Smits, P. M. (1984). Sexual disturbances arising from multiple sclerosis. *Acta Neurol. Scand.* **70**, 299–306.

87. Valleroy, M. L., and Kraft, G. H. (1984). Sexual dysfunction in multiple sclerosis. *Arch. Phys. Med. Rehabil.* **65**, 125–128.

88. Mattson, D., Petrie, M., Srivastava, D. K., and McDermott, M. (1995). Multiple sclerosis: Sexual dysfunction and its response to medication. *Arch. Neurol.* (*Chicago*) **52**, 862–868.

89. Hulter, B. M., and Lundberg, P. O. (1995). Sexual dysfunction in women with advanced multiple sclerosis. *J. Neurol., Neurosurg. Psychiatry* **59**, 83–86.

90. Joffe, R. T., Lippert, G. P., Gray, T. A., Sawa, G., and Horvath, Z. (1987). Mood disorders and multiple sclerosis. *Arch. Neurol.* (*Chicago*) **44**, 376–378.

91. Minden, S. L., Orav, J., and Reich, P. (1987). Depression in multiple sclerosis. *Gen. Hosp. Psychiatry* **9**, 424–434.

92. Whitlock, R. A., and Siskind, M. M. (1980). Depression as a major symptom of multiple sclerosis. *J. Neurol., Neurosurg. Psychiatry* **43**, 861–865.

93. Anthony, J. C., Folstein, M., Romanoski, A. J., Von Korff, M. R., Nestadt, G. R., Chahal, R., Merchant, A., Brown, C. H., Shapiro, S., Kramer, M. *et al.* (1985). Comparison of the lay diagnostic interview schedule and a standardized psychiatric diagnosis. *Arch. Gen. Psychiatry* **42**, 667–675.

94. Weissman, M. M., Bruce, M. L., Leaf, P. J., Florio, L. P., and Holger, C. (1991). Affective disorders. *In* "Psychiatric Disorders in America" (L. N. Robins and D. Regier, eds.), pp. 53–81. Free Press, New York.

95. Wells, K. B., Golding, J. M., and Burnam, M. A. (1988). Psychiatric disorders in a sample of the general population with and without chronic medical conditions. *Am. J. Psychiatry* **145**, 976–981.

96. Mohr, D. C., Goodkin, D. E., Gatto, N., and VanDer Wende, J. (1997). Depression, coping and level of neurological impairment in multiple sclerosis. *Multiple Sclerosis* **3**, 254–258.

97. Sadovnick, A. D., Remick, R. A., Allen, J., Swartz, E., Yee, M. L., Eisen, K., Farquhar, R., Hashimoto, S. A., Hooge, J., Kastrukoff, L. F., Morrison, W., Nelson, J., Oger, J., and Paty, D. W. (1996). Depression and multiple sclerosis. *Neurology* **46**, 628–632.

98. Blaivas, J. G., Holland, N. J., Giesser, B., LaRocca, N., Madonna, M., and Scheinberg, L. (1984). Multiple sclerosis bladder: Studies and care. *Ann. N.Y. Acad. Sci.* **436**, 328–346.

99. Betts, C. D., D'Mellow, M. T., and Fowler, C. J. (1993). Urinary symptoms and the neurological features of bladder dysfunction in multiple sclerosis. *J. Neurol. Neurosurg. Psychiatry* **56**, 245–250.

100. Chancellor, M. B., and Blaivas, J. G. (1994). Urological and sexual problems in multiple sclerosis. *Clin. Neurosci.* **2,** 189–195.

101. Gonor, S. E., Carroll, D. J., and Metcalf, J. B. (1985). Vesical dysfunction in multiple sclerosis. *Urology* **25,** 429–431.

102. Goldstein, I., Siroky, M. B., Sax, D. S., and Krane, R. J. (1982). Neurologic abnormalities in multiple sclerosis. *J. Urol.* **128,** 541–545.

103. Andrews, K. L., and Husmann, D. A. (1997). Bladder dysfunction and management in multiple sclerosis. *Mayo Clin. Proc.* **72,** 1176–1183.

104. Bever, C. T., Jr., Anderson, P. A., Leslie, J., Panitch, H. S., Dhib-Jalbut, S., Khan, O. A., Milo, R., Hebel, J. R., Conway, K. L., Katz, E., and Johnson, K. P. (1996). Treatment with oral 3,4-diaminopyridine improves leg strength in multiple sclerosis patients: Results of a randomized, double-blind, placebo-controlled, crossover trial. *Neurology* **47,** 1457–1462.

105. Smith, C., Birnbaum, G., Carter, J. L., Greenstein, J., Lublin, F. D., and the U.S. Tizanidine Study Group (1994). Tizanidine treatment of spasticity caused by multiple sclerosis: Results of a double-blind, placebo-controlled trial. *Neurology* **44,** S34–S43.

106. Slamovits, T. L., Rosen, C. E., Cheng, K. P., and Striph, G. G. (1991). Visual recovery in patients with optic neuritis and visual loss to no light perception. *Am. J. Ophthalmol.* **111,** 209–214.

107. Freal, J. E., Kraft, G. H., and Coryell, S. K. (1984). Symptomatic fatigue in multiple sclerosis. *Arch. Phys. Med. Rehabil.* **65,** 135–138.

108. Huber, S. J., Paulson, G. W., Shuttleworth, E. C., Chakeres, D., Clapp, L. E., Pakalnis, A., Weiss, K., and Rammohan, K. (1987). Magnetic resonance imaging correlates of dementia in multiple sclerosis. *Arch. Neurol. (Chicago)* **44,** 732–736.

109. Hohol, M. J., Guttmann, C. R. G., Orav, J., Mackin, G. A., Kikinis, R. T., Khoury, S. J., Jolesz, F. A., and Weiner, H. L. (1997). Serial neuropsychological assessment and magnetic resonance imaging analysis in multiple sclerosis. *Arch. Neurol. (Chicago)* **54,** 1018–1025.

110. Rao, S. M., Leo, G. J., Haughton, V. M., St. Aubin-Faubert, P., and Bernardin, L. (1989). Correlation of magnetic resonance imaging with neuropsychological testing in multiple sclerosis. *Neurology* **39,** 161–166.

111. Arnett, P. A., Rao, S. M., Bernardin, L., Grafman, J., Yetkin, F. Z., and Lobeck, L. (1994). Relationship between frontal lobe lesions and Wisconsin Card Sorting Test performance in patients with multiple sclerosis. *Neurology* **44,** 420–425.

112. Franklin, G. M., Heaton, R. K., Nelson, L. M., Filley, C. M., and Seibert, C. (1988). Correlation of neuropsychological and MRI findings in chronic/progressive multiple sclerosis. *Neurology* **38,** 1826–1829.

113. Runmarker, B., and Andersen, O. (1993). Prognostic factors in a multiple sclerosis incidence cohort with twenty-five years of follow-up. *Brain* **116,** 117–134.

114. Poser, S., Kreikenbaum, K., Konig, A., Poser, W., Evers, P., and Wikström, J. (1981). [Endocrinological findings in patients with multiple sclerosis (author's transl.)]. *Geburtshilfe Frauenheilkd.* **41,** 353–358.

115. Klapps, P., Seyfert, S., Fischer, T., and Scherbaum, W. A. (1992). Endocrine function in multiple sclerosis. *Acta Neurol. Scand.* **85,** 353–357.

116. Grinsted, L., Heltberg, A., Hagen, C., and Djursing, H. (1989). Serum sex hormone and gonadotropin concentrations in premenopausal women with multiple sclerosis. *J. Intern. Med.* **226,** 241–244.

117. Smith, R., and Studd, J. W. (1992). A pilot study of the effect upon multiple sclerosis of the menopause, hormone replacement therapy and the menstrual cycle. *J. R. Soc. Med.* **85,** 612–613.

118. McAlpine, D., and Compston, N. (1952). Some aspects of the natural history of disseminated sclerosis. *Q. J. Med.* **21,** 135–167.

119. McFarland, H. R. (1969). The management of multiple sclerosis. 3. Apparent suppression of symptoms by an estrogen-progestin compound. *Mo. Med.* **66,** 209–211.

120. Poser, S., Raun, N. E., Wikström, J., and Poser, W. (1979). Pregnancy, oral contraceptives, and multiple sclerosis. *Acta Neurol. Scand.* **59,** 108–118.

121. Villard-Mackintosh, L., and Vessey, M. P. (1993). Oral contraceptives and reproductive factors in multiple sclerosis incidence. *Contraception* **47,** 161.

122. Damek, D. M., and Shuster, E. A. (1997). Pregnancy and multiple sclerosis. *Mayo Clin. Proc.* **72,** 977–989.

123. Birk, K., Ford, C., Smeltzer, S., Ryan, D., Miller, R., and Rudick, R. A. (1990). The clinical course of multiple sclerosis during pregnancy and the puerperium. *Arch. Neurol. (Chicago)* **47,** 738–742.

124. Roullet, E., Verdier-Taillefer, M. H., Amarenco, P., Gharbi, G., Alperovitch, A., and Marteau, R. (1993). Pregnancy and multiple sclerosis: A longitudinal study of 125 remittent patients. *J. Neurol.* **56,** 1062–1065.

125. Worthington, J., Jones, R., Crawford, M., and Forti, A. (1994). Pregnancy and multiple sclerosis-a 3-year prospective study. *J. Neurol.* **241,** 228–233.

126. Sadovnick, A. D., Eisen, K., Hashimoto, S. A., Farquhar, R., Yee, I. M., Hooge, J., Kastrukoff, L., Oger, J. J., and Paty, D. W. (1994). Pregnancy and multiple sclerosis: A Prospective Study. *Arch. Neurol. (Chicago)* **51,** 1120–1124.

127. Runmarker, B., and Andersen, O. (1995). Pregnancy is associated with a lower risk of onset and a better prognosis in multiple sclerosis. *Brain* **118,** 253–261.

128. Duquette, P., and Girard, M. (1993). Hormonal factors in susceptibility to multiple sclerosis. *Curr. Opin. Neurol. Neurosurg.* **6,** 195–201.

129. Confavreux, C., Hutchinson, M., Hours, M. M., Cortinovis-Tourniaire, P., and Moreau, T. (1998). Rate of pregnancy-related relapse in multiple sclerosis. *N. Engl. J. Med.* **339,** 285–291.

130. Piscane, A., Impagliazzo, N., Russo, M., Valiani, R., Mandarini, A., Florio, C., and Vivo, P. (1994). Breast feeding and multiple sclerosis. *Br. Med. J.* **308,** 1411–1412.

131. Hanson, L. A., Ahlstedt, S., Andersson, B., Carlsson, B., Fallström, S. P., Mellander, L., Porras, O., Soderstrom, T., and Eden, C. S. (1985). Protective factors in milk and the development of the immune system. *Pediatrics* **75,** 172–176.

132. Weinshenker, B. G., Hader, W., Carriere, W., Baskerville, J., and Ebers, G. C. (1989). The influence of pregnancy on disability from multiple sclerosis: A Population-based Study in Middlesex County, Ontario. *Neurology* **39,** 1438–1440.

133. Stenager, E., Stenager, E. N., and Jensen, K. (1994). Effect of pregnancy on the prognosis for multiple sclerosis: A 5-Year Follow up Investigation. *Acta Neurol. Scand.* **90,** 305–308.

134. Bamford, C., Sibley, W., and Laguna, J. (1978). Anesthesia in multiple sclerosis. *Can. J. Neurol. Sci.* **5,** 41–44.

135. Bader, A. M., Hunt, C. O., Datta, S., Naulty, J. S., and Ostheimer, G. W. (1988). Anesthesia for the obstetric patient with multiple sclerosis. *J. Clin. Anesth.* **1,** 21–24.

136. Crawford, J. S. (1983). Epidural anesthesia for patients with chronic neurologic disease. *Anesth. Analg.* **62,** 621–622.

137. Stenuit, J., and Marchand, P. (1968). Les sequelles de rachia anesthisia. *Acta Neurol. Belg.* **68,** 626.

138. Birk, K., and Rudick, R. (1986). Pregnancy and multiple sclerosis. *Arch. Neurol. (Chicago)* **43,** 719–726.

139. Paty, D. W., Li, D. K. B., and the IFNB Multiple Sclerosis Study Group (1993). Interferon beta-1b is effective in relapsing-remitting multiple sclerosis: MRI results of a multicenter, randomized, double-blind, placebo-controlled trial. *Neurology* **43,** 662–667.

140. Paty, D. W. (1988). Magnetic resonance imaging in the assessment of disease activity in multiple sclerosis. *Can. J. Neurol. Sci.* **15,** 266–272.

141. McFarland, H. F., Frank, J. A., Albert, P. S., Smith, M. E., Martin, R., Harris, J. O., Patronas, N., Maloni, H., and McFarlin, D. E. (1992). Using gadolinium-enhanced magnetic resonance imaging lesions to monitor disease activity in multiple sclerosis. *Ann. Neurol.* **32,** 758–766.

142. Thompson, A. J., Kermode, A. G., MacManus, D. G., Kendall, B. E., Kingsley, D. P. E., Moseley, I. F., and McDonald, W. I. (1990). Patterns of disease activity in multiple sclerosis: clinical and magnetic resonance imaging study. *Br. Med. J.* **300,** 631–634.

143. Hafler, D. A., and Weiner, H. L. (1989). MS: A CNS and systemic autoimmune disease. *Immunol. Today* **10,** 104–107.

144. Balashov, K. E., Smith, D. R., Khoury, S. J., Hafler, D. A., and Weiner, H. L. (1997). Increased interleukin 12 production in progressive multiple sclerosis: Induction by activated CD4+ T cells via CD40 ligand. *Proc. Natl. Acad. Sci. U.S.A.* **94,** 599–603.

145. Comabella, M., Balashov, K., Smith, D., Weiner, H. L., and Khoury, S. J. (1998). "Cyclophosphamide Treatment Normalizes the Increased IL-12 Production in Patients with Chronic Progressive MS and Induces a TH-2 Cytokine Switch." American Academy of Neurology, Minneapolis, MN.

146. Chofflon, M., Juillard, C., Juillard, P., Gauthier, G., and Grau, G. (1992). Tumor necrosis factor α production as a possible predictor of relapse in patients with multiple sclerosis. *Eur. Cytokine Network* **3,** 523–531.

147. Sharief, M. K., and Hentges, R. (1991). Association between tumor necrosis factor-α and disease progression in patients with multiple sclerosis. *N. Engl. J. Med.* **325,** 467–472.

147a. Weiner, H. L., and Stazzone, L. (1998). Multiple sclerosis. *In* "Conn's Current Therapy" (R. E. Rakel, ed.) (in press).

148. The IFNB Multiple Sclerosis Study Group (1993). Interferon beta-1b is effective in relapsing-remitting multiple sclerosis: Clinical results of a multicenter, randomized, double-blind, placebo-controlled trial. *Neurology* **43,** 655–661.

149. The IFNB Multiple Sclerosis Study Group and the University of British Columbia MS/MRI Analysis Group (1995). Interferon beta-1b in the treatment of multiple sclerosis: Final outcome of the randomized controlled trial. *Neurology* **45,** 1277–1285.

150. Arnason, B. G. W. (1993). Interferon beta in multiple sclerosis. *Neurology* **43,** 641–643.

151. Hirsch, R. L., Panitch, H. S., and Johnson, K. P. (1985). Lymphocytes from multiple sclerosis patients produce elevated levels of gamma interferon in vitro. *J. Clin. Immunol.* **22,** 139.

152. Noronha, A., Toscas, A., and Jensen, M. A. (1993). Interferon beta decreases T cell activation and interferon gamma production in multiple sclerosis. *J. Neuroimmunol.* **46,** 145–153.

153. Antel, J. P., Brown-Bania, M., Reder, A., and Cashman, N. (1986). Activated suppressor cell dysfunction in multiple sclerosis. *J. Immunol.* **137,** 137–141.

154. Noronha, A., Toscas, A., and Jensen, M. A. (1990). Interferon beta augments suppressor cell function in multiple sclerosis. *Ann. Neurol.* **27,** 207–210.

155. Sibley, W. A., Bamford, C. R., and Clark, K. (1985). Clinical viral infections and multiple sclerosis. *Lancet* **1,** 1313–1315.

156. Al-Sabbagh, A., Nelson, P. A., and Weiner, H. L. (1994). Beta interferon enhances oral tolerance to MBP and PLP in experimental autoimmune encephalomyelitis. *Neurology* **44**(Suppl. 2), A242 (abstr.).

157. Smith, D. R., Balachov, K., Hafler, D. A., Khoury, S. J., and Weiner, H. L. (1997). Immune deviation following pulse cyclophosphamide/methylprednisolone treatment of multiple sclerosis: Increased interleukin-4 production and associated eosinophilia. *Ann. Neurol.* **42,** 313–318.

158. Jacobs, L. D., Cookfair, D. L., Rudick, R. A., Herndon, R. M., Richert, J. R., Salazar, A. M., Fischer, J. S., Goodkin, D. E.,

Granger, C. V., Simon, J. H. *et al.* (1996). Intramuscular interferon beta-1a for disease progression in relapsing multiple sclerosis. *Ann. Neurol.* **39,** 285–294.

159. Bornstein, M. B., Miller, A., Slagle, S., Weitzman, M., Crystal, H., Drexler, E., Keilson, M., Merriam, A., Wassertheil-Smoller, S., Spada, V., Weiss, W., Arnon, R., Jacobsohn, I., Teitelbaum, D., and Sela, M. (1987). A pilot trial of Cop 1 in exacerbating-remitting multiple sclerosis. *N. Engl. J. Med.* **41,** 533–539.

160. Bornstein, M. B., Miller, A., Slagle, S., Weitzman, M., Drexler, E., Keilson, M., Spada, V., Weiss, W., Appel, S., Rolak, L., Harati, Y., Brown, S., Arnon, R., Jacobsohn, I., Teitelbaum, D., and Sela, M. (1991). A placebo-controlled, double-blind, randomized, two-center, pilot trial of Cop 1 in chronic progressive multiple sclerosis. *Neurology* **41,** 533–539.

161. Johnson, K. P., Brooks, B. R., Cohen, J. A., Ford, C. C., Goldstein, J., Lisak, R. P., Myers, L. W., Panitch, H. S., Rose, J. W., Schiffer, R. B., Vollmer, T., Weiner, L. P., Wolinsky, J. S. and the Copolymer-1 Multiple Sclerosis Study Group (1995). Copolymer 1 reduces relapse rate and improves disability in relapsing-remitting multiple sclerosis: Results of a phase III multicenter, double-blind, placebo-controlled trial. *Neurology* **45,** 1268–1276.

162. Aharoni, R., Teitelbaum, D., Sela, M., and Arnon, R. (1997). Copolymer 1 induces T cells of the T helper type 2 that crossreact with myelin basic protein and suppress experimental autoimmune encephalomyelitis. *Proc. Natl. Acad. Sci. U.S.A.* **94,** 10821–10826.

163. Racke, M. K., Martin, R., McFarland, H., and Fritz, R. B. (1992). Copolymer-1-induced inhibition of antigen-specific T-cell activation: Interference with antigen presentation. *J. Neuroimmunol.* **37,** 75–84.

164. Teitelbaum, D., Milo, R., Arnon, R., and Sela, M. (1992). Synthetic copolymer-1 inhibits human T-cell lines specific for myelin basic protein. *Proc. Natl. Acad. Sci. U.S.A.* **89,** 137–141.

165. Kupersmith, M. J., Kaufman, D., Paty, D. W., Ebers, G., McFarland, H., Johnson, K., Reingold, S., and Whitaker, J. (1994). Megadose corticosteroids in multiple sclerosis. *Neurology* **44,** 1–4.

166. Beck, R. W., Cleary, P. A., Anderson, M. M., Jr., Keltner, J. L., Shults, W. T., Kaufman, D. I., Buckley, E. G., Corbett, J. J., Kupersmith, M. J., Miller, N. R., *et al.* (1992). A randomized, controlled trial of corticosteroids in the treatment of acute optic neuritis. *N. Engl. J. Med.* **326,** 581–588.

167. Beck, R. W., and Cleary, P. A. (1993). Optic neuritis treatment trial: One-year follow-up results. *Arch. Ophthalmol.* **111,** 773–775.

168. The Optic Neuritis Study Group (1997). The 5-year risk of MS after optic neuritis: Experience of the Optic Neuritis Treatment Trial. *Neurology* **49,** 1404–1413.

169. Gonsette, R., Demonty, L., and Delmotte, P. (1977). Intensive immunosuppression with cyclophosphamide in multiple sclerosis: Follow-up of 110 patients for 2–6 years. *J. Neurol.* **214,** 173–181.

170. Hommes, O. R., Lamers, K. J. B., and Reekers, P. (1980). Effect of intensive immunosuppression on the course of chronic progressive multiple sclerosis. *J. Neurol.* **223,** 177–190.

171. Carter, J. L., Hafler, D. A., Dawson, D. M., Orav, J., and Weiner, H. L. (1988). Immunosuppression with high-dose IV cyclophosphamide and ACTH in progressive multiple sclerosis: cummulative 6-year experience in 164 patients. *Neurology* **38,** 9–14.

172. Weiner, H. L., Mackin, G. A., Orav, E. J., Hafler, D. A., Dawson, D. M., LaPierre, Y., Herndon, R., Lehrich, J. R., Hauser, S. L., Turel, A., Fisher, M., Birnbaum, G., McArthur, J., Butler, R., Moore, M., Sigsbee, B., and Safran, A. (1993). Intermittent cyclophosphamide pulse therapy in progressive multiple sclerosis: Final report of the Northeast Cooperative Multiple Sclerosis Treatment Group. *Neurology* **43,** 910–918.

173. Hohol, M. J., Olek, M. J., Orav, E. J., Stazzone, L., Hafler, D. A., Khoury, S. J., Dawson, D. M., and Weiner, H. L. (1998). "Treat-

ment of Progressive Multiple Sclerosis with Pulse Cyclophosphamide/Methylprednisolone: Response to Therapy is Linked to the Duration of Progressive Disease. American Academy of Neurology, Minneapolis, MN.

174. Cook, S. D., Devereux, C., Troiano, R., Rohowsky-Kochan, C., Sheffet, A., Jotkowitz, A., Zito, G., and Dowling, P. C. (1992). Total lymphoid irradiation in multiple sclerosis. *In* "Treatment of Multiple Sclerosis: Trial Design, Results, and Future Perspectives" (R. A. Rudick and D. E. Goodkin, eds.), pp. 267–280. Springer-Verlag, New York.

175. The Multiple Sclerosis Study Group (1990). Efficacy and toxicity of cyclosporine in chronic progressive multiple sclerosis: A randomized, double-blinded, placebo-controlled clinical trial. *Ann. Neurol.* **27,** 591–605.

176. Rudge, P., Koetsier, J. C., Mertin, J., Mispelblom-Beyer, J. O., Van Walbeek, H. K., Clifford-Jones, R., Harrison, J., Robinson, K., Mellen, B., Poole, T., *et al.* (1989). Randomized double blind controlled trial of cyclosporin in multiple sclerosis. *J. Neurol. Neurosurg. Psychiatry* **52,** 559–565.

177. Goodkin, D. E., Rudick, R. A., Medendorp, S. V., Daughtry, M. M., Schwetz, K., Fischer, J., and Van Dyke, C. (1995). Low-dose (7.5mg) oral methotrexate reduces the rate of progression in chronic progressive multiple sclerosis. *Ann. Neurol.* **37,** 30–40.

178. Sipe, J. C., Romine, J. S., Koziol, J. A., McMillan, R., Zyroff, J., and Beutler, E. (1994). Cladribine in treatment of chronic progressive multiple sclerosis. *Lancet* **344,** 9–13.

179. Weiner, H. L., Friedman, A., Miller, A., Khoury, S. J., Al-Sabbagh, A., Santos, L., Sayegh, M., Nussenblatt, R. B., Trentham, D. E., and Hafler, D. A. (1994). Oral tolerance: Immunologic mechanisms and treatment of murine and human organ specific autoimmune diseases by oral administration of autoantigens. *Annu. Rev. Immunol.* **12,** 809–837.

180. Friedman, A., and Weiner, H. L. (1994). Induction of anergy or active suppression following oral tolerance is determined by frequency of feeding and antigen dosage. *Proc. Natl. Acad. Sci. U.S.A.* **91** (14), 6688–6692.

181. Weiner, H. L., Mackin, G. A., Matsui, M., Orav, E. J., Khoury, S. J., Dawson, D. M., and Hafler, D. A. (1993). Double-blind pilot trial of oral tolerization with myelin antigens in multiple sclerosis. *Science* **259,** 1321–1324.

182. Panitch, H. S., Francis, G., and the Oral Myelin Study Group (1997). Clinical results of a phase III trial of oral myelin in relapsing-remitting multiple sclerosis. *Ann. Neurol.* **42,** 459.

183. Durelli, L., Bongioanni, M. R., Cavallo, R., Ferrero, B., Ferri, R., Verdun, E., Bradac, G. B., Riva, A., Geuna, M., Bergamini, L., and Bergamasco, B. (1995). Interferon alpha treatment of relapsing-remitting multiple sclerosis: Long-term study of the correlations between clinical and magnetic resonance imaging results and effects on the immune function. *MS Clin. Lab. Res.* **1,** S32–37.

184. Dalakas, M. C., Illa, I., Dambrosia, J. M., Soueidan, S. A., Stein, D. P., Otero, C., Dinsmore, S. T., and McCrosky, S. (1993). A controlled trial of high-dose intravenous immune globulin infusions as treatment of dermatomyositis. *N. Engl. J. Med.* **329,** 1993–2000.

185. Achiron, A., Pras, E., Gilad, R., Ziv, I., Mandel, M., Gordon, C. R., Noy, S., Sarova-Pinhas, I., and Melamed, E. (1992). Open controlled therapeutic trial of intravenous immune globulin in relapsing-remitting multiple sclerosis. *Arch. Neurol. (Chicago)* **49,** 1233–1236.

186. Dwyer, J. M. (1992). Manipulating the immune system with immune globulin. *N. Engl. J. Med.* **326,** 107–116.

187. Achiron, A., Margalit, R., Hershkoviz, R., Markovits, D., Reshef, T., Melamed, E., Cohen, I. R., and Lider, O. (1994). Intravenous immunoglobulin treatment of experimental T-cell mediated autoimmune disease: Up-regulation of T cell proliferation and down-regulation of TNFα secretion. *J. Clin. Invest.* **93,** 600–605.

188. Fazekas, F., Deisenhammer, F., Strasser-Fuchs, S., Nahler, G., Mamoli, B., and the Austrian Immunoglobulin in Multiple Sclerosis Study Group (1997). Randomised placebo-controlled trial of monthly intravenous immunoglobulin therapy in relapsing-remitting multiple sclerosis. *Lancet* **349,** 589–593.

189. van Gelder, M., Kinwel-Bohre, E. P., and van Bekkum, D. W. (1993). Treatment of experimental allergic encephalomyelitis in rats with total body irradiation and syngeneic BMT. *Bone Marrow Transplant.* **11,** 233–241.

190. Burt, R. K., Burns, W., Ruvolo, P., Fischer, A., Shiao, C., Guimaraes, A., Barrett, J., and Hess, A. (1995). Syngeneic bone marrow transplantation eliminates V beta 8.2 T lymphocytes from the spinal cord of Lewis rats with experimental allergic encephalomyelitis. *J. Neurosci. Res.* **41,** 526–531.

191. McAllister, L. D., Beatty, P. G., and Rose, J. (1997). Allogenic bone marrow transplant for chronic myelogenous leukemia in a patient with multiple sclerosis. *Bone Marrow Transplant.* **19,** 395–397.

192. Kolar, O. J., Emanuel, D. J., Smith, F. O., Azzarelli, B., and Burt, R. K. (1997). Bone marrow transplantation (BMT) in multiple sclerosis (MS). *J. Neurol. Sci.* **150**(Suppl.), S50.

193. Burt, R. K., Traynor, A. E., Cohen, B., Karlin, K. H., Davis, F. A., Stefoski, D., Terry, C., Lobeck, L., Keever-Taylor, C., and Burns, W. H. (1997). Autologous lymphocyte depleted hematopoietic stem cell transplantation for rapidly progressive multiple sclerosis; minimal toxicity from a cyclophosphamide/total body irradiation/methylprednisolone conditioning regimen. *J. Neurol. Sci.* **150**(Suppl.), S116.

55

Systemic Lupus Erythematosus

ROSALIND RAMSEY-GOLDMAN* AND SUSAN MANZI†

*Department of Medicine, Division of Arthritis and Connective Tissue Diseases, Northwestern University, Chicago, Illinois;
†Department of Medicine, Division of Rheumatology and Clinical Immunology, University of Pittsburgh, Pittsburgh, Pennsylvania

I. Introduction

Under normal conditions, the immune system distinguishes the body's own tissues from external invaders. In autoimmune diseases the immune system inappropriately targets the body resulting in tissue injury. Autoimmune responses may be directed toward a specific organ or toward components of cells. Organ-specific autoimmune conditions include autoimmune thyroiditis and bullous dermatologic conditions, including pemphigoid. In contrast, systemic lupus erythematosus (SLE) is a disease of generalized autoimmunity because of the wide range of antigenic targets. This generalized autoreactivity in SLE results in multiorgan involvement.

Historically, the term lupus has been used in descriptions of cutaneous diseases since at least the tenth century [1]. Lupus is a Latin derivative for wolf, referring to the ulcerations around the face that eat away, bite, and destroy the flesh. For many years, lupus erythematosus was confused with lupus vulgaris (cutaneous tuberculosis). It was not until the nineteenth century that the extracutaneous (systemic) manifestations were recognized—thus, the term systemic lupus erythematosus.

SLE is a chronic inflammatory autoimmune disease that occurs predominantly in women during their childbearing years. The disease course is marked by exacerbations and remissions. This multisystem disease is characterized by the production of autoantibodies and the occurrence of tissue damage from the deposition of the antibodies in immune complexes. Immune dysregulation probably occurs from environmental triggers in the genetically susceptible host.

In this chapter we will present a brief overview of the disease manifestations and treatments for SLE. The epidemiology section discusses the evidence for the striking predilection of lupus for women. In addition, we show that as mortality from lupus is decreasing, long-term consequences of the disease and/or its treatment contribute to significant morbidity. The importance of autoantibody production and the interaction between T cells and B cells are reviewed in the pathogenesis section. Endogenous and exogenous hormone effects and pregnancy issues are discussed in the risk factor section. We suggest research directions in concluding remarks.

II. SLE Diagnosis and Management

A. Definition of SLE and Diagnosis

SLE is a chronic, inflammatory, autoimmune disease that targets various organs [2–4]. Diagnosis is generally not difficult when many typical symptoms and signs are present but may be more problematic when the disease manifests as only a few complaints or when problems occur over time. Often the diagnosis cannot be made with certainty at the onset of disease and becomes clear only after prolonged observation. The diagnosis is made largely on clinical grounds with the support of laboratory tests. In evaluating a patient with possible SLE, the physician should pay attention to family history of autoimmune disorders and to drug history. In a few circumstances, lupus and lupus-like syndromes can be precipitated by a variety of drugs. Some patients with a family history of autoimmune disease may have autoantibodies detected in their serum. However, the presence of these autoantibodies may not be clinically significant.

Criteria were developed for the purpose of disease classification in 1971 [5] and revised in 1982 [6] and 1997 (Table 55.1) [7]. These criteria are meant to ensure that large series of patients from different geographic locations are comparable to one another. Of the 11 criteria, the presence of four or more, either serially or simultaneously, is sufficient for classification of a patient as having SLE. Included are malar rash (nonscarring rash across the bridge of the nose and cheeks), discoid rash (scarring rash), photosensitivity, oral ulcers, arthritis, serositis (pleuritis or pericarditis), renal involvement, central nervous system involvement (seizures or psychosis), hematologic abnormalities (hemolytic anemia, leukopenia, thrombocytopenia), immunologic markers (antibodies to native DNA, Smith [Sm] antigen, or phospholipids including anticardiolipin IgG or IgM, lupus anticoagulant, or a false positive serologic test for syphilis), and antinuclear antibody (ANA). Although these criteria were established primarily for research purposes, they serve as useful reminders of those features that distinguish lupus from other related connective tissue diseases. However, the range of clinical manifestations in SLE is much greater than that described by the eleven classification criteria. In addition, disease severity may vary widely even in patients with the same clinical criteria.

Many patients with SLE experience fatigue, malaise, fever, anorexia, and weight loss. Disease manifestations may be cutaneous, musculoskeletal, hematologic, gastrointestinal, cardiopulmonary, renal, and neuropsychiatric [2–4,8]. Renal manifestations and neuropsychiatric disease are responsible for most of the morbidity directly related to SLE disease activity.

There is great variability in the expression, course, and histopathology of renal disease in patients with SLE [9]. Although clinically apparent nephritis develops in approximately 40–70% of patients, renal biopsies show that virtually all patients with lupus have some degree of glomerular abnormality. The most widely used histopathologic classification scheme for lupus nephritis is that derived by the World Health Organization (WHO), because there is reasonable clinical correlation between histo-

Table 55.1
1997 Revised Criteria for Classification of Systemic Lupus Erythematosus[a]

Criterion	Definition
Malar rash	Fixed erythema, flat or raised, over the malar eminences, tending to spare the nasolabial folds
Discoid rash	Erythematosus raised patches with adherent keratotic scaling and follicular plugging; atrophic scarring occurs in older lesions
Photosensitivity	Skin rash as a result of unusual reaction to sunlight, by patient history or physician observation
Oral ulcers	Oral or nasopharyngeal ulceration, usually painless, observed by a physician
Arthritis	Nonerosive arthritis involving two or more peripheral joints, characterized by tenderness, swelling, or effusion
Serositis	Pleuritis—convincing history of pleuritic pain or rub heard by a physician or evidence of pleural effusion
	Pericarditis—documented by ECG or rub or evidence of pericardial effusion
Renal disorder	Persistent proteinuria greater than 0.5 g per day or greater than 3+ if quantitation not performed
	Cellular casts—may be red cell, hemoglobin, granular, tubular, or mixed
Neurological disorder	Seizures—in the absence of offending drugs or known metabolic derangements; *e.g.,* uremia, ketoacidosis, or electrolyte imbalance
	Psychosis—in the absence of offending drugs or known metabolic derangements, *e.g.,* uremia, ketoacidosis, or electrolyte imbalance
Hematologic disorder	Hemolytic anemia—with reticulocytosis
	Leukopenia—less than 4,000/mm³ total on two or more occasions
	Lymphopenia—less than 1500/mm³ on two or more occasions
	Thrombocytopenia—less than 100,000/mm³ in the absence of offending drugs
Immunologic disorder	Anti-DNA: antibody to native DNA in abnormal titer
	Anti-Sm: presence of antibody to Sm nuclear antigen
	Positive finding of antiphospholipid antibodies based on (1) an abnormal serum level of IgG or IgM anticardiolipin antibodies, (2) a positive test result for lupus anticoagulant using a standard method, or (3) a false positive serologic test for syphilis known to be positive for at least 6 months and confirmed by *Treponema pallidum* immobilization or fluorescent treponemal antibody absorption test
Antinuclear antibody	An abnormal titer of antinuclear antibody by immunofluorescence or an equivalent assay at any point in time and in the absence of drugs known to be associated with "drug-induced lupus" syndrome

[a]The proposed classification is based on 11 criteria. For the purpose of identifying patients in clinical studies, a person shall be said to have systemic lupus erythematosus if any 4 or more of the 11 criteria are present, serially or simultaneously, during any interval of observation. These criteria have been revised based on recommendations made by the Diagnostic and Therapeutic Criteria Committee of the American College of Rheumatology [7].

pathology and prognosis [9]. Early detection of new renal involvement is essential because initiation of early therapy may prevent end-stage renal disease. Evaluation of the urine sediment is a valuable tool in assessing and monitoring lupus nephritis. A renal biopsy is generally recommended if the patient has laboratory evidence of active renal disease.

There are multiple central nervous system syndromes seen in patients with SLE [2,4,10]. Recently, the American College of Rheumatology (ACR) committee on neuropsychiatric lupus published criteria to assist with diagnosis and monitoring [10a]. Some of the most common criteria include cognitive impairment, mood disorders, psychosis, strokes, seizures, movement disorders such as chorea, and other manifestations such as headache and aseptic meningitis. Routine laboratory testing and serologies are not always helpful because of their lack of sensitivity and specificity. Some patients with neuropsychiatric disease will not have evidence of active disease elsewhere. Studies that may be helpful in evaluation include analysis of cerebrospinal fluid to rule out infection and to look for *in situ* production of immunoglobulin. Magnetic resonance imaging (MRI) can detect small reversible focal areas due to ischemia or edema as well as irreversible lesions from tissue infarction in either the white or gray matter. Single photon emission CT (SPECT) and positron emission tomography (PET) imaging have been useful in detecting abnormal cerebral metabolic function and blood flow. Investigators are hopeful that these techniques will provide increased sensitivity to microvascular abnormalities that are not detectable by MRI.

Cognitive dysfunction is commonly reported in patients with SLE but is difficult to quantitate and thus presents a diagnostic dilemma for the treating physician [10]. Specialized neuropsychological testing may be helpful in objectively documenting deficiencies in memory, problem solving, information processing, attention, and other areas. These tests may also provide baseline measurements of cognitive impairment that can be used to assess response to intervention on serial testing.

B. Management

Treatment for SLE has ranged from behavioral modification, including energy conservation and avoidance of ultraviolet light via sunscreens and protective clothing, to anti-inflammatory,

Table 55.2
Incidence of SLE by Sex/Race Group in the United States[a]

Study	Location	Date	WM	WF	BM	BF	Overall
[28]	New York, NY	1956–1965	0.3	2.5	1.1	8.1	2.0
[29]	San Francisco, CA	1965–1973	ND	ND	ND	ND	7.6
[30]	Rochester, MN	1950–1979	0.9	2.5	ND	ND	1.8
		1970–1979	0.8	3.4	ND	ND	2.2
[31]	Baltimore, MD	1970–1977	0.4	3.9	2.5	11.4	4.6
[32]	Pittsburgh, PA	1985–1990	0.4	3.5	0.7	9.2	2.4

Note. BF, Black females; BM, black males; ND, no data; WF, white females; WM, white males.
From [31] Hochberg, M. C. (1997). *Arthritis Rheum.* **28**, 80–86, with permission.
[a]Incidence rates per 100,000 persons per year.

analgesic, and immunosuppressive therapy [3,4,11,12]. Some of the most common medications used in patients with SLE include the nonsteroidal anti-inflammatory agents, low dose and topical corticosteroids, and antimalarial agents including hydroxychloroquine, chloroquine, and quinacrine. These agents may be helpful for the constitutional symptoms and cutaneous and musculoskeletal manifestations. More serious disease activity, particularly nephritis, cerebritis, and vasculitis, require higher doses of corticosteroids and/or immunosuppressive agents such as cyclophosphamide, azathioprine, methotrexate, or cyclosporin A. Traditionally, treatment for SLE was dominated by nonspecific anti-inflammatory and immunosuppressive agents. Observations of the important pathways involved in the immune dysregulation in SLE have resulted in new more targeted potential therapeutic agents. One agent is an anti-CD40 ligand antibody that partially blocks the CD40–CD40L costimulatory pathway in murine models of SLE and has been shown to prolong survival and reduce the severity of nephritis [13,14]. This antibody diminishes T-cell dependent B-cell growth and differentiation, which results in less pathogenic autoantibody production. Another agent renders the B lymphocyte unresponsive to specific immunogens, specifically DNA (dsDNA) [15]. This B cell "toleragen" results in decreased B-cell production of pathogenic antibodies to dsDNA. Clinical trials testing these agents in patients with lupus nephritis are currently underway.

There is interest in hormonal manipulation as a potentially effective treatment in SLE because of the strong evidence that hormones may induce or exacerbate disease activity. Danazol, a synthetic attenuated androgen, has been successfully used to treat immune-mediated thrombocytopenia [16] and several reports have also documented its efficacy in the treatment of autoimmune hemolytic anemia [17]. The mechanism of action of danazol is unclear. Six of eight patients tested in one series had elevated levels of IgG anti-platelet antibodies and in all six a marked decrease in these antibodies was noted following danazol treatment [18]. Clinical benefit has been noted in women, but not in men, treated with danazol for lupus-like disease associated with hereditary C1-inhibitor deficiency [18,19]. Masculinizing side effects also limit the use of danazol.

Reduced levels of androgens (androstenedione, dehydroepiandrosterone [DHEA], dehydroepiandrosterone sulfate [DHEAS], and testosterone) have been observed in women with active lupus [20,21]. These observations were extended in animal experiments that explored the effect of DHEA treatment on the development of lupus. Indeed, these studies showed delayed formation of anti-DNA antibodies and improved survival [22]. A proposed mechanism of action for DHEA is suggested by experiments regulating IL-2 secretion by activated T cells [23].

DHEA has been studied in patients with mild to moderate lupus and results of the open-label study have been published [24]. After three to six months of DHEA treatment, ten women with lupus showed an improvement in disease activity index and reduced corticosteroid requirements. Of three patients with significant proteinuria, two showed marked and one showed modest reductions in protein excretion. DHEA was well tolerated and the only frequently noted side effect was acne. On the basis of this preliminary study, two larger trials of DHEA use were initiated in multiple medical centers and practices throughout the United States and Canada. In one trial with nearly 200 patients, those with active lupus at entry into the study and who received 200 mg/day of DHEA were more often able to taper their prednisone than those who received placebo [25].

Bromocriptine, a prolactin antagonist, shows promise in suppressing active SLE. In an open-label study, bromocriptine decreased serum prolactin levels, suppressed anti-dsDNA antibodies, and significantly reduced lupus disease activity [26]. A double-blind, randomized, placebo-controlled study of bromocriptine also has been conducted [27]. The investigators reported that low-dose bromocriptine used as an adjunct to conventional treatment is a safe and effective means of reducing SLE flares. These promising studies represent the first clinical trials in lupus conducted since the 1970s.

III. SLE Incidence and Prevalence

The incidence of lupus is much higher in women than in men. The peak incidence occurs between the ages of 15 and 45, the childbearing years, when the female-to-male ratio is about 12:1. In pediatric and older-onset patients, the female-to-male ratio is closer to 2:1. SLE occurrence is three to four times higher among African-American women than Caucasian women. The average incidence of SLE in the United States has been estimated to be between 1.8 and 7.6 cases per 100,000 persons per year (Table 55.2) [28–32]. The differences in previously reported incidence rates of SLE are difficult to interpret due to limitations in study methodology. Problems that limit compara-

Table 55.3

Prevalence of SLE by Sex/Race Group in the United States[a]

Study	Location	Date	WM	WF	BM	BF	Overall
[28]	New York, NY	July 1965	3	17	3	56	14.6
[29]	San Francisco, CA	July 1973	7	71	53	283	50.8
[30]	Rochester, MN	Jan 1980	19	54	ND	ND	40.0

Note. BF, Black females; BM, black males; ND, no data; WP, white females; WM, white males.
From [31] Hochberg, M. C. (1997). *Arthritis Rheum.* **28**, 80–86, with permission.
[a]Per 100,000 persons.

bility include the lack of standardized criteria for case detection, the absence of any formal estimate of the proportion of all cases in the population actually detected, and reliance on passive case ascertainment. Many studies employed a review of inpatient medical records for case ascertainment. Although this method is efficient for large populations, patients may not be hospitalized early in the disease course. This method also tends to exclude those patients with less severe disease. Ideally, multiple sources of ascertainment should be employed. In a study examining average annual incidence of lupus from 1985 to 1990 in Allegheny County, Pennsylvania, three information sources were used to identify cases: (1) inpatient records from all 28 hospitals in the county; (2) rheumatologists in the Pittsburgh metropolitan area; and (3) the University of Pittsburgh Lupus Databank [32]. Capture-recapture methodology was used to provide a more accurate estimate of the total number of cases in the population. A total of 269 incident cases of SLE (191 definite and 78 probable by 1982 ACR criteria) were identified. The overall annual incidence rates per 100,000 person-years were 2.4 (95% Confidence Interval (CI) = 2.1–2.8) for definite SLE and 1.0 (95% CI = 0.8–1.3) for probable SLE. The crude race- and gender-specific incidence rates of definite SLE per 100,000 person-years were 0.4 (95% CI = 0.2–0.7) for white men, 3.5 (95% CI = 2.9–4.2) for white women, 0.7 (95% CI = 0.0–2.0) for African-American men, and 9.2 (95% CI = 6.8–12.5) for African-American women. These rates confirmed previous reports of an excess of SLE incidence among females compared to males and among African-Americans compared to whites.

With multiple sources of ascertainment, capture-recapture methods can be applied to provide an estimate of the total number of cases without requiring the identification of every case in the population. Using capture-recapture, the estimated overall ascertainment rate in Allegheny County was 85% (95% CI 78% to 92%). The overall ascertainment-corrected incidence rate for definite SLE per 100,000 person-years was 2.8 (95% CI = 2.6 to 3.2). This study also demonstrated the use of capture-recapture methods to potentially improve the accuracy of reported SLE incidence rates. In order to accurately interpret reported incidence rates and to evaluate comparability across studies, these authors advocate the use of capture-recapture methodology.

Prevalence of SLE in the continental United States has been reported to range from 14.6 to 50.8 cases per 100,000 people (Table 55.3) [28–30]. Similar differences in methods of case ascertainment as described for lupus incidence studies also apply for studies examining prevalence. While SLE is more com-

mon in African-American than Caucasian populations, there are no population studies of SLE prevalence in Africa. Data from case reports and series of hospitalized patients suggest that SLE is rare in Africa (Table 55.4) [33–56]. A study of females in Curacao in the West Indies reported an SLE prevalence of 83.8 per 100,000 people [56]. Prevalence of SLE in female African-Americans ranges from 17.9 to 283 per 100,000 people [29,57]. The prevalence of SLE in African-Europeans (197–207 per 100,000) appears higher than that reported in the United States [58,59]. An interesting observation in these studies is that in all cases higher prevalence of SLE in women was reported. In summary, there is a suggestion of a gradient in SLE prevalence from Africa to North America and Africa to Europe, with the lowest prevalence in Africa, although the data are meager and not directly comparable. Potential mechanisms for a prevalence gradient include genetic factors such as regional variation in gene pools caused by selection or admixture. Differences in SLE prevalence among these populations may also reflect environmental factors that trigger SLE such as nutritional, chemical, or microorganism exposures. For there to be a true gradient in SLE prevalence between Africa and North America, it is necessary to prove that the differences are not solely due to methodological differences such as criteria for diagnosis and disease severity, which could influence case ascertainment. However, as the authors of a review on prevalence gradients in SLE point out, a true geographic difference in SLE prevalence among populations of African ancestry would present an opportunity to further explore specific genetic and environmental hypotheses of SLE pathogenesis [60].

IV. Mortality and Morbidity in SLE

A. Prognosis and Factors That Affect Mortality

Studies conducted since 1955 indicate that the survival of patients with SLE has remarkably improved (Table 55.5) [61–81]. In 1955, the 5-year survival was only 50% [61], whereas 10-year survival in the 1990s approaches or exceeds 90% [73,74,78,81] and the 20-year survival approaches 70% [78]. The improved probability of survival has been attributed both to earlier diagnosis in patients with milder disease and to improved management of severe disease, including dialysis and kidney transplantation.

Host factors that may predict mortality in patients with SLE include age at onset, gender, race, and socioeconomic status. The effect of age at onset on mortality has not been unequivocally

Table 55.4
SLE Populations of African Ancestry

	Number of SLE cases	Population-based	Studied period (yr)	ARA/ACR criteria	Incidence, female	Prevalence, female
West Africa						
Ghana [33]	11	No	1983–1989	71	NA	NA
Guinea [34]	1	No	?	None	NA	NA
Nigeria [35,36]	3	No	?	None	NA	NA
Ivory Coast [37]	9	No	1972–1983	71	NA	NA
Gabon [38]	1	No	?	None	NA	NA
Zaire [39]	1	No	?	None	NA	NA
Senegal [40,41]	4	No	?	None	NA	NA
Africa, other						
Zimbabwe [40,42–46]	71	No	?,?,?,1979–1983, 1983	None, 71, 82	NA	NA
South Africa [45,47–51]	287	No	1984–1990, ?, ?, 1975–1987, 1969–1975, 1960–1972	None, 71, 82	NA	NA
Uganda [52,53]	26	No	1968–1978, ?	71, None	NA	NA
Kenya [54]	1	No	1961	None	NA	NA
Ethiopia [55]	4	No	1971–1978	None	NA	NA
Caribbean						
Curacao [56]	69	Yes	1980–1989	82	7.86 [7:1]	83.8 [10:1]
United States						
New York,	69	Yes	1956–1965	None	7.75[a] [7:1]	53.7[a] [18:1]
Alabama [57]					2.88[a] [12:1]	17.9[a]
California [29]	19	Yes	1965–1973	71	ND	283 [5:1]
Maryland [31]	223	Yes	1970–1977	71	11.4[a] [5:1]	ND
Pennsylvania [32]	48	Yes	1985–1990	82	9.0[a] [13:1]	ND
United Kingdom						
Nottingham [58]	21	Yes	1989–1990	82	31.9[a] (total)	207[a] (total)
Birmingham [59]	48	Yes	1991	82	22.8[a]	197[a] [31:1]

Note. Population-based: a study using population survey or multiple sources not limited to hospitalized patients. Incidence is per 100,000 person-years and prevalence is per 100,000 persons. Values in brackets are female to male ratios.
ARA/ACR, American Rheumatism Association/American College of Rheumatology; NA, not applicable; ND, not determined, cannot be calculated. From [60] with permission.
[a]Age-adjusted

established. An older age at presentation was found to be a risk factor for mortality in two studies [77,78]. In another study, however, the 1- to 5-year mortality rates were similar in both the adult-onset and childhood-onset SLE patient populations [81].

The role of gender in predicting mortality in SLE is also controversial. Male gender has been associated with reduced survival in some, but not all, studies [77,78,82–86]. In general, however, gender does not emerge as a significant predictor of survival when differences in age distribution between male and female SLE patients are considered [87].

In contrast, race does appear to be a significant predictor of survival, with African-American patients having a poorer prognosis than white patients in the U.S. [67,72,88] and Asians living in England having a poorer prognosis than English Caucasians [89]. Furthermore, Asian Indian, Afro–Caribbean, and Chilean patients with SLE have a poorer prognosis than either North American or European Caucasians [76,79,80].

It is difficult, however, to separate the effects of race from socioeconomic status, particularly among U.S. African-Americans and Caucasians [90]. In the Multicenter Lupus Survival Study, African-Americans exhibited a poorer prognosis than Caucasians

[67]. At study entry, African-Americans had both more severe disease and greater frequency of central nervous system involvement than Caucasians. African-Americans also were more likely to carry public rather than private medical insurance. Once these confounding variables were considered, African-American race did not significantly predict decreased survival. In contrast, even after adjustment for source of medical insurance, Reveille *et al.* [72] found that among SLE patients seen at the University of Alabama, African-American patients had a poorer prognosis than Caucasian patients. In a study of 408 SLE patients at Duke University Medical Center, Ward *et al.* [77] concluded that socioeconomic status, but not race, was associated with mortality in SLE.

Regarding SLE-related factors, renal disease, as measured by either serum creatinine or qualitative urine protein excretion, is unequivocally the most important predictor of poor outcome [66,67,72,74,78,80,90–92]. Lower than expected survival at 5, 10, and 20 years of 80%, 69%, and 53%, respectively, were reported from the Mayo Clinic for a cohort of SLE patients with lupus nephritis [93]. In Los Angeles, patients with lupus nephritis who underwent renal transplantation also had a lower than

Table 55.5
Survival Percentages for SLE over the Past Five Decades

Study	Year	Center	5 y, %	10 y, %	15 y, %	20 y, %
[61]	1955	Baltimore	50	—	—	—
[62]	1964	Cleveland	69	54	—	—
[63]	1968	New York	70	63	—	—
[64]	1971	New York	77	60	50	—
[65]	1974	Toronto	75	63	53	—
[63]	1976	Farmington	93	84	—	—
[66]	1981	Los Angeles	88	79	74	—
[67]	1982	Multicenter	86	76	—	—
[68]	1985	Sweden	97	—	—	—
[69]	1986	India	68	50	—	—
[70]	1988	Toronto	84	75	64	—
[71]	1989	Holland	92	87	—	—
[72]	1990	Alabama	89	83	79	—
[73]	1991	Finland	—	91	81	—
[74]	1991	Los Angeles	97	93	83	—
[75]	1991	Stanford	88	64	—	—
[76]	1992	India	—	—	—	—
[77]	1993	Durham	82	71	63	—
[78]	1993	Toronto	93	85	79	68
[79]	1993	Curacao	56	—	—	—
[80]	1994	Chile	92	77	66	—
[81]	1995	London	93	85	79	—

Note. From [90] Gladman, D. D. (1996). Prognosis and Treatment of systemic lupus erythematosus. *Curr. Opin. Rheumatol.* **8**, 430–437, with permission.

Table 55.6
Incidence Rates of Cardiovascular Events per 1000 Person-Years in 498 Women with Systemic Lupus Erythematosus, University of Pittsburgh, and 2208 Women, Framingham Offspring Heart Study, 1980–1993

Age (years)	SLE Rate	SLE 95% CI	Framingham Rate	Framingham 95% CI	Rate ratio	95% CI
Myocardial infarction						
15–24	6.3	0.2–35.3	0.0	0.0–11.8	∞	
25–34	3.7	0.8–10.7	0.0	0.0–1.2	∞	
35–44	8.4	4.2–15.0	0.2	0.0–0.9	52.4	21.6–98.5
45–54	4.8	1.0–14.1	2.0	0.9–3.6	2.5	0.8–6.0
55–64	8.4	1.7–24.5	2.0	0.6–4.6	4.2	1.7–7.9
65–74	8.0	1.0–28.7	0.0	0.0–17.1	∞	
Angina						
15–24	0.0	0.0–23.4	0.0	0.0–11.8	∞	
25–34	1.2	0.0–6.8	0.6	0.1–2.3	2.0	0.0–9.0
35–44	1.5	0.2–5.5	0.7	0.2–1.7	2.4	0.4–11.1
45–54	1.6	0.0–8.9	1.6	0.7–3.1	1.0	0.2–4.6
55–64	5.6	0.7–20.2	2.4	0.9–5.2	2.3	0.9–5.5
65–74	15.9	4.3–40.6	0.0	0.0–17.1	∞	

Note. SLE, systemic lupus erythematosus; CI, confidence interval. From [103] Manzi, S., *et al.* (1997). *Am. J. Epidemiol.* **145**, 408–415, with permission.

expected 5-year survival of 85.9% [94]. Race appears to be a factor in the occurrence and severity of renal disease. African-American women have the highest rates of lupus renal disease and poorer renal survival compared to Caucasian women [95].

Central nervous system disease has been associated with poor prognosis in some, but not all, cohorts in which it was studied [64,67,71,96]. It has been suggested that central nervous system involvement predicts poor outcome only in patients with active disease in other organs [97]. Greater overall disease activity has been shown to predict mortality in SLE [90]. In one study, greater disease activity at the time of first visit to a lupus clinic was a predictor of poorer survival [78].

B. Causes of Death

Although survival has greatly improved, mortality rates for patients with SLE are still three times greater than those for the general population [98]. Deaths that occur early in the course of disease are more likely to be attributable to active disease, particularly renal and central nervous system involvement, systemic vasculitis, and infections, whereas deaths that occur later tend to be the result of complications of either the disease process itself or its therapy [87,90]. Infection, which often occurs in active SLE, can also occur opportunistically in patients receiving high-dose steroids and cytotoxic therapy [99–102]. Mortality in SLE as a result of infection and other morbidity appears to have decreased over time; however, the risk of death due to active SLE has remained unchanged [90].

C. Morbidity

With the increased life expectancy of SLE patients due to improved therapy, cardiovascular disease [103–106] and osteoporosis [107–112] have emerged as significant health threats. In one study, cardiovascular events (myocardial infarction and angina pectoris) were evaluated in a cohort of 498 women with SLE seen at the University of Pittsburgh Medical Center from 1980 to 1993 (3522 person-years). The authors compared these rates with the cardiovascular event rates occurring over the same period in 2208 women of similar age participating in the Framingham Offspring Study (17,519 person-years). There were 33 first events reported in the women with SLE (11 myocardial infarction, 10 angina pectoris, and 12 angina pectoris and myocardial infarction). Two-thirds were under the age of 55 at the time of event. Age-specific rate ratios were computed to determine whether the cardiovascular events in the lupus cohort were more frequent than expected (Table 55.6) [103]. Women with lupus in the 35–44 year age group were over 50 times more likely to have a myocardial infarction than women of similar age in the Framingham Offspring Study (rate ratio = 52.4, 95% CI = 21.6–98.5). This study confirmed that premature cardiovascular disease is much more common in young premenopausal women with lupus than in a population sample. In addition, older age at lupus diagnosis, longer lupus disease duration, longer duration of corticosteroid use, hypercholesterolemia, and postmenopausal status were more common in the women with lupus who had a cardiovascular event than in those who did not have an event.

The pathogenesis of premature cardiovascular disease in women with lupus is likely multifactorial, related to the underlying vascular inflammation and arterial wall injury, adverse

Table 55.7

Standardized Morbidity Ratios for Fractures in Patients with Systemic Lupus Erythematosus Compared with United States Population Estimated from the 1994 National Health Interview Survey[a]

Age in years	Person-years at risk	Observed number of fractures	Expected number of fractures	Standardized morbidity ratio (95% CI)
<18	140	2	0.8	2.4 (0.3–8.7)
18–24	587	13	1.1	12.1 (6.4–20.7)
25–44	3546	35	10.9	3.2 (2.2–4.4)
45–64	1428	32	4.2	7.6 (5.1–10.7)
65–69	152	2	0.7	2.9 (0.4–10.5)
70+	98	2	0.4	4.9 (0.6–17.7)
Total	**5951**	**86**	**18.1**	**4.7 (3.8–5.8)**

[a]The U.S. population data sample was obtained as part of the National Health Interview Survey. These data were provided by the National Center for Health Statistics. The Center specifically disclaims responsibility for any analyses, interpretations, or conclusions.

From [112] with permission.

effects of corticosteroids, the high prevalence of renal disease and hypertension, and increased risk of thrombosis in the setting of antiphospholipid antibodies. The role of estrogen in cardiovascular disease is of particular interest in women with SLE. Estrogen replacement therapy is thought to have beneficial effects on the prevention of cardiovascular disease in postmenopausal women. In young women with lupus, the relative hyperestrogenism may have prothrombotic effects, particularly in combination with hypertension, renal disease, and antiphospholipid antibodies. The effects of estrogen on vascular disease in SLE are currently unknown.

Several studies have documented both cortical bone loss [107–109] and trabecular bone loss [107,110] in patients with lupus. Prevalence estimates of low bone mineral density (BMD) in a few studies that have measured BMD in lupus patients range from 4.5 to 40.0% [107,110,111,113–115]. In one study, the authors compared the frequency of self-reported fractures and risk factors for fracture between lupus patients and women of similar age in the general population using data from the 1994 National Health Interview Survey (NHIS) [112]. These 702 patients were seen by University of Pittsburgh Medical Center rheumatologists from 1980 to 1994, for a total of 5951 person-years of observation. Eighty-six (12.3%) of 702 women reported at least one fracture following their diagnosis of lupus. Forty-seven percent of the fractures occurred before menopause or age 50. Fracture risk was increased in the lupus cohort compared with women of similar age using NHIS data (Table 55.7); the estimated standardized morbidity ratio (SMR) was 4.7 (95% CI = 3.7–5.7). Women in the 18–24 year age group had the highest risk; they were 12 times more likely to have a fracture than the comparison group. Older age at lupus diagnosis and longer use of corticosteroids were associated with shorter time from lupus diagnosis to fracture. There are several theoretical reasons why women with lupus may have low BMD. The candidate risk factors include: (1) reproductive history (nulliparity) [116]; (2) premature menopause due to the toxic effects of im-

munosuppressive medications on ovarian function [117]; (3) avoidance of oral contraceptives and/or hormone replacement therapy because of concerns about precipitating disease flare [118,119]; (4) decreased physical activity because of fatigue [120]; (5) cytokine imbalance resulting in elevated levels of interleukin-6, which have been implicated in the pathogenesis of accelerated bone remodeling [121,122]; (6) renal disease [4]; (7) decreased vitamin D levels due to sun exposure avoidance to prevent flare of lupus [123]; and (8) corticosteroid therapy [124]. Because estrogens clearly modulate bone mass [122], the role of estrogens in osteoporosis is of particular interest in women with lupus.

There is concern about malignancy risk with and without the additional exposure to immunosuppressive drugs in patients with lupus. Most case reports show a possible relationship between lupus and non-Hodgkin's lymphoma, leukemia, and solid tumors [125–128]. The overall risk of malignancy was increased in Finnish and Danish lupus cohort studies [129,130], but not in a Canadian study [131] or in a small study in Pittsburgh, Pennsylvania [132]. Furthermore, none of these lupus cohort studies has shown a relationship between immunosuppressive drug use and cancer risk. In a study of 616 women with lupus from Chicago, Illinois, an overall increased risk of malignancy was reported (SMR = 2.0; 95% CI = 1.4–2.9) [133]. Estimated rates of cancer were calculated using age, gender, and race-adjusted cancer incidence data from Cook County, Illinois. Breast, lung, and gynecological cancer were the most common, with breast cancer significantly increased in Caucasian women. Known or suspected risk factors for breast cancer may also play a role in the putative increased risk of this malignancy in women with lupus. Elevated prolactin levels have been noted in patients with active lupus [134] and elevated prolactin levels may promote breast cancer [135]. Low parity, nulliparity, and large body size with truncal obesity [136] are risk factors that could be increased in these patients because pregnancy may be discouraged in lupus patients and they have the central obesity body habitus from treatment with corticosteroids [116]. Confirmation of this increased risk of cancer, specifically breast cancer, in women with lupus should be pursued.

V. Pathogenesis of SLE

The central immunologic disturbance in patients with SLE is autoantibody production [137,138]. These autoantibodies are directed against a host of self-antigens found in the nucleus, cytoplasm, and membranes of cells. Some of the most frequent antibodies are to nuclear antigens, which are found in over 95% of patients with SLE. Antibodies to cytoplasmic proteins associated with RNA are also a common antibody specificity in SLE. Characteristically, the autoantibodies produced in SLE are directed against molecules involved in critical cellular functions, including storage of genetic material, cell division, regulation of gene expression, RNA transcription, and RNA processing.

Autoantibody specificity may have diagnostic and prognostic implications. Recognition of the clinical features associated with specific autoantibodies has improved our understanding of the heterogeneous nature of SLE. How these autoantibodies induce pathologic disease is not completely understood. Anti-native DNA or double stranded DNA antibodies (anti-dsDNA) and antibodies to Sm, an RNA-binding protein, are highly specific for SLE. Anti-Ro antibodies bind to Ro protein, an RNA-binding protein. Although not specific for SLE, anti-Ro antibodies are frequently found in patients with a characteristic cutaneous rash (subacute cutaneous lupus) [139]. Anti-Ro antibodies are also found in the sera of women giving birth to babies with neonatal lupus syndrome including congenital heart block [140]. An increased risk of thrombosis and recurrent miscarriage have been associated with the presence of antibodies to phospholipids in patients with SLE [141].

Although the key abnormality in lupus is autoantibody production, both B and T lymphocytes are hyperactivated and evidence suggests that the autoantibodies produced in SLE are T-cell dependent. In mouse models of SLE, depletion of CD4$^+$ T cells blocks onset of disease [142] and athymic mice do not develop SLE [143]. Experimental evidence indicates that nucleosomal antigens play an important role in T-cell induction of pathogenic autoantibodies. The T-cell receptors expressed on pathogenic T cells of both murine and human SLE appear to be specific for cationic autoantigenic peptides [144,145]. In one animal model, a majority of the T-cell clones that stimulate autoantibody production by B cells are specific for nucleosomal antigens [146,147]. Furthermore, pathogenic autoantibodies to nucleosomal antigens appear first during the course of disease in humans and mice with SLE [148]. Nucleosomes circulate in normal individuals as a result of cellular apoptosis without inducing SLE. Therefore, understanding how nucleosomes function as a specific immunogen of pathogenic T cells of SLE will greatly enhance our understanding of its etiology.

Sustained B-cell and T-cell hyperactivity may be the result of reciprocal interactions. Furthermore, CD8 T cells, which could normally suppress excessive T-cell production, can supply help rather than suppression in SLE [149]. Consequently, B-cell and T-cell activation may cycle continuously. At present, it is unclear whether SLE is the result of excessive T-cell help or defective T-cell suppression [143]. Negative feedback for antibody production by circulating antibody levels is also defective in SLE.

Of particular interest is the importance of costimulatory signals between B and T cells. T cells become activated upon receiving several signals, one of which is the binding of specific antigen to the T-cell receptor. These antigens are presented to T cells by antigen presenting cells, which include B cells. Subsequently, activated T cells transiently express a ligand for CD40, referred to as CD40L. The binding of CD40L to its receptor, CD40, on antigen-specific B cells is a necessary second signal for B-cell growth and differentiation [150]. In both murine and human SLE, the interaction between CD40L on T cells and CD40 on B cells is involved in the production of pathogenic autoantibodies [13,151,152]. Under normal circumstances the immune system allows only transient expression of CD40L. However, patients with SLE express abnormally high levels of CD40L on both T and B cells and the overall number of CD40L$^+$ cells is increased [151]. Therefore, the abnormal regulation of CD40L expression may also be an important factor in the development of SLE.

Immune complexes formed by autoantibodies and their specific antigens can deposit in the blood vessels of various organs, resulting in an inflammatory response and tissue damage. Although healthy individuals can produce many, if not all, auto-

antibodies, subsets of some antibodies can be directly pathogenic. It has been suggested that cationic anti-DNA antibodies bind negatively charged components in the glomerular basement membrane, thus initiating lupus nephritis. Evidence suggests that these autoantibodies actually bind to nucleosomes and binding to glomerular basement membrane is mediated by positively charged residues in histones found in the nucleosomes [146,153,154]. Nonpathogenic antibodies can bind to tissues through electrostatic interactions or through their immune complexes [149]. It is currently of interest to determine whether there are structural differences between the autoantibodies produced by patients with SLE and those of healthy individuals. Furthermore, individuals with SLE exhibit defective mechanisms for the clearance of immune complexes, thus allowing their persistence in circulation [149]. Deficiencies of complement receptor (CR)1 and CR2 have been described in lupus patients [155]. Both inherited and acquired factors contribute to low CR levels in SLE patients [156]. Low levels of these receptors, which promote clearance of immune complexes, may result in increased levels of pathogenic immune complexes in the circulation of patients with SLE [157]. An allele of an Fc receptor (FcγRII) was identified that predisposes African-Americans to lupus nephritis [158]. Impaired Fc receptors on IgG may contribute to susceptibility to lupus due to impaired clearance of immune complexes.

SLE is thought to be caused by genetically determined immune abnormalities that are potentially triggered by hormonal factors and other environmental stimuli. An association between major histocompatibility complex (MHC) loci for human leukocyte antigens (HLA) DR2 and DR3 and SLE has been reported in some populations [159–161], but the MHC molecules themselves are also present in normal individuals. Inherited complement deficiencies, particularly homozygous C4a deficiency, confer a very high risk for SLE, and genes for C4 are located in the MHC region [162,163]. A candidate gene, 1q41-q42 region on chromosome 1, has been suggested as one genetic factor that may confer an increased risk for lupus in all ethnic groups [164]. The likely identification of multiple new candidate genes associated with predisposition to autoimmunity and SLE will dramatically increase our understanding of the pathogenesis of this disease [164–168].

VI. Hormonal and Environmental Risk Factors

A. Gender and Endogenous Hormones

Many autoimmune diseases show marked gender predilections, with women more commonly affected than men, suggesting that sex hormones play a role [169–173]. Observations in animal models and humans with autoimmune conditions indicate that sex hormones influence autoimmune reactions [169–177]. Physiological levels of estrogens are thought to stimulate the immune response and male hormones are thought to to suppress it [169,175,177]. Estrogens have been demonstrated *in vitro* to stimulate B-cell response and decrease suppressor T-cell reactivity, which can lead to an increase in autoantibody production [178]. Current evidence suggests that sex hormones influence the immune system through several mechanisms, such as via the intracellular sex steroid receptors that are found in a variety of tissues and cells, including lymphocytes [179,180]. Estrogens likely account in part for the higher immune reactivity in females against a variety of antigens and the more rapid rejection of allografts [181]. The greater immune responses in females might also increase their susceptibility to autoimmune diseases [169].

SLE is the prototypic autoimmune disease that affects women significantly more frequently than men. Much of the evidence that suggests a role for sex hormones in the development of SLE has been derived from animal models and in some of these models, estrogens accelerate the disease process. In a mouse model of SLE, the animals spontaneously develop a disease similar to human SLE, including developing antibodies to DNA and glomerulonephritis at 3 to 6 months, which progresses to death by 12 months [182]. Onset of lupus in mice is earlier and the disease is more severe in females than in males. Males of this strain have a significantly longer survival than females and castration in males accelerates mortality to the rate found in female mice [183,184]. In addition, treatment of females with androgens postpones death [22,183]. It was shown that when a lupus-like disease was induced by injecting male and female mice with procainamide-treated T cells, females developed more autoimmune disease, greater titers of anti-DNA antibodies, and accumulated more T cells in their spleens than did the male mice [185]. The authors concluded that the differences in T-cell trafficking could contribute to sex-specific disease severity. In a different murine model of lupus, investigators have shown that estrogen is capable of disrupting the balance between the expression of the proapoptotic *bax* and antiapoptotic *bl-2* genes in the mouse kidney, favoring *bax* and apoptosis [186]. The authors suggested that the enhanced apoptosis in the kidney could eventually lead to accelerated glomerulonephritis. These observations in both animal models and humans with SLE have led to intensive inquiry and investigation into the relationship between the hormonal milieu and disease state.

Both male and female patients with SLE have been found to have abnormalities in estrogen metabolism [85,170,187–189] (Fig. 55.1). Researchers have shown elevated levels of 16α-hydroxyestrone and estriol and low levels of the 2-hydroxylated estrogens in patients with SLE [174,176]. In contrast to the 16-hydroxylated metabolites, which retain significant peripheral estrogenic activity, the 2-hydroxylated compounds do not [190,191]. These observations were initially made by Lahita *et al.* when they examined urinary metabolites and later confirmed that the increased 16α-metabolites in urine were due to increased production rather than to increased excretion [189]. The elevated conversion of estradiol to 16-hydroxylated metabolites in SLE patients did not appear to be associated with age, use of corticosteroids, or obesity [189]. Lahita *et al.* failed to show an association between these metabolites and SLE disease activity. However, their study included small numbers of patients (8 men, 15 women) and lacked a quantitative measure of lupus disease activity. Patients with other illnesses, including breast cancer, endometrial cancer, chronic liver disease, or rheumatoid arthritis, failed to show elevations in the 16-hydroxylated metabolites as great as those identified in patients with SLE. There are also data suggesting that first-degree relatives of patients with SLE have similar elevations of 16-hydroxylated metabolites [188].

Lower plasma testosterone levels were demonstrated by Jungers *et al.* in women with SLE compared with age-similar healthy controls [20]. All of the SLE women in this study had regular menses and none were on oral contraceptives or had clinical liver disease. Low testosterone levels could not be explained by corticosteroid adrenal suppression because many of the SLE women had never been on corticosteroids and the others had been off corticosteroids for 6 months to 7 years prior to obtaining the hormone levels. Although testosterone levels appeared lower in 22 women with active SLE disease compared with 6 women without disease, this difference did not reach statistical significance, perhaps because of inadequate study power [21]. Testosterone is largely bound to sex hormone-binding globulin (SHBG). It is not clear whether the differences seen in testosterone levels are a reflection of lower SHBG in women with SLE. Another possible explanation for low testosterone levels in women with SLE is thought to be accelerated metabolism of testosterone via oxidation to androstenedione at C-17 [21,192]. Interestingly, men with SLE have not consistently been noted to have abnormalities in testosterone levels [192].

In a later study, 26 nulliparous women with inactive SLE who were not on corticosteroids had lower peak and 7-day postovulation serum progesterone concentrations than 21 healthy, nulliparous controls [193]. In all of the women studied, ovulation was confirmed by ultrasound. No differences were detected in serum concentrations of follicle-stimulating hormone, luteinizing hormone, prolactin, estradiol-17β, testosterone, androstenedione, DHEAS, or cortisol. Very little attention has been given to progesterone in SLE patients and thus the significance of these findings in women with presumably inactive SLE is uncertain. Interestingly, progesterone is felt to have immunosuppressive properties and has been shown to depress cell-mediated immunity and enhance suppressor cell activity [194,195].

During the normal menstrual cycle, progesterone levels rise during the luteal phase, which occurs after ovulation and prior to menses. The lower levels of progesterone during the luteal phase found in women with lupus may contribute to the increased symptoms reported by these women during the two weeks prior to menses. In addition, an inadequate production of progesterone in women may render them prone to autoimmune diseases such as SLE.

Although it is reported that SLE disease activity can be altered during pregnancy and postpartum, much less is known about the possible effects on SLE symptomatology during the normal menstrual cycle. Clinicians have recognized for some time that women with SLE often report increasing joint pain, fatigue, mouth ulcers, pleuritic chest pain, and other subjective symptoms at certain times during the menstrual cycle. Only a few investigators have actually examined this in detail [196–198]. In one report, 28 menstruating women were followed through 991 menstrual periods [196]. Increased signs or symptoms of SLE were reported in 172 (17%) of the cycles. Signs and symptoms consisted of pleurisy, pericarditis, arthritis or arthralgia, mucosal ulcers, increased rash, vasculitic lesions, and orthopnea. Most of these manifestations were confirmed by physical exam. Of the 172 periods with increased symptoms, 140 (81%) occurred in the two weeks prior to menstruation. These results suggest a possible relationship between SLE disease activity and the menstrual cycle; however, no sex hormone

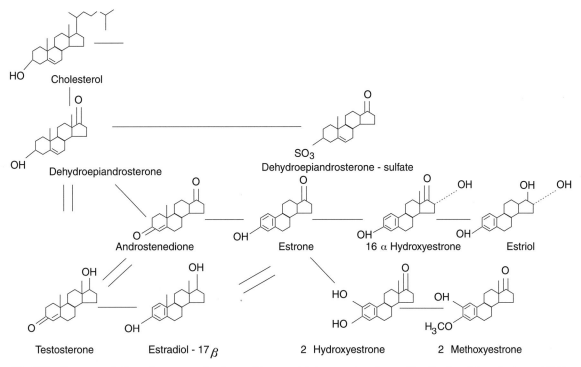

Fig. 55.1 Human metabolism of estrogen and androgen. There are data to support an abnormality of hydroxylation of estrone at C-16 to 16α-hydroxyestrone and estriol and abnormally rapid oxidation of testosterone to androstenedione at C-17. The latter finding is specific to female systemic lupus erythematosus patients. From [189a] with permission.

levels were measured. Thus, no conclusive statement about the association of sex hormones and SLE symptoms could be made. Other investigators have reported symptoms consistent with increased disease activity during the premenstrual period in 45–60% of women with SLE [197,198]. One might expect an earlier age at menarche to increase exposure to estrogen. However, there are conflicting reports concerning the impact of age at menarche on the risk of developing SLE [199]. Likewise, there is no consensus regarding the association of menstrual cycle patterns and risk of SLE [199]. These are areas that need to be examined more comprehensively.

A role for prolactin as an immunomodulator has been postulated. Prolactin is secreted by the anterior pituitary gland and regulates a variety of growth-related processes. Estrogen stimulates the production of prolactin. Prolactin receptors belong to the cytokine/growth hormone/prolactin receptor family and are expressed on immunomodulatory cells, including T and B cells [134]. Experimental evidence suggests that prolactin is involved in maintaining immunocompetence. When serum prolactin levels were artificially lowered in mice, antibody formation and cell-mediated immune responses were suppressed. Administration of prolactin reversed these effects [200–202].

In the mouse model of SLE, female mice made hyperprolactinemic had premature proteinuria, elevated serum IgG, and premature mortality. Similar mice treated with bromocriptine, which inhibits secretion of prolactin from the pituitary, had delayed elevation of serum IgG, anti-DNA antibodies, and longer survival [134,203]. Similarly, hyperprolactinemia in male mice stimulated disease activity and accelerated mortality [204].

Several studies have noted elevated prolactin levels in both male and female patients with lupus [205–208]. Although some patients with lupus have conditions that result in hyperprolactinemia, there are patients with no identifiable cause of the prolactin elevation [208–210]. The anterior pituitary cells, which express cytokine receptors, may be stimulated to produce prolactin by the elevated levels of cytokines that have been reported in patients with active SLE [211].

Physiological hyperprolactinemia, induced by pregnancy and lactation, may play a role in the lupus disease activity reported during pregnancy in some women with SLE. In one study, prolactin levels were higher in pregnant SLE patients with flare than in pregnant controls [207]. However, there has been no agreement on the relationship between prolactin levels and lupus disease activity in nonpregnant patients [209,212].

B. Fertility and Pregnancy

Fertility does not appear to be decreased in lupus patients compared with the general obstetrical population [213], except in the circumstances of very active systemic disease. There are no studies that evaluate fecundity in patients with lupus. Fetal outcome is adversely affected by maternal lupus. Only approximately 50% of pregnancies in women with lupus result in a full-term, normal birth weight infant [214–217]. Adverse fetal outcomes are common among mothers with lupus and include fetal death, preterm birth (before 37 weeks gestation), and intrauterine growth retardation. The manifestations of neonatal lupus include transient rash within the newborn period, permanent heart block, or both [140]. Fortunately, neonatal lupus is an extremely rare event. Less than 25% of infants have cutaneous neonatal lupus and less than 3% of infants have congenital heart block if their mothers with lupus also have the associated antibodies anti-SSA (Ro) or anti-SSB (La) [218]. Maternal renal disease or hypertension, previous history of fetal death, or the presence of antiphospholipid antibodies adversely influence the risk of a poor fetal outcome in women with lupus [219].

During pregnancy, estrogen and progesterone levels rise, creating a natural state in which to explore the influence of sex hormones on the course of disease. Women with lupus frequently have symptoms of their disease during pregnancy. These observations bring up several questions that have been debated in the literature since the 1950s: Does pregnancy adversely affect lupus? When does flare occur? Are flares during pregnancy more severe than those occurring when women are not pregnant?

The lack of a consensus definition for lupus flare has been a major obstacle for investigators. The diagnosis of flare in the pregnant patient is even more difficult because common pregnancy-related symptoms can also be lupus symptoms. For example, arthralgia, facial and palmar erythema, thrombocytopenia, proteinuria, and anemia all occur in women with SLE who are not pregnant. Reliable indicators of active disease in pregnant lupus patients include rising levels of anti-DNA antibody, alternative-pathway hypocomplementemia, true arthritis, true rash, mucosal ulcers, and lymphadenopathy [3]. The results of six studies are split, with three supporting the increased occurrence of lupus flares during pregnancy [220–222] and three showing no increase in the occurrence of lupus flares during pregnancy [223–225]. However, most series agree that lupus symptoms are not more severe during pregnancy than during the nonpregnant state and that disease activity can occur during any trimester and postpartum. Several methodologic issues also confound the results of these studies; these include selection of appropriate control populations for comparison, patients first diagnosed with lupus during pregnancy who may be more ill than patients with established lupus, clinical populations with a large percentage of African-American patients who have more severe lupus, counting multiple pregnancies in the same woman, and in some studies, prophylactic use of prednisone. In conclusion, lupus symptoms are present during pregnancy and after delivery. Whether flare is increased and more severe during pregnancy than during the nonpregnant state is still unclear.

C. Exogenous Hormones

The effects of oral contraceptives (OCs) and hormone replacement therapy (HRT) in women with SLE are not well characterized because these exposures are usually avoided [226]. Several retrospective studies have shown that young women with SLE used OCs less often than healthy women of similar age [227,228]. This may reflect the concerns of the treating physicians with regard to potential worsening of SLE activity with hormone use. A retrospective, multicenter survey of 404 women with SLE reported that 55 (14%) used OCs after SLE diagnosis [228]. Only seven (13%) of the 55 women reported an increase in disease activity after OC use, which predominantly took the form of musculoskeletal symptoms. Isolated case reports have described disease flares in women with SLE who are on OCs containing estrogen [229,230]. Disease flares occurred within eight weeks of instituting oral contraceptives. Other investigators have suggested that exogenous hormones may unmask subclinical SLE; they reported two cases of asymptomatic women with a biological false-positive test for syphilis (one of the serologic criteria for SLE) who developed signs and symptoms of clinical SLE three to four weeks after starting OCs [231,232]. One of the women had dramatic improvement in her symptoms within days of discontinuing the oral contraceptive [232].

Other studies have addressed the issue of the causal effects of OC use on the development of SLE. Two case-control studies have shown little or no association between the development of SLE and OC use [233,234]. A larger, prospective study of 121,645 women participating in the Nurses' Health Study reported a slightly increased relative risk of developing SLE in women with a history of OC use [235]. The women, who were classified as past users based on self-report, were followed from 1976 to 1990. During this time, 99 cases of definite SLE were confirmed. Compared with never users, the relative risk for developing SLE in women who had definite SLE was 1.4 (95% CI = 0.9–2.1) for past users of OCs. The relative risk for past users compared to never users increased to 1.9 (95% CI = 1.1–3.3) when more stringent criteria were used for case definition of SLE. The authors acknowledged the possibility that the increase in risk of SLE with more stringent case definition may be the result of including only the more severe cases and excluding the milder cases and those at an earlier stage of disease. In that case, OC use may be associated with more severe SLE manifestations rather than with an increased risk of disease development. In this same study, there was no association between duration of OC use or time since first use and the risk of developing SLE. The authors concluded that the small absolute risk of developing SLE associated with OC use in white women should not be a dominant factor in the decision to use hormonal contraception in women with SLE.

One study reported an association between OC use in healthy women seen in a birth control clinic and the presence of antinuclear antibodies (ANA) [236]. The frequency of ANA positivity was 13.4% in 210 women on OCs versus 8.5% in women who never used OCs. None of these women reported symptoms consistent with a rheumatic disease. Other studies examining healthy women attending birth control clinics have shown no association between the presence of ANA and OC use [237,238].

Researchers have compared exacerbations of SLE disease activity in women on combination oral contraceptives (estrogen–progesterone) and those on progestin-only regimens. In one study, 9 (43%) of 20 women experienced initial manifestations or flare of SLE within three months of starting the combination pills [118]. Among the 11 women on pure progestins, including five women who had previously developed lupus exacerbation while receiving estrogen, none developed signs or symptoms of lupus exacerbation during a follow-up period of 5–30 months. The authors concluded that progestin-only oral contraceptives did not have as great a potential to induce disease flares as the combination pill. However, drawing a definitive conclusion is limited by the retrospective design of the study, the underlying disease severity of SLE (all of the women had renal disease), and concomitant use of corticosteroids in these women. Other investigators have also reported a decrease in SLE disease flares with progestin-only pills, but menstrual irregularities and other intolerable side effects have dampened the enthusiasm for these oral contraceptives [229,239,240].

The administration of OCs among women with SLE has been associated with thromboembolic complications and chorea [241–244]. This risk may even be greater in patients with antiphospholipid antibodies [242,243]. Antibodies to phospholipids are found in 40–50% of patients with lupus and are associated with recurrent fetal loss and thromboembolic events [141,245,246]. Oral contraceptives are generally not recommended for those SLE women with high levels of antiphospholipid antibodies, previous thrombotic events, and those who have currently active SLE and require high doses of corticosteroids or other immunosuppressive agents.

With the increased life expectancy of women with SLE due to improved therapy, there are emerging concerns about the safety of HRT in postmenopausal women. Examples of SLE flares in postmenopausal women taking estrogen replacement for osteoporosis prevention have been published [247]. Symptoms resolved when estrogen was discontinued and promptly recurred upon reinstitution of estrogen. An increased relative risk of developing SLE was reported in postmenopausal nurses exposed to HRT [248]. In this study, 69,435 postmenopausal women aged 30–55 without a diagnosis of SLE were followed from 1976 to 1990. Forty-eight percent of participants had either been current or past users of HRT. Compared with never-users, the age-adjusted relative risk for SLE was 2.1 (95% CI = 1.1–4.0) for ever-users, 2.5 (95% CI = 1.2–5.0) for current users, and 1.8 (95% CI = 0.8–4.1) for past users. A proportional increase in risk for SLE was related to duration of HRT use. This study has several potential limitations. All hormone use was self-reported. It is possible that an increased opportunity to diagnose lupus was available to the women on estrogen replacement because of more active medical follow-up. The reasons why some women were given HRT and others were not is unclear. Subtle symptoms of SLE, including fatigue, facial flushing, myalgia, and arthralgia, may have been misinterpreted as postmenopausal symptoms leading to the initiation of HRT. Although the results of this study are intriguing, they should be interpreted with caution. In two smaller retrospective studies of postmenopausal women with SLE, HRT appeared to be well tolerated with no increase in disease flares [228,249]. Thirty of the 60 SLE women on HRT in one study had the same rate of flare as the 30 women who never used HRT [249]. Moreover, the HRT group had an improved sense of well-being, with less depression and migraine headaches [249]. In the other study, 48 (51%) of the 94 postmenopausal women surveyed had been on HRT after SLE diagnosis [228]. Only four of these women reported exacerbation of SLE disease activity.

There is some data suggesting that ovarian stimulation with fertility drugs may increase the risk of developing SLE [199]. Large clinical studies are necessary, however, to more accurately assess this association.

While there are concerns regarding the use of exogenous hormones in women with SLE, there are also many potential benefits, including effective birth control with OCs, prevention of osteoporosis, reduction in coronary artery disease, and treatment of bothersome postmenopausal symptoms, such as hot flashes, depression, headaches, and decreased libido with HRT. There are now ongoing prospective, double-blinded, placebo-controlled trials examining the risks and benefits of exogenous hormone use in women with SLE.

D. Environmental and Infectious Exposures

Although individuals may be genetically predisposed to SLE, it is likely that environmental and other exogenous triggers are involved in the initiation of disease. The effects of exogenously administered sex hormones were addressed in the previous section. Exposure to UV light can induce and exacerbate cutaneous and systemic lupus erythematosus in some individuals. The pathogenicity of UV light may be due to its ability to alter DNA as well as other autoantigens, possibly increasing their immunogenicity [149].

Drugs and medications have also been reported in association with lupus-like syndromes. The effects of procainamide, hydralazine, and various anticonvulsants and psychiatric drugs have been well established in this regard. Studies of numerous other drugs have documented an association with lupus-like syndromes [199]. Only a few, however, were prospective studies of large numbers of patients. Most were case reports of only a few individuals.

There have been some reports of an association between dietary lipids and risk of developing autoimmune diseases. It has been reported that increased consumption of omega-6 or a vegetable source of oils and decreased omega-3 intake may increase autoimmune disease because of their ability to induce the production of free radicals and proinflammatory cytokines and to suppress the production of anti-inflammatory cytokines [250]. Although dietary factors may be related to developing autoimmunity, research in this area is still inconclusive. In general, a problem facing the issue of environmental risk has been that, in contrast to the well-established criteria for SLE, there are no well-defined and accepted diagnostic criteria for drug and environmental lupus syndromes.

Infectious agents have gained increasing attention as potential etiologic agents in the development of SLE. They may initiate or flare SLE by disturbing immunoregulation, causing tissue damage leading to the release of autoantigens, or by eliciting a specific immune response by molecular mimicry. According to this mechanism, a foreign antigen having an amino acid sequence or antigenic structure similar to a self molecule might stimulate autoantibody production. This type of cross-reactivity has been suggested for SLE because of sequence similarity between certain nuclear antigens and viral or bacterial products. Of particular interest are the findings of homology between amino acid sequences of Epstein-Barr virus nuclear antigens (EBNA) and nuclear autoantigens found in SLE [251,252]. Harley and colleagues reported increased prevalence of Epstein-Barr virus infection in young patients with SLE [251]. Another group demonstrated the possibility that Epstein-Barr virus may establish a persistent infection in some SLE patients [252]. These results suggest a possible etiologic role of Epstein-Barr virus in the development of SLE. The high prevalence of Epstein-Barr virus infection in adults, however, indicates that, if Epstein-Barr virus is a causative agent, additional factors must also be required. Furthermore, the possibility of increased susceptibility to Epstein-Barr virus infection in lupus patients cannot be ruled out, particularly because B cells, which are hyperactivated in SLE, are the sites of Epstein-Barr virus infection. The dysregulation of the immune system as a result of the disease may lower the patient's intrinsic resistance to Epstein-Barr virus infection. The life-long persistence of the virus after primary infection and its tendency to reactivate as a consequence of immune dysfunction are additional factors indicating the need for caution in assigning a primary role to Epstein-Barr virus in the pathogenesis of SLE. Studies determining the temporal relationship between viral infection and onset of disease will be necessary to clarify this issue.

Herpes zoster, like Epstein-Barr virus, is a herpetovirus that persists in a latent state. In a case-control study, the occurrence of herpes zoster before the diagnosis of SLE and the use of immunosuppressive drugs were associated with the risk of developing SLE [233]. Confirmation of this finding in larger studies would support a role of herpetoviral infection in the etiology of SLE.

VII. Summary and Future Directions

SLE is a multisystem, autoimmune disease resulting in chronic inflammation. It affects women at rates nearly twelve times greater than those of men during the childbearing years. The disease is more common in American and European women of African origin, but current information suggests that SLE is rare in Africa. The contributors to overall mortality in SLE include age at onset, socioeconomic status, gender, and race. Although the impact of these factors has not been completely elucidated, it appears that renal disease has the greatest effect on mortality in African-American women. Not only is SLE more prevalent among African-American women, but these women have the highest rates of lupus renal disease and poorer renal survival compared with Caucasian women.

The pathogenesis of lupus is multifactorial and includes genetic, hormonal, and environmental factors. Associations with HLA antigens, complement deficiencies, and new candidate genes have been reported. For example, an allele of the Fc γ receptor RII has been associated with increased risk of lupus nephritis in African-Americans. Research in the genetics of lupus is accelerating, with several groups working on a particular polymorphism on chromosome 1 that may be associated with increased risk of disease.

New research on hormones in SLE animal models suggests novel hypotheses on the possible mechanisms of disease susceptibility or severity and the observed gender skew in lupus. Sex-specific severity may be related to the differences in T-cell trafficking observed in the spleens of male and female mice. Female hormones may also alter apoptosis in the kidney. The balance between the expression of the proapoptotic *bax* gene and the antiapoptotic *bcl-2* gene may be disrupted by estrogen, favoring *bax* expression and apoptosis, which could eventually lead to accelerated glomerulonephritis. Several human studies have reported increased levels of 16-hydroxylated estrogen metabolites as well as lower testosterone levels, resulting in a relative "hyperestrogenic" state. Although women with lupus report increased symptoms of their disease at certain times during the menstrual cycle (usually within two weeks of the menses), no studies have looked at the corresponding hormone levels at the time of reported symptoms. The effect of pregnancy on lupus

disease activity has been an area of some controversy. Most investigators agree that lupus disease activity can occur during any trimester and postpartum, but whether disease activity is increased and more severe during pregnancy compared with the nonpregnant state is uncertain. While there are concerns regarding the use of exogenous hormones in women with lupus because of anecdotal reports of worsening disease activity in women taking OCs and HRT, there are clearly many potential benefits, including prevention of premature cardiovascular disease and osteoporosis, which are common problems in these young women. There are ongoing, national clinical trials to examine the effects of exogenous hormone use in women with lupus.

Therapeutic options for SLE offer increasing hope. For the first time since the 1970s clinical trials in lupus are being conducted. Drugs that have been used in other diseases are finding new use in lupus. Results from clinical trials indicate that DHEA treatment improves SLE disease activity and reduces corticosteroid requirements. The prolactin antagonist bromocriptine shows promise in suppressing active SLE. Clinical studies indicate that bromocriptine decreases serum prolactin levels, suppresses anti-dsDNA antibodies, significantly reduces lupus disease activity, and reduces SLE flares. New, more targeted, therapeutic agents have also been developed based on advances in our understanding of the pathogenesis of lupus. Blocking the CD40–CD40L costimulatory pathway involved in T-cell dependent activation of B cells is the target of the anti-CD40L therapy now being evaluated. Another immunological approach involves the development of a B cell "toleragen" that renders B lymphocytes unresponsive to dsDNA, a common autoantigen in lupus. The result is decreased production of pathogenic autoantibodies to dsDNA. This agent is also currently being studied in clinical trials.

SLE is a debilitating disease that afflicts young people in the prime of life. We have seen marked improvement in survival since the 1950s, but there is still much to be done in the fight against this chronic disease. As a result of improved survival, co-morbid conditions are now apparent in patients with SLE. Identifying patients at risk for these problems should influence the development of appropriate treatment strategies. Great strides in understanding the mechanisms of immune dysregulation in lupus has resulted in the promising new targeted agents now being tested.

Acknowledgments

We extend our gratitude to Dr. Syamal Datta and Dr. Sang-Cheol Bae for their thoughtful critiques, Joan Neitznick for secretarial support, and Dr. Janice Sabatine for editorial assistance and manuscript preparation.

This work was supported by grants from the National Institutes of Health, National Institute of Arthritis and Musculoskeletal and Skin Diseases Branch (2P60 AR 30692-15, MO1 RR 0056-36, MAC P60 AR 4481101, 5R01 AL 54900-02); Commonwealth of Pennsylvania, Department of Health; Lupus Foundation of America, Illinois and Western Pennsylvania Chapters; Arthritis Foundation Clinical Science Grants, the Arthritis Foundation, National and Illinois Chapter; the American Heart Association (Grant-in-aid); and an unrestricted educational grant from Merck, Co., Inc.

References

1. Benedek, T. G. (1997). The history of lupus erythematosus. *In* "Dubois' Lupus Erythematosus" (D. J. Wallace and B. H. Hahn, eds.), 5th ed., pp. 3–16. Williams & Wilkins, Baltimore, MD.

2. Mills, J. A. (1994). Systemic lupus erythematosus. *N. Engl. J. Med.* **330,** 1871–1879.

3. Boumpas, D. T., Fessler, B. J., Austin, H. A., 3rd, Balow, J. E., Klippel, J. H., and Lockshin, M. D. (1995). Systemic lupus erythematosus: Emerging concepts. Part 2: Dermatologic and joint disease, the antiphospholipid antibody syndrome, pregnancy and hormonal therapy, morbidity and mortality, and pathogenesis. *Ann. Intern. Med.* **123,** 42–53.

4. Boumpas, D. T., Austin, H. A., 3rd, Fessler, B. J., Balow, J. E., Klippel, J. H., and Lockshin, M. D. (1995). Systemic lupus erythematosus: Emerging concepts. Part 1: Renal, neuropsychiatric, cardiovascular, pulmonary, and hematologic disease. *Ann. Intern. Med.* **122,** 940–950.

5. Cohen, A. S., Reynolds, W. E., Franklin, E. C., Kulka, J. P., Ropes, M. W., Shulman, L. E., and Wallace, S. L. (1971). Preliminary criteria for the classification of systemic lupus erythematosus. *Bull. Rheum. Dis.* **21,** 643–648.

6. Tan, E. M., Cohen, A. S., Fries, J. F., Masi, A. T., McShane, D. J., Rothfield, N. F., Schaller, J. G., Talal, N., and Winchester, R. J. (1982). The 1982 revised criteria for the classification of systemic lupus erythematosus. *Arthritis Rheum.* **25,** 1271–1277.

7. Hochberg, M. C. (1997). Updating the American College of Rheumatology revised criteria for the classification of systemic lupus erythematosus. *Arthritis Rheum.* **40,** 1725.

8. Petri, M. (1994). Clinical features of systemic lupus erythematosus. *Curr. Opin. Rheumatol.* **6,** 481–486.

9. Golbus, J., and McCune, W. J. (1994). Lupus nephritis. Classification, prognosis, immunopathogenesis, and treatment. *Rheum. Dis. Clin. North Am.* **20,** 213–242.

10. West, S. G. (1994). Neuropsychiatric lupus. *Rheum. Dis. Clin. North Am.* **20,** 129–158.

10a. ACR Ad Hoc Committee on Neuropsychiatric Lupus Nomenclature (1999). The American College of Rheumatology nomenclature and case definitions for neuropsychiatric lupus syndromes. *Arthritis Rheum.* **42,** 599–608.

11. Caccavo, D., Lagana, B., Mitterhofer, A. P., Ferri, G. M., Afeltra, A., Amoroso, A., and Bonomo, L. (1997). Long-term treatment of systemic lupus erythematosus with cyclosporin A. *Arthritis Rheum.* **40,** 27–35.

12. Fox, D. A., and McCune, W. J. (1994). Immunosuppressive drug therapy of systemic lupus erythematosus. *Rheum. Dis. Clin. North Am.* **20,** 265–299.

13. Mohan, C., Shi, Y., Laman, J. D., and Datta, S. K. (1995). Interaction between CD40 and its ligand gp39 in the development of murine lupus nephritis. *J. Immunol.* **154,** 1470–1480.

14. Kalled, S. L., Cutler, A. H., Datta, S. K., and Thomas, D. W. (1998). Anti-CD40 ligand antibody treatment of SNF1 mice with established nephritis: Preservation of kidney function. *J. Immunol.* **160,** 2158–2165.

15. Weisman, M. H., Bluestein, H. G., Berner, C. M., and de Haan, H. A. (1997). Reduction in circulating dsDNA antibody titer after administration of LJP 394. *J. Rheumatol.* **24,** 314–318.

16. Ahn, Y. S., Harrington, W. J., Simon, S. R., Mylvaganam, R., Pall, L. M., and So, A. G. (1983). Danazol for the treatment of idiopathic thrombocytopenic purpura. *N. Engl. J. Med.* **308,** 1396–1399.

17. Ahn, Y. S., Harrington, W. J., Mylvaganam, R., Ayub, J., and Pall, L. M. (1985). Danazol therapy for autoimmune hemolytic anemia. *Ann. Intern. Med.* **102,** 298–301.

18. Masse, R., Youinou, P., Dorval, J. C., and Cledes, J. (1980). Reversal of lupus-erythematosus-like disease with danazol. *Lancet* **2,** 651.

19. Fretwell, M. D., and Altman, L. C. (1982). Exacerbation of a lupus-erythematosus-like syndrome during treatment of non-C1-esterase-inhibitor-dependent angioedema with danazol. *J. Allergy Clin. Immunol.* **69,** 306–310.

20. Jungers, P., Nahoul, K., Pelissier, C., Dougados, M., Tron, F., and Bach, J. F. (1982). Low plasma androgens in women with active or quiescent systemic lupus erythematosus. *Arthritis Rheum.* **25,** 454–457.

21. Lahita, R. G., Bradlow, H. L., Ginzler, E., Pang, S., and New, M. (1987). Low plasma androgens in women with systemic lupus erythematosus. *Arthritis Rheum.* **30,** 241–248.

22. Roubinian, J. R., Talal, N., Greenspan, J. S., Goodman, J. R., and Siiteri, P. K. (1979). Delayed androgen treatment prolongs survival in murine lupus. *J. Clin. Invest.* **63,** 902–911.

23. Suzuki, T., Suzuki, N., Daynes, R. A., and Engleman, E. G. (1991). Dehydroepiandrosterone enhances IL2 production and cytotoxic effector function of human T cells. *Clin. Immunol. Immunopathol.* **61,** 202–211.

24. vanVollenhoven, R. F., Engleman, E. G., and McGuire, J. L. (1994). An open study of dehydroepiandrosterone in systemic lupus erythematosus. *Arthritis Rheum.* **37,** 1305–1310.

25. Petri, M., Lahita, R., McGuire, J., van Vollenhoven, R., Strand, V., Kunz, A., Gorelick, K., Chi, P. Y., Hsu, H., and Schwartz, K. (1997). Results of the GL701 (DHEA) multicenter steroid-sparing SLE study. *Arthritis Rheum.* **40,** S327 (abstr.).

26. McMurray, R. W., Weidensaul, D., Allen, S. H., and Walker, S. E. (1995). Efficacy of bromocriptine in an open label therapeutic trial for systemic lupus erythematosus. *J. Rheumatol.* **22,** 2084–2091.

27. Alvarez-Nemegyei, J., Cobarrubias-Cobos, A., Escalante-Triay, F., Sosa-Munoz, J., Miranda, J. M., and Jara, L. J. (1998). Bromocriptine in systemic lupus erythematosus: A double-blind, randomized, placebo-controlled study. *Lupus* **7,** 414–419.

28. Siegel, M., and Lee, S. L. (1973). The epidemiology of systemic lupus erythematosus. *Semin. Arthritis Rheum.* **3,** 1–54.

29. Fessel, W. J. (1974). Systemic lupus erythematosus in the community. Incidence, prevalence, outcome, and first symptoms; the high prevalence in black women. *Arch. Intern. Med.* **134,** 1027–1035.

30. Michet, C. J., Jr., McKenna, C. H., Elveback, L. R., Kaslow, R. A., and Kurland, L. T. (1985). Epidemiology of systemic lupus erythematosus and other connective tissue diseases in Rochester, Minnesota, 1950 through 1979. *Mayo Clin. Proc.* **60,** 105–113.

31. Hochberg, M. C. (1985). The incidence of systemic lupus erythematosus in Baltimore, Maryland, 1970–1977. *Arthritis Rheum.* **28,** 80–86.

32. McCarty, D. J., Manzi, S., Medsger, T. A., Jr., Ramsey-Goldman, R., LaPorte, R. E., and Kwoh, C. K. (1995). Incidence of systemic lupus erythematosus. Race and gender differences. *Arthritis Rheum.* **38,** 1260–1270.

33. Affram, R. K., and Neequaye, A. R. (1991). Systemic lupus erythematosus and other rheumatic disorders: Clinical experience in Accra. *Ghana Med. J.* **25,** 299–302.

34. Marton, K., Louwa, C., and Diare, M. (1974). Diseases caused by autoimmunization in Africa: Apropos of the 1st case report of acute disseminated lupus erythematosus in Guinea. *Int. J. Dermatol.* **13,** 15–19.

35. Greenwood, B. M. (1968). Autoimmune disease and parasitic infections in Nigerians. *Lancet* **2,** 380–382.

36. Adebajo, A. O. (1992). Does tumor necrosis factor protect against lupus in west Africans? *Arthritis Rheum.* **35,** 839.

37. Monnier, A., Delmarre, B., Peghini, M., Genelle, B., Dexemple, P., Lokrou, A., Niamkey, E., Soubeyrand, J., and Beda, B. Y. (1985). Acute disseminated lupus erythematosus in Ivory Coast (apropos of 9 cases). *Med. Trop. (Marseille)* **45,** 47–54.

38. Fouchet, M., Gateff, C., and Pineau, J. (1969). Acute disseminated lupus erythematosus in Africans. Apropos of the first case observed in Gabon. *Med. Trop. (Marseille)* **29,** 204–207.

39. Michaux, F. L., and Sonnet, J. (1965). Systemic lupus erythematosus in the Bantu. *Trop. Geogr. Med.* **1,** 26–39.

40. Derrien, J. P., Normand, P., Charles, D., Monnier, A., and Ducloux, M. (1976). Acute lupus erythematosus disseminatus in Black Africans. Three cases. *Bull. Soc. Med. Afr. Noire Lang. Fr.* **21,** 65–70.

41. Basset, A., Hocquet, P., Sow, A. M., and Richir, A. M. (1960). Apropos of a case of disseminated lupus erythematosus. *Bull. Soc. Med. Afr. Noire Lang. Fr.* **5**, 380–382.

42. Gelfand, M. (1969). Medical arthritis in African practice. *Cent. Afr. J. Med.* **15**, 131–165.

43. Stein, C. M., Svoren, B., Davis, P., and Blankenberg, B. (1991). A prospective analysis of patients with rheumatic diseases attending referral hospitals in Harare, Zimbabwe. *J. Rheumatol.* **18**, 1841–1844.

44. Stein, M., and Davis, P. (1990). Rheumatic disorders in Zimbabwe: A prospective analysis of patients attending a rheumatic diseases clinic. *Ann. Rheum. Dis.* **49**, 400–402.

45. Dessein, P. H., Gledhill, R. F., and Rossouw, D. S. (1988). Systemic lupus erythematosus in black South Africans. *S. Afr. Med. J.* **74**, 387–389.

46. Lutalo, S. K. (1985). Chronic inflammatory rheumatic diseases in black Zimbabweans. *Ann. Rheum. Dis.* **44**, 121–125.

47. Mody, G. M., Parag, K. B., Nathoo, B. C., Pudifin, D. J., Duursma, J., and Seedat, Y. K. (1994). High mortality with systemic lupus erythematosus in hospitalized African blacks. *Br. J. Rheumatol.* **33**, 1151–1153.

48. Tikly, M., Burgin, S., Mohanlal, P., Bellingan, A., and George, J. (1996). Autoantibodies in black South Africans with systemic lupus erythematosus: Spectrum and clinical associations. *Clin. Rheumatol.* **15**, 143–147.

49. Sutej, P. G., Gear, A. J., Morrison, R. C., Tikly, M., de Beer, M., Dos Santos, L., and Sher, R. (1989). Photosensitivity and anti-Ro (SS-A) antibodies in black patients with systemic lupus erythematosus (SLE). *Br. J. Rheumatol.* **28**, 321–324.

50. Seedat, Y. K., and Pudifin, D. (1977). Systemic lupus erythematosus in Black and Indian patients in Natal. *S. Afr. Med. J.* **51**, 335–337.

51. Jessop, S., and Meyers, O. L. (1973). Systemic lupus erythematosus in Cape Town. *S. Afr. Med. J.* **47**, 222–225.

52. Kanyerezi, B. R., Lutalo, S. K., and Kigonya, E. (1980). Systemic lupus erythematosus: Clinical presentation among Ugandan Africans. *East Afr. Med. J.* **57**, 274–278.

53. Sharp, A. G. (1961). Systemic lupus erythematosus. *East Afr. Med. J.* **38**, 134–144.

54. Olweny, C. L. (1961). Systemic lupus erythematosus in a male Kenyan African. *Makerere Med. J.* **20**, 115–118.

55. Tsega, E., Choremi, H., Bottazzo, G. F., and Doniach, D. (1980). Prevalence of autoimmune diseases and autoantibodies in Ethiopia. *Trop. Geogr. Med.* **32**, 231–236.

56. Nossent, J. C. (1992). Systemic lupus erythematosus on the Caribbean island of Curacao: An epidemiological investigation. *Ann. Rheum. Dis.* **51**, 1197–1201.

57. Siegel, M., Holley, H. L., and Lee, S. L. (1970). Epidemiologic studies on systemic lupus erythematosus. Comparative data for New York City and Jefferson County, Alabama, 1956–1965. *Arthritis Rheum.* **13**, 802–811.

58. Hopkinson, N. D., Doherty, M., and Powell, R. J. (1994). Clinical features and race-specific incidence/prevalence rates of systemic lupus erythematosus in a geographically complete cohort of patients. *Ann. Rheum. Dis.* **53**, 675–680.

59. Johnson, A. E., Gordon, C., Palmer, R. G., and Bacon, P. A. (1995). The prevalence and incidence of systemic lupus erythematosus in Birmingham, England. Relationship to ethnicity and country of birth. *Arthritis Rheum.* **38**, 551–558.

60. Bae, S.-C., Fraser, P., and Liang, M. H. (1998). The epidemiology of systemic lupus erythematosus in populations of African ancestry: A critical review of the "prevalence gradient hypothesis." *Arthritis Rheum.* **41**, 2091–2099.

61. Merrell, M., and Shulman, L. E. (1955). Determination of prognosis in chronic disease, illustrated by systemic lupus erythematosus. *J. Chronic Dis.* **1**, 12–32.

62. Kellum, R. E., and Hasericke, J. R. (1964). Systemic lupus erythematosus, a statistical evaluation of mortality based on a consecutive series of 299 patients. *Arch. Intern. Med.* **113**, 200–207.

63. Urman, J. D., and Rothfield, N. F. (1977). Corticosteroid treatment in systemic lupus erythematosus. Survival studies. *JAMA, J. Am. Med. Assoc.* **238**, 2272–2276.

64. Estes, D., and Christian, C. (1971). The natural history of systemic lupus erythematosus by prospective analysis. *Medicine (Baltimore)* **50**, 85–95.

65. Urowitz, M. B., Bookman, A. A., Koehler, B. E., Gordon, D. A., Smythe, H. A., and Ogryzlo, M. A. (1976). The bimodal mortality pattern of systemic lupus erythematosus. *Am. J. Med.* **60**, 221–225.

66. Wallace, D. J., Podell, T., Weiner, J., Klinenberg, J. R., Forouzesh, S., and Dubois, E. L. (1981). Systemic lupus erythematosus—survival patterns. Experience with 609 patients. *JAMA, J. Am. Med. Assoc.* **245**, 934–938.

67. Ginzler, E. M., Diamond, H. S., Weiner, M., Schlesinger, M., Fries, J. F., Wasner, C., Medsger, T. A., Jr., Ziegler, G., Klippel, J. H., Hadler, N. M., Albert, DA, Hess, E. V., Spencer-Green, G., Grayzel, A., Worth, D., Hahn, B. H., and Barnett, E. V. (1982). A multicenter study of outcome in systemic lupus erythematosus. I. Entry variables as predictors of prognosis. *Arthritis Rheum.* **25**, 601–611.

68. Jonsson, H., Nived, O., and Sturfelt, G. (1989). Outcome in systemic lupus erythematosus: A prospective study of patients from a defined population. *Medicine (Baltimore)* **68**, 141–150.

69. Malaviya, A. N., Misra, R., Banerjee, S., Kumar, A., Tiwari, S. C., Bhuyan, U. N., Malhotra, K. K., and Guleria, J. S. (1986). Systemic lupus erythematosus in North Indian Asians. A prospective analysis of clinical and immunological features. *Rheumatol. Int.* **6**, 97–101.

70. Stafford-Brady, F. J., Urowitz, M. B., Gladman, D. D., and Easterbrook, M. (1988). Lupus retinopathy. Patterns, associations, and prognosis. *Arthritis Rheum.* **31**, 1105–1110.

71. Swaak, A. J., Nossent, J. C., Bronsveld, W., Van Rooyen, A., Nieuwenhuys, E. J., Theuns, L., and Smeenk, R. J. (1989). Systemic lupus erythematosus. I. Outcome and survival: Dutch experience with 110 patients studied prospectively. *Ann. Rheum. Dis.* **48**, 447–454.

72. Reveille, J. D., Bartolucci, A., and Alarcon, G. S. (1990). Prognosis in systemic lupus erythematosus. Negative impact of increasing age at onset, black race, and thrombocytopenia, as well as causes of death. *Arthritis Rheum.* **33**, 37–48.

73. Gripenberg, M., and Helve, T. (1991). Outcome of systemic lupus erythematosus. A study of 66 patients over 7 years with special reference to the predictive value of anti-DNA antibody determinations. *Scand. J. Rheumatol.* **20**, 104–109.

74. Pistiner, M., Wallace, D. J., Nessim, S., Metzger, A. L., and Klinenberg, J. R. (1991). Lupus erythematosus in the 1980s: A survey of 570 patients. *Semin. Arthritis Rheum.* **21**, 55–64.

75. Seleznick, M. J., and Fries, J. F. (1991). Variables associated with decreased survival in systemic lupus erythematosus. *Semin. Arthritis Rheum.* **21**, 73–80.

76. Kumar, A., Malaviya, A. N., Singh, R. R., Singh, Y. N., Adya, C. M., and Kakkar, R. (1992). Survival in patients with systemic lupus erythematosus in India. *Rheumatol. Int.* **12**, 107–109.

77. Ward, M. M., Pyun, E., and Studenski, S. (1995). Long-term survival in systemic lupus erythematosus. Patient characteristics associated with poorer outcomes. *Arthritis Rheum.* **38**, 274–283.

78. Abu Shakra, M., Urowitz, M. B., Gladman, D. D., and Gough, J. (1995). Mortality studies in systemic lupus erythematosus. Results from a single center. II. Predictor variables for mortality. *J. Rheumatol.* **22**, 1265–1270.

79. Nossent, J. C. (1993). Course and prognostic value of Systemic Lupus Erythematosus Disease Activity Index in black Caribbean patients. *Semin. Arthritis Rheum.* **23**, 16–21.

80. Massardo, L., Martønez, M. E., Jacobelli, S., Villarroel, L., Rosenberg, H., and Rivero, S. (1994). Survival of Chilean patients with systemic lupus erythematosus. *Semin. Arthritis Rheum.* **24,** 1–11.

81. Tucker, L. B., Menon, S., Schaller, J. G., and Isenberg, D. A. (1995). Adult- and childhood-onset systemic lupus erythematosus: A comparison of onset, clinical features, serology, and outcome. *Br. J. Rheumatol.* **34,** 866–872.

82. Iseki, K., Miyasato, F., Oura, T., Uehara, H., Nishime, K., and Fukiyama, K. (1994). An epidemiologic analysis of end-stage lupus nephritis. *Am. J. Kidney Dis.* **23,** 547–554.

83. Miller, M. H., Urowitz, M. B., Gladman, D. D., and Killinger, D. W. (1983). Systemic lupus erythematosus in males. *Medicine (Baltimore)* **62,** 327–334.

84. Kaufman, L. D., Gomez-Reino, J. J., Heinicke, M. H., and Gorevic, P. D. (1989). Male lupus: Retrospective analysis of the clinical and laboratory features of 52 patients, with a review of the literature. *Semin. Arthritis Rheum.* **18,** 189–197.

85. Inman, R. D., Jovanovic, L., Markenson, J. A., Longcope, C., Dawood, M. Y., and Lockshin, M. D. (1982). Systemic lupus erythematosus in men. Genetic and endocrine features. *Arch. Intern. Med.* **142,** 1813–1815.

86. Folomeev, M., and Alekberova, Z. (1990). Survival pattern of 120 males with systemic lupus erythematosus. *J. Rheumatol.* **17,** 856–859.

87. Hochberg, M. C. (1997). The epidemiology of systemic lupus erythematosus. *In* "Dubois' Lupus Erythematosus" (D. J. Wallace and B. H. Hahn, eds.), 5th ed., pp. 49–65. Williams & Wilkins, Baltimore, MD.

88. Studenski, S., Allen, N. B., Caldwell, D. S., Rice, J. R., and Polisson, R. P. (1987). Survival in systemic lupus erythematosus. A multivariate analysis of demographic factors. *Arthritis Rheum.* **30,** 1326–1332.

89. Samanta, A., Feehally, J., Roy, S., Nichol, F. E., Sheldon, P. J., and Walls, J. (1991). High prevalence of systemic disease and mortality in Asian subjects with systemic lupus erythematosus. *Ann. Rheum. Dis.* **50,** 490–492.

90. Gladman, D. D. (1996). Prognosis and treatment of systemic lupus erythematosus. *Curr. Opin. Rheumatol.* **8,** 430–437.

91. Esdaile, J. M., Levinton, C., Federgreen, W., Hayslett, J. P., and Kashgarian, M. (1989). The clinical and renal biopsy predictors of long-term outcome in lupus nephritis: A study of 87 patients and review of the literature. *Q. J. Med.* **72,** 779–833.

92. McLaughlin, J. R., Bombardier, C., Farewell, V. T., Gladman, D. D., and Urowitz, M. B. (1994). Kidney biopsy in systemic lupus erythematosus. III. Survival analysis controlling for clinical and laboratory variables. *Arthritis Rheum.* **37,** 559–567.

93. Donadio, J. V., Jr., Hart, G. M., Bergstralh, E. J., and Holley, K. E. (1995). Prognostic determinants in lupus nephritis: A long-term clinicopathologic study. *Lupus* **4,** 109–115.

94. el Shahawy, M. A., Aswad, S., Mendez, R. G., Bangsil, R., Mendez, R., and Massry, S. G. (1995). Renal transplantation in systemic lupus erythematosus: A single-center experience with sixty-four cases. *Am. J. Nephrol.* **15,** 123–128.

95. Dooley, M. A., Hogan, S., Jennette, C., and Falk, R. (1997). Cyclophosphamide therapy for lupus nephritis: Poor renal survival in black Americans. Glomerular Disease Collaborative Network. *Kidney Int.* **51,** 1188–1195.

96. Feinglass, E. J., Arnett, F. C., Dorsch, C. A., Zizic, T. M., and Stevens, M. B. (1976). Neuropsychiatric manifestations of systemic lupus erythematosus: Diagnosis, clinical spectrum, and relationship to other features of the disease. *Medicine (Baltimore)* **55,** 323–339.

97. Sibley, J. T., Olszynski, W. P., Decoteau, W. E., and Sundaram, M. B. (1992). The incidence and prognosis of central nervous system disease in systemic lupus erythematosus. *J. Rheumatol.* **19,** 47–52.

98. Abu Shakra, M., Urowitz, M. B., and Gladman, D. D. (1994). Improved survival in a cohort of SLE patients compared to the general population over a 25-year period of observation. *Arthritis Rheum.* **37**(Suppl. 9), S216 (abstr.).

99. Duffy, K. N., Duffy, C. M., and Gladman, D. D. (1991). Infection and disease activity in systemic lupus erythematosus: A review of hospitalized patients. *J. Rheumatol.* **18,** 1180–1184.

100. Abu Shakra, M., Urowitz, M. B., Gladman, D. D., and Gough, J. (1995). Mortality studies in systemic lupus erythematosus. Results from a single center. I. Causes of death. *J. Rheumatol.* **22,** 1259–1264.

101. Hellmann, D. B., Petri, M., and Whiting-O'Keefe, Q. (1987). Fatal infections in systemic lupus erythematosus: The role of opportunistic organisms. *Medicine (Baltimore)* **66,** 341–348.

102. Nived, O., Sturfelt, G., and Wollheim, F. (1985). Systemic lupus erythematosus and infection: A controlled and prospective study including an epidemiological group. *Q. J. Med.* **55,** 271–287.

103. Manzi, S., Meilahn, E. N., Rairie, J. E., Conte, C. G., Medsger, T. A., Jr., Jansen–McWilliams, L., D'Agostino, R. B., and Kuller, L. H. (1997). Age-specific incidence rates of myocardial infarction and angina in women with systemic lupus erythematosus: Comparison with the Framingham Study. *Am. J. Epidemiol.* **145,** 408–415.

104. Shome, G. P., Sakauchi, M., Yamane, K., Takemura, H., and Kashiwagi, H. (1989). Ischemic heart disease in systemic lupus erythematosus. A retrospective study of 65 patients treated with prednisolone. *Jpn. J. Med.* **28,** 599–603.

105. Gladman, D. D., and Urowitz, M. B. (1987). Morbidity in systemic lupus erythematosus. *J. Rheumatol.* **14**(Suppl. 13), 223–226.

106. Petri, M., Perez-Gutthann, S., Spence, D., and Hochberg, M. C. (1992). Risk factors for coronary artery disease in patients with systemic lupus erythematosus. *Am. J. Med.* **93,** 513–519.

107. Kalla, A. A., Fataar, A. B., Jessop, S. J., and Bewerunge, L. (1993). Loss of trabecular bone mineral density in systemic lupus erythematosus. *Arthritis Rheum.* **36,** 1726–1734.

108. Kalla, A. A., Meyers, O. L., Parkyn, N. D., and Kotze, T. J. (1989). Osteoporosis screening—radiogrammetry revisited. *Br. J. Rheumatol.* **28,** 511–517.

109. Kalla, A. A., van Wyk Kotze, T. J., and Meyers, O. L. (1992). Metacarpal bone mass in systemic lupus erythematosus. *Clin. Rheumatol.* **11,** 475–482.

110. Formiga, F., Moga, I., Nolla, J. M., Pac, M., Mitjavila, F., and Roig Escofet, D. (1995). Loss of bone mineral density in premenopausal women with systemic lupus erythematosus. *Ann. Rheum. Dis.* **54,** 274–276.

111. Dhillon, V. B., Davies, M. C., Hall, M. L., Round, J. M., Ell, P. J., Jacobs, H. S., Snaith, M. L., and Isenberg, D. A. (1990). Assessment of the effect of oral corticosteroids on bone mineral density in systemic lupus erythematosus: A preliminary study with dual energy x ray absorptiometry. *Ann. Rheum. Dis.* **49,** 624–626.

112. Ramsey-Goldman, R., Dunn, J. E., Huang, C.-F., Dunlop, D., Raire, J., Fitzgerald, S., and Manzi, S. (1999). Frequency of fractures in women with systemic lupus erythematosus: Comparison with U.S. population data. *Arthritis Rheum.* **42,** 882–890.

113. Kipen, Y., Buchbinder, R., Forbes, A., Strauss, B., Littlejohn, G., and Morand, E. (1997). Prevalence of reduced bone mineral density in systemic lupus erythematosus and the role of steroids. *J. Rheumatol.* **24,** 1922–1929.

114. Sels, F., Dequeker, J., Verwilghen, J., and Mbuyi Muamba, J. M. (1996). SLE and osteoporosis: dependence and/or independence on glucocorticoids. *Lupus* **5,** 89–92.

115. Pons, F., Peris, P., Guanabens, N., Font, J., Huguet, M., Espinosa, G., Ingelmo, M., Munoz Gomez, J., and Setoain, J. (1995). The effect of systemic lupus erythematosus and long-term steroid therapy on bone mass in pre-menopausal women. *Br. J. Rheumatol.* **34,** 742–746.

116. Ramsey-Goldman, R. (1988). Pregnancy in systemic lupus erythematosus. *Rheum. Dis. Clin. North Am.* **14,** 169–185.

117. Boumpas, D. T., Austin, H. A., 3rd, Vaughan, E. M., Yarboro, C. H., Klippel, J. H., and Balow, J. E. (1993). Risk for sustained amenorrhea in patients with systemic lupus erythematosus receiving intermittent pulse cyclophosphamide therapy. *Ann. Intern. Med.* **119,** 366–369.

118. Jungers, P., Dougados, M., Pelissier, C., Kuttenn, F., Tron, F., Lesavre, P., and Bach, J. F. (1982). Influence of oral contraceptive therapy on the activity of systemic lupus erythematosus. *Arthritis Rheum.* **25,** 618–623.

119. Lahita, R. G. (1992). The importance of estrogens in systemic lupus erythematosus. *Clin. Immunol. Immunopathol.* **63,** 17–18.

120. Krupp, L. B., LaRocca, N. G., Muir, J., and Steinberg, A. D. (1990). A study of fatigue in systemic lupus erythematosus. *J. Rheumatol.* **17,** 1450–1452.

121. Linker-Israeli, M., Deans, R. J., Wallace, D. J., Prehn, J., Ozeri-Chen, T., and Klinenberg, J. R. (1991). Elevated levels of endogenous IL-6 in systemic lupus erythematosus. A putative role in pathogenesis. *J. Immunol.* **147,** 117–123.

122. Manolagas, S. C., and Jilka, R. L. (1995). Bone marrow, cytokines, and bone remodeling. Emerging insights into the pathophysiology of osteoporosis. *N. Engl. J. Med.* **332,** 305–311.

123. Sontheimer, R. D., and Gilliam, J. N. (1992). Systemic lupus erythematosus and the skin. *In* "Systemic Lupus Erythematosus" (R. G. Lahita, ed.), 2nd ed., pp. 657–681. Churchill-Livingstone, New York.

124. Lukert, B. P., and Raisz, L. G. (1990). Glucocorticoid-induced osteoporosis: Pathogenesis and management. *Ann. Intern. Med.* **112,** 352–364.

125. Green, J. A., Dawson, A. A., and Walker, W. (1978). Systemic lupus erythematosus and lymphoma. *Lancet* **2,** 753–756.

126. Agudelo, C. A., Schumacher, H. R., Glick, J. H., and Molina, J. (1981). Non-Hodgkin's lymphoma in systemic lupus erythematosus: Report of 4 cases with ultrastructural studies in 2. *J. Rheumatol.* **8,** 69–78.

127. Cammarata, R. J., Rodnan, G. P., and Jensen, W. N. (1963). Systemic rheumatic diseases and malignant lymphoma. *Arch. Intern. Med.* **111,** 330–337.

128. Gibbons, R. B., and Westerman, E. (1988). Acute nonlymphocytic leukemia following short-term, intermittent, intravenous cyclophosphamide treatment of lupus nephritis. *Arthritis Rheum.* **31,** 1552–1554.

129. Pettersson, T., Pukkala, E., Teppo, L., and Friman, C. (1992). Increased risk of cancer in patients with systemic lupus erythematosus. *Ann. Rheum. Dis.* **51,** 437–439.

130. Mellemkjaer, L., Andersen, V., Linet, M. S., Gridley, G., Hoover, R., and Olsen, J. H. (1997). Non-Hodgkin's lymphoma and other cancers among a cohort of patients with systemic lupus erythematosus. *Arthritis Rheum.* **40,** 761–768.

131. Abu Shakra, M., Gladman, D. D., and Urowitz, M. B. (1996). Malignancy in systemic lupus erythematosus. *Arthritis Rheum.* **39,** 1050–1054.

132. Sweeney, D. M., Manzi, S., Janosky, J., Selvaggi, K. J., Ferri, W., Medsger, T. A., Jr., and Ramsey-Goldman, R. (1995). Risk of malignancy in women with systemic lupus erythematosus. *J. Rheumatol.* **22,** 1478–1482.

133. Ramsey-Goldman, R., Mattai, S. A., Schilling, E., Chiu, Y.-L., Alo, C. J., Howe, H. L., and Manzi, S. (1998). Increased risk of malignancy in patients with systemic lupus erythematosus. *J. Invest. Med.* **46,** 217–222.

134. Walker, S. E., Allen, S. H., Hoffman, R. W., and McMurray, R. W. (1995). Prolactin: A stimulator of disease activity in systemic lupus erythematosus. *Lupus* **4,** 3–9.

135. Williams, R. R. (1976). Breast and thyroid cancer and malignant melanoma promoted by alcohol-induced pituitary secretion of prolactin, T.S.H. and M.S.H. *Lancet* **1,** 996–999.

136. Velentgas, P., and Daling, J. R. (1994). Risk factors for breast cancer in younger women. *J. Natl. Cancer Inst. Monogr.* **16,** 15–24.

137. Klinman, D. M. (1997). B-cell abnormalities characteristic of systemic lupus erythematosus. *In* "Dubois' Lupus Erythematosus" (D. J. Wallace and B. H. Hahn, eds.), 5th ed., pp. 195–206. Williams & Wilkins, Baltimore, MD.

138. Theofilopoulos, A. N., and Dixon, F. J. (1981). Etiopathogenesis of murine SLE. *Immunol. Rev.* **55,** 179–216.

139. Yamagata, H., Harley, J. B., and Reichlin, M. (1984). Molecular properties of the Ro/SSA antigen and enzyme-linked immunosorbent assay for quantitation of antibody. *J. Clin. Invest.* **74,** 625–633.

140. Lee, L. A. (1993). Neonatal lupus erythematosus. *J. Invest. Dermatol.* **100,** 9S–13S.

141. Alarcon-Segovia, D., Perez-Vazquez, M. E., Villa, A. R., Drenkard, C., and Cabiedes, J. (1992). Preliminary classification criteria for the antiphospholipid syndrome within systemic lupus erythematosus. *Semin. Arthritis Rheum.* **21,** 275–286.

142. Wofsy, D., and Seaman, W. E. (1987). Reversal of advanced murine lupus in NZB/NZW F1 mice by treatment with monoclonal antibody to L3T4. *J. Immunol.* **138,** 3247–3253.

143. Mihara, M., Ohsugi, Y., Saito, K., Miyai, T., Togashi, M., Ono, S., Murakami, S., Dobashi, K., Hirayama, F., and Hamaoka, T. (1988). Immunologic abnormality in NZB/NZW F1 mice. Thymus-independent occurrence of B cell abnormality and requirement for T cells in the development of autoimmune disease, as evidenced by an analysis of the athymic nude individuals. *J. Immunol.* **141,** 85–90.

144. Adams, S., Leblanc, P., and Datta, S. K. (1991). Junctional region sequences of T-cell receptor beta-chain genes expressed by pathogenic anti-DNA autoantibody-inducing helper T cells from lupus mice: Possible selection by cationic autoantigens. *Proc. Natl. Acad. Sci. U. S. A.* **88,** 11271–11275.

145. Desai Mehta, A., Mao, C., Rajagopalan, S., Robinson, T., and Datta, S. K. (1995). Structure and specificity of T cell receptors expressed by potentially pathogenic anti-DNA autoantibody-inducing T cells in human lupus. *J. Clin. Invest.* **95,** 531–541.

146. Mohan, C., Adams, S., Stanik, V., and Datta, S. K. (1993). Nucleosome: A major immunogen for pathogenic autoantibody-inducing T cells of lupus. *J. Exp. Med.* **177,** 1367–1381.

147. Shi, Y., Kaliyaperumal, A., Lu, L., Southwood, S., Sette, A., Michaels, M. A., and Datta, S. K. (1998). Promiscuous presentation and recognition of nucleosomal autoepitopes in lupus: Role of autoimmune T cell receptor alpha chain. *J. Exp. Med.* **187,** 367–378.

148. Burlingame, R. W., Boey, M. L., Starkebaum, G., and Rubin, R. L. (1994). The central role of chromatin in autoimmune responses to histones and DNA in systemic lupus erythematosus. *J. Clin. Invest.* **94,** 184–192.

149. Hahn, B. H. (1997). An overview of the pathogenesis of systemic lupus erythematosus. *In* "Dubois' Lupus Erythematosus" (D. J. Wallace and B. H. Hahn, eds.), 5th ed., pp. 69–75. Williams & Wilkins, Baltimore, MD.

150. Datta, S. K., and Kalled, S. L. (1997). CD40-CD40 ligand interaction in autoimmune disease. *Arthritis Rheum.* **40,** 1735–1745.

151. Desai Mehta, A., Lu, L., Ramsey–Goldman, R., and Datta, S. K. (1996). Hyperexpression of CD40 ligand by B and T cells in human lupus and its role in pathogenic autoantibody production. *J. Clin. Invest.* **97,** 2063–2073.

152. Koshy, M., Berger, D., and Crow, M. K. (1996). Increased expression of CD40 ligand on systemic lupus erythematosus lymphocytes. *J. Clin. Invest.* **98,** 826–837.

153. Berden, J. H. (1997). Lupus nephritis. *Kidney Int.* **52**, 538–558.

154. Lefkowith, J. B., and Gilkeson, G. S. (1996). Nephritogenic auto-antibodies in lupus: Current concepts and continuing controversies. *Arthritis Rheum.* **39**, 894–903.

155. Wilson, J. G., Ratnoff, W. D., Schur, P. H., and Fearon, D. T. (1986). Decreased expression of the C3b/C4b receptor (CR1) and the C3d receptor (CR2) on B lymphocytes and of CR1 on neutrophils of patients with systemic lupus erythematosus. *Arthritis Rheum.* **29**, 739–747.

156. Ross, G. D., Yount, W. J., Walport, M. J., Winfield, J. B., Parker, C. J., Fuller, C. R., Taylor, R. P., Myones, B. L., and Lachmann, P. J. (1985). Disease-associated loss of erythrocyte complement receptors (CR1, C3b receptors) in patients with systemic lupus erythematosus and other diseases involving autoantibodies and/or complement activation. *J. Immunol.* **135**, 2005–2014.

157. Wilson, J. G., and Fearon, D. T. (1984). Altered expression of complement receptors as a pathogenetic factor in systemic lupus erythematosus. *Arthritis Rheum.* **27**, 1321–1328.

158. Salmon, J. E., Millard, S., Schachter, L. A., Arnett, F. C., Ginzler, E. M., Gourley, M. F., Ramsey Goldman, R., Peterson, M. G., and Kimberly, R. P. (1996). Fc gamma RIIA alleles are heritable risk factors for lupus nephritis in African Americans. *J. Clin. Invest.* **97**, 1348–1354.

159. Reinertsen, J. L., Klippel, J. H., Johnson, A. H., Steinberg, A. D., Decker, J. L., and Mann, D. L. (1978). B-lymphocyte alloantigens associated with systemic lupus erythematosus. *N. Engl. J. Med.* **299**, 515–518.

160. Schur, P. H., Meyer, I., Garovoy, M., and Carpenter, C. B. (1982). Associations between systemic lupus erythematosus and the major histocompatibility complex: Clinical and immunological considerations. *Clin. Immunol. Immunopathol.* **24**, 263–275.

161. Gibofsky, A., Winchester, R. J., Patarroyo, M., Fotino, M., and Kunkel, H. G. (1978). Disease associations of the Ia-like human alloantigens. Contrasting patterns in rheumatoid arthritis and systemic lupus erythematosus. *J. Exp. Med.* **148**, 1728–1732.

162. Howard, P. F., Hochberg, M. C., Bias, W. B., Arnett, F. C., Jr., and McLean, R. H. (1986). Relationship between C4 null genes, HLA-D region antigens, and genetic susceptibility to systemic lupus erythematosus in Caucasian and black Americans. *Am. J. Med.* **81**, 187–193.

163. Kemp, M. E., Atkinson, J. P., Skanes, V. M., Levine, R. P., and Chaplin, D. D. (1987). Deletion of C4A genes in patients with systemic lupus erythematosus. *Arthritis Rheum.* **30**, 1015–1022.

164. Tsao, B. P., Cantor, R. M., Kalunian, K. C., Chen, C. J., Badsha, H., Singh, R., Wallace, D. J., Kitridou, R. C., Chen, S. L., Shen, N., Song, Y. W., Isenberg, D. A., Yu, C. L., Hahn, B. H., and Rotter, J. I. (1997). Evidence for linkage of a candidate chromosome 1 region to human systemic lupus erythematosus. *J. Clin. Invest.* **99**, 725–731.

165. Wakeland, E. K., Morel, L., Mohan, C., and Yui, M. (1997). Genetic dissection of lupus nephritis in murine models of SLE. *J. Clin. Immunol.* **17**, 272–281.

166. Vyse, T. J., and Kotzin, B. L. (1996). Genetic basis of systemic lupus erythematosus. *Curr. Opin. Immunol.* **8**, 843–851.

167. Kono, D. H., and Theofilopoulos, A. N. (1996). Genetic contributions to SLE. *J. Autoimmunol.* **9**, 437–452.

168. Harley, J. B., Sestak, A. L., Willis, L. G., Fu, S. M., Hansen, J. A., and Reichlin, M. (1989). A model for disease heterogeneity in systemic lupus erythematosus. Relationships between histocompatibility antigens, autoantibodies, and lymphopenia or renal disease. *Arthritis Rheum.* **32**, 826–836.

169. Cutolo, M., Sulli, A., Seriolo, B., Accardo, S., and Masi, A. T. (1995). Estrogens, the immune response and autoimmunity. *Clin. Exp. Rheumatol.* **13**, 217–226.

170. Lahita, R. G. (1996). The connective tissue diseases and the overall influence of gender. *Int. J. Fertil. Menopausal Stud.* **41**, 156–165.

171. Lahita, R. G. (1985). Sex steroids and the rheumatic diseases. *Arthritis Rheum.* **28**, 121–126.

172. Masi, A. T., Feigenbaum, S. L., and Chatterton, R. T. (1995). Hormonal and pregnancy relationships to rheumatoid arthritis: Convergent effects with immunologic and microvascular systems. *Semin. Arthritis Rheum.* **25**, 1–27.

173. Masi, A. T. (1995). Sex hormones and rheumatoid arthritis: Cause or effect relationships in a complex pathophysiology? *Clin. Exp. Rheumatol.* **13**, 227–240.

174. Schuurs, A. H., and Verheul, H. A. (1990). Effects of gender and sex steroids on the immune response. *J. Steroid Biochem.* **35**, 157–172.

175. Sthoeger, Z. M., Chiorazzi, N., and Lahita, R. G. (1988). Regulation of the immune response by sex hormones. I. In vitro effects of estradiol and testosterone on pokeweed mitogen-induced human B cell differentiation. *J. Immunol.* **141**, 91–98.

176. Ansar Ahmed, S., Penhale, W. J., and Talal, N. (1985). Sex hormones, immune responses, and autoimmune diseases. Mechanisms of sex hormone action. *Am. J. Pathol.* **121**, 531–551.

177. Grossman, C. J. (1985). Interactions between the gonadal steroids and the immune system. *Science* **227**, 257–261.

178. Carlsten, H., Nilsson, N., Jonsson, R., Backman, K., Holmdahl, R., and Tarkowski, A. (1992). Estrogen accelerates immune complex glomerulonephritis but ameliorates T cell-mediated vasculitis and sialadenitis in autoimmune MRL lpr/lpr mice. *Cell. Immunol.* **144**, 190–202.

179. Danel, L., Souweine, G., Monier, J. C., and Saez, S. (1983). Specific estrogen binding sites in human lymphoid cells and thymic cells. *J. Steroid Biochem.* **18**, 559–563.

180. Cohen, J. H., Danel, L., Cordier, G., Saez, S., and Revillard, J. P. (1983). Sex steroid receptors in peripheral T cells: Absence of androgen receptors and restriction of estrogen receptors to OKT8-positive cells. *J. Immunol.* **131**, 2767–2771.

181. Graff, R. J., Lappe, M. A., and Snell, G. D. (1969). The influence of the gonads and adrenal glands on the immune response to skin grafts. *Transplantation* **7**, 105–111.

182. Theofilopoulos, A. N. (1992). Murine models of lupus. In "Systemic Lupus Erythematosus" (R. G. Lahita, ed.), 2nd ed., pp. 121–194. Churchill-Livingstone, New York.

183. Roubinian, J. R., Papoian, R., and Talal, N. (1977). Androgenic hormones modulate autoantibody responses and improve survival in murine lupus. *J. Clin. Invest.* **59**, 1066–1070.

184. Roubinian, J. R., Talal, N., Greenspan, J. S., Goodman, J. R., and Siiteri, P. K. (1978). Effect of castration and sex hormone treatment on survival, anti-nucleic acid antibodies, and glomerulonephritis in NZB/NZW F1 mice. *J. Exp. Med.* **147**, 1568–1583.

185. Yung, R., Williams, R., Johnson, K., Phillips, C., Stoolman, L., Chang, S., and Richardson, B. (1997). Mechanisms of drug-induced lupus. III. Sex-specific differences in T cell homing may explain increased disease severity in female mice. *Arthritis Rheum.* **40**, 1334–1343.

186. Lahita, R. G., Sui, Y. P., Merrill, J., and Xie, Y. B. (1998). Sex hormone dependent apoptotic gene expression in the NXB/NZW F1 mouse kidney. *Arthritis Rheum.* **41** (Suppl.) S179 (abstr.).

187. Lahita, R. G., Bradlow, H. L., Kunkel, H. G., and Fishman, J. (1979). Alterations of estrogen metabolism in systemic lupus erythematosus. *Arthritis Rheum.* **22**, 1195–1198.

188. Lahita, R. G., Bradlow, L., Fishman, J., and Kunkel, H. G. (1982). Estrogen metabolism in systemic lupus erythematosus: Patients and family members. *Arthritis Rheum.* **25**, 843–846.

189. Lahita, R. G., Bradlow, H. L., Kunkel, H. G., and Fishman, J. (1981). Increased 16 alpha-hydroxylation of estradiol in systemic lupus erythematosus. *J. Clin. Endocrinol. Metab.* **53**, 174–178.

189a. Lahita, R. G., Bradlow, H. L., Ginzler, E., Pang, S., and New, M. (1987). Low plasma androgens in women with systemic lupus erythematosus. *Arthritis Rheum.* **30**, 241–248.

190. Martucci, C., and Fishman, J. (1977). Direction of estradiol metabolism as a control of its hormonal action—uterotrophic activity of estradiol metabolites. *Endocrinology (Baltimore)* **101**, 1709–1715.

191. Fishman, J., and Martucci, C. (1980). Biological properties of 16 alpha-hydroxyestrone: Implications in estrogen physiology and pathophysiology. *J. Clin. Endocrinol. Metab.* **51**, 611–615.

192. Lahita, R. G., Kunkel, H. G., and Bradlow, H. L. (1983). Increased oxidation of testosterone in systemic lupus erythematosus. *Arthritis Rheum.* **26**, 1517–1521.

193. Arnalich, F., Benito-Urbina, S., Gonzalez-Gancedo, P., Iglesias, E., de Miguel, E., and Gijon-Banos, J. (1992). Inadequate production of progesterone in women with systemic lupus erythematosus. *Br. J. Rheumatol.* **31**, 247–251.

194. Grossman, C. (1984). Regulation of the immune system by sex steroids. *Endocrinol. Rev.* **5**, 435–455.

195. Holdstock, G., Chastenay, B. F., and Krawitt, E. L. (1982). Effects of testosterone, oestradiol and progesterone on immune regulation. *Clin. Exp. Immunol.* **47**, 449–456.

196. Steinberg, A. D., and Steinberg, B. J. (1985). Lupus disease activity associated with menstrual cycle. *J. Rheumatol.* **12**, 816–817.

197. Rose, E., and Pillsbury, D. M. (1944). Lupus erythematosus (erythematoides) and ovarian function: Observations in a possible relationship with a report of 6 cases. *Ann. Intern. Med.* **21**, 1022–1034.

198. Lim, G. S., Petri, M., and Goldman, D. (1993). Menstruation and systemic lupus erythematosus (SLE). *Arthritis Rheum.* **36** (Suppl.), R23. (Abstr.).

199. Cooper, G. S., Dooley, M. A., Treadwell, E. L., St. Clair, E. W., Parks, C. G., and Gilkeson, G. S. (1998). Hormonal, environmental, and infectious risk factors for developing systemic lupus erythematosus. *Arthritis Rheum.* **41**, 1714–1724.

200. Hiestand, P. C., Mekler, P., Nordmann, R., Grieder, A., and Permmongkol, C. (1986). Prolactin as a modulator of lymphocyte responsiveness provides a possible mechanism of action for cyclosporine. *Proc. Natl. Acad. Sci. U. S. A.* **83**, 2599–2603.

201. Nagy, E., Berczi, I., Wren, G. E., Asa, S. L., and Kovacs, K. (1983). Immunomodulation by bromocriptine. *Immunopharmacology* **6**, 231–243.

202. Bernton, E. W., Meltzer, M. S., and Holaday, J. W. (1988). Suppression of macrophage activation and T-lymphocyte function in hypoprolactinemic mice. *Science* **239**, 401–404.

203. McMurray, R., Keisler, D., Kanuckel, K., Izui, S., and Walker, S. E. (1991). Prolactin influences autoimmune disease activity in the female B/W mouse. *J. Immunol.* **147**, 3780–3787.

204. McMurray, R., Keisler, D., Izui, S., and Walker, S. E. (1994). Hyperprolactinemia in male NZB/NZW (B/W) F1 mice: Accelerated autoimmune disease with normal circulating testosterone. *Clin. Immunol. Immunopathol.* **71**, 338–343.

205. Lavalle, C., Loyo, E., Paniagua, R., Bermudez, J. A., Herrera, J., Graef, A., Gonzalez-Barcena, D., and Fraga, A. (1987). Correlation study between prolactin and androgens in male patients with systemic lupus erythematosus. *J. Rheumatol.* **14**, 268–272.

206. Folomeev, M., Prokaeva, T., Nassonova, V., Nassonov, E., Masenko, E., and Ovtraht, N. (1990). Prolactin levels in men with SLE and RA. *J. Rheumatol.* **17**, 1569–1570.

207. Jara-Quezada, L., Graef, A., and Lavalle, C. (1991). Prolactin and gonadal hormones during pregnancy in systemic lupus erythematosus. *J. Rheumatol.* **18**, 349–353.

208. Jara, L. J., Gomez-Sanchez, C., Silveira, L. H., Martinez-Osuna, P., Vasey, F. B., and Espinoza, L. R. (1992). Hyperprolactinemia in systemic lupus erythematosus: Association with disease activity. *Am. J. Med. Sci.* **303**, 222–226.

209. Pauzner, R., Urowitz, M. B., Gladman, D. D., and Gough, J. M. (1994). Prolactin in systemic lupus erythematosus. *J. Rheumatol.* **21**, 2064–2067.

210. McMurray, R. W., Allen, S. H., Braun, A. L., Rodriguez, F., and Walker, S. E. (1994). Longstanding hyperprolactinemia associated with systemic lupus erythematosus: Possible hormonal stimulation of an autoimmune disease. *J. Rheumatol.* **21**, 843–850.

211. Walker, S. E. (1997). The importance of sex hormones in lupus. *In* "Dubois' Lupus Erythematosus" (D. J. Wallace and B. H. Hahn, eds.), 5th ed., pp. 311–322. Williams & Wilkins, Baltimore, MD.

212. McMurray, R. W., Weidensaul, D., Allen, S. H., and Walker, S. E. (1995). Efficacy of bromocriptine in an open label therapeutic trial for systemic lupus erythematosus. *J. Rheumatol.* **22**, 2084–2091.

213. Fraga, A., Mintz, G., and Orozco, J. H. (1974). Sterility and fertility rates, fetal wastage and maternal morbidity in systemic lupus erythematosus. *J. Rheumatol.* **1**, 293–298.

214. Petri, M., and Allbritton, J. (1993). Fetal outcome of lupus pregnancy: A retrospective case-control study of the Hopkins Lupus Cohort. *J. Rheumatol.* **20**, 650–656.

215. Lockshin, M. D., Harpel, P. C., Druzin, M. L., Becker, C. G., Klein, R. F., Watson, R. M., Elkon, K. B., and Reinitz, E. (1985). Lupus pregnancy. II. Unusual pattern of hypocomplementemia and thrombocytopenia in the pregnant patient. *Arthritis Rheum.* **28**, 58–66.

216. Ramsey-Goldman, R., Kutzer, J. E., Kuller, L. H., Guzick, D., Carpenter, A. B., and Medsger, T. A., Jr. (1993). Pregnancy outcome and anti-cardiolipin antibody in women with systemic lupus erythematosus. *Am. J. Epidemiol.* **138**, 1057–1069.

217. Julkunen, H., Jouhikainen, T., Kaaja, R., Leirisalo-Repo, M., Stephansson, E., Palosuo, T., Teramo, K., and Friman, C. (1993). Fetal outcome in lupus pregnancy: A retrospective case-control study of 242 pregnancies in 112 patients. *Lupus* **2**, 125–131.

218. Buyon, J. P., Winchester, R. J., Slade, S. G., Arnett, F., Copel, J., Friedman, D., and Lockshin, M. D. (1993). Identification of mothers at risk for congenital heart block and other neonatal lupus syndromes in their children. Comparison of enzyme-linked immunosorbent assay and immunoblot for measurement of anti-SS-A/Ro and anti-SS-B/La antibodies. *Arthritis Rheum.* **36**, 1263–1273.

219. Ramsey-Goldman, R., Manzi, S., Schilling, E. M., Kutzer, J. E., Luu, A. C., Medsger, T. A., Jr., for Chicago and Pittsburgh Lupus Database Participating Physicians (1995). A two center study of lupus activity in pregnant and non-pregnant women with systemic lupus erythematosus. *Lupus* **4**(Suppl. 2), 105 (abstr.)

220. Petri, M., Howard, D., and Repke, J. (1991). Frequency of lupus flare in pregnancy. The Hopkins Lupus Pregnancy Center experience. *Arthritis Rheum.* **34**, 1538–1545.

221. Wong, K. L., Chan, F. Y., and Lee, C. P. (1991). Outcome of pregnancy in patients with systemic lupus erythematosus. A prospective study. *Arch. Intern. Med.* **151**, 269–273.

222. Ruiz Irastorza, G., Lima, F., Alves, J., Khamashta, M. A., Simpson, J., Hughes, G. R., and Buchanan, N. M. (1996). Increased rate of lupus flare during pregnancy and the puerperium: A prospective study of 78 pregnancies. *Br. J. Rheumatol.* **35**, 133–138.

223. Mintz, G., Niz, J., Gutierrez, G., Garcia-Alonso, A., and Karchmer, S. (1986). Prospective study of pregnancy in systemic lupus erythematosus. Results of a multidisciplinary approach. *J. Rheumatol.* **13**, 732–739.

224. Lockshin, M. D., Reinitz, E., Druzin, M. L., Murrman, M., and Estes, D. (1984). Lupus pregnancy. Case-control prospective study demonstrating absence of lupus exacerbation during or after pregnancy. *Am. J. Med.* **77**, 893–898.

225. Urowitz, M. B., Gladman, D. D., Farewell, V. T., Stewart, J., and McDonald, J. (1993). Lupus and pregnancy studies. *Arthritis Rheum.* **36,** 1392–1397.

226. Buyon, J. P., and Wallace, D. J. (1997). The endocrine system, use of exogenous estrogens, and the urogenital tract. *In* "Dubois' Lupus Erythematosus" (D. J. Wallace and B. H. Hahn, eds.), 5th ed., pp. 817–834. Williams & Wilkins, Baltimore, MD.

227. Julkunen, H. A., Kaaja, R., and Friman, C. (1993). Contraceptive practice in women with systemic lupus erythematosus. *Br. J. Rheumatol.* **32,** 227–230.

228. Buyon, J. P., Kalunian, K. C., Skovron, M. L., Petri, M., Lahita, R. G., Merrill, J., Sammaritano, L., Yung, C., Licciardi, F., Belmont, H. M., and Hahn, B. H. (1995). Can women with systemic lupus erythematosus safely use exogenous estrogens? *J. Clin. Rheumatol.* **1,** 205–212.

229. Pimstone, B. L. (1966). Systemic lupus erythematosus exacerbated by oral contraceptives. *S. Afr. J. Obstet. Gynaecol.* **4,** 62.

230. Chapel, T. A., and Burns, R. E. (1971). Oral contraceptives and exacerbation of lupus erythematosus. *Am. J. Obstet. Gynecol.* **110,** 366–369.

231. Travers, R. L., and Hughes, G. R. (1978). Oral contraceptive therapy and systemic lupus erythematosus. *J. Rheumatol.* **5,** 448–451.

232. Garovich, M., Agudelo, C., and Pisko, E. (1980). Oral contraceptives and systemic lupus erythematosus. *Arthritis Rheum.* **23,** 1396–1398.

233. Strom, B. L., Reidenberg, M. M., West, S., Snyder, E. S., Freundlich, B., and Stolley, P. D. (1994). Shingles, allergies, family medical history, oral contraceptives, and other potential risk factors for systemic lupus erythematosus. *Am. J. Epidemiol.* **140,** 632–642.

234. Hochberg, M. C., and Kaslow, R. A. (1983). Risk factors for the development of systemic lupus erythematosus: a case-control study. *Clin. Res.* **31,** 732A (abstr.).

235. Sanchez-Guerrero, J., Karlson, E. W., Liang, M. H., Hunter, D. J., Speizer, F. E., and Colditz, G. A. (1997). Past use of oral contraceptives and the risk of developing systemic lupus erythematosus. *Arthritis Rheum.* **40,** 804–808.

236. Kay, D. R., Bole, G. G., Jr., and Ledger, W. J. (1971). Antinuclear antibodies, rheumatoid factor and C-reactive protein in serum of normal women using oral contraceptives. *Arthritis Rheum.* **14,** 239–248.

237. McKenna, C. H., Wieman, K. C., and Shulman, L. E. (1969). Oral contraceptives, rheumatic disease and autoantibodies. *Arthritis Rheum.* **12,** 313–314.

238. Tarzy, B. J., Garcia, C. R., Wallach, E. E., Zweiman, B., and Myers, A. R. (1972). Rheumatic disease, abnormal serology, and oral contraceptives. *Lancet* **2,** 501–503.

239. Mintz, G., Gutierrez, G., Deleze, M., and Rodriguez, E. (1984). Contraception with progestagens in systemic lupus erythematosus. *Contraception* **30,** 29–38.

240. Julkunen, H. A. (1991). Oral contraceptives in systemic lupus erythematosus: Side-effects and influence on the activity of SLE. *Scand. J. Rheumatol.* **20,** 427–433.

241. Pulsinelli, W. A., and Hamill, R. W. (1978). Chorea complicating oral contraceptive therapy. Case report and review of the literature. *Am. J. Med.* **65,** 557–559.

242. Asherson, R. A., Harris, E. N., Hughes, G. R., and Farquharson, R. G. (1988). Complications of oral contraceptives and antiphospholipid antibodies: Reply to the letter by Bruneau *et al. Arthritis Rheum.* **31,** 575–576.

243. Asherson, R. A., Harris, N. E., Gharavi, A. E., and Hughes, G. R. (1986). Systemic lupus erythematosus, antiphospholipid antibodies, chorea, and oral contraceptives. *Arthritis Rheum.* **29,** 1535–1536.

244. Nausieda, P. A., Koller, W. C., Weiner, W. J., and Klawans, H. L. (1979). Chorea induced by oral contraceptives. *Neurology* **29,** 1605–1609.

245. Alarcon-Segovia, D. (1992). Clinical manifestations of the antiphospholipid syndrome. *J. Rheumatol.* **19,** 1778–1781.

246. Lockshin, M. D. (1992). Antiphospholipid antibody syndrome. *JAMA, J. Am. Med. Assoc.* **268,** 1451–1453.

247. Barrett, C., Neylon, N., and Snaith, M. L. (1986). Oestrogen-induced systemic lupus erythematosus. *Br. J. Rheumatol.* **25,** 300–301.

248. Sanchez-Guerrero, J., Liang, M. H., Karlson, E. W., Hunter, D. J., and Colditz, G. A. (1995). Postmenopausal estrogen therapy and the risk for developing systemic lupus erythematosus. *Ann. Intern. Med.* **122,** 430–433.

249. Arden, N. K., Lloyd, M. E., Spector, T. D., and Hughes, G. R. (1994). Safety of hormone replacement therapy (HRT) in systemic lupus erythematosus (SLE). *Lupus* **3,** 11–13.

250. Fernandes, G. (1994). Dietary lipids and risk of autoimmune disease. *Clin. Immunol. Immunopathol.* **72,** 193–197.

251. James, J. A., Kaufman, K. M., Farris, A. D., Taylor Albert, E., Lehman, T. J., and Harley, J. B. (1997). An increased prevalence of Epstein-Barr virus infection in young patients suggests a possible etiology for systemic lupus erythematosus. *J. Clin. Invest.* **100,** 3019–3026.

252. Incaprera, M., Rindi, L., Bazzichi, A., and Garzelli, C. (1998). Potential role of the Epstein-Barr virus in systemic lupus erythematosus autoimmunity. *Clin. Exp. Rheumatol.* **16,** 289–294.

56

Asthma

MONA KIMBELL-DUNN, NEIL PEARCE, AND RICHARD BEASLEY
Wellington Asthma Research Group
Department of Medicine
Wellington School of Medicine
Wellington, New Zealand

I. Introduction

Asthma has puzzled and confused physicians and patients from the time of Hippocrates to the present day. The word "asthma" comes from a Greek word meaning "panting" [1], but reference to asthma can also be found in ancient Egyptian, Hebrew, and Indian medical writings [2,3]. There were clear observations of patients experiencing attacks of asthma in the second century and evidence of disordered anatomy in the lung as far back as the seventeenth century [4]. Despite advances in the understanding of the pathological and clinical features of asthma and increased knowledge regarding risk factors for asthma attacks, relatively little is known about why asthma prevalence is currently increasing worldwide, or why asthma is less common in women than in men before puberty but is more common in women than in men in adolescence and adult life.

In this chapter we first consider asthma diagnosis and management, including premenstrual asthma and asthma during pregnancy. We then discuss the evidence for increasing asthma prevalence worldwide and give special consideration to gender differences in the incidence, prevalence, and severity of asthma symptoms, as well as patterns of asthma death. Finally, we review what is known about risk factors for developing and/or exacerbating asthma and we discuss whether these risk factors could account for the gender differences in asthma occurrence.

II. Asthma Diagnosis and Management

A. *Asthma Definitions and Diagnosis*

The definition of asthma has become more complex as our understanding of its pathophysiology has increased. However, despite this increased complexity, the basic characteristic features of reversible airflow obstruction by which one recognizes or diagnoses the disease has changed little. In this sense, asthma could be regarded as a "condition" or "syndrome" rather than a disease [5], and it is perhaps most useful to define asthma in terms of the phenomena involved without making any etiologic implications [6]. The definition initially proposed at the Ciba Foundation conference in 1959 [7] and endorsed by the American Thoracic Society in 1962 [8] is that "asthma is a disease characterized by wide variation over short periods of time in resistance to flow in the airways of the lung." Although these features receive lesser prominence in some current definitions, as the importance of airways inflammation is appropriately rec-

ognized, they still form the basis of the recent World Health Organization/National Heart, Lung and Blood Institute definition [9] of asthma as:

> "... a chronic inflammatory disorder of the airways in which many cells play a role, in particular mast cells, eosinophils and T lymphocytes. In susceptible individuals this inflammation causes recurrent episodes of wheezing, breathlessness, chest tightness and cough, particularly at night and/or in the early morning. These symptoms are usually associated with widespread but variable airflow limitation and are at least partly reversible either spontaneously or with treatment. The inflammation also causes an associated increase in airway responsiveness to a variety of stimuli."

B. *Asthma Management*

Three components—chronic airways inflammation, reversible airflow obstruction, and enhanced bronchial reactivity—form the basis of current definitions of asthma. They also represent the major pathophysiological events leading to the symptoms of wheezing, breathlessness, chest tightness, cough, and sputum production by which physicians clinically diagnose this disorder. The diagnosis of asthma involves an overall assessment of the patient's medical history, physical examination, and laboratory test results. However, there are no universally accepted rules for combining the information from these various sources, and practice varies considerably among physicians.

The modern approach to the long-term management of asthma is based on the use of inhaled beta agonist "bronchodilator" drugs as required for symptomatic relief and inhaled anti-inflammatory "prophylactic" drugs such as inhaled corticosteroids or sodium cromoglycate to reduce underlying asthma severity. While most patients can be well controlled with such an approach, other drugs such as oral corticosteroids (*e.g.,* prednisone), methylxanthines (*e.g.,* theophylline), anticholinergic agents (*e.g.,* ipratropium bromide), long-acting beta agonists (*e.g.,* salmeterol, formoterol), or antileukotriene agents may be required in certain circumstances.

The focus of asthma management traditionally has been on relieving bronchospasm, particularly through the use of inhaled beta agonists. However, despite improved understanding of the condition, the introduction of modern asthma treatment has not reduced asthma morbidity or mortality, and some poorly selective beta agonists (isoproterenol forte and isoprenaline) have been associated with asthma mortality epidemics [10]. Thus, emphasis is shifting from the use of beta agonist drugs and oral theophylline therapy to the use of inhaled corticosteroid drugs to treat the underlying problem of inflammation. Furthermore, it has been increasingly recognized that the key person in the long-term management of asthma is the informed patient. More emphasis has therefore been given to self-management of asthma and asthma education, including the use of asthma self-management plans [11]. This trend follows the approach established for other major chronic diseases such as diabetes. However, the increasing emphasis on informed self-management raises issues of access to asthma health care and asthma education, as well as issues of communication between physicians and patients and social and psychological factors that may affect a patient's asthma self-management skills.

C. Premenstrual Asthma

There is premenstrual exacerbation in about one-third of women with asthma, particularly in those with severe asthma, although the cause is unclear [12]. For example, a U.S. survey of 57 females in a pulmonary practice found that 19 (33%) had worsening asthma premenstrually [13]. A Canadian study of asthmatic and nonasthmatic women who were not aware of premenstrual exacerbation found that the asthmatic women had a greater increase in symptoms and a small drop in morning peak expiratory flow rates before menses [14]. One study reported that all women with premenstrual asthma had some symptoms of premenstrual syndrome that could have influenced their perception of symptoms [15]. However, this is unlikely to explain the majority of cases, as studies incorporating objective measurements of lung function have shown mean falls in peak flow rate of 15 to 40 l/minute prior to menses [12]. Even greater falls in peak flow (up to 400 l/min) have been reported in individual patients [16].

The management of premenstrual asthma is the same as that for other exacerbations of asthma, with conventional antiasthma therapies recommended. Intramuscular progesterone has been shown to be effective in extreme cases of premenstrual asthma where conventional steroidal treatments failed [16], although little research has been done in this area. Other treatment modalities have been suggested as well, including oral contraceptives, diuretics, and nonsteroidal anti-inflammatory drugs (NSAIDS). These have generally been ineffective and NSAIDs may precipitate asthma exacerbations.

D. Asthma during Pregnancy

The course of asthma during pregnancy is variable [12], with some patients experiencing worsening asthma and some patients experiencing improvements. In a review of nine studies of asthma during pregnancy, Turner et al. [17] concluded that deterioration may occur in 20% of women, improvement in 30%, and no significant change in 50%. It was also observed that patients with severe asthma were likely to have worsening asthma during pregnancy, whereas patients with mild asthma experienced improvements. An individual woman was likely to repeat the same clinical pattern with each pregnancy. How-

ever, exacerbations tended to occur during the end of the second trimester with subsequent improvements. Severe asthma attacks during labor were rare [12]. Asthma severity was likely to revert to the prepregnancy level within the first 3 months post partum.

Although some studies have found higher rates of maternal complications during pregnancy in asthmatics, most studies have not found increased rates of complications but have noted a higher rate of caesarian section and use of epidural anaesthesia in asthmatic mothers [18,19]. Most studies have found that asthmatic pregnancies were not more likely to result in an increased risk of prematurity or perinatal mortality if maternal asthma was properly under control; however, infants of asthmatic mothers were more likely to be of low birth weight, and poorer asthma control led to worse fetal outcome [12]. Table 56.1 summarizes currently available information regarding the safety of asthma medications during pregnancy [20,20a].

Safety of Asthma Medications in Pregnancy

Most of the drugs being used to treat asthma are safe during pregnancy, but there are several oral corticosteroids (including prednisone and prednisolone), inhaled corticosteroids (including beclomethasone dipropionate), and beta agonists (including albuterol, isoproterenol, and metaproterenol) that have been grouped as Class C by the Food and Drug Administration (FDA). Class C indicates that the drug has been associated with adverse effects on fetal health in animals but not in humans (Table 56.1). In making decisions regarding whether to continue using asthma medications or not, the risks of taking medications should be carefully weighed against the benefits to the patient and fetus of having asthma under control.

ORAL CORTICOSTEROIDS.   While there has been some concern over the use of oral corticosteroids during pregnancy, available evidence indicates that they are safe. An increased incidence of preeclampsia in mothers taking oral corticosteroids has been reported in two studies although the mechanism for this is unknown [18,20b]. The preferred systemic corticosteroids to use during pregnancy are prednisone and methylprednisone which are likely to improve fetal outcome when used in the treatment of severe attacks of asthma during pregnancy. Dexamethasone and betamethasone cross the placenta rapidly, and for this reason dexamethasone is used for fetal indications such as hastening lung maturity or for treatment of congestive heart failure in babies with congenital heart blocks (as part of the neonatal lupus syndrome).

INHALED CORTICOSTEROIDS.   Beclomethasone dipropionate was shown to be safe to humans in two studies with no increase in perinatal morbidity or congenital malformation [20c,20d] and is considered to be the drug of choice among inhaled corticosteroids during pregnancy. A recent Swedish study involving 2014 births found no increase in the rate of congenital malformations among women taking inhaled budesonide during pregnancy [20e]. The incidence rate for congenital malformations in this study was 3.8% (95% CI 2.9, 4.6), compared to the national incidence rate of 3.5% for the same years. Triamcinolone has

Table 56.1
Safety of Selected Drugs during Pregnancy[a]

Drugs	FDA class	Safety	Adverse effects/comments
Systemic Corticosteroids[b]			
Dexamethasone[c]	C	X	Cross into fetal tissue; teratogenic in mice
Hydrocortisone	—	S	
Methyl prednisolone	—	S	Cross placenta poorly: can be used during lactation
Prednisone	C	S	Cross placenta poorly: can be used during lactation
Prednisolone	C	X	Dose-dependent; more than 10 mg/day could produce neonatal and adrenal suppression
Inhaled Corticosteroids[d]			
Beclomethasone dipropionate	C	U	Teratogenic in animal studies
Budesonide	C	U	Teratogenic in animal studies
Flunisolide	C	U	Teratogenic in animal studies
Fluticasone Propionate	C	U	Teratogenic in animal studies
Triamcinolone	C	U	Teratogenic in animal studies
Cromoglycates[e]			
Cromolyn sodium	B	S	Not known to be harmful
Nedocromil Sodium	B	S	Not known to be harmful
Beta Agonists[e]			
Albuterol	C	U	May be tocolytic (in high doses); teratogenic in animal studies; not recommended for nursing mothers
Ephedrine	—	U	Reduction in uterine blood flow (animals)
Epinephrine	C	U	Reduction in uterine blood flow and teratogenic effects in animals
Eformoterol (fermoterol fumarate)	—	—	
Fenoterol	—	U	
Isoproterenol	C	U	Animal studies have not been conducted
Metaproterenol	C	U	Teratogenic in animals, may inhibit labor
Salbutamol	C	U	
Salmeterol	C	U	Teratogenic in animals; not recommended for nursing mothers
Terbutaline	B	S	No known effects during pregnancy; contraindicated during labor
Anticholinergics			
Ipratropium bromide	B	S	No known effects
Antibiotics[f]			
Ampicillin/Penicillin	B	S	Appears in breastmilk; contraindicated for nursing
Cephalosporins	B	S	No known reproductive effects in animals or humans; appears in breastmilk
Chloramphenicol	D	X	Gray syndrome; contraindicated for pregnancy and labor
Erythromycin	B	S	Appears in breastmilk
Cotrimoxazole	C	X	Contraindicated for pregnancy, labor, and breastfeeding; hyperbilirubinemia
Tetracycline	D	X	Teeth/bone malformation; contraindicated for nursing.
Methylxanthines			
Aminophylline	C	S	May delay labor; transient fetal distress
Oxtriphylline	C	S	
Theophylline	C	U	May inhibit labor; teratogenic in animal studies; contraindicated for nursing mothers
Leukotriene Receptor			
Montelukast	—	—	Avoid during breastfeeding and pregnancy unless essential

Note. S = Safe; X = Contraindicated; U = Unknown. According to Food and Drug Administration (FDA) classification: A = drugs proved safe during pregnancy from controlled studies; B = drugs not associated with reported adverse effects in animal/human studies; C = drugs with adverse effects reported in animals but not human studies; D = drugs with proven adverse effects during human pregnancy.

[a]Sources: ''Physician's Desk Reference,'' 52nd edition (1998), Medical Economics Co., Montvale, NJ, and British Medical Association and the Royal Pharmaceutical Society of Great Britain (1998), ''British National Formulary 36,'' BMJ Books, London.

[b]Corticosteroids appear in breast milk at pharmacological doses, could suppress growth and are not advised for nursing mothers.

[c]See Lupus chapter on corticosteroid use in controlling maternal disease. Dexamethasone is specifically indicated in some rare fetal conditions.

[d]Hypoadrenalism may occur in infants of mothers taking inhaled corticosteroids during pregnancy. It is not known whether inhaled corticosteroids appear in breast milk, but caution should be used when nursing.

[e]It is not known whether these are found in breast milk.

[f]Antibiotics appear in breastmilk and caution should be exercised when deciding to take during breastfeeding.

been shown to cause teratogenicity in animal studies and fetal growth retardation in humans and as a result is not recommended (see Table 56.1).

BRONCHODILATORS. Inhaled beta agonist drugs are recommended as the bronchodilators of choice in pregnancy. Their use should obviate the need for systemic beta agonist use or nonselective beta agonists. As in the nonpregnant patient, they should be used as required for relief of symptoms rather than according to a regular scheduled regime. Some bronchodilators are not recommended for use during pregnancy, and these include adrenaline and systemic beta agonists. Theophylline is not known to have teratogenic effects when used at therapeutic doses but may prolong or complicate labor. In general, it is recommended to use inhaled rather than systemic bronchodilators during pregnancy to minimize the systemic effects of asthma medications.

III. Increasing Prevalence of Asthma

There are considerable problems in measuring asthma prevalence [21] and widely varying findings can be obtained depending on which methods are used. Standardized methods for prevalence surveys are now available for children [22] and for adults [23], but traditionally a variety of survey instruments were used, ranging from clinically diagnosed asthma to questions on "wheezing ever" or "wheezing in the last year." Nevertheless, most studies that have used the same methodology in the same population at different times have reported that the prevalence of asthma symptoms has increased in recent decades and that the magnitude of the increase has in some cases been substantial. Although methodological differences in these studies make it difficult to compare asthma prevalence among countries, the trend of increasing prevalence among populations in countries of widely differing lifestyles and ethnic groups is remarkably consistent.

Table 56.2 summarizes 24 studies of changing asthma prevalence over time [24–47]. The studies used a wide variety of questions for assessing asthma prevalence, but in each instance the same questions were used in a second "repeat" study in the same population at a later time. The table shows marked increases in reported asthma symptom prevalence in virtually all populations considered. Unfortunately, few of these studies reported findings by gender, but in those that did, the increases appeared to be occurring to a similar degree in females and in males. The reasons for these global increases in asthma prevalence are unclear, but they do not appear to be explained by "established" risk factors for asthma symptoms such as air pollution or indoor allergen exposure [48].

Because asthma prevalence depends both on asthma incidence and on average disease duration, a prevalence difference between two populations, or two time periods, could entirely depend on differences in disease duration (e.g., because of factors that prolong or exacerbate asthma symptoms) rather than on differences in incidence. Thus, the increasing asthma prevalence worldwide does not necessarily mean an increase in asthma incidence, although this is commonly assumed to be the case.

Table 56.2

Changes in Asthma Prevalence in Children and Young Adults

Country	Time period	Asthma prevalence First study	Asthma prevalence Second study	Reference
Australia	1964–1990	19.1%	46.0%	[24]
	1982–1992	12.9%	19.3%	[25]
Canada	1980–1983	3.8%	6.5%	[26]
Finland	1961–1986	0.1%	1.8%	[27]
France	1968–1982	3.3%	5.4%	[28]
Israel	1986–1990	7.9%	9.6%	[29]
Italy	1983–1993	2.9%	4.4%	[30]
Japan	1982–1992	3.3%	4.6%	[31]
Netherlands	1977–1992	19%	31%	[32]
New Zealand	1969–1982	7.1%	13.5%	[33]
	1975–1989	26.2%	34.0%	[34]
	1981–1990	22.8%	30.8%	[35]
Norway	1981–1994	3.4%	9.3%	[36]
Papua New Guinea	1973–1984	0.0%	0.6%	[37]
Sweden	1971–1981	1.9%	2.8%	[38]
Tahiti	1979–1984	11.5%	14.3%	[39]
Taiwan	1974–1985	1.3%	5.1%	[40]
UK	1956–1975	1.8%	6.3%	[41]
	1966–1990	18.3%	21.8%	[42]
	1973–1986	2.4%	3.6%	[43]
Scotland	1964–1989	10.4%	19.8%	[44]
Wales	1973–1988	4.0%	9.0%	[45]
USA	1971–1980	4.8%	7.6%	[46]
	1981–1988	3.1%	4.3%	[47]

IV. Gender Differences

A. Incidence

Few cohort studies have been conducted of asthma incidence (i.e., the percentage of new cases of asthma per year) because of the difficulties and expense of constantly monitoring a population for the occurrence of new cases. Table 56.3a shows findings from three key studies, all of which show lower asthma incidence in females than in males at younger ages but higher incidence at older ages [49–52]. This is consistent with other evidence that more girls than boys develop asthma during adolescence [53–55]. The crossover age varies among studies, but is generally assumed to be in the early teenage years (i.e., around the time of onset of puberty) [56]. All three studies show that the crossover occurs because incidence declines with age in males (although increasing again during adulthood) but declines more slowly or even increases with age during adolescence in females.

Table 56.3b shows five studies of gender differences in cumulative incidence (i.e., the percentage of people who have "ever had asthma" by a specified age [49,50,57–60]. As in Table 56.2, the estimates are not comparable among different studies because of the different asthma definitions used, thus the focus should be on gender differences within particular studies. The cumulative incidence was lower in females up to the age of 17 in all of the studies, and despite the higher incidence in

Table 56.3a

Gender Differences in Asthma Incidence (%/year) by Age Group

Country	Age group	Females	Males	Ratio	Reference
UK	0–7	2.3	2.9	0.79	[49]
	8–11	1.0	1.3	0.77	[50]
	12–16	0.6	0.9	0.68	
	17–23	0.9	0.6	1.68	
	23–33	1.6	1.3	1.22	
USA	0–4	0.9	1.4	0.64	[51]
	5–9	0.7	1.0	0.70	
	10–14	0.3	0.2	1.50	
	20–29	0.4	0.2	2.00	
	60–69	0.8	0.3	2.67	
USA	<1	2.3	3.9	0.59	[52]
	1–4	0.6	1.1	0.58	
	5–9	0.2	0.4	0.60	
	10–14	0.2	0.2	0.96	
	15–29	0.1	0.1	1.38	
	30–49	0.1	0.1	1.35	
	50+	0.1	0.1	0.81	

Table 56.3b

Gender Differences in Asthma Cumulative Incidence (%) by Age Group

Country	Age group	Females	Males	Ratio	Reference
UK	0–7	16.3	20.0	0.82	[49]
	0–11	19.6	24.0	0.82	[50]
	0–16	21.9	27.2	0.81	
	0–23	27.1	30.1	0.90	
	0–33	42.8	43.0	0.99	
Finland	18–29	0.7	0.9	0.80	[57]
	30–39	1.1	1.0	1.12	
	40–49	2.0	1.4	1.46	
	50–59	3.5	3.8	0.91	
	60–69	3.7	4.0	0.93	
	70+	5.0	4.7	1.05	
New Zealand	0–6	6.3	14.3	0.44	[58]
USA	0–9	2.4	4.6	0.52	[59]
	0–14	2.9	4.4	0.66	
USA	0–18	14.7	20.1	0.73[a]	[60]
	0–18	10.1	12.3	0.82[b]	

[a]Black children.
[b]White children.

females during adolescence, the cumulative incidence in females remained less than or equal to that in males at all ages. However, because of the different incidence patterns, the age of onset was later in females than in males. For example, Gold *et al.* [60] reported that the median age of asthma onset in whites was 6.5 years for girls and 2.5 years for boys, although there was no gender difference in median age of onset in African Americans (4.5 years for both genders).

B. Prevalence

Because of the difficulties in measuring asthma incidence in population-based studies, most comparisons of asthma occurrence in females and males have involved prevalence rather than incidence. Table 56.4 summarizes several key studies in children that involved either national samples or standardized comparisons among countries [46,49,51,61–65]. As in Table 56.2, the prevalence estimates are not comparable among different studies because of the different survey methods used, thus the focus should be on gender differences within particular studies. By far the largest study that has been conducted is the International Study of Asthma and Allergies in Childhood [22,48,64] which has involved more than 700,000 children in 155 centers in 56 countries. This study found that the 12-month period prevalence of asthma symptoms was lower in females (11.2%) than in males (14.2%) in 6–7 year olds but was slightly higher in females (15.0%) than in males (13.9%) in 13–14 year olds [61]. In fact, all of the studies in Table 56.4 showed lower prevalence in females than in males in children aged less than 12 years, whereas the patterns in adolescence were less consistent.

Table 56.4 also shows similar analyses of studies in adults. These include some smaller population studies in addition to national surveys and standardized international comparisons, because fewer studies have been done in adults. The findings are inconsistent, with some, but not all studies, showing higher prevalence in females than in males. Unfortunately, the Euro-

Table 56.4

Gender Differences in 12-Month Period Prevalence of Asthma Symptoms (%) by Age Group[a]

Country	Age group	Females	Males	Ratio	Reference
Studies in children					
Global[b]	6–7	11.2	14.2	0.79	[61]
	13–14	15.0	13.9	1.08	
UK	7	7.2	8.9	0.81	[49]
	11	3.8	5.4	0.70	
	16	2.8	3.9	0.72	
USA	<6	2.2	3.7	0.58	[62]
	6–16	2.6	3.8	0.67	[63]
USA	6–11	6.1	9.0	0.68	[46]
	12–17	4.9	8.0	0.61	
Australia	12–15	33	27	1.20	[64]
England	12–15	32	26	1.23	
Germany	12–15	21	18	1.17	
New Zealand	12–15	30	26	1.15	
Studies in adults					
New Zealand	20–44	17.0	13.2	1.29	[65]
UK	23	4.7	3.2	1.47	[49]
USA	15–19	6.8	8.9	0.76	[51]
	20–29	4.5	5.8	0.78	
	30–39	7.9	3.0	2.60	
	40–49	7.1	4.4	1.60	
	50–59	4.0	9.5	0.42	
	60–69	6.4	6.3	1.02	
	70+	6.4	7.9	0.81	
USA	17–44	2.8	2.5	1.12	[62]
	45–64	3.7	2.9	1.25	[63]
	65+	3.1	4.2	0.74	

[a]Asthma admissions per 1000 persons/year.
[b]155 centers in 56 countries.

Table 56.5
Gender Differences in Hospital Admission Rates for Asthma by Age Group[a]

Country (Years of study)	Age group	Females	Males	Ratio	Reference
Studies in children					
USA (1980–1986)	5–14	2.6	4.1	0.63	[68]
USA (1986–1989)	0–5	4.5	7.8	0.58	[69]
	6–10	2.9	4.7	0.62	
	11–20	2.3	2.2	1.05	
UK (1976–1985)	0–4	2.9	5.3	0.55	[70]
	5–14	1.4	2.3	0.61	
Studies in adults					
USA (1986–1989)	21–30	2.1	1.0	2.10	[69]
	31–40	2.3	0.9	2.70	
	41–50	2.5	1.0	2.59	
	51–60	3.1	1.4	2.14	
	61–70	2.9	1.6	1.80	
USA (1984–1989)	65–69	2.8	1.9	1.46	[71]
	70–74	3.1	2.4	1.32	
	75–79	3.0	2.5	1.21	
	80+	2.5	2.3	1.10	
UK (1976–1985)	15–44	0.7	0.5	1.51	[70]
	45–64	0.9	0.7	1.20	
	65–74	0.9	0.8	1.14	
	75+	0.6	0.6	1.10	

[a]Asthma admissions per 1000 persons/year.

Table 56.6
Gender Differences in National Asthma Mortality Rates by Age Group[a]

Country (Years of study)	Age group	Females	Males	Ratio	Reference
USA (1968–1987)	5–34 (white)	0.2	0.2	1.18	[75]
	5–34 (black)	0.8	0.8	0.96	
Denmark (1977–1993)	20+	6.7	5.2	1.29	[76]
UK (1974–1984)	1–4	0.5	0.4	1.28	[77]
	5–34	0.6	0.8	0.74	
	35–64	3.3	2.7	1.25	
Sweden[b] (1973–1988)	1–14	0.1	0.2	0.65	
	15–24	0.4	0.4	0.94	[78]

[a]Asthma deaths per 100,000 persons/year.
[b]Estimated values are reported here.

pean Community Respiratory Health Survey (ECRHS) study has not yet published gender-specific data which were reported to be inconclusive [66].

C. Severity

The concept of "asthma severity" can relate to underlying asthma severity (chronic severity) or to the severity of an acute attack (acute severity). Severe exacerbations can also be divided into those with sudden onset and those with slow onset. Although a number of studies have used detailed measures of asthma severity [67], the only data routinely available for assessing gender differences in asthma severity are those relating to asthma hospital admissions and asthma deaths. Table 56.5 lists several studies of gender differences in asthma hospital admission rates [68–71]. All studies found that hospital admission rates for asthma among children under 15 years of age were lower in females than in males, reflecting the higher asthma prevalence for boys in those age groups. In contrast, all studies in adults found higher admission rates for females. Skobeloff *et al.* [69] reported that the average length of stay was slightly higher in females than in males, indicating that the higher admission rate in females was unlikely to be due to a lower threshold for admission.

Emergency room visits are another measure of asthma severity, but these are less often reported in routine statistics. However, a study in Seattle found that adolescent girls were reported to have used the emergency room more than boys, to have used inpatient facilities less, and to have had fewer prescription

charges [72]. This is consistent with the hypothesis that asthma was more severe in females due to less access to, or utilization of, medical care [73].

Asthma severity in adolescents was measured in one study by a combination of factors that resulted in an "asthma score" [74]. These factors included the frequency of symptoms, the effect of asthma symptoms on sport, school, and/or work participation, and the use of medication and hospital facilities. Togias *et al.* found that among 151 adolescents from African-American and Caucasian backgrounds, female adolescents (aged 13–18) had more severe asthma than males (p-value = 0.01) as measured by their "asthma score," even after controlling for ethnicity, family income and education, smoking, and housing density.

D. Mortality

Asthma mortality studies have focused primarily on time trends, and few have reported gender-specific data (Table 56.6) [75–78]. In considering these studies, it is important to focus on mortality in the 5–34 year age group because the diagnosis of asthma mortality is more firmly established in this group [79] and there are major classification problems at older ages [80]. The available studies generally indicate variable gender differences in death rates from asthma (Table 56.6). For example, in the 5–34 year age group, Weiss and Wagener [75] reported slightly higher death rates in white females in the U.S., whereas Burney [77] reported higher death rates for males in the UK and a Swedish study showed little or no gender difference [78]. Among older age groups, women appear to have higher asthma mortality rates than men [76,77], although this could be complicated by diagnostic bias.

Several studies provide information on gender differences in asthma mortality without reporting the actual death rates in 5–34 year olds. Arrighi [81] reported that asthma mortality rates in the U.S. were higher in boys at ages 5–14 but were about the same for males and females among 15–34 year olds since 1980 (actual mortality rates were not reported). Robertson *et al.* [82] found no gender difference in asthma mortality in deaths in Victoria, Australia during 1986–1987, whereas Sears *et al.* [83]

reported more asthma deaths in females than in males (relative risk approximately 1.5) in 5–34 year olds in New Zealand during 1981–1983, and there were similar findings in two regions in England during 1979 [84].

Finally, several studies report gender differences in asthma mortality time trends. Mortality among 5–34 year olds in England and Wales was higher in males prior to World War I, then higher in females until the mid–1960s, but has been higher in males since 1970 [85,86]. Ehrlich and Bourne [87] reported similar death rates in white females and in white males in 5–34 year olds in South Africa during 1967–1988, with lower death rates for nonwhite females compared to nonwhite males over the same period. In the U.S., African-American males experienced higher death rates than females in 1968–1987 [75], but rates among whites were the same both for males and for females. An increase in the mortality rate of 5–34 year old American white females to nearly twice that of white males by 1994 was reported by Sly and O'Donnell [88], although these findings are difficult to interpret because they are not age-adjusted. In the same study, African-Americans showed a similar trend to whites, in that females had higher death rates than males for the same time period.

V. Risk Factors

In this section we review evidence on major risk factors for asthma. In almost all instances these factors are not gender-specific and are relevant to asthma both in males and in females. Hormonal influences on the expression of asthma and asthma symptoms are little understood and little research has been done in this area. As a result, few studies have reported gender differences in risk factor exposures, and the following discussion focuses on general rather than gender-specific risk factors. However, it is important to distinguish between risk factors for asthma symptoms and risk factors for the development of asthma itself. Many risk factors reviewed below (*e.g.,* air pollution, indoor allergen exposure) have been found to increase the frequency and severity of symptoms in people with asthma, but the evidence is much weaker as to the primary cause of asthma itself.

A. Demographic Factors

Apart from age and gender, there are a variety of other demographic factors that are associated with asthma, including birth order [89], season of birth [90], ethnicity [91], region [90], and country [48]. Studies in the 1960s and 1970s [92] suggest that asthma is more common in children in the higher social classes. However, there has been less evidence of social class differences as the diagnosis of asthma has become more widespread [93], even though diagnostic labelling of wheezing in adults differs by social class [94]. However, severe asthma appears to be more common in children in lower social classes [95] and in some disadvantaged ethnic groups [96]. Low socioeconomic status is associated with hospital admissions for asthma [97] and with reduced lung function in adults [98]. This could represent a greater prevalence of asthma in disadvantaged groups, increased severity due to environmental factors (*e.g.,* environ-

mental tobacco smoke, nutrition), or inadequate disease management and poor access to health care [95,97,99].

B. Genetic Factors and Early Life Events

Asthma appears to be multifactorial in origin and is influenced by multiple genes and environmental factors [100]. Attention has been particularly focused on chromosomes 5 and 11, both of which may contain genes relevant to asthma [101]. However, a particular genetic factor may cause susceptibility to an environmental exposure and may thereby affect one or more aspects of the complex etiologic process involved in asthma. However, whether this genetic potential is expressed will depend on whether sufficient exposure to the environmental factor occurs. In fact, the increasing prevalence of asthma indicates that genetic factors alone are unlikely to account for a substantial proportion of asthma cases [102].

It is well established that people with a family history of asthma are more likely to develop asthma themselves [58], and parental asthma is a stronger predictor of asthma in the offspring than parental atopy [103]. However, this association is not necessarily due to genetic factors, but could merely reflect similar lifestyles and exposures in family members [104]. For example, one study found that persons whose spouse developed asthma were also more likely to develop asthma [105], presumably due to common environmental exposures. Some indication of the possible contribution of genetic factors in asthma is given by studies of twins. For example, Edfors-Lubs [106] analyzed data on 7000 twin pairs from the Swedish Twin Register and found that concordance of asthma in monozygotic twins was greater than that in dizygotic twins. However, the concordance was still only 19%, and even this may, in part, be due to similar environmental exposures in monozygotic twins, including a common intrauterine environment [107].

There is now a large body of epidemiologic evidence that the intrauterine environment may have effects that increase the risk of chronic disease in adult life [108]. Airway growth is largely completed *in utero*. Nutritional deficiency *in utero* slows down weight gain and adversely affects airway growth, resulting in an increased risk of developing chronic obstructive pulmonary disease in later life. Godfrey *et al.* [107] found that a raised total serum IgE concentration in adult life (which is associated with an increased asthma risk) was associated with an increased head circumference at birth, independent of gestational age at birth, birthweight, and the mother's pelvic size and parity. They suggested that these associations reflect the long-term effects of sustaining fetal brain growth at the expense of the trunk, particularly the thymus, as a result of fetal undernutrition in late gestation. On the other hand, the association of larger head circumference at birth with subsequent risk of asthma may reflect increased maternal-fetal nutrition during the relevant stages of pregnancy, especially fatty acid status which is a major determinant of brain growth, which might also lead to programming of the developing immune system to predispose the fetus to atopic sensitization [109]. This hypothesis is more consistent with the increases in asthma prevalence in recent decades in communities in which birth weights [110], head circumference, birth length, and other markers of maternal-fetal nutrition have increased.

Early childhood infection may have a role either in promoting asthma or in protecting against sensitization [111–113]. Bronchiolitis in infancy, for example, is a risk factor for subsequent asthma [114]. On the other hand, several studies in children have found negative associations between the number of siblings in the household (and particularly with the number of older siblings) and atopy [115], hay fever [116], and asthma [89,117]. The reasons for these associations are unclear, but Strachan [116] has suggested that large family size could increase viral infections in infancy that may down regulate production of IgE [101,118].

Shaheen [111] has suggested that immunization may promote allergic sensitization. Pertussis vaccination, for example, has been known for many years to act as an adjuvant for antigen-specific responses in laboratory animals [119], and specific IgE response to pertussis toxin itself has been identified in children receiving pertussis immunization [120]. Two studies have found that infant immunization was associated with the subsequent development of asthma [121,122]. It appears, therefore, to be theoretically possible that immunization may have a contributory role in the induction of allergic disease through reducing the incidence of clinical infectious disease in early childhood, the direct IgE inducing effects of some of the infectious agents themselves, the potentiating adjuvants added to vaccines, or a combination of these.

C. Atopy

"Atopy" has previously been used as a poorly defined term to refer to allergic conditions that tend to cluster in families, including hay fever (allergic rhinitis), asthma, eczema, and other specific and nonspecific allergic states [123]. More recently, it has been characterized by the production of specific IgE in response to common environmental allergens, although total serum IgE has been found to be associated with asthma independently of specific IgE levels [124]. Prospective studies have shown that high levels of serum IgE in cord blood predict the subsequent development of asthma in childhood [125] and that levels are lower for female than for male babies [126]. IgE levels are known to gradually increase with age during early childhood, before declining again during adulthood, and adult men generally have higher levels than adult women [127]. A positive response to the application of a specific allergen in skin-prick testing reflects the production of specific IgE antibodies to the allergen and is strongly related to total serum IgE [128]. Skin-prick testing provides a convenient test for atopy in epidemiologic studies, and skin-prick testing positivity is strongly associated with asthma [129,130]. However, asthma and atopy also occur independently of each other [131]. In fact, while more than one-half of asthmatics are atopic, a substantial proportion of nonasthmatics are as well, and only about one-third of asthma in children and young adults is attributable to atopy [132].

D. Indoor Environment

Children exposed to environmental tobacco smoke are at increased risk of asthma, particularly when the mother is a smoker [133,134]. The evidence is strongest for increases in severity in children who already have asthma [135], whereas the evidence for the initial occurrence of asthma (incidence) is less conclu-sive [136]. Nevertheless, tobacco exposure *in utero* and in infancy may enhance sensitization [137].

Little is currently known about the contribution of other indoor air pollutants to the incidence and prevalence of asthma. Nitrogen dioxide from burning fossil fuels has received the most attention [138], and sulphur dioxide from burning sulphur-containing coal or gas, mosquito coil smoke [139], and formaldehyde from wood preparation [140] also have been considered. Particulates from open or closed wood and coal burning fires have received less attention in developed countries [141] but have been studied in developing countries where very high indoor levels have been encountered [142].

Dampness and humidity are not only features of old houses but may also occur in modern houses with tight insulation [143] and in modern office buildings [144]. Dampness can encourage the growth of fungal molds and house dust mites [145], as well as promote the survival of viruses in droplet spray. Cold air can itself trigger asthma attacks [146].

More than one-half of asthmatics are sensitized to at least one common allergen, but the relevant allergens may vary from country to country and culture to culture; in most Western countries house dust mites, cats, other pets, cockroaches, and molds are major sources of sensitization. Since Voorhorst and Spieksma [147] identified the house dust mite as a major source of allergen for asthma and allergic rhinitis in the 1960s, both the epidemiology and immunology of these allergens have received increasing interest [148]. Several common family pets such as cats, dogs, and rodents are also known to produce allergens that may aggravate asthma [149], and clinical studies of asthmatics have found sensitization to allergens produced by these pets [150]. Cats may be particularly important because of their widespread occurrence and the potency of their allergens [149]. Cockroach allergen may also be an important risk factor for asthma exacerbations because cockroaches are ubiquitous in inner-city areas in many countries and are highly allergenic. For example, Rosenstreich *et al.* [151] found that children who were sensitized to cockroach allergen and exposed to high levels had higher rates of asthma hospital admissions.

It should be stressed that the main evidence for the role of indoor allergens in asthma relates to the provocation of symptoms in asthmatics. The evidence is much weaker regarding the primary causation of asthma by indoor allergens, and, in fact, there is only one major study that has shown a dose-response relationship between exposure to house dust mites in infancy and the subsequent risk of developing asthma [152]. The key successful intervention study [153] found that house dust mite avoidance in infancy markedly reduced the risk of atopy, but the effect on asthma itself was less impressive and decreased with increasing age. Furthermore, there are a number of inconsistencies in international patterns of exposure to indoor allergens and patterns of asthma prevalence. For example, asthma prevalence is equally high in various centers in English-speaking countries [48], despite major differences in levels of exposure to house dust mites and other indoor allergens [154]. In fact, the epidemiologic evidence is consistent with the hypothesis that allergen exposures in infancy primarily determine which allergens susceptible persons become sensitized to, but these exposures may have little influence on whether or not these persons develop sensitization or asthma itself. Thus, although there is still considerable potential for secondary prevention of asthma through reducing indoor allergen exposures, there is at best weak and inconsistent evidence as to the potential for primary prevention [155].

E. Outdoor Environment

The role of outdoor air pollutants in asthma and other diseases has been extensively studied and debated [156]. "Conventional" air pollution includes SO_2 and airborne particulates with a size of 10 μm or less that can be inhaled into the lung. Attention also has been focused on NO_2, because its concentration has risen in recent decades due to increasing motor vehicle use. An association between traffic density on residential streets and asthma symptoms has been found in studies in Germany [157], Sweden [158], and the Netherlands [159], but not in the United Kingdom [160]. Ground level ozone is formed from nitrogen oxides, hydrocarbons, and sunlight. Ozone has been shown to increase the sensitivity to allergen challenge in asthmatics [161], and exacerbations of asthma have been reported in association with fluctuations in outdoor ozone levels. Overall, although it is well established that air pollution can exacerbate symptoms and provoke attacks in asthmatics, the weight of evidence does not support a major role for outdoor air pollution as a determinant of asthma prevalence, and there is even some evidence of a negative association with asthma prevalence internationally [48].

Airborne allergens include various pollens and spores, their derivatives, and material from other sources including plant sap, anther linings, and leaf leachate [162], as well as material of insect origin [163]. Several asthma epidemic outbreaks have been linked to thunderstorms, including studies in Birmingham [164], Melbourne [165], and London [166]. The airborne allergens involved are believed to include fungal spores [167] and pollen grains [168]. Asthma epidemic outbreaks linked to exposures from the agricultural industry include outbreaks linked to soybean dust in Barcelona [169], Cartagena [170], and New Orleans [171], and outbreaks linked to castor bean dust in Ohio [172], South Africa [173], and Brazil [174].

F. Occupational Exposures

Occupational asthma is "asthma that is caused, in whole or part, by agents encountered at work" [175]. It was documented as long ago as 1713 by Ramazzini who described grain dust asthma [176]. Nowadays, there are more than 200 known occupational causes of asthma, and asthma is the most common occupational respiratory disease in developed countries [177]. For example, asthma was the most common disease category and accounted for 28% of cases reported to the United Kingdom Surveillance of Work-related and Occupational Respiratory Diseases (SWORD) project [178]. Estimates of the proportion of adult asthma that is thought to be occupational in origin range from 2–15% in the United States, 15% in Japan [179], 5% in Spain [180], 2–3% in New Zealand [181], and 2–6% in the United Kingdom [182]. However, these studies have mainly been done in men, and there is little information available on occupational asthma in women.

G. Diet

Several dietary factors have been found to influence the development of asthma symptoms [183], including dietary sodium [184] and selenium [185,186]. Much has been written on the topic of diet as a risk factor for asthma [187]. This is not intended to be an exhaustive discussion. Specifically, we address antioxidants, fatty acids, and infant diet below.

1. Antioxidants

It has been hypothesized that changes in the diet in Western countries, particularly those involving reductions in antioxidant intake, may contribute to an increase in susceptibility to allergens and a rise in allergic disease [188,189]. A report from the Nurse's Health Study [190] found that there was a protective effect from asthma by dietary vitamin E, but not from vitamin E supplementation. Vitamin C supplementation was found to increase the risk of asthma in the same study, but this was thought to be because of use of vitamin C for the prevention of respiratory illness. In fact, serum levels of vitamin C were found to be inversely related to wheezing in the National Health and Nutrition Examination Survey II (NHANES II) [191]. In addition, it was found that intake of solid fruit (apples and pears) was inversely related to chronic, nonspecific lung disease in a study of 793 Dutch men [192]. Although other large population studies have found positive effects of dietary vitamin C in relation to lung function [193], there have been mixed results from administering vitamin C supplements to asthmatics [190].

2. Fatty Acids

The possible influence of changing patterns of consumption of dietary fat is being considered with increasing interest. In recent decades there has been an increase in the consumption of vegetable oils, accompanied by a decrease in consumption of oily fish [194–196], and it has been hypothesized that these dietary changes could have effects on subsequent generations [197–199]. Linoleic acid, a polyunsaturated fatty acid found in vegetable oils, is thought to have a proinflammatory effect through complex biomechanisms modulating the formation of eicosanoids, cytokines, and IgE. Linoleic acid intake was positively associated with chronic, nonspecific lung disease among Dutch men, as were total polyunsaturated and monounsaturated fat intakes [192]. In the Nurse's Health Study, however, asthma risk was not associated with past intake of polyunsaturated fats [190]. Omega-3 fatty acids are highly polyunsaturated fatty acids found in fish, particularly salmon, tuna, and sardines. Dry and Vincent [200] saw an improvement in lung function of 12 asthmatics after 9 months of fish oil supplementation, although their results have yet to be reproduced. In addition, current regular consumption of fresh, oily fish was found to be protective against current asthma among a group of 484 Australian children age 8 to 11 years old [201]. The Nurse's Health Study [190] found no association between fish intake in 1980 and asthma diagnosis by 1988.

3. Infant Diet

Both breastfeeding and the timing of solid food introduction have been found to be related to asthma risk in one study. Children who had been exclusively breastfed and who did not receive solid foods before the age of 15 weeks had a decreased risk of asthma diagnosis at 7 years old [202]. In a previous study, introduction of solid foods before 4 months of age was associated with higher rates for asthma diagnosis by 4 years old, whereas breastfeeding was not found to be related to asthma risk in the same group [203].

H. Viruses

Infections during infancy may protect against the development of sensitization and asthma. On the other hand, some viral infections during the first year of life may increase the risk of developing asthma [204], and infections during childhood may cause exacerbations of asthma and provoke episodes of wheezing in children who already have the condition [205,206]. In children under 5 years of age, respiratory syncytial virus (RSV) and parainfluenza (PI) are the most common pathogens, whereas in older children rhinovirus and influenza A are more important. Viral respiratory tract infections are less commonly associated with exacerbations of asthma in adults [123], although they may account for up to 40% of severe exacerbations [207]. Viral infections may provoke an IgE response and respiratory syncytial virus (RSV) produces such a response in 70% of children [208], although the persistence of the response appears to be determined by the host, and viruses do not appear to play a major part in the sensitization process [209].

I. Psychosocial Factors

There is relatively little evidence that psychosocial factors play a significant role in the development of asthma [58]. However, it has been suggested that psychosocial factors may provoke asthma attacks, and that factors such as low self-esteem among teenage girls could contribute to higher asthma rates in this age group [210,211]. Silverglade et al. [212] reported that in adolescents, although mild asthma was not associated with psychological well-being, symptoms such as irrational beliefs, anxiety, depression, and hostility were strongly associated with severe asthma (as marked by the number of attacks/year). Adolescence is usually the first time individuals make independent use of the health system [211], and there may be particular problems of overuse of reliever medications and/or underprescribing and underuse of preventer medications. For example, studies of iatrogenic asthma mortality epidemics (caused by the high-dose poorly selective beta agonists isoprenaline forte and fenoterol) have found particularly high rates of death among teenagers and in patients with psychosocial problems [10]. Thus, psychosocial factors may be important with regard to asthma severity, even if there is currently little evidence that they cause the development of asthma itself.

VI. Discussion

In summary, the gender differences in asthma prevalence and severity are not large, and asthma is generally an equally important cause of morbidity both in females and in males. Nevertheless, there are three key findings of interest with respect to gender differences: (1) asthma incidence and prevalence is consistently lower in females than in males before age 12 years; (2)

during adolescence and adulthood there is evidence of higher incidence (but not cumulative incidence) and prevalence in females; and (3) asthma severity (as indicated by hospital admissions rates and by an "asthma score") is also higher in females, but asthma mortality is not.

One possible explanation for the lower incidence and prevalence in young female children is that the average age of onset in childhood and adolescence may be later in females. For example, Horwood et al. [58] have suggested that much of childhood asthma is a genetically inherited tendency and that the age of expression is later for girls than for boys. Levels of cord blood IgE at birth are lower in girls than in boys [126], indicating a lower risk of the subsequent development of atopy and asthma. Some authors have noted that boys have smaller airways relative to lung size than girls and that this may explain the greater frequency and severity of lower respiratory tract illness in boys, even though infection rates are similar for both sexes [213,214].

Alternatively, it is possible that boys have more exposure to factors that increase asthma incidence or duration. For example, it has been suggested that boys have more lower respiratory tract infections that may be related to asthma incidence or provoke asthma attacks [59,215] or lead to asthma being incorrectly diagnosed before age 5. It has also been suggested that playing habits of boys may result in more exposure to indoor allergens, although it is unlikely that this would markedly increase the total daily allergen load. In fact, a study of 7–9 year old children in New Zealand measured higher levels of both mite and cat allergens on girl's clothes than on boy's clothes, irrespective of fabric type [216,217]. In addition to evidence of cockroach (and house dust mite) sensitivity, as shown by skin-prick tests among urban children [74,218], there have been studies that showed greater sensitivity rates to common allergens among males [219–221].

The apparently higher prevalence in adult females in some studies (Table 56.4) is of particular interest given that incidence appears to be higher in females but cumulative incidence is not (Table 56.3). This is consistent with a later age of onset of the condition in females, although this would not in itself explain the higher prevalence in female adults. Although there is evidence of higher incidence of asthma in females than in males during adolescence, there is also some evidence of poorer diagnosis and prognosis (and hence prolongation of symptoms and thereby of symptomatic asthma) in girls during adolescence [73,220,221].

Dodge and Burrows [51] suggest that women may be diagnosed with asthma more frequently than men even though they report symptoms at equal rates. In a further analysis of the Tucson cohort, Dodge et al. [222] reported that among men and women over 40 years old who presented to the physician with the same physical complaints, men were more likely to be diagnosed with emphysema and women were more likely to be diagnosed with asthma or chronic bronchitis. However, the prevalence differences shown in Table 56.4 mainly involve self-reported symptoms and are unlikely to be due to differences in diagnosis, although they could be due to differences in self-recognition and reporting of symptoms.

Alternatively, the increased prevalence (or smaller reduction in prevalence) in females after puberty could be due to hormonal influences on allergic predisposition, airway size, inflammation, and smooth muscle vascular function [60,223]. Although there is apparently no information on the relationship of testosterone to asthma occurrence [69], it is plausible that levels of some female hormones may be associated with asthma [224]. In particular, premenstrual asthma may be related to the hormonal environment which may not only cause asthma exacerbations, but may also affect the frequency and duration of asthma symptoms, resulting in an increase in the prevalence of "current asthma."

It is also possible that the higher prevalence in female compared to male adults could be due to differences in exposure to risk factors that cause the exacerbation and prolongation of symptoms. For example, it has been suggested that increased exposure to house dust mite allergen among women who primarily undertake domestic cleaning tasks could result in more asthma cases among women. However, there is currently little evidence in support of such speculations, and the evidence linking indoor allergen exposure to asthma prevalence in populations is weak and inconsistent.

Finally, the reasons for the higher rates of asthma hospital admissions in adult females are unclear, although it has been suggested that asthma may become more severe in women if physicians are reluctant to prescribe steroids out of concern for possible bone demineralization or complications during pregnancy [69]. More generally, problems of access to health care and psychosocial problems are also relevant, although it is not known whether these factors differ between female and male asthmatics.

VII. Directions for Future Research

In the context of this chapter, there are two major issues for further research: (1) why is asthma prevalence increasing worldwide? and (2) why is asthma less common in females than in males before puberty, but more common in females during adolescence and adult life?

The major focus of asthma research currently is on the causes of the global increase in prevalence. This increase does not appear to be explained by "established" asthma risk factors such as indoor allergens or air pollution [48]. Attention is, therefore, increasingly shifting to the examination of factors that may affect underlying susceptibility to the condition, including exposures *in utero* and in the first year of life. In particular, it is possible that factors associated with "Westernization" (*e.g.,* maternal diet, small family size, reduced frequency of infant infections, immunization) could increase the risk of developing asthma in childhood.

However, adult females continue to show greater asthma prevalence than adult males. The reasons for this difference are not known. Premenstrual asthma is an area deserving further research as it may have severe (in some cases) and preventable consequences, and it may give further insight into the possible role of sex hormones and asthma expression. Finally, occupational asthma is one of the most important occupational diseases, yet little research has been conducted on occupational asthma in women. However, occupational asthma is unlikely to explain the greater asthma prevalence in women.

Whatever the reasons for the gender differences in asthma incidence and prevalence, it should be stressed that they are not usually large, although some studies show markedly increased asthma severity in adult females. Furthermore, the continuing global increases in asthma prevalence are affecting both women and men, apparently with approximately equal intensity. Thus, asthma is an increasingly important cause of morbidity both in women and in men in industrialized countries and in the developing world.

Acknowledgments

The Wellington Asthma Research Group is supported by a Programme Grant from the Health Research Council of New Zealand and by a major grant from the Guardian Trust (Trustee of the David and Cassie Anderson Medical Charitable Trust).

References

1. Keeney, E. L. (1964). The history of asthma from Hippocrates to Meltzer. *J. Allergy* **35,** 215–226.
2. Ellul-Micallef, R. (1976). Asthma: A look at the past. *Br. J. Dis. Chest* **70,** 112–116.
3. Unger, L., and Harris, M. C. (1974). Stepping stones in allergy. *Ann. Allergy* **32,** 214–230.
4. Willis, T. (1678). "Practice of Physick, Pharmaceutice Rationalis or the Operations of Medicine in Humane Bodies." London.
5. Gergen, P. J., and Weiss, K. B. (1995). Epidemiology of asthma. *In* "Asthma and Rhinitis" (W. Busse and S. Holgate, eds.), pp. 15–31. Blackwell, Oxford.
6. Gross, N. J. (1980). What is this thing called love? or, defining asthma. *Am. Rev. Respir. Dis.* **121,** 203–204.
7. Ciba Foundation Guest Symposium (1959). Terminology definitions, classification of chronic pulmonary emphysema and related conditions. *Thorax* **14,** 286–299.
8. American Thoracic Society Committee on Diagnostic Standards (1962). Definitions and classification of chronic bronchitis, asthma and pulmonary emphysema. *Am. Rev. Respir. Dis.* **85,** 762–768.
9. (GINA) Global Strategy for Asthma Management and Prevention (1994). "NHLBI/WHO Workshop Report. Global Initiatives for Asthma." National Heart, Lung, and Blood Institute, Washington, DC.
10. Pearce, N. E., Beasley, R., Crane, J., and Burgess, C. (1995). Epidemiology of asthma mortality. *In* "Asthma and Rhinitis" (W. Busse and S. Holgate, eds.), pp. 58–69. Blackwell, Oxford.
11. Beasley, R., Cushley, M., and Holgate, S. T. (1989). A self management plan in the treatment of adult asthma. *Thorax* **44,** 200–204.
12. Chien, S.-M., and Mintz, S. (1993). Pregnancy and menses. *In* "Bronchial Asthma" (E. B. Weiss and M. Stein, eds.), pp. 1085–1098. Little, Brown, Boston.
13. Eliasson, O., Scherzer, H. H., and DeGraff, A. C. (1986). Morbidity in asthma in relation to the menstrual cycle. *J. Allergy Clin. Immunol.* **77,** 87–94.
14. Pauli, B. D., Reid, R. L., Hunt, P. W., Wiale, R. D., and Forkert, L. (1989). Influence of the menstrual cycle on airway function in asthmatic and normal subjects. *Am. Rev. Respir. Dis.* **140,** 358–362.
15. Rees, L. (1963). An aetiological study of premenstrual asthma. *J. Psychosom. Res.* **7,** 191–198.
16. Beynon, H. L. C., Garbett, N. D., and Barnes, P. J. (1988). Severe premenstrual exacerbations of asthma: Effect of intramuscular progesterone. *Lancet* **2,** 370–373.
17. Turner, E. S., Greenberger, P. A., and Patterson, R. (1980). Management of the pregnant asthmatic patient. *Ann. Intern. Med.* **93,** 905–918.
18. Stenius-Aarnala, B., Piirila, P., and Teramo, K. (1988). Asthma and pregnancy: A prospective study of 198 pregnancies. *Thorax* **43,** 12–18.

19. Lao, T. T., and Huengsburg, M. (1990). Labour and delivery in mothers with asthma. *Eur. J. Obstet. Gynecol. Reprod. Biol.* **35,** 183–190.

20. "Physician's Desk Reference," 52nd edition (1998). Medical Economics Company, Montvale, NJ.

20a. British Medical Association and the Royal Pharmaceutical Society of Great Britain. (1998). "British National Formulary 36." BMJ Books, London.

20b. Schatz, M., Zeiger, R. S., Harden, K., Hoffman, C. C., Chilingar, L., and Petitti, D. (1997). The safety of asthma and allergy medications during pregnancy. *J. All. Clin. Immunol.* **100,** 301–306.

20c. Stenius-Aarniala, B., Hedman, J., and Teramo, K. (1996). Acute asthma during pregnancy *Thorax* **51,** 411–414.

20d. Greenberger, P. A., and Patterson, R. (1983). Beclomethasone for severe asthma during pregnancy. *Ann. Int. Med.* **98,** 478–480.

20e. Kallen, B., Rydhstroem, H., and Anerg, A. (1999). Congenital malformations after the use of inhaled budesonide in early pregnancy. *Ob. Gynecol.* **93**(3), 392–395.

21. Pearce, N., Beasley, R., Burgess, C., and Crane, J. (1998). "Asthma Epidemiology: Principles and Methods." Oxford University Press, New York.

22. Asher, I., Keil, U., Anderson, H. R., Beasley, R., Crane, J., Martinez, F., Mitchell, E. A., Pearce, N., Sibbald, B., Stewart, A. W., Strachan, D., Weiland, S. K., and Williams, H. C. (1995). International Study of Asthma and Allergies in Childhood (ISAAC): Rationale and methods. *Eur. Respir. J.* **8,** 483–491.

23. Burney, P. G. J., Luczynska, C., Chinn, S., and Jarvis, D. (1994). The European Community Respiratory Health Survey. *Eur. Respir. J.* **7,** 954–960.

24. Robertson, C., Heycock, E., Bishop, J., Nolan, T., Olinsky, A., and Phelan, P. D. (1991). Prevalence of asthma in Melbourne schoolchildren: Changes over 26 years. *Br. Med. J.* **302,** 1116–1118.

25. Peat, J. K., van den Berg, R. H., Green, W. F., Mellis, C. M., Leeder, S. R., and Woolcock, A. J. (1994). Changing prevalence of asthma in Australian children. *Br. Med. J.* **308,** 1591–1596.

26. Infante-Rivard, C., Sukia, S. E., Roberge, D., and Baumgarten, M. (1987). The changing frequency of childhood asthma. *J. Asthma* **24,** 283–288.

27. Haahtela, T., Lindholm, H., Bjorksten, F., Koskenvuo, K., and Laitinen, L. A. (1990). Prevalence of asthma in Finnish young men. *Br. Med. J.* **301,** 266–268.

28. Perdrizet, S., Neukirch, F., Cooreman, J., and Liard, R. (1987). Prevalence of asthma in adolescents in various parts of France and its relationship to respiratory allergic manifestations. *Chest* **91,** 104S–106S.

29. Auerbach, I., Springer, C., and Godfrey, S. (1993). Total population survey of the frequency and severity of asthma in 17 year old boys in an urban area in Israel. *Thorax* **48,** 139–141.

30. Ciprandi, G., Vizzacaro, A., Cirillo, I., Crimi, P., and Canonica, G. W. (1996). Increase of asthma and allergic rhinitis prevalence in young Italian men. *Int. Arch. Allergy Immunol.* **111**(3), 278–283.

31. Nishima, S. (1993). A study of the prevalence of bronchial asthma in school children in western districts of Japan: Comparison between the studies in 1982 and in 1992 with the same methods and same districts. *Arerugi* **42,** 192–204.

32. Tirimanna, P. R., van Schayck, C. P., den Otter, J. J., van Weel, C., van Herwaarden, C. L., vanden Boom, G., van Grunsven, P. M., and van den Bosch, W. J. (1996). Prevalence of asthma and COPD in general practice in 1992: Has it changed since 1977? *Br. J. Gen. Pract.* **46,** 277–281.

33. Mitchell, E. A. (1983). Increasing prevalence of asthma in children. *N. Z. Med. J.* **96,** 463–464.

34. Shaw, R. A., Crane, J., O'Donnell, O'Donnell, T. V., Porteous, L. E., and Coleman, E. D. (1990). Increasing asthma prevalence in a rural New Zealand adolescent population: 1975–89. *Arch. Dis. Child.* **63,** 1319–1323.

35. Kljakovic, M. (1991). The change in prevalence of wheeze in 7 year old children over 19 years. *N.Z. Med. J.* **104,** 378–380.

36. Nystad, W., Magnus, P., Gulsvik, A., Skarpaas, I. J., and Carlsen, K. H. (1997). Changing prevalence of asthma in school children: Evidence for diagnostic changes in asthma in two surveys 13 yrs apart. *Eur. Respir. J.* **10,** 1046–1051.

37. Dowse, G. K., Turner, K. J., Stewart, G. A., Alpers, M. P., and Woolcock, A. J. (1985). The association between Dermatophagoides mites and the increasing prevalence of asthma in village communities within the Papua New Guinea highlands. *J. Allergy Clin. Immunol.* **75,** 75–83.

38. Alberg, N. (1989). Asthma and allergic rhinitis in Swedish conscripts. *Clin. Exp. Allergy* **19,** 59–63.

39. Liard, R., Chansin, R., Neukirch, F., Levallois, M., and Leproux, P. (1988). Prevalence of asthma among teenagers attending school in Tahiti. *J. Epidemiol. Commun. Health* **42,** 149–151.

40. Hsieh, K.-H., and Shen, J.-J. (1988). Prevalence of childhood asthma in Taipei, Taiwan and other Asian Pacific countries. *J. Asthma* **25,** 73–82.

41. Morrison Smith, J. (1976). The prevalence of asthma and wheezing in children. *Br. J. Dis. Chest* **70,** 73–77.

42. Whincup, P. H., Cook, D. P., Strachan, D. P., and Papacosta, O. (1993). Time trends in respiratory symptoms in childhood over a 24 year period. *Arch. Dis. Child.* **68,** 729–734.

43. Burney, P. G., Chinn, S., and Rona, R. J. (1990). Has the prevalence of asthma increased in children? Evidence from the national study of health and growth 1973–1986. *Br. Med. J.* **300,** 1306–1310.

44. Ninan, T. K., and Russell, G. (1992). Respiratory symptoms and atopy in Aberdeen schoolchildren: Evidence from two surveys 25 years apart. *Br. Med. J.* **304,** 873–875.

45. Burr, M. L., Butland, B. K., King, S., and Vaughan-Williams, E. (1989). Changes in asthma prevalence: Two surveys 15 years apart. *Arch. Dis. Child.* **64,** 1452–1456.

46. Gergen, P. J., Mullally, D. I., and Evans, R. (1988). National survey of prevalence of asthma among children in the United States, 1976 to 1980. *Pediatrics* **80,** 1–7.

47. Weitzman, M., Gortmaker, S. L., Sobol, A. M. (1992). Recent trends in the prevalence and severity of childhood asthma. *JAMA, J. Am. Med. Assoc.* **268,** 2673–2677.

48. ISAAC Steering Committee (Writing Committee: Beasley, R., Keil, U., von Mutius, E., and Pearce, N.) (1998). Worldwide variation in the prevalence of symptoms of asthma, allergic rhinoconjunctivitis and atopic eczema: The International Study of Asthma and Allergies in Childhood (ISAAC). *Lancet* **351,** 1225–1232.

49. Anderson, H. R., Pottier, A. C., and Strachan, D. P. (1992). Asthma from birth to age 23: Incidence and relation to prior and concurrent disease. *Thorax* **47,** 537–542.

50. Strachan, D. P., Butland, B. K., and Anderson, H. R. (1996). Incidence and prognosis of asthma and wheezing illness from early childhood to age 33 in a national British cohort. *Br. Med. J.* **312,** 1195–1199.

51. Dodge, R., and Burrows, B. (1980). The prevalence and incidence of asthma and asthma-like symptoms in a general population sample. *Am. Rev. Respir. Dis.* **122,** 567–575.

52. Yuninger, J. W., Reed, C. E., O'Connell, E. J., Melton, L. J., 3d., O'Fallon, W. M., and Silverstein, M. D. (1992). A community-based study of the epidemiology of asthma. *Am. Rev. Respir. Dis.* **146,** 888–894.

53. Montgomery, J., and Knowler, L. A. (1965). Epidemiology of asthma and allergic rhinitis. *Am. Rev. Respir. Dis.* **92,** 31–38.

54. Broder, I., Higgins, M. W., Matthews, K. P., and Keller, J. B. (1974). Epidemiology of asthma and allergic rhinitis in a total

community, Tecumseh, Michigan. IV. Natural history. *J. Allergy Clin. Immunol.* **54,** 100–110.

55. Schachter, J., and Higgins, M. W. (1976). Median age at onset of asthma and allergic rhinitis in Tecumseh, Michigan. *J. Allergy Clin. Immunol.* **57,** 342–351.

56. Balfour-Lynn, L. (1985). Childhood asthma and puberty. *Arch. Dis. Child.* **60,** 231–235.

57. Huovinen, E., Kaprio, J., Vesterinen, E., and Koskenvuo, M. (1997). Mortality of adults with asthma: A prospective cohort study. *Thorax* **52,** 49–54.

58. Horwood, L. J., Fergusson, D. M., and Shannon, F. T. (1985). Social and familial factors in the development of early childhood asthma. *Pediatrics* **75,** 859–868.

59. Schenker, M., Samet, J. M., and Speizer, F. E. (1983). Risk factors for childhood respiratory disease. *Am. Rev. Respir. Dis.* **126,** 1038–1043.

60. Gold, D. R., Wypij, D., Wang, X., Speizer, F., Pugh, M., Ware, J. H. Ferris, B. G., and Dockery, D. W. (1994). Gender- and race-specific effects of asthma and wheeze on level and growth of lung function in children in six U.S. cities. *J. Respir. Crit. Care Med.* **149,** 1198–1208.

61. Asher, M. I., Anderson, H. R., Stewart, A. W., Crane, J., Aitkhaled, N., Anabwani, G., Beasley, R., Bjorksten, B., Burr M., Clayton, T. O. et al. (1998). Worldwide variations in the prevalence of asthma symptoms: International Study of Asthma and Allergies in Childhood (ISAAC). *Eur. Respir. J.* **12,** 315–335.

62. Weiss, S. T., and Speizer, F. E. (1993). Epidemiology and natural history. *In* "Bronchial Asthma" (E. B. Weiss and M. Stein, eds.), pp. 15–25. Little, Brown, Boston.

63. Adams, P. F., and Benson, V. (1990). Current estimates from the National Health Interview Survey 1989. National Centre for Health Statistics. *Vital Health Stat. Ser.* **10,** No. 176.

64. Pearce, N., Weiland, S., Keil, U., Langridge, P., Anderson, H. R., Strachan, D., Bauman, A., Young, L., Gluyas, P., Ruffin, P. E., Crane, J., and Beasley, R. (1993). Self-reported prevalence of asthma symptoms in children in Australia, England, Germany and New Zealand: An international comparison using the ISAAC protocol. *Eur. Respir. J.* **6,** 1455–1461.

65. Lewis, S., Hales, S., Slater, T., Pearce, N., Crane, J., and Beasley, R. (1997). Geographical variation in the prevalence of asthma symptoms in New Zealand. *N. Z. Med. J.* **110,** 286–289.

66. European Community Respiratory Health Survey (ECRHS) (1996). Variations in the prevalence of respiratory symptoms, self-reported asthma attacks, and use of asthma medication in the European Community Respiratory Health Survey (ECRHS). *Eur. Respir. J.* **9,** 687–695.

67. Pearce, N., and Beasley, R. (1999). Measuring asthma morbidity. *Int. J. Tuberc. Lung Dis.* **3,** 185–191.

68. Gerstman, B. B. Bosco, L. A., and Tomita, D. K. (1993). Trends in the prevalence of asthma hospitalization in the 5–14 year-old Michigan Medicaid population, 1980 to 1986. *J. Allergy Clin. Immunol.* **91,** 838–843.

69. Skobeloff, E. M., Spivey, W. H., St. Clair, S., and Schoffstall, J. M. (1992). The influence of age and sex on asthma admissions. *JAMA, J. Am. Med. Assoc.* **268,** 3437–3440.

70. Hyndman, S. J., Williams, D. R. R., Merrill, S. L., Lipscombe, J. M., and Palmer, C. R. (1994). Rates of admission to hospital for asthma. *Br. Med. J.* **308,** 1596–1600.

71. Morris, R. D., and Munasinghe, R. L. (1994). Geographic variability in hospital admission rates for respiratory disease among the elderly in the United States. *Chest* **106,** 1172–1181.

72. Stempel, D. A., Hedblom, E. C., Durcanin-Robbins, J. P., and Sturn, L. L. (1996). Use of a pharmacy and medical claims database to document cost centers for 1993 annual asthma expenditures. *Arch. Fam. Med.* **5,** 36–40.

73. Kuhni, C. E., and Sennhauser, F. H. (1995). The Yentl Syndrome in childhood asthma: Risk factors for undertreatment in Swiss children. *Pediatr. Pulmonol.* **19,** 156–160.

74. Togias, A., Horowitz, E., Joyner, D., Guydon, L., and Malveaux, F. (1997). Evaluating the factors that relate to asthma severity in adolescents. *Int. Arch. Allergy Immunol.* **113,** 87–95.

75. Weiss, K. B., and Wagener, D. K. (1990). Changing patterns of asthma mortality: Identifying target populations at high risk. *JAMA, J. Am. Med. Assoc.* **264,** 1683–1687.

76. Prescott, E., Lange, P., Vestbo, J., and the Copenhagen City Heart Study Group (1997). Effect of gender on hospital admissions for asthma and prevalence of self-reported asthma: A prospective study based on a sample of the general population. *Thorax* **52,** 287–289.

77. Burney, P. G. J. (1986). Asthma mortality in England and Wales: Evidence for a further increase (1974–84). *Lancet* **2,** 323–326.

78. Foucard, T., and Graff-Lonnevig, V. (1994). Asthma mortality rate in Swedish children and young adults 1973–1988. *Allergy* **49,** 616–619.

79. British Thoracic Association (BTA) (1984). BTA Research Committee. Accuracy of death certificates in bronchial asthma. *Thorax* **39,** 505–509.

80. Hunt, L. W., Silverstein, M. D., Reed, C. E., O'Connell, E. J., O'Fallon, W. M., and Yunginger, J. W. (1993). Accuracy of the death certificate in a population-based study of asthmatic patients. *JAMA, J. Am. Med. Assoc.* **269,** 1947–1952.

81. Arrighi, H. M. (1995). U.S. asthma mortality: 1941–1989. *Ann. Allergy Asthma Immunol.* **74,** 321–326.

82. Robertson, C. F., Rubinfeld, A. R., and Bowes, G. (1990). Deaths from asthma in Victoria: A 12 month survey. *Med. J. Aust.* **152,** 511–517.

83. Sears, M. R., Rea, H. H., Beaglehole, R., Gillies, A. J., Holst, P. E., O'Donnell, T. V., Rothwell, R. P., and Sutherland, D. C. (1985). Asthma mortality in New Zealand: A two year national study. *N.Z. Med. J.* **98,** 271–275.

84. Sears, M. R., Rea, H. H., Rothwell, R. P. G., O'Donnell, T. V., Holst, P. E., Gillies, A. J., and Beaglehole, R. (1986). Asthma mortality: Comparison between New Zealand and England. *Br. Med. J.* **293,** 1342–1345.

85. Speizer, F. E., and Doll, R. (1968). A century of asthma deaths in young people. *Br. Med. J.* **3,** 245–246.

86. Burney, P. G. J. (1988). Asthma death in England and Wales 1931–85: Evidence for a true increase in asthma mortality. *J. Epidemiol. Commun. Health* **42,** 316–320.

87. Ehrlich, R., and Bourne, D. (1991). Asthma deaths in South Africa. *Curr. Allergy* **5,** 11–13.

88. Sly, R. M., and O'Donnell, R. (1997). Stabilization of asthma mortality. *Ann. Allergy Asthma Immunol.* **78,** 347–354.

89. Shaw, R., Woodman, K., Crane, J., Moyes, C., Kennedy, J., and Pearce, N. (1994). Risk factors for asthma in Kawerau children. *N.Z. Med. J.* **107,** 387–391.

90. Aberg, N. (1989). Birth season variation in asthma and allergic rhinitis. *Clin. Exp. Allergy* **19,** 643–648.

91. Cunningham, J., Dockery, D. W., and Speizer, F. E. (1996). Race, asthma and persistent wheeze in Philadelphia schoolchildren. *Am. J. Public Health* **86,** 1406–1409.

92. Aberg, N., Enstrom, I., and Lindberg, U. (1989). Allergic diseases in Swedish school children. *Acta Paediatr. Scand.* **78,** 246–252.

93. Mitchell, R. G., and Dawson, B. (1973). Educational and social characteristics of children with asthma. *Arch. Dis. Child.* **48,** 467–471.

94. Littlejohns, P., Ebrahim, S., and Anderson, H. R. (1989). The prevalence and diagnosis of chronic respiratory symptoms in adults. *Br. Med. J.* **298,** 1556–1560.

95. Littlejohns, P., and MacDonald, L. D. (1993). The relationship between severe asthma and social class. *Respir. Med.* **87,** 139–143.

96. Mielke, A., Reitmeir, P., and Wjst, M. (1996). Severity of childhood asthma by socioeconomic status. *Int. J. Epidemiol.* **25,** 388–393.

97. Pomare, E., Tutengaehe, H., Ramsden, I., Hight, M., Pearce, N., and Ormsby, V. (1992). Asthma in Maori people. *N.Z. Med. J.* **105,** 469–470.

98. Watson, J. P., Cowen, P., and Lewis, R. A. (1996). The relationships between asthma admission rates, routes of admission, and socioeconomic deprivation. *Eur. Respir. J.* **9,** 2087–2093.

99. Steinberg, M., and Becklake, M. R. (1986). Socio-environmental factors and lung function: A review of the literature. *S. Afr. Med. J.* **70,** 270–274.

100. Panhuysen, C. I. M., Meyers, D. A., Postma, D. S., and Bleecker, E. R. (1995). The genetics of asthma and atopy. *Allergy* **50,** 863–869.

101. Doull, I. J. M., Lawrence, S., Watson, M., Begishvili, T., Beasley, R. W., Lampe, F., Holgate, T., and Morton, N. E. (1996). Allelic association of gene markers on chromosomes 5q and 11q with atopy and bronchial hyperresponsiveness. *Am. Rev. Respir. Crit. Care Med.* **153,** 1280–1284.

102. Cullinan, P., and Newman Taylor, A. J. (1994). Asthma in children: Environmental factors. *Br. Med. J.* **308,** 1585–1586.

103. von Mutius, E., and Nicolai, T. (1996). Familial aggregation of asthma in a South Bavarian population. *Am. J. Respir. Crit. Care Med.* **153,** 1266–1272.

104. Sandford, A., Weir, T., and Paré, P. (1996). The genetics of asthma. *Am. Rev. Respir. Crit. Care Med.* **153,** 1749–1765.

105. Smith, J. M., and Knowles, L. A. (1967). Epidemiology of asthma and allergic rhinitis. II. In a university-centred community. *Am. Rev. Respir. Dis.* **92,** 31–38.

106. Edfors-Lubs, M. L. (1971). Allergy in 7000 twin pairs. *Acta Allergol.* **26,** 249–285.

107. Godfrey, K. M., Barker, D. J. P., and Osmond, C. (1994). Disproportionate fetal growth and raised IgE concentration in adult life. *Clin. Exp. Allergy* **24,** 641–648.

108. Barker, D. J. P. (1990). "Fetal and Infant Origins of Adult Disease" [editorial]. *BMJ* **301**(6761), 1111.

109. Fergusson, D., Crane, J., Beasley, R., and Horwood, L. J. (1997). Perinatal factors and atopic disease in childhood. *Clin. Exp. Allergy* **27,** 1394–1401.

110. Chike-Obi, U., David, R. J., Coutinho, R., and Wu, S.-Y. (1996). Birth weight has increased over a generation. *Am. J. Epidemiol.* **144,** 563–569.

111. Shaheen, S. O. (1995). Changing patterns of childhood infection and the rise in allergic disease. *Clin. Exp. Allergy* **25,** 1034–1037.

112. Cookson, W. O. C. M., and Moffatt, M. F. (1997). Asthma: An epidemic in the absence of infection? *Science* **275,** 41–42.

113. Holt, P. G., and Sly, P. D. (1997). Allergic respiratory disease: Strategic targets for primary prevention during childhood. *Thorax* **52,** 1–4.

114. Sigurs, N., Bjarnason, R., Sigurbergsson, F., Kjellman, B., and Bjorksten, B. (1995). Asthma and immunoglobulin E antibodies after respiratory syncytial virus bronchiolitis: A prospective cohort study with matched controls. *Pediatrics* **95,** 500–505.

115. von Mutius, E., Martinez, F. D., Fritzsch, C., Nicolai, T., Reitmeir, P., and Thiemann, H. H. (1994). Skin test reactivity and number of siblings. *Br. Med. J.* **309,** 692–695.

116. Strachan, D. P. (1989). Hay fever, hygiene, and household size. *Br. Med. J.* **299,** 1259–1260.

117. Moyes, C. D., Waldon, J., Ramadas, D., Crane, J., and Pearce, N. (1995). Respiratory symptoms and environmental factors in schoolchildren in the Bay of Plenty. *N.Z. Med. J.* **108,** 358–361.

118. Martinez, F. (1994). Role of viral infections in the inception of asthma and allergies during childhood: Could they be protective? *Thorax* **49,** 1189–1191.

119. Szentivanyi, A. (1968). The beta adrenergic theory of the atopic abnormality in bronchial asthma. *J. Allergy* **42,** 203–232.

120. Odelram, H., Ganstrom, M., Hedenskog, K., and Bjorksten, B. (1994). Immunoglobulin E and G responses to pertussis toxin after booster immunisation in relation to atopy, local reactions and aluminum content of the vaccines. *Pediatr. Allergy Immunol.* **5,** 118–123.

121. Odent, M. R., Culpin, E. E., and Kimmel, T. (1994). Pertussis vaccination and asthma: Is there a link? *JAMA, J. Am. Med. Assoc.* **272,** 592–593.

122. Kemp, T., Pearce, N., Fitzharris, P., Crane, J., Fergusson, D., St. George, I., Wickens, K., and Beasley, R. (1997). Is infant immunization a risk factor for childhood asthma and allergy? *Epidemiology* **8,** 678–680.

123. Burney, P. G. J., Anderson, H. R., and Burrows, B. (1989). Epidemiology. *In* "The role of inflammatory processes in airway hyperresponsiveness" (S. T. Holgate, J. B. L. Howell, P. G. J. Burney *et al.,* eds.), pp. 222–250. Blackwell, Oxford.

124. Sunyer, J., Antó, J. M., Castellsague, J., Soriano, J. B., and Roca, J. (1996). Total serum IgE is associated with asthma independently of specific IgE levels. *Eur. Respir. J.* **9,** 1880–1884.

125. Croner, S., Kjellman, N. I. M., Eriksson, B., and Roth, A. (1982). IgE screening in 1701 newborn infants and the development of atopic disease during infancy. *Arch. Dis. Child.* **57,** 364–368.

126. Weeke, E. (1992). Epidemiology of allergic diseases in children. *Rhinology* **30**(Suppl. 13), 5–12.

127. Barbee, R. A., Halonen, M., Lebowitz, M., and Burrows, B. (1981). Distribution of IgE in a community population sample: Correlations with age, sex and allergen skin test reactivity. *J. Allergy Clin. Immunol.* **68,** 106–111.

128. Oryszczyn, M.-P., Annesi, I., Neukirch, F., Dore, M. F., and Kauffman, F. (1991). Relationships of total IgE level, skin prick test response, and smoking habits. *Ann. Allergy* **67,** 355–358.

129. Barbee, R. A., Lebowitz, M. D., Thompson, H. C., and Burrows, B. (1976). Immediate skin test reactivity in a general population sample. *Ann. Intern. Med.* **84,** 129–133.

130. Burrows, B., Lebowitz, M. D., and Barbee, R. A. (1976). Respiratory disorders and allergy skin-test reactions. *Ann. Intern. Med.* **84,** 134–139.

131. Pearlman, D. S. (1984). Bronchial asthma: A perspective from childhood to adulthood. *Am. J. Dis. Chest* **138,** 459–466.

132. Pearce, N., Pekkanen, J., and Beasley, R. (1999) How much asthma is really attributable to atopy? *Thorax* **54,** 268–272.

133. Environmental Protection Agency (EPA) (1983). "Respiratory Health Effects of Passive Smoking: Lung Cancer and Other Disorders." EPA, Washington, D.C.

134. Department of Health and Human Services (DHHS) (1986). "The Health Consequences of Involuntary Smoking: A Report of the Surgeon General." DHHS, Washington, DC.

135. National Research Council (NRC) (1986). "Environmental Tobacco Smoke: Measuring Exposures and Assessing Health Effects." National Academy Press, Washington, DC.

136. Chen, Y., Rennie, D. C., and Dosman, J. A. (1996). Influence of environmental tobacco smoke on asthma in nonallergic and allergic children. *Epidemiology* **7,** 536–539.

137. Venables, K., Topping, M., Howe, W., Luczynska, C. M., Hawkins, R., and Taylor, A. J. (1985). Interaction of smoking, and atopy in producing IgE antibody against hapten protein conjugate. *Br. Med. J.* **290,** 201–204.

138. Neas, L., Dockery, D., Ware, J., Spengler, J. D., Speizer, F. E., and Ferris, B. G. Jr. (1991). Association of indoor nitrogen dioxide with respiratory symptoms and pulmonary function in children. *Am. J. Epidemiol.* **134,** 204–219.

139. Koo, L. C., and Ho, J. H.-C. (1994). Mosquito coil smoke and respiratory health among Hong Kong Chinese: Results of three epidemiological studies. *Indoor Environ.* **3,** 304–310.

140. Marbury, M., and Kriegler, R. (1991). Formaldehyde. *In* "Indoor Air Pollution: A Health Perspective" (J. Samet and J. D. Spengler, eds.), pp. 223–251. Johns Hopkins University Press, London.

141. Osbourne, J., and Honicky, R. (1986). Chronic respiratory symptoms in young children and indoor heating with a wood burning stove. *Am. Rev. Respir. Dis.* **133**, 300 (abstr.).

142. Anderson, H. R. (1974). The epidemiological and allergic features of asthma in the New Guinea Highlands. *Clin. Allergy* **4**, 171–183.

143. Andrae, S., Axelson, O., Bjorksten, B., Frederiksson, M., and Kjellman, N. I. (1988). Symptoms of bronchial hyperreactivity and asthma in relation to environmental factors. *Arch. Dis. Child.* **63**, 473–478.

144. Hoffman, R. E., Wood, R. C., and Kreiss, K. (1993). Building-related asthma in Denver office workers. *Am. J. Public Health* **83**, 89–93.

145. Verhoeff, A. P., Van Strien, R. T., Van Wijnen, J. H., and Brunekreef, B. (1995). Damp housing and childhood respiratory symptoms: The role of sensitization to dust mites and molds. *Am. J. Epidemiol.* **141**, 103–110.

146. Strachan, D. P., and Sanders, C. H. (1989). Damp housing and childhood asthma: Respiratory effects of indoor air temperature and relative humidity. *J. Epidemiol. Commun. Health* **43**, 7–14.

147. Voorhorst, T., Spieksma, F. M., and Varekamp, H. (1967). The house dust mite (*Dermatophagoides pteronyssinus*) and the allergens it produces: Identity with the house dust mite allergen. *J. Allergy* **39**, 325–339.

148. Platts-Mills, T., Tovey, E., Mitchell, E., and Mozarro, H. (1982). Reduction of bronchial hyperreactivity during prolonged allergen avoidance. *Lancet* **2**, 675–678.

149. Brunekreef, B., Groot, B., and Hoek, G. (1992). Pets, allergy and respiratory symptoms in children. *Int. J. Epidemiol.* **21**, 338–342.

150. Vanto, T., and Koivikko, A. (1983). Dog hypersensitivity in asthmatic children. *Acta Paediatr. Scand.* **72**, 571–575.

151. Rosenstreich, D. L., Eggleston, P., Kattan, M., Baker, D., Slavin, R. G., Gergen, P., Mitchell, H., McNiff-Mortimer, K., Lynn, H., Ownby, D., and Malveaux, F. (1997). The role of cockroach allergy and exposure to cockroach allergen in causing morbidity among inner-city children with asthma. *N. Engl. J. Med.* **336**, 1382–1384.

152. Sporik, R., Holgate, T., Platts-Mills, T., and Cogswell, J. J. (1990). Exposure to house-dust mite allergen (Der p I) and the development of asthma in childhood. *N. Engl. J. Med.* **323**, 502–507.

153. Arshad, S. H., Matthews, S., Grant, C., and Hide, D. W. (1992). Effect of allergen avoidance on development of allergic disorders in infancy. *Lancet* **339**, 1493–1497.

154. Martinez, F. D. (1997). Complexities of the genetics of asthma. *Am. J. Respir. Care Med.* **156**, S117–S122.

155. von Mutius, E. (1996). Progression of allergy and asthma through childhood to adolescence. *Thorax* **51** (Suppl. 1), S3–S6.

156. Barnes, P. J. (1994). Air pollution and asthma. *Postgrad. Med. J.* **70**, 319–325.

157. Duhme, H., Weiland, S. K., Keil, U., Kraemer, B., Schmid, M., Stender, M., and Chambless, L. (1996). The association between self-reported symptoms of asthma and allergic rhinitis and self-reported traffic density on street of residence in adolescents. *Epidemiology* **7**, 578–582.

158. Pershagen, G., Rylander, E., Norberg, S., Eriksson, M., and Nordvall, S. L. (1995). Air pollution involving nitrogen dioxin exposure and wheezing bronchitis in children. *Int. J. Epidemiol.* **24**, 1147–1153.

159. Oosterlee, A., Drijver, M., Lebret, E., and Brunekreef, B. (1995). Chronic respiratory symptoms in children and adults living along streets with high traffic density. *Occup. Environ. Med.* **53**, 241–247.

160. Livingstone, A. E., Shaddick, G., Grundy, C., and Elliott, P. (1996). Do people living near inner city main roads have more asthma needing treatment? *Br. Med. J.* **312**, 676–677.

161. Molfino, N., Wright S. C., Katz, I., Tarlo, S., Silverman, F., McClean, P. A., Szalai, J. P., Raizenne, M., Slutsky, A. S., and Zamel, N. (1991). Effects of low concentrations of ozone on inhaled allergen responses in asthmatic subjects. *Lancet* **338**, 221–222.

162. Emberlin, J. (1995). Interaction between air pollutants and aeroallergens. *Clin. Exp. Allergy* **25** (Suppl. 3), 33–39.

163. Kino, T., Chihara, J., Fukuda, K., Sasaki, Y., Shogaki, Y., and Oshima, S. (1987). Allergy to insects in Japan. III: High frequency of IgE antibody responses to insects (Moth, butterfly, caddis fly, and chironomid) in patients with bronchial asthma and immunochemical quantitation of the insect-related airborne particles smaller than 10 μm in diameter. *J. Allergy Clin. Immunol.* **79**, 857–866.

164. Packe, G. E., and Ayres, J. G. (1985). Asthma outbreak during a thunderstorm. *Lancet* **2**, 199–204.

165. Egan, P. (1985). Weather or not. *Med. J. Aust.* **142**, 330.

166. Celenza, A., Fothergill, J., Kupek, E., and Shaw, R. J. (1996). Thunderstorm associated asthma: A detailed analysis of environmental factors. *Br. Med. J.* **312**, 604–607.

167. Packe, G. E., and Ayres, J. G. (1986). Aeroallergen skin sensitivity in patients with severe asthma during a thunderstorm. *Lancet* **1**, 850–851.

168. Suphioglu, C., Singh, M. B., Taylor, P. et al. (1992). Mechanism of grass-pollen-induced asthma. *Lancet* **339**, 569–572.

169. Antó, J. M., Sunyer, J., Rodriguez-Roisin, R., Suarez-Cerrera, M., and Vazquez, L. (1989). Community outbreaks of asthma associated with inhalation of soybean dust. *N. Engl. J. Med.* **320**, 1097–1102.

170. Navarro, C., Marquez, M., Hernando, L., Galvan, F., Zapatero, L., and Caravaca, F. (1993). Epidemic asthma in Cartagena, Spain and its association with soybean sensitivity. *Epidemiology* **4**, 76–79.

171. White, M. C., Etzel, R. A., and Olson, D. R. (1993). Reexamination of epidemic asthma in New Orleans in relation to the presence of soy-carrying ships in the harbor. *Am. J. Epidemiol.* **138**, 607.

172. Figley, K. D., and Elrod, R. H. (1928). Endemic asthma due to castor bean dust. *JAMA, J. Am. Med. Assoc.* **90**, 79–82.

173. Mendes, E., and Cintra, U. (1954). Collective asthma, simulating an epidemic, provoked by castor-bean dust. *J. Allergy* **25**, 253–259.

174. Ordman, D. (1955). An outbreak of bronchial asthma in South Africa, affecting more than 200 persons, caused by castor bean dust from an oil-processing factory. *Int. Arch. Allergy* **7**, 10–24.

175. Burge, P. G. S. (1995). Occupational asthma. *In* "Respiratory Medicine" (R. A. L. Brewis, B. Corrin, G. M. Geddes, and G. J. Gibson, eds.), 2nd ed., pp. 1262–1280. Saunders, London.

176. Chan-Yeung, M., and Lam, S. (1986). Occupational asthma. *Am. Rev. Respir. Dis.* **133**, 686–703.

177. Chan-Yeung, M. (1995). Assessment of asthma in the workplace. *Chest* **108**, 1084–1117.

178. Meredith, S. K., Taylor, V. M., and McDonald, J. C. (1991). Occupational respiratory disease in the United Kingdom 1989: A report to the British Thoracic Society and the Society of Occupational Medicine by the SWORD project group. *Br. J. Ind. Med.* **48**, 292–298.

179. Chan-Yeung, M., and Malo, J.-L. (1994). Epidemiology of occupational asthma. *In* "Asthma and Rhinitis" (W. Busse and S. Holgate, eds.), pp. 44–57. Blackwell, Oxford.

180. Kogevinas, M., Antó, J. M., Soriano, J. B., Tobias, A., and Burney, P. (1996). The risk of asthma attributable to occupational exposures: A population-based study in Spain. *Am. J. Respir. Crit. Care Med.* **154**, 137–143.

181. Fishwick, D., Pearce, N., D'Souza, W., Lewis, S., Town, I., Armstrong, R., Kogevinas, M., and Crane, J. (1997). Occupational asthma in New Zealanders: A population-based study. *Occup. Environ. Med.* **54**, 301–306.

182. Meredith, S., and Nordman, H. (1996). Occupational asthma: Measures of frequency from four countries. *Thorax* **51**, 435–440.

CHAPTER 56. ASTHMA 739

183. Burney, P. G. J. (1995). Asthma: Epidemiology. *In* "Respiratory Medicine" (R. A. L. Brewis, B. Corrin, D. M. Geddes, and G. J. Gibson, eds.), 2nd ed., pp. 1098–1107. Saunders, London.

184. Burney, P. G. J. (1987). A diet rich in sodium may potentiate asthma. *Chest* **91**, 143S–148S.

185. Flatt, A., Pearce, N. E., Thomson, C. D., Sears, M. R., Robinson, M. F., and Beasley R. (1990). Reduced selenium status in asthmatic subjects in New Zealand. *Thorax* **45**, 95–99.

186. Beasley, R., Thomson, C., and Pearce, N. E. (1991). Selenium, glutathione peroxidase and asthma. *Clin. Exp. Allergy* **21**, 157–159.

187. Weiss, S. T. (1997). Diet as a risk factor for asthma. *In* "The Rising Trends in Asthma" (Ciba Foundation, Symposium 206, S. T. Holgate, chair), pp. 245–257. Wiley, West Sussex.

188. Seaton, A., Godden, D. J., and Brown, K. (1994). Increase in asthma: A more toxic environment or a more susceptible population? *Thorax* **49**, 171–174.

189. Soutar, A., Seaton, A., and Brown, K. (1997). Bronchial reactivity and dietary antioxidants. *Thorax* **52**, 166–70.

190. Troisi, R. J., Willett, W. C., Weiss, S. T., Trichopoulos, D., Rosner, B., and Speizer, F. E. (1995). A prospective study of diet and adult-onset asthma. *Am. J. Respir. Crit. Care Med.* **151**, 1401–1408.

191. Schwartz, J., and Weiss, S. T. (1990). Dietary factors and their relation to respiratory symptoms. *Am. J. Epidemiol.* **132**, 67–76.

192. Miedema, I., Feskens, E. J. M., Heederik, D., and Kromhout, D. (1993). Dietary determinants of long-term incidence of chronic nonspecific lung diseases. *Am. J. Epidemiol.* **138**, 37–45.

193. Britton, J. R., Pavord, I. D., and Richards, K. A. (1995). Dietary antioxidant vitamin intake and lung function in the general population. *Am. J. Respir. Crit. Care Med.* **151**, 1383–1387.

194. Hodge, L., Peat, J. K., and Salome, C. (1994). Increased consumption of polyunsaturated oils may be a cause of increased prevalence of childhood asthma. *Aust. N.Z. Med. J.* **24**, 727.

195. Black, P. N., and Sharpe, S. (1997). Dietary fat and asthma: Is there a connection? *Eur. Respir. J.* **10**, 1–7.

196. Fernandes, G. (1994). Dietary lipids and risk of autoimmune disease. *Clin. Immunol. Immunopathol.* **72**, 193–197.

197. Goldberg, G. R., and Prentice, A. M. (1994). Maternal and fetal determinants of adult diseases. *Nutr. Rev.* **52**, 191–200.

198. Langley-Evans, S. (1997). Fetal programming of immune function and respiratory disease. *Clin. Exp. Allergy* **27**, 1377–1379.

199. Rayon, J. I., Carver, J. D., Wyble, L. E., Weiner, D., Dickey, S. S., Benford, V. J., Chen, L. T., and Lim, D. V. (1997). The fatty acid composition of maternal diet affects lung prostaglandin E2 levels and survival from group B streptococcal sepsis in neonatal rat pups. *J. Nutr.* **127**, 1989–1982.

200. Dry, J., and Vincent, D. (1991). Effect of fish oil diet on asthma: Results of a 1-year double-blind study. *Int. Arch. Appl. Immunol.* **95**, 156–157.

201. Hodge, L., Salome, C. M., Peat, J. K., Haby, M. M., Xuan, W., and Woolcock, A. J. (1996). Consumption of oily fish and childhood asthma risk. *Med. J. Aust.* **164**, 137–140.

202. Wilson, A. C., Forsyth, J. S., Greene, S. A., Irvine, L., Hau, C., and Howie, P. W. (1998). Relation of infant diet to childhood health: Seven year follow up of cohort of children in Dundee infant feeding study. *Br. Med. J.* **316**, 21–25.

203. Fergusson, D. M., Horwood, L. J., and Shannon, F. T. (1983). Asthma and infant diet. *Arch. Dis. Child.* **58**, 48–51.

204. Pullan, C. R., and Hey, E. N. (1982). Wheezing, asthma, and pulmonary dysfunction 10 years after infection with respiratory syncytial virus in infancy. *Br. Med. J.* **284**, 1665–1669.

205. Stenius-Aarnala, B. (1987). The role of infection in asthma. *Chest* **91**, 157S–160S.

206. Welliver, R. C., Wong, D. T., Sun, M., Middleton, E., Jr., Vaughan, R. S., and Ogra, P. L. (1981). The development of respiratory syncytial virus-specific IgE and the release of histamine in nasopharyngeal secretions after infection. *N. Engl. J. Med.* **305**, 841–846.

207. Beasley, R., Coleman, E. D., Hermon, Y., Holst, P. E., O'Donnell, T. V., and Tobias M. (1988). Viral respiratory tract infection and exacerbations of asthma in adult patients. *Thorax* **43**, 679–683.

208. Burney, P. G. J. (1992). Epidemiology. *Br. Med. Bull.* **48**, 10–22.

209. Cogswell, J. J., Halliday, D. F., and Alexander, J. R. (1982). Respiratory infections in the first year of life in children at risk of developing atopy. *Br. Med. J.* **284**, 1011–1013.

210. Price, J. F. (1996). Issues in adolescent asthma: What are the needs? *Thorax* **51**(Suppl. 1), S13–S17.

211. Sweeting, H. (1995). Reversals of fortune? Sex differences in health in childhood and adolescence. *Soc. Sci. Med.* **40**, 77–90.

212. Silverglade, L., Tosi, D. J., Wise, P. S., and D-Costa, A. (1994). Irrational beliefs and emotionality in adolescents with and without bronchial asthma. *J. Gen. Psychol.* **121**, 199–207.

213. Martinez, F., Morgan, W. J., Wright, A. L., Holberg, C. J., and Taussig, L. M. (1988). Diminished lung function as a predisposing factor for wheezing respiratory illness in infants. *N. Engl. J. Med.* **319**, 1112–1117.

214. Gold, D. R., Rotnitzky, A., Damokosh, A. I., Ware, J. H., Speizer, F. E., Ferris, B. G., Jr., and Dockery, D. W. (1993). Race and gender differences in respiratory illness prevalence and their relationship to environmental exposures in children 7 to 14 years of age. *Am. Rev. Respir. Dis.* **148**, 10–18.

215. Gregg, I. (1983). Epidemiology. *In* "Asthma" (T. J. H. Clark and S. Godfrey, eds.), p. 230. Chapman & Hall, London.

216. Siebers, R., Patchett, K., Fitzharris, P., and Crane, J. (1996). Mite allergen (Der p 1) on children's clothing. *J. Allergy Clin. Immunol.* **98**, 853–854.

217. Patchett, K., Lewis, S., Crane, J., and Fitzharris, P. (1997). Cat allergen (Fel d 1) levels on school children's clothing and in primary classrooms in Wellington, New Zealand. *J. Allergy Clin. Immunol.* **100**, 755–759.

218. Malveaux, F. J., and Fletcher-Vincent, S. A. (1995). Environmental risk factors of childhood asthma in urban centers. *Environ. Health Perspect.* **103** (Suppl. 6), 59–62.

219. Kuehr, J., Frischer, T., Karmaus, W., Meinert, R., Barth R., Herrman-Kunz, E., Forster, J., and Urbanek, R. (1992). Early childhood risk factors for sensitisation at school age. *J. Allergy Clin. Immunol.* **90**, 358–363.

220. Boulet, L. P., Turcotte, H., Laprise, C., Lavertu, C., Bedard, P. M., Laroie, A., and Hebert, J. (1997). Comparative degree and type of sensitization to common indoor and outdoor allergens in subjects with allergic rhinitis and/or asthma. *Clin. Exp. Allergy* **27**, 52–59.

221. Gergen, P. J., Turkeltaub, P. C., and Kovar, M. G. (1987). A genetic-epidemiologic study of allergic skin test reactivity to eight common areoallergens in the U.S. population: Results from the National Health and Nutrition Examination Survey. *J. Allergy Clin. Immunol.* **80**, 669–679.

222. Dodge, R., Cline, M., and Burrows, B. (1986). Comparisons of asthma, emphysema, and chronic bronchitis diagnoses in a general population sample. *Am. Rev. Respir. Dis.* **133**, 981–986.

223. Redline, S., and Gold, D. (1994). Challenges in interpreting gender differences in asthma. *Am. J. Respir. Crit. Care Med.* **150**, 1219–1221.

224. Rubio, R. L., Rodriguez, G. B., and Collazo, J. J. (1988). Comparative study of progesterone, estradiol, and cortisol concentrations in asthmatic and non-asthmatic women. *Allergol. Immunopathol.* **16**, 263–266.

57

Sjögren's Syndrome

ANN L. PARKE

University of Connecticut

Farmington, Connecticut

I. Introduction and Background

A. *Description as Detailed by Henrik Sjögren*

The term syndrome as opposed to disease generally is used in medical terminology when substantive facets of an illness (*i.e.* etiology and pathology) remain poorly defined. Such is the case with Sjögren's syndrome. Although attributed to Henrik Sjögren, a Swedish ophthalmologist who in 1930 described extreme dryness of the eyes and mouth in a female patient [1], the original observation of these associated clinical complaints was first described by W. B. Hadden [2] and J. Hutchinson [3], both in 1888. Henrik Sjögren subsequently went on to present a thesis describing 19 female patients with sicca (dryness) complaints, most of whom were peri- or postmenopausal and had arthritis [4]. It was Sjögren who first appreciated the systemic nature of this illness.

B. *Primary and Secondary Sjögren's Syndrome*

Sjögren's syndrome has been defined as a triad of complaints: keratoconjunctivitis sicca (dry eyes), xerostomia (dry mouth), and the presence of a well-defined connective tissue disease such as systemic lupus erythematosus (SLE), rheumatoid arthritis (RA), or systemic sclerosis (SSc). It is then known as secondary Sjögren's syndrome. Keratoconjunctivis sicca and xerostomia occuring in the absence of a well-defined connective tissue disease is known as primary Sjögren's syndrome. The autoantibody profiles of these two categories of Sjögren's syndrome are different and patients with primary Sjögren's syndrome are more likely to produce antibodies to extractable nuclear antigen (ENA) components Ro and La [5,6] As with other connective tissue diseases, this syndrome has a predilection for women, but it occurs more commonly in postmenopausal women, unlike SLE which occurs most frequently during the childbearing years [7,8]. Patients with other connective tissue diseases usually develop their Sjögren's complaints secondary to their initial disease. It is most unusual for a primary Sjögren's patient to develop SLE or RA after the initial Sjögren's complaints, although this pattern of disease expression has been described [9,10].

C. *Evolution since 1930*

As more has been learned about this syndrome it has become apparent that it is a systemic inflammatory autoimmune disease with a predilection for the exocrine glands. Consequently, Sjögren's syndrome has also been described as an autoimmune exocrinopathy, or an autoimmune epitheliitis [11,12], and the affected exocrine glands are very active immunological sites

(see Section II.C.2). Sometimes the pathological changes in these affected glands may become malignant and a small percentage of Sjögren's patients develop lymphomas (see section V.D.) A variety of clinical complaints may be found in these patients (see Section V) but this illness consistently produces dry eyes (keratoconjunctivitis sicca) and dry mouth (xerostomia). The autoantibody profile produced by individual patients may be useful in helping to predict the spectrum of clinical complaints that may develop (see Section II.C.3).

II. Definition of Sjögren's Syndrome

A. *Criteria*

There are several sets of criteria for defining Sjögren's syndrome but many are considered to be deficient by various researchers in this field. The European criteria, developed in 1993, have been tested and overall have been found to be 92% specific and 93% sensitive for Sjögren's syndrome [13].

The European criteria were established as the result of a multicenter study, where the study protocol was subdivided into two parts. The first part was to develop a questionnaire that assessed subjective complaints of dry eyes and dry mouth and that was demonstrated to be useful for differentiating between Sjögren's patients and normal controls. The second part used the results from the first part of the study to evaluate subjective complaints and these questions were supplemented by objective tests of dryness (both keratoconjunctivitis sicca and xerostomia), a lip biopsy of the minor salivary glands from the mouth, and autoantibody production (*i.e.,* antibodies to Ro and La, rheumatoid factor, and antinuclear antibodies (ANAs)). This gave six preliminary criteria for the assessment of Sjögren's syndrome. For primary Sjögren's syndrome it was found that if four of the six criteria were positive (using antibodies to Ro and La as the only positive laboratory test), then the sensitivity of these criteria was 93.5% and the specificity was 94%, an acceptable way to diagnose primary Sjögren's syndrome. For the patients with secondary Sjögren's syndrome it was felt that the sixth criteria (autoantibody production) should not be included, leaving just five criteria. If four of these five criteria were positive (with one of the four being subjective complaints of either dry eyes or dry mouth), then the specificity was 93.9% and the sensitivity 85.1% for diagnosing secondary Sjögren's syndrome [13].

Even so, debate continues about the number and nature of objective tests to be used [13–14]. The California criteria require a positive lip biopsy for the diagnosis of definite Sjögren's syndrome and also list exclusion criteria to rule out its diagnosis [14]. It is important to remember that there are several diseases that may produce histopathological changes that are similar to

Table 57.1

Name	Test	Abnormal result
Schirmer test	Tear production measured by wetting of filter paper strip	Less than 5 mm of wetting in unanesthetized eye
Rose Bengal stain	Tear production and conjunctival health	Residual staining and presence of filaments
Fluorescein	Tear production, tear stability, and corneal health	Residual staining, dark spots where tear film broken down
Slit lamp	Examination of the anterior segment of the eye	A check of the effects of Rose Bengal + Fluorescein staining
Salivary flow rate:		
1. Unstimulated	Basal salivary production	Less than 1.5 ml per 15 minutes
2. Stimulated modified Saxon test	Chewing on weighed 4×4 gauze swab for 2 minutes	Change in weight of gauze less than 2.75 grams
Gallium and technesium scans	Gallium tests glandular inflammation Technesium tests glandular function	Marked glandular uptake of gallium Delayed uptake and discharge of technesium

those found in Sjögren's syndrome (*i.e.,* sarcoidosis, HIV disease, chronic hepatitis C, and graft versus host disease). The California criteria are more specific and less sensitive than the European criteria, which have now been modified to require either a positive lip biopsy or antibodies to Ro or La for the diagnosis of primary Sjögren's syndrome [15].

B. Definition of Normal

Many patients with Sjögren's syndrome consider that they have had their disease for many years prior to confirmation of the diagnosis [16]. This may be due to the insidious nature of the postinflammatory dryness or to the fact that some patients have become so used to living with dryness that they have forgotten what it is like to have moist mucous membranes. This phenomenon of not being aware of dryness is more common in younger patients [17].

Dryness of the eyes and mouth may not come together; one may precede the other by several years and so the diagnosis is missed. Reports have shown that a predisposition to develop Sjögren's syndrome may be suspected years before the onset of sicca complaints because of the presence of autoantibodies to Ro and La. In some women the earliest manifestation of an autoimmune diathesis and a predisposition to develop Sjögren's syndrome may be an abnormal pregnancy, producing a child with what is now known as the Neonatal Lupus Syndrome (NLS). These children may present with a transient skin rash, low white counts (cytopenias), hepatosplenomegaly, heart failure, or the most serious complaint, primary congenital complete heart block (CCHB). Some of the mothers of infants with NLS may fullfill criteria for a well-defined connective tissue disease at the time of the abnormal birth, whereas other mothers may appear to be "normal" with no clinical complaints. Long-term follow-up studies of asymptomatic mothers who gave birth to infants with NLS have determined that some of these women go on to develop the clinical features of Sjögren's syndrome, often many years after the abnormal birth [18–20].

This "subclinical" disease blurs the distinction between what is normal and what is abnormal. The maternal autoantibodies are almost certainly responsible for the fetal and neonatal pathology that occurs in NLS [21,22]. However, evidence that these autoantibodies contribute to the pathological changes found in Sjögren's patients is tenuous, even though synthesis of Ro and La antibodies by salivary and lacrimal Sjögren's glands has been demonstrated [23,24]. Indeed, some Sjögren's patients with characteristic pathological changes in salivary gland biopsies do not appear to produce any autoantibodies, making it difficult to explain glandular pathology as a consequence of autoantibody production.

C. Diagnosis

1. Abnormal Glandular Function

They are many ways to evaluate tear and saliva production. Questions pertaining to simple subjective complaints have been evaluated and some have been found to be useful for distinguishing between patients with Sjögren's syndrome and patients with sicca complaints due to other causes [13]. Objective tests of exocrine secretion must be used when evaluating patients with complaints of dryness (Table 57.1). For example, an abnormal Schirmer test, defined as <5 mm of wetting in 5 minutes in an unanesthetized eye, appears to produce an appropriate balance between specificity (76.4%) and sensitivity (75.1%), whereas the Rose Bengal test appears to be more specific (84.8%) for the diagnosis of Sjögren's syndrome [13]. It is important, however, to define exactly how these tests are done. For example, some centers take less than 10 mm of wetting as normal, and the positioning of the filter paper (*i.e.,* at the junction of the middle and nasal thirds of the lower lid) is vital if the results are to be interpretable.

Measurement of stimulated versus nonstimulated saliva production has been the center of controversy for some time. Vitali *et al.* concluded that stimulated salivary flow measurements were unacceptable in terms of specificity and sensitivity for evaluating patients with Sjögren's syndrome [13]. Salivary gland scintography is preferable to sialography [13,25] because the instillation of dye into an abnormal gland may result in long term consequences (*i.e.,* granulomatous reaction to the dye used).

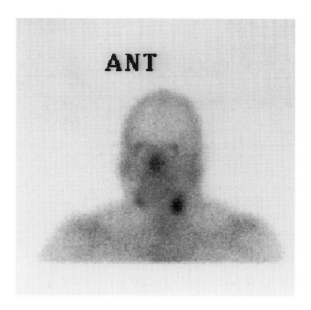 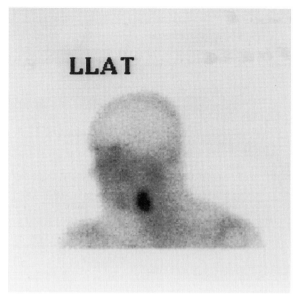

Fig. 57.1   (A) Anterior and (B) lateral views of a gallium scan showing increased uptake in the left submandibular gland of a patient with Sjögren's syndrome. Most of the right submandibular gland has been removed. Interestingly, in this patient, even though both parotid glands were swollen they did not show avidity for the gallium, but technesium scanning demonstrated poor function of both parotid and salivary glands.

Oil-based contrast medium should never be used. We prefer to use both gallium and technesium scanning for evaluation of the salivary component of Sjögren's syndrome. Gallium scanning assesses the component of glandular inflammation while a technesium scan gives information about glandular function; together they give a much more complete assessment of the nature of any glandular abnormalities (Fig. 57.1).

2. Pathology

The pathology of the affected glands is a round cell infiltrate that includes T-helper cells, B cells, and immunoglobulin-synthesizing plasma cells (Fig. 57.2) [26–28] This cellular infiltrate is focal and an aggregate of 50 or more cells is known as a focus. More than one focus of cells per 4 mm² of tissue is considered a positive biopsy [29]. A positive lip biopsy is a specific test for Sjögren's syndrome provided that: (1) the biopsy is not done in a patient with inflamed mucous membranes, and (2) the biopsy is read appropriately. Inspissated secretions can result in an inflammatory cell infiltrate and this infiltrate can be falsely interpreted as a positive biopsy [30]. Inflammatory cell infiltrates adjacent to ductal dilation, fibrosis, and atrophy can all be found in normal postmortem studies [31]. Overall, the relationship between glandular atrophy and fatty infiltration is unclear and these pathological features may be part of a separate entity or may represent end-stage postinflammatory Sjögren's syndrome [32].

Why lymphocytes infiltrate the exocrine glands, and whether this is a primary or secondary event, is open to question [33]. In addition to this cellular infiltrate, there is expression of HLA molecules, even on the ductal epithelial cells, suggesting that there is local presentation of antigen [34,35]. There is also local synthesis of immunoglobulin, including autoantibody production, making these affected glands very active immunological sites [23,28,36].

Pathological changes may be found in tissues other than exocrine glands. More than 70% of patients with primary biliary cirrhosis also have Sjögren's syndrome [37]. Chronic active hepatitis may also be found in Sjögren's syndrome patients and may be a consequence of chronic hepatitis C infection [38]. A lymphocytic interstitial pneumonitis can occur in Sjögren's patients [39], as can significant renal pathology [40] and damage to endocrine glands such as the thyroid and the pancreas [41,42]. Atrophic gastritis occurs in patients with Sjögren's syndrome [43], and patients with coeliac disease have an increased predisposition to developing autoimmune diseases including Sjögren's syndrome [44]. Vasculitis may cause end organ damage resulting in peripheral neuropathies, cutaneous lesions, and severe central nervous system disease [45]. One of the most significant pathological changes in these patients is the progression to pseudolymphoma and then to frank lymphoma [46,47] (see section VI. C) The most commonly occurring malignancy is non-Hodgkin's B-cell lymphoma. These tumors occur more frequently in patients with primary Sjögren's syndrome or in patients with glandular swelling and/or persistent lymphadenopathy [48,49]. The production of monoclonal immunoglobulins is an important step in the progression from polyclonal B cell hyperreactivity towards pseudolymphoma and eventually overt lymphoma. Any Sjögren's syndrome patient producing monoclonal immunoglobulins should be reexamined at regular intervals, at least once a year, to look for evidence confirming the progression to overt lymphoma. Persistently enlarged and firm glands must be watched carefully and biopsied if necessary.

3. Autoantibody Profile

The immunological aberration in Sjögren's syndrome involves both T cells and B cells. T-cell infiltration of exocrine glands and B-cell hyperreactivity result in autoantibody production

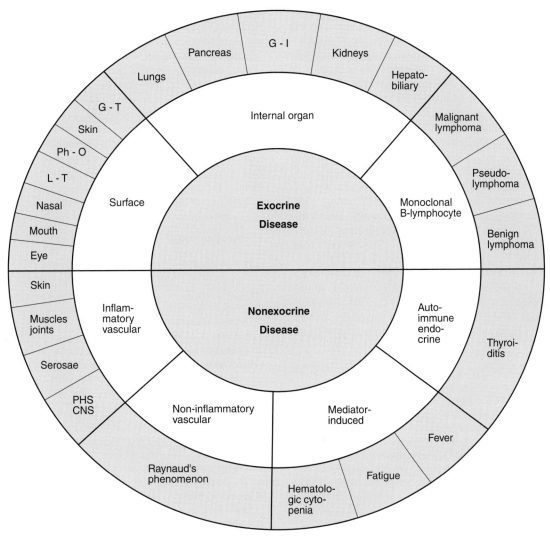

Fig. 57.2 The Sjögren's syndrome wheel was devised by Manthorpe *et al.* [16]. This figure has been published previously in *The Journal of Rheumatology* **24** (supplement 50), p. 9. Permission to reproduce this figure has been given by The Journal of Rheumatology and Professor Manthorpe.

which, in some cases, progresses to completely uncontrolled B-cell hyperreactivity and the development of B-cell lymphoma [26–28,46,47]. The autoantibody production in Sjögren's syndrome is quite nonspecific, except for antibody to the ENA component, La antigen, which seems to be quite specific for Sjögren's syndrome [50]. Antibodies to La are usually found with Ro antibodies but, even though Ro antibodies are a sensitive test for Sjögren's syndrome, they are not specific because Ro antibodies are also found in approximately 45% of SLE patients who do not have sicca complaints [50,51]. These Ro and La antibodies contribute to the antinuclear antibody (ANA) positivity that is found in approximately 80% of primary Sjögren's patients and they are responsible for a speckled or nucleolar ANA pattern. Antibodies to Ro and La antigens occur in 60–90% of Sjögren's syndrome patients depending on the methods used for the detection of the antibodies [50,52,53] and on the criteria used to define the presence of Sjögren's syndrome. Rheumatoid factors are found in 70–90% of primary Sjögren's

patients, which is a higher percentage than is found in rheumatoid patients without sicca complaints. It has been suggested that Sjögren's syndrome patients who also have rheumatoid arthritis are less likely to produce Ro and La antibodies [54].

Attempts have been made to correlate the autoantibody profile with clinical complaints in Sjögren's syndrome patients [55–57]. It has been determined that high titre Ro antibodies are associated with more severe systemic complaints, cytopenias (low white cell counts), and parotid swelling [55]. Attempts also have been made to determine if the specificity of the Ro antibody found in Sjögren's patients with primary disease is different from that found in patients with SLE and other rheumatic diseases [56,57]. Although most patients produce antibodies to more than one component of Ro antigen, some authors have concluded that patients with Sjögren's syndrome are more likely to have antibodies to the 52 KDa band of Ro antigen, whereas patients with SLE will have antibodies to both the 60 and 52 KDa bands of Ro antigen [58]. Other studies have determined that Ro antibodies are generally direct against native 60 KDa Ro and denatured 52 KDa Ro. Scofield *et al.* have demonstrated that affinity-purified antibody directed against the 52 KDa Ro leucine zipper also binds native 60 KDa Ro, suggesting that this cross-reactivity may explain the association of antibodies to both the 60 KDa and the 52 KDa Ro antigens [59].

There is considerable evidence to show that production of Ro and La antibodies is associated with a definite genetic background [60,61]. Primary Sjögren's patients who are heterozygous for DQ alleles DQ1 and DQ2 produce high levels of Ro and La antibodies [61]. Jacobsson *et al.* compared seemingly healthy mature adults who complained of mild symptoms of dryness to those without dryness and found that symptomatic subjects had elevated levels of antibodies to Ro and La compared to the asymptomatic subjects [62]. Namekawa *et al.* established T-cell lines from T cells infiltrating labial salivary glands taken from six patients with Sjögren's syndrome. These cell lines were examined for proliferative responses to recombinant 52 KDa Ro and La proteins and some did demonstrate significant proliferative responses to the Ro antigen. It was also determined that these responsive cell lines exhibited preferential usage of V β 2 and V β 3 genes, suggesting that 52 KDa Ro may be an autoantigen in patients with Sjögren's syndrome [63].

III. Epidemiology

A. Incidence and Prevalence

Very little is known about the epidemiology of Sjögren's syndrome. This is primarily because there has been a lack of consensus on the criteria required to define this syndrome and there has also being a tendency to minimize the consequences of this illness because it is not considered to be life threatening. Using the California criteria, Fox states that the prevalence of primary Sjögren's syndrome is approximately 1/250 individuals [64]. This is almost certainly an underestimate because these criteria are very specific, but not very sensitive. An epidemiologic study performed by LaLonde *et al.* involved a telephone survey of 1004 adults in Quebec (56% females and 44% males). Of the 63% who responded, 165 (16%) patients reported experiencing dry mouth frequently. Of these, 33 (19%) also complained of dry eyes and only two patients reported a diagnosis of Sjögren's syndrome. These findings suggest that as many as 3% of the population of Quebec may have symptoms of dry eyes and dry mouth, but few seek diagnosis and treatment [65].

Talal has estimated that at least four million Americans suffer from Sjögren's syndrome [66]. This estimate was based on the fact that at least 50% of patients with rheumatoid arthritis also have Sjögren's syndrome. Accepting the fact that there are many more patients who suffer from this syndrome secondary to other diseases such as SLE and SSc, as well as those who have primary disease, four million affected Americans is probably a conservative estimate. LaLonde's findings appear to support this conclusion [65].

B. Distribution in Women

Sjögren's syndrome is a disease of women, occurring in nine women to every one man. It is also a disease of middle to old age and, therefore, occurs more frequently in postmenopausal women. Sjögren's syndrome has been described in childhood and there has been some suggestion that patients with onset at an early age may experience more serious clinical disease, including the predisposition to lymphomatous change [67].

To our knowledge there have been few clinical studies evaluating the use of supplemental estrogens as a treatment for patients with Sjögren's syndrome. One longditudinal study suggested that supplemental estrogens improved salivary gland function in perimenopausal and postmenopausal women [68]. Contrary to these findings, it appears that androgens may also have a beneficial effect on Sjögren's syndrome [69,70]. The topic of the influence of sex hormones on autoimmunity and Sjögren's syndrome is discussed in more detail in section IV.B.

C. Mortality Statistics

Just as the prevalence and incidence of Sjögren's syndrome are not accurately known, mortality statistics are also obscure. However, it is known that Sjögren's syndrome patients are predisposed to develop lymphomas, and this is discussed in more detailed in Section V.D.

A study of 31 primary SSc patients followed for more than 10 years determined that the disease progressed in 71% of these patients with the development of new complaints but without any change in sicca symptoms [71]. Of 338 Sjögren's patients, also studied by Sugai *et al.*, 23 died. The major cause of death was malignancy (65%), six from malignant lymphoma and eight from other cancers (71). Other causes of death included cerebral vascular accidents (4), infection (4), hepatic failure (2), and heart failure (1) [71]. Sjögren's patients with systemic vasculitis, pneumonitis, and liver disease also have decreased survival, but the findings by Sugai *et al.* serve as an important reminder of the association of Sjögren's syndrome with malignancy.

Other long-term follow-up studies have suggested that the majority of Sjögren's syndrome patients have a comparatively benign course. Kelly *et al.* followed 1001 Sjögren's patients for a mean of 34 months. They determined that Ro antibodies were associated with more severe systemic disease, but that the syndrome appeared to remit spontaneously in some patients. Three cases of lymphoma were detected, but only one had died at the time of reporting [72]. Another long-term follow-up of patients with Sjögren's syndrome concluded that except for the risk of development of lymphoma, primary Sjögren's syndrome patients experienced a rather stable and mild clinical course and that patients with isolated keratoconjunctivitis sicca did not have an increased risk of lymphoma [73]. These three longditudinal studies were carried out in different patient populations in different parts of the world. This fact may explain the different conclusions from these studies.

IV. Host and Environmental Factors

A. Genetic Factors and Family Studies

The predisposition to autoimmunity is genetically controlled, hormonally influenced, and environmentally triggered. Family studies of Sjögren's syndrome patients have demonstrated that specific genetic markers do influence the expression of this disease, including the association of HLA DR 3 in Caucasians with primary Sjögren's [74,75]. In one family study, an attempt was made to determine if linkage disequilibrium of HLA genes was important for the development of primary Sjögren's syndrome. Using Fox's criteria to define primary Sjögren's syndrome, 16 families were analyzed and it was determined that linkage disequilibrium of HLA genes does not appear to be involved in the predisposition to Sjögren's. However, these studies did confirm an association with HLA DR 3 [76]. Wilson *et al.* have suggested that the association of HLA DR 3 is stronger with Ro positivity than with Sjögren's syndrome itself [60]. Other studies have confirmed that HLA DR 52 occurs with high frequency in patients with Sjögren's syndrome, regardless of whether it is primary or secondary [77,78], and there is now a general consensus that genes expressed at the DR loci, in particular DR3, contribute to a susceptibility to develop Sjögren's syndrome [60,61,75,79].

A discussion of the genetic aspects of a disease must include other aspects of the genetic background, including aberrations of the Fas antigen and its ligand Fas L. Fas antigen (CD95) (Fas) and tumor necrosis factor (TNF-R) receptors are part of a superfamily of receptors. The Fas antigen and its ligand are responsible for programmed cell death (apoptosis), an essential regulatory function. Induction of programmed cell death (PCD) occurs when the Fas gene is cross-linked with an antibody or with Fas L, its physiological ligand. Studies have suggested that defective Fas-mediated apoptosis may result in autoimmune disease. Animal studies have shown that mouse models with different independent autosomal recessive mutations of the Fas receptor or its ligand develop massive lymphadenopathy, autoantibodies, and a lupus-like illness [80–82]. Mutations of the Fas gene may also lead to autoimmune lymphoproliferative disorders in humans [83,84]. A study of the expression of Fas, Fas ligand, and bcl-2 (a suppressor oncogene that has been shown to suppress Fas mediated apoptosis) in salivary gland biopsies from patients with primary Sjögren's syndrome revealed that the acinar epithelial cells in these patients abnormally express Fas and Fas L. These findings suggest that Sjögren's syndrome acinar cells and ductal epithelial cells die by PCD [85]. The same study also revealed expression of Fas and bcl-2, but limited expression of Fas L, by the majority of the infiltrating lymphocytes in the salivary tissues studied from the Sjögren's syndrome patients. These findings suggest that there is limited PCD of the lymphocytes infiltrating salivary gland tissue, which may explain the predisposition to develop lymphoma found in Sjögren's syndrome patients [85]. A study of the role of the bcl-2 family of proteins in Sjögren's syndrome suggested that bcl-2 expression in Sjögren's lymphocytes may contribute to cell survival [86]. The bcl-2 transgenic mice with prolonged survival of lymphocytes developed autoantibodies and immune complex disease [87].

B. Influence of Sex Hormones and Sjögren's Syndrome

Sex hormones do influence Sjögren's syndrome [88]. Many autoimmune diseases occur more frequently in females than in males (*i.e.,* SLE, nine women to every one man), and sex hormones (estrogens) have been reported to influence disease expression both in patients with SLE (a worsening of disease) [89]

and in patients with RA (an improvement of disease) [90]. Eighty percent of RA patients clinically improve when they are pregnant [91], although this is almost certainly not just hormonally related but is also due to the altered immunological state that allows pregnancy to persist. Sjögren's syndrome is also more common in females but occurs more frequently in postmenopausal women, unlike SLE.

Studies have suggested that androgens may provide protection against Sjögren's syndrome [69,70]. Animal studies have demonstrated that testosterone improves functional activity of lacrimal glands in MRL/lpr and NZB/NZW F1 mice [92,93]. In addition to this, androgens decrease the lymphocytic infiltration of lacrimal and salivary glands but do not affect inflammation in lymphatic and splenic tissues [94]. The effect on the lacrimal tissue appears to be a hormonal effect on epithelial cells (not lymphocytes) that results in changes in cytokine production and apoptotic factors [95]. Old clinical reports suggest that androgens may be of benefit to patients with Sjögren's syndrome [96,97].

C. Environmental Factors

For many years it has been suggested that Sjögren's syndrome may be a consequence of viral infections. The viruses that have been particularly well studied have been the sialotrophic and lymphotrophic viruses and include herpes virus and retroviruses [98]. Some of the most important evidence linking retroviruses and Sjögren's syndrome comes from studies that documented that approximately 2–5% of individuals infected with HIV develop a diffuse infiltrative lymphocytic syndrome (DILS), similar clinically to Sjögren's syndrome. DILS patients may present with sicca complaints, glandular swelling and extensive extraglandular disease including interstitial pneumonitis, and lymphocytic hepatitis [98]. DILS patients generally have a more benign course than the usual HIV patient. They also differ from the idiopathic Sjögren's patient in that they are less likely to produce autoantibodies and they are more likely to be men. The histopathological changes in DILS salivary gland biopsies include lymphocytic infiltrates, but the predominant infiltrating cells are CD 8 cells, not the CD 4 cells that are found in Sjögren's patients who are not infected with HIV [99].

Other retroviruses that have been incriminated in the pathogenesis of Sjögren's syndrome include human T-cell lymphotrophic virus 1 (HTLV-1). This exogenous retrovirus becomes integrated into host DNA. If the virus becomes integrated into the host's germ line DNA, its transmission as endogenous virus to subsequent generations is a potential mechanism for autoimmunity [98]. Several animal studies have suggested a role for HTLV-1 in the pathogenesis of a Sjögren's-like syndrome [100]. HTLV-1 is endemic in Japan and in 1992, Egucki *et al.* reported that in an area endemic for HTLV-1, Japanese Sjögren's syndrome patients had a higher prevalence of antibodies for HTLV-1 compared to normal controls. These HTLV-1 patients also had more extraglandular manifestations [101]. Similar findings have been reported by Nakamura *et al.* [102]. Other Japanese studies have suggested that components of HTLV-1 proviral DNA may be found in lymphocytes infiltrating the salivary glands of HTLV-1 antibody patients, even though the clinical and laboratory findings in these HTLV-1 antibody-positive Sjögren's syndrome patients were indistinguishable from the Sjögren's syndrome patients without HTLV-1 antibodies [103]. This same research group also reported that by using PCR analysis, their studies suggested that there was restricted usage of the T-cell receptor Vβ repertoire in salivary gland and peripheral blood mononuclear cells in Sjögren's patients who were HTLV-1 antibody positive compared with Sjögren's syndrome patients who were HTLV-1 antibody negative [104]. Additional interest has been concentrated on human retrovirus 5 (HRV-5), which is probably quite ubiquitous, and may give rise to an infectiously acquired genome. Sjögren's syndrome may be an unusual autoimmune response to a common infection with HRV-5 [98]. Other reported associations between Sjögren's syndrome and viral infections include Epstein-Barr virus, human herpes virus six (HHV-6), and cytomegalovirus (CMV) [105].

It has been suggested that patients with chronic hepatitis C infection may have lymphocytic sialadenitis similar to that seen in Sjögren's syndrome patients [38]. Hepatitis C patients also have abnormal objective tests for dry mouth and dry eyes. A study of 90 Spanish patients fulfilling the European criteria for Sjögren's syndrome revealed that 13 (14%) had antibodies to hepatitis C virus (compared to 1.2% of a "normal" population from the same area). Nine of these 13 patients had clinical and laboratory markers for liver disease, while the remaining four were asymptomatic with elevated transaminase levels. Five patients agreed to liver biopsy, which showed chronic active hepatitis in all five [106].

V. Clinical Manifestations

As this is a systemic inflammatory illness, it is not surprising that the clinical manifestations are diverse. The classical complaints of Sjögren's syndrome are, of course, the dryness resulting from the autoimmune exocrinopathy, but there are other complaints that also arise from nonexocrine pathology. Because Sjögren's syndrome is associated with so many diseases, the clinical manifestations of Sjögren's syndrome sometimes can be confused with the coexisting disease. This has prompted Manthorpe et al. to design a Sjögren's syndrome wheel for describing and classifying the clinical manifestations of this syndrome (Fig. 57.2)[16]. This wheel is divided into exocrine disease and nonexocrine disease. The exocrine disease is further divided into surface and internal organ exocrine disease, with a third section allocated for monoclonal B-lymphocyte disease. The nonexocrine category is also subdivided into vascular inflammatory and noninflammatory and mediator-produced disease, with autoimmune endocrine disease as an additional subcategory.

A. Glandular Manifestations

As can be seen from the Manthorpe wheel, the exocrine manifestations are numerous. Internal organ complaints are sometimes missed until significant pathology becomes evident with pulmonary insufficiency, electrolyte and liver enzyme abnormalities, or abdominal pain. More than 70% of patients with primary biliary cirrhosis (PBC) also have symptoms consistent with Sjögren's syndrome [37]. This has prompted the concept that primary biliary cirrhosis is a dry gland syndrome with features of graft versus host disease [107]. The histopathological features of graft versus host disease resemble those of Sjögren's

syndrome, and patients with true graft versus host disease also complain of sicca symptoms. The full extent of chronic viral hepatitis in the pathogenesis of liver disease in Sjögren's patients is only just becoming apparent.

The renal manifestations of Sjögren's syndrome are diverse, the most frequent being interstitial renal tubular acidosis, which may occur in up to 30% of Sjögren's syndrome patients [40]. Patients usually have distal renal tubular acidosis; these patients have an impaired ability to transport hydrogen ion, which results in a metabolic acidosis, hypokalemia, and renal stone formation. The electrolyte disturbances cause weakness and increased debility, although in some patients the tubular abnormalities may be subclinical and are only revealed with ammonium chloride loading. It is important to note that distal renal tubular acidosis has been observed in association with various hypergammaglobulinemic states, although the precise mechanism for this association is not completely understood [108]. Erickson has studied renal disease in primary Sjögren's syndrome in detail and concludes that (1) it occurs commonly, (2) urolithiasis may precede clinical sicca complaints by many years, and (3) patients presenting with urolithiasis and distal renal tubular acidosis are more likely to have antibodies to Ro [109].

Endocrine disease may also occur in Sjögren's syndrome, and there is an increased frequency of diabetes and thyroid disease in these patients. Coll et al. studying 176 patients with autoimmune thyroid disease found that approximately one-third of these patients had features of Sjögren's syndrome [41]. In a similar study Binder et al. determined that 55% of patients with type 1 diabetes had sicca symptoms [42], which contradicted a previous report suggesting that diabetes was not more common in Sjögren's patients [110].

B. Extraglandular Disease

Manthorpe et al. classify both inflammatory and noninflammatory vascular pathology in this category [16]. It has been determined that as many as 30% of Sjögren's syndrome patients have Raynaud's phenomenon. Although this may help with our understanding of the disease process, these clinical complaints are not as worrisome as those produced by the inflammatory vascular pathology (vasculitis).

The full extent of neurological involvement in Sjögren's syndrome patients has been debated for some time. While there is little argument about peripheral neurological disease, the frequency of central nervous system involvment has been questioned [111,112]. Central nervous system disease may include unusual complaints, such as sudden neurosensory hearing loss and vestibular dysfunction [113], or the more usual manifestations of neurological deficits found in patients with vasculitis. Peripheral neurological disease includes entrapment syndromes and an unusual sensory neuropathy due to a monocytic infiltrate (lymphocytic), predominantly of the dorsal root ganglia with neuronal degeneration and fibrosis [114]. Animal model studies have suggested that a neuropathy of lacrimal glands may precede lymphocytic infiltration of these glands [115].

Traditionally little attention was given to the fatigue that occurs with this syndrome. It is one of the most disabling features of this illness. Gudbjornsson reports that many Sjögren's syndrome patients spend an extra 2 hours in bed each day [116].

This complaint may be worse for the patient than the classical exocrine symptoms [16] and this fact provides a strong argument for systemic treatment.

C. Pregnancy

Fertility and the successful completion of pregnancy are no different in Sjögren's patients, except for the neonatal lupus syndrome (NLS) (Section II.B). An abnormal pregnancy may be one of the earliest manifestations of an autoimmune diathesis. In 1977 it was recognized that women with connective tissue diseases were more likely to produce infants with primary congenital complete heart block (CCHB) than normal women [117]. Subsequent studies revealed that many mothers of children with CCHB had antibodies to Ro and La even if they were clinically normal at the time of the abnormal birth [19]. It is now known that the presence of antibodies to Ro and La predisposes mothers to have infants with NLS. The clinical features of NLS include CCHB, which is irreversible, and a transient photosensitive skin rash, which is reversible. The pathology of NLS is almost certainly due to the transplacental passage of maternal antibody, usually antibody to the Ro antigen [21,22].

Why only some Ro positive mothers have affected infants remains unclear. Whether infants develop cutaneous or cardiac disease and why the same mother, with the same antibodies present, will produce normal and abnormal children in subsequent pregnancies [19] are all questions that still need answers. The risk of having a second affected child is 26%. Children with CCHB do not do well, as more than 60% require a pacemaker and approximately 30% die [118]. Long-term follow-up of asymptomatic mothers has revealed that 48% subsequently develop symptoms of a rheumatic disease including an undifferentiated autoimmune syndrome, Sjögren's syndrome and/or SLE [19,20].

NLS is very important because it is one of the best examples of an autoantibody being directly involved in the development of pathological changes, and it has also improved our understanding of autoimmune disease by introducing the concept of subclinical autoimmunity.

D. Malignancy

Some studies have suggested that malignancy is the most frequent cause of death in Sjögren's syndrome patients [71]. An increased prevalence of non-Hodgkin's B cell lymphoma in Sjögren's syndrome patients has been known for many years [46,47]. It has even been suggested that this is a consequences of the deregulation of B-cell hyperactivity found in these patients. As B-cell hyperreactivity becomes more autonomous, the levels of autoantibodies fall, although this has not been confirmed in all studies [119]. Probably less than 10% of all Sjögren's syndrome patients develop lymphoma [49], but factors that appear to predispose patients to this complaint include primary Sjögren's syndrome, persistent parotid enlargement, an early age of onset, and monoclonal gammopathy [120,121].

The transition from polyclonal B-cell hyperactivity to monoclonal gammopathy and on to overt lymphoma may take many years [122]. There have been several documented cases of spontaneous regression of tumors in Sjögren's syndrome patients.

Frequently a watch, wait, and see strategy is adopted for patients with low grade lymphomas [119,123].

VI. Laboratory Abnormalities

A. Autoantibodies

Approximately 70% of Sjögren's syndrome patients have antinuclear antibodies (ANAs) [124]. Sixty to ninety percent of these patients produce Ro antibodies that are associated with rheumatoid factor, hypergammaglobulinemia, and cytopenias [56]. Other more specific antibodies, for example antimitochondrial antibodies, are useful for evaluating Sjögren's syndrome patients suspected of having associated liver disease (i.e., primary biliary cirrhosis or chronic hepatitis). Some patients with Sjögren's syndrome appeared to produce no autoantibodies at all, which may be beneficial because it has been suggested that the presence of Ro antibody predisposes patients to develop more severe disease. The nature of autoantibodies found in Sjögren's syndrome patients and their clinical associations are discussed in more detail in Section II.C.3.

B. Markers Indicating Glandular Abnormalities

Some simple screening tests for Sjögren's syndrome include amylase, lipase, blood sugar, BUN/creatinine, serum electrolytes, and urinary pH (looking for renal tubular acidosis). Liver function tests and screening tests for hepatitis and cryoglobulins should be ordered routinely when evaluating a patient for Sjögren's syndrome. All patients should have a mouth swab looking for candidiasis because this is a simple problem to treat and it signifies oral dryness. A chest X-ray should be done annually to screen for the development of pneumonitis and/or lymphadenopathy. A lip biopsy with the removal of 5–10 small salivary glands from the inside of the lower lip is the most definitive method for diagnosing the oral componant of Sjögren's syndrome. This simple procedure can be done in the outpatient clinic, should be performed using local anesthesia, and takes approximately 10–15 minutes.

C. Malignancy

It has been suggested that markers for malignancy in Sjögren's syndrome patients include a loss of autoantibodies and monoclonality detected by immunoglobulin electrophoresis of serum and urine [120,121]. These laboratory markers should be measured at least once a year.

VII. Management of Sjögren's Syndrome

The management of Sjögren's syndrome can be divided into the following five main categories.

A. Appropriate Diagnosis

It has been determined that it takes approximately 10 years for some Sjögren's syndrome patients to be diagnosed appropriately [16]. This is not a rare disease and it is still frequently missed.

B. Patient Education

Once the diagnosis has been made it is important to notify the patient that they have a systemic inflammatory disease that can produce numerous complaints including severe fatigue. These patients should be made aware of the need for long-term follow-up and the increased risk for lymphoma. Many patients enjoy support groups and this option should be offered to the patient when it is available.

C. Local Treatments

Local replacement of missing secretions works well for some patients with dry eyes. There are many different brands of artificial tears that are available and the patient should be encouraged to try various brands until they find the one that suits them best. Some patients are allergic to preservatives, and artificial tears with preservatives can produce painful, red eyes in some susceptible patients. The individual tear preparations without preservatives are considerably more expensive. The use of an eye ointment at night is a good idea as it gives prolonged relief and allows the patient to sleep longer. Protective glasses with covered sidearms can help to minimize evaporation of the tear film. Some patients will not be able to get relief from the use of the eyedrops alone and will require either temporary or permanent plugging of their tear ducts to relieve their dry eyes. Patients with severe eye dryness must be examined by an ophthalmolgist regularly to check for corneal abrasions and eye infections, which are complications that can lead to impaired vision.

Artificial saliva is not as well received as artificial tears. These salivary products do not seem to be able to produce the prolonged relief needed by patients, and, consequently, many patients try to stimulate saliva production by sucking candy. This is the worse thing they can do as it potentiates dental caries which are particularly prevalent in dry mouths. Patients need to see a dentist very regularly and must be educated about sore and burning mouth complaints that arise as a consequence of yeast infections, another complication of dry mouth. There is some evidence that eating live cultured yogurt that contains lactobacillus and/or acidophilus tablets can help to minimize yeast infections [125]. This is particularly important in patients on prednisone. A mouth swab taken from the circle of tissue at the back of the mouth (the fauces) and the back of the tongue may reveal a candida infection and patients should be given the appropriate treatment for at least three weeks. We use troches rather than "swish and swallow" whenever possible in an attempt to minimize the amount of glucose that the patient is taking.

Other measures that some patients find necessary include: moisturizers for the skin, KY jelly for vaginal dryness, and warm moist air for nasal and airway dryness. Some patients with exocrine pancreatic deficiency will also require replacement with pancreatic enzymes.

New oral preparations of parasympathomimetic drugs (i.e., pilocarpine hydrochloride) also have become available. The pilocarpine hydrochloride preparation appears to increase saliva production in about 60% of Sjögren's patients and is usually well tolerated [126–128]. Obviously, patients with severe cardiac or pulmonary disease should avoid treatment with parasympathomimetic agents. These oral preparations are an important step forward in managing patients appropriately.

D. Treatment of Systemic Disease

For years there has been a general trend not to treat patients with systemic agents because Sjögren's syndrome is generally not considered to be a life-threatening illness. However, patients do need treatment for the inflammatory disease because they are chronically tired, suffer with arthritis, and sometimes develop vasculitis. One of the most benign treatments is hydroxychloroquine. Fox was the first to suggest that this agent may be useful in patients with Sjögren's syndrome [129], while others have not been able to substantiate any benefit [130]. We have found this drug to be most useful in Sjögren's syndrome patients, even in patients without arthritis. Hydroxychloroquine does not produce any relief of the dryness complaints but it does help with the chronic fatigue and debility. Some patients may require additional agents such as methotrexate for arthritis and prednisone for vasculitis. Cyclosporine A is currently being evaluated, particularly as an ophthalmic solution for the benefit of dry eyes [131]. Other agents such as bromhexine [132], interferon [133], and the various preparations of gamma linoleic acid [134,135] need further evaluation.

E. Long-Term Follow-Up

Patients with Sjögren's syndrome need to be followed regularly because there is an increased risk of malignancy. Frequent assessments of persistently swollen salivary glands and clinical evaluations looking for lymphadenopathy and/or hepatosplenomegaly should be done regularly. Patients need regular dental and opthalmological examinations and they should have an annual physical examination with a chest X-ray.

VIII. Summary

Sjögren's syndrome is a chronic, inflammatory, systemic disease of women that frequently goes undiagnosed. It is a common disease that is genetically controlled and hormonally influenced (9 women to 1 man). Viral infections may be etiological agents that trigger the inflammatory process. Patients who produce a high titre of Ro antibodies appear to suffer more serious systemic complaints, while some may be clinically normal but declare their autoimmune diathesis by producing an infant with NLS.

Reluctance to treat the inflammatory process has been the standard practice for many years. This is not appropriate. Replacing or stimulating glandular secretion's is simply treating the end result of glandular destruction. Prevention of end organ damage should be the aim, just as it is in patients with systemic lupus erythematosus and rheumatoid arthritis.

Because Sjögren's syndrome is an autoimmune disease that may predispose patients to malignant change, it could be argued that Sjögren's patients, especially those with high titre Ro antibodies, should be treated sooner rather than later. With the known progression to malignancy and the chronic debility produced by inflammation and end organ damage, attempts to control the inflammatory process should be a priority.

Research efforts should be directed toward clearly defining the syndrome and determining whether the lymphocytic infiltration is a primary or secondary phenomenon. Once these basic questions have been answered it should be easier to identify the environmental triggers that stimulate the autoimmune epithelitis and the autoantibody production. This will help to identify patients destined to develop this syndrome before they become clinically dry. Prevention is the ultimate goal.

References

1. Sjögren, H. (1930). Keratoconjunctivitis Sicca. *Hygiea* **82,** 829.

2. Hadden, W. B. (1888). On "Dry Mouth" or suppression of the salivary and buccal secretions. *Trans. Clin. Soc. London* **21,** 176.

3. Hutchinson, J. (1888). A case of "Dry Mouth" (Aptyalism). *Trans. Clin. Soc. London* **21,** 180.

4. Sjögren, H. (1933). Zur Kenntnis der Keratoconjunctivitis Sicca. *Acta Ophthalmol., Suppl.* **2,** 1–151.

5. Frost-Larsen, K., Isager, H., and Manthorpe, R. (1978). Sjögren's syndrome treated with bromhexine: A randomized clinical study. *Br. Med. J.* **1,** 1579–1581.

6. Moutsopoulos, H. M., Klippel, J. H., Pavlidis, N., Wolf, R. O., Sweet, J. B., Steinberg, A. U., and Chu, F. C. (1980). Correlative histologic and serologic findings of sicca syndrome in patients with systemic lupus erythematosus. *Arthritis Rheum.* **23,** 36–40.

7. Pavlidis, N. A., Karsh, J., and Moutsopoulos, H. (1982). The clinical picture of primary Sjögren's syndrome: A retrospective study. *J. Rheum.* **9,** 685–690.

8. Dubois, E. L., and Tufanelli, D. (1964). Clinical manifestations of systemic lupus erythematosis.

9. Alarcon-Segovia, D., Ibañez, G., Hermandez-Ortiz, J., *et al.* (1974). Sjögren's syndrome in progressive systemic sclerosis (scleroderma). *Am. J. Med.* **57,** 78–85.

10. Moutsopoulos, H. M., Chused, T. M., Mann, D. L. Klippel, J. H., *et al.* (1980). Sjögren's syndrome (sicca syndrome): Current issues. *Ann. Intern. Med.* **92** (Part I), 212–226.

11. Strand, V., and Talal, N. (1980). Advances in the diagnosis and concept of Sjögren's syndrome (autoimmune exocrinopathy). *Bull. Rheum. Dis.* **30,** 1046–1052.

12. Moutsopoulos, H. M. (1994). Sjögren's syndrome: Autoimmune epithelitis. *Clin. Immunol. Immunopathol.* **72,** 162–165.

13. Vitali, C., Bombardieri, S., Moutsopoulos, H. M., Balestrieri, G., *et al.* (1993). Preliminary criteria for the classification of Sjogren's syndrome. *Arthritis Rheum.* **36,** 340–347.

14. Fox, R. I., Robinson, C. A., Curd, J. G., Kozin, F., and Howell, F. V. (1986). Sjögren's syndrome: Proposed criteria for classification. *Arthritis Rheum.* **29,** 577–585.

15. Vitali, C., Bombardieri, S., and the European Study Group on diagnostic criteria for SS. (1997). *J. Rheumatol.* **24,** 38.

16. Manthorpe, R., Asmussen, K., and Oxholm, P. (1997). Primary Sjögren's syndrome; diagnostic criteria, clinical features, and disease activity. *J. Rheumatol.* **24** (Suppl. 50), 8–11.

17. Manthorpe, R., Andersen, V., Jensen, O. A., *et al.* (1986). Editorial comments to the four sets of criteria for Sjögren's syndrome. *Scand. J. Rheumatol., Suppl* **61,** 31–35.

18. Scott, J. S., Maddison, P. J., Taylor, P. V., Esscher, E., Scott, O., and Skinner, R. P. (1983). Connective-tissue disease, antibodies to Ribonucleoprotein, and congenital heart block. *N. Engl. J. Med.* **309,** 209–212.

19. McCune, A. B., Weston, W. L., and Les, L. A. (1987). Maternal and fetal outcome in neonatal lupus erythematosus. *Ann. Intern. Med.* **106,** 518–523.

20. Waltuck, J., and Buyon, J. P. (1994). Autoantibody associated complete heart block: Outcome in mothers and children. *Ann. Intern. Med.* **120,** 544–551.

21. Reichlin, M., Brucato, A., Frank, M. B., *et al.* (1994). Concentration of autoantibodies to native 60-kd Ro/SS-A and denatured 52-kd Ro/SS-A in eluates from the heart of a child who died with congenital complete heart block. *Arthritis Rheum.* **37,** 1698–1703.

22. Alexander, E., Buyon, J. P., Provost, T. T., and Guarnieri, T. (1992). Anti-Ro/SS-A antibodies in the pathophysiology of congenital heart block in neonatal lupus syndrome, an experimental model: In vitro electrophysiologic and immunocytochemical studies. *Arthritis Rheum.* **35,** 176–189.

23. Kruize, A. A., Tilleman, E. H. B. M., Vianen, M. E., *et al.* (1997). Autoantibody synthesis in lacrimal and salivary glands of primary Sjögren's syndrome (PSS) patients. *J. Rheumatol.* **24** (Suppl. 50), Abstr. No. 28.

24. Horsfall, A. C., Rose, L. M., and Maini, R. N. (1989). Autoantibody synthesis in salivary glands of Sjögren's syndrome patients. *J. Autoimmun.* **2,** 559–568.

25. Chisholm, D. M., Blair, G. S., Low, P. S., and Whaley, K. (1971). Hydrostatic sialography as an index of salivary gland disease in Sjögren's syndrome. *Acta Radiol., Diagn.* **11,** 577–585.

26. Fox, R., Carstens, S., Fong, S., Robinson, C. A., *et al.* (1982). Use of monoclonal antibodies to analyze peripheral blood and salivary gland lymphocyte subsets in Sjögren's syndrome. *Arthritis Rheum.* **25,** 419–422.

27. Skopouli, F. N., Fox, P. C., Galanopoulou, V., Atkinson, J. C., *et al.* (1991). T cell subpopulations in the labial minor salivary gland histopathologic lesion of Sjögren's syndrome. *J. Rheum.* **18,** 210–214.

28. Talal, N., Asofsky, R., and Lighbody, P. (1970). Immunoglobulin synthesis by salivary gland lymphoid cells in Sjögren's syndrome. *J. Clin. Invest.* **49,** 49–54.

29. Chisholm, D. M., and Mason, D. K. (1968). Labial salivary gland biopsy in Sjögren's syndrome. *J. Clin. Pathol.* **21,** 656–660.

30. Daniels, T. E. (1983). Labial salivary gland biopsy in Sjögren's syndrome. Assessment as a diagnostic criterion in 362 suspected cases. *Arthritis Rheum.* **27,** 2.

31. Scott, J. (1980). Qualitative and quantitative observations on the histology of human labial salivary glands obtained postmortem. *J. Biol. Buccale* **8,** 187–200.

32. Katz, J., Yamase, H., and Parke, A. (1991). A case of Sjögren's syndrome with repeatedly negative findings on lip biopsy. *Arthritis Rheum.* **34,** 1325.

33. Buchanan, W. W. (1997). Sjögren's syndrome: The persisting conundrum. *J. Rheumatol.* **24,** 1667–1669.

34. Moutsopoulos, H. M., Hooks, J. J., Chan, C. C., *et al.* (1986). HLA-DR expression in labial salivary gland tissues in Sjögren's syndrome. *Ann. Rheum. Dis.* **45,** 677–683.

35. Lindahl, G., Hedfors, E., Klareskog, L., and Forsum, U. (1985). Epithelial HLA-DR expression and T-lymphocyte subsets in salivary glands in Sjögren's syndrome. *Clin. Exp. Immunol.* **61,** 475–482.

36. Papadopoulos, G. K., and Moutsopoulos, H. M. (1992). Slow viruses and the immune system in the pathogenesis of local tissue damage in Sjögren's syndrome. *Ann. Rheum. Dis.* **51,** 136–138.

37. Golding, P. L., Brown, R., Mason, A. M. S., and Taylor, E. (1970). "Sicca complex" in liver disease. *Br. Med. J.* **4,** 340–342.

38. Haddad, J., Deny, P., Munz-Gotheil, C., and Ambrosini, J. C. (1992). Lymphocytic sialadenitis of Sjögren's syndrome associated with chronic hepatitis C virus liver disease. *Lancet* **8,** 321–323.

39. Liebow, A. A., and Carrington, C. B. (1973). Diffuse pulmonary lymphoreticular infiltrations associated with dysproteinemia. Symposium on chronic respiratory disease. *Med. Clin. North Am.* **57,** 809–842.

40. Kassan, S. S., and Talal, N. (1987). Renal disease with Sjögren's syndrome. *Sjögren's Syndrome Clin. Immunol. Aspects* **5,** 96–101.

41. Coll, J., Anglada, J., Tomas, S., Reth, P., *et al.* (1997). High prevalence of subclinical Sjögren's syndrome features in patients with autoimmune thyroid disease. *J. Rheumatol.* **24,** 1719–1724.

42. Binder, A., Maddison, P. J., Skinner, P., Kurtz, A., and Isenberg, D. A. (1989). Sjögren's Syndrome: Association with Type-1 diabetes mellitus. *Br. J. Rheumatol.* **28,** 518–520.

43. Buchanan, W. W., Cox, A. G., Harden, R. McG., Glen, A. I. M., *et al.* (1966). Gastric studies in Sjögren's syndrome. *Gut* **7**, 351–354.

44. Parke, A. L., Fagan, E. A., Chadwick, V. S., and Hughes, G. R. V. (1984). Coelic disease and rheumatoid arthritis. *Ann. Rheum. Dis.* **43**, 378–380.

45. Alexander, E. L. (1987). Inflammatory vascular disease in Sjögren's syndrome. *Sjögren's Syndrome Clin. Immunol. Aspects* **6**, 102–125.

46. Talal, N., and Bunim, J. J. (1964). The development of malignant lymphoma in the course of Sjögren's syndrome. *Am. J. Med.* **36**, 529–540.

47. Talal, N., Sokoloff, L., and Barth, W. F. (1967). Extrasalivary lymphoid abnormalities in Sjögren's syndrome (reticulum-cell sarcoma, "pseudo-lymphoma", marcoglobulinemia). *Am. J. Med.* **43**, 50–65.

48. Anaya, J., McGuff, H., Banks, P. M., and Talal, N. (1996). Clinicopathological factors relating malignant lymphoma with Sjögren's syndrome. *Semin. Arthritis Rheum.* **25**, 337–346.

49. Kassan, S. S., Thomas, T., Moutsopoulos, H. M. *et al.* (1978). Increased risk of lymphoma in sicca syndrome. *Ann. Intern. Med.* **89**, 888–892.

50. Reichlin, M. (1979). Newly defined serologic systems in systemic lupus erythematosus and dermatomyositis. *Int. J. Dermatol.* **18**, 602–607.

51. Alexander, E. L., Hirsch, T. J., Arnett, F. C., Provost, T. T., and Stevens, M. B. (1982). Ro (SS-A) and La (SS-B) antibodies in the clinical spectrum of Sjögren's syndrome. *J. Rheumatol.* **9**, 239–246.

52. Alexander, E. L., and Provost, T. T. (1981). Ro (SS-A) and La (SS-B) antibodies. *Springer Semin. Immunopathol.* **4**, 253–273.

53. Harley, J. B., Alexander, E. L., Bias, W. B., Fox, O. F., *et al.* (1986). Anti-Ro(SS-A) and anti-La(SS-B) in patients with Sjögren's syndrome. *Arthritis Rheum.* **29**, 196–206.

54. Alspaugh, M., and Maddison, P. J. (1979). Resolution of the identity of certain antigen-antibody systems in SLE and Sjögren's syndrome: An interlaboratory collaboration. *Arthritis Rheum.* **22**, 796–798.

55. Alexander, E. L., and Provost, T. T. (1983). Cutaneous manifestations of primary Sjögren's syndrome: A reflection of vasculitis and association with anti-Ro(SSA) antibodies. *J. Invest. Dermatol.* **80**, 386–391.

56. Alexander, E. L., Arnett, F. C., Provost, T. T., and Stevens, M. B. (1983). Sjögren's syndrome: Association of anti-Ro(SS-A) antibodies with vasculitis, hematologic abnormalities, and serologic hyperreactivity. *Ann. Intern. Med.* **98**, 155–159.

57. Venables, J. P. J., Charles, P. J., Buchanan, R. C., *et al.* (1983). Quantitation and detection of isotypes of anti-SS-B antibodies by ELISA and Farr assays using affinity purified antigens. *Arthritis Rheum.* **26**, 476–477.

58. Lopez-Longo, F. J., Rodriguez-Mahou, M., Escalona, M., *et al.* (1994). Heterogeneity of the anti-Ro(SSA) response in rheumatic diseases. *J. Rheumatol.* **21**, 1450–1456.

59. Scofield, R. H., Chambers, T. L., Barber, B. D., *et al.* (1997). Autoantibody to the leucine zipper region of 52 KD RO/SSA binds native 60 KD RO/SSA: Identification of the components of the tertiary epitope in 60 KD RO/SSA. *J. Rheumatol.* **24** (Suppl. 50), Abstr. No. 27.

60. Wilson, R. W., Provost, T. T., Bias, W. B., *et al.* (1984). Sjögren's syndrome: Influence of multiple HLA-D region alloantigens on clinical and serologic expression. *Arthritis Rheum.* **27**, 1245–1253.

61. Harley, J. B., Reichlin, M., Arnett, F. C., *et al.* (1986). Gene interaction at HLA-DQ enhances autoantibody production in primary Sjögren's syndrome. *Science* **232**, 1145–1147.

62. Jacobsson, L., Hansen, B. U., Manthorpe, R., Hardgrave, K., *et al.* (1992). Association of dry eyes and dry mouth with Anti-Ro/SS-A and Anti-La/SS-B autoantibodies in normal adults. *Arthritis Rheum.* **35**, 1492–1494.

63. Namekawa, T., Kuroda, K., Kato, T., Yamamoto, K., Murata, H., Sakamaki, T., Nishioka, K., Iwamoto, I., Saitoh, Y., and Sumida, T. (1995). Identification of Ro(SSA) 52 kDa reactive T cells in labial salivary glands from patients with Sjögren's syndrome. *J. Rheumatol.* **22**, 2092–2099.

64. Fox, R. I., and Saito, I. (1994). Criteria for diagnosis of Sjögren's syndrome. *Rheum. Dis. Clin. North Am.* **20**, 391–407.

65. Lalonde, B., Lavigne, G., Goulet, J. P., and Barbeau, J. (1997). An epidemiological study of reported salivary dysfunction in an adult population of Quebec. *J. Rheumatol.* **24** Suppl. 50), (Abst. No. 51).

66. Talal, N. (1984). How to recognize and treat Sjögren's syndrome. Drug therapy. *Curr. Issues Rheumatol.*, pp. 48–54.

67. Ramos-Casals, M., Cervera, R., Font, J., *et al.* (1998). Young onset of primary Sjögren's syndrome, clinical and immunological characteristics. *Lupus* (in press).

68. Laine, M., and Leimola-Virtanen, R. (1996). Effect of hormone replacement therapy on salivary flow rate, buffer effect and pH in perimenopausal and postmenopausal women. *Arch. Oral Biol.* **41**, 91–96.

69. Sullivan, D. A., and Edwards, J. A. (1997). Androgen stimulation of lacrimal gland function in mouse models of Sjögren's syndrome. *J. Steroid Biochem. Mol. Biol.* **60**, 237–245.

70. Sullivan, D. A., Rocha, F. J., and Sato, E. H. (1995). Influence of androgen therapy on lacrimal autoimmune disease in a mouse model of Sjögren's syndrome. *Adv. Mucosal Immun.*, pp. 1199–1202.

71. Sugai, S., Cui, G., Ogawa, Y., *et al.* (1997). Analysis of the cause of death of 23 patients with SS. *Riumachi* **36**, 770–780.

72. Kelly, C. A., Foster, H., Pal, B., Gardiner, P., Malcolm, A. J., *et al.* (1991). Primary Sjögren's syndrome in north east England—A longitudinal study. *Br. J. Rheum.* **30**, 437–442.

73. Kruize, A. A., Hene, R. J., Van Der Heide, A., Bodeutsch, C., *et al.* (1996). Long-term followup of patients with Sjögren's syndrome. *Arthritis Rheum.* **39**, 297–303.

74. Reveille, J. D. (1992). The molecular genetics of systemic lupus erythematosus and Sjögren's syndrome. *Curr. Opin. Rheumatol.* **4**, 644–656.

75. Foster, H., Kelly, C., Cavanagh, G., *et al.* (1992). The association of HLA DR3 with the susceptibility and severity of primary Sjögren's syndrome in a family study. *Br. J. Rheumatol.* **31**, 309–315.

76. Foster, H., Stephenson, A., Walker, D., *et al.* (1993). Linkage studies of HLA and primary Sjögren's syndrome in multicase families. *Arthritis Rheum.* **36**, 4.

77. Manthorpe, R., Morling, N., Platz, P., Ryder, L. P., *et al.* (1981). HLA-D antigen frequencies in Sjögren's syndrome: Differences between primary and secondary forms. *Scand. J. Rheumatol.* **10**, 124–128.

78. Mann, D. L., and Moutsopoulos, H. M. (1983). HLA-DR alloantigens in different subsets of patients with Sjögren's syndrome and in family members. *Ann. Rheum. Dis.* **42**, 533–536.

79. Mann, D. L. (1987). Immunogenetics of Sjögren's syndrome. *Sjögren's Syndrome Clin. Immunol. Aspects* **3**, 235–243.

80. Elkon, K. B. (1997). Apoptosis and autoimmunity. *J. Rheumatol.* **24** (Suppl. 50), 6–7.

81. Chu, J. L., Drappa, J., Parnassa, A., and Elkon, K. (1993). The defect in fas mRNA expression in MRL/lpr mice is associated with insertion of the retrotransposon, ETn. *J. Exp. Med.* **178**, 723–730.

82. Wu, J., Zhou, T., He, J., and Mountz, J. D. (1993). Autoimmune disease in mice due to integration of an endogenous retrovirus in an apoptosis gene. *J. Exp. Med.* **178**, 461–468.

83. Fisher, G. H., Rosenberg, F. J., Straus, S., Dale, J., *et al.* (1995). Dominant interfering Fas gene mutations impair apoptosis in a

human autoimmune lymphoproliferative syndrome. *Cell* (*Cambridge, Mass.*) **81**, 935–946.

84. Rieux-Laucat, F., Le Deist, F., Hivroz, C., *et al.* (1995). Mutation in Fas associated with human lymphoproliferative syndrome and autoimmunity. *Science* **268**, 1347–1349.

85. Kong, L., Ogawa, N., Nakabayashi, T., Liu, G. T., *et al.* (1997). FAS and FAS ligand expression in the salivary glands of patients with primary Sjögren's syndrome. *Arthritis Rheum.* **40**, 1.

86. Kong, L., Robinson, C. P., Yamachika, S., Humphreys-Beher, M. G., *et al.* (1997). Salivary gland apoptosis in the absence of lymphocytes in nod mice. *J. Rheumatol.* **24** (Suppl. 50), Abstr. No. 9.

87. Strasser, A., Harris, A. W., von Boehmer, H., and Cory, S. (1994). Positive and negative selection of T cells in T cell receptor transgenic mice expressing a bcl-2 transgene. *Proc. Natl. Acad. Sci. U.S.A.* **91**, 1376–1380.

88. Sullivan, D. A. (1997). Sex hormones and Sjögren's syndrome. *J. Rheumatol.* **24** (Suppl 50), 17–32.

89. Jungers, P., Dougados, M., Pelissier, C., *et al.* (1982). Influence of oral contraceptive therapy on activity of systemic lupus erythematosus. *Arthritis Rheum.* **25**, 618–623.

90. Vandenbroucke, J. P., Valkenburg, H. A., Boersma, J. W., *et al.* (1982). Oral contraceptives and rheumatoid arthritis: further evidence for a protective effect. *Lancet* **2**, 839–842.

91. Hench, P. S. (1938). Ameliorating effect of pregnancy on chronic atrophic (infectious) rheumatoid arthritis, fibrositis and intermittent hydroarthrosis. *Proc. Staff Meet. Mayo Clin.* **13**, 161–167.

92. Ariga, H., Edwards, J., and Sullivan, D. A. (1989). Androgen control of autoimmune expression in lacrimal glands of MRL/Mp-lpr/lpr mice. *Clin. Immunol. Immunopathol.* **53**, 499–508.

93. Sato, E. H., Ariga, H., and Sullivan, D. A. (1992). Impact of androgen therapy in Sjögren's syndrome: Hormonal influence on lymphocyte populations and Ia expression in lacrimal glands of MRL/Mp-lpr/lpr mice. *Invest. Ophthalmol. Visual Sci.* **33**, 2537–2547.

94. Sato, E. H., and Sullivan, D. A. (1994). Comparative influence of steroid hormones and immunosuppressive agents on autoimmune expression in lacrimal glands of a female mouse model of Sjögren's syndrome. *Invest. Ophthalmol. Visual Sci.* **35**, 2632–2642.

95. Ono, M., Rocha, F. J., and Sullivan, D. A. (1995). Immunocytochemical location and hormonal control of androgen receptors in lacrimal tissues of the female MRL/Mp-lpr/lpr mouse model of Sjögren's syndrome. *Exp. Eye Res.* **61**, 659–666.

96. Appelmans, M. (1948). La Kerato-conjonctivite seche de Gougerot-Sjögren. (The keratoconjunctivitis sicca of Gougerot-Sjögren syndrome.) *Arch. Ophtalmol.* (*Paris*) **81**, 577–588.

97. Bruckner, R. (1945). Uber einem erfolgreich mit perandren behandelten fall von Sjogrenschem symptomen komplex. (Successful treatment of a case of Sjögren's syndrome with perandren.) *Ophthalmologica* **110**, 37–42.

98. Venables, P. J. W., and Rigby, S. P. (1997). Viruses in the etiopathogenesis of Sjögren's syndrome. *J. Rheumatol.* **24** (Suppl. 50), 3–5.

99. Itescu, S., and Winchester, R. (1992). Diffuse infiltrative lymphocytosis syndrome: A disorder occurring in human immunodeficiency virus 1 infection that may present as a sicca syndrome. *Rheum. Dis. Clin. North Am.* **18**, 683–697.

100. Green, J. E., Hinrichs, S. H., Vogel, J., and Jay, G. (1989). Exocrinopathy resembling Sjögren's syndrome in HTLV-1 *tax* transgenic mice. *Nature* (*London*) **341**, 72–74.

101. Eguchi, K., Matsuoka, N., Ida, H., *et al.* (1992). Primary Sjögren's syndrome with antibodies to HTLV-1: Clinical and laboratory features. *Ann. Rheum. Dis.* **51**, 769–776.

102. Nakamura, H., Eguchi, K., Nakamura, T., Mizokami, A., *et al.* (1997). High prevalence of Sjögren's syndrome in patients with HTLV-1 associated myelopathy. *Ann. Rheum. Dis.* **56**, 167–172.

103. Ohyama, Y., Nakamura, S., Hara, H., *et al.* (1997). Possible involvement of human T lymphotrophic virus type I in the pathogenesis of Sjögren's syndrome - Part I. *J. Rheumatol.* **24** (Suppl 50), Abstr. No. 2.

104. Sasaki, M., Nakamura, S., Ohyama, Y., Shinohara, M., *et al.* (1997). Possible involvement of human T lymphotrophic virus type I in the pathogenesis of Sjögren's syndrome - Part II. *J. Rheumatol.* **24** (Suppl. 50), Abstr. No. 1.

105. Price, E. J., and Venables, P. J. W. (1995). The etiopathogenesis of Sjögren's syndrome. *Semin. Arthritis Rheum.* **25**, 117–133.

106. Garcia-Carrasco, M., Ramos, M., Cervera, R., Font, J., *et al.* (1997). Hepatitis C virus infection in "primary" Sjögren's syndrome: Prevalence and clinical significance in a series of 90 patients. *Ann. Rheum. Dis.* **56**, 173–175.

107. Epstein, O., Thomas, H. C., and Sherlock, S. (1980). Primary biliary cirrhosis is a dry gland syndrome with features of chronic graft-versus-host disease. *Lancet* 1166–1169.

108. McCurdy, D. K., Cornwell, G. G., III, and DePratti, V. J. (1967). Hyperglobulinemic renal tubular acidosis. Report of two cases. *Ann. Intern. Med.* **67**, 110–117.

109. Eriksson, P. (1996). Renal disease in Sjögren's syndrome. Medical Dissertation, No. 495. Linkoping University.

110. Alspaugh, M. A., and Whaley, K. (1981). Sjögren's syndrome. *In* "Textbook of Rheumatology" (W. N. Kelly, E. D. Harris, S. Ruddy, and C. B. Sledge, eds.), pp. 971–979. Saunders, Philadelphia.

111. Andonopoulos, A. P., Lagos, G., Drosos, A. A., and Moutsopoulos, H. M. (1990). The spectrum of neurological involvement in Sjögren's syndrome. *Br. J. Rheumatol.* **20**, 21–23.

112. Alexander, E. L., Lijewski, J. E., Jerdan, M. S., and Alexander, G. E. (1986). Evidence of an immunopathogenic basis for central nervous system disease in primary Sjögren's syndrome. *Arthritis Rheum.* **29**, 1223–1231.

113. Alexander, E. L., (1987). Neuromuscular complications of primary Sjögren's syndrome. *Sjögren's Syndrome Clin. Immunol. Aspects* **2**, 61–82.

114. Malinow, K. L., Yannakakis, G. D., Glusman, S. M., Edlow, D. R. *et al.* (1986). Subacute sensory neuropathy secondary to dorsal root ganglionitis in primary Sjögren' syndrome. *Ann. Neurol.* **20**, 535–537.

115. Walcott, B., Claros, N., Patel, A., Peterson, E., and Brink, P. R. (1997). Changes in innervation and membrane properties of lacrimal acinar cells in the NZB/NZW F1 mouse occur before lymphocytic infiltration. *J. Rheumatol.* **24** (Suppl. 50), Abstr. No. 34.

116. Gudbjornsson, B. (1994). Clinical and experimental studies in primary Sjögren's syndrome. Thesis. Uppsala University, Uppsala, Sweden.

117. Chameides, L., Truex, RC., Vetter, V., *et al.* (1977). Association of maternal of systemic lupus erythematosus with congenital complete heart block. *N. Engl. J. Med.* **297**, 1204–1207.

118. Buyon, J. P. (1997). Autoantibodies reactive with Ro(SSA) and La(SSB) and pregnancy. *J. Rheumatol.* **24** (Suppl 50), 12–16.

119. Pavlidis, N. A., Drosos, A. A., Papadimitriou, C., Talal, N., and Moutsopoulos, H. M. (1992). Lymphoma in Sjögren's syndrome. *Med. Pediatr. Oncol.* **20**, 279–283.

120. Bridges, A. J., and England, D. M. (1989). Benign lymphoepithelial lesion: Relationship to Sjögren's syndrome and evolving malignant lymphoma. *Semin. Arthritis Rheum.* **19**, 201–208.

121. Anaya, J. M., McGuff, H. S., Banks, P. M., and Talal, N. (1996). Clinicopathological factors relating malignant lymphoma with Sjögren's syndrome. *Semin. Arthritis Rheum.* **25**, 337–346.

122. Moutsopoulos, H. M., Steinberg, A. D., Fauci, A. S., *et al.* (1983). High incidence of free monoclonal light chains in the sera of patients with Sjögren's syndrome. *J. Immunol.* **130**, 2263–2266.

123. Krikorian, J., Portlock, C., Cooney, P., and Rosenberg, S. (1980). Spontaneous regression of non-Hodgkin's lymphoma; A report of nine cases. *Cancer* (*Philadelphia*) **46**, 2093–2099.

124. Bloch, J. J., Buchanan, W. W., Wohl, M. J., and Bunim, J. J. (1965). Sjögren's syndrome. A clinical pathological, and serological study of sixty-two cases. *Medicine* (*Baltimore*) **44**, 187–231.

125. Hilton, E., Isenberg, H. D., Alperstein, P., *et al.* (1992). Ingestion of yogurt containing *Lactobacillus acidophilus* as prophylaxis for candidal vaginitis. *Ann. Intern. Med.* **116**, 353–357.

126. Fox, P. C., Atkinson, J. C., Macynski, A. A., *et al.* (1991). Pilocarpine treatment of salivary gland hypofunction and dry mouth (xerostomia). *Arch. Intern. Med.* **151**, 1149–1152.

127. Vivino, F. B., Al-Hashimi, I., Kahn, Z., *et al.* (1999). Pilocarpinc tablets for the treatment of dry mouth and dry eye symptoms in patients with Sjögren's syndrome. *Arch. Int. Med.* **159**, 174–181.

128. Khan, Z., Junidi, A., Goldlust, B., Trivedi, M., *et al.* (1997). Long term safety and efficacy of oral pilocarpine HCL for the treatment of oral symptoms associated Sjögren's syndrome. *J. Rheumatol.* **24** (Suppl. 50), Abstr. No. 41.

129. Fox, R. I., Chan, E., Benton, L., *et al.* (1988). Treatment of primary Sjögren's syndrome with hydroxychloroquine. *Am. J. Med.* **85**, 62–67.

130. Kruize, A. A., Hene, R. J., Kallenberg, C. G. M., Van Bijsterveld, *et al.* (1993) Hydroxychloroquine treatment for primary Sjögren's syndrome: A two year double blind crossover trial. *Ann. Rheum. Dis.* **52**, 360–364.

131. Donshik, P., Reis, B. L., Burk, C. T., *et al.* (1997). A dose-ranging clinical trial to assess the safety and efficacy of cyclosporine ophthalmic emulsion for the treatment of the ocular surface disease and inflammation associated with keratoconjunctivitis sicca (KCS). *J. Rheumatol.* **24** (Suppl. 50), Abstr. No. 43.

132. Manthorpe, R., Frost-Larsen, K., Hoj, L., Isager, *et al.* (1981). Bromhexine treatment of Sjögren's syndrome: Effect on lacrimal and salivary secretion, and on proteins in tear fluid and saliva. *Scand. J. Rheumatol.* **10**, 177–180.

133. Shiozawa, S., Morimoto, I., Tanaka, Y., and Shiozawa, K. (1993). A preliminary study on the interferon-α treatment for xerostomia of Sjögren's syndrome. *Br. J. Rheumatol.* **32**, 52–54.

134. Manthorpe, R., Petersen, S. H., and Prause, J. U. (1984). Primary Sjögren's syndrome treated with Efamol/Efavit. A double-blind cross-over investigation. *Rheumatol. Int.* **4**, 165–167.

135. Theander, E., Nyhagen, C., Jacobsson, L., and Manthorpe, R. (1997). Gammalinolenic acid (GLA) treatment in patients with primary Sjögren's syndrome (SS). *J. Rheumatol.* **24** (Suppl. 50), Abstr. No. 91.

Section 10

CARDIOVASCULAR DISEASE AND CARDIOVASCULAR RISK IN WOMEN

Roberta Ness

Department of Epidemiology
University of Pittsburgh
Pittsburgh, Pennsylvania

Cardiovascular disease comprises a set of arterial vascular conditions including such major conditions as coronary heart disease, stroke, and venous thromboembolism. There can be no debate that, among all of the diseases that women face, cardiovascular disease carries a uniquely heavy burden. It is the leading cause of death among women, accounting for almost half of all deaths and, although most likely to affect elderly women, cardiovascular disease is also a major killer of younger women [1]. To put this in perspective, each year the number of cardiovascular deaths among women under the age of 65 exceeds the number of deaths from ovarian cancer among women of all ages [2,3]. In contrast to this reality, women are relatively unlikely to anticipate that their death will result from cardiovascular disease. A 1995 Gallup poll indicated that only 19% of women expected to die from heart disease whereas almost two-thirds expected to die from cancer. These data strongly suggest that the perspective almost universally held a generation ago, that cardiovascular disease is not a "women's disease," continues to occupy the minds of many American women today.

Why is it important that women adopt the cause of cardiovascular disease as their own? There are many reasons, two of which are particularly important. First, this is a disease in which prevention works. That is, we understand a number of the modifiable risk factors that influence disease onset. The chapters in this section review the large observational literature that links diabetes mellitus, hypertension, hyperlipidemia, smoking, obesity, and physical inactivity to increased risk of cardiovascular disease. Many of these factors are also interlinked.

Heavier and more centrally obese middle-aged women have a constellation of adverse cardiovascular risk factors including higher systolic blood pressure and insulin levels and an adverse pattern of serum lipid levels [4,5]. Indeed, even women with body mass indices (BMIs) in the normal range can have adverse physiologic consequences if their fat distribution favors a central distribution. Hyperinsulinemia, hyperglycemia, and insulin resistance may be central to the pathophysiology linking central adiposity to hypertension and dyslipidemia [6]. Abdominal adiposity is metabolically very active and associated with (1) hepatic lipase activity and resultant decrease in high density lipoprotein2 (HDL2) and 2) peripheral release of free fatty acids which can increase hepatic synthesis and secretion of triglycerides, leading to increased very low density lipoprotein (VLDL) and subsequently to more small, dense LDL [7]. These lipid patterns (lower HDL2, higher triglycerides, and small, dense LDL) are all atherogenic. Insulin resistance in the abdominal adipose tissue diminishes the lipid metabolic activity of insulin. It also promotes systemic insulin resistance by enhancing the release of free fatty acids that inhibit glucose metabolism in the muscle and liver [8]. Further elevations in lipid concentrations occur during insulin resistance due to compromise of the complex role of insulin in regulating lipid and apolipoprotein metabolism [9]. Finally, hyperinsulinemia may increase renal salt retention, raise intramuscular calcium, and promote sympathetic nervous system activity and thus induce hypertension [10]. Thus, abdominal fat, insulin resistance, and lipid and blood pressure elevations are all highly interconnected. Dyslipidemia promotes endothelial cell dysfunction and has been hypothesized to predispose a person to coronary heart disease [11–13]. Lipids, in particular small, dense LDL, are readily oxidized and the generation of these free oxygen radicals may be toxic to tissues, thus promoting atherosclerosis [13,14].

The following chapters also catalogue the successes from clinical trials in showing that, at least in men, modification of lipids, blood pressure, and smoking behavior, as well as possibly other cardiovascular risk factors, reduces morbidity and/or mortality. In particular, the following chapters document that lowering of lipid levels, cessation of cigarette smoking, and treatment of hypertension all result in reductions in cardiovascular disease. Although clinical trial data examining real disease endpoints are lacking, there is observational evidence and a number of clinical trials using intermediate endpoints that suggest that enhanced physical activity, weight reduction, folate supplementation, and modest alcohol intake may improve cardiovascular risk.

Researchers are now addressing the fact that much of what we know about prevention has focused on trials in men. A number of ongoing, large, clinical trials have and are now examining primary and secondary prevention among women. These trials are examining treatment of cardiovascular risk factors as well as other chemopreventative approaches such as use of hormones (estrogen, progestin, and selective estrogen receptor modulators (or SERMS)) and aspirin. The results of studies emanating from this new focus on prevention of cardiovascular disease among women are just beginning to come out. For example, the first major randomized clinical trials examining (1) the impact of hormone replacement therapy on lipids [15] and (2) the impact of hormone replacement therapy on cardiovascular mortality among women with preexisting disease [16] were published since 1995. The timing of these results are most surprising given that postmenopausal estrogens have been marketed in the U.S. since the 1950s.

The second reason that it is important for women to understand the potential impact of cardiovascular disease in their lives is that it will help them to recognize early the clinical signs and symptoms of disease so that timely treatment can be administered. Holubkov and Reis (Chapter 59) discuss differences between men and women with respect to the clinical aspects of developed coronary heart disease and accessing timely medical care. Women are more likely to wait before seeking care and to have their disease missed by clinicians; they are less likely to receive aggressive intervention once disease is found. Overall, then, it is not surprising that women are more likely to die from their coronary heart disease than are men.

Can reduction of blood pressure, lipids, adiposity, sedentary lifestyle, and smoking cessation, as well as early detection and treatment of atherosclerotic plaque, eliminate cardiovascular disease among women? Clearly, death rates from heart disease have already declined dramatically among both men and women in the U.S. Age-adjusted deaths from heart disease in 1996 were 50% lower than those in 1970 among white males and 45% lower among white females over the age of 44 [17]. These re-

ductions are probably multifactoral and reflect lower smoking rates, less consumption of fat and cholesterol in the diet, and more aggressive treatment of blood pressure, diabetes, hyperlipidemia, and symptomatic coronary heart disease.

However, there is room for improvement. In this section, epidemiologic data are presented showing that a substantial number of women continue to have high blood pressure, to be hyperlipidemic, to have blood sugars that are not optimally controlled, and to live sedentary lifestyles. Women are also losing ground with respect to some important risk factors. In particular, young women are now the population group most likely to initiate smoking. In addition, both men and women are becoming more obese. Finally, some have argued that women's changing roles in society have placed them under increasing stress, resulting in more depression, anxiety, and hostility [18]. Perhaps as a result, the mortality advantage women had over men at any given age is beginning to decrease.

Beyond these general considerations, much can be learned from subgroups of women. There are specific groups of women who, by virtue of having defined physiologic aberrations, may be at high risk for cardiovascular disease. These include women with diabetes mellitus, systemic lupus erythematosis [19], polycystic ovarian syndrome [20], and preeclampsia [21–23]. In particular, Lotufo, Sabolsi, and Manson (Chapter 64) describe the excess risk of cardiovascular disease associated with both Type I and Type II diabetes. There may also be women at lower risk for cardiovascular disease due to having high endogenous or exogenous estrogen exposure during menopause. Observations like these promote the exploration of the underlying physiologic factors that account for differential distribution of cardiovascular disease. It may well be that differences in hormonal levels (including insulin, growth factors, and sex steroid hormones) and their metabolism, as well as inflammatory response, play important roles. These factors are not sufficient to explain differential patterns of health and disease but are major determinants interacting with lifestyle, environmental variables, physical and social environment, and genetic host susceptibility. By carefully examining the epidemiologic evidence within distinctive subgroups with altered risk patterns, we may be better able to understand the physiology underlying cardiovascular disease and so design better prevention and intervention strategies for both men and women. For example, both Meilahn (Chapter 60) and Evans (Chapter 66) review the data suggesting that hormone replacement therapy may create a less atherogenic lipid profile. However, despite this, estrogen and progestin taken in combination did not reduce cardiovascular events in a secondary prevention trial [16]. Is this lack of demonstrated efficacy because inflammation, which may have been enhanced by the estrogen/progestin regimen, played a larger role than lipids in short-term cardiovascular risk or because secondary prevention results cannot be extrapolated to a primary prevention context? Only future studies will tell us.

Newton, LaCroix, and Buist (Chapter 58) demonstrate differences in the patterns of both coronary heart disease and its risk factors among women in various ethnic and racial groups. Some, but not all, of the differences in disease rates are due to modifiable risk factors. Race and ethnicity may define some subsets of women who may be excellent targets for promoting prevention. As well, differences in disease frequency across racial and ethnic groups may provide clues about the genetic and environmental components of risk. Similarly, there are major geographic and socioeconomic gradients in cardiovascular disease that may both define susceptibility groups and provide opportunities to better understand the genesis of cardiovascular disease. The meaning of geographic and socioeconomic gradients is unclear but should challenge us to pursue entirely new lines of exploration. For example, diet and lifestyle differences among various populations and socioeconomic groups have long been a source of investigation. Lower socioeconomic groups are known to be at higher risk of certain infectious diseases and the potential for infection to play a role in the inflammatory component of cardiovascular disease is under intense scrutiny. Exposure to environmental factors such as pesticides that act as estrogen agonists may also vary among populations. As well, reproductive factors vary by population and socioeconomic status and may have an interesting role to play in cardiovascular risk.

Overall, then, we live in a time during which much of cardiovascular disease is already preventable given our current knowledge base. In addition, there are enticing clues that suggest we will learn more about factors such as genetics, inflammation, steroid hormones, folate metabolism, and infection that will improve our options for prevention. When it comes to reducing mortality among American women, cardiovascular disease is our greatest challenge. The stakes are high, our knowledge is already relatively substantial, but the room for improvement is great.

References

1. American Heart Association (1998). "Heart and Stroke Statistical Update. 1997." American Heart Association, Dallas, TX.
2. Centers for Disease Control and Prevention (1998). CDC wonder mortality. Available at: *http://wonder.cdc.gov/wonder*
3. Parker, S. L., Tong, T., Bolden, S., and Wingo, P. A. (1997). Cancer statistics 1997. *Ca—Cancer J. Clin.* **47,** 5–27.
4. Lyu, L., Shieh, M., Bailey, S., Dallal, G., Carrasco, W., Ordovas, J., Lichtenstein, A., and Schaefer, E. (1994). Relationship of body fat distribution with cardiovascular risk factors in healthy Chinese. *Ann. Epidemiol.* **4,** 434–44.
5. Reeder, B. A., Angel, A., Ledoux, M., Rabkin, S. W., Young, T. K., and Sweet, L. E. (1992). Obesity and its relation to cardiovascular disease risk factors in Canadian adults. *Can. Med. Assoc. J.* **146,** 2009–2019.
6. Donahue, R. P., Prineas, R. J., Bean, J. A., Donahue, R. A., Goldberg, R. B., Skyler, J. S., and Schneiderman, N. (1998). The relation of fasting insulin to blood pressure in a multiethnic population: The Miami Community Health Study. *Ann. Epidemiol.* **8,** 236–244.
7. Krauss, R. M. (1991). The tangled web of coronary risk factors. *Am. J. Med.* **90**(Suppl. 2A), 36S–41S.
8. Bjorntrop, P. (1994). Fatty acids, hyperinsulinemia, and insulin resistance: Which comes first? *Curr. Opin. Lipidol.* **5,** 166–174.
9. Taskinen, M.-R. (1995). Insulin resistance and lipoprotein metabolism. *Curr. Opin. Lipidol.* **6,** 153–160.
10. Bonner, G. (1994). Hyperinsulinemia, insulin resistance, and hypertension. *J. Cardiovasc. Pharmacol.* **24**(Suppl. 2), S39–S49.
11. Selby, J. V., Austin, M. A., Newman, B., Zhang, D., Quesenberry, C. P., Jr., Mayer, E. J., and Krauss, R. M. (1993). LDL subclass phenotypes and the insulin resistance syndrome in women. *Circulation* **88,** 381–387.
12. Witztum, J. L. (1993). Susceptibility of low-density lipoprotein to oxidative modification. *Am. J. Med.* **94,** 347–349.
13. Chait, A., Brazg, R. L., Tribble, D. L., and Krauss, R. M. (1993). Susceptibility of small, dense, low-density lipoproteins to oxidative modification in subjects with the atherogenic lipoprotein phenotype, pattern B. *Am. J. Med.* **94,** 350–356.
14. Tribble, D. L., Theil, P. M., Vandenberg, J. J. M., and Krauss, R. M. (1995). Differing 2-tocopherol oxidative lability and ascorbic acid sporing effects in byoyant and dense LDL. *Arterioscler. Thromb. Vasc. Biol.* **15,** 2025–2031.
15. Writing Group for the PEPI Trial (1995). The effects of estrogen or estrogen/progestin regimens on heart disease risk factors in postmenopausal women. *JAMA, J. Am. Med. Assoc.* **273,** 199–208.
16. Hulley, S., Grady, D., Bush, T., Furberg, C., Herrington, D., Riggs, B., Uittinghoff, E., for the HEART and Estrogen/Progestin Replacement Study (HERS) Research Group (1998). Randomized trial of estrogen plus progestin for secondary prevention of coronary heart disease in postmenopausal women. *JAMA, J. Am. Med. Assoc.* **280,** 605–613.
17. National Center for Health Statistics (1998). "Health, United States, 1998 with Socioeconomic Status and Health Chartbook." National Center for Health Statistics, Hyattsville, MD.
18. Rodin, J., and Ickovics, J. R. (1990). Women's health: Review and research agenda as we approach the 21st century. *Am. Psychol.* **45,** 1018–1034.
19. Talbott, E., Guzick, D., Kuller, L. H., and Berga, S. (1995). Coronary heart disease risk factors in women with polycystic ovary disease. *Arterioscler. Thromb. Vasc. Biol.* **15,** 821–826.
20. Manzi, S., Meilahn, E. N., Rairie, J. E., Conte, C. G., Medsger, T., Jansen-McWilliams, L., D'Agistono, R. B., and Kuller, L. H. (1997). Age-specific incidence rates of myocardial infarction and angina in women with systemic lupus erythematosus: Comparison with the Framingham Study. *Am. J. Epidemiol.* **145**(5), 408–415.
21. Chesley, L. C., Annillo, J. E., and Cosgrove, R. A. (1976). The remote prognosis of preeclamptic women: Sixth periodic report. *Am. J. Obstet. Gynecol.* **124,** 446–459.
22. Jonsdottir, L. S., Arngrimsson, R., Geirsson, R. T., Sigvaldason, H., and Sigfusson, N. (1995). Death rates from ischemic heart disease in women with a history of hypertension in pregnancy. *Acta Obstet. Gynecol. Scand.* **74,** 772–776.
23. Hannaford, P., Ferry, S., and Hirsch, S. (1997). Cardiovascular sequelae of toxaemia of pregnancy. *Heart* **77,** 154–158.

58

Overview of Risk Factors for Cardiovascular Disease

KATHERINE M. NEWTON, ANDREA Z. LACROIX, AND DIANA S.M. BUIST

Center for Health Studies, Group Health Cooperative; Department of Epidemiology, University of Washington, Seattle, Washington

I. Introduction

Despite declines in coronary heart disease (CHD) rates, CHD remains the leading cause of death in women, far exceeding death rates from cancer, pulmonary disease, and diabetes, (Fig. 58.1). Age-specific CHD death rates both in African-American and in white women lag about 10 years behind those of men. In 1995, the age-adjusted CHD death rate per 100,000 population for adults aged 45 and older was 544.5 for African-American women, 499.6 for white women, 610.3 for African-American men, and 918.8 for white men [1]. Although CHD rates increase with age, CHD is not only a disease of elderly women. Almost 20,000 women under the age of 65 die of CHD each year, and one-third of these women are under age 55. Many women with CHD continue to lead active and productive lives, but for others CHD is a chronic condition that requires medication and causes activity limitations due to ongoing angina or congestive heart failure. Annual costs attributed to CHD exceed 175 billion dollars, including health expenditures and lost productivity [2].

Until the 1990s there was a general lack of data on women and heart disease. Although women were included in many im-

portant observational studies in the United States, including the Framingham Heart Study and the Nurses Health Study, women have been excluded from almost all of the major randomized trials about CHD prevention. Fortunately the research climate is changing and women are increasingly included in, or are the primary target of, randomized trials. Results from randomized trials that include both women and men are more often reported stratified by gender. These trials and observational cohort studies are rich sources of data on the association of a wide variety of risk factors with cardiovascular disease in women.

In this chapter we briefly review the major known risk factors for cardiovascular disease in women, using data from the Third National Health and Nutrition Examination Survey (NHANES III) to demonstrate the prevalence of CHD risk factors in U.S. women [2a]. The age and race/ethnicity distribution for these data are shown in Appendix I. The association of these risk factors with cardiovascular morbidity and mortality in women are reviewed. When available, the results of intervention trials to alter CHD outcomes through risk factor modification are presented. We refer the reader to other chapters in this section for detailed discussions about the effects of hormonal milieu,

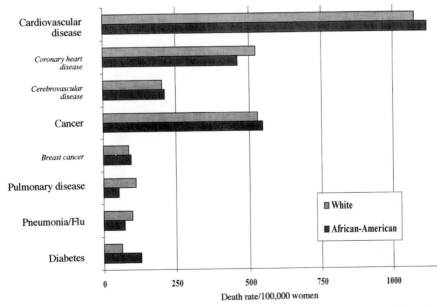

Fig. 58.1 Age-adjusted rates of death per 100,000 for common causes of death among white and African-American U.S. women, ages 45+, 1994–1995. (ICD-9-CM codes: total cardiovascular disease, 390–459.9; coronary heart disease, 410–414.9; cancer, 140–208.9; cerebrovascular disease, 430–438.9; breast cancer, 174–175.9; pulmonary disease, 490–496.9; pneumonia/flu, 480–487.9; diabetes, 250–250.9). Source: [1].

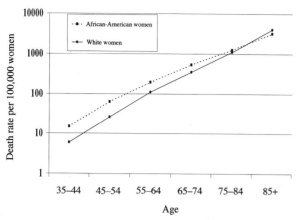

Fig. 58.2 Age-specific coronary heart disease death rates per 100,000 among white and African-American U.S. women, 1994–1995. Source: [1].

Table 58.1

Prevalence Rates of Self-Reported Prior Myocardial Infarction among Women by Age and Race/Ethnicity

Age group	Race/Ethnicity[a]			
	Caucasian	African-American	Mexican-American	Other
20–29	0.0	0.1	0.1	0.0
30–39	0.0	0.1	0.7	0.0
40–49	0.5	1.7	0.5	0.0
50–59	1.9	5.2	3.3	7.5
60–69	5.3	8.3	4.0	2.7
≥70	11.9	11.1	6.6	6.0

Source: [2a].
[a]Percentages.

thromboembolism, diabetes, obesity, and lipid disorders on CHD risk.

II. Demographic Characteristics

As shown in Figure 58.2, rates of coronary heart disease mortality among women rise exponentially with age. At ages younger than 75 years, African-American women have higher rates of coronary death than white women. At older ages the rates converge. Consistent with the trends for coronary mortality, self-reported prevalence rates of prior myocardial infarction from NHANES III also rise markedly with age and are higher for African-American women than for white women for age groups younger than 70 years (Table 58.1). Among women in their seventies, 11–12% of white and African-American women report a history of heart attack, whereas 6–7% of Latino and other minority women report having had a heart attack. Stable, recent estimates of incidence rates of coronary events by age, race, and ethnicity are difficult to come by, particularly those based on national samples. In the NHANES I Epidemiologic Follow-up Study, cumulative incidence rates of hospitalization for first myocardial infarction were 4.7% and 3.9% over 10 years of follow-up in white and African-American women, respectively [3]. These rates show a different pattern than national rates of hospital admission for acute myocardial infarction, which are higher for nonwhite women than for white women in age groups younger than 60 years [3]. Thus, differences in incidence rates of myocardial infarction among women of different races are neither consistently documented nor well understood.

In contrast, educational disparities in coronary mortality and morbidity rates are well documented, with the higher rates consistently occurring among men and women with the lowest educational attainment. Feldman [4] showed that coronary mortality rates between 1971 and 1984 were inversely associated with level of education among U.S. women. Despite the decline in coronary mortality observed during this period, women with the lowest educational attainment (<8 years) in 1984 had not yet experienced rates of mortality as low as highly educated women in 1971. In the Charleston Heart Study, years of education were inversely related to 30-year risk of coronary and all

cause mortality among African-American women but not among white women [5]. Inverse associations between educational attainment and coronary risk factors are also well documented. Thus, prevalence rates of smoking, high blood pressure, higher levels of total serum cholesterol, and lower levels of high density lipoprotein cholesterol are all more common among women with lower educational attainment compared to women with high levels of education [6,7]. These differentials in coronary risk factors typically account only partially for educational differentials in morbidity and mortality.

Thus, age, ethnicity, and education are important determinants of coronary disease risk, although the reasons for these differences are only partially understood. In the sections that follow, we examine other risk factors for coronary disease in women separately by age and ethnicity.

III. Family History of Cardiovascular Disease

Reporting of family history of CHD varies by race. In the NHANES III Survey, among women under age 50, white women are more likely than are African-American or Mexican-American women to report a family history of early myocardial infarction (before age 50). Among women age 50 years and older these racial differences in reporting of family history diminish (Table 58.2).

A history of myocardial infarction in one first-degree relative doubles, and in two or more first-degree relatives triples, myocardial infarction risk [8]. The risk for myocardial infarction is strongest when myocardial infarction in first-degree relatives occurs before ages 55 to 60 years [8,9], but risk is still elevated when myocardial infarction occurs in relatives at later ages. The risk associated with a family history of CHD is independent of other known CHD risk factors [9–11]. In the Nurses Health Study of 117,000 women aged 30 to 55 years, women with a history of parental myocardial infarction at or below age 60 years had 2.8 times the risk of nonfatal myocardial infarction, 5.0 times the risk of fatal CHD, and 3.4 times the risk for angina pectoris compared to women without a history of parental myocardial infarction [12]. For women with a history of parental myocardial infarction after age 60, there was no increase in risk for nonfatal CHD, but risk for fatal CHD was increased 2.6

Table 58.2

Family History of Myocardial Infarction in a First-Degree Relative before Age 50 among Women by Age and Race/Ethnicity

Age group	Race/Ethnicity[a]			
	Caucasian	African-American	Mexican-American	Other
20–29	23.0	13.3	11.3	10.3
30–39	24.9	13.5	11.4	17.1
40–49	23.4	14.3	16.1	8.9
50–59	16.1	15.2	17.1	4.1
60–69	17.4	11.1	10.2	8.1
≥70	9.5	4.6	10.7	6.1

Source: [2a].
[a]Percentages.

times and angina pectoris risk was increased 1.9 times compared to women with no family history of CHD [12]. Risk estimates were almost identical in a population-based study in Finland [9].

A family history of CHD puts women at increased risk for CHD events, probably due both to genetic and to environmental factors [8,13,14]. For example, siblings of persons with CHD before age 60 who are themselves less than age 60 years and free of CHD are more likely to have cholesterol levels that meet the criteria for dietary or drug therapy than the general population [15]. Women aged 18 to 39 years with a family history of CHD report less physical activity and higher BMI than women without a family history of CHD [16].

Twin studies shed further light on the influence of family history on CHD risk. In a study of female Swedish monozygotic and dizygotic twins, the relative risk of CHD for monozygotic twins was 15 and the relative risk for dizygotic twins was 2.6 when one twin died of CHD before age 55 [10]. In both mono-

zygotic and dizygotic twins, as the age at which one twin died increased, the risk for CHD among the remaining twin decreased.

Thus, family history of premature coronary disease substantially influences a woman's risk of heart disease. Such women represent a high risk group that should be the focus of intensified preventive interventions.

IV. Cigarette Smoking

In the United States, 40% of white women aged 20–29 smoke cigarettes compared with 25% of African-American women, 14% of Mexican-American women, and 7% of women in other ethnic minority groups (Fig. 58.3). Among middle-aged women, African-American women have the highest prevalence rates of cigarette smoking (32–35%). About one in five white and African-American women aged 60–69 years continue to smoke cigarettes. Since 1975, prevalence rates of smoking have declined only modestly in U.S. women [17].

Smoking is a powerful risk factor for coronary heart disease in women. In the Nurses Health Study, the relative risk of coronary disease mortality over 12 years of follow-up was 4.2 (95% confidence interval (CI) = 3.6–5.0) among current compared to never smokers. In the subgroup of women who started smoking before the age of 15, the relative risk was 9.3 compared to never smokers [18]. Even very low levels of current cigarette smoking (1 to 4 cigarettes per day) doubled the risk of coronary disease mortality compared to never smokers. Among women aged 65 and older in the Established Populations for Epidemiologic Studies of the Elderly, the age-adjusted relative risk of coronary mortality was 1.7 (95% CI = 1.3–2.4) in current compared to never smokers [19], a finding that has been reproduced in other older cohorts of women [17,20].

Smoking was also associated with early atherosclerosis, specifically more extensive fatty streaks and raised lesions in the abdominal aorta, in an autopsy study of young women decedents

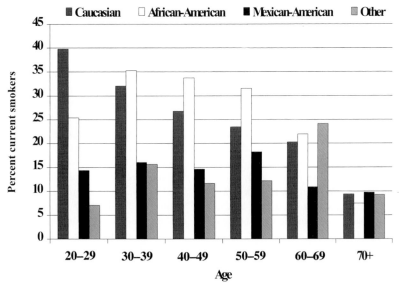

Fig. 58.3 Prevalence of current smoking among U.S. women by age and race/ethnicity. Source: [2a].

aged 15 to 34 years [21]. Nonsmoking women who live or work around smokers also incur increased coronary risk. Regular exposure to passive smoking at home or at work was associated with 1.9 times the risk of fatal and nonfatal coronary disease in the Nurses Health Study compared to those not exposed [22].

Smoking cessation dramatically reduces the risk of coronary disease events associated with cigarette smoking [17–19], even among women who have established coronary artery disease [23]. In the Nurses Health Study, within 2 years of quitting smoking there was a 24% reduction in risk of coronary events compared to current smokers, and level of risk diminished to that of never smokers within 10 to 14 years of quitting [18]. Among older women, rates of coronary death among former smokers resemble those of never smokers within 5 years of quitting [19].

Effective intervention programs exist to help women stop smoking. The best ones combine state-of-the-art behavioral strategies with physician advice, options for nicotine replacement, and consistent short- and longer-term telephone follow-up with a nurse or other health professional. Such programs can result in 1-year smoking cessation rates of 30% [24]. Given the substantial risks that cigarette smokers incur, it is imperative to have multilevel public health strategies that encourage smoking cessation in women of all ages and ethnicities.

V. High Blood Pressure

Prevalence rates of high blood pressure, defined as systolic blood pressure ≥140 mm Hg or diastolic blood pressure ≥90 mm Hg, or self-reported use of antihypertensive medications, increase markedly with age in women of all race/ethnicities (Fig. 58.4). Prevalence of high blood pressure in the U.S. is highest among African-American women, intermediate among Mexican-American women, and lowest among other minority women and white women [2a]. By their sixties, three-fourths of African-American women are classified as hypertensive by the above definition, as are one-half of women of other race/ethnicity groups. Incidence rates of high blood pressure during 9.5 years of follow-up ranged from 8% in white women aged 25–34 years to 47% in women aged 65 years and older in the NHANES I Follow-up Study [25]. At any given age, incidence rates were approximately twice as high among African-American women in that study.

High blood pressure undisputedly increases a woman's risk of stroke, coronary heart disease, and death. In the Framingham Heart Study, after 30-years of follow-up, women with definite or treated high blood pressure experienced 2–4 times the risk of these cardiovascular events compared with normotensive women [26]. In the Nurses Health Study, women who reported having hypertension had 3.5 times the risk of coronary events and 2.6 times the risk of stroke as normotensive women [27]. Relative risks relating high blood pressure to stroke and coronary disease may decline with age in women. In the Bergen Study, women with systolic blood pressure ≥160 mm Hg vs <160 mm Hg had age-specific relative risks of CHD mortality of 3.8 in 30–39 year olds, 2.9 in 50–59 year olds, 1.7 in 70–79 year olds, and 1.3 in 80–89 year olds followed for 20 years [28]. In that study, a single screening blood pressure measurement was able to predict coronary and stroke mortality risk 10 to 15 years into the future.

Risk factors for high blood pressure are similar for white and African-American women. Higher blood pressure is associated with age, obesity, weight gain, diabetes, impaired glucose tolerance, sodium intake, and increased alcohol consumption, whereas lower blood pressure is associated with smoking and potassium and protein intake [29–31].

In recent years, women have become more aware of their own hypertension diagnosis. Between NHANES II conducted in 1976–1980 and NHANES III conducted in 1988–1994, hypertension awareness increased from 77% to 79% among African-American women and from 61% to 82% among white women. Levels of hypertension treatment increased from 49% to 65% for African-American women and from 43% to 65% for white women [32]. Using the more conservative thresholds of 160/95 mm Hg for defining high blood pressure, 59% of African-American women with high blood pressure were controlled by treatment in NHANES III compared with 75% of white women [32].

In trials of older women, cardiovascular and stroke mortality were reduced substantially by active treatment among women in the European Working Party on Blood Pressure in the Elderly Study [33] and in the Systolic Hypertension in the Elderly Program (SHEP) [34]. No distinctions are made between race/ethnicity groups in current national guidelines for treatment of high blood pressure [35]. This is true despite the fact that some treatment trials have shown greater and more consistent benefits in lowering coronary and total mortality for African-American women than for white women [29]. In the SHEP trial, African-American women experienced more dramatic reductions in stroke incidence than white women (63.5% vs 26.6%), although the benefits were apparent in both race groups.

Detection and treatment of high blood pressure has major importance as a public health strategy for reducing coronary risk in women. Because short-term rates of coronary disease and stroke are low in young adult women, and because nonpharmacologic therapies including diet and exercise have shown promise in reducing high blood pressure, consideration should be given to lifestyle modification prior to initiating long-term drug therapy in younger women.

VI. Lipids and Lipoproteins

By the sixth decade of life, 29% of Hispanic women, 48% of African-American women, and 44% of white women have a total cholesterol of ≥240 mg/dl or report using a cholesterol lowering medication (Table 58.3). Elevated serum total cholesterol and low-density lipoprotein cholesterol (LDL) are associated with an increased risk of CHD in younger and middle-aged women, but to a far lesser degree in older women [36–38]. The weaker effect seen in older women may reflect confounding by the decrease in cholesterol associated with weight loss and poor health [37,39]. In the Framingham study, women with serum cholesterol concentrations >295 mg/dl had over 3 times the risk of myocardial infarction and definite coronary events than those with cholesterol concentrations <204 mg/dl [38]. In the NHANES I Epidemiologic Follow-up Study, serum cholesterol was only weakly associated with CHD risk in African-American women [40]. This weak association in African-American women may be explained by the fact that high density lipoprotein

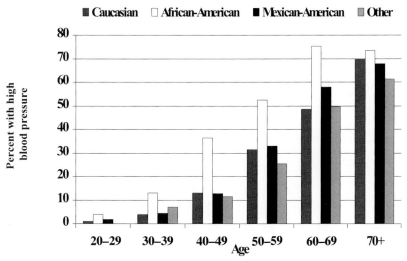

Fig. 58.4 Prevalence of high blood pressure among U.S. women by age and race/ethnicity. High blood pressure: diastolic ≥90 mm Hg, systolic ≥140 mm Hg, or self-reported use of antihypertensive medication. Source: National Center for Health Statistics. National Health and Nutrition Examination Survey III, 1988–1994.

(HDL) cholesterol levels in African-American women are generally higher than those of same-aged white women [41]. Serum cholesterol levels also influence prognosis after myocardial infarction. The risk for reinfarction is 9.2 times greater when serum cholesterol levels are 275 mg/dl or higher compared with levels <200 mg/dl [42].

Serum high-density lipoprotein cholesterol (HDL) is negatively associated with CHD. Among women, a 1 mg/dl increment in HDL is associated with a 3% decrease in total CHD risk and a 4.7% decrease in CHD mortality [43]. Furthermore, at any given level of cholesterol, higher levels of HDL confer protection against CHD [44].

Attention has been focused on subfractions of HDL and LDL: apolipoprotein AI (the major protein of LDL), apolipoprotein B (the major protein of HDL), and Lp(a) (a lipoprotein composed

of an LDL molecule bound to apolipoprotein (a)) [45]. In a study of the predictors of premature CHD at coronary arteriography, Kwiterovich found that in women ≤60 years old apolipoprotein B was associated with an increase in CHD risk (RR = 1.7, 95% CI = 1.0–2.9) [45]. Increasing levels of Lp(a) are also associated with an increase in CHD risk [46–49]. In the Framingham study, the relative risk for CHD associated with elevated Lp(a) was 1.6 (95% CI = 1.1–2.3) in women [49] and 1.9 (95% CI = 1.2–2.9) in men [48]. LDL subclass patterns also influence CHD risk. Compared with light buoyant LDL, small dense LDL is associated with a threefold increase in risk of myocardial infarction [50].

Reproductive hormones have a major impact on serum lipid levels in women. During menopause total serum cholesterol concentrations rise by an average of 19% [51]. However, use of postmenopausal estrogen replacement increases HDL cholesterol by 9–13% and lowers LDL by 10% [36]. In the Postmenopausal Estrogen/Progestin Intervention study, hormone replacement therapy was associated with more than a 15% drop in Lp(a) [52].

There is little information about the effectiveness of interventions to lower cholesterol on CHD risk in women. However, one of the few studies of the benefits of cholesterol lowering in women was the Scandinavian Simvastatin Survival Study. After a median of 5.4 years of follow-up, the relative risk for major CHD events was 0.66 (95% CI = 0.48–0.91) among women assigned to the intervention drug compared to those in the placebo group [53].

Table 58.3
Prevalence of Hypercholesterolemia[a] among Women by Age and Race/Ethnicity

Age group	Race/Ethnicity[b]			
	Caucasian	African-American	Mexican-American	Other
20–29	6.9	7.3	7.2	5.8
30–39	7.8	7.3	8.8	12.8
40–49	20.3	16.4	16.8	24.0
50–59	39.1	35.1	27.3	25.1
60–69	44.2	47.8	39.7	46.2
≥70	43.4	39.6	28.9	39.5

Source: [2a].

[a]Based on self-reported use of cholesterol lowering medication or a total serum cholesterol value ≥ 240 mg/dl.

[b]Percentages.

VII. Physical Activity, Exercise, and Sedentary Lifestyle

The 1996 Surgeon General's Report on Physical Activity and Health [54] provides a review of 36 studies published between 1953 and 1995 on the relationship between physical activity and coronary heart disease. Only four of those studies included women. The findings of all four studies showed inverse

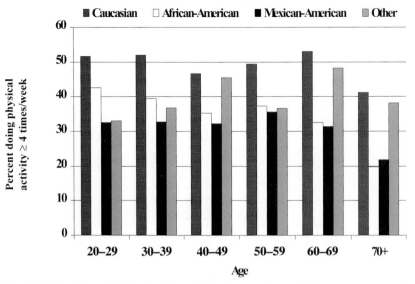

Fig. 58.5   Prevalence of physical activity ≥ 4 times per week among U.S. women by age and race/ethnicity. Physical activity includes walking, jogging or running, bicycling, swimming, aerobics, dancing, calisthenics, garden/yard work, and/or lifting weights. Source: National Center for Health Statistics. National Health and Nutrition Examination Survey III, 1988–1994.

associations between physical activity and risk of coronary disease outcomes, although definitions of physical activity varied (for example, occupational vs leisure time). Several other studies provide stronger evidence that physical activity affects coronary disease risk. A nested case-control study (50 cases, 150 controls) of acute coronary disease in women found a relative risk of 2.1 for sedentary (aerobic physical activity reported less than once per week) compared to more active women [55]. In a prospective study of 1030 women aged 65 years and older, walking 4 or more hours per week was associated with a reduced relative risk of cardiovascular hospitalization (RR = 0.63, 95% CI = 0.44–0.90) and death (RR = 0.45, 95% CI = 0.25–0.83) [56]. Among 7852 middle-aged women in the Atherosclerosis Risk in Communities Study (ARIC), indices of sports and leisure time physical activity were both similarly associated with 45–49% reduced risk of coronary heart disease incidence over 4 to 7 years [57]. Moreover, physical fitness levels measured by treadmill performance, which are partially influenced by physical activity levels in women [58], show a strong dose-response relationship with cardiovascular death [59] and with levels of several coronary risk factors [60]. Thus, although the studies remain few in number, consensus has emerged that regular physical activity is associated with reduced risk of coronary disease in women.

Current national guidelines recommend 30 minutes of moderate physical activity most days of the week [54]. When applying that definition to NHANES III data, as shown in Fig. 58.5, about half of white women reported some form of physical activity four or more times per week compared with 33–43% of African-American women and about one-third of Mexican-American women aged 20–69 years. Activity levels among women aged 70 and older were somewhat lower. Walking, the single most common form of physical activity in Americans over age 50, is reported by about one in five white and African-

American women and about one in six Mexican-American women, using the definition of walking at least one mile four or more times per week (Fig. 58.6). Thus, half or more of American women could be targeted for public health interventions aimed at increasing physical activity levels.

Long-term physical activity intervention trials with coronary endpoints have not been conducted among women. Smaller trials with intermediate risk factor endpoints such as lipid levels do not consistently show improvements in HDL-cholesterol levels [61]. However, in the Stanford Five-City Project, among 427 women aged 18–74 years, those who increased their physical activity levels over 5 years were found to have higher HDL-cholesterol levels and lower resting pulse rates than other women [62]. Perhaps the greatest challenge in thinking about the potential for physical activity to reduce rates of coronary disease among women is the development of behavioral interventions with proven effectiveness in establishing enduring, lifelong patterns of regular physical activity.

VIII. Diabetes Mellitus

The American Diabetes Association diagnostic criteria for diabetes mellitus is a random blood glucose ≥200 mg/dl or a fasting blood glucose ≥126 mg/dl [63]. In the NHANES III study, using the criteria of self-reported diabetes or a fasting plasma glucose ≥126 mg/dl, the prevalence of diabetes in African-American and Hispanic women was two to three times that of white women in the U.S. (Fig. 58.7). By age 60–69 years, about 11% of white women and 27% of African-American and Hispanic women have diabetes.

CHD and myocardial infarction rates in diabetic women approach those of nondiabetic men of similar age, diminishing, but not completely eliminating, the advantage found in nondiabetic women compared to men [64]. This is true for white [64],

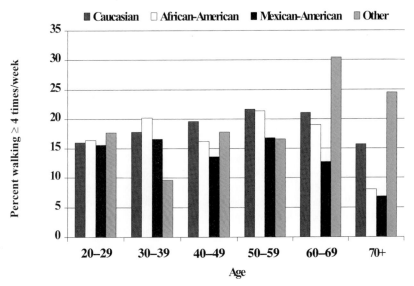

Fig. 58.6 Prevalence of walking at least one mile without stopping ≥ 4 times per week among U.S. women by age and race/ethnicity. Source: National Center for Health Statistics. National Health and Nutrition Examination Survey III, 1988–1994.

Mexican-American [65], and Japanese [66] women. Diabetes is associated with an increase in the incidence of CHD across the life span [67]. In women with diabetes, the incidence of CHD is increased five- to sevenfold [11,67], the rate of myocardial infarction is increased four- to sixfold [11,67,68], and ischemic heart disease mortality is tripled compared to rates in nondiabetic women [69]. The incidence of peripheral vascular disease is eight times higher, and the incidence of atherothrombotic stroke is 3–5 times higher, in women with diabetes compared with nondiabetics [67,70].

Diabetic women with CHD have a significantly poorer prognosis than do nondiabetic women with CHD [71,72]. After myocardial infarction, persons with diabetes are 2–4 times as likely to die in the hospital, are more likely to develop congestive heart failure and postinfarction angina pectoris, and are more likely to extend their infarct than are nondiabetics [73]. Among survivors of an initial myocardial infarction, the incidence of recurrent myocardial infarction is almost tripled, fatal CHD is doubled [74], and total mortality is 1.5–3 times that of nondiabetics [73]. Whether the degree of control of diabetes after

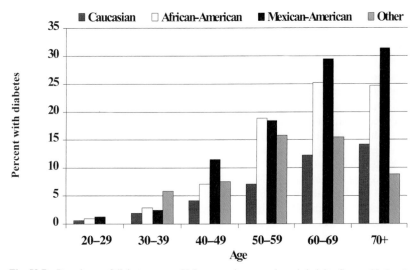

Fig. 58.7 Prevalence of diabetes among U.S. women by age and race/ethnicity. Source: National Center for Health Statistics. Based on self-report, fasting glucose levels (≥126 mg/dl), and random (nonfasting) glucose levels (≥200 mg/dl). National Health and Nutrition Examination Survey III, 1988–1994.

myocardial infarction affects survival after myocardial infarction is unknown.

Insulin resistance is also associated with CHD. Among women without diabetes in the ARIC study, the relative risks of CHD across quintiles of fasting insulin were 1.0 (<35 pmol/l), 0.76 (35.0–49.9 pmol/l), 2.08 (50.0–69.9 pmol/l), 2.08 (70.0–99.9 pmol/l), and 2.82 ($\geq$100 pmol/l) [75]. Among women in a Finnish study of factors associated with CHD, the prevalence of CHD, defined by symptoms and/or electrocardiographic changes, across quartiles of fasting plasma insulin, was 28.1% (quintiles I and II, <16.4 milliunits/l), 34.9% (III+IV, 16.4–30 milliunits/l), and 44.3% (V, >30.0 mulliunits/l), respectively [76].

The mechanisms responsible for the acceleration of myocardial dysfunction and atherosclerosis associated with diabetes are the subject of great scrutiny [77,78]. Diabetes, hyperinsulinemia, and insulin resistance are associated with higher relative weight (specifically with a central body fat distribution), higher systolic and diastolic blood pressure, lower levels of HDL, and higher total cholesterol, LDL, and triglyceride levels [67,70, 77,78]. These factors appear to be linked, possibly through a complex set of genetic and environmental factors.

The Diabetes Control and Complications Trial was the first randomized trial to show that intensive therapy to improve glycemic control reduced the risk of laser treatment of retinopathy, clinical neuropathy, and microalbuminuria in Type 1 diabetics [79]. There was also a trend towards the reduction of macrovascular events. Though clinical trials are currently underway, at this time there is no evidence that interventions to improve diabetic control in those with Type II diabetes prevents CHD events. However, because other CHD risk factors cluster in women with diabetes, particular attention should be focused on altering those risk factors where change is known to make a difference in CHD risk, including hypertension, hypercholesterolemia, and smoking.

IX. Body Weight, Weight Distribution, Obesity, and Weight Loss

Overweight is a tremendous problem among U.S. women. For women, overweight is defined as a body mass index (BMI) > 27.2 kg/m². This represents the 85th percentile of weight for women aged 20–29 in the NHANES II study or 120% of desirable weight for medium-framed women from the 1983 Metropolitan Life Insurance tables [80]. For example, a weight above 155 pounds for a woman 5 feet 3 inches tall or above 170 pounds for a woman 5 feet 6 inches tall places a woman in the category of ''overweight.'' In the National Health Examination Survey I (1960–1962), the age-adjusted prevalence of overweight was 23.6% among white women and 41.6% among African-American women [80]. By the time of the NHANES III study (1988–1994) [2a], 33.2% of white women and 48.6% of African-American women were overweight. Overweight increased with age, and by mid-life 47.6% of white women, 66.4% of African-American women, and 65.1% of Hispanic women were overweight (Fig. 58.8).

Relative weight predicts angina pectoris, CHD other than angina, CHD death, stroke, and congestive heart failure in women [81]. In the Nurses Health Study, women in the highest quartile of BMI (>29 kg/m²) had a relative risk for nonfatal myocardial

infarction and fatal CHD of 1.8 compared with women in the lowest quartile (BMI < 21 kg/m²), and their risk of angina was double that of women in the lowest quintile [82]. These relationships hold even among women of normal, or close to normal, weight; those with BMI of 25–28.9 kg/m² and 23–24.9 kg/m² had a relative risk for CHD of 2.06 and 1.45 respectively, compared with women with a BMI <21 kg/m² [83]. In the Framingham Study, obese women (Metropolitan Relative weight $\geq$ 130%) aged 50 or younger had a 2.4-fold increase in CHD risk over 26 years of follow-up compared with lean women (Metropolitan Relative Weight < 110%) [81]. Overweight also has a role in secondary prevention. Among women who have survived a first myocardial infarction, a 1-unit increase in BMI (for example, a change from 22.0 kg/m² to 23.0 kg/m²) is associated with a 3% increase in risk of reinfarction [84].

The relationship between overweight and CHD is weaker, and perhaps absent, among African-American women. In African-American women, BMI is unrelated to CHD death [5,85,86] and to all-cause mortality [85,86]. The reasons for a lack of an association between BMI and CHD risk in African-American women is not understood at this time.

Overweight continues to play a role in CHD risk among elderly women. In the NHANES I Epidemiologic Follow-up Study, BMI of 27 kg/m² or more in midlife was associated with a 70% increase in CHD risk in later life [87]. Among women 65 years of age and older, total mortality and CHD mortality were as much as doubled in those above the 70th percentile of BMI compared with those in the 10th to 29th percentile [88]. These relationships were also present, but were less strong, in women older than age 50 at entry into the study [81].

Being overweight increases a woman's risk for other important CHD risk factors. Hypertension, diabetes, and hypercholesterolemia are all more common in overweight women [89], and weight reduction is an important therapy in the management of each of these CHD risk factors.

Weight gain is associated with increased CHD risk. In the Nurses Health Study a self-reported gain of 20–34.9 kg (8 years observation) was associated with a 2.5-fold increase in risk of nonfatal myocardial infarction and fatal CHD [82]. Even among those of normal weight, those with a gain of 20 kg or more had a 25% increase in CHD risk [83]. Weight gain has also been associated with increased CHD risk in elderly women [87], but not in African-American women [85]. Again, the reasons for this lack of an association in African-American women has not been explained.

The distribution of body fat contributes to the association between overweight and CHD risk in white women. Central adiposity, often measured as waist-to-hip ratio, is positively associated with the incidence of myocardial infarction, angina pectoris, stroke, death, and CHD mortality in women [90,91]. In the Charleston Heart Study, white women in the 85th percentile of abdominal circumference had a 50% increase in all-cause mortality and in CHD mortality compared with those in the 15th percentile. After adjusting for CHD risk factors, including BMI, women in the highest tertile of waist-to-hip ratio had twice the risk of CHD mortality as women in the lowest waist-to-hip ratio tertile. However, there was no association between abdominal circumference or waist-to-hip ratio and all-cause or CHD mortality in African-American women [85].

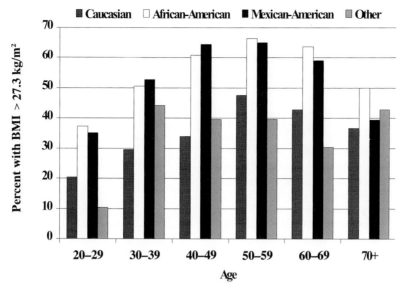

Fig. 58.8 Prevalence of overweight (body mass index (BMI) ≥ 27.3 kg/m²) among U.S. women by age and race/ethnicity. Source: National Center for Health Statistics. National Health and Nutrition Examination Survey III, 1988–1994.

We are unaware of trials of the effects of weight loss on cardiovascular disease risk in women. Nevertheless, weight loss is frequently recommended for its beneficial effects on cholesterol, blood pressure, insulin resistance, and glycemic control.

X. Folate and Homocysteine

Homocysteine is an amino acid that is an intermediate by-product of the metabolism of the dietary protein methionine. Homocysteine levels are higher in postmenopausal women than in premenopausal women, lower in women than in men, positively correlated with age, and negatively correlated with serum folate and dietary folate intake [92–94]. Methylenetetrahydrofolate (MTHFR) is an enzyme that contributes to the remethylation of homocysteine to methionine. This reaction requires folate as the substrate, thus levels of dietary and blood folate are strong determinants of homocysteine levels [95]. Normally present in only small amounts (<10 μmol/l), homocysteine is highly toxic to the vascular endothelium. Individuals with homocysteinuria, a rare autosomal recessive condition in which homocysteine levels are severely elevated, are at extremely high risk for premature atherosclerosis. A common variation in MTHFR, thermolabile MTHFR, is associated with reduced MTHFR activity and mild elevations in homocysteine [95].

Even mild elevations in homocysteine appear to increase CHD risk. In a case-control study of myocardial infarction among women younger than 45 years, those in the highest quartile (>15.6 μmol/l) of homocysteine had 2.3 times the risk of myocardial infarction compared with those in the lowest quartile (<10 μmol/l). This relationship was reversed for serum folate; women in the highest quartile of folate had half the risk of those in the lowest quartile [95]. In a study of French-Canadian women with CHD, 44% had homocysteine levels greater than the 90th percentile of control levels [94]. Mean homocysteine levels were 12.0 ± 6.3 μmol/l in women with CHD compared to 7.6 ± 4.1 μmol/l for normal controls [94]. Homocysteine levels are higher in women with angiographically confirmed obstruction of one or more major coronary arteries than in women with normal coronary arteries [96,97]. In the Framingham Heart Study, high plasma homocysteine concentrations and low concentrations of folate were associated with increased risk of extracranial carotid artery stenosis in elderly women [98]. In the ARIC study, women in the highest quintile of plasma homocysteine had a relative risk of 4.8 for carotid artery intimal-medial wall thickening compared with women in the lowest quintile [99].

Increasing homocysteine levels after menopause may partially explain the increase in CHD risk associated with aging. Postmenopausal women have an excessive rise in homocysteine after a methionine load compared to premenopausal women. Folic acid supplementation decreases this rise [100]. In a 2-year prospective study, hormone replacement therapy was associated with an 11% decrease in homocysteine. The decrease was 17% in women with high homocysteine levels, whereas levels in women with low homocysteine did not change [101]. Randomized trials of the effects of hormone replacement therapy and CHD risks may shed further light on these relationships.

XI. Alcohol

Moderate alcohol consumption has been consistently associated with a reduction in CHD risk [102–105]. In the Nurses Health Study the relative risk of severe CHD decreased as alcohol intake increased. Compared with nondrinkers, the relative risk for CHD was 40% lower among women who consumed 5.0 to 14.9 g of alcohol per day (3–9 drinks/week) and CHD risk was 60% lower in women who consumed ≥25 g/day of alcohol

[102]. Similar results have been found among women in Scotland [104], Australia [105], New Zealand [103], and Italy [106]. However, because of the health hazards associated with alcohol, it is not generally recommended that women initiate alcohol consumption to reduce CHD risk.

XII. Antioxidants

Epidemiologic findings that antioxidants decrease CHD risk are supported by evidence that oxidized LDL is present in atherosclerotic lesions [107]. Accumulation of cholesterol within the arterial intima is the hallmark of early atherosclerotic lesions, and oxidation of LDL appears to enhance the accumulation of cholesterol within atherosclerotic lesions [107].

Interest in the effects of antioxidants and CHD risk has centered on vitamins E and C and on beta-carotene. Although observational studies suggested that consumption of foods high in beta-carotene might reduce CHD risk in women [108,109], the results of randomized trials in men [110,111] and in men and women [112] show no benefit of beta-carotene supplementation on CHD risk. Similarly, evidence that vitamin C reduces CHD risk in women appears weak or lacking [108,113]. However, the evidence of cardiovascular benefit is stronger for vitamin E [114,115]. In the Nurses Health Study, the rate of major CHD (nonfatal myocardial infarction and coronary deaths) was 35% lower among women in the highest quintile of vitamin E consumption compared to women in the lowest quintile [115]. The authors found that most of the variability in vitamin E consumption was due to supplements. In the Iowa Women's Health Study, among women who did not use supplements, women in the highest quintile of Vitamin E intake had a 60% reduction in CHD death compared with women in the lowest quintile, but there was no association between vitamin A or vitamin C intake and CHD risk [113].

Randomized trials of the effect of vitamin E on the primary and secondary prevention of CHD in women are underway. Recommendations about the use of vitamin E for CHD prevention should await the results of these trials.

XIII. Summary, Conclusions, and Future Directions

Despite declines in coronary heart disease rates, CHD remains the leading cause of death in women in the United States. Family history of CHD, cigarette smoking, hypertension, diabetes mellitus, sedentary lifestyle, levels of high and low density lipoprotein cholesterol, and obesity are major risk factors for CHD in women, although observational studies support racial/ethic differences in the relative contributions of these factors to CHD risk. Low folate and high homocysteine levels are emerging as potentially important factors as well. The public health implications of these CHD risk factors in women are tremendous in their social, emotional, and fiscal impact on individuals and to the larger health care system. Randomized trials are urgently needed to develop interventions that will assist women in modifying known CHD risk factors and to test the efficacy of such changes on CHD risk. Trials are particularly crucial for minority women, where the CHD burden is greatest.

Appendix 58.1

Age and Race/Ethnicity Distribution among U.S. Women Sampled in the National Health and Nutrition Examination Survey, III 1988–1994[a]

Age group	Race/Ethnicity			
	Caucasian n (%)	African-American n (%)	Mexican-American n (%)	Other n (%)
20–29	546 (19.7)	648 (25.7)	703 (34.2)	85 (24.3)
30–39	649 (22.2)	685 (26.7)	558 (27.0)	82 (27.7)
40–49	527 (18.0)	479 (18.2)	392 (17.6)	71 (20.7)
50–59	548 (12.7)	297 (11.0)	204 (9.5)	56 (11.4)
60–69	570 (12.5)	331 (9.6)	360 (7.8)	49 (10.3)
≥70	1572 (14.9)	331 (8.8)	209 (3.9)	57 (5.7)

[a]Data displays unweighted sample size and weighted percentages. Source: [2a].

References

1. Centers for Disease Control and Prevention, Atlanta, GA. (1998). "CDC Wonder. Mortality." *http://wonder.CDC.gov/WONDER* (August, 1998).
2. American Heart Association (1997). "1998 Heart and Stroke Statistical Update." American Heart Association, Dallas, TX.
2a. National Center for Health Statistics (1988–1994). "Third National Health and Nutrition Examination." Natl. Cent. Health Stat., Washington, DC.
3. Cooper, R. S., and Ghali, J. K. (1991). Coronary heart disease: Black-white differences. *Cardiovasc. Clin.* **21,** 205–225.
4. Feldman, J. J., Makuc, D. M., Kleinman, J. C., and Cornoni-Huntley, J. (1989). National trends in educational differentials in mortality. *Am. J. Epidemiol.* **129,** 919–933.
5. Keil, J. E., Sutherland, S. E., Knapp, R. G., Lackland, D. T., Gazes, P. C., and Tyroler, H. A. (1993). Mortality rates and risk factors for coronary disease in black as compared with white men and women. *N. Engl. J. Med.* **329,** 73–78.
6. Winkleby, M. A., Jatulis, D. E., Frank, E., and Fortmann, S. P. (1992). Socioeconomic status and health: How education, income, and occupation contribute to risk factors for cardiovascular disease. *Am. J. Public Health* **82,** 816–820.
7. Kaplan, G. A., and Keil, J. E. (1993). Socioeconomic factors and cardiovascular disease: A review of the literature. *Circulation* **88,** 1973–1998.
8. Roncaglioni, M. C., Santoro, L., D'Avanzo, B., Negri, E., Nobili, A., Ledda, A., Pietropaolo, F., Franzosi, M. G., La Vecchia, C., Feruglio, G. A., and Maseri, A. (1992). Role of family history in patients with myocardial infarction. An Italian case-control study. GISSI-EFRIM Investigators. *Circulation* **85,** 2065–2072.
9. Pekka, J., Puska, P., Vartianinen, E., Pekkanin, J., and Tuomilhet, J. (1996). Parental history of premature coronary heart disease: An independent risk factor of myocardial infarction. *J. Clin. Epidemiol.* **49,** 497–503.
10. Marenberg, M. E., Risch, N., Berkman, L. F., Floderus, B., and De-Faire, U. (1994). Genetic susceptibility to death from coronary heart disease in a study of twins. *N. Engl. J. Med.* **330,** 1041–1046.
11. Stampfer, M. J., Colditz, G. A., Willett, W. C., Rosner, B., Speizer, F. E., and Hennekens, C. H. (1987). Coronary heart disease risk factors in women: The Nurse's Health Study experience. *In* "Coronary Heart Disease in Women" (E. D. Eaker, B. Packard, N. K. Wenger, T. B. Clarkson, and H. A. Tyroler, eds.), pp. 112–116, Chapter 13, Haymarket Doyma, New York.

12. Colditz, G. A., Stampfer, M. J., Willett, W. C., Rosner, B., Speizer, F. E., and Hennekens, C. H. (1986). A prospective study of parental history of myocardial infarction and coronary heart disease in women. *Am. J. Epidemiol.* **123,** 48–58.

13. Burke, G. L., Savage, P. J., Sprafka, J. M., Selby, J. V., Jacobs, D. R., Jr., Perkins, L. L., Roseman, J. M., Hughes, G. H., and Fabsitz, R. R. (1991). Relation of risk factor levels in young adulthood to parental history of disease. The CARDIA study. *Circulation* **84,** 1176–1187.

14. Nyboe, J., Jensen, G., Appleyard, M., and Schnohr, P. (1989). Risk factors for acute myocardial infarction in Copenhagen. I: Hereditary, educational and socioeconomic factors. Copenhagen City Heart Study. *Eur. Heart J.* **10,** 910–916.

15. Allen, J. K., Young D. R., Blumenthal, R. S., Moy, T. F., Yanek, L. R., Wilder, L., Becker, L. C., and Becker, D. M. (1996). Prevalence of hypercholesterolemia among siblings of persons with premature coronary heart disease, application of the second adult treatment panel guidelines. *Arch. Intern. Med.* **156,** 1654–1660.

16. Slattery, M. L., Schumacher, M. C., Hunt, S. C., and Williams, R. R. (1993). The associations between family history of coronary heart disease, physical activity, dietary intake and body size. *Int. J. Sports Med.* **14,** 93–99.

17. LaCroix, A. Z., and Omenn, G. S. (1992). Older adults and smoking. *Clin. Geriatr. Med.* **8,** 69–87.

18. Kawachi, I., Colditz, G. A., Stampfer, M. J., Willett, W. C., Manson, J. E., Rosner, B., Hunter, D. J., Hennekens, C. H., and Speizer, F. E. (1993). Smoking cessation in relation to total mortality rates in women. A prospective cohort study. *Ann. Intern. Med.* **119,** 992–1000.

19. LaCroix, A. Z., Lang, J., Scherr, P., Wallace, R. B., Cornoni-Huntley, J., Berkman, L., Curb, J. D., Evans, D., and Hennekens, C. H. (1991). Smoking and mortality among older men and women in three communities. *N. Engl. J. Med.* **324,** 1619–1625.

20. Paganini-Hill, A., and Hsu, G. (1994). Smoking and mortality among residents of a California retirement community. *Am. J. Public Health* **84,** 992–995.

21. McGill, H. C., Jr., McMahan, C. A., Malcom, G. T., Oalmann, M. C., and Strong, J. P. (1997). Effect of serum lipoproteins and smoking on atherosclerosis in young men and women. The PDAY Research Group. Pathobiological Determinants of Atherosclerosis in Youth. *Arterioscler. Thromb. Vasc. Biol.* **17,** 95–106.

22. Kawachi, I., Colditz, G. A., Speizer, F. E., Manson, J. E., Stampfer, M. J., Wallace, R. B., Wallace, R. B., Willett, W. C., and Hennekens, C. H. (1997). A prospective study of passive smoking and coronary heart disease. *Circulation* **95,** 2374–2379.

23. Hermanson, B., Omenn, G. S., Kronmal, R. A., and Gersh, B. J. (1988). Beneficial six-year outcome of smoking cessation in older men and women with coronary artery disease. Results from the CASS registry. *N. Engl. J. Med.* **319,** 1365–1369.

24. Hays, J. T., Hurt, R. D., and Dale, L. C. (1996). Smoking cessation: Evidence for or against benefit. *In* "Prevention of Myocardial Infarction" (J. E. Manson, P. M. Ridker, J. M. Gaziano, and C. H. Hennekens, eds.) pp. 99–129. Oxford University Press, Oxford and New York.

25. Cornoni-Huntley, J., LaCroix, A. Z., and Havlik, R. J. (1989). Race and sex differentials in the impact of hypertension in the United States. The National Health and Nutrition Examination Survey I Epidemiologic Follow-up Study. *Arch. Intern. Med.* **149,** 780–788.

26. Cupples, L. A., D'Agostino, R. B., and Kiely, D. (1987). Some risk factors related to the annual incidence of cardiovascular disease and death using pooled repeated biennial measurements: Framingham Heart Study, 30-Year Follow-up. *In* "The Framingham Study: An Epidemiological Investigation of Cardiovascular Disease" (W. B. Kannel, P. A. Wolf, and R. J. Garrison, eds.), pp. 1–26, Sec. 34. National Heart, Lung, and Blood Institute, Bethesda, MD.

27. Fiebach, N. H., Hebert, P. R., Stampfer, M. J., Colditz, G. A., Willett, W. C., Rosner, B., Speizer, F. E., and Hennekens, C. H. (1989). A prospective study of high blood pressure and cardiovascular disease in women. *Am. J. Epidemiol.* **130,** 646–654.

28. Selmer, R., and Tverdal, A. (1994). Mortality from stroke, coronary heart disease and all causes related to blood pressure and length of follow-up. *Scand. J. Soc. Med.* **22,** 273–282.

29. LaCroix, A. Z. (1993). Gender effects and hypertension in women. *In* "Hypertension Primer: The Essential of High Blood Pressure" (J. L. Izzo, Jr. and H. R. Black, eds.), Chapter C2, pp. 150–153. American Heart Association, Dallas, TX.

30. Haffner, S. M., Ferrannini, E., Huzuda, H. P., and Stern, M. P. (1992). Clustering of cardiovascular risk factors in confirmed prehypertensive individuals. *Hypertension* **20,** 38–45.

31. Liu, K., Ruth, K. J., Flack, J. M., Jones-Webb, R., Burke, G., Savage, P. J., and Hulley, S. B. (1996). Blood pressure in young blacks and whites: Relevance of obesity and lifestyle factors in determining differences. The CARDIA study. *Circulation* **93,** 60–66.

32. Burt, V. L., Cutler, J. A., Higgins, M., Horan, M. J., Labarthe, D., Whelton, P., Brown, C., and Roccella, E. J. (1995). Trends in the prevalence, awareness, treatment, and control of hypertension in the adult US.. population. Data from the Health Examination Surveys, 1960–1991. *Hypertension* **26,** 60–69.

33. Staessen, J., Bulpitt, C., De Leeuw, P., Fagard, R., Fletcher, A., Leonetti, G., Nissinen, A., O'Malley, K., Tuomilehto, J., Webster, J., and Williams, B. O. (1989). Relation between mortality and treated blood pressure in elderly patients with hypertension: Report of the European Working Party on High Blood Pressure in the Elderly. *Br. Med. J.* **298,** 1552–1556.

34. SHEP Cooperative Research Group (1991). Prevention of stroke by antihypertensive drug treatment in older persons with isolated systolic hypertension. Final results of the Systolic Hypertension in the Elderly Program (SHEP). *JAMA, J. Am. Med. Assoc.* **265,** 3255–3264.

35. Joint National Committee on Detection, Evaluation and Treatment of High Blood Pressure (1993). The fifth report of the Joint National Committee on Detection, Evaluation, and Treatment of High Blood Pressure (JNCV). *Arch. Intern. Med.* **153,** 154–183.

36. Bush, T. L., Fried, L. P., and Barrett-Connor, E. (1988). Cholesterol, lipoproteins and coronary heart disease in women. *Clin. Chem. (Winston-Salem, N.C.)* **34,** 660–670.

37. Manolio, T. A., Pearson, T. A., Wenger, N. K., Barrett-Conner, E., Payne, G. H., and Harlan, W. R. (1992). Cholesterol and heart disease in older persons and women: Review of an NHLBI workshop. *Ann. Epidemiol.* **2,** 161–176.

38. Wilson, P. W. F. (1990). High-density lipoprotein, low-density lipoprotein and coronary artery disease. *Am. J. Cardiol.* **66,** 7A–10A.

39. Harris, T., Kleinman, J. C., Makuc, D. M., Gillum, R., and Feldman, J. J. (1992). Is weight loss a modifier of the cholesterol-heart disease relationship in older persons. Data from the NHANES I epidemiologic follow-up study. *Ann. Epidemiol.* **2,** 35–41.

40. Gillum, R. F., Mussolino M. E., and Sempos, C. T. (1998). Baseline serum total cholesterol and coronary heart disease incidence in African-American women (the NHANES I epidemiologic follow-up study). National Health and Nutrition Examination Survey. *Am. J. Cardiol.* **81,** 1246–1249.

41. Harris-Hooker, S., and Sanford, G. L. (1994). Lipids, lipoproteins and coronary heart disease in minority populations. *Arteriosclerosis* **108,** S83–104.

42. Wong, N. D., Wilson, P. W. F., and Kannell, W. B. (1991). Serum cholesterol as a prognostic factor after myocardial infarction: The Framingham study. *Ann. Intern. Med.* **115,** 687–693.

43. Gordon, D. J., Probstfield, J. L., Garrison, R. J., Neaton, J. D., Castelli, W. P., Knoke, J. D., Jacobs, D. R., Jr., Bangdiwala, S., and Tyroler, H. A. (1989). High-density lipoprotein cholesterol and cardiovascular disease: Four prospective American studies. *Circulation* **79,** 8–15.

44. Kannel, W. B. (1987). New perspectives of cardiovascular risk factors. *Am. Heart J.* **114,** 213–219.

45. Kwiterovich, P. O., Jr., Coresh, J., Smith, H. H., Bachorik, P. S., Derby, C. A., and Pearson, T. A. (1992). Comparison of the plasma levels of apolipoproteins B and A-1, and other risk factors in men and women with premature coronary artery disease. *Am. J. Cardiol.* **69,** 1015–1021.

46. Dahlen, G. H., Guyton, J. R., Attar, M., Farmer, J. A., Kautz, J. A., and Gotto, A. M. (1986). Association of levels of lipoprotein Lp(a), plasma lipids, and other lipoproteins with coronary artery disease documented by angiography. *Circulation* **74,** 758–765.

47. Stein, J. H., and Rosenson, R. S. (1997). Lipoprotein Lp(a) excess and coronary heart disease. *Arch. Intern. Med.* **157,** 1170–1176.

48. Bostom, A. G., Cupples, L. A., Jenner, J. L., Ordovas, J. M., Seman, L. J., Wilson, P. W., Schaefer, E. J., and Castelli, W. P. (1996). Elevated plasma lipoprotein(a) and coronary heart disease in men aged 55 years and younger. A prospective study. *JAMA, J. Am. Med. Assoc.* **276,** 544–548.

49. Bostom, A. G., Gagnon, D. R., Cupples, L. A., Wilson, P. W., Jenner, J. L., Ordovas, J. M., Schaefer, E. J., and Castelli, W. P. (1994). A prospective investigation of elevated lipoprotein(a) detected by electrophoresis and cardiovascular disease in women. The Framingham Heart Study. *Circulation* **90,** 1688–1695.

50. Austin, M. A., Breslow, J. L., Hennekens, C. H., Buring, J. E., Willett, W. C., and Krauss, R. M. (1988). Low-density lipoprotein subclass patterns and risk of myocardial infarction. *JAMA, J. Am. Med. Assoc.* **260,** 1917–1921.

51. van Beresteijn, E. C., Korevaar, J. C., Huijbregts, P. C., Schouten, E. G., Burema, J., and Kok, F. J. (1993). Perimenopausal increase in serum cholesterol: A 10-year longitudinal study. *Am. J. Epidemiol.* **137,** 383–392.

52. Espeland, M. A., Marcovina, S. M., Miller, V., Wood, P. D., Wasilauskas, C., Sherwin, R., Schrott, H., and Bush, T. L. (1998). Effect of postmenopausal hormone therapy on lipoprotein(a) concentration. PEPI investigators. Postmenopausal estrogens/progestin inverventions. *Circulation* **97,** 979–986.

53. Miettinen, T. A., Pyorala, K., Olsson, A. G., Musliner, T. A., Cook, T. J., Faergeman, O., Berg, K., Pedersen, T., and Kjekshus, J. (1997). Cholesterol-lowering therapy in women and elderly patients with myocardial infarction or angina pectoris: Findings from the Scandinavian Simvastatin Survival Study (4S). *Circulation* **96,** 4211–4218.

54. U.S. Department of Health and Human Services (1996). "Physical Activity and Health: A Report of the Surgeon General." U.S. Department of Health and Human Services, Center for Disease Control and Prevention, National Center for Chronic Disease Prevention and Health Promotion, Atlanta, GA.

55. Eaton, C. B., Lapane, K. L., Garber, C. A., Assaf, A. R., Lasater, T. M., and Carleton, R. A. (1995). Sedentary lifestyle and risk of coronary heart disease in women. *Med. Sci. Sports Exercise* **27,** 1535–1539.

56. LaCroix, A. Z., Leveille, S. G., Hecht, J. A., Grothaus, L. C., and Wagner, E. H. (1996). Does walking decrease the risk of cardiovascular disease hospitalizations and death in older adults? *J. Am. Geriatr. Soc.* **44,** 113–120.

57. Folsom, A. R., Arnett, D. K., Hutchinson, R. G., Liao, F., Clegg, L. X., and Cooper, L. S. (1997a). Physical activity and incidence of coronary heart disease in middle-aged women and men. *Med. Sci. Sports Exercise* **29,** 901–909.

58. Løchen, M., and Rasmussen, K. (1992). The Tromsø study: Physical fitness, self reported physical activity, and their relationship to other coronary risk factors. *J. Epidemiol. Commun. Health* **26,** 103–107.

59. Blair, S. N., Kohl, H. W., 3rd, Paffenbarger, R. S., Jr., Clark, D. G., Cooper, K. H., and Gibbons, L. W. (1989). Physical fitness and all-cause mortality. A prospective study of healthy men and women. *JAMA, J. Am. Med. Assoc.* **262,** 2395–2401.

60. Kokkinos, P. F., Holland, J. C., Pittaras, A. E., Narayan, P., Dotson, C. O., and Papademetriou, V. (1995). Cardiorespiratory fitness and cornary heart disease risk factor association in women. *J. Am. Coll. Cardiol.* **26,** 358–364.

61. Taylor, P. A., and Ward, A. (1993). Women, high-density lipoprotein cholesterol, and exercise. *Arch. Intern. Med.* **153,** 1178–1184.

62. Young, D. R., Haskell, W. L., Jatulis, D. E., and Fortmann, S. P. (1993). Associations between changes in physical activity and risk factors for coronary heart disease in a community-based sample of men and women: The Stanford Five-City Project. *Am. J. Epidemiol.* **138,** 205–216.

63. The Expert Committee on the Diagnosis and Classification of Diabetes Mellitus (1997). Report of the Expert Committee on the Diagnosis and Classification of Diabetes Mellitus. *Diabetes Care* **20,** 1183–1197.

64. Orchard, T. J. (1996). The impact of gender and general risk factors on the occurence of atherosclerotic vascular disease in non-insulin-dependent diabetes mellitus. *Ann. Med.* **28,** 323–333.

65. Mitchell, B. D., Haffner, S. M., Huzuda, H. P., Patterson, J. K., and Stern, M. P. (1992). Diabetes and coronary heart disease risk in Mexican Americans. *Ann. Epidemiol.* **2,** 101–106.

66. Kodama, K., Sasakli, H., and Shimizu, Y. (1990). Trend of coronary heart disease and its relationship to risk factors in a Japanese population: A 26 year follow-up, Hiroshima/Nagasaki Study. *Jpn. Circ. J.* **54,** 414–421.

67. Dawber, T. R. (1980). "The Framingham Study: The Epidemiology of Atherosclerotic Disease." Harvard University Press, Cambridge, MA.

68. La Vecchia, C., Decarli, A., Franceschi, S., Gentile, A., Negri, E., and Parazzini, F. (1987a). Menstrual and reproductive factors and the risk of myocardial infarction in women under fifty-five years of age. *Am. J. Obstet. Gynecol.* **157,** 1108–1112.

69. Barrett-Connor, E., Khaw, K. T., and Wingard, D. L. (1983). Sex differential in ischemic heart disease mortality in diabetics: A prospective population-based study. *Am. J. Epidemiol.* **118,** 489–496.

70. Manson, J. E., Colditz, G. A., Stampfer, M. J., Willett, W. C., Krolewski, A. S., Rosner, B., Arky, R. A., Speizer, F. E., and Hennekens, C. H. (1991). A prospective study of maturity-onset diabetes mellitus and risk of coronary heart disease and stroke in women. *Arch. Intern. Med.* **151,** 1141–1147.

71. Khaw, K. T., and Barrett-Connor, E. (1986). Prognostic factors for mortality in a population-based study of men and women with a history of heart disease. *J. Cardiopulmonary Rehabil.* **6,** 474–480.

72. Wong, N. D., Cupples, L. A., Ostfeld, A. M., Levy, D., and Kannel, W. B. (1989). Risk factors for long-term coronary prognosis after initial myocardial infarction: The Framingham Study. *Am. J. Epidemiol.* **130,** 469–480.

73. Stone, P. H., Muller, J. E., Hartwell, T., York, B. J., Rutherford, J. D., Parker, C. B., Turi, Z. G., Strauss, H. W., Willerson, J. T., Robertson, T., Braunwald, E., Jaffe, A. S., and The MILIS Study Group (1989). The effect of diabetes mellitus on prognosis and serial left ventricular function after acute myocardial infarction: Contribution of both coronary disease and diastolic left ventricular dysfunction to the adverse prognosis. The MILIS Study Group. *J. Am. Coll. Cardiol.* **14,** 49–57.

74. Abbott, R. D., Donahue, R. P., Kannel, W. B., and Wilson, P. W. (1988). The impact of diabetes on survival following myocardial infarction in men vs women. The Framingham Study. *JAMA, J. Am. Med. Assoc.* **260**, 3456–3460.

75. Folsom, A. R., Szklo, M., Stevens, J., Liao, F., Smith, R., and Eckfeldt, J. H. (1997b). A prospective study of coronary heart disease in relation to fasting insulin, glucose, and diabetes. *Diabetes Care* **20**, 935–942.

76. Rönnemaa, T., Kaakso, M., Pyörälä, K., Kallio, V., and Puukka, P. (1991). High fasting plasma insulin is an indicator of coronary heart disease in non-insulin-dependent diabetic patients and nondiabetic subjects. *Arterioscler. Thromb.* **11**, 80–90.

77. Kaplan, N. M. (1989). The deadly quartet: Upper-body obesity, glucose intolerance, hypertriglyceridemia, and hypertension. *Arch. Intern. Med.* **149**, 1514–1520.

78. Simonson, D. C., and Dzau, V. J. (1991). Workshop IX—Lipids, insulin, diabetes. *Am. J. Med.* **90**, 2A85S–86S.

79. The Diabetes Control and Complications Trial Research Group (1993). The effect of intensive treatment of diabetes on the development and progression of long-term complications in insulin-dependent diabetes mellitus. The Diabetes Control and Complications Trial Research Group. *N. Engl. J. Med.* **329**, 977–986.

80. Kuczmarski, R. J., Flegal, K. M., Campbell, S. M., and Johnson, C. L. (1994). Increasing prevalence of overweight among U.S. adults. The National Health and Nutrition Examination Surveys, 1960 to 1991. *JAMA, J. Am. Med. Assoc.* **272**, 205–211.

81. Hubert, H. B., Feinleib, M., McNamara, P. M., and Castelli, W. P. (1983). Obesity as an independent risk factor for cardiovascular disease: A 26-year follow-up of participants in the Framingham Heart Study. *Circulation* **67**, 968–977.

82. Manson, J. E., Colditz, G. A., Stampfer, M. J., Willett, W. C., Rosner, B., Monson, R. R., Speizer, F. E., and Hennekens, C. H. (1990). A prospective study of obesity and risk of coronary heart disease in women. *N. Engl. J. Med.* **322**, 882–889.

83. Willett, W. C., Manson, J. E., Stampfer, M. J., Colditz, G. A., Rosner, B., Speizer, F. E., and Hennekens, C. H. (1995). Weight, weight changes, and coronary heart disease in women, risk within the "normal" weight range. *JAMA, J. Am. Med. Assoc.* **273**, 461–465.

84. Newton, K. M., and LaCroix, A. Z. (1996). Association of body mass index with reinfarction and survival after first myocardial infarction in women. *J. Women's Health* **5**, 433–444.

85. Stevens, J., Keil, J. E., Rust, P. F., Tyroler, H. A., Davis, C. E., and Gazes, P. C. (1992). Body mass index and body girths as predictors of mortality in black and white women. *Arch. Intern. Med.* **152**, 1257–1262.

86. Johnson, J. L., Heineman, F., Heiss, G., Hames, C. G., and Tyroler, H. A. (1986). Cardiovascular disease risk factors and mortality among black women and white women aged 40–64 years in Evans County, Georgia. *Am. J. Epidemiol.* **123**, 209–220.

87. Harris, T. B., Launer, L. J., Madans, J., and Feldman, J. J. (1997). Cohort study of effect of being overweight and change in weight on risk of coronary heart disease in old age. *Br. Med. J.* **314**, 1791–1794.

88. Harris, T., Cook, E. F., Garrison, R., Higgins, M., Kannel, W., and Goldman, L. (1988). Body mass index and mortality among nonsmoking older persons. The Framingham Heart Study. *JAMA, J. Am. Med. Assoc.* **259**, 1520–1524.

89. Van Itallie, T. B. (1985). Health implication of overweight and obesity in the United States. *Ann. Intern. Med.* **103**, 983–988.

90. Lapidus, L. and Bengtsson, C. (1987). Regional obesity as a health hazard in women—a prospective study. *Acta Med. Scand. Suppl.* **732**, 53–59.

91. Prineas, R. J., Folsom, A. R., and Kaye, S. A. (1993). Central adiposity and increased risk of coronary artery disease mortality in older women. *Am. J. Epidemiol.* **3**, 35–41.

92. Bates, C. J., Mansoor, M. A., van der Pols, J., Prentice, A., Cole, T. J., and Finch, S. (1997). Plasma total homocystein in a representative sample of 972 British men and women aged 65 and over. *Eur. J. Clin. Nutr.* **51**, 691–697.

93. Selhub, J., Jacques, P. F., Wilson, P. W. F., Rush, D., and Rosenberg, I. H. (1993). Vitamin status and intake as primary determinants of homocysteinemia in an elderly population. *JAMA, J. Am. Med. Assoc.* **270**, 2693–2698.

94. Dalery, K., Lussier-Cacan, S., Selhub, J., Davignon, J., Latour, Y., and Genest, J. (1995). Homocysteine and coronary artery disease in French Canadian Subjects: Relation with viatmins B_{12}, B_6, pyridoxal phosphate, and folate. *Am. J. Cardiol.* **75**, 1107–1111.

95. Schwartz, S. M., Siscovich, D. S., Malinow, M. R., Rosendaal, F. R., Beverly, R. K., Hess, D. L., Psaty, B. M., Longstreth, W. T., Koepsell, T. D., Raghunathan, T. E., and Reitsma, P. H. (1997). Myocardial infarction in young women in relation to plasma total homocysteine, folate, and a common variant in the methylenetetrahydrofolate reductase gene. *Circulation* **96**, 412–417.

96. Kang, S., Wong, P. W. K., Cook, H. Y., Norusis, M., and Messer, J. V. (1986). Protein-bound homocysteine, a possible risk factor for coronary artery disease. *J. Clin. Invest.* **77**, 1482–1486.

97. Robinson, K., Mayer, E. L., Miller, D. P., Green, R., van Lente, F., Gupta, A., Kottke-Marchant, K., Savon, S. R., Selhub, J., Nissen, S. E., Kutner, M., Topo, E. J., and Jacobsen, D. W. (1995). Hyperhomocysteinemia and low pyridoxal phosphate, common and independent reversible risk factors for coronary artery disease. *Circulation* **92**, 2825–2830.

98. Selhub, J., Jacques, P. F., Bostom, A. G., D'Agostino, R. B., Wilson, P. W. F., Bélanger, A. J., O'Leary, D. H., Wolf, P. A., Schaefer, E. J., and Rosenberg, I. H. (1995). Association between plasma homocysteine concentrations and extracranial carotid-artery stenosis. *N. Engl. J. Med.* **332**, 286–291.

99. Malinow, M. R., Nieto, F. J., Szklo, M., Chambless, L. E., and Bond, G. (1993). Carotid artery intimal-medial wall thickening and plasma homocysteine in asymptomatic adults. *Circulation* **87**, 1107–1113.

100. Brattstrom, L. E., Hultberg, B. L., and Hardebo, J. E. (1985). Folic acid responsive postmenopausal homocysteinemia. *Metab., Clin. Exp.* **34**, 1073–1077.

101. Van Der Mooren, M. J., Wouters, M. G., Blom, H. J., Schellekens, L. A., Eskes, T. K., and Rolland, R. (1994). Hormone replacement therapy may reduce high serum homocysteine in postmenopausal women. *Eur. J. Clin. Invest.* **24**, 733–736.

102. Colditz, G. A. (1990). A prospective assessment of moderate alcohol intake and major chronic diseases. *Ann. Epidemiol.* **1**, 167–177.

103. Jackson, R., Scragg, R., and Beaglehole, R. (1991). Alcohol consumption and risk of coronary heart disease. *Br. Med. J.* **303**, 211–216.

104. Woodward, M., and Tunstall-Pedoe, H. (1995). Alcohol consumption, diet, coronary risk factors, and prevalent coronary heart disease in men and women in the Scottish heart health study. *J. Epidemiol. Commu. Health* **49**, 354–362.

105. Simons, L. A., McCallum, J., Friedlander, Y., and Simons, J. (1996). Alcohol intake and survival in the elderly: A 77 month follow-up in the Dubbo study. *Aust. N.Z. J. Med.* **26**, 662–670.

106. Gramenzi, A., Gentile, A., Fasoli, M., Negri, E., Parazzini, F., and La Vecchia, C. (1990). Association between certain foods and risk of acute myocardial infarction in women. *Br. Med. J.* **300**, 771–773.

107. Steinbrecher, U. P. (1997). Dietary antioxidants and cardioprotection—fact or fallacy? *Can. J. Physiol. Pharmacol.* **75**, 228–233.

108. Gaziano, J. M. (1994). Antioxidant vitamins and coronary artery disease risk. *Am. J. Med.* **97,** 3A-18S–3A-28S.

109. Tavani, A., Negri, E., Avanzo, B. D., and La Vecchia, C. (1997). Beta-carotene intake and risk of nonfatal acute myocardial infarction in women. *Eur. J. Epidemiol.* **13,** 631–637.

110. Hennekens, C. H., Buring, J. E., Manson, J. E., Stampfer, M., Rosner, B., Cook, N. R., Belanger, C., LaMotte, F., Gaziano, J. M., Ridker, P. M., Willett, W., and Peto, R. (1996). Lack of effect of long-term supplementation with beta carotene on the incidence of malignant neoplasms and cardiovascular disease. *N. Engl. J. Med.* **334,** 1145–1149.

111. Alpha Tocopherol, Beta Carotenen Cancer Prevention Study Group (1994). The effect of vitamin E and beta carotene on the incidence of lung cancer and other cancers in male smokers. *N. Engl. J. Med.* **330,** 1029–1035.

112. Omenn, G. S., Goodman, G. E., Thornquist, M. D., Balmes, J., Cullen, M. R., Glass, A., Keogh, J. P., Meyskens, F. L., Valanis, B., Williams, J. H., Barnhart, S., and Hammar, S. (1996). Effects of a combination of beta carotene and vitamin A on lung cancer and cardiovascular disease. *N. Engl. J. Med.* **334,** 1150–1155.

113. Kushi, L. H., Folsom, A. R., Prineas, R. J., Mink, P. J., We, Y., and Bostick, R. M. (1996). Dietary antioxidant vitamins and death from coronary heart disease in postmenopausal women. *N. Engl. J. Med.* **334,** 1156–1162.

114. Rimm, E. B., and Stampfer, M. J. (1997). The role of antioxidants in preventive cardiology. *Curr. Opin. Cardiol.* **12,** 188–194.

115. Stampfer, M. J., Hennekens, C. H., Manson, J. E., Colditz, G. A., Rosner, B., and Willett, W. C. (1993). Vitamin E consumption and the risk of coronary disease in women. *N. Engl. J. Med.* **328,** 1444–1449.

59

Diagnosis and Treatment of Heart Disease in Women

RICHARD HOLUBKOV*,† AND STEVEN E. REIS*

*LHAS Women's Heart Center, Division of Cardiology, and †Department of Epidemiology, Graduate School of Public Health,
University of Pittsburgh, Pittsburgh, Pennsylvania

I. Introduction

Most women, as well as many of their primary health care providers, have the misconception that cardiovascular disease is of primary importance to men. In fact, cardiovascular disease is the foremost cause of death among women, accounting for twice as many deaths as all types of cancers combined. Cardiovascular disease is responsible for nearly 500,000 deaths in American women annually [1], approximately half of which may be attributed to atherosclerotic coronary artery disease (CAD).

Studies have identified gender differences in the pathophysiologic mechanisms of cardiac symptoms, outcomes following cardiac events, and responses to cardiovascular therapies. For example, women presenting with chest pain are less likely than men to have CAD as the underlying mechanism of their symptoms, whereas women with documented CAD tend to have a higher risk profile than men presenting with atherosclerosis. These observations, and others that will be discussed in this chapter, underscore the need for the development of gender-specific cardiovascular diagnostic and treatment strategies.

This chapter begins with a discussion of gender differences in the mechanisms of chest pain. After briefly addressing therapy in women with chest pain caused by abnormal coronary vasoreactivity, important risk factors for the development of atherosclerosis in women and the role of postmenopausal hormone replacement therapy in CAD prevention are discussed. The chapter concludes with a discussion of the gender-specific issues that are important in the diagnosis and treatment of coronary disease.

II. Mechanisms of Chest Pain

Obstructive atherosclerotic stenoses limit coronary blood flow during periods of increased myocardial oxygen demand, such as during exertional or emotional stress, resulting in myocardial ischemia, commonly manifested by anginal chest pain. Destabilization and rupture of atherosclerotic plaques result in coronary spasm, platelet aggregation, and thrombus formation, which cause unstable angina and/or myocardial infarction. The Framingham study demonstrated that the predominant initial manifestation of coronary atherosclerosis in women is angina, which occurred in 47% of those presenting with CAD [2]. In contrast, the predominant first presentation of men is a myocardial infarction, which occurred in 46% of men with CAD [2]. However, of patients presenting with "typical" angina (retrosternal chest pain precipitated by exertion, promptly relieved by rest, and usually described as a heavy, pressure-like, or squeezing sensation), only 58% of women versus 88% of men had angiographically-defined coronary atherosclerosis [3]. "Atypical" angina (pain not consistently caused by exertion that may not be relieved by rest and is not usually described as a pressure-like or squeezing sensation), which is more common in women than men, is associated with CAD in only 35% of women versus 67% of men. These substantial differences in presentation suggest that coronary atherosclerosis is not the underlying mechanism of chest pain in most women and that the pathophysiology of chest pain differs between women and men.

One alternative to the atherosclerosis mechanism is that chest pain may result from a dynamic reduction in myocardial perfusion caused by coronary artery spasm or attenuated coronary dilatation. This abnormal coronary vasoreactivity occurs more commonly in women and may affect either the large epicardial coronary arteries or the flow-regulating microcirculation embedded in the myocardium [4–6]. Vasospastic (Prinzmetal's) angina results from spontaneous spasm of the large epicardial coronary arteries and commonly occurs in the absence of angiographically documented atherosclerosis [4]. Clinical manifestations of vasospastic angina frequently occur at rest and include transient chest pain and electrocardiographic ST-segment elevations occasionally associated with ventricular arrhythmias or atrioventricular block [7–9]. The diagnosis of vasospastic angina may be made clinically by documenting spontaneous or acetylcholine-induced epicardial coronary constriction during coronary angiography (Fig. 59.1). Although this is most commonly seen in angiographically "normal" coronaries, studies have demonstrated that sites of coronary spasm do have pathologic evidence of atherosclerosis [10,11]. Physiologically, coronaries predisposed to vasospasm have a deficiency in endothelial production of the endogenous vasodilator nitric oxide [8,11], which is normally produced by intact vascular endothelium in response to neurohumoral stimuli (*e.g.,* acetylcholine, catecholamines), platelet products (*e.g.,* serotonin), and increased flow and shear stress (*e.g.,* hypertension). While these stimuli induce dilatation of coronaries with normal endothelial function, coronaries with endothelial dysfunction, which is an early manifestation of atherosclerosis, have impaired nitric oxide production and abnormally constrict when exposed to these stimuli. Therefore, patients with vasospastic angina may respond to pharmacologic therapies targeting normalization of endothelial function and vasodilatation.

Abnormal vasoreactivity in the coronary microvessels may also be associated with chest pain. Microvascular angina, or "Syndrome X," clinically defined as chest pain in a patient with an abnormal stress test and angiographically normal coronary arteries, is more prevalent in women and is associated with an excellent prognosis [5,6]. It is uncommon for women with this clinical syndrome to have adverse cardiac events such as myocardial infarction or cardiac death. Physiologically, microvascular angina is caused by impaired dilatation of the flow-regulating

a

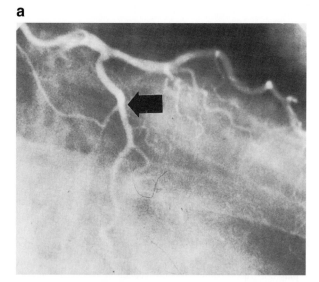

b

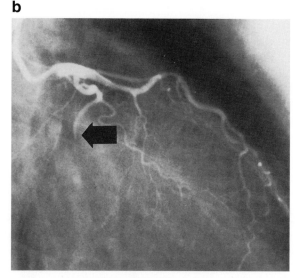

Fig. 59.1   Angiographically normal left circumflex coronary artery (a) exhibiting spontaneous vasospasm (b) in a patient with vasospastic (Prinzmetal's) angina.

coronary microvessels, which is associated with attenuated augmentation of coronary blood flow in response to increased myocardial oxygen demand [5,12]. Objective diagnosis of this syndrome is made by using intracoronary Doppler ultrasonography (Fig. 59.2) or positron emission tomography (PET) to demonstrate attenuated increases in coronary blood flow in response to vasodilators such as adenosine that induce maximal coronary hyperemia. The pathophysiologic mechanism of microvascular angina is possibly related to impaired endothelial cell nitric oxide production or to abnormal relaxation of arterial myocytes in the small coronary resistance arteries [13]. Therefore, therapies that improve coronary vasomotor tone may be beneficial in women with microvascular angina. In addition, drugs that attenuate cerebral processing of cardiac pain may also ameliorate chest pain in these women.

The prevalence of vasospastic and microvascular angina in women with typical and atypical angina and angiographically normal coronaries is not well known. Because cardiac evaluations of women with chest pain and angiographically normal coronaries do not systematically evaluate coronary physiology, these women tend to be given the diagnosis of "noncardiac" chest pain. Undoubtedly, a substantial proportion of such women have vasospastic or microvascular angina that may be treated by vasoactive pharmacologic agents. Therefore, it is incumbent upon clinicians to pursue diagnoses of disorders of coronary vasomotor tone in women with chest pain and angiographically normal coronaries. The NIH-sponsored Women's Ischemia Syndrome Evaluation (WISE) is an ongoing multicenter clinical study that is investigating the prevalence of these disorders and the optimal strategies for their diagnosis.

A. Therapy for Chest Pain Caused by Abnormal Coronary Vasoreactivity

Women with coronary heart disease must be evaluated and treated as aggressively as men; as such, the chapter turns below to a discussion of risk factors for and treatment of CAD in women. However, the sizable number of women with chest pain resulting from documented or presumed vasospastic or microvascular angina poses a therapeutic dilemma. Although these clinical syndromes are rarely associated with adverse outcomes such as myocardial infarction, their symptoms (as well as the sometimes long road to their diagnosis) can be debilitating. While traditional coronary vasodilator therapy, including nitrates and calcium channel blockers, may be used to alleviate episodes of chest pain due to abnormal coronary vasoreactivity in women, they are not universally successful. Alternative therapies may provide additional symptomatic relief. For instance, one study of postmenopausal women with "typical" angina, an abnormal exercise ECG stress test, and angiographically normal coronary arteries (i.e., microvascular angina) reported that eight weeks of cutaneous 17β-estradiol therapy significantly reduced the frequency of chest pain [14]. The results of this trial may be explained by estrogen's coronary vasodilator effects and suggest that in selected women, estrogen deficiency may be associated with microvascular angina. Thus, an hypothesized estrogen-related mechanism would explain the female predominance of patients with this disease. Similar hormonal strategies have not been systematically studied in women with vasospastic angina. Another trial demonstrated that imipramine is successful in decreasing the clinical manifestations of microvascular angina [15], likely due either to alteration of cerebral pain perception or to dilatation of the coronary microcirculation.

III. Atherosclerotic Risk Factors in Women

Chest pain due to coronary atherosclerosis is associated with an increased incidence of adverse cardiac events. Women tend to develop CAD several years later than men, resulting from cardioprotection due to circulating estrogen in their premenopausal years [16]. Nevertheless, women share with men the traditional risk factors for atherosclerosis: advanced age, diabetes, hypertension, dyslipidemia, cigarette use, and a family history

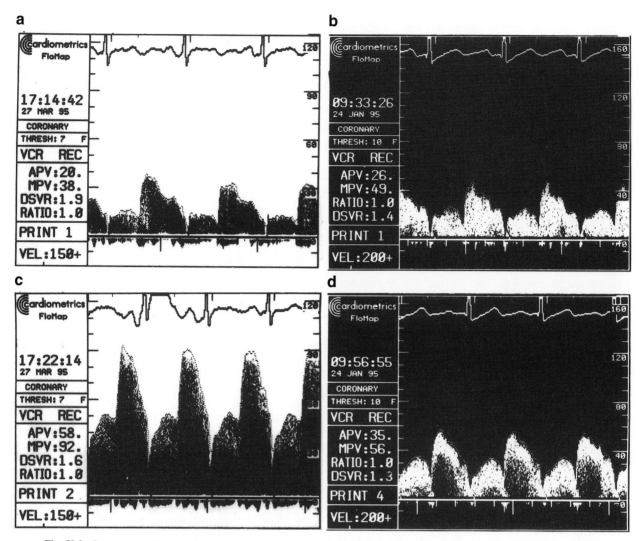

Fig. 59.2   Intracoronary Doppler ultrasound waveforms before and after intracoronary adenosine in a patient with normal coronary flow reserve (a, c) and a patient with microvascular angina manifested by attenuated flow reserve (b, d). Coronary flow reserve is the ratio of the average peak velocity (APV) of coronary blood flow after adenosine (c, d) to APV at baseline (a, b).

of premature CAD (in a first degree male relative less than 50 years old or female relative less than 60 years old).

A. Age and Menopausal Status

The incidence of CAD, which is very low in young women, increases substantially in the sixth decade of life. This correlation of CAD with the onset of natural menopause [16] has led to the hypothesis that menopause-induced hormonal changes are independently associated with acceleration of the cardiovascular atherosclerotic process. This hypothesis is supported by the finding that premature surgical menopause more than doubles CAD risk in the absence of hormonal replacement therapy [17]. Menopause may increase the risk of CAD directly by altering vascular properties or in a more indirect fashion through unfavorable alterations of the cardiovascular risk factor profile, such as increasing LDL and decreasing HDL cholesterol levels. Although coronary artery disease occurs infrequently in premeno-

pausal women, certain comorbidities such as diabetes (which is associated with impairment of estrogen binding), polycystic ovary disease, and an absence of coronary artery smooth muscle cell estrogen receptors abolish the vascular protective effects of estrogen in premenopausal women [18–21]. In postmenopausal women, hormone replacement therapy has a protective effect, as we will discuss in Section IV.

B. Diabetes Mellitus

Diabetes mellitus is a strong risk factor for coronary heart disease in women. Diabetic women have a fivefold increased risk of developing CAD compared to nondiabetic women [22]. This exceedingly high risk may be related to diabetes' direct atherosclerotic effect or to its impairment of estrogen binding [21]. Diabetes is known to compound adverse cardiovascular effects of other risk factors such as serum lipid levels [23,24], but unfavorable lipid levels only partially explain the increased risk of

CAD in women with diabetes, indicating a potent and likely multifactorial atherosclerotic effect of diabetes in women [25].

C. Hypertension

Hypertension, both moderate and severe, is associated with CAD in women [26–29]. In women with mild to moderate hypertension, a multitude of studies have demonstrated that pharmacologic lowering of blood pressure is associated with a significant reduction in cardiovascular events [26–33], particularly stroke. In women with severe hypertension, the impact of antihypertensive therapy both on CAD and on stroke has been well established.

D. Dyslipidemia

Dyslipidemia is associated with increased CAD mortality in women [34,35]. Moreover, various trials have conclusively established that pharmacologic lipid lowering therapy decreases subclinical evidence of atherosclerosis and CAD risk in women. Pharmacologic inhibitors of the enzyme 3-hydroxy-3-methyl-glutaryl-coenzyme A (HMG-CoA) reductase are now commonly prescribed as initial pharmacologic therapy for lipid lowering in postmenopausal women in order to meet the National Cholesterol Education Program's (NCEP) published guidelines. In addition, the NCEP now recommends hormone replacement therapy as an alternative to lipid lowering agents in postmenopausal women who qualify for pharmacologic intervention to lower LDL cholesterol [36]. Lipid lowering accounts for approximately half of the cardioprotective effect of postmenopausal hormone replacement therapy [35,37].

E. Cigarette Smoking

Cigarette smoking is a critically important and entirely preventable risk factor for the development of CAD, stroke, and peripheral vascular disease in women. Women who smoke one pack of cigarettes or more per day have a risk of coronary heart disease that is two to four times that of nonsmokers [38]. Smoking is also a highly significant risk factor for sudden cardiac death and for not surviving a myocardial infarction [39]. Because smoking cessation reduces CAD risk within a few years, implementation of aggressive smoking cessation strategies would be beneficial to women. Preventive strategies should target women, especially adolescent females who comprise the largest group of new smokers in the United States. The association of smoking with heart disease may be due to direct mechanisms, such as accelerated atherosclerosis due to nicotine and carbon monoxide and smoking-induced coronary artery vasoconstriction [40–42]. This association may also be due to indirect effects, given that menopause tends to occur earlier in women who smoke [43].

IV. The Role of Postmenopausal Hormone Replacement Therapy in Preventing CAD

A. Epidemiologic Evidence

It is clear that the risk of CAD in premenopausal women is low and increases significantly after menopause. As a result, many studies have examined the association of postmenopausal estrogen therapy with risk of CAD. For example, at 10 years of follow-up, the Nurses' Health Study demonstrated a 44% decrease in risk of CAD for current estrogen users versus postmenopausal women who had never used estrogen, and the Lipid Research Clinics Program follow-up study found a 66% decrease in cardiovascular mortality in users at baseline compared to nonusers [35,44]. In addition to these large prospective studies, there is a significant pool of data from numerous smaller case-control studies, cross-sectional studies involving coronary angiography, and cohort studies that supports the cardiovascular benefit of estrogen. Indeed, one pooled meta-analysis found that 15 of 16 cohort studies, 3 of 3 available angiographic studies, and 6 of 7 population-based case-control studies reported decreased risks of various cardiovascular endpoints in estrogen users compared to nonusers [45]. Combining the results of all studies yielded an estimated 44% decrease in cardiac risk (95% confidence interval (CI) = 39–50%); the estimated benefit of estrogen improved to a 50% decrease in risk when only internally controlled and prospective studies were considered.

Another analysis of all available study data since 1970 reported an estimated decrease in CAD risk of 35% (95% CI = 29–41%) for women ever using estrogen versus nonusers [46]. This study also reported that on the basis of this degree of cardioprotection and the additional assumptions of (1) a 3-fold increased risk of endometrial cancer, (2) a 1.25-fold increased risk of breast cancer, and (3) a 25% decrease in risk for hip fracture, a 50-year old white woman treated with estrogen could expect approximately one year of increased life due to the combined protective and detrimental effects of estrogen therapy. However, caution must be used in interpreting this intriguing analysis because these various assumed quantifications of the benefits and risks of estrogen therapy are not uniformly accepted.

Current clinical practice dictates the use of progestin supplementation to postmenopausal estrogen replacement regimens in women with an intact uterus to reduce the risk of endometrial neoplasia. However, progestins may reduce the cardioprotective effects of estrogen therapy because they may attenuate the beneficial mechanistic effects of estrogen on arterial dilatation and blood flow and the estrogen-related increases in HDL cholesterol levels. Although a meta-analysis of available data concluded that combined estrogen and progestin therapy may decrease the risk of endometrial carcinoma, it demonstrated that there is insufficient information to assess the effect of progestin supplementation on CAD risk [46]. Under the assumption that progestin therapy decreases risk of endometrial cancer to a normal level, the authors estimated that women at normal risk of heart disease will live longer if they use a combined hormone regimen as long as the progestin-related reduction of the beneficial estrogen effect on CAD is of a magnitude of one-third or less.

The Heart and Estrogen Replacement Study (HERS) is the first large-scale prospective placebo-controlled trial to evaluate the effect of estrogen plus progestin supplementation on prevention of cardiac death or myocardial infarction in women with established CAD [47]. Despite the reported beneficial effects of combination hormone therapy on the lipid profile, no difference was found in cardiovascular event rates at an average 4 years of follow-up in women randomized to hormone therapy and those receiving placebo. There was a significant time trend, however, with the hormone therapy group having an increased risk of early events (*i.e.,* myocardial infarction or coronary heart disease death) and a decreased cardiac event rate after the fourth year of follow-up. This late beneficial effect of combination hormone therapy is consistent with results from the Nurses' Health Study, which demonstrated a substantial decrease in CAD among women currently using estrogen with progestin compared to nonusers during 16 years of follow-up [48]. Of note, women using estrogen and progestin fared somewhat better than those using estrogen alone. It is important to note that the women in the Nurses' Health study were a self-selected group without known CAD and were substantially younger (age 30 to 55 at baseline) than participants in HERS (mean age 67 years) who had known CAD.

(The reader is referred to Chapter 93 for a more detailed discussion of the epidemiologic evidence for potential benefits and risks of hormone replacement therapy).

B. Cardioprotective Mechanisms of Postmenopausal Hormone Replacement Therapy

Chronic postmenopausal hormone therapy is associated with significant favorable changes of the lipid profile that are expected to inhibit atherosclerosis and reduce cardiovascular events. For example, the prospective placebo-controlled Postmenopausal Estrogen/Progestin Interventions (PEPI) trial demonstrated that conjugated estrogens alone (0.625 mg daily) or in combination with one of three progestin regimens (cyclic medroxyprogesterone acetate (MPA) 10 mg/day for 12 days/month; continuous MPA 2.5 mg daily; or cyclic micronized progesterone 200 mg/day for 12 days/month) is associated with a 12.7–17.7% decrease in serum LDL cholesterol and a 1.2–5.6% increase in HDL cholesterol [37; Dr. Trudy Bush, personal communication]. Other studies show that after statistical adjustment for estrogen-induced changes in HDL- and LDL cholesterol levels, the magnitude of estrogen's independent cardiovascular effect decreases significantly, suggesting that favorable alteration of lipids is a major mechanism of estrogen's cardioprotective effect. For instance, adjustment in the Lipid Research Clinics study showed that the magnitude of the coefficient of estrogen's protective effect decreased by 40%, and that the HDL level was independently associated with cardiovascular mortality [35]. Although these results suggest that a significant proportion of estrogen's cardioprotective effects are due to lipid lowering, they also demonstrate that there is an independent residual beneficial effect of estrogen. Therefore, estrogen's reductions of cardiovascular events and risk of angiographically detected coronary atherosclerosis are likely due to a combination of lipid lowering and vascular effects [49].

Estrogen's direct vascular effects are likely due to its interaction with the arterial wall. Animal studies demonstrate that chronic postmenopausal estrogen therapy directly inhibits diet-induced coronary artery intimal hyperplasia independent of lipid lowering, suggesting that estrogen has direct antiatherosclerotic properties [50]. Estrogen also accelerates functional endothelial recovery in arteries that are de-endothelialized by barotrauma, suggesting that it may limit the effects of atherogenic stimuli [51]. These direct antiatherosclerotic vascular effects are expected to be associated with lower clinical cardiovascular event rates.

Estrogen may also provide cardioprotection by mechanisms that prevent abnormal coronary vasoconstriction (*i.e.,* abnormal coronary vasoreactivity), which has been implicated as a pathophysiologic mechanism of cardiac syndromes including angina, coronary vasospasm, myocardial infarction, and sudden death. *In vitro* studies demonstrate that estrogen increases nitric oxide production by dysfunctional arterial endothelium and inhibits the production of oxidized LDL, which is a potent inhibitor of endothelium-dependent vasodilatation [52–54]. Therefore, estrogen may favorably alter the balance between myocardial supply and demand by modulating coronary vasomotor function via endothelium-dependent mechanisms. Estrogen may also increase coronary blood flow via endothelium-independent vasodilatation by directly acting on arterial myocytes, which contain estrogen receptors [18]. For example, it has been shown that estrogen induces arterial myocyte hyperpolarization, alters ATP-sensitive myocyte potassium channel kinetics, and inhibits calcium and endothelin-1 induced arterial constriction [55–58].

Estrogen's vasomotor properties favorably affect *in vivo* coronary vasoreactivity both in primates and in humans. Williams *et al.* were the first to demonstrate that estrogen reverses paradoxical acetylcholine-induced epicardial coronary constriction in ovariectomized cynomolgus monkeys, suggesting that estrogen attenuates coronary artery endothelial dysfunction, which is an early physiologic manifestation of atherosclerosis [59]. Reis *et al.* subsequently demonstrated that high dose intravenous ethinyl estradiol acutely decreases basal coronary vasomotor tone (*i.e.,* increases basal coronary blood flow) and prevents paradoxical acetylcholine-induced vasoconstriction in nonstenotic coronaries of postmenopausal women, suggesting that estrogen has direct vascular effects in women [60]. These findings were subsequently confirmed by Gilligan and Collins and their colleagues who demonstrated that *physiologic* doses of 17β-estradiol prevented acetylcholine-induced coronary constriction and potentiated acetylcholine-induced coronary flow augmentation in postmenopausal women [61,62].

As previously noted, some epidemiologic studies suggest that progestins may attenuate estrogen's cardioprotective effects, possibly by blunting favorable estrogen-induced changes in lipids. A study using an animal model showed that cyclic high-dose or continuous low-dose oral medroxyprogesterone acetate inhibits estrogen's favorable coronary vasomotor effects by 50%, suggesting another possible mechanism for progestin-related attenuation of cardioprotection [63]. However, another animal study demonstrated that progestins do not attenuate estrogen-induced inhibition of coronary atherosclerosis in ovariectomized monkeys [50]. Therefore, the overall cardiovascular impact of progestin supplementation is uncertain and needs to be determined by ongoing large-scale randomized controlled clinical trials.

V. The Diagnosis and Treatment of Coronary Artery Disease in Women

A. Noninvasive Evaluation of Chest Pain in Women

Traditional noninvasive testing strategies used to evaluate the cause of chest pain in women were formulated using the results of clinical studies performed in predominantly male study populations. However, the sensitivities and specificities of cardiovascular tests should be expected to differ between men and women because, as we have discussed, the underlying mechanisms of chest pain and the prevalence of CAD are gender-related [3]. Indeed, the Coronary Artery Surgery Study (CASS) was the first large-scale trial to demonstrate a higher false positive rate and lower specificity of the traditional exercise electrocardiographic (ECG) stress test in women compared to men [64]. This finding may have been related to the lower prevalence of CAD among women and to the observation that women were more likely to have had baseline ECG ST-segment and T-wave abnormalities that are known to be associated with an increased likelihood of a false positive exercise ECG stress test. Furthermore, the CASS study also demonstrated that an exercise ECG stress test performed in a woman with abnormal resting ST-T segments has a lower specificity compared to one performed in a man with similar ECG abnormalities, suggesting that female gender is an independent predictor of a false positive stress ECG test. Other studies have demonstrated that treadmill ECG stress tests are also less sensitive in women and that the decreased sensitivity is only partially related to the lower prevalence of severe CAD [65–68].

Due to the low sensitivity and specificity of traditional ECG stress tests among women, many clinicians now routinely perform scintigraphic stress tests. Indeed, small studies have shown that planar or SPECT imaging with Thallium-201 increase the specificity of stress testing from 36–86% to 88–97% and the sensitivity from 57–76% to 71–86% [69–71]. The use of newer perfusion agents such as Tc-99 sestamibi has also increased the accuracy of noninvasive stress tests. For instance, use of this agent with adenosine as a stressor is associated with 93% sensitivity and 69% specificity in women with nonanginal symptoms, and 92% sensitivity and 83% specificity in those with angina [72]. However, while costly scintigraphic stress tests are more accurate than exercise ECG tests, they have unique limitations in women. For instance, radioactivity emitted from the heart after radioisotope injection may be attenuated by breast tissue, resulting in a greater likelihood of a false-positive test result with fixed anterior or apical wall scintigraphic defects suggestive of myocardial infarction. In addition, women generally have smaller hearts than men do, which lessens the diagnostic accuracy of Thallium-201 SPECT imaging and may account for the reduced accuracy of this type of stress test among women [73].

The accuracy of other imaging techniques (*e.g.,* stress echocardiography) in women is still uncertain. A study of dobutamine stress echocardiography among 1209 women demonstrated a significantly lower overall sensitivity in women (78% vs 88% in men), with lower sensitivities reported in women with single-vessel CAD (72% vs 78% in men) as well as those with multi-vessel CAD (82% vs 93% in men) [74]. This study also demonstrated low specificities for this test both in women and in men (55% and 46%, respectively), which was attributed to the study's referral bias. The inherent potential limitation of this test (*i.e.,* inadequate echocardiographic imaging of the left ventricle) may be circumvented by using transesophageal echocardiography during dobutamine stress. Indeed, one small study demonstrated that this technique may accurately evaluate CAD among women presenting with chest pain [75].

The (in)accuracy of traditional noninvasive evaluations of chest pain, lower prevalence of CAD, and lesser severity of CAD in women, combined with common misperceptions about the frequency of CAD mortality in women, may bias the clinical evaluation of women with chest pain. In fact, women with CAD are diagnosed and treated less aggressively than are men. Ayanian and colleagues reported that in 82,782 hospital discharges of patients with a diagnosis of CAD, men were significantly more likely than women to have undergone diagnostic angiography and coronary revascularization [76]. Secondary analysis performed in patients with a discharge diagnosis of acute myocardial infarction demonstrated similar findings, suggesting that women with severe CAD were treated less aggressively than were their male counterparts. Similarly, the Survival and Ventricular Enlargement (SAVE) Study investigators found less frequent use of catheterization and intervention in women than men with a recent diagnosis of acute infarction and impaired left ventricular function [77].

Several studies have asserted that the less aggressive use of catheterization and intervention in women may in fact be appropriate given gender differences in other risk factors. For instance, a Beth Israel Hospital study examined rates of catheterization in patients with a primary discharge diagnosis of acute myocardial infarction and found that while overall only 22% of women had catheterization compared to 34% of men, the rates were nearly equivalent by gender within age strata (the overall difference was due to the majority of elderly patients, who are infrequently referred for catheterization postinfarction, being women). In this same study, however, there was a trend for less bypass surgery in women that remained borderline significant after adjustment for age and CAD severity [78]. A Duke University study of symptomatic outpatients undergoing exercise testing found that significantly lower rates of referral for catheterization in women could be explained by their lower physician-assessed probability of CAD (obtained before actual testing) and by less frequent positive exercise tests [79]. Additionally, a comparison of the physicians' predictions of CAD to those from validated multivariable statistical models found that the physicians tended to actually overestimate the probability of obstructive CAD in women. A second Duke study of patients catheterized and found to have significant CAD reported that women presenting with CAD were referred for coronary artery bypass graft surgery (CABG) at rates approximately equal to those of men [80]. Among high risk patients for whom CABG offered a potential survival benefit, women were referred to CABG somewhat more frequently than men, whereas among low risk patients for whom a surgical survival benefit was questionable, men were referred to CABG more frequently. The authors noted that these gender differences in referral patterns could be indicative of differential treatment and evaluation patterns earlier in the diagnostic process. A Mayo Clinic study of more than 22,000 patients undergoing angiography reported comparable use of revascularization by gender, after adjustment for baseline differences including age [81]. The results of these and other studies confirm the importance of increasing physician and patient awareness of the significance of cardiovascular disease in women and gender-related practice patterns among health care providers.

B. Prognosis after Myocardial Infarction

Women are more likely to have adverse outcomes following acute cardiac events such as myocardial infarction (MI). For instance, women more commonly die within one year following MI, despite having fewer Q-wave MIs and a better left ventricular ejection fraction at hospital discharge [16,82,83]. Furthermore, the 6-year rate of reinfarction is nearly 35% higher in women than in men. The explanation of these gender discrepancies in post-MI clinical outcome is multifactorial. For example, at the time of presentation with MI, women are approximately 7 years older than men and are more likely to have other comorbidities such as diabetes and hypertension that may adversely affect outcome [2,84]. Studies controlling for this higher risk profile in women vary in their assessment of whether female gender is independently associated with poorer post-MI outcome [2,83,85]. However, the RESCATE study reported that after a first myocardial infarction, the odds ratios for female versus male gender were 1.72 (95% CI = 1.12–2.65) and 1.73 (CI = 1.18–2.52) for 28-day and 6-month mortality, respectively, after adjustment for age, diabetes, hypertension, smoking status, and previous angina [86]. Regardless of whether poorer outcome following MI is directly related to female gender, the vast majority of studies do confirm that women have a worse prognosis than men following MI.

C. Coronary Revascularization: Coronary Artery Bypass Graft (CABG) Surgery

As is the case for MI, women presenting with CAD requiring revascularization tend to be older and present with a higher risk factor profile than men. For instance, various studies of consecutive cases in the 1970s and 1980s found that women treated with CABG were on average two to three years older than men and had more preoperative unstable angina, diabetes, hypertension, hyperlipidemia, and peripheral vascular disease [87–89]. However, while women treated with CABG had more cardiovascular risk factors, the anatomical severity of CAD tended to be markedly higher for treated men, who also had more impaired left ventricular function and a more frequent history of myocardial infarction.

Most studies have found significantly higher in-hospital CABG mortality in women than in men [87,89,90]. Several studies that performed multivariate analysis found that the increased mortality in women was explained by their smaller body surface area and vessel diameter, possibly due to increases in anastomosis difficulty or early graft closure in smaller vessels [87,89,90]. One study also reported that less frequent internal mammary artery (IMA) use as a bypass conduit in women partially explained the higher in-hospital mortality in women [90]. Less frequent use of the IMA in women may also potentially affect long-term survival, because CABG patients with IMA conduits have superior long-term results [91].

It is important to keep in mind that the case mix for surgery has changed over time. With the advent of percutaneous procedures, patients whose CAD is limited or otherwise ideal for catheter-based interventions are able to avoid or delay surgical intervention. Therefore, patients treated with CABG in recent years tend to be older and to require more urgent surgery. Despite this shift toward higher risk procedures, surgical mortality has actually decreased for both genders over time, possibly due to increased experience or improvements in cardioplegia techniques [88]. Available long-term data (5–15 years post-CABG) show that while operative mortality tends to be higher in women, survival among patients discharged alive following bypass surgery is generally comparable for men and women, with relief of angina reported less frequently in women [87,89].

Another development in coronary bypass surgery is the use of "minimally invasive" approaches. These techniques may be performed through small incisions on the beating heart without cardiopulmonary bypass using thorascopic guidance. Because technical aspects generally limit minimally invasive CABG approaches to surgical revascularization of the left anterior descending coronary artery, patients with severe multivessel coronary disease are not candidates for this emerging approach. Because women tend to have less severe CAD, they may be more likely to be eligible for this procedure, which is thought to be safer than traditional CABG. However, while this revascularization strategy is in its infancy, an analysis of 508 patients (27% women) treated in 1996–1997 with this emerging approach found female gender to be independently associated with major in-hospital events. Specifically, mortality was uniformly low among men and women, but women were at increased risk for conversion to full sternotomy and need for reintervention [92].

D. Catheter-Based Coronary Revascularization

Percutaneous transluminal coronary angioplasty (PTCA), together with other catheter-based technologies such as intracoronary stents, has become a commonly used means of coronary revascularization because it is associated with lower procedural risk, shorter in-hospital stay, and decreased cost compared with CABG. In 1993, more than 362,000 patients in the United States had PTCA and 309,000 had CABG; approximately one-third of patients receiving each procedure were women [93]. An analysis of the earliest PTCA cases enrolled in a National Heart, Lung, and Blood Institute (NHLBI) registry found that, as in the CABG experience, women presented with more unstable symptoms, more risk factors, less extensive coronary disease, and better left ventricular function compared to men [94]. Women had significantly higher rates of in-hospital mortality as well as other complications, and female gender was an independent risk factor for in-hospital death. While rates of successful reduction of stenosis were significantly lower in women, this finding is now out of date because early PTCA catheters and balloons, which had difficulty crossing the smaller target vessels that are more prevalent in women, had much higher profiles (larger diameters) than modern equipment.

The NHLBI reopened the PTCA Registry in 1985–1986 to document increased experience and significant technical improvements. Once again, women were older than men, with a higher CAD risk profile but equivalent CAD severity [95]. While angiographic success was uniformly higher and not significantly different by gender, women continued to have more untoward peri- and postprocedural events and a significantly higher in-hospital mortality rate of 2.6% versus 0.3% in men.

A very strong independent effect of gender remained after adjustment for age and other risk factors, including diabetes and history of congestive heart failure, both of which were approximately twice as prevalent in women as in men. Because of this unfavorable outcome in women, NHLBI recruited a small cohort of women in 1993–1994 and found that, despite a higher risk profile, women undergoing initial PTCA in the 1990s had significantly higher success rates, somewhat lower in-hospital mortality (1.5%), and significantly lower rates of in-hospital infarction and need for CABG compared to women in the 1985–1986 Registry [96].

At four-year follow-up of the 1985–1986 Registry, overall mortality remained significantly higher in women. However, this long-term gender effect could be accounted for by age and other baseline risk factors for mortality, including history of congestive heart failure, severe concomitant noncardiac disease, and patient classification as inoperable or high risk for surgery, all of which were twice as prevalent in women as in men. Other studies of PTCA in the late 1980s and early 1990s have also reported equivalent procedure success rates by gender and increased in-hospital mortality in women [97,98] as well as no independent long-term effect of gender on mortality [98]. An interesting long-term finding is that for catheter-based intervention, as for CABG, more women than men report recurrence of angina at follow-up [95,98].

The reported Registry series excluded patients with recent myocardial infarction. However, a study of patients treated with angioplasty for postinfarction ischemia found that women had rates of mortality and reintervention similar to those of men, both in-hospital and at an average of 3 years of follow-up [99]. Recurrence of angina was again more frequent in these women despite equivalent completeness of revascularization by gender.

Angioplasty has been supplemented with various new percutaneous technologies developed to treat difficult lesions and to reduce rates of restenosis (*e.g.*, atherectomy, intracoronary stents). When compared with men, women treated with these approaches continue to have the higher risk profile and less severe CAD seen in the days of "plain old balloon" angioplasty [100]. The NHLBI-funded New Approaches to Coronary Intervention (NACI) Registry found comparable rates of postintervention stenosis and major in-hospital events by gender in more than 2800 patients treated with new devices in the early 1990s, although women had more nonemergent CABG and somewhat longer hospital stays [100]. Studies of in-hospital complications and short-term outcome associated with these "new devices" tend to report either no difference by gender or somewhat increased immediate complications in women [101,102].

VI. Conclusions

Women with CAD continue to be treated with insufficiently aggressive strategies despite having poorer prognoses following cardiac events such as myocardial infarction. This observation may be explained, in part, by clinicians' misunderstandings about the sequelae of cardiovascular disease in women and gender-related discrepancies in cardiovascular physiology, evaluations, therapies, and outcomes. As we have pointed out, there are gender-specific issues in the mechanisms of chest pain and the likelihood that a patient with chest pain has coronary dis-

ease, and these differences and other factors affect the diagnostic ability of noninvasive evaluations. We have seen that women who are postmenopausal or who present with traditional CAD risk factors have substantially increased risk of coronary disease (and so must be evaluated aggressively), and that the cardioprotective benefits of postmenopausal estrogen have been confirmed by many studies, both epidemiologic and mechanistic. Finally, women appear to be at increased risk of adverse short-term outcome following revascularization procedures, although this gender effect is at least partly explained by the higher risk profile of women who present with CAD.

Whether or not the less aggressive evaluation of cardiovascular symptoms in women may be explained by gender differences in presentation and likelihood of coronary disease, the fact remains that current clinical strategies used for evaluation in women are based on the results of early clinical trials that focused on men. However, there has been an encouraging proliferation of cardiovascular studies focusing on women that will provide the information urgently needed to develop and refine gender-based cardiovascular diagnostic and therapeutic strategies. As previously mentioned, the NIH-sponsored WISE study is designed to optimize approaches to the diagnosis of suspected myocardial ischemia in women. Other studies including the Postmenopausal Estrogen-Progestin Intervention (PEPI) Trial, the Women's Health Initiative, the Angiographic Trial in Women, and the Estrogen Replacement and Atherosclerosis in Older Women (ERA) Study will provide further insight into the cardiovascular benefits and cancer risks of postmenopausal hormone replacement therapy. In addition to increasing awareness of the importance of coronary disease in women, the results of these studies will provide practitioners with guidance in the development of gender-specific diagnostic and therapeutic strategies in patients presenting with cardiovascular symptoms. The standard of cardiovascular care in women should improve in the next decade as the results of these studies become available.

References

1. National Center for Health Statistics (1995). "Health, United States, 1991." U.S. Public Health Service, Hyattsville, MD.
2. Murabito, J. M., Evans, J. C., Larson, M. G., and Levy, D. (1993). Prognosis after the onset of coronary heart disease. An investigation of differences in outcome between the sexes according to initial coronary disease presentation. *Circulation* **88,** 2548–2555.
3. Sox, H. C. (1983). Noninvasive testing in coronary artery disease. Selection of procedures and interpretation of results. *Postgrad. Med.* **74,** 319–336.
4. Bugiardini, R., Pozzati, A., Ottani, F., Morgagni, G. L., and Puddu, P. (1993). Vasotonic angina: A spectrum of ischemic syndromes involving functional abnormalities of the epicardial and microvascular coronary circulation. *J. Am. Coll. Cardiol.* **22,** 417–425.
5. Camici, P. G., Marraccini, P., Lorenzoni, R., Buzzigoli, G., Pecori, N., Perissinotto, A., Ferrannini, E., L'Abbate, A., and Marzilli, M. (1991). Coronary hemodynamics and myocardial metabolism in patients with syndrome X: Response to pacing stress. *J. Am. Coll. Cardiol.* **17,** 1461–1470.
6. Cannon, R. O., Bonow, R. O., Bacharach, S. L., Green, M. V., Rosing, D. R., Leon, M. B., Watson, R. M., and Epstein, S. E. (1985). Left ventricular dysfunction in patients with angina pectoris, normal epicardial coronary arteries, and abnormal vasodilator reserve. *Circulation* **71,** 218–226.

7. Onaka, H., Hirota, Y., Shimada, S., Kita, Y., Sakai, Y., Kawakami, Y., Suzuki, S., and Kawamura, K. (1996). Clinical observation of spontaneous anginal attacks and multivessel spasm in variant angina pectoris with normal coronary arteries: Evaluation by 24-hour 12-lead electrocardiography with computer analysis. *J. Am. Coll. Cardiol.* **27**, 38–44.

8. Okumura, K., Yasue, H., Matsuyama, K., Ogawa, H., Kugiyama, K., Ishizaka, H., Sumida, H., Fuji, H., Matsunaga, T., and Tsunoda, R. (1996). Diffuse disorder of coronary artery vasomotility in patients with coronary spastic angina. Hyperreactivity to the constrictor effects of acetylcholine and the dilator effects of nitroglycerin. *J. Am. Coll. Cardiol.* **27**, 45–52.

9. Egashira, K., Katsuda, Y., Mohri, M., Kuga, T., Tagawa, T., Shimokawa, H., and Takeshita, A. (1996). Basal release of endothelium-derived nitric oxide at site of spasm in patients with variant angina. *J. Am. Coll. Cardiol.* **27**, 1444–1449.

10. Yamagishi, M., Miyatake, K., Tamai, J., Nakatani, S., Koyama, J., and Nissen, S. E. (1994). Intravascular ultrasound detection of atherosclerosis at the site of focal vasospasm in angiographically normal or minimally narrowed coronary segments. *J. Am. Coll. Cardiol.* **23**, 352–357.

11. Kugiyama, K., Yasue, H., Okumura, K., Ogawa, H., Fujimoto, K., Nakao, K., Yoshimura, M., Motoyama, T., Inobe, Y., and Kawano, H. (1996). Nitric oxide is deficient in spasm arteries of patients with coronary spastic angina. *Circulation* **94**, 266–272.

12. Opherk, D., Zebe, H., Weihe, E., Mall, G., Durr, C., Gravert, B., Mehmel, H. C., Schwarz, F., and Kubler, W. (1981). Reduced coronary dilatory capacity and ultrastructural changes of the myocardium in patients with angina pectoris but normal coronary angiograms. *Circulation* **63**, 817–825.

13. Egashira, K., Inou, T., Hirooka, Y., Yamada, A., Urabe, Y., and Takeshita, A. (1993). Evidence of impaired endothelium-dependent vasodilatation in patients with angina pectoris and normal coronary angiograms. *N. Engl. J. Med.* **328**, 1659–1664.

14. Rosano, G. M. C., Peters, N. S., Lefroy, D. *et al.* (1996). 17-beta-estradiol therapy lessens angina in postmenopausal women with Syndrome X. *J. Am. Coll. Cardiol.* **28**, 1500–1505.

15. Cannon, R. O., III, Quyyumi, A. A., Mincemoyer, R., Stine, A. M., Gracely, R. H., Smith, W. B., Geraci, M. F., Black, B. C., Uhde, T. W., Waclawiw, M. A., Maher, K., and Benjamin, S. B. (1994). Imipramine in patients with chest pain despite normal coronary angiograms. *N. Engl. J. Med.* **330**, 1411–1417.

16. Lerner, D. J., and Kannel, W. B. (1986). Patterns of coronary heart disease morbidity and mortality in the sexes. A 26-year follow-up of the Framingham population. *Am. Heart J.* **111**, 383–390.

17. Colditz, G. A., Willett, W. C., Stampfer, M. J., Rosner, B., Speizer, F. E., and Hennekens, C. H. (1987). Menopause and the risk of coronary heart disease in women. *N. Engl. J. Med.* **316**, 1105–1110.

18. Losordo, D. W., Kerney, M., Kim, E. A., Jekanowski, J., and Isner, J. M. (1994). Variable expression of the estrogen receptor in normal and atherosclerotic coronary arteries of premenopausal women. *Circulation* **89**, 1501–1510.

19. Birdsall, M. A., Farquhar, C. M., and White, H. D. (1997). Association between polycystic ovaries and extent of coronary artery disease in women having cardiac catheterization. *Ann. Intern. Med.* **126**, 32–35.

20. Guzick, D. S., Talbott, E. O., Sutton-Tyrrell, K., Herzog, H. C., Kuller, L. H., and Wolfson, S. K. (1996). Carotid atherosclerosis in women with polycystic ovary syndrome: Initial results from a case-control study. *Am. J. Obstet. Gynecol.* **174**, 1224–1229.

21. Ruderman, N. B., and Haudenschild, C. (1984). Diabetes as an atherogenic factor. *Prog. Cardiovasc. Dis.* **26**, 373–412.

22. Canadian Diabetes Association (1995). "The Chronic Complications of Diabetes." Can Diabetes Assoc. (Toronto, ON, Canada). World Wide Web URL: *http://www.diabetes.ca:80/aboutdia/chronic.htm*

23. Manson, J. E., Colditz, G. A., Stampfer, M. J. *et al.* (1991). A prospective study of maturity-onset diabetes mellitus and risk of coronary heart disease and stroke in women. *Arch. Intern. Med.* **151**, 1141–1147.

24. Krolewski, A. S., Warram, J. H., Valsania, P., *et al.* (1991). Evolving natural history of coronary artery disease in diabetes mellitus. *Am. J. Med.* **90**(Suppl. 2A); 56S–61S.

25. Barrett-Connor, E. L., Cohn, B. A., Wingard, D. L. *et al.* (1991). Why is diabetes mellitus a stronger risk factor for fatal ischemic heart disease in women than in men? *JAMA, J. Am. Med. Assoc.* **265**, 627–631.

26. Kannel, W. B., McGee, D., and Gordon, T. (1976). A general cardiovascular risk profile: The Framingham Study. *Am. J. Cardiol.* **38**, 46–51.

27. Johnson, J. L., Heineman, E. F., Heiss, G., Hames, C. G., and Tyroler, H. A. (1986). Cardiovascular disease risk factors and mortality among black women and white women aged 40–64 years in Evans County, Georgia. *Am. J. Epidemiol.* **123**, 209–220.

28. Sigurdsson, J. A., Bengtsson, C., Lapidus, L., Lindquist, O., and Rafnsson, V. (1984). Morbidity and mortality in relation to blood pressure and antihypertensive treatment: A 12-year follow-up study of a population sample of Swedish women. *Acta Med. Scand.* **215**, 313–322.

29. Fiebach, N. H., Hebert, P. R., Stampfer, M. J. *et al.* (1989). A prospective study of high blood pressure and cardiovascular disease in women. *Am. J. Epidemiol.* **130**, 646–654.

30. Collins, R., Peto, R., MacMahon, S. *et al.* (1990). Blood pressure, stroke, and coronary heart disease. 2. Short-term reductions in blood pressure: Overview of randomized drug trials in their epidemiological context. *Lancet* **335**, 827–838.

31. Medical Research Council Working Party (1985). MRC trial of treatment of mild hypertension: Principal results. *Br. Med. J.* **291**, 97–104.

32. MRC Working Party (1992). Medical Research Council trial of treatment of hypertension in older adults: Principal results. *Br. Med. J.* **304**, 405–412.

33. Amery, A., Birkenhager, W., Brixko, R. *et al.* (1986). Efficacy of antihypertensive drug treatment according to age, sex, blood pressure, and previous cardiovascular disease in patients over the age of 60. *Lancet* **2**, 589–592.

34. Rich-Edwards, J. W., Manson, J. E., Hennekens, C. H., and Buring, J. E. (1995). The primary prevention of coronary heart disease in women. *N. Engl. J. Med.* **332**, 1758–1766.

35. Bush, T. L., Barrett-Connor, E., Cowan, L. D. *et al.* (1987). Cardiovascular mortality and noncontraceptive use of estrogen in women: Results from the Lipid Research Clinics Program Follow-up Study. *Circulation* **75**, 1102–1109.

36. National Cholesterol Education Program (1993). "Second Report of the Expert Panel on detection, Evaluation, and Treatment of High Blood Cholesterol in Adults." National Institutes of Health, Washington, DC.

37. The Writing Group for the PEPI Trial. (1995). Effects of estrogen or estrogen/progestin regimens on heart disease risk factors in postmenopausal women. The postmenopausal estrogen/progestin interventions (PEPI) trial. *JAMA, J. Am. Med. Assoc.* **273**, 199–208.

38. Department of Health and Human Services (1989). "Reducing the Health Consequences of Smoking: 25 Years of Progress: A Report of the Surgeon General." DHHS, Rockville, MD.

39. American Heart Association (1996). "Heart and Stroke Facts: 1997 Statistical Supplement." American Heart Association, Dallas, TX.

40. Department of Health and Human Services (1990). "The Health Benefits of Smoking Cessation. Report of the Surgeon General." DHHS, Washington, DC.

41. Celermajer, D. S., Sorenson, K. E., Georgakopoulos, D., Bull, C., Thomas, O., Robinson, J., and Deanfield, J. E. (1993). Cigarette

smoking is associated with dose-related and potentially reversible impairment of endothelium-dependent dilation in healthy young adults. *Circulation* **88**(Part 1); 2149–2155.

42. Kiowski, W., Linder, L., Stoschitzky, K., Pfisterer, M., Burckhardt, D., Burkart, F., and Buhler, F. R. (1994). Diminished vascular response to inhibition of endothelium-derived nitric oxide and enhanced vasoconstriction to exogenously administered endothelin-1 in clinically healthy smokers. *Circulation* **90**, 27–34.

43. McKinlay, S. M., Bifano, N. L., and McKinlay, J. B. (1985). Smoking and age at menopause in women. *Ann. Intern. Med.* **103**, 350–356.

44. Stampfer, M. J., Colditz, G. A., Willett, W. C. *et al.* (1991). Postmenopausal estrogen therapy and cardiovascular disease. Ten-year follow-up from the Nurses' Health Study. *N. Engl. J. Med.* **325**, 756–762.

45. Stampfer, M. J., and Colditz, G. A. (1991). Estrogen replacement therapy and coronary heart disease: A quantitative assessment of the epidemiologic evidence. *Prev. Med.* **20**, 47–63.

46. Grady, D., Rubin, S. M., Petitti, D. B. *et al.* (1992). Hormone therapy to prevent disease and prolong life in postmenopausal women. *Ann. Intern. Med.* **117**, 1016–1037.

47. Hulley, S., Grady, D., Bush, T. *et al.* (1998). Randomized trial of estrogen plus progestin for secondary prevention of coronary heart disease in postmenopausal women. *JAMA, J. Am. Med. Assoc.* **280**, 605–613.

48. Grodstein, F., Stampfer, M. J., Manson, J. E. *et al.* (1996). Postmenopausal estrogen and progestin use and the risk of cardiovascular disease. *N. Engl. J. Med.* **335**, 453–461.

49. Sullivan, J. M., Zwaag, R. V., Lemp, G. F. *et al.* (1988). Postmenopausal estrogen use and coronary atherosclerosis. *Ann. Intern. Med.* **108**, 358–363.

50. Adams, M. R., Kaplan, J. R., Manuck, S. B., Koritnik, D. R., Parks, J. S., Wolfe, M. S., and Clarkson, T. B. (1990). Inhibition of coronary artery atherosclerosis by 17-beta estradiol in ovariectomized monkeys. Lack of an effect of added progesterone. *Arteriosclerosis* **10**, 1051–1057.

51. Krasinski, K., Spyridopoulos, I., Asahara, T., van der Zee, R., Isner, J. M., and Losordo, D. W. (1997). Estradiol accelerates functional endothelial recovery after arterial injury. *Circulation* **95**, 1768–1772.

52. Van Buren, G. A., Yang, D., and Clark, K. E. (1992). Estrogen-induced uterine vasodilatation is antagonized by L-nitroarginine methyl ester, an inhibitor of nitric oxide synthesis. *Am. J. Obstet. Gynecol.* **167**, 828–833.

53. Rifici, V. A., and Khachadurian, A. K. (1992). The inhibition of low-density lipoprotein oxidation by 17-B estradiol. *Metab. Clin. Exp.* **41**, 1110–1114.

54. Subbiah, M. T. R., Kessel, B., Agrawal, M., Rajan, R., Abplanalp, W., and Rymaszewski, Z. (1993). Antioxidant potential of specific estrogens on lipid peroxidation. *J. Endocrinol. Metab.* **77**, 1095–1097.

55. Harder, D. R., and Coulson, P. B. (1979). Estrogen receptors and effects of estrogen on membrane electrical properties of coronary vascular smooth muscle. *J. Cell. Physiol.* **100**, 375–382.

56. Jiang, C., Sarrel, P. M., Poole-Wilson, P. A., and Collins, P. (1992). Acute effect of 17B-estradiol on rabbit coronary artery contractile responses to endothelin-1. *Am. J. Physiol.* **263**, H271–H275.

57. Jiang, C., Sarrel, P. M., Lindsay, D. C., Poole-Wilson, P. A., and Collins, P. (1991). Endothelium-independent relaxation of rabbit coronary artery by 17B-oestradiol *in vitro. Br. J. Pharmacol.* **104**, 1033–1037.

58. Sudhir, K., Chou, T. M., Mullen, W. L. *et al.* (1995). Mechanisms of estrogen-induced vasodilatation: In vivo studies in canine coronary conductance and resistance arteries. *J. Am. Coll. Cardiol.* **26**, 807–814.

59. Williams, J. K., Adams, M. R., and Klopfenstein, H. S. (1990). Estrogen modulates responses of atherosclerotic coronary arteries. *Circulation* **81**, 1680–1687.

60. Reis, S. E., Gloth, S. T., Blumenthal, R. S. *et al.* (1994). Ethinyl estradiol acutely attenuates abnormal coronary vasomotor responses to acetylcholine in postmenopausal women. *Circulation* **89**, 52–60.

61. Gilligan, D. M., Quyyumi, A. A., and Cannon, R. O. (1994). Effects of physiological levels of estrogen on coronary vasomotor function in postmenopausal women. *Circulation* **89**, 2545–2551.

62. Collins, P., Rosano, G. M. C., Sarrel, P. M. *et al.* (1995). 17β-estradiol attenuates acetylcholine-induced coronary arterial constriction in women but not men with coronary heart disease. *Circulation* **92**, 24–30.

63. Williams, J. K., Honore, E. K., Washburn, S. A., and Clarkson, T. B. (1994). Effects of hormone replacement therapy on reactivity of atherosclerotic coronary arteries in cynomolgus monkeys. *J. Am. Coll. Cardiol.* **24**, 1757–1761.

64. Guiteras, V. P., Chaitman, B. R., Waters, D. D. *et al.* (1982). Diagnostic accuracy of exercise ECG lead systems in clinical subsets of women. *Circulation* **65**, 1465–1474.

65. DePace, N. L., Hakki, A. H., Weinrich, D. J. *et al.* (1983). Noninvasive assessment of coronary artery disease. *Am. J. Cardiol.* **52**, 714–720.

66. Hubbard, B. L., Gibbons, R. J., Lapeyre, A. C., III, *et al.* (1992). Identification of severe coronary artery disease using simple clinical parameters. *Arch. Intern. Med.* **152**, 309–312.

67. Hlatky, M. A., Pryor, D. B., Harrell, F. E. *et al.* (1984). Factors affecting sensitivity and specificity of exercise electrocardiography. *Am. J. Med.* **77**, 64–71.

68. Chaitman, B. R., Bourassa, M. G., David, K. *et al.* (1981). Angiographic prevalence of high-risk coronary artery disease in patient subsets. (CASS). *Circulation* **64**, 360–367.

69. Friedman, T. D., Green, A. C., Iskandrian, A. S. *et al.* (1982). Exercise thallium-201 myocardial scintigraphy in women: Correlation with coronary arteriography. *Am. J. Cardiol.* **49**, 1632–1637.

70. Melin, J. A., Wiins, W., Vanbustle, R. J. *et al.* (1985). Alternative diagnostic strategies for coronary artery disease in women: Demonstration of the usefulness and efficiency of probability analysis. *Circulation* **71**, 535–542.

71. Goodgold, H. M., Rehder, J. G., Samuels, L. D. *et al.* (1987). Improved interpretation of exercise T1-201 myocardial perfusion scintigraphy in women: Characterization of breast attenuation artifacts. *Radiology* **165**, 361–366.

72. Amanullah, A. M., Berman, D. S., Erel, J., Kiat, H., Cohen, I., Germano, G., Friedman, J. D., and Hachamovitch, R. (1998). Incremental prognostic value of adenosine myocardial perfusion single-photon emission computed tomography in women with suspected coronary artery disease. *Am. J. Cardiol.* **82**, 725–730.

73. Hansen, C. L., Crabbé, D., and Rubin, S. (1996). Lower diagnostic accuracy of Thallium-201 SPECT myocardial perfusion imaging in women: An effect of smaller chamber size. *J. Am. Coll. Cardiol.* **28**, 1214–1219.

74. Secknus, M. A., and Marwick, T. H. (1997). Influence of gender and physiologic response and accuracy of dobutamine echocardiography. *Am. J. Cardiol.* **80**, 721–724.

75. Laurienzo, J. M., Cannon, R. O., III, Quyyumi, A. A., Dilsizian, V., and Panza, J. A. (1997). Improved specificity of transesophageal dobutamine stress echocardiography compared to standard tests for evaluation of coronary artery disease in women presenting with chest pain. *Am. J. Cardiol.* **80**, 1402–1407.

76. Ayanian, J. Z., and Epstein, A. M. (1991). Differences in the use of procedures between women and men hospitalized for coronary heart disease. *N. Engl. J. Med.* **325**, 221–225.

77. Steingart, R. M., Packer, M., Hamm, P., Coglianese, M. E., Gersh, B., Geltman, E. M., Sollano, J., Katz, S., Moyé, L., Basta, L. L., Lewis, S. L., Gottlieb, S. S., Bernstein, V., McEwan, P., Jacobson, K., Brown, E. J., Kukin, M. I., Kantrowitz, N. E., and Pfeffer, M. A. (1991). Sex differences in the management of coronary artery disease. *N. Engl. J. Med.* **325,** 226–230.

78. Krumholz, H. M., Douglas, P. S., Lauer, M. S., and Pasternak, R. C. (1992). Selection of patients for coronary angiography and coronary revascularization early after myocardial infarction: Is there evidence for a gender bias? *Ann. Intern. Med.* **116,** 785–790.

79. Mark, D. B., Shaw, L. K., DeLong, E. R., Califf, R. M., and Pryor, D. B. (1994). Absence of sex bias in the referral of patients for cardiac catheterization. *N. Engl. J. Med.* **330,** 1101–1106.

80. Bickell, N. A., Pieper, K. S., Lee, K. L., Mark, D. B., Golwer, G. D., Pryor, D. B., and Califf, R. M. (1992). Referral patterns for coronary artery disease treatment: Gender bias or good clinical judgment? *Ann. Intern. Med.* **116,** 791–797.

81. Bell, M. R., Berger, P. B., Holmes, D. R., Jr., Mullany, C. J., Bailey, K. R., and Gersh, B. J. (1995). Referral for coronary artery revascularization procedures after diagnostic coronary angiography: evidence for gender bias? *J. Am. Coll. Cardiol.* **25,** 1650–1655.

82. Tofler, G. H., Stone, P. H., Muller, J. E., Willich, S. N., Davis, V. G., Poole, W. K., Strauss, H. W., Willerson, J. T., Jaffe, A. S., Robertson, T. *et al.* (1987). Effects of gender and race on prognosis after myocardial infarction: Adverse prognosis for women, particularly black women. *J. Am. Coll. Cardiol.* **9,** 473–482.

83. Greenland, P., Reicher-Reiss, H., Goldbourt, U., and Behar, S. (1991). In-hospital and 1-year mortality in 1,524 women after myocardial infarction: Comparison with 4,315 men. *Circulation* **82,** 484–491.

84. Goldberg, R. J., Gorak, E. J., Yarzebski, J., Hosmer, D. W., Jr., Dalen, P., Gore, J. M., Alpert, J. S., and Dalen, J. E. (1993). A communitywide perspective of sex differences and temporal trends in the incidence and survival rates after acute myocardial infarction and out-of-hospital deaths caused by coronary heart disease. *Circulation* **87,** 1947–1953.

85. Lincoff, A. M., Califf, R. M., Ellis, S. G., Sigmon, K. N., Lee, K. L., Leimberger, J. D., and Topol, E. J. (1993). Thrombolytic therapy for women with myocardial infarction: Is there a gender gap? Thrombolysis and Angioplasty in Myocardial Infarction Study Group. *J. Am. Coll. Cardiol.* **22,** 1780–1787.

86. Marrugat, J., Sala, J., Masia, R. *et al.* (1998). Mortality differences between men and women following first myocardial infarction. *JAMA, J. Am. Med. Assoc.* **280,** 1405–1409.

87. Loop, F. D., Golding, L. R., Macmillan, J. P. *et al.* (1983). Coronary artery surgery in women compared with men: Analyses of risks and long-term results. *J. Am. Coll. Cardiol.* **1,** 383–390.

88. Mickleborough, L. L., Yasushi, T., Maruyama, H. *et al.* (1995). Is sex a factor in determining operative risk for aortocoronary bypass surgery? *Circulation* **92**(Suppl. II), II-80–II-84.

89. Davis, K. B., Chaitman, B., Ryan, T. *et al.* (1995). Comparison of 15-year survival for men and women after initial medical or sur-gical treatment for coronary artery disease: A CASS registry study. *J. Am. Coll. Cardiol.* **25,** 1000–1009.

90. O'Connor, G. T., Morton, J. R., Diehl, M. J. *et al.* (1993). Differences between men and women in hospital mortality associated with coronary artery bypass graft surgery. *Circulation* **88,** 2104–2110.

91. Cameron, A., Davis, K. B., Green, G., and Schaff, H. V. (1996). Coronary bypass surgery with internal-thoracic-artery grafts—effects on survival over a 15-year period. *N. Engl. J. Med.* **334,** 216–219.

92. Holubkov, R., Zenati, M., Akin, J. J., Erb, L., and Courcoulas, A. (1998). MIDCAB characteristics and results: The CardioThoracic Systems (CTS) Registry. *Eur. J. Cardiothorac. Surg.* **14,** S25–S30.

93. American Heart Association (1995). "Heart and Stroke Facts: 1996 Statistical Supplement." American Heart Association, Dallas, TX.

94. Cowley, M. J., Mullin, S. M., Kelsey, S. F. *et al.* (1985). Sex differences in early and long-term results of coronary angioplasty in the NHLBI PTCA registry. *Circulation* **71,** 90–97.

95. Kelsey, S. F., James, M., Holubkov, A. L. *et al.* (1993). Results of percutaneous transluminal coronary angioplasty in women. 1985–86 National Heart, Lung, and Blood Institute's coronary angioplasty registry. *Circulation* **87,** 720–727.

96. Jacobs, A. K., Kelsey, S. F., Yeh, W., Holmes, D. R., Jr., Block, P. C., Cowley, M. J., Bourassa, M. G., Williams, D. O., King, S. B., III, Faxon, D. P., Myler, R., and Detre, K. M. (1997). Documentation of decline in morbidity in women undergoing coronary angioplasty (a report from the 1993–94 NHLBI Percutaneous Transluminal Coronary Angioplasty Registry). *Am. J. Cardiol.* **80,** 979–984.

97. Bell, M. R., Holmes, D. R., Jr., Berger, P. *et al.* (1993). The changing in-hospital mortality of women undergoing percutaneous transluminal coronary angioplasty. *JAMA, J. Am. Med. Assoc.* **269,** 2091–2095.

98. Weintraub, W. S., Wenger, N. K., Kosinski, A. S., Douglas, J. S., Liberman, H. A., Morris, D. C., and King, S. B., III (1994). Percutaneous transluminal coronary angioplasty in women compared with men. *J. Am. Coll. Cardiol.* **24,** 81–90.

99. Welty, F. K., Mittleman, M. A., Healy, R. W., Muller, J. E., and Subrooks, S. J., Jr. (1994). Similar results of percutaneous transluminal coronary angioplasty for women and men with postmyocardial infarction ischemia. *J. Am. Coll. Cardiol.* **23,** 35–39.

100. Robertson, T. R., Kennard, E. D., Mehta, S., Popma, J. J., Carrozza, J. P., King, S. B., III, Holmes, D. R., Cowley, M. J., Hornung, C. A., Kent, K. M., Roubin, G. S., Litvack, F., Moses, J. W., Safian, R., Desvigne-Nickens, P., and Detre, K. M. (1997). Influence of gender on in-hospital clinical and angiographic outcomes on one-year follow-up in the New Approaches to Coronary Intervention (NACI) Registry. *Am. J. Cardiol.* **80,** 26K–39K.

101. Mehran, R., Bucher, T. A., Lansky, A. J. *et al.* (1997). Coronary stenting in women: Early in-hospital and long-term clinical outcomes. *J. Am. Coll. Cardiol.* **29**(Suppl. A), 454A (abstr.).

102. Nasser, T. K., Fry, E. T. A., Peters, T. F. *et al.* (1997). Coronary stenting in women: Clinical outcomes are equivalent to men. *J. Am. Coll. Cardiol.* **29**(Suppl. A); 71A (abstr.).

60

Hormonal Milieu and Heart Disease

ELAINE MEILAHN

Department of Epidemiology and Population Studies
London School of Hygiene and Tropical Medicine
London, England

I. Introduction

Does ovarian production of estrogen afford women cardiovascular protection, providing the biological basis for the sex difference in heart disease rates? Furthermore, can women prolong their "female advantage" after ovarian estrogen production ceases at menopause by taking supplemental estrogen therapy? The public health implications of these questions are substantial because of the potential for long-term use of pharmacological estrogen to prevent heart disease by millions of healthy women.

Belief in the cardiovascular benefit of endogenous (ovarian) and exogenous (hormone replacement therapy (HRT)) estrogen is widespread. The following chapter examines the evidence, with a focus on epidemiology, for and against the hypothesis that endogenous or exogenous estrogen provides cardioprotection for women.

Men have two to six times the risk of heart disease as compared to women, regardless of background national rates. For example, in Japan, a country with very low rates, for people 45–64 years old the male rate of heart disease is 3.2 times that of women. Similarly, the male rate is 3.7 times the female rate in England and Wales, where high rates prevail [1]. Although gradually decreasing with age, the male excess for heart disease continues until the ninth decade of life. Other manifestations of vascular disease such as stroke and peripheral vascular disease are also more common among men than women, but the excess rates among men for these conditions are not as extreme as they are for heart disease (Table 60.1) [2].

It is important to note that, despite higher rates among men and a somewhat different clinical profile, heart disease is the major cause of death for women after the age of 50. As documented by autopsy and angiography studies, the pathophysiology of coronary atherosclerosis is similar for men and women. It is a *delay* in the onset of disease among women that is the greatest contributor to the sex difference in rates; women develop heart disease as often as men, but they do so at a later age.

Heart disease has two principal components: (1) atheroma, fatty degeneration of the wall of the arteries or abnormal fatty deposits in any artery; and (2) thromboembolism, blockage of a blood vessel by a blood clot (or part of a blood clot). Atherogenesis is a tissue response to injury involving chronic inflammation and repair of the vessel wall endothelium and smooth muscle cells [3] resulting in thickened vessel walls with a narrowed opening (lumen). Narrowed arteries are more susceptible to blockage by blood clots (thromboemboli) than are healthy vessels. A high circulating level of oxidized low density lipoproteins (LDL) contributes to atherogenesis, the effects of which are partly balanced by the antioxidant activity of nitric acid in the artery wall. The clinical manifestations of heart disease not only reflect narrowing of the arteries but may also reflect unstable lesions that may induce thrombi (blood clots) and which may lead to unstable angina or a myocardial infarction. Furthermore, even intact endothelium may be dysfunctional when vessels do not vasodilate normally, contributing to acute coronary syndromes. Thus, structural changes of vessels, hemostatic factors, and cellular function contribute to the development of heart disease. If estrogen plays a role, it is likely to be via one or more of these mechanisms.

II. Do Established Risk Factors Explain the Sex Differential in Heart Disease?

First, it is important to recognize that elevated levels of the well-established risk factors for heart disease (cigarette smoking, blood pressure, and serum total cholesterol levels) result in about the same relative increase in risk for women as for men, as shown by results of the Framingham Heart Study (Table 60.2) [4]. Moreover, the fact that women in countries with high rates of heart disease have higher rates of mortality than men from countries with low rates of heart disease points to a strong impact of behavioral and/or environmental risk factors on both sexes.

Overall, men (particularly in middle age) generally smoke more than women and have higher blood pressure and serum cholesterol levels. Taking these factors into account reduces but

Table 60.1

Mortality Rates per 100,000 for Myocardial Infarction, Stroke, and Other Circulatory Disease, United States, 1995

Condition		Age			
		35–44	45–54	55–64	65–74
Myocardial infarction	Male	15.4	67.7	194.9	437.0
ICD-10 Code 270	Female	4.1	18.8	73.7	211.1
	M:F ratio	3.8	3.6	2.7	2.1
Stroke	Male	6.9	19.3	53.2	155.8
	Female	6.1	15.7	40.3	119.2
ICD-10 29	M:F ratio	1.1	1.2	1.3	1.3
Other circulatory	Male	0.4	0.6	1.7	3.3
disease	Female	0.2	0.4	1.0	2.3
ICD-10 Codes 304, 305, 309	M:F ratio	2.0	1.5	1.7	1.4

From [2].

Table 60.2
Relative Risks for Heart Disease, Framingham Heart Study, 1977–1979, Four-Year Follow-Up, Comparison of 80th versus 20th Percentile for Blood Pressure and Cholesterol Level, and for Smoker versus Nonsmoker

	Women	Men
Systolic blood pressure	1.4	1.6
Total cholesterol	1.4	1.4
HDL cholesterol	0.5	0.5
Cigarette smoking	1.2	1.1

From [4].

does not eliminate the sex differential in heart disease [5]. The factors responsible for the sex difference must be different from those responsible for geographic differences because the sex ratio is stable among countries. The factors responsible must also be independent of those related to the decline in heart disease mortality observed since the 1970s in most developed countries, because the decline has occurred for males and females and the sex ratio remains. The question of why heart disease is delayed in women relative to men can be explained only in part by behavioral and lifestyle factors.

III. Is Ovarian Hormone Production Responsible for the Lower Rate of Heart Disease in Women Compared to Men?

A. Endogenous Estrogen Levels and Heart Disease Risk

The hypothesis that endogenous estrogen protects women from heart disease is based on the very great excess risk in men compared to women, the relatively high risk observed for women who experience natural or surgical menopause at a young age, and the markedly lower risk of heart disease in women who take oral postmenopausal estrogen therapy. Direct evidence is lacking, however; studies of endogenous sex hormone concentration in men and women do not provide support for a link with heart disease risk. Although case-control studies in men have reported higher estrogen levels for cases than for controls [6,7], this has not been a consistent finding [8] and prospective studies have shown no relationship with incident heart disease [9–11]. Nor have studies of pre- and postmenopausal women found any association between endogenous estrogen or androgen levels and prevalence [12] or incidence [13–15] of heart disease (Table 60.3). Additionally, measuring the protein carriers, sex hormone binding globulin (SHBG) and corticosteroid binding globulin (CBG), as measures of androgenicity, Lapidus et al. [9] found no clear relationship with risk for myocardial infarction: the U-shaped relationship they reported is difficult to interpret with only 12 cases.

It is not clear why studies of endogenous sex hormones and risk of heart disease show no relationship. It is possible that the single measurement of sex hormone concentration used by all of these studies is insufficient to characterize individual exposure levels; sex hormone levels are difficult to characterize during premenopause because of cyclical fluctuation and also during perimenopause when intermittent ovarian production continues for several years. Moreover, postmenopausal estradiol levels tend to be so low that they are unmeasurable with available assays [12].

Estrone (not estradiol), however, is the predominant estrogen after menopause, and postmenopausal levels are generally in a measurable range. These levels exhibit stability over time similar to those of other factors shown to predict heart disease, such as cholesterol [16]. Thus, measurement difficulties should not prevent detection of a relationship between postmenopausal estrone levels and risk of heart disease. For prospective studies using a nested case-control design, sex hormone levels must be stable in frozen samples and not change while stored. Estrogen does appear to be reasonably stable while frozen [13].

Table 60.3
Endogenous Sex Hormone Levels and Heart Disease Risk in Women: Observational Epidemiological Studies

Study	Year of publication	Study population	Type of study	Hormone(s) measured	Results (age-adjusted)
[9]	1986	Random population sample of 253 Swedish women aged 54–60 (12 incident cases of myocardial infarction)	Prospective, 12 years	Serum: SHBG[a] CGB[b]	U-shaped relationship of SHBG and CBG with myocardial infarction
[12]	1994	Hospital-based angiography case-series of postmenopausal women: U.S.	Case-control, angiographically determined disease	Serum: estradiol, estrone, testosterone androstenedione	No difference in levels between cases and controls
[13] [14]	1995 1995a (DHEAS)	Retirement community in U.S., 651 postmenopausal women (82 incident cases)	Prospective, 19 years	Plasma: total and free estradiol estrone, total and free testosterone, androstenedione, DHEAS[c]	No difference in baseline levels between incident cases and noncases
[15]	1997	Cohort of premenopausal Dutch women aged 40–49 (45 incident cases, 135 controls)	Nested case-control, 5–8 year follow-up	Urinary: estrone and testosterone	No difference in baseline levels between cases and controls

[a]SHBG, sex hormone binding globulin.
[b]CBG, corticosteroid binding globulin.
[c]DHEAS, dehydroepiandrosterone.

Perhaps it is the premenopausal level of estrogen and/or androgen that confers cardioprotection. In examining this possibility, researchers have classified estrogen exposure during the reproductive years according to reproductive function, including age at menarche and at menopause, pregnancy, and menstrual history. Researchers have then related these "surrogate measures" of estrogen status to heart disease risk.

1. Age at Menarche and Menstruation

Although age at menarche has been established as a risk factor for breast cancer, limited study of its relationship to heart disease has yielded conflicting results. A case-control study of young women by La Vecchia [17] found a protective effect of later menarche (Odds Ratio (OR) = 0.49 for ages 12–14 vs <12, 95% Confidence Interval (CI) = 0.31–0.75), whereas Palmer [18] found no association between age at menarche and nonfatal myocardial infarction in a large case-control series of women aged 45–69. La Vecchia [17] also found an increased risk (OR = 1.8, 95% CI = 1.1–2.9) of myocardial infarction among women reporting a history of irregular menstrual cycling compared to those with a history of normal cycling patterns. Gorgels [15] reported a consistent finding of higher risk of heart disease among women with a history of anovulatory cycles. In an effort to link menstruation history with serum lipoprotein levels, T. McKeever (unpublished) examined whether reported history of menstrual cycle length and regularity, from age 20 to 35 years, was related to serum cholesterol levels when the subjects were in their forties and still premenopausal. Irregular cycles were associated with slightly lower high density lipoprotein (HDL) levels but with little difference in total or low density lipoprotein (LDL), while longer cycles (32–45 days) were associated with lower HDL and 8–10 mg/dl higher LDL cholesterol levels than those found among women with normal length or shorter cycles. These lipoprotein patterns are consistent with increased risk of heart disease in women with a history of irregular and/or extended menstrual cycles. Menstrual cycle patterns may be a marker for sustained abnormal sex hormone exposure [19], possibly influencing risk of heart disease (as well as breast cancer); this remains a "working hypothesis" as few studies have addressed this relationship.

2. Reproductive History and Risk of Heart Disease

Epidemiologic studies of breast cancer and osteoporosis have demonstrated the relevance of markers of duration and extent of estrogen exposure in the form of parity and age at first birth to risk of chronic disease at older ages. However, less attention has been devoted to these exposures in terms of heart disease. Of seven published cohort studies that examined parity and heart disease, all but the Nurses' Health Study [20] showed an increased risk (relative risks (RR) ranged from 1.2 to 2.5) for women with a history of five or more pregnancies compared with nulligravid women [21].

In contrast to breast cancer, for which early age at first birth is associated with reduced risk, early age at first birth appears to increase the risk for heart disease two- to threefold as reported by case-control studies of nonfatal myocardial infarction [17,18,22] or sudden death [23]. While the Harvard Nurses' Study [20] did not find this association, only 1% of study participants reported a first birth before age 20. The nurses' cohort may consist of too constrained a social class distribution. Lower

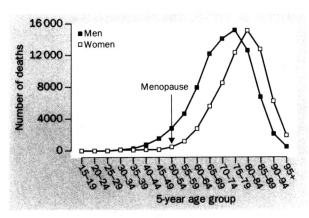

Fig. 60.1 Numbers of coronary deaths by age and sex.

socioeconomic status is associated with high parity and early childbearing and may act as a confounder, although statistical adjustment for social class and other risk factors tends to reduce but not eliminate the association between reproductive factors and heart disease [24].

One reason for an increased risk of heart disease associated with high parity and early age at first birth may be the small sustained drop in HDL cholesterol found after pregnancy, along with a positive association between parity and adiposity, particularly abdominal fat [25]. Hankinson et al. [26] measured plasma estrogen levels in a sample of 216 subjects and found that plasma estrogen levels were lower among women with high parity and those with young age at first birth, after controlling for body mass, alcohol consumption, and age. The lower endogenous estrogen level is consistent with a lower HDL cholesterol level and with higher risk of heart disease (and lower risk of breast cancer).

3. Age at Natural Menopause

Age at menopause is remarkably consistent across countries and over time with a median of 50–51 years. The cessation of menstrual cycling at menopause marks the end of the reproductive period and the decline of ovarian function. Thus, rates of heart disease should increase when the protective effect of ovarian hormones is lost; however, they do not. The rise in heart disease mortality with age is constant for women, with no acceleration after menopause, as shown by data from England and Wales for 1989–1993 [27] (Fig. 60.1). The lack of increase in rates of heart disease around age 50 years has been demonstrated repeatedly and as early as the 1950s [28], ruling out the impact of postmenopausal hormone therapy as the underlying reason [27].

Population-based studies of heart disease risk and menopausal status among women reporting menopausal age within normal limits show conflicting results. The prospective Framingham Heart Study [29] over an 18-year follow-up found a relative risk of 4.0 associated with postmenopausal status among women aged 40–54 years at study entry. In a Dutch population of the same age, Witteman [30] reported significantly greater calcification of the aorta among postmenopausal women as compared with still menstruating women. However, results from the two largest studies run counter to these findings. First, no association between carotid artery wall thickening (measured by ultrasound) and menopausal status (or estrogen therapy use) was found among over 5000 women aged 45–54 in the U.S. Atherosclerotic Risk in Communities Study [31]. In addition, the Nurses' Health Study [20] reported no excess risk of heart disease among naturally menopausal women compared with premenopausal women of the same age. Thus, vital statistics, studies of clinical endpoints, and those using subclinical measures of atherosclerotic disease have not consistently linked natural menopause with increased heart disease risk.

The failure of heart disease rates to accelerate after menopause has been referred to as a "paradox" by Tunstall-Pedoe [27], stemming from the compelling evidence that the risk of heart disease for women *should* rise after menopause. At menopause, there are marked increases in the major heart disease risk factors serum LDL cholesterol [32] and plasma fibrinogen [33], increases which are largely prevented by estrogen therapy [34]. Furthermore, women who experience natural, or surgical menopause at an early age are at higher risk of heart disease than premenopausal women of the same age [35,36].

IV. Exogenous Estrogen Therapy

It is clear that evidence for a loss of female "protection" at menopause is equivocal. Nevertheless, use of supplemental estrogen after menopause to prevent heart disease has gained widespread support in the medical and lay press. Does the scientific evidence merit such use of postmenopausal estrogen therapy? Potentially cardioprotective effects of postmenopausal estrogen therapy include favorable changes in serum lipoproteins and selected plasma hemostatic factors, antioxidant effects, augmentation of nitric oxide (important for vessel wall endothelial function), and improved vasodilation [37,38].

Moreover, over thirty observational studies have reported nearly half of the risk of coronary disease (as measured by clinical endpoints and by angiography) among women taking HRT, most of which included exposure to estrogen unopposed by progestins. Only recently have epidemiologic studies focused on newer forms of HRT in which estrogen is taken in combination with progestins. The largest series to date is based on eight deaths from heart disease among users of the combined preparation [39] and shows no reduction in benefit from the addition of progestin. Controlled trial results of exogenous estrogen in the primary prevention of heart disease are not yet available, although two very large randomized trials are currently underway in the U.S. [40] and the UK [41].

Although virtually all study results to date show reduced risk of heart disease among women who take estrogen therapy, it is particularly important to recognize the potential for bias and confounding. This has been well documented and reviewed by Barrett-Connor [42]. Most of the biases would be in the direction for overestimating favorable effects for estrogen on risk of heart disease.

Women who elect to take postmenopausal therapy tend to, before starting therapy, be thinner, be better educated, drink more alcohol, exercise more, and have lower blood pressure and fasting insulin levels and higher HDL cholesterol than women who do not take therapy [43]. Thus, a better risk factor profile among estrogen users than nonusers could bias results of observational studies that have consistently shown a markedly reduced risk of heart disease among users. Furthermore, women

with a history of chronic disease including breast cancer, diabetes mellitus, or heart disease are less likely to be prescribed HRT than women without a history of these conditions [44]. As a consequence, HRT use would be linked, both cross-sectionally and prospectively, to a lower risk of these conditions.

With respect to compliance bias, the Nurses' Health Study [45] has shown reduced risk among current (RR = 0.5) but not past (RR = 0.9) users; either the benefit is lost following cessation of use or long-term users have an underlying lower risk of heart disease than short-term users, suggesting compliance as a marker for lower risk. Evidence in support of "compliance bias" comes from results of randomized trials of lipid-lowering treatment showing consistently that, among the placebo group, compliers had about the same reduction in mortality relative to noncompliers as that found for reduced heart disease among users of estrogen therapy [46]. A large proportion of women who begin postmenopausal therapy stop using it. Even in the carefully monitored Postmenopausal Estrogen/Progestin Interventions (PEPI) Trial, about one-third of women with an intact uterus who were assigned to the therapy had stopped taking it by the end of the three-year study [34].

It is also worth noting that estrogen use is associated with reduced risk from causes of deaths not thought to be related to estrogen [47]; either estrogen has nonspecific benefits or the lower mortality rate is a consequence of the use of estrogen by healthier women [48].

Results from two reports support the need to evaluate clinical trial results before recommendations are made with respect to coronary disease prevention and estrogen therapy. A review of 22 randomized trials of short-term estrogen use [49] that tracked cardiovascular events (although not as primary trial and endpoints), found that the risk of such events among the group treated with estrogen was higher (RR = 1.39) than that for the control group. This is a surprising result if estrogen really does reduce risk by one-third to one-half. Also published are the results of the first major clinical trial of estrogen and progestin among postmenopausal women with clinical coronary disease [50] showing no benefit of therapy. The Heart and Estrogen/progestin Replacement Study (HERS) was a randomized, blinded, placebo-controlled secondary prevention trial to test whether oral conjugated estrogens plus medroxprogesterone acetate reduced nonfatal myocardial infarction and heart disease death among 2763 postmenopausal women with established heart disease. After four years of follow-up there was essentially no difference in the rate of heart disease events between the treatment and placebo groups: this was in spite of an 11% lower LDL and 10% higher HDL cholesterol in the hormone group compared with the placebo group. A striking time trend occurred during the trial with the relative hazard (relative risk) for myocardial infarction or heart disease for the hormone versus the placebo group of 1.52 in year one, 1.00 in year 2, 0.87 in year 3, and 0.67 for years 4 and 5 (p = 0.009 for trend). Thromboembolic events (deep vein thrombosis and pulmonary emboli) were also elevated among the hormone group as compared with the placebo group (RR = 2.89, 95% CI = 1.5–5.6), particularly in the first year.

The tendency for increased risk of myocardial infarction in the hormone group early in the trial with decreasing risk in subsequent years may be due to random variation, but the consistency of the trend makes this unlikely. One explanation for such a pattern is an "acute" prothrombotic effect of hormone therapy that may be particularly evident among women susceptible to arterial thrombotic events because of underlying atherosclerotic disease. Is estrogen plus progestin likely to have prothrombotic effects? Oral estrogen has consistently been shown to reduce plasma fibrinogen by about 15 mg/dl which should have favorable antithrombotic effects [34,51]. However, oral estrogen appears to be potentially procoagulatory. For example, a controlled, double-blind, cross-over study [52] of commonly used (unopposed) conjugated oral estrogen found evidence of increased thrombin generation (F1 + 2) and conversion of fibrinogen to fibrin (fibrinopeptide A), accompanied by a drop in the anticoagulatory factors anti-thrombin III and protein C. Further, Cushman *et al.* [53] found that older postmenopausal women who reported taking hormone therapy had, on average, a 59% higher level of an acute phase protein, C-reactive protein (CRP), than those not taking estrogen. This finding may be relevant to the HERS results because elevated levels of CRP have been found (in two populations of elderly women) to predict myocardial infarction and heart disease death [54]. However, results of observational studies that consistently show markedly reduced heart disease in estrogen-takers suggest that if estrogen has prothrombotic effects they may be confined to a susceptible subset of users who experience events soon after starting to use hormone therapy. Susceptible women would be unlikely to continue with hormone therapy long term and therefore would not be included as current estrogen-takers in observational studies. Thus far, estrogen use has been the focus of most observational studies. Most studies of long-term hormone therapy use have included users of unopposed estrogen (without progestin). Currently, progestins are routinely added for women who have not had a hysterectomy in order to reduce risk of endometrial cancer. The HERS trial active therapy arm included progestin in the commonly used form of medroxyprogesterone. In spite of widespread use, little is known about cardiovascular effects of progestins. Progestins may negate some of estrogen's benefit by downregulating estrogen receptors, thus making estrogen less "available." As results of the PEPI trial showed, the estrogen-related increase in HDL cholesterol was blunted by the addition of medroxyprogesterone acetate [34]. Animal experiments show medroxyprogesterone may diminish the vasodilatory effects of estrogen [55]. Furthermore, protective effects of estrogen on development of atherosclerosis in cholesterol-fed nonhuman primates were greatly diminished when medroxyprogesterone was added to estrogen [56].

Observational studies of heart disease risk among women taking estrogen plus progestins, however, have found the same low risk of heart disease among users of combined preparations as among those using estrogen alone. Users of combined preparations are subject to the same biases as studies of estrogen-only users in observational studies. Moreover, studies published to date have included too few users of combined preparations to draw conclusions about risk of heart disease associated with combined preparations.

In sum, observational studies suggest estrogen provides a high level of cardioprotection, but bias is likely to account for at least some of the observed effect. The single randomized clinical trial of estrogen plus progestin therapy for secondary prevention does not show benefit. Clinical trial results for primary prevention of heart disease will not be available until about the year 2005.

V. Comments

Heart disease among women is largely confined to older women [57] and an expected increase in incidence and prevalence of cardiovascular disease worldwide is a consequence of increasing life expectancy. Moreover, also as a consequence of greater numbers of women attaining older ages, the prevalence of risk factors such as hyperlipidemia will rise, requiring control. The assumption that heart disease in older women is related to estrogen deprivation leads to a strategy of recommending estrogen supplementation for all women, as all older women become menopausal—in spite of the conflicting evidence that estrogen protects women from heart disease.

Professor Geoffrey Rose in his book, *The Strategy of Preventive Medicine* [58], pointed out that intervention on a population basis for primary prevention should follow the evolutionary norm (*i.e.,* promote healthy behaviors to prevent chronic disease) rather than resort to mass pharmacology, which invariably carries risk. This axiom is especially pertinent to heart disease in women given the mass exposure of Western women to dietary fat and cholesterol, to cigarettes, and to a widespread lack of physical activity. The purported protection associated with estrogen therapy occurs in a "high risk milieu." Although sex hormones have an impact on coronary artery disease (CAD) risk and contribute to the gender gap in CAD, behavioral factors must play an even greater role as evidenced by (1) the higher CAD rates among women in high incidence countries relative to rates among men in low incidence countries, and (2) the declines in CAD rates among both men and women. Healthy behaviors, a diet high in fruits and vegetables and low in fat, exercise, and avoidance of tobacco have a beneficial impact on CAD for both men and women. Among women and men, risk of CAD appears to drop with improvement in risk factor profile: reduction of elevated serum LDL levels, control of hypertension, and smoking cessation [59].

A further point against use of estrogen for primary prevention of heart disease relates to the high prevalence of hypercholesterolemia among women. Experimental animal research has relied on dietary cholesterol-loading of subjects to induce atherosclerosis. In a departure from this practice, Wagner *et al.* [60] placed ovariectomized monkeys on a lipid-lowering diet with or without estrogen treatment and found that monkeys on estrogen exhibited no additional reduction in atherosclerosis over 30 months relative to diet-only monkeys in spite of lower body weight and less abdominal fat among the estrogen-fed monkeys. The point is that, in the absence of a high-cholesterol diet, atherosclerotic progression may be much less dramatic and pharmacological therapies such as estrogen may not be required. Rates of heart disease for Japanese women, who tend to eat a much lower fat/cholesterol diet than Western women, are among the lowest in the world despite their relatively low endogenous estrogen levels. Even in countries with very high rates of heart disease, such as the U.S. and Great Britain, about 2000 healthy women must take hormone therapy for a full year to prevent one heart attack; this number will be even larger for a country like France or Spain with relatively low rates of heart disease. Careful study of such low risk groups should be undertaken to further understand the role that sex hormones as well as other risk and protective factors play in coronary artery disease. Conversely, current treatment for elevated serum cholesterol levels, even among those with diagnosed disease, falls well below recommendations in the U.S. Only about one-third of women patients in one large national study were within the treatment goal of 130 mg/dl or lower for LDL cholesterol [61]. Thus, it appears that even among high-risk patients available and effective methods of lipid-lowering are underutilized.

Much work needs to be done to identify who may benefit (and who may be harmed) from use of postmenopausal estrogen therapy. The widespread elevation in serum cholesterol levels, smoking, and sedentary habits among women are points of intervention that we know will alter risk. Given the lack of clinical trial results on estrogen use and primary prevention of heart disease, general recommendations for postmenopausal estrogen therapy to prevent heart disease are premature.

References

1. Thom, T. J., Epstein, F. H., Feldman, J. J., Leaverton, P. E., and Wolz, M. (1992). "Total Mortality and Montality from Heart Disease, Cancer, and Stroke from 1950 to 1987 in 27 Countries," NIH Publ. No. 92-3088. U.S. Department of Health Human Services, Public Health Service, Washington, DC.
2. World Health Organization. (1996). World Health Statistics, 1995. WHO: Geneva, Switzerland. Table B-94.
3. Libby, P. (1996). Atheroma, more than mush. *Lancet* (Suppl. I) **348**, 1–31.
4. Castelli, W. P., Garrison, R. J., Wilson, P. W. F., Abbott, R. D., Kalovsdia, S., and Kannel, W. B. (1986). Incidence of coronary heart disease and lipoprotein levels. The Framingham Study. *JAMA, J. Am. Med. Assoc.* **256**, 2835–2838.
5. Wingard, D. L. (1984). The sex differential in morbidity, mortality and lifestyle. *Annu. Rev. Public Health* **5**, 433–458.
6. Phillips, G. B., Castelli, W. P., Abbott, R. D., and McNammara, P. M. (1983). Association of hyperestrogenaemia and coronary heart disease in men in the Framingham Cohort. *Am. J. Med.* **74**, 863–869.
7. Small, M., Lowe, G. D. O., Beastall, G. H., Beattie, J. M., McEachern, M., Hutton, I., Lorimer, A. R., and Forbes, C. D. (1985). Serum oestradiol and ischaemic heart disease: Relationship with myocardial infarction but not coronary atheroma or naemostasis. *Q. J. Med.* [*N. S.*] **57**, 775–782.
8. Heller, R. F., Jacobs, H. S., Vermeulen, A., and Deslypere, J. P. (1981). Androgens, oestrogens and coronary heart disease. *Br. Med. J.* **282**, 438–439.
9. Lapidus, L., Lindstedt, G., Lundberg, P.-A., Bengtsson, C., and Gredmark, T. (1986). Concentrations of sex-hormone binding globulin and corticosteroid binding globulin in serum in relation to cardiovascular risk factors and to 12-year incidence of cardiovascular disease and overall mortality in postmenopausal women. *Clin. Chem.* **32**, 146–152.
10. Cauley, J. A., Gutai, J. P., Kuller, L. H., and Dai, W. S. (1987). Usefulness of sex steroid hormone levels in predicting coronary artery disease in men. *Am. J. Cardiol.* **60**, 771–777.
11. Barrett-Connor, E., and Khaw, K. T. (1988). Endogenous sex hormones and cardiovascular disease in men: A Prospective Population-based Study. *Circulation* **78**, 539–545.
12. Cauley, J. A., Gutai, J. P., Glynn, N. W., Paternostro-Bales, M., Cottington, E., and Kuller, L. H. (1994). Serum estrone concentrations and coronary artery disease in postmenopausal women. *Arterioscler. Thromb.* **14**, 14–18.
13. Barrett-Connor, E., and Goodman-Gruen, D. (1995). Prospective study of endogenous sex hormones and fatal cardiovascular disease in postmenopausal women. *Br. Med. J.* **311**, 1193–1196.
14. Barrett-Connor, E., and Goodman-Gruen, D. (1995). The epidemiology of DHEAS and cardiovascular disease. *Ann. N.Y. Acad. Sci.* **774**, 259–270.
15. Gorgels, W. J. M. J., Graaf, Y. D., Blankenstein, M. A., Collette, H. J. A., Erkelens, D. W., and Banga, J. D. (1997). Urinary sex hormone excretions in premenopausal women and coronary heart disease risk: A Nested Case-referent Study in the DOM-Cohort. *Clin. Epidemiol.* **50**, 275–281.
16. Cauley, J. A., Gutai, J. P., Kuller, L. H., and Powell, J. G. (1991). Reliability and interrelations among serum sex hormones in postmenopausal women. *Am. J. Epidemiol.* **133**, 50–57.
17. LaVecchia, C., Decarli, A., Franceschi, S., Gentile, A., Negri, E., and Parazzini, F. (1987). Menstrual and reproductive factors and the risk of myocardial infarction in women under fifty-five years of age. *Am. J. Obstet. Gynecol.* **157**, 1108–1112.
18. Palmer, J. R., Rosenberg, L., and Shapiro, S. (1992). Reproductive factors and risk of myocardial infarction. *Am. J. Epidemiol.* **136**, 408–416.
19. Berge, L. N., Lyngmo, V., Svensson, B., and Nordoy, A. (1993). The bleeding time in women: An influence of sex hormones? *Acta Obstet. Gynecol. Scand.* **72**, 423–427.
20. Colditz, G. A., Willett, W. C., Stampfer, M. J., Rosner, B., Speizer, F. E., and Hennekens, C. H. (1987). A prospective study of menarche, parity, age at first birth, and coronary disease in women. *Am. J. Epidemiol.* **126**, 861–870.
21. Ness, R. B., Schotland, H. M., Flegal, K. M., and Shofer, F. S. (1994). Reproductive history and coronary heart disease risk in women. *Epidemiol. Rev.* **16**, 298–314.
22. Rosenberg, L., Miller, D. R., Kaufman, D. W., Helmrich, S. P., Van de Carr, S., Stolley, P. D., and Shapiro, S. (1983). Myocardial infarction in women under 50 years of age. *JAMA, J. Am. Med. Assoc.* **250**, 2801–2806.
23. Talbott, E. O., Kuller, L. H., Detre, K., Matthews, K., Norman, S., Kelsey, S. F., and Belle, S. (1989). Reproductive history of women dying of sudden cardiac death: A Case-control Study. *Int. J. Epidemiol.* **18**, 589–594.
24. Ness, R. B., Harris, T., Cobb, J., Flegal, K. M., Kelsey, J. L., Bélanger, A., Stunkard, A. J., and D'Agostino, R. B. (1993). Number of pregnancies and the subsequent risk of cardiovascular disease. *N. Engl. J. Med.* **328**, 1528–1533.
25. Kaye, S. A., Folsom, A. R., Princeas, R. J., Potter, J. D., and Gapstur, S. M. The association of body fat distribution with lifestyle and reproductive factors in a population study of postmenopausal women. *Int. J. Obes.* **14**, 583–591.
26. Hankinson, S. E., Colditz, G. A., Hunter, D. J., Manson, J. E., Willett, W. C., Stampfer, M. J., Longcope, C., and Speizer, F. E. (1995). Reproductive factors and family and family history of breast cancer in relation to plasma estrogen and prolactin levels in postmenopausal women in the Nurses' Health Study (United States). *Cancer Causes Control* **6**, 217–224.
27. Tunstall-Pedoe, H. (1998). Myth and paradox of coronary risk and the menopause. *Lancet* **351**, 1425–1427.
28. Tracy, R. E. (1966). Sex differences in coronary disease: Two opposing views. *J. Chronic Dis.* **19**, 1245–1251.
29. Kannel, W. B., and Wilson, P. W. F. (1995). Risk factors that alternate the female coronary disease advantage. *Arch. Intern. Med.* **155**, 57–61.
30. Witteman, J. C. M., Grobbee, D. E., Kok, F. J., Hofman, A., and Valkenburg, H. A. (1989). Increased risk of atherosclerosis in women after the menopause. *Br. Med. J.* **298**, 642–644.
31. Nabulsi, A. A., Folsom, A. R., Szklo, M., White, A., Higgins, M., and Heiss, G. (1996). No association of menopause and hormone

replacement therapy with carotid artery intima-media thickness. *Circulation* **94**, 1857–1863.

32. Matthews, K. A., Meilahn, E., Kuller, L. H., Kelsey, S. F., Caggiula, A. W., and Wing, R. R. (1989). Menopause and risk factors for coronary heart disease. *N. Engl. J. Med.* **321**, 641–646.

33. Meilahn, E. N., Kuller, L. H., Matthews, K. A., and Kiss, J. E. (1992). Hemostatic factors according to menopausal status and use of hormone replacement therapy. *Ann. Epidemiol.* **2**, 445–455.

34. Writing Group for the PEPI Trial (1995). Effects of estrogen or estrogen/progestin regimens on heart disease risk factors in postmenopausal women. *JAMA, J. Am. Med. Assoc.* **273**, 199–208.

35. Jacobson, B. K., Nilssen, S., Heuch, I., and Kvale, G. (1997). Does age at natural menopause affect mortality for ischemic heart disease? *Clin. Epidemiol.* **50**, 475–479.

36. Snowden, D. A., Kane, R. L., Beeson, W. L., Burke, G. L., Sprajka, M., Potter, J., Iso, H., Jacobs, D. A., Jr., and Phillips, R. L. (1989). Is early menopause a biologic marker of health and aging? *Am. J. Public Health* **79**, 709–714.

37. Gerhard, M., and Ganz, P. (1995). How do we explain the clinical benefits of estrogen? From bedside to bench. *Circulation* **92**, 5–8.

38. Keaney, J. F., Shwaery, G. T., Xu, A., Nicolosi, R. J., Loscalzo, J., Foxall, T. L., and Vita, J. A. (1994). 17β estradiol preserves endothelial vasodilator function and limits low-density lipoprotein oxidation in hypercholesterolemic swine. *Circulation* **89**, 2251–2259.

39. Grodstein, G., Stampfer, M. F., Manson, J. E., Colditz, G. A., Willett, W. C., Rosner, B., Speizer, F. E., and Hennekens, C. H. (1996). Postmenopausal estrogen and progestin use and the risk of cardiovascular disease. *N. Engl. J. Med.* **335**, 453–456.

40. Finnegan, L. P. (1996). The NIH Women's Health Initiative: Its evolution and expected contributions to women's health. *Am. J. Prev. Med.* **12**, 292–293.

41. Vickers, M. R., Meade, T. W., and Wilkes, H. C. (1995). Hormone replacement therapy and cardiovascular disease: The case for a randomized controlled trial. *Ciba Found. Symp.* **191**, 150–160.

42. Barrett-Connor, E., and Grady, D. (1998). Hormone replacement therapy, heart disease, and other considerations. *Annu. Rev. Public Health* **19**, 55–72.

43. Matthews, K. A., Kuller, L. H., Wing, R. R., Meilahn, E. N., and Plantinga, P. (1996). Prior to use of estrogen replacement therapy, are users healthier than nonusers? *Am. J. Epidemiol.* **143**, 971–978.

44. Moorhead, T., Hannaford, P., and Warskyj, M. (1997). Prevalence and characteristics associated with use of hormone replacement therapy in Britain. *Br. J. Obstet. Gynaecol.* **104**, 290–297.

45. Stampfer, M. J., Willet, W. C., Colditz, G. A., Rosner, B., Speizer, F. E., and Hennekens, C. H. (1985). A prospective study of postmenopausal estrogen therapy and coronary heart disease. *N. Engl. J. Med.* **313**, 1044–1049.

46. Petitti, D. B. (1994). Can compliance bias explain the results of observational studies? *Ann. Epidemiol.* **4**, 115–118.

47. Cauley, J. A., Seeley, D. G., Browner, W. S., Ensrud, K., Kuller, L. H., Lipschutz, R. C., and Hulley, S. B. (1997). Estrogen replacement therapy and mortality among older women. *Arch. Intern. Med.* **157**, 2181–2187.

48. Posthuma, W. F., Westendorp, R. G., and Nandenbrocke, J. P. (1994). Cardioprotective effect of hormone replacement therapy in postmenopausal women: Is the evidence biased? *Br. Med. J.* **308**, 1268–1269.

49. Hemminki, E., and McPherson, K. (1997). Impact of postmenopausal hormone therapy on cardiovascular events and cancer: Pooled data from clinical trials. *Br. Med. J.* **315**, 149–153.

50. Hulley, S., Grady, D., Bush, T., Furberg, C., Herrington, D., Riggs, B., Vittinghoff, E., for the HEART and Estrogen/Progestin Replacement Study (HERS) Research Group (1998). Randomized trial of estrogen plus progestin for secondary prevention of coronary heart disease in postmenopausal women. *JAMA, J. Am. Med. Assoc.* **280**, 605–613.

51. Meilahn, E. N. (1995). Hemostatic factors an ischemic heart disease risk among postmenopausal women. *J. Thromb. Thrombolysis* **1**, 125–131.

52. Caine, Y. G., Bauer, K. A., Barzager, S. M., Pen-Cate, H., Sacks, F. M., Walsh, B. W., Shiff, I., and Rosenberg, R. D. (1992). Coagulation activation following estrogen administration to postmenopausal women. *Thromb. Haemostasis* **68**, 392–395.

53. Cushman, M., Meilahn, E. N., Psaty, B. M., Kuller, L. H., Dobs, A. S., and Tracy, R. P. (1999). Hormone replacement therapy, inflammation, and hemostasis in elderly women. *Arterioscler. Thromb. Vasc. Biol.* **19**, 893–899.

54. Tracy, R. P., Lemaitre, R. N., Psaty, B. M., Ires, D. G., Evans, R. W., Cushman, M., Meilahn, E. N., and Kuller, L. H. (1997). Relationship of C-reactive protein to risk of cardiovascular disease in the elderly. Results from the Cardiovascular Health Study and the Rural Health Promotion Project. *Arterioscler. Thromb. Vasc. Biol.* **17**, 1121–1127.

55. Williams, J. K., Honore, E. K., Washburn, S. A., and Clarkson, T. B. (1994). Effects of hormone replacement therapy on reactivity of atherosclerotic coronary arteries in cynomolgus monkeys. *J. Am. Coll. Cardiol.* **24**, 1757–1761.

56. Adams, M. R., Register, T. C., Golden, D. L., Wagner, J. D., and Williams, J. R. (1997). Medroxyprogesterone acetate anagonizes inhibitory effects of conjugated aquine estrogens on coronary artery atherosclerosis. *Arterioscler. Thromb. Vasc. Biol.* **17**, 217–221.

57. Newnham, H. H., and Silberber, J. (1997). Coronary heart disease. Women's hearts are hard to break. *Lancet* **349**(Suppl. 1), 513–516.

58. Rose, G. A. (1992). Strategies of prevention: The individual and the population. In "Coronary Heart Disease Epidemiology. From Aetiology to Public Health" (M. Marmot and P. Elliot, eds.), pp. 311–324. Oxford University Press, Oxford.

59. Meilahn, E., Becker, R. C., and Corrao, J. M. (1995). Primary prevention of coronary heart disease in women. *Cardiology* **86**, 286–298.

60. Wagner, J. D., Martino, M. A., Jayo, M. J., Anthony, M. S., Clarkson, T. B., and Cefalu, W. T. (1996). The effects of hormone replacement therapy on carbohydrate metabolism and cardiovascular risk factors in surgically postmenopausal cynomolgus monkeys. *Metab. Clin. Exp.* **45**, 1254–1262.

61. Schrott, H. G., Bittner, V., Bittinghoff, E., Herrington, D. M., Hulley, S., for the HERS Research Group (1997). Adherence to National Cholesterol Education Program treatment goals in postmenopausal women with heart disease. *JAMA, J. Am. Med. Assoc.* **277**, 1281–1286.

61

Emotions and Heart Disease

GERDI WEIDNER AND HEIDI MUELLER

Department of Psychology, State University of New York at Stony Brook, Stony Brook, New York

I. Introduction

Despite the decline in heart disease in most western industrialized countries, the disease still remains a major health problem and has reached epidemic proportions in many of the newly independent countries of eastern Europe [1–6]. Typically, the category heart disease includes nonrheumatic heart disease, hypertension, and coronary heart disease (CHD). The latter comprises the major subgroup (at least 80% of heart disease [7]). The focus of this chapter will be on this major subgroup of heart disease.

Considerable progress has been made in the identification of medical risk factors for CHD both in women and in men. The major risks are elevated plasma cholesterol, hypertension, cigarette smoking, and diabetes [8]. A cluster of metabolic cardiovascular risk factor abnormalities associated with insulin resistance has been identified and labeled "syndrome X" [9]. These abnormalities include glucose intolerance, hypertriglyceridemia, low levels of high-density lipoprotein cholesterol (HDL-C), abdominal obesity, and hypertension. Syndrome X increases the risk of coronary heart disease and non-insulin-dependent diabetes mellitus both in women and in men [9]. Diabetes as a risk factor of coronary heart disease appears to be especially relevant to women, as it eliminates the 10-year advantage that women have relative to men. That is, the risk of myocardial infarction in diabetic women is equal to the risk in nondiabetic men of the same age [10].

Despite this progress, somewhere between 25 and 30% of all CHD occurs in the absence of the major risk factors [11]. Furthermore, the current epidemic of heart disease in eastern Europe is unlikely to be attributed to the traditional coronary risk factors and associated lifestyle variables, such as obesity and excessive alcohol consumption [12,13].

Several psychosocial variables reflecting personality attributes [14] and social-environmental aspects, such as social support or social integration [15,16], have emerged as independent predictors of CHD. Depression or vital exhaustion [17–19] have been added to the list of predictors. Interestingly, the upward trends in heart disease mortality in eastern Europe have paralleled increases in the prevalence of major depression [20]. For example, Kopp and Skrabski [20] report that between 1988 and 1995 the rate of persons with depressive symptomatology in the Hungarian population increased from 24.3 to 31.8%. It is conceivable that this increase in depressive symptoms in the population plays a role in the CHD epidemic in eastern European countries.

One major obstacle in the evaluation of psychosocial factors as independent predictors of CHD in women is that only limited numbers of long-term studies are available. Most of the inforation on women's traditional and psychosocial coronary risk profiles stems from a handful of pioneering studies such as the Framingham study [21], the Population Study of Women [22], and the Tecumseh Community Health Study [23]. In a review of current knowledge about coronary-prone personality attributes as independent predictors of CHD in women, the following tentative picture of a coronary-prone woman emerges: she reports symptoms of tension, suppresses anger, may be suspicious, and lacks self-assertiveness [14]. The existence of a coronary-prone situation, characterized by financial strain, social isolation, and lack of options for remedy, has also been described [14]. Thus, the above-mentioned coronary-prone personality characteristics may interact with demanding situations, resulting in an inability to cope, which in turn may increase depression and anxiety. Evidence suggests negative emotions are implicated in the development and prognosis of CHD [17,18,25]. The present chapter focuses on these emotions and their role in the etiology of CHD in initially healthy women and in women with heart disease. Studies linking emotions to CHD are discussed and directions for future research, including implications for treatment, are presented.

II. Emotions Defined

Depression is the main emotional variable being investigated with respect to CHD. Few studies evaluate anxiety and anger/hostility. The term hostility refers to outwardly expressed anger [25], rather than to hostility in the sense of suspiciousness and mistrust [26]. The link between hostility reflecting mistrust or suspiciousness (as well as the role of Type A behavior) and CHD in women has been reviewed previously [14,27,28]. The present chapter will focus on the outward expression of anger as a risk factor for CHD in women, in contrast to "anger suppression," which has also been linked to CHD in women [23] but is assessed rather infrequently. In the few studies where anger suppression is measured, it correlates highly with depression [29]. In fact, the tendency to repress anger has been linked to the development of depression [30].

Despite their dissimilar names, depression, anxiety, and anger/hostility interrelate highly and are considered aspects of one and the same construct, which Watson and Clark have termed "negative affectivity" [31]. However, each research study discussed below measures these emotions separately. Therefore, results are presented using separate labels, namely, depression, anxiety, and hostility. It should be kept in mind, however, that all three emotions tend to tap the same construct ("negative affectivity or emotionality").

III. Prospective Studies of Initially Disease-Free Community Samples

Two meta-analyses, including major prospective studies of CHD (one also included cross-sectional data), evaluated the association of personality variables and negative emotions (depression, anxiety, and hostility/anger) with CHD outcomes [32,33]. Negative emotions were indeed related to CHD outcomes in both studies. Although these early attempts to link negative emotions to CHD are promising, most of the data included in the meta-analyses came from men; when women were included, results were not reported by gender. In addition, it is often not clear to what extent the association between negative emotions and CHD was independent from the traditional coronary risk factors, such as elevated levels of plasma cholesterol, blood pressure, and smoking.

There is now mounting evidence in the United States and other countries that antecedent depression and anxiety increase the risk of heart disease [17,18,34–36] and hypertension [37,38] in population samples of women and men. In addition, excess fatigue or "vital exhaustion," which is correlated with depression, was found to predict myocardial infarction (MI) prospectively [19,39].

Most of these studies focus on depression as a predictor of CHD. The report by Anda and colleagues [34], based on NHANES 1 follow-up data, found that people with depression have a relative risk of 1.5 (95% CI = 1.0–2.3) for fatal MI compared to those without depression, even after adjusting for major coronary risk factors and sociodemographic variables. Similar results were reported in a population sample of male physicians [35].

The study by Pratt et al. [17] also supports a link between depression and CHD in community-dwelling adults. This study was based on a follow-up of the Baltimore cohort of the Epidemiologic Catchment Area Study, a survey of psychiatric disorders in the general population. A history of a major depressive episode elevated a person's risk for MI 4.5 times (95% CI, 1.65–12.44). Even a history of dysphoria (as short as two weeks) was associated with MI (odds ratio = 2.1). This elevated risk was independent of other major coronary risk factors, even those traditionally correlated with depression such as smoking, age, marital status, use of psychotropic medications, and, most importantly, gender. Thus, both women and men are at risk for developing CHD as a result of experiencing depression. Furthermore, the 20-year follow-up report of the Framingham study found that symptoms of anxiety were significant predictors of hard endpoints in cardiovascular disease (MI or coronary death) among homemakers, and depression was a significant predictor for angina pectoris (a soft endpoint) among all women [21]. Finally, the 12-year follow-up report of a population sample of elderly women and men in Denmark confirms that depression predicts mortality from all causes, including acute myocardial infarction, both for men and for women [18].

It should be noted, however, that some prospective studies do not support a link between negative emotions (depression) and heart disease [41,42]. One study among elderly hypertensive patients found no relationship between depression measured at baseline and subsequent cardiovascular outcomes [42] but did find that heart disease increased both in men and in women as their levels of depression increased over the follow-up period. This study points to an important methodological consideration when evaluating a variable for its ability to predict heat disease. Changes in emotions over time may be more sensitive as predictors of heart disease than baseline levels.

In sum, the majority of the evidence suggests that negative emotions play a role in the development of CHD both in women and in men. Considering that women consistently report more negative emotions than men [43], negative emotions may play an even greater role in coronary risk among women. Findings from some studies conducted with patients appear to support this statement (see next section).

IV. Studies of Patients with Mood Disorders and Patients with CHD

The following sections present research findings from studies of CHD among persons with mood disorders and studies of depression as a predictor of recurrent events among cardiac patients [also see 44–47 for reviews]. Demonstrating elevated coronary risk among those already depressed (but still without heart disease diagnosis) and higher likelihood of recurrent events among depressed patients may help to maximize prevention and treatment efforts.

A. Patients with Mood Disorders

In general, depressed or anxious psychiatric patients have a higher rate of general mortality than do controls [48]. Results from some small studies support the notion that depression may be a precursor of mortality due to heart disease [49,50]. For example, in a study of psychiatric patients, depression was the most common diagnosis in 21 men and 8 women who experienced MI during the study [50]. In fact, all women were depressed. A disproportionate number of both depressed male and female patients suffered an MI during the time of that study.

An association between depression and heart disease mortality in women was observed in a considerably larger prospective study of psychiatric inpatients (n = 5412) where depression was the most common diagnosis [51]. While both male and female patients had significantly higher death rates than would be expected in the general population, only the women were at increased risk for death due to heart disease. Thus, it appears that among patients with mood disorders, such as depression, women are at least as likely as men to be at risk for heart disease.

B. Patients with Heart Disease

Not surprisingly, depressive symptoms among patients with heart disease are significantly more common than among people in the general population [46]. Estimates of depression among case-series of patients with heart disease range from 15.8% to 88%, depending on the measure employed (the stricter the diagnostic criteria, the lower the rates [52–61]). For example, Vasquez-Barquero and colleagues [60] reported that almost 45% of 194 patients with CHD who were treated in an outpatient clinic were diagnosed with depressive symptoms and anxiety. Similarly, Wishnie et al. [61] found that almost all (88%) of 24 patients hospitalized for myocardial infarction (MI) rated themselves as anxious and/or depressed.

Negative emotions also emerged as a characteristic of patients with CHD in case-control studies. For example, Friedman and Booth-Kewley [62] identified a depression/anxiety cluster that was associated with the presence of CHD in men. Similarly, women who had experienced an MI 6 months prior to being studied were rated as more depressed, anxious, and angry/

hostile as compared to healthy controls [63]. One study of coronary risk factors among women in Stockholm found that women with CHD had significantly more depressive symptoms than did healthy controls [58]. When adjusting for standard risk factors, women with CHD were fourfold more likely to be in the highest versus lowest quartile of depression.

Due to the retrospective nature of these data, it is impossible to say whether depressive symptomatology preceded MI or resulted from being hospitalized or sick. Forrester and colleagues [55] found 18% of patients to have major depressive syndromes after MI; this post-MI depression was related to a history of mood disorder. For some patients, depression not only appears to precede MI, but it also increases during hospitalization and after discharge from the hospital [57]. For others, depression appears to be associated with hospitalization. Schleifer and colleagues [59] interviewed 283 women and men within two weeks after MI. The prevalence of post-MI depression was high (45%), but only 8% of the patients reported a prior psychiatric history. The findings of Cay et al. [54] suggest that depression may be related to hospitalization for approximately one-third to one-half of depressed MI patients. They found symptoms of depression and anxiety among 65% of patients hospitalized for MI. In more than 50% of patients, these symptoms were already present before admission. Similarly, Lloyd and Cawley [64] reported that nearly half of male patients with psychiatric morbidity one week following MI were rated as "psychiatrically ill" (based on patients' or relatives' responses and/or medical records) prior to the onset of MI. Travella and colleagues [53] suggest that there may be two types of depression following MI: acute depression associated with greater functional impairment and chronic or prolonged depression that may be associated with inadequate social support [59]. The first type may be more transient and related to being sick and hospitalized, whereas the second type might be more severe and play a greater role in the etiology of CHD. These results illustrate the importance of distinguishing clinical depression from transitory affective phenomena. While patients with both types of post-MI depression might benefit from treatment, prolonged treatment efforts involving the patient's social environment may be necessary only for the latter group.

Overall, there is convincing evidence that negative emotions are characteristic of patients with heart disease, and, for some, may be reactions to being sick and hospitalized. There is some indication that women may be more likely to be depressed or anxious than men following an MI [53,65], suggesting that they may be at especially high risk. The greater proportion of female patients with negative symptomatology following MI also fits in nicely with the observations that women experience greater psychosocial dysfunction than men following CHD diagnosis [66]. It is conceivable that women's worse emotional health after MI is also, in part, responsible for their poorer prognosis when compared to men [67].

The hypothesis that women may be particularly vulnerable to the effects of negative emotions on recovery from MI received support from a small study by Stern et al. [68a] of 53 men and 13 women with MI. Women's psychosocial and medical profiles were much worse than those of men, and women were half as likely as men to be alive at one year post-MI. The surviving women were about half as likely as men to be free of depression and anxiety. A recent report from the Female Coronary Risk

Study in Stockholm confirms depression as a predictor of poor prognosis of CHD in women [68b].

The reasons for women's greater risk for mortality during hospitalization following cardiac events are subject to debate. While women's greater depression may be one reason for their worse prognosis after an MI, they also may be at increased risk due to other known risk factors such as their older age and worse medical status relative to men at the time of a cardiac event.

Several larger, more methodologically rigorous studies have evaluated negative emotions as predictors of recurrent cardiac events. Ahern and colleagues [69] found that depression was an independent predictor of mortality among 335 women and men participating in the Cardiac Arrythmia Pilot Study. Similarly, Frasure-Smith et al. [70] found depression to be an independent predictor of mortality (controlling for baseline variables) 6 months post-MI in a sample of 153 men and women. These same patients were followed through 18 months post-MI. Depression continued to predict death, but the largest effects were seen during the first 6 months [24]. Increases in depressive symptoms were associated with an increased risk of stroke, MI, and/or death both in male and in female hypertensive patients [42].

There is also some indication that anxiety may predict coronary recurrence in women. An 8.5-year follow-up of 65 women participating in the Recurrent Coronary Prevention Project found high anxiety to be related to recurrent events [63].

With regard to anger, the expression of anger/hostility has been shown to be associated with the severity of coronary perfusion defects both among male and among female patients [25], and with the recurrence of life-threatening ventricular arrhythmias (and subsequent sudden cardiac death due to such events) in patients diagnosed with arrhythmia [71]. In fact, of all affective states/disorders, anger was most strongly associated with such recurrence. Frasure-Smith and colleagues [29] also found that anger increased the risk for arrhythmic events among 222 patients in the 12 months following MI.

Although there are relatively few studies on negative emotions as a predictors of recurrent cardiac events in female patients, those studies that do include women support the hypothesis that negative emotions play a role in recovery from MI among women. The study by Frasure-Smith et al. [29] reported that women had greater risk for recurrence of acute coronary syndromes following MI than did men, but after adjustment for a history of depression and anxiety, the role of gender was partially reduced. Again, this suggests that women's increased tendency to experience depression and anxiety [43] may place them at increased risk for recurrent disease following MI.

In sum, negative emotions (mostly depression) are prevalent among patients with heart disease. For some patients, these symptoms precede the disease, and for others, they appear to be more transient, resulting from acute hospitalization. It is not clear which type of depression influences cardiac events or which type increases the risk for adverse disease outcomes. In any case, depression appears to increase the risk for recurrent events both in women and in men. Its influence is largely independent of disease severity [69,70,72]. The association between negative emotions and recurrent heart disease also appears to hold across a rather wide range of measures employed in the assessment of negative emotions.

V. Future Directions

The major problem that remains is that there is insufficient research on women. When women are included, the sample sizes are generally too small to allow for the more informative multivariate statistical analyses. In addition, results are often not presented by gender. Thus, we are left to extrapolate many findings obtained from men to women. Nevertheless, based on findings from the few studies that have included women, negative emotions appear to play as powerful a role for women as they do for men. In the future, results from currently ongoing research, such as the Women's Health Initiative [73,74], a prospective follow-up of more than 160,000 postmenopausal women, and the 5-year follow-up study of 300 Swedish women with cardiovascular disease [58], will shed further light on the link between negative emotions and heart disease in women.

A. Possible Pathways Linking Negative Emotions to Heart Disease

Meanwhile, possible mechanisms have been proposed that could link heart disease risk and negative emotions. Negative emotions, resulting from inability to cope with stressful events, may increase CHD risk by directly affecting physiological parameters and/or by affecting behaviors or lifestyles associated with enhanced CHD risk. The following sections will briefly discuss these two pathways and their relationship to CHD risk.

1. Physiological Pathways

Several physiological pathways linking negative emotions to cardiac events have been proposed and reviewed previously [29,75–79]. Briefly, negative emotions play a role both in thrombogenic and in arrhythmic events. Neurotransmitters such as catecholamines and serotonin, which are considered central to dysregulation theories of affective disorders [77], can amplify platelet aggregation, facilitating thrombogenic events [75]. While the release of epinephrine in the face of physical stress or threat is adaptive (as it would protect the organism from bleeding to death during an attack by increasing platelet aggregation), the chronic release of these neurotransmitters stemming from depression, anxiety, and anger may not be advantageous.

Arrhythmic events are related to electrical instability in the myocardium facilitated by myocardial ischemia and heightened adrenergic activity [78]. Several animal models of behavioral stress, anger, and fear induction suggest that these emotions can trigger the myocardium to initiate fatal ventricular arrhythmias through increases in sympathetic activity and by upsetting the sympathetic/parasympathetic balance necessary for maintaining stable electrical stimulation of the myocardium [76,78,79]. Anger and depression have also been linked to life-threatening arrhythmias and resulting mortality in humans [24,80–83]. For example, Fraser-Smith and colleagues [24] found that post-MI mortality risk associated with depression was greatest among patients with ≥10 premature ventricular complexes per hour (PVCs). Furthermore, decreased heart rate variability (probably due to reduced parasympathetic nervous system activity [84]) among depressed patients with severe coronary artery disease has been reported [72,85]. Because the parasympathetic system acts antiarrhythmically, protecting the myocardium from electrical destabilization, decreases in activity of this system could serve to put patients at increased risk for abnormal ventricular activity.

Negative emotions also appear to enhance cardiovascular responses to stressful situations. Excessive cardiovascular reactivity to stress has been implicated in the development of hypertension [86,87] and recurrent major cardiac events in women with coronary heart disease [87b]. Siegman [25] has demonstrated that anger expression is associated with heightened cardiovascular reactivity. Shapiro and colleagues [88,89] have employed ambulatory blood pressure monitors to assess the daily fluctuations of blood pressure, which are more representative than casual measures in the assessment of overall cardiac load that is placed on the cardiovascular system. Their work has demonstrated that anger/hostility and defensiveness (indicating conflict over the expression of anger) influence ambulatory blood pressure levels. With regard to depression, Light *et al.* [90] showed that cardiovascular and sympathetic nervous system responses during mental stress were exaggerated among depressed women.

2. Behavioral Pathways

There may also be important behavioral pathways that link negative emotions and cardiac events. Negative emotions in general appear to be associated with engagement in unhealthy lifestyles. For example, students who experienced increases in negative emotions in moving from a low stress to a high stress period during the semester increased their substance use, such as alcohol consumption and smoking [91]. Similarly, in studies of stressful disruptions in social relationships, such as divorce, depression was associated with alcohol consumption and smoking [92]. The relationship between depressive symptoms and alcohol consumption appears to be especially pronounced among women. One longitudinal study of a community-based sample of adult women and men found that depressive symptoms predicted subsequent alcohol problems 4 years later in women but not in men [93]. Excessive alcohol consumption has been linked to hypertension, a major risk factor of CHD.

It is important to point out that the absence of negative emotions does not imply the presence of positive ones. Rather, the two dimensions tend to be independent from each other [94] and relate differentially to health behaviors. For example, in the Weidner *et al.* study [91], increases in negative emotions were related to substance use, but decreases in positive affect were related to decrease in exercise and self-care habits. Thus, intervention efforts aimed at health behavior change may well be directed both at decreasing negative emotions, such as depression and anxiety, **and** at increasing positive emotions, such as activity, attention, alertness, and interest in life.

Considering the association of negative emotions with unhealthy behaviors, it is not surprising that depression and anxiety are related to reduced compliance with treatment regimens [95]. Conn *et al.* [96], for example, found that depression among post-MI patients (including women) was significantly related to nonadherence to exercise, diet, medication, and the recommendation to stop smoking. Similarly, Finnegan and Suler [97] reported that depressive symptoms in post-MI women predicted poorer adherence to exercise and weight loss recommendations.

Depressive symptoms have also been associated with low levels of social support [96]. For example, in a prospective study of rural elderly people, Cerhan and Wallace [98] found that depressive symptoms predicted a decline in social support 3 years later. Social support itself is a powerful predictor of mortality due to all causes, including heart disease [15,99].

In sum, several plausible physiological and behavioral pathways linking emotions to heart disease risk have been proposed. Further research is necessary to fully elucidate these mechanisms. However, as the study of medical history reveals, diseases can be prevented or effectively treated long before causative mechanisms are understood [100]. For example, cessation of tobacco chewing to prevent oral cancer was discovered in the year 1915 [100]. However, it was not until 1974 that N'-nitrosornicotine was discovered as the causative agent of oral cancer. Thus, in spite of our incomplete knowledge of the precise mechanisms involved, efforts to lower the risk of heart disease via interventions aimed at improving negative emotions appear to be indicated.

B. Implications for Treatment

The association between negative emotions and heart disease mortality both in women and in men underlines the importance of identifying and treating such emotions in cardiac patients and among those at elevated risk for heart disease. Unfortunately, negative emotions, such as anxiety and depression, are seldom diagnosed or treated in patients with heart disease [56,101–103]. For example, Carney and colleagues [103] reported that in their sample of 77 patients undergoing elective cardiac catheterization for evaluation of coronary artery disease, 19 of the patients were found to be depressed but only five of them were diagnosed or treated for depression by their primary physician or cardiologist. Nevertheless, several promising treatment efforts in this direction have been initiated, and others are currently underway.

With regard to anger, Siegman [104] suggests that interventions aimed at reducing anger may result in improvements of health and quality of life. The goal of treatment should not be the elimination of anger, as this is not possible or desirable in a world where injustice and pain will continue, but rather on the management of anger [105]. For example, instead of expressing anger in a loud and intense voice, which is accompanied by potentially pathogenic cardiovascular reactions, it is recommended to discuss angry feelings "cooly" and rationally [26]. Siegman's work shows that communication of anger and anger-arousing experiences in a normal-to-calm, slow tone does not produce excessive cardiovascular arousal and suggests that anger management may reduce the impact of this emotion on the development of hypertension and CHD.

Similarly, treatments aimed at decreasing depressed mood either by pharmacological or by behavioral means appear to be promising. The prospective study by Pratt et al. [17] also assessed the risk associated with psychotropic medication for myocardial infarction. In spite of the concerns over potentially adverse effects of several antidepressant medications, neither benzodiazepines nor tricyclic antidepressants were related to increased coronary risk. In fact, there may be beneficial effects of pharmacological treatments for depression on heart disease outcomes [106–108]. For example, Avery and Winokur [108] found that in a sample of 519 depressed patients (62% of whom were women), those receiving adequate treatment for depression (i.e., ≥150 mg of imipramine or equivalent, 45 mg of phe-

nelzine, 30 mg of tranylcypromine, or more than five ECT treatments) were significantly less likely to experience MI during a 3-year follow-up period than those receiving inadequate treatment. A study of 473 female and 354 male patients diagnosed with affective disorders (including depression) and treated with lithium for 2 years showed that heart disease mortality was no different than what would be expected in the general population [109]. Considering that patients with affective disorders have a much higher risk for heart disease, this result is quite promising.

Behavioral interventions aimed at reducing emotional stress levels among MI patients have also been initiated [110,111]. One behavioral intervention that includes women is the "Enhancing Recovery in Coronary Heart Disease" (ENRICHD) Patients Study [66]. This study is a major multicenter randomized clinical trial that will test the effects of a psychosocial intervention, aimed at decreasing depression and increasing social support, on reinfarction and mortality in 3000 post-MI patients at high psychosocial risk (i.e., depressed and/or socially isolated patients). Fifty percent of the patients to be enrolled are expected to be women. The study will be completed in 2001 and will provide valuable information on the role of emotions in heart disease both among women and among men.

Finally, it is of interest to note that the current epidemic in heart disease in many of the newly independent countries has been linked to the relative underutilization of behavioral and pharmacological therapies for depression. Kopp and Skrabski [112], for example, speculate that the decline in cardiovascular mortality in the U.S. and Canada is associated with the greater use of psychotherapy and/or the explosion of prescriptions for antidepressants in those countries. Consequently, they have developed a very ambitious community-based program for improving mental health in Hungary that is aimed at preventing depressive symptomatology and strengthening social support [113].

Further examination of the effectiveness both of medications and of psychotherapy in reducing the risk of CHD will not only expand the range of possible interventions but will also help to solidify the causal nature of the association between these emotions and heart disease.

VI. Conclusions

As the research presented in the preceding pages indicates, there is indeed a relationship between cardiovascular disease and emotions, particularly depression and anxiety, hostility, and anger, in women as well as in men. It is both logical and poetic that that which causes us heartache may also hurt our hearts. Unfortunately, the dearth of research in this area leaves the exact nature of this relationship, as well as the mechanisms behind it, unresolved. The picture painted seems to indicate a mosaic of risk, with physiological, social, and behavioral factors relating to each of these emotions and together creating a pattern of CHD development and recurrence. One clear fact that does emerge is that the contribution of each of these emotions to the risk factor equation needs to be studied in further detail, particularly in women.

Acknowledgments

Preparation of this chapter was partially supported by a NATO Collaborative Research Grant (CRG-921325), a NATO Advanced Research Workshop Grant (ASI.ARW 972419), and an NHLBI Research Grant

(RO1-HL62156). We thank C.-W. Kohlmaun, J. R. Adiga, G. Gerhard, and Kristina Orth-Gomér for their helpful comments.

References

1. Thom, T. J., and Epstein, F. H. (1994). Heart disease, cancer, and stroke mortality trends and their interrelations: An international perspective. *Circulation* **90,** 574–582.
2. Tunstall-Pedoe, H., Kuulasmaa, K., Amouyel, P., Arveiler, D., Rajakangas, A.-M., and Pajak, A. (1994). Myocardial infarction and coronary deaths in the World Health Organization MONICA project. *Circulation* **90,** 583–612.
3. Bobak, M., and Marmot, M. (1996). East-west mortality divide and its potential explanations: proposed research agenda. *Br. Med. J.* **312,** 421–425.
4. Ginter, E. (1997). The epidemic of cardiovascular disease in eastern Europe. *N. Engl. J. Med.* **336,** 1915–1916.
5. Notzon, F. C., Komarov, Y. M., Ermakov, S. P., Sempos, C. T., Marks, J. S., and Sempos, E. V. (1998). Causes of declining life expectancy in Russia. *JAMA, J. Am. Med. Assoc.* **279,** 793–800.
6. Kristenson, M., Kucinskiene, Z., Bergdahl, B., Calkauskas, H., Urmonas, V., and Orth-Gomer, K. (1998). Increased psychosocial strain in Lithuanian versus Swedish men: The LiVicordia Study. *Psychosom. Med.* **60,** 277–282.
7. Thom, J. T. (1989). International mortality from heart disease: Rates and trends. *Int. J. Epidemiol.* **18**(Suppl. 1), S20–S28.
8. Anderson, K., Wilson, P., Odell, P., and Kannel, W. (1991). An updated coronary risk profile. A statement for health professionals. *Circulation* **83,** 356–362.
9. Sholinsky, P. D. (1998). Syndrome X. *In* "A Comprehensive Handbook of Behavioral Medicine for Women" (E. A. Blechman and K. D. Brownell, eds.), pp. 713–719. Guilford Press, New York.
10. Castelli, W. P. (1988). Cardiovascular disease in women. *Am. J. Obstet. Gynecol.* **158,** 1553–1560.
11. Jenkins, C. D. (1998). Cardiovascular disorders. *In* "A Comprehensive Handbook of Behavioral Medicine for Women" (E. A. Blechman and K. D. Brownell, eds.), pp. 604–614. Guilford Press, New York.
12. Weidner, G. (1998). Gender gap in health decline in East Europe. *Nature* **395,** 835.
13. Ginter, E. (1995). Cardiovascular risk factors in the former communist countries: Analysis of 40 European MONICA populations. *Eur. J. Epidemiol.* **11,** 199–205.
14. Weidner, G. (1995). Personality and coronary heart disease in women: Past research and future directions. *Z. Gesundheitspsycho.* **3,** 4–23.
15. House, J. S., Landis, K. R., and Umberson, D. (1988). Social relationships and health. *Science* **241,** 540–545.
16. Orth-Gomer, K. (1994). International epidemiological evidence for a relationship between social support and cardiovascular disease. *In* "Social Support and Cardiovascular Disease" (S. A. Shumaker and S. M. Czajkowski, eds.), pp. 97–117. Plenum, New York.
17. Pratt, L., Ford, D., Crum, R., Armenian, H., Gallo, J., and Eaton, W. (1996). Depression, psychotropic medication and risk of myocardial infarction: Prospective data from the Baltimore ECA follow-up. *Circulation* **94,** 3123–3129.
18. Barefoot, J. C., and Schroll, M. (1996). Symptoms of depression, acute myocardial infarction, and total mortality in a community sample. *Circulation* **93,** 1976–1980.
19. Appels, A. and Mulder, P. (1988). Excess fatigue as a precursor of myocardial infarction. *Eur. Heart J.* **9,** 758–764.
20. Kopp, M. S., and Skrabski, A. (1996). "Special Importance of Behavioral Sciences Applied to a Changing Society," pp. 142–144. Bibliotheca Septum Artium Liberalium, Budapest.
21. Eaker, E., Pinsky, J., and Castelli, W. P. (1992). Myocardial infarction and coronary death among women: Psychosocial predictors from a 20-year follow-up of women in the Framingham Study. *Am. J. Epidemiol.* **135,** 854–864.
22. Haellstroem, T., Lapidus, L., Bengtsson, C., and Edstroem, K. (1986). Psychosocial factors and risk of ischemic heart disease and death in women: A twelve-year follow-up of participants in the population study of women in Gothenburg, Sweden. *J. Psychosom. Res.* **30,** 451–459.
23. Julius, M., Harburg, E., Schork, M. A., and DiFranceisco, W. (1992). Role of suppressed anger in cause-specific deaths for married pairs (Tecumseh 1971–1988). *Pap., Int. Congr. Behav. Med. 2nd,* Hamburg, Germany.
24. Frasure-Smith, N., Lesperance, F., and Talajic, M. (1995). Depression and 18-month prognosis after myocardial infarction. *Circulation* **91,** 999–1005.
25. Siegman, A. (1993). Cardiovascular consequences of expressing, experiencing and repressing anger. *J. Behav. Med.* **16,** 539–569.
26. Cook, W. W., and Medley, D. M. (1954). Proposed hostility and pharisaic-virtue scales for the MMPI. *J. Appl. Psychol.* **38,** 414–418.
27. Weidner, G. (1994). The role of hostility and coronary-prone behaviors in the etiology of cardiovascular disease in women. *In* "Women, Behavior, and Cardiovascular Disease: Proceedings of a Conference sponsored by the National Heart, Lung, and Blood Institute" (S. M. Czajkowsi, D. R. Hill, and T. B. Clarkson, eds.), NIH. Publ. No. 94-3309, pp. 103–116. U.S. Department of Health and Human Services, Public Health Service, National Institutes of Health, Washington, DC.
28. Everson, S. A., Kauhanen, J., Kaplan, G. A., Goldberg, D. E., Julkunen, J., Tuomilehto, J., and Salonen, J. T. (1997). Hostility an increased risk of mortality and acute myocardial infarction: The mediating role of behavioral risk factors. *Am. J. Epidemiol.* **146,** 142–152.
29. Frasure-Smith, N., Lesperance, F., and Talajic, M. (1995). The impact of negative emotions on prognosis following myocardial infarction: Is it more than depression? *Health Psychol.* **14,** 388–398.
30. Riley, W. T., Treiber, F. A., and Woods, M. G. (1989). Anger and hostility in depression. *J. Nerv. Ment. Dis.* **177,** 668–674.
31. Watson, D., and Clark, L. A. (1984). Negative affectivity: The disposition to experience aversive emotional states. *Psychol. Bull.* **96,** 465–490.
32. Booth-Kewley, S., and Friedman, H. S. (1987). Psychological predictors of heart disease: A quantitative review. *Psychol. Bull.* **101,** 343–362.
33. Matthews, K. A. (1988). Coronary heart disease and Type A behaviors: Update on and alternative to the Booth-Kewley and Friedman (1987) quantitative review. *Psychol. Bull.* **104,** 373–380.
34. Anda, R., Williamson, D., Jones, D., Macera, C., Eaker, E., Glassman, A., and Marks, J. (1993). Depressed affect, hopelessness, and the risk of ischemic heart disease in a cohort of U.S. adults. *Epidemiology* **4,** 285–294.
35. Ford, D. E., Mead, L. A., Chang, P. P., Cooper-Patrick, L., Wang, N. Y., and Klag, M. J. (1998). Depression is a risk factor for coronary artery disease in men: the precursors study. *Arch. Int. Med.* **158,** 1422–1426.
36. Kawachi, I., Colditz, G. A., Ascherio, A., Rimm, E. B., Giovannucci, E., Stampfer, M. J., and Willett, W. C. (1994). Prospective study of phobic anxiety and risk of coronary heart disease in men. *Circulation* **89,** 1992–1997.
37. Jonas, B. S., Franks, P., and Ingram, D. D. (1997). Are symptoms of anxiety and depression risk factors for hypertension? *Arch. Fam. Med.* **6,** 43–49.
38. Markovitz, J. H., Matthews, K. A., Wing, R. R., Kuller, L. H., and Meilahn, E. N. (1991). Psychological, biological and health behavior predictors of blood pressure changes in middle-aged women. *J. Hypertens.* **9,** 399–406.

39. Falger, P. (1982). Factors contributing to the development of vital exhaustion and depression in male myocardial infarction patients. *Acta. Nerv. Super., Suppl.* **3,** 151–156.

40. Goldberg, E. L., Comstock, G. W., and Hornstra, R. K. (1979). Depressed mood and subsequent physical illness. *Am. J. Psychiatry* **136,** 530–534.

41. Vogt, T., Pope, C., and Hollis, J. F. (1994). Mental health status as a predictor of morbidity and mortality: A 15-year follow-up of members of a health maintenance organization. *Am. J. Public Health* **84,** 227–231.

42. Wassertheil-Smoller, S., Applegate, W. B., Berge, K., Chang, C. J., Davis, B. R., Grimm, R., Kostis, J., Pressel, S., and Schron, E. for the SHEP Cooperative Research Group (1996). Change in depression as a precursor of cardiovascular events. *Arch. Intern. Med.* **156,** 553–561.

43. Kessler, R. C., McGonagle, K. A., Zhao, S., Nelson, C. B., Hughes, M., Eshleman, S., Wittchen, H. U., and Kendler, K. S. (1994). Lifetime and 12-month prevalence of DSM-III-R psychiatric disorders in the United States. *Arch. Gen. Psychiatry* **51,** 8–19.

44. Carney R. M., Freedland, K. E., Smith, L. J., Rich, M. W., and Jaffe, A. S. (1994). Depression and anxiety as risk factors for coronary heart disease in women. *In* "Women, Behavior, and Cardiovascular Disease: Proceedings of a Conference sponsored by the National Heart, Lung, and Blood Institute" (S. M. Czajkowsi, D. R. Hill, and T. B. Clarkson, eds.), NIH Publ. No. 94-3309, pp. 117–126. U.S. Department of Health and Human Services, Public Health Service, National Institutes of Health, Washington, DC.

45. King, K. B. (1997). Psychologic and social aspects of cardiovascular disease. *Ann. Behav. Med.* **19,** 264–270.

46. Fielding, R. (1991). Depression and acute myocardial infarction: A review and reinterpretation. *Soc. Sci. Med.* **32,** 1017–1027.

47. Dalack, G. W., and Roose, S. P. (1990). Perspectives on the relationship between cardiovascular disease and affective disorder. *J. Clin. Psychiatry* **51**(7, Suppl.), 4–9.

48. Murphy, J. M., Monson, R. R., Olivier, D. C., Sobol, A. M., and Leighton, A. H. (1987). Affective disorders and mortality: A general population study. *Arch. Gen. Psychiatry* **44,** 473–480.

49. Rabins, P. V., Harvis, K., and Koven, S. (1985). High fatality rates of late life depression associated with cardiovascular disease. *J. Affective Disord.* **9,** 165–167.

50. Dreyfuss, F., Dasberg, H., and Assael, M. I. (1969). The relationship of myocardial infarction to depressive illness. *Psychother. Psychosom.* **17,** 73–81.

51. Black, D., Warrack, G., and Winokur, G. (1985). Excess mortality among psychiatric patients: The Iowa record-linkage study. *JAMA, J. Am. Med. Assoc.* **253,** 58–61.

52. Carney, R. M., Rich, M. W., Tevelde, A., Saini, J., Clark, K., and Jaffe, A. S. (1987). Major depressive disorder in coronary artery disease. *Am. J. Cardiol.* **60,** 1273–1275.

53. Travella, J. I., Forrester, A. W., Schultz, S. K., and Robinson, R. G. (1994). Depression following myocardial infarction: A one year longitudinal study. *Int. J. Psychiatry Med.* **24,** 357–369.

54. Cay, E., Vetter, N., and Philip, A. (1972). Psychological status during recovery from an acute heart attack. *J. Psychosom. Res.* **16,** 425–435.

55. Forrester, A. W., Lipsey, J. R., Teitelbaum, M. L., DePaulo, J. R., and Andrzejewski, P. L. (1992). Depression following myocardial infarction. *Int. J. Psychiatry Med.* **22,** 33–46.

56. Kurasawa, H., Shimizu, Y., Hirose, S., and Takano, T. (1983). The relationship between mental disorders and physical severities in patients with acute myocardial infarction. *Jpn. Circ. J.* **47,** 723–728.

57. Lesperance, F., Frasure-Smith, N., and Talajic, M. (1996). Major depression before and after myocardial infarction: its nature and consequences. *Psychosom. Med.* **58,** 99–110.

58. Orth-Gomer, K. (1998). Psychosocial risk factor profile in women with coronary heart disease. *In* "Women, Stress and Heart Dis-

ease" (K. Orth-Gomer, M. Chesney, and N. K. Wenger, eds.), pp. 25–38. Erlbaum, Mahwah, NJ.

59. Schleifer, S. J., Macari-Hinson, M. M., Coyle, D. A., Slater, W. R., Kahn, M., Gorlin, R., and Zucker, H. D. (1989). The nature and course of depression following myocardial infarction. *Arch. Intern. Med.* **149,** 1785–1789.

60. Vasquez-Barquero, J., Padierna Acero, J., and Ochoteco, A. (1985). Mental illness and ischemic heart disease: Analysis of psychiatric morbidity. *Gen. Hosp. Psychol.* **7,** 15–20.

61. Wishnie, H., Hackett, T., and Cassem, N. (1971). Psychological hazards of convalescence following myocardial infarction. *JAMA, J. Am. Med. Assoc.* **215,** 1292–1296.

62. Friedman, H. S., and Booth-Kewley, S. (1987). Personality, Type A behavior, and coronary heart disease: The role of emotional expression. *J. Pers. Soc. Psychol.* **53,** 783–792.

63. Graff Low, K., Thoresen, C. E., Pattillo, J. R., King, A. C., and Jenkins, C. (1994). Anxiety, depression and heart disease in women. *Int. J. Behav. Med.* **4,** 305–319.

64. Lloyd, G., and Cawley, R. (1983). Psychiatric morbidity in men one week after first myocardial infarction. *Br. Med. J.* **142,** 120–125.

65. Freedland, K. E., and Carney, R. M. (1994). Depression as a risk factor for coronary heart disease. *Pap., Int. Congr. Behav. Med. 3rd,* Amsterdam, The Netherlands.

66. Czajkowski, S. M. (1998). Psychosocial aspects of women's recovery from heart disease. *In* "Women, Stress and Heart Disease" (K. Orth-Gomer, M. Chesney, and N. K. Wenger, eds.), pp. 151–164. Erlbaum, Mahwah, NJ.

67. McKinlay, J. (1996). Some contributions from the social system to gender inequalities in heart disease. *J. Health Soc. Behav.* **37,** 1–26.

68a. Stern, J. J., Pascale, L., and Ackerman, A. (1977). Life adjustment post myocardial infarction: Determining predictive variables. *Arch. Intern. Med.* **137,** 1680–1685.

68b. Orth-Gomér, K., Horsten, M., Wamala, S. P., Schenck-Gustafsson, K., and Mittleman, M. A. (1998). Depression and social isolation in relation to prognosis of coronary heart disease in women. *Circulation* **98,** 57 (abstract, Supplement).

69. Ahern, D. K., Gorkin, L., Anderson, J. L., Tierney, C., Hallström, A., Ewart, C., Capone, R. J., Schron, E., Kornfeld, D., and Herd, J. A. (1990). Biobehavioral variables and mortality or cardiac arrest in the Cardiac Arrhythmia Pilot Study (CAPS). *Am. J. Cardiol.* **66,** 59–62.

70. Frasure-Smith, N., Lesperance, F., and Talajic, M. (1993). Depression following myocardial infarction: impact on 6-month survival. *JAMA, J. Am. Med. Assoc.* **270,** 1819–1825.

71. Reich, P., DeSilva, R., Lown, B., and Murawski, B. (1981). Acute psychological disturbances preceding life-threatening ventricular arrhythmias. *JAMA, J. Am. Med. Assoc.* **246,** 233.

72. Carney, R. M., Rich, M. W., Freedland, K. E., Saini, J., Tevelde, A., Simeone, C., and Clark, K. (1988). Major depressive disorder predicts cardiac events in patients with coronary artery disease. *Psychosom. Med.* **50,** 627–633.

73. The Women's Health Initiative Study Group (1998). Design of the Women's Health Initiative Clinical Trial and Observational Study. *Controlled Clin. Trials* **19,** 61–109.

74. Matthews, K. A., Shumaker, S. A., Bowen, D. J., Langer, R. D., Hunt, J. R., Kaplan, R. M., Klesges, R. C., and Ritenbaugh, C. (1997). Women's Health Initiative: Why now? What is it? What's new? *Am. Psychol.* **52,** 101–116.

75. Moise, A., Theroux, P., Taeymans, Y., Descoings, B., Lesperance, J., Waters, D. D., Pelletier, G. B., and Bourassa, M. G. (1983). Unstable angina and progression of coronary atherosclerosis. *N. Engl. J. Med.* **309,** 685–689.

76. Kamarck, T., and Jennings, J. R. (1991). Biobehavioral factors in sudden cardiac death. *Psychol. Bull.* **109,** 42–75.

77. Siever, L. S., and Davis, K. L. (1985). Overview: Toward a dysregulation hypothesis of depression. *Am. J. Psychiatry* **142**, 1017–1031.

78. Verrier, R. L. (1990). Behavioral stress, myocardial ischemia, and arrhythmias. *In* "Cardiac Electrophysiology: From Cell to Bedside" (D. P. Zipes and J. Jalife, eds.), pp. 343–352. Saunders, Philadelphia.

79. Lown, B., DeSilva, R. A., Reich, P., and Murawski, B. J. (1980). Psychophysiologic factors in sudden cardiac death. *Am. J. Psychiatry* **137**, 1325–1335.

80. Reich, P. (1985). Psychological predisposition to life-threatening arrhythmias. *Annu. Rev. Med.* **36**, 397–405.

81. Katz, C., Martin, R. D., Landa, B., and Chadda, K. D. (1985). Relationship of psychologic factors to frequent symptomatic ventricular arrhythmias. *Ann. Rev. Med.* **78**, 589–594.

82. Orth-Gomer, K., Edwards, M., Erhardt, L., Sjögren, A., and Theorell, T. (1980). Relation between ventricular arrhythmias and psychological profile. *Acta Med. Scand.* **207**, 31–36.

83. Kennedy, G., Hofer, M., Cohen, D., Shindledecker, M., and Fisher, J. (1987). Significance of depression and cognitive impairment in patients undergoing programmed stimulation of cardiac arrhythmias. *Psychosom. Med.* **49**, 410–421.

84. Bigger, J. T. Jr., Kleiger, R. E., Fleiss, J. L., Rolnitzky, L. M., Steinman, R. C., and Miller, J. P. (1988). Components of heart rate variability measured during healing of acute myocardial infarction. *Am. J. Cardiol.* **59**, 256–262.

85. Carney, R. M., Saunders, R. D., Freedland, K. E., Stein, P., Rich, M. W., and Jaffe, A. S. (1995). Association of depression with reduced heart rate variability in coronary artery disease. *Am. J. Cardiol.* **76**, 562–564.

86a. Light, K. C., Dolan, C. A., Davis, M. R., and Sherwood, A. (1992). Cardiovascular responses to an active coping challenge as predictors of blood pressure patterns 10–15 years later. *Psychosom. Med.* **54**, 217–230.

86b. Weidner, G., Horslen, M., Wamala, S. P., Schecnk-Gustafsson, K., Högbom, M., and Orth-Gomér, K. (1999). Prognostic value of cardiovascular reactivity to mental stress in the Stockholm Female Coronary Risk Study. *Nordiska Kongressen i. Hjärtrehabilitering,* Skellefleå, Sweden, p. 3 (abstract).

87. Weidner, G., and Messina, C. R. (1998). Cardiovascular reactivity to mental stress and cardiovascular disease. *In* "Women, Stress and Heart Disease" (K. Orth-Gomer, M. Chesney, and N. K. Wenger, eds.), pp. 219–236. Erlbaum, Mahwah, NJ.

88. Shapiro, D., Jamner, L. D., and Goldstein, I. B. (1993). Ambulatory stress psychophysiology: The study of 'compensatory and defensive counterforces' and conflict in a natural setting. *Psychosom. Med.* **55**, 309–323.

89. Shapiro, D., Goldstein, I. B., and Jamner, L. D. (1996). Effects of cynical hostility, anger out, anxiety and defensiveness on ambulatory blood pressure in black and white college students. *Psychosom. Med.* **58**, 354–364.

90. Light, K. C., Kothandapani, R. V., and Allen, M. T. (1998). Enhanced cardiovascular and catecholamine responses in women with depressive symptoms. *Int. J. Psychophysiol.* **28**, 157–166.

91. Weidner, G., Kohlmann, C.-W., Dotzauer, E., and Burns, L. R. (1996). The effects of academic stress on health behaviors in young adults. *Anxiety, Stress, Coping* **9**, 123–133.

92. Broman, C. L. (1993). Social relationships and health-related behavior. *J. Behav. Med.* **16**, 335–350.

93. Moscato, B. S., Russell, M., Zielezny, M., Bromet, E., Egri, G., Mudar, P., and Marshall, J. R. (1997). Gender differences in the relation between depressive symptoms and alcohol problems: A longitudinal perspective. *Am. J. Epidemiol.* **146**, 966–974.

94. Watson, D., Clark, L. A., and Tellegen, A. (1988). Development and validation of brief measures of positive and negative affect: The PANAS scales. *J. Pers. Soc. Psychol.* **54**, 1063–1070.

95. Chesney, M., and Darbes, L. (1998). Social support and heart disease in women: Implications for intervention. *In* "Women, Stress and Heart Disease" (K. Orth-Gomer, M. Chesney, and N. K. Wenger, eds.), pp. 165–182. Erlbaum, Mahwah, NJ.

96. Conn, V. S., Taylor, S. G., and Wiman, P. (1991). Anxiety, depression, quality of life, and self-care among survivors of myocardial infarction. *Issues Men. Health Nurs.* **12**, 321–331.

97. Finnegan, D. L., and Suler, J. R. (1985). Psychological factors associated with maintenance of improved health behaviors in postcoronary patients. *J. Psychol.* **119**, 87–94.

98. Cerhan, J. R., and Wallace, R. B. (1993). Predictors of decline in social relationships in the rural elderly. *Am. J. Epidemiol.* **137**, 870–880.

99. Shumaker, S. A., and Czajkowski, S. M. (1994). "Social Support and Cardiovascular Disease." Plenum, New York.

100. Wynder, E. L. (1994). Invited commentary: Studies in mechanisms and prevention: Striking a proper balance. *Am. J. Epidemiol.* **139**, 547–459.

101. Stern, T. A. (1985). The management of depression and anxiety following myocardial infarction. *Mt. Sinai J. Med.* **52**, 623–633.

102. Mayou, R., Foster, A., and Williamson, B. (1979). Medical care after myocardial infarction. *J. Psychosom. Res.* **23**, 23–26.

103. Carney, R. M., Rich, M. W., Tevelde, A., Saini, J., Clark, K., and Freedland, K. E. (1987). The incidence, significance and detection of depression in patients with suspected coronary artery disease. *Proc. NIMH Conf. Ment. Disord. Gen. Health Care Settings,* Seattle, WA, pp. 37–39.

104. Siegman, A. (1992). The role of expressive vocal behavior in negative emotions: Implications for stress management. *Pap., Int. Congr. Psychol. 26th,* Brussels, 1992.

105. Deffenbacher, J. (1994). Anger reduction: Issues, assessment, and intervention strategies. *In* "Anger, Hostility and the Heart" (A. Siegman and T. Smith, eds.) pp. 239–269. Earlbaum, Hillsdale, NJ.

106. Giardina E. G. V., Barnard, T., Johnson, L. L., Saroff, A. L., and Bigger, J. T., Jr. (1986). The antiarrhythmic effect of nortriptyline in cardiac patients with ventricular premature depolarization. *J. Am. Coll. Cardiol.* **7**, 1363–1369.

107. Giardina, E. G. V., and Bigger, J. T. Jr. (1982). Antiarrhythmic effect of imipramine hydrochloride in patients with ventricular premature complexes without psychological depression. *Am. J. Cardiol.* **50**, 172–179.

108. Avery, D., and Winokur, G. (1976). Mortality in depressed patients treated with electroconvulsive therapy and antidepressants. *Arch. Gen. Psychiatry* **33**, 1029–1037.

109. Ahrens, B., Muller-Oerlinghausen, B., Shou, M., Wolf, T., Alda, M., Grof, E., Grof, P., Lenz, G., Simhandl, C., Tahu, K., Vestergaard, P., Wolf, R., and Moller, H. (1995). Excess cardiovascular and suicide mortality of affective disorders may be reduced by lithium prophylaxis. *J. Affective Disord.* **33**, 67–75.

110. Frasure-Smith, N., and Prince, R. (1985). The Ischemic Heart Disease Life Stress Monitoring Program: 18-month mortality results. *Psychosom. Med.* **47**, 431–445.

111. Frasure-Smith, N., and Prince, R. (1989). Long-term follow-up of the Ischemic Heart Disease Life Stress Monitoring Program. *Psychosom. Med.* **51**, 485–513.

112. Kopp, M. S., and Skrabski, A. (1989). What does the legacy of Hans Selye and Franz Alexander mean today? The psychophysiological approach in medical practice. *Int. J. Psychophysiol.* **8**, 99–105.

113. Kopp, M. S., Skrabski, A., and Taylor, C. B. (1998). Community based programme for improving mental health in the Hungarian population: Design and methods. *Pap., Confer. Eur. Health Psychol. Soc. 12th,* Vienna, 1998.

62

Cerebrovascular Disease in Women

ROBIN L. BREY* AND STEVEN J. KITTNER[†]

*University of Texas, San Antonio, San Antonio, Texas; [†]Department of Neurology, University of Maryland School of Medicine, Department of Neurology and the Department of Epidemiology and Preventive Medicine, the Geriatrics Research, Education and Clinical Center, Baltimore Department of Veterans Affairs Medical Center, Baltimore, Maryland

I. Introduction and Scope of the Problem

Approximately 550,000 new strokes occur in the U.S. each year, with nearly half of them affecting women [1–3]. Relatively little attention has been paid to gender differences that may be important in stroke risk management and treatment [4–11]. Many stroke trials have excluded women for a variety of reasons, or included them in numbers that are too small to detect gender-specific treatment effects. Hypertension, cigarette smoking, diabetes, and hypercholesterolemia all contribute to stroke risk in women and in men [12]. There are gender differences, however, in the prevalence of these risk factors and the role they play in the etiology of stroke at various ages. Gender-specific treatment effects have been suggested for aspirin, ticlopidine, and warfarin, but definitive studies are lacking.

Ischemic cerebral infarction is the most common stroke subtype in older women and men [2,3]. In men, the predominant cause is extracranial atherosclerosis [13]. However, in women with atherosclerotic vascular disease, the intracranial vessels are most commonly affected [14]. Cardioembolic stroke is also an important cause of ischemic cerebral infarction in older women [15].

Intracerebral hemorrhage is equally common in women and men. The prevalence of subarachnoid hemorrhage is increased in women approximately 1.5-fold [16]. Furthermore, early recurrent subarachnoid hemorrhage and mortality are both substantially increased in women as compared to men [17]. The reasons for the increased risk of initial and early recurrent subarachnoid hemorrhage and mortality are not known.

II. Stroke Risk Associated with Pregnancy, Hormonal Contraception, and Postmenopausal Hormone Replacement

A. Pregnancy

Pregnancy and the postpartum period are associated with an increased risk for both ischemic stroke and intracerebral hemorrhage [18–21]. However, several published studies have suggested that the magnitude of the risk associated with pregnancy per se may not be as high as was previously thought [20,22–24]. Some potential explanations for the discrepancies between older studies that cited a high pregnancy-associated risk for stroke and these more recent studies include: failure to distinguish venous from arterial events, small numbers of end point events, and referral bias [25]. Regardless of the magnitude of risk for stroke attributed to the pregnant or postpartum state, the evaluation of a stroke in a pregnant women must include the

Table 62.1

Risk of Stroke during Pregnancy and the Postpartum Period in Three Recently Published Studies

Reference	Cerebral infarction (rate/100,000 deliveries)	Intracerebral hemorrhage (rate/100,000 deliveries)	Either type of stroke (rate/100,000 deliveries)
[19]	4.3	4.6	8.9
[15]	11	9	19
[17]	Not given	Not given	5.6
[18]	Not given	Not given	10.3

same comprehensive approach that would be used in a similarly affected nonpregnant woman.

Sharshar and colleagues [24] studied the incidence and causes of stroke associated with pregnancy and the first two postpartum weeks in 63 public maternity hospitals in the region of Ile de France between 1989 and 1992 (see Table 62.1). Because the rates of ischemic and hemorrhagic stroke were the same in this study of pregnant women and because available evidence suggests that the rates of intracerebral hemorrhage are much lower than the rates of ischemic stroke in nonpregnant women, the authors infer that pregnancy increases the risk for intracerebral hemorrhage. Their data also demonstrate that, on a per day basis, the risk of ischemic stroke or intracerebral hemorrhage is higher during the postpartum period than during any trimester of pregnancy.

Kittner and colleagues identified all female patients between the ages of 15 and 44 years in central Maryland and Washington, DC who were discharged with the diagnosis of ischemic or hemorrhagic stroke [20]. They compared the stroke risk in pregnant or postpartum women and nonpregnant women. The postpartum period was defined as up to 6 weeks after delivery, stillbirth, or induced or spontaneous abortion. For ischemic stroke, the relative risk during pregnancy, adjusted for age and race, was 0.7 (95% confidence interval (CI)-0.3–1.6), but it increased to 8.7 for the postpartum period (after a live birth or stillbirth) (95% CI = 4.6–16.7). For intracerebral hemorrhage, the adjusted relative risk was 2.5 during pregnancy (95% CI = 1.0–6.4) but 28.3 for the postpartum period (95% CI = 13.0–61.4). Overall, for either type of stroke during or within 6 weeks after pregnancy, the adjusted relative risk was 2.4 (95% CI = 1.6–3.6) and the attributable, or excess, risk was 8.1 strokes per 100,000 pregnancies (95% CI = 6.4–9.7). Three ischemic strokes and one intracerebral hemorrhage occurred in women who had undergone an induced abortion or who had delivered a stillborn child. Ten of 16 ischemic strokes (62%) occurred during the postpartum period. The risk of ischemic stroke and intracerebral hemorrhage combined were increased in the 6 weeks after delivery, but not during pregnancy itself.

Lanska and Kryscio [23] used the National Hospital Discharge survey to estimate the incidence of arterial stroke and cerebral venous thrombosis during 1979–1991. Hospital discharges were selected that met two conditions. First, there had to be a code indicating a delivery had occurred. Second, there had to be a code indicating either "cerebrovascular disorders in the puerperium" (includes acute subarachnoid hemorrhage, intracerebral hemorrhage, and ischemic arterial stroke) or "other phlebitis and thrombosis" as complications of the puerperium (including only cerebral venous thrombosis and thrombosis of intracranial sinuses). In the 280,096 sampled deliveries, there were 28 observed cases with codes for peripartum arterial stroke and 21 cases of intracranial venous thrombosis, giving observed risks of 10.0 arterial strokes and 7.5 intracranial venous thromboses per 100,000 deliveries. Only one in-hospital death occurred in a patient with ischemic or hemorrhagic arterial stroke. While the authors concluded that both arterial and venous strokes are common complications of pregnancy, the proportion of venous thromboses is higher than that reported in recent population-based series [20,22]. Because the cases were included based on discharge diagnosis codes rather than on review and adjudication by neurologists as in the other studies, the high rate of venous thromboses may be a coding artifact.

In summary, these studies suggest that the risk of ischemic stroke is not increased during pregnancy, but is increased during the postpartum period. The risk for intracerebral hemorrhage is modestly increased during pregnancy but markedly increased in the postpartum period. The rate of combined stroke in these studies ranges from 5.6 to 19 per 100,000 deliveries [20,22–24]. The *excess* combined stroke risk attributed to pregnancy and the first 6 weeks postpartum provided by Kittner's study was 8.1 strokes per 100,000 pregnancies [20]. This figure is quite helpful, as it includes data about stroke risk for women undergoing induced abortion (which was not associated with an increased risk) as well as for pregnancies resulting in stillbirths and live births.

Stroke associated with preeclampsia and eclampsia is a major contributor to stroke occurring late in pregnancy and in the postpartum period. The remainder are caused by a variety of problems and constitute the lengthy differential diagnosis considered for stroke in a young person [26]. The etiology of stroke during pregnancy in some patients could be due to physiologic changes that accompany the pregnant state. In patients without congenital heart disease or cardiomyopathy, some alteration in the coagulation and fibrinolytic systems is suspected. Fibrinogen and Factors VII, VIII, IX, X, and XII increase, whereas antithrombin III decreases in the second and third trimesters [27,28]. Platelet aggregation is also increased during pregnancy and the postpartum period. In addition, both blood glucose and plasma lipid levels are elevated during pregnancy [18], possibly contributing to an increased stroke risk as well.

Venous occlusions are not a major cause of stroke in pregnancy. They typically occur in the third trimester or postpartum period [29]. While Lanska's study [23] identified a relatively high rate of intracranial venous thrombosis during pregnancy, the number that were associated with stroke cannot be ascertained from their data.

Subarachnoid hemorrhage may be the third most common cause of nonobstetric death, with over half of these due to ruptured aneurysms or arteriovenous malformations (AVM) and about one-third related to preeclampsia or eclampsia [30]. It is sometimes difficult to determine whether an intracerebral hemorrhage is due to aneurysmal or AVM rupture or to eclampsia. All can present with headache, mental status changes, and seizures. The initial neurologic status of the patient after intracranial hemorrhage from the rupture either of an aneurysm or of an AVM was correlated with mortality and neurologic outcome. Although pregnant women may not be at increased risk for bleeding from a previously asymptomatic aneurysm, there may be a significantly increased risk for arteriovenous malformation rupture, although, as with ischemic stroke, the magnitude of the risk is controversial. Once an intracerebral hemorrhage has occurred, recurrent bleeding from an untreated aneurysm or AVM is estimated to occur in 33–50% of patients, with a maternal mortality rate of 50–60% [31].

Cerebral aneurysms in pregnancy are similar to those in the general population regarding location and association with hypertension, family history, and polycystic kidney disease. The risk of rupture increases with age in both pregnant and nonpregnant women [30]. The incidence of bleeding increases steadily throughout pregnancy and extends into the early postpartum period. Six percent of aneurysmal ruptures occur in the first trimester, 30% in the second, 55% in the third, and 9% in the first 6 weeks postpartum [32]. Initial rupture during labor and delivery is relatively rare.

In contrast to aneurysmal rupture, AVMs cause subarachnoid bleeding in younger rather than older pregnant patients. In addition, AVMs bleed with equal frequency throughout pregnancy and are less often associated with hypertension [30,31,33]. Pregnancy appears to affect the natural history of AVMs. In pregnancy, hemorrhage is much more likely to be the presenting symptom as compared to seizure in the nonpregnant patient. Labor and delivery is a high-risk period for AVM hemorrhage.

B. Parity and Subsequent Stroke Risk

An increased risk of cardiovascular disease manifestations later in life has been associated with the number of times a woman has been pregnant [34,35]. Some retrospective studies have shown an increased risk associated with parity or age at first birth [36–39] but others have not confirmed this association [40–43]. Similarly, results from prospective studies have been conflicting [44–46]. Two reports from 1993 and 1994, and the largest to date, described the findings of two studies: the Framingham Heart Study and the National Examination Survey National Epidemiologic Follow-up Study (NHEFS) [33,34]. In both of these studies, six or more pregnancies were associated with a small but consistent increase in the risk for coronary artery and cerebrovascular disease odds ratio (OR) (OR = 1.6, 95% CI = 1.1–2.2; Framingham) and (OR = 1.5, 95% CI = 1.1–1.9; NHEFS). The authors concluded that it is not clear whether gravity itself or some other unmeasured factor is actually responsible for this increased risk.

C. Hormonal Contraception

There is persuasive data from the Nurses Health Study that past use of oral contraceptives does not increase future risk of myocardial infarction or stroke [47]. With regard to current use, large case-control studies [48–51] have demonstrated that the stroke risk associated with low-estrogen (50 mcg estrogen or less) oral contraceptives is markedly decreased compared to the prior generation of high-estrogen pills. A national study of ischemic stroke in young women from Denmark [48] showed an estrogen-related dose response; progestin only pills (OR = 0.9, 95% CI = 0.4–2.4), 35–40 mcg (OR = 1.8, 95% CI = 1.1–2.9), 50 mcg (OR = 2.9, 95% CI = 1.6–5.4). A study within the Kaiser Permanente Medical Care Program of California [51] showed no evidence for an increased risk of ischemic (OR = 0.7, 95% CI = 0.3–1.7) or hemorrhagic stroke (OR = 1.0, 95% CI = 0.4–2.8), findings which were not inconsistent with the earlier Danish study [43]. The World Health Organization [50] conducted parallel studies in Europe and in the developing countries which showed statistically significant, but modest, increases in relative risk in the developing countries (ischemic stroke: OR = 3.4, hemorrhagic stroke OR = 1.7), but the lesser risks in the European centers did not achieve statistical significance. A potential explanation is that population differences and prescribing practices are important determinants of the risk of oral contraceptive use. A later report from the Danish study [49] found that the ischemic stroke risk associated with diabetes (OR = 5.4), hypertension (OR = 3.1), and migraine (OR = 2.8) were similar among oral contraceptive users and nonusers, which implies that a woman with diabetes using an oral contraceptive containing 30–40 mcg of estrogen would have an odds ratio for ischemic stroke risk of 9.9 (5.4 × 1.8). Thus, epidemiologic data relative to stroke supports continued caution in the use of oral contraceptives in women with other vascular risk factors.

The implication of migraine for oral contraceptive prescribing practices is an evolving area. At a minimum, oral contraceptives should be not be used in women with auras other than visual scintillations or prolonged auras of any type. Cessation of use is recommended in women whose headache increases in frequency or severity or who develop new aura symptoms while on oral contraceptives [52].

Combined oral contraceptives theoretically have both deleterious and protective effects on stroke risk (Table 62.2). They increase stroke risk by increasing the production of factor X,

factor II, and plasminogen, by decreasing the production of antithrombin III, and by increasing platelet aggregation [53]. Alternatively, estrogen may protect women against cardiovascular disease by a direct action on arterial walls that protects them against atheromatous injury [53]. There is also a beneficial effect of estrogen on lipoprotein metabolism, which is not substantively altered with the addition of progestin [53].

A reversible increased risk of hypertension is associated with combined oral contraceptives. Approximately 4–5% of normotensive women and 9–16% of hypertensive women have increased blood pressure with their use [53,54]. This effect is thought to be multifactorial and due to the effects of both hormones in combination with other factors such as race, family history, cigarette smoking, obesity, and the duration of oral contraceptive use. In some women, particularly if not detected early, oral contraceptive-induced hypertension could contribute to the development of stroke and heart disease. For this reason, it is important to monitor a woman's blood pressure within the first 3 months of initiating oral contraceptive therapy. Oral contraceptive therapy should be discontinued if hypertension develops, and an alternative form of contraception should be suggested.

D. Postmenopausal Estrogen Replacement

Postmenopausal estrogen replacement therapy is used by over 3 million women in the United States to treat the symptoms associated with menopause [55]. Several studies suggest that this therapy, as it is used today, extends the gender advantage for vascular disease seen in women of childbearing potential [56,57]. However, controversies exist, and the role of estrogen replacement therapy in lowering stroke risk is far from settled. Whereas the Nurses Health Study demonstrated a strong protective effect of postmenopausal estrogen replacement on risk for myocardial infarction, increased risks were seen for deep venous thrombosis and ischemic stroke [58]. Data regarding the effects of estrogen replacement therapy and stroke risk come largely from observational studies. The Women's Estrogen Stroke Trial (WEST) is an ongoing prospective randomized controlled trial of 17b-estradiol with placebo in women who have experienced a transient ischemic attack or nondisabling stroke [59]. This study will provide valuable information on the risks and benefits of estrogen replacement therapy in postmenopausal women with stroke.

As with hormonal contraceptives, there are potential effects of estrogen that may be both harmful and beneficial. Early studies, reporting on replacement therapy at doses of 2.5 mg/day, suggested an increase in cardiovascular disease risk, however. Caine and colleagues recently found a dose-dependent increase in hypercoagulability, which occurred when 0.625 and 1.25 mg of conjugated estrogen were studied in postmenopausal women [60]. These same authors had previously described favorable alterations of LDL and HDL levels with oral estrogens at both 0.625 and 1.25 mg/day doses [61]. The decrease in LDL levels resulted from accelerated LDL catabolism, and an increase in triglycerides resulted from an increased production of large, triglyceride-rich VLDL. This large VLDL fraction was cleared directly from the circulation and not converted into small VLDL or LDL. Postmenopausal estrogens could also lower the risk of

Table 62.2
Estrogen Use and Stroke Risk

Factors that increase stroke risk in patients taking estrogen	Factors that decrease stroke risk in patients taking estrogen
Increases coagulation Factors II, X and plasminogen	Has a direct antithrombotic action of the arterial wall
Decreases levels of antithrombin III	Produces favorable changes in lipoprotein metabolism
Increases platelet aggregability	
Can cause hypertension	

cardiovascular disease by beneficial effects on serum glucose and insulin levels [62] and blood pressure [63].

Many of the studies have evaluated these factors in women using unopposed estrogen replacement therapy. Unopposed estrogen replacement is associated with an increased risk for endometrial carcinoma and breast cancer and a decrease in osteoporosis [64]. Progestins appear to prevent the estrogen-induced increased risk of endometrial cancer and to have a synergistic effect with estrogen in decreasing osteoporosis. The effect on breast cancer is less clear, and, unfortunately, progestins may nullify the effects of estrogens on lipoprotein levels. The Atherosclerosis Risk in Communities Study investigators found beneficial effects on lipoprotein, fibrinogen, antithrombin III, fasting glucose, and insulin levels in women taking estrogen or estrogen plus progesterone [56]. If hormone-replacement therapy is causally related to these effects, the authors estimate that it may be associated with an approximately 42% decrease in cardiovascular disease. Although the WEST trial will provide randomized prospective data regarding the risk of recurrent stroke in women taking unopposed estrogen compared to placebo, the additional risks and benefits associated with the use of a combined estrogen-progesterone regimen must await the completion of the Women's Health Study.

III. Gender Differences in Treatable or Modifiable Stroke Risk Factors

A. Diabetes

Diabetes doubles the risk for stroke both in women and in men [65–68]. Women with both diabetes and evidence of peripheral vascular disease have a 5-fold increased risk for stroke and a 9.4-fold increased risk for coronary heart disease [65]. In diabetic men with peripheral vascular disease, only the risk for coronary heart disease is increased. Unfortunately, treatment with oral hypoglycemics may be less effective in decreasing cardiovascular risk in women with diabetes [69]. The Copenhagen Stroke Study, a community-based study of 1135 acute stroke patients, included 233 patients (105 women and 128 men) with diabetes. Diabetic patients of both genders were younger and had more frequent hypertension and less frequent intracerebral hemorrhages than nondiabetic patients [68]. Mortality was also higher in diabetic patients. Diabetic patients recovered more slowly than nondiabetic patients did. However, ultimate outcome in stroke survivors was similar between diabetic and nondiabetic patients. An older autopsy study highlights the importance of diabetes as a major stroke risk factor in women [70]. The same or an even greater degree of cerebral atherosclerosis was demonstrated in diabetic women as compared to diabetic men of similar age. These findings suggest that diabetic patients of both genders should have aggressive cardiovascular risk factor management instituted as early as possible.

B. Smoking

Several prospective studies provide strong evidence that cigarette smoking has numerous adverse health effects both in women and in men [11,71–73]. In many countries, including the U.S., women make up the majority of newly recruited smok-

ers. In addition, fewer women than men have stopped smoking since the 1960s. The association of cigarette smoking and an increased risk of stroke in young and middle-aged women was evaluated in 118,539 of the 121,700 female nurses who participated in the Nurses' Health Study Cohort [72]. Those studied were free from coronary artery disease, stroke, and cancer at entry and provided detailed information about cigarette use. During 8 years of follow-up, the investigators identified 274 strokes. The risk of stroke was increased 3.7-fold in women who smoked 25 or more cigarettes per day as compared with nonsmokers.

Lacroix and colleagues [73] found that mortality rates were twofold higher in current smokers of both genders than in those who had never smoked. Likewise, cardiovascular and cancer mortality was also increased in both genders, with a slight increase in men compared to women who were current smokers. Former smokers of both genders had cardiovascular mortality rates that were similar to people who had never smoked, regardless of the age of smoking cessation. These findings indicate that current smoking is as important a risk factor for cardiovascular mortality in elderly people as it is in young and middle-aged people of both genders. Fortunately, aggressive risk management aimed at smoking cessation appears to have the potential to powerfully influence this risk [74,75].

Smoking is thought to lead to stroke and ischemic heart disease by one of several mechanisms. Cigarette smoking has been shown to reversibly increase platelet activity even in the absence of endothelial injury both in women and in men [76]. This increase in platelet aggregability could well be related to the increased risk for cardiovascular disease seen in current smoking, and it could also explain the fairly rapid decrease in this risk with smoking cessation. In addition, the amount of smoking has been shown to independently predict the severity of carotid artery atherosclerosis [77]. The thickness of carotid plaque and the arterial wall was associated with the lifetime total number of cigarettes consumed, independent of age, hypertension, and diabetes. Future studies are needed to determine whether or not cessation of smoking has a beneficial effect on normalization or deceleration of this atherosclerotic process.

C. Hypertension

Approximately 20 million women in the United States have a blood pressure greater than 140/95 mm Hg and would be considered hypertensive [78]. With the recommendations from the JNCI IV report [79] that hypertension should be defined as a blood pressure greater than 140/90, this figure may be even higher. The importance of hypertension as a stroke risk factor has been assessed using both population-based observational studies and treatment trials. MacMahon and Rodgers performed a meta-analysis of data from seven observational studies with 6 to 25 years of follow-up involving 515,511 subjects ranging in age from 37 to 69 years [80]. Ninety-six percent of subjects were men. All subjects were free from stroke, myocardial infarction, and diabetes mellitus at the baseline evaluation. Small but prolonged changes in blood pressure had marked effects on the risk of a first stroke. Overall, in the 17 trials of antihypertensive drug treatment included in the analysis, a net blood pressure reduction of 10–12 mm Hg systolic and 5–6 mm Hg diastolic were associated with a 38% reduction in stroke *incidence*. This effect was not different between women and men, although only approximately 13,000 women were included in this analysis. These data are somewhat limited because information on death from stroke were available from all studies whereas data regarding nonfatal stroke were available from only two studies. Whether the stroke was ischemic versus hemorrhagic was not clearly differentiated in most studies. Alter and colleagues prospectively studied the role of hypertension as a risk factor for recurrent stroke in a population-based cohort of 662 patients enrolled within 1 month of ischemic or hemorrhagic stroke onset [81]. Nearly half of the cohort was made up of women, 97.2% were white, and 59.4% had hypertension at the time of study enrollment. Of the patients with ischemic stroke, 53.1% of men and 66.2% of women had a history of hypertension. Women were approximately 5 years older than men. Over approximately 2 years of follow-up, 81 of 662 patients had a second stroke. Stroke recurrence in the 621 patients with ischemic stroke was compared in the hypertensive versus normotensive groups [82]. A dose-response effect was found between the degree of blood pressure control and stroke risk, although no attempt was made to systematically treat hypertension in this study. Patients with poorly controlled diastolic hypertension had an eightfold increase (p = 0.001), those with fair control a fourfold increase (p = 0.07), and those with good control a twofold increase in stroke risk when compared to age- and gender-matched normotensive stroke patients.

MacMahon and Rodgers also performed a meta-analysis of 17 randomized trials of the effects of hypertension treatment on stroke occurrence [80]. Altogether these trials enrolled 47,653 patients and 48% were women. There were a total of 1360 strokes during follow-up in patients enrolled in these trials. Stroke incidence was decreased by 38% among patients assigned to active hypertensive treatment. The largest absolute benefits in reduction of stroke incidence were observed in trials enrolling patients with a history of cerebrovascular disease (27.3–18.8%), older patients (7.0–4.6%), and patients with more severe hypertension (8.2–4.7%). Stated slightly differently, in patients with cerebrovascular disease, one stroke would be prevented among every five patients treated for five years. For elderly patients, one stroke would be prevented among every thirty patients treated for five years, whereas for younger patients, eighty patients would need to be treated for five years to prevent one stroke. Unfortunately, the authors did not report on a gender effect.

It is true that neither observational studies nor clinical trials have enrolled sufficient numbers of women to fully explore a gender effect on the role of hypertension on stroke risk or the potential beneficial effects of treatment [83,84]. However, the large meta-analysis by MacMahon and Rodgers as well as other studies [81,82,85,86] support the protective role of lowering blood pressure on stroke risk in women as well as men. Furthermore whereas the importance of hypertension as a stroke risk factor in black women has been well recognized, these studies suggest its importance in white women as well.

D. Nonvalvular Atrial Fibrillation

Nonrheumatic atrial fibrillation is a common cardiac arrhythmia that becomes increasingly prevalent in the elderly population over 70 years of age [15]. Atrial fibrillation has been shown to be an independent risk factor for stroke both in women and in men [15,87,88]. In an analysis of data from the Framingham Study, Wolf and colleagues examined in detail the relative impacts of hypertension, coronary heart disease, cardiac failure, and atrial fibrillation on the incidence of stroke [15]. With increasing age, the effects of all but atrial fibrillation became progressively weaker. In this study, the attributable risk of atrial fibrillation for stroke increased from 1.5% in the group of subjects aged 50–59 years to 23.5% for those aged 80–89 years. When evaluating the stroke risk associated with atrial fibrillation in patients with other coexisting cardiac diseases, men with atrial fibrillation had more than a twofold, whereas women had nearly a fivefold, excess risk for stroke. Candelise and colleagues found atrial fibrillation to be a poor prognostic factor in patients hospitalized at the time of an acute stroke in 1048 consecutive stroke patients enrolled in the Italian Hemodilution Trial [89]. The adjusted risk of death associated with atrial fibrillation was significant at both 1 and 6 months (relative risk of 1.55 and 1.74, respectively). In this large retrospective case-control series, 211 patients had atrial fibrillation. Of these, 126 (60%) patients were women and 171 (88%) were over the age of 65 years. This overrepresentation of women in stroke patients with atrial fibrillation has been previously reported and likely reflects the larger percentage of women in the elderly cohort. Miller and colleagues examined the effect of aspirin in preventing cardioembolic versus noncardioembolic stroke in patients enrolled in the Stroke Prevention in Atrial Fibrillation (SPAF) Study [90]. Men with atrial fibrillation were more likely to have strokes that were classified as noncardioembolic and to benefit from aspirin. Women had a 2.5-fold increased risk for cardioembolic stroke and did not appear to benefit from aspirin.

Atrial fibrillation is clearly an important risk factor for stroke and adverse outcome after stroke in elderly women. Elderly women may not benefit from aspirin therapy because the mechanism of stroke in this group is more likely to be cardioembolic than in the group of elderly men with atrial fibrillation

E. Serum Lipids

While the relationship of serum lipids and lipoproteins to an increased risk of coronary artery and peripheral vascular disease is well established, the link to stroke risk has been less clear. A major factor slowing the understanding of the relationship of lipids to stroke has been the tendency to consider ischemic and hemorrhagic stroke together as a homogenous endpoint. There is increasing evidence that low cholesterol is associated with hemorrhagic stroke [91,92]. The inclusion of hemorrhagic with ischemic stroke in past studies has attenuated the relationship of lipoproteins to ischemic stroke. Elevated LDL and triglyceride levels have been shown to be independent risk factors in patients with atherothrombotic cerebrovascular disease [93]. There is also evidence from a prospective study of Israeli men that low HDL cholesterol is an independent predictor of mortality from ischemic stroke [94]. Similar data are not available for women.

The earliest studies evaluating the effect of diet on plasma lipid and lipoprotein levels were performed in men and the results were generalized to women. There are important differences in lipid metabolism between women and men, however, which makes these generalizations suspect. For example, before puberty the level of hepatic lipase, the enzyme of prime importance in degrading HDL cholesterol, is the same in women and men, and not surprisingly, levels of HDL are also the same [95]. After puberty, one effect of testosterone is to activate hepatic lipase activity resulting in approximately 70% lower mean HDL cholesterol levels in postpubertal men compared to women. In addition, LDL cholesterol levels are lower in women than men, likely due to an estrogen effect. These differences persist throughout most of life until, after menopause, higher LDL cholesterol levels are seen in women [96]. In a study that matched men and women for age, LDL cholesterol, triglycerides, and body mass index, a significant difference in response to dietary fat and cholesterol was seen. Women responded to a dietary increase of fat and cholesterol by further increasing mean levels of HDL cholesterol, whereas men had an increase in LDL cholesterol [97]. On the other hand, men appear to have a greater beneficial effect from modest, sustained weight loss on all cardiovascular risk factors including a decrease in serum lipids [98].

F. Alcohol

The relationship of alcohol to stroke appears to be age and dose-dependent both in women and in men [99,100]. Binge drinking has been associated with stroke in young men. Habitual heavy drinkers (more than 300 grams per week, 10 gram standard drink) have a fourfold increase in stroke risk compared to nondrinkers according to Gill and colleagues [101]. The study by Palomaki and Kaste confirmed this increased stroke risk in male heavy habitual drinkers and found a protective effect of light to moderate regular alcohol intake in men [102]. Studies evaluating populations that included both genders have also found that light to moderate alcohol consumption may decrease stroke risk by half [99,103]. A study by Rodgers and colleagues found a threefold increased stroke risk both in women and in men who were current or life-long abstainers from alcohol [99]. Based on their data, this group of investigators suggests that a maximum sensible weekly alcohol intake should be 120 grams for women and 180 grams for men.

A prospective study in 1988 that assessed the alcohol-associated stroke risk in women participants of the Nurse's Health Study has confirmed that light to moderate alcohol consumption lowers ischemic stroke risk by 50% [103]. There was, however, nearly a fourfold increase in subarachnoid hemorrhage in women who were light to moderate drinkers. The authors speculate that alcohol effects on platelets and clotting factors that could lower ischemic risk could also predispose a woman to subarachnoid hemorrhage. A report based on further prospective follow-up of that cohort confirmed the benefit of light to moderate drinking primarily in women with cardiovascular disease risk factors who were over 55 years of age [100]. A similar benefit in younger women and in women without cardiovascular disease risk factors was not seen. Similar to studies of alcohol in men, heavier drinking was associated with a substantially increased mortality.

G. Homocysteine

Accumulating epidemiologic evidence [104], supported by data from *in vitro* experiments [105] and animal models [106],

supports a causal role for homocysteine in the pathogenesis of ischemic stroke. Case-control studies of young and old adults of both genders have found homocysteine to be associated with ischemic stroke. An initial study of men and women from Finland [107] showed no relationship to stroke, and a subsequent study of men [108] showed only an equivocal relationship to stroke. However, the British Regional Heart Study [109] showed a strong, independent and graded response of homocysteine to stroke risk during a 12 year follow-up study of men 40–59 years of age. Cross-sectional data [110] show that homocysteine levels are correlated directly with other vascular risk factors, such as male gender, age, cigarette smoking, total cholesterol, and blood pressure, and inversely with factors protective against vascular disease, such as physical activity. Both in men and in women, homocysteine shows a graded independent association with carotid artery stenosis [111].

Gender differences in homocysteine levels [110], including a steeper age gradient in women, may be partly explained by sex hormones. Higher homocysteine levels after menopause were reported in some, but not in all, studies, and lower homocysteine levels have been observed with use of estrogen replacement therapy [112] and with tamoxifen [113]. This can not entirely explain the gender gradient, as it is still evident between men and women ages 65–67 [110]. Creatine-creatinine synthesis, a function of muscle mass, is a major source of homocysteine formation and may more persuasively explain the higher levels in men [114,115].

The B vitamins, folate, B6, and B12, are major determinants of population homocysteine levels [116] and supplementation with folic acid and B6, even in persons who are not vitamin deficient, is known to lower homocysteine levels. While homocysteine levels can be reduced through vitamin supplementation, it is not yet known whether this will result in a reduction in the risk of vascular disease and clinical trials to this end are in progress.

IV. Gender Differences in Other Less Common Stroke Syndromes

A. Mitral Valve Prolapse

Mitral valve prolapse is the most common cardiac valvular condition in the general population [117]. The prevalence of this condition in adults has been estimated to range from 3 to 13%, although a precise prevalence determination is hindered by a variety of technical difficulties [118,119]. While in several studies the prevalence of mitral valve prolapse appeared equal in women and men [118,119], the prevalence may be actually increased two- to threefold in young women as compared to young men [120]. A reversible prolapse of the mitral valve can occur with a decreased weight to height ratio, as is often seen in young women [121]. An age-associated increase in weight to height ratio has been suggested as an explanation for the decreased frequency of mitral valve prolapse seen in older women.

Although mitral valve prolapse is often cited as a potential cause of stroke, particularly in young people, there is controversy as to whether the presence of mitral valve prolapse in an individual patient conveys an increased risk for stroke. The frequency of stroke is not increased in neurological populations with a greater prevalence of mitral valve prolapse compared to the general population [121]. Further, in several studies of mitral valve prolapse and stroke, the incident ischemic events have been largely transient and the risk for recurrent stroke has been low [117,122]. Marks and colleagues found that patients with both mitral valve thickening and redundancy were at greater risk for infectious complications and moderate to severe mitral regurgitation often requiring mitral valve replacement than were patients without valve thickening [123]. There was no difference in the frequency of stroke between the two groups, however. In contrast, Orencia and colleagues found that stroke occurred in three of seven patients with mitral valve prolapse requiring valve replacement and was not increased relative to the general population in patients with uncomplicated mitral valve prolapse [124].

B. Migraine-Associated Stroke

The issue of migraine and stroke is of particular importance because migraine is quite prevalent in young women. There is a need for a consensus policy regarding the use of oral contraceptives among different type of migraneurs. There are several studies addressing this issue in the current low-estrogen oral contraceptive era. A study from Denmark [49] found a multiplicative relationship between the risk of oral contraceptive use and unspecified migraine. A smaller French case-control study of women under age 45 [125] found an increased ischemic stroke risk among women with migraine without aura (OR = 3.0, 95% = CI 1.5–5.8) and even greater in migraine with aura (OR = 6.2, 95% = CI = 2.1–18.0). While oral contraceptive use in the absence of migraine was associated with an odds ratio of 3.5 (95% CI = 1.5–8.3), the combination of oral contraceptive use and migraine, with or without aura, was 13.9 (95% CI = 5.5–35.1). A case-control study of transient ischemic attacks or stroke [126] found that the combination of migraine history and oral contraceptive use was reported by seven women and none of the controls (p = 0.001). Prospective studies of migraine and stroke risk have been conducted in two established cohorts. The Physicians Health Study [127] found that middle-aged and elderly men with a history of migraine had an elevated risk of ischemic stroke (OR = 2.00, 95% CI = 1.10–3.64) when compared to men without migraine. Notably, migraine was not associated with myocardial infarction. Analyses based on the National Health and Nutrition Examination Survey [128] found that both physician-diagnosed migraine and analgesic use for headaches were predictors of stroke of all types after adjustment for established stroke risk factors. Interestingly, the association was stronger at younger ages. At age 40, the estimated hazard ratio was 2.81 (95% CI = 1.45–5.43) for migraine and 4.6 (95% CI = 2.13–9.96) for headache. All cases (number unspecified) of stroke associated with migraine in women under age 45 occurred in oral contraceptive users. While the data suggest that migraine may be a risk factor for stroke and, like other risk factors, may increase the risk of oral contraceptive use, the absolute risk of ischemic stroke in young women is quite low, on the order of 10–20 per 100,000 [129,130]. Thus, recommendations regarding oral contraceptive use in migraneurs need to take into account the benefits of oral contraceptive use versus the risks.

C. Procoagulant States

Procoagulant states can contribute to the development of a variety of arterial and venous thromboses. It is difficult to estimate the prevalence of these conditions in stroke, as they are not all rigorously screened for in most studies. As many as 10% of elderly first-stroke patients [131] and possibly nearly half of stroke patients less than 50 years of age [132] harbor at least one identifiable prothrombotic condition. It should be emphasized that these conditions often occur in the setting of other recognized stroke risk factors and that a thrombotic event may result from a complex interaction of multiple pathologic features involving coagulation and abnormalities of the vessel wall or platelet function.

The changes in the coagulation system during pregnancy and with hormonal contraceptive and combined hormonal replacement use have already been discussed. There are also differences in the coagulation system between normal healthy women and men. Both fibrinogen levels and fibrinolytic activity are higher in women than in men [133]. Levels of antithrombin III begin to decline in men around the age of 40, potentially contributing to an increased risk for hypercoagulability, however no such reduction in these levels occurs in women [134].

Hart and Kanter reviewed specific prothrombotic conditions of particular importance related to stroke [134]. Most of the important procoagulant states that are associated with stroke are listed in Table 62.3. Only one additional coagulation disorder that may have special relevance to stroke in women will be mentioned here: antiphospholipid antibodies.

Antiphospholipid antibodies (aPL) are associated with thrombosis, thrombocytopenia, fetal loss, and a variety of neurological syndromes [135]. They are defined by solid-phase immunoassays and phospholipid dependent coagulation tests. When detected by solid-phase immunoassay, they may be named for the specific negatively charged phospholipid, such as cardiolipin, used as the antigen. When detected by phospholipid dependent coagulation tests, they are called lupus anticoagulants (LA). There is considerable, although incomplete, overlap between these two methods of detection and aPL is a reasonable general term for this heterogeneous antibody family.

These antibodies are highly associated with episodes of venous and arterial thrombosis, which are often recurrent. When thrombosis occurs in the arterial distribution, the brain is most frequently involved. The prevalence is probably highest in young adults and could be as high as 46% in this group [132]. The prevalence of aPL in patients with systemic lupus erythematosus (SLE), a disease which is three to five times more common in women than in men, is approximately 40%. aPL is considered to be one of the most important risk factors for thrombosis, including stroke, in patients with SLE [136]. In a case-control study by the Antiphospholipid Antibodies in Stroke Studt Group (APASS), 10% of older first ischemic stroke patients without SLE had aPL compared with only 4.3% of hospitalized controls (p = 0.02) [131]. Furthermore, this study found the presence of aPL to be an independent risk factor for stroke, equal in magnitude to diabetes mellitus or hypertension. In many studies that have studied aPL-associated stroke in non-SLE populations, the numbers of women and men with aPL and stroke in all age groups are equal. In a prospective study of

Table 62.3
Procoagulant States Associated with Stroke

A. Inherited disorders of coagulation inhibitors	Other specific disorders	Multifactorial
Protein C deficiency	High Factor V and VII levels	Pregnancy
Factor V Leiden mutation	Antiphospholipid antibodies	Physiologic stress
Protein S deficiency	Polycythemia vera	Hormonal contraceptives
Anti-thrombin III deficiency	Sickle cell disease	Nephrotic syndrome
Heparin cofactor II deficiency	Secondary polycythemias	Cancer
	Sickle cell trait	Thrombotic thrombocytopenia purpura
B. Inherited disorders of fibrinolysis	Paroxysmal noturnal hemoglobinuria	Hemolytic-uremic syndrome
Dysfibrinogenemia	Beta-thalassemia	Disseminated intravascular coagulation
Plasminogen deficiency	Essential thrombocytopenia	Lipoprotein (a)
Plasminogen activator deficiency	Secondary thrombocytosis	Macroglobulinemia
Factor XII deficiency	L-Asparaginase treatment	Cryoglobulinemia
Prekallicrein deficiency	Homocysteine	Antifibrinolytic therapy
	Activated protein C resistance without Factor V Leiden	Heparin-induced thrombocytopenia
		Severe hypoglycemia

stroke patients without SLE by Levine and colleagues, however, recurrent stroke was seen in women more commonly than in men with aPL [137]. Other features of aPL-associated stroke recurrence were the presence of multiple clinical features associated with aPL and a high titer IgG antibody.

V. Treatment

A. *Primary Prevention/Risk Reduction*

Primary stroke prevention efforts in both genders should focus on maintaining a low risk profile or beneficially altering a high risk profile, first and foremost (Table 62.4). Regardless of the gender differences in specific stroke risk factors that may exist, our largest efforts would be best spent in minimizing their effects in women and men of all ages. All people should be encouraged to stop smoking [71], eat a healthy balanced diet [138,139], and engage in regular exercise [140]. Hypertension,

Table 62.4
Stroke Risk Reduction Strategies for Women

Stop smoking
Control blood pressure
Control blood sugar
Regular exercise
Maintenance of normal weight
Use of estrogen replacement therapy
(still controversial; trials ongoing)
Use of aspirin, ticlopidine, or warfarin for secondary prevention

diabetes, and hypercholesterolemia should be sought and vigorously treated.

Postmenopausal estrogen replacement therapy appears to extend the gender advantage for vascular disease seen in women of childbearing potential [56]. Ongoing prospective studies will provide important data that should help resolve whether postmenopausal hormone replacement therapy conveys as great a benefit for stroke prevention as it does for prevention of myocardial infarction.

B. *Secondary Prevention*

1. *Gender Differences in Medical Management*

There is a controversy as to whether aspirin is equally effective in preventing stroke in women. Most trials have had insufficient numbers of women enrolled to answer this question adequately. The ISIS-2 collaborative group conducted the only major study with sufficient numbers of women for meaningful subgroup analysis of treatment effects by gender [141]. In this study all vascular endpoints as well as vascular death were considered in 4000 women and 13,000 men. Risk reduction due to aspirin therapy was approximately 20% for women and 25% for men. The SPAF investigators found that elderly women with atrial fibrillation do not benefit from aspirin in preventing a primary stroke, however, whereas men do [90]. These investigators suggest that these gender-specific effects may be explained by the different stroke mechanism in women as compared to men with atrial fibrillation.

There are many examples of *in vitro* studies that could help explain a gender-based difference in aspirin efficacy. Some studies have shown that aspirin has a greater effect in inhibiting platelet aggregation in men than in women [142]. There is also evidence for differential gender effects of aspirin on the interaction of platelets with the vessel wall subendothelium [143].

Ticlopidine, an inhibitor of platelet aggregation and thrombus formation with a mechanism distinct from that of aspirin, has been shown to be equally effective in women as in men. The Ticlopidine Aspirin Stroke Study (TASS) found that ticlopidine reduced the risk of subsequent fatal and nonfatal stroke in both genders after an initial reversible cerebrovascular ischemic event or a minor stroke [144]. The overall risk reduction for ticlopidine-treated patients was greater than that seen for aspirin-treated patients. This efficacious response in both genders was confirmed by the Canadian American Ticlopidine Study [145]. There did not appear to be any significant gender differences in the frequency or types of side effects experienced by patients taking ticlopidine. Reversible severe neutropenia, seen in about 1% of patients taking this medication, appears within the first 3 months of therapy and occurs equally in both genders.

Warfarin therapy appears to prevent strokes both in women and in men with atrial fibrillation [146–149]. The efficacy of warfarin in preventing noncardioembolic stroke is currently under investigation. The Boston Area Anticoagulation Trial for Atrial Fibrillation showed that prothrombin activation was significantly suppressed *in vivo* by warfarin but not aspirin in patients with atrial fibrillation [150]. In a study on aging and the anticoagulant response to warfarin therapy, Gurwitz and colleagues found that age, use of a medication with a potentially interactive effect with warfarin, overall medication use, and female gender all were factors associated with increased sensitivity to warfarin [151]. Thus, older women taking warfarin could be at increased risk for bleeding complications, as higher PT/INR values have been associated with increased bleeding risk [152]. This suggests the need for close and frequent monitoring of older women on warfarin therapy.

The medical treatment of stroke during pregnancy deserves special consideration. Some of the medications usually used to prevent subsequent stroke in nonpregnant patients are also used in pregnant patients. Aspirin is associated with increased fetal mortality and complications such as intrauterine growth retardation, congenital salicylate intoxication, and hemorrhage. However, available evidence from studies aimed at preventing pregnancy-induced hypertension suggests that low dose (less than 150 mg/day) aspirin during the second and third trimesters is safe for the fetus [153,154]. Other antiplatelet medications such as ticlopidine and dipyridamole have not been studied sufficiently in pregnant patients to assess their safety and should therefore not be used. In conditions where anticoagulation is required, heparin, low molecular weight heparin, and warfarin have all been used in pregnant women. Warfarin crosses the placenta and is associated with a 3–4% risk of fetal malformations that are maximal in the sixth to ninth weeks of gestation [154]. There is also the risk of fetal hemorrhage, miscarriage, stillbirth, and prematurity throughout pregnancy. Warfarin should be avoided in the first trimester but can be used after that. The fetal liver produces low levels of vitamin K-dependent coagulation factors; therefore, the risk of fetal hemorrhage may be greater with larger maternal doses. The advantage of warfarin is that it is more convenient for the patient and safer for the mother

than heparin. Fixed mini-dose warfarin (1 mg per day) has been suggested for prophylaxis of thromboembolic disease during pregnancy [155]. Apparently treatment with warfarin at this dose does not result in coagulation abnormalities in the fetus. While this has been used successfully in the prevention of venous thrombosis, it is not clear whether it would be effective in preventing stroke due to arterial thrombosis or cardiac embolism. Heparin can be used throughout pregnancy and can be given as an outpatient treatment in subcutaneous doses that increase the aPTT to about 1.5 times normal. The dose required is about 10,000 units bid for most patients. It does not cross the placenta and is therefore safe for the fetus. Thrombocytopenia and osteoporosis are both potential complications for the mother, however. Low molecular weight heparin has the advantage of not causing osteoporosis, but it still must be administered subcutaneously.

The management of the pregnant patient with an intracerebral hemorrhage also deserves special comment. Because of the high rate of rebleeding during pregnancy, authors have recommended early surgery after aneurysmal hemorrhage that occurs during pregnancy. This is supported by a study by Dias and Sekhar [30] of 106 pregnant patients who suffered an aneurysmal rupture. Fifty-five were treated with aneurysm clipping and 51 with nonoperative treatment. Maternal mortality was 11% in the operated group and 65% in the untreated group. Fetal mortality was 5% in the operated group and 27% in the untreated group. The appropriate course of treatment for patients with AVMs is less clear. In the same study by Dias and Sekhar, 13 of 36 patients with angiomatous hemorrhage were treated operatively during pregnancy and 22 were treated nonoperatively. Maternal mortality was 23% in the operated group and 32% in the untreated group. Fetal mortality was 0% in the operated group and 27% in the untreated group, but these differences were not statistically significant.

Finally, it appears that carotid endarterectomy for asymptomatic carotid stenosis (60% diameter reduction by ultrasound and/or angiography) may not be equally efficacious in women and men. Of the 1662 patients enrolled in the Asymptomatic Carotid Artery Stenosis (ACAS) study, approximately 34% were women and 95% were white [156]. The overall projected 5-year risk reduction in patients undergoing carotid endarterectomy was 53%; however, men had 66% whereas women had only 17% risk reduction. Women randomized to receive surgery had a higher perioperative complication rate than the men for unclear reasons, markedly reducing the apparent surgical benefit for women. Interestingly, no such increase in perioperative morbidity has been reported in the North American Symptomatic Carotid Endarterectomy Trial (NASET), and a larger number of women were included [157]. Information from a portion of NASET enrolling patients with carotid stenosis ranging from 30 to 69% was presented at the 23rd International Joint Conference on Stroke and Cerebral Circulation (February, 1998 in Orlando, Florida). Women with symptomatic carotid stenosis in this range enrolled in NASET did not receive a greater benefit from surgery than from medical treatment alone. Thus, carotid endarterectomy has been shown to benefit women with symptomatic carotid stenosis that is greater than 70%, whereas the benefit to women with asymptomatic carotid stenosis or symptomatic stenosis that is less than 70% is not greater than the benefit of medical management alone.

References

1. American Heart Association (1992). "Heart and Stroke Facts." American Heart Association, Dallas, TX.

2. Wolf, P. A., D'Agostino, R. B., O'Neal, M. A., Sytkowski, P., Kase, C. S., Bélanger, A. J., and Kannel, W. B. (1992). Secular trends in stroke incidence and mortality. The Framingham Study. *Stroke* **23**, 1551–1555.

3. Modan, B., and Wagener, D. K. (1992). Some epidemiological aspects of stroke: Mortality/morbidity trends, age, race, socioeconomic status. *Stroke* **23**, 1230–1236.

4. Wong, M. C., and Giuliani, M. J. (1990). Cerebrovascular disease and stroke in women. *Cardiology* **77**(Suppl. 2), 80–90.

5. Brass, L. M., Isaacsohn, J. L., Merikangas, K. R., and Robinette, C. D. (1992). A study of twins and stroke. *Stroke* **23**, 221–223.

6. Wenger, N. K., Speroff, L., and Packard, B. (1993). Cardiovascular health and disease in women. *N. Engl. J. Med.* **329**, 247–256.

7. Ellekjaer, E. F., Wyller, T. B., Sverre, J. M., and Holmen, J. (1992). Lifestyle factors and risk of cerebral infarction. *Stroke* **23**, 829–834.

8. Hershey, L. A. (1993). Gender differences in cerebrovascular disease. *Neurol. Chron.* **3**, 1–4.

9. Barnett, H. J. (1990). Stroke in women. *Can. J. Cardiol.* **6**(Suppl. B), 11B–17B.

10. Davis, P. (1994). Stroke in women. *Curr. Opin. Neurol.* **7**, 36–40.

11. Murabito, J. M. (1995). Women and cardiovascular disease: Contributions from the Framingham Heart Study. *J. Am. Med. Women's Assoc.* **50**, 35–39.

12. Wolf, P. A. (1990). An overview of the epidemiology of stroke. *Stroke* **21**(Suppl. II), II-4–II-6.

13. Flora, G. C., Baker, A. B., Loewenson, R. B., and Klassen, A. C. (1968). A comparative study of cerebral atherosclerosis in males and females. *Circulation* **38**, 859–869.

14. Caplan, L. R., Gorelick, P. B., and Hier, D. B. (1986). Race, sex and occlusive cerebrovascular disease: A review. *Stroke* **17**, 648–651.

15. Wolf, P. A., Abbott, R. D., and Kannel, W. B. (1991). Atrial fibrillation as an independent risk factor for stroke: The Framingham Study. *Stroke* **22**, 983–988.

16. Coull, B. M., Brockschmidt, J. K., Howard, G., Becker, C., Yatsu, F. M., Toole, J. F., McLeroy, K. R., Feibel, J. (1990). Community hospital-based programs in North Carolina, Oregon and New York. *Stroke* **21**, 867–873.

17. Torner, J. C., Kassell, N. F., Wallace, R. B., and Adams, H. P. (1981). Preoperative prognostic factors for rebleeding and survival in aneurysm patients receiving antifibrolytic therapy. Report of the Cooperative Aneurysm Study. *Neurosurgery* **9**, 506–513.

18. Wiebers, D. O. (1985). Ischemic cerebrovascular complications of pregnancy. *Arch. Neurol. (Chicago)* **42**, 1106–1113.

19. Wiebers, D. O., and Whisnant, J. P. (1979). The incidence of stroke among pregnant women in Rochester, Minn. 1955 through 1979. *JAMA, J. Am. Med. Assoc.* **254**, 3055–3057.

20. Kittner, S. J., Hebel, R., Feeser, B., Buchholz, D., Earley, C., Johnson, C., Macko, R., Price, T., Rosario, J., Sloan, M. A., Stern, B., Wityk, R., and Wozniak, M. (1996). Stroke risk associated with pregnancy and the post-partum period. *N. Engl. J. Med.* **335**, 768–774.

21. Simolke, G. A., Cox, S. M., and Cunningham, F. G. (1991). Cerebrovascular accidents complicating pregnancy and the puerperium. *Obstet. Gynecol.* **78**(1), 37–42.

22. Petitti, D. B., Sidney, S., Quesenberry, C. P., Jr., and Bernstein, A. (1997). Incidence of stroke and myocardial infarction in women of reproductive age. *Stroke* **28**, 280–283.

23. Lanska, D. J., and Kryscio, R. J. (1997). Peripartum stroke and intracranial venous thrombosis in the National Hospital Discharge Survey. *Gynecology* **89**, 413–418.

24. Sharshar, T., Lamy, C., Mas, J. L., for the Stroke in Pregnancy Study Group (1995). Incidence and causes of strokes associated with pregnancy and puerperium. *Stroke* **26**, 930–936.

25. Grosset, D. G., Ebrahim, S., Bone, I., and Warlow, C. (1995). Stroke in pregnancy and the puerperioum: What magnitude of risk? *J. Neurol. Neurosurg. Psychiatry* **58**, 129–131.

26. Wilterdink, J. L., and Easton, J. D. (1994). Cerebral ischemia. In Deunisky, O., Feldmann, E., and Hainline, B., Eds. *Neurological Complications of Pregnancy* Raven Press, Ltd., New York, pp. 1–11.

27. Bonnar, A. J., McNicol, G. P., and Douglas, A. S. (1970). Coagulation and fibrinolytic mechanisms during and after normal childbirth. *Br. Med. J.* **2**, 200–203.

28. Meher-Homji, N. J., Montemagno, R., Thilaganathan, B., and Nicolaides, K. H. (1994). Platelet size and glycoprotein Ib and IIIa expression in normal fetal and maternal blood. *Am. J. Obstet. Gynecol.* **171**(3), 791–796.

29. Donaldson, J. O., and Lee Nora, S. (1994). Arterial and venous stroke associated with pregnancy. *Neurol. Clin.* **12**, 3.

30. Dias, M. S. (1994). Neurovascular emergencies in pregnancy. *Clin. Obstet. Gynecol.* **37**(2), 337–354.

31. Wilterdink, J. L., and Feldmann, E. (1994). Cerebral hemorrhage. In Deunisky, O., Feldmann, E., and Hainline, B., Eds. *Neurological Complications of Pregnancy* Raven Press, Ltd., New York, pp. 13–23.

32. Hunt, H., Schifrin, B., and Suzuki, K. (1974). Ruptured berry aneurysms and pregnancy. *Obstet. Gynecol.* **43**, 827–836.

33. Lanzino, G., Jensen, M. E., Cappelletto, B., and Kassell, N. F. (1994). Arteriovenous malformations that rupture during pregnancy: A management dilemma. *Acta Neurochir.* **126**, 102–106.

34. Ness, R. B., Harris, T., Cobb, J., Flegal, K. M. F., Kelsey, J. L., Balanger, A., Stunkard, A. J., and D'Agostino, R. B. (1993). Number of pregnancies and the subsequent risk of cardiovascular disease. *N. Engl. J. Med.* **328**, 1528–1533.

35. Qureshi, A. I., Giles, W. H., Croft, J. B., and Stern, B. J. (1997). Number of pregnancies and risk for stroke and stroke subtypes. *Arch. Neurol. (Chicago)* **54**, 203–206.

36. Winkelstein, W., and Rekate, A. C. (1964). Age trend of mortality from coronary artery disease in women and observations on the reproductive patterns of those affected. *Am. Heart J.* **67**, 481–488.

37. Winkelstein, W., Jr., Stenchever, M. A., and Lilienfeld, A. M. (1958). Occurrence of pregnancy, abortion, and artificial menopause among women with coronary artery disease: A preliminary study. *J. Chronic Dis.* **7**, 273–286.

38. Bengtsson, C., Rybo, G., and Westerberg, H. (1973). Number of pregnancies, use of oral contraceptives and menopausal age in women with ischaemic heart disease, compared to a population of sample women. *Acta Med. Scand. Suppl.* **549**, 75–81.

39. Beral, V. (1985). Long term effects of childbearing on health. *J. Epidemiol. Commun. Health* **39**, 343–346.

40. Ritterband, A. B., Jaffe, I. A., Densen, P. M., Magagna, J. F., and Reed, E. (1963). Gonadal function and the development of coronary heart disease. *Circulation* **27**, 237–251.

41. Mann, J. I., Doll, R., Thorogood, M., Vessey, M. P., and Waters, W. E. (1976). Risk factors for myocardial infarction in young women. *Br. J. Prev. Soc. Med.* **30**, 94–100.

42. Mann, J. I., Vessey, M. P., Thorogood, M., and Doll, R. (1975). Myocardial infarction in young women with special reference to oral contraceptive practice. *Br. Med. J.* **2**, 241–245.

43. Oliver, M. F. (1974). Ischaemic heart disease in young women. *Br. Med. J.* **4**, 253–259.

44. Colditz, G. A., Willett, W. C., Stampfer, M. J., Rosner, B., Speizer, F. E., and Hennekens, C. H. (1987). A prospective study of age at menarche, parity, age at first birth, and coronary heart disease in women. *Am. J. Epidemiol.* **126**, 861–870.

45. Kannel, W. B., Hortland, M. C., McNamara, P. M., and Gordon, T. (1976). Menopause and risk of cardiovascular disease: The Framingham Study. *Ann. Intern. Med.* **85**, 447–452.

46. Haynes, S. G., and Feinleib, M. (1980). Women, work and coronary heart disease: Prospective finding from the Framingham Heart Study. *Am. J. Public Health* **70**, 133–141.

47. Stampfer, M. J., Willett, W. C., Colditz, G. A., Speizer, F. E., and Hennekens, C. H. (1988). A prospective study of past use of oral contraceptive agents and risk of cardiovascular disease. *N. Engl. J. Med.* **319**, 1313–1317.

48. Lidegaard, O. (1993). Oral contraceptives and risk of a cerebral thromboembolic attack: Result of a case-control study. *Br. Med. J.* **306**, 956–963.

49. Lidegaard, O. (1995). Oral contraceptives, pregnancy and the risk of cerebral thromboembolism: The influence of diabetes, hypertension, migraine and previous thrombotic disease. *Br. J. Obstet. Gynaecol.* **102**, 153–159.

50. WHO Collaborative Study of Cardiovascular Disease and Steroid Hormone Contraception (1996). Ischaemic stroke and combined oral contraceptives: Results of an international, multicentre, case-control study. *Lancet* **348**, 498–505.

51. Petitti, D. B., Sidney, S., Bernstein, A., Wolf, S., Quesenberry, C., and Ziel, H. K. (1996). Stroke in users of low-dose oral contraceptives. *N. Engl. J. Med.* **335**, 8–15.

52. Becker, W. J. (1997). Migraine and oral contraceptives. *Can. J. Neurol. Sci.* **24**, 16–24.

53. Baird, D. T., and Glasier, A. F. (1993). Hormonal contraception. *N. Engl. J. Med.* **328**, 1543–1549.

54. American College of Obstetrics and Gynecology, Technical Bulletin (1995). Hormonal contraception. *Int. J. Gynecol. Obstet.* **48**, 115–126.

55. Kenndey, D. L., and Forbes, M. B. (1985). Noncontraceptive estrogens and progestins: Use patterns over time. *Obstet. Gynecol.* **65**, 441–446.

56. Nabulsi, A. A., Folsom, A. R., White, A., Patsch, W., Heiss, G., Wu, K. K., and Szklo, M. (1993). Association of hormone-replacement therapy with various cardiovascular risk factors in postmenopausal women. *N. Engl. J. Med.* **328**, 1069–1076.

57. Sitruk-Ware, R. (1995). Cardiovascular risk at the menopause-role of sexual steroids. *Horm. Res.* **43**, 58–63.

58. Stampfer, M. J., Coldwitz, G. A., Willett, W. C., Manson, J. E., Rosner, B., Speizer, F. E., and Hennekens, C. H. (1991). Postmenopausal estrogen therapy and cardiovascular disease. Ten-year follow-up from the nurses' health study. *N. Engl. J. Med.* **325**, 756–762.

59. Kernan, W. N., Brass, L. M., Viscoli, C. M., Sarrel, P. M., Makuch, R., and Horwitz, R. I. (1998). Estrogen after ischemic stroke: Clinical basis and design of the womens' estrogen for stroke trial. *J. Stroke Cerebrovasc. Dis.* **7**, 85–95.

60. Caine, Y. G., Bauer, K. A., Barzegar, S., ten Cate, H., Sacks, F. M., Walsh, B. W., Schiff, I., and Rosenberg, R. D. (1992). Coagulation activation following estrogen administration to postmenopausal women. *Thromb. Haemostasis.* **4**, 392–395.

61. Walsh, B. W., Schiff, I., Rosner, B., Greenberg, L., Ravnikar, V., and Sacks, F. M. (1991). Effects of postmenopausal estrogen replacement on the concentrations and metabolism of plasma lipoproteins. *N. Engl. J. Med.* **325**, 1196–1204.

62. Barrett-Connor, E. (1990). Putative complications of estrogen replacement therapy: Hypertension, diabetes, thrombophlebitis, and gallstones. *In* "The Menopause: Biological and Clinical Consequence of Ovarian Failure; Evolution and Management" (S. G. Korenman, ed.), pp. 199–209. Serono Symp., Norwell, MA.

63. Mashchak, C. A., and Lobo, R. A. (1985). Estrogen replacement therapy and hypertension. *J. Reproductive Med* **30**, 805–810.

64. Notelovitz, M. (1989). Estrogen replacement therapy: Indications, contraindications, and agent selection. *Am. J. Obstet. Gynecol.* **161**, 1832–1841.

65. Abbott, R. D. (1990). Epidemiology of some peripheral arterial findings in diabetic men and women: Experiences from the Framingham Study. *Am. J. Med.* **88**, 376–381.

66. Manson, J. E., Colditz, G. A., Stampfer, M. J., Willett, W. C., Krolewski, A. S., Rosner, B., Arky, R. A., Speizer, F. E., and Hennekens, C. H. (1991). A prospective study of maturity-onset diabetes mellitus and risk of coronary heart disease and stroke in women. *Arch. Intern. Med.* **151**(6), 1141–1147.

67. Kuusisto, J., Mykkanen, L., Pyorala, K., and Laakso, M. (1994). Non-insulin-dependent diabetes and its metabolic control are important predictors of stroke in elderly subjects. *Stroke* **25**(6), 1157–1164.

68. Jørgensen, H. S., Nakayama, H., Raaschou, H. O., and Olsen, T. S. (1994). Stroke in patients with diabetes. The Copenhagen Stroke Study. *Stroke* **25**(10), 1977–1984.

69. Brand, F. N., Abbott, R. D., and Kannel, W. B. (1989). Diabetes, intermittent claudication, and the risk of cardiovascular events: The Framingham Study. *Diabetes* **38**, 504–509.

70. Flora, G. C., Baker, A. B., Loewenson, R. B., and Klassen, A. C. (1968). A comparative study of cerebral atherosclerosis in males and females. *Circulation* **38**, 859–869.

71. Kawachi, I., Colditz, G. A., Stampfer, M. J., Willett, W. C., Manson, J. E., Rosner, B., Speizer, F. E., and Hennekens, C. H. (1993). Smoking cessation and decreased risk of stroke in women. *JAMA, J. Am. Med. Assoc.* **269**, 232–236.

72. Colditz, G. A., Bonita, R., Stampfer, M. J., Willett, W. C., Rosner, B., Speizer, F. E., and Hennekens, C. H. (1988). Cigarette smoking and risk of stroke in middle-aged women. *N. Engl. J. Med.* **318**, 937–941.

73. Lacroix, A. Z., Lang, J., Scherr, P., Wallace, R. B., Huntley, J. C., Berkman, L., Curb, J. D., Evans, D., and Hennekens, C. H. (1991). Smoking and mortality among older men and women in three communities. *N. Engl. J. Med.* **324**, 1619–1625.

74. Fielding, J. E. (1987). Smoking and women: Tragedy of the majority. *N. Engl. J. Med.* **317**, 13343–1345.

75. Bergman, A. M. (1989). Facts on women smoking. *J. Am. Med. Women's Assoc.* **44**, 55–59.

76. Rangemark, C., Ciabattoni, G., and Wennmalm, A. (1993). Excretion of thromboxane metabolites in healthy women after cessation of smoking. *Arterioscler. Thromb.* **12**, 777–782.

77. Dempsey, R. J., and Moore, R. W. (1992). Amount of smoking independently predicts carotid artery atherosclerosis severity. *Stroke* **23**, 693–696.

78. Hall, P. M. (1990). Hypertension in women. *Cardiology* **77**(Suppl. 2), 25–30.

79. Joint National Committee (1997). The sixth report of the Joint National Committee on prevention, detection, evaluation and treatment of high blood pressure. *Arch. Intern. Med.* **157**, 2413–2446.

80. MacMahon, S., and Rodgers, A. (1994). Blood pressure, antihypertensive treatment and stroke risk. *J. Hypertens.* **12**, S5–S14.

81. Alter, M., Friday, G., Lai, S., O'Connell, J., and Sobel, E. (1994). Hypertension and risk of stroke recurrence. *Stroke* **25**, 1605–1610.

82. Lai, S., Alter, M., Friday, G., and Sobel, E. (1994). A multifactorial analysis of risk factors for recurrence of ischemic stroke. *Stroke* **25**, 958–962.

83. Anastos, K., Charney, P., Charon, R. A., Cohen, E., Jones, C. Y., Marte, C., Swiderski, D. M., Wheat, M. E., and Williams, S.

(1991). Hypertension in women: What is really known? The Women's Caucus, Working Group on Women's Health of the Society of General Internal Medicine. *Ann. Intern. Med.* **115,** 287.

84. Kaplan, N. M. (1995). The treatment of hypertension in women. *Arch. Intern. Med.* **155,** 563–567.

85. Cook, N. R., Cohen, J., Hebert, P. R., Taylor, J. O., and Hennekens, C. H. (1995). Implications of small reductions in diastolic blood pressure for primary prevention. *Arch. Intern. Med.* **155,** 701–709.

86. Fiebach, N. H., Hebert, P. R., Stampher, M. J., Colditz, G. A., Willett, W. C., Rosner, B., Speizer, F. E., and Hennekens, C. H. (1989). A prospective study of high blood pressure and cardiovascular disease in women. *Am. J. Epidemiol.* **130,** 646–654.

87. Flegel, K. M., and Hanley, J. (1989). Risk factors for stroke and other embolic events in patients with nonrheumatic atrial fibrillation. *Stroke* **20,** 1000–1004.

88. Noel, P., Grégoire, F., Capon, A., and Lehert, P. (1991). Atrial fibrillation as a risk factor for deep venous thrombosis and pulmonary emboli in stroke patients. *Stroke* **22,** 760–762.

89. Candelise, L., Pinardi, G., and Morabito, A. (1991). Mortality in acute stroke with atrial fibrillation. *Stroke* **22,** 169–174.

90. Miller, V. T., Rothrock, J. F., Pearce, L. A., Feinberg, W. M., Hart, R. G., and Anderson, D. C. (1993). Ischemic stoke in patients with atrial fibrillation: Effect of aspirin according to stroke mechanism. *Neurology* **43,** 32–36.

91. Iso, H., Jacobs, D. R., Wentworth, D., Neaton, J., Cohen, J. D., for the MRFIT Research Group (1989). Serum cholesterol levels and 6-year mortality from stroke in 350,977 men screened for the Multiple Risk Factor Intervention Trial. *N. Engl. J. Med.* **320,** 904–910.

92. Abbott, R. D., Sharp, D. S., Burchfiel, C. M., Carb, J. D., Rodriguez, B. L., Hakim, A. A., and Yano, K. (1997). Cross-sectional and longitudinal changes in total and high-density-lipoprotein cholesterol levels over a 20-year period in elderly men: The Honolulu Heart Program. *Am. Epidemiol.* **7,** 417–424.

93. Hachinsk, V., Graffagnino, C., Beaudry, M., Bernier, G., Buck, C., Donner, A., Spence, J. D., Gordon, D., and Wolfe, B. M. J. (1996). Lipids and stroke: A paradox resolved. *Arch. Neurol.* (*Chicago*) **53,** 303–308.

94. Tanne, D., Yaari, S., and Goldbourt, U. (1997). High-density lipoprotein cholesterol and risk of ischemic stroke mortality. A 21-year follow-up of 8586 men from the Israeli Ischemic Heart Disease Study. *Stroke* **28,** 83–87.

95. Sorva, R., Kuusi, T., Dunkel, L., and Taskinen, M. R. (1987). Effects of endogenous sex steroids on serum lipoproteins and most heparin plasma lipolytic enzymes. *J. Clin. Endocrinol. Metab.* **66,** 408–413.

96. The Lipid Research Clinics Program Epidemiology Committee (1979). Plasma lipid distribution in selected North American populations: The Lipid Research Clinics Program Prevalence Study. *Circulation* **60,** 427–429.

97. Clifton, P. M., and Nestel, P. J. (1992). Influence of gender, body mass index, and age on response of plasma lipids to dietary fat plus cholesterol. *Arterioscler. Thromb.* **12,** 955–992.

98. Wing, R. R., and Jeffery, R. W. (1995). Effect of modest weight loss on changes in cardiovascular risk factors: Are there differences between men and women or between weight loss and maintenance? *Int. J. Obes. Relat. Metab. Disord.* **19,** 67–73.

99. Rodgers, H., Aitken, P. D., French, J. M., Curless, R. H., Bates, D., and James, O. F. W. (1993). Alcohol and stroke: A case study of drinking habits past and present. *Stroke* **24,** 1473–1477.

100. Fuchs, C. S., Stampfer, M. J., Colditz, G. A., Giovannucci, E. L., Manson, J. E., Kawachi, I., Hunter, D. J., Hankinson, S. E., Hennekens, C. H., Rosner, B., Speizer, F. E., and Willett, W. C. (1995).

Alcohol consumption and mortality among women. *N. Engl. J. Med.* **332,** 1245–1250.

101. Gill, J. S., Zezulka, A. V., Shipley, M. J., Gill, S. K., and Beevers, D. G. (1986). Stroke and alcohol consumption. *N. Engl. J. Med.* **315,** 1041–1046.

102. Palomaki, H., and Kaste, M. (1993). Regular light-to-moderate intake of alcohol and the risk of ischemic stroke. Is there a beneficial effect? *Stroke* **24,** 1828–1832.

103. Stampfer, M. J., Colditz, G. A., Willett, W. C., Speizer, F. E., and Hennekens, C. H. (1988). A prospective study of moderate alcohol consumption and the risk of coronary disease and stroke in women. *N. Engl. J. Med.* **319,** 267–273.

104. Boushey, C. J., Beresford, S. A. A., Omenn, G. S., and Motulsky, A. G. (1995). A quantitative assessment of plasma homocysteine as a risk factor for vascular disease. *JAMA, J. Am. Med. Assoc.* **274,** 1049–1057.

105. Lentz, S. R., and Sadler, J. E. (1991). Inhibition of thrombomodulin surface expression and protein C activation by the thrombogenic agent homocysteine. *J. Clin. Invest.* **88,** 1906–1914.

106. Rolland, P. H., Friggi, A., Bartlatier, A., Piquet, P., Latrille, V., Faye, M. M., Guillou, J., Charpiot, P., Bodard, H., Gihiringhelli, O., Calaf, R., Luccioni, R., and Garcon, D. (1995). Hyperhomocysteinemia-induced vascular damage in the minipig: Captopril-hydrochlorothiazide combination prevents elastic alterations. *Circulation* **91,** 1161–1174.

107. Alfthan, G., Pekkanen, J., Jauhiainen, M., Pitkaniemi, J., Karvonen, M., Tuomilehto, J., Salonen, J. T., and Ehnholm, C. (1994). Relation of serum homocysteine and lipoprotein(a) concentrations to atherosclerotic disease in a prospective Finnish population based study. *Atherosclerosis* **106,** 9–19.

108. Verhoef, P., Hennekens, C. H., Malinow, M. R., Kok, F. J., Willett, W. C., and Stampfer, M. J. (1994). A prospective study of plasma homocyst(e)ine and risk of ischemic stroke. *Stroke* **25,** 1924–1930.

109. Perry, I. F., Fefsum, H., Morris, R. W., Ebrahim, S. B., Ueland, P. M., and Shaper, A. G. (1995). Prospective study of serum total homocysteine concentration risk of stroke in middle-aged British men. *Lancet* **346,** 1395–1398.

110. Nygard, O., Vollset, S. E., Refsum, H., Stensvold, I., Tverdal, A., Nordrehaug, J. E., Ueland, P. M., and Kvale, G. (1995). Total plasma homocysteine and cardiovascular risk profile: The Hordaland Homocysteine Study. *JAMA, J. Am. Med. Assoc.* **274,** 1526–1533.

111. Selhub, J., Jacques, P. F., Bostom, A. G., D'Agostino, R. B., Wilson, P. W. F., Bélanger, A. J., O'Leary, D. H., Wolf, P. A., Schaefer, E. J., and Rosenberg, I. H. (1995). Association between plasma homocysteine concentrations and extracranial carotid-artery stenosis. *N. Engl. J. Med.* **332,** 286–291.

112. Van der Mooren, M. J., Wouters, M. G. A. J., Blom, H. J., Schellekens, L. A., Eskes, T. K. A. B., and Rolland, R. (1994). Hormone replacement therapy may reduce high serum homocysteine in postmenopausal women. *Eur. J. Clin. Invest.* **24,** 733–736.

113. Anker, G., Lonning, P. E., Ueland, P. M., Refsum, H., and Lien, E. A. (1995). Plasma levels of the atherogenic amino acid homocysteine in post-menopausal women with breast cancer treated with tamoxifen. *Int. J. Cancer* **60,** 1–4.

114. Wu, L. L., Wu, J., Hunt, S. C., James, B. C., Vincent, G. M., Williams, R. R., and Hopkins, P. N. (1994). Plasma homocysteine as a risk factor for early familial coronary artery disease. *Clin. Chem. (Winston-Salem, N.C.)* **40,** 552–561.

115. Brattstrom, L., Lindgren, A., Israelsson, B., Andersson, A., and Hultberg, B. (1994). Homocysteine and cysteine: Determinants of plasma levels in middle aged and elderly subjects. *J. Intern. Med.* **236,** 631–641.

116. Selhub, J., Jacques, P. F., Wilson, P. W. F., Rush, D., and Rosenberg, I. H. (1993). Vitamin status and intake as primary determi-

nants of homocystinemia in an elderly population. *JAMA, J. Am. Med. Assoc.* **270**, 2693–2698.

117. Wolf, P. A., and Sila, C. A. (1987). Cerebral ischemia with mitral valve prolapse. *Am. Heart J.* **113**, 1308–1315.

118. Darsee, J. R., Mikolich, J. R., Nicloff, N. B., and Lesser, L. E. (1979). Prevalence of mitral valve prolapse in presumably healthy young men. *Circulation* **59**, 619–622.

119. Hickey, A. J., Wolfers, J., and Wilcken, D. E. L. (1976). Mitral valve prolapse. *Med. J. Aust.* **1**, 31–33.

120. Devereux, R. B., Perloff, J. K., Reichek, N., and Josephson, M. E. (1976). Mitral valve prolapse. *Circulation* **54**, 3–14.

121. Heck, A. F. (1989). Neurologic aspects of mitral valve prolapse. *Angiology* **40**, 743–751.

122. Jones, H. R., Jr., Naggar, C. Z., Seljan, M. P., and Downing, L. L. (1982). Mitral valve prolapse and cerebral ischemic events. A comparison between a neurology population with stroke and a cardiology population with mitral valve prolapse observed for five years. *Stroke* **13**, 451–453.

123. Marks, A. R., Choong, C. Y., Chir, M. B. B., Sanfilippo, A. J., Ferre, M., and Weyman, A. E. (1989). Identification of high-risk and low-risk subgroups of patients with mitral-valve prolapse. *N. Engl. J. Med.* **320**, 1031–1036.

124. Orencia, A. J., Petty, G. W., Khandheria, B. K., Annegers, J. F., Ballard, D. J., Sicks, J. D., O'Fallon, W. M., and Whisnant, J. P. (1995). Risk of stroke with mitral valve prolapse in population-based cohort study. *Stroke* **26**(1), 7–13.

125. Tzourio, C., Tehindrazanarivelo, A., Inglesias, S., Alperovitch, A., Chedru, F., d'Anglejan-Chatillon, J., and Bousser, M. G. (1995). Case-control study of migraine and risk of ischaemic stroke in young women. *Br. Med. J.* **310**, 830–833.

126. Carolei, A., Marini, C., and DeMatteis, G. (1996). History of migraine and risk of cerebral ischaemia in young adults. The Italian National Research Council Study Group on Stroke in the Young. *Lancet* **347**, 1503–1506.

127. Buring, J. F., Hebert, P., Romero, J., Kittross, A., Cook, N., Manson, J., Peto, R., and Hennekens, C. (1995). Migraine and subsequent risk of stroke in the Physicians' Health Study. *Arch. Neurol. (Chicago)* **52**, 129–134.

128. Merikangas, K. R., Fenton, B. T., Cheng, S. H., Stolar, M. J., and Risch, N. (1997). Association between migraine and stroke in a large-scale epidemiological study of the United States. *Arch. Neurol. (Chicago)* **54**, 362–368.

129. Kittner, S. J., McCarter, R. J., Sherwin, R. W., Sloan, M. A., Stern, B. J., Johnson, C. J., Buchholz, D., Seipp, M. J., and Price, T. R. (1993). Black-white differences in stroke risk among young adults. *Stroke* **24**(12 Suppl.), I13–I15.

130. Petitti, D. B., Sidney, S., Quesenberry, C. P., Jr., and Bernstein, A. (1997). Incidence of stroke and myocardial infarction in women of reproductive age. *Stroke* **28**, 280–283.

131. Antiphospholipid Antibodies in Stroke Study group (1993): Anticardiolipin antibodies are an increase risk for first ischemic stroke. *Neurology* **43**, 2069–2072.

132. Brey, R. L., Hart, R. G., Sherman, D. G., and Tegeler, C. H. (1990). Antiphospholipid antibodies and cerebral ischemia in young people. *Neurology* **40**, 1190–1196.

133. Peberdy, M. A., and Ornato, J. P. (1992). Coronary artery disease in women. *Heart Dis. Stroke* **1**, 315–319.

134. Hart, R. G., and Kanter, M. C. (1990). Hematologic disorders and ischemic stroke. *Stroke* **21**, 1111–1121.

135. Brey, R. L. (1992). Antiphospholipid antibodies and ischemic stroke. *Heart Dis. Stroke* **1**, 379–382.

136. Love, P. E., and Santoro, S. A. (1990). Antiphospholipid antibodies and the lupus anticoagulant in systemic lupus erythematosus (SLE) and in non-SLE disorders. *Ann. Intern. Med.* **112**, 682–698.

137. Levine, S. R., Brey, R. L., Sawaya, K. L., Salowich-Palm, L., Kokkinos, J., Kostrzema, B., Perry, M., Havstad, S., and Carey, J. (1995). Recurrent stroke and thrombo-occlusive events in the antiphospholipid syndrome. *Ann. Neurol.* **38**, 119–124.

138. Dennis, K. E., and Goldberg, A. P. (1993). Differential effects of body fatness and body fat distribution on risk factors for cardiovascular disease in women. Impact of weight loss. *Arterioscler. Thromb.* **13**, 1487–1494.

139. Leenen, R., van der Kooy, K., Droop, A., Seidell, J. C., Deurenberg, P., Weststrate, J. A., and Hautvast, J. G. A. J. (1993). Visceral fat loss measured by magnetic resonance imaging in relation to changes in serum lipid levels of obese men and women. *Arterioscler. Thromb.* **13**, 487–494.

140. Colantonio, A., Kasl, S. V., and Ostfeld, A. M. (1992). Level of function predicts first stroke in the elderly. *Stroke* **23**, 1355–1357.

141. ISIS-2 Collaborative Group (1988). Randomized trial of intravenous streptokinase, oral aspirin, boh or neither among 17,187 cases of myocardial infarction: ISIS-2. *Lancet* **2**, 349–360.

142. Harrison, M. J. G., and Weisblatt, E. (1983). A sex difference in the effect of aspirin on "spontaneous' platelet aggregation in whole blood. *Thromb. Haemostasis.* **50**, 771–774.

143. Escolar, G., Bastida, E., Garrido, M., Rodriguez-Gomez, J., Castillo, R., and Ordinas, A. (1986). Sex-related differences in the effects of aspirin on the interaction of platelets with subendothelium. *Thromb. Res.* **44**, 837–884.

144. Hass, W. K., Easton, J. D., Adams, H. P., Jr., Pryse-Phillips, W., Molony, B. A., Anderson, S., Kamm, B., for the Ticlopidine Aspirin Stroke study group (1989). A randomized trial comparing ticlopidine hydrochloride with aspirin for the prevention of stroke in high-risk patients. *N. Engl. J. Med.* **321**, 501–507.

145. Gent, M., Blakely, J. A., Easton, J. D., Ellis, D. J., Hachinski, V. C., Harbison, J. W., Panak, E., Roberts, R. S., Sicurella, J., Turpie, A. G. G., and the CATS Group (1989). The Canadian American Ticlopidine Study (CATS) in thromboembolic stroke. *Lancet* **1**, 1215–1220.

146. Boston Area Anticoagulation Trial for Atrial Fibrillation Investigators (1991). The effect of low dose warfarin on the risk of stroke in patients with non-rheumatic atrial fibrillation. *N. Engl. J. Med.* **323**, 1505–1511.

147. Peterson, P., Godtfredson, J., Boysen, G., Andersen, E. E., and Andersen, B. (1989). Placebo-controlled, randomized trial of warfarin and aspirin for prevention of thromboembolic complications in chronic atrial fibrillation: The Copenhagen AFASAK Study. *Lancet* **1**, 175–179.

148. Stroke Prevention in Atrial Fibrillation Investigators (1991). Stroke prevention in atrial fibrillation study: Final results. *Circulation* **89**, 527–539.

149. Veterans Affairs Stroke Prevention in Nonrheumatic Atrial Fibrillation Investigators (1992). Warfarin in the prevention of stroke associated with nonrheumatic atrial fibrillation. *N. Engl. J. Med.* **327**, 1406–1412.

150. Kistler, J. P., Singer, D. E., Millenson, M. M., Bauer, K. A., Gress, D. R., Barzegar, S., Hughes, R. A., Sheehan, M. A., Maraventano, S. W., Oertel, L. B., Rosner, B., and Rosenberg, R. D. (1993). Effect of low-intensity warfarin anticoagulation on level of activity of the hemostatic system in patients with atrial fibrillation. *Stroke* **24**, 1360–1365.

151. Gurwitz, J. H., Avorn, J., Ross-Degnan, D., Choodnovskiy, I., and Ansell, J. (1992). Aging and the anticoagulant response to warfarin therapy. *Ann. Intern. Med.* **116**(11), 901–904.

152. Fihn, S. D., McDonell, M., Martin, D., Henikoff, J., Vermes, D., Kent, D., and White, R. H. (1993). Risk factors for complications of chronic anticoagulation. A multicenter study. Warfarin Optimized Outpatient Follow-up Study Group. *Ann. Intern. Med.* **119**(9), 957.

153. CLASP Collaborative Group, CLASP Coordinating Centre, Radcliffe Infirmary, Oxford (1995). Low dose aspirin study in pregnancy. *Br. J. Obstet. Gynaecol.* **102,** 861–868.

154. Oakley, C. M. (1995). Clinical perspective: anticoagulation and pregnancy. *Eur. Heart J.* **16,** 1317–1319.

155. Porreco, R. P., McDuffie, R. S., Jr., and Peck, S. D. (1993). Fixed mini-dose warfarin for prophylaxis of thromboembolic disease in pregnancy: A safe alternative for the fetus. *Obstet. Gynecol.* **81,** 806–807.

156. Executive Committee for the Asymptomatic Carotid Atherosclerosis Study (1995). Endarterectomy for asymptomatic carotid artery stenosis. *JAMA, J. Am. Med. Assoc.* **273,** 1421–1428.

157. North American Symptomatic Carotid Endarterectomy Trial Collaborators (1991). Beneficial effect of carotid endarterectomy in symptomatic patients with high-grade carotid stenosis. *N. Engl. J. Med.* **325,** 445–453.

63

Venous Thromboembolism

LYNN ROSENBERG
Slone Epidemiology Unit
Boston University School of Medicine
Brookline, Massachusetts

I. Introduction

Thrombi that have formed in the deep venous system often resolve spontaneously, but deep vein thrombosis (DVT) is nonetheless a serious medical concern because 5–20% of patients with a deep calf vein thrombus develop pulmonary embolism (PE), for which the fatality rate is 10% [1,2]. More than 95% of PEs are thought to result from DVT, and it has been estimated that PE may be responsible for more than 50,000 deaths per year in the U.S. [1,2]. Superficial vein thrombosis—of the greater or lesser saphenous veins and their tributaries—does not have important clinical consequences. Only the deep venous forms of venous thromboembolism (VTE) are considered here.

II. Symptoms, Diagnosis, Natural History, Prophylaxis, and Treatment

A. Deep Vein Thrombosis (DVT)

The clinical diagnosis of DVT is unreliable. Symptoms of thrombi in the deep veins of the legs are diffuse—pain, heat, swelling, and red blotches on the skin. DVT often occurs in the absence of symptoms. Conversely, when symptoms do occur, DVT frequently is not the cause [1,2]. The condition occurs most frequently in the legs, but occurrences in the arms are increasing because of the increased use in the hospital of subclavian (upper extremity) catheters that can result in local clot formation. Symptoms in the arms are similar to those in the legs.

The most reliable diagnostic test for DVT, ascending contrast venography, is invasive and its use is limited. Noninvasive tests, which include impedance plethysmography and doppler ultrasound, have poor sensitivity for deep calf thrombi. Imprecision in diagnostic criteria may account for varying estimates of incidence rates across studies.

Deep venous thrombi often develop near a venous valve. Growth may continue from continued platelet and fibrin accretion. The body's fibrinolysis system may be sufficient to dissolve all or most of the thrombus. The remainder undergoes organization, which leaves behind a fibrotic zone. This process may render valves incompetent and result in luminal narrowing.

As prophylaxis against the development of deep venous thrombi, patients who have DVT, or who are at high risk, may receive external pneumatic compression of the veins or be treated with parenteral or oral anticoagulants. These measures are often considered for patients with prolonged bed rest or who are undergoing thoracic, abdominal, or orthopedic surgery, such as hip replacement. Parenteral and oral anticoagulants also are used to treat DVT.

B. Pulmonary Embolism (PE)

Pleuritic chest pain and shortness of breath are among the common symptoms of PE [1,2]. As is the case for DVT, PE symptoms are diffuse and nondefinitive and may be caused by a variety of conditions. If PE is suspected, the key issue is to differentiate between it and other potential serious causes of chest pain, such as coronary artery disease.

Diagnostic tools for the documentation of PE include lung scintigraphy, pulmonary arteriography, echocardiography, computed tomography, and magnetic resonance imaging. An invasive test, pulmonary arteriography, is definitive.

PE usually arises from thrombi in the legs. It can cause mechanical obstruction of blood flow to the lungs, vasoconstriction, pulmonary hypertension, and pulmonary infarction. Occasionally, right-sided heart failure occurs, and the acute fall in cardiac output can lead to sudden death.

The risk of PE is greatest during the first few days after the formation of a thrombus in the deep venous system. Efforts to reduce the risk of PE focus on reducing the likelihood of DVT. Thus, postoperative patients may be treated with anticoagulants, mechanical means to augment venous return in the legs, leg exercises, and efforts to promote early ambulation.

III. Risk Factors Shared by Women and Men

Over a century ago, Virchow proposed that venous thrombosis tends to occur when any of a triad of conditions is present: stasis, abnormalities in the venous wall, and alterations in blood coagulability in the venous system [1–3]. A large quantity of evidence has since supported this proposition.

Conditions or procedures that lead to stasis, such as surgery requiring general anesthesia, prolonged bed rest, major surgery to the lower extremities (*e.g.,* hip replacement), and fractures and injuries to the lower extremities are associated with increased risk of venous thrombosis [2,4,5]. Prolonged immobilization of the legs, such as during long plane flights, is a risk factor even in healthy persons. Obesity is also associated with increased risk of VTE, possibly because of increased stasis.

Abnormalities in the endothelium of the vein increase the risk of VTE [1,2]. These abnormalities arise from a variety of causes, including cancer, long-standing cardiovascular disease, or indwelling catheters used pre- or postoperatively.

Coagulation alterations associated with hypercoagulability in the venous system include deficiencies in antithrombin III, protein C, and components of the fibrinolytic system [1,2]. A relatively common abnormality, factor V Leiden mutation, present in about 5% of the populations studied to date, is associated with a sevenfold increase in the risk of VTE [6]. Because specific coagulation abnormalities are uncommon, routine testing is not conducted and these abnormalities are usually discovered after the event. For this reason, the prediction of the risk of VTE is almost always based on clinical characteristics. Thus, family history of VTE has often been used in epidemiologic studies as a surrogate, albeit an imperfect one, for underlying coagulation defects.

Venous and arterial thrombosis share two major risk factors, older age and obesity, but have few others in common. For example, while cigarette smoking is a powerful risk factor for arterial thrombosis, it has appeared to be unrelated to VTE. Similarly, hypertension and diabetes are strongly related to increased risk of arterial thrombosis, but they are weakly related, if at all, to venous thrombosis [7].

IV. Pregnancy

During pregnancy, changes occur in coagulation and fibrinolytic factors that favor increased venous thrombosis. In addition, the iliac veins may be compressed by the enlarged uterus, giving rise to stasis [3]. The increase in the risk of thrombosis persists into the early postpartum period.

Farmer estimated the incidence of VTE to be 60 per 100,000 pregnancies, based on data from British general medical practices [8]. In contrast, the incidence was 10 per 100,000 women per year among women who were not pregnant and who were not using oral contraceptives. Given the problems in diagnosing DVT and PE, these estimates are imprecise. Nonetheless, they clearly indicate that pregnancy is a strong risk factor for VTE. Heparin therapy may be used in pregnant women with DVT to prevent PE because it is thought to not cross the placenta. Coumadin is not used.

V. Oral Contraceptive Use

A. Early Epidemiologic Studies

Soon after the first generation of combined oral contraceptives (formulations which contained an estrogen together with a progestagen) was marketed in the early 1960s, case reports linked their use to increased risks of VTE, myocardial infarction, and stroke. Case-control and follow-up studies conducted in the 1960s and 1970s assessed the influence of the first and second generations of oral contraceptives [9,10]. First-generation formulations contained 50 μg or more of estrogen combined with a first-generation progestagen, such as norethisterone; second-generation pills contained less than 50 μg of estrogen combined with a second-generation progestagen, such as levonorgestrel. The studies established that the incidence of VTE was increased in current users, and that past use was unrelated to risk. The relative risk estimates for VTE ranged from about 2 to 11 for current use relative to nonuse, and higher estrogen doses were associated with higher risks [9–14]. Based largely on British data, the incidence of VTE per 100,000 women at risk per year was estimated to be 110 for current oral contraceptive users and 30 for nonusers, with a case fatality rate of 1–2% [9,10,15–17]. It should be noted that diagnostic criteria for VTE were less precise than they are now.

B. Recent Epidemiologic Studies

1. Background

Third-generation formulations containing desogestrel, gestodene, or norgestimate were commonly used in the UK and in Europe before 1995. Formulations containing desogestrel are used by about 15% of oral contraceptive users in the U.S. [18]. Second-generation formulations account for most of the remaining use. In October 1995, the UK Committee on Safety of Medicines [19], the British equivalent of the U.S. Food and Drug Administration, issued a warning that the risk of VTE among users of formulations containing desogestrel or gestodene was two times greater than that among users of second-generation oral contraceptives. The warning was based on three then-unpublished studies [20–23]. (It should be noted that the studies provided no information on formulations containing norgestimate.) Those studies, now published, and several subsequent studies that included data on both second- and third-generation oral contraceptives, are considered below.

2. Epidemiologic Findings

The three studies considered by the Committee on Safety of Medicines included two hospital-based case-control studies: The World Health Organization (WHO) Collaborative Study [20,21] and the Transnational Study [22]. The Committee also considered a study based on computerized medical records from UK general practices [23], the General Practice Research Database. Subsequently, data were published from the Leiden Thrombophilia Study [24], in which ascertainment of the cases was population-based; from studies derived from another computerized general practice database, the Mediplus database [25,26]; and from reanalyses of some of the published data [27]. All of the studies were conducted after 1987, and all considered idiopathic VTE. The definitions varied as did the diagnostic criteria, but generally women with major risk factors such as a previous VTE, cancer, or recent surgery were excluded. The data were collected either by personal interview [20–22,24] or from computerized records [23,25,26].

Overall, the studies found relative risks of VTE of about 3 to 4 for current use of second-generation oral contraceptives relative to never use. Past use was unrelated to VTE risk. The increase in risk for current use of third-generation preparations relative to never use was greater than that for second-generation use in several studies; when third-generation use was compared to second-generation use, the relative risk estimate was significantly increased by 1.5- to 2-fold. In particular, the WHO study compared use of third-generation preparations to use of preparations containing levonorgestrel, the most commonly used second-generation pill; the relative risk estimate was 2.2 (95% confidence interval (CI) = 1.1–4.2) [21]. In this comparison, the few users of norgestimate-containing preparations were included with the levonorgestrel users, the reasoning being that norgestimate is metabolized to levonorgestrel and its metabolites [28,29]. When the Transnational Study repeated the WHO comparison of third-generation formulations to formulations containing levonorgestrel or norgestimate, the relative risk estimate was 1.5 (95% CI = 1.0–2.2) [22]. Based on the data of the General Practice Research Database, Jick *et al.* estimated the relative risk to be 2 for the comparison of third-generation preparations containing desogestrel or gestodene to preparations containing levonorgestrel [23]. In the Leiden study, only third-generation preparations containing desogestrel had been used; the relative risk estimate for the comparison of those preparations to levonorgestrel use was 2.2 [24]. By contrast, one study from the Mediplus database suggested no difference in the risk for second- and third-generation products [25], and another suggested that the risk was elevated only for the most recently introduced third-generation formulations (which had lower doses of estrogen than those on the market longer) [26]. In some studies [22,23], the risk was greater for shorter duration than for longer duration use.

Further analyses of the Transnational Study data were conducted in an attempt to identify biases that may have influenced the results [27]. Individual formulations were compared separately to oral contraceptives containing levonorgestrel, whereas previously they had been compared to formulations containing levonorgestrel or norgestimate [22]. The relative risk estimate was increased for every third-generation oral contraceptive formulation, including preparations containing norgestimate; the estimates ranged from 1.57 to 1.86. Among women aged 25–44, the relative risk was greater the more recently the formulation had been introduced (p < 0.001 for trend).

In summary, the relative risk estimates for VTE in recent studies of second and third generation oral contraceptives have been similar to or smaller than those for first-generation pills [9,10]. Several studies suggested that third-generation oral contraceptives are associated with a risk of VTE 1.5 to 2 times greater than that associated with second-generation formulations. Some data suggested that the risk was greater for women who had been using oral contraceptives for shorter durations and for women who have used the most recently introduced formulations.

3. Potential Biases

Information bias is unlikely to explain the results of the various positive studies because findings were similar in studies based on prescription records and on self-report. Although uncontrolled confounding is possible, a variety of major risk factors were controlled.

Diagnostic bias cannot be ruled out. Because it is known that oral contraceptives cause VTE, oral contraceptive users with symptoms suggestive of VTE, such as swelling in the leg, may more often have been referred and worked up for VTE than nonusers [30,31]. This would have resulted in overestimation of the relative risk of VTE for users relative to nonusers. If such selective diagnosis was related to the specific oral contraceptive formulation used, comparisons between formulations also could have been biased. In this regard, German physicians surveyed after the issuance of the advisory by the UK Committee on Safety of Medicines reported that they would be more likely to investigate the symptoms of women thought to be at higher risk (*e.g.,* because of obesity or family history of thrombosis) [31]. They also reported that they were more likely to prescribe third-generation oral contraceptives to such women. These practices would result in third-generation users having a higher likelihood of referral for the investigation of symptoms of thrombosis, which could result in a positive bias in the comparison of third- to second-generation use.

The comparison of third- to second-generation oral contraceptive use also could have been biased if the users of the two generations of pills differed in their susceptibility to VTE. This difference could have occurred through the differential depletion of susceptible women from the pool of users. For example, women who had a genetic susceptibility to VTE (*e.g.,* those with factor V Leiden mutation) might have developed VTE soon after starting oral contraceptive use and they would have been removed from the pool of users. A greater number of susceptible women could have been removed in this way from the pool of users of preparations long on the market than from the pool of users of newer formulations, resulting in an upward bias in the comparison of newer to older formulations. In the Transnational Study, in an analysis of specific formulations relative to formulations with levonorgestrel (which was on the market longest), the relative risk estimate was elevated and similar in magnitude for all third-generation formulations, including those containing norgestimate [27]. Because norgestimate is converted to levonorgestrel and its metabolites [28,29], there is no compelling biologic credibility for a difference in risk between levonorgestrel and norgestimate. In addition, among women 25–44 years of age, the relative risk estimate was greater the more recently the formulation had been marketed, which is consistent with greater attrition of women susceptible to VTE among users of older formulations [27]. The higher relative risk estimates in the WHO study [21], the Transnational Study [22], and the most recent Mediplus study [26] for more recently introduced formulations containing 20 µg of estrogen than for older formulations containing 30 µg of estrogen also are compatible with greater attrition of susceptible women among users of formulations on the market longer. The finding that durations of use were shorter among third-generation users than among second-generation users is also compatible [22,23].

Differential prescribing could have resulted in differences in susceptibility among users. If particular oral contraceptives were perceived as being safer, women thought to be at greater risk of VTE could have been selectively prescribed those formulations, resulting in an upward bias in the relative risk estimates for those oral contraceptives. Indeed, the oral contraceptive manufacturers promoted third-generation formulations as being safer. A tendency to prescribe third-generation formulations to women at higher risk was reported in surveys of British [32], German [31,32], French [33], and Dutch [34] physicians.

In summary, there is evidence that biases may have inflated the relative risk estimate for the comparison of third- to second-generation oral contraceptive use.

4. Effects of Oral Contraceptives on Biologic Parameters

The epidemiologic findings of increased risks of VTE in current users of oral contraceptives, but not in past users, suggested an acute mechanism. Pharmacologic evidence indicated that oral contraceptives increased the tendency for both arterial and venous thrombosis, and it also suggested that the effects were related to the dose of estrogen [9,10]. Based on that evidence, manufacturers of oral contraceptives have progressively lowered the doses of estrogen and progestagen; current doses are about one-fifth and one-tenth of early doses, respectively [35,36].

The third-generation progestagens—desogestrel, gestodene, and norgestimate—are less androgenic than their predecessors and have little effect on lipid metabolism or glucose tolerance [35–39]. However, it appears that effects that may influence thrombosis in the venous system are of greater importance in the occurrence of VTE. Both second- and third-generation oral contraceptives have small effects on the hemostatic system [36,37]. They increase the tendency to coagulation, but they also increase fibrinolysis. The general effect in the natural anti-coagulation system is a reduction in factors that inhibit coagulation. The effects of first-generation oral contraceptives were greater than those of later generations, because the magnitude of the effects on hemostatic variables is related to the dose of estrogen. Even after the changes induced by oral contraceptives, however, the values of the hemostatic values tend to be within normal ranges, and the relevance of the changes to clinical events is unknown.

In 1989, the results from a small pharmacokinetic study [40] indicated that serum levels of estrogen from oral contraceptives that women had taken were considerably higher in gestodene users than in desogestrel users, raising the possibility that gestodene-containing formulations might carry a greater risk of estrogen-related diseases such as VTE. However, the findings were not confirmed in subsequent studies [41–46].

It has been established that people with a genetic predisposition to VTE are the group most susceptible to the effects of oral contraceptives. Those with the factor V Leiden mutation, a defect in the body's natural anticoagulation system, have a risk of VTE appreciably greater than that of persons without the defect. For example, in the Leiden Thrombophilia Study, the increase in risk was sevenfold [6]. That study also found an interaction of oral contraceptive use with the mutation: the risk of VTE was increased 35-fold among oral contraceptive users who had the mutation [47]. The latter finding, while biologically plausible, requires confirmation. Users of third-generation oral contraceptives were found to have developed a resistance to the blood's anticoagulation system that was similar to the degree of activated protein C resistance observed in persons with the factor V Leiden mutation; the effect was significantly higher than that among women using second-generation formulations [48]. The clinical import of this change has not been established.

In summary, biologic credibility for a higher risk of VTE for users of third-generation oral contraceptives remains to be established. Evidence indicating that third-generation formulations have a greater effect than second-generation formulations on a component of the natural anticoagulation system are suggestive of a higher risk of VTE for third-generation users [6]. However, the finding has not yet been confirmed, nor have the changes been linked to the occurrence of clinical VTE. There is no other evidence to suggest that third-generation oral contraceptives affect hemostasis or other biologic systems in a way that would increase VTE risk relative to second-generation formulations. There also has not been a convincing biologic explanation for the greater increase in risk associated with formulations containing 20 μg of estrogen compared to those containing 30 μg of estrogen, as has been observed in several studies.

5. Safety of Second- and Third-Generation Oral Contraceptives

The incidence of idiopathic VTE per 100,000 women-years at risk was estimated to be 3.9 for nonusers, 10.5 for users of oral contraceptives containing levonorgestrel, and 21.3 for users of formulations containing desogestrel or gestodene, based on the UK data in the WHO study [20,21]. Estimates based on the Mediplus database [25,26] have been higher (e.g., about 35 per 100,000 women years for users of oral contraceptives with levonorgestrel), but both the WHO and Mediplus estimates are an order of magnitude lower than the estimates from epidemiologic studies conducted in the 1960s and 1970s [9]. It may be that part of the difference is due to stricter diagnostic criteria in recent studies. Whether or not this is so, the incidence of VTE associated with the use of second- and third-generation oral contraceptives is lower than that estimated for earlier formulations. The incidence of VTE in pregnant women has been estimated to be about 2–3 times that among women who were using oral contraceptives, and 6 times that of women who were not pregnant and not using oral contraceptives [8]. Thus, the use of second- or third-generation oral contraceptives carries a risk of VTE no greater than, and probably less than, that of pregnancy.

In assessing the overall safety of second- and third-generation oral contraceptives, it is important to assess all potential adverse outcomes. Two reports have addressed the relation of second- and third-generation oral contraceptive use to the risk of myocardial infarction [49,50]. Both suggested a lower incidence of myocardial infarction among users of third-generation pills than among users of second-generation pills. Based on the results of those studies and the studies of VTE, and using estimates of incidence and mortality taken from U.S. hospital discharge data and national mortality statistics, Schwingl and Shelton estimated age- and formulation-specific incidence rates of VTE and myocardial infarction [51]. They estimated the annual incidence of VTE per 100,000 nonusers of oral contraceptives to be 2.2, 4.8, and 4.7 at ages 15–24, 25–34, and 35–44, respectively. The corresponding mortality rates were 0.1, 0.1, and 0.2 per 100,000 nonusers. The annual incidence of myocardial infarction per 100,000 nonusers was estimated to be 1.3, 2.4, and 14.7 at ages 15–24, 25–34, and 35–44; the mortality rates per 100,000 nonusers were 0.1, 0.4, and 2.4, respectively. They concluded that there would be little difference in the burden of these diseases between users of second- and third-generation formulations who were under the age of 35. For women aged 35–44, however, the potential decrease in the risk of myocardial infarction more than offset the potential increase in the risk of VTE in third-generation users. It should be noted that all the differences in risk, even at ages 35–44, were relatively small, usually less than 1 event per 10,000 women per year. It also should be noted that the increased risk of myocardial infarction among women who use oral contraceptives is concentrated among cigarette smokers [9,52]. It may be that, among nonsmokers, the increased risk of VTE is not offset by the decreased risk of myocardial infarction. However, Schwingl and Shelton did not estimate smoking-specific risks.

VI. Postmenopausal Use of Female Hormone Supplements

A. Background

Until recently, there has been relatively little interest in the relationship of use of postmenopausal female hormone supplements (hormone replacement therapy (HRT)) to the risk of VTE.

Most of the studies of this relationship were small and had little statistical power [53]. The lack of interest may seem surprising in view of the fact that oral contraceptives increase the risk of VTE. However, the estrogen dose in postmenopausal female hormone supplements is lower than that in oral contraceptives [53,54]. Women who use postmenopausal female hormone supplements take daily doses of 1.25 mg or less [55–58].

B. Recent Findings

1. Epidemiologic Studies of Postmenopausal Female Hormone Use and VTE

In 1996 and 1997, four observational studies were published, all of which indicated an increased risk of VTE among women who were using female hormone supplements. One was a follow-up study [55] and three were case-control studies [56–58]. As in the studies of oral contraceptive use, the focus was on idiopathic disease.

Grodstein et al. studied 27 cases of PE that occurred among postmenopausal women in the Nurses' Health Study [55]. The hazard ratio for PE among current users of postmenopausal female hormone supplements (mostly unopposed estrogen) was 2.1 (95% of CI = 1.2–3.8). Past use and the duration of current and past use were unrelated to risk. There was no trend of increasing risk with increasing estrogen dose among current users, but numbers for subanalyses were small.

Jick et al. studied 31 cases of DVT and 11 cases of PE that occurred among women aged 50–74 in a group health plan in Seattle, Washington [56]. Four controls from the plan were age-matched to each case. Information on female hormone use was obtained from computerized pharmacy records. The relative risk estimate for current use was 3.6 (95% CI = 1.6–7.8) overall, 4.0 (95% CI = 1.6–9.7) for DVT, and 2.5 (95% CI = 0.5–12.2) for PE. Past use was unrelated to risk. The relative risk increased with increasing daily dose of estrogen, but numbers were small. Point estimates were elevated both for estrogen taken alone and for estrogen taken with a progestagen.

Daly et al. studied 104 cases of VTE and 178 age-matched hospital controls from the Oxford region of the U.K. [57]. Information on medication use and medical history was obtained by interview. For current use relative to nonuse, the odds ratio was 3.5 (95% CI = 1.8–7.0). There was a statistically significant trend for the risk to decrease as the duration of use increased. The point estimate was greater for higher dose preparations but the estimates for all doses were compatible with each other. There were no significant differences between estimates for unopposed and combined therapy, or between transdermal and oral therapy. Past use was unrelated to risk.

Pérez Gutthann et al. compared 195 cases of DVT and 97 cases of PE with 10,000 randomly selected controls, based on data from the General Practice Research Database [58]. For current use of HRT, the adjusted odds ratio was 2.1 (95% CI = 1.4–3.2). Past used was unrelated to risk.

In summary, in the observational studies the risk of VTE increased from two- to fourfold among women using postmenopausal female hormone supplements, whether estrogen alone or estrogen with a progestagen. Past use had no effect on risk. There is some evidence that the risk may be greatest for women on HRT for short durations [57], compatible with the idea that

susceptible women may succumb to the disease early on. However, the evidence on the duration of use, and evidence suggesting a greater risk for higher doses [56], is scanty and needs confirmation.

In 1998, results from a four-year randomized trial of estrogen plus progestin for the secondary prevention of coronary heart disease were published [59]. Among 2763 postmenopausal women under age 80 who had been randomized to 0.625 mg of conjugated estrogen daily together with 2.5 mg of medroxyprogesterone acetate or to placebo, more women in the hormone group than in the placebo group experienced VTE. Among hormone users, there were 25 cases of DVT and 11 cases of PE, 2 of which were fatal; among placebo users, there were 8 cases of DVT and 4 of PE. The relative risk estimate for VTE was 2.89 (95% CI = 1.50–5.58). The relative risk of VTE was greater in the earlier years of the study than in the later years; the rate per 1000 user-years was 9.6 in year 1 and 4.0 in years 4 and 5. These results confirm the findings of the observational studies and also are consistent with the idea that women susceptible to VTE may be affected early after they begin use of female hormone supplements.

2. Potential Biases

The similarity of findings from the observational studies to those from a randomized clinical trial indicates that information bias and confounding do not explain the associations. Diagnostic bias is an unlikely explanation of the randomized trial results because investigators were blinded to the exposure status of the participants. However, diagnostic bias is a concern in the observational studies because women who use HRT are under greater medical surveillance than those who do not. If users were referred and worked up for symptoms of VTE more often than nonusers, this would have resulted in an upward bias in the relative risk estimates. PE has a more dramatic presentation than DVT of the legs and, therefore, diagnostic bias might be less of a problem for this condition than for DVT. That the relative risk estimates for PE were smaller than those for DVT in the observational studies is compatible with less diagnostic bias.

3. Biologic Mechanism

HRT has effects on hemostasis similar to, but smaller than, those of oral contraceptives [60–64]. The overall effect is to increase the tendency to clot formation. It is not known how any of these changes are related to the occurrence of clinical events.

4. Safety of Postmenopausal Female Hormone Supplements

The incidence of idiopathic PE per 100,000 nonusers per year was estimated to be 6 among women 50–59 years of age and 16 at ages 60 or older among American nurses [55]. The annual incidence of idiopathic VTE among women aged 50–74 who did not use supplements was estimated to be 9 per 100,000 based on data from Seattle [56]. Daly et al. estimated the annual incidence of idiopathic VTE per 100,000 nonusers of HRT among women 45–64 to be 10.9, based on UK data [57]. Thus, the incidence of idiopathic VTE among current users of HRT may be 20–40 per 100,000 users per year. It is worth noting again, however, that estimates of the incidence of VTE are imprecise because of problems in the diagnosis (both false positives and false negatives).

The incidence of VTE attributable to use of postmenopausal female hormone supplements is small in comparison with the increase in the incidence of breast cancer or the reduction in the incidence of myocardial infarction estimated to be attributable to such use [65]. However, virtually all of the evidence concerning breast cancer and myocardial infarction is from observational studies and the estimates may have been affected to varying degrees by bias. For example, the magnitude of reduction in the risk of myocardial infarction among female hormone users may have been overestimated because of the tendency of healthier women to use HRT [66]. Evidence from randomized trials of HRT for the primary prevention of coronary heart disease will not be available for at least a decade. Results from the secondary prevention trial [59] showed that HRT did not reduce the overall rate of coronary heart disease events in postmenopausal women with previous coronary disease; there was a suggestion, however, of possible protection after longer durations of use. These results do not agree with those of observational studies that have observed both primary and secondary protection. In any event, for women at high risk of VTE because of coagulation defects, obesity, serious cardiovascular disease, cancer, or other risk factors, the incidence of HRT-associated VTE will be greater than that among low risk women, and this should be weighed when use of female hormone supplements is considered.

References

1. Creager, M. A., and Dzau, V. J. (1994). Vascular diseases of the extremities. *In* "Harrison's Principles of Internal Medicine" (K. J. Isselbacher, E. Braunwald, J. D. Wilson, J. B. Martin, A. S. Fauch, and D. L. Kasper, eds.), 13th ed., pp. 1135–1143. McGraw-Hill, New York.

2. Moser, K. M. (1994). Pulmonary thromboembolism. *In* "Harrison's Principles of Internal Medicine" (K. J. Isselbacher, E. Braunwald, J. D. Wilson, J. B. Martin, A. S. Fauch, and D. L. Kasper, eds.), 13th ed., pp. 1214–1220. McGraw-Hill, New York.

3. Ferris, T. F. (1994). Medical disorders during pregnancy. *In* "Harrison's Principles of Internal Medicine" (K. J. Isselbacher, E. Braunwald, J. D. Wilson, J. B. Martin, A. S. Fauch, and D. L. Kasper, eds.), 13th ed., pp. 18–23. McGraw-Hill, New York.

4. Thromboembolic Risk Factors (THRIFT) Consensus Group (1992). Risk of and prophylaxis for venous thromboembolism in hospital patients. *Br. Med. J.* **305,** 567–574.

5. Lowe, G. D., Osborne, D. H., McArdle, B. M., Smith, A., Carter, D. C., Forbes, C. D., McClaren, D., Prentice, C. R. (1982). Prediction and selective prophylaxis of venous thrombosis in elective gastrointestinal surgery. *Lancet* **1,** 409–412.

6. Koster, T., Rosendaal, F. R., de Ronde, H., Brët, E., Vandenbroucke, J. P., and Bertina, R. M. (1993). Venous thrombosis due to poor anticoagulant response to activated protein C: Leiden Thrombophilia Study. *Lancet* **342,** 1503–1506.

7. Manson, J. E., Tosteson, H., Ridker, P. M., Satterfeld, S., Herbert, P., O'Connor, G. T., Buring, J. E., and Hennekens, C. H. (1992). The primary prevention of myocardial infarction. *N. Engl. J. Med.* **326,** 1406–1416.

8. Farmer, R. D. T., and Preston, T. D. (1995). The risk of venous thromboembolism associated with low oestrogen oral contraceptives. *J. Obstet. Gynaecol.* **15,** 195–200.

9. Stadel, B. V. (1981). Oral contraceptives and cardiovascular disease. *N. Engl. J. Med.* **305,** 612–617, 672–677.

10. Vessey, M. P. (1980). Female hormones and vascular disease: Epidemiologic overview. *Br. J. Fam. Plann.* **6**(Suppl.), 1–12.

11. Royal College of General Practitioners' Oral Contraception Study (1978). Oral contraceptives, venous thrombosis, and varicose veins. *J. R. Coll. Gen. Pract.* **28,** 393–399.

12. Stolley, P. D., Tonascia, J. A., Tockman, M. S., Sartwell, P. E., Rutledge, A. H., and Jacobs, M. P. (1975). Thrombosis and low-estrogen oral contraceptives. *Am. J. Epidemiol.* **102,** 197–208.

13. Inman, W. H. W., Vessey, M. P., Westerholm, B., and Engelund, A. (1970). Thromboembolic diseases and the steroidal content of oral contraceptives: A report to the Committee on Safety of Drugs. *Br. Med. J.* **2,** 203–209.

14. Meade, T. W., Greenberg, G., and Thompson, S. G. (1980). Progestogens and cardiovascular reactions associated with oral contraceptives and a comparison of the safety of 50- and 30-μg oestrogen preparations. *Br. Med. J.* **280,** 1157–1161.

15. Royal College of General Practitioners' Oral Contraception Study (1977). Mortality among oral-contraceptive users. *Lancet* **2,** 727–731.

16. Vessey, M. P. (1978). Steroid contraception, venous thromboembolism, and stroke: Data from countries other than the United States. *In* "Risks, Benefits, and Controversies in Fertility Control" (J. J. Sciarra, G. I. Zatuchni, and J. J. Speidel, eds.), pp. 113–121. Harper & Row, Hagerstown, MD.

17. Maguire, M. G., Tonascia, J., Sartwell, P. E., Stolley, P. D., and Tockman, M. S. (1979). Increased risk of thrombosis due to oral contraceptives: A further report. *Am. J. Epidemiol.* **110,** 188–195.

18. Food and Drug Administration (1995). "FDA Talk Paper. Oral Contraceptives and Risk of Blood Clots." FDA, Washington, DC.

19. U.K. Committee on Safety of Medicines (1995). Letter, October 18.

20. Poulter, N. R., Chang, C. L., Farley, T. M. M., Meirik, O., and Marmot, M. G. (1995). Venous thromboembolic disease and combined oral contraceptives: Results of international multicenter case-control study. World Health Organization Collaborative Study of Cardiovascular Disease and Steroid Hormone Contraception. *Lancet* **346,** 1575–1582.

21. Farley, T. M. M., Meirik, O., Chang, C. L., Marmot, M. G., and Poulter, N. R. (1995). Effect of different progestagens in low oestrogen oral contraceptives on venous thromboembolic disease. *Lancet* **346,** 1582–1588.

22. Spitzer, W. O., Lewis, M. A., Heinemann, L. A. J., Thorogood, M., and MacRae, K. D., on behalf of Transnational Research Group on Oral Contraceptives and the Health of Young Women (1996). Third generation oral contraceptives and risk of venous thrombolic disorders: An International Case-control Study. *Br. Med. J.* **312,** 83–88.

23. Jick, H., Jick, S. S., Gurewich, V., Myers, M. W., and Vasilakis, C. (1995). Risk of idiopathic cardiovascular death and nonfatal venous thromboembolism in women using oral contraceptives with differing progestagen components. *Lancet* **346,** 1589–1593.

24. Bloemenkamp, K. W. M., Rosendaal, F. R., Helmerhorst, F. M., Buller, H. R., and Vandenbroucke, J. P. (1995). Enhancement of factor V Leiden mutation of risk of deep-vein thrombosis associated with oral contraceptives containing third-generation progestagen. *Lancet* **346,** 1593–1596.

25. Farmer, R. (1996). Safety of modern oral contraceptives. *Lancet* **347,** 259.

26. Farmer, R. D. T., Lawrenson, R. A., Thompson, C. R., Kennedy, J. G., and Hambleton, I. R. (1997). Population-based study of risk of various thromboembolism associated with various oral contraceptives. *Lancet* **349,** 83–88.

27. Lewis, M. A., Heinemann, L. A. J., Mac Rae, K. D., Bruppacher, R., and Spitzer, W. O., with the Transnational Research Group on Oral Contraceptives and the Health of Young Women (1996). The increased risk of venous thromboembolism and the use of third generation progestagens: Role of bias in observational research. *Contraception* **54,** 5–13.

28. Stanczyk, F. Z., and Roy, S. (1990). Metabolism of levonorgestrel, norethindrone and structurally related contraceptive steroids. *Contraception* **42**, 67–96.

29. Shenfield, G. M., and Griffin, J. M. (1991). Clinical pharmacokinetics of contraceptive steroids: An update. *Clin. Pharmacokinet.* **20**, 15–37.

30. Realini, J. P., and Goldzieher, J. W. (1985). Oral contraceptives and cardiovascular disease: A critique of epidemiologic studies. *Am. J. Ostet. Gynecol.* **152**, 729–798.

31. Heinemann, L. A. J., Lewis, M. A., Assmann, A., Gravens, L., and Guggenmoos-Holzmann, I. (1996). Could preferential prescribing and referral behavior of physicians explain the elevated thrombotic risk found to be associated with third generation oral contraceptives? *Pharmacoepidemiol. Drug Saf.* **5**, 285–294.

32. Dunn, N., Heinemann, L. A. J., and Mann, R. D. (1996). Are third-generation oral contraceptives really more thrombotic than second-generation? Some potential biases examined and compared in German and English doctors. *Int. Conf. Pharmacoepidemiol., 12th,* Amsterdam, The Netherlands, *1996.*

33. Jamin, C., and de Mouzon, J. (1996). Selective prescribing of third-generation oral contraceptives (OCs). *Contraception* **54**, 55–56.

34. Herings, R. M. C., de Boer, A., Urquhart, J., and Leufkens, H. G. M. (1996). Non-causal explanations for the increased risk of venous thrombembolism among users of third-generation oral contraceptives. *Pharmacoepidemiol. Drug Saf.* **5**, 588.

35. Fotherby, K., and Caldwell, A. D. S. (1994). New progestogens in oral contraception. *Contraception* **49**, 1–32.

36. Speroff, L., De Cherney, A., and the Advisory Board for the New Progestins (1993). Evaluation of a new generation of oral contraceptives. *Obstet. Gynecol.* **81**, 1034–1047.

37. Robinson, G. E. (1994). Low-dose combined oral contraceptives. *Br. J. Obstet. Gynaecol.* **101**, 1036–1041.

38. Newton, J. R. (1995). Classification and comparison of oral contraceptives containing new generation progestogens. *Hum. Reprod. Update* **1**, 231–263.

39. Godsland, I. F., Crook, D., Simpson, R., Proudler, T., Felton, C., Lees, B., Anyaoku, V., Devenport, M., and Wynn, V. (1990). The effect of different formulations of oral contraceptives on lipid and carbohydrate metabolism. *N. Engl. J. Med.* **323**, 1375–1381.

40. Jung-Hoffmann, C., and Kuhl, H. (1989). Interaction with the pharmacokinetics of ethinyl estradiol and progestogens contained in oral contraceptives. *Contraception* **40**, 299–312.

41. Himpel, M., Täuber, U., Kuhnz, W., Pfeffer, M., Brill, K., Heitbecker, R., Louton, T., and Steinberg, B. (1990). Comparison of serum ethinyl estradiol, sex-hormone-binding globulin, corticoid-binding globulin and cortisol levels in women using two low-dose combined oral contraceptives. *Horm. Res.* **33**, 35–39.

42. Kuhnz, W. (1990). Pharmacokinetics of the contraceptive steroids levonorgestrel and gestodene after single and multiple oral administration to women. *Am. J. Obstet. Gynecol.* **163**, 2120–2127.

43. Kuhnz, W., Sostarek, D., Gansau, C., Louton, T., and Mahler, M. (1991). Single and multiple administration of a new triphasic oral contraceptive to women: Pharmacokinetics of ethinyl estradiol and free and total testosterone levels in serum. *Am. J. Obstet. Gynecol.* **165**, 596–602.

44. Back, D. J., Ward, S., and Orme, M. L. E. (1991). Recent pharmacokenitic studies of low-dose oral contraceptives. *Adv. Contraception* **7**(Suppl. 3), 164–179.

45. Dibbelt, L., Knuppen, R., Jütting, G., Heinmann, S., Klipping, C. O., and Parikka-Olexik, H. (1991). Group comparison of serum ethinyl estrndiol, SHBG and CBG levels in 83 women using two low-dose combination oral contraceptives for three months. *Contraception* **43**, 1–21.

46. Orme, M., Back, D. J., Ward, S., and Green, S. (1991). The pharmacokenitics of ethinylestradiol in the presence and absence of gestodene and desogestrel. *Contraception* **43**, 305–316.

47. Vandenbroucke, J. P., Koster, T., Briet, E., Reitsma, P. H., Bertina, R. M., and Rosendaal, F. R. (1994). Increased risk of venous thrombosis in oral contraceptive users who are carriers of factor V Leiden mutation. *Lancet* **344**, 1453–1457.

48. Rosing, J., Tans, G., Nicolaes, G. A., Thomassen, M. C., van Oerde, R., van der Ploeg, P. M., Heijnen, P., Hamulyak, K., and Hemker, A. C. (1997). Oral contraceptives and venous thrombosis: Different sensitivities to activated protein C in women using second- and third-generation oral contraceptives. *Br. J. Pharmacol.* **97**, 233–238.

49. Lewis, M. A., Spitzer, W. O., Heinemann, L. A. J., MacRae, K. D., Bruppacher, R., and Thorogood, M., on behalf of Transnational Research Group on Oral Contraceptives and Health of Young Women (1996). Third generation oral contraeptives and risk of myocardial infarction: an international case-control study. *Br. Med. J.* **312**, 88–90.

50. Lidegaard, O., and Edström, B. (1996). Oral contraceptives and acute myocardial infarction: A case-control study. *Eur. J. Contraception Reprod. Health Care* **1**, 74.

51. Schwingl, P. J., and Shelton, J. (1997). Modeled estimates of myocardial infarction and venous thromboembolic disease in users of second and third generation oral contraceptives. *Contraception* **55**, 125–129.

52. Slone, D., Shapiro, S., Kaufman, D. W., Rosenberg, L., Miettinen, O. S., and Stolley, P. D. (1981). Risk of myocardial infarction in relation to current and discontinued use of oral contraceptives. *N. Engl. J. Med.* **305**, 420–424.

53. Vandenbroucke, J. P., and Helmerhorst, F. M. (1996). Risk of venous thrombosis with hormone replacement therapy. *Lancet* **349**, 972.

54. Labo, R. A. (1992). Estrogen and risk of coagulopathy. *Am. J. Med.* **92**, 283–285.

55. Grodstein, F., Stampfer, M. J., Goldhaber, S. Z., Manson, J. E., Colditz, G. A., Speizer, F. E., Willett, W. C., and Hennekens, C. H. (1996). Prospective study of exogenous hormones and risk of pulmonary embolism in women. *Lancet* **348**, 983–987.

56. Jick, H., Derby, L. E., Myers, M. W., Vasilakis, C., and Newton, K. M. (1996). Risk of hospital admission for idiopathic venous thromboembolism among users of postmenopausal oestrogens. *Lancet* **348**, 381–383.

57. Daly, E., Vessey, M. P., Hawkins, M. M., Carson, J. L., Gough, P., and Warsh, S. (1996). Risk of venous thromboembolism in users of hormone replacement therapy. *Lancet* **348**, 977–980.

58. Pérez Gutthann, S., García Rodríguez, L. A., Castellsague, J., and Duque Oliart, A. (1997). Hormone replacement therapy and risk of venous thromboembolism: Population based case-control study. *Br. Med. J.* **314**, 796–800.

59. Hully, S., Grady, D., Bush, T., Furberg, C., Herrington, D., Riggs, B., and Vittinghoff, E., for the Heart and Estrogen/progestin. Replacement Study (HERS) Research Group (1998). *JAMA, J. Am. Med. Assoc.* **280**, 605–613.

60. Poller, L., Thomson, J. M., and Coope, J. (1977). Conjugated equine estrogens and blood clotting: a follow-up report. *Br. Med. J.* **1**, 935–936.

61. Stangel, J. J., Innerfield, I., Reyniak, J., and Stone, M. L. (1977). The effect of conjugated estrogens on coagulability in menopausal women. *Obstet. Gynecol.* **49**, 314–316.

62. Notelovitz, M., Kitchens, C. S., and Ware, M. D. (1984). Coagulation and fibrinolysis in estrogen-treated surgically menopausal women. *Obstet. Gynecol.* **63**, 621–625.

63. Notelovitz, M., Kitchens, C., Ware, M. D., Hirschberg, K., and Coone, L. (1983). Combination estrogen and progestin replacement

therapy does not adversely affect coagulation. *Obstet. Gynecol.* **62,** 596–600.

64. Postmenopausal Estrogen/Progestin Interventions Trial Writing Group (1995). Effects of estrogen/progestin regimens on heart disease risk factors in postmenopausal women. *JAMA, J. Am. Med. Assoc.* **273,** 199–208.

65. Grady, D., Rubin, S. M., Petitti, D. B., Fox, C. S., Black, D., Ettinger, B., Ernster, V. L., and Cummings, S. R. (1992). Hormone therapy to prevent disease and prolong life in postmenopausal women. *Ann. Intern. Med.* **117,** 1016–1037.

66. Rosenberg, L. (1993). Hormone replacement therapy: The need for reconsideration. *Am. J. Public Health* **83,** 1670–1673.

64

Diabetes in Women

PAULO A. LOTUFO,*,† MARY SABOLSI,* AND JOANN E. MANSON*

*Division of Preventive Medicine, Brigham and Women's Hospital/Harvard Medical School, Boston, Massachusetts; †University of Sao Paulo, School of Medicine, Sao Paulo, Brazil

I. Introduction

Diabetes mellitus is a major cause of morbidity and mortality in the United States and many other developed countries. The overall prevalence of diabetes is somewhat greater in women than in men, and some of the risk factors for diabetes and the health consequences of this disease are more pronounced in women. There are also issues unique to women, including pregnancy complications in women with preexisting diabetes and the development of gestational diabetes. This chapter provides an overview of the prevalence of diabetes, risk factors for its development, health consequences of the disease, and treatment and prevention strategies, with particular emphasis on the impact of diabetes in women.

II. Definition

Until 1997, the definition of diabetes was based on elevated fasting and/or postglucose load levels of plasma glucose. In 1997, the *Expert Committee on the Diagnosis and Classification of Diabetes Mellitus* issued new guidelines endorsing the measurement of fasting plasma glucose based on its greater convenience, acceptability to patients, and lower cost [1]. The committee also recommended using a lower glycemic level ($\geq$126 mg/dl) as the cutpoint for diabetes diagnosis than the threshold previously endorsed by the World Health Organization ($\geq$140 mg/dl).

The nomenclature for diabetes types has also been revised. A lack of insulin hormone production, previously termed insulin-dependent diabetes mellitus (IDDM) or juvenile-onset diabetes, is now referred to as type 1 diabetes mellitus. Type 2 diabetes, formerly termed noninsulin dependent diabetes mellitus (NIDDM) or adult-onset diabetes, occurs when circulating insulin is available but cells do not use it efficiently. The primary defect of type 2 diabetes is insulin resistance, which is strongly associated with hypertriglyceridemia, low levels of high density lipoprotein cholesterol (HDL), and hypertension, all of which play a role in the development of coronary heart disease events and, consequently, higher mortality rates. Gestational diabetes mellitus (GDM) is defined as glucose intolerance with the onset of pregnancy, although there is a possibility that unrecognized glucose intolerance may have antedated the pregnancy. Finally, a new condition called impaired glucose tolerance (or impaired fasting glucose) has been defined. This refers to individuals whose fasting plasma glucose levels are in between those used to define normoglycemia and frank diabetes ($\geq$110 to <126 mg/dl). Individuals with this condition are at high risk for developing diabetes and cardiovascular complications (Table 64.1).

Table 64.1

Definition of Diabetes Mellitus (DM) and Gestational Diabetes Mellitus (GDM)[a]

Definition

Diabetes can be diagnosed in any one of the following three ways, by any of these three tests (should be confirmed by retesting on a different day)

a fasting plasma glucose (FPG) $\geq$126 mg/dl after no caloric intake for at least 8 hours

a casual plasma glucose $\geq$200 mg/dl taken at any time of day without regard to time of the last meal with diabetes symptoms of increased urination, increased thirst, and unexplained weight loss

a two-hour oral glucose tolerance test (OGTT) value of $\geq$200 mg/dl

Classification

Diabetes mellitus according to the above classification

Impaired fasting glucose (IFG), a new category, when fasting plasma glucose $\geq$110 mg/dl but lower than 126 mg/dl

Impaired glucose tolerance (IGT), an existing category, when result of the oral glucose tolerance test is $\geq$140 mg/dl but lower than 200 mg/dl at 2 hours

Gestational Diabetes

Gestational diabetes mellitus (GDM) is defined as any degree of glucose intolerance with onset or first recognition during pregnancy, although there is a possibility that unrecognized glucose intolerance may have antedated or begun concomitantly with the pregnancy. Six weeks or more after delivery, the woman should be reclassified, as in the following categories: (1) diabetes, (2) IFG, (3) IGT, or (4) normoglycemia. The definition of GDM applies when the condition persists after pregnancy but in the majority of cases of GDM glucose regulation will return to normal after delivery.

[a]Source: [1].

III. Prevalence, Public Health Burden, and Distribution of Diabetes

In 1997, it was estimated that there were 10.3 million U.S. residents with diagnosed diabetes, with 90–95% of these being type 2 diabetes and 5–10% being type 1 diabetes [2–5]. Based on oral glucose tolerance testing in representative samples, however, it is estimated that there are approximately 5.4 million additional cases of undiagnosed diabetes in the U.S.

Diabetes was the seventh leading cause of death listed on U.S. death certificates in 1995. Based on death certificate data, diabetes contributed to 187,800 deaths in 1995. Approximately 595,000 new cases of type 2 diabetes and 30,000 new cases of type 1 diabetes are diagnosed each year. The economic costs of diabetes are substantial, with direct health care costs estimated to be $44.1 billion annually. There are an additional $54.1 billion in indirect costs, including lost wages and productivity [6].

The overall age-adjusted prevalence of physician-diagnosed diabetes is the same in women and in men (5.2% vs 5.3%). A sex differential in diabetes, however, is expressed in racial minority groups. The gender gap is significant among non-Hispanic blacks (women 9.5%, men 7.3%) and Mexican-Americans (women 10.9%, men 7.7%) [2]. It remains unclear, however, whether gender independently affects the risk of developing diabetes or whether disparities in prevalence and incidence merely reflect gender differences in risk factors for diabetes, such as body mass index, physical activity, and other variables [3,7].

For type 1 diabetes, which accounts for only a small proportion of total diabetes cases, there is a slight excess of male cases among white populations and an excess of female cases among nonwhite populations [4]. Age-adjusted incidence rates of type 1 diabetes (100,000 person/year) are higher in white (13.3 to 20.6) than in Hispanic (4.1 to 15.2), African-American (5.6 to 11.0), and Oriental (6.4) population groups [7a].

Prevalence of diabetes in adults worldwide was estimated to be 4.0% (135 million) in 1995 and is expected to rise to 5.4% (300 million people) by the year 2025 [7b]. India, China, and the United States are the countries with the largest number of people with diabetes. In developing countries, the majority of people with diabetes are in the age range of 45–64 years; in developed countries the majority of diabetics are older than 65 years [7b].

There is great international variation in the incidence of diabetes. A diabetes registry of 15 countries reported substantial differences in the incidence of type 1 diabetes, with age-adjusted rates among those younger than 15 years of age ranging from 1.7 per 100,000 person-years in Hokkaido, Japan to 29.5 per 100,000 person-years in Finland [8]. Migrant studies and cross-sectional comparisons of members of a given ethnic group living in different regions provide valuable information concerning the likely role of environmental factors as determinants of diabetes risk. A study of Asians who migrated to the United Kingdom showed that rates of type 1 diabetes in children increased steadily over a period of 12 years, reaching a level comparable to that of non-Asian children in the U.K. [9]. As for type 2 diabetes, a study of blacks in Nigeria, the Caribbean, the United States, and the United Kingdom suggested that the self-reported prevalence of diabetes varied greatly, with an age-adjusted prevalence of 2% in Nigeria, 9% in the Caribbean, and 11% among blacks in the U.S. and the U.K. Increasing body mass index was correlated strongly with greater prevalence of type 2 diabetes, suggesting that the balance of energy intake and expenditure is an important determinant of type 2 diabetes among those individuals with similar genetic background [10].

IV. Risk Factors for Diabetes

Just as the two main forms of diabetes vary in their prevalence and typical age of onset, each form has its own distinct set of risk factors.

A. Risk Factors for Type 1 Diabetes

Genetic susceptibility plays an important role in the risk of type 1 diabetes. However, the observation of concordance of type 1 disease in only slightly more than one-third of monozygous twin pairs [11] suggests a strong influence of environmental cofactors. The development of type 1 diabetes appears to be strongly immune-mediated. Nearly 90% of individuals diagnosed with type 1 disease have immune markers of beta-cell destruction, the most common being islet cell autoantibodies, but anti-insulin autoantibodies and autoantibodies against specific enzymes are also commonly detected [12].

Data from both the northern and southern hemispheres demonstrate a seasonal variation in the onset of type 1 diabetes, with a reduction in the number of cases diagnosed during warm summer months [13]. These findings have led to speculation that infection may be an important risk factor of type 1 diabetes. Seasonal variation has also been reported in the birth months of children with type 1 diabetes, with more relative births of diabetic children from March to August in three diabetes registries in the United Kingdom [14]. This observation has prompted speculation that maternal infection during pregnancy with a virus or viruses with a similar seasonal variation may be related to risk of type 1 diabetes in their offspring [14]. In a nested case-control study utilizing frozen sera collected from mothers at the time of delivery, maternal enteroviral infection with specific IgM antibodies for viruses such as echo 9 and 30 and coxsackie B5 was associated with a relative risk of subsequent type 1 diabetes in their offspring of 3.2 (95% CI = 1.4–7.3) [15].

Nutritional factors may also play a role in the development of type 1 diabetes. The most widely studied exposures have been breastfeeding and exposure to cow's milk protein. A number of studies have shown decreased risks of type 1 diabetes in those who were breastfed until at least three months of age, while the early introduction of cow's milk into an infant's diet has been associated with greater subsequent risk of diabetes [13,16,17]. In studies assessing cell-mediated immunity to cow's milk components in patients with recent onset of type 1 diabetes, findings were positive for beta-casein [18], negative for beta-lactoglobin [19], and divergent concerning bovine serum albumin [20,21]. Based on the totality of available evidence, the American Academy of Pediatrics has stated that breastfeeding is strongly recommended as the primary source of nutrition during the first year of life. In families with a history of type 1 diabetes, particularly in a sibling of the infant, breastfeeding and avoidance of commercially-available cow's milk and products containing intact cow's milk protein during the first year of life are strongly encouraged [22].

B. Risk Factors for Type 2 Diabetes

Genetics as well as perinatal factors, in this case principally low birth weight, have been correlated with greater insulin resistance and subsequent increased risk of type 2 diabetes [23–25]. However, lifestyle habits and health variables in adulthood have been shown to exert a far greater influence on the risk of developing type 2 diabetes. Indeed, the prominent role of modifiable risk factors as determinants of type 2 diabetes is such that this disease should be regarded as a largely preventable condition.

Obesity, regional fat distribution, and weight gain during adulthood are the most important constellation of modifiable risk factors for type 2 diabetes. This assumes particular importance in women, who suffer disproportionately from obesity, and in black and Hispanic women, in whom the prevalence of obesity is substantially higher than that in white women. In the First National Health and Nutrition Examination Survey (NHANES I), 68% of prevalent cases of diabetes were attributable to excess body mass index (BMI) (weight in kilograms, divided by height in meters, squared) [26]. In the Nurses' Health Study, a prospective cohort study of U.S. women, 90% of type 2 diabetes cases in white women were estimated to be attributable to BMIs higher than the Metropolitan Insurance Companies "recommended" weights (BMI > 22). There was a linear relationship with no apparent threshold between body mass index and the development of diabetes [27]. The risk of developing diabetes for women with BMIs > 35 was 93.2 times (95% CI = 81.4–106.6) that of women with BMI < 22 [27]. However, even women with body mass indices between 23 and 25, a range that corresponds with "normal" weight according to most guidelines, had a greater than threefold (95% CI = 2.0–6.6) elevation in age-adjusted risk of diabetes compared to women with BMIs less than 22. These findings, like those from studies of BMI and risks of cardiovascular disease and total mortality, suggest that average or "normal" weights may not be optimal and provide further evidence for the need to reevaluate accepted standards for desirable weight. Marked increases in the prevalence of obesity in the U.S. during the 1980s portend increases in the incidence of obesity-related type 2 diabetes.

Independent of overall BMI, regional fat distribution, typically measured as the ratio of waist-to-hip circumference, is also strongly associated with risk of developing type 2 diabetes. In the Nurses' Health Study, after controlling for BMI and other potential confounding variables, there was a threefold greater risk of developing diabetes among those in the 90th percentile of waist-to-hip ratio (0.86) compared with those in the 10th percentile (0.70) (Relative Risk (RR) = 3.1; 95% CI = 2.3–4.1) [28]. For waist measurement alone, the relative risk of diabetes for women in the 90th percentile of waist circumference (92 cm/ 36.2 inches) versus the 10th percentile (67 cm/26.2 inches) was 5.1 (95% CI = 2.9–8.9) [28].

Weight gain during adult life is also a risk factor for type 2 diabetes, independent of level of adiposity or family history. In the Nurses' Health Study, women who had gained 5.0 to 7.9 kg after age 18 had a relative risk of diabetes of 1.9 (95% CI = 0.5–2.3), in comparison to women with stable weight during this period [29]. Conversely, women who lost more than 5.0 kg reduced their risk by at least 50%. In another study of men and women, one in four new cases of type 2 diabetes was attributable to weight gain during a 9-year follow-up period [26].

Physical inactivity, which is closely correlated with obesity, is also positively associated with the risk of developing type 2 diabetes. Conversely, physical activity increases insulin sensitivity and loss of adipose mass with preservation of lean body mass and is strongly associated with decreased risk of diabetes. In Nurses' Health Study analyses, based on 8 years of follow-up, women who engaged in vigorous exercise at least once per week had an age-adjusted relative risk of type 2 diabetes of 0.67 (95% CI = 0.60–0.75) compared with less active women [30]. This apparent protective effect of exercise was attenuated by adjustment for body mass index but remained statistically significant (RR = 0.84, 95% CI = 0.74–0.95). When analysis was restricted to the 2 years following ascertainment of physical activity level and to symptomatic cases of type 2 diabetes, the age-adjusted relative risk for active women was 0.50 (95% CI = 0.35–0.71), and the age- and body mass index-adjusted RR was 0.69 (95% CI = 0.48–1.0).

In a population-based cohort among 5000 men and women in Finland, physical activity was associated with decreased risk of type 2 diabetes in women but not in men [31]. The benefits of physical activity on diabetes risk appear to extend to older ages as well. In a Massachusetts cohort of men and women aged 65 and older (in which women made up two-thirds of the study population), those with low levels of activity had a greater risk of type 2 diabetes than those with moderate activity levels [32].

With respect to dietary factors, higher glycemic load with increased intake of rich, refined carbohydrates and lower intake of fiber and minerals such as potassium, magnesium, and calcium, are all related to increased risks of type 2 diabetes [33]. The role of dietary fat and fatty acids in the etiology of type 2 diabetes requires further study.

Cigarette smoking is a strong, independent risk factor for coronary heart disease (CHD). Evidence that smoking decreases two-hour insulin tolerance test results and causes a transient increase in serum glucose also raised the possibility of a causal role in diabetes [34]. In the Nurses' Health Study, women who reported smoking in excess of 25 cigarettes per day had a 42% greater risk of type 2 diabetes than nonsmokers [35]. Similar findings have been observed in other cohorts [36].

An association between oral contraceptive (OC) use and risk of diabetes emerged from cross-sectional data of the Second National Health and Nutrition Examination Survey (NHANES II). In this survey, age-adjusted rates of impaired glucose tolerance were 15.4% among current OC users and 6.3% among nonusers during the period of 1976–1980 [37]. In contrast, prospective data from the Nurses' Health Study indicated no increased risk of diabetes among OC users during 10 years of follow-up [38]. Many of the studies examining this relationship, however, were carried out during the period when only high-dose, combined estrogen and progestin preparations were available. Since the mid-1980s, newer OC agents with reduced doses of estrogen and progestin have become widely used, as have triphasic OC agents. Studies indicate that users of these newer low-dose agents have lower levels of glucose intolerance than levels seen among users of high-dose preparations [39]. In the Nurses' Health Study II (NHS II), current users of low-dose OC preparations had no appreciable increased risk of diabetes during 4 years of follow-up in the early 1990s [40]. A nonsignificant increased risk observed for current OC users in NHS II (RR = 1.6; 95% CI = 0.9–3.1) was attenuated when analyses were restricted to symptomatic type 2 diabetes (RR = 1.3; 95% CI = 0.6–2.8), suggesting that differential surveillance may explain all or part of any observed excess diabetes risk.

Two large-scale cohorts, the Nurses' Health Study [41] and the Rancho Bernardo Study [42], have reported no increased diabetes risk among users of postmenopausal hormone replacement therapy.

V. Health Consequences of Diabetes

The incidence and severity of acute and chronic complications associated with diabetes are often greater in patients with type 1 diabetes than in those with type 2 diabetes. However, because type 2 diabetes accounts for the overwhelming proportion of all cases of diabetes, the population-wide impact of diabetes-related illness is greater from type 2 disease.

A. Health Consequences of Type 1 Diabetes

Patients with type 1 diabetes are at substantially increased risk of a range of serious health conditions. Acute metabolic complications of type 1 diabetes include diabetic ketoacidosis, hyperosmolar nonketotic coma, lactic acidosis, and hypoglycemia. Diabetic retinopathy is the leading cause of new cases of blindness in adults in the United States. In a U.S. study, 97% of patients with type 1 diabetes had evidence of retinopathy within 15 years of diagnosis [43]. Patients with type 1 diabetes are also at markedly increased risk of neuropathy, end stage renal disease, peripheral vascular disease, CHD, and stroke. In females with type 1 diabetes, menarche typically occurs an average of one year later than in unaffected females, and they experience increased dysmenorrhea and irregular menstrual cycles [44].

In the Pittsburgh Insulin-Dependent Diabetes Mellitus Morbidity and Mortality Study, total mortality in type 1 diabetes patients aged 25–45 years was 20 times that of the general population [45]. In an earlier Danish study, nearly half of all type 1 diabetes patients died after an illness duration of 35 years, regardless of the age of onset [46]. As with type 2 diabetes, CHD is the predominant cause of death in type 1 patients. In a cohort of type 1 diabetes patients followed by the Joslin Clinic, cumulative CHD mortality by age 55 was 35% in men and in women [47]. CHD mortality in nondiabetic men and women of corresponding age are 8% and 4%, respectively.

B. Health Consequences of Type 2 Diabetes

1. Coronary Heart Disease

Retinopathy, nephropathy, and neuropathy are also complications of type 2 diabetes, and a substantially elevated risk of CHD is one of the most serious consequences of this condition. Numerous prospective cohort studies have demonstrated that diabetes is a stronger risk factor for CHD in women compared to men; compared with nondiabetics, age-adjusted CHD mortality rates are 2 to 3 times higher in diabetic men and 3 to 7 times higher in diabetic women [48–52] (Table 64.2) [52–52b]. This more pronounced adverse effect of diabetes in women appears to abolish completely the usual female advantage in CHD mortality. The reasons for the sex differential in CHD risk from diabetes are not completely understood but may relate to particularly deleterious effects of diabetes on lipids and blood pressure in women [53]. In the Framingham Heart Study, diabetic women had lower levels of high-density lipoprotein cholesterol (HDL-C) than diabetic men [54]. The relative risk of coronary events in this population for diabetic versus nondiabetic subjects was 5.1 in women and 2.4 in men. In cross-sectional data from the United Kingdom, women with type 2 diabetes had significantly higher levels of total and low-density lipoprotein cholesterol (LDL-C) than age-matched male counterparts with diabetes [55]. In the San Antonio Heart Study, the size of LDL particles (a characteristic associated with greater atherogenicity)

Table 64.2

Prospective Population-Based Studies of Coronary Heart Disease Mortality in Predominantly Middle-Aged Diabetic Patients

Study Location	Year	Age	Follow-up (years)	Mortality risk ratio[a] Males	Mortality risk ratio[a] Females
DuPont Company	1970	<20–64[b]	10	2.9[c,d]	
Israel	1977	≥40	5	3.4[c,d]	
Framingham	1979	45–74	20	1.7	3.3[e]
Evans County	1980	≥22	4.5	1.0	2.8
Rancho Bernardo	1983	40–79	7	2.4	3.5
Warsaw	1984	18–68	9.5	1.3	1.7[c]
Tecumseh	1985	40–54	20	7.9	7.7[e]
		55–69		2.6	0.6[e]
		≥70		1.9	5.3[c,e]
Whitehall	1985	40–64	10	2.5[e]	
Chicago	1986	35–64	9	3.8	4.7[e]
Finland	1986	40–69	11	2.0	4.1[c,d]
Nurses' Health Study	1991	30–35	8		3.0[e]

Note. Modified from [52,52a,52b] with permission.

[a]Rancho Bernardo and Warsaw identified ischemic heart disease; Israel identified myocardial infarction; all others identified coronary heart disease.

[b]Among a total of 370 diabetic patients, only 9 patients under 40 years old.

[c]Relative risk (observed/expected death proportion, standardized mortality ratio, Cox standardized risk ratio) Evans County and DuPont Company not age-adjusted; all others age-adjusted.

[d]Newly diagnosed diabetic patients.

[e]Multiple adjusted risk, including age and major coronary heart disease risk factors.

was significantly smaller (<235.5 Å) in diabetic men and women than in nondiabetic subjects, but only in women did this remain the case after adjustment for triglyceride and HDL-C levels [56]. The stronger association between LDL size and diabetes in women compared to men may also be a factor in the greater relative risk of coronary disease in women with diabetes.

Traditional risk factors for CHD are highly prevalent in individuals diagnosed with type 2 diabetes. In addition to the previously described associations between obesity, physical inactivity, and smoking and risk of developing diabetes, hypertension is twice as prevalent among diabetic women aged 45–64 years (49%) compared to the general U.S. adult population (26%) [57]. In addition, hypertension is independently associated with many of the common clinical sequelae of diabetes, including CHD, stroke, renal failure, and peripheral arterial disease. In relation to nondiabetic individuals, those with type 2 diabetes also have higher mean total cholesterol, LDL-C, and triglycerides, and lower mean HDL-C [57].

The increased risk of CHD among diabetics is certainly due, at least in part, to factors strongly correlated with diabetes, including hypertension, hypercholesterolemia, obesity, and sedentary lifestyle, which are independently and causally related to CHD. However, it is clear that traditional risk factors alone do not fully account for the increase in CHD among diabetics. In

Table 64.3a

Prevalence of Atherogenic and Potentially Atherogenic Factors in Type 1 and Type 2 Diabetes

	Type I		
Factor	Normal renal function	Impaired renal function	Type 2
Hypertension	0	+++	++
Hypercholesterolemia	0−+	+++	+
Hypertriglyceridemia	0−+	+++	+++
Remnant particles	?	+++	++
Decreased HDL cholesterol	0−+	+++	++
Obesity	0	0	+++
Hyperinsulinemia	+	++	++
Immune complexes	++	?	0

Note. The symbol (+) indicates increased prevalence, with (+++) signifying the greatest increase. A prevalence not different from that of the general population is indicated by 0. Lipid and lipoprotein abnormalities in type I diabetics with normal renal function are usually associated with poor control. Source: Modified from [61a].

Table 64.3b

Alterations in Plasma Lipoproteins in Patients with Types 1 and 2 Diabetes

Lipoprotein	Alteration
LDL	Increased plasma level
	Glycosylation
	Modified in arterial wall to enhance uptake by macrophage
	Triglyceride-enriched
HDL	Decreased plasma level
	Glycosylation
	Triglyceride-enriched
VLDL	Increased plasma level
	Impaired degradation leading to remnant accumulation
	Altered composition → endothelial toxicity; increased lipid uptake by macrophages
Chylomicra	Impaired degradation causing remnant accumulation
Remnants	Increased levels in some type II diabetics

the Nurses' Health Study, diabetes was associated with a markedly increased risk of CHD (nonfatal MI plus fatal CHD) (age-adjusted RR = 6.7; 95% CI = 5.3–8.4) and total cardiovascular mortality (age-adjusted RR = 6.3; 95% CI = 4.6–8.6) [52]. Although these relative risks were attenuated in multivariate analyses controlling for known coronary risk factors, a substantial independent effect of diabetes persisted, with relative risks for CHD of 3.1 (95% CI = 2.3–4.2) and total cardiovascular mortality of 3.0 (95% CI = 1.9–4.8), respectively. With regard to potential explanations for the residual independent effect of diabetes in the etiology of thrombotic vascular events [58], accumulating evidence points toward a role of hemostatic risk factors, which are highly prevalent in diabetics. These factors include hyperaggregable platelets [59], impaired fibrinolytic activity [60], increased oxidative stress, and glycosylation of tissue proteins [60]. In 16-year follow-up data in the Framingham cohort, increasing levels of fibrinogen were correlated with increasing blood glucose concentrations, an association that was particularly pronounced in women [61]. Mechanisms that may mediate the increased risk of CHD in type 1 and type 2 diabetes are summarized in Table 64.3a and 64.3b [61a].

Diabetes is also associated with a range of poor outcomes following myocardial infarction (MI), and this appears to be more pronounced in women. In the WHO-MONICA study, the age-adjusted 28-day case fatality rate for acute MI was 25% in diabetic women compared with 16% in nondiabetic women [62]. In the Framingham study, among women survivors of MI, diabetics developed congestive heart failure four times as often as nondiabetics (16% vs 3.8%). Diabetes was associated with a doubling of risk of reinfarction in women but had no significant independent effect on reinfarction in men [63].

2. Stroke

Diabetes is associated with an increased risk of stroke. In the Nurses' Health Study, type 2 diabetes was associated with a more than fivefold increase in risk of ischemic stroke (age-adjusted RR = 5.4; 95% CI = 3.3–9.0). The relative risk was attenuated, but remained significant, after adjustment for hypertension and other cardiovascular risk factors (RR = 3.0; 95% CI = 1.6–5.7) [52]. As with CHD, there is some evidence of a sex differential in the diabetes-stroke relationship. In the Framingham Study, the age-adjusted relative risk of brain infarction among diabetics compared with nondiabetics was 2.7 in men and 3.8 in women [61]. In a cohort study in Northern Sweden, there was also a higher relative risk of stroke among diabetic women than their male counterparts [64].

3. Total Mortality

Diabetes is clearly associated with an increased risk of total mortality [65]. In the Nurses' Health Study, women with diabetes had an age-adjusted risk of total mortality three times greater than that of nondiabetic women (RR = 3.0; 95% CI = 2.5–3.7) [52]. This was attenuated, but remained significant, in multivariate analyses adjusting for known coronary risk factors (RR = 1.9; 95% CI = 1.4–2.4). More than two dozen analytic epidemiologic studies have examined the relationship between diabetes and total mortality, and after adjusting for a range of coronary risk factors, the risk of mortality in persons with diabetes is generally twice that of nondiabetic subjects [65].

4. Depression

Diabetic patients also appear to be at increased risk for depression, although it is unclear whether they are more prone to depressive illness than those with other chronic diseases [66]. In several studies, the prevalence of major depression has been found to be approximately three times higher in people with diabetes than among the general population [66]. Although data on gender differences are limited, the occurrence of depression among diabetics appears to follow the same pattern seen in the general population, with greater prevalence in women than in men [67].

Health care providers should be attentive to signs of depression in patients with diabetes. However, the diagnosis must be made particularly carefully and should utilize clinical tools such the Beck Inventory and other reliable indices of depression, because malaise, fatigue, eating disorders, and sleep disturbances may be due to hyperglycemia or other metabolic abnormalities in diabetic patients [68]. Depression is not only highly prevalent among people with diabetes, but it may adversely affect the management of this condition if the depression causes a patient to be less compliant with his or her dietary, drug, and exercise programs. Thus, accurate diagnosis and treatment of depression in diabetic patients is crucial to the overall management of the disease.

Two randomized trials have evaluated pharmacologic antidepressant therapy in diabetic patients. In one, tricyclic antidepressant therapy, which produced remission of pain and sleep disturbances from diabetic neuropathies, was also associated with relief of depression. In the second trial, nortriptyline yielded substantial mood improvement in diabetics [69]. Selective serotonin reuptake inhibitors may be more appropriate for diabetes patients with eating disorders such as binging and carbohydrate cravings. However, because all pharmacologic treatments of depression have side effects, some of which may be of potentially greater relevance in diabetes, cognitive-behavioral therapy (CBT) approaches may be particularly appropriate in diabetic patients with depression. A randomized controlled trial comparing CBT (n = 25) and a control group using antidepressants (n = 26) during 10 weeks showed in an intention-to-treat analysis that remission of depression was greater in the CBT group (70% or 14 of 20) than in the control group (33.3% or 7 of 21) [69a]. However posttreatment glycosylated hemoglobin levels were not different in the two groups.

5. Quality of Life Considerations

Diabetes can have a profound impact on the overall quality of life of affected individuals, family members, and others in their social support network. Diabetes is associated with increased risks of several major disabilities, including blindness and lower-extremity amputations. There are also a number of minor symptoms that are more prevalent in diabetics, including impaired sensation in the feet and hands and delayed stomach emptying. Women with diabetes are more prone to vaginal pruritus and infections and may also experience sexual dysfunction due to neuropathy-related loss of sensation in the genital area [70].

6. Osteoporosis

Osteoporosis, a condition characterized by decreased bone mass density (BMD) and increased risk of bone fractures, is common in postmenopausal women. Both obesity and type 2 diabetes are associated with preservation of BMD in women [71,72]. In a cohort of 25,298 women followed in Norway for 12 to 16 years, those with diabetes were at increased risk of hip fracture (RR = 5.81, 95% CI = 2.15–15.71) [72a]. However, this result was not adjusted for weight change (loss or gain) that could be a counfounding factor for diabetes and bone fractures. In the Rancho Bernardo Study population in California, type 2 diabetes was associated with higher BMD levels in women but not in men. This association in women persisted after control for a range of risk factors, including obesity, suggesting that there may be an independent effect of diabetes on bone density. Increased fasting insulin levels have also been shown to be cor-

related with increased BMD in women, but not in men, and it has been hypothesized that the association between diabetes and BMD may reflect the greater androgenicity reported in women with hyperglycemia and hyperinsulinemic conditions [73]. The current evidence does not permit us to judge if osteoporosis and hip fractures are positively or negatively associated with diabetes.

VI. Unique Issues in Women

While most risk factors for, and the health consequences of, diabetes are similar in men and women there are several issues unique to women with diabetes.

A. Gestational Diabetes

Gestational diabetes mellitus (GDM), defined as glucose intolerance first recognized during pregnancy, occurs in 1–14% of pregnancies in the United States, with most studies reporting a prevalence of GDM of 2–5% [74]. Women who develop GDM are at increased risk of subsequent development of type 2 diabetes; approximately 40% of former gestational diabetic women develop diabetes within 20 years [75]. The effects of GDM on pregnancy outcome are not clear. While most studies have reported no increased perinatal mortality in the offspring of women with GDM, these reports have come from studies in which GDM was carefully monitored, with women receiving dietary or insulin therapy, as appropriate. There also may be increased surveillance of women with GDM, and the increased level of monitoring and medical care that may result could offset the hazards of this condition to the fetus. In studies where GDM screening has been carried out but caregivers have remained blinded to the results or not used them in the management of the pregnancy, excess perinatal mortality has been reported in women with elevated plasma glucose levels [70]. As regards perinatal morbidity in GDM, the most widely reported potentially adverse outcome is macrosomia, defined variously as birth weight greater than 4000 grams or 4500 grams.

B. Pregnancy in Women with Preexisting Diabetes

Pregestational diabetes is present in approximately 0.1–0.3% of pregnancies in the United States. In cases where no special preconception diabetes management has been established, spontaneous abortions occur in 7–17% of clinically recognized pregnancies and major fetal malformations occur in 7–13% [76]. Both outcomes are related to degree of maternal hyperglycemia in the first trimester. Indeed, several series have reported comparable rates of spontaneous abortion in nondiabetic women and diabetic women whose condition is well-controlled. The principal fetal malformations believed to be associated with pregestational diabetes are anomalies of the central nervous and cardiovascular systems. Findings from several studies indicate that rates of fetal malformations are lower among women who take part in preconceptional diabetes management programs compared to usual care [76]. As with GDM, an excess of macrosomia is seen in offspring of women with pregestational diabetes. Based on U.S. data from 1980, overall perinatal mortality rates (from 28 weeks of gestation onward and neonatal deaths within less than 7 days of birth) associated with pregestational diabetes (30–40 per 1000 live births) are approximately 2–3 times those associated with nondiabetic pregnancies (15 per 1000 live births) [76].

Overall maternal mortality rates in pregnancy, however, appear comparable in diabetic and nondiabetic women, although the lack of adequate population-based data precludes detection of possible modest differences. Women with prior vascular disease appear to be at increased risk of mortality during pregnancy. Pregnancy-induced hypertension and preeclampsia occur more frequently in diabetic than in nondiabetic women, possibly due to the effects of diagnosed and incipient diabetic nephropathy as well as the overall higher prevalence of hypertensive disorders in diabetic populations, particularly those with type 2 disease.

VII. Treatment and Prevention of Diabetes

A. Treatment of Diabetes

The treatment of diabetes includes strategies to directly improve glycemic control as well as interventions to reduce risks of cardiovascular disease and other clinical consequences of diabetes. Dietary treatment plus modest to moderate weight loss can yield normalization or near normalization of fasting postprandial glycemia in most overweight patients with type 2 diabetes [77]. In a study of type 2 diabetes patients in Northern Ireland, 80% of patients achieved glycemic control (fasting blood glucose <180 mg/dl) with dietary therapy alone [78]. A range of dietary regimens, including low-fat/high-carbohydrate diets, very-low-calorie diets, as well as increases in fiber and monounsaturated fat, may beneficially influence glycemic control and blood lipids in patients with type 2 diabetes. Newer oral hypoglycemic agents appear to provide effective glycemic control in appropriate type 2 diabetes patients and may also be associated with reductions in triglyceride and total cholesterol and increased levels of HDL cholesterol.

Strict glycemic control with intensified insulin treatment appears to favorably affect risk of microvascular complications in type 1 diabetes, including retinopathy, nephropathy, and neuropathy. In the Diabetes Control and Complications Trial (DCCT) [79], conducted among 1441 type 1 diabetes patients with and without retinopathy at baseline, intensive therapy was associated with significant reductions in the development of retinopathy by 76% (95% CI = 62%–82%), of retinopathy progression among already affected patients by 54% (95% CI = 19%–74%), in occurrence of microalbuminuria by 39% (95% CI = 21%–52%), and in clinical neuropathy by 60% (95% CI = 38%–74%) [79]. Available evidence remains inconclusive, however, as to whether there are similar benefits in type 2 diabetes or on risks of macrovascular disease in patients with either type 1 or type 2 disease [77]. In the DCCT, intensive insulin therapy reduced the development of hypercholesterolemia in type 1 diabetes patients [79]. Rates of cardiovascular disease events were lower in the intensive therapy group, although differences were not significant, perhaps due to small numbers. Indirect evidence concerning glycemic control and CHD comes from the Framingham Heart Study, in which there was an independent association in women (but not in men) between CHD events and chronic hyperglycemia, as assessed by glycosylated hemoglobin levels

[80]. Despite the paucity of direct evidence concerning glycemic control and macrovascular complications of diabetes, the available data suggest that such benefits are likely in both type 1 and type 2 diabetes, and that glycemic control is also likely to reduce microvascular complications in type 2 diabetes patients.

Patients with diabetes may benefit from interventions used among the broader population of patients at increased risk of cardiovascular disease. Dyslipidemia is highly correlated with type 2 diabetes, and 3-hydroxy-3-methylglutaryl coenzyme A (HMG-CoA) reductase inhibitors are well-tolerated and highly effective in reducing lipid levels in diabetic patients [77,81].

Hypertension is also highly prevalent in patients with type 1 and type 2 diabetes, as well as among individuals with impaired glucose tolerance. Lifestyle modifications, including weight control, exercise, smoking cessation, and dietary changes are all appropriate components of the management of hypertensive diabetes patients. Although limited randomized trial data are available that directly quantify the benefits of drug treatment of hypertension in diabetes, diabetic patients appear likely to accrue benefits to cardiovascular disease risk comparable to those seen in drug trials in general hypertensive populations, including clear reductions in risk of stroke and CHD [82]. For diabetic patients requiring pharmacologic therapy, current guidelines recommend use of angiotensin-converting enzyme inhibitors, calcium channel blockers, alpha-adrenergic blockers, or low-dose diuretics (although diuretics may worsen glycemic control in some patients). These agents are generally preferable to beta-blockers, which can have adverse effects on peripheral blood flow and may also prolong or mask symptoms of hypoglycemia. However, beta-blocker use by people with diabetes reduces CHD risk as much as, or more than, it does in people without diabetes [83]. Diabetics who have suffered a myocardial infarction have a real benefit from using beta-blockers. In a large cohort of 201,752 patients (46% women), post-MI use of beta-blockers was associated with a 2-year reduction in overall mortality of 36% (95% CI = 31%–40 [83a].

Low-dose aspirin therapy, which has been demonstrated to reduce risks of subsequent vascular events in secondary prevention [84] and risk of a first myocardial infarction in men [85], may also be a valuable adjunct in the treatment of diabetes. In the Early Treatment Diabetic Retinopathy Study, a randomized trial among 3711 diabetic men and women, aspirin treatment was associated with a 22% reduction in risk of nonfatal and fatal myocardial infarction (RR = 0.78; 95% CI = 0.55–0.95), with no evidence of adverse effects [86]. In the Antiplatelet Trialists' Collaboration overview of secondary prevention trials, aspirin treatment was associated with a 19% reduction in risk of subsequent vascular events (nonfatal myocardial infarction, nonfatal stroke, or vascular death) in the subgroup of diabetic patients, a benefit comparable to the overall reduction among secondary prevention patients [84]. The American Diabetes Association has recommended that low-dose aspirin therapy (81–325 mg/day) be initiated in diabetic men and women with prior occlusive vascular disease and that such therapy be considered in primary prevention among type 1 and type 2 diabetes patients at elevated risk due to family history, cigarette smoking, hypertension, obesity, albuminuria, or dyslipidemia [87].

Postmenopausal hormone therapy is associated with a reduction in risk of CHD in epidemiologic studies [88]. In the Nurses'

Health Study, there were comparable reductions in CHD risk associated with postmenopausal hormone therapy use among nondiabetic and diabetic women [89]. This finding, in conjunction with the lack of evidence that postmenopausal estrogen therapy adversely affects glucose tolerance or risk of developing type 2 diabetes [41], suggests that estrogen therapy may be appropriate in those diabetic patients for whom the benefit-to-risk assessment otherwise favors treatment.

B. Prevention of Diabetes

As described in the earlier section on risk factors for diabetes, there are a number of modifiable risk factors for type 2 diabetes. Modifiable determinants of risk include obesity, physical inactivity, and dietary factors. Cigarette smoking also appears to be related to risk of type 2 diabetes; its avoidance or cessation, therefore, would likely decrease risk. Evidence that low-birth weight may be associated with future risk of type 2 diabetes suggests that improved perinatal care and maternal nutrition and health might also lead to decreases in future cases of diabetes. Less evidence is available concerning potentially modifiable determinants of risk of type 1 diabetes. However, data suggest a potential role of breastfeeding and avoidance of cow's milk in the early months of life in prevention of type 1 diabetes in offspring. Based on these data, it is reasonable to conclude that a substantial proportion of type 2 diabetes cases may be preventable.

VIII. Conclusions

Diabetes is one of the most prevalent chronic diseases in the United States as well in other countries. The overwhelming majority of cases in the U.S. are type 2 diabetes. While rates of type 2 diabetes are similar in men and women, the relative risk of coronary heart disease—one of the leading and most serious health consequences of diabetes—is greater in women than in men. In addition, women with preexisting diabetes are at increased risk for complications during pregnancy, and pregnant women without diagnosed diabetes are at risk of developing gestational diabetes. Despite the clear role of genetics, a growing body of evidence indicates that modifiable risk factors account for a substantial proportion of cases of type 2 diabetes. Improvements in the population-wide prevalence of these factors, including obesity, physical inactivity, dietary intake patterns, and cigarette smoking, will contribute to the prevention of diabetes. Similarly, favorable modification of these risk factors by women with diabetes—in conjunction with good glycemic control and appropriate additional pharmacologic therapy (e.g., aspirin, antihypertensive therapy, cholesterol-lowering therapy, postmenopausal estrogen therapy in selected subgroups of patients)—should reduce the risk of adverse health consequences of this disease.

References

1. The Expert Committee on the Diagnosis and Classification of Diabetes Mellitus (1997). Report of the expert committee on the diagnosis and classification of diabetes mellitus. *Diabetes Care* **20**, 1183–1197.

2. Harris, M., Flegal, K., Cowie, C., Eberhardt, M., Goldstein, D., Little, R., Wiedmayier, H., and Byrd-Holt, D. (1998). Prevalence of diabetes, impaired fasting glucose, and impaired glucose tolerance in U.S. adults. The Third National Health and Nutrition Examination Survey, 1988–1994. *Diabetes Care* **21**, 518–524.

3. LaPorte, R. E., Matsushima, M., and Chang, Y.-F. (1995). Prevalence and incidence of insulin-dependent diabetes mellitus. *In* "National Diabetes Data Group. Diabetes in America," 2nd ed., NIH Publ. No. 95-1468. National Institutes of Health, National Institute of Diabetes and Digestive and Kidney Diseases, Bethesda, MD.

4. Kenny, S. J., Aubert, R. E., and Geiss, L. S. (1995). Prevalence and incidence of non-insulin-dependent diabetes mellitus. *In* "National Diabetes Data Group. Diabetes in America," 2nd ed., NIH Publ. No. 95-1468. National Institutes of Health, National Institute of Diabetes and Digestive and Kidney Diseases, Bethesda, MD.

5. Centers for Disease Control and Prevention (1997). "National Diabetes Fact Sheet: National Estimates and General Information on Diabetes in the United States." U.S. Department of Health and Human Services, Centers for Disease Control and Prevention, Atlanta, GA.

6. American Diabetes Association (1998). Economic consequences of diabetes in the U.S. in 1997. *Diabetes Care* **21**, 296–309.

7. Rewers, M., and Hamman, R. D. (1995). Risk factors for non-insulin-dependent diabetes. *In* "National Diabetes Data Group. Diabetes in America," 2nd ed., NIH Publ. No. 95-1468. National Institutes of Health, National Institute of Diabetes and Digestive and Kidney Diseases, Bethesda, MD.

7a. LaPorte, R., Matsushima, M., and Chang, Y. (1995). Prevalence and incidence of insulin-dependent diabetes. *In* "National Diabetes Data Group. Diabetes in America," 2nd ed., NIH Publ. No. 95-1468. National Institutes of Health, National Institute of Diabetes and Digestive and Kidney Diseases, Bethesda, MD.

7b. King, H., Aubert, R. E., and Herman, W. H. (1998). Global burden of diabetes, 1995–2025. *Diabetes Care* **21**, 1414–1431.

8. Diabetes Epidemiology Research International Group (DERI) (1988). Geographic patterns of childhood insulin-dependent diabetes mellitus. *Diabetes* **37**, 113–119.

9. Bodansky, H., Staines, A., Stephenson, C., Haigh, D., and Cartwright, R. (1992). Evidence for an environmental effect in the aetiology of insulin dependent diabetes in a transmigratory population. *Br. Med. J.* **304**, 1020–1022.

10. Cooper, R. S., Rotimi, C. N., Kaufman, J. S., Owoaje, E. E., Fraser, H., Forrester, T., Wilks, R., Riste, L. K., and Cruickshank, J. K. (1997). Prevalence of NIDDM among populations of the African diaspora. *Diabetes Care* **20**, 343–348.

11. Olmos, P., A'Hern, R., Heaton, D. A., Millward, B. A., Risley, D., Pyke, D. A., and Leslie, R. D. (1988). The significance of concordance rate for type 1 (insulin-dependent) diabetes in identical twins. *Diabetologia* **31**, 747–750.

12. Atkinsons, M. A., MacLaren, N. K., and Riley, W. J. (1986). Are insulin autoantibodies markers for insulin-dependent mellitus? *Diabetes* **35**, 894–899.

13. Dorman, J. S., McCarthy, B. J., O'Leary, L. A., and Koehler, A. N. (1995). Risk factors for insulin-dependent diabetes. *In* "National Diabetes Data Group. Diabetes in America," 2nd ed., NIH Publ. No. 95-1468. National Institutes of Health, National Institute of Diabetes and Digestive and Kidney Diseases, Bethesda, MD.

14. Rothwell, P. M., Staines, A., Smail, P. *et al.* (1996). Seasonality of birth of patients with childhood diabetes in Britain. *Br. Med. J.* **312**, 1456–1457.

15. Dahlquist, G. G., Ivarsson, S., Lindberg, B., and Forsgren, M. (1995). Maternal enteroviral infection during pregnancy as a risk factor for childhood IDDM. A Population-based Case-control Study. *Diabetes* **44**, 408–413.

16. Borch-Johnsen, K., Joner, G., Mandrup-Poulsen, T., Christy, M., Zachau-Christiansen, B., Kastrup, K., and Nerup, J. (1984). Rela-

tionship between breast feeding and incidence rates of insulin-dependent diabetes mellitus: A hypothesis. *Lancet* **2,** 1083–1086.

17. Gimeno, S. G., and Souza, J. M. (1997). IDDM and milk consumption. *Diabetes Care* **20,** 1256–1260.

18. Cavallo, M. G., Fava, D., Monetini, L., Barone, F., and Pozzilli, P. (1996). Cell-mediated immune response to β casein in recent onset insulin-dependent diabetes: Implications for disease pathogenesis. *Lancet* **348,** 926–928.

19. Vaarala, O., Klemetti, P., Savilahti, E., Reijonen, H., Ilonen, J., and Akerblom, H. K. (1996). Cellular immune response to cow's milk β-lactoglobulin in patients with newly diagnosed IDDM. *Diabetes* **45,** 178–182.

20. Karjalainen, J., Martin, J. M., Knip, M., Ilonen, J., Robinson, B. H., Savilahti, E., Akerblom, H. K., and Dosch, H. M. (1992). A bovine albumin peptide as a possible trigger of insulin-dependent diabetes mellitus. *N. Engl. J. Med.* **327,** 302–307.

21. Atkinson, M. A., Bowman, M. A., Kao, K. J., Campbell, L., Dush, P. J., Shah, S. C., Simell, O., and MacLaren, N. K. (1994). *N. Engl. J. Med.* **329,** 1853–1858.

22. Work Group on Cow's Milk Protein and Diabetes Mellitus (1990). Infant feeding practices and their possible relationship to the etiology of diabetes mellitus. *Pediatrics* **94,** 752–754.

23. Barker, D. J., Hales, C. N., Fall, C. H., Osmond, C., Phipps, K., and Clark, P. M. (1993). Type 2 (non-insulin-dependent) diabetes mellitus, hypertension and hyperlipidaemia (syndrome X): Relation to reduced fetal growth. *Diabetologia* **36,** 62–67.

24. Phillips, D. I., Barker, D. J., Hales, C. N., Hirst, S., and Osmond, C. (1994). Thinness at birth and insulin resistance in adult life. *Diabetologia* **37,** 150–154.

25. Poulsen, P., Vaag, A. A., Kyvik, K. O., Moller Jensen, D., and Beck-Nielsen, H. (1997). Low birth weight is associated with NIDDM in discordant monozygotic and dizygotic twin pairs. *Diabetologia* **40,** 439–446.

26. Ford, E. S., Williamson, D. F., and Liu, S. (1997). Weight change and diabetes incidence: Findings from a national cohort of U.S. adults. *Am. J. Epidemiol.* **146,** 214–222.

27. Colditz, G. A., Willett, W. C., Stampfer, M. J., Manson, J. E., Hennekens, C. H., Arky, R. A., and Speizer, F. E. (1990). Weight as a risk factor for clinical diabetes in women. *Am. J. Epidemiol.* **132,** 501–513.

28. Carey, V. J., Walters, E. E., Colditz, G. A., Solomon, C. G., Willett, W. C., Rosner, B. A., Speizer, F. E., and Manson, J. E. (1997). Body fat distribution and risk of non-insulin-dependent diabetes mellitus in women. The Nurses' Health Study. *Am. J. Epidemiol.* **145,** 614–619.

29. Colditz, G. A., Willett, W. C., Rotnitzky, A., and Manson, J. E. (1995). Weight gain as a risk factor for clinical diabetes mellitus in women. *Ann. Intern. Med.* **122,** 481–486.

30. Manson, J. E., Rimm, E. B., Stampfer, M. J., Colditz, G. A., Willett, W. C., Krolewski, A. S., Rosner, B., Hennekens, C. H., and Speizer, F. E. (1991). Physical activity and incidence of non-insulin-dependent diabetes mellitus in women. *Lancet* **338,** 774–778.

31. Haapanen, N., Miilunpalo, S., Vuori, I., Oja, P., and Pasanen, M. (1997). Association of leisure time physical activity with the risk of coronary heart disease, hypertension and diabetes in middle-aged men and women. *I. J. Epidemiol.* **26,** 739–747.

32. Gurwitz, J. H., Field, T. S., Glynn, R. J., Manson, J. E., Avorn, J., Taylor, J. O., and Hennekens, C. H. (1994). Risk factors for non-insulin-dependent diabetes mellitus requiring treatment in the elderly. *J. Am. Geriatr. Soc.* **42,** 1235–1240.

33. Salmeron, J., Manson, J. E., and Stampfer, M. J. (1997). Dietary fiber, glycemic load, and the risk of non-insulin-dependent diabetes mellitus in women. *JAMA, J. Am. Med. Assoc.* **277,** 142, 147.

34. Janzon, L., Berntorp, K., Hanson, M., Lindell, S. E., and Trell, E. (1983). Glucose tolerance and smoking: A population study of oral

and intravenous glucose tolerance test in middle-aged men. *Diabetologia* **25,** 86–88.

35. Rimm, E. B., Manson, J. E., Stampfer, M. J., Colditz, G. A., Willett, W. C., Rosner, B., Hennekens, C. H., and Speizer, F. E. (1993). Cigarette smoking and the risk of diabetes in women. *Am. J. Public Health* **83,** 211–214.

36. Manson, J. E., and Spelsberg, A. (1994). Primary prevention of non-insulin-dependent diabetes mellitus. *Am. J. Prev. Med.* **10,** 172–184.

37. Russel-Briefel, R., Ezzati, T. M., Perlman, J. A., and Murphy, R. S. (1987). Impaired glucose tolerance in women using oral contraceptives: United States, 1976–1980. *J. Chronic Dis.* **40,** 3–11.

38. Rimm, E. B., Manson, J. E., Stampfer, M. J., Colditz, G. A., Willett, W. C., Rosner, B., Hennekens, C. H., and Speizer, F. E. (1992). Oral contraceptive use and the risk of type 2 (non-insulin-dependent) diabetes mellitus in a large prospective study of women. *Diabetologia* **35,** 967–972.

39. Harvengt, C. (1992). The effect of oral contraceptive use on the incidence of impaired glucose tolerance and diabetes mellitus. *Diabetes Metab.* **18,** 71–77.

40. Chasan-Taber, L., Willett, W. C., Stampfer, M. J., Hunter, D. J., Colditz, G. A., Spiegelman, D., and Manson, J. E. (1997). A prospective study of oral contraceptives and NIDDM among U.S. women. *Diabetes Care* **20,** 330–335.

41. Manson, J. E., Rimm, E. B., Colditz, G. A., Willett, W. C., Nathan, D. M., Arky, R. A., Rosner, B., Hennekens, C. H., Speizer, F. E., and Stampfer, M. J. (1992). A prospective study of postmenopausal estrogen therapy and subsequent incidence of non-insulin-dependent diabetes mellitus. *Ann. Epidemiol.* **2,** 665–673.

42. Gabal, L. L., Goodman-Gruen, D., and Barrett-Connor, E. (1997). The effect of postmenopausal estrogen therapy on the risk of non-insulin-dependent diabetes mellitus. *Am. J. Public Health* **87,** 443–445.

43. Harris, M. I. (1995). Summary. *In* "National Diabetes Data Group. Diabetes in America, 2nd ed., NIH Publ. No. 95-1468. National Institutes of Health, National Institute of Diabetes and Digestive and Kidney Diseases, Bethesda, MD.

44. Kjaer, K., Hagen, C., Sando, S. H., and Eshoj, O. (1992). Epidemiology of menarche and menstrual disturbances in an unselected group of women with insulin-dependent diabetes mellitus compared to controls. *J. Clin. Endocrinol. Metab.* **75,** 524–529.

45. Dorman, J. S., Laporte, R. E., Kuller, L. H., Cruickshanks, K. J., Orchard, T. J., Wagener, D. K., Becker, D. J., Cavender, D. E., and Drash, A. L. (1984). The Pittsburgh Insulin-Dependent Diabetes Mellitus (IDDM) Morbidity and Mortality Study: Mortality results. *Diabetes* **33,** 271–276.

46. Deckert, T., Poulsen, J. E., and Larsen, M. (1978). Prognosis of diabetics with diabetes onset before age thirty-one. I. Survival, causes of death, and complications. *Diabetologia* **14,** 363–370.

47. Krolewski, A. S., Kosinski, E. J., Warram, J. H., Leland, O. S., Busick, E. J., Asmal, A. C., Rand, L. L., Christlieb, A. R., Bradley, R. F., and Kahn, C. R. (1987). Magnitude and determinants of coronary artery disease in juvenile-onset, insulin-dependent diabetes mellitus. *Am. J. Cardiol.* **59,** 750–755.

48. Barrett-Connor, E., and Wingard, D. L. (1983). Sex differential in ischemic heart disease mortality in diabetics: A prospective population-based study. *Am. J. Epidemiol.* **118,** 489–496.

49. Heyden, S., Heiss, G., Bartel, A. G., and Hames, C. G. (1980). Sex differences in coronary mortality among diabetics in Evans County, Georgia. *J. Chronic Dis.* **33,** 265–273.

50. Kannel, W. B., and McGee, D. L. (1979). Diabetes and glucose intolerance as risk factors for cardiovascular disease: The Framingham Study. *Diabetes Care* **2,** 120–126.

51. Krolewski, A. S., Warram, J. H., and Christlieb, A. R. (1985). Onset, course, complications, and prognosis of diabetes mellitus. *In*

"Joslin's Diabetes Mellitus" (A. Marble, L. P. Krall, R. F. Bradley, A. R. Christlieb, and J. S. Soeldner, eds.). Lea & Febiger, Philadelphia.

52. Manson, J. E., Colditz, G. A., Stampfer, M. J., Willett, W. C., Krolewski, A. S., Rosner, B., Arky, R. A., Speizer, F. E., and Hennekens, C. H. (1991). A prospective study of maturity-onset diabetes mellitus and risk of coronary heart disease and stroke in women. *Arch. Intern. Med.* **151,** 1141–1147.

52a. Barrett-Connor, E., and Orchard, T. (1985). *In* "National Diabetes Group, Diabetes in America," 2nd ed., NIH Publ. 85-1468. Vol. 16, pp. 1–41. National Institutes of Health, National Institute of Diabetes and Digestive and Kidney Diseases, Bethesda, MD.

52b. Penzram, G. (1987). Mortality and survival in type 2 diabetes mellitus. *Diabetologia* **30,** 123–131.

53. Mosca, L., Manson, J. E., Sutherland, S. E., Langer, R. D., Manolio, T., and Barrett-Connor E. (1997). Cardiovascular disease in women. A statement for healthcare professionals from the American Heart Association. *Circulation* **96,** 2468–2482.

54. Kannel, W. B. (1985). Lipids, diabetes, and coronary heart disease: Insights from the Framingham Study. *Am. Heart J.* **110,** 1100–1107.

55. UK Prospective Diabetes Study Group (1997). UK Prospective Diabetes Study 27: Plasma lipids and lipoproteins at diagnosis of NIDDM by age and sex. *Diabetes Care* **20,** 1683, 1687.

56. Haffner, S. M., Mykkanen, L., Stern, M. P., Paidi, M., and Howard, B. V. (1994). Greater effect of diabetes on ldl size in women than in men. *Diabetes Care* **17,** 1164–1171.

57. Cowie, C. C., and Harris, M. I. (1995). Physical and metabolic characteristics of persons with diabetes. *In* "National Diabetes Data Group. Diabetes in America," 2nd ed., NIH Publ. No. 95-1468. National Institutes of Health, National Institute of Diabetes and Digestive and Kidney Diseases, Bethesda, MD.

58. Ridker, P. M., and Hennekens, C. H. (1991). Hemostatic risk factors for coronary heart disease. *Circulation* **83,** 1098–1100.

59. Halushka, P. V., Lurie, D., and Colwell, J. A. (1977). Increased synthesis of prostaglandin E-like material by platelets from patients with diabetes mellitus. *N. Engl. J. Med.* **297,** 1306.

60. Juhan-Vague, I., Vague, P., Alessi, M. C., Badier, C., Valadier, J., Aillaud, M. F., and Atlan, C. (1987). Relationship between plasma insulin, triglyceride, body mass index, and plasminogen activator inhibitor 1. *Diabetes Metab.* **13,** 331–336.

61. Kannel, W. B., D'Agostino, R. B., Wilson, P. W. F. *et al.* (1990). Diabetes, fibrinogen, and risk of cardiovascular disease: The Framingham Study. *Am. Heart J.* **120,** 672–679.

61a. Ruderman, and Haudenschild, (1984). Diabetes as an atherogenic factor. *Prog. Cardiovasc. Dis.* **26,** 373–412.

62. Chun, B. Y., Dobson, A. J., and Heller, R. F. (1997). The impact of diabetes on survival among patients with first myocardial infarction. *Diabetes Care* **20,** 704–708.

63. Abbott, R. D., Donahue, R. P., Kannel, W. B., and Wilson, P. W. (1989). The impact of diabetes on survival following myocardial infarction in men vs. women: The Framingham Study. *JAMA, J. Am. Med. Assoc.* **261,** 3409–3410.

64. Stegmayr, B., and Asplund, K. (1995). Diabetes as a risk factor for stroke. A population perspective. *Diabetologia* **38,** 1061–1068.

65. Geiss, L. S., Herman, W. H., and Smith, P. J. (1995). Mortality in non-insulin-dependent diabetes. *In* "National Diabetes Data Group. Diabetes in America," 2nd ed., NIH Publ. No. 95-1468. National Institutes of Health, National Institute of Diabetes and Digestive and Kidney Diseases, Bethesda, MD.

66. Lustman, P. J., and Gavard, J. A. (1995). Pyschosocial aspects of diabetes in adult populations. *In* "National Diabetes Data Group. Diabetes in America," 2nd ed., NIH Publ. No. 95-1468. National Institutes of Health, National Institute of Diabetes and Digestive and Kidney Diseases, Bethesda, MD.

67. Griffith, L. S., and Lustman, P. J. (1997). Depression in women with diabetes. *Diabetes Spectrum* **10,** 216–223.

68. Lustman, P. J., Clouse, R. E., Griffith, L. S., Carney, R. M., and Freedland, K. E. (1997). Screening for depression in diabetes using the Beck Depression Inventory. *Psychosom. Med.* **59,** 24–31.

69. Lustman, P. J., Griffith, L. S., Clouse, R. E., Freedland, K. E., Eisen, S. A., Rubin, E. H., Carney, R. M., and McGill, J. B. (1997). Effects of nortriptyline on depression and glycemic control in diabetes: Results of a double-blind placebo-controlled trial. *Psychosom. Med.* **59,** 241–250.

69a. Lustman, P. J., Griffith, L. S., Freedland, K. E., Kissel, S. S., and Clouse, R. E. (1998). Cognitive behavior therapy for depression in type 2 diabetes mellitus. A randomized, controlled trial. *Ann. Intern. Med.* **129,** 613–621.

70. Veciana, M. (1998). Diabetes and female sexuality. *Diabetes Rev.* **6,** 54–64.

71. Edelstein, S. L., and Barrett-Connor, E. (1993). Relationship between body size and bone mineral density in elderly men and women. *Am. J. Epidemiol.* **138,** 160–169.

72. Barrett-Connor, E., and Holbrook, T. L. (1992). Sex differences in osteoporosis in older adults with non-insulin-dependent diabetes mellitus. *JAMA, J. Am. Med. Assoc.* **268,** 3333–3337.

72a. Meyer, H. E., Tverdal, A., and Falch, J. A. (1993). Risk factors for hip fracture in middle-aged Norwegian women and men. *Am. J. Epidemiol.* **137,** 1203–1211.

73. Barrett-Connor, E., and Kritz-Silverstein, D. (1996). Does hyperinsulinemia preserve bone? *Diabetes Care* **19,** 1388–1392.

74. Coustan, D. R. (1995). Gestational diabetes. *In* "National Diabetes Data Group. Diabetes in America," 2nd ed., NIH Publ. No. 95-1468. National Institutes of Health, National Institute of Diabetes and Digestive and Kidney Diseases, Bethesda, MD.

75. O'Sullivan, J. B. (1984) Subsequent morbidity among GDM women. *In* "Carbohydrate Metabolism in Pregnancy and the Newborn" (H. W. Sutherland and J. M. Stowers, eds.), pp. 174–180. Churchill-Livingstone, New York.

76. Buchanan, T. A. (1995). Pregnancy in preexisting diabetes. *In* "National Diabetes Data Group. Diabetes in America, 2nd ed., NIH Publ. No. 95-1468. National Institutes of Health, National Institute of Diabetes and Digestive and Kidney Diseases, Bethesda, MD.

77. Manson, J. E., and Spelsberg, A. (1996). Risk modification in the diabetic patient. *In* "Prevention of Myocardial Infarction" *In* (J. E. Manson, P. M. Ridker, J. M. Gaziano, and C. H. Hennekens, eds.). Oxford University Press, New York.

78. Hadden, D. R., Blair, A. L. T., Wilson, E. A. *et al.* (1986). Natural history of diabetes presenting age 40–69: A prospective study of the influence of intensitive dietary therapy. *Q. J. Med.* **230,** 579–598.

79. The Diabetes Control and Complications Trial Research Group (1993). The effect of intensive treatment of diabetes on the development and progression of long-term complications of insulin-dependent diabetes mellitus. *N. Engl. J. Med.* **329,** 977–986.

80. Singer, D. E., Nathan, D. M., Anderson, K. M., Wilson, P. W., and Evans, J. C. (1992). Association of HbA$_{1c}$ with prevalent cardiovascular diseaes in the original cohort of the Framingham Heart Study. *Diabetes* **41,** 202–208.

81. American Diabetes Association (1998). Management of dyslipidemia among adults with diabetes. *Diabetes Care* **21,** 179–182.

82. Collins, R., Peto, R., MacMahon, S., Hebert, P., Fiebach, N. H., Eberlein, K. A., Godwin, J., Qizilbash, N., Taylor, J. O., and Hennekens, C. H. (1990). Blood pressure, stroke, and coronary heart disease. Part 2. Short-term reductions in blood pressure: Overview of randomised drug trials in their epidemiologic context. *Lancet* **335,** 825–838.

83. National High Blood Pressure Education Program Working Group (1997). "The Sixth Report of Joint National Committee on Prevention,

Detection, Evaluation, and Treatment of High Blood Pressure,'' NIH Publ. No. 98-4080. National Institutes of Health, National Heart, Lung, and Blood Institute, Bethesda, MD.

83a. Gottlieb, S. S., McCarter, R. J., and Vogel, R. A. (1998). Effect of beta-blockade on mortality among high-risk and low-risk patients after myocardial infarction. *N. Engl. J. Med.* **339,** 489–497.

84. Antiplatelet Trialists' Collaboration (1994). Collaborative overview of randomized trials of antiplatelet treatment. Part I: Prevention of vascular death, myocardial infarction and stroke by prolonged antiplatelet therapy in different categories of patients. *Br. Med. J.* **308,** 81–106.

85. Steering Committee of the Physicians' Health Study Research Group (1989) Final report on the aspirin component of the ongoing Physicians' Health Study. *N. Engl. J. Med.* **321,** 129–135.

86. ETDRS Investigators (1992). Aspirin effects on mortality and morbidity in patients with diabetes mellitus. Early Treatment Diabetic Retinopathy Study Report 14. *JAMA, J. Am. Med. Assoc.* **268,** 1292–1300.

87. American Diabetes Association (1998). Aspirin therapy in diabetes. *Diabetes Care* **21,** S45–S46.

88. Stampfer, M. J., and Colditz, G. A. (1991). Estrogen replacement therapy and coronary heart disease: A quantitative assessment of the epidemiologic evidence. *Prev. Med.* **20,** 47–63.

89. Stampfer, M. J., Colditz, G. A., Willett, W. C., Manson, J. E., Rosner, B., Speizer, F. E., and Hennekens, C. H. (1991). Postmenopausal estrogen therapy and cardiovascular disease. Ten-yearfollow-up from the nurses' health study. *N. Engl. J. Med.* **325,** 756–762.

65

Obesity

KATHERINE M. FLEGAL

National Center for Health Statistics
Centers for Disease Control and Prevention
Hyattsville, Maryland

I. Introduction

Obesity, or excess body fat, appears to be a highly prevalent and increasing condition in the U.S. and elsewhere in the world. Concerns about body weight are high, particularly among women. Obesity is associated with excess mortality and excess risk of coronary heart disease, hypertension, hyperlipidemia, diabetes, gallbladder disease, certain cancers, and osteoarthritis, although some features of these relationships remain uncertain or controversial. The causes of obesity are poorly understood, and both the prevention and the treatment of obesity remain elusive goals.

II. Definitions and Measurement

The human body can be described on the atomic, molecular, cellular, tissue-system, or whole body levels [1]. On the molecular level the main components are lipid, water, protein, and mineral. The lipid component consists both of essential membrane and nervous tissue lipids and of nonessential lipids consisting principally of triglycerides (triacylglycerols) stored in adipose tissue cells known as adipocytes. Body fat may be defined as consisting of these stored triglycerides. On the tissue-system level, the main components are skeletal muscle, adipose tissue, bone, blood and visceral organs, plus a residue. Adipose tissue is a structure that includes adipocytes and vascular, neural, and other tissues.

Women are characterized by a higher percentage of body fat than men, in part because of lower muscle and bone mass. Studies of younger populations suggest that on average women's body fat content is approximately 10 percentage points greater than men's [2]. The reference woman described by Forbes has a body fat content of 28% vs 15% for the reference man [3]. The anatomical location of adipose tissue also tends to differ between men and women [2,4–6]. Women are characterized by obligatory fat deposits in the breasts, hips, and thighs [2,6]. These sex-specific fat stores appear to be beneficial in terms of survival and reproduction and are sometimes also considered essential lipids for women [6]. Men lack these obligatory fat deposits and tend to store fat abdominally [4]. Relative to men, women tend to have a higher proportion of body fat stored in subcutaneous rather than visceral adipose tissue [5].

Obesity is generally defined as the condition of excess body fat. However, there is no precise definition of excess. The degree of adiposity is a continuous trait not marked by any clear division into normal and abnormal. In addition, it is difficult to measure body fat content. Thus, for practical purposes, obesity is often defined as excess body weight rather than as excess fat. In epidemiologic studies, body mass index (BMI) or Quetelet's index, calculated as weight (kg) divided by height (m) squared is often used to express weight adjusted for height [7].

The interpretation of BMI both in terms of body fatness and in terms of comparison to a weight standard varies by gender, age, and other factors [8]. Because of the differences in body composition between men and women, women will tend to have a considerably higher percentage of body fat than men at the same BMI. Because of the changes in body composition with age, older persons will tend to have a higher percentage of body fat than younger people at the same BMI. A particular value of BMI generally is associated with somewhat different risks for men and women or for people of different ages. In addition, only if the same body weight standards are considered to be appropriate for both men and women does a given value of BMI have the same meaning in terms of relative weight. Thus, a given value of BMI may be numerically the same for men and women and for people of different ages but may not represent the same percentage body fat, the same degree of risk, or even necessarily the same degree of overweight relative to a weight standard.

The terms overweight and obesity have been used with a wide variety of meanings. Definitions of overweight and obesity are usually based on measurements of body weight rather than of body fat. In 1959 and in 1983, the life insurance industry produced height and weight tables based on the mortality experience of policy holders that indicated by height and frame size the range of weights at which mortality was lowest for men and women ages 25–59 years [9,10]. One convention has been to define overweight as a body weight 20% or more above the midpoint of the weight range for a medium frame size from these Metropolitan tables [11].

A variety of definitions using BMI have also been proposed. In 1981, Garrow proposed a classification of grades I, II, and III obesity, defined by BMI categories of 25–29.9, 30–40, and >40, for both men and women [12]. The value of 25 was selected in part because it was approximately equivalent to the upper end of the weight range for large frame sizes from the 1959 Metropolitan tables. Similar BMI classifications, but with different terminology, were also used by Bray [13] and more recently by a 1995 World Health Organization (WHO) expert committee [14]. In 1985, a WHO expert consultation defined obesity as a BMI of 30 or more for men and 28.6 or more for women [15]. Also in 1985, a National Institutes of Health (NIH) Consensus Conference supported a definition of overweight as a BMI ≥ 27.8 for men and ≥27.3 for women based on the

Table 65.1
Classification of Overweight and Obesity in Adults According to BMI

Classification	BMI level	Risk of comorbidities
Underweight	<18.5	Low
Normal range	18.5–24.9	Average
Overweight	≥25	
Pre-obese	25.0–29.9	Increased
Obese Class I	30.0–34.9	Moderate
Obese Class II	35.0–39.9	Severe
Obese Class III	≥40.0	Very severe

Source: [16].

U.S. population distributions because these values corresponded approximately to 120% of the midpoint of the 1983 Metropolitan tables [11]. A classification similar to the one proposed by Garrow, but using different terminology and with an additional cutpoint of a BMI of 35, was accepted as part of the 1997 WHO consultation on obesity [16]. This WHO classification defined overweight as a BMI of 25 or greater and obesity as a BMI of 30 or greater, with some additional subdivisions. The full classification from WHO, which is currently recommended for international use, is presented in Table 65.1. A BMI of 25 was also used as the upper limit of the range of healthy weights for adults from the 1995 Dietary Guidelines for Americans [17].

Data comparing several of these proposed standards for women are presented in Table 65.2, which shows selected values from the 1959 and 1983 Metropolitan tables and also shows the weights that are equivalent to BMI values of 25 and 30. The

value of a BMI of 25 is close to the upper range of the large frame size from the 1959 tables, on which it was originally based. The differences among these criteria vary considerably by height.

III. Overweight and Obesity—Prevalence and Trends

Systematic data on obesity cannot generally be gathered from medical records or vital statistics; thus, prevalence estimates are usually derived from surveys or population studies. Because of the logistical difficulties involved in making measurements of body fat in population studies, virtually all data on prevalence and trends are based on measurements of weight and height rather than of body fat.

Estimates of the prevalence of overweight (BMI ≥ 25.0) and obesity (BMI ≥ 30.0) from the National Health and Nutrition Examination Survey (NHANES) program of successive cross-sectional national health examination surveys are shown in Tables 65.3 and 65.4 for the United States from 1960 to 1994 [18]. The prevalence of overweight (BMI ≥ 25.0) was high in 1960–1962 (48% for men and 39% for women) but changed relatively little over the time period 1960–1980, as shown in Table 65.3. However, between NHANES II (1976–1980) and NHANES III (1988–1994), the prevalence of BMI ≥ 25.0 increased to 59% for men and 50% for women. The prevalence of BMI ≥ 25.0 increased with age for both men and women, reaching a maximum in the age range 50–59 years and then declining somewhat. However, data from the 1988–1994 survey (NHANES III) show that for men and women 80 years old and above, the prevalence of BMI ≥ 25.0 exceeded 50%.

The prevalence of obesity (BMI ≥ 30.0) was 10% for men and 15% for women in 1960–1962, as shown in Table 65.4. Like the prevalence of BMI ≥ 25.0, the prevalence of BMI ≥

Table 65.2
Values of Weight (lbs) for Height Corresponding to Specified Standards and to Specified Values of Body Mass Index (BMI)

| Height | | Metropolitan tables[a] | | | | | | Weight equivalent to | | | |
| | | Range of weights for medium frame size | | 120% of midpoint of range for medium frame size | | Maximum of range for large frame size | | BMI = 25 | | BMI = 30 | |
Ft, in	Cm	1959	1983	1959	1983	1959	1983	Lbs	(Kg)	Lbs	(Kg)
4'9"	145	94–106	106–118	120	134	118	128	115	(52)	138	(63)
4'10"	147	97–109	108–120	124	137	121	131	119	(54)	143	(65)
4'11"	150	100–112	110–123	127	140	124	134	124	(56)	148	(67)
5'0"	152	103–115	112–126	131	143	127	137	128	(58)	153	(70)
5'1"	155	106–118	115–129	134	146	130	140	132	(60)	158	(72)
5'2"	157	109–122	118–132	139	150	134	144	136	(62)	164	(75)
5'3"	160	112–126	121–135	143	154	138	148	141	(64)	169	(77)
5'4"	163	116–131	124–138	148	157	142	152	145	(66)	174	(79)
5'5"	165	120–135	127–141	153	161	146	156	150	(68)	180	(82)
5'6"	168	124–139	130–144	158	164	150	160	155	(70)	186	(85)
5'7"	170	128–143	133–147	163	168	154	164	159	(72)	191	(87)
5'8"	173	132–147	136–150	167	172	159	167	164	(75)	197	(90)
5'9"	175	136–151	139–153	172	175	164	170	169	(77)	203	(92)
5'10"	178	140–155	142–156	177	179	169	173	174	(79)	209	(95)

[a]From [9] and [10].

Table 65.3

Prevalence (%) of Overweight (BMI ≥ 25) by Age, Sex, and Survey: United States, 1960–1994

Survey[a]	20–29 y	30–39 y	40–49 y	50–59 y	60–69 y	70–79 y[b]	80+ y[c]	Total (20+)	Age-adjusted total (20–74)[d]
Men									
NHES I (1960–1962)	39.9	49.6	53.6	54.1	52.9	36.0	—	48.9	48.2
NHANES I (1971–1974)	38.6	58.1	63.6	58.4	55.6	52.7	—	53.5	52.9
NHANES II (1976–1980)	37.0	52.6	60.3	60.8	57.4	53.3	—	51.5	51.4
NHANES III (1988–1994)	43.1	58.1	65.5	73.0	70.3	63.1	50.6	59.4	59.3
Women									
NHES I (1960–1962)	17.0	32.8	42.3	55.0	63.1	57.4	—	41.5	38.7
NHANES I (1971–1974)	23.2	35.0	44.6	52.2	56.2	55.9	—	41.0	39.7
NHANES II (1976–1980)	25.0	36.8	44.4	52.8	56.5	58.2	—	41.6	40.8
NHANES III (1988–1994)	33.1	47.0	52.7	64.4	64.0	57.9	50.1	50.7	49.6

[a]NHES I: National Health Examination Survey, Cycle I; NHANES I, II, III: first; second, and third National Health and Nutrition Examination Survey.

[b]For NHANES I and NHANES II, the estimates in this category are for persons ages 70–74 only because the upper age limit for the survey was 74 years.

[c]Data for this age group only available from NHANES III.

[d]Age-adjusted to the 1980 U.S. population.

30.0 also showed a large increase between NHANES II and NHANES III, to 20% for men and 25% for women. The pattern of changes with age in the prevalence of BMI ≥ 30.0 was also similar to that seen for the prevalence of BMI ≥ 25.0.

Data from other areas of the world show great variability in the prevalence of overweight and obesity, although data are not available for many countries and the many methodological differences among studies make precise comparisons difficult [16]. In China, the estimated prevalence of BMI ≥ 25 is relatively low both for men (9%) and for women (13%) [19]. Data from Brazil, Australia, and from European countries show prevalences of BMI ≥ 30 somewhat lower than those in the U.S. but

of a similar order of magnitude [20–22]. For example, the WHO MONICA data suggest that the prevalence of BMI ≥ 30 for Europe as a whole would be in the range of 10–20% for men and 15–25% for women [23]. The prevalence of BMI ≥ 30 in the U.S. falls at the upper end of this range. Data from several Pacific Island populations show high prevalences of BMI ≥ 30. For example, in three areas of Western Samoa, the prevalence of BMI ≥ 30 ranged from 56 to 74% for women and from 36 to 57% for men [24].

Increases in the prevalence of obesity similar to those seen in the United States have been reported from a number of other countries and regions of the world [16]. For example, in Britain,

Table 65.4

Prevalence (%) of Overweight (BMI ≥ 30) by Age, Sex, and Survey: United States, 1960–1994

Survey[a]	20–29 y	30–39 y	40–49 y	50–59 y	60–69 y	70–79 y[b]	80+ y[c]	Total (20+)	Age-adjusted total (20–74)[d]
Men									
NHES I (1960–1962)	9.0	10.4	11.9	13.4	7.7	8.6	—	10.5	10.4
NHANES I (1971–1974)	8.0	13.3	14.2	15.3	10.3	11.1	—	12.0	11.8
NHANES II (1976–1980)	8.1	12.1	16.4	14.3	13.5	13.6	—	12.3	12.3
NHANES III (1988–1994)	12.5	17.2	23.1	28.9	24.8	20.0	8.0	19.5	19.9
Women									
NHES I (1960–1962)	6.1	12.1	17.1	20.4	27.2	21.9	—	16.2	15.1
NHANES I (1971–1974)	8.2	15.1	17.6	22.0	24.0	21.9	—	16.7	16.1
NHANES II (1976–1980)	9.0	16.8	18.1	22.6	22.0	19.4	—	16.8	16.5
NHANES III (1988–1994)	14.6	25.8	26.9	35.6	29.8	25.0	15.1	25.0	24.9

[a]NHES I: National Health Examination Survey, Cycle I; NHANES I, II, III: first; second, and third National Health and Nutrition Examination Survey.

[b]For NHANES I and NHANES II, the estimates in this category are for persons ages 70–74 only because the upper age limit for the survey was 74 years.

[c]Data for this age group only available from NHANES III.

[d]Age-adjusted to the 1980 U.S. population.

the prevalence of BMI > 30 doubled over the time period 1980 to 1991, even though the prevalence of BMI > 30 in Britain still remained considerably below the prevalence in the U.S. [25].

IV. Host and Environmental Determinants

The answer to the question of whether the prevalence of overweight or obesity is higher for men or for women is unclear given the differences in body composition between men and women and the issues in defining obesity. The answer also varies according to the definition used. Generally, both in the U.S. and in Europe, the prevalence of overweight, defined as a BMI ≥ 25, is higher for men than for women, but the prevalence of obesity, defined as BMI ≥ 30, is higher for women than for men. This difference arises because the prevalence of BMI 25–29 is considerably higher for men than for women [23].

In Europe and the United States, differences in the prevalence of overweight or obesity between countries or between race–ethnic groups tend to be more pronounced for women than for men [18,23]. For example, in the WHO MONICA study in Europe, which gathered data from 39 sites in 18 countries, the prevalence of BMI ≥ 30 was similar for men across all sites. However, for women there were marked differences between sites, with higher values for women from Eastern Europe. Similarly, in the United States, for women, but not for men, there are marked differences in the prevalence of overweight (BMI ≥ 25.0) and obesity (BMI ≥ 30.0) by race–ethnic group, as shown in Table 65.5 for non-Hispanic whites, non-Hispanic blacks, and Mexican-Americans.

Tobacco use is widespread and could be considered a major environmental determinant of weight and overweight. Smoking and tobacco use are associated with lower body weights and a lower prevalence of overweight or obesity [26]. Although the precise mechanisms are not clear, this appears to be related to a direct effect of nicotine on metabolism rather than to reduced energy intake among smokers [27].

V. Influence of Women's Social Roles or Context

Body size is often associated with socioeconomic status. However, the magnitude and the direction of the association tend to differ both by level of economic development and by gender [28,29]. In less developed countries, higher weight may be associated with wealth and prosperity, and there may be a positive association between socioeconomic status and body size for both men and women. Historically in many contexts, greater body size, including tallness, increased muscularity, and increased fatness, has symbolized power, dominance, wealth, or high social standing.

For men in developed countries, height is positively associated with socioeconomic status but weight and BMI tend to be weakly, if at all, associated with socioeconomic status. For women in developed countries, however, weight and BMI have a strong inverse association with socioeconomic status. The slender body that in the past might have reflected economic deprivation, limited access to food, or the necessity for hard physical labor now may require expenditures of time, money, and effort to achieve. The finding in several studies that obesity

Table 65.5

Age-Adjusted Prevalence (%) of Overweight (BMI ≥ 25.0) and Obesity (BMI ≥ 30.0) in the United States by Race-Ethnic Group for Men and Women ages 20–74 y: NHANES III (1988–1994)[a]

	Men		Women	
	BMI ≥25.0	BMI ≥30.0	BMI ≥25.0	BMI ≥30.0
Non-Hispanic white	59.6	20.0	45.5	22.4
Non-Hispanic black	57.5	21.3	66.5	37.4
Mexican-American	67.1	23.1	67.6	34.2

[a]Age-adjusted to the 1980 U.S. population.

is predictive of subsequent education and earnings for women but not for men may reflect the stronger association between obesity and low socioeconomic status at baseline for women than for men [30,31]. Weight appears to vary considerably more by socioeconomic status, race–ethnicity, and nationality for women than for men, suggesting that body weight may be more closely associated with social and cultural roles for women than for men.

Body size and weight are important aspects of appearance and are highly valued culturally for women in many countries. In the U.S., women are often more concerned about weight for reasons of appearance than for reasons of fitness [32]. Concern about weight is high even among women who are not overweight and has been termed a "normative discontent" [33].

VI. Risk Factors

The human body can metabolize protein, carbohydrate, and fat to meet energy needs. The principal energy storage is in the form of fat, which, unlike protein or carbohydrate, can be stored in the body in relatively large amounts. Adipose tissue can expand to accommodate increasing amounts of triacylglycerols by increasing fat cell weight or fat cell number. This ability to store fat allows energy stores to be mobilized in times of famine or food deprivation. From an historical and evolutionary perspective, starvation is a greater danger than overabundance. Famine and starvation still occur in the world today in wartime and in other adverse political and economic conditions [34].

Throughout history, considerable effort has been devoted to finding ways to improve the adequacy and stability of the food supply and to reduce the energy expenditure required for work. As a result, an organism adapted for a situation in which food was limited and physical exertion was required is now often confronted with an environment in which palatable energy-dense foods are easily obtained with minimal physical activity. Increased modernization and a Westernized diet and lifestyle are associated with an increased prevalence of overweight in many developing countries. This has been conceived of as part of a transition to modernity, sometimes referred to as the nutrition transition [35]. To date, however, there is little evidence of a point at which this process ceases.

The question of what are the risk factors leading to obesity may be restated as the more general question of what factors determine body weight and body composition in the absence of major environmental constraints. These factors are complex, and many aspects are not well understood [36,37]. From the point of view of energy balance, the initial development of overweight is due to energy intake that exceeds energy expenditure. However, the interrelationships among energy expenditure, adipose tissue, and energy intake suggest some degree of physiological regulation of body weight or body fat content [36]. Advances, such as the identification of leptin, the product of the *ob* gene, may lead to a better understanding of these interrelationships [38].

Energy intake and energy expenditure, the ability to store excess fat under conditions of overfeeding, and the ability to lose fat under conditions of underfeeding all appear to have genetic elements [39–42]. Twin studies and adoption studies show a high heritability of body mass index [39,41,42]. When young healthy male identical twins were overfed under controlled conditions in an experimental intervention, weight gains varied widely, from 4.3 kg to 13.3 kg, even though all subjects consumed identical amounts of excess energy [40]. However, members of the same twin pair were significantly more alike than were unrelated individuals for changes in weight, body composition, subcutaneous fat distribution, and abdominal visceral fat. Similar experimental studies of negative energy balance also showed similar results for weight loss [40]. Nonexperimental studies of female twins also show genetic effects on weight and on weight change [41,42].

Genetic factors, however, are unlikely to explain the current increases in the prevalence of overweight and obesity occurring in the U.S., the U.K. and many other countries. Clearly, individual behaviors and social, cultural, and environmental contexts must also play an important role. Surprisingly little is known about these increases in overweight and obesity. Studies in the U.S. and elsewhere show that although smoking cessation might account for some of the increase in overweight, it does not explain the majority of the increase and cannot be the sole contributing factor [43]. Reductions in physical activity, often of a type that is difficult to measure, may account for some of these trends [25,44]. Changes in energy intake may also contribute but are difficult to measure [45].

VII. Clinical Issues

Although there are a wide variety of treatments designed to induce individual weight loss, almost all rely on some combination of restricted dietary intake, surgical intervention, or pharmacological treatment. Dietary treatments include very-low-calorie diets (VLCDs) of <800 kcal/day (<3350 kJ/day) provided in the context of a comprehensive treatment program and standard low-calorie diets (LCDs) of 800–1200 kcal/day (3350–5020 kJ/day) that form part of many clinical, commercial, and self-help programs. VLCDs are considered generally safe used under proper medical supervision and appear to be more effective for short-term weight loss than LCDs [46]. Behavior modification that addresses both eating and exercise patterns is a more comprehensive approach to the treatment of obesity that has also proved successful for short-term weight loss [47]. Long-term maintenance of weight loss tends to be unsatisfactory with dietary and behavioral interventions [46,47].

The goal of surgical intervention is anatomical reconfiguration designed to reduce food intake or absorption [48,49]. Earlier procedures involving intestinal bypass were associated with severe complications and unacceptable sequelae. The most common methods of bariatric surgery at present are gastroplasties and gastric bypass surgery. In vertical banded gastroplasty, the upper portion of the stomach is stapled to create a small upper pouch, which limits the amount of food that can be ingested but does not affect absorption. The Roux-en-Y gastric bypass creates a larger gastric pouch that empties directly into the jejunum and bypasses the duodenum. This both limits the amount of food that can be ingested and causes some degree of malabsorption. Surgical intervention is often an effective mode of weight loss for severely overweight patients. During the first 1–2 years after the surgery, 60–75% of the initial excess weight may be lost. On a long-term basis there tends to be some partial regain [49].

Drugs are generally considered not a sole source of treatment but one component of a comprehensive weight-loss program that includes dietary restriction and behavior modification [50–52]. A new phase in the pharmacological treatment of obesity began in 1991 with the publication of the results of a trial of phentermine and fenfluramine used in combination [53]. Although the combination had not been approved by the U.S. Food and Drug Administration (FDA), in 1996 the total number of prescriptions in the United States for fenfluramine and phentermine exceeded 18 million. Dexfenfluramine, developed during the 1970s and licensed in Europe in the 1980s, was approved by the FDA in 1996 for use in the United States, where it was marketed under the name of Redux. Controversy arose over the effect of dexfenfluramine on the risk of developing primary pulmonary hypertension [54,55]. However, a greater concern arose when valvular heart disease was reported in 24 women treated with fenfluramine-phentermine who had no history of cardiac disease [56]. Because of additional reports of fenfluramine- or dexfenfluramine-associated valvulopathy, the FDA then requested withdrawal of these drugs from the market [57].

Several new drugs have also been developed. Sibutramine (Meridia) is an antiobesity drug that works to suppress appetite primarily by inhibiting the reuptake of the neurotransmitters norepinephrine and serotonin [58]. Orlistat (Xenical) is the first of a new class of nonsystemically acting antiobesity drugs called lipase inhibitors that act in the gastrointestinal tract to block the absorption of fat by about 30% [59].

Public health and clinical guidelines on treatment for obesity generally include degree of overweight, health risk factors, and comorbid conditions as aspects to consider when making decisions on treatment and identifying the most appropriate treatment for an individual [14,17,60,61] An Institute of Medicine committee has recommended criteria for evaluating weight-management programs [62]. Evidence suggests that even without reaching "ideal" weight, a moderate amount of weight loss can be beneficial in terms of reducing levels of some risk factors for mortality or morbidity, such as blood pressure [63].

Long-term weight loss has proven difficult to achieve. Most studies of dietary and behavioral treatments show that after 5 years most of the weight lost will have been regained [32]. The lack of long-term efficacy of treatments for obesity may be due to compensatory changes in energy expenditure that oppose the maintenance of a weight different from usual weight [64].

There have been some controversies regarding the value and appropriate uses of treatments for obesity [65–68]. Intensive treatments to achieve weight loss, including VLCDs, surgery, and pharmacological treatments, are not without risks. Weight loss has been shown to improve numerous health risk factors. However, the benefits of weight loss on long-term health outcomes, such as disease incidence and mortality, have been difficult to demonstrate. Some observational studies suggest that weight loss might actually be disadvantageous [69]. The frequent regain of lost weight has also raised concerns over the effects of cycles of loss and regain [70–72]. A number of large population-based observational studies have suggested increased risk of mortality and morbidity associated with fluctuations in body weight. Methodological issues with these studies include the wide variety of definitions of weight cycling used, the difficulty of distinguishing between voluntary and involuntary weight loss, the lack of association of weight cycling with changes in cardiovascular risk factors or other explanatory factors, and the possibility of different effects in different subgroups. Because of limited and inconsistent evidence, it has been suggested that there should be clinical trials of the effects of weight loss [73,74].

VIII. Epidemiologic Issues, Including Methodology for Study and Public Health Impact

Overweight or obesity, variously defined, clearly affects large numbers of people. Even though the prevalence depends on the exact definition chosen, nonetheless obesity appears to be a highly prevalent and increasing condition in the U.S. and in many other parts of the world. Little is known about the causes of these increases. Interventions at the community or workforce level have not prevented increases [75]. Little research has been done on the prevention of obesity, which appears to be a formidable task requiring some innovative public health strategies [44,76].

The net effect of overweight and obesity on morbidity and mortality is difficult to quantify. The figure of 300,000 deaths per year in the U.S. is an estimate for mortality related to any aspect of dietary intake and physical activity and is not specific to obesity [77]. Higher body weight is associated with increased incidence and prevalence of numerous conditions, including hypertension, diabetes mellitus, hyperlipidemia, certain cancers, musculoskeletal disorders, and cardiovascular disease, and with increased risk of disability [78,79]. Higher body weights among women are associated with increased risks of cardiovascular mortality and morbidity [80,81]. On the other hand, higher body weight is also associated with some benefits, including increased bone density and a lower prevalence of osteoporosis and of hip fracture [6]. In the older age groups, which tend to have the highest mortality and morbidity, there appears to be less of an association of weight with mortality than in younger groups [82,83]. It has been suggested that at older ages, the negative aspects of obesity may to some extent be counterbalanced by the positive aspects [83].

Many aspects of the relationship between weight and weight change and health remain controversial, with findings remarkably unresolved. Different studies have produced a wide variety of results regarding the issue of what weights are associated with lowest mortality and morbidity at different ages [81,84–87]. Some studies have shown little effect of BMI on all-cause mortality for women [88]. The effects of weight loss and weight cycling on mortality and morbidity remain controversial [69].

The costs of obesity are high, although few true economic evaluations of obesity have been carried out [89]. The direct medical costs of obesity in the U.S. have been estimated to be over $50 billion in 1995 dollars, with almost the same amount of indirect costs [90]. The costs of treatments for obesity are also high. For example, it has been estimated that in 1989 alone, Americans spent over $30 billion trying to lose weight [46].

Considerable attention has been paid to body weight and body mass index in epidemiological studies. Body weight and height can be easily and accurately measured. In addition, most adults know their height and weight fairly accurately; thus, many epidemiologic studies, such as the Nurses Health Study, use data on self-reported height and weight [91]. However, self-reported weight is characterized by systematic underreporting that is greater at higher weight levels [92]. This systematic bias could affect the relationship between BMI and mortality in studies that use self-reported weight and height.

Although obesity is considered a multifactorial condition, nonetheless it is often viewed unidimensionally and described and studied as a simple issue of body weight. However, some data suggest that body weight, which is only an indirect measure of body fat content, and even body fat content itself may only be indirectly related to health risk. For example, some research suggests that physical fitness may be a more significant concern than weight per se [93,94]. One important risk factor appears to be the distribution of body fat [95,96]. In particular, excess visceral abdominal fat has been suggested as part of what has been variously called the insulin resistance syndrome, syndrome X, or the deadly quartet—a cluster of metabolic abnormalities related to hypertension, hyperinsulinemia, hypertriglyceridemias, and perhaps also to hyperlipidemia, diabetes, and cardiovascular disease [97–100]. It has even been suggested that some individuals whose weight is normal may be considered "metabolically obese" [101].

IX. Summary and Conclusions/Future Directions

Body weight is salient and a matter of extreme social concern. Body weight is also easily measured and can be obtained through self-report, thus making it feasible for large-scale studies. The emphasis on body weight in epidemiologic studies is consistent with contemporary cultural preoccupations but may be somewhat misleading. Health risk factors may not be adequately captured by simple measurements or reports of body weight, particularly when weight is only measured at a single point in time. An individual's body weight reflects body composition and adipose tissue distribution that in turn reflect a combination of genetic factors, physiological status, individual behaviors, and environmental and social influences. The real health risk factors are difficult to identify in this complex situation.

Obesity has increasingly been medicalized, but this medicalization should not necessarily be accepted uncritically [65,102, 103]. Defining obesity as a chronic disease in itself, rather than a risk factor for other conditions, has important implications for treatment decisions and for health care reimbursements, particularly considering that at present our knowledge regarding the effects of weight loss in obesity on health outcomes is limited. Should otherwise healthy individuals be treated for obesity and if so at what level of body weight? How vigorous a treatment is appropriate, and should treatment be life-long? These questions may be particularly pertinent for women for a number of reasons. Women tend to have not only higher prevalences of obesity but also high levels of concern about body weight even at lower levels of BMI. On the other hand, women may be at somewhat lower risk than men for adverse health outcomes, in part because women tend to have lower-body obesity and adipose tissue located subcutaneously rather than viscerally [104].

Many important basic questions remain to be answered and may need to be reformulated. Can obesity be prevented? What are the causes of obesity? What are the health consequences of obesity? What are the effects of weight loss on health? Controversies over weight loss could potentially be clarified by clinical trials. Several ventures are currently in progress that will help to provide answers to some of these questions. The Swedish Obese Subjects study includes surgical intervention for a cohort of severely obese individuals with follow-up data being collected on quality of life and on subsequent mortality and morbidity [105]. Plans are currently underway in the U.S. for a clinical trial of weight loss, the SHOW trial, sponsored by the NIH and the Centers for Disease Control and Prevention [74].

The social costs of obesity are high, and the social costs of attempts to prevent or to treat obesity are high as well. The prevalence of obesity is increasing in most parts of the world and appears likely to continue to increase in the future. The health risks associated with these increases and the risks and benefits of prevention and treatment strategies, including long-term use of pharmacologic treatment, need to be evaluated objectively. The challenge is to identify realistic goals and strategies to maintain and improve health in the context of increasing environmental pressure towards higher weights and higher levels of overweight and obesity.

References

1. Wang, Z. M., Pierson, R. N., Jr., and Heymsfield, S. B. (1992). The five-level model: A new approach to organizing body-composition research. *Am. J. Clin. Nutr.* **56,** 19–28.
2. Vogel, J. A., and Friedl, K. E. (1992). Body fat assessment in women. Special considerations. *Sports Med.* **13,** 245–269.
3. Forbes, G. B. (1988). Body composition: Influence of nutrition, disease, growth, and aging. *In* "Modern Nutrition in Health and Disease" (M. E. Shils and V. R. Young, eds.), 7th ed., pp. 533–556. Lea & Febiger, Philadelphia.
4. Ley, C. J., Lees, B., and Stevenson, J. C. (1992). Sex- and menopause-associated changes in body-fat distribution. *Am. J. Clin. Nutr.* **55,** 950–954.
5. Lemieux, S., Prud'homme, D., Bouchard, C., Tremblay, A., and Despres, J. P. (1993). Sex differences in the relation of visceral adipose tissue accumulation to total body fatness. *Am. J. Clin. Nutr.* **58,** 463–467.
6. Norgan, N. G. (1997). The beneficial effects of body fat and adipose tissue in humans. *Int. J. Obes. Relat. Metab. Disord.* **21,** 738–746.
7. Keys, A., Fidanza, F., Karvonen, M. J., Kimura, N., and Taylor, H. L. (1972). Indices of relative weight and obesity. *J. Chronic Dis.* **25,** 329–343.
8. Baumgartner, R. N., Heymsfield, S. B., and Roche, A. F. (1995). Human body composition and the epidemiology of chronic disease. *Obes. Res.* **3,** 73–95.
9. Metropolitan Life Insurance Company (1959). New weight standards for men and women. *Stat. Bull.* **40** (Nov–Dec), 1–4.
10. Metropolitan Life Insurance Company (1983). 1983 Metropolitan height and weight tables. *Stat. Bull.* **64** (Jan–June), 2–9.
11. National Institutes of Health Consensus Development Panel on the Health Implications of Obesity (1985). Health implications of obesity: National Institutes of Health Consensus Development Conference Statement. *Ann. Intern. Med.* **103,** 1073–1077.
12. Garrow J. S. (1981). "Treat Obesity Seriously: A Clinical Manual." Churchill-Livingstone, Edinburgh.
13. Bray, G. A. (1985). Complications of obesity. *Ann. Intern. Med.* **103,** 1052–1062.
14. World Health Organization Expert Committee on Physical Status (1995). "Physical Status: The Use and Interpretation of Anthropometry. Report of a WHO Expert Committee," Tech. Rep. Ser. 854. WHO, Geneva.
15. World Health Organization (1985). "Energy and Protein Requirements. Report of a Joint FAO/WHO/UNU Expert Consultation," Tech. Rep. Ser. 724. WHO, Geneva.
16. World Health Organization (1998). "Obesity: Preventing and Managing the Global Epidemic. Report of a WHO Consultation on Obesity. WHO, Geneva.
17. U.S. Department of Agriculture (1995). "Report of the Dietary Guidelines Advisory Committee on the Dietary Guidelines for Americans, 1995, to the Secretary of Health and Human Services and the Secretary of Agriculture." USDA, Washington, DC.
18. Flegal, K. M., Carroll, M. D., Kuczmarski, R. J., and Johnson, C. L. (1998). Overweight and obesity in the United States: Prevalence and trends, 1960–1994. *Int. J. Obes. Relat. Metab. Disord.* **22,** 39–47.
19. Popkin, B. M., Paeratakul, S., Ge, K., and Zhai, F. (1995). Body weight patterns among the Chinese: Results from the 1989 and 1991 China Health and Nutrition Surveys. *Am. J. Public Health* **85,** 690–694.
20. Sichieri, R., Coitinho, D. C., Leao, M. M., Recine, E., and Everhart, J. E. (1994). High temporal, geographic, and income variation in body mass index among adults in Brazil. *Am. J. Public Health* **84,** 793–798.
21. Bennett, S. A., and Magnus, P. (1994). Trends in cardiovascular risk factors in Australia. Results from the National Heart Foundation's Risk Factor Prevalence Study, 1980–1989. *Med. J. Aust.* **161,** 519–527.
22. Pietinen, P., Vartiainen, E., and Mannisto, S. (1996). Trends in body mass index and obesity among adults in Finland from 1972 to 1992. *Int. J. Obes. Relat. Metab. Disord.* **20,** 114–120.
23. Seidell, J. C., and Flegal, K. M. (1997). Assessing obesity: Classification and epidemiology. *Br. Med. Bull.* **53,** 238–252.
24. Hodge, A. M., Dowse, G. K., Toelupe, P., Collins, V. R., Imo, T., and Zimmet, P. Z. (1994). Dramatic increase in the prevalence of obesity in western Samoa over the 13 year period 1978–1991. *Int. J. Obes. Relat. Metab. Disord.* **18,** 419–428.
25. Prentice, A. M., and Jebb, S. A. (1995). Obesity in Britain: Gluttony or sloth? *Br. Med. J.* **311,** 437–439.
26. Klesges, R. C., Meyers, A. W., Klesges, L. M., and LaVasque, M. E. (1989). Smoking, body weight, and their effects on smoking behavior: A comprehensive review of the literature. *Psychol. Bull.* **106,** 204–230.

27. Wack, J. T., and Rodin, J. (1982). Smoking and its effects on body weight and the systems of caloric regulation. *Am. J. Clin. Nutr.* **35,** 366–380.

28. Cassidy, C. M. (1991). The good body: When big is better. *Med. Anthropol.* **13,** 181–213.

29. Sobal, J., and Stunkard, A. J. (1989). Socioeconomic status and obesity: A review of the literature. *Psychol. Bull.* **105,** 260–275.

30. Sargent, J. D., and Blanchflower, D. G. (1994). Obesity and stature in adolescence and earnings in young adulthood. Analysis of a British birth cohort. *Arch. Pediatr. Adolesc. Med.* **148,** 681–687.

31. Gortmaker, S. L., Must, A., Perrin, J. M., Sobol, A. M., and Dietz, W. H. (1993). Social and economic consequences of overweight in adolescence and young adulthood. *N. Engl. J. Med.* **329,** 1008–1012.

32. NIH Technology Assessment Conference Panel (1992). Methods for voluntary weight loss and control. *Ann. Intern. Med.* **116,** 942–949.

33. Rodin, J., Silberstein, L., and Striegel-Moore, R. (1984). Women and weight: A normative discontent. *Nebr. Symp. Motiv.* **32,** 267–307.

34. Latham, M. C. (1997). "Human Nutrition in the Developing World." Food and Agriculture Organization of the United Nations, Rome.

35. Popkin, B. M. (1994). The nutrition transition in low-income countries: An emerging crisis. *Nutr. Rev.* **52,** 285–298.

36. Rosenbaum, M., Leibel, R. L., and Hirsch, J. (1997). Obesity. *N. Engl. J. Med.* **337,** 396–407.

37. Saltzman, E., and Roberts, S. B. (1995). The role of energy expenditure in energy regulation: Findings from a decade of research. *Nutr. Rev.* **53,** 209–220.

38. Auwerx, J., and Staels, B. (1998). Leptin. *Lancet* **351,** 737–742.

39. Stunkard, A. J., Harris, J. R., Pedersen, N. L., and McClearn, G. E. (1990). The body-mass index of twins who have been reared apart. *N. Engl. J. Med.* **322,** 1483–1487.

40. Bouchard, C., and Tremblay, A. (1997). Genetic influences on the response of body fat and fat distribution to positive and negative energy balances in human identical twins. *J. Nutr.* **127**(5 Suppl.), 943S–947S.

41. Austin, M. A., Friedlander, Y., Newman, B., Edwards, K., Mayer-Davis, E. J., and King, M. C. (1997). Genetic influences on changes in body mass index: A longitudinal analysis of women twins. *Obes. Res.* **5,** 326–331.

42. Korkeila, M., Kaprio, J., Rissanen, A., and Koskenvuo, M. (1995). Consistency and change of body mass index and weight. A study on 5967 adult Finnish twin pairs. *Int. J. Obes. Relat. Metab. Disord.* **19,** 310–317.

43. Flegal, K. M., Troiano, R. P., Pamuk, E. R., Kuczmarski, R. J., and Campbell, S. M. (1995). The influence of smoking cessation on the prevalence of overweight in the United States. *N. Engl. J. Med.* **333,** 1165–1170.

44. James, W. P. T. (1995). A public health approach to the problem of obesity. *Int. J. Obes. Relat. Metab. Disord.* **19**(Suppl. 3), S37–S45.

45. Briefel, R. R., Sempos, C. T., McDowell, M. A., Chien, S., and Alaimo, K. (1997). Dietary methods research in the third National Health and Nutrition Examination Survey: Underreporting of energy intake. *Am. J. Clin. Nutr.* **65**(4 Suppl.), 1203S–1209S.

46. National Task Force on the Prevention and Treatment of Obesity (1993). Very low-calorie diets. *JAMA, J. Am. Med. Assoc.* **270,** 967–974.

47. Foreyt, J. P., and Goodrick, G. K. (1993). Evidence for success of behavior modification in weight loss and control. *Ann. Intern. Med.* **119**(7, Pt. 2), 698–701.

48. Mason, E. E., and Doherty, C. (1993). Surgery. *In* "Obesity: Theory and Therapy" (A. J. Stunkard and T. A. Wadden, eds.), 2nd ed., pp. 313–325. Raven Press, New York.

49. Kolanowski, J. (1997). Surgical treatment for morbid obesity. *Br. Med. Bull.* **53,** 433–444.

50. Bray, G. A. (1995). Evaluation of drugs for treating obesity. *Obes. Res.* (3, Suppl. 4), 425S–434S.

51. Blundell, J. E., and Halford, J. C. (1995). Pharmacological aspects of obesity treatment: Towards the 21st century. *Int. J. Obes. Relat. Metab. Disord.* **19**(Suppl. 3), S51–S55.

52. Finer, N. (1997). Present and future pharmacological approaches. *Br. Med. Bull.* **53,** 409–432.

53. Weintraub, M. (1992). Long-term weight control: The National Heart, Lung, and Blood Institute funded multimodal intervention study. *Clin. Pharmacol. Ther.* **51,** 581–585; *erratum:* **52**(3), 323.

54. Abenhaim, L., Moride, Y., Brenot, F., Rich, S., Benichou, J., Kurz, X., Higenbottam, T., Oakley, C., Wouters, E., Aubier, M., Simonneau, G., and Begaud, B. (1996). Appetite-suppressant drugs and the risk of primary pulmonary hypertension. International Primary Pulmonary Hypertension Study Group. *N. Engl. J. Med.* **335,**609–616.

55. Manson, J. E., and Faich, G. A. (1996). Pharmacotherapy for obesity—do the benefits outweigh the risks? *N. Engl. J. Med.* **335,** 659–660.

56. Connolly, H. M., Crary, J. L., McGoon, M. D., Hensrud, D. D., Edwards, B. S., Edwards, W. D., and Schaff, H. V. (1997). Valvular heart disease associated with fenfluramine-phentermine. *N. Engl. J. Med.* **337,** 581–588.

57. Centers for Disease Control and Prevention. (1997). Cardiac valvulopathy associated with exposure to fenfluramine or dexfenfluramine. U.S. Department of Health and Human Services Interim Public Health Recommendations, November 1997. *Morbid. Mortal. Wkly. Rep.* **46,** 1061–1066.

58. Ryan, D. H., Kaiser, P., and Bray, G. A. (1995). Sibutramine: A novel new agent for obesity treatment. *Obes. Res.* **3** (Suppl. 4), 553S–559S.

59. Drent, M. L., and van der Veen, E. A. (1995). First clinical studies with orlistat: A short review. *Obes. Res.* **3** (Suppl. 4), 623S–625S.

60. Brownell, K. D., and Wadden, T. A. (1991). The heterogeneity of obesity: Fitting treatments to individuals. *Behav. Ther.* **22,** 153–177.

61. Yanovski, S. Z. (1993). A practical approach to treatment of the obese patient. *Arch. Fam. Med.* **2,** 309–316.

62. Stern, J. S., Hirsch, J., Blair, S. N., Foreyt, J. P., Frank, A., Kumanyika, S. K., Madans, J. H., Marlatt, G. A., St. John, S. T., and Stunkard, A. J. (1995). Weighing the options: Criteria for evaluating weight-management programs. The Committee to Develop Criteria for Evaluating the Outcomes of Approaches to Prevent and Treat Obesity. *Obes. Res.* **3,** 591–604.

63. Goldstein, D. J. (1992). Beneficial health effects of modest weight loss. *Int. J. Obes. Relat. Metab. Disord.* **16,** 397–415.

64. Leibel, R. L., Rosenbaum, M., and Hirsch, J. (1995). Changes in energy expenditure resulting from altered body weight. *N. Engl. J. Med.* **332,** 621–628.

65. Kassirer, J. P., and Angell, M. (1998). Losing weight—an ill-fated New Year's resolution. *N. Engl. J. Med.* **338,** 52–54.

66. Garrow, J. S. (1994). Should obesity be treated? Treatment is necessary. *Br. Med. J.* **309,** 654–655.

67. Wooley, S. C., and Garner, D. M. (1994). Dietary treatments for obesity are ineffective. *Br. Med. J.* **309,** 655–656.

68. Brownell, K. D., and Rodin, J. (1994). The dieting maelstrom. Is it possible and advisable to lose weight? *Am. Psychol.* **49,** 781–791.

69. Williamson, D. F., and Pamuk, E. R. (1993). The association between weight loss and increased longevity. A review of the evidence. *Ann. Intern. Med.* **119**(7, Pt. 2), 731–736.

70. National Task Force on the Prevention and Treatment of Obesity (1994). Weight cycling. *JAMA, J. Am. Med. Assoc.* **272**, 1196–1202.

71. Brownell, K. D., and Rodin, J. (1994). Medical, metabolic, and psychological effects of weight cycling. *Arch. Intern. Med.* **154**, 1325–1330.

72. Lissner, L., Odell, P. M., D'Agostino, R. B., Stokes, J., Kreger, B. E., Bélanger, A. J., and Brownell, K. D. (1991). Variability of body weight and health outcomes in the Framingham population. *N. Engl. J. Med.* **324**, 1839–1844.

73. Stern, M. P. (1995). The case for randomized clinical trials on the treatment of obesity. *Obes. Res.*(3, Suppl. 2), 299s–306s.

74. Yanovski, S. Z., Bain, R. P., and Williamson, D. F. (1999). Report of a National Institutes of Health—Centers for Disease Control and Prevention workshop on the feasibility of conducting a randomized clinical trial to estimate the long-term health effects of intentional weight loss in obese persons. *Am. J. Clin. Nutr.* **69**, 366–372.

75. Jeffrey, R. W. (1998). Prevention of obesity. *In* "Handbook of Obesity" (G. A. Bray, C. Bouchard, and W. P. T. James, eds.), pp. 819–829. Dekker, New York.

76. Gill, T. P. (1997). Key issues in the prevention of obesity. *Br. Med. Bull.* **53**, 359–388.

77. McGinnis, J. M., and Foege, W. H. (1993). Actual causes of death in the United States. *JAMA, J. Am. Med. Assoc.* **270**, 2207–2212.

78. Pi-Sunyer, F. X. (1993). Medical hazards of obesity. *Ann. Intern. Med.* **119**(7, Pt. 2), 655–660.

79. Bray, G. A. (1996). Health hazards of obesity. *Endocrinol. Metab. Clin. North Am.* **25**, 907–919.

80. Rissanen, A., Knekt, P., Heliovaara, M., Aromaa, A., Reunanen, A., and Maatela, J. (1991). Weight and mortality in Finnish women. *J. Clin. Epidemiol.* **44**, 787–795.

81. Manson, J. E., Willett, W. C., Stampfer, M. J., Colditz, G. A., Hunter, D. J., Hankinson, S. E., Hennekens, C. H., and Speizer, F. E. (1995). Body weight and mortality among women. *N. Engl. J. Med.* **333**, 677–685.

82. Stevens, J., Cai, J., Pamuk, E. R., Williamson, D. F., Thun, M. J., and Wood, J. L. (1998). The effect of age on the association between body-mass index and mortality. *N. Engl. J. Med.* **338**, 1–7.

83. Diehr, P., Bild, D. E., Harris, T. B., Duxbury, A., Siscovick, D., and Rossi, M. (1998). Body mass index and mortality in nonsmoking older adults: The Cardiovascular Health Study. *Am. J. Public Health* **88**, 623–629.

84. Troiano, R. P., Frongillo, E. A., Jr., Sobal, J., and Levitsky, D. A. (1996). The relationship between body weight and mortality: A quantitative analysis of combined information from existing studies. *Int. J. Obes. Relat. Metab. Disord.* **20**, 63–75.

85. Durazo-Arvizu, R., Cooper, R. S., Luke, A., Prewitt, T. E., Liao, Y., and McGee, D. L. (1997). Relative weight and mortality in U.S. blacks and whites: Findings from representative national population samples. *Ann. Epidemiol.* **7**, 383–395.

86. Solomon, C. G., and Manson, J. E. (1997). Obesity and mortality: A review of the epidemiologic data. *Am. J. Clin. Nutr.* **66**(4 Suppl.), 1044S–1050S.

87. Andres, R., Elahi, D., Tobin, J. D., Muller, D. C., and Brant, L. (1985). Impact of age on weight goals. *Ann. Intern. Med.* **103**, 1030–1033.

88. Seidell, J. C., Verschuren, W. M., van Leer, E. M., and Kromhout, D. (1996). Overweight, underweight, and mortality. A prospective study of 48,287 men and women. *Arch. Intern. Med.* **156**, 958–963.

89. Hughes, D., and McGuire, A. (1997). A review of the economic analysis of obesity. *Br. Med. Bull.* **53**, 253–263.

90. Wolf, A. M., and Colditz, G. A. (1998). Current estimates of the economic cost of obesity in the United States. *Obes. Res.* **6**, 97–106.

91. Troy, L. M., Hunter, D. J., Manson, J. E., Colditz, G. A., Stampfer, M. J., and Willett, W. C. (1995). The validity of recalled weight among younger women. *Int. J. Obes. Relat. Metab. Disord.* **19**, 570–572.

92. Plankey, M. W., Stevens, J., Flegal, K. M., and Rust, P. F. (1997). Prediction equations do not eliminate systematic error in self-reported body mass index. *Obes. Res.* **5**, 308–314.

93. Blair, S. N., Kohl, H. W., and Barlow, C. E. (1993). Physical activity, physical fitness, and all-cause mortality in women: Do women need to be active? *J. Am. Coll. Nutr.* **12**, 368–371.

94. Barlow, C. E., Kohl, H. W., 3rd, Gibbons, L. W., and Blair, S. N. (1995). Physical fitness, mortality and obesity. *Int. J. Obes. Relat. Metab. Disord.* **19**(Suppl. 4), S41–S44.

95. Despres, J. P. (1993). Abdominal obesity as important component of insulin-resistance syndrome. *Nutrition* **9**, 452–459.

96. Abate, N., and Garg, A. (1995). Heterogeneity in adipose tissue metabolism: Causes, implications and management of regional adiposity. *Prog. Lipid Res.* **34**, 53–70.

97. DeFronzo, R. A., and Ferrannini, E. (1991). Insulin resistance. A multifaceted syndrome responsible for NIDDM, obesity, hypertension, dyslipidemia, and atherosclerotic cardiovascular disease. *Diabetes Care* **14**, 173–194.

98. Reaven, G. M. (1993). Role of insulin resistance in human disease (syndrome X): An expanded definition. *Annu. Rev. Med.* **44**, 121–131.

99. Ferrannini, E., and Natali, A. (1991). Essential hypertension, metabolic disorders, and insulin resistance. *Am. Heart J.* **121**(4, Pt. 2), 1274–1282.

100. Kaplan, N. M. (1989). The deadly quartet. Upper-body obesity, glucose intolerance, hypertriglyceridemia, and hypertension. *Arch. Intern. Med.* **149**, 1514–1520.

101. Ruderman, N., Chisholm, D., Pi-Sunyer, X., and Schneider, S. (1998). The metabolically obese, normal-weight individual revisited. *Diabetes* **47**, 699–713.

102. Ritenbaugh, C. (1982). Obesity as a culture-bound syndrome. *Cult. Med. Psychiatry* **6**, 347–361.

103. Sobal, J. (1995). The medicalization and demedicalization of obesity. *In* "Eating Agendas: Food and Nutrition as Social Problems" (D. Maurer and J. Sobal, eds.), pp. 67–90. de Gruyter, New York.

104. Stevens, J. (1995). Obesity, fat patterning and cardiovascular risk. *Adv. Exp. Med. Biol.* **369**, 21–27.

105. Sjöström, L., Larsson, B., Backman, L., Bengtsson, C., Bouchard, C., Dahlgren, S., Hallgren, P., Jonsson, E., Karlsson, J., Lapidus, L., Lindroos, A., Lindstedt, S., Lissner, L., Narbro, K., Naslund, I., Olbe, L., Sullivan, M., Sylvan, A., Wedel, H., and Agren, G. (1992). Swedish obese subjects (SOS). Recruitment for an intervention study and a selected description of the obese state. *Int. J. Obes. Relat. Metab. Disord.* **16**, 465–479.

66

Coronary Heart Disease, Lipid Metabolism, and Steroid Hormones in Women

RHOBERT W. EVANS
Department of Epidemiology
University of Pittsburgh
Pittsburgh, Pennsylvania

I. Introduction

Coronary Heart Disease (CHD) is the leading cause of death among women. In the United States the annual average number of women who died of CHD during 1988 to 1992 was 362,000; in contrast, an average of 234,000 died from malignant neoplasms, the second leading cause of death. Despite the customary tendency to associate CHD with men, the proportion of women, 35.1%, who die of CHD is marginally higher than that for men, 32.8% [1]. The major risk factors for CHD are cigarette smoking, diabetes, family history of CHD, hypertension, hyperlipidemia, obesity, and physical inactivity, and they are similar for men and women [2,3]. In this chapter, the main focus will be on the association between lipids and CHD with an emphasis on the role of steroid sex hormones in lipoprotein metabolism. The major pathways of lipid metabolism are outlined and the targets for hormone action noted. In addition to lipoproteins, reference will be made to the role of eicosanoids, oxidized fatty acids, in CHD. The effects of oral contraceptives (OC), hormone replacement therapy (HRT), and phytoestrogens on lipoprotein metabolism, as well as the modifications in lipid levels observed during puberty, the menstrual cycle, pregnancy, menopause, and polycystic ovary syndrome, are reported. Changes in lipoprotein concentrations with age in women are related to incidence of CHD and the results contrasted with the values obtained for men. The influence of other CHD risk factors such as parity, smoking, obesity, physical inactivity, and diabetes on lipoprotein profiles is also cited. Finally, the effect of dietary, pharmacological, and hormonal therapy on lipoprotein levels and CHD rates are presented.

II. Metabolism

In this section the fundamental pathways of lipid metabolism are presented and the actions of steroid hormones noted. A broad review of lipids and womens' health was published in 1991 [4].

A. Dietary Lipids

Approximately 90% of the fat consumed per day is triacylglycerol (TG). The remainder includes glycolipids, phospholipids (PL), nonesterified fatty acids (NEFA), and about 100–700 mg of cholesterol. (Unless specifically noted the term cholesterol in this chapter refers to both esterified [CE] and free cholesterol [C]). About 50–70% of the dietary cholesterol is absorbed by what is generally considered to be a diffusion process. However, the observation that less than 5% of plant sterols are absorbed suggests that a transport protein is also involved that may serve as a target for manipulating the efficiency of cholesterol absorption [5]. Dietary cholesterol and saturated fats may raise low density lipoprotein cholesterol (LDLc) concentrations by increasing the rate of LDL synthesis and by reducing LDL receptor activity [6].

B. Fatty Acids

Saturated fatty acids, whether synthesized or of dietary origin, can be further elongated and desaturated (Fig. 66.1). Little conversion of 16:0 (palmitic) to 16:1n7 (palmitoleic) takes place but 18:0 (stearic) is efficiently desaturated to 18:1n9 (oleic). This may explain why 18:0, unlike other saturated fatty acids (12:0, 14:0, 16:0), does not raise TC (total cholesterol), LDLc,

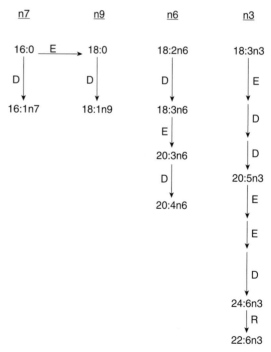

Fig. 66.1 Simplified representation of fatty acid synthesis in mammalian cells. D, desaturation; E, two carbon elongation; R, retroconversion (two carbon cleavage).

or high density lipoprotein cholesterol (HDLc) levels. Feeding 18:0 is effectively giving the individual 18:1n9 [7].

About 4% of the fatty acids ingested possess a trans, instead of a cis, double bond. These trans fatty acids are enriched in dairy and hydrogenated food products. The trans monounsaturated fatty acids, primarily elaidic (18:1n9t), may lower HDLc and raise lipoprotein(a) (Lp[a]) and LDLc levels, thus eliciting an atherogenic lipoprotein profile [8].

Mammalian cells cannot synthesize n6 and n3 fatty acids. These are essential fatty acids that must be ingested in the diet. Mammalian cells can elongate and desaturate dietary precursors, which are mainly linoleic acid (18:2n6) and linolenic acid (18:3n3), but there is no interconversion between the n6 and n3 families (Fig. 66.1). Linoleic acid (18:2n6) is widespread in the diet whereas arachidonic acid (20:4n6) is mainly found in meat products. Arachidonic acid is the major precursor for eicosanoids which have broad physiological effects, including the prothrombotic and antithrombotic roles of thromboxane (TxA_2) and prostacyclin (PGI_2), respectively. The n6 fatty acids are associated with reduced levels of TC and LDLc. The suggested mechanisms for this include diminished synthesis of very low density lipoprotein (VLDL), enhanced LDL receptor activity, and less cholesterol per LDL particle. Similarly, w6 acids may elicit small reductions in HDLc levels by inhibiting the synthesis of apoprotein A-I (apoA-I) [9].

Eicosapentaenoic acid (20:5n3) is also a precursor of eicosanoids, including TxA_3 and PGI_3, but it is present at far lower concentrations than 20:4n6 and is less efficiently metabolized. The inactivity of TxA_3 may partly explain the reduced rate of CHD among populations that eat large amounts of fish, a source that is rich in 20:5n3 and 22:6n3. However, the competition between n6 and n3 fatty acids for synthetic enzymes and their interactions at the eicosanoid level are complex and, in general, poorly understood. Linolenic acid (18:3n3) is plentiful in vegetable sources but its conversion to 20:5n3 and 22:6n3 in mammalian cells is inefficient. Adequate levels of 20:5n3 and 22:6n3 require intake of the preformed acids that are found in abundance among marine organisms. The elongated n3 fatty acids have little effect on TC but lower TG concentrations. The n3 acids stimulate the transcription of peroxisomal oxidizing enzymes, thus fatty acids are directed to oxidation rather than to TG and lipoprotein synthesis [10]. The n3 acids have little effect on HDLc levels [11].

C. Chylomicron (Exogenous) Pathway

The main function of the chylomicrons is to deliver dietary TG to adipose tissue, muscles, and the liver. The lipids after entering the mucosal cells are converted back into TG, CE, and PL and then packaged with endogenous lipids and apoproteins into nascent chylomicrons (Fig. 66.2). These large, TG-rich lipoproteins (TGRL) are secreted into the lymph and then migrate via the thoracic duct into the superior vena cava. The nascent chylomicrons contain $apoB_{48}$ and the soluble apolipoproteins (apoA-I, apoA-II, and apoA-IV). Once in the circulation the chylomicrons transfer apoA-I, apoA-II, apoA-IV, and PL to HDL, and receive apoC-II, apoE, and CE from HDL. Exchange of surface lipids and apoproteins occurs extensively between lipoproteins. The apoC-II is essential for activating LPL (lipoprotein lipase), present in the capillary beds, which, together with hepatic lipase (HL), hydrolyses the TG of the chylomicrons. Glycerol, monoacylglycerol (MG), and NEFA are released into the circulation and as the chylomicrons decrease in size they become comparatively TG-poor and eventually form chylomicron remnants (CR). These CR spare the apoC-II by returning it to the HDL before they are taken up by the liver via the CR and LDL receptors which recognize apoE moieties. The half-life of the chylomicrons within the circulation is about 10 min [12].

Estrogens primarily influence the chylomicron pathway by promoting the clearance of CR from the circulation [13]. This is beneficial as prolonged postprandial elevation of TG has been postulated to be thrombogenic. The levels of the procoagulation factors, factor VII and plasminogen activator inhibitor-1 (PAI-1), have been correlated with the concentration of TG and, *in vitro*, elevation of TGRL levels results in enhanced release of PAI-1 by endothelial cells [14,15]. Estrogens may elicit their effect by enhancing the activity of LDL receptors which facilitates removal of the CR from the circulation [16].

D. Very Low Density Lipoprotein (VLDL) (Endogenous) Pathway

The major role of VLDL is to deliver endogenous TG to adipose tissue and muscles. The VLDL are synthesized in the liver and the main apoproteins present in the mature VLDL are $apoB_{100}$, apoC-II, and apoE. The lipids in VLDL can be the result of *de novo* synthesis, come from the diet via CR, or be returning to the liver in intermediate density lipoproteins (IDL) or LDL (Fig. 66.3). NEFA, transported by albumin, are also obtained from adipose tissue. The VLDL pathway is similar to the chylomicron pathway. The TG of VLDL are hydrolyzed by lipoprotein lipase and the particles gradually become smaller, denser, and enriched in cholesterol. VLDL have a half-life of about 2 hours. At an intermediate stage IDL are present that have a half-life of a few minutes. The IDL contain $apoB_{100}$ and multiple copies of apoE which are recognized by the remnant and LDL receptors. The IDL remaining in the blood are subjected to further hydrolysis by lipoprotein lipase and eventually transfer all their apoEs to HDL. Ultimately, small cholesterol-rich LDL particles are formed. The LDL contain one copy of $apoB_{100}$ that is poorly recognized by the LDL receptor, thus the LDL have a comparatively long half-life of about three days. About 70% of the LDL return to the liver and the remainder deliver cholesterol to extrahepatic tissue.

Estrogens affect the VLDL pathway by stimulating the synthesis of TG and $apoB_{100}$ in the liver, thus providing more substrate for VLDL synthesis. This change would favor an increase in VLDL and LDL levels. However, estrogens also inhibit HL and LPL activities, thus again promoting accumulation of VLDL but reducing LDL formation. The role of estrogens in increasing the number of LDL receptors exposed, enhancing the uptake of IDL and LDL, further diminishes LDL accumulation.

The removal of IDL and LDL by the liver is also influenced by the particular apoE isoforms present. ApoE exists as three common alleles, apoE2, apoE3, and apoE4, of which apoE3 is

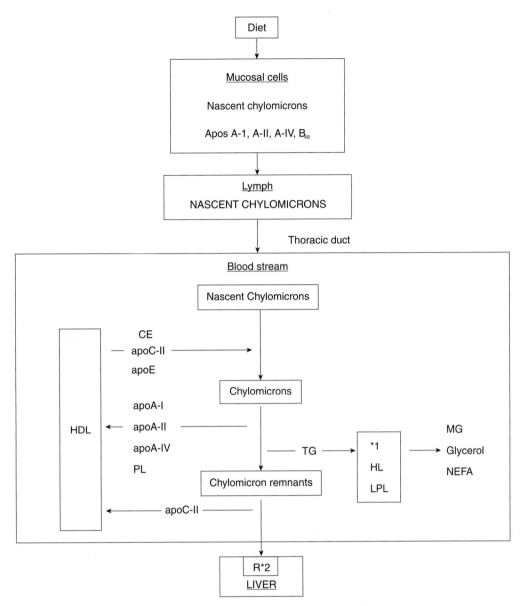

Fig. 66.2 Metabolism of chylomicrons. *1, hepatic (HL) and lipoprotein lipases (LPL) activity inhibited by estrogens and promoted by progestogens and androgens; *2, expression of receptors (R) stimulated by estrogens.

the most common. The frequencies of the apoE alleles are similar among men and women (apoE3, 82%; apoE4, 12%; apoE2, 6%) [17,18]. A meta-analysis indicated that in comparison to subjects with the apoE3/3 phenotype, TC levels are higher in subjects with apoE4/4 or apoE4/3 phenotypes but lower in those with apoE3/2 or apoE2/2 phenotypes. All ApoE phenotypes except apoE4/4 raised TG concentrations compared to the apoE3/3 set. ApoE phenotype has little effect on HDLc values although the apoE4/3 group has lower HDLc levels than the apoE3/3 subjects [19].

The elevated TC and LDLc levels associated with apoE4 are due to enhanced efficiency of intestinal cholesterol absorption and the more efficient clearance of chylomicron remnants (order

of efficiencies: apoE4 > apoE3 > apoE2). This results in decreased numbers of LDL receptors and elevated levels of TC and LDLc. The lipid alterations associated with the apoE2 reflect the inefficient conversion of VLDL to LDL and the delayed clearance of CR. The apoE2 binds poorly (<2% compared to apoE3) to the LDL and apoE receptors. The decreased delivery of cholesterol to the liver increases the number of LDL receptors resulting in reduced levels of TC and LDLc [17,20].

The presence of apoE4 is associated with an enhanced risk of CHD but this increased risk is only partially explained by the altered lipoprotein profile. The relationship between apoE2 and CHD is unclear as the low frequency of this isoform hinders investigation [21,22].

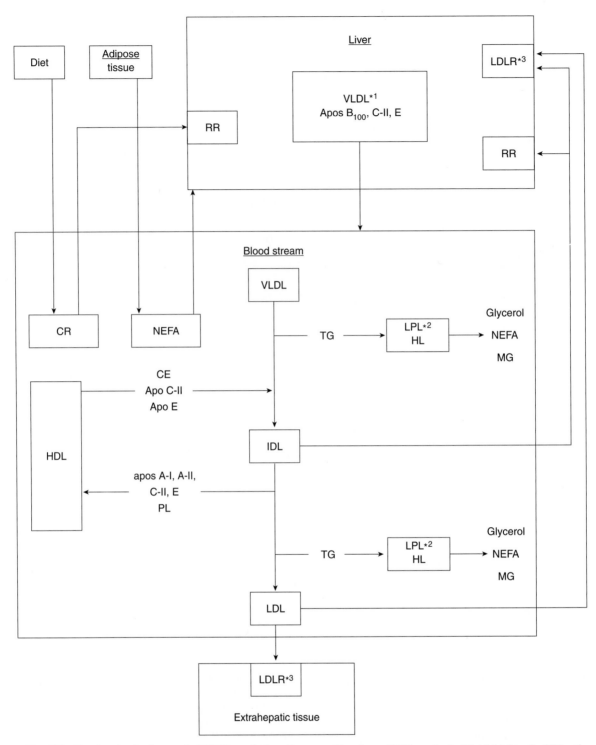

Fig. 66.3   Very low density lipoprotein (VLDL) metabolism. Estrogens: *1, enhance VLDL synthesis; *2, inhibit hepatic (HL) and lipoprotein lipase (LPL) activity; *3, promote LDL receptor activity. LDLR, LDL receptor; RR, remnant receptor.

In a separate protective role, estrogens may inhibit LDL oxidation, thus reducing the uptake of modified LDL by scavenger receptors and retarding subsequent foam cell formation. The mechanisms by which estrogens slow lipoprotein oxidation are unclear but may include effects on the redox status of iron [23].

E. High Density Lipoprotein (HDL) Metabolism

HDL play an important role in delivering apoproteins E and C-II to chylomicrons and VLDL. However, HDL metabolism is complex and poorly understood. HDL is synthesized both in the

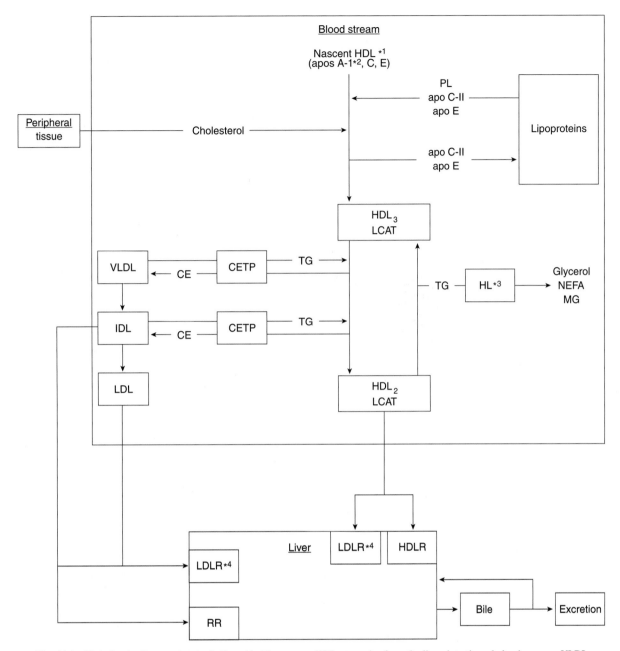

Fig. 66.4 High density lipoprotein metabolism. *1, The nascent HDL may arise from the liver, intestine, chylomicrons, or VLDL. Estrogens: *2, promote apo A-1 synthesis; *3, inhibit hepatic lipase (HL); *4, promote LDL receptors. Testosterone increases HL activity [34]. HDLR, HDL receptor; LDLR, LDL receptor; RR, remnant receptor; LCAT, lecithin cholesterol acyl transferase; CETP, cholesterol ester transfer protein.

liver, with apoE predominating, and in the intestine, with apoA-I predominating. HDL may also arise from the surface remnants generated during VLDL and chylomicron lipolysis, with apoC predominating (Fig. 66.4). The nascent HDL are discoid in shape and rapidly exchange lipids and apoproteins with other lipoproteins, eventually forming HDL_3. Unesterfied cholesterol is readily exchangeable and partially originates from peripheral tissue. Within the HDL_3, LCAT (lecithin-cholesterol acyl transferase) catalyses the transfer of the fatty acid at the 2-position

of phosphatidylcholine, usually linoleic acid, to cholesterol. The resulting lysophosphatidylcholine diffuses into the plasma whereas the CE enters the core of the HDL_3 and is resistant to diffusion. However, a cholesterol ester transfer protein (CETP) catalyses the exchange of CE within HDL_3 for the TG present in chylomicrons, VLDL and IDL. Gradually the HDL_3 is converted into the larger and less dense HDL_2. The TG in HDL_2 can then be hydrolyzed by hepatic lipase and reform HDL_3. Unlike LPL, hepatic lipase is much more active towards HDL than it is against apoB-containing lipoproteins (VLDL, LDL). Alternatively the HDL_2 is removed by the liver, possibly via the LDL receptor or by a poorly described HDL receptor. Within the liver the HDL is degraded and the cholesterol may enter the bile as cholesterol or as bile salts. Some of the bile is excreted, providing a route for elimination of cholesterol from the body. Mammalian cells do not have enzymes capable of degrading the cholesterol ring structure and no specific pathway for excretion of cholesterol exists. As chylomicrons VLDL and IDL exchange triglycerides for cholesterol esters in HDL, the uptake of chylomicron remnants, IDL and LDL, by the liver also contributes to the delivery of extrahepatic cholesterol to the liver. This combined activity of HDL and the other lipoproteins to transport cholesterol from extrahepatic tissue to the liver is known as reverse cholesterol transport and may partly explain the protective role of HDL against CHD [24]. HDL may also limit CHD by transporting paraoxonase, an enzyme that inhibits LDL oxidation [25].

The half-life of HDL in the circulation is about 2 days and it should be noted that circulating HDLc and TG concentrations are inversely related. This reflects reduced LPL activity releasing less substrate from TGRL (triglyceride rich lipoproteins) for HDL synthesis and excessive transfer of CE from HDL to TGRL. Estrogens elevate HDL_2 concentrations by promoting synthesis of apoA-1 and by inhibiting hepatic lipase [12]. In general, it is HDL_2 that may be associated with protection against CHD, and the higher HDLc values found in women are primarily due to increased levels of HDL_2 [26]. In a more refined analysis, five HDL subspecies (2a, 2b, 3a, 3b, 3c) can be separated based on size and apoA-I and apoA-II content. CHD risk is associated with increased proportions of HDL_{3b} and HDL_{3c} relative to HDL_{2b}. Compared to men, but not prepubertal boys, women have higher levels of HDL_{2a}, HDL_{2b}, and HDL_{3a} and lower concentrations of HDL_{3b} and HDL_{3c}. Cross-sectional analysis of adults, adolescents, boys, and girls indicated that the divergence in HDL subclasses occurs during puberty [27].

F. Lipoprotein(a)

Lp(a) is synthesized in the liver and is similar in structure to LDL but contains an apo(a) moiety bound to the $apoB_{100}$ via a disulfide bridge. The apo(a) contains a Kringle domain with 11 distinct Kringle types. One bears a strong homology to the Kringle 5 of plasminogen whereas the others are similar to the Kringle 4 of plasminogen. All of the Kringle moieties are present as single copies except for Kringle 4 type 2, which is repeated 3 to 40 times. The larger isoforms are synthesized at reduced rates (glycosylation in the Golgi is the limiting factor), resulting in an inverse relationship between the molecular weight of Lp(a) and its circulating levels. The number of Kringle repeats may influence the thrombogenic activity of the particle. It is suggested that Lp(a) promotes atherogenesis via the same mechanism as LDL, by delivering, possibly after oxidation, cholesterol to macrophages which leads to the formation of foam cells. Lp(a), however, may also have a thrombogenic role. Lp(a) inhibits fibrolysis by competing with plasminogen for receptors on fibrin and cell surfaces and by interfering with the activity of plasminogen activator. The large number of isoforms complicates the study of Lp(a) and many reports are contradictory.

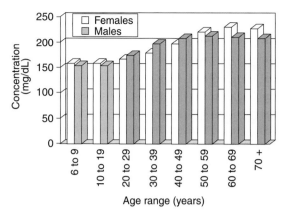

Fig. 66.5 Mean total cholesterol concentration by age among white populations. Data from [31].

However, some studies indicate that the smaller forms of Lp(a) are more atherogenic. Kark *et al.* [28] noted an association between small apo(a) isoforms and CHD among women but not among men. However, Wild *et al.* [29] observed the association for men only. Estrogens may have a protective role by reducing Lp(a) levels [30].

III. Lipid Distributions

In this section, lipid distributions by age for men and women are presented separately. The influences on lipoprotein profiles of puberty, the menstrual cycle, polycystic ovary syndrome, pregnancy, and menopause are noted, as are the impact of HRT, oral contraceptives, and phytoestrogens.

A. Age

Lipid distributions by age, for white men and women, are shown in Figures 66.5–66.8 [31]. Fewer data are available for other ethnic groups, although the trends for African Americans are comparable. Data for adults aged 20 and over from the

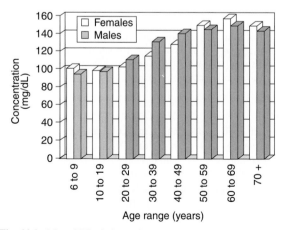

Fig. 66.6 Mean LDL cholesterol concentrations by age among white populations. Data from [31].

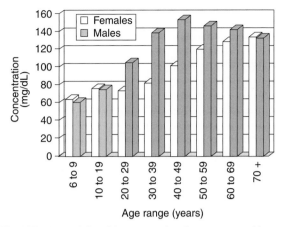

Fig. 66.7 Mean triglyceride concentrations by age among white populations. Data from [31].

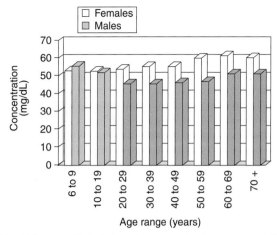

Fig. 66.8 Mean HDL cholesterol concentrations by age among white populations. Data from [31].

NHANES III study exhibit similar trends [32]. The levels for total cholesterol among men and women are comparable for the first two decades of life (Fig. 66.5). Subsequently, TC concentrations increase, but more rapidly in men than in women aged 20 to 50 years. TC levels then plateau for men but continue to rise among women, leading to the observation that women over age 50 have higher TC values than men. LDLc values exhibit similar trends (Fig. 66.6). The delay in the increase in TC and LDLc concentrations among women compared to men parallels the later manifestation of CHD in women. What Figure 66.6 does not show is that as LDLc concentration increases with age in women, the LDL also tends to form smaller particles and these are associated with greater atherogenic risk [33]. TG concentrations are similar in men and women during the first 20 years of life but the subsequent increase in concentration among men is much more rapid than that in women, reaching a peak in the fifth decade of life before gradually decreasing. The increase in TG levels among women is gradual but eventually the values reach those of men in the 70 plus age group (Fig. 66.7).

The levels of HDLc may be greater among males during the first decade of life but thereafter the values are higher for adolescent girls and women. HDLc concentrations tend to decrease in boys during the first 20 years of life and do not subsequently increase until the seventh decade is reached (Fig. 66.8). The increase in testosterone associated with puberty reduces HDLc concentration by promoting hepatic lipase activity [34]. HDLc levels in women tend to gradually rise with age and a pronounced difference between men and women is apparent after the age of 20 [31].

TC and TG trends with age are similar for black and white women, although TG concentrations are consistently higher among white women (Fig. 66.9) [31]. Few data are available for Lp(a), but the levels increase with age in women but not among men. The rise at the age of menopause is particularly dramatic for black women (Fig. 66.10) [35].

B. Menstrual Cycle

Superimposed on these lipid distributions in women are the effects of the menstrual cycle. During the follicular phase, estro-

gens are dominant, but during the luteal phase the levels of progesterone rise markedly, and in the luteal phase there are small reductions in the concentrations of TC and LDLc [36].

C. Polycystic Ovary Syndrome (PCOS)

Women with polycystic ovary syndrome (PCOS) suffer disturbed menstrual cycles and tend to be obese. Metabolically they show an increased prevalence for glucose intolerance and insulin resistance, syndromes associated with disruption of lipoprotein metabolism. Their elevated level of androgens would be expected to reduce their concentrations of HDLc because androgens enhance hepatic lipase activity, thus promoting the catabolism of HDL. Women with PCOS have been reported to exhibit a modified lipoprotein pattern, but when controlling for body mass index (BMI), the lipoprotein profile of the PCOS women is not significantly different from that of the controls. This observation may be explained by the off-setting effects of androgens and estrogens. Among women with PCOS, not only are androgen levels elevated, but the total estrogen effect may also be increased. Estrone synthesis is promoted by peripheral

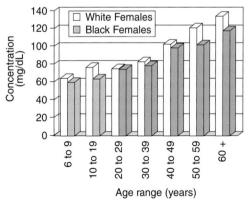

Fig. 66.9 Mean triglyceride concentrations by age among black/white females. Data from [31].

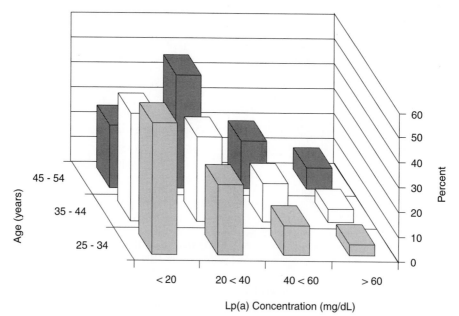

Fig. 66.10   Effect of age on the frequency distribution of Lp(a) in a female Nigerian population. Data from [35].

aromatization in fat tissue and concentrations of progesterone and of sex hormone binding globulin are low, thus increasing the concentration of free estradiol [37].

D. Pregnancy

During pregnancy when estriol and progesterone levels rise substantially, the concentrations of all lipids increase dramatically despite the associated expansion in plasma volume. The HDLc and Lp(a) values subsequently decline during the third trimester [38,39]. Elevations in TG and free fatty acids (FFA) are further exaggerated during preeclampsia, but all lipoprotein levels decrease significantly within 48 hours postpartum [40]. The increases in TC and LDLc values during pregnancy are surprising. Estrogen levels rise during pregnancy, yet estrogens administered during HRT are known to reduce TC and LDLc concentrations. However, progesterone levels also increase during pregnancy and it is the ratio of the hormones that may be paramount. It is also possible that during pregnancy the effect of estrogens in stimulating VLDL synthesis and inhibiting LPL may be more prominent than their role in promoting LDL receptor activity. Herrera *et al.* suggested that only the very high concentrations of estrogens observed during pregnancy inhibit LPL [41].

E. Menopause

Elevated levels of TC, LDLc, TG, and Lp(a) are associated with menopause when concentrations of estrogen and progesterone are very low. HDLc levels fall during menopause. The main estrogen in circulation during menopause is estrone, a much less

potent estrogen than estradiol, the main circulating estrogen prior to menopause [42].

F. Hormone Replacement Therapy

Hormone replacement therapy (HRT) involving oral estrogens, but not systemically administered natural estrogens, limits the increases in TC and, particularly, in LDLc, observed during menopause, (Figs. 66.11 and 66.12) [31]. HRT is also effective in raising the concentration of HDLc, mainly HDL$_2$ (HDL$_{3a}$ and HDL$_{2a}$ in an alternative analytical procedure) (Fig. 66.13) [27,31,43]. HRT raises TG among premenopausal women but has little influence on TG levels in postmenopausal women, although it may lower TG in women over age 70 (Fig. 66.14) [31]. Studies in the U.S. indicate that among women aged 40 to 60, 35% use HRT, but the percentage decreases with age: 15% and 7% for women over 65 and 80, respectively [44]. HRT must be maintained to sustain a modified lipoprotein profile.

The effects of estrogens on lipoproteins depend on the dose, schedule, and route of administration of the estrogens. Naturally occurring estrogens are effective when given orally, a result of the "first pass" effect, whereby the portal system efficiently delivers estrogens absorbed from the intestine to the liver. Synthetic estrogens (*e.g.,* ethinyl estradiol) are very potent and can elicit alterations in lipoprotein levels even when given transdermally or vaginally [45].

Historically, most HRT protocols involved unopposed estrogen therapy. However, because estrogens used as a single agent increase the risk of endometrial cancer, progestogens were introduced into HRT. Progesterone, the only naturally occurring progestogen, is not believed to influence lipoprotein levels, but synthetic progestogens do modify lipoprotein levels as a result

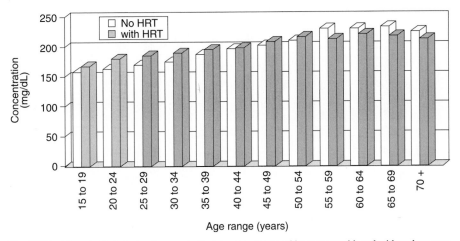

Fig. 66.11   Mean total cholesterol concentrations by age among white women with and without hormone replacement therapy. Data from [31].

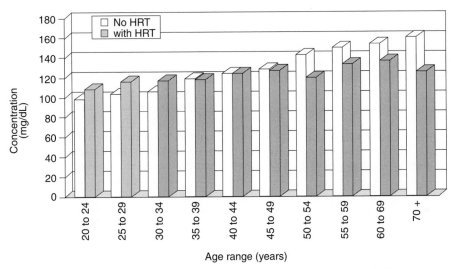

Fig. 66.12   Mean LDL cholesterol concentrations by age among white women with and without hormone replacement therapy. Data from [31].

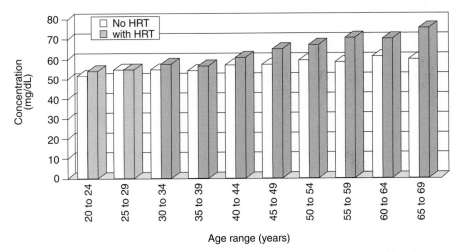

Fig. 66.13   Mean HDL cholesterol concentrations by age among white women with and without hormone replacement therapy. Data from [31].

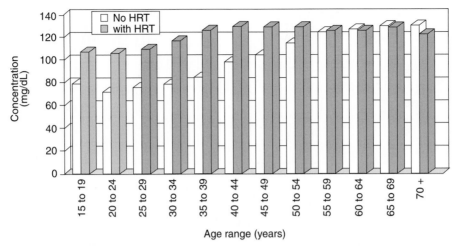

Fig. 66.14   Mean triglyceride concentrations by age among white women with and without hormone replacement therapy. Data from [31].

of their androgenic activity. The results for combined HRT suggest that levels of progestogens can be selected that protect the endometrium but do not greatly alter the modification in lipoprotein levels elicited by estrogens except for lowering HDLc. The latter effect is most apparent for HDL$_2$, the protective lipoprotein [46].

G. Oral Contraceptives (OCs)

As is the case for HRT, evaluation of OC effects is complicated by the many regimens employed. Oral contraceptives contain high concentrations of both synthetic estrogens and progestogens. Estrogens raise HDLc and TG concentrations whereas they lower LDLc levels. The synthetic progestogens, except the gonanes (third generation progestogens), have opposing effects. All OC formulations increase concentrations of TG. Similarly, LDLc levels are raised by OCs, except those containing desogestrel, a gonane, which lowers LDLc. HDLc concentrations vary inversely with the degree of androgenicity of the OC formulation. Desogestrel is an exception and raises HDLc levels [47,48]. Overall, published reports indicate that past use of oral contraceptives is not associated with an increased risk of CHD. However, several studies of the older formulations suggest that current use of oral contraceptives, particularly by smokers, raises the risk of CHD. It is unclear whether this enhanced risk is due to a thromboembolic effect of estrogens or to the reduction in HDLc levels elicited by progestogens [49]. Although not conclusive, evidence suggests that this elevated risk is diminished when modern OCs, those containing <50 μg estrogen and a third generation progestogen, are employed [50].

H. Phytoestrogens

Phytoestrogens are nonsteroidal estrogens that are abundant in semivegetarian diets typical of Asian countries. There are two main groups, isoflavonoids and lignans, which occur at particularly high concentrations in soy and flax products, respectively. The phytoestrogens have both estrogenic and antiestrogenic ac-

tivities, the latter the result of competing with 17β-estradiol for estrogen receptors. Although phytoestrogens exhibit only weak estrogenic activity, their potentially high concentrations, over 100-fold greater than endogenous estrogens, can elicit significant effects. It has also been proposed that phytoestrogens can modulate biological effects via a broad range of properties that are independent of estrogen receptors [51].

Phytoestrogens have been reported to reduce the levels of TC and LDLc but the results are not uniform. The suggested mechanisms include the inhibition of cholesterol absorption and synthesis and the upregulation of LDL receptors [52,53]. The above actions of phytoestrogens plus an antithrombotic activity [54] are the basis for proposing that phytoestrogens have a beneficial role in CHD. The epidemiologic evidence emphasizes the low rates of CHD and high intakes of phytoestrogens among Asian populations. However, this benefit can also be attributed to the low fat and high fiber content of the Asian diets.

IV. Lipoproteins as Risk Factors for CHD

In this section, the association between lipoprotein levels and CHD is presented and the effects of dietary, hormonal, and pharmaceutical interventions are noted. The implications of altering lipoprotein levels are significant. A meta-analysis concluded that lowering TC by 10% reduces coronary risk by 54% when therapy is initiated at age 40, but the risk is only attenuated by 27% if start of treatment is delayed until age 60 [55]. Furthermore, it is estimated that only half of the women in the U.S. aged 30–79 have desirable (<130 mg/dl) circulating levels of LDLc [56]. Blacks have a disproportionately higher percentage of individuals with elevated LDLc and depressed HDLc concentrations.

Lipoproteins are risk factors for CHD both among men and among women, but the relationship between lipoproteins and CHD is somewhat different for each gender. The association between TC or LDLc and CHD is less consistent for women than it is for men [57–59]. This may partly reflect the earlier exposure of men to elevated TC and LDLc levels. It also results from the higher concentrations of HDLc found among women,

Table 66.1
NCEP Treatment Recommendations Based on LDLc Concentrations[a]

Patient category	Dietary therapy		Drug treatment	
	Initiation level (mg/dl)	LDL goal (mg/dl)	Initiation level (mg/dl)	LDL goal (mg/dl)
Without CHD and with fewer than two risk factors[b]	≥160	<160	≥190	<160
Without CHD and with two or more risk factors[b]	≥130	<130	≥160	<130
With CHD	>100	≤100	≥130	≤100

[a]NCEP = National Cholesterol Educational Program.

[b]Risk factors: age (≥ 45 for men; ≥ 55 for women or premature menopause without estrogen replacement therapy); male sex; cigarette smoking; diabetes mellitus; family history of premature CHD; hypertension; low HDLc (<35mg/dl) and obesity.

thus, for women, HDLc is a greater contributor to TC than it is among men. Women have more HDL to offset the effects of the LDL. It has been suggested that the protective role of HDLc on CHD risk is at least as powerful as the atherogenic effect of LDLc [60]. Several studies have shown that HDLc levels are inversely correlated with CHD. The risk is independent of TG concentrations, which are negatively related to HDLc values [61–63]. A 1 mg/dl increase in HDLc concentration is associated with a 2–3% decline in risk [63]. The TC:HDLc ratio is a strong predictor of CHD but its significance differs between men and women. At values below 7.5, men face a greater risk of CHD than women, but at ratios above 7.5 the risk of CHD is similar for men and women. Values of about 3.5 are optimal [64].

TG are much stronger predictors of CHD among women than they are among men [57,58]. In a prospective study of Norwegian women the association was observed even when nonfasting blood samples were analyzed [65]. The deleterious action of TG may include effects on coagulation and on fibrinolytic factors, particularly factor VII and plasminogen activator inhibitor-1 [66–70].

The suggestion that TG are a risk factor for CHD among women indicated that the use of HRT, which raises TG levels among hypertriglyceridemic women, must be done cautiously. Similarly, when evaluating interventions that reduce LDLc concentrations, attention must be given to the HDLc values. In general, interventions involving LDLc levels are more appropriate for postmenopausal women or those suffering from severe hypercholesterolemia [59].

Despite these gender differences in CHD risk factors, the National Cholesterol Education Program (NCEP) guidelines for lipoprotein levels are similar for men and women, although male sex is considered a separate risk factor. The concentrations of TC are classified as: <200 mg/dl, desirable; 200–239 mg/dl, borderline risk; ≥240 mg/dl, high risk. Even if TC is below 200 mg/dl intervention may be needed if the HDLc is below 35 mg/dl [71]. The NCEP treatment recommendations, based on LDLc values, are shown in Table 66.1. Drug treatment is attempted

after failure of diet modification. The other cardiovascular risk factors indicated are age (≥45 for men; ≥55 for women or premature menopause without estrogen replacement therapy), male sex, cigarette smoking, diabetes mellitus, family history of premature CHD, hypertension, low HDLc (<35 mg/dl) and obesity. HDLc ≥60 mg/dl is considered protective.

V. Related Risk Factors

Several CHD risk factors, including pregnancy for women, involve aspects of estrogen and lipoprotein metabolism.

A. Parity

Studies of CHD risk by parity are inconclusive and may be confounded by smoking, social class, and number of spontaneous abortions. Interpretations of the modifications that occur in lipoprotein levels during pregnancy are muddled by the changes that occur afterwards. The increase in HDLc during pregnancy would be expected to be protective against CHD but subsequently, women who have had more pregnancies have lower HDLc concentrations than those who had none or fewer pregnancies [49].

B. Smoking

Smoking is a strong risk factor for CHD, which modifies estrogen metabolism and leads to premature menopause. Compared to nonsmokers, cigarette smokers have higher circulating levels of LDLc and TG and a reduced concentration of HDLc [72]. Despite the increase in weight associated with cessation of smoking, women who stop smoking greatly decrease their risk of CHD. In societies where the prevalence of smoking among women is very low compared to that of men, the rates of CHD are also disproportionately reduced among the women [73].

C. Obesity

The relationship between obesity and CHD among women is inconsistent, but abdominal obesity may be a greater risk factor for CHD than peripheral obesity [74]. Similarly, the detrimental lipoprotein profile associated with obesity is more pronounced among those with abdominal obesity [75,76]. Any benefit of weight loss in women may be more complicated than that observed for men. Postmenopausal obese women have higher levels of estrone than lean women due to its formation during aromatization of androstenedione in peripheral fat. Loss of fat may reduce estrone levels and subsequently diminish the concentration of HDLc [59].

D. Physical Activity

Reduced fat tissue and subsequent loss of estrone synthesis is also a factor in evaluating the benefits of physical exercise for women, but, in general, physical activity leads to healthier profiles for women. Lipoprotein levels improve unless the reduction in fat tissue proceeds to the extent that estrone synthesis is compromised and detrimental modifications to lipoprotein concentrations occur [77]. The relationship between CHD and physical activity are inconclusive [78–80]. The research is hampered by the few women who exercise regularly and the fact that these women are more likely to be in low risk groups and less likely to benefit from exercise than women in high risk groups. However, in a prospective study employing cardiovascular fitness as a surrogate measure for physical activity, quintile analysis indicated a very high age-adjusted relative risk for cardiovascular disease (CVD) of 8, for both men and women, when comparing the lowest quintile to quintiles 4 and 5 [81].

E. Diabetes

Diabetic patients experience an increased risk of CHD, and, unlike the general population, premenopausal women do not exhibit a protective effect. Diabetic patients do manifest adverse lipoprotein profiles but their extent does not explain the increased risk for CHD or the absence of a sex differential [82]. However, glycosylation of LDL may make it more prone to oxidation [83], and a subsequent study suggested that diabetic women, but not diabetic men, are under greater oxidative stress than are nondiabetics and this may contribute to the increased risk of CHD observed among diabetic women [84].

VI. Intervention

A. Dietary Intervention

CHD risk among both men and women has been linked to dietary intake of cholesterol and saturated fatty acids. The rationale for dietary therapy of lipoprotein levels is partly based on the work of Keys *et al.* [85]. The group studied the response of total cholesterol (TC) levels in men to changes in dietary fat intake and developed an empirical formula:

$$\Delta TC \text{ (mg/dl)} = 2.7\Delta S - 1.3\Delta P + 1.5\Delta Z$$

where ΔS and ΔP represent the percentage difference in saturated and polyunsaturated fat between two diets, respectively, and Z is the square root of the change in dietary cholesterol in units of milligrams per 1000 kcal. Other formulae have been proposed and are reviewed by Kris-Etherton and Yu [7]. In the Keys *et al.* formula [85], alterations in saturated fat have twice the importance of the changes in polyunsaturated fat and this weighting is reflected in epidemiologic studies. Caggiula and Mustad [86], in a critical review of studies involving dietary fat and lipoproteins, found that a positive association between blood TC and saturated fat intake was regularly observed but that the correlation of blood TC with the other dietary acids was inconsistent. Similarly, they concluded that only dietary saturated fatty acids and cholesterol were consistently correlated (positively) with CHD.

Dietary Guidelines of the NCEP recommend restrictions in fat intake. For step 1, the recommendations are: cholesterol, under 300 mg/day; total fat, under 30% of total calories; polyunsaturated fat, up to 10% of total calories; monounsaturated fat, 10–15% of total calories; saturated fat, under 10% of total calories. If this approach fails to elicit satisfactory changes in the lipoprotein profile then a more rigorous step 2 diet is proposed. The step 2 recommendations are similar to those of step 1 except for cholesterol: under 200 mg/day of cholesterol and

saturated fat intake under 7% of total calories [71]. A major concern involving nutritional intervention is the degree of subject adherence to the recommended diet. Among U.S. women, dietary intake of cholesterol is modest at about 230 mg/day but average intake of total and saturated fat remains high. In the 1970s, average intake among adult women of total and saturated fat was 40% and 14% of daily calories, respectively. In the late 1980s the values were 34% and 12%, respectively [87]. Complicating the activity of dietary interventions is the ability of mammalian cells to synthesize both cholesterol and saturated fats from acetate. Endogenous rates of cholesterol synthesis (8–10 mg/kg . d) are high compared to the amounts of cholesterol ingested [88].

B. Pharmaceutical Intervention

Several classes of drugs are available for modifying lipoprotein levels, but currently emphasis is placed on the statins, drugs that act primarily by inhibiting 3-hydroxy-3-methyl glutaryl CoA reductase (HMG-CoA reductase), the rate limiting enzyme in cholesterol biosynthesis. These inhibitors can lower circulating TG, TC, and LDLc levels by 30%, 40%, and 60%, respectively. Most studies of the clinical significance of these drugs involve men, but Scandinavian and American studies indicated 33% and 46% reductions in major coronary events among women receiving Simvastatin [89] or Pravastatin [90], respectively. The equivalent percentage reductions among men were 30% and 20%, respectively.

C. Hormonal Intervention

A review of the research involving HRT and CHD indicated a broad agreement among the investigators for a protective effect for the therapy, although the experimental designs involved a wide range of regimens and populations [91]. However, the studies were generally observational in design and a beneficial effect for HRT was not supported by a clinical intervention trial [92]. During 4.1 years of follow-up in this secondary prevention trial of postmenopausal women with an intact uterus, no benefit of an estrogen and medroxyprogesterone mix was observed. Examination of the first year data indicated that estrogen was detrimental due to an elevated incidence of thromboemboli. The subjects receiving supplements also suffered increased risk of gallbladder disease. Estrogens promote hepatic lipoprotein uptake and inhibit bile acid synthesis, thus raising biliary cholesterol and causing cholelithiasis [93]. During the last two years of the study estrogen therapy did tend to protect against CHD. This may reflect the improved lipoprotein profile of the subjects observed after one year of therapy.

D. Selective Estrogen Receptor Modulators (SERMs)

The wide-ranging effects of estrogen have prompted the search for selective estrogen receptor modulators. Estrogens may benefit bone and the heart but be detrimental to the breast and the endometrium. SERMs, it is hoped, would retain the beneficial aspects of estrogens but not exhibit their negative effects and would ideally inhibit the deleterious actions of estrogens. The search for SERMS takes advantage of the intricate activities of estrogens. In addition to modifying lipoprotein levels, estrogens influence blood pressure, carbohydrate metabolism, cytokines, endothelial cells, and hemostatic factors [94–96]. The actions of estrogens involve two receptor subtypes, the differential distribution and activation of which determine the response elicited [97].

One SERM, tamoxifen, may be useful in the treatment of breast cancer, osteoporosis, and CHD. However, it may promote uterine cancer. Raloxifene may also possess these useful effects but may not promote uterine cancer. However, both tamoxifen and raloxifene increase the risk of thromboemboli.

Administration of either tamoxifen or raloxifene elicits favorable changes in lipoprotein levels, although the effects may be limited to postmenopausal women. Tamoxifen reduces TC, LDLc, and Lp(a) concentrations but its effects on HDLc and TG are unclear [98–100]. Raloxifene reduces LDLc and Lp(a) levels and does not influence TG values. Raloxifene also does not alter HDLc concentrations, although it may increase the level of HDL_2, the beneficial component [101]. Preliminary evidence suggests that tamoxifen may be protective against CHD but that it also increases the incidence of endometrial cancer [102,103].

VII. Conclusions

Men and women share many risk factors for CHD, including hyperlipidemia. Nevertheless, CHD among women raises specific questions. It is unclear why a first myocardial infarction (MI) among women is associated with greater morbidity and mortality than is evident for men. The greater severity of MI among women, which is similarly not understood, only partially accounts for the differing outcomes [104]. Further information is also needed on the mechanisms whereby estrogens reduce CHD risk. The benefit is incompletely explained by the changes in lipoprotein levels induced by estrogens. Interpreting the influence of estrogens on lipoprotein profiles is complex. Possible confounders are the biological and analytical variations that are superimposed on the estrogen effects. This produces contradictions in the literature, particularly when small differences are involved, as is often the case for HDLc.

HRT may reduce the risk of CHD by 50% in postmenopausal women, and for selected groups, gains in life expectancy may exceed 3 years [49,105]. Nevertheless, further information is required on the implication for women of therapies that modify TG and HDLc concentrations. Most prominently, results are needed from clinical trials on the benefits and disadvantages of long-term combined HRT, difficult studies that are further complicated by the wide variety of regimens possible.

Acknowledgments

I thank Drs. D. Kelley, L. Kuller, and R. Ness for critically reviewing this manuscript and I thank Ms. M. J. Snyder for preparing the figures and typing this manuscript.

References

1. Pickle, L. W., Mungiole, M., Jones, G. K., and White, A. A. (1996). "Atlas of United States Mortality," DHHS Publ. No. (PHS) 97-1015. U.S. Department of Health and Human Services, Centers for Disease Control and Prevention, National Center for Health Statistics, Hyattsville, MD.

2. Kuller, L. H., Meilahn, E., Bunker, C., Yong, L. C., Sutton-Tyrrell, K., and Matthews, K. (1995). Development of risk factors for cardiovascular disease among women from adolescence to older ages. Am. J. Med. Sci. 310(Suppl. 1), S91–S100.

3. Kuller, L. H., and Meilahn, E. N. (1996). Risk factors for cardiovascular disease among women. Curr. Opin. Lipidol. 7, 203–208.

4. Redmond, G. P., ed. (1991). "Lipids and Women's Health." Springer-Verlag, New York.

5. Thurnhofer, H., and Hauser, H. (1990). Uptake of cholesterol by small intestinal brush border membrane is protein-mediated. Biochemistry 29, 2142–2148.

6. Daumerie C. M., Woollett L. A., and Dietschy J. M. (1992). Fatty acids regulate hepatic low density lipoprotein receptor activity through redistribution of intracellular cholesterol pools. Proc. Natl. Acad. Sci. U.S.A. 89, 10797–10801.

7. Kris-Etherton, P. M., and Yu, S. (1997). Individual fatty acid effects on plasma lipids and lipoproteins: Human studies. Am. J. Clin. Nutr. 65(Suppl.), 1628S–1644S.

8. Mensink, R. P., and Katan, M. B. (1993). Trans monounsaturated fatty acids in nutrition and their impact on serum lipoprotein levels in man. Prog. Lipid Res. 32(1), 111–122.

9. Grundy, S. M., and Denke, M. A. (1990). Dietary influences on serum lipids and lipoproteins. J. Lipid Res. 31, 1149–1172.

10. Clarke, S. D., and Jump, D. B. (1993). Regulation of gene transcription by polyunsaturated fatty acids. Prog. Lipid Res. 32(2), 139–149.

11. Harris, W. S. (1997). n-3 fatty acids and serum lipoproteins: Human studies. Am. J. Clin. Nutr. 65(Suppl.), 164S–154S.

12. Gevers Leuven, J. A. (1994). Sex steroids and lipoprotein metabolism. Pharmacol. Ther. 64, 99–126.

13. Westerveld, H. T., Meyer, E., de Bruin, T. W. A., and Erkelens, D. W. (1997). Oestrogens and postprandial lipid metabolism. Biochem. Soc. Trans. 25(1), 45–49.

14. Mitropoulos, K. A., Miller, G. J., Watts, G. F., and Durrington, P. N. (1992). Lipolysis of triglyceride-rich lipoproteins activates coagulant factor XII: A study in familial lipoprotein-lipase deficiency. Atherosclerosis 95, 119–125.

15. Hoffman, C. J., Miller, R. H., and Hultin, M. B. (1992). Correlation of factor VII activity and antigen with cholesterol and triglycerides in healthy young adults. Arterioscler. Thromb. 12, 267–270.

16. Urabe, M., Yamamoto, T., Kashiwagi, T., Okubo, T., Tsuchiya, H., Iwasa, K., Kikuchi N., Yakota, K., Hosokawa, K., and Honjo, H. (1996). Effect of estrogen replacement therapy on hepatic triglyceride lipase, lipoprotein lipase and lipids including apolipoprotein E in climacteric and elderly women. Endoc. J. 43, 737–742.

17. Weintraub, M. S., Eisenberg, S., and Breslow, J. L. (1987). Dietary fat clearance in normal subjects is regulated by genetic variation in apolipoprotein E. J. Clin. Invest. 80, 1571–1577.

18. Xhignesse M., Lussier-Cacan S., Sing, C. F., Kessling, A. M., and Davignon, J. (1991). Influences of common variants of apolipoprotein E on measures of lipid metabolism in a sample selected for health. Arterioscler. Thromb. 11, 1100–1110.

19. Dallongeville, J., Lussier-Cacan, S., and Davignon, J. (1992). Modulation of plasma triglyceride levels by apoE phenotype: A meta-analysis. J. Lipid Res. 33, 447–454.

20. Kesaniemi, Y. A., Ehnholm, C., and Miettinen, T. A. (1987). Intestinal cholesterol absorption efficiency in man is related to apoprotein E phenotype. J. Clin. Invest. 80, 578–581.

21. Davignon, J., Greg R. E., and Sing C. F. (1988) Apolipoprotein E polymorphism and atherosclerosis. Arteriosclerosis 8, 1–21.

22. Eichner, J. E., Kuller, L. H., Orchard T. J., Grandits G. A., McCallum, L. M., Ferrell, R. E., and Neaton J. D. (1993). Relation of apolipoprotein E phenotype to myocardial infarction and mortality from coronary artery disease. Am. J. Cardiol. 71, 160–165.

23. Lacort, M., Leal, A. M., Liza, M., Martin, C., Martinez, R., and Ruiz-Larrea, M. B. (1995). Protective effect of estrogens and catecholestrogens against peroxidative membrane damage in vitro. *Lipids* **30**(2), 141–146.

24. Schmitz, G., and Williamson, E. (1991). High-density lipoprotein metabolism, reverse cholesterol transport and membrane protection. *Curr. Opin. Lipidol.* **2**, 177–189.

25. Mackness, B., Hunt, R., Durrington, P. N., and Mackness, M. I. (1997). Increased immunolocalization of paraoxonase, clustering, and apolipoprotein A-I in the human artery wall with the progression of atherosclerosis. *Arterioscler. Thromb. Vasc. Biol.* **17**(7), 1233–1238.

26. Krauss, R. M. (1982). Regulation of high density lipoprotein levels. *Med. Clin. North Am.* **66**, 403–430.

27. Williams, P. T., Vranizan, K. M., Austin, M. A., and Krauss, R. M. (1993). Associations of age, adiposity, alcohol intake, menstrual status, and estrogen therapy with high-density lipoprotein subclasses. *Arterioscler. Thromb.* **13**, 1654–1661.

28. Kark, J. D., Sandholzer, C., Friedlander, Y., and Utermann, G. (1993). Plasma Lp(a), apolipoprotein(a) isoforms and acute myocardial infarction in men and women: A case-control study in the Jerusalem population. *Atherosclerosis* **98**, 139–151.

29. Wild, S. H., Fortmann, S. P., and Marcovina, S. M. (1997). A prospective case-control study of lipoprotein(a) levels and Apo(a) size and risk of coronary heart disease in Stanford five-city project participants. *Arterioscler. Thromb. Vasc. Biol.* **17**, 239–245.

30. Soma, M. R., Meschia, M., Bruschi, F., Morrisett, J. D., Paoletti, R., Fumagalli, R., and Crosignani, P. (1992). Hormonal agents used in lowering lipoprotein(a). *Chem. Phys. Lipids* **67/68**, 345–350.

31. Hainline, A., Karon, J., and Lippel, K., eds. (1982). "Manual of Laboratory Operations, Lipid Research Clinics Program, Lipid and Lipoprotein Analysis," 2nd ed. U.S. Department of Health and Human Services, Bethesda, MD.

32. Johnson, C. L., Rifkind, B. M., Sempos, C. T., Carroll, M. D., Bachorik, P. S., Briefel, R. R., Gordon, D. J., Burt, V. L., Brown, C. D., Lippel, K., and Cleeman, J. I. (1993). Declining serum total cholesterol levels among U.S. adults. The National Health and Nutrition Examination Surveys. *JAMA, J. Am. Med. Assoc.* **269**, 3002–3008.

33. Gardner, C. D., Fortmann, S. P., and Krauss, R. M. (1996). Association of small low-density lipoprotein particles with the incidence of coronary artery disease in men and women. *JAMA, J. Am. Med. Assoc.* **276**, 875–881.

34. Bagatell, C. J., and Bremner, W. J. (1995). Androgen and progestagen effects on plasma lipids. *Prog. Cardiovasc. Dis.* **38**(3), 255–271.

35. Evans, R. W., Bunker, C. H., Ukoli, F. A. M., and Kuller, L. H. (1997). Lipoprotein (a) distribution in a Nigerian population. *Ethnicity Health* **2**(1/2), 47–58.

36. Hemer, H. A., Valles de Bourges, V., Ayala, J. J., Brito, G., Diaz-Sanchez, V., and Garza-Flores, J. (1985). Variations in serum lipids and lipoproteins throughout the menstrual cycle. *Fertil. Steril.* **44**, 80–84.

37. Holte, J., Bergh, T., Berne, C., and Lithell, H. (1994). Serum lipoprotein lipid profile in women with the polycystic ovary syndrome: Relation to anthropometric, endocrine and metabolic variables. *Clin. Endocrinol.* **41**, 463–471.

38. Knopp, R. H., Magee, M. S., Bonet, B., and Gomez-Coronado, D. (1991). Lipid metabolism in pregnancy. *In* "Principles of Perinatal-Neonatal Metabolism" (R. M. Cowett, ed.), pp. 177–203. Springer-Verlag, New York.

39. Zechner, R., Desoye, G., Schweditsch, M. O., Pfeiffer, K. P., and Kostner, G. M. (1986). Fluctuations of plasma lipoprotein-A concentrations during pregnancy and post partum. *Metab. Clin. Exp.* **35**, 333–336.

40. Hubel, C. A., McLaughlin, M. K., Evans, R. W., Hauth, B. A., Sims, C. J., and Roberts, J. M. (1996). Fasting serum triglycerides, free fatty acids, and malondialdehyde are increased in preeclampsia, are positively correlated, and decrease within 48 hours post partum. *Am. J. Obstet. Gynecol.* **174**, 975–982.

41. Herrera, E., Lasuncion, M. A., Gomez-Coronado, D., Aranda, P., Lopez-Luna, P., and Maier, I. (1988). Role of lipoprotein lipase activity on lipoprotein metabolism and the fate of circulating triglycerides in pregnancy. *Am. J. Obstet. Gynecol.* **158**, 1575–1583.

42. MacDonald, P. C., Edman, C. D., Hemsell, D. L., Porter, J. C., and Siiteri, P. K. (1978). Effect of obesity on conversion of plasma androstenedione to estrone in postmenopausal women with and without endometrial cancer. *Am. J. Obstet. Gynecol.* **130**, 448.

43. Cauley, J. A., LePorte, R. E., Kuller, L. H., Bates M., and Sandler R. B. (1983). Menopausal estrogen use, high-density lipoprotein cholesterol subfractions and liver function. *Atherosclerosis* **49**, 31–39.

44. Carr, B. R. (1996). HRT management: The American experience. *Eur. J. Obstet. Gynecol. Reprod. Biol.* **64**(Suppl.), S17–S20.

45. Goebelsmann, U., Maschak, C. A., and Mishell, D. R. (1985). Comparison of hepatic impact of oral and vaginal administration of ethinyl estradiol. *Am. J. Obstet. Gynecol.* **151**, 868–877.

46. Miller, V. T., Muesing, R. A., LaRosa, J. C., Stoy, D. B., Phillips, E. A., and Stillman, R. J. (1991). Effects of conjugated equine estrogen with and without three different progestogens on lipoproteins, high-density lipoprotein subfractions, and apolipoprotein A-I. *Obstet. Gynecol.* **77**(2), 235–240.

47. Miller, V. T., and LaRosa, J. C. (1991). Sex steroids and lipoproteins. *In* "Lipids and Women's Health" (G. P. Redmond, ed.), pp. 48–65. Springer-Verlag, New York.

48. Harvengt, C., Desager, J. P., Gaspard, U., and Lepot, M. (1988). Changes in lipoprotein composition in women receiving two low-dose oral contraceptives containing ethinyl estradiol and gonane progestins. *Contraception* **37**, 565–575.

49. Barrett-Connor, E., and Bush, T. L. (1991). Estrogen and coronary heart disease in women. *JAMA, J. Am. Med. Assoc.* **265**, 1861–1867.

50. Rosenberg, L., Palmer, J. R., Sands, M. I., Grimes, D., Bergman, U., Daling, J., and Mills, A. (1997). Modern oral contraceptives and cardiovascular disease. *Am. J. Obstet. Gynecol.* **177**, 707–715.

51. Tham, D. M., Gardner, C. D., and Haskell, W. L. (1998). Potential health benefits of dietary phytoestrogens: A review of the clinical, epidemiological and mechanistic evidence. *J. Clin. Endocrinol. Metab.* **83**, 2223–2235.

52. Hirose, N., Inoue, T., Nishihara, K., Sugano, M., Akimoto, K., Shimizu, S., and Yamada, H. (1991). Inhibition of cholesterol absorption and synthesis in rats by sesamin. *J. Lipid Res.* **32**, 629–638.

53. Sirtori, C. R., Lovati, M. R., Manzoni, C., Monetti, M., Pazucconi, F., and Gatti, E. (1995). Soy and cholesterol reduction: Clinical experience. *J. Nutr.* **125**, 598S–605S.

54. Wilcox, J. N., and Blumenthal, B. F. (1995). Thrombotic mechanisms in atherosclerosis: Potential impact of soy proteins. *J. Nutr.* **125**, 631S–638S.

55. Law, M. R., Wald, N. J., and Thompson, S. G. (1994). By how much and how quickly does reduction in serum cholesterol concentration lower risk of ischaemic heart disease? *Br. Med. J.* **308**, 367–373.

56. Posner, B. M., Cupples, A. D., Gagnon, D., Wilson, P. W. F., Chetwynd, K., and Felix, D. (1993). The rationale and potential efficacy of preventive nutrition in heart disease: The Framingham Offspring-Spouse Study. *Arch. Intern. Med.* **153**, 1549–1556.

57. Bass, C. M., Newschaffer, M. S., Klag, M. J., and Bush, T. L. (1993). Plasma lipoprotein levels as predictors of cardiovascular death in women. *Arch. Intern. Med.* **153**, 2209–2216.

58. Bengtsson, C., Bjorkelund, C., Lapidus, L., and Lissner, L. (1993). Associations of serum lipid concentrations and obesity with mortality in women: 20 year follow up of participants in prospective population study in Gothenburg, Sweden. *Br. Med. J.* **307,** 1385–1388.

59. Meilahn, E. N., Becker, R. C., and Corrao, J. M. (1995). Primary prevention of coronary heart disease in women. *Cardiology* **86,** 286–298.

60. Kannel, W. B. (1983). High-density lipoproteins: Epidemiologic profile and risks of coronary artery disease. *Am. J. Cardiol.* **52,** 9b–12b.

61. Castelli, W. P., Garrison, R. J., Wilson, P. W. F., Abbott, R. D., Kalousdian, S., and Kannel, W. B. (1986). Incidence of coronary heart disease and lipoprotein cholesterol levels. The Framingham Study. *JAMA, J. Am. Med. Assoc.* **256,** 2835–2838.

62. Brunner, D., Weisbort, M. D., Meshulam, N., Schwartz, S., Gross, J., Saltz-Rennert, H., Altman, S., and Loebl, K. (1987). Relation of serum total cholesterol and high-density lipoprotein cholesterol percentage to the incidence of definite coronary events: Twenty-year follow-up of the Donolo-Tel Aviv Prospective coronary artery disease study. *Am. J. Cardiol.* **59,** 1271–1276.

63. Gordon, D. J., Probstfield, J. L., Garrison, R. J., Neaton, J. D., Castelli, W. P., Knoke, J. D., Jacobs, D. R., Bangdiwala, S., and Tyroler, H. A. (1989). High-density lipoprotein cholesterol and cardiovascular disease. Four prospective American studies. *Circulation* **79,** 8–15.

64. Kannel, W. B. (1997). Metabolic risk factors for coronary heart disease in women: Perspective from the Framingham Study. *Am. Heart J.* **114**(2), 413–419.

65. Stensvold, I., Tverdal, A., Urdal, P., and Graff-Iversen, S. (1993). Nonfasting serum triglyceride concentration and mortality from coronary heart disease and any cause in middle aged Norwegian women. *Br. Med. J.* **307,** 1318–1322.

66. Miller, G. J., Martin, J. C. Mitropoulos, K. A., Reeves, B. E., Thompson, R. L., Meade, T. W., Cooper, J. A., and Cruickshank, J. K. (1991). Plasma factor VII is activated by postprandial triglyceridaemia, irrespective of dietary fat composition. *Atherosclerosis* **86**(2–3), 163–171.

67. Mussoni, L., Mannucci, L., Sirtori, M., Camera, M., Maderna, P., Sironi, L., and Tremoli, E. (1992). Hypertriglyceridemia and regulation of fibrinolytic activity. *Arteriosclerosis Thromb.* **12**(1), 19–27.

68. Camera, M., Mussoni, L., Maderna, P., Sironi, L., Prati, L., Colli, S., Bernini, F., Corsini, A., and Tremoli, E. (1993). Effect of atherogenic lipoproteins on PAI-1 synthesis by endothelial cells. *Cytotechnology,* **11**(Suppl. 1), S144–S146.

69. Mitropoulos, K. A. (1994). Lipoprotein metabolism and thrombosis. *Curr. Opin. Lipidol.* **5**(3), 227–235.

70. Bradley, W. A., Booyse, F. M., and Giaturco, S. H. (1994). Fibrinolytic and thrombotic factors in atherosclerosis and IHD: The influence of triglyceride rich lipoproteins (TGRLP). *Atherosclerosis* **108**(Suppl.), S31–S39.

71. National Cholesterol Education Program (NCEP) Expert Panel (1993). Summary of the second report of the National Cholesterol Education Program (NCEP) Expert Panel on detection, evaluation, and treatment of high blood cholesterol in adults (Adult Treatment Panel II). *JAMA, J. Am. Med. Assoc.* **269,** 3015–3023.

72. Craig, W. Y., Palomaki, G. E., and Haddow, J. E. (1989). Cigarette smoking and serum lipid and lipoprotein levels: An analysis of published data. *Br. Med. J.* **298,** 784–788.

73. Rosenberg, L., Palmer, J. R., and Shapiro, S. (1990). Decline in the risk of myocardial infarction among women who stop smoking. *N. Engl. J. Med.* **322,** 213–217.

74. Lapidus, L., Bengstsson, C., Larsson, B., Pennert, K., Tybo, E., and Sjostrom, L. (1984). Distribution of adipose tissue and risk of cardiovascular disease and death: 12- year follow-up of participants in the population study of women in Gothenburg, Sweden. *Br. Med. J.* **289,** 1257–1261.

75. Wing, R. R., Matthews, K. A., Kuller, L. H., Meilahn, E. N., and Plantinga, P. (1992) Waist-to-hip ratio in middle aged women. Associations with behavior and psychosocial factors with changes in cardiovascular risk factors and with changes in cardiovascular risk factors. *Arterioscler. Thromb.* **11,** 1250–1257.

76. Soler, J. T., Kushi, L. H., Prineas, R. J., and Seal, U. S. (1988). Association of body fat distribution with plasma lipids, lipoproteins, apolipoproteins A1 and B in postmenopausal women. *J. Clin. Epidemiol.* **41,** 1075–1081.

77. Owens, J. F., Matthews, K. A., Wing, R. R., and Kuller, L. H. (1990). Physical activity and cardiovascular risk: A cross-sectional study of middle-aged premenopausal women. *Prev. Med.* **19,** 147–157.

78. Kannel, W. B., and Sorlie, P. (1986). Some health benefits of physical activity. The Framingham Study. *Arch. Intern. Med.* **139,** 857–861.

79. Salonen, J. T., Puska, P., and Tuomilehto, J. (1982). Physical activity and risk of myocardial infarction, cerebral stroke and death: A longitudinal study in eastern Finland. *Am. J. Epidemiol.* **115,** 526–537.

80. Lapidus, L., and Bengtsson, C. (1986). Socioeconomic factors and physical activity in relation to cardiovascular disease and death: A 112-year follow-up of participants in a population study of women in Gothenberg, Sweden. *Br. Heart J.* **55,** 295–301.

81. Blair, S. N., Kohl, H. W., and Pffenbarger, R. S. (1989). Physical fitness and all-cause mortality. A prospective study of healthy men and women. *JAMA, J. Am. Med. Assoc.* **262,** 2395–2401.

82. Orchard, T. J. (1991). Dyslipoproteinemia and diabetes. *Endocrinol. Metab. Clin. North Am.* **19,** 361–380.

83. Bowie, A., Owens, D., Collins, P., Johnson, A., and Tomkin, G. H. (1993). Glycosylated low density lipoprotein is more sensitive to oxidation: Implications for the diabetic patient? *Atherosclerosis* **102,** 63–67.

84. Evans, R. W., and Orchard, T. J. (1994). Oxidized lipids in insulin-dependent diabetes mellitus: A sex-diabetes interaction? *Metab. Clin. Exp.* **43**(9), 1196–1200.

85. Keys, A., Anderson, J. T., and Grande, F. (1965). Serum cholesterol response to changes in the diet II. The effect of cholesterol in the diet. *Metab. Clin. Exp.* **14**(7), 759–765.

86. Caggiula, A. W., and Mustad, V. A. (1997). Effects of dietary fat and fatty acids on coronary artery disease risk and total and lipoprotein cholesterol concentrations: Epidemiologic studies. *Am. J. Clin. Nutr.* **65**(Suppl.), 1597S–1610S.

87. Life Sciences Research Office (1995). "Third Report on Nutrition Monitoring in the United States," Vol. 2. Federation of American Societies for Experimental Biology, Interagency Board for Nutrition Monitoring and Related Research, U.S. Government Printing Office, Washington, DC.

88. Osono, Y., Woollett, L. A., Herz, J., and Dietschy, J. M. (1995). Role of the low density lipoprotein receptor in the flux of cholesterol through the plasma and across the tissues of the mouse. *J. Clin. Invest.* **95,** 1124–1132.

89. Scandinavian Simvastain Survival Study Group (1994). Randomised trial of cholesterol lowering in 4444 patients with coronary heart disease: The Scandinavian Simvastatin Survival Study (4S). *Lancet* **344,** 1383–1389.

90. Sacks, F. M., Pfeffer, M. A., Moye, L. A., Rouleau, J. L., Rutherford, J. D., Cole, T. G., Brown, L., Warnica, J. W., Arnold, J. M. O., Wun, C. C., Davis, B. R., Braunwald, E., for the Cholesterol and Recurrent Events Trial Investigators (1996). The effect of pravastatin on coronary events after myocardial infarction in patients with average cholesterol levels. *N. Engl. J. Med.* **335,** 1001–1009.

91. Grady, D., Rubin, S. M., Pettiti, D. B., Fox, C. S., Black, D., Ettinger, B., Ernster, V. L., and Cummings, S. R. (1992). Hormone therapy to prevent disease and prolong life in postmenopausal women. *Ann. Inter. Med.* **117,** 1016–1037.

92. Hulley, S., Grady, D., Bush, T., Furberg, C., Herrington, D., Riggs, B., Vittinghoff, E., for the Heart Estrogen/progestin Replacement Study (HERS) Research Group (1998). Randomized trial of estrogen plus progestin for secondary prevention of coronary heart disease in postmenopausal women. *JAMA, J. Am. Med. Assoc.* **280,** 605–613.

93. Everson, G. T., McKinley, C., and Kern, F., Jr. (1991). Mechanisms of gallstone formation in women. *J. Clin. Invest.* **87,** 237–246.

94. Wong, M., Thompson, T. L., and Moss, R. L. (1996). Nongenomic actions of estrogen in the brain: Physiological significance and cellular mechanisms. *Crit. Rev. Neurobiol.* **10**(2), 189–203.

95. Hales, A. M., Chamberlain, C. G., Murphy, C. R., and McAvoy, J. W. (1997). Estrogen protects lenses against cataract induced by transforming growth factor-beta (TGFbeta). *J. Exp. Med.* **185**(2), 273–280.

96. Gaspard, U. J., Gottal, J. M., and van den Brule, F. A. (1995). Postmenopausal changes of lipid and glucose metabolism: a review of their main aspects. *Maturitas* **21**(3), 171–178.

97. Kuiper, G. G. J. M., Carlsson, B., Grandien, K., Enmark, E., Haggblad, J., Nilsson, S., and Gustafsson, J.-A. (1997). Comparison of the ligand binding specificity and transcript tissue distribution of estrogen receptors α and β. *Endocrinology* (*Baltimore*) **138,** 863–870.

98. Ilanchezhian S., Thangaraju M., and Sachdanandam P. (1995). Plasma lipids and lipoprotein alterations in tamoxifen-treated breast cancer women in relation to the menopausal status. *Cancer Biochem. Biophys.* **15,** 83–90.

99. Grey, A. B., Stapleton, J. P., Evans, M. C., and Reid, I. R. (1995). The effect of the anti-estrogen tamoxifen on cardiovascular risk factors in normal postmenopausal women. *J. Clin. Endocrinol. Metab.* **80,** 3191–3195.

100. Saarto, T., Blomqvist, C., Ehnholm, C., Taskinen, M. R., and Elomaa, I. (1996). Antiatherogenic effects of adjuvant antiestrogens: A randomized trial comparing the effects of tamoxifen and toremifene on plasma lipid levels in postmenopausal women with node-positive breast cancer. *J. Clin. Oncol.* **14,** 429–433.

101. Walsh, B. W., Kuller, L. H., Wild, R. A., Paul, S., Farmer, M., Lawrence, J. B., Shah, A. S., and Anderson, P. W. (1998). Effects of raoxifene on serum lipids and coagulation factors in healthy postmenopausal women. *JAMA, J. Am. Med. Assoc.* **279,** 1445–1451.

102. McDonald, C. C., Stewart, H. J., for the Scottish Breast Cancer Committee (1991). Fatal myocardial infarction in the Scottish adjuvant tamoxifen trial. *Br. Med. J.* **303,** 435–437.

103. Rutqvist, L. E., Mattsson, A., for the Stockholm Breast Cancer Study Group (1993). Cardiac and thromboembolic morbidity among postmenopausal women with early stage breast cancer in a randomized trial of adjuvant tamoxifen. *J. Natl. Cancer Inst.* **85,** 1398–1406.

104. Marrugat, J., Sala, J., Masia, R., Pavesi, M., Sanz, G., Valle, V., Molina, L., Seres, L., Roberto, E., for the RESCATE Investigators (1998). Mortality differences between men and women following first myocardial infarction. *JAMA, J. Am. Med. Assoc.* **280,** 1405–1409.

105. Eckman, N. F., Karas, R. H., Pauker, S. G., Goldberg, R. J., Ross, E. M., Orr, R. K., and Wong, J. B. (1997). Patient-specific decisions about hormone replacement therapy in postmenopausal women. *JAMA, J. Am. Med. Assoc.* **277**(14), 1140–1147.

Section 11

CANCER

Louise A. Brinton

Environmental Epidemiology Branch
National Cancer Institute
Bethesda, Maryland

I. Introduction

Cancer in women is still very much a serious public health concern given that approximately 40% of American women will be diagnosed at some time in their life with a cancer other than a superficial skin cancer and 20% will die from cancer (Chapter 67). Most of the cancers that occur in American women have incidence rates that are among the highest in the world. Although we hear much about the occurrence of breast cancer, with figures indicating that one in every eight women will develop the disease, statistics indicate that lung cancer is a more serious public health concern in terms of disability and number of deaths. The major cause of this disease is well known, namely cigarette smoking. While this should enable effective prevention strategies, necessary behavior modification efforts are often unsuccessful. Thus, there are serious obstacles to overcoming the increasing incidence of this fatal disease. There have also been challenges to counteracting the occurrence of many of the other cancers, whose incidences continue to remain high and essentially unchanged over time.

In approaching the prevention of cancers in women, special emphasis must be placed on understanding exposures that are unique to women and that merit special consideration, because of either changing prevalences or unique interactions with other risk factors. Studies have provided many new insights regarding possible effects of a variety of potential etiologic agents. These efforts have been greatly aided by technologic advances that have expanded our understanding of underlying biologic processes and the natural history of these lesions. This is perhaps best exemplified by the tremendous advances that have been made in understanding the genetic etiology of some cancers. In addition, technologic advances that have enabled more precise categorization of diseases as well as more reliable measures of exposure have also been beneficial. These efforts have been essential towards developing effective preventive approaches. Many of the exciting epidemiologic observations regarding cancers in women are elaborated in this overview. However, it is clear that we are only beginning to embark on an era of discovery and a fuller appreciation of the complexities involved in both the etiology and prevention of cancers in women, and that there is much opportunity for new advances in the coming years.

II. Unique Exposures among Women

In studying cancer in women, a number of unique exposures must be considered, including effects of reproductive and menstrual factors. These factors have been of historic interest to breast and gynecologic cancers, but it has been recognized that they may also affect other cancers, including those of the lung and colorectum. Although much has been learned by studying relationships with such factors as parity, age at first birth, age at menarche, and type of menopause, we still have much that can be clarified. This includes our better understanding the biologic underpinnings of many of these factors. For instance, the biologic rationale remains unclear for both the well-recognized beneficial effects of sterilization and tubal ligation on ovarian cancer risk [1,2] as well as of early ages at first birth on breast cancer risk. This latter relationship has been recognized since the turn of the twentieth century, but whether the relationship derives from pregnancy-induced endogenous hormonal changes or breast tissue structural alterations remains to be clarified [3].

Another exposure unique to women is exogenous hormone use, including oral contraceptives and menopausal replacement hormones. These increasingly prescribed medications have received widespread attention in terms of their effects on cancer risk. Much has been clarified, but prescribing recommendations are complicated by the extremely divergent effects of these hormones in different target tissues. For instance, oral contraceptives substantially reduce the risk of ovarian [4] and endometrial cancers [5] but increase the risk of early onset breast cancers [6]. Effects of menopausal estrogens are even more controversial because they offer substantial protection against a number of important chronic diseases, including cardiovascular diseases and osteoporosis, as well as probably against colorectal cancers. (Chapter 76). However, the increased risk of endometrial cancer and possibly of breast cancer is often viewed with much more alarm, even though it is apparent that the overall benefits on mortality of hormone use in most instances far outweigh the risks [7]. Despite considerable research attention, this topic remains one that engenders considerable confusion and controversy. Further complicating an assessment of ultimate risks versus benefits is that most epidemiologic literature relates to use of estrogens alone. Increasingly, however, estrogens are being prescribed in conjunction with progestins, given the well-recognized benefit of this regimen against estrogen-induced endometrial hyperplasia and associated endometrial cancer (Chapter 72). Given the newness of this therapeutic regimen, epidemiologic data are only beginning to emerge regarding effects on cancer risk. Of concern is the observation that progestins appear to have very different effects on endometrial tissue compared to breast tissue. In endometrial tissue, progestogens counteract the epithelial proliferative effects of estrogens, but in breast tissue progestogens appear to act as mitogens, inducing cellular proliferation [8]. This has led to the hypothesis that the combined therapy may be more hazardous to breast cancer risk than estrogens alone, a notion that has received some support from epidemiologic investigations [9–11]. However, findings regarding this relationship have been conflicting, leading to difficulties in determining how this prescribing approach should be viewed, especially in terms of how it affects the risk-benefit balance of hormone replacement therapy.

In terms of cancer risk, there are a number of other concerns regarding possible iatrogenic agents. Ovulation-stimulating drugs have been hypothesized to lead to a possible increase in the risk of ovarian cancer, although effects remain unresolved [12; Chapter 71]. Trends towards delayed childbearing and declining fecundity with age have led to substantial increases in usage of these medications over time. Technologic advances, such as in vitro fertilization, have also been associated with different therapeutic approaches, with unknown implications for cancer risk [13]. However, given the recency of these approaches and the fact that they are used among young women, it may be years before we fully understand their implications for cancer risk. In addition, new hypotheses have arisen about other drugs, including tranquilizers, whose usage in women far exceeds that in men. Observations of an increase in ovarian cancer risk associated with these drugs merit further attention [14]. Fortunately, not all news about drugs is bad, with some medications having been linked with reductions in cancer risk. This includes possi-

ble reductions in colorectal, ovarian, and breast cancer with aspirin and other nonsteroidal anti-inflammatory drugs [15,16].

III. Exposures Requiring Special Consideration in Women

A number of exposures require special consideration in women. These include factors that either have different prevalences in women as compared to men or exposures that have special effects in women, often reflecting interactions with other hormonal risk factors. The range of such factors is wide, encompassing general environmental exposures (*e.g.*, pesticides), occupational factors, cigarette smoking, nutritional intake, anthropometric measures, and physical activity. Many new hypotheses have emerged, some reflective of changing exposures.

The effects of environmental agents have come under intense scrutiny, particularly given the widespread geographic variation in cancers, including a concentration of many cancers in highly industrialized areas [17]. There are a number of lines of evidence that lung cancer risk may be influenced by both outdoor and indoor air pollution [18,19], but the precise nature of the relationship bears further investigation. Potential effects of radon exposure have received intense public attention, but it remains to be clarified to what extent this exposure relates to cancer risk, especially in women [20]. For most other cancers in women, there is little biologic evidence to support the notion that general environmental factors play a major etiologic role. However, observations have emphasized a possible etiologic role in various hormonally related cancers for a class of agents known as endocrine disruptors, including DDT, PCBs, and possibly some other chemicals. The relationship of these agents to cancer risk in women is currently unresolved, although it is certainly a topic of intense investigation. Although most attention has focused on breast cancer [21], effects on other cancers, including endometrial cancer, have also been examined [22]. Some of the uncertainty in the relationships may derive from methodologic difficulties in assessing changing exposures over time [23,24], a problem that plagues most environmental exposures. Thus, whether there are chemicals that are carcinogenic in women remains uncertain. In men, major clues have been derived from studies of occupational exposures. However, in women, few studies have been undertaken regarding occupational predictors of cancer risk. This undoubtedly reflects the limited numbers of women that were actively employed in industrial occupations in past eras. However, given that increasing numbers of women are entering the workforce, studies are beginning to emerge suggesting that certain exposures merit attention, including possible effects of asbestos and metals on lung cancer risk [25,26], pharmaceutical and selected chemicals on breast cancer risk [27,28] and herbicides on ovarian cancer risk [29]. Although it may still be years before we are fully able to resolve effects in this country, research opportunities exist in many other countries where women have been actively employed for many years. A number of such investigations are currently underway, and results are awaited with great interest.

The effects of cigarette smoking on cancer risk in women are of obvious concern. The increasing prevalence of smoking in women has led to major increases in lung cancer incidence (Chapter 75). Of concern is that the risk of lung cancer associated with smoking may be greater in women than men for comparable amounts of smoking. This suggests a possible hormonal etiology for lung cancer. Although cigarette smoking is best recognized for its effects on lung cancer risk, results are emerging that it may also increase the risk of several other cancers in women, including cancers of the colorectum and cervix. Underlying biologic mechanisms remain less clear than those for lung cancer, although for cervical cancer an enhancement of the effects of the human papilloma viruses must be considered [30], especially given the recognized immunosuppressive effects of smoking. In addition to active smoking, passive smoking effects have gained recognition, not only for lung cancer but also for other cancers. The effects of cigarette smoking on breast cancer risk remain controversial, with divergent results from various studies. However, it has been suggested that active smoking effects may have been obscured by the failure to account for passive smoking effects. Notably, in several studies, when active smokers were compared to individuals unexposed to either active or passive smoking, an increased risk of breast cancer associated with active smoking was observed [31,32]. This observation requires further confirmation, but, if true, could have major implications for breast cancer and possibly also for other cancers. At least one cancer, endometrial cancer, has been shown to be reduced among cigarette smokers [33−35]. Although several biologic mechanisms have been proposed, including the alteration of the absorption, distribution, or metabolism of hormones, the precise mode of action remains unclear. Further pursuit of this anomalous finding may provide etiologic insights regarding endometrial cancers as well as smoking-related pathology in general.

Although it has been postulated that many of the cancers in women are influenced by dietary factors, the individual components of diet that contribute to risk remain relatively undefined. Many studies have focused extensive attention on determining the relationship of dietary fat to breast cancer risk, an issue that remains quite controversial [36,37]. A better understanding of how dietary fat interrelates with other breast cancer risk factors, including endogenous hormones, may clarify effects [38]. A broadening of approach to include the role of other dietary components, for breast as well as for other cancers, also appears necessary. Thus, of interest are studies that have noted possible effects of lactose intolerance on ovarian cancer risk [39], and iron intake on colorectal cancer risk [40]. Studies have also attempted to address the role of a variety of potential protective factors. Studies to date appear to indicate that there may be substantial beneficial effects for a variety of cancers for diets that are high in intake of fruits or vegetables, or, in some cases, in specific micronutrients. Classically, studies have focused on effects of vitamins A, C, and E, but it has become apparent that attention should also be given to other dietary constituents, including calcium, vitamin D, selenium, phytochemicals, fiber, and folate, because findings are beginning to emerge suggesting that these may be equally important for several cancers (e.g., colorectal cancer). Ways in which foods are prepared are also of concern, including potential effects of heterocyclic aromatic amines produced in meats cooked at high temperatures [41]. Increasingly, studies are attempting to relate dietary constituents to other risk factors to enhance our understanding of carcinogenic mechanisms. For instance, it has been postulated that folate deficiencies may be involved in both the increased risks of cervical cancer associated with multiple pregnancies and of colorectal cancer associated with alcohol consumption. Studies in the 1990s have also stressed the importance of a more global approach to dietary assessment, focusing simultaneously on a variety of nutritional components that may be highly correlated with each other and with other lifestyle factors, including general health status. The need for a concomitant assessment of energy expenditure seems indicated. It now appears that physical activity may have definite advantages for several cancers of importance to women, including cancers of the breast, endometrium, and colorectum [42]. Future studies clarifying the respective roles of diet, anthropometry, and energy expenditure on risk of these cancers will be important because this is one area that has significant potential in terms of possible preventive approaches.

Alcohol consumption has been recognized as having a variety of health implications, including some related to cancer risk. Despite early studies that dismissed the association between alcohol consumption and breast cancer risk as reflective of other lifestyle factors, other studies support the notion of a biologic effect [43]. The relationship is biologically plausible given the recognized effects of alcohol on a variety of endogenous hormones. Other mechanisms of action, however, are possible, including increased cellular permeability, various hepatobiliary-associated changes, and direct effects of nitrosamines or other constituents of alcoholic beverages. It remains unclear whether there are critical time periods that affect risk more than others and what levels of consumption are most hazardous. The relationship of alcohol consumption to other cancer sites in women remains less clear, although there is support for further investigation into possible effects on cancers of the lung and colorectum.

While we still have much to learn about dietary relationships with cancer risk, there is clear evidence that body size has an effect on a number of cancers in women, including endometrial cancer, postmenopausal breast cancer, colorectal cancer, and possibly ovarian cancer [44]. The potential impact of this risk factor on cancer occurrence is emphasized by the widespread prevalence of obesity among American women. In addition to body size, the distribution of body fat also appears to relate to risk of many of these female cancers. For example, women whose weight distributes abdominally have higher risks of breast cancer than those with fat that distributes primarily peripherally [45]. Further study of the predictors and correlates of body fat distribution may provide clues to the etiology of some of these cancers. Preliminary evidence indicates that there may be genetic as well as lifestyle predictors. In addition, anthropometric factors may modulate the effects of other factors. Of interest in this regard is an investigation showing that the effects of body size on breast cancer risk were most apparent in women who had not taken menopausal hormones [46]. Thus, future efforts to elucidate mechanisms related to diet, anthropometry, and physical activity will need to have a comprehensive perspective and one that considers a wide variety of underlying biologic mechanisms, including endogenous hormones.

IV. Technologic Advances in Exposure Assessment

Foremost among the genetic advances that have been made are the identification of some major genes involved in the etiology of several cancers in women. The identification of BRCA1 and subsequently BRCA2, the colorectal adenomatous polyposis gene (APC), and a variety of tumor suppressor genes (including p53) have revolutionized ideas about how to approach the prevention of various cancers, including breast, ovarian, and colorectal cancers. Although several of these genes are highly penetrant (*i.e.,* individuals with the genes have a high likelihood of developing disease), it appears that the genes are relatively rare and thus explain only a relatively small proportion of cancer cases [47]. At least for breast cancer, it is also clear that these high-penetrant genes do not entirely explain familial effects, with many women who have a family history of the disease showing no evidence of the specific mutations (Chapter 69). This suggests that other genes may be identified in the future that could also play a major role in predicting these cancers. Thus, future efforts will undoubtedly focus on the identification of additional genetic markers to enhance efforts to develop better preventive approaches.

In addition to the identification of highly penetrant genes, advances have been made toward identifying a number of more common genetic polymorphisms involved in either carcinogen or hormone metabolism. Population studies evaluating the role of common polymorphisms are only beginning to emerge and it is unclear whether findings will have practical preventive implications. Nonetheless, it appears to be an exciting period of discovery and one that holds great promise for advancing our understanding of carcinogenic mechanisms. Particularly exciting are studies that attempt to integrate genetic markers with other identified disease risk factors (*i.e.,* the assessment of gene–environment interactions). The chapters that follow illustrate a number of examples where such studies may be insightful, including examining effects of cigarette smoking and consumption of meat cooked at high temperatures by N-acetyltransferase (NAT1 and NAT2) genotype, other polycyclic aromatic hydrocarbon exposures (including cigarette smoking) by glutathione S-transferase (GST), alcohol consumption by methylenetetrahydrofolate (MTHRF), and various reproductive and menstrual parameters by genes involved in hormone metabolism, including the cytochrome P450 P450c17α (CYP17) and catechol O-methyltransferase (COMT) enzymes. These studies present a number of unique challenges. As with any interdisciplinary study, careful exposure assessment is necessary. This is not only true for the genetic markers but also for the environmental agents, which can be difficult to assess given changing exposures over time and influences of factors that can affect measurement levels, including disease onset and type of treatment. In order to be informative, gene–environment studies must be large, especially if the prevalence of either the genetic marker or the environmental exposure of interest is low [48] or if either factor is misclassified [49]. It is important that the driving force behind these assessments be biologic, as opportunities for chance findings are high. Despite the challenges involved in assessing gene–environment interactions, it is clear that this line of inquiry will be increasingly used to better understand etiologic processes. However, we still have much to learn, including how genes relate to each other (*i.e.,* gene–gene interactions) as well as how the genes are functionally expressed. Studies such as one that addressed the relationship between the CYP17 polymorphism and plasma hormone levels [50] exemplify an important needed approach in elaborating on the functional effects of genetic polymorphisms.

Technologic advances have also assisted in more precisely measuring a number of other biomarkers of interest for cancers in women. This has included the assessment of endogenous hormones, whose relationship to most cancers has remained enigmatic. Thus, a number of methodologic evaluations to assess the reliability and validity of hormone measurements have been vital towards advancing progress in this area [51,52]. Early studies were usually unable to relate endogenous estrogens to breast cancer risk, but more recent studies, which have benefitted from both increased assay precision as well as better methodologic approaches, have provided support to the notion that estrogens are predictive of subsequent breast cancer risk [53,54]. Also enhancing our understanding of the hormonal etiology of female cancers has been a broadening of the outlook to include not only estrogens but also other hormones that could have an impact on risk. Of note are several studies that have related androgen levels to breast cancer risk [53,55,56]. Other studies are beginning to emerge that are assessing other hormonal influences, including a variety of growth factors, such as insulin-like growth factors [57]. Furthermore, our knowledge is being expanded by investigations of hormonal influences not only on breast cancer but also on other cancers suspected of being hormonally related, including ovarian and endometrial cancer, where relationships with endogenous hormones have only recently been explored. Thus, of interest are findings that endometrial cancer appears to be influenced by estrogen levels [58,59] and that ovarian cancer is not related to gonadotrophin levels [60]. Future studies that focus on additional hormonal correlates for various cancers will undoubtedly increase our knowledge of endocrine mechanisms. This will derive not only from a broadening of the number of hormonal influences assessed but also through an examination of their intercorrelated effects. In addition, emerging investigations that relate hormone levels to risk factors may enhance our understanding of when in the carcinogenic process these influences are most important. Studies that are assessing hormone levels in target tissue (*e.g.,* breast tissue) rather than in blood or urine will also be essential towards advancing our understanding of hormonal carcinogenesis.

V. Disease Heterogeneity

Some of the lack of progress in understanding the etiology of cancers in women may reflect the persistence of the notion that cancers are single disease entities. It is becoming progressively clear that significant disease heterogeneity exists for a number of cancers. Thus, studies that attempt to distinguish between etiologically distinct subtypes of tumors may be important in our understanding of etiologic processes. Although it has been known for some time that risk factors differ between squamous cell and adenocarcinomas of the lung, efforts are now being made to define subgroups of other cancers that might be etiologically distinct. For instance, studies of endometrial cancer have shown that serous carcinomas demonstrate quite different risk profiles from other epithelial tumors [61]. It appears that similar approaches for distinguishing subsets of ovarian cancer might prove useful. Furthermore, studies of breast cancer that account for hormone receptor status indicate that this may distinguish etiologically distinct subtypes. Important advances in more precisely measuring hormone receptor status may help in clarifying etiologic differences between tumor subgroups. It is apparent that attempts should be made to consider combined effects of both estrogen and progesterone hormone receptor status in these studies as this may assist in more precisely defining etiologically distinct disease entities [62]. It is possible to use a variety of other molecular probes to distinguish tumor types, including the assessment of various tumor suppressor genes (such as p53) and oncogenes. Finally, studies should not dismiss the importance of stage of disease at detection because it appears that a number of factors may operate quite differently for early as opposed to late stage tumors. The differing effects of exogenous hormones by stage of endometrial cancer at presentation are well recognized [63], but it is increasingly being appreciated that other factors may also vary (*e.g.,* an enhanced effect of alcohol on later stage breast cancers) [64].

VI. Recommended Study Approaches

Although we have struggled to advance our understanding of most of the major cancers in women, substantial progress has been made in understanding the etiology of cervical cancer (Chapter 73). Notably, it is now recognized that an essential etiologic agent for this disease is the human papilloma virus [65]. Although a number of issues remain unclear, including what factors might promote the progression of low-grade lesions to neoplastic conditions, we are well on our way to obtaining sufficient knowledge to launch efforts to eradicate this disease through vaccinology efforts [66]. Much of what we have learned about this disease has derived from careful natural history studies that could serve as a model for other cancer sites where much less is known about what factors are involved in the progression of precursor conditions. In particular, it would appear advantageous to apply some of these approaches to the studies of breast, endometrial, and ovarian neoplasms, where our understanding of biologic processes is still limited. Studies of factors involved in the progression of endometrial hyperplasias, benign ovarian tumors, proliferative breast diseases, and *in situ* cancers may be especially useful. In addition, investigations to enhance our understanding of biologic correlates of early disease markers appear warranted. For instance, it is now well recognized that the extent of mammographic breast density is highly predictive of subsequent breast cancer, yet only a limited number of studies have attempted to clarify the hormonal and histologic correlates of this early endpoint biomarker (Chapter 68).

Although much research has focused on the integration of molecular probes into well-designed analytic studies, descriptive studies should not be overlooked as they also can provide important clues regarding the etiology of diseases. In particular, most cancers in women show tremendous geographic variation, oftentimes as much as five- to tenfold differences in incidence and/or mortality across countries. Studies in migrants, whose disease rates are in fluctuation, may be especially useful in enhancing our understanding of elusive associations. In particular, the extent to which rates change in migrating populations and how quickly they change may give clues to which risk factors are involved and when in a woman's life they are most important. Risks in fact can vary quite substantially by migrant status, as demonstrated in a study of breast cancer [67]. Thus, additional studies in migrant populations, as well as in other nonimmigrant minority populations, should be pursued. In particular, unusual patterns of cancer or interrelationships with risk factors in these populations may be informative. For instance, it remains enigmatic why African-American women experience un-

usually high risks of premenopausal breast cancer compared to white women or why African-American women experience higher smoking-associated risks of lung cancer. Studies to resolve these issues may assist in furthering our understanding of these tumors and in providing clues to genetic mechanisms that should be pursued. Studies that include minorities may provide data that will assist in achieving sufficient heterogeneity of exposures or in providing wider ranges over which exposures can be examined. These studies may be especially useful in clarifying effects of dietary or genetic factors.

Finally, future studies may benefit from careful assessments of the timing of exposures. Studies have clarified that weight gain late in life appears to be most predictive of breast cancer risk [45], but the timing of other exposures remains less clear, including dietary patterns (including alcohol consumption) and physical activity. The adolescent period has come under scrutiny as a possible window of opportunity, although the challenges in how to accurately measure such exposures for cancers that generally occur late in life are obvious [68]. Even more challenging is how to account for pre- and postnatal exposures, which have been postulated to also affect risk of some cancers, including breast cancer [69,70].

VII. Prevention

Given what we have learned about the etiology of cancer in women, what are the prospects for prevention? For some cancers, such as cervical and possibly colorectal cancers, the prospects are good. For others, we still need to make serious advances (Chapter 77). Fortunately, we are currently in an era where biologic advances are enabling us to approach epidemiologic investigations with much more precision and insight. Sufficient knowledge has been gained to warrant the initiation of several chemoprevention trials. The benefits of tamoxifen in reducing the incidence of breast cancer in high-risk women has been demonstrated [71]. However, given the recognition that tamoxifen continues to be associated with certain risks, including an increased incidence of endometrial cancer, attempts are being made to identify alternative chemopreventive agents [72]. Currently underway are several trials to assess the effectiveness of alternatives, including various Selective Estrogen Receptor Modulators (SERMs). Many clinical trials of other agents are underway, and it remains to be demonstrated which of these will have a significant impact in reducing risk of cancers in women. Although several appear promising based on results from either analytic or clinical studies, expectations may not always be borne out by randomized clinical trials, as evidenced by results from beta-carotene trials that showed higher lung cancer risks in both men and women [73]. In addition, the large Women's Health Initiative, currently underway [74], should contribute a wealth of data regarding potential effects on cancer risk of dietary interventions (as well as exogenous hormones). This investigation should also provide extremely useful information for assessing compliance to a variety of recommended behavioral modifications.

Although our ultimate goal is for primary prevention, we fortunately have also seen a number of advances in secondary prevention. Screening techniques have improved considerably for a number of cancers in women. The efficacy of Pap smears in reducing the incidence of cervical disease is obvious, with new technologies on the horizon that may be even more effective in reducing disease occurrence. However, challenges still exist to assure that women most in need of screening will avail themselves of the technology and receive adequate follow-up when needed (Chapter 74). This is a particular challenge in areas of the world where the rates of disease are highest; in these areas, the extent to which technologic advances in screening modalities will be feasible is unclear. The benefit of mammographic screening for detection of breast cancer in older women is well recognized, but for younger women it is more controversial (Chapter 70). Given the importance of early detection of this disease, this topic will continue to receive much attention. Screening policies for colorectal cancer also continue to be reviewed and revised. Fewer advances have been made in early detection of ovarian and lung cancers, although many efforts are currently being expended to improve diagnostic accuracy.

VIII. Summary

It is clear that most cancers in women have a multifactorial etiology which complicates our understanding of their occurrence and presents difficulties for developing preventive strategies. Fortunately, we have learned much about the etiology of many cancers in women, primarily through multidisciplinary efforts. New molecular probes are being discovered each day, enabling many different directions for future research. The next decade promises to be an important one for clarifying appropriate approaches to these tumors. Given the complex etiologies of these cancers and the changing prevalence of many of the postulated risk factors, it is obvious that prevention must be approached from a number of different perspectives. This includes the need for epidemiologists to collaborate with others in different disciplines in order to clarify biologic mechanisms. Although there are many challenges ahead, it is apparent that many new insights will be gained by current research directions. Particularly encouraging has been progress in defining the epidemiology of cervical cancer. If this model of research can be extended, we may be well on our way to significantly reducing the incidence of some of these major cancers in women.

References

1. Green, A., Purdie, D., Bain, C. et al. (1997). Tubal sterilisation, hysterectomy and decreased risk of ovarian cancer. Int. J. Cancer 71, 948–951.
2. Kreiger, N., Sloan, M., Cotterchio, M. et al. (1997). Surgical procedures associated with risk of ovarian cancer. Int. J. Epidemiol. 26, 710–715.
3. Russo, J., and Russo, I. H. (1994). Toward a physiologic approach to breast cancer prevention. Cancer Epidemiol., Biomarkers Prev. 3, 353–364.
4. Risch, H. A. (1998). Hormonal etiology of epithelial ovarian cancer, with a hypothesis concerning the role of androgens and progesterone. J. Natl. Cancer Inst. 90, 1774–1786.
5. Voigt, L. F., Deng, Q., and Weiss, N. S. (1994). Recency, duration, and progestin content of oral contraceptives in relation to the incidence of endometrial cancer (Washington, USA). Cancer Causes Control 5, 227–233.
6. Collaborative Group on Hormonal Factors in Breast Cancer (1996). Breast cancer and hormonal contraceptives: Collaborative reanalysis of individual data on 53,297 women with breast cancer and

100,239 women with breast cancer from 54 epidemiological studies. Lancet 347, 1713–1727.
7. Grodstein, F., Stampfer, M. J., Colditz, G. A. et al. (1997). Postmenopausal hormone therapy and mortality. N. Engl. J. Med. 336, 1769–1775.
8. Key, T. J. A., and Pike, M. C. (1988). The role of oestrogens and progestagens in the epidemiology and prevention of breast cancer. Eur. J. Cancer Clin. Oncol. 24, 29–43.
9. Bergkvist, L., Adami, H.-O., Persson, I. et al. (1989). The risk of breast cancer after estrogen and estrogen-progestin replacement. N. Engl. J. Med. 321, 293–297.
10. Colditz, G. A., Hankinson, S. E., Hunter, D. J. et al. (1995). The use of estrogens and progestins and the risk of breast cancer in postmenopausal women. N. Engl. J. Med. 332, 1589–1593.
11. Schairer, C., Byrne, C., Keyl, P. M. et al. (1994). Menopausal estrogen and estrogen-progestin replacement therapy and risk of breast cancer (United States). Cancer Causes Control 5, 491–500.
12. Glud, E., Kjaer, S. K., Troisi, R. et al. (1998). Fertility drugs and ovarian cancer. Epidemiol. Rev. 20, 237–257.
13. Venn, A., Watson, L., Lumley, J. et al. (1995). Breast and ovarian cancer incidence after infertility and in vitro fertilisation. Lancet 346, 995–1000.
14. Harlow, B. L., Cramer, D. W., Baron, J. A. et al. (1998). Psychotropic medication use and risk of epithelial ovarian cancer. Cancer Epidemiol, Biomarkers Prev. 7, 697–702.
15. Berkel, H. J., Holcombe, R. F., Middlebrooks, M. et al. (1996). Nonsteroidal antiinflammatory drugs and colorectal cancer. Epidemiol. Rev. 18, 205–217.
16. Cramer, D. W., Harlow, B. L., Titus-Ernstoff, L. et al. (1998). Over-the-counter analgesics and risk of ovarian cancer. Lancet 351, 104–107.
17. Sturgeon, S. R., Schairer, C., Gail, M. et al. (1995). Geographic variation in mortality from breast cancer among white women in the United States. J. Natl. Cancer Inst. 87, 1846–1853.
18. Ko, Y. C., Lee, C. H., Chen, M. J. et al. (1997). Risk factors for primary lung cancer among non-smoking women in Taiwan. Int. J. Epidemiol. 26, 24–31.
19. Xu, Z. Y., Blot, W. J., Xiao, H. P. et al. (1989). Smoking, air pollution and the high rates of lung cancer in Shenyang, China. J. Natl. Cancer Inst. 81, 1800–1806.
20. Lubin, J. H., and Boice, J. D. (1997). Lung cancer risk from residential radon. Meta-analysis of eight epidemiological studies. J. Natl. Cancer Inst. 89, 49–57.
21. Hoyer, A. P., Grandjean, P., Jorgensen, T., et al. (1998). Organochlorine exposure and risk of breast cancer. Lancet 352, 1816–1820.
22. Sturgeon, S. R., Brock, J. W., Potischman, N. et al. (1998). Serum concentrations of organochlorine compounds and endometrial cancer risk (United States). Cancer Causes Control 9, 417–424.
23. Gammon, M. D., Wolff, M. S., Neugut, A. I. et al. (1997). Temporal variation in chlorinated hydrocarbons in healthy women. Cancer Epidemiol., Biomarkers Prev. 6, 327–332.
24. Gammon, M. D., Wolff, M. S., Neugut, A. I. et al. (1996). Treatment for breast cancer and blood levels of chlorinated hydrocarbons. Cancer Epidemiol., Biomarkers Prev. 5, 467–471.
25. Brownson, R. C., Alavanja, M. C. R., and Chang, J. C. (1993). Occupational risk factors for lung cancer among nonsmoking women, a case-control study in Missouri (United States). Cancer Causes Control 4, 449–454.
26. Wu-Williams, A. H., Xu, Z. Y., Blot, W. J. et al. (1993). Occupation and lung cancer risk among women in northern China. Am. J. Ind. Med. 24, 67–79.
27. Blair, A., Hartge, P., Stewart, P. A. et al. (1998). Mortality and cancer incidence of aircraft maintenance workers exposed to trichloroethylene and other organic solvents and chemicals: Extended follow-up. Occup. Environ. Med. J. 55, 161–171.

28. Goldberg, M. S., and Labreche, F. (1996). Occupational risk factors for female breast cancer: A review. *J. Occup. Environ. Med.* **53,** 145–156.

29. Donna, A., Crosignani, P., Robutti, F. *et al.* (1989). Triazine herbicides and ovarian epithelial neoplasms. *Scand. J. Work Environ. Health* **15,** 47–53.

30. Burger, M. P. M., Hollema, H., Gouw, A. S. H. *et al.* (1993). Cigarette smoking and human papillomavirus in patients with reported cervical cytological abnormality. *Br. Med. J.* **306,** 749–752.

31. Lash, T. L., and Aschengrau, A. (1999). Active and passive cigarette smoking and the occurrence of breast cancer. *Am. J. Epidemiol.* **149,** 5–12.

32. Morabia, A., Bernstein, M., Heritier, S. *et al.* (1996). Relation of breast cancer with passive and active exposure to tobacco smoke. *Am. J. Epidemiol.* **143,** 918–928.

33. Austin, H., Drews, C., and Partridge, E. E. (1993). A case-control study of endometrial cancer in relation to cigarette smoking, serum estrogen levels, and alcohol use. *Am. J. Obstet. Gynecol.* **169,** 1086–1091.

34. Brinton, L. A., Barrett, R. J., Berman, M. L. *et al.* (1993). Cigarette smoking and the risk of endometrial cancer. *Am. J. Epidemiol.* **137,** 281–291.

35. Parazzini, F., La Vecchia, C., Negri, E. *et al.* (1995). Smoking and risk of endometrial cancer: Results from an Italian case-control Study. *Gynecol. Oncol.* **56,** 195–199.

36. Howe, G. R., Hirohata, T., Hislop, T. G. *et al.* (1990). Dietary factors and risk of breast cancer: Combined analysis of 12 case-control studies. *J. Natl. Cancer Inst.* **82,** 561–569.

37. Hunter, D. J., Spiegelman, D., Adami, H.-O. *et al.* (1996). Cohort studies of fat intake and the risk of breast cancer—a pooled analysis. *N. Engl. J. Med.* **334,** 358–361.

38. Wu, A. H., Pike, M. C., and Stram, D. O. (1999). Meta-analysis: Dietary fat intake, serum estrogen levels, and the risk of breast cancer. *J. Natl. Cancer Inst.* **91,** 529–534.

39. Cramer, D. W., Harlow, B. L., Willett, W. C. *et al.* (1989). Galactose consumption and metabolism in relation to the risk of ovarian cancer. *Lancet* **2,** 66–71.

40. Wurzelmann, J. I., Silver, A., Schreinemachers, D. M. *et al.* (1996). Iron intake and the risk of colorectal cancer. *Cancer Epidemiol., Biomarkers Prev.* **5,** 503–507.

41. Zheng, W., Gustafson, D. R., Sinha, R. *et al.* (1998). Well-done meat intake and risk of breast cancer. *J. Natl. Cancer Inst.* **90,** 1724–1729.

42. Moore, M. A., Park, C. B., and Tsuda, H. (1998). Physical exercise: A pillar for cancer prevention. *Eur. J. Cancer Prev.* **7,** 177–193.

43. Longnecker, M. P. (1994). Alcoholic beverage consumption in relation to risk of breast cancer: Meta-analysis and review. *Cancer Causes Control* **5,** 73–82.

44. Carroll, K. K. (1998). Obesity as a risk factor for certain types of cancers. *Lipids* **33,** 1055–1059.

45. Ballard-Barbash, R. (1994). Anthropometry and breast cancer. Body size—a moving target. *Cancer (Philadelphia)* **74,** 1090–1100.

46. Huang, Z., Hankinson, S. E., Colditz, G. A. *et al.* (1997). Dual effects of weight and weight gain on breast cancer risk. *JAMA, J. Am. Med. Assoc.* **278,** 1407–1411.

47. Streuwing, J. P., Hartge, P., Wacholder, S. *et al.* (1997). The risk of cancer associated with mutations of BRCA1 and BRCA2 among Ashkenazi Jews. *N. Engl. J. Med.* **336,** 1401–1408.

48. Garcia-Closas, M., and Lubin, J. H. (1999). Power and sample size calculations in case-control studies of gene-environment interactions: Comments on different approaches. *Am. J. Epidemiol.* **149,** 689–692.

49. Garcia-Closas, M., Rothman, N., Stewart, W. F. *et al.* (1999). Impact of misclassification on sample size in case-control studies of gene-environment interactions. *Cancer Epidemiol. Biomarkers Prev* (in press).

50. Haiman, C. A., Hankinson, S. E., Spiegelman, D. *et al.* (1999). The relationship between a polymorphism in *CYP17* with plasma hormone levels and breast cancer. *Cancer Res.* **59,** 1015–1020.

51. Falk, R. T., Dorgan, J. F., Kahle, L. *et al.* (1997). Assay reproducibility of hormone measurements in postmenopausal women. *Cancer Epidemiol., Biomarkers Prev.* **6,** 429–432.

52. Hankinson, S. E., Manson, J. E., Spiegelman, D. *et al.* (1995). Reproducibility of plasma hormone levels in postmenopausal women over a 2–3 year period. *Cancer Epidemiol., Biomarkers Prev.* **4,** 649–654.

53. Hankinson, S. E., Willett, W. C., Manson, J. E. *et al.* (1998). Plasma sex steroid hormone levels and risk of breast cancer in postmenopausal women. *J. Natl. Cancer Inst.* **90,** 1292–1299.

54. Thomas, H. V., Reeves, G. K., and Key, T. J. (1997). Endogenous estrogen and postmenopausal breast cancer: A quantitative review. *Cancer Causes Control* **8,** 922–928.

55. Dorgan, J. F., Stanczyk, F. Z., Longcope, C. *et al.* (1997). Relationship of serum dehydroepiandrostrone (DHEA), DHEA sulfate and 5-androstene-3β, 17β-diol to risk of breast cancer in postmenopausal women. *Cancer Epidemiol., Biomarkers Prev.* **6,** 177–181.

56. Zeleniuch-Jacquotte, A., Bruning, P. F., Bonfrer, J. M. G. *et al.* (1997). Relation of serum levels of testosterone and dehydroepiandrosterone sulfate to risk of breast cancer in postmenopausal women. *Am. J. Epidemiol.* **145,** 1030–1038.

57. Hankinson, S. E., Willett, W. C., Colditz, G. A. *et al.* (1998). Circulating concentrations of insulin-like growth factor-I and risk of breast cancer. *Lancet* **351,** 1393–1396.

58. Nyholm, H. C. J., Nielsen, A. L., Lyndrup, J. *et al.* (1993). Plasma oestrogens in postmenopausal women with endometrial cancer. *Br. J. Obstet. Gynaecol.* **100,** 1115–1119.

59. Potischman, N., Hoover, R. N., Brinton, L. A. *et al.* (1996). Case-control study of endogenous steroid hormones and endometrial cancer. *J. Natl. Cancer Inst.* **88,** 1127–1135.

60. Helzlsouer, K. J., Alberg, A. J., Gordon, G. B. *et al.* (1995). Serum gonadotropins and steroid hormones and the development of ovarian cancer. *JAMA, J. Am. Med. Assoc.* **274,** 1926–1930.

61. Sherman, M. E., Sturgeon, S., Brinton, L. A., *et al.* (1997). Risk factors and hormone levels in patients with serous and endometrioid uterine carcinomas. *Mod. Pathol.* **10,** 963–968.

62. Potter, J. D., Cerhan, J. R., Sellers, T. A. *et al.* (1995). Progesterone and estrogen receptors and mammary neoplasia in the Iowa Women's Health Study: How many kinds of breast cancer are there? *Cancer Epidemiol., Biomarkers Prev.* **4,** 319–326.

63. Shapiro, S., Kelly, J. P., Rosenberg, L. *et al.* (1985). Risk of localized and widespread endometrial cancer in relation to recent and discontinued use of conjugated estrogens. *N. Engl. J. Med.* **313,** 969–972.

64. Swanson, C. A., Coates, R. J., Malone, K. E., *et al.* (1997). Alcohol consumption and breast cancer risk among women under age 45 years. *Epidemiology* **8,** 231–237.

65. Schiffman, M. H., Bauer, H. M., Hoover, R. N. *et al.* (1993). Epidemiologic evidence showing that human papillomavirus infection causes most cervical intraepithelial neoplasia. *J. Natl. Cancer Inst.* **85,** 958–964.

66. Hildesheim, A. (1997). Human papillomavirus variants: Implication for natural history studies and vaccine development efforts. *J. Natl. Cancer Inst.* **89,** 752–753.

67. Ziegler, R. G., Hoover, R. N., Pike, M. C. *et al.* (1993). Migration patterns and breast cancer risk in Asian-American women. *J. Natl. Cancer Inst.* **85,** 1819–1827.

68. Colditz, G. A., and Franzier, A. L. (1995). Models of breast cancer show that risk is set by events of early life: Prevention efforts must shift focus. *Cancer Epidemiol., Biomarkers Prev.* **4,** 567–571.

69. Hilakivi-Clarke, L., Clarke, R., and Lippman, M. E. (1994). Perinatal factors increase breast cancer risk. *Breast Cancer Res. Treat.* **31,** 273–284.

70. Trichopoulos, D. (1990). Is breast cancer initiated in utero? *Epidemiology* **1**, 95–96.

71. Fisher, B., Costantino, J. P., Wickerham, D. L. *et al.* (1998). Tamoxifen for prevention of breast cancer: Report of the National Surgical Adjuvant Breast and Bowel Project P-1 Study. *J. Natl. Cancer Inst.* **90**, 1371–1388.

72. Jordan, V. C. (1998). Antiestrogenic action of raloxifene and tamoxifen: Today and tomorrow. *J. Natl. Cancer Inst.* **90**, 967–971.

73. Omenn, G. S., Goodman, G. E., Thornquist, M. D. *et al.* (1996). Risk factors for lung cancer and for intervention effects in CARET, the Beta-Carotene and Retinol Efficacy Trial. *J. Natl. Cancer Inst.* **88**, 1550–1559.

74. Women's Health Initiative Study Group (1998). Design of the women's health initiative clinical trial and observational study. *Controlled Clin. Trials* **19**, 61–109.

67

Cancers in Women

SUSAN S. DEVESA

Biostatistics Branch
Division of Cancer Epidemiology and Genetics
National Cancer Institute, National Institutes of Health
Bethesda, Maryland

I. Impact of Cancer

In the United States, cancer is the second leading cause of death both among men and among women, following only deaths due to heart disease [1]. The estimated number of deaths due to cancer during 1998 exceeded 564,000, with more than 294,000 occurring among men and 270,000 among women. Cancer was the leading cause of death among women aged 35–74 years.

Although there are at least 40 forms of cancer, six sites accounted for more than 60% of all deaths due to cancer among American women: breast, lung, colorectum, cervix uteri, corpus uteri, and ovary (Table 67.1). It was estimated that during 1998, lung cancer would be the most frequent form of cancer death among American women, accounting for almost 25% of all their cancer deaths. It also was estimated that breast and colorectal cancers would account for 16% and 11% of cancer deaths among women, respectively. The same six cancers accounted for almost 67% of all newly-diagnosed cases, excluding the common superficial skin cancers that generally are caught early, are very treatable, and usually are nonfatal. By far the most commonly diagnosed cancer was breast cancer, which alone accounted for almost 30% of all cases. Lung and colorectal cancers each accounted for more than 10% of the cases.

Based on data from the Surveillance, Epidemiology, and End Results (SEER) program incidence files and the National Center for Health Statistics mortality files, the lifetime probability for an American woman of being diagnosed with any form of non-superficial skin cancer was estimated to be 38%, or almost 2 in 5

(Table 67.2) [2]. For the specific cancers, the lifetime risk of ever being diagnosed varied from 12.5% for breast cancer (or 1 in 8), to about 6% for lung or colorectal cancer, to less than 1% for invasive cervix uteri cancer. The lifetime risk of dying from any form of cancer was 20.5%, or 1 in 5, with the risks ranging from 4.5% for lung cancer to 0.3% for cervix uteri cancer.

II. Trends in Cancer Rates among American Women

Mortality rates, age-adjusted using the 1970 U.S. standard, varied considerably according to cancer, race, gender, and over time (Fig. 67.1). Rates spanning 1950–1954 through 1990–1994 and 1995 were available for the racial categories white and nonwhite, and since 1970–1974 specifically for blacks. Rates per 100,000 person-years for breast cancer have been relatively stable among white women, whereas rates have increased among nonwhites, surpassing those among whites very recently; rates were higher among blacks than nonwhites since at least 1970, and exceeded those among whites around 1980. In contrast, lung cancer rates in the past were lower than the breast cancer rates among all race/gender groups, but rapid increases have altered the picture. Over the entire time period, rates rose by about 600% among women, by 169% among white men, and by 302% among black men. Rates always were considerably higher among men than among women, and rates among men peaked during the 1980s. Increases among women occurred later but accelerated during the late 1960s through the late 1990s, with some hint of less rapid increases during the 1990s. Lung cancer rates in the 1950s were higher among whites than nonwhites; in recent years, there has been little difference in the

Table 67.1
Estimates of Cancer among American Women, 1998

Cancer	Numbers	
	Cases	Deaths
All cancers[a]	600,700	270,600
Breast	178,700	43,500
Lung	80,100	67,000
Colorectum	67,000	28,600
Cervix uteri	13,700	4,900
Corpus uteri	36,100	6,300
Ovary	25,400	14,500

Data from [1].
[a]Excluding superficial skin cancers.

Table 67.2
Lifetime Risk (Percent) for American Women of Being Diagnosed with Cancer or Dying from Cancer

Cancer	Being diagnosed	Dying
All cancers[a]	38.0	20.5
Breast	12.5	3.4
Lung	5.6	4.5
Colorectum	5.6	2.5
Cervix uteri	0.8	0.3
Corpus uteri	2.7	0.5
Ovary	1.8	1.1

Data from [2].
[a]Excluding superficial skin cancers.

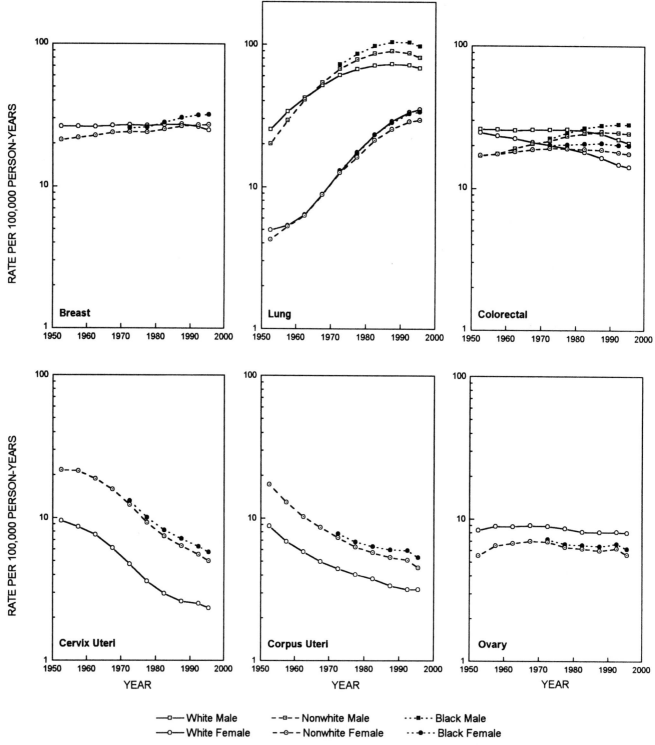

Fig. 67.1 Age-adjusted (1970 U.S. standard) mortality trends in the United States for six cancers by race and sex, 1950–1954 to 1990–1994, 1995.

rates for blacks and whites among women, in contrast to markedly higher rates for blacks among men. The trends for colorectal cancer have varied. Declines were consistent among white women but were apparent only recently among white men. Non-

whites showed early increases in colorectal cancer rates, but the increases were less rapid among women than among men, and the rates have plateaued in recent years. Rates for cancer of the cervix uteri declined more than 75%, and those for cancer of the

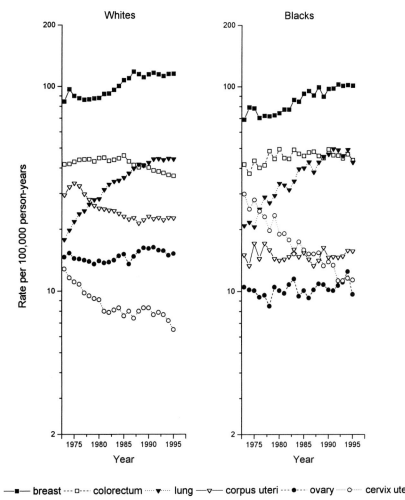

Fig. 67.2 Age-adjusted (1970 U.S. standard) incidence trends for six cancers among women by race, SEER program, 1973–1995.

corpus uteri (including cancer of the uterus not otherwise specified) declined more than 60%; rates were consistently higher among blacks and nonwhites than among whites. Ovarian cancer mortality rates did not change greatly over time, and rates among whites always were higher than those among nonwhites or blacks.

Incidence data based on newly-diagnosed cases are available since the early 1970s from nine population-based registries in five states and four metropolitan areas participating in the SEER program; about 10% of the U.S. population live in these areas [2]. Figure 67.2 presents the age-adjusted (1970 U.S. standard) incidence trends for the six common cancers among white and black women. The diagnosis of breast cancer rose in the early 1970s, which was coincident with the increased awareness and publicity associated with the diagnosis in the wives of the president and vice-president. Increases during the 1980s are largely attributable to the rising use of mammography and the detection of smaller tumors. As with the mortality data, lung cancer incidence rose rapidly among both races and surpassed colorectal cancer among whites in 1990. During the 1970s the incidence of corpus uteri cancer rose and subsequently fell in conjunction

with the use of unopposed estrogens; decreases during the 1980s may be related to the rising use of opposed estrogens. Cervix uteri cancer incidence declined substantially among both races, which was related to the use of Pap smears for the detection of premalignant disease.

III. Racial/Ethnic Variation in Rates among American Women

Cancer mortality rates among women varied considerably according to racial/ethnic group (Table 67.3) [3]. Total cancer rates were highest among Alaska Natives, blacks, and Hawaiians, and were lowest among Filipinos, Hispanics, Chinese, and Japanese. Lung and colorectal cancer mortality rates also were highest among Alaska Natives; there were too few deaths to calculate rates for the other four cancers. Among the other racial/ethnic groups, black women had the highest mortality rates from breast, colorectal, and cervix uteri cancers, and the second highest rate for corpus uteri cancer. Non-Hispanic whites had the highest rate of ovarian cancer mortality and the second highest rate for breast, lung, and colorectal cancers. The second

Table 67.3

Racial/Ethnic Variation in Cancer Mortality Rates[a] among U.S. Women, 1988–1992

Cancer	All cancers	Breast	Lung	Colorectum	Cervix uteri	Corpus uteri	Ovary
Alaska Native	179	—	45.3	24.0	—	—	—
American Indian[b]	99	—	—	—	—	—	—
Black	168	31.4	31.5	20.4	6.7	6.0	6.6
Chinese	86	11.2	18.5	10.5	2.6	2.2	4.0
Filipino	63	11.9	10.0	5.8	2.4	1.3	3.4
Hawaiian	168	25.0	44.1	11.4	—	8.4	7.3
Japanese	88	12.5	12.9	12.3	1.5	1.9	5.0
White	140	27.0	31.9	15.3	2.5	3.2	8.1
Hispanic (Total)	85	15.0	10.8	8.3	3.4	2.3	4.8
White Hispanic	89	15.7	11.2	8.6	3.6	2.4	5.1
White Non-Hispanic	143	27.7	32.9	15.6	2.5	3.3	8.2

Note.—rate not calculated when fewer than 25 deaths. Data from [3].

[a]Per 100,000 woman-years, age-adjusted using the 1970 U.S. standard.

[b]New Mexico.

highest cervix uteri cancer rate occurred among white Hispanics. Hawaiian women experienced the highest mortality rate from lung and corpus uteri cancers. Rates for most cancers were low among Filipino women, although the lowest breast cancer rate was among Chinese women.

Cancer incidence patterns were somewhat different (Table 67.4). The rate for all cancers combined was highest among non-Hispanic whites, largely due to elevated breast cancer rates. Relative to non-Hispanic whites, Hispanic whites were at lower cancer risk overall and had the lowest rates for all of the cancers shown except for cervix uteri; invasive cervical cancer incidence among Hispanics was more than twice that among non-Hispanic whites. As in the mortality data, Alaska Natives and blacks had relatively high incidence overall because they had the highest rates of lung and colorectal cancers. Compared to whites, total incidence rates and rates for most cancers shown were considerably lower among the Chinese, Filipino, and Jap-

anese, except for cancers of the colorectum among the Chinese- and Japanese, and cervix uteri among the Chinese and Filipinos. Rates among Hawaiian women were relatively high for cancers of the breast, lung, and especially corpus uteri. In addition to the racial/ethnic groups for whom mortality data were available, Vietnamese and Koreans are represented in the incidence data. Vietnamese women experienced relatively low rates of breast and corpus uteri cancers but they had the highest invasive cervix uteri cancer rate. The lowest total cancer incidence rates occurred among American Indians and Koreans and were related to their notably low breast and colorectal cancer rates.

IV. International Variation in Cancer Rates

International incidence data for 1988–1992 were available from the most recent volume of Cancer Incidence in Five Con-

Table 67.4

Racial/Ethnic Variation in Cancer Incidence Rates[a] among Women, SEER Program, 1988–1992

Cancer	All cancers	Breast	Lung	Colorectum	Cervix uteri	Corpus uteri	Ovary
Alaska Native	348	78.9	50.6	67.4	15.8	—	—
American Indian[b]	180	31.6	—	15.3	9.9	10.7	17.5
Black	326	95.4	44.2	45.5	13.2	14.4	10.2
Chinese	213	55.0	25.3	33.6	7.3	11.6	9.3
Filipino	224	73.1	17.5	20.9	9.6	12.1	10.2
Hawaiian	321	105.6	43.1	30.5	9.3	23.9	11.8
Japanese	241	82.3	15.2	39.5	5.8	14.5	10.1
Korean	180	28.5	16.0	21.9	15.2	3.8	7.0
Vietnamese	273	37.5	31.2	27.1	43.0	8.4	13.8
White	346	111.8	41.5	38.3	8.7	22.3	15.8
White Non-Hispanic	354	115.7	43.7	39.2	7.5	23.0	16.2
White Hispanic	256	73.5	20.4	25.9	17.1	14.5	12.1
Hispanic (Total)	243	69.8	19.5	24.7	16.2	13.7	11.4

Note.—rate not calculated when fewer than 25 cases. Data from [3].

[a]Per 100,000 woman-years, age-adjusted using the 1970 U.S. standard.

[b]New Mexico.

Table 67.5
International Variation in Cancer Incidence Rates among Women, 1988–1992

Cancer	All cancers	Breast	Lung	Colorectum	Cervix uteri	Corpus uteri	Ovary
United States							
U.S., Los Angeles: Hispanic White	199.4	57.4	14.5	18.2	17.9	11.5	9.0
U.S., SEER: White	280.9	90.7	33.8	29.5	7.5	18.2	11.9
U.S., SEER: Black	271.7	79.4	38.5	35.3	12.0	11.4	8.1
Asia							
China, Shanghai	153.2	26.5	18.2	18.1	3.3	3.7	5.8
India, Bombay	125.4	28.2	3.7	5.6	20.2	2.5	7.2
Israel: All Jews	239.3	77.4	9.2	31.3	5.3	10.8	11.6
Japan, Osaka	154.8	24.3	12.4	20.0	9.2	3.0	5.6
Europe							
Denmark	261.6	73.3	25.4	30.3	15.2	14.7	14.0
France, Calvados	195.9	76.3	4.6	24.2	10.4	8.4	9.0
Italy, Varese	226.5	73.5	8.5	26.9	6.4	12.6	10.2
Sweden	228.5	72.9	10.9	24.1	8.0	13.2	13.2
UK, England and Wales	225.5	68.8	22.8	23.7	12.5	8.7	12.4
Oceania							
Australia, New South Wales	240.9	67.2	14.9	30.9	9.9	8.8	8.4
New Zealand: Non-Maori	274.6	77.2	18.2	40.8	11.9	9.4	11.0

Note. Rates per 100,000 person-years, age-adjusted using world population standard. Data from [4].

tinents (Table 67.5) [4]. These rates were age-adjusted using the world standard, which has a relatively younger age distribution than the 1970 U.S. standard, so these rates are lower than those shown in Table 67.4. Among these 14 registries, the highest total cancer incidence occurred among U.S. SEER whites, followed by New Zealand non-Maoris and U.S. SEER blacks. The high U.S. rates were due to the high rates of breast and lung cancers, whereas colorectal cancer rates were notably elevated in New Zealand. The total cancer incidence among U.S. white Hispanics was relatively low; rates for most of the individual cancers also were low, except for cervix uteri cancer. Within Europe, the total cancer rate was highest in Denmark because of elevated rates of ovarian and corpus uteri cancers. The lowest rates were in Bombay, India (where the rates for all cancers except cervix uteri were low), in Shanghai, China (although the lung cancer rate there was among the top third), and in Osaka, Japan (where the lowest rates for breast and ovarian cancers occurred). From a different perspective, breast cancer rates were highest among U.S. whites and blacks (91 and 79 per 100,000 person-years, respectively), more than three times those in most of Asia. The high U.S. rates for lung cancer, exceeding 33, were more than eight times those in Bombay (3.7). The colorectal cancer rates in New Zealand and U.S. blacks (35–41) were about six times those in Bombay (5.6). Cervix uteri cancer rates in Bombay and U.S. Hispanic whites (18–20) were more than four times those in Shanghai (3.3). The corpus uteri cancer rates varied about sevenfold (from 2.5 to 18.2), and those for ovarian cancer more than twofold (from 5.6 to 14).

V. Age-Specific Cancer Rates among American Women

Age-specific mortality curves in the United States for these six cancers among white and black women during 1991–1995

are presented in Figure 67.3. Rates for all six cancers rose rapidly with age. Among women under age 50, breast cancer was the leading cause of cancer death. Lung cancer was the dominant form among women ages 55 to 80, and colorectal cancer predominated among the oldest women. Lung cancer mortality rates peaked among women in their 70s, reflecting the cohort-specific histories of cigarette smoking. Deaths due to uterine cancer were more likely cervical in origin among women under age 60 and corpus in origin at older ages.

Breast cancer was the most frequent form of cancer diagnosed across virtually the entire age range (Fig. 67.4). Only at the very young ages were ovarian and cervix uteri cancers more common. Cervix uteri cancer was unique in that incidence rates increase until about age 40 and then plateau thereafter, more convincingly among whites than among blacks. During the early 1990s, incidence rates both for lung cancer and for corpus uteri cancer peaked among women in their 70s. At the oldest ages, colorectal cancer was diagnosed as frequently as breast cancer.

Although difficult to discern from these figures, age-specific mortality and incidence rates were higher among blacks than among whites at virtually all ages for colorectal and cervix uteri cancers, and were higher among whites than among blacks for ovarian cancer [2]. Breast cancer incidence rates were higher among whites than among blacks only at ages 50 and older, in contrast to higher incidence rates among blacks at younger ages and higher mortality rates across virtually all ages. Corpus uteri cancer (including cancer of uterus not otherwise specified) mortality rates also were consistently higher among blacks than among whites, in contrast to higher incidence rates among whites. Lung cancer incidence and mortality rates were higher among blacks up to ages 65 or 70, with higher rates among whites at older ages.

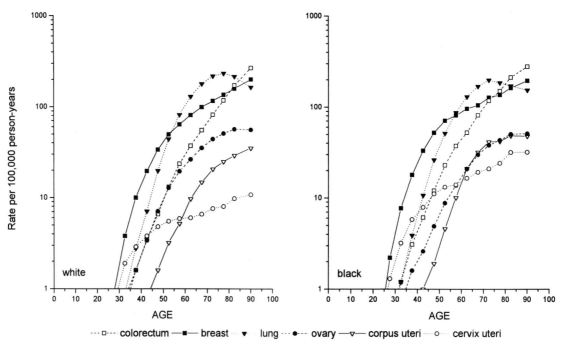

Fig. 67.3   Age-specific mortality curves in the United States for six cancers among white and black women, 1991–1995.

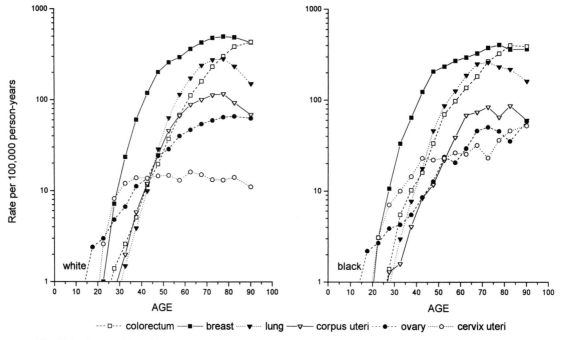

Fig. 67.4   Age-specific incidence curves for six cancers among white and black women, SEER program, 1991–1995.

VI. Stage at Diagnosis of Cancer and Patient Survival

Information regarding the stage of disease at the time of diagnosis is available from the SEER program [2]. During the early 1990s, more than 80% of lung cancers and about 90% or more of the five other cancers were staged (Table 67.6). More than half of ovarian cancers and lung cancers were not diagnosed until there was distant spread, with 24% or fewer diagnosed at a localized stage. In contrast, half or more of breast cancers and cervical cancers were diagnosed while still localized, at least among whites, and 10% or fewer had spread to distant sites. The largest proportion localized at diagnosis occurred for corpus

Table 67.6
Distribution of Major Cancers among Women by Stage at Diagnosis,
SEER Program, 1989–1994

Cancer	Race	Total	Localized	Regional	Distant	Unstaged
				%		
Breast	White	100%	62	29	6	3
	Black	100%	50	35	9	5
Lung	White	100%	15	23	47	15
	Black	100%	12	24	52	12
Colorectum	White	100%	37	37	20	6
	Black	100%	32	35	25	7
Cervix uteri	White	100%	55	31	7	7
	Black	100%	43	37	10	11
Corpus uteri	White	100%	75	12	9	4
	Black	100%	50	22	20	8
Ovary	White	100%	24	9	61	6
	Black	100%	24	10	58	8

Data from [2].

uteri cancers among white women (75%), although the proportion was only 50% among blacks. The distribution was more even for colorectal cancers, with 32–37% each for localized and regional stages and about 20–25% for distant stage at diagnosis.

Five-year relative survival rates, adjusted for general population mortality experience, were highest among women with breast cancer and tended to be somewhat higher among whites than among blacks (Table 67.7). For all stages combined, the

Table 67.7
Five-Year Relative Survival Rates by Stage at Diagnosis among Women with Diagnoses
of Cancer, SEER Program, 1989–1994

Cancer	Race	Total	Localized	Regional	Distant	Unstaged
				%		
Breast	White	86.7	98.1	78.3	23.2	52.2
	Black	70.6	88.9	62.4	14.4	47.3
Lung	White	14.5	50.8	20.5	2.1	7.6
	Black	11.2	43.6	15.2	2.2	8.3
Colorectum	White	63.0	92.5	66.7	8.6	34.6
	Black	52.5	83.8	61.2	6.5	37.0
Cervix uteri	White	71.5	92.5	49.4	12.9	60.9
	Black	59.0	85.5	38.0	8.2	65.9[†]
Corpus uteri	White	86.5	97.2	68.6	29.9	54.8
	Black	54.4	79.3	39.4	12.3	37.3[a]
Ovary	White	50.1	95.9	79.6	28.1	29.0
	Black	46.3	91.0	78.0[a]	24.3	33.2[b]

Data from [2].
[a]The standard error of the survival rate is between 5 and 10 percentage points.
[b]The standard error of the survival rate is greater than 10 percentage points.

largest differences were for patients with breast or corpus uteri cancer, with smaller but still notable differences for women with colorectal or cervical cancer. Racial differences for lung or ovarian cancer patients were minimal, even though the overall survival rate was 46–50% for ovarian cancer and less than 15% for lung cancer. Survival rates varied substantially by stage at diagnosis, from 80% or greater for all the localized cancers shown except lung cancer to less than 30% for those with distant disease. Only about half of patients diagnosed with localized lung cancer survived 5 years, and the rate declined to 2% for those with distant disease. Although the racial differences in stage-specific survival rates tended to be smaller than those for all stages combined, blacks still generally had a less favorable prognosis than whites. The more advanced stages at diagnosis and poorer survival rates within stage contributed to the higher mortality rates for breast and corpus uteri cancers among blacks versus whites, in the face of lower incidence rates.

VII. Conclusions

Although cancers at various anatomic sites share the common characteristics of malignant behavior, including uncontrolled proliferation and invasion of other tissues, it is clear that the patterns of occurrence in the population, time trends, and survival rates vary considerably. Risk factors and causes of some cancers have been identified, discussed in other chapters in this volume, but additional research is still warranted. Primary prevention, when feasible, is the most efficient method to avoid the development of cancer. As is evident from the variation in survival rates by stage of disease, one of the most important factors influencing outcome is early diagnosis.

Acknowledgments

I appreciate the sustained high quality registry operations of the SEER program participants, the dedication of the NCI SEER staff, and the computer programming and figure development by John Lahey of IMS, Inc.

References

1. Landis, S. H., Murray, T., Bolden, S., and Wingo, P. A. (1998). Cancer statistics, 1998. *Ca—Cancer J. Clin.* **48,** 6–29.
2. Ries, L. A. G., Kosary, C. L., Hankey, B. F., and Edwards, B. K., eds. (1998). "SEER Cancer Statistics Review, 1973–1995." National Cancer Institute, Bethesda, MD.
3. Miller, B. A., Kolonel, L. N., Bernstein, L., Young, J. L., Jr., Swanson, G. M., West, D. W., Key, C. R., Liff, J. M., Glover, C. S., and Alexander, G. A., eds. (1996). "Racial/Ethnic Patterns of Cancer in the United States 1988–1992," NIH Publ. No. 96-4104, pp. 1–A9. National Cancer Institute, Bethesda, MD.
4. Parkin, D. M., Whelan, S. L., Ferlay, J., Raymond, L., and Young, J., eds. (1997). "Cancer Incidence in Five Continents," Vol. 7, IARC Sci. Publ. No. 143. IARC, Lyon, France.

68

Breast Cancer Epidemiology, Treatment, and Prevention

GISKE URSIN,* DARCY V. SPICER,† AND LESLIE BERNSTEIN*

*Department of Preventive Medicine and †Department of Medicine, University of Southern California School of Medicine, Norris Comprehensive Cancer Center, Los Angeles, California

I. Introduction

Breast cancer is the most common cancer among women worldwide [1] and continues to be a major cause of cancer deaths [2]. It is estimated that nearly 180,000 women in the United States (U.S.) will be diagnosed with invasive breast cancer in 1998 and that more than 43,000 women will die of breast cancer [3].

This chapter describes the epidemiology of breast cancer, including trends in incidence and mortality and risk factors for the disease, as well as some clinical issues. Breast cancer screening and genetics are discussed in other chapters in this volume.

II. Trends in Incidence and Mortality

The incidence rates of female breast cancer are highest among North American and northern European women, intermediate among women living in southern Europe and Central and South America, and lowest among women living in Asia and Africa (Table 68.1) [1]. The largest increases in incidence rates in recent years have occurred in some of the Asian countries; for example, rates more than doubled in Singapore and Japan between 1970 and 1990 (Fig. 68.1) [1,4,5]. Mortality rates have also increased since 1970 in a number of countries with low incidence rates, such as Japan, Singapore, Columbia, Spain, and Yugoslavia [6]. However, in the U.S. and Great Britain, the two countries with the highest incidence rates in 1985, mortality rates declined between 1985 and 1990 (Fig. 68.2) [7,8]. Incidence rates vary within the U.S. by racial ethnic group for both invasive and *in situ* breast cancer (Fig. 68.3) [9]. Non-Hispanic whites have the highest rates of both invasive and *in situ* cancer, whereas Korean and American Indian women have the lowest rates. Although the incidence rates are highest among non-Hispanic white women, breast cancer mortality rates are highest among African Americans (Fig. 68.4) [9].

The epidemic of breast cancer throughout the world has caused much concern and increased awareness of the disease. Several factors have likely contributed to the observed increases in incidence rates, including changes in the prevalence of breast cancer risk factors, increased frequency of mammographic screening, and improved methods of ascertainment by cancer registries in some countries. The increases in breast cancer mortality rates observed in many areas of the world may be the result of true increases in breast cancer incidence without improvements in survival or to changes in cause of death coding on death certificates [6]. The declining breast cancer mortality rates observed in the presence of stable incidence rates, as seen in many industrialized countries, may be due to improvements in breast cancer treatment per se, better survival due to earlier

Table 68.1

Age-Adjusted Incidence Rates of Female Breast Cancer for Women Aged 35–74 (Per 100,000 Women in the Population) from Selected Cancer Registries around the World and in the United States during the Time Period 1986–1992[a]

Registry	Years of diagnosis	Age-adjusted incidence rates per 100,000 women
U.S., Connecticut	1988–1992	225.7
England and Wales	1988–1990	168.4
Norway	1988–1992	130.2
Yugoslavia (Slovenia)	1988–1992	114.9
Spain (Zaragoza)	1986–1990	97.9
Colombia (Cali)	1987–1991	87.6
Singapore (Chinese)	1988–1992	98.5
Japan (Miyagi)	1988–1992	79.1
India (Bombay)	1988–1992	68.5
China (Shanghai)	1988–1992	65.8

Note. Incidence rates are adjusted to the world standard population.
[a]From [1].

cancer diagnosis, or the diagnosis of cancers that are histologically malignant but biologically benign. Data from Los Angeles, one of the 11 national Surveillance, Epidemiology, and End Results (SEER) registries, indicate that breast cancer rates have plateaued after a peak in 1988 [L. Bernstein, 1998, unpublished data].

III. Role of Endogenous Hormones

The substantial body of experimental, clinical, and epidemiologic evidence on breast cancer indicates that hormones play a major role in its etiology [10]. A majority of the known nondemographic risk factors for breast cancer discussed in this chapter may be interpreted as measures of the cumulative exposure of the breast to estrogen, and perhaps, to progesterone. The actions of these ovarian hormones and their exogenous formulations (used in combination oral contraceptives and hormone replacement therapy) on the breast do not appear to be genotoxic, but do affect the rate of cell division [11]. Studies of breast epithelial cell division rates show that proliferation rates are low during the follicular phase of the menstrual cycle, when estradiol and progesterone levels are low, and are higher during the luteal phase of the cycle when levels of these ovarian hormones are higher [12]. Furthermore, well designed seroepidemiologic studies of hormone levels of women at high and at low risk of breast cancer, as well as studies of hormone levels of breast cancer

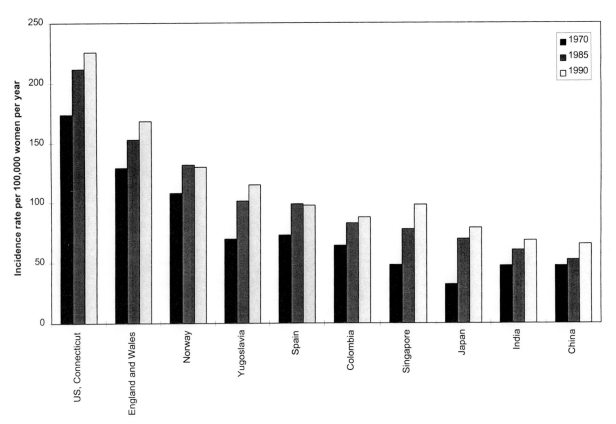

Fig. 68.1   Invasive breast cancer incidence rates in women aged 35–74 over time [4,5].

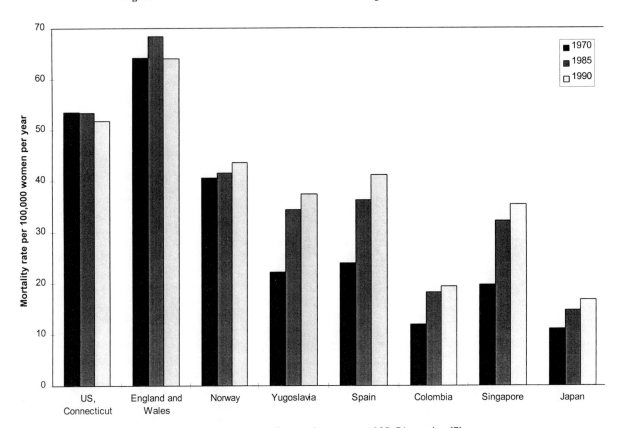

Fig. 68.2   Breast cancer mortality rates in women aged 35–74 over time [7].

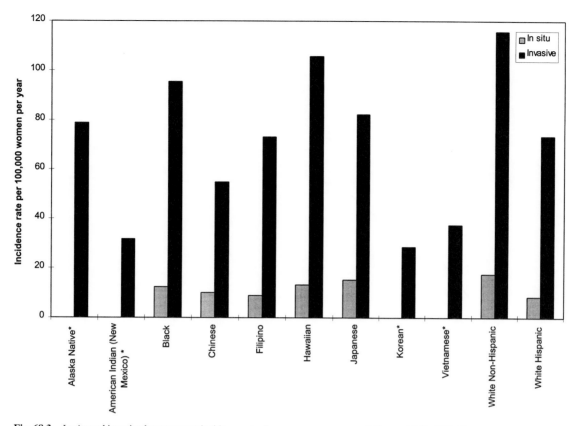

Fig. 68.3 *In situ* and invasive breast cancer incidence rates by race among women of all ages, 1988–1992 (data from the Surveillance, Epidemiology, and End Results (SEER) registries) [9]. In situ rates not calculated when fewer than 25 cases exist.

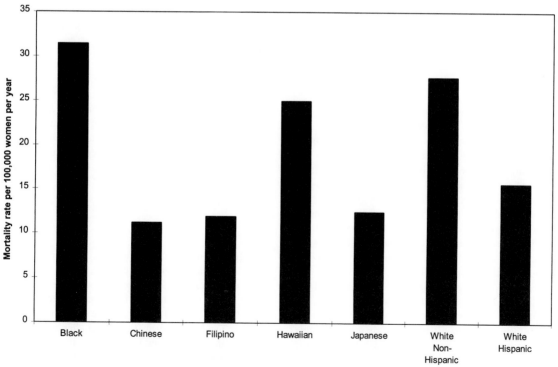

Fig. 68.4 Rates of breast cancer mortality by race among women of all ages in the U.S., 1988–1992 [9]. Fewer than 25 deaths were recorded in the interval for Native Americans, and a rate was not recorded. Mortality data were not available for Koreans and Vietnamese.

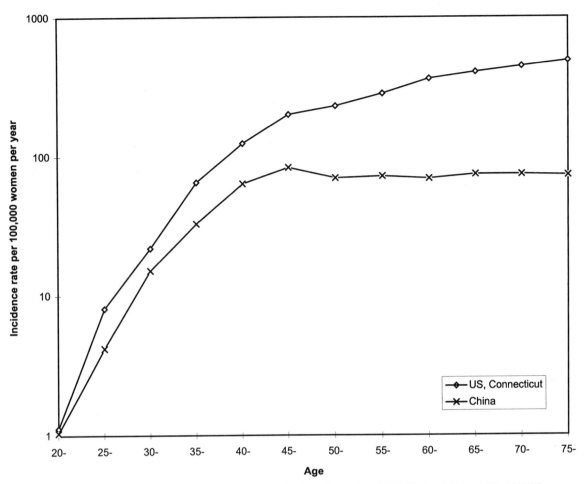

Fig. 68.5 Invasive breast cancer incidence rates by age; data from Connecticut, 1988–1992, and China, 1988–1992 [1].

patients and controls, show that differences in estrogen levels can account for differences in risk [13].

Seroepidemiologic studies of premenopausal women are complicated by the need to study all women at the same time during the menstrual cycle; when comparing luteal phase hormone levels, the appropriate day to collect biologic specimens can only be determined retrospectively by counting backwards from the first day of the next menstrual cycle (*i.e.,* the first day of menstrual bleeding). A further complication of these studies is the practical necessity of using only one or a few spot samples from one woman to represent the overall hormonal milieu to which she is exposed. Because of these factors, study results have been more consistent for postmenopausal women than for premenopausal women. Comparisons of postmenopausal breast cancer patients and controls demonstrate that patients have higher levels of circulating estrogens than controls [13].

The available evidence regarding endogenous progesterone levels is confusing and inconsistent because studies of this hormone must be restricted to premenopausal women. Most studies that have evaluated progesterone levels have not reported whether women were measured during ovulatory cycles. During anovulatory menstrual cycles, progesterone levels remain low throughout the cycle. Furthermore, it is impractical in

epidemiologic studies to make certain that the progesterone measurement occurs on the day of the progesterone peak; this would require daily measurements during the luteal phase of the cycle. Androgens may also play a role in breast cancer development. Testosterone has been associated with breast cancer risk both in case-control [13] and in cohort studies [14,15]. The role is less clear for androstenedione, DHEA, and DHEAS [13].

IV. Risk Factors

A. Age

Age is an important predictor of breast cancer risk, as it is for other epithelial cancers (Fig. 68.5). Among women in high-risk countries, incidence rates increase with age, but the upward slope of the age-specific incidence curve declines after about 45–50 years of age. In the past, little or no increase in the age-specific incidence rates was observed after age 45–50 years in those countries with low breast cancer incidence. This is still the case for some low-risk countries such as China (Fig. 68.5). However, as rates have increased in many other low-risk countries, the age-specific incidence curves have changed, reflecting a gradual increase in incidence after menopause similar to that seen in the U.S. This change in the slope of breast cancer incidence rates around age 45–50 years suggests the involvement of reproductive hormones in the etiology of breast cancer.

B. Age at Menarche

Modest elevations in breast cancer risk are associated with early age at menarche. In general, breast cancer risk declines 10–20% with each year that menarche is delayed. In a study of young women, Henderson and colleagues [16] showed that breast cancer risk was associated not only with the age at onset of menstruation but also with the age when "regular" or predictable menstruation was first established. Breast cancer risk was more than two times greater among women whose menstrual cycles became regular within one year of their first menstrual period than among women with a 5-year or longer delay in the onset of regular cycles. Furthermore, among women with early menarche (age 12 years or younger) who had regular cycles within their first menstrual year, breast cancer risk was more than four times greater than among women with late menarche (age 13 years or older) and a long duration of irregular cycles.

These observations suggest that regular ovulatory menstrual cycles increase a woman's risk of breast cancer [17]. This concept is supported by studies documenting the frequency of ovulation in relation to age at menarche and years since first menstrual period (gynecologic age). Among girls 15–19 years of age, selected from populations with differing levels of breast cancer risk, those with later menarche were more likely to have anovulatory menstrual cycles than those with early menarche, given the same gynecologic age [18]. After adjusting for gynecologic age, the frequency of ovulatory menstrual cycles across the different populations varied according to their risk of breast cancer. Apter and Vihko [19], in a longitudinal study of Finnish schoolgirls, also found that those with early menarche established ovulatory cycles more quickly than girls with later onset of menstruation.

C. Age at Menopause

Breast cancer incidence rates increase more slowly or plateau after menopause (Fig. 68.5). Analytic epidemiologic studies have consistently demonstrated that late age at menopause is associated with greater breast cancer risk. Among women whose menstrual periods stop naturally, breast cancer risk is about two times greater for those who experienced menopause at age 55 years or later than for those who underwent menopause at age 45 or younger [20]. Artificial menopause by bilateral oophorectomy also markedly reduces breast cancer risk and the effect is somewhat greater than that for natural menopause [20,21]. This differential effect can be explained by the fact that ovarian function does not stop at the time of menopause among women with intact ovaries, but declines over a period of a few months or years. Hysterectomy without oophorectomy does not affect breast cancer risk [21].

D. Parity and Age at First Full-Term Pregnancy

Two of the earliest known and most reproducible features of breast cancer epidemiology are the decreased risk associated

with increased parity and the increased risk of single women. MacMahon and colleagues [22] advanced the understanding of the role of pregnancy in altering breast cancer risk, showing that the risk of breast cancer among single women and nulliparous married women was similar (approximately 1.4 times the risk of parous married women). Furthermore, they showed that the protective effect of parity was due to a protective effect of a young age at first birth; women who had a first birth at age 20 years or younger had about half the breast cancer risk of women who completed their first full-term pregnancy at age 30 years or older. In their study, after adjusting for age at first full-term pregnancy, subsequent births after the first had no influence on the risk of developing breast cancer. Studies in other populations have observed a small additional protective effect of an increasing number of births, suggesting that, under certain circumstances, multiparity does offer some further protection [23].

Married women who have a late first full-term pregnancy (generally taken to be after the age of 35 years) are actually at an elevated risk of breast cancer, compared with nulliparous women [22]. This paradoxical effect of a late first full-term pregnancy has been confirmed repeatedly.

The immediate effect of a full-term pregnancy appears to be one of temporarily increasing risk; Bruzzi and colleagues [24] showed that breast cancer risk was greater among women whose last birth was within the past 3 years than among those of equivalent age whose most recent birth was at least 10 years earlier. Therefore, it appears that pregnancy has two contradictory effects on breast cancer risk. During the first trimester of pregnancy there is a rapid rise in the level of "free" estradiol (that which is available physiologically), an effect that is more apparent in the first than in subsequent pregnancies [25]. The net effect of this exposure in the early part of pregnancy is to expose the breast to the equivalent of several ovulatory menstrual cycles over a short period of time. Consistent with this theory is the observation that women who experienced nausea and vomiting of pregnancy serious enough to require treatment have somewhat greater breast cancer risk than those not experiencing this condition [26]. This effect was most extreme among women with full-term pregnancies within the past 5 years. A possible explanation for this finding is that women with serious nausea and vomiting of pregnancy (that is, those diagnosed with hyperemesis gravidarum) have higher levels of estradiol during the first trimester of pregnancy than those without nausea or vomiting [27].

In the long run, the negative effect of the first part of a pregnancy on breast cancer risk can be overridden by two beneficial consequences of a completed pregnancy. Prolactin levels are substantially lower in parous than in nulliparous women [28,29]. Secondly, parous women have lower levels of circulating estradiol and higher levels of sex-hormone binding globulin, which reduce the level of bioavailable or free estradiol [28].

A number of studies have examined the risk of breast cancer associated with spontaneous and induced abortions. Wingo and colleagues [30] reviewed the published literature on this topic, describing the inconsistent findings across studies and the difficulties in evaluating these associations. They concluded that breast cancer risk does not appear to be associated with an increased number of either spontaneous or induced abortions.

E. Lactation

Although the evidence is not entirely consistent, the number of studies showing that lactation protects against breast cancer development has increased substantially. Breastfeeding is a potentially modifiable behavior, and thus, its impact on breast cancer risk is extremely important. Evidence is fairly convincing that lactation reduces the risk of breast cancer among premenopausal women [26,31]. Evidence is less certain with regard to the risk of postmenopausal breast cancer [31,32]. In two publications, Enger and colleagues showed that the risk of premenopausal breast cancer (among women 40 years of age or younger) was nearly 35% lower among those who breastfed for more than 15 months relative to those not breastfeeding their children [26] and that postmenopausal women who breastfed more than 15 months had nearly a 30% lower risk than women who never breastfed their children [32]. The reason some studies have not observed protective effects of lactation may be due to the small proportion of women with sufficient lactation experience. Variations in the time when supplementary feeding is introduced and in the frequency and duration of each breastfeeding episode may also contribute to the inconsistent findings. An explanation for the protection afforded by lactation is that the cumulative number of ovulatory menstrual cycles a woman experiences will be lower among women with substantial lactation experience because breastfeeding results in a substantial delay in reestablishing ovulation following a completed pregnancy.

F. Weight Gain and Obesity

Heavy body weight increases the risk for breast cancer in postmenopausal women [33]. Most studies report relative risks of 1.5 to 2.0 for the heaviest group compared with the lightest group. The increased risk in heavy postmenopausal women can be attributable to the higher levels of circulating estrogen in these women, since the main source of endogenous estrogen after menopause is the conversion of the androgen precursor, androstenedione, to estrone in adipose tissue. In a study of 95,256 nurses, weight gain during adulthood was related to breast cancer risk in postmenopausal women who had never used hormonal replacement therapy; women who had gained 20 kg or more since age 18 years were at almost twice the risk of breast cancer compared to women who had not gained weight [34]. In contrast, heavy body weight may be associated with a slightly decreased risk in premenopausal women, with relative risk estimates in the range of 0.7 to 0.9 [35]. This possible protective effect of obesity in the premenopausal years may result from obese women having fewer ovulatory cycles than thin women [36]. In several studies, greater height has been associated with a modest elevation in risk, particularly in postmenopausal women. One hypothesis is that high caloric intake during childhood and adolescence is associated both with being tall and with increased breast cancer risk [33].

I. Physical Activity

Strenuous physical activity may delay menarche. Girls who engage in ballet dancing, swimming, or running have a considerable delay in the onset of menses [37]. Moderate physical activity during adolescence also can lead to anovulatory men-

strual cycles. In a study of high school girls, those who engaged in regular, moderate physical activity (averaging at least 600 Kcal of energy expended per week) were 2.9 times more likely to experience anovulatory cycles than were girls who engaged in lesser amounts of physical activity [38].

Early epidemiologic studies of the relationship between physical activity and breast cancer risk used surrogate measures of activity, such as occupation or participation in college sports [39]. Despite this indirect approach, many of these studies have shown a lower breast cancer risk among more active women. A study conducted among women 40 years of age or younger demonstrated that adolescent and adult participation in physical exercise activities significantly reduces breast cancer risk [40]. The risk for women who averaged 4 or more hours of exercise activity per week during their reproductive years was nearly 60% lower than that of inactive women. These results were later confirmed in a cohort study of Norwegian women [41]. Whether the reduction in breast cancer risk associated with physical activity extends to postmenopausal women is still uncertain [41]. Because exercise offers a feasible opportunity for breast cancer risk reduction, more research is needed to evaluate the relationship and to establish the mechanisms by which the possible protection might operate [42].

J. Benign Breast Disease

Certain types of benign breast disease are associated with breast cancer risk. Compared to women with no proliferative disease (or epithelial hyperplasia), women with proliferative disease without atypia have up to twice the risk of breast cancer, while women with atypical hyperplasia are at two to five times greater risk [43]. Atypial lobular hyperplasia is associated with four- to fivefold increased risk of breast cancer, while the estimates for atypical ductal disease vary from two to four [44–46]. Fibroadenoma without proliferative disease is not a strong risk factor for breast cancer, whereas cystic disease (with cysts ≥ 3mm) may be associated with a two- to fourfold increased risk of breast cancer [43].

K. Exogenous Hormones

Several studies have shown that the long-term use of oral contraceptives increases the risk for breast cancer diagnosed before age 35 and possibly up to age 45, ages at which breast cancer is relatively uncommon. Three studies conducted in Los Angeles County are typical of the range of effects observed for the relationship between oral contraceptives and breast cancer risk. In a study of women aged 37 or younger that was completed in 1983, breast cancer risk was strongly related to a woman's history of oral contraceptive use [47]. In a study of women diagnosed with breast cancer more recently (1983–1987), risk was not related to oral contraceptive use, possibly reflecting changes in pill formulation over time [48]. In this second study, only women under age 35 who had used oral contraceptives at a young age (before age 18) appeared to be at slightly greater risk of breast cancer than women who had not used oral contraceptives. The third study showed a marked increase in the risk of premenopausal bilateral breast cancer with increasing duration of oral contraceptive use [49]. The Collaborative Group on Hormonal Factors in Breast Cancer published a reanalysis of data collected from 54 breast cancer studies, conducted in 25 countries, that specifically collected detailed information on oral contraceptive use [50,51]. In this summary, a history of recent oral contraceptive use, rather than duration of use, was more predictive of breast cancer risk. The effect of recent oral contraceptive use was strongest among those women who first used oral contraceptives before age 20. The greatest increase in breast cancer risk was observed among women who were youngest at the time of their diagnoses. In this pooled analysis, the breast cancers diagnosed among oral contraceptive users were less advanced clinically than those diagnosed among women who had never used oral contraceptives.

The risk for breast cancer diagnosed after age 45 does not appear to be affected by oral contraceptive use, but this situation needs to be closely monitored as substantial numbers of women with a history of long-term use have recently entered this age group. Progestin-only oral contraceptives and the long-acting injectable progestogen contraceptive, depot-medroxyprogesterone acetate (DMPA), do not appear to affect breast cancer risk [52].

Most studies that have included sufficiently large numbers of women who have used estrogen replacement therapy for extended periods of time (*e.g.*, for more than 10 years) indicate a modest increase in breast cancer risk among exposed women, with risk increasing approximately 3% per year of use [53]. Among studies conducted in the United States, where the use of conjugated equine estrogens is the norm, breast cancer risk increases about 2.2% per year of use of a standard dose regimen (0.625 mg/day). This translates into increases in risk of 10% after 5 years, 20% after 10 years, and 40% after 15 years of use.

Data are just becoming available on whether a combined replacement therapy regimen (with both estrogen and progestin) affects breast cancer risk. In combination hormone therapy, progestins may enhance the proliferative effects of estrogen, but to date no firm conclusions can be reached regarding the effects of such regimens on risk.

Other sources of exogenous hormones have also been evaluated with regard to breast cancer risk. The use of diethylstilbestrol (DES) by pregnant women in the late 1940s to the 1960s to reduce the risk of fetal loss has been found in most studies to be associated with a modest, approximately 50%, increase in the risk of breast cancer [54].

Studies of chemically induced mammary tumor induction in a rat model show that administration of human chorionic gonadotrophin (hCG) reduces tumor incidence in a manner comparable to that of a completed pregnancy [55]. In the Los Angeles case-control study of women aged 40 years or younger, those who had received hCG injections as part of a weight loss regimen or as a component of infertility treatment had lower risk of breast cancer [56].

L. Mammographic Densities

The radiographic appearance of the breast is determined by the relative amounts of fat, connective tissue, and epithelial tissue. On mammograms, fat appears as radiologically lucent areas, whereas connective and epithelial tissue appear as areas of high radiologic density. These radiodense areas are referred to as mammographic densities.

Wolfe described four "parenchymal" patterns (N1, P1, P2, DY) of increasing densities [57,58]. In his two original studies,

Wolfe reported that the risk of incident breast cancer was much greater in women with the DY pattern than in women with the N1 pattern [58]. However, this classification scheme was found to have substantial interobserver variation [59], and subsequently Wolfe and other investigators classified the breast in terms of the percentage of the breast showing densities [60–62].

The results relating mammographic densities to breast cancer risk have been reviewed in detail by Saftlas and Szklo [63] and Oza and Boyd [64], and have been summarized in a meta-analysis by Warner and colleagues [65]. These reviews have concluded that the risk of breast cancer associated with mammographic densities remains after adjusting for other risk factors, with increased densities associated with greater risk [62–64]. The magnitude of the risk is greater than that associated with most other known breast cancer risk factors. In fact, two studies suggest that women with 75% or more densities have more than a fourfold greater risk of breast cancer than women who have no mammographic densities [66,67]. The results are similar for premenopausal and postmenopausal women of all ages [67].

M. Diet

Interest in the relationship between diet and breast cancer risk has been based largely on the observed international variation in incidence rates and on animal experiments showing that a high fat diet increases the risk of chemically induced mammary tumors. Although results from epidemiologic studies have been conflicting, most cohort studies do not indicate that a high-fat diet in adulthood increases the risk of breast cancer [68]. It is possible, however, that a high-fat diet in childhood and adolescence could affect breast cancer risk. It has also been suggested that low caloric intake during childhood and adolescence may reduce risk [33], but to date evidence for this hypothesis is only circumstantial. Some studies suggest that vitamin A and other antioxidant vitamins slightly decrease risk; in fact, studies are currently being conducted to determine whether treatment with fenretinide, a synthetic derivative of retinoic acid, reduces the risk of breast cancer recurrence and mortality [69].

Finally, it has been suggested that consumption of foods with high phytoestrogen content, such as soybean products, reduces breast cancer risk. Phytoestrogens are weakly estrogenic and could compete with more potent human endogenous estrogens at estrogen binding sites, thereby reducing the possible carcinogenic effects of these stronger estrogens. In a case-control study of women newly diagnosed with early breast cancer (who were studied prior to initiation of treatment) and population controls, breast cancer risk was substantially reduced among those with higher urinary levels of two phytoestrogen compounds, the isoflavone equol and the lignan enterolactone [70]. It is proposed that phytoestrogens may affect the uptake and metabolism of sex hormones by contributing to the regulation of sex hormone binding globulin [71]; they also appear to have antioxidant properties [72].

N. Alcohol Consumption

Most epidemiologic studies of the relationship of alcohol consumption to breast cancer risk have suggested a positive association. A meta-analysis of 38 such studies concluded that a modest, linear, statistically significant dose-response relationship exists [73]. In relation to nondrinkers, the breast cancer risks of women consuming one, two, and three alcoholic drinks daily were increased 11, 24, and 38%, respectively. A meta-analysis of six cohort studies involving more than 322,000 women confirmed these results [74]. Both meta-analyses concluded that no particular type of alcoholic beverage was more predictive of increased risk than any other. No completely satisfactory biologic explanation of the alcohol-breast cancer relationship has been identified, although several possible mechanisms have been proposed [75]. The most plausible of these is a possible mediating effect of reproductive steroid hormones. Although a causal relationship is possible, confounding cannot be ruled out as an explanation for the alcohol-breast cancer association.

O. In utero *Factors*

Intrauterine exposure to high concentrations of pregnancy estrogens during gestation may be important in the etiology of breast cancer. Pre-eclampsia and eclampsia as well as small fetal size are negatively related to estrogen levels [76] and are associated with lower subsequent risk of breast cancer [76,77].

P. *Radiation*

Among women younger than 40 years of age, exposure to ionizing radiation in moderate to high doses is known to increase breast cancer risk, with risk increasing as a function of dose [78]. A variety of sources of radiation have been implicated, including the atomic bomb explosions in Japan in 1945, and repeated exposure of the chest in the course of medical treatment for such conditions as tuberculosis, postpartum mastitis, benign breast conditions, Hodgkin's disease, ankylosing spondylitis, scoliosis, tinea capitis, enlarged thymus, skin hemangioma, and childhood cancer. Little is known about the effect of very low radiation doses on risk, such as might occur during certain diagnostic procedures or occupational exposures.

Q. *Other Environmental Exposures*

It has been speculated that certain chemicals in the environment mimic the action of estrogen or alter the activity of estrogen by increasing the production of 16-alpha-hydroxyestrone. Although 16-alpha-hydroxyestrone has been reported to enhance breast cell proliferation and perhaps damage DNA [79], it is controversial whether this estrogen metabolite plays a role in breast cancer [80–82]. There has been concern about the potential estrogenic effect of organochlorine pesticides. One study reported a positive association between levels of these pesticides in the blood and breast cancer risk [83], but a second study by some of the same investigators [84] and a large cohort study [85] have not supported this finding.

V. Clinical Issues

A. *Prognostic Factors*

Breast tumor size and lymph node involvement are the two major prognostic factors for the risk of breast cancer recurrence and death. Many other possible prognostic factors have been suggested. These include age, nuclear and histologic grade, estrogen and progesterone receptor status (ER and PR) [86,87], tumor necrosis factor, micrometastasis, proliferative indices of the tumor, mutations in the P53 tumor suppressor gene, and HER2-neu expression [88,89]. Although these factors have proven useful in identifying subgroups of women who are likely to have better or poorer survival, their positive predictive value is still relatively low.

VI. Treatment for Breast Cancer

A woman's prognosis following breast cancer diagnosis is better when the cancer is detected at an early stage. The treatment options available to patients depend on tumor characteristics as well as patient variables. The alternatives available to patients are described briefly below. This discussion emphasizes the treatment of stage I and II cancer and includes a short additional comment on treatment of *in situ* cancers as well as more advanced cancers.

A. In situ *Breast Cancer*

Mastectomy has traditionally been the treatment of choice for ductal carcinoma *in situ*. However, local excision and radiation therapy may be a good alternative [90]. Although recurrence rates are higher with this latter approach to treatment, any invasive cancer occurring at a later time can be treated with mastectomy, and the resulting survival rates are comparable to primary treatment of the *in situ* carcinoma with mastectomy [91]. Currently about half of all ductal carcinomas *in situ* are treated with lumpectomy [92].

B. *Stage I and II Breast Cancer*

1. Surgery

Breast-conserving surgery combined with radiotherapy represents a standard mode of treatment for most women with stage I and stage II breast cancer and is associated with survival rates that are equivalent to those achieved for mastectomy patients [93,94]. Clinical practice varies substantially across the U.S., with breast-conserving therapy being much more common at teaching hospitals than at nonteaching hospitals, and a more common treatment for white women than for African-American women [95].

Axillary lymph node dissection has long been recommended as a staging procedure to identify patients who have histologic involvement of the lymph nodes and could benefit from systemic adjuvant therapy (see section 3). However, axillary lymph node removal is associated with considerable morbidity, including arm swelling (lymphadema) and discomfort that can be long-lasting. The extent to which such dissection is appropriate is controversial. Some investigators recommend biopsy of the first node in the chain that receives the lymphatic drainage (the sentinel node), reserving axillary dissection for patients with a positive sentinel node [96]. Because study results suggest that both node-positive and node-negative patients may benefit from adjuvant chemotherapy, axillary dissections may provide little information to influence therapy [97].

Whether the timing of surgery during the menstrual cycle is important remains controversial. Several studies suggest that women who have their surgery during the luteal phase of the menstrual cycle have longer disease-free survival than those who have surgery in the follicular phase of the menstrual cycle. Other studies have failed to confirm this finding, or have found the opposite result [98].

2. Radiotherapy

Radiation therapy is conventionally administered to women who have received breast-conserving surgery. Standard radiation therapy consists of external beam radiation to the entire breast, with or without a boost (given with an interstitial radioactive implant or by external (electron) beam radiation). Several trials have demonstrated that breast conserving surgery without radiotherapy results in higher breast cancer recurrence rates [94,99]. Studies suggest that delaying radiotherapy until after completion of chemotherapy may be safe and, arguably, preferable (see section 3).

3. Adjuvant Systemic Therapy—Chemotherapy

In 1976, Bonadonna and colleagues reported clinical trial results that showed that treating node-positive women with a combination of cyclophosphamide, methothrexate, and fluorouracil (CMF) lowered the rate of breast cancer recurrence [100]. Many trials have been performed since that time, and adjuvant chemotherapy regimens incorporating doxorubicin are frequently used. Recent studies suggest a role for the addition of taxanes. The side effects of the various chemotherapeutic agents can be substantial. Common side effects include nausea, vomiting, alopecia (hair loss), and weight gain. Less common side effects include suppression of the bone marrow, heart failure (anthracycline), thromboembolic events, and premature menopause [101,102].

Adjuvant chemotherapy was first used successfully in breast cancer patients with affected lymph nodes. However, evidence suggests that both node-negative and node-positive patients may benefit. The Early Breast Cancer Trialists' Collaborative Group (EBCTCG) overview published in 1992 suggested that adjuvant chemotherapy reduces recurrence rates and increases survival rates in both premenopausal and postmenopausal women with stage I (node-negative) and stage II breast cancer [103]. The results from the National Surgical Adjuvant Breast Project (NSABP) B-20 trial [97] suggest that all women with node-negative ER-positive tumors benefit from chemotherapy. However, given the side effects of chemotherapy, the extent to which all node-negative patients should be treated with chemotherapy in order to prevent relapse in the few women who would otherwise have had one is controversial. Ongoing clinical trials should soon identify molecular markers that can be used to predict which patients with early stage cancer are most likely to have it recur and hence to benefit the most from chemotherapy.

The duration of most chemotherapeutic regimens is 6 months or less. The optimal therapy for patients at risk of distant metastasis may be to undergo chemotherapy first, delaying radiotherapy until chemotherapy has been completed (this increases the chances of successfully completing the chemotherapy with the required doses) [104]. Current trials are exploring chemotherapy prior to surgery (so called neoadjuvant chemotherapy). Histologic markers following neoadjuvant chemotherapy obtained at time of surgery may then indicate whether further chemotherapy will be of benefit to the patient [105].

4. Adjuvant Systemic Therapy—Tamoxifen

Tamoxifen decreases the risk of relapse and death in node-positive breast cancer patients aged 50 and over and in receptor-positive node-negative patients [106]. While the optimal duration of tamoxifen therapy for node-positive cancers remains controversial, data from the NSABP B-14 trial suggest that 5 years of treatment is at least as good as 10 years of treatment. Possible side effects of tamoxifen therapy include hot flashes, menstrual irregularities, skin rash, vaginal discharge or irritation, development of benign ovarian cysts, and cataracts. More serious side effects include the possibility of thromboembolic events and the development of endometrial cancer [107]. Combined tamoxifen and chemotherapy may result in increased toxicity, particularly an increased risk of thromboembolic events [97,102].

C. Advanced Breast Cancer

The primary therapeutic options available for the management of advanced breast cancer include hormonal manipulations and systemic chemotherapy. Radiotherapy and occasionally surgery are appropriate for the treatment of painful or compressive sites of disease, such as spinal cord compression, and for the treatment of brain metastases. In addition, interventions such as pleuroedesis for pleural effusions, intrathecal chemotherapy for meningeal carcinomatosis, and internal fixation for metastases to weight-bearing bones may be important in management.

Hormonal manipulations are an important therapeutic option for patients with advanced breast cancer. Generally accepted predictors for response to hormonal therapies include hormone receptor status, disease-free interval, pace of metastasis, and response to prior hormonal manipulations. The antiestrogen tamoxifen is the most frequently used agent. However, alternate antiestrogens including toremifene have been developed and demonstrate similar antitumor activity in postmenopausal patients [108]. Agonists of gonadotrophin releasing hormone (GnRH), when given continuously, suppress the release of gonadotrophins among premenopausal women and reduce ovarian sex-steroid production to postmenopausal levels. Goserelin, a potent GnRH agonist, is an effective alternative to surgical oophorectomy in the management of advanced premenopausal hormone-sensitive breast cancer [109]. The progestin megestrol acetate is a useful hormonal agent that is frequently employed as a second line hormonal therapy. New safer aromatase inhibitors, including anastrozole [110] and letrozole [111], have replaced aminoglutethimide as second or third line hormonal therapies for advanced breast cancer in postmenopausal women. The sequential use of these and less frequently used agents (estrogens and androgens) may result in protracted control of advanced breast cancer in some patients.

Whether bone marrow transplantation in combination with high dose chemotherapy in advanced breast cancer is beneficial is currently not clear. Published data from small uncontrolled trials have proved the feasibility [112] but not the effectiveness of such treatment [113].

VII. Prospects for Prevention

Knowledge of the epidemiology of breast cancer continues to evolve but more remains to be learned. For women, breast cancer risk appears to be determined in large part by the cumulative exposure of breast epithelium to estrogen and progesterone. Most of this exposure is accumulated during the years of active ovarian function. Possible approaches to primary prevention require a detailed understanding of the factors that influence the onset, regularity, and quality of ovarian activity. Chemoprevention of breast cancer via hormonal manipulation has received much attention [114]. Possible approaches include the use of a luteinizing hormone-releasing hormone agonist combined with low-dose add-back hormones to reduce exposure to ovarian hormones. Tamoxifen, a hormonal therapy widely used in the treatment of breast cancer, was found to reduce the risk of breast cancer in a clinical trial of women at high risk of breast cancer. However, the drug also increased the risk of endometrial cancer, pulmonary embolism, stroke, and deep vein thrombosis [115]. Two European trials [116,117], although smaller than the U.S. Study [115], found no difference between tamoxifen treated and control women. Unfortunately, on the basis of the currently established risk factors, there is little practical advice that can be given to most women about how to reduce their risk, although lactation, limitation of alcohol intake, participation in physical activity, and prevention of postmenopausal obesity offer possible approaches. Ongoing and future epidemiologic research, such as studies of dietary intake of phytoestrogens and those of interactions between breast cancer susceptibility genes and certain exposures, may offer further prospects for breast cancer prevention. With various preventive approaches and further data on risk factors, it is hoped that sufficient knowledge will be gained to substantially reduce breast cancer incidence.

References

1. Parkin, D. M., Whelan, S. L., Ferlay, J., Raymond, L., and Young, J., eds. (1997). "Cancer Incidence in Five Continents," Vol. 7, IARC Sci. Publ. No. 143. IARC, Lyon, France.
2. Boring, C. C., Squires, T. S., and Tong, T. (1993). Cancer Statistics 1993. Ca—Cancer J. Clin. **43,** 7–26.
3. American Cancer Society, Surveillance Research (1998). Available at: *http://www.cancer.org/bcn/info/brstats.html*
4. Waterhouse, J., Muir, C., Shanmugaratnam, K., and Powell, J., eds. (1982). "Cancer Incidence in Five Continents," Vol. 4, IARC Sci. Publ. No. 42. IAEA, Lyon, France.
5. Parkin, D. M., Muir, C. S., Whelan, S. L., Gao, Y. T., Ferlay, J., and Powell, J. (1992). "Cancer Incidence in Five Continents," Vol. 6, IARC Sci. Publ. No. 120. IARC, Lyon, France.
6. Ursin, G., Bernstein, L., and Pike, M. C. (1994). Breast cancer. *Cancer Surv.* **19–20,** 241–264.
7. World Health Organization (1972–1993). "World Health Statistics Annual 1971–1992." WHO, Geneva.
8. Wingo, P. A., Ries, L. A., Rosenberg, H. M., Miller, D. S., and Edwards, B. S. (1998). Cancer incidence and mortality 1973–1995. *Cancer (Philadelphia)* (in press).
9. Miller, B. A., Kolonel, L. N., Bernstein, L., *et al.,* eds. (1996). "Racial/Ethnic Patterns of Cancer in the United States 1988–

1992," NIH Publ. No. 96-4104. National Cancer Institute, Bethesda, MD.
10. Henderson, B. E., Ross, R. K., Pike, M. C., and Casagrande, J. T. (1982). Endogenous hormones as a major factor in human cancer. *Cancer Res.* **42,** 3232–3239.
11. Preston-Martin, S., Pike, M. C., Ross, R. K., Jones, P. A., and Henderson, B. E. (1990). Increased cell division as a cause of human cancer. *Cancer Res.* **50,** 7415–7421.
12. Pike, M. C., Spicer, D. V., Dahmoush, L., and Press, M. F. (1993). Estrogens, progestogens, normal breast cell proliferation and breast cancer risk. *Epidemiol. Rev.* **15,** 17–35.
13. Bernstein, L., and Ross, R. K. (1993). Endogenous hormones and breast cancer risk. *Epidemiol. Rev.* **15,** 48–65.
14. Dorgan, J. F., Longcope, C., Stephenson, H. E., Jr., Falk, R. T., Miller, R., Franz, C., Kahle, L., Campbell, W. S., Tangrea, J. A., and Schatzkin, A. (1996). Relation of prediagnostic serum estrogen and androgen levels to breast cancer risk. *Cancer Epidemiol., Biomarkers Prev.* **5,** 533–539.
15. Zeleniuch-Jacquotte, A., Bruning, P. F., Bonfrer, J. M. G., Koenig, K. L., Shore, R. E., Kim, M. Y., Pasternack, B. S., and Toniolo, P. (1997). Relation of serum levels of testosterone and dehydroepiandrosterone sulfate to risk of breast cancer in postmenopausal women. *Am. J. Epidemiol.* **145,** 1030–1038.
16. Henderson, B. E., Ross, R. K., and Bernstein, L. (1988). Estrogens as a cause of human cancer: The Richard and Hinda Rosenthal Foundation Award Lecture. *Cancer Res.* **48,** 246–253.
17. Henderson, B. E., Ross, R. K., Judd, H. L., Krailo, M. D., and Pike, M. C. (1985). Do regular ovulatory cycles increase breast cancer risk? *Cancer (Philadelphia)* **56,** 1206–1208.
18. MacMahon, B., Trichopoulos, D., Brown, J., Andersen, M. P., Aoki, K., Cole, P., deWaard, F., Kauraniemi, T., *et al.* (1982). Age at menarche, probability of ovulation and breast cancer risk. *Int. J. Cancer* **29,** 13–16.
19. Apter, D., and Vihko, R. (1983). Early menarche, a risk factor for breast cancer, indicates early onset of ovulatory cycles. *J. Clin. Endocrinol. Metab.* **57,** 82–86.
20. Trichopoulos, D., MacMahon, B., and Cole, P. (1972). The menopause and breast cancer risk. *J. Natl. Cancer Inst. (U.S.)* **48,** 605–613.
21. Brinton, L. A., Schairer, C., Hoover, R. N. *et al.* (1988). Menstrual factors and risk of breast cancer. *Cancer Invest.* **6,** 245–254.
22. MacMahon, B., Cole, P., Lin, M. T., Lowe, C. R., Mirra, A. P., Ravnihar, B., Salber, E. J., Valaoras, V. G., and Yuasa, S. (1970). Age at first birth and cancer of the breast. A summary of an international study. *Bull. W.H.O.* **43,** 209–221.
23. Yuan, J. M., Yu, M. C., Ross, R. K., Gao, Y. T., and Henderson, B. E. (1988). Risk factors for breast cancer in Chinese women in Shanghai. *Cancer Res.* **48,** 949–1953.
24. Bruzzi, P., Negri, E., La Vecchia, C., Decarli, A., Palli, D., Parazzini, F., and Del Turco, M. R. (1988). Short term increase in risk of breast cancer after full term pregnancy. *Br. Med. J.* **297,** 1096–1098.
25. Bernstein, L., Depue, R. H., Ross, R. K., Judd, H. L., Pike, M. C., and Henderson, B. E. (1986). Higher maternal levels of free estradiol in first compared to second pregnancy: A study of early gestational differences. *JNCI, J. Natl. Cancer Inst.* **75,** 1035–1039.
26. Enger, S. M., Ross, R. K., Henderson, B., and Bernstein, L. (1997). Breast feeding history, pregnancy experience and risk of breast cancer. *Br. J. Cancer* **76,** 118–123.
27. Depue, R. H., Bernstein, L., Ross, R. K., Judd, H. L., Pike, M. C., and Henderson, B. E. (1987). Hyperemesis gravidarum in relation to estradiol levels, pregnancy outcome, and other maternal factors: a sero-epidemiologic study. *Am. J. Obstet. Gynecol.* **156,** 1137–1141.
28. Bernstein, L., Pike, M. C., Ross, R. K., Judd, H. L., Brown, J. B., and Henderson, B. E. (1985). Estrogen and sex hormone-binding

globulin levels in nulliparous and parous women. *JNCI, J. Natl. Cancer Inst.* **74,** 741–745.

29. Musey, V. C., Collins, D. C., Musey, P. I., Martino-Saltzman, D., and Preedy, J. R. (1987). Long-term effect of a first pregnancy on the secretion of prolactin. *N. Engl. J. Med.* **316,** 229–234.

30. Wingo, P. A., Newsome, K., Marks, J. S., Calle, E. E., and Parker, S. L. (1997). The risk of breast cancer following spontaneous and induced abortion. *Cancer Causes Control* **8,** 93–108.

31. Newcomb, P. A., Storer, B. E., Longnecker, M. P., Mittendorf, R., Greenberg, E. R., Clapp, R. W., Burke, K. P., Willett, W. C., and MacMahon, B. (1994). Lactation and a reduced risk of premenopausal breast cancer. *N. Engl. J. Med.* **330,** 81–86.

32. Enger, S. M., Ross, R. K., Paganini-Hill, A., and Bernstein, L. (1998). Breastfeeding experience and breast cancer risk among postmenopausal women. *Cancer Epidemiol., Biomarkers Prev.* **7,** 365–369.

33. Hunter, D. J., and Willett, W. C. (1993). Diet, body size, and breast cancer. *Epidemiol. Rev.* **15,** 110–132.

34. Huang, Z., Hankinson, S. E., Colditz, G. A., Stampfer, M. J., Hunter, D. J., Manson, J. E., Hennekens, C. H., Rosner, B., Speizer, F. E., and Willett, W. C. (1997). Dual effects of weight and weight gain on breast cancer risk. *JAMA, J. Am. Med. Assoc.* **278,** 1407–1411.

35. Ursin, G., Longnecker, M. P., Haile, R. W., and Greenland, S. (1995). A meta-analysis of body mass index and risk of premenopausal breast cancer. *Epidemiology* **6,** 137–141.

36. Pike, M. C. (1990). Reducing cancer risk in women through lifestyle-mediated changes in hormone levels. *Cancer Detect. Prev.* **14,** 595–607.

37. Frisch, R. E., Gotz-Welbergen, A. V., McArthur, J. W., Albright, T., Witschi, J., Bullen, B., Birnholz, J., Reed, R. B., and Hermann, H. (1981). Delayed menarche and amenorrhea of college athletes in relation to age at onset of training. *JAMA, J. Am. Med. Assoc.* **246,** 1559–1563.

38. Bernstein, L., Ross, R. K., Lobo, R., Hanisch, R., Krailo, M. D., and Henderson, B. E. (1987). Effects of moderate physical activity on menstrual cycle patterns in adolescence: Implications for breast cancer prevention. *Br. J. Cancer* **55,** 681–685.

39. Friedenreich, C. M., and Rohan, T. E. (1995). A review of physical activity and breast cancer. *Epidemiology* **6,** 311–317.

40. Bernstein, L., Henderson, B. E., Hanisch, R., Sullivan-Halley, J., and Ross, R. K. (1994). Physical exercise and reduced risk of breast cancer in young women. *J. Natl. Cancer Inst.* **86,** 1403–1408.

41. Thune, I., Brenn, T., Lund, E., and Gaard, M. (1997). Physical activity and the risk of breast cancer. *N. Engl. J. Med.* **336,** 1269–1275.

42. Gammon, M. D., Britton, J. A., and Teitelbaum, S. L. (1996). Does physical activity reduce the risk of breast cancer. A review of the epidemiologic literature. *Menopause* **3,** 172–180.

43. Bodian, C. A. (1993). Benign breast diseases, carcinoma in situ, and breast cancer risk. *Epidemiol. Rev.* **15,** 177–187.

44. Page, D. L., Vander Zwaag, R., Rogers, L. W., Williams L. T., Walker, W. E., and Hartmann, W. H. (1978). Relation between component parts of fibrocystic disease complex and breast cancer. *JNCI, J. Natl. Cancer Inst.* **61,** 1055–1063.

45. Page, D. L., Dupont, W. D., Rogers, L. W., and Rados, M. S. (1985). Atypical hyperplastic lesions of the female breast. *Cancer (Philadelphia)* **5,** 2698–2708.

46. Marshall, L. M., Hunter, D. J., Connolly, J. L., Schnitt, S. J., Byrne, C., and London, S. J. (1997). Risk of breast cancer associated hyperplasia of lobular and ductal types. *Cancer Epidemiol., Biomarkers Prev.* **6,** 297–301.

47. Bernstein, L., Pike, M. C., Krailo, M., and Henderson, B. E. (1990). Update of the Los Angeles Study of oral contraceptives and breast cancer: 1981 and 1983. *In* "Oral Contraceptives and Breast Cancer" (R. D. Mann, ed.), pp. 169–180. Parthenon Publishing, Park Ridge, NJ.

48. Ursin, G., Ross, R. K., Sullivan-Halley, J., Hanisch, R., Henderson, B., and Bernstein, L. (1998). Use of oral contraceptives and risk of breast cancer in young women. *Breast Cancer Res. Treat.* **50,** 175–184.

49. Ursin, G., Aragaki, C. A., Paganini-Hill, A., Siemiatycki, J., Thompson, W. D., and Haile, R. W. (1992). Oral contraceptives and premenopausal bilateral breast cancer, a case-control study. *Epidemiology* **3,** 414–419.

50. Collaborative Group on Hormonal Factors in Breast Cancer (1996). Breast cancer and hormonal contraceptives. *Lancet* **347,** 1713–1727.

51. Collaborative Group on Hormonal Factors in Breast Cancer (1996). Breast cancer and hormonal contraceptives: Further results. *Contraception* **54S,** 1–106.

52. Stanford, J. L., and Thomas, D. B. (1993). Exogenous progestins and breast cancer. *Epidemiol. Rev.* **15,** 98–107.

53. Pike, M. C., Bernstein, L., and Spicer, D. V. (1993). The relationship of exogenous hormones to breast cancer risk. *In* "Current Therapy in Oncology" (J. E. Niederhuber, ed.), pp. 292–303. Mosby, St. Louis, MO.

54. Malone, K. E. (1993). Diethylstilbestrol (DES) and breast cancer. *Epidemiol. Rev.* **15,** 108–109.

55. Russo, J., and Russo, I. H. (1994). Toward a physiological approach to breast cancer prevention. *Cancer Epidemiol., Biomarkers Prev.* **3,** 353–364.

56. Bernstein, L., Hanisch, R., Sullivan-Halley, J., and Ross, R. K. (1995). Treatment with human chorionic gonadotropin and risk of breast cancer. *Cancer Epidemiol., Biomarkers Prev.* **4,** 437–440.

57. Wolfe, J. N. (1976). Breast patterns as an index of risk for developing breast cancer. *AJR, Am. J. Roentgenol.* **126,** 1130–1137.

58. Wolfe, J. N. (1976). Breast parenchymal patterns and their changes with age. *Radiology* **121,** 545–552.

59. Boyd, N. F., Wolfson, C., Moskowitz, M., Carlile, T., Petitclerc, C., Ferri, H. A., Fishell, E., Gregire, A., Kiernan, M., Longley, J. D., Simor, I. S., and Miller, A. B. (1986). Observer variation in the classification of mammographic parenchymal pattern. *J. Chronic Dis.* **39,** 465–72.

60. Brisson, J., Sadowsky, N. L., Twaddle, J. A., Morrison, A. S., Cole, P., and Merletti, F. (1982). The relation of mammographic features of the breast to breast cancer risk factors. *Am. J. Epidemiol.* **115,** 438–443.

61. Wolfe, J. N., Saftlas, A. F., and Salane, M. (1987). Mammographic parenchymal patterns and quantitative evaluation of mammographic densities: A case-control study. *Am. J. Radiol.* **148,** 1087–1092.

62. Saftlas, A. F., Hoover, R. N., Brinton, L. A., Szklo, M., Olson, D. R., Salane, M., and Wolfe, J. N. (1991). Mammographic densities and risk of breast cancer. *Cancer (Philadelphia)* **67,** 2833–2838.

63. Saftlas, A. F., and Szklo, M. (1987). Mammographic parenchymal patterns and breast cancer risk. *Epidemiol. Rev.* **9,** 146–174.

64. Oza, A. M., and Boyd, N. F. (1993). Mammographic parenchymal patterns: A marker of breast cancer risk. *Epidemiol. Rev.* **15,** 196–208.

65. Warner, E., Lockwood, G., Math, M., Tritchler, D., and Boyd, N. F. (1992). The risk of breast cancer associated with mammographic parenchymal patterns: A meta-analysis of the published literature to examine the effect of method of classification. *Cancer Detect. Prev.* **16,** 67–72.

66. Boyd, N. F., Byng, J., Jong, R., Fishell, E., Little, L., Miller, A. B., Lockwood, G., Tritchler, D., and Yaffe, M. (1995). Quantitative classification of mammographic densities and breast cancer risks: Results from the Canadian National Breast Screening Study. *J. Natl. Cancer Inst.* **87,** 670–675.

67. Byrne, C., Schiarer, C., Wolfe, J., Parekh, N., Salane, M., Brinton, L. A., Hoover, R., and Haile, R. (1995). Mammographic features and breast cancer risk: Effects with time, age and menopause status. *J. Natl. Cancer Inst.* **87,** 1622–1629.

68. Hunter, D. J., Spiegelman, D., Adami, H. O., Beeson, L., van den Brandt, P. A., Folsom, A. R., Fraser, G. E., Goldbohm, R. A., Graham, S., Howe, G. R., *et al.* (1995). Cohort studies of fat intake and the risk of breast cancer: A pooled analysis. *N. Engl. J. Med.* **334,** 356–361.

69. Decensi, A., Formelli, F., Torrisi, R., and Costa, A. (1993). Breast cancer chemoprevention: Studies with 4-HPR alone and in combination with tamoxifen using circulating growth factors as potential surrogate endpoints. *J. Cell. Biochem.* **17G** (Suppl.), 226–233.

70. Ingram, D., Lopez, D., Kolybaba, M., and Sanders, K. (1997). Case-control study of phytooestrogens and breast cancer. *Lancet* **350,** 990–994.

71. Adlercreutz, H., Mousavi, Y., Clark, J., Hockerstedt, K., Hamalainen, E., Wahala, K., Makela, T., and Hase, T. (1992). Dietary phytoestrogens and cancer: In vitro and in vivo studies. *J. Steroid Biochem. Mol. Biol.* **41,** 331–337.

72. Messina, M., and Messina, V. (1991). Increasing use of soyfoods and their potential role in cancer prevention. *J. Am. Diet. Assoc.* **91,** 836–840.

73. Longnecker, M. P. (1994). Alcoholic beverage consumption in relation to risk of breast cancer: meta-analysis and review. *Cancer Causes Control* **5,** 73–82.

74. Smith-Warner, S. A., Spiegelman, D., Yaun, S. S., van den Brandt, P. A., Folsom, A. R., Goldbohm, R. A., Graham, S., Holmberg, L., Howe, G. R., Marshall, J. R., Miller, A. B., Potter, J. D., Speizer, F. E., Willett, W. C. Wolk, A., and Hunter, D. J. (1998). Alcohol and breast cancer in women: A pooled analysis of cohort studies. *JAMA, J. Am. Med. Assoc.* **279,** 535–540.

75. Schatzkin, A., and Longnecker, M. P. (1994). Alcohol and breast cancer: Where are we now and where do we go from here? *Cancer (Philadelphia)* **74,** 1101–1110.

76. Ekbom, A., Trichopoulos, D., Adami, H. O., Hsieh, C. C., and Lan, S. J. (1992). Evidence of prenatal influences on breast cancer risk. *Lancet* **340,** 1015–1018.

77. Michels, K. B., Trichopoulos, D., Robins, J. M., Rosner, B. A., Manson, J. E., Hunter, D. J., Colditz, G. A., Hankinson, S. E., Speizer, F. E., and Willett, W. C. (1996). Birthweight as a risk factor for breast cancer. *Lancet* **348,** 1542–1546.

78. John, E. M., and Kelsey, J. L. (1993). Radiation and other environmental exposures and breast cancer. *Epidemiol. Rev.* **15,** 157–162.

79. Davis, D. L., and Bradlow, H. L. (1995). Can environmental estrogens cause breast cancer? *Sci. Am.* **273,** 166–172.

80. Schneider, J., Kinne, D., Fracchia, A., Pierce, V., Bradlow, H. L., and Fishman, J. (1982). Abnormal oxidative metabolism of estradiol in women with breast cancer. *Proc. Natl. Acad. Sci. U.S.A.* **79,** 3047–3051.

81. Adlercreutz, H., Fotsis, T., Höckerstedt, K., Hämäläinen, E., Bannwart, C., Bloigu, S., Valtonen, A., and Ollus, A. (1989). Diet and urinary estrogen profile in premenopausal omnivorous and vegetarian women and in premenopausal women with breast cancer. *J. Steroid Biochem.* **34,** 527–530.

82. Ursin, G., London, S., Stanczyk, F. Z., Gentzschein, E., Paganini-Hill, A., Ross, R. K., and Pike, M. C. (1997). A pilot study of urinary estrogen metabolites (16α-OHE1 and 2-OHE1) in postmenopausal women with and without breast cancer. *Environ. Health Perspect.* **105**(Suppl.), 601–605.

83. Wolff, M. S., Toniolo, P. G., Lee, E. W., Rivera, M., and Dubin, N. (1993). Blood levels of organochlorine residues and risk of breast cancer. *J. Natl. Cancer Inst.* **85,** 648–652.

84. Krieger, N., Wolff, M. S., Hiatt, R. A., Rivera, M., Vogelman, J., and Orentreich, N. (1994). Breast cancer and serum organochlo-

rines: A prospective study among white, black, and Asian women. *J. Natl. Cancer Inst.* **86,** 589–599.

85. Hunter, D. J., Hankinson, S. E., Laden, F., Colditz, G. A., Manson, J. E., Willett, W. C., Speizer, F. E., and Wolff, M. S. (1997). Plasma organochlorine levels and the risk of breast cancer. *N. Engl. J. Med.* **337,** 1253–1258.

86. Fisher, B., Fisher, E. R., Redmond, C., and Brown, A. (1986). Tumor nuclear grade, estrogen receptor, and progesterone receptor: Their value alone or in combination as indicators of outcome following adjuvant therapy for breast cancer. *Breast Cancer Res. Treat.* **7,** 147–160.

87. Fisher, E. R., Redmond, C., Fisher, B., Bass, G., and Contributing NSABP Investigators (1990). Pathologic findings from the National Surgical Adjuvant Breast and Bowel Projects (NSABP): Prognostic discriminants for 8-year survival for node-negative invasive breast cancer patients. *Cancer (Philadelphia)* **65**(Suppl.), 2121–2128.

88. Wenger, C. R., Beardslee, S., Owens, M. A., *et al.* (1993). DNA ploidy, S-phase, and steroid receptors in more than 127,000 breast cancer patients. *Breast Cancer Res. Treat.* **28,** 9–20.

89. Allred, D. C., Clark, G. M., Tandon, A. K., *et al.* (1992). HER-2/neu in node-negative breast cancer: Prognostic significance of overexpression influenced by the presence of in situ carcinoma. *J. Clin. Oncol.* **10,** 599–605.

90. Fisher, E. R., Costantino, J., Fisher, B., *et al.* (1995). Pathologic findings from the National Surgical Adjuvant Breast Project (NSABP) protocol B-17: Intraductal carcinoma (ductal carcinoma in situ). *Cancer (Philadelphia)* **75,** 1310–1319.

91. Solin, L. J., Fourquet, A., McCormick, B., *et al.* (1994). Salvage treatment for local recurrence following breast-conserving surgery and definitive irradiation for ductal carcinoma in situ (intraductal carcinoma) of the breast. *Int. J. Radiat. Oncol., Biol., Phy.* **30,** 3–9.

92. Ernster, V. L., Barclay, J., Kerlikowske, K., Grady, D., and Henderson, I. C. (1996). Incidence of and treatment for ductal carcinoma in situ of the breast. *JAMA, J. Am. Med. Assoc.* **275,** 913–918.

93. Fisher, B., Anderson, S., Redmond, C. K., *et al.* (1995). Reanalysis and results after 12 years of follow-up in a randomized clinical trial comparing total mastectomy with lumpectomy with or without irradiation in the treatment of breast cancer. *N. Engl. J. Med.* **333,** 1456–1461.

94. Veronesi, U., Salvadori, B., Luini, A., *et al.* (1995). Breast conservation is a safe method in patients with small cancer of the breast: Long-term results of three randomised trials on 1,973 patients. *Eur. J. Cancer* **31A,** 1574–1579.

95. Ayanian, J. Z., and Guadagnoli, E. (1996). Variations in breast cancer treatment by patient and provider characteristics. *Breast Cancer Res. Treat.* **40,** 65–74.

96. Veronesi, U., Paganelli, G., Galimberti, V., *et al.* (1997). Sentinel-node biopsy to avoid axillary dissection in breast cancer with clinically negative lymph-nodes. *Lancet* **349,** 1864–1867.

97. Fisher, B., Dignam, J., Wolmark, N., DeCillis, A., Emir, B., Wickerham, D. L., Bryant, J., Dimitrov, N. V., Abramson, N., Atkins, J. N., Shibata, H., Deschenes, L., and Margolese, R. G. (1997). Tamoxifen and chemotherapy for lymph node-negative, estrogen receptor-positive breast cancer. *J. Natl. Cancer Inst.* **89,** 1673–1682.

98. McGuire, W. L., Hilsenbeck, S., and Clark, G. M. (1992). Optimal mastectomy timing. *J. Natl. Cancer Inst.* **84,** 346–348.

99. Clark, R. M., McCulloch, P. B., Levine, M. N., *et al.* (1992). Randomized clinical trial to assess the effectiveness of breast irradiation following lumpectomy and axillary dissection for node-negative breast cancer. *J. Natl. Cancer Inst.* **84,** 683–689.

100. Bonadonna, G., Brusamolino, E., Valagussa, P., Rossi, A., Brugnatelli, L., Brambilla, C., DeLena, M., Tancini, G., Bajetta, E., Musumeci, R., and Veronesi, U. (1976). Combination chemother-

apy as an adjuvant treatment in operable breast cancer. *N. Engl. J. Med.* **294,** 405–410.

101. Fisher, B., Brown, A. M., Dimitrov, N. V., *et al.* (1990). Two months of doxorubicin-cyclophosphamide with and without interval reinduction therapy compared with 6 months of cyclophosphamide, methotrexate, and fluorouracil in positive-node breast cancer patients with tamoxifen-nonresponsive tumors: Results from the National Surgical Adjuvant Breast and Bowel Project B-15. *J. Clin. Oncol.* **8,** 1483–1496.

102. Pritchard, K. I., Paterson, A. H., Paul, N. A., *et al.* (1996). Increased thromboembolic complications with concurrent tamoxifen and chemotherapy in a randomized trial of adjuvant therapy for women with breast cancer. *J. Clin. Oncol.* **14,** 2731–2737.

103. Early Breast Cancer Trialists' Collaborative Group (1992). Systemic treatment of early breast cancer by hormonal, cytotoxic, or immune therapy: 133 randomised trials involving 31,000 recurrences and 24,000 deaths among 75,000 women. Part II. *Lancet* **339,** 71–85.

104. Recht, A., Come, S. E., Henderson, I. C., *et al.* (1996). The sequencing of chemotherapy and radiation therapy after conservative surgery for early-stage breast cancer. *N. Engl. J. Med.* **334,** 1356–1361.

105. Makris, A., Powles, T. J., Dowsett, M., Osborne, C. K., Trott, P. A., Fernando, I. N., Ashley, S. E., Ormerod, M. G., Titley, J. C., Gregory, R. K., and Allred, D. C. (1997). Prediction of response to neoadjuvant chemoendocrine therapy in primary breast carcinomas. *Clin. Cancer Res.* **3,** 593–600.

106. Early Breast Cancer Trialists' Collaborative Group (1992). Systemic treatment of early breast cancer by hormonal, cytotoxic, or immune therapy: 133 randomised trials involving 31,000 recurrences and 24,000 deaths among 75,000 women. Part I. *Lancet* **339,** 1–15.

107. van Leeuwen, F. E., Benraadt, J., Coebergh, J. W., *et al.* (1994). Risk of endometrial cancer after tamoxifen treatment of breast cancer. *Lancet* **343,** 448–452.

108. Hayes, D. F., Van Zyl, J. A., Hacking, A., *et al.* (1995). Randomized comparison of tamoxifen and two separate doses of toremifene in postmenopausal patients with metastatic breast cancer. *J. Clin. Oncol.* **13,** 2556–2566.

109. Blamey, R. W., Jonat, W., Kaufmann, M., Bianco, A. R., and Namer, M. (1992). Goserelin depot in the treatment of premenopausal advanced breast cancer. *Eur. J. Cancer* **28A,** 810–814.

110. Jonat, W., Howell, A., Blomqvist, C., Eiermann, W., Winblad, G., *et al.* (1996). A randomised trial comparing two doses of the new selective aromatase inhibitor anastrozole (Arimidex) with megestrol acetate in postmenopausal patients with advanced breast cancer. *Eur. J. Cancer* **32A,** 404–412.

111. Dombernowsky, P., Smith, I., Falkson, G., *et al.* (1998). Letrozole, a new oral aromatase inhibitor for advanced breast cancer: Double-blind randomized trial showing a dose effect and improved efficacy and tolerability compared with megestrol acetate. *J. Clin. Oncol.* **16,** 453–461.

112. Ayash, L. J., Elias, Ibrahim, J., Schwartz, G., Wheeler, C., Reich, E., Lynch, C., Warren, D., Shapiro, C., Richardson, P., Hurd D., Schnipper, L., Frei, E., III, and Antman, K. (1998). High-dose multimodality therapy with autologous stem-cell support for stage IIIb breast carcinoma. *J. Clin. Oncol.* **16,** 1000–1007.

113. Canellos, G. P. (1997). Selection bias in trials of transplantation for metastatic breast cancer: Have we picked the apple before it was ripe? *J. Clin. Oncol.* **15,** 3169–3170.

114. Henderson, B. E., Ross, R. K., and Pike, M. C. (1993). Hormonal chemoprevention of cancer in women. *Science* **259,** 633–638.

115. Fisher, B., Costantino, J. P., Wickerham, D. L., Redmond, C. K., *et al.* (1998). Tamoxifen for prevention of breast cancer: Report of the National Surgical Adjuvant Breast and Bowel Project P-1 Study. *J. Natl. Cancer Inst.* **90,** 1371–1388.

116. Powles, T., Eeles, R., Ashley, S., Easton, D., Chang, J., Dowsett, M., Tidy, A., Viggers, J., and Davey, J. (1998). Interim analysis of the incidence of breast cancer in the Royal Marsden Hospital tamoxifen randomised chemoprevention trail. *Lancet* **352,** 98–101.

117. Veronesi, U., Maisonneuve, P., Costa, A., *et al.* (1998). Prevention of breast cancer with tamoxifen: Preliminary findings from the Italian randomized trial among hysterectomised women. *Lancet* **352,** 93–97.

69

Inherited Genetic Susceptibility and Breast Cancer

BETH NEWMAN[1]

Department of Epidemiology and Lineberger Comprehensive Cancer Center
University of North Carolina
Chapel Hill, North Carolina

I. Introduction

Many people now assume that whether or not they get cancer has something to do with their genes. However, that is only part of the story. During the 1990s, the potential to identify genes that increase risk of breast cancer due to inherited susceptibility has been realized. Arising from a desire to explain the unusually high frequency of breast cancer that occurs in some families, it was also thought that once we found the genes for inherited breast cancer, we would know more about sporadic (noninherited) forms of the disease as well. A number of genes related to breast cancer now have been described. Not too surprisingly, the more we learn, the more aware we become of the complexity that characterizes breast cancer genetics. This chapter provides an overview of our current knowledge about inherited genetic susceptibility to breast cancer, with particular emphasis on BRCA1 and BRCA2. In addition, it addresses some of the clinical and public health implications, including prospects for prevention and treatment of breast cancer, and issues that we, as a society, must deal with as a result of this increasing focus on genetics. But first, a little background.

II. Background

A. *Brief Orientation to Cancer Genetics*

Each of us inherits two copies (called alleles) of every gene—one from our mother's egg and one from our father's sperm. With a few exceptions, all cells in our bodies contain these two alleles. Each allele is composed of a sequence of chemical units known as deoxyribonucleic acid (DNA). The DNA sequence serves as code for a protein, which plays some role in our bodies. However, the exact DNA sequence, which potentially influences the structure and/or function of the encoded protein, varies among individuals and can be altered. Some alleles, therefore, may contribute to disease because of their particular DNA sequence [1,2].

Most diseases, including breast cancer, require a number of factors before they occur. Breast cancer is thought to arise from a series of genetic alterations (mutations) that accumulate in a breast cell during a woman's (or man's) life. It may take up to five or six of these alterations. Some of these may be inherited at conception and therefore appear in virtually all cells of the body (germline mutations), but most occur specifically in the breast cell(s) later in life (somatic mutations). Depending on the particular genes and mutations involved, whether the muta-

tions are repaired or otherwise corrected, and whether only one or both copies of the gene are affected, the breast cell may escape its usual controls and become a cancer. When one or more mutations is passed down from parent to offspring over the generations, families may experience many cases of breast cancer among their members.

B. *Familial Aggregation of Breast Cancer*

The first published account of familial breast cancer appeared in the medical literature in 1866, when a physician named Broca described the breast cancer experiences of his wife's family [3]. A more systematic investigation of the causes of breast cancer mortality published in 1926 confirmed that a family history of the disease was one of the best predictors; women who died of breast cancer were more likely than others to have family members who had been diagnosed with breast cancer as well [4]. That is as far as science got in understanding familial aggregation of breast cancer for several decades, although numerous epidemiologic studies addressed the issue with increasing sophistication, substituting incidence (disease occurrence) data for mortality data, moving from clinical populations to the general population, and analyzing separately premenopausal and postmenopausal disease [5–8].

The overall impression remained that a family history of breast cancer was associated with increased risk of breast cancer occurrence [5–8]. Across all ages, women with breast cancer have been, on average, two to three times more likely than women without breast cancer to report a family history of the disease in a mother or a sister. When the family history of breast cancer involved multiple first-degree relatives, the observed risk of breast cancer was even more elevated, at 4-fold or greater [5,8], but when restricted to more distant relatives, such as aunts or grandmothers, the family history was associated with a lower, 1.5- to 2-fold, risk of breast cancer [9]. Attempts to explain these associations solely on the basis of familial aggregation of breast cancer risk factors or as a result of greater recall among women with breast cancer compared with women without breast cancer have failed [7]. Moreover, family history associations of this modest magnitude have been shown statistically to be consistent with strong genetic effects, at least in a subset of families [7]. Eventually the suspicion that familial aggregation of breast cancer was due to an underlying genetic transmission of susceptibility motivated scientists to begin approaching the disease from a more biological perspective.

During the 1970s, physicians and scientists described large, extended family trees containing multiple relatives with breast cancer [10,11], similar to Broca's 1866 account [3]. These fam-

[1]Present address: School of Public Health, Queensland University of Technology, Brisbane, Australia.

ily trees were quite exceptional and generally reflected cancers among family members across generations far in excess from that expected in the general population. The researchers recognized that the pattern of who got cancer and who did not was consistent with what might be expected of a genetically inherited condition. In some branches of these families, almost half of the women were diagnosed with breast cancer sometime during their adult years, suggesting that inheritance of a single susceptibility allele (autosomal dominant transmission) might be influencing disease occurrence in the family. In addition to breast cancer, other cancers, including ovarian, endometrial, prostate, and colon cancers, were seen in family members as well.

These descriptive reports were followed in the 1980s by formal statistical evaluations of the pattern of disease occurrence in families using a method called segregation analysis. The results across studies were remarkably consistent, mostly indicating that the inheritance of even one copy of a rare, variant sequence of DNA (the susceptibility allele), from either the mother or father, might confer greatly increased risk of developing breast cancer later in life. Although the first studies were conducted on select samples of more high-risk families [7,12], similar findings were obtained in subsequent analyses on samples of families from the general public [7,8]. These latter analyses provided more generalizable results and estimated that between 1 in 150 to 1 in 800 people in the United States may have inherited one of these breast cancer susceptibility alleles [13,14]. The results from segregation analyses, again suggesting autosomal dominant transmission, thus provided another important clue that susceptibility to breast cancer due to inheritance of a germline genetic mutation might be operating in at least some families. The rapid development of molecular biology as a field during the 1980s and 1990s then enhanced the feasibility of identifying specific disease susceptibility genes. In fact, breast cancer is the first relatively common form of cancer for which a susceptibility gene has been identified.

III. Inherited Susceptibility to Breast Cancer

Several genes now have been implicated in inherited susceptibility to breast cancer (Table 69.1). When particular alleles can be traced within families and occur in the women (and possibly men) with breast cancer, while being absent in most individuals without breast cancer, there is strong circumstantial evidence for an influence on disease development. Conceptually, this is the basis of genetic linkage analysis. This method is particularly good at identifying genes with alleles that confer greatly increased risk of breast cancer, so-called high-penetrance or high-risk alleles. Generally speaking, the mutations that distinguish these alleles significantly alter the protein's normal structure or function, but they are relatively uncommon. As a consequence, their impact on breast cancer in the general population, measured by attributable fraction, is likely to be low.

In contrast, other alleles of the same or different genes may increase risk of breast cancer to a lesser degree, but because they are more common, they may have a bigger impact on breast cancer in the general population. This kind of genetic influence is more difficult to identify, in part because it is weaker and in part because it may require the presence of additional genetic or nongenetic factors to affect the development of breast cancer.

Table 69.1
Relative Importance of Susceptibility Genes for Breast Cancer

Gene	Prevalence of susceptibility alleles[a]	Relative risk[b]	Proportion of breast cancer[c]
BRCA1	<1%	~4–10	3–7%
BRCA2	<1%	~4–10	2–7%
p53	Rare	~4–5	<1%
PTEN	Rare	?	Rare
AR[d]	~75%	~1	0%
ER	?	?	?
ATM	0.5–1.5%	~4	~1–11%
HRAS1	6%	~2	~9%
Metabolism gene[e]	10–70%	~1.5–2.5	~5–50%

[a]Proportion of the general population inheriting at least one copy of a susceptibility allele at the gene noted.

[b]Risk of breast cancer among women inheriting a susceptibility allele compared to women who did not or to women in the general population; estimates reflect overall relative risks rather than age-specific relative risks.

[c]Proportion of women with breast cancer from the general population who have inherited a susceptibility allele; some estimates derived directly from genotyping breast cancer patients unselected for family history and others derived from formulae using susceptibility allele prevalence and relative risk.

[d]Susceptibility allele defined as <22 CAG repeats in exon 3 of androgen receptor (AR) gene. Results presented for women; may differ for men. A relative risk of 1 means women who inherited an AR susceptibility allele are at no different risk of breast cancer than women lacking an AR susceptibility allele.

[e]This category includes a variety of genes encoding enzymes involved in metabolism of carcinogens or hormones (e.g., NAT2, COMT, CYP17, GSTM1, etc.); range for proportion of breast cancer attributed to susceptibility allele(s) at genes like these is estimated first from the low values and then from the high values of prevalence and relative risk.

Usually research on such lower penetrance alleles requires some prior knowledge either about the function of the protein encoded by the gene or about the gene's involvement in disease susceptibility. Knowledge of the interacting factors, including other genes or particular environmental or lifestyle exposures, may be necessary as well.

Table 69.1 lists the genes currently being studied for inherited susceptibility to breast cancer and provides some estimates of prevalence as well as relative risks and attributable fractions. Prevalence refers to the frequency of the susceptibility allele(s). Relative risk refers to the risk of breast cancer among women inheriting the susceptibility allele(s) (i.e., penetrance) compared to women in the general population. Attributable fraction here refers to the proportion of women with breast cancer from the general population who have inherited the susceptibility allele(s), which depends both on their penetrance and on their prevalence. Many of these estimates are based on strong assumptions and/or limited research and hence are most useful as preliminary indicators of the relative importance of these genes. As more is learned about these genes and their relationships to breast cancer development, the estimates of prevalence, relative risk, and attributable fraction are likely to change.

Table 69.2
Characteristics of BRCA1 and BRCA2

	BRCA1	BRCA2
Size		
Gene	~84,000 base pairs 22 coding exons	~100,000 base pairs 27 coding exons
Protein	1863 amino acids	3418 amino acids
Mutation characteristics		
Types of mutations	Substitutions, deletions, insertions	Deletions most frequently, also insertions, substitutions
Number reported	>500	>300
Mutations reported only once	~57%	~64%
Disease-related mutations	Mostly protein truncating mutations	Mostly protein truncating mutations
Founder effects	Yes	Yes
Associated cancers		
Characteristic cancers	Breast and ovary	Breast (male and female) and ovary
Other possible cancers	Prostate, colon	Pancreas, prostate, ocular melanoma

A. BRCA1 and BRCA2

1. Description

The two most well-known susceptibility genes for breast cancer are BRCA1 on chromosome 17q21 [15] and BRCA2 on chromosome 13q12–13 [16]. The DNA sequences of these two genes were characterized within months of each other [17,18] and share a number of similarities (see Table 69.2). Structurally, both are large genes with numerous coding regions (exons) interspersed by noncoding regions (introns) and a large, central exon (exon 11 in both BRCA1 and BRCA2). Originally, neither bore much similarity to other, known genes, and there were few, readily recognizable functional domains. Now it appears that both may contain DNA sequences that resemble genes involved in regulation of DNA transcription and possibly in control of the cell cycle, and both may play important roles in growth and development and in the repair of DNA damage [19,20]. Susceptibility alleles appear to increase risk of both breast and ovarian cancer; however, the risk of ovarian cancer appears higher for BRCA1, and the risk of breast cancer among men is much higher for BRCA2 [21]. In addition, BRCA1 alleles may elevate risks of prostate and colon cancers [22–24], and BRCA2 alleles may influence the risks of prostate and pancreatic cancers [25,26] as well as ocular melanoma [27].

2. Mutational Spectrum

As of mid-1998, over 500 different mutations of all types were reported for BRCA1 and over 300 for BRCA2 [28]. The mutational spectra, including nucleotide substitutions, insertions, and deletions, reveal no "hotspots" (regions that are more prone to alteration). Rather, alterations occur throughout the DNA sequence of both genes, with only a few mutations recur-

ring with appreciable frequency (see Section III.A.4. on Founder Effects). In fact, approximately 60% of all BRCA1 and BRCA2 mutations have been reported only once. This poses a challenge for identification of susceptibility alleles in genetic testing. For both genes, most alterations known to increase disease risk (referred to here as "susceptibility alleles" but often called simply "mutations") result in truncated protein, presumably with reduced or absent biological activity. In addition, other alterations have been observed, including some presumed to be neutral with respect to disease risk (called polymorphisms) and some of unknown functional significance. Many of these introduce subtle changes in the DNA sequence, are relatively common, and/or fail to cosegregate with disease in family members. Whether or not such alterations nevertheless increase susceptibility to breast cancer, possibly to a lesser degree than the protein truncation mutations, is under investigation [29–31].

3. Prevalence

How common are these BRCA1 and BRCA2 mutations that increase susceptibility to breast cancer? Our estimates of the prevalence of these alleles have been profoundly influenced by the specific study designs used [7]. It now appears that susceptibility alleles at either BRCA1 or BRCA2 account for virtually all so-called high-risk, breast–ovarian cancer families (i.e., families reporting at least one relative with ovarian cancer and a total of four or more cases of breast or ovarian cancer) [21,32,33]. In contrast, only about half (35–80% depending on the study) of families with more than four cases of breast cancer in the absence of ovarian cancer test positive for a BRCA1 or BRCA2 susceptibility allele [21,33,34]. Still lower prevalence estimates of 13–16% for BRCA1 susceptibility alleles are obtained from studies of women selected generally for family history or young age at diagnosis [7,35,36], whereas only 3–7% of women with breast cancer from the general population have tested positive for a susceptibility allele at BRCA1 [31,37,38]. Research on BRCA2 is more limited, but approximately 2–7% of female [7,39; B. Newman, unpublished findings] and 4–14% of male [40,41] breast cancer patients have tested positive for a BRCA2 susceptibility allele. Hence, around 90% of breast cancer occurs in the absence of a BRCA1 or BRCA2 mutation because these susceptibility alleles are uncommon, even among women with breast cancer. In the general population, roughly 1 in 300 to 1 in 800 women have inherited a BRCA1 mutation [42,43], and BRCA2 mutations are likely just as rare. However, variations in these prevalences do occur.

4. Founder Effects

Not all populations have the same frequency of susceptibility allele(s). The prevalence of inherited breast cancer, and the specific mutations involved, have their origins in the history and geography of human populations. Until recently, marriage (or more accurately mating) worldwide largely took place within extended communities, providing the conditions for local differences, known in genetics as founder effects. A founder effect results from a single, mutational event in the distant past being passed on through the generations within a particular group [44]. In Europe, these differences largely parallel geography [45]. For example, in Iceland a five base-pair deletion in exon 9 of BRCA2 (known as 999del5) occurs in 0.4% of the general population and in 8% of women and 40% of men diagnosed with breast cancer [25,46]. In the U.S., Canada, and Australia, differences largely reflect ancestry prior to immigration but are being influenced by population dynamics since relocation [47].

One of the most striking examples of genetic founder effects on breast cancer occurs among Ashkenazi Jewish women both in the U.S. and in Israel. Among people with this ancestry, the combined frequency of three founder alleles at BRCA1 and BRCA2 is approximately 2%, with approximately 10% of breast cancer overall attributable to these particular mutations [22]. A single BRCA1 mutation—the two base-pair deletion in exon 2 (denoted 185delAG)—is found in approximately 1% of individuals, but up to 20–30% of breast cancer patients, of Ashkenazi Jewish descent [48]. Genetic analysis of markers near the 185delAG variant in unrelated individuals claiming Ashkenazi Jewish descent suggests a common ancestor in whom the original mutation occurred hundreds of years ago, although there is also evidence suggesting that this same mutation has occurred independently more than once in the past. Thus, the prevalence of susceptibility allele(s), and therefore the likelihood of inheriting such an allele, depends in part on ancestry, defined on the basis of geography, ethnicity, religion, skin color, or other attributes that influenced marriage and mating in the distant and recent past.

5. Penetrance

Despite their low prevalence overall, susceptibility alleles at BRCA1 and BRCA2 are important contributors to breast cancer. Those women who inherit at least one susceptibility allele from either gene are at increased risk of developing the disease. Studies based on the high-risk families with four or more affected relatives estimate that cumulative risk of breast cancer by age 70 among women carrying a susceptibility allele may be as high as 87% for BRCA1 [49] and 84% for BRCA2 [21]. Comparable figures for ovarian cancer by age 70 are 44% for BRCA1 and 27% for BRCA2 [21,49]. However, estimates from three studies of Ashkenazi Jewish families, not selected for a family history of breast or ovarian cancer, suggest that cumulative risk by age 70 may be as low as 40–60% for breast cancer and less than 20% for ovarian cancer among those carrying one of the BRCA1 founder alleles [48,50]. It is possible that these alleles have lower penetrance (i.e., risk) than other BRCA1 or BRCA2 susceptibility alleles. Alternatively, these lower estimates of penetrance may have been observed because the study did not rely on high-risk families who themselves may have been selected inadvertently for other genetic or environmental factors that influenced development of disease among susceptibility allele carriers. The possibility that penetrance of BRCA1 or BRCA2 susceptibility alleles is lower than previously estimated also is consistent with observations from studies on breast cancer patients from referral clinics and the general population that indicate that many women inheriting a BRCA1 or BRCA2 susceptibility allele report little or no family history of disease [31,35,37,38].

Thus, contrary to some earlier conceptions of genetic conditions, inheritance of a BRCA1 or BRCA2 susceptibility allele does not mean a woman necessarily will be diagnosed with breast cancer. The relative risk of breast cancer, comparing women who inherited a BRCA1 or BRCA2 mutation to those

who did not, is higher at younger ages than at older ages (possibly 20-fold or higher instead of the 4- to 10-fold relative risk overall), in part because breast cancer is relatively more common among older women in general. Moreover, among younger women, the relative risk associated with BRCA2 mutations may be lower than the relative risk for BRCA1 mutations [21,39]. However, the diagnosis of breast cancer may occur anytime during adulthood [21,49]. Hence other factors may influence whether or not a woman inheriting a BRCA1 or BRCA2 susceptibility allele ultimately develops breast cancer. These same factors, or different ones, may influence the age at which she is diagnosed. If some of these factors can be modified, then the risk of breast cancer among women inheriting a susceptibility allele may be reduced.

6. Modifiers

The genetic or environmental/behavioral factors that influence disease development among women inheriting a BRCA1 or BRCA2 susceptibility allele are not yet known. Two reports have been published describing genes that may influence risk of disease among BRCA1 or BRCA2 mutation carriers. In the first, women who inherited rare alleles at the HRAS1 locus (described further in Section III.B.5.) in addition to a BRCA1 susceptibility allele were more likely to develop ovarian cancer than women who inherited only the BRCA1 mutation [51]. However, no such association was observed for breast cancer. A study of women of Ashkenazi Jewish descent has suggested that a particular mutation in the adenomatous polyposis coli (APC) gene on chromosome 5q21–22 increases risk of breast cancer particularly among those who inherited one of the founder mutations in BRCA1 or BRCA2 [52]. Neither of these observations has yet been replicated by other researchers; however, scientists have begun to propose potential mechanisms to explain how the genes, or the proteins they encode, interact to increase risk of breast or ovarian cancer.

Research on nongenetic modifiers of BRCA1 and BRCA2 is not much further along and is characterized by conflicting findings. Two studies have reported an increased risk of breast cancer among BRCA1 mutation carriers born since 1930 or 1940 [53,54], suggesting that environmental exposures or lifestyle behaviors may modify disease development by influencing age at diagnosis or overall risk. Reproductive factors, including earlier age at first menstrual period, later age at first full-term pregnancy, and fewer births, may explain some, but not all, of this trend. Two additional exposures that have become more common during recent decades also have been implicated as risk factors for breast cancer among women with susceptibility alleles. Use of oral contraceptives was associated with increased risk of breast cancer among women with a BRCA1 or BRCA2 susceptibility allele in one study [55]. In contrast, women with a BRCA1 or BRCA2 mutation who used oral contraceptives were less likely to develop ovarian cancer than those who did not [56]. In another study, women with a susceptibility allele at BRCA1 or BRCA2 who smoked cigarettes were less likely to be diagnosed with breast cancer [57]. The possibility of a chemical in cigarette smoke that might delay or prevent breast cancer among genetically susceptible women is provocative. However, these observations suggesting that BRCA1- or BRCA2-related cancer risk may be modified by use of oral contraceptives or cigarette smoking are preliminary; further evaluation is warranted, and studies have been initiated. Knowledge about factors that influence breast cancer risk among genetically susceptible women, especially if they are modifiable, may provide opportunities for prevention. Despite the obvious importance of learning more about such factors, this research has progressed slowly, in large part because of the expense and difficulty in testing for susceptibility alleles and the low frequency of those alleles.

B. Other Susceptibility Genes

1. BRCA3

Susceptibility alleles at either BRCA1 or BRCA2 account for a substantial proportion of familial breast cancer. Nevertheless, most researchers studying families with multiple cases of breast cancer have at least a few that do not appear to have inherited a susceptibility allele in BRCA1 or BRCA2 [21,33,34]. Several genome-wide searches are currently underway on these high-risk families, in an attempt to localize additional breast cancer susceptibility gene(s). However, it is also possible that no other genes account for a significant proportion of the remaining familial breast cancer. Instead, the clustering of breast cancer in these families may represent common environmental exposures or behaviors or the chance co-occurrence of a relatively common disease. Alternatively, a variety of other genes may each contribute in a minor way, several of which are described in the remainder of this section.

2. p53 and PTEN

A number of rare genetic disorders have been associated with increased risk of breast cancer, with two of particular interest. The Li-Fraumeni syndrome is characterized by a diverse array of childhood and adult tumors, including most prominently breast cancer, bone and soft tissue sarcomas, brain tumors, leukemia, and adrenocortical carcinoma. Inherited mutations in the p53 tumor suppressor gene on chromosome 17p13 have been found in about 50% of families with the Li-Fraumeni syndrome [58,59]. In addition, tissue-specific mutations in p53 that occur after conception and probably later in life (i.e., somatic mutations) have been reported in a wide variety of cancers, including 15–50% of breast cancers, depending on the stage of the tumor [60]. Somatic p53 mutations are considered a relatively common occurrence during the carcinogenic process. However, inherited p53 mutations account for little breast cancer overall; fewer than 1% of breast cancer patients lacking a family history consistent with the Li-Fraumeni syndrome carry an inherited p53 mutation [61].

A similar situation exists for Cowden disease, a rare, autosomal dominant, genetic disorder characterized by lesions of the skin and mucous membranes and benign tumors in a variety of tissues, including the skin, thyroid, colon, endometrium, brain, and breast. In addition, women with Cowden disease are at elevated risk of breast cancer. One gene for Cowden disease has been localized to chromosome 10q22–23 and is called PTEN, among other names [62]. Like p53, PTEN is a tumor suppressor gene, and noninherited, somatic mutations in the PTEN gene have been seen in cancers of the breast, brain, and prostate [63].

Inherited mutations in this gene have been identified in many, but not in all, families with Cowden disease [64]; however, once again, they are extremely rare in women with breast cancer who do not have Cowden disease [64,65].

3. ER and AR

Other candidate genes for inherited susceptibility to breast cancer code for hormone receptors. One is the estrogen receptor (ER) gene on chromosome 6p24–27, or a nearby gene, which may be linked to breast cancer in some families [66; M.-C. King, personal communication]. However, no particular inherited mutations in the ER gene have been associated consistently with breast cancer risk [67,68]. Another candidate is the androgen-receptor gene (AR) on chromosome Xq11. Inherited mutations in exon 3 have been implicated in the breast cancers of three men with androgen insufficiency, including two brothers [69,70]. Exon 1 of the AR gene is characterized by a region of DNA composed of a variable number of CAG sequence repeats. This region influences the DNA transcription function of the gene, with longer repeat lengths correlated with decreased androgen activity [71]. Men with fewer repeats have increased risk of prostate cancer [72], but studies evaluating the same variations in relation to breast cancer risk in women have observed no such association [B. Newman, unpublished findings; A. Spurdle, personal communication].

4. ATM

Ataxia telangiectasia (AT) is an autosomal recessive (i.e., requiring two susceptibility alleles) disorder characterized by severe neurologic symptoms, unusual sensitivity to ionizing radiation, and a 100-fold elevated risk of cancer, particularly leukemia and lymphoma. Individuals who inherit only one AT susceptibility allele lack clinical signs of the disease but may be at increased risk of cancer. Excess breast cancers have been observed among mothers of AT patients, who are obligate carriers of a susceptibility allele [73], and risk may be influenced by exposure to ionizing radiation from diagnostic, therapeutic, or occupational sources [74]. It has been estimated that approximately 1% of people in the general population have inherited one AT susceptibility allele [74,75]. Assuming modest estimates of prevalence (0.5–1.0%) and relative risks ranging from 2–7, between 1 and 11% of breast cancers may be attributable to an AT susceptibility allele [75].

The ATM gene is a major susceptibility gene for AT, which is mutated in some but not in all individuals with the disease. It is located on chromosome 11q22–23, and direct genotyping is now possible though difficult [76]. A study of AT patients and their families found that female relatives with an ATM mutation had a fourfold elevated risk of breast cancer [77], similar to results from studies prior to availability of genotype information [73]. However, a second study [78] of breast cancer patients diagnosed prior to age 40 reported a prevalence of ATM mutations of 0.5%, lower than the prevalence observed among controls (1.0%). This apparent conflict in results has been contested on a number of methodologic and theoretical grounds [76,79]. The most important criticism points out that the elevation in breast cancer risk among ATM mutation carriers would not be seen until older ages, if, as hypothesized, cumulative exposure to ionizing radiation is a necessary modifying factor. None of the studies have addressed this or other interactions with ATM susceptibility alleles.

5. HRAS1

The Harvey-ras (HRAS1) oncogene is located on chromosome 11p15, adjacent to a region comprised of 30–100 repeated copies of a 28 base-pair sequence of DNA [80]. The biological function of these variable number of repeats is not yet known, but they may influence transcription of DNA. There are several alleles, based on repeat length, that are fairly common, as well as multiple, rare alleles. The risk of breast cancer among women who inherit one rare HRAS1 allele may be increased approximately twofold. Elevated risks of similar magnitude have been observed for other cancers as well. Risk among individuals inheriting two rare alleles was at least double that of carriers of one rare allele. The association between breast cancer and HRAS1 rare alleles may be stronger among African-American women [81], but this requires additional study because the definition of a rare allele is based on frequencies found among whites. Approximately 6% of the population has inherited at least one rare HRAS1 allele, and it has been estimated that approximately 1 in 11 cancers of the breast, colorectum, or bladder may be due to inheritance of a susceptibility allele at HRAS1 [80].

6. Metabolism Genes

A number of genes have been identified that code for enzymes involved in metabolism [82]. These enzymes help our bodies process foods, beverages, medications, substances like tobacco, and workplace exposures, and they also influence the synthesis of hormones and other important compounds produced internally. Many of the genes have variant DNA sequences that occur in 10–50% or more of the population. Because these genetic variants are relatively common and do not cosegregate with disease in families, they are referred to as polymorphisms. However, these polymorphisms often are associated with variations in the biological activity of the enzymes for which they code, making them more or less efficient at activating, detoxifying, or synthesizing substances during the metabolic process. Hence, they operate as potential genetic modifiers of exogenous exposures to, and endogenous levels of, chemicals. Unfortunately, their relationships with respect to breast cancer susceptibility have proven difficult to establish.

One example is a low-activity allele of the COMT gene, which codes for catechol-O-methyltransferase, an enzyme that inactivates potentially carcinogenic forms of estrogen known as catechol estrogens. Allele frequencies differ somewhat by ethnicity, but approximately 50–80% of the population have inherited one or two copies of the low-activity allele [83–85], which is associated with a reduction in the methylation activity of the enzyme. Women who have inherited at least one copy of the low-activity COMT allele have been reported to be at increased risk of postmenopausal breast cancer and decreased risk of premenopausal disease in one study [83] and to be at increased risk of premenopausal breast cancer and reduced risk of postmenopausal disease in another [84]. A third, much larger study found no association between the COMT low-activity allele and either pre- or postmenopausal breast cancer, with relative risk estimates that were roughly equal to the weighted averages of the

other two studies [85]. Thus, the relationship between the COMT low-activity allele and breast cancer remains unclear. A similar pattern of contradictory findings has been reported for glutathione-S-transferase (GST) enzymes [86] and for CYP17 [87,88]. Research is currently underway on other genes coding for cytochrome P450 enzymes, including CYP1A1 and CYP19.

An alternative approach to evaluating the role of metabolism genes in susceptibility to breast cancer is to study the at-risk genotype in conjunction with the environmental exposure(s) that the enzyme metabolizes. A good example is cigarette smoking, for which the association with breast cancer has been inconsistent [89]. It has been hypothesized that cigarette smoking may increase risk of breast cancer only among genetically susceptible women. N-acetyltransferases (NAT) are enzymes that help process compounds in tobacco smoke. Some studies of breast cancer have examined interactions between cigarette smoking and alleles at the NAT2 gene. The first study found that smoking cigarettes was associated with a strong increased risk of breast cancer among postmenopausal women with NAT2 alleles conferring slow acetylator status [90]. No increased risk was seen among postmenopausal women who inherited the alleles for rapid acetylator status or among premenopausal women. In contrast, two subsequent studies [91,92] found little evidence for an association between cigarette smoking and breast cancer among women with NAT2-slow alleles, although one observed an increased risk of postmenopausal disease among women inheriting alleles for rapid NAT2 activity [92]. Again, the inconsistency across studies may be due, at least in part, to relatively small sample sizes which are insufficient for rigorous evaluation of the many different subgroups involved in statistical analysis of gene–environment interactions. Currently there are no metabolism genes that have been associated consistently with breast cancer risk, with or without inclusion of the relevant environmental exposures.

IV. Future Work

Inherited alterations in additional genes are currently being studied in relation to breast cancer, and results should become available in the early 2000s. These include the gene for the vitamin D receptor, genes related to repair of DNA damage, and genes coding for enzymes that metabolize alcohol. Others of interest include genes coding for additional cytochrome P450 and related metabolic enzymes, as well as genes that code for insulin-like growth factors and other hormones or their receptors. As new candidate genes are identified and potential susceptibility alleles are characterized, these will be evaluated as well. In addition to assessing their influence on the development of breast cancer, studies are beginning to evaluate possible effects of inherited genetic alterations on the natural history of the disease, including the pace of cancer progression and the likelihood of metastasis and recurrence [93]. While potentially revealing additional genetic contributions to breast cancer, this research also may provide insights into relevant, and possibly avoidable, environmental exposures that influence the disease process.

Breast cancer research also continues to focus on somatic genetic alterations (*i.e.,* those mutations that occur in DNA of breast cells after conception). Contrary to expectation, somatic mutations in BRCA1 or BRCA2 have not been observed in the tumors of breast cancer patients lacking an inherited BRCA1 or BRCA2 mutation [19]. Although beyond the scope of this review, noninherited alterations have been identified in other genes, some of which have been associated with prognosis or responsiveness to particular treatments or with distinct risk factor profiles [7].

The interrelationships between inherited mutations in susceptibility genes, like BRCA1 and BRCA2, and somatic genetic alterations or other "downstream" events during tumor development comprise another new area of inquiry. Researchers are beginning to identify the additional changes that accumulate in breast epithelial cells prior to the diagnosis of breast cancer. For example, it has been recognized for some time that tumors with inherited mutations in BRCA1 or BRCA2 exhibit somatic loss of the other, wild-type allele (i.e., the gene copy not associated with breast cancer susceptibility) [19]. These tumors have been described as having histologic features usually associated with aggressive disease and poorer prognosis [94]. Nevertheless, early studies, still considered inconclusive, have reported that disease-free or overall survival may not be much different among breast cancer patients with BRCA1 or BRCA2 mutations than that experienced by breast cancer patients lacking inherited predisposition [94]. Similar research on tumor characteristics and prognosis for women with inherited mutations in other susceptibility genes is lacking.

V. Clinical and Social Implications

Our knowledge regarding the biology of inherited susceptibility to breast cancer is evolving rapidly. With time, understanding the genetic changes that influence development of breast cancer may lead to opportunities for more targeted treatments or even prevention. For example, when we know the functions of the BRCA1 or BRCA2 proteins, it may be possible to develop drugs that compensate for defective genes. Even in the absence of this information, an early trial to assess the feasibility, toxicity, and possible outcome of BRCA1 gene replacement for the treatment of ovarian cancer has been initiated, and a similar study for breast cancer is being developed [95]. In gene therapy, the genetic sequence itself, rather than its protein product, is introduced directly into the tumor or surrounding area. However, the effective use of such gene therapy or chemotherapy is likely years away.

An alternative application, known as predictive genetic testing, is being marketed to identify women at high risk of developing breast cancer. There are a number of practical issues that complicate this enterprise: the test can be quite expensive (currently around $2500 U.S. for both BRCA1 and BRCA2); we do not know how to interpret some results of the tests (i.e., the biological or clinical significance of some variants is still unknown); even when a known disease-related mutation is identified, we are not exactly sure how likely it is that the woman will get breast cancer (i.e., penetrance estimates cover a range of risks); and a negative test does not mean the woman will escape breast cancer (because she might have an undetected susceptibility allele in the tested gene or another gene, or she may develop disease in the absence of inherited susceptibility). In addition, it is challenging to inform those interested in genetic

testing about the potential risks and benefits to ensure that they can make meaningful decisions about testing [96]. A number of difficulties occur, including a lack of familiarity with genetics on the part of the general public and many health care providers, a preference for assessments that are more definitive (i.e., yes/no) rather than being probabilistic in nature, and the potential for psychological distress during the consent and testing process as well as after results are received. Genetic testing therefore tends to be most effective when it is done in the context of a family history of breast cancer (or related conditions) and several relatives are tested. Under these circumstances, the results of the test can be more informative, and the potential risks and benefits may take on different weights given the family's past experiences with the disease.

One of the most problematic aspects of the debate surrounding clinical use of predictive genetic testing is that we lack effective interventions to prevent breast cancer. Most of the current options differ little from medical approaches for treating the disease. Although it seems intuitively appealing to assume that removal of the target organs will reduce risk of subsequent cancer, there are anecdotal reports of women undergoing prophylactic surgery who subsequently develop breast cancer (manifested as a mass on the chest wall) or ovarian cancer (manifested as a mass in the peritoneum) in the absence of breasts and ovaries. Preliminary findings from formal statistical evaluations of the probability of developing breast cancer after prophylactic mastectomy suggest that women who undergo these operations are at lower risk [94]; however, assessments among women with inherited genetic susceptibility have not yet been done. Pharmacologic agents, such as tamoxifen, which has recently been approved by the FDA for chemoprevention of breast cancer, as well as other medications, are also under investigation [94]. Yet again, the effectiveness specifically among women who carry a susceptibility gene mutation is not known, nor have the long-term consequences of these interventions been evaluated for women's risks of other hormone-related diseases, such as cardiovascular disease or osteoporosis. Increased screening with mammography, viewed as the least invasive medical alternative, would not prevent breast cancer, but by aiding early detection of disease, it might improve survival after diagnosis. However, the benefits of mammographic screening in women younger than 50 years remain controversial [97]. Moreover, if some susceptibility genes are involved in DNA repair, as appears possible for BRCA1, BRCA2, and ATM, then mutation carriers may have impaired function. More frequent exposure to ionizing radiation, a known breast carcinogen, which would occur with increased mammography starting at younger ages, may actually increase breast cancer risk.

Some of the personal and social dilemmas raised by the availability of predictive genetic testing are summarized in Table 69.3, in an attempt to focus attention on unintended, but potential, consequences of genetic research. First, we need to ask whether the rational desire of women to manage their risk with respect to breast cancer merely leads to earlier medicalization of their lives. Given the trend in breast cancer treatment toward less invasive procedures, it is ironic that one option commonly considered for disease prevention among women with inherited susceptibility is removal of both breasts, particularly if progno-

Table 69.3
Trade-Offs for Predictive Genetic Testing

Groups involved	Reasonable goals	Possible consequences?
Individuals	Decrease risk	Premature medicalization
Scientists, media	Report results	Unrealistic expectations
Clinicians/hospitals/ biotechnology companies	Provide services	Indiscriminant testing
Insurers/employers	Reduce expenses	Loss of coverage, jobs
Relatives	Plan for future	Intrafamily conflict
Society	Health for all	Eugenics

sis after diagnosis is better for these women. Two breast cancer survivors have written about their concern that this approach may more accurately reflect a treatment for the widespread fear of breast cancer [98]. This anxiety, along with the well-meaning intentions of scientists and the media to share scientific and technological developments with each other and the public, may be contributing to common misconceptions regarding what genetic testing can offer. There is ample documentation that individuals tend to overestimate the benefits of genetic testing relative to the limitations and risks [99], thereby helping to create the market forces that encourage biotechnology firms, hospitals, and physicians to provide genetic testing services. The increasing availability of such services, in turn, may be fueling less discriminating use of genetic tests than is generally recommended.

Methods are now being developed to help women and their clinicians assess the probability of an inherited genetic susceptibility to breast cancer based on family history to facilitate the process of informed decision making about testing [100]. This is important, because the knowledge of carrier status alone can constitute a real risk. It is the objective of most insurance companies and employers to contain costs, and excluding individuals who test positive for inherited predisposition to costly and life-threatening disease from coverage or jobs is a strategy that has been exercised in some situations [101,102]. Additionally, because genetic information has implications not only for the individual being tested but for family members as well, one family member's desire to plan for the future may generate intrafamily conflict and discord. At the societal level, programs designed to enhance the health and well-being of people are already beginning to explore the utility of using genetics to help target so-called "high-risk" subgroups [103]. Although surprisingly little has been published regarding potential eugenic applications of genetic testing [104], the potential for this remains [103].

These concerns suggest that the social issues surrounding predictive genetic testing are likely to be far-reaching and difficult to resolve, and highlight the need for us, as a society, to deal directly with this powerful new technology and approach to health care. As a start, legislation to protect privacy and prevent discrimination, at both the federal and state levels, has begun to be passed [105]. However, legislation alone is no guarantee, and many ethical issues are still under debate or just now being articulated. The women's movement has prepared many of us to be more discriminating in evaluating the costs and benefits of new developments aimed at women and to recognize that unanticipated, and sometimes negative, repercussions follow the introduction of new information and technology against a background of old attitudes. These experiences may be very valuable, as one of the first test cases for integrating genetics with medical policy involves a prominent women's health problem.

VI. Conclusions

Despite the gradual accumulation of knowledge, the emerging picture of breast cancer genetics is still characterized by uncertainty and complexity. As such, it serves as a model of what might be expected for other common cancers and chronic diseases. We do not yet know how many genes confer increased susceptibility to breast cancer, nor do we have accurate measures of the prevalence of susceptibility alleles or the risk or proportion of breast cancer due to currently recognized susceptibility genes. Our understanding of the biological functions of the known susceptibility genes, although rapidly evolving, is still limited. The potential modifiers of gene expression and the subsequent molecular events during carcinogenesis are not yet characterized for women who inherit a susceptibility allele. The clinical implications of inherited susceptibility to breast cancer, including prognosis and potential interventions, are only now being evaluated. Many of these questions are likely to be answered over the coming years, as a consequence of the intensive research efforts underway. We can look forward to increased understanding of breast cancer causation, including contributions of environmental and behavioral risk factors, and progress in prevention and treatment of breast cancer is likely. However, the societal repercussions of dealing with health and disease in relation to inherited genetic predispositions will likely take longer to appreciate and resolve. These factors represent some of the biggest challenges in this new era of molecular genetics.

References

Frequent use of secondary sources, such as textbooks, reviews, and commentaries, is made to limit the list of references; the reader is encouraged to consult bibliographies in these sources for specific, primary reports of research findings.

1. Lewin, B. (1997). "Genes VI." Oxford University Press, Oxford and New York.
2. Ruddon, R. W. (1995). "Cancer Biology," 3rd ed. Oxford University Press, Oxford and New York.
3. Broca, P. P. (1866). "Traites des tumeurs," Vol. 1. P. Asselin, Paris.
4. Lane-Claypon, J. E. (1926). A further report on cancer of the breasts, with special reference to its associated antecedent conditions. *Rep. Minist. Health,* No. 32.
5. Kelsey, J. L. (1979). A review of the epidemiology of human breast cancer. *Epidemiol. Rev.* **1,** 74–109.
6. Kelsey, J. L., and Gammon, M. D. (1990). Epidemiology of breast cancer. *Epidemiol. Rev.* **12,** 228–240.
7. Newman, B., Millikan, R. C., and King, M.-C. (1997). Genetic epidemiology of breast and ovarian cancers. *Epidemiol. Rev.* **19,** 69–79.

8. Eby, N., Chang-Claude, J., and Bishop, D. T. (1994). Familial risk and genetic susceptibility for breast cancer. *Cancer Causes Control* **5,** 458–470.

9. Slattery, M. L., and Kerber, R. A. (1993). A comprehensive evaluation of family history and breast cancer risk: The Utah Population Database. *JAMA, J. Am. Med. Assoc.* **270,** 1563–1568.

10. Anderson, D. E. (1974). Genetic study of breast cancer: Identification of a high risk group. *Cancer (Philadelphia)* **34,** 1090–1097.

11. Lynch, H. T., Guirgis, H. A., Albert, S., *et al.* (1974). Familial association of carcinoma of the breast and ovary. *Surg., Gynecol. Obstet.* **138,** 717–724.

12. Bailey-Wilson, J., Cannon, L. A., and King, M.-C. (1986). Genetic analysis of human breast cancer: A synthesis of contributions to GAW IV. *Genet. Epidemiol.,* Suppl. **1,** 15–35.

13. Claus, E. B., Risch, N., and Thompson, W. D. (1991). Genetic analysis of breast cancer in the Cancer and Steroid Hormone Study. *Am. J. Hum. Genet.* **48,** 232–242.

14. Newman, B., Austin, M. A., Lee, M., and King, M.-C. (1988). Inheritance of human breast cancer: Evidence for autosomal dominant transmission in high-risk families. *Proc. Natl. Acad. Sci. U.S.A.* **85,** 3044–3048.

15. Hall, J., Lee, M., Newman, B., *et al.* (1990). Linkage of early onset familial breast cancer to chromosome 17q21. *Science* **250,** 1684–1689.

16. Wooster, R., Neuhausen, S. L., Mangion, J., *et al.* (1994). Localization of a breast cancer susceptibility gene, BRCA2, to chromosome 13q12-13. *Science* **265,** 2088–2090.

17. Miki, Y., Swenson, J., Shattuck-Eidens, D., *et al.* (1994). A strong candidate for the breast and ovarian cancer susceptibility gene BRCA1. *Science* **266,** 66–71.

18. Wooster, R., Bignell, G., Lancaster, J., *et al.* (1995). Identification of the breast cancer susceptibility gene BRCA2. *Nature (London)* **378,** 789–792.

19. Lee, W.-H., Chew, H. K., Farmer, A. A., and Chen, P.-L. (1998). Biological functions of the BRCA1 protein. *Breast Dis.* **10,** 11–22.

20. Wong, A. K. C., Pero, R., Ormonde, P. A., *et al.* (1997). RAD51 interacts with the evolutionarily conserved BRC motifs in the human breast cancer susceptibility gene Brca2. *J. Biol. Chem.* **272,** 31941–31944.

21. Ford, D., Easton, D. F., Stratton, M., *et al.* (1998). Genetic heterogeneity and penetrance analysis of the BRCA1 and BRCA2 genes in breast cancer families. *Am. J. Hum. Genet.* **62,** 676–689.

22. Struewing, J. P., Hartge, P., Wacholder, S., *et al.* (1997). The risk of cancer associated with specific mutations of BRCA1 and BRCA2 among Ashkenazi Jews. *N. Engl. J. Med.* **336,** 1401–1408.

23. Langston, A. A., Stanford, J. L., Wicklund, K. G., *et al.* (1996). Germ-line BRCA1 mutations in selected men with prostate cancer. *Am. J. Hum. Genet.* **58,** 881–885.

24. Ford, D., Easton, D. F., Bishop, D. T., *et al.* (1994). Risks of cancer in BRCA1-mutation carriers. *Lancet* **343,** 692–695.

25. Thorlacius, S., Olafsdottir, G., Tryggvadottir, L., *et al.* (1996). A single BRCA2 mutation in male and female breast cancer families from Iceland with varied cancer phenotypes. *Nat. Genet.* **13,** 117–119.

26. Phelan, C. M., Lancaster, J. M., Tonin, P., *et al.* (1996). Mutation analysis of the BRCA2 gene in 49 site-specific breast-cancer families. *Nat. Genet.* **13,** 120–122.

27. Easton, D. F., Steele, L., Gields, P., *et al.* (1997). Cancer risks in two large breast cancer families linked to BRCA2 on chromosome 13q12-13. *Am. J. Hum. Genet.* **61,** 120–128.

28. Breast Cancer Information Core (1998). World Wide Web, available at: *http://www.nhgri.nih.gov/Intramural_research/Lab_transfer/Bic*

29. Barker, D. F., Almeida, E. R. A., Casey, G., *et al.* (1996). BRCA1 R841W: A strong candidate for a common mutation with moderate phenotype. *Genet. Epidemiol.* **13,** 595–604.

30. Dunning, A. M., Chiano, M., Smith, N. R., *et al.* (1997). Common BRCA1 variants and susceptibility to breast and ovarian cancer in the general population. *Hum. Mol. Genet.* **6,** 285–289.

31. Newman, B., Mu, H., Butler, L. M., *et al.* (1998). Frequency of breast cancer attributable to BRCA1 in a population-based series of American women. *JAMA, J. Am. Med. Assoc.* **279,** 915–921.

32. Narod, S., Ford, D., Devilee, P., *et al.* (1995). Genetic heterogeneity of breast-ovarian cancer revisited. *Am. J. Hum. Genet.* **57,** 957–958.

33. Schubert, E., Lee, M., Mefford, H., *et al.* (1997). BRCA2 in American families with four or more cases of breast or ovarian cancer: Recurrent and novel mutations, variable expression, penetrance, and characteristics of families not attributable to BRCA1 or BRCA2. *Am. J. Hum. Genet.* **60,** 1031–1040.

34. Serova, O. M., Mazoyer, S., Puget, N., *et al.* (1997). Mutations in BRCA1 and BRCA2 in breast cancer families: Are there more breast cancer-susceptibility genes? *Am. J. Hum. Genet.* **60,** 486–495.

35. Couch, F. J., DeShano, M. L., Blackwood, A., *et al.* (1997). BRCA1 mutations in women attending clinics that evaluate the risk of breast cancer. *N. Engl. J. Med.* **336,** 1409–1415.

36. Shattuck-Eidens, D., Oliphant, A., McClure, M., *et al.* (1997). BRCA1 sequence analysis in women at high risk for susceptibility mutations: Risk factor analysis and implications for genetic testing. *JAMA, J. Am. Med. Assoc.* **278,** 1242–1250.

37. Malone, K. E., Daling, J. R., Thompson, J. D., *et al.* (1998). BRCA1 mutations and breast cancer in the general population. *JAMA, J. Am. Med. Assoc.* **279,** 922–929.

38. Southey, M. C., Tesoriero, A. A., Andersen, C. R., *et al.* (1998). BRCA1 mutations and other sequence variants in a population-based sample of Australian women with breast cancer. *Br. J. Cancer* **79,** 34–39.

39. Krainer, M., Silva-Arrieta, S., FitzGerald, M. G., *et al.* (1997). Differential contributions of BRCA1 and BRCA2 to early-onset breast cancer. *N. Engl. J. Med.* **336,** 1416–1421.

40. Couch, F. J., Farid, L. M., DeShano, M. L., *et al.* (1996). BRCA2 germline mutations in male breast cancer cases and breast cancer families. *Nat. Genet.* **13,** 123–125.

41. Friedman, L. S., Gayther, S. A., Kurosaki, T., *et al.* (1997). Mutation analysis of BRCA1 and BRCA2 in a male breast cancer population. *Am. J. Hum. Genet.* **60,** 313–319.

42. Whittemore, A. S., Gong, G., and Itnyre, J. (1997). Prevalence and contribution of BRCA1 mutations in breast cancer and ovarian cancer: Results from three U.S. population-based case-control studies of ovarian cancer. *Am. J. Hum. Genet.* **60,** 496–504.

43. Ford, D., Easton, D. F., and Peto, J. (1995). Estimates of the gene frequency of BRCA1 and its contribution to breast and ovarian cancer incidence. *Am. J. Hum. Genet.* **57,** 1457–1462.

44. Cavalli-Sforza, L. L., Menozzi, P., and Piazza, A. (1994). "The History and Geography of Human Genes." Princeton University Press, Princeton, NJ.

45. Szabo, C. I., and King, M.-C. (1997). Population genetics of BRCA1 and BRCA2. *Am. J. Hum. Genet.* **60,** 1013–1020.

46. Johannesdottir, G., Gudmundsson, J., Bergthorsson, J. T., *et al.* (1996). High prevalence of the 999del5 mutation in Icelandic breast and ovarian cancer patients. *Cancer Res.* **56,** 3663–3665.

47. Simard, J., Tonin, P., Durocher, F., *et al.* (1994). Common origins of BRCA1 mutations in Canadian breast and ovarian cancer families. *Nat. Genet.* **8,** 392–398.

48. Struewing, J. P. (1998). BRCA1 in special populations. *Breast Dis.* **10,** 71–75.

49. Easton, D. F., Ford, D., Bishop, D. T., *et al.* (1995). Breast and ovarian cancer incidence in BRCA1-mutation carriers. *Am. J. Hum. Genet.* **56,** 265–271.

50. Fodor, F. H., Weston, A., Bleiweiss, I. J., *et al.* (1998). Frequency and carrier risk associated with common BRCA1 and BRCA2

mutations in Ashkenazi Jewish breast cancer patients. *Am. J. Hum. Genet.* **63,** 45–51.

51. Phelan, C. M., Rebbeck, T. R., Weber, B. L., *et al.* (1996). Ovarian cancer risk in BRCA1 carriers is modified by the HRAS1 variable number of tandem repeat (VNTR) locus. *Nat. Genet.* **12,** 309–311.

52. Redston, M., Nathanson, K. L., Yuan, Z. Q., *et al.* (1998). The APCI1307K allele and breast cancer risk. *Nat. Genet.* **20,** 13–14.

53. Narod, S. A., Goldgar, D., Cannon-Albright, L., *et al.* (1995). Risk modifiers in carriers of BRCA1 mutations. *Int. J. Cancer* **64,** 394–398.

54. Chang-Claude, J., Becher, H., Eby, N., *et al.* (1997). Modifying effect of reproductive risk factors on the age at onset of breast cancer for German BRCA1 mutation carriers. *J. Cancer Res. Clin. Oncol.* **123,** 272–279.

55. Ursin, G., Henderson, B. E., Haile, R. W., *et al.* (1997). Does oral contraceptive use increase the riskof breast cancer in women with BRCA1/BRCA2 mutations more than in other women? *Cancer Res.* **57,** 3678–3681.

56. Narod, S. A., Risch, H., Moslehi, R., *et al.* (1998). Oral contraceptives and the risk of hereditary ovarian cancer. *N. Engl. J. Med.* **339,** 424–428.

57. Brunet, J.-S., Parviz, G., Rebbeck, T. R., *et al.* (1998). Effect of smoking on breast cancer in carriers of mutant BRCA1 or BRCA2 genes. *J. Natl. Cancer Inst.* **90,** 761–766.

58. Birch, J. M., Hartley, A. L., Tricker, K. J., *et al.* (1994). Prevalence and diversity of constitutional mutations in the p53 gene among 21 Li-Fraumeni families. *Cancer Res.* **54,** 1298–1304.

59. Frebourg, T., Barbier, N., Yan, Y., *et al.* (1995). Germ-line p53 mutations in 15 families with Li-Fraumeni Syndrome. *Am. J. Hum. Genet.* **56,** 608–615.

60. Elledge, R. M., and Allred, D. C. (1994). The p53 tumor suppressor gene in breast cancer. *Breast Cancer Res. Treat.* **32,** 39–47.

61. Sidransky, D., Tokino, T., Helzlsouer, K., *et al.* (1992). Inherited p53 gene mutations in breast cancer. *Cancer Res.* **52,** 2984–2986.

62. Nelen, M. R., Padberg, G. W., Peeters, E. A. J., *et al.* (1996). Localization of the gene for Cowden disease to chromosome 10q22-23. *Nat. Genet.* **13,** 114–116.

63. Li, J., Yen, C., Liaw, D., *et al.* (1997). PTEN, a putative protein tyrosine phosphatase gene mutated in human brain, breast, and prostate cancer. *Science* **275,** 1943–1946.

64. Tsou, H. C., Teng, D. H.-F., Ping, X. L., *et al.* (1997). The role of MMAC1 mutations in early-onset breast cancer: Causative association with Cowden Syndrome and excluded in BRCA1-negative cases. *Am J. Hum. Genet.* **61,** 1036–1043.

65. Lynch, E. D., Ostermeyer, E. A., Lee, M. K., *et al.* (1997). Inherited mutations in PTEN that are associated with breast cancer, cowden disease, and juvenile polyposis. *Am. J. Hum. Genet.* **61,** 1254–1260.

66. Zuppan, P., Hall, J. M., Lee, M. K., *et al.* (1991). Possible linkage of the estrogen receptor gene to breast cancer in a family with late-onset disease. *Am. J. Hum. Genet.* **48,** 1065–1068.

67. Dowsett, M., Daffada, A., Chan, C. M. W., and Johnston, S. R. D. (1997). Oestrogen receptor mutants and variants in breast cancer. *Eur. J. Cancer* **33,** 1177–1183.

68. Southey, M. C., Batten, L. E., McCredie, M. R. E., *et al.* (1998). Estrogen receptor polymorphism at codon 325 and risk of breast cancer in women before age forty. *J. Natl. Cancer Inst.* **90,** 532–536.

69. Wooster, R., Mangion, J., Eeles, R., *et al.* (1992). A germline mutation in the androgen receptor gene in two brothers with breast cancer and Reifenstein syndrome. *Nat. Genet.* **2,** 132–134.

70. Lobaccaro, J.-M., Lumbroso, S., Belon, C., *et al.* (1993). Male breast cancer and the androgen receptor gene. *Nat. Genet.* **5,** 109–110.

71. Chamberlain, N. L., Driver, E. D., and Miesfeld, R. L. (1994). The length and location of CAG trinucleotide repeats in the androgen receptor N-terminal domain affect transactivation function. *Nucleic Acids Res.* **22,** 3181–3186.

72. Giovannucci, E., Stampfer, M. J., Drithivas, K., *et al.* (1997). The CAG repeat within the androgen receptor gene and its relationship to prostate cancer. *Proc. Natl. Acad. Sci. U.S.A.* **94,** 3320–3323.

73. Morrell, D., Cromartie, E., and Swift, M. (1986). Mortality and cancer incidence in 263 patients with ataxia-telangiectasia. *J. Natl. Cancer Inst.* **77,** 89–92.

74. Swift, M., Morrell, D., Massey, R. B., and Chase, C. L. (1991). Incidence of cancer in 161 families affected by Ataxia-Telangiectasia. *N. Engl. J. Med.* **325,** 1831–1836.

75. Easton, D. F. (1994). Cancer risks in A-T heterozygotes. *Int. J. Radiat. Biol.* **66,** S177–S182.

76. Bebb, G., Glickman, B., Gelmon, K., and Gatti, R. (1997). "AT risk" for breast cancer. *Lancet* **349,** 1784–1785.

77. Athma, P., Rappaport, R., and Swift, M. (1996). Molecular genotyping shows that ataxia-telangiectasia heterozygotes are predisposed to breast cancer. *Cancer Genet. Cytogenet.* **92,** 130–134.

78. FitzGerald, M. G., Bean, J. M., Hegde, S. R., *et al.* (1997). Heterozygous ATM mutations do not contribute to early onset of breast cancer. *Nat. Genet.* **15,** 307–310.

79. Bishop, D. T., and Hopper, J. (1997). AT-tributable risks? *Nat. Genet.* **15,** 226–227.

80. Krontiris, T. G., Devlin, B., Karp, D. D., *et al.* (1993). An association between the risk of cancer and mutations in the HRAS1 minisatellite locus. *N. Engl. J. Med.* **329,** 517–523.

81. Garrett, P. A., Hulka, B. S., Kim, Y. L., and Farber, R. A. (1993). HRAS protooncogene polymorphism and breast cancer. *Cancer Epidemiol., Biomarkers Prev.* **2,** 131–138.

82. Smith, G., Stanley, L. A., Sim, E., *et al.* (1995). Metabolic polymorphisms and cancer susceptibility. *Cancer Surv.* **25,** 27–65.

83. Lavigne, H., Helzlsouer, K., Huang, H.-Y., *et al.* (1997). An association between the allele coding for a low activity variant of Catechol-O-methyltransferase and the risk for breast cancer. *Cancer Res.* **57,** 5493–5497.

84. Thompson, P. A., Shields, P. G., Freudenheim, J. L., *et al.* (1998). Genetic polymorphisms in Catechol-O-Methyltransferase, menopausal status, and breast cancer risk. *Cancer Res.* **58,** 2107–2110.

85. Millikan, R. C., Pittman, G., Tse, C.-K. J., *et al.* (1998). Catechol-O-Methyltransferase (COMT) and breast cancer risk. *Carcinogenesis (London)* **19,** 1943–1947.

86. Kelsey, K. T., and Wiencke, J. K. (1998). Growing pains for the environmental genetics of breast cancer: Observations on a study of the Glutathione-S-Transferases. *J. Natl. Cancer Inst.* **90,** 484–485.

87. Feigelson, H. S., Coetzee, G. A., Kolonel, L. N., *et al.* (1997). A polymorphism in the CYP17 gene increases the risk of breast cancer. *Cancer Res.* **57,** 1063–1065.

88. Dunning, A. M., Healey, C. S., Pharoah, P. D., *et al.* (1998). No association between a polymorphism in the steroid metabolism gene CYP17 and risk of breast cancer. *Br. J. Cancer* **77,** 2045–2047.

89. Palmer, J., and Rosenberg, L. (1993). Cigarette smoking and the risk of breast cancer. *Epidemiol. Rev.* **15,** 145–156.

90. Ambrosone, C. B., Freudenheim, J. L., Graham, S., *et al.* (1996). Cigarette smoking, N-acetyltransferase genetic polymorphisms, and breast cancer risk. *JAMA, J. Am. Med. Assoc.* **276,** 1494–1501.

91. Hunter, D., Hankinson, S., Hough, H., *et al.* (1997). A prospective study of NAT2 acetylation genotype, cigarette smoking and risk of breast cancer. *Carcinogenesis (London)* **18,** 2127–2132.

92. Millikan, R. C., Pittman, G. S., Newman, B., *et al.* (1998). Cigarette smoking, N-acetyltransferases 1 and 2, and breast cancer risk. *Cancer Epidemiol., Biomarkers Prev.* **7,** 371–378.

93. Narod, S. A. (1998). Host susceptibility to cancer progression. *Am. J. Hum. Genet.* **63,** 1–5.

94. Osborne, C. K., Elledge, R. M., Brown, P. H., and Hilsenbeck, S. G. (1998). BRCA1 in clinical breast cancer. *Breast Dis.* **10,** 77–88.

95. Tait, D. L., Jensen, R. A., Holt, J. T., *et al.* (1998). Gene therapy for breast and ovarian cancer with BRCA1. *Breast Dis.* **10,** 89–98.

96. Rimer, B. K., Sugarman, J., Winer, E., *et al.* (1998). Informed consent for BRCA1 and BRCA2 testing. *Breast Dis.* **10,** 99–114.

97. Kerlickowske, K., Grady, D., Rubin, S., *et al.* (1995). Efficacy of screening mammography: A meta-analysis. *JAMA, J. Am. Med. Assoc.* **273,** 149–154.

98. Barr, P., and Reese-Coulbourne, J. (1998). Commentary from the patient's perspective. *Breast Dis.* **10,** 137–140.

99. Freedman, T. G. (1997). Genetic susceptibility testing: A therapeutic illusion? *Cancer (Philadelphia)* **79,** 2063–2064.

100. Berry, D. A., and Parmigiani, G. (1998). Assessing the benefits of testing for breast cancer susceptibility genes: A decision analysis. *Breast Dis.* **10,** 115–125.

101. Rothenberg, K., Fuller, B., Rothstein, M., *et al.* (1997). Genetic information and the workplace: Legislative approaches and policy challenges. *Science* **275,** 1755–1757.

102. Rothenberg, K. H. (1995). Genetic information and health insurance: State legislative approaches. *J. Law Med. Ethics* **23,** 312–319.

103. Hubbard, R., and Wald, E. (1993). "Exploding the Gene Myth." Beacon Press, Boston.

104. Lancaster, J. M., Wiseman, R. W., and Berchuck, A. (1996). An inevitable dilemma: Prenatal testing for mutations in the BRCA1 breast-ovarian cancer susceptibility gene. *Obstet. Gynecol.* **87,** 306–309.

105. Dressler, L. (1998). Genetic testing for the BRCA1 gene and the need for protection from discrimination: An evolving legislative and social issue. *Breast Dis.* **10,** 127–135.

70

Breast Cancer Screening

KARLA KERLIKOWSKE

Department of Epidemiology and Biostatistics, Department of Medicine,
and General Internal Medicine Section at Department of Veterans Affairs
University of California, San Francisco
San Francisco, California

I. Introduction

There are very few preventive measures that reduce an average-risk woman's risk of breast cancer. Strenuous exercise, maintaining ideal body weight, minimizing alcohol intake, breastfeeding, and avoidance of long-term hormone replacement therapy are a few potential modifiable risk factors [1–9]. Thus, screening for early stage disease is the principal means of reducing breast cancer mortality. Because mass screening for breast cancer involves primarily healthy women, it is important for women and health practitioners to understand the potential benefits as well as the harms and limitations of screening for breast cancer.

II. Goal of Screening

The goal of screening is to avert death from breast cancer. In order for that to occur, breast cancer must be identified in the preclinical phase and be biologically significant, treatment must be more effective in the preclinical phase than in the symptomatic phase, the screening test must have a high sensitivity and specificity, and it must be widely applied in the target population. For a screening to be cost-effective, early detection must not only reduce the rate of death from breast cancer but the number of false positive screening tests should be relatively low and the screening test inexpensive.

Breast cancer has a detectable preclinical phase that can be identified before women are symptomatic. Survival is over 90% when small breast tumors (less than 10 mm) are identified and treated before women become symptomatic. Mammography, clinical breast exam (CBE), and breast self-examination (BSE) are the three screening tests that have been evaluated for early detection of breast cancer. The potential benefits, harms, limitations, and cost-effectiveness of these tests are described below.

III. Efficacy and Accuracy of Screening Mammography

A. *Randomized Controlled Trials*

The randomized controlled trial is the most unbiased means to assess whether a screening test reduces the likelihood of death in a person who has the disease and is considered the gold standard when evaluating the efficacy of screening tests. There have been ten randomized controlled trials conducted to determine whether undergoing screening mammography decreases the chance of dying from breast cancer (Table 70.1). These trials differ in type of randomization (cluster, individual), type of intervention (screening intervals from 12 to 33 months, single-view or two-view mammography, screening with or without clinical breast examination), and age of the study population. There have been several meta-analyses [10–14] published that combine data from the randomized controlled trials of screening mammography in order to quantify the overall impact of screening on breast cancer mortality and to obtain a more stable estimate of the effect of screening according to age.

1. Efficacy by Age

a. WOMEN AGES 40 TO 49 YEARS. Pooled results from randomized controlled trials have not demonstrated that screening mammography significantly reduces breast cancer mortality in women ages 40 to 49 years within the first 7 to 9 years following the initiation of screening (Table 70.1) [10,12]. There is a trend toward a significant reduction in breast cancer mortality 10 to 14 years after the start of screening (Table 70.1) [11,14]. Based on a subgroup analysis of women ages 39 to 49 years, the Gothenburg trial is the only individual study to report a statistically significant reduction in breast cancer mortality 11 years after the initiation of screening [15]. This is in contrast to the Canadian National Breast Screening Study, the only study designed specifically for women ages 40 to 49 years, that found no reduction in breast cancer mortality after 10.5 years [16]. It has been estimated that to demonstrate a 25% reduction in breast cancer mortality at 5 years with 80% statistical power would require enrollment of 500,000 women ages 40 to 49 years in a randomized control trial [17]. However, only 25,945 women were enrolled in the Gothenburg study [15], compared with 50,430 in the Canadian trial [16]. It is unclear why the Gothenburg trial found a statistically significant reduction in breast cancer mortality when other trials have failed to do so. One explanation is that the Gothenburg results are a chance finding, given the study was not designed specifically for women ages 40 to 49 years. Another possible explanation is that the Gothenburg finding is real but results from the fact that the mortality in the control group was more than twice as high as the mortality rate in the control group in the Canadian trial, suggesting that women in the control group in the Canadian study presented with earlier stage disease compared with the control group in the Gothenburg study [18]. The high mortality rate in the control group in the Gothenburg study may reflect the twofold higher rate of lymph node positive tumors in the control group [15] compared with the rate in the Canadian study [16,19]. Lastly, there has been concern about data quality [20,21] in the Gothenburg trial

Table 70.1

Randomized Controlled Trials of Screening Mammography Included in Meta-Analysis of Women Aged 40 to 74 Years[a]

Study	Screening interval (mo.)	Annual clinical breast exam	Reduction in breast cancer mortality		
			Women 40 to 49 7–9 yr from first screen	Women 40 to 49 10–14 yr from first screen	Women 50 to 74 7–9 yr from first screen
Gothenburg	18	no	27%[b]	44%	9%[b]
Stockholm[c]	24–28[d]	no	None	None	35%[b]
HIP	12	yes	19%[b]	23%[b]	35%
Canadian I[c]	12	yes	None	None	NA[e]
Canadian II	12	yes	NA[e]	NA[e]	None
Ostergotland	24	no	None	None	31%
Kopparberg	24	no	24%[b]	27%[b]	39%
Malmo I[c]	21	no	None	33%[b]	21%[b]
Malmo II[c]	21	no	NA[f]	31%[b]	NA[e]
Edinburgh[c]	24	yes	None	27%[b]	20%[b]
Meta-analysis results			+2% (95% CI = −18 − +27%)	16% (95% CI = 1 − 29%)	27% (95% CI = 16 − 37%)

[a]Adapted from [10], *JAMA*, 1995, **273**, 149–154, copyright 1993–1996, American Medical Association, and [11], *Journal of the National Cancer Institute*, 1997, **22**, 79–86, by permission of Oxford University Press.

[b]Findings do not achieve statistical significance.

[c]Data presented but unpublished in peer-reviewed journal.

[d]First round 28 months after baseline exam, second round 24 months after first round.

[e]Not applicable.

[f]Not available.

due to discrepancies among published reports from this trial [15,22,23].

Among 10,000 women 40 years old, 150 will be diagnosed with invasive breast cancer in the next ten years, and of these, 37 will die of the disease. Using results from the pooled analysis of all randomized controlled trials (Table 70.1), if 10,000 women get routine mammography over the next 10 years, 4 of the 37 breast cancer deaths may be averted (Table 70.2). This means that for women in their forties, mammography prevents one breast cancer death for every 2500 women screened annually for 10 years, or one death averted per 25,000 mammograms

Table 70.2

Annual Mammography in 10,000 40-Year-Old Women for 10 Years Compared with Biennial Mammography in 10,000 50-Year-Old Women for 20 Years[a]

	Age (y)	
	40 to 49	50 to 69
Abnormal result	3,000	2,500
Biopsy	750	1,000
Breast cancer		
Invasive	150	580
DCIS	50	220
Die of breast cancer	37	260
Breast cancer deaths averted from screening	4	37
Mammograms performed per breast cancer		
death averted	25,000	2,700
Cost per year of life saved	$150,000	$21,000

[a]Adapted from [24], *Ann. Intern. Med.*, 1997, **127**, 955–965, with permission.

performed [24]. More deaths from breast cancer are not averted because many breast cancers detected by mammography can be diagnosed later and still be cured. Also, some cancers detected on mammography are already too advanced at the time of detection to make a difference. Lastly, because breast cancer is less common in younger than in older women, there are fewer potential breast cancer deaths to avert [25].

Interventions that cost less than $50,000 per life-year saved are generally viewed favorably. The incremental cost-effectiveness of screening women ages 40 to 49 years annually for ten years is $150,000 per year of life saved [24]. Three factors make screening mammography less cost-effective in younger women than in older women: (1) the incidence of breast cancer begins to rise around age 40 yet remains two- to threefold lower than in women ages 50 to 69 years; 2) the long delay until the onset of a reduction in breast cancer mortality among younger screened women (Table 70.1); and 3) the lower relative risk reduction for mammography making screening less efficacious among younger compared to older women.

b. WOMEN AGES 50 TO 69 YEARS. Screening mammography has been shown to reduce mortality from breast cancer 27% among women ages 50 and older 7 to 9 years after the initiation of screening (Table 70.1) [10,26]. The mortality reduction begins to appear as early as 4 to 5 years after the initiation of screening [27,28]. Among 10,000 women 50 years old, 580 will be diagnosed with invasive breast cancer in the next 20 years, and of these, 260 will die of the disease. If 10,000 women get routine mammography over the next 20 years, 37 of the 260 breast cancer deaths may be averted. This means that for women ages 50 and older, routine mammography averts one breast cancer death for every 270 women screened regularly for 20 years, or one death averted per 2700 examinations performed (Table

70.2) [24]. It is more cost-effective to screen women ages 50 to 69 years then to screen women ages 40 to 49 years. The cost-effectiveness ratio of screening women ages 50 to 69 years biennially for twenty years is $21,000 per year of life saved (Table 70.2), a ratio comparable to other recommended screening interventions [24].

c. WOMEN AGES 70 YEARS AND OLDER.   There are inadequate data from randomized controlled trials to draw a conclusion regarding the benefit of screening mammography in women ages 70 and older. Data from the combined Swedish trials reported a relative risk of 0.78 (95% CI = 0.53–1.20) at 13 years of follow-up [29]. Small numbers limit the statistical power of this analysis to provide meaningful results. A decision analysis of the utility of screening for breast cancer in women ages 65 to 85 years reported that, on average, life expectancy would be extended about 2 days for women ages 65 to 74 years and 1 day for women ages 75 to 85 years in the screened population. For women diagnosed with breast cancer by screening mammography, life-expectancy is increased by 617 days for women ages 65 to 69 years and 178 days for women ages 85 and older. Life expectancy is extended less for women with breast cancer detected by screening mammography who also have heart failure, with increases in life expectancy of 311 days and 126 days for women ages 65 to 69 years and 85 years or more, respectively [30]. Thus, screening mammography may benefit some elderly women through detection of early breast cancers, especially if they do not have comorbid conditions. However, if elderly women have three or more comorbid conditions (i.e., hypertension, diabetes, arthritis, history of myocardial infarction, stroke, respiratory disease, or other types of cancer), regardless of the stage at diagnosis of breast cancer, death from causes other than breast cancer is 20-fold more likely within 3 years [31]. Given this, performing screening mammography when an elderly woman's life expectancy is less than 5 years will not likely impact her overall mortality but may influence her quality of life if she has to live with the knowledge she has cancer, be subjected to unnecessary diagnostic evaluations of abnormal mammographic results (the vast majority (86–92%) of which do not represent cancer), and be exposed to surgical treatment of clinically insignificant lesions [32,33].

2. *Efficacy According to Length of Screening Interval and Whether CBE is Performed*

a. SCREENING INTERVAL.   There are only two clinical trials that screened women ages 40 to 49 years annually and the results are conflicting. The HIP trial showed a nonsignificant 23% reduction in breast cancer mortality 10 years after screening began [28] and the Canadian trial showed a nonsignificant 14% increase in breast cancer mortality after 11 years [16]. Of note, the subgroup analysis of women ages 40 to 49 years in the Gothenburg trial showed a significant decrease in breast cancer mortality when women were screened every 18 months. The lead time[1] for development of a mammographically detectable tumor is thought to be 1.25 years in women ages 35 to 49 years compared with 3.5 years for women ages 50 and older [34,35].

[1]Lead time is the amount of time between detection of an occult lesion on a mammogram and when the same lesion would cause symptoms.

This suggests that a greater proportion of invasive breast cancers grow more rapidly in younger women and require a shorter interval between screening mammographic examinations to detect small occult tumors as early as possible. This theory is supported by the observation that the sensitivity of screening mammography decreases with increasing size of tumor. That is, tumors that are not detected by mammography are larger at presentation than tumors that are mammographically detected. A lower sensitivity for detecting large tumors is more marked in younger than in older women, suggesting that tumors in younger women are especially rapid-growing [36]. Also in support of more rapid tumor growth rates in younger women is the finding that the sensitivity of mammography decreases rapidly as the length of time between screening examinations increases [36]. This is further supported by the observation that among women ages 40 to 49 years, a greater proportion of small tumors detected by screening mammography are associated with positive lymph nodes when compared with older women [37,38]. Taken together, these findings suggest that the tumor biology is different in younger than in older women. Given the indirect evidence that rapid tumor growth rates may be the major contributing factor to the lower sensitivity of mammography, if women ages 40 to 49 years request screening mammography, they should be screened every 12 to 18 months.

Among women ages 50 and older, screening every 18 to 33 months results in a 23% (95% CI = 12–32%) reduction in breast cancer mortality. Screening annually results in a similar reduction (23%; 95% CI = 0–41%). The lead time for development of a mammographically detectable tumor is thought to be about 3.5 years in women ages 50 and older [34]. Thus, screening biennially allows sufficient time to detect breast cancer at a curable stage. Screening more frequently than biennially in this age group does not result in a higher reduction in breast cancer mortality, but it does increase the cost of screening [24]. Screening triennially results in unacceptable rates of interval cancers [39,40].

b. CLINICAL BREAST EXAM.   Screening mammography results in a significant reduction in breast cancer mortality regardless of whether clinical breast examination is performed in conjunction with mammography [10]. Among women ages 50 and older, breast cancer mortality was decreased 24% among those who did not receive CBE and 20% among those who did undergo CBE in conjunction with mammography [10]. The U.S. Preventive Services Task Force (USPSTF) recommends screening mammography every one to two years for women ages 50 to 69 years with or without CBE [41].

3. *Efficacy According to Family History of Breast Cancer*

Other than female gender and older age, having a family history of breast cancer, defined as having a first-degree relative with breast cancer, is one of the strongest risk factors for breast cancer [42]. There are no clinical trials or subgroup analyses evaluating the efficacy of screening mammography in women who have a family history of breast cancer. Among women less than 50 years of age, those who had a relative diagnosed with breast cancer before age 50 have a five times greater risk of dying of breast cancer than those without a family history [43]. This finding is supported by a study that also shows that women who were younger than 50 years when diagnosed with breast

Table 70.3
Performance of First Screening Mammography by Decade of Age

	Age (y)			
	40 to 49	50 to 59	60 to 69	≥ 70
Abnormal exams %	5.8–6.4	5.6–6.8	4.6–8.0	3.7–7.4
PPV mammography[a]				
Invasive cancer only %	2.6–4.4	6.3–7.9	9.3–12.2	14.8–15.1
All breast cancer %	4.6–6.2	9.0–10.0	12.0–14.9	19.0–20.1
Sensitivity[b]				
Invasive cancer only %	78.0	90.9	89.8	87.0
DCIS	100	100	100	100
All breast cancer %	84.7	93.0	91.2	91.2

[a]Data from University of California Mobile Mammography Screening Program, 1985–1996 (adapted from [52], *Journal of the National Cancer Institute,* 1997, **22,** 105–111, by permission of Oxford University Press) and National Breast and Cervical Cancer Early Detection Program 1991–1995 (adapted from [53]).

[b]Data from University of California Mobile Mammography Screening Program, 1985–1996 (adapted from [52]).

cancer were at increased risk of dying of breast cancer if they had a family history of breast cancer [44]. Among women ages 50 and older, the effect of family history was much smaller, regardless of the relative's age at diagnosis [43].

The positive predictive value (PPV)[2] of screening mammography is increased two to threefold in women ages 40 to 59 years with a family history of breast cancer because of the higher prevalence of disease in these women [45]. However, the sensitivity of mammography is primarily influenced by age, not by family history status. The sensitivity of mammography is similar or slightly lower for women younger than 50 with a family history compared to women without a family history, even though women with a family history are at higher risk of breast cancer [36]. Although studies confirming a benefit from screening high-risk young women are lacking, recommendations for screening such women have been made on other grounds, including a high burden of suffering (increased risk of disease and possibly death from breast cancer) and a PPV of mammography similar to that of women ages 50 to 69 years [41].

4. Efficacy According to Menopausal Status or Hormone Replacement Therapy

There are no clinical trials or subgroup analyses evaluating the efficacy of screening mammography according to menopause status or hormone replacement therapy. One study reported the sensitivity of mammography was lower for premenopausal women compared with postmenopausal women [36]. Another study reported that the sensitivity of mammography is lower among women who use hormone replacement therapy [46], although the confidence intervals around the estimate of sensitivity were wide. It is not known whether the sensitivity of mammography among hormone users was lower because radiologists missed cancers visible on films, because cancers were obscured by dense breast tissue, or because the cancer went from undetectably small on mammography to very large in a

[2]The PPV of screening mammography is calculated as the percentage of women with abnormal screening results who are subsequently diagnosed with breast cancer.

short amount of time. Of note, the specificity of mammography has also been reported to be significantly lower among women who used hormone replacement therapy compared with nonusers [46,47]. The lower specificity among hormone users may be because hormone replacement therapy increases breast density in about 25% of users [48]. Hormone replacement therapy may slightly increase a woman's risk of breast cancer [9], but it is unclear whether hormone-induced cancers are rapid-growing. On the one hand, it has been reported that women who have ever used hormone replacement tend to have more *in situ* or localized tumors at detection, possibly because of earlier detection by mammography [9,49]. On the other hand, two small studies have reported that the extent of disease among women using hormone replacement therapy is the same as that among nonusers [50] or possibly greater, with more stage II tumors at diagnosis [51]. Lastly, it has been reported among current or recent users of hormone replacement that increased duration of use may increase the risk of disease spread [9]. Given the conflicting and limited data regarding hormone replacement therapy and tumor growth rates and extent of disease, there is insufficient evidence at this time to recommend that postmenopausal women who take hormone replacement therapy undergo screening mammography more often than those who do not.

B. Accuracy of Screening Mammography

The percentage of first screening mammographic examinations with abnormal results increases with age (Table 70.3). The PPV of mammography also increases with age with women ages 50 to 59 years, having about a twofold higher PPV of mammography than women ages 40 to 49 years (Table 70.3). This means for every 100 forty-year-old women with abnormal mammography, about 3.5 will have invasive cancer compared with 7, 11, and 15 per 100 women in their fifties, sixties, and seventies or older, respectively [52,53]. The PPV of mammography is somewhat higher for all ages of women when all breast disease outcomes (invasive cancer and ductal carcinoma *in situ*) are considered, but it still remains low for women ages 40 to 49 years. The incidence of breast cancer increases about 1.5-fold every 10 years starting at age 40 up to age 70, with approximately 76% of all invasive breast cancers diagnosed after age 50 [54]. The observed increase in PPV with increasing age is most likely due to the higher prevalence of breast cancer in older women.

Studies of modern screening mammography [55–60] report overall sensitivities of screening mammography (71.1–91.5%) similar to those published for randomized controlled trials [26]. Two studies report the sensitivity of mammography by age and show that sensitivity is lower for women younger than 50 years (63% and 80%) compared to women ages 50 and older (89% and 94%) [55,57]. A study that evaluated the sensitivity of modern screening mammography by decade of age shows that the sensitivity of mammography to detect invasive breast cancer is lower among women ages 40 to 49 years compared with women ages 50 and older (78% versus 92%; Table 70.3) [36,52]. The sensitivity of mammography to detect invasive cancer is lowest for women ages 30 to 39 years [36]. The sensitivity of mammography is primarily influenced by the ability of radiologists to identify breast cancers on mammography and by the rate at which breast cancers double in size between screening examinations. Consequently, a false negative examination can occur when a radiologist does not identify a breast lesion that is visible on mammography or when an undetectable breast cancer grows quickly and is discovered clinically before the next screening examination. It has been hypothesized that the lower sensitivity of mammography in younger women is due to the lower fat content of young women's breasts which makes them less radiolucent on film screen mammography (thus, obscuring small tumors). However, two studies have shown that the sensitivity of mammography does not vary according to breast density among younger women but rather that the lower sensitivity is more likely a result of rapid tumor growth rates [36,61].

IV. Harms and Limitations of Screening Mammography

Screening mammography may harm women through additional diagnostic evaluations following an abnormal mammography result with associated morbidity and anxiety, the potential detection and surgical treatment of clinically insignificant lesions which may have no impact on mortality [62], and false reassurance resulting from having a normal examination. In addition, a large proportion (up to 91%) of women report having some degree of pain during mammography, with a small proportion of women (less than 15%) reporting intense pain [63]. Women who are annoyed by medical tests and visits to doctors, bothered by the discomfort of undergoing mammography and unnecessary additional tests, or who experience anxiety waiting for test results might defer screening, even if the weight of the evidence suggests a benefit. For these reasons, women should be informed of the potential harms and limitations of undergoing screening mammography, as well as the potential benefits, and be allowed to participate in the decision whether to be screened.

A. Diagnostic Evaluations and Associated Morbidity and Anxiety

One consequence of the low PPV of mammography (Table 70.3) is the high number of diagnostic evaluations. On average,

approximately 1.5 to 2 additional diagnostic tests are performed per abnormal screening examination [41,64]. Because the PPV of mammography is low in women ages 40 to 49 years, these women have the potential to be subjected to the greatest harm because they will undergo the greatest number of diagnostic tests to find the fewest cancers. For example, among 100 average-risk women ages 40 to 49 years with an abnormal first screening examination, about 94 do not have cancer (Table 70.3) and must undergo further diagnostic evaluation that may include tests such as clinical breast examination, additional mammography, ultrasound, needle aspiration, or excisional biopsy. Women 40 to 49 years of age undergo approximately 45 diagnostic tests for every cancer detected by screening mammography compared to 15 for every cancer detected in women ages 50 and older [41]. The yield of invasive cancer diagnosed per breast biopsy increases with age from 11–15% in women ages 40 to 49 years to 38–39% in women ages 70 and older [52,53]. Therefore, for women less than age 50, only 1 in 7 biopsies will have cancer whereas 1 in 3 will have cancer in older women. The lower yield of cancer per breast biopsy and higher number of diagnostic tests per cancer detected in younger women is due to the lower incidence of breast cancer in these women.

Because most mammographic abnormalities are nonpalpable, needle localization biopsy or core biopsy is often required. Although risk is low, there are complications associated with biopsies, such as hematomas, infection, and scarring, and from wire localization itself, complications include vasovagal reactions (7%) and rarely, prolonged bleeding (1%) and extreme pain (1%) [65]. In addition, a substantial proportion of women have increased anxiety about breast cancer compared to women with normal mammographic results, even after learning they do not have cancer [66–69]. Twenty-nine percent have persistent anxiety 18 months after an abnormal mammographic result compared to women with normal results (13%), especially women who undergo breast biopsies [66]. However, such anxiety does not appear to interfere with subsequent adherence to screening. In contrast, women who have decreased anxiety about breast cancer after undergoing screening mammography are less likely to obtain subsequent routine mammography [66]. Lastly, some women may be wrongly labeled as being at higher risk of breast cancer as a result of having a false-positive mammographic examination, which may affect recommendations for subsequent screening and insurance status.

The risk of at least one abnormal mammographic exam, false-positive exam, and breast biopsy in women screened annually for 10 years is high for all ages (Table 70.4). If a 40 year old woman elects to be screened annually for ten years (i.e., ten mammographic examinations in ten years), she should be informed she has a 30% chance of having at least one abnormal screening examination that will require a diagnostic work-up, a 28% chance of at least one false-positive examination, and a 7% chance of undergoing at least one breast biopsy (Table 70.4). A 50 year old woman who elects to be screened annually for ten years should be informed she has a 26% chance of having at least one abnormal screening examination that will require a diagnostic work-up, a 23% chance of at least one false-positive examination, and a 10% chance of undergoing at least one breast biopsy. For all women irrespective of age, the chance of an abnormal test and false-positive test is greater than the risk of breast cancer (Table 70.4). However, for younger women the

Table 70.4

Risk of at Least One Abnormal Mammographic Exam, False-Positive Exam, and Breast Biopsy if Screened Annually for Ten Years[a]

	Age (y)			
Risk	40	50	60	>70
Abnormal exam	30%	26%	23%	26%
False-positive exam	28%	23%	20%	22%
Biopsy	7.5%	10.4%	10.4%	10%
Invasive breast cancer[b]	1.5%	2.4%	3.4%	3.5%
DCIS[b]	0.5%	1.0%	1.2%	1.1%

[a]Adapted from [52], *Journal of the National Cancer Institute,* **22**, 105–111, by permission of Oxford University Press, and [54].

[b]Risk of invasive breast cancer or ductal carcinoma *in situ* in the next ten years.

risk of a false-positive test is the highest because the incidence of breast cancer is lower in these women. It is important to emphasize that these numbers are based on abnormal rates for first screening and subsequent screening assuming high quality screening mammography and may be a conservative estimate of the risk of at least one abnormal mammographic exam, false-positive exam, and breast biopsy [52]. The estimated cumulative risk of a false-positive examination after 10 mammograms may be as high as 49% [70].

B. Increased Detection of Ductal Carcinoma In Situ (DCIS)

DCIS is a breast lesion that is contained within the milk ducts of the breast. DCIS lesions contain some cells with malignant features but not all such lesions behave as cancer (*i.e.,* they will not spread outside the ducts and invade surrounding breast tissue, nor will they be life threatening). In other words, only some DCIS lesions will eventually become invasive cancer.

It is thought that 11–32% of DCIS lesions progress to invasive cancer over 18 to 30 years [71,72], and up to 13% recur as invasive cancer over 8 years if treated by wide excision alone [73]. Of breast cancers detected by screening mammography in average-risk women ages 40 to 49 years, approximately 30–44% are DCIS compared to 20–30% of those detected by mammography in women ages 50 and older [41,52,53]. Data from the Surveillance Epidemiology and End Results (SEER) program depict over a 300% increase in DCIS from 1983 to 1992 for women ages 40 years and older, with the greatest number of DCIS cases detected in women ages 50 and older [62]. In 1992, there were an estimated 23,438 cases of DCIS compared with 4901 in 1983 [62,74]. Because the vast majority of DCIS is nonpalpable and, therefore, detected by screening mammography, the increase use of mammography is the primary reason for the increased incidence of DCIS [75].

Given that the natural history of DCIS is unknown, particularly the natural history of mammographically detected DCIS, the clinical dilemma lies in not being able to distinguish which lesions will progress to invasive cancer. This results in the vast majority of women with DCIS receiving some surgical treatment to prevent progression of the relatively few DCIS lesions

that have the potential to progress to invasive cancer. Almost all women who have DCIS detected are currently treated either by mastectomy or lumpectomy with or without radiation, with less than 3% receiving no treatment [62,74]. In 1992, it was estimated that 41.3% (or a total of 2529) of DCIS cases in women ages 70 and older in the U.S. were treated by mastectomy, which was similar to the 45.3% of DCIS cases treated by mastectomy in women ages 40 to 49 years (2250 mastectomies) [62]. Mortality from breast cancer is low among women diagnosed with DCIS. Only 1.0–3.4% die of breast cancer within 8 to 10 years of diagnosis [73,76,77]. Whether the low risk of death from breast cancer is due to effective treatment or the fact that the majority of DCIS are relatively benign is not known. Thus, screening mammography may be benefiting some women whose DCIS would become invasive cancer. However, it is potentially harming other women whose DCIS would never become invasive cancer, who, for lack of good prognostic indicators, are almost always treated surgically. Whether or not detection of DCIS by mammography averts breast cancer deaths is unknown.

C. False Reassurance

Of 100 women ages 40 to 49 years with invasive breast cancer, about 22 will go undetected by screening mammography, compared with 9 of 100 women ages 50 to 59 years with invasive cancer (Table 70.3). This means that potentially 22 women ages 40 to 49 years with invasive breast cancer will be told their screening examination is normal and may be falsely reassured that they do not have breast cancer and therefore will not seek medical attention for breast symptoms. Women who have a normal result and do not have breast cancer also may be reassured by having a normal screening examination that they do not have breast cancer. For example, the annual risk of invasive breast cancer for a 40 year old woman is about 1 in 625 [54]. Having a normal screening examination decreases her risk to about 1 in 2500 [78]. Although the very low risk of breast cancer after a normal screening examination may reassure women that they do not have breast cancer, the risk of breast cancer *before* mammography is already quite low. The need for reassurance from mammography might not be necessary if women understood that the risk of breast cancer prior to mammography is already very low [79].

V. Cost of Screening Mammography

High volume screening mammography programs (greater than 20 to 35 mammograms per day) offer screening examinations at $60–75 per screen. Facilities that operate at a lower utilization (less than six a day) need to charge at least $100 to break even and usually charge $120 to $250 to make a profit [81]. There has been an explosion of dedicated mammography machines in the U.S., with a projected 10,000 machines installed as of 1990, a 369% surplus [80]. Based on guidelines that recommend mammography screening for women ages 50 to 69 years every 1 to 2 years, the demand for mammography only requires 2600 mammography machines. Clinicians should refer patients to accredited high volume mammography programs with well-trained and experienced personnel to ensure that patients undergo high quality mammography at a low cost.

Table 70.5

Frequency of Mammographic Results in a First Screened Population and Risk of Breast Cancer Based on Mammographic Result

Mammographic result	Frequency[a]	Risk of breast cancer[a]	Likelihood ratio[b]
Normal or benign	87–93%	0.05–0.1%	0.1
Probably benign	1–2%	0.3–2%	NA
Incomplete, need additional imaging evaluation	4–8%	2–10%	7
Suspicious	0.3–1.4%	10–55%	125
Malignant	0.1%	60–100%	2200

[a]Data from University of California Mobile Mammography Screening Program, 1985–1992 (adapted from [45], *JAMA*, 1993, **270**, 2444–2450, copyright 1993–1996. American Medical Association) and the New Mexico Mammography Project, 1991–1993 (adapted from [60], *Cancer* **78**, 1996, 1731–1739. Copyright © 1996 American Cancer Society. Reprinted by permission of Wiley-Liss, Inc., a subsidiary of John Wiley & Sons, Inc.).

[b]Data from University of California Mobile Mammography Screening Program, 1985–1991 (adapted from [79], *JAMA*, 1996, **276**, 39–43, copyright 1993–1996. American Medical Association). Likelihood ratios are the ratio of diseased to nondiseased persons for a given test result.

VI. Interpreting Mammographic Results

The most common (and most worrisome) mammographic abnormalities are masses and calcifications. Radiologists generally describe both masses and calcifications in terms of location, size, and other characteristics (such as shape, borders, pattern). In addition to describing findings, radiologists should make an *assessment and recommendation* [81]. The American College of Radiology recommends one of six assessments for interpretation of a screening mammographic examination (Table 70.5). Good quality mammography facilities generally call 5–10% of all *screening* films abnormal. A higher rate should be considered unacceptable because it results in a large number of healthy women undergoing additional diagnostic evaluation.

Likelihood ratios[3] associated with screening mammography interpreted as "suspicious" and "malignant" are associated with a substantial increase in the risk of breast cancer, irrespective of age (Table 70.5) [78]. However, these interpretations only account for about 2% of all abnormal mammographic results.

VII. Efficacy and Accuracy of Clinical Breast Exam

There are no studies that compare the effectiveness of CBE alone compared to screening mammography or CBE alone to no screening. Although mammography plus CBE will detect more breast cancer than clinical breast examination alone [82], mammography plus CBE does not decrease breast cancer mortality beyond the reduction achieved by mammography alone for women ages 50 to 69 years [10]. In the Canadian trial of

---

[3]Likelihood ratios are the ratio of diseased to nondiseased persons for a given test result.

women ages 50 to 59 years [83], mammography and CBE did not decrease breast cancer mortality beyond the reduction achieved by CBE alone. These results can be interpreted in one of two ways: mammography failed to decrease breast cancer mortality or CBE was just as effective as mammography plus CBE, suggesting that mammography adds little to decreasing breast cancer mortality if CBE is performed by a trained practitioner. The fact that the number of node-positive tumors was similar in the mammography plus CBE group and CBE alone group [83] and that the sensitivity of mammography in the Canadian trial was similar to or better than that in other randomized controlled trials [26] would suggest that mammography did not fail to decrease breast cancer mortality. Rather, when performed by skilled practitioners, clinical breast exam can be as efficacious as mammography among older women.

It has been suggested that two-modality screening may be more advantageous in women ages 40 to 49 years because clinical breast examination is the primary mode of cancer detection in 24% of two-modality screened cases compared with 12% for older women [82]. There are only two clinical trials that screened women ages 40 to 49 years with mammography plus CBE and the results are conflicting. The HIP trial showed a nonsignificant 23% reduction in breast cancer mortality 10 years after screening began [28] and the Canadian trial showed a nonsignificant 14% increase in breast cancer mortality [16]. Five Swedish randomized controlled trials (Table 70.1) used only mammography and four of five showed a trend toward a reduction in breast cancer mortality 10 years after the initiation of screening. As with women ages 50 and older, it does not appear that mammography plus CBE decreases breast cancer mortality beyond the reduction achieved by mammography alone.

The sensitivity of CBE is highest for lesions 1.0 cm or larger (87–88%) and lower for lesions smaller than 1.0 cm (34–55%). Compared with mammography, the sensitivity of CBE (57–83%) and the PPV of CBE (1.5–4.0%) are lower [84]. As with any screening test, there are potential harms. The rate of false positive CBEs is highest for women ages 40 to 49 years (6%) who undergo screening CBE and declines with age to 3.5% for women ages 50 to 59 years, 2.5% for women ages 60 to 69 years, and 2.2% for women ages 70 to 79 years [70]. If a woman elects to undergo annual CBE over 10 years (*i.e.,* ten screening CBEs in 10 years), she has a 13% chance of having at least one false-positive examination that will require a diagnostic work-up and an estimated cumulative risk of a false-positive CBE of 22% [70]. The risk of having at least one biopsy as a result of a false-positive test is 6% after ten CBEs [70]. CBE alone may be most advantageous for women ages 70 and older because CBE results in a lower number of breast biopsies (0.5%) per woman screened compared with mammography (1.0–2.5%) and in lower rates of detection of DCIS. However, in proficient hands CBE may detect a high proportion of clinically important lesions that impact on breast cancer mortality [83].

VIII. Efficacy and Accuracy of Breast Self-Exam

BSE has an overall sensitivity of 26%, which decreases with age from 41% for women ages 35 to 39 years old to 21% for women ages 60 to 74 years [85]. In the United Kingdom Trial of Early Detection of Breast Cancer, a nonrandomized community trial, there was no reduction in breast cancer mortality in the BSE communities compared to communities that did not perform BSE [86]. There have been two randomized controlled trials, one in Leningrad of women ages 40 to 64 years [87] and one in Shanghai of women ages 31 to 64 years [88], that have directly tested the effectiveness of BSE to reduce breast cancer mortality. In the Leningrad trial, all women also underwent yearly CBE. After 8 years of follow-up, the Leningrad study reported no difference in the number of breast cancers diagnosed in the BSE group versus the control group and no difference with regard to the size of primary tumors or incidence of metastasis or regional lymph nodes [89]. The Shanghai study reported similar results after 5 years of follow-up. In addition, the authors reported no difference in breast cancer mortality between the BSE trained group and the control group [89]. However, in both studies [87,88] there were increases in physician visits, referrals for further diagnostic evaluations, and excisional biopsies among women in the BSE group compared with those in the control group. Both studies reported a twofold greater number of benign breast lesions identified in the BSE group than in the control group, and the Leningrad study reported 50% more excisional biopsies in the BSE group.

In summary, although observational studies have shown that women who report regular practice of BSE present with smaller tumors than women who do not practice BSE, when the gold standard for evaluation of screening, the randomized controlled trial, is used to determine the efficacy of BSE, there is no substantial stage shift from late to earlier stage disease in women who report regular practice of BSE and no reduction in breast cancer mortality. In addition, from the Leningrad trial it appears that BSE plus CBE does not result in a reduction in breast cancer mortality compared with women who only receive annual CBE. Finally, performing BSE results in additional physician visits and diagnostic tests without identifying more breast cancers or earlier stage disease.

IX. Breast Cancer Screening Recommendations

Baseline mammography at age 35 is no longer recommended by any organization. The question is, at what age should routine screening mammography start? There is disagreement concerning whether the potential mortality benefit of mammography screening in women ages 40 to 49 years outweighs the known associated harms, which is reflected in differing guidelines about what age routine screening mammography should start. The USPSTF [41] and the American College of Physicians, as well as most European countries, recommend mammography every 1–2 years for women beginning at age 50. A National Institutes of Health (NIH) consensus panel concluded in January of 1997 that the data currently available do not warrant a universal recommendation for mammography for women in their forties, but that women should be fully informed of the risks and benefits of screening [90]. This view has been supported by others who have suggested that because there is uncertainty as to whether the potential benefits of screening on a regular basis outweigh the known harms for women ages 40 to 49 years, the choice to screen or not screen is a woman's personal decision [91–94]. On the other hand, the American Cancer Society recommends that mammography screening begin at age 40 and be

performed annually [95]. The National Cancer Institute (NCI) recommended that women ages 40 to 49 years undergo screening every 1–2 years and that women at higher risk of breast cancer discuss with their physicians not only how frequently they should be screened, but whether they should begin screening *before* age 40 [96]. At the same time, the NCI did recommend that women be fully informed of the benefits and risks of mammography before deciding to undergo screening. Given these conflicting recommendations and the lack of compelling evidence that the benefits of mammography screening outweigh the known risks for women ages 40 to 49 years, women considering mammography screening should be informed of the potential benefits, harms, and limitations of the test (see Section X Informed Decision Making for Breast Cancer Screening) so that they can make informed decisions based on their personal risk status and utility for the associated harms and potential benefits of screening.

It has been suggested that individualized risk be assessed and used to determine who should undergo screening mammography and that women at higher than average risk for breast cancer, such as those who have a family history of breast cancer, should start screening at an earlier age [97]. Unfortunately, there are no data to support this recommendation. All available evidence shows that the accuracy and efficacy of screening mammography is primarily related to age and the biology of disease, not to the absolute risk of disease [10–14,26,27,36,41,52,53]. For example, the sensitivity of mammography is similar or slightly lower for women younger than 50 with a family history of breast cancer compared to women who do not have a family history [36], (*i.e.,* the sensitivity of mammography is not higher just because women with a family history are at higher risk of disease). Although studies confirming a benefit from screening in young women with a family history of a first-degree relative with breast cancer are lacking, recommendations for screening such women have been made on other grounds, including a high burden of suffering (increased risk of disease and possibly death from breast cancer) and a PPV of mammography similar to that of women ages 50 to 69 years. There is no evidence to recommend for or against CBE alone in women ages 40 to 49 years.

For women ages 50 to 69 years there is universal agreement that they should undergo screening mammography every 1–2 years. There is no evidence that screening annually is more effective than screening biennially, but it is more costly. The USPSTF recommends that screening mammography be performed with or without CBE because there is no evidence that CBE increases the benefit beyond that achieved by mammography alone. The age at which to stop performing routine screening mammography has not been determined. A decision regarding screening women beyond age 69 years should be based on a woman's general health, the presence of comorbid conditions, and her willingness to undergo additional tests to find and treat breast lesions that may have no impact on her mortality.

The USPSTF leaves screening with BSE up to the discretion of the practitioner and woman because there is no evidence to suggest that performing BSE decreases breast cancer mortality. However, the American Cancer Society recommends routine BSE starting at age 20.

Recommendations for breast cancer screening by four organizations are summarized in Table 70.6. Opinions of organizations

Table 70.6
Recommendations for Breast Cancer Screening

Screening modality	U.S. Preventive Services Task Force (USPSTF)	Canadian Task Force on the Period Health Exam	American Cancer Scoiety (ACS)	National Cancer Institute (NCI)
	Frequency of screening			
Mammography				
40 to 49	0[a]	0[a]	Annual	Every 1 to 2 years
50 to 69	Every 1 to 2 years	Annual	Annual	Every 1 to 2 years
70 and older[b]	0[a]	0[a]	Annual	Every 1 to 2 years
Clinical breast exam				
40 to 49	0[a]	0[a]	Annual	Every 1 to 2 years
50 to 69	Optional every 1 to 2 years with mammography	Annual	Annual	Every 1 to 2 years
Breast self-exam	0[a]	0[a]	Every month[c]	No recommendation

[a]0 = Does not recommend for or against screening.

[b]USPSTF recommends that women ages 70 and older who have a reasonable life expectancy may consider screening past age 69 years. ACS and NCI recommend screening women ages 40 years and older with no upper age limit on when to stop screening.

[c]ACS recommends monthly breast self-examination starting at age 20 years without an upper age limit of when to stop screening.

differ because the process by which organizations establish guidelines varies. The USPSTF and Canadian Task Force on the Periodic Health Exam adhere to a high standard of evidence in recommending guidelines for screening and develop such guidelines by conducting an exhaustive review of the literature with explicit linking of the quality of the data to the strength of the recommendation. The American Cancer Society and NIH recommendations are primarily established by group consensus based on expert opinion.

X. Informed Decision Making for Screening

All women who request or are offered screening mammography should be informed of the chance of an abnormal result, the chance of a false-positive examination, the chance of undergoing a breast biopsy, the chance of finding breast cancer, and their age-specific risk of breast cancer (Table 70.4). Women should also be informed of the available evidence that screening mammography reduces breast cancer mortality for women in their age group. This is especially important for (1) women ages 40 to 49 years because the absolute benefit of screening mammography is small, and (2) women ages 70 and older in whom the benefits of screening mammography may not outweigh the costs of screening and treatment of early lesions that may have no impact on mortality. In addition, women should be informed that (1) BSE has a very low sensitivity to detect breast cancer, does not decrease breast cancer mortality, and may result in additional diagnostic procedures including breast biopsies, and that (2) CBE alone has a lower sensitivity than mammography to detect breast cancer, there is insufficient evidence to determine whether CBE decreases breast cancer mortality, and CBE may result in additional diagnostic procedures. Health practitioners need to assist women in understanding what factors might influence their choice to undergo or not undergo screening, such as their attitude toward pain, risk, and inconvenience,

and assist them in understanding the potential benefits, harms, and limitations of mammography, CBE, and BSE [93].

Acknowledgments

This work was supported by an NCI-funded Breast Cancer Surveillance Consortium cooperative agreement (1 U01 CA 63740) and NCI-funded Breast Cancer SPORE grant (P50 CA58207).

References

1. Huang, Z., Hankinson, S. E., Colditz, G. A., et al. (1997). Dual effects of weight and weight gain on breast cancer risk. JAMA, J. Am. Med. Assoc. **278**, 1407–1411.
2. Swanson, C. A., Coates, R. J., Malone, K. E., et al. (1997). Alcohol consumption and breast cancer risk among women under age 45 years. Epidemiology **8**, 231–237.
3. Bernstein, L., Henderson, B. E., Hanisch, R., Sullivan-Halley, J., and Ross, R. K. (1994). Physical exercise and reduced risk of breast cancer in young women. J. Natl. Cancer Inst. **86**, 1403–1408.
4. Thune, I., Brenn, T., Lund, E., and Gaard, M. (1997). Physical activity and the risk of breast cancer. N. Engl. J. Med. **336**, 1269–1275.
5. Smith-Warner, S. A., Spiegelman, D., Yaun, S. S., et al. (1998). Alcohol and breast cancer in women. JAMA, J. Am. Med. Assoc. **279**, 535–540.
6. Freudenheim, J. L., Marshall, J. R., Vena, J. E., et al. (1997). Lactation history and breast cancer risk. Am. J. Epidemiol. **146**, 932–938.
7. Michels, K. B., Willett, W. C., Rosner, B. A., et al. (1996). Prospective assessment of breastfeeding and breast cancer incidence among 89,887 women. Lancet **347**, 431–436.
8. Newcomb, P. A., Storer, B. E., Longnecker, M. P., et al. (1994). Lactation and a reduced risk of premenopausal breast cancer. N. Engl. J. Med. **330**, 81–87.
9. Beral, V. (1997). Breast cancer and hormone replacement therapy: Collaborative reanalysis of data from 51 epidemiological studies of 52 705 women with breast cancer and 108 411 women without breast cancer. Lancet **350**, 1047–1059.
10. Kerlikowske, K., Grady, D., Rubin, S. M., Sandrock, C., and Ernster, V. (1995). Efficacy of screening mammography: a meta-analysis. JAMA, J. Am. Med. Assoc. **273**, 149–154.

11. Kerlikowske, K. (1997). Efficacy of screening mammography among women aged 40 to 49 years and 50 to 69 years: Comparison or relative and absolute benefit. *J. Natl. Cancer Inst.* **22,** 79–86.

12. Elwood, J. M., Cox, B., and Richardson, A. K. (1993). The effectiveness of breast cancer screening by mammography in younger women. *Online J. Curr. Clin. Trials (Ser. Online)* **2,** Doc. No. 32.

13. Glasziou, P. P., Woodward, A. J., and Mahon, C. M. (1995). Mammographic screening trials for women aged under 50: A quality assessment and meta-analysis. *Med. J. Aust.* **162,** 625–629.

14. Glasziou, P., and Irwig, L. (1997). The quality and interpretation of mammographic screening trials for women ages 40–49. *J. Natl. Cancer Inst.* **22,** 73–77.

15. Bjurstam, N., Björneld, L., Duffy, S. W., *et al.* (1997). The Gothenberg Breast Screening Trial. *Cancer* **80,** 2091–2099.

16. Miller, A. B., To, T., Baines, C. J., and Wall, C. (1997). The Canadian National Breast Screening Study: Update on breast cancer mortality. *J. Natl. Cancer Inst.* **22,** 37–41.

17. Kopans, D. B., Halpern, E., and Hulka, C. A. (1994). Statistical power in breast cancer screening trials and mortality reduction among women 40–49 with particular emphasis on the National Breast Screening Study of Canada. *Cancer (Philadephia)* **74,** 1196–1203.

18. Narod, S. A. (1997). On being the right size: a reappraisal of mammography trials in Canada and Sweden. *Lancet* **349,** 1846.

19. Miller, A. B., Baines, C. J., To, T., and Wall, C. (1992). Canadian National Breast Screening Study: Breast cancer detection and death rates among women aged 40 to 49 years. *Can. Med. Assoc. J.* **147,** 1459–1476.

20. Kerlikowske, K. M., Grady, D., and Ernster, V. (1995). Benefit of mammography screening in women ages 40 to 49 years: Current evidence from randomized controlled trials. *Cancer (Philadelphia)* **76,** 1679–1680.

21. Berry, D. A. (1998). Benefits and risks of screening mammography for women in their forties: A statistical appraisal. *J. Natl. Cancer Inst.* **90,** 1431–1439.

22. Tabár, L., Larsson, L.-G., Andersson, I., *et al.* (1996). Breast cancer screening with mammography in women aged 40–49 years: Report of the Organizing Committee and Collaborators, Falun Meeting, Sweden, March 21–22, 1996. *Int. J. Cancer.* **68,** 693–699.

23. Bjurstam, N., Björneld, L., Duffy, S. W., *et al.* (1997). The Gothenburg Breast Cancer Screening Trial: Preliminary results on breast cancer mortality for women aged 39–49. *J. Natl. Cancer Inst.* **22,** 53–55.

24. Salzmann, P., Kerlikowske, K., and Phillips, K. (1997). Cost-effectiveness of extending screening mammography guidelines to include women 40 to 49 years of age. *Ann. Intern. Med.* **127,** 955–965.

25. Esserman, L., and Kerlikowske, K. (1996). Should we recommend screening mammography for women aged 40 to 49 years. *Oncology* **10,** 357–364.

26. Fletcher, S. W., Black, W., Harris, R., Rimer, B., and Shapiro, S. (1993). Report of the international workshop on screening for breast cancer. *J. Natl. Cancer Inst.* **85,** 1644–1656.

27. Nystrom, L. *et al.* (1993). Breast cancer screening with mammography: Overview of the Swedish randomized trials. *Lancet* **341,** 973–978.

28. Shapiro, S. (1988). "Periodic Screening for Breast Cancer: The Health Insurance Plan Project and its Sequelae, 1963–1986." Johns Hopkins University Press, Baltimore, MD.

29. Chen, H. H., Tabar, L., Fagerberg, G., and Duffy, S. W. (1995). Effect of breast cancer screening after age 65. *J. Med. Screen.* **2,** 10–14.

30. Mandelblatt, J. S., Wheat, M. E., Monane, M., Moshief, R. D., Hollengerg, J. P., and Tang, J. (1992). Breast cancer screening for elderly women with and without comorbid conditions. *Ann. Intern. Med.* **116,** 722–730.

31. Satariano, W. A., and Ragland, D. R. (1994). The effect of comorbidity on 3-year survival of women with primary breast cancer. *Ann. Intern. Med.* **120,** 104–111.

32. Welch, H. G., and Fisher, E. S. (1998). Diagnostic testing following screening mammography in the elderly. *J. Natl. Cancer Inst.* **90,** 389–392.

33. Smith-Bindman, R., and Kerlikowske, K. (1998). Is there a downside to elderly women undergoing screening mammography? *J. Natl. Cancer Inst.* **90,** 1322–1323.

34. Moskowitz, M. (1986). Breast cancer: Age-specific growth rates and screening strategies. *Radiology* **161,** 37–41.

35. Tabar, L., Faberberg, G., Day, N. E., *et al.* (1987). What is the optimum interval between mammographic screening examinations? An analysis based on the latest results of the Swedish two-county breast cancer screening trial. *Br. J. Cancer* **55,** 547–551.

36. Kerlikowske, K., Grady, D., Barclay, J., Sickles, E. A., and Ernster, V. (1996). Effect of age, breast density, and family history on the sensitivity of first screening mammography. *JAMA, J. Am. Med. Assoc.* **276,** 33–38.

37. Peer, P. G. M., Holland, R., Hendriks, J. H. C. L., Mravunac, M., and Verbeek, A. L. M. (1994). Age-specific effectiveness of the Nijmegen population-based breast cancer-screening program: Assessment of early indicators of screening effectiveness. *J. Natl. Cancer Inst.* **86,** 436–441.

38. Peer, P. G. M., Verbeek, A. L. M., Mravunac, M., Hendriks, J. H. C. L., and Holland, R. (1996). Prognosis of younger and older patients with early breast cancer. *Br. J. Cancer* **73,** 382–385.

39. Threlfall, A. G., Woodman, C. B. J., and Prior, P. (1997). Breast screening programme: Should the interval between tests depend on age? *Lancet* **349,** 472.

40. Asbury, D., Boggis, C. R. M., Sheals, D., Threlfall, A. G., and Woodman, C. B. J. (1996). NHS breast screening programme: is the high incidence of interval cancers inevitable? *Lancet* **313,** 1369–1370.

41. U.S. Preventive Services Task Force (1996) "Guide to Clinical Preventive Services," 2nd ed. Williams & Wilkins, Baltimore, MD.

42. Harris, J. R., Lippman, M. E., Veronesi, U., and Willett, W. (1992). Breast cancer. *N. Engl. J. Med.* **327,** 319–328.

43. Calle, E. E., Martin, L. M., Thun, M. J., Miracle, H. L., and Heath, C. W. (1993). Family history, age, and risk of fatal breast cancer. *Am. J. Epidemiol.* **138,** 675–681.

44. Slattery, M. L., Berry, D., and Kerber, R. A. (1993). Is survival among women diagnosed with breast cancer influenced by family history of breast cancer? *Epidemiology* **4,** 543–548.

45. Kerlikowske, K., Grady, D., Barclay, J., Sickles, E. A., Eaton, A., and Ernster, V. (1993). Positive predictive value of screening mammography by age and family history of breast cancer. *JAMA, J. Am. Med. Assoc.* **270,** 2444–2450.

46. Laya, M. B., Larson, E. B., Taplin, S. H., and White, E. (1996). Effect of estrogen replacement therapy on the specificity and sensitivity of screening mammography. *J. Natl. Cancer Inst.* **88,** 643–649.

47. Thurfjell, E. L., Holmberg, L. H., and Persson, I. R. (1997). Screening mammography: Sensitivity and specificity in relation to hormone replacement therapy. *Radiology* **203,** 339–341.

48. Leung, W., Goldberg, F., Zee, B., and Sterns, E. (1997). Mammographic density in women on postmenopausal hormone replacement therapy. *Surgery* **122,** 669–674.

49. Schairer, C., Byrne, C., Keyl, P. M., Brinton, L. A., Sturgeon, S. R., and Hoover, R. N. (1994). Menopausal estrogen and estrogen-progestin replacement therapy and risk of breast cancer (United States). *Cancer Causes Control* **5,** 491–500.

50. Sellers, T. A., Mink, P. J., Cerhan, J. R., *et al.* (1997). The role of hormone replacement therapy in the risk for breast cancer and total

mortality in women with a family history of breast cancer. *Ann. Intern. Med.* **127,** 973–980.

51. Bonnier, P., Romain, S., Giacalone, P. L., Laffargue, F., Martin, P. M., and Piana, L. (1995). Clinical and biologic prognostic factors in breast cancer diagnosed during postmenopausal hormone replacement therapy. *Obstet. Gynecol.* **85,** 11–17.

52. Kerlikowske, K., and Barclay, J. (1997). Outcomes of modern screening mammography. *J. Natl. Cancer Inst.* **22,** 105–111.

53. May, D. S., Lee, N. C., Nadel, M. R., Henson, R. M., and Miller, D. S. (1998). The National Breast and Cervical Cancer Early Detection Program: Report on the first 4 years of mammography provided to medically underserved women. *AJR, Am. J. Roentgenol.* **170,** 97–104.

54. Kosary, C. L., Ries, L. A. G., Miller, B. A., Hankey, B. F., Harras, A., and Edwards, B. K. (1995). "SEER Cancer Statistics Review, 1973–1992: Tables and Graphs," NIH Publ. No. 96-2789. National Cancer Institute, Bethesda, MD.

55. Burhenne, H. J., Burhenne, L. W., Goldberg, F., *et al.* (1994). Interval breast cancers in the screening mammography program of British Columbia: Analysis and classification. *AJR, Am. J. Roentgenol.* **162,** 1067–1071.

56. Bird, R. E. (1989). Low-cost screening mammography: Report on finances and review of 21,716 consecutive cases. *Radiology* **171,** 87–90.

57. Linver, M. N., Paster, S. B., Rosenberg, R. D., Key, C. R., Stidley, C. A., and King, W. V. (1992). Improvement in mammography interpretation skills in a community radiology practice after dedicated teaching courses: 2-year medical audit of 38,633 cases. *Radiology* **184,** 39–43.

58. Robertson, C. L. (1993). A private breast imaging practice: Medical audit of 25,788 screening and 1,077 diagnostic examinations. *Radiology* **187,** 75–79.

59. Sienko, D. G., Hahn, R. A., Mills, E. M., *et al.* (1993). Mammography use and outcomes in a community: The Greater Lansing Area Mammography Study. *Cancer* **71,** 1801–1809.

60. Rosenberg, R. D., Lando, J. F., Hunt, W. C., *et al.* (1996). The New Mexico mammography project: Screening mammography performance in Albuquerque, New Mexico, 1991 to 1993. *Cancer* **78,** 1731–1739.

61. Tabar, L., Fagerberg, G., Chen, H. *et al.* (1995). Efficacy of breast cancer screening by age: New results from the Swedish two-county trial. *Cancer* **75,** 2507–2517.

62. Ernster, V. L., Barclay, J., Kerlikowske, K., Grady, D., and Henderson, I. C. (1996). Incidence of and treatment for ductal carcinoma in situ of the breast. *JAMA, J. Am. Med. Assoc.* **275,** 913–918.

63. Kornguth, P. J., Keefe, F. J., and Conaway, M. R. (1996). Pain during mammography: Characteristics and relationship to demographic and medical variables. *Pain* **66,** 187–194.

64. Chang, S. W., Kerlikowske, K., Napoles-Springer, A., Posner, S. F., Sickles, E. A., and Pérez-Stable, E. J. (1996). Racial differences in timeliness of follow-up after abnormal screening mammography. *Cancer* **78,** 1395–1402.

65. Dixon, J., Chetty, U., and Forrest, A. (1988). Wound infection after breast biopsy. *Br. J. Surg.* **75,** 918–919.

66. Lerman, C., Tock, B., Rimer, B., *et al.* (1991). Psychological and behavioral implications of abnormal mammograms. *Ann. Intern. Med.* **114,** 657–661.

67. Cockburn, J., Staples, M., Hurley, S. F., *et al.* (1994). Psychological consequences of screening mammography. *J. Med. Screen.* **1,** 7–12.

68. Ellman, R., Angeli, N., Christians, A., *et al.* (1989). Psychiatric morbidity associated with screening for breast cancer. *Br. J. Cancer* **60,** 781–784.

69. Gram, I. T., Lund, H. E., and Slenker, S. E. (1990). Quality of life following a false positive mammogram. *Br. J. Cancer* **62,** 1018–1022.

70. Elmore, J. G., Barton, M. B., Moceri, V. M., and Fletcher, S. W. (1998). Ten-year risk of false positive screening mammograms and clinical breast examinations. *N. Engl. J. Med.* **338,** 1089–1096.

71. Page, D. L., Dupont, W. D., Rogers, L. W., *et al.* (1982). Intraductal carcinoma of the breast: Follow-up after biopsy only. *Cancer* **49,** 751–758.

72. Page, D. L., Dupont, W. D., Rogers, L. W., Jensen, R. A., and Schuyler, P. A. (1995). Continued local recurrence of carcinoma 15–25 years after a diagnosis of low grade ductal carcinoma in situ of the breast treated only by biopsy. *Cancer* **76,** 1197–1200.

73. Fisher, B., Dignam, J., Wolmark, N., *et al.* (1998). Lumpectomy and radiation therapy for the treatment of intraductal breast cancer: Findings from national surgical adjuvant breast and bowel project B-17. *J. Clin. Oncol.* **16,** 441–452.

74. Ernster, V. L., and Barclay, J. (1997). Increases in ductal carcinoma in situ (DCIS) of the breast in relation to mammography: A dilemma. *J. Natl. Cancer Inst.* **22,** 151–156.

75. White, E., Lee, C. Y., and Kristal, A. R. (1990). Evaluation of the increase in breast cancer incidence in relation to mammography use. *J. Natl. Cancer Inst.* **82,** 1546–1552.

76. Bradley, S. J., Weaver, D. W., and Bouwman, D. L. (1990). Alternative in the surgical management of in situ breast cancer: A meta-analysis of outcome. *Am. Surg.* **56,** 428–432.

77. Ernster, V. L., Barclay, J., Kerlikowske, K., Wilkie, H., and Ballard-Barbash, R. (1999). Mortality among women with ductal carcinoma in situ of the breast in the population-based SEER program. *Arch. Intern. Med.* In press.

78. Kerlikowske, K., Grady, D., Barclay, J., Sickles, E. A., and Ernster, V. (1996). Likelihood ratios for modern screening mammography. *JAMA, J. Am. Med. Assoc.* **276,** 39–43.

79. Black, W. C., Nease, R. F., and Tosteson, A. N. A. (1995). Perceptions of breast cancer risk and screening effectiveness in women younger than 50 years of age. *J. Natl. Cancer Inst.* **87,** 720–731.

80. Brown, M. L., Kessler, L. G., and Rueter, F. G. (1990). Is the supply of mammography machines outstripping need and demand? *Ann. Intern. Med.* **113,** 547–552.

81. Olson, L. K. (1993). Interpreting the mammogram report. *Am. Fam. Physician* **47,** 396–403.

82. Baines, C. J. (1997). Mammography versus clinical breast examination. *J. Natl. Cancer Inst.* **22,** 125–129.

83. Miller, A. B., Baines, C. J., To, T., and Wall, C. (1992). Canadian National Breast Screening Study: 2. Breast cancer detection and death rates among women aged 50 to 59 years. *Can. Med. Assoc. J.* **147,** 1477–1488.

84. Baines, C. J., Miller, A. B., and Bassett, A. A. (1989). Physical examination: Its roles as a single screening modality in the Canadian National Breast Screening Study. *Cancer* **63,** 1816–1822.

85. O'Malley, M. S., and Fletcher, S. W. (1987). Screening for breast cancer with breast self-examination: A critical review. *JAMA, J. Am. Med. Assoc.* **257,** 2196–2203.

86. United Kingdom Trial of Early Detection of Breast Cancer Group (1993). Breast cancer mortality after 10 years in the UK Trial of Early Detection of Breast Cancer. *Breast* **2,** 13–20.

87. Semiglazov, V. F., Moiseyenko, V. M., Bavli, J. L., *et al.* (1992). The role of breast self-examination in early breast cancer detection (results of the 5-years USSR/WHO randomized study in Leningrad). *Eur. J. Epidemiol.* **8,** 498–502.

88. Thomas, D. B., Gao, D. L., Self, S. G., *et al.* (1997). Randomized trial of breast self-examination in Shanghai: Methodology and preliminary results. *J. Natl. Cancer Inst.* **89,** 355–365.

89. Semiglazov, V. F., Sagaidak, V. N., Moiseyenko, V. M., and Mikhailov, E. A. (1993). Study of the role of breast self-examination in the reduction of mortality from breast cancer. *Eur. J. Cancer* **29A,** 2039–2046.

90. National Institutes of Health Consensus Development Conference Panel (1997). National Institutes of Health Consensus Development Conference Statement: Breast cancer screening for women ages 40–49, January 21–23, 1997. *J. Natl. Cancer Inst.* **89,** 1015–1020.

91. Woolf, S. H., and Lawrence, R. S. (1997). Preserving scientific debate and patient choice: Lessons from the consensus panel on mammography screening. *JAMA, J. Am. Med. Assoc.* **278,** 2105–2108.

92. Pauker, S. G., and Kassirer, J. P. (1997). Contentious screening decisions: Does the choice matter? *N. Engl. J. Med.* **336,** 1243–1244.

93. Ernster, V. L. (1997). Mammography screening for women aged 40 through 49—a guidelines saga and a clarion call for informed decision making. *Am. J. Public Health* **87,** 1103–1106.

94. Eddy, D. M. (1997). Breast cancer screening in women younger than 50 years of age: What's next? *Ann. Intern. Med.* **127,** 1035–1036.

95. Mettlin, C., and Smart, C. R. (1994). Breast cancer detection guidelines for women aged 40 to 49 years: Rationale for the American Cancer Society reaffirmation of recommendations. *Ca—Cancer J. Clin.* **44,** 248–255.

96. National Cancer Advisory Board (1997). NCAB endorses mammograms for "average risk" women 40–49; screening schedules may vary among individuals. *Cancer Lett.* **23,** 4–7.

97. Gail, M., and Rimer, B. (1998). Risk-based recommendations for mammographic screening for women in their forties. *J. Clin. Oncol.* **16,** 3105–3114.

71

Ovarian Cancer

REBECCA TROISI AND PATRICIA HARTGE
Division of Cancer Epidemiology and Genetics
National Cancer Institute
National Institutes of Health
Bethesda, Maryland

I. Introduction

Ovarian cancer poses a significant threat to women's health, especially in the western hemisphere. A white woman in the United States has a 2% chance of developing ovarian cancer and a 1% chance of dying from it [1]. Some patterns of risk for ovarian cancer parallel those for breast and endometrial cancer, but many do not. Existing theories for the development of ovarian cancer include epithelial trauma from repeated ovulations, exposure to elevated gonadotropin levels, and exposure to exogenous carcinogenic agents that enter the peritoneal cavity via the vagina. Extensive investigation of the epidemiology of this disease has provided a consistent picture of the major risk factors but has not yet revealed the fundamental etiology. Examination of the genetics, cytology, and pathology of ovarian cancer has shed light on the natural history but has not yet illuminated the sequence of changes from earliest genetic abnormality to overt disease. Detection of some early lesions by ultrasound or by serum assays has raised the prospect of screening but has not yet been established as effective in experimental trials. At this writing, the investigation of ovarian cancer proceeds on many fronts. Investigators hope for a convergence of findings in the near future that will substantially advance our understanding of this feared disease.

II. Biology of Ovarian Cancer

The ovaries are almond-sized organs covered with a thin layer of coelomic epithelium. Cancers may arise in any of the three basic cell types of the ovary: the surface epithelium, the germ cells, and the stroma [2]. Germ cell tumors are rare, occur at young ages, and include dysgerminomas, teratomas, and choriocarcinomas. About 90% of ovarian cancers are epithelial, and are further characterized as serous, endometrioid, or mucinous. Some evidence of etiologic heterogeneity among the epithelial cancer cell types has been presented, but the differences are uncertain and the epithelial cancers typically are grouped together in epidemiologic investigation. Similarly, borderline tumors, or cancers of low malignant potential, exhibit distinct clinical behavior but not distinct etiology and are generally included with invasive disease.

After menarche, each ovary typically ovulates on alternating months. With ovulation, follicular and luteal cysts occur commonly and appear not to affect risk of carcinoma. Microscopic inclusion cysts have been suggested to give rise to cancer but the evidence is conflicting [3].

At menopause, ovarian function ceases and the organs decrease in size. The age-specific risk of ovarian cancer parallels, but lags behind, the age-specific measures of normal ovarian function. Women with ovarian cancer typically come to diagnosis with nonspecific symptoms such as constipation, urinary frequency, or pelvic pressure [4]. Transvaginal [5] and abdominal ultrasound and serum level of the cancer antigen CA 125 [6] are among the tests used to make a diagnosis. By extension, these techniques are being evaluated as screening methods for early asymptomatic ovarian cancer.

III. Ovarian Cancer in Populations

A. Age-Specific Incidence and Mortality in U.S. Black and White Women

The age-adjusted incidence of ovarian cancer is 15 cases per 100,000 woman-years in the United States [1]. Incidence rates are low before the age of 40, increase exponentially until age 65, and plateau thereafter (Fig. 71.1). The age-adjusted incidence rates are 10.2/100,000 for women younger than 65 years of age and 57.9/100,000 for women 65 years of age and older. Mortality rates follow a similar pattern. The age-adjusted mortality rates are 7.8/100,000 for all ages, 3.9/100,000 for women younger than 65 years of age, and 43.6/100,000 for women 65 years of age and older. The overall five-year survival rate is 46.4% but is highly dependent on age and stage. Five-year survival rates are higher among younger women (78.1% for women <45 years of age vs 23.5% for women 75+), and for localized (92.6%) compared with distant (25.3%) disease. Most disease (57%), however, is diagnosed as distant rather than localized (24%) or regional (13%) stage. The stage distribution also depends on age, with younger women more likely to be diagnosed with earlier stage and older women with later stage disease.

Black women in the United States face somewhat lower risks of developing or dying from ovarian cancer (Fig. 71.1). The lifetime risk of being diagnosed with ovarian cancer is 1.86% among whites and only 1.15% among blacks, and risk of dying from ovarian cancer is 1.21% and 0.74%, respectively. The age-adjusted incidence rate in blacks is 10.9/100,000 compared with 15.6/100,000 in whites, and age-adjusted mortality rates are 6.6/100,000 and 8.0/100,000, respectively. Survival is similar in black and white women younger than 50 years of age but is slightly worse in older black women than in older white women (27.6% vs 36.6%). The stage distribution at diagnosis is similar for whites and blacks.

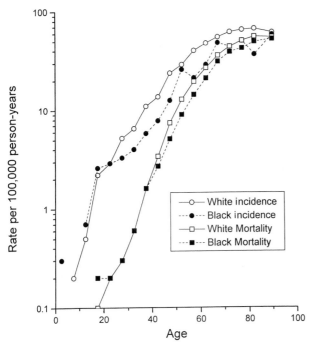

Fig. 71.1 Age-specific ovarian cancer incidence and mortality rates in U.S. black and white women, 1990–1994.

B. International Variation

Ovarian cancer develops two to three times more often in northern European women than in Japanese women, as shown in Figure 71.2. Incidence rates are highest in the northern European countries, U.S. whites, Canada, Israel and New Zealand's non-Maori population. Rates in the Far East, the Caribbean, and Central and South America rank lowest. Such variation suggests lifestyle effects, but these effects are less marked than those for breast cancer.

C. Time Trends in Age-Specific Incidence and Mortality

Figure 71.3 presents time trends in the U.S. age-adjusted ovarian cancer mortality rates by race. Overall, mortality rates have decreased slightly since the late 1960s. White mortality rates have been consistently higher than nonwhite rates for decades. Mortality rates converged slightly in the 1950s with rates among nonwhites approaching those of whites. Since the 1950s, time trends by race have appeared to be similar, although an increase among blacks in recent years is suggested.

IV. Influences on Individual Risk

A. Genetics and Host Influences

Inherited ovarian cancers may account for 3–5% of all cases, with a larger fraction among the younger cases. Case-control studies suggest a three- or fourfold risk of ovarian cancer in women who have a first-degree relative diagnosed with ovarian cancer [7]. In families with multiple ovarian cancers, breast cancers also occur more often than expected. In addition, primary ovarian cancer develops more often than expected in women who have survived breast cancer [8].

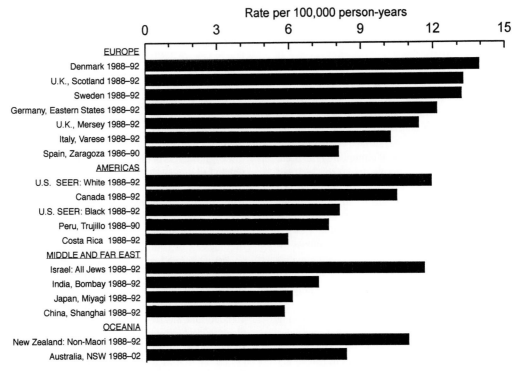

Fig. 71.2 International variation in ovarian cancer incidence rates by continent.

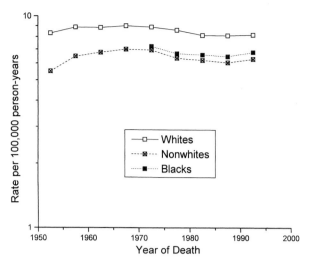

Fig. 71.3 Age-adjusted ovarian cancer mortality rates in the U.S., by race, 1950–1954 to 1990–1994.

In a fraction of families with unusual numbers of breast and ovarian cancers, perhaps 50%, it is possible to detect a mutation in BRCA1 [9] or BRCA2 [10]. Among women who carry a mutation in either of the genes, the risk of developing ovarian cancer is estimated to be 17% [11]. The proportion of disease associated with BRCA1 or BRCA2 is greater at younger ages, much as other genetically related cancers are most evident at younger ages. Some preliminary investigations suggest that the relative effect of a mutation on breast versus ovarian cancer may vary between BRCA1 and BRCA2 or vary depending on the locus of the mutation, but no firm conclusions can yet be drawn.

B. Reproductive Factors

Table 71.1 describes selected case-control and cohort studies of ovarian cancer and Table 71.2 summarizes the literature on the associations of the major reproductive factors with ovarian cancer risk. Several reproductive factors have been firmly associated with a reduced risk of ovarian cancer. Women who have had a full-term pregnancy have a lower risk of ovarian cancer than nulliparous women, and each subsequent birth confers additional protection [7]. In the analysis of combined U.S. case-control studies [7], risk of ovarian cancer was reduced by 13–19% with each live birth. The effect of pregnancy to reduce risk of ovarian cancer is independent of age at which a woman's first pregnancy occurs [7]. Incomplete pregnancies, due to spontaneous or induced abortion, may also decrease risk, although the protection afforded appears to be less than that associated with full-term pregnancies [7]. Another factor associated with a lower risk of ovarian cancer is oral contraceptive pill use. In virtually all studies, women who have used oral contraceptive pills are at lower risk of ovarian cancer than women who have not used them, and risk decreases by approximately 10% for every year of use [38]. Parous and nulliparous women probably experience similar protection from oral contraceptives [38,39] and are protected for at least 10 years after cessation of use [38]. Several types of gynecologic surgery also decrease risk of ovar-

ian cancer. Women who have had a tubal ligation [7,40–43] or hysterectomy (with preservation of at least one ovary) [7,15,40–43] have a lower risk of ovarian cancer, and the reduction in risk remains up to 15 years or more after surgery [41,43].

Other reproductive factors either do not appear to be related to ovarian cancer risk or are inconclusive. Most studies indicate that risk of ovarian cancer, unlike breast cancer, does not increase with an early age at menarche [7,23] or a late age at first birth [7,23]. Regularity of menstrual cycles does not appear to affect risk [18,19,26], and, though less studied, risk does not seem to vary by cycle length [18], amenorrhea [18], or premenstrual symptoms (*e.g.,* irritability, breast tenderness, edema) [19]. Women who have breastfed are at lower risk of ovarian cancer in studies conducted in the U.S. [7] and Australia [31], but most studies in other low-risk countries show no association with breastfeeding [17,19,30].

A relationship of infertility with ovarian cancer risk seems likely but has not been firmly established or clearly explained. Risk has been shown to increase with increasing number of contraceptive-free years of marriage [7,47] and is slightly elevated among nulligravida, but not gravid women who have a physician's diagnosis of infertility [7]. In the analysis of the combined U.S. case-control studies [7], risk was elevated among infertile women reporting ovulatory abnormalities, but not among women with other or unspecified types of infertility, compared with women reporting no history of infertility. The elevated risk, however, was restricted to women with a history of infertility who used fertility drugs. Studies of cancer risk in populations of infertile women [32,34,35,48] have generally had better information than case-control studies on type of infertility but have lacked sufficient numbers of ovarian cancer cases to assess risk. In a U.S. study of infertile women [35], no overall increase in ovarian cancer risk was observed, although an increase in risk was suggested for women with progesterone deficiency/oligoovulation/anovulation compared with other causes of infertility. In studies conducted in Israel [34,37], no overall association was found with infertility, but infertile women with adequate estrogen and progesterone production had a higher than expected frequency of ovarian cancer compared with population rates. In an Australian cohort of women referred for in-vitro fertilization (IVF) [33], risk associated with unexplained infertility was nearly twenty times that of infertility due to known causes after taking into account whether women were treated with fertility drugs. The results, however, were based on a small number of cases. Thus, it is possible that only certain types of infertility are associated with an excess risk of ovarian cancer.

Even if higher risk among infertile women is likely, the relationship of fertility drugs and ovarian cancer risk is inconclusive at present. Combined data from three U.S. case-control studies with information on fertility drug use revealed an increased risk of ovarian cancer among the infertile women who had used fertility drugs [7]. While risk of ovarian cancer was similar among fertile and infertile women, infertile women who had used fertility drugs had three times the risk of women lacking a history of infertility. Most of the increase in risk associated with fertility drug use was observed among infertile women who remained nulligravid. In a cohort study of U.S. infertile women [32], prolonged use of clomiphene citrate, a common fertility drug, was associated with 11 times the risk of ovarian cancer

Table 71.1

Selected Epidemiologic Studies of Ovarian Cancer

Geographic area	Years	Subjects	Age
Case-control studies[a]			
London and Oxford, U.K. [12]	1978–1983	235/451 (hospital)	<65
Milan, Italy [13,14]	1983–1987	634/1,626 (hospital)	22–74
	1979–1980	161/561 (hospital)	19–69
Athens, Greece [15,16]	1989–1991	189/200 (hospital visitors)	<75
	1980–1981	150/250 (hospital)	all
Hokkaido, Japan [17]	1980–1981	110/220 (hospital, outpatient)	all
	1985–1986		
Shanghai, China [18]	1984–1986	229/229 (population)	18–70
Beijing, China [19]	1984–1986	112/224 (community)	all
U.S. Collaborative Study [20]	1956–1986	2,197/8,893 (hospital, community)	all
WHO Collaborative Study [21]	1979–1986	368/2,397 (hospital)	all
(Australia, Chile, China, Israel,			
Mexico, Philippines, Thailand)			
Cohort studies[b]			
Norway [22]	1953–1991	1,694/1,145,076	20–56
U.S. Nurses' Health Study [23]	1976–1988	260/121,700	30–67

[a]Subjects: # cases/# controls (type).
[b]Subjects: # cases/# subjects.

Table 71.2

The Associations of Several Major Reproductive Factors with Ovarian Cancer Risk

Factors	Comparison	Association	Comments
Age at menarche	Early vs late	None	No association [7,14,15,23,24] Inverse association [18,25] Positive association [19]
Menstrual patterns	Irregular vs regular periods	None	No association [18,19,26] Inverse association [13]
Parity	Parous vs nulliparous	Reduced risk	Inverse association [7,15,17–19,22,23,27]
	Number of children	Inverse trend	[7]
Age at first birth	Late vs early	None	No association [7,19,23] Positive association [15,27 (no trend)] Inverse association (among uniparous women only) [22]
Ovulatory years	Increasing number	Increased risk	Positive association [14,17,19,26,28,29]
Breastfeeding	Ever vs never	Inconclusive	No association Japanese [17], Chinese [19], WHO International studies [30] Inverse association U.S. [7], Australian studies [31]
Infertility	History vs no history	Inconclusive	Positive association with ovulatory abnormalities [7] (but no increase overall in women without drug use) [32], unexplained infertility [33], and nonhormonal infertility (suggested) [34] No association [35–37]
	Treated vs untreated	Inconclusive	Positive association [7,32] No association [18,33,34,36]
Oral contraceptive use	Ever vs never	Reduced risk	Inverse association [7,23(≥5 years of use),25,38,39] No association [15,19]
	Number of years	Inverse trend	[38]
Tubal ligation	Ever vs never	Reduced risk	Inverse association [7,40–43]
Hysterectomy	Ever vs never	Reduced risk	Inverse association [7,15,40–43]
Age at natural menopause	Late vs early	Inconclusive	No association [7,18,19,23] Positive association [15,24 (suggested)]
Noncontraceptive estrogen use	Ever vs never	Inconclusive	No association [7,16,44] Positive association [12,45,46 (fatal cancer)]

compared with no use. This association held regardless of gravidity, although use of human chorionic gonadotropin (hCG) was not associated with risk. In the cohort of Australian women referred for IVF [33], risk was similar for women who had undergone IVF compared with those who had not. In extended follow-up of the Israeli cohort [37], ovarian cancer risk among women treated with clomiphene was slightly elevated but not statistically different from risk among women who were not treated. Other studies also have not found an association between fertility drugs and risk of ovarian cancer [18,36]. All of these studies have lacked sufficient numbers of ovarian cancer cases that were exposed to fertility drugs and details of fertility drug use. Further research is underway to determine whether ovarian cancer risk is elevated in women who have taken fertility drugs [49].

In the age-specific incidence data, risk declines after menopause. This pattern might suggest that late natural menopause is associated with an increased risk of ovarian cancer, although, in general, the epidemiologic data are not supportive [7,23]. Similarly, most studies of noncontraceptive hormone use and ovarian cancer have not provided strong evidence of an association [7].

C. Nonreproductive Factors

Several potential nonreproductive risk factors for ovarian cancer have been assessed, although no strong or consistent associations have emerged.

1. Diet

An influence of dietary intake on ovarian cancer risk has been proposed but has not been extensively evaluated [50]. Cramer and colleagues hypothesized that galactose consumption in combination with low levels of its key metabolic enzyme, galactose-1-phosphate uridyltransferase, may be involved in ovarian pathogenesis by increasing gonadotropin levels [51]. They showed that women who had lower enzyme activity and higher consumption of galactose were at higher risk of ovarian cancer, but several other studies have failed to confirm this finding [52,53]. Current research is assessing whether diets high in total or saturated fat may increase risk of ovarian cancer, possibly through a hormonal mechanism [50,54–56]. Positive associations with animal fat intake were demonstrated in several [54–56], but not in all [57,58], studies. While effects of other types of fat, macronutrients, and cholesterol have been investigated, the data available are too scant to conclude whether they are related to risk [50,57–59]. There is some evidence that diets high in fruits and vegetables [50,54,55,57] and in carotenoids such as beta-carotene [58] decrease ovarian cancer risk, but more studies are needed.

Various, largely inconsistent, associations have been reported for coffee and alcohol consumption, which are studied more easily than other dietary factors. Moderate to heavy coffee consumption has been associated with an increased risk of ovarian cancer in some studies [60], but most studies indicate no increase in risk [15,16,56,57,61,62]. A positive association with alcohol consumption has been suggested in some studies [16,63,64] but has not been confirmed in others [15,56,57,60,65].

2. Body Size

The relationship of body size with ovarian cancer risk is quite unclear. Results from the combined U.S. studies [7] found conflicting results that depended on the source of controls; relative weight was inversely associated with risk when cases were compared with hospital controls and positively associated with risk when population controls were used. Additional studies show some increase in risk with increasing levels of weight adjusted for height [29,66], but other data indicate no association [16,18,58]. One study found that risk decreased with increasing weight after adjustment for height [57]. Height does not appear to be related to risk [16,18,26].

3. Smoking

Smoking does not appear to be an important risk factor for ovarian cancer [15,16,56,57,60,63].

4. Talc

An excess risk of ovarian cancer mortality observed among occupational cohorts of women exposed to asbestos (reviewed in Weiss *et al.* [67]) raised concern over whether talc exposure might also increase risk. Talc and asbestos share certain physical properties, and talc is sometimes contaminated with asbestos particles. Talc applied to undergarments, sanitary napkins, diaphragms, or directly to the perineal area may subsequently migrate to the peritoneal cavity and ovaries. Most [12,19,41, 60,68–70], but not all [71,72], studies show an increase in risk of ovarian cancer with talc use. Risk appears to be associated primarily with the direct application of talc to the perineum and not with other sources of exposure. Trends with duration [41,60,69], frequency [60,70], and age at first use [41,70] have generally not been observed, however, raising doubt about whether the association with talc is real.

5. Antidepressant and Analgesic Drugs

Antidepressants have been speculated to increase ovarian cancer risk through possible effects on estrogen and gonadotropin levels. An early Greek study [16] found that women with ovarian cancer were more likely to have used psychotropic drugs than were controls, but a subsequent study of Greek women [72] failed to confirm those findings. A U.S. case-control study [73] found that self-reported prior use for 1–6 months of both antidepressants and benzodiazepine tranquilizers was associated with an increased risk. Elevated risks were confined to first use that occurred before age 50 and that occurred at least 10 years prior to diagnosis. These findings were corroborated in another case-control study [74] that showed that the association was primarily found with psychotropic drugs operating through dopaminergic, or gabaergic mechanisms rather than through sertoninergic pathways.

A U.S. study [75] found a significant decrease in ovarian cancer risk among women who used paracetamol (acetaminophen) and the effect was most marked for frequent and long-term use. Risk associated with aspirin use was also reduced, although the reduction was not statistically significant, and no association was found for ibuprofen or prescribed analgesics. Some findings suggest that the effect of paracetamol on ovarian cancer risk may be mediated through lowered gonadotropin levels [76].

6. Mumps

Mumps and other viruses typically encountered in childhood appear not to be related to risk. The mumps virus has been investigated because of its potential to damage the ovaries and cause premature menopause [77]. Several early studies suggested that history of mumps infection was less common in women with ovarian cancer [78] and that subclinical cases of mumps were more likely [79]. Cramer [77] reported that postmenopausal, but not premenopausal, women with ovarian cancer were less likely to have a history of mumps. Other studies, however, have found little difference in history of mumps infection between cases and controls as assessed by self-report [63,80,81] and by serological evidence [82].

7. Occupational and Other Environmental Exposures

Women employed as hairdressers or cosmetologists may have a small excess risk of ovarian cancer [83]. An elevated ovarian cancer risk among women who frequently dye their hair was also noted in a recent Greek study [72] but not in an earlier one [16].

While data on atomic bomb survivors demonstrate some excess risk of ovarian cancer in women who were less than 50 at the time of exposure, studies seem to indicate that there should be little concern over the amount of radiation to which women are commonly exposed (reviewed in Weiss *et al.* [67]). Likewise, women occupationally exposed to high levels of asbestos through employment in the assembly of gas masks or in an asbestos factory have been shown to have an excess risk of ovarian cancer, but exposure to lower levels is probably not an important determinant of risk [67].

D. Possible Common Pathways

The etiology of ovarian cancer is not established but several theories have been offered. The "incessant ovulation hypothesis," first suggested by Fathalla [84], posits that epithelial neoplasms result from repeated ovulation, causing minor trauma to the ovarian epithelium and predisposing it to malignant transformation. By this theory, periods of anovulation should be protective. Indeed, risk of ovarian cancer is lower in parous women and women who have used oral contraceptives, and has been shown, with some exceptions, to increase with various markers of ovulatory time (roughly calculated as age at menopause less age at menarche, and total time pregnant, breastfeeding, or using oral contraceptives) [14,17,19,26,28,29]. However, the magnitude of diminished risk associated with the individual components of the composite measure of anovulatory time appears to be greater than that predicted solely by anovulatory periods, suggesting that other mechanisms may also be important [85]. The ovulation hypothesis predicts that women who ovulate infrequently should have a decreased risk of ovarian cancer; however, either no difference or *increased* risks of ovarian cancer have been observed among infertile women in previous studies.

One proposed mechanism to explain the protection afforded by tubal ligation and hysterectomy is the prevention of ovarian exposure to exogenous carcinogenic agents (*e.g.*, talc) that enter the peritoneal cavity through the vagina. Another explanation is that surgery compromises blood flow to the ovaries, leading to involution, fewer cell divisions, and reduced risk. A third explanation is that surgery is accompanied by surveillance and early removal of at-risk ovaries [86].

A hypergonadotropic state (elevated gonadotropin levels) also has been suggested as an etiologic factor for ovarian cancer pathogenesis based on experimental studies in animals relating high gonadotropin production with ovarian tumors [87]. Cramer [88] proposed a model in which persistent gonadotropin stimulation of the ovary predisposes it to the development of malignancy. Few studies have assessed directly the relation of gonadotropin levels to risk, although in one study, *low* serum gonadotropin levels were associated with an increased risk of ovarian cancer [89]. Decreased risks associated with pregnancy, oral contraceptive use, and possibly breastfeeding—factors associated with lower gonadotropin levels—provide some support for this hypothesis. However, these factors are also associated with temporary cessation of ovulation and are consistent with the hypothesis that anovulatory periods are protective against risk. The gonadotropin hypothesis would predict that women with specific types of infertility associated with gonadotropin levels should have an altered risk, and women who took fertility drugs (which either stimulate endogenous gonadotropin release or are gonadotropins themselves) might also be at increased risk. As stated earlier, the effects of specific types of infertility and ovulation induction drugs on ovarian cancer risk are inconclusive as yet. A reduced risk of ovarian cancer would be expected with use of noncontraceptive exogenous hormones because the normally elevated gonadotropin levels associated with the peri- and postmenopausal period are decreased with use [90]. The data on noncontraceptive hormones at present, however, do not consistently support this hypothesis.

V. Future Research Directions

Primary prevention requires better understanding of the etiology of ovarian cancer. Three well-established epidemiologic observations suggest promising strategies for prevention and require further epidemiologic investigation in the near term. Oral contraceptive pills reduce risk, at least in the formulations most commonly used in past decades. A family history of ovarian cancer and, to a lesser extent, breast cancer identifies women at moderately increased risk. Presence of a mutation in BRCA1 or BRCA2 identifies women at markedly increased risk. Thus, one clear priority for epidemiologic research is to assess the effect of oral contraceptives in high-risk women. In addition, further epidemiologic and pharmacologic research is needed to formulate agents like oral contraceptives to reduce risk of ovarian cancer while not increasing risk of other diseases. Continuing genetic studies in women in high-risk families are needed to discover whether genes other than BRCA1 and BRCA2 can be linked to ovarian cancer. When the normal functions of BRCA1 and BRCA2 are known, the mechanisms that underlie the nongenetic risk factors (*e.g.*, repeated ovulation) also will be clearer. Thus, genetic epidemiology will raise many new hypotheses in the next few years. In general, numerous interdisciplinary and translational studies can be expected, even in the absence of a unified etiologic understanding.

Additional priorities for epidemiologic research are suggested by several new leads and several older, unresolved issues. The effects of infertility and of therapies used to treat infertility will require large and detailed studies. Resolution of the possible link to talcum powder and other physical contaminants in the reproductive tract may require data collected before diagnosis, that is, from cohorts, because the reported effects are modest, inconsistent, and uncertain because of possible recall errors. The investigation of hormonal influences will proceed partly in association with related work in breast and endometrial cancer, partly for practical reasons and partly because the similarities and differences among the three cancers will aid in the interpretation of measured hormonal effects. Even though few strong or clear effects of diet on ovarian cancer have emerged, diet will draw attention if a clean link to breast cancer is defined. Dietary effects and hormonal effects may be related, via body mass index, and various genetic polymorphisms influence metabolism of food and of hormones.

One of the major hypotheses for ovarian carcinogenesis involves persistent gonadotropin stimulation of the ovary, yet little work has been done in directly assessing the relation of gonadotropin levels to ovarian cancer risk. Research in this area has been hindered by the potential for "disease effects" whereby changes in hormone levels are a consequence of the tumor itself. This is particularly problematic for a disease in which a high proportion of cases are diagnosed at late stage. Ideally, hormone levels would be measured several years in advance of the advent of cancer. This may only be achieved in a prospective study with a very large number of subjects. Alternatively, hormone measurements could be made in women with a marker highly predictive of future ovarian cancer development. At present, however, such a marker has not been identified. In the meantime, studies are limited to factors associated with gonadotropin levels such as pregnancy and oral contraceptive use.

Secondary prevention, through detection of early cancers, seems particularly desirable because ovarian cancers are often detected late in the course of development when therapy is less effective. Ongoing trials of screening for ovarian cancer may reveal whether ultrasound and serum assays, alone or in combination [6], can reduce ovarian cancer mortality in postmenopausal women in general or in those at highest risk [91]. Various serum markers are being investigated as prognostic tools, most successfully CA-125 [92].

References

1. Ries, L. A. G., Kosary, C. L., Hankey, B. F., Miller, B. A., Harras, A., and Edwards, B. K., eds. (1997). "SEER Cancer Statistics Review, 1973–1994," NIH Publ. No. 97-2789. National Cancer Institute, Bethesda, MD.
2. Kurman, R. J., ed. (1994). "Blaustein's Pathology of the Female Genital Tract," 4th ed. Springer-Verlag, New York.
3. Scully, R. E. (1995). Pathology of ovarian cancer precursors. *J. Cell. Biochem., Suppl.* **23,** 208–218.
4. Gershenson, D. M., Tortolero-Luna, G., Malpica, A., Baker, V. V., Whittaker, L., Johnson, E., and Mitchell, M. F. (1996). Ovarian intraepithelial neoplasia and ovarian cancer. *Obstet. Gynecol. Clin. North Am.* **23,** 475–543.
5. Bourne, T. H., Campbell, S., Reynolds, K. M., Whitehead, M. I., Hampson, J., Royston, P., Crayford, T. J. B., and Collins, W. P. (1993). Screening for early familial ovarian cancer with transvaginal ultrasonography and colour blood flow imaging. *Br. Med. J.* **306,** 1025–1029.
6. Jacobs, I., Davies, A. P., Bridges, J., Stabile, I., Fay, T., Lower, A., Grudzinskas, J. G., and Oram, D. (1993). Prevalence screening for ovarian cancer in postmenopausal women by CA 125 measurement and ultrasonography. *Br. Med. J.* **306,** 1030–1034.
7. Whittemore, A. S., Harris, R., Itnyre, J., and the Collaborative Ovarian Cancer Group (1992). Characteristics relating to ovarian cancer risk: Collaborative analysis of 12 U.S. case-control studies. II. Invasive epithelial ovarian cancers in white women. *Am. J. Epidemiol.* **136,** 1184–1203.
8. Harvey, E. B., and Brinton, L. A. (1985). Second cancer following cancer of the breast in Connecticut, 1935–82. *In* "Multiple Primary Cancers in Connecticut and Denmark" (J. D. Boice, R. E. Curtis, R. A. Kleinerman, H. H. Storm, O. M. Jensen, H. S. Jensen, J. T. Flannery, and J. F. Fraumeni, eds.), NIH Publ. No. 85-2714. National Cancer Institute, Bethesda, MD.
9. Miki, Y., Swensen, J., Shattuck-Eidens, D., *et al.* (1994). Isolation of BRCA1, the 17q-linked breast and ovarian cancer susceptibility gene. *Science* **266,** 66–71.
10. Wooster, R., Neuhausen, S. L., Mangion, J., Quirk, Y., Ford, D., Collins, N., Nguyen, K., Seal, S., Tran, T., and Averill, D. (1994). Localization of a breast cancer susceptibility gene, BRCA2, to chromosome 13q12-13. *Science* **265,** 2088–2090.
11. Struewing, J. P., Hartge, P., Wacholder, S., Baker, S. M., Berlin, M., McAdams, M., Timmerman, M. M., Brody, L. C., and Tucker, M. A. (1997). The risk of cancer associated with specific mutations of BRCA1 and BRCA2 among Ashkenazi Jews. *N. Engl. J. Med.* **336,** 1401–1408.
12. Booth, M., Beral, V., and Smith, P. (1989). Risk factors for ovarian cancer: A case-control study. *Br. J. Cancer* **60,** 592–598.
13. Parazzini, F., La Vecchia, C., Negri, E., and Gentile, A. (1989). Menstrual factors and the risk of epithelial ovarian cancer. *J. Clin. Epidemiol.* **42,** 443–448.
14. Franceschi, S., La Vecchia, C., Helmrich, S. P., Mangioni, C., and Tognoni, G. (1982). Risk factors for epithelial ovarian cancer in Italy. *Am. J. Epidemiol.* **115,** 714–719.
15. Polychronopoulou, A., Tzonou, A., Hsieh, C.-C., Kaprinis, G., Rebelakos, A., Toupadaki, N., and Trichopoulos, D. (1993). Reproductive variables, tobacco, ethanol, coffee and somatometry as risk factors for ovarian cancer. *Int. J. Cancer* **55,** 402–407.
16. Tzonou, A., Day, N. E., Trichopoulos, D., Walker, A., Saliaraki, M., Papapostolou, M., and Polychronopoulou, A. (1984). The epidemiology of ovarian cancer in Greece: A case-control study. *Eur. J. Cancer Clin. Oncol.* **20,** 1045–1052.
17. Mori, M., Harabuchi, I., Miyake, H., Casagrande, J. T., Henderson, B. E., and Ross, R. K. (1988). Reproductive, genetic and dietary risk factors for ovarian cancer. *Am. J. Epidemiol.* **128,** 771–777.
18. Shu, X.O., Brinton, L. A., Gao, Y. T., and Yuan, J. M. (1989). Population-based case-control study of ovarian cancer in Shanghai. *Cancer Res.* **49,** 3670–3674.
19. Chen, Y., Wu, P.-C., Lang, J.-H., Ge, W.-J., Hartge, P., and Brinton, L. A. (1992). Risk factors for epithelial ovarian cancer in Beijing, China. *Int. J. Epidemiol.* **21,** 23–29.
20. Whittemore, A. S., Harris, R., Intyre, J., Halpern, J., and the Collaborative Ovarian Cancer Group (1992). Characteristics relating to ovarian cancer risk: Collaborative analysis of 12 U.S. case-control studies. I. Methods. *Am. J. Epidemiol.* **136,** 1175–1183.
21. The WHO Collaborative Study of Neoplasia and Steroid Contraceptives (1989). Epithelial ovarian cancer and combined oral contraceptives. *Int. J. Epidemiol.* **18,** 538–545.
22. Albrektsen, G., Heuch, I., and Kvåle, G. (1996). Reproductive factors and incidence of epithelial ovarian cancer: A Norwegian prospective study. *Cancer Causes Control* **7,** 421–427.
23. Hankinson, S. E., Colditz, G. A., Hunter, D. J., Willett, W. C., Stampfer, M. J., Rosner, B., Hennekens, C. H., and Speizer, F. E.

(1995). A prospective study of reproductive factors and risk of epithelial ovarian cancer. *Cancer (Philadelphia)* **76,** 284–290.

24. Franceschi, S., La Vecchia, C., Booth, M., Tzonou, A., Negri, E., Parazzini, F., Trichopoulos, D., and Beral, V. (1991). Pooled analysis of 3 European case-control studies of ovarian cancer: II. Age at menarche and at menopause. *Int. J. Cancer* **49,** 57–60.

25. Tavani, A., Negri, E., Franceschi, S., Parazzini, F., and La Vecchia, C. (1993). Risk factors for epithelial ovarian cancer in women under age 45. *Eur. J. Cancer* **29A,** 1297–1301.

26. Hildreth, N. G., Kelsey, J. L., LiVolsi, V. A., Fischer, D. B., Holford, T. R., Mostow, E. D., Schwartz, P. E., and White, C. (1981). An epidemiologic study of epithelial carcinoma of the ovary. *Am. J. Epidemiol.* **114,** 398–405.

27. Negri, E., Franceschi, S., Tzonou, A., Booth, M., La Vecchia, C., Parazzini, F., Beral, V., Boyle, P., and Trichopoulos, D. (1991). Pooled analysis of 3 European case-control studies of ovarian cancer: I. Reproductive factors and risk of epithelial ovarian cancer. *Int. J. Cancer* **49,** 50–56.

28. Wu, M. L., Whittemore, A. S., Paffenbarger, R. S., Sarles, D. L., Kampert, J. B., Grosser, S., Jung, D. L., Ballon, S., Hendrickson, M., and Mohle-Boetani, J. (1988). Personal and Environmental characteristics related to epithelial ovarian cancer. I. Reproductive and menstrual events and oral contraceptive use. *Am. J. Epidemiol.* **128,** 1216–1227.

29. Casagrande, J. T., Pike, M. C., Ross, R. K., Louise, E. W., Roy, S., and Henderson, B. E. (1979). "Incessant ovulation" and ovarian cancer. *Lancet,* July 28, pp. 170–172.

30. Rosenblatt, K. A., Thomas, D. B., and The World Health Organization Collaborative Study of Neoplasia and Steroid Contraceptives (1993). Lactation and the risk of epithelial ovarian cancer. *Int. J. Epidemiol.* **22,** 192–197.

31. Siskind, V., Green, A., Bain, C., and Purdie, D. (1997). Breastfeeding, menopause and epithelial ovarian cancer. *Epidemiology* **8,** 188–191.

32. Rossing, M. A., Daling, J. R., Weiss, N. S., Moore, D. E., and Self, S. G. (1994). Ovarian tumors in a cohort of infertile women. *N. Engl. J. Med.* **331,** 771–776.

33. Venn, A., Watson, L., Lumley, J., Giles, G., King, C., and Healy, D. (1995). Breast and ovarian cancer incidence after infertility and in vitro fertilisation. *Lancet* **346,** 995–1000.

34. Ron, E., Lunenfeld, B., Menczer, J., Blumstein, T., Katz, L., Oelsner, G., and Serr, D. (1987). Cancer incidence in a cohort of infertile women. *Am. J. Epidemiol.* **125,** 780–790.

35. Brinton, L. A., Melton, J., Malkasian, G. D., Bond, A., and Hoover, R. (1989). Cancer risk after evaluation for infertility. *Am. J. Epidemiol.* **129,** 712–722.

36. Franceschi, S., La Vecchia, C., Negri, E., Guarneri, S., Montella, M., Conti, E., and Parazzini, F. (1994). Fertility drugs and risk of epithelial ovarian cancer in Italy. *Hum. Reprod.* **9,** 1673–1675.

37. Modan, B., Ron, E., Lerner-Geva, L., Blumstein, T., Menczer, J., Rabinovici, J., Oelsner, G., Freedman, L., Mashiach, S., and Lunenfeld, B. (1998). Cancer incidence in a cohort of infertile women. *Am. J. Epidemiol.* **147,** 1038–1042.

38. Hankinson, S. E., Colditz, G. A., Hunter, D. J., Spencer, T. L., Rosner, B., and Stampfer, M. J. (1992). A quantitative assessment of oral contraceptive use and risk of ovarian cancer. *Obstet. Gynecol.* **80,** 708–714.

39. Rosenberg, L., Palmer, J. R., Zauber, A. G., Warshauer, M. E., Lewis, J. L., Strom, B. L., Harlap, S., and Shapiro, S. (1994). A case-control study of oral contraceptive use and invasive epithelial ovarian cancer. *Am. J. Epidemiol.* **139,** 654–661.

40. Rosenblatt, K. A., Thomas, D. B., and The World Health Organization Collaborative Study of Neoplasia and Steroid Contraceptives (1996). Reduced risk of ovarian cancer in women with a tubal li-

gation or hysterectomy. *Cancer Epidemiol., Biomarkers Prev.* **5,** 933–935.

41. Green, A., Purdie, D., Bain, C., Siskind, V., Russell, P., Quinn, M., Ward, B., and the Survey of Women's Health Study Group (1997). Tubal sterilisation, hysterectomy and decreased risk of ovarian cancer. *Int. J. Cancer* **71,** 948–951.

42. Krieger, N., Sloan, M., Cotterchio, M., and Parsons, P. (1997). Surgical procedures associated with risk of ovarian cancer. *Int. J. Epidemiol.* **26,** 710–715.

43. Hankinson, S. E., Hunter, D. J., Colditz, G. A., Willett, W. C., Stampfer, M. J., Rosner, B., Hennekens, C. H., and Speizer, F. E. (1993). Tubal ligation, hysterectomy, and risk of ovarian cancer. A prospective study. *J. Am. Med. Assoc.* **270,** 2813–2818.

44. Annegers, J. F., O'Fallon, W., and Kurland, L. T. (1977). Exogenous oestrogens and ovarian cancer. *Lancet,* October 22, pp. 869–870.

45. Hoover, R., Gray, L. A., and Fraumeni, J. F. (1977). Stilboestrol (diethylstillbestrol) and the risk of ovarian cancer. *Lancet,* September 10, pp. 533–534.

46. Rodriguez, C., Calle, E. E., Coates, R. J., Miracle-McMahill, H. L., Thun, M. J., and Health, C. W. (1995). Estrogen replacement therapy and fatal ovarian cancer. *Am. J. Epidemiol.* **141,** 828–835.

47. Nasca, P. C., Greenwald, P., Chorost, S., Richart, R., and Caputo, T. (1984). An epidemiologic case-control study of ovarian cancer and reproductive factors. *Am. J. Epidemiol.* **119,** 705–713.

48. Coulam, C. B., Annegers, J. F., and Kranz, J. S. (1983). Chronic anovulation syndrome and associated neoplasia. *Obstet. Gynecol.* **61,** 403–407.

49. Kaufman, S. C., Spirtas R., and Alexander, N. J. (1995). Do fertility drugs cause ovarian tumors? *J. Women's Health* **4,** 247–258.

50. World Cancer Research Fund/ American Institute for Cancer Research (1997). "Food, Nutrition and the Prevention of Cancer: A Global Perspective." American Institute for Cancer Research, Washington, DC.

51. Cramer, D. W., Harlow, B. L., Willett, W. C., Welch, W. R., Bell, D. A., Scully, R. E., Ng, W. G., and Knapp, R. C. (1989). Galactose consumption and metabolism in relation to the risk of ovarian cancer. *Lancet,* July 8, pp. 66–71.

52. Mettlin, C. J., and Piver, M. S. (1990). A case-control study of milk-drinking and ovarian cancer risk. *Am. J. Epidemiol.* **132,** 871–876.

53. Herrinton, L. J., Weiss, N. S., Beresford, S. A. A., Stanford, J. L., Wolfla, D. M., Feng, Z., and Scott, C. R. (1995). Lactose and galactose intake and metabolism in relation to the risk of epithelial ovarian cancer. *Am. J. Epidemiol.* **141,** 407–416.

54. La Vecchia, C., Decarli, A., Negri, E., Parazzini, F., Gentile, A., Cecchetti, G., Fasoli, M., and Franceschi, S. (1987). Dietary factors and the risk of epithelial ovarian cancer. *J. Natl. Cancer Inst.* **79,** 663–669.

55. Shu, X. O., Gao, Y. T., Yuan, J. M., Ziegler, R. G., and Brinton, L. A. (1989). Dietary factors and epithelial ovarian cancer. *Br. J. Cancer* **59,** 92–96.

56. Cramer, D. W., Welch, W. R., Hutchison, G. B., Willett, W., and Scully, R. E. (1984). Dietary animal fat in relation to ovarian cancer risk. *Obstet. Gynecol.* **63,** 833–838.

57. Byers, T., Marshall, J., Graham, S., Mettlin, C., and Swanson, M. (1983). A case-control study of dietary and nondietary factors in ovarian cancer. *J. Natl. Cancer Inst.* **71,** 681–686.

58. Slattery, M. L., Schuman, K. L., West, D. W., French, T. K., and Robison, L. M. (1989). Nutrient intake and ovarian cancer. *Am. J. Epidemiol.* **130,** 497–502.

59. Tzonou, A, Hsieh, C. C., Polychronopoulou, A., Kaprinis, G., Toupadaki, N., Trichopoulou, A., Karakatsani, A., and Trichopoulos, D. (1993). Diet and ovarian cancer: A case-control study in Greece. *Int. J. Cancer* **55,** 411–414.

60. Whittemore, A. S., Wu, M. L., Paffenbarger, R. S., Sarles, D. L., Kampert, J. B., Grosser, S., Jung, D. L., Ballon, S., and Hendrickson, M. (1988). Personal and environmental characteristics related to epithelial ovarian cancer. II. Exposures to talcum powder, tobacco, alcohol, and coffee. *Am. J. Epidemiol.* **128**, 1228–1240.

61. Hartge, P., Lesher, L. P., McGowan, L., and Hoover, R. (1982). Coffee and ovarian cancer. *Int. J. Cancer* **30**, 531–532.

62. Leviton, A. (1990). Methylxanthine consumption and the risk of ovarian malignancy. *Cancer Lett.* **51**, 91–101.

63. Hartge, P., Schiffman, M. H., Hoover, R., McGowan, L., Lesher, L., and Norris, H. J. (1989). A case-control study of epithelial ovarian cancer. *Am. J. Obstet. Gynecol.* **161**, 10–16.

64. La Vecchia, C., Negri, E., Franceschi, S., Parazzini, F., Gentile, A., and Fasoli, M. (1992). Alcohol and epithelial ovarian cancer. *J. Clin. Epidemiol.* **45**, 1025–1030.

65. Gwinn, M. L., Webster, L. A., Lee, N. C., Layde, P. M., Rubin, G. L., and the Cancer and Steroid Hormone Study Group (1986). Alcohol consumption and ovarian cancer risk. *Am. J. Epidemiol.* **123**, 759–766.

66. Farrow, D. C., Weiss, N. S., Lyon, J. L., and Daling, J. R. (1989). Association of obesity and ovarian cancer in a case-control study. *Am. J. Epidemiol.* **129**, 1300–1304.

67. Weiss, N. S., Cook, L. S., Farrow, D. C., and Rosenblatt, K. A. (1996). Ovarian cancer. *In* "Cancer Epidemiology and Prevention" (D. Schottenfeld and J. F. Fraumeni, eds.), 2nd ed., pp 1040–1057. Oxford University Press, New York.

68. Cramer, D. W., Welch, W. R., Scully, R. E., and Wojciechowski, C. A. (1982). Ovarian cancer and talc. *Cancer (Philadelphia)* **50**, 372–376.

69. Cook, L. S., Kamb, M. L., and Weiss, N. S. (1997). Perineal powder exposure and the risk of ovarian cancer. *Am. J. Epidemiol.* **145**, 459–465.

70. Harlow, B. L., Cramer, D. W., Bell, D. A., and Welch, W. R. (1992). Perineal exposure to talc and ovarian cancer risk. *Obstet. Gynecol.* **80**, 19–26.

71. Hartge, P., Hoover, R., Lesher, L. P., and McGowan, L. (1983). Talc and ovarian cancer. *J. Am. Med. Assoc.* **250**, 1844.

72. Tzonou, A., Polychronopoulou, A., Hsieh, C.-C., Rebelakos, A., Karakatsani, A., and Trichopoulos, D. (1993). Hair dyes, analgesics, tranquilizers and perineal talc application as risk factors for ovarian cancer. *Int. J. Cancer* **55**, 408–410.

73. Harlow, B. L., and Cramer, D. W. (1995). Self-reported use of antidepressants or benzodiazepine tranquilizers and risk of epithelial ovarian cancer: Evidence from two combined case-control studies (Massachusetts, United States). *Cancer Causes Control* **6**, 130–134.

74. Harlow, B. L., Cramer, D. W., Baron, J. A., Titus-Ernstoff, L., and Greenberg, E. R. (1998). Psychotropic medication use and risk of epithelial ovarian cancer. *Cancer Epidemiol., Biomarkers Prev.* **7**, 697–702.

75. Cramer, D. W., Harlow, B. L., Titus-Ernstoff, L., Bohlke, K., Welch, W. R., and Greenberg, E. R. (1998). Over-the-counter analgesics and risk of ovarian cancer. *Lancet* **351**, 104–107.

76. Cramer, D. W., Liberman, R. F., Hornstein, M. D., McShane, P., Powers, D., Li, E. Y., and Barbieri, R. (1998). Basal hormone levels in women who use acetaminophen for menstrual pain. *Fertil. Steril.* **70**, 371–373.

77. Cramer, D. W., Welch, W. R., Cassells, S., and Scully, R. E. (1983). Mumps, menarche, menopause, and ovarian cancer. *Am. J. Obstet. Gynecol.* **147**, 1–6.

78. Newhouse, M. L., Pearson, R. M., Fullerton, J. M., Boesen, E. A. M., and Shannon, H. S. (1977). A case control study of carcinoma of the ovary. *Br. J. Prev. Soc. Med.* **31**, 148–153.

79. Menczer, J., Modan, M., Ranon, L., and Golan, A. (1979). Possible role of mumps virus in the etiology of ovarian cancer. *Cancer (Philadelphia)* **43**, 1375–1379.

80. McGowan, L., Parent, L., Lednar, W., and Norris, H. J. (1979). The woman at risk for developing ovarian cancer. *Gynecol. Oncol.* **7**, 325–344.

81. Schiffman, M. H., Hartge, P., Lesher, L. P., and McGowan, L. (1985). Mumps and postmenopausal ovarian cancer. *Am. J. Obstet. Gynecol.* **152**, 116–117.

82. Golan, A., Joosting, A. C. C., and Orchard, M. E. (1979). Mumps virus and ovarian cancer. *S. Afr. Med. J.* **56**, 18–20.

83. Boffetta, P., Andersen, A., Lynge, E., Barlow, L., and Pukkala, E. (1994). Employment as hairdresser and risk of ovarian cancer and non-hodgkin's lymphomas among women. *J. Occup. Environ. Med.* **36**, 61–65.

84. Fathalla, M. F. (1971). Incessant ovulation—A factor in ovarian neoplasia? *Lancet,* July 17, p. 163.

85. Risch, H. A., Weiss, N. S., Lyon, J. L., Daling, J. R., and Liff, J. M. (1983). Events of reproductive life and the incidence of epithelial ovarian cancer. *Am. J. Epidemiol.* **117**, 128–139.

86. Weiss, N. S., and Harlow, B. L. (1986.) Why does hysterectomy without bilateral oophorectomy influence the subsequent incidence of ovarian cancer? *Am. J. Epidemiol.* **124**, 856–858.

87. McGowan, L., and Davis, R. H. (1970). Intrasplenic ovarian grafts in Syrian hamsters and peritoneal fluid cellular distribution. *Proc. Soc. Exp. Biol. Med.* **134**, 507.

88. Cramer, D. W., and Welch, W. R. (1983). Determinants of ovarian cancer risk. II. Inferences regarding pathogenesis. *J. Natl. Cancer Inst.* **71**, 717–721.

89. Helzlsouer, K. J., Alberg, A. J., Gordon, G. B., Longcope, C., Bush, T. L., Hoffman, S. C., and Comstock, G. W. (1995). Serum gonadotropins and steroid hormones and the development of ovarian cancer. *J. Am. Med. Assoc.* **274**, 1926–1930.

90. Hartge, P., Hoover, R., McGowan, L., Lesher, L., and Norris, H. J. (1988). Menopause and ovarian cancer. *Am. J. Epidemiol.* **127**, 990–998.

91. National Institutes of Health Consensus Development Conference Statement (1994). Ovarian cancer: Screening, treatment, and follow-up. *Gynecol. Oncol.* **55**, S4–S14.

92. Schutter, E. M., Sohn, C., Kristen, P., Mobus, V., Crombach, G., Kaufmann, M., Caffier, H., Kreienberg, R., Verstraeten, A. A., and Kenemans, P. (1998). Estimation of probability of malignancy using a logistic model combining physical examination, ultrasound, serum CA 125, and serum CA 72-4 in postmenopausal women with a pelvic mass: An international multicenter study. *Gynecol. Oncol.* **69**, 56–63.

72

Endometrial Cancer

LINDA S. COOK*,† AND NOEL S. WEISS†,‡

*Department of Community Health Sciences, University of Calgary, Calgary, Alberta, Canada; †Fred Hutchinson Cancer Research Center, Seattle, Washington; ‡Epidemiology Department, University of Washington, Seattle, Washington

I. Introduction

Endometrial cancer is a relatively common gynecologic cancer; diagnosis generally occurs after abnormal uterine bleeding or spotting [1]. The five-year relative survival following diagnosis is greater than 84% overall [2], but it varies by the size, spread, and morphology of the cancer. Five-year relative survival is 96 to 100% for women with tumors confined to the uterine body (*i.e.,* Stage I) or highly differentiated tumors (*i.e.,* low grade), but it falls to 27 to 47% with tumors that have spread beyond the pelvis (Stage IV) or are poorly differentiated (*i.e.,* high grade) [2,3].

Current evidence indicates that exposure of the endometrium to high circulating levels of estrogens, especially from exogenous sources, increases the likelihood of developing this disease. The causal role of exogenous estrogens is highlighted by the sharp rise in endometrial cancer incidence in the early 1970s in the U.S. that followed an increase in the use of unopposed estrogens among postmenopausal women. There is also evidence that progestogens (both endogenous and exogenous) have a beneficial effect on the endometrium. The actions of many of the other known or suspected factors that alter endometrial cancer risk, such as obesity, reproductive characteristics, certain medical conditions, and cigarette smoking, may be explained at least in part by their influence on estrogen and progestogen activity.

II. Demographic Patterns of Incidence and Mortality

The body (corpus) of the uterus contains several different types of tissue: the endometrium (the inner mucosal layer); the myometrium (the thick, middle, muscular layer); and the serosa (the thin external coat). The cervix is the lower portion of the uterus that lies below the uterine body. Endometrial cancer occurrence has often been described through the proxy measure of cancer of the corpus uteri. This measure is a relatively good proxy for endometrial cancer; among African-Americans, 75% of cancers of the corpus uteri are endometrial in nature, and among white Americans, the figure is more than 90% [2]. Mortality from endometrial cancer can be estimated from rates for cancer of the uterine corpus or through a combination of rates of cancer of the corpus uteri and uterine cancer, not otherwise specified (NOS). The latter measure is probably the more useful but is nonetheless somewhat inaccurate because a substantial proportion of deaths in the "NOS" category may be due to cervical cancer. This proportion has decreased over time due to diminishing proportions of cervical cancers included in the uterine cancer, NOS category [4] as well as to decreases in cervical cancer mortality [2]. This reduction in cervical cancer mortality complicates interpretation of time trends in mortality for the combined category of uterine cancers of the corpus plus uterine cancer NOS. Both incidence and mortality rates will also be artificially low, because corrections are not made in routinely available data for the proportion of women who have had their uterus removed and are no longer at risk for developing endometrial cancer. Nonetheless, even with these limitations, we can draw some conclusions about changes and patterns in endometrial cancer incidence and mortality.

A. Time

Rapid changes have occurred in the incidence of endometrial cancer since the 1960s in the United States. The incidence began to rise in the 1960s and reached a peak in the mid-1970s [2,5]. The increase in incidence was experienced primarily by postmenopausal women and was generally greater in the western than in the eastern part of the country (Table 72.1) [5a–5c]. The increase shown in the table is actually an underestimate of the true increase; the rate of hysterectomy for reasons other than

Table 72.1

Annual Incidence of Invasive Carcinoma of the Uterine Corpus by Age: Connecticut and Alameda County, California, 1960–1994[a]

Time period	Connecticut: age (years)		Alameda County[b]: age (years)	
	30–49	50–69	30–49	50–69
1960–1964	15.4	66.5	11.0	70.6
1965–1969	13.7	67.9	15.7	109.8
1970–1971	11.6	77.1	17.0	135.6
1972–1973	16.7	84.4	23.2	195.4
1974–1975	14.9	96.2	19.3	186.6
1976–1977	12.6	84.0	12.6	177.6
1978–1979	11.3	79.9	6.7	127.2
1980–1981	11.7	74.2	10.1	110.2
1982–1983	8.6	73.5	9.3	96.8
1984–1985	10.4	69.5	7.6	105.8
1986–1987	11.2	66.3	8.3	88.0
1988–1989	8.7	64.7	8.0	89.3
1990–1991	9.7	69.0	9.9	76.0
1992–1993	12.5	75.9	7.0	62.5
1994	11.8	72.5	8.1	65.3

Personal communications and National Cancer Institute.

[a]Rate per 100,000 adjusted, within the broad age groups shown, to a uniform standard (10-year age groups for Connecticut, 5-year age groups for Alameda County). Data compiled from [5a,5c].

[b]Whites only.

Table 72.2
Cancer Incidence of the Uterine Corpus in Selected Populations

Continent	Country	Area	Race	"Truncated" incidence 1969–1972[b]	1973–1977[c]	1978–1982[d]	1983–1987[e]
Africa	Nigeria	Ibadan	All	4.2	—	—	—
	Senegal	Dakar	All	—	3.8	—	—
	Gambia	All	All	—	—	—	2.6
South America	Brazil	Sao Paulo	All	18.1	25.7	20.3	—
		Porto Alegre	All	—	—	15.5	13.2
North America	U.S.	Detroit	White	44.1	56.5	38.9	42.4
			Black	20.8	18.6	15.6	18.6
		New Mexico	Spanish	17.9	17.5	18.8	—
			Other White	39.0	53.7	30.9	—
			Am. Indian	11.4	9.2	9.7	—
Asia	India	Bombay	All	3.1	3.0	3.9	4.6
	Israel	All	Jews	22.9	21.4	16.6	21.0
			Non-Jews	3.1	1.7	6.0	6.8
	China	Shanghai	All	—	10.4	6.5	7.4
	Japan	Miyagi	All	3.2	4.6	5.9	8.4
Europe	Norway	All	All	23.2	25.3	23.9	28.2
	U.K.	Oxford	All	20.7	22.7	18.2	21.9
	Spain	Navarra	All	—	27.6	23.7	23.3
	Yugoslavia	Slovenia	All	21.2	23.4	23.1	25.2
Oceania	U.S.	Hawaii	White	71.4	76.7	36.7	34.8
			Japanese	40.7	46.7	31.4	28.2
			Chinese	49.0	66.9	37.1	32.6
			Hawaiian	63.8	75.3	51.7	38.2
	New Zealand	All	Maori	58.1	33.4	31.5	38.9
			Non-Maori	22.8	24.3	17.6	18.7

[a]Annual rate per 100,000, ages 35–64, standardized to the age distribution of the World Standard Population.
[b][16] Depending on population, rates apply to a part of the period 1969–1972.
[c][17] Depending on population, rates apply to a part of the period 1973–1977.
[d][18] Depending on population, rates apply to a part of the period 1978–1982.
[e][19] Depending on population, rates apply to a part of the period 1983–1987.

cancer rose rapidly during the same time period and the rates presented have not been corrected for this change [6–9]. Following the peak in the mid-1970s, the incidence of endometrial cancer steadily declined in the United States until the 1990s (Tables 72.1, 72.2) [2,16–19]. In contrast to these changes, the incidence during the same time period outside of North America generally remained stable (Table 72.2) or increased only a small degree [10].

The increased incidence in the United States during the 1970s does not appear to be an artifact of changes in diagnostic criteria or classifying estrogen-induced hyperplasia as cancer due to similarities in morphology [11]. Based on histologic reviews of reported endometrial cancers that were conducted using conservative criteria to identify cancer, the incidence of endometrial cancer during the mid-1970s was unequivocally greater than that of earlier years [4,12]. The increased incidence in the 1960s to 1970s and the ensuing decline in rates do, however, parallel patterns of postmenopausal unopposed estrogen use [13]. Replacement estrogens were introduced into medical practice in the 1930s, but they were not widely taken by postmenopausal women in this country until the 1960s. Estrogen use was greater in the western United States than in other regions (Table 72.1)

[14], and outside of North America use was generally uncommon [15]. With increasing evidence of an association between estrogens and endometrial cancer, the United States Food and Drug Administration issued a warning to physicians in 1976 [20]. A steady decline in the proportion of American women who used unopposed estrogens ensued, initially due to a reduction in use of unopposed estrogen and later to an increase in use of combined estrogen-progestogen hormone replacement therapy [13,21–25].

In the United States, age-standardized mortality due to corpus cancer and uterine cancer NOS decreased by 60% between 1950 (8.9 per 100,000) and 1985 (3.6 per 100,000) [26] and has remained relatively stable since that time [2]. The magnitude of this decrease is clearly an overestimate of any true decrease in corpus cancer mortality; over the same time period, cervical cancer mortality decreased by 75% [26] and the hysterectomy rate rose by as much as 60% [6,7,27]. After adjustment for the declining proportion of women with an intact uterus and the contribution of cervical cancer deaths included in uterine cancer NOS mortality, an estimated 14% decrease in corpus cancer mortality occurred between 1960 and 1970 in the United States [28]. Likewise, age-standardized mortality rates from cancer of the body of the uterus and cancer of uterus, part unspecified, also show a downward trend between 1955 and 1990 in England and Wales, Hungary, Italy, Japan, France, Canada, Norway, Australia, and Yugoslavia [29]. Adjustment for the rising hysterectomy rate [30] diminished, but did not eliminate, the overall downward trend in mortality in England and Wales [31].

B. Nationality/Geographic Area

Even prior to the estrogen-stimulated increase in the incidence of uterine corpus cancer, rates among U.S. white women were greater than rates among European women, as were the rates among African-American and Asian-American women relative to their counterparts in Africa and Asia, respectively (Table 72.2).

C. Age

Starting at about 50 years of age, uterine cancer mortality steadily increases with increasing age (Fig. 72.1) [2,5a]. In comparison, the incidence of uterine cancer rises rapidly in late reproductive life, peaks between 60 and 70 years of age, and plateaus or slightly declines in later life (Fig. 72.1). Exposure of the postmenopausal female population to unopposed estrogens affects the relationship between endometrial cancer incidence and age. In populations in which estrogen use is widespread among postmenopausal women, the peak incidence is accentuated (Table 72.1).

D. Race/Ethnicity

Maoris in New Zealand experience a rate of uterine corpus cancer approximately twice that of non-Maoris (Table 72.2). The high rate among Maoris is not due to an unusually high incidence of corpus cancer that is nonendometrial in nature (F. Foster, personal communication). Hawaiian women also have a relatively high incidence of uterine corpus cancer that rivals the rate among U.S. whites (Table 72.2). This suggests that some feature of their shared Polynesian heritage or some common characteristic such as obesity in Maoris and Hawaiian women contributes to a relatively high incidence of endometrial cancer.

Worldwide, the incidence of corpus cancer among whites exceeds that among women of other races (Table 72.2). In the United States, differences in incidence between whites and African-Americans depends on age (Fig. 72.1), with the greatest disparity in the postmenopausal years between the ages of 50 and 75. In contrast, uterine cancer mortality is higher among African-Americans than among whites beginning at about 50 years of age.

III. Hormonal Risk Factors

A. Endogenous Estrogens

Estrogens are the primary stimulants of endometrial proliferation. Because cellular proliferation is a prerequisite for carcinogenesis [32], and unchecked proliferation can lead to malignant transformation [33], it follows that estrogens may be a necessary cause for the development of at least some endometrial car-

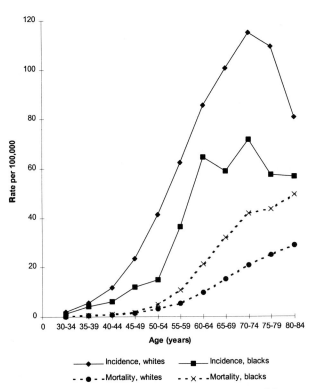

Fig. 72.1 Average annual age-specific incidence of cancer of the uterine corpus and mortality from cancer of the uterine corpus and uterus, NOS: 1990–1994. Sources: [2,5a].

cinomas. Indeed, risk factors for endometrial cancer include medical conditions that are known to result in relatively high endogenous estrogen levels.

1. Estrogen-Secreting Ovarian Tumors

Among women with estrogen-secreting ovarian tumors (*e.g.,* granulosa-theca cell tumors), the prevalence of endometrial carcinoma at the time of oophorectomy ranges from 6 to 21% [34–37]. It is likely that this observed frequency of endometrial tumors is considerably greater than expected even beyond possible biases in the selection of women and the diagnoses of uterine cancer.

2. Polycystic Ovaries

A number of case series have suggested that a high proportion of young women with endometrial cancer [38,39] and hyperplasia [40] have polycystic ovaries and amenorrhea or oligomenorrhea (the polycystic ovary syndrome or Stein-Leventhal syndrome). One follow-up study indicated that women with polycystic ovaries, Stein-Leventhal syndrome, or a clinically diagnosed ovulatory disorder had a three-fold elevated risk for endometrial cancer, although this was based on only five women who developed cancer [41]. Due to chronically elevated levels of luteinizing hormone, women with polycystic ovaries secrete abnormally large quantities of androstenedione. This is metabolized peripherally by the enzyme aromatase to estrone (a potent estrogen) and results in levels of estrone similar to those found at the peak of the normal ovulatory cycle [42].

3. Weight or Body Mass

Consistently, studies have found that both pre- and postmenopausal women with endometrial cancer are more likely to be overweight than other women [43–66]. Depending on the measures used to identify excess weight or body mass, the risk for overweight women relative to women of normal weight is roughly 2 to 20. While some studies report that risk increases with incremental increases in weight or body mass [50,54,63,66], others find a strong elevation in risk only among obese women [45,46,49,56,64]. Whether a low level of physical activity plays a role in obesity-related endometrial cancer is unclear at the present time. Among the studies that adjusted for body mass, the results of three suggest there is an elevated risk for relatively inactive women [67–69], whereas another does not [66]. Obesity leads to a net increase in the amount of endogenous estrogens, both through increased conversion of androstenedione to estrogen [70–72] and through decreased circulating levels of sex hormone-binding globulin [73–75].

B. Endogenous Progesterone

The surge of luteal progesterone secretion that begins just prior to ovulation each month in premenopausal women arrests endometrial proliferation, promotes secretory differentiation of the endometrium, and initiates endometrial sloughing in the absence of fertilization [76]. These actions decrease the likelihood of developing hyperplastic or neoplastic lesions because cell differentiation alone can arrest or reverse neoplastic cellular transformation [33]. Thus, conditions that are characterized by low endogenous progesterone levels, particularly when coupled with relatively high estrogen levels, are generally associated with higher rates of endometrial cancer. This probably is the explanation for the peak incidence of endometrial cancer that occurs after menopause (Fig. 72.1), when limited estrogen production continues (through peripheral conversion of adrenal androgens) without any cyclic progesterone production or substantial peripheral production of progesterone. Similarly, women with the polycystic ovary syndrome not only have excessive production of estrogen but also lack cyclic progesterone secretion [39,77,78]. Lastly, the increased risk of endometrial cancer in premenopausal obese women may be related not only to a hyperestrogenic state but also to decreased progesterone levels. A study of six obese oligomenorrheic premenopausal women found consistently subnormal levels of serum progesterone relative to ten nonobese controls, even though the luteal phase in all subjects lasted at least ten days [79]. Estrogen levels did not differ between the two groups.

C. Exogenous Estrogens

1. Postmenopausal Estrogens

a. HYPERPLASIA.   It is well established that women who take estrogens unopposed by progestogens develop adenomatous hyperplasia of the uterus more commonly than do other women [80–82]. For example, among the women assigned to receive unopposed estrogen therapy (conjugated equine estrogen, 0.625 mg/day) in the Postmenopausal Estrogen/Progestin Interventions (PEPI) trial, 22% developed adenomatous hyperplasia and 12% developed atypical hyperplasia after three years of follow-up [81]. Less than 1% of women in the placebo arm and only 1% of other women who received various combinations of estrogen and progestogens in the trial developed hyperplasia. In another randomized trial, the Continuous Hormones as Replacement Therapy (CHART) Study, endometrial hyperplasia occurred in 55.6% of women receiving 10 μg of unopposed ethinyl estradiol daily compared with 1.7% of the women receiving placebo [82]. The credibility of this association is strengthened by the knowledge that some hyperplastic lesions regress when estrogens are discontinued [83] and that most hyperplastic lesions can be successfully treated with progestogens [84]. One or more forms of hyperplasia probably represent an early stage in the development of endometrial carcinoma, as they frequently coexist with endometrial cancer in estrogen users and have been observed to precede the appearance of carcinoma [85–87]. A follow-up study of 170 women (mean follow-up: 13.4 years) with untreated hyperplasia found that 1.6% of the women with simple hyperplasia and 23% of the women with atypical hyperplasia developed endometrial cancer [88].

b. ENDOMETRIAL CANCER.   Nearly all epidemiologic studies of the question have observed unopposed estrogen use to be associated with an increase of endometrial cancer [47–50,52, 54,56,89–119]. This literature has been reviewed in detail elsewhere [120–122]. In the interest of brevity, the following discussion will provide a general summary of specific features of estrogen use.

   i. Duration/Recency   Most studies have found an elevation in risk beginning after 1 to 5 years of use, the size of which increases further with increasing duration of use [47–49,52,56, 89–94,99,101–104,110,113,114,116,119,123]. Relative to the incidence among current users, risk of endometrial cancer decreases after cessation of estrogen therapy [48,52,89,92,96,115, 116,123–125]. In three studies there remained little or no residual excess above the rate in women who never used hormones after 2 drug-free years had elapsed [52,125,126], whereas in other studies some excess risk persisted for up to 5 years [92,101] and even up to 10 years [94,113,115]. The results of two studies suggest that following cessation, a particularly large excess risk remains for women with the longest durations of estrogen therapy [56,123].

   ii. Type   Use of conjugated estrogens, the type of estrogen most commonly prescribed in the U.S., has consistently been related to an increased endometrial cancer risk of roughly 2- to 20-fold [52,89,91–93,96,99,101,102,105,114,127]. Similar results have been reported for other nonconjugated estrogens (e.g., stilbestrol or ethinyl estradiol) in most studies [89,96,99, 102,114], but not in all [92]. Among users of vaginal hormone creams, one study reported an elevated risk of endometrial cancer [49], whereas two others did not [102,116]. Transdermal patches and intramuscular injections of estrogen have been evaluated in too few women to adequately address their influence on risk.

   iii. Dosage   Although no increase in the incidence of hyperplasia was noted among women treated with 0.3 mg of esterified estrogen relative to the incidence in women treated with placebo [128], the risk of endometrial cancer appears to be elevated with all commonly prescribed dosages of conjugated estrogens (0.3 mg to 1.25 mg per day or the equivalent amount of other estrogens), and it may rise with increasing dose [48,52,96,99,101,102,106,127]. Women who take estrogens orally usually do so either on a daily basis or with a break of 5 to 7 days each month. Each of these regimens are associated with an elevated risk both for endometrial hyperplasia [129] and for endometrial cancer [48,52,89,91,96,99,101].

iv. Stage of Cancer While both higher and lower grades of endometrial cancer are found in excess among unopposed estrogen users, the highest risks are seen for the less invasive and more highly differentiated tumors [49,94,96,101,130]. This may be due, in part, to atypical hyperplasia being misdiagnosed as early endometrial cancer in some instances [4,11] and to earlier detection of cancer in estrogen users compared to nonusers [89,131]. Whatever the reason(s), among women who develop endometrial cancer, postmenopausal estrogen users have a more favorable survival experience than do nonusers [132–135].

v. Influence of Other Risk Factors on the Estrogen/Endometrial Cancer Association Four studies have noted an elevated relative risk of endometrial cancer among estrogen-users relative to nonusers both in smokers and in nonsmokers [46,56,115,116]. The elevated relative risk was greater among smokers in one of these studies [116] but greater in nonsmokers in another [46]. Most studies have reported similar elevations in relative risk irrespective of parity among estrogen-users relative to nonusers [56,103,104,116], although one noted a particularly high relative risk in women of low parity [46]. In both users and nonusers of oral contraceptives, the use of estrogen replacement therapy was associated with an elevated relative risk of endometrial cancer in two studies [115,116], whereas in another there was no elevation in relative risk among oral contraceptive users who later used estrogen [56]. Almost all studies have reported an elevation in relative risk among estrogen-users compared to nonusers across all categories of weight or body mass [46,49,54,56,103,104,115], although the results of some suggest that the relative risk is greater among women of lower weight or body mass [49,115] or is even restricted to this group [52,116]. Even if the relative risk for endometrial cancer is somewhat greater for women of lower body mass, the absolute increase in the incidence of endometrial cancer associated with estrogen use may be quite similar among light and heavy women.

2. Other Unopposed Estrogen Use

Further information about the long-term effects of estrogen supplementation comes from a series of young women with gonadal agenesis (undeveloped ovaries) who received diethylstibestrol for prolonged periods. These women had an usually high incidence of endometrial cancer [136]. Additionally, a group of patients with breast cancer who had received estrogenic hormones (primarily stilbestrol) as a cancer treatment had about twice the incidence of endometrial cancer than that expected for other breast cancer patients [98].

3. Tamoxifen

Tamoxifen, a nonsteroidal hormone used in the treatment of breast cancer, has been shown to be estrogenic in the human uterus [137,138] and it appears to increase the risk of endometrial cancer, particularly when given at a relatively high daily dosage (30 to 40 mg) or for relatively long periods of time (5 or more years). In randomized controlled trials, relative risks of 6.4 (95% CI = 1.4–28.0), 7.5 (95% CI = 1.7–33.0), and 3.3 (95% CI = 0.6–32.4) for endometrial cancer have been associated with tamoxifen regimens of 40 mg/day for 2 years, 20 mg/day for 5 years, and 30 mg per/day for approximately 1 year, respectively [139–141]. Two case-control studies also found the suggestion of an elevated risk with 5 or more years of tamoxifen use [142,143]. No elevation was noted in smaller trials for regimens of 20 mg/day for 1 year and 20 mg/day for 5 years [144,145], as well as in a case-control study conducted in a population in which tamoxifen use tended to be of relatively short duration [146]. A tamoxifen-associated increase in endometrial cancer is of particular concern given the proposed prophylactic use of tamoxifen in the primary prevention of breast cancer among healthy women [147]. Another nonsteroidal hormone, raloxifene, is being developed as a treatment for osteoporosis in postmenopausal women [148]. In contrast to tamoxifen, raloxifene does not appear to stimulate proliferation of the endometrium [149].

D. Exogenous Progestogens

Exogenous progestogens mimic the actions of luteal progesterone in promoting differentiation and arresting proliferation of endometrial tissue. These beneficial effects are present when continuous low-dose progestogens are given with estrogens, a combination of exogenous hormones that results in no endometrial sloughing [150–152].

1. Postmenopausal Estrogens–Progestogens

a. HYPERPLASIA. As mentioned previously, several studies have found that estrogen-stimulated hyperplasia reverts to normal endometrium after administration of exogenous progestogens [83,84,153]. Several studies have found that an estrogen–progestogen regimen is associated with a lower occurrence of hyperplasia (0–4%) [84,154–157] than is estrogen alone, particularly when the progestogens are used for more than 10 days each month [84,155,156]. The occurrence of hyperplasia in women randomized to progestogen supplementation (either continuous or for 12 days each month) was found to be very low (<2%) relative to women receiving unopposed estrogen (44–56%), and it was no different than that among women receiving placebo [81,82]. To counter the often unwanted monthly menses in women receiving progestogen supplementation of estrogen each month, a "long-cycle" regimen has been proposed where progestogen supplementation is given for 10 to 14 days once every 3 months. In one study, only 1 of the 219 women followed for 1 year and none of the 132 women followed for 2 years had evidence of hyperplasia with a long-cycle regimen that included 20 mg per day of medroxyprogesterone acetate for 14 days [158]. In another study that included 10 mg per day of medroxyprogesterone acetate for 14 days, 1.5% of the 214 women given a long-cycle regimen developed hyperplasia after one year of follow-up, which was similar to the baseline prevalence of 0.9% [159]. However, a Scandinavian trial was recently discontinued after 3 years of follow-up due to an unacceptably high occurrence of hyperplasia in the long-cycle group (6%) relative to a monthly-cycle group (<1%); the long-cycle group received 1 mg per day of norethindrone for 10 days [160].

Table 72.3

Summary of Studies Evaluating Postmenopausal Estrogen Plus Progestogen Therapy and Endometrial Cancer

Reference	Study design	Number of exposed cases	Type/measure of combined therapy	RR	95% CI
[108]	Cohort	8	Rate in estrogen users vs nonusers	1.9	NP
		2[a]	Rate in E+P users vs nonusers	0.2	NP
[114]	Cohort	48	Rate in estrogen users vs general population	1.4	(1.1, 3.2)
		7[b]	Rate in cyclic E+P users vs general population	0.9	(0.4, 2.0)
[162]	Case-control	54	Estrogen users vs nonusers	3.1	(1.6, 5.8)
		18[a]	Any use of E+P vs nonuse	1.3	(0.6, 2.8)
		11[a]	E+P with progestogen for <10 days/month	2.0	(0.7, 5.3)
		7[a]	E+P with progestogen for ≥10 days/month	0.9	(0.3, 2.4)
[117]	Case-control	32	Current/recent estrogen users vs nonusers	6.5	(3.1, 13.3)
		18[a]	Current/recent use of E+P vs nonusers	1.9	(0.9, 3.8)
		6[a]	Past use of E+P vs nonuse	0.9	(0.3, 3.4)
[116]	Case-control	60	Estrogen users vs nonusers	3.4	(1.8, 6.3)
		11[a]	Any use of E+P vs nonuse	1.8	(0.6, 4.9)
[118]	Case-control	324	Estrogen users vs nonusers	4.0	(3.1, 5.1)
		67[b]	Any use of E+P vs nonuse	1.4	(1.0, 1.9)
		25[b]	E+P with progestogen for <10 days/month	3.1	(1.7, 5.7)
		25[b]	E+P with progestogen for ≥10 days/month	1.3	(0.8, 2.2)
[119]	Case-control	422[c]	For each 5 years estrogen use vs nonuse	2.2[d]	(1.9, 2.5)
			For each 5 years E+P use vs nonuse		
		74[c]	With progestogen <10 days/month	1.9[d]	(1.3, 2.7)
		79[c]	With progestogen ≥10 days/month	1.1[d]	(0.8, 1.4)
		94[c]	With continuous progestogen	1.1[d]	(0.8, 1.4)

Note. RR, relative risk; E+P, estrogen plus progestogen; NP, information not provided in published paper.

[a]Women with prior unopposed estrogen use included.

[b]Women with prior unopposed estrogen use excluded.

[c]Categories not mutually exclusive; 509 cases had used one or more types of hormone replacement therapy.

[d]Adjusted for use of other types of hormone replacement therapy.

b. ENDOMETRIAL CANCER. Compared to the epidemiologic information on unopposed estrogen therapy and endometrial cancer, information on combined estrogen-progestogen therapy and endometrial cancer is relatively sparse. In a small randomized controlled trial of institutionalized women, no cases of endometrial cancer were observed in the combined therapy group, and only one was observed in the placebo group [161]. A small cohort study also observed no cases of endometrial cancer among women who received estrogen-progestogen therapy, whereas three cases were observed among the nonusers [107]. Other studies that have assessed the impact of progestogen supplementation of postmenopausal estrogens on the risk of endometrial cancer are summarized in Table 72.3. As predicted from the studies of hyperplasia, all observed a lower risk of endometrial cancer associated with the use of combined therapy compared to therapy with unopposed estrogens. However, compared with women who did not use hormones, only one study found a reduction in endometrial cancer risk with any combined therapy use [108], whereas the remainder found no difference or a modest elevation in risk [114,116–119,162]. Two studies report that endometrial cancer risk with combined hormone therapy is lower for all stages of disease relative to that for unopposed estrogen users [119,163].

Studies that were large enough to address the question reported a two- to fourfold elevated risk associated with short cyclic progestogen use (*i.e.,* for less than 10 days each month) [118,119,162], particularly when this regimen was used for 5 years or longer [118,119]. No elevation in risk has been noted with a regimen of longer cyclic progestogen use (*i.e.,* for 10 or more days per month) when taken for less than 5 years [118,119], whereas more than 5 years use of this regimen has been associated with an elevated risk in one study [118] but not in another [119]. Only one study has evaluated continuous combined therapy; it found no elevation in endometrial cancer risk relative to that in nonusers of hormones [119].

The results of studies that have investigated the impact of using a relatively high progestogen dosage (*e.g.,* 10 mg/day of methoxyprogesterone acetate) are not consistent. In one study women given a higher dosage of progestogen for more than 10 days per month for 5 or more years still had an elevated endometrial cancer risk compared with nonusers of hormones [118], whereas another study reported no elevation in risk associated with the same regimen [119]. Another study reported that, in general, a lower proportion of women with endometrial cancer (17%) compared to other women (29%), had used high-dosage progestogen in combined therapy [117].

In summary, while available evidence suggests that the occurrence of endometrial cancer is lower in combined estrogen-progestogen users than in estrogen-only users, the long-term risks associated with different dosages and monthly durations of progestogen supplementation remain unclear at present.

2. Hormonal Contraception

Women are also exposed to exogenous progestogens through use of hormonal contraceptive agents. Since the 1970s this has included sequential and combination oral contraceptives, progestogens-only pills (minipills), and the long-acting injectable or implanted progestogens (*e.g.,* medroxyprogesterone). While a fair amount of study has been devoted to oral contraceptive use in relation to endometrial cancer risk, little information is available concerning the impact of the other hormone contraceptives. One U.S. study reported no increase in the incidence of uterine cancer after 13 years of follow-up among African-American women receiving depot medroxyprogesterone acetate (DMPA) injections relative to African-American women in general [164]. A study in Thailand reported that endometrial cancer incidence was approximately 80% (95% CI = 0.1–0.8) lower among DMPA users than among other women [165].

Initially, oral contraceptive pills (OCPs) were available as sequential preparations (estrogen only followed by a short course of estrogen plus progestogen) or as combination preparations (concurrent estrogen and progestogen). Starting in the mid-1970s, there were reports of endometrial abnormalities [166,167] and an increased risk of endometrial cancer [50,166,168–170] among women who used sequential preparations, particularly Oracon (Mead Johnson, Evansville, IL), which contained a relatively potent estrogen (ethinyl estradiol) followed by a weak progestogen (dimethisterone). Two studies observed no excess risk with use of sequential OCPs or Oracon [49,171], but they included a very small number of women who used sequential OCPs. Sequential preparations were removed from the consumer market in the United States and Canada in 1976.

In contrast, women who have taken combination OCPs have about one-half the risk of endometrial cancer as do nonusers [49,50,117,126,169,171–180]. The reduced risk first appears to be evident after about 2 to 5 years of use [50,117,171,173,174, 177,178] and continues to decrease as the duration of OCP use increases [49,50,126,171,173,174,177]. Some studies report a greater reduction in risk with more recent use [117,177,178], but another study with enough former OCP users to address this issue reported no difference [174]. When duration and recency of use were evaluated jointly, longer durations of use (>5 years), but not shorter durations of use (<5 years), were associated with a reduced endometrial cancer risk irrespective of recency [179]. Some studies report that the reduction in risk may be greatest with OCPs in which progestogen effects predominate [173] or that contain higher dosages of progestogen [181], but another study found that a longer duration of use (>5 years), and not progestogen dosage, was most predictive of a reduced risk [179].

In joint evaluation with other risk factors for endometrial cancer, no reduction in relative risk was found among combination OCP users in the heaviest category of body weight in two studies [50,178], whereas a reduced relative risk was found among OCP users regardless of weight or body mass in two others [174,177]. Three studies found that relative risk reductions with combination OCP use were strongest among parous women [177] or women of higher parity (>5 births) [175,178], but another noted a reduced relative risk for endometrial cancer only

among OCP users who were nulliparous [174]. No reduction in relative risk among OCP users who later used estrogen replacement therapy (ERT) for 3 or more years was noted in two studies [178,179], whereas a reduced relative risk among OCP users who ever used ERT was noted in four others [171,173,174,177]. The inclusion of women with brief use (less than 2 or 3 years) of ERT in the latter studies could have obscured the attenuation of the association between combined OCPs and endometrial cancer that appears to be present among women with longer durations of ERT use.

IV. Other Risk Factors

A. Events of Reproductive Life

Evidence for a relationship of age at menarche to endometrial cancer is mixed, with some studies reporting an earlier menarche among cases than among controls [45,49,54,65,182–185], whereas others do not [44,47,64,89,126]. A more consistent finding is that women who experience a natural menopause at a relatively late age are at greater risk of endometrial cancer than are other women [45,47,49,54,63,65,126,182,183,185].

Nulliparity has consistently been observed to be associated with an increased risk of endometrial cancer, and among women with at least one birth, the probability of developing endometrial cancer is inversely related to the number of births [45,49,50,56, 59,63–65,126,183,184,186,187]. Each additional childbirth also has been related to a 9.5% reduction in mortality from cancer of the uterine corpus [188]. In contrast to findings on breast cancer, no consistent relationship has been found between age at first birth and the incidence of endometrial cancer after adjustment for parity [50,57,182,184,185,187,189]. Evidence for a reduced risk of endometrial cancer associated with an older age at last pregnancy is also mixed [57,182,184,185,187,189]. An international study observed that the longer a woman has breastfed over her reproductive life, the lower her risk of endometrial cancer [190]. In contrast, two studies in the U.S. and Japan found no difference in risk between women who had and had not breastfed [64,184].

After accounting for parity, impaired fertility measured in various ways (*e.g.,* 3 or more years of unsuccessfully attempting pregnancy, seeking medical advice for infertility, or self-reported infertility) has been associated with an increased risk of endometrial cancer in most [50,170,184,191,192], but not in all [185], studies. In addition, the occurrence of endometrial cancer among cohorts of women either referred for fertility treatment or diagnosed with infertility was roughly three to five times that expected based on the rate in the general population [192,193]. In one of these cohorts, a ninefold elevation in risk was noted among women whose infertility was characterized by normal estrogen production but progesterone deficiency [193].

B. Intrauterine Contraceptive Devices (IUDs)

IUDs evoke alterations in the surface morphology of cells [194] as well as acute and chronic endometrial inflammation [195], but it is unclear how these actions may influence the development of endometrial cancer. Seven epidemiologic studies have investigated this relationship and none report any elevation in risk. The results of five showed a 40–50% reduction in endometrial cancer risk among women who ever used an IUD compared to never users [174,196–199], and the reduced risk persisted for many years after use [196–199]. The remaining two studies found no alteration in risk [59,200].

C. Abnormal Glucose Tolerance and Diabetes Mellitus

In two cohorts of women with diabetes mellitus, the incidence of endometrial cancer was 40–80% higher than that among women in general [201,202]. Even after exclusion of women with reported obesity, modest elevations (20–70%) remained. Two other cohort studies of diabetic women found no elevation in endometrial cancer risk [203,204], but both studies lacked statistical power to detect a relatively small excess. A higher frequency of diabetes mellitus has been reported among women with endometrial cancer in many [45,47,52,56,92,102–104,184,186,205–208], but not in all [49,51,66,89,105], case-control studies. Among those studies that in some way accounted for body weight, an important confounding variable, at least a twofold elevation in risk remained in most [45,184, 186,207], but not in all, of them [49,66].

D. Hypertension

Although a relationship between hypertension and endometrial cancer has long been claimed, results of the epidemiologic studies that accounted for age and some measure of weight, factors associated both with endometrial cancer incidence and with hypertension [209], have been inconsistent. Two studies reported a 70–100% elevation in risk [45,46], two others found a suggestion of an elevation [51,186], but three others found no alteration in risk [49,66,184] for hypertensive relative to normotensive women.

E. Gallbladder Disease

Gallbladder disease, like hypertension, has been more commonly reported among women with endometrial cancer than among other women, but only a few studies have accounted for important confounding factors. One study found that a history of gallbladder disease was related to an elevated endometrial cancer risk among women with and without exposure to exogenous estrogens [89], whereas another study that adjusted for use of hormone replacement therapy and oral contraceptive use, as well as some proxy measures of endogenous estrogen exposure (e.g., weight and parity) [184], found no elevation in risk. Thus, it is possible that the reported link between gallbladder disease and endometrial cancer is due to the association of endogenous estrogens with both conditions.

F. Diet, Alcohol Consumption

On average, women who reside in countries in which high-fat diets are prevalent generally have relatively high rates of corpus cancer [210], but evidence of such an association on an individual level is mixed. Using various indices of dietary fat intake, several studies report that endometrial cancer risk is increased with higher levels of fat intake after adjustment for weight or body mass [66,211–214], whereas others do not [215–217]. While most studies find at least the suggestion of a

reduced risk with higher consumption of vegetables [66,212, 214,216], fruits [66,212,214], or vegetable- and fruit-related micronutrients [217], this is not entirely consistent [211,213]. In general, other evidence for a relationship between types of food, macronutrients, and micronutrients in the diet and endometrial cancer risk is inconsistent and these relationships are presently unresolved. The European Prospective Investigation into Cancer and Nutrition (EPIC) study may help to clarify some of these potential relationships [218]; cancer incidence is being monitored in over 400,000 people in nine countries who are providing detailed information on diet as well as providing biological samples (plasma, serum, lymphocytes, and erythrocytes) for future biochemical and molecular studies.

Apart from one study that found an elevated risk for women who drank more than two alcoholic beverages a day relative to nondrinkers [219], most studies have found that alcohol consumption is unrelated to endometrial cancer risk, especially among postmenopausal women [60,64,66,220–222]. The results of four studies that evaluated younger women (approximately 55 years of age or less) suggest that those with the highest alcohol consumption (ranging from 4 to >14 drinks/ week) may have a modest reduction in risk compared with nondrinkers [60,219,222,223].

G. Cigarette Smoking

Most case-control studies of endometrial cancer that have ascertained cigarette smoking behavior have noted a lower proportion of smokers among cases than among controls (Table 72.4 and [55,64,208,220,224,225]), especially among postmenopausal women [46,53,183,226–233]. Thus, if cigarette smoking does prevent some women from developing endometrial cancer,

Table 72.4

Epidemiologic Studies of Cigarette Smoking and Endometrial Cancer by Menopausal Status

		Relative risk	
Reference	Measure of smoking	Premenopausal women	Postmenopausal women
[46]	Ever vs never	0.5[a]	0.4
[226]	Ever vs never	1.3	0.4
[227][b]	Ever vs never	1.1	0.8
[228][b]	Postmenopausal smoking vs never	—	0.5
[229]	≥25 cigarettes/day vs never	0.9	0.5
[230]	>40 cigarettes/day vs never	1.9[c]	0.4[c]
[231]	≥20 cigarettes/day vs never	0.6	0.6
[53]	Current vs never	0.6	0.6
[183]	Current vs never	2.3	0.2
[232]	Current vs never	0.5	0.4
[233]	Current vs never	2.8	0.8

[a]Noel Weiss and Lynda Voigt, personal communication.

[b]Centers for Disease Control Cancer and Steroid Hormone Study; the study by Franks et al. [228] is a subset of the study of Tyler et al. [227].

[c]Women < 50 and ≥ 50 years of age.

it is likely to do so in ways other than by reducing ovarian production of estrogens.

Although not entirely consistent [231], the negative association has been particularly strong in recent smokers [53,55,220, 225,227–230,232,233], which would be predicted if smoking were to exert an effect by interfering in some way with the action or availability of hormones involved in endometrial carcinogenesis. Most studies have not found differences in blood levels of estradiol or estrone associated with smoking [220,234,235], but there may be differential metabolism of these compounds among smokers that favors the 2-hydroxyestrone pathway which produces a metabolite of low estrogenic activity [236,237]. Similarly, higher levels of circulating progesterone, as well as a higher ratio of progesterone to estrogen, among smokers [235] may also act to diminish net estrogenic effects. These findings are consistent with the observations that smoking is related to an increased risk of osteoporosis [238,239] and an earlier age at menopause [240,241], both of which are associated with decreased estrogenic activity.

H. Genetic Predisposition

Only a handful of studies have evaluated the risk of endometrial cancer associated with a history of endometrial cancer in one or more first-degree relatives. Two studies that included women up to 75 years of age reported modest (50–90%) elevations in risk with a positive family history [49,242]. The results of two other studies restricted to women less than 55 years of age observed a stronger association [50,243]. At the present time, it is unknown if these associations are due more to shared genes or to the shared environment among female family members.

It is also possible that germline allelic variants in oncogenes, tumor-suppressor genes, genes encoding DNA repair or methylation enzymes, or genes encoding enzymes that metabolize carcinogens or procarcinogens (such as compounds found in food or tobacco smoke) may place women at greater or lesser risk of endometrial cancer. An elevated endometrial cancer risk has been reported for women with a family history of colorectal cancer [243]. One explanation might be that the same germline mutations in genes encoding DNA mismatch repair enzymes important in the familial aggregation of colorectal cancer [244] may also influence the development of endometrial carcinoma. Although this possibility is just beginning to be explored, the results of one study suggest that women carrying a rare p53 allele, a rare P450 CYP1A1 allele, or a rare methylenetetrahydrofolate reductase gene allele have an elevated risk of endometrial cancer relative to women with the respective wild type genes [245,246]. The activity and function of these allelic variants in the development of endometrial cancer remains to be elucidated; such knowledge would help identify a genetic basis for altered cancer susceptibility.

V. Disease Prevention

Our knowledge of identified risk factors for endometrial cancer indicates that an impact can be made in the primary prevention of this disease by minimizing the use of high dose unopposed estrogens. Knowledge of this risk factor, and action upon it, have already had a favorable impact on the occurrence of endometrial cancer in the U.S. This was accomplished in two steps: an initial reduction in unopposed estrogen use (*i.e.,* a reduction in any type of exogenous hormone use for replacement therapy) and then increases in combined estrogen-progestogen use. Depending on the duration and type of estrogen-progestogen regimen used, endometrial cancer risk among combined users is either intermediate to that for unopposed estrogen users (higher risk) and nonhormone users (lower risk) or similar to that for nonhormone users. In terms of other modifiable risk factors, a decrease in the prevalence of extreme obesity in women would also reduce the occurrence of endometrial cancer (in addition to having many other health benefits).

VI. Future Research Directions

The reasons for the international variation in the occurrence of endometrial cancer among women remain unclear, apart from differences in the use of postmenopausal hormones and in the prevalence of obesity. A likely explanation is that lifestyle differences are responsible. A challenge for future investigations will be to determine if distinct aspects of lifestyle (*e.g.,* dietary factors, exercise, or occupation) account for all or part of this variation beyond any influence they may have on the prevalence of obesity.

The incidence of endometrial cancer associated with hormone replacement therapy among postmenopausal women will continue to be an active area of research as the means of providing replacement therapy continue to change. This research will include continued evaluation of estrogens supplemented with cyclic or continuous progestogens, as well as the use of low dosage estrogens alone (*e.g.,* 0.3 mg/day of esterified estrogen). Long-term evaluations of selective estrogen receptor modulators (SERMs), such as raloxifene, that may provide some of the benefits of estrogen replacement therapy without inducing endometrial proliferation are also needed. Finally, as we learn more about the genetic susceptibilities for endometrial cancer, it may ultimately be possible to identify those women who can safely take unopposed estrogens without sustaining an elevated risk for this disease.

References

1. Burke, T. W., Tortolero-Luna, G., Malpica, A., Baker, V. V., Whittaker, L., Johnson, E., and Follen Mitchell, M. (1996). Endometrial hyperplasia and endometrial cancer. *Obstet. Gynecol. Clin. North Am.* **23,** 411–456.
2. Ries, L. A. G., Kosary, C. L., Hankey, B. F., Miller, B. A., Harras, A., and Edwards, B. K., eds. (1997). "SEER Cancer Statistics Review, 1973–1994," NIH Publ. No. 97-2789. National Cancer Institute, Bethesda, MD.
3. Kodama, S., Kase, H., Tanaka, K., and Matsui, K. (1996). Multivariate analysis of prognostic factors in patients with endometrial cancer. *Int. J. Gynaecol. Obstet.* **53,** 23–30.
4. Szekely, D. R., Weiss, N. S., and Schweid, A. I. (1978). Incidence of endometrial carcinoma in King County, Washington: A standardized histologic review. *J. Natl. Cancer Inst. (U.S.)* **60,** 985–989.
5. Weiss, N. S., Szekely, D. R., and Austin, D. F. (1976). Increasing incidence of endometrial cancer in the United States. *N. Engl. J. Med.* **294,** 1259–1262.
5a. Marrett, L. D., Elwood, J. M., Epid, S. M., *et al.* (1978). Recent trends in the incidence and mortality of cancer of the uterine corpus in Connecticut. *Gynecol. Oncol.* **6,** 183.

5b. State of California, Department of Health (1978). "Uterine Cancer Incidence," Vol. V, No. 1. State of California, Dept. of Health, Sacramento.

5c. National Cancer Institute, DCPC, Surveillance Program, Cancer Statistics Branch (1997). "Surveillance, Epidemiology, and End Results (SEER) Program Public Use CD-ROM (1973–1994)." National Cancer Institute, Bethesda, MD. Based on the August 1996 submission.

6. Lyon, J. L., and Gardner, J. W. (1977). The rising frequency of hysterectomy: Its effect on uterine cancer rates. *Am. J. Epidemiol.* **105,** 439–443.

7. Koepsell, T. D., Weiss, N. S., Thompson, D. J., and Martin, D. P. (1980). Prevalence of prior hysterectomy in the Seattle-Tacoma area. *Am. J. Public Health* **70,** 40–47.

8. Dicker, R. C., Scally, M. J., Greenspan, J. R., Layde, P. M., Ory, H. W., Maze, J. M., and Smith, J. C. (1982). Hysterectomy among women of reproductive age: Trends in the United States, 1970–1978. *JAMA, J. Am. Med. Assoc.* **248,** 323–327.

9. Howe, H. L. (1984). Age-specific hysterectomy and oophorectomy prevalence rates and the risks for cancer of the reproductive system. *Am. J. Public Health* **74,** 560–563.

10. Pagel, J., and Bock, J. E. (1984). Endometrial cancer. A review. *Dan. Med. Bull.* **31,** 333–345.

11. Gordon, M. D., and Ireland, K. (1994). Pathology of hyperplasia and carcinoma of the endometrium. *Semin. Oncol.* **21,** 64–70.

12. Gordon, J., Reagan, J. W., Finkle, W. D., and Ziel, H. K. (1977). Estrogen and endometrial carcinoma. An independent pathology review supporting original risk estimate. *N. Engl. J. Med.* **297,** 570–571.

13. Austin, D. F., and Roe, K. M. (1982). The decreasing incidence of endometrial cancer: Public health implications. *Am. J. Public Health* **72,** 65–68.

14. Jick, H., Walker, A. M., and Rothman, K. J. (1980). The epidemic of endometrial cancer: A commentary. *Am. J. Public Health* **70,** 264–267.

15. Doll, R., Kinlen, L. J., and Skegg, D. C. (1976). Letter: Incidence of endometrical carcinoma. *Br. Med. J.* **1,** 1071–1072.

16. Waterhouse, J., Muri, C., Correa, P., and Powell, J. (1976). "Cancer Incidence in Five Continents," Vol. 3 IARC, Lyon, France.

17. Waterhouse, J., Muri, C., Shanmugaratnam, K., and Powell, J. (1982). "Cancer Incidence in Five Continents," Vol. 4. IARC, Lyon, France.

18. Muir, C., Waterhouse, J., Mack, T., Powell, J., and Whelan, S. (1987). "Cancer Incidence in Five Continents," Vol. 5. IARC, Lyon, France.

19. Parkin, D. M., Muir, C., Whelan, S., Gao, Y.-T., Ferlay, J., and Powell, J. (1992). "Cancer Incidence in Five Continents," Vol. 6. IARC, Lyon, France.

20. Food and Drug Administration (1976). Estrogens and endometrial cancer. *FDA Drug Bull.* **6,** 18–20.

21. Standeven, M., Criqui, M. H., Klauber, M. R., Gabriel, S., and Barrett-Connor, E. (1986). Correlates of change in postmenopausal estrogen use in a population-based study. *Am. J. Epidemiol.* **124,** 268–274.

22. Gruber, J. S., and Luciani, C. T. (1986). Physicians changing postmenopausal sex hormone prescribing regimens. *Prog. Clin. Biol. Res.* **216,** 325–335.

23. Ross, R. K., Paganini-Hill, A., Roy, S., Chao, A., and Henderson, B. E. (1988). Past and present preferred prescribing practices of hormone replacement therapy among Los Angeles gynecologists: Possible implications for public health. *Am. J. Public Health* **78,** 516–519.

24. Kennedy, D. L., Baum, C., and Forbes, M. B. (1985). Noncontraceptive estrogens and progestins: Use patterns over time. *Obstet. Gynecol.* **65,** 441–446.

25. Wysowski, D. K., Golden, L., and Burke, L. (1995). Use of menopausal estrogens and medroxyprogesterone in the United States, 1982–1992. *Obstet. Gynecol.* **85,** 6–10.

26. Division of Cancer Prevention and Control, National Cancer Institute (1988). "1987 Annual Cancer Statistics Review: Including Cancer Trends 1950–1985." National Institutes of Health, Bethesda, MD.

27. U.S. Department of Health and Human Services (1981). Hysterectomy in women aged 15–44, United States, 1970–1978. *Morbid. Mortal. Wkly. Rep.* **30,** 173–176.

28. Weiss, N. S. (1978). Assessing the risks from menopausal estrogen use: What can we learn from trends in mortality from uterine cancer. *J. Chronic Dis.* **31,** 705–708.

29. Mant, J. W., and Vessey, M. P. (1994). Ovarian and Endometrial Cancers, *Cancer Surveys* **19–20,** 287–307.

30. Coulter, A., McPherson, K., and Vessey, M. (1988). Do British women undergo too many or too few hysterectomies? *Soc. Sci. Med.* **27,** 987–994.

31. Villard, L., and Murphy, M. (1990). Endometrial cancer trends in England and Wales: A possible protective effect of oral contraceptives. *Int. J. Epidemiol.* **19,** 255–258.

32. Ryser, H. J. (1971). Chemical carcinogenesis. *N. Engl. J. Med.* **285,** 721–734.

33. Pitot, H. C. (1986). "Fundamentals of Oncology." Dekker, New York.

34. Diddle, A. W. (1952). Granulosa and theca cell ovarian tumors: Prognosis. *Cancer (Philadelphia)* **5,** 215–228.

35. Larson, J. A. (1954). Estrogens and endometrial cancer. *Obstet. Gynecol.* **3,** 551–572.

36. Salerno, L. J. (1962). Feminizing mesenchymomas of the ovary—An analysis of 28 granulosa-theca cell tumors and their relationship to coexistent carcinoma. *Am. J. Obstet. Gynecol.* **84,** 731–738.

37. Gusberg, S. B., and Kardon, P. (1971). Proliferative endometrial response to theca-granulosa cell tumors. *Am. J. Obstet. Gynecol.* **111,** 633–643.

38. Dockerty, M. B., Lovelady, S. B., and Foust, G. T., Jr. (1951). Carcinoma of the corpus uteri in young women. *Am. J. Obstet. Gynecol.* **61,** 966–981.

39. Farhi, D. C., Nosanchuk, J., and Silverberg, S. G. (1986). Endometrial adenocarcinoma in women under 25 years of age. *Obstet. Gynecol.* **68,** 741–745.

40. Chamlian, D. L., and Taylor, H. B. (1970). Endometrial hyperplasia in young women. *Obstet. Gynecol.* **36,** 659–666.

41. Coulam, C. B., Annegers, J. F., and Kranz, J. S. (1983). Chronic anovulation syndrome and associated neoplasia. *Obstet. Gynecol.* **61,** 403–407.

42. Siiteri, P. K., and MacDonald, P. C. (1973). The role of extraglandular estrogen in human endocrinology. *Handb. Physiol.* **2,** 615.

43. Damon, A. (1960). Host factors in cancer of the breast and uterine cervix and corpus. *J. Natl. Cancer Inst. (U.S.)* **24,** 483–516.

44. Wynder, E. L., Escher, G. C., and Mantel, N. (1966). An epidemiological investigation of cancer of the endometrium. *Cancer (Philadelphia)* **19,** 489–520.

45. Elwood, J. M., Cole, P., Rothman, K. J., and Kaplan, S. D. (1977). Epidemiology of endometrial cancer. *J. Natl. Cancer Inst. (U.S.)* **59,** 1055–1060.

46. Weiss, N. S., Farewall, V. T., Szekely, D. R., English, D. R., and Kiviat, N. (1980). Oestrogens and endometrial cancer: Effect of other risk factors on the association. *Maturitas* **2,** 185–190.

47. Spengler, R. F., Clarke, E. A., Woolever, A., Newman, A. M., and Osborn, R. W. (1981). Exogenous estrogens and endometrial cancer: A case-control study and assessment of potential bias. *Am. J. Epidemiol.* **114,** 497–506.

48. Stavraky, K. M., Collins, J. A., Donner, A., and Wells, G. A. (1981). A comparison of estrogen use by women with endometrial

cancer, gynecologic disorders, and other illnesses. *Am. J. Obstet. Gynecol.* **141,** 547–555.

49. Kelsey, J. L., LiVolsi, V. A., Holford, T. R., Fischer, D. B., Mostow, E. D., Schwartz, P. E., O'Connor, T., and White, C. (1982). A case-control study of cancer of the endometrium. *Am. J. Epidemiol.* **116,** 333–342.

50. Henderson, B. E., Casagrande, J. T., Pike, M. C., Mack, T., Rosario, I., and Duke, A. (1983). The epidemiology of endometrial cancer in young women. *Br. J. Cancer* **47,** 749–756.

51. Pettersson, B., Adami, H. O., and Bergström, R. (1985). Obesity, hypertension and diabetes as risk factors for endometrial cancer. *In* "Risk Factors for Endometrial Carcinoma." Acta Universitatis Upsaliensis, Uppsala, Comprehensive Summaries of Uppsala Dissertations from the Faculty of Medicine, Vol. 6, 42 pages.

52. Hulka, B. S., Fowler, W. C., Jr., Kaufman, D. G., Grimson, R. C., Greenberg, B. G., Hogue, C. J., Berger, G. S., and Pulliam, C. C. (1980). Estrogen and endometrial cancer: Cases and two control groups from North Carolina. *Am. J. Obstet. Gynecol.* **137,** 92–101.

53. Lawrence, C., Tessaro, I., Durgerian, S., Caputo, T., Richart, R., Jacobson, H., and Greenwald, P. (1987). Smoking, body weight, and early-stage endometrial cancer. *Cancer (Philadelphia)* **59,** 1665–1669.

54. Ewertz, M., Schou, G., and Boice, J. D., Jr (1988). The joint effect of risk factors on endometrial cancer. *Eur. J. Cancer Clin. Oncol.* **24,** 189–194.

55. Folsom, A. R., Kaye, S. A., Potter, J. D., and Prineas, R. J. (1989). Association on incident carcinoma of the endometrium with body weight and fat distribution in older women: Early findings of the Iowa Women's Health Study. *Cancer Res.* **49,** 6828–6831.

56. Rubin, G. L., Peterson, H. B., Lee, N. C., Maes, E. F., Wingo, P. A., and Becker, S. (1990). Estrogen replacement therapy and the risk of endometrial cancer: Remaining controversies. *Am. J. Obstet. Gynecol.* **162,** 148–154.

57. Parazzini, F., La Vecchia, C., Negri, E., Fedele, L., and Balotta, F. (1991). Reproductive factors and risk of endometrial cancer. *Am. J. Obstet. Gynecol.* **164,** 522–527.

58. Austin, H., Austin, J. M., Partridge, E. E., Hatch, K. D., and Shingleton, H. M. (1991). Endometrial cancer, obesity, and body fat distribution. *Cancer Res.* **51,** 568–572.

59. Shu, X.-O., Brinton, L. A., Zheng, W., Gao, Y. T., Fan, J., and Fraumeni, J. F. (1991). A population-based case-control study of endometrial cancer in Shanghai, China. *Int. J. Cancer* **49,** 38–43.

60. Swanson, C. A., Wilbanks, G. D., Twiggs, L. B., Mortel, R., Berman, M. L., Barrett, R. J., and Brinton, L. A. (1993). Moderate alcohol consumption and the risk of endometrial cancer. *Epidemiology* **4,** 530–536.

61. Tornberg, S. A., and Carstensen, J. M. (1994). Relationship between Quetelet's index and cancer of breast and female genital tract in 47,000 women followed for 25 years. *Br. J. Cancer* **69,** 358–361.

62. Olson, S. H., Trevisan, M., Marshall, J. R., Graham, S., Zielezny, M., Vena, J. E., Hellmann, R., and Freudenheim, J. L. (1995). Body mass index, weight gain, and risk of endometrial cancer. *Nutr. Cancer* **23,** 141–149.

63. Baanders-van Halewyn, E. A., Blankenstein, M. A., Thijssen, J. H., de Ridder, C. M., and de Waard, F. (1996). A comparative study of risk factors for hyperplasia and cancer of the endometrium. *Eur. J. Cancer Prev.* **5,** 105–112.

64. Hirose, K., Tajima, K., Hamajima, N., Takezaki, T., Inoue, M., Kuroishi, T., Kuzuya, K., Nakamura, S., and Tokudome, S. (1996). Subsite (cervix/endometrium)-specific risk and protective factors in uterus cancer. *Jpn. J. Cancer Res.* **87,** 1001–1009.

65. Kalandidi, A., Tzonou, A., Lipworth, L., Gamatsi, I., Filippa, D., and Trichopoulos, D. (1996). A case-control study of endometrial cancer in relation to reproductive, somatometric, and life-style variables. *Oncology* **53,** 354–359.

66. Goodman, M. T., Hankin, J. H., Wilkens, L. R., Lyu, L. C., McDuffie, K., Liu, L. Q., and Kolonel, L. N. (1997). Diet, body size, physical activity, and the risk of endometrial cancer. *Cancer Res.* **57,** 5077–5085.

67. Shu, X. O., Hatch, M. C., Zheng, W., Gao, Y. T., and Brinton, L. A. (1993). Physical activity and risk of endometrial cancer. *Epidemiology* **4,** 342–349.

68. Levi, F., La Vecchia, C., Negri, E., and Franceschi, S. (1993). Selected physical activities and the risk of endometrial cancer. *Br. J. Cancer* **67,** 846–851.

69. Sturgeon, S. R., Brinton, L. A., Berman, M. L., Mortel, R., Twiggs, L. B., Barrett, R. J., and Wilbanks, G. D. (1993). Past and present physical activity and endometrial cancer risk. *Br. J. Cancer* **68,** 584–589.

70. MacDonald, P. C., and Siiteri, P. K. (1974). The relationship between the extraglandular production of estrone and the occurrence of endometrial neoplasia. *Gynecol. Oncol.* **2,** 259–263.

71. MacDonald, P. C., Edman, C. D., Hemsell, D. L., Porter, J. C., and Siiteri, P. K. (1978). Effect of obesity on conversion of plasma androstenedione to estrone in postmenopausal women with and without endometrial cancer. *Am. J. Obstet. Gynecol.* **130,** 448–455.

72. Edman, C. D., and MacDonald, P. C. (1978). Effect of obesity on conversion of plasma androstenedione to estrone in ovulatory and anovulatory young women. *Am. J. Obstet. Gynecol.* **130,** 456–461.

73. Davidson, B. J., Gambone, J. C., Lagasse, L. D., Castaldo, T. W., Hammond, G. L., Siiteri, P. K., and Judd, H. L. (1981). Free estradiol in postmenopausal women with and without endometrial cancer. *J. Clin. Endocrinol. Metab.* **52,** 404–408.

74. Kaye, S. A., Folsom, A. R., Soler, J. T., Prineas, R. J., and Potter, J. D. (1991). Association of body mass and fat distribution with sex hormone concentrations in postmenopausal women. *Int. J. Epidemiol.* **20,** 151–156.

75. Nyholm, H. C., Nielsen, A. L., Lyndrup, J., Dreisler, A., Hagen, C., and Haug, E. (1993). Plasma oestrogens in postmenopausal women with endometrial cancer. *Br. J. Obstet. Gynaecol.* **100,** 1115–1119.

76. King, R. J. B., and Whitehead, M. I. (1983). Estrogen and progestin effects on epithelium and stroma from pre- and postmenopausal endometria: Application to clinical studies of the climacteric syndrome *In* "Steroids and Endometrial Cancer" (V. M. Jasonni, ed.), pp. 105–115. Raven Press, New York.

77. Lucas, W. E. (1974). Causal relationships between endocrine-metabolic variables in patients with endometrial carcinoma. *Obstet. Gynecol. Surv.* **29,** 507–528.

78. Yen, S. S. (1980). The polycystic ovary syndrome. *Clin. Endocrinol. (Oxford)* **12,** 177–207.

79. Sherman, B. M., and Korenman, S. G. (1974). Measurement of serum LH, FSH, estradiol and progesterone in disorders of the human menstrual cycle: The inadequate luteal phase. *J. Clin. Endocrinol. Metab.* **39,** 145–149.

80. Gusberg, S. B. (1947). Precursors of corpus carcinoma estrogens and adenomatous hyperplasia. *Am. J. Obstet. Gynecol.* **54,** 905–927.

81. The Writing Group for the PEPI Trial (1995). Effects of estrogen or estrogen/progestin regimens on heart disease risk factors in postmenopausal women: The Postmenopausal Estrogen/Progestin Interventions (PEPI) Trial. *J. Am. Med. Assoc.* **273,** 199–208.

82. Speroff, L., Rowan, J., Symons, J., Genant, H., and Wilborn, W. (1996). The comparative effect on bone density, endometrium, and lipids of continuous hormones as replacement therapy (CHART Study). *JAMA, J. Am. Med. Assoc.* **276,** 1397–1403.

83. Kistner, R. W. (1973). Endometrial alterations associated with estrogen and estrogen-progestin combinations. *In* "The Uterus" (H. J. Norris, A. T. Hertig, and M. R. Abell, eds.), pp. 227–254. Williams & Wilkins, Baltimore, MD.

84. Thom, M. H., White, P. J., Williams, R. M., Sturdee, D. W., Paterson, M. E., Wade-Evans, T., and Studd, J. W. (1979). Prevention and treatment of endometrial disease in climacteric women receiving oestrogen therapy. *Lancet* 2, 455–457.

85. Gusberg, S. B., and Hall, R. E. (1961). Precursors of corpus cancer III. The appearance of cancer of the endometrium in estrogenically conditioned patients. *Obstet. Gynecol.* 17, 397–412.

86. Pettersson, B., Adami, H. O., Lindgren, A., and Hesselius, I. (1985). Endometrial polyps and hyperplasia as risk factors for endometrial carcinoma. A case-control study of curettage specimens. *Acta Obstet. Gynecol. Scand.* 64, 653–659.

87. Deligdisch, L., and Holinka, C. F. (1987). Endometrial carcinoma: Two diseases? *Cancer Detect. Prev.* 10, 237–246.

88. Kurman, R. J., Kaminski, P. F., and Norris, H. J. (1985). The behavior of endometrial hyperplasia. A long-term study of "untreated" hyperplasia in 170 patients. *Cancer (Philadelphia)* 56, 403–412.

89. Mack, T. M., Pike, M. C., Henderson, B. E., Pfeffer, R. I., Gerkins, V. R., Arthur, M., and Brown, S. E. (1976). Estrogens and endometrial cancer in a retirement community. *N. Engl. J. Med.* 294, 1262–1267.

90. La Vecchia, C., Franceschi, S., DeCarli, A., *et al.* (1984). Risk factors for endometrial cancer at different ages. *NCI, J. Natl. Cancer Inst.* 73, 667–671.

91. McDonald, T. W., Annegers, J. F., O'Fallon, W. M., Dockerty, M. B., Malkasian, G. D., Jr., and Kurland, L. T. (1977). Exogenous estrogen and endometrial carcinoma: Case-control and incidence study. *Am. J. Obstet. Gynecol.* 127, 572–580.

92. Shapiro, S., Kaufman, D. W., Slone, D., Rosenberg, L., Miettinen, O. S., Stolley, P. D., Rosenshein, N. B., Watring, W. G., Leavitt, T., Jr., and Knapp, R. C. (1980). Recent and past use of conjugated estrogens in relation to adenocarcinoma of the endometrium. *N. Engl. J. Med.* 303, 485–489.

93. Ziel, H. K., and Finkle, W. D. (1975). Increased risk of endometrial carcinoma among users of conjugated estrogens. *N. Engl. J. Med.* 293, 1167–1170.

94. Shapiro, S., Kelly, J. P., Rosenberg, L., Kaufman, D. W., Helmrich, S. P., Rosenshein, N. B., Lewis, J. L., Jr., Knapp, R. C., Stolley, P. D., and Schottenfeld, D. (1985). Risk of localized and widespread endometrial cancer in relation to recent and discontinued use of conjugated estrogens. *N. Engl. J. Med.* 313, 969–972.

95. Smith, D. C., Prentice, R., Thompson, D. J., and Herrmann, W. L. (1975). Association of exogenous estrogen and endometrial carcinoma. *N. Engl. J. Med.* 293, 1164–1167.

96. Weiss, N. S., Szekely, D. R., English, D. R., and Schweid, A. I. (1979). Endometrial cancer in relation to patterns of menopausal estrogen use. *JAMA, J. Am. Med. Assoc.* 242, 261–264.

97. Hunt, K., Vessey, M., McPherson, K., and Coleman, M. (1987). Long-term surveillance of mortality and cancer incidence in women receiving hormone replacement therapy. *Br. J. Obstet. Gynaecol.* 94, 620–635.

98. Hoover, R., Fraumeni, J. F., Everson, R., and Myers, M. H. (1976). Cancer of the uterine corpus after hormonal treatment for breast cancer. *Lancet* 1, 885–887.

99. Antunes, C. M., Strolley, P. D., Rosenshein, N. B., Davies, J. L., Tonascia, J. A., Brown, C., Burnett, L., Rutledge, A., Pokempner, M., and Garcia, R. (1979). Endometrial cancer and estrogen use. Report of a large case-control study. *N. Engl. J. Med.* 300, 9–13.

100. Vakil, D., Morgan, R., and Haliday, M. (1983). Exogenous estrogens and development of breast and endometrial cancer. *Cancer Detect. Prev.* 6, 415–424.

101. Buring, J. E., Bain, C. J., and Ehrmann, R. L. (1986). Conjugated estrogen use and risk of endometrial cancer. *Am. J. Epidemiol.* 124, 434–441.

102. Gray, L. A., Sr., Christopherson, W. M., and Hoover, R. N. (1977). Estrogens and endometrial carcinoma. *Obstet. Gynecol.* 49, 385–389.

103. Hoogerland, D. L., Buchler, D. A., Crowley, J. J., and Carr, W. F. (1978). Estrogen use—risk of endometrial carcinoma. *Gynecol. Oncol.* 6, 451–458.

104. Jelovsek, F. R., Hammond, C. B., Woodard, B. H., Draffin, R., Lee, K. L., Creasman, W. T., and Parker, R. T. (1980). Risk of exogenous estrogen therapy and endometrial cancer. *Am. J. Obstet. Gynecol.* 137, 85–91.

105. Horwitz, R. I., and Feinstein, A. R. (1978). Alternative analytic methods for case-control studies of estrogens and endometrial cancer. *N. Engl. J. Med.* 299, 1089–1094.

106. Jick, H., Watkins, R. N., Hunter, J. R., Dinan, B. J., Madsen, S., Rothman, K. J., and Walker, A. M. (1979). Replacement estrogens and endometrial cancer. *N. Engl. J. Med.* 300, 218–222.

107. Hammond, C. B., Jelovsek, F. R., Lee, K. L., Creasman, W. T., and Parker, R. T. (1979). Effects of long-term estrogen replacement therapy. I. Metabolic effects. *Am. J. Obstet. Gynecol.* 133, 525–536.

108. Gambrell, R. D., Massey, F. M., Castaneda, T. A., Ugenas, A. J., and Ricci, C. A. (1979). Reduced incidence of endometrial cancer among postmenopausal women treated with progestogens. *J. Am. Geriatr. Soc.* 27, 389–394.

109. Lafferty, F., and Helmuth, D. (1985). Postmenopausal estrogen-replacement: The prevention of osteoporosis and systemic effects. *Maturitas* 7, 147–159.

110. Stampfer, M., Colgitz, G., Willet, W., Rosner, B., Hennekens, C., and Speizer, F. (1986). A prospective study of exogenous hormones and risk of endometrial cancer. *Am. J. Epidemiol.* 124, 520 (abstr.).

111. Petitti, D. B., Perlman, J. A., and Sidney, S. (1987). Noncontraceptive estrogens and mortality: Long-term follow-up of women in the Walnut Creek Study. *Obstet. Gynecol.* 70, 289–293.

112. Ettinger, B., Golditch, I. M., and Friedman, G. (1988). Gynecologic consequences of long-term, unopposed estrogen replacement therapy. *Maturitas* 10, 271–282.

113. Paganini-Hill, A., Ross, R. K., and Henderson, B. E. (1989). Endometrial cancer and patterns of use of oestrogen replacement therapy: A cohort study. *Br. J. Cancer* 59, 445–447.

114. Persson, I., Adami, H., Bergkvist, L., Lindgren, A., Pettersson, B., Hoover, R., and Schairer, C. (1989). Risk of endometrial cancer after treatment with oestrogens alone or in conjunction with progestogens: Results of a prospective study. *Br. Med. J.* 298, 147–151.

115. Levi, F., La Vecchia, C., Gulie, C., Franceschi, S., and Negri, E. (1993). Oestrogen replacement treatment and the risk of endometrial cancer: An assessment of the role of covariates. *Eur. J. Cancer* 29A, 1445–1449.

116. Brinton, L. A., Hoover, R. N., and the Endometrial Cancer Collaborative Group (1993). Estrogen replacement therapy and endometrial cancer risk: Unresolved issues. *Obstet. Gynecol.* 81, 265–271.

117. Jick, S. S., Walker, A. M., and Jick, H. (1993). Estrogens, progesterone, and endometrial cancer. *Epidemiology* 4, 20–24.

118. Beresford, S. A. A., Weiss, N. S., Voigt, L. F., and McKnight, B. (1997). Risk of endometrial cancer in relation to use of oestrogen combined with cyclic progestagen in postmenopausal women. *Lancet* 349, 458–461.

119. Pike, M. C., Peters, R. K., Cozen, W., Probst-Hensch, N. M., Felix, J. C., Wan, P. C., and Mack, T. M. (1997). Estrogen-progestin replacement therapy and endometrial cancer. *J. Nat. Cancer Inst.* 89, 1110–1116.

120. Grady, D., and Ernster, V. L. (1996). Endometrial cancer. *In* "Cancer Epidemiology and Prevention" (D. Schottenfeld and J. F. Fraumeni, eds.), pp. 1058–1089. Oxford University Press, New York.

121. Grady D., Gebretsadik, T., Kerlikowske K., Ernster, V., and Petitti, D. (1995). Hormone replacement therapy and endometrial cancer risk: A meta-analysis. *Obstet. Gynecol.* 85, 304–313.

122. Herrington, L. J., and Weiss, N. S. (1993). Postmenopausal unopposed estrogens: Characteristics of use in relation to the risk of endometrial carcinoma. *Ann. Epidemiol.* **3,** 308–318.

123. Green, P. K., Weiss, N. S., McKnight, B., Voigt, L. F., and Beresford, S. A. (1996). Risk of endometrial cancer following cessation of menopausal hormone use (Washington, United States). *Cancer Causes Control* **7,** 575–580.

124. Pettersson, B., Adami, H. O., Persson, I., Bergstrom, R., Lindgren, A., and Johansson, E. D. (1986). Climacteric symptoms and estrogen replacement therapy in women with endometrial carcinoma. *Acta Obstet. Gynecol. Scand.* **65,** 81–87.

125. Finkle, W. D., Greenland, S., Miettinen, O. S., and Ziel, H. K. (1995). Endometrial cancer risk after discontinuing use of unopposed conjugated estrogens (California, United States). *Cancer Causes Control* **6,** 99–102.

126. Pettersson, B., Adami, H. O., Bergstrom, R., and Johansson, E. D. (1986). Menstruation span—a time-limited risk factor for endometrial carcinoma. *Acta Obstet. Gynecol. Scand.* **65,** 247–255.

127. Cushing K. L., Weiss, N. S., Voigt, L. F., McKnight, B., and Beresford, S. A. A. (1998). Risk of endometrial cancer in relation to use of low-dose, unopposed estrogens. *Obstet. Gynecol.* **91,** 35–39.

128. Genant, H. K., Lucas, J., Weiss, S., Akin, M., Emkey, R., McNaney-Flint, H., Downs, R., Mortola, J., Watts, N., Yang, H. M., Banav, N., Brennan, J. J., and Nolan, J. C. (1997). Low-dose esterified estrogen therapy. *Archiv. Intern. Med.* **157,** 2609–2615.

129. Schiff, I., Sela, H. K., Cramer, D., Tulchinsky, D., and Ryan, K. J. (1982). Endometrial hyperplasia in women on cyclic or continuous estrogen regimens. *Fertil. Steril.* **37,** 79–82.

130. Hulka, B. S., Kaufman, D. G., Fowler, W. C., Jr., Grimson, R. C., and Greenberg, B. G. (1980). Predominance of early endometrial cancers after long-term estrogen use. *JAMA, J. Am. Med. Assoc.* **244,** 2419–2422.

131. Weiss, N. S. (1978). Noncontraceptive estrogens and abnormalities of endometrial proliferation. *Ann. Intern. Med.* **88,** 410–412.

132. Robboy, S. J., and Bradley, R. (1979). Changing trends and prognostic features in endometrial cancer associated with exogenous estrogen therapy. *Obstet. Gynecol.* **54,** 269–277.

133. Elwood, J. M., and Boyes, D. A. (1980). Clinical and pathological features and survival of endometrial cancer patients in relation to prior use of estrogens. *Gynecol. Oncol.* **10,** 173–187.

134. Collins, J., Donner, A., Allen, L. H., and Adams, O. (1980). Oestrogen use and survival in endometrial cancer. *Lancet* **2,** 961–964.

135. Chu, J., Schweid, A. I., and Weiss, N. S. (1982). Survival among women with endometrial cancer: A comparison of estrogen users and nonusers. *Am. J. Obstet. Gynecol.* **143,** 569–573.

136. Cutler, B. S., Forbes, A. P., Ingersoll, F. M., and Scully, R. E. (1972). Endometrial carcinoma after stilbestrol therapy in gonadal dysgenesis. *N. Engl. J. Med.* **287,** 628–631.

137. Gorodeski, G. I., Beery, R., Lunenfeld, B., and Geier, A. (1992). Tamoxifen increases plasma estrogen-binding equivalents and has estradiol agonistic effect on histologically normal premenopausal and postmenopausal endometrium. *Fertil. Steril.* **57,** 320–327.

138. Satyaswaroop, P. G., Zaino, R. J., and Mortel, R. (1984). Estrogen-like effects of tamoxifen on human endometrial carcinoma transplanted into nude mice. *Cancer Res.* **44,** 4006–4010.

139. Fornander, T., Rutqvist, L. E., Cedermark, B., Glas, U., Mattsson, A., Silfversward, C., Skoog, L., Somell, A., Theve, T., Wilking, N., Askergren, J., and Hjalmar, M.-L. (1989). Adjuvant tamoxifen in early breast cancer: occurrence of new primary cancers. *Lancet* **1,** 117–120.

140. Fisher, B., Costantino, J. P., Redmond, C. K., Fisher E. R., Wickerham, D. L., and Cronin, W. M. (1994). Endometrial cancer in tamoxifen-treated breast cancer patients: Findings from the National Surgical Adjuvant Breast and Bowel Project (NSABP) B-14. *J. Nat. Cancer Inst.* **86,** 27–37.

141. Andersson, M., Storm, H. H., and Mouridsen, H. T. (1991). Incidence of new primary cancers after adjuvant tamoxifen therapy and radiotherapy for early breast cancer. *J. Nat. Cancer Inst.* **83,** 1013–1017.

142. van Leeuwen, F. E., Benraadt, J., Coebergh, J. W. W., Kiemeney, L. A. L. M., Gimbrere, C. H. F., Otter, R., Schouten, L. J., Damhui, R. A. M., Bontenbal, M., Diepenhorst, F. W., van den Belt-Dusebout, A. W., and van Tinteren, H. (1994). Risk of endometrial cancer after tamoxifen treatment of breast cancer. *Lancet* **343,** 448–452.

143. Sasco, A. J., Chaplin, G., Amoros, E., and Saez, S. (1996). Endometrial cancer following breast cancer: effect of tamoxifen and castration by radiotherapy. *Epidemiology* **7,** 9–13.

144. Ribeiro, G., and Swindell, R. (1992). The Christie Hospital Adjuvant Tamoxifen Trial. *J. Nat. Cancer Inst. Monog.* **11,** 121–125.

145. Stewart, H. J., and The Scottish Cancer Trials Breast Group (1992). The Scottish trial of adjuvant tamoxifen in node-negative breast cancer. *J. Natl. Cancer Inst. Monogr.* **11,** 117–120.

146. Cook, L. S., Weiss, N. S., Schwartz, S. M., White, E., McKnight, B., Moore, D. E., and Daling, J. R. (1995). A population-based study of tamoxifen therapy and subsequent ovarian, endometrial, and breast cancers. *J. Nat. Cancer Inst.* **87,** 1359–1364.

147. Fisher, B., and Redmond, C. (1991). New perspective on cancer of the contralateral breast: A marker for assessing tamoxifen as a preventive agent. *J. Nat. Cancer Inst.* **83,** 1278–1280.

148. Jordan, V. C. (1995). Tamoxifen: Toxicities and drug resistance during the treatment and prevention of breast cancer. *Annu. Rev. Pharmacol. Toxicol.* **35,** 195–211.

149. Boss, S. M., Huster, W. J., Neild, J. A., Glant, M. D., Eisenhut, C. C., and Draper, M. W. (1997). Effects of raloxifene hydrochloride on the endometrium of postmenopausal women. *Am. J. Obstet. Gynecol.* **177,** 1458–1464.

150. Staland, B. (1985). Continuous treatment with a combination of estrogen and gestagen—a way of avoiding endometrial stimulation. Clinical experiences with Kliogest. *Acta Obstet. Gynecol. Scand., Suppl.* **130,** 29–35.

151. Mattsson, L. A., and Samsioe, G. (1985). Estrogen-progestogen replacement in climacteric women, particularly as regards a new type of continuous regimen. *Acta Obstet. Gynecol. Scand., Suppl.* **130,** 53–58.

152. Magos, A. L., Brincat, M., Studd, J. W., Wardle, P., Schlesinger, P., and O'Dowd, T. (1985). Amenorrhea and endometrial atrophy with continuous oral estrogen and progestogen therapy in postmenopausal women. *Obstet. Gynecol.* **65,** 496–499.

153. Whitehead, M. I., McQueen, J., Beard, R. J., Minardi, J., and Campbell, S. (1977). The effects of cyclical oestrogen therapy and sequential oestrogen/progestogen therapy on the endometrium of postmenopausal women. *Acta Obstet. Gynecol. Scand., Suppl.* **65,** 91–101.

154. Whitehead, M. I., King, R. J., McQueen, J., and Campbell, S. (1979). Endometrial histology and biochemistry in climacteric women during oestrogen and oestrogen/progestogen therapy. *J. R. Soc. Med.* **72,** 322–327.

155. Paterson, M. E., Wade-Evans, T., Sturdee, D. W., Thom, M. H., and Studd, J. W. (1980). Endometrial disease after treatment with oestrogens and progestogens in the climacteric. *Br. Med. J.* **280,** 822–824.

156. Sturdee, D. W., Wade-Evans, T., Paterson, M. E., Thom, M., and Studd, J. W. (1978). Relations between bleeding pattern, endometrial histology, and oestrogen treatment in menopausal women. *Br. Med. J.* **1,** 1575–1577.

157. Clisham, P. R., de Ziegler, D., Lozano, K., and Judd, H. L. (1991). Comparison of continuous versus sequential estrogen and pro-

gestin therapy in postmenopausal women. *Obstet. Gynecol.* **77,** 241–246.

158. Hirvonen, E., Salmi, T., Puolakka, J., Heikkinen, J., Grnfors, E., Hulkko, S., Makarainen, L., Nummi, S., Pekonen, F., Rautio, A.-M., Sunderstrom, H., Telimaa, S., Wilen-Rosenqvist, G., Virkkunen, A., and Wahlstrom, T. (1995). Can progestin be limited to very third month only in postmenopausal women taking eestrogen? *Maturitas* **21,** 39–44.

159. Ettinger, B., Selby, J., Citron, J. T., Vangessel, A., Ettinger, V. M., and Hendrickson, M. R. (1994). Cyclic hormone replacement therapy using quarterly progestin. *Obstet. Gynecol.* **83,** 693–700.

160. Cerin, A., Heldaas, K., Moeller, B., and the Scandinavian LongCycle Study Group (1996). Adverse endometrial effects of long-cycle estrogen and progestogen replacement therapy (letter). *N. Engl. J. Med.* **334,** 668–669.

161. Nachtigall, L. E., Nachtigall, R. H., and Nachtigall, R. D. (1979). Estrogen replacement therapy. II. A prospective study in the relationship to carcinoma and cardiovascular and metabolic problems. *Obstet. Gynecol.* **54,** 74–79.

162. Voigt, L. F., Weiss, N. S., Chu, J., Daling, J. R., Mcknight, B., and Van Belle, G. (1991). Progestagen supplementation of exogenous oestrogens and risk of endometrial cancer. *Lancet* **338,** 274–277.

163. Shapiro, J. A., Weiss, N. S., Beresford, S. A. A., and Voigt, L. F. (1998). Menopausal hormone use and endometrial cancer, by tumor grade and invasion. *Epidemiology* **9,** 99–101.

164. Liang, A. P., Levenson, A. G., Layde, P. M., Shelton, J. D., Hatcher, R. A., Potts, M., and Michelson, M. J. (1983). Risk of breast, uterine corpus, and ovarian cancer in women receiving medroxyprogesterone injections. *JAMA, J. Am. Med. Assoc.* **249,** 2909–2912.

165. WHO Collaborative Study of Neoplasia and Steroid Contraceptives (1991). Depot medroxyprogesterone acetate (DMPA) and risk of endometrial cancer. *Int. J. Cancer* **49,** 186–190.

166. Lyon, F. A., and Frisch, M. J. (1976). Endometrial abnormalities occurring in young women on long-term sequential oral contraceptives. *Obstet. Gynecol.* **47,** 639–643.

167. Kaufman, R. H., Reeves, K. O., and Dougherty, C. M. (1976). Severe atypical endometrial changes and sequential contraceptive use. *JAMA, J. Am. Med. Assoc.* **236,** 923–926.

168. Silverberg, S. G., Makowski, E. L., and Roche, W. D. (1977). Endometrial carcinoma in women under 40 years of age: Comparison of cases in oral contraceptive users and nonusers. *Cancer (Philadelphia)* **39,** 592–598.

169. Weiss, N. S., and Sayvetz, T. A. (1980). Incidence of endometrial cancer in relation to the use of oral contraceptives. *N. Engl. J. Med.* **302,** 551–554.

170. Centers for Disease Control, Cancer and Steroid Hormone Study (1983). Oral contraceptive use and the risk of endometrial cancer. *JAMA, J. Am. Med. Assoc.* **249,** 1600–1604.

171. Kaufman, D. W., Shapiro, S., Slone, D., Rosenberg, L., Miettinen, O. S., Stolley, P. D., Knapp, R. C., Leavitt, T., Jr., Watring, W. G., Rosenshein, N. B., Lewis, J. L., Schottenfeld, D., and Engle, R. L., Jr. (1980). Decreased risk of endometrial cancer among oral-contraceptive users. *N. Engl. J. Med.* **303,** 1045–1047.

172. La Vecchia, C., Decarli, A., Fasoli, M., Franceschi, S., Gentile, A., Negri, E., Parazzini, F., and Tognoni, G. (1986). Oral contraceptives and cancers of the breast and of the female genital tract. Interim results from a case-control study. *Br. J. Cancer* **54,** 311–317.

173. Hulka, B. S., Chambless, L. E., Kaufman, D. G., Fowler, W. C., Jr., and Greenberg, B. G. (1982). Protection against endometrial carcinoma by combination-product oral contraceptives. *JAMA, J. Am. Med. Assoc.* **247,** 475–477.

174. Centers for Disease Control and the National Institute of Child Health and Human Development, Cancer and Steroid Hormone

Study (1987). Combination oral contraceptive use and the risk of endometrial cancer. *JAMA, J. Am. Med. Assoc.* **257,** 796–800.

175. Armstrong, B. K., Ray, R. M., and Thomas, D. B. (1988). Endometrial cancer and combined oral contraceptives. The Who Collaborative Study of Neoplasia and Steroid Contraceptives. *Int. J. Epidemiol.* **17,** 263–269.

176. Beral V., Hannaford, P., and Kay, C. (1988). Oral contraceptive use and malignancies of the genital tract. *Lancet* **2,** 1331–1335.

177. Levi, F., La Vecchia, C., Gulie, C., Negri, E., Monnier, V., Franceschi, S., Delaloye, J.-F., and De Grandi, P. (1991). Oral contraceptives and the risk of endometrial cancer. *Cancer Causes Control* **2,** 99–103.

178. Stanford, J. L., Brinton, L. A., Berman, M. L., Mortel, R., Twiggs, L. B., Barrett, R. J., Wilbanks, G. D., and Hoover, R. N. (1993). Oral contraceptives and endometrial cancer: Do other risk factors modify the association? *Int. J. Cancer* **54,** 243–248.

179. Voigt, L. F., Deng, Q., and Weiss, N. S. (1994). Recency, duration, and progestin content of oral contraceptives in relation to the incidence of endometrial cancer (Washington, USA). *Cancer Causes Control* **5,** 227–233.

180. Vessey, M. P., and Painter, R. (1995). Endometrial and ovarian cancer and oral contraceptives—findings in a large cohort study. *Br. J. Cancer* **71,** 1340–1342.

181. Rosenblatt, K. A., Thomas, D. B., and the WHO Collaborative Study of Neoplasia and Steroid Contraceptives (1991). Hormonal content of combined oral contraceptives in relation to the reduced risk of endometrial cancer. *Int. J. Cancer* **49,** 870–874.

182. Kvale, G., Heuch, I., and Ursin, G. (1988). Reproductive factors and risk of cancer of the uterine corpus: A prospective study. *Cancer Res.* **48,** 6217–6221.

183. Koumantaki, Y., Tzonou, A., Koumantakis, E., Kaklamani, E., Aravantinos, D., and Trichopoulos, D. (1989). A case-control study of cancer of endometrium in Athens. *Int. J. Cancer* **43,** 795–799.

184. Brinton, L. A., Berman, M. L., Mortel, R., Twiggs, L. B., Barrett, R. J., Wilbanks, G. D., Lannom, L., and Hoover, R. N. (1992). Reproductive, menstrual, and medical risk factors for endometrial cancer: Results from a case-control study. *Am. J. Obstet. Gynecol.* **167,** 1317–1325.

185. McPherson, C. P., Sellers, T. A., Potter, J. D., Bostick, R. M., and Folsom, A. R. (1996). Reproductive factors and risk of endometrial cancer. The Iowa Women's Health Study. *Am. J. Epidemiol.* **143,** 1195–1202.

186. Inoue, M., Okayama, A., Fujita, M., Enomoto, T., Tanizawa, O., and Ueshima, H. (1994). A case-control study on risk factors for uterine endometrial cancer in Japan. *Jpn. J. Cancer Res.* **85,** 346–350.

187. Albrektsen, G., Heuch, I., Tretli, S., and Kvale, G. (1995). Is the risk of cancer of the corpus uteri reduced by a recent pregnancy? A prospective study of 765,756 Norwegian women. *Int. J. Cancer* **61,** 485–490.

188. Lochen, M. L., and Lund, E. (1997). Childbearing and mortality from cancer of the corpus uteri. *Acta Obstet. Gynecol. Scand.* **76,** 373–377.

189. Lesko, S. M., Rosenberg, L., Kaufman, D. W., Stolley, P., Warshauer, M. E., Lewis, J. L., Jr., and Shapiro, S. (1991). Endometrial cancer and age at last delivery: Evidence for an association. *Am. J. Epidemiol.* **133,** 554–559.

190. Rosenblatt, K. A., and Thomas, D. B. (1995). Prolonged lactation and endometrial cancer. WHO Collaborative Study of Neoplasia and Steroid Contraceptives. *Int. J. Epidemiol.* **24,** 499–503.

191. Escobedo, L. G., Lee, N. C., Peterson, H. B., and Wingo, P. A. (1991). Infertility-associated endometrial cancer risk may be limited to specific subgroups of infertile women. *Obstet. Gynecol.* **77,** 124–128.

192. Venn, A., Watson, L., Lumley, J., Giles, G., King, C., and Healy, D. (1995). Breast and ovarian cancer incidence after infertility and in vitro fertilization. *Lancet* **346**, 995–1000.

193. Modan, B., Ron, E., Lerner-Geva, L., Blumstein, T., Menczer, J., Rabinovici, J., Oelsner, G., Freedman, L., Mashiach, S., and Lunenfeld, B. (1998). Cancer incidence in a cohort of infertile women. *Am. J. Epidemiol.* **147**, 1038–1042.

194. Shaw, S. T., and Macaulay, L. K. (1979). Morphorlogic studies on IUD-induced metrorrhagia. II. Surface changes of the endometrium and microscopic localization of bleeding sites. *Contraception* **19**, 63–81.

195. Moyer, D. L., and Mishell, D. R. (1971). Reactions of human endometrium to the intrauterine foreign body. II. Long-term effects on the endometrial histology and cytology. *Am. J. Obstet. Gynecol.* **111**, 66–80.

196. Hill, D. A., Weiss, N. S., Voigt, L. F., and Beresford, S. A. (1997). Endometrial cancer in relation to intra-uterine device use. *Int. J. Cancer* **70**, 278–281.

197. Castellsague, X., Thompson, W. D., and Dubrow, R. (1993). Intra-uterine contraception and the risk of endometrial cancer. *Int. J. Cancer* **54**, 911–916.

198. Parazzini, F., La Vecchia, C., and Moroni, S. (1994). Intrauterine device use and risk of endometrial cancer. *Br. J. Cancer* **70**, 672–673.

199. Sturgeon, S. R., Brinton, L. A., Berman, M. L., Mortel, R., Twiggs, L. B., Barrett, R. J., Wilbanks, G. D., and Lurain, J. R. (1997). Intrauterine device use and endometrial cancer risk. *Int. J. Epidemiol.* **26**, 496–500.

200. Rosenblatt, K. A., Thomas, D. B., and the WHO Collaborative Study of Neoplasia and Steroid Contraception (1996). Intrauterine devices and endometrial cancer. *Contraception* **54**, 329–332.

201. Wideroff, L., Gridley, G., Mellemkjaer, L., Chow, W.-H., Linet, M., Keehn, S., Borch-Johnsen, K., and Olsen, J. H. (1997). Cancer incidence in a population-based cohort of patients hospitalized with diabetes mellitus in Denmark. *J. Nat. Cancer Inst.* **89**, 1360–1365.

202. Weiderpass, E., Gridley G., Persson, I., Nyren, O., Ekbom, A., and Adami, H.-O. (1997). Risk of endometrial and breast cancer in patients with diabetes mellitus. *Int. J. Cancer* **71**, 360–363.

203. Ragozzino, M., Melton, L. J., 3rd, Chu, C. P., and Palumbo, P. J. (1982). Subsequent cancer risk in the incidence cohort of Rochester, Minnesota, residents with diabetes mellitus. *J. Chronic Dis.* **35**, 13–19.

204. Kessler, I. I. (1970). Cancer mortality among diabetics. *J. Natl. Cancer Inst. (U.S.)* **44**, 673–680.

205. O'Mara, B. A., Byers, T., and Schoenfeld E. (1985). Diabetes mellitus and cancer risk: A multisite case-control study. *J. Chronic Dis.* **38**, 435–441.

206. Maatela, J., Aromaa, A., Salmi, T., Pohja, M., Vuento, M., and Gronroos, M. (1994). The risk of endometrial cancer in diabetic and hypertensive patients: A nationwide record-linkage study in Finland. *Ann. Chir. Gynaecol., Suppl.* **208**, 20–24.

207. La Vecchia, C., Negri, E., Franceschi, S., D'Avanzo, B., and Boyle, P. (1994). A case-control study of diabetes mellitus and cancer risk. *Br. J. Cancer* **70**, 950–953.

208. Olson, S. H., Vena, J. E., Dorn, J. P., Marshall, J. R., Zielezny, M., Laughlin, R., and Graham, S. (1997). Exercise, occupational activity, and risk of endometrial cancer. *Ann. Epidemiol.* **7**, 46–53.

209. Ascherio, A., Hennekens, C., Willet, W. C., Sacks, F., Rosner, B., Manson, J., Witteman, J., and Stampfer, M. J. (1996). Prospective study of nutritional factors, blood pressure, and hypertension among U.S. women. *Hypertension* **27**, 1065–1072.

210. Armstrong, B., and Doll, R. (1975). Environmental factors and cancer incidence and mortality in different countries, with special reference to dietary practices. *Int. J. Cancer* **15**, 617–631.

211. Potischman, N., Swanson, C. A., Brinton, L. A., McAdams, M., Barrett, R. J., Berman, M. L., Mortel, R., Twiggs, L. B., Wilbanks, G. D., and Hoover, R. N. (1993). Dietary associations in a case-control study of endometrial cancer. *Cancer Causes Control* **4**, 239–250.

212. La Vecchia, C., Decarli, A., Fasoli, M., and Gentile, A. (1986). Nutrition and diet in the etiology of endometrial cancer. *Cancer (Philadelphia)* **57**, 1248–1253.

213. Shu, X. O., Zheng, W., Potischman, N., Brinton, L. A., Hatch, M. C., Gao, Y. T., and Fraumeni, J. F., Jr (1993). A population-based case-control study of dietary factors and endometrial cancer in Shanghai, People's Republic of China. *Am. J. Epidemiol.* **137**, 155–165.

214. Levi, F., Franceschi, S., Negri, E., and La Vecchia, C. (1993). Dietary factors and the risk of endometrial cancer. *Cancer (Philadelphia)* **71**, 3575–3581.

215. Zheng, W., Kushi, L. H., Potter, J. D., Sellers, T. A., Doyle, T. J., Bostick, R. M., and Folsom, A. R. (1995). Dietary intake of energy and animal foods and endometrial cancer incidence. The Iowa Women's Health Study. *Am. J. Epidemiol.* **142**, 388–394.

216. Tzonou, A., Lipworth, L., Kalandidi, A., Trichopoulou, A., Gamatsi, I., Hsieh, C. C., Notara, V., and Trichopoulos, D. (1996). Dietary factors and the risk of endometrial cancer: A case-control study in Greece. *Br. J. Cancer* **73**, 1284–1290.

217. Barbone, F., Austin, H., and Partridge, E. E. (1993). Diet and endometrial cancer: A case-control study. *Am. J. Epidemiol.* **137**, 393–403.

218. International Agency for Research on Cancer and World Health Organization (1997). "Biennial Report 1996/1997." IARC, Lyon, France.

219. Parazzini, F., La Vecchia, C., D'Avanzo, B., Moroni, S., Chatenoud, L., and Ricci, E. (1995). Alcohol and endometrial cancer risk: Findings from an Italian case-control study. *Nutr. Cancer* **23**, 55–62.

220. Austin, H., Drews, C., and Partridge, E. E. (1993). A case-control study of endometrial cancer in relation to cigarette smoking, serum estrogen levels, and alcohol use. *Am. J. Obstet. Gynecol.* **169**, 1086–1091.

221. Gapstur, S. M., Potter, J. D., Sellers, T. A., Kushi, L. H., and Folsom, A. R. (1993). Alcohol consumption and postmenopausal endometrial cancer: Results from the Iowa Women's Health Study. *Cancer Causes Control* **4**, 323–329.

222. Newcomb, P. A., Trentham-Dietz, A., and Storer, B. E. (1997). Alcohol consumption in relation to endometrial cancer risk. *Cancer Epidemiol. Biomarkers Prev.* **6**, 775–778.

223. Webster, L. A., Weiss, N. S., and the Cancer and Steroid Hormone Study Group (1989). Alcoholic beverage consumption and the risk of endometrial cancer. *Int. J. Epidemiol.* **18**, 786–791.

224. Baron, J. A., Byers, T., Greenberg, E. R., Cummings, K. M., and Swanson, M. (1986). Cigarette smoking in women with cancers of the breast and reproductive organs. *J. Natl. Cancer Inst.* **77**, 677–680.

225. Elliott, E. A., Matanoski, G. M., Rosenshein, N. B., Grumbine, F. C., and Diamond, E. L. (1990). Body fat patterning in women with endometrial cancer. *Gynecol. Oncol.* **39**, 253–258.

226. Smith, E. M., Sowers, M. F., and Burns, T. L. (1984). Effects of smoking on the development of female reproductive cancers. *JNCI, J. Natl. Cancer Inst.* **73**, 371–376.

227. Tyler, C. W., Jr, Webster, L. A., Ory, H. W., and Rubin, G. L. (1985). Endometrial cancer: how does cigarette smoking influence the risk of women under age 55 years having this tumor? *Am. J. Obstet. Gynecol.* **151**, 899–905.

228. Franks, A. L., Kendrick, J. S., and Tyler, C. W., Jr. (1987). Postmenopausal smoking, estrogen replacement therapy, and the risk of endometrial cancer. *Am. J. Obstet. Gynecol.* **156**, 20–23.

229. Lesko, S. M., Rosenberg, L., Kaufman, D. W., Helmrich, S. P., Miller, D. R., Strom, B., Schottenfeld, D., Rosenshein, N. B., Knapp, R. C., Lewis, J., and Shapiro, S. (1985). Cigarette smoking and the risk of endometrial cancer. *N. Engl. J. Med.* **313**, 593–596.

230. Stockwell, H. G., and Lyman, G. H. (1987). Cigarette smoking and the risk of female reproductive cancer. *Am. J. Obstet. Gynecol.* **157,** 35–40.

231. Parazzini, F., La Vecchia, C., Negri, E., Moroni, S., and Chatenoud, L. (1995). Smoking and risk of endometrial cancer: Results from an Italian case-control study. *Gynecol. Oncol.* **56,** 195–199.

232. Brinton, L. A., Barrett, R. J., Berman, M. L., Mortel, R., Twiggs, L. B., and Wilbanks, G. D. (1993). Cigarette smoking and the risk of endometrial cancer. *Am. J. Epidemiol.* **137,** 281–291.

233. Weir, H. K., Sloan, M., and Kreiger, N. (1994). The relationship between cigarette smoking and the risk of endometrial neoplasms. *Int. J. Epidemiol.* **23,** 261–266.

234. Jensen, J., Christiansen, C., and Rodbro, P. (1985). Cigarette smoking, serum estrogens, and bone loss during hormone-replacement therapy early after menopause. *N. Engl. J. Med.* **313,** 973–975.

235. Friedman, A. J., Ravnikar, V. A., and Barbieri, R. L. (1987). Serum steroid hormone profiles in postmenopausal smokers and nonsmokers. *Fertil. Steril.* **47,** 398–401.

236. Michnovicz, J. J., Hershcopf, R. J., Naganuma, H., Bradlow, H. L., and Fishman, J. (1986). Increased 2-hydroxylation of estradiol as a possible mechanism for the anti-estrogenic effect of cigarette smoking. *N. Engl. J. Med.* **315,** 1305–1309.

237. Key, T. J. A., Pike, M. C., Brown, J. B., Hermon, C., Allen, D. S., and Wang, D. Y. (1996). Cigarette smoking and urinary oestrogen excretion in premenopausal and post-menopausal women. *Br. J. Cancer* **74,** 1313–1316.

238. Daniell, H. W. (1976). Osteoporosis of the slender smoker. Vertebral compression fractures and loss of metacarpal cortex in relation to postmenopausal cigarette smoking and lack of obesity. *Arch. Intern. Med.* **136,** 298–304.

239. Williams, A. R., Weiss, N. S., Ure, C. L., Ballard, J., and Daling, J. R. (1982). Effect of weight, smoking, and estrogen use on the risk of hip and forearm fractures in postmenopausal women. *Obstet. Gynecol.* **60,** 695–699.

240. Kaufman, D. W., Slone, D., Rosenberg, L., Miettinen, O. S., and Shapiro, S. (1980). Cigarette smoking and age at natural menopause. *Am. J. Public Health* **70,** 420–422.

241. Willett, W., Stampfer, M. J., Bain, C., Lipnick, R., Speizer, F. E., Rosner, B., Cramer, D., and Hennekens, C. H. (1983). Cigarette smoking, relative weight, and menopause. *Am. J. Epidemiol.* **117,** 651–658.

242. Parazzini, F., La Vecchia, C., Moroni, S., Chatenoud, L., and Ricci, E. (1994). Family history and the risk of endometrial cancer. *Int. J. Cancer* **59,** 460–462.

243. Gruber, S. B., Thompson, W. D., and the Cancer and Steroid Hormone Study Group (1996). A population-based study of endometrial cancer and familial risk in younger women. *Cancer Epidemiol., Biomarkers Prev.* **5,** 411–417.

244. Marra, G., and Boland, C. R. (1995). Hereditary nonpolyposis colorectal cancer: The syndrome, the genes, and historical perspectives. *J. Nat. Cancer Inst.* **87,** 1114–1125.

245. Esteller, M., Garcia, A., Martinez-Palones, J. M., Xercavins, J., and Reventos, J. (1997). Germ line polymorphisms in cytochrome-P450 1A1 (C4887 CYP1A1) and methylenetetrahydrofolate reductase (MTHFR) genes and endometrial cancer. *Carcinogenesis* (*London*) **18,** 2307–2311.

246. Esteller, M., Garcia, A., Martinez-Palones, J. M., Xercavins, J., and Reventos, J. (1997). Susceptibility to endometrial cancer: Influence of allelism at p53, glutathione S-transferase (GSTm1 and GSTt1) and cytochrome P-450 (CYP1A1) loci. *Br. J. Cancer* **75,** 1385–1388.

73

Cervical Cancer

F. XAVIER BOSCH* AND NUBIA MUÑOZ†

*Institut Català d'Oncologia, L'Hospitalet del Llobregat, Barcelona, Spain; †International Agency for
Research on Cancer, Lyon, France

I. Introduction

Cancer of the cervix is the second most common cancer in women after cancer of the breast and is the commonest type of cancer in developing countries. In developed countries, cervical cancer incidence ranks fifth and clusters among women in the lower and marginal socioeconomic groups, largely as a consequence of isolation from preventive programs. For more than a century, epidemiologic observations had pointed out the parallels between sexual behavior and cervical cancer, leading to the hypothesis that one (or several) sexually transmissible agent(s) had to be a major cause of cervical cancer. However, only in the 1990s has the etiology of cervical cancer been established. It is now recognized that over 99% of the cases in all countries are related to some types of Human Papillomaviruses (HPV). The best marker of exposure is currently the detection of type-specific HPV DNA (Deoxyribonucleic acid) in cancer cells. These infections are common in the young age groups (*i.e.,* prevalence as high as 30–50%) and resolve spontaneously in most instances. The typical HPV DNA prevalence after the third decade lies between 5 and 20% among women in most populations and these probably form the true high-risk group for cervical cancer. These findings offer new opportunities for improving screening and primary prevention of cervical cancer through HPV testing and vaccination. However, they also bring to the screening scenario the troubles and emotional issues associated with the management of a potentially oncogenic sexually transmitted disease.

II. What Is Cervical Cancer?

Cervical cancer develops in the epithelia of the neck of the uterus from a single cell undergoing neoplastic transformation. Most of the time, the process is initiated in a peculiar anatomical region known as transitional or transformation zone (TZ) where the columnar epithelium of the inner part of the cervix, the endocervix, merges with the nonkeratinizing squamous epithelium of the external part of the cervix, the exocervix. It has been established that invasive cancer is the final stage of a continuous process historically described as a series of discrete stages (*i.e.,* Papanicolau PAP I to V; mild to severe dysplasia and carcinoma *in situ* [CIS]; cervical intraepithelial neoplasia [CIN] I to III; low and high grade squamous intraepithelial neoplasia [LGSIL and HGSIL]). All of these varying nomenclatures try to describe in different frames a unique morphologic process with overlapping boundaries. For example, the severe dysplasia/carcinoma *in situ* (Dysplasia classification, 1973) largely corresponds to the CIN III lesions (Richard classification, 1973) and to HGSIL (Bethesda classification, 1988).

The early stages of cervical cancer are often diagnosed by examining exfoliated cells from the cervix taken by gently scraping the surface with a wooden spatula (or other more evolved instruments). This popular procedure is known as "the Pap smear" in recognition of the work of Dr. Papanicolaou who first described the procedure. The Pap smear is the test currently recommended for screening the general population and is part of the diagnostic procedures to evaluate women with genital symptoms. If early diagnosis and treatment is not performed, cervical cancer expands locally to invade the rest of the cervix, the vagina, and laterally to the parametrium and the pelvic walls. Lymph node metastases occur in the pelvis and the retroperitoneum and eventually distant metastases develop. The progressive clinical stages have also been categorized by the International Federation of Obstetrics and Gynecology (FIGO) into four stages (I–IV) and 15 substages with marked differences in prognosis. Since the 1980s, morphologic changes in exfoliated cells related to HPV infection (*i.e.,* koilocytotic changes) have been described and incorporated as a component of the earliest stage of the neoplastic process (LGSIL).

III. Assessing Early Cancer and Persistent Infection with Human Papillomavirus

As in any screening scheme, a major determinant of the success in reducing mortality is the ability to reach the population at risk. Most deaths to cervical cancer in developed countries occur in the social groups or in individuals that have never or rarely been screened. However, screening for cervical cancer with cytology has been the paradigm of cancer prevention for over 50 years. Organized screening has proven in some countries to contribute to a significant reduction in the incidence and mortality of cervical cancer [1].

However, the cytology test has some recognized limitations attributable in broad categories to (i) poor smear taking and preparation, (ii) poor reading and reporting of the smears, and (iii) poor follow-up of women with abnormalities in the smear. There is a reasonable agreement that, in routine screening, an unavoidable error rate of 5–10% is to be expected. In some countries, false negative errors (women given a diagnosis of normal smear who subsequently develop cervical cancer) are subject to intense litigation, promoting better quality control procedures and the adoption of additional screening methods (with subsequent cost increases) to avoid both false negatives and overtreatment (see Chapter 76).

The best measurement of exposure to HPV requires testing for HPV DNA in the exfoliated cervical cells or in cervical biopsies. The methods have been considerably developed and standardized assays are available for use in routine practice.

Women and Health   932

Copyright © 2000 by Academic Press.
All rights of reproduction in any form reserved.

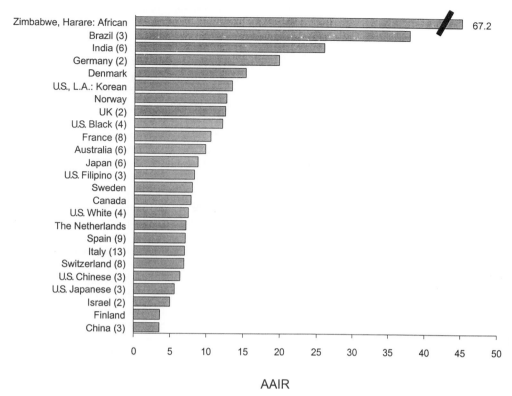

Fig. 73.1 Age-adjusted incidence rates of cervical cancer in selected cancer registries. Numbers in parentheses indicate the number of registries included. AAIR: age-adjusted incidence rates per 100,000 per year.

Sampling from the penis/urethra to detect HPV DNA carriers among men is feasible, although the yield of cells is lower and acceptance of the procedure by men is still in its infancy.

Serologic assays to characterize the natural history of HPV infections are under active development. Transient HPV infections usually escape immunological response and, with current methods, serum antibodies are undetectable. Persistent infections trigger some response and antibodies to the late capsid proteins (L1) are detectable in about half of infected patients. Invasive disease sharply increases the production of antibodies to viral genes E6 and E7 while decreasing the presence of anti-L1 antibodies. Serologic assays may in the future increase our ability to conduct larger surveys that include men.

IV. Incidence and Mortality of Cervical Cancer: Worldwide Perspective

Excluding nonmelanocytic skin cancers, cancer of the cervix is the second most common cancer in women worldwide and is the commonest type of cancer in developing countries. The World Health Organization (WHO) estimated that the number of new cases of invasive cancer of the cervix in 1996 was 525,000 (about 12% of new cases of cancer in women), with over 240,000 deaths from the disease [2]. When men and women are considered together, cancer of the cervix ranks as the fifth most common cancer worldwide, after cancers of the lung, stomach, breast, and large bowel, and it accounts for

an estimated 5% of all human cancers and 3.4% of all cancer deaths. The highest incidence rates occur in the developing areas of the world, particularly parts of Asia, South America, and Africa, where it accounts for 20–30% of all cancers among women. The lowest rates are reported from Australia and New Zealand, Southern Europe, North America, and Western Asia (the Middle East) [3].

Figure 73.1 shows a selection of the age-adjusted incidence rates of cancer of the cervix broadly representing the variation across continents [4]. Figure 73.2 shows the range of variation in the age-adjusted incidence rates in different regions within countries. This variation should not be attributed to differences in diagnostic capabilities across regions or to major differences in the ability to access medical care. Figures 73.1 and 73.2 also show that the striking differences in the incidence of cervical cancer observed between developed and developing countries are also reflected by race within developed countries. In the U.S., black populations show two-fold increased incidence rates as compared to whites. In the U.S. population, strong gradients have been reported among cervical cancer incidence, education, and income [5]. Figure 73.2 illustrates that, in developed countries, the variation in cervical cancer incidence must be also determined by factors that are only partially described. These factors may refer to differences in the prevalence of HPV, the efficacy of the local screening strategies, or the distribution of individuals with particular genetic traits. Survival from cervical cancer is closely related to stage at diagnosis and strong differences by socioeconomic status are also observed. In the U.S., for example,

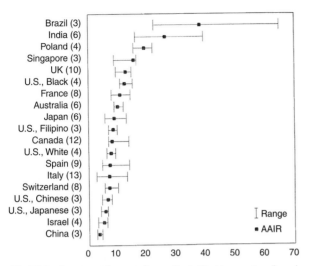

Fig. 73.2 Average and range of age-adjusted incidence rates of cervical cancer in all cancer registries within countries. Numbers in parentheses indicate the number of registries included. AAIR: age-adjusted incidence rates per 100,000 per year.

black women have consistently higher stages at diagnosis and lower survival rates than white women [6].

V. Time Trends in Cervical Cancer

Incidence and mortality due to cervical cancer have been declining in most developed countries since at least the 1960s. Part of the reduction is generally attributed to the widespread screening practices as well as to the increasing level of obstetrical and gynecological care offered to women. Correspondingly, in countries where such developments have not been implemented, the scarce available data suggest that incidence and mortality have remained constant or have even increased. Using data from 60 population-based Cancer Registries, trends in incidence were evaluated in 32 defined populations within the interval 1962 to 1992. The study included close to 175,000 cervical cancer cases and close to 20,000 cervical adenocarcinomas [7]. A steady decline in incidence appeared generally in the 1940s and 1950s. However, some relevant exceptions in some developed countries occurred, including the UK, New Zealand, Australia, and some areas of the U.S. In these populations, the younger generations (*i.e.,* women born after 1930) showed a reverse in the trend, with increases in incidence and mortality. The reverse trend is generally attributed to the generalized changes in sexual behavior occurring in the 1960s and thereafter in developed countries, leading to an increase in the prevalence of infections with HPVs. The related increase in cervical cancer expected in the 1980s and thereafter is only partially observed because of the competing effects of screening. It has been shown that the predominance of adenocarcinomas observed in the cases that escape screening (sometimes referred to as rapidly progressing cervical cancer) is at least in part due to the limitations of the current methods of specimen collection, which are relatively less efficient in sampling from the endocervical canal [8,9].

VI. Environmental and Host Determinants of Cervical Cancer

Cervical cancer is a rare and late outcome of a common sexually acquired infection with some strains of Human Papillomavirus (HPV). At least 70 types together with several subtypes of HPV have been described. Traditionally, types are considered to be high- or low-risk, taking into account their prevalence in cervical cancer specimens. In some cases an intermediate risk group is also described that largely includes types for which evidence is scanty and uncertain. HPV types 6 and 11 are related to the development of external genital warts and are rarely found in cervical cancer specimens (*i.e.,* <1 in 1000). These are genuine low-risk HPV types. Close to 50% of cervical cancer cases worldwide are related to HPV 16. HPV 18 accounts for an additional 10–15% and is particularly common in cervical adenocarcinomas (40–50%). Types 45 and 59 are found in about 5–10% of the cases and other types are present in lower proportions (*i.e.,* <5%).

Most HPV infections first occur in the young age groups and clear spontaneously. However, a fraction of HPV-exposed women remain persistent carriers and constitute the high-risk group for cervical cancer. Some of the known determinants of HPV infection are the common risk factors for any other Sexually Transmitted Disease (STD) and for cervical cancer, namely number of different sexual partners, number of recent new partners, and history of other STDs [10,11]. These factors apply also to the sexual behavior of the husband and to other surrogates of HPV infection such as the presence of other STDs in the husband's medical history.

However, having established that only a fraction of HPV infections will evolve to invasive cancer it should be anticipated that other determinants of the evolution of HPV infections are in force. In addition to viral determinants such as viral type, viral load, or perhaps genetic variants of the common HPV types, some host factors such as genetic susceptibility or p53 polymorphisms and other environmental factors such as smoking and hormonal treatments are being intensively studied and will be briefly discussed in the section on risk factors.

VII. Risk Factors for Cervical Cancer

A. Human Papillomaviruses and Host Factors

The evidence relating HPV infections to cervical cancer includes an impressive and consistent body of studies indicating a strong and specific role of the viral infection in all countries where investigations have taken place. The association has been recognized as causal in nature by a number of international review parties since the early 1990s [12–14]. For illustration, we will briefly review some of the most recent and relevant studies.

The International Biological Study on Cervical Cancer (IBSCC) is an International Agency for Research on Cancer (IARC) coordinated survey that included over 1000 biopsy specimens of invasive cervical cancer from 22 countries. A Polymerase Chain Reaction (PCR)-based system including 26 type-specific probes and a generic probe was used. This large study identified HPV DNA in 93% of the specimens, half of which were of the type

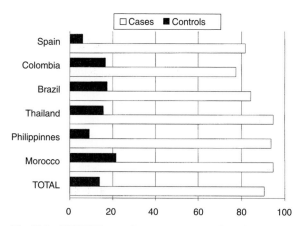

Fig. 73.3 HPV DNA prevalence among cases of cervical cancer and controls in the IARC multicentric study (preliminary results).

HPV 16. Cervical adenocarcinomas showed a similar HPV DNA prevalence, but HPV 18 was the most common type. Only one putative new viral type was identified, suggesting that most of the HPV types related to cervical cancer have been identified [15]. Subsequently, the 7% of specimens that remained HPV DNA-negative in this study were retested using a different HPV detection system and using a sandwich method (verifying in multiple histological cuts that recognizable neoplastic cells were used in the HPV testing procedure) to ensure the quality of the materials processed for HPV. As a result, the estimated prevalence of HPV DNA in cervical cancer biopsies increased from 93.0 to 99.8% [16].

The IARC research program on HPV also organized a series of case-control studies in different countries, mostly in areas at high risk for cervical cancer. Some results concerning HPV are summarized in Fig. 73.3 and Table 73.1. The table includes the adjusted odds ratios (OR) (the factor by which the risk of cervical cancer of a given woman is multiplied if HPV DNA is detected) as well as the estimates of the attributable fraction (AF, the proportion of disease in %,) that is related to HPV DNA.

The magnitude of the ORs and AFs shown in Table 73.1 suggests that the association is one of the strongest identified for any human cancer and, along with the consistent experimen-

tal and animal data, makes the case for claiming a necessary cause of the disease (*i.e.*, that HPV-negative cervical cancer cases are extremely rare) [17–21].

In developed countries, invasive cervical cancer is, however, a relatively rare disease, and most diagnoses occur at earlier stages (*i.e.*, Carcinoma *in Situ*, HGSIL or CIN III). Several studies in both developed and developing countries have also shown that HPV is related to these precursor lesions with the same strength (as measured by the magnitude of the ORs) as the more advanced invasive cancers [22–27].

Most HPV infections are transient and self-limiting. However, in some instances these viral infections persist for prolonged periods of time. Most of the time persistency is related to the high-risk viral types and efforts are now being devoted to describing which factors determine the outcome of the infection to either resolution or persistency and progression.

Prospective studies are beginning to describe factors associated with persistence of HPV and progression to cervical neoplasia following infection. High-risk viral type, increased viral load (the estimates of the amount of viral DNA per cell), and advanced age have been found to be related to progression in some studies. Less consistently, infection with multiple types has also been reported [28].

An intratypic variant of HPV is a type whose L1 coding region (L stands for "Late" and denotes a highly conserved region in the envelope of the virus; there are two of such regions, L1 and L2) differs by less than 2% when compared to the "prototypic" genomic sequence. It has been postulated that certain variants could have different biologic characteristics that could result in enhanced oncogenicity. A prospective study indicated a 6.5-fold higher risk of CIN 2–3 among women with nonprototypic variants of HPV 16 compared to women with prototypic variants. These results require confirmation in larger studies [29].

In addition to viral-related factors, some host factors are being unveiled that may modulate the outcome of a given HPV infection. Among these, certain alleles and haplotypes of the human leukocyte antigen (HLA), which influence the immune response of individuals to particular antigens, have been proposed. Several studies have detected either increased or decreased risk of CIN or invasive cancer in relation to HLA, but no striking association with any HLA type has been found [30]. Genetic or acquired immunosuppression has clearly been demonstrated

Table 73.1

HPV DNA Prevalence among Cases of Cervical Cancer and Controls and Risk Estimates in the IARC Multicentric Study (Preliminary Results)

Study	Cases (HPV + %)	Controls (HPV + %)	ORa	(95% CL)	AF (%)
Spain	81.7	5.9	72.4	(31.5–166.6)	81
Colombia	77.3	16.7	17.2	(8.2–35.8)	73
Brazil	84.4	17.4	25.9	(15.0–44.7)	81
Thailand	94.7	15.7	104.9	(58.6–187.5)	94
Philippines	93.7	9.2	156.0	(89.1–273.2)	93
Morocco	94.6	21.6	59.0	(27.5–126.6)	93
Total	90.4	13.8	65.2	(50.5–84.1)	90

Note. ORa, odds ratios adjusted for age; ORa Total, odds ratios adjusted for age and country.

as a cofactor for progression of HPV-induced lesions. An example of genetic immune deficiency is provided by the epidermodysplasia verruciformis model seen in the skin, and the two examples of acquired deficiency come from the observation of more frequent and more serious HPV-associated disease among renal transplant patients [13] and patients infected with the Human Immunodeficiency Virus (HIV) [31].

One report indicates that certain genetic variants of the tumor suppressor gene p53 are more susceptible to degradation of the p53 protein by the E6 oncoprotein of high risk HPVs. According to this model, women who are homozygous for arginine at position 72 of the p53 gene appear to be at higher risk of HPV-associated cervical cancer than women who are heterozygous [32]. This or other polymorphisms of the tumor suppressor genes could be important markers of susceptibility to HPV-associated neoplasia.

B. Other Environmental Risk Factors for Cervical Dysplasia and Cervical Cancer

Before HPV was investigated, epidemiologic studies identified a series of factors as being more prevalent in cases of cervical cancer than in their control groups. This was the case for different sexual and reproductive behavioral traits, use of Oral Contraceptives (OC), smoking, or history of venereal infections, typically Herpes virus type 2. The ORs observed for such associations were in the range of 1 to 3 (as compared to ORs in the hundreds for some HPV types) and the results were inconsistent across studies. Having unveiled the very strong associations with HPV, all of these putative additional factors require reevaluation. In general terms, what remains to be established is if and which additional environmental factors result in persistence of HPV infection and its progression to cervical cancer. Most

investigators have attempted such reevaluation by restricting the comparison of the relevant exposures in cases of cervical cancer, most of which were shown to be HPV-positive, with their HPV-positive controls (women in the same age groups from the same underlying population, with HPV infection but without cancer). The most relevant of these studies will be briefly reviewed.

1. Sexual Behavior of Women

Analyses among HPV-positive women in studies of invasive cancer and CIN III have shown that the number of sexual partners, the key risk factor for cervical cancer, is no longer related to the disease among women who are HPV-positive. This finding clearly suggests that number of partners is a surrogate measure of HPV infection. It also suggests that among women who are HPV-positive, further increasing the number of partners would not add to the risk of developing cervical cancer because the key exposure has already occurred [18,22,27,33,34]. Interestingly, age at first sexual intercourse or age at first marriage (both interpreted as surrogate measures of age at first HPV infection) remain as independent factors in several studies that also considered HPV as a risk factor. This repeatedly reported finding suggests the existence of a period of increased susceptibility to HPV of the transitional zone of the cervix in the perimenarchy. A study in Rabat, Morocco, where tradition favors that very young women marry older men, showed an increased risk of cervical cancer in relation to the age difference between spouses at the time of the women's first marriage [21].

2. The Role of Men's Sexual Behavior in the Epidemiologic Chain Leading to Cervical Cancer

Fig. 73.4 shows the correlation of the average number of lifetime sexual partners reported by men and women and the incidence rates (age-adjusted) of cervical cancer in the Cancer

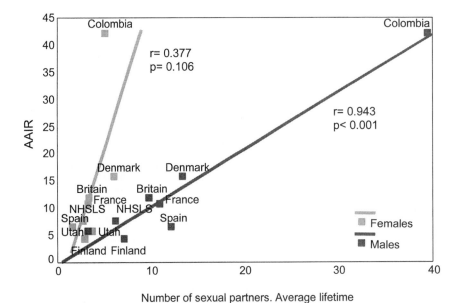

Fig. 73.4 Average number of lifetime sexual partners and incidence of cervical cancer: population-based surveys. AAIR: age-adjusted incidence per 100,000 per year. Males in Spain and Colombia are the husbands of a random sample of women.

Registries in their countries. All of the estimates of the number of partners are based on population surveys [35–40].

Although the correlation is not adjusted by the penetrance of screening programs in the populations (a confounder that tends to decrease the true incidence rates of cervical cancer), the difference between the slopes of the two correlations is striking. The range of the reported average number of sexual partners of men in eight populations is wide (3 to 40) and the correlation with the incidence of cervical cancer is statistically significant. In contrast, the variability in the average number of partners reported by women in the same countries is narrower (range 1.6 to 6.1) and insufficient to show a significant correlation with the incidence of cervical cancer. This crude analysis suggests that the greatest proportion of the geographic variability in cervical cancer incidence is attributable to differences in men's sexual behavior rather than to women's.

Formal case-control studies relating men's sexual promiscuity and cervical cancer in their spouses have clearly shown an increased risk with the number of additional sexual partners and the number of sexual contacts of the husband (or current partner) with prostitutes. Furthermore, husbands of women with cervical cancer more often showed HPV DNA in cell scrapes from their penis's than husbands of control women, although temporality and direction of the infection cannot be established from case-control studies [41]. In high-risk countries in Latin America, the reported number of partners by men is very high and the differences in sexual behavior between husbands of cervical cancer cases and husbands of controls is less obvious. These findings suggest that in these populations HPV infections are widespread and thus the number of partners poorly reflects the probability of infection by any given partner [42].

Not much is known about the distribution and the dynamics of HPV infections in populations. It has been proposed that some groups are at particularly high risk and could be considered reservoirs of the infection. These would include women (and men) practicing prostitution, carriers of the HIV infection, and persons with a high number of new sexual partners.

Recognition of the venereal nature of cervical cancer, as well as of some other cancers of the genital tract, particularly anal cancer [43], may raise some social and ethical issues that are new to the cancer field. Women with cervical cancer and their husbands may suffer the additional burden of social discrimination and feelings of personal guilt in communities and among individuals that enforce restrictive patterns of sexual behavior. Furthermore, the increasing practice of adding HPV tests as adjuvants to screening for cervical cancer will identify men and women who are HPV-carriers (some in the young age groups) to whom, at the present time, little can be offered as treatment. These women and their partners may undergo severe damage to their self-esteem and sexual life. Gynecologists are likely to be in the forefront of such problems and should probably seek appropriate support as well as conduct research into new treatments for the HPV infections that they are now able to diagnose.

3. Other Sexually Transmitted Diseases

Antibodies to other sexually transmitted diseases (STDs) have been investigated and occasionally found associated with cervical cancer. It is likely that these are merely surrogates of HPV infections because they share the basic epidemiologic

traits (i.e., a women with multiple partners is at increased risk of exposure to any given STD. Therefore, some weak and inconsistent associations are to be expected between HPV and any other venereal infection without necessarily having any specific biological implication). Alternatively, coexistence of other sexually transmitted diseases may facilitate implantation of HPV infections because of epithelial erosions. Studies restricted to HPV-positive women have not clearly identified other STDs as independent factors for cervical cancer. *Chlamidia trachomatis* has been singularized in several studies as related to cervical cancer after adjustment for other factors, but in the opinion of several investigators the issue still requires further research [27,44,45].

A particularly significant interaction has been repeatedly reported between HPV and HIV. Women (and men) who are exposed to HIV are more likely to be HPV-positive and to develop HPV-related neoplasia. Further, the stage of the neoplasia (cervical or anal) is related to the intensity of the immunosupression as measured by the level of $CD4^+$ counts [13,31,43]. The presence of cervical cancer was recommended in 1992 as an independent criterion to establish the diagnosis of AIDS in women with HIV infection.

4. Oral Contraceptives and Parity

A role of external hormones in cervical cancer has been suspected for some time and several studies concluded that long-term use of Oral Contraceptives (OCs) was a moderate risk factor [46]. Studies in Colombia, Spain, and Brazil found that the only significant differences among HPV-positive cases and controls were the use and the duration of use of OCs. In these three countries, the analysis by duration of use of OCs suggests a positive trend, although the conclusion was limited by the small number of control women that were HPV DNA positive [18,22,27]. The finding was not confirmed in studies on preinvasive neoplasia in Denmark and in the U.S. [22,27].

Parity showed inconsistent results in two countries in Latin America, both at high risk for cervical cancer and both with high parity rates. In Colombia there was no association with invasive cancer but a moderate association with CIN III, and in Brazil there was a strong association with invasive cancer. Parity was associated with cervical cancer but not with CIN III in Spain. Among HPV-positive women there was no association of invasive cancer with parity in the Spain and Colombia studies, whereas a significant trend with increasing parity was observed in Brazil [18,33,34]. A moderate effect of parity in the risk of preinvasive neoplasia was observed in the Spain and Colombia studies, in Denmark, and in the U.S. [22,27].

The hypothesis that hormonal factors may play a role as a cofactor in the acquisition of the HPV chronic carrier state and/or in the progression from HPV to HPV-related neoplasia has inconsistent results in experimental work and needs to be reassessed by studies in populations that include large number of cases and of HPV-positive controls.

5. Tobacco Smoking and Cervical Cancer

Cigarette smoking showed inconsistent and weak associations in the studies in Spain, Colombia, and Brazil that did not support an independent effect of smoking. However, another PCR-based study reported a two-fold increased risk of squamous intraepithelial lesion (SIL) for HPV-positive women who were current smokers as compared to HPV-positive women who were nonsmokers [27]. Cigarette smoke contains powerful nonorgan-specific carcinogens and smoke metabolites which have been detected in the cervical mucosa [47]. Therefore, the hypothesis of a carcinogenic effect in the cervix cannot be conclusively ruled out.

6. Diet

Dietary factors are related to epithelial cancers and fresh fruits and vegetables are strong protective factors for cancers of the oral cavity, esophagus, stomach, and colorectum. Some protection is also claimed for cancers of the larynx, pancreas, breast, and urinary bladder. Consequently, several studies have evaluated food and nutrients as possible cofactors in cervical cancer. In general, associations between foods or nutrients with cervical cancer are of low magnitude and inconsistent. Low intake of vitamin C and carotenoids seems a fairly consistent finding among cases [48]. An international review board has evaluated the role of dietary factors and cancer. In relation to cervical cancer it was concluded that intake of vegetables and fruits, carotenoids, vitamin C, and vitamin E could possibly reduce the incidence of cervical cancer. The panel found lack of evidence for a protective effect of folate and retinol [49]. Methodological issues involved in nutritional case-control studies (i.e., difficulties in recalling diets) and the scarcity of studies assessing dietary factors and its interactions with HPV contribute to the uncertainty of the associations. The issue clearly demands further research.

VIII. Clinical Issues: Early Diagnosis, Prognosis, Treatment, and Prevention

A. Early Diagnosis

HPV testing fulfills most of the requirements for a screening test: it aims at detecting a viral infection that precedes neoplasia by some years and the viral DNA can be detected with high sensitivity and reproducibility. In the age groups 30–35 and above, HPV tests are predictive of the risk of high-grade neoplastic lesions. Furthermore, taking cell specimens for HPV testing is a noninvasive procedure and should be socially acceptable. This new technology is slowly being introduced because some aspects of the natural history are still poorly understood and currently no clinical treatment is available if infection without neoplasia is identified. This latter issue is particularly relevant when HPV DNA is detected among men. Finally, the cost efficiency of HPV testing as a screening tool is under evaluation. The earliest studies using HPV as part of the screening procedures have suggested that in the presence of a cytological diagnosis of uncertain significance (AGUS/ASCUS), HPV tests show higher predictive values for high-grade lesions than does repeated cytology [50].

A large population study in Amsterdam is evaluating a screening strategy by which the risk of advanced neoplasia is predicted either by an abnormal cytology or by the presence of a high-risk HPV type. The study aims at proving that normalcy in both assays could reliably rule out neoplasia for an extended period of time (i.e., 8 to 10 years) and prolong screening intervals. If

proven, the strategy could, at least theoretically, liberate resources to increase surveillance of women who are not regularly screened and concentrate diagnostic and treatment resources in women at high risk for cervical cancer [51].

B. HPV as a Prognostic Factor in Invasive Cervical Cancer

Some early studies discussed the value of the presence/absence of HPV type as an indicator of the prognosis in cervical cancer patients with inconsistent findings. The clinical follow-up of the 471 population-based cases recruited in the Colombia and Spain study was completed after a median time interval of 5 years. HPV DNA was identified by PCR in 79.1% of the cases. Presence/absence of HPV DNA, HPV type, or some crude estimates of the HPV viral load were not related to survival. However, cases related to HPV 18 and of unspecified type had (nonstatistically significant) poorer survival than HPV 16 related cases [52]. Other investigations have also suggested worse prognosis linked to HPV 18 related cervical cancer [53]. These are areas where further clinical research is warranted.

C. Treatment of Cervical Cancer

In early stages (FIGO stage I), cervical cancer is a curable disease (i.e., 5 year survival rates are as high as 80–90%), requiring in most instances limited surgery or radiotherapy. Advanced stages (stage II and above) require major surgery and/or radiotherapy with a more somber prognosis (i.e., the 5 year survival rate for stage II is around 50–60% and for stage III around 25–30%). Advanced or recurrent cervical cancer is usually treated with palliative chemotherapy. End-stage cervical cancer is a most disturbing condition that usually concurs with severe pain and distress.

D. Prevention: HPV Vaccines

The majority, if not all, cases of cervical cancer are related to HPV infections. The majority of the cases of cancer of the anal canal and half of the cancers of the vulva, vagina, and penis are also related to HPV, and there are indications that other cancers may be partially involved (i.e., the oral cavity, the esophagus, the skin, and the urinary bladder) [13,54,55]. Definite evidence for these cancer sites is still lacking, but given the impact of cervical cancer alone, it can be claimed that of the known human carcinogens, HPV may be the first in importance among women. HPV vaccines are thus of great potential interest.

Three types of HPV vaccines are being developed: prophylactic to prevent HPV infection and its associated neoplastic lesions; therapeutic to induce regression of HPV-associated lesions; and chimeric that combine prophylactic and therapeutic vaccines to induce both effects. Prophylactic HPV vaccines are based on structural proteins or late antigens in the L1 or L2 regions synthetically produced as virus-like particles (VLPs) assembled by overexpression of the major capsid protein (L1) in various expression vectors. VLPs have the same surface topography as infectious viral particles and present the conformational epitopes required for generating high titers of neutralizing antibodies but do not contain the potentially oncogenic viral DNA. Vaccination experiments have shown the protective efficacy of VLP-based vaccines in cattle, rabbits, and dogs against challenge with bovine papillomavirus [56], cottontail rabbit papillomavirus [57], and canine oral papillomavirus [58], respectively. Therapeutic vaccines target the E6 and E7 oncoproteins of HPV. Various recombinant HPV vaccines are being developed and some of them are already being tested in small-scale phase I trials [59]. Combined prophylactic and therapeutic vaccine are based on full-length HPV 16 L2–E7 chimeric proteins that are incorporated into L1 VLPs [57]. Women with normal or abnormal cytology positive for high-risk HPVs could be a target population for this vaccine.

IX. Epidemiologic Research and Public Health Issues

In spite of the relevance of recent findings, several areas requiring further research can still be identified:

1. Studies on the epidemiology and natural history of HPV infections in the community. These studies should address issues such as the incidence of HPV infection following exposure, the determinants of persistency and progression as opposed to spontaneous regression of the HPV infection, the role of additional routes of HPV infection such as mother to child transmission, and the role of men as vectors and occasionally as victims of the infection.

2. Studies on the biological specificity of the transformation zone of the cervix to identify the traits that make it highly vulnerable to HPV. Of particular interest would be to describe the changes that occur with age from menarchy.

3. Technology development should ensure that HPV testing is made available to the clinical laboratory in a reliable and standardized manner. Some of these assays are commercially available and are actively being tested.

4. Ongoing research should provide additional evidence of the benefits and the costs of adding HPV testing to screening strategies in different settings and in the triage of borderline cytological lesions.

5. Vaccine trials that are now in Phase I and II should lead to Phase III trials involving samples from the general population. These studies should provide data on the efficacy of the vaccination products in preventing HPV infections and neoplasia and ultimately in preventing mortality due to cervical cancer.

6. Clinical research should investigate new treatments for HPV infections that may be of use in subclinical lesions in men and women. Psychological support has to be foreseen for individuals identified as HPV carriers.

Finally, it has to be of major concern that cervical cancer is a lethal and frequent disease among young and middle-aged women in the third world. In most of these countries, new vaccines are prohibitively expensive for a number of years after successful production and testing. Strategies must soon be developed to ensure that costs do not unnecessarily delay the worldwide introduction of a prophylactic HPV vaccine.

X. Summary and Conclusions/Future Directions

Several studies have convincingly shown that cervical cancer is a long-term sequela of certain unresolved HPV infections of the uterine cervix. As observed for other sexually transmitted diseases, HPV infections and cervical cancer are most common in the poorest countries and among the most deprived social groups in affluent societies. The strongest known determinants of the geographical and social variation in the incidence of cervical cancer are related to the sexual behavior patterns of the population and to lack of screening or to reduced participation of the population to the available screening programs. Screening programs are likely to benefit from the addition of HPV testing to the standard cytology. HPV testing may also prove to be beneficial in the triage of lesions of uncertain nature (ASCUS/AGUS) and in the definition of the appropriate referrals to colposcopy. The prospects for developing HPV vaccines are advanced. Phase II–III trials involving large numbers of participants from the general population should be initiated relatively soon, with the aim of demonstrating a reduction in mortality due to cervical cancer in the vaccinated group. The completion of this work may offer a real preventive option to third world countries.

Acknowledgments

This review was partially supported by grants from the Fondo de Investigaciones Sanitarias (FIS) of the Spanish Government (FIS98/0699 and FIS98/0646) and from the Comisión Interministerial de Ciencia y Tecnología of the Spanish Government (SAF96-0323).

References

1. Miller, A. B. (1992). Control of carcinoma of cervix by exfoliative cytology screening. In "Gynecologic Oncology" (M. Coppleson, ed.), Vol. 1, pp. 543–555. Churchill Livingstone, Edinburgh.
2. World Health Organization (1997). "The World Health Report." WHO, Geneva.
3. Parkin, D. M., Pisani, P., and Ferlay, J. (1993). Estimates of the worldwide incidence of eighteen major cancers in 1985. Int. J. Cancer 54, 1–13.
4. World Health Organization (1997). "Cancer Incidence in Five Continents," Vol. 7. IARC, Lyon, France.
5. Baquet, C. R., Horm, J. W., Gibbs, T., and Greenwald, P. (1991). Socioeconomic factors and cancer incidence among blacks and whites. J. Natl. Cancer Inst. 83, 551–557.
6. Kosary, C. L., Ries, L. A. G., Miller, B. A., Hankey, B. F., Harras, A., and Edwards, B. K., eds. (1995). "SEER Cancer Statistics Review, 1973–1992." National Institutes of Health, Bethesda, MD.
7. Vizcaino, A. P., Moreno, V., Bosch, F. X., Muñoz, N., Barros-Dios, X. M., and Parkin, D. M. (1998). International trends in the incidence of cervical cancer: I. Adenocarcinoma and adenosquamous cell carcinomas. Int. J. Cancer 75, 536–545.
8. Schwartz, P. E., Hadjimichael, O., Lowell, D. M., Merino, M. J., and Janerich, D. (1996). Rapidly progressive cervical cancer: The Connecticut experience. Am. J. Obstet. Gynecol. 175(4, Pt. 2), 1104–1109.
9. DeMay, R. M. (1996). Cytopathology of false negatives preceding cervical carcinoma. Am. J. Obstet. Gynecol. 175(4), 1110–1114.
10. Muñoz, N., Kato, I., Bosch, F. X., Eluf-Neto, J., De Sanjosé, S., Ascunce, N., Gili, M., Izarzugaza, I., Viladiu, P., Tormo, M. J., Moreo, P., Gonzalez, L. C., Tafur, L., Walboomers, J. M. M., and Shah, K. V. (1996). Risk factors for HPV DNA detection in middle-aged women. Sex. Transm. Dis. 23, 504–510.
11. Kjaer, S. K., Van den Brule, A. J. C., Bock, J. E., Poll, P. A., Engholm, G., Sherman, M. E., and Meijer, C. J. L. M. (1997). Determinants for genital human papillomavirus (HPV) infection in 1000 randomly chosen young danish women with normal Pap smear: Are there different risk profiles for oncogenic and nononcogenic HPV types? Cancer Epidemiol., Biomarkers Prev. 6, 799–805.

12. Muñoz, N., Bosch, F. X., Shah, K. V., and Meheus, A. (1992). "The Epidemiology of Human Papillomavirus and Cervical Cancer," IARC Sci. Publ. No. 119. IARC, Lyon, France.

13. International Agency for Research on Cancer (1995). "Human Papillomaviruses—IARC Monographs on the Evaluation of Carcinogenic Risks to Humans," Vol. 64. IARC, Lyon, France.

14. National Institutes of Health Consensus Development Panel (1996). NIH Consensus Development Conference Statement: Cervical cancer. *J. Natl. Cancer Inst. Monogr.* **21,** vii–xiii

15. Bosch, F. X., Manos, M., Muñoz, N., Sherman, M., Jansen, A., Peto, J., Schiffman, M., Moreno, V., Shah, K. V., and the IBSCC Study Group (1995). Prevalence of human papillomavirus in cervical cancer: A worldwide perspective. *J. Natl. Cancer Inst.* **87,** 796–802.

16. Walboomers, J. M. M., Jacobs, M. V., Manos, M. M., Bosch, F. X., Kumer, J. A., Shah, K. V., Snijders, P. J. F., Meijer, C. J. L. M., and Muñoz, N. (1999). Human papillomavirus, a necessary cause of invasive cervical cancer worldwide. *J. Pathol.* (in press).

17. Muñoz, N., Bosch, F. X., De Sanjosé, S., Tafur, L., Izarzugaza, I., Gili, M., Viladiu, P., Navarro, C., Martos, C., Ascunce, N., Gonzalez, L. C., Kaldor, J. M., Guerrero, E., Lorincz, A. T., Santamaria, M., Alonso de Ruiz, P., Aristizabal, N., and Shah, K. V. (1992). The causal link between human papillomavirus and invasive cervical cancer: A population-based case-control study in Colombia and Spain. *Int. J. Cancer* **52,** 743–749.

18. Eluf-Neto, J., Booth, M., Muñoz, N., Bosch, F. X., Meijer, C. J. L. M., and Walboomers, J. M. M. (1994). Human papillomavirus and invasive cervical cancer in Brazil. *Br. J. Cancer* **69,** 114–119.

19. Chichareon, S., Herrero, R., Muñoz, N., Bosch, F. X., Jacobs, M. V., Deacon, J., Santamaria, M., Chongsuvivatwong, V., Meijer, C. J. L. M., and Walboomers, J. M. M. (1998). Risk factors for cervical cancer in Thailand: A case-control study. *J. Natl. Cancer Inst.* **90** (1), 50–57.

20. Ngelangel, C., Muñoz, N., Bosch, F. X., Limson, G. M., Festin, M. R., Deacon, J., Jacobs, M. V., Santamaria, M., Meijer, C. J. L. M., and Walboomers, J. M. M. (1998). Causes of cervical cancer in the Philippines: A case-control study. *J. Natl. Cancer Inst.* **90**(1), 43–49.

21. Chaouki, N., Bosch, F. X., Muñoz, N., Meijer, C. J. L. M., El Gueddari, B., El Ghazi, A., Deacon, J., Castellsagué, X., and Walboomers, J. M. M. (1998). The viral origin of cervical cancer in Rabat, Morocco. *Int. J. Cancer* **75**(4), 546–554.

22. Schiffman, M. H., Bauer, H. M., Hoover, R. N., Glass, A. G., Cadell, D. M., Rush, B. B., Scott, D. R., Sherman, M. E., Kurman, R. J., Wacholder, S., Stanton, C. K., and Manos, M. M. (1993). Epidemiologic evidence showing that human papillomavirus infection causes most cervical intraepithelial neoplasia. *J. Natl. Cancer Inst.* **85,** 958–964.

23. Bosch, F. X., Muñoz, N., De Sanjosé, S., Navarro, C., Moreo, P., Ascunce, N., Gonzalez, L. C., Tafur, L., Gili, M., Larrañaga, I., Viladiu, P., Daniel, R. W., Alonso de Ruiz, P., Aristizabal, N., Santamaria, M., Guerrero, E., and Shah, K. V. (1993). HPV and CIN III: A case-control study in Spain and Colombia. *Cancer Epidemiol., Biomarkers Prev.* **2,** 415–422.

24. Olsen, A. O., Gjoen, K., Sauer, T., Orstavik, I., Naess, O., Kierulf, K., Sponland, G., and Magnus, P. (1996). Human papillomavirus and cervical intraepithelial neoplasia grade II–III: A population-based case-control study. *Int. J. Cancer* **61,** 312–315.

25. Moreno, V., Muñoz, N., Bosch, F. X., De Sanjosé, S., Gonzalez, L. C., Tafur, L., Gili, M., Izarzugaza, I., Navarro, C., Vergara, A., Viladiu, P., Ascunce, N., and Shah, K. (1995). Risk factors for progression of cervical intraepithelial neoplasm grade III to invasive cervical cancer. *Cancer Epidemiol., Biomarkers Prev.* **4,** 459–467.

26. Liaw, K., Hsing, A. W., Chen, C.-J., Schiffman, M. H., Zhang, T. Y., Hsieh, C. Y., Greer, C. E., You, S. L., Huang, T. W., Wu, T. C.,

O'Leary, T. J., Seidman, J. D., Blot, W. J., Meinert, C. L., and Manos, M. M. (1995). Human papillomavirus and cervical neoplasia: a case-control study in Taiwan. *Int. J. Cancer* **62,** 565–571.

27. Kjaer, S. K., Van den Brule, A. J. C., Bock, J. E., Poll, P. A., Engholm, G., Sherman, M. E., Walboomers, J. M. M., and Meijer, C. J. L. M. (1996). Human papillomavirus—The most significant risk determinant of cervical intraepithelial neoplasia. *Int. J. Cancer* **65,** 601–606.

28. Ho, G. Y. F., Bierman, R., Beardsley, L., Chang, C. J., and Burk, R. D. (1998). Natural history of cervicovaginal papillomavirus infection in young women. *N. Engl. J. Med.* **338,** 423–428.

29. Xi, L. F., Koutsky, L. A., Galloway, D. A., Kuypers, J., Hughes, J. P., Wheeler, C. M., Holmes, K. K., and Kiviat, N. B. (1997). Genomic variation of human papillomavirus type 16 and risk for high grade cervical intraepithelial neoplasia. *J. Natl. Cancer Inst.* **89,** 796–802.

30. Davies, D. H., and Stauss, H. J. (1997). The significance of human leukocyte antigen associations with cervical cancer. *Papillomavirus Rep.* **8,** 43–49.

31. Sun, X. W., Kuhn, L., Ellerbrock, T. V., Chiasson, M. A., Busch, T. J., and Wright, T. C. (1997). Human papillomavirus infection in women infected with the human immunodeficiency virus. *N. Engl. J. Med.* **337,** 1343–1349.

32. Storey, A., Thomas, M., Kalita, A., Harwood, C., Gardiol, D., Mantovani, F., Breuer, J., Leigh, I. M., Maltashewski, G., and Banks, L. (1998). Role of a p53 polymorphism in the development of human papillomavirus-associated cancer. *Nature (London)* **393,** 229–234.

33. Bosch, F. X., Muñoz, N., De Sanjosé, S., Izarzugaza, I., Gili, M., Viladiu, P., Tormo, M. J., Moreo, P., Ascunce, N., Gonzalez, L. C., Tafur, L., Kaldor, J. M., Guerrero, E., Aristizabal, N., Santamaria, M., and Alonso de Ruiz, P. (1992). Risk factors for cervical cancer in Colombia and Spain. *Int. J. Cancer* **52,** 750–758.

34. Muñoz, N., Bosch, F. X., De Sanjosé, S., Vergara, A., Del Moral, A., Muñoz, M. T., Tafur, L., Gili, M., Izarzugaza, I., Viladiu, P., Navarro, C., Alonso de Ruiz, P., Aristizabal, N., Santamaria, M., Orfila, J., Daniel, R. W., Guerrero, E., and Shah, K. V. (1993). Risk factors for cervical intraepithelial neoplasia grade III/carcinoma in situ in Spain and Colombia. *Cancer Epidemiol. Biomarkers Prev.* **2,** 423–431.

35. Bosch, F. X., Muñoz, N., De Sanjosé, S., Guerrero, E., Ghaffari, A. M., Kaldor, J., Castellsagué, X., and Shah, K. V. (1994). Importance of human papillomavirus endemicity in the incidence of cervical cancer: An extension of the hypothesis on sexual behavior. *Cancer Epidemiol., Biomarkers Prev.* **3,** 375–379.

36. Johnson, A. M., and Wadsworth, J. (1992). Sexual lifestyles and HIV risk. *Nature (London)* **360,** 410–412.

37. ACFS (1992). AIDS and sexual behaviour in France. *Nature (London)* **360,** 407–409.

38. Melbye, M., and Biggar, R. J. (1992). Interactions between persons at risk for AIDS and the general population in denmark. *Am. J. Epidemiol.* **135,** 593–602.

39. Slattery, M. L., Overall, J. C., Abbott, T. M., French, T. K., Robison, L. M., and Gardner, J. (1989). Sexual activity, contraception, genital infections and cervical cancer: Support for a sexually transmitted disease hypothesis. *Am. J. Epidemiol.* **130,** 248–258.

40. Laumann, E. O., Gagnon, J. H., Michael, R. T., and Michaels, S. (1994). "The Social Organization of Sexuality." University of Chicago Press, Chicago and London.

41. Bosch, F. X., Castellsagué, X., Muñoz, N., De Sanjosé, S., Ghaffari, A. M., Gonzalez, L. C., Gili, M., Izarzugaza, I., Viladiu, P., Navarro, C., Vergara, A., Ascunce, N., Guerrero, E., and Shah, K. V. (1996). Male sexual behavior and human papillomavirus DNA: Key risk factors for cervical cancer in Spain. *J. Natl. Cancer Inst.* **88**(15), 1060–1067.

42. Muñoz, N., Castellsagué, X., Bosch, F. X., Tafur, L., De Sanjosé, S., Aristizabal, N., Ghaffari, A. M., and Shah, K. V. (1996). Difficulty in elucidating the male role in cervical cancer in Colombia, a high-risk area for the disease. *J. Natl. Cancer Inst.* **88**(15), 1068–1075.

43. Frisch, M., Glimelius, B., Van den Brule, A. J. C., Wohlfahrt, J., Meijer, C. J. L. M., Walboomers, J. M. M., Goldman, S., Svensson, C., Adami, H. O., and Melbye, M. (1997). Sexually transmitted infection as a cause of anal cancer. *N. Engl. J. Med.* **337**(19), 1350–1358.

44. De Sanjosé, S., Muñoz, N., Bosch, F. X., Reimann, K., Pedersen, N. S., Orfilia, J., Ascunce, N., Gonzalez, L. C., Tafur, L., Gili, M., Lette, I., Viladiu, P., Tormo, M. J., Moreo, P., Shah, K., and Wahren, B. (1994). Sexually transmitted agents and cervical neoplasia in Colombia and Spain. *Int. J. Cancer* **56**, 358–363.

45. Muñoz, N., Kato, I., Bosch, F. X., De Sanjosé, S., Sundquist, V.-A., Izarzugaza, I., Gonzalez, L. C., Tafur, L., Gili, M., Viladiu, P., Navarro, C., Moreo, P., Guerrero, E., Shah, K. V., and Wahren, B. (1995). Cervical cancer and Herpes Simplex Virus type 2: Case-control studies in Spain and Colombia, with special reference to immunoglobulin-G sub-classes. *Int. J. Cancer* **60**, 438–442.

46. World Health Organization (1993). Invasive squamous-cell cervical carcinoma and combined oral contraceptives: Results from a multinational study. *Int. J. Cancer* **55**, 228–236.

47. Schiffman, M. H., Haley, N. J., Felton, J. S., Andrews, A. W., Kaslow, R. A., Lancaster, W. D., Kurman, R. J., Brinton, L. A., Lannom, L. B., and Hoffmann, D. (1987). Biochemical epidemiology of cervical neoplasia: Measuring cigarette smoke constituents in the cervix. *Cancer Res.* **47**, 3886–3888.

48. Potischman, N., and Brinton, L. A. (1996). Nutrition and cervical neoplasia. *Cancer Causes Control* **7**, 113–126.

49. World Cancer Research Fund and American Institute for Cancer Research (1997). Cervix. *In* "Food, Nutrition and the Prevention of Cancer: A Global Perspective," pp. 301–308. WCRF & AICR, Washington, DC.

50. Cuzick, J., Szarewski, A., Terry, G., Ho, L., Hanby, A., Maddox, P., Anderson, M., Kocjan, G., Steele, S. T., and Guillebaud, J. (1995). Human papillomavirus testing in primary cervical screening. *Lancet* **345**, 1533–1536.

51. Meijer, C. J. L. M., Van den Brule, A. J. C., Snijders, P. J. F., Helmerhorst, T., Kenemans, P., and Walboomers, J. M. M. (1992). Detection of human papillomavirus in cervical scrapes by the polymerase chain reaction in relation to cytology: Possible implications for cervical cancer screening. *In* "The Epidemiology of Human Papillomavirus and Cervical Cancer" (N. Muñoz, F. X. Bosch, K. V. Shah, and A. Meheus, eds.) IARC Sci. Publ. No. 119, pp. 271–281. IARC, Lyon, France.

52. Viladiu, P., Bosch, F. X., Castellsagué, X., Muñoz, N., Escribà, J. M., Hamsikova, E., Hofmannova, V., Guerrero, E., Izquierdo, A., Navarro, C., Moreo, P., Izarzugaza, I., Ascunce, N., Gili, M., Muñoz, M. T., Tafur, L., Shah, K. V., and Vonka, V. (1997). HPV DNA and antibodies to HPV 16 E2, L2 and E7 peptides as predictors of survival in patients with squamous cell cervical cancer. *J. Clin. Oncol.* **15**, 610–619.

53. Burger, R. A., Monk, B. J., Kurosaki, T., Anton-Culver, H., Vasilev, S. A., Berman, M. L., and Wilczynski, S. P. (1996). Human papillomavirus type 18: Association with poor prognosis in early stage cervical cancer. *J. Natl. Cancer Inst.* **88**, 1361–1368.

54. Zur Hausen, H. (1997). Human papillomaviruses and cancer: A retrospective. *In* "Papillomaviruses in Human Cancer: The Role of E6 and E7 Oncoproteins" (M. Tommasino, ed.), pp. 1–24. Springer, Heidelberg.

55. Franceschi, S., Muñoz, N., Bosch, F. X., Snijders, P. J. F., and Walboomers, J. M. M. (1996). Human papillomavirus and cancers of the upper aero-digestive tract: A review of epidemiological and experimental evidence. *Cancer Epidemiol., Biomarkers Prev.* **5**, 567–575.

56. Jarrett, W. F., O'Neil, B. W., Gaukroger, J. M., Smith, K. T., Laird, H. M., and Campo, M. S. (1990). Studies on vaccination against papillomaviruses: The immunity after infection and vaccination with bovine papillomaviruses of different types. *Vet. Rec.* **126**, 473–475.

57. Breitburd, F., Kirnbauer, R., Hubbert, N. L., Nonnenmacher, B., Trin-Dinh-Desmarquet, C., Orth, G., Schiller, J. T., and Lowy, D. R. (1995). Immunization with viruslike particles from cottontail rabbit papillomavirus (CRPV) can protect against experimental CRPV infection. *J. Virol.* **69**(6), 3959–3963.

58. Suzich, J. A., Ghim, S. J., Palmer-Hill, F. J., White, W. I., Tamura, J. K., Bell, J. A., Newsome, J. A., Jenson, A. B., and Schlegel, R. (1995). Systemic immunization with papillomavirus L1 protein completely prevents the development of viral mucosal papillomas. *Proc. Natl. Acad. Sci. U.S.A.* **92**, 11553–11557.

59. Borysiewicz, L. K., Fiander, A., Nimako, M., Man, S., Wilkinson, G. W. G., Westmoreland, D., Evans, A. S., Adams, A. S., Stacey, S. N., Boursnell, M. E. G., Rutherford, E., Hickling, J. K., and Inglis, S. C. (1996). A recombinant vaccinia virus encoding human papillomavirus types 16 and 18, E6 and E7 proteins as immunotherapy for cervical cancer. *Lancet* **347**, 1523–1527.

74

Cervical Cancer Screening

DIANE SOLOMON* AND MARK SCHIFFMAN†

*Division of Cancer Prevention and †Division of Cancer Epidemiology and Genetics, National Cancer Institute,
Bethesda, Maryland

I. Introduction

Cervical cancer mortality in the United States has decreased since the 1950s by over 70% [1]. The decrease is attributed largely to the introduction of the Papanicolaou test in the 1940s. Cervical cancer, once the number one cancer killer of women, now ranks tenth in cancer deaths for women in the U.S. An estimated 15,000 women are still diagnosed each year with cervical cancer and approximately 4800 will die of their disease. However, worldwide, cervical cancer is the third most common cancer in women behind breast and colon cancer, and it ranks first in many developing countries that lack screening programs [2].

The accessibility of the cervix to direct examination and the relatively slow progression to cervical cancer from recognized and treatable precursor lesions make cervical neoplasia an ideal target for screening and prevention efforts. The success of screening has been demonstrated most directly and convincingly in Scandinavia. Countries with formal screening programs with wide population coverage experienced substantial drops in incidence and mortality while neighboring countries with limited population screening did not [3,4].

An empirical evaluation of screening programs in eight countries [5] as well as a mathematical model developed by Eddy [6] found that screening every 3 years affords appreciably more protection compared with screening every 5 or 10 years. However, in this evaluation, little protection was gained by screening annually compared with every three years.

In Canada, Great Britain, and many European countries, screening recommendations range from every 3–5 years. In the U.S., consensus recommendations adopted by the American Cancer Society, National Cancer Institute, American College of Obstetricians and Gynecologists, and others call for three consecutive annual screening tests for women who have initiated sexual activity or have reached age 18. If the results of these three tests are negative, the screening interval may be extended at the discretion of the clinician.

Historically, unscreened subpopulations of women in the U.S. include older women, uninsured and impoverished women, minority women (particularly Hispanic and older African-American women), and women residing in rural areas [7]. Some of these patterns are changing whereas others are not. In the 1994 National Health Interview Survey of the U.S. population [8], 77% of women reported having had a Pap test in the past 3 years. Age remains a factor; screening was higher among women 18–44 (82%) compared to women 65 and older (57%). However, there were no marked differences between African-Americans, Hispanic whites, and non-Hispanic whites, or metropolitan versus nonmetropolitan residents in the 18–44 age group. Socioeconomic measures continue to show significant differences in screening coverage; women who did not complete high school and whose family income was less than $20,000 reported lower rates of screening compared to women with education beyond high school or family income exceeding $20,000 [8].

In many developing countries, screening is available to only a small segment of the population through urban clinics or hospitals, or not at all [9]. Obstacles to comprehensive cervical cancer screening include lack of public and clinician awareness of cervical cancer as a health problem, lack of awareness of the benefits of screening, inadequate numbers of trained clinicians, inadequate supplies, inadequate laboratory facilities and personnel to evaluate specimens, loss to follow-up, and inadequate treatment facilities [10]. For such countries, comprehensive cytologic screening performed at regular intervals is unattainable at the present time. Other approaches to screening must be considered, such as limiting the age range for screening, limiting screening to a single test for women at the maximally beneficial age (*e.g.,* between the ages of 30–35), or utilizing noncytologic approaches to screening that do not require an extensive infrastructure of trained personnel.

II. Papanicolaou Test

The Papanicolaou (Pap) test is currently the most widely utilized cervical cancer screening technique in the U.S. as well as internationally. Named for George Papanicolaou, one of the originators of cervical cytologic diagnosis, the test involves gently scraping cells from the surface of the cervix and evaluating the fixed and stained sample microscopically to detect abnormal morphologic cell changes. Although there are new technologies currently available and others in development that may dramatically alter screening in the future, the Papanicolaou test is still the standard of care and serves as a paradigm to discuss components of a screening process.

A. *Specimen Collection*

Obtaining an adequate specimen is an essential step that requires some training and experience. The clinician should visually inspect the cervix and identify the "squamocolumnar junction" where the smooth squamous surface of the ectocervix changes to the cobblestone-like glandular lining of the endocervix which leads into the uterine cavity. Sampling should be directed to this ring of tissue, as this is the region where the majority of cervical lesions arise. In comparison to a spatula alone, it has been demonstrated that use of either a combination of a spatula and a cervical brush or a broom-shaped device that samples both the ectocervix and endocervix simultaneously results in increased detection of abnormalities [11].

Table 74.1
Cervical Diagnostic Terminology

Dysplasia	Atypia	HPV	Mild dysplasia	Moderate dysplasia	Severe dysplasia	CIS
CIN	Atypia	HPV	CIN 1	CIN 2	CIN 3	
Bethesda	ASCUS	LSIL		HSIL		

In the conventional Pap "smear," the cellular sample collected on the instrument(s) is spread over the surface of a glass slide. The object is to quickly but evenly spread the material over the slide, thinning out large clumps but avoiding excessive manipulation that can damage cells. Studies have shown that more than half of the material collected on the sampling instrument is not transferred to the glass slide but remains on the device and is therefore lost for microscopic analysis [12]. After smearing, rapid fixation of the specimen by alcohol immersion or spray is essential to preserve morphologic detail. Air-drying of the sample may limit the interpretability of the specimen.

B. Laboratory Evaluation and Diagnosis

Once accessioned in the laboratory, the slides are stained using a polychrome process that was developed by Papanicolaou and bears his name. When optimally performed, it results in excellent nuclear detail and cytoplasmic transparency that allows visualization through areas of overlapping cells.

Specimen adequacy is assessed microscopically based on a number of parameters including number and types of epithelial cells present, morphologic preservation, and presence of obscuring factors, such as blood, inflammation, or air-drying, that may limit microscopic visualization of the cells [13]. An "adequate" specimen consists of well-preserved, evenly distributed squamous and glandular cells. The presence of both epithelial cell types provides indirect evidence that the squamocolumnar junction has been sampled.

The process of diagnostic evaluation of a Pap test is highly labor-intensive and subjective. A cervical specimen may consist of over 100,000 cells of which only a small number may be abnormal. The process of microscopic screening is performed by trained cytotechnologists who must be able to detect the rare abnormal cell amidst thousands of cytologically normal cells. Any identified abnormal or questionable cytologic changes are then referred to a pathologist for diagnostic interpretation.

Pap test results may be reported using a variety of terminology systems. A translation table [Table 74.1] is helpful to convert from one nomenclature to another. At the time of the emergence of cytology as a diagnostic discipline in the 1940s–1950s, Dr. George Papanicolaou devised a numeric classification (I-V) to communicate the degree of confidence that cancer cells were present in a specimen. As used initially by Papanicolaou, the numeric designations represented the following: Class I—benign; Class II—minor cellular abnormalities considered benign; Class III—cells suspicious for but not diagnostic of cancer; Class IV—cells fairly conclusive for malignancy; and Class V—cells diagnostic of cancer.

As the field of cytology expanded, numeric designations largely gave way to terminology systems that included a designation of the degree of abnormality identified, for example the four grades of dysplasia (mild, moderate, severe, and carcinoma-*in situ* (CIS)). Richart introduced the term cervical intraepithelial neoplasia (CIN), grades 1, 2 and 3, to promote the concept of a disease continuum of precursors to invasive cancer [14]. The morphologic criteria for the three grades of CIN are based on tissue architecture: the proportional thickness of the epithelium involved by disorderly growth and cytologic atypia. Mild and moderate dysplasia roughly correspond to CIN 1 and CIN 2, respectively. However, CIN 3 encompasses severe dysplasia and CIS, thus eliminating a difficult and sometimes arbitrary diagnostic distinction between almost vs complete full-thickness abnormality.

Koilocytosis, a descriptive diagnostic term indicating cellular changes of perinuclear cytoplasmic cavitation, was recognized by Meisels to be a manifestation of genital human papillomavirus (HPV) infection [15]. Initially, HPV cellular changes were considered distinct from "true" dysplasia or CIN and not part of the precursor pathway to cervical cancer. However, as techniques for identifying HPV became more sensitive, HPV DNA was found in the vast majority of cervical neoplasias studied [16]. The pathogenesis of cervical neoplasia and cervical cancer is now known to be due to HPV, based on epidemiologic, virologic, and experimental evidence. Therefore, isolation of "koilocytotic atypia" or "HPV effect" as a separate distinct entity from dysplasia/CIN is no longer biologically valid.

The Bethesda System, developed at a National Cancer Institute workshop in 1988 [17] and refined in 1991 [18], collapses the cytologic diagnostic subcategories of intraepithelial lesions into low- and high-grade squamous intraepithelial lesions, abbreviated as LSIL and HSIL, respectively. This division is based on the concept of HPV-induced cellular changes as discrete processes of (1) LSIL as acute infection with any HPV type resulting in mild, usually transient cytologic effects, and (2) HSIL as the result of persistent infection with predominantly oncogenic HPV types and the interplay of a variety of factors, including host immune response, that poses a substantial risk of invasion [19]. While the CIN classification remains widely used in cervical histopathology, the Bethesda System (TBS) is more commonly used to report Pap test results.

The Bethesda System also introduced the term "atypical squamous cells of undetermined significance" (ASCUS) to reflect equivocal, abnormal changes that are quantitatively or qualitatively insufficient to establish a definitive diagnosis of SIL. ASCUS is not a single diagnostic entity and is therefore associated with highly variable clinical outcomes. It does represent an improvement, however, over older classifications that used "atypia" to encompass reactive changes and HPV-associated cell changes in addition to equivocal findings. In the Bethesda System, reactive changes are categorized as "benign" and HPV changes are subsumed under SIL.

Abnormal Pap test results are not evenly distributed among the diagnostic categories described above. Rather, in a screened population such as that in the U.S., the distribution of abnormalities resembles a pyramid with relatively few cancers at the top and millions of low grade and equivocal diagnoses comprising the very broad base (Fig. 74.1). In the U.S., cancers represent

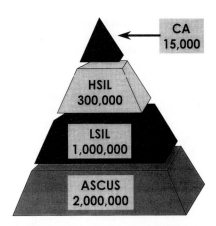

Fig. 74.1 Cervical lesions: pyramid of diagnoses.

far less that one-tenth of 1% of diagnoses and high-grade lesions constitute approximately six-tenths of 1% [20,21]. By contrast, LSIL and ASCUS account for an estimated 6% of all Pap test results, which translates to 3 million women in the U.S. annually.

C. Screening Test Characteristics

A screening test can be evaluated by several parameters. In the 2 × 2 screening table (Table 74.2), the results of a dichotomous screening test are presented compared to the disease state of the population screened: for example, HSIL or above (HSIL+) versus not. Four test outcomes are possible: A represents true-positives (positive results in individuals with HSIL+); D is the number of true-negatives (negative results in individuals without HSIL+); B reflects false-positives (positive results in individuals without HSIL+); and C corresponds to false-negatives (negative results in individuals with HSIL+).

Sensitivity [A / (A + C)] is the proportion of diseased individuals correctly detected by a positive test. Specificity [D/ (B + D)] is the proportion of disease-free individuals who receive a negative test result. Positive predictive value [A/(A + B)] indicates the percentage of positive test results that correctly identify the presence of disease. Negative predictive value [D/ (C + D)] reflects the percentage of negative tests correctly indicating the absence of disease (i.e., the reassurance provided by a negative test).

The sensitivity of the Papanicolaou test for high-grade lesions or cancer is estimated to be up to 70–80% (specificity 94–97%) [22,23]. Test sensitivity must be distinguished from program sensitivity. The former is a measure of the sensitivity of a single test at one point in time. The latter is the sensitivity of a series of tests at intervals determined by the screening program to detect an abnormality at any single test event. Repeat screening at regular intervals therefore compensates somewhat for the limitations of the sensitivity of the technique.

In the context of cervical screening, two main types of error contribute to lower sensitivity. Sampling error occurs when a cervical lesion is present but cells representative of the abnormality are not present on the glass slide specimen. Sampling error may occur if either the lesion is not sampled or if abnormal cells collected on the sampling implement are not transferred to

Table 74.2
Schematic Outcomes of a Diagnostic Test

Test result	Disease		Total
	Present	Absent	
Positive	A	B	A + B
Negative	C	D	C + D
Total	**A + C**	**B + D**	

the slide. Factors that contribute to sampling error include small size of the lesion, inaccessible location of the lesion (high in the endocervical canal, for example), or inappropriate sampling technique. Laboratory error occurs when cells diagnostic of an intraepithelial lesion or carcinoma are present in the specimen but are not identified as abnormal when the result is reported. Factors that may contribute to laboratory error include presence of only a few abnormal cells, small size of the abnormal cells, presence of inflammation or blood obscuring cells, or diagnostic misinterpretation of the significance of identified cell abnormalities. Even under optimal screening conditions, sampling and laboratory error cannot be entirely eliminated.

D. Threshold for Further Follow-up

The objective of cervical cancer screening is to prevent the development of invasive cervical cancer, ideally by effectively identifying and treating the minimal number of women with histologically confirmed precursor lesions. In order to identify such women, the screening test threshold for further evaluation could be set at a cytologic diagnosis of "high-grade lesion." This cut-point would yield a high percentage of confirmed high-grade tissue lesions among the women evaluated (a high positive predictive value of a positive result). However, many of the women who harbor a true high-grade tissue lesion may be "missed" because the severity of a lesion may be undercalled on the screening test. In two large studies, one-fourth to over one-half of prevalent cases of high-grade neoplasia were associated with cytologic diagnoses of ASCUS or LSIL [24,25].

Lowering the test threshold for further evaluation, to LSIL or ASCUS for example, improves sensitivity and NPV (reassurance of a negative result) of the screening process at the expense of a loss of specificity. As can be seen from the pyramid of abnormal cytology diagnoses in Figure 74.1, lowering the cut-point for further evaluation from HSIL to ASCUS increases by a factor of ten the number of women referred to follow-up.

The screening test threshold for follow-up and/or treatment of cervical abnormalities will vary depending on prevailing management paradigms, medicolegal issues, economic factors, and societal expectations. In countries where sensitivity is emphasized over specificity, lesser abnormalities will trigger additional follow-up compared to countries that favor a more cost-effective approach to screening.

The trade-off of sensitivity and specificity for a given test is graphically depicted by Receiver Operating Characteristic (ROC) curves that plot sensitivity and (1-specificity) along the Y- and X-axis respectively, as the test cut-points change. ROC curves are useful tools to compare the performance of a given test at different thresholds, as well as to compare different tests or combinations of tests [26].

E. Follow-up and Management of Abnormalities

Screening cannot be effective without follow-up of abnormal results and treatment of lesions as appropriate. Loss to follow-up is a significant problem. In two studies [27,28] 13 and 15% of cervical cancers that occurred in women who had ever had a Pap test were attributed to lack of either patient notification or patient compliance with recommended treatment. ''One-stop'' screening, diagnosis, and treatment clinics have been established in a few high-risk areas to address this problem [29,30]. However, this is a labor-intensive approach to screening that cannot feasibly be applied as yet on a large scale.

In the U.S., high-grade cytologic lesions are managed by visual evaluation of the cervix with magnification (colposcopy), directed biopsy, and—in women with histologically-confirmed HSIL—destruction or removal of the lesion and transformation zone of the cervix. However, there is currently no consensus as to the appropriate management of women with LSIL or ASCUS, which comprise the vast majority of abnormal Pap smear results. Options include immediate colposcopy and directed biopsy as with high-grade lesions versus follow-up with repeat cytology every 4–6 months, with colposcopy indicated only if an abnormality persists.

Most low-grade changes regress spontaneously; only a minority of such lesions would progress without treatment. However, currently there is no way to determine morphologically which patients are at risk. The available data indicate that for many specimens demonstrating ASCUS, patients do not have a significant lesion and follow-up smears will be normal. In 25–60% of patients, however, further evaluation will detect a squamous intraepithelial lesion. The majority of these lesions are low grade: only 15–30% are high-grade [31,32]. Of note, the yield of high-grade lesions is increased in cases of ASCUS involving atypical metaplastic cells [33].

As discussed in the next section, future studies may provide a molecular basis to distinguish true precursors of neoplasia from minor lesions of no significant clinical import; this would allow a more coherent and rational approach to diagnosis and management of women. For example, HPV testing may have utility to identify which women with equivocal Pap test results are at greatest risk of a significant lesion.

III. New Cervical Cancer Screening Techniques

The cervical/vaginal Pap smear has been tremendously successful in reducing the death rate from cervical cancer. However, as with any medical test, the Pap smear has limitations, particularly with respect to false-negative screening results.

Interest has focused on development of technologies to enhance the accuracy of cervical cancer screening. Some of these techniques are directed at improving the sampling and specimen quality, others are focused on improving the laboratory microscopic screening process, and some techniques are visual or molecular rather than microscopic.

A. Techniques to Improve Sampling and Specimen Quality

As noted earlier, with conventional smear techniques only a fraction of the cellular material collected from the cervix is transferred to the glass slide. By contrast, liquid-based collection, in which the implement is vigorously rinsed in a vial of preservative/fixative, recovers much more of the cervical sample [12]. The vial is then transported to the laboratory where the specimen is agitated to disaggregate cell clumps and a subsample of cell material is deposited on a glass slide in a fairly uniform, thin layer. The method of subsampling and cell transfer to the glass slide varies depending on the particular device. ThinPrep 2000 (Cytyc, Boxborough, MA) is a semiautomated single-sample processor that uses suction filtration. CytoRich (AutoCyte Inc., Elon College, NC) utilizes centrifugation, sedimentation through a density gradient, and filtration to process multiple samples at a time. These techniques cost $10–20 more than a conventional Pap smear.

Clinical trials comparing conventional smears and ''residual-to-vial'' (after a smear has been made) liquid-based preparations have shown equal or greater sensitivity of the thin-layer preparations in detection of low-grade lesions [34,35]. Some studies also claim equal or increased sensitivity for high-grade lesions [24,36,37].

Liquid-based collection eliminates vagaries of collector-operator errors of uneven or incomplete transfer of cellular material or air-drying artifact. Improved fixation and presentation of the even distribution of material in a more uniform fashion may make detection of abnormal cells easier. Another theoretical advantage of liquid-based collection is that residual specimen would be available for additional testing as may be appropriate. Several studies have noted the potential advantage of ''reflex'' HPV testing in the setting of a low-grade or equivocal cytology result [38,39]. The ability to test for HPV from the same cytologic sample would eliminate the need for an additional patient visit to collect a separate sample. Ongoing studies are evaluating the cost-effectiveness of such a management approach.

B. Computerized Screening Technologies

These devices utilize computer image analysis technology to screen cervical cytology specimens in an effort to reduce false-negative results. Two instruments have received FDA approval for rescreening (secondary screening) of previously evaluated specimens determined to be negative by routine manual screening: Papnet (AutoCyte) and AutoPap (NeoPath Inc., Redmond, WA). Only AutoPap is also FDA approved for use in initial (primary) screening of cervical specimens.

Used in a rescreening mode, AutoPap identifies approximately 20% of previously diagnosed ''negative'' cases as most likely to contain an abnormality. These cases undergo repeat manual screening by the cytotechnologist. Papnet analysis generates a digital tape of images of the most abnormal cells or cell groups as identified by the computer. The tape is reviewed by a cytotechnologist at a computer monitor workstation. In some cases, an abnormality may be diagnosed based on review of the digital images; in other cases, an abnormality may be suspected and the glass slide may be selected for manual microscopic review.

Compared to random 10% quality control rescreening, both computer rescreening devices increase detection of abnormalities in cases previously diagnosed as negative. Using 100% rescreening as the reference standard, AutoPap, set to select 20% of slides for review, identified 77% of LSIL and above [40] (7.7 times more than a random 10% review). Papnet assisted rescreening sensitivity for LSIL and above is estimated to be in the range of 85% [41] (8.5 times more that a random 10% review). However, this increased sensitivity is primarily for AS-CUS and LSIL diagnoses and comes at significant cost. Used in a secondary screening mode, these technologies are cost-effective only if incorporated into a less frequent screening strategy [41].

Operating as a primary screener, the AutoPap computer identifies approximately 25% of cases—those with the lowest rank score—as least likely to contain an abnormality; these slides are not reviewed by a cytotechnologist. The remaining 75% of specimens undergo manual microscopic screening. In addition, of those cases reviewed as "negative" by the cytotechnologist, a subset with the highest rank score as determined by the computer are then subjected to a second round of manual screening.

Although not FDA-approved for use in primary screening, a study comparing primary Papnet screening to conventional microscopic screening showed promising results [22].

C. Nonmicroscopic Screening Technologies

In addition to the efforts to improve cytologic screening, several adjunctive screening technologies are being considered in an effort to improve cervical screening sensitivity. These can be categorized as: (1) visual evaluation techniques; (2) electro-optical probes; and (3) testing for molecular markers.

1. Visual Evaluation Technologies

The colposcope was first developed in 1925 in Germany to visually evaluate the cervical epithelium for abnormal changes. Five percent acetic acid is usually applied to the cervix during the examination to enhance the contrast between normal and abnormal tissue. Although colposcopy is used for primary screening in some countries such as Switzerland, in most areas coloposcopy is not an economically feasible primary screening tool due to the relatively high price of the procedure which requires highly trained colposcopists. In certain high-risk populations, colposcopy may be used as a screening technique. However, colposcopy is generally restricted to patients with a previously identified abnormality, to direct tissue biopsy to the most abnormal area of the cervix, and to visually evaluate the location and extent of a lesion prior to therapy. It is worth noting that the practice of colposcopy has not been standardized, leading to uncertain intercolposcopist variability.

Cervicography utilizes a 35-mm camera with a fixed focal-distance telephoto macrolens to take a photographic image of the cervix that can be sent off-site for diagnostic interpretation by expert readers. Several early studies evaluating the sensitivity of cervicography plus cytology in comparison to cytology alone found that the addition of cervicography increases screening sensitivity, primarily for low-grade lesions, but with unacceptable loss of screening specificity [42–44].

In a large population-based screening study in Costa Rica, Schneider et al. compared the results of cytology, cervicography, and HPV testing, alone and in combination, for 8460 women [45]. If women with positive screening cervicography were referred to colposcopy, all 11 invasive cancers would have been detected. However, the sensitivity for high-grade intraepithelial lesions was only 48%. The technique was logistically feasible and inexpensive but had limited utility in postmenopausal women due to the migration of the squamocolumnar junction into the endocervical canal beyond the visual range of cervicography.

Visual inspection (VI) is a very low-cost approach to screening that may be an option for areas that do not have access to comprehensive cervical cytological screening [46]. VI, at its most basic, consists of looking at the untreated cervix for visual signs of a high-grade lesion or cancer. VI may be enhanced by cervical application of acetic acid (termed VIA) and the use of low-power magnification (termed VIAM) to detect acetowhite lesions. Use of VI or VIAM alone, without cytologic screening, is unlikely to achieve the accuracy of the Pap test or even cervicography. However, VI may prove to be a cost-effective approach to decrease cervical mortality in countries that cannot afford a comprehensive cytological screening program [47].

2. Electro-optical Probe Devices

Several electro-optical probe devices are currently under development or in clinical trials; none, at this writing, are FDA approved. This technology is based on measurable differences in the physical properties of electrical decay and light scatter of normal and abnormal epithelium [48]. The devices typically consist of a small desktop processor and a sterilizable or disposable fiber-optic probe that is inserted into the vagina. The probe emits a mild electrical and/or optical stimulus to the surface of the cervix and then measures the voltage decay or light transmission and scatter properties of the tissue. Immediate results are available to the operator and theoretically could be used either in a primary screening mode to select women to be evaluated by colposcopy or in a triage mode to direct colposcopic biopsy to the most abnormal areas of the cervix.

3. Molecular Markers: HPV Testing

As mentioned earlier, the pathogenesis of cervical neoplasia and cervical cancer is known to be due to HPV, based on epidemiologic, virologic, and experimental evidence. While there are over 90 HPV types, including those that cause cutaneous warts, approximately 30 types infect the anogenital tract. About half of the anogenital HPV types have been identified in cervical cancers and are termed "oncogenic" or "cancer associated" types; the remainder are classified as "low risk." Although HPV infection of sexually active women is common, cervical cancer is not; therefore, other factors, including the host immune response, determine the course of infection and the potential to develop significant cervical disease.

HPV testing is based on detection of HPV DNA in cervical specimens, as clinically useful serologic assays for the full range of oncogenic HPV types have not yet been fully validated. Advances in HPV DNA testing methodology now allow testing directly from residual liquid-based cytology specimens

(PreservCyt, Cytyc) or a separately collected sample. Studies comparing polymerase chain reaction (PCR)-based and Hybrid Capture (HC)-based HPV detection systems show excellent agreement between PCR results and the newest generation HC test (HC II) using a 1.0-pg/ml cutoff [49].

Several screening strategies utilizing HPV testing may be considered. These scenarios can be categorized broadly as: (1) primary screening, in addition to, or as a substitute for cervical cytology; or (2) secondary testing following an ASCUS cytologic abnormality, to clarify the cytologic diagnosis or to triage women for further colposcopic evaluation.

In young sexually active women, the high prevalence of HPV infection, often not associated with significant cervical disease, probably precludes use of HPV testing as a primary screening strategy in this setting. However, HPV DNA prevalence declines sharply with age while the sensitivity of HPV DNA for cervical neoplasia remains high. Therefore, the positive predictive value of finding HPV DNA rises with age. Moreover, the accuracy of Pap smear cytology declines with age due to sampling false negatives (as the squamocolumnar junction migrates into the endocervical canal) and false positives (associated with atrophic (estrogen-depleted) cellular changes). Thus, HPV testing may potentially be a cost-effective, primary screening strategy in older women [50]. Evaluation of the strategy is underway in at least two European countries.

In well-screened populations, the number of ASCUS and LSIL cases detected by cytology will greatly outnumber high-grade lesions and cancer. While there is general consensus (in the U.S.) that such diagnoses warrant increased monitoring, it is not clear whether colposcopy and biopsy or more frequent cytologic sampling represents optimal management. There is a trade-off between aggressive follow-up of cytologic changes that, in the majority of women, would regress spontaneously and underdiagnosis of the minority of women at risk for a significant HSIL or cancer. Sherman *et al.* have demonstrated that HPV testing may clarify inconclusive cytologic diagnoses by separating true lesions from mimics unrelated to cervical neoplasia [51]. HPV testing may also help to identify which women harbor occult HSIL at the time of an ASCUS diagnosis [52], or possibly predict which low grade lesions will progress over time to HSIL. Currently, a multicenter, prospective clinical trial sponsored by the National Cancer Institute is underway to evaluate the potential role of HPV testing in the management of ASCUS and LSIL.

D. Evaluation of New Technologies

Many of these new technologies are well beyond the financial capabilities of developing countries that are seeking to establish or improve existing screening programs. However, using ROC curve and cost-effectiveness analyses, one might develop a more rationally based screening program that may improve sensitivity at no/little extra cost. Using a new technology or a combination of technologies will increase the cost of a screening event; however, the gain in sensitivity may allow less frequent screening that theoretically could result in cost-neutral implementation.

At this writing it is not clear which technology or combination of methodologies will emerge as new standards of care in the U.S. or internationally. A cost-effectiveness analysis of the

three FDA-approved technologies to improve the accuracy of cytology screening finds that although each enhancement increases the sensitivity of screening, the marginal benefit in terms of lives saved compared to conventional annual screening is small, and the costs are relatively high [41]. However, the study points out that as technologies evolve and/or less frequent screening strategies are considered, the cost-effectiveness ratios may shift in favor of new approaches to screening.

References

1. Landis, S. H., Murray, T., Bolden, S., and Wingo, P. A. (1999). Cancer statistics 1999. *C—Cancer J. Clin.* **49,** 8–32.
2. Parkin, D. M., Pisani, P., and Ferlay, J. (1999). Global cancer statistics. *Ca—Cancer J. Clin.* **49,** 33–64.
3. Johannessson, G., Geirsson, G., and Day, N. (1978). The effect of mass screening in Iceland, 1965–74, on the incidence and mortality of cervical carcinoma. *Int. J. Cancer* **21,** 418–425.
4. Hakama, M., Magnus, K., Pettersson, F., Storm, H., and Tulinius, H. (1991). Effect of organized screening on the risk of cervical cancer. *In* "Cancer Screening. UICC Project on Evaluation of Screening for Cancer" (A. B. Miller, J. Chamberlain, N. E. Day, M. Hakama, and P. C. Prorock, eds.) pp. 153–162. International Union Against Cancer, Cambridge, UK.
5. Day, N. E., for IARC Working Group (1986). Screening for squamous cervical cancer: Duration of low risk after negative results of cervical cytology and its implications for screening policies. *Br. Med. J.* **293,** 659–664.
6. Eddy, D. M. (1990). Screening for cervical cancer. *Ann. Intern. Med.* **113,** 214–226.
7. Brown, C. L. (1996). Screening patterns for cervical cancer: How best to reach the unscreened population. *Natl. Cancer Inst. Monogr.* **21,** 7–11.
8. Cucinelli, J., programmer (1996). 1994 National Health Interview Survey public use data file. *Morbid. Mortal. Wkly. Rep.* **45,** 57–61.
9. Lunt, R. (1984). Worldwide early detection of cervical cancer. *Obstet. Gynecol.* **63,** 708–713.
10. Bishop, A., Wells, E., Sherris, J., Tsu, V. D., and Crook, B. (1995). Cervical cancer: Evolving prevention strategies for developing countries. *Reprod. Health Matters* **6,** 60–70.
11. Boon, M. E., de Graaff Guilloud, J. C., and Rietveld, W. J. (1989). Analysis of five sampling methods for the preparation of cervical smears. *Acta Cytol.* **33,** 843–848.
12. Hutchinson, M. L., Isenstein, L. M., Goodman, A., Hurley, A. A., Douglass, K. L., Jui, K. K., Patten, F. W., and Zahniser, D. J. (1994). Homogeneous sampling accounts for the increased diagnostic accuracy using the ThinPrep processor. *Am. J. Clin. Pathol.* **101,** 215–219.
13. Solomon, D., and Henry, M. (1995). Specimen adequacy. *In* "Compendium on Quality Assurance, Proficiency Testing and Workload Limitations in Clinical Cytology," pp. 90–94. Tutorials of Cytology, Chicago.
14. Richart, R. M. (1968). Natural history of cervical intraepithelial neoplasia. *Clin. Obstet. Gynecol.* **5,** 748.
15. Meisels, A., and Fortin, A. (1976). Condylomatous lesions of the cervix and vagina. I. Cytologic patterns. *Acta Cytol.* **20,** 505–509.
16. Schiffman, M. H. (1992). Recent progress in defining the epidemiology of human papillomavirus infection and cervical neoplasia. *J. Natl. Cancer Inst.* **84,** 394–398.
17. Solomon, D. (1989). The 1988 Bethesda system for reporting cervical/vaginal cytologic diagnoses. *JAMA, J. Am. Med. Assoc.* **262,** 931–934.
18. Broder, S. (1992). The Bethesda system for reporting cervical/vaginal cytologic diagnoses—report of the 1991 Bethesda workshop. *JAMA, J. Am. Med. Assoc.* **267,** 1892.

19. Schiffman, M. H., Liaw, K. L., Herrero, R., Sherman, M. E., Hildesheim, A., and Solomon, D. (1998). Epidemiologic support for a simplified view of cervical carcinogenesis. *Eur. Bull.* **1,** 2–6.

20. College of American Pathologists (1994). "Interlaboratory Comparison Program in Cervicovaginal Cytology." College of American Pathologists.

21. Liaw, K. L., Glass, A. G., Manos, M. M., Greer, C., Scott, D. R., Sherman, M. E., Burk, R. D., Kurman, R. J., Wacholder, S., Rush, B. B., Cadell, D. M., Lawler, P., Tabor, D., and Schiffman, M. (1999). HPV testing of cytologically-normal women predicts the risk of cervical squamous intraepithelial lesions. *J. Natl. Cancer Inst.* **91,** 954–960.

22. Sherman, M. E., Schiffman, M., Herrero, R., Kelly, D., Bratti, C., Mango, L. J., Alfaro, M., Hutchinson, M. L., Mena, F., Hildesheim, A., Morales, J., Greenberg, M. D., Balmaceda, J., and Lorincz, A. T. (1998). Performance of a semiautomated Papanicolaou smear screening system. *Cancer Cytopathol.* **84,** 273–280.

23. Myers, E. R. *et al.* "Evaluation of Cervical Cytology," Technol. Assess. No. 5, 99-E010. Agency for Health Care Policy and Research (AHCPR).

24. Hutchinson, M., Zahniser, D., Sherman, M. E., Herrero, R., Alfaro, M., Bratti, C., Hildesheim, A., Greenberg, M., Lorincz, A., Morales, J., and Schiffman, M. (1999). Utility of liquid-based cytology for cervical cancer screening: Results of a population-based study conducted in a high cervical cancer incidence region of Costa Rica. *Cancer Cytopathol.* **87,** 48–55.

25. Kinney, W. K., Manos, M. M., Hurley, L. B., and Ransley, J. E. (1998). Where's the high-grade cervical neoplasia? The importance of minimally abnormal papanicolaou diagnoses. *Obstet. Gynecol.* **91,** 973–976.

26. Fahey, M. T., Irwig, L., and Macaskill, P. (1995). Meta-analysis of Pap test accuracy. *Am. J. Epidemiol.* **141,** 680–689.

27. Janerich, D. T., Hadjimichael, O., Schwartz, P. E., Lowell, D. M., Meigs, J. W., Merino, M. J., Flannery, J. T., and Polednak, A. P. (1995). The screening histories of women with invasive cervical cancer, Connecticut. *Am. J. Public Health* **85,** 791–794.

28. Carmichael, J. A., Jeffrey, J. F., Steele, H. D., and Ohlke, I. D. (1984). The cytologic history of 245 patients developing invasive cervical carcinoma. *Am. J. Obstet. Gynecol.* **148,** 685–690.

29. Burger, R. A., Monk, B. J., Van Nostrand, K. M., Greep, N., Anton-Culver, H., and Manetta, A. (1995). Single-visit program for cervical cancer prevention in a high-risk population. *Obstet. Gynecol.* **86**(4, Pt. 1), 491–498.

30. Megevand, E., Van Wyk, W., Knight, B., and Bloch, B. (1996). Can cervical cancer be prevented by a see, screen, and treat program? A pilot study. *Am. J. Obstet. Gynecol.* **174,** 923–928.

31. Davey, D. D., Naryshkin, S., Nielsen, M. L., and Kline, T. S. (1994). Atypical squamous cells of undetermined significance: Interlaboratory comparison and quality assurance monitors. *Diagn. Cytopathol.* **11,** 390–396.

32. Williams, M. L., Rimm, D. L., Pedigo, M. A., and Frable, W. J. (1997). Atypical squamous cells of undetermined significance: Correlative histologic and follow-up studies from an academic medical center. *Diagn. Cytopathol.* **16,** 1–7.

33. Sheils, L. A., and Wilbur, D. C. (1997). Atypical squamous cells of undetermined significance: Stratification of the risk of association with, or progression to, squamous intraepithelial lesions based on morphologic subcategorization. *Acta Cytol.* **41,** 1065–1072.

34. Wilbur, D., Cibas, E., Merritt, S., James, L. P., Berger, B., and Bonfiglio, T. (1994). ThinPrep Processor clinical trials demonstrate an increased detection rate of abnormal cervical cytologic specimens. *Am. J. Clin. Pathol.* **101,** 209–214.

35. Lee, K. R., Ashfaq, R., Birdsong, G. G., Corkill, M. E., McIntosh, K. M., and Inhorn, S. L. (1997). Comparison Of conventional Papanicolaou smears and a fluid-based, thin-layer system for cervical cancer screening. *Obstet. Gynecol.* **90,** 278–284.

36. Papillo, J. L., Zarka, M. A., and St. John, T. L. (1998). Evaluation of the ThinPrep Pap test in clinical practice. *Acta Cytol.* **42,** 203–208.

37. Roberts, J. M., Gurley, A. M., Thurloe, J. K., Bowditch, R., and Laverty, C. R. A. (1997). Evaluation of the ThinPrep Pap test as an adjunct to the conventional Pap smear. MJA *Med. J. Aust.* **167,** 466–469.

38. Ferenczy, A., Franco, E., Arseneau, J., Wright, T. C., and Richart, R. M. (1996). Diagnostic performance of hybrid capture human papillomavirus deoxyribonucleic assay combined with liquid-based cytologic study. *Am. J. Obstet. Gynecol.* **175,** 651–656.

39. Sherman, M. E., Schiffman, M. H., Lorincz, A. T., Herrero, R., Hutchinson, M. L., Bratti, C., Zahniser, D., Morales, J., Hildesheim, A., Helgesen, K., Kelly, D., Alfaro, M., Mena, F., Balmaceda, I., Mango, L., and Greenberg, M. (1997). Cervical specimens collected in liquid buffer are suitable for both cytologic screening and ancillary human papillomavirus testing. *Cancer* (*Philadelphia*) **81,** 89–97.

40. Colgan, T. J., Patten, S. F., and Lee, J. S. J. (1995). A clinical trial of the AutoPap 300 QC system for quality control of cervicovaginal cytology in the clinical laboratory. *Acta Cytol.* **39,** 1191–1198.

41. Brown, A. D., and Garber, A. (1999). Cost-effectiveness of 3 methods to enhance the sensitivity of Papanicolaou testing. *JAMA, J. Am. Med. Assoc.* **281,** 347–353.

42. Coibion, M., Autier, P., Vandam, P., Delobelle, A., Huet, F., Hertens, D., Vosse, M., Andry, M., DeSutter, P., Heimann, R., and Miloiu, A. (1994). Is there a role for cervicography in the detection of premalignant lesions of the cervix uteri? *Br. J. Cancer* **70,** 125–128.

43. Baldauf, J.-J., Dreyfus, M., Ritter, J., Meyer, P., and Philippe, E. (1997). Cervicography Does it improve cervical cancer screening? *Acta Cytol.* **41,** 295–301.

44. Ferris, D. G., Payne, P., Frisch, L. E., Milner, F. H., diPaola, F. M., and Petry, L. J. (1993). Cervicography: Adjunctive cervical cancer screening by primary care clinicians. *J. Fam. Pract.* **37,** 158–164.

45. Schneider, D. L., Herrero, R., Bratti, C., Greenberg, M., Hildesheim, A., Sherman, M. E., Morales, J., Hutchinson, M. L., Sedlacek, T. V., Lorincz, A., Mango, L., Wacholder, S., Alfaro, M., and Schiffman, M. (1999). *Am. J. Obstet. Gynecol.* **180,** 290–298.

46. Singh, V., Sehgal, A., and Luthra, U. K. (1992). Screening for cervical cancer by direct inspection. *Br. Med. J.* **304,** 534–535.

47. Sankaranarayanan, R., Wesley, R., Somanathan, T., Dhakad, N., Shyamalakumary, B., Amma, N. S., Parkin, D. M., and Nair, M. K. (1998). *Cancer* (*Philadelphia*) **83**(10), 2150–2156.

48. Wunderman, I., Coppleson, M., Skladnev, V. N., and Reid, B. L. (1995). Polarprobe: A precancer detection instrument. *J. Gynecol. Technol.* **1,** 105–110.

49. Peyton, C. L., Schiffman, M., Lorincz, A. T., Hunt, W. C., Mielzynska, J., Bratti, C., Eaton, S., Hildesheim, A., Morera, L. A., Rodriguez, A. C., Herrero, R., Sherman, M. E., and Wheeler, C. M. (1998). Comparison of PCR- and Hybrid Capture-based human papillomavirus detection systems using multiple cervical specimen collection strategies. *J. Clin. Microbiol.* **36,** 3248–3254.

50. Schiffman, M. H., and Sherman, M. E. (1994). HPV testing to improve cervical cancer screeing. *In* "Early Detection of Cancer: Molecular Markers" (S. Srivastava, S. M. Lippman, W. K. Hong, and J. L. Mulshine, eds.), Futura Publ. Co., Armonk, NY.

51. Sherman, M. E., Schiffman, M. H., Lorincz, A. T., Manos, M. M., Scott, D. R. *et al.* (1994). Toward objective quality assurance in cervical cytopathology. *Am. J. Clin. Pathol.* **102,** 182–187.

52. Cox, J. T., Lorincz, A. T., Schiffman, M. H., Sherman, M. E., Cullen, A., and Kurman, R. J. (1995). Human papillomavirus testing by hybrid capture appears to be useful in triaging women with a cytologic diagnosis of atypical squamous cells of undetermined significance. *Am. J. Obstet. Gynecol.* **172,** 946–954.

75

Epidemiology of Lung Cancer in Women

ANNA H. WU

Norris Comprehensive Cancer Center
University of Southern California
Los Angeles, California

I. Introduction

Lung cancer is the second most common cancer and the leading cause of cancer mortality in women in the U.S., accounting for 13% of all malignancies and 25% of all cancer deaths [1]. This chapter reviews the changing demographic pattern of lung cancer in women. This is followed by a discussion of the causes of this cancer, with emphasis on active and passive smoking, host susceptibility and dietary factors, occupational hazards, and air pollution.

II. Magnitude of Problem and Descriptive Epidemiology

A. *International Variation*

Lung cancer is the fifth most frequent cancer of women worldwide [2]. International and racial/ethnic comparisons of the age-standardized incidence rates of female lung cancer show wide variations ranging from 62 per 100,000 among the Maori in New Zealand to less than 2 per 100,000 in Madras, India [3] (Table 75.1). The incidence rates among white and black women in the U.S. are among the highest in the world. In Europe, the rates are highest in the United Kingdom and Denmark,

whereas rates in Mediterranean Europe are still relatively low. In Asia, with the exception of low rates in India, rates on the international scale are intermediate in Japan and the Philippines but are high among Chinese. Lung cancer incidence and mortality in women, as in men, show strong cohort patterns related to age at smoking initiation and subsequent smoking prevalence. Although standards of ascertainment and diagnosis may account for some of the observed differences in rates internationally, differences in smoking patterns likely explain most of this variation. One exception is the high lung cancer rate among Chinese females in Asia, where the prevalence of smoking in women has historically been low [2,4,5].

B. *Trends in the United States*

1. *Time Trends*

In the U.S., female lung cancer incidence has increased 600% during the past five decades, from 7 per 100,000 person in the late 1940s [6] to 42 per 100,000 in 1988–1992 [7]. Increases in mortality are comparable to those seen for incidence. Lung cancer surpassed breast cancer as the leading cause of cancer death in U.S. women in 1987. The average annual increase in female lung cancer incidence peaked between 1969 and 1974 (10.1% per year) [8] but declined to 6.3% for 1975–1979 and to just 1.8% for 1987–1992 [7]. Concurrently, lung cancer death rates have also declined among white females under age 45 between 1973 and 1986 [9]. The expectation, though, is that overall lung cancer incidence and mortality rates in U.S. women will continue to rise until the early twenty-first century.

2. *Race*

There is considerable racial/ethnic variation in female lung cancer incidence in the U.S. Incidence rates (per 100,000) are high in Alaska Natives, blacks, non-Hispanic whites, and Hawaiians (43 to 51); intermediate in Vietnamese and Chinese (25 to 31); and comparatively low in Hispanics, Filipinos, Koreans, and Japanese (15 to 20). Racial/ethnic differences in lung cancer mortality rates follow a similar pattern [10].

3. *Histology*

Increases in female lung cancer incidence have been accompanied by shifts in the relative proportions of the four major histologic types of lung cancer (adenocarcinoma, squamous cell, small or oat cell, and large cell carcinomas). Based on national Surveillance, Epidemiology, End Results (SEER) data between 1969 and 1986, the greatest percentage increase in women occurred for small cell carcinoma (320%) and adenocarcinoma

Table 75.1

Age-Standardized Incidence Rates (World Standard) per 100,000 for Females, 1988–1992

United States/Canada		South America	
U.S. SEER, white	29.9	Cuba	15.7
U.S. SEER, black	28.1	Brazil, Porto Alegre	16.8
Canada	23.9	Brazil, Goiania	11.6
United Kingdom/Europe		Colombia, Cali	9.8
UK, England and Wales	20.5	Costa Rica	4.7
UK, Scotland	30.5	Paraguay	3.6
Denmark	23.1	Peru	4.2
Poland, Warsaw City	16.4	Asia	
Poland, Warsaw Rural	7.0	Hong Kong	32.6
Switzerland, Basel	10.3	China, Tianjin	33.2
Norway	9.7	China, Shanghai	18.1
Netherlands	7.9	Japan, Osaka	11.7
Italy, Trieste	13.4	Philippines, Manila	16.3
Italy, Florence	8.9	India, Bombay	3.0
Italy, Ragusa	3.0	India, Madras	1.4
Portugal	5.9	Maori, New Zealand	62.2
France, Bas Rhin	5.6		
Spain, Basque Country	3.8		
Spain, Tarragona	3.0		

(220%), followed by increases in squamous cell carcinoma (182%) [11]. Subsequent SEER incidence data between 1989 and 1991 showed continued increases in these three histologic types but reductions in large cell carcinoma [12,13]. In white and black women, the absolute rates are highest for adenocarcinoma, followed by squamous cell and small cell carcinoma histologies [12]. Adenocarcinoma is the predominant cell type in women and accounts for 43% of the lung cancer in white and black women [12]. It has been suggested that secular trends in the increase in adenocarcinoma are compatible with the widespread adoption of filtered cigarettes in the 1950s and the accompanying changes in smoking behavior [14–17]. (see Section III.A.3).

4. Survival Rates

Since the 1980s, the 5-year relative survival rate for lung cancer in women has remained at 15% [7]. The survival rate for cancers localized at diagnosis is some 25 times higher than that for metastatic disease, but only 15% of women are diagnosed at the localized stage. Survival rates also differ by race, age, and cell type of lung cancer. Overall survival rates are considerably better in white (16%) than in black women (11%), for those diagnosed under 65 (14%) than in those 65 or older (11%), and for women with non-small cell carcinomas (18%) than for those with small cell carcinomas (6%).

III. Risk Factors

A. Active Smoking

In 1964 the first Surgeon General Report declared cigarette smoking to be a major cause of lung cancer. In the ensuing 30 years, evidence accumulated to clarify in detail the relationship between lung cancer risk in women and various aspects of smoking behavior, including number of cigarettes smoked per day, duration of smoking, age at starting, inhalation patterns, effect of cessation, and types of tobacco products smoked [18–26].

Several conclusions can be summarized from this wealth of data. First, the dominant reason for the lung cancer epidemic in U.S. women is the widespread adoption of smoking in women beginning in the 1930s. Two prospective studies initiated by the American Cancer Society (ACS) in 1959 and 1982 (*i.e.,* Cancer Prevention Study I (CPS I) and Cancer Prevention Study II (CPS II)) have been particularly informative. Between the two study periods, the age-adjusted female lung cancer death rates for current smokers increased almost sixfold (from 26.1 to 154.6 per 100,000) whereas the death rates in never smokers remained essentially constant (9.6 to 12.0 per 100,000). Over 90% of the female lung cancer deaths in the CPS II were directly attributable to cigarette smoking compared to some 60% in CPS I [27,28]. Second, there is ample evidence from cohort and case-control studies that documented exposure-response relationships between lung cancer risks in women and measures of exposure to tobacco smoke. Overall, female smokers carry a 5- to 15-fold increase in lung cancer risk relative to lifelong nonsmokers [18–22,27,28]. Differences in risk between studies can be explained by differences in lifetime cumulative smoke exposure related to age at starting, duration of smoking, amount of cigarettes smoked, and inhalation patterns in different populations and during different calendar periods.

1. Various Parameters of Tobacco Exposure

The collective evidence confirms that lung cancer risk in women increases with increasing number of cigarettes smoked; most studies conducted in North America have shown a 10- to 20-fold increased risk associated with smoking a pack or more of cigarettes per day relative to nonsmokers [22]. Similarly, the risk increases with increasing duration of smoking independent of the number of cigarettes smoked per day. Studies that have investigated intensity versus duration of smoking have demonstrated independent effects [29–32]. Age at starting is highly correlated with duration of smoking, but one study also showed an independent effect with age at starting after adjustment for duration of smoking [33]. Women who inhale deeply or most of the time [34–37] show at least a twofold higher risk of lung cancer than those who inhale less deeply or less frequently.

Both case-control and cohort studies have shown increasing lung cancer risk in relation to increasing proportion of "tar" yield of cigarettes [38–40]. In some studies, women who smoked only filtered cigarettes experienced a 20–30% lower risk than lifelong nonfilter smokers [30,36]. Some data suggest that not all lung cancer cell types show similar benefits associated with use of filtered cigarettes [16] (see Section III.A.3). Reduction in risk relative to nonsmokers results from cessation of smoking. The risk of lung cancer among exsmokers decreases with increasing years of abstinence; the reduced risk is usually evident after 5 years of cessation compared with continued smoking [21]. Risk reduction is also influenced by characteristics of active smoking (*e.g.,* duration and amount smoked) prior to smoking cessation [21]. Data from the CPS II showed that although smoking cessation is beneficial at any age, those who quit at younger ages derive greater benefits [41]. Lung cancer risk always remains higher for former compared to lifelong nonsmokers [21].

2. Gender Difference in Risk

Three cohort studies with small numbers of women [42–44] and a growing number of case-control studies [36,40,41,45–47] have reported that risk of lung cancer associated with smoking may be greater in women than in men for comparable amounts of smoking. The largest and most consistent finding is a female excess in risk for small cell carcinomas, whereas less consistent or smaller female excesses in risks for squamous cell and adenocarcinoma have been observed (Table 75.2). Although the epidemiologic evidence is not yet conclusive that women are, in fact, more susceptible to smoking-induced lung cancer than men, there are potential biologic explanations. The greater risk in women may relate to estrogen and/or other hormonal receptors in the lung epithelium that are sex-specific [48]. Some studies also show slower nicotine metabolism in women than in men [49]. Differences in the activity of various cytochrome P-450 enzymes that metabolically activate or detoxify smoke carcinogens may also render women more susceptible [50–52].

3. Histology

Smoking clearly increases the risk of all histologic types of lung cancer but the strength of the smoking association differs by histology. The association is strongest for small or oat cell

Table 75.2

Smoking and Risk[a] of Lung Cancer by Cell Type; Females (F) and Males (M) in Selected Case-Control Studies

| | | Cell type | | | | | |
| | | Adenocarcinoma | | Squamous cell | | Oat/small cell | |
Reference	Exposure	F	M	F	M	F	M
[36]	Cigarettes/day						
	1–9	1.0	1.0	1.0	1.0	1.0	1.0
	10–19	2.0	1.9	2.4	1.5	2.4	1.8
	20–29	1.1	2.5	5.3	2.1	6.2	2.5
	30+	2.3	3.5	4.2	3.1	5.6	3.2
[29]	Any smoking	3.6	4.8	10.6	18.9	59.0	22.9
	<20/day, <35 years	1.8	1.1	2.5	4.8	18.0	4.1
	20+/day, <35 years	3.0	2.8	7.1	12.4	36.8	16.5
	<20/day, 35+ years	3.6	5.2	10.9	18.0	57.6	20.4
	20+/day, 35+ years	6.5	7.4	21.1	29.0	132.0	35.8
[46]	Ever	20.1	11.1	20.1	11.1	37.6	11.4
	Former	19.2	8.7	19.2	8.7	7.9	29.8
	Current	20.6	13.7	20.6	13.7	42.5	15.1
	<20 cig/day	11.7	7.6	11.7	7.6	25.6	7.3
	20+ cig/day	26.1	17.2	26.1	17.2	53.1	19.2
[45]	Ever	9.5	17.9	26.4	36.1	86.0	37.5
	Former	5.8	13.1	13.5	22.9	43.3	14.0
	Current	11.6	21.7	35.2	49.3	119.1	62.0
	<2 pack/day	8.8	16.5	24.0	35.3	76.7	27.6
	2+ pack/day	24.2	37.5	72.3	76.0	316.1	95.3
[47]	40 pack-years	8.8	5.4	101	15.5	87.3	14.9
[40]	1–19 pack-years	6.8	2.4	11.9	6.5	NA	NA
	20–39 pack-years	11.2	5.8	26.4	24.1	NA	NA
	40–49 pack-years	21.4	11.6	48.8	48.9	NA	NA
	50+ pack-years	32.7	13.8	95.2	82.1	NA	NA

[a]With the exception of the results reported by Lubin and Blot [36], the risks shown are computed relative to nonsmokers.

NA, not available.

carcinoma, intermediate for squamous cell carcinoma, and weakest for adenocarcinoma [5,15,23,25]. There is a well-established predilection of adenocarcinoma among nonsmokers and ex-smokers. Some 60–80% of lung cancers in nonsmokers and long-term former smokers are adenocarcinomas compared to only 40% of lung cancer in women who are active smokers [53,54].

Reasons for the relative increases in adenocarcinoma in women and in men over time are not known. In one study, both women and men who were lifetime filtered cigarette smokers compared to lifetime nonfiltered cigarette smokers showed a significantly reduced risk of squamous cell carcinoma odds ratio (OR) = 0.4, 95% CI = 0.2–0.8) but not of adenocarcinoma (OR = 0.9, 95% CI = 0.5–1.7) [16]. These investigators interpreted these and their previous findings [17] to support the hypothesis that use of filtered cigarettes may help explain the temporal increase of adenocarcinoma and the decrease of squamous cell carcinoma of the lung. Their rationale is that deep inhalation of filtered products increases the deposition of small particles in the periphery of the lung. The association between risk of different lung cancer cell types and types of tobacco products warrants further investigation.

4. Biomarkers of Exposure

A follow-up study of prediagnostic urinary nicotine and cotinine corroborated the strong association between tobacco exposure and risk of lung cancer in women [55,56]. Based on concentrations of total nicotine, women who were biochemically proven active smokers showed a significant eight-fold higher risk than nonsmokers. Risk of lung cancer also increased as a function of increasing concentrations of total nicotine metabolites. In a subset of subjects with data on histologic type, significant associations for all histologic types of lung cancer were found but the risks were two to four times higher for non-adenocarcinoma than for adenocarcinoma.

5. Racial/Ethnic Variation

There is general agreement that the risk for lung cancer associated with smoking is higher for black women than for white women for comparable amounts smoked [42,57–59]. The black–white differences in risk of lung cancer are more marked for squamous cell carcinoma than for adenocarcinoma of the lung [58,59] and in younger than in older women [59]. The lung cancer risk associated with smoking may be lower for

Asian-American women than for whites [42,60] but risks appear to be similar for Hispanic and non-Hispanic white women [61]. Racial/ethnic differences in lung cancer risk may relate to differences between ethnic groups in the intensity of smoking, type of tobacco products smoked, exposure to other risk factors such as diet or occupational exposures, and/or differences in susceptibility to the carcinogenic effects of tobacco smoke.

Smoking associated risks also differ in studies conducted in different countries. Compared to a 5- to 15-fold increased risk in studies conducted in North America, a 2- to 4-fold increased risk of lung cancer for smokers compared to never smokers has been reported in Greece [62], Spain [63], Germany [64], and France [65]. Similar or even lower risk estimates associated with smoking have been reported among Chinese women in Asia [5,66]. The attributable risk for lung cancer associated with smoking is also lower in these studies, ranging from less than 10% in Taiwan to about 45% in Hong Kong and China compared to approximately 90% in the U.S. The lower attributable risk associated with smoking among women in these studies compared to studies in the U.S. is due to: (1) the lower magnitude of risk associated with smoking; and (2) the lower percentage of women in these areas who are smokers (10–30%), including the lower percentage of heavy smokers (<10%) and the lower cumulative tobacco exposure among women in these areas. After accounting for these differences, the effect of cigarette smoking on lung cancer risk among women in these studies is qualitatively and quantitatively similar to its effects on lung cancer risk in studies conducted in North America [5,62,65].

B. Environmental Tobacco Smoke (ETS)

Since 1981, nearly 40 epidemiologic studies have evaluated the role of ETS, based on spouses' smoking habits, as a risk factor for lung cancer. Most of these studies have been case-control studies in women because lung cancer is very rare among never-smokers and a higher percentage of women than men have never smoked [67,68]. Because most of the studies had small sample sizes that lacked statistical power to detect weak associations, a meta-analysis approach has been used to calculate a pooled estimate. Based on an evaluation of the weight of the evidence from human and animal studies, the U.S. Environmental Protection Agency classified ETS as a class A carcinogen [69], thus strengthening previous conclusions reached by the Surgeon General [70] and the National Research Council [71].

The most recent meta-analysis on lung cancer and ETS [68], published in 1997, covered over 4600 lung cancers in female lifetime never smokers from 37 epidemiologic studies (four prospective and 33 case-control studies) in nine countries. A pooled relative risk of 1.24 (95% CI = 1.13–1.36) was obtained for female nonsmokers married to smokers compared to nonsmokers married to nonsmokers. Based on much smaller numbers of lung cancer in men who have never smoked, a pooled relative risk of 1.34 (95% CI = 0.97–1.84) was found for men married to smokers compared to those married to nonsmokers. In a subset of studies that had information on number of cigarettes smoked by husbands and years having lived with a smoker, there were significant dose–response trends with either measure of exposure. The pooled risk estimate associated with ETS was stronger for squamous and small cell carcinoma (OR = 1.58,

95% CI = 1.14–2.19) than for adenocarcinoma (OR = 1.25, 95% CI = 1.07–1.46). Although data on other sources of ETS exposure (i.e., at work or social setting) are less consistent, they also point to ETS as deleterious [72]. The attributable risk for lung cancer associated with ETS in lifetime female nonsmokers in the U.S. is less than 10% [73].

Because there are limitations with the questionnaire approach, there is much interest in identifying sensitive and specific biomarkers of exposure to ETS. Cotinine, the major metabolite of nicotine, is elevated in biological fluids of passive smokers [67–69]. In the follow-up study of prediagnostic urinary cotinine described earlier, there was a twofold increased risk of lung cancer associated with increasing urinary cotinine excretion in nonsmokers who were passively exposed [55]. However, cotinine is probably not a good marker of the total carcinogenic exposure. There are a number of carcinogenic biomarkers of tobacco exposure (e.g., serum PAH-albumin adduct or hemoglobin adduct of 4-aminobiphenyl) with higher relative levels in nonsmokers exposed to ETS compared to cotinine [72].

C. Susceptibility

Among cigarette smokers, only one in ten will develop lung cancer. In addition to chance and competing causes of death, host susceptibility factors are likely to affect an individual's risk of developing lung cancer. Tobacco carcinogens, including polycyclic aromatic hydrocarbons (PAH), arylamines, alkyl halides, and nitrosamines, require metabolic activation to transform into fully carcinogenic agents. It is likely that genetic differences in the activation and detoxification of tobacco smoke-derived procarcinogens lead to individual variation in susceptibility to lung cancer [74]. The influence of various genes involved in metabolic activation or detoxification of carcinogens contained in cigarette smoke has been investigated extensively in relation to risk of lung cancer. Although results to date are inconsistent and show associations in only some ethnic groups or only when several genes are considered jointly, many published studies have been small. Moreover, few of these studies included women and, even then, most did not present separate results in women.

1. Tobacco Metabolism Genes

The cytochrome P4502D6 (CYP2D6 gene), which genetically controls the ability to metabolize the antihypertensive agent debrisoquine, has been investigated as a risk factor for lung cancer because it is suspected that differential drug metabolism may reflect individual variation in carcinogen metabolism. Despite initial reports of a strong association between CYP2D6 phenotype and risk of lung cancer [75], subsequent studies on the risk of lung cancer and CYP2D6 genotype and/or phenotype have been inconsistent [76,77].

The CYP1A1 gene codes for aryl hydrocarbon hydroxylase (AHH), an enzyme that plays an important role in the activation of PAHs. Two polymorphisms of the CYP1A1 gene, the MspI and the Ile/Val, have been investigated extensively. The homozygous variant has been associated with increased risk of lung cancer in Japanese populations [78,79] but not among whites in the U.S. [80], Swedes [81], Finns [82], or Germans [83]. In a study of whites in Boston, CYP1A1 MspI heterozygous genotype was associated with a modest increased risk (OR = 1.5,

95% CI = 1.0–2.3) that was evident only after adjustment for smoking habits [84]. There is no consensus whether the association between CYP1A1 genotype and lung cancer risk differs by levels of tobacco exposure [78,84].

The GSTM1 gene codes for a member of a family of enzymes that detoxify reactive chemical entities by promoting their conjugation to glutathione [85]. Known substrates for the GSTM1 protein products include PAHs, specifically the metabolically generated epoxide intermediates of benzo(a)pyrene. The polymorphism in the GSTM1 gene is due to a complete deletion of the gene that results in a lack of enzyme activity. A meta-analysis of 12 case-control studies on GSTM1 and lung cancer suggested a small increased risk (OR = 1.4, 95% CI = 1.2–1.6) associated with the GSTM1 null genotype [86]. When studies are stratified by race, the main support for a GSTM1 lung cancer association comes from studies conducted in Japan, whereas the evidence is weak in western countries. Results from large studies in Los Angeles and Boston also show little association between GSTM1 null genotype and risk of lung cancer in white and black males and females [84,87]. Studies that have evaluated the interaction between GSTM1 and cumulative cigarette smoking have also obtained contradictory results, with about half of the studies finding a positive association only among smokers with high cumulative cigarette exposure.

The role of GSTT1, another member of the glutathione S-transferases which catalyzes the detoxification of tobacco smoke constituents such as alkyl halides, has been investigated. Although there is no evidence of an association between GSTT1 genotype and risk of lung cancer [88–90], individuals with both GSTT1 null and GSTM1 null genotypes experienced a nearly threefold increased risk after adjustment for smoking habits [90]. This finding suggests that deficiencies in both enzymes may be needed to alter risk.

Several studies have shown significantly increased risk among individuals with both GSTM1 null and CYP1A1 Msp I heterozygous genotypes [81,84] or GSTM1 null and CYP1A1 Msp I homozygous genotypes [78,79]. Thus, host susceptibility to lung cancer may depend on the metabolic balance between various genes involved in the detoxification and activation of tobacco carcinogens.

2. Genetic Alterations

The DNA-mutating agents in tobacco smoke produce mutations both in oncogenes and in tumor suppressor genes. Among the genes commonly altered in lung cancers are K-ras and p53. The K-ras mutations occur frequently in adenocarcinoma of the lung in men and women [91,92].

p53 mutations are common in lung cancer, with the highest prevalence (70–90%) in small cell carcinomas and the lowest (30–50%) in adenocarcinoma [93,94]. The frequency of p53 mutations is positively correlated with lifetime cigarette consumption [51,95,96]. Only a few studies of p53 mutations and lung cancers have included females. In a study conducted in the United Kingdom, the frequency of p53 mutations was lower in women (38%) compared to men (53%) [51]. However, in a study at the Mayo Clinic including mostly smokers, mutation frequency was comparable in females (48%) and males (49%) [52]. Lower frequencies of p53 mutation have been reported in a study of lung cancer among Chinese women in Hong Kong

(20%) and women in Japan (31%), presumably because the majority of these subjects were nonsmokers with adenocarcinoma of the lung [97].

D. Family History

The relationship of family history to risk of lung cancer has been investigated in case-control studies [32,34,98–107] and through segregation analysis [108–110]. In lung cancer studies of mostly smokers [32,34,98–101,106,107], family history of lung cancer has been associated with a 60% to five-fold increased risk of lung cancer after controlling for personal smoking habits and other risk factors. Information on smoking habits by family members was generally absent in these studies. It has been difficult to determine the extent to which familial susceptibility to the development of lung cancer can be attributed to aggregation of smoking habits in families versus shared genetic susceptibility.

Studies of lung cancer in nonsmokers can provide insights regarding this question. Three large studies of family history in lifetime nonsmokers or long-term exsmokers have been conducted in the U.S. In a multicenter study of lung cancer in nonsmoking women, lung cancer in all first-degree relatives was associated with a 29% increased risk that did not achieve statistical significance. However, a significant threefold increased risk of lung cancer was observed if lung cancer occurred in mothers and sisters [103]. In a study in Detroit, nonsmoking lung cancer patients were also more likely than nonsmoking controls to have a first-degree relative with lung cancer (OR = 1.4). This effect was due entirely to an increased risk among nonsmoking females (OR = 1.9, 95% CI = 0.9–3.3) [102]. In a study in Missouri, family history of lung cancer was associated with a significant increased risk among long-term exsmokers (OR = 2.9) but there was no such association in nonsmokers (OR = 1.1) [104]. A study of lung cancer in nonsmoking women in Xuan Wei County, China showed a statistically significant fourfold increased risk due to family history of lung cancer [105]. These authors noted that the strong finding may be due to shared exposure to indoor air pollution from "smoky" coal combustion that has been implicated as the main cause of lung cancer in this region (see Section I.2.).

There is also support that risk associated with a family history of lung cancer tends to be stronger among subjects with early-onset (<60 years of age) lung cancer [102,103,108]. It is unclear whether there are differences in familial risk by histologic type [101–103,111].

E. Previous Lung Diseases

Increased lung cancer risk has been reported among individuals with impaired pulmonary function [112] or with history of various nonmalignant lung diseases [26,113]. Although these studies have adjusted for the effects of smoking, it is difficult to be certain that there is an independent effect of previous lung diseases on risk of lung cancer because smoking also increases the risk of many nonmalignant lung diseases. Residual confounding by active smoking was eliminated in two large case-control studies of previous lung diseases among female nonsmokers in the U.S. In both studies, there was a statistically significant

40–60% increased risk associated with history of previous lung disease [114,115].

History of asthma emerged as a significant risk factor in both studies of nonsmokers [114,115]. Increased risk associated with pneumonia has been reported both for adenocarcinoma and for nonadenocarcinomas [31,32,107,114–116], while a history of chronic bronchitis and emphysema has been reported, but mainly among smokers [99], for squamous or small cell carcinomas [31,32,115]. Associations between risk of lung cancer and history of tuberculosis have been found in some cohort and case-control studies [117]; the strongest evidence comes from studies conducted among Chinese in the U.S. [118] and in Asia [66,119]. In most studies, the risks associated with tuberculosis were more apparent for adenocarcinoma than squamous cell carcinomas [114,115,119]. Tuberculosis and pneumonia are known to result occasionally in lung scars, which have been associated with adenocarcinomas [120], but it is controversial whether scarring precedes the development of cancer. The previous lung diseases did not appear to be early manifestations of lung cancer. Significant increased risks for lung cancer were found for those with early-onset (i.e., lung diseases first diagnosed before age 21 years) of asthma [114] or tuberculosis [107,114] in some studies.

A prospective serologic study of male smokers [121] offered additional support that previous lung diseases may have an independent effect on risk of lung cancer. In this study, markers of chronic *Chlamydia pneumonia* infection, based on stable elevated IgA antibody and immune complexes, were positively associated with risk of lung cancer (OR = 1.6, 95% CI = 1.0–2.3) after adjustment for amount and duration of smoking.

F. Endocrine Factors

The role of endocrine factors, in particular estrogens, has been investigated in an attempt to explain the higher prevalence of adenocarcinoma of the lung among women. There are no consistent patterns between risk of lung cancer and factors including age at menarche, parity, and breastfeeding. However, in a study in China, short menstrual cycles (<26 days) were associated with a statistically significant twofold increased risk of all lung cancer [31], but this was not confirmed in a U.S. study [122]. There is some suggestion of higher risks of lung cancer in association with late age at menopause [31,32,122] and history of surgical menopause [107,117,122]. Three studies also suggest a 20–70% increased risk associated with use of hormone replacement therapy [107,122,123]. Thus, further investigation of the role of menopausal events in lung cancer is warranted, with careful consideration of the interrelationships between smoking habits, age at menopause, type of menopause, and use of menopausal hormones [124].

G. Dietary Factors

Since the first report of a possible role of diet and lung cancer in men [125], some twenty studies that have included only women or large numbers (at least 50) of women have been published [31,32,107,116,126–143]. Most were case-control studies and four were cohort or nested case-control studies. A food frequency questionnaire was typically used to assess diet but the number of food items included ranged from a few [126] to over a hundred [127]. The specific dietary component

under investigation has also varied. Studies have reported results on specific vegetables or fruits, vegetables and fruits as a group, specific micronutrients, or some combination of the above. Five studies included only female nonsmokers and long-term exsmokers [128–132], whereas the other studies included mostly smokers and adjusted for smoking in various ways in the analyses.

1. Vegetables, Fruits, and Specific Micronutrients

The majority of cohort and case-control studies in women show that diets high in vegetables and fruits are associated with lower risk of lung cancer. Cohort studies conducted in a retirement community in Southern California [133], in Iowa [134], and in Japan [135] reported that high intake of vegetables was associated with a 30–50% lower risk of lung cancer in women (in a New York State cohort [136], intake of vegetables and fruits was not presented).

In case-control studies of female smokers and nonsmokers combined, the strongest evidence of a protective effect of vegetables and fruits came from studies conducted in Hawaii [127,137], New Jersey [138], and Spain [139]. Lower risks of lung cancer in association with high intake of beta-carotene rich foods including carrots and dark leafy greens were also reported in studies conducted in Los Angeles County [107], Canada [140], and Singapore [126]. On the other hand, three other studies conducted in the U.S. [141–143] and two studies in China [31,32] found little evidence of any association between risk of lung cancer in women and intake of vegetables, carotene, vitamin C, or other components in fruits and vegetables.

Of the five case-control studies conducted in nonsmokers, studies in Florida [128] and New York [129] reported that high intake of vegetables was associated with a statistically significant reduced risk. In Hong Kong [130] and Greece [131], high intake of fruits was associated with a significantly reduced risk, but intake of vegetables was only weakly [130] or not related to risk [131]. In Missouri [132], risk of lung cancer was not associated with intake of vegetables and fruits; only high intake of beans and peas was associated with a statistically significant lower risk.

Despite support for vegetables and fruits playing a protective role, the specific component of vegetables and fruits that protect against lung cancer is unclear. A lower risk associated with carotenoid intake has been reported in some [107,116,127–129,137,138,140] but not in other studies in women [130–134,136,141–143]. The strongest evidence for a protective effect of vitamin C came from three studies in female nonsmokers [128,130,131] but no association was reported in other studies [127,132–134,141,142]. There is also little consistent evidence of an association between risk and reduced intake of lutein, lycopene, cryptoxanthin [128,134,137], vitamin E, or folate [127,132,136,141]. Other possible protective constituents in vegetables and fruits (e.g., indoles, isothiocyanates, lignans) are poorly studied. Studies in neither women nor men show clear differences in the diet associations by histologic type or smoking status [144,145].

In addition to the observational studies, the association between lung cancer risk and serological concentrations of several antioxidants has been investigated, mainly in men. Six of seven studies of prediagnostic serological beta-carotene support an inverse relationship with lung cancer, including studies with results pertaining to women [144,146]. However, results on lung cancer risk and prediagnostic serum ascorbic acid, alpha-tocopherol, and selenium are less consistent [144,146]. Despite support from observational and serological studies for a protective role of beta-carotene, three chemoprevention trials have either shown no benefit or a possible harmful effect of beta-carotene supplementation on lung cancer mortality. Only men were included in the Finnish [147] and the Physicians' Health Study [148]. Some 6000 heavy female smokers were included in the Beta-Carotene and Retinol Efficacy Trial. Results in women, as in men, showed an excess risk of lung cancer in association with supplementation [149].

2. Fat and Cholesterol

A diet high in fat or cholesterol has been associated with increased risk of lung cancer in men but the evidence in women is weak. The strongest evidence in women was based on a case-control study in Toronto [140], whereas no association was observed in case-control studies conducted in Florida [150], Hawaii [151], and in cohort studies in Iowa [152] or New York State [136]. A study of lung cancer in female nonsmokers in Missouri reported a strong association between intake of saturated fat and risk of lung cancer [132]. This association was not confirmed in a subsequent study of lung cancer in women smokers in the same study area [153]. At this time, it is not known whether findings of a diet high in fat may be a marker of a role of heterocyclic amines, which has been implicated in various cancers including colon and breast [154]. However, the considerable inconsistencies in the cholesterol/fat association by gender, lung cancer cell type, smoking status, and respondent type thus weaken the hypothesis that cholesterol/fat has a major etiologic role in lung cancer.

3. Alcohol

A relationship between alcohol intake and lung cancer has been found in some studies of men, but few studies have included women [155]. In a New York State cohort, there was no association between alcohol intake and risk in women, but there was an association in men, particularly among heavy smokers [136]. In a study of adenocarcinoma of the lung in women [107], the increased risk associated with high intake of beer and total ethanol diminished and was no longer statistically significant after adjustment for smoking habits. In the Iowa cohort study in women, there was some suggestion of an association between alcohol intake and risk that was due primarily to beer drinking. However, intake of beer was low in this population [156]. The close correlation between alcohol intake and smoking makes it difficult to identify the separate effects of alcohol.

H. Occupation

Various exposures at the workplace, including asbestos, arsenic, and uranium, have been associated with increased risk of lung cancer [26]. Occupational exposures have been estimated to account for 5% of lung cancer in women and 15% in men [157]. Studies in women have generally confirmed that women employed in potentially hazardous occupations appear to experience increased risks of lung cancer that are similar to those in men. Some of the studies in women have included large numbers of nonsmokers and have thus enabled the investigation of

occupational exposures in the absence of tobacco smoking, an important potential confounder in the relationship between occupational exposures and lung cancer.

In the U.S., a study of lung cancer among women in Texas found that employment of women in a high-risk industry or occupation results in consistently increased smoking-adjusted risk [158]. In Missouri, among female nonsmokers, work in the dry-cleaning business and self-reported exposure to asbestos and pesticide resulted in a two- to threefold increased risk. Slight elevations (ORs were 1.2 to 2.1) were noted for job categories including shoemaking, shipbuilding, foundry work, rubber industry, housing construction, and beautician [159]. In a study of lung cancer among nonsmokers in Illinois, females who were currently employed in eating and drinking places showed a statistically significant 90% elevated risk of lung cancer [160]. Because information on occupation and smoking was abstracted from medical records, one can only surmise that exposure to ETS may be common at such eating places and may help explain this excess.

Additional occupations and exposures were implicated [161, 162] in two large case-control studies of lung cancer among women in China. In Shanghai, after controlling for smoking, risk was elevated for employment in the manufacture of non-metallic mineral products and in metallurgical, shipbuilding, and construction work [161]. In Northern China, increased risks were associated with occupations involved in the manufacture of transportation equipment, sewing, metal smelting and treatment, and workplace exposures to coal dusts and smoke from burning fuel. In the Northern China study, most of the associations were similar for smokers and nonsmokers and for adenocarcinoma and nonadenocarcinomas [162].

I. Air Pollution

1. Urban Pollution

Atmospheric pollution has been implicated to play a role in the etiology of lung cancer. Analytic epidemiologic studies have been limited by difficulties in defining and measuring air pollution and other exposures that may confound the association. Several case-control studies that included women have been conducted in areas considered to be heavily industrialized or polluted to assess the role of air pollution after adjustment for smoking and other relevant risk factors. In two studies that used residence near an industrial site or factories as a proxy measure of exposure to air pollution, there was some suggestion of increased risk [66,163]. Female nonsmokers in Kaohsiung, Taiwan who resided near a heavy industrial site for over 20 years showed nearly a twofold significant increased risk [66]. Similarly, in Shenyang, China, significantly more female lung cancer cases than controls reported living in a residence within 200 meters of any type of factory after adjusting for smoking and other factors [163]. However, studies conducted in Greece [62], Poland [164], and Japan [165] that used different definitions of pollution showed no clear associations.

2. Indoor Air Pollution

Although the effect of outdoor air pollution may be small or present in only subgroups of women, there is accumulating evidence that indoor air pollution from cooking and heating devices may play an important role in female lung cancer, particularly in some areas of China and elsewhere in Asia. In Xuan Wei County, China, the high rates of lung cancer in women and men were largely explained by living in homes without chimneys and indoor use of smoky coal [105,166]. Analysis of indoor air during cooking indicated that residents who used smoky coal were exposed to occupational levels of PAHs [167]. In northeast China, indoor pollution from the use of "kang" and other coal-burning heating devices has been suggested to contribute to the areas' high rates of lung cancer [32,163]. When an overall indoor air pollution index was used (based on use of coal central heating, coal cooking fuel, and location of cook stove), there was an increasing trend in risk, particularly for squamous cell carcinoma [163]. Levels of both indoor and outdoor wintertime benzo(a)pyrene in this part of northern China exceeded standards for cities in the U.S. by more than 60-fold. In addition, studies in China and Taiwan have reported higher risks for women who prepared meals in a kitchen without a fume extractor [66], who lived in homes without a separate kitchen or with poor air circulation [168], who had prolonged exposure to fumes and oil volatiles from cooking, particularly rapeseed oil [31,32], or who reported smokiness in the house and experienced eye irritation during cooking [31,32]. Condensates of the volatile emissions from rapeseed and soybean oil have been found to be mutagenic [169].

3. Radon

Studies of underground miners exposed to radioactive radon gas and its alpha-particle emitting decay products have found that exposure increases the risk of lung cancer. Radon in underground mines can reach levels up to 10,000 working-level months (WLMs). A significant excess risk has been found for cumulative exposure as low as 50 to 100 WLMs. Based on extrapolation from studies of underground miners, increases in lung cancer risk of about twofold have been projected at the lower doses associated with long-term residential exposures to radon. There are concerns about extrapolating risk from miners because exposure in a typical U.S. home is substantially lower (10–20 WLM) than that experienced by miners and because of qualitative differences between exposure in an underground mine and a domestic residence [170].

In the 1990s, eight large case-control studies, including five studies in women only [171–175], have been conducted to assess directly lung cancer risk from indoor radon. A meta-analysis of the eight studies produced an estimated relative risk of 1.14 (95% CI = 1.0–1.3) at 150 Bq/m3 (4.05pCi/l). However, the evidence from the studies in women is inconsistent and weak. Studies in China [171] and Canada [173] showed no association with exposure, both overall and by subgroup analysis, by smoking status, and histologic type of lung cancer. In Missouri, there was also no association in all subjects combined, by age or smoking status. A suggestive association was reported for adenocarcinoma of the lung and among subjects who were interviewed directly [174]. In New Jersey, lung cancer risk increased more than twofold among those living in homes with radon levels of ≥4 versus <1 pCi/l, but the group exposed to ≥4 pCi/l consisted of only six cases and two controls. The radon association was apparent only among light smokers and for large cell carcinomas [172]. In the Stockholm, Sweden study,

there was a significant trend in risk with radon level but this trend disappeared when exposure was adjusted for recent residential occupancy. In this study, the radon association was observed only among nonsmokers and heavy smokers and was more apparent for small cell and squamous cell carcinoma [175]. A second Swedish study that included over 1300 males and females with lung cancer provided strong evidence of a trend of increasing risk with increasing indoor radon exposure. The findings were similar by smoking status and by histological type. The risk estimates from this study were consistent with projections from the studies of lung cancer in miners [176].

Lubin and Boice [177] explained that the mixed results may be the consequence of inherent limitations including: (1) the extreme uncertainty in estimating accurately individuals' past exposure based on current radon measurements; (2) the low expected relative risks even at high radon levels (150 Bq/m³); and (3) that less than 10% of U.S. houses have these high radon levels. They calculated that a single case-control study with thousands of subjects would be required to determine the risk of indoor radon.

IV. Summary

Cigarette smoking is the most important risk factor for lung cancer in women and is responsible for some 90% of the disease in women. There is general consensus that exposure to ETS increases risk of lung cancer in nonsmoking women but the attributable risk is low (<10%). The role of tobacco-metabolizing genes in modifying the risk of smoking-induced lung cancer is presently unclear and data specific to women are largely absent. There is accumulating evidence that family history of lung cancer and previous lung diseases are associated with increased risk in smokers and nonsmokers. The evidence for a role of endocrine factors, especially female hormones, in the etiology of lung cancer is sparse and noncompelling. Diets high in vegetables and fruits are associated with lower risk of lung cancer but the specific mechanism of action and the relevant anticarcinogenic agent have not been clearly identified. Few studies in women have data on specific occupational exposures and they generally point to comparable high risks in women and in men. Small increased risks are also associated with exposure to various sources of air pollution including urban pollution and radon.

References

1. Anonymous (1997). "Cancer Facts and Figures—1997." American Cancer Society, Atlanta, GA.
2. Parkin, D. M., Pisani, P., and Ferlay, J. (1993). Estimates of the worldwide incidence of eighteen major cancers in 1985. *Int. J. Cancer* **54,** 594–606.
3. Parkin, D. M., Muir, C. S., Whelon, S. L., Gao, Y. T., Ferlay, J., and Powell, J. (1992). "Cancer Incidence in Five Continents," Vol. 6, IARC Sci. Pub. No. 120. IARC, Lyon, France.
4. Koo, L. C., and Ho, J. H. C. (1990). Worldwide epidemiological patterns of lung cancer in nonsmokers. *Int. J. Epidemiol.* **19,** (Suppl. 1), S14–S23.
5. Wu-Williams, A. H., and Samet, J. M. (1994). Lung cancer and cigarette smoking. *In* "Epidemiology of Lung Cancer" (J. M. Samet, ed.), pp. 71–108. Dekker, New York.

6. Devesa, S. S., and Silverman, D. T. (1978). Cancer incidence and mortality trends in the United States. *J. Natl. Cancer Inst. (U.S.)* **60,** 545–571.

7. Kosary, C. L., Ries, L. A. G., Miller, B. A., Hankey, B. F., Harras, A., and Edwards, B. K., eds. (1995). "SEER Cancer Statistics Review, 1973–1992," Tables and Graphs, NIH Publ. No. 96-2789, National Cancer Institute, Bethesda, MD.

8. Pollack, E. S., and Horm, J. W. (1980). Trends in cancer incidence and mortality in the United States, 1969–76. *JNCI, J. Natl. Cancer Inst.* **64,** 1091–1103.

9. Devesa, S. S., Blot, W. J., and Fraumeni, J. F. (1989). Declining lung cancer rates among young men and women in the United States, a cohort analysis. *J. Natl. Cancer Inst.* **81,** 1568–1571.

10. Miller, B. A., Kolonel, L. N., Bernstein, L., Young, J. L., Jr., Swanson, G. M., West, D., Key, C. R., Liff, J. M., Glover, C. S., Alexander, G. A., eds. (1996). "Racial/Ethnic Patterns of Cancer in the United States 1988–1992," NIH Publ. No. 96-4104. National Cancer Institute, Bethesda, MD.

11. Devesa, S. S., Shaw, G. L., and Blot, W. J. (1991) Changing patterns of lung cancer incidence by histologic type. *Cancer Epidemiol., Biomarkers Prev.* **1,** 29–34.

12. Travis, W. D., Travis, L. B., and Devesa, S. S. (1995). Lung cancer. *Cancer (Philadelphia)* **75,** 191–202.

13. Travis, W. D., Lubin, J. Ries, L., and Devesa, S. (1996). United States lung carcinoma incidence trends. *Cancer (Philadelphia)* **77,** 2464–2470.

14. Zheng, T., Holford, T. R., Boyle, P., Chen, Y., Ward, B. A., Flannery, J., and Mayne, S. T. (1994). Time trend and the age-period-cohort effect on the incidence of histologic types of lung cancer in Connecticut 1960–1989. *Cancer (Philadelphia)* **74,** 1556–1567.

15. Thun, M. T., Lally, C. A., Flannery, J. T., Calle, E. E., Flanders, W. D., and Health, C. W. (1997). Cigarette smoking and changes in the histopathology of lung cancer. *J. Natl. Cancer Inst.* **89,** 1580–1586.

16. Stellman, S. D., Muscat, J. E., Thompson, S., Hoffmann, D., and Wynder, E. L. (1997). Risk of squamous cell carcinoma and adenocarcinoma of the lung in relation to lifetime filter cigarette smoking. *Cancer (Philadelphia)* **80,** 382–388.

17. Wynder, E. L., and Kabat, G. C. (1988). The effect of low-yield cigarette smoking on lung cancer risk. *Cancer (Philadelphia)* **62,** 1223–1230.

18. U.S. Public Health Services (1980). "The Health Consequences of Smoking for Women. A Report of the Surgeon General." U.S. Department of Health and Human Services, Office on Smoking and Health, Rockville, MD.

19. U.S. Department of Health and Human Services (1982). "The Health Consequences of Smoking, Cancer. A Report of the Surgeon General," DHHS Publ. No (PHS) 82-50179. Public Health Service, Office on Smoking and Health, Rockville, MD.

20. U.S. Department of Health and Human Services (1989). "Reducing the Health Consequences of Smoking, 25 Years of Progress. A report of the Surgeon General," DHHS Publ. No. (CDC) 89-8411. U.S. Govt. Printing Office, Washington, DC.

21. U.S. Department of Health and Human Services (1990). "The Health Benefits of Smoking Cessation." U.S. Department of Health and Human Services, Centers for Disease Control, Center of Chronic Disease Prevention and Health Promotion, Office of Smoking and Health, Rockville, MD.

22. U.S. Department of Health and Human Services (1997). "Women and Smoking." U.S. Department of Health and Human Services, Centers for Disease Control, Center of Chronic Disease Prevention and Health Promotion, Office of Smoking and Health, Rockville, MD.

23. International Agency for Research on Cancer (1986). "IARC Monographs on the Evaluation of the Carcinogenic Risk of Chemicals to Humans, Tobacco Smoking," Vol. 38. World Health Organization, Lyon, France.

24. Ernster, V. L. (1994). The epidemiology of lung cancer in women. *Ann. Epidemiol.* **4,** 102–110.

25. Ernster, V. L. (1996). Female lung cancer. *Annu. Rev. Public Health* **17,** 97–114.

26. Blot, W. J., and Fraumeni, J. F., Jr. (1996). Cancers of the lung and pleura. *In* "Cancer Epidemiology and Prevention" (D. Schottenfeld and J. F. Fraumeni, eds.), pp 637–665. Oxford University Press, New York.

27. Thun, M. J., Day-Lally, C., Myers, D. G., Calle, E. E., Flanders, D., Zhu, B., Namboodiri, M. M., and Health, C. W., Jr. (1997). Trends in tobacco smoking and mortality from cigarette use in Cancer Prevention Studies I (1959 through 1965) and II (1982 through 1988). *In* "Changes in Cigarette-related Disease Risks and Their Implication for Prevention and Control," Smoking and Tobacco Control Monogr. No. 8. U.S. Department of Health and Human Services, Public Health Service, National Institutes of Health, National Cancer Institute, Bethesda, MD.

28. Thun, M. J., Myers, D. G., Day-Lally, C., Namboodiri, M. M., Calle, E. E., Flanders, D., Adams, S. L., and Health, C. W., Jr. (1997). Age and the exposure-response relationships between cigarette smoking and premature death in Cancer Prevention II. *In* "Changes in Cigarette-related Disease Risks and Their Implication for Prevention and Control," Smoking and Tobacco Control Monogr. No. 8. U.S. Department of Health and Human Services, Public Health Service, National Institutes of Health, National Cancer Institute, Bethesda, MD.

29. Schoenberg, J. B., Wilcox, H. B., Mason, T. J., Bill, J., and Stemhagen, A. (1989). Variation in smoking-related risk among New Jersey women. *Am. J. Epidemiol.* **130,** 688–695.

30. Garfinkel, L., and Stellman, S. D. (1988). Smoking and lung cancer in women, findings in a prospective study. *Cancer Res.* **48,** 6951–6955.

31. Gao, Y. T., Blot, W. J., Zheng, W., Ershow, A. G., Hsu, C. W., Levin, L. I., Zhang, R., and Fraumeni, J. F. (1987) Lung cancer among Chinese women. *Int. J. Cancer* **40,** 604–609.

32. Wu-Williams, A. H., Dai, X. D., Blot, W., Xu, Z. Y., Sun, X. W., Xiao H. P., Stone, B. J., Yu, S. F., Feng, Y. P., Ershow, A. G., Sun, J., Fraumeni, J. F., and Henderson, B. E. (1990). Lung cancer among women in north-east China. *Br. J. Cancer* **62,** 982–987.

33. Hegmann, K. T., Fraser, A. M., Keaney, R. P., Moser, S. E., Nilasena, D. S., Sedlars, M., Higham-Gren, L., and Lyon, J. L. (1993). The effect of age at smoking initiation on lung cancer risk. *Epidemiology* **4,** 444–448.

34. Osann, K. E. (1991). Lung cancer in women. The importance of smoking, family history of cancer, and medical history of respiratory disease. *Cancer Res.* **51,** 4893–4897.

35. Joly, O. G., Lubin, J. H., and Caraballoso, M. (1983). Dark tobacco and lung cancer in Cuba. *JNCI, J. Natl. Cancer Inst.* **70,** 1033–1039.

36. Lubin, J. H., and Blot, W. (1984). Assessment of lung cancer risk factors by histologic category. *JNCI, J. Natl. Cancer Inst.* **73,** 383–389.

37. Sidney, S., Tekawa, I. S., and Friedman, G. D. (1993). A prospective study of cigarette tar yield and lung cancer. *Cancer Causes Control* **4,** 3–10.

38. Kaufman, D. W., Palmer, J. R., Rosenberg, L., Stolley, P., Warshauer, E., and Shapiro, S. (1989). Tar content of cigarettes in relation to lung cancer. *Am. J. Epidemiol.* **129,** 703–711.

39. Zang, E. A., and Wynder, E. L. (1992). Cumulative tar exposure. A new index for estimating lung cancer risk among cigarette smokers. *Cancer (Philadelphia)* **70,** 69–76.

40. Zang, E. A., and Wynder, E. L. (1996). Differences in lung cancer risk between men and women, examination of the evidence. *J. Natl. Cancer Inst.* **88,** 183–192.

41. Halpern, M. T., Gillespie, B. W., and Warner, K. E. (1993). Patterns of absolute risk of lung cancer mortality in former smokers. *J. Natl. Cancer Inst.* **85**, 457–464.

42. Friedman, G. D., Tekawa, I., Sadler, M., and Sidney, S. (1997). Smoking and mortality, The Kaiser Permanente Experience. *In* "Changes in Cigarette-related Disease Risks and Their Implication for Prevention and Control," Smoking and Tobacco Control Monogr. No. 8. U.S. Department of Health and Human Services, Public Health Service, National Institutes of Health, National Cancer Institute, Bethesda, MD.

43. Engeland, A., Haldorsen, T., Andersen, A., and Tretli, S. (1996). The impact of smoking habits on lung cancer risk, 28 years' observation of 26,000 Norwegian men and women. *Cancer Causes Control* **7**, 366–376.

44. Cohn, B. A., Wingard, D. L., Cirillo, P. M., Cohen, R. D., Reynoids, P., and Kaplan, G. A. (1996). Re, differences in lung cancer risk between men and women, examination of the evidence. *J. Natl. Cancer Inst.* **88**, 1867.

45. Osann, K. E., Anton-Culver, H., Kurosaki, T., and Taylor, T. (1993). Sex differences in lung-cancer risk associated with cigarette smoking. *Int. J. Cancer* **54**, 44–48.

46. Brownson, R. C., Chang, J. C., and Davis, J. R. (1992). Gender and histologic type variations in smoking-related risk of lung cancer. *Epidemiology* **3**, 61–64.

47. Risch, H. A., Howe, G. R., Jain, M., Burch, J. D., Holowaty, E. J., and Miller, A. B. (1993). Are female smokers at higher risk for lung cancer than male smokers? A case-control analysis by histologic type. *Am. J. Epidemiol.* **138**, 281–293.

48. Chaudhuri, P. K., Thomas, P. A., Walker, M. J., Briele, H. A., Tapas, K., Gupta, D., and Beattie, C. W. (1984). Steroid receptors in human lung cancer cytosols. *Cancer Lett.* **16**, 327–332.

49. Benowitz, N. L., and Jacob, P. (1984). Daily intake of nicotine during cigarette smoking. *Clin. Pharmacol. Ther.* **35**, 499–504.

50. Ryberg, D., Hewer, A., Phillips, D. H., and Haugen, A. (1994). Different susceptibility to smoking-induced DNA damage among male and female lung cancer patients. *Cancer Res.* **54**, 5801–5803.

51. Kure, E. H., Ryberg, D., Hewer, A., Phillips, D. H., Skaug, V., Baera, R., and Haugen, A. (1996). P53 mutations in lung tumours, relationship to gender and lung DNA adduct levels. *Carcinogenesis* (*London*) **17**, 2201–2205.

52. Guinee, D. G., Travis, W. D., Trivers, G. E., De Beneditti, V. M. G., Cawley, H., Welsch, J. A., Bennett, W. P., Jett, J., Colby, T. V., Tazelaar, H., Abbondanzo, S. L., Pairolero, P., Trastek, V., Caporaso, N. E., Liotta, L. A., and Harris, C. C. (1995). Gender comparisons in human lung cancer, analysis of p53 mutations, anti-p53 serum antibodies and C-erb B-2 expression. *Carcinogenesis* (*London*) **16**, 993–1002.

53. Muscat, J. E., and Wynder, E. L. (1995). Lung cancer pathology in smokers, ex-smokers and never smokers. *Cancer Lett.* **88**, 1–5.

54. Stockwell, H. G., Armstrong, A. W., Leaverton, P. E. (1990). Histopathology of lung cancer among smokers and nonsmokers in Florida. *Int. J. Epidemiol.* **19**(Suppl. 1), S48–S52.

55. de Waard, F., Kemmeren, J. M., van Ginkel, L. A., and Stolker, A. A. M. (1995). Urinary cotinine and lung cancer risk in a female cohort. *Br. J. Cancer* **72**, 784–787.

56. Ellard, G. A., de Waard, F., and Kemmeren, J. M. (1995). Urinary nicotine metabolite excretion and lung cancer risk in a female cohort. *Br. J. Cancer* **72**, 788–791.

57. Correa, P., Pickle, L. W., Fontham, E., Dalager, N., Lin, Y., Haenszel, W., and Johnson, W. (1984). The causes of lung cancer in Louisiana. *In* "Lung Cancer, Causes and Prevention," pp. 73–99. Verlag Chemie International, New York.

58. Harris, R. E., Zang, E. A., Anderson, J. I., and Wynder, E. L. (1993). Race and sex differences in cancer risk associated with cigarette smoking. *Int. J. Epidemiol.* **22**, 592–599.

59. Schwartz, A. G., and Swanson, G. M (1997). Lung carcinoma in African-Americans and whites. A population-based study in metropolitan Detroit, Michigan. *Cancer* (*Philadelphia*) **79**, 45–52.

60. Le Marchand, L, Wilkens, L. R., and Kolonel, L. N. (1992). Ethnic differences in the lung cancer risk associated with smoking. *Cancer Epidemiol., Biomarkers Prev.* **1**, 103–107.

61. Humble, C. G., Samet, J. M., Pathak, D. R., and Skipper, B. J. (1985). Cigarette smoking and lung cancer in "Hispanic' whites and other whites in New Mexico. *Am. J. Public Health* **75**, 145–148.

62. Katsouyanni K, Trichopoulos D., Kalandidi A, Tomos, P., and Riboli, E. (1991). A case-control study of air pollution and tobacco smoking in lung cancer among women in Athens. *Prev. Med.* **20**, 271–78.

63. Agudo, A., Barnadas, A., Pallares, C., Martinez, I., Fabregat, X., Rosello, J., Estape, J., Planas, J., and Gonzalez, C. A. (1994). Lung cancer and cigarette smoking in women. A case-control study in Barcelone (Spain). *Int. J. Cancer* **59**, 165–169.

64. Becher, H., Jockel, K. H., Timm, J., Wichmann, H. E., and Drescher, S. (1991). Smoking cessation and non-smoking intervals, effect of different smoking patterns on lung-cancer risk. *Cancer Causes Control* **2**, 381–387.

65. Benhamou, E., Benhamou, S., and Flamant, R. (1987). Lung cancer and women. Results of a French case-control study. *Br. J. Cancer* **55**, 91–95.

66. Ko, Y. C., Lee, C. H., Chen, M. J., Huang, C. C., Chang, W. Y., Lin, H. J., Wang, H. Z., and Chang, P. Y. (1997). Risk factors for primary lung cancer among non-smoking women in Taiwan. *Int. J. Epidemiol.* **26**, 24–31.

67. Dockery, D. W., and Tichopoulos, D. (1997). Risk of lung cancer from environmental exposures to tobacco smoke. *Cancer Causes Control* **8**, 333–345.

68. Hackshaw, A. K., Law, M. R., and Wald, N. J. (1997). The accumulated evidence on lung cancer and environmental tobacco smoke. *Br. Med. J.* **315**, 980–988.

69. U.S. Environmental Protection Agency (1992). "Respiratory Health Effects of Passive Smoking, Lung Cancer and Other Disorders," EPA/600/6-90/006F. EPA, Washington, DC.

70. U.S. Department of Health and Human Services (1986). "The Health Consequences of Involuntary Smoking," A Report of the Surgeon General, DHHS Publ. No (PHS) 87-8398. DHHS, Washington, DC.

71. National Research Council, Committee on Passive Smoking (1986). "Environmental Tobacco Smoke, Measuring Exposures and Assessing Health Effects." National Academy Press, Washington, DC.

72. Wu, A. H. (1997). Environmental tobacco smoke. II. Lung cancer. *In* "Topics in Environmental Epidemiology" (K. Steenland and D. Savitz, eds.), pp. 227–255. Oxford University Press, New York.

73. Alavanja, M. C. R., Brownson, R. C., Benichou, J., Swanson, C., and Boice, J. D., Jr. (1995). Attributable risk of lung cancer in lifetime nonsmokers and long-term ex-smokers (Missouri, United States). *Cancer Causes Control* **6**, 209–216.

74. Perera, F. P. (1996). Insights into cancer susceptibility, risk assessment, and prevention. *J. Natl. Cancer Inst.* **88**, 496–509.

75. Caporaso, N. E., Hayes, R. B., Dosemeci, M., Hoover, R., Ayesh, R., Hetzel, M., and Idle, J. (1989). Lung cancer risk, occupational exposure, and the debrisoquine metabolic phenotype. *Cancer Res.* **49**, 3675–3679.

76. Caporaso, N., Debaun, M. R., and Rothman, N. (1995). Lung cancer and CYP2D6 (the debrisoquine polymorphism), sources of heterogeneity in the proposed association. *Pharmacogenetics* **5**, S129–S134.

77. Shaw, G. L., Falk, R. T., Deslauriers, J., Frame, J. N., Nesbitt, J. C., Pass, H. I., Issaq, H. J., Hoover, R. N., and Tucker, M. A.

(1995). Debrisoquine metabolism and lung cancer risk. *Cancer Epidemiol. Biomarkers Prev.* **4,** 41–48.

78. Nakachi, K., Imai, K., Hayashi, S., and Kawajiri, K. (1993). Polymorphisms of the CYP1A1 and glutathione S-transferases genes associated susceptibility of lung cancer in relation to cigarette dose in Japanese populations. *Cancer Res.* **53,** 2994–2999.

79. Kihara, M., Kihara, M., and Noda, K. (1995). Risk of squamous and small cell carcinomas of the lung modulated by combinations of CYP1A1 and GSTM1 gene polymorphisms in a Japanese population. *Carcinogenesis* (*London*) **16,** 2331–2336.

80. Shields, P. G., Caporoso, N. E., Falk, R. T., Sugimura, H., Trivers, G. E., Trump, B. F., Hoover, H. N., Weston, A., and Harris, C. C. (1993). Lung cancer, race, and a CYP1A1 genetic polymorphism. *Cancer Epidemiol., Biomarkers Prev.* **2,** 481–485.

81. Alexandrie, A., Sundberg, M. I., Seidegard, J., Tornling, G., and Rannug, A. (1994). Genetic susceptibility to lung cancer with special emphasis on CYP1A1 and GSTM1, a study of host factors in relation to age at onset, gender and histological cancer types. *Carcinogenesis* (*London*) **15,** 1785–1790.

82. Hirvonen, A., Husgafvel-Pursiainen, K., Karjalainen, A, Anttila, S., and Vainio, H. (1992). Point-mutational MspI and Ile-Val polymorphisms closely linked to the CYP1A1 gene, lack of association with susceptibility to lung cancer in a Finnish study population. *Cancer Epidemiol., Biomarkers Prev.* **1,** 485–489.

83. Drakoulis, N., Cascorbi, I., Brockmoller, J., Gross, C. R., and Roots, I. (1994). Polymorphisms in the human CYP1A1 gene as susceptibility factors for lung cancer: Exon-7 mutation (4889 A to G), and a T to C mutation in the 3'-flanking region. *Clin. Invest.* **72,** 240–248.

84. Closas-Garcia, M., Kelsey, K. T., Wiencke, J. K., Xu, X., Wain, J. C., and Christiani, D. C. (1997). A case-control study of cytochrome P450 1A1, glutathione S-transferase M1, cigarette smoking and lung cancer susceptibility (Massachusetts, United States). *Cancer Causes Control* **8,** 544–553.

85. Board, P., Coggan, M., Johnston, P. *et al.* (1990). Genetic heterogeneity of the human glutathione transferases, a complex of gene families. *Pharmacol. Ther.* **48,** 357–369.

86. McWilliams, J. E., Sanderson, B. J. S., Harris, E. L., Richert-Boe, K. E., and Henner, W. D. (1995). Glutathione S-transferase M1 (GSTM1) deficiency and lung cancer risk. *Cancer Epidemiol., Biomarkers Prev.* **4,** 589–594.

87. London, S. J., Daly, A. K., Cooper, J., Navidi, W. C., Carpenter, C. L., and Idle, J. R. (1995). Polymorphism of glutathione S-transferase M1 and lung cancer risk among African-Americans and Caucasians in Los Angeles County, California. *J. Natl. Cancer Inst.* **87,** 1246–1253.

88. Rebbeck, T. R. (1997). Molecular epidemiology of the human glutathione S-transferase genotypes gstml and GSTTl in cancer susceptibility. *Cancer Epidemiol., Biomarkers Prev.* **6,** 733–743.

89. Deakin, M., Elder, J., Hendrickse, C., *et al.* (1996). Glutathione S-transferase GSTT1 genotypes and susceptibility to cancer, studies of interactions with GSTM1 in lung, oral, gastric and colorectal cancers. *Carcinogenesis* (*London*) **17,** 881–884.

90. Kelsey, K. L., Spitz, M. R., Zuo, Z. F., and Wiencke, J. K. (1997). Polymorphisms in the glutathione S-transferase class mu and theta genes interact and increase susceptibility to lung cancer in minority populations (Texas, United States). *Cancer Causes Control* **8,** 554–559.

91. Rodenjuis, S., and Slebos, R. J. C. (1992). Clinical significance of ras oncogene activation in human lung cancer. *Cancer Res.* **52,** (Suppl.) 2665s–2669s.

92. Westra, W. H., Slebos, R. J. C., Offerhaus, G. H. A., Goodman, S. N., Evers, S. G., Kensler, T. W., Askin, F. B., Rodenhuis, S., and Hruben, R. H. (1993). K-ras oncogene activation in lung adenocarcinomas from former smokers. *Cancer* (*Philadelphia*) **72,** 432–438.

93. Greenblatt, M. S., Bennett, W. P., Hollstein, M., and Harris, C. C. (1994). Mutations in the p53 tumor suppressor gene, clues to cancer etiology and molecular pathogenesis. *Cancer Res.* **54,** 4855–4878.

94. Harris, C. C. (1996). Structure and function of the p53 tumor suppressor gene, clues for rational cancer therapeutic strategies. *J. Natl. Cancer Inst.* **88,** 1442–1455.

95. Suzuki, H., Takahashi, T., Kuroishi, T., Suyama, M., Ariyoshi, Y., and Takahashi, T. (1992). P53 mutations in non-small cell lung cancer in Japan, association between mutations and smoking. *Cancer Res.* **52,** 734–736.

96. Kondo, K., Tsuzuki, H., Sasa, M., Sumitomo, M., Uyama, T., and Monden, Y. (1996). A dose-response relationship between the frequency of p53 mutations and tobacco consumption in lung cancer patients. *J. Surg. Oncol.* **61,** 20–26.

97. Takagi, Y., Koo, L. C., Osada, H., Ueda, R., Kyaw, K., Ma, C. C., Suyama, M., Saji, S., Takahashi, T., Tominaga, S., and Takahashi, T. (1995). Distinct mutational spectrum of the p53 gene in lung cancers from Chinese women in Hong Kong. *Cancer Res.* **55,** 5354–5357.

98. Tokuhata, G. K., and Lilienfeld, A. M. (1963). Familial aggregation of lung cancer in humans. *J. Natl. Cancer Inst.* (*U.S.*) **30,** 289–312.

99. Samet, J. M., Humble, C. G., and Pathak, D. R. (1986). Personal and family history of respiratory disease and lung cancer risk. *Am. Rev. Respir. Dis.* **134,** 466–670.

100. Ooi, W. L., Elston, R. C., Chen, V. W., Wilson-Bailey, J. E., and Rothschild, H. (1986). Increased familial risk of lung cancer. *JNCI, J. Natl. Cancer Inst.* **76,** 217–222.

101. Shaw, G. L., Falk, R. T., Pickle, L. W., Mason, T., and Buffler, P. A. (1991). Lung cancer risk associated with cancer in relatives. *J. Clin. Epidemiol.* **44,** 429–437.

102. Schwartz, A. G., Yang, P., and Swanson, G. M. (1996). Familial risk of lung cancer among nonsmokers and their relatives. *Am. J. Epidemiol.* **144,** 554–562.

103. Wu, A. H., Fontham, E. T. H., Reynolds, P., Greenberg, R. S., Buffler, P., Liff, J., Boyd, P., and Correa, P. (1996). Family history of cancer and risk of lung cancer among lifetime nonsmoking women in the United States. *Am. J. Epidemiol.* **143,** 535–542.

104. Brownson, R. C., Alavanja, M. C. R., Caporaso, N., Berger, E., and Chang, J. C. (1997). Family history of cancer and risk of lung cancer in lifetime non-smokers and long-term ex-smokers. *Int. J. Epidemiol.* **26,** 256–263.

105. Liu, Z., He, X., and Chapman, R. S. (1991). Smoking and other risk factors for lung cancer in Xuanwei, China. *Int. J. Epidemiol.* **20,** 26–31.

106. McDuffie, H. H. (1991). Clustering of cancer in families of patients with primary lung cancer. *J. Clin. Epidemiol.* **44,** 69–76.

107. Wu, A. H., Yu, M. C., Thomas, D. C., Pike, M. C., and Henderson, B. E. (1988). Personal and family history of lung disease as risk factors for adenocarcinoma of the lung. *Cancer Res.* **48,** 7279–7284.

108. Sellers, T. A., Bailey-Wilson, J. E., Elston, R. C., Wilson, A. F., Elston, G. Z., Ooi, W. L., and Rothschild, H. (1990). Evidence for mendelian inheritance in the pathogenesis of lung cancer. *J. Natl. Cancer Inst.* **82,** 1272–1279.

109. Gauderman, W. J., Morrison, J. L., Carpenter, C. L., and Thomas, D. C. (1997). Analysis of gene-smoking interaction in lung cancer. *Genet. Epidemiol.* **14,** 199–214.

110. Yang, P., Schwartz, A. G., McAllister, A. E., Aston, C. E., and Swanson, G. M. (1997). Genetic analysis of families with nonsmoking lung cancer probands. *Genet. Epidemiol.* **14,** 181–197.

111. Sellers, T. A., Elston, R. C., Atwood, L. D., and Rothschild, H. (1992). Lung cancer histologic type and family history of cancer. *Cancer* (*Philadelphia*) **69,** 86–91.

112. Tockman, M. S., Anthonisen, N. R., Wright, E. C, and Donithan, M. G. (1987). Airways obstruction and the risk for lung cancer. *Ann. Intern. Med.* **106,** 512–518.

113. Cohen, B., Diamond, E. L., Graves, C. G., *et al.* (1977). A common familial component in lung cancer and chronic obstructive pulmonary disease. *Lancet* **2,** 523–526.

114. Wu, A. H., Fontham, E. T. H., Reynolds, P., Greenberg, R. S., Buffler, P., Liff, J., Boyd, P., Henderson, B. E., and Correa, P. (1995). Previous lung disease and risk of lung cancer among lifetime nonsmoking women in the United States. *Am. J. Epidemiol.* **141,** 1023–1032.

115. Alavanja, M. C. R., Brownson, R. C., Boice, J. D., and Hock, E. (1992) Preexisting lung disease and lung cancer among nonsmoking women. *Am. J. Epidemiol.* **136,** 623–632.

116. Wu, A. H., Henderson, B. E., Pike, M. C., and Yu, M. C. (1985). Smoking and other risk factors for lung cancer in women. *JNCI, J. Natl. Cancer Inst.* **74,** 747–751.

117. Aoki, K. (1993). Excess incidence of lung cancer among pulmonary tuberculosis patient. *Jpn. J. Clin. Oncol.* **23,** 205–220.

118. Hinds, M. W., Cohen, H. I., and Kolonel, L. N. (1982). Tuberculosis and lung cancer risk in nonsmoking women. *Am. Rev. Respir. Dis.* **125,** 776–778.

119. Zheng, W., Blot, W. J., Liao, M. L., Wang, Z. W., Levin, L. I., Zhao, J. J., Fraumeni, J. F., and Gao, Y. T. (1987). Lung cancer and prior tuberculosis infection in Shanghai. *Br. J. Cancer* **56,** 501–504.

120. Bakris, G. L., Mulopulos, G. P., Korchik, R., Ezdinli, E. Z., Ro, J., and Yoon, B. (1983). Pulmonary scar carcinoma. A clinicopathologic analysis. *Cancer (Philadelphia)* **52,** 493–497.

121. Laurila, A. L., Anttila, T., and Laara, E. (1997). Serological evidence of an association between chlamydia pneumoniae infection and lung cancer. *Int. J. Cancer* **74,** 31–34.

122. Taioli, E., and Wynder, E. L. (1994). Endocrine factors and adenocarcinoma of the lung in women. *J. Natl. Cancer Inst.* **86,** 869–870.

123. Adami, H. O., Persson, I., Hoover, R., Schairer, C., and Bergkvist, L. (1989). Risk of cancer in women receiving hormone replacement therapy. *Int. J. Cancer* **44,** 833–839.

124. Derby, C. A., Hume, A. L., Barbour, M. M., McPhillips, J. B., Lasater, T. M., and Carleton, R. A. (1993). Correlates of postmenopausal estrogen use and trends through the 1980s in two southeastern New England communities. *Am. J. Epidemiol.* **137,** 1125–1135.

125. Bjelke, E. (1975). Dietary vitamin A and human lung cancer. *Int. J. Cancer* **15,** 561–565.

126. MacLennan, R., Costa, J. D., Day, N. E., Law, C. H., Ng, Y. K., and Shanmugaratnam, K. (1977). Risk factors for lung cancer in Singapore Chinese, a population with high female incidence rates. *Int. J. Cancer* **20,** 854–860.

127. Le Marchand, L., Yoshizawa, C. N., Kolonel, L. N., Hankin, J. H., and Goodman, M. T. (1989). Vegetable consumption and lung cancer risk, a population-based case-control study in Hawaii. *J. Natl. Cancer Inst.* **81,** 1158–1164.

128. Candelora, E. C., Stockwell, H. G., Armstrong, A. W., and Pinkham, P. A. (1992). Dietary intake and risk of lung cancer in women who never smoked. *Nutr. Cancer* **17,** 263–270.

129. Mayne, S. T., Janerich, D. T., Greenwald, P., Chorost, S., Tucci, C., Zaman, M. B., Melamed, M. R., Kiely, M., and McKneally, M. F. (1994). Dietary beta carotene and lung cancer risk in U.S. nonsmokers. *J. Natl. Cancer Inst.* **86,** 33–38.

130. Koo, L. C. (1988). Dietary habits and lung cancer risk among Chinese females in Hong Kong who never smoked. *Nutr. Cancer* **11,** 155–172.

131. Kalandidi, A., Katsouyanni, K., Voropoulou, N., Bastas, G., Saracci, R., and Trichopoulos, D. (1990). Passive smoking and diet in the etiology of lung cancer among non-smokers. *Cancer Causes Control* **1,** 15–21.

132. Alavanja, M. C. R., Brown, C. C., Swanson, C., and Brownson, R. C. (1993). Saturated fat intake and lung cancer risk among nonsmoking women in Missouri. *J. Natl. Cancer Inst.* **85,** 1906–1916.

133. Shibata, A., Paganini-Hill, A., Ross, R. K., and Henderson, B. E. (1992). Intake of vegetables, fruit, beta carotene, vitamin C and vitamin supplements and cancer incidence among the elderly, a prospective study. *Br. J. Cancer* **66,** 673–679.

134. Steinmetz, K. A., Potter, J. D., and Folsom, A. R. (1993). Vegetables, fruit, and lung cancer in the Iowa women's health study. *Cancer Res.* **53,** 536–543.

135. Hirayama, T. (1979). Diet and cancer. *Nutr. Cancer* **1,** 67–81.

136. Bandero, E. V., Freudenheim, J. L., Marshall, J. R., Zielezny, R., Priore, R. L., Brasure, J., Baptiste, M., and Graham, S. (1997). Diet and alcohol consumption and lung cancer risk in the New York State Cohort (United States). *Cancer Causes Control* **8,** 828–840.

137. LeMarchand, L., Hankin, J. H., Kolonel, L. N., Beecher, G. R., Wilkens, L. R. and Zhao, L. P. (1992). Intake of specific carotenoids and lung cancer risk. *Cancer Epidemiol., Biomarkers Prev.* **2,** 183–187.

138. Dorgan, J. F., Ziegler, R. G., Schoenberg, J. B., Hartge, P., McAdams, M. J., Falk, R. T., Wilcox, H. B., and Shaw, G. L. (1993). Race and sex differences in associations of vegetables, fruits, and carotenoids, with lung cancer risk in New Jersey (United States). *Cancer Causes Control* **4,** 273–281.

139. Agudo, A., Esteve, M. G., Pallares, C., Martinez-Ballarin, I., Fabregat, X., Malats, N., Machengs, I. Badia, A., and Gonzalez, C. A. (1997). Vegetable and fruit intake and the risk of lung cancer in women in Barcelona, Spain. *Eur. J. Cancer* **33,** 1256–1261.

140. Jain, M., Burch, J. D., Howe, G. R., Risch, H. A., and Miller, A. B. (1990). Dietary factors and risk of lung cancer, results from a case-control study, Toronto, 1981–1985. *Int. J. Cancer* **45,** 287–293.

141. Byers, T. E., Graham, S., Haughey, B. P., Marshall, J. R., and Swanson, M. K. (1987). Diet and lung cancer risk, findings from the Western New York Diet Study. *Am. J. Epidemiol.* **125,** 351–363.

142. Hinds, M. W., Kolonel, L. N., Hankin, J. H., and Lee, J. (1984). Dietary vitamin A, carotene, vitamin C, and risk of lung cancer in Hawaii. *Am. J. Epidemiol.* **119,** 227–237.

143. Fontham, E. T. H., Pickle, L. W., Haenszel, W., Correa, P., Lin, Y., and Falk, R. T. (1988). Dietary vitamins A and C and lung cancer risk in Louisiana. *Cancer (Philadelphia)* **62,** 2267–2273.

144. Ziegler, R. G., Mayne, S. T., and Swanson, C. A. (1995). Nutrition and lung cancer. *Cancer Causes Control* **7,** 157–177.

145. Koo, L. C. (1997). Diet and lung cancer 20+ years later, More questions than answers? *Int. J. Cancer. Suppl.* **10,** 22–29.

146. Comstock, G. W., Alberg, A. J., Huang, H. Y., Wu, K., Burke, A. E., Hoffman, S. C., Norkus, E. P., Gross, M., Cutler, R. G., Morris, S., Spate, V. L., and Helzlsouer, K. J. (1997). The risk of developing lung cancer associated with antioxidants in the blood, ascorbic acid, carotenoids, a-tocopherol, selenium, and total peroxyl radical absorbing capacity. *Cancer Epidemiol., Biomarkers Prev.* **6,** 907–916.

147. Albanes, D., Heinonen, O. P., Taylor, P. R., *et al.* (1996). A-Tocopherol and b-carotene supplements and lung cancer incidence in the Alpha-Tocopherol, Beta-Carotene Cancer Prevention Study, effects of base-line characteristics and study compliance. *J. Natl. Cancer Inst.* **99,** 1560–1570.

148. Hennekens, C. H., Buring, J. E., Manson, J. E., *et al.* (1996). Lack of effect of long-term supplementation with beta carotene on the incidence of malignant neoplasms and cardiovascular disease. *N. Engl. J. Med.* **334,** 1145–1149.

149. Omenn, G. S., Goodman, G. E., Thornquist, M. D., *et al.* (1996). Risk factors for lung cancer and for intervention effects in CARET, the Beta-Carotene and Retinol Efficacy Trial. *J. Natl. Cancer Inst.* **88,** 1550–1559.

150. Stockwell, H. C., and Candelora, E. C. (1992). Dietary cholesterol and incidence of lung cancer. The Western Electric Study (Letters to the Editor). *Am. J. Epidemiol.* **136,** 1167.

151. Goodman, M. T., Kolonel, L. N., Yoshizawa, C. N., and Hankin, J. H. (1988). The effect of dietary cholesterol and fat on the risk of lung cancer in Hawaii. *Am. J. Epidemiol.* **128,** 1241–1255.

152. Wu, Y., Zheng, W., Sellers, T. A., Kushi, L. H., Bostick, R. M., and Potter, J. D. (1994). Dietary cholesterol, fat, and lung cancer incidence among older women: The Iowa Women's Health Study (United States). *Cancer Causes Control* **5,** 395–400.

153. Swanson, C. A., Brown, C. C., Sinha, R., Kulldorff, M., Brownson, R. C., and Alavanja, M. C. R. (1997). Dietary fats and lung cancer risk among women: The Missouri Women's Health Study (United States). *Cancer Causes Control* **8,** 883–893.

154. Sinha, R., and Capruaso, N. (1997). Heterocyclic amines, cytochrome P4501A2, N-acetyl-transferase: Issues involved in incorporating putative genetic susceptibility markers into epidemiologic studies. *Ann. Epidemiol.* **7,** 350–356.

155. Carpenter, C. L., Morgenstern, H., and London, S. J. (1998). Alcoholic beverage consumption and lung cancer risk among residents of Los Angeles County. *J. Nutr.* **128,** 694–700.

156. Potter, J. D., Sellers, T. A., Folsom, A. R., and McGovern, P. G. (1992). Alcohol, beer, and lung cancer in postmenopausal women. The Iowa Women's Health Study. *Ann. Epidemiol.* **2,** 587–595.

157. Doll, R., and Peto, R. (1981). The causes of cancer, quantitative estimates of avoidable risks of cancer in the United States today. *JNCI, J. Natl. Cancer Inst.* **66,** 1192–1308.

158. Ives, J. C., Buffler, P. A., Selwyn, B. J., Hardy, R. J., and Decker, M. (1988). Lung cancer mortality among women employed in high-risk industries and occupations in Harris County, Texas, 1977–1990. *Am. J. Epidemiol.* **127,** 65–74.

159. Brownson, R. C., Alavanja, M. C. R., and Chang, J. C. (1993). Occupational risk factors for lung cancer among nonsmoking women, a case-control study in Missouri (United States). *Cancer Causes Control* **4,** 449–454.

160. Keller, J. E., and Howe, H. L. (1993). Risk factors for lung cancer among nonsmoking Illinois residents. *Environ. Res.* **60,** 1–11.

161. Levin, L. I., Zheng, W., Blot, W. J., Gao, Y. T., and Fraumeni, J. F. (1988). Occupation and lung cancer in Shanghai, a case-control study. *Br. J. Ind. Med.* **45,** 450–458.

162. Wu-Williams, A. H, Xu, Z. Y., Blot, W. J., Dai, D. X., Louie, R., Xiao, H. P., Stone B. J., Sun, X. W., Yu, S. F., Feng, Y. P., Fraumeni, J. F., and Henderson, B. E. (1993). Occupation and lung cancer risk among women in northern China. *Am. J. Ind. Med.* **24,** 67–79.

163. Xu, Z. Y., Blot, W. J., Xiao, H. P., Wu, A., Feng, Y. P., Stone, B. J., Jie, S., Ershow, A. G., Henderson, B. E., and Fraumeni, J. F. (1989). Smoking, air pollution and the high rates of lung cancer in Shenyang, China. *J. Natl. Cancer Inst.* **81,** 1800–1806.

164. Jedrychowski, W., Becher, H., Wahrendorf, J., and Basa-Cierpialek, Z. (1990). A case-control study of lung cancer with special reference to the effect of air pollution in Poland. *J. Epidemol. Commun. Health* **44,** 114–120.

165. Hitosugi, M. (1968). Epidemiological study of lung cancer with special reference to the effect of air pollution and smoking habits. *Inst. Public Health Bull.* **17,** 237–256.

166. Mumford, J. L., He, X. Z., Chapman, R. S., *et al.* (1987). Lung cancer and indoor air pollution in Xuan Wei, China. *Science* **235,** 217–220.

167. Mumford, J. L., Li, X., Hu, F., Lu, X. B., and Chuang, J. C. (1995). Human exposure and dosimetry of polycyclic aromatic hydrocarbons in urine from Xuan Wei, China with high lung cancer mortality associated with exposure to unvented coal smoke. *Carcinogenesis (London)* **16,** 3031–3036.

168. Liu Q., Sasco, A. J., Riboli, E., and Hu, M. X. (1993). Indoor air pollution and lung cancer in Guangzhou, People'' Republic of China. *Am. J. Epidemiol.* **137,** 145–154.

169. Shields, P. G., Xu, G. X., Blot, W. J., Fraumeni, J. F., Trivers, G. E., Pellizzari, E. D., Qu, Y. H., Gao, Y. T., and Harris, C. C. (1995). Mutagens from heated Chinese and U.S. cooking oils. *J. Natl. Cancer Inst.* **87,** 836–841.

170. Darby, S. C., and Samet, J. M. (1994). Radon. *In* "Epidemiology of Lung Cancer" (J. M. Samet, ed.), pp. 219–243. Dekker, New York.

171. Blot, W. J., Xu, Z. Y., Boice, J., John, D., Zhao, D. Z., Stone, B. J., Sun, J., Jing, L. B., and Fraumeni, J. F., Jr. (1990). Indoor radon and lung cancer in China. *J. Natl. Cancer Inst.* **82,** 1025–1030.

172. Schoenberg, J. B., Klotz, J. B., Wilcox, H. B., Nicholls, G. P., Gil-del-Real, M. T., Stemhagen, A., and Mason, T. J. (1990). Case-control study of residential radon and lung cancer among New Jersey women. *Cancer Res.* **50,** 6520–6524.

173. Letourneau, E., Krewski, D., Choi, N., Goddard, M., McGregor, R., Zielinski, J. M., and Du, J. (1994). Case-control study of residential radon and lung cancer in Winnipeg, Manitoba, Canada. *Am. J. Epidemiol.* **140,** 310–322.

174. Alavanja, M. C., Brownson, R. C., Lubin J. H., Berger, R., Chang, J., and Boice, J. D., Jr. (1994). Residential radon exposure and lung cancer among nonsmoking women. *J. Natl. Cancer Inst.* **86,** 1829–1837.

175. Pershagen, G., Liang, Z. H., Hrubec, Z., Svensson, C., and Boice, J. D., Jr. (1992). Residential radon exposure and lung cancer in Swedish women. *Health Phys.* **63,** 179–186.

176. Pershagen, G., Akerblom, G., Axelson, O., Clavensjo, B., Damber, L., Desai, G., *et al.* (1994). Residential radon exposure and lung cancer in Sweden. *N. Engl. J. Med.* **330,** 159–164.

177. Lubin, J. H., and Boice, J. D. (1997). Lung cancer risk from residential radon, meta-analysis of eight epidemiologic studies. *J. Natl. Cancer Inst.* **89,** 49–57.

76

Colon and Rectal Cancer in Women

ELIZABETH A. PLATZ*,† AND EDWARD GIOVANNUCCI†,‡

Departments of *Epidemiology and †Nutrition, Harvard School of Public Health, Boston, Massachusetts; ‡Channing Laboratory,
Department of Medicine, Harvard Medical School and Brigham and Women's Hospital, Boston, Massachusetts

I. Introduction

Risk of colorectal cancer (CRC) is largely determined by life-style, including diet, obesity, physical inactivity, smoking, and alcohol consumption; a strong genetic predisposition accounts for only a small proportion of cases [1]. Unlike most other cancer sites, CRC affects women and men almost equally [2]. However, differences in CRC risk or strengths of association for these factors with CRC between men and women should not be unexpected as biological variation exists in large bowel physiology, including bile acid profiles, lumenal pH, and transit time [3]. In this chapter, we describe clinical characteristics, descriptive and analytic epidemiology, and prevention of CRC (specifically adenocarcinoma) and adenoma, the precursor for CRC, in women. As several articles have comprehensively reviewed the etiology of CRC and adenoma [4,5], we focus here on those factors that are unique to women, such as aspects of reproductive life, or those for which evidence supports variation in the CRC association between the sexes.

A. Clinical Characteristics of CRC

The colon may be divided into several anatomic regions: cecum, ascending colon, hepatic flexure, transverse colon, splenic flexure, descending colon, sigmoid colon, and the rectum [6]. The right colon (cecum through the mid-transverse colon) is responsible for digestion and water resorption, while the left colon stores the end products of digestion. The colonic wall is composed of four histologically distinct layers. The lumenal layer is the mucosum, which consists of tall columnar epithelial cells in a single layer lining the crypts of Lieberkühn. The epithelial cells at the bottom two-thirds of the crypt maintain their proliferative capacity, while those in the upper third are well-differentiated. The rate of turnover of the epithelial cell layer is six to eight days. Goblet cells, which produce mucus, are interspersed among the epithelial cells. Neuroendocrine cells are found at the base of the crypt. The mucosum is anchored by a basement membrane and is supported by the lamina propria, which includes blood and lymph vessels and nerve fibers. The muscularis mucosae separates the mucosum and submucosum. The submucosum is composed of connective tissue and contains arteries, veins, and lymph vessels. The muscle coat and serosa are the third and fourth layers.

The majority of CRCs are adenocarcinomas, which arise from the colorectal epithelium [7] typically from an existing benign adenomatous polyp (also called adenoma) [8]. Adenomas may develop 10 or more years before CRC, although only about one-tenth of adenomas are thought to progress to malig-

nancy [9]. Adenomas are neoplastic growths of the colorectal epithelium that extend into the lumen. Other nonadenomatous polyps, such as the common hyperplastic polyp, are not neoplastic and are not considered to be cancer precursors [10].

Adenomas show perturbed crypt epithelial cell proliferation and differentiation. Proliferation may extend into the upper third of the crypt and onto the mucosal surface, and the epithelial cells lose features of differentiation [11]. With continued hyperproliferation, the usual glandular structure is distorted and a raised growth develops. Degree of dysplasia correlates with adenoma size. Hyperproliferation plays a key role in the probability of transformation to malignancy. Because growth of a cell entails DNA replication and there is an inherent error rate during replication, the more frequently a cell undergoes replication, the greater the probability that mutations will arise, possibly in genes that when altered increase the risk of transformation. Also, faster growing cells may have a shorter time during which to repair DNA damage resulting from endogenous or exogenous agents. Without repair, a genetic alteration becomes a heritable change (if the cell survives), and if the mutation is in a growth control gene, then the probability of transformation increases.

The natural history of the transitions from normal colorectal mucosum to CRC has been studied extensively and genetic alterations accompanying these transitions are being identified [12,13]. The model of colorectal carcinogenesis proposed by Vogelstein illustrates the progression from normal epithelium, hyperplasia, early to intermediate to late adenoma, carcinoma, and finally metastasis [13]. In this model an accumulation of mutations in four to five genes is necessary for malignant transformation of normal epithelium [13]. Genetic events observed in sporadic and familial colorectal adenomas and adenocarcinomas have included deactivating mutations in or loss of the adenomatous polyposis coli gene (APC, a tumor suppressor gene, maps to 5q), perturbation in DNA methylation (methylation plays a role in regulating transcription), activating mutations in the K-ras gene (an oncogene, maps to 12p), loss of the deleted in colon cancer gene (DCC, a tumor suppressor gene, maps to 18q), and loss of the p53 gene (a tumor suppressor gene, maps to 17p). Changes in APC tend to occur early in the pathway, while those in p53 occur later, although the order in which these genetic changes arise does not appear to be critical. Vogelstein's model may now be updated to include the addition of new genes, such as mutations in DNA mismatch repair genes, which result in DNA hypermutability; overexpression of COX-2, the gene encoding the inducible form of cyclooxygenase [14] and which may curb apoptosis [15]; and novel changes in already identified genes, such as the T-to-A transversion at nucleotide 3920 in *APC*, which does not directly affect *APC* gene product function but which results in a region of hypermutability within the gene [16]. Some studies have demonstrated a varying array of genetic changes in proximal and distal colon and rectal cancers, suggesting that factors affecting initiation and progression may differ by site within the large bowel [17,18].

Individuals with a first-degree relative with CRC consistently have been shown to be at an increased risk for the disease, with a relative risk (RR) around two [4]. This finding was also reported in prospective studies [19] in which differential recall of family history between cases and noncases is unlikely because report preceded diagnosis, and in multiethnic populations [20]. Some of these cases represent clustering within families due to shared dietary and lifestyle risk factors, while others represent common genetic constitution. Several inherited CRC susceptibility syndromes have been described, including familial adenomatous polyposis (FAP) [21] and hereditary nonpolyposis CRC (HNPCC) [22]. The inheritance of the predisposition to CRC does not appear to vary by gender. FAP, characterized by innumerable adenomas that are histologically indistinct sporadic adenoma throughout the colon, is autosomal dominantly inherited, with virtually 100% penetrance by age 50. Affected patients inherit a germline mutation in the *APC* gene, a tumor suppressor gene mapping to the q21 region of chromosome 5 [8,23]. Gardner's syndrome, which co-occurs with osteomas, epidermal cysts, and dental anomalies, and Turcot's syndrome, which co-occurs with central nervous system cancers, are FAP variants.

HNPCC, also known as Lynch syndrome, is characterized by an early age of CRC development (mean 40–45 years), primarily in the proximal colon (70%), with multiple primary concurrent (synchronous) and later developing (metachronous) tumors, and without numerous adenomas [22,24]. HNPCC may be divided into Lynch syndrome I and II, where families with II have histories of cancer of extracolonic sites, especially of the endometrium, ovary, stomach, small intestine, ureter, and renal pelvis. HNPCC patients carry a germline mutation in DNA mismatch repair genes, most commonly *hMSH2* (chromosome 2p21–22) and *hMLH1* (3p21), and less frequently *hPMS1* (2q31–33) and *hPMS2* (7p22). In these individuals, mismatch repair is not appreciably affected until the unaffected copy is inactivated through somatic mutation. Cells with inactivation of both copies show an increase in uncorrected errors with each round of replication, particularly in microsatellites (regions of tandem repeats), and acquire a hypermutable phenotype [24]. The penetrance is 75–80% [22], and 1–5% of CRC cases may be due to HNPCC [24].

Patients with inflammatory bowel disease, in particular extensive ulcerative colitis of 10 or more years duration, have 20 times the risk of CRC compared to the general population [25]. During the active phase of the condition, the colorectal mucosum becomes increasingly dysplastic [8]. Adenocarcinomas from patients with these conditions show a similar spectrum of genetic alterations as those found in sporadic CRCs [8].

B. Detection, Treatment, and Prognosis of CRC

Rectal bleeding and abdominal, pelvic, or back pain are frequent symptoms of later-stage CRC [7,8], while adenomas are generally asymptomatic and are detected through routine endoscopic screening or work-up for unrelated gastrointestinal conditions [26]. Changes in bowel habits and abdominal discomfort are typically indicative of bowel obstruction that occurs late in disease progression. Sites of metastasis typically include the lymph nodes, liver, peritoneum, and lung, but not bone.

Adenomas may be found throughout the colorectum. They range in size from millimeters to centimeters, and are classified by the type of glandular distortion as tubular, villous, or tubulovillous, of which tubular adenomas are the most prevalent and villous have the greatest potential for malignant transformation [11]. Pedunculated adenomas are attached to the colonic wall by a stalk, while sessile adenomas are stalkless, broad, and relatively flat. CRCs are categorized histologically by grade (describing the extent of loss of glandular structure and nuclear orientation) and by stage at presentation (describing the invasiveness through the colonic wall and spread of the tumor to regional and distant sites) [7].

Polyps are generally removed during endoscopy. Large polyps may be biopsied and later removed with surgery. Treatment for colon cancer is generally radical resection of half or more of the large bowel depending on the affected region [7]. Following surgery, chemotherapy may be used to eliminate micrometastases. For rectal tumors with invasion localized to the submucosum and that are smaller than 3 cm, local excision is used. Because removal of adequate margins may not be possible for larger rectal tumors, combinations of chemo- and radiotherapy have been investigated for control of recurrence and metastasis. Metastatic CRC has not been successfully treated with chemotherapy. Carcinoembryonic antigen (CEA) is used in conjunction with frequent physical examination and colonoscopy to detect recurrence [7].

II. Descriptive Epidemiology of CRC

A. CRC Incidence and Mortality in the U.S.

From the U.S. Surveillance, Epidemiology, and End Results Program (SEER), the American Cancer Society estimates that in 1999, 94,700 colon and 34,700 rectum cancer cases will be newly diagnosed, and of these 54% and 44%, respectively, will be diagnosed in women [2]. 56,600 people will die from CRC, 51% of them women. After cancers of the breast and lung/bronchus, CRC is the third most commonly occurring nonskin cancer site and cancer cause of death in women, accounting for 11% of incident cancers and 11% of cancer deaths. By subsite within the colon and rectum, the most frequent locations are the sigmoid, rectum, and cecum [27]. Age-adjusted rates are similar between men and women in the right colon, but they are somewhat lower for women in the left colon, with a male to female ratio of 1.12 for the cecum, 1.41 for the sigmoid, and 1.71 for the rectum [27]. Invasive CRC is rarely diagnosed in those under age 40 (women: 1 in 1871), but increases dramatically in middle age (women: 1 in 150) and elderly years (women: 1 in 32). The lifetime risk (*i.e.,* cumulative incidence) of invasive CRC is 1 in 18 both in men and in women [2]. In women 80 years and older, the number of CRC deaths exceeds those from breast cancer [2].

Because adenomas are usually asymptomatic, their detection may occur years after onset; thus, the appropriate measure of their frequency is prevalence. The prevalence of adenomas increases with age and is greater in men than in women [26]. Autopsy studies suggest that one- to three-fifths of individuals have prevalent adenomas, and screenings of average risk populations have reported that one-fourth to two-fifths of individuals have adenomas [26].

B. Time Trends

Annual age-adjusted CRC mortality rates peaked in the 1940s at about 25/100,000 women and have steadily fallen through 1990 to around 15/100,000 in the U.S. [2]. The pattern in men is somewhat different, with the age-adjusted mortality rates remaining steady since 1950 until a slight decline in the mid-1980s. Although the rate of left colon cancer remains higher, rates of right colon cancer increased more rapidly than did left colon cancer during 1976–1987 [27].

C. Racial/Ethnic and International Variation in CRC Incidence, Mortality, and Survival Rates

There is some racial/ethnic variation in colon cancer mortality rates in the U.S. However, among African Americans, Asians, Native Americans, Hispanics, and whites, CRC similarly accounts for about 10% of cancer deaths [2]. By subsite within the colon, there is little racial variation in incidence rates (1976–1987) of cancer of the cecum and ascending colon [27]. African Americans have higher rates for the transverse and descending colon, whereas whites have higher rates for the sigmoid, rectosigmoid, and rectum.

Five-year colon cancer survival rates have increased from 50 to 64% in whites and 46 to 52% in African Americans between the mid-1970s and the late 1980s and mid-1990s in the U.S. [2]. Survival is approximately 90% for early stage CRC [2]. Differences in overall CRC survival rates in these two racial groups appear to be due to a greater proportion of African Americans diagnosed at later stages (25% versus 20% with distant disease) as well as poorer survival at a given stage [2], which may possibly reflect reduced access to and quality of medical care [8]. In 30 to 50% of CRC cases who have undergone resection the disease recurs within two years [7]. Stage at diagnosis is a predictor of likelihood of recurrence [7].

Internationally, marked variation exists in CRC mortality. In women, 1990–1993 colorectal mortality rates, adjusted to the world standard age distribution, are highest in Hungary (19.0/100,000 women), the Czech Republic (18.2), and New Zealand (17.6), and lowest in Albania (2.1), Mexico (3.1), and Uzbekistan (3.8) [2a]. The U.S. ranks 22nd (10.9) for women and 27th (16.0) for men. In general, rates are highest in Northern Europe and in countries populated by those of Northern European descent. In past decades, Asian and African countries had the lowest rates. However, with the introduction of Western diet and lifestyle into Asian society, rates have begun to increase. A now classic descriptive study by Haenszel and Kurihara [28] showed that Japanese migrants living in the U.S., a region with a higher CRC mortality rate, have a higher rate than Japanese living in Japan. Among countries, adenoma prevalence corresponds to the incidence rate of CRC [11].

III. Analytic Epidemiology of CRC and Polyps

The wide international variation in CRC incidence rates and the changing rates among migrants moving from low to high risk countries suggest that some feature of diet or lifestyle influences CRC occurrence. In addition, the differences in rates of CRC by subsite between men and women suggest the contribution of hormones and reproductive history. Here we summarize the epidemiologic findings of the relation of dietary, lifestyle, and reproductive factors on CRC risk. In particular we provide greater detail on findings from large, comprehensive studies of women and on those factors that are specific to females. Finally we describe the gene–environment interactions in sporadic CRC.

A. Diet

Doll and Peto used international comparisons of exposure prevalences and disease rates to estimate that up to 90% of colon cancers may have a dietary contribution [29]. The central dietary hypotheses that have been explored for CRC are that high fiber, fruit and vegetable, folate, methionine, calcium, and vitamin D intakes are protective, whereas high animal fat, red meat, and alcohol consumption are adverse. The bulk of the studies examining these relationships have been case-control studies, which may be limited by the potential for differential recall of past diet between cases and controls. For this reason, prospective studies with adequate size, length of follow-up, and dietary assessment methods are preferred for examining the contribution of diet to cancer risk. Also, it is important to consider the impact of measurement error in assessing long term dietary patterns, which are most likely to be relevant for colorectal cancer, and the difficulty in attributing effects to one nutrient among many intercorrelated nutrients [30].

1. Fruits, Vegetables, and Fiber

Many case-control studies have pointed to the beneficial effect of fruits and vegetables, total dietary fiber, or fiber from fruits or vegetables on CRC incidence [31–40]. A meta-analysis of case-control studies published through 1988 found an odds ratio (OR) for CRC of 0.58 for fiber and 0.48 for vegetables when comparing extreme quantiles (*e.g.*, quartiles or quintiles) [41]. However, prospective studies generally have produced more modest [42–44] or not supportive [45] results. Similarly for adenomas, some [46–53], but not all, studies [54–56] have supported the role of fruits and vegetables or fiber. Nonetheless, across all studies the most consistent dietary finding for colorectal carcinogenesis is for a protective effect of vegetable consumption [4].

Burkitt first proposed the fiber hypothesis based on his observation of low colon cancer rates in regions of Africa where fiber intake is high [57]. Dietary fiber consists of plant cell wall polysaccharides and lignin, which are not hydrolyzable by human digestive enzymes [58], and includes pectin, cellulose, hemicellulose, and lignin [59]. Several plausible mechanisms for the beneficial effect of dietary fiber have been proposed. For example, fiber may increase fecal bulk, possibly diluting carcinogens [57,59–61], or may be broken down to butyrate [62], which induces colon cell differentiation. In addition, secondary bile acids, which promote the neoplastic effect of experimental co-

Ionic carcinogens in animal models [63,64], may adsorb to some types of dietary fiber or may be sequestered in the viscous fecal mass, thus limiting reabsorption and enhancing their excretion in the feces [65]. The influence of fiber may depend on fiber type and source and may be modulated by food preparation and cooking methods and by mastication [66].

In many studies, high fiber intake may indicate a generally healthy lifestyle. For example, in early follow-up of Health Professionals Follow-up Study members, high fiber intake was strongly related to low adenoma risk [67]; however, with additional follow-up and control for recently identified CRC risk factors, including physical activity, smoking, and folate, the effect of fiber was substantially weaker [53]. Fiber from grains and cereals frequently has not been shown to be protective against CRC and adenoma, suggesting that fiber type (i.e., soluble, insoluble) is important or that the findings for fruits and vegetables may not be entirely due to their fiber component, but instead to other nutrients in those foods [31]. For example, in a case-control study conducted in Kaiser-Permanente members undergoing a screening sigmoidoscopy, the inverse relation between adenomas and high carotenoid vegetables and cruciferous vegetables was only modestly attenuated when controlling for hypothesized protective components of vegetables (i.e., fiber, folate, β-carotene, and vitamin C) and other factors, suggesting that constituents of these vegetables other than fiber may be acting [52].

Fruits and vegetables also contain folate, antioxidants, and other phytochemicals. Antioxidants found in fruits and vegetables include carotenoids and vitamins A, C, and E and the mineral selenium. These nutrients have been shown to be inversely related to CRC risk, although findings are not always the same for women and men. For example, in a prospective study of California retirees, CRC was inversely related to β-carotene intake in men only, to vitamin C intake in women only, and unrelated to vitamin A total intake or from supplements in both sexes [42]. Use of multivitamins and vitamin C and E supplements in both women and men, and use of vitamin A supplements in women, were inversely related to CRC risk in a population-based case-control study in Washington State [68]. Plasma levels of carotenoids have not been observed to be related to adenoma risk [69].

Fruits and vegetables also contain plant-specific phytochemicals such as allium in garlic and glucosinolates and isothiocyanates in cruciferous vegetables. Garlic has frequently been found to be inversely related to CRC risk [44]. Some phytochemicals, including indoles and isothiocyanates, induce P450 metabolic enzymes that detoxify carcinogens. Brassica vegetables (e.g., cabbage, broccoli, and brussels sprouts), a genus within the Cruciferae family, appear to be inversely related to CRC and adenoma in case-control studies [70], although prospective studies may not support this conclusion [44]. Flavonoids, which are plant phenols with antioxidant properties, are not clearly related to CRC risk [71].

2. Animal Fat and Red Meat

Across countries, age-adjusted incidence and mortality of CRC is correlated with national per capita disappearance of animal fat and meat [72,73]. However, per capita fat and meat intake may be a marker of potential deleterious correlates of Western lifestyle [43], such as physical inactivity and low intake

of fruits and vegetables; these are difficult to exclude as confounders in ecologic studies. Although numerous case-control studies show a positive association between CRC and red meat intake (a major source of animal fat in Western diets) and CRC and fat consumption, the relation of fat composition with CRC risk does not appear to be independent of total energy balance [43,74]. In addition, findings from the case-control studies may have been limited by differential accuracy of recall of past diet between cases and controls. Two large prospective studies of women [75] and men [76] showed positive associations for CRC and red meat. Other prospective cohort studies of women showed no [77] or equivocal [45] relationships between red meat or total and animal fat and colon and/or rectal cancer. No association between total protein and CRC risk has been detected in case-control studies, whereas a possible inverse relation, especially from nonred meat (e.g., poultry, fish), has been suggested in prospective studies of women [45,75,76,78].

One hypothesis to explain the red meat effect independent of its fat content is the presence of mutagenic heterocyclic amines [79] produced by the pyrolization of amino acids and creatinine during high temperature cooking [80]. Some epidemiologic studies have found that individuals who frequently consume heavily browned fried meat have a higher risk of CRC, especially rectal cancer, compared to those consuming lightly browned fried meat, even when controlling for total energy and physical activity [81]. Other studies have shown no association between fried meat and CRC risk [82].

3. Folate, Methionine, and Alcohol

The folate class of compounds provides methyl moieties for a variety of biochemical pathways, including thymine synthesis from uracil and the conversion of homocysteine to methionine, the major methyl donor for DNA methylation of cytosine in CpG nucleotides [83]. The adverse effects of reduced methyl status (i.e., enhanced DNA misincorporation of uridylate for thymidylate and subsequent damage from attempted repair, and DNA hypomethylation leading to perturbed expression of proto-oncogenes and tumor suppressor genes) suggest that folate and methionine play a role early in the carcinogenic pathway. A major metabolite of alcohol, acetaldehyde, may affect several steps in the methyl transfer pathway, notably inhibition of the enzymes methionine synthase [84] and DNA methylase [85], and inactivation of N^5-methyltetrahydrofolate [86].

As described earlier, fruits and vegetables and nonred meat protein were inversely related to CRC and adenoma in several epidemiologic studies. These findings may point to the folate and methionine contents of these foods, respectively. Most studies that estimated intake of these nutrients have shown a reduction in CRC and adenoma risk with increasing intake of folate [33,34,38,49,68,87–92] and methionine [88,90]. In two large prospective cohort studies, the relative risk (RR) for adenoma comparing the highest to lowest quintiles of folate was 0.66 for women and 0.63 for men, which appeared to be independent of other vitamins found in fruits and vegetables [88]. Comparing extreme quintiles of methionine, the RR for large adenoma (≥1 cm) was 0.62 in men and women in the same two cohorts [88]. Folate from supplements was inversely related to CRC risk in men (≥400 μg/day versus none: OR = 0.59) and women (OR = 0.44) in a population-based case-control study in

Washington state [68]. Erythrocyte and plasma folate concentrations were inversely related to adenoma risk in men, but not in women, among Kaiser-Permanente members undergoing sigmoidoscopy [93].

Although not entirely consistent, epidemiologic studies [94], in particular prospective cohort studies [42,88,90,95], support a positive association between alcohol consumption and CRC and adenoma. In a large prospective study of men, comparing more than 2 drinks/day to 0.25 or fewer, the RR for CRC was 2.07, and past drinkers remained at an elevated CRC risk [90]. Among prospectively followed residents of a California retirement community consuming at least 31 ml per day of alcohol compared to those who drank less than daily, men and women had an RR of CRC of 2.42 and 1.45, respectively [42]. In a cohort of Japanese men in Hawaii, alcohol consumption was positively related both to colon (comparing at least 24 ounces/month to 0: RR = 1.39) and to rectal (RR = 2.30) cancers, but was notably stronger for the rectum [95].

Adenoma risk did not appear to vary among beer, wine, or liquor sources of alcohol in a large cohort study [88], whereas rectal cancer risk was greater for beer consumption in a case-control study [96]. The combined state of low folate, low methionine, and high alcohol intake is strongly (RR > 2) associated with an increased risk of CRC in men [90,91] and of adenoma in women and men [88]. Individuals with higher intakes of folate and methionine were not at an increased risk of CRC among those consuming more than two drinks daily [90]. In a French case-control study, alcohol was positively related to large adenoma, but not to small adenoma or to CRC, among men, but not among women, possibly because alcohol intake was substantially lower in women [97].

4. Calcium, Vitamin D, and Dairy Products

In vitro and *in vivo* studies suggest a beneficial effect of calcium on colonic cell proliferation and differentiation and tumor incidence, perhaps mediated through calcium's role as a second messenger in the signal transduction cascade or as a binder of unconjugated bile acids and free fatty acids [98], which may act as promoters. Although many epidemiologic studies have evaluated the calcium-CRC relationship, few provide statistically significant evidence of a protective effect [36,38,40,68,89,99–104]. Most studies have not observed a relationship between calcium intake and adenoma [46,49,55,105,106]. A meta-analysis of cohort and case-control studies published between 1980 and 1994 did not show a notable protective effect of calcium on CRC or on adenoma, and it found large heterogeneity in estimates not accounted for by differences in endpoint, subsite, gender, fat intake, calcium source and dose, or study design [107].

Experimental studies have demonstrated that $1,25(OH)_2D$, the active metabolite of vitamin D, inhibits proliferation and induces differentiation of human colorectal cells [108–111] and reduces tumor growth [112]. Ecologic studies showed that areas with higher sunlight exposure (ultraviolet light photoconverts skin precursors to vitamin D) had lower rates of CRC incidence and death [113–116]. In those studies where the RR for the contrast of extreme quantiles of vitamin D intake or the trend across quantiles was statistically significant, the RR for colon or CRC, comparing the upper and lower quantiles, ranged from 0.5 to 0.7. Suggestions of inverse relations for plasma 25(OH)D concentration and CRC were observed in nested studies in the Washington County, Maryland cohort [117] and in the Finnish Alpha-Tocopherol, Beta-Carotene Cancer Prevention Study [118], whereas no relationship was observed for 25(OH)D in a later group of CRC cases in the Washington County study [119]. Neither of the studies that measured $1,25(OH)_2D$ level observed a relationship with CRC [118,119]. Few studies have evaluated the relationship of vitamin D and adenoma, and in the even fewer studies that observed inverse relationships [105,106], findings were not consistent across gender or location within the colon.

Dairy products are a major source of calcium and vitamin D. Only a small proportion of the numerous epidemiologic studies of dairy products and CRC show statistically significant inverse associations [45,100,101,120,121]. Inverse relationships have not been observed for adenoma and milk and other dairy products [46,48,54,105,106]. Fermented dairy products (*e.g.*, yogurt and cheese) also contain lactic acid bacteria and in some studies they are inversely related to CRC and adenoma in at least one site or gender [104,106,121–123].

5. Sucrose, Glycemic Load, and Insulin

Another etiologic hypothesis for CRC and adenoma relates to insulin control [124]. Insulin is a growth factor, which has been shown to be mitogenic for malignant colonic cells *in vitro* [124]. *In vivo* insulin delivered via injection increases the incidence of azoxymethane-initiated colon tumors in rats, supporting its role as a colon cancer promotor [125]. A preliminary report suggested that individuals with adenoma or adenocarcinoma have higher insulin and triglyceride concentrations and have greater waist-to-hip ratios, a marker of central adiposity and a correlate of insulin resistance, than comparison individuals who had a negative endoscopy [126]. High consumption of simple sugars, easily digestible starch, and highly processed foods, and low intake of water-soluble fiber result in postprandial hyperinsulinemia. Conversely, unrefined diets with a high fiber content, particularly water-soluble fiber, reduce the rate of glucose absorption from the upper gastrointestinal tract and may increase insulin sensitivity [127].

A cohort of 35,216 Iowa women, among whom 212 cases of colon cancer were identified, noted a suggestion of a positive relationship between colon cancer and servings of sucrose-containing foods or daily intake of sucrose, but not for fructose or total carbohydrate intake [77]. In another cohort study of 14,727 women living in New York and Florida, among whom 100 CRC cases were ascertained, no relationship with carbohydrate was observed [45].

6. Iron

Two mechanisms for a potential adverse effect of iron on CRC risk have been postulated [128]. First, free radicals may be generated by iron, which in turn may cause DNA damage, a potential initiating event, and lipid peroxidation, which may result in increased cellular proliferation to replace damaged cells, a potential promotional event. Second, iron is a necessary nutrient in culture media for cellular growth and, thus, may participate in proliferative control.

An increased risk of CRC with higher intake of iron was observed in a follow-up of the National Health and Nutrition Examination Survey I (NHANES I) population, with the elevation being greatest for the proximal colon and for women [128].

Case-control studies generally have not shown a positive relationship between iron intake and CRC [34,104,120,129] or adenoma [46,49,92], although one case-control study found a strong positive relationship for rectal cancer in men (extreme quartiles: RR = 2.15) but not in women (extreme tertiles: RR = 0.84) [130]. Transferrin saturation, a measure of the proportion of transferrin binding sites that are filled by iron, was inversely related to colon and rectal cancer in men but not in women in a follow-up study of Kaiser-Permanente members [131]. Plasma ferritin, a marker of total iron stores, was weakly positively related to adenoma risk among those with a previous negative sigmoidoscopy [132]. In the NHANES I cohort, serum iron, which represents both dietary iron intake as well as binding protein equilibria, was related only to distal colon cancer in women [128].

B. Lifestyle Factors

The international variation in CRC rates also suggests that lifestyle may be an important determinant of risk. Hypotheses explored to date include obesity, physical inactivity, and smoking as factors that increase CRC and adenoma risk, while aspirin has been hypothesized to reduce risk. As with diet, prospective cohort studies with adequate exposure assessment and duration of follow-up are less likely to be subject to the selection and observation biases that limit the findings from case-control studies.

1. Obesity

Energy restriction has long been known to reduce tumor incidence in rodents [133]. The mechanism underlying this observation may be related to the reduction in level of insulin, a normal and malignant colonic cell growth factor *in vitro*, with calorie restriction [134]. Some epidemiologic studies, especially cohort studies [42,77,95,135,136], have noted that higher body mass index (BMI = weight in kilograms divided by the square of height in meters), a measure of obesity, is related to an elevated risk of CRC. This relationship is stronger in men than in women, which may be indicative of the higher prevalence of central obesity and accompanying hyperinsulinemia in men compared to women [134,135]. In the Iowa Women's Health Study, those in the upper three quintiles of baseline BMI had about a 50% higher colon cancer risk, while for waist-to-hip ratio no risk gradient was observed [77]. In the Nurses' Health Study, an increasing risk gradient with BMI was detected for distal (BMI $\geq$ 29 versus < 21 kg/m^2 RR = 1.96) but not for proximal colon cancer [136]. Women with high central obesity had a greater risk of CRC than women with a high BMI but a low waist-to-hip ratio. No relationship of CRC to BMI was noted in the New York University Women's Health Study; however, the follow-up was relatively short and the number of CRC cases was small [45]. In the Nurses' Health Study, for large adenoma, the RR comparing the top to bottom quintiles of BMI was 2.21 [134]. The relationship with BMI was weaker for small adenoma and nonexistent for rectal adenoma.

2. Physical Activity

Physical activity may affect CRC risk through several routes. First is maintenance of body weight through metabolic use of excess caloric intake, which influences hyperinsulinemia [134]. Second, activity directly reduces insulin levels and heightens

response of muscle to insulin [137]. Third, physical activity is important for bowel motility [138]. More than 40 case-control and cohort studies have examined the relationship of occupational and recreational activities as markers of physical activity to CRC [139]. These have consistently shown a decreased risk of CRC with higher activity on the order of a 50% reduction when comparing highest and lowest activity levels [139]. The effect of physical activity appears to be independent of obesity (BMI).

In the Nurses' Health Study, women whose leisure-time physical activity exceeded 21 MET-hours/week (*e.g.,* running 2 hours/week) had about half the risk of total colon cancer and a third of the risk of distal colon cancer as those exercising under 2 MET-hours/week [136]. A MET-hour is the metabolic equivalent of sitting at rest for one hour [140]. Among 11,888 retirees prospectively followed, among whom 58 men and 68 women were diagnosed with CRC, those exercising more than 2 hours/day compared to those exercising under 1 hour had a RR of 0.40 for men, but only 0.89 for women [42]. Although not statistically significant, physical activity in women did appear to be more strongly related to left colon cancer (RR = 0.68). Physical activity is also related to a reduced risk of adenoma [134]. For example, in the Nurses' Health Study, the RR for large colorectal adenoma, comparing the top to bottom quintiles of leisure-time activities, was 0.57; the effect was weaker for small and rectal adenoma [134]. The stronger association with large compared to small adenoma suggests that physical activity affects adenoma growth [134].

3. Smoking

Many studies have not shown a relationship between cigarette smoking and CRC [4]. In the Iowa Women's Health Study, the RR for colon cancer for current and former smokers were 1.09 and 0.92, respectively, and there was no relationship between CRC and packyears smoked [77]. In the New York University Women's Health Study, the RR for CRC was 0.97 for current and 0.99 for past smoking relative to nonsmokers [45]. When the time since starting smoking and smoking duration are considered, smoking does appear to be related CRC risk, however. In the Nurses' Health Study (586 CRC cases among 118,334 women), an increased risk of CRC was only observed among women who had begun smoking at least 10 cigarettes/day more than 35 years before diagnosis, with the RR increasing from 1.47 for 35–39 years ago to 2.00 for more than 45 years ago [141]. Similarly, in a large cohort of men, smoking more than 35 years ago was related to CRC risk [142]. A moderately sized prospective study of Japanese men in Hawaii with long-term follow-up found an increased risk of colon (RR = 1.48) and rectal (RR = 1.92) cancers with at least 31 pack-years of cigarette smoking [95]. An effect of cigarette smoking in women is supported by a case-control study in Wisconsin for both colon (current versus never: OR = 1.33) and rectal (OR = 1.70) cancers [143]. CRC risk increased with number of cigarettes smoked daily, duration of smoking, and younger age at starting to smoke, and it was also elevated among former smokers.

Several studies have noted a positive relationship between cigarette smoking and adenoma [144]. In the Nurses' Health Study (564 adenoma cases among 12,143 women who underwent endoscopy), smoking in the immediate past 20 years was positively related more strongly to small adenoma than to large [141]. In a large cohort of men, recent smoking was related to small polyps, and smoking more than 20 years ago was related to large polyps [142].

Tobacco smoke constituents are likely initiators in the carcinogenesis pathway. This hypothesis is supported by the observation that the occurrence of small adenomas is influenced by recent smoking, large adenomas by more distant, and CRC by even more remote smoking.

4. Aspirin Use

Sulindac, a nonsteroidal anti-inflammatory drug (NSAID), decreased the number and size of adenomas in patients with polyposis in several small nonrandomized studies [15]. Case-control and cohort studies have suggested that regular use of aspirin and other NSAIDs may reduce the incidence of CRC and adenoma or death from CRC, including in women [145–151]. A large randomized, placebo-controlled trial investigating the benefit of aspirin on cardiovascular disease in male physicians did not observe an inverse association between aspirin and CRC, possibly because of the low dose of aspirin used or the relatively short duration follow-up [152]. From large prospective studies, the observed reduction in risk with regular aspirin use was 30% for total and 50% for metastatic or fatal CRC in women [150] and in men [153], with the reduction being greatest for long-term consistent use. It is unlikely that the inverse relationship of aspirin to CRC risk is a result of heightened detection and removal of adenoma due to work-up for gastrointestinal bleeding caused by aspirin among users [153].

Three mechanisms have been proposed to account for the benefit of NSAIDs on CRC risk [15], two of which are related to the ability of NSAIDs to reduce the production of the prostaglandin cascade from arachidonic acid by inhibiting the enzyme cyclooxygenase. First, lowered prostaglandin production limits the inhibition of immune surveillance for malignant cells induced by prostaglandin E_2. Second, lower prostaglandin production limits prostaglandin-enhanced tumor growth and invasion. Third, NSAIDs appear to inhibit progression through the cell cycle, evidenced by an accumulation of cells at G_0/G_1 of the cell cycle and induction of apoptosis.

C. Reproductive Factors

Early reports suggested that unmarried women [154], including elderly nuns [155], were at an increased risk of CRC compared to married women, where marital status was a marker of (nulli)parity. In 1980, McMichael and Potter hypothesized that pregnancy and other reproductive factors that increase steroid hormone levels, such as use of oral contraceptives (OC) and postmenopausal hormone (PMH) replacement, would protect against CRC by reducing flow of bile acids, which may be cancer promoters in the colon [156]. Other proposed effects of reproductive factors may be related to estrogen's effect of reducing colonic motility [157] and/or may be mediated by receptors for estrogen and other hormones that are present in normal and in malignant colonic epithelium [158,159].

1. Age at Menarche

There is little evidence for a relationship between age at menarche and either CRC or adenoma. A few studies have suggested positive relations for CRC [160,161], some noted inverse relations [162–167], and others found no important relationship [45,168–173]. For adenoma, a statistically significant 40% decreased risk of colorectal adenoma among women with menarche before age 13 years was observed among 128 cases and 283 colonoscopy negative controls in New York [174]. Two other studies did not show a relationship between adenoma and age at menarche [166,175].

2. Age at First Birth

There is no consistent evidence that age at first birth affects CRC or adenoma risk. No relationship was observed in several studies [161,163–165,169,170,172,176–181], positive relations were suggested in some [162,167,171,182–186], and inverse relationships were found in others [45,160,173,187]. For adenoma and age at first birth, a cohort [175] and case-control [174] study did not show a relationship, whereas another case-control study found a nonstatistically significant inverse association [188].

3. Parity

The majority of the case-control studies examining parity and CRC have found evidence of a modest inverse relationship [45,161,162,164,165,168,169,171,176,178,179,182,183, 185,187,189,190], although only half were statistically significant [164,171,176,178,179,182,185,190]. Only one cohort study found a lower risk of CRC among women who had three or more children compared to none, which was not statistically significant [42]. Conversely, other cohort and population-based case-control studies suggest a positive relationship. For example, an analysis in the Nurses' Health Study with 12 years of follow-up suggested an increased risk among women with a parity of more than four [167]. In the prospective Iowa Women's Health Study, a positive association with colon cancer was observed, with a RR of 1.80 among those with three or more live births compared to none [77]. A registry linkage study in Norway noted an inverse association between parity and cancer in the ascending colon, no association for transverse or descending colon, and a U-shaped relationship for the sigmoid colon, rectosigmoid junction, and rectum [186]. In a population-based case-control study, a U-shaped relationship between colon cancer and parity was observed; risk decreased with up to four pregnancies then increased with each additional pregnancy both for incomplete and complete pregnancies [169]. The relationship was stronger for cancer of the distal colon than for the proximal colon. No relationship between parity and adenoma was observed in three case-control studies [166,174,188], whereas in the prospective Nurses' Health Study, an elevated risk of distal colon adenoma was seen for three through six or more term pregnancies [175]. Although nulliparous women appear to be at an increased risk for CRC relative to having had one or more births in some of the above indicated studies, women diagnosed with infertility, some of whom may have underlying endocrine dysfunction, do not appear to be at an elevated risk of CRC [191].

During pregnancy, estrogens are tenfold higher due to fetal–placental contribution [157]. Competing hypotheses for the effect of elevated estrogen on CRC risk can be described, where parity is a marker for duration of exposure to elevated pregnancy estrogens. The estrogen-associated change in the bile acid profile hypothesis could explain an inverse relationship between CRC and parity. At least two possible explanations could account for a positive association between parity and CRC. First, estrogen reduces bowel motility [157]; it is commonly held that decreased transit time and increased bowel motility reduce risk by minimizing contact between lumen carcinogens and the colonic epithelium. Second, pregnancy perturbs carbohydrate metabolism which results in decreased glucose tolerance and increased secretion of insulin [157], a possible colon tumor promotor [124,192]. All three of these possible estrogen-mediated effects, however, would be more likely to have promotional roles and would act close in time to polyp development, and thus are not entirely consistent with the generally long span between the reproductive years and the detection of polyps. A possible explanation for the positive relationship between parity and adenoma in the cohort study independent of estrogen is folate inadequacy, because during pregnancy folate demand increases due both to fetal uptake and to increased catabolism [193].

4. Oral Contraceptive Use

The evidence for a relationship between OC use and CRC is equivocal, with studies finding no [162,169,170,187], positive [178], or inverse [161,166,167,182,184] associations, of which only two [166,167] were statistically significant. In a small Danish case-control study, only current OC users showed the reduction in risk [166], and in the Nurses' Health Study, the inverse relationship between OC use and CRC was duration-dependent [167]. An inverse association between current or ever users of OC and adenoma was suggested in case-control studies [166,174], but not in another case-control [188] or a cohort study [175].

It is biologically plausible that OC use would be protective for CRC and adenoma because, like pregnancy, it diminishes flow of bile acids [157], which again may be promoters of carcinogenesis. However, OCs also impede the conversion of the folate isomer found in foods to the form that is absorbable in the gastrointestinal tract [157]; in extreme cases it results in malabsorption [194]. Again, it has been previously shown in the Nurses' Health Study cohort that higher intake of folate is moderately protective against CRC and adenoma [88]. In addition, like pregnancy, OCs perturb glucose control [157].

5. Age at Menopause

Age at menopause does not appear to be related to risk of CRC or adenoma [166,175]. A Greek case-control study found an inverse relationship with CRC [160], a case-control study in the Netherlands found a nonstatistically significant positive relationship with CRC [162], and a study of Chinese women living in North America and China saw a positive relationship, but only for rectal cancer in North America [164].

6. Postmenopausal Hormone Use (PMH)

Unlike the other reproductive factors examined, PMH use fairly consistently has been found to be inversely related to CRC [161,165,170,173,184,187,195–197], with more recent use being more protective than past use. A few studies did not observe a relationship [169,178,182,198], and one case-control study saw a positive relationship with current use [166]. Consistent with CRC findings, for adenomas three case-control studies noted an inverse association for PMH use [166,174,188], and a weak effect on adenoma risk was seen in a cohort study [197]. The sites within the large bowel for which a reduced risk was detected with PMH use has varied between studies, however.

The mechanism of the potential benefit of PMH remains un-

known, but proposed hypotheses include effects on bile acid profiles, the diacylglycerol second messenger, or estrogen receptor gene methylation [199]. PMHs also improve insulin sensitivity [200,201], which might also modify CRC risk. An alternative explanation for the generally observed inverse relationship is bias. Women who are prescribed replacement hormones have greater frequency of medical contact during their use, and in general, these women tend to be more proactive in seeking screening. Women taking PMHs may be more likely to have undergone endoscopy and to have had polyps removed, and thus would be at lower risk of developing CRC. Conversely, women with undiagnosed CRC who are experiencing early bowel symptoms may be more likely to stop taking PMHs and would be former users at the time of diagnosis, thus biasing the results towards current use appearing protective.

D. Other Inherited Cancer Susceptibility Genes and Gene–Environment Interactions

For the vast majority of CRCs, cases do not have a strong family history of the disease and have not inherited germline mutations in major cellular growth control or repair genes. However, genetic variation between individuals at loci encoding detoxification enzymes, other metabolic enzymes, or growth factor/hormone receptors may lead to differential inherited cancer susceptibility. The prevalence of these variants is generally high enough in the population to be considered an alternative allele for the gene rather than a mutation in the normal allele. For the detoxification enzymes, differences in cancer susceptibility attributable to variants may not manifest themselves unless exposed to a carcinogen that the enzyme plays a role in activating or detoxifying. Thus, allelic variation at these loci may be viewed as modifiers of an exposure's carcinogenicity. Generally, the relative risk associated with the riskier allele is modest.

Several groups have investigated the relationship between adenoma and CRC risk with polymorphisms in phase II biotransformation enzymes, which are involved in rendering potentially toxic lipophilic compounds water soluble and, thus, more readily excretable. The N-acetyltransferases, NAT1 and NAT2, acetylate aromatic amines such as those that are found in cigarette smoke and meat cooked at high temperatures, which in some instances results in bioactivation instead of detoxification of these compounds. Using phenotypic assessment of acetylation genotype, the fast acetylator phenotype was associated with an almost twofold increased risk of CRC, and risk of CRC or adenoma increased with increasing meat consumption among fast acetylators, but not among slow acetylators [202]. Two studies did not observe a relationship between the fast acetylation NAT2 genotypes and CRC [203] or adenoma [204] risk overall. One study observed a twofold increased risk of CRC associated with the fast acetylator NAT1*10 allele, which was even greater among those with higher stage disease or the NAT2 fast acetylator genotype [203], whereas a second study did not observe this relationship for distal adenomas and NAT1*10 or the interaction of NAT*10 and NAT2 fast acetylator genotype [205].

The family glutathione S-transferases (GST) are phase II enzymes that conjugate glutathione to compounds such as polycyclic aromatic hydrocarbons, which are found in cigarette smoke, producing a more water soluble metabolite. Some individuals lack specific GST subtypes and are called GST null. The

GSTM1 null genotype was associated with a twofold higher risk of distal CRC, although there was no variation by smoking status [206], whereas in two other studies there was no association between this genotype and adenoma [207] or CRC [208]. The GSTT1 null genotype was associated with a twofold increased risk of CRC in a study in England [208] but not in a study in Japan [206]. Differences in findings between these studies of phase II biotransformation enzymes and colorectal neoplasia may be due to varying linkage disequilibria among the racial and ethnic groups included in these studies for alleles encoding two or more enzymes acting in concert to modify potential toxicants. These populations may be exposed to differing arrays and levels of procarcinogens, for example through differences in amounts of meat consumed, cooking methods, numbers of cigarettes smoked, and exposure to other agents in the environment or workplace that induce these enzymes.

Because of the importance of folate and methionine for competing precursors for DNA methylation and production of the nucleotide thymidine, investigators have examined a lower activity variant in the enzyme 5,10-methylenetetrahydrofolate reductase (MTHFR), which catalyzes conversion between 5,10-methylenetrahydrofolate and 5-methyltetrahydrofolate, CRC risk, and MTHFR's interaction with alcohol. In two nested case-control studies, men who were homozygous for the variant MTHFR allele had about half the risk of CRC compared to men who were homozygous or heterozygous for the more common allele [209,210]. Higher alcohol consumption reduced or eliminated the apparent protective effect of the MTHFR polymorphism in both studies.

The main methodologic issues in the planning and conduct of these heterogeneity studies are adequate sample size (more specifically, sufficient prevalence of the rarest allele to be evaluated), adequate dietary or environmental exposure gradient, confounding, and, importantly, the ethics of genotyping individuals and relating genotype to cancer and other disease risks [211–213]. Prospective collection of blood or other sources of DNA is preferable because variants may also influence case survival.

IV. Primary and Secondary Prevention

A. Screening Methods and Recommended Frequency

Early detection and removal of colorectal adenoma reduces risk of CRC [8,26], and early diagnosis and treatment of CRC likely increases survival. Several early detection tests are available. The fecal occult blood test (FOBT) involves collection of three consecutive stools after consuming a specified diet and testing for the presence of blood, which may indicate either a colorectal adenoma or cancer. Sigmoidoscopy involves cleansing the descending and sigmoid colon and directly visualizing the lower half or third of the colorectum using a 60 cm endoscope. Colonoscopy involves cleansing the colorectum and directly visualizing the entire colon and rectum using a colonoscope. A double-contrast barium enema (DCBE) involves barium and air to visualize the colorectal mucosal profile using X-rays.

The American Cancer Society (ACS) updated their recommendations for CRC screening to encompass three risk level groups [214]. The guidelines for individuals at average risk and who are older than 50 years include a FOBT annually, sigmoidoscopy every 5 years, colonoscopy every 10 years, or DCBE every 5 to 10 years, along with a digital rectal examination at the time of endoscopy or DCBE. Individuals with a family history of FAP or HNPPC, or who have inflammatory bowel disease, including ulcerative colitis and Crohn's disease, are considered to be at high risk. Individuals at moderate risk are those with a previous small or large adenoma or who have a family history of colorectal adenoma or cancer in first-degree relatives. Recommendations for high and moderate risk individuals have been tailored to their expected risk of CRC and vary in age at which to begin screening, frequency of screening, and screening methods used [214]. Those who have had resection for CRC should have either a colonoscopy or DCBE within 1 year of resection, and, if normal, in 3 years, and if normal again, then in 5 years.

B. CRC Prevention Trials

Several ongoing trials are assessing dietary and nutrient interventions in the prevention of CRC. The Polyp Prevention Trial is a randomized placebo-controlled intervention study examining the effect of low-fat, high fiber, high vegetable and fruit diet pattern on adenoma recurrence [215]. A randomized, double blinded, placebo-controlled trial of folate supplementation (1 mg/day) on adenoma recurrence is being conducted among participants in the Nurses' Health Study I and II and in the Health Professionals Follow-up Study. The Women's Health Initiative is investigating, among other hypotheses, the benefit of a low fat diet on CRC risk in postmenopausal women [216].

Several dietary randomized intervention trials have been completed. β-carotene did not appear to influence CRC incidence in the Physicians' Health Study [217] or in the Alpha-Tocopherol, Beta Carotene Cancer Prevention Study [218]. Similarly, no effect of β-carotene or of vitamins C and E supplementation on adenoma recurrence was observed in the Polyp Prevention Study [219]. β-carotene supplementation for 3 months in 101 participants, some of whom had a history of adenoma or CRC and some with no history, did not alter markers of cellular proliferation [220]. The Australian Polyp Prevention Project observed nonstatistically significant decreases in large adenoma recurrence with a low fat diet or supplementation with wheat bran and no difference with β-carotene supplementation in 411 participants [221]. No effect of wheat bran and calcium supplementation for 9 months was found on markers of rectal epithelial proliferation among 100 participants with a prior resected adenoma [222].

A double-blind intervention study demonstrated that individuals supplemented with selenium (200 μg/day) had a 60% reduction in colorectal cancer risk [223]. The study, however, was composed mostly of men (about 75%), and there were insufficient numbers of women to indicate whether they would benefit. An observational study based on toenail selenium level as an integrated marker of intake did not support a benefit [224], so further study in women, as well as in men, is warranted.

Because CRC is rare in the context of clinical trials, two approaches have been used to increase the number of endpoints for evaluation of interventions. The first is to enroll subjects at increased risk for CRC, including those with a previous adenoma or colon cancer or those with a strong family history. The second is to employ intermediate markers of CRC risk [225]. An increase in crypt cell proliferative activity and proliferative zone extension up the crypt appears to correspond with progression through the adenoma to carcinoma sequence [226–228]. Several methods have been used to detect proliferation, including incorporation of tritiated thymidine or 5-bromodeoxyuridine, a thymidine analog, into DNA of replicating cells, or by immunohistochemical detection of S-phase-associated proliferating cell nuclear antigen [225].

V. Summary

Diet and lifestyle clearly are major contributors to CRC and adenoma risk. Although identification of the specific influential factors and the mechanisms by which they exert their effect on CRC development have not been fully elucidated, certainly a diet high in fruits and vegetables, low in animal fat, red meat, and refined flours and sugars, coupled with ample physical activity and avoidance of smoking and alcohol, is prudent for decreasing CRC and adenoma risk. With dietary and lifestyle modifications, as well as adherence to screening recommendations, it is very likely that CRC is largely preventable in both genders.

Several areas of research warrant additional investigation to understand the etiology and prevention of CRC in general, and specifically for women. Efforts should be focused on further uncovering the mechanisms underlying the adverse effect of combined central adiposity, physical inactivity, and possibly a high glycemic index diet, and combined low folate, low methionine, and high alcohol diets. Substantial epidemiologic evidence has been gathered to support the theory that these two modifiable constellations of factors increase CRC and adenoma risk and plausible, although as yet unproven, modes of action have been postulated. Identification of polymorphic genes that influence susceptibility to endogenous and exogenous carcinogens and procarcinogens and quantification of variation in risk of CRC and adenoma to an exposure with a given polymorphism should be pursued, most feasibly in large, well-characterized cohorts with archival DNA and long-term follow-up.

Although the role of reproductive factors in CRC etiology is not entirely clear, one factor, PMH use, deserves further consideration. Besides usefulness in ameliorating peri- and postmenopausal symptoms, PMH use has already acknowledged benefits for cardiovascular disease and osteoporosis and risks for endometrial and breast cancers [229]. Because its use is widespread and the benefit in Western societies (e.g., those consuming higher fat, low fruit and vegetable diets, and are inactive) may be far reaching, it is important to uncover whether the observed association is causal or merely represents bias. Even if replacement hormones are conclusively shown to reduce risk of CRC, each women and her physician must weigh the risks and benefits of use based on her personal and family history of disease.

Among promising CRC and adenoma chemopreventive agents that require additional work are folate, selenium, and NSAIDs. The mechanism underlying the role of folate in CRC risk is rapidly emerging. Dose and timing of supplementation relative to adenoma formation, likely critical determinants of its action, as well as folate congeners with optimal bioavailability, must be

further explored. Similarly for NSAIDs and other compounds that inhibit cyclooxygenase, determination of dose and timing of use, and refinement so that these compounds do not have the gastrointestinal side effects of aspirin, are needed [230].

Acknowledgments

Elizabeth Platz was supported by a National Research Service Award (T32 CA 09001).

References

1. Lynch, H. T., and Smyrk, T. C. (1993). Genetics, natural history, tumor spectrum and pathology of hereditary nonpolyposis colorectal cancer: An updated review. *Gastroenterology* **104**, 1535–1549.
2. Landis, S. H., *et al.* (1999). Cancer Statistics, 1999. *Ca—Cancer J. Clin.* **49**, 8–31.
2a. Landis, S. H., *et al.* (1998). Cancer Statistics, 1998. *Ca—Cancer J. Clin.* **48**, 6–29.
3. Lampe, J. W., *et al.* (1993). Sex differences in colonic function: A randomised trial. *Gut* **34**, 531–536.
4. Potter, J. D., *et al.* (1993). Colon cancer: A review of the epidemiology. *Epidemiol. Rev.* **15**, 499–545.
5. Peipins, L. A., and Sandler, R. S. (1994). Epidemiology of colorectal adenoma. *Epidemiol. Rev.* **16**, 273–297.
6. Haubrich, W. S. (1995). Anatomy of the colon. *In* "Bockus Gastroenterology" (W. Haubrich, F. Schaffner, and J. Berk, eds.), 5th ed., pp. 1573–1591. Saunders, Philadelphia.
7. Levin, B., and Raijman, I. (1995). Malignant tumors of the colon and rectum. *In* "Bockus Gastroenterology" (W. Haubrich, F. Schaffner, and J. Berk, eds.), 5th ed., pp. 1744–1772. Saunders, Philadelphia.
8. Jessup, J. M., *et al.* (1997). Diagnosing colorectal carcinoma: Clinical and molecular approaches. *Ca—Cancer J. Clin.* **47**, 70–92.
9. Lev, R. (1990). "Adenomatous Polyps of the Colon." Springer-Verlag, New York.
10. Jass, J. R. (1991). Nature and clinical significance of colorectal hyperplastic polyp. *Semin. Colon Rectal Surg.* **2**, 246–252.
11. Lee, R. G. (1995). Benign tumors of the colon. *In* "Bockus Gastroenterology" (W. Haubrich, F. Schaffner, and J. Berk, eds.), 5th ed., pp. 1715–1730. Saunders, Philadelphia.
12. Vogelstein, B., *et al.* (1988). Genetic alterations during colorectal-tumor development. *N. Engl. J. Med.* **319**, 525–532.
13. Fearon, E. R., and Vogelstein, B. (1990). A genetic model for colorectal tumorigenesis. *Cell (Cambridge, Mass.)* **61**, 759–767.
14. Oshima, M., *et al.* (1996). Suppression of intestinal polyposis in $Apc^{\Delta 716}$ knockout mice by inhibition of cyclooxygenase 2 (COX-2). *Cell (Cambridge, Mass.)* **87**, 803–809.
15. Berkel, H., *et al.* (1996). Nonsteroidal antiinflammatory drugs and colorectal cancer. *Epidemiol. Rev.* **18**, 205–217.
16. Laken, S. J., *et al.* (1997). Familial colorectal cancer in Ashkenazim due to a hypermutable tract in *APC. Nat. Genet.* **17**, 79–83.
17. Delattre, O., *et al.* (1989). Multiple genetic alterations in distal and proximal colorectal cancer. *Lancet* **2**, 353–356.
18. Watatani, M., *et al.* (1996). Allelic loss of chromosome 17p, mutation of the p53 gene, and microsatellite instability in right- and left-sided colorectal cancer. *Cancer (Philadelphia)* **77**, 1688–1693.
19. Fuchs, C. S., *et al.* (1994). A prospective study of family history and the risk of colorectal cancer. *N. Engl. J. Med.* **331**, 1669–1674.
20. Le Marchand, L., *et al.* (1996). Family history and risk of colorectal cancer in the multiethnic population of Hawaii. *Am. J. Epidemiol.* **144**, 1122–1128.
21. Lynch, P. M. (1995). Polyposis syndromes. *In* "Bockus Gastroenterology" (W. Haubrich, F. Schaffner, and J. Berk, eds.), 5th ed., pp. 1731–1743. Saunders, Philadelphia.

22. Lynch, H. T., and Smyrk, T. (1996). Hereditary nonpolyposis colorectal cancer (Lynch Syndrome): An updated review. *Cancer (Philadelphia)* **78,** 1149–1167.

23. Groden, J., *et al.* (1991). Identification and characterization of the familial adenomatous polyposis coli gene. *Cell (Cambridge, Mass.)* **66,** 589–600.

24. Marra, G., and Boland, C. R. (1995). Hereditary nonpolyposis colorectal cancer: The syndrome, the genes, and historical perspectives. *J. Natl. Cancer Inst.* **87,** 1114–1125.

25. Korelitz, B. I. (1995). Inflammatory bowel disease: Cancer. *In* "Bockus Gastroenterology" (W. Haubrich, F. Schaffner, and J. Berk, eds.), 5th ed., pp. 1364–1373. Saunders, Philadelphia.

26. Markowitz, A. J., and Winawer, S. J. (1997). Management of colorectal polyps. *Ca—Cancer J. Clin.* **47,** 93–112.

27. Devesa, S. S., and Chow, W. H. (1993). Variation in colorectal cancer incidence in the United States by subsite of origin. *Cancer (Philadelphia)* **71,** 3819–3826.

28. Haenszel, W., and Kurihara, M. (1968). Studies of Japanese migrants. I. Mortality from cancer and other diseases among Japanese in the United States. *J. Natl. Cancer Inst. (U.S.)* **40,** 43–68.

29. Doll, R., and Peto, R. (1981). The causes of cancer: Quantitative estimates of avoidable risks of cancer in the United States today. *JNCI, J. Natl. Cancer Inst.* **66,** 1191–1308.

30. Willett, W. C. (1990). "Nutritional Epidemiology." Oxford University Press, New York.

31. Slattery, M. L., *et al.* (1988). Diet and colon cancer: Assessment of risk by fiber type and food source. *J. Natl. Cancer Inst.* **80,** 1474–1480.

32. Freudenheim, J. L., *et al.* (1990). Risks associated with source of fiber and fiber components in cancer of the colon and rectum. *Cancer Res.* **50,** 3295–3300.

33. Benito, E., *et al.* (1990). A population-based case-control study of colorectal cancer in Majorca. I. Dietary factors. *Int. J. Cancer* **45,** 69–76.

34. Benito, E., *et al.* (1991). Nutritional factors in colorectal cancer risk: A case-control study in Majorca. *Int. J. Cancer* **49,** 161–167.

35. Hu, J., *et al.* (1991). Diet and cancer of the colon and rectum: A case-control study in China. *Int. J. Epidemiol.* **20,** 362–367.

36. Arbman, G., *et al.* (1992). Cereal fiber, calcium, and colorectal cancer. *Cancer (Philadelphia)* **69,** 2042–2048.

37. Iscovich, J. M., *et al.* (1992). Colon cancer in Argentina. I: Risk from intake of dietary items. *Int. J. Cancer* **51,** 851–857.

38. Meyer, F., and White, E. (1993). Alcohol and nutrients in relation to colon cancer in middle-aged adults. *Am. J. Epidemiol.* **138,** 225–236.

39. Zaridze, D., *et al.* (1993). Diet and colorectal cancer: Results of two case-control studies in Russia. *Eur. J. Cancer* **29A,** 112–115.

40. Ghadirian, P., *et al.* (1997). Nutritional factors and colon carcinoma: A case-control study involving French Canadians in Montréal, Quebec, Canada. *Cancer (Philadelphia)* **80,** 858–864.

41. Trock, B., Lanza, E., and Greenwald, P. (1990). Dietary fiber, vegetables, and colon cancer: Critical review and meta-analyses of the epidemiologic evidence. *J. Natl. Cancer Inst.* **82,** 650–661.

42. Wu, A. H., *et al.* (1987). Alcohol, physical activity and other risk factors for colorectal cancer: A prospective study. *Br. J. Cancer* **55,** 687–694.

43. Giovannucci, E., and Willett, W. C. (1994). Dietary factors and risk of colon cancer. *Ann. Med.* **26,** 443–452.

44. Steinmetz, K. A., *et al.* (1994). Vegetables, fruit, and colon cancer in the Iowa Women's Health Study. *Am. J. Epidemiol.* **139,** 1–15.

45. Kato, I., *et al.* (1997). Prospective study of diet and female colorectal cancer: The New York University Women's Health Study. *Nutr. Cancer* **28,** 276–281.

46. Macquart-Moulin, G., *et al.* (1987). Colorectal polyps and diet: A case-control study in Marseilles. *Int. J. Cancer* **40,** 179–188.

47. Kato, I., *et al.* (1990). A comparative case-control study of colorectal cancer and adenoma. *Jpn. J. Cancer Res.* **81,** 1101–1108.

48. Kune, G. A., *et al.* (1991). Colorectal polyps, diet, alcohol, and family history of colorectal cancer: A case-control study. *Nutr. Cancer* **16,** 25–30.

49. Benito, E., *et al.* (1993). Diet and colorectal adenomas: A case-control study in Majorca. *Int. J. Cancer* **55,** 213–219.

50. Sandler, R. S., *et al.* (1993). Diet and risk of colorectal adenomas: Macronutrients, cholesterol, and fiber. *J. Natl. Cancer Inst.* **85,** 884–891.

51. Smith, S. A., *et al.* (1995). Vegetable and fruit consumption and adenomatous polyps: The University of Minnesota Cancer Prevention Research Unit Case-control Study. *Proc. Am. Assoc. Cancer Res.* **36,** 286.

52. Witte, J. S., *et al.* (1996). Relation of vegetable, fruit, and grain consumption to colorectal adenomatous polyps. *Am. J. Epidemiol.* **144,** 1015–1025.

53. Platz, E. A., *et al.* (1997). Dietary fiber and distal colorectal adenoma in men. *Cancer Epidemiol., Biomarkers Prev.* **6,** 661–670.

54. Kono, S., *et al.* (1993). Relationship of diet to small and large adenomas of the sigmoid colon. *Jpn. J. Cancer Res.* **84,** 13–19.

55. Little, J., *et al.* (1993). Colorectal adenomas and diet: A case-control study of subjects participating in the Nottingham faecal occult blood screening programme. *Br. J. Cancer* **67,** 177–184.

56. Neugut, A. I., *et al.* (1993). Dietary risk factors for the incidence and recurrence of colorectal adenomatous polyps. *Ann. Intern. Med.* **118,** 91–95.

57. Burkitt, D. P. (1971). Epidemiology of cancer of the colon and rectum. *Cancer (Philadelphia)* **28,** 3–13.

58. Trowell, H. (1974). Definitions of fibre. *Lancet* **1,** 503.

59. Lanza, E., and Butrum, R. R. (1986). A critical review of food fiber analysis and data. *J. Am. Diet. Assoc.* **86,** 732–740.

60. Cranston, D., McWhinnie, D., and Collin, J. (1988). Dietary fibre and gastrointestinal disease. *Br. J. Surg.* **75,** 508–512.

61. Eastwood, M. (1987). Physiological properties of dietary fibre. *Mol. Aspects Med.* **9,** 31–40.

62. Augeron, C., and Laboisse, C. L. (1984). Emergence of permanently differentiated cell clones in a human colonic cancer cell line in culture after treatment with sodium butyrate. *Cancer Res.* **44,** 3961–3969.

63. Chomchai, C., Bhadrachari, N., and Nigro, N. D. (1974). The effect of bile on the induction of experimental intestinal tumors in rats. *Dis. Colon Rectum* **17,** 310–312.

64. Narisawa, T., *et al.* (1974). Promoting effect of bile acids on colon carcinogenesis after intrarectal instillation of N-methyl-N'-nitro-N-nitrosoguanidine in rats. *J. Natl. Cancer Inst. (U.S.)* **53,** 1093–1097.

65. Jenkins, D. J. A., *et al.* (1993). Effect on blood lipids of very high intakes of fiber in diets low in saturated fat and cholesterol. *N. Engl. J. Med.* **329,** 21–26.

66. Eastwood, M. A. (1992). The physiological effect of dietary fiber: An update. *Annu. Rev. Nutr.* **12,** 19–35.

67. Giovannucci, E., *et al.* (1992). Relationship of diet to risk of colorectal adenoma in men. *J. Natl. Cancer Inst.* **84,** 91–98.

68. White, E., Shannon, J. S., and Patterson, R. E. (1997). Relationship between vitamin and calcium supplement use and colon cancer. *Cancer Epidemiol., Biomarkers Prev.* **6,** 769–774.

69. Shikany, J. M., *et al.* (1997). Plasma carotenoids and the prevalence of adenomatous polyps of the distal colon and rectum. *Am. J. Epidemiol.* **145,** 552–557.

70. Verhoeven, D. T. H., *et al.* (1996). Epidemiological studies on brassica vegetables and cancer risk. *Cancer Epidemiol., Biomarkers Prev.* **5,** 733–748.

71. Knekt, P., *et al.* (1997). Dietary flavonoids and the risk of lung cancer and other malignant neoplasms. *Am. J. Epidemiol.* **146,** 223–230.

72. Armstrong, B., and Doll, R. (1975). Environmental factors and cancer incidence and mortality in different countries, with special reference to dietary practices. *Int. J. Cancer* **15**, 617–631.

73. Rose, D., Boyar, A., and Wynder, E. (1986). International comparisons of mortality rates for cancer of the breast, ovary, prostate, and colon, and *per capita* food consumption. *Cancer (Philadelphia)* **58**, 2363–2371.

74. Willett, W., *et al.* (1990). Relation of meat, fat, and fiber intake to the risk of colon cancer in a prospective study among women. *N. Engl. J. Med.* **323**, 1664–1672.

75. Giovannucci, E., *et al.* (1994). Intake of fat, meat, and fiber in relation to risk of colon cancer in men. *Cancer Res.* **54**, 2390–2397.

76. Bostick, R., *et al.* (1994). Sugar, meat, and fat intake, and nondietary risk factors for colon cancer incidence in Iowa women (United States). *Cancer Causes Control* **5**, 38–52.

77. Goldbohm, R., *et al.* (1994). A prospective cohort study on the relation between meat consumption and the risk of colon cancer. *Cancer Res.* **54**, 718–723.

78. Vineis, P., and McMichael, A. (1996). Interplay between heterocyclic amines in cooked meat and metabolic phenotype in the etiology of colon cancer. *Cancer Causes Control* **7**, 479–486.

79. Nagao, M., and Sugimura, T. (1993). Carcinogenic factors in food with relevance to colon cancer development. *Muta. Res.* **290**, 43–51.

80. Gerhardsson de Verdier, M., *et al.* (1991). Meat, cooking methods and colorectal cancer: A case-referent study in Stockholm. *Int. J. Cancer* **49**, 520–525.

81. Knekt, P., *et al.* (1994). Intake of fried meat and risk fo cancer: A follow-up study in Finland. *Int. J. Cancer* **59**, 756–760.

82. Mason, J. (1994). Folate and colonic carcinogenesis: Searching for a mechanistic understanding. *J. Nutr. Biochem.* **5**, 170–175.

83. Barak, A., *et al.* (1987). Effects of prolonged ethanol feeding on methionine metabolism in rat liver. *Biochem. Cell Biol.* **65**, 230–233.

84. Garro, A., *et al.* (1991). Ethanol consumption inhibits fetal DNA methylation in mice: Implications for the fetal alcohol syndrome. *Alcohol.: Clin. Exp. Res.* **15**, 395–398.

85. Garro, A. J., *et al.* (1991). Ethanol consumption inhibits fetal DNA methylation in mice: implications for the fetal alcohol syndrome. *Alcohol.: Clin. Exp. Res.* **15**, 395–398.

86. Shaw, S., *et al.* (1989). Cleavage of folates during ethanol metabolism. *Biochem. J.* **257**, 277–280.

87. Freudenheim, J. L., *et al.* (1991). Folate intake and carcinogenesis of the colon and rectum. *Int. J. Epidemiol.* **20**, 368–374.

88. Giovannucci, E., *et al.* (1993). Folate, methionine, and alcohol intake and risk of colorectal adenoma. *J. Natl. Cancer Inst.* **85**, 875–884.

89. Ferraroni, M., *et al.* (1994). Selected micronutrient intake and the risk of colorectal cancer. *Br. J. Cancer* **70**, 1150–1155.

90. Giovannucci, E., *et al.* (1995). Alcohol, low-methionine—low-folate diets, and risk of colon cancer in men. *J. Natl. Cancer Inst.* **87**, 265–273.

91. Glynn, S. A., *et al.* (1996). Colorectal cancer and folate status: A nested case-control study among male smokers. *Cancer Epidemiol., Biomarkers Prev.* **5**, 487–494.

92. Tseng, M., *et al.* (1996). Micronutrients and the risk of colorectal adenomas. *Am. J. Epidemiol.* **144**, 1005–1014.

93. Bird, C. L., *et al.* (1995). Red cell and plasma folate, folate consumption, and the risk of colorectal adenomatous polyps. *Cancer Epidemiol., Biomarkers Prev.* **4**, 709–714.

94. Kune, G. A. (1996). Alcohol consumption. *In* "Causes and Control of Colorectal Cancer," pp. 117–138. Kluwer Academic Publishers, Boston.

95. Chyou, P. H., Nomura, A. M. Y., and Stemmermann, G. N. (1996). A prospective study of colon and rectal cancer among Hawaii Japanese men. *Ann. Epidemiol.* **6**, 276–282.

96. Kune, S., Kune, G. A., and Watson, L. F. (1987). Case-control study of alcoholic beverages as etiological factors: The Melbourne Colorectal Cancer Study. *Nutr. Cancer* **9**, 43–56.

97. Boutron, M.-C., *et al.* (1995). Tobacco, alcohol, and colorectal tumors: A multistep process. *Am. J. Epidemiol.* **141**, 1038–1046.

98. Newmark, H. L., Wargovich, M. J., and Bruce, W. R. (1984). Colon cancer and dietary fat, phosphate, and calcium: A hypothesis. *JNCI, J. Natl. Cancer Inst.* **72**, 1323–1325.

99. Garland, C., *et al.* (1985). Dietary vitamin D and calcium and risk of colorectal cancer: A 19-year prospective study in men. *Lancet* **1**, 307–309.

100. Kune, S., Kune, G. A., and Watson, L. F. (1987). Case-control study of dietary etiological factors: The Melbourne Colorectal Cancer Study. *Nutr. Cancer* **9**, 21–42.

101. Slattery, M. L., Sorenson, A. W., and Ford, M. H. (1988). Dietary calcium intake as a mitigating factor in colon cancer. *Am. J. Epidemiol.* **128**, 504–514.

102. Stemmermann, G. N., Nomura, A., and Chyou, P.-H. (1990). The influence of dairy and nondairy calcium on subsite large-bowel cancer risk. *Dis. Colon Rectum* **33**, 190–194.

103. Whittemore, A. S., *et al.* (1990). Diet, physical activity, and colorectal cancer among Chinese in North America and China. *J. Natl. Cancer Inst.* **82**, 915–926.

104. Peters, R. K., *et al.* (1992). Diet and colon cancer in Los Angeles County, California. *Cancer Causes Control* **3**, 457–473.

105. Kampman, E., *et al.* (1994). Calcium, vitamin D, dairy foods, and the occurrence of colorectal adenomas among men and women in two prospective studies. *Am. J. Epidemiol.* **139**, 16–29.

106. Boutron, M.-C., *et al.* (1996). Calcium, phosphorus, vitamin D, dairy products and colorectal carcinogenesis: A French case-control study. *Br. J. Cancer* **74**, 145–151.

107. Bergsma-Kadijk, J. A., *et al.* (1996). Calcium does not protect against colorectal neoplasia. *Epidemiology* **7**, 590–597.

108. Lointier, P., *et al.* (1987). The role of vitamin D_3 in the proliferation of a human colon cancer cell line *in vitro*. *Anticancer Res.* **7**, 817–822.

109. Thomas, M. G., Tebbutt, S., and Williamson, R. C. N. (1992). Vitamin D and its metabolites inhibit cell proliferation in human rectal mucosa and a colon cancer cell line. *Gut* **33**, 1660–1663.

110. Cross, H. S., Farsoudi, K. H., and Peterlik, M. (1993). Growth inhibition of human colon adenocarcinoma-derived Caco-2 cells by 1,25-dihydroxyvitamin D_3 and two synthetic analogs: Relation to in vitro hypercalcemic potential. *Naunyn-Scyhmiedeberg's Arch. Pharmacol.* **347**, 105–110.

111. Shabahang, M., *et al.* (1994). Growth inhibition of HT-29 human colon cancer cells by analogues of 1,25-dihydroxyvitamin D_3. *Cancer Res.* **54**, 4057–4064.

112. Eisman, J. A., Barkla, D. H., and Tutton, P. J. M. (1987). Suppression of *in vivo* growth of human cancer solid tumor xenografts by 1,25-dihydroxyvitamin D_3. *Cancer Res.* **47**, 21–25.

113. Garland, C. F., and Garland, F. C. (1980). Do sunlight and vitamin D reduce the likelihood of colon cancer? *Int. J. Epidemiol.* **9**, 227–231.

114. Spitz, M. R., Paolucci, M. J., and Newell, G. R. (1988). Occurrence of colorectal cancer: Focus on Texas. *Cancer Bull.* **40**, 187–191.

115. Emerson, J. C., and Weiss, N. S. (1992). Colorectal cancer and solar radiation. *Cancer Causes Control* **3**, 95–99.

116. Gorham, E. D., Garland, C. F., and Garland, F. C. (1989). Acid haze air pollution and breast and colon cancer mortality in 20 Canadian cities. *Can. J. Public Health.* **80**, 96–100.

117. Garland, C. F., *et al.* (1989). Serum 25-hydroxyvitamin D and colon cancer: Eight-year prospective study. *Lancet* **2**, 1176–1178.

118. Tangrea, J., *et al.* (1997). Serum levels of vitamin D metabolites and the subsequent risk of colon and rectal cancer in Finnish men. *Cancer Causes Control* **8**, 615–625.

119. Braun, M. M., *et al.* (1995). Colon cancer and serum vitamin D metabolite levels 10–17 prior to diagnosis. *Am. J. Epidemiol.* **142,** 608–611.

120. Macquart-Moulin, G., *et al.* (1986). Case-control study on colorectal cancer and diet in Marseilles. *Int. J. Cancer* **38,** 183–191.

121. Shannon, J., *et al.* (1996). Relationship of food groups and water intake to colon cancer risk. *Cancer Epidemiol., Biomarkers Prev.* **5,** 495–502.

122. Miller, A. B., *et al.* (1983). Food items and food groups as risk factors in a case-control study of diet and colo-rectal cancer. *Int. J. Cancer* **32,** 155–161.

123. Young, T. B., and Wolf, D. A. (1988). Case-control study of proximal and distal colon cancer and diet in Wisconsin. *Int. J. Cancer* **42,** 167–175.

124. Giovannucci, E. (1995). Insulin and colon cancer. *Cancer Causes Control* **6,** 164–179.

125. Tran, T. T., Medline, A., and Bruce, W. R. (1996). Insulin promotion of colon tumors in rats. *Cancer Epidemiol., Biomarkers Prev.* **5,** 1013–1015.

126. McKeown-Eyssen, G. E., and the Toronto Polyp Prevention Group (1996). Insulin resistance and the risk of colorectal neoplasia. *Cancer Epidemiol., Biomarkers Prev.* **5,** 235.

127. Riccardi, G., and Rivellese, A. A. (1991). Effects of dietary fiber and carbohydrate on glucose and lipoprotein metabolism in diabetic patients. *Diabetes Care* **14,** 1115–1125.

128. Wurzelmann, J. I., *et al.* (1996). Iron intake and the risk of colorectal cancer. *Cancer Epidemiol., Biomarkers Prev.* **5,** 503–507.

129. Tuyns, A. J., Haelterman, M., and Kaaks, R. (1987). Colorectal cancer and the intake of nutrients: Oligosaccharides are a risk factor, fats are not. A case-control study in Belgium. *Nutr. Cancer* **10,** 181–196.

130. Freudenheim, J. L., *et al.* (1990). A case-control study of diet and rectal cancer in western New York. *Am. J. Epidemiol.* **131,** 612–624.

131. Herrinton, L. J., *et al.* (1995). Transferrin saturation and risk of cancer. *Am. J. Epidemiol.* **142,** 692–698.

132. Bird, C. L., *et al.* (1996). Plasma ferritin, iron intake, and the risk of colorectal polyps. *Am. J. Epidemiol.* **144,** 34–41.

133. Kritchevsky, D. (1993). Colorectal cancer: The role of dietary fat and caloric restriction. *Muta. Res.* **290,** 63–70.

134. Giovannucci, E., *et al.* (1996). Physical activity, obesity and risk of colorectal adenoma in women. *Cancer Causes Control* **7,** 253–263.

135. Giovannucci, E., *et al.* (1995). Physical activity, obesity, and risk for colon cancer and adenoma in men. *Ann. Intern. Med.* **122,** 327–334.

136. Martínez, M. E., *et al.* (1997). Leisure-time physical activity, body size, and colon cancer in women. *J. Natl. Cancer Inst.* **89,** 948–955.

137. Koivisto, V. A., Yki-Jarvinen, H., and DeFronzo, R. A. (1988). Physical training and insulin sensitivity. *Diabetes Metab. Rev.* **1,** 445–481.

138. Sarna, S. K. (1991). Physiology and pathophysiology of colonic motor activity. II. *Dig. Dis. Sci.* **36,** 998–1018.

139. Colditz, G. A., Cannuscio, C. C., and Frazier, A. L. (1997). Physical activity and reduced risk of colon cancer: Implications for prevention. *Cancer Causes Control* **8,** 649–667.

140. Chasen-Taber, S., *et al.* (1996). Reproducibility and validity of a self-administered physical activity questionnaire for male health professionals. *Epidemiology* **7,** 1–6.

141. Giovannucci, E., *et al.* (1994). A prospective study of cigarette smoking and risk of colorectal adenoma and colorectal cancer in U.S. men. *J. Natl. Cancer Inst.* **86,** 183–191.

142. Giovannucci, E., *et al.* (1994). A prospective study of cigarette smoking and risk of colorectal adenoma and colorectal cancer in U.S. women. *J. Natl. Cancer Inst.* **86,** 192–199.

143. Newcomb, P. A., Storer, B. E., and Marcus, P. M. (1995). Cigarette smoking in relation to risk of large bowel cancer in women. *Cancer Res.* **55,** 4906–4909.

144. Martinez, M. E., *et al.* (1995). Cigarette smoking and alcohol consumption as risk factors for colorectal adenomatous polyps. *J. Natl. Cancer Inst.* **87,** 274–279.

145. Rosenberg, L., *et al.* (1991). A hypothesis: Nonsteroidal anti-inflammatory drugs reduce the incidence of large-bowel cancer. *J. Natl. Cancer Inst.* **83,** 355–358.

146. Thun, M. J., Namboodiri, N. M., and Heath, C. W., Jr. (1991). Aspirin use and reduced risk of fatal colon cancer. *N. Engl. J. Med.* **325,** 1593–1596.

147. Suh, O., Mettlin, C., and Petrelli, N. J. (1993). Aspirin use, cancer, and polyps of the large bowel. *Cancer (Philadelphia)* **72,** 1171–1177.

148. Greenberg, E. R., *et al.* (1993). Reduced risk of large-bowel adenomas among aspirin users. *J. Natl. Cancer Inst.* **85,** 912–916.

149. Muscat, J. E., Stellman, S. D., and Wynder, E. L. (1994). Nonsteroidal antiinflammatory drugs and colorectal cancer. *Cancer (Philadelphia)* **74,** 1847–1854.

150. Giovannucci, E., *et al.* (1995). Aspirin use and the risk of colorectal cancer in women. *N. Engl. J. Med.* **333,** 609–614.

151. Reeves, M. J., *et al.* (1996). Nonsteroidal anti-inflammatory drug use and protection against colorectal cancer in women. *Cancer Epidemiol., Biomarkers Prev.* **5,** 955–960.

152. Gann, P. H., *et al.* (1993). Low-dose aspirin and incidence of colorectal tumors in a randomized trial. *J. Natl. Cancer Inst.* **85,** 1220–1224.

153. Giovannucci, E., *et al.* (1994). Aspirin use and the risk of colorectal cancer and adenoma in male health professionals. *Ann. Intern. Med.* **121,** 241–246.

154. Ernster, V. L., *et al.* (1979). Cancer incidence by marital status: U.S. Third National Cancer Survey. *JNCI, J. Natl. Cancer Inst.* **63,** 567–585.

155. Fraumeni Jr, J. F., *et al.* (1969). Cancer mortality among nuns: Role of marital status in etiology of neoplastic disease in women. *J. Natl. Cancer Inst. (U.S.)* **42,** 455–468.

156. McMichael, A. J., and Potter, J. D. (1980). Reproduction, endogenous and exogenous sex hormones, and colon cancer: A review and hypothesis. *JNCI, J. Natl. Cancer Inst.* **65,** 1201–1207.

157. Goldfien, A., and Monroe, S. E. (1994). Ovaries. In "Basic and Clinical Endocrinology" (F. S. Greenspan and J. D. Baxter, eds.), 4th ed., pp. 419–470. Appleton & Lange, Norwalk, CT.

158. Alford, T. C., *et al.* (1979). Steroid hormone receptors in human colon cancers. *Cancer (Philadelphia)* **43,** 980–984.

159. Francavilla, A., *et al.* (1987). Nuclear and cytosolic estrogen receptors in human colon carcinoma and in surrounding noncancerous colonic tissue. *Gastroenterology* **93,** 1301–1306.

160. Papadimitriou, C., *et al.* (1984). Biosocial correlates of colorectal cancer in Greece. *Int. J. Epidemiol.* **13,** 155–159.

161. Kampman, E., *et al.* (1997). Hormone replacement therapy, reproductive history, and colon cancer: A multicenter, case-control study in the United States. *Cancer Causes Control* **8,** 146–158.

162. Kampman, E., *et al.* (1994). Reproductive and hormonal factors in male and female colon cancer. *Eur. J. Cancer Prev.* **3,** 329–336.

163. Negri, E., *et al.* (1989). Reproductive and menstrual factors and risk of colorectal cancer. *Cancer Res.* **49,** 7158–7161.

164. Wu-Williams, A. H., *et al.* (1991). Reproductive factors and colorectal cancer risk among Chinese females. *Cancer Res.* **51,** 2307–2311.

165. Gerhardsson de Verdier, M., and London, S. (1992). Reproductive factors, exogenous female hormones, and colorectal cancer by subsite. *Cancer Causes Control* **3,** 335–360.

166. Olsen, J., Olsen, J. S., and Kronborg, O. (1994). Risk factors for cancers and adenomas of the large intestine. An analysis of food items and reproductive factors. *Cancer J.* **7,** 103–107.

167. Martínez, M. E., *et al.* (1997). A prospective study of reproductive factors, oral contraceptive use, and risk of colorectal cancer. *Cancer Epidemiol., Biomarkers Prev.* **6,** 1–5.

168. Haenszel, W., Locke, F. B., and Segi, M. (1980). A case-control study of large bowel cancer in Japan. *JNCI, J. Natl. Cancer Inst.* **64,** 17–22.

169. Peters, R. K., *et al.* (1990). Reproductive factors and colon cancers. *Br. J. Cancer* **61,** 741–748.

170. Chute, C. G., *et al.* (1991). A prospective study of reproductive history and exogenous estrogens on the risk of colorectal cancer in women. *Epidemiology* **2,** 201–207.

171. Franceschi, S., *et al.* (1991). Colorectal cancer in northeast Italy: Reproductive, menstrual and female hormone-related factors. *Eur. J. Cancer* **27,** 604–608.

172. Kvle, G., and Heuch, I. (1991). Is the incidence of colorectal cancer related to reproduction? A prospective study of 63,000 women. *Int. J. Cancer* **47,** 390–395.

173. Marcus, P. M., *et al.* (1995). The association of reproductive and menstrual characteristics and colon and rectal cancer risk in Wisconsin women. *Ann. Epidemiol.* **5,** 303–309.

174. Jacobson, J. S., *et al.* (1995). Reproductive risk factors for colorectal adenomatous polyps (New York City, NY, United States). *Cancer Causes Control* **6,** 513–518.

175. Platz, E. A., *et al.* (1997). Parity and other reproductive factors and risk of adenomatous polyps of the distal colorectum (United States). *Cancer Causes Control* **8,** 894–903.

176. Dales, L. G., *et al.* (1979). A case-control study of relationships of diet and other traits to colorectal cancer in American blacks. *Am. J. Epidemiol.* **109,** 132–144.

177. Miller, A. B., *et al.* (1980). A study of cancer, parity and age at first pregnancy. *J. Chronic. Dis.* **33,** 595–605.

178. Weiss, N. S., Daling, J. R., and Chow, W. H. (1981). Incidence of cancer of the large bowel in women in relation to reproductive and hormonal factors. *JNCI, J. Natl. Cancer Inst.* **67,** 57–60.

179. Cantor, K. P., Lynch, C. F., and Johnson, D. (1993). Reproductive factors and risk of brain, colon, and other malignancies in Iowa (United States). *Cancer Causes Control* **4,** 505–511.

180. La Vecchia, C., *et al.* (1993). Long-term impact of reproductive factors on cancer risk. *Int. J. Cancer* **53,** 215–219.

181. Slattery, M. L., Mineau, G. P., and Kerber, R. A. (1995). Reproductive factors and colon cancer: The influences of age, tumor site, and family history on risk (Utah, United States). *Cancer Causes Control* **6,** 332–338.

182. Potter, J. D., and McMichael, A. J. (1983). Large bowel cancer in women in relation to reproductive and hormonal factors: A Case-control Study. *JNCI, J. Natl. Cancer Inst.* **71,** 703–709.

183. Howe, G. R., Craib, K. J. P., and Miller, A. B. (1985). Age at first pregnancy and risk of colorectal cancer: A Case-control Study. *JNCI, J. Natl. Cancer Inst.* **74,** 1155–1159.

184. Furner, S. E., *et al.* (1989). A case-control study of large bowel cancer and hormone exposure in women. *Cancer Res.* **49,** 4936–4940.

185. Kune, G. A., Kune, S., and Watson, L. F. (1989). Children, age at first birth, and colorectal cancer risk. *Am. J. Epidemiol.* **129,** 533–542.

186. Kravdal, Ø., *et al.* (1993). A sub-site-specific analysis of the relationship between colorectal cancer and parity in complete male and female Norwegian birth cohorts. *Int. J. Cancer* **53,** 56–61.

187. Jacobs, E. J., White, E., and Weiss, N. S. (1994). Exogenous hormones, reproductive history, and colon cancer (Seattle, Washington, USA). *Cancer Causes Control* **5,** 359–366.

188. Potter, J. D., *et al.* (1996). Hormone replacement therapy is associated with lower risk of adenomatous polyps of the large bowel: The Minnesota Cancer Prevention Research Unit Case-control Study. *Cancer Epidemiol., Biomarkers Prev.* **5,** 779–784.

189. Bjelke, E. (1975). Colorectal cancer: Clues from epidemiology. *Int. Congr. Ser. 354—Excerpta Med.* **6,** 324–330.

190. McMichael, A. J., and Potter, J. D. (1984). Parity and death from colon cancer in women: A Case-control Study. *Commun. Health Stud.* **8,** 19–25.

191. Brinton, L. A., *et al.* (1989). Cancer risk after evaluation for infertility. *Am. J. Epidemiol.* **129,** 712–722.

192. McKeown-Eyssen, G. (1994). Epidemiology of colorectal cancer revisited: Are serum triglycerides and/or plasma glucose associated with risk? *Cancer Epidemiol., Biomarkers Prev.* **3,** 687–695.

193. McPartlin, J., *et al.* (1993). Accelerated folate breakdown in pregnancy. *Lancet* **341,** 148–149.

194. Necheles, T. F., and Snyder, L. M. (1970). Malabsorption of folate polyglutamates associated with oral contraceptive therapy. *N. Engl. J. Med.* **282,** 858–859.

195. Calle, E. E., *et al.* (1995). Estrogen replacement therapy and risk of fatal colon cancer in a prospective cohort of postmenopausal women. *J. Natl. Cancer Inst.* **87,** 517–523.

196. Newcomb, P. A., and Storer, B. E. (1995). Postmenopausal hormone use and risk of large-bowel cancer. *J. Natl. Cancer Inst.* **87,** 1067–1071.

197. Grodstein, F., *et al.* (1998). Postmenopausal hormone use and risk for colorectal cancer and adenoma. *Ann. Intern. Med.* **128,** 705–712.

198. Risch, H. A., and Howe, G. R. (1995). Menopausal hormone use and colorectal cancer in Saskatchewan: A Record Linkage Cohort Study. *Cancer Epidemiol., Biomarkers Prev.* **4,** 21–28.

199. Potter, J. D. (1995). Hormones and colon cancer. *J. Natl. Cancer Inst.* **87,** 1039–1040.

200. Barrett-Connor, E., and Laakso, M. (1990). Ischemic heart disease risk in postmenopausal women: Effects of estrogen use on glucose and insulin levels. *Arteriosclerosis* **10,** 531–534.

201. The Writing Group for the PEPI Trial (1995). Effects of estrogen or estrogen/progestin regimens on heart disease risk factors in postmenopausal women: The Postmenopausal Estrogen/Progestin Interventions (PEPI) Trial. *JAMA, J. Am. Med. Assoc.* **273,** 199–208.

202. Roberts-Thomson, I. C., *et al.* (1996). Diet, acetylator phenotype, and risk of colorectal neoplasia. *Lancet* **347,** 1372–1374.

203. Bell, D. A., *et al.* (1995). Polyadenylation polymorphism in the acetyltransferase 1 gene (*NAT1*) increases risk of colorectal cancer. *Cancer Res.* **55,** 3537–3542.

204. Probst-Hensch, N. M., *et al.* (1995). Acetylation polymorphism and prevalence of colorectal adenomas. *Cancer Res.* **55,** 2017–2020.

205. Probst-Hensch, N. M., *et al.* (1996). Lack of association between the polyadenylation polymorphism in the *NAT1* (acetyltransferase 1) gene and colorectal adenomas. *Carcinogenesis (London)* **17,** 2125–2129.

206. Katoh, T., *et al.* (1996). Glutathione S-transferase M1 (GSTM1) and T1 (GSTT1) genetic polymorphism and susceptibility to gastric and colorectal adenocarcinoma. *Carcinogenesis (London)* **17,** 1855–1859.

207. Lin, H. J., *et al.* (1995). Glutathione transferase (GSTM1) null genotype, smoking, and prevalence of colorectal adenomas. *Cancer Res.* **55,** 1224–1226.

208. Deakin, M., *et al.* (1996). Glutathione S-transferase GSTT1 genotypes and susceptibility to cancer: Studies of interactions with GSTM1 in lung, oral, gastric and colorectal cancers. *Carcinogenesis (London)* **17,** 881–884.

209. Chen, J., *et al.* (1996). A methylenetetrahydrofolate reductase polymorphism and the risk of colorectal cancer. *Cancer Res.* **56,** 4862–4864.

210. Ma, J., *et al.* (1997). Methylenetetrahydrofolate reductase polymorphism, dietary interactions, and risk of colorectal cancer. *Cancer Res.* **57,** 1098–1102.

211. Foppa, I., and Spiegelman, D. (1997). Power and sample size calculations for case-control studies of gene-environment interactions

with a polytomous exposure variable. *Am. J. Epidemiol.* **146,** 596–604.

212. Yang, Q., and Khoury, M. J. (1997). Evolving methods in genetic epidemiology. III. Gene-environment interaction in epidemiologic research. *Epidemiol. Rev.* **19,** 33–43.

213. Holtzman, N. A., and Andrews, L. B. (1997). Ethical and legal issues in genetic epidemiology. *Epidemiol. Rev.* **19,** 163–174.

214. Byers, T., *et al.* (1997). American Cancer Society guidelines for screening and surveillance for early detection of colorectal polyps and cancer: Update 1997. *Ca—Cancer J. Clin.* **47,** 154–160.

215. Schatzkin, A., *et al.* (1996). The Polyp Prevention Trial I: Rationale, design, recruitment, and baseline participant characteristics. *Cancer Epidemiol., Biomarkers Prev.* **5,** 375–383.

216. Chlebowski, R. T., and Grosvenor, M. (1994). The scope of nutrition intervention trials with cancer-related endpoints. *Cancer (Philadelphia)* **74,** 2734–2738.

217. Hennekens, C. H., *et al.* (1996). Lack of effect of long-term supplementation with beta carotene on the incidence of malignant neoplasms and cardiovascular disease. *N. Engl. J. Med.* **334,** 1145–1149.

218. The Alpha-Tocopherol Beta Carotene Cancer Prevention Study Group (1994). The effect of vitamin E and beta carotene on the incidence of lung cancer and other cancers in male smokers. *N. Engl. J. Med.* **330,** 1029–1035.

219. Greenberg, E. R., *et al.* (1994). A clinical trial of antioxidant vitamins to prevent colorectal adenoma. *N. Engl. J. Med.* **331,** 141–147.

220. Frommel, T. O., *et al.* (1995). Effect of β-carotene supplementation on indices of colonic cell proliferation. *J. Natl. Cancer Inst.* **87,** 1781–1787.

221. MacLennan, R., *et al.* (1995). Randomized trial of intake of fat, fiber, and beta carotene to prevent colorectal adenomas. *J. Natl. Cancer Inst.* **87,** 1760–1766.

222. Alberts, D. S., *et al.* (1997). The effect of wheat bran fiber and calcium supplementation on rectal mucosal proliferation rates in patients with resected adenomatous colorectal polyps. *Cancer Epidemiol., Biomarkers Prev.* **6,** 161–169.

223. Clark, L. C., *et al.* (1996). Effects of selenium supplementation for cancer prevention in patients with carcinoma of the skin. A randomized controlled trial. Nutritional Prevention of Cancer Study Group. *JAMA, J. Am. Med. Assoc.* **276,** 1957–1963.

224. Garland, M., *et al.* (1995). Prospective study of toenail selenium levels and cancer among women. *J. Natl. Cancer Inst.* **87,** 497–505.

225. Einspahr, J. G., *et al.* (1997). Surrogate end-point biomarkers as measures of colon cancer risk and their use in cancer chemoprevention trials. *Cancer Epidemiol., Biomarkers Prev.* **6,** 37–48.

226. Terpstra, O. T., *et al.* (1987). Abnormal pattern of cell proliferation in the entire colonic mucosa of patients with colon adenoma or cancer. *Gastroenterology* **92,** 704–708.

227. Risio, M., *et al.* (1988). Immunohistochemical study of epithelial cell proliferation in hyperplastic polyps, adenomas, and adenocarcinomas of the large bowel. *Gastroenterology* **94,** 899–906.

228. Risio, M., Candelaresi, G., and Rossini, F. P. (1993). Bromodeoxyuridine uptake and proliferating cell nuclear antigen expression throughout the colorectal tumor sequence. *Cancer Epidemiol., Biomarkers Prev.* **2,** 363–367.

229. Grady, D., *et al.* (1992). Hormone therapy to prevent disease and prolong life in postmenopausal women. *Ann. Intern. Med.* **117,** 1016–1037.

230. Hong, W. K., and Sporn, M. B. (1997). Recent advances in chemoprevention of cancer. *Science* **278,** 1073–1077.

77

Cancer Prevention for Women

KAREN GLANZ AND THOMAS M. VOGT
Cancer Research Center of Hawaii
University of Hawaii
Honolulu, Hawaii

I. Introduction

Half of the cancer deaths in the nation can be prevented if current knowledge about prevention is applied [1]. Cigarette smoking is the leading cause of cancer death [2]. There is a growing consensus that diets high in fat and low in fiber, fruits, and vegetables raise cancer risk and risk for many other diseases as well. Other effective strategies for reducing cancer morbidity and mortality include appropriate screening for breast, cervical, and colorectal cancers, exercise, avoidance of excessive sun exposure, and avoidance of heavy alcohol consumption [3]. Because screening has been discussed in several earlier chapters, this chapter focuses on behavioral and chemopreventive prevention approaches.

In the United States, breast cancer accounts for 30% of all female cancer cases, lung cancer accounts for 13%, and colorectal cancer accounts for 11%. However, lung cancer causes 25% of cancer deaths among women (66,000 per year), breast cancer causes 17% of cancer deaths (43,900 per year), and colorectal cancer causes about 10% of female cancer deaths (27,900 per year) [4,5]. Cancers of the uterus and ovary account for 6% and 4% of cases and 2% and 5% of female cancer deaths, respectively. The remaining 40% of cancer deaths are caused by a wide variety of cancers affecting many different organ systems.

Since 1960, age-adjusted cancer death rates among women have fallen for cancers of the stomach, pancreas, endometrium, and uterus. They have fallen slightly for breast and colorectal cancers and have risen rapidly for lung cancer. Other major forms of cancer have shown little change in recent decades. Because lung, breast, and colorectal cancers account for more than half of all cancer cases and cancer deaths among women, factors contributing to morbidity and mortality from cancers of these sites deserve special emphasis in prevention programs.

This chapter will present a brief overview of the variety of actions that can be taken by women to reduce their cancer risk and discuss strategies for encouraging their adoption. The most remarkable feature of these recommendations is that they are mostly neither new nor technologic. Certain behaviors such as smoking, poor diets, sedentary lifestyle, excessive drinking, and prolonged sun exposure raise cancer risk. Cancer risk can be substantially reduced by modifying these behaviors. Chemopreventive and hormonal strategies also hold promise for preventing some major female cancers. While new findings about cancer biology may one day allow successful treatment of even advanced cancer, at present, prevention holds the greatest promise for avoiding illness, suffering, unnecessary medical costs, and premature death.

II. Cancer Prevention: The Evidence and Status

This section summarizes the evidence for cancer prevention through nutrition and weight control, avoidance of tobacco, regular physical activity, limited consumption of alcohol, sun protection, chemoprevention, and hormone replacement therapy. Each of these types of actions or preventive factors offers opportunity for primary prevention of cancer among women, and many are particularly promising for preventing female cancers. Table 77.1 summarizes the preventive actions/factors, major affected cancer sites, and some special issues for women.

A. Nutrition and Weight Control

Each year, evidence of the influence of nutrition on health is growing [6,7]. The cumulative body of research, which includes animal, laboratory, clinical, and epidemiologic findings, provides compelling evidence of the association of dietary excesses and imbalances with chronic disease [6]. Five of the ten leading causes of death for Americans are associated with dietary practices: heart disease, some cancers, stroke, diabetes, and atherosclerosis [8].

Doll and Peto estimated that 35% of all cancer mortality in the United States is related to diet [9], and increasing evidence indicates that nutrition plays an important role in the initiation, promotion, and progression of cancer. Nutritional factors affect the incidence of cancers of the breast, colon and rectum, lung, and prostate—the leading causes of cancer mortality in the United States both for women and for men [10]. High intake of dietary fat, especially saturated fat, and obesity among postmenopausal women are associated with increased risk of breast and colorectal cancers [11–15]. Adequate intake of fruits, vegetables, and fiber-rich foods appear to have protective effects [16,17]. Fat intake and obesity are also associated with cancers of the endometrium, cervix, pancreas, kidney, and other sites, and fruits and vegetables have been found to reduce risk of cancers of the cervix, oral cavity, bladder, and stomach [6]. Genetics, physical activity, and alcohol consumption are factors that may modulate the effects of nutrition on disease occurrence and progression [10].

The National Cancer Institute has established dietary goals that include a reduction in average consumption of fat to no more than 30% of energy, an increase in fiber consumption to 20–30 grams per day, and increased intake of fruits and vegetables [18,19]. Other, similar dietary guidelines have been released by the U.S. Departments of Agriculture and Health and Human Services [20] and by the American Institute for Cancer Research [21].

Table 77.1

Preventing Cancer among Women: Affected Cancer Sites and Special Issues

Preventive factor or action	Affected cancer site(s)[a]	Special issues for women
Nutrition	Breast	Women's roles in food purchase and preparation
Fat intake (especially saturated fat)	Colorectal	Work roles and eating out
Overweight/obesity	Lung	Risk reduction for other chronic diseases
Fruits and vegetables	Cervix	Interaction with genetics, activity, and alcohol
Fiber		
Tobacco use	Lung	Weight gain due to cessation
	Cervix	Stress management tool, depression self-medicating
		Reproductive effects
		Second-hand smoke and child health
Physical activity	Colon	Role in energy balance and weight control
	Breast	Barriers: time, convenience, social support, self-efficacy, health problems
		Realistic goals of moderate activity now an option
Alcohol consumption	Breast	Balance with possible cardiovascular benefit of
	Colorectal	moderate drinking
Sun protection and skin examination	Skin	Motivation to reduce photoaging or wrinkling
		Skin self-examination may be paired with BSE
		Protecting children may enhance self-protection
Chemoprevention	Breast	Risk/benefit ratio
Tamoxifen	Endometrium	Appropriate prescribing decisions
Hormone replacement therapy (HRT)	Breast	Relief from menopausal symptoms
	Endometrium	Risk/benefit (CVD, osteoporosis) and duration of use

[a]Only major cancer sites for women and female cancers are shown.

During the 1990s, dietary fat intake as a percentage of energy has decreased in the U.S., although concomitant increases in intake of fiber and micronutrients from fruits and vegetables have not occurred [22]. Though women are more likely than men to consume diets that meet cancer prevention guidelines [23], women's eating patterns vary considerably by demographic subgroups and by patterns of eating away from home [24].

The prevalence of overweight among American adults increased by 19% between the 1970s and the early 1990s [25]; more than 50% of all women are now considered overweight or obese, defined as a body mass index (BMI) greater than 25 kg/m². While the prevalence of overweight is higher in black women, the rate of increase is more rapid among white women. The reasons for these disturbing trends are not entirely clear, though sedentary lifestyle along with nutrition appears to play an important role. Indeed, evidence suggests that weight loss maintenance requires a remarkably high level of routine physical activity [26]. The epidemic of obesity has led to some efforts to better address this problem in health care, including an emphasis on treating those with other health risks such as high cholesterol or hypertension and efforts to prevent weight gain that often accompanies aging [27].

Although there remains no proven strategy for preventing breast and colorectal cancers, healthful nutritional practices hold great promise for prevention. Furthermore, dietary recommendations for breast and colorectal cancer prevention are consistent with current population-wide dietary guidelines [18–21]. Thus, the direction of research and health programs has begun to shift into intervention trials that attempt to encourage women to adhere to dietary advice and test the efficacy of good nutrition for prevention. Women remain the principal food shoppers and preparers, making behavior change programs for women all the more promising. Prevention efforts directed toward women, in a variety of research settings, will be discussed further in section IV.

B. Tobacco Use

Tobacco is responsible for about two million deaths each year in developed nations, making it the leading cause of premature death [28]. In the United States, about 66,000 women die annually of tobacco-induced lung cancer [4,5], and 1500 women die from the effects of passive smoking [29]. Smoking cessation reduces the risk of dying from tobacco-related disease [30]. Between 1959 and 1988, the proportion of all deaths among women attributable to tobacco rose 256%, from 18.7% to 47.9% of all deaths [31]. Although the prevalence of smoking declined by about 0.5% per year from 1965 to 1987, the rate of decline in tobacco use in the United States was four times greater among men as compared to women [32]. As a consequence, while lung cancer rates have stabilized for men, they are increasing rapidly among women. Lung cancer is now the leading cause of cancer death among women [30] and is the only common form of cancer that is increasing in women. Indeed, if not for lung cancer, total cancer death rates among women would have declined significantly in the last twenty years.

Smoking is a women's issue. This is underscored not only by the dramatic trends in lung cancer among women, but also by advertising targeted at women, weight-related issues, reproductive issues, and methods for controlling stress.

In most countries, tobacco use began among men. Once that market was saturated, tobacco companies began to target women and children, to expand the current market and assure a future one. Women are more likely than men to use tobacco to cope with stress and because they are concerned about a slender appearance [33]. Tobacco companies recognized these differences years ago, responding with brands like Virginia Slims and ad campaigns portraying female smokers as glamorous, and showing pastoral scenes in Salem and Newport commercials.

Concern about weight gain is an important cause of smoking initiation among young women and of relapse among women trying to stop smoking. Both men and women gain an average of 10 to 12 pounds after quitting, but this gain is more likely to lead to relapse among women [32]. Tobacco companies actually market cigarettes as "weight loss sticks" in some countries, and in the United States, their models clearly convey this message.

Women are more likely than men to say that they use cigarettes to control stress, emotional upsets, loneliness, and depression [34]. Effective alternative coping and stress management strategies are important to helping women quit smoking.

Cigarettes retard fetal growth and development, are associated with miscarriage and other pregnancy problems, and affect a variety of other health problems related to women's reproductive systems, including earlier menopause, excess stroke risk by contraceptive users, and greater risk of osteoporosis [35]. Despite these problems, many women continue to smoke through pregnancy [36], and the majority of those who quit during a pregnancy resume smoking afterwards [37]. Because smoking increases the risk and severity of respiratory ailments among children exposed to second hand smoke, the consequences of maternal smoking are multiplied beyond the mother to her family.

Quitting smoking has powerful and important benefits for women. Smoking cessation has been shown to reduce the risk of heart disease, cancer, and total mortality among women smokers who quit [31]. For example, a group of 1000 women smokers aged 40–44 would expect to experience 30 deaths during the next 14 years if they keep smoking. If they all quit, only 17 deaths would occur, which is a reduction of 43% in risk of dying. Opportunities to assess and intervene with tobacco use are readily available, both within medical systems and in community settings. To achieve significant progress against lung cancer and other tobacco-induced illnesses among women, public policy along with physician and patient expectations and medical support systems need to vigorously address the challenges of smoking prevention and cessation.

C. Physical Activity

Physical activity has numerous beneficial physiologic and psychological effects. Public health guidelines recommend at least a moderate amount of physical activity on a regular basis (*i.e.,* activity that burns about 150 calories/day or about 30 minutes of moderate physical activity per day), which can be achieved through any of a variety of activities [38]. High levels of regular physical activity are associated with lower mortality rates for adults, and even those who are moderately active have lower mortality rates than those who are sedentary [38]. Regular

physical activity lowers the risk of cardiovascular disease and non-insulin-dependent diabetes mellitus, can attenuate or prevent osteoporosis, and helps maintain healthy weight [38,39]. In addition, regular exercise is associated with a decreased risk of colon cancer [40] and breast cancer [41–46]. A possible mechanism that has been suggested for the role of exercise in protection against cancers of the breast and reproductive organs is through reduction of body fat and suppression of circulating hormones [47].

In addition to its well-established impact on physical health, regular exercise can relieve symptoms of depression and anxiety and mood and appears to improve health-related quality of life [38,48,49]. Despite the recognized benefits of physical activity, more than 60% of American adults do not achieve the recommended amount of regular physical activity and 25% are not active at all. Inactivity is more prevalent in women than in men, increases with age, and is inversely associated with education and income levels [38]. Female adolescents are much less physically active than male adolescents [38], which is particularly troubling given that physical activity levels during adolescence and young adulthood are associated with reduced risk of breast cancer [44].

The promulgation of guidelines that advise people to engage in moderate physical activities may make it easier for women to increase their activity levels because they can meet these exercise goals through short bouts of activity throughout the day [50]. Some issues that may deter women's participation in physical activity include lack of time, inconvenience of exercise facilities, lack of social support for women to exercise, low self-confidence for exercise, and, for many older women, health problems [50]. Motivating factors may include enjoyment of exercise and/or sport, weight control, cardiovascular risk reduction, and improved mood [38,51].

For many older women who may not have been encouraged to undertake physically active lifestyles when they were younger, the adoption of regular moderate physical activity may be an attainable and realistic goal. For future generations of women, special initiatives within and outside school may set the stage for better overall health and reduce cancer incidence later in life. Programs targeting both individual factors (cognitions, skills, social support) and environmental factors (accessibility, safety, and natural opportunities to be active) hold promise in promoting physical activity among women [38,50,51].

D. Alcohol Consumption

There is convincing evidence that alcoholic drinks increase the risk of cancers of the mouth, pharynx, larynx, esophagus, and liver, and that smoking multiplies this risk [21]. Alcohol consumption also appears to increase the risk for cancers of the colon, rectum, and breast [6]. Of particular interest to women is the significant research on the possible association between alcohol and breast cancer that has been underway since the early 1980s [52–54]. Although the findings are not completely consistent, positive associations between moderate alcohol intake and breast cancer have been observed in many studies [53–55]. The evidence for moderate to heavy drinking seems stronger; Howe *et al.* performed a combined analysis of six studies and found consistently increased breast cancer risk associated with consuming about four drinks (40 g of alcohol) per day [56]. The plausibility of the alcohol–breast cancer link is supported by two possible mechanisms: (1) alcohol produces acetaldehyde, a mutagen and cocarcinogen, and (2) moderate alcohol consumption increases levels of several hormones linked with breast cancer [57]. The latter hypothesis is consistent with a dose-response relationship of alcohol and breast cancer incidence.

The implications of advising women to limit alcohol consumption to help prevent breast cancer are not simple, because both research and media attention have suggested that drinking alcoholic beverages (especially wine) can reduce heart disease risk. These recommendations are not necessarily contradictory. Women should simply avoid excess alcohol intake. We are not aware of any targeted program or intervention efforts to prevent excess drinking for women or to control alcohol intake for the sake of cancer prevention. A good place to start might be to include alcohol education and prevention in nutrition interventions.

E. Skin Cancer Prevention and Sun Protection

Skin cancer is among the most common cancers in the United States. An estimated 34,100 cases of melanoma and 900,000 cases of nonmelanoma skin cancer (NMSC) were diagnosed in 1998 [5,58], and approximately 9430 persons died from skin cancer [59]. The rate of increase of melanomas leads that of all other cancers [60], and the overall percentage increase in death rate of 47.9% among men was the highest for all cancers [61]. If the current trend continues, the lifetime risk of developing melanoma will increase from 1 in 100 to 1 in 75 by the year 2000 [62]. The financial costs and impact on the health care system are substantial: in 1992, the cost of skin cancer treatment in the United States exceeded $ 500 million [63,64]. The total public health burden of nonmelanoma skin cancers, while also great, is not well quantified [58].

While skin cancer is among the most common of cancers, it is also one of the most preventable [59] and survivable if detected early [65]. Risk factors for skin cancer include excess sun exposure, male gender, family history, personal history of skin cancer or precancerous lesions, and physical characteristics [5,58,66–68]. Geographic and climatic factors also play a role [62,68]. Some risk indicators are useful in identifying persons at higher risk for certain types of skin cancer. Others, such as sun exposure and dysplastic nevi, are modifiable or treatable.

Increased cumulative sun exposure is a risk factor for NMSC, and sunburns, especially severe blistering sunburns, are associated with higher incidence of malignant melanoma [63,67,68]. The causal relationship between damage to skin cells by ultraviolet radiation from sunlight and almost all NMSCs and melanomas has now been confirmed at the molecular level [69]. The amount of ultraviolet-B (UV-B) radiation appears to have a dose-response relationship both to melanoma and to NMSC [70], and depletion of the earth's protective ozone layer leads directly to increased UV-B exposure [71].

Prevention guidelines include limiting sun exposure, using sunscreens and protective clothing, performing regular skin self-examination, and seeking professional evaluation of suspicious skin changes. Children, adults, and persons at high risk can all benefit substantially from adopting preventive practices. Nevertheless, awareness, knowledge, concern, and practice of prevention and early detection are relatively low in the United States [72], although these factors are somewhat higher among women, whites, and people with higher education and incomes [72]. Also, women, whites, older persons, persons with more formal education, people who perceive themselves to be at greater risk, and those who had relatives or friends with skin cancer are found to be more likely to perform skin self-examination or obtain physician skin examination [73–76].

Although there appears to be increased interest in promoting skin cancer prevention in the United States since the mid-1990s, there are few reports of controlled intervention studies. Most interventions have been targeted at children, adolescents, or outdoor workers. Some intervention trials that are underway are examining the efficacy of preventive education in a wide variety of settings including pharmacies, health maintenance organizations, and outdoor recreation settings. Others are targeting persons at moderate to high risk for melanoma and evaluating the impact of tailored messages provided by telephone or through the mail.

A special consideration for women may be the motivational effect of sun protection for its benefits in reducing the effects of wrinkling, or photoaging. They also might be encouraged to perform skin self-examinations in conjunction with monthly breast self-examinations for breast cancer. Because of their concern about protecting their children's health, the importance of sun protection for children may also encourage women to be more attentive to skin cancer prevention.

F. Chemoprevention

Epidemiologic, in vitro, and animal studies confirm that cancer can be promoted or inhibited by various chemicals [77]. This knowledge has led to research to identify agents that inhibit one or more steps in the development of cancer in hopes of finding ways to reduce the incidence of cancer by the prophylactic use of such compounds. This approach is termed "chemoprevention."

Many chemopreventive agents are derived from foods. Epidemiologic studies of plant-derived chemicals are complicated by the extraordinary complexity of diet. For example, fruits and vegetables contain many different compounds, and because they are consumed together in foods, it is difficult to separate their effects and to determine whether more than one compound is required to achieve a benefit.

Persons with a high intake of Vitamin A have a reduced risk of lung cancer [78–80]. Nevertheless, three large trials of vitamin A/beta-carotene have all failed to show any reduction in lung cancer risk [81–83]. The difficulty of applying and testing the findings from descriptive epidemiologic studies in clinical trials may explain the failure of these vitamin A/beta-carotene studies.

Allium (onions, garlic), soy beans, and brassica vegetables (cauliflower, broccoli, cabbage) all produce a variety of compounds that appear to inhibit the development of cancer [84]. These compounds include phytoestrogens (plant compounds with mild antiestrogenic activity), flavinoids (potent antioxidants), protease inhibitors (may block invasive potential of cancer cell), glutathione (antioxidant), and others. The phytoestrogens in soy, berries, tomatoes, onions, squash, and other vegetables may be especially important for women because they can block the effects of estrogen, a potential contributing factor to breast cancer risk. Many of these compounds impair carcinogenesis in animals, but human trials are not near at hand. However, these foods are all included in the nutritional recommendations for cancer prevention through diet [19].

The antiestrogen, tamoxifen, is routinely prescribed as a means of preventing breast cancer recurrence. It may also have benefit for osteoporosis and cardiovascular disease. A study has been underway to evaluate the benefits of tamoxifen [85], and the trial was stopped in light of the early finding of reduced rates of breast cancer by approximately 38% in the tamoxifen group compared to a placebo [86].

Aspirin and other nonsteroidal anti-inflammatory drugs may reduce the risk of bowel cancer [87]. Vitamins C and E may [88] or may not [89] reduce progression of premalignant colon polyps. Wheat fiber and calcium supplementation are also under study as potential chemopreventive agents for colorectal cancer.

The problems of preventing cancer on a large scale with a chemopreventive strategy have been extensively discussed [85,90]. Widely used chemopreventive agents must be very safe, inexpensive, and well understood to have public health implications. Thus, their identification and testing will require many years, and, with a few exceptions such as tamoxifen and related drugs, are unlikely to lead to widely applicable preventive measures in the near future.

G. Hormone Replacement Therapy (HRT)

Few drug therapies have remained so controversial for so long as has hormone replacement therapy (HRT). Since the 1950s when it was introduced, postmenopausal estrogens have undergone wide swings in popularity. First hailed as a panacea to reduce menopausal symptoms, HRT became less popular when studies showed that it raised the risk of endometrial cancer. Early enthusiasm was rekindled once endometrial cancer risk was shown to be reduced by the addition of progesterone [91]. Epidemiologic studies also suggested that HRT protects against heart disease and osteoporosis [92,93]. The benefits on bone density have since been confirmed in clinical studies.

Several large, case-control and longitudinal studies have suggested that a modest increase in breast cancer observed in case control studies may actually underestimate the true magnitude of risk [94,95]. New drugs such as the antiestrogens tamoxifen and raloxifene may actually reduce risk of cancer and heart disease and may be more effective than estrogen at increasing bone density among osteoporotic women. However, further studies are required to confirm the balance of risk and benefit from these drugs.

Many of the benefits of HRT require long-term use. Lack of compliance to HRT is high. About three-fourths of women are not taking HRT regularly a year after it was first prescribed [96]. Consequently, the relative benefits and risks of HRT use remain in question [97]. Certainly, there are women who would benefit (*e.g.,* those with significant menopausal symptoms) and others who should probably not take HRT (*e.g.,* those with breast cancer), although some experts recommend it for all and others for no one. The risks of HRT have generally been shown to be dependent on long-term use. Short-term use probably confers little risk and, except for relief of menopausal symptoms, little benefit. The balance of risk and benefit for long-term use is by

no means clear and women should discuss this with their physicians before beginning use. The lack of clarity on this issue is not likely to be resolved until a randomized clinical trial of the effects of HRT on morbidity and mortality is completed. This is one of the aims of the large, decade-long Women's Health Initiative study that is currently underway.

H. Limits of Current Knowledge

The ideal of prevention would be a solution that removes the need to worry about cancer. Public health and medicine have produced a few answers like these, and those that have been found are mainly for infectious rather than for chronic diseases. Simple solutions for cancer prevention are unlikely in the near future. As our understanding of cancer biology improves, so too does our recognition that cancer is not one, but many, diseases, and that these diseases arise as the result of varying mixtures of genetic and environmental interactions.

The easy part of learning to prevent cancer is that we know some ways to reduce the chances that cancers will develop: do not smoke; eat a low fat diet high in whole grains, fruits, and vegetables; get at least a moderate amount of exercise; do not drink too much; and limit sun exposure. Following through with these practices is not always easy, of course. Other prevention measures remain less certain and often are controversial. The role of postmenopausal estrogens in cancer, the benefits and risks of various specific food supplements and additives, incremental cancer risk from asbestos in buildings, or chemicals in the food and water supplies are unclear. At present, adopting and sustaining healthy behaviors relating to diet, smoking, exercise, sun exposure, and alcohol use are the surest way to reduce cancer risk.

III. What Is Unique to Women about Cancer Prevention?

Except for breast cancer and reproductive cancers, cancer risk factors appear similar for women and men. Diet and tobacco use are implicated in the major cancers both for women and for men. The dilemma over hormone replacement therapy is the single most critical cancer risk issue that is unique to women. The debate about the relative benefits and risks of postmenopausal estrogens is marked by diverse and rapidly changing views, and is not likely to be resolved in the near future. Cancer screening programs are more likely to be gender-directed than are primary prevention programs because of the efficacy of early detection technologies for sex-specific cancer sites (*i.e.,* breast, cervix, and ovary for women and prostate for men).

However, there may be considerable gender differences in the approaches to, and success in achieving, behavior changes designed to reduce risk of cancer. The challenge of changing behaviors relating to cancer risk differs among men and women in part because of differences in biological and social roles (see Table 77.1). Weight gain associated with childbearing can increase the problems of overweight for women, and this in turn may increase risk for some cancers. Weight gain as a result of menopause may also increase risk. Even though more women are in the workforce than ever before in industrialized nations, women are still more likely to purchase and prepare food than are men. This situation provides both opportunities (*e.g.,* knowledge, skill, access to foods) and challenges (*e.g.,* having to prepare to suit the tastes of family members).

Although smoking issues differ for men and women also, success rates by gender within programs do not differ greatly. While women may approach smoking cessation differently than men, they succeed at similar rates [98]. Consequently, smoking programs probably do not need to be specifically tailored for women in order to be effective. One study suggested that adjusting dates of smoking withdrawal to the follicular phase (days 1 to 15) of the menstrual cycle may improve success rates [99], but this strategy remains unproven at present.

In the area of primary prevention of cancer, most programs are designed to reach both men and women rather than being single-sex oriented. Typically, programs in worksites, community, and medical care settings attract more women than men as participants. Nevertheless, women's prevention programs for changes in diet, smoking, moderate alcohol consumption, physical activity, and sun protection are seldom segregated from such programs for men. Thus, our discussion in the next section is not narrowly focused on women because there are insufficient data and case examples to warrant this.

IV. Strategies and Programs for Changing Cancer Risk Behaviors

Behavior change is not easy. Large scientific literatures have accumulated that deal with changes in risk behaviors. The literature confirms the difficulty of maintaining behavior changes over long periods of time and the importance of repeated efforts and high levels of commitment. Few people succeed in making desired behavior changes the first time they try, so persistence and preparation are important contributors to success. Success is enhanced by attempting change in small steps, by obtaining support and encouragement from friends, family, and clinicians, and by reducing or eliminating environmental cues and temptations that trigger undesirable behaviors. Brief advice from medical care workers, enrollment in intensive smoking cessation or weight loss programs, and individual counseling have all been shown to enhance self-change efforts.

A broad range of settings are feasible and have been found effective for reaching women with behavior change programs. In community settings, programs to help highly motivated individuals to change risk behaviors such as smoking, overweight, and lack of exercise are available from the volunteer associations such as the American Lung Association, the American Cancer Society, and the YMCA/YWCA, and from commercial providers. Such intensive programs are also offered by many medical systems and employers. While they can be useful for those willing to attend them, only a fraction of those who want to quit smoking, lose weight, or enter exercise programs are willing to invest the required time and/or money.

Drugs are available to assist both weight loss and smoking cessation efforts, but they are much more effective when offered along with conscientious advice and counseling than as stand-alone prescriptions. Many people make changes in their smoking, exercise, and dietary habits on their own, with or without books, pamphlets, tapes, and other supporting materials.

Social policies and mass media are important influences on risk behaviors and can promote cessation of harmful practices or adoption of desirable behaviors. At the societal level, increases in tobacco taxes help to reduce the uptake of smoking by adolescents, and well-crafted media campaigns can both increase smoking cessation and reduce uptake. Media may send mixed messages, however. Media images promote the desirability of being thin but also depict the pleasures of consuming large quantities of rich foods that are high in fat and low in important nutrients. Effective public health approaches to reducing tobacco use [30–32], promoting healthful dietary change [100–102], and increasing physical activity [38,50] offer promise, much of which remains to be fulfilled.

V. Future Directions

Half of all cancer deaths in the United States can be prevented if current knowledge about prevention is applied [1]. At present, prevention holds the greatest promise for avoiding illness, suffering, unnecessary medical costs, and premature death. Avoiding tobacco, eating healthful diets, making regular physical activity a part of our lives, and avoiding excessive sun exposure and heavy alcohol consumption are the central known behavioral strategies for preventing cancer.

Because lung, breast, and colorectal cancers account for more than half of all cancer cases and cancer deaths among women, factors contributing to morbidity and mortality from cancers of these sites deserve special emphasis in prevention programs. In identifying strategies for promoting cancer prevention among women, it is valuable to consider issues of special concern to women: reproductive issues, factors associated with appearance such as weight and wrinkling, child health, and women's multiple social roles.

Cancer risk behaviors are also associated with risk factors for other chronic diseases, and people's patterns of multiple behaviors warrant attention beyond that which is given to single behavioral or chemopreventive prevention strategies. Multiple risk factor interventions are especially appealing because of the interrelationship among behaviors [103] and the shared determinants across various distinct behaviors such as smoking, physical activity, and diet [104,105]. Preventive programs with multiple foci can be efficient and targeted for women's needs [106] and should be further developed and evaluated for their possible role in preventing cancer among women.

References

1. Harvard Center for Cancer Prevention (1996). Harvard report on cancer prevention. Vol. 1: Causes of human cancer. Summary. *Cancer Causes Control* **7,** S55–S58.
2. Harvard Center for Cancer Prevention (1996). Harvard report on cancer prevention. Vol. 1: Causes of human cancer. Smoking. *Cancer Causes Control* **7,** S5–S6.
3. Osborne, M., Boyle, P., and Lipkin, M. (1997). Cancer prevention. *Lancet* **349** (Suppl. II), 27–30.
4. Parker, S. L., Tong, T., Bolden, S., and Wingo, P. (1997). Cancer Statistics, 1997. *Ca—Cancer Jo. Clini.* **47,** 5–27.
5. American Cancer Society (1998). "Cancer Facts and Figures—1998." American Cancer Society, Atlanta, GA.
6. National Research Council, National Academy of Sciences (1989). "Diet and Health: Implications for Reducing Chronic Disease Risk." National Academy Press, Washington, DC.

7. P. Thomas, ed. (1991). "Improving America's Diet and Health: From Recommendations to Action." National Academy Press, Washington, DC.

8. Department of Health and Human Services (1988). "The Surgeon General's Report on Nutrition and Health." U.S. Govt. Printing Office, Washington, DC.

9. Doll, R., and Peto, R. (1981). The causes of cancer: Quantitative estimates of avoidable risks of cancer in the United States today. *JNCI, J. Natl. Cancer Inst.* **66**, 1191–1308.

10. Glanz, K. (1997). Behavioral research contributions and needs in cancer prevention and control: Dietary change. *Prev. Med.* **26**, S43–S55.

11. Prentice, R. L., and Sheppard, L. (1990). Dietary fat and cancer: Consistency of the epidemiologic data, and disease prevention that may follow from a practical reduction in fat consumption. *Cancer Causes Control* **1**, 81–97.

12. Hankin, J. H. (1993). Role of nutrition in women's health: Diet and breast cancer. *J. Am. Dietetic Assoc.* **93**, 994–999.

13. Folsom, A. R., Kaye, S. A., Prineas, R. J., *et al.* (1990). Increased incidence of carcinoma of the breast associated with abdominal adiposity in post-menopausal women. *Am. J. Epidemiol.* **131**, 794–803.

14. Le Marchand, L. (1992). Anthropometry, body composition, and cancer. *In* "Macronutrients: Investigating Their Role in Cancer" (M. Micozzi and T. Moon, eds.), pp. 321–342. Dekker, New York.

15. Lubin, F., Ruder, A. M., Wax, Y., and Modan, B. (1985). Overweight and changes in weight throughout adult life in breast cancer etiology. *Am. J. Epidemiol.* **122**, 579–588.

16. Hunter, D. J., Manson, J. E., and Colditz, G. A. (1993). A prospective study of the intake of vitamins C, E, and A and the risk of breast cancer. *N. Engl. J. Med.* **329**, 234–240.

17. Rohan, T. E., Howe, G. R., and Friedenreich, C. M. (1993). Dietary fiber, vitamins A, C, and E, and risk of breast cancer: A cohort study. *Cancer Causes Control* **4**, 29–37.

18. P. Greenwald and E. Sondik, eds. (1986). "NCI Monographs: Cancer Control Objectives for the Nation: 1985–2000," NIH Publ. No. 86-2880, No. 2. National Cancer Institute, Bethesda, MD.

19. Butrum, R., Clifford, C. K., and Lanza, E. (1988). NCI dietary guidelines rationale. *Am. J. Clin. Nutr.* **48**, 888–895.

20. Kennedy, E., Meyers, L., and Layden, W. (1996). The 1995 Dietary Guidelines for Americans: An overview. *J. Am. Dietet. Assoc.* **96**, 234–237.

21. American Institute for Cancer Research (1998). "Diet and Health Recommendations for Cancer Prevention." American Institute for Cancer Research, Washington, DC.

22. Norris, J., Harnack, L., Carmichael, S., Pouan, T., Wakmoto, P., and Block, G. (1997). U.S. trends in nutrient in- take: The 1987 and 1992 National Health Interview Surveys. *Am. J. Public Health* **87**, 740–746.

23. Morris, D. H., Sorensen, G., Stodddard, A. M., and Fitzgerald, G. (1992). Comparison between food choices of working adults and dietary patterns recommended by the National Cancer Institute. *J. Am. Diet. Assoc.* **92**, 1272–1274.

24. Haines, P. S., Hungerford, D. W., Popkin, D., and Guilkey, D. K. (1992). Eating patterns and energy and nutrient intakes of U.S. women. *J. Am. Diet. Assoc.* **92**, 698–704, 707.

25. NHLBI, National Institutes of Health (1998). "Clinical Guidelines on the Identification, Evaluation, and Treatment of Overweight and Obesity in Adults. The Evidence Report." National Institute of Health, Bethesda, MD.

26. Schoeller, D. A., Shay, K., and Kushner, R. F. (1997). How much physical activity is needed to minimize weight gain in previously obese women? *Am. J. Clin. Nutr.* **66**, 551–556.

27. St. Jeor, S. T. (1997). New trends in weight management. *J. Am. Dietetic Assoc.* **97**, 1096–1098.

28. Peto, R., Lopez, A., Boreham, J., Thun, M., and Heath, C., Jr. (1995). Health effects of tobacco use. *In* "Tobacco and Health" (K. Slama, ed.). Plenum, New York.

29. U.S. Department of Health and Human Services (1989). "Reducing the Health Consequences of Smoking: 25 Years of Progress. A Report of the Surgeon General," DHHS Publ. No. (CDC) 89-8411. U.S. Govt. Printing Office, Washington, DC.

30. U.S. Department of Health and Human Services (1990). "The Health Benefits of Smoking Cessation. A Report of the Surgeon General," DHHS Publ. No. (CDC) 90-8416. U.S. Govt. Printing Office, Washington, DC.

31. Kawachi, I., Colditz, G. A., Stampfer, M. J., Willett, W. C., Manson, J. E., Rosner, B., Hunter, D. J., Hennekens, C. H., and Speizer, F. E. (1997). Smoking cessation and decreased risks of total mortality, stroke, and coronary heart disease incidence among women: A prospective cohort study. *J. Natl. Cancer Inst. Monog.* **8.**

32. Gritz, E., Brooks, L., and Nielsen, I. (1995). Gender differences in smoking cessation. *In* "Tobacco and Health" (K. Slama, ed.). Plenum Press, New York.

33. Waldron, I. (1991). Patterns and causes of gender differences in smoking. *Soc. Sci. Med.* **32**, 989–1005.

34. Solomon, L. J., and Flynn, B. S. (1993). Women who smoke. *In* "Nicotine Addiction: Principles and Management" (C. T. Orleans and J. Slade, eds.). Oxford University Press, New York.

35. Berman, B. A., and Gritz, E. R. (1991). Women and smoking: Current trends and issues for the 1990s. *J. Substance Abuse* **3**, 221–238.

36. Williamson, D. F., Serdula, M. K., Kendrick, J. S., and Binkin, N. J. (1985). Comparing the prevalence of smoking in pregnant and nonpregnant women, 1985 to 1986. *JAMA, J. Am. Med. Assoc.* **261**, 70–74.

37. Fingerhut, L. A., Kleinman, J. C., and Kendrick, J. S. (1990). Smoking before, during, and after pregnancy. *Am. J. Public Health* **80**, 541–544.

38. U.S. Department of Health and Human Services (1996). "Physical Activity and Health: A Report of the Surgeon General." USDHHS, Centers for Disease Control and Prevention, U.S. Govt. Printing Office, Washington, DC.

39. Blair, S., Horton, E., Leon, A., Lee, I. M., Drinkwater, B. L., Dishman, R. K., Mackey, M., and Kienholz, M. L. (1996). Physical activity, nutrition, and chronic disease. *Med. Sci. Sports Exercise* **28**, 335–349.

40. Colditz, G. A., Cannuscio, C. C., and Frazier, A. L. (1997). Physical activity and reduced risk of colon cancer: Implications for prevention. *Cancer Causes Control* **8**, 649–667.

41. Kramer, M. M., and Wells, C. L. (1996). Does physical activity reduce risk of estrogen-dependent cancer in women? *Med. Sci. Sports Exercise* **28**, 322–334.

42. Friedenreich, C. M., and Rohan, T. E. (1995). A review of physical activity and breast cancer. *Epidemiology* **6**, 311–317.

43. D'Avanzo, B., Nanni, O., La Vecchia, C., Franceschi, S., Negri, E., Giacosa, A., Conti, E., Montella, M., Talamini, R., and Decarli, A. (1996). Physical activity and breast cancer risk. *Cancer Epidemiol., Biomarkers Prev.* **5**, 155–160.

44. Mittendorf, R., Longnecker, M. P., Newcomb, P. A., Dietz, A. T., Greenberg, R., Bogdan, G. R., Clapp, R. W., and Willett, W. C. (1995). Strenuous physical activity in young adulthood and risk of breast cancer. *Cancer Causes Control* **6**, 347–353.

45. Pinto, B. M., and Marcus, B. H. (1994). Physical activity, exercise, and cancer in women. *Med. Exercise, Nutr. Health* **3**, 102–111.

46. Bernstein, L., Henderson, B. E., Hanisch, R., Sullivan-Halley, J., and Ross, R. K. (1994). Physical exercise and reduced risk of breast cancer in young women. *J. Natl. Cancer Inst.* **86**, 1403–1408.

47. Shepard, R. J. (1996). Exercise and cancer: Linkage with obesity? *Crit. Rev. Food Sci. Nutr.* **36**, 321–339.

48. King, A., Taylor, C. B., Haskell, W. L., and DeBusk, R. F. (1989). Influence of regular aerobic exercise on psychological health: A randomized, controlled trial of healthy middle-aged adults. *Health Psychol.* **8,** 305–324.

49. Wilcox, S., and Storandt, M. (1996). Relationship among age, exercise, and psychological variables in a community sample of women. *Health Psychol.* **15,** 110–113.

50. Pinto, B. M., Marcus, B. H., and Clark, M. M. (1996). Promoting physical activity in women: The new challenges. *Am. J. Prev. Med.* **12,** 395–400.

51. Marcus, B. H., Dubbert, P. M., King, A. C., and Pinto, B. M. (1995). Physical activity in women: Current status and future directions. *In* "Women's Health" (A. Stanton and S. Gallant, eds.), pp. 349–379. American Psychological Association, Washington, DC.

52. Longnecker, M. P., Berlin, J. A., Ozra, M. J., *et al.* (1988). A meta-analysis of alcohol consumption in relation to risk of breast cancer. *JAMA, J. Am. Med. Assoc.* **260,** 652–656.

53. Lowenfels, A. B., and Zevola, S. A. (1989). Alcohol and breast cancer: An overview. *Alcohol. Clin. Exp. Res.* **13,** 109–111.

54. Rosenberg, L., Metzger, L. S., and Palmer, J. R. (1993). Alcohol consumption and risk of breast cancer: A review of the epidemiologic evidence. *Epidemiol. Rev.* **15,** 133–144.

55. Gapstur, S. M., Potter, J. D., Sellers, T. A., *et al.* (1992). Increased risk of breast cancer with alcohol consumption in postmenopausal women. *Am. J. Epidemiol.* **136,** 1221–1231.

56. Howe, G. H., Rohan, T., Decarli, A., *et al.* (1991). The association between alcohol and breast cancer risk: Evidence from the combined analysis of six dietary case-control studies. *Int. J. Cancer* **47,** 707–710.

57. Longnecker, M. P. (1993). Do hormones link alcohol with breast cancer? *J. Natl. Cancer Inst.* **85,** 692–693.

58. Miller, D., and Weinstock, M. A. (1994). Nonmelanoma skin cancer in the United States: Incidence. *J. Am. Acad. Dermatol.* **30,** 774–778.

59. U.S. Department of Health and Human Services (1996). "National Skin Cancer Prevention Education Program, At-A-Glance 1996," World Wide Web pages. U.S. Public Health Service, Washington, DC.

60. Fraser, M. C., and Hartge, P. (1996). Melanoma of the skin. *In* "Cancer Rates and Risk" (A. Harras, ed.), NIH Publ. No. 96-691, pp. 163–166. U.S. Department of Health and Human Services, National Institutes of Health, Cancer Statistics Branch, Division of Cancer Prevention and Control, Public Health Service, Washington, DC.

61. Morbidity and Mortality Weekly Report (1996). Dramatic rise in incidence and deaths from melanoma in the United States, 1973 to 1992. *Primary Care Cancer* **16,** 5–6.

62. U.S. Department of Health and Human Services (1996). *In* "Cancer Rates and Risk" (A. Harras, ed.), NIH Publ. No. 96-691. National Institutes of Health, Cancer Statistics Branch, Division of Cancer Prevention and Control, Public Health Service, Washington, DC.

63. National Cancer Institute (1994). Prevention of skin cancer. *In* "PDQ Cancer Screening/Prevention Summary." NCI, NIH, Bethesda, MD.

64. Brown, M. L. (1990). The national economic burden of cancer: An update. *J. Natl. Cancer Inst.* **82,** 1811–1814.

65. Wingo, P. A., Tong, T., and Bolden, S. (1995). "Cancer Statistics." American Cancer Society, Atlanta, GA.

66. Preston, D. S., and Stern, R. S. (1992). Nonmelanoma cancers of the skin. *N. Engl. J. Med.* **327**(23), 1649–1662.

67. Author (1996). Ultraviolet light. *Cancer Causes Control* **7,** S39–S40.

68. Koh, H. K., Lew, R. A., Geller, A. C., Miller, D. R., and Davis, B. E. (1995). Skin cancer: Prevention and control. *In* "Cancer

Prevention and Control" (P. Greenwald, B. S. Kramer, and D. L. Weed, eds.), pp. 611–640. Dekker, New York.

69. Ziegler, A., Jonason, A. S., Leffell, D. J., *et al.* (1994). Sunburn and p53 in the onset of skin cancer. *Nature (London)* **372,** 773–776.

70. Scotto, J. (1996). Skin (nonmelanoma). *In* "Cancer Rates and Risk" (A. Harras, ed.), NIH Publ. No. 96–691, pp. 188–190. U.S. Department of Health and Human Services, National Cancer Institute, Division of Cancer Prevention and Control, Public Health Service, Washington, DC.

71. Morison, W. L. (1996). The effect of environmental changes. *Skin Cancer Found. J.* **14,** 21–23.

72. Miller, D. R., Geller, A. C., Wyatt, S. W., *et al.* (1996). Melanoma awareness and self-examination practices: Results of a United States survey. *J. Am. Acad. Dermatol.* **34,** 962–970.

73. Arthey, S., and Clarke, V. A. (1995). Suntanning and sun protection: A review of the psychological literature. *Soc. Sci. Med.* **40**(2), 265–274.

74. Girgis, A., Campbell, E. M., Redman, S., and Sanson-Fisher, R. (1991). Screening for melanoma: A community survey of prevalence and predictors. *Med. J. Aust.* **154,** 338–343.

75. Anderson, P., Lowe, J., Stanton, W., and Balanda, K. (1994). Skin cancer prevention: A link between primary prevention and early detection? *Aust. J. Public Health* **18.**

76. Hill, D., White, V., Marks, R., Theobald, T., Borland, R., and Roy, C. (1992). Melanoma prevention: Behavioral and nonbehavioral factors in sunburn in an Australian urban population. *Prev. Med.* **21,** 654–669.

77. Mettlin, C. (1997). Chemoprevention: Will it work? *Int. J. Cancer, Suppl.* **10,** 18–21.

78. Hirayama, T. (1979). Diet and cancer. *Nutr. Cancer* **1,** 67–72.

79. Mettlin, C. (1989). Milk drinking, other beverage habits and lung cancer risk. *Int. J. Cancer* **43,** 608–612.

80. Mettlin, C., Graham, S., and Swanson, M. (1979). Vitamin A and lung cancer. *JNCI, J. Natl. Cancer Inst.* **62,** 1435–1438.

81. ATBC (Alpha-Tocopherol, Beta Carotene) Prevention Study Group (1994). The effect of vitamin E and beta carotene on the incidence of lung cancer and other cancers in male smokers. *N. Engl. J. Med.* **330,** 1029–1035.

82. Hennekens, C. H., Buring, J. E., Manson, J. E., Stampfer, M., Rosner, B., Willett, W., and Peto, R. (1996). Lack of effect of long-term supplementation with beta carotene on the incidence of malignant neoplasms and cardiovascular disease. *N. Engl. J. Med.* **334,** 1145–1149.

83. Omenn, G., Goodman, G., Thornquist, M., Balmes, J., Cullen, M., Glass, A., Keough, J., Meyskens, F., Valanis, B., Williams, J., Barnhart, S., and Hammar, S. (1996). Effects of a combination of beta carotene and vitamin A on lung cancer and cardiovascular disease. *N. Engl. J. Med.* **334,** 1150–1155.

84. World Cancer Research Fund (W.C.R.F.) (1997). "Food, Nutrition and the Prevention of Cancer: A Global Perspective." American Institute for Cancer Research, Washington, DC.

85. Greenwald, P. (1995). Preventive clinical trials: An overview. *Ann. N.Y. Acad. Sci.* **768,** 129–140.

86. Various news reports (not yet published in a journal), April 1998.

87. Suh, O., Mettlin, C., and Petrelli, N. (1993). Aspirin use, cancer and polyps of the large bowel. *Cancer (Philadelphia)* **72,** 1171–1177.

88. DeCosse, J. J., Miller, H. H., and Lesser, M. L. (1989). Effect of wheat fiber and vitamins C and E on rectal polyps in patients with familial adenomatous polyposis. *J. Natl. Cancer Inst.* **81,** 1290–1300.

89. McKeown-Eyssen, G., Holloway, C., Jazmaji, V., Bright-See, E., Dion, P., and Bruce, W. (1988). A randomized trial of vitamins C and E in the prevention of recurrence of colorectal polyps. *Cancer Res.* **48,** 4701–4705.

90. Nixon, D. W. (1994). Special aspects of cancer prevention trials. *Cancer (Philadelphia)* **74,** 2683–2686.

91. Sherman, M. E., Sturgeon, S., Brinton, L., and Kurman, R. J. (1995). Endometrial cancer chemoprevention: Implications of diverse pathways of carcinogenesis. *J. Cell. Biochem. Suppl.* **23,** 160–164.

92. Matthews, K. A., Kuller, L. H., Wing, R. R., Meilahn, E. N., and Plantinga, P. (1996). Prior to use of estrogen replacement therapy, are users healthier than nonusers? *Am. J. Epidemiol.* **143,** 971–984.

93. Sturgeon, S. R., Schairer, C., Binton, L. A., Pearson, T., and Hoover, R. N. (1995). Evidence of a healthy estrogen user survivor effect. *Epidemiology* **6,** 227–231.

94. Kuller, L. H., Cauley, J. A., Lucas, L., Cummings, S., and Browner, W. S. (1997). Sex steroid hormones, bone mineral density, and risk of breast cancer. *Environ. Health Perspect.* **105** (Suppl. 3), 593–599.

95. Persson, I., Yuen, J., Bergkvist, L., and Schairer, C. (1996). Cancer incidence and mortality in women receiving estrogen and estrogen-progestin replacement therapy—long-term follow-up of a Swedish cohort. *Int. J. Cancer* **67,** 327–332.

96. Berman, R. S., Epstein, R. S., and Lydick, E. G. (1996). Compliance of women in taking estrogen replacement therapy. *J. Women's Health* **5,** 213–220.

97. Jacobs, H. (1996). Hormone replacement therapy for all? Not for everybody. *Br. Med. J.* **313,** 351–352.

98. Whitlock, E. P., Vogt, T. M., Hollis, J. F., and Lichtenstein, E. (1997). Does gender affect responses to a brief clinic-based smoking intervention? *Am. J. Prev. Med.* **13,** 159–166.

99. O'Hara, P., Porter, S. A., and Anderson, B. P. (1989). The influences of menstrual cycle changes on the tobacco withdrawal syndrome in women. *Addict. Behav.* **14,** 595–600.

100. Glanz, K., Lankenau, B., Foerster, S., Temple, S., Mullis, R., and Schmid, T. (1995). Environmental and policy approaches to cardiovascular disease prevention through nutrition: opportunities for state and local action. *Health Educ. Q.* **22,** 512–527.

101. Brunner, E., White, I., Thorogood, M., Bristow, A., Curle, D., and Marmot, M. (1997). Can dietary interventions change diet and cardiovascular risk factors? A meta-analysis of randomized clinical trials. *Am. J. Public Health* **87,** 1415–1422.

102. Glanz, K. (1994). Reducing breast cancer risk through changes in diet and alcohol intake: From clinic to community. *Ann. Behav. Med.* **16,** 334–346.

103. Emmons, K., Marcus, B., Linnan, L., Rossi, J. S., and Abrams, D. B. (1994). Mechanisms in multiple risk factor interventions: Smoking, physical activity, and dietary fat intake among manufacturing workers. *Prev. Med.* **23,** 481–489.

104. Boyle, R., O'Connor, P., Pronk, N., and Tan, A. (1998). Stages of change for physical activity, diet, and smoking among HMO members with chronic conditions. *Am. J. Health Promotion* **12,** 170–75.

105. King, N. A., Trembla, A., and Blundell, J. E. (1997). Effects of exercise on appetite control: iplications for energy balance. *Med. Sci. Sports Exercise* **29,** 1076–1089.

106. Simkin-Silverman, L., Wing, R. R., Hansen, D. H., Klem, M. L., Pasagian-Macaulay, A., Meilahn, E. N., and Kuller, L. H. (1995). Prevention of cardiovascular risk factor elevations in healthy premenopausal women. *Prev. Med.* **24,** 509–517.

Section 12

MENTAL DISORDERS

Evelyn J. Bromet

Department of Psychiatry and Behavioral Science
State University of New York at Stony Brook
Stony Brook, New York

Psychiatric disorders encompass a wide range of health conditions, including childhood disorders such as attention deficit-hyperactivity disorder; adult disorders, such as anxiety and mood disorders, substance abuse, and psychotic disorders; and disorders arising primarily at older ages, such as dementia. A study of the lifetime prevalence of mental illness concluded that close to 50% of people 15–54 years of age had an episode of mental illness at some point in their lives, and 29% experienced an episode of mental disorder in the year prior to assessment [1]. Not only are psychiatric disorders a leading cause of morbidity in the population, but they have also been linked to excess mortality from suicide and other causes [2]. While some disorders are time-limited or less severe and do not require professional intervention, others may exert a profound short- or long-term impact on quality of life. Current efforts are underway to define the subgroups who are most in need of treatment by virtue of their associated level of impairment or disability [3].

The chapters in this section provide both an overview of women's mental health (Chapter 78) and an in-depth analysis of five common conditions: affective disorders (depression and manic-depressive illness) (Chapter 79), anxiety disorders (both general anxiety disorder and specific diagnoses such as panic disorder and phobia) (Chapter 80), posttraumatic stress disorder (or PTSD) (Chapter 81), eating disorders (Chapter 82), and substance use disorders (Chapter 83). The timeliness of these discussions is underscored by the dramatic progress in our ability to diagnose and treat these conditions and the explosion of descriptive epidemiologic information.

Several themes run through these chapters. One is the progress that has been made in our ability to diagnose psychiatric

disorders. Clinical and epidemiologic research have been aided by the development of specific criteria for formulating a differential diagnosis and assessment tools that can elicit the cardinal symptoms and histories that are needed to make these diagnoses. In the U.S., the Diagnostic and Statistical Manual 4th revised edition, or DSM-IV, is the official classification system of the American Psychiatric Association [4]. Second, the risk factors for mental disorders are multifactorial and can include genetic inheritance, biologic factors, and/or environmental adversities. Similarly, once diagnosed with a mental illness, the prognostic factors can include both innate or historical characteristics of the individual as well as extrinsic or environmental influences, such as social stress or support. Of course, while these causal agents may act upon disease independently, evidence is accumulating to suggest that they interact in complex ways to increase the risk of disease onset. Thus, for example, life events explain less than 10% of risk for the onset of depression, but adding knowledge of vulnerability (*e.g.,* loss of a parent in childhood; absence of a confiding relationship; family history of depression) can increase the explanatory power beyond 50% [5]. Third, across the disorders described in this section, most women (or men) who meet diagnostic criteria never seek treatment from a mental health professional; rather, if treatment is sought, the source of care is most frequently the general medical sector [6]. This continues to be the case for these disorders even though there have been exciting advances in the treatment of psychiatric disorders involving both refinements in psychotherapy and the development of new medications. There are other conditions, such as schizophrenia and severe cognitive disorders, for which most individuals do receive professional mental health treatment. In addition, besides severity of illness, utilization is influenced by accessibility of care, insurance coverage, and other personal and demographic characteristics. The fourth theme in these chapters is that comorbidity (*i.e., having a history of more than one disorder*) is the rule rather than the exception. This means that women who experience a single disorder have an elevated risk for developing one or more additional psychiatric disorders. For some conditions, like PTSD, the rate of comorbidity is as high as 80%. Thus, the risk factors for different psychiatric disorders may overlap as well.

Many adult disorders have their roots in childhood. For example, women who develop schizophrenia more often report adjustment problems when they were very young than women who are free of this disease. Women who behave in antisocial ways or abuse drugs or alcohol often had personality difficulties when they were young. Women who develop PTSD after a catastrophic life event have a greater likelihood of childhood exposure to sexual or physical abuse.

Physical illness is a major risk factor for mental illness, and vice versa. In addition, many of the high risk groups for psychiatric disorder and many of the risk factors for these disorders cut across psychiatric and medical conditions. Examples include homelessness, exposure to neurotoxic substances in the workplace, and malnutrition—all of which serve as risk factors for both psychiatric and medical conditions. Unfortunately, too often, researchers artificially separate physical and mental disorders. Similarly, screening programs focus on one aspect of health to the exclusion of all others. Thus, to date, most epidemiologic studies focus either on psychiatric disorders with insufficient and unsystematic information on physical illness or on physical disorders with inadequate and nonstandardized information on mental health. It is our hope that this section of the book will encourage women's health researchers to develop conceptual models of illness that take both mental and physical health characteristics into account. Finally, it is important to recognize that this section is not inclusive of all psychiatric disorders. We have focused on those disorders that are most commonly found in adult women in the population.

References

1. Kessler, R. C., McGonagle, K. A., Zhao, S., Nelson, C. B., Hughes, M., Eshleman, S., Wittchen, H. U., and Kendler, K. S. (1994). Lifetime and 12-month prevalence of DSM-III-R psychiatric disorders in the United States: Results from the National Comorbidity Survey. *Arch. Gen. Psychiatry* **51,** 8–19.
2. Bruce, M., Leaf, P., Rozal, G., Florio, L., and Hoff, R. (1994). Psychiatric status and 9-year mortality data in the New Haven Epidemiologic Catchment Area study. *Am. J. Psychiatry* **151,** 716–721.
3. Regier, D. A., Kaelber, C. T., Rae, D. S., Farmer, M. E., Knauper, B., Kessler, R., and Norquist, G. (1998). Limitations of diagnostic criteria and assessment instruments for mental disorders. *Arch. Gen. Psychiatry* **55,** 109–115.
4. American Psychiatric Association (1994). "Diagnostic and Statistical Manual of Mental Disorders," 4th ed. American Psychiatric Association, Washington, DC.
5. Mann, A. (1997). The evolving face of psychiatric epidemiology. *Br. J. Psychiatry* **171,** 314–318.
6. Kaelber, C. T., and Regier, D. A. (1995). Directions in psychiatric epidemiology. *Curr. Opin. Psychiatry* **8,** 109–115.

78

Mental Illness in Women

MARY V. SEEMAN
University of Toronto
Centre for Addiction and Mental Health
Toronto, Ontario, Canada

I. Introduction

Men and women differ in the prevalence and severity of mental illnesses. The validity of this finding is undermined by the fact that a diagnosis of mental illness is primarily based on reported symptoms and rarely on objective signs, biopsies, blood tests, or indices of structural abnormality of the brain or specific neuronal circuit dysfunction. To compound the diagnostic problem, the nature and range of symptoms experienced during an episode of mental illness overlap considerably with short-term mood alterations and cognitive distortions that are ubiquitous in the general population. Given these constraints, it has become necessary to establish more or less arbitrary thresholds for the symptoms that define a specific illness. In addition, it is acknowledged that psychiatric symptoms fluctuate, especially in the degree to which they impede functioning, that some people seek help more readily than others and are therefore more readily identifiable, and also that different interviewing styles elicit greater or lesser disclosure of private concerns.

The American Psychiatric Association's current Diagnostic and Statistical Manual [1] establishes thresholds for symptoms, assesses functional impairment, and bases its categories of illness, as much as possible, on demographic studies that use objective, standardized diagnostic tools. It must nevertheless be recognized that it is not yet possible to categorize mental illnesses into etiologically discrete entities and it is therefore likely that new discoveries and changes in psychiatric nomenclature will, in the future, dramatically alter what are understood today to be male/female differences in the expression of the various psychiatric diseases.

II. Historical Trends

The interpretation of results of studies emanating from any specific decade of history has to be tempered by understanding the historical context of the era in which the study was conducted and reported. It has been variously popular to either emphasize or negate any substantial differences between men and women. The historical context of the 1970s was to expose the false ideology of built-in male/female differences and to attribute all apparent disparities to economic conditions and to social conditioning. The trend shifted in the 1980s, with more attention paid to the positive attributes of women, such as the importance of interpersonal connection in the lives of women. Study results from this period need, therefore, to be read and understood from this new standpoint. Inherent difference between men and women, whether innate or learned, was taken as fact in the 1980s, although the dichotomy was sometimes understood as the difference between masculinity and femininity, a social concept (*i.e.,* gender, rather than a biological concept, *i.e.,* sex). For a review of these changing historical trends, see Hare-Mustin and Marecek [2].

The 1990s have brought a new evolution of thinking about male and female health. Although it is always difficult to define prevailing winds of influence, there seem to be two prominent currents of thought at present. The first is an outgrowth of a culture of "perceived victimization" where increasing attention is being paid to the experience of trauma in all its forms—homophobia, racism, physical, sexual, mental abuse, harassment in the workplace, role conflict, inequitable opportunity—as explanatory of difference, notably between women and men [3]. The second is the deconstruction of the nature/nurture distinction—the increasing recognition of the complex interaction of genes and environment and the appreciation that individuals, through their genetic endowment, create their own individual environments (the nature of nurture) while, at the same time, modulating their gene expression through experience (the nurturing of nature) [4,5].

Subsequent chapters in this section on mental disorders will single out for gender analysis the affective or mood disorders, the anxiety disorders (with a special focus on posttraumatic disorders), the eating disorders, and the addictive disorders. Alzheimer's disease will be covered in the section on aging. This introductory mental health chapter, while recognizing that mental illnesses are distinct and varied, will attempt to discuss sexually dimorphic biological, interpersonal, and social determinants of mental health in a global overview and will also address gender-specific treatment.

Although this chapter will use as examples only "mainstream" psychiatric diagnoses, it is, of course, important to note that, as psychiatric categories evolve, what is considered mainstream changes and gender ratios change accordingly. This may be because etiologic agents are discovered and illnesses previously prevalent are consequently eliminated. Such is the case for neurosyphilis, which accounted for 19% of male psychiatric admissions versus only 3% of female admissions in the years 1908–1912 in the United Kingdom, but virtually disappeared from that country by 1950–1952 [6]. It also may be that new categories arise out of large scale social upheavals such as wars, new ideologies, and new epidemics. Post world war neuroses, the Gulf War syndrome, neurasthenia, fugue, dissociative identity disorder, environmental hypersensitivity syndrome, hypoglycemia, candidiasis hypersensitivity, adult attention deficit disorder, and chronic fatigue disorder are examples of reconfigurations of psychological symptoms into new syndromes that often affect men and women to significantly different degrees [7].

III. Demographics of Mental Health and Illness

Although boys have more psychiatric problems than girls [8], after puberty there is an abrupt change in psychiatric statistics [9], and virtually all of the major disorders, including mood disorders, anxiety disorders, somatoform disorders [10], dissociative disorders, and eating disorders, become substantially more common in women than in men, at least in North America. Addictive disorders, sexual disorders, and impulse-control disorders continue to be more prevalent in men [11–13]. In an epidemiologic study of mental health in Canada conducted in 1991, The Ontario Mental Health Supplement [14], there were wide sex variations in the rates of specific types of mental disorders. Depending on age, the ratio of women to men with respect to anxiety disorders ranged from 3:2 to 5:2. The ratio for mood disorders ranged from 2:1 to 3:1. The ratio for substance abuse ranged from 2:1 to 6:1 in favor of men. Among the personality disorders, only antisocial personality disorder is diagnosed more frequently in men than in women. Women are significantly more likely than men to meet criteria for borderline personality diagnosis, but the categorization of personality disorder remains a controversial area [15–20]. Adjustment disorders and schizophrenia are essentially equally prevalent in the two sexes. Schizophrenia is a relatively rare disorder (lifetime morbidity risk is under 1%) but it is regarded by many as the most serious of mental illnesses.

IV. Issues Related to Ascertainment of Mental Illness

Before community survey methodology permitted accurate general population sampling [21], mental health prevalence figures came, to a large extent, from treatment registers. Male/female differences in help-seeking and male/female differences in access to health care in certain social classes and geographic areas have rendered these results suspect. The Ontario Mental Health Supplement study showed clearly that *medical* mental health service delivery to women was twice that of men even though the rates of total psychiatric illness in the two sexes were not far apart. Either women seek help, at least from physicians, more readily than men or the nature of their illness (depression, anxiety) necessitates more medical treatment than does the kind of illness more common to men (*i.e.,* addictive disorders) [22]. These differences need to be kept in mind when extrapolating from clinical samples.

V. Thresholds for Defining Mental Illness

Women always report more symptoms than men on epidemiologic surveys of mental health and it has been variously speculated that they must have a lower threshhold for reporting. In a Canadian study designed to test this hypothesis, Toussignant and colleagues [23] interviewed over 200 men and 200 women and concluded that the women were not more prone to confide, nor did they report less serious symptoms than the men,

but they did report more of them. Please see Chapter 79 for further discussion of this issue.

VI. Male/Female Differences in Attributing Distress to Emotion and Acknowledging Emotional Ill Health

Since the landmark paper by Phillips and Segal [24] claimed that the higher rate of symptoms reported by women in community surveys did not reflect psychological distress but was a consequence of it being less stigmatizing for women than for men to admit to such symptoms, many sociologists have investigated these issues. Several alternative hypotheses have been tested in an attempt to understand why women report more symptoms.

The impetus behind this large field of inquiry has been a series of influential papers by the sociologist Walter Gove, who attributed the sex differences to the possibly more problematic role of marriage for women than for men [25–28]. Some of the issues raised by Gove follow. Because interviewers in most surveys tend to be women, it is possible that people confide more to persons of their own sex. Women may have a tendency toward yea-saying (acquiescence), or a tendency to respond according to perceived cultural norms (a quest for social approval), or a tendency to endorse socially desirable items. Time-frames may also influence gender responsiveness (*e.g.,* men hypothetically living more in the present than women and, therefore, forgetting symptoms unless they are being currently experienced). It was also argued that items on most surveys were geared more to women's problems (psychophysiological symptoms) than to men's problems (*i.e.,* impulsivity questions). Other related issues had to do with self-awareness. Women may be more aware of their internal sensations and better able to translate them into a psychological vocabulary; they may hold a higher ideal of health than men and consequently find their own level of health wanting. In confirmation of Gove's earlier work, the Montreal-based study by Toussignant *et al.* tested for all these potential biases and found no evidence for any of them [23]. Subsequently, in a sample based on over 2000 first-degree relatives of affectively disordered probands of a National Institute of Mental Health (NIMH) collaborative study of depression, the lack of sex bias in reporting was once again supported [29]. Please see Chapter 79 for further discussion of this issue.

VII. How Does Women's Changing Social Role Impact on Rates and Severity of Disease?

Wilhelm and Parker designed a prospective longitudinal study to answer this challenging question with respect to depression. In 1978, they began following a cohort of 114 female and 56 male trainee teachers in Australia who initially, when they were in college, reported high rates of depression but no gender difference. The investigators' hypothesis was that the usual female over male preponderance would begin to show itself as the cohort got married, had children, and gender social roles began to diverge. At the fifteen year follow-up [30], the women's rates for depression (especially when combined with anxiety disorders) were just beginning to exhibit significant differences over those of men. Significantly fewer women than men were in full-time employment after 15 years, but there is thus far no indication that employment status or marital status has played any role in the development of depressive and/or anxiety syndromes.

VIII. Host/Environment Issues

Culture, class, occupation, and age norms interact with sex in complex ways to produce differing rates of illness. For instance, somatoform disorder, which is much more common in women than in men in North America, is more often reported in Greece by men than by women [1]. Chronic pain conditions, more common in women than men, occur very frequently among male manual laborers. Among Jews, depression may be as common in men as in women, attributable in part, perhaps, to relative alcohol abstinence [31]. It has often been speculated that many men vulnerable to depression abuse alcohol in an effort at self-medication and are consequently given a diagnosis of substance abuse that masks their underlying depression.

Whereas 90% of eating disorders are diagnosed in women, these disorders occur almost exclusively in industrialized societies where food is abundant. That societal pressures toward thinness are an important part of the pathogenesis of eating disorders is supported by the high ratio of homosexuality in eating disordered men, because the gay subculture places the same premium on appearance as most industrialized societies place preferentially on women [32]. With respect to data from nonindustrialized countries, however, it must be recognized that women in these societies may have proportionally less access to health care and, thus, to identification.

In the Ontario Mental Health Supplement study, age played a significant role in mental health. Men in the 15–19 age group showed four times the rate of mental disorder as men between the ages of 45 and 64. Women in the younger age group had twice the rate of mental illness as women in the older age category. In the younger age group, men reported higher rates of illness than women (35% vs 29%); the rate was essentially the same in the age group 20–44 years but, in the age group 45–64, almost twice as many women as men reported symptoms meeting thresholds for mental illness (15% vs 8%) [14]. The explanation seems to be that certain disorders (substance abuse, eating disorder) diminish with age, while others (depression, anxiety) do not [33].

IX. Risk Factors for Women

Is it possible that collective and conflicting social expectations (to be attractive and nurturant, to be agreeable and autonomous, to be socially interactive and competitive) impact so much on women that they explain the differential sex prevalence of eating disorders, depression, agoraphobia, and social phobia, as well as "borderline" personality disorder? Silverstein and Perlick [34], in a review of the depression literature, suggest that gender differences occur only at particular historical times (*i.e.*, those in which adolescence is reached during periods of increasing opportunities for female achievement).

Child-rearing has been implicated as the vehicle through which social pressures are communicated to offspring. However, if this were the main source of expectations resulting in illness, why would the adult differences in prevalence rates not become apparent until adolescence? Perhaps child-rearing prac-

tices inculcate personality traits in women (amiability, eagerness to please, field dependence) that make them vulnerable to later pressure from peers and from the ambient popular culture as transmitted via music, film, and print media.

Peer expectations are among the most powerful shapers of ideology and behavior. Girls reinforce in each other the wish to starve and purge in order to be thin; boys reinforce substance use and other forms of risk taking. The precise nature of these pressures is influenced by socioeconomic class, ethnicity, and locale.

A. The Role of Economics, Social Status, and Social Role

Mental illness is more prevalent in the context of poverty and economic hardship, and women all over the globe are financially disadvantaged relative to men. This is seen most dramatically (and is of most concern) in relation to single mothers who live in financially constrained circumstances, who suffer high rates of mental illnesses, and who, at the same time, bear sole responsibility for the upbringing of a new generation of disadvantaged children [35]. Eleven percent of Canadian families are headed by single mothers. These women are at special risk for the development of mood disorders, anxiety disorders, and addictive disorders [35,36].

B. Social Supports

In polar opposite to the impact of poverty, the quality of social relationships has been shown to protect against psychiatric symptoms and to buffer the effects of stress [37]. The relative lack of perceived social supports (generally a lesser problem for women than for men) can be considered as an independent risk factor for mental illness or it can be seen as arising secondarily from a genetic incapacity to form or benefit from social networks. In an elegant population study sampling approximately 2000 Caucasian female twins (854 pairs of whom 497 were monozygotic and 354 were dizygotic) and using the liability threshold model of analysis, Kendler and colleagues concluded that between 40 and 80% of the temporally stable variance in dimensions of social support was due to genetic differences [4,38]. This suggests that social supports develop in part as a result of an active effort to develop and sustain them [39].

C. Stressful Life Events

Exposure to stressful life events has been postulated as the vehicle through which genetically vulnerable individuals eventually succumb to psychiatric disease [5]. Of the various factors studied as putative precursors to depression (family history of depression, temperament, social support, lifetime traumas, recent interpersonal difficulties, and recent stressful life events), Kendler *et al.* [40] showed that recent stressful life events were the single most important predictor of major depression in women. Genetic factors came second. The advent of stressful life events, they postulated, is a combination of bad luck and genetically modulated exposure to the environment. Because their sample consisted only of women, it is not possible to compare the sexes either in terms of the frequency of "bad luck" or on temperamental differences that might induce "bad luck," but both sociological (economic disadvantage, role conflict) and biological (temperamental irritability induced by hormone fluxes) factors could work to the disadvantage of women. It has also been stated that women pay the "price of caring" in that their extended friendships and nurturant social roles put them into close contact with relatively large networks of people and that life events occurring in any part of their social network impact on them to a greater extent than would be the case for men.

D. Childhood Trauma

Whereas the Kendler study [4] showed proximal stressful events to be more important for mood disorders than earlier, more distal events, the opposite may be true for the development of borderline personality disorder. There is, however, an overlap between this diagnostic category and that of mood disorder which may explain the continuing contradictions in the literature with respect to the etiologic role of childhood trauma [41–43]. It is currently thought, but this may simply reflect the world view of the 1990s, that physical, and especially sexual, abuse in children is an important risk factor for adult emotional impairment, but that it does not constitute a diagnostically-specific risk factor.

X. Evolutionary Aspects of Social Roles

Biological givens shape social expectations and continue to do so in the present, despite widespread beliefs in equality of the sexes and gender-neutral child-rearing. In general, men after puberty are taller and stronger than their female same-age peers. After puberty, lipid stores in women's bodies increase; women's bodies become round and breasts appear. Women begin to menstruate in preparation for reproduction. As adults, they carry each fetus for nine months prior to labor and delivery; they give birth and nurse their infants. The lengthy proximity with their young as well as powerful hormonal secretions create unique mother-child bonds. From an evolutionary viewpoint, with respect to the biological imperative of ensuring the survival of one's DNA, in most species, males and females use different behavioral strategies. Females know that their infants have inherited their genetic material and have a selfish interest in ensuring that their babies survive. Depending on the species and on the relative impact of role socialization, ensuring survival of the young includes maintaining an outward appearance and attitude designed to promote the continuing loyalty of protective, but potentially aggressive and faithless, males. It also includes an emphasis on nurturing skills. Males, on the other hand, never absolutely certain that their mate's child carries their genes, must evolve different selfish strategies. It can be argued that the male's best bet is to attract partners who appear to be healthy enough to weather repeated pregnancy and childbirth and to impregnate them as frequently as possible. Moreover, just in case, the male may be wise to impregnate other females as well, in order to increase the chances of survival of his own selfish genes. In the light of selection pressures, with males being obligated to be on continuous courtship alert, is it any wonder that they fall prey to the lure of alcohol and mind numbing substances that help them overcome their sexual inhibitions? Does this help to explain the male/female ratio in substance abuse? According to evolutionary theory, females prone to depression and anxiety may be especially attractive to males; their likelihood

to stay close to home lulls the male's fear of potential infidelity. Does this help to explain the male/female ratio of depressive and anxiety disorders? Like any grand theory, evolutionary theory can explain essentially all behavior and couch it in the language of biological imperativeness. Experimental data in humans, however, do not support any of these speculations.

XI. Hormonal Effects on Brain Function

Genetic mechanisms other than hormonal ones may determine sexual dimorphism, but not much is known about them yet. Hormonal effects, however, have been well studied. When testes develop in male fetuses and begin to produce androgens, males and female brains begin to differentiate. Maximum central nervous system sensitivity to the organizational effects of gonadal steroids is presumed to occur between human gestational weeks 14 and 16, when peak concentrations of testosterone are present in fetal serum. Testosterone enters the male brain and attaches to hormone receptors, setting off a cascade of effects that forever shapes the male brain into an organ whose structure and function and pace of development is somewhat different from the female's [44,45]. During development (and also during adult life) gonadal steroid receptors are expressed in several areas of the cerebral cortex that mediate cognition and affect: the nuclei of the septum and the diagonal band of Broca, the hippocampus, the allocortex, isocortex, and amygdaloid complex [46]. The best studied receptors are the ones for estrogens, known to regulate neuronal function in a number of important ways, essentially to prevent cell death and to promote growth of cell connections and, thus, to enhance neural communication [47].

The schizophrenia and the Alzheimer's literature both provide evidence for estrogen's protection of the integrity of neuronal circuits. The coming on line of gonadal hormones at puberty may delay the onset of schizophrenia in women; their withdrawal after menopause produces cognitive change and may increase women's risk for Alzheimer's disease [48].

It is important to know more about what happens in the brain of both men and women at the time of puberty because this is the period when sex ratios for symptoms of depression and anxiety dramatically alter. Primate and human data suggest that certain neurons (not all) undergo active growth and/or pruning during this stage and that the pace of these events and their final steady state may well be affected by the activational action of pubertal hormones [49]. Men's brains, as well as women's, are bathed in estrogen during this time period because a percentage of testosterone, upon entering the brain, is enzymatically converted to estradiol. In women, the cyclic monthly withdrawal of hormones may now begin to interfere with one of the natural functions of estrogen—its ability to neutralize the effects of glucocorticoids released in response to stress—thus, perhaps, rendering women more vulnerable than men to stress hormones and, consequently (although this is highly speculative), to depression and anxiety [50].

XII. Brain Maturation

Male brains develop more slowly than female brains so that, at birth, they are less mature than weight-adjusted female brains. There is also a difference in lateralization, with the two hemispheres in girls being more symmetrical [51]. The process of maturation is completed faster in girls than it is in boys, both with respect to progressive myelination and to volumetric increases. To generalize from relatively few replicated studies, this may mean that boys' neuronal circuitry is more vulnerable than that of girls at early ages because it lags behind in protective shielding such as that provided by myelin sheaths. Boys' hemispheric functions are, early on, more specialized than those of girls. Girls' brains are more plastic (*i.e.,* they are able to recover function after one-sided brain injury). Girls may learn faster than boys in their younger years because the pace of development of their brains is faster. At some point in adolescence, women's brains complete the process of final connectivity and pruning of unwanted redundant neurons. Men's brains continue for some time later to form and lose neuronal connections and to become increasingly specialized in their functions. Women's brains show stronger interhemispheric connections [52]. Adult male brains facilitate specialized functions; female brains preserve plenipotentiality. A vivid demonstration of this are Positron Emission Tomography (PET) scans illustrating areas in the brain used in a rhyming task. Only left language areas light up in male subjects whereas bilateral language areas shine in female participants [53].

In light of the preceding discussion, it is possible that stressors impairing specific brain circuitry in early years have more severe impact on boys, with girls being protected by the presence of a second well-functioning brain hemisphere (much as the presence of a healthy allele on the female's second X chromosome protects her from X-linked genetic disease). But with a longer time for reparation (via pruning of unwanted synapses and connections), boys may emerge from adolescence with lesser vulnerability to repeated stress than girls, despite female built-in protection. Women's stronger interhemispheric connections may generalize past stressors and invoke past memories with a concomitant release of stress hormones and resultant symptoms of depression and anxiety. This is, of course, speculation.

XIII. Women and Pharmacological Response

Serious psychiatric illness is treated with psychopharmaceuticals which are becoming more specific, freer of adverse effects, and more effective over time. For reasons of convenience and of safety, drug development has traditionally been based on experiments with male animals and on early human screening in mainly male populations [54]. But it cannot be assumed that women and men respond in the same way to the various psychopharmacological agents. One important difference is that women's bodies contain far more adipose tissue than men's per unit of body weight. Because antipsychotic drugs, antidepressants, and anxiolytics are lipophilic (bind to fat molecules), they will be retained longer in women's bodies after drug discontinuation; they can also be released unexpectedly from fat stores during rapid weight loss, causing untoward side-effects. Blood flow to the brain (taking the drugs along with it) is under hormonal control. It is more rapid in women than in men [55] which means that psychotropic agents reach their targets faster. This probably varies with the time of the month.

There are sex differences in the activity of liver enzymes that degrade drugs and turn them into other molecules (sometimes chemically active, sometimes not) prior to their elimination from the body. Alcohol, for instance, is much less efficiently detoxified by the female liver than by the male liver so that the same initial concentration of blood alcohol is more toxic for women than it is for men. Increasingly, the prescription of psychopharmacological agents is complicated by one agent inducing enzymes that either enhance or reduce the activity of a second agent being used for a comorbid condition. Women, more than men, are likely to be receiving adjunctive medication because they suffer more often than men from allergies, from thyroid conditions, from arthritic conditions, from depressions and anxieties, and from insomnia. They are more likely to be taking contraceptives, hormone replacement therapies, pain pills, anti-inflammatory pills, thyroid pills, and diuretics. Drug interactions are, therefore, more frequently encountered in women and, correspondingly, so are adverse drug reactions. The concomitant use of nicotine and alcohol with prescribed medication, as well as age effects, are equally important when evaluating the safety and efficacy of a drug regimen [56].

Men and women differ in which side-effects are least tolerable and which ones are likely to reduce adherence to a prescribed regimen. Because appearance is of importance to most women, drugs that induce weight gain are especially problematic. Sedating drugs interfere with childcare and drugs that cause blood pressure drops and secondary falls are particularly dangerous for older women at risk for osteoporosis and hip fracture [57].

In treating psychosis, it is useful to know that estrogens block dopamine and that, all things being equal, women need lower doses of antipsychotic dopamine blockers than men do [58]. The older antipsychotics, by increasing prolactin levels, correspondingly lowered estrogen secretion, so that women's requirement of lower dosing was not apparent. The newer generation antipsychotics have less effect on prolactin and women should be expected to respond to them at lower doses than men.

XIV. Other Clinical Issues

Clinically important gender issues can interfere with accurate diagnosis and management of specific psychiatric illness. For instance, depression and anxiety are more prevalent among women than among men in the general population and will also be relatively more prevalent in the population who suffer from major psychiatric illness such as schizophrenia. The recognition of prominent mood symptoms may mask the underlying disorder and may delay appropriate treatment. This overlay of mood lability in women may partially account for what is usually considered their later onset age for schizophrenia [59]. If treatment delay adversely affects long-term outcome, as is being currently hypothesized, it is important to keep in mind that gender (as well as culture, personality, and social class) can influence the expression of illness and its recognition.

Even more critical to effective outcome is the development and maintenance of a therapeutic alliance between patient and care provider. This is crucial for serious illness with propensity for chronicity and frequent relapse. Women are considered more affiliative than men, form intimate relationships more readily, and sustain them more easily. Perhaps because of this, the outcome for schizophrenia in women is superior to that in men along many outcome dimensions [60] for the first 15 years following a first episode of psychosis. In later years, the outcome advantage is lost [61]. It is possible that hormonal changes at menopause alter affiliative tendencies (which may be mediated through female hormones) or, perhaps, that strong therapeutic alliances are good in the short-run but are inherently risky in the long-run because they undermine autonomy or expose women to repeated emotional losses as care providers come and go.

Family contexts, long suspected to be critical in the triggering and the buffering of psychiatric symptoms, are perceived and experienced differently by men and women. There is a rich and complex literature in this area. Boys appear to be more sensitive than girls to early family upheavals such as death and divorce. Girls, on the other hand, are more likely to be victims of sexual abuse at the hands of family members. In adolescence, boys perceive their families as more demanding and less supportive than do girls. Married men are better protected against psychiatric disease than single men; the opposite is true for women. Women are the caretakers of children and of elderly parents; the emotional burdens of family life fall more severely on their shoulders than on those of men. Women suffer more than men from domestic abuse but, when ill with schizophrenia, are less exposed to hostility and criticism on the part of family members, and, therefore, are better protected against relapse. Perhaps for this reason, family therapies in schizophrenia work better for women than they do for men. In summary, it appears as if women's superior affiliative skills operate in the family context as they do elsewhere, on one hand shielding women from some of the pressures experienced by men and on the other hand exposing them to the toll of caring, exacted in a relative loss of autonomy [62].

Other important clinical issues for women with psychiatric disease are their cycling hormones and the need to recognize that menstrual fluctuations, pregnancies, postpartum periods, and menopause have both a biological and a psychological impact on symptoms. Most women feel subjectively different at different times of the month and respond differently to perceived slights as well as to alcohol or to therapeutic medications. In schizophrenia, for example, symptoms get worse at low estrogen times, are considerably dampened during pregnancy, but return postpartum and are usually increased at menopause [63]. Menopausal worsening of schizophrenia in women may be due to estrogen withdrawal but may also be due to the age-related loss of social supports. Such losses are less important in schizophrenic men who, for the most part, have isolated themselves from others since before the onset of illness.

XV. Impact

The greatest impact of psychiatric disease in women, in human terms, is on their children. Psychiatric disorders have been shown to lead to teenage parenthood in both sexes, but especially in women [64]. The ratio of teenage parenthood is increasing in the United States because of a decrease in births among older women. The consequences of this for infants are many and varied: teenage mothers are less responsive to their infants' needs than are older mothers, they display less positive attitudes toward motherhood, their expectations of their children are unrealistic, and they are more likely to abuse their children. The children born to young mothers weigh less at birth, show a delay in cognition, and have more school and conduct problems [65].

Women with major mental illness are becoming more sexually active [66] which means that health services have to learn how to intervene more quickly and more effectively to prevent unwanted pregnancies, to care for the mother-infant dyad during pregnancy and postpartum, and to provide mentally ill mothers with opportunities to learn optimal parenting skills. It has been estimated that 50% of children of mothers with schizophrenia will develop some type of psychiatric disorder [67,68]. Contributory factors may be a relative lack of prenatal care, prenatal exposure to drugs and alcohol, obstetric complications, socioeconomic problems, and mothering issues secondary to the cognitive deficits associated with schizophrenia or to the secondary effects of neuroleptic medication. Once recognized, many of these issues can be addressed by appropriate service provision.

XVI. Summary and Conclusions

This overview discusses factors that impinge on men and women without sufficiently emphasizing that specific factors may be more important or perhaps exclusively important in specific diseases. For instance, the conclusions reached in a well-argued paper by Cloninger *et al.* [69] were that while sex differences in alcoholism were due to both familial and non-familial factors, in women the nonfamilial environmental factors exerted most of the influence. In contrast, the data showed that familial factors (both genetic and environmental) relevant to the development of antisocial personality were largely the same for women and men. The authors argued that inferences from prevalence data to the etiology of sex differences may be spurious unless all of the information is considered separately for each disease. Ideally, that would mean knowing the genetic transmission mechanism of each specific illness and assessing the early family environment relevant to each illness, the role of birth trauma, of physical and sexual abuse, the nature of familial stressors, and the relevant cultural-ecological influences. These would need to include diagnostic and judicial procedures, response biases in reporting and help-seeking, adult social role divergence, exposure to stressful life events, and extent and quality of social supports. It would also mean testing for neuroendocrine differences in level, rate of change, cyclicity, and timing of peak effects, as well as understanding gross and microscopic central nervous system differences that mediate perception, cognition, affect regulation, and behavior.

To any initial difference in response bias attributable to early brain differentiation is added early experience, which further magnifies sex differences not only because boys and girls are dealt somewhat different environmental cards but also because, from birth, their (on average) disparate temperaments and skill sets elicit different responses from those around them [70].

Women and men are more alike than not. The overlap between cognitive traits that are considered most divergent (such as visuospatial memory, for instance) is considerable. The range of differences within each sex is certainly larger than mean differences between the sexes. Nevertheless, the differences that do exist may offer clues to the origin and perpetuation of certain psychiatric diseases and, for that reason, it is important to identify them and investigate them.

References

1. American Psychiatric Association (1994). "Diagnostic and Statistical Manual of Mental Disorders," 4th ed. American Psychiatric Association, Washington, DC.

2. Hare-Mustin, R., and Marecek, J. (1988). The meaning of difference: Gender theory, postmodernism, and psychology. *Am. Psychol.* **43**, 455–464.

3. Herman, J. L. (1992). "Trauma and Recovery." Basic Books, New York.

4. Kendler, K. S. (1997). Social support: A genetic-epidemiologic analysis. *Am. J. Psychiatry* **154**, 1398–1404.

5. Post, R. M. (1992). Transduction of psychosocial stress into the neurobiology of recurrent affective disorder. *Am. J. Psychiatry* **149**, 999–1010.

6. Busfield, J. (1982). Gender and mental illness. *Int. J. Ment. Health* **11**, 46–66.

7. Stewart, D. E. (1990). The changing face of somatization. *Psychosomatics* **31**, 153–158.

8. Earls, F. (1987). Sex differences in psychiatric disorders: Origins and developmental influences. *Psychiatr. Dev.* **1**, 1–23.

9. Almqvist, F. (1986). Sex differences in adolescent psychopathology *Acta Psychiat. Scand.* **73**, 295–306.

10. Wool, C. A., and Barsky, A. J. (1994). Do women somatize more than men? *Psychosomatics* **35**, 445–452.

11. Bland, R. C., Newman, S. C., and Orn, H. (1988). Epidemiology of psychiatric disorders in Edmonton. *Acta Psychiatr. Scand.* **77**(Suppl. 338), 1–80.

12. Kessler, R. C., McGonagle, K. A., Zhao, S., Nelson, C. B., Hughes, M., Eshleman, S., Wittchen, H., and Kendler, K. S. (1994). Lifetime and 12-month prevalence of DSM-III-R psychiatric disorders in the United States: Results from the National Comorbidity Survey. *Arch. Gen. Psychiatry* **51**, 8–19.

13. Robins, L. N., and Regier, D. A., eds. (1991). "Psychiatric Disorders in America: The Epidemiologic Catchment Area Study." Free Press, New York.

14. Offord, D. R., Boyle, M. H., Campbell, D., Goering, P., Lin, E., Wong, M., and Racine, Y. (1996). One-year prevalence of psychiatric disorder in Ontarians 15 to 64 years of age. *Can. J. Psychiatry* **41**, 559–563.

15. Bardenstein, K. K., and McGlashan, T. H. (1988). The natural history of a residentially treated borderline sample: Gender differences. *J. Pers. Disord.* **2**, 69–83.

16. Corbitt, E. M., and Widiger, T. A. (1995). Sex differences among the personality disorders: An exploration of the data. *Clin. Psychol.: Sci. Pract.* **2**, 225–238.

17. Golomb, M., Fava, M., Abraham, M., and Rosenbaum, J. F. (1995). Gender differences in personality disorders. *Am. J. Psychiatry* **152**, 579–582.

18. Grilo, C. M., Becker, D. F., Fehon, D. C., Walker, M. L., Edell, W. S., and McGlashan, T. H. (1996). Gender differences in personality disorders in psychiatrically hospitalized adolescents. *Am. J. Psychiatry* **153**, 1089–1091.

19. Gunderson, J. G., Zanarini, M. C., and Kisiel, C. L. (1991). Borderline personality disorder: A review of data on DSM-III-R descriptions. *J. Pers. Disord.* **5**, 340–352.

20. Loranger, A. W., Lenzenweger, M. F., Gartner, A. F., Susman, V. L., Herzig, J., Zammit, G. K., Gartner, J. D., Abrams, R. C., and Young, R. C. (1991). Trait-state artifacts and the diagnosis of personality disorders. *Arch. Gen. Psychiatry* **48**, 720–728.

21. Bland, R. C. (1988). Investigations of the prevalence of psychiatric disorders. *Acta Psychiatr: Scand.* **77**(Suppl. 338), 7–16.

22. Rhodes, A. R., and Goering, P. (1994). Gender differences in the use of outpatient mental health services. *J. Ment. Health Admin.* **21**, 338–345.

23. Tousignant, M., Brosseau, R., and Tremblay, L. (1987). Sex biases in mental health scales: Do women tend to report less serious symptoms and confide more than men? *Psychol. Med.* **17**, 203–215.

24. Phillips, D. L., and Segal, B. E. (1969). Sexual status and psychiatric symptoms. *Am. Sociol Rev.* **34**, 58–72.

25. Gove, W. (1978). Sex differences in mental illness among adult men and women: An examination of four questions raised whether or not women actually have higher rates. *Soc. Sci. Med.* **12**, 187–198.

26. Gove, W. (1980). Mental illnes and the psychiatric treatment of women. *Psychol. Women's Q.* **4**, 345–362.

27. Gove, W. (1984). Gender differences in mental and physical illness: The effects of fixed roles and nurturant roles. *Soc. Sci. Med.* **19**, 77–84.

28. Gove, W. R., and Tudor, J. F. (1973). Adult sex roles and mental illness. *Am. J. Sociol.* **78**, 812–835.

29. Young, M. A., Fogg, L. F., Scheftner, W. A. *et al.* (1990). Sex differences in the lifetime prevalence of depression: Does varying the diagnostic criteria reduce the female/male ratio? *J. Affective Disord.* **18**, 187–192.

30. Wilhelm, K., Parker, G., and Hadzi-Pavlovic, D. (1997). Fifteen years on: Evolving ideas in researching sex differences in depression. *Psychol. Med.* **27**, 875–883.

31. Levav, I., Kohn, R., Golding, J. M., and Weissman, M. M. (1997). Vulnerability of Jews to affective disorders. *Am. J. Psychiatry* **154**, 941–947.

32. Carlat, D. J., Camargo, C. A., and Herzog, D. B. (1997). Eating disorders in males: A report on 135 patients. *Am. J. Psychiatry* **154**, 1127–1132.

33. Bland, R. C., Newman, S. C., and Orn, H. (1997). Age and remission of psychiatric disorders *Can. J. Psychiatry* **42**, 722–729.

34. Silverstein, B., and Perlick, D. (1991). Gender differences in depression: historical changes. *Acta Psychiatr. Scand.* **84**, 327–331.

35. Lipman, E. L., Offord, D. R., and Boyle, M. H. (1997). Single mothers in Ontario: Sociodemographic, physical and mental health characteristics. *Can. Med. Assoc. J.* **156**, 639–645.

36. Avison, W. R. (1997). Single motherhood and mental health: implications for primary prevention. *Can. Med. Assoc. J.* **56**, 661–663.

37. Cohen, S., and Wills, T. A. (1985). Stress, social support, and the buffering hypothesis. *Psychol. Bull.* **98**, 310–357.

38. Kessler, R. C., Kendler, K. S., Heath, A. C., Neale, M. C., and Eaves, L. J. (1992). Social support, depressed mood, and adjustment to stress: A genetic epidemiologic investigation. *J. Pers. Soc. Psychol.* **62**, 257–272.

39. Monroe, S. M., and Steiner, S. C. (1986). Social support and psychopathology: Interrelations with preexisting disorder, stress, and personality. *J. Abnorm. Psychol.* **95**, 29–39.

40. Kendler, K. S., Kessler, R. C., Neale, M. C., Heath, A. C., and Eaves, L. J. (1993). The prediction of major depression in women: Toward an integrated etiologic model. *Am. J. Psychiatry* **150**, 1139–1148.

41. Gabbard, G. O. (1994). "Psychodynamic Psychiatry in Clinical Practice. The DSM-IV Edition," Chapter 15, pp. 449–496. American Psychiatric Press, Washington, DC.

42. Paris, J., and Zweig-Frank, H. (1992). A critical review of the role of childhood sexual abuse in the etiology of borderline personality disorder. *Can. J. Psychiatry* **37**, 125–128.

43. Zanarini, M. C., Williams, A. A., Lewis, R. E., Reich, R. B., Vera, S. C., Marino, M. F., Levin, A., Yong, L., and Frankenburg, F. R. (1997). Reported pathological childhood experiences associated with a development of borderline personality disorder. *Am. J. Psychiatry* **154**, 1101–1106.

44. Breedlove, S. M. (1994). Sexual differentiation of the human nervous system. *In* "Annual Review of Psychology" (L. W. Porter and M. R. Rosenzweig, eds.), pp. 389–418. Annual Reviews, Palo Alto, CA.

45. Pilgrim, C., and Reisert, I. (1992). Differences between male and female brains—Developmental mechanisms and implications. *Horm. Metab. Res.* **24,** 353–359.

46. Miranda, R. C., and Sohrabji, F. (1996). Gonadal steroid receptors: Possible roles in the etiology and therapy of cognitive and neurological disorders. *Annu. Rep. Med. Chem.* **31,** 11–20.

47. Toran-Allerand, C. D. (1996). The estrogen/neurotrophin connection during neural development: Is co-localization of estrogen receptors with the neurotrophins and their receptors biologically relevant? *Dev. Neurosci.* **18,** 36–41.

48. Wickelgren, I. (1997). Estrogen stakes claim to cognition. *Science* **276,** 675–678.

49. Lewis, D. A. (1997). Development of the prefrontal cortex during adolescence: Insights into vulnerable neural circuits in schizophrenia. *Neuropsychopharmacology* **16,** 385–398.

50. Sapolsky, R. M. (1996). Stress, glucocorticoids, and damage to the nervous system: The current state of confusion. *Stress* **1,** 1–19.

51. Lewis, D. W., and Diamond, M. C. (1995). The influence of gonadal steroids on the asymmetry of the cerebral cortex. *In* "Brain Asymmetry" (R. J. Davidson and K. Hugdahl, eds.), pp. 31–50. MIT Press, Cambridge, MA.

52. Nopoulos, P. C., Schultz, S. K., and Andreasen, N. C. (1998). Brain and behavior. *In* "Textbook of Women's Health" (L. A. Wallis, ed.), Chapter 98, pp. 795–802. Lippincott-Raven Press, Philadelphia.

53. Shaywitz, B., Shaywitz, S., Pugh, K. *et al.* (1995). Sex difference in the functional organization of the brain for language. *Nature (London)* **373,** 607–609.

54. Seeman, M. V. (1997). CNS clinical drug trials and women. *In* "The Handbook of Psychopharmacology Trials" (M. Hertzman and D. E. Feltner, eds.), pp. 100–122. New York University Press, New York.

55. Gur, R. C., Gur, R. E., Obrist, W. D., Hungerbuhler, J. P., Younkin, D., Rosen, A. D., Skolnik, B. E., and Reivich, M. (1982). Sex and handedness differences in cerebral blood flow during rest and cognitive activity. *Science* **217,** 659–661.

56. Hamilton, J. A., and Yonkers, K. A. (1996). Sex differences in pharmacokinetics of psychotropic medications. Part I. Physiological basis for effects. *In* "Psychopharmacology and Women: Sex, Gender, and Hormones" (M. J. Jensvold, U. Halbreich, and J. A.

Hamilton, eds.), Chapter 2, pp. 11–42. American Psychiatric Press, Washington, DC.

57. Seeman, M. V. (1994). Sex differences in the prediction of neuroleptic response. *In* "Prediction of Neuroleptic Treatment Outcome in Schizophrenia" (W. Gaebel and A. G. Awad, eds.), pp. 51–64. Springer-Verlag, Wien.

58. Seeman, M. V. (1989). Neuroleptic prescription for men and women. *Soc. Pharmacol.* **3,** 219–236.

59. Häfner, H., and Heiden, W. (1997). Epidemiology of schizophrenia. *Can. J. Psychiatry* **42,** 139–151.

60. Harrison, G., Croudace, P., Mason, C., Glazebrook, C., and Medley, I. (1996). Predicting the long-term outcome of schizophrenia. *Psychol. Med.* **26,** 697–705.

61. Opjordsmoen, S. (1991). Long-term clinical outcome of schizophrenia with special reference to gender differences. *Acta Psychiatr. Scand.* **83,** 307–313.

62. Seeman, M. V. (1997). Psychopathology in women and men: focus on female hormones. *Am. J. Psychiatry* **154,** 1641–1647.

63. Seeman, M. V. (1996). The role of estrogen in schizophrenia. *J. Psychiatry Neurosci.* **21,** 123–127.

64. Kessler, R. C., Berglund, P. A., Foster, C. L., Saunders, W. B., Stang, P. E., and Walters, E. E. (1997). Social consequences of psychiatric disorders. II: Teenage parenthood. *Am. J. Psychiatry* **154,** 1405–1411.

65. McLanahan, S. S. (1994). The consequence of single motherhood. *Am. Prospect* **18,** 48–58.

66. Miller, L. J. (1997). Sexuality, reproduction, and family planning in women with schizophrenia. *Schizophr. Bull.* **23,** 623–635,

67. Raine, A., Brennan, P., and Mednick, S. A. (1997). Interaction between birth complications and early maternal rejection in predisposing individuals to adult violence: Specificity to serious, early-onset violence *Am. J. Psychiatry* **154,** 1265–1271.

68. Silverman, M. M. (1989). Children of psychiatrically ill parents: A prevention perspective. *Hosp. Commun. Psychiatry* **40,** 1257–1265.

69. Cloninger, R. C., Christiansen, K. O., Reich, T., and Gottesman, I. I. (1978). Implications of sex differences in the prevalences of antisocial personality, alcoholism, and criminality for familial transmission. *Arch. Gen. Psychiatry* **35,** 941–951.

70. Kimura, D. (1992). Sex differences in the brain. *Sci. Am.* **31,** 119–125.

79

Gender and Mood Disorders

RONALD C. KESSLER
Department of Health Care Policy
Harvard Medical School
Boston, Massachusetts

I. Introduction and Overview

This chapter reviews research on gender differences in mood disorders. Data are reported on prevalence, correlates, help-seeking, and treatment response. The main results are fairly easy to summarize. In community epidemiologic surveys, women consistently have a higher rate of mood disorders than men. The course of mood disorders does not appear to differ by gender. Risk factors for the onset of mood disorders are generally the same for women and men. Mood disorders have a wide variety of adverse consequences, most of which do not differ by gender. There is some evidence that women with a mood disorder are more likely than comparable men to seek professional help and that treatment response is also linked to gender.

II. Mood Disorders Defined

Mood disorders are psychiatric disorders in which mood disturbance is the predominant clinical feature. Modern diagnostic systems including the American Psychiatric Association (APA) Diagnostic and Statistical Manual of Mental Disorders [1] and the World Health Organization (WHO) International Classification of Disease [2] divide mood disorders into two types: depression and mania. Mania is defined as a period lasting a minimum of several days in which the main clinical feature is uncontrollable euphoria or irritability accompanied by a number of other cognitive and behavioral symptoms and serious impairment in functioning. Patients who have manic episodes usually also have intermittent episodes of depression [3], or "bipolar disorder" (*i.e.*, moving back and forth between manic and depressive episodes). Such a high proportion of manic patients develop subsequent depression that the term bipolar disorder is used to describe patients with mania even if they have never been depressed.

Depression is further divided into major depressive episodes (a period lasting a minimum of 2 weeks characterized by persistent depressed mood or anhedonia accompanied by a number of associated cognitive and behavioral symptoms) and dysthymia (a period lasting a minimum of 2 years characterized by

[1]Address comments to the author at the Department of Health Care Policy, Harvard Medical School, 180 Longwood Avenue, Boston, MA 02115-5899, 617-432-3587 (phone), 617-432-3588 (fax), Kessler@ hcp.med.harvard.edu (e-mail). Portions of this chapter previously appeared in the *Journal of Affective Disorders* (**30**, 15–26), the *Journal of American Medical Women's Association* (in press), the book chapter *Sex, Society, and Madness* (in press), and the book chapter *Mood Disorders in Women* (in press) and are reproduced here with permission of the publishers.

depressed mood more days than not and accompanied by a number of symptoms similar to those for a major depressive episode, but not with the persistence of a major depressive episode). Provisional diagnoses proposed for further testing are minor depression (similar to major depression, but with a smaller number of associated symptoms) and brief recurrent depression (similar to minor depression, but with duration of episodes as short as 2–3 days and recurrence at least once a month over a period of at least one year).

III. Issues Related to Ascertainment

Ascertainment of mood disorders is typically based on patient self-reports, although information obtained from clinical observations and from informants should be included in a thorough diagnostic evaluation. No definitive medical tests are available to diagnose mood disorders. Semistructured research diagnostic interviews have been developed for use by clinicians to standardize psychiatric diagnoses [4]. Diagnoses based on these clinical interviews represent the gold standard in the field. However, more typically general population epidemiologic studies base diagnoses on less expensive fully structured interviews administered by trained interviewers who are not clinicians [5]. Methodological research has documented good concordance between diagnoses of depression based on these fully structured interviews and follow-up semistructured clinical validation interviews [6,7], although fully structured diagnostic interviews appear to overdiagnose mania [8].

Concerns have been raised that gender differences in estimated prevalence of depression are due to self-report bias. Societal norms and sex-role socialization experiences may selectively predispose women to admit depression in epidemiologic surveys [9,10]. However, the available evidence is inconsistent with this hypothesis in two important ways.

First, a number of methodological studies have been carried out on this type of response bias in community surveys of nonspecific psychological distress [11–13]. These studies used standard psychometric methods to assess potential biasing factors such as social desirability, expressivity, lying, and yeasaying/ naysaying. No evidence was found in any of these studies that the significantly higher levels of self-reported distress found among women compared to men were due to these biasing factors.

Second, a more indirect, but in some ways more revealing, evaluation of the impact of response bias on the gender difference in major depression was carried out by Young *et al.* [10]. They hypothesized that lower reluctance to admit depression should lead women to be more likely than men to admit ever having a period lasting 2 weeks or longer of being sad, blue, or

Table 79.1

Gender Ratio of Lifetime Diagnosis of DSM-II-R Major Depressive Episodes (MDE) in the National Comorbidity Survey by Varying Diagnostic Criteria

	Percentage meeting MDE lifetime diagnosis			
Diagnostic criteria	Male	Female	Female/Male	Average
Stem[a]	45.7	57.6	1.3	51.7
Stem plus 1 or more symptoms	25.2	35.9	1.4	30.6
Stem plus 2 or more symptoms	20.9	32.0	1.5	26.5
Stem plus 3 or more symptoms	16.6	26.9	1.6	21.8
Stem plus 4 or more symptoms	12.7	21.3	1.7	17.1
Stem plus 5 or more symptoms	8.5	16.6	2.0	12.6
Stem plus 6 or more symptoms	5.7	11.0	1.9	8.4
Stem plus 7 or more symptoms	3.3	6.1	1.9	4.8
Stem plus all 8 symptoms	1.2	3.0	2.5	2.1

Note. Reproduced with permission from an earlier NCS report [16].

[a]The "stem" refers to having two or more weeks of persistent depressed mood or anhedonia.

depressed, but should not affect reports of the less stigmatizing symptoms associated with depressed mood to constitute a major depressive episode, such as sleep disturbance, eating disturbance, and lack of energy. However, the opposite was found by Young and colleagues in an analysis of a representative sample of nonpatient relatives of depressed probands. The female-to-male ratio of a 2-week period of depressed mood or anhedonia (1.3) was much smaller than the ratio of less stigmatizing associated symptoms (1.7).

As shown in Table 79.1, we found very similar results in the nationally representative general population data collected in the U.S. National Comorbidity Survey (NCS) [14]. The female versus male (F:M) ratio for the lifetime prevalence of DSM-III-R [15] major depression in the NCS increased steadily from 1.3 for endorsement of the major depression diagnostic stem question to 2.5 for endorsement of the stem question plus all eight of the other A Criteria stipulated in DSM-III-R [16]. It is implausible that a pattern of this sort would be due to greater reluctance to admit depression on the part of women.

Before concluding that this pattern argues against a differential reporting bias, it is important to recognize that this is exactly what one would expect from invalidity due to differential forgetting. The notion here is that men might remember the vague outlines of past depressive episodes as well as women but they might be more likely than women to minimize in memory the impairment and constellation of symptoms associated with these episodes. Angst and Dobler-Mikola [17] reported a pattern of this sort in a longitudinal study of young adults. Prospective data in their study showed no gender difference in the period prevalence of depression even though retrospective data over the same recall period suggested that women had higher rates than men. Wilhelm and Parker [18] subsequently reported a similar result regarding male underreporting in a separate sample and added the important finding that women were more likely to overreport depression than men (*i.e.,* retrospectively to report a previously reported subclinical episode as having met full criteria). These results argue persuasively that the gender difference in reported lifetime depression is, at least in part, due to a gender difference in accuracy of retrospective reporting. As

noted by Ernst and Angst [19], it is unlikely that this differential retrospective reporting bias completely explains the gender difference in reported depression. The most important evidence in this regard comes from studies of current or recent prevalence, where recall becomes unimportant. Women consistently have higher rates of current depression than men in these studies [20,21]. This finding, of course, is not definitive. It could be that a higher proportion of depressed men consistently refuse to admit their depression to survey interviewers. However, this possibility is inconsistent with the fact that a higher prevalence of current depression among women than among men is found not only in studies that rely on self-report, but also in those that use informant reports. For example, adult informants in family studies report higher rates of depression among their mothers than among their fathers [22].

Another methodological possibility is that men are as likely to be depressed as women but that some sort of recognition bias (as opposed to bias due to differential recall failure or differential willingness to report) leads to more underreporting among men than women. This would come about if men were more likely than women to mask their depression so completely that neither they nor those close to them are aware of it. It would also come about if depression among men was more likely than among women to manifest itself as irritability rather than dysphoria or anhedonia. Current diagnostic systems allow for this possibility among children and adolescents, but not among adults. It would be informative to explore this possibility in future epidemiologic studies.

The evaluation of irritability also raises questions about the accuracy of the assessment of mania. In community epidemiologic surveys that use fully structured diagnostic interviews, the majority of respondents who are classified as manic meet criteria by reporting periods of irritability, but deny ever having euphoria. This is very different from the typical symptom profile seen clinically, where a lifetime history of euphoria characterizes the vast majority of manic individuals [3]. Research aimed at exploring the validity of diagnostic assessment in community surveys shows that the respondents with irritability, but not euphoria, are very often incorrectly diagnosed as manic [8].

As these people are more likely to be men than women, a more rigorous assessment of mania in the general population that excludes such false positives might find that women have a higher prevalence of true mania than men.

IV. Historical Trends

There is evidence from epidemiologic surveys carried out in a number of different countries that the prevalence of major depression has increased dramatically over the past few decades [23]. Estimates based on retrospective age of onset reports obtained in the NCS [24] suggest that the lifetime prevalence of major depression among men in the age range 20–24 increased from 3.7% in the early 1960s to 17.8% in the early 1990s, and among women in the same age range over this same time period prevalence increased from 5.7 to 28.1%.

Neither the population distribution of sex hormones nor the gene pool changes this quickly, which means that the influence of changing environmental conditions must account for these dramatic increases. Before considering what such environmental conditions might be, it is important to recognize that controversy exists about the meaning of the cohort effect. This is because the strongest evidence for this effect comes from cross-sectional surveys that use retrospective age of onset reports to carry out synthetic cohort analyses. Recall failure and reluctance to admit depression might both increase with age and, if so, would create the false impression that the prevalence of depression had increased in recent cohorts. Another possibility is that selective mortality or other forms of sample censoring due to depression might increase with age.

We do not know if these influences are at work. There is at least some indirect evidence consistent with the possibility of age-related recall bias [25]. Furthermore, evidence from simulations suggests that it would not be difficult to produce results regarding cohort effects similar to those found in retrospective surveys by assuming plausible levels of censoring and recall biases [26]. However, other evidence suggests that the cohort effect might represent a real temporal increase and a narrowing of the sex difference; this evidence is based on several long-term prospective epidemiologic surveys [27–29] and archival time-series data documenting increases in suicide rates similar to the estimated increases in depression.

Additional data supporting the claim that the cohort effect is real come from specifications in cross-sectional epidemiologic data. Perhaps the most intriguing of these is the finding that the cohort effect for major depression is much more pronounced for secondary than for primary disorders [30]. Specifically, disaggregated analysis of the NCS data suggests that the lifetime prevalence of major depression in people without a history of any prior psychiatric disorder (i.e., "primary" major depression) has not changed substantially since the 1960s. The prevalence of major depression subsequent to other disorders (i.e., "secondary" depression), in comparison, has increased substantially over this same period of time.

Such a finding is unlikely to be caused by recall bias. Indeed, in light of the fact that comorbid depression is generally more severe than pure depression [30] and the plausible assumption that severe disorders are less likely to go unreported than non-severe disorders, we might expect the opposite pattern to have

occurred if recall error explained the cohort effect. Another finding that we would not expect if recall error was the primary reason for the observed intercohort difference in the survey data is that the concentration of the cohort effect in secondary depression has been due to increases in the prevalence of other primary disorders such as anxiety and drug addiction, not to increases in the transition probabilities from these primary disorders to secondary depression.

V. Distribution in Women

A. Prevalence

A higher prevalence of depression among women than among men is one of the most widely documented findings in psychiatric epidemiology. This gender difference has been found throughout the world using a variety of diagnostic schemes and interview methods [31–36]. The prevalence of major depression among women in these studies has typically been between 1.5 to 3 times that of men. Despite this consistency, there has been enormous variation in the estimated total population prevalence of major depression, with lifetime prevalence estimates ranging between 6 [20] and 17% [21].

A substantial number of epidemiologic studies have also examined gender differences in dysthymia [37–40]. These studies have consistently found that dysthymia is more prevalent among women than men, with female:male (F:M) prevalence ratios generally around 2:1. Total population lifetime prevalence estimates of dysthymia are typically in the range of 6–8% when diagnostic hierarchy rules are not made. However, when the definition of dysthymia excludes respondents with an intercurrent major depressive episode over a 2-year period, the total population lifetime prevalence estimate of dysthymia drops to less than 3% [24]. Smaller numbers of epidemiologic studies have examined gender differences in minor depression [41] or brief recurrent depression [42]. These types of subthreshold depression are consistently found more among women than men.

Epidemiologic studies have not found meaningful gender differences, in comparison, in the prevalence of mania [Epidemiologic Catchment Area (ECA)] [30]. It is important to remember in this regard, as noted earlier in the chapter, that the evaluation of mania has low validity in community epidemiologic surveys [6,7]. This low validity, in conjunction with the low total population prevalence of true mania (1–2% lifetime prevalence), introduces considerable uncertainty into the evaluation of gender differences. However, treatment studies have also failed to find a gender difference in mania [3], adding support to this finding in general population surveys.

B. Age of Onset Distributions

A question that can be raised is whether lifetime F:M prevalence ratios are the same regardless of the age of respondents. This question is motivated by the possibility that the rates of depression in women and men might be getting more similar in recent years [27,43,44]. We are aware of only one direct examination of this possibility [24]. This study made use of NCS respondents' retrospective reports of the age when they first experienced their depression. By comparing these reports across

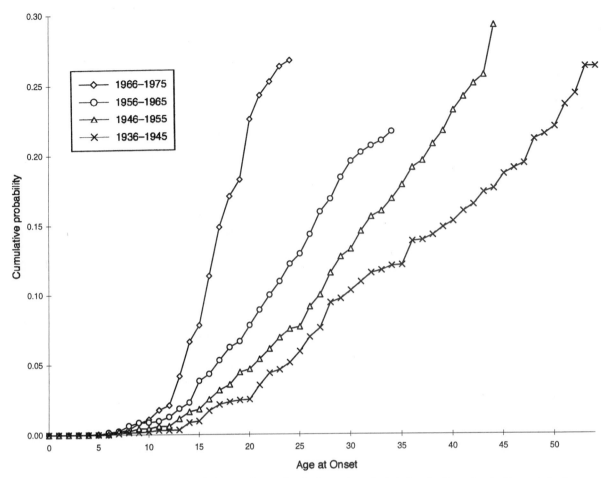

Fig. 79.1   Cohort differences in the cumulative age of onset of major depressive episode in females. Reproduced with permission from an earlier NCS report [52].

subsamples who differed in age at interview but focusing on the same recalled ages, it was possible to generate synthetic age of onset curves for each sample cohort to see whether there has been a change over time.

These curves are presented in Figures 79.1 and 79.2 for the cumulative lifetime (LT) prevalence of ever having a major depressive episode. Separate curves are presented for each of four NCS 10-year birth cohorts, beginning with the oldest respondents, who were born in the decade before the end of World War II (1936–1945), and ending with the youngest respondents, who were born during the decade spanning the second half of the 1960s and the first half of the 1970s (1966–1975). The first of the two figures shows the estimated cumulative LT prevalence of depression among women in these four cohorts, while the second shows parallel results for men. There is a consistent trend for LT prevalence to be higher at all ages in successively younger cohorts in both figures. This means that LT depression is becoming more common in recent cohorts. Similar analyses were carried out for bipolar disorder, but no significant cohort effect was found [8].

Is the cohort effect for depression associated with a change in the F:M LT prevalence ratio? The answer is presented in

Table 79.2, where we see NCS cohort-specific cumulative ratios of this sort by 5-year age intervals. An elevated female ratio can be observed by age 14 in the youngest cohort and by age 9 in the next two older cohorts, but not until age 24 in the oldest cohort. This later emergence of the gender difference in the oldest cohort could be due to a true cohort effect or to less accurate recall of early-onset depression in the oldest cohort. We would not expect this cohort effect to exist if biological factors were largely responsible for the gender difference in depression (*e.g.,* if onset of puberty played a powerful part in causing the emergence of depression among girls). Therefore, if this cohort effect is genuine and not a result of bias, it is likely due to environmental rather than biological risk factors.

Despite this cohort difference in the age at which the gender difference first emerges, the F:M ratio does not differ greatly across cohorts by age 24, the oldest age of onset at which we can compare all four NCS cohorts. The ratio is 1.9 in the youngest cohort and decreases monotonically to 1.6 in the oldest cohort, a difference that is not statistically significant. There is a trend for the ratio at older ages to be larger in successively older cohorts. However, only one of the 14 possible pairwise intercohort comparisons in Table 79.2 beginning at age 24 is statisti-

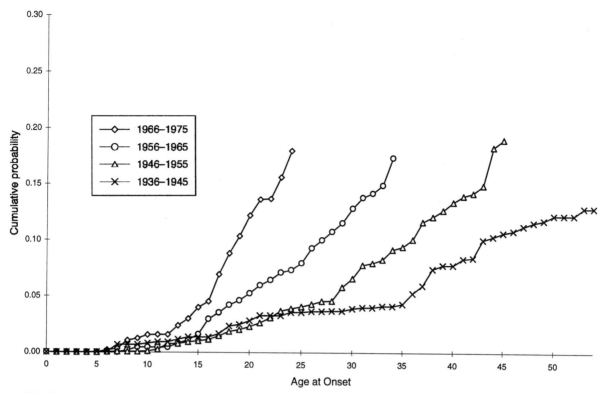

Fig. 79.2 Cohort differences in the cumulative age of onset of major depressive episode in males. Reproduced with permission from an earlier NCS report [52].

cally significant. The general similarity of the ratios across cohorts beginning at age 24 is much more impressive than this one significant difference. This similarity across cohorts is particularly striking in light of the finding in Figures 79.1 and 79.2 that there is a roughly fivefold increase in the lifetime prevalence of depression across these cohorts.

C. The Effects of Prior Psychopathology

Breslau [45] proposed the existence of an interesting specification—that the gender difference in depression is partly due to a difference in prior anxiety. She supported this claim by showing that the odds ratio (OR) of gender predicting major depression substantially attenuates when controls are introduced for the prior existence of anxiety. A similar result was reported by Wilhelm, *et al.* [46]. This finding is indirectly consistent with the result, reported above, that the cohort effect for major depression is largely confined to secondary depression. However, this finding is limited in that it focuses on a predictor that is characteristic of women (*i.e.*, anxiety) while ignoring other comparable predictors that are more characteristic of men (*e.g.*, alcohol/drug abuse, conduct disorder, antisocial personality disorder).

In an effort to investigate this issue, Breslau's result was replicated in the NCS by estimating a pair of survival models to predict first onset of major depression [47]. The only predictor in the first model was gender. The F:M OR in this model was 1.9. The second model introduced a series of controls for prior

anxiety disorders. Consistent with Breslau, the introduction of these controls was found to reduce the F:M OR to 1.6.

However, a third model was also estimated that controlled for prior substance use disorders and conduct disorder rather than prior anxiety disorders. The F:M OR in this third model increased to 2.4. This means that the higher risk of depression among women compared to men would be even greater than it is were it not for the fact that men are more likely than women to have prior substance and conduct problems.

Finally, a model was estimated that controlled simultaneously for prior anxiety disorders, substance use disorders, and conduct disorder. The F:M OR in this final model was 1.9—exactly what it was in the model that had no controls. Furthermore, in the subsample of respondents with no history of any of these prior disorders, the F:M OR was 2.2. These results show clearly, contrary to Breslau's conclusion, that history of prior psychiatric disorders does not play an important part in explaining the observed gender difference in onset risk of major depression.

D. Illness Course

There are some theories about the reasons for gender differences in depression that emphasize the importance of differential persistence. For example, sex-role theories suggest that the chronic stresses and lack of access to effective coping resources associated with traditional female roles lead to a higher prevalence of depression among women [48,49]. Rumination theory suggests that women are more likely than men to dwell on

Table 79.2

Cumulative Female:Male Odds Ratios for Lifetime Major Depressive Episode[a]

Age of onset	Cohort							
	1966–1975		1956–1965		1946–1955		1936–1945	
	OR	(95% CI)	OR	(95% CI)	OR	(95% CI)	OR	(95% CI)
9	0.7	(0.2–1.9)	1.9	(0.4–7.9)	1.7	—	0.2	—
14	2.0[b]	(1.0–3.8)	2.4[b]	(1.3–4.3)	1.5	(0.8–3.2)	0.6	(0.1–2.8)
19	2.0[b]	(1.3–3.2)	1.6[b]	(1.2–2.3)	1.5	(0.9–2.6)	0.8	(0.3–2.2)
24	1.9[b]	(1.3–2.9)	1.8[b]	(1.4–2.3)	1.6[b]	(1.1–2.4)	1.6	(0.7–3.4)
29			1.6[b]	(1.3–2.1)	1.7[b]	(1.2–2.4)	2.6[b]	(1.5–4.5)
34			1.5[b]	(1.2–2.0)	1.9[b]	(1.4–2.5)	2.8[b]	(1.7–4.5)
39					1.8[b]	(1.4–2.3)	2.0[b]	(1.3–3.0)
44					1.7[b]	(1.4, 2.2)	1.8[b]	(1.2–2.7)
49							1.9[b]	(1.2–2.8)
54							2.0[b]	(1.3–3.0)

Note. Reproduced with permission from an earlier NCS report [24]. OR, odds radio; 95% CI, 95% confidence interval; —, low precision.

[a]The results are based on a series of discrete-time survival models in which gender and age (person-year) were used to predict onset of major depressive episodes in subsamples defined by the cross-classification of age (person-year) and cohort. The ORs are gender effects.

[b]Statistically significant at the 0.05 level (two-tailed test).

problems and, because of this, to let transient symptoms of dysphoria grow into clinically significant episodes of depression [50]. Both of these perspectives imply that the higher point prevalence of clinically significant depression among women is due, at least in part, to a higher persistence than that found among men.

There is some evidence from retrospective studies that women do, in fact, have a more chronic course of depression than men [19]. However, the differential recall bias discussed earlier in the chapter appears to explain this effect. Indeed, Ernst and Angst [19] found that evidence for a higher recurrence risk of major depression among women disappears entirely when adjustments are made for differential recall bias. Furthermore, in the NCS, where special procedures were used to stimulate active memory search [6], no gender difference in recurrence risk was found [24]. The NCS also failed to yield evidence of a gender difference either in speed of episode recovery or in chronicity. Furthermore, a report from the NIMH Collaborative Program on the Psychobiology of Depression reported prospective data from a multisite patient sample that showed no gender difference in the course of depression [51].

Additional information about gender differences in the course of mood disorders comes from the same epidemiologic studies that documented a higher lifetime prevalence of major depression among women compared to men. These studies also show that women are more likely than men to have a recent episode of depression. Between one-third and more than half of the respondents in these surveys who were estimated ever to experience depression are classified in these studies as having an episode of depression during the year prior to the interview, while up to 80% of those with history of mania are estimated to have a past year episode of either mania or depression. However, although no explicit investigations of course were carried out in most of these studies, inspection of the prevalence estimates shows that the ratios of current to lifetime prevalence, a rough

indicator of persistence, do not differ consistently by gender.

These sorts of indirect results are not definitive because they fail to adjust for the possibility of gender differences either in age of onset or in time since onset. A more formal analysis of persistence with cross-sectional data is possible by controlling for age of onset and years since onset. Results of such an analysis carried out in the NCS [52] are presented in Table 79.3, where the F:M OR of 12-month prevalence among people with a LT disorder, controlling for age of onset and years since onset of that disorder are shown. There is no evidence in these data for a gender difference in the course of mood disorders.

The one exception to this general pattern is the finding in many clinical studies that women with bipolar disorder have a higher ratio of depressive to manic episodes than men with bipolar disorder [53–55]. I am aware of only one attempt to verify this pattern in a general population sample [8]. No evidence was

Table 79.3

Gender Differences (Female:Male) in Persistence of Mood Disorders[a]

Disorder	OR	(95% CI)
Major Depressive Episode	1.0	(0.7–1.4)
Manic Episode	0.5	(0.2–1.3)
Dysthymia	0.8	(0.5–1.3)

Note. Reproduced with permission from an earlier NCS report [52]. OR, odds ratio; 95% CI, 95% confidence interval.

[a]Persistence is defined as 12-month prevalence among respondents with a lifetime history of the disorder and age of onset more than one year prior to the time of interview. The ORs are the effects of gender (female:male) in a multivariate logistic regression equation controlling for age of onset and number of years since onset.

found in this one study of a gender difference in the ratio of manic to depressive episodes among bipolar individuals.

VI. Influences of Women's Social Roles

It has been known since the 1970s [56] that women report higher levels of depressed mood than men and that this gender difference is stronger for married people than for the unmarried. More recent data have shown that the same specification exists for major depression. This specification was the main empirical basis for the sex-role theory of female depression put forth originally in the early 1970s [57] and subsequently elaborated by a number of feminist scholars in both the scientific [48,58] and popular [59,60] literature. The basic claim of this theory is that women are more depressed then men because of the higher levels of stress and lower levels of fulfillment in female compared to male sex roles. The specification by marital status, according to this account, is due to the fact that married women are more strongly exposed to traditional sex-role experiences than single women.

However, the data are inconsistent with this line of thinking. Epidemiologic data show that the gender difference in risk of first onset of major depression is the same among married as among never married or previously married women. Two other processes lead to a stronger gender difference in depression among married compared to not married people. First, depression has different effects on marital stability for women. Second, although there is no gender difference in the chronicity or recurrence of major depression, the environmental experiences that are associated with chronicity and recurrence are different for men and women. For example, financial pressures are more depressogenic for men than women whereas family problems are more depressogenic for women than men [61]. Together, these two processes create a stronger association between gender and depression among married people.

VII. Risk Factors

Most research on risk factors for mood disorders has focused on the predictors of episode onset of major depression. A number of consistently significant risk factors have been found, including family history of psychiatric disorder [62,63], childhood adversities [64,65], various aspects of personality [66,67], social isolation [68,69], and exposure to stressful life experiences [70,71]. It is important to note that the family history results have generally shown little specificity; that is, family histories of anxiety or alcoholism or other psychiatric disorders have often been just as important as a family history of mood disorders in predicting depression [22,72]. The same has been true for the effects of childhood adversities, as the intercorrelations among the many different types of adversity that predict depression are so strong that it is impossible to pinpoint any individual types of adversity as being especially important [22,73]. Some specificity has been found, in comparison, in the effects of stressful life experiences, with stressors involving loss strongly related to risk of depression, stressors involving danger related to risk of anxiety, and stressors involving a combination of danger and loss related to risk of mixed anxiety-depression [74–76].

Despite the considerable research carried out on risk factors for episode onset of depression, the confusion noted in an earlier section of the chapter between first episodes and recurrences makes it impossible to draw firm conclusions about the importance of these risk factors. A good example of this confusion can be found in the work of Nazroo et al. [77] on the differential effects of stress on episode onset of major depression among women and men. Nazroo and colleagues argued that the gender difference in risk of past-year episodes of major depression is due in large part to a greater vulnerability to the effects of stressful life events and difficulties among women compared to men. However, given that the vast majority of past-year episodes of major depression in the adult general population are recurrences rather than first onsets, this argument is misguided. Nazroo et al. failed to distinguish first onsets from recurrences in the analysis of past-year episode onset and so confounded three separable processes: (1) the significant gender difference in risk of first onset in the subsample of respondents who had never been depressed prior to the past year; (2) the insignificant gender difference in recurrence risk in the subsample of respondents who were asymptomatic at the beginning of the past-year but who had a subsequent recurrence of major depression; and (3) the significant gender difference in the proportion of respondents who have an elevated risk of episode recurrence due to the fact that they have a history of depression.

Furthermore, exposure to stressful life experiences is higher among people with a history of depression compared to those without such a history [78]. The impact of stress on the onset of depressive episodes is also stronger among people with a history of depression than among those without such a history. Given that women are more likely than men to have a history of depression, these differences in exposure and impact will create the impression that the higher rate of past-year episode onset among women is due to a combination of differential exposure and differential reactivity to stress when, in fact, it is largely due to a higher proportion of women having a history of depression.

Our work with the NCS has shown that a number of putative risk factors for a gender difference in depression work in exactly the same way. One of the most important of these is personality. There has been considerable discussion of the possibility that personality differences account for the higher prevalence of depression among women than among men. Low self-esteem, interpersonal dependency, pessimistic attributional style, low perceived control, expressiveness, and instrumentality have all been implicated in this way [31,79–82]. However, few of the empirical studies that have documented these effects controlled for prior history of depression and those that did failed to find powerful effects of personality on the gender difference in depression [83,84]. When we carried out similar analyses in the NCS we found that personality differences appeared to explain part of the association between gender and current depression in the total sample when we did not control for prior history of depression. When we controlled for history of depression, however, all such significant associations disappeared.

VIII. Biological Markers of Susceptibility and Exposure

The possibility that sex hormones are involved in the gender difference in depression is raised by the finding that the

Table 79.4

Probability of 12-Month Service Use for Psychiatric Problems in Separate Service Sectors by 12-Month NCS/DSM-III-R Disorders by Gender

Disorder	Service use for psychiatric problems during the past 12 months[a]											
	MHS		GM		HCS		HS		SH		ANY	
	%	(se)	%	(se)	%	(se)	%	(se)	%	(se)	%	(se)
Any mood disorder												
Female	19.8	(2.2)	14.5[b]	(2.2)	29.0	(2.7)	14.4	(1.5)	7.5	(1.0)	36.8	(2.8)
Male	23.2	(3.2)	8.6	(2.3)	26.0	(3.5)	12.8	(2.6)	10.1	(2.0)	35.6	(4.0)
Any 12-month NCS disorder												
Female	13.4	(1.2)	10.7[b]	(1.5)	20.3[b]	(1.8)	10.5[b]	(0.9)	6.4	(0.7)	27.7[b]	(1.7)
Male	11.2	(1.4)	4.5	(1.0)	13.7	(1.6)	6.9	(1.0)	8.6	(1.2)	21.1	(1.9)
No 12-month NCS disorder												
Female	4.2[b]	(1.0)	2.2	(0.5)	5.5	(1.0)	4.3	(1.1)	1.7	(0.5)	9.8[b]	(1.4)
Male	1.5	(0.4)	2.0	(1.0)	3.5	(1.1)	2.5	(0.7)	1.0	(0.2)	6.5	(1.1)

Note. Reproduced with permission from an earlier NCS report [52].

[a]MHS, mental health specialty; GM, general medical; HCS, health care system (either MHS or GM); HS, human services; SH, self-help; ANY, any services; se, standard error.

[b]Statistically significant gender difference at the 0.05 level (two-tailed test).

observed gender difference emerges at about the time of puberty [85]. Angold *et al.* [86] are currently conducting an important study that investigates this issue by following cohorts of prepubescent boys and girls through puberty using direct measures of sex hormones from blood samples. Preliminary results suggest that the increase in depression among girls relative to boys occurs sharply at mid-puberty (Tanner stage III) and that pubertal status is more important than age in predicting the gender difference in depression.

However, an intriguing specification in earlier work must be considered when evaluating these findings. A gender difference in low self-esteem, which often accompanies depression, emerges at different ages depending on the timing of the transition from primary to secondary school and the beginning of girls' exposure to the pressures associated with dating older boys. Simmons and Blyth [87] found that the gender difference in low self-esteem emerges in the seventh grade when students are situated in a school system that has grades seven through nine in a middle school, but not until the ninth grade when students hail from a system that has a K to eighth grade primary school and a four-year high school. Clearly, even if sex hormones are involved, this specification strongly suggests that there is an interaction with environmental stresses related to sex roles in potentiating this differential vulnerability.

It is also possible to search for a biological basis of the gender difference in depression at the genetic level. The twin research of Kendler *et al.* [88] has shown clearly that there is a strong genetic basis for major depression among women. Until recently this work was confined exclusively to female-female twin pairs. More recently, though, male-male pairs and male-female pairs have been added to the Kendler twin dataset and a separate nationally representative sample of both same-sex and opposite-sex twin-pairs has been collected [89]. These new studies, results of which have not yet been reported, will soon provide important information on the extent to which genes are involved in the sex difference in depression.

Related research on bipolar disorder has already advanced beyond this point by documenting not only family aggregation [90], but also by finding susceptibility genes [91,92]. There is considerable controversy about these results, however, due to concern about bias in ascertainment and lack of replication [93]. Given the progress in genetic mapping and advances in the automation of genotyping, it is likely that this controversy will be resolved in the near future.

IX. Clinical Issues

A. *Gender Differences in Patterns of Help-Seeking*

Research on help-seeking consistently finds that women with psychiatric disorders are significantly more likely than comparable men to seek treatment [94,95]. International research shows that this pattern is true in countries throughout the world [96]. The gender difference in help-seeking persists after adjusting for differential need for services. Is it implausible to argue that women have more ready access to services than men, as available evidence is clear in showing that both financial [97] and nonfinancial [98] barriers to seeking care are greater for women than men.

The most recent general population data on gender and help-seeking in the U.S., from the NCS, are reported in Table 79.4. As shown in the first two rows of the table, women with a 12-month mood disorder in this survey were significantly more likely than their male counterparts to seek help from the general medical sector (14.5% vs 8.6%). Due to slightly higher rates of services use in the mental health speciality and self-help sectors among men, though, there is no gender difference in the overall proportion of respondents who sought some type of services for a mood disorder (36.8% vs 35.6%). The average number of visits in each service sector among respondents in treatment for a mood disorder was also examined and no significant gender difference was found.

The third and fourth rows of Table 79.4 show that, unlike the situation for mood disorders, women with one of the other disorders assessed in the NCS (including anxiety disorders, substance use disorders, and nonaffective psychosis) were significantly more likely than comparable men to seek help (27.7% vs 21.1%). The last two rows, finally, show that women without any of the disorders assessed in the NCS were significantly more likely than comparable men to seek some type of services for a psychiatric problem (9.8% vs. 6.5%). It is unclear why the situation here is different than it is for mood disorders.

B. Gender Differences in Treatment Response

Effective treatments are available for mood disorders, with rates of treatment response that compare favorably to those for most chronic physical disorders. There is emerging evidence, however, that optimal treatments for nonbipolar depression are different for women than for men [99]. Focusing first on psychotherapy, women with severe depression are less responsive to cognitive behavioral therapy (CBT) than are severely depressed men, whereas there is no gender difference in response to CBT among patients with less severe depression [100]. There is no gender difference in response to interpersonal psychotherapy (IPT) at any level of depression severity [101]. Thase *et al.* [99] speculated that the greater success of IPT compared to CBT among women with severe depression might be due to the fact that the focus of IPT on roles, emotions, and interpersonal processes is more consistent with the coping styles used by women than the focus of CBT on structure, homework, and rationality.

Optimal pharmacologic treatments for depression also differ by gender. Clinical trials that report results by gender have consistently found that response to imipramine is greater among men than among women [102]. Research suggests that this may be due to the fact that depressed women are more likely than depressed men to have reversed vegetative symptoms (*i.e.,* increases in appetite and energy compared to the more typical decreases seen in depression) and that patients with these symptoms have more difficulty tolerating the side effects of imipramine and also are less responsive than more typical depressives to imipramine and other tricyclic antidepressants (TCAs) [99]. Interestingly, these gender differences seem to disappear after menopause. Women have fewer side effects and greater treatment response to selective serotonin reuptake inhibitors (SSRIs) compared to TCAs, while differences between TCAs and SSRIs among men are much more modest [99].

X. Epidemiologic Issues

The WHO Global Burden of Disease (GBD) Study estimated that nonbipolar major depression was the fourth leading cause of disease-related disability in the world as of 1990 and will become the second leading cause by the year 2020 [103]. This extremely high ranking is due to the fact that depression is a very common and chronic disease, that it has a much earlier age of onset than most other chronic diseases, and that it is associated with serious role impairment. Research has shown that the impact of depression on role functioning and quality of life is comparable to that of serious chronic physical conditions [104]. Bipolar disorder, while much lower in overall burden from a

public health perspective due to its comparatively low prevalence, is a severely disabling disorder that blights the lives of many patients and their families [3].

One might expect that the global burden of depression would be much greater among women than among men. The GBD Study concluded that this is true. In fact, major depression was estimated to be the number one cause of disease-related disability in the world among women. In drawing this conclusion, though, the GBD investigators made the untested assumption that the impact of depression on role functioning is the same for men and women. As it turns out, research confirms the plausibility of this assumption. Two lines of this research are worthy of mention: one on depression and role transitions and the other on depression and role performance.

The first of these two has been studied largely in the NCS. The basic logic was to use retrospectively reported age of onset data to classify respondents according to the existence of psychiatric disorders prior to the age of a particular life course transition and then to use survival analysis to determine whether these disorders significantly predicted the transitions. Analyses of this sort were carried out for a range of outcomes [52,105–109]. Early-onset mood disorders were found to predict truncated educational attainment, teen childbearing, marital instability, occupational instability, and low socioeconomic status. Importantly, all of these effects were as strong among women as among men.

Information about role performance also comes from epidemiologic surveys. A number of such surveys have documented that psychiatric disorders are associated with increases in sickness absence from work as well as with self-reported increases in work cutback and reductions in quality of work [110,111]. Analyses of this sort in the NCS [112] found that depression is associated with substantial work loss and role cutback and that the magnitudes of these effects are very similar for women and men.

XI. Conclusions and Future Directions

Mood disorders are among the most commonly occurring and seriously impairing diseases to affect women. Major depression, the most prevalent of the mood disorders, is arguably the number one disease among women in the world today in terms of total disease burden. Fortunately, effective treatments exist and knowledge is accumulating about treatment strategies that are especially effective among women.

However, only a minority of depressed women seek professional treatment (typically between one-third and one-fourth across studies), leading to a prolongation of unnecessary suffering. Several efforts are underway in the U.S. to increase the proportion of women and men with depression who obtain treatment, including special training programs for primary care physicians and obstetricians/gynecologists to increase recognition of depression [113] and a Depression Screening Day to increase public awareness of depression [114]. Evaluations and refinements of these programs are being carried out, and international dissemination efforts coordinated by the WHO have also been initiated.

There is also a problem of maintaining patients in treatment for depression. Dropout from pharmacotherapy is common because of adverse side effects. Fortunately, newer medications have fewer side effects, promising to address dropout associated with treatment tolerance. However, many other people both in psychotherapy and in pharmacotherapy for depression quit prematurely because, according to their reports, they want to manage on their own as soon as their symptom relief is adequate. This kind of premature termination leads to incomplete episode resolution and increased recurrence risk. We strongly suspect that the stigma of being treated for a mental illness plays a role in patients' responses, but our knowledge is woefully incomplete because this has been a neglected area of research. A new emphasis on premature termination dropout is needed with the aim of developing innovative approaches to reduce dropout.

Documentation of dramatically increasing rates of depression in recent cohorts has led to a call for a new focus on preventive interventions [115]. Given that depression often begins early in life and that the gender difference in depression emerges with puberty, such interventions have so far focused on children and adolescents [116,117]. Although a great deal of risk factor research has been carried out in an effort to inform these interventions, the confounding of information about prior history with information about first onset hampers the development of evidence-based preventive interventions. More focused risk factor research carried out in conjunction with intervention planning teams is needed to overcome these limitations.

Finally, in order for progress to be made on all the fronts reviewed, health policymakers need to recognize the importance of depression. The research funds available to study depression continue to lag far behind those available to study diseases that more typically occur in men. Health care systems continue to impose cost-control mechanisms, such as copayments and caps on number of visits, for the treatment of depression and other emotional disorders that do not exist for the treatment of physical disorders. New initiatives are needed to facilitate, rather than impede, efforts to seek treatment for these disorders in the context of systems designed to assure quality of cost-effective provision of established therapies.

Acknowledgments

Preparation of this chapter was partially supported by grants R01 MH41135, R01 MH46376, R01 MH49098, and K05 MH00507 from the U.S. National Institute of Mental Health with supplemental support from the National Institute of Drug Abuse (through a supplement to MH46376) and the WT Grant Foundations (Grant 90135190).

References

1. American Psychiatric Association (1994). "Diagnostic and Statistical Manual of Mental Disorders," 4th ed. APA, Washington, DC.
2. World Health Organization (1992). "The ICD-10 Classification of Mental and Behavioral Disorders." WHO, Geneva.
3. Goodwin, F. K., and Jamison, K. J. (1990). "Manic-Depressive Illness." Oxford University Press, New York.
4. Spitzer, R. L., Williams, J. B. W., Gibbon, M., and First, M. B. (1992). The structured clinical interview for DSM-III-R (SCID). I. History, rationale, and description. *Arch. Gen. Psychiatry* **49**, 624–629.
5. World Health Organization (1997). "Composite International Diagnostic Interview (CIDI, Version 2.1)." WHO, Geneva.
6. Kessler, R. C., Wittchen, H.-U., Abelson, J. M., Kendler, K. S., Knäuper, B., McGonagle, K. A., Schwarz, N., and Zhao, S. (1998). Methodological studies of the Composite International Diagnostic Interview (CIDI) in the U.S. National Comorbidity Survey. *Int. J. Methods Psychiatr. Res.* **7**, 33–55.
7. Wittchen, H.-U. (1994). Reliability and validity studies of the WHO-Composite International Diagnostic Interview (CIDI): A critical review. *J. Psychiatr. Res.* **28**, 57–84.
8. Kessler, R. C., Rubinow, D. R., Holmes, C., Abelson, J. M., and Zhao, S. (1997). The epidemiology of DSM-III-R bipolar I disorder in a general population survey. *Psychol. Med.* **27**, 1079–1089.
9. Phillips, D., and Segal, B. (1969). Sexual status and psychiatric symptoms. *Am. Sociol. Rev.* **34**, 58–72.
10. Young, M. A., Fogg, L. F., Scheftner, W. A., Keller, M. B., and Fawcett, J. A. (1990). Sex differences in the lifetime prevalence of depression. *J. Affective Disord.* **18**, 187–192.
11. Clancy, K., and Gove, W. (1972). Sex differences in mental illness: An analysis of response bias in self-reports. *Am. J. Clin. Hypnosis* **14**, 205–216.
12. Gove, W. R., and Geerken, M. R. (1977). Response bias in surveys of mental health: An empirical investigation. *Am. J. Sociol.* **82**, 1289–1317.
13. Gove, W. R. (1978). Sex differences in mental illness among adult men and women: An evaluation of four questions raised regarding the evidence on the higher rates of women. *Soc. Sci. Med.* **12B**, 187–198.
14. Kessler, R. C., McGonagle, K. A., Zhao, S., Nelson, C. B., Hughes, M., Eshleman, S., Wittchen, H.-U., and Kendler, K. S. (1994). Lifetime and 12-month prevalence of DSM-III-R psychiatric disorders in the United States: Results from the National Comorbidity Survey. *Arch. Gen. Psychiatry* **51**, 8–19.
15. American Psychiatric Association (1987). "Diagnostic and Statistical Manual of Mental Disorders," 3rd rev. ed. APA, Washington, DC.
16. Kessler, R. C., McGonagle, K. A., Swartz, M. S., Blazer, D. G., and Nelson, C. B. (1993). Sex and depression in the National Comorbidity Survey I: Lifetime prevalence, chronicity and recurrence. *J. Affective Disord.* **29**, 85–96.
17. Angst, J., and Dobler-Mikola, A. (1984). Do the diagnostic criteria determine the sex ration in depression? *J. Affective Disord.* **7**, 189–198.
18. Wilhelm, K., and Parker, G. (1994). Sex differences in lifetime depression rates: Fact or artefact? *Psychol. Med.* **24**, 97–111.
19. Ernst, C., and Angst, J. (1992). The Zurich study: XII. Sex difference in depression. Evidence from longitudinal epidemiological data. *Eur. Arch. Psychiatry Clin. Neurosci.* **241**, 222–230.
20. Weissman, M. M., Bruce, M. L., Leaf, P. J., Florio, L. P., and Holzer, C., III (1991). Affective disorders. *In* "Psychiatric Disorders in America: The Epidemiologic Catchment Area Study" (L. N. Robins and D. A. Regier, eds.), pp. 53–80. Free Press, New York.
21. Blazer, D. G., Kessler, R. C., McGonagle, K. A., and Swartz, M. S. (1994). The prevalence and distribution of major depression in a national community sample: The National Comorbidity Survey. *Am. J. Psychiatry* **151**, 979–986.
22. Kendler, K. S., Davis, C. G., and Kessler, R. C. (1997). The familial aggregation of common psychiatric and substance abuse disorders in the National Comorbidity Survey: A family history study. *Br. J. Psychiatry* **170**, 541–548.
23. Cross-National Collaborative Group (1992). The changing rate of major depression: Cross-national comparisons. *JAMA, J. Am. Med. Assoc.* **268**, 3098–3105.
24. Kessler, R. C., McGonagle, K. A., Nelson, C. B., Hughes, M., Swartz, M. S., and Blazer, D. G. (1994). Sex and depression in the National Comorbidity Survey II: Cohort effects. *J. Affective Disord.* **30**, 15–26.

25. Simon, G. E., and VonKorff, M. (1995). Recall of psychiatric history in cross-sectional surveys: Implications for epidemiologic research. *Epidemiol. Rev.* **17,** 221–227.

26. Giuffra, L. A., and Risch, N. (1994). Diminished recall and the cohort effect of major depression: A simulation study. *Psychol. Med.* **24,** 375–383.

27. Kessler, R. C., and McRae, J. A., Jr. (1981). Trends in the relationship between sex and psychological distress: 1957–1976. *Am. Sociol. Rev.* **46,** 443–452.

28. Srole, L., and Fisher, A. K. (1980). The Midtown Manhattan Longitudinal Study vs. *The Mental Paradise Lost Doctrine:* A controversy joined. *Arch. Gen. Psychiatry* **37,** 209–221.

29. Hagnell, O., Lanke, J., Rorsman, B., and Ojesjo, L. (1982). Are we entering an age of melancholy? Depressive illness in a prospective epidemiologic study over 25 years: The Lunby Study, Sweden. *Psychol. Med.* **12,** 279–289.

30. Kessler, R. C., Nelson, C. B., McGonagle, K. A., Liu, J., Swartz, M. S., and Blazer, D. G. (1996). Comorbidity of DSM-III-R major depressive disorder in the general population; Results from the U.S. National Comorbidity Survey. *Br. J. Psychiatry* **168,** 17–30.

31. Bebbington, P. E. (1988). The social epidemiology of clinical depression. *In* "Handbook of Social Psychiatry" (A. S. Henderson, and G. D. Burrows, eds.), pp. 87–102. Elsevier, Amsterdam.

32. Nolen-Hoeksema, S. (1987). Sex differences in unipolar depression: Evidence and theory. *Psychol. Bull.* **101,** 259–282.

33. Weissman, M. M., and Klerman, J. K. (1977). Sex differences and the epidemiology of depression. *Arch. Gen. Psychiatry* **34,** 98–111.

34. Weissman, M. M., and Klerman, J. K. (1985). Gender and depression. *Trends Neurosci.* **8,** 416–420.

35. Weissman, M. M., and Klerman, G. L. (1992). Depression: Current understanding and changing trends. *Annu. Rev. Public Health* **13,** 319–339.

36. Weissman, M. M., Leaf, P. J., Holzer, C. E., III, Myers, J. K., and Tischler, G. L. (1984). The epidemiology of depression. An update on sex differences in rates. *J. Affective Disord.* **7,** 179–188.

37. Bland, R. C., Newman, S. C., and Orn, H. (1988). Period prevalence of psychiatric disorders in Edmonton. *Acta Psychiatr Scand.* **77,** 3–42.

38. Bland, R. C., Orn, H., and Newman, S. C. (1988). Lifetime prevalence of psychiatric disorders in Edmonton. *Acta Psychiatr Scand.* **77,** 24–32.

39. Canino, G. J., Bird, H. R., Shrout, P. E., Rubio-Stipec, M., Bravo, M., Martinez, R., Sesman, M., and Guevara, L. M. (1987). The prevalence of specific psychiatric disorders in Puerto Rico. *Arch. Gen. Psychiatry* **44,** 727–735.

40. Wells, J. E., Bushnell, J. A., Hornblow, A. R., Joyce, P. R., and Oakley-Browne, M. A. (1989). Christchurch Psychiatric Epidemiology Study. Part I: Methodology and lifetime relevance for specific psychiatric disorders. *Aust. N. Z. J. Psychiatry* **23,** 315–326.

41. Kessler, R. C., Zhao, S., Blazer, D. G., and Swartz, M. S. (1997). Prevalence, correlates and course of minor depression and major depression in the NCS. *J. Affective Disord.* **45,** 19–30.

42. Angst, J., and Merikangas, K. (1997). The depressive spectrum: Diagnostic classification and course. *J. Affect Disord.* **45**(1–2), 31–39.

43. Murphy, J. M. (1986). Trends in depression and anxiety: Men and women. *Acta Psychiatr. Scand.* **73**(2), 113–127.

44. Weissman, M. M., Bland, R., Joyce, P. R., and Newman, S. (1993). Sex differences in rates of depression: Cross-national perspectives. Special Issue: Toward a new psychobiology of depression in women. *J. Affective Disord.* **29**(2–3), 77–84.

45. Breslau, N. (1995). Sex differences in depression: A role for preexisting anxiety. *Psychiatry Res.* **58,** 1–12.

46. Wilhelm, K., Parker, G., and Hadzi-Pavlovic, D. (1997). Fifteen years on: Evolving ideas in researching sex differences in depression. *Psychol. Med.* **27,** 875–883.

47. Kessler, R. C. (1999). Sex differences in major depression: Epidemiologic findings. *In* "Sex, Society, and Madness" (E. Frank, ed.). American Psychiatric Press, Washington, DC (in press).

48. Barnett, R. C., Biener, G. K., and Baruch, G. K. (1987). "Gender and Stress." Free Press, New York.

49. Mirowsky, J., and Ross, C. E. (1989). "Social Causes of Psychological Distress." de Gruyter, New York.

50. Nolen-Hoeksema, S. (1990). "Sex Differences in Depression." Stanford University Press, Palo Alto.

51. Simpson, H. B., Nee, J. C., and Endicott, J. (1997). First-episode major depression: Few sex differences in course. *Arch. Gen. Psychiatry* **54,** 633–639.

52. Kessler, R. C. (1998). Sex differences in DSM-III-R psychiatric disorders in the United States: Results from the National Comorbidity Survey. *J. Am. Med. Women's Assoc.* **53,** 148–158.

53. Dunner, D. L., and Hall, K. S. (1980). Social adjustment and psychological precipitants in mania. *In* "Mania: An Evolving Concept" (R. H. Belmaker and H. M. van Praag, eds.), pp. 337–347. Spectrum Publ., Jamaica, NY.

54. Taylor, M. A., and Abrams, R. (1981). Gender differences in bipolar affective disorder. *J. Affective Disord.* **3,** 261–277.

55. Roy-Byrne, P., Post, R. M., Uhde, T. W., Porcu, T., and Davis D. (1985). The longitudinal course of recurrent affective illness: Life chart data from research patients at the NIMH. *Acta Psychiatr. Scand.* **317,** 5–34.

56. Gove, W. R. (1972). The relationship between sex roles, marital status, and mental illness. *Soc. Forces* **51,** 34–44.

57. Gove, W. R., and Tudor, J. F. (1973). Adult sex roles and mental illness. *Am. J. Sociol.* **78,** 812–835.

58. Belle, D. (1982). "Lives in Stress: Women and Depression." Sage Publ., Thousand Oaks, CA.

59. Scarf, M. (1980). "Unfinished Business: Pressure Points in the Lives of Women." Doubleday, Garden City, NY.

60. Faludi, S. (1991). "Backlash: The Undeclared War Against American Women." Crown, New York.

61. Kessler, R. C., and McLeod, J. D. (1984). Sex differences in vulnerability to undesirable life events. *Am. Sociol. Rev.* **49,** 620–631.

62. Gershon, E. S., Hamovit, J., Guroff, J. J., Dibble, E., Leckman, J. F., Sceery, W., Targum, S. D., Nurnberger, J. I., Jr., Goldin, L. R., and Bunney, W. E., Jr. (1982). A family study of schizoaffective, bipolar I, bipolar II, unipolar and normal control probands. *Arch. Gen. Psychiatry* **39,** 1157–1167.

63. Weissman, M. M., Gershon, E. S., Kidd, K. K., Prusoff, B. A., Leckman, J. F., Dibble, E., Hamovit, J., Thompson, W. D., Pauls, D. L., and Guroff, J. J. (1984). Psychiatric disorders in the relatives of probands with affective disorders: The Yale University National Institute of Mental Health Collaborative Study. *Arch. Gen. Psychiatry* **41,** 13–21.

64. Oakley-Browne, M. A., Joyce, P. R., Wells, J. E., Bushnell, J. A., and Hornblow, A. R. (1995). Adverse parenting and other childhood experience as risk factors for depression in women aged 18–44 years. *J. Affective Disord.* **34,** 13–23.

65. Rodgers, B. (1994). Pathways between parental divorce and adult depression. *J. Child Psychol. Psychiatry* **35,** 1289–1308.

66. Abrahams, B., Feldman, S., and Nash, S. C. (1978). Sex-role self concept and sex-role attitudes: Enduring personality characteristics or adaptations to changing life situations? *Dev. Psychol.* **14,** 393–400.

67. Klerman, G. L., and Hirshfeld, R. M. A. (1988). Personality as a vulnerability factor: With special attention to clinical depression. *In* "Handbook of Social Psychiatry" (A. S. Henderson and G. D. Burrows, eds.), pp. 41–53. Elsevier, New York.

68. Brown, G. W., Bifulco, A., and Harris, T. O. (1987). Life events, vulnerability and onset of depression: some refinements. *Br. J. Psychiatry* **140,** 30–42.

69. Gotlib, I. H., and Hammen, C. (1992). "Psychological Aspects of Depression: Toward a Cognitive-Interpersonal Integration." Wiley, New York.

70. Dohrenwend, B. P., Shrout, P. E., Link, B. G., Skodol, A. E., and Stueve, A. (1995). Life events and other possible psychosocial risk factors for episodes of schizophrenia and major depression: A case-control study. In "Does Stress Cause Psychiatric Illness?" (C. Mazure, ed.), pp. 43–66. American Psychiatric Press, Washington, DC.

71. Kessler, R. C. (1997). The effects of stressful life events on depression. In "Annual Review of Psychology," pp. 191–214. Annual Reviews, Palo Alto, CA.

72. Merikangas, K. R., Risch, N. J., and Weissman, M. M. (1994). Comorbidity and co-transmission of alcoholism, anxiety and depression. Psychol. Med. 24, 69–80.

73. Mullen, P. E., Martin, J. L., Anderson, J. C., Romans, S. E. and Herbison, G. P. (1996). The long-term impact of the physical, emotional, and sexual abuse of children: A community study. Child Abuse Neglect 20, 7–21.

74. Finlay-Jones, R., and Brown, G. W. (1981). Types of stressful life events and the onset of anxiety and depressive disorders. Psychol. Med. 11, 803–815.

75. Finlay-Jones, R., and Brown, G. W. (1989). Types of stressful life event and onset of anxiety and depressive disorders. Psychol. Med. 11, 803–815.

76. Brown, G. W., Harris, T. O., and Eales, M. J. (1993). Aetiology of anxiety and depressive disorders in an inner-city population: II. Comorbidity and adversity. Psychol. Med. 23(1), 155–165.

77. Nazroo, J. Y., Edwards, A. C., and Brown, G. W. (1997). Gender differences in the onset of depression following a shared life event: A study of couples. Psychol. Med. 27, 9–19.

78. Kessler, R. C., and Magee, W. J. (1993). Childhood adversities and adult depression: Basic patterns of association in a U.S. National Survey. Psychol. Med. 23, 679–690.

79. Abramson, L. Y., and Andrews, D. E. (1982). Cognitive models of depression: Implications for sex differences in vulnerability to depression. Int. J. Ment. Health 11, 77–94.

80. Bassoff, E. S., and Glass, G. V. (1982). The relationship between sex roles and mental health: A meta-analysis of twenty-six studies. Couns. Psychol. 10, 105–112.

81. Baucom, D. H., and Danker-Brown, P. (1984). Sex role identity and sex stereotyped tasks in the development of learned helplessness in women. J. Pers. Soc. Psychol. 46, 422–430.

82. Whiteley, B. E., Jr. (1985). Sex role orientation and psychological well-being: Two meta-analyses. Sex Roles 12, 207–225.

83. Hirshfeld, R. M. A., Klerman, G. L., Clayton, P. J., Keller, M. B., McDonald-Scott, P., and Larkin, B. H. (1983). Assessing personality: Effects of the depressive state on trait measurement. Am. J. Psychiatry 140, 695–699.

84. Hirshfeld, R. M. A., Klerman, G. L., Clayton, P. J., Keller, M. B., and Andreasen, N. C. (1984). Personality and gender-related differences in depression. J. Affective Disord. 7, 211–221.

85. Angold, A., and Worthman, C. W. (1993). Puberty onset of gender differences in rates of depression: A developmental, epidemiologic and neuroendocrine perspective. J. Affective Disord. 29, 145–158.

86. Angold, A., Costello, E. J., and Worthman, C. W. (1998). Puberty and depression: The roles of age, pubertal status, and pubertal timing. Psychol. Med. 28, 51–61.

87. Simmons, R. G., and Blyth, D. A. (1987). "Moving into Adolescence: The Impact of Pubertal Change and School Context." de Gruyter, New York.

88. Kendler, K. S., Neale, M. C., Kessler, R. C., Heath, A. C., and Eaves, L. J. (1993). The lifetime history of major depression in women. Arch. Gen. Psychiatry 50, 863–870.

89. Brim, O. G., and Featherman, D. (1999). Surveying midlife development in the United States. Aging Soc. (in press).

90. Winokur, G. (1991). "Mania and Depression: A Classification of Syndrome and Disease." Johns Hopkins University Press, Baltimore, MD.

91. Blackwood, D. H. R., He, L., Morris, S. W., McLean, A., Whitton, C., Thomson, M., Walker, M. T., Woodburn, K., Sharp, C. M., Wright, A. F., Shibasaki, Y., St. Clair, D. M., Porteous, D. J., and Muir, W. J. (1996). A locus for bipolar affective disorder on chromosome 4p. Nat. Genet. 12, 427–430.

92. Freimer, N. B., Reus, V. I., Escamilla, M. A., McInnes, L. A., Spensy, M., Leon, P., Service, S. K., Smith, L. B., Silva, S., Rojas, E., Gallegos, A., Meza, L., Fournier, E., Baharloo, S., Blankenship, K., Tyler, D. J., Batki, S., Vinogradov, S., Weissenbach, J., Barondes, S. H., and Sandkuijl, L. A. (1996). Genetic mapping usage haplotype, association and linkage methods suggests a locus for severe bipolar disorder (BPI) at 18q22-q23. Nat. Genet. 12, 436–441.

93. Morell, V. (1996). Manic-depression findings spark polarized debate. Science 272, 31–32.

94. Kessler, R. C., Brown, R. L., and Broman, C. L. (1981). Sex differences in psychiatric help-seeking: Evidence from four large-scale surveys. J. Health Soc. Behav. 22, 49–64.

95. Leaf, P. J., and Bruce, M. L. (1987). Gender differences in the use of mental health-related services: A reexamination. J. Health Soc. Behav. 28, 171–183.

96. Alegria, M., Kessler, R. C., Bijl, R., Lin, E., Heeringa, S., Takeuchi, D. T., and Kolody, B. (1999). Comparing mental health service use data across countries. In "Unmet Need in Mental Health Service Delivery" (G. Andrews, ed.). Cambridge University Press, Cambridge, UK.

97. Wilensky, G. R., and Cafferata, G. L. (1983). Women and the use of health services. Women's Health 73, 128–133.

98. Gove, W. R. (1984). Gender differences in mental and physical illness: The effects of fixed roles and nurturant roles. Soc. Sci. Med. 19, 77–91.

99. Thase, M. E., Frank, E., Kornstein, S. G., and Yonkers, K. A. (1999). Sex-related differences in response to treatments of depression. In "Sex, Society, and Madness: Gender and Psychopathology" (E. Frank, ed.), American Psychiatric Press, Washington, DC.

100. Thase, M. E., Reynolds, C. F., III, Frank, E., Simons, A. D., McGeary, J., Fasiczka, A. L., Garamoni, G. G., Jennings, J. R., and Kupfer, D. J. (1994). Do depressed men and women respond similarly to cognitive behavior therapy? Am. J. Psychiatry 151, 500–505.

101. Thase, M. E., Buysse, D. J., Frank, E., Cherry, C. R., Cornes, C. L., Mallinger, A. G., and Kupfer, D. J. (1997). Which depressed patients will respond to interpersonal psychotherapy? The role of abnormal EEG sleep profiles. Am. J. Psychiatry 154, 502–509.

102. Hamilton, J. A., Grant, M., and Jensvold, M. F. (1996). Sex and treatment of depression. In "Psychopharmacology and Women. Sex, Gender, and Hormones" (M. F. Jensvold, U. Halbreich, and J. A. Hamilton, eds.), pp. 241–260. American Psychiatric Press, Washington, DC.

103. Murray, C. J. L., and Lopez, A. D. (1996). Alternative visions of the future: Projecting mortality and disability, 1990–2020. In "The Global Burden of Disease: A Comprehensive Assessment of Mortality and Disability from Diseases, Injuries, and Risk Factors in 1990 and Projected to 2020" (C. J. L. Murray and A. D. Lopez, eds.), pp. 325–395. Harvard University Press, Boston.

104. Wells, K. B., Sturm, R., Sherbourne, C. D., and Meredith, L. (1996). "Caring for Depression." Harvard University Press, Cambridge, MA.

105. Kessler, R. C., Foster, C. L., Saunders, W. B., and Stang, P. E. (1995). The social consequences of psychiatric disorders. I. Educational attainment. *Am. J. Psychiatry* **152,** 1026–1032.

106. Kessler, R. C., Berglund, P. A., Foster, C. L., Saunders, W. B., Stang, P. E. and Walters, E. E. (1997). The social consequences of psychiatric disorders: II. Teenage parenthood. *Am. J. Psychiatry* **154,** 1405–1411.

107. Kessler, R. C. and Forthofer, M. S. (1999). The effects of psychiatric disorders on family formation and stability. *In* "Conflict and Cohesion in Families: Causes and Consequences" (J. Brooks-Gunn and M. Cox, eds.), pp. 301–320. Cambridge University Press, New York.

108. Jayakody, R., Danziger, S., and Kessler, R. C. (1998). Early onset psychiatric disorders and male socioeconomic status. *Soc. Sci. Res.* **27,** 371–387.

109. Ettner, S. L., Frank, R. G., and Kessler, R. C. (1997). The impact of psychiatric disorders on labor market outcomes. *Ind. Labor Relations Rev.* **51,** 64–81.

110. Kouzis, A. C., and Eaton, W. W. (1994). Emotional disability days: Prevalence and predictors. *Am. J. Public Health* **84,** 1304–1307.

111. Berndt, E. R., Ernst, R., Stan, N., Finkelstein, S. N., Greenberg, P. E., Keith, A., and Bailit, H. (1997). "Illness and Productivity: Objective Workplace Evidence," Working Paper 42-97. MIT Program on the Pharmaceutical Industry, MIT, Boston.

112. Kessler, R. C., and Frank, R. G. (1997). The impact of psychiatric disorders on work loss days. *Psychol. Med.* **27,** 861–873.

113. Coyne, J. C., Schwenk, T. L., and Fechner-Bates, S. (1995). Nondetection of depression by primary care physicians reconsidered. *Gen. Hosp. Psychiatry* **17**(1), 3–12.

114. Greenfield, S. F., Reizes, J. M., Magruder, K. M., Muenz, L. R., Kopans, B., and Jacobs, D. G. (1997). Effectiveness of community-based screening for depression. *Am. J. Psychiatry* **154,** 1391–1397.

115. Institute of Medicine (1994). "Reducing Risks for Mental Disorders: Frontiers for Preventive Intervention Research." National Academy Press, Washington, DC.

116. Dryfoos, J. G. (1990). "Adolescents at Risk: Prevalence and Prevention." Oxford University Press, New York.

117. Hamburg, D. A. (1992). "Today's Children: Creating a Future for a Generation in Crisis." Times Books, New York.

80

Anxiety Disorders in Women

KATHLEEN RIES MERIKANGAS* AND RACHEL A. POLLOCK†

*Departments of Epidemiology and Public Health, Psychiatry, and Psychology, Yale University School of Medicine; †Department of Psychology, Yale University, New Haven, Connecticut

I. Introduction

During the 1990s there has been an increasing focus on the anxiety disorders, which have been shown to be the most prevalent psychiatric syndromes in the U.S. population. Nearly one-third of women report a lifetime history of one of the major anxiety disorders. Anxiety is associated with intense subjective distress, social impairment, and medical and psychiatric comorbidity. Although there is abundant literature on sex differences in depression, there is far less empirical research investigating the male–female differences in anxiety disorders. This chapter provides an overview of anxiety disorders in women and systematically explores possible explanations for the female preponderance. Explanations that may falsely elevate the sex ratio for anxiety disorders include sampling bias, reporting differences, artifacts of the classification system, and confounding by other clinical or demographic correlates. The major classes of possible mechanisms for the sex difference, including historical (evolutionary), demographic, developmental, genetic, biologic, and psychosocial risk factors, are reviewed. Finally, the authors suggest future areas of research that may not only further elucidate the relationship between women and manifest anxiety but also provide impetus for the identification of risk factors and early intervention.

II. Background

Anxiety is an ubiquitous human emotion. Its expression can range from a normal reaction to an acutely threatening stimulus to an anxiety attack with multiple physical sensations and fear of impending doom in response to an unknown stimulus. There are two properties that underlie the definitions of anxiety [1]. It is generally unpleasant and future-oriented. It is distinguished from fear in that it either has no discernible source of danger or the emotion is disproportionate to the fear stimulus. Lader [1] defines pathologic anxiety by subjective assessment of the patient that the symptoms are more frequent, more severe, or more persistent than that to which he/she is accustomed or is able to tolerate.

Prior to the current classification of anxiety states, disorders of anxiety were described by a variety of terms such as effort syndrome, neurocirculatory asthenia, soldier's heart, cardiovascular neurosis, and hyperdynamic beta-adrenergic circulatory state. That these disorders were often assumed to be psychogenic in origin was challenged by Cohen and coworkers [2] who could find no evidence to support this assumption.

Despite dramatic advances in our understanding of genetics, neurobiology, and the links between the environment and central nervous system functioning, the etiology of the anxiety disorders is still relatively unknown. There remain no pathognomonic markers with which a presumptive diagnosis of an anxiety disorder may be made in the absence of the clinical interview determining the cognitive and physiologic response of an individual to purported specific or nonspecific environmental stimuli. This highlights the importance of the empirical epidemiologic approach to investigating the definitions and risk factors for the expression of anxiety across the life course. Although the high magnitude of anxiety in females has been well-documented, our understanding of the causes of the sex difference in anxiety is quite limited.

III. Definitions of the Anxiety Disorders

In this chapter, we consider the anxiety disorders as defined by the American Psychiatric Association's Diagnostic and Statistical Manual for Psychiatric Disorders, 4th edition (DSM-IV) [3] criteria including panic, phobias, and generalized anxiety. The major subtypes of anxiety states include panic disorder (with or without agoraphobia), specific phobia, social phobia, and generalized anxiety disorder. The key phenomenologic features of the major anxiety disorders are presented in Appendix A. Obsessive compulsive disorder and posttraumatic stress disorder are also included as anxiety disorders in the DSM-IV; obsessive compulsive disorder will not be considered in the present chapter because the base rates and risk factors differ from those of the other anxiety disorders. Posttraumatic stress disorder is discussed elsewhere in this volume. Despite the progress in nomenclature represented by the definitions of subtypes of anxiety states, the distinctions are still somewhat blurred. The cooccurrence of more than one anxiety state within individuals and the lack of longitudinal stability of specific disorders suggest that further research is necessary to validate the overlap between the anxiety disorders and the boundaries between anxiety and depression.

IV. Assessment

Because of the broad meaning of the term anxiety, numerous measures of anxiety have been employed. The most commonly used assessments include self-report checklists of both state and trait anxiety (*e.g.,* Beck Anxiety Inventory [4], State–Trait Anxiety Inventory [5], fears (Fear Survey Schedule-Revised [6]; Anxiety Sensitivity Index [7]), phobias (*e.g.,* Fear Questionnaire [8], avoidance (*e.g.,* Mobility Inventory [9]), as well as clinician-administered symptom checklists (*e.g.,* Hamilton Anxiety Rating Scale [10]). In addition, there are several structured and semistructured diagnostic interviews designed for administration either by lay interviewers (*e.g.,* Diagnostic Interview Schedule

Table 80.1
Sex-Specific Lifetime Prevalence of Anxiety Disorders in Community Surveys in the U.S.

Anxiety disorders	Epidemiologic catchment area study[a]			National comorbidity survey[b]		
	Males (%)	Females (%)	Sex ratio (F:M)	Males (%)	Females (%)	Sex ratio (F:M)
Total	1.8	10.3	5.7	19.2	30.5	1.6
Generalized anxiety disorder	4.3	6.8	1.6	3.6	6.6	1.8
Panic disorder	0.9	2.0	2.2	2.0	5.0	2.5
Phobic disorders						
Agoraphobia without panic	3.2	7.9	2.5	3.5	7.0	2.0
Simple phobia	7.8	14.5	1.9	6.7	15.7	2.3
Social phobia	2.5	2.9	1.2	11.1	15.5	1.4

[a][19].
[b][20].

(D.I.S.) [11]) or by experienced clinicians (*e.g.,* Structured Clinical Interview for DSM-III (S.C.I.D.) [12,13], Anxiety Disorders Interview Schedule, Revised (ADIS-R) [14]). The ADIS-R [14,15] is the most comprehensive diagnostic interview for the assessment of the diagnostic criteria of anxiety and affective disorders.

Several psychophysiologic indicators of anxiety have been used to examine both human and animal anxiety. Experimental models that induce stress and measure autonomic output to test the human "fight or flight" response to threat have been used to study the range of triggers, correlates, and responses to fear-provoking situations. Behavioral tasks, such as giving a speech, or receiving the threat of shock have been used to experimentally induce anxiety states in normal individuals as well as in those with ongoing anxiety disorders. Measures of changes in pulse, galvanic skin response, heart rate, and temperature regulation, as well as observations of facial expression, blushing, and other overt signs of anxiety are presumed to provide a more accurate depiction of anxiety than are self-reports or interviews about typical response patterns to stress.

V. Historical Evolution

There has been abundant literature regarding the evolutionary significance of anxiety. Both human and nonhuman primates have been evolutionarily predisposed to acquire fears and phobias to objects or situations that may have posed a threat to the survival of the species. Selective advantage would have been conferred to individuals who easily acquired fears of dangerous objects or situations that might have had selective advantage in the struggle for existence. Biological preparedness is postulated to be responsible for the rapid acquisition, irrationality, and high resistance to extinction that are characteristic of phobias [16].

VI. Evidence for Sex Differences in the Prevalence of Anxiety Disorders

A large body of systematic evidence on the prevalence of anxiety states across diverse settings has revealed that anxiety states are more common in women. In contrast to research on

depression, however, there is some evidence that the preponderance of women with anxiety states is greater in general practice and community settings than in inpatient facilities [17,18]. Thus, it is unlikely that the female excess of anxiety can be attributed to an increased tendency for women to be hospitalized for anxiety. The following section provides a review of sex differences in anxiety disorders in community surveys of adults and children.

A. Sex Differences in Anxiety Disorders in Community Surveys

1. Adults

a. RATES OF ANXIETY DISORDERS. The results of the two large-scale community-based surveys of psychiatric disorders of adults in the United States, the Epidemiological Catchment Area study (ECA) [19] and the National Comorbidity Study (NCS) [20], confirm the sex differences in anxiety disorders that have been reported in earlier epidemiologic studies across the world since the 1980s [21–30]. The ECA is a large-scale study of a probability sample of adults in five sites in the United States. Diagnostic information necessary for ascertaining the DSM-III criteria for the major psychiatric disorders was collected via a structured diagnostic instrument, the D.I.S. [31], administered by lay interviewers. The NCS, which followed the ECA by almost a decade, is based on a national stratified multistage area probability sample of noninstitutionalized adults. The Composite International Diagnostic Interview (CIDI) [32], a fully structured diagnostic interview, was used to collect the diagnostic information. The major differences between these two surveys are: a decade in time between the dates of the surveys; a national sampling frame of the NCS compared to a nonrandom five-site (based on catchment areas) design in the ECA; the use of DSM-III-R in the NCS rather than DSM-III criteria; different diagnostic instruments; and the age range of the sample.

Table 80.1 presents the sex-specific lifetime prevalence and sex ratios for the major subtypes of anxiety disorders assessed in the ECA and NCS. Although the magnitude of the rates of anxiety disorders varies substantially between the two studies,

Table 80.2
Sex-Specific Prevalence of Anxiety Disorders in Youth: Surveys

Study		Anxiety disorders, total			Overanx/ GAD		Separation anxiety		Simple phobia		Social phobia	
References	Age	M %	F %	Sex ratio	M %	F %	M %	F %	M %	F %	M %	F %
[35][a]	8–17	13.3	28.6	2.1	9.5	15.2	4.8	21.0	1.0	5.7	1.0	1.0
[36][a]	14–17	5.6	11.7	2.1	0.7	1.8	2.4	5.8	1.1	2.8	0.5	2.4
[37][a]	14–24	8.3	20.3	2.4	—	—	—	—	1.2	2.3	2.2	3.5
[38][b]	9–13	4.5	7.0	1.6	1.0	2.4	2.7	4.3	0.3	0.8	0.3	0.8
[39][b,c]	8–16	0.1	0.4	2.9	0.31	0.6	0.1	0.2	0.3	0.6	0.2	0.3

[a]Lifetime prevalence.
[b]3-month prevalence.
[c]Impairment required.

the sex ratio is strikingly similar: in both studies women have an approximately twofold elevation in lifetime rates of panic, generalized anxiety disorder, agoraphobia, and simple phobia than do males. In contrast, there is a nearly equal sex ratio for the lifetime prevalence of social phobia. In both studies, the sex differences are constant across all other demographic factors including race, urban–rural residence, education, and social class.

2. Gender Differences in Anxiety in Youth

a. SEX-SPECIFIC RATES OF ANXIETY IN CHILDREN AND ADO-LESCENTS. Similar to the findings for adults, there is generally a lack of gender differences in treated samples of children and adolescents [33]. In contrast, studies of community settings reveal that the sex difference in anxiety disorders is already apparent at ages 9–12. A comprehensive review of epidemiologic studies of anxiety disorders by Orvaschel and Weissman [34] revealed that fears tended to be more common in girls than in boys across all ages but that worries showed no particular sex difference.

Table 80.2 presents the sex-specific prevalence of anxiety disorders in community or school-based surveys of children and adolescents as defined by the DSM-III, DSM-III-R, or DSM-IV criteria [35–39]. Similar to the sex ratio for adults, girls tend to have more of all subtypes of anxiety disorders, irrespective of the age composition of the sample. Several other community-based studies of children and adolescents that examined anxiety

symptoms or disorders in general also revealed higher rates of anxiety in girls across all ages [40–43]. These findings are consistent in concluding that the sex difference in anxiety begins early in life and persists across all developmental phases.

b. AGE OF ONSET AND COURSE. Epidemiologic surveys of adults reveal that the female preponderance of anxiety disorders is present across all stages of life but is most pronounced throughout early and mid-adulthood. Figures 80.1 and 80.2 shows the sex differences in the prevalence of phobias and panic by current age and gender. The rates of both subtypes of anxiety disorders in males are rather stable throughout adult life, whereas the rates in females peak in the fourth and fifth decades of life and decrease thereafter. The increased rates in females are present across all ages and do not diminish as the rates of anxiety decrease in late life.

Evidence regarding the evolution of sex differences in anxiety disorders in community samples of children and adolescents is quite sparse. A notable exception is the prospective longitudinal study by Lewinsohn et al. [36] of adolescents aged 14–18 from the general community. In this study, the Oregon Adolescent Depression Project, sex differences in the prevalence,

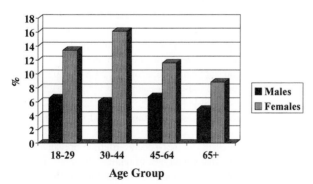

Fig. 80.1 One-year prevalence of phobia by age and gender (ECA).

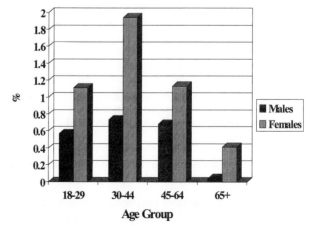

Fig. 80.2 One-year prevalence of panic by age and gender (ECA).

course, risk factors, onset, and comorbidity of anxiety disorders were examined. As compared to males, females had greater prevalences of current anxiety disorders (12.2% vs 8.5%) and past anxiety disorders (5.2% vs 2.7%), as well as anxiety symptom scores on a dimensional rating ($M = 1.9$ vs 0.9).

Epidemiologic studies of anxiety disorders in childhood and adolescence may provide clues regarding possible explanations for the sex difference in anxiety by examining the age at which the sex difference becomes apparent [44]. Inspection of incidence curves reveals a sharp increase in girls beginning as early as age 5, with a continuously increasing slope throughout adolescence. Although rates of anxiety among males also increase throughout childhood and adolescence, the rise is far more gradual than that of females and begins to level in late adolescence. Thus, by age 6, females have significantly greater rates of anxiety than males. Despite the far more rapid increase with age in girls than in boys, there was no sex difference in the mean age at onset of anxiety or in its duration [44].

Several follow-up studies of children and adolescents have shown that there is a high level of stability of anxiety disorders in general, with some switching between categories of anxiety disorders over time [45]. The stability of internalizing disorders in girls has been reported to be greater than that of boys in two prospective longitudinal follow-up studies [41,46,47]. However, the results of other studies suggest a sex difference in the continuity of emotional problems. Bolognini and colleagues [48] found that girls without emotional problems in childhood who developed new onset emotional problems in adolescence tended to have persistent problems into adulthood, whereas boys with emotional disorders in childhood tended to remit in adulthood. A recent 8-year follow-up study of a community sample of youth ages 9–18 at study entry provides compelling evidence for the stability of the subtypes of anxiety disorders and yields some interesting sex differences in the stability of anxiety over time [49]. The stability of both social phobia and simple phobia was highly specific over time, whereas overanxious disorder was associated with major depression, social phobia, and generalized anxiety in early adulthood. However, no gender differences were found in the stability of the anxiety states.

VII. Artifactual Explanations for Sex Differences in Anxiety

Because the sex difference in anxiety states does not appear to be an artifact of the sampling source or developmental stage, there are a host of other possible explanations that may be invoked to explain possible methodologic explanations for the female excess of anxiety.

A. Sex Differences in Reporting of Anxiety

In assessing possible explanations for sex differences in depression and anxiety among adolescents, Lewinsohn *et al.* [44] concluded that the sex differences were real rather than caused by reporting bias because there was no gender difference in measures of social desirability. However, this still did not rule out possible gender differences in reporting of anxiety symptoms because of other reasons such as differential awareness or cognitive interpretation between males and females. Although little systematic evidence is available on sex differences in reporting of symptoms of anxiety among adults, studies of the reliability of telephone interviewing have yielded some evidence that males are more likely to report anxiety syndromes by telephone than by direct interview, whereas females have equal rates by both methods [50,51]. Thus, it is unlikely that reporting bias can explain the large differences in prevalence of anxiety among females; however, there is suggestive evidence that some tendency for underreporting by males at direct interview may increase the magnitude of the sex ratio for anxiety states. Similar reviews of artifactual explanations for the sex difference in depression have also concluded that females truly manifest increased rates of depression, irrespective of the sampling, definition of depression, and methods of assessment [52].

Some of the major impediments to assessment of anxiety in young children are the lack of stability of reporting symptoms and the lack of development at the cognitive level to judge the meaning of physical symptoms and signs as well as the significance of fears and worries. The use of a maternal report to provide context for symptoms in children has helped to determine the significance of self-reported symptoms of anxiety in children. Maternal interpretation may underestimate the extent to which children suffer from anxiety. However, systematic underreporting by parents regarding boys is unlikely because Rapee [53] found that parent–child agreement was greater for boys than girls for anxiety disorders. In contrast, these authors found that inter-rater agreement between diagnostic interviewers was nearly identical for boys and for girls. Direct interviews of older children suggest that boys begin to deny fears because of cultural conditioning regarding the masculine sex role [54].

B. Sex Differences in the Clinical Components and Correlates of Anxiety

The female preponderance of anxiety is also found for anxiety symptoms and subthreshold definitions of anxiety [37], as well as for several of the components of the diagnostic categories of anxiety disorders, including severity, recurrence, impairment, and course. Data from both the ECA and NCS reveal that the female preponderance of anxiety was observed across all prevalence periods, and with increasingly strict definitions of anxiety. However, females also may be more likely to suffer from the consequences of anxiety symptoms than males. This interpretation is supported by the finding in the Yale Family Study that a significant proportion of male relatives of probands with anxiety disorders did not meet the impairment and/or avoidance criteria for anxiety, yet manifested full symptomatic expression of the symptoms of panic and/or social phobia [51]. Thus, women may exhibit more impairment and/or consequences of anxiety than males who may have better coping skills at the same symptom level as females. In contrast, there is some evidence that males suffer more subjective distress from social phobia than females [55]. The authors attributed this difference to the increased role demands of males, who are expected to exhibit greater levels of initiative and assertiveness, particularly in the occupational sphere.

The results of prospective follow-up of clinical samples have yielded inconsistent results regarding sex differences in the course of anxiety. Whereas some studies report greater impairment and recurrence of anxiety symptoms in women [56,57], others found no sex differences upon follow-up [56]. The discrepant findings are likely to be attributable to differences in the samples themselves, which all derive from clinical settings.

In contrast, studies of children and adolescents suggest that there are few gender differences in either age at onset or short-term course of anxiety disorders. Lewinsohn *et al.* [44] reported that despite the greater rates of anxiety in girls across all ages, there was no difference between boys and girls in the average age at onset of anxiety (mean for girls = 8.0 ± 3.9; mean for boys = 8.5 ± 3.8). With respect to course, two follow-up studies of adolescents found no gender difference in either the stability or the recovery rates from anxiety (or depressive disorders) in clinical [58] and community samples followed for 3–4 years [59].

C. Comorbidity

1. Patterns of Psychiatric Comorbidity of Anxiety in Adults

Comorbidity between anxiety disorders and other psychiatric disorders may provide one explanation for the sex difference in anxiety. Sex differences in the patterns of comorbidity between depression and specific subtypes of anxiety states were observed in the Zurich Cohort Study [60]. Whereas comorbidity between depression and panic or generalized anxiety was independent of sex, comorbidity of phobic states and depression was limited to females and was called "sex-dependent comorbidity" by the authors. Family and twin studies also show common underlying etiologic factors for the broad diagnostic categories of anxiety and depression [61–64], but longitudinal studies of comorbidity tend to reveal that anxiety disorders, particularly phobic disorders, tend to have far more stability over time than does depression [65].

Likewise, there is fairly compelling evidence that comorbidity between anxiety and substance disorders is not an explanation for increased rates of anxiety among females. This explanation would posit that the higher rates of substance disorders in males would result from the male tendency to use alcohol or drugs to self-medicate underlying anxiety states. However, the results of several epidemiologic studies have found that comorbidity between anxiety disorders and either alcohol or drug dependence tends to be either more common in females [65,66] or equal across gender [67]. Despite their lower rates of substance abuse and dependence, females are even more likely to exhibit both anxiety and substance disorders. This suggests that the sex difference in anxiety states cannot be explained by the increased frequency of substance disorders in males resulting from masked expression of underlying anxiety states.

Evidence from family study data also tends to rule out comorbidity between substance and anxiety disorders as an artifactual explanation for the lower rates of anxiety in males. The results of a family study of sex-specific patterns and pathways of comorbidity of anxiety disorders and alcoholism revealed that the associations between alcohol dependence and anxiety disorders were primarily a function of the effect in female relatives. However, there was a more consistent association between alcoholism and social phobia across sexes than between alcoholism and panic disorder, for which there was no evidence for comorbidity in males. However, when repetitive spontaneous panic

attacks were considered, a strong association emerged in both genders. This suggests that sex differences in the manifestations of anxiety states may derive in part from the classification system.

2. Patterns of Psychiatric Comorbidity with Anxiety Disorders in Youth

Comorbidity between anxiety disorders and other DSM-III-R or -IV disorders are even more common in adolescents than in adults [50]. Anxiety disorders are associated with all of the other major classes of disorders including depression, disruptive behaviors, eating disorders, and substance use. In the Virginia Twin Study of Adolescent Behavioral Development (VTSABD), there was a significant degree of overlap within the subtypes of anxiety disorders as well as with depression, which was stronger in girls than in boys. However, there was little overlap between anxiety disorders and behavior disorders [39]. In older adolescents, anxiety disorders are more strongly associated with regular substance use, including cigarettes, alcohol, and illicit substances, in girls than in boys [68]. A review of comorbidity of anxiety and depression by Brady and Kendall [69] suggests that anxiety and depression may be part of a developmental sequence in which anxiety is expressed earlier in life than is depression. Thus, although comorbidity between anxiety and both depression and substance problems is quite common in children and adolescents of both genders, further research on the mechanisms for links between specific disorders both across and within genders is necessary.

3. Medical Comorbidity

Evidence suggests that women tend to utilize treatment for psychopathology with greater frequency than men. Most individuals will initially present with symptoms to a primary health care provider. It may be that referral to a mental health provider is increased when women present with physiological and somatic complaints, whereas attention to rule out organic etiology may occur more frequently when a male presents with similar symptoms. Social biases not withstanding, anxiety disorders are common psychiatric disorders among women. Correct diagnostic classification via appropriate assessment demands attention to differential diagnosis, which considers organic factors that may be confounded with, for example, panic disorder and generalized anxiety disorder. A variety of medical conditions may be associated with significant anxiety and distress. Hyperthyroidism, cardiac and respiratory disease, systemic lupus erythematosis, anemia, asthma, complex partial seizures, and late-onset diabetes are highly prevalent in female populations. These disorders may not only complicate the treatment of the anxious patient, but they also may elicit and worsen their anxiety symptoms [70]. Furthermore, certain treatment modalities, such as systemic corticosteroids and bronchodilators, may mimic or exacerbate symptoms of anxiety, complicating the clinical profile. Because many of these disorders are more common in females, they may provide a partial explanation for the increased prevalence of anxiety symptoms in women.

VIII. Explanations for Sex Differences in Anxiety

The lack of compelling evidence for male–female differences in either the definitions or the reporting of anxiety symptoms and syndromes suggests that there is a true female preponderance of anxiety disorders. In this section, we review the possible explanations for the increased risk of anxiety in females.

A. Genetic/Familial Factors

1. Family and Twin Studies

The familial aggregation of all of the major subtypes of anxiety disorders has been well established [71]. The results of twin studies of anxiety symptoms and disorders reveal that familial clustering of anxiety can be attributed to a large extent to genetic factors [61–64]. Despite this growing body of genetic epidemiologic research on anxiety, there are nearly no data on sex-specific transmission of the anxiety disorders from either family or twin study data. The characteristic sex differences in the anxiety states observed in community studies are also found among the relatives in family studies, yet none of these studies has systematically investigated the role of the sex of the affected proband in familial aggregation. There is some suggestive evidence that male relatives may require greater genetic loading for the expression of anxiety states than females, but the small number of affected relatives in most of these studies diminishes their power to investigate the role of sex-specific transmission of anxiety disorders [72]. Evidence from the Yale Family Study of Comorbidity of Anxiety Disorders and Substance Disorders suggests that there may be increased familial aggregation of anxiety disorders among female probands. Preliminary evidence suggests that this may be attributable in part to methodologic factors including underreporting among male probands [73].

Likewise, the twin studies of anxiety disorders have either been conducted with exclusively female samples [61–64] or have not investigated sex differences in concordance rates because of small numbers of twins with specific subtypes of anxiety disorders. Torgersen [74] reported greater familial tendencies among males than females for phobic fears, confirming some earlier twin studies of clinical samples [75]. Reich and colleagues [76] conclude that these findings suggest that males may require greater genetic loading than females to manifest anxiety states.

Although the above review does not provide evidence that genetic factors underlie the sex differences in anxiety states, investigation of the possible components of anxiety that are under genetic control may provide insight into sex differences in their mediation or expression. In a comprehensive consideration of what may be inherited, Marks [75] reviews the components of anxiety that have been investigated in both human and animal studies. Evidence from twin studies has indicated that somatic manifestations of anxiety are under some degree of genetic control. These studies demonstrate that physiologic responses such as pulse, respiration rate, and galvanic skin response are more alike in monozygotic than in dizygotic twin pairs. Furthermore, twin studies of personality factors have shown high heritability of anxiety reaction. Finally, the results of animal studies have suggested that anxiety or emotionality is under genetic control. Selective breeding experiments with mammals have demonstrated that emotional activity analogous to anxiety is controlled by multiple genes [75]. These findings suggest that anxiety and fear states are highly heterogeneous and that future studies need to investigate the extent to which the components of anxiety result from common versus unique genetic factors and the role of environmental factors, either biologic or social, in either potentiating or suppressing their expression.

2. High Risk Studies

Studies of children of parents with anxiety have become an increasingly important source of information on the premorbid risk factors and early forms of expression of anxiety. Increased rates of anxiety symptoms and disorders among offspring of parents with anxiety disorders have been demonstrated [77–82]. However, despite their potential to elucidate some of the possible mechanisms for sex differences in anxiety, there has been little application of such studies to investigate this issue. Beidel and Turner [82] reported no sex difference in the rates of anxiety disorders among offspring of parents with anxiety compared to those of parents with depression or anxiety plus depression. The results of a study of offspring of parents of probands with anxiety disorders compared to psychiatric and normal controls revealed that both male and female offspring of parents with anxiety disorders had significantly greater rates of anxiety disorders than those of either psychiatric or normal controls [73]. Whereas the magnitude of the association between parental and child anxiety was stable across childhood for girls, there was a large increase with increasing age among male offspring. However, sex differences did emerge when considering anxiety sensitivity, a suspected cognitive predictor of anxiety disorders [7]. Anxiety sensitivity scores were greater among girls when compared with boys [83].

The elevation in rates of the childhood anxiety disorders in this study suggests that both the cognitive and somatic manifestations of these conditions may occur in childhood but that changes associated with adolescence, including increased interpersonal social demands and/or biologic maturation, are necessary for expression of the full diagnostic entity.

With respect to parental anxiety, no differences were found in the rates of anxiety among offspring according to the sex of the affected parent. The rate of anxiety disorders among offspring of couples where *both* parents were affected with anxiety and/or affective disorders was significantly elevated over that in which only one parent had an anxiety disorder, providing further evidence for specificity of familial aggregation of the anxiety disorders. This strong degree of specificity of transmission of anxiety disorders, coupled with the finding that anxiety/affective families did not show dysregulation in either the structure (*i.e.,* rates of divorce and single parenting) or functioning (problems in parent–child interactions) of the family constellation, may suggest the presence of an underlying temperamental vulnerability for anxiety disorders.

B. Temperament

Sex differences are not only apparent in anxiety disorders but also in their underlying temperamental components. In an early study of sex differences in the development of fear (Jersild and Holmes, 1935, in Rothbart [84]), extended latency to approach was demonstrated in girls on multiple variables. Control and inhibition may be associated with the development of language and socialization efforts in girls [85]. It has been suggested that girls may be more influenced by the potential threat of punishment for approach behavior (which may be misinterpreted for aggressive behavior) and hence react with compliance and inhibition [85–87].

One of the earliest indicators of vulnerability to the development of anxiety is behavioral inhibition, characterized by increased physiological reactivity or behavioral withdrawal in the face of novel stimuli or challenging situations [88]. Behavioral inhibition may be a manifestation of a biological predisposition characterized by both overt behavioral (*e.g.,* cessation of play, latency to interact in the presence of unfamiliar objects and people) and physiologic indicators (*e.g.,* low heart rate variability, accelerated heart rate, increased salivary cortisol level, pupillary dilation, increased cortisol level). There is an increased frequency of behavioral inhibition among children of parents with anxiety disorders compared to those of normal controls [77,78,80,82,89–92].

Although the literature on familial risk factors associated with behavioral inhibition is expanding, attention to gender differences at these young children at risk is sparse. Sex differences in measures of inhibition (motor hesitation during reaches [84,93,94]) were observed during the second six months of an infant's life. Additional reports, including cross-cultural investigation have indicated that male infants are more likely to approach novelty in contrast to females who are predisposed to withdrawal behaviors [95–97]. Similarly, Reznick and colleagues [93] demonstrated sex differences at both 14 and 21 months, with the latter manifesting significant differences in inhibition for girls. In follow-up studies of behavioral inhibition, girls who had been categorized as behaviorally inhibited were more likely than boys to be stably inhibited both at 7.5 and at 13 years of age [94]. Similar sex differences were found in the Colorado twin study.

The stability of behavioral inhibition across early childhood has been found to be far greater among girls than among boys [98]. Few studies have evaluated the differences in manifest inhibition and approach/avoidance in both clinical and nonclinical samples, leaving gaps in the conceptualization of the construct of inhibition. The expression of behavioral inhibition studied prospectively may reveal patterns of anxiety symptomatology similar to those endorsed in adult populations.

C. Biologic Vulnerability Risk Factors

1. Neurobiologic Factors

The three major neurotransmitter systems that have been investigated etiologically in anxiety include the gamma-aminobutyric acid (GABA)-benzodiazepine system, the noradrenergic system, and the serotonin system [99]. Experimental provocation of these systems has elucidated possible neurobiologic mechanisms that underlie anxiety states [100]. Aside from a limited number of studies that have demonstrated differential prolactin response to thyroid releasing hormone among women [101–103], the lack of systematic investigation of sex differences in neurobiologic function is striking. There is also some preliminary evidence that women with anxiety respond better to the serotonin-uptake inhibitors than men. One possible explanation is that estrogens increase the receptor binding of serotonin [104]. Likewise, the greater sensitivity of women to environmental light

suggests that women may be more sensitive to circadian shifts than men.

2. Neuroendocrine Factors

Based on a review of sex differences in stress in humans and in animals, Young [105] suggested that greater stress responsiveness in women may be influenced by modulation of HPA-axis activity by ovarian steroids. She concluded that gonadal steroid antagonism of glucocorticoids and the increased stress responsiveness of females may lead to increased anxiety disorders in women.

Indirect evidence of the involvement of neuroendocrine factors in gender-specific processes of anxiety disorders in females is also suggested by changes in the expression of anxiety associated with the menstrual cycle, pregnancy, and the postpartum period. The menstrual cycle has been the subject of several investigations of endocrine function and anxiety. Studies of premenstrual mood changes have neglected the significance of anxiety symptoms and disorders in women. In a community study of young adults, Merikangas and colleagues [106] reported that premenstrual anxiety symptoms were far more frequent and incapacitating than were depressive symptoms. Moreover, in addition to depression, comorbidity with premenstrual syndrome was found for all of the major anxiety syndromes as well.

Research suggests that ovarian hormones may mediate the expression of anxiety symptomatology, particularly in those women at risk for anxiety disorders. For example, the premenstrual phase was found to be associated with the onset or worsening of panic symptoms in panic disorder patients [107]. Perna and colleagues [108] demonstrated that women with panic disorder experienced heightened sensitivity to biological provocation (*e.g.,* CO_2) during the early follicular phase of the menstrual cycle compared with normal controls. Similarly, samples of women without anxiety diagnoses experiencing PMS had an increase in panic-like symptoms and hyperreactivity in response to biological challenge [109]. Generalized anxiety disorder (GAD) has also demonstrated worsening during the premenstrual period [110]. These studies further implicate the association of gonadal hormones and anxiety in women. A finding that shed significant light on the role of sex hormones is the work of Hayward *et al.* [111], which revealed that after controlling for age, pubertal stage predicted the onset of panic attacks in young adolescent girls selected from a community sample.

Pregnancy and the postpartum period have also been studied in their relationship to anxiety disorders; however, the literature is limited by few investigations of the specific subtypes of anxiety. Pregnancy and the puerperium are unquestionably periods of physiologic, hormonal, and neurochemical change. The behavioral impact of such neuroendocrine alteration is widespread (*e.g.,* aggressive, emotional, maternal, and feeding behaviors) and is likely influenced by multiple neuroanatomical pathways [112,113]. Unfortunately, rigorous investigations are sparse. There has been a consistent trend in the literature suggesting that pregnancy may in fact be protective against the development of panic disorder or reduce the frequency of panic attacks. In two studies, the clinical course of anxiety disorders during pregnancy was obtained through retrospective self-report data. These findings, though highly variable, at least substantiate to some extent the notion of biologic protection from panic disorder during pregnancy [114,115]. Villeponteaux and colleagues [114] indicate that although panic disorder symptoms improved during pregnancy, agoraphobic avoidance was either maintained or exacerbated. The purported mechanisms underlying this "protective" phenomenon include the following: blunting of the sympatho-adrenal response in pregnancy may act to stabilize autonomic reactivity; elevated gonadal hormones (*e.g.*, progesterone metabolites) may have an inhibitory function at receptor sites (*e.g.*, GABA) to provide antianxiety effects; and psychosocial vulnerability factors contributing to anxiety may be lessened as pregnancy serves to increase self-esteem and purposefulness [116]. Klein and colleagues [117] hypothesize that lactation protects against panic disorder. This may be related to the secretion of oxytocin, which has been linked to attachment behavior [118]. In addition, progesterone may be involved in decreasing blood CO_2 levels, hence increasing the tolerance threshold in CO_2-sensitive individuals, which is suggested in the etiology and maintenance of panic disorder [117].

The postpartum period is considered a risk factor for the onset or exacerbation of anxiety psychopathology mediated by both biological and psychosocial factors [112,117,119–121]. Case studies have revealed panic symptom remission during pregnancy and/or relapse or onset in the postpartum period [116,122]. Studies by Sholomskas and colleagues [119] and Cohen *et al.* [121] showed that there was an increased risk associated with the onset of panic disorder at the postpartum period. The latter study also revealed that patients who received pharmacotherapy by the third trimester were significantly less likely to have experienced postpartum worsening of their anxiety symptoms. Anxiolytic administration prophylactiacally may decrease relapse rates in these patients.

Further research attention to hormonally-mediated biological components is needed to understand gender differences predisposing women to experience decreased resiliency to fear-provoking stimuli. Data on the impact of pregnancy on the course and outcome of anxiety disorders are limited and inconsistent, with reports indicating both increases and decreases in anxiety symptomatology during pregnancy.

3. Psychophysiologic Vulnerability Markers

The association between a variety of psychophysiologic markers and anxiety has been studied both in adults and in children. Most research fails to report patterns of psychophysiologic functioning by sex. In a study of high-risk adolescent offspring of parents with anxiety disorders compared to psychiatric and normal controls, the startle reflex and its potentiation by aversive states was used as a possible marker of vulnerability to anxiety disorders [123,124]. Startle was found to discriminate between children at high- and low-risk for anxiety disorders. However, different abnormalities for high-risk males and females were observed. Startle levels were elevated among high-risk females, whereas high-risk males exhibited greater magnitude of startle potentiation during aversive anticipation. Two possible explanations for the gender differences in the high-risk groups were suggested by the authors: 1) differential sensitivity among males and females to explicit threat versus the broader contextual stimuli which are mediated by different neurobiologic pathways; and 2) different developmental levels in males

and females in which the vulnerability to anxiety may be physiologically expressed earlier in females.

D. Psychosocial Risk Factors and Impact

1. Social Roles

A question of equal importance to address is what psychosocial factors, cultural influences, and gender stereotypes may be responsible for many of the differences in the prevalence rates of anxiety disorders? Sex differences in social roles may be involved in either the development or consequence of anxiety states. Bruch and Cheek [125] cite evidence that shyness, inhibition, and nervousness interfere more with male socialization compared to that of females because of societal expectations that males have more initiative, adventurousness, and self-confidence. This may in part explain the more balanced sex ratio for social phobia in contrast to that of other subtypes of anxiety. Social learning has also been implicated in the development and maintenance of low assertiveness and low risk taking associated with women and linked to anxiety [99]. Shear promotes the theory that sex roles, specifically, orginate as individuals learn gender-stereotyped behavior from their interpersonal interactions, culture, and by modeling the behaviors of others [99]. With the changing social role expectations during the 1990s, anxiety states would be expected to have a greater impact on women as they assume a more equal balance in occupational and familial role functions.

In terms of social consequences of anxiety, there is some evidence that women with phobic states may have little impairment in either the marital or parental role. There is little evidence for spouse concordance for anxiety states [71]. Couples with a female agoraphobic have been found to have comparable levels of marital adjustment compared to those of normals, whereas couples with females with depression or substance abuse had high levels of marital impairment [126]. Negative consequences of anxiety in the parenting role include decreased emotional availability, overprotection, and modeling of maladaptive coping strategies. However, the perception of care generated by anxious mothers, even in the presence of overprotection, may actually be beneficial to the development of self efficacy. Thus, anxiety disorders may not be associated with impairment in the marital or parental role despite the impairment in external social relationships and occupational function reported by individuals with anxiety disorders. Indeed, agoraphobia may have had adaptive value in reproduction and child-rearing during the early evolutionary phase of human existence [127]. For example, postpartum women may experience increased anxiety in response to novel situations that surfaced with the birth of a child (*e.g.*, decreased time, career limitation, marital and intrafamilial conflict). A new environment stipulates the acquisition of skills to adapt to new social demands. Cognitive–behavioral models (although not systematically investigated in the postpartum period) may prove effective in the control of fear, avoidance, and negative automatic thoughts should they arise during this stressful time period.

One of the major consequences of social inhibition in childhood stems from distress secondary to speaking in class, which is often an evaluative component in academic settings. Occupational impairment is also a serious consequence of social phobia in adulthood because many individuals with phobias must alter their educatinoal or occupational aspirations because of their inability to perform particular activities required for specific tasks such as public speaking, air travel, eating in public, or use of elevators. This may in part explain the lack of sex differences in the prevalence of social phobia. Agoraphobia, often the most extreme form of phobic states, may often leave a person housebound for years, leading to pervasive role impairment. However, most of the association between anxiety and social roles appears to be causal, with anxiety leading to social role impairment rather than the reverse, as is often reported for depression. Thus, it is unlikely that the sex difference in anxiety states results from differential role stress among females.

2. Life Events, Stress Reactivity, and Coping Styles

Life events have often been designated as a causal role in the onset of anxiety states, particularly phobias. Life stressors and other triggers that to some extent threaten an individual's notion of safety and security in the world are often, at least retrospectively, perceived to predate the onset of anxiety disorders. Adverse developmental experiences have been associated with the onset of anxiety disorders [128]. While it is likely that life stress may exacerbate phobic and anxiety states, Marks [129] concludes that phobic states resulting from exposure are far more rare than those that emerge with no apparent exposure.

Although there are few sex differences in the frequency of stressful life events [100], women may exhibit greater levels of emotional upheaval in the face of environmental challenge. Supportive evidence is derived from experimentally induced stress [130,131]. For example, Frankenhaeuser and colleagues [132] demonstrated interesting sex differences in biological and psychological responses to examination stress. Although both males and females demonstrated an increased excretion of cortisol, adrenaline, noradrenaline, and MHPG from baseline, males secreted proportionately more than females. Whereas self-rated discomfort correlated with poor performance in males, the opposite was found in females; the greater the self-rated discomfort, the better the performance. Moreover, females were far more likely to report feelings of discomfort and failure than were males. Although these findings do not have direct relevance for underlying neurobiologic differences in anxiety between men and women, they do suggest a mismatch between physiologic and psychologic interpretation of anxiety states under stress.

Sex differences in coping styles have also been implicated as a possible explanation for the increased vulnerability to anxiety in women [33]. Citing Nolen-Hoeksma [133] Craske notes that girls tend to employ a coping style characterized by worry and rumination, whereas boys use more active problem-solving strategies. In this respect, differential coping styles would also provide an explanation for sex differences in depression.

One of the most comprehensive investigations of the role of psychosocial risk factors explaining gender difference in anxiety was conducted by Lewinsohn et al. [44] in his longitudinal community survey of adolescents. The psychosocial correlates of anxiety and sex-specific associations with anxiety disorders and symptoms were similar to those previously examined for depression. Females exhibited more life events, higher self-consciousness, lower self-esteem, higher social competence, greater emotional reliance, more social support from friends,

more physical illness, and poorer self-rated physical health. Males and females reported no differences in social role expectations associated with anxiety. However, the gender difference in anxiety was not attributable to these psychosocial risk factors because the gender difference persisted after adjustment for gender differences in the psychosocial factors. The authors conclude that these findings support biologic or genetic vulnerability rather than psychosocial factors as the explanatory mechanism for the female preponderance of anxiety in youth [44].

IX. Treatment of Anxiety

Standard treatments for anxiety include both pharmacological and psychological interventions. The pharmacopoeia for anxiety disorders is diverse. Agents demonstrating clinical efficacy include monoamine oxidase inhibitors (MAOIs: e.g., phenelzine), tricyclic antidepressants (TCAs: e.g., imipramine, desipramine), benzodiazepines (BZDs: e.g., clonazepam, lorazepam), beta-blockers (e.g., propranolol, atenolol), selective serotonin reuptake inhibitors (SSRIs: e.g., fluoxetine, paroxetine, sertraline), and other medications (e.g., buspirone). The efficacy of nearly all of these agents has been evaluated in controlled clinical trials [134]. Unique risks exist when deciding an appropriate course of treatment in women of childbearing age. Because the long-term impact of psychotropic medication during childbearing years is relatively unknown and the potential risk of teratogenicity remains a concern, clinicians should be particularly sensitive to the dosing and careful monitoring of pharmacological agents used in this population.

Cognitive–behavioral therapy (CBT) has shown clinical efficacy in a wide range of anxiety disorders and is often indicated as a first-line or adjunct treatment. CBT integrates psychoeducation about the nature of anxiety and the fear-of-fear cycle, relaxation and breathing retraining, cognitive restructuring to address catastrophic misinterpretation of bodily sensations and/or negative automatic thoughts, and exposure aimed at feared sensations and situational avoidance. In controlled trials for panic disorder with or without agoraphobia, GAD, and social phobia, CBT demonstrated success rates greater than placebo and equal to or greater than pharmacotherapy alone. Follow-up studies indicate these treatment gains are often maintained over time [135].

Despite the wealth of available literature on treatment efficacy for anxiety disorders, such investigations have failed to scrutinize gender differences in treatment outcome. Beyond gaining a more comprehensive understanding of the etiology and course of anxiety presentation across gender, it is also critical to understand these differences when considering therapeutic options. Some preliminary evidence suggests that there are no gender differences in baseline symptom severity or treatment outcome to either CBT therapy or medication [112].

There is very limited empirical research and literature regarding pharmacotherapy and psychotherapy in the treatment of anxiety disorders in children and adolescents [136]. Studies describe many of the same agents used in adults with anxiety disorders; however, information about therapeutic dose as well as short- and long-term efficacy remains unstandardized. Similar to the adult literature, specific attention to gender differences in medication treatment outcome trials of anxiety disorders in children and adolescents remains unreported. CBT in this younger population has been steadily gaining attention as a beneficial treatment option. Systematic desensitization, *in vivo* exposure, skill building, and cognitive coping strategies have been incorporated into protocol-driven treatments for anxiety disorders in children. Barrett *et al.* [137] have investigated the integration of a familial component of treatment for the parents of anxious children, involving skills building and education about the manifested signs and symptoms of childhood anxiety. Anxiety management skills for the anxious parent have also been added. Family-based treatment outcomes indicate a benefit above and beyond CBT alone.

X. Summary and Conclusions

In summary, there is consistent evidence that women have higher rates of anxiety disorders and symptoms than men across the age span. None of the possible artifactual explanations, including sampling, demographic, or social risk factors, appear to convincingly explain the preponderance of anxiety in women. However, there is some evidence that males are less likely to report anxiety symptoms and to suffer from a lower magnitude of impairment and distress than women, with the possible exception of social phobia.

In a review of the multiple possible explanations for the higher rates of anxiety in women, the most compelling explanation is that women are more likely to have higher levels of arousal, psychophysiologic response to stress, and somatic symptoms, as well as greater awareness of somatic anxiety, than men. Increased social sensitivity may also lead to increased awareness of anxiety symptoms in women. There is also substantial evidence that the oscillation of female reproductive hormones across the life cycle may be associated with greater levels of expression of underlying vulnerability for anxiety in women, as well as increased severity and frequency of anxiety symptoms among ongoing cases.

The widespread lack of consideration of sex differences in anxiety across all domains of research is striking. The major conclusion of this work is the need for research that systematically examines sex differences in vulnerability, manifestation, causes, and consequences of anxiety disorders. Future research should address the following issues:

1. Studies designed to develop more accurate and developmentally sensitive methods of assessment of anxiety with a focus on developing objective measures of the components of anxiety.
2. Follow-up studies of community-based samples of individuals with anxiety to examine the longitudinal stability of anxiety, the overlap within the anxiety disorders and between the anxiety disorders and other types of psychiatric and medical disorders, and risk factors and correlates associated with chronicity and remission.
3. Additional research on the role of comorbidity of anxiety and depression in the sex differences for emotional disorders.
4. Following up the intriguing evidence that the onset of panic attacks is associated with specific stages of pubertal development, further research attention to hormonally mediated neurobiological function is needed to understand gender differences predisposing women to experience decreased resiliency to fear-provoking stimuli.

5. Systematic study of the role of sex-specific transmission in family and twin studies of anxiety disorders in order to identify the extent to which genetic and environmental factors contribute to the etiology of anxiety disorders.

In addition to research implications of this work, there are several clinical issues of relevance to anxiety disorders in women. First, the sheer magnitude of anxiety states, affecting up to one-third of women across the life span, warrants intensive focus on increased education and treatment of anxiety in the general population. Second, the focus on links between depression and the female reproductive cycle have failed to consider the importance of anxiety symptoms and disorders that may overshadow those of depression, particularly in younger women. Third, the importance of early intervention in anxiety disorders is highlighted by the evidence of its temperamental underpinnings and chronicity across the life span.

The dramatic increase in rates of onset of anxiety disorders at mid-adolescence provides an important target for prevention of the crystallization of these disorders and subsequent impairment in the accomplishment of both education and social life goals. The overwhelming focus on disruptive behavior problems in youth should be balanced by new initiatives to identify those who suffer quietly from anxiety. Lack of adequate identification of those with anxiety limits the opportunities for application of the intervention and treatment strategies that have been well developed for anxiety disorders in clinical settings. Increased focus on research regarding the causes of sex differences in anxiety, as well as better recognition and more effective treatment of anxiety in community settings, particularly in young women, may not only benefit the large number of persons suffering from the consequences of anxiety, but will also bring us closer to understanding the elusive etiology of the anxiety disorders.

Appendix A

Panic disorder (± agoraphobia)

Recurrent unexpected panic attacks, characterized by four or more of the following:
Palpitations
Sweating
Trembling or shaking
Shortness of breath
Feeling of choking ("air hunger")
Chest pain or discomfort
Nausea or abdominal distress
Feeling dizzy, lightheaded, or faint
Derealization/depersonalization
Fear of losing control or going crazy
Fear of dying
Numbness or tingling
Chills or hot flashes

Persistent concern of future attacks

Worry about the meaning of or consequences of the attacks (*e.g.,* heart attack, stroke)

Significant change in behavior related to the attacks (*e.g.,* avoiding places where panic attacks have occurred)
± the presence of agoraphobia

Agoraphobia

Fear of being in places or situations from which escape might be difficult, embarrassing, or where help may be unavailable in the event of having a panic attack

Often results in avoidance of the feared places or situations, for example:

Crowds
Stores
Bridges
Tunnels
Traveling on a bus, train, airplane
Theaters
Standing in a line
Small enclosed rooms

Generalized anxiety disorder

Excessive anxiety and worry about a number of events or activities (future-oriented), occurring more days than not for at least 6 months

Worry is difficult to control

Worry is associated with at least three of the following symptoms:

Restlessness or feeling keyed up or on edge
Easily fatigued
Difficulty concentrating
Irritability
Muscle tension
Sleep disturbance

Anxiety and worry causes significant distress and impairment in social, occupational, or other daily functioning

Specific phobia

Marked and persistent fear that is excessive, unreasonable, cued by the presence or anticipation of a specific object or situation, for example:

Flying
Enclosed spaces
Heights
Storms
Animals (*e.g.*, snakes, spiders)
Receiving an injection
Blood

Provokes an immediate anxiety response

Recognition that the fear is excessive or unreasonable

Avoidance, anticipatory anxiety, or distress is significantly impairing

Social phobia

Marked and persistent fear of one or more social or performance situations where the person is concerned with negative evaluation or scrutiny by others, for example:

Public speaking
Writing, eating, drinking in public
Initiating or maintaining conversations

Fears humiliation or embarrassment, perhaps by manifesting anxiety symptoms (*e.g.*, blushing, sweating)

Feared social or performance situations are avoided or endured with intense anxiety or distress

Recognition that the fear is excessive or unreasonable

Avoidance, anticipatory anxiety, or distress is significantly impairing

Acknowledgments

This work was supported primarily by grant DA05348 and in part by grants AA07080, AA09978, DA09055, MH36197, and by Research Scientist Development Award K02 DA00293 to Dr. Merikangas from the Alcohol, Drug Abuse and the Mental Health Administration of the United States Public Health Service.

References

1. Lader, M. (1972). The nature of anxiety. *Brit. J. Psychiatry* **121**, 481–491.
2. Cohen, M., Badal, D., Kilpatrick, A., Reed, E., and White, P. (1951). The high familial prevalence of neuro-circulatory asthenia (anxiety neurosis effort syndrome). *Am. J. Hum. Genet.* **3**, 126–158.
3. American Psychiatric Association (1994). "Diagnostic and Statistical Manual for Psychiatric Disorders," 4th ed., APA, Washington, DC.
4. Beck, A., Epstein, N., Brown, G., and Steer, R. (1988). An inventory for measuring clinical anxiety: Psychometric properties. *J. Consult. Clin. Psychiatry* **4**, 561–571.
5. Spielberger, C., Gorsuch, R., Lushene, R., Vagg, P., and Jacobs, G. (1983). "Manual for the State-Trait Anxiety Inventory." Consulting Psychologists Press, Palo Alto, CA.
6. Wolpe, J., and Lang, P. A. (1964). A Fear Survey Schedule for use in behaviour therapy. *Behav. Res.* **2**, 27–30.
7. Reiss, S., Peterson, R., Gursky, D., and McNally, R. (1986). Anxiety sensitivity, anxiety frequency, and the prediction of fearlessness. *Behav. Res. Ther.* **24**, 1–8.
8. Marks, I., and Mathews, A. (1979). Brief standard self-rating for phobic patients. *Behav. Res. Ther.* **17**, 263–267.
9. Chambless, D., Caputo, C., Jason, S., Gracely, E., and Williams, C. (1985). The mobility inventory for agoraphobia. *Behav. Res. Ther.* **23**, 35–44.
10. Hamilton, M. (1959). The assessment of anxiety states by rating. *Bri. J. Med. Psychol.* **32**, 50–55.
11. Robins, L., Helzer, J., Croughan, J., and Ratcliff, K. (1981). The National Institute of Mental Health Diagnostic Interview Schedule: Its history, characteristics, and validity. *Arch. Gen. Psychiatry* **38**, 381–389.
12. Spitzer, R., Williams, J., Gibbon, M., and First, M. (1992). The Structured Clinical Interview for DSM-III-R (SCID). I: History, rationale, and description. *Arch. Gen. Psychiatry* **49**, 624–629.
13. Spitzer, R., Williams, J., Gibbon, M., and First, M. (1990). "Structured Clinical Interview for DSM-III-R—Patient Edition (SCID-P)." American Psychiatric Press; Washington, DC.
14. Di Nardo, P., and Barlow, D. (1988). "Anxiety Disorders Interview Schedule—Revised (ADIS-R)." Phobia and Anxiety Disorders Clinic, Albany, NY.
15. Di Nardo, P., Moras, K., Barlow, D., Rapee, R., and Brown, T. (1993). Reliability of the DSM-III-R anxiety disorder categories using the Anxiety Disorders Interview Schedule—Revised (ADIS-R). *Arch. Gen. Psychiatry* **50**, 251–256.
16. Seligman, M. (1971). Preparedness and phobias. *Behav. Ther.* **2**, 307–320.
17. Marks, I., and Lader, M. (1973). Anxiety states (anxiety neurosis): A review. *J. Nerv. Ment. Dis.* **156**, 3–18.
18. Schneier, F. R., Liebowitz, M., and Beidel, D. (1995). "Social Phobia," 2 vols. American Psychiatric Press, Washington, DC.
19. Robins, L., Helzer, J., Weissman, M. *et al.* (1984). Lifetime prevalence of specific psychiatric disorders in three sites. *Arch. Gen. Psychiatry* **41**, 949–958.

20. Kessler, R., McGonagle, K., Zhao, S. *et al.* (1994). Lifetime and 12-month prevalence of DSM-III-R psychiatric disorders in the United States: Results from the National Comorbidity Survey. *Arch. Gen. Psychiatry* **51**, 8–19.

21. Merikangas, K., and Weissman, M. (1985). Epidemiology of anxiety disorders in adulthood. *In* "Epidemiologic Psychiatry Psychiatry" (M. M. Weissman, R. Michels, and J. O. Cavenor, eds.), pp. 1–12. Lippincott, Philadelphia.

22. Canino, G., Bird, H., Shrout, P. *et al.* (1987). The prevalence of specific psychiatric disorders in Puerto Rico. *Arch. Gen. Psychiatry* **44**, 727–735.

23. Bland, R., Orn, H., and Newman, S. (1988). Lifetime prevalence of psychiatric disorders in Edmonton. *Acta Psychiatr. Scand., Suppl.* **338**, 24–32.

24. Faravelli, C., Guerrini Degl'Innocenti, B., and Giardnelli, L. (1989). Epidemiology of anxiety disorders in Florence. *Acta Psychiatr. Scand.* **79**, 308–312.

25. Lepine, J., Pariente, P., Boulenger, J. *et al.* (1989). Anxiety disorders in a French general psychiatric outpatient sample. Comparison between DSM-III and DSM-III-R criteria. *Soc. Psychiatry Psychiatr. Epidemiol.* **24**, 301–308.

26. Wells, J., Bushnell, J., Hornblow, A., Joyce, P., and Oakley-Browne, M. (1989). Christchurch psychiatric epidemiology study: Methodology and lifetime for specific psychiatric disorders. *Aust. N. Z. J. Psychiatry* **23**, 315–326.

27. Angst, J. (1993). Comorbidity of anxiety, phobia, compulsions, and depression. *Int. Clin. Psychopharmacol.* **8**, 21–25.

28. Lee, C., Kwak, Y., Yamamototo, J. *et al.*, (1990). Psychiatric epidemiology in Korea: Part 1. Gender and age differences in Seoul. *J. Nerv. Ment. Dis.* **178**, 242–246.

29. Wacker, H., Mullejans, R., Klein, K., and Battegay, R. (1992). Identification of cases of anxiety disorders and affective disorders in the community according to the ICD-10 and DSM-III-R by using the Composite International Diagnostic Interview (CIDI). *Int. J. Methods Psychiatr. Res.* **2**, 91–100.

30. Wittchen, H., Essau, C., von Zerssen, D., Krieg, J., and Zaudig, M. (1992). Lifetime and six-month prevalence of mental disorders in the Munich Follow-Up. *Eur. Arch. Psychiatry Clin. Neurosc.* **241**, 247–258.

31. Robins, L., Helzer, J., Croughan, J., and Ratcliff, K. (1981). The National Institute of Mental Health Diagnostic Interview Schedule: Its history, characteristics, and validity. *Arch. Gen. Psychiatry* **38**, 381–389.

32. Robins, L., Wing, J., Wittchen, H. *et al.* (1988). The Composite International Diagnostic Interview. *Arch. Gen. Psychiatry* **45**, 1069–1077.

33. Craske, M. (1997). Fear and anxiety in children and adolescents. *Bull. Menninger Clin.* **61**, 4–36.

34. Orvaschel, H., and Weissman, M. (1986). Epidemiology of anxiety disorders in children: A review. *In* "Anxiety Disorders of Childhood" (R. Gittelman, ed.), pp. 58–72. Guilford, New York.

35. Kashani, J. H., and Orvaschel, H. (1990). A community study of anxiety in children and adolescents. *Amer. J. Psychiatry* **147**(3), 313–318.

36. Lewinsohn, P., Hops, H., Roberts, R., Seeley, J., and Andrews, J. (1993). Adolescent psychopathology: I. Prevalence and incidence of depression and other DSM-III-R disorders in high school students. *J. Abnorm. Psychol.* **102**, 133–144.

37. Wittchen, H., Nelson, C., and Lachner, G. (1998). Prevalence of mental disorders and psychosocial impairments in adolescents and young adults. *Psychol. Med.* **28**, 109–126.

38. Costello, J., Angold, A., Burns, B. *et al.* (1996). The Great Smoky Mountains Study of Youths: Goals, design, methods, and the prevalence of DSM-III-R Disorders. *Arch. Gen. Psychiatry* **53**, 1129–1136.

39. Simonoff, E., Pickles, A., Meyer, J. *et al.* (1997). The Virginia Twin Study of Adolescent Behavioral Development: Influences of age, sex, and inpairment on rates of disorders. *Arch. Gen. Psychiatry* **47**, 487–496.

40. Whitaker, A., Johnson, J., Shaffer, D. *et al.* (1990). Uncommon troubles in young people: Prevalence estimates of selected psychiatric disorders in a nonreferred population. *Arch. Gen. Psychiatry* **47**, 487–496.

41. Feehan, M., McGee, R., and Williams, S. (1993). Mental health disorders from age 15 to 18 years. *J. Am. Acad. Child Adolesc. Psychiatry* **32**, 1118–1126.

42. Fergusson, D., Horwood, L., and Lynskey, M. (1993). Prevalence and comorbidity of DSM-III-R diagnoses in a birth cohort of 15 year olds. *J. Am. Acad. Child Adolesc. Psychiatry* **32**, 1127–1134.

43. Verhulst, F., van der Ende, J., Ferdinand, R., and Kasius, M. (1997). The prevalence of DSM-III-R diagnoses in a national sample of Dutch adolescents. *Arch. Gen. Psychiatry* **54**, 329–336.

44. Lewinsohn, P., Lewinsohn, M., Gotlib, I., Seeley, J., and Allen, N. (1998). Gender differences in anxiety disorders and anxiety symptoms in adolescents. *J. Abnorm. Psychol.* **107**, 109–117.

45. Cantwell, D., and Baker, L. (1989). Stability and natural history of DSM-III childhood diagonses. *J. Am. Acad. Child Adolesc. Psychol.* **28**, 691–700.

46. McGee, R., Feehan, M., Williams, S., and Anderson, J. (1992). DSM-III disorders from age 11 to age 15 years. *J. Am. Acad. Child Adolesc. Psychiatry* **31**(1), 50–59.

47. Ferdinand, R., and Verhulst, F. (1996). Psychopathology from adolescence into young adulthood: An 8-year follow-up study. *Am. J. Psychiatry* **152**, 1586–1594.

48. Bolognini, M., Bettschart, W., Plancherel, B., and Rossier, L. (1989). From the child to the young adult: Sex differences in the antecedents of psychological problems. A retrospective study over ten years. *Soc. Psychiatry Psychiatr. Epidemiol.* **24**, 179–186.

49. Pine, D., Cohen, P., Gurley, D., Brook, J., and Ma, Y. (1998). The risk for early adulthood anxiety and depressive disorders in adolescents with anxiety and depressive disorders. *Arch. Gen. Psychiatry* **55**, 56–64.

50. Rhode, P., Lewinsohn, P., and Seeley, J. (1997). Comparability of telephone and face-to-face interviews in assessing axis I and axis II disorders. *Am. J. Psychiatry* **154**, 1593–1598.

51. Merikangas, K., Stevens, D., Fenton, B. *et al.* (1998). Comorbidity of alcoholism and anxiety disorders: The role of family studies. *Psychol. Med.* **28**, 773–788.

52. Merikangas, K., Weissman, M., and Pauls, D. (1985). Genetic factors in the sex ratio of major depression. *Psychol. Med.* **15**, 63–69.

53. Rapee, R., Barrett, P., Dadds, M., and Evans, L. (1994). Reliability of the DSM-III-R childhood anxiety disorders using structured interview: Inter-rater and parent-child agreement. *J. Am. Acad. Child Adolesc. Psychiatry* **154**, 1593–1598.

54. Bauer, D. (1976). An exploratory study of developmental changes in children's fears. *J. Child Psychol. Psychiatry* **17**, 69–74.

55. Pollard, C., and Henderson, J. (1988). Four types of social phobia in a community sample. *J. Nerv. Ment. Dis.* **176**, 440–445.

56. Noyes, R. R. J., Christiansen, J., Suelzer, M., Pfhol, B., and Coreil, W. A. (1990). The outcome of Panic Disorder: relationship to diagnostic subtypes and comorbidity. *Arch. Gen. Psychiatry* **47**, 808–818.

57. Yonkers, K. A. Z. C., Allsworth, J., Warshaw, M., Shea, T., and Keller, M. B. (1998). Is the course of panic disorder the same in women and men? *Am. J. Psychiatry* **155**, 596–602.

58. Last, C., Perrin, S., Hersen, M., and Kazdin, A. (1996). A prospective study of childhood anxiety disorders. *J. Am. Acad. Child Adolesc. Psychiatry* **35**, 1502–1510.

59. Lewinsohn, P., Lewinsohn, M., Gotlib, I., Seeley, J., and Allen, N. (1998). Gender differences in anxiety disorders and anxiety symptoms in adolescents. *J. Abnorm. Psychol.* **107**, 109–117.

60. Ernst, C., and Angst, J. (1992). The Zurich Study: XII. Sex difference in depression. Evidence from longitudinal epidemiologic data. *Eur. Arch. Psychiatry Clin. Neurosci.* **241,** 222–230.

61. Kendler, K., Heath, A., Martin, N., and Eaves, L. (1987). Symptoms of anxiety and symptoms of depression: Same genes, different environments? *Arch. Gen. Psychiatry* **44,** 451–457.

62. Kendler, K., Neale, M., Kessler, R., Heath, A., and Eaves, L. (1992). The genetic epidemiology of phobias in women: The interrelationship of agoraphobia, social phobia, situational phobia, and simple phobia. *Arch. Gen. Psychiatry* **49,** 273–281.

63. Kendler, K., Walters, E., Neale, M. *et al.* (1995). The structure of the genetic and environmental risk factors for six major psychiatric disorders in women: Phobia, generalized anxiety disorder, panic disorder, bulimia, major depression, and alcoholism. *Arch. Gen. Psychiatry* **52,** 374–383.

64. Kendler, K., Neale, M., Kessler, R., Heath, A., and Eaves, L. (1993). Major depression and phobias: The genetic and environmental sources of comorbidity. *Psychol. Med.* **23,** 361–371.

65. Angst, J., Vollrath, M., Merikangas, K., and Ernst, C. (1990). Comorbidity of anxiety and depression in the Zurich Cohort Study of young adults. *In* "Comorbidity of Mood and Anxiety Disorders" (J. Maser and C. Cloninger, eds.). American Psychiatric Press, Washington, DC.

66. Kessler, R. (1995). The National Comorbidity Survey: Preliminary results and future directions. *Int. J. Methods Psychiatr. Res.* **5,** 139–151.

67. Schneier, F., Johnson, J., Hornig, C., Liebowitz, M., and Weissman, M. (1992). Social phobia: Comorbidity and morbidity in an epidemiologic sample. *Arch. Gen. Psychiatry* **49,** 282–288.

68. Kandel, D., Johnson, J., Bird, H. *et al.* (1997). Psychiatric disorders associated with substance use among children and adolescents: Findings from the Methods for the Epidemiology of Child and Adolescent Mental Disorders (MECA) Study. *J. Abnorm. Child Psychol.* **25,** 121–132.

69. Brady, E., and Kendall, P. (1992). Comorbidity of anxiety and depression in children and adolescents. *Psychol. Bull.* **111,** 244–255.

70. Pollock, R., Rosenbaum, J., Marrs, A., Miller, B., and Biederman, J. (1995). Anxiety disorders of childhood: Implications for adult psychopathology. *Psychiatr. Clin. North Am.* **18,** 745–766.

71. Merikangas, K., and Swendsen, J. (1997). Genetic epidemiology of psychiatric disorders. *Epidemiol. Rev.* **19,** 1–12.

72. Hopper, J., Judd, F., Derrick, P., and Burrows, D. (1987). A family study of panic disorder. *Genet. Epidemiol.* **4,** 33–41.

73. Merikangas, K., Stevens, D., Fenton, B. *et al.* (1998). Comorbidity and familial aggregation of alcoholism and anxiety disorders. *Psychol. Med.* **28,** 773–788.

74. Torgersen, S. (1983). Genetic factors in anxiety disorders. *Arch. Gen. Psychiatry* **40,** 1085–1089.

75. Marks, I. (1986). Genetics of fear and anxiety disorders. *Br. J. Psychiatry* **149,** 406–418.

76. Reich, T., James, J., and Morris, C. (1972). The use of multiple thresholds in determining the mode of transmission of semicontinuous traits. *Ann. Hum. Genet.* **36,** 163–186.

77. Turner, S., Beidel, D., and Costello, A. (1987). Psychopathology in the offspring of anxiety disorders patients. *J. Consult. Clin. Psychol.* **55,** 229–235.

78. Biederman, J., Rosenbaum, J., Bolduc, E., Faraone, S., and Hirshfeld, D. (1991). A high risk study of young children of parents with panic disorders and agoraphobia with and without comorbid major depression. *Psychiatry Res.* **37,** 333–348.

79. Last, C., Hersen, M., Kazdin, A., Orvaschel, H., and Perrin, S. (1991). Anxiety disorders in children and their families. *Arch. Gen. Psychiatry* **48.**

80. Sylvester, C., Hyde, T., and Reichler, R. (1988). Clinical psychopathology among children of adults with panic disorder. *In* "Rel-

atives at Risk for Mental Disorder" (D. Dunner, E. Gershon, and J. Barrett, eds.). Raven Press, New York.

81. Warner, V., Mufson, L., and Weissman, M. (1995). Offspring at high risk for depression and anxiety: Mechanisms of psychiatric disorder. *J. Am. Acad. Child Adolesc. Psychiatry* **34,** 786–797.

82. Beidel, D., and Turner, S. (1997). At risk for anxiety: I. Psychopathology in the offspring of anxious parents. *J. Am. Acad. Child Adolesc. Psychiatry* **36,** 918–924.

83. Pollock, R., Carter, A., Dierker, I., and Merikangas, K. (1998). Anxiety sensitivity in children at familial risk for psychopathology. *Banff Int. Conf. Behav. Sci., 30th,* Banff, Alberta, Canada, *1998.*

84. Rothbart, M. (1989). Behavioral approach and inhibition. *In* "Perspectives on Behavioral Inhibition" (J. Reznick, ed.). University of Chicago Press, Chicago.

85. Minton, C., Kagan, J., and Levine, J. (1971). Maternal control and obedience in the two-year-old. *Child Dev.* **42,** 1873–1894.

86. Smith, P., and Dagliersh, L. (1977). Sex differences in parent and infant behavior in the home. *Child Dev.* **48,** 1250–1254.

87. Pederson, F., and Bell, R. (1970). Sex differences in preschool children without histories of complications of pregnancy and delivery. *Dev. Psychol.* **3,** 10–15.

88. Kagan, J., and Reznick, S. (1986). Shyness and temperament. *In* "Shyness: Perspectives on Research and Treatment" (W. Jones, J. Cheek, and S. Briggs, ed.). Plenum, New York.

89. Biederman, J., Rosenbaum, J., Bolduc-Murphy, E. *et al.* (1993). A three-year follow-up of children with and without behavioral inhibition. *J. Am. Acad. Child Adolesc. Psychiatry* **32,** 814–821.

90. Rosenbaum, J., Biederman, J., Hirshfeld, D., Bolduc, E., and Chaloff, J. (1991). Behavioral inhibition in children: A possible precursor to panic disorder or social phobia. *J. Clin Psychiatry* **52,** 5–9.

91. Rosenbaum, J., Biederman, J., Bolduc-Murphy, E. *et al.* (1993). Behavioral inhibition in childhood: A risk factor for anxiety disorders. *Harv. Rev. Psychiatry* **1,** 2–16.

92. Rosenbaum, J., Biederman, J., and Gersten, M. (1988). Behavioral inhibition in children of parents with panic disorder and agoraphobia: A controlled study. *Arch. Gen. Psychiatry* **45,** 463–470.

93. Reznick, J., Gibbons, J., Johnson, M., and McDonough, P. (1989). Behavioral inhibition in a normative sample. *In* "Perspectives on Behavioral Inibition" (J. Reznick, ed.) University of Chicago Press, Chicago.

94. Kagan, J. (1994). "Galen's Prophecy: Temperment in Human Nature." Basic Books, New York.

95. Maziade, M. *et al.* (1984). Infant temperament: SES and gender differences and reliability of measurement in a large Quebec sample. *Merril-Palmer Q.* **30,** 213–216.

96. Carey, W., and McDevitt, S. (1978). Revision of the Infant Temperament Questionnaire. *Pediatrics* **61.**

97. Hsu, C., Soong, W., Stigler, J., Hong, C., and Liang, C. (1981). The temperamental characteristics of Chinese babies. *Child Dev.* **52,** 1337–1340.

98. Hirshfeld, D., Rosenbaum, J., Biederman, J. *et al.* (1992). Stable behavioral inhibition and its association with anxiety disorder. *J. Am. Acad. Child Adolesc. Psychiatry* **31,** 103–111.

99. Shear, K. (1997). Anxiety disorders in women: Gender-related modulation of neurobiology and behavior. *Semin. Reprod. Endocrinol.* **15,** 69–76.

100. McNally, R. (1994). "Panic Disorder: A Critical Analysis." Guilford Press, New York.

101. Castellani, S., Quillen, M., Vaughn, D. *et al.* (1988). TSH and catecholamine response to TRH in panic disorder. *Biol. Psychiatry* **24,** 87–90.

102. Roy-Byrne, P., Mellman, T., and Uhde, T. (1988). Biologic findings in panic disorder: Neuroendocrine and sleep-related abnormalities. *J. Anxiety Disord.* **2,** 17–29.

103. Roy-Byrne, P., Uhde, T., and Rubinow, D. (1986). Reduced TSH and prolactin responses to TRH in patients with panic disorder. *Am. J. Psychiatry* **143**, 503–507.

104. Steiner, M., Steinberg, S., Stewart, D. *et al.* (1995). Fluoxetine in the treatment of premenstrual dysphoria. *N. Engl. J. Med.* **332**, 1529–1534.

105. Young, E. (1998). Sex differences and the HPA Axis: Implications for psychiatric disease. *J. Gender-Specific Med.* **1**, 21–27.

106. Merikangas, K., Foeldenyi, M., and Angst, J. (1993). The Zurich Study: Patterns of menstrual distrubances in the community: Results of the Zurich Cohort Study. *Eur. Arch. Psychiatry Clin. Neurosci.* **243**, 23–32.

107. Stein, M., Schmidt, P., Rubinow, D., and Uhde, T. (1989). Panic disorder and the menstrual cycle: Panic disorder patients, health control subjects, and patients with premenstrual syndrome. *Am. J. Psychiatry* **146**, 1299–1303.

108. Perna, G., Brambilla, F., Arancio, C., and Bellodi, L. (1995). Menstrual cycle related sensitivity to 35% CO2 in panic patients. *Biol. Psychiatry* **37**, 528–532.

109. Cook, B., Noyes, R., Garvey, M., Beach, V., Sobotka, J., and Chaudry, D. (1990). Anxiety and the menstrual cycle in panic disorder. *J. Affective Disord.* **19**, 221–226.

110. McLeod, D., Hoehm-Saric, R., Foster, G., and Hipsley, P. (1993). The influence of premenstrual syndrome on ratings of anxiety in women with generalized anxiety disorder. *Acta Psychiatr. Scand.* **88**, 248–251.

111. Hayward, C., Killen, J., Hammer, L. *et al.* (1992). Pubertal stage and panic attack history in sixth- and seventh-grade girls. *Am. J. Psychiatry* **149**, 1239–1243.

112. Shear, M., and Mammen, O. (1995). Anxiety disorders in pregnant and post-partum women. *Psychopharmacol. Bull.* **31**, 693–703.

113. Deakin, J. (1988). Relevance of hormone-CNS interactions to psychological changes in the puerperium. *In* "Motherhood and Mental Illness" (R. Kumar and I. F. Brockington, eds.), pp. 113–132. Wright, London.

114. Villeponteaux, V., Lydiard, R., Larais, M., Stuart, G., and Ballenger, J. (1992). The effects of pregnancy on preexisting panic disorder. *J. Clin. Psychiatry* **53**, 201–203.

115. Northcott, C., and Stein, M. (1994). Panic disorder in pregnancy. *J. Clin. Psychiatry* **55**, 539–542.

116. George, D., Ladenheim, J., and Nutt, D. (1987). Effect of pregnancy on panic attacks. *Am. J. Psychiatry* **144**, 1078–1079.

117. Klein, D. (1994). Pregnancy and panic disorder. *J. Clin. Psychiatry* **55**, 293–294.

118. Insel, T. R. (1992). Oxytocin: A neuropeptide for affiliation: Evidence from behavioral receptor autoradiographic, and comparative studies. *Psychoneuroendocrinology* **17**, 35.

119. Sholomskas, D., Wickarmaratne, P., Dogolo, L., O'Brian, D., Leaf, P., and Woods, S. (1993). Postpartum onset of panic disorder: A coincidental event? *J. Clin. Psychiatry* **54**, 156–159.

120. Cohen, L., Sichel, D., Dimmock, J., and Rosenbaum, J. (1994). Impact of pregnancy on panic disorder: A case series. *J. Clin. Psychiatry* **55**, 284–288.

121. Cohen, L., Sichel, D., Dimmock, J., and Rosenbaum, J. (1994). Postpartum course in women with preexisting panic disorder. *J. Clin. Psychiatry* **55**, 289–292.

122. Metz, A., Sichel, D., and Goff, D. (1988). Postpartum panic disorder. *J. Clin. Psychiatry* **49**, 278–279.

123. Grillon, C., Dierker, I., and Merikangas, K. (1997). Startle modulation in children at risk for anxiety disorders and/or alcoholism. *J. Am. Acad. Child Adolesc. Psychiatry* **36**, 925–932.

124. Grillon, C. D. L., and Merikangas, K. R. (1998). Fear-potentiated startle in adolescent offspring of parents with anxiety disorders. *Biol. Psychiatry* 990–997.

125. Bruch, M., and Cheek, J. (1995). Developmental factors in childhood and adolescent shyness. *In* "Social Phobia: Diagnosis, Assessment and Treatment" (RG. Heimberg, M. R. Liebowitz, D. A. Hope, and F.R. S., eds.), pp. 163–182. Guilford Press, New York.

126. Arrindell, W., and Emmelkamp, P. (1986). Marital adjustment, intimacy and needs in female agoraphobics and their partners: A controlled study.

127. Marks, I. (1969). "Fears and Phobias." Academic Press, New York.

128. Raskin, M., Peeke, H., Dickman, W., and Pinkster, H. (1982). Panic and generalized anxiety disorders: Develolpmental antecedents and precipitants. *Arch. Gen. Psychiatry* **39**, 687–689.

129. Marks, I. (1987). The development of normal fear: A review. *J. Child Psychol. Psychiatry* **28**, 667–697.

130. Frankenhaeuser, M., Dunne, E. U. L. (1976). Sex differences in sympathetic-adrenal medullary reactions induced by different stressors. *Psychopharmacology* **47**, 1–5.

131. Johansson, G., and Post, B. (1974). Catecholamine output of males and females over a one-year period. *Acta Physiol. Scand.* **92**, 557–565.

132. Frankenhaeuser, M., von Wright, M., Collins, A., von Wright, J., Sedvall, G., and Swahn, C. (1978). Sex differences in psychoneuroendocrine reactions to examination stress. *Psychosom. Med.* **40**, 334–343.

133. Nolen-Hoeksema, S. (1987). Sex differences in unipolar depression: Evidence and theory. *Psychol. Bull.* **101**, 259–282.

134. Rosenbaum, J., Pollock, R., Otto, M., and Pollack, M. (1995). Integrated treatment of panic disorder. *Bull. Menninger Clin.* **59**, 4–26.

135. Otto, M., Peneva, S., Pollock, R., and Smoller, J. (1995). Cognitive-behavioral and pharmacologic perspectives on the treatments of post-traumatic stress disorder. *In* Clinical "Challenges in Psychiatry: Pharmacologic and Psychosocial Perspectives on the Treatment of Post-Traumatic Stress Disorders" (O. Pollack and J. Rosenbaum, eds.). Guilford Press, New York.

136. Albano, A. M. CB. (1995). "Treatment of Anxiety Disorders of Childhood," 18 vols. Saunders, Philadelphia.

137. Barrett, P. M., and Rapee, R. M. (1996). Family treatment of childhood anxiety: A controlled trial. *J. Consult. Clin. Psychol.* **64**, 333–342.

81

Posttraumatic Stress Disorder (PTSD)

NAOMI BRESLAU

Departments of Psychiatry, Biostatistics, and Epidemiology, Henry Ford Health System, Detroit, Michigan; Department of
Psychiatry, Case Western Reserve University School of Medicine, Cleveland, Ohio; Department of Psychiatry, University of
Michigan School of Medicine, Ann Arbor, Michigan

I. Introduction

The introduction of posttraumatic stress disorder (PTSD) in
1980 into the official nosologic classification, the DSM-III [1],
marked the beginning of contemporary research on the psychiatric response of victims of severely traumatic events, such as
military combat, rape, or disaster. The definition of PTSD in the
DSM-III and in subsequent DSM editions [2,3] is based on a
conceptual model that brackets traumatic events from other
stressful experiences and PTSD from other responses to stress.
Traumatic (catastrophic) events, in contrast with "ordinary"
stressful experiences, have been linked etiologically in the DSM
to a specific syndrome, that of PTSD. Criterial symptoms are
defined in terms of their connection, temporally and in content,
with a distinct traumatic event: reexperiencing the event in intrusive thoughts and dreams, avoidance of stimuli that symbolize the event, numbing of general responsiveness, and increased
arousal not present before the event. It is this connection with a
distinct event that renders the list of 17 symptoms—many of
them are among the characteristic features of other psychiatric
disorders—a specific syndrome that is the signature malady in
victims of trauma.

Research on PTSD since 1980 has focused primarily on
Vietnam War veterans and, to a lesser extent, on victims of
specific traumatic events, such as disasters, rape, and other criminal assault. A handful of epidemiologic studies of PTSD in the
general population have been conducted in the 1990s [4–10].
These studies describe the prevalence of traumatic events and
PTSD and their distributions across subgroups of the population
and suggest risk factors for exposure to trauma and PTSD. They
describe aspects of the natural history of PTSD, including the
duration of symptoms and comorbidity with other psychiatric
disorders, and examine hypotheses about causal pathways between PTSD and the disorders with which strong lifetime associations have been observed. This chapter reviews the evidence
from these studies. Reports on special populations exposed to a
distinct trauma are cited with respect to risk factors for PTSD.
Evidence from clinical samples is not included, because studies
based on such samples may confound characteristics of care-
seeking with those of the disorder. Excluded from this review
are studies on the health effects of violence against women.
There is a growing literature on the physical effects of intentional violence experienced by women, including rape, physical
assault, and childhood sexual abuse. Several studies have described mental health effects other than PTSD, including symptoms of anxiety and depression, substance abuse, and sexual
dysfunction [11,12]. While this literature suggests that violence
is a serious public health problem, the topic is outside the
scope of this review and is covered in Chapter 42, Violence
Against Women.

II. The Definition of Trauma and PTSD

The core features of PTSD in the current nosology, the DSM-
IV [3], consist of (1) the "stressor criterion" that defines the
etiologic event in PTSD; and (2) the configuration of symptoms
that defines the characteristic PTSD syndrome. Three symptom
groups constitute the PTSD syndrome: (a) reexperiencing the
trauma in nightmares, intrusive memories, or "flashbacks," (b)
numbing of affect and avoidance of thoughts, acts, and situations that symbolize the trauma, and (c) symptoms of excess
arousal. The diagnosis requires the persistence of symptoms for
at least one month and a clinically significant distress or impairment. This definition of the syndrome of PTSD in DSM-IV has
changed little from earlier DSM editions. However, the stressor
criterion has changed materially. The new definition broadens
the range of "qualifying" traumatic events. In addition to the
types of events that represent the core category of traumas that
have initially been used to define PTSD (*i.e.,* military combat,
disaster, and criminal violence), the expanded DSM-IV definition attempts to capture the aggregate clinical experience regarding the types of events that might culminate in PTSD.
Although stressors defined as less extreme are explicitly excluded (*e.g.,* "spouse leaving," "being fired" [3, p. 427], the
revised definition of traumatic events is clearly more inclusive.
For the first time in DSM-IV, death of a loved one from any
cause, including natural causes, qualifies as a stressor, as long
as it was "sudden and unexpected." Being diagnosed with a
life-threatening illness is another example of the range of traumatic events that are included in the new definition. The revised
definition has introduced a subjective component, requiring
that "the person's response involved intense fear, helplessness, or horror," although the timing of this response has not
been specified. The revision in DSM-IV—the broader range
of qualifying traumatic events and the added criterion of a specific emotional response—aligns the stressor criterion with ongoing clinical practice, which is guided by the principle that
people may perceive and respond differently to outwardly similar events.

The DSM-IV definition of PTSD requires, for the first time,
that the syndrome cause significant distress or impairment, an
addition that renders the diagnostic definition of PTSD more
stringent, particularly in epidemiologic surveys of the general

population, as opposed to clinical practice, where impairment or distress is generally the reason for seeking treatment.

III. Measurement Issues

A. Assessing Exposure to Traumatic Events

The standard measurement procedure in contemporary epidemiologic studies of psychiatric disorders, including PTSD, has been the National Institute of Mental Health-Diagnostic Interview Schedule (NIMH-DIS) [13,14]. The NIMH-DIS is a structured interview designed to be administered by experienced interviewers without clinical training. The PTSD section inquires about lifetime history of traumatic events and then elicits information about PTSD symptoms in connection with an event nominated by the respondents as the worst that they had ever experienced. The inquiry about PTSD symptoms continues for up to three traumatic events, two additional events apart from the worst. Modifications to the NIMH-DIS have been introduced in some epidemiologic studies, chiefly in the approach to eliciting information about exposure to traumatic events [4,6].

Estimates of the prevalence of exposure to traumatic events vary by the inclusiveness of the stressor criterion and the methods used to measure exposure to qualifying stressors. In previous editions of the DSM (i.e., DSM-III and -IIIR), qualifying stressors were defined as events that would be "distressing to almost everyone" and as "generally outside the range of usual human experience." Typical PTSD traumas were military combat, rape, physical assault, natural disaster, witnessing violence, and learning about violent injury or a violent death of a loved one. DSM-IV has broadened the stressor criterion beyond the earlier definition, as described earlier. As a result, estimates of the prevalence of exposure to traumatic events according to DSM-IV can be expected to be higher than those based on earlier definitions. Evidence indicates that the subjective component of the stressor criterion in DSM-IV does little to offset the broadened range of qualifying stressors. Approximately 90% of those who have ever been exposed to one or more stressors have responded with "intense fear, helplessness, or horror" to one (the worst) of their stressors [15].

Differences in the estimates of the prevalence of exposure to traumatic events across epidemiologic studies that predated the publication of the DSM-IV in 1994, when the stressor criterion was revised, reflect differences in measurement approaches. The key difference is between studies that used the revised NIMH-DIS [14], which elicits history of exposure to traumatic events with a single question that incorporates examples of typical PTSD events, and studies that used a list of events and inquired about each event separately. The two Epidemiologic Catchment Area (ECA) sites that gathered data on PTSD used the original NIMH-DIS [13], which inquired only about history of stressors that resulted in PTSD symptoms [16,17]. The use of a list of events and the number of events included in the list have important implications for the estimates of prevalence of traumatic events, as will be summarized in section IVA. The use of a list of events has become the standard measurement procedure, with the incorporation of the approach in the NIMH-DIS revised for DSM-IV [18]. There still remains a potential for variability in the construction of such a list across studies.

B. Estimating the Conditional Risk of PTSD

The probability of PTSD given exposure to traumatic events is referred to as the conditional risk (or conditional probability) of PTSD. Estimates of the conditional risk of PTSD are derived from information on the prevalence of exposure to one or more qualifying events and the proportion of those exposed who meet criteria for PTSD in connection with one of their events. Estimates can be calculated for the *overall* conditional risk of PTSD and for the conditional risk associated with specific types of traumatic events. Estimates of the overall conditional risk of PTSD depend in part on the stressor definition. A narrow definition that includes only rare and highly traumatic events would yield a lower prevalence of exposure and a higher conditional risk of PTSD, compared with an inclusive definition that encompasses a wide range of events. Even when the same stressor criterion is used, a measurement procedure that yields a lower prevalence of exposure, as when a single question is used, is likely to result in a higher estimate of the conditional risk of PTSD compared to a procedure that yields a higher prevalence of exposure, as when a list of events is used. The use of a list, instead of a single question, enhances recall of traumatic events that are less memorable but also less likely to have led to PTSD [19].

An important methodologic issue in previous epidemiologic studies that estimated the conditional risk of PTSD has been identified by Kessler et al. [7,20]. Previous studies have examined the PTSD effects of traumatic events nominated by the respondents as the worst or the most upsetting they had ever experienced. The approach is efficient for estimating the prevalence of PTSD because events nominated by respondents as the worst are the events most likely to lead to PTSD [19]. However, the conditional risk of PTSD from the worst events is higher than the conditional risk of PTSD associated with the entire category of traumatic events defined in the DSM as potentially leading to PTSD. The worst events represent the extreme end of the distribution of traumatic events considered as potential causes of PTSD. Furthermore, it is reasonable to assume that traumatic events that have led to marked and enduring psychologic distress would be more likely to be selected by respondents as the worst. A contamination of the selection of traumatic events by their psychologic sequelae, including PTSD symptoms, would result in a spuriously strong association between exposure to traumatic events and PTSD. A selection of a random event from the complete list of traumatic events reported by each respondent avoids this contamination and provides a representative sample of qualifying events. Evidence supporting the bias in estimates of the risk of PTSD based on the worst traumas has been reported [10].

In the 1996 Detroit Area Survey of Trauma [10], we applied this method to estimate the conditional risk of PTSD, as defined in the DSM-IV. The survey was conducted on a representative sample of 2181 persons 18–45 years of age in the Detroit primary metropolitan statistical area (PMSA). A random digit dialing method [21] was used to select the sample and a computer-assisted telephone interview was used to obtain the

Table 81.1
Estimates of Lifetime Prevalence of Exposure to Trauma and PTSD (%)

Study	Exposure		PTSD	
	Male	Female	Male	Female
[4]	43.0	36.7	6.0	11.3
[5]	73.6	64.8	—	—
[6]	—	69.0	—	12.3
[7]	60.7	51.2	5.0	10.4
[8]	—	40.0	—	13.8
[9]	81.3	74.2	—	—
[10]	92.2	87.1	10.2	18.3

Note. Studies 4–8 are based on DSM-III and -III-R; studies 9 and 10 are based on DSM-IV, although only 10 covered the entire range of traumas in DSM-IV. Studies 4 and 8 are based on a single question; other studies used a list of traumas. Studies 5 and 9 evaluated only current prevalence of PTSD.

data. Screening was completed in 76.2% of households and cooperation rate in eligible households was 86.8%. The sample was weighted to adjust for the sampling design and to approximate the sample distribution on key sociodemographic factors to the population of the geographic area in the 1990 U.S. Census [22,23]. A comparison of the distribution of the sample with the distribution of the population of the PMSA suggests that we succeeded in getting a representative sample of the population with respect to key characteristics [10].

The computer-assisted telephone interview took 30 minutes to complete. It began with an enumeration of 19 types of traumatic events that operationalize the DSM-IV stressor criterion. An endorsement of an event type was followed by questions on the number of times an event of that type had occurred and the respondent's age at each time. A computer selected one random event out of the list of events of each respondent for the evaluation of PTSD symptoms, using the DIS-IV, with slight modifications (some items were adapted from the WHO Composite International Diagnostic Interview 2.1 [24]). Responses were used to diagnose PTSD based on DSM-IV criteria. A validation study in a subset of the sample (n = 53) documented high agreement between the telephone administered DIS and independent clinical interviews conducted on the telephone by trained psychiatric social workers (sensitivity = 95% and specificity = 71%) [25].

IV. Sex Differences in Exposure to Traumatic Events

A. Sex Differences in Overall Exposure to Traumatic Events

Table 81.1 presents estimates of lifetime exposure to traumatic events in males and females reported in epidemiologic studies of the general population. The lifetime prevalence of exposure to traumatic events is lower in females than in males. Estimates of the lifetime prevalence of exposure in studies that preceded the publication of DSM-IV range from 36.7 to 69.0% in females and from 43.0 to 73.6% in males [4–8]. Studies that used the revised NIMH-DIS, in which information on exposure

was elicited by a single question, yielded lower estimates than studies that used a list of events [4,8]. Two studies that applied the DSM-IV definition and used lists of events yielded higher estimates of exposure in both sexes [9,10]. The first epidemiologic survey of DSM-IV PTSD, conducted in Winnipeg, Canada, did not cover the entire range of traumatic events in the expanded definition, but nonetheless reported very high rates of exposure: 74.2% in females and 81.3% in males [9]. In the 1996 Detroit Area Survey of Trauma, which used a comprehensive list of DSM-IV traumatic events, the lifetime prevalence of traumatic events in females was 87.1% and in males was 92.2% [10]. While the overall lifetime prevalence of exposure in the Detroit Area Survey of Trauma cannot be directly compared with other estimates because of differences in the stressor criteria, the prevalence estimates of specific trauma types are comparable to those from the National Comorbidity Survey (NCS) [7]. For example, the prevalence of rape in the NCS was 9.2% in females and 0.7% in males, compared to 9.4% and 1.1%, respectively, in the Detroit Area Survey of Trauma. The prevalence of exposure to disaster in the NCS was 15.3% in females and 18.9% in males, compared to 15.3% and 17.9%, respectively, in the Detroit Area Survey.

Although the sex difference in lifetime exposure across these epidemiologic studies is small (male: female ratio <1.2), it is consistent. Epidemiologic surveys also reported that the proportion of females with a history of multiple traumas is lower than the proportion of males [7] and that the mean number of traumatic events reported by females is lower than that reported by males [10]. Thus the overall burden of trauma, in terms of lifetime prevalence and the average number of traumatic events, is lower in females than in males.

B. Sex Differences across Types of Traumas

The overall pattern of a lower burden of trauma in females vs males obscures an important variation across types of traumatic events. Table 81.2 illustrates the variation in the sex differences in exposure across event types observed in the Detroit Area Survey of Trauma [10]. Females had significantly higher rates of rape and other sexual assault. In contrast, males had significantly higher rates of other types of assaultive violence, including being shot or stabbed, mugged or threatened with a weapon, and having been badly beaten. The overall lifetime prevalence of assaultive violence, as a category, was significantly higher in males than females, a difference that persisted when race, education, income, and marital status were controlled. As can be seen in Table 81.2, males had significantly higher lifetime rates of serious accidents, both motor vehicle and other types, and of witnessing acts of violence. The prevalence of exposure to natural disaster did not vary between the sexes. For all of the events subsumed under the category of other injury or shocking experiences, males had a significantly higher rate than females, even when key sociodemographic characteristics were controlled. In contrast with assaultive violence and other injury or shocking event, the two categories of events experienced directly, sex differences in learning about trauma to a close relative or friend or a sudden unexpected death of a loved one were small. Other epidemiologic studies have also reported higher rates of rape and sexual assault in females, higher rates of accidents and wit-

Table 81.2

Lifetime Prevalence of Exposure to Traumatic Events by Sex

	Females		Males	
	% Exposed	(se)	% Exposed	(se)
Assaultive violence	**32.4**	**(1.7)**	**43.3**	**(1.7)**
Military combat	0.2	(0.2)	2.8	(0.6)
Rape	9.4	(1.1)	1.1	(0.4)
Held captive/tortured/ kidnapped	2.0	(0.5)	1.7	(0.5)
Shot/stabbed	1.8	(0.5)	8.2	(1.0)
Sexual assault other than rape	9.4	(1.1)	2.8	(0.6)
Mugged/held-up/ threatened with weapon	16.4	(1.3)	34.0	(1.6)
Badly beaten-up	9.8	(1.1)	13.1	(1.2)
Other injury or shocking event	**52.0**	**(1.8)**	**68.0**	**(1.6)**
Serious car accident	23.5	(1.6)	32.8	(1.6)
Other serious accident	9.5	(1.1)	18.5	(1.3)
Natural disaster	15.3	(1.3)	17.9	(1.3)
Life-threatening illness	5.9	(0.8)	3.6	(0.6)
Child's life-threatening illness	3.5	(0.7)	2.6	(0.5)
Witnessed killing/serious injury	18.6	(1.4)	40.1	(1.7)
Discovering a dead body	6.2	(0.9)	9.1	(1.0)
Learning of traumas to others[a]	**61.8**	**(1.8)**	**63.1**	**(1.7)**
Sudden unexpected death of relative/friend	**59.0**	**(1.8)**	**61.1**	**(1.7)**
Any trauma	**87.1**	**(1.2)**	**92.2**	**(1.0)**

Note. Based on the total pool of traumatic events compiled from the complete lists of events reported by the respondents. Se, Standard error. From the Detroit Area Survey of Trauma.

[a]The 4 specific event types under this category are not displayed.

nessing acts of violence in males, and trivial sex differences in exposure to disaster and learning about a traumatic event of a loved one [5,7,9].

Lifetime prevalence of assaultive violence in the Detroit Area Survey of Trauma was higher in males and females who were black and in lower social classes. In contrast, other categories of trauma were unrelated (or only weakly related) to race and social class. We also found that exposure to assaultive violence in both sexes peaked at 16–20 years of age and fell precipitously after age 20. The rate in males was higher than in females across the age range covered in the study (*i.e.,* up to age 45). A wide sex difference occurred in the decade from 31 to 40 years of age due to a continued decline in the rate of experiencing assaultive violence in the fourth decade in females but not in males [10].

V. Sex Differences in PTSD

A. Sex Differences in the Conditional Risk of PTSD

Epidemiologic surveys that reported on the lifetime prevalence of DSM-IIIR PTSD yielded population estimates that fall within a narrow range: 5–6% in males and 10–14% in females

(Table 81.1). The lifetime prevalence of DSM-IV PTSD estimated in the Detroit Area Survey of Trauma was 10.2% in males and 18.3% in females. The higher prevalence estimates in this study are probably due to exposure to new event types included in the DSM-IV definition, which are prevalent in the community, even though they confer a relatively low risk for PTSD. (The two ECA sites that measured the lifetime prevalence of DSM-III PTSD yielded lower estimates, approximately 1%, but a similar sex difference [16,17]). The sex difference in PTSD is not due to females' higher lifetime prevalence of exposure to traumatic events. The lifetime prevalence of exposure is higher in males, as is the proportion with a history of multiple events. Females' higher prevalence of PTSD reflects their overall higher conditional risk of PTSD, that is, their higher probability of experiencing PTSD following exposure to traumatic events. However, there is consistent evidence of sex differences in the distributions of exposure across specific event types. A key issue concerns the extent to which females' higher prevalence of PTSD is accounted for by their higher rates of certain types of trauma, primarily rape, which confer a high risk of PTSD.

B. The Detroit Area Survey of Trauma

To address this question, results from the Detroit Area Survey of Trauma are presented. The survey provides estimates of the conditional risk of PTSD across event types, free of the bias in previous studies that focused on traumatic events nominated by the respondents as the worst they had ever experienced. Sex-specific estimates of the conditional risk of PTSD, based on the randomly selected traumatic events, appear in Table 81.3. The table also displays the distributions of the randomly selected events across types. The conditional risk of PTSD associated with exposure to any trauma was 13.0% in females and 6.2% in males (p < 0.001; female to male ratio = 2.1). Adjusting for key sociodemographic characteristics does not alter these results. The overall sex difference in the conditional risk of PTSD was due primarily to females' greater risk of PTSD following exposure to assaultive violence (35.7% vs 6.0%; p < 0.001; female to male ratio = 6.0). Sex differences in the three other categories of traumatic events were not significant. Specifically, the conditional risk of PTSD associated with other injury or shocking experience was 5.4% in females and 6.6% in males, with learning about traumas to a close relative or friend, was 3.2% in females and 1.4% in males, and with sudden unexpected death of a loved one was 16.2% in females and 12.6% in males.

The distributions across types of events involving assaultive violence varied between the sexes. Most important, rape was not a randomly selected event for any male, a reflection of the low prevalence of rape in males (*i.e.,* 1.1%), whereas rape was the randomly selected event in 4% of females. The absence of males whose randomly selected event was rape precludes a sex comparison of the conditional risk of PTSD associated with rape. Furthermore, because rape conferred a relatively high risk of PTSD in females (49%), the sex inequality in the exposure to rape is likely to have influenced the sex ratio of the overall conditional risk of PTSD and the conditional risk associated with the category of assaultive violence. With the exception of rape and military combat, on which sex comparisons cannot

Table 81.3
The Risk of PTSD by Type of Traumatic Event (n = 1957)

	Females		Males	
	% Exposed	% PTSD	% Exposed	% PTSD
Assaultive violence	**19.7**	**35.7**	**15.8**	**6.0**
Military combat	0	—	1.2	0
Rape	4.0	49.0	0	—
Held captive/tortured/ kidnapped	0.7	78.2	0.2	0
Shot/stabbed	0.6	0	2.7	18.1
Sexual assault other than rape	4.0	24.4	0.3	15.7
Mugged/held-up/threatened with weapon	5.9	17.5	8.0	2.4
Badly beaten-up	4.4	56.2	3.4	6.4
Other injury or shocking event	**28.6**	**5.4**	**38.3**	**6.6**
Serious car accident	6.7	3.6	9.4	1.6
Other serious accident	3.8	28.3	5.5	10.4
Natural disaster	6.5	0	5.6	7.3
Life-threatening illness	1.7	1.0	1.2	1.2
Child's life-threatening illness	0.8	0	0.9	17.8
Witnessed killing/serious injury	6.5	2.8	13.6	9.1
Discovering a dead body	2.6	0.5	2.3	0
Learning of traumas to others[a]	**30.4**	**3.2**	**26.8**	**1.4**
Sudden unexpected death of relative/friend	**21.3**	**16.2**	**19.1**	**12.6**
Any trauma	**100.0**	**13.0**	**100.0**	**6.2**

Note. The 1st column for each sex displays the distribution of the randomly selected traumas across event types. From the Detroit Area Survey of Trauma.

[a]The 4 specific event types under this category are not displayed.

be made, females' higher risk of PTSD applied to all but one event type subsumed under assaultive violence. Specifically, females had a higher risk of PTSD in connection with having been held captive/tortured/kidnapped, sexual assault other than rape, mugged/held-up/or threatened with a weapon, and having been badly beaten (Table 81.3).

To address the disparity between the sexes in the percentage whose randomly selected event was rape, we compared the conditional risk of PTSD between the sexes, excluding rape. The conditional risk of PTSD associated with assaultive violence in the remaining sample (n = 1925) was 32.3% in females and 6.0% in males (p < 0.001; female to male ratio = 5.4). The overall conditional risk of PTSD associated with any trauma (excluding rape) was 11.5% in females and 6.2% in males (p < 0.01; female to male ratio = 1.9).

Similar results were observed in an epidemiologic study that measured current prevalence of full and partial PTSD [9]. A sex difference in the conditional risk of PTSD was estimated, excluding those with lifetime sexual trauma, by a composite variable that includes child sexual abuse and adult forced sex, including rape. The relative risk of current PTSD, full and partial combined, in females vs males who were exposed to other types of traumas was 3.5 (95% Confidence Interval = 1.2–10.1) (Dr. Murray Stein, written communication of unpublished data, February 24, 1998).

C. The Most Frequent Precipitating Traumas among PTSD Cases

The distribution of PTSD cases across the trauma types that precipitated the disorder was estimated in the Detroit Area Survey of Trauma based on the randomly selected traumatic events. In both sexes the single most frequent cause of PTSD was sudden unexpected death of a loved one, with 27% of female cases and 38% of male cases attributable to an event of this type. A marked sex difference was observed in the proportion of PTSD cases attributable to assaultive violence: 54% of female cases vs 15% of male cases. The sex difference is in large part due to females' higher risk for PTSD following assaultive violence, and the sex difference in the distribution of trauma types contributes to this pattern as well.

VI. Sex Differences in the Course of PTSD

A. Number of Symptoms

Two features in the course of PTSD associated with the randomly selected events were examined in the Detroit Area Survey of Trauma: the frequencies of criterial symptoms of PTSD and the duration of the disorder. Among those with PTSD, significant sex differences were observed in the prevalence of intense psychological reactivity to stimuli that symbolize the trauma (restricted affect and exaggerated startle response), with the females' rates exceeding the males'. The mean number of PTSD symptoms was significantly higher in females than in males (12 vs 11; p < 0.05). The sex difference in the mean number of PTSD symptoms is due almost entirely to the higher proportion of female PTSD cases attributable to assaultive violence, a category of traumatic events that is associated with more severe PTSD in terms of number of symptoms.

B. Duration of PTSD

Duration of PTSD varied markedly between the sexes. The median time to remission of symptoms was 48 months in females vs 12 months in males. This overall fourfold sex difference in duration is a function of two factors. First, PTSD persisted longer in cases that resulted from traumas experienced directly (*i.e.,* assaultive violence and other injury or shocking experience), compared to learning about a trauma to a loved one or a sudden unexpected death of a loved one. Second, independent of trauma type, PTSD persisted longer in females than males. The median duration of PTSD due to traumas experienced directly was 72 months in females vs 24 months in males, whereas the median duration of PTSD due to learning about a trauma to a loved one or sudden unexpected death of a loved one was 15 months in females vs 6.5 months in males. Estimated in a Cox proportional hazards model, the likelihood of remission in females was approximately half the likelihood in males after being adjusted for the effects of trauma type [10]. A higher risk for chronicity in females with PTSD had been previously reported [26].

VII. Risk Factors for Exposure and PTSD

A. Risk Factors for Exposure to Traumatic Events

As summarized in Section IVB, exposure to traumatic events that involve assaultive violence is higher in males than in females and varies by race, social class, and age. Exposure to other types of traumatic events is more equally distributed across subgroups of the population. Apart from these factors, several personal risk factors have been shown to be associated with exposure to traumatic events, although there is evidence to suggest that these risk factors might not apply uniformly across categories of trauma [27]. Early conduct problems, family history of psychiatric disorder, and personality traits have been reported to be associated with exposure [4,16,28]. Another important risk factor for exposure, revealed in an analysis of prospective data of the Detroit area longitudinal study of young adults, is prior history of trauma: Persons with a history of trauma at baseline were nearly twice as likely to experience a new trauma during the follow-up interval compared to those with no history of trauma [28]. Similar results were reported by others [29]. Evidence from two studies indicates that major depression predicted subsequent exposure to traumatic events [8,27]. This finding is consistent with reports of an increased risk of ordinary stressful life events (as distinguished from traumatic events in PTSD) in persons with prior depression [30,31]. A weaker relationship with preexisting anxiety disorders has also been observed [27]. A role of illicit drug use disorder in predicting exposure to traumatic events has been observed in some, but not all, studies [8,32,33].

B. Risk Factors for PTSD Following Exposure

The risk for PTSD in persons exposed to traumatic events varies by type of trauma and by sex, with events involving assaultive violence conferring a higher risk than other event types and females having a higher risk than males. The sex difference in the vulnerability to the PTSD effects of trauma is unlikely to be influenced by differences in the rate of childhood trauma or preexisting psychiatric disorders [19]. Personal characteristics found to be associated with the risk of PTSD include personality traits, history of psychiatric disorder, and family history of psychiatric disorder [34,35]. Among persons exposed to traumatic events, those with high neuroticism scores have been reported to be at increased risk for PTSD [4]. Preexisting major depression increases the risk of PTSD following exposure, as does family history of major depression [4,8,20,35–38].

The effects of history of prior trauma on the PTSD effects of subsequent trauma have been examined. Higher rates of childhood trauma have been reported in Vietnam combat veterans with PTSD [17,39–41]. Prior victimization in women who have been raped and prior combat stress in men fighting a subsequent war have been reported to influence the psychological sequelae of the later trauma [42,43]. Results from the Detroit Area Survey of Trauma suggest that history of prior exposure to assaultive violence is associated with increased vulnerability to the

PTSD effects of subsequent traumas, a vulnerability that persists for years. The effect of prior exposure to other traumatic events on the PTSD effects of subsequent traumas fades with the passage of time.

A report that examined risk factors for males and females separately shows that sex differences in the associations between risk factors and either exposure or PTSD were minor, suggesting similar relationships in both sexes (i.e., no interactions between risk factors and sex). The study also demonstrated the role of preexisting mental disorders and parental psychiatric history as risk factors for exposure and for PTSD [29]. Preexisting substance use disorder was associated with a significant increase in the risk of exposure in both sexes but was unrelated to the risk of PTSD in either sex. None of the factors subsumed under childhood family stress (e.g., parental aggression, nonconfiding relationship with mother, divorce) were associated with a significant increase in the risk of PTSD in either sex, whereas parental aggression toward the respondent and divorce were significantly associated with an increased risk for exposure in females and males, respectively.

In sum, traumatic events are not random phenomena. Exposure to assaultive violence is higher in persons in lower social classes, in blacks, and in the young. While it is generally higher in males, exposure to some types of assaultive violence, namely rape and other sexual assault, is higher in females. Other risk factors for exposure include preexisting major depression and possibly substance use disorders, family history of psychiatric disorder, personality characteristics, and history of conduct disorder and family adversity in childhood. Risk factors for PTSD following trauma include female sex, personality characteristics, preexisting major depression, and history of trauma in childhood.

VIII. Conclusions

Estimates of the lifetime prevalence of PTSD in females are relatively high, ranging between 10–14% in studies preceding DSM-IV and over 18% in a study that used a comprehensive list of traumatic events as defined in DSM-IV. These estimates are approximately twice as high as those in males. Furthermore, the duration of PTSD symptoms in females with the disorder is nearly four times longer than in male counterparts: 48 months vs 12 months, on average. Thus the burden of PTSD in females, in terms of the lifetime prevalence and the persistence of the disorder, is far greater than in males.

At the core of the heavy burden of PTSD in females is the unique role of assaultive violence. Females are at a far greater risk for PTSD when exposed to traumatic events involving assaultive violence than are males. With respect to other categories of traumatic events, sex differences are small. Although females' higher vulnerability to the PTSD effects of assaultive violence is in part attributable to the higher prevalence of rape among them, the sex difference persists when this is taken into account. Another aspect of the burden is the longer duration of PTSD in females than males. The difference is due to the higher proportion of female PTSD cases attributable to assaultive violence, coupled with the fact that the disorder persists longer in females regardless of cause. When PTSD in females is the result of experiencing assaultive violence, the duration of the disorder is more than twice as long as in male counterparts.

The heaviest burden of PTSD is likely to be borne by young women, those in late adolescence and early adulthood, when females' risk for exposure to assaultive violence is at its peak. The risk of exposure to assaultive violence falls precipitously after age 25 and, in women, it continues to decline in the fourth decade of life. However, past exposure to assaultive violence increases the vulnerability to the PTSD effects of subsequent traumas of less magnitude for many years.

It should be noted that data on the PTSD effects of trauma summarized here are based on epidemiologic studies conducted in U.S. communities in peacetime and not under special circumstances of disaster. The PTSD effects of military combat in men have been well documented, chiefly in Vietnam War veterans. By applying to community residents the criteria of a disorder that was introduced primarily to capture the psychiatric sequelae of Vietnam military combat, these epidemiologic studies have shown that, under peacetime conditions, PTSD is relatively common among community residents and that the burden of the disorder is heavier in females than in males. Research that replicates findings concerning the role of assaultive violence in the sex difference in PTSD is needed.

References

1. American Psychiatric Association (1980). "Diagnostic and Statistical Manual of Mental Disorders," 3rd ed. American Psychiatric Press, Washington, DC.
2. American Psychiatric Association (1987). "Diagnostic and Statistical Manual of Mental Disorders," 3rd rev. ed. American Psychiatric Press, Washington, DC.
3. American Psychiatric Association (1994). "Diagnostic and Statistical Manual of Mental Disorders," 4th ed. American Psychiatric Press, Washington, DC.
4. Breslau, N., Davis, G. C., Andreski, P., and Peterson, E. (1991). Traumatic events and posttraumatic stress disorder in an urban population of young adults. *Arch. Gen. Psychiatry* **48,** 216–222.
5. Norris, F. H. (1992). Epidemiology of trauma: Frequency and impact of different potentially traumatic events on different demographic groups. *J. Consult. Clin. Psychol.* **60,** 409–418.
6. Resnick, H. S., Kilpatrick, D. G., Dansky, B. S., Saunders, B. E., and Best, C. L. (1993). Prevalence of civilian trauma and posttraumatic stress disorder in a representative national sample of women. *J. Consult. Clin. Psychol.* **61,** 984–991.
7. Kessler, R. C., Sonnega. A., Bromet, E., and Nelson, C. B. (1995). Posttraumatic stress disorder in the National Comorbidity Survey. *Arch. Gen. Psychiatry* **52,** 1048–1060.
8. Breslau, N., Davis, G. C., Peterson, E., and Schultz, L. (1997). Psychiatric sequelae of posttraumatic stress disorder in women. *Arch. Gen. Psychiatry* **54,** 81–87.
9. Stein, M. B., Walker, J. R., Hazen, A. L., and Forde, D. R. (1997). Full and partial posttraumatic stress disorder: Findings from a community survey. *Am. J. Psychiatry* **154,** 114–119.
10. Breslau, N., Kessler, R. C., Chilcoat, H. D., Schultz, L. R., Davis, G. C., and Andreski, P. (1998). Trauma and posttraumatic stress disorder in the community: The 1996 Detroit area survey of trauma. *Arch. Gen. Psychiatry* **55,** 626–632.
11. Koss, M. P., and Heslet, L. (1992). Somatic consequences of violence against women. *Arch. Fam. Med.* **1,** 53–59.
12. Goodman, L. A., Koss, M. P., and Russo, N. F. (1993). Violence against women: Physical and mental health effects. Part I: Research findings. *App. Prev. Psychol.* **2,** 79–89.
13. Robins, L. N., Helzer, J. E., Croughland, J. L., Williams, J. B. W., and Spitzer, R. L. (1981). "NIMH Diagnostic Interview Schedule,

Version III," Publ. ADM-T-42-3 (5-8-81). Public Health Service (PHS), NIMH, Rockville, MD.
14. Robins, L. N., Helzer, J. E., Cottler, L., and Golding, E. (1989). "NIMH Diagnostic Interview Schedule, Version III Revised." Washington University, St. Louis, MO.
15. Breslau, N., Chilcoat, H. D., Kessler, R. C., and Davis, G. C. (1999). Previous exposure and the PTSD effects of a subsequent trauma: Results from the Detroit Area Survey of Trauma. *Am. J. Psychiatry* **156,** 902–907.
16. Helzer, J. E., Robins, L. N., and McEvoy, L. (1987). Post-traumatic stress disorder in the general population: Findings of the Epidemiologic Catchment Area Survey. *N. Engl. J. Med.* **317,** 1630–1634.
17. Davidson, J. R. T., Hughes, D., and Blazer, D. G. (1991). Posttraumatic stress disorder in the community: An epidemiological study. *Psychol. Med.* **21,** 713–721.
18. Robins, L., Cottler, L., Bucholz, K., and Compton, W. (1995). "Diagnostic Interview Schedule for DSM-IV." Washington University, St. Louis, MO.
19. Breslau, N., Davis, G. C., Andreski, P., Peterson, E., and Schultz, L. R. (1997). Sex differences in posttraumatic stress disorder. *Arch. Gen. Psychiatry* **54,** 1044–1048.
20. Kessler, R. C., Sonnega, A., Bromet, E., Hughes, M., Nelson, C. B., and Breslau, N. (1999). Epidemiologic risk factors for trauma and PTSD. *In* "Risk Factors for Posttraumatic Stress Disorder" (R. Yehuda, ed.). Am. Psychiatric Press Assoc., Washington, DC 23–59.
21. Survey Sampling (1996). "Random Digit Telephone Sampling Methodology." Survey Sampling, Fairfield, CT.
22. Potthoff, R. F. (1994). Telephone sampling in epidemiologic research: To reap the benefits, avoid the pitfalls. *Am. J. Epidemiol.* **10,** 967–978.
23. Kessler, R. C., Little, R. J. A., and Groves, R. M. (1995) Advances in strategies for minimizing and adjusting for survey nonresponse. *Epidemiology* **17,** 192–204.
24. World Health Organization (1997). "Composite International Diagnostic Interview (CIDI), Version 2.1." WHO, Geneva.
25. Breslau, N., Kessler, R., and Peterson, E. L. (1998). PTSD assessment with a structured interview: Reliability and concordance with a standardized clinical interview. *Int. J. Methods Psychiatry Res.* 7:121–127.
26. Breslau, N., and Davis, G. C. (1992). Posttraumatic stress disorder in an urban population of young adults: Risk factors for chronicity. *Am. J. Psychiatry* **149,** 671–675.
27. Breslau, N., Davis, G. C., Andreski, P., Federman, B. and Anthony, J. C. (1998). Epidemiologic findings on PTSD and comorbid disorders. *In* "Adversity, Stress, and Psychopathology" (B. Dohrenwend, ed.), pp. 319–330. Oxford University Press, New York.
28. Breslau, N., Davis, G. C., and Andreski, P. (1995). Risk factors for PTSD related traumatic events: A prospective analysis. *Am. J. Psychiatry* **152,** 529–535.
29. Bromet, E., Sonnega, A., and Kessler, R. C. (1998). Risk factors for DSM-III-R Posttraumatic Stress Disorder: Findings from the National Comorbidity Survey. *Am. J. Epidemiol.* **147,** 353–361.
30. Kendler, K. S., Kessler, R. C., Neal, M. C., Heath, A. C., and Eaves, L. J. (1993). The prediction of major depression in women: Toward an integrated etiologic model. *Am. J. Psychiatry* **150,** 1139–1148.
31. Brown, G. W., Harris, T. O., and Eales, M. J. (1993). Aetiology of anxiety and depressive disorders in an inner-city population. 1. Early adversity. *Psychol. Med.* **23,** 143–154.
32. Cottler, L. B., Compton, W. M., Mager, D., Spitznagel, E. L., Janca, A. (1992). Posttraumatic stress disorder among substance users from the general population. *Am. J. Psychiatry* **149,** 664–670.
33. Chilcoat, H. D., and Breslau, N. (1998) PTSD and drug use disorders: Testing causal pathways. *Arch. Gen. Psychiatry* **55,** 913–917.
34. Breslau, N. (1999). Epidemiology of trauma and post-traumatic stress disorder. *In* "Psychological Trauma" (R. Yehuda, ed.), *Rev.*

Psychiatry, (J. M. Oldham and M. B. Riba, eds.), Vol. 17, pp. 1–29. American Psychiatric Press, Washington, DC.

35. Green, B. L., Grace, M. C., Lindy, J. D., Gleser, G. C., and Leonard, A. (1990). Risk factors for PTSD and other diagnoses in the general sample of Vietnam veterans. *Am. J. Psychiatry* **147,** 729–733.

36. Davidson, J., Swartz, M., Storck, M., Krishnan, R. R., and Hammett, E. (1985). A diagnostic and family study of posttraumatic stress disorder. *Am. J. Psychiatry* **142,** 90–93.

37. McFarlane, A. C. (1989). The aetiology of post-traumatic morbidity: Predisposing, precipitating and perpetuating factors. *Br. J. Psychiatry* **154,** 221–228.

38. McFarlane, A. C. (1990). Vulnerability to Posttraumatic Stress Disorder. *In* "Posttraumatic Stress Disorder: Etiology, Phenomenology, and Treatment" (M. E. Wold and A. D. Mosnaim, eds.), vol. 1, pp. 3–17. American Psychiatric Press, Washington, DC.

39. Kulka, R. A., Schlenger, W. E., Fairbank, J. A., Hough, R. L., Jordan, B. K., Marmar, C. R., and Weiss, D. S. (1990). "Trauma and the Vietnam War Generation." Brunner/Mazel, New York.

40. Zaidi, L. Y., and Foy, D. W. (1994). Childhood abuse experiences and combat-related PTSD. *J. Traumatic Stress* **7,** 33–42.

41. Bremner, J. D., Southwick, S. M., Johnson, D. R., Yehuda, R., and Charney, D. S. (1993). Childhood physical abuse and combat-related posttraumatic stress disorder in Vietnam veterans. *Am. J. Psychiatry* **150,** 235–239.

42. Foa, E. B., and Riggs, D. S. (1993). Posttraumatic stress disorder and rape. *In* "American Psychiatric Press Review of Psychiatry" (J. M. Oldham, M. B. Riba, and A. Tasman, eds.), Vol. 12. American Psychiatric Press, Washington, DC.

43. Solomon, Z., Mikulincer, M., and Jakob, B. R. (1987). Exposure to recurrent combat stress: Combat stress reactions among Israeli soldiers in the Lebanon War. *Psychol. Med.* **17,** 433–440.

82

Eating Disorders

KATHERINE A. HALMI
Cornell University Medical College,
Eating Disorder Program New York Hospital—Westchester Division
White Plains, New York

I. Definition and Diagnosis

The eating disorders are syndromes that are classified on the basis of the clusters of symptoms with which they present. They are not specific diseases with a single common cause, course, or pathophysiology. This chapter will focus predominately on the well-defined eating disorders of anorexia nervosa and bulimia nervosa. Other variants of these disorders and binge eating disorder have not been so thoroughly investigated over a long period of time and less information is available on them.

A. Criteria for Anorexia Nervosa

The three criteria proposed by Russell [1] in 1970 are still imbedded in the latest Diagnostic and Statistical Manual of Mental Disorders Fourth Edition (DSM-IV) criteria [2]. The former criteria pertain to weight loss, a morbid fear of becoming fat, and evidence of an endocrine disorder (amenorrhea). Because there is no specific amount of weight loss associated with the other symptoms of anorexia nervosa, the current DSM-IV first criterion simply states "Refusal to maintain body at or above a minimally normal weight for age and height." In the absence of other physical illness, anyone maintaining a weight less than 85% of that expected should be thoroughly interviewed for other symptomatology of anorexia nervosa. The second criterion, "intense fear of gaining weight," has remained unchanged through the decades. Anorectic patients' fear of gaining weight exists even when they are emaciated. Although these patients are disinterested in and even resistant to treatment, they often deny this fear, which must then be inferred by observation of their behavior. They will develop rigorous exercise programs and severely restrict their total food intake in order to prevent weight gain.

The third criterion pertaining to body image disturbance has evolved since the 1980s into a more complex concept. The initial definition of body image disturbance was a narrow definition of body image related to visual self-perception. Because studies have shown that many anorectics do not visually overestimate their sizes and that overestimation is not unique to those with anorexia nervosa [3], the criterion was then reworded in DSM-III-R to focus on attitudinal and affective dimensions of body image. The newly worded criterion, "overconcern with body size and shape," also did not distinguish sufficiently between anorexia nervosa patients and the general female population. Thus it was revised in DSM-IV to emphasize the central concern of weight and shape in the evaluation of the self, in addition to a reference to the denial of the serious consequences of weight loss.

The fourth criterion for diagnosis of anorexia nervosa in the DSM-IV is amenorrhea. Amenorrhea can occur before noticeable weight loss has occurred [4]. Russell [5] suggested that the amenorrhea may be caused by a primary disturbance of hypothalamic function and that the full expression of this disturbance is induced by psychological stress. He thought that the malnutrition of anorexia nervosa perpetuated the amenorrhea but was not primarily responsible for the endocrine disorder.

In the current DSM-IV, anorexia nervosa is divided into two subtypes: the restrictor type and the binge-purger type. Studies have consistently demonstrated that impulsive behaviors including stealing, drug abuse, suicide attempts, self-mutilations, and mood lability are more prevalent in anorectic-bulimics compared with anorectic restrictors. The anorectic-bulimics also have a higher prevalence of premorbid obesity, familial obesity, and debilitating personality traits [6–9].

B. Criteria for Bulimia Nervosa

Bulimia nervosa was identified as a separate entity in 1979 by Russell [10]. He proposed three criteria: a powerful and intractable urge to overeat, resulting in episodes of overeating; avoidance of "fattening" effects of food by inducing vomiting or abusing purgatives or both; and a morbid fear of becoming fat. These criteria do not distinguish bulimia nervosa patients from anorexia nervosa patients who have the binge-purge subtype. Because there are physiological differences between bulimic people who lose large amounts of weight and have anorexia nervosa and those who never lose weight, it became necessary to give a more precise definition to bulimia nervosa. In the current DSM-IV, the first criterion is recurrent episodes of binge eating with "a sense of lack of control over eating" during these episodes. The second criterion is the recurrent use of inappropriate compensatory behaviors to avoid weight gain (self-induced vomiting). The third criterion, designed to address chronicity and frequency, is a minimum average of two episodes of binge eating and two inappropriate compensatory behaviors a week for at least three months. This criterion is not grounded in specific research. The fourth criterion states that self-evaluation is unduly influenced by body shape and weight and more accurately describes the preoccupation of bulimics with their physical appearance than the previous "morbid fear of becoming fat." The fifth and final criterion of bulimia nervosa is that the disturbance does not occur exclusively during episodes of anorexia nervosa.

The diagnosis of bulimia nervosa is also subtyped into a purging type for those who regularly engage in self-induced vomiting or the use of laxatives or diuretics and a nonpurging type for those who use strict dieting, fasting, or vigorous exercise, but do not regularly engage in purging. Bulimic people who purge differ from binge eaters who do not purge in that the latter tend to have less body image disturbance and less anxiety concerning eating [11,12]. Bulimic people who do not purge tend to be obese.

C. Binge Eating Disorder and Eating Disorder Not Otherwise Specified

Binge eating disorder is listed in the "not otherwise specified" category of the DSM-IV for eating disorders. Field trials are being conducted to provide evidence as to whether binge eating disorder should be a specific diagnostic category. Based on their clinical work and a search of the literature, Devlin *et al.* [13] came to the conclusion that a significant number of people had problems with binge eating but did not meet criteria for bulimia nervosa. People with this disorder lack the compensatory weight-control behaviors and the overconcern with weight and shape. Spitzer *et al.* [14] proposed criteria for this syndrome and termed it "binge eating disorder." It is distinguished from bulimia nervosa nonpurge type by the lack of any compensatory behavior to avoid weight gain and the necessity of bingeing twice a week for a six-month period.

There are many variations of anorexia nervosa and bulimia nervosa that do not meet criteria for these disorders. Unfortunately, there is not enough information on response to treatment and temporal course to specifically identify them. An example is the individual of normal body weight who regularly engages in self-induced vomiting after eating small amounts of food. Prospective longitudinal studies should provide more information on the course of these maladaptive eating behaviors in order to help us determine if they can be identified as a specific diagnostic entity.

II. Historical Context and Controversial Issues

A. Saints of the Middle Ages

Disturbances of eating behavior are not just a twentieth century invention; they go far back in history. It is fashionable to think of the eating disorders, anorexia nervosa and bulimia nervosa, as culture-bound syndromes. The well-documented cases of irreversible self-starvation in the fasting female saints of the Middle Ages [15] describe typical signs and symptoms of anorexia nervosa. An example is Princess Margaret of Hungary who lived from 1242–1271 [16]. She was the daughter of a king and was raised in a Dominican convent where she excelled in all of her studies and in all of the undesirable chores of the monastery. She was described as practicing the austerities of fasting, deprivation of sleep, exhausting menial work, and other bodily penances to a heroic degree. She was never idle and allowed all the food at the table to pass her untasted and would often slip out to pray while her sisters ate. She intensified her dieting when King Bela confronted her with suitors and died at the age of 28 with a mind that was clear and alert and a poor wasted body.

Whether dieting for sainthood or dieting for thinness produces the same determination not to gain weight is still a matter of controversy. Most likely the psychobiological vulnerability

factors that induce the development of irreversible starvation in Medieval saints are similar to those inciting the emergence of anorexia nervosa and bulimia nervosa in twentieth century young women [16].

B. Seventeenth and Eighteenth Century Case Descriptions

In the second half of the seventeenth century, case descriptions of self-starvation began to appear in the form of printed pamphlets. One of these cases was an 18 year old English girl named Martha Taylor who had lost her period and, after a siege of vomiting, stopped taking all solid food. Over the following year, she became emaciated. Martha was examined by physicians, teachers, and theologians. John Reynolds [18] wrote "most of these damsels fall into this abstinence between the age of 14 and 20 years. It's probable that the feminale humours in these virgins may by a long abode in their vessels grow acid. Her age confirms the probability of a ferment in the feminals." Twenty-two years later Richard Morton [19] described two cases of typical anorexia nervosa symptomatology and distinguished them from consumption. In the eighteenth century Robert Whytt described a case of "nervous atrophy" in a young man of 14 [20]. The description was that of typical anorectic symptomatology.

C. Nineteenth Century Victorian Case Descriptions

In 1860, Dr. Louis-Victor Marcé of Paris [21] described several cases of "young girls, who at the period of puberty, and after a precocious development, become subject to inappetency carried to the utmost limits. What ever the duration of their abstinence, they experience a distaste for food, which the most pressing want is unable to overcome. These patients arrive at a delirious conviction that they can not or ought not to eat. In one word, the gastric nervous disorder becomes cerebro-nervous."

Often the modern history of anorexia nervosa is regarded as starting in 1873 when the London physician, Sir William Gull, and the Paris neuropsychiatrist, Ernest Charles Lasequé, almost simultaneously published papers on the description and treatment of "hysterical anorexia" [22,23]. Lasequé recognized the emotional turmoil of the patients and Gull recommended a treatment: "the patient should be fed at regular intervals, and surrounded by persons who would have moral control over them; relations and friends being the generally the worst attendants" [22].

In the United States, very little attention was paid to these cases of fasting women. However, William Chipley, chief medical officer of the Eastern Lunatic Asylum of Kentucky, did publish an article in 1859 on the causes and treatment of sitomania (intense dread of eating or aversion to food) [24].

A book by William Alexander Hammond, a New York neurologist, published in 1879 described fasting girls and, in particular, the well publicized Brooklyn fasting girl, Molly Fancher [25]. Another American neurologist, Silas Weir Mitchell [26], described cases of great self-starvation. The first Canadian publication on anorexia nervosa occurred in 1895 when P. R. Inches [27] read a paper on anorexia nervosa before the St. John Medical Society in Halifax, Nova Scotia. After Gull's and Lasequé's publications, the name anorexia nervosa generally took hold. By the twentieth century, it was commonly used in reference to cases of self-induced starvation.

D. Emergence of Twentieth Century Eating Disorders

The first major publication on anorexia nervosa in the twentieth century was a book by Bliss and Branch [28] that summarized the current endocrine studies as well as the history and psychological descriptions of the disorder. Hilda Bruch [29] carefully articulated the psychological turmoil of these patients. She coined the phrase "the relentless pursuit of thinness" and described the "paralyzing sense of ineffectiveness which pervades all thinking and activities."

In 1979, Gerald Russell [10] described bulimia nervosa and reported on thirty patients (28 females, 2 males), discussing how they differed from typical anorexia nervosa patients. All of these publications spurred the interest in eating disorders and led to increased recognition, improved diagnosis, and more research on this previously ignored area of dysfunction.

III. Epidemiology

A. Incidence

Incidence rates, the number of new cases in the population in a specified period of time, are commonly expressed with eating disorders as the rate per one hundred thousand population per year. A major difficulty in studying the epidemiology of eating disorders has been the changes in criteria for these illnesses over time. Despite this problem, research suggests that the overall incidence of both anorexia nervosa and bulimia nervosa has increased since the 1950s. There was a consistent increase in the registered incidence of anorexia nervosa from 1931 to 1986 in cases presenting to the health care system in several industrialized countries (*e.g.,* Sweden, United States, England, Switzerland, Scotland) in the numerous studies reviewed by Hoek [30]. For example, a study conducted in northeastern Scotland [31] showed that between 1965 and 1991 there was almost a sixfold increase in the incidence of anorexia nervosa (from 3/100,000 to 17/100,000 cases).

In a study conducted in southwest London [32] between the period of July 1991 and June 1992, the incidence of anorexia was found to be 2.7 cases per 100,000 total population. In females aged 15–29 years, the incidence was 19.2 cases per 100,000 women. Overall, an average estimate of the incidence of anorexia nervosa in the 1990s in industrialized countries is about 8 per 100,000 people per year. Incidence rates of anorexia nervosa have increased since the 1950s both in the United States and in western Europe. This is especially true for females 15 to 24 years old.

The only large incidence study of bulimia nervosa was in Holland. It reported an annual incidence of bulimia nervosa of 19.9 per 100,000 people during 1985–1986 and 11.4 per 100,000 people during 1985–1989 [32].

B. Prevalence and Gender Distributions

The prevalence of anorexia nervosa among young females in community populations has been estimated in a number of survey studies in western European countries. A summary of these studies reported an average prevalence of anorexia nervosa, using strict diagnostic criteria, in 0.28% of young females [30]. In Holland in the 1980s, the one-year prevalence in females aged 15–29 in primary care was 0.16% [32]. In another study conducted with primary care physicians in the U.K., the prevalence of anorexia nervosa was 20.2 cases per 100,000 people (0.02% of the total population) [33]. In Minnesota, the point prevalence of anorexia nervosa was 0.2% for females and 0.02% for males in January 1980. In January 1985, the point prevalence in the same community was 0.48% of girls aged 15–19 years [34]. A review of more than fifty prevalence studies conducted between 1981 and 1989 showed a remarkably consistent prevalence of bulimia nervosa of about 1% [35]. The prevalence of bulimia nervosa in a Canadian community sample was also 1% [36]. In summary, the prevalence of anorexia nervosa is estimated to be 0.28% in industrialized countries, and prevalence is about 1% for bulimia nervosa.

C. Lifetime Expectancy

In a study conducted in the United States [37], lifetime prevalence for anorexia nervosa was found to be 0.51% (narrowly defined) and 3.7% (broadly defined). A questionnaire-based study of women in Norway reported a lifetime prevalence for all eating disorders of 8.7% [38]. When divided into subgroups, the lifetime prevalence was 3.2% for binge-eating disorder, 1.6% for bulimia nervosa, 0.4% for anorexia nervosa, and 3.0% for eating disorders not otherwise specified. In the Canadian study [36], the lifetime prevalence for bulimia nervosa was 1.1% for females and 0.1% for males. In England, the lifetime prevalence was found to be 2.6% for bulimia nervosa in women aged 18–34 [39].

D. Sociocultural Differences

Transcultural studies show that anorexia nervosa is rare in nonwestern, poorly industrialized countries [40]. When nonwesterners are exposed to western ideals of thinness, they are significantly affected. For example, a study of the prevalence of eating disorders among Greek and Turkish girls who remained in their homeland and those who were living in Germany showed twice the prevalence in the latter compared to the former group [41]. In a study in Holland, the prevalence of bulimia nervosa was highest in large cities, intermediate in urbanized areas, and lowest in rural areas [42]. No differences were found in prevalence of anorexia nervosa in urban versus rural areas. In studies conducted in the early 1970s, there was some evidence that the prevalence of eating disorders was higher in higher social economic groups [43,44]. However, a review of eight studies in the 1980s [45,45a] failed to support that relationship. Social and cultural factors have significantly influenced the prevalence and nature of eating disorders. In the twentieth century, it appears that the powerful effect of the western ideal of a slender body type has been influential in producing dieting behavior and thus eating disorders.

IV. Risk Factors

A. Sociological Factors

At present there are no data derived from methodologically sound, controlled studies to support the theories on risk for developing eating disorders pertaining to women's social roles or cultural factors. Only hypotheses and theories can be mentioned.

1. Influence of Women's Social Roles

The common theme for the social influence on the development of eating disorders in women is that eating disorders are complex expressions of resistance to or subversion of demands that women conform to in the context of impossible and oppressive social expectations. The relative lack of power held by women is considered responsible for making them easily shaped and influenced by social expectations. Women are indoctrinated into a belief system that overvalues female beauty and, in particular, overvalues thinness. Women cannot achieve satisfactory self-esteem without attaining ideals that are impossible to fulfill. The eating disorder has an adaptive function in that it is a reflection of the woman's need to assert her autonomy and strength in a culture that rejects such qualities in females.

2. Cultural Factors

Weight patterns and physical appearance expectations vary from one society to another [46]. As noted in the epidemiology section, industrialized societies have higher rates of eating disorders than nonwestern societies. This has been attributed to values held by western, technologically advanced societies; in particular, thinness and weight control are highly esteemed.

There are suggestions that cultural differences in dietary habits, patterns of parent-child interactions, and value orientation in family structure may reveal more about the societal impact on the development of eating disorders [47,48]. At this time, more refinement in research methodology is needed to accurately assess the influence of cultural differences on the development of eating disorders.

B. Psychological Factors

1. Personality Characteristics

The premorbid personality of the individual with anorexia nervosa-restricting type is described in retrospective clinical reports as obsessional, socially inhibited, compliant, and emotionally restrained. Anorexia nervosa patients often perceive themselves as ineffectual and, in fact, have lacked the experiences necessary to foster a sense of competency and control. Because it is almost impossible to study patients before they develop anorexia nervosa, the next best technique is to study them after recovery from the illness. One study compared long-term anorectics to controls and found that anorectics had greater risk avoidance, restraint in emotional expression, and greater conformance to authority. Recovered anorectics, compared to their sisters, also displayed greater self and impulse control, industriousness, responsibility, interpersonal insecurity, excessive conformance to rules and standards, and highly regimented behavior [49]. In another study [50], anorectic patients showed a decrease in obsessive-compulsive symptoms but no decrease in trait obsessiveness after weight gain. Their trait obsessiveness was defined as social introversion, overly compliant behavior, an inflexible thinking style, limited social spontaneity, and reduced affect. A study comparing anorectic subtypes found that the binge-purge type of anorectic was more similar to normal weight bulimics than to anorectic restrictors in personality characteristics [51]; that is, bulimic anorectics, like normal weight bulimics, tended to be more impulsive, emotionally distraught, extroverted, and sexually active. A review of personality dis-

order studies in anorectic subtypes showed that about one-fourth of restricting anorectics have cluster C (anxious-fearful) personality disorder diagnoses. Anorectic bulimics, surprisingly, have an equal likelihood to receive a cluster C diagnosis but, in addition, about one-third will have a cluster B (dramatic-erratic) personality disorder diagnosis [52]. Multiple patterns of impulsivity are very strongly associated with the presence of binge-eating [53]. An example is the combination of bulimia with substance abuse disorders, kleptomania, and self-mutilation. No fundamental personality trait or traits have been identified as risk factors for the development of bulimia nervosa. Studies have produced heterogeneous personality profiles in the bulimia nervosa patients.

2. Family Functioning

It is unclear whether patterns of interaction in eating disorder families reflect a causal connection in the pathogenesis of eating disorders or an untoward consequence of such illnesses on family life. Studies of anorectic families show that they have more rigidity in their family organization, less clear interpersonal boundaries, and tend to avoid open discussions of disagreements among themselves compared with control families [54]. Parents of bulimia nervosa patients were found to be insufficiently nurturing and lacking in empathy to their daughters [55]. In addition, the families have been described as having a lack of parental affection; negative, hostile and disengaged interactions within the family; parental impulsivity; and family alcoholism and obesity.

3. Stressful Events

Studies investigating the relationship between sexual abuse and eating disorders have produced highly discrepant results [56]. Overall, approximately 30% of eating disorder patients have been sexually abused in childhood. This figure is comparable to rates found in normal populations. Another study showed that childhood psychological abuse in multiple forms increased the likelihood of lifetime comorbid Axis I disorders and personality pathology in bulimic patients [57]. It is probably reasonable to assume that for some patients there may be a direct link between sexual trauma and disturbed eating behavior. However, in general, sexual abuse may best be considered a risk factor for developing a wide range of psychological and psychiatric problems. A surprisingly low rate of sexual abuse has been reported among anorectic restrictors compared to bulimic anorectics or normal weight bulimics [58].

Case studies have shown that normative developmental events, such as the onset of puberty, leaving home, or beginning a new school, can precipitate the onset of an eating disorder in premorbidly vulnerable individuals [59]. Adverse life events, such as the death of a close relative, the breakup of a relationship, or illness, may also be a precipitant for the development of an eating disorder in individuals with premorbid vulnerability.

V. Clinical Issues

A. Psychiatric Comorbidity

In the 1990s, a substantial number of studies were produced on the psychiatric comorbidity of Axis I, clinical syndromes, and Axis II, developmental and personality disorder psychopathology, with eating disorders. For Axis I diagnoses, the highest lifetime comorbidity with both anorexia nervosa and bulimia nervosa are affective disorders, especially major depressive disorder. For anorexia nervosa, the second highest comorbid disorders are the anxiety disorders, in particular obsessive-compulsive disorder. Substance abuse ranks highest in bulimia nervosa but is also frequently present in anorectic-bulimics.

Three studies administered the Diagnostic Interview Schedule (DIS) to anorexia nervosa patients to establish comorbid DSM-III Axis I diagnoses. These studies, all of which had adequate control groups, found a lifetime prevalence for major depression of 68%, 36%, and 38%, respectively [60–62]. Another interview study found that 54% of anorectic patients had a lifetime diagnosis of major affective disorder [63]. A study of 105 inpatients that used the Structured Clinical Interview for DSM-III-R (SCID) reported that the lifetime prevalence of any affective disorder was 41.2% in anorectic restrictors, 82.0% in anorectic bulimics, 64.5% in bulimia nervosa patients, and 78.0% in bulimia nervosa patients with a past history of anorexia nervosa [64]. The anorectic restrictor subgroup was significantly more likely than the other groups to have had no affective disorder. The majority (64%) of the anorectic restrictors, but only 33% of the bulimia nervosa patients, developed their eating disorder as their first Axis I disorder. Three other studies found the incidence of lifetime major depression to be higher in the anorectic bulimic group than in the anorectic restricting group [65–67]. In a large study of bulimia nervosa patients, 73% had a diagnosis of "any affective disorder" and 63% had major depressive disorder [68].

Two studies used the DIS interview and found a lifetime prevalence of 65% and 60% for anxiety disorders [60,62]. The two most prevalent of the anxiety disorders in both studies were social phobia and obsessive-compulsive disorder (OCD). The rates of lifetime OCD in anorectic patients were 34% and 26%, respectively.

In a study by Braun et al. [64], 18.3% of the bulimic subtypes had a social phobia compared with 2.9% of the anorectic restrictors. A comparison of the age at onset of the social phobia and the eating disorder showed that 78% had the social phobia earlier and 21% had the eating disorder earlier. Anxiety disorders obviously contribute considerable comorbidity to patients with eating disorders.

In a review of 51 studies, Holderness et al. [69] concluded that the relationship between substance abuse and bulimia nervosa was far stronger than that of substance abuse and anorexia nervosa. In another study Selby and Moreno [70], compared the personal and family history of substance use problems in women with anorexia nervosa, bulimia nervosa, obesity, and depression. The bulimic subjects reported a greater frequency of both personal and family substance abuse problems compared to the other groups. The bulimic subjects with and without substance abuse problems reported similar frequencies of family substance abuse problems.

To date, there are no adequately designed longitudinal studies to determine whether the existence of a comorbid condition has an impact on the outcome of the eating disorder. It is also important to determine the influence of comorbid psychiatric disorders as risk factors for the development of an eating disorder.

In one large interview study [64], 69% of the patients had at least one personality disorder (Personality disorders are dis-

cussed in section IVB1). Of the patients who had personality disorders, 93% also had an Axis I comorbidity. Cluster A disorders were diagnosed in only 4.8% of the sample and cluster B disorders were diagnosed in 2.0% of the overall sample. Thirty-one percent of the bulimic subgroups and none of the anorectic restrictors had cluster B disorders. Borderline personality disorder was present in 25% of the bulimic subgroups and was the most common cluster B condition. Cluster C personality disorders were present in 29.5% of the sample. Avoidant personality disorder was the most common (14.3%), followed by dependent (10.5%), obsessive-compulsive (6.7%), and passive aggressive (4.8%). The prevalence of cluster C personality disorders did not vary according to eating disorder subtype. In this study, patients with no personality disorder were significantly more likely to have no affective disorder and no substance dependence disorder compared with the sample as a whole.

The questions still to be answered are: (1) Is a woman with a cluster B personality disorder at risk for developing bulimia, and (2) Does the development of an eating disorder, especially during adolescent years, have a formative effect on personality?

B. Medical and Physiological Factors

Emaciation can produce abnormalities in hematopoiesis, such as leucopenia and relative lymphocytosis. In patients who engage in self-induced vomiting or abuse laxatives and diuretics, hypokalemic alkalosis may develop. These patients often have elevated serum bicarbonate, hypochloremia, and hypokalemia. Patients with electrolyte disturbances have physical symptoms of weakness, lethargy, and, at times, cardiac arrhythmias. The latter condition may result in sudden cardiac arrest, a cause of death in patients who purge. Elevation of serum enzymes reflects fatty degeneration of the liver and is present both in the emaciated anorectic phase and during refeeding. Elevated serum cholesterol levels seem to occur more frequently in younger patients and return to normal with weight gain. Hypercarotenemia is often observed in malnourished anorectic patients [4].

Patients who purge can have severe erosion of the enamel of their teeth, pathologic pulp exposures, loss of integrity of the dental arches, diminished masticatory ability, and an obvious unasthetic appearance of their teeth.

Parotid gland enlargement is associated with elevated serum amylase levels in bulimics who binge and vomit. The serum amylase level can be used to follow reduction of vomiting in purging patients who deny their vomiting episodes. Acute dilatation of the stomach is a rare emergency condition for patients who binge. Esophageal tears usually accompanied by shock can also occur in the process of self-induced vomiting. Severe abdominal pain in the bulimic patient should alert the physician to a diagnosis of gastric dilatation and the need for nasal gastric suction, x-rays, and surgical consultation.

Cardiac failure may be caused by cardiomyopathy from Ipecac intoxication. (Ipecac may be used to induce vomiting.) This is a medical emergency that usually results in death. Symptoms of precordial pain, dyspnea, and generalized muscle weakness associated with hypotension, tachycardia, and electrocardiogram abnormalities should alert one to a possible Ipecac intoxication [4].

Neurotransmitter and neuroendocrine functioning can be affected in anorexia nervosa and bulimia nervosa. These disturbances generally reverse with weight gain and cessation of purging behaviors. For example, thyroid changes found in anorexia nervosa are similar to those found in starvation and are reversible with refeeding. The increased corticotropin-releasing hormone (CRH) secretion in underweight anorexic patients returns to normal with weight restoration [71]. These findings suggest that CRH hypersecretion is due to weight loss alone. However, CRH, which is a potent anorectic hormone, may have a role in maintaining anorectic behaviors and initiating a relapse.

Amenorrhea is an essential clinical feature in the diagnosis of anorexia nervosa. In underweight anorectics, basal levels of leutinizing hormone (LH), follicle-stimulating hormone, and estrogen are decreased. With weight restoration, the menstrual cycle usually returns and normal LH secretion patterns are reestablished in some, but not all, patients [71]. Dysfunction present in neurotransmitter systems that influence gonadotrophin releasing hormone release may be present in anorexia nervosa. Neuroendocrine disturbances in bulimia are highly variable. Studies have shown no consistent findings.

Animal studies have shown that several neurotransmitters and neuropeptides modulate eating behavior. One of special interest is serotonin, which produces fullness or satiety. Serotonin is also implicated in obsessive-compulsive disorder and affective disorder, both of which are frequently present in eating disorders. Thus far, only rather crude measures of serotonin function can be made in humans. Kaye et al. [73] found elevated cerebrospinal fluid levels of 5-hydroxyindoleacetic acid (5-HIAA), a metabolite of serotonin in long-term recovered anorectic women compared with control subjects. This may indicate a trait abnormality of serotonin functioning that contributes to the development and maintenance of anorexia nervosa.

Several studies have reported impaired serotonergic function in bulimic patients. Jimerson et al. [74] showed that more severely binge-eating patients had lower cerebrospinal fluid 5-HIAA levels compared with control subjects. Brewerton et al. [75] showed that bulimic patients had a blunted prolactin response to m-chlorophenylpiperazine (m-CPP), and McBride et al. [76] showed that these patients have a reduced prolactin response to the serotonin agonist fenfluramine. Aberrations of serotonergic function in anorexia nervosa and bulimia nervosa provide the pharmacological rationale for pharmacotherapy of these disorders (see section VIA1). Because pharmacotherapy is not based on some of the other neuroendocrine and neuropeptide aberrations found during the acute illness states of these eating disorders, these aberrations will not be discussed here but are reviewed in Halmi [77].

C. Course of Illness

Long-term studies have been conducted on patients who presented for treatment. Onset of anorexia is usually between the ages of fourteen and fifteen or at age eighteen [78]. The course subsequently can vary from a single episode with recovery to weight recovery with repeated relapses or a chronic course that results in death. In two very large sample studies, mortality rates at ten years after presentation were 6.6% [60] and 18% at thirty years follow-up [79]. In the same ten-year follow-up study, about one-fourth of the patients had recovered and an additional one-third were functioning but still had anorectic symptomatology. The others remained in poor condition [80]. The optimistic

news in the thirty-year follow-up study [79] is that a substantial number of chronically ill patients gradually improved to the "good" category over the years. By the thirty-year follow-up, 75% of the patients were in the good category.

There are no large sample size follow-up studies of bulimia nervosa. However, a review of 88 articles [81], with follow-up periods ranging from six months to ten years, showed that the typical age at onset for bulimia was between eighteen and nineteen years. Mortality rates varied from 0 to 3%. Between five and ten years of follow-up, about 50% of the bulimics were fully recovered, while 20% continued to meet diagnostic criteria for bulimia nervosa. About one-third of recovered bulimics relapsed within four years of presentation.

D. Predictors of Outcome

In a ten-year follow-up study, predictors of good outcome were early age at onset of illness (under the age of eighteen) and the absence of vomiting, laxative abuse, and repeated hospitalizations [80]. In another follow-up study, purging predicted poor outcome. Those with purging behavior in conjunction with social impairment were likely to have a particularly unfavorable course [82]. A review of bulimia nervosa studies [81] reported that bulimia nervosa patients with personality disorders marked by problems with impulse control had a worse prognosis than similar patients without such problems.

VI. Treatment of Anorexia Nervosa and Bulimia Nervosa

A. Anorexia Nervosa

1. Medical Management and Pharmacotherapy

The more severely ill anorectic patient can present an extremely difficult medical-management challenge. It is best if these patients are hospitalized in a specialized inpatient treatment setting that has a team of individuals highly skilled in the multidisciplinary management of anorectic patients. Medical management involves weight restoration, nutritional rehabilitation, rehydration, and correction of serum electrolytes. Inpatient hospitalization should include daily monitoring of weight, food, calorie intake, and urine output. Patients should be monitored closely for attempts to vomit. If patients are in outpatient therapy, they should be weighed weekly in the physician's office with periodic physical examinations.

Medications can be useful adjuncts in the treatment of anorexia nervosa. Cyproheptadine in high doses (up to 28 mg per day) can facilitate weight gain in anorectic restrictors and also has an antidepressant effect [83]. There is a suggestion from several small sample size, open studies that both fluoxetine and clomipramine (serotonin reuptake inhibitors) may be effective in preventing relapse in anorexia nervosa [84,85].

2. Cognitive Behavioral Therapy (CBT)

Cognitive and behavior therapy principles can be applied in both inpatient and outpatient settings. Behavior therapy has been found to be effective in inducing weight gain [86]. There are no large sample size, controlled studies of cognitive therapy with behavior therapy in anorexia nervosa patients. Because anorexia nervosa patients are very resistant and unmotivated for treatment, the first step of establishing a therapeutic relationship can be very difficult. Monitoring is an essential component of CBT. Patients are taught to monitor their food intake, their feelings and emotions, their bingeing and purging behaviors, and their problems in interpersonal relationships. Cognitive restructuring is a method patients are taught in order to identify automatic thoughts and challenge their core beliefs. Problem solving is a specific method whereby patients learn how to think through and devise strategies to cope with their food-related and/or interpersonal problems. If the patient learns to use these techniques effectively, it reduces her reliance on anorectic behaviors as a means of coping [87].

3. Family Therapy

A family analysis should be done on all anorexia nervosa patients who are living with their families. On the basis of this analysis a clinical judgment can be made as to the advisability of family therapy or counseling. There will be some cases in which family therapy is not possible. In other cases brief counseling sessions with immediate family members may be the extent of the family therapy required. A controlled family study showed that anorectic patients under the age of eighteen benefited from family therapy, whereas patients over the age of eighteen did worse in family therapy compared with the controlled therapy [88]. In actual practice most clinicians find it necessary to combine individual therapy with some sort of family counseling in managing their anorexia nervosa patients.

B. Bulimia Nervosa

1. Cognitive Behavioral Therapy

Treatment outcome studies in bulimia nervosa have proliferated in the 1990s in contrast to the relatively few treatment studies of anorexia nervosa. CBT has been found to be the most effective treatment in over 35 controlled studies [89,90]. A summary of those studies shows that about 40–50 of patients are abstinent from both binge eating and purging at the end of treatment (16–20 weeks). Another 30–40% show a marked improvement in the reduction of bingeing and purging. Of those 30–40% who do not show improvement immediately posttreatment, most show improvement to full recovery one year after treatment. It is important to note that the data supporting the efficacy of CBT in treating bulimia nervosa are based on strict adherence to rigorously implemented, highly detailed manual-guided treatments that include about 18 to 20 sessions over 5 to 6 months. In the treatment of bulimia nervosa, CBT uses a number of cognitive and behavioral procedures for the purpose of interrupting the self-maintaining behavioral cycle of bingeing and dieting and altering the individual's cognitions and beliefs about food, weight, body image, and overall self-concept.

2. Pharmacotherapy

In the 1990s, over a dozen double-blind, placebo-controlled trials of antidepressants, including desipramine, imipramine, amitriptyline, nortriptyline, phenelzine, and fluoxetine, have been conducted in normal weight outpatients with bulimia nervosa [91,92]. In all trials, the antidepressants were significantly more effective than placebo in reducing binge eating. However, the average abstinence rate from bingeing and purging in these studies was 22%, indicating that the majority of patients were still symptomatic at the end of treatment. The few long-term studies to evaluate maintenance of change showed disappointing results, with over 80% of patients relapsing [93,94]. There is some indication now that if medications are used the treatment should be continued for six months to prevent relapse. However, studies are needed to assess the adequate length and time for pharmacotherapy.

One large multicenter collaborative study showed that fluoxetine in doses of 60 mg per day was effective in reducing binge/purge episodes. Controlled double-blind placebo trials are needed for other serotonin-selective reuptake inhibitors (SSRI) medications [95]. With respect to combining CBT and pharmacotherapy, one study found that group CBT was superior to imipramine in decreasing binge eating and purging, but the combined treatment demonstrated no additive effects to those of the CBT group alone [96]. Another study [97] compared individual CBT, desipramine, or the combination and found that all groups were similar at sixteen weeks. At thirty-two weeks, only the combined treatment given for twenty-four weeks was superior to medication given for sixteen weeks.

3. Family Therapy

Family therapy is not widely used in the treatment of bulimia nervosa, mainly because most patients with bulimia nervosa are in their twenties or older and living away from their family of origin. While there is consensus that families of younger patients should be involved with their treatment, controlled studies have not been conducted to demonstrate this.

4. Interpersonal Therapy

Interpersonal therapy (IPT) focuses on interpersonal functioning and does not target eating behaviors. In the one controlled study comparing IPT to CBT [98], it was found that IPT was not as effective as CBT at the end of treatment, but at the end of two years and six years it had produced the same reduction of binge/purge behavior. These findings suggest that IPT has a delayed but powerful effect. This study needs to be replicated before definitive statements can be made concerning the effectiveness of IPT in the treatment of bulimia nervosa.

5. Group Therapy and Guided Self-Help

Although widely practiced using CBT techniques, group therapy for bulimia has not been studied in a controlled assessment to compare it with individual CBT. One study showed group CBT was more effective than imipramine [96]. CBT group therapy is especially cost efficacious both for patients and for treatment centers.

Guided self-help has been assessed with two specifically designed manuals [99,100]. However, adequately designed controlled studies for the efficacy of self-help have not been conducted.

VII. Model for Conceptualizing the Eating Disorders

Eating disorders do not have a single cause or a predictable course. They begin with dieting or restrained eating behavior. Often the dieting is for the purpose of becoming thinner and more attractive. However, dieting can also follow a severe stress or physical illness. Behaviors and influences antecedent to the dieting experience can be categorized into problems of biological vulnerability, psychological predispositions, and societal influences, all of which have been discussed in this chapter. It is the integrative effect of the disturbances in these categories on the dieting behavior that propels the individual person into developing an eating disorder. As the dieting continues, starvation effects, weight loss, nutritional effects, and psychological changes occur. A sustaining cycle of core dysfunctional eating behaviors develops with both psychological and physiological reinforcement. At the present time, no treatment modality can predict recovery in the specific patient.

References

1. Russell, G. F. M., and Beardwood, C. J. (1970). Amenorrhea in the eating disorders: Anorexia nervosa and obesity. *Psychother. Psyschosom.* **18,** 358–364.

2. American Psychiatric Association (1994). "Diagnostic and Statistical Manual of Mental Disorders," 4th ed. APA, Washington, DC.

3. Lindhohn, L., and Wilson, G. T. (1988). Body image assessment in patients with bulimia nervosa and nonnal controls. *Int. J. Eat. Disord.* **7,** 527–539.

4. Halmi, K. A., and Falk, J. R. (1981). Common physiological changes in anorexia nervosa. *Int. J. Eating Disord.* **1,** 16–27.

5. Russell, G. F. M. (1969). Metabolic, endocrine and psychiatric aspects of anorexia nervosa. *Sci. Basis Med.* **15,** 236–240.

6. Casper, R., Eckert E., and Halmi, K. A. (1980). Bulimia: Its incidence and clinical importance in patients with anorexia nervosa. *Arch. Gen. Psychiatry* **37,** 1030–1035.

7. Garfinkel, P. E., Moldofsky, H., and Gardner, D. M. (1980). The heterogeneity of anorexia nervosa: Bulimia as a distinct subgroup. *Arch. Gen. Psychiatry* **37,** 1036–1040.

8. Strober, M., Salkin, B., and Burroughs, J. (1982). Validity of the bulimia—restrictor distinctions in anorexia nervosa: Parental personality characteristics and familial psychiatric morbidity. *J. Nerv. Men. Dis.* **170,** 345–351.

9. Eckert, E., Halmi, K. A., and Marchi, P. (1987). Comparison of bulimic and nonbulimic anorexia nervosa patients during treatment. *Psychol. Med.* **17,** 891–898.

10. Russell, G. F. M. (1979) Bulimia nervosa; and aminious variant of anorexia nervosa. *Psychol. Med.* **9,** 492–448.

11. Davis, C. G., Williamson, D. A., and Gorecmy, T. (1986). Body image distortion in bulimia: An important distinction between binge purgers and binge eaters. *Paper Annu. Conv. Assoc. Adv. Behav. Ther.,* Chicago, *1997.*

12. Duchman, E. G., Williamson, D. A., and Strickler, P. M. (1986). Dietary restraint in bulimia. *Paper Annu. Conv. Asso. Adv. Behav. Ther.,* Chicago, *1986.*

13. Devlin, M. J., Walsh, B. T., and Spitzer, R. L. (1992). Is there another binge eating disorder? A review of the literature on overeating in the absence of bulimia nervosa. *Int. J Eat. Disord.* **11,** 341–350.

14. Spitzer, R., Devlin, M. J., and Walsh, B. T. (1992). Binge eating disorder: Multi site field trial of the diagnostic criteria. *Int. J. Eat. Disord.* **11,** 191–203.

15. Bell, R. M. (1985). "Holy Anorexia." University of Chicago Press, Chicago.

16. Halmi, K. A. (1994). Images in psychiatry. *Am. J. Psychiatry* **151,** 1216.

17. Johnston, N. (Letter in Latin dated June 29, 1669). To Timothy Clark "Concerning the young fasting women in Derbyshire, named Martha Taylor, together with his apprehension of some impostor in the affair." *J. R. Soc. London* **3,** 389–392 (1667–1671).

18. Reynolds, J. A. (1669). "Discourse upon Prodigious Abstinence: Occasion by the 12 Months Fasting of Martha Taylor, the Fained Derbyshire Dawnosell." London. RW.

19. Morton, R. (1689). "Phthisiologia, Seu Exercitationes de Phtisi." S. Smith, London.

20. Whytt, R. (1764). "Observations on the Nature, Causes and Cure of those Disorders which have been Commonly Called Nervous, Hypochondriac or Hysteric." Becket, DeHondt and Balfour, Edinburgh.

21. Marcé, L. V. (1860). On a form of hypocondriacle delirum occurring consecutive to dyspepsia, and characterized by refusal of food. *J. Psychol. Med Ment. Pathol.* **13,** 264–266.

22. Gull, W. (1888). Anorexia nervosa. *Lancet* **1,** 516–517.

23. Lasequé, E. C. (1873). On hysterical anorexia. *Med. Times Gaz.* **2,** 265–266.

24. Chipley, W. S. (1859). Sitomania: It's causes and treatment. *Am. J. Insanity* **16,** 1–42.

25. Hammond, W. A. (1879). "Fasting Girls: Their Physiology and Pathology." Putnam, New York.

26. Mitchell, S. W. (1881). "Lectures on the Diseases of the Nervous System, Especially in Women." Henry C. Lea's, Philadelphia.

27. Inches, P. R. (1895). Anorexia nervosa. *Marit. Med News (Halifax)* **7,** 73–75.

28. Bliss, E. L., and Branch, C. H. H. (1960). "Anorexia nervosa: It's History, Psychology and Biology." Hoeber, New York.

29. Bruch, H. (1973). "Eating Disorders: Obesity, Anorexia nervosa, and the Person Within." Basic Books, New York.

30. Hoek, H. W. (1993). Review of the epidemiological studies of eating disorders. *Int. Rev. Psychiatry* **5,** 61–74.

31. Eagles, R., Johnston, M., Hunter, D., Lobban, M, and Millar, H. (1995). Increasing incidences of anorexia nervosa in the female population of northeast Scotland. *Am. J. Psychiatry* **152,** 1266–1271.

32. Hoek, H. W. (1991). The incidence and prevalence of anorexia nervosa and bulimia nervosa in primary care. *Psychol. Med.* **21,** 455–460.

33. Rooney, B., McClelland, L., Chrisp, A. H., and Sedgwick, P. M. (1995). The incidence and prevalence of anorexia nervosa in three suburban health districts in southwest London, UK. *Int. J. Eat. Disord.* **18,** 299–307.

34. Lucas, A. R., Beard, C. M., O'Fallon, W. M., and Kurland, L. T. (1991). Fifty-year trends in the incidence of anorexia nervosa in Rochester, Minnesota: A population-base study. *Am. J. Psychiatry* **148,** 917–922.

35. Fairburn, C. G., and Beglin, S. J. (1990). Studies of the epidemiology of bulimia nervosa. *Am. J. Psychiatry* **147,** 104–108.

36. Garfinkel, P. E., Lin, E., Goering, P., Spegg, C., Goldbloom, D. S., Kennedy, S., Kaplan, A. S., and Woodside, D. B. (1995). Bulimia nervosa in a Canadian community sample: Prevalence and comparison of subgroups. *Am. J. Psychiatry* **152,** 1052–1058.

37. Walters, E., and Kendler, K. (1995). Anorexia nervosa and anorectic like syndromes in a population-based female twin sample. *Am. J. Psychiatry* **152,** 64–71.

38. Gotestam, K. G., Erickson, L., and Hagen, H. (1995). An epidemiological study of eating disorder in Norwegian psychiatric institutions. **18,** 263–268.

39. Bushnell, J. A., Wells, J. W., Hornblow, A. R., Oakley-Browne, M. A., and Joyce, P. (1990). Prevalence of three bulimia syndromes in the general population. *Psychol. Med.* **20,** 671–680.

40. Lee, S., Leung, T., Lee, A. M., Yu, H., and Lunge, C. M. (1996). Body dissatisfaction among Chinese undergraduates and it's implications for eating disorders in Hong Kong. *Int. J. Eat. Disord.* **20,** 77–84.

41. Fichter, M. M., Elton, M., Sourdi, L., Weyerer, S., and Coptagel-Ilal, L. G. (1988). Anorexia nervosa in Greek and Turkish adolescents. *Eur. Arch. Psychiatry Neurol. Sci.* **237,** 200–208.

42. Hoek, H. W., Bartelds, A. I, and Bosveld, J. J. F. (1995). The impact of urbanization on detection rates of eating disorders. *Am J. Psychiatry* **152,** 1272–1278.

43. Crisp, A. H., Palmer, R. L., and Kaluchy, R. S. (1976). How common is anorexia nervosa? A prevalence study. *B. J. Psychiatry* **128,** 549–554.

44. Kendell, R. E., Hall, D. J., Bailey, and Barbgian, H. N. (1973). The epidemiology of anorexia nervosa. *Psychol. Med.* **3,** 53–61.

45. Gard, M. C., and Freeman, C. P. (1996). The dismanteling of a myth: A review of eating disorders and social economic status. *Int. J. Eat. Disord.* **20,** 1–12.

45a. Striegel-Moore, R., Schreiber, G. B., Pike, K. M., Wilfley, D. E., and Rodin, J. (1995). Drive for thinness in black and white preadolescent girls. *Int. J. Eat. Disord.* **18,** 59–69.

46. DiNicola, F. A. (1990). Anorexia multiforme: Self-starvation in historical and cultural context. Part 1: Self-starvation as a historical chamelun. *Transcult. Psychiatr. Res. Rev.* **27,** 165–196.

47. Pate, J. D., Pumariega, A. G., Hester, C., and Gemer, D. M. (1992). Cross-cultural patterns in eating disorders: A review. *J. Am. Acad. Child Adolesc. Psychiatry* **31,** 802–809.

48. Yates, A. (1990). Current propectives on eating disorders: H. Treatment, outcome and research directions. *J. Am. Acad. Child Adolesc. Psychiatry* **29,** 1–9.

49. Casper, R. C. (1990). Personality features of women with good outcome from restricting anorexia nervosa. *Psycholsom. Med.* **52,** 156–170.

50. Strober, M. (1980). Personality and symptomalogical features in young, nonchronic anorexia nervosa patients. *J. Psycholsom. Res.* **24,** 353–359.

51. DaCosta, and Halmi, K. A. (1992). Classifications of anorexia nervosa: Question of subtypes. *Int. J. Eat. Disord.* **11,** 305–313.

52. Halmi, K. A. (1997). Comorbidity of the eating disorders. *In* "Baillière's Clinical Psychiatry Series" (D. Jimmerson and W. Kaye, eds.), Vol. 39, pp. 291–302. Baillière Tindall, London.

53. Lacey, J. H., and Evans, C. D. H. (1986). The impulsivist: A multi-impulsive personality disorder. *Br. J. Addict.* **81,** 641–649.

54. Strober, M., and Humphrey, L. L. (1997). Familial contributions to the etiology and course of anorexia nervosa. *J. Consult. Clin. Psychol.* **55,** 654–659.

55. Humphrey, L. L. (1986). Structural analysis of parent-child relationships in eating disorders. *J. Abnorm. Psychol.* **95,** 395–402.

56. Connors, M. E., and Morse, W. (1993). Sexual abuse and eating disorders: A review. *Int. J. Eat. Disord.* **13,** 1–11.

57. Rorty, M., Yager, J., and Rossotto, E. (1994). Childhood sexual, physical and psychological abuse and the relationship to comorbid psychopathology and bulimia nervosa. *Int. J. Eat. Disord.* **16,** 317–334.

58. Loller, G., Halek, C., and Crisp, A. H. (1993). Sexual abuse as a factor in anorexia nervosa: Evidence from two separate case series. *J. Psychosom. Res.* **37,** 873–879.

59. Cooper, Z. (1995). The development and maintenance of eating disorders. *In* "Eating Disorders and Obesity: A Comprehensive Handbook" (K. D. Braunell and C. G. Fairburn, eds.), pp. 199–206. Guilford Press, New York and London.

60. Halmi, K. A., Eckert, E., Mark, E. P., Sampugnaro, R., Apple, R., and Cohen, J. (1991). Comorbidity of psychiatric diagnosis in anorexia nervosa. *Arch. Gen. Psychiatry* **48,** 712–718.

61. Laessle, R. G., Kittl, S., Fichter, M. M., Wittchen, H., and Pirke, K. M. (1987). Major affective disorder in anorexia nervosa and bulimia nervosa: A descriptive diagnostic study. *Br. J. Psychiatry* **151,** 785–789.

62. Toner, B., Garfinkel, P., and Garner, D. (1988). Affective and anxiety disorders in the long term follow-up of anorexia nervosa. *Int. J. Psychiatry Med.* **18,** 357–364.

63. Gershon, E., Schreiber, J. L., and Hamovit, J. R. (1984). Clinical findings in patients with anorexia nervosa and affective illness in their relatives. *Am. J. Psychiatry* **141,** 1419–1422.

64. Braun, D. L., Sunday, S. R., and Halmi, K. A. (1994). Psychiatric comorbidity in patients with eating disorders. *Psychol. Med.* **24,** 859–867.

65. Herzog, D. B., Keller, M. S., and Lavori, P. M. (1992). The prevalence of personality disorders in 210 women with eating disorders. *J. Clin. Psychiatry* **53,** 147–152.

66. Fornari, V., Kaplan, M., and Sanberg, D. E. (1992). Depressive and anxiety disorders in anorexia nervosa and bulimia nervosa. *Int. J. Eat. Disord.* **12,** 21–29.

67. Hudson, J. I., Pulp, H. G., and Jonas, J. M. (1983). Phenomenological relationship of eating disorders to major affective disorder. *Psychiatry Res.* **9,** 345–354.

68. Brewerton, T. D., Lydiard, R. B., and Herzog, D. B. (1995). Comorbidity of Axis I psychiatric disorders in bulimia nervosa. *J. Clin. Psychiatry* **56,** 77–80.

69. Holdemess, C. C., Brooks-Gun, J., and Warren, M. P. (1994). Comorbidity of eating disorders and substance abuse review of the literature. *Int. J. Eat. Disord.* **16,** 1–34.

70. Selbey, M. J., and Moreno, J. K. (1995). Personal and familial substance misuse patterns among eating disorder and depressed subjects. *Int. J. Addict.* **30,** 1169–1176.

71. Hotta, M., Shibasaki, T., and Masuda, A. (1986). The responses of plasma adrenal corticotropin and cortisol to corticotropin-releasing hormone and cerebral spinal fluid immunoreactive CRH in anorexia nervosa patients. *J. Clin. Endocrinol. Metab.* **62,** 319–324.

72. Weiner, H. (1989). Psychoendocrinology of anorexia nervosa. *Psychiatr. Clin. North Am.* **12,** 187–206.

73. Kaye, W. H., George, D. T., and Gwirtsman, H. E. (1991). Altered serotonin activity in anorexia nervosa after long term weight restoration. *Arch. Gen. Psychiatry* **48,** 556–562.

74. Jimerson, D. C., Lesem, M. D., and Kaye, W. H. (1992). Low serotonin and dopamine metabolite concentrations in cerebral spinal fluid from bulimic patients with frequent bingeing episodes. *Arch. Gen. Psychiatry* **49,** 132–138.

75. Brewerton, T. D., Brandt, H. A., and Lesem, M. D. (1990). Serotonin in eating disorders. *In* "Serotonin in Major Psychiatric Disorders" (E. F. Coccaro and D. L. Murphy, eds.), pp. 153–184. American Psychiatric Press, Washington, DC.

76. McBride, P. A., Anderson, G. M., and Khait, V. D. (1991). Serotonergic responsiveness in eating disorders. *Psychopharm. Bull.* **27,** 365–372.

77. Halmi, K. A. (1995). Basic biological overview of eating disorders. *In* "Psychopharmacology; The Fourth Generation of Progress" (F. E. Bloom and D. J. Kupfer, eds.), pp. 1609–1616. Raven Press, New York.

78. Halmi, K. A., Casper, R. C., and Eckerd, E. D. (1979). Unique features associated with age of onset of anorexia nervosa. *Psychiatry Res.* **1,** 209–215.

79. Theander, S. (1985). Outcome and prognosis in anorexia nervosa and bulimia: Some results of previous investigations compared with those of a Swedish long term study. *J. Psychiatr. Res.* **19,** 493–508.

80. Eckert, E. D., Halmi, K. A., and Mark, E. P. (1995). Ten year follow-up of anorexia nervosa: Clinical course and outcome. *Psychol. Med.* **25,** 143–156.

81. Kell, P. K., and Mitchell, J. E. (1997). Outcome and bulimia nervosa. *Am. J. Psychiatry* **154,** 313–321.

82. Herzog, W., Schellberg, D., and Deter, H. C. (1997). First recovery in anorexia nervosa patients in the long-term course: A discrete time survival analysis. *J. Consult. Clin. Psychol.* **65,** 169–177.

83. Halmi, K. A., Eckert, E., and Ladu, T. J. (1986). Anorexia nervosa: Treatment efficacy of cyproheptadine and amitriptyline. *Arch. Gen. Psychiatry* **48,** 177–181.

84. Crisp, A. H., Lacey, J. H., and Crutchfield, M. (1987). Clomipramine and drive in people with anorexia nervosa: An inpatient study. *Br. J. Psychiatry* **144,** 238–246.

85. Kaye, W. H., Weltzin, T. E., and Hsul, K. J. (1991). An open trial of fluoxetine in patients with anorexia nervosa. *J. Clin. Psychiatry* **52,** 464–471.

86. Wulliemier, F., Rossel, F., and Sinclair, K. (1975). Behavior therapy in anorexia nervosa. *J. Psychosom. Res.* **19,** 267–272.

87. Kleifield, E., Wagner, S., and Halmi, K. A. (1986). Cognitive-behavioral treatment of anorexia nervosa. *Psychiatr. Clin. North Am.* **19,** 715–737.

88. Russell, G. F. M., Svmukler, G. I., and Dare, C. (1987). An evaluation of family therapy in anorexia nervosa and bulimia nervosa. *Arch. Gen. Psychiatry* **44,** 1047–1056.

89. Mitchell, J. E., and Raymond, N. C. (1992). Cognitive-behavioral therapy in treatment of bulimia nervosa. *In* "The Psychobiology and Treatment of Anorexia Nervosa and Bulimia Nervosa" (K. A. Halmi, ed.). American Psychiatric Press, Washington, DC.

90. Wilson, G. T., and Fairburn, C. G. (1993). Cognitive treatments for eating disorders. *J. Consult. Clin. Psychol.* **61,** 261–269.

91. Mitchell, J. E., and De Zwan, M. (1993). Pharmacological treatments of binge eating. *In* "Binge Eating: Nature, Assessment and Treatment" (C. G. Fairburn and G. T. Wilson, eds.). Guilford Press, New York.

92. Welch, B. T., and Devlin, M. J. (1992). The pharmacological treatment of eating disorders. *In* "Psychiatric Clinics of North America" (D. Shafer, ed.), pp. 149–160. Saunders, Philadelphia.

93. Pyle, R. L., Mitchell, J. E., and Eckert, E. D. (1990). Maintenance treatment and six month outcome of bulimia nervosa patients who respond to initial treatment. *Am. J. Psychiatry* **147,** 871–875.

94. Welch, B. T., Hadigan, C. M., and Devlin, M. J. (1991). Long term outcome of antidepressant treatment for bulimia nervosa. *Am. J. Psychiatry* **148,** 1206–1212.

95. Fluoxetine Bulimia Nervosa Collaborative Study Group (1992). Fluoxetine in the treatment of bulimia nervosa. *Arch. Gen. Psychiatry* **49,** 139–143.

96. Mitchell, J. E., Pyle, R. L., and Eckert, E. D. (1990). A comparison study of antidepressants and structured group psychotherapy in the treatment of bulimia nervosa. *Arch. Gen. Psychiatry* **47,** 149–157.

97. Agras, W. S., Rossiter, E. M., and Arnow, B. (1992). Pharmacologic and cognitive-behavioral treatment for bulimia nervosa: A controlled comparison. *Am. J. Psychiatry* **149,** 82–87.

98. Fairburn, C. G., Joan, R., and Pelevar, R. C. (1992). Three psychological treatments for bulimia nervosa: A comparative trial. *Arch. Gen. Psychiatry* **48,** 463–469.

99. Cooper, P. J. (1993). "Bulimia Nervosa: A Guide to Recovery." Robinson, London.

100. Fairburn, C. G. (1995). "Overcoming Binge Eating." Guilford Press, New York.

83

Addictive Disorders

SHIRLEY Y. HILL
Department of Psychiatry
University of Pittsburgh Medical Center
Pittsburgh, Pennsylvania

I. Introduction and Background

Population survey data are consistent in showing that women are at greater risk for affective and anxiety disorders than men but are at lower risk for substance abuse/dependence [1–3]. The first large scale multisite survey, the Epidemiological Catchment Area (ECA) program [1], also found rates of drug abuse/dependence to be higher in males than in females (7.8% versus 4.8%), as was the case for alcohol abuse/alcohol dependence (23.8% versus 4.6%). More recent data from the National Institute on Alcoholism and Alcohol Abuse (NIAAA) National Longitudinal Alcohol Epidemiologic Survey indicated that men are significantly more likely to have a history of alcohol dependence than are women (18.6% versus 8.4%) [3]. Data from the National Comorbidity Survey [2] also indicated that alcohol dependence was lower in women (8.2% versus 20.1%), as was substance dependence (which includes both drug and alcohol dependence; 17.9% versus 35.4%), an approximately 0.5 times lower rate for women. Unlike the case for substance dependence, the National Comorbidity Survey [2] found that for affective disorders, the ratio of affected cases (lifetime diagnosis) among women was approximately 1.5 times higher than it was for men (23.9% versus 14.7%).

Because rates of alcohol and drug dependence are consistently lower in women than in men, one may question why the physical and mental health consequences of these disorders should be considered a major focus of public health policy. This chapter will argue that while exposure to drugs and alcohol may be lower in women than in men, there is reason to believe that the health consequences may be greater for women.

Alcohol is the most widely used psychoactive drug in the United States among women and girls over the age of 12. However, only a minority of women who drink become alcoholic. Nevertheless, there are known risks and some possible benefits of consuming alcohol in quantities well below those typically consumed by alcoholic women. In addition to the direct effects of alcohol on the cardiovascular system, liver, brain, and gastrointestinal tract, there are indirect effects on women's health. For example, women who drink excessively may be more prone to alcohol-induced depressive disorders. Also, women who drink excessively are more likely to associate with others who use alcohol and drugs excessively, thereby increasing the likelihood of their being a victim of domestic violence. Association with alcohol and drug dependent partners clearly brings greater risk for exposure to HIV, both because the woman who associates with drug users may risk exposure to intravenous users and because frequent drug/alcohol intoxication for the woman may

mean a lower likelihood of protected sex. Also, cocaine appears to be growing in popularity at a faster rate among women than men. This drug brings not only direct health effects [4], but also greater exposure to prostitution for drugs, particularly among young women and girls living on the street, and exposure to intravenous drug using partners.

In summary, though women may be at lower risk overall for alcohol and drug dependence, there is evidence, which the following review will provide, that the health consequences of this dependence may, in some instances, be greater for women than for men.

II. Definition of the Topic

The term addictive disorder has been used to describe such diverse behaviors as pathological gambling, eating disorders, and sexual addiction, as well as alcohol and drug dependence. The focus of this chapter will be exclusively on alcohol and drug dependence. It may be useful to consider the various definitions of alcohol and drug dependence that have been used.

The essential feature of substance dependence, according to the authors of DSM-IV, is a "cluster of cognitive, behavioral and physiological symptoms indicating the individual continues use of the substance despite significant substance-related problems. A diagnosis of Substance Dependence can be applied to every class of substances except caffeine." Substance abuse is distinguished from substance dependence in DSM-IV. The individual with substance abuse must exhibit a pattern of substance use that leads to impairment or distress during one 12 month period. These include failure to fulfill obligations and exposure to hazards (*e.g.*, driving while intoxicated (DWI), legal, social, or interpersonal problems). An individual who meets criteria for substance dependence must meet criteria for any three symptoms from a list of seven items covering tolerance; withdrawal; loss of control; inability to cut down; excessive time procuring the drug of choice; reduction in social, occupational, or recreational activities due to drug seeking; and use of the substance in spite of knowledge that the drug causes physical or psychological problems.

The interesting aspect about the current definition of substance abuse and substance dependence outlined in DSM-IV is that more options are listed for dependence (seven) than for abuse (four) and fewer items need to be met (one or more) for abuse, than for dependence (three or more). Thus, there is no qualitative difference between the two forms. The earlier DSM-III version required either tolerance or dependence to be present for substance dependence to be diagnosed. Thus, the newer version does not require either of these but assumes a linear scale of severity with abuse requiring fewer symptoms.

Other definitions that have been used widely include Research Diagnostic Criteria (RDC) and the Feighner Criteria [5]. These older definitions have similar elements in common with the newer DSM-IV definitions but differ in the conceptualization of how symptoms must be distributed in order to meet full criteria for either abuse or dependence. For example, the RDC definition of alcoholism was conceived as a low threshold definition (any three or more symptoms out of 20 possible symptoms) so that persons meeting criteria would typically have milder forms than those seen in treatment settings. The Feighner Criteria distinguished between "probable" and "definite" alcoholism, the former requiring two positive categories of symptoms while the latter required three. Unlike other systems that counted symptoms, this diagnostic system utilizes groups of symptoms. These can be broadly categorized as: (1) manifestations of withdrawal, (2) inability to stop or control drinking, (3) legal and/or social difficulties due to drinking, and (4) individual or others believe the individual drinks too much. As can readily be seen, in order for an individual to have a definite diagnosis of alcoholism by Feighner Criteria, the person would need to demonstrate evidence of either withdrawal or inability to stop or control drinking. Thus, the earlier DSM-III and Feighner Criteria have much greater similarity in conceptual orientation, requiring that the presence of withdrawal, tolerance, or failure to stop be present.

How do these varying definitions of alcohol or drug dependence affect our notions about gender differences in prevalence? The varying definitions offered by DSM-III, DSM-IV, RDC, and Feighner systems reflect the degree to which each attempts to embody what Edwards and Gross termed the alcohol dependence syndrome [6]. This syndrome included such symptoms as tolerance, withdrawal, drinking to avoid withdrawal, and impaired control. This definition, in contrast to social/occupational/family functioning definitions, is more than just a semantic difference. In the U.S. population, alcohol-related social and/or occupational problems appear somewhat transient [7] whereas the alcohol dependence syndrome appears to be prognostic of later problems [8]. With respect to gender differences in prevalence, definitions may make a crucial difference in detecting cases. For women who work only in the home, detecting occupational and social impairment may be comparatively more difficult than for men. Also, women are less likely than men to experience legal difficulties due to substance abuse. Additionally, because of smaller body size and differences in alcohol metabolism [9], there is reason to believe that women may develop alcohol tolerance and dependence more rapidly than men do.

III. Issues Related to Ascertainment

A. Sources of Heterogeneity

It is critically important to our understanding of gender differences in patterns of substance use, abuse, and dependence to specify the degree of familial/genetic loading of the cases, the severity (early onset or late onset), the source of the sample (clinical or general population), and degree of comorbidity present. All of these factors will affect case finding, presentation, and outcome. Consequently, meaningful comparisons by gender cannot be made without consideration of these factors.

1. Familial/Genetic Factors

Transmission of alcoholism within families occurs at a higher rate than most psychiatric disorders, including the affective disorders [10]. Transmission of drug dependence within families also occurs at rates exceeding population prevalence [11]. Adoption studies both in Sweden [12] and in the U.S. [11] point to genetic factors in the etiology of alcohol and drug dependence in women.

Further evidence for familial/genetic mediation of alcoholism in women can be gained from studies of adult twin pairs. Typically, monozygotic (MZ) and dizygotic (DZ) pairs are compared with respect to concordance for drinking or drinking problems. Both concordance for drinking [13] and alcohol dependence [14–18] have been investigated. While some studies [14,15,17] did not find MZ/DZ rates suggestive of genetic mediation in female twin pairs, recent data [16,18] are quite convincing. Pickens and colleagues [16], studying 114 male and 55 female twin pairs, concluded that a genetic component was operating in female alcoholism as well as in male alcoholism. Kendler et al. [18], studying 1030 MZ and DZ twin pairs, found substantially higher correlations in MZ than DZ twins, with over 50% of the variance in alcoholism risk explained by genetic factors.

2. Age of Onset as a Marker of Severity

Cloninger et al. [19] have described the characteristics of male Type I alcoholics, in contrast to Type II alcoholics, as those whose likelihood of drinking depends much more heavily on the environmental milieu in which the individual resides. The latter, or Type II males, tend to be of the familial form and display earlier onset. Previously we have noted [20] that there appear to be two forms of alcoholism in women. One is generally less severe and often arises because of environmental pressures to drink (e.g., divorce, children have grown up and left the home). This form has a later onset corresponding to one of the peaks in heavy drinking identified in women between the ages of 35 and 49 [1]. The other form is more severe and has an early onset. This form appears to correspond to an additional peak of heavy drinking that occurs between ages 18–24 [1]. While many young women drink heavily, often with problems that would allow them to meet criteria for alcohol dependence, many will "mature out." However, one subgroup of women has been identified that consists of women who show excessive drinking in young adulthood but who do not appear to change their pattern of drinking and persist in this abusive style of drinking. It is proposed that this form is most often a part of a familial alcoholism diathesis, often being part of an intergenerational pattern of substance abuse that reappears in multiple generations. Accordingly, this form has greater severity and is present even in families where the environment is relatively favorable (the family has a higher socioeconomic status, is intact, and subscribes to traditional values).

Further evidence for two types of alcoholism in women comes from studies by Glenn and Nixon [21] and Lex et al.

[22]. Glenn and Nixon [21] showed that when women were classified by age of onset, several significant differences emerged with respect to the severity of alcoholism exhibited. In general, the early onset alcoholism group displayed greater severity of symptoms and more affected relatives. Lex *et al.* [22], studying women who came to the attention of the courts because of their drinking (DWI offenses), found these women to have an earlier onset of alcohol problems and a greater density of familial alcoholism than those without court involvement.

3. Alcoholism: Comorbidity with Other Psychiatric Disorders

a. CLINICAL POPULATIONS.   Numerous studies of clinical populations have pointed to elevated rates of depressive symptoms among alcoholics [23–25]. Ross and colleagues [24], studying 260 men and 241 women seeking treatment for alcohol and drug dependence, evaluated the rates of comorbidity for other psychiatric disorders and found exceptionally high rates of other psychiatric disorders. Among persons seeking treatment for either alcohol or drug dependence, 78% were found to have a lifetime diagnosis of another psychiatric disorder, with 65% currently having another disorder. Those individuals meeting criteria for both alcohol and drug dependence had the highest rates of comorbidity. The affective disorders were among the most common, with 24% of the sample meeting lifetime criteria for major depression. Higher rates of major depression have also been reported for treatment-seeking opiate-dependent individuals [26].

Examining the patterns of comorbidity, particularly by gender, is useful with regard to etiology, health consequences, and treatment. If substance-dependent women show greatly elevated rates of depression compared to substance-dependent men, then the etiology of substance dependence in women can be said to be more clearly based on an affective diathesis. This would suggest, perhaps, a better treatment response to antidepressant medication for women. On the other hand, the excess cases of depression seen in women may merely reflect the gender ratio observed in the general population. In this case, the greater likelihood that women will carry a diagnosis of depression compared to men is noteworthy, but it suggests little about etiology or treatment of the underlying substance dependence.

b. POPULATION SURVEYS.   The ECA survey found that alcoholics had higher rates for every psychiatric disorder examined than did nonalcoholics [27]. Substance-abusing women were found to be depressed more often than substance-abusing men. Further examination of these data show, however, that this is due to the greater female/male ratio for depression seen in the general population. When the prevalence ratios by gender for the co-occurrence of major depression and alcoholism were calculated, quite similar ratios were found [28]. The ratio for women was 2.7 whereas that for men was 2.4. Thus, the study authors concluded that while the association between antisocial personality disorder, other substance abuse, and mania were quite strong, the association between alcoholism and depressive disorder, though positive, was not.

Therefore, it is commonly observed in general population surveys that rates of major depression are elevated among both male and female alcoholics relative to nonalcoholics. How-

ever, this trend is not nearly as impressive as it is in clinical samples. Investigations of clinical samples tend to report much higher rates of depression among alcoholics than nonalcoholics [29], particularly for women. Further examination of clinical samples reveals, however, that the proportion of women treated for alcohol dependence who carry a lifetime diagnosis of major depression may actually be representative of the proportion one would expect in the general population. For example, Hesselbrock [29] studied the admission and one-year follow-up diagnosis of 197 male and 69 female alcoholics admitted to three inpatient treatment centers in the greater Hartford, Connecticut area. Approximately one-third of the males had an admission diagnosis of major depression with or without antisocial personality disorder and one-half of the women alcoholics carried a diagnosis of depression (with or without ASP). Thus, 1.5 times as many women as men carried an admission diagnosis of depression. This is the approximate ratio seen in the general population, 15% and 23% for males and females, respectively [2]. In other words, though the process of having alcoholism elevates rates of depression, it would appear the relative risk for developing depression by gender is not altered by this process. Moreover, even in clinical samples, the increased risk that female alcoholics in comparison to male alcoholics have for carrying a diagnosis of depression, is the risk that all women in the general population share, which is approximately 1.5 times the relative risk seen in men.

IV. Historical Evolution/Trends

Understanding the historical trends in women's drinking may shed some light on why men more often tend to become alcoholics than do women. Because alcohol use is often the "port of entry" for other drug use/dependence, review of historical trends in the drinking practices of women may also provide insight regarding historical trends in drug use among women.

In Roman times women were not allowed to become intoxicated. To do so resulted in death, usually by stoning. In Victorian times alcohol and drugs were to be used sparingly, lest the women be considered a "fallen angel." This attitude persisted up until WWII in the U.S. With the entry into factories and other workplaces, women adopted dress and behavior more similar to those of men. Because of this trend, some have argued that work stress and multiple role stress have contributed to greater problems with alcohol and drug use among women. However, there seems to be little evidence that this is so. The most likely explanation can be found in greater accessibility [30]. At any rate, there is clearly a trend for greater use of alcohol since WWII. This trend is most visible among younger aged (18–24) women, where the major peak in heavy drinking and alcohol dependence occurs. Of some concern is the fact that the age of onset to begin drinking has become earlier. Grant [3] compared three cohorts for their onset of drinking. With each cohort examined, from the earliest (1894–1937) to the most recent (1968–1974), onset of drinking was earlier, which in turn reflected the cumulative probability of developing alcohol dependence. Thus, the historical trend has been for women to use alcohol more freely, and with this greater exposure, they began drinking at an earlier age.

V. Distribution of Alcohol and Drug Dependence by Gender

A. Lifetime Prevalence of Alcohol Dependence in the U.S.

Among lifetime users of alcohol, both men and women exhibit peak levels of alcohol dependence between the ages of 18 and 24, with 35.4% of men and 24.6% of women affected [3]. Recent surveys [2,3] provided data by age, ethnicity, and socioeconomic status, but they did not provide information about gender. However, in these surveys greater odds for developing alcohol abuse were found for the lowest income group (less than $20,000 per year) and for those with the least education (less than high school). It is reasonable to assume that the same relationship would hold if analyses were restricted to women. However, it is noteworthy that during the young adult years (ages 18–24), rates of alcohol dependence in men and women show the greatest convergence, with rates for women not far behind those for men.

B. Mortality Statistics

Description of the mortality rates for substance-dependent women is complicated by the fact that women abuse and become dependent on a wide variety of psychoactive drugs (e.g., alcohol, cocaine, opioids, and sedative–hypnotic drugs). The acute and chronic effects of these drugs vary widely, particularly with respect to the particular organ system affected (e.g., cardiovascular, neuroendocrine). Moreover, the risks vary with the age of the woman being considered, as patterns of drug use vary by age. For example, women over the age of 65 are much more likely to receive prescriptions for psychoactive medication than are men [31]. These specific effects have been reviewed elsewhere [32]. For the present discussion a comparison of mortality due to alcoholism by gender will be discussed. Also, because our focus is on mental health issues, special emphasis will be placed on suicide mortality by gender with reference to both drug and alcohol dependence.

A number of adverse health consequences are associated with use of alcohol among persons drinking at levels high enough to cause interpersonal, legal, and social problems and who would be classified as alcoholic by most diagnostic systems. It is not surprising that an early review of alcohol-related mortality by gender [33] found that when mortality ratios were compared for alcoholics and nonalcoholics, an excess mortality for women was found in all studies examined (mortality ratios of 2.0–7.0 times higher for alcoholics). One study illustrates these findings quite dramatically. Smith et al. [34] conducted an 11 year follow-up of 100 alcohol-dependent women with a mean age of 44 years and found that over one-third were deceased.

In the general population, women have higher rates of attempted suicide [35], whereas men have higher rates of completed suicide [36]. Using age- and gender-appropriate controls, one sees a different pattern for alcoholic men and women. In four of five studies in which sex-specific rates were determined, women were found to have higher observed-to-expected ratios than men [33]. The pattern appears to be similar for other drugs. The incidence density of suicide was found to be higher among drug addicts than controls in Norway [37], with excess mortality

by suicide being higher among women and among the youngest groups examined. These data suggest that females with serious alcohol or drug problems are not reticent about using injurious methods with clear lethality.

There is evidence that adolescent girls who abuse drugs may be at especially high risk for suicide. Among adolescents who attempt suicide, those who use illicit drugs outnumber controls by 8:1 [38]. Another study of adolescent outpatient substance abusers found a three-fold excess in risk for suicide, with significantly more females at risk than males [39]. Also, there is accumulating evidence that substance abuse among adolescents is related to greater frequency and repetitiveness of suicide attempts with more medically lethal attempts (e.g., firearms) and greater seriousness of intention [40,41].

The presence of a depressive illness is also a significant factor in attempted suicide, particularly among youth. Clarkin [42], studying admissions to a hospital unit specializing in depression, noted that among adolescents who attempted suicide within three months of admission, both substance abuse and depression were prominent features of their presentation (33% were drug dependent, 25% were alcohol dependent, and 71% met criteria for major affective disorder).

In summary, substance use affects the health of women, especially adolescent females, by elevating their rates of suicide attempts and completed suicides. Clinical depression also plays a role, amplifying the impact of substance use and dependence. Because females of all ages appear to be at greater risk for affective disorders, substance dependence among adolescent females is especially worrisome. Additionally, other reckless behavior (driving under the influence of alcohol and/or drugs, unprotected sexual activity) may be more likely for the individual who is depressed, impulsive, and intoxicated, or otherwise feels life is not worth living.

VI. Host and Environmental Determinants

In addition to the obvious differences in host characteristics (age, gender, ethnicity) that affect the individual's likelihood of drinking or using drugs, there are some less obvious host characteristics (temperament, psychiatric status, family history, of psychiatric disorders) that must be discussed in the context of environmental determinants. For example, Millstein et al. [43] studied risk taking behavior in young adolescents from an inner city middle school serving grades 6–8 and found that those children who engaged in any given risk behavior were more likely to engage in other risk behaviors as well. Adolescents who used tobacco, alcohol, or street drugs were significantly more likely to be sexually active and drive or ride in a car while they or the driver was under the influence of a substance. In that study, one-fifth of these young adolescents reported being sexually active (50% of the Black adolescents, 23% of the Hispanics, and 11% of the Whites). A third said they did not take precautions against acquiring sexually transmitted diseases.

The tendency to take risks has been described as "sensation seeking." This temperament characteristic has been thought to be an enduring characteristic of the individual and may be related to inherited variations in one or more neurotransmitters. Cloninger has suggested that alterations in dopamine activity might account for differences in sensation seeking [44]. Some, but not all, laboratories have found evidence for this [45,46]. Other temperament characteristics such as negative affectivity, have been found to be genetically linked to risk for alcohol dependence [47]. Some individuals may be genetically predisposed to risk taking behavior, including use of alcohol and drugs.

Clearly, family history of alcohol and drug dependence confers an increased risk to offspring. Early adoption data for alcoholism showed a four-fold increase in risk to offspring even when the child did not live with the alcoholic parent [48]. Drug dependence has also been shown to be elevated in adopted offspring of biological parents who are drug dependent [11]. Moreover, exposure to alcoholic or drug abusing parents alters the risk to offspring. In a study of the biological offspring of alcoholic mothers from high density for alcoholism families, children were found to have increased risk for psychopathology. This risk was increased by the presence of an alcoholic father in the home, whether biological or surrogate [49].

Family history of alcoholism appears to exert an effect through gender. Bohman *et al.* [12] found a threefold excess of alcohol abusers among adopted daughters of biological alcoholic mothers compared to controls. No excess was found for the daughters of male offspring. Whether this is due to greater saliency of same-sexed role models when one has a genetic susceptibility or to a gender-specific genetic form of alcoholism is currently unknown.

VII. Women's Social Roles

A. Employment

Early surveys of women's drinking and its relationship to employment tended to emphasize the adverse effect of employment on drinking [50]. Because married, employed women tended to report higher rates of heavy drinking than single employed women, the cause of the increased level of heavy drinking was assumed to be the role conflict between the demands of work and home. Based on data from a 1981 U.S. national sample, Wilsnack *et al.* [51] concluded that women employed full-time had only slightly higher rates of heavy drinking than did full-time homemakers. Several studies suggest that the influence of employment on rates of heavy drinking has more to do with the type of employment than whether the woman is employed outside the home.

Employment in male-dominated occupations was reported to be associated with greater frequency of drinking by Norwegian women [30]. This trend appears to have been substantiated in other samples, including some in Czechoslovakia [52] and the U.S. [53]. Also, a survey of 3000 women in Helsinki, Finland [54] found that the proportion of co-workers who are women in the workplace influences women's drinking. Controlling for socioeconomic status, income, and education, this study found that those work places with the highest proportion of female co-workers tended to have the lowest rates of drinking for women.

B. Marital Status

Early studies consistently showed that married women were less likely to be heavy drinkers than separated or divorced women [55]. Subsequent longitudinal studies appear to confirm

this. Young women who were divorced during the period of observation showed increased drinking, whereas those who remarried showed decreased drinking [56]. Also, Wilsnack *et al.* [57] found that women who lived with partners to whom they were not married had the highest rates of heavy drinking when compared to either divorced, separated, or married women. Most surveys do not analyze data with respect to whether the woman was a heavy drinker before the marital status change, nor do they assess whether the partner was a heavy drinker or alcohol dependent. However, when results are analyzed by husbands' drinking, clear influences do emerge [30]; that is, women living with substance-dependent persons tend to use more substances while living with this partner.

C. Sexual Orientation

Early studies reported higher rates of alcoholism and problem drinking in gay samples than in the general population [58–60]. For example, Saghir and Robins [58] reported that 25% of a sample of gay women were excessive drinkers, with 10% classified as alcoholic. This is in comparison to 5% of heterosexual women being excessive, and none being reported as alcoholic in that study. However, these early studies had distinct limitations that alter the conclusions that can be reached. Some did not analyze data for women [60], while others obtained samples from gay-oriented bars [59,60]. More recently there have been a few large scale surveys [61,62] initiated that did not depend on the more biased samples that would be obtained utilizing bars as sources for case finding. The McKirnan and Peterson [61] survey of 3400 gay women and men suggests that while reported rates of heavy drinking were not higher among gays, rates of problem drinking were significantly higher. This was due to two factors. First, in general population samples, men and women differ considerably in problem rates, while among gays, men and women do not differ in problem rates. Second, the commonly observed decline in alcohol problems usually seen with age among women in general population surveys does not apply to gay women. The National Lesbian Health Care Survey, which involved 1925 lesbians, described the frequency of drinking and drug use for these women but did not focus on problems related to use. However, questions were asked concerning "worry about use." For alcohol, 14% worried about their own use, while 7% worried about marijuana and 2% worried about cocaine use. These rates appear quite low.

While the larger surveys were an improvement over those that obtained cases through gay bars, there are some major limitations to the available studies. First, the response rates were quite low (42%) in the Bradford *et al.* study [62] and in the McKirnan and Peterson study (16%) [61]. Also, the method of distribution of questionnaires was limited; distribution was through personal networks in many cities. Additionally, further distortion may have been introduced by participation being offered by some gay groups but not others, so that the particular composition of groups responding to the questionnaires could have influenced the outcome of the survey. Moreover, neither study described characteristics of nonresponders compared to responders, further introducing potential bias in the obtained results. In other words, a representative sample of gays has yet to be collected in which alcohol and drug dependence have been thoroughly assessed. Clearly, more research is needed to determine the prevalence of alcohol and drug use in gay populations. With this information in hand, more informed decisions about treatment and intervention can be made.

VIII. Risk Factors

A. Social Factors

Cross-cultural variation will be examined with respect to alcohol use because availability across cultures is probably more similar than are specific drugs of abuse. Those cultures that have low availability of particular drugs because of more intense policing of drug trafficking will often have alcohol readily available because of locally made alcohol products. In fact, women are often a part of the manufacture of alcohol products in third-world countries.

In most cultures more men than women become alcoholic. This has led to the misconception that women have reduced access to alcohol in most cultures. However, Child et al. [63] examined records concerning drinking in 139 societies and found adequate data for analysis from 113; only four did not allow women to drink. Only in one society was drunkenness forbidden in women. Moreover, in some cultures women actually have greater access to alcohol than do men. As noted by Heath [64], in Latin America and parts of Africa women are the primary producers and distributors of grain or vegetable-based home brews that are used both as foods and as intoxicants. In some of these societies, women own and operate public drinking establishments where they market the alcohol they produce. In some tribal societies the home brew business is the focus of the economy.

Discussion of social factors that influence women's drinking must be evaluated from the premises we hold concerning the greater or lesser genetic vulnerability that women are thought to have. One early misconception was that there is a greater genetic (biological) vulnerability to alcoholism for men than for women [65]. However, formal tests of genetic heterogeneity among women alcoholics were not performed in that family study of women [65] so that the possible existence of both genetic and nongenetic forms could not be determined directly. Therefore, we suspect that the more genetic form of alcoholism is as likely to occur in women as it is in men, but the higher rates seen overall in men are largely due to the fact that environmental pressures to drink heavily are much greater for males. This conclusion is, in part, based on analysis of data from two ongoing high-density family studies in which families are selected either on the basis of the presence of two alcoholic brothers (NIAAA Grant AA 05909) or on the basis of two alcoholic sisters (NIAAA Grant AA 08082). In these families, the age of onset for females who become alcoholic is as early as it is for males who become alcoholic [20]. Also, families selected through female probands for their high density of alcoholism and multigenerational affectation show that female alcoholism can be as severe as the most familial form identified in men. The median age of onset for alcoholism in these women is only 16 years of age. Thus, comparison of measures of severity such as age of onset show that when ascertainment criteria are equated for familial density, women show equal severity. These results are consistent with reports using twin samples of a substantial heritability of alcohol dependence in women [18].

If the discrepancy between alcoholism rates among men and women is largely due to variations in environmental pressures to not drink heavily in the first place, one should find widely varying ratios between male and female rates from society to society. This is exactly what one sees in reviewing the ratio of male to female alcoholism across those societies that permit women to drink, which, as noted previously, includes the majority of societies. For example, reanalysis of data provided by epidemiological surveys in Israel and in the U.S. [66] reveals a six-month prevalence male:female ratio for alcoholism of approximately 6:1 in the U.S., versus 14:1 in Israel. Another cross-national study using the Diagnostic Interview Schedule, the instrument originally developed for the U.S. Epidemiological Catchment Area study, compared the lifetime prevalence of alcohol abuse and alcohol dependence in the U.S., Canada, Puerto Rico, Taiwan, and South Korea [67]. This study found male: female ratios of 5.4:1, 4.7:1, 9.8:1, 29:1, and 20:1, respectively. Clearly, cultural variation changes the likelihood that individuals will become alcoholic.

Cultural variation in attitudes towards drinking can change the risk for developing alcoholism in a given population by increasing overall levels of exposure. Interestingly, these attitudes towards drinking are gender dependent and appear to change with the sociocultural environment. This observation is illustrated in an intriguing study in which the Korean version of the DIS was administered to Korean men and women living in two sites, Kangwha, Korea and Yanbian, China [68]. All were native Koreans but lived in different sociocultural environments, with those living in China subject to more traditional and conservative attitudes towards drinking. A significant difference in lifetime prevalence of alcohol abuse and alcohol dependence was reported by the authors. What is of interest with respect to sociocultural influences on female drinking and alcoholism was the widely varying male to female ratios in alcohol abuse. For Kangwha, the ratio was 17.5:1, whereas the ratio for Yanbian, China, where traditional societal values are more salient, was 115:1!

In contrast, in those societies that are more permissive of drinking, the male: female ratio is much less discrepant. For example, Brown et al. [69] examined alcoholism rates in Southern Cheyenne Indians and found a male to female ratio of 1.7:1. Moreover, the majority of both the male and female alcoholics studied had an early onset (<25 years) of the disorder. In fact, there are reports of females drinking more than males in Native American samples, as noted by Weibel-Orlando [70]. She found that Sioux women drank more frequently than Sioux men and appeared to consume more alcohol per occasion than did the men.

Because alcohol use is influenced by cultural restraints, it is not surprising that women are less likely to drink abusively and to consume less alcohol overall than men. However, if one form of alcohol dependence shows greater genetic mediation, then one might expect that the manifestations of alcoholism would be equivalent in men and women once they become alcoholic. Several studies appear to support this conclusion. The *World Health* collaborative project has been collecting data concerning the prevalence of harmful levels of alcohol consumption in six countries: Australia, Bulgaria, Kenya, Mexico, Norway, and the

U.S. For all national groups the level of consumption for men is significantly higher than that for women. However, among alcoholics the average daily intake for women was 74% of the intake for men (176 g for women and 200 g for men). Correcting for body weight differences indicates that women alcoholics may be drinking more in terms of gram per kilogram intake than are alcoholic men.

Further, Kawakami *et al.* [71] found that among a sample of over 2500 employees of a computer factory in Tokyo, 15% of the men and 6% of the women could be classified as having alcohol problems based on an alcoholism screening test (Kurihama Alcoholism Screening Test) designed for use among Japanese. The test embodies similar concepts to those used in DSM-III definitions of alcoholism. Importantly, there was no significant gender difference in the prevalence of alcohol-related problems for a given amount of alcohol consumption. These results support the conclusions reached by Edwards *et al.* [72] that women have the same susceptibility for alcohol-related problems when they drink the same amount as males.

B. Genetic Factors

1. Family History of Substance Abuse

As noted previously, women with a family history of alcohol dependence are at greater risk for developing alcohol problems (See Section III). Among the more intriguing notions concerning genetic effects is the possibility that individuals with genes predisposing them to certain temperaments or clinical conditions may choose their environment. For example, while it is well known that peer pressure to use alcohol and drugs is a significant factor in the development of substance use among youth, there is some evidence that adolescents predisposed to such use may select their friends on this basis. Thus, environmental effects are not random. Women who select substance abusing partners are also changing their environment so that such effects must not be considered random. Thus, the genotype by environment interactions (G X E) for women may be a critical aspect of whether they begin drinking heavily or go on to alcohol dependence.

2. Risk Markers for Susceptibility to Substance Dependence in Women

Several research efforts have been initiated since the early 1980s in an attempt to identify reliable and valid markers for alcoholism risk. Broadly defined, these markers appear to belong within the domains of personality or temperament, on the one hand, and cognition on the other. A search for risk markers for drug dependence in women has not been as intense. The following review will focus, therefore, on alcohol dependence in women.

a. EVENT-RELATED POTENTIALS AND RISK FOR DEVELOPING ALCOHOL AND DRUG DEPENDENCE. Event-related potential (ERP) characteristics, including particular components of the waveform (*e.g.,* P300), are of interest for several reasons. P300 is a scalp-positive wave that occurs after an informative event. These components are of particular interest because, first, long-latency components of ERPs, including P300, are associated with particular sensory and cognitive aspects of information processing. Second, the ERP waveform appears to be under genetic control [73,74]. Although there has been some controversy surrounding whether P300 is a risk marker for later development of alcoholism and/or drug dependence, it should be noted that there are now several laboratories besides our own [77–79] showing consistent results [75,76]. Decrements in P300 are usually seen when high-risk minor children are tested with sufficiently difficult paradigms, and when the density of the alcoholism in the high-risk group is sufficient to allow the variation between high-risk (HR) and low-risk (LR) groups to be explored (see Hill and Steinhauer [78] and Steinhauer and Hill [79], for discussion).

b. RISK TO CHILDREN OF FEMALE ALCOHOLICS VERSUS CHILDREN OF MALE ALCOHOLICS. Our laboratory has been in a unique position to compare children of alcoholics from male alcoholism families with children from female alcoholism families. As part of a large scale family study of alcoholism in which families are identified through multigenerational alcoholism and the presence of at least two alcoholic brothers per family, we are studying children of male alcoholics prospectively (NIAAA Grant AA 05909-15). We have observed remarkable consistency in our findings of reduced P300 in these HR children, using both auditory and visual oddball paradigms [77–79].

Using a similar ascertainment scheme requiring multiple alcoholic family members, we have also been studying families of alcoholic women (NIAAA Grant AA 08082-08). We have chosen this design because we believe that in those families where female alcoholism is multigenerational, with multiple affected relatives, one might expect an early onset of the problem. These families are much more likely to transmit alcoholism to the next generation, thereby increasing risk to offspring. Also, because this subtype is more severe, a minority of affected individuals will achieve remission simply as part of a "maturing out" process.

Analysis of data obtained from offspring of female proband alcoholism families indicated, once again, that the amplitude of the P300 component is smaller in HR children (p = 0.02) when compared with control children [80]. Because of the potential importance of these findings in identifying the highest risk children for possible intervention, we looked for other explanations for why children from female alcoholism families would show reduced P300. The variables we considered were: (1) drinking during pregnancy; (2) presence of "other psychopathology" in mothers, especially antisocial personality disorder that has been linked to P300 reduction by some, but not all, laboratories; and (3) lower socioeconomic status. None of these possible sources of variance contributed significantly to the results obtained, leading us to conclude that a diathesis for alcoholism was a strong predictor of P300 reduction in children from these female alcoholism families.

Establishing that P300 deficits occur in female alcoholics is another important step. Our laboratory has an extensive data base containing information on adult relatives of male proband alcoholics (both alcoholic and nonalcoholic) and low-risk subjects [81,82]. A smaller data set of high-risk women has enabled us to determine that those who develop alcoholism exhibit a 50% reduction in the amplitude of P300 in comparison to their nonalcoholic sisters and to controls in response to information processing demands [83]. Recent history of drinking (past 7 and past 30 days) was entered into these analyses as covariants without significant effect. Thus, absence of a P300 decrement might be protective for the high-risk nonalcoholic sister. Alternatively, exposure to alcohol or drugs during adulthood may alter cognitive functioning. (The contaminating factor of alcohol and drug exposure is removed in studies of high-risk children who have not begun to use alcohol or drugs.)

3. Metabolism

Alcohol metabolism may play a role in the development of alcoholism. Women may be at greater risk for developing alcohol dependence than men because they produce less alcohol dehydrogenase (ADH) in the gastrointestinal tract relative to men [9]. Although most of the alcohol a woman consumes is metabolized in the liver, nevertheless, variation in ADH in the gastrointestinal tract has implications for how quickly she will become intoxicated.

4. Temperament and Personality

There is ample evidence that personality traits are genetically mediated [84]. Although there has been the suggestion that an "alcoholic personality" is a predisposing factor for development of alcoholism, solid evidence for this is difficult to obtain. The long term effects of chronic use of alcohol and drugs can alter personality so one does not know if the assessed personality of the chronic alcohol dependent women is cause or consequence. One strategy for uncovering possibly important temperament characteristics is to assess the nonalcoholic young offspring of alcoholics or drug dependent women, or their nondependent relatives for clues regarding important temperament characteristics. Work in our laboratory suggests that traits of alienation, or the tendency to believe negative outcomes are the result of "bad luck" and to view others' motives with suspicion, may be a characteristic that is heritable within alcoholic families. Negative affectivity has also emerged as an important characteristic within families with multiple cases of alcohol dependence, a trait that appears to be linked to one of the dopamine receptor genes [47].

IX. Clinical Issues

A. Diagnosis and Case Finding

Epidemiological studies find cases of alcoholism and problem drinking in women through interviewing representative samples of the general population. To the extent that reliable estimates of problems are revealed by such surveys, comparison with other case-finding methods (e.g., treatment facilities) can be made. There are four common methods for finding cases: drinking driver rehabilitation programs, public inebriate programs, employee assistance programs (EAPs), and medical visits. All four of these case-finding methods, which could provide opportunities for intervention for women, have a common problem; they have difficulty reaching women. The drinking driver rehabilitation program and the employee assistance program fail to identify women with problems more often than is the case with men [85]. The public inebriate programs are not as effec-

tive with women because public drunkenness occurs more frequently among male than female alcoholics [86]. Even among heavy drinkers, gender differences in legal problems have been found. Perkins [87], in a study of a college-age population, found that property damage, injury to others, fighting, and impaired driving were reported substantially more often by male than female heavy drinkers.

The opportunity for case finding and intervention in medical settings would appear to be great. Two studies have estimated the prevalence of alcohol abuse or dependence in U.S. community hospitals to be between 20–26% of all admissions [88,89]. Unfortunately, though alcohol-related medical problems (e.g., cirrhosis) facilitate the identification of problem drinkers and alcoholics [90], women with alcohol problems are less likely to be identified in a medical setting [91].

B. Treatment Issues and Outcome

1. The Proportion of Women Treated in Comparison to Those Affected

The ECA study provided estimates of the likelihood that individuals who met criteria for specific psychiatric disorders would receive treatment [92]. Results of that survey showed that a number of psychiatric disorders (e.g., schizophrenia) were brought to the attention of a physician more frequently than were alcohol or drug problems. Data for both sexes were combined for the category "ever mention symptom to a doctor," with 15% and 18% reporting a positive response for alcohol abuse and drug abuse, respectively. Furthermore, 14% of women with alcohol abuse received some form of treatment whereas 9% of men did. For drug abuse, 7% of women and 11% of men received treatment. Thus, it appears that only a minority of women with drug or alcohol abuse problems receive treatment.

2. Types of Treatment

A significant portion of women with alcohol or drug dependence problems are treated through self-help groups such as Alcoholics Anonymous (AA) or Narcotics Anonymous (NA). Although both AA and NA are quite popular, minimal study of their effectiveness has been undertaken. However, there has been a systematic evaluation of AA by Walsh and colleagues [93] in the context of a multisite collaborative study [94] sponsored by the National Institute on Alcoholism and Alcohol Abuse. Comparison of three psychosocial treatments (cognitive-behavioral therapy, motivational enhancement treatment, and AA) showed that all were effective in increasing the number of days of abstinence in the first year and in reducing the number of drinks on any day they did drink. Both Project MATCH [94] and the Walsh et al. study [93] support the efficacy of the AA approach, which also has a reduced cost. Also, the study by Walsh and colleagues [93] showed the best treatment outcomes in patients assigned to inpatient treatment. They found that over 60% of AA clients ultimately required inpatient care. However, few women were included in this study, precluding data analysis by gender. One of the traditional barriers to treatment, which affects women more than men, is the inability to enter inpatient facilities because of issues of childcare responsibility [95]. Without a reliable adult to take over child-rearing responsibilities while the woman is in treatment, inpatient treatment is often rejected.

3. Outcome of Treatment for Women

Few studies have compared the treatment outcomes of men and women [96]. Rarely have treatment modalities or intensity of treatment (inpatient versus outpatient) been studied in the context of gender differences. However, one large study that compared 1675 men and 684 women [97] found no difference in outcome at one year. More research is needed to determine the most effective methods of treatment for women. However, there are known personal factors that appear to be reasonably good predictors of favorable and unfavorable outcomes for women. Smith and Cloninger [98] found that a history of loss of control drinking, delirium tremens, early onset alcoholism, lack of employment, diagnosis of antisocial personality disorder, and being unmarried or separated were all predictors of poor outcome.

X. Epidemiological Issues

A. Heterogenity of the Disorder

It is becoming clear that alcohol and drug dependent individuals frequently meet criteria for concomitant psychiatric disorders [2,27]. For this reason, there is substantial heterogeneity among both substance abusing women and men. Our notions about the prototypic characteristics of drug- and alcohol-dependent women and the most efficacious methods for case finding and intervention may be related to the specific subtype of substance abuse a particular woman may have. For example, we know that women with depressive symptoms are more likely to seek treatment [99]. Also, the effect of having concomitant depression has differential effects in women compared to men. Rounsaville *et al.* [100] found that depression in male alcoholics was associated with poorer outcome whereas for female alcoholics the reverse was true; those with depression had better outcomes. Thus, the conclusions that can be reached concerning these and other gender issues are limited by the more modest available literature on substance abusing women compared to men.

B. Public Health Implications

There is a growing literature on prevention of alcohol and drug problems, some of which discuss methods for reaching women and young girls [101,102]. The common theme among most of those targeting women has been the issue of whether special programs are needed for women, and if so, are there techniques that work best with women? Also, the issue arises about which subpopulations should be targeted. Experiences in the treatment of alcohol dependence would suggest that in the area of prevention, special programs focused on women would be most effective. Dahlgren and Willander [103] conducted a controlled study of 200 alcoholic women randomly assigned either to a speciality treatment program for women or to the standard alcoholism treatment at the same facility. After 2 years of follow-up, the women attending the specialized treatment program showed greater improvement in drinking outcome, job stability, and relationships with children. Thus, it may be more appropriate to develop prevention programs specifically aimed at women.

Also, of considerable importance is the need to break the intergenerational transmission of alcohol and drug problems among the offspring of alcohol- and drug-dependent women. The observation that offspring of alcoholic women have greater psychopathology than do children of alcoholic men, as ongoing research in our laboratory has found [49], suggests that intervening in the lives of offspring of substance abusing women is especially important.

XI. Summary and Conclusions/Future Directions

This review has summarized the literature concerning the prevalence of alcohol and drug dependence in women in comparison to men, showing that men outnumber women for both. It is argued that while the number of women who are affected is fewer, the societal costs of women's alcohol and drug problems may be greater. Women appear to incur greater health consequences, including higher rates of mortality, if they become alcoholic than do alcoholic men [104]. Because women provide most of the child rearing, even in contemporary society, impaired performance of this role will have a significant impact on the likelihood that substance abuse behaviors will be transmitted to offspring. Also, offspring of alcoholic mothers show increased rates of both internalizing and externalizing behaviors during childhood and adolescence.

Evidence has been presented which suggests that societal pressures may reduce the number of heavy drinking women in any particular society, but once the threshold for heavy drinking is crossed, women appear to be as likely to have alcohol dependence problems as men. Furthermore, there is evidence that, like the situation with male alcoholics, there are at least two types of alcohol dependence: one that is primarily nongenetic in origin, arising out of late-life stressors, and another that appears to depend on the woman's family history of alcohol dependence and is more likely genetically mediated. Of course, the existence of these two types is for heuristic purposes; in fact, risk is undoubtedly on a continuum. Nevertheless, recognition of these two types in women enhances our understanding that women can have severe forms of alcohol dependence with an early onset (often in late adolescence or early adulthood) and with severe physical and psychological consequences. Less is known about drug dependence in women from a genetic/familial point of view. However, data from the Iowa adoption study suggest that genetic factors play an important role in the etiology of drug dependence in women as well.

Acknowledgments

Supported by grants from the National Institute on Alcohol Abuse and Alcoholism AA05909-14, AA08082-08, and AA11304-01.

References

1. Robins, L. N., Helzer, J. E., Pryzbeck, T. R., and Regier, D. A. (1988). Alcohol disorders in the community: A report from the Epidemiological Catchment Area. *In* "Alcoholism: Origins and Outcome" (R. M. Rose and J. E. Barrett, eds.), pp. 15–30. Raven Press, New York.
2. Kessler, R. C., McGonagle, K. A., Zhao, S., Nelson, C. B., Hughes, M., Eshleman, S., Wittchen, H. U., and Kendler, K. S. (1994). Lifetime and 12-month prevalence of DSM-III-R psychi-

atric disorders in the United States. *Arch. Gen. Psychiatry* **51**, 8–19.

3. Grant, B. F. (1997). Prevalence and correlates of alcohol use and DSM-IV alcohol dependence in the United States: Results of the national longitudinal alcohol epidemiologic survey. *J. Stud. Alcohol* **58**, 464–473.

4. Kaufman, M. J., Levin, J. M., Ross, M. H., Lange, N., Rose, S. L., Kukes, T. J., Mendelson, J. H., Lukas, S. E., Cohen, B. M., and Renshaw, P. F. (1998). Cocaine-induced cerebral vasoconstriction detected in humans with magnetic resonance angiography. *JAMA, J. Am. Med. Assoc.* **279**, 376–380.

5. Feighner, J. P., Robins, E., Guze, S. B., Woodruff, R. A., Jr., Winokur, G., and Munoz, R. (1972). Diagnostic criteria for use in psychiatric research. *Arch. Gen. Psychiatry* **26**, 57–63.

6. Edwards, G., and Gross, M. M. (1976). Alcohol dependence: Provisional description of a clinical syndrome. *Br. Med. J.* **1**, 1058–1061.

7. Clark, W. B., and Cahalan, D. (1976). Changes in problem drinking over a four-year span. *Addict. Behav.* **1**, 251–259.

8. Report of a WHO Group of Investigators (1977). On criteria for identifying and classifying disabilities related to alcohol consumption, in alcohol-related disabilities. *In* "World Health Organization" (G. Edwards, M. M. Gross, M. Keller, *et al.*, eds.), p. 10. WHO, Geneva.

9. Frezza, M., di Padova, C., Pozzato, G., Terpin, M., Baraona, E., and Lieber, C. S. (1990). High blood alcohol levels in women. The role of decreased gastric alcohol dehydrogenase activity and first-pass metabolism. *N. Engl. J. Med.* **322**, 95–99.

10. Merikangas, K. R., Leckman, J. F., Prusoff, B. A., *et al.* (1985). Familial transmission of depression and alcoholism. *Arch. Gen. Psychiatry* **42**, 367–372.

11. Cadoret, R. J., Troughton, E., O'Gorman, T. W., and Heywood, E. (1986). An adoption study of genetic and environmental factors in drug abuse. *Arch. Gen. Psychiatry* **43**, 1131–1136.

12. Bohman, M., Sigvardsson, S., and Cloninger, C. R. (1981). Maternal inheritance of alcohol abuse: Cross-fostering analysis of adopted women. *Arch. Gen. Psychiatry* **38**, 965–969.

13. Heath, A. C., Jardine, R., and Martin, N. G. (1989). Interactive effects of genotype and social environment on alcohol consumption in female twins. *J. Stud. Alcohol* **50**, 38–48.

14. Gurling, H. M. D., Murray, R. M., and Clifford, C. A. (1981). Investigations into the genetics of alcohol dependence and into its effects on brain function. *In* "Twin Research 3: Epidemiological and Clinical Studies" (L. Gedda, P. Parisi, and W. E. Nance, eds.). Liss, New York.

15. McGue, M., Pickens, R. W., and Svikis, D. S. (1992). Sex and age effects on the inheritance of alcohol problems: A twin study. *J. Abnorm. Psychol.* **101**, 3–17.

16. Pickens, R. W., Svikis, D. S., McGue, M., Lykken, D. T., Heston, L. L., and Clayton, P. J. (1991). Heterogeneity in the inheritance of alcoholism: A study of male and female twins. *Arch. Gen. Psychiatry* **48**, 19–28.

17. Pickens, R. W., and Svikis, D. S. (1988). The twin method in the study of vulnerability to drug abuse. *Bio. Vulnerabil. Drug Abuse, Res. Monog.* **89**, 41–51.

18. Kendler, K. S., Heath, A. C., Neale, M. C., Kessler, R. C., and Eaves, L. J. (1992). A population-based twin study of alcoholism in women. *JAMA, J. Am. Med. Assoc.* **268**, 1877–1882.

19. Cloninger, C. R., Bohman, M., and Sigvardsson, S. (1981). Inheritance of alcohol abuse: Cross-fostering analysis of adopted men. *Arch. Gen. Psychiatry* **38**, 861–868.

20. Hill, S. Y., and Smith, T. R. (1991). Evidence for genetic mediation of alcoholism in women. *J. Subst. Abuse* **3**, 159–174.

21. Glenn, S. W., and Nixon, S. J. (1991). Applications of Cloninger's subtypes in a female alcoholic sample. *Alcohol. Clin. Exp. Res.* **15**, 851–857.

22. Lex, B. W., Sholar, J. W., Bower, T., and Mendelson, J. H. (1991). Putative type II alcoholism characteristics in female third DUI offenders in Massachusetts: A pilot study. *Alcohol* **8**, 283–287.

23. Schuckit, M. (1983). Alcoholic patients with secondary depression. *Am. J. Psychiatry* **140**, 711–714.

24. Ross, H. E., Glaser, F. B., and Germanson, T. (1988). The prevalence of psychiatric disorders in patients with alcohol and other drug problems. *Arch. Gen. Psychiatry* **45**, 1023–1031.

25. Merikangas, K. R., and Gelernter, C. S. (1990). Comorbidity for alcohol and depression. *Psychiatr. Clin. North Am.* **13**, 613–632.

26. Rounsaville, B. J., and Kleber, H. D. (1985). Untreated opiate addicts: How do they differ from those seeking treatment? *Arch. Gen. Psychiatry* **42**, 1072–1077.

27. Helzer, J. E., and Pryzbeck, T. R. (1988). The co-occurrence of alcoholism with other psychiatric disorders in the general population and its impact on treatment. *J. Stud. Alcohol* **49**, 219–224.

28. Helzer, J. E., Burnam, A., and McEvoy, L. T. (1991). Alcohol abuse and dependence. *In* "Psychiatric Disorders in America: The Epidemiological Catchment Area Study" (L. N. Robins and D. A. Regier, eds.), pp. 81–115. Free Press, New York.

29. Hesselbrock, M. N. (1991). Gender comparison of antisocial personality disorder and depression in alcoholism. *J. Subst. Abuse* **3**, 205–219.

30. Hammer, T., and Vaglum, P. (1989). The increase in alcohol consumption among women: A phenomenon related to accessibility or stress? A general population study. *Br. J. Addict.* **84**, 767–775.

31. Pakesch, G., Loimer, N., Rasinger, E., Tutsch, G., and Katschnig, H. (1989). The prevalence of psychoactive drug intake in a metropolitan population. *Pharmacopsychiatry* **22**, 61–65.

32. Hill, S. Y. (1993). The health implications of substance abuse in women. *In* "Women and Substance Abuse: A Gender Analysis and Review of Health and Policy Implications." World Health Organization, Geneva.

33. Hill, S. Y. (1984). Vulnerability to the biomedical consequences of alcoholism and alcohol-related problems among women. *In* "Alcohol Problems in Women" (S. Wilsnack and L. Beckman, eds.), pp. 121–154. Guilford Press, New York.

34. Smith, E. M., Cloninger, C. R., and Bradford, S. (1983). Predictors of mortality in alcoholic women: A prospective follow-up study. *Alcohol.: Clin. Exp. Res.* **7**, 237–243.

35. Bratfos, O. (1971). Attempted suicide. *Acta Psychiatr. Scand.* **47**, 38–56.

36. Gibbs, J. P. (1966). Suicide. *In* "Contemporary Social Problems" (R. K. Merton and R. A. Nisbet, eds.), 2nd ed., pp. 281–321. Harcourt, Brace and World, New York.

37. Rossow, I. (1994). Suicide among drug addicts in Norway. *Addiction* **89**, 1667–1673.

38. Kirkpatrick-Smith, J., Rich, A., and Bonner, R. (1989). Alcohol abuse and suicide ideation in adolescents. *Am. Assoc. Suicidol.,* San Diego, CA.

39. Berman, A. L., and Schwartz, R. H. (1990). Suicide attempts among adolescent drug users. *Am. J. Dis. Child.* **144**, 310–314.

40. Crumley, F. E. (1990). Substance abuse and adolescent suicidal behavior. *JAMA, J. Am. Med. Assoc.* **263**, 3051–3056.

41. Brent, D. A., Perper, J. A., and Allman, C. J. (1987). Alcohol, firearms, and suicide among youth. Temporal trends in Allegheny County, Pennsylvania, 1960 to 1983. *JAMA, J. Am. Med. Assoc.* **257**, 3369–3372.

42. Clarkin, J. F., Friedman, R. C., Hurt, S. W., Corn, R., and Aronoff, M. (1984). Affective and character pathology of suicidal adolescent and young adult inpatients. *J. Clin. Psychiatry* **45**, 19–22.

43. Millstein, S. G., Irwin, C. E., Jr., Adler, N. E., Cohn, L. D., Kegeles, S. M., and Dolcini, M. M. (1992). Health-risk behaviors and health concerns among young adolescents. *Pediatrics* **89**, 422–428.

44. Cloninger, C. R. (1987). A systematic method for clinical description and classification of personality variants: A proposal. *Arch. Gen. Psychiatry* **44,** 573–588.

45. Ebstein, R. P., Novick, O., Umansky, R., Priel, B., Osher, Y., Blaine, D., Bennett, E. R., Nemanov, L., Katz, M., and Belmaker, R. H. (1996). Dopamine D4 receptor (D4DR) exon III polymorphism associated with the human personality trait of novelty seeking. *Nat. Genet.* **12,** 78–80.

46. Benjamin, J., Li, L., Patterson, C., Greenberg, B. D., Murphy, D. L., and Hamer, D. H. (1996). Population and familial association between the D4 dopamine receptor gene and measures of novelty seeking. *Nat. Genet.* **12,** 81–84.

47. Hill, S. Y., Xu, J., Zezza, N., Wipprecht, G., Locke, J., and Neiswanger, K. (1997). Personality traits and dopamine: Linkage studies in families of alcoholics for D2 and D4 dopamine receptors and the DAT1 gene. *Proc. Am. Coll. Neuropharmacol.* **36,** 96.

48. Goodwin, D. W., Schulsinger, F., Hermansen, L., Guze, S. B., and Winokur, G. (1973). Alcohol problems in adoptees raised apart from alcoholic biological parents. *Arch. Gen. Psychiatry* **28,** 238–243.

49. Hill, S. Y., and Muka, D. (1996). Childhood psychopathology in children from families of alcoholic female probands. *J. Am. Acad. Child Adolesc. Psychiatry* **35,** 725–733.

50. Johnson, P. B. (1982). Sex differences, women's roles and alcohol use: Preliminary national data. *J. Soc. Issues* **2,** 93–116.

51. Wilsnack, S. C., Klassen, A. D., and Wright, S. I. (1986). Gender-role orientations and drinking among women in a U.S. national survey. *In* "Proceedings of the 34th International Congress on Alcoholism and Drug Dependence," pp. 242–255. International Council on Alcohol and Addictions, Calgary, Alberta, Canada.

52. Kubicka, L., Csemy, L., and Kozeny, J. (1991). The sociodemographic, microsocial, and attitudinal context of Czech women's drinking. *Pap., Symp. Alcohol, Fam. Significant Others, Soc. Res. Inst. Alcohol Stud. Nord. Counc. Alcohol Drug Res.,* Helsinki, Finland.

53. Wilsnack, R. W., and Wright, S. I. (1991). Women in predominantly male occupations: Relationships to problem drinking. *Pap., Annu. Meet. Soc. Study Soc. Probl.,* Cincinnati, OH.

54. Haavio-Mannila, E. (1991). Impact of colleagues and family members on female alcohol use. *Pap., Symp. Alcohol, Fam. Significant Others, Soc. Res. Inst. Alcohol Stud. Nord. Counc. Alcohol Drug Res.,* Helsinki, Finland.

55. Clark, W. B., and Midanik, L. (1982). Alcohol use and alcohol problems among U.S. adults: Results of the 1979 national survey. *In* "National Institute on Alcohol Abuse and Alcoholism, Alcohol Consumption and Related Problems," Alcohol and Health Monogr. No. 1, DHHS Publ. No. ADM 82-1190, pp. 3–52. U.S. Gov. Printing Office, Washington, DC.

56. Hanna, E., Faden, V., and Harford, T. (1993). Marriage: Does it protect young women from alcoholism? *J. Subst. Abuse* **5,** 1–14.

57. Wilsnack, S. C., Klassen, A. D., Schur, B. E., and Wilsnack, R. W. (1991). Predicting onset and chronicity of women's problem drinking: A five-year longitudinal analysis. *Am. J. Public Health* **81,** 305–318.

58. Saghir, M. T., and Robins, E. (1973). "Male and Female Homosexuality: A Comprehensive Investigation." Williams & Wilkins, Baltimore, MD.

59. Fifeld, L., DeCrscenzo, T. A., and Latham, J. D. (1975). "On My Way to Nowhere: Alienated, Isolated, Drunk." Gay Community Center, Los Angeles.

60. Lohrenz, L., Connely, J., Coyne, L., and Spare, L. (1978). Alcohol problems in several midwest homosexual populations. *J. Stud. Alcohol* **39,** 1959–1963.

61. McKirnan, D. J., and Peterson, P. L. (1989). Alcohol and drug use among homosexual men and women: Epidemiology and population characteristics. *Addict. Behav.* **14,** 545–553.

62. Bradford, J., Ryan, C., and Rothblum, E. D. (1994). National lesbian health care survey: Implications for mental health care. *J. Consult. Clin. Psychol.* **62,** 228–242.

63. Child, I. L., Barry, H., and Bacon, M. K. (1965). A cross-cultural study of drinking: III. Sex differences. *Q. J. Stud. Alcohol* **3,** 49–61.

64. Heath, D. B. (1991). Women and alcohol: Cross-cultural perspectives. *J. Subst. Abuse* **3,** 175–185.

65. Gilligan, S. B., Reich, T., and Cloninger, C. R. (1987). Etiologic heterogeneity in alcoholism. *Genet. Epidemiol.* **4,** 395–414.

66. Levav, I., Kohn, R., Dohrenwend, B. P., *et al.* (1993). An epidemiological study of mental disorders in a 10-year cohort of young adults in Israel. *Psychol. Med.* **23,** 691–707.

67. Helzer, J. E., Canino, G. J., Yeh, E. K., *et al.* (1990). Alcoholism—North America and Asia: A comparison of population surveys with the diagnostic interview schedule. *Arch. Gen. Psychiatry* **47,** 313–319.

68. Namkoong, K., Lee, H. Y., Lee, M. H., *et al.* (1991). Cross-cultural study of alcoholism: Comparison between Kangwha, Korea and Yanbian, China. *Yonsei Med. J.* **32,** 319–325.

69. Brown, G. L., Albaugh, B. J., Robin, R. W., *et al.* (1993). Alcoholism and substance abuse among selected Southern Cheyenne Indians. *Cult. Med. Psychiatry* **16,** 531–542.

70. Weibel-Orlando, J. (1996). Women and alcohol: Special populations and cross-cultural variations. *In* "Women and Alcohol: Health Related Issues," pp. 161–187. National Institute on Alcohol Abuse and Alcoholism, Rockville, MD.

71. Kawakami, N., Haratani, T., Hemmi, T. *et al.* (1992). Prevalence and demographic correlates of alcohol-related problems in Japanese employees. *Soc. Psychiatry Psychiatr. Epidemiol.* **27,** 198–202.

72. Edwards, G., Chandler, J., Hensman, C., *et al.* (1972). Drinking in a London suburb: II. Correlates of trouble with drinking among men. *Q. J. Stud. Alcohol* **6**(Suppl.), 94–119.

73. Aston, C. E., and Hill, S. Y. (1990). A segregation analysis of the P300 component of the event-related potential. *Am. J. Hum. Genet.* **47**(Suppl.), A127.

74. van Beijsterveldt, C. E. M. (1996). "The Genetics of Electrophysiological Indices of Brain Activity: An EEG Study in Adolescent Twins." University of Amsterdam.

75. Begleiter, H., Porjesz, B., Bihari, B., *et al.* (1984). Event-related brain potentials in boys at risk for alcoholism. *Science* **225,** 1493–1496.

76. Berman, S. M., Whipple, S. C., Fitch, R. J., *et al.* (1993). P300 in young boys as a predictor of adolescent substance use. *Alcohol* **10,** 69–76.

77. Hill, S. Y., Steinhauer, S. R., Park, J., *et al.* (1990). Event-related potential characteristics in children of alcoholics from high density families. *Alcohol.: Clin. Exp. Res.* **14,** 6–16.

78. Hill, S. Y., and Steinhauer, S. R. (1993). Assessment of prepubertal and postpubertal boys and girls at risk for developing alcoholism with P300 from a visual discrimination task. *J. Stud. Alcohol* **54,** 350–358.

79. Steinhauer, S. R., and Hill, S. Y. (1993). Auditory event-related potentials in children at high risk for alcoholism. *J. Stud. Alcohol* **54,** 408–421.

80. Hill, S. Y., Muka, D. H., Steinhauer, S. R., and Locke, J. (1995). P300 amplitude decrements in children from families of alcoholic female probands. *Biol. Psychiatry* **38,** 622–632.

81. Hill, S. Y., Steinhauer, S. R., Zubin, J., *et al.* (1988). Event-related potentials as markers for alcoholism risk in high density families. *Alcohol.: Clin. Exp. Res.* **12,** 545–554.

82. Hill, S. Y., Steinhauer, S. R., and Locke, J. (1995). Event-related potentials in alcoholic men, their high-risk male relatives and low-risk male controls. *Alcohol.: Clin. Exp. Res.* **19,** 567–576.

83. Hill, S. Y., and Steinhauer, S. R. (1993). Event-related potentials in women at risk for alcoholism. *Alcohol* **10,** 349–354.

84. Eaves, L. J., Eysenck H. J., and Martin, N. G. (1989). "Genes, Culture and Personality: An Empirical Approach." Academic Press, San Diego, CA.

85. Reichman, W. (1983). Affecting attitudes and assumptions about women and alcohol problems. *Alcohol Health Res. World* **7**, 6–10.

86. Nirenberg, T. D., and Gomberg, E. S. L. (1993). Antecedents and consequences. *In* "Women and Substance Abuse" (E. S. L. Gomberg and T. D. Nirenberg, eds.), pp. 118–141. Ablex, Norwood, NJ.

87. Perkins, H. W. (1991). Gender patterns in consequences of collegiate alcohol abuse: A 10 year study of trends in an undergraduate population. *J. Stud. Alcohol* **52**, 458–462.

88. Bush, B., Shaw, S., Cleary, P., Delbanco, T. L., and Aronson, M. D. (1987). Screening for alcohol abuse using the CAGE questionnaire. *Am. J. Med.* **82**, 231–235.

89. Muller, A. (1996). Alcohol consumption and community hospital admissions in the United States: A dynamic regression analysis, 1950–1992. *Addiction* **91**, 321–342.

90. Moore, R. D., Bone, L. R., Geller, G., Mamon, J. A., Stokes, E. J., and Levine, D. M. (1989). Prevalence, detection, and treatment of alcoholism in hospitalized patients. *JAMA, J. Am. Med. Assoc.* **261**, 403–407.

91. Amodei, N., Williams, J. F., Seale, J. P., and Alvarado, M. L. (1996). Gender differences in medical presentation and detection of patients with a history of alcohol abuse or dependence. *J. Addict. Dis.* **15**, 19–31.

92. Robins, L. N., Locke, B. Z., and Regier, D. A. (1991). An overview of psychiatric disorders in America. *In* "Psychiatric Disorders in America: The Epidemiologic Catchment Area Study" (L. N. Robins and D. A. Regier, eds.), pp. 328–366. Free Press, New York.

93. Walsh, D. C., Hingson, R. W., Merrigan, D. M., *et al.* (1991). A randomized trial of treatment options for alcohol-abusing workers. *N. Engl. J. Med.* **325**, 775–782.

94. Project MATCH Research Group (1997). Matching alcoholism treatments to client heterogeneity: Project MATCH posttreatment outcomes. *J. Stud. Alcohol* **58**, 7–29.

95. Beckman, L. J., and Amaro, H. (1984). Patterns of women's use of alcohol treatment agencies. *In* "Alcohol Problems in Women" (S. Wilsnack and L. J. Beckman, eds.), pp. 319–348. Guilford Press, New York.

96. McCrady, B. S., and Raytek, H. (1993). Women and substance abuse: Treatment modalities and outcomes. *In* "Women and Substance Abuse" (E. S. L. Gomberg and T. D. Nirenberg, eds.), pp. 314–338. Ablex, Norwood, NJ.

97. Filstead, W. J. (1990). "Treatment Outcome: An Evaluation of Adult and Youth Treatment Services." Parkside Medical Services Corporation, Park Ridge, IL.

98. Smith, E. M., and Cloninger, C. R. (1984). A prospective twelve-year follow-up of alcoholic women: A prognostic scale for long-term outcome. *NIDA Res. Monogr.* **55**, 245–251.

99. Vaglum, S., Vaglum, P., and Larson, O. (1987). Depression and alcohol consumption in nonalcoholic and alcoholic women. *Acta Psychiatr. Scand.* **75**, 577–584.

100. Rounsaville, B., Dolinsky, Z., Babor, T., and Meyer, R. (1987). Psychopathology as a predictor of treatment outcome in alcoholics. *Arch. Gen. Psychiatry* **44**, 505–513.

101. Bry, B. H. (1984). Substance abuse in women: Etiology and prevention. *In* "Social and Psychological Problems of Women: Prevention and Crisis Intervention" (A. U. Rickel, M. Gerrard, and I. Iscoe, eds.), pp. 253–272. Hemisphere Publishing, Washington, DC.

102. Morrisey, E. R. (1986). Of women, by women or for women? Selected issues in the primary prevention of drinking problems. *In* "Women and Alcohol: Health-Related Issues," Res. Monogr. No. 16, DHHS Publ. No. ADM 86-1139, pp. 226–259. National Institute on Alcohol Abuse and Alcoholism, Washington, DC.

103. Dahlgren, L., and Willander, A. (1989). Are special treatment facilities for female alcoholics needed? A controlled 2-year study from a specialized female unit (EWA) versus a mixed male/female treatment facility. *Alcohol.: Clin. Exp. Res.* **13**, 499–504.

104. Hill, S. Y. (1995). Mental and physical health consequences of alcohol use in women. *Recent Dev. Alcohol.* **12**, 181–197.

Section 13

POORLY UNDERSTOOD CONDITIONS

Karen B. Schmaling
Department of Psychiatry and Behavioral Sciences
University of Washington
Seattle, Washington

In this section, experts on chronic fatigue syndrome (CFS), fibromyaglia (FM), multiple chemical sensitivity (MCS), headache, temporomandibular disorders (TMD), irritable bowel syndrome (IBS), and interstitial cystitis (IC) provide up-to-date reviews of their topics. The reviews address issues of definition and diagnosis, pathogenesis, epidemiology, and treatment, with a special emphasis on the health effects for women. These conditions share a number of common characteristics: they are chronic, ascertained by patient report, lack objective diagnostic markers, have symptoms out of proportion to demonstrable physiological abnormalities, occur more frequently among women than men, occur frequently in conjunction with psychiatric disorders, are associated with significant functional impairment, and lack effective treatment strategies.

I. Issues with Case Definition and Ascertainment

There is variability with the method with which case definitions have been developed among the different disorders. For example, the development of the case definitions for CFS, FM, TMD, and headache were sponsored by professional organizations (the Centers for Disease Control and Prevention, the American College of Rheumatology, the American Dental Association, and the International Headache Society, respectively). By contrast, the case definitions of other disorders, such as IBS, have been developed by groups of interested individuals not convened under the aegis of a professional organization. Some definitions are based more on empirical data than others; for example, the case definition for CFS makes no reference to supporting data, whereas the FM case definition is more strongly

rooted in empirical knowledge. Finally, some conditions lack an accepted case definition, such as MCS. More work is needed to define reliably applicable case definitions. The development of case definition subtypes (*e.g.,* as has been done for headache) and multiaxial systems (*e.g.,* for TMD) that code both clinical diagnoses and psychosocial profile would be useful for both clinicians and researchers to decrease heterogeneity and help place these syndromes in broader contexts.

In addition to shared characteristics, these disorders have shared, or overlapping, symptoms. The degree of symptom overlap is so extensive among some of these conditions that the diagnosis may depend on which symptom is emphasized the most by the patient and on the specialty of the physician. Fatigue, joint pain, muscle pain, and cognitive difficulties are criteria for CFS [1] and also are common among persons with MCS and FM [2]. Pain complaints are common to all disorders: pain is a central feature for headache, FM, TMD, IBS, and IC, and a common, if not defining, feature of CFS and MCS.

That the case definitions of these syndromes are in various states of evolution (with the case definition for headache representing perhaps the most sophisticated level of development among these conditions and for MCS the least) presents challenges for the conduct of appropriate studies of the epidemiology of the syndromes. The extensive reliance on patient perceptions also creates difficulties in case ascertainment. Patients with these conditions typically seek care from specialty physicians; proper epidemiology studies depend on the identification of unbiased samples. Few epidemiology studies of any of these conditions have been conducted utilizing appropriate methodology. These conditions occur more frequently among women than men, averaging a 2:1 female-to-male ratio that tends to be more female predominant among certain conditions (headache, FM) and when nonpopulation-based samples or sampling techniques are used. Future studies must attend to the role of health care seeking behavior in case ascertainment.

II. The Role of Perceptual Factors

Patient reports of symptoms are strongly influenced by perceptual factors in addition to cultural, socioenvironmental, and other factors. Perception of any stimulus depends on the intensity of the stimulus and on person characteristics including cognitive (*e.g.,* the person's attentional focus; interpretive style) and affective factors (*e.g.,* mood). Several of the chapters in this section describe research efforts that involve the presentation of a standardized stimulus and measurement of subjects' responses to the stimulus. For example, there is evidence that patients with IBS have lower pain thresholds than healthy control subjects in response to bowel distention [3] but do not differ in their pain thresholds to peripheral stimuli [4]. This pattern of results may suggest a symptom-specific response threshold. However, among patients with other illnesses, there is evidence of symptom amplification. For example, complaints of poor memory and concentration are common among patients with CFS; however, standardized neuropsychological tests generally reveal normal ranging functioning except for subtle difficulties in complex information processing [5]. A similar pattern has been noted during exercise: perceived exertion among patients with CFS is greater than that among sedentary controls, despite similar exercise capacities [6]. Furthermore, research does not support a more sensitive olfactory threshold among patients with MCS compared with healthy control subjects [7]. Patients with FM have lower thresholds to pain, an illness-specific stimulus, but also to noise, a nonspecific stimulus [8], suggesting generalized sensitivity or hypervigilance. Further research efforts along these lines are needed, not to question the patient's experience but to better understand the covariates of these important perceptual processes. For example, the systematic examination of gender differences in perception, while also examining cognitive and affective factors, is needed.

III. Common Illness Risk Factors and Consequences and Their Implications for Treatment

The examination of risk factors associated with the syndromes is in an early stage of investigation. Commonly recognized comorbid conditions, such as psychiatric disorders, are neither sufficient to explain the illnesses nor present among all patients with these illnesses. Several of the reviews in this section do suggest that trauma and victimization may be related to the disorders. For example, a history of sexual abuse may be associated with sensitization among patients with MCS and may be more common among patients with IBS, FM [9], and CFS [10]. Women are more likely to report illnesses, including those reviewed in this section, seek medical care, and be victimized [11]. Victimization has been associated repeatedly with neuroendocrine dysregulation [12] that in turn is an area of active investigation among several of these illnesses. Further identification of potential common pathways is needed.

A common consequence of the conditions in this section is avoidance: patients with pain problems naturally tend to avoid triggers that may result in pain (*e.g.,* exercise); chemically sensitive patients may avoid strong smells and odors; patients with IBS and TMD may avoid certain foods. Avoidance probably contributes to the functional limitations and lifestyle restrictions observed among these patients. The recommended treatment approaches to the conditions in this section generally combine conservative medical treatment and nonmedication treatments with the goal of enhancing functional status, not necessarily symptom relief. Nonmedication treatment approaches typically involve gradually reintroducing that which has been avoided (*e.g.,* enhancing the patient's level of activity through a graded exercise program). Desensitization and the enhancement of hardiness are other ways to conceptualize the nonmedication treatment approaches recommended for these illnesses. Multidisciplinary treatment is generally indicated to best address the different components of illness-related disability.

IV. Future Directions

Coherent explanatory models are needed for the conditions reviewed in this section. Areas of investigation common to these poorly understood conditions include the interaction of HPA axis function (see Chapter 84), exogenous and endogenous hormones (see Chapters 86, 88, and 89), and sensitization processes (see Chapter 90). For example, exogenous hormone use (birth control pills and hormone replacement therapy) may enhance the sensitization process (see Chapter 89), thereby also partially accounting for the predominance of women with these disorders. The association between stress and illness has been well established. Stress should be broadly defined to include physical and emotional challenges that were remote compared to illness onset (*e.g.,* childhood sexual abuse), more proximal events, and also potential perpetuating factors. The physiological and emotional effects of stress also must be considered in future research efforts to advance our understanding of these illnesses and to help patients cope more effectively with symptom complaints.

References

1. Fukuda, K., Straus, S. E., Hickie, I., Sharpe, M. C., Dobbins, J. G. H., Komaroff, A., and The International Chronic Fatigue Syndrome Study Group (1994). The chronic fatigue syndrome: A comprehensive approach to its definition and study. *Ann. Intern. Med.* **121,** 953–959.
2. Buchwald, D., and Garrity, D. (1994). Comparisons of patients with chronic fatigue syndrome, fibromyalgia, and multiple chemical sensitivities. *Arch. Intern. Med.* **154,** 2049–2053.
3. Munakata, J., Naliboff, B., Harraf, F., Kodner, A., Lembo, T., Chang, L., Silverman, D. H. S., and Mayer, E. A. (1997). Repetitive sigmoid stimulation induces rectal hyperalgesia in patients with irritable bowel syndrome. *Gastroenterology* **112,** 55–63.
4. Cook, I. J., van Eeden, A., and Collins, S. M. (1987). Patients with irritable bowel syndrome have greater pain tolerance than normal subjects. *Gastroenterology* **93,** 727–733.
5. Tiersky, L. A., Johnson, S. K., Lange, G., Natelson, B. H., and DeLuca, J. (1997). Neuropsychology of chronic fatigue syndrome: A critical review. *J. Clin. Exp. Neuropsychol.* **19,** 560–586.
6. Gibson, H., Carroll, N., Clague, J. E., and Edwards, R. H. T. (1993). Exercise performance and fatiguability in patients with chronic fatigue syndrome. *J. Neurol., Neurosurg. Psychiatry* **5,** 993–998.

7. Doty, R. L., Deems, D. A., Frye, R. E., Pelberg, R., and Shapiro, A. (1988). Olfactory sensitivity, nasal resistance, and autonomic function in patients with multiple chemical sensitivities. *Arch. Otolarngol. Head Neck Surg.* **114,** 1422–1427.

8. McDermid, A. J., Rollman, G. B., and McCain, G. A. (1996). Generalized hypervigilance in fibromyalgia: Evidence of perceptual amplification. *Pain* **66,** 133–144.

9. Walker, E. A., Keegan, D., Gardner, G., Sullivan, M., Bernstein, D., and Katon, W. J. (1997). Psychosocial factors in fibromyalgia compared with rheumatoid arthritis: II. Sexual, physical, and emotional abuse and neglect. *Psychosom. Med.* **59,** 572–577.

10. Schmaling, K. B., and DiClementi, J. D. (1995). Interpersonal stressors in chronic fatigue syndrome: A pilot study. *J. Chronic Fatigue Syndr.* **3/4,** 153–158.

11. Koss, M. P., Koss, P. G., and Woodruff, W. J. (1991). Deleterious effects of criminal victimization on women's health and medical utilization. *Arch. Intern. Med.* **151,** 342–347.

12. Lemieux, A. M., and Coe, C. L. (1995). Abuse-related posttraumatic stress disorder: Evidence for chronic neuroendocrine activation in women. *Psychosom. Med.* **57,** 105–115.

84

Chronic Fatigue Syndrome

TOM REA AND DEDRA BUCHWALD
Department of Medicine
University of Washington
Seattle, Washington

I. Introduction

Fatigue is a common symptom in the community [1–4] and a frequent complaint in clinical practice, being reported by at least 20% of patients seeking medical care [5–9]. Typically the fatigue is transient, self-limiting, and explained by prevailing circumstances. However, a minority of persons experience persistent and disabling fatigue. In some cases, such fatigue may result from medical diseases such as anemia or hypothyroidism. However, in other individuals, chronic fatigue cannot be readily explained by organic illness and may represent chronic fatigue syndrome (CFS).

CFS is an illness characterized by profound disabling fatigue lasting at least 6 months accompanied by symptoms of sleep disturbance, musculoskeletal pain, and neurocognitive impairment [10]. As its name implies, CFS is a symptom-based diagnosis without distinguishing physical examination findings or routine laboratory findings. Although infectious, immunological, neuroendocrine, sleep disorder, and psychiatric mechanisms have been investigated, a unifying etiology for CFS has yet to emerge. Regardless of the pathogenesis, those with CFS have substantially impaired functional status similar to that of other chronic illnesses, resulting in significant personal and economic morbidity [11–13]. This chapter will detail the issues of CFS diagnosis, epidemiology, pathogenesis, treatment, and prognosis.

II. History and Case Definition

Although the Centers for Disease Control did not publish the first case definition of CFS until 1988, the medical literature is replete with descriptions of similar fatiguing conditions throughout history. As early as 1750 Sir Richard Manningham coined the term "febricula," a condition characterized by "low fever, listlessness with great lassitude and weariness all over the body . . . (and) sometimes the patient is a little delirious and forgetful" [14,15]. In the nineteenth century, "nervous exhaustion" or "neurasthenia" was popularized by the writings of neurologist George Beard. Beard described an illness of "profound physical and mental fatigue, nervous dyspepsia, . . . and protracted postexertional muscle weakness." Persons with "neurasthenia" were "likely female and victims of nervous system overload caused by environmental stressors" [16,17]. During the twentieth century, several geographically distinct outbreaks of unexplained fatigue were reported. These illnesses, termed "epidemic neuromyasthenia," were characterized by (1) "an acute phase of low-grade fever, headache, sore throat, myalgia, and malaise; (2) persistence of debilitating physical and mental fatigue, mood changes, and sleep disturbances; and (3) the ab-

sence of both laboratory abnormalities consistent with the degree of symptom severity and mortality" [17]. More recently, descriptions of such fatiguing conditions have reflected attempts to invoke a specific pathogenesis. Hence, chronic fatigue was termed "chronic brucellosis," "chronic candidiasis," and "chronic Epstein Barr virus (EBV) infection," but none of those pathologies proved to be the unifying etiology [18–20].

Indeed, in light of the heterogeneity of those with "chronic EBV infection" and the overall lack of research in this area, a group of epidemiologists, researchers, and clinicians was convened by Centers for Disease Control and Prevention (CDC) in 1987 to develop a consensus statement describing the key clinical features of what is now known as CFS. The CFS case definition highlighted the syndrome's primary feature, removed the implication that Epstein Barr virus was the etiologic agent, and provided a construct to study this condition [21]. Similar working formulations were also developed in England and Australia [22,23]. In 1994, the CDC revised its original definition, thereby establishing the current international criteria for CFS [10]. As shown in Table 84.1, the new criteria eliminated the previous physical exam criteria, reduced the number of required symptoms from 8 to 4, and clarified exclusionary conditions. The most notable modification was the inclusion of persons with most nonpsychotic psychiatric disorders.

III. Epidemiology

The epidemiology of CFS varies considerably depending on which definition is employed, the particular population evaluated, and the study's methodology [24]. In early studies, the prevalence of CFS was estimated at 0.004% to 0.130%. [22,25,26]. These estimates relied on physician report and referral (sentinel physician) or physician recall—methodologies that are open to ascertainment bias. In population-based primary care samples, the prevalence of CFS in United Kingdom adults was 2.6%, using the more inclusive 1994 international CDC case definition [27], and 0.3% and 1.2% in American and English patients, respectively, employing the more restrictive 1988 CDC criteria [5,27]. In a population-based study of a community sample in the United States, the prevalence of CFS was estimated to be between 0.075% and 0.270% [28]. Ascertainment bias likely accounts for the significantly lower prevalence rates of CFS reported in older studies [22,25,26,29]. CFS also occurs in children and adolescents, but seemingly at a lower rate [30].

Historically, early reports in the lay press and from tertiary clinics suggested that CFS primarily affected young, Caucasian, socioeconomically successful women [22,25]. Indeed, most persons diagnosed with CFS are 30–40 years of age. However,

Table 84.1

1994 CDC Case Definition of Chronic Fatigue Syndrome

In a patient with severe fatigue that persists or relapses for 6 months. Exclude if patient is found to have:

1. Active medical condition that may explain the chronic fatigue, such as untreated hypothyroidism, sleep apnea, or narcolepsy;
2. Previously diagnosed medical conditions that have not fully resolved, such as previously treated malignancies or unresolved cases of hepatitis B or C virus infection;
3. Any past or current major depressive disorder with psychotic or melancholic features, including bipolar affective disorders, schizophrenia, delusional disorders, dementias, anorexia nervosa, or bulimia nervosa;
4. Alcohol or other substance abuse within two years before the onset of chronic fatigue and at any time afterward.

Classify as chronic fatigue syndrome if:

Fatigue is sufficiently severe and of new or definite onset (not lifelong), not substantially alleviated by rest, results in substantial reduction in previous levels of occupational, educational, social or personal activities; and

Four or more of the following symptoms (all of which must have started with or after the onset of the fatigue) are concurrently present for 6 months:

1. Substantially impaired memory or concentration
2. Sore throat
3. Tender cervical or axillary lymph nodes
4. Muscle pain
5. Multi-joint pain
6. New headaches
7. Unrefreshing sleep
8. Post-exertional malaise

Table 84.2

Frequency of Symptoms Reported in CFS

Symptom	Frequency (%)
Fatigue	100
Impaired cognition	50–85
Depression	50–85
Sore throat	50–75
Anxiety	50–70
Postexertional malaise	50–60
Premenstrual worsening	50–60
Stiffness	50–60
Visual blurring	50–60
Nausea	50–60
Muscle weakness	40–70
Arthralgias	40–50
Tachycardia	40–50
Headaches	35–85
Dizziness	30–50
Parasthesias	30–50
Dry eyes	30–40
Dry mouth	30–40
Diarrhea	30–40
Anorexia	30–40
Cough	30–40
Finger swelling	30–40
Night sweats	30–40
Painful lymph nodes	30–40
Rash	30–40
Low-grade fever	20–95
Myalgias	20–95
Sleep disturbance	15–90

From [70].

studies in community and primary-care samples reveal that CFS and the symptoms of chronic fatigue have only a modest predilection for women, with an odds ratio of 1.3 to 1.7 for females compared to males [4,7,22,27,28]. Additionally, very few gender-specific differences appear to be evident in CFS. Among a tertiary care population with CFS, women did have a higher frequency of tender or enlarged lymph nodes and coexisting fibromyalgia, as well as lower scores on a survey of physical functioning. Men, on the other hand, more often possessed pharyngeal inflammation and a higher lifetime prevalence of alcoholism. All other measures, however, including symptoms, physical exam findings, laboratory results, health-related functional status, and psychosocial health, were similar between male and female groups [31]. More research is required to elucidate gender characteristics in primary and community populations. Finally, although study is limited, CFS does not appear to necessarily favor upper-class Caucasians but rather affects all groups, regardless of socioeconomic class or ethnicity [22,27,28,32–34].

IV. Symptoms, Signs, and Laboratory Findings

Fatigue is the hallmark of CFS. The fatigue itself is exceptional, causing significant functional impairment. Nearly all those with CFS note a decrease in social relationships and other unwanted consequences of the illness [23]. Furthermore, one-third are unable to work and another third can only work part-time [11]. Often, patients with CFS will recall that the initial onset of fatigue was abrupt and dramatic, frequently accompanied by other symptoms suggestive of acute infection [35,36]. Interestingly, many enjoyed excellent physical fitness and energy premorbidly [37]. After the onset of illness however, patients often note that physical exertion typically exacerbates the fatigue, a phenomenon not readily exhibited in the research setting [38]. In addition, as shown in Table 84.2, numerous symptoms are reported by patients with CFS [39]. However, the symptoms that comprise the CFS case definition are all reported more frequently in CFS than in other chronic illnesses including depression [40]. Finally, a history of atopy is noted more often in those with CFS than in the general population [41].

As reflected by its absence as a criterion in the case definition, the physical examination in CFS is generally unremarkable. Although occasional pharyngitis, fever, or lymphadenopathy may be present, such findings provide little diagnostic value and were thus not included in the 1994 CDC definition [10]. The most common examination abnormality is the presence of specific musculoskeletal sites that elicit tenderness on palpation. These tender points are characteristic of fibromyalgia, which occurs in up to 70% of CFS patients [39,42]. Likewise, there are no consistent routine laboratory abnormalities in patients

Table 84.3
NIH-Recommended Laboratory Tests in Patients with Debilitating Chronic Fatigue

Standard tests
  Complete blood count with white blood cell differential
  Erythrocyte sedimentation rate
  Urinalysis
  Blood urea nitrogen, creatinine, and electrolytes
  Glucose
  Calcium, phosphorous
  Thyroid stimulating hormone
  Alanine aminotransferase
  Alkaline phosphatase
  Total protein, albumin, and globulin

Optional tests, as clinically indicated
  Antinuclear antibodies
  Serum cortisol
  Rheumatoid factor
  Immunoglobulin levels
  Tuberculin skin testing
  Lyme serology
  HIV serology

with CFS. Although some studies have found increased frequencies of elevated alkaline phosphatase, cholesterol, and circulating autoantibodies, these deviations are typically mild and not useful in the diagnosis or care of CFS patients [43–45].

V. Evaluation of the Chronically Fatigued Patient

Although chronic fatigue is a common concern in primary care, only a small minority (2–5%) of patients will actually fulfill criteria for CFS. A similarly small proportion will have organic diseases such as anemia, hypothyroidism, or occult malignancy. By far, the majority of patients presenting with chronic fatigue will have depression or anxiety disorders [8,46].

As no specific test is diagnostic of CFS, the workup of the chronically-fatigued patient should be aimed at detecting underlying organic diseases and psychiatric disorders, remembering that nonpsychotic psychiatric illness (see Table 84.1) does not exclude a diagnosis of CFS. A thorough history and physical examination provide the cornerstone of the evaluation and should direct the subsequent workup. In screening for organic diseases in the chronically fatigued patient, the National Institutes of Health (NIH) has recommended a battery of standard laboratory tests, shown in Table 84.3 [36], although they uncommonly identify a cause of the fatigue [47–49]. Other investigations, including central nervous system imaging, formal neuropsychological evaluation, or tilt table testing, have yet to establish their usefulness in the diagnosis or management of CFS and should typically be reserved for research purposes.

VI. Pathophysiology

Despite extensive investigation, a cause of CFS has not been demonstrated. One explanation may be that a specific unifying pathogenesis has yet to be identified. More likely, CFS represents heterogeneous factors rather than a lone etiology. Indeed,

some elements may predispose a person to CFS, others may precipitate the illness, while still others might perpetuate the disorder [50,51].

A. Psychiatric Illness

Persons with CFS have an increased prevalence of current and lifetime affective disorders, primarily depression, compared to other chronically-ill populations or healthy controls [46,52–55]. In CFS, 50–75% of patients have either a lifetime history or are currently experiencing depression. Generalized anxiety disorder and somatoform disorders also occur at greater rates in patients with CFS than in the general population [55–58]. In most [52,53,59], but not all cases [5,60], the affective disorder exists prior to the emergence of CFS. In light of these findings, some observers have felt that CFS may be an atypical presentation of depression.

However, several factors do not support such a simple explanation. First, a significant proportion of patients with CFS do not have a current or lifetime affective disorder. Second, patients with major depression often have a central upregulation of the hypothalamic-pituitary-adrenal (HPA) axis, resulting in mild hypercortisolism [61–64]. Conversely, patients with CFS often display a central downregulation of the HPA axis, resulting in mild hypocortisolism [65]. Third, the typical sleep abnormalities of major depression, reduced REM latency and increased REM density [66], are not typically evident in CFS. Finally, initial trials with antidepressant doses of fluoxetine, a selective serotonin reuptake inhibitor, failed to improve patients with CFS [67]. Thus, although affective disorders occur more frequently in CFS and should be treated, their precise mechanistic role or contribution to its symptoms and morbidity has yet to be completely defined.

B. Central Nervous System (CNS)

Several of the symptoms of CFS, including fatigue, impaired concentration or memory, and headache, suggest a CNS role in the syndrome's pathophysiology. Indeed, in an effort to gain insight into the disorder, researchers have investigated a CNS link to CFS by means of neuroimaging, cognitive testing, autonomic assessment, and the study of neuropeptides. Neuroimaging research in CFS has primarily entailed magnetic resonance imaging (MRI) and single-photon emission computed tomography (SPECT). Some studies using MRI revealed punctate areas of high signal in the white matter, often in the subcortical areas, more frequently in CFS patients than in healthy controls [68–70]. However, other studies found no increase in white matter disease or other CNS pathology on MRI in CFS patients when compared to either healthy or depressed controls [71,72]. In addition, the MRI abnormalities were not associated with neurocognitive performance [71]. Abnormalities in CNS perfusion, typically hypoperfusion, have been found more often on SPECT testing in CFS patients than in healthy or depressed controls, although no specific anatomic pattern has emerged and the effect of comorbid depression is difficult to ascertain [73,74]. Although intriguing, the functional significance and clinical utility of these findings remain uncertain and await further clarification [75].

Despite the considerable frequency of subjective cognitive complaints among those with CFS, results of formal neuropsychological testing of cognition have been variable and sometimes contradictory. The weight of the evidence suggests a modest but significant deficit in information processing. Such a shortcoming could account for the poorer performance on complex attention and information-processing tasks [76]. Coexisting psychological distress or psychiatric disease may contribute in part to these deficits. In general, however, persons with CFS appear to possess normal cognitive, memory, and global intellectual abilities [76,77].

Autonomic dysfunction has been implicated in the pathophysiology of CFS by some but not all investigations [78–81]. This dysfunction has been demonstrated by tilt-table testing and manifests itself by hypotension with bradycardia (vasovagal reaction) or hypotension with tachycardia (vasodepressor reaction) upon vertical tilting. Furthermore, anecdotal reports suggest that CFS patients with symptoms indicative of this neurally-mediated hypotension often improve (but do not recover entirely) with fluid, salt, or fludrocortisone therapy [82]. However, this therapy is unlikely to be useful for all patients with CFS [83]. Again, the precise nature and extent of autonomic system involvement in CFS are still undetermined.

Finally, abnormalities in the HPA axis and serotonin pathways have been identified in CFS patients. As previously discussed, patients with CFS exhibit hypocortisolism while patients with depression, in contrast, typically manifest CNS-mediated hypercortisolism. The hypocortisolism of CFS appears to originate from a CNS source of HPA dysfunction rather than from a primary adrenal site [65]. Some [84–86], but not all [87,88], studies have demonstrated abnormalities of CNS serotonin physiology in patients with CFS. Specifically, administration of serotonin agonists causes a significant increase in serum prolactin levels in CFS patients relative to depressed and healthy controls, suggesting a CNS upregulation of the serotonergic system in CFS [84–86]. Conversely, patients with depression have a suppressed serotonin-mediated prolactin response [84,89]. Of interest, serotonin pathways supply input to the HPA axis [90,91]. In this regard, some evidence indicates a blunted adrenocorticotropic hormone (ACTH) response to serotonin in CFS patients [84,92]. Such an altered neuropeptide state could provoke the symptoms of CFS. Although the precise interaction of serotonin and the HPA axis and its pathophysiological significance are still undetermined, these findings are provocative and deserve further study.

C. Infection

Viral infection has been proposed as a cause of CFS for several reasons. First, many persons with CFS recall that the syndrome began with a flu-like illness [35,36]. Second, several of the CFS minor criteria, including sore throat, myalgias, and tender cervical or axillary lymph nodes, resemble symptoms of an acute viral infection. Third, numerous geographical clusters of CFS-like illnesses have been reported, suggesting a possible "epidemic" or infectious component [93]. Finally, viral infection provokes the immune system, which produces cytokines and alters the HPA axis, which in turn induce fatigue and alter sleep patterns [94–96]. Indeed, Epstein-Barr virus, human her-

pes virus 6, group B coxsackie virus, human T lymphotropic virus II, hepatitis C, enteroviruses, and retroviruses, among others, have all been proposed as etiologic agents in CFS [97]. However, with rare exception [98], there has been no consistent evidence to date that CFS results from a specific infection [99]. In fact, some CFS patients have no clinical or laboratory evidence of viral infection [100]. Furthermore, therapy for CFS with antiviral agents such as acyclovir or alpha interferon provides no benefit in CFS [101,102]. Therefore it is improbable that a single infectious agent causes CFS. Rather, a heterogeneous group of infections may act as a trigger, inciting initial immune and/or neuroendocrine events with consequent fatigue. This fatigue may subsequently be perpetuated by an individual's poor biological and psychological adaptive skills [50].

D. Sleep

Despite the frequency of subjective sleep disturbances in CFS [103] and the similarities between CFS patients and sleep-deprived healthy subjects [104], surprisingly few studies have investigated the relationship between CFS and sleep. Those with CFS frequently note more difficulty falling asleep, more interrupted sleep, and more daytime napping than healthy or chronically-ill controls [105–108]. Interestingly, sleep disturbance does not appear to correlate with fatigue severity [108, 109]. Polysomnography in CFS subjects has yielded variable results. Some studies revealed a characteristic "alpha intrusion" during non-REM sleep [110], although other trials have not verified this result [107,108,111]. Consequently, in contrast to depression, which typically manifests shortened REM-sleep latency [66], polysomnography has yet to yield a consistent diagnostic abnormality specific to CFS. Finally, individuals with presumptive CFS are occasionally discovered to have significant sleep apnea or narcolepsy upon sleep testing [108,112], diagnoses that exclude CFS and are readily treatable. Thus, whether sleep is an etiologic, perpetuating, or complicating factor in CFS awaits clarification.

E. Immune System

Dysregulation of the immune system has been proposed as an underlying cause of CFS. Indeed, cytokines such as interleukin 2 and interferons can cause the symptoms characteristic of CFS [113,114]. However, very few consistent relationships have emerged from extensive study of the immune system in patients with CFS. Research does support depressed function of natural killer cells and increased expression of activation markers on the surface of T lymphocytes in patients with CFS [115–118]. Investigations of humoral immunity and cytokines have produced conflicting results, possibly due in part to varying laboratory techniques and the heterogenity of study populations. Regardless, the clinical importance of immune system abnormalities, even when consistently established, remains uncertain in CFS.

VII. Treatment

Because the cause of CFS is still uncertain and few well-designed trials have been conducted to evaluate treatment modalities, therapy should be directed toward relieving symptoms and improving function. Initial therapy should always include education. The physician should offer supportive counseling, symptom acceptance, and information about the current understanding of CFS [102]. Specifically, the physician should provide reassurance that CFS carries no excessive mortality, symptoms often improve with time although relapse may occur, and that, even though there is no specific cure, several therapeutic options can provide benefit [119].

Treatments to consider include antidepressant pharmacotherapy, nonsteroidal anti-inflammatories, cognitive behavioral therapy (CBT), and graded-exercise programs. Tricyclic antidepressant medications are administered at bedtime to improve sleep and diminish pain. Because patients appear particularly sensitive to the adverse effects of tricyclics, low doses (e.g., 10–30 mg of nortriptyline) are typically used in CFS treatment [102,120]. Although the benefit of this practice has not been conclusively demonstrated in controlled trials, its success in the related disorder of fibromyalgia [121,122] makes it a reasonable intervention, especially in those with sleep and pain symptoms. Response to selective serotonin reuptake inhibitors such as fluoxetine has been minimal, possibly because of the aforementioned serotonergic hypersensitivity demonstrated in CFS [67,84,85]. Monoamine oxidase inhibitors have demonstrated some early modest promise, especially in populations with significant vegetative symptoms [123–125]. A trial of acetominophen or nonsteroidal anti-inflammatory agents may be worthwhile in patients with prominent musculoskeletal pain complaints [120].

The use of CBT derives from the beliefs that CFS may be perpetuated by ineffective coping and unhelpful health beliefs and its success in other illnesses such as depression, chronic low back pain, and atypical chest pain [126,127]. Controlled trials have found that approximately 70% of patients receiving several months of weekly CBT versus only 20% of the placebo group demonstrated functional improvement [128,129]. Although encouraging, the exact content and duration of the CBT require careful scrutiny; CBT was associated with little benefit in earlier studies that used shorter duration and different composition of therapy [130,131] than in the recent trials conducted in the United Kingdom.

CFS patients without psychiatric illness and sleep disturbances appear to receive short- and long-term subjective and objective functional benefits from a graded aerobic exercise program [132,133]. The program involved initial endurance conditioning such as walking and later incorporated strengthening exercises. The program ran for several months and combined supervised and independent physiotherapy [134]. It remains to be seen whether these results can be generalized to those with comorbid affective and sleep disorders.

Multiple other modalities including intravenous immunoglobulin, acyclovir, vitamin B_{12}, intramuscular dialyzable leukocyte extract, intramuscular magnesium, and essential fatty acid therapy have not shown consistent benefit in the treatment of CFS [135]. Many patients use alternative medicine treatments with unknown outcome [135]. Finally, it is essential to recognize and treat comorbid illness such as depression and sleep apnea. In summary, successful therapy for CFS must be built on a foundation of patient–physician respect and advo-

cacy. Specific treatment regimens should be individualized, reflecting the heterogeneity of the CFS population.

VIII. Prognosis

CFS has a variable course. Improvement occurs in 20–70% of patients depending on the study group and duration of follow-up. Less than 10%, however, recover fully to their premorbid state [136]. Another 10–20% may worsen during follow-up [137]. Older age, more chronic illness, comorbid psychiatric disease, and attribution of illness to a physical cause such as infection have consistently predicted poor recovery in CFS [136]. Conversely, children and adolescents appear to recover more readily [138]. Female gender does not appear to influence outcome [139]. As these findings generally reflect "untreated" CFS, specific therapies may significantly improve outcome and should offer optimism for CFS patients.

IX. Health Care Utilization and Economics

Similar to other chronic illnesses, the morbidity of CFS impacts societal and health care resources. Persons with CFS are much more likely to access health care than the general population, averaging between 20 and 30 total care visits per year [11,140]. Such persons often employ varied means of care including allopathic generalists and specialists, chiropractors, naturopaths, and acupuncturists, usually in an uncoordinated fashion [11]. Consequently, the direct expense of medical care for individuals with CFS is at least twice that of the general population [11]. Furthermore, the indirect costs of CFS appear significant. Upwards of one-third of CFS patients are unable to work, which is a fraction similar to that of individuals with rheumatoid arthritis or severe osteoarthritis [11]. A subset of these unemployed persons collect disability compensation for their CFS. Although preliminary study indicates a comparable degree of employment disability in CFS [140], more research is needed to determine the effect of gender on health care use and employment. In practice, disability in CFS should not be viewed as permanent and efforts to return to the work force should proceed as the illness dictates.

X. Related Disorders

The symptoms of chronic fatigue and CFS often cooccur with other so-called "functional illnesses" such as fibromyalgia, multiple chemical sensitivities, irritable bowel syndrome, and temperomandibular joint syndrome. This relationship has been best studied in fibromyalgia, a syndrome of characteristic tender points and chronic diffuse body pain [141]. Despite their contrasting definitions, 20–70% of patients with fibromyalgia also meet criteria for CFS [142–144], and conversely, between 35–70% of those with CFS-like illnesses have concurrent fibromyalgia [31,42]. Variable expressions of a common shared pathophysiology may explain the extensive overlap among these syndromes [145]. In the clinical setting, an appreciation of the coexistence of these disorders will help the physician and patient to achieve satisfactory care for these challenging illnesses.

XI. Conclusions

CFS is an illness characterized by extraordinary fatigue, along with subjective cognitive, musculoskeletal, and sleep symptoms. The syndrome lacks a specific diagnostic test or identifying physical examination finding. Whether CFS represents a discrete disorder is still unknown. Those individuals who fulfill the criteria of CFS suffer striking functional impairment in their personal and professional relationships. Currently, treatment is largely empiric and rarely curative. Improving upon this outcome requires a better understanding of CFS pathophysiology. Evidence to date points to the importance of biological and psychological contributions. Future study which simultaneously assesses both of these factors may provide the insight needed to unravel these complexities and consequently enhance the care of patients with CFS.

References

1. Chen, M. (1986). The epidemiology of self-perceived fatigue among adults. *Prev. Med.* **15,** 74–81.

2. Derogatis, L. R., Lipman, R. S., Rickels, K., Uhlenhuth, E. H., and Covi, L. (1974). The Hopkins Symptom Checklist (HSCL): A self report symptom inventory. *Behav. Sci.* **19,** 1–15.

3. Kellner, R., and Sheffield, B. (1973). The one week prevalence of symptoms in neurotic patients and normals. *Am. J. Psychiatry* **130,** 102–105.

4. Pawlikowska, T., Chadler, T., Hirsch, S. R., Wallace, P., Wright, D. J., and Wessely, S. C. (1994). Population based study of fatigue and psychological distress. *Br. Med. J.* **308,** 763–766.

5. Bates, D., Schmitt, W., Buchwald, D., Ware, N. C., Lee, J., Thoyer, E., Kornish, R. J., and Komaroff, A. L. (1993). Prevalence of fatigue and chronic fatigue syndrome in a primary care practice. *Arch. Intern. Med.* **153,** 2759–2765.

6. Cath'ebras, P. J., Robbins, J. M., Kirmayer, L. J., and Hayton, B. C. (1992). Fatigue in primary care: Prevalence, psychiatric comorbidity, illness behavior, and outcome. *J. Gen. Intern. Med.* **7,** 276–286.

7. David, A., Pelosi, A., McDonald, E., Stephens, D., Ledger, D., Rathbone, R., and Mann, A. (1990). Tired, weak, or in need of rest: A profile of fatigue among general practice attenders. *Br. Med. J.* **301,** 1199–1202.

8. Kroenke, K., Wood, D. R., Mangelsdorff, D., Meier, N. J., and Powell, J. B. (1988). Chronic fatigue in primary care: Prevalence, patient characteristics and outcome. *JAMA, J. Am. Med. Assoc.* **260,** 929–934.

9. McDonald, E., David, A. S., Pelosi, A. J., and Mann, A. H. (1993). Chronic fatigue in primary care attenders. *Psychol. Med.* **23,** 987–998.

10. Fukuda, K., Straus, S. E., Hickie, I., Sharpe, M. C., Dobbins, J. G., and Komaroff, A. (1994). The chronic fatigue syndrome: A comprehensive approach to its definition and study. *Ann. Intern. Med.* **121,** 953–959.

11. Bombardier, C. H., and Buchwald, D. (1996). Chronic fatigue, chronic fatigue syndrome, and fibromyalgia: Disability and healthcare use. *Med. Care* **34,** 924–930.

12. Buchwald, D., Pearlman, T., Umali, J., Schmaling, K., and Katon, W. (1996). Functional status in patients with chronic fatigue syndrome, other fatiguing illnesses, and healthy individuals. *Am. J. Med.* **101,** 364–370.

13. Komaroff, A. L., Fagioli, L. R., Doolittle, T. H., Gandek, B., Gleit, M. A., Guerriero, R. T., and Kornish, R. J. (1996). Health status in patients with chronic fatigue syndrome and in general population and disease comparison groups. *Am. J. Med.* **101,** 281–290.

14. Manningham, R. (1750). "The Symptoms, Nature, Causes and Cure of the Febricula or Little Fever: Commonly Called the Nervous or Hysteric Fever; the Fever on the Spirits; Vapours, Hypo, or Spleen," 2nd ed., pp. 52–53. J Robinson, London.

15. Straus, S. E. (1991). History of chronic fatigue syndrome. *Rev. Infect. Dis.* **13**(Suppl. 1), S2–S7.

16. Beard, G. (1869). Neurasthenia, or nervous exhaustion. *Boston Med. Surg. J.* **3,** 217–220.

17. Kim, E. (1994). A brief history of chronic fatigue syndrome. *JAMA, J. Am. Med. Assoc.* **272,** 1070–1071.

18. Evans, A. C. (1947). Brucellosis in the United States. *Am. J. Public Health* **37,** 139–151.

19. Truss, C. O. (1981). The role of candida albicans in human illness. *J. Orthomol. Psychiatry* **10,** 228–238.

20. Straus, S. E. (1998). The chronic mononucleosis syndrome. *J. Infect. Dis.* **157,** 405–412.

21. Holmes, G. P., Kaplan, J. E., Gantz, N. M., Komaroff, A. L., Schonberger, L. B., Straus, S. E., Jones, J. F., Dubois, R. E., Cunningham-Rundles, C., and Pahwa, S. (1988). Chronic fatigue syndrome: A working case definition. *Ann. Intern. Med.* **108,** 387–389.

22. Lloyd, A. R., Hickie, I., Boughton, C. R., Spencer, O., and Wakefield, D. (1990). Prevalence of the chronic fatigue syndrome in an Australian population. *Med. J. Aust.* **153,** 522–528.

23. Sharpe, M. C., Archard, L. C., Banatvala, J. E., Borysiewicz, L. K., Clare, A. W., David, A., Edwards, R. H., Hawton, K. E., Lambert, H. P., and Lane, R. J. (1991). A report—chronic fatigue syndrome: Guidelines for research. *J. R. Soc. Med.* **84,** 118–121.

24. Richman, J., Flaherty, J., and Rospenda, K. (1994). Chronic fatigue syndrome: Have flawed assumptions been derived from treatment-based studies? *Am. J. Public Health* **84,** 282–284.

25. Gunn, W., Connell, D., and Randall, B. (1993). Epidemiology of chronic fatigue syndrome: The Centers for Disease Control Study. *Ciba Found. Symp.* **173,** 83–101.

26. Ho-Yen, D., and McNamara, I. (1991). General practitioners' experience of the chronic fatigue syndrome. *Br. J. Gen. Pract.* **41,** 324–326.

27. Wessely, S., Chadler, T., Hirsch, S., Wallace, P., and Wright, D. (1997). The prevalence and morbidity of chronic fatigue and the chronic fatigue syndrome: A prospective primary care study. *Am. J. Public Health* **87,** 1449–1455.

28. Buchwald, D., Umali, P., Umali, J., Kith, P., Pearlman, T., and Komaroff, A. L. (1995). Chronic fatigue and the chronic fatigue syndrome in a Pacific Northwest health care system. *Ann. Intern. Med.* **123,** 81–88.

29. Wessely, S. (1995). The epidemiology of chronic fatigue syndrome. *Epidemiol. Rev.* **17,** 139–151.

30. Jordan, K. M., Landis, D. A., Downey, M. C., Osterman, S. L., Thurm, A. E., and Jason, L. A. (1998). Chronic fatigue syndrome in children and adolescents: A review. *J. Adolesc. Health* **22,** 4–18.

31. Buchwald, D., Pearlman, T., Kith, P., and Schmaling, K. (1994). Gender differences in patients with chronic fatigue syndrome. *J. Gen. Intern. Med.* **9,** 397–401.

32. Buchwald, D., Manson, S. M., Pearlman, T., Umali, J., and Kith, P. (1996). Race and ethnicity in patients with chronic fatigue. *J. Chronic Fatigue Syndr.* **2,** 53–66.

33. Minowa, M., and Jiamo, M. (1996). Descriptive epidemiology of chronic fatigue syndrome based on a nationwide survey in Japan. *J. Epidemiol.* **6,** 75–80.

34. Euba, R., Chadler, T., Deale, A., and Wessely, S. (1996). A comparison of the characteristics of chronic fatigue syndrome in primary and tertiary care. *Br. J. Psychiatry* **168,** 121–126.

35. Salit, I. E. (1997). Precipitating factors for the chronic fatigue syndrome. *J. Psychiatr. Res.* **31,** 59–65.

36. Schlueiderberg, A., Straus, S. E., Peterson, P., Blumenthal, S., Komaroff, A. L., Spring, S. B., Landay, A., and Buchwald, D. (1992). Chronic fatigue syndrome research: Definition and medical outcome assessment. *Ann. Intern. Med.* **117,** 325–331.

37. MacDonald, K. L., Osterholm, M. T., LeDell, K. H., White, K. E., Schenck, C. H., Chao, C. C., Persing, D. H., Johnson, R. C., Barker, J. M., and Peterson, P. K. (1996). A case-control study to assess possible triggers and cofactors in chronic fatigue syndrome. *Am. J. Med.* **100,** 548–554.

38. Sisto, S. A., LaManca, J., Cordero, D. L., Bergen, M. T., Ellis, S. P., Drastal, S., Boda, W. L., Tapp, W. N., and Natelson, B. H. (1996). Metabolic and cardiovascular effects of a progressive exercise test in patients with chronic fatigue syndrome. *Am. J. Med.* **100,** 634–640.

39. Komaroff, A. L., and Buchwald, D. (1991). Symptoms and signs of chronic fatigue syndrome. *Rev. Infect. Dis.* **13**(Suppl. 1), S8–S11.

40. Komaroff, A. L., Fagioli, L. R., Geiger, A. M., Doolittle, T. H., Lee, J., Kornish, R. J., Gleit, M. A., and Guerriero, R. T. (1996). An examination of the working case definition of chronic fatigue syndrome. *Am. J. Med.* **100,** 56–64.

41. Strauss, S. E., Dale, J. K., Wright, R., and Metcalfe, D. D. (1988). Allergy and the chronic fatigue syndrome. *J. Allergy Clin. Immunol.* **81,** 791–795.

42. Goldenberg, D. L., Simms, R. W., Geiger, A., and Komaroff, A. L. (1990). High frequency of fibromyalgia in patients with chronic fatigue seen in primary care practice. *Arthritis Rheum.* **33,** 381–387.

43. Bates, D. W., Buchwald, D., Lee, J., Kith, P., Doolittle, T., Rutherford, C., Churchill, W. H., Schur, P. H., Wener, M., and Wybenga, D. (1995). Laboratory abnormalities in patients with chronic fatigue syndrome. *Arch. Intern. Med.* **155,** 97–103.

44. Buchwald, D., and Komaroff, A. L. (1991). Laboratory findings in chronic fatigue syndrome. *Rev. Infect. Dis.* **13**(Suppl. 1), S12–S18.

45. Calabrese, L. H., Davis, M. E., and Wilke, W. S. (1994). Chronic fatigue syndrome and a disorder resembling Sjögren's syndrome: Preliminary report. *Clin. Infect. Dis.* **18**(Suppl. 1), S28–S31.

46. Manu, P., Matthews, D. A., Lane, T. J., Tennen, H., Hesselbrock, V., Mendola, R., and Affleck, G. (1988). The mental health of patients with a chief complaint of chronic fatigue: A prospective evaluation and follow-up. *Arch. Intern. Med.* **148,** 2213–2217.

47. Risdale, L., Evans, A., Jerrett, W., Mandalia, S., Osler, K., and Vora, H. (1993). Patients with fatigue in general practice: A prospective study. *Br. Med. J.* **307,** 103–106.

48. Valdini, A., Steinhardt, S., and Feldman, E. (1989). Usefulness of a standard battery of laboratory tests in investigating chronic fatigue in adults. *Fam. Pract.* **6,** 286–291.

49. Lane, T., Matthews, D., and Manu, P. (1990). The low yield of physical examinations and laboratory investigations of patients with chronic fatigue. *Am. J. Med. Sci.* **299,** 313–318.

50. White, P. D. (1997). The relationship between infection and fatigue. *J. Psychosom. Res.* **43,** 345–350.

51. Sharpe, M. (1996). Chronic fatigue syndrome. *Psychiatr. Clin. North Am.* **19,** 549–573.

52. Wood, G. C., Bentall, R. P., Gopfert, M., and Edwards, R. H. T. (1991). A comparative assessment of patients with chronic fatigue syndrome and muscle disease. *Psychol. Med.* **21,** 618–628.

53. Wessely, S., and Powell, R. (1991). Fatigue syndromes: A comparison of chronic "postviral' fatigue with neuromuscular and affective disorders. *J. Neurol., Neurosurg. Psychiatry* **52,** 940–948.

54. Katon, W. J., Buchwald, D., Simon, G., Russo, J. E., and Mease, P. J. (1991). Psychiatric illness in patients with chronic fatigue and those with rheumatoid arthritis. *J. Gen. Intern. Med.* **6,** 277–285.

55. Wessely, S., Chadler, T., Hirsch, S., Wallace, P., and Wright, D. (1996). Psychological symptoms, somatic symptoms, and psychiatric disorder in chronic fatigue and chronic fatigue syndrome: A prospective study in the primary care setting. *Am. J. Psychiatry* **153,** 1050–1059.

56. Fischler, B., Cluydts, R., DeGucht, V., Kaufman, L., and DeMeirleir, K. (1997). Generalized anxiety disorder in chronic fatigue syndrome. *Acta Psychiatr. Scand.* **95,** 405–413.

57. Kruesi, M. J. P., Dale, J., and Straus, S. E. (1989). Psychiatric diagnoses in patients who have the chronic fatigue syndrome. *J. Clin. Psychiatry* **50,** 53–56.

58. Lane, T. J., Manu, P., and Matthews, D. A. (1991). Depression and somatization in the chronic fatigue syndrome. *Am. J. Med.* **91,** 335–344.

59. Manu, P., Matthews, D. A., and Lane, T. J. (1989). Depression among patients with a chief complaint of chronic fatigue. *J. Affective Disord.* **17,** 165–172.

60. Hickie, I., Lloyd, A., Wakefield, D., and Parker, G. (1990). The psychiatric status of patients with chronic fatigue syndrome. *Br. J. Psychiatry* **156,** 534–540.

61. Dinan, T. G. (1994). Glucocorticoids and the genesis of depressive illness. *Br. J. Psychiatry* **164,** 365–372.

62. Checkley, S. A. (1992). Neuroendocrine mechanisms and the precipitation of depression by life events. *Br. J. Psychiatry* **160**(Suppl. 15), 7–17.

63. Carroll, B. J., Feinberg, M., Greden, J. F., Tarika, J., Albala, A. A., Haskett, R. F., James, N. M., Kronfol, Z., Lohr, N., Steiner, M., deVigne, J. P., and Young, E. (1981). A specific laboratory test for the diagnosis of melancholia: standardization, validation, and clinical utility. *Arch. Gen. Psychiatry* **38,** 15–22.

64. Carroll, B. J., Curtis, G. C., and Mendels, J. (1976). Neuroendocrine regulation in depression. I. Limbic system-adrenocortical dysfunction. *Arch. Gen. Psychiatry* **33,** 1039–1044.

65. Demitrack, M. A., Dale, J. K., Straus, S. E., Laue, L., Listwak, S. J., Kruesi, M. J., Chrousos, G. P., and Gold, P. W. (1991). Evidence for impaired activation of the hypothalamic-pituitary-adrenal axis in patients with chronic fatigue syndrome. *J. Clin. Endocrinol. Metab.* **73,** 1224–1234.

66. Gillin, J. C., Sitaram, N., and Wehr, T. (   ). Sleep and affective illness. *In* "Neurobiology of Mood Disorder" (R. M. Post and J. C. Ballenger, eds.), pp. 157–189. Williams & Williams, New York.

67. Vercoulen, J. H. M. M., Swanink, C. M. A., Zitman, F. G., Vreden, S. G., Hoofs, M. P., Fennis, J. F., Galama, J. M., van der Meer, J. W., and Bleijenberg, G. (1996). Randomized, double-blind, placebo-controlled study of fluoxetine in chronic fatigue syndrome. *Lancet* **347,** 858–861.

68. Buchwald, D., Cheney, P. R., Peterson, D. L., Henry, B., Wormsley, S. B., Geiger, A., Ablashi, D. V., Sasahuddin, S. Z., Saxinger, C., and Biddle, R. (1992). A chronic illness characterized by fatigue, neurologic and immunologic disorders, and active human herpesvirus type 6 infection. *Ann. Intern. Med.* **116,** 103–113.

69. Schwartz, R. B., Garada, B. M., Komaroff, A. L., Tice, H. M., Gleit, M., Jolesz, F. A., and Holman, B. L. (1994). Detection of intracranial abnormalities in patients with chronic fatigue syndrome: Comparison of MR imaging and SPECT. *AJR. Am. J. Roentgenol.* **162,** 935–941.

70. Natelson, B. H., Cohen, J. M., Brassloff, I., and Lee, H. J. (1993). A controlled study of brain magnetic resonance imaging in patients with the chronic fatigue syndrome. *J. Neurol. Sci.* **120,** 213–217.

71. Cope, H., Pernet, A., and Kendall, B. (1995). Cognitive functioning and magnetic resonance imaging in chronic fatigue. *Br. J. Psychiatry* **176,** 86–94.

72. Greco, A., Tannock, C., and Brostoff, J. (1997). Brain MR in chronic fatigue syndrome. *Am. J. Neuroradiol.* **18,** 1265–1269.

73. Ichise, M., Salit, I. E., Abbey, S. E., Chung, D. G., Gray, B., Kirsh, J. C., and Freedman, M. (1992). Assessment of regional cerebral

perfusion by 99Tcm-HMPAO SPECT in chronic fatigue syndrome. *Nucl. Med. Commun.* **13,** 767–772.

74. Schwartz, R. B., Komaroff, A. L., Garada, B. M., Gleit, M., Doolittle, T. H., Bates, D. W., Vasile, R. G., and Holman, B. L. (1994). SPECT imaging of the brain: Comparison of findings in patients with chronic fatigue syndrome, AIDS dementia complex, and major unipolar depression. *AJR, Am. J. Roentgenol.* **162,** 943–951.

75. Cope, H., and David, A. S. (1996). Neuroimaging in chronic fatigue syndrome. *J. Neurol. Neurosurg. Psychiatry* **60,** 471–473.

76. Moss-Morris, R., Petrie, K. J., Large, R. G., and Kydd, R. R. (1996). Neuropsychological deficits in chronic fatigue syndrome: Artifact or reality? *J. Neurol. Neurosurg. Psychiatry* **60,** 474–477.

77. Tiersky, L. A., Johnson, S. K., Lange, G., Natelson, B. H., and DeLuca, J. (1997). Neuropsychology of chronic fatigue syndrome: A critical review. *J. Clin. Exp. Neuropsychol.* **19,** 560–586.

78. Rowe, P. C., Bou-Holaigah, I., Kan, J. S., and Calkins, H. (1995). Is neurally mediated hypotension an unrecognized cause of chronic fatigue? *Lancet* **345,** 623–624.

79. Bou-Holaigah, I., Rowe, P. C., Kan, J., and Calkins, H. (1995). The relationship between neurally mediated hypotension and the chronic fatigue syndrome. *JAMA, J. Am. Med. Assoc.* **274,** 961–967.

80. Lapp, C. W., Glenn, F., and Davis, P. (1996). Neurally mediated hypotension and symptomatic orthostatic tachycardia in chronic fatigue syndrome. *Proc. Am. Assoc. Chronic Fatigue Syndr. Res. Conf.,* Abstr., p. 23.

81. Freeman, R., and Komaroff, A. L. (1997). Does the chronic fatigue syndrome involve the autonomic nervous system? *Am. J. Med.* **102,** 357–364.

82. Wilke, W. S., Fouad-Tarazi, F. M., Cash, J. M., and Calabrese, L. H. (1998). The connection between chronic fatigue syndrome and neurally mediated hypotension. *Cleveland Clin. J. Med.* **65,** 261–266.

83. Peterson, P. K., Pheley, A., Schroeppel, J., Schenck, C., Marshall, P., Kind, A., Haugland, J. M., Lambrecht, L. J., Swan, S., and Goldsmith, S. (1998). A preliminary placebo-controlled crossover trial of fludrocortisone for chronic fatigue syndrome. *Arch. Intern. Med.* **158,** 908–914.

84. Cleare, A. J., Bearn, J., Allain, T., McGregor, A., Wessely, S., Murray, R. M., and O'Keane, V. (1995). Contrasting neuroendocrine responses in depression and chronic fatigue syndrome. *J. Affective Disord.* **35,** 283–289.

85. Bakheit, A. M. O., Behan, P. O., Dinan, T. G., Gray, C. E., and O'Keane, V. (1992). Possible upregulation of hypothalamic 5-hydroxytryptamine receptors in patients with postviral fatigue syndrome. *Br. Med. J.* **304,** 1010–1012.

86. Sharpe, M., Hawton, K., Clements, A., and Cowen, P. J. (1997). Increased brain serotonin function in men with chronic fatigue syndrome. *Br. Med. J.* **315,** 164–165.

87. Bearn, J., Allain, T., Coskeran, P., Munro, N., Butler, J., McGregor, A., and Wessely, S. (1995). Neuroendocrine responses to d-fenfluramine and insulin-induced hypoglycemia in chronic fatigue syndrome. *Biol. Psychiatry* **37,** 245–252.

88. Yatham, L. N., Morehouse, R. L., Chisholm, B. T., Haase, D. A., MacDonald, D. D., and Marrie, T. J. (1995). Neuroendocrine assessment of serotonin function in chronic fatigue syndrome. *Can. J. Psychiatry* **40,** 93–96.

89. O'Keane, V., and Dinan, T. G. (1991). Prolactin and cortisol responses to d-fenfluramine in major depression: Evidence for diminished responsivity of central serotonergic function. *Am. J. Psychiatry* **148,** 1009–1015.

90. Dinan, T. G. (1996). Serotonin and the regulation of hypothalamic-pituitary-adrenal axis function. *Life Sci.* **58,** 1683–1690.

91. Gilbert, F., Brazell, C., Tricklebank, M. S., and Stahl, S. M. (1988). Activation of the 5HT1a receptor subtype increases rat plasma ACTH concentrations. *Eur. J. Pharm.* **147,** 431–437.

92. Dinan, T. G., Majeed, T., Lavelle, E., Scott, L. V., Berti, C., and Behan, P. (1997). Blunted serotonin-mediated activation of the hypothalamic-pituitary-adrenal axis in chronic fatigue syndrome. *Psychoneuroendocrinology* **22,** 261–267.

93. Briggs, N. C., and Levine, P. H. (1994). A comparative review of systemic and neurological symptomatology in 12 outbreaks collectively described as chronic fatigue syndrome, epidemic neuromyasthenia, and myalgic encephalomyelitis. *Clin. Infect. Dis.* **18**(Suppl. 1), S32–S42.

94. Dunn, A. J. (1993). Infection as a stressor: A cytokine-mediated activation of the hypothalamo-pituitary-adrenal axis? *Ciba Found. Symp.* **173,** 226–239.

95. Pollmacher, T., Mullington, J., Korth, C., and Hinze, S. D. (1995). Influence of host defense activation on sleep in humans. *Adv. Neuroimmunol.* **5,** 155–169.

96. Chrousos, G. P. (1998). Integration of the immune and endocrine systems by interleukin-6. P-134. In Papanicolaou DA, moderator. The pathophysiologic roles of interleukin-6 in human disease. *Ann. Intern. Med.* **128,** 127–137.

97. Ablashi, D. V. (1994). Viral studies of CFS. *Clin. Infect. Dis.* **18**(Suppl. 1), 130–132.

98. White, P. D., Thomas, J. M., Amess, J., Grover, S. A., Kangro, H. O., and Clare, A. W. (1995). The existence of a fatigue syndrome after glandular fever. *Psychol. Med.* **25,** 907–916.

99. Hotopf, M. H., and Wessely, S. (1994). Viruses, neurosis and fatigue. *J. Psychosom. Res.* **38,** 499–514.

100. Farrar, D. J., Locke, S. E., and Kantrowitz, F. G. (1995). Chronic fatigue syndrome 1: Etiology and pathogenesis. *Behav. Med.* **21,** 5–16.

101. Straus, S. E., Dale, J. K., Tobi, M., Lawley, T., Preble, O., Blaese, R. M., Hallahan, C., and Henle, W. (1988). Acyclovir treatment of the chronic fatigue syndrome: Lack of efficacy in a placebo-controlled trial. *N. Engl. J. Med.* **319,** 1692–1698.

102. Wilson, A., Hickie, I., Lloyd, A., and Wakefield, D. (1994). The treatment of chronic fatigue syndrome: Science and speculation. *Am. J. Med.* **96,** 544–550.

103. Farmer, A., Jones, I., Hillier, J., Llewelyn, M., Borysiewicz, L., and Smith, A. (1995). Neurasthenia revisited: ICD-10 and DSM-IIIR psychiatric syndromes in chronic fatigue patients and comparison subjects. *Br. J. Psychiatry* **167,** 503–506.

104. Neyta, N., and Horne, J. A. (1990). Effects of sleep extension and reduction on mood in healthy adults. *Hum. Psychopharmacol.* **6,** 173–188.

105. Morriss, R. K., Wearden, A. J., and Battersby, L. (1997). The relation of sleep difficulties to fatigue, mood, and disability in chronic fatigue syndrome. *J. Psychosom. Res.* **42,** 597–605.

106. Sharpley, A., Clements, A., Hawton, K., and Sharpe, M. (1997). Do patients with "pure" chronic fatigue syndrome (neurasthenia) have abnormal sleep? *Psychosom. Med.* **59,** 592–596.

107. Morris, R., Sharpe, M., Sharpley, A., Cowen, P. J., Hawton, K., and Morris, J. (1993). Abnormalities of sleep in patients with chronic fatigue syndrome. *Br. Med. J.* **306,** 1161–1164.

108. Krupp, L. B., Jandorf, L., Coyle, P. K., and Mendelson, W. B. (1993). Sleep disturbance in chronic fatigue syndrome. *J. Psychosom. Res.* **37,** 325–331.

109. Vercoulen, J. H. M. M., Swanink, C. M. A., Fennis, J. F. M., Galama, J. M., van der Meer, J. W., and Bleijenberg, G. (1994). Dimensional assessment of chronic fatigue syndrome. *J. Psychosom. Res.* **38,** 383–392.

110. Whelton, C. L., Salit, I., and Moldofsky, H. (1992). Sleep, Epstein-Barr virus infection, musculoskeletal pain, and depressive symptoms in chronic fatigue syndrome. *J. Rheumatol.* **19,** 939–943.

111. Manu, P., Lane, T. J., Mathews, D. A., Castriotta, R. J., Watson, R. K., and Abeles, M. (1994). Alpha-delta sleep patterns in patients with chief complaint of chronic fatigue. *South. Med. J.* **87,** 1289–1290.

112. Buchwald, D., Pascualy, R., Bombardier, C., and Kith, P. (1994). Sleep disorders in patients with chronic fatigue. *Clin. Infect. Dis.* **18**(Suppl. 1), S68–S72.

113. Denicoff, K. D., Durkin, T. M., Lotze, M. T., Quinlan, P. E., Davis, C. L., Listwak, S. J., Rosenberg, S. A., and Rubinow, D. R. (1989). The neuroendocrine effects of interleukin-2 treatment. *J. Clin. Endocrinol. Metab.* **69**, 402–410.

114. McDonald, L., Mann, A., and Thomas, H. C. (1987). Interferons and mediators of psychiatric morbidity: An investigation in a trial of recombinant alpha interferon in hepatitis B carriers. *Lancet* **2**, 1175–1178.

115. Landay, A. L., Jessop, C., Lennette, E. T., and Levy, J. A. (1991). Chronic fatigue syndrome: Clinical condition associated with immune activation. *Lancet* **338**, 707–712.

116. Caligiuri, M., Murray, C., Buchwald, D., Levine, H., Cheney, P., Peterson, D., Komaroff, A. L., and Ritz, J. (1987). Phenotypic and functional deficiency of natural killer cells in patients with chronic fatigue syndrome. *J. Immunol.* **139**, 3306–3313.

117. Barker, E., Fujimura, S. F., Fadem, M. B., Landay, A. L., and Levy, J. A. (1994). Immunologic abnormalities associated with chronic fatigue syndrome. *Clin. Infect. Dis.* **18**(Suppl. 1), S136–S141.

118. Morrison, L. J. A., Behan, W. M. H., and Behan, P. O. (1991). Changes in natural killer cell phenotype in patients with post-viral fatigue syndrome. *Clin. Exp. Immunol.* **83**, 441–446.

119. Kantrowitz, F. G., Farrar, D. J., and Locke, S. E. (1995). Chronic fatigue syndrome 2: Treatment and future research. *Behav. Med.* **21**, 17–24.

120. Komaroff, A. L., and Buchwald, D. S. (1998). Chronic fatigue syndrome: An update. *Annu. Rev. Med.* **49**, 1–13.

121. Goldenberg, D. L., Felson, D. T., and Dinerman, H. (1986). A randomized, controlled trial of amitryptiline and naproxen in the treatment of patients with fibromyalgia. *Arthritis Rheum.* **29**, 1371–77.

122. Goodnick, P. J., and Sandoval, R. (1993). Psychotropic treatment of chronic fatigue syndrome and related disorders. *J. Clin. Psychiatry* **54**, 13–20.

123. White, P. D., and Cleary, K. J. (1997). An open study of the efficacy and adverse effects of moclobemide in patients with the chronic fatigue syndrome. *Int. Clin. Psychopharmacol.* **12**, 47–52.

124. Natelson, B. H., Cheu, J., Hill, N., Bergen, M., Korn, L., Denny, T., and Dahl, K. (1998). Single-blind, placebo phase-in trial of two escalating doses of selegiline in the chronic fatigue syndrome. *Neuropsychobiology* **37**, 150–154.

125. Natelson, B. H., Cheu, J., Pareja, J., Ellis, S. P., Policastro, T., and Findley, T. W. (1996). Randomized, double-blind, controlled placebo-phase trial of low dose phenelzine in the chronic fatigue syndrome. *Psychopharmacology* **124**, 226–230.

126. Wessely, S., David, A., Butler, S., and Chadler, T. (1989). The management of the chronic "post-viral" fatigue syndrome. *J. R. Coll. Gen. Pract.* **39**, 26–29.

127. Surawy, C., Hackman, A., Hawton, K., and Sharpe, M. (1995). Chronic fatigue syndrome: A cognitive approach. *Behav. Res. Ther.* **33**, 534–544.

128. Sharpe, M., Hawton, K., Simkin, S., Surawy, C., Hackmann, A., Klimes, I., Peto, T., Warrell, D., and Seagroatt, V. (1996). Cognitive behaviour therapy for the chronic fatigue syndrome: A randomised controlled trial. *Br. Med. J.* **312**, 22–26.

129. Deale, A., Chadler, T., Marks, I., and Wessely, S. (1997). Cognitive behavior therapy for chronic fatigue syndrome: A randomized controlled trial. *Am. J. Psychiatry* **154**, 408–414.

130. Friedberg, F., and Krupp, L. B. (1994). A comparison of cognitive behavioral treatment for chronic fatigue syndrome and primary depression. *Clin. Infect. Dis.* **18**(Suppl. 1), S105–S110.

131. Lloyd, A., Hickie, I., Brockman, A., Hickie, C., Wilson, A., Dwyer, J., and Wakefield, D. (1993). Immunologic and psychologic therapy for patients with chronic fatigue syndrome: A double-blind, placebo-controlled trial. *Am. J. Med.* **94**, 197–203.

132. Fulcher, K. Y., and White, P. D. (1997). Randomised controlled trial of graded exercise in patients with the chronic fatigue syndrome. *Br. Med. J.* **314**, 1647–1652.

133. Weardon, A., Morris, R., and Mullis, R. (1996). A double blind placebo controlled trial of fluoxetine and graded exercise for chronic fatigue syndrome. *Eur. Psychol.* **11**(Suppl. 4), 273 (abstr.).

134. Fulcher, K. Y., and White, P. D. (1998). Chronic fatigue syndrome: A description of graded exercise treatment. *Physiotherapy* **84**, 1–4.

135. Kantrowitz, F. G., Farrar, D. J., and Locke, S. E. (1995). Chronic fatigue syndrome 2: Treatment and future research. *Behav. Med.* **21**, 17–24.

136. Joyce, J., Hotopf, M., and Wessely, S. (1997). The prognosis of chronic fatigue and chronic fatigue syndrome: A systematic review. *Q. J. Med.* **90**, 223–233.

137. Vercoulen, J. H. M. M., Swanink, C. M. A., Fennis, J. F. M., Galama, J. M., van der Meer, J. W., and Bleijenberg, G. (1996). Prognosis in chronic fatigue syndrome: A prospective study of the natural course. *J. Neurol. Neurosurg. Psychiatry* **60**, 489–494.

138. Krilov, L. R., Fisher, M., Friedman, S. B., Reitman, D., and Mandel, F. S. (1998). Course and outcome of chronic fatigue in children and adolescents. *Pediatrics* **102**, 360–366.

139. Sharpe, M., Hawton, K., Seagrott, V., and Pasvol, G. (1992). Follow up of patients presenting with fatigue to an infectious diseases clinic. *Br. Med. J.* **305**, 147–152.

140. Lloyd, A. R., and Pender, H. (1992). The economic impact of chronic fatigue syndrome. *Med. J. Aust.* **157**, 599–601.

141. Wolfe, F., Smythe, H. A., Yunus, M. B., Bennett, R. M., Bombardier, C., Goldenberg, D. L., Tugwell, P., Campbell, S. M., Abeles, M., and Clark, P. (1990). The American College of Rheumatology 1990 criteria for the classification of fibromyalgia. *Arthritis Rheum.* **33**, 160–172.

142. Buchwald, D., and Garrity, D. (1994). Comparisons of patients with chronic fatigue syndrome, fibromyalgia, and multiple chemical sensitives. *Arch. Intern. Med.* **154**, 2049–2053.

143. Hudson, J. I., Goldenberg, D. L., Pope, H. G., Keck, P. E., and Schlesinger, L. (1992). Comorbidity of fibromyalgia with medical and psychiatric disorders. *Am. J. Med.* **92**, 363–367.

144. Norregaard, J., Bulow, P. M., Prescott, E., Jacobsen, S., and Danneskiold Samsoe, B. (1993). A 4 year follow-up study in fibromyalgia. Relationship to chronic fatigue syndrome. *Scand. J. Rheumatol.* **22**, 35–38.

145. Hudson, J. I., and Pope, H. G. (1990). Affective spectrum disorder: Does antidepressant response identify a family of disorders with a common pathophysiology? *Am. J. Psychiatry* **147**, 552–564.

85

Fibromyalgia

DONNA J. HAWLEY* AND FREDERICK WOLFE[†]

*School of Nursing, Wichita State University, Wichita, Kansas; [†]University of Kansas School of Medicine-Wichita, and Wichita
Arthritis Research and Clinical Centers, Wichita, Kansas

I. Introduction/Background

Fibromyalgia is a controversial syndrome. It is a descriptive syndrome but is often treated as a disease. Its diagnosis as well as its severity depends almost entirely on self-report [1]. It seems to be associated with increased psychosocial distress as manifested by depression, somatization or somatoform disorders, anxiety problems [2–5], and, in some studies, increased rates of divorce, smoking, physical and sexual abuse, and alcohol–drug use [6–11]. It is associated with high rates of work disability and service utilization [12]. In population surveys it does not appear to be a distinct syndrome but rather the end of a spectrum of distress-related symptoms [9,13]. Finally, it has been suggested that fibromyalgia may be a byproduct of the later twentieth Century Zeitgeist, a creation by society and the medical system [14]. However, in spite of these problems, many persons seeking medical care clearly "have" fibromyalgia.

II. Definition

Fibromyalgia is a clinical syndrome characterized by widespread pain and decreased pain threshold that is associated with a series of somatic symptoms that may include sleep disturbance, fatigue, psychological distress, headaches, and irritable bowel syndrome, among other symptoms (Table 85.1). The disorder affects women eight to nine times more often than men and has an overall prevalence in U.S. women of about 2%. While it may begin in childhood, its major prevalence occurs after the age of 50. (Fig. 85.1).

III. Historical Evolution and Trends

Fibromyalgia, originally called fibrositis, may be traced to descriptions of tender point sites in the early part of the nineteenth Century [15]. Fibrositis was coined in 1904 by Gowers to describe a syndrome characterized by pain in the neck, shoulder, arm, back, and legs [16]. Although he and others that followed him recognized that there was no inflammation involved, the term persisted until 1990 [1,16]. Hench, in 1974, was the first to suggest that fibromyalgia was a better descriptive term for a condition involving "muscle as well as ligamentous and tendinous connective tissue" [17].

The modern era of fibromyalgia began in the late 1970s when Smythe and Moldofsky first published criteria for the syndrome [18]. The disorder found a convenient first home within rheumatology because so many rheumatic disease patients had generalized musculoskeletal pain that could not be adequately explained by the then current diagnostic paradigms. The early years of fibromyalgia were spent describing the fundamental

Table 85.1

Prevalence of Pain and Symptoms in the 1990 American College of Rheumatology Study of Criteria for the Classification of Fibromyalgia

Criterion	% Positive	Classification accuracy
Pain symptoms		
Pain posterior thorax	72.3	73.9
15+ painful sites	55.6	70.6
Neck pain	85.3	67.5
Low back pain	78.8	66.6
Widespread pain	97.6	65.9
Symptoms		
Sleep disturbance	74.6	73.8
"Pain all over"	67.0	73.6
Fatigue	81.4	71.7
Morning stiffness >15 minutes	77.0	67.2
Paresthesias	62.8	63.6
Anxiety	47.8	62.9
Headache	52.8	62.3
Prior depression	31.5	58.0
Irritable bowel syndrome	29.6	57.1
Sicca symptoms	35.8	55.4
Urinary urgency	26.3	54.2
Dysmenorrhea history	40.6	53.4
Raynaud's phenomenon	16.7	51.6
Modulating factors		
Noise	24.0	68.5
Cold	79.3	66.6
Poor sleep	76.0	65.2
Anxiety	69.0	63.7
Humidity	59.6	63.6
Stress	63.0	60.4
Fatigue	76.7	60.3
Weather change	66.1	60.3
Warmth	78.0	50.8

Modified from Wolfe *et al.* [1] with permission.

features—pain, tender points, psychological distress, fatigue, and sleep disturbance, establishing the reliability of diagnostic assessments, and developing criteria for diagnosis and classification. With the publication of the American College of Rheumatology (ACR) criteria [1], fibromyalgia became a describable and identifiable syndrome with readily applicable criteria.

With the establishment of fibromyalgia criteria [1,18,19], links to a series of syndromes were established, including irritable bowel and bladder syndrome, chronic fatigue syndrome, interstitial cystitis, depression, panic disorders, physical and

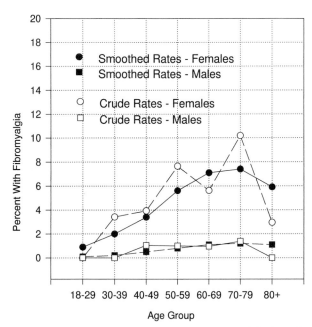

Fig. 85.1 Prevalence for age and sex for fibromyalgia from the population-based study conducted in Wichita, Kansas for people 18 years and older. Circles represent women and squares men. From Wolfe *et al.* [9] with permission.

sexual abuse, paresthesias, and noncardiac chest pain [2,3,6,20–26], and Hudson labeled the syndrome as part of an affective spectrum disorder [27]. Epidemiology studies next suggested that the signs and symptoms of fibromyalgia were not unique but instead represented the end of a spectrum of distress [9,13].

While these investigations of fibromyalgia were defining the syndrome, others using physical, biochemical, and neuroimaging methods were making efforts to find causes and mechanisms for the clinical manifestations. Macro- and micro-abnormalities of muscle, hypothalamic-pituitary-adrenal (HPA) abnormalities, brain flood flow changes, and serotonin and other central nervous system amino acids were all proposed as causative or contributing factors [28–34].

A wide series of treatment options developed, including drugs, vitamins, herbs, physical therapy, exercise, electrical stimulation, and acupuncture. Whole clinics devoted to fibromyalgia developed. An environment of Internet sites devoted to fibromyalgia developed, and the number of patient support groups, often linking fibromyalgia to chronic fatigue syndrome and chemical sensitivities syndrome, grew rapidly. Although sporadic reports of meaningful improvement appeared, most therapies did not produce positive results or were only of marginal benefit [12,35,36].

Fibromyalgia has grown to have important societal consequences. It began to be a recognized cause for disability pensions [37], and in a number of countries litigation regarding fibromyalgia and its relationship to trauma consumed significant resources [38]. In addition, there seemed to be overdiagnosis of the syndrome both by patients and by physicians [39]. It grew to be an easy catchall for difficult to categorize musculoskeletal problems or for those with psychological distress [38]. Patients with

greater distress and psychological abnormalities began to dominate those seen in clinical settings [12,35], although symptoms and outcomes may be less severe for community populations [7].

Thus the desirable aspects of identifying and understanding a syndrome began to clash with the undesirable consequences of the syndrome, and serious direct and indirect criticism arose. What, after all, was fibromyalgia? Was it a useful construct? In many respects the criticism came back to the issue that had haunted the syndrome all along. Wasn't it really just distress or a psychological disorder? Wasn't it a problem of adequate coping and of behavioral expressions of pain?

IV. Epidemiology: Prevalence of Fibromyalgia in the Clinic and Community

Fibromyalgia and its counterpart, chronic widespread pain, are commonly seen in clinical situations and in the community. Both are more common in women than in men, with women outnumbering men eight or nine to one [9,13]. Gender differences in a variety of pain conditions including musculoskeletal pain are well-documented and have been reviewed extensively [40]. Women generally report greater responses to pain induced in experimental situations [41] as well as more chronic widespread pain, a lower pain threshold, fatigue, self-reported sleep disturbances, and irritable bowel symptoms than men [42]. Gender differences extend to studies of children with chronic pain and fibromyalgia, although the differences are not consistent, especially prior to adolescence [43].

A. Clinical Studies

In a U.S. general medical clinic, fibromyalgia was noted in 3.7% of the patients and 73% were women [44]. Prevalence estimates in family practice clinics were noted at 1.6% in the Netherlands and 2.1% in a U.S. study [45,46]. In rheumatology clinics, fibromyalgia is only exceeded by rheumatoid arthritis and osteoarthritis in the frequency of disorders seen, occurring with a prevalence of 12% [47] to 20% [19].

B. Community Studies

The only North American study to date noted the prevalence of fibromyalgia to be 0.5% (95% Confidence Interval (CI)-0.0–1.0) and 3.7% (95% CI-2.3–4.6) in men and women, respectively, and 2.0% (95% CI-1.4–2.7) for both sexes combined [9]. European prevalence estimates were as high as 10.5% (95% CI-6.4–14.6) when studying women only [48]. Extrapolating from a major population-based study of chronic widespread pain in the United Kingdom, the prevalence for both sexes was about 4% [13]. Prevalence studies outside the industrialized countries are not available, but two investigations provide some interesting information for consideration. Lyddell and Meyers studied essentially all residents (84%) of a small South African village, finding that 3.2% met ACR tender point criteria, of whom 100% were women [49]. A study of Pima Indians in the United States, ages 35 to 64 years (N=105), found no evidence of widespread pain, with over 90% reporting that they had never experienced regional pain for over 3 months [50]. In summary, population-based studies conducted to date indicate the adult prevalence (both sexes) is about 2–3% when all studies are considered, with prevalence in women several percentage points higher than that in men [9,13,51,52].

C. Prevalence in Children

There have been only a few prevalence studies of children with fibromyalgia or widespread pain. Buskila and colleagues reported a prevalence of 3.9% in boys and 8.8% in girls ages 9 to 15 years in one school in Israel [53]. Mikkelsson *et al*, in a population-based study of slightly younger children (9–12 years) in Finland, reported a prevalence of 7.5% for widespread pain occurring at least once per week during the last 3 months; however, they reported no significant difference between the sexes [54,55]. Cicuttini, and Littlejohn, reporting from a referral rheumatology practice, found that 19 of 60 consecutively referred adolescent women met criteria for fibromyalgia [56].

V. Epidemiology: Demographic Characteristics

Although community prevalence studies have been done [13,9], the majority of studies addressing demographic characteristics have been conducted in special clinical referral centers and may not reflect the characteristics of persons in the community or those seen in primary care settings.

Mäkelä and Heliövaara [57] performed an evaluation on 7217 Finns aged 30 years or older. Using the Yurus Criteria for fibromyalgia [19], they found a strong association between fibromyalgia and lower education levels in multivariate analyses, but not with body mass index, physical stress at work, or smoking. The presence of heavy labor and osteoarthritis in participants may have confounded the results. Wolfe *et al*. in their large community prevalence study, also noted an association of lower education level with fibromyalgia [9]. They reported that not completing high school was more common among those with fibromyalgia, with an odds ratio of 3.52 (95% CI-1.04–11.90) [9].

In this same study, Wolfe *et al*. reported a crude divorce rate of 20.3% [9]. Persons with fibromyalgia were 4.32 (95% CI-1.03–18.12) times more likely to be divorced than those in the general population without fibromyalgia. Hawley *et al*. studied marital status and divorce in a clinical sample [11]. Among 7293 consecutive patients seen in a rheumatology practice, 15.4% of 792 rheumatoid arthritis patients, 12.3% of those with osteoarthritis of the knee or hip, and 25.7% of patients with fibromyalgia had ever been divorced. The odds of ever being divorced were 2.2 (95% CI-1.50–2.99) times greater for patients with fibromyalgia compared with rheumatoid arthritis patients. However, current divorce status did not differ among any patient diagnostic group. Middleton *et al*. [10], in a study of fibromyalgia as part of systemic lupus erythematosus, found that the divorce rate was 10.9% among those without fibromyalgia and approximately 38% in those with fibromyalgia.

In a separate study, Wolfe and colleagues examined rheumatoid arthritis, osteoarthritis, and fibromyalgia patients [14] and by cluster analysis classified patients into groups of "coping," "stressed," and "dysfunctional" patients without regard to diagnosis. Increasing patterns of distress were associated with increased smoking, BMI, divorce rate, and lower income and education regardless of diagnosis. Approximately 15% of rheumatoid arthritis and 15% of osteoarthritis patients were in the "dysfunctional cluster," but 51% of those with fibromyalgia were in that cluster; however, only 18.5% of the original sample had the diagnosis of fibromyalgia.

Fibromyalgia in diverse ethnic groups has not been studied extensively. With the exception of the study in a small village in South Africa [49] and one with Pima Native Americans in the U.S. [50], ethnic distribution is either not reported or approaches 90% Caucasian. Although not yet supported by published studies, there is some evidence that fibromyalgia may occur less often in blacks than in Caucasians. Calvo-Alen and colleagues [58] at the University of Alabama retrospectively reviewed charts of 149 patients with systemic lupus erythematosus that had positive antinuclear antibodies and found that 37 could be classified as "fibromyalgia-like." Two (5%) of the "fibromyalgia-like" patients and almost 60% (N = 71) of the remaining patients were African American.

Prevalence of physical and sexual abuse determined through self-report varies from 53 to 76% in fibromyalgia patients seen in rheumatic disease clinics [6,8,20,59]. Rates for fibromyalgia patients were 10% greater than those for other female rheumatic disease patients [8], healthy controls [8], and rheumatoid arthritis patients [6].

VI. Diagnosis and Methodologies for Diagnosis

Fibromyalgia may be defined operationally as (1) widespread pain complaints, (2) decreased pain threshold, as identified by the presence of *many tender points,* and (3) the presence of characteristic symptoms such as fatigue, sleep disturbance, headaches, irritable bowel syndrome, and psychological distress, among other symptoms. The first two items of this definition form the criteria requirement of the ACR (*i.e.,* widespread pain and pain on palpation at a minimum of 11 of 18 tender point sites). The complete criteria are listed in Table 85.2 and depicted in Figure 85.2.

Fibromyalgia has been characterized as a disorder of "pain modulation" [60–63], "pain amplification" [64], and as an "irritable everything syndrome" [64]. It may be conceptualized as a syndrome in which the threshold for all stimuli appears to be decreased and/or that stimuli appear to be "amplified."

A. Pain Amplification

Nociceptive stimuli, which are in most individuals below the pain threshold, are perceived as pain in persons with fibromyalgia. Mild nociceptive stimuli are perceived as more than mild and often as severe pain. People with fibromyalgia appear to feel, interpret or amplify stimuli to the point of pain when those without the syndrome may experience little if any distress [65]. Figure 85.3 displays a typical pain drawing from a patient with fibromyalgia. The figure emphasizes the characteristic distribution of fibromyalgia pain: widespread, axial, radiating, and joint associated. The extent of the pain is often surprising and, in and of itself, may suggest the diagnosis.

B. Pain Threshold

The major observable manifestation of the reduced pain threshold is seen with the palpation of tender points. Tender points are central to the diagnosis of fibromyalgia and are the

Table 85.2

The 1990 American College of Rheumatology Criteria for the Classification of Fibromyalgia

1. History of widespread pain.

    Definition: Pain is considered widespread when all of the following are present: pain in the left side of the body, pain in the right side of the body, pain above the waist and pain below the waist. In addition, axial skeletal pain (cervical spine or anterior chest or thoracic spine or low back) must be present. In this definition shoulder and buttock pain is considered as pain for each involved side. "Low back" pain is considered lower segment pain.

2. Pain in 11 of 18 tender point sites on digital palpation.

    Definition: Pain, on digital palpation, must be present in at least 11 of the following 18 tender point sites:

    Occiput: bilateral, at the suboccipital muscle insertions.

    Low cervical: bilateral, at the anterior aspects of the inter-transverse spaces at C5-C7.

    Trapezius: bilateral, at the midpoint of the upper border.

    Supraspinatus: bilateral, at origins, above the scapula spine near the medial border.

    2nd rib: bilateral, at the second costochondral junctions, just lateral to the junctions on upper surfaces.

    Lateral epicondyle: bilateral, 2 cm distal to the epicondyles.

    Gluteal: bilateral, in upper outer quadrants of buttocks in anterior fold of muscle.

    Greater trochanter: bilateral, posterior to the trochanteric prominence.

    Knees: bilateral, at the medial fat pad proximal to the joint line.

    Digital palpation should be performed with an approximate force of 4 kg/mm².

    For a tender point to be considered "positive" the subject must state that the palpation was painful. "Tender" is not to be considered painful.

Note. For classification purposes patients will be said to have fibromyalgia if both criteria are satisfied. Widespread pain must have been present for at least 3 months. The presence of a second clincal disorder does not exclude the diagnosis of fibromyalgia. From Wolfe *et al.* [1] with permission.

Fig. 85.2 The 18 tender point sites from the 1990 American College of Rheumatology (ACR) criteria for the classification of fibromyalgia. Eleven of 18 tender points satisfy the tenderness criterion. From Wolfe *et al.* [1] with permission.

method that demonstrates lowered pain threshold [18]. There are specific anatomical areas in muscles, tendon, fat pads, or over bone where the pain threshold to palpation is ordinarily reduced even in people who do not have fibromyalgia. In fibromyalgia these points are quite painful. Many such sites exist throughout the body [66,67]. For use in diagnosis, a number of sites have been identified that tend to be painful in those with fibromyalgia and nonpainful in those without fibromyalgia. The most commonly used map of tender point sites is that proposed by Wolfe *et al.* [1]. See Fig. 85.2 and Table 85.2.

In fibromyalgia patients, pain is elicited by palpating the tender point site with approximately 4 kg/mm² of force. This is equivalent to the force necessary to produce blanching in the nail of the palpating finger. Palpation is usually performed using the second and third fingers or the thumb, and often with a

rolling motion. A positive tender point is one in which the patient states that the examination caused pain. Tender point sites may be tender in those without fibromyalgia; therefore, a response indicating "pain" or "hurt" is necessary for a tender point to be considered positive [1]. The patient determines the pain, not the examiner.

Tenderness can be quantified somewhat more objectively by the use of pressure algometers or dolorimeters. Algometers remove the variability of the palpating force during digital examination as well as the need to interpret the patient's response. A value equal to or less than 4 kg/1.77 cm² has been used as approximately equivalent to a positive tender point [1,68–70].

Trigger point is a term that is sometimes confused with tender point. Trigger points are pressure-sensitive, tender areas in muscle. The concept has been used as a diagnostic feature and a treatment focus area for myofascial pain syndromes [71,72]. Trigger points have not been well studied within controlled situations and controversy concerning their pathological significance remains [73]. A review of trigger points is beyond the purview of this chapter and the concept is addressed only to make the point that trigger points and tender points are different. Generally speaking, fibromyalgia tenderness, as evidenced by a high count of tender points, is a reflection or manifestation of the decreased pain threshold, while the trigger points of the

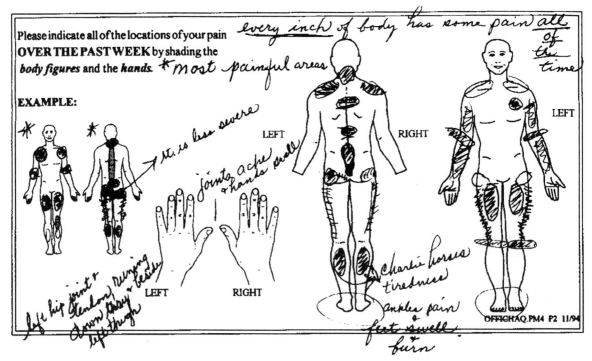

Fig. 85.3 This figure is a typical pain drawing from a patient with fibromyalgia. The figure emphasizes the characteristic distribution of fibromyalgia pain: widespread, axial, radiating, and joint associated. The hand-written comments are illustrative of the distress experienced by persons with fibromyalgia. The extent of the pain is often surprising and, in and of itself, may suggest the diagnosis.

myofascial pain syndrome do not reflect a generally low pain threshold but refer to tender muscle areas. There is nothing within these two constructs that prevents a tender point from also being a trigger point or, stated another way, fibromyalgia and the myofascial pain syndrome can coexist.

C. Symptoms

The symptoms of fibromyalgia (Table 85.1) have been studied extensively [19,44,74–78]. These symptoms may be used with varying accuracy to classify persons as having or not having fibromyalgia. The classification accuracy of these symptoms when compared to rheumatic disease control patients is shown in Table 85.1 [1]. Sleep disturbance, fatigue, and morning stiffness are present in 75–81% of patients, while "pain all over," paresthesia, headache, and anxiety occur in 49–67% of patients [1]. Symptom criteria and widespread pain alone may allow identification of persons with fibromyalgia; however, symptoms alone will fail to identify many patients with fibromyalgia. Symptoms have a high false-positive rate, particularly in those with other rheumatic conditions, as indicated in Table 85.1. In the 1990 ACR classification criteria study, the addition of symptom criteria to the tender point count and the presence of widespread pain did not improve criteria sensitivity and specificity [1].

D. Clinical Presentations: Diagnosis and Differential Diagnosis

Fibromyalgia is frequently misdiagnosed either as rheumatoid arthritis or as osteoarthritis because peripheral joint pain in the knees, shoulders, hands, and wrists are common presenting symptoms and the patient often complains of joint swelling [79]. The physical symptoms may also suggest systemic lupus erythematosus, especially when [19] a positive antinuclear antibody is identified [80,81]. Pain predominating in the axial skeletal area may mimic compressive nerve root and disk syndromes, and the prevalence of axial skeletal surgery is high among those with the syndrome [12,82]. Childhood fibromyalgia may be misdiagnosed as juvenile or adult rheumatoid arthritis for reasons similar to those in adults: joint pain and reported swelling. Chronic fatigue syndrome (CFS) shares most features in common with fibromyalgia [25]. The main difference between the syndromes is the predominance of the incapacitating nature of the fatigue in CFS and the predominance of pain in fibromyalgia.

VII. Clinical Syndrome

A. Subtypes and Characteristics

Describing, defining, and then developing acceptable diagnostic criteria for fibromyalgia has presented challenges for clinicians and researchers alike. Until 1990, three types were recognized [1]. *Primary* fibromyalgia was said to exist when no other disease or pathology was present that could explain the symptoms or cause the disorder. *Secondary* fibromyalgia existed where another condition such as sleep apnea, osteoarthritis, or even trauma might have caused the disorder. *Concomitant* fibromyalgia was present when another disorder such as rheumatoid arthritis (RA), systemic lupus erythematosus (SLE), or osteoarthritis (OA) coexisted with the fibromyalgia without a causal inference being made.

Two different processes led to the evolution of this set of terms. First, in order to study fibromyalgia "cleanly" (primary fibromyalgia), other disorders whose symptoms and physical findings *might* overlap with fibromyalgia needed to be excluded. The second logical process assumed that fibromyalgia could be caused by illnesses such as hypothyroidism, sleep apnea, or trauma; however, the causal link was based on anecdotal reports. These causal connections still remain largely unsubstantiated and *concomitant* fibromyalgia is a more accurate term that expresses this uncertainty.

In 1990, the ACR criteria for the classification of fibromyalgia [1] "abolished the distinction between primary and secondary or concomitant fibromyalgia" for diagnostic purposes because no clinical or diagnostic differences between the various types of fibromyalgia could be found, nor were the subclassifications useful as part of the diagnostic criteria [1]. Thus, fibromyalgia ceased to be a disorder of exclusion. Instead it became a disorder that could be identified by specific criteria and could exist concurrently with a variety of rheumatic and nonrheumatic conditions.

B. Onset Types and Characteristics

Fibromyalgia may begin in childhood or in the elderly but most commonly it begins in midlife. A review of types and characteristics of onset may provide clues to risk factors, although at this point data are limited and depend on the memory and self-report of those with the condition.

Several reports document the presence and characteristics of fibromyalgia in children [43,53–55,83–85]; however, these reports do not address age when these symptoms began. Persistence of symptoms one year to two years later was noted in 27% and 29.7% in studies from one school in Israel and a large community in Finland, respectively [54,84]. These persistence rates are very similar to the 35% persistence rate observed by Kennedy and Felson in their community based follow-up study, but considerably lower than rates in patients seen in clinics [86–89].

Adult-onset fibromyalgia has several patterns that may be classified as follows: childhood onset of symptoms, gradual onset of symptoms in adult life, fibromyalgia in association with various (often) chronic musculoskeletal problems, fibromyalgia following trauma, and fibromyalgia following apparent viral infection.

Patients presenting as young and middle-aged adults often describe a lifelong history of mild musculoskeletal complaints, usually beginning in adolescence. Difficulty with athletic activities in school, "growing pains," and "leg aches" are described frequently [90]. This group may have irritable bowel syndrome and migraine headaches that also develop in adolescence. Yunus reported 14–28% [19] with childhood onset while Wolfe reported a rate of 16% [90]. A gradual onset pattern also has been described when an otherwise healthy adult experiences gradually increasing musculoskeletal pain [90].

Post-surgical or "viral" onset usually occurs in a generally previously healthy individual. Fibromyalgia symptoms may follow a viral or "flu-like" syndrome. Depending on the sample and the specific question asked, this type of onset has been reported in 6% [91], 18% [90,92], 22% [93], and 55% [93,94] of patients.

Fibromyalgia also occurs as part of other chronic musculoskeletal conditions, including two identifiable subgroups. In the first, fibromyalgia develops as a concomitant of other musculoskeletal problems, particularly osteoarthritis complaints of the axial skeleton and peripheral joints. Patients complain more than would be expected about their "arthritic" problems and about more arthritic problems than are expected. Goldenberg reported that 15% of patients seen in a referral rheumatology clinic had this type of insidious onset [92].

In the second subgroup, fibromyalgia is diagnosed in a background of major musculoskeletal medical complaints, multiple surgical interventions, and is associated with marked pain behavior. Individuals frequently have long surgical and medical histories, significant anxiety, and aspects of somatization disorder [5,82]. The specific date of onset generally cannot be determined.

"Posttraumatic" fibromyalgia may include both physical and emotional trauma as self-reported precipitating events for fibromyalgia. Physical trauma represents injuries from a motor vehicle accident and work injuries, with some investigators including surgery and medical illness as part of physical trauma. Again, varying with the type of physical trauma included in the definition, patients report this type of onset 11–33% of the time [90,92,95,96]. Wolfe and colleagues, in a large study of over 800 patients, found that those receiving social security disability payments were more than two times as likely to report onset of symptoms following trauma than those not receiving such benefits [95]. Similarly, Greenfield [96] reported that patients whose fibromyalgia may be traced to a trauma, surgery, or a medical illness are more likely to report limited physical activity, be unemployed, or be receiving disability payments than those whose fibromyalgia began through a different mechanism.

Emotional trauma such as death in the family, divorce, combat, certain types of surgery with an emotional overlay, or other traumatic emotional events may also be perceived by patients as the triggering event. Percentages vary from 14% reported by Goldenberg in Boston, to 26% from Aaron and colleagues in Alabama, to 36% by Burckhardt in Oregon [92,97,98].

Retrospective reports of an association with antecedent events and the subsequent development of fibromyalgia should be weighed with considerable care and skepticism because association is not causation. There is considerable noise in these data. The issue of causality is complex, and an "injury" is not always an injury. The link between trauma and fibromyalgia on a pathophysiological basis is often very tenuous and needs further investigation [92,97,99–102]. There is no general consensus that trauma causes fibromyalgia, and further studies are needed. Compensation issues remain unresolved and vary with the social system of the country [96,97,100–102].

VIII. Clinical Symptoms

As previously described, the main clinical feature of fibromyalgia is widespread pain. Other clinical symptoms are illustrated on Table 85.1. Fatigue, sleep disturbances, irritable bowel syndrome, irritable bladder syndrome, and psychological distress warrant additional consideration.

A. Fatigue

Fatigue is a common symptom in the general population with a prevalence of 18–33%. It is 1–2% higher in women than in men. Fatigue is also a major symptom in the rheumatic disorders, with over 90% of patients reporting some degree of fatigue [103]. In a study of almost 1500 rheumatic disease patients, 76% of the fibromyalgia patients reported substantial fatigue while only 40% of the rheumatoid arthritis and osteoarthritis patients reported such a fatigue level [104]. (Substantial fatigue was a score $\geq$ 2 on a 3 point visual analogue scale.) In a general population survey, approximately 66% of subjects with fibromyalgia complained of fatigue [9]. The suggestion of Campbell et al. that fatigue be qualified by the definition of (usually or often) being "too tired during the day to do what I want to do" adds an appropriate method of assessment [44]. Visual analog and other scales may add additional standardized methods of assessment [105,106].

B. Sleep Disturbance

Moldofsky originally described the sleep disturbance of fibromyalgia in terms of "alpha delta intrusion" observable in electro encephalographs (EEGs) during sleep studies [107]. Studies of sleep architecture have found increases in Stage I, a reduction in delta waves, more frequent and longer awakenings, and reduced sleeping time compared to healthy controls [108]. Sleep disturbance as a cause or consequence of fibromyalgia remains under study. Nonetheless, sleep problems cause major distress for patients. "Awaking unrefreshed" is often used clinically to evaluate sleep complaints; however, the relationship between symptomatology and sleep physiology has not been confirmed [44,78,109,110].

C. Irritable Bowel Syndrome (IBS)

IBS has been reported in 34–53% of patients with fibromyalgia [1,19,44,47]. Using a validated self-administered questionnaire, Triadafilopoulus reported "altered bowel function" in 73% of patients; 64% reported "abdominal pain" [111].

D. Irritable Bladder Syndrome and Interstitial Cystitis

Although bladder symptomatology has not been studied extensively in people with fibromyalgia [74], the relationship between various genitourinary symptoms, disorders, and pain syndromes is well recognized [112]. In a survey of over 2000 individuals with interstitial cystitis, fibromyalgia was diagnosed in 17%. This prevalence was the third most common accompanying diagnosis, with allergies (42%) the most common and irritable bowel syndrome next (30%) [26].

E. Psychological Distress

Since 1985, psychological distress in patients with fibromyalgia has been examined by numerous investigators in many countries using a variety of instruments and interview procedures with remarkably uniform results. While there are exceptions [77,113], the majority of studies have demonstrated a consistent pattern of emotional and psychological distress in fibromyalgia patients. Studies have centered on depression and depressive symptoms, anxiety, and somatization.

Depressive symptoms, lifetime depression, and current depression are more frequent and serious in fibromyalgia patients when compared to rheumatoid arthritis patients, other rheumatic disease patients, and healthy controls. Lifetime depression rates as determined by standard interview techniques have ranged from 20 [5] to 86% [114], with a median of about 58% in 12 studies [3]. In four studies comparing lifetime depression in patients with rheumatoid arthritis (RA) to fibromyalgia patients, RA lifetime depression rates ranged from 5 [115] to 39% [113], while rates for healthy controls in two studies varied from 9 to 26% [113,116]. Correspondingly, depressive symptoms and current depression, measured through various self-report questionnaires and interviews, have been consistently higher in fibromyalgia patients compared to rheumatoid arthritis patients [117–119], other rheumatoid disease patients [118], and healthy controls [116,120].

High levels of anxiety have been reported for persons with fibromyalgia using self-report scales and structured interviews. Using the same self-report questionnaire, Burckhardt and colleagues [117] and Hawley and Wolfe [118] reported higher anxiety levels for fibromyalgia compared to rheumatoid arthritis patients. Similar findings were reported by Walker and colleagues using the structured Diagnostic Interview Schedule (DIS) [2]. Krag, in Denmark, also found higher levels of anxiety in fibromyalgia patients compared to rheumatoid arthritis and low back pain patients [121].

Somatization rates that exceed those found in rheumatoid arthritis patients have been documented through the Diagnostic Interview Schedule [2,4,5] and also on questionnaires [122]. Self-reports by fibromyalgia patients of physical symptoms and comorbid conditions relating to both numbers of symptoms and perceived severity exceed those of other rheumatic disease patients, illustrating further somatization as a feature of this disorder [2,123]. In fact, as shown in Table 85.1, very high rates of symptom reporting are strongly associated with fibromyalgia.

While the evidence for psychological distress as a component of fibromyalgia is extensive, one must also recognize that the issue is complex. Pain is an important stressor and not all patients with fibromyalgia have antecedent psychological abnormality or react to pain in an inappropriate way. This observation was underscored by Bennett when he said "You don't have to be psychologically disturbed to have fibromyalgia" [124].

IX. Additional Physical Findings

A. Swelling

Patients frequently complain of swelling, often in characteristic locations—in the hands, around the elbows, and in the medial aspect of the knees. Yunus and colleagues referred to this phenomenon as "subjective swelling" because actual swelling could not be observed [19]. Physicians have observed, and patients have described, a tightness of their wristwatch band or rings and the loss of the skin lines over the metacarpal phalangeal joints. Swelling is nonarticular and often improves during the day. The swelling may be related to complaints of carpal tunnel symptoms. In an early study, 32% of fibromyalgia patients seen in a referral center reported such nonarticular swelling [19]. Wolfe and colleagues in their community study of fibromyalgia found that 71% of those with fibromyalgia reported "joint swelling," although such swelling was not observed by the study investigators [9,52].

B. Neurogenic Inflammation

Reactive hyperemia was described early in the descriptions of fibromyalgia [18,19]. Quantification of the presence of erythema in the skin following mechanical (palpation) or chemical (capsaicin) stimulation was demonstrated by Littlejohn and associates who showed increased erythema in those with fibromyalgia compared with controls [125] and normals [126]. Common reactive hyperemia, as seen in a clinical examination, lacks sufficient specificity to be used for diagnosis because it is seen (in lesser proportion) in others with rheumatic diseases [1].

C. Tissue Compliance

Tissue compliance, or the firmness or consistency of the tissue [127,128], is significantly lower in patients with fibromyalgia than in pain-free controls and has good diagnostic specificity [126].

1. Skin Fold Tenderness

Skin fold tenderness is evaluated by grasping and rolling between the fingers a major muscle. In the pioneering study by Granges and Littlejohn, skin fold tenderness was present in 95% of fibromyalgia patients, in 33% of pain-free control subjects, and was absent in all the exercising, physically fit control participants [126]. Skin fold tenderness is another indicator of pain threshold.

D. Laboratory Studies

Routine laboratory studies are generally normal in fibromyalgia patients unless an additional disorder is present that is associated with laboratory abnormalities. A small subset of fibromyalgia patients (20.5%) who had features suggestive of a connective tissue disorder, including Raynaud's phenomenon (30.5%), sicca syndrome (14%), low C3 (17%), IgG deposition at the dermal epidermal junction (14%), and positive antinuclear antibodies (23%), was identified by Dinerman et al. [76]. Enestrom et al. have shown in skin biopsies that fibromyalgia patients had significantly higher IgG deposits in the dermis, higher number of mast cells, and higher reactivity for collagen III than healthy controls, rheumatoid arthritis patients, and those with chronic pain from a whiplash injury [129]. This finding has not been replicated and must be viewed with caution [129].

A series of antibody-related findings have been reported. Antibodies to serotonin receptor have been reported in fibromyalgia [130]. Anti-striated muscle antibodies and anti-smooth muscle antibodies (but not other auto-antibodies) were found in 40–55% of fibromyalgia patients but not in controls in a single study [131]. However, other reports found no evidence of increases in antibodies [132]. Another study found that the prevalence of thyroid microsomal antibodies was significantly higher in persons with fibromyalgia compared with those without chronic widespread musculoskeletal complaints (16.0% versus 7.3%); however, the fibromyalgia diagnosis was not confirmed by examination [133]. In addition, sophisticated studies of cytokines and immune regulation in fibromyalgia disclosed no abnormalities [134].

X. Pathophysiology

Investigators have examined muscle structure and metabolism, neurochemicals, and mechanism of central processing of stimuli to determine possible pathophysiological mechanisms that could explain fibromyalgia. These mechanisms are still not well understood and controversies remain. Specific biological markers, anatomical changes, or chemical abnormalities have been postulated and identified in a few studies. Disturbances of muscle function and structure, nociception, HPA axis, central neurotransmission, and psychological functioning have all been implicated. Which factors might be causative or contributory and which are epiphenomena remain open for further study.

Substance P is a neuropeptide that seems to facilitate nociception [31]. Substance P in the cerebrospinal fluid (though not in plasma) [31,135] or urine [31] appears to be higher in those with fibromyalgia than in normals [136–138]. Whether elevated levels of substance P are unique to fibromyalgia or are found in other painful conditions is still not determined [31,139].

Studies using single photon emission computed tomography (SPECT) and positron emission tomography (PET) have suggested decreases in regional cerebral flow in the thalamus and the caudate nucleus [140]. Hypotheses that decreased regional cerebral blood flow may be related to alterations of pain thresholds seen in fibromyalgia are being suggested and tested [29,140]. Additional studies in this area are ongoing.

Other generalized measures of central nervous system (CNS) dysfunction have been described, including abnormal cortisol diurnal response [141], increased prolactin levels [142,143], decreased somatomedin-C levels [144], abnormal response to thyroid stimulating hormone [145,146], low levels of serotonin or its metabolites centrally and in the CNS [137] [147], and abnormal central levels of amino acids [34,148].

Serum muscle enzymes and electromyography are normal [75]. Some histological studies have found structural abnormalities in muscle fibers of fibromyalgia patients compared to controls [149–152] while others have found little if any difference morphologically between patients and controls [153]. The specificity and importance of the findings relative to pathogenesis of fibromyalgia remain areas of debate [30,32,154]. Magnetic resonance spectroscopy (MRS) has been performed to evaluate metabolic activities in muscles of fibromyalgia patients both under resting and under exercising conditions [154,155]. Several studies have failed to find serious metabolic abnormalities [156–158]; however, one study identified such a metabolic abnormality [159].

The effects of hormonal changes on pain and other symptoms of fibromyalgia during both the menstrual cycle and menopause may be important; however, the data are very limited. Crofford points out that the high prevalence of fibromyalgia and other stress-related syndromes in women "... supports examination of the reproductive axis and its effects on other neuroendocrine systems" [160]. A retrospective study that involved interviews of fibromyalgia patients and controls found that symptoms were worse in the premenstrual period and that pregnancy and the postpartum period tended to exacerbate symptoms [161]. Hapidou and Rollman reported that tender point counts increased during the follicular or postmenstrual phase compared to the luteal or intermenstural phase of college-age women with normal cycles compared to those taking oral contraceptives [162]. Certainly this area warrants increased attention and controlled clinical investigations of the changes in pain perceptions before, during, and following the menstrual period.

XI. Treatment

Treatment of fibromyalgia, when considered as a whole, has not been successful. Short-term improvements are noted from various interventions; however, long-term effects are not promising [12,35,163,164]. Common treatment modalities are discussed below.

A. Treatment of Musculoskeletal Problems Concurrent with Fibromyalgia

Population data as well as clinical data suggest that most persons with fibromyalgia are above age 50; therefore, they commonly have accompanying musculoskeletal disorders. Treatment of these disorders may improve fibromyalgia symptoms. In patients with inflammatory arthritis who are treated with corticosteroids during an acute flare, important reductions in tender point counts and pain scores have been reported [165]. Nonsteroidal anti-inflammatory drug (NSAIDs) therapy in primary fibromyalgia has not been shown to be of benefit in several controlled trials [166–168], but it can be of value in some patients with osteoarthritis and with degenerative disorders of the axial skeleton.

B. Exercise

An early finding by Moldofsky [169] that sedentary individuals deprived of sleep developed fibromyalgia-like symptoms while trained athletes did not, and the observations by Bennett and colleagues that over 80% of fibromyalgia patients were "deconditioned," have led to studies of various exercise programs [170]. McCain and colleagues performed the first major study of aerobic exercise. Cardiovascular fitness improved in an indoor bicycle aerobic group but not in those undergoing stretching [171]. Pain threshold scores, patient and physician global assessment scores, and pain scores improved. Although subsequent controlled trials have shown aerobic exercise programs to be of benefit in fibromyalgia [172–176], improvement in aerobic capacity and physical fitness have been minimal [171, 174–177]. Some studies report improvement in self-efficacy, generally feeling better, and improved quality of life [174]. Recruitment of participants [177] and high numbers of withdrawal from exercise programs remain problematic [175,177].

C. Psychopharmacological Agents

Amitriptyline, and to a lesser extent cyclobenzaprine, are commonly prescribed in treatment of fibromyalgia. Short-term clinical trials [110,168,178–182] have demonstrated benefits from both drugs, including lower pain scores. Effects are short-lived, with only about one-third of patients starting therapy continuing for as long a year. Symptoms may revert to pretreatment levels a few weeks after therapy begins [178,183]. Originally these drugs were thought to affect a non-REM sleep problem in fibromyalgia; however, this hypothesis has not been validated [109,110]. Aprazolam, an agent with effects on anxiety and depression in the doses employed, appears to have a very marginal benefit in the syndrome in one study [166] and no benefit in another [121].

Fluoxetine, an antidepressant that selectively inhibits the reuptake of serotonin, has been shown to have marginal or uncertain benefits [184,185]. Goldenberg and colleagues [186] found in a double-blind crossover trial that fluoxetine and amitriptyline combined was more effective than either drug alone, with significant improvements in pain, global well-being, and sleep. Changes in depression and fatigue were not significant.

Weighing the hazards and benefits of using tricyclics, other antidepressants, and anxiolytics in persons with long-standing complex problems warrants careful consideration by health care providers. Their prescription alone may continue a failing pattern of one drug after another.

D. Cognitive Behavioral Therapy and Other Psychoeducational Interventions

Various psychoeducational interventions, including cognitive behavioral therapy [187–189], and comprehensive interdisciplinary programs have been used for the treatment of fibromyalgia [190,191]. Both approaches are conceptually appealing and have demonstrated success in other rheumatic disorders [192,193]. The comprehensive multidisciplinary approaches may include medications, exercise training, relaxation training, electromyographic (EMG) biofeedback, cognitive restructuring, aerobic and stretching exercise, biomechanics training, pacing, and family education. Evaluation of such programs is challenging because it is difficult to develop adequate control treatments [136] because the selection processes, compliance, and drop-outs influence results, and because long-term follow-up is rarely reported. In addition, the milieu in which treatment is administered may be more important than the treatment itself. The expense of such programs remains a major consideration [190,194] and prohibits attendance by many people.

Two groups have reported follow-up from comprehensive programs. Bennett and colleagues found that about one-third of participants in their comprehensive program continued to show improvement at two years [190]. White and Nelson found that only three of their ten "target variables" continued at significantly improved levels compared to pretreatment at 30 months of follow-up [188]. Two well-designed studies comparing various cognitive behavioral treatments with attention placebo/ educational interventions or exercise failed to demonstrate any significant differences between treatment approaches [187,195].

E. Pain

Surprisingly, analgesic efficacy has not been studied extensively, although about a third of fibromyalgia patients report using over-the-counter analgesics [196]. Russell and colleagues have studied tramadol and have found it provides modest benefits [197].

While generally narcotics are not considered appropriate for use in fibromyalgia [36], guidelines suggest some indications for use in chronic nonmalignant pain [198]. Controlled studies of narcotic use in fibromyalgia should be considered. Sorenson *et al.* [199] in Sweden identified "responders" and "nonresponders" in a double-blind crossover study of intravenous morphine, lidocain, ketamine, and saline. Thirteen of eighteen participants responded to one or more drugs (but not to the saline), two responded to all four substances, and three were nonresponders. Plasma concentrations were similar both in responders and in nonresponders. These authors suggest additional studies of the use of narcotics in fibromyalgia using a division of patients into responders and nonresponders [199].

F. Other Treatment Interventions

A number of other therapies have not been shown to be effective. Prednisone is not effective in the treatment of fibromyalgia [200], nor is somatostatin [201]. Other pharmacological therapies with serotonin receptor (S2) blockers [202], 5-hydroxy-L-tryptophan [203,204], muscle relaxants [205], calcitonin, and S-adenyslmethionine [206] have not been beneficial. One study of growth hormone indicated overall improvement in symptomatology for women with low levels of insulin-like growth factor (IGF-1) in a nine-month trial. When the injections of growth hormone were stopped, ". . . patients experienced a worsening of symptoms" [207].

XII. Issues and Future Directions

Twenty-two years after fibromyalgia criteria were first proposed, fibromyalgia is a widely diagnosed syndrome with a broad literature that has expanded from rheumatology into general medicine. Fibromyalgia has become a convenient and often useful diagnosis. Even so, it remains a controversial syndrome. Some argue that it is primarily a manifestation of abnormal pain behavior, appears to be part of an effective spectrum disorder, has its major etiology in psychiatric illness, and is not, after all, a real disease, but only the morbid end of a distress spectrum. It is additionally criticized because of high rates of disability, based primarily on self-report. Finally, it has also been suggested that fibromyalgia may be, in part, a creation of our society and culture at one moment in time and that by invoking the syndrome and its treatments we continue it and expand it. Regardless of the truth of these concerns, there are clearly patients who fit the syndrome and fit no other illness, and not all patients with fibromyalgia have psychiatric illness.

Fibromyalgia has prompted physicians to investigate the issues of fatigue, sleep disturbance, pain, and psychosocial distress. These key features of common illnesses have been generally ignored in the traditional biomedical model of disease, but they are at the heart of fibromyalgia.

Most therapies have little effectiveness, and it is not likely that any breakthrough is imminent. Thought should be given to diagnosing fibromyalgia less frequently and to treating it less frequently with medications.

The current trend in research regarding fibromyalgia relates to mechanisms of pain and distress expression. It is likely to teach us more about how people become ill, but it does not seem likely to unravel the multifarious threads that drive the etiology of the syndrome. If this research leads to the investigation of the greater spectrum of distress rather than the extreme end, it may yield important data that will be useful in fibromyalgia and in other pain and distress-related conditions.

References

1. Wolfe, F., Smythe, H. A., Yunus, M. B., Bennett, R. M., Bombardier, C., Goldenberg, D. L., Tugwell, P., Abeles, M., Campbell, S. M., Clark, P., Fam, A. G., Farber, S. J., Fiechtner, J. J., Franklin, C. M., Gatter, R. A., Hamaty, D., Lessard, J., Lichtbroun, A. S., Masi, A. T., McCain, G. A., Reynolds, W. J., Romano, T. J., Russell, I. J., and Sheon, R. P. (1990). The American College of Rheumatology 1990 Criteria for the Classification of Fibromyalgia: Report of the Multicenter Criteria Committee. *Arthritis Rheum.* **33,** 160–172.

2. Walker, E. A., Keegan, D., Gardner, G., Sullivan, M., Katon, W. J., and Bernstein, D. (1997). Psychosocial factors in fibromyalgia compared with rheumatoid arthritis: I. Psychiatric diagnoses and functional disability. *Psychosom. Med.* **59,** 565–571.

3. Hudson, J. I., and Pope, H. G. J. (1996). The relationship between fibromyalgia and major depressive disorder. *Rheum. Dis. Clin. North Am.* **22,** 285–303.

4. Hudson, J. I., Goldenberg, D. L., Pope, H. G. J., Keck, P. E. J., and Schlesinger, L. (1992). Comorbidity of fibromyalgia with medical and psychiatric disorders. *Am. J. Med.* **92,** 363–367.

5. Kirmayer, L. J., Robbins, J. M., and Kapusta, M. A. (1988). Somatization and depression in fibromyalgia syndrome. *Am. J. Psychiatry* **145,** 950–954.

6. Walker, E. A., Keegan, D., Gardner, G., Sullivan, M., Bernstein, D., and Katon, W. J. (1997). Psychosocial factors in fibromyalgia compared with rheumatoid arthritis: II. Sexual, physical, and emotional abuse and neglect. *Psychosom. Med.* **59,** 572–577.

7. Macfarlane, G. J., Thomas, E., Papageorgiou, A. C., Schollum, J., Croft, P. R., and Silman, A. J. (1996). The natural history of chronic pain in the community: A better prognosis than in the clinic? *J. Rheumatol.* **23,** 1617–1620.

8. Boisset-Pioro, M. H., Esdaile, J. M., and Fitzcharles, M. A. (1995). Sexual and physical abuse in women with fibromyalgia syndrome. *Arthritis Rheum.* **38,** 235–241.

9. Wolfe, F., Ross, K., Anderson, J., Russell, I. J., and Hebert, L. (1995). The prevalence and characteristics of fibromyalgia in the general population. *Arthritis Rheum.* **38,** 19–28.

10. Middleton, G. D., Mcfarlin, J. E., and Lipsky, P. E. (1994). The prevalence and clinical impact of fibromyalgia in systemic lupus erythematosus. *Arthritis Rheum.* **37,** 1181–1188.

11. Hawley, D. J., Wolfe, F., Cathey, M. A., and Roberts, F. K. (1991). Marital status in rheumatoid arthritis and other rheumatic disorders: A study of 7,293 patients. *J. Rheumatol.* **18,** 654–660.

12. Wolfe, F., Anderson, J., Harkness, D., Bennett, R. M., Caro, X. J., Goldenberg, D. L., Russell, I. J., and Yunus, M. B. (1997). A prospective, longitudinal, multicenter study of service utilization and costs in fibromyalgia. *Arthritis Rheum.* **40,** 1560–1570.

13. Croft, P., Rigby, A. S., Boswell, R., Schollum, J., and Silman, A. J. (1993). The prevalence of widespread pain in the general population. *J. Rheumatol.* **20**, 710–713.

14. Wolfe, F. (1998). Fibromyalgia in a grand bio-psycho-social model—Reply. *J. Rheumatol.* **25**(5), 1029–1030.

15. Goldenberg, D. L. (1987). Fibromyalgia syndrome. An emerging but controversial condition. *JAMA, J. Am. Med. Assoc.* **257**, 2782–2787.

16. Smythe, H. A. (1989). Fibrositis syndrome: A historical perspective. *J. Rheumatol.* **16**(Suppl. 19), 2–6.

17. Hench, P. K. (1986). Secondary fibrositis. *Am. J. Med.* **81**, 60–62.

18. Smythe, H. A., and Moldofsky, H. (1977). Two contributions to understanding of the "fibrositis" syndrome. *Bull. Rheum. Dis.* **28**, 928–931.

19. Yunus, M. B., Masi, A. T., Calabro, J. J., Miller, K. A., and Feigenbaum, S. L. (1981). Primary fibromyalgia (fibrositis): Clinical study of 50 patients with matched normal controls. *Semin. Arthritis Rheum.* **11**, 151–171.

20. Taylor, M. L., Trotter, D. R., and Csuka, M. E. (1995). The prevalence of sexual abuse in women with fibromyalgia. *Arthritis Rheum.* **38**, 229–234.

21. Mukerji, B., Mukerji, V., Alpert, M. A., and Selukar, R. (1995). The prevalence of rheumatologic disorders in patients with chest pain and angiographically normal coronary arteries. *Angiology* **46**, 425–430.

22. Goldenberg, D. L. (1997). Fibromyalgia, chronic fatigue syndrome, and myofascial pain syndrome. *Curr. Opin. Rheumatol.* **9**, 135–143.

23. Donald, F., Esdaile, J. M., Kimoff, J. R., and Fitzcharles, M. A. (1996). Musculoskeletal complaints and fibromyalgia in patients attending a respiratory sleep disorders clinic. *J. Rheumatol.* **23**, 1612–1616.

24. Clauw, D. J., and Chrousos, G. P. (1997). Chronic pain and fatigue syndromes: Overlapping clinical and neuroendocrine features and potential pathogenic mechanisms. *Neuroimmunomodulation* **4**, 134–153.

25. Buchwald, D. (1996). Fibromyalgia and chronic fatigue syndrome: Similarities and differences. *Rheum. Dis. Clin. North Am.* **22**, 219–243.

26. Alagiri, M., Chottiner, S., Ratner, V., Slade, D., and Hanno, P. M. (1997). Interstitial-cystitis: Unexplained associations with other chronic disease and pain syndromes. *Urology* **49**, 52–57.

27. Hudson, J. I., and Pope, H. G. (1994). The concept of affective spectrum disorder: Relationship to fibromyalgia and other syndromes of chronic fatigue and chronic muscle pain. *Bailliere Clin. Rheumatol.* **8**, 839–856.

28. Demitrack, M. A., and Crofford, L. J. (1998). Evidence for and pathophysiologic implications of hypothalamic-pituitary-adrenal axis dysregulation in fibromyalgia and chronic fatigue syndrome. *Ann. N.Y. Acad. Sci.* **840**, 684–697.

29. Mountz, J. M., Bradley, L. A., and Alarcon, G. S. (1998). Abnormal functional activity of the central nervous system in fibromyalgia syndrome. *Am. J. Med. Sci.* **315**, 385–396.

30. Olsen, N. J., and Park, J. H. (1998). Skeletal muscle abnormalities in patients with fibromyalgia. *Am. J. Med. Sci.* **315**, 351–358.

31. Russell, I. J. (1998). Advances in fibromyalgia: Possible role for central neurochemicals. *Am. J. Med. Sci.* **315**, 377–384.

32. Simms, R. W. (1998). Fibromyalgia is not a muscle disorder. *Am. J. Med. Sci.* **315**, 346–350.

33. Johansson, G., Risberg, J., Rosenhall, U., Orndahl, G., Svennerholm, L., and Nystrom, S. (1995). Cerebral dysfunction in fibromyalgia: Evidence from regional cerebral blood flow measurements, otoneurological tests and cerebrospinal fluid analysis. *Acta Psychiatr. Scand.* **91**, 86–94.

34. Yunus, M. B., Dailey, J. W., Aldag, J. C., Masi, A. T., and Jobe, P. C. (1992). Plasma tryptophan and other amino acids in primary fibromyalgia—A controlled study. *J. Rheumatol.* **19**, 90–94.

35. Wolfe, F., Anderson, J., Harkness, D., Bennett, R. M., Caro, X. J., Goldenberg, D. L., Russell, I. J., and Yunus, M. B. (1997). Health status and disease severity in fibromyalgia: Results of a six-center longitudinal study. *Arthritis Rheum.* **40**, 1571–1579.

36. Alarcon, G. S., and Bradley, L. A. (1998). Advances in the treatment of fibromyalgia: Current status and future directions. *Am. J. Med. Sci.* **315**, 397–404.

37. Csillag, C. (1992). Conference—Fibromyalgia—The Copenhagen Declaration. *Lancet* **340**, 663–664.

38. Wolfe, F. (1997). The fibromyalgia problem. *J. Rheumatol.* **24**, 1247–1249.

39. Fitzcharles, M. A., and Esdaile, J. M. (1997). The overdiagnosis of fibromyalgia syndrome. *Am. J. Med.* **103**(1), 44–50.

40. Unruh, A. M. (1996). Gender variations in clinical pain experience. *Pain* **65**(2–3), 123–167.

41. Fillingim, R. B., Maixner, W., Kincaid, S., and Silva, S. (1998). Sex differences in temporal summation but not sensory-discriminative processing of thermal pain. *Pain* **75**(1), 121–127.

42. Wolfe, F., Ross, K., Anderson, J., and Russell, I. J. (1995). Aspects of fibromyalgia in the general population: Sex, pain threshold, and fibromyalgia symptoms. *J. Rheumatol.* **22**, 151–156.

43. Siegel, D. M., Janeway, D., and Baum, J. (1998). Fibromyalgia syndrome in children and adolescents: Clinical features at presentation and status at follow-up. *Pediatrics* **101**(3), 377–382.

44. Campbell, S. M., Clark, S., Tindall, E. A., Forehand, M. E., and Bennett, R. M. (1983). Clinical characteristics of fibrositis. I. A "blinded," controlled study of symptoms and tender points. *Arthritis Rheum.* **26**, 817–824.

45. Bazelmans, E., Vercoulen, J. H., Galama, J. M., van Weel, C., van der Meer, J. W., and Bleijenberg, G. (1997). [Prevalence of chronic fatigue syndrome and primary fibromyalgia syndrome in The Netherlands]. *Ned. Tijdschr. Geneeskd.* **141**, 1520–1523.

46. Hartz, A., and Kirchdoerfer, E. (1987). Undetected fibrositis in primary care practice. *J. Fam. Pract.* **25**, 365–369.

47. Wolfe, F., and Cathey, M. A. (1983). Prevalence of primary and secondary fibrositis. *J. Rheumatol.* **10**, 965–968.

48. Forseth, K. O., and Gran, J. T. (1992). The prevalence of fibromyalgia among women aged 20–49 years in Arendal, Norway. *Scand. J. Rheumatol.* **21**, 74–78.

49. Lyddell, C., and Meyers, O. L. (1992). The prevalence of fibromyalgia in a South African community. *Scand. J. Rheumatol.* **Suppl.** **94**, S143 (abstr.).

50. Jacobsson, L. T., Nagi, D. K., Pillemer, S. R., Knowler, W. C., Hanson, R. L., Pettitt, D. J., and Bennett, P. H. (1996). Low prevalences of chronic widespread pain and shoulder disorders among the Pima Indians. *J. Rheumatol.* **23**, 907–909.

51. Prescott, E., Kjoller, M., Jacobsen, S., Bulow, P. M., Danneskioldsamsoe, B., and Kamperjorgensen, F. (1993). Fibromyalgia in the adult Danish population.1. A prevalence study. *Scand. J. Rheumatol.* **22**, 233–237.

52. Raspe, H. H., Baumgartner, C., and Wolfe, F. (1993). The prevalence of fibromyalgia in a rural German community: How much difference do different criteria make? *Arthritis Rheum.* **36**(Suppl. 9), S48.

53. Buskila, D., Press, J., Gedalia, A., Klein, M., Neumann, L., Boehm, R., and Sukenik, S. (1993). Assessment of nonarticular tenderness and prevalence of fibromyalgia in children. *J. Rheumatol.* **20**, 368–370.

54. Mikkelsson, M., Salminen, J. J., and Kautiainen, H. (1997). Nonspecific musculoskeletal pain in preadolescents. prevalence and 1- year persistence. *Pain* **73**(1), 29–35.

55. Mikkelsson, M., Sourander, A., Piha, J., and Salminen, J. J. (1997). Psychiatric symptoms in preadolescents with musculoskeletal pain and fibromyalgia. *Pediatrics* **100**(2), 220–227.

56. Cicuttini, F., and Littlejohn, G. O. (1989). Female adolescent rheumatological presentations: The importance of chronic pain syndromes. *Aust. Paediatr. J.* **25**, 21–24.

57. Mäkelä, M., and Heliövaara, M. (1991). Prevalence of primary fibromyalgia in the Finish population. *Br. Med. J.* **303**, 216–219.

58. Calvo-Alen, J., Bastian, H. M., Straaton, K. V., Burgard, S. L., Mikhail, I. S., and Alarcon, G. S. (1995). Identification of patient subsets among those presumptively diagnosed with, referred, and/or followed up for systemic lupus erythematosus at a large tertiary care center. *Arthritis Rheum.* **38**, 1475–1484.

59. Alexander, R. W., Bradley, L. A., Alarcon, G. S., Trianaalexander, M., Aaron, L. A., Alberts, K. R., Martin, M. Y., and Stewart, K. E. (1998). Sexual and physical abuse in women with fibromyalgia: Association with outpatient healthcare utilization and pain medication usage. *Arthritis Care Res.* **11**(2), 102–115.

60. Goldstein, J. A. (1994). Fibromyalgia syndrome: A pain modulation disorder related to altered limbic function? *Bailliere's Clin. Rheumatol.* **8**, 777–800.

61. Yunus, M. B. (1992). Towards a model of pathophysiology of fibromyalgia—Aberrant central pain mechanisms with peripheral modulation. *J. Rheumatol.* **19**, 846–850.

62. Moldofsky, H. (1982). Rheumatic pain modulation syndrome: The interrelationships between sleep, central nervous system serotonin, and pain. *Adv. Neurol.* **33**, 51–57.

63. Smythe, H. A. (1979). Fibrositis as a disorder of pain modulation. *Clin. Rheum. Dis.* **5**, 823–832.

64. Smythe, H. A. (1985). "Fibrositis" and other diffuse musculoskeletal syndromes. *In* "Textbook of Rheumatology" (W. N. Kelley, E. D. Harris, Jr., S. Ruddy, and C. B. Sledge, eds.), Saunders, Philadelphia.

65. McDermid, A. J., Rollman, G. B., and McCain, G. A. (1996). Generalized hypervigilance in fibromyalgia: Evidence of perceptual amplification. *Pain* **66**(2–3), 133–144.

66. Lautenschläger, J., Bruckle, W., Seglias, J., and Müller, W. (1989). [Localized pressure pain in the diagnosis of generalized tendomyopathy (fibromyalgia)]. *Z. Rheumatol.* **48**, 132–138.

67. Simms, R. W., Goldenberg, D. L., Felson, D. T., and Mason, J. H. (1988). Tenderness in 75 anatomic sites. Distinguishing fibromyalgia patients from controls. *Arthritis Rheum.* **31**, 182–187.

68. Croft, P. R., Nahit, E. S., Macfarlane, G. J., and Silman, A. J. (1996). Interobserver reliability in measuring flexion, internal rotation, and external rotation of the hip using a plurimeter. *Ann. Rheum. Dis.* **55**(5), 320–323.

69. Smythe, H. A., Buskila, D., Urowitz, S., and Langevitz, P. (1992). Control and "fibrositic" tenderness: Comparison of two dolorimeters. *J. Rheumatol.* **19**, 768–771.

70. Smythe, H. A., Gladman, A., Dagenais, P., Kraishi, M., and Blake, R. (1992). Relation between fibrositic and control site tenderness—Effects of dolorimeter scale length and footplate size. *J. Rheumatol.* **19**, 284–289.

71. Travell, J. G., and Simons, D. G. (1992). "Myofascial Pain and Dysfunction: The Trigger Point Manual," Vol. 2. Williams & Wilkins, Baltimore, MD.

72. Simons, D. G. (1987). "Myofascial Pain Syndrome Due to Trigger Points." Mosby, St. Louis, MO.

73. Wolfe, F., Simons, D. G., Fricton, J. R., Bennett, R. M., Goldenberg, D. L., Gerwin, R., Hathaway, D., McCain, G. A., Russell, I. J., Sanders, H. O., and Skootsky, S. A. (1992). The fibromyalgia and myofascial pain syndromes—A preliminary study of tender points and trigger points in persons with fibromyalgia, myofascial pain syndrome and no disease. *J. Rheumatol.* **19**, 944–951.

74. Wallace, D. J. (1990). Genitourinary manifestations of fibrositis: An increased association with the female urethral syndrome. *J. Rheumatol.* **17**, 238–239.

75. Bengtsson, A., Henriksson, K. G., Jorfeldt, L., Kagedal, B., Lennmarken, C., and Lindstrom, F. (1986). Primary fibromyalgia. A clinical and laboratory study of 55 patients. *Scand. J. Rheumatol.* **15**, 340–347.

76. Dinerman, H., Goldenberg, D. L., and Felson, D. T. (1986). A prospective evaluation of 118 patients with the fibromyalgia syndrome: Prevalence of Raynaud's phenomenon, sicca symptoms, ANA, low complement, and Ig deposition at the dermal-epidermal junction. *J. Rheumatol.* **13**, 368–373.

77. Clark, S., Campbell, S. M., Forehand, M. E., Tindall, E. A., and Bennett, R. M. (1985). Clinical characteristics of fibrositis. II. A "blinded," controlled study using standard psychological tests. *Arthritis Rheum.* **28**, 132–137.

78. Wolfe, F., Hawley, D. J., Cathey, M. A., Caro, X., and Russell, I. J. (1985). Fibrositis: Symptom frequency and criteria for diagnosis. An evaluation of 291 rheumatic disease patients and 58 normal individuals. *J. Rheumatol.* **12**, 1159–1163.

79. Reilly, P. A., and Littlejohn, G. O. (1992). Peripheral arthralgic presentation of fibrositis/fibromyalgia syndrome. *J. Rheumatol.* **19**, 281–283.

80. Tan, E. M., Cohen, A. S., Fries, J. F., Masi, A. T., McShane, D. J., Rothfield, N. F., Schaller, J. G., Talal, N., and Winchester, R. (1982). The 1982 revised criteria for the classification of systemic lupus erythematosus. *Arthritis Rheum.* **25**, 1271–1271.

81. Hochberg, M. C. (1997). Updating the American College of Rheumatology revised criteria for the classification of systemic lupus erythematosus. *Arthritis Rheum.* **40**(9), 1725–1725.

82. Cathey, M. A., Wolfe, F., Kleinheksel, S. M., and Hawley, D. J. (1986) Socioeconomic impact of fibrositis. A study of 81 patients with primary fibrositis. *Am. J. Med.* **81**, 78–84.

83. Reid, G. J., Lang, B. A., and McGrath, P. J. (1997). Primary juvenile fibromyalgia: Psychological adjustment, family functioning, coping, and functional disability. *Arthritis Rheum.* **40**, 752–760.

84. Buskila, D., Neumann, L., Hershman, E., Gedalia, A., Press, J., and Sukenik, S. (1995). Fibromyalgia syndrome in children—an outcome study. *J. Rheumatol.* **22**, 525–528.

85. Yunus, M. B., and Masi, A. T. (1985). Juvenile primary fibromyalgia syndrome. A clinical study of thirty-three patients and matched normal controls. *Arthritis Rheum.* **28**, 138–145.

86. Kennedy, M., and Felson, D. T. (1996). A prospective long-term study of fibromyalgia syndrome. *Arthritis Rheum.* **39**, 682–685.

87. Granges, G., Zilko, P., and Littlejohn, G. O. (1994). Fibromyalgia syndrome—Assessment of the severity of the condition 2 years after diagnosis. *J. Rheumatol.* **21**, 523–529.

88. Hawley, D. J., Wolfe, F., and Cathey, M. A. (1988). Pain, functional disability, and psychological status: A 12-month study of severity in fibromyalgia. *J. Rheumatol.* **15**, 1551–1556.

89. Felson, D. T., and Goldenberg, D. L. (1986). The natural history of fibromyalgia. *Arthritis Rheum.* **29**, 1522–1526.

90. Wolfe, F. (1986). The clinical syndrome of fibrositis. *Am. J. Med.* **81**, 7–14.

91. Ledingham, J., Doherty, S., and Doherty, M. (1993) Primary fibromyalgia syndrome—an outcome study. *Br. J. Rheumatol.* **32**, 139–142.

92. Goldenberg, D. L. (1993). Do infections trigger fibromyalgia? *Arthritis Rheum.* **36**, 1489–1492.

93. Burckhardt, C. S., Clark, S. R., Campbell, S. M., O'Reilly, C. A., Wiens, A. N., and Bennett, R. M. (1992). The onset of fibromyalgia: an analysis of early symptoms and initiating events. *Arthritis Rheumat. Suppl.* **53**, 524 (abstr.).

94. Buchwald, D., Sullivan, J. L., and Komaroff, A. L. (1987). Frequency of "chronic active Epstein-Barr virus infection' in a general medical practice. *JAMA, J. Am. Med. Assoc.* **257,** 2303–2307.

95. Wolfe, F., Anderson, J., Harkness, D., Bennett, R. M., Caro, X. J., Goldenberg, D. L., Russell, I. J., and Yunus, M. B. (1997). Work and disability status of persons with fibromyalgia. *J. Rheumatol.* **24,** 1171–1178.

96. Greenfield, S., Fitzcharles, M. A., and Esdaile, J. M. (1992). Reactive fibromyalgia syndrome. *Arthritis Rheum.* **35,** 678–681.

97. Aaron, L. A., Bradley, L. A., Alarcon, G. S., Triana-Alexander, M., Alexander, R. W., Martin, M. Y., and Alberts, K. R. (1997). Perceived physical and emotional trauma as precipitating events in fibromyalgia. Associations with health care seeking and disability status but not pain severity. *Arthritis Rheum.* **40,** 453–460.

98. Burckhardt, C. S., Clark, S. R., Campbell, S. M., O'Reilly, C. A., Wiens, A. N., and Bennett, R. M. (1992). The onset of fibromyalgia: An analysis of early symptoms and initiating events. *Arthritis Rheum.* **53,** S241(abstr.).

99. Buskila, D., Neumann, L., Vaisberg, G., Alkalay, D., and Wolfe, F. (1997). Increased rates of fibromyalgia following cervical spine injury. A controlled study of 161 cases of traumatic injury. *Arthritis Rheum.* **40,** 446–452.

100. Hadler, N. M. (1991). When is an "idiopathic' rheumatic disease a personal injury? *Occup. Probl. Med. Pract.* **6**(1), 1–8.

101. Hadler, N. M. (1989). Work-related disorders of the upper extremity. Part I: Cumulative trauma disorders—a critical review. *Occup. Probl. Med. Pract.* **4,** 1–8.

102. Littlejohn, G. O. (1989). Medicolegal aspects of fibrositis syndrome. *J. Rheumatol., Suppl.* **19,** 169–173.

103. Hawley, D. J., and Wolfe, F. (1997). Fatigue and musculoskeletal pain. *Phys. Med. Rehabil. Clin. North Am.* **8,** 101–111.

104. Wolfe, F., Hawley, D. J., and Wilson, K. (1996). The prevalence and meaning of fatigue in rheumatic disease. *J. Rheumatol.* **23,** 1407–1417.

105. Wolfe, F. (1994). Data collection and utilization: A methodology for clinical practice and clinical research. *In* "Rheumatoid Arthritis: Pathogenesis, Assessment, Outcome, and Treatment" (F. Wolfe and T. Pincus, eds.), pp. 463–514. Dekker, New York.

106. Burckhardt, C. S., Clark, S. R., and Bennett, R. M. (1991). The fibromyalgia impact questionnaire: Development and validation. *J. Rheumatol.* **18,** 728–733.

107. Moldofsky, H., Scarisbrick, P., England, R., and Smythe, H. A. (1975). Musculoskeletal symptoms and non-REM sleep disturbance in patients with "fibrositis syndrome" and healthy subjects. *Psychosom. Med.* **37,** 341–351.

108. Harding, S. M. (1998). Sleep in fibromyalgia patients: Subjective and objective findings. *Am. J. Med. Sci.* **315,** 367–376.

109. Carette, S., Oakson, G., Guimont, C., and Steriade, M. (1995). Sleep electroencephalography and the clinical response to amitriptyline in patients with fibromyalgia. *Arthritis Rheum.* **38,** 1211–1217.

110. Reynolds, W. J., Moldofsky, H., Saskin, P., and Lue, F. A. (1991). The effects of cyclobenzaprine on sleep physiology and symptoms in patients with fibromyalgia. *J. Rheumatol.* **18,** 452–454.

111. Triadafilopoulos, G., Simms, R. W., and Goldenberg, D. L. (1991). Bowel dysfunction in fibromyalgia syndrome. *Dig. Dis. Sci.* **36,** 59–64.

112. Wesselmann, U., Burnett, A. L., and Heinberg, L. J. (1997). The urogenital and rectal pain syndromes. *Pain* **73**(3), 269–294.

113. Ahles, T. A., Khan, S. A., Yunus, M. B., Spiegel, D. A., and Masi, A. T. (1991). Psychiatric status of patients with primary fibromyalgia, patients with rheumatoid arthritis, and subjects without pain—A blind comparison of DSM-III diagnoses. *Am. J. Psychiatry* **148,** 1721–1726.

114. Tariot, P. N., Yocum, D., and Kalin, N. H. (1986). Psychiatric disorders in fibromyalgia. *Am. J. Psychiatry* **143,** 812–813.

115. Alfici, S., Sigal, M., and Landau, M. (1989). Primary fibromyalgia syndrome—a variant of depressive disorder? *Psychother. Psychosom.* **51,** 156–161.

116. Aaron, L. A., Bradley, L. A., Alarcon, G. S., Alexander, R. W., Triana-Alexander, M., Martin, M. Y., and Alberts, K. R. (1996). Psychiatric diagnoses in patients with fibromyalgia are related to health care-seeking behavior rather than to illness. *Arthritis Rheum.* **39,** 436–445.

117. Burckhardt, C. S., Clark, S. R., and Bennett, R. M. (1993). Fibromyalgia and quality of life: A comparative analysis. *J. Rheumatol.* **20,** 475–479.

118. Hawley, D. J., and Wolfe, F. (1993). Depression is not more common in rheumatoid arthritis: A 10 year longitudinal study of 6,608 rheumatic disease patients. *J. Rheumatol.* **20,** 2025–2031.

119. Piergiacomi, G., Blasetti, P., Berti, C., Ercolani, M., and Cervini, C. (1989). Personality pattern in rheumatoid arthritis and fibromyalgic syndrome. Psychological investigation. *Z. Rheumatol.* **48,** 288–293.

120. Celiker, R., Borman, P., Oktem, F., Gokce-Kutsal, Y., and Basgoze, O. (1997). Psychological disturbance in fibromyalgia: Relation to pain severity. *Clin. Rheumatol.* **16,** 179–184.

121. Krag, N. J., Norregaard, J., Larsen, J. K., and Danneskioldsamsoe, B. (1994). A blinded, controlled evaluation of anxiety and depressive symptoms in patients with fibromyalgia, as measured by standardized psychometric interview scales. *Acta Psychiatr. Scand.* **89,** 370–375.

122. Robbins, J. M., Kirmayer, L. J. and Kapusta, M. A. (1990). Illness worry and disability in fibromyalgia syndrome. *Int. J. Psychiatry Med.* **20,** 49–63.

123. Wolfe, F. (1997). Evidence for disordered symptom appraisal in fibromyalgia: Increased rates of reported comorbidity and comorbidity severity. *Arthritis Rheum.* **40,** S117 (abstr.).

124. Bennett, R. M. (1987). Personal communication (Unpublished).

125. Littlejohn, G. O., Weinstein, C., and Helme, R. D. (1987). Increased neurogenic inflammation in fibrositis syndrome. *J. Rheumatol.* **14,** 1022–1025.

126. Granges, G., and Littlejohn, G. O. (1993). A comparative study of clinical signs in fibromyalgia/fibrositis syndrome, healthy and exercising subjects. *J. Rheumatol.* **20,** 344–351.

127. Fischer, A. A. (1987). Tissue compliance meter for objective, quantitative documentation of soft tissue consistency and pathology. *Arch. Phys. Med. Rehabil.* **68,** 122–125.

128. Fischer, A. A. (1987). Muscle tone in normal persons measured by tissue compliance. *J. Neurol. orthop. Med. Surg.* **8,** 227–233.

129. Enestrom, S., Bengtsson, A., and Frodin, T. (1997). Dermal IgG deposits and increase of mast cells in patients with fibromyalgia— Relevant findings or epiphenomena? *Scand. J. Rheumatol.* **26**(4), 308–313.

130. Klein, R., and Berg, P. A. (1994). A comparative study on antibodies to nucleoli and 5- hydroxytryptamine in patients with fibromyalgia syndrome and tryptophan-induced eosinophilia-myalgia syndrome. *Clin. Invest.* **72,** 541–549.

131. Jacobsen, S., Höyer-Madsen, M., Danneskiold-Samsoe, B., and Wiik, A. (1990). Screening for autoantibodies in patients with primary fibromyalgia syndrome and a matched control group. *APMIS* **98,** 655–658.

132. Bengtsson, A., Ernerudh, J., Vrethem, M., and Skogh, T. (1990). Absence of autoantibodies in primary fibromyalgia. *J. Rheumatol.* **17,** 1682–1683.

133. Aarflot, T., and Bruusgaard, D. (1996). Association between chronic widespread musculoskeletal complaints and thyroid auto-

immunity. Results from a community survey. *Scand. J. Primary Health Care* **14**, 111–115.

134. Wallace, D. J., Bowman, R. L., Wormsley, S. B., and Peter, J. B. (1989). Cytokines and immune regulation in patients with fibrositis. *Arthritis Rheum.* **32**, 1334–1335, *erratum*: p. 1607.

135. Reynolds, W. J., Chiu, B., and Inman, R. D. (1988). Plasma substance P levels in fibrositis. *J. Rheumatol.* **15**, 1802–1803.

136. Vaeroy, H., Helle, R., Frre, O., Kass, E., and Terenius, L. (1988). Elevated CSF levels of substance P and high incidence of Raynaud phenomenon in patients with fibromyalgia: New features for diagnosis. *Pain* **32**, 21–26.

137. Russell, I. J., Vaeroy, H., Javors, M., and Nyberg, F. (1992). Cerebrospinal fluid biogenic amine metabolites in Fibromyalgia/Fibrositis Syndrome and rheumatoid arthritis. *Arthritis Rheum.* **35**, 550–556.

138. Russell, I. J., Orr, M. D., Littman, B., Vipraio, G. A., Alboukrek, D., Michalek, J. E., Lopez, Y., and Mackillip, F. (1994). Elevated cerebrospinal fluid levels of substance P in patients with the fibromyalgia syndrome. *Arthritis Rheum.* **37**, 1593–1601.

139. Lindh, C., Liu, Z., Lyrenas, S., Ordeberg, G., and Nyberg, F. (1997). Elevated cerebrospinal fluid substance P-like immunoreactivity in patients with painful osteoarthritis, but not in patients with rhizopatic pain from a herniated lumbar disc. *Scand. J. Rheumatol.* **26**(6), 468–472.

140. Mountz, J. M., Bradley, L. A., Modell, J. G., Alexander, R. W., Triana-Alexander, M., Aaron, L. A., Stewart, K. E., Alarcon, G. S., and Mountz, J. D. (1995). Fibromyalgia in women. Abnormalities of regional cerebral blood flow in the thalamus and the caudate nucleus are associated with low pain threshold levels. *Arthritis Rheum.* **38**, 926–938.

141. Crofford, L. J., Engleberg, N. C., and Demitrack, M. A. (1996). Neurohormonal perturbations in fibromyalgia. *Bailliere's Clin. Rheumatol.* **10**, 365–378.

142. Jara, L. J., Gomez-Sanchez, C., and Espinoza, L. R. (1991). Prolactin in primary fibromyalgia and rheumatoid arthritis. *J. Rheumatol.* **18**, 480–481.

143. Ferraccioli, G. F., Cavalieri, F., Salaffi, F., Fontana, S., Scita, F., Nolli, M., and Maestri, D. (1990). Neuroendocrinologic findings in primary fibromyalgia (soft tissue chronic pain syndrome) and in other chronic rheumatic conditions (rheumatoid arthritis, low back pain). *J. Rheumatol.* **17**, 869–873.

144. Bennett, R. M., Clark, S. R., Campbell, S. M., and Burckhardt, C. S. (1992). Low levels of somatomedin C in patients with the fibromyalgia syndrome—A possible link between sleep and muscle pain. *Arthritis Rheum.* **35**, 1113–1116.

145. Carette, S., and Lefrancois, L. (1988). Fibrositis and primary hypothyroidism. *J. Rheumatol.* **15**, 1418–1421.

146. Wilke, W. S., Sheeler, L. R., and Makarowski, W. S. (1981). Hypothyroidism with presenting symptoms of fibrositis. *J. Rheumatol.* **8**, 626–631.

147. Houvenagel, E., Forzy, G., Cortet, B., and Vincent, G. (1990). 5-Hydroxy indol acetic acid in cerebrospinal fluid in fibromyalgia. *Arthritis Rheum.* **33**, S55.

148. Russell, I. J. (1989). Neurohormonal aspects of the fibromyalgia syndrome. *Rheum. Dis. Clin. North Am.*, **15**, 149–168.

149. Bengtsson, A., Henriksson, K. G., and Larsson, J. (1986). Muscle biopsy in primary fibromyalgia. Light-microscopical and histochemical findings. *Scand. J. Rheumatol.* **15**, 1–6.

150. Drewes, A. M., Andreasen, A., Schroder, H. D., Hogsaa, B., and Jennum, P. (1993). Pathology of skeletal muscle in fibromyalgia—A histo- immuno- chemical and ultrastructural study. *Br. J. Rheumatol.* **32**, 479–483.

151. Lindh, M., Johansson, G., Hedberg, M., Henning, G. B., and Grimby, G. (1995). Muscle fiber characteristics, capillaries and

enzymes in patients with fibromyalgia and controls. *Scand. J. Rheumatol.* **24**, 34–37.

152. Lindman, R., Hagberg, M., Angqvist, K. A., Soderlund, K., Hultman, E., and Thornell, L. E. (1991). Changes in muscle morphology in chronic trapezius myalgia. *Scand. J. Work Environ. Health* **17**, 347–355.

153. Yunus, M. B., Kalyan Raman, U. P., Kalyan Raman, K., and Masi, A. T. (1986). Pathologic changes in muscle in primary fibromyalgia syndrome. *Am. J. Med.* **81**, 38–42.

154. Simms, R. W. (1996). Is there muscle pathology in fibromyalgia syndrome? *Rheum. Dis. Clin. North Am.* **22**, 245–266.

155. Lund, N., Bengtsson, A., and Thorborg, P. (1986). Muscle tissue oxygen pressure in primary fibromyalgia. *Scand. J. Rheumatol.* **15**, 165–173.

156. Jacobsen, S., Jenson, K. E., Thomsen, C., Danneskiold-Samsoe, B., and Henriksen, O. (1992). 31P magnetic resonance spectroscopy of skeletal muscle in patients with fibromyalgia. *J. Rheumatol.* **19**, 1600–1603.

157. Jubrias, S. A., Bennett, R. M., and Klug, G. A. (1994). Increased incidence of a resonance in the phosphodiester region of P-31 nuclear magnetic resonance spectra in the skeletal muscle of fibromyalgia patients. *Arthritis Rheum.* **37**, 801–807.

158. Simms, R. W., Roy, S. H., Hrovat, M., Anderson, J. J., Skrinar, G., LePoole, S. R., Zerbini, C. A. F., DeLuca, C., and Jolesz, F. (1994). Lack of association between fibromyalgia syndrome and abnormalities in muscle energy metabolism. *Arthritis Rheum.* **37**, 794–800.

159. Park, J. H., Phothimat, P., Oates, C. T., Hernanzschulman, M., and Olsen, N. J. (1998). Use of P-31 magnetic resonance spectroscopy to detect metabolic abnormalities in muscles of patients with fibromyalgia. *Arthritis Rheum.* **41**(3), 406–413.

160. Crofford, L. J. (1998). Neuroendocrine abnormalities in fibromyalgia and related disorders. *Am. J. Med. Sci.* **315**(6), 359–366.

161. Ostensen, M., Rugelsjoen, A., and Wigers, S. H. (1997). The effect of reproductive events and alterations of sex hormone levels on the symptoms of fibromyalgia. *Scand. J. Rheumatol.* **26**(5), 355–360.

162. Hapidou, E. G., and Rollman, G. B. (1998). Menstrual cycle modulation of tender points. *Pain* **77**(2), 151–161.

163. Wolfe, F. (1997). The fibromyalgia problem. *J. Rheumatol.* **24**(7), 1247–1249.

164. Carette, S. (1995). What have clinical trials taught us about the treatment of fibromyalgia? *J. Musculoskeletal Pain* **3**, 133–140.

165. Wolfe, F., Cathey, M. A., and Kleinheksel, S. M. (1984). Fibrositis (fibromyalgia) in rheumatoid arthritis. *J. Rheumatol.* **11**, 814–818.

166. Russell, I. J., Fletcher, E. M., Michalek, J. E., McBroom, P. C., and Hester, G. G. (1991). Treatment of primary fibrositis/fibromyalgia syndrome with ibuprofen and alprazolam—A double-blind, placebo-controlled study. *Arthritis Rheum.* **34**, 552–560.

167. Yunus, M. B., Masi, A. T., and Aldag, J. C. (1989). Short term effects of ibuprofen in primary fibromyalgia syndrome: A double blind, placebo controlled trial. *J. Rheumatol.* **16**, 527–532; *erratum* p. 855.

168. Goldenberg, D. L., Felson, D. T., and Dinerman, H. (1986). A randomized, controlled trial of amitriptyline and naproxen in the treatment of patients with fibromyalgia. *Arthritis Rheum.* **29**, 1371–1377.

169. Moldofsky, H., and Scarisbrick, P. (1976). Induction of neurasthenic musculoskeletal pain syndrome by selective sleep stage deprivation. *Psychosom. Med.* **38**, 35–44.

170. Bennett, R. M., Clark, S. R., Goldberg, L., Nelson, D., Bonafede, R. P., Porter, J., and Specht, D. (1989). Aerobic fitness in patients with fibrositis: A controlled study of respiratory gas exchange and

[133]xenon clearance from exercising muscle. *Arthritis Rheum.* **32**, 454–460.

171. McCain, G. A., Bell, D. A., Mai, F. M., and Halliday, P. D. (1988). A controlled study of the effects of a supervised cardiovascular fitness training program on the manifestations of fibromyalgia. *Arthritis Rheum.* **31**, 1135–1141.

172. Mengshoel, A. M., Komnaes, H. B., and Forre, O. (1992). The effects of 20 weeks of physical fitness training in female patients with fibromyalgia. *Clin. Exp. Rheumatol.* **10**, 345–349.

173. Isomeri, R., Mikkelsson, M., and Latikka, P. (1992). Effects of amitripyline and cardiovascular fitness training on the pain of fibromyalgia patients. *Scand. J. Rheumatol., Suppl.* **21**, 47 (abstr.).

174. Burckhardt, C. S., Mannerkorpi, K., Hedenberg, L., and Bjelle, A. (1994). A randomized, controlled clinical trial of education and physical training for women with fibromyalgia. *J. Rheumatol.* **21**, 714–720.

175. Verstappen, F. T. J., van Santen-Hoeufit, H. M. S., Bolwijn, P. H., van der Linden, S., and Kuipers, H. (1997). Effects of a group activity program for fibromyalgia patients on physical fitness and well being. *J. Musculoskeletal Pain* **5**(4), 17–28.

176. Martin, L., Nutting, A., MacIntosh, B. R., Edworthy, S. M., Butterwick, D., and Cook, J. (1996). An exercise program in the treatment of fibromyalgia. *J. Rheumatol.* **23**, 1050–1053.

177. Norregaard, J., Lykkegaard, J. J., Mehlsen, J., and Danneskiold-samsoe, B. (1997). Exercise training in treatment of fibromyalgia. *J. Musculoskeletal Pain* **5**(1), 71–79.

178. Carette, S., McCain, G. A., Bell, D. A., and Fam, A. G. (1986). Evaluation of amitriptyline in primary fibrositis. A double- blind, placebo-controlled study. *Arthritis Rheum.* **29**, 655–659.

179. Bennett, R. M., Gatter, R. A., Campbell, S. M., Andrews, R. P., Clark, S. R., and Scarola, J. A. (1988). A comparison of cyclobenzaprine and placebo in the management of fibrositis: A double-blind controlled study. *Arthritis Rheum.* **31**, 1535–1542.

180. Quimby, L. G., Gratwick, G. M., Whitney, C. D., and Block, S. R. (1989). A randomized trial of cyclobenzaprine for the treatment of fibromyalgia. *J. Rheumatol., Suppl.* **19**, 140–143.

181. Jaeschke, R., Adachi, J. D., Guyatt, G., Keller, J., and Wong, B. (1991). Clinical usefulness of amitriptyline in fibromyalgia: The results of 23 N-of-1 randomized controlled trials. *J. Rheumatol.* **18**, 447–451.

182. Santandrea, S., Montrone, F., Sarziputtini, P., Boccassini, L., and Caruso, I. (1993). A double-blind crossover study of 2 cyclobenzaprine regimens in primary fibromyalgia syndrome. *J. Int. Med. Res.* **21**, 74–80.

183. Carette, S., Bell, M. J., Reynolds, W. J., Haraoui, B., McCain, G. A., Bykerk, V. P., Edworthy, S. M., Baron, M., Koehler, B. E., Fam, A. G., Bellamy, N., and Guimont, C. (1994). Comparison of amitriptyline, cyclobenzaprine, and placebo in the treatment of fibromyalgia—A randomized, double- blind clinical trial. *Arthritis Rheum.* **37**, 32–40.

184. Wolfe, F., Cathey, M. A., and Hawley, D. J. (1994). A double-blind placebo controlled trial of fluoxetine in fibromyalgia. *Scand. J. Rheumatol.* **23**, 255–259.

185. Geller, S. A. (1989). Treatment of fibrositis with fluoxetine hydrochloride (Prozac). *Am. J. Med.* **87**, 594–595.

186. Goldenberg, D., Mayskiy, M., Mossey, C., Ruthazer, R., and Schmid, C. (1996). A randomized, double-blind crossover trial of fluoxetine and amitriptyline in the treatment of fibromyalgia. *Arthritis Rheum.* **39**, 1852–1859.

187. Vlaeyen, J. W., Teeken-Gruben, N. J., Goossens, M. E., Rutten-van Molken, M. P., Pelt, R. A., van Eek, H., and Heuts, P. H. (1996). Cognitive-educational treatment of fibromyalgia: A randomized clinical trial. I. Clinical effects. *J. Rheumatol.* **23**, 1237–1245.

188. White, K. P., and Nielson, W. R. (1995). Cognitive behavioral treatment of fibromyalgia syndrome: A followup assessment. *J. Rheumatol.* **22**, 717–721.

189. Flor, H., and Birbaumer, N. (1993). Comparison of the efficacy of electromyographic biofeedback, cognitive-behavioral therapy, and conservative medical interventions in the treatment of chronic musculoskeletal pain. *J. Consult. Clin. Psychol.* **61**, 653–658.

190. Bennett, R. M., Burckhardt, C. S., Clark, S. R., O'Reilly, C. A., Wiens, A. N., and Campbell, S. M. (1996). Group treatment of fibromyalgia: A 6 month outpatient program. *J. Rheumatol.* **23**, 521–528.

191. Goldenberg, D. L., Kaplan, K. H., Nadeau, M. G., Brodeur, C., Smith, S., and Schmid, C. H. (1994). A controlled study of a stress-reduction, cognitive-behavioral treatment program in fibromyalgia. *J. Musculoskeletal Pain* **2**, 67–66.

192. Hawley, D. J. (1995). Psycho-educational interventions in the treatment of arthritis. *Bailliere's Clin. Rheumatol.* **9**, 803–823.

193. Bradley, L. A. (1989). Cognitive-behavioral therapy for primary fibromyalgia. *J. Rheumatol., Suppl.* **19**, 131–136.

194. Goossens, M. E., Rutten-van Molken, M. P., Leidl, R. M., Bos, S. G., Vlaeyen, J. W., and Teeken-Gruben, N. J. (1996). Cognitive-educational treatment of fibromyalgia: A randomized clinical trial. II. Economic evaluation. *J. Rheumatol.* **23**, 1246–1254.

195. Nicassio, P. M., Radojevic, V., Weisman, M. H., Schuman, C., Kim, J., Schoenfeld-Smith, K., and Krall, T. (1997). A comparison of behavioral and educational interventions for fibromyalgia. *J. Rheumatol.* **24**, 2000–2007.

196. Cathey, M. A., Wolfe, F., Roberts, F. K., Bennett, R. M., Caro, X., Goldenberg, D. L., Russell, I. J., and Yunus, M. B. (1990). Demographic, work disability, service utilization and treatment characteristics of 620 fibromyalgia patients in rheumatologic practice. *Arthritis Rheum.* **33**, S10 (abstr.).

197. Russell, I. J., Kamin, M., Sager, D., Bennett, R. B., Schnitzer, T., Green, J. A., and Katz, W. A. (1997). Efficacy of Ultram™ (Tramadol HCL) treatment of fibromyalgia syndrome: Preliminary analysis of a multi-center, randomized, placebo-controlled study. *Arthritis Rheum.* **40**(Suppl.), S117.

198. The Federation of State Medical Boards of the United States, Inc. (1998). "Model Guidelines for the Use of Controlled Substances for the Treatment of Pain," Adopted, May 2, 1998. Federation of State Medical Boards of the United States, Euless, Texas, pp. 76039–3855.

199. Sorensen, J., Bengtsson, A., Ahlner, J., Henriksson, K. G., Ekselius, L., and Bengtsson, M. (1997). Fibromyalgia—are there different mechanisms in the processing of pain? A double blind crossover comparison of analgesic drugs. *J. Rheumatol.* **24**, 1615–1621.

200. Clark, S., Tindall, E., and Bennett, R. M. (1985). A double blind crossover trial of prednisone versus placebo in the treatment of fibrositis. *J. Rheumatol.* **12**, 980–983.

201. Wolfe, F., Mullis, M., and Cathey, M. A. (1991). A double blind placebo controlled trial of somatostatin in fibromyalgia. *Arthritis Rheum.* **34**, S188 (abstr.).

202. Stratz, T., Mennet, P., Benn, H. P., and Müller, W. (1991). Blockade of S2 receptors—A new approach to therapy of primary fibromyalgia syndrome. *Z. Rheumatol.* **50**, 21–22.

203. Puttini, P. S., and Caruso, I. (1992). Primary fibromyalgia syndrome and 5-Hydroxy-L-Tryptophan—A 90-day open study. *J. Int. Med. Res.* **20**, 182–189.

204. Caruso, I., Sarzi Puttini, P., Cazzola, M., and Azzolini, V. (1990). Double-blind study of 5-hydroxytryptophan versus placebo in the treatment of primary fibromyalgia syndrome. *J. Int. Med. Res.* **18**, 201–209.

205. Vaeroy, H., Abrahamsen, A., Frre, O., and Kass, E. (1989). Treatment of fibromyalgia (fibrositis syndrome): A parallel double blind trial with carisoprodol, paracetamol and caffeine (Somadril comp) versus placebo. *Clin. Rheumatol.* **8,** 245–250.

206. Bessette, L., Carette, S., Fossel, A. H., and Lew, R. A. (1998). A placebo controlled crossover trial of subcutaneous salmon calci-

tonin in the treatment of patients with fibromyalgia. *Scand. J. Rheumatol.* **27**(2), 112–116.

207. Bennett, R. M., Clark, S. C., and Walczyk, J. (1998). A randomized, double-blind, placebo-controlled study of growth hormone in the treatment of fibromyalgia. *Am. J. Med.* **104**(3), 227–231.

86

Epidemiology of Headache in Women: Emphasis on Migraine

SANDA W. HAMELSKY,* WALTER F. STEWART,† AND RICHARD B. LIPTON‡

*New Jersey Graduate Program in Public Health, Rutgers University/University of Medicine and Dentistry of New Jersey, New Brunswick, New Jersey; *Bristol-Myers Squibb, New York, New York; †Department of Epidemiology, The Johns Hopkins University, Baltimore, Maryland; †Innovative Medical Research, Towson, Maryland; ‡Departments of Neurology, Epidemiology, and Social Medicine, Albert Einstein College of Medicine, Bronx, New York; ‡Headache Unit, Montefiore Medical Center, Bronx, New York; ‡Innovative Medical Research, Stamford, Connecticut

I. Introduction

The epidemiology of headache varies with headache type and gender. Migraine prevalence is generally higher among women than among men. The female to male gender prevalence ratio for migraine is about 2.8 [1], while the corresponding ratio for episodic tension-type headache is close to 1.0 [2]. Cluster headache, on the other hand, shows a clear male preponderance, with a gender ratio range of 1 to 6 [3].

Headache, especially migraine, has a substantial impact on the individual and society. Migraine causes sufferers to miss work, social activities, and time with their families. Furthermore, migraine causes reduced productivity in the workplace and increased health care utilization. The estimated annual lost productivity cost for the employed migraine population in the United States has ranged from $1.4–17 billion [4–6], with a 1999 estimate of about $13 billion [7].

Because migraine is much more common in women and because the burden of migraine is so great, this chapter focuses primarily on the epidemiology of migraine in women. We first consider the diagnostic criteria, which is followed by discussions of the incidence and prevalence of migraine. Next, we discuss female sex hormones and headache, as well as genetics. We close with a discussion of the public health and economic impact of headache.

II. Diagnostic Criteria

In 1988, the International Headache Society (IHS) established diagnostic criteria [8] for headache that provided the basis for most subsequent epidemiologic research. These criteria have also been widely used in biologic and genetic research as well as in clinical trials. The IHS criteria define four major categories of primary headache disorders and eight major categories of secondary headache disorders. Secondary headaches are outside the scope of the present discussion. Though seven subtypes of migraine are defined, the two most common are migraine without aura and migraine with aura. The IHS also provides case definitions of episodic and chronic tension-type headache. The case definitions for these disorders are presented in Tables 86.1 and 86.2.

The IHS classification system greatly facilitated the study of headache by establishing standard diagnostic criteria for most headache disorders. The diagnostic criteria were developed through expert consensus and represent a marked improvement over previous diagnostic criteria. They are written in uncomplicated, operational terms and give explicit detail regarding which headache features are necessary and which features preclude each diagnosis. The methodologic issues associated with the validation and implementation of the IHS criteria have been discussed [9–11].

III. Incidence

Few prospective studies of migraine have been conducted to estimate incidence rates. Breslau et al. [12] estimated the incidence of migraine in a random sample of members of a large Health Maintenance Organization ranging in age from 21 to 30 years. A total of 1007 participants were initially interviewed. Follow-up interviews were conducted among 972 (97.5%) of the participants 3.5 and 5.5 years later. The 5-year incidence of migraine was calculated among the 848 persons in the panel of 972 who had not met the criteria for migraine at baseline. A total of 71 (8.4%) cases of migraine were identified (female, 60; male, 11).

Two population-based studies used the reported age of migraine onset to estimate the incidence of migraine. Stewart et al. [13] conducted telephone interviews among 10,169 residents of Washington County, Maryland who were between the ages of 12 and 29. In total, 392 males and 1018 females were identified as migraine sufferers. This study is unique because it adjusted for telescoping, a type of recall bias. Telescoping is the tendency to report the occurrence of past events at times closer to the present [14]. Thus, studies estimating the age-specific incidence of migraine based on recall would likely be biased toward older ages of headache onset [14,15]. To minimize the effect of telescoping, Stewart et al. estimated the age-specific incidence rates after adjusting for the time interval between the reported age of onset and the age at interview. Individuals between the ages of 12 and 30 were interviewed, limiting the generalizability of results.

Below the age of 10, the incidence of migraine, especially migraine without aura, was higher among males than among females. For example, among females between 6 and 7 years old, the incidence was 6.3 per 1000 person-years for migraine with aura and 5.4 per 1000 person-years for migraine without aura. For males in the same age group, the incidences for migraine with aura and migraine without aura were 5.6 and 8.3 per 1000 person-years, respectively. However, during the adolescent years, the incidence of migraine was more common among

Table 86.1
International Headache Society (IHS) Diagnostic Criteria for Migraine

Migraine without aura

Description: Idiopathic, recurring headache disorder manifesting in attacks lasting 4–72 hours. Typical characteristics of headache are unilateral location, pulsating quality, moderate or severe intensity, aggravation by routine physical activity, and association with nausea, photo-, and phonophobia.

Diagnostic criteria:

A. At least five attacks fulfilling B–D
B. Headache attacks lasting 4–72 hours (untreated or unsuccessfully treated)
C. Headache has at least two of the following characteristics:
    1. Unilateral location
    2. Pulsating quality
    3. Moderate or severe intensity inhibits or prohibits daily activities
    4. Aggravation by walking stairs or similar routine physical activity
D. During headache at least one of the following:
    1. Nausea and/or vomiting
    2. Photophobia and phonophobia
E. At least one of the following:
    1. History, physical and neurological examinations do not suggest symptomatic headache due to an underlying condition
    2. History and/or physical- and/or neurological examinations do suggest such a disorder, but it is ruled out by appropriate investigations
    3. Such a disorder is present, but migraine attacks do not occur for the first time in close temporal relation to the disorder

Migraine with aura

Description: Idiopathic, recurring disorder manifesting with attacks of neurological symptoms unequivocally localizable to cerebral cortex or brain stem, usually gradually developed over 5–20 minutes and usually lasting less than 60 minutes. Headache, nausea, and/or photophobia usually follow neurological aura symptoms directly or after a free interval of less than an hour. The headache usually lasts 4–72 hours, but may be completely absent.

Diagnostic Criteria:

A. At least two attacks fulfilling B
B. At least three of the following four characteristics:
    1. One or more fully reversible aura symptoms indicating focal cerebral cortical and/or brain stem dysfunction
    2. At least one aura symptom develops gradually over more than 4 minutes or 2 or more symptoms occur in succession
    3. No aura symptom lasts more than 60 minutes. If more than one aura symptom is present, accepted duration is proportionally increased
    4. Headache follows aura with a free interval of less than 60 minutes (It may also begin before or simultaneously with aura)
C. At least one of the following:
    1. History, physical, and neurological examinations do not suggest symptomatic headache due to an underlying condition
    2. History and/or physical- and/or neurological examinations do suggest such a disorder, but it is ruled out by appropriate investigations
    3. Such a disorder is present, but migraine attacks do not occur for the first time in close temporal relation to the disorder

Table 86.2
International Headache Society (IHS) Diagnostic Criteria for Tension-Type Headaches

Episodic tension-type headache

Description: Recurrent episodes of headache lasting minutes to days. The pain is typically pressing/tightening in quality, of mild or moderate intensity, bilateral in location, and does not worsen with routine physical activity. Nausea is absent, but photophobia or phonophobia may be present.

Diagnostic criteria:

A. At least ten previous headache episodes fulfilling criteria B–D. Number of days with such headache < 180/year (< 15/month).
B. Headache lasting from 30 minutes to 7 days
C. At least two of the following pain characteristics:
    1. Pressing/tightening (nonpulsating) quality
    2. Mild or moderate intensity (may inhibit, but does not prohibit, activities)
    3. Bilateral location
    4. No aggravation by walking stairs or similar routine physical activity
D. Both of the following:
    1. No nausea or vomiting (anorexia may occur)
    2. Photophobia and phonophobia are absent, or one but not the other is present
E. At least one of the following:
    1. History, physical-, and neurological examinations do not suggest symptomatic headache due to an underlying condition.
    2. History and/or physical and/or neurological examinations do suggest such a disorder, but it is ruled out by appropriate investigations.
    3. Such a disorder is present, but tension-type headache does not occur for the first time in close temporal relation to the disorder.

Chronic tension-type headache

Description: Headache present for at least 15 days a month during at least 6 months. The headache is usually pressing/tightening in quality, mild or moderate in severity, bilateral, and does not worsen with routine physical activity. Nausea, photophobia, or phonophobia may occur.

Diagnostic criteria:

A. Average headache frequence ≥ 15 days/month (180 days/year) for ≥ 6 months fulfilling criteria B–D
B. At least two of the following pain characteristics:
    1. Pressing/tightening quality
    2. Mild or moderate intensity (may inhibit, but does not prohibit activities)
    3. Bilateral location
    4. No aggravation by walking stairs or similar routine physical activity
C. Both of the following:
    1. No vomiting
    2. No more than one of the following:
        Nausea, photophobia, or phonophobia
D. At least one of the following:
    1. History, physical, and neurological examinations do not suggest symptomatic headache due to an underlying condition
    2. History and/or physical and/or neurological examinations do suggest such a disorder, but it is ruled out by appropriate investigations
    3. Such a disorder is present, but tension-type headache does not occur for the first time in close temporal relation to the disorder

females than among males. For example, among females between 12 and 13 years old, the incidence was 14.1 per 1000 person-years for migraine with aura and 17.3 per 1000 person-years for migraine without aura. The corresponding values for males were 3.5 and 9.9 per 1000 person-years. Most of the new onset cases among females were cases of migraine without aura. In addition, the peak incidence of migraine was higher among females (migraine without aura: females, 18.9; males, 10.1; migraine with aura: females, 14.1; males, 6.6), but occurred earlier in males (migraine without aura: females, 14–17 years; males, 10–11 years; migraine with aura: females, 12–13 years; males, 4–5 years).

Rasmussen [16] reported that the age-adjusted annual incidence of migraine was 3.7 per 1000 person-years (female, 5.8; male, 1.6). This study did not estimate age-specific incidence, nor did it report incidence by migraine subtypes.

Stang *et al.* [17] used linked medical records to estimate the incidence of migraine in Olmstead County, Minnesota. As a consequence, only medically recognized headaches were included. Of the 6400 patient records reviewed, 629 fulfilled the IHS criteria for migraine. The reported incidences were lower than in any of the aforementioned studies. Among females, the incidence of migraine peaked between 20 and 24 years, at 6.9 per 1000 person-years. In males, the highest incidence, 2.5 per 1000 person-years, occurred between 10 and 14 years of age. The decreased incidences and later peaks in incidence relative to the study of Stewart *et al.* [13] may reflect the study methodology. Only individuals who consulted a health care provider for headache were identified in the study Stang *et al.*

IV. Prevalence

Many studies have estimated the prevalence of migraine. These prevalence estimates vary widely, from 2 to 57% in females and 0 to 46% in males, and differ according to age, gender, race, geography, and socioeconomic status [18,19]. Much of the variation in prevalence estimates is due to differences in case definition, especially for studies conducted prior to the publication of the IHS criteria [18,19]. Furthermore, because migraine prevalence varies by sociodemographic factors, study samples that differ according to these factors vary in estimated prevalence. Table 86.3 [19–36] presents the age- and gender-specific migraine prevalence estimates from 18 population-based studies that used the IHS criteria [19]. Two published meta-analyses that help explain the variation in prevalence estimates will be reviewed in this section [18,19].

Stewart *et al.* [18] conducted a meta-analysis of 24 population-based studies of migraine prevalence published prior to 1994 to describe the variation in reported prevalence estimates. At that time, only five studies used the IHS criteria. The authors reported that approximately 70% of the variation was accounted for by gender (14.5%), age (age 2.9%; age^2 13.5%), and case definition (36.1%). Migraine was more prevalent among females than among males and varied with age, peaking between 35 and 45 years of age both in males and in females. The most consistent estimates of migraine prevalence were found among studies that used case definitions that included any two of the following characteristics: unilateral head pain, nausea, and visual aura. Other factors, including method of selecting the study population, the source of the population, the response rate, and whether diagnoses were confirmed by a clinical examination were not found to significantly affect the variation in prevalence.

Another meta-analysis included 18 reports that used the IHS criteria for migraine [19]. Separate gender-specific models were developed. Consistent with the previous meta-analysis [18], migraine was more prevalent among females than among males; prevalence also varied by age, peaking between 35 and 45 years in both genders. The age distribution of the population and geographic location of the study explained 74% of the variation in migraine prevalence among females and 58% of the variation among males. The geographic differences may reflect, at least in part, the influence of race. Methodological factors (*e.g.,* sampling method, response method, response rate, recall period) did not explain the variation in migraine prevalence among studies.

These two meta-analyses provide important information regarding the variation in migraine prevalence, but some of the variability remains unexplained. The authors suggest that unmeasured factors such as socioeconomic status or cultural differences in symptom reporting may be important predictors of variation in migraine prevalence [19].

A. Prevalence by Age

Most studies of migraine prevalence have reported variation by age. Figure 86.1 from the American Migraine Study [1,37] demonstrates that migraine prevalence is generally highest between 25 and 55 years, typically the most productive years of life. Migraine prevalence peaks in the late 30s to early 40s and declines during the fifth decade [1,37]. Similarly, the prevalence of episodic tension-type headache peaks during the third decade in both men (42.3%) and women (46.9%), and then declines thereafter [2].

B. Prevalence by Gender

Several studies indicate that prior to puberty, the prevalence of migraine is slightly higher in boys than in girls; however, as adolescence approaches, migraine becomes more prevalent among females [20,38–42]. In the American Migraine Study [1,37], a study of individuals 12 and older, the average female to male migraine prevalence ratio was 2.8, with a peak of 3.3 between 40 and 45 years. The ratio remained above 2.0 even after the age of menopause, suggesting that constitutional changes after menarche play a role in the excess prevalence in females.

C. Prevalence by Race and Geographic Region

Migraine prevalence may vary by race and geographic region. In one study conducted in the United States, the lowest prevalence was observed among Asian-Americans, intermediate estimates were reported in African-Americans, and the highest prevalence was observed among Caucasians, before and after adjusting for sociodemographic covariates [26]. Consistent with this study, another meta-analysis found that prevalence was lowest in Africa and Asia and higher in Europe and Central/South America. The highest prevalence was found in North America [19]. Figure 86.2 presents the age-adjusted migraine prevalence by geographic region. Similarly, the prevalence of episodic tension-type headache is higher among Caucasians than among African-Americans in both men and women [2].

Table 86.3

Gender- and Age-Specific Migraine Prevalence as Reported in 18 Population-Based Studies of Migraine That Used International Headache Society (IHS) Diagnostic Criteria

Author	Age range		Female	Male	Author	Age range		Female	Male
[20]	5	15	11.5	9.7	[21]	11	11	1.1	4.2
[22]	5	15	4.4	2.6		12	12	4.8	3.3
	16	25	11.3	8.1		13	13	3.8	2.1
	26	35	11.5	6.7		14	14	5.2	1.6
	36	45	20.3	17.7	[23]	25	34	18.0	5.0
	46	55	22.7	7.7		35	44	14.0	7.0
	56	65	6.7	0.0		45	54	12.0	6.0
	66	87	5.3	0.0		55	64	19.0	7.0
[24]	0	9	0.0	0.8	[25][a]	15	19	11.5	2.0
	10	19	3.7	1.5		20	29	13.0	7.0
	20	29	8.8	0.0		30	39	20.0	6.0
	30	39	8.5	0.0		40	49	18.0	2.5
	40	49	18.0	2.3		50	59	11.0	3.0
	50	59	8.1	3.1		60	69	9.0	0.5
	60	69	4.0	0.0		70	79	3.0	0.0
	0	9	4.8	0.7	[26][b]	18	25	22.7	10.0
	10	19	20.1	8.1		26	30	22.1	11.5
	20	29	16.8	5.0		31	35	19.0	8.2
	30	39	22.1	10.0		36	40	21.2	9.8
	40	49	21.5	16.1		41	45	21.0	9.1
	50	59	38.1	12.0		46	55	20.9	6.5
	60	69	15.4	0.0		56	65	9.8	3.2
[27]	10	18	10.2	9.6	[1][b]	12	19	8.1	4.3
[28]	21	30	12.9	3.4		20	29	19.8	7.1
[29]	0	9	1.4	1.5		30	39	28.7	8.9
	10	19	8.2	8.4		40	49	24.4	7.4
	20	29	9.7	6.2		50	59	18.7	6.0
	30	39	10.1	5.8		60	69	10.5	3.4
	40	49	13.0	6.6		70	85	6.6	2.5
	50	59	10.3	7.1	[30]	20	29	4.8	1.1
	60	69	8.1	6.5		30	39	5.6	4.3
[31]	65	84	2.0	0.0		40	49	3.8	1.4
[32][b]	15	24	6.0	2.0		50	59	3.4	1.6
	25	34	12.0	3.9		60	69	1.9	0.6
	35	49	12.0	2.4		70	79	0.6	0.6
	50	64	6.7	2.2		80	89	0.0	0.0
	65	75	2.8	0.9	[33][a]	17	29	15.5	13.4
[34][b]	12	17	10.2	5.4		30	49	14.4	10.7
	18	23	13.1	5.2		50	70	12.1	5.3
	24	29	18.8	5.4	[35]	15	24	0.8	0.4
[36]	18	24	18.3	10.9		25	34	2.6	0.9
	25	34	23.0	9.1		35	44	2.7	1.0
	35	44	33.3	8.7		45	54	1.4	0.2
	45	54	32.3	5.9		55	64	0.6	0.4
	55	64	28.7	5.4		65	74	0.4	0.2
	65	74	11.7	1.9					

[a]Data were estimated from published information.
[b]Data were provided directly by the author.
From [19].

Several factors may explain the international variation in migraine prevalence. The hypothesis of race-related differences in genetic susceptibility to migraine is supported by the findings that migraine prevalence is low in Africa and Asia and remains low among African-Americans and Asians in the United States. However, because the prevalence among Asians living in Asia is even lower than that among Asians living in the United States,

environmental or cultural risk factors may also explain the international variation.

D. Prevalence by Socioeconomic Status

Migraine was once thought to be a disease of the affluent. This perception may be explained by patterns of consultation

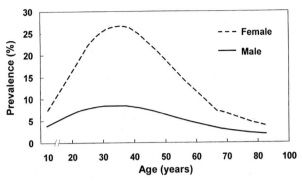

Fig. 86.1 Age- and sex-specific prevalence of migraine. From [37]; reproduced with permission from Advanstar Communications Inc. as reprinted from *Neurology*, 1993, Vol. 43, Supplement 13, and pages S6–S10. Neurology is a registered trademark of the American Academy of Neurology.

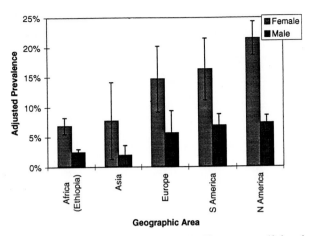

Fig. 86.2 Prevalence of migraine by geographic area at age 40, based on 18 population-based studies that used International Headache Society (IHS) diagnostic criteria. Estimates are model-based and adjusted for age and geographic area. 95% confidence intervals are shown. From [19].

for medical care. Because medical diagnosis of migraine is more common among high income groups [43], migraine may appear to be a disease of high income in the physician's office. However, several population-based studies from North America have demonstrated that in the community, migraine prevalence is inversely related to household income or education [1,26,37,44]. As income or education increased, migraine prevalence declined. The National Health Interview Survey (NHIS) demonstrated lower prevalence in the low compared with middle income group; however, prevalence was highest among the highest income group [5]. Because this study relied on self-reported medical diagnosis of migraine, and medical diagnosis of migraine rises with income, differential ascertainment by income may explain the association in the highest income group. Studies conducted outside the U.S. have not consistently supported the inverse relationship between migraine prevalence and household income [20,36,45,46]. It is unclear what factors may account for the international variation in findings.

Though migraine prevalence is inversely related to measures of socioeconomic status, the prevalence of episodic tension-type headache shows a direct relationship with education level. The prevalence of episodic tension-type headache increases with increasing education level [2]. The contrasting epidemiologic profiles of migraine and tension-type headache support the notion that these disorders are biologically distinct.

V. Female Hormones and Headache

Some of the age and gender variation in migraine prevalence may be accounted for by changes in levels of the female sex hormones, estrogen and progesterone [47,48]. Female sex hormone levels increase during puberty—the time that the female to male gender prevalence ratio of migraine increases. The female to male ratio continues to increase until about age 42 and then declines, perhaps coinciding with decreasing estrogen levels in anticipation of menopause [1,37]. Changing hormone levels during pregnancy may also affect the incidence and severity of migraine. Finally, the use of supplemental hormones such as oral contraceptives may affect the occurrence of migraine. Although hormonal fluctuations are important triggers of mi-

graine, hormonal factors alone do not account for the female preponderance of migraine, in general, and the persistently elevated gender ratios in later life. Other factors, yet unidentified, are likely to be involved.

Somerville's estrogen withdrawal hypothesis helps explain the relationship between fluctuating hormone levels and migraine. Somerville [48] studied six women during the premenstrual and menstrual phases of two menstrual cycles. During the first cycle, plasma levels of estradiol and progesterone were recorded a few days before and then during menstruation. Somerville reported that migraine typically occurred a few days before, or during, menstruation, a time when hormone levels are the lowest. Because a previous study failed to prove that declining progesterone levels caused migraine [49], Somerville hypothesized that estradiol withdrawal induced migraine. During the second cycle, women were given supplemental estrogen to maintain high plasma levels of estradiol during the premenstrual and menstrual phases. Menstruation was not influenced by estrogen supplements; however, the migraine attack was postponed three to nine days in five of the six women. Thus, Somerville concluded that falling estradiol levels played an important role in precipitating migraine. Two additional studies supported Somerville's conclusions [50,51].

A. Menstrual Migraine

Many women report an association between migraine and menstruation. Both doctors and patients commonly refer to migraine attacks that occur before, during, or after menstruation as "menstrual" migraine. Despite its wide use, the term has no standard definition. Consequently, prevalence estimates for "menstrual" migraine are highly variable, ranging from 4 to 73% [52]. Because women with menstrual migraine may be more likely to consult physicians, selection bias in clinic-based samples may also contribute to the variability [53,54].

The IHS has not agreed on a definition for "menstrual" migraine but provides the following comment: "Migraine without aura may occur almost exclusively at a particular time of the menstrual cycle—so-called menstrual migraine. Generally accepted criteria for this entity are not available. It seems reasonable to demand that 90 percent of attacks should occur between 2 days before menses and the last day of menses, but further epidemiological knowledge is needed" [8]. Several investigators have proposed definitions based on the timing of attacks, type of migraine, and the proportion of attacks that occur within a defined period of the menstrual cycle [52,55]. The term "true menstrual migraine" (TMM) usually refers to migraine headaches that occur almost exclusively with menses and not at any other time of the month. Migraine that occurs during menstruation is often associated with dysmenorrhea (difficult and painful menstruation) and is said to be more difficult to treat.

Several studies suggest that menses is associated with migraine without aura rather than migraine with aura [53,54,56] Johannes et al. [53] conducted a population-based diary study of women with migraine with aura. Daily diaries were used to collect information on the characteristics of the headache attacks as well as menstrual cycle events over a four-month time period. Headaches were classified as migraine with or without aura, episodic tension-type headache, or other. Subjects reported headaches on 29% of the study days, with a marked increase during the first three days of the menstrual period (odds ratio (OR) = 1.45, p < 0.002) (Fig. 86.3). This excess risk was attributed to an increase in attacks of migraine without aura (OR = 1.66, p < 0.002).

Consistent with this population-based study [53], in a clinic-based study of female migraine sufferers, MacGregor et al. [54] reported that migraine attacks occur at varying times throughout the month but with increased occurrence around the time of the menstrual period. In this study, the incidence of migraine headaches peaked on the first and second days of the menstrual period.

In both of these studies, the timing of increased risk for migraine coincides with the time of estrogen withdrawal, supporting Somerville's estrogen withdrawal hypothesis. Although circulating hormone levels are an important precipitating factor in headache attacks, they are not an exclusive trigger. In the Johannes et al. [53] study, only 33.9% of headache attacks occurred during the first three days of menstruation. Similarly, MacGregor et al. [54] reported that only 34.5% of women in the study had an increased number of attacks during the "menstrual" migraine period (days −2 to +3 of the menstrual cycle). Thus, headache attacks often occur on other days of the month. Additional population-based diary studies and clinical trials that collect daily blood levels of estrogen and progesterone would enhance our understanding of menstrual migraine.

B. Migraine, Oral Contraceptive Use, and Stroke

Evidence for an association between migraine and oral contraceptive (OC) use is inconsistent. (See Silberstein [52] and Becker [57] for comprehensive reviews). Initial use of OCs may provoke migraine attacks for the first time; alternatively, existing migraine may become more severe or frequent, the pattern of symptoms may change, or attacks may improve [58]. Becker [57] added that there may be no change in status at all. This variation in results among studies may be explained by methodological weaknesses in the research reported to date. Results may vary over time because the dosages of estrogen and progestin contained in OCs have decreased.

One concern with OC use is that it may increase the risk of stroke among migraine sufferers. There is support for an association between migraine and stroke in women under the age of 45 years [59]. In two independent case-control studies, Tzourio et al. found that the risk of ischemic stroke was substantially greater among young women with IHS migraine than in young women without migraine (OR = 4.3, 95% Confidence Interval (CI) = 1.2–16.3 [60] and OR = 3.5, 95% CI = 1.8–6.4 [61]). Furthermore, the risk associated with migraine with aura (OR = 6.2; 95% CI = 2.1–18.0) was considerably higher than that for migraine without aura (OR = 3.0; 95% CI = 1.5–5.8) [61].

Tzourio et al. also explored the relationship between OCs, migraine, and stroke in women under 45 years old [61]. Compared to women without migraine who were not using OCs, young women with migraine who used OCs had almost a 14-fold increased risk of stroke (OR = 13.9; 95% CI = 5.5–35.1); young women without migraine who were using OCs had about a 4-fold increased risk of stroke (OR = 3.5, 95% CI = 1.5–8.3). The authors concluded that there was no evidence of statistical interaction between migraine and OC use. The increased risk of stroke among OC users with migraine is inconsistent with the higher risk of stroke among migraine sufferers compared to nonmigraine sufferers.

The strength of these associations must be put into context. Assuming that the incidence of ischemic stroke in young women is 10 in 100,000 woman-years, the absolute risk of ischemic stroke would be 19 per 100,000 women per year for women with migraine [61]. Thus, while migraine and the combination of migraine and OC use are significant risk factors for stroke in young women, the absolute risk is still relatively low.

Migrainous women who smoke are also at increased risk of stroke. Tzourio et al. [61] found that compared to women without migraine who never smoked, the odds ratio among migrainous women who were heavy smokers was 10.2 (95% CI = 3.5–29.9). No statistical interaction between smoking and migraine was observed.

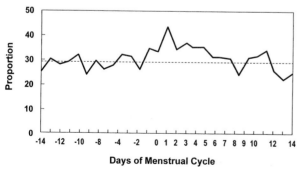

Fig. 86.3 Proportion of person-days on which headache was reported for each day of the menstrual cycle among 74 diary study participants. Washington County, Maryland, 1987 to 1988. Solid line indicates proportion of person-days with any type of headache. Dotted line indicates overall average proportion of headaches. From [53].

C. Migraine and Pregnancy

Evidence regarding the relationship between migraine and pregnancy is also inconsistent. Pregnancy has been reported to induce, improve, worsen, or cause no change in migraine [52]. Ratinahirana *et al.* [62] interviewed 116 women with IHS diagnosed migraine 1–3 days after delivery to assess the relationship between migraine and pregnancy. They reported that migraine improved or disappeared in 102 of 147 (69.4%) pregnancies. Results were similar for women with or without aura. Interestingly, improvement was more common among women who had previously suffered from menstrual migraine. For these women, improvement in migraine during pregnancy may be explained by the consistently elevated estrogen levels that occur during pregnancy, thus supporting Somerville's estrogen withdrawal hypothesis [48].

Marcus and Scharff [63] conducted the first prospective study to measure changes in headache that occur during pregnancy. In a study of 46 women with chronic headache (migraine, episodic tension-type, and combined migraine and episodic tension-type headache), there was no significant change in any headache type over the course of the pregnancy. These study results contrast with the significant reduction in headaches reported in many retrospective studies [64].

D. Migraine, Menopause, and Estrogen Replacement Therapy

Although migraine prevalence decreases after the fourth decade of life, migraine may worsen or regress during menopause [65]. Neri *et al.* [66] reported an improvement in symptoms after menopause. Neri *et al.* studied 556 postmenopausal women and found that headache was present in 13.7%. Two-thirds of the women who experienced migraine prior to menopause reported improvement in their headaches after the onset of menopause. Surgical menopause usually worsened migraine. The estrogen withdrawal hypothesis would predict migraine exacerbation during menopause due to fluctuating hormone levels, with improvement after menopause when estrogen levels are more consistent and low.

Hormonal replacement therapy is often used to treat the symptoms of menopause and to prevent osteoporosis [67,68]. This treatment may be helpful in alleviating menopausal symptoms; however, studies have found that estrogen replacement therapy may exacerbate migraine [69], or, in some instances, relieve it [70]. Hormonal replacement therapy that is administered in a cyclical fashion to mimic the natural menstrual cycle may trigger migraine in women sensitive to estrogen withdrawal. In this instance, changing to continuous estrogen administration may alleviate the migraine attacks [52].

VI. Genetics

Familial aggregation of migraine has been recognized for several hundred years. Although all reported studies found an increased risk of migraine among family members of migraine probands, the relative risks varied from 1.5 to 19.3 [71]. Lack of a standard definition of migraine and highly selected study samples explain much of the variation. In addition, some of the studies interviewed the migraine proband to determine the migraine status of the relatives. This interview method has been demonstrated to result in an overestimate of familial aggregation because probands with migraine are more likely to be aware of migraine in their relatives [72]. Directly interviewing family members reduces the potential for bias [72].

Two population-based studies that do not suffer from these methodologic weaknesses examined the association between the risk of migraine among relatives of migraine probands and nonmigrainous controls. Russell and Olesen [73] studied 378 probands, 1109 first-degree relatives, and 229 spouses and found that first-degree relatives of probands with migraine without aura had a significantly increased risk of having migraine either with or without aura. In addition, first-degree relatives of probands with migraine with aura had nearly four times the risk of having migraine with aura, but no increased risk of migraine without aura. Finally, the spouses of probands with migraine without aura had 1.4 times the risk of migraine without aura, suggesting that the environment may be an important risk factor for migraine without aura. Thus, according to the results of this study, migraine without aura has both genetic and environmental risk factors, whereas migraine with aura is associated almost exclusively with genetic factors.

In a second population-based family study, Stewart *et al.* [74] found that the risk of migraine was 50% higher in relatives of migraine probands than in relatives of controls, thus supporting familial aggregation of migraine. In addition, migraine risk was higher among relatives of disabled migraine sufferers compared with nondisabled migraine sufferers. The familial aggregation and disability association may depend on the gender of the family member. Among migraine probands without disability, familial aggregation was statistically significant in male, but not in female, relatives. Alternatively, among disabled migraine probands, familial aggregation was observed in both male and female relatives. The authors hypothesized that these results may suggest that the expression of a less severe variant of migraine in females may be caused, in part, by a gender-related nonfamilial factor such as the acute and long-term effects of sex hormones.

Both clinic-based [75] and population-based [76–78] twin studies demonstrate that monozygotic twins are more concordant for migraine than dizygotic twins, providing additional support for a genetic basis for migraine. The three twin studies establish that nongenetic factors are important because many monozygotic twin pairs are discordant for migraine, again highlighting the importance of environmental factors.

A genetic locus for familial hemiplegic migraine (FHM), a rare subtype of migraine with aura, was found on chromosome 19 [79], with a second locus on chromosome 1 [80]. These discoveries have increased the hope that the genetic bases of more common forms of migraine may be determined. Peroutka *et al.* [81] analyzed the dopamine D_2 receptor (DRD2) as a candidate gene for migraine with or without aura. The investigators found that individuals with migraine with aura had a significantly increased frequency of the DRD2 NcoI C allele compared with controls. There was no increased frequency among individuals with migraine without aura. Thus, dopamine receptor genes may play a role in the pathophysiology of migraine with aura.

VII. Risk Factors and Comorbidities

In addition to the demographic (age, gender, sociodemographic status, and geographic region) and genetic risk factors, migraine is comorbid with a number of other disorders; that is, certain disorders occur more frequently among individuals with migraine than in the general population. Population-based studies demonstrate that migraine is comorbid with depression, anxiety disorders, manic depressive illness [28,82–90] and epilepsy [91–93]. Migraine is also associated with stroke in women below the age of 45 years [60,61] and with Raynaud's syndrome [94].

Certain foods, beverages, or circumstances have been reported to "trigger" migraine attacks, although clinical trials and analytic epidemiologic studies are generally lacking. Migraine triggers vary by individual and may include chocolate, cheeses, alcoholic beverages (especially red wine), citrus fruits, and foods containing monosodium glutamate, nitrates, and aspartame [95]. While some controlled clinical trials of the association between dietary triggers and migraine exist [96–98], most of the information about triggers has been obtained through patient reports collected in clinic-based samples [99–102]. These patient reports represent beliefs about causes of migraine rather than proven causes of migraine.

VIII. Treatment

Both pharmacological and nonpharmacological methods are used to treat headaches. Pharmacological treatment may be acute (abortive) or preventive (prophylactic). Acute treatments are given at the time of the attack to relieve pain and other symptoms. Prophylactic treatments are taken on a daily basis whether or not headache is present to prevent migraine attacks and reduce their severity. Migraine treatment has been extensively reviewed elsewhere [95,103]. The development of sumatriptan and related compounds (e.g., naratriptan, zolmitriptan, and rizatriptan) as well as novel delivery systems (e.g., injection, intranasal, and tablets that dissolve in the oral cavity) have dramatically improved therapeutic options and patient outcomes. Additionally, the approval of a nonprescription treatment for migraine in the United States [104] provides for increased awareness of treatment for the many migraine sufferers who do not consult physicians.

In a study of patterns of medication use, Celentano et al. [105] found that among female migraine sufferers, approximately 57% treat their migraine attacks exclusively with nonprescription drugs, 40% use prescription drugs, and 3% do not use medication. The corresponding male percentages are 67%, 28%, and 5%.

Nonpharmacological treatments such as relaxation and biofeedback are helpful for some headache sufferers. Behavioral interventions such as maintaining a regular schedule, getting adequate sleep and exercise, and giving up tobacco may also be beneficial. These methods have been reviewed elsewhere [95,106].

IX. Public Health Significance

Headache has a substantial impact on the individual and society. The majority of studies that assess the impact of headache have focused specifically on the impact of migraine. The individual impact of migraine has traditionally been measured in terms of the frequency and severity of attacks, which vary considerably by gender. Quality of life studies have been conducted to complement the existing knowledge about the individual impact of headache. The societal impact of headache is often measured through its economic terms. Economic costs include both direct as well as indirect costs. Direct costs include health care utilization figures such as rates of outpatient visits, hospitalization, the use of emergency department services, and the cost of prescriptions. Indirect costs include work loss and absenteeism due to headache.

A. Individual Impact—Frequency and Severity of Attacks

The individual impact of migraine is a function of the level of pain intensity, the presence of associated symptoms such as nausea, photophobia, and phonophobia, and the frequency and duration of the attacks. Migraine attacks vary among individuals in frequency, pain intensity, and disability. In general, women report greater individual impact from their headaches than men.

On average, women report higher levels of pain intensity with their headaches than men. Celentano et al. [107] conducted a population-based study to describe gender differences in the pain, symptom frequency, and duration of headache attacks. The study evaluated the most recent headache in 6347 respondents between 12 and 29 years of age who had a headache attack within 4 weeks of the interview. The focus on the most recent headache was intended to maximize the accuracy of symptom reporting. The estimated pain associated with each subject's headache was consistently greater for women than for men. In a review of population-based studies, Stewart et al. [108] confirmed that women report more pain with their migraine attacks than their male counterparts.

Women also suffer from more frequent headache attacks and attacks of longer duration. Celentano et al. [107] reported that while 47.6% of males and 36.7% of females reported only one headache during a four week period, almost twice as many females as males reported four or more headaches during the same period (females 18.4%; males 10.6%). Stewart et al. found similar results among migraine sufferers [108]. Women were significantly more likely to report headache-related disability and to seek health care services for their headaches, even after adjusting for headache severity [108,109]. Women were also more likely to report a medical diagnosis of migraine, even when the analysis was restricted to individuals who sought medical care [43,109]. Finally, women report more severe and more frequent disability from their migraine attacks [108].

A combination of physiological and social factors may explain the gender differences in the experience of headache [107]. The socialization process may lead women to report symptoms more readily than men [110]. In addition, women may be more attentive to their headache symptoms [111]. Hormonal factors may modify the experience of headache. Additional studies are needed to explain the interplay between social and physiological influences in the experience of headache.

B. Quality of Life Studies

Quality of life instruments assess patients' perceptions of physical and role functioning and well-being during or between attacks. Both generic and disease-specific instruments have been used to measure health-related quality of life in headache. Generic quality of life instruments assess quality of life associated with physical, social, psychological, and behavioral life domains and are designed to measure the impact of a range of illnesses using a common scale. They facilitate comparisons of quality of life across diagnostic categories. Commonly used generic measures of health-related quality of life are the Medical Outcomes Study instruments: Short Form 20 (SF20) and Short Form 36 (SF36) [112].

Disease-specific measures are designed to target the quality of life impact of specific illnesses. These measures are often designed to be sensitive to change over time [113]. Many of these measures were developed for use in studies designed to assess improvement in quality of life while taking a specific treatment [114–118].

Most quality of life studies have been conducted in clinic-based samples. Quality of life in these samples may differ from comparable measures in the population, limiting generalizability. In addition, many clinic-based studies of quality of life lack a contemporaneous control group and do not control for comorbid conditions. In studies that assess response to treatment over time, the lack of a control group may overestimate benefits of treatment because of regression toward the mean.

Studies conducted in clinic-based samples demonstrate that migraine sufferers have lower quality of life than control subjects from the United States population [119,120]. Osterhaus et al. [120] found that the quality of life of migraine sufferers was similar to patients with depression but lower than that of other patients with hypertension, osteoarthritis, or diabetes. Another clinic-based study found that migraine sufferers experience poorer quality of life, even between attacks [121].

One study estimated the impact of migraine on quality of life using the SF-36 in a population-based sample [122]. A total of 618 migraine sufferers and 5421 controls participated in the study. Migraine sufferers, regardless of the presence of aura, reported significantly lower levels of functioning and well-being compared with the control population. In addition, quality of life and frequency of attacks were inversely correlated; as the frequency of attacks increased, quality of life decreased.

C. Societal Impact

1. Direct Costs

Physician consultation for headache varies based on gender, age, and severity. In a population-based study of headache sufferers (without specific diagnostic requirements) conducted from 1986 to 1987, 13.9% of women and 5.6% of men were current consulters (physician consultation within 1 year of interview), 12.8% of women and 8.0% of men were lapsed consulters (physician consultation more than 1 year before interview), and 86.4% of men and 73.3% of women had never consulted a physician [123]. Physician consultation is more common among migraine sufferers. Data from the American Migraine Study indicate that 68% of women and 57% of men have ever consulted a physician for migraine [109].

Several studies have examined the factors associated with consultation among migraine sufferers and found that consultation is more common among females than males [109,123]. Among both males and females, consultation was more likely with increasing age [109]. Migraine sufferers with severe attacks are more likely to see a physician [109]. Furthermore, Lipton et al. [109] reported that certain headache characteristics are associated with medical consultation for headache. Among females, these characteristics were higher pain intensity, number of migraine symptoms, attack duration, and disability. Migraine sufferers who had never consulted a physician for headache reported that their symptoms were not severe enough to warrant consultation and/or that they were able to treat their headaches effectively with nonprescription medications.

As demonstrated by Clouse and Osterhaus [124], patients with migraine have higher consultation rates than the general population. In this study, 1336 migraine sufferers enrolled in a United Health Care Corporation-affiliated plan were matched on the basis of age, sex, duration of enrollment, and subscriber/dependent status to a nonmigraine sufferer. During the 18-month study period, migraine sufferers made 2616 physician visits for migraine and 19,971 visits not related to migraine. Nonmigraine sufferers, in contrast, made a total of 13,072 physician visits during the same period. The evidence indicates that a small percentage of migraine sufferers account for the majority of physician visits. Clouse and Osterhaus [124] reported that while most patients reported 1–4 visits per year, 7.6% reported more than 12 visits.

Population-based studies indicate that emergency department (ED) visits occur among 14 [124,125] to 20% [105] of migraine sufferers. ED use is more common among females than among males (20% of females vs 13% of males) [105] and varies with medication use. In one study [105], the highest utilization rates were observed among subjects using prescription medication (33% of females and 27% of males); subjects using nonprescription drugs used the ED at one-third the rate of those using prescription drugs. Subjects who did not use medication had intermediate ED utilization rates (20% for men and 15% for women). The investigators were unable to determine whether the high rate of ED use among prescription drug users was due to the severity of the subject's headaches, medication failure, or side effects associated with the medication.

Hospitalization is not common among headache sufferers; however, overall hospitalization rates among migraine sufferers are reported to be approximately twice as high as those in a healthy population [124]. Lifetime hospitalization rates for migraine vary between 6% [5] and 11% [124] and are slightly higher in females than in males [5].

Few studies have translated health care utilization patterns into monetary costs. Clouse and Osterhaus [124] estimated the average claim cost per member per month ($145); however, costs were not separated by diagnosis. Among migraine sufferers who have comorbid conditions, these costs reflect treatment for migraine as well as the comorbid conditions. Osterhaus et al.,

[4] estimated that yearly direct medical costs were $817 per migraine sufferer. This figure is probably overestimated because the investigators studied a sample of clinical trial participants.

Hu *et al.* [7], on the other hand, estimated over $1 billion in annual treatment costs, about $100 per migraine sufferer per year, in a population-based study of migraine sufferers. Female sufferers accounted for approximately 80% of the total costs, slightly out of proportion to their representation in the population. Physician visits (60%) and prescription drugs (30%) accounted for the majority of treatment related costs. Emergency department costs accounted for less than 1% of total costs.

2. Indirect Costs

Work-related disability may be the most important component of the economic impact of headache. Work-related disability is usually measured through lost work days, but many headache sufferers stay at work during attacks and work with reduced effectiveness. Thus, it is also important to measure work loss due to reduced productivity as well as to absenteeism. Most studies of work loss due to headache focus on migraine, perhaps because migraine is more likely to interfere with a person's ability to work [126]. Many studies examine actual days of missed work, time at work with headache, and percent effectiveness while at work with a headache. The components are sometimes combined in an index termed "lost work day equivalents" (LWDE), which equals actual days of missed work plus days at work with headache times one minus percent effectiveness while at work with headache [127].

Many of the early estimates of work loss due to migraine may be inaccurate because studies included highly selected samples and incomplete ascertainment of migraine [4,5]. Later population-based studies included improved migraine ascertainment but asked migraine sufferers to recall missed days of work and reduced effectiveness on days over an extended period of time [127–129]. This approach is limited by uncertain accuracy of recall. To address these limitations, Von Korff *et al.* [126] estimated the number of lost work days and lost workday equivalents in a population-based sample of employed migraine sufferers who completed a daily diary for 3 months. This method was designed to reduce the influence of recall on estimates of work loss. During this 3-month period, migraine sufferers missed an average of 1.1 days per month due to headache, of which 0.7 lost work days were due to migraine, 0.3 to migrainous headaches (headache meeting some, but not all, of the IHS criteria for migraine), and 0.1 to other headaches (nonmigrainous). Migraine sufferers experienced an average of three lost work day equivalents, of which 1.4 were due to migraine and an additional 0.7 were due to migrainous headaches. The number of lost work days and LWDEs reported by von Korff *et al.* [126] are higher than those reported in other population-based studies [5,127,130]. Part of the difference may be accounted for by improved measurement of lost work time through the use of a daily diary to improve recall.

The most severely affected migraine sufferers accounted for most of the work loss and reduced work performance. Figure 86.4 from the von Korff *et al.* [126] study demonstrates that the most disabled 20% of the participants accounted for 50% of the LWDE and 77% of the lost work days. Thirty percent of the migraine sufferers accounted for 64% of the LWDE and 90% of

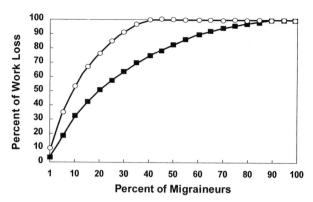

Fig. 86.4   Percent of employed migraineurs accounting for the indicated percentage of lost work days (line with circles) and lost work day equivalents (line with squares), for all types of headaches. Cumulative percentage of total disability was estimated by ordering subjects from highest to lowest on lost work days and lost work day equivalents. From [126].

the lost work days. Forty percent of subjects accounted for 75% of the lost work day equivalents and 100% of the lost work days. Similarly, Stewart *et al.* [127], found that a total of 51.1% of females and 38.1% of males experienced six or more LWDEs per year, accounting for more than 90% of all LWDEs in this population. Thus, to substantially reduce work loss due to migraine, the most severely affected sufferers must be targeted for treatment.

While many studies have estimated lost workdays and reduced effectiveness at work, only a few studies have estimated the cost associated with productivity loss. These studies had significant methodological weaknesses. One study probably underestimated the cost because it was based on self-identified migraine sufferers, thus excluding those sufferers who did not know they had migraine [5]. Another study generalized from a sample of clinical trials participants; this sample overestimates severity of disease in the general population [4].

Hu *et al.* [7] studied a population-based sample of migraine sufferers to estimate the indirect costs of migraine using the human capital method. Prevalence estimates were derived from the American Migraine Study [1], a national probability sample of over 20,000 individuals residing in the United States. Work loss data were derived from the Baltimore county survey [26], a sample of over 13,000. The authors estimate that migraine costs American employers about $13 billion per year due to absenteeism and reduced effectiveness at work. Approximately 62% of the cost was due to absenteeism. Importantly, the greatest indirect costs were found among middle-aged (30–49 years) males and females. Human capital studies assume that the economic value of a missed day of work is equivalent to the wages for that day. If coworkers or the employee with migraine compensate, this may be an overestimate. On the other hand, if costly errors are made during an attack, this may be an underestimate.

Furthermore, human capital studies only account for the value of time lost by the working population. It does not assess the value of time lost by a care giver or homemaker. Because migraine prevalence peaks during the childbearing and family care-giving years, it can have a substantial impact on family functioning when a care giver's role is repeatedly disrupted. Smith [131] surveyed a population-based sample of migraine sufferers about how migraine affected their family life. Sixty-one percent of the 350 migraine sufferers (269 females, 81 males) reported that their migraine attacks had a significant impact on their families. The majority of respondents (>76%) delayed or postponed household duties and activities with children or a spouse when they had a migraine attack. Social activities were also delayed or postponed, but not as often as household duties or time with children or a spouse. Thus, migraine not only has an impact on the individual and society, but on family functioning as well.

X. Conclusions

Headache is a very common condition, especially in women. The prevalence of migraine is greater among females than among males after about age 12 and peaks during the late 30s and early 40s, typically the most productive years of life [1,37]. The average female to male migraine prevalence ratio is 2.8, with a peak of 3.3 between 40–45 years of age [1,37].

The female preponderance of migraine may be due, at least in part, to the influence of female sex hormones. Migraine incidence, severity, and frequency is reported to vary with menstruation, oral contraceptive use, pregnancy, and menopause. However, because the greater prevalence of migraine continues into late life, female hormones are unlikely to account for all of the variation in migraine prevalence. Other factors, thus far unidentified, may be involved.

Headache has a substantial impact on the individual and society. Women are particularly affected by their headaches, reporting greater pain and disability than their male counterparts. Women also report more frequent attacks and attacks of longer duration. Headache also has a large economic impact on society. Health care costs related to migraine are estimated at over $1 billion per year, 80% of which are generated by female sufferers [7]. In addition, migraine costs employers an estimated $13 billion each year in lost and reduced productivity [7].

Public health interventions should be aimed at understanding the female preponderance of headache beyond the influence of hormones. Additional research is needed to understand the gender differences in headache experience. Understanding how headaches impact women's lives in areas other than work, such as when taking care of children or participating in social activities, will help further understanding of the impact of headache.

References

1. Stewart, W. F., Lipton, R. B., Celentano, D. D., and Reed, M. L. (1992). Prevalence of migraine headache in the United States: Relation to age, income, race, and other sociodemographic factors. *JAMA, J. Am. Med. Assoc.* **267,** 64–69.
2. Schwartz, B. S., Stewart, W. F., Simon, D., and Lipton, R. B. (1998). Epidemiology of tension-type headache. *JAMA, J. Am. Med. Assoc.* **279,** 381–383.
3. Raskin, N. H. (1988). "Headache." Churchill-Livingstone, New York.
4. Osterhaus, J. T., Gutterman, D. L., and Plachetka, J. R. (1992). Healthcare resource and lost labour costs of migraine headache in the U.S. *PharmacoEconomics* **1,** 67–76.
5. Stang, P., and Osterhaus, J. (1993). Impact of migraine in the United States: Data from the National Health Interview Survey. *Headache* **33,** 29–35.
6. de Lissovoy, G., and Lazarus, S. (1994). The economic cost of migraine. Present state of knowledge. *Neurology* **44**(Suppl. 4), S56–S62.
7. Hu, X., Markson, L., Lipton, R., Stewart, W., Berger, M. (1999). Disability and economic costs of migraine in the United States: A population-based approach. *Arch. Intern. Med.* **159,** 813–818.
8. Headache Classification Committee of the International Headache Society (1988). Classification and diagnostic criteria for headache disorders, cranial neuralgias and facial pain. *Cephalalgia* **8**(Suppl. 7), 1–96.
9. Merikangas, K., Whitaker, A., and Angst, J. (1993). Validation of diagnostic criteria for migraine in the Zurich Longitudinal Cohort Study. *Cephalalgia* **13**(Suppl. 12); 47–53.
10. Lipton, R., Stewart, W., and Merikangas, K. (1993). Reliability in headache diagnosis. *Cephalalgia* **13**(Suppl. 12), 29–33.
11. Lipton, R., and Stewart, W. (1997). Prevalence and impact of migraine. *Neuro. Clin.* **15,** 1–13.
12. Breslau, N., Chilcoat, H., and Andreski, P. (1996). Further evidence on the link between migraine and neuroticism. *Neurology* **47,** 663–667.
13. Stewart, W. F., Linet, M., Celentano, D., Van, N., and Ziegler, D. (1991). Age- and sex-specific incidence rates of migraine with and without visual aura. *Am. J. Epidemiol.* **134,** 1111–1120.
14. Brown, N., Rips, L., and Shevell, S. (1985). The subjective dates of natural events in very long-term memory. *Cognit. Psychol.* **17,** 139–177.
15. Cummings, R., Kelsey, J., and Nevitt, M. (1990). Methodologic issues in the study of frequent and recurrent health problems. *Ann. Epidemiol.* **1,** 49–56.
16. Rasmussen, B. (1995). Epidemiology of headache. *Cephalalgia* **15,** 45–68.
17. Stang, P., Yanagihara, T., Swanson, J., Beard, C., O'Fallon, W., *et al.* (1992). Incidence of migraine headache: A Population-based Study in Olmstead County, Minnesota. *Neurology,* **42,** 1657–1662.
18. Stewart, W., Simon, D., Schechter, A., and Lipton, R. (1995). Population variation in migraine prevalence: A meta-analysis. *J. Clin. Epidemiol.* **48,** 269–280.
19. Scher, A., Stewart, W., and Lipton, R. (1999). Epidemiology of migraine: A meta-analytic approach. Submitted for publication.
20. Abu-Arefeh, I., and Russell, G. (1994). Prevalence of headache and migraine in schoolchildren. *Br. Med. J.* **309,** 765–769.
21. Raieli, V., Raimondo, D., Cammalleri, R., and Camarda, R. (1995). Migraine headaches in adolescents: A Student Population-based Study in Monreale. *Cephalalgia* **15,** 5–12.
22. Alders, E., Hentzen, A., and Tan, C. (1996). A community-based prevalence study on headache in Malaysia. *Headache* **36,** 379–384.
23. Rasmussen, B., Jensen, R., Schroll, M., and Olesen, J. (1991). Epidemiology of headache in a general population—A Prevalence Study. *J. Clin. Epidemiol.* **44,** 1147–1157.
24. Arregui, A., Cabrera, J., Leon-Velarde, F., Paredes, S., Viscarra, D., *et al.* (1991). High prevalence of migraine in a high-altitude population. *Neurology* **41,** 1668–1669
25. Sakai, F., and Igarashi, H. (1997). Prevalence of migraine in Japan: A Nationwide Survey. *Cephalalgia* **17,** 15–22.
26. Stewart, W. F., Lipton, R. B., and Liberman, J. (1996). Variation in migraine prevalence by race. *Neurology* **16,** 231–238.
27. Barea, L., Tannhauser, M., and Rotta, N. (1996). An epidemiologic study of headache among children and adolescents of southern Brazil. *Cephalalgia* **16,** 545–549.
28. Breslau, N., Davis, G., and Andreski, P. (1991). Migraine, psychiatric disorders, and suicide attempts: An epidemiologic study of young adults. *Psychiatry Res.* **37,** 11–23.

29. Cruz, M., Cruz, I., Preux, P., Schantz, P., and Dumas, M. (1995). Headache and cysticercosis in Ecuador, South America. *Headache* **35,** 93–97.

30. Tekle Haimanot, R., Seraw, B., Forsgren, L., Ekbom, K., and Ekstedt, J. (1995). Migraine, chronic tension-type headache, and cluster headache in an Ethiopian rural community. *Cephalalgia* **15,** 482–488.

31. Franceschi, M., Colombo, B., Rossi, P., and Canal, N. (1997). Headache in a population-based elderly cohort. An ancillary study to the Italian Longitudinal Study of Aging (ILSA). *Headache* **37,** 79–82.

32. Henry, P., Michel, P., Brochet, B., Dartigues, J., Tison, S., *et al.* (1992). A nationwide survey of migraine in France: Prevalence and clinical features in adults. *Cephalalgia* **12,** 229–237.

33. Thomson, A., White, G., and West, R. (1993). The prevalence of bad headaches including migraine in a multiethnic community. *N.Z. Med. J.* **106,** 477–480.

34. Linet, M. S., Stewart, W. F., Celentano, D. D., Ziegler, D., and Sprecher, M. (1989). An epidemiologic study of headache among adolescents and young adults. *JAMA, J. Am. Med. Assoc.* **261,** 2211–2216.

35. Wong, T., Wong, K., Yu, T., and Kay, R. (1995). Prevalence of migraine and other headaches in Hong Kong. *NeuroEpidemiology* **14,** 82–91.

36. O'Brien, B., Goeree, R., and Streiner, D. (1994). Prevalence of migraine headache in Canada: A population-based survey. *Int J Epidemiol* **23,** 1020–1026.

37. Lipton, R. B., and Stewart, W. F. (1993). Migraine in the United States: A review of epidemiology and health care use. *Neurology* **43**(Suppl. 3), S6–S10.

38. Bille, B. (1962). Migraine in school children. *Acta Paediat.* **51**(Suppl. 136), 1–151.

39. Bille, B. (1989). Migraine in children: Prevalence, clinical features, and a 30-year follow up. *In* "Migraine and Other Headaches" (M. Ferrari and X. Lataste, eds.). Parthenon, NJ.

40. Sillanpaa, M., Piekkala, P., and Kero, P. (1991). Prevalence of headache at preschool age in an unselected child population. *Cephalalgia* **11,** 239–242.

41. Sillanpaa, M. (1994). Headache in children. *In* "Headache Classification and Epidemiology" (J. Olesen, ed.), pp. 273–281. Raven Press, New York.

42. Lipton, R. (1997). Diagnosis and epidemiology of pediatric migraine. *Curr. Opin. Neurol.* **10,** 231–236.

43. Lipton, R., Stewart, W., Celentano, D., and Reed, M. (1992). Undiagnosed migraine headaches: A comparison of symptom-based and reported physician diagnosis. *Arch. Intern. Med.* **152,** 1273–1278.

44. Kryst, S., and Scherl, E. (1994). A population-based survey of the social and personal impact of migraine. *Headache* **34,** 344–350.

45. Gobel, H., Petersen-Braun, M., and Soyka, D. (1994). The epidemiology of headache in Germany: A nationwide survey of a representative sample on the basis of the headache classification of the International Headache Society. *Cephalalgia* **14,** 97–106.

46. Rasmussen, B. (1992). Migraine and tension-type headache in a general population: Psychosocial factors. *Int. J. Epidemiol.* **21,** 1138–1143.

47. Silberstein, S., and Merriam, G. (1991). Estrogens, progestins, and headache. *Neurology* **41,** 786–793.

48. Somerville, B. W. (1972). The role of estradiol withdrawal in the etiology of menstrual migraine. *Neurology* **22,** 355–365.

49. Somerville, B. (1971). The role of progesterone in menstrual migraine. *Neurology* **21,** 853–859.

50. de Lignieres, B., Vincens, M., Mauvais-Jarvis, P., Mas, J., Touboul, P., *et al.* (1986). Prevention of menstrual migraine by percutaneous oestradiol. *Br. Med. J.* **293,** 1540

51. Dennerstein, L., Morse, C., Burrows, G., Oats, J., Brown, J., *et al.* (1988). Menstrual migraine: A double-blind trial of percutaneous estradiol. *Gynecol. Endocrinol.* **2,** 113–120.

52. Silberstein, S., and Merriam, G. (1997). Sex hormones and headache. In "Blue Books of Practical Neurology—Headache" (P. Goadsby and S. Silberstein eds.). Butterworth-Heinemann, Boston

53. Johannes, C., Linet, M., Stewart, W., Celentano, D., Lipton, R., *et al.* (1995). Relationship of headache to phase of the menstrual cycle among young women: A Daily Diary Study. *Neurology* **45,** 1076–1082.

54. MacGregor, E., Chia, H., Vohrah, R., and Wilkinson, M. (1990). Migraine and menstruation: A Pilot Study. *Cephalalgia* **10,** 305–310.

55. MacGregor, E. (1996). "Menstrual" migraine: Towards a definition. *Cephalalgia* **16,** 11–21.

56. Cupini, L., Matteis, M., Troisi, E., Calabresi, P., Bernardi, G., *et al.* (1995). Sex-hormone related events in migrainous females. *Cephalalgia* **15,** 140–144.

57. Becker, W. (1997). Migraine and oral contraceptives. *Can. J. Neurol. Sci.* **24,** 16–21.

58. Bickerstaff, E. (1975). "Neurological Complications of Oral Contraceptives." Oxford University Press, Oxford.

59. Dayno, J., Silberstein, S., and Lipton, R. (1996). Migraine comorbidity: Epilepsy and stroke. *Adv. Clin. Neurosci.* **6,** 365–385.

60. Tzourio, C., Iglesias, S., Hubert, J., Visy, J., Alperovitch, A., *et al.* (1993). Migraine and risk of ischaemic stroke: A Case-control Study. *Br. Med. J.* **307,** 289–292.

61. Tzourio, C., Tehindrazanarivelo, A., Iglesias, S., Alperovitch, A., Chedru, F., *et al.* (1995). Case-control study of migraine and risk of ischaemic stroke in young women. *Br. Med. J.* **310,** 830–833.

62. Ratinahirana, H., Darbois, Y., and Bousser, M. (1990). Migraine and pregnancy: A prospective study in 703 women after delivery. *Neurology* **40**(Suppl. 1), 437.

63. Marcus, D., and Scharff, L. (1998). Headache during pregnancy and in the post-partum: A Prospective Study. *Headache* **38,** 392–393.

64. Silberstein, S. (1997). Migraine and pregnancy. *Neurol. Clinics* **15,** 209–231.

65. Whitty, C., and Hockaday, J. (1968). Migraine: A follow-up study of 92 patients. *Br. Med. J.* **1,** 735–736.

66. Neri, I., Granella, F., Nappi, R., Manzoni, G., Facchinetti, F., *et al.* (1993). Characteristics of headache at menopause: A Clinico-epidemiologic Study. *Maturitas* **17,** 31–37.

67. Shoemaker, E., Forney, J., and MacDonald, P. (1977). Estrogen treatment of postmenopausal women. *JAMA, J. Am. Med. Assoc.* **238,** 1524–1530.

68. LaRosa, J. (1995). Has ERT come of age? *Lancet* **345,** 76–77.

69. Kudrow, L. (1975). The relationship of headache frequency to hormone use in migraine. *Headache* **15,** 36–49.

70. Martin, P., Burnier, A., Segré, E., and Huix, F. (1971). Graded sequential therapy in the menopause: A Double-blind Study. *Am. J. Obstet. Gynecol.* **111,** 178–186.

71. Merikangas, K. (1990). Genetic epidemiology of migraine. *In* "Migraine: A Spectrum of Ideas" (M. Sandler and G. Collins, eds.), pp. 40–47. Oxford University Press, New York.

72. Ottman, R., Hong, S., and Lipton, R. B. (1993). Validity of family history data on severe headache and migraine. *Neurology* **43,** 1954–1960.

73. Russell, M. B., and Olesen, J. (1995). Increased familial risk and evidence of genetic factor in migraine. *Br. Med. J.* **311,** 541–544.

74. Stewart, W. F., Staffa, J., Lipton, R. B., and Ottman, R. (1997). Familial risk of migraine: A Population-based Study. *Ann. Neurol.* **41,** 166–172.

75. Merikangas, K. R. (1996). Genetics of migraine and other headache. *Curr. Opin. Neurol.* **9,** 202–205.

76. Honkasalo, M., Kaprio, J., Winter, T., Heikkila, K., Sillanpaa, M., *et al.* (1995). Migraine and concomitant symptoms among 8167 adult twin pairs. *Headache* **35,** 70–78.

77. Larsson, B., Bille, B., and Pederson, N. (1995). Genetic influence in headaches: A Swedish Twin Study. *Headache* **35,** 513–519.

78. Merikangas, K., Tierney, C., Martin, N., Heath, A., and Risch, N. (1994). Genetics of migraine in the Australian twin registry. *In* "New Advances in Headache Research" (F. Rose, ed.), pp. 27–28. Smith-Gordon, London.

79. Joutel, A., Bousser, M. G., Biousse, V., Labauge, P., Chabriat, H., *et al.* (1993). A gene for familial hemiplegic migraine maps to chromosome 19. *Nat. Genet.* **5,** 40–45.

80. Gardner, K., Barmada, M., Ptacek, L., and Hoffman, E. (1997). A new locus for hemiplegic migraine maps to chromosome 1q31. *Neurology* **49,** 1193–1195.

81. Peroutka, S., Wilhoit, T., and Jones, K. (1997). Clinical susceptibility to migraine with aura is modified by dopamine D2 receptor (DRD2) Ncol alleles. *Neurology* **49,** 201–206.

82. Silberstein, S., Lipton, R., and Breslau, N. (1995). Migraine: Association with personality characteristics and psychopathology. *Cephalalgia* **15,** 358–369.

83. Merikangas, K., Angst, J., and Isler, H. (1990). Migraine and psychopathology. Results of the Zurich Cohort Study of Young Adults. *Arch. Gen. Psychiatry* **47,** 849–853.

84. Stewart, W., Linet, M., and Celentano, D. (1989). Migraine headaches and panic attacks. *Psychosom. Med.* **51,** 559–569.

85. Stewart, W., Schechter, A., and Liberman, J. (1992). Physician consultation for headache pain and history of panic: Results from a Population-based Study. *Am. J. Med.* **92,** 35S–40S.

86. Merikangas, K., Stevens, D., and Angst, J. (1993). Headache and personality: Results of a community sample of young adults. 18th Collegium Internationale Neuro-Psychopharmacologicum Congress: Migraine: The interface between neurology and psychiatry (1991, Nice, France). *J. Psychiatr. Res.* **27,** 187–196.

87. Breslau, N. (1992). Migraine, suicidal ideation, and suicide attempts. *Neurology* **42,** 392–395.

88. Breslau, N., and Davis, G. (1992). Migraine, major depression and panic disorder: A prospective epidemiologic study of young adults. *Cephalalgia* **12,** 85–90.

89. Breslau, N., and David, G. (1993). Migraine, physical health and psychiatric disorder: A prospective epidemiologic study in young adults. *J. Psychiatr. Res.* **27,** 211–221.

90. Breslau, N., Davis, G., Schultz, L., and Peterson, F. (1994). Joint 1994 Wolff Award Presentation. Migraine and major depression: A Longitudinal Study. *Headache* **34,** 387–393.

91. Andermann, E., and Andermann, F. (1987). Migraine-epilepsy relationships: Epidemiological and genetic aspects. *In* "Migraine and Epilepsy" (F. Andermann and E. Lugaresi, eds.), p. 281. Butterworth, Boston.

92. Lipton, R., Ottman, R., Ehrenberg, B., and Hauser, W. (1994). Comorbidity of migraine: The connection between migraine and epilepsy. *Neurology* **44,** 2105–2110.

93. Ottman, R., and Lipton, R. (1996). Is the comorbidity of epilepsy and migraine due to a shared genetic susceptibility? *Neurology* **47,** 918–924.

94. Schechter, A., Rozen, T., Silberstein, S., Stewart, W., and Lipton, R. (1998). Raynaud's phenomenon and migraine: Are they co-morbid? A Population-based Case-control Study. *Neurology* **50** (Suppl. 4), A435.

95. Silberstein, S., Lipton, R., and Goadsby, P. (1998). "Headache in Clinical Practice" Isis Medical Media Ltd., Oxford.

96. Littlewood, J., Gibb, C., Glover, V., Sandler, M., Davies, P., *et al.* (1988). Red wine as a cause of migraine. *Lancet* **1,** 558–559.

97. Koehler, S., and Glaros, A. (1988). The effect of aspartame on migraine headache. *Headache* **28,** 10–14.

98. Schiffman, S., Buckley, I. C., Sampson, H., Massey, E., Baraniuk, J., *et al.* (1987). Aspartame and susceptibility to headache. *N. Eng. J. Med.* **317.**

99. Robbins, L. (1993). Precipitating factors in migraine: A retrospective review of 494 patients. *Headache* **34,** 214–216.

100. Peatfield, R. (1995). Relationships between food, wine, and beer-precipitated migrainous headaches. *Headache* **35,** 355–357.

101. Peatfield, R., Glover, V., Littlewood, J., Sandler, M., and Rose, C. (1984). The prevalence of diet-induced migraine. *Cephalalgia* **4,** 179–183.

102. Selby, G., and Lance, J. (1960). Observations on 500 cases of migraine and allied vascular headache. *J. Neurol. Neurosurg. Psychiatry* **23,** 23–32.

103. Olesen, J., and Tfelt-Hansen, P. (1997). "Headache Treatment: Trial Methodology and New Drugs." Lippincott-Raven, Philadelphia.

104. Lipton, R., Stewart, W., Ryan, R., Saper, J., Silberstein, S., *et al.* (1998). Efficacy and safety of acetaminophen, aspirin, and caffeine in alleviating migraine headache pain. Three double-blind, randomized placebo-controlled trials. *Arch. Neurol. (Chicago)* **55,** 210–217.

105. Celentano, D., Stewart, W., Lipton, R., and Reed, M. (1992). Medication use and disability among migraineurs: A National Probability Sample Survey. *Headache* **32,** 223–228.

106. Silberstein, S., and Lipton, R. (1994). Overview of diagnosis and treatment of migraine. *Neurology* **44,** 6–16.

107. Celentano, D. D., Linet, M. S., and Stewart, W. F. (1990). Gender differences in the experience of headache. *Soc. Sci. Med.* **30,** 1289–1295.

108. Stewart, W., Schechter, A., and Lipton, R. (1994). Migraine heterogeneity. Disability, pain intensity, and attack frequency and duration. *Neurology* **44**(Suppl. 4), S24–S39.

109. Lipton, R., Stewart, W., and Simon, D. (1998). Medical consultation for migraine: Results of the American Migraine Study. *Headache* **38,** 87–90.

110. Lewis, C., and Lewis, M. (1977). The potential impact of sexual equality on health. *N. Engl. J. Med.* **297,** 863–869.

111. Hibbard, J., and Pope, C. (1983). Gender role, illness orientation, and the use of medical services. *Soc. Sci. Med.* **17,** 129–137.

112. Tarlov, A., Ware, J., Greenfield, S., Nelson, E., Perrin, E., *et al.* (1989). The Medical Outcomes Study: An application of methods for monitoring the results of medical care. *JAMA, J. Am. Med. Assoc.* **262,** 925–930.

113. Hurst, B., Macclesfield, U., and Patrick, D. (1998). Assessing outcomes of treatment for migraine headache using generic and specific measures. *Neurology* **50**(Suppl. 4), A180–A181.

114. Hurst, B., and Patrick, D. (1998). Quality of life improvement in responders to long-term treatment with zomig. *Headache* **38,** 385–386.

115. Saper, J., Burke, T., Wong, W., and Batenhorst, A. (1998). Improvements in health related quality of life with long-term use of sumatriptan therapy for migraine. *Headache* **38,** 400.

116. Lofland, J., Johnson, N., Nash, D., and Batenhorst, A. (1998). Improvements in managed care patients' health-related quality of life after sumatriptan (Imitrex). *Headache* **38,** 391.

117. Patrick, D., and Hurst, B. (1998). Consistency of meaningful migraine relief and quality of life in patients treated with Zomig. *Headache* **38,** 397.

118. Santanello, N., Davies, G., Kramer, M., Matzura-Wolfe, D., and Lipton, R. (1998). Determinants of migraine-specific quality of life. *Headache* **38,** 400.

119. Solomon, G., Skobieranda, F., and Gragg, L. (1993). Quality of life and well-being of headache patients: Measurement by the medical outcomes study instrument. *Headache* **33,** 351–358.

120. Osterhaus, J., Townsend, R., Gandek, B., and Ware, J. (1994). Measuring functional status and well-being of patients with migraine headache. *Headache* **34,** 337–343.

121. Dahlof, C., and Dimenas, E. (1995). Migraine patients experience poorer subjective well-being/quality of life even between attacks. *Cephalalgia* **15,** 31–36.

122. Terwindt, G., Launer, L., and Ferrari, M. (1998). The impact of migraine on quality of life in the general population: The GEM Study. *Neurology* **50**(Suppl. 4), A434.

123. Linet, M., Celentano, D., and Stewart, W. (1991). Headache characteristics associated with physician consultation: A Population-based Survey. *Am. J. Prev. Med.* **7,** 40–46.

124. Clouse, J., and Osterhaus, J. (1994). Healthcare resource use and costs associated with migraine in a managed healthcare setting. *Ann. Pharmacother.* **28,** 659–664.

125. Edmeads, J., Findlay, H., Tugwell, P., Pryse-Phillips, W., Nelson, R., *et al.* (1993). Impact of migraine and tension-type headache on life-style, consulting behavior, and medication use: A Canadian Population Survey. *Can. J. Neurol. Sci.* **20,** 131–137.

126. von Korff, M., Stewart, W. F., Simon, D. S., and Lipton, R. B. (1998). Migraine and reduced work performance: A Population-based Diary Study. *Neurology* **50,** 1741–1745.

127. Stewart, W., Lipton, R., and Simon, D. (1996). Work-related disability: Results from the American Migraine Study. *Cephalalgia* **16,** 231–238.

128. Rasmussen, B., Jensen, R., and Olesen, J. (1992). Impact of headache on sickness absence and utilization of medical services: A Danish Population Study. *J. Epidemiol Commun. Health* **46,** 443–446.

129. Schwartz, B., Stewart, W., and Lipton, R. (1997). Lost workdays and decreased work effectiveness associated with headache in the workplace. *J. Occup. Environ. Med.* **39,** 320–327.

130. van Roijen, L., Essink-Bot, M., Koopmanschap, M., Michel, B., and Rutten, F. (1995). Societal perspective on the burden of migraine in the Netherlands. *PharmacoEconomics* **7,** 170–179.

131. Smith, R. (1998). Impact of migraine on the family. *Headache* **38,** 423–426.

87

Irritable Bowel Syndrome

NICHOLAS J. TALLEY AND NATASHA KOLOSKI

Department of Medicine
University of Sydney
Nepean Hospital
Sydney, Australia

I. Introduction

The irritable bowel syndrome (IBS) represents a constellation of gastrointestinal symptoms that currently have no structural or biochemical explanation. Rigorous clinical studies suggest that IBS is indeed a disease because it has a significant impact on quality of life, although an exact pathophysiological explanation remains elusive. Equally unexplained is why IBS affects women in Western countries more frequently than men. In this chapter, the definition, natural history, pathophysiology, and management of IBS will be reviewed, and relevant gender differences will be highlighted where applicable.

II. Definition

Early on, IBS was very vaguely defined as any abdominal pain or bowel disturbance that did not have a clear explanation [1]. Subsequently, a number of diseases were identified that indeed did appear to explain the symptoms, such as lactose intolerance and bloating [2] (although this is overdiagnosed, as discussed later) or collagenous colitis and diarrhea (an inflammatory disease of the colon with clear-cut histological changes) [3]. However, such findings only applied to relatively small subsets of these cases. It also became increasingly recognized that other groups of these patients had specific pathophysiological abnormalities. For example, in subsets of women with very severe constipation, colonic inertia (very slow colonic transit due to a motor disorder) or pelvic floor dysfunction (sometimes linked to childbirth injury) have been identified to be important in accounting for the complaints [4,5]. Thus, selected subgroups of patients with painless chronic constipation or painless chronic diarrhea seemed to be distinct from patients who reported abdominal pain linked to a disturbance of bowel dysfunction (*i.e.*, IBS).

IBS is only one of many functional gastrointestinal disorders. By definition all these disorders comprise three months or more of chronic or recurrent gastrointestinal complaints where there is no structural or biochemical explanation for the symptoms using currently available tests (6). IBS is now considered to be characterized by abdominal pain or discomfort associated with a change in bowel function. Specific internationally accepted symptom criteria have been developed to help positively identify cases for both research and clinical practice [6]. Other functional gastrointestinal disorders include nonulcer (or functional) dyspepsia, aerophagia, rumination, functional constipation, functional diarrhea, and functional abdominal pain, which are all comprehensively reviewed elsewhere [6].

III. Issues Related to Ascertainment

A comprehensive Medline and Current Contents search was conducted to identify relevant articles on IBS and womens' health using the MeSH terms irritable bowel syndrome, "women," and "gender" from 1993 to 1998. A manual search of the reference lists of these articles and searches of references in review articles were used to identify other relevant citations. Only references relevant to each of the major chapter headings are included in this chapter.

IV. Historical Evolution

The recognition of IBS has been relatively recent. Here the history of the condition will be summarized and an overview of how the current diagnostic criteria evolved will be presented.

A. History of Irritable Bowel Syndrome

Symptoms of abdominal pain and altered bowel habit in the absence of organic pathology have been described in the literature since the first part of the nineteenth century [1,7]. The first case was probably reported by Powell in the UK in 1820 [8]. While the passage of large volumes of mucus was a major feature in early reports, this is rare today. The term irritable bowel syndrome was probably first used in 1944 by Peters and Bargen [8]. Over the decades, numerous other terms have been used to denote this condition, including passing "membranes" per rectum, pseudo-membranous enteritis, mucous colitis, myxoneurosis, chronic catarrhal colitis, unstable colon, and, in the 1930s, "irritable colon" [9–11]. Chaudhary and Truelove in 1962 subdivided the irritable colon into "spastic colon" and "painless diarrhoea," providing the first systematic description of the condition [1]. The Rome classification of the functional gastrointestinal disorders has refined the definition and provided internationally accepted diagnostic criteria [6].

B. Development of the Current Diagnostic Criteria for Irritable Bowel Syndrome

One of the first attempts to try to objectively identify the key symptoms in IBS was undertaken by a research group from Bristol [12]. They measured 15 gastrointestinal symptoms in a cohort of 65 patients eventually diagnosed as having IBS or structural disease in an outpatient clinic. Six symptoms (now called the Manning criteria after the first author) were identified as being more common in patients with IBS than in those with

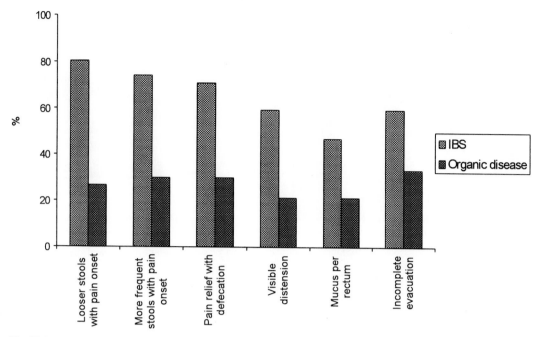

Fig. 87.1 Symptoms more likely to be found in the irritable bowel syndrome (IBS) than in organic gastrointestinal disease. Data from [12].

structural disease (Fig. 87.1). Among IBS cases, 94% had two or more of these six symptoms compared with 46% of patients with organic disease [12]. Other researchers have confirmed that the greater the number of these six symptom criteria that are present, the greater the probability that the diagnosis is IBS. This especially applies in women as well as in younger patients

Table 87.1
Rome Definitions of Irritable Bowel Syndrome

ROME I

At least 3 months of continuous or recurrent symptoms of
1. Abdominal pain or discomfort that is
   a. Relieved with defecation
   b. Associated with a change in frequency of stool; and/or
   c. Associated with a change in consistency of stool; and
2. Two or more of the following, at least on one-fourth of occasions or days:
   a. Altered stool frequency (defined as more than three bowel movements each day or less than three bowel movements each week);
   b. Altered stool form (lumpy/hard or loose/watery stool);
   c. Altered stool passage (straining, urgency, or feeling of incomplete evacuation);
   d. Passage of mucus; and/or
   e. Bloating or feeling of abdominal distension

ROME II

Twelve weeks or more in the past 12 months of abdominal pain or discomfort that has two out of three of the following features:
• Relieved with defecation
• Associated with a change in frequency of stool
• Associated with a change in the consistency of stool

[13–17]. A German study also showed that the symptoms of abdominal pain, flatulence and bowel irregularity, pellet-like stools or mucus, diarrhea and constipation, and symptoms for more than two years were of value in identifying IBS, although negative basic laboratory tests had more diagnostic weight [18].

Based on these clinical studies, and fanned by the need to standardize diagnostic criteria for IBS for clinical research purposes, a group of experts met in Rome to formulate consensus-based criteria using a Delphi approach [19]. There have been subsequent revisions of the internationally accepted Rome criteria based on new data, as summarized in Table 87.1. The current criteria are supported by clinical experience and factor analysis, which have independently shown that an ''IBS factor'' exists in nonpatients [20,21]. However, the Rome criteria must be considered diagnostic criteria in evolution because firm validity data continue to be lacking, although major studies are in progress. A key feature of IBS in all recent definitions has been the absolute requirement for the presence of abdominal pain associated with bowel dysfunction and the absence of any definite structural or biochemical explanation for these symptoms [19].

V. Distribution in Women

A. Prevalence

IBS is remarkably common in the general population. In the United States, the prevalence has ranged from 9 to 20% depending on the definition applied [22–25]. If a broader classification approach is used (*e.g.,* the Manning criteria), the prevalence is higher than if restrictive criteria such as the Rome diagnostic criteria are applied. IBS is more common in women than in men

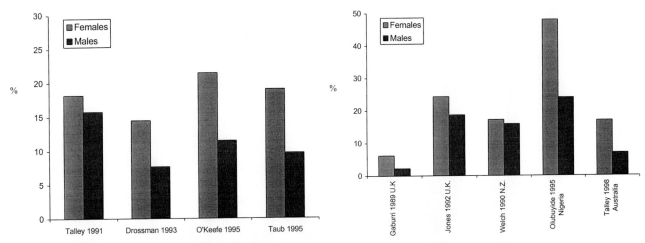

Fig. 87.2 Prevalence of irritable bowel syndrome (IBS) in males and females in the United States. Data from [20,22,24,26].

Fig. 87.3 Prevalence of irritable bowel syndrome (IBS) in males and females in international studies. Data from [27–31].

in the U.S. (Fig. 87.2) [20,22,24,26]. Similar results have been reported from other western countries and Africa (Fig. 87.3) [27–31]. The prevalence of IBS symptoms appears to be similar in pre- and postmenopausal women, although bowel dysfunction was more common in premenopausal cases in family practice in one study [32].

B. Natural History

IBS does not impair survival. For example, a retrospective 30-year cohort study of subjects in the Olmsted County community in Minnesota showed that those with IBS had a survival rate almost identical to those without the condition (indeed, the IBS cohort lived a median of a few days longer) [33].

The rate of development of IBS is much lower than the prevalence [34,35]. A U.S. population study showed that 9% of persons developed a new onset of IBS over a 12–20 month follow-up period [35]. However, and importantly, a similar number lost their symptoms over the same time period, so overall the prevalence in the population was unchanged from year to year [35]. In those with absolutely no prior history of gastrointestinal symptoms, lower rates of symptom onset have been observed [34]. Therefore, the true annual incidence of IBS is likely to be about 1–2% [34]. An explanation for these fluctuations in symptom patterns is not available and does not appear to be related to gender, age, or other sociodemographic factors.

The majority of outpatients with a diagnosis of IBS (*i.e.,* health care-seeking persons) continue to be symptomatic over time [33,35–46]. The identification of prognostic features has been inconsistent in the studies to date, but those with more anxiety or a longer duration of complaints may have a worse prognosis.

C. Health Care-Seeking Behavior

Overall, IBS is one of the most common diagnoses in gastroenterology practice (although clinicians often use a rather broad

definition) [47]. IBS accounts for at least 3.5 million physician visits in the U.S. annually [48,49].

Although IBS is common, the number of people who present for medical care is remarkably variable [50]. Studies from the United States and Europe have reported that only a minority with IBS (ranging from 10 to 50%) had seen a physician or alternative health care provider for this problem in the past [27,29,51,52]. This behavior, however, may be culturally determined. For example, a population-based study from Australia, where health care provided by family practitioners is essentially free, reported much higher health care-seeking rates (70% for IBS) [50]. This suggests that access to the health care system may be one important force in promoting or retarding health care seeking.

Other major reasons why only some patients present for care in the U.S. remain inadequately understood. Pain severity is one clear factor and this is true for both men and women, but those with more severe pain only represent a minority seeking care [24,25,52]. Women may also seek health care more often than men for the symptoms of bloating, mucus, looser stools with pain onset, and incomplete rectal evacuation, based on a British report [52]. Psychological distress has been shown to be higher in patients with IBS compared with nonconsulters who have similar symptoms, suggesting that psychological distress may drive some to seek health care [53,54]. Somatization may also drive some IBS sufferers to more frequently seek medical attention [24], but not all studies agree [25,55]. Gender differences in the health perceptions of IBS patients have also been observed; males appear to have better health perceptions in general but demonstrate more concern regarding bowel habits than females [23]. Some patients with IBS may learn a pattern of behavior that leads them to seek care more frequently (abnormal illness behavior), which may result from childhood experiences [56].

D. Symptom Subgroups Based on Bowel Disturbance

It has been suggested that patients with IBS can be subdivided into symptom subgroups based on the bowel habit disturbance that predominates [1,57]. The validity, however, of this subgroup classification remains to be confirmed. One U.S. population-based study subdivided IBS subjects into those with predominant constipation, predominant diarrhea, alternating diarrhea and constipation, and unclassifiable [35]. These subgroups had an approximately even prevalence in this study. Interestingly, the gender ratios were similar in these subgroups except in the constipation-predominant IBS subgroup, where women were more commonly affected [35]. Why symptoms of constipation are more common in women with IBS remains to be adequately explained.

VI. Host and Environmental Determinants

In most studies, women have had a higher prevalence of IBS than men across all adult age groups including the elderly. The female to male ratio in IBS is usually in the range of 2:1 to 4:1 [20,24,26,51,58]. It remains to be explained why women are more affected. None of the current pathogenic models account for this gender difference, although genetic factors or possibly hormonal factors could be important.

In the U.S., Hispanics and African-Americans appear to have a prevalence similar to Caucasians [22,23], while rates may be lower in Asians in California [59], although adequate data are lacking. If correct, an explanation for the racial differences is not available.

VII. Influence of Women's Social Roles

Clinical studies have confirmed that the Manning symptom criteria were useful discriminators of IBS versus organic disease. However, the discriminant value of these symptoms in men was substantially less than that found in women. Thompson evaluated 156 patients with IBS [60]. He reported that mucus, feelings of incomplete evacuation, abdominal distension, and scybala were reported significantly more frequently by women than men with IBS. In contrast, pain relieved by defecation and pain followed by a change in stool consistency or frequency were similar in both women and men [60]. Smith *et al.* rated the Manning criteria in 36 men and 61 women with abdominal pain or disturbed defecation or both; 65% had IBS, but the Manning criteria only correlated significantly with IBS in women [61]. Overall, it appears that the diagnostic value of the Rome criteria in men is less than it is in women, although not all studies agree [15]. This may be because men report fewer nonpainful IBS-type symptoms than women [60].

In most studies from Western countries, women are not only more likely than men to have IBS, but they are also more likely to seek care for the condition (Fig. 87.4) [23,27,29]. However, this behavior may also be culturally determined, as it does not appear to be true in India and Sri Lanka. In these two countries, males predominantly present for medical care with IBS [62–67]. One possible explanation may be social differences in expectation of illness. Differentiated risks acquired from roles, stress, lifestyles, and preventative health practices may promote sex differences in health care-seeking behavior in IBS [68]. Psychosocial factors that modulate how men and women perceive and evaluate symptoms may also play a role [68]. Whether IBS is truly more common in men than women in the Indian or Sri

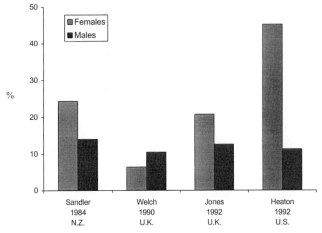

Fig. 87.4 Proportion of males and females with irritable bowel syndrome (IBS) who sought health care. Data from [27,29,51,52].

Lankan community is unknown, as population-based data are unavailable.

VIII. Risk Factors

A. Psychological Distress

There remains controversy about the link between psychological factors and IBS. However, there is growing evidence that psychological factors are important based on unbiased population-based studies and the response to psychological interventions in IBS.

Overall, psychiatric diagnoses and psychological distress are more common in patients with IBS who present for care to a specialist center compared with controls, based on several studies [58,69–72]. However, these findings may in part reflect selection bias, as disturbed patients may be more likely to seek care, artificially enriching this population. In primary care, the prevalence of psychological disturbance is likely to be substantially lower, but good data are unavailable. In two studies of volunteers, nonpatients (nonconsulters) with IBS had personality profiles very similar to symptom-free controls, whereas patients (consulters) with IBS had greater psychosocial disturbances (although these were much less than the disturbances observed in tertiary care) [53,54]. Hence, a primary role for psychosocial factors in the etiology of IBS remains in dispute.

Up to one-fifth of IBS patients can date their symptoms to an acute gastroenteritis-type illness due to *Salmonella*. In a prospective study, those with increased levels of neuroticism, anxiety, and depression had a greater risk of developing IBS after infection than those who had lower levels [73]. These data strongly suggest an interrelationship between psychological factors and infection-induced intestinal dysfunction [73]. This is key evidence that psychological factors can be causal and not just associated with health care-seeking in IBS. Other population-based data suggest that neuroticism is an important risk factor for IBS [31].

A relationship between stress and symptoms in IBS also remains controversial. However, two prospective studies of women

Table 87.2
Abuse and Functional Gastrointestinal Disorders

			Compared groups (%)		
Study year	Sample	Type of abuse	Functional gastrointestinal disorders	Organic disease	Healthy controls
[76]	Female gastroenterology patients	Any sexual abuse	53.0	37.0	
[59]	Health examinees	Any sexual abuse	35.6 (IBS) (severe) 21.0 (IBS) (less severe)		10.2
[80]	Outpatients and spouses	Childhood abuse	31.0 (IBS)		16.0
[77]	Population	Any abuse	50.0 (IBS)		23.3
[78]	Outpatients	Any abuse	22.1	16.2	
[81]	Outpatients	Any sexual assault	54.0 (IBS)	42.0	
[79]	Hospital outpatients	Sexual abuse			
	Females		34.4 (IBS)	21.7	9.9
	Males		19.4 (IBS)	8.0	4.9
[31]	Population	Any childhood sexual abuse	27.1 (IBS)		20.4
		Any adulthood sexual abuse	23.0 (IBS)		14.1

have shed some new light on this area. Levy *et al.* found that there was a positive relationship between daily stress and symptoms in both patients and nonpatients with IBS over two months [74]. Similarly, Whitehead *et al.,* in a study of 383 women followed over a year, observed that the group with IBS had higher levels of stress and stress level correlated with bowel symptoms [75]. While stress may therefore aggravate IBS complaints, it is unknown whether stress can induce the development of IBS.

B. Sexual and Physical Abuse in IBS

The high frequency of sexual and physical abuse in gastrointestinal patients was first reported in the U.S. in 1990, and these observations have subsequently been largely confirmed [76,77]. In a consecutive sample of women outpatients, 44% of 206 patients had a history of sexual or physical abuse in childhood or adult life. Importantly, patients with functional gastrointestinal disorders including IBS reported a higher rate of abuse (31%) than patients with organic disease (18%). It is relevant that 70% of these patients had not previously disclosed any history of abuse to anyone [76].

In a population-based U.S. study, 30% of the population identified a history of abuse using a self-report questionnaire similar to the one applied in the outpatient study [77]. Patients with a history of sexual abuse (both women and men) were nearly three times as likely to have a functional bowel disorder [77]. Other studies have similarly identified a link between abuse and IBS, although not all studies have agreed (Table 87.2) [31,59,78–81]. Not only has abuse been linked to the syndrome itself, but it was also associated with higher rates of health care-seeking behavior in a U.S. study [77]. Other population-based data have suggested that abuse may predispose a person to the development of neuroticism, which in turn may be a predisposing factor for IBS, although this remains to be established [31].

IX. Biological Markers of Susceptibility or Exposure

There are no accepted biological markers of disease, but new insights into the condition have been derived from physiological studies and twin studies. It is beyond the scope of this chapter to review all potential pathophysiological mechanisms [47].

A. Motility and Visceral Hypersensitivity

Usually a disturbance of gut motility is blamed for the symptoms of IBS, but the evidence currently suggests that motor disturbances in the gut are not primarily the cause. While abnormal colonic myoelectric activity at three cycles per minute was reported in the past [82,83], this is now generally believed to represent either an epiphenomenon or a recording artifact [84]. Basal colonic motility is normal in IBS [85]. On the other hand, the colon in IBS may be more responsive than normal to acute stress and hormonal factors [85,86]. A few specific small intestinal motor patterns have been shown to be associated with abdominal pain in patients with IBS. Discrete clustered contractions in the jejunum and prolonged propagated contractions in the ileum coincide with abdominal pain in some patients [87,88]. However, not all studies have confirmed that these findings occur more frequently in IBS patients than in healthy subjects [89].

Perhaps one of the key observations in IBS is the finding that many, if not most, patients have gut pain thresholds that are significantly lower than healthy controls (*i.e.,* visceral hypersensitivity) [47,57,90,91]. Various methods have been used to distend the bowel in order to assess pain thresholds. New studies suggest that rapid high distension pressure applied in the sigmoid can induce rectal hypersensitivity in IBS but not in controls [92]. Laxative-induced diarrhea has been shown to induce rectal sensitivity in women but not in men [93]. Sexual abuse in the past, however, has not been associated with abnormal rectal pain sensitivity [94]. Importantly, in contrast to the gut, peripheral pain thresholds, such as pain induced by hand immersion in iced water or electrical stimulation, are no different in IBS patients and healthy controls [95]. These findings suggest that these patients are not just complainers with a generally low pain threshold.

Abnormalities in urinary bladder function, including detrusor instability [96], gallbladder dysfunction [97], and asthma [98], have been reported in patients with IBS, suggesting that there may be a generalized disturbance of autonomic function. Increased morning and evening urine norepinephrine and cortisol levels have also been observed in women with IBS [99].

B. Brain Imaging

Sleep disturbance is common in IBS [100,101]. Moreover, the multisymptom nature of IBS suggests that processing of gut signals centrally may be important in the pathogenesis of IBS. Studies applying PET scanning indicate that there are indeed differences in regional cerebral blood flow between patients with and without IBS [102]. Moreover, women may have different patterns from men [103], suggesting that gender-related differences in IBS could be in part centrally mediated.

C. Genetics

Clinically, IBS is often reported to run in families [104]. However, there are few data on heritability in IBS. One study reported that of 686 individual twins from same-sex pairs, 5% had a medical diagnosis of a functional gastrointestinal disorder. Statistical modeling suggested that 57% of the variance in this group was attributable to genetic variance while the remainder was due to the individual's unique environment [105]. Whether this genetic component in functional bowel disease, assuming other studies confirm its existence, is separate from other polygenetic inherited conditions such as neuroticism (also linked to IBS) remains to be determined.

X. Clinical Issues

A. Diagnosis

In IBS, abdominal pain may occur anywhere in the abdomen but is classically present in the lower abdomen [47]. The pain of IBS may often move to different sites in the abdomen. The pain is related to disturbed defecation, and often pain relief occurs with defecation or there is a change in stool frequency or stool form associated with the onset of the pain [47]. Some bowel symptoms are exacerbated by menses, especially increased bowel gas in IBS [106]. Bowel disturbances are classically described as alternating between constipation and diarrhea, although about half the patients have a predominant pattern of constipation or diarrhea and a quarter cannot be subtyped [47]. Postpartum women with IBS have been observed to have greater rectal urgency and loss of control of flatus but not incontinence of stool compared to women without IBS [107].

IBS patients often report nongastrointestinal symptoms as well. These include headaches, backaches, lethargy, and urinary and gynecological symptoms including dysmenorrhea and dyspare-

nuria [108–110]. Women with IBS have been observed to report symptoms such as headaches, nausea, and tiredness more often than their male counterparts [111]. Prior [112] reported that half of women presenting to a gynecology clinic in the UK had IBS and only 8% had gynecological pathology. Similar results have been reported by others in the UK and in the U.S. [113,114]. Many women with IBS have undergone hysterectomy, laparoscopy, or other gynecological procedures, probably unnecessarily, because IBS was not recognized initially. Prior *et al.* reported that symptoms suggestive of IBS arose in 10% of women within six months after hysterectomy [112].

Obtaining a menstrual history in premenopausal women is essential to distinguish IBS from other conditions. For example, pelvic inflammatory disease can present with lower abdominal pain and a vaginal discharge. Endometriosis may cause midmenstrual cycle pain that can be associated with bowel disturbance.

A diagnosis of IBS should be a positive one rather than just a diagnosis of exclusion [47]. The Rome criteria provide useful guidance for positive symptom features. The physical examination is usually normal. Sometimes there is nonspecific abdominal tenderness. There may be scars from previous surgery such as hysterectomy, cholecystectomy, or appendectomy [115].

If the clinical picture indicates a diagnosis of IBS, then the diagnostic evaluation can be limited to excluding large bowel cancer and inflammatory bowel disease [47]. If the findings are atypical, then a more extensive evaluation may be indicated depending on the clinical picture. Hematology and chemistry group, erythrocyte sedimentation rate or C-reactive protein, stool examination for occult blood, and flexible sigmoidoscopy should be performed in all cases if not previously undertaken. If there is predominant diarrhea, stool studies must be obtained and thyroid function should be tested. If the patient is 50 years of age or older, then a complete colonic evaluation by colonoscopy (or barium enema and sigmoidoscopy) is recommended unless an adequate diagnostic evaluation has already been performed.

Lactose intolerance should be suspected, particularly in certain ethnic groups such as persons of Jewish descent, blacks, native Americans, and Asians [116]. Bloating and diarrhea are often the predominant symptoms. It is now clear that most patients with lactose malabsorption do not have an increase in bloating, diarrhea, or abdominal discomfort after ingestion of a cup of conventional milk daily or its lactose equivalent [117]. Therefore, many patients with a diagnosis of lactose intolerance in fact have IBS that has been mislabeled. A two-week lactose-free diet is a practical test, but hydrogen breath testing is useful in equivocal cases [118].

B. Treatment

1. General Measures

Reassurance and advice remain key parts of the management that must be initiated by the physician. The patient should be told about the nature of IBS in terms he or she can understand and advised that it never leads to cancer or reduces life expectancy [47,118]. Physicians' perceptions should also be recognized. There is evidence that physicians perceive the frequency of abdominal pain more accurately in males than in females [111].

The initial treatment of IBS is generally agreed to be dietary intervention [47,118]. The average American diet contains only approximately 12 g of fiber per day. For IBS patients, it is recommended that dietary fiber be increased very slowly to about 20 to 30 g of fiber per day. Unprocessed wheat bran remains an inexpensive and convenient dietary fiber source. Alternatively, commercially available fiber supplements are available. These are safe and often better tolerated than other sources. Although the available clinical trial data overall are flawed regarding the value of fiber and fiber supplements, the studies suggest that abdominal pain, diarrhea, and bloating are no more likely to respond to fiber supplements than placebo but constipation may improve [119]. However, anecdotally, some patients with diarrhea do very well on fiber because it firms up the stools. Notably, bloating, distension, and rectal flatus are exacerbated in a subset of patients with IBS who increase their dietary fiber or use fiber supplements [120], but these symptoms usually subside within a few weeks if patients can be convinced to continue with a high fiber diet and only increase the intake very slowly.

Modification of certain dietary components may assist some patients with IBS. Excess gas (rectal flatus) may occur with ingestion of food such as beans, legumes, lentils, and cabbage because their complex carbohydrate content is not absorbed by the small bowel but is fermented by bacteria in the colon; avoidance may help reduce flatus [121]. Bloating and flatulence may respond to reduction or avoidance of carbonated beverages to reduce sorbitol ingestion [122]. Alpha-D-galactosidase (Beano) in healthy volunteers reduces gas production after black bean ingestion [123], but the value of this compound in IBS is unknown.

2. Drugs

In clinical trials, the placebo response ranges from 20 to 70% [124,125]. Unfortunately, there is no unequivocal evidence that any single agent is truly superior to placebo in IBS. However, based on clinical experience and selected trial data some drugs have a place in the management of IBS.

a. ABDOMINAL PAIN AND BLOATING. Anticholinergics are most often prescribed for pain in U.S. IBS patients [118]. An anticholinergic drug taken 30–45 minutes before a meal can temporarily reduce postprandial pain or diarrhea. The major anticholinergic agents used are belladonna, dicyclomine, hyoscyamine, and propantheline [120,126,127]. Sublingual preparations (e.g., levsin) are available but are unlikely to have a more rapid onset of action. Suppository preparations (e.g., hyoscyamine) are also available but are generally less popular with patients. Despite their widespread use, the efficacy of anticholinergics generally remains controversial. Peppermint oil is another available antispasmodic but the results of trials have been mixed [128]. Poynard et al., in a meta-analysis of antispasmodic drugs, identified 148 trials published between 1959 and 1992, but only 82 were randomized studies [129]. Overall, there was a significant improvement in global outcome with antispasmodics (62% improvement) compared with placebo (35% improvement). There was, however, no significant improvement detected for abdominal distension or constipation.

Cisapride is a serotonin (5HT$_4$-type) receptor agonist that induces motor stimulation of the bowel [130], implying it might be of value in constipation-predominant IBS. However, randomized double-blind placebo-controlled trials have been disappointing [131,132]. The value of this prokinetic is thus likely to be minimal.

Serotonin is a mediator of visceral gut sensation via 5HT$_3$ receptors. In IBS, the 5HT$_3$ receptor antagonist ondansetron has been shown to improve stool consistency in those with diarrhea, but some individuals had an exacerbation of abdominal pain [133]. Granisetron, another selective 5HT$_3$ antagonist, was more promising [134]. Most interest, however, has focused on another agent, alosetron, which is not yet available. Interestingly, this drug was shown to be superior to placebo in women but not in men with IBS in one large trial [135], based on a post hoc analysis. These results therefore remain to be confirmed in adequate trials that set out to test gender-specific differences prospectively.

b. BOWEL DISTURBANCE. Stimulant laxatives for constipation include bisacodyl, danthron, phenolphthalein, and senna [136]. They are not recommended in IBS because of the risk of myenteric plexus damage, although the data on which these concerns are based are not adequate [137]. The osmotic laxatives including milk of magnesia or lactulose are often useful and are generally safe [138]. They represent second line treatment for constipation-predominant IBS patients who fail to respond to dietary fiber and fiber supplements. Suppositories or enema preparations are sometimes useful for bowel retraining in patients with IBS associated with severe constipation [138].

The opioid antidiarrheal agents (loperamide, diphenoxylate) inhibit intestinal secretion and increase fluid and electrolyte absorption by prolonging intestinal transit time [139]. Loperamide may also increase anal sphincter tone in patients with diarrhea, which may be helpful in the small subset of patients with IBS who also have fecal incontinence [47]. Diphenoxylate is combined with atropine because this reduces the problem of excessive self-medication. Codeine phosphate should generally be avoided because it has a high risk of inducing dependence.

A subset of patients with diarrhea appears to respond to cholestyramine, although it has not been subjected to double-blind controlled trials in IBS. The mechanism of effect is presumed to be through drug absorption of bile salts that spill over from the terminal ileum into the colon, inducing diarrhea [140,141]. Cholestyramine is also safe and therefore sometimes warrants a trial.

c. INTRACTABLE SYMPTOMS. In patients with more difficult-to-control symptoms of IBS, antidepressant agents have a useful place. The tricyclic agents have a peripheral anticholinergic action as well as a central analgesic action [142,143]. It is recommended that the tricyclic antidepressants be started at a low dose (e.g., amitriptyline 10–25 mg at night). It may take weeks for the drug to produce any benefit. This class of medications may make constipation worse. Selective serotonin reuptake inhibitors have a presumed central action producing analgesia [142]. Serotonin reuptake inhibitors such as fluoxetine, paroxetine, and sertraline should be given in the morning initially. There is no controlled trial evidence with these agents, but anecdotally they are of value in IBS and they have less side effects than tricyclics [144].

Benzodiazepines should be avoided in IBS because of their potential for habituation and interaction with other drugs and alcohol. They have no established efficacy in IBS.

Following continuous administration of gonadotropin releasing hormone, there is a reduction in the synthesis of gonadotropins as well as follicle stimulating hormone and luteinizing hormone. Luprolide is a gonadotropin releasing hormone analog that must be given subcutaneously. A randomized controlled trial demonstrated significant improvement in women with severe functional gastrointestinal complaints who were administered luprolide [145,146], but the trial had flaws and the results remain to be confirmed. As luprolide induces chemical castration, side effects are numerous. It therefore is not currently recommended.

3. Psychotherapy and Behavioral Therapy

The results from the studies of psychological treatments in IBS have been limited by methodological problems. A systematic review identified 14 controlled trials of which eight reported that a psychological treatment was superior to control [147]. Stress reduction incorporating education and breathing exercises, relaxation therapy, and cognitive behavioral therapy have been tested with good results. Dynamically orientated psychotherapy and hypnotherapy appear to be most promising.

In patients with resistant symptoms, a psychological treatment approach should therefore be considered [148]. Patients who appear to respond best to psychological treatment are those with intermittent symptoms that they clearly link to stress or emotional difficulties [147]. Patients who are resistant to the idea that psychological factors are important in their illness or those with constant pain, distension, or constipation are least likely to respond [147].

XI. Epidemiologic and Public Health Issues

Rigorous evaluation of the validity of the Rome criteria for IBS is still required. However, physiological studies that have applied the criteria have produced promising results regarding potential biological markers and the current classification is supported by epidemiologic and factor analysis studies.

The public health impact of IBS is large because of the high prevalence of the condition. The costs of IBS in a community setting have been estimated. Using a billing database with linkages to medical care, a community study estimated that the median annual health service charges (in 1992 U.S. dollars) for subjects with IBS were $742 compared with $429 for asymptomatic persons; overall, the odds of incurring charges were 1.6 times greater in IBS sufferers [149]. Extrapolating these results to the U.S. population, the excess charges for IBS in the United States are at least 8 billion dollars annually [149,150]. This figure, however, is probably an underestimate because outpatient drug costs and indirect costs were not measured. Data from the U.S. National Ambulatory Care Surveys have shown that drugs were prescribed for approximately 75% of patients presenting with IBS and account for at least 2.2 million prescriptions annually [49]. Over-the-counter drugs also need to be added into cost calculations; in the U.S., laxative sales over-the-counter alone are approximately $348 million annually [150]. The application of unnecessary tests and surgery also must substantially increase the total costs [47,151].

IBS is a major cause of absenteeism from work. The U.S. householder survey estimated that the adjusted mean number of days of work missed over a one year period was 13.4 for those

with IBS compared to 4.9 for persons without functional gastrointestinal complaints [22].

The impact of IBS on individual quality of life can be great [26,152,153]. O'Keefe et al. showed that among 520 elderly community subjects, those with IBS had a significantly lower quality of life [26]. IBS symptom severity has been linked to reduced productivity such as time lost from work [153]. Similarly, it has been observed that undergraduate students who sought health care for IBS experienced a reduction in the quality of life compared with nonconsulters with IBS, even when the effects of psychological distress and neuroticism were controlled [152].

In recognition of the importance of IBS, lay groups around the U.S. have been formed by patients with the condition. Patient support groups have proved extremely popular because they can help an individual to develop a sense of control over their problem. The International Foundation for Bowel Dysfunction (IFBD) is a nationwide lay organization, founded by Nancy Norton, with links to academia such as the Functional Brain Gut Research Group (FBGRG). The IFBD has had a major impact in educating the public and Congress regarding the magnitude and importance of the problem of IBS so that solutions can be found.

XII. Conclusions

IBS is no longer considered to be an unimportant affliction. Its social and economic costs, and the recognition that it is not just a disorder of the mind, have contributed to the change in viewpoint. Studies of gender and cultural issues should help to shed more light on the impact of IBS in the community. Careful pathophysiological and treatment studies, especially in primary care patients and in nonpatients in the community, are needed to help identify the most appropriate management. There remain many unanswered questions, but eventually a full explanation of this still mysterious syndrome will surely be found.

References

1. Chaudhary, N. A., and Truelove, S. C. (1962). The irritable colon syndrome. Q. J. Med. 31, 307–322.
2. Vesa, T. H., Seppo, L. M., Marteau, P. R., Sahi, T., and Korpela, R. (1998). Role of irritable bowel syndrome in subjective lactose intolerance. Am. J. Clin. Nut. 67, 710–715.
3. Delarive, J., Saraga, E., Dorta, G., and Blum, A. (1998). Budesonide in the treatment of collagenous colitis. Digestion 59, 364–366.
4. Camilleri, M., and Ford, M. J. (1998). Review article: Colonic sensorimotor physiology in health, and its alteration in constipation and diarrhoeal disorders. Aliment. Pharmacol. Ther. 12, 287–302.
5. Bharucha, A. E. (1998). Obstructed defecation: Don't strain in vain. Am. J. Gastroenterol. 93, 1019–1020.
6. Drossman, D. A., Richter, J. E., Talley, N. J., Thompson, G. W., Corazziari, E., and Whitehead, W. E., eds. (1994). "The Functional Gastrointestinal Disorders. Diagnosis, Pathophysiology and Treatment—A Multinational Consensus." Little, Brown, Boston.
7. Howship, J. (1930). "Practical Remarks on the Discrimination and Successful Treatment of Spasmodic Stricture of the Colon." Burgess & Hill, London.
8. Peters, G. A., and Bargen, J. A. (1944). The irritable bowel syndrome. Gastroenterology 3, 399–402.
9. Fielding, J. F. (1977). The irritable bowel syndrome. Part 1: Clinical spectrum. Clin. Gastroenterol. 6, 607–622.

10. Almy, T. P., Kern, F., and Tulin, M. (1949). Alterations in colonic function in man under stress II. Experimental production of sigmoid spasm in healthy persons. *Gastroenterology* **12,** 425–436.

11. Cumming, W. (1849). Electro-galvanism in a peculiar affliction of the mucous membrane of the bowels. *London Med Gaz.* [N.S.] **9,** 969–973.

12. Manning, A. P., Thompson, W. G., Heaton, K. W., and Morris, A. F. (1978). Towards positive diagnosis of the irritable bowel. *Br. Med. J.* **2,** 653–654.

13. Thompson, W. G. (1984). Gastrointestinal symptoms in the irritable bowel compared with peptic ulcer and inflammatory bowel disease. *Gut* **25,** 1089–1092.

14. Talley, N. J., Phillips, S. F., Melton, L. J., Mulvihill, C., Wiltgen, C., and Zinsmeister, A. R. (1990). Diagnostic value of the Manning criteria in irritable bowel syndrome. *Gut* **31,** 77–81.

15. Poynard, T., Couturier, D., Frexinos, I., Bommelaer, G., Hernandez, M., Dapoigny, M., Buscoil, L., Benand-Agostini, H., Chaput, C., and Rheims, N. (1992). French experience of Manning's criteria in irritable bowel syndrome. *Eur. J. Gastroenterol. Hepatol.* **4,** 747–752.

16. Rao, K. P., Gupta, S., Jain, A. K., Agrawal, A. K., and Gupta, J. P. (1993). Evaluation of Manning criteria in the diagnosis of irritable bowel syndrome. *J. Assoc. Physicians India* **41,** 357–358.

17. Jeong, H., Lee, H. R., Yoo, B. C., and Park, S. M. (1993). Manning criteria in irritable bowel syndrome: Its diagnostic significance. *Korean J. Intern. Med.* **8,** 34–39.

18. Kruis, W., Thieme, C., Weinzierl, M., Schussler, P., Holl, J., and Paulus, W. (1984). A diagnostic score for the irritable bowel syndrome. Its value in the exclusion of organic disease. *Gastroenterology* **87,** 1–7

19. Drossman, D. A., Thompson, W. G., Talley, N. J., Funch-Jensen, P., Janssens, J., and Whitehead, W. E. (1990). Identification of sub-groups of functional gastrointestinal disorders. *Gastroenterol. Int.* **4,** 159–172.

20. Taub, E., Cuevas, J. L., Cook, E. W., III, Crowell, M., and Whitehead, W. E. (1995). Irritable bowel syndrome defined by factor analysis. Gender and race comparisons. *Dig. Dis. Sci.* **40,** 2647–2655.

21. Talley, N. J., Holtmann, G., Zinsmeister, A. R., and Jones, M. (1997). Gastrointestinal symptoms cluster into distinct upper and lower groupings consistent with the Rome classification: A Three Country Population-based Study. *Gastroenterology* **112,** A834.

22. Drossman, D. A., Li, Z., Andruzzi, E., Temple, R. D., Talley, N. J., Thompson, W. G., Whitehead, W. E., Hanssens, J., Funch-Jensen, P., Corazziari, E., Richter, J. E., and Koch, G. G. (1993). U.S. householder survey of functional gastrointestinal disorders: Prevalence, sociodemography and health impact. *Dig. Dis. Sci.* **38,** 1569–1580.

23. Zuckerman, M. J., Guerra, L. G., Drossman, D. A., Foland, J. A., and Gregory, G. G. (1996). Health care seeking behaviors related to bowel complaints. Hispanics versus non-Hispanic whites. *Dig. Dis. Sci.* **41,** 77–82.

24. Talley, N. J., Zinsmeister, A. R., Van Dyke, C., and Melton, L. J. (1991). Epidemiology of colonic symptoms and the irritable bowel syndrome. *Gastroenterology* **101,** 927–934.

25. Hyams, J. S., Burke, G., Davis, P. M., Rzepski, B., and Andrulonis, P. A. (1996). Abdominal pain and irritable bowel syndrome in adolescents: A Community-based Study. *J. Pediatr.* **129,** 220–226.

26. O'Keefe, E. A., Talley, N. J., Zinsmeister, A. R., and Jacobsen, S. J. (1995). Bowel disorders impair functional status and quality of life in the elderly: A Population-based Study. *J. Gerontol. Med. Sci.* **50A,** M184–M189.

27. Jones, R., and Lydeard, S. (1992). Irritable bowel syndrome in the general population. *Br. Med. J.* **304,** 87–90.

28. Gaburri, M., Bassotti, G., Bacci, G., Cinti, A., Bosso, R., Ceccarelli, P., Paolocci, N., Pelli, M. A., and Morelli, A. (1989). Functional gut disorders and health care seeking behavior in an Italian non-patient population. *Rec. Prog. Med.* **80,** 241–244.

29. Welch, G. W., and Pomare, E. W. (1990). Functional gastrointestinal symptoms in a Wellington community sample. *N.Z. Med. J.* **103,** 418–420.

30. Olubuyide, L. O., Olawuyi, F., and Fasanmade, A. A. (1995). A study of irritable bowel syndrome diagnosed by Manning criteria in an African population. *Dig. Dis. Sci.* **40,** 983–985.

31. Talley, N. J., Boyce, P. M., and Jones, M. (1998). Is the association between irritable bowel syndrome and abuse explained by neuroticism? A Population Based Study. *Gut* **42,** 47–53.

32. Triadafilopoulos, G., Finlayson, M., and Grellet, C. (1998). Bowel dysfunction in postmenopausal women. *Women's Health* **27,** 55–66.

33. Owens, D. M., Nelsons, D. K., and Talley, N. J. (1995). Irritable bowel syndrome: Long-term prognosis and the physician-patient interaction. *Ann. Intern. Med.* **122,** 107–112.

34. Agreus, L., Svardsudd, K., Nyren, O., and Tibblin, G. (1994). Irritable bowel syndrome and dyspepsia in the general population: Overlap and lack of stability over time. *Gastroenterology* **109,** 671–680.

35. Talley, N. J., Weaver, A. L., Zinsmeister, A. R., and Melton, L. J., III (1992). Onset and disappearance of gastrointestinal symptoms and functional gastrointestinal disorders. *Am. J. Epidemiol.* **136,** 165–167.

36. Waller, S. L., and Misiewicz, J. J. (1969). Prognosis in the irritable-bowel syndrome. A Prospective Study. *Lancet* **2,** 754–756.

37. Holmes, K. M., and Salter, R. H. (1982). Irritable bowel syndrome—a safe diagnosis? *Br. Med. J.* **285,** 1533–1534.

38. Svendsen, J. H., Munch, L. K., and Andersen, J. R. (1985). Irritable bowel syndrome—prognosis and diagnostic safety. A 5 Year Follow-up Study. *Scand. J. Gastroenterol.* **20,** 415–418.

39. Harvey, R. F., Mauad, E. C., and Brown, A. M. (1987). Prognosis in the irritable bowel syndrome: A 5 Year Prospective Study. *Lancet* **1,** 963–965.

40. Lembo, T., Fullerton, S., Diehl, D., Raeen, H., Munakata, J., Naliboff, B., and Mayer, E. A. (1996). Symptom duration in patients with irritable bowel syndrome. *Am. J. Gastroenterol.* **91,** 898–905.

41. Prior, A., and Whorwell, P. J. (1989). Gynaecological consultation in patients with the irritable bowel syndrome. *Gut* **30,** 996–998.

42. Otte, J. J., Larsen, L., and Andersen, J. R. (1986). Irritable bowel syndrome and symptomatic diverticular disease—different diseases? *Am. J. Gastroenterol.* **81,** 529–531.

43. Fowlie, S., Eastwood, M. A., and Ford, M. J. (1992). Irritable bowel syndrome: The influence of psychological factors on the symptom complex. *J. Psychosom. Res.* **36,** 169–173.

44. Danquechin Dorval, E., Delvaux, M., Allemand, H., Allouche, S., Van Egroo, L. D., and Lepen, C. (1994). Profile and evolution of irritable bowel syndrome. Prospective national epidemiological study of 1301 patients followed-up for 9 months in gastroenterology. *Gastroenterol. Clin. Biolo.* **18,** 145–150.

45. Blewett, A., Allison, M., Calcraft, B., Moore, R., Jenkins, P., and Sullivan, G. (1996). Psychiatric disorder and outcome in irritable bowel syndrome. *Psychosomatics* **37,** 155–160.

46. Hahn, B., Watson, M., Yan, S., Gunput, D., and Heuijerjans, J. (1998). Irritable bowel syndrome symptom patterns: Frequency, duration, and severity. *Dig. Dis. Sci.* **43,** 2715–2718.

47. Drossman, D. A., Whitehead, W. E., and Camilleri, M. (1997). Irritable bowel syndrome—A technical review for practice guideline development. *Gastroenterology* **112,** 2120–2137.

48. Sandler, R. S. (1990). Epidemiology of irritable bowel syndrome in the United States. *Gastroenterology* **99,** 409–415.

49. Everhart, J. E., and Renault, P. F. (1991). Irritable bowel syndrome in office-based practice in the United States. *Gastroenterology* **100,** 998–1005.

50. Talley, N. J., Boyce, P. M., and Jones, M. (1997). Predictors of health care seeking for irritable bowel syndrome: A population Based Study. *Gut* **41,** 394–398.

51. Sandler, R. S., Drossman, D. A., Nathan, H. P., and McKee, D. C. (1984). Symptom complaints and health care seeking behavior in subjects with bowel dysfunction. *Gastroenterology* **87**, 314–318.

52. Heaton, K. W., O Donnell, L. J. D., Braddon, F. E. M. *et al.* (1992). Symptoms of irritable bowel syndrome in a British urban community: Consulters and non-consulters. *Gastroenterology* **102**, 1962–1967.

53. Drossman, D. A., McKee, D. C., Sandler, R. S., Mitchell, C. M., Cramer, E. M., Lowman, B. C., and Burger, A. L. (1988). Psychosocial factors in the irritable bowel syndrome. A multivariate study of patients and nonpatients with irritable bowel syndrome. *Gastroenterology* **95**, 701–708.

54. Whitehead, W. E., Bosmajian, L., Zonderman, A. B., Costa, P. T., Jr., and Schuster, M. M. (1988). Symptoms of psychologic distress associated with irritable bowel syndrome. Comparison of community and medical clinical samples. *Gastroenterology* **95**, 709–714.

55. Gick, M. L., and Thompson, W. G. (1997). Negative affect and the seeking of medical care in university students with irritable bowel syndrome: A Preliminary Study. *J. Psychosom. Res.* **43**, 535–540.

56. Whitehead, W. E., Winget, C., Fedoravicius, A. S., Wooley, S., and Blackwell, B. (1982). Learned illness behavior in patients with irritable bowel syndrome and peptic ulcer. *Dig. Dis. Sci.* **27**, 202–208.

57. Whitehead, W. E., Engel, B. T., and Schuster, M. M. (1980). Irritable bowel syndrome: Physiological and psychological differences between diarrhea-predominant and constipation-predominant patients. *Dig. Dis. Sci.* **25**, 404–413.

58. Young, S. J., Alpers, D. H., Norland, C. C., and Woodruff, R. A. (1976). Psychiatric illness and the irritable bowel syndrome. Practical implications for the primary physician. *Gastroenterology* **70**, 162–166.

59. Longstreth, G. F., and Wolde-Tsadik, G. (1993). Irritable bowel-type symptoms in HMO examinees: Prevalence, demographics and clinical correlates. *Dig. Dis. Sci.* **38**, 1581–1589.

60. Thompson, W. G. (1997). Gender differences in irritable bowel symptoms. *Eur. J. Gastroenterol. Hepatol.* **9**, 299–302.

61. Smith, R. C., Greenbaum, D. S., Vancouver, J. B., Henry, R. C., Reinhart, M. A., Greenbaum, R. B., Dean, H. A., and Mayle, J. E. (1991). Gender differences in Manning criteria in the irritable bowel syndrome. *Gastroenterology* **100**, 591–595.

62. Jain, A. A., Gupta, O. P., Jajoo, U. N., and Sidhwa, H. K. (1991). Clinical profile of irritable bowel syndrome at a rural based teaching hospital in central India. *J. Assoc Physicians India* **39**, 385–386.

63. Pimparkar, B. D. (1970). Irritable colon syndrome. *J. Indian Med. Assoc.* **54**, 95–105.

64. Mendis, B. L. J., Wijesiriwardena, B. C., Sheriff, J. H. R., and Dharmadasa, K. (1982). Irritable bowel syndrome. *Ceylon Med. J.* **27**, 171–181.

65. Mathur, A., Tandon, B. N., and Prakkash, O. M. (1966). Irritable colon syndrome. *J. Indian Med. Assoc.* **46**, 651–655.

66. Bordie, A. K. (1972). Functional disorders of the colon. *J. Indian Med. Assoc.* **58**, 451–456.

67. Kapoor, K. K., Nigam, P., Rastogi, C. K., Kumar, A., and Gupta, A. K. (1985). Clinical profile of the irritable bowel syndrome. *Indian J. Gastroenterol.* **4**, 15–16.

68. Verbrugge, L. M. (1985). Gender and health: An update on hypotheses and evidence. *J. Health Soc. Behav.* **26**, 156–182.

69. Drossman, D. A., and Thompson, W. G. (1992). The irritable bowel syndrome: Review and a graduated multicomponent treatment approach. *Ann. Intern. Med.* **116**, 1009–1016.

70. Toner, B. B., Garfinkel, P. E., and Jeejeebhoy, K. N. (1990). Psychological factors in the irritable bowel syndrome. *Can. J. Psychiatry* **35**, 158–161.

71. Creed, F. H., and Guthrie, E. (1987). Psychological factors in the irritable bowel syndrome. *Gut* **28**, 1307–1318.

72. Esler, M. D., and Goulston, K. J. (1973). Levels of anxiety in colonic disorders. *N. Engl. J. Med.* **288**, 16–20.

73. Gwee, K. A., Graham, J. C., McKendrick, M. W., Collins, S. M., Marshall, J. S., Walters, S. J., and Read, N. W. (1996). Psychometric scores and persistence of irritable bowel after infectious diarrhoea. *Lancet* **247**, 150–153.

74. Levy, R. L., Cain, K. C., Jarrett, M., and Heitkemper, M. M. (1997). The relationship between daily life stress and gastrointestinal symptoms in women with irritable bowel syndrome. *J. Behav. Med.* **20**, 177–193.

75. Whitehead, W. E., Crowell, M. D., Robinson, J. C., Heller, B. R., and Schuster, M. M. (1992). Effects of stressful life events on bowel symptoms: Subjects with irritable bowel syndrome compared with subjects without bowel dysfunction. *Gut* **33**, 825–830.

76. Drossman, D. A., Leserman, J., Nachman, G., Li, Z., Gluck, H., Tommey, T. C., and Mitchell, M. (1990). Sexual and physical abuse in women with functional or organic gastrointestinal disorders. *Ann. Intern. Med.* **113**, 828–833.

77. Talley, N. J., Fett, S. L., Zinsmeister, A. R., and Melton, L. J., III (1994). Gastrointestinal tract symptoms and self reported abuse: A Population-based study. *Gastroenterology* **107**, 1040–1049.

78. Talley, N. J., Fett, S. L., and Zinsmeister, A. R. (1995). Self-reported abuse and gastrointestinal disease in outpatients: Association with irritable bowel-type symptoms. *Am. J. Gastroenterol.* **90**, 366–371.

79. Delvaux, M., Denis, P., Allemand, H., and the French Club of Digestive Motility (1997). Sexual abuse is more frequently reported by IBS patients than by patients with organic diseases or controls. Results of a multicentre inquiry. *Eur. J. Gastroenterol. Hepatol.* **9**, 345–352.

80. Talley, N. J., Kramlinger, K. G., Burton, M. C., Colwell, L. J., and Zinsmeister, A. R. (1993). Psychiatric disorders and childhood abuse in the irritable bowel syndrome. *Eur. J. Gastroenterol. Hepatol.* **5**, 647–654.

81. Walker, E. A., Gelfand, A. N., Gelfand, M. D., and Katon, W. J. (1995). Psychiatric diagnoses, sexual and physical victimization, and disability in patients with irritable bowel syndrome or inflammatory bowel disease. *Psychol. Med.* **25**, 1259–1267.

82. Latimer, P. R., Sarna, S. K., Campbell, D., Latimer, M. R., Waterfall, W. E., and Daniel, E. E. (1981). Colonic motor and myoelectrical activity: A comparative study of normal patients, psychoneurotic patients and patients with irritable bowel syndrome (IBS). *Gastroenterology* **80**, 893–901.

83. Snape, W. J., Carlson, G. M., and Cohen, S. (1976). Colonic myoelectric activity in the irritable bowel syndrome. *Gastroenterology* **70**, 326–330.

84. McKee, D. P., and Quigley, E. M. (1993). Intestinal motility in irritable bowel syndrome: Is IBS a motility disorder? Part 1. Definition of IBS and colonic motility. *Dig. Dis. Sci.* **38**, 1761–1772.

85. Vassallo, M. J., Camilleri, M., Phillips, S. F., Steadman, C. J., Talley, N. J., Hanson, R. B., and Haddad, A. C. (1992). Colonic tone and motility in patients with irritable bowel syndrome. *Mayo Clin. Proc.* **67**, 725–731.

86. Harvey, R. F., and Read, A. E. (1973). Effect of cholecystokinin on colon motility and symptoms in patients with the irritable bowel syndrome. *Lancet* **1**, 1–3.

87. Kellow, J. E., Gill, R. C., and Wingate, D. L. (1990). Prolonged ambulant recordings of small bowel motility demonstrate abnormalities in the irritable bowel syndrome. *Gastroenterology* **98**, 1208–1218.

88. Kellow, J. E., and Phillips, S. F. (1987). Altered small bowel motility in irritable bowel syndrome is correlated with symptoms. *Gastroenterology* **92**, 1885–1893.

89. Gorard, D. A., Libby, G. W., and Farthing, M. J. G. (1994). Ambulatory small intestinal motility in diarrhea-predominant irritable bowel syndrome. *Gut* **35**, 203–210.

90. Mertz, H., Naliboff, B., Munakata, J., Niazi, N., and Mayer, E. A. (1995). Altered rectal perception is a biological marker of patients with irritable bowel syndrome. *Gastroenterology* **109**, 40–52.

91. Zighelboim, J., Talley, N. J., Phillips, S. F., Harmsen, W. S., and Zinsmeister, A. R. (1995). Visceral perception in irritable bowel syndrome. Rectal and gastric responses to distension and serotonin type 3 antagonism. *Dig. Dis. Sci.* **40**, 819–827.

92. Munakata, J., Naliboff, B., Harraf, F., Kodner, A., Lembo, T., Chang, L., Silverman, D. H. S., and Mayer, E. A. (1997). Repetitive sigmoid stimulation induces rectal hyperalgesia in patients with irritable bowel syndrome. *Gastroenterology* **112**, 55–63.

93. Houghton, L. A., Wych, J., and Whorwell, P. J. (1995). Acute diarrhoea induces rectal sensitivity in women but not men. *Gut* **37**, 270–273.

94. Whitehead, W. E., Crowell, M. D., Davidoff, A. L., Palsson, O. S., and Schuster, M. M. (1997). Pain from rectal distension in women with irritable bowel syndrome: Relationship to sexual abuse. *Dig. Dis. Sci.* **42**, 796–804.

95. Cook, I. J., van Eeden, A., and Collins, S. M. (1987). Patients with irritable bowel syndrome have greater pain tolerance than normal subjects. *Gastroenterology* **93**, 727–733.

96. Whorwell, P. J., Lupton, E. W., Erduran, D., and Wilson, K. (1986). Bladder smooth muscle dysfunction in patients with irritable bowel syndrome. *Gut* **27**, 1014–1017.

97. Sood, G. K., Baijal, S. S., Lahoti, D., and Broor, S. L. (1993). Abnormal gallbladder function in patients with irritable bowel syndrome. *Am. J. Gastroenterol.* **88**, 1387–1390.

98. White, A. M., Stevens, W. H., Upton, A. R., O'Byrne, P. M., and Collins, S. M. (1991). Airway responsiveness to inhaled methacholine in patients with irritable bowel syndrome. *Gastroenterology* **100**, 68–74.

99. Heitkemper, M., Jarrett, M., Cain, K., Shaver, J., Bond, E., Woods, N. F., and Walker, E. (1996). Increased urine catecholamines and cortisol in women with irritable bowel syndrome. *Am. J. Gastroenterol.* **91**, 906–912.

100. Orr, W. C., Crowell, M. D., Lin, B., Harnish, M. J., and Chen, J. D. (1997). Sleep and gastric function in irritable bowel syndrome: Derailing the brain-gut axis. *Gut* **41**, 390–393.

101. Mertz, H., Fass, R., Kodner, A., Yan-Go, F., Fullerton, S., and Mayer, E. A. (1998). Effect of amitriptyline on symptoms, sleep, and visceral perception in patients with functional dyspepsia. *Am. J. Gastroenterol.* **93**, 160–165.

102. Silverman, D. H., Munakata, J. A., Ennes, H., Mandelkern, M. A., Hoh, C. K., and Mayer, E. A. (1997). Regional cerebral activity in normal and pathological perception of visceral pain. *Gastroenterology* **112**, 64–72.

103. Chang, L., Mayer, E. A., Munakata, J. *et al.* (1998). Difference in left prefrontal activation to visceral and somatic stimuli assessed by 0-15-water PET in female patients with irritable bowel syndrome (IBS) and fibromyalgia. *Gastroenterology* **114**, A732.

104. Woodman, C. L., Breen, K., Noyes, R., Moss, C., Fagerholm, R., Yagla, S. J., and Summers, R. (1998). The relationship between irritable bowel syndrome and psychiatric illness. A Family Study. *Psychosomatics* **39**, 45–54.

105. Morris-Yates, A. D., Talley, N. J., Boyce, P., and Andrews, G. (1998). Evidence of a genetic contribution to functional bowel disorder. *Am. J. Gastroenterol.* **93**, 1311–1317.

106. Whitehead, W. E., Cheskin, L. J., Heller, B. R., Robinson, J. C., Crowell, M. D., Benjamin, C., and Schuster, M. M. (1990). Evidence for exacerbation of irritable bowel syndrome during menses. *Gastroenterology* **98**, 1485–1489.

107. Donnelly, V. S., O'Herlihy, C., Campbell, D. M., and O'Connell, P. R. (1998). Postpartum fecal incontinence is more common in women with irritable bowel syndrome. *Dis. Colon Rectum* **41**, 586–589.

108. Whorwell, P. J., McCallum, M., Creed, F. H., and Roberts, C. T. (1986). Non-colonic features of irritable bowel syndrome. *Gut* **27**, 37–40.

109. Kamm, M. A. (1997). Chronic pelvic pain in women—gastroenterological, gynaecological or psychological? *Int. J. Colorectal Dis.* **12**, 57–62.

110. Fass, R., Fullerton, S., Naliboff, B., Hirsh, T., and Mayer, E. P. (1998). Sexual dysfunction in patients with irritable bowel syndrome and non-ulcer dyspepsia. *Digestion* **59**, 79–85.

111. Van Dulmen, A. M., Fennis, J. F., Mokkink, H. G., van der Velden, H. A., and Bleijenberg, G. (1994). Doctors' perception of patients' cognitions and complaints in irritable bowel syndrome at an outpatient clinic. *J. Psychosom. Res.* **38**, 581–590.

112. Prior, A., Stanley, K. M., Smith, A. R., and Read, N. W. (1992). Relation between hysterectomy and the irritable bowel: A Prospective Study. *Gut* **33**, 814–817.

113. Longstreth, G. F., Preskill, D. B., and Youkeles, L. (1990). Irritable bowel syndrome in women having diagnostic laparoscopy or hysterectomy. Relation to gynecologic features and outcome. *Dig. Dis. Sci.* **35**, 1285–1290.

114. Heaton, K. W., Parker, D., and Cripps, H. (1993). Bowel function and irritable bowel symptoms after hysterectomy and cholecystectomy: A Population Based Study. *Gut* **34**, 1108–1111.

115. Burns, D. G. (1986). The risk of abdominal surgery in irritable bowel syndrome. *S. Afr. Med. J.* **70**, 91.

116. Rao, D. R., Bello, H., Warren, A. P., and Brown, G. E. (1994). Prevalence of lactose maldigestion. Influence and interaction of age, race, and sex. *Dig. Dis. Sci.* **39**, 1519–1524.

117. Suarez, F. L., Savaiano, D. A., and Levitt, M. D. (1995). A comparison of symptoms after the consumption of milk or lactose-hydrolyzed milk by people with self-reported severe lactose intolerance. *N. Engl. J. Med.* **333**, 1–4.

118. Camilleri, M., and Prather, C. M. (1992). The irritable bowel syndrome: Mechanisms and a practical approach to management. *Ann. Intern. Med.* **116**, 1001–1008.

119. Lambert, J. P., Brunt, P. W., Mowat, N. A., Khin, C. C., Lai, C. K., Morrison, V., Dickerson, J. W., and Eastwood, M. A. (1991). The value of prescribed "high-fibre" diets for the treatment of irritable bowel syndrome. *Eur. J. Clin. Nutr.* **45**, 601–609.

120. Francis, C. Y., and Whorwell, P. J. (1994). Bran and irritable bowel syndrome: Time for reappraisal. *Lancet* **344**, 39–40.

121. Levitt, M. D., Furne, J., and Olsson, S. (1996). The relation of passage of gas and abdominal bloating to colonic gas production. *Ann. Intern. Med.* **124**, 422–424.

122. Fernandez-Banares, F., Esteve-Pardo, M., de Leon, R., Humbert, P., Cabre, E., Llovet, J. M., and Gassull, M. A. (1993). Sugar malabsorption in functional bowel disease: Clinical implications. *Am. J. Gastroenterol.* **88**, 2044–2050.

123. Ganiats, T. G., Norcross, W. A., Halverson, A. L., Burford, P. A., and Palinkas, L. A. (1994). Does Beano prevent gas? A double-blind crossover study of oral alpha-galactosidase to treat dietary oligosaccharide intolerance. *J. Fam. Pract.* **39**, 441–445.

124. Klein, K. B. (1988). Controlled treatment trials in the irritable bowel syndrome: A critique. *Gastroenterology* **95**, 232–241.

125. Talley, N. J., Nyren, O., Drossman, D. A., Heaton, K. W., Veldhuyzen van Zanten, S. J. O., Koch, M. M., and Ransohoff, D. F. (1993). The irritable bowel syndrome: Toward optimal design of controlled treatment trials. *Gastroenterol. Int.* **6**, 189–211.

126. Ritchie, J. A., and Truelove, S. C. (1979). Treatment of irritable bowel syndrome with lorazepam, hyoscine butylbromide, and ispaghula husk. *Br. Med. J.* 376–378.

127. Page, J. G., and Dirnberger, G. M. (1981). Treatment of the irritable bowel syndrome with bentyl (Dicyclomine Hydrochloride). *J. Clin. Gastroenterol.* **3**, 153–156.

128. Liu, J. H., Chen, G. H., Yen, H. Z., Huang, C. K., and Poon, S. K. (1997). Enteric-coated peppermint oil capsules in the treatment of irritable bowel syndrome: A Prospective, Randomized Trial. *J. Gastroenterol.* **32**, 765–768.

129. Poynard, T., Naveau, S., Mory, B., and Chaput, J. C. (1994). Meta-analysis of smooth muscle relaxants in the treatment of irritable bowel syndrome. *Aliment. Pharmacol. Ther.* **8**, 499–510.

130. Reynolds, J. C. (1989). Prokinetic agents: A key in the future of gastroenterology. *Gastroenterol. Clin. North Am.* **18**, 437–457.

131. Farup, P. G., Hovdenak, N., Wetterhus, S., Lange, O. J., Hovde, O., and Trondstad, R. (1998). The symptomatic effect of cisapride in patients with irritable bowel syndrome and constipation. *Scand. J. Gastroenterol.* **33**, 128–31.

132. Schutze, K., Brandstatter, G., Dragosics, B., Judmaier, G., and Hentschel, (1997). Double-blind study of the effect of cisapride on constipation and components of the irritable bowel syndrome. *Aliment. Pharmacol. Ther.* **11**, 387–394.

133. Steadman, C. J., Talley, N. J., Phillips, S. F., and Zinsmeister, A. R. (1992). Selective 5-hydroxytryptamine type 3 receptor antagonism with ondansetron as treatment for diarrhoea-predominant irritable bowel syndrome. *Mayo Clin. Proc.* **67**, 732–738.

134. Prior, A., and Read, N. W. (1993). Reduction of rectal sensitivity and post-prandial motility by granisetron, a 5 HT3-receptor antagonist, in patients with irritable bowel syndrome. *Aliment. Pharmacol. Ther.* **7**, 175–180.

135. Northcutt, A. R., Camilleri, M., Mayer, E. A. *et al.* (1998). Alosetron, a 5HT$_3$ receptor antagonist, is effective in the treatment of female irritable bowel syndrome patients. *Gastroenterology* **114**, A812.

136. Petticrew, M., Watt, I., and Sheldon, T. (1997). Systematic review of the effectiveness of laxatives in the elderly. *Health Technol. Assess.* **1**, 1–52.

137. Joo, J. S., Ehrenpreis, E. D., Gonzalez, L., Kaye, M., Breno, S., Wexner, S. D., Zaitman, D., and Secrest, K. (1998). Alterations in colonic anatomy induced by chronic stimulant laxatives: The cathartic colon revisited. *J. Clin. Gastroenterol.* **26**, 283–286.

138. Prather, C. M., and Ortiz-Camacho, C. P. (1998). Evaluation and treatment of constipation and fecal impaction in adults. *Mayo Clin. Proc.* **73**, 881–886.

139. Efskind, P. S., Bernklev, T., and Vatn, M. H. (1996). A double-blind placebo-controlled trial with loperamide in irritable bowel syndrome. *Scand. J. Gastroenterol.* **31**, 463–468.

140. Luman, W., Williams, A. J., Merrick, M. V., and Eastwood, M. A. (1995). Idiopathic bile acid malabsorption: Long-term outcome. *Eur. J. Gastroenterol. Hepatol.* **7**, 641–645.

141. Williams, A. J., Merrick, M. V., and Eastwood, M. A. (1991). Idiopathic bile acid malabsorption—A review of clinical presentation, diagnosis, and response to treatment. *Gut* **32**, 1004–1006.

142. Clouse, R. E., Lustman, P. J., Geisman, R. A., and Alpers, D. H. (1994). Antidepressant therapy in 138 patients with irritable bowel syndrome: A five-year clinical experience. *Aliment. Pharmacol. Ther.* **8**, 409–416.

143. Clouse, R. E. (1994). Antidepressants for functional gastrointestinal syndromes. *Dig. Dis. Sci.* **39**, 2352–2363.

144. Gram, L. F. (1994). Fluoxetine. *N. Engl. J. Med.* **20**, 1354–1361.

145. Mathias, J. R., Clench, M. H., Roberts, P. H., and Reeves-Darby, V. G. (1994). Effect of leuprolide acetate in patients with functional bowel disease. Long-term follow-up after double-blind, placebo-controlled study. *Dig. Dis. Sci.* **39**, 1163–1170.

146. Mathias, J. R., Clench, M. H., Reeves-Darby, Fox, L. M., Hsu, P. H., Roberts, P. H., Smith, L. L., and Stiglich, N. J. (1994). Effect of leuprolide acetate in patients with moderate to severe functional bowel disease. Double-blind, Placebo-controlled Study. *Dig. Dis. Sci.* **39**, 1151–1162.

147. Talley, N. J., Owen, B. K., Boyce, P., and Paterson, K. (1996). Psychological treatments for irritable bowel syndrome: A critique of controlled treatment trials. *Am. J. Gastroenterol.* **91**, 277–283.

148. Drossman, D. A. (1995). Diagnosing and treating patients with refractory functional gastrointestinal disorders. *Ann. Intern. Med.* **123**, 688–697.

149. Talley, N. J., Gabriel, S. E., Harmsen, W. S., Zinsmeister, A. R., and Evans, R. W. (1995). Medical costs in community subjects with irritable bowel syndrome. *Gastroenterology* **109**, 1736–1741.

150. Longstreth, G. F. (1995). Irritable bowel syndrome—a multibillion-dollar problem. *Gastroenterology* **109**, 2029–2031.

151. Jones, R. H. (1996). Clinical economics review—gastrointestinal disease in primary care. *Aliment. Pharmacol. Ther.* **10**, 233–239.

152. Whitehead, W. E., Burnett, C. K., Cook, E. W., III, and Taub, E. (1996). Impact of irritable bowel syndrome on quality of life. *Dig. Dis. Sci.* **41**, 2248–2253.

153. Hahn, B. A., Kirchdoerfer, L. J., Fullerton, S., and Mayer, E. (1997). Patient perceived severity of irritable bowel syndrome in relation to symptoms, health resource utilization and quality of life. *Aliment. Pharmacol. Ther.* **11**, 553–559.

88

Interstitial Cystitis

C. LOWELL PARSONS, MEHDI KAMAREI, AND MANOJ MONGA
Division of Urology
University of California, San Diego
San Diego, California

I. Introduction

Interstitial cystitis (IC) is a severe and debilitating chronic pain syndrome that afflicts the bladder. It is characterized by severe urinary frequency, urgency, and/or lower abdominal or perineal pain. IC is an affliction whose etiology and pathogenesis remain obscure, complex, and controversial. It is a disease that typically has a gradual onset with an insidious progression. However, the variability of clinical onset can range from patients suffering from mild symptoms over a long period to others with a short and more severe disease course. This variability has led to the assumption that IC is a syndrome with a heterogeneous etiology. Based on these facts and with its unpredictable natural history, the diagnosis of IC is a problem—clinically and morphologically—especially in the early phases.

IC is usually misdiagnosed initially as chronic bacterial cystitis in females or perhaps called urethral syndrome in other cases. In men, it is often confused with either bladder outlet obstruction due to prostate enlargement or "prostatitis." As the disease progresses, it is the appearance of pain that is the major impetus to seek medical attention. Pain is the most disabling part of the syndrome and the most difficult for the physician to treat. This chapter will review the definition, epidemiology, pathogenesis, diagnosis, and treatment of IC.

II. Definition of Interstitial Cystitis

The definition of IC has changed significantly in recent years, secondary to the marked interest in studying this disease. The diagnosis of interstitial cystitis has traditionally been a diagnosis of exclusion. In 1987 a group of researchers met at the National Institutes of Health (NIH) to establish strict clinical criteria for characterizing the IC syndrome, so as to standardize patient selection for research studies [1]. The criteria include the presence of two or more of the following: pain on bladder filling relieved by emptying, pain (suprapubic, pelvic, urethral, vaginal, or perineal), glomerulations on endoscopy, or decreased compliance on cystometrogram; Hunner's ulcer automatically includes a patient as meeting case criteria. A number of exclusionary conditions and criteria also were specified, such as being less than 18 years of age, or a waking urinary frequency of less than five times in 12 hours. Patients meeting these criteria have advanced disease. It was well accepted by the IC researchers that they would identify only a portion of the patients (probably less than one-third) and that many people with the IC syndrome would not have all the published parameters of IC. In reality, these criteria would exclude those patients with milder, inter-

mittent symptoms who may benefit from early identification and intervention.

The *sine qua non* of the definition of IC is that the patient has significant urinary urgency, frequency, and/or bladder pain. Upon clinical evaluation, these patients should have no other definable pathology such as urinary infections, carcinoma, radiation, or medication-induced cystitis. This syndrome may encompass a number of different etiologies that cause a bladder insult, ultimately resulting in urinary frequency and urgency, the bladder's only clinical response to noxious stimuli. The expansion of the definition of IC is controversial, but as clinical and epidemiologic data accumulate, it is steadily becoming accepted. An important purpose for utilizing a broad definition of the syndrome is that IC may be underdiagnosed. Many patients with milder forms could readily benefit from therapy if the diagnosis is considered. Treatment may be inappropriately withheld if the physician reserves the diagnosis of IC for patients with more severe symptoms. In this review of IC, this more liberal definition will be used.

III. Epidemiology of Interstitial Cystitis

A. Incidence and Prevalence

Although IC was first identified in 1907 by Nitze, few epidemiologic studies have been reported. Oravisto [2] studied one million Caucasians in Finland and found prevalences of 10.1 cases/100,000 men and 18.1/100,000 women. Based on this population prevalence of IC, the annual incidence of IC was estimated to be 1.2/100,000 females in Finland. The prevalence of IC in the Netherlands was estimated to be 8–16 per 100,000 women, using a questionnaire sent to urologists [3].

The prevalence of IC in the United States was estimated using three different sources: urologists, IC patients, and a sample from the general population. As a result, estimates ranged from 19,400 to 90,700 women diagnosed with IC. Using a weighted average of the three estimates there are at least 43,500 women with IC in the United States (estimated annual incidence of 2.6/100,000 women) [4]. Jones *et al.* [4a] reported that 501/100,000 people (865/100,000 females) of the U.S. civilian noninstitutionalized population had a diagnosis of IC, based on self-reports of a previous diagnosis of IC in 20,561 adults interviewed. Held *et al.* also estimated a worst case scenario prevalence in the U.S. to be 450,000 people [4].

All studies have their limitations, either due to bias produced by using self-reported questionnaires, by investigating selected subpopulations, or due to the different definitions of IC. In a large study in England of 1000 successive patients presenting to outpatient centers for a clinical trial, 50% of the patients with signs and symptoms of urinary tract infection (urgency/frequency) had negative cultures. The researchers concluded that these patients had "the urethral syndrome," which they defined as having signs and symptoms of infection but negative cultures [5]. These patients may have a mild form of IC that gradually escalates over the years until symptoms become severe enough and they are identified as having IC. All patients with urinary urgency, frequency, or bladder pain may represent the same disease process but are early or late in this spectrum of time in terms of having mild or severe disease. Generally the disease appears to incubate slowly over the years, and with progressive increase in symptoms and decrease in bladder function, the disease takes its toll.

B. Public Health Impact

In 1987, the individual costs for the U.S. American health system were calculated to be $3870 per patient per year, and an additional economic burden of $4039 per patient per year was placed on the economy system due to the inability to work. Thus, the overall costs can be extrapolated to about $348 million per year for about 44,000 IC patients in the United States [4].

C. Impact on Functional Status

The difficulty patients with IC face is that they have an elusive disease about which little is known and for which diagnostic and therapeutic directions are under discussion. The impact of IC on the patients' quality of life is significant. The problems in delayed diagnosis may lead to inappropriate suggestions of a psychological etiology rather than a physiological abnormality. A high percentage of patients are limited in their daily and leisure activities and experience disruptions of their occupational, social, familial, and sexual lives; they suffer from sleeplessness, lack of concentration, and some may need psychological support [4,6,7].

IV. Demographic Risk Factors

The demographic characteristics of adult IC patients can be summarized as middle-aged, female, and white. The mean age of IC patients is between 42 and 53 years [4,8,9]. The median age of diagnosis of IC is between 42 and 46 years and the average duration of symptoms prior to diagnosis is 3–4 years [4,10]. More than 30% of patients are younger than 30 years when IC symptoms start [6]. Moreover, there are several reports of children with IC [6,11,12] and patients older than 80 years with the syndrome [6,11]. Conditions such as nocturnal enuresis and urinary frequency of childhood may represent early forms of IC. According to the NIH research criteria, however, an age less than 18 years is an exclusion criteria for IC. Females suffer more often from IC than males. The female/male ratio in most studies is 9–10:1 [3,6,8,9,13,14]. However, in different subpopulations, such as in children, a ratio of 2:3 females/males is reported [12].

There are only a few studies evaluating ethnic factors in IC. Most studies report that the prevalence of IC in African-Americans is lower than that of the general American population. There are several reports that Africans may experience IC at a lower frequency than African-Americans but this finding may reflect a difference in access to medical care rather than a true difference in the pathophysiology of IC. A 400% increased incidence was noted in people of Jewish descent in a University of California, San Diego study compared with the general population [4].

Held et al. [4] reported that the educational level, family income per year, household size, and marital status, as well as the number of sexual partners, were not statistically different in IC patients compared to the general population. However, childhood bladder difficulties increase the risk for IC 12-fold, while prior urinary tract infections increases the risk for IC 2-fold [4].

Potential risk factors, symptoms, and psychological aspects of IC were investigated in IC patients (300 patients from the National Institute of Arthritis, Diabetes, Digestive and Kidney Diseases, La Jolla, California and 246 patients from the Division of Urology, University of California, San Diego) using a 200-item evaluation and were compared to controls with non-IC related bladder disorders (n = 171) [6]. This study found an increased risk for IC in patients with a history of allergic reactions (medication-related, hay fever, or food allergies), in patients with chronic or frequent upper respiratory tract infections (including sinusitis), and patients with autoimmune diseases or irritable bowel syndrome. Significantly more IC patients had positive Epstein-Barr virus titers than controls, and hysterectomies were significantly more common in IC patients than in female controls. No differences between IC patients and controls were seen in the prevalence of coexisting thyroid disease, diabetes, radiotherapy, chemotherapy, tuberculosis, or urinary calculi. No increased intrafamilial risk was identified for IC. For a variety of malignancies (cervical, renal, bladder, urethral), insufficient control data were available for meaningful comparisons [6].

V. Pathogenesis

Since the original description of the "elusive ulcer" of Hunner in 1915, there has been slow progress in defining the etiology of IC [15,16]. In part, because the severe form of IC is relatively rare, the identification of a large study population is difficult. Lymphatic, infectious, neurologic, psychologic, autoimmune, and vasculitic etiologies have been proposed for IC [13,17–23]. Most of the proposed etiologies are hypothetical, with little data to either substantiate or refute their role. For example, Oravisto [22] demonstrated that there are increased antinuclear antibodies with IC; however, there is no obvious association with systemic autoimmune phenomena in these patients. Modest rises in autoantibodies may occur in many chronic illnesses, not necessarily inducing the illness but as a nonspecific response to it.

Currently, several factors seem to play an etiologic role in IC. One theory is that there is a defective bladder epithelium with loss of the "blood–urine" barrier resulting in a leaky membrane [24,25]. An epithelium permeable to small molecules could explain many of the symptoms associated with the complex. A chronic leak of small molecules into the bladder interstitium could induce sensory nerves to depolarize, resulting in urgency–frequency and pain [26,27]. In particular, diffusion of potassium across the membrane could trigger the sensory nerve endings [28]. A well-controlled study in 56 patients supported the hypothesis that the bladder surface in many patients with IC may leak [25]. These investigators have developed a sensitive leak assay and found that about 70% of patients with IC have a "leaky epithelium" while 30% do not [29]. Those patients who do not leak may suffer from an alternative etiologic process such as a neurologic abnormality or inflammation. These findings support the concept that several etiologic mechanisms contribute to the development of IC and it may be possible to stratify patients by etiology (epithelial leakers vs nonleakers).

The other etiologic theory for IC of bladder inflammation has little support. Most of the patients do not have significant signs of inflammatory responses in their bladder muscle or serum [30]. Investigations have been unable to isolate any white cells, cytokines, or other inflammatory mediators in urine in 90–95% of patients with the IC syndrome [31].

The role of mast cells in this disorder is not understood. The central point of contention is whether mast cells play a causative or secondary role in the pathogenesis of IC. A causal relationship has been proposed based on the theory that degranulation of mast cells may produce symptoms of IC. Alternatively, the presence of mast cells in the interstitium may be a response to a primary disorder in IC such as an epithelial leak, and mast cells may be part of a defense mechanism that may ultimately become part of the problem. The conflicting reports in the literature indicate that mast cells are not uniformly increased in IC. This may be due to suboptimal staining techniques, although the large standard deviations suggest subgroups of patients in whom bladder mastocytosis is prevalent [32]. Mast cells are prominent in 30% of patients and degranulation may be important in the provocation of symptoms, especially in atopic people [31,33–37]. Control of this degranulation may be beneficial therapy in some patients. It is important to reemphasize the potential for multiple etiologies of IC. The presence of mast cells in some patients may prove to be important for diagnosis and therapy in this select subset.

A proliferation of sympathetic bladder nerve fibers occurs in IC [38]. Substance P-containing nerve fibers lying adjacent to mast cells are increased in IC and a close anatomical relationship between bladder neurons and mast cells has been demonstrated [39]. IC patients report symptom exacerbation by stress [9]. Stress affects the development of inflammation, and the contribution of the nervous system to the regulation of immune and inflammatory processes is now accepted, possibly through corticotrophin-releasing hormone (CRH), which has peripheral proinflammatory actions that include mast cell activation. The stress of restraints in animal models results in bladder and intestinal mast cell activation [40]. The relationship of mast cells, substance P, and stress exacerbation may identify a link in the pathogenesis of irritable bowel syndrome (which cooccurs frequently in IC patients) and IC [39].

Premenopausal IC patients frequently report symptom exacerbation perimenstrually or at the time of ovulation and this may lead to a gynecologic evaluation and a diagnosis of endome-

triosis. Estradiol augments bladder mast cell histamine secretion and proliferation of bladder mast cells in animal models and the response of bladder mast cells to substance P *in vitro*. Bladder mast cells in IC have increased expression of high affinity estrogen receptors but only few progesterone receptors [41]. The effects of the female sex hormones on mast cells may explain the high incidence of IC in women and the worsening of symptoms perimenstrually. Thus, IC has many of the characteristics of a neuroimmunoendocrine disorder. It may be important to address mast cell abnormalities during therapy for IC, and this multifactorial etiology for IC supports the need for a combinational therapeutic approach to control the disease.

It has been suggested that IC is a psychological disorder. Many patients with IC are depressed [8]. Those suffering from severe nocturia will exhibit even more profound depression due to sleep deprivation. Our experience and that of earlier researchers suggest that no one has been cured of IC by psychotherapy [13]. It is important to emphasize that treating depression can improve overall sense of well-being and help patients cope with their disease, but it will not cure the IC or reduce the number of daily voids. Acute stress may flare IC symptoms [9] and stress reduction may improve symptoms. Stress factors could be emotional or physical, such as viral infections, exercise, travel in a car or plane, or jogging. It is important for the patient that the physician emphasizes that IC is not a psychological disorder.

VI. Diagnosis

A. History

The cornerstone symptoms of interstitial cystitis are significant sensory urgency and urinary frequency with no identifiable cause. The symptoms of interstitial cystitis are usually attributed to bacterial infection; however, only 30% of women have positive bacterial cultures, and therefore over 70% of women diagnosed with cystitis have a nonbacterial problem [5]. A history of pelvic radiation or surgery should be elicited. Risk factors for bladder carcinoma including tobacco use and occupational exposure to carcinogens (industrial dyes, etc.) should be noted. Any episodes of gross hematuria should be documented.

Most patients will have associated bladder pain. One study of over 200 patients reported that 15% of IC patients had little to no bladder pain whereas 85% of patients presented with significant pain [42]. It is important, but often difficult, to determine whether or not the pain is of bladder origin. Ask the patient if the pain, despite being constantly present, worsens if the bladder is not emptied and diminishes in intensity with voiding. The bladder pain of IC may be experienced in the suprapubic region, perineum, vagina, lower back, or medial thigh [13]. Two-thirds of patients do not experience dysuria.

Nocturia is variable, but in general, 90% of patients complain of voiding at least one to two times per night [42]. Nocturia increases with the severity and duration of the disease. The average patient voids approximately 16 times per day; a minimum for diagnostic purposes is considered to be eight voids per day [1]. The average voided volume is 75 ml.

Between 85 and 90% of individuals with IC are female. Of those who are sexually active, the majority (75%) will complain of exacerbation of the symptom complex associated with sexual intercourse [42]. The increase in symptoms may be felt during sexual activity, immediately after, or within 24 hours. In addition, most women who are still menstruating will complain of a flare of symptoms several days prior to the onset of the menstrual cycle [13,42].

Duration of symptoms helps to define patients with IC versus those with urgency–frequency syndrome (UFS). The diagnosis of IC is more likely if the individual has had the presence of continuous symptoms for at least six months. Differentiating between IC and UFS is worthwhile because UFS may need little or no therapy and the prognosis for the patient is good.

B. Voiding Diary

An accurate assessment of the number of daily voids and average volume is determined from a 3-day voiding log where each void is measured and recorded by the patient at home. It was found that the average IC patient voids 16 times per day with a capacity of 73 ml compared to age-matched controls, who void an average of 270 ml 6 times per day; patients averaged five episodes of nocturia per night [1]. The voiding profile is a useful method to help establish the diagnosis of IC and may be used subsequently to create a therapeutic plan and to monitor progress in therapy. As might be anticipated, patients with a longer disease history have a smaller functional bladder capacity as reflected in a lower average voided volume and a higher number of daily voids.

C. Physical Examination

There is one important part of the examination that helps confirm the diagnosis of IC. On physical examination, over 95% of patients will complain of a tender bladder base during the pelvic examination. This discomfort is easily demonstrated by palpation of the anterior vaginal wall.

D. Laboratory Tests

Urine analysis on voided specimens is not useful in IC patients because their low voided volumes make midstream collection impossible. One sees only vaginal secretions unless a catheterized specimen is obtained. A catheterized specimen examined under the microscope should show no bacteria, and most will show no red or white blood cells. A urine culture should be sent if the urinalysis suggests infection. Urine should be sent for cytological evaluation to exclude carcinoma *in situ*. Patients presenting with hematuria are rare but require a full urological work-up to exclude malignancy.

E. Urodynamic Testing

The cystometrogram (CMG) is a valuable study to perform in patients with IC because a normal study essentially excludes the diagnosis of IC. Significant urinary urgency can usually be documented with cystometry. If gas cystometry is performed, a sensation of significant urgency will be experienced at <125 ml, and with water cystometry at <150 ml. If this portion of the CMG is normal, patients may not have IC or only have a mild form. In 75 patients with cystometrograms reported by Parsons [42], the average bladder capacity was 220 ml, with over 90% of patients having a functional volume of less than 350 ml.

Bladder pain may be provoked by the CMG, and this is an important diagnostic tool. However, there is an important caveat relative to maximum bladder capacity. A small group of patients with significant "end-stage" IC will develop detrusor myopathy (about 5%) [42,43]. Individuals with this complication will have large atonic bladders with little muscle present. They have moderate to severe sensory urgency, large bladder capacities (>1000 ml), and usually carry residual urine (>100 ml). Detrusor function is poor or absent. In fact, many patients with IC have poor muscle function in their bladder and empty only with difficulty. Because most of the patients are female, they are able to void, but primarily with a valsalva maneuver. Due to the detrusor myopathy and generalized atrophy of bladder muscle with this disease, males with low voiding pressure may require a program of clean intermittent catheterization (CIC).

F. Cystoscopic Evaluation

Cystoscopic evaluation of the bladder under anesthesia is primarily important as a therapeutic maneuver. It is not necessary for diagnosis. Examination under local anesthesia is to be discouraged because it offers little help in diagnosis and causes the patient severe discomfort. It is recommended that when IC is suspected, a cystoscopy be performed under anesthesia. It is not recommended that all patients have a cystoscopy. Most patients do not need this procedure unless severe symptoms are present. It is best to omit cystoscopy on milder patients and proceed with other therapies. The diagnosis is confirmed by one of two findings: a Hunner's ulcer or the presence of diffuse glomerulations or petechial hemorrhages. A classic Hunner's ulcer, a velvety-red patch of urothelium, is identified in only 6–8% of IC patients. In healthy individuals, the anesthetized bladder capacity is 900 ml, while in IC patients it is reduced to 550–650 ml.

Since the initial report by Bumpus in 1930, bladder hydrodistention has been a mainstay of therapy of IC [44]. Few would question the activity of hydrodistention in ameliorating the symptoms in the majority of IC patients. The procedure must be performed under anesthesia because it is not possible to dilate a painful bladder without anesthesia. Pressure dilatation of the bladder using a syringe should not be done because it can result in bladder rupture. A maximum hydrostatic pressure of 80–100 cm H_2O pressure is recommended.

The mechanism by which hydrodistention improves symptoms is unknown; several theories have been postulated. Neuropraxis induced by mechanical trauma may occur in some individuals. However, few patients awaken with decreased pain, which contradicts the neuropraxis concept. Most patients awaken from anesthesia with significantly worse pain that slowly improves over 2–3 weeks. This pain usually requires narcotic analgesia. Remission will occur in the majority of patients over the course of several weeks. As a result of the increased pain, it is recommended that all patients receive belladonna and opium rectal suppositories immediately in the recovery room, or, better yet, instillation of 10 ml of 2% viscous xylocaine into the bladder at the end of hydrodistention. In addition, they should be discharged with narcotic medication to control the increased pain.

We believe that the exacerbation in patient symptoms with hydrodistention is due to epithelial damage from the mechanical trauma. The disruption in the integrity of the mucosal cells increases the epithelial leak, causing symptoms to flare. Healing may occur over the next several weeks which correlates with the time of clinical remission. Perhaps the epithelium regenerates and for a period of time is "healthy" and impermeable. Then, whatever events initiated the disease resurface to cause a relapse.

Remissions may persist between 4 and 12 months, following which hydrodistention may be repeated as needed. If no remission is obtained, hydrodilation should be repeated at least two more times because frequently, in our experience, patients respond to a subsequent dilation.

Random bladder biopsies are performed at the completion of cystoscopy. Biopsies should not be done before distension so as to avoid bladder perforation. The biopsy itself is not diagnostic for IC but can rule out other diseases such as carcinoma *in situ*. The findings on pathologic examination may include the presence of mast cells on toluidine blue staining and a thinned mucosa [45,46]. A normal biopsy does not exclude IC and should not be so utilized in diagnosis. Conversely, no pathologic findings specifically make a diagnosis of IC.

While diagnosis of IC depends in part on abnormal cystoscopic findings, one cannot arbitrarily rule out the disease purely by the endoscopic findings. There are many patients who have IC without such findings who will benefit from therapy. We emphasize that the primary manifestation of this disease complex is significant urinary urgency or frequency and perhaps few or no other findings.

G. The Potassium Test

A simple method has been devised by Parsons (the Parsons Test) to measure epithelial permeability. The test is based on the hypothesis that if one places a solution of KCl into a normal bladder, it provokes no symptoms of urgency or pain. On the other hand, if placed into a bladder that has an impaired mechanism to maintain the impermeable epithelium, then the potassium diffuses across the transitional cells to stimulate sensory nerves and cause urgency or pain.

To perform the test, two solutions are placed into the bladder for 3–5 minutes each using a regular Foley catheter. Solution 1 is 40 ml of sterile water and Solution 2 is 40 ml of KCl (40 mEq in 100 ml water). The total volume of 40 ml is used to reduce stimulation due to volume sensitivity. Solution 1 is then drained and replaced by Solution 2. Symptoms of pain and urgency are graded on a scale of 1–5 before and after instillation of each solution.

If the patient does not respond to water and states the KCl solution is causing their symptoms of pain and/or urgency to increase by an increment of 2 or more on the graded symptom scale, this is considered a positive test. In one study, 70% of patients [29] had provocation of symptoms while only 4% of control subjects responded. It is a useful test for an abnormal permeability barrier, as separation of epithelial leakers from nonleakers may help direct therapy.

The test may be positive in patients with radiation cystitis or bacterial cystitis. It will also be positive in 25% of patients with

detrusor instability and 4% of normal individuals. Response to potassium instillation may be muted by recent hydrodilation or instillation therapy.

VII. Therapy

No therapy is uniformly successful in IC. There are three basic approaches to treat IC. One utilizes antidepressants, antispasmodics, and antihistamines. A second utilizes cytodestructive techniques including dilation, DMSO, chlorpactin, silver nitrate, and BCG. Last, cytoprotective techniques utilize heparin and pentosanpolysulfate. Frequently these therapies must be combined; partial remissions are often achieved with pentosanpolysulfate, heparin, and intravesical dimethylsulfoxide (DMSO).

Few advances had been made in therapy for IC until recently. Most medications were employed empirically, and all were studied without controls. When discussing therapy with the individual patient, it is important for the physician to emphasize to the patient that if the symptoms have been present for more than a year, no particular therapy is likely to be curative. While he or she may have a significant remission of symptoms, in all probability, relapse will occur. If patients are prepared for this eventuality, they are much less distressed when symptoms return and cope better with their disease. The physician–patient relationship is strengthened in terms of credibility if this area is addressed prior to initiating treatment. Patients readily accept this explanation and overall appear to adjust to their disorder when their outlook is realistic.

A. Oral Therapy

1. Antihistamines

Antihistamines are critical to managing IC in people with hay fever, sinusitis, or food allergies. Patients in good control of their symptoms will breakthrough in allergy season. Antihistamines have been evaluated in IC but without controlled studies. Antihistamines were chosen because of the possible role of mast cells in the pathogenesis of the disease [46–49]. While most patients may not respond to antihistamines, subsets of patients seem to have major benefit, especially when combined with other therapy. The beneficial effect of antihistamines is more pronounced in IC patients with a history of allergies, increased numbers of bladder mast cells, or elevated urine levels of mast cell mediators [35].

The heterocyclic piperazine H1-receptor antagonist, hydroxyzine, has considerable benefit when given at 25–75 mg/day for more than 3 months [50]. Beneficial effects appear 2–3 months after start of treatment and patients are urged to stay on medication for at least 3 months to determine effectiveness. The beneficial effects of hydroxyzine are attributed to its ability to block neurogenic stimulation of bladder mast cells, in addition to anticholinergic, anxiolytic, and sedative properties.

The anti-allergic mast cell "stabilizer" drug disodium cromoglycate (cromolyn) had been tried in chemical cystitis and pretreatment appeared to reduce the inflammatory response. Unfortunately, intravesical cromolyn does not have significant therapeutic efficacy in IC, probably because it is not very soluble and does not inhibit mucosal mast cells [51].

2. Antidepressants

Chronic pain and sleep loss can cause depression. Thus, it is valuable to place most IC patients with moderate or severe symptoms on antidepressant medications. Tricyclic antidepressants have several modes of action that are beneficial. They have side effects of drowsiness (to aid sleep), increased pain thresholds, and elevation of mood. If tricyclic antidepressants are used, one starts with low doses and warns patients that they will be fatigued for 12–15 hours per day for the first 2–3 weeks of therapy. Once they become tolerant to this side effect, the dosage is increased if needed.

Amitriptyline or imipramine can be prescribed in doses of 10–25 mg one hour before bedtime [52]. The tricyclic antidepressant amitriptyline reduces the symptoms of IC in about 30% of patients, an action which may be due to its ability to inhibit histamine secretion from mast cells [51]. In an uncontrolled trial, amitriptyline was reported by Hanno *et al.* [52] to ameliorate the symptoms of IC. Patients were treated with 25 mg of amitriptyline one hour before bedtime for one week; the dose was then increased weekly by 25 mg to 75 mg. Fifty percent of patients responded to this medication. The exact mechanism of action of amitriptyline is unknown, although it may block H1 histamine receptors and perhaps mast cell degranulation. More likely the drug raises pain tolerance due to its antidepressant activity. Fluoxetine (Prozac) 20–40 mg per day or sertraline (Zoloft) 50–100 mg per day may also be prescribed.

Antidepressant therapy is an important adjunct to treatment. It does not cure IC, but patients function much better with their disabling symptoms if not depressed. In essence, they "feel better" even if they still void 20 times per day. Epithelial nonleakers respond best to antidepressants. Surprisingly, many patients (about 25–30%) improve dramatically with antidepressant monotherapy. We place all moderately or severely symptomatic patients on antidepressants until other treatment modalities such as heparin or pentosanpolysulfate have been successfully initiated.

3. Alkalinizers

Polycitra is an oral agent that not only alkalinizes the urine but also binds potassium. Both effects may be beneficial in IC and it is recommended that patients receive a trial of therapy for 3–6 months. Employing two doses a day of medication appears to be sufficient. In general, this drug should be combined with other treatments such as heparin, pentosanpolysulfate, or amitriptyline to obtain the best effect.

4. Arginine

Smith and associates reported the use of L-arginine for patients with IC with beneficial effects [53]. It was used in a limited open-phase study and its true activity is unknown.

5. Pentosanpolysulfate

Parsons *et al.* [54,55] first reported pentosanpolysulfate as active in ameliorating the symptoms of IC. Because pentosanpolysulfate (Elmiron) is a sulfated polysaccharide, theoretically it may augment the bladder surface defense mechanism or detoxify urine agents such as quaternary amines that may disrupt the bladder epithelial surface. In a controlled clinical study, 42% of patients were shown to have their symptoms controlled versus 20% for placebo [54]. This finding has been borne out in several subsequent studies, including a five-center trial where 28% of patients on Elmiron compared to 13% on placebo improved [55]. In a seven-center study of 150 patients there was a 32% patient improvement on drug versus 15% on placebo [56]. Additionally, an English–Danish study also found a significant reduction of pain in patients on Elmiron compared to placebo [33]. It is employed in an oral dose of 100 or 200 mg, three times a day. Males respond to higher dosages. Responses may require 3–6 months of therapy, and the medication is well-tolerated. In patients with moderate disease, it appears to have about 40–50% activity. In the controlled clinical trials that were done on patients with severe disease, its activity was lower. Continued use of Elmiron for several years leads to long-term disease control in most of the initial responders. This pattern has not been previously found with any other therapy except heparin.

6. Corticosteroids

Proponents of an inflammatory etiology for IC support the therapeutic role of steroids. Badenoch found significant improvement in 19 of 25 patients treated with prednisone [57]. However, all were treated following hydrodistention under anesthesia which may have been responsible for most of the benefit. As with most drugs, there have been no controlled clinical trials conducted on the efficacy of steroids in the treatment of IC.

B. Intravesical Therapy

1. Cytodestructive

a. DIMETHYL SULFOXIDE. Dimethylsulfoxide (DMSO) was approved for use in IC in 1977 [58]. While no controlled clinical trials ever were conducted with DMSO, it does appear to induce remission in 34–40% of IC patients. The difficulty with DMSO is that it may induce an excellent remission in the first one to three cycles of therapy, but as an individual relapses and requires subsequent treatment, progressive resistance to its beneficial effects is seen.

For treatment, 50 ml of 50% DMSO are instilled into the bladder for 5–10 minutes. Longer periods are unnecessary because DMSO rapidly absorbs into the bloodstream. Instillations are performed on an outpatient basis or the patient can be taught self-instillation. Patients receive 6–8 weekly treatments to determine whether a therapeutic response is achieved. If the patient has moderate or severe symptoms, the therapy is continued for an additional 4–6 months on alternative weeks. Once DMSO therapy is discontinued, the patient will likely become resistant to its use.

Some patients will experience a flare of symptoms when DMSO is placed into the bladder. This phenomenon may be related to DMSO's ability to degranulate mast cells and may occur primarily in patients who have significant bladder mastocytosis. Nonetheless, DMSO may be very effective in treating these patients. Should the patient experience pain with DMSO, it is recommended that he or she receive 10 ml of intravesical 2% viscous xylocaine jelly 15 minutes before instilling DMSO.

If this is not successful, then the use of an injectable narcotic or Toradol 60 mg intramuscularly before the intravesical instillation is indicated. The flare of symptoms associated with DMSO usually disappears over 24 hours. As these patients receive subsequent treatments, the pain tends to diminish.

Patients may receive indefinite therapy using DMSO. As originally reported by Stewart [58], patients have used DMSO weekly for several years without difficulty. DMSO has been reported to be associated with cataracts in animals; however, this complication has not been reported in humans. If the patient is maintained on chronic therapy, it is recommended that he or she have a slit lamp evaluation at 3–6 month intervals.

b. SILVER NITRATE.   Intravesical silver nitrate was first reported in 1926 by Dodson [59]. Pool used a treatment regimen of bladder irrigations that were begun under anesthesia with 1:5000 concentration and gradually increased on a daily basis to a 1% solution [60]. This study was uncontrolled, with concurrent hydrodilation of the bladder under anesthesia. Pool reported good results in 89% of patients. There have been other uncontrolled studies reporting that this compound is helpful; nevertheless, it is not widely used today. One caution is that silver nitrate should never be instilled after a bladder biopsy, as intraperitoneal or extraperitoneal extravasation could result in serious complications.

c. SODIUM OXYCHLOROSENE (CLORPACTIN).   Clorpactin is a highly reactive chemical compound that is a modified derivative of hypochlorous acid in a buffered base. Its activity depends on the liberation of hypochlorous acid with resulting detergent and oxidizing effects [61]. Wishard [61] treated 20 patients with five weekly instillations of 0.2% Clorpactin WCS-90 under local anesthesia. Improvement was reported in 14 of the 20 patients, but follow-up was brief. Messing and Stamey [62] treated 38 patients with 0.4% Clorpactin and reported significant improvement in 72%. Ureteral reflux is a contraindication to the use of Clorpactin. It is recommended that the compound usually be instilled under anesthesia.

d. BCG (BACILLUS CALMETTE-GUERIN).   BCG therapy for IC has been reported to be efficacious [63,64]. However, because of its potential infectious side-effects, it is recommended as a second-line agent. BCG causes an intense desquamation of the bladder mucosa.

2. Cytoprotective

One major breakthrough in therapy is the use of heparin-like drugs (heparin, pentosanpolysulfate) which, when effective, will reverse the course of the disease by restoring the glycosaminoglycan layer. Patients rarely become resistant to their use. Heparin-like drugs may take 3–4 months for a response, depending on how advanced the disease is. However, if patients respond initially, they usually continue to experience prolonged remissions and progressive improvement in symptoms and bladder function. A badly damaged bladder may take three years to recover adequate function.

a. HEPARIN.   Parenteral heparin has been reported to alleviate the symptoms of IC in an uncontrolled study [65]. Chronic sys-temic heparin therapy cannot be employed in most individuals as it results in osteoporosis in 100% of patients who use it for 26 weeks.

Intravesical heparin has a significant activity in approximately 50% of patients; however, we have noted a placebo effect of approximately 20% [54,66]. Heparin (20,000 IU) is initially instilled intravesically on a daily basis for 3–4 months. The frequency is then reduced to 3–4 times per week. This treatment can be carried on indefinitely. Oral pentosanpolysulfate may be given simultaneously. It may take 2–4 months before improvements are noticed, but patients should be encouraged to continue therapy for at least 6 months. The best improvements are noted after 1–2 years, and long-term therapy is recommended for patients with moderate or worse disease who respond to its use. Serum prothrombin time (PT) and partial thromboplastin time (PTT) and platelets are monitored for several weeks after therapy begins to rule out the formation of an unusual antibody to heparin or systemic absorption. Patients are instructed in self-catheterization so this therapy can be performed at home.

C. Surgical Therapy

Approximately 2% of patients presenting with IC to the University of California, San Diego Medical Center have ultimately undergone some type of surgery for disease that is severe and refractory to all medical treatment. A variety of surgical procedures may be considered.

1. Bladder Deinnervation (Ingelman-Sanberg Procedure)

Attempts at surgical ablation of bladder innervation by cystolysis are to be discouraged because most patients will fail this and develop a neurogenic bladder with significant urinary pain, frequency, and perhaps urinary retention and the need for intermittent catheterization.

2. Bladder Augmentation

IC patients were once thought to have small bladders that lead to frequent voiding, but the reverse is true. Sensory urgency stimulates frequent voiding which subsequently leads to the development of a small dysfunctional bladder. Hence, attempts to augment the bladder with a patch of bowel are likely to fail. Patients will then have a capacity that is perhaps large, have more difficulty emptying (usually requiring intermittent catheterization), but still retain all their sensory urgency and pain [67].

3. Urinary Diversion

There are no controlled studies evaluating diversion alone, but studies suggest it is not effective [68]. We have encountered two patients who required secondary cystectomy following urinary diversions due to persistent pelvic pain. The pain was eliminated by removal of the bladder. In counseling, one should tell patients that diversion alone may not be sufficient to control their pain and they may subsequently require a cystectomy. The patient can then decide whether or not they want to risk more than one surgery. Few patients elect the potential of two surgeries.

4. Cystectomy

Cystectomy is the mainstay of therapy for patients with "end-stage bladder." Patient acceptance has increased with the current enthusiasm for continent urinary diversions. Pelvic pain will present after the procedure in 5% of patients. In general, if the patients have classic IC symptoms and the usual stigmata of IC under anesthesia, they are likely to have relief of their symptoms by cystectomy. Those individuals with severe pelvic pain not associated with classic parameters of IC and particularly not exacerbated by bladder filling will be unlikely to have their pain alleviated.

When continent diversion is performed, 20–30% of patients will develop pouch pain 6–36 months after surgery. This pain can be managed successfully by having the patient instill 10,000 units of heparin in 10 ml water into the pouch after each catheterization.

D. Biofeedback and Bladder Training

Whatever therapy is successful at alleviating the pain and sensory urgency of IC, the individual afflicted with the chronic form of the disorder will have a small capacity bladder that is in part the result of sensory urgency and in part due to frequent low-volume voiding. In controlled clinical trials, it has been reported that even with good remission of pain and urgency, there is almost no change in urinary frequency over a 12-week period [54–56]. This issue must be addressed in order to obtain a functional recovery of the bladder. Persistent urinary frequency from a small bladder can be reversed after therapy has controlled urgency and pain. This result is accomplished by training the patients to undergo a program of progressively holding their urine to gradually increase their bladder capacity and voiding interval [69]. This therapy can be directed by a urological nurse.

To begin this treatment, obtain a 3-day voiding diary from the patient (to include time of voiding and a measurement of volume). Determine the average time interval between voids, and gradually increase this interval monthly. For example, if the patient voids every hour, it is recommended that he or she attempt to void every hour and a quarter and at the end of one month increase that to an hour and a half. The patient should never progress too quickly because they will become discouraged and compliance will decrease. It takes 3–5 months of this protocol to start to achieve results, at which time the bladder capacity will increase approximately 2.5 times and there will be a corresponding reduction in urgency and the number of voids per day. In patients who have minimal or no pain associated with their urinary frequency, bladder training may be the only therapy indicated to improve their symptoms.

E. Dietary and Sexual Behavior Modifications

Many IC patients find that modifying their diet helps them control symptoms and avoid periods of symptom exacerbations. Diet does not cause the illness but may exacerbate it. Determining which foods may cause problems requires perseverance. In one survey, 50% of patients with IC reported that acidic, alcoholic, or carbonated beverages, and coffee or tea increased their IC pain [70].

Some of the suggested items to restrict or avoid include aged cheeses, sour cream, yogurt, chocolate, fava beans, lima beans, onions, tofu, tomatoes, apples, apricots, avocados, bananas, cantaloupes, citrus fruits, cranberries, grapes, nectarines, peaches,

Table 88.1
Symptom-Stratified Therapy

Mild patients:
 Polycitra
 Elmiron
 DMSO for 3 months may induce long remission
 If unsuccessful, heparin
Moderate-severe patients:
 Polycitra
 Elavil, Elmiron, and/or daily intravesical heparin
 When improved, slowly taper heparin
Severe patients:
 Daily intravesical heparin, Elavil, Elmiron

pineapples, plums, pomegranates, rhubarb, strawberries, rye and sourdough bread, most nuts, alcoholic beverages, carbonated drinks, coffee, tea, mayonnaise, miso, spicy foods, soy sauce, vinegar, salad dressing, smoked meats, smoked fish, canned or processed meats and fish, corned beef, citric acid, monosodium glutamate, Nutrasweet, saccharine, tobacco, diet pills, recreational drugs, and cold and allergy medications containing ephedrine and pseudoephedrine. Although the list of "forbidden" foods may appear daunting, remember that there are many foods that can be enjoyed. Many IC patients report little trouble with rice, potatoes, pasta, vegetables, meat, or chicken. The patients are instructed to add some of the forbidden items every 5 to 7 days to their diet and determine if their symptoms are exacerbated.

IC can also have a disruptive effect on sexuality and relationships [71]. Sometimes, IC patients and their partners avoid sex entirely because they are afraid of causing pain or an exacerbation of their symptoms. Good partner communication is essential for solving sexual problems. Otherwise, partners may be unable to make the necessary adjustments in their sexual activities. By far the most common sexual problem among IC patients, whether they are heterosexual or homosexual, is pain before, during, or after sexual activity. Intercourse seems to cause the most discomfort. Experimenting with sexual positions may reduce the pain. In some IC patients, sexual arousal may intensify symptoms. Cuddling, sensate focusing, meditation, and caressing may be helpful.

VIII. Summary

A. Treatment Algorithm

Treatment selection can be stratified based on either the response to the potassium test and/or the severity of symptoms. Potassium-positive patients respond best to heparin, pentosanpolysulfate, Polycitra, hydrodilation, and DMSO instillation therapy. Potassium-negative patients respond best to antidepressants and possibly pentosanpolysulfate; however, they represent the major challenge for the future as they respond poorly to intravesical therapy. Nerve stimulation and acupuncture may be alternative routes to relief for these patients. Hydroxyzine is added in those patients with a history of allergies. Symptom directed stratification is listed in Table 88.1.

IX. Conclusions

Interest in interstitial cystitis has increased in recent years. Advances in our understanding and management of the illness along with increased public and physician awareness will hopefully encourage more patients to seek help. The apparent preponderance of women with IC requires further study. One possible explanatory model involves the effects of female sex hormones on mast cells, and mast cells appear to be increased in a subset of patients with IC. By utilizing the suggested treatment modalities, 75–85% of patients with moderate to severe IC can experience significant indefinite remissions with conservative therapy and avoid the need for extirpative surgery. Early recognition and intervention are crucial to achieve positive results for patients' physical, psychological, and social well-being.

References

1. Gillenwater, J. Y., and Wein, A. J. (1988). Summary of the National Institute of Arthritis, Diabetes, Digestive and Kidney Diseases Workshop on Interstitial Cystitis, National Institutes of Health, Bethesda, Maryland, August 28–29, 1987. *J. Urol.* **140,** 203–206.
2. Oravisto, K. J. (1975). Epidemiology of interstitial cystitis. *Ann. Chir. Gynaecol. Fenn.* **64,** 75–77.
3. Bade, J. J., Rijcken, B., and Mensink, H. J. (1995). Interstitial cystitis in the Netherlands: Prevalence, diagnostic criteria, and therapeutic preferences. *J. Urol.* **154,** 2035–2037.
4. Held, P. J., Hanno, P. M., Wein, A. J., Pauly, M. V., and Cahn, M. A. (1990). Epidemiology of interstitial cystitis: 2. *In* "Interstitial Cystitis" (P. M. Hanno, D. R. Staskin, R. J. Krane, and A. J. Wein, eds.), pp. 29–48. Springer-Verlag, London.
4a. Jones, C. A., and Nyberg, L. (1997). Epidemiology of interstitial cystitis. *Urology* **49** (Suppl. 5A), 2–9.
5. Hamilton-Miller, J. M. T. (1994). The urethral syndrome and its management. *J. Antimicrob. Chemo.* **33**(Suppl. A), 63–73.
6. Koziol, J. A. (1994). Epidemiology of interstitial cystitis. *Urol. Clin. North Am.* **21,** 7–20.
7. Ratner, V., Slade, D., and Greene, G. (1994). Interstitial cystitis—a patient's perspective. *Urol. Clin. North Am.* **21,** 1–5.
8. Hanno, P., Levin, R. M., Monson, F. C., Tenscher, C., Zhou, Z. Z., Ruggieri, M., Whitmore, K., and Wein, A. J. (1990). Diagnosis of interstitial cystitis. *J. Urol.* **143,** 278–281.
9. Koziol, J. A., Clark, D. C., Gittes, R. F., and Tan, E. M. (1993). The natural history of interstitial cystitis: A survey of 374 patients. *J. Urol.* **149,** 465–469.
10. El-Mansoury, M., Boucher, W., Sant, G. R., and Theoharides, T. C. (1994). Increased urine histamine and methylhistamine in interstitial cystitis. *J. Urol.* **152,** 350–353.
11. Oravisto, K. J. (1990). Epidemiology of interstitial cystitis 1. *In* "Interstitial Cystitis" (P. M. Hanno, D. R. Staskin, R. J. Krane, and A. J. Wein, eds.), pp. 25–28. Springer-Verlag, London.
12. Geist, R. W., and Antolak, S. J. (1970). Interstitial cystitis in children. *J. Urol.* **104,** 922–925.
13. Hand, J. R. (1949). Interstitial cystitis, a report of 223 cases. *J. Urol.* **61,** 291.
14. Wein, A. J., and Broderick, G. A. (1994). Interstitial cystitis: Current and future approaches to diagnosis and treatment. *Urol. Clin. North Am.* **21,** 153–161.
15. Hunner, G. L. (1915). A rare type of bladder ulcer in women: Report of cases. *Boston Med. Surg. J.* **172,** 660–664.
16. Hunner, G. L. (1918). Elusive ulcer of the bladder: Further notes on a rare type of bladder ulcer with a report of 25 cases. *Am. J. Obstet.* **78,** 374.

17. Oravisto, K. J., Alfthan, O. S., and Jokinen, E. J. (1970). Interstitial cystitis. Clinical and immunological findings. *Scand. J. Urol. Nephrol.* **4,** 37–42.
18. Hanash, K. A., and Pool, T. L. (1970). Interstitial and hemorrhagic cystitis: Viral, bacterial and fungal studies. *J. Urol.* **104,** 705–706.
19. Oravisto, K. J., and Alfthan, O. S. (1976). Treatment of interstitial cystitis with immunosuppression and chloroquine derivatives. *Eur. Urol.* **2,** 82–84.
20. Silk, M. R. (1970). Bladder antibodies in interstitial cystitis. *J. Urol.* **103,** 307–309.
21. Holm-Bentzen, M., and Lose, G. (1987). Pathology and pathogenesis of interstitial cystitis. *Urology* **29**(4, Suppl.), 8–13.
22. Oravisto, K. J. (1990). Interstitial cystitis as an autoimmune disease. A review. *Eur. Urol.* **6,** 10–13.
23. Weaver, R. G., Dougherty, T. F., and Natoli, C. (1963). Recent concepts of interstitial cystitis. *J. Urol.* **89,** 377.
24. Eldrup, J., Thorup, J., Nielsen, S. L., Hald, T., and Hainau, B. (1983). Permeability and ultrastructure of human bladder epithelium. *Br. J. Urol.* **55,** 488–492.
25. Parsons, C. L., Lilly, J. D., and Stein, P. (1991). Epithelial dysfunction in non-bacterial cystitis (interstitial cystitis). *J. Urol.* **145,** 732–735.
26. Lilly, J. D., and Parsons, C. L. (1990). Bladder surface glycosaminoglycans: A human epithelial permeability barrier. *Surg., Gynecol. Obstet.* **171,** 493–496.
27. Parsons, C. L., Boychuk, D., Jones, S., Hurst, R., and Callahan, H. (1990) Bladder surface glycosaminoglycans: An epithelial permeability barrier. *J. Urol.* **143,** 139–142.
28. Hohlbrugger, G., and Lentsch, P. (1985). Intravesical ions, osmolality and pH influence the volume pressure response in the normal rat bladder, and this is more pronounced after DMSO exposure. *Eur. Urol.* **11,** 127–130.
29. Parsons, C. L., Stein, P. C., Bidair, M., and Lebow, D. (1994). Abnormal sensitivity to intravesical potassium in interstitial cystitis and radiation cystitis. *Neurourol. Urodyn.* **13,** 515–520.
30. MacDermott, J. P., Miller, C. H., Levy, N., and Stone, A. R. (1991). Cellular immunity in interstitial cystitis. *J. Urol.* **145,** 274–278.
31. Holm-Bentzen, J., Halt, T., and Sondergaard, I. (1986). Urinary excretion of a metabolite of histamine (1,4-methyl-imidazole-acetic-acid). *J. Urol.* **135,** 187.
32. Theoharides, T. C., Sant, G. R., El-Mansoury, M., Letourneau, R. I., Ucci, A. A., Jr., and Meares, E. M., Jr. (1995). Activation of bladder mast cells in interstitial cystitis: A light and electron microscopic study. *J. Urol.* **153,** 629–636.
33. Holm-Bentzen, M., Jacobsen, F., Nerstrom, B., Lose, G., Kristensen, J. K., Pedersen, R. H., Krarup, T., Feggether, J., Bates, P., Barnard, R., Larsen, S., and Hald, T. (1987). Painful bladder disease: Clinical and pathoanatomical differences in 115 patients. *J. Urol.* **138,** 500.
34. Lotz, M., Villiger, P. M., Hugli, T., Koziol, J., and Zuraw, B. L. (1994). Interleukin-6 and interstitial cystitis. *J. Urol.* **152,** 869–873.
35. Kastrup, J., Hald, J., and Larsen, L. (1983). Histamine content and mast cell count of detrusor muscle in patients with interstitial cystitis and other types of chronic cystitis. *Br. J. Urol.* **55,** 495–500.
36. Sant, G. R. (1989). Interstitial cystitis: Pathophysiology, clinical evaluation and treatment. *Ann. Urol.* **3,** 172–179.
37. Sant, G. R., Kalaru, P., and Ucci, A. A., Jr. (1988). Mucosal mast cell (MMC) contribution to bladder mastocytosis in interstitial cystitis. *J. Urol.* **139,** 276A.
38. Thompson, A. C., and Christmas, T. J. (1996). Interstitial cystitis—an update. *Br. J. Urol.* **78,** 813–820.
39. Pang, X., Boucher, W., Triadafilopoulos, G., Sant, G. R., and Theoharides, T. C. (1996). Mast cell and substance P-positive nerve involvement in a patient with both irritable bowel syndrome and interstitial cystitis. *Urology* **47,** 436–438.

40. Spanos, C., Pang, X., Ligris, K., Letourneau, R. J., Alferes, L., Alexacos, N., Sant, G. R., and Theoharides, T. C. (1997). Stress-induced bladder mast cell activation: Implications for interstitial cystitis. *J. Urol.* **157,** 669–672.

41. Pang, X., Cotreau-Bibbo, M. M., Sant, G. R., and Theoharides, T. C. (1995). Bladder mast cell expression of high affinity estrogen receptors in patients with interstitial cystitis. *Br. J. Urol.* **75,** 154–161.

42. Parsons, C. L. (1990). Interstitial cystitis: Clinical manifestations and diagnostic criteria in over 200 cases. *Neurourol. Urodyn.* **9,** 241–250.

43. Holm-Bentzen, M., Larsen, S., Hainau, B., and Hald, T. (1985). Non-obstructive detrusor myopathy in a group of patients with chronic bacterial cystitis. *Scand. J. Urol. Nephrol.* **19,** 21.

44. Bumpus, H. C. (1930). Interstitial cystitis. *Med. Clin. North Am.* **13,** 1495.

45. Theoharides, T. C., and Sant, G. R. (1991). Bladder mast cell activation in interstitial cystitis. *Semin. Urol.* **9,** 74–87.

46. Larsen, S., Thompson, S. A., Hald, T., Barnard, R. J., Gilpin, C. J., Dixon, J. S., and Gosling, J. A. (1982). Mast cells in interstitial cystitis. *Br. J. Urol.* **54,** 283.

47. Smith, B. H., and Dehner, L. P. (1972). Chronic ulcerating interstitial cystitis (Hunner's ulcer). *Arch. Pathol.* **93,** 76–81.

48. Bohne, A. W., Hodson, J. M., Rebuck, J. W., and Reinhard, R. E. (1962). An abnormal leukocyte response in interstitial cystitis. *J. Urol.* **88,** 387.

49. Simmons, J. L. (1961). Interstitial cystitis: An explanation for the beneficial effect of an antihistamine. *J. Urol.* **85,** 149.

50. Theoharides, T. (1994). Hydroxyzine in the treatment of interstitial cystitis. *Urol. Clin. North Am.* **21,** 113–119.

51. Theoharides, T. C. (1996). The mast cell: A neuroimmunoendocrine master player. *Int. J. Tissue React.* **18,** 1–21.

52. Hanno, P. M., Buehler, J., and Wein, A. J. (1989). Use of amitriptyline in the treatment of interstitial cystitis. *J. Urol.* **141,** 846–848.

53. Smith, S. D., Wheeler, M. A., Foster, H. E., Jr., and Weiss, R. M. (1997). Improvement in interstitial cystitis symptom scores during treatment with oral L-arginine. *J. Urol.* **158**(3, Pt. 1), 703–708.

54. Parsons, C. L., and Mulholland, S. (1987). Successful therapy of interstitial cystitis with pentosanpolysulfate. *J. Urol.* **138,** 513–516.

55. Mulholland, S. G., Hanno, P., Parsons, C. L., Sant, G. R., and Staskin, D. R. (1990). Pentosan polysulfate sodium for therapy of interstitial cystitis: A double-blind placebo-controlled clinical study. *Urology* **35,** 552–558.

56. Parsons, C. L., Benson, G., Childs, S. J., Hanno, P., Sant, G. R., and Webster, G. (1993). A quantitatively controlled method to prospectively study interstitial cystitis and which demonstrates the efficacy of pentosanpolysulfate. *J. Urol.* **150,** 845–848.

57. Badenoch, A. W. (1971). Chronic interstitial cystitis. *Br. J. Urol.* **43,** 718.

58. Stewart, B. H., Persky, L., and Kiser, W. S. (1968). The use of dimethylsulfoxide (DMSO) in the treatment of interstitial cystitis. *J. Urol.* **98,** 671.

59. Dodson, A. I. (1926). Hunner's ulcer of the bladder: a report of 10 cases. *Va. Med. Mon.* **53,** 305.

60. Pool, T. L. (1967). Interstitial cystitis: Clinical considerations and treatment. *Clin. Obstet. Gynecol.* **10,** 185–191.

61. Wishard, W. N., Nourse, M. H., and Mertz, J. H. O. (1957). Use of clorpactin wcs90 for relief of symptoms due to interstitial cystitis. *J. Urol.* **77,** 420.

62. Messing, E. M., and Stamey, T. A. (1978). Interstitial cystitis: Early diagnosis, pathology, and treatment. *Urology* **12,** 381.

63. Peters, K., Diokno, A., Steinert, B., Yuhico, M., Mitchell, B., Khrota, S., Gillette, B., and Gonzalez, J. (1997). The efficacy of intravesical Tice strain bacillus Calmette-Guerin in the treatment of interstitial cystitis: A double-blind, prospective, placebo controlled trial. *J. Urol.* **157,** 2090–2094.

64. Peters, K. M., Diokno, A. C., Steinert, B. W., and Gonzalez, J. A. (1998). The efficacy of intravesical bacillus Calmette-Guerin in the treatment of interstitial cystitis: Long-term follow up. *J. Urol.* **159**(5), 1483–1486; discussion: pp. 1486–1487.

65. Lose, G., Frandsen, B., and Hojensgard, J. C. (1983). Chronic interstitial cystitis: Increased levels of eosinophil cationic protein in serum and urine and an ameliorating effect of subcutaneous heparin. *Scand. J. Urol. Nephrol.* **17,** 159.

66. Parsons, C. L., Housley, T., Schmidt, J. D., and Lebow, D. (1994). Treatment of interstitial cystitis with intravesical heparin. *Br. J. Urol.* **73,** 504–507.

67. Nielsen, K. K., Kromann-Andersen, B., Steven, K., and Hald, T. (1990). Failure of combined supratrigonal cystectomy and Mainz ileocecocystoplasty in intractable interstitial cystitis: Is histology and mast cell count a reliable predictor for the outcome of surgery? *J. Urol.* **144**(2, Pt. 1), 255–258.

68. Eigner, E. G., and Freiha, F. S. (1990). The fate of the remaining bladder following supravesical diversion. *J. Urol.* **144,** 31–33.

69. Parsons, C. L., and Koprowski, P. (1991). Interstitial cystitis: Successful management by a pattern of increasing urinary voiding interval. *Urology* **37,** 207–212.

70. Interstitial Cystitis Association (1993). "IC and Diet," pamphlet. Interstitial Cystitis Association, Rockville, Maryland.

71. Interstitial Cystitis Association (1993). "IC and Sex," pamphlet. Interstitial Cystitis Association, Rockville, Maryland.

89

Temporomandibular Disorders

LINDA LeRESCHE* AND MARK DRANGSHOLT*,†

*Department of Oral Medicine, University of Washington, Seattle, Washington; †Departments of Oral Medicine, Dental Public
Health Sciences and Epidemiology, University of Washington, Seattle, Washington

I. Background

Temporomandibular disorders are a collection of common conditions affecting the temporomandibular joint (TMJ) and the muscles of mastication [1] (Fig. 89.1). These disorders are principally characterized by pain in the temporomandibular region, but the global term "temporomandibular disorders" (TMD) also covers functional problems such as limitations in jaw opening or deviation of the jaw to one side upon opening, as well as joint sounds during jaw function [2]. Frequently, both pain and functional problems are present, but the link between pain and functional problems is not clear. Because the vast majority of patients with TMD who seek care do so because of the pain they experience [3,4], this chapter will focus primarily on temporomandibular pain, rather than the nonpainful conditions of the TMJ, such as TMJ clicking, locking, or limitation of jaw opening.

A number of terms have been used to describe problems in the temporomandibular region, including TMJ syndrome (or simply TMJ), myofascial pain dysfunction syndrome (MPD), craniomandibular disorders (CMD), and temporomandibular pain dysfunction syndrome (TMPDS). This plethora of terminology may stem from the fact that, historically, signs and symptoms in the temporomandibular region have been considered a syndrome. Depending on the practitioner's view regarding which symptoms are the most important elements of the syndrome, as well as hypotheses regarding the origin of the symptoms and philosophy regarding how to best treat the problem, different terminology would be applied. In 1983, the President of the American Dental Association convened a conference on the examination, diagnosis, and management of temporomandibular disorders [2]. This body articulated the idea that symptoms in the temporomandibular region are the result of a variety of disorders (including, for example, displacements of the temporomandibular joint disc and problems with the muscles of mastication) rather than a single syndrome with a single etiology. They coined the term temporomandibular disorders (TMD) to cover all of these conditions. Temporomandibular disorders have historically been considered the province of dental specialists. Physicians typically receive little or no training regarding TMD problems [5], and these disorders are frequently not well understood either by primary care physicians or by general dentists [6]. The research literature on TMD is almost exclusively confined to dental journals. A systematic review of TMD literature published in the past thirty years found that only five substantive articles (out of over 12,400 publications on TMD) had been published in major medical journals [5]. Ironically, most people with TMD pain problems first seek care from either a physician or a general dentist [7,8].

TMD pain typically runs a recurrent or chronic course over several years, without deterioration of structure [9]. Approximately half of those who seek care for TMD and are treated conservatively are pain-free at five year follow-up [10]. The side effects of invasive or irreversible treatments (including open joint surgery, jaw repositioning, and use of TMJ implants) may represent the greatest biological risks associated with these conditions [11,12]. Women are at greater risk than men for developing TMD pain, with a prevalence ratio of approximately 2:1 in population-based studies. However, the female-to-male ratio in tertiary care clinics is much higher, ranging from about 5:1 to 9:1 [13]. Furthermore, among persons seeking care for TMD, women are more likely than men to receive surgical treatments [14]. Thus, both the prevalence and treatment patterns for TMD appear to place women at higher risk for negative consequences.

II. Definitions

Temporomandibular disorders are best categorized as musculoskeletal pain conditions. Almost all persons who seek care for these conditions do so because of pain [3,4]. Pain may be located in the temporomandibular joint region (arthralgia), in the masticatory muscles (particularly the masseter and temporalis) and associated soft tissues (myofascial pain), or, more frequently, in both muscle and joint structures.

In general, the intensity and temporal characteristics of TMD are similar to those of back pain [15]. On average, persons with TMD identified in the community rate the usual intensity of

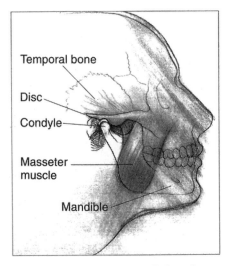

Fig. 89.1 The temporomandibular region.

Labels: Temporal bone, Disc, Condyle, Masseter muscle, Mandible

their pain as 4.3 on a 10-point scale (vs 4.7 for back pain), with 16% of those with TMD and 15% of those with back pain rating the pain as "severe." The percentage who report experiencing pain on more than half the days in the past 6 months is 28% for TMD and 29% for back pain.

Diagnostic criteria are available for differentiating subtypes of TMD (*e.g.*, myofascial pain, arthralgia) [16,17] based on presenting signs and symptoms. However, as with back pain, specific structural abnormalities are often not identifiable in TMD, and even when identifiable abnormalities exist, their relationship to pain is not clear [18]. Thus, treatment for most cases is appropriately directed toward management of pain and other symptoms rather than toward correction of specific structural abnormalities. Again, analogous to back pain, surgery may be indicated in a small percentage of cases. For TMD, these cases are usually those that involve radiographically documented structural disorders of the jaw joint that have not responded to conservative therapy [17].

As mentioned previously, certain signs, including joint sounds, limited ability to open the mandible, and deviation of the jaw on opening frequently accompany TMD pain. However, even though these signs are more common among persons with TMD pain, they are also frequently present in persons in the population without pain in the temporomandibular region [3]. Thus, no individual sign has high specificity for diagnostic purposes. Furthermore, there are no prospective studies indicating that the presence of any of these signs increases the risk of onset for TMD pain. Early theories concerning the etiology of TMD focused primarily on the structural and functional relationships between the upper and lower teeth and jaws, or the dental occlusion [19,20]. These theories were based primarily on clinical observations rather than on epidemiologic studies. Case-control studies have generally found that occlusal variables differ little in cases and controls, and the few differences found could be the consequences, rather than the cause, of the condition [21]. Thus, although some dental practitioners still hold occlusally based theories of TMD, few researchers would define TMD as a disorder of the occlusion.

III. Issues Related to Ascertainment

A. Case Definition

The International Association for the Study of Pain defines pain as "an unpleasant sensory and emotional experience associated with actual or potential tissue damage, or described in terms of such damage." [22]. Because pain is defined as an experience, and because no "objective" tests for pain exist, the presence of specific types of pain must be ascertained by self-report. This situation is not unique to pain but also holds for many other symptoms such as fatigue, dizziness, and depressed mood. In epidemiologic studies, TMD pain is generally defined by its location (*i.e.*, pain in the "jaw joint" [23] or "the muscles of your jaw or your jaw joint" [24] or "the muscles of your face . . . the joint in front of your ear" [15]) and by some temporal information, based either on time period (*e.g.*, "in the last week") or frequency of pain (*e.g.*, "often or very often"). Additional factors that are sometimes used in case definition include whether the pain occurs at rest (ambient), with jaw

movement (functional), or is unspecified, and pain severity (*e.g.*, mild, moderate, or severe). The quality of the pain (*e.g.*, "ache") is less frequently specified.

B. Prevalence Measures

One important issue in measuring the prevalence of TMD is how long a recall period should be used in asking subjects about pain. Because the course of TMD pain is recurrent for many people, a respondent who typically experiences pain could be pain-free at the time of survey and might not be counted as a case. Thus, point prevalence measures may be inappropriate for this type of pain condition, with period prevalence representing a better measure. Given that period prevalence is used, Von Korff has pointed out [25] that there are trade-offs for short versus long reporting intervals. All else being equal, a short recall period should minimize the problem of respondents forgetting a pain condition they may not be experiencing at the time of survey. However, persons with acute or recurrent pain may be underrepresented relative to those with chronic pain if the reporting interval is very short. That is, among all persons experiencing TMD pain during a specific interval (*e.g.*, 1 year), those with persistent pain would be more likely to be identified than persons with acute or recurrent pain if the reporting interval were very short (*e.g.*, 1 week). In addition, a short reporting period might not be enough to capture important information regarding characteristics of pain and pain-related disability (*e.g.*, average intensity, pain-related disability days). Data are available to support the validity of a 6-month reporting interval [25]. Intervals from 3 months to 1 year are typically used in prevalence surveys.

C. Severity

Because pain is a very common human experience, epidemiologic studies of pain often employ severity thresholds as part of the case definition in order to identify cases with clinically significant pain. Thus, in epidemiologic studies of pain, including TMD pain, respondents are sometimes asked not to report fleeting or minor pain, or to report only pain that is significant enough to interfere with activity. Of course, different prevalences are found depending on the severity criterion employed. For example, in one study [24], jaw pain that was "severe" was reported by 10% of respondents, but jaw pain of any intensity was reported by over a third of the sample.

D. Indices of TMD

Quite a number of epidemiologic studies of TMD have used as their case definition the Helkimo Indices [26]. These indices (one derived from self-report and one derived from clinical findings) are based on combinations of signs and symptoms of temporomandibular disorders, including both pain and nonpain complaints. Symptoms thought to indicate greater pathology are given greater weight. Using these methods, prevalence is defined as the number of people in the population meeting certain cut-offs for severity. The rationale for using these indices is that, as mentioned earlier, many of the signs thought to be characteristic of temporomandibular disorders (*e.g.*, joint clicking) are quite common in the general population and may actually represent normal variation rather than pathology [27]. Because many people are not significantly bothered by their symptoms, analyses that simply count the number of people with a particular sign or symptom may not provide a good estimate of the burden of these disorders in the population. Thus, severity indices were devised in an attempt to better document the public health significance of these problems. Unfortunately, there are a number of psychometric problems with these indices [28]. Furthermore, while this approach can document the overall burden of signs and symptoms in the population, organizing data in this way may not be helpful for other uses of epidemiology [29], such as attempts to understand the etiology of specific TMD conditions. An additional disadvantage of these indices is that it is not possible to extract data on the prevalence of pain (or any other individual symptom) if the only data presented relate to the indices.

IV. Historical Trends

Pain and dislocation in the jaw region were described and treated in humans as early as 3000 BC [30]. Descriptions of syndromes including some symptoms now considered characteristic of TMD were first reported in the medical and dental literature in the 1930s [19]. To our knowledge, no studies have been published that have specifically looked at trends in TMD incidence or prevalence over time. There have been a few investigations that repeated surveys several years apart in the same community to compare the prevalence of TMD pain in different cohorts [31–34]. These studies were consistent in finding a trend toward decreased prevalence with time, although the differences found were sometimes quite small. Unfortunately, all of these surveys involved only elderly persons, and most were concerned specifically with joint pain. No studies could be located that focused on masticatory muscle pain in children or younger and middle-aged adults, as opposed to these studies which are based on older persons and may reflect pain associated with osteoarthritis. There is evidence that other head pain conditions, such as migraine [35] and tension-type headache in children [36], have shown increases in incidence or prevalence over the last few decades, while another series of repeated cross-sectional studies showed decreases in reported musculoskeletal pain [37].

Many TMD researchers share the impression that temporomandibular pain occurred at a low rate in the early part of the twentieth century, became more prevalent in the 1960s, and that prevalence is now stable or possibly declining. However, these impressions come largely from experience with clinic populations and could simply reflect clinicians' abilities to identify TMD, their willingness to treat these conditions, changes in reimbursement for TMD treatment, or changes in the public perception such that TMD symptoms came to be viewed as a disorder that could be helped by medical or dental treatment. There is clearly now a greater awareness of TMD than previously, as evidenced by an increase in the number of published studies [38], the amount of funded research, and interest expressed in the lay press. However, the population most at risk for TMD in the U.S.—women aged 20 to 45 years—also increased substantially in recent decades because of the "baby

boom" after World War II. Thus, any perceived increase could be simply a reflection of this population bulge [39].

Because well designed epidemiologic surveys were conducted for several Scandinavian populations in the 1970s and several North American populations in the 1980s and 1990s, in the future we should be better able to document historical trends in the prevalence of TMD [40].

V. Distribution in Women

A. Prevalence

We conducted a comprehensive review of the literature to assess the prevalence of TMD pain in both genders [5]. We will draw on that review as we discuss the prevalence of TMD pain in women. In order for epidemiologic studies to be considered useful for the purposes of our review, we required, at a minimum, that these studies be conducted within a population-based sample and use a pain definition of either self-reported ambient or functional pain in or around the TMJ. We also required that gender-specific estimates be provided. Studies meeting these criteria come primarily from Scandinavia, other European countries, North America, and Japan. There are few studies of the prevalence of TMD pain in developing countries.

We located 34 articles written in English and published from 1965 to 1998 that were population-based studies of the prevalence of TMD pain in adults. Thirteen of these studies reported either ambient or unspecified TMD pain with separate gender-specific prevalence estimates. Although the definitions and questions to assess pain varied considerably, the overall population prevalence of reported TMD pain in women was in a relatively narrow range, with the lowest estimate being 4.9% for "headaches in or near the ear" [41] to 14–15% for "facial and jaw pain" [26] or "pain in the muscles of the face, the joint in front of the ear, or inside the ear" [15]. In the 11 studies that included both genders, the point estimates for women were roughly double the estimates for men. The female-to-male gender prevalence ratio ranged from 1.2 to 2.6, with 6 of the 11 studies reporting female-to-male prevalence ratios in the range of 1.8–2.0.

B. Age Distribution

With a couple of exceptions [26,41], population-based studies of adults have found the prevalence of TMD pain to be lower in the elderly than among younger people. Some studies have found that rates decline steadily with age in adults [42,43] whereas others have found a bell-shaped pattern, with the highest prevalence among those aged 25–54 [15,24,33]. There are few studies of TMD pain in children and adolescents, but those studies that are available suggest that prevalence is quite low (2–6%) in both sexes before puberty [5,40]. One study of Swedish adolescents that focused specifically on TMD pain found that prevalence increased with age after age 13, but much more so for girls than for boys, such that the prevalence among girls aged 14–18 was roughly 28%—much higher than the rate among boys the same age (about 8–9%) [44].

Differing patterns in the ratios of female- and male-specific prevalence with age may also be apparent. However, there are presently too few studies with the necessary detail to make any conclusions except that the proportion of males with pain is lower at all ages for adults. If the bell-shaped pattern for TMD prevalence with age is found to be correct, the age- and gender-specific prevalence patterns of TMD pain appear to be very similar to that of migraine [45], another pain condition with a female predominance, especially from ages 15 to 50.

C. Incidence and Duration

There are almost no longitudinal studies that assess the incidence of TMD pain, and of the three studies identified, only two allowed calculation of gender-specific incidence rates. One of these studies [46] followed a cohort of 361 Japanese school girls from age 12 to age 16 and examined pain occurring in response to opening or forward movement of the jaw during a clinical examination. Rates of onset for this type of pain ranged from 2.4 per 100 person-years to 3.9 per 100 person-years. The second study examined rates of onset of TMD pain over three years in an adult HMO population [47], originally aged 18–75. The onset rate among women was 2.6 per 100 person-years, about 1.5 times the onset rate in men. Thus, from the few studies available, incidence rates for TMD pain appear to be fairly low. However, interpretation of incidence rates for TMD pain is difficult, as is the case for other episodic and recurrent disorders [48], due to problems in retrospectively dating onset and the fact that offsets occur that may or may not be followed by recurrences. Although the data for TMD pain are few, the available information provides hints that the age- and gender-specific incidence curves for TMD pain may be similar to those of migraine—incidence is greater for females than for males and it peaks in the teens and declines slowly with increasing age [35].

There are almost no published data on the duration of TMD episodes or total duration of these disorders over time. One study found that 49% of a cohort of community cases and TMD patients treated conservatively were pain free (*i.e.*, had not experienced pain in the prior 6 months) five years after the baseline assessment [10]. Data from a second sample of treated cases indicated that, compared with men, women are more likely to describe their TMD pain as persistent and recurrent [49]. The rates of offset over a 3-year period (*i.e.*, no pain in the prior 2 years) were also significantly lower for women (18%) than for men (42%). Thus, from the very minimal evidence available, it appears likely that the higher prevalence of TMD in women is associated with both higher rates of onset and longer duration of the condition.

D. Morbidity

Like other pain conditions, TMD can interfere with ability to work and to concentrate. For example, one study found that about one-quarter of people referred to a craniofacial pain unit reported that they were unable to work to full capacity due to pain [50]. In another study, 64% of a small, selected group of patients with myofascial TMD pain reported decreased efficiency at work, compared to 8% of subjects who reported clenching or grinding their teeth but no pain [51].

In a large study of persons seeking treatment for TMD pain in primary care, about 16% were found to experience moderate or severe levels of pain-related disability [52] (compared with about 30% of headache patients and 37% of back pain patients). As with other pain conditions, women with TMD pain are significantly more likely than men to experience such pain-related disability [53]. Persons with TMD who experience pain-related disability are at increased risk for unemployment, frequent use of opioid medications, and frequent use of the health care system. Among persons whose TMD pain is associated with high levels of interference and disability days (nearly all of whom are women), over half (57%) experience elevated depression. Even among those who are not disabled but experience high levels of TMD pain, about 30% experience depression [52]. Thus, from a public health perspective, the pain itself, as well as depression and other aspects of pain-related disability, represent the major aspects of morbidity associated with TMD.

For a small but significant number of individuals, primarily women, devastating physical impacts of TMD have also occurred, due to irreversible treatments for TMD problems. For example, from the 1970s to the early 1990s, proplast-teflon implants were sometimes inserted into temporomandibular joints to replace temporomandibular joint discs. Approximately 26,000 of these devices were released by the manufacturer to hospitals and clinicians, but data are not available on the exact number of implants actually placed in human joints. Failure rates of 50–100% have been reported for these devices, with some severe outcomes including deformity, autoimmune reactions, and severe, unremitting, debilitating pain [11,12].

VI. Host and Environmental Determinants

A. Gender and Age

Rates of TMD pain are consistently found to be higher in women than in men [5,13,40], with a gender ratio of about two women for every man. Virtually all studies find the highest prevalence for both sexes to be between the ages of 15 and 50 years. As yet, it remains unclear what aspects of women's biology, psychology, or social roles predispose them to suffer these conditions more often than men, although some of the risk factor studies suggest that hormonal factors may play some role.

B. Race and Ethnicity

We know of only one population-based study of TMD that examined rates by race. Using data from the National Health Interview Survey, Lipton *et al.* [42] found a slightly lower prevalence of jaw joint and facial pain among African-Americans when compared to whites, although this estimate was not adjusted for socioeconomic status (SES) or gender.

C. Geographic Variation

Nearly all of the population-based studies of TMD pain prevalence have been carried out in Europe, North America, and Japan. Within this set of studies, no systematic variation of prevalence by geographical area is apparent. One national study in the United States [42] examined the prevalence of jaw joint and facial pain by geographic region. Rates of jaw joint pain were highest in the West, lowest in the Northeast, and intermediate in the Midwest and South. Rates of facial pain were also highest in the West, but were lowest in the South and intermediate in the Midwest and Northeast. It is quite possible that these rates are influenced by the awareness of specific aspects of TMD in each of the regions and by variations in practice patterns and access to care, as well as by other environmental factors.

D. Family History

To our knowledge, there are only two studies investigating familial factors in TMD, and both compared the occurrence of TMD in dizygotic (DZ) and monozygotic (MZ) twins. One small study of 94 twins [54] did not show any genetic influence, whereas a study of 146 MZ and 96 DZ twins reported a modest, nonsignificant heritability estimate of 12–24% for TMD pain and other symptoms [55]. One of the authors of this paper estimated that 1000 twin pairs would be needed to demonstrate that the contribution from a genetic effect was not due to chance alone because the prevalence of TMD pain was relatively low (8.7%) in their twin population [J. Hodges, personal communication, 1998]. Based on these two studies, it appears that TMD pain does not have a large heritable component; however, much more work is needed in this area before firm conclusions can be drawn.

We know of no research that has investigated specifically whether a history of TMD pain in parents may be risk factor for TMD in their children. For chronic pain in general, the empirical evidence (reviewed by Turk *et al.* [56]) seems to indicate that both children and adults with chronic pain are likely to have individuals in their families with chronic pain. It is frequently speculated that these individuals may serve as models for how to behave when in pain. However, as Turk *et al.* point out, most of the reported studies of chronic pain in families suffer from one or more significant methodological problems, including reliance on retrospective self-report of patients and families, lack of control groups, and potential for recall bias.

E. Income

Although TMD has been stereotyped as an ailment of upper middle class suburbanites, it is likely that this stereotype is based on impressions from early studies in which women of higher SES were shown to seek care for TMD pain at higher rates than those of low SES [57,58]. It is possible, however, that these studies reflect referral bias, because much of TMD care is not covered by insurance in the U.S. The few population-based studies that have measured income have shown a slightly greater prevalence of low SES patients in population-based samples [3,59] or in clinics that take patients with state-supported care [60]. Thus, it appears that the relationship between TMD pain and income is not clear. Further population-based studies are needed to explore this relationship.

VII. Influence of Women's Social Roles

The data available from population-based studies reveal no apparent relationship between TMD pain and marital status, family composition, or employment status. Although some studies have reported high rates of TMD in specific occupational groups (*e.g.,* musicians), these investigations rarely have

control groups. Overall, there does not appear to be a relationship between TMD pain and occupation or social role. We know of no studies investigating the relationship between TMD and sexual orientation.

VIII. Risk Factors

Until recently, the major focus in TMD epidemiology has been on documenting the burden of these conditions rather than on attempting to understand their etiology. Nevertheless, some studies have been undertaken specifically to identify risk factors, and additional information can be obtained from several investigations not explicitly designed for this purpose. Because of the clear association of TMD pain with female gender, and because of prevalence differences with age, studies that attempt to assess risk factors other than age and gender should control for these factors. Unfortunately, not all studies have used such controls; in fact, there have been only a handful of cohort studies and a small number of case-control studies that employed appropriate methods for assessing risk.

The only cohort study that controlled for the possible confounding effects of age, gender, and education [47] showed that, in an adult population of HMO enrollees, high depression scores (RR = 1.6) and the existence of other pain conditions at baseline (RR = 3.7) were associated with the onset of TMD pain in the next three years. Among the depressed, the risk of onset was highest for those with severe depression and those whose depression was present both at baseline and at follow up. However, only the risk for existing pain was statistically significant. These findings are important because they demonstrate a temporal relationship between existing somatic complaints and the onset of pain. A number of cross-sectional studies have found TMD patients to be more psychologically and somatically distressed than those in control groups, but these studies have not been able to demonstrate whether psychological distress preceded or followed the onset of pain symptoms.

In a cohort study of adolescent Japanese girls, Kitai *et al.* [46] found no association between several measures of malocclusion and the onset of TMD pain on function during a 5-year follow-up period. These two investigations are the only cohort studies that have measured the association between the onset of TMD pain and pre-existing exposure measures. All other longitudinal studies located that investigated putative risk factors for TMD pain failed to assess the temporal sequence of the onset of the disorder and the occurrence of the purported risk factor, a major criterion for demonstrating cause and effect.

The remaining information on risk factors for TMD pain comes from case-control studies. Although over 100 case-control studies of TMD have been conducted, many did not attempt to control for gender or age or did not have pain as a major aspect of the case definition. A clinic-based study of 133 TMD cases and 133 controls who were seeking dental care but did not have TMD showed strong associations between TMD pain and self-reported bruxism (clenching the jaw or grinding the teeth), as well as migraine, mixed, and tension-type headaches [61]. Cases and controls were matched by age and gender. Unfortunately, this study excluded controls who used medications, which likely elevated associations of TMD and headache.

In an attempt to discern what aspects of female gender might increase the risk of TMD pain, our research group investigated the effects of exogenous hormone use in women in two large population-based case-control studies with analyses that controlled for age and health care utilization [62]. We found small but significant associations between seeking care for TMD pain and use of oral contraceptives (odds ratio [OR] = 1.19, 95% confidence interval (CI) 1.01–1.40) [62]. In a similar analysis among older women, there was also a significant effect of postmenopausal estrogen use (OR = 1.32, 95% CI 1.10–1.57). A dose-response relationship was found between cumulative estrogen dose in the prior year and risk of TMD.

Two small clinic-based case-control studies [63,64] have investigated the relationship between forward head posture and TMD pain. One found significantly elevated odds ratios for forward head posture with TMD pain, while the other failed to find a relationship. Another small case-control study found significant relationships with several novel exposures, including chronic respiratory infections, gastrointestinal infections, and chronic pain in long-term partners [65].

Several other putative risk factors, including occlusal factors, trauma, generalized joint hypermobility, dental extractions, and orthodontic treatment have engendered interest among clinicians. However, although there have been some case-control studies investigating these putative risks, all the studies we located that investigated these factors either did not include TMD pain as a major aspect of the case definition or did not attempt to control for confounding by gender or other factors. Thus, the role, if any, that these factors might play in TMD pain remains speculative and unknown.

In summary, there are surprisingly few analytic studies of TMD pain that fulfill basic epidemiologic standards. Most have uncontrolled confounding, are too small, have inadequate analyses, or have other problems, such as lack of validity or reliability of the measures. In addition, the few well-designed studies have investigated a wide variety of risk factors, with few potential risk factors for TMD pain being evaluated in more than one study. Clearly more research is needed in this area to confirm the findings of the few existing studies and to investigate additional potential risk factors.

IX. Biologic Markers of Susceptibility or Exposure

Currently, there are no known biologic markers of susceptibility to TMD pain. There are also no known biologic markers of exposure to the condition or its risk factors.

X. Clinical Issues

A. Diagnostic Criteria

Although there are a number of different diagnostic classification systems for TMD available for research and clinical use, the state of knowledge in this field is such that the available classification systems for TMD, like those for headache and for back pain, are descriptive and empirically based rather than based on etiology. In an attempt to produce a standardized set of criteria for use in research, a group of investigators from several institutions developed the Research Diagnostic Criteria for TMD (RDC/TMD) in the early 1990s [16]. The RDC/TMD is a multiaxial system in that it classifies subjects both by their clinical (Axis I) diagnoses and by their psychosocial (Axis II)

profile. Specifications are provided for conducting a standardized clinical examination, and algorithms are used to assign clinical diagnoses to subjects based on examination findings and a few self-report questions referring primarily to presence and location of pain. Classification of the subject's psychosocial status is based on standardized psychometric instruments. The RDC/TMD has been tested for reliability and for clinical utility of the multiaxial system [10,66], and it provides a common language for describing research populations. However, the RDC/TMD was not designed specifically for clinical use and it is not used extensively in clinical settings. In fact, there is no generally accepted approach to diagnostic classification of TMD in clinical settings [1]. Thus, patients still receive a range of diagnostic labels that may carry different meanings for different practitioners.

B. Therapies

There have been few controlled trials of TMD therapies, although the number of such trials is growing [38]. Conservative treatments such as occlusal splints ("night guards"), heat and cold application, jaw exercises, and relaxation have all been shown in clinical trials to be effective in reducing pain [23,67–70]. However, in some studies of occlusal splints, placebo-controlled interventions have shown results roughly equivalent to the active treatments [23]. Long-term follow-up studies of both conservative and nonconservative therapies show equivalent results [71], and randomized clinical trials of irreversible treatments are lacking. In the absence of controlled trials indicating the success of invasive or irreversible treatments, and recognizing that such treatments can have significant side effects, the American Dental Association's President's Conference on Temporomandibular Disorders recommended in 1983 that only conservative reversible therapies be used as initial treatments for TMD [2]. Given the current state of knowledge in the field, this advice remains appropriate today.

C. Implications of TMD as a Chronic or Recurrent Condition

Because TMD pain is typically chronic or recurrent, it can be associated with significant psychosocial impact. Treatment approaches that do not systematically consider the psychosocial impact are likely to address only part of the problem, whereas management approaches that include consideration of the patient's psychological state and the impact of the pain on the patient's life are more likely to meet with success [72,73]. A further implication of the chronic/recurrent nature of TMD pain is that, like other chronic conditions such as diabetes or heart disease, its management requires that patients assume major responsibility for their own care, in consultation with a health care provider who offers a diagnosis, information on treatment options, monitoring, and reinforcement for the patient's long-term management efforts [74].

D. Gender Issues in TMD Treatment

Although the female-to-male prevalence ratio for TMD in the community is about 2:1, in some tertiary care clinics, as many as 90% of the patients are women. The gender ratios of TMD patients in primary dental and medical care are not known, but it is clear that many more women than men are treated for TMD [14]. Although women now constitute about one-third of dental students, the vast majority of practicing dentists are men [75]. As a result of these disease patterns and practice patterns, the vast majority of female TMD patients are currently treated by male providers. To our knowledge there are no studies of patient preferences regarding care for TMD, or whether the gender of the provider influences the type of treatment offered and patient satisfaction with care. However, this would appear to be an important area for future research.

XI. Epidemiologic Issues

Because TMD pain is quite common, it may have significant public health impact in terms of personal suffering and diminished quality of life. Perhaps because it is not a life-threatening condition and does not have as significant an impact on work attendance as do some other pain conditions, the public health significance of TMD may not have been recognized. Additional studies are needed to accurately document the burden that this condition places on society in general and women in particular.

Methodologic issues that need to be addressed in epidemiologic studies include the issues of case ascertainment described earlier. Specifically, attention should be paid to case definition, including not only pain location, but also temporal aspects of pain and pain severity. A standardized case definition based on the presence of pain of specific severity and duration would greatly advance research. In addition, studies assessing risk factors need to be conducted. Epidemiologic studies of particular subtypes of TMD are also needed in order to determine whether the population distribution and risk factors for specific subtypes of TMD are similar or distinct.

Finally, in order to discern whether the differentially high rates of TMD pain in women are attributable to biological or sociocultural factors, prevalence surveys and studies of risk factors for TMD pain should be carried out in a range of cultures in developing as well as in developed countries.

XII. Summary and Conclusions/Future Directions

TMD pain is a common problem in the population. Its prevalence is highest among women of reproductive age. The course of TMD pain is typically chronic or recurrent. Although not life-threatening, TMD pain can be extremely disabling for some people and is associated with work loss, excessive use of medications and health care, and high levels of psychological distress. Women are more likely than men to experience TMD pain and more likely to become disabled if pain is present.

Future research is needed on risk and etiologic factors for TMD pain, and particularly on reasons for the increased risk of TMD pain in women. Specifically, the possible role of hormones as etiologic or exacerbating factors in these disorders is intriguing and should be further investigated. In addition to research on TMD in particular, research on gender and pain, more broadly, may also provide some insights.

Currently, most TMD treatment is empirically based. Choice of treatment modalities is often more highly influenced by the background, biases, and clinical experience of the clinician than by the scientific literature. Approaches to the management of TMD need to be tested in randomized clinical trials. Both conventional clinic-based treatments and self-management approaches for controlling pain should be assessed.

Finally, practitioners need to be educated about TMD pain and its similarity to other pain conditions, as well as appropriate palliative management approaches for TMD. By decreasing misdiagnoses, unnecessary tests and procedures, and unproven irreversible therapies that cause iatrogenic disorders, much of the disability that accompanies TMD pain could be prevented.

References

1. National Institutes of Health (1996). Management of temporomandibular disorders: National Institutes of Health Technology Assessment Conference statement. *J. Am. Dent. Assoc.* **127,** 1595–1606.
2. Laskin, D., Greenfield, W., Gale, E., Rugh, J., Neff, P., Alling, C., and Ayer, W. A., eds. (1983). "The President's Conference on the Examination, Diagnosis and Management of Temporomandibular Disorders." American Dental Association, Chicago.
3. Dworkin, S. F., Huggins, K. H., LeResche, L., Von Korff, M., Howard, J., Truelove, E., and Sommers, E. (1990). Epidemiology of signs and symptoms in temporomandibular disorders: Clinical signs in cases and controls. *J. Am. Dent. Assoc.* **120,** 273–281.
4. Motegi, E., Miyazaki, H., Ogura, I., Konishi, H., and Sebata, M. (1992). An orthodontic study of temporomandibular joint disorders: Part I: Epidemiological research in Japanese 6–18 year olds. *Angle Orthodontist* **62,** 249–256.
5. Drangsholt, M., and LeResche, L. (1999). Temporomandibular disorder pain. *In* "Epidemiology of Pain" (I. K. Crombie, P. R. Croft, S. J. Linton, L. LeResche, and M. Von Korff, eds.), pp. 203–233. IASP Press, Seattle, WA.
6. LeResche, L., Truelove, E., and Dworkin, S. F. (1993). Dentists' knowledge and beliefs concerning temporomandibular disorders. *J. Am. Dent. Assoc.* **124,** 90–106.
7. Turp, J. C., Kowalski, C. J., and Stohler, C. S. (1998). Treatment-seeking patterns of facial pain patients: Many possibilities, limited satisfaction. *J. Orofacial Pain* **12,** 61–66.
8. Glaros, A. G., Glass, E. G., and Hayden, W. J. (1995). History of treatment received by patients with TMD: A preliminary investigation. *J. Orofacial Pain* **9,** 147–151.
9. Dworkin, S. F., LeResche, L., Mancl, L., Ohrbach, R., Truelove, E., Huggins, K., and Von Korff, M. (1996). Longitudinal course of TMD-related pain in clinic and community cases. *In* "Abstracts, 8th World Congress on Pain." IASP Press, Seattle, WA.
10. Ohrbach, R., and Dworkin, S. F. (1998). Five-year outcomes in TMD: Relationship of changes in pain to changes in physical and psychological variables. *Pain* **74,** 315–326.
11. Milam, S. B. (1997). Failed implants and multiple operations. *Oral Surg. Oral Med. Oral Pathol. Oral Radiol. Endodontics* **83,** 156–162.
12. Spagnoli, D., and Kent, J. N. (1992). Multicenter evaluation of temporomandibular joint Proplast-Teflon disk implant. *Oral Surg. Oral Med. Oral Pathol. Oral Radiol. Endodontics* **74,** 411–421.
13. Carlsson, G. E., and LeResche, L. (1995). Epidemiology of temporomandibular disorders. *In* "Progress in Pain Research and Management" (B. J. Sessle, P. S. Bryant, and R. A. Dionne, eds.), pp. 211–226. IASP Press, Seattle, WA.
14. Marbach, J. J., Ballard, G. T., Frankel, M. R., and Raphael, K. G. (1997). Patterns of TMJ surgery: Evidence of sex differences. *J. Am. Dent. Assoc.* **128,** 609–614.
15. Von Korff, M., Dworkin, S. F., LeResche, L., and Kruger, A. (1988). An epidemiologic comparison of pain complaints. *Pain* **32,** 173–183.
16. Dworkin, S. F., and LeResche, L. (1992). Research Diagnostic Criteria for Temporomandibular Disorders: Review, criteria, examina-

tions and specifications, critique. *J. Craniomandib. Disord. Facial Oral Pain* **6,** 301–355.

17. American Academy of Orofacial Pain (1996). Differential diagnosis and management considerations of temporomandibular disorders. *In* "Orofacial Pain: Guidelines for Assessment, Diagnosis, and Management" (J. P. Okeson, ed.), pp. 113–184. Quintessence Publishing Co., Chicago.

18. Jensen, M. C., Brant-Zawadzki, M. N., Obuchowski, N., Modic, M. T., Malkasian, D., and Ross, J. S. (1994). Magnetic resonance imaging of the lumbar spine in people without back pain. *N. Engl. J. Med.* **331,** 69–73.

19. Costen, J. B. (1997). A syndrome of ear and sinus symptoms dependent upon disturbed function of the temporomandibular joint [originally published 1934]. *Ann. Otol. Rhinol. Laryngol.* **43,** 1–15.

20. Ramfjord, S. P., and Ash, M., Jr. (1966). Functional disturbances of temporomandibular joints and muscles. *In* "Occlusion," pp. 160–178. Saunders, Philadelphia.

21. Pullinger, A. G., Seligman, D. A., and Gornbein, J. A. (1993). A multiple logistic regression analysis of the risk and relative odds of temporomandibular disorders as a function of common occlusal features. *J. Dent. Res.* **72,** 968–979.

22. H. Merskey and N. Bogduk, eds. (1994). Classification of Chronic Pain: Descriptions of Chronic Pain Syndromes and Definitions of Pain Terms, Second Edition, IASP Press, Seattle, WA.

23. Dao, T. T. T., Lavigne, G. J., Charbonneau, A., Feine, J. S., and Lund, J. P. (1994). The efficacy of oral splints in the treatment of myofascial pain of the jaw muscles: A controlled clinical study. *Pain* **56,** 85–94.

24. Goulet, J.-P., Lavigne, G. J., and Lund, J. P. (1995). Jaw pain prevalence among French-speaking Canadians in Quebec and related symptoms of temporomandibular disorders. *J. Dent. Res.* **74,** 1738–1744.

25. Von Korff, M. (1992). Epidemiologic and survey methods: Chronic pain assessment. *In* "Handbook of Pain Assessment" (D. C. Turk and R. Melzack, eds.), pp. 391–408. Guilford Press, New York.

26. Helkimo, M. (1974). Studies on function and dysfunction of the masticatory system: IV. Age and sex distribution of symptoms of dysfunction of the masticatory system in Lapps in the north of Finland. *Acta Odontol. Scand.* **32,** 255–267.

27. Goyer, R. A., and Rogan, W. J. (1986). When is biologic change an indicator of disease? *In* "New and Sensitive Indicators of Health Impacts of Environmental Agents" (D. W. Underhill and E. P. Radford, eds.), pp. 17–25. University of Pittsburgh Graduate School of Public Health Center for Environmental Epidemiology, Pittsburgh, PA.

28. Van der Weele, L. T., and Dibbets, J. M. H. (1987). Helkimo's index: A scale or just a set of symptoms? *J. Oral Rehabil.* **14,** 229–237.

29. Morris, J. N. (1975). "Uses of Epidemiology." Churchill-Livingstone, Edinburgh.

30. McNeill, C. (1997). History and evolution of TMD concepts. *Oral Surg. Oral Med. Oral Pathol. Oral Radiol. Endodontics* **83,** 51–60.

31. Sato, H., Osterberg, T., Ahlqwist, M., Carlsson, G. E., Grondahl, H.-G., and Rubinstein, B. (1996). Association between radiographic findings in the mandibular condyle and temporomandibular dysfunction in an elderly population. *Acta Odontol. Scand.* **54,** 384–390.

32. Osterberg, T., Carlsson, G. E., Wedel, A., and Johansson, U. (1992). A cross-sectional and longitudinal study of craniomandibular dysfunction in an elderly population. *J. Craniomandib. Disord. Facial Oral Pain* **6,** 237–246.

33. Locker, D., and Slade, G. (1988). Prevalence of symptoms associated with temporomandibular disorders in a Canadian population. *Commun. Dent. Oral Epidemiol.* **16,** 310–313.

34. Locker, D., and Miller, Y. (1994). Subjectively reported oral health status in an adult population. *Commun. Dent. Oral Epidemiol.* **22,** 425–430.

35. Lipton, R. B., and Stewart, W. F. (1997). Prevalence and impact of migraine. *Neurol. Clin.* **15,** 1–13.

36. Sillanpaa, M., and Anttila, P. (1996). Increasing prevalence of headache in 7-year-old schoolchildren. *Headache* **36,** 466–470.

37. Manninen, P., Riihimaki, H., and Heliovaara, M. (1996). Has musculoskeletal pain become less prevalent? *Scand. J. Rheumatol.* **25,** 37–41.

38. Antczak-Bouckoms, A. (1995). Epidemiology of research for temporomandibular disorders. *J. Orofacial Pain* **9,** 226–234.

39. Rugh, J. D., and Solberg, W. L. (1985). Oral health status in the United States: Temporomandibular disorders. *J. Dent. Educ.* **49,** 398–405.

40. LeResche, L. (1997). Epidemiology of temporomandibular disorders: Implications for the investigation of etiologic factors. *Crit. Rev. Oral Biol. Med.* **8,** 291–305.

41. Agerberg, G., and Bergenholtz, A. (1989). Craniomandibular disorders in adult populations of West Bothnia, Sweden. *Acta Odontol. Scand.* **47,** 129–140.

42. Lipton, J. A., Ship, J. A., and Larach-Robinson, D. (1993). Estimated prevalence and distribution of reported orofacial pain in the United States. *J. Am. Dent. Assoc.* **124,** 115–121.

43. Matsuka, Y., Yatani, H., Kuboki, T., and Yamashita, A. (1996). Temporomandibular disorders in the adult population of Okayama City, Japan. *J. Craniomandib. Pract.* **14,** 159–162.

44. List, T., Wahlund, K., Wenneberg, B., and Dworkin, S. F. (1999). Temporomandibular disorders in children and adolescents: Prevalence of pain, gender differences, and perceived treatment need. *J. Orofacial Pain* **13,** 9–20.

45. Lipton, R. B., Stewart, W. F., and Von Korff, M. (1997). Burden of migraine: Societal costs and therapeutic opportunities. *Neurology* **48,** S4–S9.

46. Kitai, N., Takada, K., Yasuda, Y., Verdonck, A., and Carels, C. (1997). Pain and other cardinal TMJ dysfunction symptoms: A longitudinal survey of Japanese female adolescents. *J. Oral Rehabil.* **24,** 741–748.

47. Von Korff, M., LeResche, L., and Dworkin, S. F. (1993). First onset of common pain symptoms: A prospective study of depression as a risk factor. *Pain* **55,** 251–258.

48. Lawrence, R. C., Helmick, C. G., Arnett, F. C., Deyo, R. A., Felson, D. T., Giannini, E. H., Heyse, S. P., Hirsch, R., Hochberg, M. C., Hunder, G. G., Liang, M. H., Pillemer, S. R., Steen, V. D., and Wolfe, F. (1998). Estimates of the prevalence of arthritis and selected musculoskeletal disorders in the United States. *Arthritis Rheum.* **41,** 778–799.

49. Dworkin, S. F., LeResche, L., Truelove, E., and Saunders, K. (1997). Gender differences in onset and duration of TMD and headache. *J. Dent. Res.* **76,** 148 (Abst. No. 1078).

50. Murray, H., Locker, D., Mock, D., and Tenenbaum, H. C. (1996). Pain and the quality of life in patients referred to a craniofacial pain unit. *J. Orofacial Pain* **10,** 316–323.

51. Dao, T. T. T., Lund, J. P., and Lavigne, G. J. (1994). Comparison of pain and quality of life in bruxers and patients with myofascial pain of the masticatory muscles. *J. Orofacial Pain* **8,** 350–356.

52. Von Korff, M., Ormel, J., Keefe, F. J., and Dworkin, S. F. (1992). Grading the severity of chronic pain. *Pain* **50,** 133–149.

53. Von Korff, M., Dworkin, S. F., and LeResche, L. (1990). Graded chronic pain status: An epidemiologic evaluation. *Pain* **40,** 279–291.

54. Heiberg, A., Heloe, B., Heiberg, A. N., Heloe, L. A., Magnus, P., Berg, K., and Nance, W. E. (1980). Myofascial pain dysfunction (MPD) syndrome in twins. *Commun. Dent Oral Epidemiol.* **8,** 434–436.

55. Michalowicz, B., Pihlstrom, B., Hodges, J., and Bouchard, T., Jr. (1998). Genetic and environmental influences on signs and symptoms of TMD. *J. Dent. Res.* **77,** 919 (abstr.).

56. Turk, D. C., Flor, H., and Rudy, T. E. (1987). Pain and families. I. Etiology, maintenance, and psychosocial impact. *Pain* **30,** 3–27.

57. Franks, A. S. T. (1964). The social character of temporomandibular joint dysfunction. *Dent. Pract.,* November, pp. 94–100.

58. Heloe, B., Heloe, L. A., and Heiberg, A. (1977). Relationship between sociomedical factors and TMJ-symptoms in Norwegians with myofascial pain-dysfunction syndrome. *Commun. Dent. Oral Epidemiol.* **5,** 207–212.

59. Von Korff, M. (1995). Health services research and temporomandibular pain. *In* "Temporomandibular Disorders and Related Pain Conditions" (B. J. Sessle, P. S. Bryant, and R. A. Dionne, eds.), pp. 227–236. IASP Press, Seattle, WA.

60. Smith, J. A., and Syrop, S. (1994). TMD incidence and socioeconomic standing: Eradicating the myth of the upper class patient. *N. Y. State Dent. J.* **60,** 36–39.

61. Molina, O. F., dos Santos, J., Jr., Nelson, S. J., and Grossman, E. (1997). Prevalence of modalities of headaches and bruxism among patients with craniomandibular disorder. *J. Craniomandib. Pract.* **15,** 314–325.

62. LeResche, L., Saunders, K., Von Korff, M., Barlow, W., and Dworkin, S. F. (1997). Use of exogenous hormones and risk of temporomandibular disorder pain. *Pain* **69,** 153–160.

63. Hackney, J., Bade, D., and Clawson, A. (1993). Relationship between forward head posture and diagnosed internal derangement of the temporomandibular joint. *J. Orofacial Pain* **7,** 386–390.

64. Lee, W.-Y., Okeson, J. P., and Lindroth, J. (1995). The relationship between forward head posture and temporomandibular disorders. *J. Orofacial Pain* **9,** 161–167.

65. McGregor, N. R., Butt, H. L., Zerbes, M., Klineberg, I. J., Dunstan, R. H., and Roberts, T. K. (1996). Assessment of pain (distribution and onset), symptoms, SCL-90-R Inventory responses, and the association with infectious events in patients with chronic orofacial pain. *J. Orofacial Pain* **10,** 339–350.

66. Hayward, C., Killen, J. D., Wilson, D. M., Hammer, R. D., Litt, I. F., Kraemer, H. C., Haydel, F., Varady, A., and Taylor, C. B. (1997). Psychiatric risk associated with early puberty in adolescent girls. *J. Am. Acad. Child Adolesc. Psychiatry* **36,** 255–262.

67. Truelove, E. L., Huggins, K. H., Dworkin, S. F., Mancl, L., Sommers, E., and LeResche, L. (1998). RCT splint treatment outcomes in TMD: Initial self-report findings. *J. Dent. Res.* **77**(Spec. Issue A), 143 (Abstr. No. 304).

68. Burgess, J. A., Sommers, E., Truelove, E. L., and Dworkin, S. F. (1988). Short-term effect of two therapeutic methods on myofascial pain and dysfunction of the masticatory system. *J. Prosthet. Dent.* **60,** 606–610.

69. Dworkin, S. F., Turner, J. A., Wilson, L., Massoth, D., Whitney, C., Huggins, K. H., Burgess, J., Sommers, E., and Truelove, E. (1994). Brief group cognitive-behavioral intervention for temporomandibular disorders. *Pain* **59,** 175–187.

70. Dworkin, S. F. (1997). Behavioral and educational modalities. *Oral Surg. Oral Med. Oral Pathol. Oral Radiol. Endodontics* **83,** 128–133.

71. Greene, C. L., and Marbach, J. J. (1982). Epidemiologic studies of mandibular dysfunction: A critical view. *J. Prosthet. Dent.* **48,** 184–190.

72. McCreary, C. P., Clark, G. T., Oakley, M. E., and Flack, V. (1992). Predicting response to treatment for temporomandibular disorders. *J. Craniomandib. Disord. Facial Oral Pain* **6,** 161–169.

73. Dworkin, S. F. (1996). The case for incorporating biobehavioral treatments into TMD management. *J. Am. Dent. Assoc.* **127,** 1607–1610.

74. Von Korff, M., Gruman, J., Schaefer, J., Curry, S. J., and Wagner, E. H. (1997). Collaborative management of chronic illness. *Ann. Intern. Med.* **127,** 1097–1102.

75. LeResche, L., Truelove, E. L., and Dworkin, S. F. (1993). Temporomandibular disorders: A survey of dentists' knowledge and beliefs. *J. Am. Dent. Assoc.* **124,** 90–106.

90

Multiple Chemical Sensitivity

CAROL M. BALDWIN,* IRIS R. BELL,† MERCEDES FERNANDEZ,‡ AND GARY E. R. SCHWARTZ§

*Departments of Medicine and Psychology, and Respiratory Sciences Center, University of Arizona, and the Department of Psychiatry, Tucson Veterans Affairs Medical Center, Tucson, Arizona; †Departments of Psychiatry, Psychology, and Family and Community Medicine, University of Arizona, and the Department of Psychiatry, Tucson Veterans Affairs Medical Center, Tucson, Arizona; ‡Department of Psychology, University of Arizona, and the Department of Psychiatry, Tucson Veterans Affairs Medical Center, Tucson, Arizona; §Departments of Psychiatry, Psychology, Neurology and Medicine, University of Arizona, Tucson, Arizona

I have gout, asthma, and seven other maladies,
but am otherwise very well
Lady Holland's Memoir, Vol. 1, Ch. 10

I. Introduction

Chemical sensitivity is a chronic condition involving the subjective experience of illness due to low levels of environmental chemical odors that most people find neutral or nontoxic [1–3]. People who suffer from chemical sensitivity usually have multiple symptoms in multiple systems, most notably central nervous, musculoskeletal, respiratory, and gastrointestinal. Many people with chemical sensitivity report illness from not one but numerous different, chemically unrelated substances that culminates in a clinical condition termed multiple chemical sensitivity (MCS). Most MCS patients report intolerance to multiple drugs and foods as well as chemicals, and these different triggering agents can all mobilize the same symptoms in a given individual [4,5]. A majority of people with MCS are women and, like Lady Holland, have polysymptomatic complaints. Unlike Lady Holland, MCS patients may appear well clinically yet have poorer health outcomes. For example, MCS patients report high rates of disability [3] and extremely poor quality of life [6]. Their MCS requires numerous lifestyle changes to accommodate their illness [7], and these restrictions impinge markedly on their ability to function in everyday activities at home, work, school, and recreation.

A. Multiple Chemical Sensitivity Defined

MCS does not have a generally accepted case definition for research or clinical purposes. Most proposed definitions require that acute low-level chemical exposures cause symptom flares and the avoidance of exposure to resolve symptoms. Cullen [8] has provided the most consistently used definition—"an acquired disorder of recurrent symptoms, referable to multiple organ systems, occurring in response to chemically unrelated compounds at doses far below those established in the general population to cause harmful effects" [p. 655]. Ashford and Miller [9] emphasize the importance of a two-step process in MCS involving (i) initiation and (ii) elicitation. They indicate that it may require a relatively smaller number of specific chemicals (*e.g.,* certain pesticides or solvents) at relatively higher doses for initiation, whereas elicitation typically involves a broader spectrum of agents (*e.g.,* perfumes, cleaning solutions, fresh newspapers) at low, nontoxic levels. Cullen's definition requires a

history of an identifiable initiating chemical exposure for the diagnosis of MCS. However, emerging research suggests that while a large proportion of people with MCS and related conditions can identify specific chemicals as elicitors, they cannot necessarily identify the initiating exposure [10] (see Section VIII.F).

B. Chemical Odor Intolerance, the Focal Symptom of MCS

The single symptom that all MCS patients share is that of chemical odor intolerance (CI), or the subjective complaint of feeling ill on exposure to everyday chemicals, such as perfume, fresh paint, or car exhaust [11]. On a questionnaire for chemical sensitivity (synonymous with CI), Kipen *et al.* [12] found that scores higher than 23 out of 122 common substances distinguished MCS patients (n = 39) from non-MCS patients (n = 221) and normals (n = 436). Prevalence rates of daily symptoms from CI or physician diagnosis of MCS fall in the range of 4–6% [1] (see Section VIII.F). Among the U.S. general population, 15–30% report less severe degrees of chemical intolerance [13–15].

C. Distinguishing MCS from Sick Building Syndrome

MCS and sick building syndrome (SBS) have often been spoken of as synonymous entities. However, MCS is differentiated from SBS by the occurrence of symptoms in any indoor or outdoor environment in which the affected person encounters chemical exposures. In contrast, building-related illnesses occur only in a specific indoor setting, such as an office building, and can result not only from volatile chemical exposure but also from aeroallergens (*e.g.,* fungi), perhaps interacting with workplace stress [16]. SBS can precede the development of MCS [17]. Rates of SBS, as in MCS (see Section II.A), are reportedly higher in women compared to men [16,18]. Exact gender ratios for SBS are not known.

D. Overlap with Chronic Fatigue Syndrome, Fibromyalgia, and Gulf War Syndrome

Of note, a number of patients with various unexplained illnesses such as chronic fatigue syndrome (CFS) and fibromyalgia

Table 90.1

Heritable and Environmental Risk Factors Implicated in Sensitization and in the Phenomenology of CI, a Focal Symptom of MCS

Sensitizability	Chemical intolerance	Literature (human and animal studies)
Female gender	Female gender	[9,15,31–36]
Genetics	Family history of substance abuse	[14,32,37,38]
Increased sucrose intake	Carbohydrate craving	[32,39]
Hyperreactivity to novelty	Increased reports of shyness	[14,32,34,40]
Lateral asymmetry (leftward rotation)	Increased reports of left-handedness	[41–43]
Industrial chemicals (*e.g.*, toluene, formaldehyde, organophosphates)	"Feeling ill" on exposure to common odorants (*e.g.*, perfume, pesticide, fresh paint, new carpet, auto exhaust)	[11,12,14,23,32,44–48]
Drugs (*e.g.*, narcotics, stimulants, antidepressants, anxiolytics, neuroleptics, alcohol)	Family history of drug/alcohol abuse, but CI themselves cannot tolerate even small amounts	[14,32,37,49–52]
Stressful events (physical, psychological, and/or socioemotional)	Increased reports of sexual, physical, and emotional abuse, particularly in early life	[1,14,27,33,53–55]

Note. See Bell *et al.* [1] for additional discussion of endogenous mediators implicated in sensitization.

(FM) [19], as well as a subset of chronically ill Persian Gulf veterans (GWS) [20], share many of the polysymptomatic complaints of MCS patients and people with CI. Severe fatigue, muscle and joint pain, irritable bowel syndrome, headaches, and memory problems are some of the more common overlapping complaints. Chemical intolerance is frequently reported by these patients as well [19,21,22].

II. Epidemiology

A. Demographic Variables

Until recently, MCS had engendered a great deal of controversy (see Section VI) but little data. While we still know relatively little about the condition, we now have some data on its epidemiology and on patient characteristics [1,23–29]. In both population- and clinically-based studies, more women than men are affected by MCS and CI, with female:male ratios ranging from 1.6:1 to 3:1 [13–15,30]. The average age at which MCS patients present is between 30 and 40 years in most studies. Many MCS samples have people with higher levels of education than do healthy control groups. To date, little is known about ethnic characteristics, but most of the clinical- and population-based samples with MCS have been non-Hispanic white.

B. Health Profiles

Patients with MCS and/or CI are more likely to report personal histories of rhinitis, sinusitis, bronchitis, migraine headache, irritable bowel, food intolerances, arthritis, ovarian cysts, menstrual dysfunction, depression, anxiety, and panic disorders [1,13,31]. Miller and Mitzel [3] found that the most frequently reported symptoms in MCS patients included feeling unreal/

spacey (often a symptom of limbic nervous system dysfunction), memory difficulties, dizziness/lightheadedness, problems focusing eyes, muscle aches, tingling in fingers/toes, fatigue, headache, depressed feelings, chest discomfort, feeling irritable/edgy, shortness of breath, problems digesting food, eye burning/irritation, and loss of motivation. Their family histories are notable for increased rates of rhinitis, heart disease, diabetes, and substance abuse [1,13,31]. Studies using objective measures support the self-reported data in certain areas. For example, Doty *et al.* [30] demonstrated that MCS patients have increased nasal resistance, even at rest (*cf.*, rhinitis, sinusitis histories). Elderly people who have increased CI exhibit poorer total sleep time and sleep efficiency on polysomnographic recordings during a milk-containing diet as compared with a soy-based diet (*cf.*, insomnia symptoms, food intolerances; milk is a common food intolerance) [4,5]. These same elderly people with CI have higher waking supine blood pressures and heart rates in the laboratory (*cf.*, family hypertension histories) [26].

III. Risk Factors

A. Susceptibility to Sensitization

In both animals and human subjects, certain risk factors (Table 90.1) [31–55] have been observed that suggest susceptibility to neural sensitization (see Section V.A.1). These suspected risk factors include female gender, genetics, increased spontaneous intake of sucrose (a positive hedonic substance, whose intake is regulated in part by the mesolimbic pathways), heightened behavioral reactivity to novelty, and asymmetry of dopaminergic function in mesolimbic structures (as demonstrated in animal models by consistent preference for leftward turning after test dose delivery) [1,28,56]. In large surveys, Bell *et al.* [4,32,57]

found parallels between all of these risk factors and characteristics of persons with CI (*i.e.,* more common in women than men, increased family histories of substance abuse, elevated carbohydrate craving, greater shyness in community CI but not in MCS (shyness is a marker of novelty responses by a hyperreactive limbic nervous system) [33], and higher rates of left-handedness among persons who endorse a screening question on chemical sensitivity ("I consider myself to be especially sensitive to certain chemicals") [41].

B. Sexual Abuse and Susceptibility to Sensitization

In addition, a history of trauma, including sexual abuse, may be a risk factor for sensitization [27,53,58]. In a laboratory study of CI women without a history of sexual abuse, sexually abused women without CI, and controls with no history of CI or sexual abuse, Fernandez *et al.* [54] observed increases in electroencephalograph (EEG) alpha frequency bands over time in both the CI and abused women compared to the normal controls. Notably, they found that women with sexual abuse histories, but not CI, were sensitized, as shown by low and high EEG alpha activity, nonspecifically to every odorant stimulus, including room air sham. However, women with CI, but no sexual abuse histories, showed sensitization to particular odorants, *i.e.,* propylene glycol or peppermint, in the low alpha frequency band over sessions but did not show sensitization to others (*i.e.,* vanilla or room air sham). Notably, the low alpha band findings in the CI women in this study replicate findings in an animal study of sensitization [59]. Normal women habituated or showed no change over time to the same stimuli.

IV. Diagnosing MCS

Although diagnostic criteria have been established for chronic fatigue syndrome (CFS) and fibromyalgia (FM), there are still no clear-cut diagnostic guidelines for MCS [60]. Clinically, MCS patients report heightened awareness of odors in their environment. Despite this report, systematic tests of olfactory sensory ability in several different laboratories have shown that MCS patients and others with CI do not differ from normals on olfactory threshold or identification tests [30]. One study suggests that MCS patients may experience standardized odor stimuli on the University of Pennsylvania Smell Identification Test (UPSIT) as more unpleasant and/or more intense [61], but another study indicates that persons with less severe degrees of CI do not differ from normals on hedonic or strength ratings for odors [62].

A. Differentiating MCS from CI

The lack of case definition for MCS has lead our research team to use a chemical odor intolerance index (CII) [11] and an extended chemical sensitivity questionnaire [12] (both validated measures), in combination with a four-item lifestyle change report [7], in our studies to assist in: (a) differentiating MCS from CI, (b) determining the prevalence rates of MCS and CI in as consistent a manner as possible (see Section VIII.F), and (c) investigating patterns of illness/disability in MCS and/or CI individuals who have or have not made lifestyle changes. In addition,

Table 90.2

Selected Subjective Measures Used to Assess Differences between Groups in Studies of MCS and CI

Subjective measures	Reference(s)
Bell Chemical Odor Intolerance Index (CII)	[11]
Kipen Chemical Sensitivities Questionnaire	[12]
Barsky Somatic Symptom Amplification	[14]
Cheek-Buss Shyness Scale/self-reported shyness	[4,14,33,41]
Early Life Stress Reports	[1,27,33]
Food and/or drug intolerance reports	[1,4,5]
Handedness	[40,41]
Marlow-Crowne Social Desirability	[41]
McLean Limbic Symptom Checklist	[32,40]
M.D.-diagnosed depression, or reported health problems	[13,31,33,57,63]
Minnesota Multiphasic Personality Inventory (MMPI)-2	[10,64]
Pennebaker Symptom Inventory	[14]
Personal and/or family health histories	[13–15,31,32,41]
Profile of Moods States Scale (POMS)	[29]
Simon Lifestyle Change Questionnaire	[27,41,63]
Sleep quality	[24,57]
Symptom Checklist 90 (revised) (SCL-90-R)	[14,29,41,63]
Toronto Alexithymia Scale	[10]

we use a battery of subjective measures of personal and family health/psychiatric histories and personality characteristics, such as shyness (Table 90.2) [63,64], in conjunction with psychophysiological tests (Table 90.3) [65–67], such as cognitive tasks and electroencephalograph activity (qEEG), in order to determine further any distinguishing characteristics within and between MCS and CI individuals. By analogy, researchers [68,69] are beginning to find differences in subsets of the poorly understood condition, chronic fatigue syndrome.

Another possible difference between MCS and CI is that the nonolfactory neurobehavioral rather than the olfactory sensory functions of limbic areas receiving olfactory information are

Table 90.3

Objective Testing Used to Assess Differences between Groups in Studies of MCS and CI

Objective measures	Reference(s)
Beta-endorphin levels	[23]
Blood pressure/Heart rate measures	[1,26,27,56]
Cognitive Tasks (CVMT[a], DAT[b])	[10,65]
Immune profiles	[10,66]
Neopterin levels (serum)	[29]
Polysomnography	[24]
qEEG	[25,54,63,67]

[a]Continuous visual memory task.
[b]Divided attention task.

selectively disturbed in persons with CI. Olfactory stimuli, as well as other classes of limbic-relevant stimuli, would then activate abnormal responses mediated by these common pathways [70]. Although MCS patients and persons with less severe CI share many symptoms, they may differ in important, albeit as yet undiscovered, ways. Thus, it is unclear whether or not the findings reported earlier on olfactory perception reflect inconsistencies between studies or true differences between subsets of the chemically sensitive (*i.e.,* those who are sufficiently ill to receive a label of "MCS patient" and those who are simply chemically intolerant and higher functioning).

1. Trait Shyness in CI

Studies in our own laboratory suggest the possibility that community-based persons with CI may be different in a number of respects from MCS patients. Healthy persons with CI, many of whom cannot identify an initiating chemical, are significantly more shy than are MCS patients [14]. Shyness is a risk factor for anxiety disorders [71]. Notably, Fiedler *et al.* [10] have shown that it is the persons with CI and no initiating chemical exposure who have a higher rate of psychiatric pathology, including anxiety disorders, compared with MCS patients with an initiating chemical history. It is possible that different individuals with chemical sensitivity develop the problem from different origins, which converge on a final common pathway of manifestations. In other words, psychopathology can be a correlated, but not necessarily a causal feature, of chemical sensitivity, and it may be more important in some individuals than in others.

2. CI with and without Lifestyle Changes and Cognitive Task Skills

Fiedler *et al.* [10] found that MCS patients with a chemical initiator exhibited difficulties with a particular aspect of visual memory, a problem not seen in chemically sensitive persons without a chemical initiator. We [28] have a parallel set of findings that persons with CI who have made lifestyle changes (similar to MCS patients) have abnormalities in visual memory performance similar to those of the MCS patients of Fiedler *et al.* However, those individuals with CI who had not made lifestyle changes had an entirely different pattern of cognitive findings, with normal visual memory performance but slower reaction time performance on a visual divided attention test. Thus, the type of cognitive dysfunction may differ between MCS patients and persons with less disabling CI.

3. CI with and without Lifestyle Changes and qEEG Differences

We have found differential susceptibility to sensitization of waking electroencephalograph (qEEG) delta activity during repeated musk odor exposures in people with CI with lifestyle changes versus people with CI without such changes [72]. Surprisingly, the lifestyle change CI patients (who would be most similar to MCS patients) were less sensitizable to repeated low-level exposures in the laboratory than were the non-lifestyle change CI patients. One possibility in the latter situation is that the MCS-like group, in contrast with the nonlifestyle change CI patients, are already at ceiling for reactivity and cannot show further change in the laboratory [73]. Taken together, the findings suggest that there may indeed be meaningful differences

between subsets of chemically sensitive individuals. However, it is presently unknown whether or not the differences in psychological profiles, cognitive performance, and psychophysiological response patterns between people with CI with and without lifestyle changes reflect premorbid risk factors for development of clinical disorder, results of chronic illness, different conditions, and/or epiphenomena.

V. Possible Mechanisms

A. Neural Sensitization

Various investigators and clinicians have proposed numerous mechanisms for MCS. These range from psychogenic to immunological to neurological in focus [2]. At this time, no mechanism has been definitively ruled in or out, but research has shifted to an emphasis on possible central and peripheral nervous system mechanisms (*i.e.,* limbic/mesolimbic sensitization [1,28,29,74] and neurogenic inflammation [75,76]). The neural sensitization and neurogenic inflammation hypotheses together could account in large part for: (a) susceptibility to low level chemicals, (b) the variety of psychiatric, endocrine, and immune-related dysfunctions reported by women with MCS by way of low-threshold activation of the hypothalamic-pituitary-adrenal (HPA) axis, and (c) peripheral somatic symptoms, such as pain and inflammatory phenomena. We will emphasize neural sensitization because of our own work in this area [1,24,28,29,34, 56,77] and the relevance of sensitization to many clinical features of MCS and lesser degrees of CI [35,74].

1. Neural Sensitization Defined

Neural sensitization is the progressive amplification of a given response over time to repeated intermittent exposures of a stimulus. Animal studies [44] and human functional neuroimaging studies [78] suggest that chronic repeated intermittent solvent exposures, for example, can alter dopaminergic receptors and/or dopamine synthesis, thereby facilitating mesolimbic sensitization. Neural sensitization has been most well documented in animal models, primarily rodents [35,49,50]. Greater vulnerability to sensitization has been reported in female and in neutered adult male animals [35,36], suggesting that gonadal hormones, particularly estrogen, influence sensitizability. Classes of stimuli that can initiate sensitization in these animal studies include various drugs [45,50,51], certain environmental chemicals (*e.g.,* solvents, volatiles, pesticides) [44], odorants such as peppermint [79], endogenous mediators, including substance P analogs (a major mediator of neurogenic inflammation) [55], and physical or psychological stress [50]. Kay [79] reported that toluene and other odorants with both olfactory and trigeminal features were better at inducing a sensitization of paroxysmal EEG beta activity in animals than were more purely olfactory odorants such as vanilla.

2. Neural Sensitization, Limbic Kindling, and Neuropsychiatric Disorders

Sensitization in animals has also been proposed as a model for the long-term course of a number of different neuropsychiatric disorders (*e.g.,* depression, panic disorder, posttraumatic stress disorder, chronic pain, somatoform disorders) that some investigators have found in excess in the personal histories [74,80], or family histories (*e.g.,* drug craving in substance abuse), of persons with CI and MCS [1]. Limbic kindling is a special case of neural sensitization that serves as an animal model for temporal lobe epilepsy [74]. Typically, in animal models, repeated electrical or chemical stimuli induce a progressive lowering of firing threshold until a stimulus that caused little or no initial response later elicits a full seizure. Many pesticides facilitate limbic kindling [81]. They can elevate cholinergic function and/or increase excitatory amino acid activity, both of which can increase CNS excitability and favor limbic nervous system sensitization. The amygdala has modulatory input into the activity of the mesolimbic pathways [51]. Even subconvulsive kindling of the amygdala can induce lasting increases in anxious behavior in animals [52]. An implication of these observations is that a psychological symptom being exhibited by a woman with MCS could reflect dysfunctional neural substrates, not necessarily that the current chemical sensitivity is a psychogenic problem, a misattribution, or imagination. This model also predicts that, in large epidemiologic surveys, we should find not only comorbid psychiatric disorders, but also an excess of documented temporal lobe epilepsy and its variants in MCS patients and people with CI as compared to controls without CI.

3. Animal and Human Outcome Variables Related to Neural Sensitization

Diverse outcome variables in rodent models that can show sensitization include psychomotor activity levels, various central neurotransmitters and hormones, blood pressure and heart rate, electroencephalograph (EEG) activity, especially in the alpha frequency band, and pain behavior. Healy *et al.* [82] have shown that effects of amygdaloid kindling on vagal tone influenced abnormal cardiac patterns in animals. Research in human subjects has shown that it is possible to sensitize a range of outcome measures, including autonomic nervous system variables such as cardiovascular responses [1,26], EEG alpha [*e.g.,* 54,63] and delta frequency activity [72], and spontaneous blink rate, which is controlled by CNS dopamine, a neurotransmitter that plays a key role in mesolimbic sensitization [83]. Subsequently, the size of the host response elicited by the same level or a lower intensity stimulus grows with the passage of time alone. A one-time low-level exposure may not initiate or elicit hyperreactivity that a series of intermittent exposures might initiate, but the amplified response may not be elicitable until several days to weeks of no exposure have elapsed (*i.e.,* single session studies cannot reliably test for sensitizability).

B. Neurogenic Inflammation

Neurogenic inflammation is a form of inflammation (redness, heat, swelling, pain) initiated by activation of peripheral nervous system c-fiber neurons rather than by immunological events [75]. The neuronal activity leads to neuropeptide release and inflammation at sites different from the original stimulus. Substance P is one of several peptides that mediate this process.

Many different chemicals, including formaldehyde, ether, and cigarette smoke, can exert an irritant effect by stimulating these c-fibers in humans. Individuals differ in the susceptibility

of their c-fibers to this form of chemical reactivity. The CNS integrates the information from such peripheral events and may upregulate or modulate responsivity of the local tissues to the mediators. Meggs [84] has pointed out that neurogenic inflammation may play a role in certain asthmas, rhinitis, and migraine headache, all conditions commonly reported in MCS and CI. He has preliminary supporting evidence for inflammatory changes from lymphocytic infiltrates of nasal mucosal tissue in MCS patients, findings consistent with neurogenic inflammatory events [76]. This work is limited, however, by the lack of matched control groups.

VI. Skepticism Surrounding MCS

A. Ambiguous Exposure-Response Relationships, Chemophobia, Litigation Confounds

MCS fosters intense debate, often characterized by irrational and illogical arguments. Skeptics [85,86] insist that MCS is not a new condition, but rather that it represents a new label for age-old problems such as hysteria or somatization disorder and other psychiatric diagnoses. Some individuals suggest that MCS patients are simply mistaken; low level chemicals are not causing their symptoms. This issue is confounded by the fact that traditional exposure-response relationships are ambiguous; subthreshold exposures trigger large responses. Chemophobia, as well as occupationally-based litigation, further confound MCS. Skeptics do not offer alternative mechanistic explanations for the inflammatory conditions that often accompany MCS. They also overlook the fact that in clinical psychiatry, conditions that begin after the usual age of onset typically have biological rather than psychogenic origins (*e.g.,* somatization disorder begins before age 30, whereas MCS usually begins between age 30–40) [77]. Trying to stop the debate by conferring psychiatric diagnoses is incorrect; arguing over labels will not resolve the question of etiology [34,77,87,88].

B. Psychogenic versus Psychobiological Underpinnings

Psychiatric research is at the threshold of many breakthroughs with new tools to examine brain chemistry and brain function *in vivo* [89,90]. Convergent bodies of evidence indicate that all of the major psychiatric disorders have important biological mechanisms as underpinnings. Patients who meet criteria for somatization disorder also have high rates of comorbid lifetime diagnoses of these other major psychiatric disorders, the same disorders that have biological mechanisms [91]. It is illogical to assume that their somatic symptoms would be figments of their imagination while the rest of their psychopathology has biological roots. Stress and adverse life events can initiate, trigger, and exacerbate medical conditions [92]. Even in posttraumatic stress disorder [90,93], symptoms derive from acquired neurobiological dysfunction, not from a patient's imagination or misattribution. An overwhelming body of data demonstrates that exogenous or endogenous mediators, drugs, and chemicals can activate the same pathways as can stress and adverse life events [94].

At the same time, skeptics continue to point to a relative lack of well-controlled data on MCS. However, some data are finally beginning to emerge. We are at a point in the field when professionals on all sides of the issue should take a rational look at all of the available evidence and undertake difficult but manageable scientific studies before entering into further debate. All of us need to acknowledge that the data may not support some firmly held beliefs about MCS of both proponents and skeptics. If we wish to help the patients who are suffering, we need to pursue more research and follow the data where they lead.

VII. Treatment

Treatment for MCS remains as elusive as a clear-cut diagnosis for the disorder. Anecdotally, patients who seek assistance for MCS and the accompanying polysymptomatic complaints have reported a great deal of frustration with the traditional medical community. Gibson et al. [6] examined psychosocial and quality of life data from 305 individuals with self-identified MCS (80.3% women). Lifestyle disruption was pervasive; work and financial problems, reduced mobility, reduced access to resources and public space, little medical or social support, and quality of medical services all exacerbated the considerable personal distress experienced by MCS patients. Standard medical tests covered by health maintenance organizations are usually negative, and nonstandard testing is either capitated, or not covered, due to the ''experimental'' nature of the testing. Many of these patients are referred to psychiatrists, particularly if the referring physician believes the disorder to be primarily psychogenic, if established medical treatments fail to relieve the chronic complaints, or both. Eventually, a number of these patients become increasingly more isolated in rural areas in specially modified ''sterile'' environments in order to reduce their chances of being exposed to the variety of chemicals that trigger their symptoms. This isolation also results in reduced social support. Some patients have developed grass-roots organizations in order to disseminate information regarding the condition and, in part, in response to the lack of assistance they have received from their health care providers. This networking has also provided MCS patients with names of physicians who specialize in treating ''environmental illness.''

A. Traditional and Nontraditional Interventions

Relevant to the polysymptomatic presenting problems, there are no universal treatment guidelines for MCS. Traditionally, treatment approaches have included psychotropic medications for psychiatric comorbidities, and selective serotonin reuptake inhibitors (SSRIs), such as fluoxetine, for pain relief. However, it is not uncommon for MCS patients to react adversely to the very interventions that are intended to reduce their symptoms. There is a growing body of literature to suggest that patients with unexplained illnesses, such as MCS, FM, and CFS, are turning to alternative therapies for amelioration of symptoms [19,95]. In an interviewer-based survey of 301 patients attending a rheumatology clinic, Pioro-Boisset et al. [95] reported alternative medicine approaches used extensively by individuals with rheumatologic conditions, particularly FM patients. Given the overlap between FM and MCS, it is probable that MCS patients are also more likely to turn to these alternative approaches, such as spiritual practices (meditation, prayer, self-help groups), alternative practitioners (acupuncturists, chiropractors,

homeopaths, massage therapists), botanicals (herbal preparations, vitamins), and dietary modifications. Allopathic health care providers should be aware of these approaches as potentially palliative options for MCS patients. Palliative treatment can be guided by the MCS patient in terms of what works best for her with the least amount of discomfort and/or exacerbation of symptoms.

B. Need for Health Care Provider Education

In order to provide quality patient care, it is also important for health care providers to receive education regarding MCS and its hallmark symptom of chemical odor intolerance. Whether skeptical of the disorder or not, medical education regarding health care delivery to MCS patients should include training in pertinent documentation of personal and family history, including early life stress, odors known to elicit symptoms, and any initiating events. Thorough history-taking (e.g., differentiating between CI with and without lifestyle changes and/or an initiating odorant) can become an important component in determining appropriate treatment guidelines specific to the patient, including behavioral modifications at home and at work. Active listening in an open and caring manner and familiarity with organizations that provide networking could help the MCS patient feel less socially isolated and more amenable to traditional medical care.

C. Discrepancy in Diagnosing Women's Health Complaints?

Finally, the preponderance (up to 80%) of women with MCS in population-based samples could be clinically relevant to women's health and health care delivery. There has been a tradition in allopathic medicine of attributing women's symptomatic complaints to psychogenic origins, while men's health complaints (even if they parallel their female cohorts) have generally been given a medical diagnosis with referral to a medical specialist [96]. Again, polysymptomatic complaints are a feature of MCS, as in other poorly understood conditions such as CFS and FM, and a majority of these patients are women. While skeptics [85,86] would suggest that women with MCS are exhibiting a variation of somatoform disorder, neuroimmune and/or neurohormonal dysregulations can present as major psychiatric disorders, and somatization may be symptomatic of such dysregulations [77]. For instance, serum neopterin levels, a nonspecific marker for inflammation [97], showed significant correlations with three scores measuring different aspects of ''somatization'' in women with CI, but not in depressed women without CI or normals [29]. These measures were the Symptom Checklist 90 (revised) somatization subscale [98], McLean Limbic Symptom Checklist somatic subscale (somatic symptoms of temporal lobe epilepsy) [99], and the Profile of Mood States Scale fatigue subscale [POMS—Educational and Testing Service, San Diego, CA]. Such findings offer preliminary evidence in support of biological mediation, possibly inflammatory in nature, for the multiple system symptoms of at least some women with CI. Even in a general medical population, Kroenke and Spitzer [91] recommended that evaluation and management of physical symptomatology become a focal issue in women's health; they found that physical and somatoform symptoms were reported 50% more often by women compared to men even without psychiatric comorbidity. Clearly, there is a need for further research in primary health care of women in general, and the diagnosis, mechanisms, and optimization of clinical management of women with MCS and CI in particular.

VIII. Future Research in MCS

A. Autonomic Measures

Laboratory studies in human subjects offer considerations for the design of future research in MCS. Careful choice of study subjects and controls and the use of multiple sessions separated by enough time (in days) are crucial. For example, Newlin and Thomson [37] showed that sons of alcoholics can sensitize autonomic measures over two weeks to repeated ethanol ingestion in the laboratory, while sons of nonalcoholics habituate to the same procedures. Morrow and Steinhauer [46] found that solvent-exposed workers sensitize and/or fail to habituate heart rate and pupil responses to a stressful complex cognitive task over weeks to months in comparison with unexposed workers. Such findings certainly indicate the need to (a) select control groups who do not overlap the characteristics of patients under study, and (b) evaluate subjects over more than one session. Many earlier investigations in MCS failed to screen comparison groups for CI and for acute, peak, or cumulative higher level chemical exposure histories, let alone other stimuli (e.g., past physical or psychological trauma) also capable of initiating sensitization. In addition, most studies of MCS patients have used only one session to compare with controls.

B. Reducing False Negative Outcomes

In drug studies, the data suggest that persons who are naïve to the test agent in the laboratory will exhibit sensitization to low-dose repeated exposures, whereas persons who are chronic drug addicts will not. Gorelick and Rothman [73] have proposed that the explanation for this discrepancy may lie in a ceiling effect (i.e., drug addicts are already maximally sensitized and showing their peak responsiveness to any given dose, which is perhaps also attenuated by concomitant tolerance processes). Naïve subjects are less likely to have preexisting sensitization or tolerance to the test agent in a laboratory situation. Thus, knowing the prior history of the subjects with the test agent or cross-sensitizing agents at the time of the laboratory sessions is crucial. Recency of last exposure is a key variable; in animal studies, it takes up to 5 days after the initiating agent to see evidence of a sensitized response to the next exposure [50]. An alternative explanation is that some persons with CI are more vulnerable to context-dependent than to context-independent sensitization. In context dependency, it would be less possible to elicit sensitization unless the test substance is given in the physical setting in which the sensitization was initiated [56,100]. Finally, because of the importance of the intensity of the initiating stimulus, it may require a convergence of weaker stimuli, such as mixed rather than single solvent exposures or chemicals plus life stress, to set sensitization in motion. Studies on chemical sensitization in human subjects will need to take all of these

potential factors into consideration in designing appropriate studies. Failure to do so may lead to false negative outcomes.

C. Capturing Susceptibility and Sensitization with qEEG

These research design issues could pertain to MCS in a variety of ways. For the clinical situation, we have both a likely hierarchy of potential neural sensitizers as initiators (e.g., pesticides and solvents more than other classes of chemicals) [3] and individual variability in susceptibility (i.e., different degrees of sensitizability from different factors) to being sensitized. One of several promising markers of sensitization may be spectral electroencephalograph (EEG) activity, especially in view of evidence of increased EEG delta and alpha activity in solvent-exposed workers [101], a population who also have an elevated prevalence of CI in other studies [102]. As in animal studies, persons who have already instituted chemical avoidance for extended periods may pass through a period of heightened reactivity, but they eventually may show less sensitizability.

D. Testing Categories of MCS Patients with Appropriate Controls

The data imply that abuse history may be a particularly powerful sensitizing experience for women. An important next step would be to study MCS patients with and without sexual abuse history compared with appropriate controls. It may be that the clinical picture of some MCS patients is complicated by an interaction with prior abuse and chemical exposures. Prior abuse may heighten the capacity to sensitize to many different chemicals and stressors (sham), with widespread hyperreactivity as a consequence of stimulus generalization (many odorants; not only stress, but also foods, drugs). It is important to note, however, that these findings suggest past abuse is not necessary for vulnerable people to have current CI from specific chemicals, especially substances with some trigeminal stimulus properties.

E. Cross-Sensitization and Research Design Issues

The nature of sensitization requires further attention to design issues seemingly far removed from a question focused on chemicals or odorants. In animals, physical or psychological stress cross-sensitizes with drugs [49]. In other words, studies of chemical sensitization in people must take into account the history of the subjects and controls with regard to a full range of possible cross-sensitizing categories of initiating stimuli, including life stress, as well as the specific nature and intensity of the initiating and eliciting stimuli. Women who currently see themselves as sensitive to chemicals could have arrived at that state from past life traumas alone, past chemical exposures alone, past drug use alone, past infections alone (which activated sensitizing cytokines and other mediators), or some combination of all of these factors [1].

F. Controlling for Potential Sources of Bias in Prevalence Studies

The 4–6% prevalence rates we cited for MCS appear high and requires commentary. Lack of adequate case definition (see

Section I.A) for either research or clinical purposes makes it difficult to ascertain the prevalence rates of MCS in the general population. We derived a rough estimate of MCS in the general public from three different surveys using three different questions: (1) members of community-living elderly in the Southwest with physician-diagnosed chemical sensitivity (4% of 192 subjects) [33]; (2) a random-dial telephone survey of self-reported chemical sensitivity in a rural Southern population in which 4–5% of 1027 respondents reported daily or almost daily symptoms of chemical sensitivity [15]; and (3) 6.3% of 4046 randomly-selected individuals in a state-wide California survey, who reported being diagnosed with MCS by a physician [103]. We recognize that the use of telephone surveys in the recruitment of participants for these types of studies is subject to selection bias as well as questions of the validity of self-reported illness. In addition, problems related to diagnosis of MCS (information bias) must be considered. Our research group uses two validated measures [11,12] and a lifestyle change report [7] to define our MCS and CI groups (see Section IV.A). Until a case definition is established for the disorder, the prevalence rates of MCS will be difficult to determine. It is essential that researchers studying MCS and CI make explicit their criteria for subject selection, data obtained by self-report, and any potential for bias in selection and recruitment. It is of interest to note, however, that despite variations in the ways in which these different studies derived their chemically sensitive samples, the percentages fall within similar numerical parameters.

IX. Conclusions

Only a truly multidisciplinary research approach involving basic science research in sensitization and clinical research can address the complex multifactorial problems of women with MCS and related CI. Emerging data suggest the likelihood of subtypes with differing clinical histories, disability, and psychopathology. Research on neural sensitization and neurogenic inflammation is only a starting point for understanding the many contributing factors and mediators of the clinical disorder. These models also facilitate animal studies on initiation processes that are not possible in human subjects [44,45,47,79]. MCS may be a heterogeneous group of related conditions with different initiating etiologies and overlapping endpoint phenomenologies. Improved understanding of MCS may finally advance research on the nature of and treatment for illnesses in a subset of women previously dismissed by physicians as having somatoform disorders.

Acknowledgments

The authors acknowledge Margaret Kurzius-Spencer, M.S., MPH, Biostatistics Core, Respiratory Sciences Center, University of Arizona, for her thoughtful review and suggestions, as well as the guidance and feedback provided by Section Editor Karen B. Schmaling, Ph.D., ABPP. We also offer our appreciation to all the participants, their families, the research team members, and funding agencies (Environmental Health Foundation; NIH/NHLBI SCOR Grant HL14136; Wallace Genetic Foundation) who made our studies possible.

References

1. Bell, I. R., Baldwin, C. M., and Schwartz, G. E. (1998). Illness from low levels of environmental chemicals: Relevance to chronic fatigue syndrome and fibromyalgia. *Am. J. Med.* **105**(3A), 74S–82S.

2. Fiedler, N., and Kipen, H. (1997). Chemical sensitivity: The scientific literature. *Environ. Health Perspect.* **105**(Suppl. 2), 409–415.

3. Miller, C. S., and Mitzel, H. (1995). Chemical sensitivity attributed to pesticide exposure vs. remodeling. *Arch. Environ. Health* **50**, 119–129.

4. Bell, I. R., Schwartz, G. E., Peterson, J. M., and Amend, D. (1993). Symptom and personality profiles of young adults from a college student population with self-reported illness from foods and chemicals. *J. Am. Coll. Nutr.* **12**, 693–702.

5. Bell, I. R., Schwartz, G. E., Peterson, J. M., Amend, D., and Stini, W. A. (1993). Possible time-dependent sensitization to xenobiotics: Self-reported illness from chemical odors, foods, and opiate drugs in an older adult population. *Arch. Environ. Health* **48**, 315–327.

6. Gibson, P. R., Cheavens, J., and Warren, M. L. (1996). Chemical sensitivity/chemical injury and life disruption. *Women Ther.* **19**, 63–79.

7. Simon, G. E., Katon, W. J., and Sparks, P. J. (1990). Allergic to life: Psychological factors in environmental illness. *Am. J. Psychiatry* **147**, 901–906.

8. Cullen, M. R. (1987). The worker with multiple chemical sensitivities: An overview. *Occup Med: State of the Art Rev.,* pp. 655–662.

9. Ashford, N. A., and Miller, C. S. (1998). ''Chemical Exposures. Low Levels and High Stakes,'' 2nd ed., Van Nostrand-Reinhold, New York.

10. Fiedler, N., Kipen, H. M., DeLuca, J., Kelly-McNeil, K., and Natelson, B. (1996). A controlled comparison of multiple chemical sensitivities and chronic fatigue syndrome. *Psychosom. Med.* **58**, 38–49.

11. Szarek, M. J., Bell, I. R., and Schwartz, G. E. (1997). Validation of a brief screening measure of environmental chemical sensitivity: The chemical odor intolerance index. *J. Environ. Psychol.* **17**, 345–351.

12. Kipen, H. M., Hallman, W., Kelly-McNeil, K., and Fiedler, N. (1995). Measuring chemical sensitivity prevalence: A questionnaire for population studies. *Am. J. Public Health* **85**, 574–577.

13. Baldwin, C. M., Bell, I. R., O'Rourke, M. K., and Lebowitz, M. D. (1997). The association of respiratory problems in a community sample with self-reported chemical intolerance. *Eur. J. Epidemiol.* **13**, 547–552.

14. Bell, I. R., Peterson, J. M., and Schwartz, G. E. (1995). Medical histories and psychological profiles of middle-aged women with and without self-reported illness from environmental chemicals. *J. Clin. Psychiatry* **56**, 151–60.

15. Meggs, W. J., Dunn, K. A., Bloch, R. M., Goodman, P. E., and Davidoff, A. L. (1996). Prevalence and nature of allergy and chemical sensitivity in a general population. *Arch. Environ. Health* **51**, 275–282.

16. Kreiss, K. (1989). The epidemiology of building-related complaints and illness. *Occup. Med.: State of the Art Rev.* **4**, 575–592.

17. Welch, L. S., and Sokas, R. (1992). Development of multiple chemical sensitivity after an outbreak of sick building syndrome. *Toxicol. Ind. Health* **8**, 47–50.

18. Wallace, L., Nelson, C. J., Kollander, M., Leaderer, B., Bascom, R., and Dunteman, G. (1991). Indoor air quality and work environment study: Multivariate statistical analysis of health, comfort, and odor perceptions as related to personal and workplace characteristics. *U.S. Environ. Prot. Agency* **4**, 32–33 (EPA Headquarters Buildings, Atmospheric Research and Exposure Assessment Laboratory, 21M-3004).

19. Buchwald, D., and Garrity, D. (1994). Comparison of patients with chronic fatigue syndrome, fibromyalgia, and multiple chemical sensitivities. *Arch. Intern. Med.* **154**, 2049–2053.

20. Fiedler, N., Kipen, H. M., Natelson, B., and Ottenweller, J. (1996). Chemical sensitivities and the Gulf War: Dept. of Veterans Affairs

Research Center in basic and clinical science studies of environmental hazards. *Regul. Toxicol. Pharmacol.* **24,** S129–S138.

21. Bell, I. R., Warg-Damiani, L., Baldwin, C. M., Walsh, M., and Schwartz, G. E. R. (1998). Self-reported chemical sensitivity and wartime chemical exposures in Gulf War veterans with and without decreased global health ratings. *Mil. Med.* **163,** 725–732.

22. Slotkoff, A. T., Radulovic, D. A., and Clauw, D. J. (1997). The relationship between fibromyalgia and the multiple chemical sensitivity syndrome. *Scand. J. Rheumatol.* **26,** 364–367.

23. Bell, I. R., Bootzin, R. R., Davis, T. P., Hau, V., Ritenbaugh, C., Johnson, K. A., and Schwartz, G. E. (1996). Time-dependent sensitization of plasma beta-endorphin in community elderly with self-reported environmental chemical odor intolerance. *Biol. Psychiatry* **40,** 134–143.

24. Bell, I. R., Bootzin, R. R., Ritenbaugh, C., Wyatt, J. K., De-Giovanni, G., Kulinovich, T., Anthony, J. L., Kuo, T. F., Rider, S. P., Peterson, J. M., Schwartz, G. E., and Johnson, K. A. (1996). A polysomnographic study of sleep disturbance in community elderly with self-reported environmental chemical odor intolerance. *Biol. Psychiatry* **40,** 123–133.

25. Bell, I. R., Kline, J. P., Schwartz, G. E., and Peterson, J. M. (1997). Quantitative EEG patterns during nose versus mouth inhalation of filtered room air in young adults with and without self-reported chemical odor intolerance. *Int. J. Psychophysiol.* **59,** 144–149.

26. Bell, I. R., Schwartz, G. E., Bootzin, R. R., and Wyatt, J. K. (1997). Time-dependent sensitization of heart rate and blood pressure over multiple laboratory sessions in elderly individuals with chemical odor intolerance. *Arch. Environ. Health* **52,** 6–17.

27. Bell, I. R., Baldwin, C. M., Russek, L. G. S., Schwartz, G. E. R., and Hardin, E. E. (1999). Early life stress, negative paternal relationships, and chemical intolerance in middle-aged women: Support for a neural sensitization model. *J. Women's Health* **7,** 1135–1147.

28. Bell, I. R., Bootzin, R. R., Schwartz, G. E., Baldwin, C. M., and Ballesteros, F. (1999). Differing patterns of cognitive dysfunction and heart rate reactivity in chemically-intolerant individuals with and without lifestyle changes. *J. Chron. Fatigue Synd.* **5,** 4–25.

29. Bell, I. R., Patarca, R., Baldwin, C. M., Klimas, N. G., Schwartz, G. E., and Hardin, E. E. (1998). Serum neopterin and somatization in women with chemical intolerance, depressives, and normals. *Neuropsychobiology* **38,** 13–18.

30. Doty, R. L., Deems, D. A., Frye, R. E., Pelberg, R, and Shapiro, A. (1988). Olfactory sensitivity, nasal resistance, and autonomic function in patients with multiple chemical sensitivities. *Arch. Otolaryngol. Head Neck Surg.* **114,** 1422–1427.

31. Baldwin, C. M., and Bell, I. R. (1998). Increased cardiopulmonary disease risk in a community-based sample with chemical odor intolerance: Implications for women's health and health care utilization. *Arch. Environ. Health* **53,** 347–353.

32. Bell, I. R., Hardin, E. E., Baldwin, C. M., and Schwartz, G. E. (1995). Increased limbic system symptomatology and sensitizability of young adults with chemical and noise sensitivities. *Environ. Res.* **70,** 84–97.

33. Bell, I. R., Schwartz, G. E., Amend, D., Peterson, J. M., and Stini W. A. (1994). Sensitization to early life stress and response to chemical odors in older adults. *Biol. Psychiatry* **35,** 857–863.

34. Bell, I. R. (1994). Neuropsychiatric aspects of sensitization to low level chemicals: A neural sensitization model. *Toxicol. Ind. Health* **10,** 277–312.

35. Antelman, S. M. (1994). Time-dependent sensitization in animals: A possible model of multiple chemical sensitivity in humans. *Toxicol. Ind. Health* **10,** 335–342.

36. Robinson, T. E., Becker, J. B., and Presty, S. K. (1982). Long-term facilitation of amphetamine-induced rotational behavior and striatal dopamine release produced by a single exposure to amphetamine: Sex differences. *Brain Res.* **253,** 231–241.

37. Newlin, D. B., and Thomson, J. B. (1991). Chronic tolerance and sensitization to alcohol in sons of alcoholics. *Alcohol.: Clin. Exp. Res.* **15,** 399–405.

38. Tolliver, G. E., Belknap, J. K., Woods, W. E., and Carney, J. M. (1994). Genetic analysis of sensitization and tolerance to cocaine. *J. Pharmacol. Exp. Ther.* **270,** 1230–1238.

39. Sills, T. L., and Vaccarino, F. J. (1994). Individual differences in sugar intake predict the locomotor response to acute and repeated amphetamine administration. *Psychopharmacology* **116,** 1–8.

40. Hooks, M. S., Jones, G. H., Neill, D. B., and Justice, J. B. (1991). Individual differences in amphetamine sensitization: dose-dependent effects. *Pharmacol. Biochem. Behav.* **41,** 203–210.

41. Bell, I. R., Miller, C. S., Schwartz, G. E., Peterson, J. M., and Amend, D. (1996). Neuropsychiatric and somatic characteristics of young adults with and without self reported chemical odor intolerance and chemical sensitivity. *Arch. Environ. Health* **51,** 275–282.

42. Baldwin, C. M., Bell, I. R., Schwartz, G. E., and Quan, S. F. (1997). Asthma, handedness, neurodevelopmental disorder reports, and limbic system reactivity in a college-based sample. *Ann. Behav. Med.* **17,** S115.

43. LaHoste, G. J., Mormede, P., Rivet, J. M., and LeMoal, M. (1988). Differential sensitization to amphetamine and stress responsivity as a function of inherent laterality. *Brain Res.* **453,** 381–384.

44. von Euler, G., Ogren, S., Eneroth, P., Fuxe, K., and Gustafsson, J. A. (1994). Persistent effects of 80 ppm toluene on dopamine-regulated locomotor activity and prolactin secretion in the male rat. *Neuro Toxicology* **15,** 621–624.

45. Sorg, B. A., Willis, J. R., See, R. E., Hopkins, B., and Westberg, H. H. (1998). Repeated low-level formaldehyde exposure produces cross-sensitization to cocaine: Possible relevance to chemical sensitivity in humans. *Neuropsychopharmacology* **18,** 385–394.

46. Morrow, L. A., and Steinhauer, S. R. (1995). Alterations in heart rate and pupillary response in persons with organic solvent exposure. *Biol. Psychiatry* **37,** 721–730.

47. Sorg, B. A., Willis, J. R., Nowatka, T. C., Ulibarri, C., See, R. E., and Westberg, H. H. (1996). A proposed animal neurosensitization model for multiple chemical sensitivity in studies with formalin. *Toxicology* **111,** 135–145.

48. Ryan, C. M., Morrow, L. A., and Hodgson, M. (1988). Cacosmia and neurobehavioral dysfunction associated with occupational exposure to mixtures of organic solvents. *Am. J. Psychiatry* **145,** 1442–1445.

49. Antelman, S. M. (1988). Time-dependent sensitization as the cornerstone for a new approach to pharmacotherapy: drugs as foreign/stressful stimuli. *Drug Dev. Res.* **14,** 1–30.

50. Kalivas, P. W., Sorg, B. A., and Hooks, M. S. (1993). The pharmacology and neural circuitry of sensitization to psychostimulants. *Behav. Pharmacol.* **4,** 315–334.

51. Kalivas, P. W., and Alesdatter, J. E. (1993). Involvement of N-methyl-D-aspartate receptor stimulation in the ventral tegmental area and amygdala in behavioral sensitization to cocaine. *J. Pharmacol. Exp. Ther.* **267,** 486–495.

52. Adamec, R. (1994). Modeling anxiety disorders following chemical exposures. *Toxicol. Ind. Health* **10,** 391–420.

53. Staudenmayer, H., Selner, M. E., and Selner, J. C. (1993). Adult sequelae of childhood abuse presenting as environmental illness. *Ann. Allergy* **71,** 538–546.

54. Fernandez, M., Bell, J. R., and Schwartz, G. E. R. (1999). EEG sensitization during chemical exposures in women with and without chemical sensitivity of unknown etiology. *Toxicol. Indust. Health* **15,** 305–312.

55. Kalivas, P. W., and Stewart, J. (1991). Dopamine transmission in the initiation and expression of drug- and stress-induced sensitization of motor activity. *Brain Res. Rev.* **16,** 223–244.

56. Bell, I. R., Schwartz, G. E., Baldwin, C. M., Hardin, E. E., Klimas, N., Kline, J. P., Patarca, R., and Song, Z. Y. (1997). Individual differences in neural sensitization and the role of context in illness from low level environmental chemical exposures. *Environ. Health Perspect.* **105**(Suppl. 2), 457–466.

57. Bell, I. R., Schwartz, G. E., Peterson, J. M., and Amend, D. (1993). Self reported illness from chemical odors in young adults without clinical syndromes or occupational exposures. *Arch. Environ. Health* **48**, 6–13.

58. Yehuda, R., and Antelman, S. M. (1993). Criteria for rationally evaluating animal models of posttraumatic stress disorder. *Biol. Psychiatry* **33**, 479–486.

59. Ferger, B., Stahl, D., and Kuschinsky, K. (1996). Effects of cocaine on the EEG power spectrum of rats are significantly altered after its repeated administration: Do they reflect sensitization phenomena? *Arch. Pharmacol.* **353**, 545–551.

60. Nethercott, J. R., Davidoff, L. L., Curbow, B., and Abbey, H. (1993). Multiple chemical sensitivities syndrome: toward a working case definition. *Arch. Environ. Health* **48**, 19–26.

61. Kurtz, D., White, T., and Belknap, E. (1993). Perceived odor intensity and hedonics in subjects with multiple chemical sensitivity. *Chem. Senses* **18**, 584.

62. Fernandez, M., Schwartz, G. E. R., and Bell, I. R. (1998). Intensity and pleasantness ratings of odorants by women with ideopathic chemical sensitivity. submitted for publication.

63. Bell, I. R., Schwartz, G. E., Baldwin, C. M., Hardin, E. E., and Kline, J. P. (1998). Differential resting qEEG alpha patterns in women with environmental chemical intolerance, depressives, and normals. *Biol. Psychiatry* **43**, 376–388.

64. Fernandez, M., Bell, I. R., Herring, A. M., Baldwin, C. M., McCormick, L. M., and Schwartz, G. E. S. (1998). Similarities of MMPI profiles in chemically sensitive subjects and chronically medically ill patients. Submitted for publication.

65. Bell, I. R., Wyatt, J. K., Bootzin, R. R., and Schwartz, G. E. (1996). Slowed reaction time performance on a divided attention task in elderly with environmental chemical odor intolerance. *Int. J. Neurosci.* **84**, 127–134.

66. Kipen, H. M., Fiedler, N., Maccia, C., Yurkow, E., Todaro, J., and Laskin, D. (1992). Immunologic evaluation of chemically sensitive patients. *Toxicol. Ind. Health* **8**, 125–135.

67. Schwartz, G. E., Bell, I. R., Dikman, Z. V., Fernandez, M., Kline, J. P., Peterson, J. M., and Wright, K. P. (1994). EEG responses to low-level chemicals in normals and cacosmics. *Toxicol. Ind. Health* **10**, 633–643.

68. DeLuca, J., Johnson, S. K., Ellis, S. P., and Natelson, B. H. (1997). Cognitive functioning is impaired in patients with chronic fatigue syndrome devoid of psychiatric disease. *J. Neurol. Neurosurg. Psychiatry* **62**, 151–155.

69. Hickie, I., Lloyd, A., Hadzi-Pavlovic, D., Parker, G., Bird, K., and Wakefield, D. (1995). Can the chronic fatigue syndrome be defined by distinct clinical features? *Psychol. Med.* **25**, 925–935.

70. Locatelli, M., Bellodi, L., Perna, G., and Scarone S. (1993). EEG power modifications in panic disorder during a temporolimbic activation task: Relationships with temporal lobe clinical symptomatology. *J. Neuropsychiatry Clin. Neurosci.* **5**, 409–414.

71. Rosenbaum, J. F., Biederman, J., Bolduc-Murphy, F. A., Faraone, S. B., Chaloff, J., Hirshfeld, D. R., and Kagan, J. (1993). Behavioral inhibition in childhood: A risk factor for anxiety disorders. *Harv. Rev. Psychiatry* **1**, 2–16.

72. Bell, I. R., Szarek, M. J., DiCenso, D. R., Baldwin, C. M., Schwartz, G. E., and Bootzin, R. R. (1999). Patterns of waking EEG spectral power in chemically intolerant individuals during repeated chemical exposures. *Intl. J. Neurosci.* **97**, 41–59.

73. Gorelick, D. A., and Rothman, R. B. (1997). Stimulant sensitization in humans. *Biol. Psychiatry* **42**, 230–231.

74. Bell, I. R., Miller, C. S., and Schwartz, G. E. (1992). An olfactory-limbic model of multiple chemical sensitivity syndrome: Possible relationships to kindling and affective spectrum disorders. *Biol. Psychiatry* **32**, 218–242.

75. Bascom, R., Meggs, W. J., Frampton, M., Hudness, K., Kilburn, K., Kobal, G., Medinsky, M., and Rea, W. (1997). Neurogenic inflammation: With additional discussion of central and perceptual integration of nonneurogenic inflammation. *Environ. Health Perspect.* **105**(Suppl. 2), 531–537.

76. Meggs, W. J. (1997). Hypothesis for induction and propagation of chemical sensitivity based on biopsy studies. *Environ. Health Perspect.* **105**(Suppl. 2), 473–478.

77. Bell, I. R. (1994). Somatization disorder: Health care costs in the decade of the brain. *Biol. Psychiatry* **35**, 81–83.

78. Edling, C., Hellman, B., Arvidson, B., Andersson, J., Hartvig, P., Lilja, A., Valind, S., and Langstrom, B. (1997). Do organic solvents induce changes in the dopaminergic system? Positron emission tomography studies of occupationally exposed subjects. *Int. Arch. Occup. Environ. Health* **70**, 180–186.

79. Kay, L. M. (1996). Support for the kindling hypothesis in multiple chemical sensitivity syndrome (MCSS) induction. *Soc. Neurosci.* **22**, 1825.

80. Ursin, H. (1997). Sensitization, somatization, and subjective health complaints. *Int. J. Behav. Med.* **4**, 105–116.

81. Gilbert, M. E. (1995). Repeated exposure to lindane leads to behavioral sensitization and facilitates electrical kindling. *Neurotoxicol. Teratol.* **17**, 131–141.

82. Healy, B., Peck, J., and Healy, M. R. (1995). The effect of amygdaloid kindling on heart period and heart period variability. *Epilepsy Res.* **21**, 109–114.

83. Strakowski, S. M., Sax, K. W., Setters, M. J., and Keck, P. E. (1996). Enhanced response to repeated d-amphetamine challenge: Evidence for behavioral sensitization in humans. *Biol. Psychiatry* **40**, 872–880.

84. Meggs, W. J. (1993). Neurogenic inflammation and sensitivity to environmental chemicals. *Environ. Health Perspect.* **101**, 234–238.

85. Gots, R. E. (1996). Multiple chemical sensitivities: Distinguishing between psychogenic and toxicodynamic. *Regul. Toxicol. Pharmacol.* **24**, S8–S15.

86. Terr, A. I. (1993). Multiple chemical sensitivities. *Ann. Intern. Med.* **199**, 163–164.

87. Hudson, J. I., Geldenberg, D. L., Pope, H. G., Keck, P., and Schlesinger, L. (1992). Comorbidity of fibromyalgia with medical and psychiatric disorders. *Am. J. Med.* **92**, 363–367.

88. Johnson, S. K., DeLuca, J., and Natelson, B. H. (1996). Assessing somatization in the chronic fatigue syndrome. *Psychosom. Med.* **58**, 50–57.

89. Kandel, E. R. (1998). A new intellectual framework for psychiatry. *Am. J. Psychiatry* **155**, 457–469.

90. Rauch, S. L., van der Kolk, B. A., Fisler, R. E., Alpert, N. M., Orr, S. P., Savage, C. R., Fischman, A. J., Jenike, M. A., and Pitman, R. K. (1996). A symptom provocation study of posttraumatic stress disorder using positron emission tomography and script-driven imagery. *Arch. Gen. Psychiatry* **53**, 380–387.

91. Kroenke, K., and Spitzer, R. L. (1998). Gender differences in the reporting of physical and somatoform symptoms. *Psychosom. Med.* **60**, 150–155.

92. Pike, J. L., Smith, T. L., Hauger, R. L., Nicassio, P. M., Patterson, T. L., McClintick, J., Costlow, C., and Irwin, M. R. (1997). Chronic life stress alters sympathetic, neuroendocrine, and immune responsivity to an acute psychological stressor in humans. *Psychosom. Med.* **59**, 447–457.

93. van der Kolk, B. A., Greenberg, M. S., Orr, S. P., and Pitman, R. K. (1989). Endogenous opioids, stress induced analgesia, and posttraumatic stress disorder. *Psychopharmacol. Bull.* **25**, 417–421.

94. Leserman, J., Li, Z., Hu, Y. J. B., and Drossman, D. A. (1998). How multiple types of stressors impact on health. *Psychosom. Med.* **60,** 175–181.

95. Pioro-Boisset, M., Esdaile, J. M., and Fitzcharles, M. A. (1996). Alternative medicine use in fibromyalgia syndrome. *Arthritis Care Res.* **9,** 13–17.

96. Maynard, C., Beshansky, J. R., Griffith, J. L., and Selker, H. P. (1996). Influence of sex on the use of cardiac procedure in patients presenting to the emergency department. A prospective multicenter study. *Circulation* **94,** 1193–1198.

97. Patarca, R., Bell, I. R., and Fletcher, M. A. (1997). Pteridines and neuroimmune function and pathology. *J. Chronic Fatigue Syndr.* **3,** 69–86.

98. Derogatis, L. R. (1977). "Symptom Checklist-90 (revised). Administration, Scoring and Procedures Manual I." Clinical Psychometrics Research, Baltimore, MD.

99. Teicher, M. H., Glod, C. A., Surrey, J., and Swett, C. (1993). Early childhood abuse and limbic system ratings in adult psychiatric outpatients. *J. Neuropsychiatry Clin. Neurosci.* **5,** 301–306.

100. Shors, T. J., and Servatius, R. J. (1997). The contribution of stressor intensity, duration, and context to the stress-induced facilitation of associative learning. *Neurobiol. Learn. Mem.* **68,** 92–96.

101. Muttray, A., Lang, J., Mayer-Popken, O., and Konietzko, J. (1995). Acute changes in the EEG of workers exposed to mixtures of organic solvents. *Int. J. Occup. Med. Environ. Health* **8,** 131–137.

102. Morrow, L. A., Ryan, C. M., Hodgson, M. J., and Robin, N. (1990). Alterations in cognitive and psychological functioning after organic solvent exposure. *J. Occup. Med.* **32,** 444–450.

103. Kreutzer, R., Neutra, R. R., and Lashuay, N. (1999). The prevalence of people reporting sensitivities to chemicals in a population based survey. *Am. J. Epidemiol.* **150,** 1–12.

Part V

GERIATRIC HEALTH

Section 14

AGING

Jack M. Guralnik

Epidemiology, Demography, and Biometry Program
National Institute on Aging
Bethesda, Maryland

Population aging in both developed and developing countries will have an ever increasing impact not only on the medical and long-term care systems but also on the entire social structure. At the close of the twentieth century it is instructive to take a broad overview of the demographic changes that have occurred in the past 100 years. For women, life expectancy in the U.S. rose from 48.3 years in 1900 to 79.1 years in 1996 [1]. Much of the increase in life expectancy in the first half of the century was related to improvements in early life mortality rates, but since the 1950s there have been rapid mortality declines at older ages as well. The 30% decline seen in mortality rates in white and black women aged 65–74 between 1960 and 1991 was particularly impressive [2].

Along with these changes in mortality and life expectancy, the number of old and very old persons, particularly women, has risen dramatically. In 1940 there were only 211,000 women aged 85 and older in the U.S. This number increased 14-fold, to nearly 2.9 million, by the end of the century. To appreciate just how profound the changes in women's aging have been, one need look only at the remarkable increase in the proportion of middle aged women who survive to very old age that occurred over the course of the twentieth century. In 1900, only 3% of 50-year-old women could expect to live to age 90. By 1990, this number had risen to over 22%, considerably higher than the 10% of 50-year-old men expected to live to age 90 [3]. The rapidly growing number of older women and the high proportion of women living to very old age make strong arguments for the importance of studying the aging process and diseases of aging in women.

It is of interest that although much of the demographic change that has led to a rapidly expanding population of older women has involved declines in the major causes of death, the chapters in this section are not about those diseases. In essence, it is the escape from early mortality that has made the conditions described here ever more important. The three leading causes of death, heart disease, cancer, and stroke, account for nearly 70% of deaths in older people and also have a substantial impact on loss of functional status and need for long-term care. At the same time, in a long-living population, conditions that may not lead to mortality but have major impacts on health status, functional outcomes, and quality of life become equally important. These conditions, discussed in the following chapters, include changes related to menopause, osteoporosis, osteoarthritis, incontinence, sensory impairment, and cognitive impairment. The prevalences of all of these conditions rise steeply with increasing age, making them highly relevant for the population of women that is increasingly likely to survive to very old age.

Because older people frequently have multiple chronic conditions it is useful to go beyond the assessment of a single disease at a time and study the overall impact of a person's total burden of disease. The examination of functional status, disability, and level of independence has proven to be an important way to represent the overall impact of disease on a person's health status. Although a single condition such as a hip fracture may cause catastrophic disability, many older people suffer from progressive decline in functional abilities due to the increasing impact of several conditions. Chapter 91, on morbidity, disability, and mortality, outlines the approaches to measuring functional loss and disability and demonstrates the higher prevalence of disability in older women compared to men. The rapid rise in disability with increasing age is compelling. Only 8% of women aged 65–74 need help at home or live in a nursing home, but this number rises to 55% in those aged 85 and over [3]. This chapter's presentation of disability as the cumulative effect of physiologic decline and chronic disease sets the stage for the more in-depth presentations of the other chapters in this section.

Although the emphasis of most aging research is on the conditions and disabilities that have a negative impact on older people's lives, new research on successful aging has helped to reduce some of the stereotypes of aging as a time of inevitable frailty and cognitive decline. Chapter 99 shows that when the total older population living in the community is studied we see a very different picture than the one that arises from studying older patients in hospitals and clinics. Large, community-based epidemiologic studies have revealed that many older women maintain high levels of physical functioning, cognitive abilities, and well-being for most of their older years. Work that has begun to uncover the determinants of successful aging is reviewed in Chapter 99. Successful aging can be maintained both by preventing the onset of disease and by minimizing the functional consequences of diseases that people develop. In this regard, improvement in behavioral risk factors, such as increased physical activity, maintenance of appropriate weight, and cessation of smoking, can have positive impacts on successful aging both through reduced risk of disease onset and through direct beneficial physiologic effects. This chapter makes a convincing case that successful aging is a definable, relevant medical and public health outcome and that we should evaluate the impact of promising interventions, such as estrogen therapy, exercise, and psychosocial interventions, on the maintenance of successful aging in women.

Because much of the emphasis of biomedical research since the 1950s has been on the so-called killer diseases, the conditions presented in this section have not traditionally received the level of attention that leads to rapid progress. Fortunately, there has been an increased appreciation for just how great an impact these conditions have on older women's lives. When we finally dedicated resources to studying these conditions, gains in our understanding of them have been remarkable and the findings detailed in these chapters show an exponential growth in the body of knowledge we have acquired since the 1980s.

Menopause is perhaps the best example of an aging process that traditionally received relatively little research attention. Chapter 92 describes the stages of menopause and their associated hormonal changes. It demonstrates that we now have a better understanding of the symptoms of menopause, with some insights developed into the relationship of symptoms to both extrinsic factors and intrinsic characteristics, particularly changes in hormone status. The author of the chapter reminds us that there is still a great deal to be learned about this transition period, which can occupy 5–10% of a woman's lifetime. Of particular importance is the issue of the potential consequences of the menopausal transition on both current and long-term disease development and progression. It is known that the incidence of most chronic diseases goes up after menopause and that estrogen loss plays a role in many of these. It is unclear, and needs to be more thoroughly investigated, whether the metabolic, immunologic, and body composition changes during the menopausal transition itself might have an effect on the increased risk of disease after menopause.

Estrogen loss and the risks and benefits of hormone replacement therapy (HRT) represent a complex area in which much has been learned but many questions remain. Chapter 93 describes a large body of evidence demonstrating that women taking HRT have decreased risk of mortality, coronary heart disease, osteoporosis, and possibly colon cancer, but probable increased risk of breast cancer, endometrial cancer, deep vein thrombosis, and pulmonary embolism. The chapter describes the critical methodologic issues that prevent the existing observational studies from providing a definitive answer to these questions and makes the case for clinical trials such as the Women's Health Initiative. Further research is also needed to fully evaluate the potential benefits of combination HRT, which contains progestins, and selective estrogen receptor modulator drugs (SERMS), which may have beneficial effects on the cardiovascular system and bone while at the same time being protective against breast cancer.

Musculoskeletal diseases, including osteoporosis and osteoarthritis, are among the most disabling conditions affecting older women. Osteoporosis has important implications for fracture risk, functional outcomes, and use of medical and long-term care services. As reviewed in Chapter 94, 40% of women will, in their lifetimes, have a fracture of one or more of the three most common fracture sites: the hip, vertebrae, or distal forearm. This is perhaps not surprising considering that 70% of women aged 80 and over have osteoporosis [4]. Hip fractures frequently lead to permanent loss of functioning, institutionalization, and death. Low bone mineral density (BMD) is a major risk factor for fracture, and Chapter 94 reviews findings on the biology of bone growth, modeling, and remodeling. Osteoporosis is determined by a combination of peak bone mass at skeletal maturity and the amount of bone loss later in life, particularly during and after menopause. Findings on genetic and environmental determinants, especially calcium intake, exercise, and HRT, provide strategies throughout a woman's life to reduce late-life osteoporosis.

Arthritis is the most commonly reported chronic condition in the U.S. and its burden falls most heavily on older persons. Arthritis is more common in women than men after the age of 45 and the majority of women aged 65 and older report arthritis in health surveys. Clinical evaluation of older populations confirms that the most common form of arthritis, osteoarthritis (OA), is highly prevalent. For example, it has been shown that one-third of older persons have radiographic OA of the knees, with 10% having symptomatic knee OA [5]. Chapter 95 describes the work that has been done to develop standardized diagnostic criteria for OA which will be valuable for continued research on this condition. The chapter provides a comprehensive review of the risk factors for OA, particularly weight and weight gain, occupation, high level physical activity, and joint injury. It then reviews the clinical presentation, factors related to progression, and current approaches to treatment.

Chapter 96, on urinary incontinence, summarizes a wide variety of research being done in this area. The problem is of particular concern to women, with about one in three older women having problems with incontinence at some time and up to 14% having daily incontinence [6]. The psychological, social, and physical impact of urinary incontinence can be considerable. A substantial amount of work has been done on the descriptive epidemiology and the search for risk factors for this condition, but the authors point out various methodologic limitations that have led to inconsistent results across studies. Progress in treating urinary incontinence has been made with behavioral strategies, pharmacologic interventions, and surgery.

Chapter 97 describes the high rates of sensory impairment at advanced ages in vision, hearing, taste, and smell. Vision loss has the greatest impact on global functioning and quality of life and is caused by a variety of ailments in older persons. Refractive error and cataracts are very common and easily correctable, whereas more serious vision impairment can result from age-related macular degeneration, diabetic retinopathy, and glaucoma. The authors estimate that two-thirds of vision impairment can be successfully treated, with the remaining one-third usually resulting in permanent vision loss. Older people with vision loss have been shown to have increased disability, reduced physical activity, greater social isolation, poorer quality of life, and higher mortality rates. Low-vision rehabilitation services can offer benefits to older people with permanent vision loss but they are greatly underutilized. The authors make a strong case that in research estimating visual problems and their impact in older persons it is necessary to go beyond the simple measurement of visual acuity and to include measures of contrast and glare sensitivity and visual field and stereoacuity measures. Hearing impairment can have a large impact on an older person's quality of life. The prevalence of hearing impairment rises rapidly with increasing age and although women have somewhat less hearing impairment than men at younger ages, by the time men and women pass 80 years of age more than 80% have hearing loss of greater than 25 decibels in either ear [7]. The loss of taste and smell in old age has less impact on quality of life and function than loss of vision and hearing, but the full measure of the impact of their loss has not been determined. When these two senses are impaired, food is less desirable and the resulting decrease in caloric intake and subsequent weight loss may result in many adverse outcomes in older persons. The chapter closes by reviewing emerging methods for evaluating taste and smell and describing the small amount of epidemiologic information on the prevalence and risk factors for impairments in these chemical senses.

With the increasing number of women living to very old age, dementia, and particularly Alzheimer's disease, has a major impact on the health of the population of older women. Women are no more likely than men to get Alzheimer's disease, but because the incidence rate climbs so steeply with age, and be-

cause there are so many more very old women than men, there are many more women suffering from the disease. Furthermore, among those with the disease, women live longer with their symptoms before they are diagnosed, are institutionalized earlier in the course of the disease, spend substantially more time in institutions than men with the disease, and, overall, live longer with the disease [8]. Chapter 98 summarizes the rapid growth in our knowledge about Alzheimer's disease that has come from epidemiologic studies, basic research, and clinical investigations. Of particular relevance to women are the findings that use of postmenopausal HRT may be associated with decreased risk of Alzheimer's disease and the research that sheds light on how estrogen might affect the pathophysiology of this disease.

Taken as a group, the diverse chapters in this section provide a comprehensive review of many conditions that have a profound impact on the health of older women. Each of these chapters makes a strong case for the importance of understanding, preventing, and more effectively treating the conditions they review. These are conditions that are not responsible for the majority of deaths but strongly influence well-being, quality of life, and the ability of older women to lead independent productive lives.

References

1. Peters, K. D., Kochanek, K. D., and Murphy, S. L. (1998) "Deaths: Final Data for 1996," Nat. Vital Stat. Rep., Vol. 47, No. 9. National Center for Health Statistics, Hyattsville, Md.
2. U.S. Bureau of the Census (1996). "65+ in the United States, Current Population Reports, Special Studies, P23-190." U.S. Govt. Printing Office, Washington, DC.
3. Guralnik, J. M., Leveille, S. G., Hirsch, R., Ferrucci, L., and Fried, L. P. (1997). The impact of disability in older women. *J. Am. Med Women's Assoc.* **52,** 113–129.
4. Melton, L. J., Chrischilles, E. A., Cooper, C., Lane, A. W., and Riggs, B. L. (1992). How many women have osteoporosis? *J. Bone Miner. Res.* **7,** 1005–1010.
5. Felson, D. T., Naimark, A., Anderson, J., Kazis, L., Castelli, W., and Meenan, R. F. (1987). The prevalence of knee osteoarthritis in the elderly. *Arthritis Rheum.* **30,** 914–918.
6. Thom, D. (1998). Variation in estimates of urinary incontinence prevalence in the community: Effects of differences in definition, population characteristics, and study type. *J. Am. Geriatr. Soc.* **46,** 473–480.
7. Cruickshanks, K. J., Wiley, T. L., Tweed, T. S., Klein, B. E., Klein, R., Mares-Perlman, J. A., and Nondahl, D. M. (1998). Prevalence of hearing loss in older adults in Beaver Dam, Wisconsin: The Epidemiology of Hearing Loss. *Am. J. Epidemiol.* **148,** 879–886.
8. Jost, B. C., and Grossberg, G. T. (1995). The natural history of Alzheimer's disease: A brain bank study. *J. Am. Geriatr. Soc.* **43,** 1248–1255.

91

Morbidity, Disability, and Mortality

SUZANNE G. LEVEILLE AND JACK M. GURALNIK

Epidemiology, Demography and Biometry Program
National Institute on Aging
Bethesda, Maryland

I. Introduction

Older women live longer, develop more chronic disease, and experience more prolonged and severe disability than older men. Although women have lower mortality rates than men, the prevalence of nonlethal chronic diseases, such as arthritis, osteoporosis, and depression, is generally higher in older women. With more comorbidity, women have greater risk for disability. Nearly twice as many women as men over the age of 65 are living in nursing homes (5.3% versus 2.5%, 1995 National Nursing Home Survey, National Center for Health Statistics [NCHS]). These differences cannot be fully explained based on current evidence, but in this chapter we will summarize our knowledge to date related to trends in health, disease, and mortality in older women. Gender, racial, and geographic variations will be presented to help portray health, illness, and death in the later years of women's lives.

Among persons aged 65 and older in the United States in 1996, almost 60% were women. The proportion of women in the older population climbs dramatically with age to over 70% in those aged 85 and older. These gender differences are expected to lessen as we enter the twenty-first century. However, a major shift is expected to occur in the age distribution of women and men in the entire population that is related to the aging of the "baby boom" generation born between 1946 and 1964. In 1994, older women represented 15% of the female population. This figure is expected to grow to 18% by 2020 and 22% by 2040 (Fig. 91.1) [1–4]. Comparable figures for men are 11, 15, and 19%, respectively. These dramatic changes are

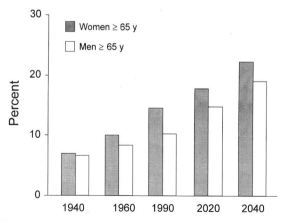

Fig. 91.1 Percentage of the U.S. population age 65 years and older, 1940 to 2040. Sources: [1–4].

largely due to reduced mortality at older ages coupled with declining birth rates. In 1994, there were 20 million women aged 65 and older in the U.S. compared to 13.5 million men. By 2050, the older female population is expected to grow to 42 million, and the older male population to 35 million.

The United States is not alone in this demographic transition; most developed countries are experiencing similar age shifts. As of 1992, the U.S. ranked twentieth among developed nations in the percentage of the population aged 65 and above. The higher proportions of women in the older populations are also evident across the developed world, where women typically outlive men by 5–9 years. Japan has the fastest growing aged population [5]. Older adults comprised 7% of Japan's population in 1970 and grew to 14% in just 26 years. Alternatively, Sweden has a much slower rate of change, taking 85 years (from 1890 to 1975) to achieve the same change, but it has the highest proportion of older adults in its population (17.9% in 1992). Even with a slower rate of change, the number of older adults in Sweden is expected to increase by 33% by the year 2025.

In the coming decades, the increase in the size of the older population is expected to be highest in developing countries, generally in the range of 200–400% between 1990 and 2025 [5]. As of 1996, 17.9% of the world's population aged 75 and older lived in China, a higher proportion than in any other country (U.S. Bureau of the Census, International Programs Center, International Data Base), 1996, http://www.census.gov/ipc/www/idbnew.html). The United States had the next highest proportion, at 11.8%. Because of the large total populations in many developing countries, the majority of the 1 billion people aged 60 and above in 2020 will be living in the developing world. In summary, these changing demographics highlight the urgent need for improving our understanding of health problems of older people, and of older women in particular.

II. Health Status

Although women face many risks related to chronic disease and disability as they age, the majority of women are healthy well into their later years. In 1994, 75% of women aged 65–74 years reported their health to be good, very good, or excellent [6]. Among women aged 75 and older, 69% rated their health as good to excellent. Many factors influence self-assessed health, including income and race. Persons with family incomes of less than $14,000 per year were five times more likely to report their health as fair or poor compared to those with incomes of $50,000 or more [6]. Among blacks aged 75 and older, only 53% reported their health as good to excellent, compared to 70% of whites. It is important to note that most older women rate their health as good or better even though they are living with chronic conditions. Nearly half of women aged 75 and older reported activity limitations caused by chronic conditions, although two-thirds reported good to excellent health.

III. Chronic Conditions

The prevalence of chronic conditions increases with age. Arthritis is the most prevalent chronic condition affecting older women, followed by high blood pressure and hearing impairment (Table 91.1). National estimates show that more women than men aged 65 and older have arthritis (55.4% versus 42.9%, respectively), and prevalence of musculoskeletal joint symptoms such as pain and stiffness are typically higher in women [7]. In addition, women with arthritis are more likely to report activity limitations than are men (23% versus 18%) [8]. Although more research is needed in this area, it is estimated that, in persons over age 65, incidence of knee osteoarthritis is approximately 2% per year in women and 1% per year in men [9]. These rates are high for either sex and point to an urgent need for the development of new strategies and treatments to prevent or control this very common and disabling disease.

Rates of hypertension increase markedly with age and are somewhat higher in older women than men (39.6% versus 32.0%, respectively). Among persons aged 75 and older, 77% of women and 64% of men report high blood pressure (National Health and Nutrition Examination III (NHANES), NCHS, 1988–1994). African-Americans and persons with low incomes are at higher risk for developing hypertension. In the decade from 1980 to 1990, there were major advances in increasing public awareness of hypertension. However, more than a third of persons who have the disease are unaware of it. Death rates in 1995 from hypertension were more than four times higher in African-American women than in white women (22.2 versus 5.0 per 100,000 women, respectively) [10]. Treatment of hypertension in older adults has been shown to reduce risks for stroke, cardiovascular disease, and mortality.

Ischemic heart disease prevalence is somewhat higher in older men than women; however, the numbers in both genders are very high (16.9% versus 10.8%, Table 91.1). Rates of heart disease are strongly linked to age in women. About one in nine women aged 45 to 64 has clinical evidence of heart disease, but this number jumps to one in three women aged 65 and above [11]. In the older ages, myocardial infarctions are twice as likely to be fatal in the first few weeks following the event in women compared to men. Data from the Cardiovascular Health Study showed that the rates of new or recurrent heart attacks increase from 7.8 per thousand in nonblack women aged 65–74 to 21.0 per thousand at ages 75–84, to 24.2 per thousand in women aged 85 and older [10]. In black women, the rates are higher in the 65–74 year age group, but the rates show a modest decline in the oldest ages (rates were 13.3, 18.3, and 14.1, respectively). Activities of daily living (ADL) disability related to heart disease increases markedly with age, occurring in 12.5% of women aged 65–74 with heart disease and up to 40% in those over 85 years who have the disease [13].

Women have higher risks for osteoporotic fractures than men, although gender differences in fracture rates vary by fracture site. Women have two to three times the lifetime risk of hip fracture

Table 91.1
Prevalence (Percent) of Chronic Conditions in Women and Men, Aged 65 Years and Older: National Health Interview Survey, 1994, United States

Chronic conditions	Women		Men	
	Percent	Rank	Percent	Rank
Arthritis	55.4	1	42.9	1
High blood pressure	39.6	2	32.0	3
Hearing impairment	23.8	3	35.4	2
Cataracts	19.2	4	13.0	5
Chronic sinusitis	17.6	5	11.7	6
Ischemic heart disease[a]	10.8	6	16.9	4
Varicose veins	9.8	7	4.2	>10
Diabetes	9.7	8	10.7	8
Deformity/orthopedic impairment–back	9.3	9	8.3	10
COPD[b]	8.9	10	11.0	7
Visual impairment	7.6	>10	9.2	9

Source: [7].

[a]Based on 12% estimated overlap in prevalence of myocardial infarction and angina.

[b]Chronic Obstructive Pulmonary Disease includes emphysema and chronic bronchitis; estimated 8% overlap in disease prevalence.

of men. Age-specific incidence of vertebral fractures does not appear to be higher in women than in men, but women have higher lifetime risks of spine fractures related to living longer [14]. Lifetime risk of wrist fracture is 16% in 50-year-old women versus only 2.5% in 50-year-old men [15]. African-Americans have substantially lower risks of osteoporotic fractures compared to whites; overall, the incidence of hip fractures in blacks in the United States is about half that of whites. Studies of younger women show that blacks attain higher peak bone mass and have slower bone loss at menopause than white women [16]. Cross-national research in African populations has shown that fracture rates are lower among black African women compared to white populations, even when their bone mineral density was found to be lower than that of white women [17]. Risk of osteoporotic fractures is generally lower in Asian and Hispanic women than in whites; however, Asian women have comparable risks to white women for vertebral fracture [14].

Cancer is a disease of older adults, with 60% of all cancers diagnosed in persons aged 65 and older. More than two-thirds of cancer deaths occur in those over the age of 65. Cancer incidence rates are similar in older white and black women: 1600–1700 per 100,000 per year, as of 1995 (SEER Cancer Statistics, National Cancer Institute). The incidence for all cancers in women peak at 1973 per 100,000 (approximately 2% per year) in the 80–84 year old age group [18]. These rates are about half those of comparably aged men. Among cancers, breast cancer is the most common in older women and increases with age until age 75, then the rate flattens and decreases slightly after age 80. The next most common cancers in older women are colorectal and lung, followed by uterine and ovarian cancers and lymphomas. Lung cancer is the leading cause of cancer mortality in older women and is followed by breast and colorectal cancers.

Cognitive impairment related to Alzheimer's disease and other dementias is difficult to measure in large national surveys, but a number of smaller population-based studies have found cognitive impairment to be among the top ten chronic conditions affecting older adults. Alzheimer's disease represents more than half of the age-associated dementias. Differences between studies in the classification of dementias and Alzhemer's disease have contributed to broad ranges of prevalence estimates across populations. Prevalence of Alzheimer's disease increases exponentially with age from about 1–2% of those aged 65–74 to 7% of persons aged 75–85 to 25% of those aged 85 and older [19,20]. Prevalence and incidence of Alzheimer's disease appear to be slightly higher in older women than in comparably aged men. Research that is currently underway, including the development of Alzheimer's disease registries, will hopefully lead to a clearer understanding of the epidemiology of senile dementias.

In addition to the most disabling and life-threatening chronic diseases, there are a number of common chronic conditions that disproportionately affect older women, including thyroid disorders, cataracts, migraine headaches, anemias, gastrointestinal problems, bladder conditions, varicose veins, and foot deformities [7]. Another common problem affecting many older women is chronic pain, which is often related to musculoskeletal disease. Prevalence of chronic pain varies with how pain is defined and with the population studied. Research has found that more than half of older women and about one-third of older men report chronic pain, often at multiple sites [21,22]. Reported pain intensity is also typically higher in women than in men.

Obesity is a problem for many women throughout their lives and it poses health hazards particularly in the older ages. According to national surveys conducted from 1988 to 1994, the proportion of women who were overweight was highest (48%) in the 55–64 year old age group. In the older ages, the proportions were somewhat lower: 42% of women aged 65–74 and 35% of those aged 75 and older [6]. More than half of black women and Mexican-American women aged 20–74 were overweight (body mass index > 27.3 kg/m²) between 1988 and 1994, versus about one-third of white women. Several major chronic conditions are strongly associated with obesity including osteoarthritis, heart disease, diabetes, and low back pain. Related to higher risks of diseases and the impact of a less active lifestyle, overweight individuals are more prone to disability in late life. In a large epidemiologic study of mobility loss in older adults, women in the top 20% of body mass index (weight in kilograms divided by height in squared meters) had a 40% higher risk of losing mobility during the four years of follow-up than those in the mid-range of body mass index [23]. This is a very high figure considering that more than a third of the total study population lost mobility during the follow-up period.

Considering the many chronic conditions affecting the lives of older women, it is not surprising that comorbidity, or the presence of cooccurring chronic diseases, poses a substantial hazard in this population. National survey data showed that 45% of women aged 60–69 had two or more chronic conditions, and this figure jumped to 70% in those aged 80 and older [24]. Comparable figures in men were 35% percent and 53%, respectively. There is evidence suggesting that comorbidity is more prevalent in African-Americans and Native Americans than in

whites [25]. More research is needed to understand the impact of comorbidity and to develop strategies to prevent diseases and provide more effective treatments for the multiple chronic conditions affecting older adults.

IV. Disability

A. Measurement and Prevalence

Recent decades have heralded major research efforts to study the causes of disability in older adults. It has long been known that women experience a higher burden of disability in old age. In an early epidemiologic study of disability, Nagi reported that nearly twice as many women as men aged 75 and older needed assistance with mobility and self-care activities [26]. Later reports continued to describe the differences in disability rates between women and men, both in terms of needing assistance with daily activities in the home and in rates of institutionalization. Presently, there are a variety of measures used to assess daily functioning and disability levels in older populations. A common measure referred to as activities of daily living (ADLs) includes basic self-care tasks such as bathing, dressing, transferring from a bed to a chair, eating, and using the toilet. Instrumental activities of daily living (IADLs) include more complex tasks such as performing light housework, shopping, managing money, getting around inside the home, using the telephone, and preparing meals. National data from 1994 and 1995 showed that 19% of women aged 65 and older were either receiving help with IADLs or ADLs at home or were living in a nursing home, compared to 12% of men (Fig. 91.2) [27]. In the older age groups the gender differences are more marked, with 55% of women aged 85 and older needing help or living in an institution compared to 37% of men the same age. These striking gender differences are related to higher incidence of disability in women coupled with longer survival of disabled women compared to disabled men.

Prevalence of disability varies by race and socioeconomic status. National estimates from 1994 showed that among older white women, 14% of those living in the community required assistance with ADLs or IADLs compared to 20% of older black women [7]. Education is strongly associated with lower risk for disability. Ten percent of older women who had a high school education or greater were disabled in ADLs or IADLs versus 20% of those who did not complete high school.

A number of large prospective studies have identified several risk factors for disability (Table 91.2). Prevalent chronic conditions that have been found to predict disability include arthritis, heart and lung disease, stroke, diabetes, and cognitive impairment. Other factors that predispose older adults to disability include sociodemographic characteristics, obesity, and health behaviors such as smoking and inactivity.

Another approach to evaluating function is to use tests of physical performance. These include measures of gait speed, balance, strength, and other tests. Studies have shown that decrements in physical performance predict subsequent loss of independence in daily activities among those who are initially nondisabled. Specifically, slow walking speed has been shown to be a strong predictor of loss of mobility [28]. This important new information can be used to identify persons most at risk for

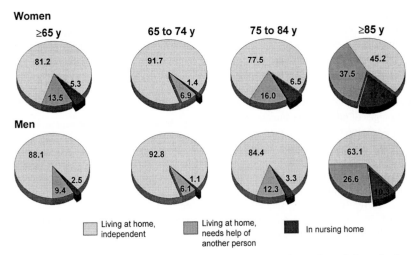

Fig. 91.2 Percentage of older women and men by age group that lives at home independently, lives at home but requires help or supervision from another person, or lives in a nursing home. Independence was measured by not needing help with one or more of the following: bathing, dressing, eating, transferring from bed to chair, using the toilet, getting around inside the home, cooking, shopping, using the telephone, and doing light housework. From J. M. Guralnik, S. G. Leveille, R. Hirsch, L. Ferrucci, and L. P. Fried (1997) [27]. The impact of disability in older women. *J. Am. Med. Women's Assoc.* **52,** 113–120 with permission.

developing disability and to develop and test interventions to prevent functional loss in vulnerable seniors.

B. Active Life Expectancy

A newer method that is being used to measure functional independence in older adults is the calculation of active life expectancy and disabled life expectancy, terms that describe expected number of years of life without and with disability. These are two components of total life expectancy. Considering population dynamics using these concepts allows researchers to monitor mortality and disability trends in a combined measure. Active life expectancy at age 65 has been shown to vary with

Table 91.2
Risk Factors for Disability in Older Adults from Epidemiologic Studies

Sociodemographic characteristics	Medical conditions
Older age	Arthritis
Female	Heart disease
Lower education	Lung disease
Lower income	Stroke
Health behaviors	Cognitive impairment
Physical inactivity	Hip fracture
Current smoking	Cancer
High alcohol intake	Depression
Health characteristics	High blood pressure
Obesity	Vision impairment
Fair or poor self-rated health	
Number of chronic conditions	
Use of psychoactive medications	

sex, race, and educational status (Fig. 91.3) [29]. Longer active life expectancy is typically associated with longer disabled life expectancy, an unfortunate consequence of living to older ages. At age 65, black women have longer active life expectancy and longer total life expectancy compared to black men and whites of both sexes in the same education level. Higher education is associated with 2–4 years of longer disability-free life across sex and race groups and is also associated with somewhat longer expected disabled life, particularly for women. The challenge of aging research is to postpone mortality and also disability, thus increasing active life expectancy and shortening the period of disability prior to death in late life.

Reports have presented encouraging findings indicating that modifiable risk factors may postpone mortality as well as the development of disability in late life. In a study of university alumni, disability was postponed by more than five years among those with low health risks compared to those with high health risks, based on a summary score of smoking status, body mass index, and vigorous activity [30]. Other studies using population-based data have presented evidence suggesting that moderate physical activity can reduce the likelihood of disability prior to death in advanced old age and increase active life expectancy in older adults [31,32].

V. Mortality

The mortality rates of older women have decreased dramatically since the 1950s and have generally paralleled mortality reductions in older men. Between 1960 and 1991, there was a 30% decline in death rates for both black and white women aged 65–74 [33]. Although white men in this age group experienced similar reductions, death rates in black men of the same age declined by only 16%. In the entire population across sex and race groups,

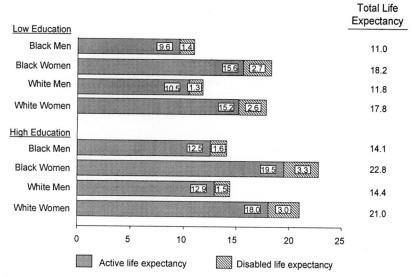

Fig. 91.3 Active life expectancy, disabled life expectancy, and total life expectancy in older women and men by by race and education level. Figure adapted from data published in [29].

the most dramatic decreases in death rates were related to heart disease mortality that peaked in the 1960s and has been on the decline since that time. There have been similar declines in death rates due to strokes, although the decreases in death rates due to noncardiovascular diseases such as cancer showed more modest declines and have begun to increase in recent years.

In consort with declining mortality, life expectancy has changed dramatically for women in the course of the twentieth century. The greatest gains were made in the first half of the century, primarily due to reductions in premature deaths related to infections and maternal complications (Fig. 91.4) [6]. Although life expectancy for older adults rose prior to 1950, the latter half of the twentieth century has seen continued increases in life expectancy from the age of 65. Older white women have

experienced the greatest gains in life expectancy, which, for those aged 65, increased from 15 years in 1950 to 19.1 years in 1995. In 65-year-old black women, the gains were less dramatic: from 14.9 years to 17.1 years over the same time period [6]. Comparable estimates of changes in life expectancies in 65-year-old men were 1–2 years less than in women (increasing from 12.8 to 15.3 years from 1950 to 1995). These trends are largely attributed to lifestyle changes and improvements in the medical management of chronic illnesses.

As of 1995, life expectancy at birth for women in the U.S. was 6.4 years longer than for men (78.9 years versus 72.5 years). At age 65, the female to male difference in life expectancy narrows to just under four years. As mentioned earlier, total life expectancy varies sharply by race, with whites overall

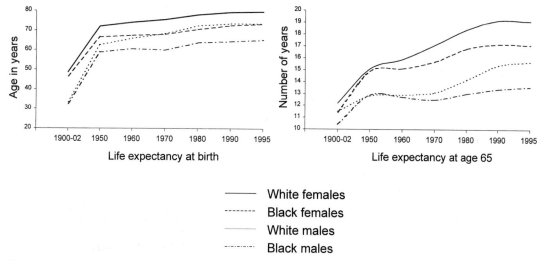

Fig. 91.4 Trends in life expectancy from birth and from age 65 years according to sex and race, from 1900 to 1995, United States. Source: [6].

Table 91.3

Number of Deaths and Death Rates (per 100,000) for the Ten Leading Causes of Death in Persons Aged 65 Years and Older, by Race and Sex: United States, 1995

Causes of death (ICD - 9 Code)	White female		Black female		White male		Black male	
	Number	Rate	Number	Rate	Number	Rate	Number	Rate
All causes	823,925	4,643.9	83,632	5,024.1	693,659	5,633.5	70,808	6,713.0
Diseases of heart (390-398, 402, 404–429)	304,076	1,713.9	31,125	1,870.4	249,444	2,025.8	23,428	2,221.1
Malignant neoplasms, including neoplasms of lymphatic and hematopoietic tissue (140-208)	162,820	917.7	16,522	992.5	176,481	1,433.3	20,139	1,909.3
Cerebrovascular diseases (430-438)	78,371	441.7	8,214	493.4	44,756	363.5	5,181	491.2
Chronic obstructive pulmonary diseases and allied conditions (490-496)	39,880	224.8	1,831	110.0	43,074	349.8	2,772	262.8
Pneumonia and influenza (480-487)	38,237	215.5	2,946	177.0	29,221	237.3	2,659	252.1
Diabetes mellitus (250)	21,354	120.4	4,406	264.7	15,454	125.5	2,315	219.5
Accidents, motor vehicle crashes, and adverse effects (E810-E825)	13,383	75.4	1,117	67.1	12,756	103.6	1,365	129.4
Alzheimer's disease (331.0)	12,635	71.2			6,428	52.2		
Atherosclerosis (440)	9,309	52.5						
Nephritis, nephrotic syndrome, and nephrosis (580-589)	8,876	50.0	1,716	103.1	8,050	65.4	1,238	117.4
Septicemia (038)			1,518	91.2	5,584	45.3	1,064	100.9
Hypertension with or without renal disease (401, 403)			1,208	72.6			742	70.3
All other causes (residual)	134,984	760.8	13,019	782.1	102,411	831.7	9,905	939.0

Source: [34].

having a 6.9 year advantage over blacks from birth and white females having a 5.7 year advantage over black females [34]. These race differences narrow to 1–2 years in those over age 65, and there is little difference by the age of 85. Cross-nationally, life expectancy at birth is highest for women in Japan (82.5 years), France (81.8 years), and Switzerland (81.0 years) [33]. As of 1990, the United States ranked fourteenth in life expectancy at birth for women. However, a cross-national comparison of mortality in 1987 showed that white women and men at the age of 80 in the United States had the greatest life expectancy (9.1 years and 7.0 years, respectively) compared to their age peers in Japan, Sweden, France, and England [35].

The three leading causes of death in persons aged 65 and older—heart disease, cancer, and cerebrovascular disease—are the same in women and men, regardless of race (Table 91.3). In 1995, these three conditions accounted for nearly 70% of deaths in older adults. Women have lower death rates than men for each of the leading causes of death. However, in the case of heart disease, men have higher death rates but more women than men die from this condition because there are so many more older women in the population. In 1995, 500,883 women and 450,523 men died from cardiovascular disease [34]. The death rate due to heart disease was 6% higher in black women compared to white women. Among white women, Chronic Obstructive Pulmonary Disease (COPD) and pneumonia/influenza are the fourth and fifth most common causes of death (Table 91.3). In black women, diabetes is the fourth leading cause of death, followed by pneumonia/influenza.

VI. Ascertainment

In each of the areas described in this chapter, there are methodologic considerations. The ascertainment of chronic conditions continues to be problematic. Traditional reliance on self-report in national surveys has been shown to have its limitations, resulting in over- or underreporting biases. Certain diseases, such as osteoporosis, must be clinically evaluated using new diagnostic procedures that have not typically been utilized in primary care. This results in a substantial underreporting. Similarly, vertebral fractures are often not radiographically assessed and may be underreported. The prevalence of cognitive impairment is difficult to determine because many persons with dementias are living in institutions and are often left out of national surveys. In addition, persons with mild cognitive impairments may not be identified in surveys and their responses may not be reliable.

Describing the prevalence of disability has a number of limitations. Definitions of disability, which may be based on functional difficulty or incapacity in a range of activities, may vary substantially among surveys and subsequent reports. It has also been shown that caregivers and the persons they assist may differ markedly in their assessment of the individual's disability levels. The cognitive component of disability has not received the amount of attention that physical functional losses have received. As mentioned earlier, cognitive impairment is likely a more substantial problem in older adults than has been characterized in national reports to date.

There are limitations in the methods used for determining cause of death, especially in the very old. Attributing death to a single cause in persons afflicted by multiple serious chronic conditions can be a questionable practice. For example, when a person dies in the later stages of dementia related to Alzheimer's disease or cerebrovascular disease but also has heart disease, the underlying cause of death could be any of these conditions. This is further complicated by the fact that incomplete medical evaluations in the very old may lead to inaccurate diagnoses of causes of death. Better methods are needed for documenting causes of death, including allowing for the identification of multiple major causes of death. In addition, autopsies could provide more specific information about critical events leading to death in persons with multiple diseases.

VII. Summary

In our aging nation, the health problems of older adults, and older women in particular, will become magnified in the first half of the twenty-first century, when more than one in five women will be age 65 or older. More and more women will be facing risks related to living to very old ages. Although new treatments are being discovered that provide relief from several chronic diseases, major efforts are needed to effectively prevent or reduce disablement from the most prevalent chronic conditions such as arthritis, heart disease, and dementia. Better identification of persons at highest risk for disease and disability will lead to effective prevention strategies. Public health efforts to promote widespread implementation of healthy behaviors to reduce health risks could serve a twofold purpose of preventing disease and reducing risks for disability in older adults. It is the successes in public health and medicine in the twentieth century that have given rise to the new challenges we face today. The health of older women during the coming decades will largely be determined by progress in preventing or delaying the onset of chronic diseases and reducing the period of disability prior to death in very old age.

References

1. U.S. Bureau of the Census (1965). "Public Use Data files, PE-11," Curr. Popul. Rep., Ser. P-25, No. 311. U.S. Govt. Printing Office, Washington, DC.
2. U.S. Bureau of the Census (1974). "Public Use Data files, PE-11," Curr. Popul. Rep., Ser. P-25, No. 519. U.S. Govt. Printing Office, Washington, DC.
3. Byerly, E. R., and Deardorff, K. (1995). "National and State Population Estimates: 1900 to 1994," Curr. Popul. Rep., P25-1127. U.S. Bureau of the Census, U.S. Govt. Printing Office, Washington, DC.
4. Day, J. C. (1993). "Population Projections of the United States, by Age, Sex, Race, and Hispanic Origin: 1993 to 2050," Curr. Popul. Rep., P25-1104. U.S. Bureau of the Census, U.S. Govt. Printing Office, Washington, DC.
5. U.S. Bureau of the Census (1992). "An Aging World II," Int. Popul. Rep., P25, 92–3. U.S. Govt. Printing Office, Washington, DC.
6. National Center for Health Statistics (1997). "Health, United States 1996–97 and Injury Chartbook," DHHS Publ. No. (PHS)97-1232. National Center for Health Statistics, Hyattsville, MD.
7. Adams, P. F., and Marano, M. A. (1995). Current estimates from the National Health Interview Survey, 1994. National Center for Health Statistics. *Vital Health Stat.* 10(193).
8. Collins, J. G. (1997) Prevalence of selected chronic conditions: United States, 1990–1992. National Center for Health Statistics. *Vital Health Stat.* 10(194).
9. Felson, D. T., Zhang, Y., Hannan, M. T., Naimark, A., Weissman, B. N., Aliabadi, P., and Levy, D. (1995). The incidence and natural history of knee osteoarthritis in the elderly: The Framingham Osteoarthritis Study. *Arthritis Rheum.* 38, 1500–1505.
10. American Heart Association (1998). "Heart and Stroke A-Z Guide. Women and Cardiovascular Diseases," Biostat. Fact Sheets. (Available at website: *www.americanheart.org/Heart_and_Stroke_A_Z_Guide/*)
11. Wenger, N. K. (1997). Coronary heart disease: An older woman's major health risk. *Br. Med. J.* 315, 1085–1090.
12. Reference deleted in proof.
13. Bild, D. E., Fitzpatrick, A., Fried, L. P., Wong, N. D., Haan, M. N., Lyles, L., Bovill, E., Polak, J. F., and Schulz, R. (1993). Age-related trends in cardiovascular morbidity and physical functioning in the elderly: The Cardiovascular Health Study. *J. Am. Geriatr. Soc.* 41, 1047–1056.
14. Melton, L. J. (1997). Epidemiology of spinal osteoporosis. *Spine* 22, 2s–11s.
15. Melton, L. J. (1995). How many women have osteoporosis now? *J. Bone Miner. Res.* 10, 175–177.
16. Luckey, M. M., Wallenstein, S., Lapinski, R., and Meier, D. (1996). A prospective study of bone loss in African-American and white women—A Clinical Research Center Study. *J. Clin. Endocrinol. Metab.* 81, 2948–2956.
17. Aspray, T. J., Prentice, A., Cole, T. J., Sawo, Y., Reeve, J., and Francis, R. M. (1996). Low bone mineral content is common but osteoporotic fractures are rare in elderly rural Gambian women. *J. Bone Miner. Res.* 11, 1019–1025.
18. Yancik, R. (1997). Epidemiology of cancer in the elderly, current status and projections for the future. *Rays* 22S, 3–9.
19. White, L., Petrovich, H., Ross, W., Masaki, K., Abbott, R. D., Teng, E. L., Rodriguez, B. L., Blanchette, P. L., Havlik, R. J., Werowske, G., Chiu, D., Foley, D. J., Murdaugh, C., and Curb, J. D. (1996). Prevalence of dementia in older Japanese-American men in Hawaii: The Honolulu-Asia Study. *JAMA, J. Am. Med. Assoc.* 276, 955–960.
20. Bachman, D. L., Wolf, P. A., Linn, R., Knoefel, J. E., Cobb, J., Bélanger, A., D'Agostino, R. B., and White, L. R. (1992). Prevalence of dementia and probable senile dementia of the Alzheimer type in the Framingham Study. *Neurology* 42, 115–119.
21. Brattberg, G., Parker, M. G., and Thorslund, M. (1997). A longitudinal study of pain: Reported pain form middle age to old age. *Clin. J. Pain* 13, 144–149.
22. Woo, J., Ho, S. C., Lau, J., and Leung, P. C. (1994). Musculoskeletal complaints and associated consequences in elderly Chinese aged 70 years and over. *J. Rheumatol.* 21, 1927–1931.
23. LaCroix, A. Z., Guralnik, J. M., Berkman, L. F., Wallace, R. B., and Satterfield, S. (1993). Maintaining mobility in late life. II. Smoking, alcohol consumption, and body mass index. *Am. J. Epidemiol.* 137, 858–869.
24. Guralnik, J. M., LaCroix, A. Z., Everett, D. F., and Kovar, M. G. (1989). "Aging in the Eighties: The Prevalence of Comorbidity and its Association with Disability," Adv. Data, Vital Health Stat. 170. National Center for Health Statistics, Hyattsville, MD.
25. Chapleski, E. E., Lichtenberg, P. A., Dwyer, J. W., Youngblade, L. M., and Tsai, P. F. (1997). Morbidity and comorbidity among Great Lakes American Indians: Predictors of functional ability. *Gerontologist* 37, 588–597.
26. Nagi, S. Z. (1976). An epidemiology of disability among adults in the United States. *Milbank Mem. Fund Q. Health Soc.* 54, 439–467.
27. Guralnik, J. M., Leveille, S. G., Hirsch, R., Ferrucci, L., and Fried, L. P. (1997). The impact of disability in older women. *J. Am. Med. Women's Assoc.* 52, 113–120.
28. Guralnik, J. M., Ferrucci, L., Simonsick, E. M., Salive, M. E., and Wallace, R. B. (1995). Lower-extremity function in persons over the age of 70 years as a predictor of subsequent disability. *N. Engl. J. Med.* 332, 556–561.
29. Guralnik, J. M., Land, K. C., Blazer, D., Fillenbaum, G. G., and Branchi, L. G. (1993). Educational status and active life expectancy among older blacks and whites. *N. Engl. J. Med.* 329, 110–116.

30. Vita, A. J., Terry, R. B., Hubert, H. B., and Fries, J. F. (1998). Aging, health risks, and cumulative disability. *N. Engl. J. Med.* **338,** 1035–1041.

31. Leveille, S. G., Guralnik, J. M., Ferrucci, L., and Langlois, J. (1999). Aging successfully until death in old age: Opportunities for increasing active life expectancy. *Am. J. Epidemiol.* **149,** 654–664.

32. Ferrucci, L., Izmirlian, G., Leveille, S. G., Phillips, C. L., Corti, M. C., Brock, D. B., and Guralnik, J. M. (1999). Smoking, physical activity and active life expectancy. *Am. J. Epidemiol.* **149,** 645–653.

33. U. S. Bureau of the Census (1996). "Current Population Reports, Special Studies, P23-190, 65+ in the United States." U.S. Govt. Printing Office, Washington, DC.

34. Anderson, R. N., Kochanek, K. D., and Murphy, S. L. (1997). "Report of Final Mortality Statistics, 1995," Mon. Vital Stat. Rep., Vol. 45, No. 11, Suppl. 2. National Center for Health Statistics, Hyattsville, MD.

35. Manton, K. G., and Vaupei, J. W. (1995). Survival after the age of 80 in the United States, Sweden, France, England, and Japan. *N. Engl. J. Med.* **333,** 1232–1235.

92

Menopause: Its Epidemiology

MARYFRAN R. SOWERS
Department of Epidemiology
University of Michigan
Ann Arbor, Michigan

I. Introduction

In women, the midlife, which includes the menopausal transition period from active reproductive capacity to reproductive senescence, has been less well studied than any other period of the life span except extreme old age. Understanding and defining the attributes of this transitional period is challenging because of its complexities. There is the challenge of addressing the overlay of ovarian change on normal aging. There is the challenge of its variation in presentation, including alterations in bleeding patterns, hormone patterns, and physical and psychosocial characteristics. The complexities in characterizing midlife in women also make it difficult to evaluate the potential social and medical morbidities that are thought to be associated with the transitional stages of the menopause.

This chapter will primarily focus on the perimenopausal experience, beginning with an operational definition of the menopause and the scope of the transition in terms of age and ethnicity. We will characterize the current understanding of the relationship between bleeding characteristics and the attending hormonal variation around the transition. Thereafter, we will address potential morbidity, which has been suggested as having its origin, directly or indirectly, in the events of the menopausal transition. The chapter will close with a summary of the theories and conceptual models that influence the nature and interpretation of the research and its ultimate implementation in health practice.

II. Definition of the Menopause

The most frequently used definition of the menopause is that of the World Health Organization [1]. The *menopause* is defined as the permanent cessation of menstruation resulting from loss of ovarian follicular activity. By convention, this attribution is made in retrospect following 12 months of amenorrhea not due to other factors such as pregnancy or lactation. The *perimenopause* (or climacteric) includes the period prior to the menopause when endocrinological, clinical, and biological changes associated with the menopause are occurring, as well as the first year following the menopause. The *postmenopause* is defined as the period after the menopause and begins after 12 months of spontaneous amenorrhea has been observed.

III. Hormonal Events and the Menopausal Transition

Studies of the processes associated with ovarian aging have generally involved clinical studies conducted in a limited number

of women over a relatively brief period of time relative to the length of the transitional period, which, in its entirety, is probably 8–12 years in duration. Alternatively, studies have contrasted the characteristics of women who are clearly postmenopausal with those characteristics of women who are not yet menopausal (as typically defined by age) but for whom the proportion of pre- vs perimenopause status is not really known. Study subjects have been predominantly Caucasian, yet there are intriguing reports that suggest that menopausal presentation is different among varied ethnic groups. Based on the human studies and a number of animal studies, a working paradigm of the sequence of hormonal events associated with the menopause has been developed. There appears to be a three-phase evolutionary process that describes the transitions that occur along a time continuum.

Characteristics thought to be associated with *the first phase* of menopause include the following: there is an increase in follicle stimulating hormone (FSH) secretion, without concomitant increase in the luteinizing hormone (LH) level [2]; a reduction in the secretion of inhibin and follistatin, peptides that inhibit the release of FSH from the pituitary gland (in a closed feedback system with FSH); a reduction in the number of follicles; an increasing likelihood of premature ovulation; more frequent decreased progesterone production with luteal insufficiency [2–4]; hyperestrogenism in ovulatory cycles [5,6]; and an excess of estrogen relative to progesterone. However, it is important to recognize that these events are on a continuum and may not represent discrete phases with a well-defined onset. For example, Klein [7] reported that women of advanced reproductive age (aged 40–45) did have evidence of poorer quality of the aging oocytes, but there was little evidence of compromise in follicle development or granulosa cell function.

The second phase of the menopausal transition is characterized by instability in estrogen levels and an increasing probability of anovulatory cycles [8], with an apparent failure to mount an LH surge in response to an estrogen stimulus [3]. Abnormalities of the gonadotrophin releasing hormone (GnRH) pulse generator, including rapid follicular phase [7], luteal phase [9], and pulsatile LH secretion have also been reported. Gonadal hormone levels are erratic such that cycles continue to reflect hyperestrogenism relative to progesterone [10]. These fluctuating changes in the gonadal hormone concentrations, in turn, should be reflected in altered presentation of menstrual cycle lengths and bleeding patterns [8,11–13]. This may include an increased frequency of dysfunctional uterine bleeding (which may be attributable to anovulation).

In *the third phase* of the menopausal transition, the ovarian follicles no longer respond to FSH and LH. Plasma estradiol levels drop below 20 pg/ml while progesterone production is nonmeasurable. Menstrual bleeding ceases. FSH and LH levels are consistently elevated, remaining so for an extended period of time [14,15], with an FSH/LH ratio greater than one [10].

This working paradigm of the hormonal events of the menopausal transition has not concurrently been related to bleeding patterns and symptom presentation. The paradigm also provides no index of the social, psychological, or physical characteristics that might influence the variation in hormone concentrations, an individual woman's responsiveness to these hormone concentrations, or the sequence of events.

One might argue that there is evidence of a fourth phase in this transitional process, at least in some women. Cross-sectional studies of serum hormone levels suggest that 20–40% of women have estrogen levels indicative of some follicular activity during the first 6–12 months following the last menstrual period [16,17]. Additional evidence also suggests that the amount of estradiol secreted by the ovary in the postmenopause is correlated with the degree of ovarian stromal hyperplasia [18], and it is estimated that approximately 10% of women [19] will retain an estrogenic cervical smear up to 10 years beyond the last menstrual period [20].

While the major estrogen prior to the menopause is estradiol, the primary estrogen following the menopause is estrone [21]. The average estrogen production in the postmenopause is estimated to be 40 μg/day for estrone (primarily generated by the peripheral aromatization of androstenedione) and 6 μg/day of estradiol. This is in comparison with productions of 80–500 μg/day of estradiol and 80–300 μg/day for estrone in the premenopause [22]. Longcope [23] reported that declining concentrations of estrone, estradiol, and estrone sulfate stabilized within 12 months after the last menses.

The role of other hormones, apart from the sex hormones, in the menopausal transition is relatively unexplored. For example, Ballinger [24] found no difference in thyroxine level according to menopause status; however, triiodothyronine and TSH were higher in early premenopausal women than in later premenopausal women. Postmenopausal women had the lowest levels of both triiodothyronine and TSH. Administration of estrogen formulations to postmenopausal women has resulted in reductions of insulin-like growth factor-I and elevations in growth hormone and growth hormone binding protein, a system that has been associated with the aging process and decreased muscle mass [25].

IV. Placing the Menopausal Transition in the Context of Age and Time

The menopausal transition is a function of age and time. The universal questions include: at what age does the menopausal transition begin? How long does the transition last? Are the phases (as described in the working paradigm) predictable in length? Are the lengths of the phases, indeed the total transition time, related to the chronological age at which the transition commences?

The menopausal transition transects the aging process. The degree to which these aging-related changes in the endocrine environment influence the menopausal process is uncertain. As an example, aging can have an impact upon hepatic metabolism,

CNS feedback sensitivity or secretory activity, or hormone bio-activity, all of which will impact the functional presentation associated with hormone behavior.

Metcalf and colleagues [2,8,26] attempted to distinguish the effects of aging from the effect of approaching menopause on hormone levels. No significant changes were observed in the magnitude of the pregnanediol peak in 21–35 day menstrual cycles as women aged. However, women aged 40–55 who had experienced anovulatory menstrual cycles had significantly lower pregnanediol levels in their ovulatory cycles. FSH and, to a lesser extent, LH may exhibit age-dependent changes that are unrelated to the proximity of menopause; however, among women over age 40, change in gonadotropin levels was considerably more marked in women who had begun to experience menstrual irregularity than in women who had not. The characteristic low estrogen and high gonadotropin levels of older women appears to evolve gradually beginning up to ten years before, and continuing at least one year after, menopause [27].

A. Predicting Age at Menopause

With the difficulty in characterizing actual hormone patterns, menopause has been most frequently characterized in studies of populations in relation to the age of onset of menopause. The average age of menopause in populations ranges from 48 to 52 years with indications that there is a trend toward an increasingly older age at onset [28]. It is not clear in much of the epidemiologic literature whether the typically reported age of menopause represents the age of the last menstrual period (LMP) or the LMP plus one year, to assure consistency with the WHO definition. Furthermore, many of the studies fail to identify whether the age of menopause being reported also reflects the comingling of information contributed from women with surgical menopause. This is an important consideration as the proportions of populations with surgical menopause have substantial variation according to ethnicity and geography.

Characteristics that predispose a woman to an earlier onset of menopause are poorly understood. It has been suggested that the younger age at menopause observed in women living at higher altitudes is a reflection of relative hypoxia. This hypoxia results in lower serum progesterone and estradiol concentrations during the luteal phase [28]. Studies conducted in Papua, New Guinea indicated that malnourished women experienced menopause approximately 4 years earlier than better-nourished women [29]. Body size may be an important factor, albeit there are relatively few studies. It has been suggested that greater weight and height [30,31] predispose a woman to a later age at menopause.

Consistently, cigarette smoking has been the environmental characteristic most frequently associated with an earlier age of onset of menopause. Women who smoke have an age of onset 1–2 years earlier than women who do not smoke [32–40]. There have been multiple mechanisms proposed to explain this observation. These potential mechanisms include speculation that smokers have lower estrogen levels [40], metabolize estrogen more rapidly [41], or have altered estrogen receptor binding [42] relative to nonsmokers. Alternatively, aromatic hydrocarbons may act on the ovary and accelerate follicle aging [40]. Alcohol consumption, another lifestyle behavior, has also been examined relative to age at menopause. Torgerson et al. [43] have argued that moderate alcohol consumption is associated with a later age at menopause and that this later age contributes to the protective effect of alcohol on cardiovascular risk in women.

Scientists have sought explanations for age at menopause in other reproductive characteristics. For example, Whelan *et al.* [44] reported that women with menstrual cycles less than 26 days duration were more like to have an earlier age at menopause. Khaw [45] reported that reproductive history, including earlier menarche, greater parity, and oral contraceptive use were associated with later age at menopause. A partial explanation for these observations may be associated with the potential role of FSH. There is evidence that FSH influences the rate of follicular depletion. It has been hypothesized that the *longer FSH concentrations are suppressed* (as would occur with pregnancy or oral contraceptive use), the greater likelihood that menopause is delayed [44,46]. However, these studies have not adjusted for the potential confounding effects of body size and cigarette smoking. Furthermore, these findings are inconsistent with a knowledge of the change at FSH levels, which, on average, begin to rise at approximately age 40. An alternative theory suggests that growth retardation during gestation may be associated with a reduced number of primordial follicles which, in turn, lead to an earlier age at menopause. However, studies have yet to associate age at menopause with birth weight [47].

Other work suggests that there is a familial association in the menopausal age between mothers and daughters [48,49]. Current studies, however, have been unable to distinguish whether the familiality is genetic in origin or whether the association is a function of a shared environment or is artifactual.

In summary, it appears that the average age of menopause is occurring at a later chronological age. Studies have not addressed whether there is also increasing variation of age at menopause within populations showing an increasingly older average age at onset. Behaviors that appear to support the secular trend in age at menopause include the reduction in smoking frequency, increase in average body size, improved nutritional status, and, potentially, use of oral contraceptives.

V. "Symptomatology" of the Menopause

A. The Social/Physical Context of the Menopause

There is controversy about whether the events of menopause should be interpreted within a medical context of "symptomatology," particularly because there appears to be a differential presentation of menopausal "symptoms" among different cultures. Social scientists have argued that differential presentation of physical or psychological symptoms are, at least in part, a result of the social expectations of a particular sociocultural group. Thus, a particular culture could contribute to the label of dysfunctionality that might be linked to menopause. It is speculated that there will be more complaints about "symptoms" in a youth-oriented culture where fertility is valued and aging is feared. Another culture in which attaining menopause is associated with greater status and greater freedom than during the time of reproduction might be associated with fewer complaints of "symptoms." For example, Flint [50] reported that wealthy Indian (south Asian) women (n = 483) experienced few symptoms and looked forward to menopause when they would be

released from purdah (being veiled and secluded). Maoz [51] reported similar findings in a study of Arabian women who were one of five cultural groups studied in Israel. These intriguing investigations have not been accompanied by hormone studies to ascertain whether duration and intensity of endocrine characteristics are different between cultures.

There is mixed evidence as to what social and economic environments may be associated with menopausal symptomatology. Severne [52] reported that during the climacteric, work outside the home had a positive effect on women with higher socioeconomic status (SES). However, the converse was true in women with lower SES. Hunter [53] reported that lower social class was predictive of depressed mood, anxiety, cognitive difficulties, and sleep problems. Several studies have suggested that fewer menopausal symptoms were observed among women with a well-integrated social network or a strong social support network [54].

At issue is also whether change in physical or behavioral characteristics should be attributed to the menopause, specifically, or more generally to aging. Utian [55] has maintained that true menopausal symptoms are only those that are estrogen dependent, these being hot flashes, atrophic vaginitis, and osteoporosis. In contrast, several menopause studies have utilized lists of characteristics (such as the widely used 41-item list by Neugarten [56]), which include characteristics that also reflect the ongoing aging process. This mixed characterization of aging with menopause may help explain why studies report no correlation between "symptoms" and the degree of estrogen deficiency as judged by plasma estrogen levels [24,57]. Likewise, these lists may not be adequate to discern if characteristics have really changed over time or if the woman would respond similarly to the list even if she were younger and premenopausal or older and truly postmenopausal.

The following sections will address individual "symptoms" commonly associated with the menopausal transition. We will also attempt to integrate information about social and cultural information linked to the particular perimenopausal characteristics.

B. Dysfunctional Uterine Bleeding during the Perimenopause

As previously described, characteristics of the perimenopause include a decline in the number of ovarian follicles (to a depleted state in the postmenopause) and a decrease in the proportion of ovulatory cycles. Likewise, there are changes in menstrual cycle length and bleeding patterns associated with the menopausal transition [8,11–13]. There is speculation that dysfunctional uterine bleeding (DUB) may be attributable to this anovulatory pattern. It is suggested that dysfunctional uterine bleeding may arise from two scenarios associated with intermittent anovulation preceding actual menopause. In the first scenario, the oocyte may respond to FSH stimulation with the formation of a follicle; however, the level of estrogen is insufficient to induce the LH surge and the attendant ovulation. The follicle continues to supply the endometrium with estrogen and the endometrium continues to grow. However, the lack of progesterone impedes the normal secretory response, the sloughing of the endometrium. Eventually, estrogen levels drop, hormonal support for endometrial growth is lost, and bleeding occurs. In a second scenario, the estrogen levels build slowly over an extended period of time, allowing for significant proliferation of endometrium. Eventually, estrogen levels are of a magnitude to induce the LH surge and ovulation. However, in the interim, substantial endometrial tissue is accumulated for sloughing and the resulting menstrual flow is extremely heavy and prolonged.

There was equal distribution of menorrhagia (very heavy menstrual bleeding) between those classified as *early* and *late premenopause*. Ballinger [24] reported no difference in hormone profiles during these events; however, it is extremely difficult to predict when a DUB event will occur and the monitoring required to evaluate the hormone state of a woman undergoing DUB is problematic to implement.

C. Hot Flashes and Flushes

A "hot flash" is defined as the increase and perception of heat within or on the body. Typically, the length of a flash ranges from 5 to 12 minutes in length [58]. The prevalence of hot flashes is estimated to be 50–85% in perimenopausal women [56,59,60]; however, in 10–15% of women, hot flash activity is of sufficient intensity as to disrupt daily activities [61]. The prevalence of hot flashes is greater in the first two years after menopause and declines with increased years postmenopause. Kronenberg [61,62] reviewed our understanding of the central and peripheral thermoregulatory physiology in hot flashes and summarized the observed changes in reproductive hormones, catecholamines, and opioid peptides.

Women with surgical menopause tend to have a greater prevalence of hot flashes, at least in the year following oophorectomy [62]. Circadian rhythms in hot flash presentation have been described, with increased presentation, either in the morning or the evening, suggesting component rhythms at 8 or 12 hours in addition to the 24-hour rhythms. The rhythms observed in hot flash presentation do not appear to be related to whether women are experiencing natural or surgically induced menopause [63].

The prevalence of hot flashes varies greatly among the populations where it has been studied. This variation may arise for multiple reasons, including type of study population, climate, and culture. First, one would expect to see more concern about hot flash frequency and intensity in studies of clinical populations as compared to community- or population-based studies. Furthermore, how one elicits information and the case definitions may not be equivalent in the different studies. Second, it has been suggested that incidence of hot flashes is lower in warmer climates, leading some investigators to suggest that hot flashes are not being differentiated from fevers of local origin, particularly malaria [59]. Unfortunately, these climatic and geographic characteristics contribute to the definition of hot flashes in different cultures. Thus, it becomes difficult to discern whether differences are real or artifactual or whether the differences are cultural or climatic. Third, all cultures do not share a similar presentation of hot flashes, suggesting that culture has a role in defining the expression of and attitudes towards menopause. While hot flashes have been reported in a variety of cultures, Japanese and Indonesian women report fewer hot flashes than do women in studies located in western cultures. Beyene [64] reported that Mayan women in Mexico do not include hot flashes in their responses about menopausal transition. Thus,

investigators debate as to whether cultural variability should be interpreted as differences in physiology or differences in perceptions of characteristics. Cultural interpretations are also more complex because these interpretations of cultural differences represent geographic differences and differences in environment factors (*i.e.,* diet, physical activity, and smoking), as well as cultural definitions of social roles for women. In western cultures, there have been attempts to identify those factors that are associated with hot flashes. To date, no association has been found with employment status, social class, age, marital status, or parity [60].

Differences in cultural presentation of the menopause transition is an area of active investigation, spurred in part by studies that have shown ethnic differences in hormonal levels during the menopausal transition. For example, lower serum estradiol and testosterone levels have been reported in Chinese as compared to Western women [65]. Studd [66] has argued that hot flashes are due to the process of estrogen withdrawal rather than to the deficiency of estrogens.

Sherman *et al.* [67] reported no difference in women with and without hot flashes relative to age at menarche, age at menopause, number of pregnancies, height, or coexisting medical conditions. Weight was negatively correlated with hot flash frequency; women with a lower weight reported a greater frequency of hot flashes, presumably because of the higher estrone concentrations in heavier women.

There have been relatively few investigations of the metabolic consequences being reflected by hot flash frequency or intensity. However, clinical studies have suggested that neuroendocrinologic changes contribute to hyperglycemia and hypertriglyceridemia in women with severe and frequent vasomotor episodes [68].

D. Depression

Depression is a major public health problem with a preponderance of expression among women. As reviewed by McKinlay [69], the excess risk for depression among women appears to be real as opposed to being attributable to gender differences in use of health care, symptom recognition, experience with and reaction to life events, or measurement artifact.

The prevalence of depression has been evaluated in large sample surveys in populations located in Sweden, the United Kingdom, Canada, and the U.S. In these studies, there appears to be no excess risk for depressive disorder associated with the menopause [70–72]. McKinlay *et al.* [69] reach the same conclusion in their review of two decades of epidemiologic studies. Other studies have found no greater frequency of admission to psychiatric hospitals [73] or regional mental health centers [74] among women in the menopausal age range. Ballinger [24] also reported no difference in reproductive hormone levels in perimenopausal women who scored higher on depression scales versus those who scored lower, although there was a significantly higher level of both triiodothyronine and TSH in women classified as having lower depression scores.

In a prospective study of 2500 women, McKinlay [69] reported that although high depressive symptom scores (CES-D) were more frequent in those with surgical menopause, among the remainder of the cohort menopausal status was not predictive of depression. Higher depression scores were more likely to be observed in widowed, divorced, or separated women with less than 12 years of education. Limitations in physical activity or recently diagnosed chronic disease conditions, along with worry about others, were also more frequent among those with the highest depression scores.

E. Urinary Tract Infections

An increase in asymptomatic bacteriuria with age [75–78], has been attributed, in part, to the expected hormonal changes with the menopause. Vaginal walls become thinner and lose their ability to lubricate quickly so that the vagina is less moist. Decreased lubrication can increase the mechanical irritation associated with sexual intercourse. It has been suggested that mechanical irritation can facilitate invasion by uropathogens or that declines in urethral closure pressure make it easier for perineal pathogens to gain entry into the bladder. With menopause, the vaginal pH increases, allowing growth of many bacterial species, including known uropathogens [79,80].

Estrogen replacement therapy is hypothesized to modify risk of UTI both positively and negatively [80,81]. The literature describing the effects of estrogen replacement therapy on UTI risk is inconclusive and contributes little to understanding the role of menopause in UTI. A clinical series [80] and a crossover experiment [79] suggested that the restoration of premenopausal pH and mucosa with estrogen replacement therapy prevented colonization by gram-negative bacteria and, thus, UTI in postmenopausal women. A clinical trial found administration of estriol reduced recurrent urinary tract infections in postmenopausal women, probably by modifying the vaginal flora [82]. However, a large case-control study (3616 cases and 19,162 controls) conducted using an automated data base found a twofold increased risk of UTI in older women (aged 50–69 years) with intact uteri who used estrogen [81]. Results were adjusted for atrophic vaginitis, neurologic deficit, diabetes mellitus, incontinence, and age. The investigators had no measure of the indications for estrogen use among women with uteri, such as presence of menopausal symptoms.

F. Urinary Incontinence

Urinary incontinence has long been associated both with aging and with menopause in women [83,84], although examination of the literature suggests that for menopause, the associations are far from conclusive. Here is an example of the inconsistencies in this area. Approximately 20% of 45-year-old Dutch and Swedish women suffer some degree of urinary incontinence [85,86], in contrast to approximately 7–9% of premenopausal women [86–88] and 10–12% of postmenopausal women. Hording [85] could not identify an increased likelihood of incontinence among age-matched (45-year old) pre- and postmenopausal women, an observation supported by Hagstad [89] and Milsom [90].

While it is recognized that the bladder and urethral tissues are estrogen-sensitive, a clinical trial of 49 nonsupplemented women compared to 23 estrogen-supplemented women showed only marginal improvement in urethral function [91]. Thus, to address the question of the menopausal transition as a risk factor

for urinary incontinence, it would appear that future study designs need to more specifically characterize the estrogen status of women in relation to urinary continence and need a clearly defined and implementable case definition of continence. Further, the designs need to be prospective in nature in order to evaluate the relative impacts of age and estrogen status.

G. Menopause and the Epithelium: Impact on the Vagina

Within the constellation of characteristics associated with the menopause, attenuation of sexual behavior may result, in part, because of the atrophic endothelial changes in the vagina. Atrophic changes of the vagina include thinning of the endometrial endothelium, loss of rugation, and decrease in the lubrication accompanying sexual arousal. These changes are, in turn, associated with dyspareunia.

Hallström [92] has reported significant attenuation of sexual activity in a cross-sectional study of Swedish women, but this observation is not consistently reported [93]. Furthermore, there have been few longitudinal studies in this area. In one longitudinal, albeit short-term, study, McCoy et al. [94] reported that hot flash ratings were negatively and significantly associated with frequency of intercourse. They observed no association between hormone levels and frequency of sexual activity.

Potentially, one limitation in this area is the focus on estrogen alone. Studies have shown sexual interest and motivation are more strongly influenced by androgen (testosterone) as opposed to estrogen [95]. Much of the clinical literature suggests that estrogen administration alleviates symptomatology associated with atrophic vagination and dyspareunia; however, estrogen levels appear to be less related to motivation [96]. Much of the classic menopause literature does not address the role of testosterone levels or the rapidity of change in testosterone level. If libido and arousal are constituents of good mental health during menopause, then studies need to expand their evaluation of hormone levels beyond that of estrogen.

H. Summary

The relationship of hormonal changes to changes in bleeding or other symptoms remains underexplored. Issues remain about the centrality of the role of estrogen versus the more aggregated role of estrogen and other hormones including the androgens. Issues remain as to whether the nature of "symptom" presentation is based on actual hormone levels, the rapidity of change in these levels, or both. Issues remain in terms of understanding how hormone concentrations or changes in those concentrations are modulated by the social and physical characteristics of individual women. Certainly, some women experience the menopausal transition as an abrupt transition and other women experience a more gradual change. Some women report severe symptoms, while others experience few or very mild symptoms. It is important to understand how the hormone paradigm plays out in healthy, free-living populations and to describe the range of normal variation across the population and between population subgroups in terms of the ultimate functional presentation.

It is equally important to expand the characterization of the social environment (including employment, social role identification, and economic status) to determine whether that social environment affects the hormone paradigm. Conversely, it is also important to determine if there are unique social environments that are more likely to sensitize women to marked changes in hormone status that occur during the midlife.

VI. Is the Menopausal Transition a Keystone in Subsequent Chronic Disease?

A number of investigations have noted that the midlife (age approximately 50 years) is associated with a significant rise in the prevalence of chronic disease among women. There is speculation as to whether metabolic changes associated with the menopause, interacting with or accelerating events of normal aging, promote the increased incidence of heart disease, diabetes, hypertension, breast cancer, osteoarthritis, and autoimmune disease [97].

A. Bone Density and Calcium Mobilization

Osteoporosis is considered the classic estrogen deficiency disease, and is classified by Utian [55] as a menopausally-related disease. Many investigators believe the estrogen loss of menopause is a major factor in bone mass loss, a loss which will, in some women, eventually become sufficient to lead to osteoporotic fracture. Most researchers agree that there is a period of rapid bone loss immediately following the menopause, [57,98–105]. It is thought that continued low levels of estrogen after menopause are largely responsible for the decrease in bone mass [103,106–108]. One study, after comparing women with oophorectomy with naturally menopausal women, suggested that 52–66% of bone loss is caused by estrogen deficiency [104].

Some studies have reported evidence of some bone loss in the perimenopause [98,99,109–112]. However, other studies have detected no significant changes preceding the menopause [104,113,114]. Change in bone mass prior to menopause may also be bone site specific. One study showed a significant decrease in the lumbar spine beginning in the third decade but no substantial loss in the peripheral skeleton before the menopause [102,107,115,116]. Population studies suggest that loss in spinal bone may occur faster around the perimenopause than after menopause [98,117,118].

Bone loss in women with oophorectomy as compared to women experiencing natural menopause has not been sufficiently studied. One investigation of 64 oophorectomized women and 309 naturally menopausal women found a greater percentage of bone loss over time in the natural group, but this could be confounded by age, as the women with surgical menopause were significantly younger [104].

Studies of women with early natural menopause (before 45 years) differ greatly in their conclusions as to the subsequent impact on osteoporosis and fracture. Some have found evidence that low bone mineral density and high subsequent fracture rates and osteoporosis risk can be correlated with early menopause [33,119–121]. However, others have reported no significant association [106,122], leading some researchers to hypothesize that if early menopause is a risk for osteoporosis and fracture, it is likely due only to a longer period of low bone mass [122]. No studies have examined the intensity of menopausal symptomatology in relation to bone density.

There are an increasing number of studies relating menopause to biochemical markers of bone turnover [123]. One study of 178 women aged 29–78 found that the estrogen decline after menopause occurred simultaneously with changes in biochemical indices of bone turnover [113]. Many researchers have observed a rise in serum osteocalcin levels, which has been positively correlated with bone loss of menopause [101,103, 107,113,124–126]. However, some studies have found different menopause phase-specific patterns, such as a continuous increase in bone turnover marker levels from premenopause to postmenopause [125], or a decline before menopause, a rise at menopause, and a fall in the seventh and eighth decades [124].

Pyridinoline cross-link components of collagen, a marker of bone resorption, have been associated with bone changes in the menopause. A study of 68 aged 50–76 who had experienced natural menopause reported that deoxy-pyridinoline *decreased* with age and slightly with age at menopause [127]. In this study, 58% of the variation in the rate of loss of bone mineral content could be explained by pyridinoline and estradiol glucuronide assays and body mass index. Other studies have shown a rise in hydroxyproline, another indicator of total bone resorption, accompanying the menopause [57,101,113,124]. One study emphasized the greater rise in hydroxyproline relative to the rise in alkaline phosphatase, a difference that decreases gradually with time and diminishing rate of bone loss [57].

Urinary calcium has been found to increase at the menopause [57,113,124–126] and after the menopause [57,113]. One study found that plasma and urinary calcium levels fell after a peak at menopause [57]. The increased calcium excretion found in the menopause has been shown to be negatively associated with bone density and bone mineral content [57,106].

The responsiveness of the vitamin D system to estrogen may also contribute to the bone loss observed with menopause. 1,25-dihydroxyvitamin D is an integral part of the hormonal environment that maintains calcium balance by regulating calcium absorption from the intestine, calcium resorption from the skeleton, and calcium retention in the kidney. Sowers [112] has shown that there is a significantly greater 1,25-dihydroxyvitamin D level in premenopausal women versus postmenopausal women. Furthermore, in women using HRT, 1,25-dihydroxyvitamin D levels were similar to those of premenopausal women.

B. Obesity

Obesity is defined by body composition proportions; that is, there is an excess of fat tissue in relation to the amount of lean mass. Measures of body topology, including waist-to-hip ratio (WHR), describe the distribution of fat mass in body segments. It is increasingly well appreciated that body composition, including topology, is related to multiple disease conditions in women, including heart disease, diabetes, gall bladder disease, certain cancers, osteoporosis, and arthritis [128]. Additionally, these measures are also associated with a number of undesirable metabolic characteristics, including elevated triglycerides, low HDL cholesterol concentrations, glucose intolerance, and hyperinsulinemia, and potentially plasma leptin concentrations that are believed to be risk factors for cardiovascular disease and diabetes [129].

The change in body composition that culminates in menopause or the final menstrual period (FMP) has not been well characterized. Furthermore, existing data are limited predominantly to studies of Caucasian women and use methodological approaches that preclude comparison between ethnic groups. Limited evidence suggests that body composition changes with ovarian aging and/or chronologic aging. Sowers has shown that in the perimenopausal period, on average 1 kg gain in total weight annually is associated with a 1.4 kg increase in the fat compartment and a 0.5 kg decrease in the lean compartment [128]. Aloia et al. [130] suggested that natural menopause is associated with an acceleration of loss of fat-free (muscle) tissue as well as skeletal mass. The rate of loss of total body potassium, a marker of muscle mass, was greatest in the first three years following menopause, although these investigators noted a slow (and statistically insignificant) loss in the fat-free mass in the perimenopause.

There is greater evidence of association between ovarian aging and body composition topology as described with WHR [131]. With the transition to menopause, the relative adipose distribution in women changes in the direction of that observed in men [132]. WHR ratios have been reported to change slightly from premenopause (0.73 ± 0.05, n = 550, aged 43.4 ± 3.9 years) to perimenopause (0.74 ± 0.06, n = 168, aged 49.2 ± 3.8 years) but to increase significantly after menopause (0.78 ± 0.06, n = 1133, aged 58.6 ± 4.7 years) [133]. When dual energy X-ray absorptiometry technology was used to determine degree of central fat mass, women transitioning to both perimenopause and postmenopause had greater central adiposity than premenopausal women [134].

With menopause, an increase in visceral fat mass is preventable by hormonal replacement therapy [135,136]. It is noteworthy that HRT has been reported to minimize the shift from gynoid to android fat distribution, but does not influence BMI or fat mass [137].

There is some evidence of enlargement of femoral subcutaneous adipocytes to support the demands of pregnancy and lactation, and this enlargement is reported to decline with menopause [138]. Furthermore, progesterone competes with the glucocorticoid receptor (GR) [139,140] and may protect from glucocorticoid effects during the late luteal phase of the menstrual cycle, a protection which would be lost with menopause.

A very limited number of studies with uniquely selected populations have provided a preliminary characterization of the midlife transitional period. However, currently no single study has provided an integrated picture of body composition (including topology) and its change during the menopausal transition. This would entail making simultaneous and repeated measures of ovarian aging, body composition and other health-related outcomes (i.e., menopausal "symptoms," bone mineral density, lipids, glucose) throughout the transition with a sample size sufficient to characterize any menopausally-related patterns.

C. Cardiovascular Disease

Cardiovascular disease (CVD) is the major cause of death in U.S. women, accounting for more death in women (45%) than men (39%), considering all ages [141]. Coronary heart disease (CHD) accounts for 34% of all deaths in women.

Menopause is thought to be a risk factor for heart disease based on two types of studies. First, following oophorectomy, women have been shown to have an increased incidence of atherosclerosis [142]. Second, studies of exogenous estrogen use in the postmenopause indicate that women have fewer CVD events [143]. The mechanisms by which estrogens affect coronary heart disease appear to fall into two major categories. Stevenson [144] suggests that benefits include both a direct effect on the arteries as well as an effect on the metabolic risk factors for CHD. The latter would include improvement in the lipoprotein and lipid profile, and attenuation of disturbances of glucose and insulin metabolism, as well as stabilization of the hemostatic factors which would favor fibrinolysis rather than coagulation. The direct action on arteries suggests improvement in arterial compliance. Primarily, this information is from studies of women using hormone replacement therapy. However, the body of information about the actual menopausal transition and mechanisms associated with CVD is surprisingly limited.

1. Lipids

It is well established that total, LDL, and HDL cholesterol are important risk factors for CHD. Some evidence suggests that low levels of HDL cholesterol and high levels of triglycerides are particularly important risk factors for CHD in women. Epidemiologic studies show that the change from pre- to postmenopausal status is not associated with increases in blood pressure, blood glucose, or body weight [145–147]. However, change in menopausal status is associated with increases in total cholesterol and LDL cholesterol and with a decline in HDL during the menopause due to the decline in HDL_2 [148]. Studies with Pima and African-American women, however, show no variability in lipids according to menopausal status [149,150].

2. Clotting System

The majority of clinical vascular events (myocardial infarction, unstable angina, and stroke) are associated with the formation of a platelet-rich fibrin clot at the site of prior atherosclerotic damage of a coronary or cerebral vessel. The potential for arterial thrombosis is a function of platelet adhesion and aggregability, the state of the clotting system, and the capacity for clot dissolution or fibrinolysis.

Fibrinogen, Factor VII, and PAI-1 levels are higher in postmenopausal women as compared to premenopausal women. Furthermore, hormone replacement therapy appears to curtail these increases in fibrinogen [151–153], in Factor VIIc [151], and in both tPA and PAI-1 [154,155]. However, there are no specific data on a direct relationship between hormone levels and clotting factor changes.

3. Blood Pressure

The majority of studies fail to find an association between menopausal status or hormone use and systolic blood pressure (SBP) or diastolic blood pressure (DBP) [147,148,156–159]. However, there are some studies reporting positive associations between ovarian aging and blood pressure changes [146,160,161].

4. Insulin Resistance

Changes in insulin and insulin sensitivity at the menopause may play a role in women's increased risk for coronary heart disease at older ages. However, the findings have been conflicting. One population-based study did not show an increase in fasting insulin with change in menopausal status [147]. A cross-sectional study showed a fall in insulin levels in 426 pre- and postmenopausal Italian women [162], and a Turkish study [163] showed no difference in the insulin: glucose ratio among women with premature ovarian failure and natural menopause. In contrast, other studies suggested that menopause was associated with increased fasting insulin concentration and suggested that the loss of ovarian function may explain this increase, [164–166]. Furthermore, studies have shown that 2-hour insulin secretion in postmenopausal women did not differ among those who received hormone replacement, but increasing insulin resistance thereafter brings about an increase in circulating insulin concentrations.

Rather consistently, studies have shown that abdominal obesity is associated with alterations in glucose-insulin homeostasis, while other studies have shown that age and menopausal status are known correlates of visceral adiposity. It also appears that menopause per se may not be associated with changes in glucose tolerance; however, there may be a reduction in pancreatic insulin secretion and a compensatory reduction in insulin secretion. Studies are needed to address the connectivity of these events, that is, to show that with the onset of menopause (and accompanying age), insulin resistance increases as central adiposity increases. Supportive information in the form of measured concentrations of estradiol, estrone, testosterone, progesterone, cortisol, insulin, and glucose throughout this process will help establish the temporal sequence of these events.

D. Rheumatoid Arthritis (RA)

The potential for changes in hormone concentrations to influence the initiation or progression of autoimmune diseases in women is not well appreciated. Two studies have described the increased incidence of RA at the age of the menopausal transition [167,168]. Considering the apparent disease remission with pregnancy (a high estrogen state) or with the use of oral contraceptives [169,170], it has been speculated that the decline in the incidence of rheumatoid arthritis is attributable to the use of oral contraceptive and postmenopausal hormone replacement [171,172]. In that scenario, the decline in estrogen levels might be a component of the presentation of autoimmune diseases. Not all studies support considering only estrogen levels around the menopausal transition [173]. Other investigators have noted that testosterone levels are associated with an increased incidence level of RA in men [174]; however, the role of testosterone (and its change in level with the menopause) does not appear to have been extensively explored in women [175].

Two mechanisms are potential explanations for the apparently protective effect of estrogen in autoimmune diseases including rheumatoid arthritis, Crohn's disease, and ulcerative colitis [176]. Estrogen has importance as an inhibitor of cell-mediated immunity and in the development of certain T lymphocytes. The first mechanism for the effect of estrogen on the immune system is via estrogen receptors on the T lymphocytes. There appears to be a dose-dependent response of lymphocytic proliferation to the addition of estradiol. Low concentrations act in a stimulatory fashion whereas higher concentrations act in an inhibitory fashion. Low concentrations have been reported to stimulate CD8+ suppressor cells [177] or to stimulate CD4+ helper cells [178].

A second mechanism is via a permissive role where cytokines and other growth factors take on a second messenger role. The cytokine interleukin-1 (IL-1), associated with the inflammatory response, is stimulated by low concentrations of estradiol and progesterone [179]. Studies indicate that a loss of estrogen (estradiol) allows peripheral blood monocytes to secrete more IL-1 (i.e., estrogen inhibits IL-1 production) and TNF-α, a response also associated with greater bone turnover. Those events that are catabolic to bone are also a part of the cytokine-mediated pathway through which estrogen depletion is associated in the inflammatory response.

E. Osteoarthritis (OA)

Based on their landmark publication in 1925, Cecil and Archer are credited with first suggesting that the presentation of osteoarthritis increases remarkably around the menopause. Though bone and joint pain are commonly accepted "symptoms" of menopause, there are surprisingly few studies that specifically address the issue of menopause and arthritis. In the Tecumseh Community Health Study, the incidence of osteoarthritis of the hand (OAH) in premenopausal women is less than that of men. After age 50, the incidence in women is higher than that in men and continues to be greater until age 80 [180,181]. It is acknowledged that osteoarthritis more commonly affects more joints and is usually more severe in women than in men [182].

Spector et al. [183,184] observed an association between osteoarthritis and hysterectomy and, based on a review of evidence from human and animal studies, have suggested that OA might be hormonally mediated. Schouten et al. [185] reported that there is a positive association between frequency of Heberden's nodes and later age at menopause, but they could not identify the "menopause-related osteoarthritis" suggested by Spector and Campion [184]. Other studies have also failed to demonstrate a positive association between the menopausal transition and either the initiation or progression of OA [186,187]; however, studies that follow OA status in women experiencing the menopausal transition have not been undertaken.

F. Menopause and the Dementias

There is a significant interest in defining the role of the menopause in relation to the dementias, particularly Alzheimer's disease (AD). In experimental animal studies, oophorectomy is associated with decreased neuronal regeneration and synaptic function [188]. Likewise, the gonadal hormones have been related to learning and memory function [189] in animal studies. A number of trials have shown improvements in memory and/or cognition with the administration of hormones [190–193], however improvements have not been universally noted in other trials [194,195].

There has been no evidence that women with an early age at menopause have an increased risk of Alzheimer's disease [196,197]. However, van Duijn [197] suggests that in those with a family history of Alzheimer's disease, age at menopause was significantly earlier in Alzheimer's patients. This is an area that is greatly underexplored and of primary importance.

In summary, it appears that there is substantial opportunity for changes occurring during the menopausal transition to have an impact on subsequent presentation with chronic diseases. However, the following are not clear in the present body of work:

- that estradiol is the central factor in this impact as opposed to estrogenic, androgenic, and adrenal hormone interplay
- that menopausal changes have a major role as opposed to a more minor role
- that the relative importance of these menopausal changes are equal in each chronic disease
- that a single mechanism could be expected to account for the impact of menopausal changes
- that menopausal changes would account for an impact independent of the social and physical environment of the individual woman

VII. Theories and Conceptual Models of Menopause

As indicated in this chapter, there is much to be learned about the menopausal transition. However, the theories and conceptual models surrounding the menopause shape the nature of the questions to be addressed as well as the implementation of interpretation and that research. Barile [198] summarizes four theoretical approaches to considering the menopause. The *biological theory* ascribes the experience of the menopause to biology, particularly within the framework of alterations in metabolism and endocrine status. Typically, there is a pronounced focus on ovarian function, and in some, attendant focus on hormone replacement. The *psychological/psychosocial* school of thought highlights the importance of stressors and losses as catalysts for symptoms. The attendant focus then suggests the need for social supports and coping skills. The *sociocultural/environmental theory* indicates that cultural constructs define our behaviors toward the menopause and its potential symptoms. Finally, the *feminist theory* views the menopause as a normal developmental stage. Women are urged to achieve control of the experience and to be an active participant in addressing the challenges of symptoms. Barile [198] argues for a more inclusive approach that acknowledges the need for and contributed value of each of these theories, minimizing the reductionist approach that is often associated with the menopause debate. It is argued that the menopause is such a complex process that there is a reasonable opportunity for each approach to contribute to the knowledge base.

The Study of Women's Health Across the Nation (SWAN) represents an important step in understanding women's health and subscribes to this more inclusive approach to considering the menopausal transition. It is the first national study to describe women at the midlife. The study encompasses unprecedented breadth in carefully characterizing social, psychological, and physiologic status, with measures of cardiovascular status, bone mineral density, depression, sociodemographic characteristics, symptomatology, use of health care, and a myriad of other factors. It will act as the landmark study to evaluate these factors in relation to serially-measured hormone levels. Furthermore, SWAN uniquely addresses these questions, longitudinally, in five major racial/ethnic groups: African Americans, Caucasians, Chinese, Japanese, and Hispanics, offering the opportunity to consider the contributions of culture and biology.

VIII. Summary

More information about the menopausal transition period is necessary to promote health in women. This transition period

may encompass 5–10% of the lifetime of women. The menopausal transition period is the portal through which women will approach one-third of the life span. However, it is a period about which relatively little is known with regard to the development processes and metabolic changes that may influence subsequent health status. Examples of changes might include impact on body composition and nutritional requirements.

The transition experience for women is highly variable, ranging from an unmarked transition to a pronounced physical and emotional experience. While there are indications that the decline in estradiol levels with menopause has important ramifications for the expression of subsequent disease in women, the nature of these associations is not clearly characterized, particularly relative to the variability of functional expression seen among women. The lack of well-characterized associations of menopause with chronic diseases is undoubtedly due to difficulty in defining the menopausal stages and difficulty in doing serum hormone sampling. Added to these limitations is another impediment. Chronic diseases will not be expressed within a few short months of the menopause; therefore, the temporal sequence of hormone change with clinical disease expression is difficult to establish.

Future studies of the menopause need to consider similar issues that have arisen relative to studies in symptomatology. There needs to be a more concise definition of menopausal status. Potential confounders, including ethnicity, body size, and socioeconomic status, must be considered both in design and in analysis. Finally, women with medically induced (*i.e.,* oophorectomy and radiation therapy) menopause should be evaluated separately from those with natural menopause.

References

1. World Health Organization. Report of a WHO Scientific Group (1981). Research on the menopause. WHO Tech. Rep. Ser. **670,** 8–12.
2. Metcalf, M. G., Donald, R. A., and Livesey, J. H. (1982). Pituitary-ovarian function before, during and after the menopause: A longitudinal study. *Clin. Endocrinol. (Oxford)* **17,** 489–494.
3. van Look, P. F., Lothian, H., Hunter, W. M., Michie, E. A., and Baird, D. T. (1977). Hypothalamic-pituitary-ovarian in perimenopausal women. *Clin. Endocrinol. (Oxford)* **7,** 13–31.
4. Reyes, F. I., Winter, J. S. D., and Faiman, C. (1977). Pituitary-ovarian relationships preceding the menopause. *Am. J. Obstet. Gynecol.* **129,** 557–564.
5. Santoro, N., Brown, J. R., Adel, T., and Skurnick, J. H. (1996). Characterization of reproductive hormonal dynamics in the perimenopause. *J. Clin. Endocrinol. Metab.* **81**(4); 1495–1501.
6. Shideler, S. E., DeVane, G. W., Kaira, P. S., Benirschke, K., and Lasley, B. L. (1989). Ovarian-pituitary hormone interactions during the perimenopause. *Maturitas* **11,** 331–339.
7. Klein, N. A., Battaglia, D. E., Miller, P. B., Branigan, E. F., Giudice, L. C., and Soules, M. R. (1996). Ovarian follicular development and the follicular fluid hormones and growth factors in normal women of advanced reproductive age. *J. Clin. Endocrinol. Metab.* **81,** 1946–1951.
8. Metcalf, M. G. (1979). Incidence of ovulatory cycles in women approaching the menopause. *J. Biosoc. Sci.* **11,** 39–48.
9. Reame, N. E., Kelch, R. P., Beitins, I. Z., Zawacki, C. M., and Padmanabhan, V. (1996). Age effects of FSH and pulsatile LH secretion across the menstrual cycles of premenopausal women. *J. Clin. Endocrinol. Metab.* **81,** 1512–1518.

10. Vagenakis, A. G. (1989). Endocrine aspects of menopause. *Clin. Rheumatol.* **8,** 48–51.

11. Treloar, A. E., Boynton, R. E., Behn, B. G., and Brown, B. W. (1967). Variation of the human menstrual cycle through reproductive life. *Int. J. Fertil.* **12,** 77–127.

12. Vollman, R. F. (1977). The menstrual cycle. *In* "Major Problems in Obstetrics and Gynecology," Vol. 7. Saunders, Philadelphia.

13. Rutherford, A. M. (1978). The menopause. *N. Z. Med. J.* **87,** 251–253.

14. Chakravarti, S., Collins, W. P., Forecast, J. D., Newton, J. R., Oram, D. H., and Studd, J. W. W. (1976). Hormonal profiles after the menopause. *Br. Med. J.* **2,** 784–787.

15. Scaglia, H., Medina, M., Pinto-Ferreira, A. L, Vazques, G., Gual, C., and Perez-Palacios, G. (1976). Pituitary LH and FSH secretion and responsiveness in women of old age. *Acta Endocrinol. (Copenhagen)* **81,** 673–679.

16. Rannevik, G., Carlstrom, K., Jeppsson, S., Bjerre, B., and Svanberg, J. (1986). A prospective long-term study of women from pre-menopause to post-menopause: Changing profiles of gonadotropins, oestrogens and androgens. *Maturitas* **8,** 297–307.

17. Trevoux, R., DeBrux, J., Castanier, M., Nahoul, K., Soule, J.-P., and Scholler, R. (1986). Endometrium and plasma hormone profile in the peri-menopause and post- menopause. *Maturitas* **8,** 309–326.

18. Lucisano, A., Russo, N., Acampora, M. G., Fabrino, A., Fattibene, M., Parlati, E., Maniccia, E., and Dell'Acqua, M. (1986). Ovarian and peripheral androgen and oestrogen levels in post-menopausal women: Correlations with ovarian histology. *Maturitas* **8,** 57–65.

19. De Waard, F., Pot, H., Tonckens-Nanniga, N. E., Baanders-van Halewin, E. A., and Thijssen, J. H. H. (1972). Longitudinal studies on the phenomenon of postmenopausal estrogen. *Acta Cytol.* **16,** 273–278.

20. Grattarola, R., Secreto, G., and Recchione, C. (1975). Correlation between urinary testosterone or estrogen excretion levels and interstitial cell-stimulating hormone concentrations in normal post-menopausal women. *Am. J. Obstet. Gynecol.* **121,** 380–381.

21. Baird, D. T., and Guevara, A. (1969). Concentration of unconjugated estrone and estradiol in peripheral plasma in non-pregnant women throughout the menstrual cycle, castrate and postmenopausal women and men. *J. Clin. Endocrinol. Metab.* **29,** 149–156.

22. Longcope, G. (1971). Metabolic clearance rates and blood production rates of estrogens in postmenopausal women. *Am. J. Obstet. Gynecol.* **111,** 778–781.

23. Longcope, C., Franz, C., Morello, C., Baker, R., and Johnston, C. C., Jr. (1986). Steroid and gonadotropin levels in women during the peri-menopausal years. *Maturitas* **8,** 189–196.

24. Ballinger, C. B., Browning, M. C. K., and Smith, A. H. W. (1987). Hormone profiles and psychological symptoms in peri-menopausal women. *Maturitas* **9,** 235–251.

25. Kelly, J. J., Rajkovic, I. A., O'Sullivan, A. J., Sernia, C., and Ho, K. K. Y. (1993). Effects of different oral oestrogen formulations on insulin-like growth factor-I, growth hormone and growth hormone binding protein in post-menopausal women. *Clin. Endocrinol. (Oxford)* **39,** 561–567.

26. Metcalf, M. G., and Livesey, J. H. (1985). Gonadotropin excretion in fertile women: Effect of age and the onset of the menopausal transition. *J. Endocrinol.* **105,** 357–362.

27. Longcope, C. (1990). Hormone dynamics at the menopause. *Ann. N.Y. Acad. Sci.* **592,** 21–30.

28. Flint, M. P. (1997). Secular trends in menopause age. *J. Psychosom. Obstet. Gynecol.* **18,** 65–72.

29. Scragg, R. F. R. (1973). Menopause and reproductive span in rural Niugini. *Proc. Annu. Symp. Papau New Guinea Med. Soc.,* Port Moresby, pp. 126–144.

30. MacMahon, B., and Worcester, J. (1966). Age at menopause, United States 1960–1962. *Vital Health Stat. Ser. 11* No. 19.

31. Brand, P. C., and Lehert, P. H. (1978). A new way of looking at environmental variables that may affect the age at menopause. *Maturitas* **1,** 121–132.

32. Baron, J. A. (1984). Smoking and estrogen-related disease. *Am. J. Epidemiol.* **119,** 9–22.

33. Lindquist, O., and Bengtsson, C. (1979). The effect of smoking on menopausal age. *Maturitas* **1,** 191–199.

34. Jick, H., and Porter, J. (1977). Relation between smoking and age of natural menopause. *Lancet* **2,** 1354–1355.

35. Andersen, F. S., Transbol, I., and Christiansen, C. (1982). Is cigarette smoking a promotor of the menopause? *Acta Med. Scand.* **212,** 137–139.

36. Adena, M. A., and Gallagher, H. G. (1982). Cigarette smoking and the age at menopause. *Ann. Hum. Biol.* **9,** 121–130.

37. McKinlay, S. M., Bifano, N. L., and McKinlay, J. B. (1985). Smoking and age at menopause in women. *Ann. Intern. Med.* **103,** 350–356.

38. Brambilla, D. J., and McKinlay, S. M. (1989). A prospective study of factors affecting age at menopause. *J. Clin. Epidemiol.* **42,** 1031–1039.

39. Willet, W., Stampfer, M. J., Bain, C., Lipnick, R., Speizer, F. E., Rosner, B., Cramer, D., and Hennekens, C. H. (1983). Cigarette smoking, relative weight, and menopause. *Am. J. Epidemiol.* **117,** 651–658.

40. Mattison, D. R., and Thorgeirsson, S. S. (1978). Smoking and industrial pollution, and their effects on menopause and ovarian cancer. *Lancet* **1,** 187–188.

41. Michnovicz, J. J., Hershcopf, R. J., Naganuma, H., Bradlow, H. L., and Fishman, J. (1986). Increased 2-hydroxlation of estradiol as a possible mechanism for the anti-estrogenic effect of cigarette smoking. *N. Engl. J. Med.* **315,** 1151–1154.

42. Longcope, C., and Johnston, C. C. (1988). Androgen and estrogen dynamics in pre- and postmenopausal women: A comparison between smokers and nonsmokers. *J. Clin. Endocrinol. Metab.* **67,** 379–383.

43. Torgerson, D. J., Thomas, R. E., Campbell, M. K., and Reid, D. M. (1997a). Alcohol consumption and age of maternal menopause are associated with menopause onset. *Maturitas* **26,** 21–25.

44. Whelan, E. A., Sandler, D. P., McConnaughey, D. R., and Weinberg, C. R. (1990). Menstrual and reproductive characteristics and age at natural menopause. *Am. J. Epidemiol.* **131,** 625–632.

45. Khaw, K-T. (1992). Epidemiology of the menopause. *Br. Med. J.* **48,** 249–261.

46. Richardson, S. J., and Nelson, J. F. (1990). Follicular depletion during the menopausal transition. *Ann. N.Y. Acad. Sci.* **592,** 13–20.

47. Cresswell, J. L., Egger, P., Fall, C. H. D., Osmond, C., Fraser, R. B., and Barker, D. J. P. (1997). Is the age of menopause determined in-utero? *Early Hum. Dev.* **49,** 143–148.

48. Torgerson, D. J., Thomas, R. E., and Reid, D. M. (1997b). Mothers and daughters menopausal ages: Is there a link? *Eur. J. Obstet. Gynecol. Reprod. Biol.* **74,** 63–66.

49. Cramer, D. W., Xu, H., and Harlow, B. L. (1995). Family history as a predictor of early menopause. *Fertil. Steril.* **64,** 740–745.

50. Flint, M. P. (1979). Sociology and anthropology of the menopause. *In* "Female and Male Climacteric" (P. A. van Keep, and D. M. Serr, eds.), 1–146. MTP Press, Lancaster, England.

51. Maoz, B. (1973). The perception of menopause in five ethnic groups in Israel. Thesis, Ph.D. dissertation. Leiden, the Netherlands.

52. Severne, L. (1985). Coping with life events and stress at the climacteric. *In* "Proceedings of the 4th International Congress on the Menopause" (N. Notelovitz and P. A. van Keep, eds.), pp. 200–217. MTP Press, Lancaster, England.

53. Hunter, M., Battersby, R., and Whitehead, M. (1986). Relationships between psychological symptoms, somatic complaints and menopausal status. *Maturitas* **8,** 217–228.

54. van Keep, P. A. (1983). The menopause, part B: Psychosomatic aspects of the menopause. *In* "Handbook of Psychosomatic Obstetrics and Gynaecology" (L. Dennerstein and G. Burrows, eds.), 313–325. Elsevier, New York.

55. Utian, W. H. (1980). The place of oestriol therapy after menopause. *Acta Endocrinol. (Copenhagen), Suppl.* **233,** 51–56.

56. Neugarten, B. L., and Kraines, R. J. (1965). Menopausal symptoms in women of various ages. *Psychosom. Med.* **27,** 266–273.

57. Nordin, B. E. C., and Polley, K. J. (1987). Metabolic consequences of the menopause. A cross-sectional, longitudinal, and intervention study on 557 normal postmenopausal women. *Calcif. Tissue Int.* **41,**(Suppl. 1), S1–S59.

58. Voda, A. M., Feldman, B. M., and Gronseth, E. (1984). Description of the hot flash: Sensations, meaning and change in frequency across time. *In* "The Climacteric in Perspective" Chapter 23, pp. 259–269. MTP Press, Hingham, MA.

59. Thompson, B., Hart, S. A., and Durno, D. (1973). Menopausal age and symptomatology in a general practice. *J. Biosoc. Sci.* **5,** 71–82.

60. McKinlay, S. M., and Jefferys, M. (1974). The menopausal syndrome. *Br. J. Prev. Soc. Med.* **28,** 108–115.

61. Kronenberg, F., and Downey, J. A. (1987). Thermoregulatory physiology of menopausal hot flashes: A review. *Can. J. Physiol. Pharmacol.* **65,** 1312–1324.

62. Kronenberg, F. (1990). Hot flashes: Epidemiology and physiology. *Ann. N.Y. Acad. Sci.* **592,** 52–86; discussion: pp. 123–133.

63. Albright, D. L., Voda, A. M., Smolensky, M. H., Hsi, B., and Decker, M. (1989). Circadian rhythms in hot flashes in natural and surgically-induced menopause. *Chronobiol. Int.* **6,** 279–284.

64. Beyene, Y. (1986). Cultural significance and physiological manifestations of menopause. A biocultural analysis. *Cult. Med. Psychiatry* **10,** 47–71.

65. Bungay, G. T., Vessey, M. P., and McPherson, C. K. (1980). Study of symptoms in middle life with special reference to the menopause. *Br. Med. J.* **287,** 181–183.

66. Studd, J. (1992). Complications of hormone replacement therapy in post-menopausal women. *J. R. Soc. Med.* **85,** 376–378.

67. Sherman, B. M., Wallace, R. B., Bean, J. A., Chang, Y., and Schlabaugh, L. (1981). The relationship of menopause to medical and reproductive experience. *J. Gerontol.* **36,** 306–309.

68. Cignarelli, M., Cicinelli, E., Corso, M., Cospite, M. R., Garruti, G., Tafaro, E., Giorgino, R., and Schonauer, S. (1989). Biophysical and endocrine-metabolic changes during menopausal hot flashes: Increase in plasma free fatty acid and norepinephrine levels. *Gynecol. Obstet. Invest.* **27,** 34–37.

69. McKinlay, J. B., McKinlay, S. M., and Brambilla, D. (1987). The relative contributions of endocrine changes and social circumstances to depression in mid-aged women. *J. Health Soc. Behav.* **28,** 345–363.

70. Hallstrom, T., and Samuelsson, S. (1985). Mental health in the climacteric. The longitudinal study of women in Gothenburg. *Acta Obstet. Gynecol. Scand. Suppl.* **130,** 13–18.

71. Gath, D., Osborn, M., Bungay, G., Iles, S., Day, A., Bond, A., and Passingham, C. (1987). Psychiatric disorder and gynaecological symptoms in middle aged women: A community survey. *Br. Med. J.* **294,** 213–218.

72. Kaufert, P., Lock, M., McKinlay, S., Beyenne, Y., Coope, J., Davis, D., Eliasson, M., Gognalons-Nicolet, Goodman, M., and Holte, A. (1986). Menopause research: The Korpilsmip workshop. *Soc. Sci. Med.* **15E,** 1285–1289.

73. Winokur, G. (1973). Depression in the menopause. *Am. J. Psychiatry* **130,** 92–93.

74. Smith, W. G. (1971). Critical life-events and prevention strategies in mental health. *Arch. Gen. Psychiatry* **25,** 103–109.

75. Kunin, C. M., and McCormack, R. C. (1968). An epidemiologic study of bacteriuria and blood pressure among nuns and working women. *N. Engl. J. Med.* **278,** 635–642.

76. Evans, D. A., Williams, D. N., Laughlin, L. W., Miao, L., Warren, J. W., Hennekens, C. H., Shimada, J., Chapman, W. G., Rosner, B., and Taylor, J. O. (1978). Bacteriuria in a population-based cohort of women. *J. Infect. Dis.* **138,** 768–773.

77. Bengtsson, C., Bengtsson, U., and Lincoln, K. (1980). Bacteriuria in a population sample of women. *Acta Med. Scand.* **208,** 417–423.

78. Vorland, L. H., Carlson, K., and Aalen, O. (1985). An epidemiological survey of urinary tract infections among outpatients in northern Norway. *Scand. J. Infect. Dis.* **17,** 277–283.

79. Larsen, B., Goplerud, C. P., Petzold, C. R., Ohm-Smith, M. J., and Galask, R. P. (1982). Effect of estrogen treatment on the genital tract flora of postmenopausal women. *Obstet. Gynecol.* **60,** 20–24.

80. Parsons, C. I. (1987). Lower urinary tract infections in women. *Urol. Clin. North Am.* **14,** 247–250.

81. Orlander, J. D., Jick, S. S., Dean, A. D., and Jick, H. (1992). Urinary tract infections and estrogen use in older women. *J. Am. Geriatr. Soc.* **40,** 817–820.

82. Raz, R., and Stamm, W. E. (1993). A controlled trial of intravaginal estriol in postmenopausal women with recurrent urinary tract infections. *N. Engl. J. Med.* **329,** 753–756.

83. Miodrag, A., Castleden, C. M., and Vallance, T. R. (1988). Sex hormones and the female urinary tract. *Drugs* **36,** 491–504.

84. Versi, E. (1990). Incontinence in the climacteric. *Clin. Obstet. Gynecol.* **33,** 392–398.

85. Hording, U., Pedersen, K. H., Sidenius, K., and Hedegaard, L. (1986). Urinary incontinence in 45-year-old women. An epidemiological survey. *Scand. J. Urol. Nephrol.* **20,** 183–186.

86. Iosif, S., Henriksson, L., and Ulmsten, U. (1981). The frequency of disorders of the lower urinary tract, urinary incontinence in particular, as evaluated by a questionnaire survey in a gynaecological health control population. *Acta. Obstet. Gynecol. Scand.* **60,** 71–76.

87. Feneley, R. C. L., Sheperd, A. M., Powell, P. H., and Blannin, J. (1979). Urinary incontinence: Prevalence and needs. *Br. J. Urol.* **51,** 493–496.

88. Thomas, T. M., Plymat, K. R., Blannin, J., and Meade, T. W. (1980). Prevalence of urinary incontinence. *Br. Med. J.* **281,** 1243–1245.

89. Hagstad, A., and Janson, P. O. (1986). The epidemiology of climacteric symptoms. *Acta Obstet. Gynecol. Scand., Suppl.* **134,** 59–65.

90. Milsom, I., Ekelund, P., Molander, U., Arvidsson, L., and Areskoug, B. (1993). The influence of age, parity, oral contraception, hysterectomy and menopause on the prevalence of urinary incontinence in women. *J. Urol.* **249,** 1459–1462.

91. Fantl, A., Wyman, J. F., Anderson, R. L., Matt, D. W., and Bump, R. C. (1988). Postmenopausal urinary incontinence: comparison between non-estrogen-supplemented and estrogen-supplemented women. *Obstet. Gynecol.* **71,** 823–828.

92. Hallström, T. (1977). Sexuality in the climacteric. *Clin. Obstet. Gynecol.* **4,** 227–237.

93. Davidson, J. M., Gray, G. D., and Smith, E. R. (1983). The sexual psychoendocrinology of aging. *In* "Neuroendocrinology of Aging" (J. Meites, ed.), 261–276. Plenum, New York.

94. McCoy, N., Cutler, W., and Davidson, J. M. (1985). Relationships among sexual behavior, hot flashes, and hormone levels in perimenopausal women. *Arch. Sex. Behav.* **14,** 385–394.

95. Sherwin, B. B., Gelfand, M. M., and Brender, W. (1985). Androgen enhances sexual motivation in females: A prospective, crossover study of sex steroid administration in the surgical menopause. *Psychosom. Med.* **47,** 339–351.

96. Myers, J. K., Lindenthal, J. J., and Pepper, M. P. (1975). Life events, social integration, and psychiatric symptomatology. *J. Heal. Soc. Behav.* **16**, 421–427.

97. Bjorntorp, P. (1988). The associations between obesity, adipose tissue distribution and disease. *Acta Med. Scand., Suppl.* **723**, 121–134.

98. Elders, P. J., Netelenbos, J. C., Lips, P., van Ginkel, F. C., and van der Stelt, P. F. (1988). Accelerated vertebral bone loss in relation to the menopause: A cross-sectional study on lumbar bone density in 286 women of 46 to 55 years of age. *Bone Miner.* **5**, 11–19.

99. Elders, P. J., Netelenbos, J. C., Lips, P., Khoe, E., van Ginkel, F. C., Hulshof, K. F., and van der Stelt, P. F. (1989). Perimenopausal bone mass and risk factors. *Bone Miner.* **7**, 289–299.

100. Nilas, L., and Christiansen, C. (1988). Rates of bone loss in normal women: Evidence of accelerated trabecular bone loss after the menopause. *Eur. J. Clin. Invest.* **18**, 529–534.

101. Mazzuoli, G., Minisola, S., Valtorta, C., Antonelli, R., Tabolli, S., and Bigi, F. (1985). Changes in mineral content and biochemical bone markers at the menopause. *Isr. J. Med. Sci.* **21**, 875–877.

102. Geusens, P., Dequecker, J., Verstraeten, A., and Nijs, J. (1986). Age-, sex-, and menopause-related changes of vertebral and peripheral bone: Population study using dual and single photon absorptiometry and radiogrammetry. *J. Nucl. Med.* **27**, 1540–1549.

103. Slemenda, C., Hui, S. L., Longscope, C., and Johnston, C. C. (1987). Sex steroids and bone mass. A study of changes about the time of menopause. *J. Clin. Invest.* **80**, 1261–1269.

104. Gnudi, S., Mongiorgi, R., Figus, E., and Bertocchi, G. (1990). Evaluation of the relative rates of bone mineral content loss in postmenopause due to both estrogen deficiency and ageing. *Boll. Soc. Ital. Biol. Sper.* **66**, 1153–1159.

105. Nilas, L., and Christiansen, C. (1989). The pathophysiology of peri- and postmenopausal bone loss. *Br. J. Obstet. Gynecol.* **96**, 580–587.

106. Reid, I. R., Ames, R., Evans, M. C., Sharpe, S., Gamble, G., France, J. T., Lim, T. M., and Cundy, T. F. (1992). Determinants of total body and regional bone mineral density on normal postmenopausal women—a key role for fat mass. *J. Clin. Endocrinol. Metab.* **75**, 45–51.

107. Johnston, C. C. Jr, Hui, S. L., Witt, R. M., Appledorn, R., Baker, R. S., and Longcope, C. (1985). Early menopausal changes in bone mass and sex steroids. *J. Clin. Endocrinol. Metab.* **61**, 905–911.

108. Steinberg, K. K., Freni-Titulaer, L. W., DePuey, E. G., Miller, D. T., Sgoutas, D. S., Coralli, C. H., Phillips, D. L., Rogers, T. N., and Clark, R. V. (1989). Sex steroids and bone density in premenopausal and perimenopausal women. *J. Clin. Endocrinol. Metab.* **69**, 533–539.

109. Ribot, C., Tremollieres, F., Pouilles, J. M., Louvet, J. P., and Guiraud, R. (1988). Influence of the menopause and aging on spinal density in French women. *Bone Miner.* **5**, 89–97.

110. Hedlund, L. R., and Gallagher, J. C. (1989). The effect of age and menopause on bone mineral density of the proximal femur. *J. Bone Miner. Res.* **4**, 639–642.

111. Stevenson, J. C., Lees, B., Devenport, M., Cust, M. P., and Ganger, K. F. (1989). Determinants of bone density in normal women: Risk factors for osteoporosis? *Br. Med. J.* **298**, 924–928.

112. Sowers, M-F., Clark, M. K., Hollis, B., Wallace, R. B., and Jannausch, M. (1992). Radial bone mineral density in pre- and perimenopausal women: A prospective study of rates and risk factors for loss. *J. Bone Miner. Res.* **7**, 647–657.

113. Nilas, L., and Christiansen, C. (1987). Bone mass and its relationship to age and the menopause. *J. Clin. Endocrinol. Metab.* **65**, 697–702.

114. Recker, R. R., Lappe, J. M., Davies, K. M., and Kimmel, D. B. (1992). Changes in bone mass immediately before menopause. *J. Bone Miner. Res.* **7**, 857–862.

115. Ohta, H., Makita, K., Suda, Y., Ikeda, T., Masuzawa, T., and Nozawa, S. (1992). Influence of oophorectomy on serum levels of sex steroids and bone metabolism and assessment of bone mineral density in lumbar trabecular bone by QCT-C value. *J. Bone Miner. Res.* **7**, 659–665.

116. Ohta, H., Ikeda, T., Masuzawa, T., Makita, K., Suda, Y., and Nozawa, S. (1993). Differences in axial bone mineral density, serum levels of sex steroids, and bone metabolism between postmenopausal and age- and body size-matched premenopausal subjects. *Bone* **14**, 111–116.

117. Block, J. E., Smith, R., Glueer, C. C., Steiger, P., Ettinger, B., and Genant, H. K. (1989). Models of spinal trabecular bone loss as determined by quantitative computed tomography. *J. Bone Miner. Res.* **4**, 249–257.

118. Hui, S. L., Slemenda, C. W., Johnston, C. C., and Appledorn, C. R. (1987). Effects of age and menopause on vertebral bone density. *Bone Miner.* **2**, 141–146.

119. Bagur, A. C., and Mautalen, C. A. (1992). Risk for developing osteoporosis in untreated premature menopause. *Calcif. Tissue Int.* **51**, 4–7.

120. Johnell, O., and Nilsson, B. E. (1984). Life-style and bone mineral mass in perimenopausal women. *Calcif. Tissue Int.* **36**, 354–356.

121. Gardsell, P., Johnell, O., and Nilsson, B. E. (1991). The impact of menopausal age on future fragility fracture risk. *J. Bone Miner. Res.* **6**, 429–433.

122. Seeman, E. Cooper, M. E., Hooper, J. L., Parkinson, E., McKay, J., and Jerums, G. (1988). Effect of early menopause on bone mass in normal women and patients with osteoporosis. *Am. J. Med.* **85**, 213–216.

123. Falch, J. A., and Gautvik, K. M. (1988). A longitudinal study of pre- and postmenopausal changes in calcium metabolism. *Bone* **9**, 15–19.

124. Kelly, P. J., Pocock, N. A., Sambrook, P. N., and Eisman, J. A. (1989). Age and menopause-related changes in indices of bone turnover. *J. Clin. Endocrinol. Metab.* **69**, 1160–1165.

125. Pansini, F., Bonaccorsi, G., Calisesi, M., Farina, A., Levato, F., Mazzotta, D., Bagni, B., and Mollica, G. (1992). Evaluation of bone metabolic markers as indicators of osteopenia in climacteric women. *Gynecol. Obstet. Invest.* **33**, 231–235.

126. Ribot, C., Tremolliere, F., Pouilles, J. M., Bonneu, M., Germain, F., and Louvet, J. P. (1987). Obesity and postmenopausal bone loss: The influence of obesity on vertebral density and bone turnover in postmenopausal women. *Bone* **8**, 327–331.

127. Mole, P. A., Walkinshaw, M. H., Robins, S. P., and Paterson, C. R. (1992). Can urinary pyridinium crosslinks and urinary oestrogens predict bone mass and rate of bone loss after the menopause? *Eur. J. Clin. Invest.* **22**, 767–771.

128. Sowers, M.-F., Crutchfield, M., Jannausch, M. L., and Russell-Aulet, M. (1996). Longitudinal changes in body composition in women approaching the midlife. *Ann. Hum. Biol.* **23**, 253–265.

129. Nicklas, B. J., Toth, M. J., Goldberg, A. P., and Poehlman, E. T. (1997). Racial differences in plasma leptin concentrations in obese postmenopausal women. *J. Clin. Endocrinol. Metab.* **82**, 315–317.

130. Aloia, J. F., McGowan, D. M., Vaswani, A. N., Ross, P., and Cohn, S. H. (1991). Relationship of menopause to skeletal and muscle mass. *Am. J. Clin. Nutr.* **53**, 1378–1383.

131. Tremollieres, F. A., Pouilles, J.-M., and Ribot, C. A. (1996). Relative influence of age and menopause on total and regional body composition changes in postmenopausal women. *Am. J. Obstet. Gynecol.* **175**, 1594–1600.

132. Ley, C. J., Lees, B., and Stevenson, J. C. (1992). Sex- and menopause-associated changes in body-fat distribution. *Am. J. Clin. Nutr.* **55**, 950–954.

133. Sonnenschein, E. G., Kim, M. Y., Pasternack, B. S., and Toniolo, P. G. (1993). Sources of variability in waist and hip measurements in middle-aged women. *Am. J. Epidemiol.* **138**, 301–309.

134. Panotopoulos, G., Ruiz, J. C., Raison, J., Guy-Grand, B., and Basdevant, A. (1996). Menopause, fat and lean distribution in obese women. *Maturitas* **25,** 11–19.

135. Haarbo, J., Marslew, U., Gotfredsen, A., and Christiansen, C. (1991). Postmenopausal hormone replacement therapy prevents central distribution of body fat after menopause. *Metab., Clin. Exp.* **40,** 1323–1326.

136. Rebuffé-Scrive, M., Eldh, J., Hafström, L.-O., and Björntorp, P. (1986). Metabolism of mammary, abdominal and femoral adipocytes in women before and after menopause. *Metab. Clin. Exp.* **35,** 792–797.

137. Reubinoff, B. E., Wurtman, J., Rojansky, N., Adler, D., Stein, P., Schenker, J. G., and Brzezinski, A. (1995). Effects of hormone replacement therapy on weight, body composition, fat distribution and food intake in early postmenopausal women: A prospective study. *Fertil. Steril.* **64,** 963–968.

138. Rebuffé-Scrive, M., Enk, L., Crona, N., Lonnroth, P., Abrahamsson, L., Smith, U., and Björntorp, P. (1985). Fat cell metabolism in different regions in women: Effect of menstrual cycle, pregnancy, and lactation. *J. Clin. Invest.* **75,** 1973–1976.

139. Rebuffé-Scrive, M., Lundholm, K., and Björntorp, P. (1985). Glucocorticoid hormone binding to human adipose tissue. *Eur. J. Clin. Invest.* **15,** 267–272.

140. Xu, X., Hoebeke, J., and Björntorp, P. (1990). Progestin binds to the glucocorticoid receptor and mediates antiglucocorticoid effect in rat adipose precursor cells. *J. Steroid Biochem.* **36,** 465–471.

141. Kochanek, K. D., Maurer, J. D., and Rosenberg, H. M. (1994). Why did black life expectancy decline from 1984 through 1989 in the United States? *Am. J. Public Health* **84,** 938–944.

142. Parrish, H. M., Carr, C. A., Hall, D. G., and King, T. M. (1967). Time interval from castration in premenopausal women to development of excessive coronary atherosclerosis. *Am. J. Obstet. Gynecol.* **99,** 155–162.

143. van der Graaf, Y., de Kleijn, M. J. J., and van der Schouw, Y. T. (1997). Menopause and cardiovascular disease. *J. Psychosomatic Obstet. Gynecol.* **18**(2), 113–120.

144. Stevenson, J. C. (1996). Mechanisms whereby oestrogens influence arterial health. *Eur. J. Obstet. Gynecol. Reprod. Biol.* **65,** 39–42.

145. Hjortland, M. C., McNamara, P. M., and Kannel, W. B. (1976). Some atherogenic concomitants of menopause: The Framingham Study. *Am. J. Epidemiol.* **103,** 304–311.

146. Lindquist, O. (1982). Influence of the menopause on ischaemic heart disease and its risk factors and on bone mineral content. *Acta Obstet. Gynecol. Scand.* **110,** 1–32.

147. Matthews, K. A., Meilahn, E. N., Kuller, L. H., Kelsey, S. F., Caggiula, A. W., and Wing, R. R. (1989). Menopause and risk factors for coronary heart disease. *N. Engl. J. Med.* **321,** 641–646.

148. Matthews, K. A., Wing, R. R., Kuller, L. H., Meilahn, E. N., and Plantinga, P. (1994). Influence of the perimenopause on cardiovascular risk factors and symptoms in middle-aged healthy women. *Arch. Intern. Med.* **154,** 2349–2355.

149. Baird, D. D., Tyroler, H. A., Heiss, G., Chambless, L. E., and Hames, C. G. (1985). Menopausal change in serum cholesterol. Black/white differences in Evans County, Georgia. *Am. J. Epidemiol.* **122,** 982–993.

150. Hamman, R. F., Bennett, P. H., and Miller, M. (1975). The effect of menopause on serum cholesterol in American (Pima) Indian women. *Am. J. Epidemiol.* **102,** 164–169.

151. Scarabin, P-Y., Plu-Bureau, G., Bara, L., Bonithon-Kopp, B., Guize, L., and Samama, M. M. (1993). Haemostatic variables and menopausal status: Influence of hormone replacement therapy. *Thromb. Haemostasis* **70,** 584–587.

152. Lee, A. J., Lowe, G. D. Smith, W. C., and Tunstall-Pedoe, H. (1993). Plasma fibrinogen in women: Relationships with oral con-

traception, the menopause and hormone replacement therapy. *Br. J. Haematol.* **83,** 616–621.

153. Salomaa, V., Stinson, V., Kark, J. D., Folsom, A. R., David, C. E., and Wu, K. K. (1995). Association of fibrinolytic parameters with early atherosclerosis. The ARIC Study. *Circulation* **91,** 284–290.

154. Shahar, E., Folsom, A. R., Salomaa, V. V., Stinson, V. L., McGovern, P. G., Shimakawa, T., Chambless, L. E., and Wu, K. K. (1996). Relation of hormone-replacement therapy to measures of plasma fibrinolytic activity. Atherosclerosis Risk in Communities (ARIC) Study Investigators. *Circulation* **93,** 1970–1975.

155. Katz, R. J., Hsia, J., Walker, P., Jacobs, H., and Kessler, C. (1996). Effects of hormone replacement therapy on the circadian pattern of atherothrombotic risk factors. *Am. J. Cardiol.* **78,** 876–880.

156. Casiglia, E., d'Este, D., Ginocchio, G., Colangeli, G., Onesto, C., Tramontin, P., and Ambrosio, G. B. (1996). Lack of influence of menopause on blood pressure and cardiovascular risk profile: A 16-year longitudinal study concerning a cohort of 568 women. *J. Hyperten.* **14,** 729–736.

157. Portaluppi, F., Pansini, F., Manfredini, R., and Mollica, G. (1997). Relative influence of menopausal status, age, and body mass index on blood pressure. *Hypertension* **29,** 976–979.

158. Markovitz, J. H., Matthews, K. A., Wing, R. R., Kuller, L. H., and Meilahn, E. N. (1991). Psychological, biological and health behavior predictors of blood pressure changes in middle-aged women. *J. Hypertens.* **9,** 399–406.

159. The Writing Group for the PEPI Trial (1995). Effects of estrogen or estrogen/progestin regimens on heart disease risk factors in postmenopausal women. *JAMA, J. Am. Med. Assoc.* **273,** 199–208.

160. Dallongeville, J., Marecaux, N., Isorez, D., Zylbergberg, G., Fruchart, J. C., and Amouyel, P. (1995). Multiple coronary heart disease risk factors are associated with menopause and influenced by substitutive hormonal therapy in a cohort of French women. *Atherosclerosis* **118,** 123–133.

161. Lindquist, O., Bengtsson, C., and Lapidus, L. (1985). Relationships between the menopause and risk factors for ischaemic heart disease. *Acta Obstet. Gynecol. Scand., Suppl.* **130,** 43–47.

162. Pasquali, R., Casimirri, F., Pascal, G., Tortelli, O., Morselli Labate, O., Bertazzo, D., Vicennati, V., and Gaddi, A. (1997). Influence of menopause on blood cholesterol levels in women: The role of body composition, fat distribution and hormonal milieu. The Virgilio Menopause Health Group. *J. Intern. Med.* **24,** 195–203.

163. Senoz, S., Direm, B., Gulekli, B., and Gokmen, O. (1996). Estrogen deprivation, rather than age, is responsible for the poor lipid profile and carbohydrate metabolism in women. *Maturitas* **25,** 107–114.

164. Berger, G. M., Maidoo, J., Gounden, N., and Gouws, E. (1995). Marked hyperinsulinaemia in postmenopausal, healthy Indian (Asian) women. *Diabetic Med.* **12,** 788–795.

165. Folsom, A. R., Burke, G. I., Ballew, C., Jacobs, D. R., Jr., Haskell, W. L., Donahue, R. P., Liu, K. A., and Hilner, J. E. (1989). Relation of body fatness and its distribution to cardiovascular risk factors in young blacks and whites. *Am. J. Epidemiol.* **130,** 911–924.

166. Lindheim, S. R., Presser, S. C., Ditkoff, E. C., Vijod, M. A., Stanczyk, F. Z., and Lobo, R. A. (1993). A possible bimodal effect of estrogen on insulin sensitivity in postmenopausal women and the alternating effect of added progestin. *Fertil. Steril.* **60,** 664.

167. Goemaere, S., Ackerman, C., Goethals, K., De Keyser, F., Van Der Straeten, C., Verbruggen, G., Mielants, H., and Veys, E. M. (1990). Onset of symptoms of rheumatoid arthritis in relation to age, sex and menopausal transition. *J. Rheum.* **17,** 1620–1622.

168. Vermeulen, A. (1965). Behandeling van reumatoide artritis. [Treatment of rheumatoid arthritis]. *Belg. Tijdschr. Geneeskd.* **21,** 1137–1141.

169. Perselin, R. H. (1976). The effect of pregnancy on rheumatoid arthritis. *Bull. Rheum. Dis.* **27,** 922–927.

170. Wingrave, S. J., and Kay, C. R. (1978). Reduction in incidences of rheumatoid arthritis associated with oral contraceptives. *Lancet* **1**, 569–571.

171. Linos, A., Worthington, J. W., O'Fallon, W. M., and Kurland, L. T. (1980). The epidemiology of rheumatoid arthritis in Rochester, Minnesota: A study of incidence, prevalence and mortality. *Am. J. Epidemiol.* **111**, 87–98.

172. Royal College of General Practitioners' Oral Contraception Study (1978). Reduction in incidence of rheumatoid arthritis associated with oral contraceptives. *Lancet* **1**, 569–571.

173. del Junco, D. J., Annegers, J. F., Luthra, H. S., Coulam, C. B., and Kurland, L. T. (1985). Do oral contraceptives prevent rheumatoid arthritis? *JAMA, J. Am. Med. Assoc.* **254**, 1938–1941.

174. Spector, T. D., Perry, L. A., Tubb, G., Silman, A. J., and Huskisson, E. C. (1988). Low free testosterone levels in rheumatoid arthritis. *Ann. Rheum. Dis.* **47**, 65–68.

175. Cutolo, M., Balleari, E., Giusti, M., Monachesi, M., and Accardo, S. (1986). Sex hormone status in women suffering from rheumatoid arthritis. *J. Rheum.* **13**, 1019–1023.

176. Samisoe, G. (1987). Sexual endocrinology in older women. *Acta Obstet. Gynecol. Scand., Suppl.* **140**, 23–27.

177. Stimson, W. H. (1988). Estrogen and human T-lymphocytes: Presence of specific receptors in T-suppressor/cytotoxic subset. *Scand. J. Immunol.* **28**, 345–350.

178. Novotny, E. A., Raveche, E. S., Sharrow, S., Ottinger, O., and Steinberg, A. D. (1983). Analysis of thymocyte subpopulations following treatment with sex hormones. *Clin. Immunol. Immunopathol.* **28**, 205–217.

179. Polan, M. L., Daniele, A., and Juo, A. (1988). Gonadal steroids modulate human monocyte interleukin-1 (IL-1) activity. *Fertil. Steril.* **49**, 964–968.

180. Mikkelsen, W. M., and Duff, I. F. (1970). Age-specific prevalence of radiographic abnormalities of the joints of the hands, wrists and cervical spine of adult residents of the Tecumseh, Michigan, Community Health Study area, 1962–1965. *J. Chronic Dis.* **23**, 151–159.

181. Roberts, J., and Burch, T. A. (1966). Osteoarthritis in adults. By age, sex, race and geographic area: United States 1960–1962. *Vital Health Stat., Ser. 11* No. 15.

182. Lawrence, J. S. (1977). "Rheumatism in Populations." Heinemann, London.

183. Spector, T. D., Brown, G. C., and Silman, A. J. (1988). Increased rates of previous hysterectomy and gynaecological operations in women with osteoarthritis. *Br. Med. J.* **297**, 899–900.

184. Spector, T. D., and Campion, G. D. (1989). Generalized osteoarthritis: A hormonally mediated disease. *Ann. Rheum. Dis.* **48**, 523–527.

185. Schouten, J. S. A. G., van den Ouweland, F. A., and Valkenburg, H. A. (1992). Natural menopause, oophorectomy, hysterectomy and the risk of osteoarthritis of the dip joints. *Scand. J. Rheum.* **21**, 196–200.

186. Hannan, M. T., Felson, D. T., Anderson, J. J., Naimark, A., and Kannel, W. B. (1990). Estrogen use and radiographic osteoarthritis of the knee in women. *Arthritis Rheum.* **33**, 525–532.

187. Samanta, A., Jones, A., Regan, M., Wilson, S., and Doherty, M. (1993). Is osteoarthritis in women affected by hormonal changes or smoking? *Br. J. Rheum.* **32**, 366–370.

188. Wong, M., and Moss, R. L. (1992). Long-term and short-term electrophysiological effects of estrogen on the synaptic properties of hippocampal CA 1 neurons. *J. Neurosci.* **12**, 3217–3225.

189. van Harren, F. A., van Hest, A., Heinsbroeck, R. P. (1988). Behavioral differences between male and female rats. Effects of gonadal hormones on learning and memory. *Neurosci. Biobehav. Rev.* **14**, 23–33.

190. Campbell, S., and Whitehead, M. (1977). Oestrogen therapy and the menopausal syndrome. *Clin. Obstet. Gynecol.* **4**, 31–47.

191. Fedor-Freybergh, P. (1977). The influence of oestrogen on the well being and mental performance in climacteric and postmenopausal women. *Acta Obstet. Gynecol. Scand., Suppl.* **64**, 5–69.

192. Sherwin, B. B. (1988). Estrogen and/or androgen replacement therapy and cognitive functioning in surgically menopausal women. *Psychoneuroendocrinology* **13**, 345–357.

193. Robinson, D., Friedman, L., Marcus, R., Tinklenberg, J., and Yesavage, J. (1994). Estrogen replacement therapy and memory in older women. *J. Am. Geriatr. Soc.* **42**, 919–922.

194. Rauramo, L., Lagerspetz, K., Engblom, P., and Punnonen, R. (1975). The effect of castration and peroral estrogen therapy on some psychological functions. *Front. Horm. Res.* **3**, 94–104.

195. Ditkoff, E. C., Crary, W. G., Cristo, M., and Lobo, R. A. (1991). Estrogen improves psychological function in asymptomatic postmenopausal women. *Obstet. Gynecol.* **78**, 345–357.

196. Paganini-Hill, A., and Henderson, V. W. (1994). Estrogen deficiency and risk of Alzheimer's disease in women. *Am. J. Epidemiol.* **140**, 256–261.

197. van Duijn, C. M. (1997). Menopause and the brain. *J. Psychosomatic Obstet. Gynecol.* **18**(2), 121–125.

198. Barile, L. A. (1997). Theories of menopause. Brief comparative synopsis. *J. Psychosoc. Nurs.* **35**(2), 36–39.

93

The Risks and Benefits of Hormone Replacement Therapy— Weighing the Evidence

PHYLLIS A. WINGO* AND ANNE McTIERNAN†

*American Cancer Society, Atlanta, Georgia; †Fred Hutchinson Cancer Research Center, Seattle, Washington

I. Introduction

In 1997, an estimated 45.1 million prescriptions for Premarin were dispensed, more than for any other prescription drug in the United States [1]. On average, about one in five menopausal women in the United States were using noncontraceptive hormones in 1992 [2], and an estimated 71% of women who had had a bilateral oophorectomy during 1990–1992 reported having used hormone replacement therapy (HRT) at some time [3]. Because so many peri- and postmenopausal women use hormones for relief of vasomotor symptoms and to treat urogenital atrophy, the balance of health risks and benefits related to this common exposure needs careful consideration.

II. Historic Evolution and Trends in Use

HRT has been available since the 1940s [3], although the specific hormone combinations and delivery systems have changed over time. The earliest approved hormones were administered in oral pills, by injection, and in vaginal creams or suppositories; all of these routes of administration are still in use. In 1974,

similar percentages of physician office visits for treatment of menopausal symptoms resulted in oral estrogen prescriptions (37%) and injected estrogens (36%) [4]. However, by 1986, the mix of oral and injected estrogens had changed to 52 and 14%, respectively [4].

In 1986, transdermal estradiol was first marketed under the brand name Estraderm. Between 1987 and 1992, the number of prescriptions for Estraderm increased threefold from 1.5 to 4.7 million [2]. In Britain, subdermal implant systems for delivering postmenopausal hormones have been available since the 1980s [5]. Similar delivery systems are not approved for use in the United States [Lisa Rarick, Food and Drug Administration, personal communication, September 1998].

Although conjugated estrogens have been available in the United States since the 1940s [3], they were not widely used until the 1960s [3]. In general, the use of noncontraceptive estrogens has waxed and waned depending on the perceptions of risks and benefits [2,4,6]. During the 1960s and 1970s, estrogen use increased to about 28 million prescriptions in 1975 [6] when several studies reported an association with endometrial cancer (Fig. 93.1) [7,8]. Over the next several years, prescriptions for

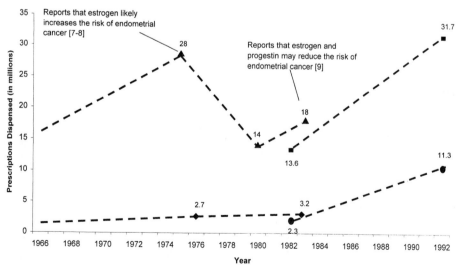

Fig. 93.1 Estimated number of prescriptions dispensed for noncontraceptive estrogen and progestins, 1966–1992.

noncontraceptive estrogens decreased by half [4,6]. Then, in the early 1980s, use began to increase again, and by 1992, the annual number of prescriptions dispensed for oral menopausal estrogens had increased to previously high levels of 31.7 million [2]. This increase in the use of estrogens in the 1980s actually reflected an increase in the use of combination HRT—estrogen plus progestin therapy. The addition of progestin to estrogen for treating symptoms of menopause ameliorates most of the adverse effects of unopposed estrogen on the uterus [9,10].

III. Characteristics of Women Who Use HRT

Because women who use HRT visit their physician regularly to get their hormone prescriptions refilled and to have checkups, they may get an earlier diagnosis and receive treatment for cancer or heart disease [11]. Such differences are important and need to be considered when evaluating the HRT literature. The protective effects of HRT use on cardiovascular disease and perhaps colon cancer and the possible adverse effects on other cancers may be influenced by selection, surveillance, diagnostic, and other biases.

Women who use HRT have different sociodemographic characteristics, may be more likely to practice healthy behaviors, and may have a better cardiovascular risk factor profile than women who have never used HRT [3,11–13]. The Epidemiologic Followup Study to the First National Health and Nutrition Examination Survey was used to examine patterns of HRT over time and to describe the characteristics of menopausal women who had ever used this treatment and who continued to use it for at least five years [3]. In this study, women who had ever used HRT were more likely to have undergone hysterectomy or bilateral oophorectomy, to be white, to live in the western United States, to have taken oral contraceptives in the past, to be lean, and to consume more alcohol than women who had never used HRT. Smoking, physical activity, and self-reported health status were not related to HRT use in this cohort. Similarly, serum cholesterol and health conditions such as a history of diabetes mellitus or hypertension were not related to the use of HRT.

Other studies described the relationship between HRT use and several of the same factors, including surgical menopause, leanness, and white race [11–13]. A study of upper-middle class menopausal women ages 50 to 79 years in California examined the relationship between hormone use and healthy changes in lifestyle (e.g., decreased salt and fat intake, increased fiber intake, increased levels of exercise, weight loss, stress reduction, etc.) [11]. Study participants were asked about physician office visits that resulted in blood pressure checks, cholesterol measurement, screening examinations for cancer, and other prevention activities. All healthy behaviors examined, except for decreased fat and salt intake and use of blood pressure screening, were related to estrogen use. Current estrogen users had the highest prevalence of healthy behaviors, followed by past users and never users.

A prospective investigation found a more favorable profile of cardiovascular risk factors before using estrogen replacement therapy among women who subsequently used hormones than among those who did not [12]. Prior to using estrogen replacement therapy, women who subsequently used it had higher levels of HDL cholesterol, higher levels of physical activity and

alcohol consumption, and lower levels of apolipoprotein B, systolic and diastolic blood pressure, weight, and fasting insulin than nonusers.

IV. Contraindications to HRT Use

Estrogen replacement therapy may be contraindicated in some women. Exogenous estrogens raise triglyceride levels [14]; women with hypertriglyceridemia (levels above 500 mg/dl) can experience elevations of triglycerides to 1000 mg/dl or more, and cases of hypertriglyceride-associated pancreatitis have been reported [15]. Oral estrogens are metabolized through the liver with a first-pass mechanism [16] and thus can affect clotting factors that are formed in the liver. Women with a history of deep vein thrombosis or pulmonary embolism are at increased risk of recurrence if they use estrogen replacement therapy [17–20] and so are advised to avoid HRT. Some women who have periods of immobilization as a result of trauma or surgery are advised to discontinue HRT temporarily [21].

Estrogens can be carcinogenic to various tissues, including the endometrium [10] and breast [22]. Thus, women who have been diagnosed with endometrial cancer are usually advised not to take estrogen therapy, although those with hysterectomy-treated, well-differentiated, stage I carcinoma can probably be safely treated with estrogen plus progesterone therapy [23]. Women with adenomatous endometrial hyperplasia are advised not to use estrogen therapy if they have not undergone hysterectomy, but they are often prescribed progestin therapy to counteract endogenous estrogen stimulation of the endometrium [24]. Use of HRT is contraindicated for women with a history of breast cancer, although some clinicians treat menopausal symptoms with short courses of estrogens [25]. One study is testing HRT in 1100 hormone receptor-negative breast cancer patients in a randomized controlled clinical trial [26]. It is likely that long-duration use of HRT poses too large a risk of recurrence or new cancer development to be acceptable to most patients.

Estrogen exacerbates endometriosis; thus, women with a history of severe endometriosis might have recurrence of endometriosis-related symptoms on estrogen therapy that is not opposed by progesterone [27]. Exogenous estrogen may be contraindicated in women with active gallbladder disease [28,29]. The Heart and Estrogen/Progestin Replacement Study (HERS) trial found almost a 40% increased risk (p = 0.05) of developing gallbladder disease after 4–5 years of follow-up in women with coronary disease randomized to combined estrogen–progestin therapy compared with placebo [20]. Estrogen therapy may also cause exacerbation of symptoms related to migraine headaches [30] and may increase risk of lupus flares [31].

V. Epidemiologic Studies

HRT has been implicated as both beneficial and harmful to the health and quality of life for menopausal women. Many studies have examined the role of exogenous hormones in the development of hormone-dependent cancers, cardiovascular disease, and osteoporosis. As each relationship is summarized in the following sections, the decades during which women were using HRT in each study should be kept in mind. Use that occurred before the mid-1980s was almost exclusively estrogen replacement therapy. Since then, the standard treatment for symptoms of menopause has been combination HRT for women with intact uteri. Use of combination therapy has increased substantially since the 1980s [1,2] even though about 15 years have elapsed since combination HRT was first recommended for women who still have a uterus.

A. Cancer

1. Breast Cancer

Early epidemiologic studies pointed to a role for hormones in the etiology of breast cancer; women who had early bilateral oophorectomy, a late age at menarche, or an early age at menopause were less likely to develop breast cancer than women who did not have these factors [22,32]. A combined analysis of six prospective studies supports the role of endogenous estrogens in the development of postmenopausal breast cancer [33]. The mean concentration of serum estradiol collected from postmenopausal women was 15% higher in those who subsequently developed breast cancer than in those who did not develop disease. Findings from the combined analysis of the case-control data were similar, but the test for heterogeneity across studies was statistically significant. Newer hypotheses focus on how specific reproductive events and exogenous and endogenous hormones affect breast cell proliferation and the accumulation of DNA damage [22].

There is an extensive epidemiologic literature regarding the relationship between exogenous menopausal hormone use and the risk of developing breast cancer. The Collaborative Group on Hormonal Factors in Breast Cancer combined original data from 51 epidemiologic studies to study the relationship between HRT use and breast cancer [34]. This collaborative investigation was estimated to include about 90% of the world's epidemiologic data on 52,705 women with breast cancer and 108,411 women without breast cancer. Because the average year of diagnosis for the women included in the collaborative analysis was 1985, the reported HRT and breast cancer associations were based primarily on the use of unopposed estrogen. Among women who were currently using HRT or who had stopped using HRT during the previous 1–4 years, the risk of breast cancer increased 2.3% for each year of HRT (95% confidence interval (CI) = 1.1–3.6). The increase in risk for each year of HRT use was hypothesized to be similar in magnitude to the effects of delaying menopause each year among women who have never used HRT. Among current and recent HRT users, the trend of increasing risk with increasing duration of use was significant, and women who had been using HRT for 15 or more years had a significantly elevated risk of 1.56 of developing breast cancer. The increase in risk was more pronounced among lean women than among women with high body mass index. However, women who used HRT had tumors that were less likely to have spread to axillary nodes and to distant sites than women who had never used HRT.

The increased risk found for current and recent users was not seen among women who had stopped using HRT five or more years in the past, regardless of duration of use [34]. Among past users, risk estimates were 1.12 for women who used HRT for less than one and 1–4 years, 0.90 for women who used HRT for 5–9 years, and 0.95 for women who used HRT for 10 or more years. None of these estimates was statistically significant, and there was no trend in risk by time since last use.

Only a few studies have examined the effects of estrogen plus progestin on breast cancer risk and, in aggregate, they do not provide sufficient evidence for making a judgment about a positive relationship [35–42]. In the collaborative analysis, risk estimates for combination therapy were elevated and were slightly greater than those for unopposed estrogen. However, none of the comparisons were significantly different than the risk for unopposed estrogen, and none differed significantly from one. Animal and in vitro studies have yielded conflicting results regarding the role for progestins in the development of breast cancer [22]. Animal studies suggest that addition of progestin to estrogen replacement therapy results in greater breast cell proliferation than estrogens alone whereas in vitro studies suggest an inhibitory effect of progestins.

In summary, the collaborative study provides strong evidence for a relationship between HRT use, more specifically estrogen replacement therapy, and the risk of developing breast cancer [34]. The nonsignificant χ^2 tests for heterogeneity suggested that there was no variation in findings across studies and that no one study dominated the overall results. Risk increased with longer duration of use and decreased after stopping use; the increased effect was almost entirely absent about five years after stopping use.

Additional studies are needed to address the long-term effects of combination HRT. Although the relationship between combination HRT use and breast cancer risk has been examined in a few studies, the existing evidence suggests that estrogen plus progestin use is unlikely to reduce risk and thus leaves open the question about whether combination HRT use increases risk or has no effect. In addition, because HRT use was associated with tumors that were less likely to have spread to axillary nodes and distant sites [34], more data are needed to examine the influence of HRT use on the virulence of breast tumors and on breast cancer survival.

2. Colorectal Cancer

Colorectal cancer is the third leading cause of cancer and cancer death in women after breast and lung cancer [43]. Several biologic mechanisms have been proposed for a possible protective effect of HRT on the risk of developing colon cancer: (1) a reduction in the concentration of secondary bile acids that have been hypothesized to initiate or promote malignancy in the colon [44] and (2) a loss of the tumor suppressor function on the estrogen receptor (ER) gene in epithelial colorectal tissue [45,46].

The preponderance of the epidemiologic literature available through 1994 as reviewed by Calle and colleagues [47,48] and several more recent studies [49–53] suggested an inverse relationship between HRT and the risk of developing cancer of the colon/rectum. Some studies found a relationship with duration of use [47,51,54], and other studies found that the reduction in risk was most apparent for women who were currently using HRT or had recently stopped using HRT [47,49,50,52,54,55]. Some epidemiologic studies reported reduced risks only for colon cancer, some reported reduced risks only for cancer of the rectum, and others found that risk was reduced for both sites [47]. However, not all studies found an inverse association (see review by Calle et al. [47]) [56–60].

A meta-analysis concluded that HRT use reduced the risk of developing colon cancer in women [61]. The random effects model yielded a significantly reduced risk of colon cancer for women who had ever used HRT (0.86, 95% CI = 0.73–0.99) and for women who had used HRT for more than five years (0.73, 95% CI = 0.53–1.02). The largest reduction in risk was for women who were current or recent HRT users (0.69, 95% CI = 0.52–0.91). Reduced risks appeared more pronounced in studies published since 1990 than in those published earlier.

In order to refine our understanding of the relationship between HRT and cancers of the colon/rectum, laboratory investigations are needed to investigate the possible biologic mechanisms, and epidemiologic studies are needed to confirm the current hypothesis of a protective effect. The number of studies that examined exposure to HRT in detail and included dosage of HRT and multiple levels of duration of use and time since first or last use was limited. Future studies should include a complete assessment of potentially confounding factors and of the various measures of HRT in relation to the occurrence of colon cancer and, separately, of rectal cancer.

3. Endometrial Cancer

Endometrial cancer, or cancer of the lining of the uterus, is the fourth leading cause of cancer in women, accounting for about 36,100 new diagnoses each year, and it ranks seventh among the top ten cancer mortality sites [43]. Beginning with two reports in the New England Journal of Medicine in 1975, a substantial epidemiologic literature supporting an association between the development of endometrial cancer and use of estrogen replacement therapy has accumulated [7,8,62,63]. The risk related to unopposed estrogen use is at least twofold for ever use, increases with duration and dose, is consistent across many studies, and involves the direct effects of estrogen on hyperplastic changes to the endometrium [62,63]. Even low dose unopposed estrogen therapy (0.3 mg per day) carries some increased risk of this malignancy [64]. The increased risk of endometrial cancer associated with use of unopposed estrogens appears to persist for at least five years after stopping use, and according to a meta-analysis, women who used HRT had nearly three times the risk of dying from endometrial cancer as women who had never used HRT (2.7, 95% CI = 0.9–8.0) [63].

Findings from the Postmenopausal Estrogen/Progestin Interventions (PEPI) Trial strongly support the use of progestin in combination with estrogen for women who still have a uterus [10]. In this randomized clinical trial, the histologic changes to the endometrium were examined among 596 postmenopausal women aged 45–64 years who received the placebo, 0.625 mg/day conjugated equine estrogens (CEE) alone, CEE 0.625 mg/day plus cyclic medroxyprogesterone acetate (MPA) 10 mg/day for 12 days/month, CEE 0.625 mg/day plus continuous MPA 2.5 mg/day, or CEE 0.625 mg/day plus cyclic micronized progesterone 200 mg/day for 12 days/month. Women who took only CEE were more likely to develop hyperplastic changes to the endometrium than women who used an estrogen plus progestin regimen. CEE plus cyclic or continuous MPA or CEE plus cyclic micronized progesterone provided protection against hyperplastic changes to the endometrium.

Use of combination oral contraceptives provides protection from developing endometrial cancer that persists for at least 15 years after stopping use [65]. One study has examined whether use of oral contraceptives in the past results in different risks of developing endometrial cancer for women who have also used

estrogen replacement therapy [66]. In this study, women who used estrogen replacement therapy for at least six years had a sevenfold statistically significant increased risk of developing endometrial cancer. When risks were examined by whether or not women had used oral contraceptives, risks were largest for women who had never used the pill. Although risks were lower for women who had used oral contraceptives, estrogen replacement therapy still appeared to increase their risk about twofold. These findings should be interpreted with caution due to the small number of cases who used both estrogen replacement therapy and oral contraceptives and require replication in other investigations.

The existing literature with information about combination therapy and endometrial cancer includes only limited results regarding duration of use, cyclic versus continuous therapy, and specific doses of estrogen and progestin [67–73]. One study in Los Angeles found no increased risk of endometrial cancer among women who used continuous combination therapy [67]. In general, risk estimates for using estrogen plus progestin were lower than risks for using unopposed estrogen [67–69,71,72]. Of note, the lower estimates of risk for combination therapy were still elevated risks, particularly for cyclic progestin use of fewer than 10 days per month [67,69,71]. For example, in the case-control study conducted in western Washington State, the risk for using estrogen plus progestin therapy for less than 10 days per month increased from 2.1 (95% CI = 0.9–4.7) for 6 to 35 months use to 3.7 (95% CI = 1.7–8.2) for at least 60 months use [69]. The risks of using estrogen plus progestin therapy for 10 or *more* days per month for these same durations were 0.8 (95% CI = 0.4–1.8) and 2.5 (95% CI = 1.1–5.5), respectively. The study from Los Angeles reported significantly elevated risks for sequential combination therapy taken for fewer than 10 days per month and no increased risk for therapy of 10 or more days per month [67].

In summary, the positive relationship between unopposed estrogen therapy and the risk of endometrial cancer in women with a uterus is well established. The unanswered questions involve the long-term effects of combination therapy, the optimal regimen for progestin therapy (cyclic versus continuous, number of days, dosage, and type of progestin [MPA or micronized progesterone]), and any possible modification by prior oral contraceptive use or other risk factors for endometrial cancer. While the effects of estrogen plus progestin use on the endometrium appear to be different than those of estrogen alone, estrogen plus progestin therapy may also increase risk.

4. Ovarian Cancer

Ovarian cancer is less common than the other hormone-related cancers in women but contributes disproportionately to cancer mortality [43]. Several hypotheses provide biologic plausibility for a relationship between ovarian cancer and HRT [74–76]. One hypothesis postulated that estrogen use might reduce the risk of ovarian cancer because estrogen lowers the production of pituitary gonadotropins [74,75]. Another theory posited that ovarian tumorigenesis occurred in two stages: (1) formation of inclusion cysts entrapped by ovarian surface epithelium into the ovarian stroma, followed by (2) estrogen-stimulated proliferation and malignant transformation of the inclusion cysts [76].

According to a collaborative analysis [77], two reviews that included cohort and case-control analyses of this relationship [78,79], and two more studies [80,81], the epidemiologic evidence for a relationship between ovarian cancer and HRT is unclear. Many studies showed no appreciable relationship but reported only on ever use of estrogen replacement therapy [81]. Some studies suggested the possibility of a slightly increased risk of ovarian cancer with longer duration of estrogen replacement therapy, but they generally did not yield a statistically significant risk estimate for the longest duration nor a significant trend [80,81]. A prospective study of fatal ovarian cancer and estrogen replacement therapy found a significant trend of increasing risk with increasing duration of use [80]. The pooled analyses of the hospital- and population-based case-control studies available before 1992, which included large numbers of cases (406 and 824 cases respectively), did not find a trend of increasing risk of epithelial ovarian cancer with increasing duration of use [77]. These findings did not differ by menopausal status or histologic diagnosis.

A meta-analysis of the relationship between HRT and the risk of developing ovarian cancer found that women who had ever used HRT had a slightly elevated risk of developing ovarian cancer (1.15, 95% CI = 1.05–1.27), and women with more than ten years of use had the highest risk (1.27, 95% CI = 1.00–1.61) [82]. At the p = 0.10 level of statistical significance specified by the investigators for assessing the consistency of findings across studies, the tests for heterogeneity for ever use and for long duration of use were statistically significant.

Three studies reported a relationship between noncontraceptive estrogen use and specific histologic subtypes of epithelial ovarian tumors [78,83,84]. These studies reported two- to threefold statistically significant or marginally significant increased risks for endometrioid tumors among women who had ever used noncontraceptive estrogens. In the Canadian study [78], women who used estrogen replacement therapy for at least five years had the following risks of developing specific types of tumors: 2.03 (95% CI = 1.04–3.97) for serous ovarian tumors, 2.81 (95% CI = 1.15–6.89) for endometrioid tumors, and 0.58 (95% CI = 0.08–4.21) for mucinous tumors. Another study did not confirm different risks by histologic subtype [85].

Women who have used oral contraceptives have a reduced risk of ovarian cancer that persists for at least 15 years after stopping use [86]. Two studies examined whether the relationship between ovarian cancer risk and use of noncontraceptive estrogen varied by prior use of oral contraceptives [66,78]. Women who had never used oral contraceptives but had used HRT had a threefold significantly elevated risk of epithelial ovarian cancer; by comparison, the risk associated with HRT use among women who had used oral contraceptives in the past was reduced [66]. In the Canadian study, the trend for using estrogen replacement therapy increased significantly among women who had never used oral contraceptives [78]. The trend for estrogen replacement therapy among women who had used oral contraceptives in the past was in the same direction as for never users but was not significant. The findings from both studies were based on small numbers of cases exposed to both estrogen replacement therapy and oral contraceptive use and should be interpreted with caution.

In summary, the question of an association between estrogen replacement therapy and ovarian cancer remains unanswered, although HRT does not appear to reduce the risk of ovarian cancer. There are several important research questions that deserve further investigation in addition to the basic question. First, what is the relationship between ovarian cancer and combination therapy (estrogen plus progestin use) for women who still have a uterus? Second, does risk vary by histologic diagnosis, specifically between mucinous versus endometrioid or serous ovarian tumors? Finally, is risk different for women who have used oral contraceptives in the past or for women who have other long-lasting risk factors?

B. Cardiovascular Disease

1. Coronary Heart Disease

Risk of atherosclerosis increases after menopause or bilateral oophorectomy [87]. The majority of epidemiologic observational studies note a benefit from estrogen against coronary heart disease [88–90], coronary death [90–92], and all-cause mortality [93,94]. Overall, epidemiologic evidence indicates a 30–50% reduction in risk of coronary heart disease among postmenopausal women who use estrogen therapy [88].

In the Nurses' Health Study, women who used estrogen alone had a 40% reduction (95% CI = 17%–57%) in risk of major coronary heart disease (nonfatal myocardial infarction, coronary death); the comparable risk reduction for users of combined therapy was 61% (95% CI = 22–81%) [90]. This effect was attenuated, however, in long-term users and was not present in past users. Furthermore, the reduction in risk was greater in younger than in older women. Coronary revascularization procedures were equally common in current users compared to never-users of postmenopausal estrogen therapy, which could indicate a healthy user effect. Thus, women using HRT might have more access to early diagnosis and invasive therapy of coronary disease compared with nonusers.

A prospective cohort study in Sweden involved the linkage of hormone prescriptions dispensed during 1977–1980 to central health records (1977–1983) of the entire Uppsala female population of 1.4 million women aged 35 and older [95]. The age-adjusted relative risk of myocardial infarction in hormone users was 0.81 (95% CI = 0.7–0.92). Women who used the more potent estrogens (CEE or estradiol) had lower risks than women who had used other estrogens. The addition of the progestin levonorgestrel to estradiol did not alter the protective effect of estradiol. As with the Nurses' Health Study, the protective effect for cardiovascular disease associated with hormone use was greater in younger than in older women.

The mechanisms for the hypothesized improvement in cardiovascular disease risk with estrogen therapy may be multifactorial. Exogenous estrogen increases HDL and lowers LDL cholesterol [14,96]. In an analysis of the PEPI Trial, serum measures of cardiovascular disease were examined in 875 postmenopausal women who were randomized to placebo or one of four HRT regimens, described previously. After three years of treatment, investigators found mean increases in plasma HDL cholesterol of 1.2–5.8 mg/dl in women randomized to combined hormone therapy or unopposed estrogen compared with a decrease of 1.2 mg/dl in placebo-treated women, with the largest benefit observed for unopposed estrogen. Significant lowering of LDL and total cholesterol was also observed in the active-treatment groups. The beneficial effects of estrogen were blunted with the addition of progestins.

Decreases in fibrinogen and antithrombin III, and increases in plasminogen, factor X, and factor VII, have been found in various randomized trials of HRT [14,97,98]. Estrogen also has direct beneficial effects on the coronary arterial wall in humans; it improves blood flow [99–102] and has antioxidant properties that may slow early stages of atherosclerosis [103]. Conversely, progestins may reverse the direct benefits of estrogen on coronary vasculature [104,105].

The effect of HRT on recurrent coronary heart disease and other cardiovascular disease was tested in 2763 postmenopausal women (mean age 66.7 years) with a documented history of coronary disease in the Heart and Estrogen/Progestin Replacement Study (HERS) [20]. HERS was a randomized, blinded, placebo-controlled clinical trial of an average of 4.1 years treatment with either combined 0.625 mg CEE plus 2.5 mg MPA daily or placebo medication. Pre-existing coronary disease was defined as evidence of one or more of myocardial infarction, coronary artery bypass graft surgery, percutaneous coronary revascularization, or angiographic evidence of $\geq$50% occlusion of one or more major coronary arteries. Despite a net 11% lower LDL cholesterol and 10% higher HDL cholesterol level in the HRT arm compared with placebo ($p < 0.001$), there was no significant difference between groups in occurrence of myocardial infarction or coronary death (relative hazard 0.99, 95% CI = 0.80–1.22). There were no significant differences between groups in incidence of other cardiovascular outcomes (coronary revascularization, unstable angina, congestive heart failure, resuscitated cardiac arrest, stroke or transient ischemic attack, and peripheral arterial disease). There was a statistically significant time trend; more coronary events were observed in the HRT arm compared to the placebo arm in year 1 of follow-up, and fewer events were observed in the HRT arm in years 4 and 5 of follow-up.

2. Stroke

The data on HRT effect on stroke incidence are unclear. Among the published observational cohort studies of HRT use and stroke incidence or death [106], three studies found a statistically significant decrease in risk of stroke among HRT users [107–109], and one found a small but statistically significant increase in risk [110]. In the Nurses' Health Study, little association between HRT use and stroke occurrence was observed for ever use of estrogens (1.27, 95% CI = 0.95–1.69). Most of the increased risk was in ischemic stroke (1.4, 95% CI = 1.02–1.92) [90]. In the HERS trial, the incidence of stroke did not differ significantly between women randomized to HRT and those assigned to placebo (relative hazard 1.13, 95% CI = 0.85–1.48) [20].

It is difficult to predict the effect of HRT, either unopposed estrogen or combination therapy, on coronary or other cardiovascular diseases. Some lipid and coagulation factor changes (LDL-C, HDL-C, fibrinogen, plasminogen) could reduce risk of cardiovascular disease, while others (triglycerides, factor VII, factor X, antithrombin III) could increase cardiovascular disease

risk [106]. Beneficial effects of estrogen on the cardiovascular system may be lessened by the addition of progestins [106]. Finally, the degree of protection afforded to intermediate outcomes such as lipids and coagulation factors may not translate into that degree of protection against cardiovascular disease occurrence [111].

There are several major unresolved issues surrounding the association between HRT use and risk of cardiovascular disease. The effect of lipid and coagulation factor changes with HRT use on clinical endpoints is unknown for healthy women. It does not necessarily follow that improvement in lipid fractions will translate into improvement in clinical disease risk. As described in Section III, women who use HRT are more likely to follow health practices that protect against atherosclerosis and its sequelae compared with women who do not use HRT. The need for large, controlled, randomized clinical trials of HRT and clinical endpoints is supported by results from the HERS trial. Observational data indicated that HRT would be protective against cardiovascular disease recurrence and mortality in women with preexisting coronary heart disease [94,112–114]. The HERS trial found no benefit, and considerable risk, for HRT use among women at high risk for coronary events and death [20]. The HERS results, however, should not be extrapolated to unopposed estrogen therapy and HRT used for primary prevention [115].

3. Venous Thromboembolic Disease

There is evidence that oral estrogen therapy increases risk of pulmonary embolism and deep vein thrombosis in postmenopausal women. Three large prospective studies found that current, but not past, HRT users had a statistically significant two- to threefold increased risk for pulmonary embolism and deep vein thrombosis [17–19]. The risk was highest in the first few years of use and was increased for both unopposed estrogen therapy and combination therapy. This is consistent with a procoagulant mechanism of estrogen that is not counteracted by progestins. Other exposures, such as smoking, did not affect the relationship between HRT and risk of thromboembolic disease. The HERS clinical trial found that HRT use increased risk of venous thromboembolic disease nearly threefold (relative hazard 2.89, 95% CI = 1.50–5.58). The risk was greatest in the first year of follow-up (relative hazard 3.29, 95% CI = 1.07–10.08), but the relative hazard did not go below 2.05 in 4–5 years of follow-up.

It has been postulated that transdermal and other nonoral HRT preparations will pose less risk of thromboembolic disease because of bypassing the liver pathway [116], largely because of small trial evidence that transdermal estrogen does not elevate renin substrate or change levels of the clotting factors fibrinopeptide A, high-molecular weight fibrinogen, antithrombin III activity, or antithrombin III antigen [117]. There has not been a clinical trial of transdermal estrogen of sufficiently large scale to assess effects on clinical thromboembolic disease clinical endpoints, nor are there large-scale, long-term observational data to support this. It is also not clear if the increased risk of thromboembolic disease associated with oral HRT applies to all women, or to specific subgroups with other risk factors such as obesity, history of clotting disorders, or a family history of thromboembolic disease.

C. Osteoporosis

On average, women lose 1% of bone mass per year in the first few years after menopause, and the risk of fracture increases fourfold for each 1 standard deviation decline in bone density at the hip [118]. This bone loss is thought to result from the loss of ovarian estrogens at menopause. Estrogen stabilizes bone mineral density [107,119,120], and many observational studies support a 30–50% reduction in fracture rates in women who use postmenopausal estrogen long-term compared with nonusers [121]. Estrogen is one of the drugs approved by the Food and Drug Administration (FDA) for the prevention and treatment of osteoporosis [122]. The timing of estrogen use for optimal protection is important. Observational data suggest that use during the period of risk for fractures (e.g., over age 75) is most protective, regardless of whether estrogen was used earlier at the time of menopause [121,123,124]. In the Study of Osteoporotic Fractures, current estrogen users who had started estrogen within five years of menopause had a 70% decreased risk of hip fracture and wrist fracture and a 50% decreased risk of all nonspinal fractures [121]. Previous estrogen use for more than ten years and use begun soon after menopause did not confer this protection if the woman was not a current estrogen user. Analysis of data from 670 women who had bone mineral density measurements in the Framingham cohort indicated that women should take estrogen therapy for at least seven years after menopause to receive benefit to bone density [123]. The data also indicated that duration of therapy of this length may be insufficient to protect the oldest, highest risk women over age 75 years. The HERS trial found no reduction in incidence of hip or other fracture in women randomized to HRT compared with those randomized to placebo, although the study may not have had sufficient power to detect meaningful differences in fracture rates between the two arms [20].

D. Menopausal Symptoms

Approximately 20% of women seek medical care for symptoms of menopause [125,126]. As a result of declining circulating levels of estrogen, women may experience vasomotor symptoms that consist of hot flashes, night sweats, irritability and other emotional symptoms, vaginal symptoms including dryness and difficulty with sexual functioning, and urinary tract symptoms. Most menopausal symptoms have been shown in short-term (one year or less) randomized placebo-controlled clinical trials to be dramatically relieved with estrogen or combined estrogen plus progestin [127]. Acute relief of vasomotor symptoms has been demonstrated in laboratory settings [127].

The use of HRT, however, presents its own set of symptoms that may or may not be tolerated by women. Women using either estrogen alone or in combination with progestin can be subject to vaginal spotting or bleeding. Use of a cyclic progestin preparation results in regular bleeding similar to that of regular menstrual cycling [24]. Use of unopposed estrogen or estrogen plus continuous progestin preparations often results in unscheduled vaginal spotting or bleeding [24,128]. Other symptoms may include breast tenderness, headache, bloating, and depression (from progestins) [129]. The most commonly reported reasons for women stopping hormone therapy include vaginal bleeding and breast tenderness, in addition to concerns about cancer risk [130,131].

E. Cognitive and Memory Dysfunction

Alzheimer's disease is the most common dementia of older women, affecting 3–11% of community residents over age 65, with the highest prevalence in the oldest old [132,133]. Most studies of dementia and HRT have been observational in design, which can lead to biased conclusions as women with higher levels of cognition may be more likely to take or be prescribed HRT [134]. In these observational studies, HRT users have been found to have lower risk of Alzheimer's disease [135], fewer deaths related to Alzheimer's disease or related dementia [109], higher scores on the Mini-Mental State Examination [135], and delayed onset of Alzheimer's disease [136]. In the Rancho Bernando Study, no effect of postmenopausal estrogen therapy on cognitive performance was observed [137]. In a short-term clinical trial, women on placebo showed a decrease in immediate and delayed recall of paired-associates following surgical menopause, compared with no change in estrogen-treated women [138]. Another placebo-controlled clinical trial failed to demonstrate cognitive benefits from low-dose transdermal estradiol [139]. Several small estrogen treatment trials (most open-label) report a significant improvement of Alzheimer's symptoms among estrogen-treated women [140,141]. Estrogen might protect against dementia through several mechanisms. Estrogen suppresses levels of apolipoprotein E, a risk factor for late-onset Alzheimer's disease [142], favorably regulates the production of Alzheimer amyloid precursor protein [142], and has anti-inflammatory properties that could protect brain tissue [143]. It stimulates production of basal forebrain cholinergic neurons [144,145], may work in concert with neurotrophins to regulate protein synthesis necessary for neuronal differentiation [146], and may improve cerebral perfusion [147].

F. All Cause Mortality

Numerous studies that have assessed the effects of HRT on all-cause mortality have consistently found a protective effect [53,73,91–93,148–152]. In general, the risk of death from all causes was lower by 20–50% for women who had used HRT compared to women who had never used menopausal hormones. The reduction in risk has been attributed primarily to the reduced risk of dying from coronary heart disease. However, because lower mortality rates have also been reported for conditions that are generally not believed to be associated with hormone use, other explanations need to be considered [153]. Possible explanations include: (1) multifactorial biologic effects of estrogen and (2) bias introduced through the selective use of estrogen by women who have a low cardiovascular risk profile and who practice healthy behaviors [153].

VI. Selective Estrogen Receptor Modulator Drugs (SERMS)

Estrogen enters target cells by diffusing through the outer cell membrane, then binds to the nuclear estrogen receptor (ER), and thus gains entry to the cell nucleus where it has direct effects on cell transcription [154]. Several pharmaceutical preparations, manufactured chemicals, and naturally occurring substances act at the level of the ER as either estrogen agonists or antagonists [155]. Tamoxifen, a first-generation SERM, has been successfully used to treat estrogen-receptor-positive breast cancer and has been found to reduce risk of death by 25% in node-positive, ER-positive breast cancer patients using the medication for five years after diagnosis [156]. Breast cancer patients using Tamoxifen were also noted to have about a one-third lower incidence of new primary breast cancers [157], which led to the hypothesis that Tamoxifen might be useful in the primary prevention of breast cancer in high-risk women [158]. The National Surgical Adjuvant Breast and Bowel Project (NSABP) Breast Cancer Prevention Trial released main trial results, which showed that high-risk women randomized to Tamoxifen for five years had a 45% decrease in incidence of breast cancer compared with women randomized to placebo [159]. Two smaller European clinical trials (7879 women) found no difference in rates of breast cancer occurrence between women randomized to Tamoxifen and women randomized to placebo [160,161].

Results of a two-year randomized placebo-controlled trial of Raloxifene (a SERM) versus placebo in 7705 postmenopausal women at normal risk for developing breast cancer showed a 66% reduced risk for breast cancer in the group randomized to Raloxifene [162]. Tamoxifen has been found to be proestrogenic at the level of the cardiovascular system [163], bone [164,165], and endometrium [166,167]. This action is reflected in the reduced incidence of cardiovascular disease and osteoporotic fractures and the increased incidence of endometrial cancer in women taking Tamoxifen either for breast cancer treatment [168] or in the prevention trials [159,163]. Raloxifene may have similar effects on the cardiovascular and bone systems [169,170] but appears to be antiestrogenic at the endometrium [171] and so theoretically should not increase risk for endometrial cancer. Raloxifene has been found to increase bone density in postmenopausal women [169] and is now FDA approved for the treatment and prevention of osteoporosis. The NSABP Study of Tamoxifen and Raloxifene (STAR) will compare these two SERMS on the incidence of primary breast cancer in 22,000 high risk postmenopausal (natural or surgical) women who will be randomized to Tamoxifen or Raloxifene for five years, and followed for at least two additional years [162]. This study will also assess the effect on risk of cardiovascular disease, fractures, toxicity, and adverse events. Several other SERM preparations are in Phase I or II trial testing [172].

VII. Summary and Future Research

A. Conclusions and Effects of Bias

HRT is used extensively in the United States [1,2] and has been prescribed for some women for the rest of their lives to prevent or treat specific conditions [173]. In addition to the beneficial influence of HRT on treating symptoms of menopause, estrogen replacement therapy has been associated with lower mortality and longer survival and may decrease the risk of coronary heart disease, osteoporosis, and possibly colon cancer. On the negative side, estrogen replacement therapy likely increases

the risk of breast cancer, endometrial cancer, deep vein thrombosis, and pulmonary embolism. The effects of estrogen replacement therapy on ovarian cancer, stroke, and cognitive and memory dysfunction seem less clear.

Selection biases need to be considered in the interpretation of observational studies about HRT. The possibility that women who take estrogen have fewer cardiovascular risk factors and are more likely to practice healthy dietary and exercise behaviors raises concerns that the beneficial effects on the coronary heart disease as seen in observational studies may be overestimated [12,153,174]. Similarly, women who suffer from hot flashes and other symptoms of menopause likely have lower levels of endogenous estrogens and hence may have a lower risk of developing breast cancer [22]. Because these women are also more likely to use menopausal estrogens, their inclusion in studies of HRT and breast cancer could lead to an underestimate of the adverse effects of HRT on this disease. In addition, HRT prescribing practices may have further complicated the ability to address this concern because the various contraindications over time suggested prudence in prescribing HRT for women who had risk factors or were not in good health [174]. The resolution of such issues can only be addressed in a randomized trial.

B. Unanswered Questions

Because the use of combination HRT increased substantially only since the 1980s [2], the published literature does not provide information about the long-term effects of combination HRT. Future studies should examine the effects of combination HRT on the occurrence, mortality, and survival from the cardiovascular diseases, hormone-dependent cancers, and selected other conditions. The minimum effective dose is not known [173]. In order to specify the optimum duration of therapy required to maximize benefits and to decrease risks, future investigations should include data about the length of time from starting use to an increased/decreased risk and about the length time that risk/benefit persists after stopping use. Little is known about how much progestin to take each month and whether to use progestin sequentially or at the same time as estrogen. There are only limited data about the number of progestin pills to take each month (e.g., 10 pills, more than 10 pills, or every day). Additional data are needed to clarify which subgroups of women are most likely to benefit from using HRT. Before ages 55–59 years, breast cancer mortality is of greater concern than heart disease (Fig. 93.2). At older ages, cardiovascular mortality overwhelms mortality from other hormone-dependent conditions.

C. The Women's Health Initiative (WHI)

The many open questions regarding the risks and benefits of long-term use of HRT prompted the funding in 1992 of the WHI Clinical Trial [122]. The WHI comprises three clinical intervention trials: HRT, calcium/vitamin D, and low-fat dietary modification. Women were recruited during 1993–1998 from diverse racial, ethnic, and geographic groups to 40 clinical centers across the United States. In the HRT trial, 27,500 postmenopausal women aged 50–79 years have been randomized to HRT or placebo. Women with a uterus are randomized in equal proportions to either placebo or combined estrogen plus progestin

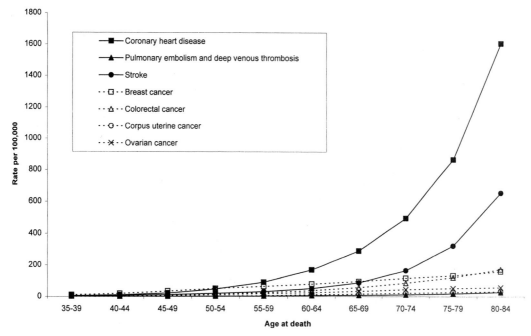

Fig. 93.2 Age-specific cardiovascular and cancer mortality rates, females, U.S., 1995.

(CEE 0.625 mg/day plus MPA 2.5 mg/day). Women who have previously undergone hysterectomy (45% of the cohort) are randomized in equal proportions to either placebo or unopposed estrogen (CEE 0.625 mg/day). Women are followed semiannually for 9–12 years for the occurrence of specific endpoints (coronary heart disease, coronary deaths, other cardiovascular disease, breast and other cancers, and osteoporotic fractures). The WHI will provide definitive answers about the effect of HRT on preventing coronary heart disease and osteoporotic fractures and the potential adverse effects on risk of breast and other cancers, thromboembolic disease, and other measures of health and disease. It will provide these answers without the bias that is present in observational studies from health, lifestyle, and socioeconomic differences between HRT users and nonusers.

The WHI Memory Study, an ancillary study to the WHI HRT trial, is designed to assess the effects of HRT on the development and progression of symptoms associated with dementia [134]. This study will assess cognition at baseline and follow-up (9–12 years) in participants aged 65 and older at randomization using the Mini-Mental State Examination. Participants with low scores will undergo more intensive neuropsychological testing and neuropsychiatric examinations to classify women to dementia status. It is hypothesized that dementia-related symptoms will be delayed in women who are on active (estrogen or combined estrogen plus progestin) therapy as compared to placebo.

In summary, only a randomized trial such as the WHI can provide assurances about the long-term effects of HRT. In the meantime, a woman and her health care provider should discuss her family history of diseases and other individual risk factors and weigh her possible risks and benefits of HRT, SERMS, and other treatment modalities, including but not limited to taking low dose aspirin, controlling coronary risk factors (*e.g.,* blood pressure, diabetes), taking calcium/vitamin D or alendronate for

prevention of osteoporotic fractures [175], and adopting healthy behaviors such as dietary modification, regular physical activity, avoidance of tobacco and excess alcohol, and maintenance of a healthy body weight.

References

1. IMS America (1998). Top 200 drugs of 1997. *Pharm. Times* **64**(4), 31–49.
2. Wysowski, D. K., Golden, L., and Burke, L. (1995). Use of menopausal estrogens and medroxyprogesterone in the United States. *Obstet. Gynecol.* **85**(1), 6–10.
3. Brett, K. M., and Madans, J. H. (1997). Use of postmenopausal hormone replacement therapy: Estimates from a Nationally Representative Cohort Study. *Am. J. Epidemiol.* **145**(6), 536–545.
4. Hemminki, E., Kennedy, D. L., Baum, C., and McKinlay, S. M. (1988). Prescribing of noncontraceptive estrogens and progestins in the United States, 1974–86. *Am. J. Public Health* **78**(11), 1479–1481.
5. Ellerington, M. C, and Whitehead, M. I. (1991). Hormone replacement therapy and its current use in British specialist clinical practice. *In* "Hormone Replacement Therapy and Breast Cancer Risk" (R. D. Mann, ed.), pp. 18–20. The Parthenon Publishing Group, Park Ridge, NJ.
6. Kennedy, D. L., Baum, C., and Forbes, M. B. (1985). Noncontraceptive estrogens and progestins: Use patterns over time. *Obstet. Gynecol.* **65**(3), 441–446.
7. Smith, D. C., Prentice, R., Thompson, D. J., and Herrmann, W. L. (1975). Association of exogenous estrogen and endometrial carcinoma. *N. Engl. J. Med.* **293**, 1164–1167.
8. Ziel, H. K., and Finkle, W. D. (1975). Increased risk of endometrial carcinoma among users of conjugated estrogens. *N. Engl. J. Med.* **293**, 1167–1170.
9. Gambrell, R. D. (1982). The menopause: Benefits and risks of estrogen-progestogen replacement therapy. *Fertil. Steril.* **37**, 457–474.

10. The Writing Group for the PEPI Trial (1996). Effects of hormone replacement therapy on endometrial histology in postmenopausal women: The Postmenopausal Estrogen/Progestin Interventions (PEPI) Trial. *JAMA, J. Am. Med. Assoc.* **275**, 370–375.

11. Barrett-Conner, E. (1991). Postmenopausal estrogen and prevention bias. *Ann. Intern. Med.* **115**(6), 455–456.

12. Matthews, K. A., Kuller, L. H., Wing, R. R., Meilahn, E. N., and Plantinga, P. (1996). Prior use of estrogen replacement therapy, are users healthier than nonusers? *Am. J. Epidemiol.* **143**, 971–978.

13. Harris, R. B., Laws, A., Reddy, V. M., King, A., and Haskell, W. L. (1990). Are women using postmenopausal estrogens? A Community Survey. *Am. J. Public Health* **80**(10), 1266–1268.

14. The Writing Group for the PEPI Trial (1995). Effects of estrogen or estrogen/progestin regimens on heart disease risk factors in postmenopausal women: The Postmenopausal Estrogen/Progestin Interventions (PEPI) Trial. *JAMA, J. Am. Med. Assoc.* **273**, 199–208.

15. Glueck, C. J., Lang, J., Hamer, T., and Tracy, T. (1994). Severe hypertriglyceridemia and pancreatitis when estrogen replacement therapy is given to hypertriglyceridemic women. *J. Lab. Clin. Med.* **123**, 59–64.

16. Jaffe, R. B. (1996). The menopause and perimenopausal period. *In* "Yen SSC, Reproductive Endrocrinology" (R. B. Jaffe, ed.), pp. 389–408. Saunders, Philadelphia.

17. Daly, E., Vesey, M. P., Hawkins, M. M., Carson, J. L., Gough, P., and Marsh S. (1996). Risk of venous thromboembolism in users of hormone replacement therapy. *Lancet* **348**, 977–980.

18. Grodstein, F., Stampfer, M. J., Goldhaber, S. Z., Manson, J. E., Colditz, G. A., Speizer, F. E., Willett, W. C., and Hennekens, C. H. (1996). Prospective study of exogenous hormones and risk of pulmonary embolus in women. *Lancet* **348**, 983–987.

19. Jick, H., Derby, L. E., Myers, M. W., Vasilakis, C., and Newton, K. M. (1996). Risk of hospital admission of idiopathic venous thromboembolism among users of postmenopausal oestrogens. *Lancet* **348**, 981–983.

20. Hulley, S., Grady, D., Bush, T., Furberg, C., Herrington, D., Riggs, B., and Vittinghoff, E. (1998). Randomized trial of estrogen plus progestin for secondary prevention of coronary heart disease in postmenopausal women. *JAMA, J. Am. Med. Assoc.* **280**, 605–613.

21. Johnson, S. R. (1998). Menopause and hormone replacement therapy. *Women's Health Issues* **82**, 297–320.

22. Colditz, G. A. (1998). Relationship between estrogen levels, use of hormone replacement therapy, and breast cancer. *J. Natl. Cancer Inst.* **90**, 814–823.

23. Belchetz, P. E. (1994). Hormonal treatment of postmenopausal women. *N. Engl. J. Med.* **330**, 1062–1071.

24. Whitehead, M. I., Hillard, T. C., and Crook, D. (1990). The role and use of progestogens. *Obstet. Gynecol.* **75**, 59s–76s.

25. Creaseman, W. T. (1991). Estrogen replacement therapy: Is previously treated cancer a contraindication? *Obstet. Gynecol.* **77**, 308–312.

26. Vassilopoulou-Sellin, R., and Theriault, R. L. (1994). Randomized prospective trial of estrogen-replacement therapy in women with a history of breast cancer. *Natl. Cancer Inst. Monogr.* **16**, 153–159.

27. Varner, R. E. (1990). Hormone replacement therapy. *In* "Postreproductive Gynecology" (H. M. Shingleton and W. G. Hurt, eds.), pp. 143–169. Churchill-Livingstone, New York.

28. The Boston Collaborative Drug Surveillance Project (1974). Surgically confirmed gallbladder disease, venous thromboembolism, and breast tumors in relation to postmenopausal estrogen therapy. *N. Engl. J. Med.* **290**, 15–19.

29. Grodstein, F., Colditz, G. A., and Stampfer, M. J. (1994). Postmenopausal hormone use and cholecystectomy in a large prospective study. *Obstet. Gynecol.* **83**, 5–11.

30. Edelson, R. N. (1985). Menstrual migraine and other hormonal aspects of migraine. *Headache* **25**, 376–379.

31. Sanchez-Guerrero, J., Liang, M. H., Karlson, E. W., Hunter, D. J., and Colditz, G. A. (1995). Postmenopausal estrogen therapy and the risk for developing systemic lupus erythematosus. *Ann. Intern. Med.* **122**, 430–433.

32. Trichopoulos, D., MacMahon, B., and Cole, P. (1972). Menopause and breast cancer risk. *J. Natl. Cancer Inst. (U.S.)* **48**, 605–613.

33. Thomas, H. V., Reeves, G. K., and Key, T. J. (1997). Endogenous estrogen and postmenopausal breast cancer: A quantitative review. *Cancer Causes Control* **8**, 922–928.

34. Collaborative Group on Hormonal Factors in Breast Cancer (1997). Breast cancer and hormone replacement therapy: Collaborative reanalysis of data from 51 epidemiological studies of 52,702 women with breast cancer and 108,411 women without breast cancer. *Lancet* **350**, 1047–1059.

35. Bergkvist, L., Adami, H. O., Persson, I., Hoover, R., and Schairer, C. (1989). The risk of breast cancer after estrogen and estrogen-progestin replacement. *N. Engl. J. Med.* **321**(5), 293–297.

36. Colditz, G. A., Stampfer, M. J., Willett, W. C., Hunter, D. J., Manson, J. E., Hennekens, C. H., Rosner, B. A., and Speizer, F. E. (1992). Type of postmenopausal hormone use and risk of breast cancer: 12-year follow-up from the Nurses' Health Study. *Cancer Causes Control* **3**, 433–439.

37. Colditz, G. A., Hankinson, S. E., Hunter, D. J., Willett, W. C., Manson, E., Stampfer, M. J., Hennekens, C., Rosner, B., and Speizer, F. E. (1995). The use of estrogens and progestins and the risk of breast cancer in postmenopausal women. *N. Engl. J. Med.* **332**(24), 1589–1593.

38. Colditz, G., and Rosner, B. (1998). Use of estrogen plus progestin is associated with greater increase in breast cancer risk than estrogen alone for the Nurses' Health Study Research Group. *Am. J. Epidemiol.* **147**(11), S64–S254.

39. Risch, H. A., and Howe, G. R. (1994). Menopausal hormone usage and breast cancer in Saskatchewan: A record-linkage Cohort Study. *Am. J. Epidemiol.* **139**(7), 670–683.

40. Schairer, C., Byrne, C., Keyl, P. M., Brinton, L. A., Sturgeon, S. R., and Hoover, R. N. (1994). Menopausal estrogen and estrogen-progestin replacement therapy and risk of breast cancer (United States). *Cancer Causes Control* **5**, 491–500.

41. Stanford, J. L., Weiss, N. S., Voigt, L. F., Daling, J. R., Habel, L. A., and Rossing, M. A. (1995). Combined estrogen and progestin hormone replacement therapy in relation to risk of breast cancer in middle-aged women. *JAMA, J. Am. Med. Assoc.* **274**(2), 137–142.

42. Newcomb, P. A., Longnecker, M. P., Storer, B. E., Mittendorf, R., Baron, J., Clapp, R. W., Bogdan, G., and Willett, W. C. (1995). Long-term hormone replacement therapy and risk of breast cancer in postmenopausal women. *Am. J. Epidemiol.* **142**(8), 788–795.

43. Landis, S. H., Murray, T., Bolden S., and Wingo, P. A. (1998). Cancer statistics, 1998. *Ca—Cancer J. Clin.* **48**, 6–29.

44. McMichael, A. J., and Potter, J. D. (1980). Reproduction, endogenous and exogenous sex hormones, and colon cancer: A review and hypothesis. *JNCI, J. Natl. Cancer Inst.* **65**, 1201–1207.

45. Potter, J. D., Bostick, R. M., Grandits, G. A., Fosdick L., Elmer P., Wood, J., Grambsch, P., and Louis, T. A. (1996). Hormone replacement therapy is associated with lower risk of adenomatous polyps of the large bowel: The Minnesota Cancer Prevention Research Unit Case-Control Study. *Cancer Epidemiol, Biomarkers Prev.* **5**(10), 779–784.

46. Issa, J. P., Ottaviano, Y. l., Celano, P., Hamilton, S. R., Davidson, N. E., and Baylin S. B. (1994). Methylation of the estrogen receptor CpG island links aging and neoplasia in human colon. *Nat. Genet.* **7**, 536–540.

47. Calle, E. E., Miracle-McMahill, H. L., Thun, M. J., and Heath, C. W. (1995). Estrogen replacement therapy and risk of fatal colon cancer in a prospective cohort of postmenopausal women. *J. Natl. Cancer Inst.* **87**, 517–523.

48. Calle, E. E. (1997). Hormone replacement therapy and colorectal cancer: Interpreting the evidence. *Cancer Causes Control* **8**, 127–129.

49. Grodstein, F., Martinez E., Platz, E. A., Giovannucci, E., Colditz, G. A., Kautzky, M., Fuchs, C., and Stampfer, M. J. (1998). Postmenopausal hormone use and risk for colorectal cancer and adenoma. *Ann. Intern. Med.* **128**, 705–712.

50. Kampman, E., Potter, J. D., Slattery, M. L., Caan, B. J., and Edwards, S. (1997). Hormone replacement therapy, reproductive history, and colon cancer: A Multi-center, Case-control Study in the United States. *Cancer Causes Control* **8**(2), 146–158.

51. Fernandez, E., LaVecchia, C., D'Avanzo, B., Franceschi, S., Negri, E., and Parazzini, F. (1996). Oral contraceptives, hormone replacement therapy, and the risk of colorectal cancer. *Br. J. Cancer* **73**, 1431–1435.

52. Newcomb, P. A., and Storer, B. E. (1995). Postmenopausal hormone use and risk of large-bowel cancer. *J. Natl. Cancer Inst.* **87**, 1067–1071.

53. Folsom, A. R., Mink, P. J., Sellers, T. A., Hong, C. P., Zheng, W., and Potter, J. D. (1995). Hormone replacement therapy and morbidity and mortality in a prospective study of postmenopausal women. *Am. J. Public Health* **85**, 1128–1132.

54. Jacobs, E. J., White E., and Weiss, N. S. (1994). Exogenous hormones, reproductive history, and colon cancer. *Cancer Causes Control* **5**, 359–366.

55. Troisi, R., Schairer, C., Chow, W. H., Schatzkin, A., Brinton, L. A., and Fraumini, J. F. (1997). A prospective study of menopausal hormones and risk of colorectal cancer. *Cancer Causes Control* **8**, 130–138.

56. Wu, A. H., Paganini-Hill, A., Ross, R. K., and Henderson, B. E. (1987). Alcohol, physical activity, and other risk factors for colorectal cancer: A prospective study. *Br. J. Cancer* **55**, 687–694.

57. Wu-Williams, A. J., Lee, M., Whittemore, A. S., Gallagher, R. P., Jiao, D.-A., Zheng, S., Zhou, L., Wang, X.-H., Chen, K., Jung, D., Chong-Ze, T., Ling, C., Xu, J. Y., Paffenbarger, R. S., and Henderson, B. E. (1991). Reproductive factors and colorectal cancer risk among Chinese females. *Cancer Res.* **51**, 2307–2311.

58. Risch, H. A., and Howe, G. R. (1995). Menopausal hormone use and colorectal cancer in Saskatchewan: A Record Linkage Cohort Study. *Cancer Epidemiol., Biomarkers Prev.* **4**, 21–28.

59. Weiss, N. S., Daling, J. R., and Chow, W. H. (1981). Incidence of cancer of the large bowel in women in relation to reproductive and hormonal factors. *JNCI, J. Natl. Cancer Inst.* **71**, 703–709.

60. Peters, R. K., Pike, M. C., Chang, W. W., and Mack, T. M. (1990). Reproductive factors and colon cancers. *Br. J. Cancer* **61**, 741–748.

61. Hebert-Croteau, N. (1998). A meta-analysis of hormone replacement therapy and colon cancer in women. *Cancer Epidemiol., Biomarkers Prev.* **7**, 653–659.

62. Rose, P. G. (1996). Endometrial carcinoma. *N. Engl. J. Med.* **335**(9), 640–649.

63. Grady, D., and Ernster, V. L. (1997). Hormone replacement therapy and endometrial cancer: Are current regimens safe? *J. Natl. Cancer Inst.* **89**(15), 1088–1089.

64. Cushing, K. L., Weiss, N. S., Voigt, L. F., McKnight, B., and Beresford, S. A. (1998). Risk of endometrial cancer in relation to use of low-dose unopposed estrogens. *Obstet. Gynecol.* **91**(1), 35–39.

65. CDC-NICHD Cancer and Steroid Hormone Study Group (1987). Combination oral contraceptive use and the risk of endometrial cancer. *JAMA, J. Am. Med. Assoc.* **257**, 796–800.

66. Lee, N. C., Wingo, P. A., Peterson, H. B., Rubin, G. L., and Sattin, R. W. (1986). Estrogen therapy and the risk of breast, ovarian, and endometrial cancer. *In* "Aging, Reproduction, and the Climacteric" (L. Mastroianni and C. A. Paulsen, eds.), pp. 287–303. Plenum, New York.

67. Pike, M. C., Peters, R. K., Cozen, W., Probst-Hensch, N. M., Felix, J. C., Wan, P. C., and Mack, T. M. (1998). Estrogen-progestin replacement therapy and endometrial cancer. *J. Natl. Cancer Inst.* **90**(2), 164–166.

68. Shapiro, J. A., Weiss, N. S., Beresford, S. A., and Voigt, L. F. (1998). Menopausal hormone use and endometrial cancer by tumor grade and invasion. *Epidemiology* **9**(1), 99–101.

69. Beresford, S. A., Weiss, N. S., Voigt, L. F., and McKnight, B. (1997). Risk of endometrial cancer in relation to use of oestrogen combined with cyclic progestagen therapy in postmenopausal women. *Lancet* **349**, 458–461.

70. Grady, D., Gebretsadik, T., Kerlikowske, K., Ernster, V., and Petitti, D. (1995). Hormone replacement therapy and endometrial cancer risk: A meta-analysis. *Obstet. Gynecol.* **85**(2), 304–314.

71. Voigt, L. F., Weiss, N. S., Chu, J., Daling, J. R., McKnight, B., and VanBelle, G. (1991). Progestagen supplementation of exogenous oestrogens and risk of endometrial cancer. *Lancet* **338**, 274–277.

72. Persson, I. R., Adami, H. O., Bergkvist, L., Lindgren, A., Pettersson, B., Hoover, R., and Schairer, C. (1989). Risk of endometrial cancer after treatment with oestrogens alone or in conjunction with progestogens: Results of a prospective study. *Br. Med. J.* **298**, 147–151.

73. Hunt, K., Vessey, M., McPherson, K., and Coleman, M. (1987). Long-term surveillance of mortality and cancer incidence in women receiving hormone replacement therapy. *Br. J. Obstet. Gynaecol.* **94**, 620–635.

74. Stadel, B. V. (1975). The etiology and prevention of ovarian cancer. *Am. J. Obstet. Gynecol.* **123**, 772–773.

75. McGowan, L., Parent, L., Lednar, W., and Norris, H. J. (1979). The woman at risk for developing ovarian cancer. *Gynecol. Oncol.* **7**, 325–344.

76. Cramer, D. W., Hutchison, G. B., Welch, W. R., Scully, R. E., and Ryan, K. J. (1983). Determinants of ovarian cancer risk. I. Reproductive experiences and family history. *JNCI, J. Natl. Cancer Inst.* **71**, 711–716.

77. Whittemore, A. S., Harris, R., Intyre, J., and the Collaborative Ovarian Cancer Group (1992). Characteristics relating to ovarian cancer risk: Collaborative analyses of 12 case-control studies. II. Invasive epithelial ovarian cancers in white women. *Am. J. Epidemiol.* **136**, 1184–1203.

78. Rodriguez, C., Calle, E. E., Coates, R. J., Miracle-McMahill, H. L., Thun, M. J., and Heath, C. W. (1995). Estrogen replacement therapy and fatal ovarian cancer. *Am. J. Epidemiol.* **141**, 828–835.

79. Brewer-Newsome, K. (1995). Ovarian cancer risk and estrogen replacement therapy. A report submitted to the Emory University School of Public Health, Atlanta, GA, in partial fulfillment of the degree of master of public health.

80. Risch, H. A. (1996). Estrogen replacement therapy and risk of epithelial ovarian cancer. *Gynecol. Oncol.* **63**, 254–257.

81. Purdie, D., Green, A., Bain, C., Siskind, V., Ward, B., Hacker, N., Quinn, M., Wright, G., Russell, P., and Susil, B. (1995). Reproductive and other factors and risk of epithelial ovarian cancer: An Australian Case-control Study. Survey of Women's Health Study Group. *Int. J. Cancer* **62**(6), 678–684.

82. Garg, P. P., Kerlikowske, K., Subak, L., and Grady, D. (1998). Hormone replacement therapy and the risk of epithelial ovarian carcinoma: A meta-analysis. *Obstet. Gynecol.* **92**, 472–479.

83. Weiss, N. S., Lyon, J. L., Krishnamurthy, S., Dietert, S. E., Liff, J. M., and Daling, J. R. (1981). Noncontraceptive estrogen use and the occurrence of ovarian cancer. *JNCI, J. Natl. Cancer Inst.* **68**(1), 95–98.

84. LaVecchia, C., Liberati, A., and Franceschi, S. (1982). Noncontraceptive estrogen use and the occurrence of ovarian cancer. *JNCI, J. Natl. Cancer Inst.* **69**(6), 1207.

85. Hempling, R. E., Wong, C., Piver, M. S., Natarajan, N., and Mettlin, C. J. (1997). Hormone replacement therapy as a risk factor for epithelial ovarian cancer: Results of a Case-control Study. *Obstet. Gynecol.* **89**(6), 1012–1016.

86. CDC-NICHD Cancer and Steroid Hormone Study Group (1987). The reduction in risk of ovarian cancer associated with oral contraceptive use. *N. Engl. J. Med.* **316**(11), 650–655.

87. Colditz, G. A., Willett, W. C., Stampfer, M. J., Rosner, B., Speizer, F. E., and Hennekens, C. H. (1987). Menopause and risk of coronary heart disease in women *N. Engl. J. Med.* **316**, 1105–1110.

88. Stampfer, M. J., and Colditz, G. A. (1991). Estrogen replacement therapy and coronary heart disease: A quantitative assessment of the epidemiologic evidence. *Prev. Med.* **20**, 47–63.

89. Grady, D., Rubin, S. M., Petitti, D. B., Fox, C. S., Black, D., Ettinger, B., Ernster, V. L., and Cummings, S. R. (1992). Hormone therapy to prevent disease and prolong life in postmenopausal women. *Ann. Intern. Med.* **117**(12), 1016–1037.

90. Grodstein, F., Stampfer, M. J., Manson, J. A., Colditz, G. A., Willett, W. C., Rosner, B., Speizer, F. E., and Hennekens, C. H. (1996). Postmenopausal estrogen and progestin use and the risk of cardiovascular disease. *N. Engl. J. Med.* **335**(7), 453–461.

91. Ettinger, B., Friedman, G. D., Bush, T., and Quesenberry, C. P., Jr., (1996). Reduced mortality associated with long-term postmenopausal estrogen therapy. *Obstet. Gynecol.* **87**, 6–12.

92. Criqui, M. H., Suarez, L., Barrett-Connor, E., McPhillips, J., Wingard, D. L., and Garland, C. (1988). Postmenopausal estrogen use and mortality: Results from a prospective study in a defined, homogeneous community. *Am. J. Epidemiol.* **128**, 606–614.

93. Henderson, B. E., Paganini-Hill, A., and Ross, R. K. (1991). Decreased mortality in users of estrogen replacement therapy. *Arch. Intern. Med.* **151**, 75–78.

94. Bush, T. L., Barrett-Connor, E., Cowan, L. D., Criqui, M. H., Wallace, R. B., Suchindran, C. M., Tyroler, H. A., and Rifkind, B. M. (1987). Cardiovascular mortality and noncontraceptive use of estrogen in women: Results from the Lipid Research Clinics Program Follow-up Study. *Circulation* **75**, 1102–1109.

95. Falkeborn, M., Persson, I., Adami, H. O., Bergstrom R., Eaker, E., Lithell, H., Mohsen, R., and Naessen, T. (1992). The risk of acute myocardial infarction after estrogen and estrogen-progestogen replacement. *Br. J. Obstet. Gynaecol.* **99**, 821–828.

96. Walsh, B. W., Schiff, I., Rosner, B., Greenberg, L., Ravnihar, V., and Sacks, F. M. (1991). Effects of postmenopausal estrogen replacement on the concentrations and metabolism of plasma lipoproteins. *N. Engl. J. Med.* **325**(17), 1196–1204.

97. Lobo, R. A., Pickar, J. H., Wild, R. A., Walsh, B., and Hirvonen, E. (1994). Metabolic impact of adding medroxyprogesterone acetate to conjugated estrogen therapy in postmenopausal women. *Obstet. Gynecol.* **84**, 987–995.

98. Medical Research Council's General Practice Research Framework (1996). Randomized comparison of estrogen vs. estrogen plus progestogen hormone replacement therapy in women with hysterectomy. *Br. Med. J.* **3**, 473–478.

99. Lieberman, E. H., Gerhard, M. D., Uehata, A., Walsh, B. W., Selwyn, A. P., Ganz, P., Yeung, A. C., and Creager, M. A. (1994). Estrogen improves endothelium-dependent flow-mediated vasodilatation in postmenopausal women. *Ann. Intern. Med.* **121**, 936–941.

100. Giraud, G. D., Morton, M. J., Wilson, R. A., Burry, K. A., and Speroff, L. (1996). Effects of estrogen and progestin on aortic size and compliance in postmenopausal women. *Am. J. Obstet. Gynecol.* **174**, 1708–1717.

101. Sullivan, J. M. (1995). Coronary arteriography in estrogen-treated postmenopausal women. *Prog. Cardiovasc. Dis.* **38**, 211–222.

102. Rosano, G. M. C., Sarrel, P. M., Poole-Wilson, P. A., and Collins, P. (1993). Beneficial effect of estrogen on exercise-induced myocardial ischaemia in women with coronary artery disease. *Lancet* **342**, 133–136.

103. Rossouw, J. E. (1998). Estrogens and cardiovascular disease. *In* "Evidence Based Cardiology" (S. Yusuf, J. A. Cairns, J. A. Camm, J. A. Fallen, E. L. Gersh, and B. London, eds.), pp. 315–328. *Br. Med. Assoc.,* London.

104. Sullivan, J. M., Shala, B. A., Miller, L. A., Lerner, J. L., and McBrayer, J. D. (1995). Progestin enhances vasoconstrictor responses in postmenopausal women receiving estrogen replacement therapy. *Menopause* **2**, 193–199.

105. Levine, R. L., Chen, S. J., Durand, J., Chen, Y. F., and Oparil, S. (1996). Medroxyprogesterone attenuates estrogen-mediated inhibition of neointima formation after balloon injury of the rat carotid artery. *Circulation* **94**, 2221–2227.

106. Rossouw, J. E. (1996). Estrogens for prevention of coronary heart disease. Putting the brakes on the bandwagon. *Circulation* **94**, 2982–2985.

107. Ettinger, B., Genant, H. K., Steiger, P., and Madvig, P. (1992). Low-dosage micronized 17 beta-estradiol prevents bone loss in postmenopausal women. *Am. J. Obstet. Gynecol.* **166**, 479–488.

108. Falkeborn, M., Persson, I., Terent, A., Adami, H. O., Lithell, H., and Bergstrom, R. (1993). Hormone replacement therapy and the risk of stroke: Follow-up of a Population-based Cohort in Sweden. *Arch. Intern. Med.* **153**, 1201–1209.

109. Paganini-Hill, A., and Henderson, V. W. (1994). Estrogen deficiency and risk of Alzheimer's disease in women. *Am. J. Epidemiol.* **140**, 256–261.

110. Wilson, P. W., Garrison, R. J., and Castelli, W. P. (1985). Postmenopausal estrogen use, cigarette smoking, and cardiovascular morbidity in women over 50: The Framingham Study. *N. Engl. J. Med.* **313**(17), 1038–1043.

111. Fleming, T. R., and DeMets, D. L. (1996). Surrogate end points in clinical trials: Are we being misled? *Ann. Intern. Med.* **125**, 605–613.

112. Newton, K. M., LaCroix, A. Z., McKnight, B., Knopp, R. H., Siscovick, D. S., Heckbert, S. R., and Weiss, N. S. (1997). Estrogen replacement therapy and prognosis after first myocardial infarction. *Am. J. Epidemiol.* **145**, 269–277.

113. Sullivan, J. M., El-Zeky, F., VanderZwaag, R., and Ramanathan, K. B. (1997). Effect on survival of estrogen replacement therapy after coronary artery bypass grafting. *Am. J. Cardiol.* **79**, 847–850.

114. O'Keefe, J. H., Kim, S. C., Hall, R. R., Cochran, V. C., Lawhorn, S. L., and McCallister, B. D. (1997). Estrogen replacement therapy after coronary angioplasty in women. *J. Am. Coll. Cardiol.* **29**, 1–5.

115. Petitti, D. B. (1998). Hormone replacement therapy and heart disease prevention: Experimentation trumps observation. *JAMA, J. Am. Med. Assoc.* **280**, 650–651.

116. Baker, V. L. (1994). Alternatives to oral estrogen replacement, transdermal patches, percutaneous gels, vaginal creams and rings, implants and other methods of delivery *Obstet. Gynecol. Clin. North Am.* **21**(2), 271–297.

117. Chetkowski, R. J., Meldrum, D. R., Steingold, K. A., Randle, D., Lu, J. K., Eggena, P., Hersham, J. M., Alkjaersig, N. K., Fletcher, A. P., and Judd, H. L. (1986). Biologic effects of transdermal estradiol. *N. Engl. J. Med.* **314**(25), 1615–1620.

118. Purdie, D. W., and Horsman, A. (1990). Population screening and the prevention of osteoporosis. *In* "HRT and Osteoporosis" (J. O. Drife and J. W. W. Studd, eds.), pp. 251–264. Springer-Verlag, London.

119. The Writing Group for the PEPI Trial (1996). Effects of hormone therapy on bone mineral density: Results from the Postmenopausal Estrogen/Progestin Interventions (PEPI) Trial. *JAMA, J. Am. Med. Assoc.* **276**, 1389–1396.

120. Speroff, L., Rowan, J., Symons, J., Genent, H., and Wilborb, W. (1996). The comparative effect on bone density, endometrium, and lipids of continuous hormones as replacement therapy (CHART study). A Randomized Controlled Trial. *JAMA, J. Am. Med. Assoc.* **276**, 1397–1403.

121. Cauley, J. A., Seeley, D. G., Ensrud, K., Ettinger, B., Black, D., and Cummings, S. R. (1995). Estrogen replacement therapy and fractures in older women: Study of Osteoporotic Fractures Research Group. *Ann. Intern. Med.* **122**, 9–16.

122. The Women's Health Initiative Study Group (1998). Design of the Women's Health Initiative Clinical Trial and Observational Study. *Controlled Clin. Trials* **19**, 61–109.

123. Felson, D. T., Zhang, Y., Hannan, M. T., Kiel, D. P., Wilson, P. W., and Anderson, J. J. (1993). The effect of postmenopausal estrogen therapy on bone density in elderly women. *N. Engl. J. Med.* **329**, 1141–1146.

124. Schneider, D. L., Barrett-Connor, E. B., and Morton, D. J. (1997). Timing of postmenopausal estrogen for optimal bone mineral density: The Rancho Bernardo Study. *JAMA, J. Am. Med. Assoc.* **277**, 543–547.

125. Avis, N. E., McKinley, S. M. (1995). The Massachusetts Women's Health Study: An epidemiologic investigation of the menopause. *J. Am. Med. Women's Assoc.* **50**, 45–49.

126. Morse, C. A., Smith, A., Dennerstein, L., Green, A., Hopper, J., and Burger, H. (1994). The treatment-seeking women at menopause. *Maturitas* **18**, 161–173.

127. Reis, S. E., Gloth, S. T., Blumenthal, R. S., Resar, J. R., Zacur, H. A., Gerstenblith, G., and Brinker, J. A. (1994). Ethinyl estradiol acutely attenuates abnormal coronary vasomotor responses to acetylcholine in postmenopausal women. *Circulation* **89**, 52–60.

128. Marslew, U., Riis, B. J., and Christiana, C. (1991). Bleeding patterns during continuous combined estrogen-progestogen therapy. *Am. J. Obstet. Gynecol.* **164**, 1163–1168.

129. Sherwin, B. B. (1997). The impact of different doses of estrogen and progestin on mood and sexual behavior in postmenopausal women. *J. Clin. Endocrinol. Metab.* **72**, 336–343.

130. Hahn, R. (1989). Compliance considerations with estrogen replacement: Withdrawal bleeding and other factors. *Am. J. Obstet. Gynecol.* **161**, 1854–1858.

131. Newton, K. M., LaCroixz, A. Z., Leveille, S. G., Rutter, C., Keenan, N. L., and Anderson, L. A. (1997). Women's beliefs and decision about hormone replacement therapy. *J. Women's Health* **6**, 459–465.

132. Hebert, L. E., Scherr, P. A., Beckett, L. A., Albert, M. S., Pilgrim, D. M., Chown, M. H., Funkenstein, H. H., and Evans, D. A. (1995). Age-specific incidence of Alzheimer's disease in a community population. *JAMA, J. Am. Med. Assoc.* **273**, 1354–1359.

133. Bachman, D. L., Wolf, P. A., Linn, R. T., Knoefe, J. E., Xobb, J. L., Bélanger, A. J., White, L. R., and D'Agostino, R. B. (1993). Incidence of dementia and probable Alzheimer's disease in a general population. The Framingham Study. *Neurology* **43**, 515–519.

134. Shumaker, S. A., Reboussin, B. A., Espeland, M. A., Rapp, S. R., McBee, W. L., Kailey, M., Bowen, D., Terrell, T., and Jones, B. N., for the WHIMS Investigators (1998). The women's health initiative memory study (WHIMS). A trial of the effect of estrogen therapy in preventing and slowing the progress of dementia. *Controlled Clin. Trials* (in press).

135. Henderson, V. W., Paganini-Hill, A., Emanual, C. K., Dunn, M. E., and Buckwalter, J. G. (1994). Estrogen replacement therapy in older women: Comparisons between Alzheimer's disease cases and nondemented control subjects. *Arch. Neurol. (Chicago)* **51**, 896–900.

136. Tang, M-X., Jacobs, D., Stern, Y., Marder, K., Schofield, P., Gurland, B., Andrews, H., and Mayeux, R. (1996). Effect of oestrogen during menopause on risk and age at onset of Alzheimer's disease. *Lancet* **348**, 429–432.

137. Barrett-Connor, E., and Kirtz-Silverstein, D. (1993). Estrogen replacement therapy and cognitive function in older women. *JAMA, J. Am. Med. Assoc.* **269**, 2637–2641.

138. Phillips, S. M., and Sherwin, B. B. (1992). Effects of estrogen on memory function in surgically menopausal women. *Psychoneuroendocrinology* **17**, 485–495.

139. Fillit, H. (1994). Estrogen in the pathogenesis and treatment of Alzheimer's disease in postmenopausal women. *Ann. N.Y. Acad. Sci.* **743**, 233–238.

140. Honjo, H., Ogino, Y., Tanaka, K., *et al.* (1993). An effect of conjugated estogen to cognitive impairment in women with senile dementia–Alzheimer's type: A placebo-controlled Double-blind Study. *J. Jpn. Menopause Soc.* **1**, 167–171.

141. Ohkura, T., Isse, K., Akazawa, K., Hammamoto, M., Yaoi, Y., and Hagino, N. (1994). Low-dose estrogen replacement therapy for Alzheimer disease in women. *Menopause* **1**, 125–130.

142. Jaffe, A. B., Toran-Allerand, C. D., Greengard, P., and Gandy, S. E. (1994). Estrogen regulates metabolism of Alzheimer amyloid B precursor protein. *J. Biol. Chem.* **269**, 13065–13068.

143. Josefsson, E., Tarkowski, A., and Carlstein, H. (1992). Anti-inflammatory properties of estogen in vivo suppression of leukocyte production in bone marrow and redistribution of peropheral blood neutrophils. *Cell. Immunol.* **142**, 67–78.

144. Singh, M., Meyer, E. M., Millard, W. J., and Simpkins, J. W. (1994). Ovarian steroid deprivation results in a reversible learning impairment and compromised cholinergic function in female Sprague-Dawley rats. *Brain Res.* **644**, 305–312.

145. Gibbs, R. B., Wu, D., Hersh, L. B., and Pfaff, D. W. (1994). Effects of estrogen replacement on the relative levels of choline acetyltransferase, trkA, and nerve growth factor messenger RNAs in the basal forebrain and hippocampal formation of adult rats. *Exp. Neurol.* **129**, 70–80.

146. Toran-Allerand, C. D. (1996). Mechanisms of estrogen action during neural development: Mediation by interaction with the neurotrophins and their receptors. *J. Steroid Biochem. Mol. Biol.* **56**, 169–178.

147. Funk, J. L., Mortel, K. F., and Meyer, J. S. (1991). Effects of estrogen replacement therapy of cerebral perfusion and cognition among postmenopausal women. *Dementia* **2**, 268–272.

148. Bush, T. L., Cowan, L. D., Barrett-Connor, E., Criqui, M. H., Karon, J. M., Wallace, R. B., Tyroler, H. A., and Rifkind, B. M. (1983). Estrogen use and all-cause mortality. *JAMA, J. Am. Med. Assoc.* **249**, 903–906.

149. Petitti, D. B., Perlman, J. A., and Sidney, S. (1987). Noncontraceptive estrogens and mortality: Long-term follow-up of women in the Walnut Creek Study. *Obstet. Gynecol.* **70**, 289–293.

150. Hunt, K., Vessey, M., McPherson, K., and Coleman, M. (1990). Mortality in a cohort of long-term users of hormone replacement therapy: An updated analysis. *Br. J. Obstet. Gynaecol.* **97**, 1080–1086.

151. Grodstein, F., Stampfer, M. J., Colditz, G. A., Willett, W. C., Manson, J. E., Joffe, M., Rosner, B., Fuchs, C., Hankinson, S. E., Hunter, D. J., Hennekens, C. H., and Speizer, F. E. (1997). Postmenopausal hormone therapy and mortality. *N. Engl. J. Med.* **336**, 1769–1775.

152. Schairer, C., Adami, H. O., Hoover, R., and Perrson I. (1997). Cause-specific mortality in women receiving hormone replacement therapy. *Epidemiology* **8**, 59–65.

153. Barrett-Conner, E. (1998). Fortnightly review: Hormone replacement therapy. *Br. Med. J.* **317**, 457–461.

154. Dauvois, S., and Parker, M. G. (1996). Nucleocytoplasmic shuttling of estrogen receptors is blocked by "Pure Anti-Estrogens."

In "Hormonal Carcinogenesis" (J. J. Li, S. A. Li, J. A. Gustaffson, S. Nandi, S., and L. I. Sekely, eds.), pp. 79–85. Springer, New York.

155. Jordan, V. C. (1998). Antiestrogenic action of Raloxifene and Tamoxifen: Today and tomorrow. *J. Natl. Cancer Inst.* **90,** 967–971.

156. Early Breast Cancer Trialists' Collaborative Group (1992). Systemic treatment of early breast cancer by hormonal, cytotoxic, or immune therapy. 133 randomized trials involving 31,000 recurrences and 24,000 deaths among 75,000 women. *Lancet* **339,** 1–15.

157. Early Breast Cancer Trialists' Collaborative Group (1998). Tamoxifen for early breast cancer: An overview of the randomized trials. *Lancet* **351,** 1451–1467.

158. Fisher, B., and Redmond, C. (1991). New perspective on cancer of the contralateral breast: A marker for assessing Tamoxifen as a preventing agent. *J. Natl. Cancer Inst.* **83,** 1278–1280.

159. Stat Bite (1998). Breast cancer incidence in the breast cancer prevention trial. *J. Natl. Cancer Inst.* **90,** 648.

160. Powles, T., Eeles, R., Ashley, S., Easton, D., Chang, J., Dowsett, M., Tidy, A., Viggers, J., and Davey, J. (1998). Interim analyses of the incidence of breast cancer in the Royal Marsden Hospital tamoxifen randomised chemoprevention trial. *Lancet* **352,** 98–101.

161. Veronesi, U., Mainonneuve, P., Costa, A., Sacchini, V., Maltoni, C., Robertson, C., Rotmensz, N., and Boyle, P. (1998). Italian Tamoxifen Prevention Study. Prevention of breast cancer with tamoxifen: Preliminary findings from the Italian randomised trial among hysterectomized women. *Lancet* **352,** 93–97.

162. News (1998). In search of the perfect SERM: Beyond Tamoxifen and Raloxifene. *J. Natl. Cancer Inst.* **90,** 956–957.

163. Powles, T., Hardy, J., Ashley, S. E, Farrington, F. M., Cosgrove, D., Davey, J. B., Dowsett, M., McKinna, J. A., Nash, A. G., and Sinnett, H. D. (1989). A pilot trial to evaluate the acute toxicity and feasibility of tamoxifen for prevention of breast cancer. *Br. J. Cancer* **60,** 126–131.

164. Love, R. R., Mazess, R., Tormey, D. C., Barden, H. S., Newcomb, P. A., and Jordan, V. C. (1988). Bone mineral density in women with breast cancer treated with adjuvant tamoxifen for at least two years. *Breast Cancer Res. Treat.* **12,** 297–302.

165. Turken, S., Siris, E., Selden, D., Flaster, E., Hyman, G., and Lindsay, R. (1989). Effects of tamoxifen on spinal bone density in women with breast cancer. *J. Natl. Cancer Inst.* **81,** 1086–1088.

166. Creaseman, W. T. (1997). Endometrial cancer: Incidence, prognostic factors, diagnosis and treatment. *Semin. Oncol.* **24,** sl-140–sl-150.

167. MacMahon, B. (1997). Overview of studies on endometrial cancer and other types of cancer in humans: Perspectives of an epidemiologist. *Semin. Oncol.* **24,** sl-122–sl-139.

168. Rutqvist, L., Johansson, H., Signomklao, T., Jahansson, U., Fornander, T., and Wilking, N. (1995). Adjuvant tamoxifen therapy for early stage breast cancer and second primary malignancies. Stockholm Breast Cancer Study Group. *J. Natl. Cancer Inst.* **87,** 645–651.

169. Delmas, P. D., Bjarnason, N. H., Mitlak, B. H., Ravou, A. C., Shah, A. S., Huster, W. J., Draper, M., and Christiansen, C. (1997). Effects of raloxifene on bone mineral density, serum cholesterol concentrations, and uterine endometrium in postmenopausal women. *N. Engl. J. Med.* **337,** 1641–1647.

170. Draper, M. W., Flowers, D. E., Huster, W. J., Neild, J. A., Hooper, K. D., and Arnaud, C. (1996). A controlled trial of raloxifene (LY139481) HCI: Impact on bone turnover and serum lipid profile in healthy postmenopausal women. *J. Bone Miner. Res.* **11,** 835–842.

171. Gottardis, M. M., Ricchio, M. E. Satyaswaroop, P. G., and Jordan, V. C. (1990). Effect of steroidal and nonsteroidal antiestrogens on the growth of a tamoxifen stimulated human endometrial carcinoma (EnCa101) in athymic mice. *Cancer Res.* **50,** 3189–3192.

172. Jordan, V. C., MacGregor, J. I., and Tonetti, D. A. (1997). Tamoxifen: From breast cancer therapy to the design of a postmenopausal prevention maintenance therapy. *Osteoporosis Int.* **7,** S52–S57.

173. Rosenberg, L. (1993). Hormone replacement therapy: The need for reconsideration. *Am. J. Public Health* **83**(12), 1670–1673.

174. Hemminki, E., and Sihvo, S. (1993). A review of postmenopausal hormone therapy recommendations: Potential for selection bias. *Obstet. Gynecol.* **82,** 1021–1028.

175. Black, D. M., Cummings, S. R., Karpf, D. B., Cauley, J. A., Thompson, D. E., Nevitt, M. C., Bauer, D. C., Genant, H. K., Haskell, W. L., Marcus, R., Ott, S. M., Torner, J. C., Quandt, S. A., Reiss, T. F., and Ensrud, K. E. (1996). Randomised trial of effect of alendronate on risk of fracture in women with existing vertebral fractures. Fracture Intervention Trial Research Group. *Lancet* **348,** 1535–1541.

94

Osteoporosis

JANE A. CAULEY AND MICHELLE E. DANIELSON
University of Pittsburgh
Department of Epidemiology
Pittsburgh, Pennsylvania

I. Definition

Osteoporosis is a systemic skeletal disease characterized by low bone mass and microarchitectural deterioration of bone tissue with a consequent increase in bone fragility and fractures after minimal trauma [1]. An expert panel convened by the World Health Organization (WHO) developed a definition of osteoporosis based on bone densitometry: "osteopenia" is a bone mineral density (BMD) measurement of the hip, spine, or distal forearm of > 1.0 but ≤ 2.5 standard deviations (SD) below the young normal mean; "osteoporosis" is a BMD at one of the three sites of >2.5 SD below the young normal mean [2]. This operational definition of osteoporosis was developed to estimate the worldwide prevalence of osteoporosis. It is currently used clinically to identify women in need of treatment or prevention of osteoporosis. Of importance is the fact that the WHO definition relies solely on BMD measurements, with no consideration of the microarchitectural deterioration in bone tissue that accompanies the loss of bone mass. Nevertheless, it is believed that BMD measurements do provide some information on the microarchitectural properties of bone, especially if the problems are severe [3].

II. Public Health Impact

Osteoporosis is a major public health problem. Of the 1.3 million fractures that occur in the United States each year, 70% of all fractures among individuals age 45+ are attributable to osteoporosis [4]. Estimates of the proportion of Caucasian women who have osteoporosis, as defined by the WHO, show dramatic increases with age (Fig. 94.1) [5]. About 15% of women aged 50–59 have osteoporosis at any BMD site, while 70% of women age 80 or greater are considered osteoporotic. The three most common fractures associated with osteoporosis are hip, vertebral, and distal forearm, but data from the Study of Osteoporotic Fractures have shown that most fractures, at least among women 65 years or older are, indeed, due to osteoporosis, thus widening the public impact of osteoporosis [5].

A. Hip Fractures

In the United States, 250,000 individuals over age 65 fracture their hips each year; by 2040, over 650,000 hip fractures will occur each year [6,7]. One in six (17.5%) white women will fracture their hips in their lifetime [5]. Worldwide, the total number of hip fractures in women in 1990 was estimated at 917,000. This number is expected to double to 1,821,000 in the year 2025 and increase to 3,112,000 in the year 2050, even assuming no change in the age- and sex-specific incidence. The increase reflects the dramatic demographic shifts in the age structure of the population [8]. The impact of this marked increase in hip fractures will be greatest in Asia and in developing countries. In 1990, about one-quarter of all hip fractures occurred in Asia; by 2050, almost half of all hip fractures worldwide will occur in Asia [9].

Hip fractures have major consequences, including 18–33% mortality within the first year of the hip fracture [10,11]. Most of this excess mortality occurs in the first six months after the fracture in individuals with the greatest comorbidity [12]. However, data suggest that the excess mortality continues for as long as 5 years after the fracture even among individuals who were in relatively good health at the time of their hip fracture [13].

Among hip fracture survivors, only 21% regained their pre-fracture functioning in six instrumental activities of daily living [14]. Hip fractures are also a major cause of institutionalization, with 45% of those who were living independently in their community discharged to a nursing home after their hip fracture hospitalization and 15–25% remaining institutionalized for a year after their fracture [15]. The direct and indirect cost of osteoporosis in the United States in 1995 approached $14 billion [16].

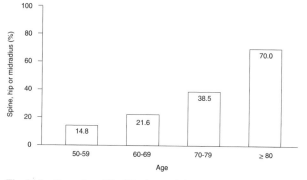

Fig. 94.1 Proportion (%) of Rochester, Minnesota, women with bone mineral measurements more than 2.5 standard deviations below the mean for young normal women. Mean is from 48 subjects under age 40 who were randomly sampled from Rochester, Minnesota population. None of them were known to have any disorder that might influence bone metabolism. Age-adjusted to the population structure of 1990 United States, white women 50 years of age or older. From [5], with permission.

B. *Vertebral Fractures*

The lifetime risk of clinical vertebral fractures approaches that of hip fractures, with a risk of 15.6% among women and 5.0% among men [5]. Only one-third of vertebral fractures reach clinical attention and estimates of vertebral fracture prevalence that rely on morphometric definitions (measurement of vertebral heights) suggest that the overall prevalence of vertebral fractures is actually much higher (about 20%) and is similar in men and women [17,18]. Traditionally, it was not believed that vertebral fractures were associated with a significant mortality. However, data from the Mayo Clinic demonstrated almost a 20% decreased survival 5 years after the diagnosis, especially among individuals over age 75 years [13].

Vertebral fractures also have a significant impact on the quality of life and functional status of older individuals. Among women aged 65–70 years, the most severe vertebral deformities were associated with moderate to severe back pain (odds ratio [OR] = 1.9; 95% confidence intervals (CI) = 1.5–2.4) and a higher risk of disability involving the back (OR = 2.6; CI = 1.7–3.9) [19]. Incident vertebral fractures, even those not recognized clinically, are associated with substantial increases in back pain and functional limitation due to back pain [20]: among women with at least one incident vertebral fracture, the OR of experiencing increased back pain was 2.4 (95% CI = 1.7–3.3); the OR for back disability was 2.6% (95% CI = 1.9–3.7).

C. *Distal Forearm*

The lifetime risk of distal forearm fractures is 2.5% in men and 16% in women [5]. Although there is no excess mortality risk associated with distal forearm fractures, about 50% of subjects reported fair or poor functional status six months after the fracture [21].

D. *Summary*

In summary, osteoporotic fractures are relatively common events. Forty percent of women and 13% of men will experience a hip, vertebral, or distal forearm fracture in their lifetime [5]. Although hip fracture incidence has stabilized in the United States, the number of fractures will increase because of the relative increase in the numbers of persons over age 65. The impact of hip fracture on mortality and morbidity is greatest among individuals with the poorest health status and is likely to increase because the fastest growing segment of our older population is individuals aged 85 and older. It is these individuals who are most likely to die or to become institutionalized or disabled from their hip fracture. Preventive efforts are urgently needed to diminish the societal impact of hip fractures.

III. Evolution of Bone

Three basic processes govern the dynamic nature of bone: (1) growth, (2) modeling, and (3) remodeling. During the skeletal life span, each plays an important role, operating both alone and simultaneously with other processes depending on the state of flux of the skeleton. Some have referred to "osteoporosis" as a pediatric disease with adult consequences. Therefore, in order to have a full understanding of osteoporosis, it is important to consider the lifetime evolution of bone.

A. *Bone Growth and Modeling*

Bone formation begins *in utero*. The skeletal outline is clearly apparent and the future shape and proportion of the long bones is probably set by 26 weeks of gestation. From conception to epiphyseal closure, there is a progressive increase in both cortical and trabecular bone. Bone formation exceeds bone resorption throughout the growth period resulting in dramatic linear growth at the epiphyseal plate. Upon cessation of linear growth, modeling continues to increase the mass and improve the architecture of the bone in response to mechanical loading, thus improving overall strength. Only slight gender differences in bone mass are apparent before puberty. The accretion of bone mineral during adolescence parallels pubertal changes, including the increase in gonadal hormone levels, and is most pronounced between Tanner stages 2 and 5 [22]. While girls achieve their peak bone mass at an earlier age than boys, the pubertal growth spurt is longer and faster in boys, resulting in a greater peak bone mass at skeletal maturity [23]. It is estimated that up to 90% of adult height is achieved [24] and nearly one-half of adult bone mass accrued [25] during puberty. It is generally accepted that peak bone mass occurs during the late teens or early twenties [26,27].

B. *Bone Remodeling: Combined Processes of Formation and Resorption*

Bone remodeling is the dominant process in adulthood whereby "old" bone is replaced by "new" bone in response to fatigue damage. It is estimated that approximately 25% of trabecular bone and 3% of cortical bone are replaced annually in adults [28]. This process does not occur randomly but rather at sites of previous resorption in a programmed series of events known as the activation–resorption–formation (ARF) sequence. In the ARF sequence, small units of bone, called basic multicellular units (BMUs), are removed and replaced in a cyclic fashion over multiple bone surfaces such that a coupling of formation and resorption is maintained. Osteoblasts are the bone forming cells that later become osteocytes during the process of calcification. Osteoclasts are the bone resorbing cells. In simple terms, the ARF sequence proceeds as follows: (1) the *activation* phase is characterized by mobilization of osteoclasts that attach to the bone surface and form a sealed compartment; (2) during the *resorption* phase, lysosomal enzymes synthesized by the osteoclasts are secreted via the ruffled border into the compartment, dissolving the bone mineral and leaving an area of excavation. At the end of this phase, there is a brief reversal period during which formation becomes coupled with resorption, and (3) finally, during the *formation* phase osteoblasts form osteoid, which is subsequently mineralized. Remodeling occurs primarily at the endosteal surface and one cycle takes approximately 3–6 months to complete [28].

Bone loss occurs when there is an uncoupling of resorption and formation. The timing, extent, and exact nature of bone loss continues to be an area of intense research. Bone loss may begin as early as the third decade of life and is universal, occurring in both genders and across racial and ethnic groups. In women, a biphasic pattern of loss occurs: an early, slower phase that is age-related superimposed by one that accompanies menopause and is significantly more rapid. Age-related bone loss is probably due to decreased bone formation, while menopause-related bone loss is attributed to increased bone resorption. Because of its high metabolic activity, trabecular bone loss begins prior to menopause in the late perimenopausal period. Loss of both cortical and trabecular bone accelerates during menopause and continues over the next 5–10 years. Following a short period of slow bone loss, bone loss appears to accelerate after the age of 75. As a result, in her lifetime, a woman will experience a global bone loss of up to 50% relative to her peak mass. Because postmenopausal bone loss varies widely between women and from site to site, it has been proposed by some [29] but not all researchers [30] that 25–30% of women could be classified as "fast bone losers" who may be more vulnerable to osteoporotic fracture later in life.

A number of local and systemic factors regulate bone modeling and remodeling. A detailed description of these factors is outside the scope of this review [31]. Local factors that affect skeletal growth include insulin-like growth factor (IGF), transforming growth factor-β (TGF-β) including bone morphogenetic proteins (BMPs), fibroblast growth factors (FGF), platelet-derived growth factors (PDGF), and select cytokines (*e.g.,* interleukin, tumor necrosis factor, and colony-stimulating factor). Systemic factors that regulate bone remodeling through their influence on either formation or resorption include a number of different polypeptide (parathyroid, calcitonin, insulin, growth), steroid (1,25-dihydroxyvitamin D_3, glucocorticoid, sex), and thyroid hormones.

IV. Genetic Aspects of Bone Mass and Osteoporosis

Family and twin studies provide consistently strong evidence for a significant genetic component to BMD and perhaps osteoporotic risk, with as much as 80% of the variability in BMD attributed to heritable factors [32]. Other indicators of bone status, such as biochemical markers of bone turnover, calcium absorption, and quantitative ultrasound parameters, also appear to be heritable, although these findings are just beginning to be reported in the literature. Heritability can be broadly defined as the ratio of the genetic variance, or that which is attributable to genotypic or allelic differences among individuals, to the total phenotypic variance in the population. Several methods employed to examine the heritability of and genetic contribution to bone mass are discussed below.

A. Family Studies

Analyses of parent–offspring and sibling data have demonstrated significant familial correlations in bone mass, expressed as both bone mineral content (BMC) and BMD (r = 0.20–0.70) [33–37]. Heritability estimates have ranged from 0.34 to 0.72 [36–38]. Women with a family history of osteoporosis, defined by either low bone mass or fracture, have significantly lower bone mass (as much as 1 SD difference) compared to women without such a history [38,39]. However, it is important to remember that other factors related to BMD and risk for osteo-

porosis, such as body composition [40], alcohol intake [41], and physical activity [42], also show a familial tendency and may contribute to the heritability seen in BMD. Consequently, bone mass and osteoporotic risk are most likely a function of gene–environment interaction, but more research is needed to characterize the exact nature of such interaction.

B. Twin Studies

The classic twin model is a powerful tool for examining the genetic contribution to quantitative physical traits because it controls for the potential confounding of differences in age, generation, or lifestyle/environmental factors that can occur in family studies. Comparison of the intraclass correlation or within-pair differences in monozygotic (MZ) and dizygotic (DZ) twins can be used to calculate heritability which in turn gives a reasonable estimate of the genetic control of a trait. Consistent with a genetic contribution to bone mass, MZ correlations are generally greater than DZ correlations and heritability estimates for morphometric measures, BMC, and BMD are reported to be in the range of 0.30–0.90 [43–45]. Other indicators of bone status, including biochemical markers of bone turnover [46], rates of bone loss [47], microarchitectural properties as measured by quantitative ultrasound [48], and geometric parameters (*e.g.,* hip axis length) [48] have also been found to have a significant genetic component in twin studies.

C. Genetic Markers

With the emergence of more sophisticated methods of genetic mapping, several gene polymorphisms have been identified that demonstrate an association with BMD and possibly fracture risk in both men and women. (Table 94.1) [49]. Because effects on BMD cannot be used to predict fracture results, more studies of fracture outcomes are needed.

Table 94.1
Summary of Potential Genetic Polymorphisms That Contribute to Osteoporosis

Gene name	Gene symbol
α$_2$HS-glycoprotein	AHSG
Estrogen receptor	ESR
Interleukin 6	IL6
Collagen type Iα2	COL1A2
Vitamin D receptor	VDR
Collage type Iα1	COL1A1
Transforming growth factor β1	TGFB1
Apolipoprotein E	APOE
Calcitonin receptor	CTR
Interleukin 1 receptor antagonist	IL1RN
Insulin-like growth factor-I	IGF-I
Human leucocyte antigen	HLA-A, HLA-B (class I); HLA-DR (class II)
Osteocalcin	OC

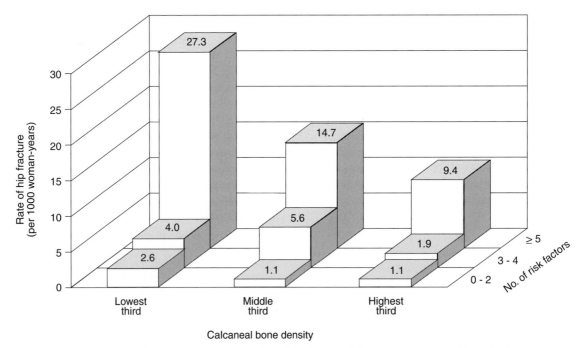

Fig. 94.2 Annual risk of hip fracture according to the number of risk factors and the age-specific calcaneal bone density. (From [54]. Copyright © 1995, Massachusetts Medical Society. All rights reserved)

D. Summary

In summary, unequivocal evidence exists for a significant genetic effect on both appendicular and axial bone mass. Several gene polymorphisms have been identified that demonstrate an association with bone mass and research continues to uncover new genetic links at a rapid pace. Continued research is needed to expand our knowledge about the complex relationship between our genes and the environment and how they interact and affect skeletal health throughout the life span.

V. Risk Factors for Fracture

Most fractures in older women are associated with low bone mass and, hence, may be considered osteoporotic. Indeed, the only fractures that appear to be unrelated to osteoporosis are fractures of the toes, fingers, and face. However, despite this link with low bone mass, risk factors for different fractures are heterogeneous.

A. Hip Fracture

Established risk factors for hip fracture have been summarized in several reviews [50,51]. Hip fractures increase exponentially with age. The incidence of hip fractures in white women (per 1000 person-years) is 2.2 (ages 65–69), 4.4 (ages 70–74), 9.5 (ages 75–79), 16.9 (ages 80–84), 27.9 (ages 85–90), and 34.2 (age 90+) [52]. The incidence of hip fracture is higher for white women than for African-American, Asian, and Hispanic women. In a California hospital discharge study of hip fracture, the age-adjusted incidence (per 1000 person-years) of hip fracture was 6.2 among white women, 3.8 among Asian

women, 2.4 among African-American women, and 2.2 among Hispanic women (53).

In addition to older age, white race, and low BMD, risk factors for hip fracture include falling on the hip, neuromuscular impairment, low bone quality as assessed by ultrasound attenuation, history of falls, never use of hormone replacement therapy (HRT), low muscle strength, family history of fracture, low body weight, self-reported poor health status, weight loss, low physical activity, and Parkinson's disease. Weaker and more inconsistent associations have been found with increased caffeine intake, low calcium intake, fracture history, greater height, longer hip axis length, psychotropic drug use, smoking, thiazide diuretic use, and visual impairment. There is even greater uncertainty about whether alcohol, protein intake, antioxidant vitamins, home hazards, water fluoride, and markers of bone turnover are risk factors for hip fracture [51].

Of importance, however, women with multiple risk factors and low bone density have an especially elevated risk of hip fracture (Fig. 94.2) (54). Information on many established risk factors for hip fracture can be easily ascertained within a clinical setting and several are modifiable. Assessment of the total number of risk factors may lead to identification of high risk women who may be in need of aggressive prophylaxis.

B. Wrist and Proximal Humerus Fractures

Wrist fractures tend to occur in relatively younger postmenopausal women who are healthier and more physically active than women who fracture their hip [55,56]. Independent predictors of wrist fracture were decreased BMD at the distal radius (relative risk [RR] = 1.8; 95% CI = 1.6–2.1), a history of recurrent falls (RR = 1.6; 95% CI = 1.2–2.0), and having had a previous fracture since age 50 years (RR = 1.3, 95% CI = 1.1–1.6). Current use of HRT was protective (RR = 0.6; 95% CI = 0.4–0.8). For women over age 75, poor cognitive function was also associated with an increased risk [56].

Proximal humerus fractures, on the other hand, tend to occur in older women who have poorer neuromuscular function [55]. Factors associated with an increased risk of proximal humerus fractures include a recent decline in health status, insulin dependent diabetes mellitus, infrequent walking, and poor neuromuscular function [55].

C. Vertebral Fractures

Only one-third of vertebral fractures come to clinical attention. Studies that rely on morphometric definitions of vertebral fractures (based on measurement of vertebral heights) often refer to vertebral fractures as vertebral deformities.

Women with vertebral fractures are at an increased risk of other osteoporotic fractures including hip fractures [57]. Most studies of risk factors for vertebral fractures have focused on comparisons of women with and without vertebral fractures.

Several reports from the European Vertebral Osteoporosis Study have identified risk factors for prevalent vertebral fractures including positive family history [58], reduced physical activity [59] and several reproductive factors including late menarche and early menopause. Use of oral contraceptives was associated with a 25% reduction in the risk of vertebral deformity [60]. Women who reported drinking alcohol on more than five days per week had a reduced risk of vertebral deformity [61]. Finally, there was a significant trend of decreasing risk with increasing degree of obesity [62]. The Dubbo Epidemiology Study reported that cigarette smoking increased this risk of vertebral deformity [63].

There are few studies of risk factors for incident vertebral fractures. Low bone mass and the presence of two or more prevalent vertebral fractures increased the risk of incident vertebral fractures 75-fold, relative to women with the highest BMD and no fractures [64]. In the Study of Osteoporotic Fractures, women with at least one new incident vertebral fracture had a lower body weight, lower body mass index (BMI), greater decreases in grip strength, greater height loss, and were less likely to report current estrogen use [20].

D. Ankle and Foot Fractures

The relationship between BMD and ankle and foot fractures is somewhat weaker than that for BMD and other types of fractures [6]. Risk factors for ankle fractures include history of falling, vigorous physical activity, more weight gain since age 25, self-reported osteoarthritis, sister's history of hip fracture, and out of the house ≤ 1 week. Factors associated with foot fractures include insulin dependent diabetes, use of seizure medications, or benzodiazepines, hyperthyroidism, and poor depth perception.

VI. Risk Factors and BMD

Many risk factors for fracture are believed to influence fracture risk because of an underlying association with BMD. For example, obese individuals have greater BMD and a lower risk of fracture. If a risk factor decreases the risk of fracture by preserving bone mass, then adjustment for bone mass should substantially decrease the association between the factor and the risk of fracture. However, we have shown that inclusion of BMD in multivariate models has only modest effects on the relative risk estimates [54]. For example, the multivariate adjusted relative risk of wrist fracture among current estrogen users was 0.39 (95% CI = 0.24–0.64). Adjustment for distal radius BMD had only a modest effect on the risk associated with estrogen use (RR = 0.50; 95% CI = 0.31–0.83). Estrogen may therefore prevent fractures in other ways besides improvement in bone mass [65]. Similarly, the relative risk of vertebral fracture in the Dubbo Study associated with cigarette smoking was 3.14 (95% CI = 1.27–7.7), which was independent of BMD [63]. The effect of smoking on fracture may also be mediated by changes in bone strength that are not measured by traditional densitometry methods.

VII. Prevention of Osteoporosis

Early intervention is a key to averting the development of osteoporosis and a number of risk factors for fracture and osteoporosis are amenable to change. There are two primary approaches to prevention of osteoporosis: maximization of peak bone mass and prevention of fractures.

The risk for osteoporosis is determined both by the amount of peak bone mass attained at skeletal maturity and by the amount and rate of bone loss that occurs during the menopause in women and with aging in general. Optimizing the genetically programmed peak skeletal mass is considered to be a key factor in the prevention of osteoporosis later in life.

Nutrition plays a significant role in the attainment of peak skeletal mass, in particular calcium intake. In the U.S. alone, adolescent intake of calcium has been on the decline, with the NHANES III data indicating that as many as 93% of girls and 53% of boys ages 9–13 are consuming < 77% of the recommended calcium intake of 1300 mg/day [66]. Carefully controlled calcium balance studies and calcium supplementation trials have clearly demonstrated the importance of adequate calcium intake for full skeletal maturity [67]. Young girls may be particularly vulnerable to calcium inadequacy due to their increased dieting behavior during adolescence and the common misconception they have that milk products are necessarily high in fat, thus resulting in decreased consumption of calcium-dense foods. Other nutrients of concern among youth include high sodium and caffeine intakes, which may alter calcium homeostasis via higher urinary excretion of calcium [68].

Physical activity is another lifestyle factor that may play a significant role with regard to the attainment of peak bone mass, but less is known about the type, intensity, and duration most likely to benefit the growing skeleton. The mechanical loading and altered strain patterns that accompany weight-bearing activity are thought to be potent osteogenic stimulators in the growing skeleton, particularly when they occur prior to and around the time of puberty [69]. In general, children who have higher levels of habitual activity tend to have higher bone mass than children who are inactive [70]. More importantly, the relationship between activity and bone mass is further mediated by the loading properties of the activity engaged in. Physical activity that involves weight-bearing and variable loading of the bone are associated with greater bone mass as compared to non-weight-bearing activities and those that involve primarily active loading (muscular contraction, *e.g.,* resistance training) [71].

Excessive weight loss via extreme dieting behavior or excessive exercise or clinically diagnosed disordered eating (*e.g.,* anorexia nervosa) can have a deleterious effect on the growing skeleton [72]. Disturbances in the menstrual cycle, such as oligomenorrhea and amenorrhea, which often accompany excessive weight loss, intense exercise training, or disordered eating, can lead to low bone mass which may or may not be reversible with proper nutrition, weight gain, and return of menses [73,74].

Fractures may be prevented through both pharmacologic and lifestyle interventions aimed at preserving bone mass and bone strength. Many risk factors for fracture are modifiable, such as increasing physical activity and improving muscle strength. Most fractures occur because of a fall, and hence, prevention of falls and attention to risk factors for falls such as use of long-acting benzodidzepines are key to the prevention of fractures.

A national panel convened by the National Osteoporosis Foundation made the following public health recommendations for individuals of all ages: ensure adequate calcium intake, between 1000 mg and 1500 mg per day; ensure adequate vitamin D, 400 IU to 800 IU daily; encourage exercising and stopping smoking [75]. In these guidelines, the panel identified five risk factors for fractures that could identify women who are at high risk of fracture. These include: low BMD; low body weight (< 127 pounds), cigarette smoking; history of fracture of the hip, wrist, or vertebrae in a first-degree relative; and history of prior fracture after age 40. Women with these risk factors could be targeted for BMD testing and, if warranted, pharmacologic interventions.

VIII. Osteoporosis and Other Diseases

BMD may be a useful marker of a woman's cumulative exposure to estrogen. It is this cumulative or lifetime exposure to estrogen that could determine a woman's risk of developing several diseases, all of which may be related to estrogens.

As part of the Study of Osteoporotic Fractures, we have examined the relationship between BMD and a number of important conditions that are common in older women (Table 94.2). Low BMD is associated with an increased risk of experiencing a hip [76], wrist [56], or any clinical fracture [77]. The association between BMD and fracture is consistent across all BMD sites [76].

Low BMD was also associated with aortic calcification, although the magnitude of the association was weak and confined to the radial BMD sites [78]. Decreased vascular flow to the lower extremities as measured by the Ankle/Arm Index (the ratio of the posterior tibial and brachial systolic blood pressures) was associated with increased rate of bone loss at the hip and calcaneus [79]. This suggests that there may be some link between certain manifestations of cardiovascular disease and osteoporosis.

Low BMD is also associated with an increased risk of nontrauma mortality, especially deaths due to stroke [80]. Results of one study of BMD and incident stroke are consistent with the

Table 94.2

Relationship between Bone Mineral Density and Selected Outcomes in Women

	RR	95% CI
Hip fracture [78]	1.5	(1.2–1.9)
Wrist fractures [57]	1.9[a]	(1.7–2.2)
All fractures [79]	1.32[a]	(1.14–1.53)
Aortic calcification [80]	1.20[a]	(1.0–1.3)
Nontrauma mortality [82]	1.22[a]	(1.08–1.37)
Coronary heart disease mortality [82]	1.17[a]	(0.92–1.51)
Stroke mortality [82]	1.75[a]	(1.15–2.65)
Stroke incidence [83]	1.26[a]	(1.00–1.58)
Hip osteoarthritis [84]	2.7[b]	(0.7–4.6)[b]
Breast cancer [85]	1.34[c]	(1.09–1.62)

[a]RR for one standard deviation *decrease* in BMD.
[b]Percent increase in BMD.
[c]RR for one standard deviation *increase* in BMD.

mortality data [81]. This association is probably not causal; BMD may be a marker of comorbidity and general health status. The association with stroke mortality is intriguing, however, because it is stronger than the association between systolic blood pressure and stroke. Several possible mechanisms operating through estrogen, homocysteine, and vitamin D may account for this observation and are currently being explored. Radiographic hip osteoarthritis (OA) was associated with higher BMD, consistent with the role of elevated BMD in the pathogenesis of hip OA [82].

Finally, one standard deviation increase in BMD was associated with 30% increase in the risk of breast cancer, consistent with the hypothesis that endogenous estrogens are major etiologic factors in breast cancer in older women [83].

We tested the hypothesis in a case-cohort study, that the underlying mechanism for this association between breast cancer and high BMD reflected elevated endogenous estrogens. We compared sex hormones in 97 cases of incident breast cancer and 244 women who were a random subset of the cohort. Increased endogenous estrogens and androgens were associated with an increased risk of breast cancer (Table 94.3) [84]. Of importance, however, inclusion of either non-SHBG bound estradiol or free testosterone in the models did not attenuate the relationship between BMD and breast cancer. In fact, the association between BMD and breast cancer was strongest in the multivariate model including both hormones as well as adjustments for many risk factors for breast cancer.

Across each tertile of non-SHBG bound estradiol and free testosterone, the relative hazard of breast cancer increased with increasing BMD. (Figs. 94.3 and 94.4). Women with the highest BMD were at a higher risk of breast cancer compared to women with the lowest BMD. Similarly, within each tertile of BMD, the risk of breast cancer increased with increasing levels of sex steroids. Of note, the relative risk of breast cancer was greatest for women with the highest BMD and the highest non-SHBG bound estradiol (RR = 3.6; 95% CI = 1.4–9.3 (Fig. 94.3)) or the highest free testosterone (RR = 3.4; 95% CI = 1.0–9.1 (Fig. 94.4). There was, however, no significant interaction between hormone levels, BMD, and risk of breast cancer.

Table 94.3

Bone Mineral Density, Non-SHBG Bound Estradiol, Free Testosterone, and the Risk of Breast Cancer[a]

Variable (unit)	Hazard	Relative (95% CI)	p
Model 1			
Distal radius BMD (0.08 g/cm²)[b]	1.20	(0.94–1.53)	0.15
Model 2			
Non-SHBG bound E2[c]	3.70	(1.72–7.69)	0.0007
Model 3			
Free testosterone[c]	3.03	(1.49–6.25)	0.0023
Model 4			
Distal radius BMD/(0.08 g/cm²)[b]	1.21	(0.94–1.29)	0.14
Non-SHBG bound E2 (Q2-4)[c]	3.57	(1.67–7.69)	0.001
Model 5			
Distal radius BMD/(0.08 g/cm²)[b]	1.24	(0.96–1.60)	0.10
Free testosterone (Q2-4)[c]	3.13	(1.54–6.25)	0.002
Model 6			
Distal radius BMD/(0.08 g/cm²)[b]	1.25	(0.96–1.63)	0.10
Non-SHBG bound E2 (Q2, 3, or 4)[c]	2.70	(1.23–5.88)	0.013
Free testosterone (Q2, 3, or 4)[c]	2.50	(1.18–5.26)	0.018
Model 7 (multivariate)[d]			
Distal radius BMD/(0.08 g/cm²)[b]	1.32	(0.98–1.78)	0.07
Non-SHBG bound E2 (Q2-4)[c]	2.50	(1.09–5.89)	0.03
Free testosterone (Q2-4)[c]	2.86	(1.23–6.67)	0.01

[a]All models adjusted for age and body weight.

[b]One standard deviation increase in BMD equals 0.08 g/cm².

[c]Quartile 2, 3, or 4 versus quartile 1.

[d]Multivariate model. Also adjusted for current smoking, clinic, walk for exercise, alcohol intake (drinks per week), family history of breast cancer, age at menarche, age at menopause.

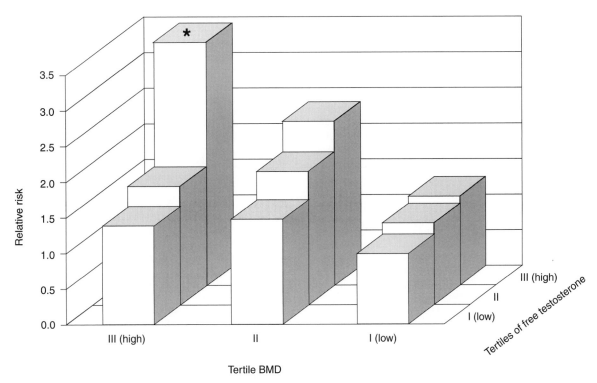

Fig. 94.3 Relative risk (age-adjusted) of breast cancer according to distal radius bone mineral density (tertile of BMD) and non-SHBG bound estradiol (tertile). *, $p < 0.05$.

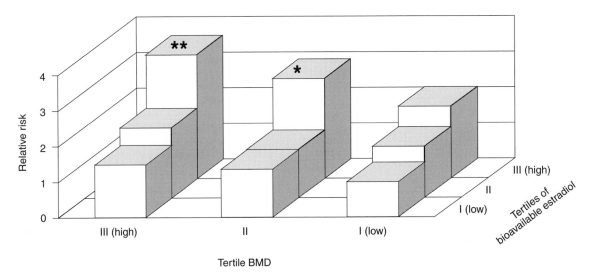

Fig. 94.4 Relative risk (age-adjusted) of breast cancer according to distal radius bone mineral density (tertile of BMD) and free testosterone (tertile). *, $p < 0.05$, **$p < 0.01$

These analyses suggest that there may be other mechanisms for an association between BMD and breast cancer and illustrate how improving our understanding of the association between BMD and other diseases in women could lead to new and important hypotheses about their etiology.

Further understanding of whether these associations reflect common pathways in the aging process, shared risk factors, endogenous estrogen levels, or other potential mechanisms could substantially expand our understanding of the etiology of all of these conditions.

IX. Future Directions

Epidemiologic studies have made a number of important advances over the course of the 1990s. Important risk factors for hip fracture have been delineated, although there still remain several important questions about the role of certain factors (*e.g.,* antioxidant use). Nevertheless, the main challenge in the future will be development of effective prevention strategies that could focus on BMD, fractures, and falls. It is likely that there may be genetic–environment interaction on risk of fracture. Identification of which genes place individuals at high risk and how these genetic factors interact with lifestyle/environmental factors will be essential.

References

1. Anonymous. (1993). Consensus development conference: diagnosis, prophylaxis, and treatment of osteoporosis. *Am. J. Med.* **94**(6), 646–650.

2. Kanis, J. A., Melton, J., Christiansen, C., Johnston, C. C., and Khaltaev, N. (1994). The diagnosis of osteoporosis. *J. Bone Miner. Res.* **9**, 1137–1141.

3. Marcus, R. (1996). The nature of osteoporosis. *In* "Osteoporosis" (R. Marcus, D. Feldman, and J. Kelsey, eds.), pp. 647–659. Academic Press, San Diego, CA.

4. Iskrant, A. P., and Smith, R. W. (1969). Osteoporosis in women 45 years and over related to subsequent fractures. *Public Health Rep.* **84**, 33–38.

5. Melton, L. J., Chrischilles, E. A., Cooper, C., Lane, A. W., and Riggs, B. L. (1992). How many women have osteoporosis? *J. Bone Miner. Res.* **7**, 1005–1010.

6. Seeley, D. G., Browner, W. S., Nevitt, M. C., Genant, H. K., Scott, J. C., and Cummings, S. R. (1991). Which fractures are associated with low appendicular bone mass in elderly women? The Study of Osteoporotic Fractures Research Group. *Ann. Intern. Med.* **155**(11), 837–842.

7. National Center for Health Statistics (1985). "Utilization of Short Stay Hospitals." U.S. Govt. Printing Office, Washington, DC.

8. Schneider, E. L., and Guralnik, J. M. (1990). The aging of America—Impact of health care costs. *JAMA, J. Am. Med. Assoc.* **263**, 2335–2340.

9. Gullberg, B., Johnell, O., and Kanis, J. A. (1997). World-wide projections for hip fracture. *Osteoporosis Int.* **7**, 407–413.

10. Magaziner, J., Simonsick, E. M., Kashner, M., Hebel, J. R., and Kenzora, J. E. (1989). Survival experience of aged hip fracture patients. *Am. J. Public Health* **79**, 274–278.

11. Jacobsen, S. J., Goldberg, J., Milist, P., Brody, J. A., Stiers, W., and Rimm, A. A. (1992). Race and sex differences in mortality following fracture of the hip. *Am. J. Public Health* **82**, 1147–1150.

12. Marottoli, R. A., Berkman, L. F., Leo-Summers, L., and Cooney, L. M. (1994). Predictors of mortality and institutionalization after hip fracture: The New Haven EPESE Cohort. *Am. J. Public Health* **84**, 1807–1812.

13. Cooper, C., Alkinson, E. J., Jacobsen, S. J., O'Fallon, W. M., and Melton, L. J. (1993). Population based study of survival following osteoporotic fractures. *Am. J. Epidemiol.* **137**, 1001–1005.

14. Jette, A. M., Harris, B. A., Cleary, P. D., and Campion, E. W. (1987). Functional recovery after hip fracture. *Arch. Phys. Med. Rehabil.* **68**, 735–740.

15. Magaziner, J., Simonsick, E. M., Kashner, T. M., Hebel, J. R., and Kenzora, J. E. (1990). Predictors of functional recovery one year following hospital discharge for hip fracture: A Prospective Study. *J. Gerontol. Med. Sci.* **45**, M101–M107.

16. Ray, N. F., Chan, J. K., Thaner, M. and Melton, L. J. (1997). Medical expenditures for the treatment of osteoporotic fractures in the United States in 1995: Report from the National Osteoporosis Foundation. *J. Bone Miner. Res.* **12**, 24–35.

17. O'Neill, T. W., Felsenberg, D., Varlow, J., Cooper, C., Kanis, J. A., and Silman, A. J. (1997). The prevalence of vertebral deformity in

European men and women: The European Vertebral Osteoporosis Study. *J. Bone Miner. Res.* **11**, 1010–1018.

18. Davies, K. M., Stegman, M. R., Heaney, R. P., and Recker, R. R. (1996). Prevalence and severity of vertebral fracture: The Saunders County Bone Quality Study. *Osteoporosis Int.* **6**, 160–165.

19. Ettinger, B., Black, D. M., Nevitt, M. C., Rundle, A. C., Cauley, J. A., Cummings, S. R., and Genant, H. K. (1992). Contribution of vertebral deformities to chronic back pain. *J. Bone Miner. Res.* **7**, 449–456.

20. Nevitt, M. C., Ettinger, B., Black, D., Stone, K., Jamal, S., Ensrud, K., Segal, M., Genant, H. K., and Cummings, S. R. (1998). The association of radiographically detected vertebral fractures with back pain and function: A Prospective Study. *Ann. Intern. Med.* **128**(10), 793–800.

21. Kaukonen, J. P., Karaharju, E. O., Porras, M., Luthje, P., and Jakobsson, A. (1988). Functional recovery after fractures of the distal forearm. *Ann. Chir. Gynaecol.* **77**, 27–31.

22. Ilich, J. Z., Badenhop, N. E., and Matkovic, V. (1996). Primary prevention of osteoporosis: Pediatric approach to disease of the elderly. *Women's Health Issues* **6**, 194–203.

23. Bonjour, J. P., and Rizzoli, R. (1996). Bone acquisition in adolescence. *In* "Osteoporosis" (R. Marcus, D. Feldman, and J. Kelsey, eds.), pp. 465–476. Academic Press, San Diego, CA.

24. Matkovic, V., Fontana, D., Tominac, C., Goel, P., and Chestnut, C. H. (1990). Factors that influence peak bone mass formation: A study of calcium balance and the inheritance of bone mass in adolescent females. *Am. J. Clin. Nutr.* **52**, 878–888.

25. Kreipe, R. E. (1995). Bone mineral density in adolescents. *Pediatr. Ann.* **24**, 308–315.

26. Teegarden, D., Proulx, W. R., Martin, B. R., Zhao, J. McCabe, G. P., Lyle, R. M., Peacock, M., Slemenda, C., Johnston, C. C., and Weaver, C. M. (1995). Peak bone mass in young women. *J. Bone Miner. Res.* **10**, 711–715.

27. Matkovic, V., Jelic, V., Wardlaw, G. M., Ilich, J. Z., Goel, P. K., Wright, J. K., Andon, M. B., Smith, K. T., and Heaney, R. P. (1994). Timing of peak bone mass in caucasian females and its implication for the prevention osteoporosis. *J. Clin. Invest.* **93**, 799–808.

28. Dempster, D. W. (1995). Bone remodeling. *In* "Osteoporosis: Etiology, Diagnosis, and Management" (B. L. Riggs and L. J. Melton, eds.), pp. 67–91. Lippincott-Raven, Philadelphia.

29. Hansen, M. A., Overgaard, K., Riis, B. J., and Christiansen, C. (1991). Role of peak bone mass and bone loss in postmenopausal osteoporosis: 12 Year Study. *Br. Med. J.* **303**, 961–964.

30. Nordin, B. E. C., Need, A. G., Prince, R. L., Horowitz, M., Gutteridge, D. H., and Papapoulos, S. E. (1993). Osteoporosis. *In* "Metabolic Bone and Stone Disease" (B. E. C. Nordin, A. G. Need, and H. A. Morris, eds.), pp. 1–82. Churchill-Livingstone, New York.

31. Canalis, E. (1996). Regulation of bone remodeling. *In* "Primer on the Metabolic Bone Diseases and Disorders of Mineral Metabolism" (M. J. Favus, ed.) 3rd ed., pp. 29–34. Lippincott-Raven, Philadelphia.

32. Pocock, N. A., Eisman, J. A., Hopper, J., Yeates, M., Sambrook, P. N., and Eberl, S. (1987). Genetic determinants of bone mass in adults. *J. Clin. Invest.* **80**, 706–710.

33. Hansen, M. A., Hassager, C., Jensen, S. B., and Christiansen, C. (1992). Is heritability a risk factor for postmenopausal osteoporosis? *J. Bone Miner. Res.* **7**, 1037–1043.

34. Lutz, J., and Tesar, R. (1990). Mother-daughter pairs: Spinal and femoral bone densities and dietary intakes. *Am. J. Clin. Nutr.* **52**, 872–877.

35. Matkovic, V., Fontana, D., Tominac, C., Goel, P., and Chestnut, C. H., III (1990). Factors that influence peak bone mass formation: A study of calcium balance and the inheritance of bone mass in adolescent females. *Am. J. Clin. Nutr.* **52**, 878–888.

36. Krall, E. A., and Dawson-Hughes, B. (1993). Heritable and life-style determinants of bone mineral density. *J. Bone Miner. Res.* **8**, 1–9.

37. Lutz, J. (1986). Bone mineral, serum calcium, and dietary intakes of mother/daughter pairs. *Am. J. Clin. Nutr.* **44**, 99–106.

38. Dorman, J. S., Towers, J. D., and Kuller, L. H. (1999). Familial resemblance of bone mineral density (BMD) and calcaneal ultrasound attenuation: the BMD in mothers and daughters study. *J. Bone Miner. Res.* **14**, 102–110.

39. Soroko, S. B., Barrett-Connor, E., Edelstein, S. L., and Kritz-Silverstein, D. (1994). Family history of osteoporosis and bone mineral density at the axial skeleton: The Rancho Bernardo Study. *J. Bone Miner. Res.* **9**, 761–769.

40. Nguyen, T. V., Howard, G. M., Kelly, P. J., and Eisman, J. A. (1998). Bone mass, lean mass, and fat mass: Same genes or same environments? *Am. J. Epidemiol.* **147**, 3–16.

41. Gottlieb, N. H., and Baker, J. A. (1986). The relative influence of health beliefs, parental and peer behaviors and exercise program participation on smoking, alcohol use and physical activity. *Soc. Sci. Med.* **22**, 915–927.

42. Freedson, P. S., and Evenson, S. (1991). Familial aggregation in physical activity. *Res. Q. Exercise Sport* **62**, 384–389.

43. Slemenda, C. W., Christian, J. C., Williams, C. J., Norton, J. A., and Johnston, C. C. (1991). Genetic determinants of bone mass in adult women: A reevaluation of the twin model and the potential importance of gene interaction on heritability estimates. *J. Bone Miner. Res.* **6**, 561–567.

44. Flicker, L., Hopper, J. L., Rodgers, L., Kaymakci, B., Green, R. M., and Wark, J. D. (1995). Bone density determinants in elderly women: A Twin Study. *J. Bone Miner. Res.* **10**, 1607–1613.

45. Arden, N. K., and Spector, T. D. (1997). Genetic influences on muscle strength, lean body mass, and bone mineral density: A Twin Study. *J. Bone Miner. Res.* **12**, 2076–2081.

46. Kelly, P. J., Hopper, J. L., Macaskill, G. T., Pocock, N. A., Sambrook, P. N., and Eisman, J. A. (1991). Genetic factors in bone turnover. *J. Clin. Endocrinol. Metab.* **72**, 808–813.

47. Kelly, P. J., Nguyen, T., Hopper, J., Pocock, N., Sambrook, P., and Eisman, J. (1993). Changes in axial bone density with age: A Twin Study. *J. Bone Miner. Res.* **8**, 11–17.

48. Arden, N. K., Baker, J., Hogg, C., Baan, K., and Spector, T. D. (1996). The heritability of bone mineral density, ultrasound of the calcaneus and hip axis length: A Study of Postmenopausal Twins. *J. Bone Miner. Res.* **11**, 530–534.

49. Zmuda, J. M., Cauley, J. A., and Ferrell, R. E. (1999). Recent progress in understanding the genetic susceptibility to osteoporosis. *Genet. Epidemiol.* **16**, 356–367.

50. Cummings, S. R., Kelsey, J. L., Nevitt, M. C., and O'Dowd, K. J. (1985). Epidemiology of osteoporosis and osteoporotic fractures. *Epidemiol. Rev.* **7**, 178–208.

51. Cumming, R. G., Nevitt, M. C., and Cummings, S. R. (1997). Epidemiology of hip fractures. *Epidemiol. Rev.* **19**(2), 244–257.

52. Baron, J. A., Barrett, J., Malenka, D., Fisher, E., Kniffin, W., Bubolz, T., and Tosteson, T. (1994). Racial differences in fracture risk. *Epidemiology* **5**(1), 42–47.

53. Silverman, S. L., and Madison, R. E. (1988). Decreased incidence of hip fracture in Hispanics, Asians, and Blacks: California Hospital Discharge Data. *Am. J. Public Health* **78**, 1482–1483.

54. Cummings, S. R., Nevitt, M. C., Browner, W. S., Stone, K., Fox, K. M., Ensrud, K. E., Cauley, J., Black, D., and Vogt, T. M. (1995). Risk factors for hip fracture in white women. *N. Engl. J. Med.* **332**(12), 767–773.

55. Kelsey, J. L., Browner, W. S., Seeley, D. G., Nevitt, M. C., and Cummings, S. R. (1992). Risk factors for fractures of the distal forearm and proximal humerus. *Am. J. Epidemiol.* **135**(5), 477–489.

56. Vogt, M. T., Cauley, J. A., Stone, K., Williams, J. R., and Herndon, J. H. (1999). Distal radius fracture among elderly women: a 10-year

follow-up study of descriptive characteristics and risk factors. Submitted for publication.

57. Black, D. M., Palermo, L., Nevitt, M. C., Genant, H. K., Epstein, R., San Valentine, R., and Cummings, S. R. (1995). Comparisons of methods for defining prevalent vertebral deformities: The Study of Osteoporotic Fractures. *J. Bone Miner. Res.* **10**(6), 890–902.

58. Diaz, M. N., O'Neill, T. W., and Silman, A. J. (1997). The influence of family history of hip fracture on the risk of vertebral deformity in men and women: The European Vertebral Osteoporosis Study. *Bone* **20**, 145–149.

59. Silman, A. J., O'Neill, T. W., Cooper, C., Kanis, J., Felsenberg, D., and the European Vertebral Osteoporosis Study Group (1997). Influence of physical activity on vertebral deformity in men and women: Results from the European Vertebral Osteoporosis Study. *J. Bone Miner. Res.* **12**, 813–819.

60. Silman, A. J., O'Neill, T. W., Cooper, C., Kanis, J., Felsenberg, D., and the European Vertebral Osteoporosis Study Group (1997). Influence of hormonal and reproductive factors on the risk of vertebral deformity in European women. *Osteoporosis Int.* **7**, 72–78.

61. Diaz, M. N., O'Neill, T. W., and Silman, A. J. (1997). The influence of alcohol consumption on the risk of vertebral deformity. *Osteoporosis Int.* **7**, 65–71.

62. Johnell, O., O'Neill, T., Felsenberg, D., Kanis, J., Cooper, C., and Silman, A. J. (1997). Anthropometric measurements and vertebral deformities. *Am. J. Epidemiol.* **146**, 287–293.

63. Jones, G., White, C., Nguyen, T., Sambrook, P., Kelly, P., and Eisman, J. (1995). Cigarette smoking and vertebral body deformity. *JAMA, J. Am. Med. Assoc.* **274**, 1834.

64. Ross, P. D., Davis, J. W., Epstein, R. S., and Wasnich, R. D. (1991). Pre-existing fractures and bone mass predict vertebral fracture incidence in women. *Ann. Intern. Med.* **114**, 919–923.

65. Cauley, J. A., Seeley, D. G., Ensrud, K., Black, D., Cummings, S. R., for the SOF Research Group (1995). Estrogen replacement therapy and fractures in older women. *Ann. Intern. Med.* **122**(1), 9–16.

66. Dietary Reference Intakes: Calcium, Phosphorus, Magnesium, Vitamin D, and Fluoride. (1997). Standing Committee on the Scientific Evaluation of Dietary Reference Intakes. Food & Nutrition Board. National Academy, Washington, DC.

67. Lloyd, T., Martel, J. K., Rollings, N., Andon, M. B., Kulin, H., Demers, L. M., Eggli, D. F., Kieselhorst, K., and Chinchilli, V. M. (1996). The effect of calcium supplementation and tanner stage on bone density, content and area in teenage women. *Osteoporosis Int.* **6**, 276–283.

68. Matkovic, V., and Ilich, J. Z. (1993). Calcium requirements for growth: Are current recommendations adequate? *Nut. Rev.* **51**, 171–180.

69. Morris, F. L., Naughton, G. A., Gibbs, J. L., Carlson, J. S., and Wark, J. D. (1997). Prospective ten-month exercise intervention in premenarcheal girls: Positive effects on bone and lean mass. *J. Bone Miner. Res.* **12**, 1453–1462.

70. Ruiz, J. C., Mandel, C., and Garabedian, M. (1995). Influence of spontaneous calcium intake and physical activity on the vertebral and femoral bone mineral density of children and adolescents. *J. Bone Miner. Res.* **10**, 675–682.

71. Chilibeck, P. D., Digby, G. S., and Webber, C. E. (1995). Exercise and bone mineral density. *Sports Med.* **19**, 103–122.

72. Bachrach, L. K., Guido, D., Katzman, D., Litt, I. F., and Marcus, R. (1990). Decreased bone density in adolescent girls with anorexia nervosa. *Pediatrics* **86**, 440–447.

73. Drinkwater, B. L., Bruemner, B., and Chestnut, C. H. (1990). Menstrual history as a determinant of current bone density in young athletes. *JAMA, J. Am. Med. Assoc.* **263**, 545–548.

74. Keen, A. D., and Drinkwater, B. L. (1997). Irreversible bone loss in former amenorrheic athletes. *Osteoporosis Int.* **7**, 311–315.

75. Eddy, D. M., Johnston, C. C., Cummings, S. R., Dawson-Hughes, B., Lindsay, R., Melton, L. J., and Slemenda, C. W. (1998). Osteoporosis. Review of the evidence for prevention, diagnosis and treatment and cost-effectiveness analysis. Osteoporosis Int. **8**(S4), S1–S88.

76. Cummings, S. R., Black, D. M., Nevitt, M. C., Browner, W., Cauley, J., Ensrud, K., Genant, H. K., Palermo, L., Scott, J., and Vogt, T. M. (1993). Bone density at various sites for prediction of hip fractures: The Study of Osteoporotic Fractures. *Lancet* **341**, 72–75.

77. Black, D. M., Cummings, S. R., Genant, H. K., Nevitt, M. C., Palermo, L., and Browner, W. (1992). Axial and appendicular bone density predicts fractures in older women. *J. Bone Miner. Res.* **7**, 633–638.

78. Vogt, M. T., Valentin, R. S., Forrest, K. Y., Nevitt, M. C., and Cauley, J. A. (1997). Bone mineral density and aortic calcification: The Study of Osteoporotic Fractures. *J. Am. Geriatr. Soc.* **45**(2), 140–145.

79. Vogt, M. T., Cauley, J. A., Kuller, L. H., and Nevitt, M. C. (1997). Bone mineral density and blood flow to the lower extremities. *J. Bone Miner. Res.* **12**(2), 283–289.

80. Browner, W. S., Seeley, D. G., Vogt, T. M., and Cummings, S. R. (1991). Non-trauma mortality in elderly women with low bone mineral density. *Lancet* **338**, 355–358.

81. Browner, W. S., Pressman, A. R., Nevitt, M. C., Cauley, J. A., and Cummings, S. R. (1993). Association between low bone density and stroke in elderly women: The Study of Osteoporotic Fractures. *Stroke* **24**, 940–946.

82. Nevitt, M. C., Lane, N. E., Scott, J., Hochberg, M., Genant, H. K., and Cummings, S. R. (1995). Radiographic osteoarthritis of the hip and bone mineral density. *Arthritis Rheum.* **38**(7), 907–916.

83. Cauley, J. A., Lucas, F. L., Kuller, L. H., Vogt, M. T., Browner, W. S., and Cummings, S. R. (1996). Bone mineral density and risk of breast cancer in older women. *JAMA, J. Am. Med. Assoc.* **276**(17), 1404–1408.

84. Cauley, J. A., Lucas, F. L., Kuller, L. H., Stone, K., Browner, W., and Cummings, S. R. (1999). Elevated serum estradiol and testosterone concentrations are associated with a high risk of breast cancer. *Ann. Intern. Med.* **130**, 270–277.

95

Osteoarthritis and Other Musculoskeletal Diseases

MARC C. HOCHBERG,*,†,‡ JEAN C. SCOTT,*,† AND MARGARET LETHBRIDGE-CEJKU*

*Division of Rheumatology and Clinical Immunology, Department of Medicine, and †Department of Epidemiology and Preventive Medicine, University of Maryland School of Medicine, and ‡Geriatric Research, Education, and Clinical Center, Veterans Affairs Maryland Health Care System, Baltimore, Maryland

I. Introduction

Arthritis and musculoskeletal diseases are the most common chronic diseases and causes of physical disability in the United States [1]. Based on data from the 1989–1991 National Health Interview Survey (NHIS), 15.1% of the civilian, noninstitutionalized population of the United States reported the presence of a musculoskeletal condition that was classified as arthritis by the National Arthritis Data Work Group; the overall age-adjusted prevalence in women was 21.9% [2,3]. Applying these prevalence ratios to 1990 population census figures, an estimated 38 million persons aged 15 and above in the United States, including almost 23 million women, are affected by arthritis [2,3]. In women, the prevalence of self-reported arthritis increased with increasing age (Fig. 95.1); the majority of women aged 65 and above reported the presence of an arthritis diagnosis. Age-adjusted prevalence was higher in women of non-Hispanic than of Hispanic ethnicity; there was no difference in age-adjusted prevalence between whites and blacks, but prevalence was lower in Asians [3,4]. Other factors associated with the presence of self-reported arthritis, analyzed in a subset of women aged 15 and above, included being overweight, defined as having a body mass index (BMI) of 27.3 kg/m^2 or greater, having a lower level of formal education, defined as not being a high school graduate, and living in a household with an annual income less than $20,000 [3,5].

The impact of arthritis in women in the United States can be estimated by examining its relationship with disability and its associated economic costs. In the NHIS, disability was defined as some difficulty in performing one or more activities of daily living or instrumental activities of daily living. Overall, almost 5% of women reported that they had an activity limitation attributable to arthritis [3]. Disability attributed to arthritis increased with increasing age (Fig. 95.2); age-adjusted rates of activity limitation attributed to arthritis were higher in black and Native American women than in white women, in overweight women, as well as in women with lower levels of formal education, and women with lower incomes [2,3]. In an examination of baseline data from the NHIS Longitudinal Supplement on Aging, Verbrugge and colleagues noted that women with other chronic diseases or illnesses in addition to arthritis, and women who were either underweight (BMI < 20 kg/m^2) or severely overweight (BMI > 30 kg/m^2), were also more likely to have arthritis resulting in disability [6]. In an examination of baseline data from the Women's Health and Aging Study, Hochberg and colleagues noted that women with self-reported physician-diagnosed arthritis had significantly greater odds of reporting difficulty with task performance than women without arthritis, even after adjustment for age, race, education level, mini-mental state score, and presence of other chronic conditions; odds ratios ranged from a low of 1.7 for using the toilet to a high of 3.7 for grasping or handling [7].

Musculoskeletal diseases such as arthritis exact a heavy economic burden in the United States [8–10] and in other developed countries [11]. The total cost of arthritis to the U.S. economy in 1988 (including inpatient and outpatient care, nursing home care, medications, and lost productivity) was estimated at $54.6 billion; in 1992 dollars, this figure was $64.8

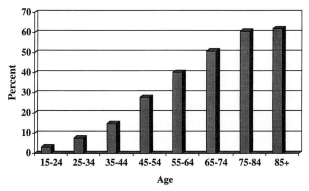

Fig. 95.1 Age-specific average annual prevalence of self-reported arthritis in women. Data derived from the National Health Interview Survey—United States, 1989–1991 [2,3].

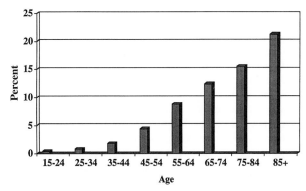

Fig. 95.2 Age-specific average annual prevalence of activity limitation attributed to arthritis in women. Data derived from the National Health Interview Survey—United States, 1989–1991 [2,3].

billion [10]. Less than a quarter of these costs are due to direct medical care, including inpatient and outpatient hospital and physician charges, and costs of pharmaceuticals. The vast majority of costs are attributable to indirect costs, predominantly due to lost wages.

More than 100 diseases make up the spectrum of arthritis and musculoskeletal disorders. Most of these diseases are uncommon, are of unknown cause, and allow little opportunity for primary or secondary prevention in the general population [12]. One disorder, osteoarthritis (OA), makes up the vast majority of the disability and economic costs of arthritis and is also subject to primary and secondary prevention initiatives. The remainder of this chapter focuses on OA; other chapters in this volume cover the disorders of rheumatoid arthritis (Chapter 53), systemic lupus erythematosus (Chapter 55), Sjogren's syndrome (Chapter 57), fibromyalgia (Chapter 85), and osteoporosis (Chapter 94).

II. Definitions of Osteoarthritis

Prior to 1986, no standard definition of OA, formerly known as degenerative joint disease, existed. In 1986, the Subcommittee on Osteoarthritis of the American College of Rheumatology Diagnostic and Therapeutic Criteria Committee proposed the following definition of OA: "A heterogeneous group of conditions that lead to joint symptoms and signs which are associated with defective integrity of articular cartilage, in addition to related changes in the underlying bone at the joint margins" [13].

A more comprehensive definition of OA was developed at a conference on the Etiopathogenesis of Osteoarthritis sponsored by the National Institute of Arthritis, Diabetes, Digestive and Kidney Diseases, National Institute on Aging, American Academy of Orthopaedic Surgeons, National Arthritis Advisory Board, and Arthritis Foundation [14]. This definition summarized the clinical, pathophysiologic, biochemical, and biomechanical changes that characterize OA:

> "Clinically, the disease is characterized by joint pain, tenderness, limitation of movement, crepitus, occasional effusion, and variable degrees of local inflammation, but without systemic effects. Pathologically, the disease is characterized by irregularly distributed loss of cartilage more frequently in areas of increased load, sclerosis of subchondral bone, subchondral cysts, marginal osteophytes, increased metaphyseal blood flow, and variable synovial inflammation. Histologically, the disease is characterized early by fragmentation of the cartilage surface, cloning of chondrocytes, vertical clefts in the cartilage, variable crystal deposition, remodeling, and eventual violation of the tidemark by blood vessels. It is also characterized by evidence of repair, particularly in osteophytes, and later by total loss of cartilage, sclerosis, and focal osteonecrosis of the subchondral bone. Biomechanically, the disease is characterized by alteration of the tensile, compressive, and shear properties and hydraulic permeability of the cartilage, increased water, and excessive swelling. These cartilage changes are accompanied by increased stiffness of the subchondral bone. Biochemically, the disease is characterized by reduction in the

proteoglycan concentration, possible alterations in the size and aggregation of proteoglycans, alteration in collagen fibril size and weave, and increased synthesis and degradation of matrix macromolecules."

The most recent definition of OA was developed in 1994 at a workshop titled "New Horizons in Osteoarthritis" sponsored by the American Academy of Orthopaedic Surgeons, National Institute of Arthritis, Musculoskeletal and Skin Diseases, National Institute on Aging, Arthritis Foundation, and Orthopaedic Research and Education Foundation [15]. This definition underscores the concept that OA may not represent a single disease entity:

> "Osteoarthritis is a group of overlapping distinct diseases, which may have different etiologies but with similar biologic, morphologic, and clinical outcomes. The disease processes not only affect the articular cartilage, but involve the entire joint, including the subchondral bone, ligaments, capsule, synovial membrane, and periarticular muscles. Ultimately, the articular cartilage degenerates with fibrillation, fissures, ulceration, and full thickness loss of the joint surface."

III. Classification of Osteoarthritis

A. Classification Schema

Osteoarthritis is a disorder of diverse etiologies. A classification schema for OA developed at the "Workshop on Etiopathogenesis of Osteoarthritis" is shown in Table 95.1 [14]. Idiopathic osteoarthritis is divided into two forms: localized and generalized. Generalized OA represents the form of OA described by Kellgren and Moore involving three or more joint groups [16]. Patients with an underlying disease that appears to have caused their OA are classified as having secondary OA. Some forms of secondary OA (e.g., that due to chronic trauma from leisure and/or occupational activities) may be considered as risk factors for idiopathic OA; conversely, risk factors for idiopathic OA (e.g., overweight) may be considered as causes of secondary OA.

B. Radiographic Criteria

Classically, the diagnosis of OA in epidemiologic studies has relied on the characteristic radiographic changes described by Kellgren and Lawrence in 1957 [17] and illustrated in the Atlas of Standard Radiographs [18]. These features include: (1) formation of osteophytes on the joint margins or in ligamentous attachments; (2) periarticular ossicles, chiefly in relation to distal and proximal interphalangeal joints; (3) narrowing of joint space associated with sclerosis of subchondral bone; (4) cystic areas with sclerotic walls situated in the subchondral bone; and (5) altered shape of the bone ends, particularly the head of the femur. Combinations of these changes considered together led to the development of an ordinal grading scheme for severity of radiographic features of OA: 0 = normal, 1 = doubtful, 2 = minimal, 3 = moderate, and 4 = severe.

Table 95.1
Classification of Osteoarthritis[a]

I. Idiopathic
  A. Localized
    1. Hands
    2. Feet
    3. Knee
    4. Hip
    5. Spine
    6. Other single sites
  B. Generalized
II. Secondary
  A. Traumatic
  B. Congenital or developmental diseases
  C. Metabolic diseases
    1. Ochronosis
    2. Hemochromatosis
    3. Wilson's disease
    4. Gaucher's disease
  D. Endocrine diseases
    1. Acromegaly
    2. Hyperparathyroidism
    3. Diabetes mellitus
    4. Hypothyroidism
  E. Calcium deposition disease
    1. Calcium pyrophosphate dihydrate deposition disease
    2. Apatite arthropathy
  F. Other bone and joint diseases
  G. Neuropathic (Charcot) arthropathy
  H. Endemic disorders
  I. Miscellaneous conditions

[a]Modified from [14].

Potential limitations of the use of the Kellgren-Lawrence grading scales, as illustrated in the Atlas on Standard Radiographs, have been noted [19]. Radiographic grading scales that focus on individual radiographic features of OA at specific joint groups have now been published for the hand [20], hip [21,22], and knee [23,24], as well as for all three peripheral joint groups [25].

C. Clinical Criteria

There are potential limitations to the use of purely radiographic criteria for case definition, especially in clinical studies of OA. At the Third International Symposium on Population Studies of the Rheumatic Diseases in 1966, the Subcommittee on Diagnostic Criteria for Osteoarthrosis recommended that future population-based studies should investigate the predictive value of certain historical, physical, and laboratory findings for the typical radiographic features of OA on a joint-by-joint basis [26]. Such historical features included pain on motion, pain at rest, nocturnal joint pain, and morning stiffness. Features on physical examination included bony enlargement, limitation of motion, and crepitus. Laboratory features included erythrocyte sedimentation rate, tests for rheumatoid factor, serum uric acid, and appropriate analyses of synovial fluid.

The Subcommittee on Osteoarthritis of the American College of Rheumatology's Diagnostic and Therapeutic Criteria Committee proposed sets of clinical criteria for the classification of OA of the knee [13,27], hand [28], and hip [29]; these criteria sets have been amplified into algorithms by Altman for ease of use in clinical research and population-based studies [30]. These criteria sets identify patients with clinical OA, as the major inclusion parameter is joint pain for most days of the prior month. This contrasts with the use of radiographic features alone wherein many, if not most, subjects do not report joint pain. Thus, prevalence estimates using different case definitions will likely be systematically lower when based on the American College of Rheumatology classification criteria as opposed to traditional radiographic criteria [31]; readers need to be aware of this when reviewing published studies.

IV. Descriptive Epidemiology of Osteoarthritis

A. Prevalence

The prevalence of OA has been estimated in two U.S. national studies: the National Health Examination Survey (NHES), conducted from 1960 to 1962 [32], and the First National Health and Nutrition Examination Survey (NHANES-I), conducted from 1971 to 1975 [33]. The case definition of OA was based on radiographic changes in the hands and feet in the NHES and on radiographic changes in knees and hips in NHANES-I. In addition, a physician's clinical diagnosis of OA was also available in NHANES-I. These data were summarized by the National Arthritis Data Work Group in 1989 and again in 1998 [34,35].

Overall, about one-third of adults aged 25–74 years have radiographic evidence of OA involving at least one site. Specifically, 33% had changes of definite OA of the hands, 22% of the feet, and 4% of the knee [34]. Among persons aged 55–74, corresponding prevalence ratios were 70% for the hands, 40% for the feet, 10% for the knees, and 3% for the hips [35]. Overall, a clinical diagnosis of OA, based on symptoms and physical findings, was made by the examining physician in 12% of 6913 examinees aged 25–74 years in NHANES-I [33]; using estimates for the 1990 U.S. Population, the National Arthritis Data Work Group estimated that over 20 million adults have physician-diagnosed OA [35].

Population-based data from the Framingham Osteoarthritis Study, a prevalence survey of radiographic knee OA in white elders aged 63–93 years, suggest that one-third of both men and women in this age group have evidence of definite radiographic OA of the knees [36]. These results are similar to those from the Baltimore Longitudinal Study on Aging [37].

The prevalence of symptomatic knee OA can also be estimated from NHANES-I and the Framingham Osteoarthritis Study. Subjects are considered to be symptomatic if they report pain in or around their knees on most days of at least one month. The prevalence of symptomatic knee OA was 1.6% in adults aged 25–74 years using data from NHANES-I [33] and 9.5% among adults aged 63–93 years in the Framingham Osteoarthritis Study [36].

B. Incidence

The incidence of radiographic OA of the knee was examined in surviving members of the Framingham Osteoarthritis Study who underwent repeat standing knee radiographs in 1992–1993; these individuals had a mean age of 71 years at time of baseline radiographs in 1983–1985 [38]. In women, the cumulative incidence of radiographic knee OA was 18.1% and the cumulative incidence of symptomatic knee OA was 8.1%. These rates did not differ between women below age 70 or age 70 and above; rates were 80 to 100% higher in women than in men, after adjusting for age.

The incidence of radiographic OA of the hand was also examined in surviving members of the Framingham Osteoarthritis Study who underwent repeat hand radiographs in 1992–1993; these individuals had a mean age of 55 years at time of baseline radiographs in 1967–1969 [39]. The cumulative incidence of radiographic hand OA was higher in women than in men; the most common joints affected were the distal and proximal interphalangeal joints and joints at the thumb base.

The incidence of clinically diagnosed, symptomatic OA of the hand, hip, and knee has been estimated in two studies [40,41]. The most recent data, based on persons who were members of the Fallon Community Health Plan in Massachusetts between 1988 and 1992, showed that the incidence of symptomatic hand, hip, and knee OA increased with increasing age in women. Incidence reached maximum values of 529, 583, and 1082 cases per 100,000 women-years in the 70–79 year age group for symptomatic hand, hip, and knee OA, respectively [41]. Incidence declined slightly in women aged 80–89 years. No cases of symptomatic OA occurred in women below age 30, and only one case of symptomatic hip and four cases of symptomatic knee OA occurred in approximately 47,000 women-years of follow-up in women below age 40 years.

C. Demographic Factors

Prevalence of OA, as well as the proportion of cases with moderate or severe disease, increase with increasing age at least through age 65–74 years. Osteoarthritis is more common among men than women under age 45, comparable between ages 45 and 54, and more common among women than men over age 54. Clinically, patterns of joint involvement also demonstrate sex differences, with women having on average more joints involved and more frequent complaints of morning stiffness and joint swelling than men. Among adults in the United States who participated in NHANES-I, radiographic knee OA was more common among black women than white women [42]. No differences were noted in the prevalence of hip OA between black and white women in the U.S. [43] or between Native American women and U.S. white women [44]. It has been suggested that the prevalence of hip OA is lower in women of Asian ethnicity; this inference is derived from comparative rates of total hip replacement for coxarthrosis in San Francisco, CA and may be subject to several limitations, including cultural and referral biases [45]. Studies in Asian women living in Asia, however, have consistently shown a lower prevalence of hip OA compared to that expected in Caucasian women [46].

D. Geographic Distribution

Data from U.S. national surveys suggest that arthritis and associated disability are more prevalent in the South and least

prevalent in the Northeast [2]. Possible explanations for these patterns include lower socioeconomic status and greater proportion of the population employed in manual labor in the South. The Johnston County Osteoarthritis Project, being conducted by investigators from the University of North Carolina at Chapel Hill and funded by the Centers for Disease Control and Prevention and the National Institutes of Health, is currently examining reasons for these regional variations as well as possible black–white differences among residents of the southeastern United States [47].

E. Time Trends

No data are available on time trends in the prevalence of OA in the United States. The National Center for Health Statistics recently released the clinical data from NHANES-III, conducted from 1988 to 1993; this study also included radiographs of the hands and knees in older persons. These radiographic data should become available in 1999; they will be useful for comparing the prevalence of OA with the prevalence based on earlier national surveys.

F. Mortality

Few data are available on mortality associated with OA [48]. The relationship between radiographically defined OA and survival was examined in a cohort of 296 women who underwent full-body radiographs when first employed in the radium dial-painting industry [49]. Women with an increasing number of joints affected with OA had decreased survival; the age-adjusted hazard ratio was 1.45 for each 3-unit increase in the number of joints involved. These results were largely unchanged after adjustment for BMI, smoking, and comorbid conditions.

V. Risk Factors for the Development of Osteoarthritis

Risk factors for the development of OA have been the subject of many studies; most of these studies, until recently, have been cross-sectional or retrospective in design. During the 1990s results of longitudinal studies have become available; largely, these results confirm and extend findings from older cross-sectional studies.

Risk factors for OA can be divided into genetic and nongenetic host factors and environmental factors (Table 95.2). This section focuses on several of these risk factors specifically associated with the development of OA of the knee and hip: overweight, occupation and physical activity, and joint injury/trauma. The reader is referred to additional reviews for a detailed discussion of the role of other factors in the development of OA at these joint groups as well as at other sites, including the hand and spine [50–53].

A. Overweight

1. Knee Osteoarthritis

Overweight is clearly the most important modifiable risk factor for the development of knee OA in women; however, its role as a risk factor for the development of OA of the hip remains

Table 95.2

Factors Associated with the Presence of Osteoarthritis

Host factors
  Genetic
    Female sex
    Inherited disorders of type II collagen gene (*e.g.,* Stickler's syndrome)
    Other inherited disorders of bones and joints
    Race/ethnicity
    Higher bone mass (?)
  Nongenetic host factors
    Increasing age
    Overweight
    Depletion of female sex hormones (i.e., postmenopausal state)
    Developmental and acquired bone and joint diseases
    Previous joint surgery (e.g., meniscectomy)
Environmental factors
  Selected occupations and greater physical demands of work
  Major trauma to joints
  Leisure and/or sports activities

controversial. Many cross-sectional epidemiologic studies have found an association between obesity and radiographically-defined knee OA [42,54–58]; this was confirmed in a twin study [59] and in several prospective cohort studies [60–64].

Is the association of overweight with OA of the knee related to the distribution of body weight and/or the amount of body fat? Davis and colleagues examined the relationship between body fat distribution and OA of the knees in subjects aged 45–74 years in NHANES-I [65]. There was no association of either subscapular skinfold thickness or triceps skinfold thickness with either unilateral or bilateral OA of the knees in women after adjustment for age, race, and BMI. BMI remained significantly associated with bilateral but not unilateral knee OA in women. Hochberg and colleagues examined the relationship between body fat distribution and percent body fat in 275 Caucasian women participants in the Baltimore Longitudinal Study of Aging; 99 women had radiographic features of definite knee OA [66]. Neither a central body fat distribution, defined as a higher waist–hip ratio, nor higher percent body fat were associated with bilateral knee OA in women after adjustment for BMI. Thus, based on these two studies, it appears that being overweight is the important factor related to the presence of knee OA, not whether the weight is predominantly fat versus lean body weight or where the weight is distributed.

Is the relationship of overweight with knee OA due to systemic effects of obesity and possibly mediated by metabolic factors related to obesity? There are conflicting results regarding this question. Davis and colleagues examined the role of metabolic factors in subjects participating in NHANES-I and found that the strength of the association between overweight and the presence of knee OA was not diminished by adjustment for potential confounding variables including blood pressure, serum cholesterol, serum uric acid, body fat distribution, and history of diabetes [67]. Similar results were noted from analysis of data from the Baltimore Longitudinal Study of Aging which failed to demonstrate significant confounding or effect modification of the association of BMI with definite knee OA by blood

pressure, fasting and 2-hour serum glucose and insulin levels, and fasting serum lipid levels, including cholesterol, triglycerides, and HDL-cholesterol [68]. Hart and colleagues, however, did find an association between knee OA and elevated serum glucose, serum cholesterol, and use of diuretics after adjustment for age and weight in women participating in the baseline evaluation in the Chingford Study [69].

Does prevention of weight gain or weight loss among those overweight result in a decreased risk of developing knee OA? Felson and colleagues analyzed data from the Framingham study to examine the effect of weight change from examination 1 on the incidence of symptomatic knee OA in women [70]. After adjusting for baseline BMI, they found a relationship between weight change from 6 to 12 years prior to the radiographic examination; women whose weight increased by 2 kg/m^2 had between a 25 and 35% increased risk of having current symptomatic knee OA compared to women without weight change, while those whose weight decreased by the same magnitude had a reduced risk of developing current symptomatic knee OA. Focusing on the 10-year interval prior to the radiographic examination, the authors showed that the odds of having current symptomatic knee OA were reduced by 50% for a loss of every 2 kg/m^2. These findings were confirmed in the longitudinal component of the Framingham Osteoarthritis Study [61]. Weight change, measured as a continuous variable, was strongly associated with the development of knee OA; the odds ratio (OR) for every 10 pound weight gain from baseline was 1.6 (95% confidence interval (CI) = 1.2–2.3) in women. Furthermore, when subjects were categorized into three group (gained 5 pounds or more, lost 5 pounds or more, or all others), those who gained 5 pounds or more had an increased risk of developing knee OA, while those who lost 5 pounds or more had a reduced risk of developing knee OA compared to the remaining subjects (OR = 3.8 95% CI = 0.7–20.7) and 0.5 (95% CI = 0.2–1.1), respectively. Thus, these data strongly suggest that weight loss can reduce the risk of developing both radiographic and symptomatic knee OA.

2. Hip Osteoarthritis

The results of cross-sectional studies examining the association of obesity with hip OA suggest that the relationship is of lower strength than that with OA of the knee [43,54,56,71–73]. Several of these studies have demonstrated that the association with overweight is stronger for bilateral hip OA than unilateral hip OA [43,71,72].

B. Occupation and Physical Activity

Certain occupations that require repetitive use of particular joints over long periods of time have been associated with the development of site-specific OA. Specific occupational groups with increased risks of OA include miners, who have an excess of knee and lumbar spine disease; dockers and shipyard workers, who have an excess of hand and knee OA; cotton and mill workers, who have an excess of hand OA involving specific finger joints; pneumatic tool operators, who have an excess of elbow and wrist OA; concrete workers and painters, who have an excess of knee OA; and farmers, who have an excess of hip OA. The relationship between occupation and OA has been the subject of several reviews [74–76]; in addition, Buckwalter has reviewed the biomechanical mechanisms underlying the relationship between abuse of joints and development of OA [77].

1. Knee Osteoarthritis

Anderson and Felson examined data from NHANES-I to study the relationship between occupation and knee OA in the general population [42]. In multiple logistic regression models adjusted for race, BMI, and education levels, women aged 55–64 years who worked in jobs with either increasing strength demands or higher knee-bending demands had greater odds of radiographic knee OA. Cooper and colleagues tested the hypothesis that specific occupational physical activities were risk factors for knee OA in a population-based case-control study in Bristol, England [78]. Subjects with moderate-to-severe knee OA were significantly more likely than controls to report occupations that required 30 minutes per day or more of squatting or kneeling, or climbing more than ten flights of stairs.

2. Hip Osteoarthritis

Occupational physical activity has also been associated with hip OA in men in several studies conducted in Europe. In a case-control study in England, the authors failed to demonstrate an association between occupational lifting and hip OA in women; however, only 15% of the 401 women studied had been in paid employment for 90% or more of their working lives [79]. No studies examining the relationship between occupation and hip OA in women have been conducted in the U.S.

C. Sports and Exercise

Many studies have been conducted to examine the relationship between regular physical activity and OA of the hip and knee; most of the European studies have included elite athletes, particularly football players, runners, and soccer players. Panush and Lane [80] and Lane and Buckwalter [81] reviewed these studies and concluded that individuals who participate in sports at a highly competitive level (i.e., elite athletes) or who have abnormal or injured joints appear to be at increased risk of developing OA as compared to persons with normal joints who participate in low impact activities. This was confirmed in a retrospective cohort study of 81 female former elite athletes that found a two- to three-fold excess odds of radiographic hip and knee OA compared to age-matched female population controls [82]. Running as a recreational activity does not appear to be a risk factor for the development of OA of the knee in the absence of knee injury [83–85].

Habitual physical activity also appears to be a risk factor for knee and hip OA. In the Framingham Osteoarthritis Study, women who had a score in the highest quartile of the Framingham Physical Activity Index had three-fold greater odds of developing incident knee OA compared to women in the lowest quartile of this measure [61]. In a case-control study of hip OA in England, there was a modest but significant association of sporting activity (i.e., tennis and swimming) and disease [72]. Analysis of data from the Study of Osteoporotic Fractures also suggests a modest but significant association between severe hip OA in elderly women and lifetime participation in recreational physical activity [86].

D. Joint Injury/Trauma

As noted above, leisure physical activity is associated with a modest increase in the risk of both knee and hip OA in the absence of joint injury. Knee injury, especially tears of the medial meniscus and rupture of the anterior cruciate ligament, is strongly associated with an increased risk of knee OA in elite male soccer players [87,88]. Is joint injury associated with knee and hip OA in women in the general population?

1. Knee Osteoarthritis

Davis and colleagues studied the association between knee injury and unilateral and bilateral radiographic OA of the knee in 3885 subjects aged 45–74 years who participated in NHANES-I, of whom 226 (4.9%) had bilateral and 75 (1.8%) had unilateral OA [89]. Overall, a history of right knee injury was present in 5.8% of subjects with bilateral knee OA, 15.8% of 37 subjects with right knee OA, and 1.5% of controls, while a history of left knee injury was present in 4.6% of those with bilateral knee OA, 27% of subjects with left knee OA, and 1.8% of controls. In multiple polychotomous logistic regression analysis adjusting for age, sex, and BMI, the odds ratio for the association of knee injury with bilateral knee OA was 3.51 (95% CI = 1.80–6.83) as compared with right knee and left knee OA of 16.30 (95% CI = 6.50–40.9) and 10.90 (95% CI = 3.72–31.93), respectively. Cooper and colleagues examined the relationship of knee injury to symptomatic knee OA involving the tibiofemoral and/or patellofemoral component [90]. In multiple logistic regression models adjusted for age, BMI, family history of knee arthritis, and presence of Heberden's nodes, knee injury was associated with significantly higher odds of radiographic knee OA in both sexes (OR = 3.4; 95% CI = 1.7–6.7). In the Framingham Osteoarthritis Study, women with an interim knee injury had a twofold greater odds of developing knee OA; this did not reach statistical significance, however, because of small numbers of women with such injuries [61].

2. Hip Osteoarthritis

Tepper and Hochberg studied the association between hip injury and radiographic OA of the hip in 2358 subjects aged 55–74 years who participated in NHANES-I, of whom only 73 (3.1%) had definite OA of one or both hips [43]. In multiple logistic regression models adjusted for age, race, and education, a history of hip injury was significantly associated with higher odds of hip OA (OR = 7.84; 95% CI = 2.11–29.10). When the analysis was performed examining the relationship between hip injury and either unilateral or bilateral hip OA, the odds ratio for the association of hip injury with unilateral hip OA was 24.2 (95% CI = 3.84–153) as compared with bilateral hip OA of 4.17 (95% CI = 0.50–34.7). Heliovaara and colleagues studied the relationship between history of lower limb injury and clinically diagnosed hip OA in a population study in Finland [71]. In multiple logistic regression models adjusted for age, gender, BMI, and physical stress at work, a history of lower limb injury was associated with OA in both sexes. As in the NHANES-I dataset, the association was stronger with unilateral than bilateral hip OA. These results were confirmed in a case-control study of 401 women undergoing total hip arthroplasty for hip OA in England; previous hip injury was more strongly associated with unilateral as compared to bilateral hip OA; the odds ratio for the association of hip injury with unilateral hip OA was 10.6 (95% CI = 2.4–47.4) as compared with bilateral hip OA of 2.3 (95% CI = 1.0–5.1) [72]. These data suggest that hip and knee injury are important risk factors for hip and knee OA, respectively, especially unilateral hip and knee OA.

VI. Clinical Features and Diagnosis [91]

Typically, the patient with OA is a middle-aged or elderly woman who presents with pain and stiffness accompanied by loss of function. Pain, gradual or insidious in onset, is usually mild in intensity, worsened by use of involved joints, and improved or relieved with rest. Pain at rest or nocturnal pain are features of severe disease. The mechanism of pain in patients with OA is multifactorial; pain may result from periostitis at sites of bony remodeling, subchondral microfractures, capsular irritation from osteophytes, periarticular muscle spasm, bone angina due to decreased blood flow and elevated intraosseus pressure, and synovial inflammation accompanied by the release of prostaglandins, leukotrienes, and various cytokines, including interleukin-1.

Morning stiffness is common in patients with OA; the duration of morning stiffness, however, is significantly shorter, often less than 30 minutes, than in patients with active rheumatoid arthritis. Gel phenomenon, or stiffness after inactivity, is also common and resolves within several minutes. Both pain and stiffness are modified by weather changes in the majority of patients; symptoms worsen in damp, cool, and rainy weather and improve after the weather improves. Patients with OA of the knees may also complain of a sense of instability or buckling, especially when descending stairs or curbs.

On examination, findings are usually localized to symptomatic joints. Bony enlargement with tenderness at the joint margins and at attachments of the joint capsule and periarticular tendons is common. Limitation of motion of the affected joint is usually related to osteophyte formation and/or severe cartilage loss. Signs of local inflammation may be present, including warmth and soft-tissue swelling due to joint effusion. Instability can be detected by excess joint motion; locking of a joint during range of motion is likely due to loose bodies. Crepitus, felt on passive range of joint motion and due to irregularity of the opposing cartilage surfaces, is present in over 90% of patients with OA of the knee. Malalignment is present in almost 50% of patients with OA of the knee; a varus deformity due to loss of articular cartilage in the medial compartment is more common than a valgus deformity.

The clinical diagnosis of OA is usually confirmed with radiographs. Routine laboratory tests, including complete blood count, erythrocyte sedimentation rate, chemistry panels, and urinalysis, are usually normal. These studies, especially the complete blood count and chemistry panel, should be obtained at the time of baseline evaluation of patients with OA prior to instituting therapy with nonsteroidal anti-inflammatory drugs and at 3–6 month intervals during therapy. Tests for rheumatoid factor, commonly performed in the evaluation of patients with arthritis, may be positive in low titer in up to 20% of elderly women; therefore, the presence of rheumatoid factor does not exclude a diagnosis of OA if the clinical picture and radiographic changes are most consistent with this disorder. Synovial fluid analysis usually re-

Table 95.3

Factors Associated with Progression of Osteoarthritis of the Knee

Older age
Female sex
Overweight
Generalized osteoarthritis
    Heberden's nodes
Low dietary intake of antioxidants
Low dietary intake (serum levels) of vitamin D

veals Type I noninflammatory fluid. The finding of an inflammatory fluid with an elevated white cell count suggests either a superimposed microcrystalline process (*e.g.*, gout, pseudogout, or basic calcium apetite crystals) or septic arthritis.

VII. Risk Factors for the Progression of Osteoarthritis

Factors associated with the progression of OA are listed in Table 95.3 [92,93]. Secondary prevention strategies directed towards weight loss through dietary instruction and regular low intensity aerobic physical exercise, and muscle strengthening through either aerobic exercise or resistance training have been shown to decrease symptoms and functional limitation in patients with knee OA [94,94a]; it is possible that this intervention may slow progression of knee OA and reduce the need for total joint replacement. In addition, data from the Framingham Osteoarthritis Study suggest that elders with a low dietary intake of antioxidants, especially vitamin C and vitamin E, as well as those with low serum levels of vitamin D, have higher rates of progression of radiographic changes of knee OA [95–97]. Hence, dietary supplementation with vitamin C 1000 mg/day, vitamin D 400 IU/day, and vitamin E 400 IU/day may be indicated in older persons with knee OA.

VIII. Management of Patients with Osteoarthritis

As there is no known cure for OA at present, current medical treatment of patients with OA is symptomatic (Table 95.4). Systematic reviews of randomized clinical trials of psychoeducational

Table 95.4

Options for Medical Management of Patients with Osteoarthritis

Nonpharmacologic modalities
    Patient education
    Cognitive-behavioral interventions
    Social support via telephone contact
    Physical therapy, including muscle strengthening
    Aerobic exercise
    Weight loss
Pharmacologic therapy
    Nonopioid analgesics (*e.g.*, acetaminophen)
    Nonsteroidal anti-inflammatory drugs
    Topical analgesics (*e.g.*,capsaicin cream)
    Intra-articular corticosteroid injections
    Intra-articular hyaluronic acid injections (limited to knees)
    Opioid analgesics
    Nutritional supplements (*e.g.*, glucosamine and chondroitin sulfate)

interventions [98], other nonpharmacologic modalities [99], and pharmacologic therapy of hip and knee OA [100,101], as well as guidelines for the medical management of hip and knee OA, have been published [102,103]. These guidelines, as well as the results of a survey of community-based practicing rheumatologists in the United States [104], led to the development of an approach to the management of patients with OA recommended by the Arthritis Foundation [105]. A detailed discussion of specific options in the management of patients with OA is beyond the scope of this chapter; the interested reader is referred elsewhere for such material [91,106].

References

1. Kelsey, J. L., and Hochberg, M. C. (1988). Epidemiology of chronic musculoskeletal disorders. *Annu. Rev. Public Health* **9**, 379–401.

2. Centers for Disease Control (1994). Arthritis prevalence and activity limitations—United States, 1990. *Morbid. Mortal. Wkly. Rep.* **43**, 433–438.

3. Centers for Disease Control (1995). Prevalence and impact of arthritis among women—United States, 1989–1991. *Morbid. Mortal. Wkly. Rep.* **44**, 329–334.

4. Centers for Disease Control (1996). Prevalence and impact of arthritis by race and ethnicity—United States, 1989–1991. *Morbid. Mortal. Wkly. Rep.* **45**, 373–378.

5. Centers for Disease Control (1966). Factors associated with prevalent self-reported arthritis and other rheumatic conditions—United States, 1989–1991. *Morbid. Mortal. Wkly. Rep.* **45**, 487–491.

6. Verbrugge, L. M., Gates, D. M., and Ike, R. W. (1991). Risk factors for disability among U.S. adults with arthritis. *J. Clin. Epidemiol.* **44**, 167–182.

7. Hochberg, M. C., Williamson, J., Kasper, J. D., and Fried, L. P. (1997). Musculoskeletal disability in elderly community-dwelling women: Data from the Women's Health and Aging Study. *In* "Proceedings of the 19th *ILAR Congress of Rheumatology*" (P. H. Feng, ed.), pp. 206–208. Communication Consultants, Singapore.

8. Felts, W., and Yelin, E. (1989). The economic impact of the rheumatic diseases in the United States. *J. Rheumatol.* **16**, 867–884.

9. Yelin, E. H., and Felts, W. R. (1990). A summary of the impact of musculoskeletal conditions in the United States. *Arthritis Rheum.* **33**, 750–755.

10. Yelin, E., and Callahan, L. F., for the National Arthritis Data Work Group (1995). The economic cost and social and psychological impact of musculoskeletal conditions. *Arthritis Rheum.* **38**, 1351–1362.

11. March, L. M., and Bachmeier, C. J. (1997). Economics of osteoarthritis: A global perspective. *Bailliere's Clin. Rheumatol.* **11**, 817–834.

12. Hochberg, M. C., and Flores, R. H. (1995). Arthritis and connective tissue diseases. *In* "Physician's Guide to Rare Diseases" (J. Thoene, ed.), 2nd ed., pp. 745–786. Dowden, Montvale, NJ.

13. Altman, R., Asch, E., Bloch, D., Bole, G., Borenstein, D., Brandt, K. *et al.* (1986). Development of criteria for the classification and reporting of osteoarthritis: Classification of osteoarthritis of the knee. *Arthritis Rheum.* **29**, 1039–1049.

14. Brandt, K. D., Mankin, H. J., and Shulman, L. E. (1986). Workshop on etiopathogenesis of osteoarthritis. *J. Rheumatol.* **13**, 1126–1160.

15. Keuttner, K., and Goldberg, V. M., eds. (1995). Osteoarthritic Disorders," pp. xxi–xxv. American Academy of Orthopaedic Surgeons, Rosemont, IL.

16. Kellgren, J. H., and Moore, G. (1952). Generalized osteoarthritis and Heberden's nodes. *Br. Med. J.* **1**, 181–187.

17. Kellgren, J. H., and Lawrence, J. S. (1957). Radiologic assessment of osteoarthrosis. *Ann. Rheum. Dis.* **16**, 494–501.

18. University of Manchester, Department of Rheumatology and Medical Illustration (1973). "The Epidemiology of Chronic Rheumatism," Vol. 2, pp. 1–15. Davis, Philadelphia.

19. Spector, T. D., and Hochberg, M. C. (1994). Methodological problems in the epidemiological study of osteoarthritis. *Ann. Rheum. Dis.* **53**, 143–146.

20. Kallman, D. A., Wigley, F. M., Scott, W. W., Jr., Hochberg, M. C., and Tobin, J. D. (1989). New radiographic grading scales for osteoarthritis of the hand. *Arthritis Rheum.* **32**, 1584–1591.

21. Croft, P., Cooper, C., Wickham, C., and Coggon, D. (1990). Defining osteoarthritis of the hip for epidemiologic studies. *Am. J. Epidemiol.* **132**, 514–522.

22. Lane, N. E., Nevitt, M. C., Genant, H. K., and Hochberg, M. C. (1993). Reliability of new indices of radiographic osteoarthritis of the hand and hip and lumbar disc degeneration. *J. Rheumatol.* **20**, 1911–1918.

23. Spector, T. D., Cooper, C., Cushnaghan, J., Hart, D. J., and Dieppe, P. A. (1992). "A Radiographic Atlas of Knee Osteoarthritis." Springer-Verlag, London.

24. Scott, W. W., Jr., Lethbridge-Cejku, M., Reichle, R., Wigley, F. M., Tobin, J. D., and Hochberg, M. C. (1993). Reliability of grading scales for individual radiographic features of osteoarthritis of the knee: The Baltimore Longitudinal Study of Aging Atlas of Knee Osteoarthritis. *Invest. Radiol.* **28**, 497–501.

25. Altman, R. D., Hochberg, M. C., Murphy, W. A., Jr., Wolfe, F., and Lequesne, M. (1995). Atlas of individual radiographic features in osteoarthritis. *Osteoarthritis Cartilage* **3**(Suppl. A), 3–70.

26. Bennett, P. H., and Wood, P. H. N., eds. (1968). "Population Studies of the Rheumatic Diseases," Int. Cong. Ser. No. 148, pp. 417–419. Excerpta Med. Found., Amsterdam.

27. Altman, R. D., Meenan, R. F., Hochberg, M. C., Bole, G. G., Jr., Brandt, K., Cooke, T. D. V. *et al.* (1983). An approach to developing criteria for the clinical diagnosis and classification of osteoarthritis: a status report of the American Rheumatism Association Diagnostic Subcommittee on Osteoarthritis. *J. Rheumatol.* **10**, 180–183.

28. Altman, R., Alarcon, G., Appelrough, D., Bloch, D., Borenstein, D., Brandt, K. *et al.* (1990). The American College of Rheumatology criteria for the classification and reporting of osteoarthritis of the hand. *Arthritis Rheum.* **33**, 1601–1610.

29. Altman, R., Alarcon, G., Appelrough, D., Bloch, D., Borenstein, D., Brandt, K. *et al.* (1991). The American College of Rheumatology criteria for the classification and reporting of osteoarthritis of the hip. *Arthritis Rheum.* **34**, 505–514.

30. Altman, R. (1991). Classification of disease: Osteoarthritis. *Semin. Arthritis Rheum.* **20**(6, Suppl. 2), 40–47.

31. Hart, D. J., Leedham-Green, M., and Spector, T. D. (1991). The prevalence of knee osteoarthritis in the general population using different clinical criteria: The Chingford Study. *Br. J. Rheumatol.* **30**(S2), 72.

32. Engle, A. (1966). Osteoarthritis in adults by selected demographic characteristics, United States—1960–1962. *Vital Health Stat., Ser.* **11**, No. 20.

33. Maurer, K. (1979). Basic data on arthritis knee, hip and sacroiliac joints in adults ages 25–74 years, United States, 1971–1975. *Vital Health Stat., Ser.* **11**, No. 213.

34. Lawrence, R. C., Hochberg, M. C., Kelsey, J. L. *et al.* (1989). Estimates of the prevalence of selected arthritic and musculoskeletal diseases in the United States. *J. Rheumatol.* **16**, 427–441.

35. Lawrence, R. C., Helmick, C. G., Arnett, F. C. *et al.* (1998). Prevalence estimates of arthritis and selected musculoskeletal diseases in the United States. *Arthritis Rheum.* **41**, 778–799.

36. Felson, D. T., Naimark, A., Anderson, J., Kazis, L., Castelli, W., and Meenan, R. F. (1987). The prevalence of knee osteoarthritis in the elderly. *Arthritis Rheum.* **30,** 914–918.

37. Lethbridge-Cejku, M., Tobin, J. D., Scott, W. W., Jr., Reichle, R., Plato, C. C., and Hochberg, M. C. (1994). The relationship of age and gender to prevalence and pattern of radiographic changes of osteoarthritis of the knee: Data from the Baltimore Longitudinal Study of Aging. *Aging Clin. Exp. Res.* **6,** 353–357.

38. Felson, D. T., Zhang, Y., Hannan, M. T., Naimark, A., Weissman, B. N., Aliabadi, P., and Levy, D. (1995). The incidence and natural history of knee osteoarthritis in the elderly. *Arthritis Rheum.* **38,** 1500–1505.

39. Chaisson, C. E., Zhang, Y., McAlindon, T. E., Hannan, M. T., Aliabadi, P., Naimark, A., Levy, D., and Felson, D. T. (1997). Radiographic hand osteoarthritis: Incidence, patterns, and influence of pre-existing disease in a population based sample. *J. Rheumatol.* **24,** 1337–1343.

40. Wilson, M. G., Michet, C. J., Ilstrup, D. M., and Melton, L. J., III (1990). Idiopathic symptomatic osteoarthritis of the hip and knee: A Population-based Incidence Study. *Mayo Clin. Proc.* **65,** 1214–1221.

41. Oliveria, S. A., Felson, D. T., Reed, J. I., Cirillo, P. A., and Walker, A. M. (1995). Incidence of symptomatic hand, hip, and knee osteoarthritis among patients in a health maintenance organization. *Arthritis Rheum.* **38,** 1134–1141.

42. Anderson, J. J., and Felson, D. T. (1988). Factors associated with osteoarthritis of the knee in the First National Health and Nutrition Examination Survey (NHANES-I): Evidence for an association with overweight, race, and physical demands of work. *Am. J. Epidemiol.* **128,** 179–189.

43. Tepper, S., and Hochberg, M. C. (1993). Factors associated with hip osteoarthritis: Data from the National Health and Nutrition Examination Survey (NHANES-I). *Am. J. Epidemiol.* **137,** 1081–1088.

44. Hirsch, R., Fernandes, R. J., Pillemer, S. R., Hochberg, M. C., Lane, N. E., Altman, R. D., Bloch, D. A., Knowler, W. C., and Bennett, P. H. (1998). Hip osteoarthritis prevalence estimates by three radiographic scoring systems. *Arthritis Rheum.* **41,** 361–368.

45. Hoaglund, F. T., Oishi, C. S., and Gialamas, G. G. (1995). Extreme variations in racial rates of total hip arthroplasty for primary coxarthrosis: A Population-based Study in San Francisco. *Ann. Rheum. Dis.* **54,** 107–110.

46. Lau, E. M., Symmons, D. P., and Croft, P. (1996). The epidemiology of hip osteoarthritis and rheumatoid arthritis in the Orient. *Clin. Orthop.* **323,** 81–90.

47. Jordan, J. M., Linder, G. F., Renner, J. B., and Fryer, J. G. (1995). The impact of arthritis in rural populations. *Arthritis Care Res.* **8,** 242–250.

48. Callahan, L. F., and Pincus, T. (1995). Mortality in the rheumatic diseases. *Arthritis Care Res.* **8,** 229–241.

49. Cerhan, J. R., Wallace, R. B., el-Khoury, G. Y., Moore, T. E., and Long, C. R. (1995). Decreased survival with increasing prevalence of full-body, radiographically defined osteoarthritis in women. *Am. J. Epidemiol.* **141,** 225–234.

50. Hochberg, M. C. (1991). Epidemiologic considerations in the primary prevention of osteoarthritis. *J. Rheumatol.* **18,** 1438–1440.

51. Silman, A. J., and Hochberg, M. C. (1993). "Epidemiology of the Rheumatic Diseases," pp. 257–288. Oxford University Press, Oxford.

52. Creamer, P., and Hochberg, M. C. (1997). Osteoarthritis. *Lancet* **350,** 503–508.

53. Felson, D. T., and Zhang, Y. (1998). An update on the epidemiology of knee and hip osteoarthritis with a view to prevention. *Arthritis Rheum.* **41,** 1343–1355.

54. Hartz, A. J., Fischer, M. E., Brill, G. *et al.* (1986). The association of obesity with joint pain and osteoarthritis in the HANES data. *J. Chronic Dis.* **39,** 311–319.

55. Davis, M. A., Ettinger, W. H., Neuhaus, J. M., and Hauck, W. W. (1998). Sex differences in osteoarthritis of the knee: the role of obesity. *Am. J. Epidemiol.* **127,** 1019–1030.

56. van Saase, J. L. C. M., Vandenbroucke, J. P., van Romunde, L. K. J., and Valkenburg, H. A. (1988). Osteoarthritis and obesity in the general population: A relationship calling for an explanation. *J. Rheumatol.* **15,** 1152–1158.

57. Bagge, E., Bjelle, A., Eden, S., and Svanborg, A. (1991). Factors associated with radiographic osteoarthritis: Results from a population study 70-year-old people in Göteborg. *J. Rheumatol.* **18,** 1218–1222.

58. Hart, D. J., and Spector, T. D. (1993). The relationship of obesity, fat distribution and osteoarthritis in women in the general population: The Chingford Study. *J. Rheumatol.* **20,** 331–335.

59. Cicuttini, F. M., Baker, J. R., and Spector, T. D. (1996). The association of obesity with osteoarthritis of the hand and knee in women: A Twin Study. *J. Rheumatol.* **23,** 1221–1226.

60. Felson, D. T., Anderson, J. J., Naimark, A. A., Walker, A. M., and Meenan, R. F. (1988). Obesity and knee osteoarthritis: The Framingham Study. *Ann. Intern. Med.* **109,** 18–24.

61. Felson, D. T., Zhang, Y., Hannan, M. T., Naimark, A., Weissman, B., Aliabadi, P., and Levy, D. (1997). Risk factors for incident radiographic knee osteoarthritis in the elderly: The Framingham Osteoarthritis Study. *Arthritis Rheum.* **40,** 728–733.

62. Lethbridge-Cejku, M., Creamer, P., Wilson, P. D., Hochberg, M. C., Scott, W. W., Jr., and Tobin, J. D. (1998). Risk factors for incident knee osteoarthritis: Data from the Baltimore Longitudinal Study on Aging. *Arthritis Rheum.* **41**(9, Suppl.), S182.

63. Spector, T. D., Hart, D. J., and Doyle, D. V. (1994). Incidence and progression of osteoarthritis in women with unilateral knee disease in the general population: The effect of obesity. *Ann. Rheum. Dis.* **53,** 565–568.

64. Manninen, P., Riihimaki, H., Heliovaara, M., and Mäkelä, P. (1996). Overweight, gender and knee osteoarthritis. *Int. J. Obes. Relat. Metab. Disord.* **20,** 595–597.

65. Davis, M. A., Neuhaus, J. M., Ettinger, W. H., and Mueller, W. H. (1990). Body fat distribution and osteoarthritis. *Am. J. Epidemiol.* **132,** 701–707.

66. Hochberg, M. C., Lethbridge-Cejku, M., Scott, W. W., Jr., Reichle, R., Plato, C. C., and Tobin, J. D. (1995). The association of body weight, body fatness and body fat distribution with osteoarthritis of the knee: Data from the Baltimore Longitudinal Study of Aging. *J. Rheumatol.* **22,** 488–493.

67. Davis, M. A., Ettinger, W. H., and Neuhaus, J. M. (1988). The role of metabolic factors and blood pressure in the association of obesity with osteoarthritis of the knee. *J. Rheumatol.* **15,** 1827–1832.

68. Martin, K., Lethbridge-Cejku, M., Muller, D., Elahi, D., Andres, R., Plato, C. C., Tobin, J. D., and Hochberg, M. C. (1997). Metabolic correlates of obesity and radiographic features of knee osteoarthritis: Data from the Baltimore Longitudinal Study of Aging. *J. Rheumatol.* **24,** 702–707.

69. Hart, D. J., Doyle, D. V., and Spector, T. D. (1995). Association between metabolic factors and knee osteoarthritis in women: The Chingford Study. *J. Rheumatol.* **22,** 1118–1123.

70. Felson, D. T., Zhang, Y., Anthony, J. M., Naimark, A., and Anderson, J. J. (1992). Weight loss reduces the risk for symptomatic knee osteoarthritis in women: The Framingham Study. *Ann. Intern. Med.* **116,** 535–539.

71. Heliovaara, M., Mäkelä, M., Imprivaara, O., Knekt, P., Aromaa, A., and Sievers, K. (1993). Association of overweight, trauma and

workload with coxarthrosis: A health study of 7,217 persons. *Acta Orthop. Scand.* **64,** 513–518.

72. Cooper, C., Inskip, H., Croft, P., Campbell, L., Smith, G., McLaren, M., and Coggon, D. (1998). Individual risk factors for hip osteoarthritis: Obesity, hip injury, and physical activity. *Am. J. Epidemiol.* **147,** 516–522.

73. Nevitt, M. C., Cummings, S. R., Lane, N. E., Hochberg, M. C., Scott, J. C., Pressman, A. R., Genant, H. K., and Cauley, J. A., for the Study of Osteoporotic Fractures Research Group (1996). Association of estrogen replacement therapy with the risk of osteoarthritis of the hip in elderly white women. *Arch. Intern. Med.* **156,** 2073–2080.

74. Cooper, C. (1995). Occupational activity and the risk of osteoarthritis. *J. Rheumatol.* **22**(Suppl. 43), 10–12.

75. Felson, D. T. (1994). Do occupation-related physical factors contribute to arthritis? *Bailliere's Clin. Rheumatol.* **8,** 63–77.

76. Maetzel, A., Mäkelä, M., Hawker, G., and Bombardier, C. (1997). Osteoarthritis of the hip and knee and mechanical occupational exposure: A systematic reviw of the evidence. *J. Rheumatol.* **24,** 1599–607.

77. Buckwalter, J. A. (1995). Osteoarthritis and articular cartilage use, disuse, and abuse: Experimental studies. *J. Rheumatol.* **22**(Suppl. 43), 13–15.

78. Cooper, C., McAlindon, T., Coggon, D., Egger, P., and Dieppe, P. (1994). Occupational activity and osteoarthritis of the knee. *Ann. Rheum. Dis.* **53,** 90–93.

79. Coggon, D., Kelingray, S., Inskip, H., Croft, P., Campbell, L., and Cooper, C. (1998). Osteoarthritis of the hip and occupational lifting. *Am. J. Epidemiol.* **147,** 523–528.

80. Panush, R. S., and Lane, N. E. (1994). Exercise and the musculoskeletal system. *Bailliere's Clin. Rheumatol.* **8,** 79–102.

81. Lane, N. E., and Buckwalter, J. A. (1993). Exercise: A cause of osteoarthritis? *Rheum. Dis. Clin. North Am.* **19,** 617–633.

82. Spector, T. D., Harris, P. A., Hart, D. J., Cicuttini, F. M., Nandra, D., Etherington, J., Wolman, R. L., and Doyle, D. V. (1996). Risk of osteoarthritis associated with long-term weight-bearing sports: A radiologic survey of the hips and knees in female ex-athletes and population controls. *Arthritis Rheum.* **39,** 988–995.

83. Lane, N. E., Bloch, D. A., Jones, H. H., Marshall, W., Jr., Wood, P. D., and Fries, J. F. (1986). Long distance running, bone density and osteoarthritis. *JAMA, J. Am. Med. Assoc.* **255,** 1147–1151.

84. Lane, N. E., Bloch, D. A., Hubert, H. B., Jones, H., Simpson, U., and Fries, J. F. (1990). Running, osteoarthritis, and bone density: Initial 2-year Longitudinal Study. *Am. J. Med.* **88,** 452–459.

85. Lane, N. E., Michel, B., Bjorkengren, A., Oehlert, J., Shi, H., Bloch, D. A., and Fries, J. F. (1993). The risk of osteoarthritis with running and aging: A 5-year Longitudinal Study. *J. Rheumatol.* **20,** 461–468.

86. Lane, N. E., Nevitt, M. C., Pressman, A., Cummings, S. R., Hochberg, M. C., and Scott, J. (1996). The relationship of physical activity and osteoarthritis of the hip in elderly white women. *Arthritis Rheum.* **39**(Suppl. 9), S309.

87. Neyret, P., Donell, S. T., DeJour, D., and DeJour, H. (1993). Partial meniscectomy and anterior cruciate ligament rupture in soccer players: A study with a minimum 20-year followup. *Am. J. Sports Med.* **21,** 455–460.

88. Roos, H., Lindberg, H., Gardsell, P., Lohmander, L. S., and Wingstrand, H. (1994). The prevalence of gonarthrosis and its relation to meniscectomy in former soccer players. *Am. J. Sports Med.* **22,** 219–222.

89. Davis, M. A., Ettinger, W. H., Neuhaus, J. M., Cho, S. A., and Houck, W. W. (1989). The association of knee injury and obesity with unilateral and bilateral osteoarthritis of the knee. *Am. J. Epidemiol.* **130,** 278–288.

90. Cooper, C., McAlindon, T., Snow, S., Vines, K., Young, P., Kirwan, J., and Dieppe, P. (1994). Mechanical and constitutional risk factors for symptomatic knee osteoarthritis: Differences between medial tibiofemoral and patellofemoral disease. *J. Rheumatol.* **21,** 307–313.

91. Hochberg, M. C. (1997). Osteoarthritis: Clinical features and treatment. *In* "Primer on the Rheumatic Diseases" J. H. Klippel, C. M. Weyand, and R. L. Wortmann, eds. 11th ed., pp. 218–221. Arthritis Foundation, Atlanta, GA.

92. Felson, D. T. (1993). The course of osteoarthritis and factors that affect it. *Rheum. Dis. Clin. North Am.* **19,** 607–615.

93. Hochberg, M. C. (1996). Progression of osteoarthritis. *Ann. Rheum. Dis.* **55,** 685–688.

94. Martin, K., Nicklas, B. J., Bunyard, L. B., Tretter, L. D., Dennis, K. E., Goldberg, A. P., and Hochberg, M. C. (1996). Weight loss and walking improve symptoms of knee osteoarthritis. *Arthritis Rheum.* **39**(Suppl. 9); S225.

94a. Ettinger, W. H., Jr., Burns, R., Messier, S. P., Applegate, W., Rejeski, W. J., Morgan, T., Shumaker, S., Berry, M. J., O'Toole, M., Monu, J., and Craven, T. (1997). A randomized trial comparing aerobic exercise and resistance exercise with a health education program in older adults with knee osteoarthritis. The Fitness Arthritis and Seniors Trial. *JAMA, J. Am. Med. Assoc.* **277,** 25–31.

95. McAlindon, R. E., Jacques, P., Zhang, Y., Hannan, M. T., Aliabadi, P., Weissman, B., Rush, D., Levy, D., and Felson, D. T. (1996). Do antioxidant micronutrients protect against the development and progression of knee osteoarthritis? *Arthritis Rheum.* **39,** 648–656.

96. McAlindon, R. E., Felson, D. T., Zhang, Y., Hannan, M. T., Aliabadi, P., Weissman, B., Rush, D., Wilson, P. W., and Jacques, P. (1996). Relation of dietary intake and serumn levels of vitamin D to progression of osteoarthritis of the knee among participants in the Framingham Study. *Ann. Intern. Med.* **125,** 353–359.

97. McAlindon, T., and Felson, D. T. (1997). Nutrition: Risk factors for osteoarthritis. *Ann. Rheum. Dis.* **56,** 397–402.

98. Hawley, D. J. (1995). Psycho-educational interventions in the treatment of arthritis. *Bailliere's Clin. Rheumatol.* **9,** 803–823.

99. Puett, D. W., and Griffin, M. R. (1994). Published trials of nonmedicinal and noninvasive therapies for hip and knee osteoarthritis. *Ann. Intern. Med.* **121,** 133–140.

100. Towheed, T. E., and Hochberg, M. C. (1997). A systematic review of randomized controlled trials of pharmacologic therapy in osteoarthritis of the hip. *J. Rheumatol.* **24,** 549–557.

101. Towheed, T. E., and Hochberg, M. C. (1997). A systematic review of randomized controlled trials of pharmacological therapy in osteoarthritis of the knee, with an emphasis on trial methodology. *Semin. Arthritis Rheum.* **26,** 755–770.

102. Hochberg, M. C., Altman, R. D., Brandt, K. D. *et al.* (1995). Guidelines for the medical management of osteoarthritis. Part I: Osteoarthritis of the hip. *Arthritis Rheum.* **38,** 1535–1540.

103. Hochberg, M. C., Altman, R. D., Brandt, K. D. *et al.* (1995). Guidelines for the medical management of osteoarthritis. Part II: Osteoarthritis of the knee. *Arthritis Rheum.* **38,** 1541–1546.

104. Hochberg, M. C., Perlmutter, D. L., Hudson, J. I., and Altman, R. D. (1996). Preferences in the management of osteoarthritis of the hip and knee: Results of a survey of community-based rheumatologists in the United States. *Arthritis Care Res.* **9,** 170–176.

105. Hochberg, M. C. (1996). "The Osteoarthritis Action Plan: Guidelines for the Medical Management of Osteoarthritis." Arthritis Foundation, Atlanta, GA.

106. Creamer, P., Flores, R. H., and Hochberg, M. C. (1998). Management of osteoarthritis in older adults. *Clin. Geriatr. Med.* **14**(3), 435–454.

96

Urinary Incontinence

NANCY H. FULTZ* AND A. REGULA HERZOG*,†

*Institute for Social Research, University of Michigan, Ann Arbor, Michigan; †Institute of Gerontology, Department of Psychology,
University of Michigan, Ann Arbor, Michigan

I. Introduction

Urinary incontinence is a health concern of particular relevance to women. The prevalence rate among women is at least twice as high as the rate among men [1]. About one in four younger women and one in three older women experience this condition [2]. Moreover, with the aging of the American population, greater numbers of women will be at risk.

This trend is significant because the consequences of incontinence can be debilitating. Urinary incontinence has been associated with psychosocial distress, with incontinent persons reporting diminished well-being and reduced quality of life as a result of the condition [3,4]. There are accounts of incontinent individuals curtailing social activities and sexual relationships because of concerns about odor and leakage [3,5]. The physical sequelae of constant wetness can include rashes, pressure ulcers, and urinary tract infections [6,7]. Falls (and, possibly, fractures) may result from a dash to the bathroom, a slip on wet floors, or the need for a late-night change of bedding [6,7]. In addition, the economic impact of urinary incontinence has been estimated at more than $26 billion dollars (1995 dollars), or over $3500 per incontinent person [8]. This includes the cost of diagnosis, treatment, routine care, and consequences such as longer hospitalization periods and additional admissions to institutions [8]. As suggested by the latter, incontinence has been cited as a major precipitating factor for nursing home admission [9]. Indeed, Thom *et al.* found that urinary incontinence increased the risk of hospital and nursing home admissions independent of age, sex, and several comorbid conditions [10].

Given the possibility of these outcomes, it is not surprising that urinary incontinence is an area of active research. In this chapter, we summarize current knowledge about incontinence and offer some guidance for future research. Several reviews of the literature on urinary incontinence are available and are cited in the following relevant sections.

II. Definition

In 1979 the International Continence Society (ICS) defined urinary incontinence as involuntary loss of urine that presents a social or hygienic problem and is objectively demonstrable [11]. Several aspects of this definition have been challenged and the resulting discussions have made it clear that the usefulness of a particular definition may depend on its purpose. For example, Foldspang and Mommsen [12] have argued that the part of the definition about urine loss presenting a problem confounds more subjective with more objective aspects and thus is less useful for epidemiologic descriptions and investigations of risk factors than an objective definition would be. The part of the definition that requires objective demonstration of loss limits the ICS definition for community-based epidemiologic investigations because objective demonstration of UI is difficult to achieve outside of the clinical setting; even in the clinic, self-reports of loss cannot always be demonstrated by clinical or urodynamic tests. Other important aspects of UI are lacking from the ICS definition. Thus, it is unclear whether any involuntary loss of urine constitutes UI or whether only a certain level of severity, as defined by frequency or quantity of loss, should be so considered. Likewise, the time frame over which occurrence of involuntary urine loss should be considered is left unspecified in the ICS definition. Because UI symptoms can occur irregularly, a time frame for measurement, such as a month or year, must be established.

Holtedahl and Hunskaar [13] calculated prevalence estimates using different definitions of UI for the same sample of 50 to 70 year old women. The prevalence of any self-reported leakage was 47%. Self-reported regular UI (two or more episodes of incontinence per month) with or without objective demonstration was found for 31% of women, and regular incontinence according to the full ICS definition was found for 19%. The results indicate that the ICS definition is rather restrictive and yields prevalence estimates that are lower than many other definitions used in epidemiologic studies.

A further factor complicating the conceptualization and measurement of UI in epidemiologic studies lies in the nature of the condition. UI is a chronic condition (or set of conditions) that often starts slowly and comes and goes for a considerable time period before it becomes fully established [14]. Moreover, over time people get used to their UI and notice it less. These aspects of the condition can interfere with valid assessment.

UI is defined by symptoms that are caused by a number of underlying pathologies. In recognition of the two major types of UI, the ICS definition also includes definitions of stress and urge types of incontinence. Symptoms of urge incontinence are UI associated with a strong and sudden desire to void. Urge incontinence often results from involuntary contractions of the smooth muscle in the bladder wall that can be identified through urodynamic testing. Symptoms of stress urinary incontinence are UI associated with physical activities such as coughing, sneezing, and laughing that increase intra-abdominal pressure. Stress urinary incontinence typically is caused by a weak sphincter or by displacement of the urethra and bladder neck during exertion.

III. Measurement

A. General Self Reports

Self reports of UI are widely used. They are elicited through standardized questions that translate medical symptomatology, definitions, and criteria into common language. Typically, respondents are asked about involuntary loss of urine for measuring any UI, about loss when sneezing or coughing for stress UI, and/or about loss without warning for urge UI. A scale of urinary and incontinence symptoms, termed the Urogenital Distress Inventory, has been developed to collect self-reported information [15]. This scale measures primarily symptoms of obstruction and irritation and thus cannot be considered a full measurement of UI and its types.

Responses to self-report questions reflect to some degree medical conditions and symptoms such as urinary incontinence, but they are also influenced by cognitive and motivational processes of the responding individual. In other words, respondents might be unwilling to report an embarrassing condition like incontinence and thereby underreport the condition [1]. They also might be unable to report specific frequencies of urine loss within given time frames because UI episodes can represent fairly frequent but not very salient events and thus be difficult to remember or estimate accurately [16].

Information about the psychometric properties of general self reports of urinary incontinence is limited. Resnick and his colleagues [17] have reported low test-retest reliability of reports of incontinence over the time period of two weeks, whereas we [1] found reasonable test-retest reliability between the beginning and the end of the questionnaire. A few validation investigations have been reported. Diokno and his colleagues invited both continent and incontinent respondents from a community survey for extensive clinical investigations. They found good agreement between self reports of UI and the clinical assessment (reported in Herzog and Fultz [1]). They also reported satisfactory agreement (a) between self reports of stress UI and a provocative stress test and urethral competence established by urodynamics and (b) between self reports of urge UI and uninhibited detrusor contractions established by urodynamics [18]. One Swedish study reported that only a small percentage of self-reported UI could not be verified in the clinic [19]. Another study [20] revealed less than satisfactory predictive validity of self-reported types of UI compared with urodynamic investigations. When there is disagreement between measures, it is not obvious which measure is lacking validity. Clinical and even urodynamic investigations have their own characteristic errors and therefore should be regarded as other measures, rather than as gold standards. More research on the quality of the available survey questions and scales is needed.

B. Diaries

Diaries of urine loss episodes and urinary symptoms have been used to reduce the cognitive burden of reporting because of their shorter reporting period. However, diaries have their own limitations, including the need to keep them over extended periods of time in order to avoid undue influence of short-term variation and the resulting burden to respondents. A few studies examined the test-retest reliability of the diary method for the measurement of incontinence and its validity against standard self-report measures or urodynamic findings and reported satisfactory correlations [21–23], although measures of agreement would have been more appropriate statistical measures.

C. Proxy Reports

Reports by staff or next of kin often are used for the assessment of institutionalized or impaired populations. An example is the Resident Assessment Protocol developed for the Minimum Data Set that is based on staff observations and chart reviews. The reliability and validity of the protocol for assessment of UI are encouraging [24].

D. Clinical Evaluation

In the clinic, symptoms of UI can be elicited during a medical history-taking. The questions that are asked are similar to those used in self-reported survey instruments but typically are less standardized. For the evaluation of stress incontinence, a provocative stress test provides direct observation of leakage when coughing with a full bladder, preferably in a standing position. Such a test shows satisfactory agreement with the self report of stress UI [18]. A number of more specialized urodynamic tests of bladder and urethral dysfunctions are also available to aid diagnosis but are not recommended for standard assessment of UI and its types.

IV. Historical Evolution/Trends

In our opinion, advances in the diagnosis and treatment of urinary incontinence have been associated with a gradual change in attitudes about the condition. Although there continue to be gaps in the public's knowledge about incontinence [25], we detect a greater openness about the condition and a lessening of the stigma that has surrounded it. The 1990s may have brought urinary incontinence into the mainstream of public awareness and acceptance. It would be interesting to verify or refute this observation through a content analysis of articles, programs, and advertisements from print and broadcast media. It also would be interesting to replicate earlier measures of attitudes and beliefs about incontinence to document the presence or absence of change. In one early study, Mitteness used qualitative methods to assess beliefs and attitudes about incontinence [26]. The Medical, Epidemiologic, and Social Aspects of Aging (MESA) project took a more quantitative approach [27,28]. To our knowledge, the most recent major review of the literature on beliefs about incontinence was prepared in the late 1980s [29].

V. Distribution of Urinary Incontinence among Women

A. Prevalence

Prevalence—the probability of being incontinent within a defined population and at a defined time point—is important for establishing the distribution of the condition in the population and for projecting the need for health and medical services. Several reviews of UI prevalence are available [1,2,6,30–32].

1. All-Cause UI

In a 1988 review of several European and American epidemiologic studies of older women, we identified a 10–40% range of prevalence estimates of any UI and suggested a preliminary UI prevalence of 40% among older women as not improbable

[1]. Whereas the wide range had been noted before [32], we systematically reviewed the evidence and argued that the variability cannot be attributed solely to variations in the definition of UI or in study samples, as is usually done. We argued that differential underreporting, attributable to variation in survey procedures such as topic introduction, may also be important. Thus, we placed the prevalence estimate at the higher end of the range. Since 1988, a number of additional epidemiologic studies—many of them European—have been published; we reviewed some of them in 1996 [30]. Another review [31] applied meta-analytical techniques to review 48 available European and North American prevalence studies of UI and argued for overreporting of UI in epidemiologic surveys without clinical or urodynamic confirmation. At this point we are not convinced by their argument and see no reason to reconsider the conclusions from our two previous reviews, which attested to the wide range of UI prevalence estimates and placed the likely prevalence of any incontinence among older women around 40%. This conclusion is supported by a 1998 U.S. study that reports a 49% prevalence of any UI among women 50 years of age or older [33] and by a descriptive review of major UI studies by Thom [2] that places prevalence of any UI among older women at about 35% and among younger women at about 28%.

2. Severe UI

In our previous review [1] we also noted that the prevalence estimate for severe UI (typically defined as urine loss at least weekly, "regularly," or "most of the time") among older women is lower than the prevalence of any incontinence and that the available estimates are less variable, ranging from 5 to 15% and centering around 7%. Likewise, Thom's [2] review places daily UI among older women at about 14%. We suggested that the lesser variation might be due to the fact that a severe condition cannot be as easily denied and is therefore less sensitive to variation in question wording, survey design, and survey procedures.

3. Types of UI

In our previous review we suggested that older women suffer primarily from simple stress or mixed incontinence [1]. Thom's [2] review shows that the proportion of incontinent women suffering from simple stress to those suffering from urge and mixed forms declines with age; whereas about half of all middle-aged and younger incontinent women suffer from simple stress UI, only about one quarter of all older incontinent women suffer from simple stress UI.

B. Incidence and Remission of UI

Incidence—the probability of becoming incontinent over a defined period of time—is the appropriate measure for studying the onset of the condition and its risk factors. Because of the formidable data requirements of incidence studies, little information is available about the incidence and the remission of UI and about related risk factors.

We know of only a few studies that have investigated incidence and remission of UI among older women and most were reviewed earlier [1,30,34]. While the reported incidence rates differ considerably, it is clear that they are substantial. Likewise,

rates of remission—the probability among previously incontinent persons of becoming continent—while varying across studies, are not negligible. Incidence and remission rates are confounded by unreliability of the measures. As noted, some of the UI measures have limited reliability. Future research must address the issue of how much of the observed onset and remission of UI is attributable to measurement error.

C. Sociodemographic Differences in UI

Evidence about age differences in UI among women is somewhat conflicting. Thom [2] indicates a modest trend for higher prevalence among older women than among middle-aged and younger women.

Few studies have examined racial differences. In a nationwide survey of Americans 70 years old and older, we observed a significant racial difference in the prevalence of UI; Caucasian women indicated a higher prevalence of all-cause UI than did African-American women [35]. Likewise, Brown and her colleagues [36] found a higher prevalence of stress incontinence for white women than for African-American women. Two other studies reported similar results [37,38].

Surprisingly, very little research has focused on socioeconomic status and UI, despite the fact that socioeconomic status has turned out to be an important correlate in many other aspects of health. For a review of work in this area, see the article by Adler *et al.* [39].

VI. Influence of Women's Social Roles or Context

Wells [40] noted the importance of gender differences in norms regarding urination and urine control. She suggested that men are allowed greater freedom to urinate whenever the need arises, whereas women are more encumbered by culturally prescribed rules of modesty. The design and location of public toilets has reflected and reinforced these gender differences in attitudes and behaviors [40].

It also is likely that women and men interpret and respond to the symptom of urine loss differently. Women's experiences with incontinence during pregnancy or postpartum may make urine loss appear to be a normal part of being female [41]. Similarly, experience with menstruation may make it easier for women to adapt to urinary leakage [14,40]. These processes of normalization and adaptation can deter seeking treatment for incontinence [41]. Moreover, physicians, traditionally, have been male, and embarrassment about discussing urine loss with someone of the opposite sex may prevent women from seeking treatment [40]. One study found that only 13% of incontinent women had sought care for urinary symptoms, compared with 29% of incontinent men [33].

VII. Risk Factors

The literature on risk factors for urinary incontinence is relatively young, and knowledge in this area is incomplete. In our attempts to summarize findings, we have been frustrated by the inconsistency of results across studies. We suspect that the lack of consensus regarding risk factors is due, at least in part, to variations in sample design, specificity of measurement, and analytic approaches. This issue is addressed in greater detail in the Epidemiologic Issues section.

A. Health

1. Poor Health

Urinary incontinence is associated with a wide range of medical conditions, suggesting its multifactorial origin [42]. Self-rated health measures often are used to collect survey respondents' subjective evaluations or summaries of their total health situation. Using data from the East Boston component of the Established Populations for Epidemiologic Study of the Elderly, Wetle and her colleagues found that poor self-rated health was related to difficulty holding urine [43]. Similarly, Brown *et al.* [44] reported that the prevalence of daily incontinence increased significantly with poor self-rated health. In multivariate analyses of a national sample of community-dwelling older adults, the relationship between incontinence and poor self-rated health was significant for certain subgroups of the sample [45]. When incontinence status was used as a predictor of four health concepts, however, it was found to relate to perceived limitations in usual role activities because of physical health, but not to perceived limitations in usual role activities because of emotional problems, to perceived limitations in social activities, or to respondents' overall perceptions of their health [46].

2. Urinary and Bowel Symptoms

Women who experience other urinary or bowel problems may be predisposed to incontinence. Data from the MESA project showed a significant relationship between urine loss and symptoms of bladder irritation/infection or difficulty emptying the bladder [42]. Constipation and fecal incontinence also were identified as correlates of urinary incontinence in the MESA data [42]. Similarly, a study of older East Boston residents found that fecal incontinence was related to difficulty holding urine [43]. A study of older women in the Netherlands reported that urinary frequency, urinary urgency (for ages 60–84), and nocturia (for ages 85+) were independently associated with urinary incontinence, although fecal incontinence was not [47].

3. Functional Impairments

Functional impairments, particularly mobility limitations, are associated with difficulty holding urine [43] and with urine loss [42,48,49]. Indicators of mobility limitations include a diagnosis of arthritis, the use of equipment such as a cane or walker, and several measures of lower body physical performance. Two causal paths have been proposed to interpret the bivariate relationship [6]. One possibility is that incontinence is the direct consequence of problems with accessing a toilet and adjusting clothing. Alternatively, both incontinence and mobility limitations may be due to an underlying condition such as dementia, Parkinsonism, or stroke.

4. Cognitive Impairment

Several studies have reported a relationship between cognitive impairment and urinary incontinence. Campbell *et al.* [50] studied older adults living in New Zealand and concluded that those with dementia were more likely to be incontinent than

were those with normal cognitive function. Similar results were reported by Ekelund and Rundgren [51] on admissions to a long-term-care institution and by Ouslander and his colleagues [49], who found a relationship between mental function measures and continence status among VA nursing home residents. No relationship between mental status and difficulty holding urine was found in a general community sample, however [43].

B. Reproductive Factors

A comprehensive review of the clinical and epidemiologic research literature on reproductive and hormonal risk factors for urinary incontinence is provided by Thom and Brown [52].

1. Childbirth

Childbirth is one of the most heavily researched factors in women's incontinence. It is reasonable to suppose that the extra weight of pregnancy, followed by the straining and stretching of labor and birth, would weaken and/or damage the physical structures that support urine control. Several studies found a relationship between pregnancy or childbirth and incontinence [37,53–56], a few did not [42,57,58], and one reported the relationship only for white women [38]. Studies vary greatly in the extent to which specifics about labor and delivery are provided. In their review of this literature, Thom and Brown concluded that vaginal delivery is a risk factor for both transient postpartum incontinence and incontinence in later life, with the strongest association found for stress incontinence [52].

2. Menopause/Hormone Replacement

With menopause, atrophy of the urogenital tissues increases susceptibility to urinary tract infections and to urinary frequency and urgency. Estrogen replacement can reverse these atrophic changes. Moreover, estrogen therapy for urinary incontinence appears to subjectively improve this condition in postmenopausal women [59]. Taken together, this evidence suggests that estrogen loss at menopause contributes to incontinence.

Findings have not confirmed this relationship, however. Rekers and colleagues reported somewhat mixed results about the relationship of menopause and incontinence [60]. Milsom *et al.* [54] reported no significant difference in the prevalence of incontinence between pre- and postmenopausal women from two birth cohorts, despite excluding women who were taking estrogen or had had surgical removal of the ovaries. Separate studies of middle-aged women [57], 45-year-old women [58], and women aged 25 and over (plus those under 21 taking oral contraceptives) [55] also failed to document higher rates of incontinence among those who were postmenopausal. Moreover, three studies that included estrogen replacement as a correlate of urinary incontinence found an unexpected significant positive relationship between these two variables [37,42,44].

3. Hysterectomy

Hysterectomy has been investigated as a risk factor for incontinence, both because of its impact on hormone levels and because of the possibility of neurological damage to the pelvic floor during surgery. Two studies that examined the relationship between prior hysterectomy and urinary incontinence found a positive association [44,54]. A third study found a similar (but not statistically significant) relationship [58], and a fourth reported that the association between hysterectomy and incontinence became insignificant after adjustment for estrogen replacement [37]. Thom and Brown [52] noted that findings of a relationship between hysterectomy and incontinence typically emerge from retrospective epidemiologic studies such as these, rather than from prospective studies of outcomes in the first years after surgery. This pattern is consistent with the hypothesis of nerve damage, which, while initially compensated, could increase the risk of incontinence in later life [52].

C. Behaviors

1. Smoking

Cigarette smoke appears to have a harmful effect on the bladder. For example, many patients with interstitial cystitis believe that smoking worsens their symptoms [61]. Thus, it seems reasonable to suspect that substances in cigarette smoke could lead to incontinence. In addition, a chronic cough due to smoking might exert pressure on pelvic floor muscles, resulting in urine loss. Bump and McClish [62] found a strong statistical association between smoking and incontinence in a case control study of over 600 women. Similarly, Tampakoudis *et al.* [63] reported a statistically significant relationship between smoking and incontinence. Smoking was not related to continence status, however, in a survey of healthy, middle-aged women conducted by Burgio and her colleagues [57]. Cigarette smoking also was not associated with daily urinary incontinence in a large well-controlled study of community-dwelling older women [44].

2. Exercise

Speculations about the relationship between exercise and incontinence have suggested both a positive and a negative relationship. Immobility is associated with incontinence, which raises the possibility that increased mobility might promote continence. A study of improvements in urine control gained through a nursing home walking program supports this hypothesis [64]. Exercise that involves impact, however, such as aerobics or running, could increase intra-abdominal pressure enough to cause incontinence during those activities. Findings that about one-third of women who exercise experience incontinence during repetitive bouncing activities confirm this [65]. Of course, the two arguments are not mutually exclusive, and women need not forgo exercise due to concerns about urine loss. Exercise-induced incontinence can be successfully prevented with simple mechanical devices such as a tampon [66]. Moreover, strenuous physical exercise at younger ages does not appear to predispose women to incontinence in later life [67]. Of concern are those women who discontinue physical activities secondary to UI, thus missing the health benefits of regular exercise [65].

3. Obesity

Obesity is commonly thought to contribute to incontinence, and a number of studies that include body mass index as a predictor find a significant association [36,37,44,57,67,68]. For example, multivariate analysis of survey data from female members of a large HMO showed that the likelihood of incontinence was significantly higher for respondents in the fourth quartile of BMI than for those in the first [37]. Moreover, Bump and his colleagues reported improvements in lower urinary tract function for obese women following surgery to reduce weight [69]. One study that included waist/hip ratio found central obesity to be a significant predictor for stress incontinence, independent of BMI [36]. Kölbl and Riss [70], however, did not find a significant difference between continent and incontinent women for several indices of relative weight, although markedly high BMI was associated with a positive clinical stress test.

VIII. Clinical Issues (Diagnosis, Treatment)

Given findings that a large percentage of people with urinary incontinence never report it to a health care professional [33,53,57,71,72], it is incumbent upon clinicians to raise the topic with their patients. This can be done in a supportive and nonjudgmental way through simple open-ended questions such as "Tell me about any problems you are having with your bladder" [73]. Observed wetness, skin irritation, or a detectable odor of urine are other clues to the presence of incontinence, as are comments from caregivers. Any indications of urinary incontinence should lead to further diagnosis and treatment.

The *Clinical Practice Guideline Update* [73] published by the Agency for Health Care Policy and Research is a comprehensive reference on the diagnosis and treatment of urinary incontinence in adults. This 1996 document reflects the work of a large panel of prominent gynecologists, urologists, nurses, gerontologists, and other specialists. We refer interested readers to its useful text and extensive bibliography. The writing of the following sections on diagnosis and treatment was informed largely by this reference.

A. Diagnosis

Urinary incontinence can be a symptom or consequence of numerous internal (physiologic) and/or external factors. Therefore, a careful evaluation is necessary to confirm the presence of the condition, identify possible contributing factors, differentiate patients who require further testing from those who can begin initial treatment, and develop a diagnosis. The basic evaluation should include a history, physical examination, estimation of postvoid residual volume, and urinalysis [73]. For older women, age-related changes to the lower urinary tract and the likelihood of additional pathologic, physiologic, or pharmacologic factors must be considered [7]. When necessary, additional information can be collected through specialized diagnostic tests, including urodynamic tests, endoscopic tests, and imaging tests [73].

B. Treatment

1. Behavioral

Behavioral interventions are frequently selected as a first-line treatment for incontinence because they are often effective, carry little risk of side effects, and do not limit future treatment options. Included in this category are toileting assistance, bladder retraining, and pelvic muscle rehabilitation. The optimal technique depends on the cognitive and physical capacities of the incontinent person, the type of incontinence that she experiences, and the setting (home or long-term care facility) in which she resides [74].

Toileting assistance comprises routine or scheduled toileting, habit training, and prompted voiding [73]. These interventions establish a pattern for urination, which may reduce incontinent episodes. With routine or scheduled toileting, the goal is to keep the patient dry by offering the opportunity to void every two to four hours. Habit training modifies that approach by taking the incontinent person's natural voiding pattern into account when scheduling toileting. This technique is particularly appropriate for people living at home with a caregiver [73]. Prompted voiding is used as a supplement to habit training for individuals who have the cognitive capacity to respond to bladder fullness or to prompts to void. The three major elements of prompted voiding are monitoring (to check for incontinence and to reinforce the ability to discriminate between wetness and dryness), prompting (to teach appropriate responses to requests to use the toilet or to bladder fullness), and praising (to reward continence and attempts to toilet) [73].

Bladder retraining involves cognitive and physiologic suppression of the urge to void in order to prolong the periods between voids. This is a particularly effective strategy for cognitively aware patients with urge or mixed incontinence, although it also has been found to be valuable in the management of stress incontinence [73]. In a bladder retraining program, the patient learns about lower urinary tract function and about relaxation, pelvic muscle contraction, and distraction techniques for inhibiting urinary urgency. She is then given a timetable for urination that gradually increases the length of time between scheduled voids. The efficacy of this method was evaluated by Fantl and his colleagues in a clinical trial of 123 women 55 years of age or older [75]. They found that bladder training significantly reduced the number of incontinent episodes per week in the treatment group [75].

Rehabilitation of pelvic muscles can be achieved through pelvic muscle exercises. The strengthening of periurethral and perivaginal muscles through repeated contractions and relaxation has been advocated as a treatment for women with stress incontinence [73,76]. The use of a planned contraction that is timed to precede an increase in intra-abdominal pressure has demonstrated reductions in stress urine loss [77]. Biofeedback may be of assistance in learning and practicing pelvic muscle exercises [78]. Vaginal weight training has also been used to augment these exercises [79].

Electrical stimulation of the pelvic floor is another approach to pelvic muscle rehabilitation. Electrical pulses are transmitted through vaginal or anal sensors or surface electrodes to induce contraction of the levator ani, external urethral and anal sphincters [73]. Interestingly, electrical stimulation appears to contract the pelvic structures *and* to inhibit detrusor overactivity, suggesting its effectiveness for treating urge and mixed incontinence as well as the pure stress type [73,80,81].

2. Pharmacologic

Pharmacotherapy can be a beneficial approach to treating urinary incontinence [6,73,82,83]. For urge incontinence, medications with anticholinergic and smooth muscle relaxant properties are effective. Oxybutynin has both properties and has been ap-

proved specifically for the control of uninhibited bladder contractions. It has been considered the anticholinergic agent of choice, with propantheline as the second-line option [73]. A new drug, tolterodine, is now available for the treatment of urgency symptoms. In clinical trials, tolterodine was found to be effective and better tolerated than oxybutynin [84]. Side effects of anticholinergic agents include dry mouth, blurred vision, and constipation. In addition, these drugs should be used with caution in cognitively impaired patients because of the potential for worsening mental status. Anticholinergic medications should not be used for patients with documented narrow angle glaucoma. Tricyclic antidepressants such as imipramine may be useful for decreasing bladder contractility but should be reserved for carefully evaluated patients because of adverse cardiac and anticholinergic effects.

Pharmacotherapy for stress urinary incontinence includes alpha-adrenergic agonists and estrogen. The sustained release form of phenylpropanolamine is the first-line alpha-adrenergic agonist. This drug works to increase outlet resistance by enhancing urethral smooth muscle tone. It is contraindicated in patients with coronary heart disease, hypertension, or thyroid disease. Side effects include dry mouth, nausea, insomnia, rash, itching, and restlessness. For postmenopausal women with stress incontinence, estrogen therapy—either alone or in combination with phenylpropanolamine—may be appropriate, although the evidence is inconsistent (for a full discussion, see Thom and Brown [52]). Estrogen appears to have a beneficial effect on the urethra, as reflected in subjective improvement of symptoms; the exact mechanism through which it might promote continence is not known, however, and its efficacy has yet to be conclusively demonstrated in randomized controlled trials. This therapy is contraindicated in women with known or suspected breast cancer or uterine cancer, and estrogen therapy should include a progestin if a woman's uterus is intact.

3. Surgical

Surgery can be an appropriate treatment for selected incontinent women (and men) for whom more conservative measures are not feasible or were not successful. It is generally reserved for patients with obstruction or with stress incontinence due to urethral hypermobility and/or intrinsic sphincter deficiency [82]. Surgery for the management of urge incontinence is not common and is usually considered only for highly symptomatic patients for whom other options have failed [73].

Surgical procedures designed to stabilize the urethra in the correct position include needle suspension of the bladder neck and retropubic suspension of the urethrovesical junction [73,82]. These techniques appear to be more efficacious than anterior vaginal repair for hypermobility [73]. For the management of intrinsic sphincter deficiency, options include placing a sling under the urethrovesical junction, injecting a bulking material such as collagen into the periurethral area, or implanting an artificial sphincter [73,82]. Reviews of the literature on surgical techniques for incontinence have noted the difficulty of evaluating and comparing procedures because of omissions or variations across the relevant studies [6,73,85].

4. Self-Care Behaviors

Self-care behaviors are widely used to manage involuntary urine loss [72,86,87]. Among the most popular of these methods are frequent voiding, using protective products or garments, locating or staying near a toilet, and changing volume or type of beverages. Although the majority of women report satisfaction with them, most self-care behaviors will do little or nothing to improve bladder control over time [86]. Some, in fact, may be counterproductive [86]. Limiting fluid intake, for example, results in highly concentrated and more irritating urine.

IX. Epidemiologic Issues

In addition to the definition and measurement issues, study design, sample design, mode of data collection, and analytic methods are important issues in epidemiologic studies of UI. Variability in these methodological features of past studies are likely to account for some of the variability of estimates of prevalence, incidence, and risk factors reported in the literature. We will comment on each of these issues briefly.

A. Design

A rather large number of studies on the prevalence of UI have been published, but the studies use almost exclusively cross-sectional designs. Cross-sectional designs can provide prevalence estimates and document correlations with risk factors but allow only weak inferences about causation. Longitudinal or prospective designs are required to estimate incidence rates of UI, to describe the course of the condition according to types and severity, and to provide the basis for strong causal inferences between the condition and its risk factors. Such designs are still rare in UI research and should be given high priority.

B. Sample Design and Response Rate

Many samples in UI research are small, and small samples cannot give reliable results because of lack of power. This is especially problematic when comparing findings across race, ethnicity, or other demographic characteristics that subdivide the population. Moreover, even large samples may not be representative of the population to which generalizations are to be made. For example, clinical groups do not represent the entire adult population and may provide misleading findings regarding disease severity or treatment utilization. Findings from samples of older women cannot be generalized to younger women (and vice versa) because of the possibility that risk factors or other correlates change with age. Further, population samples often include only noninstitutionalized respondents and so omit the oldest and frailest individuals. Finally, even large representative samples are unbiased only if response rates are high or nonresponse is not systematic. Not all studies in UI research obtain high response rates or evaluate the nature of the nonrespondents.

C. Data Collection Mode

Many studies collect data on UI by mailed self-administered questionnaires; fewer studies have used face-to-face or telephone interviews. Interviews allow more detailed explorations of issues and they generally achieve higher response rates than do mail questionnaires. But it is also possible that responses elicited by interview are more susceptible to social desirability bias than those elicited by mail questionnaires. The only available study of this issue, by Thom [2], suggests little effect of the mode of data collection on estimates of the prevalence of urinary incontinence.

D. Analytical Methods

Bivariate statistical analysis procedures such as tables or correlations do not control for confounding variables, which are commonplace in nonexperimental research. When confounding variables are not controlled, the analysis may lead to erroneous conclusions due to spurious correlations. The fact that some studies present only bivariate relationships, whereas others present multivariate models, undoubtedly explains some of the inconsistency of findings on risk factors for incontinence.

The failure to control for type of incontinence when examining the relationship between incontinence and risk factors contributes to the inconsistency of findings. Because involuntary urine loss can be a symptom of different underlying conditions, it could be expected that a single set of risk factors might not apply to both urge and stress incontinence. Studies, however, have not always had sufficiently detailed measures of incontinence and sufficiently large sample sizes to do more than compare those who report urine loss with those who do not. A priority should be placed on research that improves our ability to accurately predict who is at risk for incontinence by delineating the risk factors by type of UI.

X. Conclusion

Significant gains have been made in the epidemiology, diagnosis, and treatment of urinary incontinence. This is good news for the many people who have endured involuntary urine loss. A need for additional work is evident, however, from the number of inconsistent findings and unresolved issues that remain. In this regard, the lack of comparability across studies of urinary incontinence has been a major hindrance. There have been strong calls for greater uniformity of measurement and design, including a standardized definition of the condition. We endorse these calls, with the caveat that any choice of standards must be guided by principles of scientific rigor. Given the importance of research on incontinence, we are confident that this need will be addressed and that further advances are forthcoming.

Acknowledgments

The authors thank Jeanette Brown, Ananias Diokno, and Carolyn Sampselle for helpful comments on an earlier version of this chapter. Funding for the writing of this chapter was provided, in part, by grants R37 AG08511 from the National Institute on Aging and R01 DK47543 from the National Institute of Diabetes and Digestive and Kidney Diseases.

References

1. Herzog, A. R., and Fultz, N. H. (1990). Prevalence and incidence of urinary incontinence in community-dwelling populations. *J. Am. Geriatr. Soc.* **38**, 273–281.
2. Thom, D. (1998). Variation in estimates of urinary incontinence prevalence in the community: Effects of differences in definition, population characteristics, and study type. *J. Am. Geriatr. Soc.* **46**, 473–480.

3. Naughton, M. J., and Wyman, J. F. (1997). Quality of life in geriatric patients with lower urinary tract dysfunction. *Am. J. Med. Sci.* **314,** 219–227.

4. Wyman, J. F., Harkins, S. W., and Fantl, J. A. (1990). Psychosocial impact of urinary incontinence in the community-dwelling population. *J. Am. Geriatr. Soc.* **38,** 282–288.

5. Lam, G. W., Foldspang, A., Elving, L. B., and Mommsen, S. (1992). Social context, social abstention, and problem recognition correlated with adult female urinary incontinence. *Dan. Med. Bull.* **39,** 565–570.

6. Herzog, A. R., Diokno, A. C., and Fultz, N. H. (1989). Urinary incontinence: Medical and psychosocial aspects. *Annu. Rev. Gerontol. Geriatr.* **9,** 74–119.

7. Resnick, N. M. (1996). Geriatric incontinence. *Urol. Clin. North Am. Geriatr. Urol.* **23,** 55–74.

8. Wagner, T. H., and Hu, T. W. (1998). Economic costs of urinary incontinence in 1995. *Urology* **51,** 355–361.

9. Smallegan, M. (1985). There was nothing else to do: Needs for care before nursing home admission. *Gerontologist* **25,** 364–369.

10. Thom, D. H., Haan, M. N., and Van Den Eeden, S. K. (1997). Medically recognized urinary incontinence and risks of hospitalization, nursing home admission and mortality. *Age Ageing* **26,** 367–374.

11. Bates, P., Bradley, W. E., Glen, E., Griffiths, D., Melchior, H., Rowan, D., Sterling, A., Zinner, N., and Hald, T. (1979). The standardization of terminology of lower urinary tract function. *J. Urol.* **121,** 551–554.

12. Foldspang, A., and Mommsen, S. (1997). The International Continence Society (ICS) incontinence definition: Is the social and hygienic aspect appropriate for etiologic research? *J. Clin. Epidemiol.* **50,** 1055–1060.

13. Holtedahl, K., and Hunskaar, S. (1998). Prevalence, 1-year incidence and factors associated with urinary incontinence: A population based study of women 50–74 years of age in primary care. *Maturitas* **28,** 205–211.

14. Fultz, N. H., and Herzog, A. R. (1993). Measuring urinary incontinence in surveys. *Gerontologist* **33,** 708–713.

15. Shumaker, S. A., Wyman, J. F., Uebersax, J. S., McClish, D., and Fantl, J. A. (1994). Health-related quality of life measures for women with urinary incontinence: The Incontinence Impact Questionnaire and the Urogenital Distress Inventory. Continence Program in Women (CPW) Research Group. *Qual. Life Res.* **3,** 291–306.

16. Menon, G. (1994). Judgements of behavioral frequencies: Memory search and retrieval strategies. *In* "Autobiographical Memory and the Validity of Retrospective Reports" (N. Schwarz and S. Sudman, eds.), pp. 161–172. Springer, New York.

17. Resnick, N. M., Beckett, L. A., Branch, L. G., Scherr, P. A., and Wetle, T. (1994). Short-term variability of self report of incontinence in older persons. *J. Am. Geriatr. Soc.* **42,** 202–207.

18. Diokno, A. C., Normolle, D. P., Brown, M. B., and Herzog, A. R. (1990). Urodynamic tests for female geriatric urinary incontinence. *Urology* **36,** 431–439.

19. Molander, U., Milsom, I., Ekelund, P., and Mellström, D. (1990). An epidemiologic study of urinary incontinence and related urogenital symptoms in elderly women. *Maturitas* **12,** 51–60.

20. Kirschner-Hermanns R., Scherr, P. A., Branch, L. G., Wetle, T., and Resnick, N. M. (1998). Accuracy of survey questions for geriatric urinary incontinence. *J. Urol.* **159,** 1903–1908.

21. Diokno, A. C., Wells, T. J., and Brink, C. A. (1987). Urinary incontinence in elderly women: Urodynamic evaluation. *J. Am. Geriatr. Soc.* **35,** 940–946.

22. Elser, D. M., Fantl, J. A., and McClish, D. K., and the Continence Program for Women Research Group. (1995). Comparison of "subjective" and "objective" measures of severity of urinary incontinence in women. *Neurourol. Urodyn.* **14,** 311–316.

23. Wyman, J. F., Choi, S. C., Wilson, M. S., and Fantl, J. A. (1988). The urinary diary in evaluation of incontinent women: A test-retest analysis. *Obstet. Gynecol.* **71,** 812–817.

24. Resnick, N. M., Brandeis, G. H., Baumann, M. M., and Morris, J. N. (1996). Evaluating a national assessment strategy for urinary incontinence in nursing home residents: Reliability of the Minimum Data Set and validity of the Resident Assessment Protocol. *Neurourol. Urodyn.* **15,** 583–598.

25. Branch, L. G., Walker, L. A., Wetle, T. T., DuBeau, C. E., and Resnick, N. M. (1994). Urinary incontinence knowledge among community-dwelling people 65 years of age and older. *J. Am. Geriatr. Soc.* **42,** 1257–1262.

26. Mitteness, L. S. (1987). So what do you expect when you're 85?: Urinary incontinence in late life. *Res. Sociol. Health Care* **6,** 177–219.

27. Herzog, A. R., and Fultz, N. H. (1988). Urinary incontinence in the community: Prevalence, consequences, management, and beliefs. *Top. Geriatr. Rehabil.* **3,** 1–12.

28. Herzog, A. R., and Fultz, N. H. (1990). Epidemiology of urinary incontinence: Prevalence, incidence, and correlates in community populations. *Urology, Suppl.* **36,** 2–10.

29. Mitteness, L. S. (1990). Knowledge and beliefs about urinary incontinence in adulthood and old age. *J. Am. Geriatr. Soc.* **38,** 374–378.

30. Fultz, N. H., and Herzog, A. R. (1996). Epidemiology of urinary symptoms in the geriatric population. *Urol. Clin. North Am.: Geriatr. Urol.* **23,** 1–10.

31. Hampel, C., Wienhold, D., Benken, N., Eggersmann, C., and Thüroff, J. W. (1997). Prevalence and natural history of female incontinence. *Eur. Urol.* **32** (suppl. 2), 3–12.

32. Mohide, E. A. (1986). The prevalence and scope of urinary incontinence. *Clin. Geriatr. Med.* **2,** 639–659.

33. Roberts, R. O., Jacobsen, S. J., Rhodes, T., Reilly, W. T., Girman, C. J., Talley, N. J., and Lieber, M. M. (1998). Urinary incontinence in a community-based cohort: Prevalence and healthcare-seeking. *J. Am. Geriatr. Soc.* **46,** 467–472.

34. Nygaard, I. E., and Lemke, J. H. (1996). Urinary incontinence in rural older women: Prevalence, incidence and remission. *J. Am. Geriatr. Soc.* **44,** 1049–1054.

35. Fultz, N. H., Herzog, A. R., Raghunathan, T. E., Wallace, R. B., and Diokno, A. C. (1999). Prevalence and severity of urinary incontinence in older African American and Caucasian women. *J. Gerontol. Med. Sci.* **54A,** M299–M303.

36. Brown, J. S., Grady, D., Ouslander, J. G., Herzog, A. R., Varner, R. E., and Posner, S. F. (1999). Prevalence of urinary incontinence and associated risk factors in postmenopausal women. Heart & Estrogen/Progestin Replacement Study (HERS). *Obstet. Gynecol.* **94,** 66–70.

37. Thom, D. H., Van Den Eeden, S. K., and Brown, J. S. (1997). Evaluation of parturition and other reproductive variables as risk factors for urinary incontinence in later life. *Obstet. Gynecol.* **90,** 983–989.

38. Burgio, K. L., Locher, J. L., Zyczynski, H., Hardin, J. M., and Singh, K. (1996). Urinary incontinence during pregnancy in a racially mixed sample: Characteristics and predisposing factors. *Int. Urolgynecol. J.* **7,** 69–73.

39. Adler, N. E., Boyce, W. T., Chesney, M. A., Cohen, S., Folkman, S., Kahn, R. L., and Syme, S. L. (1994). Socioeconomic status and health: The challenge of the gradient. *Am. Psychol.* **49,** 15–24.

40. Wells, T. J. (1984). Social and psychological implications of incontinence. *In* "Urology in Old Age" (J. C. Brocklehurst, ed.), pp. 107–126. Churchill Livingstone, New York.

41. Umlauf, M. G., Goode, P. S., and Burgio, K. L. (1996). Psychosocial issues in geriatric urology. *Urol. Clin. North Am.: Geriatr. Urol.* **23,** 127–136.

42. Diokno, A. C., Brock, B. M., Herzog, A. R., and Bromberg, J. (1990). Medical correlates of urinary incontinence in the elderly. *Urology* **36**, 129–138.

43. Wetle, T., Scherr, P., Branch, L. G., Resnick, N. M., Harris, T., Evans, D., and Taylor, J. O. (1995). Difficulty with holding urine among older persons in a geographically defined community: Prevalence and correlates. *J. Am. Geriatr. Soc.* **43**, 349–355.

44. Brown, J. S., Seeley, D. G., Fong, J., Black, D. M., Ensrud, K. E., and Grady, D. (1996). Urinary incontinence in older women: Who is at risk? *Obstet. Gynecol.* **87**, 715–721.

45. Johnson, T. M., Kincade, J. E., Bernard, S. L., Busby-Whitehead, J., Hertz-Picciotto, I., and DeFries, G. H. (1998). The association of urinary incontinence with poor self-rated health. *J. Am. Geriatr. Soc.* **46**, 693–699.

46. Kutner, N. G., Schechtman, K. B., Ory, M. G., Baker, D. I., and the FICSIT Group (1994). Older adults' perceptions of their health and functioning in relation to sleep disturbance, falling, and urinary incontinence. *J. Am. Geriatr. Soc.* **42**, 757–762.

47. Kok, A. L. M., Voorhorst, F. J., Burger, C. W., Van Houten, P., Kenemans, P., and Janssens, J. (1992). Urinary and faecal incontinence in community-residing elderly women. *Age Ageing* **21**, 211–215.

48. Tinetti, M. E., Inouye, S. K., Gill, T. M., and Doucette, J. T. (1995). Shared risk factors for falls, incontinence, and functional dependence—unifying the approach to geriatric syndromes. *J. Am. Med. Assoc.* **273**, 1348–1353.

49. Ouslander, J. G., Uman, G. C., Urman, H. N., and Rubenstein, L. Z. (1987). Incontinence among nursing home patients: Clinical and functional correlates. *J. Am. Geriatr. Soc.* **35**, 324–330.

50. Campbell, A. J., Reinken, J., and McCosh, L. (1985). Incontinence in the elderly: Prevalence and prognosis. *Age Ageing* **14**, 65–70.

51. Ekelund, P., and Rundgren, A. (1987). Urinary incontinence in the elderly with implications for hospital care consumption and social disability. *Arch. Gerontol. Geriatr.* **6**, 11–18.

52. Thom, D. H., and Brown, J. S. (1998). Reproductive and hormonal risk factors for urinary incontinence in later life: A review of the clinical and epidemiologic literature. *J. Am. Geriatr. Soc.* **46**, 1411–1417.

53. Holst, K., and Wilson, P. D. (1988). The prevalence of female urinary incontinence and reasons for not seeking treatment. *N. Z. Med. J.* **101**, 756–758.

54. Milsom, I., Ekelund, P., Molander, U., Arvidsson, L., and Areskoug, B. (1993). The influence of age, parity, oral contraception, hysterectomy and menopause on the prevalence of urinary incontinence in women. *J. Urol.* **149**, 1459–1462.

55. Jolleys, J. V. (1988). Reported prevalence of urinary incontinence in women in a general practice. *Br. Med. J.* **296**, 1300–1302.

56. Sommer, P., Bauer, T., Nielsen, K. K., Kristensen, E. S., Hermann, G. G., Steven, K., and Nordling, J. (1990). Voiding patterns and prevalence of incontinence in women. A questionnaire survey. *Br. J. Urol.* **66**, 12–15.

57. Burgio, K. L., Matthews, K. A., and Engel, B. T. (1991). Prevalence, incidence and correlates of urinary incontinence in healthy, middle-aged women. *J. Urol.* **146**, 1255–1259.

58. Hørding, U., Pedersen, K. H., Sidenius, K., and Hedegaard, L. (1986). Urinary incontinence in 45-year-old women: An epidemiological survey. *Scand. J. Urol. Nephrol.* **20**, 183–186.

59. Fantl, J. A., Cardozo, L., McClish, D. K., and the Hormones and Urogenital Therapy Committee (1994). Estrogen therapy in the management of urinary incontinence in postmenopausal women: A meta-analysis. First report of the Hormones and Urogenital Therapy Committee. *Obstet. Gynecol.* **83**, 12–18.

60. Rekers, H., Drogendijk, A. C., Valkenburg, H. A., and Riphagen, F. (1992). The menopause, urinary incontinence and other symptoms of the genito-urinary tract. *Maturitas* **15**, 101–111.

61. Harris, M. M. (1994). "Interstitial Cystitis," NIH Publ. No. 94-3220. National Institute of Diabetes and Digestive and Kidney Diseases, National Kidney and Urologic Diseases Information Clearinghouse, Bethesda, MD.

62. Bump, R. C., and McClish, D. K. (1992). Cigarette smoking and urinary incontinence in women. *Am. J. Obstet. Gynecol.* **167**, 1213–1218.

63. Tampakoudis, P., Tantanassis, T., Grimbizis, G., Papaletsos, M., and Mantalenakis, S. (1995). Cigarette smoking and urinary incontinence in women—a new calculative method of estimating the exposure to smoke. *Eur. J. Obstet., Gynecol. Reprod. Biol.* **63**, 27–30.

64. Jirovec, M. M. (1991). The impact of daily exercise on the mobility, balance and urine control of cognitively impaired nursing home residents. *Int. J. Nurs. Stud.* **28**, 145–151.

65. Nygaard, I., DeLancey, J. O. L., Arnsdorf, L., and Murphy, E. (1990). Exercise and incontinence. *Obstet. Gynecol.* **75**, 848–851.

66. Nygaard, I. (1995). Prevention of exercise incontinence with mechanical devices. *J. Reprod. Med.* **40**, 89–94.

67. Nygaard, I. E. (1997). Does prolonged high-impact activity contribute to later urinary incontinence? A retrospective cohort study of female Olympians. *Obstet. Gynecol.* **90**, 718–722.

68. Dwyer, P. L., Lee, E. T. C., and Hay, D. M. (1988). Obesity and urinary incontinence in women. *Br. J. Obstet. Gynaecol.* **95**, 91–96.

69. Bump, R. C., Sugerman, H. J., Fantl, J. A., and McClish, D. K. (1992). Obesity and lower urinary tract function in women: Effect of surgically induced weight loss. *Am. J. Obstet. Gynecol.* **167**, 392–399.

70. Kölbl, H., and Riss, P. (1988). Obesity and stress urinary incontinence: Significance of indices of relative weight. *Urol. Int.* **43**, 7–10.

71. Burgio, K. L., Ives, D. G., Locher, J. L., Arena, V. C., and Kuller, L. H. (1994). Treatment seeking for urinary incontinence in older adults. *J. Am. Geriatr. Soc.* **42**, 208–212.

72. Herzog, A. R., Fultz, N. H., Normolle, D. P., Brock, B. M., and Diokno, A. C. (1989). Methods used to manage urinary incontinence by older adults in the community. *J. Am. Geriatr. Soc.* **37**, 339–347.

73. Fantl, J. A., Newman, D. K., Colling, J., DeLancey, J. O. L., Keeys, C., Loughery, R., McDowell, B. J., Norton, P., Ouslander, J., Schnelle, J., Staskin, D., Tries, J., Urich, V., Vitousek, S. H., Weiss, B. D., and Whitmore, K. (1996). "Urinary Incontinence in Adults: Acute and Chronic Management," Clin. Pract. Guideline, No. 2 1996 Update, AHCPR Pub. No. 96-0682. U.S. Department of Health and Human Services, Public Health Service, Agency for Health Care Policy and Research, Rockville, MD.

74. Burgio, K. L., and Burgio, L. D. (1986). Behavior therapies for urinary incontinence in the elderly. *Clin. Geriatr. Med.* **2**, 809–827.

75. Fantl, J. A., Wyman, J. F., McClish, D. K., Harkins, S. W., Elswick, R. K., Taylor, J. R., and Hadley, E. C. (1991). Efficacy of bladder training in older women with urinary incontinence. *J. Am. Med. Assoc.* **265**, 609–613.

76. Kegel, A. H. (1951). Physiologic therapy for urinary stress incontinence. *J. Am. Med. Assoc.* **146**, 915–917.

77. Miller, J. M., Ashton-Miller, J. A., and DeLancey, J. O. L. (1998). A pelvic muscle precontraction can reduce cough-related urine loss in selected women with mild SUI. *J. Am. Geriatr. Soc.* **46**, 870–874.

78. Burns, P. A., Pranikoff, K., Nochajski, T. H., Hadley, E. C., Levy, K. J., and Ory, M. G. (1993). A comparison of effectiveness of biofeedback and pelvic muscle exercise treatment of stress incontinence in older community-dwelling women. *J. Gerontol. Med. Sci.* **48**, M167–M174.

79. Peattie, A. B., Plevnik, S., and Stanton, S. L. (1988). Vaginal cones: A conservative method of treating genuine stress incontinence. *Br. J. Obstet. Gynaecol.* **95**, 1049–1053.

80. Sand, P. K., Richardson, D. A., Staskin, D. R., Swift, S. E., Appell, R. A., Whitmore, K. E., and Ostergard, D. R. (1995). Pelvic floor electrical stimulation in the treatment of genuine stress incontinence: A multicenter, placebo-controlled trial. *Am. J. Obstet. Gynecol.* **173,** 72–79.

81. Brubaker, L., Benson, J. T., Bent, A., Clark, A., and Shott, S. (1997). Transvaginal electrical stimulation for female urinary incontinence. *Am. J. Obstet. Gynecol.* **177,** 536–540.

82. Busby-Whitehead, J., and Johnson, T. M. (1998). Urinary incontinence. *Clin. Geriatr. Med.* **14,** 285–296.

83. Wein, A. J. (1998). Pharmacologic options for the overactive bladder. *Urology* **51** (Suppl. 2A), 43–47.

84. Appell, R. A. (1997). Clinical efficacy and safety of tolterodine in the treatment of overactive bladder: A pooled analysis. *Urology* **50** (Suppl. 6A), 90–96.

85. Black, N. A., and Downs, S. H. (1996). The effectiveness of surgery for stress incontinence in women: A systematic review. *Br. J. Urol.* **78,** 497–510.

86. Engberg, S. J., McDowell, B. J., Burgio, K. L., Watson, J. E., and Belle, S. (1995). Self-care behaviors of older women with urinary incontinence. *J. Gerontol. Nurs.* **21,** 7–14.

87. Mitteness, L. S. (1987). The management of urinary incontinence by community-living elderly. *Gerontologist* **27,** 185–193.

97

Sensory Impairment

GARY S. RUBIN* AND KAREN J. CRUICKSHANKS†

*Lions Vision Center, Wilmer Eye Institute, Baltimore, Maryland; †Department of Ophthalmology and Vision Sciences, University of Wisconsin—Madison, Madison, Wisconsin

I. Introduction

While sensory impairment can occur at any age, it becomes much more common in later life. The purpose of this chapter is to review the epidemiology of age-related sensory impairment in women and to indicate areas where additional research is needed. Separate sections are devoted to vision impairment, hearing impairment, and impairment of the chemical senses (taste and smell). In each section we provide a brief overview of the causes of impairment and a review of important issues in the assessment and measurement of sensory loss. This is followed by a summary of the literature concerning prevalence and risk factors for the impairment.

II. Vision Impairment

A. Introduction

Vision impairment can lead to physical disability and decreased quality of life. Several epidemiologic studies have demonstrated that vision impairment is associated with dependency in daily activities, reduced physical activity, social isolation, and mortality in older individuals [1–10]. One of the leading causes of vision impairment is uncorrected refractive error. Eyeglasses and contact lenses are such simple and effective treatments that discussions of vision impairment often exclude refractive error. However, it is estimated that as much as one-third of all vision impairment in the U.S. adult population is caused by uncorrected refractive error [11,12].

Another leading cause of vision impairment is cataract. The lens consists of a central nucleus surrounded by layers of cortex enclosed by an elastic capsule. Cataracts are typically classified according to location as nuclear, [anterior] cortical, or [posterior] subcapsular. In nuclear cataracts there is an accelerated yellowing of the lens and an increase in diffuse light scatter throughout the nucleus. Cortical cataracts tend to be more localized with areas of turbidity sequestered in annular rings or wedges of the lens cortex. Posterior subcapsular cataracts are the most highly localized and typically appear as a small opaque region near the center of the posterior pole of the lens. It is estimated that over one million cataract surgeries are performed in the U.S. every year, making it the most frequent surgical procedure [13]. The vast majority of these surgeries are extremely successful, restoring vision to normal or near-normal levels. Nevertheless, untreated cataract accounts for nearly one-third of the cases of vision impairment observed in population-based studies [14,15].

Most of the remaining cases of vision impairment in later life are caused by one of three age-related eye diseases: age-related macular degeneration (AMD), diabetic retinopathy, or glaucoma. AMD comes in two forms. In the more common "dry" (atrophic) form, there is a gradual atrophy of the retinal cells that transduce the light stimulus. The atrophy often follows a characteristic pattern beginning as a horseshoe that surrounds the macula. The horseshoe gradually coalesces into a ring then spreads both in towards the fovea and out towards the periphery. The less common "wet" (exudative) form is characterized by the growth of new blood vessels from behind the retina that break though into the retinal tissues. These vessels also leak blood and eventually scar over. In both types of macular degeneration there is a complete loss of visual sensitivity in the affected area, resulting in a blind spot or scotoma. Exudative AMD can be treated with laser photocoagulation which limits the spread of new vessels. However, the treatment itself creates a scotoma wherever it is applied. There is no treatment at present for atrophic AMD.

Diabetic retinopathy is an eye disease affecting diabetic individuals in which the blood vessels inside the retina proliferate and break through the retinal surface. These vessels are fragile and tend to leak blood into the gelatinous vitreous in front of the retina. Scar tissue may form which detaches the retina from the back of the eye. Prior to the proliferative stage just described there may a buildup of fluid that elevates and distorts the retinal surface. The disease is treated by applying a grid or ring of small laser burns to the retina. This slows the proliferation of new blood vessels and reduces the buildup of fluid in the retina. Leaky blood vessels can also be sealed with laser photocoagulation.

Diabetic retinopathy and macular degeneration are most disabling when their effects encroach upon the foveal region in the very center of the retina. The fovea is the area of best spatial resolution. However, foveal scotomas disrupt many visual tasks to an extent that far outweighs the loss of resolution (acuity). For example, individuals with macular degeneration seldom read faster than about 50 words/minute whereas those with equivalent acuity loss caused by cataract may reach 150 words/minute or more with adequate magnification [16]. One explanation for the disproportionate effect of foveal scotomas is that loss of foveal vision disrupts eye movement patterns and this results in inefficient processing of visual information [17]. A second possibility is that peripheral vision may be less well suited for processing text and other images because of its relative insensitivity to spatial position information and increased sensitivity to crowding from neighboring images [18].

Glaucoma is an insidious disease that destroys the optic nerve and may go undetected for many years. If left untreated, glaucoma will cause peripheral vision loss that progresses slowly towards the fovea, culminating in blindness. The treatment for glaucoma is to lower the pressure inside the eye, although the mechanism by which elevated eye pressure destroys the optic nerve is not well understood. Glaucoma is classically described as causing a peripheral field impairment; however, evidence suggests that central vision is also affected early in the disease. The central vision problem specifically affects the processing of dynamic stimuli such as flickering patterns or moving objects [19]. These dynamic vision deficits are believed to contribute to the difficulties with orientation, mobility, and navigation that accompany glaucoma [20].

Although approximately two-thirds of the cases of vision impairment can be successfully treated with eyeglasses or cataract surgery, the remaining one-third usually result in permanent vision loss. Many of these individuals can benefit from special optical and electronic devices such as magnifiers and closed-circuit televisions, and from low-vision rehabilitation services [21,22]. Low-vision devices are not covered by many insurance plans or Medicare and the coverage for low-vision rehabilitation services is not consistent. Moreover, the availability of the services tends to be limited to larger urban areas and many individuals with vision impairment are not made aware of low-vision rehabilitation by their physicians.

Vision impairment is seldom life-threatening. Nevertheless, vision impairment is among the most disabling of chronic medical conditions; it is more disabling than arthritis, pulmonary disease, or hearing impairment in young adults, and more disabling than all conditions except diabetes and cancer in the oldest adults [23]. The limitations in employment, education, and recreation imposed by vision loss are disproportionate to health care expenditures for vision impairment assessment, research, and rehabilitation.

B. Measurement Issues

Visual acuity is the standard test for assessing visual function and most population-based studies have used visual acuity loss as the sole determinant of vision impairment. While good acuity is necessary for some activities, such as reading fine print, it is only weakly associated with one's ability to see large low-contrast objects, such as nearby faces [24], or to navigate safely and independently in unfamiliar environments [25,26]. Other measures may provide important information about visual function that cannot be captured by acuity alone. Contrast sensitivity is one such measure that has received considerable attention. In a healthy human eye, contrast sensitivity and visual acuity are highly correlated. However, contrast sensitivity may be markedly reduced despite near-normal visual acuity [27]. Contrast sensitivity has been shown to be important for predicting reading speed in patients with severe vision impairment [28,29] and in older individuals who are free from obvious ocular pathology [30]. Low-contrast sensitivity is also associated with postural instability [31] and mobility difficulty [26] in low-vision patients. Older observers require higher contrast to recognize "real-world" images such as traffic signs [32] and faces [33].

Some individuals with excellent visual acuity report particular difficulty seeing objects in the presence of glare. "Disability glare" refers to the reduced visibility of a target due to the presence of a light source elsewhere in the visual field. Any disorder that increases intraocular light scatter, such as lens opacity, may cause problems due to disability glare. For people with normal vision, glare sensitivity measurements are correlated with simulated nighttime driving performance and correspond to subjective complaints about glare from oncoming headlights [34]. However, other studies of disability glare in patients with mild to moderate cataracts failed to detect an association between glare symptoms and scores on disability glare tests [35,36].

Studies of visually impaired patients have shown that both central and lower mid-peripheral visual fields are important for mobility performance [25,37]. The relationship between visual field loss and driving skills has received considerable attention [38–40]. In a California study of 10,000 drivers, binocular field loss was associated with increased crash and driving conviction rates, although only 4% of those with visual field loss reported coexisting loss of visual acuity [38].

It is often presumed that loss of stereoscopic depth perception should be related to performance of everyday tasks. While there is a vast array of research on the underlying mechanisms of stereopsis and on its relation to other aspects of visual function, there is almost no work on the implications of poor stereopsis on daily activities. We previously reported [41] that stereoacuity was unrelated to self-reported difficulty with daily activities in a sample of 220 older adults. However, Nevitt and coworkers [42] found that stereoacuity was a significant risk factor for recurrent falls in the elderly.

Given the mounting evidence for the importance of multiple measures of visual function, epidemiologic studies have begun to incorporate contrast and glare sensitivity, visual field, and stereoacuity measurements [12,43,44].

C. Descriptive Epidemiology of Vision Impairment

There have been dozens of reports of the prevalence and incidence of vision impairment throughout the world. This section focuses on studies that used objective vision measures (i.e., visual acuity) rather than subjective questionnaire data. In addition, we have limited our consideration to vision impairment in industrialized nations where the causes of vision loss differ dramatically from those in developing nations. Unless otherwise noted, vision impairment is defined as best-corrected acuity worse than 20/40 in the better-seeing eye.

The Framingham Eye Study was one the first population-based assessments of visual acuity and eye disease in the U.S. [45]. Survivors of the original Framingham Heart Study cohort, aged 52–85, were examined in 1973–1975. The overall prevalence of vision impairment was 2.7% for men and 3.5% for women; however, age-specific gender differences were significant only for ages 65–74 (1.6% for men vs 3.2% for women). The age-specific prevalence of cataract and AMD were higher for women than for men, whereas the prevalence of diabetic retinopathy and glaucoma were higher for men than for women [46,47].

The Baltimore Eye Survey [48] was conducted in 1985–1988 in an urban setting designed to sample an equal number of white and African-American individuals age 40 and over. There was no difference between the rates of vision impairment between men and women after age adjustment. However, the prevalence of vision impairment among blacks (3.3%) was 1.75 times greater than the prevalence among whites (2.7%). Although vision impairment was associated with socioeconomic status, as measured by educational level, income, and employment status, there remained an excess prevalence of impairment in African-Americans after adjustment for these factors. Cataract was twice as likely to be the cause of vision impairment in blacks compared to whites, diabetic retinopathy five times more likely, and glaucoma six times more likely. In contrast, AMD was 1.5 times more likely to be the cause of impairment in whites compared to blacks [14].

As a complement to the urban population-based data from Baltimore, a study of vision impairment in a rural population was conducted in 1988 in the Mud Creek Valley of southeastern Kentucky [49]. The area is described as one of the poorest in the country, with 30% of the population below the poverty level. The prevalence of binocular blindness (acuity worse than 20/400 in the better eye) was 0.44%, which was twice the national rate. There was no difference in blindness or vision impairment rates by gender; however, cataract was the leading cause of vision loss in women, whereas AMD was the leading cause in men.

The Beaver Dam Eye Study was initiated in 1988–1990, with a follow-up study conducted between 1993–1995. Residents of Beaver Dam, WI who were 44–85 years of age were invited to participate in a study of age-related eye diseases. The prevalence of vision impairment was 6.5% in women and 3.6% in men. Vision impairment was more common in women than in men at all ages [50]. This study also included a measure of visual sensitivity, or the dimmest light that could be detected within the central 25° region of the visual field [44]. Women had lower visual sensitivity than men at all ages, but the average difference was small (less than 0.05 log unit). The prevalence and severity of cortical cataract and the incidence of nuclear cataract were greater in women than in men. The prevalence and incidence of AMD were also greater in women than in men, but there was no gender difference for glaucoma [51]. The Beaver Dam Eye Study looked at the relationship between visual function and mobility as assessed by a history of falls and hip fracture and by measured walking speed. Acuity, contrast sensitivity, and visual sensitivity to light were associated with mobility. Although women were more likely to report falls and hip fractures and to walk more slowly, there was no evidence that the relationships of these variables to vision impairment differed by gender [52].

Acuity and visual field tests were administered to subjects 55 and older in Rotterdam, the Netherlands, during 1990–1993. Visual field results were used to determine the prevalence of blindness, but not lesser degrees of vision impairment, following definitions set forth by the World Health Organization (WHO) [53]. The prevalence of vision impairment increased dramatically with advancing age, as in all studies, from 0.1% at age 55 to 9.0% at age 85. The prevalence of vision impairment and blindness was slightly higher in women than in men within 10-year age strata; however, the authors found no statistically significant difference after adjustment for age within age strata. A similar study conducted in Melbourne, Australia, the Melbourne Vision impairment Project, also used acuity and fields to assess visual function in adults aged 40 and over [54]. Vision impairment was defined as acuity worse than 20/60 or visual field constriction to 20° or less. The age-standardized prevalence of visual acuity impairment was higher for women than for men (0.8% vs 0.5% using the more stringent WHO definition), but there was no gender difference for visual field constriction (prevalences not given).

The Blue Mountain Eye Study was conducted in an urban setting west of Sydney, Australia in 1992–1993. Visual acuity and eye disease were evaluated in a population-based sample that was 49–97 years of age. The investigators reported a significantly higher prevalence of vision impairment (5.7% vs 3.2%), especially severe impairment (20/200 or worse acuity, 0.9% vs 0.3%), in women [55]. These differences were observed in all age strata and were attributable to higher rates of cortical cataract, AMD, and glaucoma in women [56–58]. There was no gender difference in the age-adjusted prevalence of diabetic retinopathy [59]. Vision-impaired individuals were three times more likely to use community support services and six times less likely to go out alone than those with normal vision [60]. The association of vision impairment with reliance on community support was more evident in women than in men. As in Beaver Dam, the Blue Mountain study found a positive relationship between vision impairment and a self-reported history of falls. In addition to visual acuity loss, reduced contrast sensitivity and restricted visual fields also contributed to the association of vision and falls [61].

The most recent population-based study of vision impairment is the Salisbury Eye Evaluation (SEE) project [62]. Baseline testing was conducted in 1993–1995 in Salisbury, MD, with follow-up testing two years later. Participants were 65–84 years old at the time of enrollment and the African-American population was deliberately oversampled so that cross-racial comparisons could be made. The SEE study included the most comprehensive vision assessment performed in any epidemiologic study, with tests of visual acuity at normal and low luminance, contrast and glare sensitivity, stereoacuity, and visual fields [12]. In age-specific comparisons there were no consistent gender differences in prevalence of vision impairment in contrast to the findings in the Blue Mountain Study (Fig. 97.1). The overall rate of visual acuity impairment for white SEE participants was somewhat lower than that reported by other studies with predominately white populations (3.5%), but it was twice as high among African-American participants (7.2%). Black participants were also twice as likely as white participants to have a contrast sensitivity impairment (letter contrast sensitivity < 1.35 log units, OR (odds ratio) = 2.2, 95% CI (confidence interval) = 1.79–4.6). After adjustment for age and race there were no differences in prevalence rates of acuity or contrast sensitivity impairment according to gender. The SEE study included performance-based measures of mobility, reading speed, face recognition, and simulated instrumental activities of daily living, and a self-reported assessment of falls and difficulty with daily activities. Visual acuity was strongly associated with reading speed and to a lesser degree with face recognition [63]. Acuity and contrast sensitivity were independently associated with mobility performance (timed walks and stair climb). Visual fields were associated with a history of falls [64] and all of the vision measures were associated with low scores on the daily activity questionnaire [65]. There were no consistent gender differences in the relationships between vision variables and disability measures.

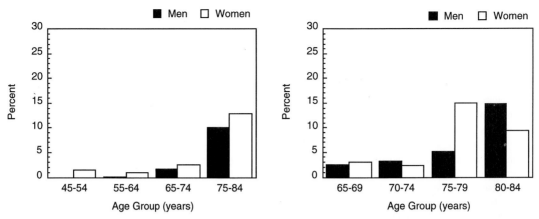

Fig. 97.1 Prevalence of vision impairment by age and gender in two recent studies. Blue Mountain Eye Study, 1992–1993, Blue Mountains, Australia (left) and Salisbury Eye Evaluation, 1993–1995, Salisbury, MD (right).

D. Risk Factors

All of the population-based studies of vision impairment have demonstrated a pronounced increase in the prevalence of vision loss with advancing age. Nevertheless, it is not unusual to find individuals who maintain 20/20 visual acuity even into their eighties or nineties. Given the wide variability of visual function in later life, there is great interest in determining the influence of suspected risk factors on the development of vision-threatening eye conditions.

Socioeconomic factors such as education and income were shown to be associated with vision impairment in the relatively homogenous Beaver Dam population [66] and the racially diverse Baltimore and Salisbury populations [12,67]. Adjusting for socioeconomic status reduces but does not eliminate the excess prevalence of vision impairment among African-Americans compared to whites.

Other factors that have been implicated in the etiology of vision loss are sunlight exposure, cigarette smoking, alcohol consumption, cardiovascular risk factors, hormone status, and micronutrients. There is strong evidence for a link between sunlight exposure and the development of cortical cataracts. Until recently, the association had been established for men only [68,69]. However, data from the SEE study extended these findings to women [70]. Participants in SEE provided retrospective data on their occupational and recreational exposure to sunlight and usage of hats and glasses. Experiments were done to determine how hats and sunglasses affected ocular exposure. A model was then developed that estimated each participant's cumulative UV-B exposure. Cortical cataracts were more likely to occur in those with greater UV-B exposure (OR = 1.57, CI = 1.04–2.38 comparing those in the highest exposure quartile to those in the lowest quartile). The effect of sunlight exposure was similar for men and women and for white and African-American participants. Most importantly, the authors found that wearing a hat with a brim reduced ocular exposure by 30–50% and ordinary plastic glasses or sunglasses were effective in blocking virtually all of the UV-B.

Cigarette smoking has also been shown to increase the risk of cataract, but in this case the association is with nuclear and possibly posterior subcapsular, not cortical lens, opacities. Initial cross-sectional data from Maryland [71], Wisconsin [72], and Australia [73] demonstrated a modest increase in risk of cataract for both male and female smokers. Subsequent longitudinal data [74] reinforced this finding and showed that the risk of cataract progression was significantly higher in current smokers than nonsmokers or exsmokers (OR = 2.4; CI = 1.0–6.0). Smoking is also associated with AMD, particularly the vision-threatening severe forms of the disease. Data from the Rotterdam study demonstrated a dose-response relationship between smoking and AMD, with current smokers having a 6.6-fold higher risk while former smokers had a 3.2-fold higher risk than those who had never smoked [75]. The Beaver Dam study found a stronger association for men than for women (OR = 3.3 for men; 2.5 for women who smoked compared to nonsmokers or exsmokers [76]), but the Blue Mountain study found a stronger association for women than for men [77]. Longitudinal data from Beaver Dam indicated that the incidence of AMD was higher in current smokers than in those who had never smoked or who had stopped smoking before the beginning of the study [78]. An important implication of these data is that quitting smoking is an effective way to reduce the risk of cataract and AMD.

Data concerning the relationship between alcohol use and age-related eye disease are complicated. The Beaver Dam study reported that heavy drinking was associated with all types of cataract, whereas moderate alcohol use was protective for nuclear lens opacities [79]. The Blue Mountain study found that moderate alcohol use decreased the risk of cortical opacities, whereas heavy use increased the risk of nuclear cataract in smokers only. The picture is equally confusing for AMD. The Beaver Dam study found beer consumption to be associated with early AMD in men and late-stage AMD in men and women. There was no association with consumption of wine or other liquors [80,81].

Several studies have looked at cardiovascular risk factors in age-related eye disease. In the Beaver Dam Study, early AMD was associated with serum cholesterol in both men and women, although the contribution of high-density lipoproteins (HDL) and low-density lipoproteins (LDL) differed by gender [82]. High levels of HDL cholesterol reduced the risk of cortical cataract in women, and high levels of LDL cholesterol increased the risk of posterior subcapsular cataract in men. The Blue Mountain study found no association of serum lipids with AMD, but it did report an association of late-stage AMD with plasma fibrinogen level [83]. Cardiovascular disease increased the risk of AMD in the Rotterdam study [84] but was not associated with AMD in the Beaver Dam or the Blue Mountain studies [82,83]. Hypertension has also been associated with some forms of glaucoma [85].

Data from a variety of sources suggest that there is a relationship between estrogen and eye disease. The Beaver Dam study found that older age at menopause was associated with lower rates of cortical cataract and hormone replacement therapy reduced the risk of nuclear cataracts [86]. In the Blue Mountain study, women who took hormone replacement therapy had a reduced risk of nuclear cataract (OR = 0.4; CI = 0.2–0.8), and greater years from menarche to menopause was associated with lower prevalence of AMD [87]. Together, these studies suggest that estrogen exposure may reduce the risk of some types of age-related eye disease.

The popular press has been filled with reports of the beneficial effects of micronutrients and dietary supplements for reducing the risk of vision loss in later life. This has lead to a thriving commercial market for vitamin supplements whose names allude to supposed sight-saving properties. However, the support for such claims in the published literature is scarce and often conflicting. For example, one study that included men only found that higher levels of serum antioxidants were associated with reduced risk of AMD, but vitamin supplement intake was not [88]. A case-control study of 356 patients with advanced AMD showed a protective effect of dietary antioxidant intake [89]. Population-based studies have not found such a relationship. Antioxidants and zinc were investigated in the Beaver Dam study [90] which showed modest inverse associations between these micronutrients and specific retinal changes that often precede AMD, but there was no overall or consistent protective effect for any of these substances for early AMD. The Blue Mountain study failed to find an association between serum antioxidants and AMD [91]. Most researchers agree that the resolution of these inconsistencies will depend on the results of ongoing clinical trials such as the Age-Related Eye Disease Study (AREDS) sponsored by the National Eye Institute.

Despite the uncertain and inconsistent data regarding alcohol use, cardiovascular risk factors, and nutrition, there is compelling evidence that cigarette smoking and sunlight exposure increase the risk of age-related eye disease. Simple prevention strategies such as smoking cessation and the wearing of glasses and hats can reduce the risk of cataract and AMD, two of the most common causes of vision impairment in later life.

E. Summary

The main causes of vision impairment in the developed world have been well established. Uncorrected refractive error and cataract each account for approximately one-third of cases of vision loss in people over age 65. Because effective treatments are available for both of these conditions, fully two-thirds of vision impairment can be considered reversible. The reasons why individuals fail to take full advantage of available treatments is not fully understood and requires further study. The remaining one-third of vision impairment is caused by diseases that may be treatable but for which the vision loss can not be fully restored. Therefore, individuals with these irreversible vision impairments must receive low-vision rehabilitation if they are to take maximum advantage of their residual visual function. However, low-vision rehabilitation services are woefully underfunded and underutilized.

Vision impairment affects both men and women in approximately equal proportions, although there is some evidence that cataract and AMD are more common in women. Smoking and sunlight exposure increase the risk of cataract and AMD and it appears that both men and women would benefit from lifestyle choices that reduce exposure to these risk factors. It was pointed out that most of the literature on vision impairment has concentrated on the loss of visual resolution ability (acuity). However, evidence suggests that other aspects of visual function, such as contrast sensitivity and peripheral visual fields, may also be important predictors of difficulty with everyday activities. Further study is needed to determine how these other visual functions change with age and the onset of eye disease.

III. Hearing Impairment

A. Introduction

Hearing loss in adulthood can cause difficulties in communication that may affect the quality of life. While hearing loss can occur at any age and may affect many components of the auditory pathway, most hearing impairment in older adults is sensorineural (noise-induced hearing loss or presbycusis (age-related)) and is thought to primarily reflect alterations in hair cell function within the cochlea and changes in auditory neural function [92,93]. With sensorineural hearing loss, older people may experience difficulty understanding speech when there are challenging listening conditions such as background noise or multiple speakers (as at a party or in a restaurant). People with age-related hearing loss generally have similar degrees of loss in each ear with greater reductions in sensitivity to high-pitched sounds than to low-pitched sounds. With increasing severity of loss, people are unable to hear sounds unless they are very loud. Hearing loss in older adults has been associated with a greater likelihood of depression, disability, and residence in nursing homes [4,7,94–99]. It can be frustrating for family members to communicate with the hearing impaired individual and frustrating for the person with the impairment. There is concern that hearing impaired older adults may become more socially isolated [99].

There is no cure for age-related hearing loss, although some benefit may be obtained from using hearing aids or other amplification devices and/or adopting new listening strategies such as minimizing background noise, facing the speaker, and learning speech reading techniques [94,100–103]. Hearing aids and assistive devices are not covered by many insurance plans or Medicare, may be difficult to adjust to, and do not overcome many problems with speech understanding in background noise [102]. Few older people with hearing loss use hearing aids, which may reflect access/cost issues, reluctance to acknowledge the hearing impairment, or the technological limitations of the devices

[103]. Many people with hearing impairment are unaware of listening strategies and less costly devices that can provide some help in specific listening situations.

B. Measurement Issues

Pure-tone air- and bone-conduction audiometry is the standard method to clinically assess hearing loss [93,104]. During this test, tones are presented at various intensities to determine, with a bracketing procedure, the minimum volume detectable (threshold). Hearing loss is measured in decibels with respect to normative data; a 25 dB HL loss means the person does not hear a sound until it is presented at a volume 25 dB louder than that detectable by a person with normal hearing. Usually a series of frequencies is tested from 250 Hz to 8000 Hz, but thresholds for very high frequency sound (9000 Hz and above) can also be measured. The range most important for understanding speech is 500–2000 Hz, with higher frequencies (3000–8000 Hz) contributing to distinguishing some consonant sounds.

Other forms of hearing testing include speech understanding tasks that test the ability of the listener to correctly interpret complex sound (speech) [105,106]. These tests usually adjust the presentation level for the person's hearing level in order for the volume to be perceived as similar across subjects. Listening tasks may be made more challenging by degrading the sound of the recording or presenting a competing message to ignore. These tests generally are used in conjunction with air- and bone-conduction audiometry to assess the potential benefit from hearing aids.

In epidemiologic studies, audiometric testing requires a sound-treated room to be in compliance with standards for maximum permissible ambient noise levels, calibrated audiometers, and highly trained personnel. The involvement of experienced, certified audiologists in designing testing facilities, training and supervising technicians, and conducting periodic calibrations (every six months) is essential for high quality data collection. Standard testing techniques such as those recommended by the American Speech Hearing and Language Association are designed to generate reliable and consistent thresholds (within ± 5 dB HL) [104].

Although there are reliable standard assessments of hearing sensitivity, epidemiologic studies have been hampered by the lack of a consistent definition of hearing loss. Studies have reported mean (or median) thresholds by frequency, by the percentage of people with abnormal thresholds for each frequency tested using various standards for abnormal (>25 dB HL, >40 dB HL, etc.), or they have applied a cut-point to determine if the average of the pure tone thresholds (PTA) at several frequencies is abnormal [107–112]. While some studies used a three frequency average of thresholds at 500, 1000, and 2000 Hz, and others included 3000 Hz or 4000 Hz, a PTA above 25 dB HL has usually been considered abnormal. Some studies have classified an individual as affected if the better ear PTA was abnormal and some studies have classified people as hearing impaired if the PTA was abnormal in either ear.

Few epidemiologic studies in this country have been able to use audiometric testing to obtain performance-based measures of hearing. Some epidemiologic studies have relied on questionnaires to determine prevalence of hearing loss. Unlike some other areas in chronic disease epidemiology, there have been few attempts to employ validated and standardized questions of hearing loss. Questions have varied from ascertaining mild degrees of hearing loss to profound deafness. One study did compare the sensitivity and specificity of several questionnaire approaches to audiometric measures of hearing loss and found that a simple question, "Do you feel you have a hearing loss," was the best self-reported measure of hearing loss [113].

The screening versions of the Hearing Handicap Inventory (versions exist for adults and for the elderly), developed by Weinstein and Ventry, are in wide use clinically to detect people who might benefit from amplification [114–116]. This questionnaire was designed to measure self-perceived handicap from hearing loss and has been validated with audiometric data.

C. Descriptive Epidemiology of Hearing Impairment

A review of the epidemiology of hearing loss is hampered by the methodological issues described in the previous section. Audiometers and reference standards have changed over the years and few studies have reported results in consistent ways to enable comparisons. This section focuses primarily on reports of the prevalence and incidence of hearing loss or impairment in the United States.

The Health Interview Survey, a periodic interview of samples of the U.S. population, has included questions about trouble hearing normal speech [117]. The prevalence of hearing trouble in women age 65 years or older was 24.2% in 1990–1991 compared to 35.9% among men. These rates appeared to represent an increase in reported hearing trouble since 1971 when the rates were 23.6% and 32.6% for women and men, respectively. However, it is well recognized that self-reported measures probably underestimate the true prevalence of hearing loss in the population. Nonetheless, hearing loss ranks as one of the most common chronic conditions affecting older adults in the United States.

A report from the Alameda County Study also suggests that the prevalence of hearing impairment has increased since the 1970s [118]. In this longitudinal study, participants were asked a simple question about trouble hearing. The age-adjusted prevalence of reported hearing impairment was 11.7% for men and 7.3% for women in 1965. In 1994, the age-adjusted prevalence had increased to 23.7% among men and 13.7% among women. Thus, among both men and women, the prevalence of hearing loss had doubled during the study period.

Hearing testing was conducted as part of the 1971 Health and Nutrition Examination Survey in the United States [109,112]. Hearing thresholds appeared to be better in women than in men but no sex-specific prevalence rates of hearing loss have been published from this study. Overall, among adults 65 years of age and older, 30% had a hearing loss based on an average PTA for 500, 1000, and 2000 Hz. Blacks appeared to have slightly worse hearing thresholds than whites, but again, age-adjusted, sex-specific, and race-specific rates were not presented.

In the Hispanic Health and Nutrition Examination Survey, hearing testing was conducted on adults who were 20–74 years of age [111]. Cuban American women appeared to have higher rates of hearing loss ($PTA_{0.5-2 \text{ kHz}} > 25$ dB HL) than Mexican American or Puerto Rican women, similar to the pattern seen among men. Mexican American or Puerto Rican women, similar to the pattern seen among men. Mexican American women

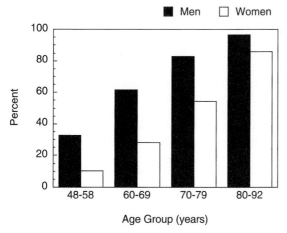

Fig. 97.2 Prevalence of hearing loss by age and gender. Epidemiology of Hearing Loss Study, 1993–1995, Beaver Dam, WI.

were less likely to have hearing loss than Mexican American men (10.9% vs 15.8%), but there were no significant gender differences for the other two Hispanic groups.

In 1978–1979, hearing testing was added to examination 15 of the Framingham Heart Study [110]. Participants were aged 57–89 years (n = 935 men and 1358 women) at the time of the audiometric evaluation. The authors found that women had better hearing at higher frequencies (2000–8000 Hz) but did not differ from men in the lower frequencies. After adjusting for age and other potential confounders (including noise exposure), women were 37% less likely to have a hearing loss in the speech frequencies (OR = 0.63).

Hearing was tested again in this cohort in 1983–1985 (n = 1662 adults) [107]. At this examination, 26.7% of women and 32.5% of men had a hearing loss ($PTA_{0.5-2\ kHz} \geq 26$ dB HL). Data on the incidence of hearing decline were available for 1475 people tested twice approximately six years apart [119]. Women seemed to experience greater decline than men in the lower frequencies, with no gender difference in the rate of decline at higher frequencies. Sex-specific incidence rates were not presented. Overall, the incidence of hearing loss was 8.4% for right ears and 13.7% for left ears.

The Baltimore Longitudinal Study of Aging (BLSA) followed people who were aged 30–80 years with no evidence of hearing loss, middle ear disease, or occupational noise exposure. Follow-up time varied from 2 to 23 years and the majority of participants were men. During the follow-up, 6% developed a hearing loss [120]. Rates of change varied by age and were greater for men than for women [121].

Hearing thresholds also have been measured with pure-tone audiometry in a population-based study of age-related hearing loss in Beaver Dam, WI [108]. Participants in the 1988–1990 Beaver Dam Eye Study (described earlier) were asked to participate in the hearing study (n = 4541). Of those eligible, 3753 (82.6%) participated in the hearing study. The prevalence of hearing loss ($PTA_{0.5-4\ k\ Hz} > 25$ dB HL, either ear) was 36.2% among women and 58.6% among men. As shown in Figure 97.2, the prevalence of hearing loss increased with age among women but was greater among women than men in each age group. After adjusting for age, men were more than four times as likely to have a hearing loss as women (OR = 4.42, 95% CI = 3.73–5.24).

It is interesting to note in Figure 97.2 that the prevalence in women seems to lag about 10 years behind the prevalence in men. For example, the prevalence of hearing loss among women aged 60–69 years was 28%, which is similar to the rate for men 48–59 years (33%). However, even in the oldest group (those 80–92 years of age), about 14% of women did not have a hearing loss, suggesting that hearing loss is not an inevitable part of aging.

The overall prevalence of hearing loss in Beaver Dam women was lower than that reported in a study of 267 Iowa women aged 60–85 years. In the Iowa study, 60% of women had a hearing loss ($PTA_{1-4\ kHz} \geq 25$ dB HL) [122]. This higher rate may reflect the older age distribution of the Iowa women or the effect of excluding thresholds at 500 Hz from the calculation of the pure tone average. (Hearing is better at lower frequencies).

D. Risk Factors

Although most clinical studies have demonstrated a clear decline in hearing with age and better hearing in women than in men, these studies have focused on industrialized Western societies. There are insufficient data to determine if prevalence varies by race and ethnicity. In an early study by Rosen in a rural African tribe, there was no gender difference in hearing thresholds across the age spectrum, nor did hearing decline with age [123]. Rosen speculated that there were environmental and lifestyle differences beyond noise exposure that explained the preservation of good hearing in older men and women.

The causes of hearing loss in older adults are unknown. It is clear that exposure to loud noises for long periods of time can result in permanent hearing loss through destruction of the outer hair cells [124,125]. However, it is unclear what proportion of hearing loss in older populations can be attributed to noise exposure in occupational settings, and the contributions of exposures during leisure time or military service have not been described in population-based studies. It is not known if the female hearing advantage is attributable to less exposure to noise.

In the Beaver Dam study, education and income level were inversely associated with the prevalence of hearing loss [108]. After adjusting for age and gender, people who had not completed high school were about 2.4 times as likely to have a hearing loss as those with a college a education. People who earned less than $30,000 were about twice as likely to have a hearing loss as those with incomes of $60,000 or more per year. Occupational exposure to noise was associated with an increased likelihood of having a hearing loss (OR = 1.31, 95% CI = 1.10–1.56). People in service, production, and operations occupations (jobs more often associated with noise exposure) were more likely to have a hearing loss than those in management positions. However, the association of socioeconomic status indicators with hearing loss was not explained by the type of occupation or occupational noise exposure. After adjusting for age, occupation, noise exposure, and education, women remained less likely to have a hearing loss than men (OR = 0.27, 95% CI = 0.22–0.34), suggesting that other risk factors are important contributors to the gender difference in risk.

Other factors that have been implicated in the etiology of hearing loss are cardiovascular disease (CVD), cigarette smoking, alcohol consumption, elevated cholesterol, diabetes, and high blood pressure. These factors may affect blood supply to the cochlea and, therefore, the health of the hair cells [126,127]. Cardiovascular disease or atherosclerosis has been a hypothesized risk factor for hearing loss since the 1970s. Rosen hypothesized that the absence of an age-related decline in hearing among the Mabaans in Sudan was due to their low exposure to noise, heart disease, hypertension, and diabetes, as well as their healthy lifestyle [123]. They consumed diets high in fruits and fiber, did not smoke, and were very physically active. Rosen's results in the Sudan tribe were replicated in a study of Kalahari Bushmen, where hearing did not dramatically decline with age [128]. Results from other rural or geographically isolated populations, however, have shown age-related declines in hearing that may reflect their greater exposure to noise, Western lifestyles, or higher fat diets [129–131].

Ecologic studies have shown that the prevalence of hearing loss is higher in areas with high rates of CVD compared with areas with low rates of CVD [132,133]. Results from some animal studies [134] and pathology series [126,127,135] have demonstrated that atherosclerosis can affect blood vessels supplying the inner ear and that reduction in the blood supply may lead to degeneration of hair cells and other components of the auditory system. A case-control study of women with ischemic heart disease and a control group thought to have normal coronary arteries found an adjusted odds ratio of 8.0 for hearing loss [136].

In Framingham, with audiometric measures of hearing and objective assessment of the 30-year occurrence of CVD, cardiovascular disease was associated with hearing loss [137]. Risk of hearing loss associated with CVD appeared to be higher for women than for men (OR = 3.06 for women vs OR = 1.75 for total CVD). Among women, coronary heart disease and intermittent claudication was also strongly associated with hearing loss. Gates speculated that microvascular disease may affect the blood supply of the stria vascularis, that estrogen may explain the overall female advantage in part, and that hormone replacement therapy may have protective effects for hearing. A study by Clark *et al.* did not find a protective effect of estrogen use among women [138].

Cigarette smoking has been reported to be associated with hearing loss in adults in some clinical studies [139–142]. Weiss found that men who smoked more than one pack a day were more likely to have poorer hearing thresholds at 250–1000 Hz than nonsmokers and "light" smokers, but there was no difference at higher frequencies [140]. Siegelaub *et al.* [142] reported on a large study of 33,146 men and women seen at Kaiser-Permanente. Among men without a history of noise exposure, current smokers were more likely to have a hearing loss at 4000 Hz than nonsmokers, but the size of the effect was small and there was no association among women. The Baltimore Longitudinal Study of Aging found no association between cigarette smoking and the development of a hearing loss in 531 white, upper-middle class men [120].

There have been few population-based studies of smoking and hearing. In the Health Interview Survey, men who smoked two or more packs per day were more likely to report having a hearing loss than nonsmokers [143]. In the Framingham study,

which tested hearing with audiometry, there was no association between cigarette smoking and hearing loss [137].

Among participants in the Beaver Dam study, current smokers were 1.7 times as likely to have a hearing loss as nonsmokers [144]. There was a slight suggestion of increasing risk with increasing exposure, based on reported pack-years of smoking. In a small subset analysis, nonsmokers who lived with current smokers were more likely to have a hearing loss than nonsmokers who lived alone or with a nonsmoker, suggesting that passive exposure to environmental tobacco smoke may also play a role. A longitudinal follow-up of this cohort is currently underway to determine if baseline smoking patterns predict incidence of hearing loss.

Some studies of chronic alcoholics suggest that alcohol abuse may have an adverse effect on hearing thresholds [145–147]. In the Baltimore Longitudinal Study of Aging (n = 531), there was a slight but nonsignificant inverse association between moderate alcohol consumption and hearing loss [120]. Among participants in the Beaver Dam Epidemiology of Hearing Loss Study, those who reported consuming moderate amounts of alcohol were less likely to have hearing loss than nondrinkers [148]. These patterns are consistent with an atherosclerotic process and are similar to those reported for alcohol and cardiovascular disease.

A number of clinical studies have investigated the possible association of diabetes and hearing loss [149–151]. However, there was no association between diabetes and hearing thresholds in the Framingham study [137]. Among women, the PTA for 0.25–1 kHz did worsen as blood glucose levels increased, which may suggest a link. In the Beaver Dam Study, there was a weak association between Type 2 diabetes and hearing loss when controlling for potential confounders (OR = 1.41, 95%, CI = 1.05–1.88), but there was no association between glycemic control and hearing loss [152].

The evidence from the Framingham and Beaver Dam studies suggests that many modifiable lifestyle factors may contribute to the etiology of hearing loss in older adults. Longitudinal data are needed to confirm these associations and determine their relative contributions in order to develop prevention strategies beyond noise abatement.

E. Summary

The factors that contribute to hearing loss remain to be determined. Women appear to be less likely to have hearing problems than men and this gender difference does not appear to be completely explained by differences in noise exposure. The studies showing relationships with cardiovascular disease, smoking, and other factors suggest that hearing loss may be preventable. Studies are needed to determine the descriptive epidemiology of this disorder in order to understand the magnitude of this public health problem and to identify effective prevention strategies. Most estimates of the prevalence of hearing loss suggest that at least 30% of women over age 65 years have a hearing loss. In the Beaver Dam Study, 86% of women 80 years of age and older had a hearing loss. With the aging of the population and the improved survival for women, large numbers of older women will be faced with adjusting to their own hearing loss and to the hearing impairments of those around them. There is some evidence to suggest that hearing loss is associated with other sensory impairments, functional disabilities, and depression [4,7,94–99,153]. Improved access to hearing-related health care services and increased awareness of available assistance may greatly improve the quality of life for older women while advances are being made to reduce the burden of this important public health problem affecting the lives of many older women.

IV. Olfaction and Taste Disorders

A. Introduction

The chemical senses of taste and olfaction may play important roles in our perception of foods' desirability and the ability to identify potentially dangerous exposures and hazards (to food that has spoiled, natural gas leaks, smoke from fires, etc.). Genetic factors, medications, illnesses, and head injuries may lead to a decreased ability to detect odors and tastes or diminished sensitivity to these stimulants [154–161]. It is well recognized that older people are more likely to experience impairments in these sensory systems, similar to the age pattern described for vision and hearing, yet epidemiologic data on the prevalence, incidence, or natural history of these disorders are remarkably limited.

Most clinical and basic research suggests that women in general are less likely to be affected by smell and taste disorders than men [154,155,159–170]. However, these studies have usually failed to adequately control for potential confounding effects of smoking, occupational exposures, and other factors, and most rely on volunteers or clinic-based patients. Women seem to experience a reduction in smell and taste sensitivity at older ages, which may be similar to the decline seen among men [170]. Because of the roles of taste and smell in food selection and appetite, there is great concern among chemosensory scientists that aging-related deficits may increase the risk of malnutrition among the elderly. This would decrease the ability of older adults to ward off infection and recover from illnesses or surgical procedures, and it may diminish overall quality of life [153,171,172].

B. Measurement Issues

As with vision and hearing, scientists are refining their ability to measure these impairments and capture the many dimensions of these sensory functions. People may be unable to detect a single group of odors or tastes (specific anosmia or ageusia), unable to taste or smell anything, able to taste or smell only strong (intense) stimulants (hyposmia or hypogeusia), have an increased sensitivity to odors or tastes (hyperosmia or hypergeusia), or have a distorted perception (phantom smells, unpleasant tastes in the mouth, etc.). Defining the cut-points in the spectrum of function that defines abnormal is a difficult task for these complex sensory processes.

The primary limitations in studies of taste and olfaction have been the difficulties of standardizing objective measures of disease, defining the range of disorders, and the absence of epidemiologic study designs. In general, investigators have used either questionnaires or testing to identify disorders in detection, sensitivity (threshold), or classification (identification) of odors and tastes. Chemosensory disorders may represent a defect in the receptor cells, neural pathways, and/or cerebral function. Test performance may be influenced by both the strength (intensity) and duration of exposure. There are few data on the comparability of these various testing strategies [173].

Tests of taste disorders have focused on the four basic tastes (sweet, salt, sour, and bitter) as well the ability to detect certain compounds (phenylthiocarbamide) [154,160,164,174,175]. These tests generally require cleansing the tongue and applying a solution to an area or multiple areas of the tongue surface, chewing a treated piece of paper or wafer [174], or swishing a liquid around the mouth. The concentration of the stimulus depends on the preparation technique, freshness of the solution, physical environment (temperature and humidity), application method, and preparation of the tongue or mouth. Acute illnesses, medications, allergy symptoms, smoking, and the presence of dentures or other dental appliances may interfere with perception [156,160,164]. Cognitive impairments may also influence performance on identification tasks, although there are some testing strategies that can minimize the effect for those with mild impairments.

Olfaction tests have generally consisted of threshold tests where the concentration required to detect the odor is assessed (butanol, ethanol, acetone, etc.) or detection and/or identification tasks where the participant is asked whether or not he/she smells anything (detection) and what the odor is (identification) [154–159,162–168,173,175–181]. The most commonly used test in clinical studies has been the University of Pennsylvania Smell Identification test (UPSIT) which consists of a 40-item series of microencapsulated odorants [164,178]. The participant is asked to scratch and sniff the sample then respond to a question asking if he/she can smell anything and if yes to identify the smell from an item-specific list of choices. Raw scores or scores adjusted to published norms have been used for analysis purposes.

The San Diego Odor Identification Test (SDOIT) was developed to be suitable for children and older adults [177]. It uses a picture board to facilitate the naming task and then presents a series of eight common odorants (cinnamon, peanut butter, mustard, coffee, chocolate, baby powder, bubble gum, and Playdoh) for identification. Missed items are presented in a second trial and the score is the total number correctly identified in the combined trials.

Standardization of the stimulus can be a problem for each of these types of tests. The absolute concentration of the odorant that gets to a person's nose is not measured. While the UPSIT has demonstrated good shelf life [178], there are few data on the test kit variability or stimulus strength. The SDOIT relies on a detailed protocol about brands, containers, wrapping methods, and replacement calendar to ensure adequate odorant strength [177]. The physical environment (temperature and humidity) and the subject's familiarity with the odor to be named may influence test performance for each of these methods. As with tests of taste, acute illnesses, medications, allergy symptoms, dentures, and smoking may interfere with perception [161, 164,165]. Older people with early cognitive disorders may also have difficulty with naming tasks and some studies have reported an association between Alzheimer's disease and olfactory deficits [157].

One article reports on a new clinical screening evaluation that may be feasible for epidemiologic studies as well [182]. In this test, the Alcohol Sniff Test (AST), an alcohol pad is presented to the participant at a fixed distance (30 cm) from the person's nares. The pad is moved closer in standard increments until the person detects the odor of the isopropyl alcohol. That distance is measured four times, with the mean of the four trials considered the detection threshold.

C. Descriptive Epidemiology of Smell and Taste Disorders

There are no population-based data on the prevalence or incidence of smell or taste disorders. As part of the Baltimore Longitudinal Study of Aging, the participants selected for the oral physiology substudy (n = 387) were asked questions about their sense of smell and taste [155]. Odor identification was assessed with the 40-item UPSIT. Participants (n = 221 men and 166 women) were between the ages of 19–95 years, community-dwelling, and generally healthy. Most participants were non-Hispanic White (96%) and middle class. This study solicited volunteers primarily by word-of-mouth, so the prevalence data are not generalizable to populations.

The prevalence of self-reported changes in smell varied with age. Among women aged 19–59, 10% reported a change in smell (most reported a worsening) while 15% of those 60–79 years of age and 19% of those 80–95 years of age reported changes in smell. A similar pattern with age was found for men. In each age group, men were more likely to report changes than women. Older women were more likely to report changes in taste than younger women (reports ranged from 8–17% among women less than 80 years of age, whereas 35% of women in the oldest age group reported changes in taste). UPSIT scores declined and the prevalence of anosmia increased with age for both men and women. Women generally had higher UPSIT scores than men, although age-specific differences were not always statistically significant. Overall, 5.9% of the participants were anosmic. There were no significant associations between medical problems, medication use, or smoking and anosmia. Given the small sample size, the power to detect effects was quite limited.

A three-year follow-up study of 85 males and 76 women who participated in the first olfaction testing was conducted [159]. UPSIT scores declined with age during the three-year follow-up for both men and women. Women had better scores than men (greater sensitivity). The average rate of decline was one point per year among those 85 years of age or older. Men's scores declined at an earlier age than women's (decline evident at age 55 for men but not until age 75 for women), which may suggest some female advantage associated with estrogens [159].

As part of the BLSA, a small subset of men and women (n = 81, aged 23–88 years) also underwent taste detection thresholds using four series of solutions [169]. There was a correlation between taste threshold and age (thresholds declining with age) for salt and bitter but no association between age and sweet or sour thresholds. Men were more likely to have elevated thresholds for the sour stimulus than women.

A large clinical series of 750 consecutive patients from the University of Pennsylvania Smell and Taste Center were also studied with the UPSIT and a health questionnaire [164]. This group was predominately non-Hispanic white, with a mean age

of 49 years for men (n = 336) and 52 years for women (n = 414). In this clinical setting, 20.4% of the participants reported a loss of olfaction, 8.7% a taste dysfunction, and 57.7% reported problems with both taste and smell.

The largest survey to date, and the source of many reports about the patterns of olfaction function, was conducted in 1986 by Gilbert and Wysocki in collaboration with the National Geographic Society [162,166–168,171]. In this survey, a questionnaire about olfaction and six "scratch and sniff" samples (androstenone [sweat], isoamyl acetate [banana], galaxolide [musk], eugenol [cloves], mercaptan [natural gas], and rose) were included in an issue of the magazine. The ability to detect the odor and the ability to recognize the odor were measured. Responses were obtained from over 1.5 million people for a response rate of 13%. Of course, readers of this specialized magazine who responded were likely to represent a very biased sample of individuals, so these cannot be considered prevalence data. Among Caucasian respondents, 1.2% could not smell anything at all. The ability to detect an odor varied by country of residence, type of work, smoking, and age. Women were more likely to correctly identify odors than men.

Olfaction testing is included as part of an ongoing population-based study in Beaver Dam, WI. During the five-year follow-up examination for the Epidemiology of Hearing Loss Study, participants (n = 3753) are being asked to complete a taste and smell questionnaire (by interview) and the SDOIT [108,177]. Participants were all residents of Beaver Dam at identification during a private census in 1987–1988 and were aged 48–92 years at the initial hearing study (1993–1995). The examination is scheduled to be completed in June of 2000 and will provide the first population-based data on the prevalence of anosmia and odor identification disorders.

D. Risk Factors

With the dearth of epidemiologic data, there is little evidence for risk factors for olfaction and taste disorders. Clinical studies have suggested that geographic location, occupation, medications, illnesses, allergies, and genetic factors may contribute to olfaction and taste disorders. Longitudinal epidemiologic studies are needed to evaluate these potential risk factors in populations to determine which are major contributors to chemosensory deficits and to identify ways to prevent or delay their onset and progression.

E. Summary

Basic epidemiologic data are needed for these chemosensory disorders. There are no reliable population-based data to determine the magnitude of the impact of these impairments or to identify risk factors for these disorders.

References

1. Havlik, R. J. (1986). "Aging in the Eighties, Impaired Senses for Sound and Light in Persons Age 65 Years and Older: Preliminary Data from the Supplement on Aging to the National Health Interview Survey: United States, January–June 1984, "Advance Data from Vital and Health Statistics, pp. 1–7, National Center for Health Statistics, Washington, DC.

2. Jette, A. M., and Branch, L. G. (1985). Impairment and disability in the aged. *J. Chronic Dis.* **38**, 59–65.

3. LaForge, R. G., Spector, W. D., and Sternberg, J. (1992). The relationship of vision and hearing impairment to one-year mortality and functional decline. *J. Aging Health* **4**, 126–148.

4. Carabellese, C., Appollonio, I., Rozzini, R., Bianchetti, A., Frisoni, G. B., Frattola, L., and Trabucchi, M. (1993). Sensory impairment and quality of life in a community elderly population. *J. Am. Geriatr. Soc.* **41**, 401–407.

5. Rudberg, M. A., Furner, S. E., Dunn, J. E., and Cassel, C. K. (1993). The relationship of visual and hearing impairments to disability: An analysis using the longitudinal study of aging. *J. Gerontol.* **48**, M261–M265.

6. Salive, M. E., Guralnik, J., Glynn, R. J., Christen, W., Wallace, R. B., and Ostfeld, A. M. (1994). Association of visual impairment with mobility and physical function. *J. Am. Geriatr. Soc.* **42**, 287–292.

7. Appollonio, I., Carabellese, C., Magni, E., Frattola, L., and Trabucchi, M. (1995). Sensory impairments and morality in an elderly community population: A Six-year Follow-up Study. *Age Ageing* **24**, 30–36.

8. Hakkinen, L. (1984). Vision in the elderly and its use in the social environment. *Scand. J. Soc. Med.* **35**(Suppl.), 5–60.

9. Thompson, J. R., Gibson, J. M., and Jagger, C. (1989). The association between visual impairment and mortality in elderly people. *Age Aging* **18**, 83–88.

10. Klein, R., Klein, B. E., and Moss, S. E. (1995). Age-related eye disease and survival. The Beaver Dam Eye Study. *Arch. Ophthalmol.* **113**, 333–339.

11. Tielsch, J. M., Sommer, A., Witt, K., Katz, J., and Royall, R. M. (1990). Blindness and visual impairment in an American urban population. The Baltimore Eye Survey. *Arch. Ophthalmol.* **108**, 286–290.

12. Rubin, G., West, S., Muñoz, B., Bandeen-Roche, K., Zeger, S., Fried, L., and the SEE Project Team (1997). A comprehensive assessment of visual impairment in an older American population: SEE Study. *Invest. Ophthalmol. Visual Sci.* **38**, 557–568.

13. Steinberg, E. P., Tielsch, J. M., Schein, O. D., Javitt, J. C., Sharkey, P., Cassard, S. D., Legro, M. W., Diener-West, M., Bass, E. B., Damiano, A. M., Steinwachs, D. M., and Sommer, A. (1994). National study of cataract surgery outcomes. *Ophthalmology* **101**, 1131–1141.

14. Rahmani, B., Tielsch, J. M., Katz, J., Gottsch, J., Quigley, H., Javitt, J., and Sommer, A. (1996). The cause-specific prevalence of visual impairment in an urban population. The Baltimore Eye Survey. *Ophthalmology* **103**, 1721–1726.

15. Klaver, C. C., Wolfs, R. C., Vingerling, J. R., Hofman, A., and de Jong, P. T. (1998). Age-specific prevalence and causes of blindness and visual impairment in an older population: The Rotterdam Study. *Arch. Ophthalmol.* **116**, 653–658.

16. Legge, G. E., Rubin, G. S., Pelli, D. G., and Schleske, M. M. (1985). Psychophysics of reading. II. Low vision. *Vision Res.* **25**, 253–266.

17. Rubin, G. S., and Turano, K. (1994). Low vision reading with sequential word presentation. *Vision Res.* **34**, 1723–1733.

18. Bouma, H. (1970). Interaction effects in parafoveal letter recognition. *Nature* (*London*) **226**, 177–178.

19. Bullimore, M. A., Wood, J. M., and Swenson, K. (1993). Motion perception in glaucoma. *Invest. Ophthalmol. Visual Sci.* **34**, 3526–3533.

20. Geruschat, D., Turano, K., and Stahl, J. (1998). Traditional measures of mobility performance and retinitis pigmentosa. *Optom. Visual Sci.* **75**, 525–537.

21. Leat, S. J., Fryer, A., and Rumney, N. J. (1994). Outcome of low vision aid provision: The effectiveness of a low vision clinic. *Optom. Visual Sci.* **71**, 199–206.

22. Raasch, T. W., Leat, S. J., Kleinstein, R. N., Bullimore, M. A., and Cutter, G. R. (1997). Evaluating the value of low-vision services. *J. Am. Optom. Assoc.* **68**, 287–295.

23. Verbrugge, L. M., and Patrick, D. L. (1995). Seven chronic conditions: Their impact on U.S. adults' activity levels and use of medical services. *Am. J. Public Health* **85**, 173–182.

24. Rubin, G. S., and Schuchard, R. A. (1990). Does Contrast sensitivity predict face recognition performance in low-vision observers? "Noninvasive Assessment of the Visual System." Optical Society of America, Washington, DC.

25. Brown, B., Brabyn, J., Welch, L., Haegerstrom-Portnoy, G., and Colenbrander, A. (1986). The contribution of vision variables to mobility in age-related maculopathy patients. *Am. J. Optom. Physiol. Opt.* **63**, 733–739.

26. Marron, J. A., and Bailey, I. L. (1982). Visual factors and orientation-mobility performance. *Am. J. Optom. Physiol. Opt.* **59**, 413–426.

27. Rubin, G. S. (1989). Assessment of visual function in eyes with visual loss. *In* "Ophthalmology Clinics of North America: Assessment of Visual Function for the Clinician" (D. Fuller and D. G. Birch, eds.), pp. 357–367. Saunders, Philadelphia.

28. Rubin, G. S. (1986). Predicting reading performance in low-vision patients with age-related maculopathy. *In* "Low Vision: Principles and Applications" (G. C. Woo, ed.). Springer-Verlag, New York.

29. Rubin, G. S., and Legge, G. E. (1989). Psychophysics of reading. VI. The role of contrast in low vision. *Vision Res.* **29**, 79–91.

30. Akutsu, H., Legge, G. E., Ross, J. A., and Schuebel, K. J. (1991). Psychophysics of reading. X. Effects of age-related changes in vision. *J. Gerontol.* **46**, 325–331.

31. Turano, K. A., Dagnelie, G., and Herdman, S. J. (1996). Visual stabilization of posture in persons with central visual field loss. *Invest. Ophthalmol. Visual Sci.* **37**, 1483–1491.

32. Owsley, C., and Sloane, M. E. (1987). Contrast sensitivity, acuity, and the perception of "real-world' targets. *Br. J. Ophthalmol.* **71**, 791–796.

33. Owsley, C., Sekuler, R., and Boldt, C. (1981). Aging and low-contrast vision: Face perception. *Invest. Ophthalmol. Visual Sci.* **21**, 362–365.

34. Pulling, N. H., Wolf, E., Sturgis, S. P., Vaillancourt, D. R., and Dolliver, J. J. (1980). Headlight glare resistance and driver age. *Hum. Factors* **22**, 103–112.

35. Adamsons, I., Vitale, S., Stark, W., and Rubin, G. (1996). The association of post-operative subjective visual function with acuity, glare, and contrast sensitivity in subjects with early cataract. *Arch. Ophthalmol.* **114**, 529–536.

36. Elliott, D. B., Hurst, M. A., and Weatherill, J. (1990). Comparing clinical tests of visual function in cataract with the patient's perceived visual disability. *Eye* **4**, 712–717.

37. Lovie-Kitchin, J., Mainstone, J., Robinson, J., and Brown, B. (1990). What areas of the visual field are important for mobility in low vision patients? *Clin. Vision Sci.* **5**, 249–263.

38. Johnson, C. A., and Keltner, J. L. (1983). Incidence of visual field loss in 20,000 eyes and its relationship to driving performance. *Arch. Ophthalmol.* **101**, 371–375.

39. Fishman, G., Anderson, R., and Stinson, L. (1981). Driving performance of retinitis pigmentosa patients. *Br. J. Ophthalmol.* **65**, 122–126.

40. Hofstetter, H. (1976). Visual acuity and highway accidents. *J. Am. Optom. Assoc.* **47**, 887–893.

41. Rubin, G. S., Bandeen-Roche, K., Prasada-Rao, P., and Fried, L. P. (1994). Visual impairment and disability in older adults. *Optom. Visual Sci.* **71**, 750–760.

42. Nevitt, M. C., Cummings, S. R., Kidd, S., and Black, D. (1989). Risk factors for recurrent nonsyncopal falls: A Prospective Study. *JAMA, J. Am. Med. Assoc.* **261**, 2663–2668.

43. Dargent-Molina, P., Hays, M., and Bréart, G. (1996). Sensory impairments and physical disability in aged women living at home. *Int. J. Epidemiol.* **25,** 621–629.

44. Klein, B. E., Klein, R., and Jensen, S. C. (1996). Visual sensitivity and age-related eye diseases. The Beaver Dam Eye Study. *Ophthalmic Epidemiol.* **3,** 47–55.

45. Kahn, H. A., Leibowitz, H. M., Ganley, J. P., Kini, M. M., Colton, T., Nickerson, R. S., and Dawber, T. R. (1977). The Framingham Eye Study. I. Outline and major prevalence findings. *Am. J. Epidemiol.* **106,** 17–32.

46. Kahn, H. A., and Milton, R. C. (1980). Revised Framingham eye study prevalence of glaucoma and diabetic retinopathy. *Am. J. Epidemiol.* **111,** 769–776.

47. Kini, M. M., Leibowitz, H. M., Colton, T., Nickerson, R. J., Ganley, J., and Dawber, T. R. (1978). Prevalence of senile cataract, diabetic retinopathy, senile macular degeneration, and open-angle glaucoma in the Framingham Eye Study. *Am. J. Ophthalmol.* **85,** 28–34.

48. Tielsch, J. M., Sommer, A., Witt, K., Katz, J., and Royall, R. M. (1990). Blindness and visual impairment in an American urban population. The Baltimore Eye Survey. *Arch. Ophthalmol.* **108,** 286–290.

49. Dana, M. R., Tielsch, J. M., Enger, C., Joyce, E., Santoli, J. M., and Taylor, H. R. (1990). Visual impairment in a rural Appalachian community. Prevalence and causes. *JAMA, J. Am. Med. Assoc.* **264,** 2400–2405.

50. Klein, R., Klein, B. E., Linton, K. L., and De Mets, D. L. (1991). The Beaver Dam Eye Study: Visual acuity. *Ophthalmology* **98,** 1310–1315.

51. Klein, R., Wang, Q., Klein, B. K. E., Moss, S. E., and Meuer, S. M. (1995). The relationship of age-related maculopathy, cataract, and glaucoma to visual acuity. *Invest. Ophthalmol. Visual Sci.* **36,** 182–191.

52. Klein, B. E. K., Klein, R., Lee, K. E., and Cruickshanks, K. J. (1998). Performance-based and self-assessed measures of visual function as related to history of falls, hip fractures, and measured gait time. *Ophthalmology* **105,** 160–164.

53. World Health Organization (1980). "International Classification of Impairments, Disabilities, and Handicaps: A Manual of Classification Relating to the Consequences of Disease. Introduction," pp. 7–19, WHO, Geneva.

54. Taylor, H. R., Livingston, P. M., Stanislavsky, Y. L., and McCarty, C. A. (1997). Visual impairment in Australia: Distance visual acuity, near vision, and visual field findings of the Melbourne Visual Impairment Project. *Am. J. Ophthalmol.* **123,** 328–337.

55. Attebo, K., Mitchell, P., and Smith, W. (1996). Visual acuity and the causes of visual loss in Australia. The Blue Mountains Eye Study. *Ophthalmology* **103,** 357–364.

56. Mitchell, P., Smith, W., Attebo, K., and Healey, P. R. (1996). Prevalence of open-angle glaucoma in Australia. The Blue Mountains Eye Study. *Ophthalmology* **103,** 1661–1669.

57. Mitchell, P., Smith, W., Attebo, K., and Wang, J. J. (1995). Prevalence of age-related maculopathy in Australia. The Blue Mountains Eye Study. *Ophthalmology* **102,** 1450–1460.

58. Mitchell, P., Cumming, R. G., Attebo, K., and Panchapakesan, J. (1997). Prevalence of cataract in Australia: The Blue Mountains Eye Study. *Ophthalmology* **104,** 581–588.

59. Mitchell, P., Smith, W., Wang, J. J., and Attebo, K. (1998). Prevalence of diabetic retinopathy in an older community. The Blue Mountains Eye Study. *Ophthalmology* **105,** 406–411.

60. Wang, J. J., Mitchell, P., Smith, W., Cumming, R. G., and Attebo, K. (1999). Impact of visual impairment on use of community support services by elderly persons: The Blue Mountains Eye Study. *Invest. Ophthalmol. Visual Sci.* **40,** 12–19.

61. Ivers, R. Q., Cumming, R. G., Mitchell, P., and Attebo, K. (1998). Visual impairment and falls in older adults: The Blue Mountains Eye Study. *J. Am. Geriatr. Soc.* **46,** 58–64.

62. West, S. K., Munoz, B., Rubin, G. S., Schein, O. D., Bandeen-Roche, K., Zeger, S., German, S., and Fried, L. P. (1997). Function and visual impairment in a population-based study of older adults. The SEE Project. Salisbury Eye Evaluation. *Invest. Ophthalmol. Visual Sci.* **38,** 72–82.

63. Rubin, G. S., Muñoz, B., Bandeen-Roche, K., Fried, L., West, S. K., and the SEE Project Team (1996). The impact of visual impairment on objective and subjective measures of physical disability: SEE Project. *Invest. Ophthalmol. Visual Sci.* **37,** S184.

64. Friedman, S. M., Rubin, G. S., West, S. K., Muñoz, B., Bandeen-Roche, K., Zeger, S. L., Schein, O. D., Fried, L. P. and the SEE Project Team (1996). The relationship between visual function and balance: The SEE Project. *Invest. Ophthalmol. Visual Sci.* **37,** S184.

65. Valbuena, M., Bandeen-Roche, K., Rubin, G. S., Muñoz, B., and West, S. K. (1999). Self-reported assessment of visual function in a population-based study: The SEE Project. Salisbury Eye Evaluation. *Invest. Ophthalmol. Visual Sci.* **40,** 280–288.

66. Klein, R., Klein, B. E., Jensen, S. C., Moss, S. E., and Cruickshanks, K. J. (1994). The relation of socioeconomic factors to age-related cataract, maculopathy, and impaired vision. The Beaver Dam Eye Study. *Ophthalmology* **101,** 1969–1979.

67. Tielsch, J. M., Sommer, A., Katz, J., Quigley, H., and Ezrine, S. (1991). Socioeconomic status and visual impairment among urban Americans. Baltimore Eye Survey Research Group. *Arch. Ophthalmol.* **109,** 637–641.

68. Taylor, H., West, S., Rosenthal, F., Muñoz, B., Newland, H., Abbey, H., and Emmett, E. (1988). Effect of ultraviolet radiation on cataract formation. *N. Engl. J. Med.* **319,** 1429–1433.

69. Cruickshanks, K. J., Klein, B. E., and Klein, R. (1992). Ultraviolet light exposure and lens opacities: The Beaver Dam Eye Study. *Am. J. Public Health* **82,** 1658–1662.

70. West, S. K., Duncan, D. D., Muñoz, B., Rubin, G. S., Fried, L. P., Bandeen-Roche, K., and Schein, O. D. (1998). Sunlight exposure and risk of lens opacities in a population-based study: The Salisbury Eye Evaluation Project. *JAMA, J. Am. Med. Assoc.* **280,** 714–718.

71. West, S., Muñoz, B., Emmett, E. A., and Taylor, H. R. (1989). Cigarette smoking and risk of nuclear cataracts. *Arch. Ophthalmol.* **107,** 1166–1169.

72. Klein, B. E., Klein, R., Linton, K. L., and Franke, T. (1993). Cigarette smoking and lens opacities: The Beaver Dam Eye Study. *Am. J. Prev. Med.* **9,** 27–30.

73. Cumming, R. G., and Mitchell, P. (1997). Alcohol, smoking, and cataracts: The Blue Mountains Eye Study. *Arch. Ophthalmol.* **115,** 1296–1303.

74. West, S., Muñoz, B., Schein, O. D., Vitale, S., Maguire, M., Taylor, H. R., and Bressler, N. M. (1995). Cigarette smoking and risk for progression of nuclear opacities. *Arch. Ophthalmol.* **113,** 1377–1380.

75. Vingerling, J. R., Hofman, A., Grobbee, D. E., and de Jong, P. T. (1996). Age-related macular degeneration and smoking. The Rotterdam Study. *Arch. Ophthalmol.* **114,** 1193–1196.

76. Klein, R., Klein, B. E., Linton, K. L., and DeMets, D. L. (1993). The Beaver Dam Eye Study: The relation of age-related maculopathy to smoking. *Am. J. Epidemiol.* **137,** 190–200.

77. Smith, W., Mitchell, P., and Leeder, S. R. (1996). Smoking and age-related maculopathy. The Blue Mountains Eye Study. *Arch. Ophthalmol.* **114,** 1518–1523.

78. Klein, R., Klein, B. E., and Moss, S. E. (1998). Relation of smoking to the incidence of age-related maculopathy. The Beaver Dam Eye Study. *Am. J. Epidemiol.* **147,** 103–110.

79. Ritter, L. L., Klein, B. E., Klein, R., and Mares-Perlman, J. A. (1993). Alcohol use and lens opacities in the Beaver Dam Eye Study. *Arch. Ophthalmol.* **111,** 113–117.

80. Moss, S. E., Klein, R., Klein, B. E., Jensen, S. C., and Meuer, S. M. (1998). Alcohol consumption and the 5-year incidence of age-related maculopathy: The Beaver Dam Eye Study. *Ophthalmology* **105,** 789–794.

81. Ritter, L. L., Klein, R., Klein, B. E., Mares-Perlman, J. A., and Jensen, S. C. (1995). Alcohol use and age-related maculopathy in the Beaver Dam Eye Study. *Am. J. Ophthalmol.* **120,** 190–196.

82. Klein, R., Klein, B. E., and Franke, T. (1993). The relationship of cardiovascular disease and its risk factors to age-related maculopathy. The Beaver Dam Eye Study. *Ophthalmology* **100,** 406–414.

83. Smith, W., Mitchell, P., Leeder, S. R., and Wang, J. J. (1998). Plasma fibrinogen levels, other cardiovascular risk factors, and age-related maculopathy: The Blue Mountains Eye Study. *Arch. Ophthalmol.* **116,** 583–587.

84. Vingerling, J. R., Dielemans, I., Bots, M. L., Hofman, A., Grobbee, D. E., and de Jong, P. T. (1995). Age-related macular degeneration is associated with atherosclerosis. The Rotterdam Study. *Am. J. Epidemiol.* **142,** 404–409.

85. Dielemans, I., Vingerling, J. R., Algra, D., Hofman, A., Grobbee, D. E., and de Jong, P. T. (1995). Primary open-angle glaucoma, intraocular pressure, and systemic blood pressure in the general elderly population. The Rotterdam Study. *Ophthalmology* **102,** 54–60.

86. Klein, B. E., Klein, R., and Ritter, L. L. (1994). Is there evidence of an estrogen effect on age-related lens opacities? The Beaver Dam Eye Study. *Arch. Ophthalmol.* **112,** 85–91.

87. Smith, W., Mitchell, P., and Wang, J. J. (1997). Gender, oestrogen, hormone replacement and age-related macular degeneration: Results from the Blue Mountains Eye Study. *Aust. N. Z. J. Ophthalmol.* **25,** S13–S15.

88. West, S., Vitale, S., Hallfrisch, J., Muñoz, B., Muller, D., Bressler, S., and Bressler, N. M. (1994). Are antioxidants or supplements protective for age-related macular degeneration? *Arch. Ophthalmol.* **112,** 222–227.

89. Seddon, J. M., Ajani, U. A., Sperduto, R. D., Hiller, R., Blair, N., Burton, T. C., Farber, M. D., Gragoudas, E. S., Haller, J., Miller, D. T., Yanuzzi, L. A., and Willet, W. (1994). Dietary carotenoids, vitamins A, C, and E, and advanced age-related macular degeneration. Eye Disease Case-Control Study Group. JAMA, *J. Am. Med. Assoc.* **272,** 1413–1420, *erratum: Ibid.* **273**(8), 622 (1995).

90. VandenLangenberg, G. M., Mares-Perlman, J. A., Klein, R., Klein, B. E., Brady, W. E., and Palta, M. (1998). Associations between antioxidant and zinc intake and the 5-year incidence of early age-related maculopathy in the Beaver Dam Eye Study. *Am. J. Epidemiol.* **148,** 204–214.

91. Smith, W., Mitchell, P., and Rochester, C. (1997). Serum beta carotene, alpha tocopherol, and age-related maculopathy: The Blue Mountains Eye Study. *Am. J. Ophthalmol.* **124,** 838–840.

92. Davis, A. C., (1991). Epidemiological profile of hearing impairments: The scale and nature of the problem with special reference to the elderly. *Acta Otolaryngol., Suppl.* **476,** 23–31.

93. Martin, F. N. (1991). "Introduction to Audiology." Prentice-Hall, Englewood Cliffs, NJ.

94. Mulrow, C. D., Aguilar, C., Endicott, J. E., Tuley, M. R., Velez, R., Charlip, W. S., Rhodes, M. C., Hill, J. A., and DeNino, L. A. (1990). Quality-of-life changes and hearing impairment. A randomized trial. *Ann. Intern. Med.* **113,** 188–194.

95. Strawbridge, W. J., Cohen, R. D., Shema, S. H., and Kaplan, G. A. (1996). Successful aging: Predictors and associated activities. *Am. J. Epidemiol.* **44**(2), 135–141.

96. Bess, F. H., Lichtenstein, M. J., Logan, S. A., Burger, M. C., and Nelson, E. (1989). Hearing impairment as a determinant of function in the elderly. *J. Am. Geriatr. Soc.* **37,** 123–128.

97. Gerson, L. W., Jarjoura, D., and McCord, G. (1989). Risk of imbalance in elderly people with impaired hearing or vision. *Age Ageing* **18,** 31–34.

98. Gilhiome Herbst, K. R., Meredith, R., and Stephens, S. D. (1990). Implications of hearing impairment for elderly people in London and in Wales. *Acta Otolaryngol.* **476**(Suppl.), 209–214.

99. Weinstein, B. E., and Vestry, I. M. (1982). Hearing impairment and social isolation in the elderly. *J. Speech Hear. Res.* **25,** 593–599.

100. Jerger, J., Chmiel, R., Wilson, N., and Luchi, R. (1995). Hearing impairment in older adults: New concepts. *J. Am. Geriatr. Soc.* **43,** 928–935.

101. Andersson, G., Green, M., and Melin, L. (1997). Behavioural hearing tactics: A controlled trial of a short treatment program. *Behav. Res. Ther.* **35,** 523–530.

102. Walden, B. E., Busacco, D. A., and Montgomery, A. A. (1993). Benefit from visual cues in auditory-visual speech recognition by middle-aged and elderly persons. *J. Speech Hear. Res.* **36,** 431–436.

103. Popelka, M. M., Cruickshanks, K. J., Wiley, T. L., Tweed, T. S., Klein, B. E. K., and Klein, R. (1998). Low prevalence of hearing aid use among older adults with hearing loss: The Epidemiology of Hearing Loss Study. *J. Am. Geriatr. Soc.* **46,** 1075–1078.

104. American Speech-Language- Hearing Association (1978). Guidelines for manual pure-tone threshold audiometry. *Am. Speech Hear. Assoc.* **20,** 297–301.

105. Wilson, R. H., Zizz, C. A., Shanks, J. E., and Causey, J. D. (1990). Normative data in quiet, broadband noise and competing message for Northwestern University Test #6 by a female speaker. *J. Speech Hear. Dis.* **55,** 771–778.

106. Wiley, T. L., Cruickshanks, K. J., Nondahl, D. M., Klein, R., and Klein, B. E. K. (1998). Aging and word recognition in competing messages. *J. Am. Acad. Audiol.* **9,** 191–198.

107. Gates, G. A., Cooper, J. C., Jr., Kannel, W. B., and Miller, N. J. (1990). Hearing in the elderly: The Framingham Cohort, 1983– 1985. *Ear Hear.* **11,** 247–256.

108. Cruickshanks, K. J., Wiley, T. L., Tweed, T. S., Klein, B. E. K., Klein, R., Mares-Perlman, J. A., and Nondahl, D. M. (1998). Prevalence of hearing loss in older adults in Beaver Dam, WI: The Epidemiology of Hearing Loss Study. *Am. J. Epidemiol.* **148**(9), 879–886.

109. Ries, P. W. (1985). Demography of hearing loss. *In* "Adjustment to Adult Hearing Loss" (H. Orlans, (ed.), pp. 3–20. College Hill Press, San Diego, CA.

110. Moscicki, E. K., Elkins, E. F., Baum, H. M., and McNamara, P. M. (1985). Hearing loss in the elderly: An Epidemiologic Study of the Framingham Heart Study Cohort. *Ear Hear.* **6**(4), 184–190.

111. Lee, D. J., Carlson, D. L., Lee, H. M., Ray, L. A., and Markides, K. S. (1991). Hearing loss and hearing aid use in Hispanic adults: Results from the Hispanic Health and Nutrition Examination Survey. *Am. J. Public Health* **81,** 1471–1474.

112. Rowland, M. (1980). "Basic Data on Hearing Levels of Adults 25–74 Years: United States—1971-75," Publ. No. (PHS) 80-1663. Vital and Health Statistics 1980; Department of Health, Education and Welfare, Washington, DC.

113. Nondahl, D. M., Cruickshanks, K. J., Wiley, T. L., Tweed, T. S., Ershler, W., Klein, R., and Klein, B. E. K. (1998). Accuracy of self-reported hearing loss. *Audiology* **37,** 295–301.

114. Ventry, I. M., and Weinstein, B. E. (1983). Identification of elderly people with hearing problems. *Am. Soc. Hear. Assoc.* **25,** 37–47.

115. Lichtenstein, M. J., Bess, F. H., and Logan, S. A. (1988). Diagnostic performance of the Hearing Handicap Inventory for the Elderly

(Screening Version) against differing definitions of hearing loss. *Ear Hear.* 9(4), 208–211.

116. Mulrow, C. D., Tuley, M. R., and Aguilar, C. (1990). Discriminating and responsiveness abilities of two hearing handicap scales. *Ear Hear.* 11(3), 176–180.

117. Ries, P. W. (1994). Prevalence and characteristics of persons with hearing trouble: United States 1990–91. National Center for Health Statistics. *Vital & Health Statistics* 10, 1–75.

118. Wallhagen, M. I., Strawbridge, W. J., Cohen, R. D., and Kaplan, G. A. (1997). An increasing prevalence of hearing impairment and associated risk factors over three decades of the Alameda County Study. *Am. J. Public Health* 87, 440–442.

119. Gates, G. A., and Cooper, J. C. (1991). Incidence of hearing decline in the elderly. *Acta Otolaryngol.* 111, 240–248.

120. Brant, L. J., Gordon-Salant, S., Pearson, J. D., Klein, L. L., Morrell Metter, E. J., and Fozard, J. L. (1996). Risk factors related to age-related hearing loss in the speech frequencies. *J. Am. Acad. Audiol.* 7, 152–160.

121. Pearson, J. D., Morrell, C. H., Gordon-Salant, S., Brant, L. J., Metter, E. J., Klein, L. L., and Fozard, J. L. (1995). Gender differences in a longitudinal study of age-associated hearing loss. *J. Access. Soc. Am.* 97(2), 1196–1205.

122. Clark, K., Sowers, M., Wallace, R. B., and Anderson, C. (1991). The accuracy of self-reported hearing loss in women aged 60–85 years. *Am. J. Epidemiol.* 134(7), 704–708.

123. Rosen, S., Bergman, M., Plester, D., El-Mofty, A., and Satti, M. H. (1962). Presbycusis study of a relatively noise-free population in the Sudan. *Ann. Otol. Rhinol. Laryngol.* 71, 727–743.

124. Ewertson, H. W. (1973). Epidemiology of professional noise-induced hearing loss. *Audiology* 12, 453–458.

125. OSHA (1983). Occupational noise exposure: Hearing conservation amendment: Final rule. *Fed. Regist.* 48, 9737–9785.

126. Johnsson, L. G., and Hawkins, J. E. (1972). Vascular changes in the human inner ear associated with aging. *Ann. Otol.* 81, 364–376.

127. Makishima, K. (1978). Arteriolar sclerosis as a cause of presbycusis. *Otolaryngology* 86, 322–326.

128. Jarvis, J., and vanHeerden, H. G. (1962). The acuity of hearing in the Kalahari Bushmen. A Pilot Study. *J. Laryngol. Otol.* 81, 63–68.

129. Goycoolea, M. V., Goycoolea, H. G., Farfan, C. R., Rodriquez, L. G., Martinez, G. C., and Vidal, R. (1986). Effect of life in industrialized societies on hearing in natives of Easter Island. *Laryngoscope* 96, 1391–1396.

130. Counter, S. A., and Klareskov, B. (1990). Hypoacusis among the Polar Eskimos of Northwest Greenland. *Scanning Audiol.* 19, 149–160.

131. Bois, E., Bonaiti, C., Lallemont, M., Moatti, L., Feingold, N., Mayer, F. M., and Feingold, J. (1987). Studies on an isolated West Indies population. III. Epidemiologic study of sensorineural hearing loss. *NeuroEpidemiology* 6, 139–149.

132. Rosen, S. (1966). Hearing studies in selected urban and rural populations. *Trans. N. Y. Acad. Sci.* 29, 9–21.

133. Rosen, S., Preobrajensky, N., Tbilisi, S. K., Glazunov, I., Tbilisi, N. K., and Rosen, H. V. (1970). Epidemiologic hearing studies in the USSR. *Arch. Otolaryngol.* 91, 424–428.

134. Morizono, T., and Paparell, M. M. (1978). Hypercholesterolemia and auditory dysfunction. *Ann. Otol.* 87, 804–814.

135. Belal, A. (1980). Pathology of vascular sensorineural hearing impairment. *Laryngoscope* 90, 1831–1839.

136. Susmano, A., and Rosenbush, S. W. (1988). Hearing loss and ischemic heart disease. *Am. J. Otol.* 9(5), 403–408.

137. Gates, G. A., Cobb, J. L., D'Agostino, R. B., and Wolff, P. A. (1993). The relation of hearing in the elderly to the presence of cardiovascular disease and cardiovascular risk factors. *Arch. Otolaryngol. Head Neck Surg.* 119, 156–161.

138. Clark, K., Sowers, M. R., Wallace, R. B., Jannausch, M. L., Lemke, J., and Anderson, C. V. (1995). Age-related hearing loss and bone mass in a population of rural women aged 60 to 85 years. *Ann. Epidemiol.* 5, 8–14.

139. Weston, T. E. T. (1964). Presbyacusis. A clinical study. *J. Laryngol. Otol.* 78, 273–286.

140. Weiss, W. (1970). How smoking affects hearing. *Med. Times* 98(11), 84–89.

141. Zelman, S. (1973). Correlation of smoking history with hearing loss. *JAMA, J. Am. Med. Assoc.* 223(8), 920.

142. Siegelaub, A. B., Friedman, G. D., Adour, K., and Seltzer, C. C. (1974). Hearing loss in adults. Relation to age, sex, exposure to loud noise and cigarette smoking. *Arch. Environ. Health* 29, 107–109.

143. National Center for Health Statistics (1967). "Data from the National Health Survey: Cigarette Smoking and Health Characteristics, July 1964–June 1965," Vital Health Stat., Publ. Ser. 10, No. 34, pp. 11 and 14. U.S. Public Health Service, Washington, DC.

144. Cruickshanks, K. J., Klein, R., Klein, B. E. K., Wiley, T. L., Nondahl, D. M., and Tweed, T. S. (1998). Cigarette smoking and hearing loss: The Epidemiology of Hearing Loss Study. *JAMA, J. Am. Med. Assoc.* 279(21), 1715–1719.

145. Golabek, W., and Niedzielska, G. (1984). Audiologic investigation of chronic alcoholics. *Clin. Otolaryngol.* 9(5), 257–261.

146. Rosenhall, U., Sixt, E., Sundh, V., and Svanborg, A. (1993). Correlations between presbycusis and extrinsic noxious factors. *Audiology* 32(4), 234–243.

147. Spitzer, J. B. (1981). Auditory effects of chronic alcoholism. *Drug Alcohol Depend.* 8(4), 317–335.

148. Popelka, M. M., Cruickshanks, K. J., Wiley, T. L., Tweed, T. S., Klein, B. E. K., and Klein, R. (1999). Moderate alcohol consumption and hearing loss; A protective effect. *J. Am. Geriatr. Soc.* (submitted for publication).

149. Kurien, M., Thomas, K., and Bhanu, T. S. (1989). Hearing threshold in patients with diabetes mellitus. *J. Laryngol. Otol.* 103, 164–168.

150. Parving, A., Elberling, C., Balle, V., Parbo, J., Dejgaard, A., and Parving, H. H. (1990). Hearing disorders in patients with insulin dependent diabetes mellitus. *Audiology* 29, 113–121.

151. Axelsson, A., Sigroth, K., and Vertes, D. (1978). Hearing in diabetics. *Acta Oto-Laryngol., Suppl.* 356, 3–23.

152. Dalton, D. S., Cruickshanks, K. J., Klein, R., Klein, B. E. K., and Wiley, T. L. (1998). The association of non- insulin-dependent diabetes and hearing loss. *Diabetes Care* 21(9), 1540–1544.

153. Ives, D. G., Bonino, P., Traven, N. D., and Kuller, L. H. (1995). Characteristics and comorbidities of rural older adults with hearing impairment. *J. Am. Geriatr. Soc.* 43, 803–806.

154. Schiffman, S. S. (1997). Taste and smell losses in normal aging and disease. *JAMA, J. Am. Med. Assoc.* 278(16), 1357–1362.

155. Ship, J. A., and Weiffenbah, J. M. (1993). Age, gender, medical treatment, and medication effects on smell identification. *J. Gerontol.* 48(1), M26–M32.

156. Moll, B., Klimek, L., Eggers, G., and Mann, W. (1998). Comparison of olfactory function in patients with seasonal and perennial allergic rhinitis. *Allergy* 53, 297–301.

157. Moberg, P. J., Doty, R. L., Mahr, R. N., Mesholam, R. J., Arnold, S. E., Turetsky, B. I., and Gur, R. E. (1997). Olfactory identification in elderly schizophrenia and Alzheimer's disease. *Neurobiol. Aging* 18(2), 163–167.

158. Doty, R. L., Yousem, D. M., Pham, L. T., Kreshak, A. A., Geckle, R., and Lee, W. W. (1997). Olfactory dysfunction in patients with head trauma. *Arch. Neurol.* (*Chicago*) 54, 1131–1140.

159. Ship, J. A., Pearson, J. D., Cruise, L. J., Brant, L. J., and Metter, E. F. (1996). Longitudinal changes in smell identification. *J. Gerontol.* **51A**(2), M86–M91.

160. Baker, K. A., Didcock, E. A., Kemm, J. R., and Patrick, J. M. (1983). Effect of age, sex, and illness on salt taste detection thresholds. *Age Ageing* **12**, 159–165.

161. Griep, M. L., Collys, K., Mets, T. F., Slop, D., Laska, M., and Massart, D. L. (1996). Sensory detection of food odour in relation to dental status, gender and age. *Gerontology* **13**(1), 56–62.

162. Gilbert, A. N., and Wysocki, C. J. (1987). The Smell Survey. *Natl. Geogr.* **72**, 514–525.

163. Doty, R. L., Gregor, T., and Monroe, C. (1986). Quantitative assessment of olfactory function in an industrial setting. *J. Occup. Med.* **28**(6), 457–460.

164. Deems, D. A., Doty, R. L., Settle, G., Moore-Gillon, V., Shaman, P., Mester, A. F., Kimmelman, C. P., Brightman, V. J., and Snow, J. B. (1991). Smell and taste disorders, a study of 750 patients from the University of Pennsylvania Smell and Taste Center. *Arch. Otolaryngol. Head Neck Surg.* **117**, 519–528.

165. Doty, R. L., Shaman, P., Applebaum, S. L., Giberson, R., Siksorski, L., and Rosenberg, L. (1984). Smell identification ability: Changes with age. *Science* **226**, 1441–1442.

166. Corwin, J., Loury, M., and Gilbert, A. N. (1995). Workplace, age, and sex as mediators of olfactory function: Data from the National Geographic Smell Survey. *J. Gerontol.* **50B**(4), 179–186.

167. Wysocki, C. J., and Gilbert, A. N. (1989). National Geographic Smell Survey. Effects of age are heterogenous. *Ann. N.Y. Acad. Sci.* **561**, 12–28.

168. Barber, C. E. (1997). Olfactory acuity as a function of age and gender: A comparison of African and American samples. *Int. J. Aging Hum. Dev.* **44**(4), 317–334.

169. Weiffenbach, J. M., Baum, B. J., and Burghauser, R. (1982). Taste thresholds: Quality specific variation with human aging. *J. Gerontol.* **37**(3), 372–377.

170. Fikentscher, R., Roseburg, B., Spinar, H., and Bruchmller, W. (1977). Loss of taste in the elderly: Sex differences. *Clin. Otolaryngol.* **2**, 183–189.

171. Wysocki, C. J., and Pelchat, M. L. (1993). The effects of aging on the human sense of smell and its relationship to food choice. *Crit. Rev. Food Sci. Nutr.* **33**(1), 63–82.

172. Duffy, V. B., Backstrand, J. R., and Ferris, A. M. (1995). Olfactory dysfunction and related nutritional risk in free-living, elderly women. *J. Am. Diet. Assoc.* **95**, 879–884.

173. Doty, R. L., McKeown, D. A., Lee, W. W., and Shaman, P. (1995). A study of test-retest reliability of ten olfactory tests. *Chem. Senses* **20**, 645–656.

174. Hummel, T., Erras, A., and Kobal, G. (1997). A test for the screening of taste function. *Rhinology* **35**, 146–148.

175. Odeigah, P. G. C. (1994). Smell acuity for acetone and its relationship to taste ability to phenylthiocarbamide in a Nigerian population. *East Afr. Med. J.* **71**(7), 462–466.

176. Murphy, C., Gilmore, M. M., Seery, C. S., Salmon, D. P., and Lasker, B. R. (1990). Olfactory thresholds are associated with degree of dementia in Alzheimer's disease. *Neurobiol. Aging* **11**, 465–469.

177. Murphy, C., Anderson, J. A., and Markison, S. (1986). Psychophysical assessment of chemosensory disorders in clinical populations. *In* "Olfaction and Taste XI" (K. Kurihara, N. Suszuki, and H. Ogawa, eds.), pp. 609–613. Springer, Tokyo.

178. Doty, R. L., and Agrawal, U. (1989). The shelf life of the University of Pennsylvania Smell Identification Test (UPSIT). *Laryngoscope* **99**, 402–404.

179. Doty, R. L., Marcus, A., and Lee, W. W. (1996). Development of the 12-item Cross-Cultural Smell Identification Test (CC-SIT). *Laryngoscope* **106**, 353–356.

180. Richman, R. A., Wallace, K., and Sheehe, P. R. (1995). Assessment of an abbreviated odorant identification task for children: A rapid screening device for schools and clinics. *Acta Paediatr.* **84**, 434–437.

181. Liu, H., Wang, S., Lin, K., Lin, K., Fuh, J., and Teng, E. L. (1995). Performance on a smell screening test (the MODSIT): A study of 510 predominantly illiterate Chinese subjects. *Physiol. Behav.* **58**, 1251–1255.

182. Davidson, T. M., and Murphy, C. (1997). Rapid clinical evaluation of anosmia. The Alcohol Sniff Test. *Arch. Otolaryngol. Head Neck Surg.* **123**, 591–594.

98

Alzheimer's Disease

RICHARD MAYEUX* AND SAMUEL GANDY†

*Gertrude H. Sergievsky Center and Taub Alzheimer's Disease Research Center at Columbia University, College of Physicians and Surgeons, New York, New York; †Dementia Research Program, The Nathan S. Kline Institute for Psychiatric Research, New York University, Orangeburg, New York

I. Introduction

Around the seventeenth century, "dementia" was characterized as the "extinction of the imagination and judgement," but the clinical entity as it is now conceived was developed over the seventeenth and eighteenth centuries [1]. Thomas Willis, known for his observations on the circulatory system, also wrote on the potential causes of impaired cognition in his *Practice of Physik* [2]. Over the years, various organizations have contributed to the definition of dementia and attempted to develop standardized criteria; those developed for the Diagnostic and Statistical Manual of the American Psychiatric Association [3] remain the most widely accepted.

There are over 60 different causes of dementia [4–6], but Alzheimer's disease is the most frequent. Depending on the age of the population under consideration, Alzheimer's disease accounts for up to 60% of cases of dementia. When considered as a coexisting condition, it is present in up to 80% of dementias. Among patients with dementia, the proportion of those with Alzheimer's disease increases with increasing age.

This review will cover what is known about the illness in terms of the clinical and pathological characteristics, diagnosis, current treatments, and genetic and environmental risk factors. We include separate sections on the epidemiology of Alzheimer's disease among women and a very detailed discussion of the putative role of estrogen.

II. The Clinical Syndrome of Alzheimer's Disease

The manifestations of Alzheimer's disease evolve rather typically from the earliest signs of impaired memory to severe cognitive loss, inanition, and death [7]. Gradual and progressive decline with occasional plateaus is most typical. Memory impairment is the usual presenting manifestation [8–10]. Difficulty recalling people's names, telephone numbers, and the details of events of the day or conversations may be seen as insignificant early on but can represent the initial signs of disease. In contrast, memory for remote events of life experiences and for some previously learned facts remains relatively unimpaired at the beginning of the illness.

Word-finding difficulty often parallels the onset of memory problems, and impaired reading comprehension and written language may also be seen early in Alzheimer's disease [11]. Changes in visual perception can also result in difficulties with activities of daily living, such as in dressing oneself or getting lost in familiar surroundings. Depressed mood at some stage of the illness has been reported [12,13]. Major depression with

insomnia or anorexia occurs in 5–8% of patients with Alzheimer's disease regardless of severity [14]. Depressed mood can occur prior to or coincident with the onset of dementia in up to 25% of patients [15]. Delusions and psychotic behavior increase with progression of Alzheimer's disease and remain persistent in 20% of patients. Agitation exists in up to 20% of patients, increasing during the later stages of disease [16,17]. Hallucinations occur with similar frequency and may be either visual or auditory. No sex differences in clinical course have been reported.

Except for the mental state, the neurological examination is usually normal. Extrapyramidal features, including rigidity, bradykinesia, shuffling gait, and postural change, are relatively common in patients with Alzheimer's disease [18,19]. Primary motor and sensory functions are otherwise spared. Oculomotor, cerebellar, or peripheral nerve abnormalities on physical examination strongly raise the possibility of other forms of dementia.

Analysis of 100 autopsy-confirmed cases with Alzheimer's disease [20] indicated that the average duration of symptoms until death may be ten years, with a range of 4 to 16 years. Disease duration tends to be longer for women than for men.

Criteria for the clinical diagnosis were established by a joint effort of the National Institute of Neurological and Communicative Disorders and Stroke and the Alzheimer's Disease and Related Disorders Association in 1984 and are referred to as the NINCDS-ADRDA criteria [21]. These criteria include a history of progressive deterioration in cognitive ability in the absence of other known neurological or medical problems. Psychological testing, brain imaging, and other recommendations establish three levels of diagnostic certainty. Patients without associated illnesses are termed *probable,* while *possible* refers to patients meeting criteria, with other illnesses that may contribute, such as hypothyroidism or cerebrovascular disease. Using the NINCDS-ADRDA criteria, the sensitivity of the clinical diagnosis of probable or possible Alzheimer's disease is over 90% and the specificity is approximately 60% [22]. The designation of *definite* Alzheimer's disease has been reserved for autopsy-confirmed disease.

At postmortem examination, the brain in Alzheimer's disease shows dense plaques containing the β-amyloid peptide (Aβ). A 700-amino acid integral membrane precursor, the amyloid precursor protein (APP), undergoes proteolysis resulting in the accumulation of Aβ in brain which is a key step in the pathogenesis of the disease. Evidence in support of this causal link between APP and Alzheimer's disease comes from the discovery that mutations within the APP gene coding sequence lead to an autosomal dominant form of Alzheimer's disease.

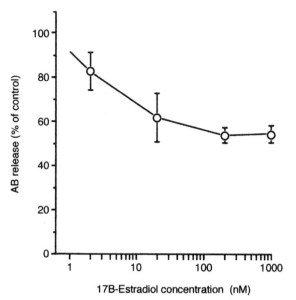

Fig. 98.1 Deposits of amyloid-β peptide (Aβ) in the cerebral cortex. (Photo by Gunnar Gouras; digitized image by Cory Peterhoff.)

Gandy and Greengard [23] reviewed the role of amyloid in Alzheimer's disease. Amyloid is a generic description applied to a heterogeneous class of tissue protein precipitates that have the common feature of beta-pleated sheet secondary structure, a characteristic that confers affinity for the histochemical dye Congo red. Amyloids may be deposited in a general manner throughout the body (systemic amyloids) or confined to a particular organ (*e.g.,* cerebral amyloid, renal amyloid), but in Alzheimer's disease amyloid is deposited around meningeal and cerebral vessels and in gray matter. The gray matter deposits are multifocal, coalescing into military structures known as plaques (Fig. 98.1). Parenchymal amyloid plaques are distributed in the brain in a characteristic fashion, differentially affecting the various cerebral and cerebellar lobes and cortical laminae.

Aβ may be a by-product of APP metabolism, but APP function may not be intimately involved in the cause of Alzheimer's disease. It is possible that the proteolytic processing step for APP involves cleavage within the Aβ domain (Fig. 98.2) from which a soluble form of APP (sAPP) is released. This pathway is designated the α-secretory cleavage or release processing pathway for APP because the enzyme(s) that perform(s) this nonamyloidogenic cleavage has been designated α-secretase. In this pathway the Aβ domain is destroyed. Alternatively, Aβ production begins by cleavage at the Aβ amino-terminus (Fig. 98.2) by another enzyme, designated β-secretase, probably beginning in the constitutive secretory pathway. This provides a codistribution opportunity involving APP and β-secretase prior to the encounter between APP and the Aβ-destroying α-secretase, which appears to act primarily at the plasma membrane. At least some Aβ arises from APP molecules following their residence at the cell surface. Presumably these APP molecules have escaped cell-surface α-secretase and have encountered β-secretase on the cell surface or else have been internalized intact by β-secretase in an endosome.

Of particular interest has been the elucidation of the molecular and cellular mechanisms by which the carboxyl-terminus of Aβ is generated, because this region of the APP molecule resides within an intramembranous domain (Fig. 98.2). The protease(s) responsible for that cleavage, designated the γ-secretase(s) (Fig. 98.2 lower panel), is particularly interesting not only because of the novelty of the active site–substrate reaction within a membranous domain, but also in some familial forms of Alzheimer's disease mutations in the presenilin genes 1 and 2 appear to control γ-secretase function [24,25]. Wolfe *et al.* [26], have suggested that presenilin 1 and γ-secretase might be the same protein. The generation of highly aggregatable peptides are believed to initiate Aβ accumulation in all forms of the disease [27,28].

Other pathological features include elements of degenerating neurons and neurofibrillary tangles composed of abnormally phosphorylated *tau* protein. There is substantial loss of neurons and synapses throughout the neocortex. Degeneration in the basal forebrain profoundly reduces acetylcholine and the enzymatic activities of cholineacetyltransferase and acetylcholinesterase [29]. These changes occur early and account for the initial impairment in memory [30].

III. Frequency of Alzheimer's Disease

The proportion of individuals with Alzheimer's disease increases, particularly after the age of 65. Five to ten percent of 65 year olds have Alzheimer's disease, and this increases to as high as 30–40% at age 85 or older. Jorm *et al.* [31] demonstrated the relationship between prevalence and age to be consistent across studies, with rates of Alzheimer's disease doubling every five years to age 95. In East Boston, the prevalence of probable Alzheimer's disease was 47.2% among subjects older than 85 years and 18.7% in those aged 75–84 [32]. A similar exponential increase in the prevalence of dementia with age was also found in a study of dementia in Stockholm [33]. In the Bronx Aging study, the age-specific incidence, or the number of new cases arising over a specific period of time, rose steeply over three age intervals from 1.3% per year between ages 75 and 79 to 6.0% per year for individuals aged 85 and older [34]. Similar increases in the age-specific incidence have been reported for East Boston [35], Great Britain [36], Southwest France [37], and Nottingham [38], all suggesting exponential increases to a peak of around 8% per year.

IV. Gender and the Frequency of Alzheimer's Disease

Several studies have suggested that the prevalence of Alzheimer's disease may be higher in women than in men [39–41], although data are conflicting [34,42,43]. Women with Alzheimer's disease generally survive longer with the illness than do men [44,45]. Thus, the apparent increased risk of Alzheimer's disease among women, based on prevalence studies, might simply reflect this differential mortality. Others have argued that the risk of Alzheimer's disease is greater in women than men based on studies of the incidence rate [46,47]. Among individuals aged 85 years or more, Gussekloo *et al.* [47] found the risk for women to be 8.9% per year and that for men to be 2.7% per year. However, some studies find no differences in rates for men and women [36,37,48,49]. Aronson *et al.* [34] noted higher

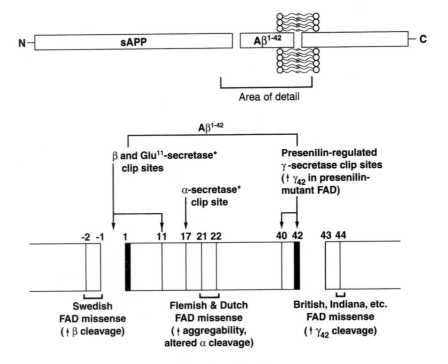

Fig. 98.2 Complete structure of the Alzheimer β-amyloid precursor protein (APP) (upper panel), with finer structure displayed below. FAD = familial Alzheimer's disease. The extracellular/ intraluminal domain of APP is directed leftward in these images.

incidence rates for women up to the age of 84, but for older women the incidence rates were similar to those for men. In families with at least one affected individual, women who are first-degree relatives have a higher lifetime risk of developing Alzheimer's disease than men [50,51].

V. Inheritance and Genes

Not only have a large number of multigenerational families with autosomal dominant Alzheimer's disease been described, but siblings of patients have twice the expected lifetime risk of developing the disease [53,54], and concordance of Alzheimer's disease between monozygotic twins is significantly greater than that observed in dizygotic twins [55].

Mutations in three genes, the APP gene on chromosome 21, the presenilin 1 (PS1) on chromosome 14, and the presenilin 2 (PS2) on chromosome 1, result in an autosomal dominant form of the disease beginning as early as the third decade of life [56]. The existence of over 50 mutations in PS1 suggests that this may be the most common form of familial early-onset Alzheimer's disease. Studies of the clinically relevant mutant APP molecules from families have indicated that these APP mutations can lead to enhanced generation or aggregability of Aβ, suggesting a pathogenic role. On the other hand, PS1 and PS2 are distinct from the immediate regulatory and coding regions of the APP gene, indicating that defects in molecules other than

APP can also lead to cerebral amyloidogenesis and familial Alzheimer's disease. To date, all APP and non-APP mutations can be demonstrated to have the common feature of promoting amyloidogenesis of Aβ.

In contrast, the ε4 polymorphism of the apolipoprotein-E (APOE) gene on chromosome 19 has been associated with both sporadic and familial disease with onset usually after age 65 [57]. In some families with late-onset Alzheimer's disease, each APOE-ε4 allele lowers the age at onset [58]. This is also true in some families with mutations in the amyloid precursor protein [56] and in Down syndrome [59]. The association between the APOE-ε4 allele and Alzheimer's disease is weaker among African-Americans and Caribbean Hispanics [60–62]. Polymorphisms in regulatory genes may affect APOE expression and may also account for some of the variability in disease risk [63,64]. These regulatory polymorphisms appear to produce an allele-specific alteration in APOE transcription. Consistent with other genes involved in Alzheimer's disease, APOE may also act through a complex and poorly understood relationship with Aβ [65]. APOE is an obligatory participant in Aβ accumulation [66,67], and postmortem [68,69] and cell culture [70] data indicate that APOE isoforms exert at least some of their effects via controlling Aβ accumulation, perhaps at the stage of clearance of Aβ peptides. APOE-isoform-specific synaptic remodelling also appears to be a consistent and potentially relevant property of these molecules [71,72].

APOE genotyping for the diagnosis of Alzheimer's disease has been suggested because of the strong association with the APOE-ε4 allele. In a few small postmortem studies [73–75], the presence of an APOE-ε4 allele provided a sensitivity for the clinical diagnosis of Alzheimer's disease that varied from 46 to 78%, while specificity was nearly 100%. However, the patients represented a highly selective group of individuals who had met criteria for probable Alzheimer's disease. In a collaborative study [22] involving over 2000 patients with dementia who had come to autopsy, the sensitivity and specificity of the clinical diagnosis were 93 and 55%, while the sensitivity and specificity of the APOE-ε4 allele was 65 and 68%, respectively. The APOE genotype improved the overall specificity to 84% in patients first meeting clinical criteria for Alzheimer's disease, although sensitivity decreased. The APOE genotype, when used alone as a diagnostic test for Alzheimer's disease, was considered inadequate in terms of sensitivity or specificity. However, this study did suggest that when used in combination with clinical criteria, the APOE genotype would significantly improves the specificity of the diagnosis by decreasing the rate of false positive diagnoses.

A new locus on chromosome 12 has pointed to the possibility of yet another genetic susceptibility locus for Alzheimer's disease [76,77]. Wu *et al.* [78] did not confirm the linkage using the National Institute of Mental Health Alzheimer's Disease-Sibpairs cohort, but Blacker *et al.* [79] reported an association with a deletion polymorphism in the alpha-2-macroglobulin (α2M) gene on chromosome 12. An association with a different polymorphism in this gene suggested that the gene may well be involved in Alzheimer's disease risk, but the exact site of the polymorphism may not have been identified [80]. Nevertheless, α2M protein has been implicated in Aβ clearance [81,82].

Based on putative biological roles, several genes have been proposed as putative candidates involved in disease risk. An intron in the PS1 gene [83–85], the alpha-1-antichymotrypsin gene [86–88], HLA-A2 [89], very low density lipoprotein (VLDL) [90,91], low density receptor lipoprotein (LRP) [92], and butyrylcholinesterase [93] may also harbor polymorphisms altering risk independently or by interaction with APOE. While both the LRP and intronic PS1 polymorphisms remain reasonable candidates, neither alters protein function, as they are not at splice sites, nor do they change the predicted amino acid sequences. A polymorphism in the angiotensin-converting enzyme that has been associated with longevity [94] has been associated with risk of Alzheimer's disease [95], but the association remains unconfirmed.

VI. Other Factors That Alter the Risk of Alzheimer's Disease

A. Down Syndrome

Adults with Down Syndrome develop the neuropathological changes of Alzheimer's disease by age 40, but not all patients become demented. The risk of Alzheimer's disease associated with a family history of Down Syndrome is increased two to threefold [96]. Schupf *et al.* [97] found the risk of Alzheimer's disease among mothers who were young (before age 35) when their child with Down syndrome was born was five times that of mothers who had children with other types of mental retar-

dation, while the risks of dementia among mothers who were older (over age 35) at the proband's birth was comparable to that of mothers of children with other types of mental retardation. Risk of dementia was not increased for fathers of patients with Down syndrome. These finding have suggested that there may be shared genetic susceptibility to developing Alzheimer's disease and having a child with Down syndrome before the age of 35. The authors proposed that this association might implicate a form of accelerated aging in the mothers.

B. Depression

A prior history of depression has been associated with Alzheimer's disease [98]. Even a prior history of "medically treated" depression can be associated with a threefold increase in risk of Alzheimer's disease. Whether depression represents incipient disease or is an early manifestation remains to be determined, but even when depression occurred ten years earlier the risk remained significant in some studies. Devanand *et al.* [15] found that even a persistently depressed mood might increase in parallel with cognitive failure and that depressed mood alone was associated with increased risk of incident dementia in a study of elderly individuals. However, the presence of a depressed mood also increased with increasing cognitive difficulty, suggesting that it was probably an early manifestation of disease.

C. Education

Educational achievement has profound influence on the frequency of Alzheimer's disease [41,99,100] which has led some to propose that a "cognitive reserve" develops in direct response to the increase in educational experience. In support of this is the observation that illiteracy and the lack of formal education have been associated with Alzheimer's disease among Chinese [41]. In the U.S., Stern and associates [100] found that the cumulative risk of Alzheimer's disease was significantly increased among elderly with less than eight years of education and lesser degrees of occupational achievement compared to those of a similar age with more education and with greater occupational achievement. Whether these effects are direct or mediated through genetic or environmental effects remains unknown. For example, Snowdon *et al.* [101] found that linguistic ability during the second decade of life might predict cognitive impairment and Alzheimer's disease during later years. However, while exposure to solvents has been associated with Alzheimer's disease, the exposure occurred among individuals with few years of education [102].

D. Anti-inflammatory Agents

Use of anti-inflammatory agents was found to be less frequent among patients with Alzheimer's disease than among controls [103]. This is consistent with the known pathogenesis as inflammation is a component of amyloid deposition that activates complement. Because chronic inflammation has been associated with amyloid deposition and because amyloid deposition may actually activate the complement cascade, anti-inflammatory agents could play an important role in the disease by slowing or inhibiting the pathogenesis.

E. Smoking

Cigarette smoking was once purported to be a protective factor in several cross-sectional studies [104,105]. However, a prospective study of dementia and Alzheimer's disease from Rotterdam [106] indicated that smokers had a statistically significant two to fourfold increase in risk of Alzheimer's disease, but only among those individuals without an APOE-ε4 allele. Similar results were found by Merchant *et al.* [107] in the Washington Heights study in New York City. In these studies, smoking was thought to increase the risk of dementia through a complex interaction with the cerebral vessels. In fact, an association between smoking and dementia associated with cerebrovascular disease has also been established [108].

F. Antioxidants

Though attractive as protective factors, no antioxidant has proven to be a significant protective factor against Alzheimer's disease. Alpha tocopherol (vitamin E) limits free radical formation, oxidative stress, and lipid peroxidation [109,110] and promotes survival of neurons in cell culture exposed to β-amyloid [111]. In a double-blind, placebo-controlled, multicenter trial [112], patients with Alzheimer's disease were randomized to either alpha tocopherol 2000 international units per day, selegiline 10 mg per day, both, or placebo. Those receiving both drugs fared better in terms of survival than those on placebo but slightly poorer than those on either agent alone, indicating that the effects of selegiline and alpha tocopherol were not additive. Neither drug was associated with an improvement in cognitive function.

Ginko biloba, an extract from the leaves of a subtropical tree, like other nootropics affect the presynaptic cholinergic system in animals [113]. Ginko biloba was found to provide a small benefit on cognitive testing in patients with Alzheimer's disease, but in that trial a large number of dropouts has raised concerns about the validity of the investigations [114].

G. Traumatic Head Injury

Several studies suggest an association between head injury and Alzheimer's disease [115,116], but these observations have remained inconsistent because most studies have been cross-sectional. Furthermore, the effect of head injury on the risk of Alzheimer's disease may be restricted to individuals with APOE genotypes containing one or more copies of ε4 [117,118]. At least two prospective studies have found an association between prior head injury and the age-at-onset of Alzheimer's disease, suggesting a complex interaction possibly related to amyloid metabolism [119,120].

H. Estrogen

The use of estrogen by postmenopausal women may also be associated with a decreased risk of Alzheimer's disease. Women who took estrogen had about a 50% reduction in occurrence of Alzheimer's disease in a prospective study of mortality rates [121]. Women who began menstruation at later ages had higher risk of Alzheimer's disease compared with women who had

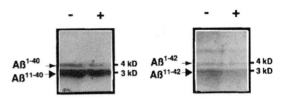

Fig. 98.3 Generation and regulation of Aβ formation by rodent and human neurons in cell culture [127]. Representative autoradiographs (left) show levels of various Aβ species (Aβ 1-40, 11-40, 1-42 and 11-42) generated by cultured rodent neurons in the absence (−) or presence (+) of 17β-estradiol. Dose–response curve (right) illustrates the generation of total Aβ species by cultured human neurons, as a function of 17β-estradiol concentration.

earlier onset of menstruation. The age-at-onset for Alzheimer's disease was significantly later and the relative risk was significantly lower for women who took, compared to women who did not take, estrogen, even after adjustment for differences in education, ethnic group, age, and APOE genotype in a population-based study in New York City [122]. This finding was confirmed in two additional prospective investigations in Baltimore [123] and in Italy [124].

A meta-analysis of published studies of the effects of estrogen on cognitive function in women with dementia [125] revealed primarily positive results but concluded that prospective placebo-controlled double-blind evaluations were required to assess the effectiveness of estrogen replacement in delaying or preventing Alzheimer's disease. Such studies are now in progress.

It is attractive to consider that estradiol might lower disease risk, or possibly age-at-onset of disease, by inhibiting Aβ accumulation which may initiate the pathogenesis of Alzheimer's disease. Steroids hormone such as 17β-estradiol [126,127] and dihydroepiandrostenedione (DHEA) [128] are examples of signal transduction compounds that may regulate APP metabolism *in vitro*. Xu [127] and Gridley *et al.* [129] demonstrated that estradiol diminishes Aβ generation (Fig. 98.3), while others [128,130] observed accumulation of sAPP in the conditioned media of cultured cells treated with estradiol or DHEA. The accumulation of sAPP resulting from steroids may be caused by their ability to stimulate the amount or the activity of α-secretase. The level or activity of protein kinase C (PKC) increases, augmenting "regulated APP cleavage" along the α-secretase pathway described earlier. Estrogen has been shown to regulate PKC both in normal and in neoplastic tissue [131]. The effects of estrogen on PKC may also be mediated by polypeptide growth factors [132–135], many of which enhance the accumulation of sAPP in the conditioned media of cultured cells [136,137]. The colocalization of estrogen receptors in the basal forebrain, cerebral cortex, and hippocampus with the receptors for nerve growth factor, as well as with nerve growth factor itself, is consistent with the idea that growth factors might play a role in mediating the effects of estrogen in the central nervous system [138]. In addition, steroid hormones such as estrogen and neurotrophins have been demonstrated to potentiate the ability of incoming neurotransmitters to activate the "regulated α-secretase cleavage of APP," as measured by sAPP release [139–142].

From a therapeutic standpoint, altering APP metabolism might be beneficial in individuals with, or at risk for, Alzheimer's disease, and it is along this line that 17β-estradiol—via its ability to elevate PKC activity—has been identified as a potential modulator of Aβ metabolism. This phenomenon is now under investigation in living animals. Such investigations should begin to clarify whether estradiol can indeed regulate Aβ generation and/or accumulation *in vivo*. Aside from the possible effects of estradiol on Aβ generation, the hormone might also (or instead) modify other factors known to contribute to Aβ accumulation, including the processing of soluble Aβ into an aggregated form [143,144] or the association of Aβ with other molecules, such as a-1-antichymotrypsin [145], heparan sulfate proteoglycan [146], and APOE [147]. Heparan sulfate proteoglycans in particular appear to play a role in plaque persistence. Because it is now possible to model brain Aβ accumulation and neurodegeneration *in vivo* [148], these possibilities can be assessed directly. It will also be possible to apply animal models of Aβ deposition and neurodegeneration [148] to efforts at elucidating whether the apparent effects of estrogen in delaying or preventing Alzheimer's disease in postmenopausal women are due to one or a combination of its various documented activities relevant to the pathogenesis of the disease, (*i.e.*, by lowering Aβ load (by altering Aβ metabolism), by sustaining the basal forebrain cholinergic system [149], by modulating interactions with growth factors or their receptors [150,151], by supporting neuritic plasticity [152,153], or by serving as an antioxidant [154]). One cannot yet exclude the possibility that estradiol plays several of these roles in its apparent ability to delay or prevent Alzheimer's disease.

VII. Treatment of Alzheimer's Disease

Improvement of memory and cognition has been the primary strategy in Alzheimer's disease, but treatments that maintain independent function, slow disease progression, and prevent disease are also in development. Anticholinesterases decrease the hydrolysis of acetylcholine released from the presynaptic neuron into the synaptic cleft by inhibition of the enzyme acetylcholinesterase, resulting in stimulation of the cholinergic receptor, and are the only approved treatments for Alzheimer's disease. While there are small yet consistent benefits, they must be viewed as palliative treatments. Only subtle differences exist between these drugs, but the adverse effects differ considerably. Tacrine (tetrahydroaminoacridine), a reversible acridine-based anticholinesterase, was the first approved drug for Alzheimer's disease, but its plasma half-life of 2–4 hours requires four daily doses [155]. Donepezil, the second drug approved for Alzheimer's disease by the Food and Drug Administration, became available in 1996. It is a piperdine-based molecule unrelated to tacrine, with high specificity for acetylcholinesterase and no hepatic toxicity. Compared with tacrine and physostigmine, donepezil has minimal peripheral anticholinesterase activity and has a plasma half-life of 70 hours, allowing for a single daily dose [156]. At the maximum dosages, tacrine improves cognitive test performance slightly more than donepezil. However, donepezil was rated slightly better by the clinician's rating scales, produces few side effects and no change in liver transaminases, and can be given once a day [155–157]. Though currently unavailable in the U.S., rivastigmine and metrifonate compare with donepezil on cognitive performance, but adverse effects have been slightly more frequent.

Both alpha tocopherol and selegiline delay the later stages of Alzheimer's disease [112], but it is difficult to know whether the reported difference is clinically meaningful. Unlike selegiline, alpha tocopherol does not interact with other medication and can be used in the majority of patients without concern.

Psychotropic drugs for Alzheimer's disease warrant more attention because they become important in the management of the later stages of disease. For depression, nearly all approaches are similar in efficacy, but there are only a handful of randomized, controlled studies upon which to make therapeutic decisions. Effective means to diminish depression as well as delusions, agitation, and other difficult behaviors without serious adverse effects are clearly needed.

VIII. Conclusions

Alzheimer's disease is a "complex" disorder that clearly increases with advancing age in men and women of all ethnic groups. It is likely that the etiology of Alzheimer's disease is both genetic and environmental. The way in which genes and environmental factors interact in the pathogenesis of Alzheimer's disease will be crucial to developing effective interventions. Hormone replacement therapy that includes estrogen may well be an important first step, but one that desperately needs the results of rigorous clinical trials before recommendations can be made.

Acknowledgments

The work was supported by United States Federal Grants (AG08702, AG07232, AG09464, AG10491) and the New York State Office of Mental Health.

References

1. Berrios, G. E. (1987). Dementia during the seventeenth and eighteenth centuries, a conceptual history. *Psychol. Med.* **17,** 829–837.
2. Berrios, G. E. (1990). Alzheimer's disease, a conceptual history. *Int. J. Geriatr. Psychiatry* **5,** 355–365.
3. American Psychiatric Association (1994). "Diagnostic and Statistical Manual of Mental Disorders," Rev. 3rd ed. American Psychiatric Press, Washington, DC.
4. Katzman, R. (1986). Alzheimer's disease. *N. Engl. J. Med.* **314,** 964–973.
5. Friedland, R. P. (1994). Epidemiology and neurobiology of the multiple determinants of Alzheimer's disease. *Neurobiol Aging* **15,** 239–241.
6. Mayeux, R., and Chun, M. R. (1994). Acquired and hereditary dementias. *In* "Merritt's Textbook of Neurology" (L. P. Rowland, ed.), pp. 677–685. Williams & Wilkins, Baltimore, MD.
7. Cummings, J. L., and Benson, D. F. (1992). "Dementia, a Clinical Approach." Butterworth-Heinemann, London.
8. Price, B. H., Gurvit, H., Weintraub, S., Geula, C., Leimkuhler, E., and Mesulam, M. (1993). Neuropsychological patterns and language deficits in 20 consecutive cases of autopsy-confirmed Alzheimer's disease. *Arch. Neurol. (Chicago)* **50,** 931–937.
9. McCormick, W. C., Kukull, W. A., Van Belle, G., Bowen, J. D., Teri, L., and Larson, E. B. (1994). Symptom patterns and comorbidity in the early stages of Alzheimer's disease. *J. Am. Geriatr. Soc.* **42,** 517–521.

10. Schofield, P. W., Jacobs, D., Marder, K., Sano, M., and Stern, Y. (1997). The validity of new memory complaints in the elderly. *Arch. Neurol. (Chicago)* **54**, 756–759.

11. Faber-Langendoen, K., Morris, J. C., Knesevich, J. W., LaBarge, E., Miller, J. P., and Berg, L. (1988). Aphasia in senile dementia of the Alzheimer type. *Ann. Neurol.* **23**, 365–370.

12. Reichman, W. E., and Coyne, A. C., (1995). Depressive symptoms in Alzheimer's disease and multi-infarct dementia. *J. Geriatr. Psychiatry Neurol.* **8**, 96–99.

13. Migliorelli, R., Tesón, A., Sabe, L., Petracchi, M., Leiguarda, R., and Starkstein, S. E. (1995). Prevalence and correlates of dysthymia and major depression among patients with Alzheimer's disease. *Am. J. Psychiatry* **152**, 37–44.

14. Rovner, B. W., Broadhead, J. H., Spencer, M., Carson, K., and Folstein, M. F. (1989). Depression and Alzheimer's disease. *Am. J. Psychiatry* **146**, 350–353.

15. Devanand, D. P., Sano, M., Tang, M. X., Taylor, S., Gurland, B. J., Wilder, D., Stern, Y., and Mayeux, R. (1996). Depressed mood and the incidence of Alzheimer's disease in the community elderly. *Arch. Gen. Psychiatry* **53**, 175–182.

16. Devanand, D. P., Brockington, C. D., Moody, B. J., Brown, R. P., Mayeux, R., Endicott, J., and Sackeim, H. A. (1992). Behavioral syndromes in Alzheimer's disease. *Int. Psychogeriatr.* **4**, 161–184.

17. Devanand, D. P., Jacobs, D. M., Tang, M. X., Del Castillo-Castaneda, C., Sano, M., Marder, K., Bell, K., Bylsma, F. W., Brandt, J., Albert, M., and Stern, Y. (1997). The course of psychopathologic features in mild to moderate Alzheimer's disease. *Arch. Gen. Psychiatry* **54**, 257–263.

18. Stern, Y., Mayeux, R., Sano, M., Hauser, W. A., and Bush, T. (1987). Predictors of disease course in patients with probable Alzheimer's disease. *Neurology* **37**, 1649–1653.

19. Funkenstein, H. H., Albert, M. S., Cook, N. R., West, C. G., Scherr, P. A., Chown, M. J., Pilgrim, D., and Evans, D. A. (1993). Extrapyramidal signs and other neurologic findings in clinically diagnosed Alzheimer's disease, a community-based study. *Arch. Neurol. (Chicago)* **50**, 51–56.

20. Jost, B. C., and Grossberg, G. T. (1995). The natural history of Alzheimer's disease, a brain bank study. *J. Am. Geriatr. Soc.* **43**, 1248–1255.

21. McKhann, G., Drachman, D., Folstein, M., Katzman, R., Price, D., and Stadian, E. M. (1984). Clinical Diagnosis of Alzheimer's disease, report on the NINCDS-ADRDA work group under the auspices of the Dept of Health and Human Services Task Force on Alzheimer's disease. *Neurology* **34**, 939–944.

22. Mayeux, R., Saunders, A. M., Shea, S., Mirra, S., Evans, D., Roses, A. D., Hyman, B. T., Crain, B., Tang, M.-X., Phelps, C. H., and the Alzheimer's Disease Centers Consortium on APOE and Alzheimer's Disease (1998). Utility of the apolipoprokinE genotype in the diagnosis of Alzheimer's disease. *N. Engl. J. Med.* **338**, 506–512.

23. Gandy, S., and Greengard, P. (1994). Processing of Alzheimer Aβ-amyloid precursor protein: cell biology, regulation, and role in Alzheimer disease. *Int. Rev. Neurobiol.* **36**, 29–50.

24. Scheuner, D., Eckman, C., Jensen, M., Song, X., Citron, M., Suzuki, N., Bird, T. D., Hardy, J., Hutton, M., Kukull, W., Larson, E., Levy-Lahad, E., Viitanen, M., Peskind, E., Poorkaj, P., Schellenberg, G., Tanzi, R., Wasco, W., Lannfelt, L., Selkoe, D., and Younkin, S. (1996). Secreted amyloid beta-protein similar to that in the senile plaques of Alzheimer's disease is increased in vivo by the presenilin 1 and 2 and APP mutations linked to familial Alzheimer's disease. *Nat. Med.* **2**(8), 864–870.

25. Borchelt, D. R. Thinakaran, G., Eckman, C. B., Lee, M. K., Davenport, F., Ratovitsky, T., Prada, C. M., Kim, G., Seekins, S., Yager, D., Slunt, H. H., Wang, R., Seeger, M., Levey, A. I., Gandy, S. E., Copeland, N. G., Jenkins, N. A., Price, D. L., Younkin, S. G., and Sisodia, S. S. (1996). Familial Alzheimer's disease-linked presenilin 1 variants elevate Abeta1-42/1-40 ratio in vitro and in vivo. *Neuron* **17**(5), 1005–1013.

26. Wolfe, M. E., Xia, W., Ostaszewski, B., Diehl, T. S., Kimberly, W. T., and Selkoe, D. J. (1999). Two transmembrane aspartates in presenilin-1 required for presenilin endoproteolysis and γ-secretase activity *Nature (London)* **398**, 513–517.

27. Iwatsubo, T., Odaka, A., Suzuki, N., Mizusawa, H., Nukina, N., and Ihara, Y. (1994). Visualization of A beta 42(43) and A beta 40 in senile plaques with end-specific A beta monoclonals, evidence that an initially deposited species is A beta 42(43). *Neuron* **12**(1), 45–53.

28. Lemere, C. A., Blusztajn, J. K., Yamaguchi, H., Wisniewski, T., Saido, T. C., and Selkoe, D. J. (1996). Sequence of deposition of heterogeneous amyloid beta-peptides and *APOE* in Down syndrome, implications for initial events in amyloid plaque formation. *Neurobiol. Dis.* **3**(1), 16–32.

29. Price, D. L., Thinakaran, G., Borchelt, D. R., Martin, L. J., Crain, B. J., Sisodia, S. S., and Troncoso, J. C. (1998). Neuropathology of Alzheimer's disease and animal models. In "Neuropathology of Dementing Disorders" (W. R. Markesbery, ed.), pp. 121–141. Oxford University Press, London.

30. Winkler, J., Thal, L. J., Gage, F. H., and Fisher, L. J. (1998). Cholinergic strategies for Alzheimer's disease. *J. Mol. Med.* **76**, 555–567.

31. Jorm, A. E., Korten, A. E., and Henderson, A. S. (1987). The prevalence of dementia, a quantitative integration of the literature. *Acta Psychiatr. Scand.* **76**, 465–479.

32. Evans, D. A., Funkenstein, H. H., Albert, M. S., Scherr, P. A., Cook, N. R., Chown, M. J., Hebert, L. E., Hennekens, C. H., and Taylor, J. O. (1989). Prevalence of Alzheimer's disease in a community population of older persons. Higher than previously reported. *JAMA, J. Am. Med. Assoc.* **262**, 2551–2556.

33. Fratiglioni, L., Grut, M., Forsell, Y., Viitanen, M., Grafstrom, M., Holmen, K., Ericsson, K., Backman, L., Ahlbom, A., and Winblad, B. (1991). Prevalence of Alzheimer's disease and other dementias in an elderly urban population, relationship with age, sex and education. *Neurology* **41**, 1886–1892.

34. Aronson, M. K., Ooi, W. L., Geva, D. L., Masur, D., Blau, A., and Frishman, W. (1991). Age-dependent incidence, prevalence, and mortality in the old old. *Arch. Intern. Med.* **151**, 989–992.

35. Hebert, L. E., Scherr, P. A., Beckett, L. A., Albert, M. S., Pilgrim, D. M., Chown, M. J., Funkenstein, H. H., and Evans, D. A. (1995). Age-specific incidence of Alzheimer's disease in a community population. *JAMA, J. Am. Med. Assoc.* **273**, 1354–1359.

36. Paykel, E. S., Brayne, C., Huppert, F. A., Gill, C., Barkley, C., Gehlhaar, E., Beardsall, L., Girling, D. M., Pollitt, P., and O'Connor, D. (1994) Incidence of dementia in a population older than 75 years in the United Kingdom. *Arch. Gen. Psychiatry* **51**, 325–332.

37. Letenneur, L., Commenges, D., Dartigues, J. F., and Barberger-Gateau, P. (1994). Incidence of dementia and Alzheimer's disease in elderly community residents of South-Western France. *Int. J. Epidemiol.* **23**, 1256–1261.

38. Morgan, K., Lilley, J. M., Arie, T., Byrne, E. J., Jones, R., and Waite, J. (1993). Incidence of dementia in a representative British sample. *Br. J. Psychiatry* **163**, 467–470.

39. Sulkava, R., Wikstrom, J., Aromaa, A., Raitasalo, R., Lehtinen, V., Lahtela, K., and Palo, J. (1985). Prevalence of severe dementia in Finland. *Neurology* **35**, 1025–1029.

40. Rocca, W. A., Bonaiuto, S., Lippi, A., Luciani, P., Turtu, F., Cavarzeran, F., and Amaducci, L. (1990) Prevalence of clinically diagnosed Alzheimer's disease and other dementing disorders, a

door-to-door survey in Appignano, Macerata Province, Italy. *Neurology* **40,** 626–631.

41. Zhang, M., Katzman, R., Salmon, D., Jin, H., Cai, G., Wang, Z., Qu, G., Grant, I., Yu, E., Levy, P., Klauber, M. R., and Liu, W. T. (1990). The prevalence of dementia and Alzheimer's disease in Shanghai, China, impact of age, gender and education. *Ann. Neurol.* **27,** 428–437.

42. Fichter, M. M., Meller, I., Schröppel, H., and Steinkirchner, R. (1995). Dementia and cognitive impairment in the oldest old in the community. Prevalence and comorbidity. *Br. J. Psychiatry* **166,** 621–629.

43. Rocca, W. A., Cha, R. H., Waring, S. C., and Kokmen, E. (1998). Incidence of dementia and Alzheimer's disease, a reanalysis of data from Rochester, Minnesota, 1975–1984. *Am. J. Epidemiol.* **148,** 51–62.

44. Jagger, C., Clarke, M., and Stone, A. (1995). Predictors of survival with Alzheimer's disease, a community-based study. *Psychol. Med.* **25,** 171–177.

45. Corder, E. H., Saunders, A. M., Strittmatter, W. J., Schmechel, D. E., Gaskell, P. C., Jr., Rimmler, J. B., Locke, P. A., Conneally, P. M., Schmader, K. E., and Tanzi, R. E. (1995). Apolipoprotein E, survival in Alzheimer's disease patients, and the competing risks of death and Alzheimer's disease. *Neurology* **45,** 1323–1328.

46. Katzman, R., Aronson, M., Fuld, P., Kawas, C., Brown, T., Morgenstern, H., Frishman, W., Gidez, L., Eder, H., and Ooi, W. L. (1989). Development of dementing illnesses in an 80-year-old volunteer cohort. *Ann. Neurol.* **25,** 317–324.

47. Gussekloo, J., Heeren, T. J., Izaks, G. J., Ligthart, G. J., and Rooijmans, H. G. M. (1995). A community based study of the incidence of dementia in subjects aged 85 years and over. *J. Neurol., Neurosurg. Psychiatry* **59,** 507–510.

48. Liu, H. C., Lin, K. N., Teng, E. L., Wang, S. J., Fuh, J. L., Guo, N. W., Chou, P., Hu, H. H., and Chiang, B. N. (1995). Prevalence and subtypes of dementia in Taiwan, a community survey of 5297 individuals. *J. Am. Geriatr. Soc.* **43,** 144–149.

49. Bachman, D. L., Wolf, P. A., Linn, R., Knoefel, J. E., Cobb, J., Bélanger, A., D'Agostino, R. B., and White, L. R. (1992). Prevalence of dementia and probable senile dementia of the Alzheimer type in the Framingham study. *Neurology* **42,** 115–119.

50. Farrer, L. A., Cupples, L. A., van Duijn, C. M., Kurz, A., Zimmer, R., Muller, U., Green, R. C., Clarke, V., Shoffner, J., and Wallace, D. C. (1995). Apolipoprotein E genotype in patients with Alzheimer's disease, implications for the risk of dementia among relatives. *Ann. Neurol.* **38,** 797–808.

51. Devi, G., Ottman, R., Tang, M.-X., Marder, K., Stern, Y., Tycko, B., and Mayeux, R. (1999). Influence of Apolipoprotein E genotype on familial aggregation of Alzheimer's disease in an urban population. *Neurology* (in press).

52. Breitner, J. C., Silverman, J. M., Mohs, R. C., and Davis, K. L. (1988). Familial aggregation in Alzheimer's disease, comparison of risk among relatives of early- and late-onset cases, and among male and female relatives in successive generations. *Neurology* **38,** 207–212.

53. Mayeux, R., Sano, M., Chen, J., Tatemichi, T., and Stern, Y. (1991). Risk of dementia in first-degree relatives of patients with Alzheimer's disease and related disorders. *Arch. Neurol. (Chicago)* **48,** 269–273.

54. Farrer, L. A., Myers, R. H., Cupples, L. A., St. George-Hyslop, P. H., Bird, T. D., Rossor, M. N., Mullan, M. J., Polinsky, R., Nee, L., and Heston, L. (1990). Transmission and age-at-onset patterns in familial Alzheimer's disease, evidence for heterogeneity. *Neurology* **40,** 395–403.

55. Breitner, J. C., Welsh, K. A., Gau, B. A., McDonald, W. M., Steffens, D. C., Saunders, A. M., Magruder, K. M., Helms, M. J.,

Plassman, B. L., and Folstein, M. F. (1995). Alzheimer's disease in the National Academy of Sciences-National Research Council Registry of Aging Twin Veterans. III. Detection of cases, longitudinal results, and observations on twin concordance. *Arch. Neurol. (Chicago)* **52,** 763–771.

56. Levy-Lahad, E., and Bird, T. D. (1996). Genetic factors in Alzheimer's disease, a review of recent advances. *Ann. Neurol.* **40,** 829–840.

57. Saunders, A. M., Strittmatter, W. J., Schmechel, D., George-Hyslop, P. H., Pericak-Vance, M. A., Joo, S. H., Rosi, B. L., Gusella, J. F., Crapper-MacLachlan, D. R., Alberts, M. J., Haines, J. L., and Roses, A. D. (1993). Association of apolipoprotein E allele epsilon 4 with late-onset familial and sporadic Alzheimer's disease. *Neurology* **43,** 1467–1472.

58. Corder, E. H., Saunders, A. M., Strittmatter, W. J., Schmechel, D. E., Gaskell, P. C., Small, G. W., Roses, A. D., Haines, J. L., and Pericak-Vance, M. A. (1993). Gene dose of apolipoprotein E type 4 allele and the risk of Alzheimer's disease in late onset families. *Science* **261,** 921–923.

59. Schupf, N., Kapell, D., Zigman, W., Canto, B., Tycko, B., and Mayeux, R. (1996). Onset of dementia is associated with apolipoprotein ε4 in Down syndrome. *Ann. Neurol.* **40,** 799–801.

60. Maestre, G., Ottman, R., Stern, Y., Gurland, B., Chun, M., Tang, M.-X., Shelanski, M., Tycko, B., and Mayeux, R. (1995). Apolipoprotein-E and Alzheimer's disease, ethnic variation in genotypic risks. *Ann. Neurol.* **37,** 254–259.

61. Tang, M.-X., Stern, Y., Marder, K., Bell, K., Gurland, B., Lantingua, R., Andrews, H., Tycko, B., and Mayeux, R. (1998). APOE risks and the frequency of Alzheimer's disease among African-Americans, Caucasians and Hispanics. *JAMA, J. Am. Med. Assoc.* **279,** 751–755.

62. Farrer, L. A., Cupples, L. A., Haines, J. L., Hyman, B., Kukull, W. A., Mayeux, R., Myers, R. H., Pericak-Vance, M. A., Risch, N., and van Duijn, C. M. (1997). Effects of age, gender and ethnicity on the association between apolipoprotein-E genotype and Alzheimer's disease. *JAMA, J. Am. Med. Assoc.* **278,** 1349–1356.

63. Lambert, J. C., Perez-Tur, J., Dupire, M. J., Galasko, D., Mann, D., Amouyel, P., Hardy, J., Delacourte, A., and Chartier-Harlin, M. C. (1997). Distortion of allelic expression of apolipoprotein E in Alzheimer's disease. *Hum. Mol. Genet.* **6,** 2151–2154.

64. Bullido, M. J., Artiga, M. J., Recuero, M., Sastre, I., Garcia, M. A., Aldudo, J., Lendon, C., Han, S. W., Morris, J. C., Frank, A., Vazquez, J., Goate, A., and Valdivieso, F. (1998). A polymorphism in the regulatory region of APOE associated with risk of Alzheimer's dementia. *Nat. Genet.* **18,** 69–71.

65. Selkoe, D. J. (1997). Alzheimer's disease, genotypes, phenotype and treatments. *Science* **275,** 630–631.

66. Strittmatter, W. J., Saunders, A. M., Schmechel, D., Pericak-Vance, M., Enghild, J., Salvesen, G. S., and Roses, A. D. (1993). Apolipoprotein E, High-avidity binding to β-amyloid and increased frequency of type 4 allele in late-onset familial Alzheimer disease. *Proc. Natl. Acad. Sci. U.S.A.* **90,** 1977–1981.

67. Bales, K. R., Verina, T., Dodel, R. C., Du, Y., Altstiel, L., Bender, M., St. George-Hyslop, P., Johnstone, E. M., Little, S. P., Cummins, D. J., Piccardo, P., Ghetti, B., and Paul, S. M. (1997). Lack of apolipoprotein E dramatically reduces amyloid beta-peptide deposition. *Nat. Genet.* **17,** 263–264.

68. Rebeck, G. W., Reiter, J. S., Strickland, D. K., and Hyman, B. T. (1993). Apolipoprotein E in sporadic Alzheimer's disease, allelic variation and receptor interactions. *Neuron* **11**(4), 575–580.

69. Polvikoski, T., Sulkava, R., Haltia, M., Kainulainen, K., Vuorio, A., Verkkoniemi, A., Niinisto, L., Halonen, P., and Kontula, K. (1995). Apolipoprotein E, dementia, and cortical deposition of beta-amyloid protein. *N. Engl. J. Med.* **333**(19), 1242–1247.

70. Yang, D. S. (1999). Apolipoprotein E promotes the association of amyloid-β with Chinese Hamster ovary cells in an isoform-specific manner. *Neuroscience* (in press).

71. Arendt, T., Schindler, C., Bruckner, M. K., Eschrich, K., Bigl, V., Zedlick, D., and Marcova, L. (1997). Plastic neuronal remodeling is impaired in patients with Alzheimer's disease carrying apolipoprotein epsilon 4 allele. *J. Neurosci.* **17,** 516–529.

72. Sun, Y., Wu, S., Bu, G., Onifade, M. K., Patel, S. N., LaDu, M. J., Fagan, A. M., and Holtzman, D. M. (1998). Glial fibrillary acidic protein-apolipoprotein E (apoE) transgenic mice, astrocyte-specific expression and differing biological effects of astrocyte-secreted apoE3 and apoE4 lipoproteins. *J. Neurosci.* **18**(9), 3261–3272.

73. Saunders, A. M., Hulette, O., Welsh-Bohmer, K. A., Schmechel, D. E., Crain, B., Burke, J. R., Alberts, M. J., Strittmatter, W. J., Breitner, J. C., and Rosenberg, C. (1996). Specificity, sensitivity and predictive value of apolipoprotein-E genotyping for sporadic Alzheimer's disease. *Lancet* **348,** 90–93.

74. Kakulas, B. A., Wilton, S. D., Fabian, V. A., and Jones, T. M. (1996). Apolipoprotein E genotyping in the diagnosis of Alzheimer's disease in an autopsy confirmed series. *Lancet* **348,** 483–484.

75. Smith, A. D., Jobst, K. A., Johnston, C., Joachim, C., and Nagy, Z. (1996). Apolipoprotein E genotyping in the diagnosis of Alzheimer's disease. *Lancet* **348,** 483–484.

76. Pericak-Vance, M. A., Bass, M. P., Yamaoka, L. H., Gaskell, P. C., Scott, W. K., Terwedow, H. A., Menold, M. M., Conneally, P. M., Small, G. W., Vance, J. M., Saunders, A. M., Roses, A. D., and Haines, J. L. (1997). Complete genomic screen in late-onset familial Alzheimer disease. Evidence for a new locus on chromosome 12. *JAMA, J. Am. Med. Assoc.* **278,** 1237–1241.

77. Rogaeva, E., Premkumar, S., Song, Y., Sorbi, S., Brindle, N., Paterson, A., Duara, R., Levesque, G., Yu, G., Nishimura, M., Ikeda, M., O'Toole, C., Kawarai, T., Jorge, R., Vilarino, D., Bruni, A. C., Farrer, L. A., and St. George-Hyslop, P. H. (1998). Evidence for an Alzheimer disease susceptibility locus on chromosome 12 and for further locus heterogeneity. *JAMA, J. Am. Med. Assoc.* **280,** 614–618.

78. Wu, W. S., Holmans, P., Wavrant-De Vrieze, F., Shears, S., Kehoe, P., Crook, R., Booth, J., Williams, N., Perez-Tur, J., Roehl, K., Fenton, I., Chartier-Harlin, M. C., Lovestone, S., Williams, J., Hutton, M., Hardy, J., Owen, M. J., and Goate, A. (1998). Genetic studies on chromosome 12 in late onset Alzheimer disease. *JAMA, J. Am. Med. Assoc.* **280,** 619–622.

79. Blacker, D., Wilcox, M. A., Laird, N. M., Rodes, L., Horvath, S. M., Go, R. C., Perry, R., Watson, B., Jr., Bassett, S. S., McInnis, M. G., Albert, M. S., Hyman, B. T., and Tanzi, R. E. (1998). Alpha-2-macroglobulin is genetically associated with Alzheimer disease. *Nat. Genet.* **19,** 357–360.

80. Liao, A., Nitsch, R. M., Greenberg, S. M., Finckh, U., Blacker, D., Albert, M., Rebeck, G. W., Gomez-Isla, T., Clatworthy, A., Binetti, G., Hock, C., Mueller-Thomsen, T., Mann, U., Zuchowski, K., Beisiegel, U., Staehelin, H., Growdon, J. H., Tanzi, R. E., and Hyman, B. T. (1998). Genetic association of an alpha2-macroglobulin (Val1000lle) polymorphism and Alzheimer's disease. *Hum. Mol. Genet.* **7,** 1953–1956.

81. Hughes, S. R., Khorkova, O., Goyal, S., Knaeblein, J., Heroux, J., Riedel, N. G., and Sahasrabudhe, S. B. (1998). Alpha2-macroglobulin associates with beta-amyloid peptide and prevents fibril formation. *Proc. Natl. Acad. Sci. U.S.A.* **95,** 3275–3280.

82. Trommsdorff, M., Borg, J. P., Margolis, B., and Herz, J. (1998). Interaction of cytosolic adaptor proteins with neuronal apolipoprotein E receptors and the amyloid precursor protein. *J. Biol. Chem.* **273,** 33556–33560.

83. Wragg, M., Hutton, M., and Talbot, C. (1996). Genetic association between intronic polymorphism in presenilin-1 gene and late-onset Alzheimer's disease. Alzheimer's Disease Collaborative Group. *Lancet* **347,** 509–512.

84. Higuchi, S., Muramatsu, T., Matsushita, S., Arai, H., and Sasaki, H. (1996). Presenilin-1 polymorphism and Alzheimer's disease. *Lancet* **347,** 1186.

85. Kehoe, P., Williams, J., Lovestone, S., Wilcock, G., and Owen, M. J. (1996). Presenilin-1 polymorphism and Alzheimer's disease. The UK Alzheimer's Disease Collaborative Group. *Lancet* **347,** 1185.

86. Kamboh, M. I., Sanghera, D. K., Ferrell, R. E., and DeKosky, S. T. (1995). APOE*4-associated Alzheimer's disease risk is modified by alpha 1-antichymotrypsin polymorphism. *Nat. Genet.* **10,** 486–488.

87. Muller, U., Bodeker, R. H., Gerundt, I., and Kurz, A. (1996). Lack of association between alpha 1-antichymotrypsin polymorphism, Alzheimer's disease, and allele epsilon 4 of apolipoprotein E. *Neurology* **47,** 1575–1577.

88. Haines, J. L., Pritchard, M. L., Saunders, A. M., Schildkraut, J. M., Growdon, J. H., Gaskell, P. C., Farrer, L. A., Auerbach, S. A., Gusella, J. F., Locke, P. A., Rosi, B. L., Yamaoka, L., Small, G. W., Conneally, P. M., Roses, A. D., and Pericak-Vance, M. (1996). No association between alpha 1-antichymotrypsin and familial Alzheimer's disease. *Ann. N. Y. Acad. Sci.* **802,** 35–41.

89. Payami, H., Schellenberg, G. D., Zareparsi, S., Kaye, J., Sexton, G. J., Head, M. A., Matsuyama, S. S., Jarvik, L. E., Miller, B., McManus, D. Q., Bird, T. D., Katzman, R., Heston, L., Norman, D., and Small, G. W. (1997). Evidence for association of HLA-A2 allele with onset age of Alzheimer's disease. *Neurology* **49,** 512–518.

90. Okuizumi, K., Onodera, O., Seki, K., Tanaka, H., Namba, Y., Ikeda, K., Saunders, A. M., Pericak-Vance, M. A., Roses, A. D., and Tsuji, S. (1996). Lack of association of very low density lipoprotein receptor gene polymorphism with Caucasian Alzheimer's disease. *Ann. Neurol.* **40,** 251–254.

91. Okuizumi, K., Onodera, O., Seki, K., Tanaka, H., Namba, Y., Ikeda, K., Saunders, A. M., Pericak-Vance, M. A., Roses, A. D., and Tsuji, S. (1995). Genetic association of very low density lipoprotein (VLDL) receptor gene locus with sporadic Alzheimer's disease. *Nat. Genet.* **11,** 207–209.

92. Wavrant-DeVrieze, F., Perez-Tur, J., Lambert, J. C., Frigard, B., Pasquier, F., Delacourte, A., Amouyel, P., Hardy, J., and Chartier-Harlin, M. C. (1997). Genetic association studies between the low-density lipoprotein receptor-related protein gene (LRP) the Alzheimer's disease. *Neurosci. Lett.* **222,** 187–190.

93. Lehmann, D. J., Johnston, C., and Smith, A. D. (1997). Synergy between the genes for butyrylcholinesterase K variant and apolipoprotein E4 in late-onset confirmed Alzheimer's disease. *Hum. Mol. Genet.* **6,** 1933–1936.

94. Schachter, F., Faure-Delanef, L., Guenot, F., Rouger, H., Froguel, P., Lesueur-Ginot, L., and Cohen, D. (1994). Genetic associations with human longevity at the APOE and ACE loci. *Nat. Genet.* **6,** 29–32.

95. Kehoe, P. G., Mcllroy, S., Williams, H., Holmans, P., Holmes, C., Liolitsa, D., Vahidassr, D., Powell, J., Liddell, M., Plomin, R., Dynan, K., Williams, N., Neal, J., Cairns, N. J., Wilcock, G., Passmore, P., Lovestone, S., Williams, J., and Owen, M. J. (1999). Variation in DCP1, encoding ACE, is associated with susceptibility to Alzheimer disease. *Nat. Genet.* **21,** 71–72.

96. Heyman, A., Wilkinson, W. E., Stafford, J. A., Helms, M. J., Sigmon, A. H., and Weinberg, T. (1984). Alzheimer's disease, A study of epidemiological aspects. *Ann. Neurol.* **15,** 335–341.

97. Schupf, N., Dapell, D., Lee, J., Ottman, R., and Mayeux, R. (1994). Increased risk for Alzheimer's disease in mothers of adults with Down syndrome. *Lancet* **344,** 353–356.

98. Jorm, A. F., van Duijn, C. M., Chandra, V., Fratiglioni, L., Graves, A. D., Heyman, A., Kokmen, E., Kondo, K., Mortimer, J. A., Rocca, W. A., Shalat, S. L., Soininen, H., and Hofman, A. (1991). Psychiatric history and related exposures as risk factors for Alz-

heimer's disease, a collaborative re-analysis of case-control studies. *Int. J. Epidemiol.* **20**(2), S43–S47.

99. Kittner, S. J., White, L. R., Farmer, M. E., Wolz, M., Kaplan, E., Moes, E., Brody, J. A., and Feinleib, M. (1986). Methodological issues in screening for dementia, the problem of education adjustment. *J. Chronic Dis.* **39**, 163–174.

100. Stern, Y., Gurland, B., Tatemichi, T. K., Tang, M.-X., Wilder, D., and Mayeux, R. (1994). Influence of education and occupation on the incidence of dementia. *JAMA, J. Am. Med. Assoc.* **271**, 1004–1010.

101. Snowdon, D. A., Greiner, L. H., Mortimer, J. A., Riley, K. P., Greiner, P. A., and Markesbery, W. R. (1996). Linguistic ability in early life and cognitive function and Alzheimer's disease in late life. *JAMA, J. Am. Med. Assoc.* **275**, 528–532.

102. The Canadian Study of Health and Aging (1994). Risk factors for Alzheimer's disease in Canada. *Neurology* **44**, 2073–2080.

103. Stewart, W., Kawas, C., Corrada, M., and Metter, E. (1997). The risk of Alzheimer's disease and duration of NSAID use. *Neurology* **48**, 626–632.

104. Graves, A. B., van Duijn, C. M., Chandra, V., Fratiglioni, L., Heyman, A., Jorm, A. F., Kokmen, E., Kondo, K., Mortimer, J. A., Rocca, W. A., Shalat, S. L., Soininen, H., and Hofman, A. (1991). Alcohol and tobacco consumption as risk factors for Alzheimer's disease. A collaborative re-analysis of case-control studies. *Int. J. Epidemiol.* **20**, S48–S57.

105. van Duijn, C. M., and Hofman, A. (1991). Relation between nicotine intake and Alzheimer's disease. *Br. Med. J.* **302**, 1491–1494.

106. Ott, A., Slooter, A. J. C., Hofman, A., van Harskap, F., Wittemen, J. C. M., Van Broeckhoven, C., van Duijn, C. M., and Breteler, M. M. B. (1998). Smoking and risk of dementia and Alzheimer's disease in a population-based cohort study. *Lancet* **351**, 1840–1843.

107. Merchant, C., Tang, M.-X., Albert, S., Manly, J., Stern, Y., and Mayeux, R. (1999). The influence of smoking on the risk of Alzheimer's disease. *Neurology* **52**, 1408–1412.

108. Gorelick, P. B., Brody, J., Cohen, D., Freels, S., Levy, P., Dollear, W., Forman, H., and Harris, Y. (1993). Risk factors for dementia associated with multiple cerebral infarcts. A case-control analysis in predominantly African-American hospital-based patients. *Arch. Neurol. (Chicago)* **50**, 714–720.

109. Halliwell, B., and Gutteridge, J. M. C. (1985). Oxygen radicals in the nervous system. *Trends Neurosci.* **8**, 22–26.

110. Yoshida, S., Busto, R., Watson, B. D., Santiso, M., and Gindsberg, M. D. (1985). Post ischemic cerebral lipid peroxidation in vitro, modification by dietary vitamin E. *J. Neurochem.* **44**, 1593–1601.

111. Behl, C., and Sagara, Y. (1997). Mechanism of amyloid beta protein induced neuronal cell death, current concepts and future perspectives. *J. Neural Transm. Suppl.* **49**, 125–134.

112. Sano, M., Ernesto, C., Thomas, R. G., Klauber, M. R., Schafer, K., Grundman, M., Woodbury, P., Growdon, J., Cotman, C. W., Pfeiffer, E., Schneider, L. S., and Thal, L. J. (1997). A controlled trial of selegiline, alpha-tocopherol, or both as treatment for Alzheimer's disease. *N. Engl. J. Med.* **336**, 1216–1222.

113. Kanowski, S., Herrmann, W. M., Stephan, K., Wierich, W., and Horr, R. (1996). Proof of efficacy of the ginkgo biloba special extract EGb 761 in outpatients suffering from mild to moderate primary degenerative dementia of the Alzheimer type or multi-infarct dementia. *Pharmacopsychiatry* **29**, 47–56.

114. Le Bars, P. L., Katz, M. M., Berman, N., Itil, T. M., Freedman, A. M., and Schatzberg, A. F. (1997). A placebo-controlled, double-blind, randomized trial of an extract of ginkgo biloba for dementia. North American EGb Study Group. *JAMA, J. Am. Med. Assoc.* **278**, 1327–1332.

115. van Duijn, C. M., Tanja, T. A., Haaxma, R., Schulte, W., Saan, R. J., Lameris, A. J., Antonides-Hendriks, G., and Hofman, A. (1992). Head trauma and the risk of Alzheimer's disease. *Am. J. Epidemiol.* **135**, 775–782.

116. Mayeux, R., Ottman, R., Tang, M.-X., Noboa-Bauza, L., Marder, K., Gurland, B., and Stern, Y. (1993). Genetic susceptibility and head injury as risk factors for Alzheimer's disease among community-dwelling elderly persons and their first-degree relatives. *Ann. Neurol.* **33**, 494–501.

117. Mayeux, R., Ottman, R., Maestre, G., Ngai, C., Tang, M.-X., Ginsberg, H., Chun, M., Tycko, B., and Shelanski, M. S. (1995). Synergistic effects of traumatic head injury and apolipoprotein-ε4 in patients with Alzheimer's disease. *Neurology* **45**, 555–557.

118. Jordan, B. D., Relkin, N. R., Ravdin, L. D., Jacobs, A. R., Bennett, A., and Gandy, S. (1997). Apolipoprotein E epsilon4 associated with chronic traumatic brain injury in boxing. *JAMA, J. Am. Med. Assoc.* **278**, 136–140.

119. Schofield, P. W., Tang, M.-X., Marder, K., Bell, K., Dooneief, G., Chun, M., Sano, M., Stern, Y., and Mayeux, R. (1997). Alzheimer's disease following remote head injury, an incidence study. *J. Neurol., Neurosurg. Psychiatry* **62**, 119–124.

120. Nemetz, P. N., Leibson, C., Naessens, J. M., Beard, M., Kokmen, E., Annegers, J. F., and Kurland, L. T. (1999). Traumatic brain injury and time to onset of Alzheimer's disease, a population-based study. *Am. J. Epidemiol.* **149**, 32–40.

121. Paganini-Hill, A., and Henderson, V. W. (1994). Estrogen deficiency and risk of Alzheimer's disease in women. *Am. J. Epidemiol.* **140**, 256–261.

122. Tang, M.-X., Jacobs, D., Stern, Y., Marder, K., Schofield, P., Andrews, H., Gurland, B., and Mayeux, R. (1996). Effect of oestrogen during menopause on risk and age-at-onset of Alzheimer's disease. *Lancet* **348**, 429–432.

123. Kawas, C., Resnick, S., Morrison, A., Brookmeyer, R., Corrada, M., Zonderman, A., Bacal, C., Donnel Lingle, D., and Metter, E. (1997). A prospective study of estrogen replacement therapy and the risk of developing Alzheimer's disease. The Baltimore Longitudinal Study of Aging. *Neurology* **48**, 1517–1521.

124. Baldereschi, M., Di Carlo, A., Lepore, V., Bracco, L., Maggi, S., Grigoletto, E., Scarlato, G., and Amaducci, L. (1998). Estrogen-replacement therapy and Alzheimer's disease in the Italian Longitudinal Study on Aging. *Neurology* **50**, 996–1002.

125. Yaffe, K., Sawaya, G., Lieburg, I., and Grady, D. (1998). Estrogen therapy in postmenopausal women. *JAMA, J. Am. Med. Assoc.* **279**, 688–695.

126. Jaffe, A. B., Toran-Allerand, C. D., Greengard, P., and Gandy, S. E. (1994). Estrogen regulates metabolism of Alzheimer amyloid beta precursor protein. *J. Biol. Chem.* **269**(18), 13065–13068.

127. Xu, H. (1998). Estrogen reduces neuronal generation of Alzheimer beta-amyloid peptides. *Nat. Med.* **4**(4), 447–451.

128. Danenberg, H. D. (1995). Dehydroepiandrosterone augments M1-muscarinic receptor-stimulated amyloid precursor protein secretion in desensitized PC12M1 cells. *Ann. N.Y. Acad. Sci.* **774**, 300–303.

129. Chang, D., Kwan, J., and Timiras, P. S. (1997). Estrogens influence growth, maturation and amyloid beta-peptide production in neuroblastoma cells and in a beta-APP transfected kidney 293 cell line. *Adv. Exp. Med. Biol.* **429**, 261–271.

130. Danenberg, H. D., Haring, R., Fisher, A., Pittel, Z., Gurwitz, D., and Heldman, E. (1996). Dehydroepiandrosterone (DHEA) increases production and release of Alzheimer's amyloid precursor protein. *Life Sci.* **59**(19), 1651–1657.

131. Maizels, E. T., Miller, J. B., Cutler, R. E., Jr., Jackiw, V., Carney, E. M., Mizuno, K., Ohno, S., and Hunzicker-Dunn, M. (1992). Estrogen modulates Ca(2+)-independent lipid-stimulated kinase in the rabbit corpus luteum of pseudopregnancy. Identification of luteal estrogen-modulated lipid-stimulated kinase as protein kinase C delta. *J. Biol. Chem.* **267**, 17061–17068.

132. Dickson, R. B., McManaway, M. E., and Lippman, M. E. (1986). Characterization of estrogen responsive transforming activity in human breast cancer cell lines. *Cancer Res.* **46**, 1707–1713.

133. Dickson, R. B., Huff, K. K., Spencer, E. M., and Lippman, M. E. (1986). Induction of epidermal growth factor-related polypeptides by 17 beta-estradiol in MCF-7 human breast cancer cells. *Endocrinology (Baltimore)* **118,** 138−142.

134. Bates, S. E., Davidson, N. E., Valverius, E. M., Freter, C. E., Dickson, R. B., Tam, J. P., Kudlow, J. E., Lippman, M. E., and Salomon, D. S. (1988). Expression of transforming growth factor alpha and its messenger ribonucleic acid in human breast cancer, its regulation by estrogen and its possible functional significance. *Mol. Endocrinol.* **2,** 543−555.

135. Lippman, M. E., Dickson, R. B., Gelmann, E. P., Rosen, N., Knabbe, C., Bates, S., Bronzert, D., Huff, K., and Kasid, A. (1988). Growth regulatory peptide production by human breast carcinoma cells. *J. Steroid Biochem.* **30,** 53−61.

136. Fukuyama, R., Chandrasekaran, K., and Rapoport, S. I. (1993). Nerve growth factor-induced neuronal differentiation is accompanied by differential induction and localization of the amyloid precursor protein (APP) in PC12 cells and variant PC12S cells. *Brain Res. Mol. Brain Res.* **17,** 17−22.

137. Refolo, L. M., Salton, S. R., Anderson, J. P., Mehta, P., and Robakis, N. K. (1989). Nerve and epidermal growth factors induce the release of the Alzheimer amyloid precursor from PC 12 cell cultures. *Biochem. Biophys. Res. Commun.* **164,** 664−670.

138. Miranda, R. C., Sohrabji, F., and Toran-Allerand, C. D. (1993). Neuronal colocalization of mRNAs for neurotrophins and their receptors in the developing central nervous system suggests a potential for autocrine interactions. *Proc. Natl. Acad. Sci. U.S.A.* **90,** 6439−6443.

139. Fisher, A. (1996). M1 agonists for the treatment of Alzheimer's disease. Novel therapeutic properties and clinical update. *Ann. N.Y. Acad. Sci.* **777,** 189−196.

140. Haring, R. (1998). Mitogen-activated protein kinase-dependent and protein kinase C-dependent pathways link the m1 muscarinic receptor to β-amyloid precursor protein secretion. *J. Neurochem.* **71,** 2094−2103.

141. Rossner, S., Ueberham, U., Schliebs, R., Perez-Polo, J. R., and Bigl, V. (1998). The regulation of amyloid precursor protein metabolism by cholinergic mechanisms and neurotrophic receptor signalling. *Prog. Neurobiol.* **56,** 541−569.

142. Rossner, S., Ueberham, U., Schliebs, R., Perez-Polo, J. R., and Bigl, V. (1998). p75 and trkA receptor signalling independently regulate amyloid precursor protein mRNA expression, isoform composition and protein secretion in PC12 cells. *J. Neurochem.* **71,** 757−766.

143. Burdick, D., Soreghan, B., Kwon, M., Kosmoski, J., Knauer, M., Henschen, A., Yates, J., Cotman, C., and Glabe, C. (1992). Assembly and aggregation properties of synthetic Alzheimer's A4/β amyloid peptide analogs. *J. Biol. Chem.* **267,** 546−554.

144. Pike, C. J., Burdick, D., Walencewicz, A. J., Glabe, C. G., and Cotman, C. W. (1993). Neurodegeneration induced by β-amyloid peptides in vitro. The role of peptide assembly state. *J. Neurosci.* **13,** 1676−1687.

145. Abraham, C. R., Selkoe, D. J., and Potter, H. (1988). Immunochemical identification of the serine protease inhibitor α-1-antichymotrypsin in the brain amyloid deposits of Alzheimer's disease. *Cell (Cambridge, Mass.)* **52,** 487−501.

146. Snow, A. D., Sekiguchi, R., Nochlin, D., Fraser, P., Kimata, K., Mizutani, A., Arai, M., Schreier, W. A., and Morgan, D. G. (1994). An important role of heparan sulfate proteoglycan (perlecan) in a model system for the deposition and persistence of fibrillar Aβ amyloid in rat brain. *Neuron* **12,** 219−234.

147. Wisniewski, T., and Frangione, B. (1992). Apolipoprotein E, A pathological chaperone in patients with cerebral and systemic amyloid. *Neurosci. Lett.* **135,** 235−238.

148. Calhoun, M. E. (1998). Neuron loss in APP transgenic mice. *Nature (London)* **395,** 755−756.

149. Luine, V. N. (1985). Estradiol increases choline acetyltransferase activity in specific basal forebrain nuclei and projection areas of female rats. *Exp. Neurol.* **89**(2), 484−490.

150. Miranda, R. C., Sohrabji, F., and Toran-Allerand, C. D. (1993). Interactions of estrogen with the neurotrophins and their receptors during neural development. *Mol. Cell. Neurosci.* **4,** 510−525.

151. Sohrabji, F., Greene, L. A., Miranda, R. C., and Toran-Allerand, C. D. (1994). Reciprocal regulation of estrogen and NGF receptors by their ligands in PC12 cells. *J. Neurobiol.* **25**(8), 974−988.

152. Woolley, C. S. (1998). Estrogen-mediated structural and functional synaptic plasticity in the female rat hippocampus. *Horm. Behav.* **34**(2), 140−148.

153. Stone, D. J., Rozovsky, I., Morgan, T. E., Anderson, C. P., and Finch, C. E. (1998). Increased synaptic sprouting in response to estrogen via an apolipoprotein E-dependent mechanism, Implications for Alzheimer's disease. *J. Neurosci.* **18**(9), 3180−3185.

154. Gridley, K. E., Green, P. S., and Simpkins, J. W. (1998). A novel, synergistic interaction between 17β-estradiol and glutathione in the protection of neurons against Aβ 25-35-induced toxicity In vitro. *Mol. Pharmacol.* **4**(5), 874−880.

155. Knapp, M. J., Knopman, D. S., Solomon, P. R., Pendlebury, W. W., Davis, C. S., and Gracon, S. I. (1994). A 30-week randomized controlled trial of high-dose tacrine in patients with Alzheimer's disease. *JAMA, J. Am. Med. Assoc.* **271,** 985−991.

156. Geldmacher, D. S. (1997). Donepezil (Aricept) therapy for Alzheimer's disease. *Compr. Ther.* **23,** 492−493.

157. Rogers, S. L., and Friedhoff, L. T. (1996). The efficacy and safety of donepezil in patients with Alzheimer's disease: Results of a U.S. multicenter, randomized, double-blind, placebo-controlled trial. The Donepezil Study Group. *Dementia* **7,** 293−303.

99

Successful Aging

JENNIFER B. UNGER* AND TERESA E. SEEMAN†

*Institute for Health Promotion and Disease Prevention Research, University of Southern California School of Medicine, Los Angeles, California; †Department of Geriatrics, University of California, Los Angeles School of Medicine, Los Angeles, California

I. Introduction

Early conceptualizations of the aging process characterized aging as a period of declining health, increasing dependence on assistance from others, and gradual disengagement from previously enjoyed activities and relationships. In a sense, aging was viewed as the preparation for death. Recently, however, researchers have abandoned this pessimistic view and adopted a more positive view of aging. The concept of successful aging posits that aging can be a time of continued independence, enjoyment of activities and relationships, and good health.

The concept of successful aging is particularly important for women. Women have a longer life expectancy than men do, but they tend to spend a larger proportion of their old age living with disability. Therefore, interventions to promote successful aging among women have the potential to affect public health substantially by enhancing physical functioning and quality of life among the increasing number of older women.

To promote successful aging among older women, it is important to understand the biological, behavioral, and social factors that make successful aging most likely. If interventions can be developed to increase these protective factors, more older women may be able to enjoy more successful aging.

II. Background

Improvements in living conditions, health habits, and medical care in the twentieth century have resulted in a dramatic increase in life expectancy for older women. A woman's life expectancy at age 65 currently is nearly 20 years [1]. If these extra years are years of health and vitality, they provide an opportunity for older women to participate in social, physical, and cultural activities and to interact with family and friends. In fact, a growing trend among older women is to embark on entirely new occupations, interests, and hobbies in their later years, after having completed a successful career and raised a family [2]. As the Baby Boom generation ages, this generation of women may challenge the stereotype of older women as frail, inactive, and lonely. New careers, new romantic relationships, and new interests may become the norm for healthy, active, older women.

Unfortunately, for many women, the older years are characterized by disease and disability, hindering their attempts to remain active and productive in old age. The average 70-year-old woman can expect to live 49% of her remaining life with slight impairments in physical functioning and an additional 20% of her remaining life with a chronic disability that renders her unable to perform the functions necessary for personal care and independent living [3]. At every age after 65 years, women have a longer life expectancy than men do, but a larger number of those remaining years will be lived with physical disability [4]. Although absolutely perfect physical health is not essential for a rewarding, productive old age, disease and disability can hinder an older woman's attempts to enjoy her old age.

Many older women are able to lead healthy productive lives relatively free from disease, disability, and cognitive impairment. What factors determine which older women will avoid functional decline in old age? Successful aging is a term used to describe those older adults who manage to negotiate the process of aging without succumbing to physical and cognitive declines.

III. Definition

Successful aging has been defined in different ways by different researchers. Most definitions emphasize the maintenance of physical, cognitive, and social functioning. Physical functioning includes the maintenance of independence in activities of daily living, instrumental activities of daily living, and more advanced physical mobility. Cognitive functioning includes the ability to react appropriately to environmental stimuli, recall recent and past events, and solve abstract and concrete problems. Social functioning includes involvement in a supportive social network and continued participation in activities such as hobbies, social events, and volunteer work.

Rowe and Kahn [5] list three components of successful aging: avoiding disease and disability, maintaining high cognitive and physical function, and remaining actively engaged with life. According to Rowe and Kahn, the combination of these three factors provides the most complete representation of successful aging. Success in each of these areas depends on success in the other two areas. In order to maintain active engagement in life, a person must have high cognitive and physical function. In order to maintain high cognitive and physical function, a person must avoid chronic disease. Therefore, in order to be considered a successful ager, a person must "succeed" within each of these three areas.

Baltes and Baltes [6] list seven criteria for successful aging: length of life, biological health, mental health, cognitive efficacy, social competence and productivity, personal control, and life satisfaction. Success in each of these seven domains contributes to the underlying concept of successful aging, but these seven criteria do not necessarily occur together or occur in a particular order. They represent different facets of a state that is not easily defined. In addition, successful aging may have different meanings to different people, so Baltes and Baltes stress the importance of assessing the older person's perception of successful aging in addition to the definition imposed on the older person by the researcher.

Fries [7] provides a medical perspective on successful aging. According to Fries, the life expectancy of humans has a fixed upper limit. Improvements in medical technology can increase the proportion of humans who can survive to this age limit, but it will not be possible to extend the human life span beyond this limit. Therefore, advances in medicine will allow humans to live relatively disease-free until they reach the biological limit of life expectancy, and then they will die after a relatively short period of impairment. Viewed from this perspective, successful aging represents the compression of morbidity into a very short period just before death, thereby allowing humans to remain healthy until just before death.

IV. Issues Related to Ascertainment

Because successful aging is a multidimensional concept, it is difficult to measure objectively. There are no strict criteria for classifying a person as a successful ager; successful aging is a general term used to describe an older person who functions well across multiple domains. Unfortunately, there is no universally-accepted tool to measure successful aging, such as a "successful aging scale" or a "successful aging test." Researchers tend to infer that an older woman is aging successfully if she lacks evidence of physical, cognitive, or psychological impairment. Although successful aging is more than merely the absence of measurable impairment, most existing measures of successful aging tend to assess impairment across multiple domains and use this information to draw conclusions about successful aging. Measures of successful aging typically include objective and subjective indicators of physical and cognitive functioning.

A. Measuring Physical Functioning

Physical functioning typically is measured either with objective physical performance measures or with subjective self-reports of activity limitations.

1. Objective Physical Performance Measures

Objective measures of physical performance typically involve asking an older woman to perform specific physical tasks. In the MacArthur Study of Successful Aging [8], a trained interviewer visited the respondents in their homes and administered physical performance measures. These included measures of balance (e.g., standing with the feet in a tandem, semi-tandem, or side-by-side position), gait (e.g., amount of time needed to walk across a room), lower extremity strength (e.g., ability to rise from a chair five times in a row), and upper extremity range of motion (e.g., ability to rotate shoulders).

2. Subjective Self-Report Measures of Activities of Daily Living

Subjective measures of physical functioning typically ask the respondent to indicate whether she is able to perform specific tasks. These tasks are divided into Activities of Daily Living (ADL) and Instrumental Activities of Daily Living (IADL). Severe impairment is indicated by respondents' reported inability to perform ADL, such as bathing, dressing, transferring from bed to chair, using the toilet, grooming, and eating [9]. Moderate impairment is indicated by respondents' reported inability to

perform IADL, such as using a telephone, shopping, preparing meals, housekeeping, taking medication, and handling one's own finances and transportation [10]. Before an older woman becomes unable to perform ADL or IADL, she may experience slight decrements in physical ability. Scales have been developed to detect minor disabilities in physical mobility, such as the inability to do heavy housework, walk up and down stairs, push or pull heavy objects, stoop, crouch, kneel, or handle small objects [11,12]. Using a combination of these self-report scales, it is possible to determine a woman's level of impairment in performing various tasks that are necessary for independent living.

3. Reliability and Validity of Self-Report Measures

Subjective self-report measures frequently are used in large-scale population surveys because they can be administered with telephone or written surveys. Objective physical performance measures are more expensive and time-consuming because they typically require a visit to the woman's home by a trained interviewer. However, the disadvantage of using a self-report measure rather than an objective performance measure is that the self-report measure may be biased by the respondent's over- or underestimation of her physical ability. A woman may report that she is unable to perform a task simply because she lacks the confidence to attempt it. Conversely, if a woman fears the loss of independence associated with functional disability, she may claim to be able to perform tasks that she actually cannot perform independently. Comparisons of subjective and objective measures of physical functioning [13,14] have found that self-report measures are quite accurate. However, there were gender differences in inaccurate reporting; women showed a greater tendency to underestimate their ability, whereas men showed a greater tendency to overestimate their ability. This suggests that some portion of the gender difference in self-reported physical functioning may be an artifact of self-report bias. Although self-reports of functional impairment tend to fluctuate over long time intervals, test–retest reliability of these measures tends to be quite high over short time intervals [15].

B. Measuring Cognitive Function

Like physical functioning, cognitive functioning consists of several dimensions. Perlmutter [16] has proposed a three-tiered model of cognitive functioning. Tier 1, "fluid intelligence," represents the basic mechanisms or processing resources used to solve problems. These abilities are most likely to decline with age. Tier 2, "crystallized intelligence," is the knowledge base of facts and skills learned through formal education and informal experience. Tier 3 contains thinking strategies for processing knowledge and coping with one's own cognitive resources. This "metacognition" or "metamemory" can continue to grow throughout the life span.

The AHEAD study [17] contains several cognitive measures designed to assess these dimensions of cognitive functioning. These include immediate and delayed free-recall of lists of nouns, serial subtraction (starting at 100 and subtracting 7 repeatedly), knowledge tests (naming the day of the week, the date, the current President, and common objects), tests of abstract reasoning, and self-ratings of memory skills.

Physicians typically use the Mini-Mental State Examination (MMSE) to assess cognitive function in older adults [18]. The MMSE is a short, structured, 30-item examination that takes 5–10 minutes to administer, and it has been shown to have good sensitivity and specificity for distinguishing patients with dementia from normal controls [19].

Although physical functioning and cognitive functioning appear to represent very different domains, they tend to be interrelated. In a study of older adults living in Kungsholmen, Stockholm [20], dementia was the strongest predictor of development of ADL dependence over a three-year period. Therefore, physical decline and cognitive decline may represent different symptoms or manifestations of a general syndrome of unsuccessful aging. Measuring both physical and cognitive functioning is one way to triangulate on the difficult-to-measure concept of successful aging.

V. Historical Evolution/Trends

A. Activity Theory vs Disengagement Theory

Before the concept of successful aging was introduced, the prevailing view among researchers and practitioners in gerontology was that functional decline and cessation of activities was a normal and inevitable part of aging. Disengagement theory [21], for example, states that individuals tend to withdraw from their previous activities and social relationships as they grow older, in preparation for the ultimate separation of death.

Other theorists have challenged this pessimistic view of aging. Activity theory [22], for example, states that withdrawal from activities and social relationships is not an inevitable consequence of aging. This theory asserts that healthy aging involves the continuation of relationships and activities enjoyed during middle age. According to activity theory, most older people should be able to remain independent and socially integrated throughout most of their old age. Although aging is associated with definite physical and social changes, Havighurst [23] asserted that many older people are able to adapt to these changes without sacrificing their physical and emotional well-being. According to Havighurst [23], this successful adjustment to these changes defines successful aging.

During the 1980s and 1990s, research in successful aging focused on identifying the genetic, biomedical, behavioral, and social factors that enable certain older adults to retain or even enhance their physical and mental functioning late in life [5]. In longitudinal studies such as the MacArthur Study of Successful Aging [8], initially high-functioning older adults were studied extensively and followed over time to determine who would age successfully and who would succumb to physical or cognitive disability. Findings from studies such as MacArthur have contributed a great deal of knowledge about the predictors of successful aging.

B. A Broader Definition of Successful Aging

Early definitions of successful aging emphasized the lack of physical disability. This implies that an older person with a physical disability cannot be classified as a successful ager. However, researchers [24,25] have expanded the definition of

successful aging to include those older people who are able to maintain a sense of personal mastery and life satisfaction and avoid depression and anxiety despite a physical limitation or chronic illness. People who are able to cope effectively with the challenge of a health problem may be even more resilient than are those who have had the good fortune of avoiding disabling illnesses. Although many studies have measured successful aging as a lack of physical or cognitive disability, this definition may exclude the large number of older women who are functioning effectively despite the challenge of health problems. Curb et al. [24] introduced the term "effective aging" to describe older adults who either maintain high levels of functioning or maximize their remaining functional capacity despite the physiological declines, risk factors, and clinically diagnosed diseases that are a reality of life for many older people.

VI. Distribution in Women, Including Incidence, Prevalence, and Mortality Statistics

A. Prevalence of Physical Impairment in Older Women

Physical impairment is more prevalent among older women than among older men [15,26–28]. Figure 99.1, adapted from Crimmins et al. [29], shows the prevalence of impairment among community-dwelling men and women 70 years of age and over in the 1993 National Health Interview Survey.

Part of the reason for the gender difference in the prevalence of physical impairment is the fact that men and women are at risk for different medical conditions. Older men are at greater risk for the diseases that are the major causes of death among older adults, such as cardiovascular disease and stroke. However, older women are at greater risk for the diseases that result in chronic impairment, such as arthritis and osteoporosis. Therefore, older men are more likely to die, but older women are more likely to become disabled and live many years with a chronic disability [30,31]. As a consequence, the older population consists of a large number of women (of whom many are disabled) and a smaller number of men (of whom few are disabled).

Figure 99.2 shows the percentage of noninstitutionalized older adults in the United States in 1995 who had activity limitations. As shown in the figure, women ages 75 years and older had the highest prevalence of physical impairment.

The gender difference in functioning is not consistent across domains of functioning. In a study of Canadian older adults [28], women were more likely than men to have limitations in mobility and vision, but men were more likely than women to have limitations in hearing and speech.

Figure 99.3 shows the percentage of older adults in the United States in 1995 who were in nursing homes [32]. Although the risk of nursing home placement is low for both men and women until 75 years of age, after age 75 the risk begins to increase dramatically, especially for women. This is an indicator of the proportion of older adults who are unable to live independently and do not have social network members willing or able to care for them.

B. Prevalence of Cognitive Impairment in Older Women

Cognitive impairment is more prevalent among older women than among older men [33], and older women are more likely

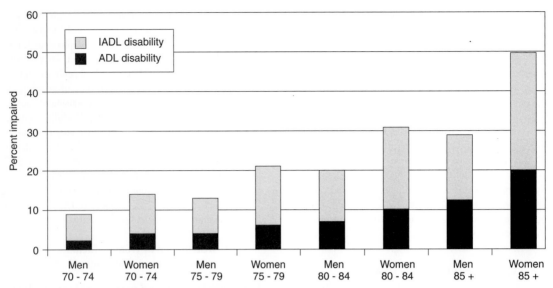

Fig. 99.1 Prevalence of impairment among community-dwelling men and women 70 years of age and over, National Health Interview Survey, 1993. Adapted from [29].

than older men to experience cognitive declines over time [34]. Dementia is defined as global impairment of cognitive function that interferes with normal activities [35]. The prevalence of dementia increases steadily with age, approximately doubling every five years [36]. Studies of community-dwelling older persons in North America have reported dementia in 0.8–1.6% of persons 65–74 years old, 7–8% of persons 75–84 years old, and 18–32% of persons over 85 [37].

VII. Host and Environmental Determinants

A. Demographic Determinants

1. Gender

Women have a longer life expectancy than do men, but they also are more likely to suffer from physical and/or cognitive impairment during those extra years of life.

2. Race/Ethnicity

Race/ethnicity is another important correlate of successful aging among women. Most studies of ethnicity and aging have

focused on differences between whites and African-Americans. Few studies have had sufficient statistical power to examine differences in successful aging among other ethnic groups such as Hispanics or Asian-Americans, although analyses of the older respondents in the Hispanic HANES survey [38,39] have provided descriptive information about the health status of older Hispanic women. Successful aging appears to be more common among white women than among African-American women. White women have a lower prevalence of physical impairment [40] and cognitive impairment [41] than do African-American women. Although this effect may be explained partially by differences in socioeconomic status between whites and African-Americans, this racial difference persists even after controlling for income and education [42]. Racial/ethnic differences in successful aging may reflect a variety of social and physical differences between African-Americans and whites. The prevalence of cardiovascular disease, hypertension, diabetes, obesity, and several types of cancer is higher among African-Americans than among whites [27,43,44]. These diseases may increase the risk

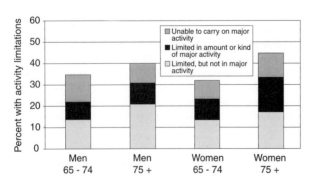

Fig. 99.2 Percentage of older adults with activity limitations, United States, 1995 [32].

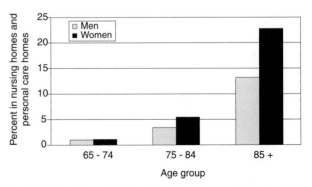

Fig. 99.3 Percentage of older adults who are in nursing homes, 1995 [32].

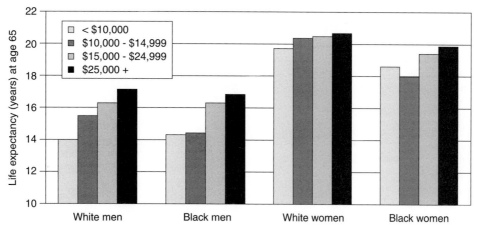

Fig. 99.4 Life expectancy at age 65 by gender, ethnicity, and annual household income [32].

of functional impairment among African-Americans relative to whites. The prevalence of osteoporosis, however, is lower among African-Americans than among whites [45]. There are also psychosocial differences between African-American and white women. African-American women tend to be embedded in stronger and more durable kin networks than whites [46]. Experience with large and dense social support networks may make African-American older women more comfortable relying on others for assistance in performing tasks. Although supportive social networks generally are beneficial for health, excessive reliance on social network members for instrumental support may cause minority older women to become overly dependent on assistance from others, thus decreasing their ability to function independently in the environment [47]. In addition, the adverse social conditions experienced more frequently by minorities, including racial discrimination, poverty, and unsafe neighborhoods, may increase their risk for injuries that may lead to functional decline [48].

3. Socioeconomic Status

Higher income is associated with a lower prevalence of physical impairment and a lower risk of functional decline [40,49,50]. Unfortunately, older women are more likely to have lower income and thus may be at increased risk for negative outcomes [51]. Older women with low income may have greater exposure to health threats, including inadequate or unsafe housing conditions, crime-ridden neighborhoods, and inadequate nutrition [52]. These conditions may increase their risk of injury and decrease their probability of obtaining adequate medical care, which may make them less likely to age successfully.

Figure 99.4 shows life expectancy at age 65 by gender, ethnicity, and annual household income [32]. Although women have a higher life expectancy than do men, life expectancy is lower among black women than among white women, and it is lower among women with low household income.

4. Education

Higher education is associated with lower prevalence of physical impairment [3,40,42,53,54]. Education also appears to protect against or compensate for cognitive declines [55,56].

The prevalence of dementia and Alzheimer's disease is lower among older people with higher levels of education [57–59]. It is not clear whether educational level reflects an extra cognitive capacity that is either innate or is acquired through the educational process. Evidence suggests that the protective effect of education on cognitive functioning may be due to the fact that people who acquire more education tend to be more intelligent. Their superior cognitive processing skills may enable them to maintain a high level of cognitive performance even after suffering an insult to the brain [60].

5. Depression

The prevalence of depression among older women is significantly higher than among older men [61–63]. The association between depression and physical functioning appears to be bi-directional; depression is a risk factor for subsequent decline in physical functioning [64,65], while physical impairment is a risk factor for subsequent increases in depression [66,67]. Depression also is associated with poor performance on memory tests and lower self-reports of cognitive performance [68].

B. Health Behaviors

1. Physical Activity

Physical activity may prevent or lessen age-related declines in physical functioning [69–71]. In the Longitudinal Study of Aging, physical activity was associated with the slope of decline in physical functioning over a six-year period; sedentary women experienced greater functional decline than did those who were more active [72]. Among men and women ages 70–74 with no functional limitations at baseline, those who were inactive had a 50% greater risk of losing functional status in the next two years [73]. Older people who reported high levels of recreational physical activity also were significantly less likely than sedentary individuals to have functional impairments three years later [74]. Older subjects in the 1993 Behavioral Risk Factor Surveillance System who reported engaging in any type of physical activity were significantly less likely to report health-related ADL difficulties than were subjects who did not engage in physical activity [75]. In one study, the number of

minutes per week that subjects exercised was significantly negatively correlated with the number of days that they were sick
in bed in the preceding six months [76]. Because older women
are more likely than older men to be sedentary [77], encouraging older women to participate in physical activity could greatly
enhance successful aging among women.

2. Obesity/Malnutrition

Maintenance of an appropriate weight is associated with successful aging. Obesity and underweight each have been associated with increased risk of physical impairment, as well as poor
subjective health and well-being [69,78]. Older women, especially those who live alone, are at high risk for malnutrition,
which can lead to physical disability [79]. This association is
likely bidirectional (*i.e.,* malnutrition increases the risk of disability, and disability increases the risk of malnutrition). Obesity
is a risk factor for diabetes, cardiovascular disease, and cancer,
so obese women may be at higher risk for disease-related disability. However, obesity decreases the risk of osteoporosis, a
major cause of disability among older women [80].

3. Alcohol Use

Approximately 5–6% of community-living older people experience problems with alcoholism [81]. The prevalence of alcohol use is lower among older women than among older men.
In the MacArthur Study of Successful Aging [T. E. Seeman,
unpublished data, 1997], 57% of the older women reported no
alcohol intake as compared with 37% of the men.

Although alcohol use has not been directly associated with
successful aging, excessive alcohol use can increase the risk of
falls and poor self-care, which can lead to functional decline
[81]. Even moderate intake of alcohol among older adults is
associated with increased risk of immune system impairment,
hypertension, cardiac arrhythmia, myocardial infarction, cardiomyopathy, stroke, alcohol dementia, cancer, malnutrition,
cirrhosis and other liver diseases, decreased nutritional intake
and absorption, depression, suicidal tendency, anxiety, delusions, and sleep disturbances, all of which may have implications for successful aging [82]. Conversely, some studies [83]
have found that older adults who abstain completely from alcohol use are at higher risk of mortality. This may indicate that
people who already are ill tend to abstain from alcohol use, or it
may indicate a protective effect of small amounts of alcohol [83].

4. Smoking

Smoking is a risk factor for the two most common causes of
death among women—cardiovascular disease and cancer [84,85].
The prevalence of smoking is lower among older adults than
among younger individuals, and it is lower among older women
than among older men. Therefore, few studies of the effects of
smoking have focused on older women. Nevertheless, 11% of
women age 65 and older are current smokers [86]. Among older
women, smoking has been associated with increased risk of
poor self-rated health, respiratory problems, medication use, un-
happiness, and dissatisfaction with social relationships [87].
Even after a lifetime of smoking, older women still can benefit
from smoking cessation. After quitting smoking, a woman's risk
of mortality, coronary events, smoking-related cancers, and
chronic obstructive pulmonary disease will begin to improve

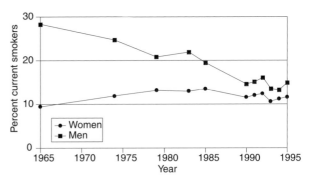

Fig. 99.5 Percentage of people ages 65 and older who were current
cigarette smokers [32].

immediately and will even approach the risk of a never-smoker
after 15–20 years of abstinence [88]. In addition, smoking cessation can reduce the prevalence of respiratory symptoms, slow
the rate of decline in pulmonary function, reduce functional
impairment, increase exercise tolerance, slow osteoporosis, and
reduce the risk of hip fractures [69].

Smoking as a risk factor for unsuccessful aging among
women has not been studied extensively. In the past, the prevalence of smoking among older women has been quite low. However, smoking initiation among young women showed a sharp
increase in the 1960s, coincident with the increase in advertising
of women's cigarette brands [89]. Since that time, rates of smoking cessation have been significantly lower among women than
among men. As the cohort of women who began smoking as
adolescents in the 1960s ages, the effects of smoking on successful aging may become increasingly salient.

Figure 99.5 shows the trends in prevalence of current smoking among older adults from 1965 to 1995 [32]. Although smoking was primarily a men's activity in the 1970s years ago, there
now is very little difference between older men and older
women in smoking prevalence. If this trend continues, the prevalence of tobacco-related disease among elderly women will
continue to increase.

C. Psychological Factors

1. Self-Efficacy

Self-efficacy appears to be an important determinant of successful aging. Self-efficacy, or one's confidence in one's ability
to perform tasks [90], is a key factor in determining whether an
older person will attempt a task. Older women with low self-
efficacy may cease attempting to perform complex tasks, even
though their actual physical ability to perform these tasks is not
impaired [91]. If they stop trying to perform complex tasks,
they may lose the ability to perform the tasks and they may
become dependent on others for assistance. Efforts to promote
self-efficacy may be especially valuable for women because
older women are more likely than are men to underestimate their
functional capacity [14] and to have lower self-efficacy beliefs
in general [92]. Self-efficacy may be especially important for
older people who experience declines in physical ability. In
an 18-month study of older men and women [93], among the

respondents who experienced declines in objectively measured physical performance, those with high self-efficacy were less likely to experience declines in subjectively-rated functional status. This suggests that self-efficacy may help an older woman to persevere in trying to maintain her functional independence, even if she experiences health problems that threaten her ability to function independently.

Older people with high self-efficacy also are more likely to practice healthy behaviors, including exercise, low dietary fat intake, weight control, low alcohol intake, and avoidance of smoking [94]. These health behaviors are likely to enhance health and functioning. Although self-efficacy is associated with maintenance of cognitive performance among men [92], this association was not observed among women.

VIII. Influence of Women's Social Roles or Context

A. Social Support/Social Networks/Social Isolation

Social integration is another important factor in successful aging. Social integration describes the extent to which an individual is embedded in a social network (a web of social relationships) that provides social support resources, such as empathy, love, trust, caring, tangible aid, service, advice, suggestions, information, and feedback for self-evaluation [95]. Because older women are likely to outlive their spouses, social support from relatives, friends, neighbors, and service providers may be especially important for successful aging among women.

Social support is associated with better physical functioning in older adults, and the absence of social support is associated with declines in functional status. People without a supportive social network are at increased risk of functional decline and are less likely to improve their physical status [96–100]. In general, the associations between social integration and health are stronger among men than among women, but several studies have reported significant protective effects of social support among older women [72,101].

Several types of social support may affect successful aging in older women. *Emotional support* involves the provision of empathy, love, trust, and caring [95]. High levels of emotional support are associated with decreased risk of functional decline [8]. *Instrumental support* involves the provision of tangible aid and services that assist a person in need [95]. High levels of instrumental support have been associated with negative health outcomes among older women [102], although this effect was not significant in other studies [47]. High instrumental support may be an early marker for medical conditions that may soon result in physical impairment [103]. Older women who marshal social network members to assist them may be doing so because they are experiencing the beginnings of functional decline. Therefore, instrumental support may be a result, rather than a cause, of physical disability. *Informational support* involves the provision of advice, suggestions, and information [95]. Although informational support has not been associated directly with physical or cognitive functioning in older adults, it has been associated with recovery from stroke [104], and informational support groups have been shown to be as effective as emotional support groups in helping patients adjust to cancer [105]. *Appraisal support* involves the provision of constructive feedback, affirmation, and social comparison that helps the person to make self-evaluations [95]. Older women with high levels of appraisal support did not show the age-related elevations in blood pressure that were evident among women with low appraisal support [106], suggesting that appraisal support may be associated with successful aging.

B. Marital Status

The majority of older women in the United States are unmarried (59%), while only 25% of older men are unmarried [1]. The gender difference in marital status is due to women's greater longevity, their tendency to marry older men, and their tendency not to remarry after becoming widowed.

Married individuals are less likely to experience age-related or disease-related declines in physical functioning than those who are single, divorced, or widowed [98,107]. Studies of the effects of widowhood [108–110] generally have found that widowhood is associated with an increased risk of functional decline, mortality, and illness. However, these effects tend to be stronger among men than among women. The excess mortality associated with widowhood was only 6% in women as compared with 17% in men [109]. Women who become widowed typically show short-term psychological disturbance, but they tend to return to their previous levels of psychological well-being within a few years [111]. Although the negative effects of widowhood may not be as severe for women as for men, the public health impact of widowhood is considerable because of the large proportion of older women who experience widowhood.

C. Activities

1. Employment/Volunteerism

Employment, whether paid or unpaid, may enhance life satisfaction and successful aging [112]. Older people who work at paying jobs express higher levels of life satisfaction than do those who are not employed [113]. Approximately one-half of retired people in the United States engage in some type of volunteer work [114]. Retirees who engaged in volunteer work for ten or more hours per week reported greater feelings of purpose in life [115]. Involvement in work activities enables an older woman to use her physical and cognitive skills; therefore, these abilities may be less likely to deteriorate. In addition, work usually involves contact with a social network, and the social interactions with people at work may enhance successful aging. Therefore, continued involvement with work or volunteer activities may increase the likelihood of successful aging among women.

D. Living Conditions

Older women are significantly more likely to live alone than are older men. Of the 9.4 million persons age 65 years or older who lived alone in 1993, eight out of ten were women [1]. Older women who are able to live alone typically have higher levels of functioning than do those women who live with others; they report fewer medical conditions and less functional and cognitive impairment [116]. However, this probably does not indicate that living alone enhances health and successful aging. Rather, it may indicate that the older women who do experience functional limitations are more likely to be institutionalized. Older women who are unable to live alone independently and do not have a spouse, child, or sibling to provide aid are at high risk of nursing home admission [117,118]. Because older women are likely to outlive their spouses, they may be especially at risk of nursing home admission.

E. Sexual Orientation

Little research has focused on the aging process among older lesbian women. The few studies that have examined aging among lesbians [119,120] generally have found that older lesbian women report good physical and mental health and positive attitudes toward aging. There is no direct evidence that older lesbian women age more or less successfully than do heterosexual women. In fact, their daily lives and concerns appear to be similar to those of heterosexual older women [120]. However, researchers have hypothesized that lesbian and gay older women who are able to challenge traditional sex-role assumptions may age more successfully because they may be better prepared to cope with role transitions and stressful life events that typically occur with age [121].

IX. Biologic Markers of Susceptibility or Exposure

A. Allostatic Load

The cumulative effects of stress over a lifetime may affect health and longevity, decreasing the probability of successful aging [122]. Stress is associated with impairments in immune and cardiovascular functioning, but the underlying biological mechanism is not understood completely. One theory of the association between stress and disease [123] posits that the body's physiologic systems fluctuate to meet demands from external forces, a state termed allostasis. Allostatic load, the strain on the body caused by this repeated fluctuation, is the physiological cost of chronic exposure to fluctuating or heightened neural or neuroendocrine response resulting from repeated or chronic environmental challenge [123].

Allostatic load has been operationalized as the sum of physiological parameters for which the respondent is in the highest-risk quartile [124]. These measures include systolic and diastolic blood pressure, waist-to-hip ratio, serum high-density lipoprotein (HDL) and total cholesterol levels, blood plasma levels of total glycosylated hemoglobin, serum dehydroepiandrosterone sulfate (DHEA-S), and urinary cortisol, norepinephrine, and epinephrine excretion levels. In a 2.5-year longitudinal study, Seeman *et al.* [124] found that allostatic load was associated with poor physical functioning at baseline, decline in physical functioning over time, poor cognitive performance at baseline, and decline in cognitive performance over time. Newer data [T. E. Seeman, unpublished data, 1997] have shown similar results over a seven-year period. These findings suggest that allostatic load is a risk factor for less successful aging. In general, allostatic load is lower among older women than among older men [122], but even small increases in allostatic load may affect successful aging among women.

B. Cortisol

Cortisol is one of the primary hormones produced during times of stress. Over time, repeated excretion of cortisol in response to stress may have harmful effects on successful aging. In the MacArthur Study of Successful Aging, for example, increases in cortisol over a 2.5-year period were associated with an increased risk of fractures among initially high-functioning older women [125]. Also in the MacArthur study [126], women with greater cortisol excretion had poorer baseline memory performance, and women who exhibited increases in cortisol excretion over a 2.5-year period were more likely to show declines in memory performance. Fortunately, women who experienced declines in cortisol exhibited improvements in memory performance. These results indicate that cortisol excretion in response to stress can lead to less successful aging, and reductions in cortisol may be associated with maintaining functional status or even regaining lost functional ability.

X. Intervention Programs to Promote Successful Aging

Several types of interventions have been shown to promote successful aging. These include pharmacologic interventions, such as estrogen replacement therapy, nutritional interventions, such as calcium supplementation, and behavioral interventions, such as physical activity programs.

A. Pharmacologic and Nutritional Interventions

Estrogen replacement therapy may be useful in preventing age-related cognitive declines among older women. A meta-analysis of observational studies and clinical trials [127] found that estrogen replacement therapy was associated with a 29% decreased risk of developing dementia among postmenopausal women. Estrogen replacement therapy may prevent or slow cognitive declines in women by promoting cholinergic and serotonergic activity in specific brain regions, maintaining neural circuitry, altering lipoprotein levels, and preventing cerebral ischemia. Estrogen replacement therapy also has physiological benefits that may promote maintenance of physical functioning. In the Postmenopausal Estrogen/Progestin Interventions (PEPI) Trial, estrogen replacement therapy was associated with decreased LDL cholesterol, increased HDL cholesterol, and maintenance of bone mineral density [128,129].

Dietary supplementation with calcium and vitamin D also may enhance successful aging. Several studies [130–133] have shown that calcium and vitamin D supplementation are associated with reduced bone loss, which may decrease the risk of debilitating fractures.

B. Physical Activity Interventions

Because physical activity is such an important determinant of maintenance of physical functioning, exercise programs may play a significant role in increasing the likelihood of successful aging. Several studies have demonstrated beneficial effects of strength and endurance training on health and functional status among older adults. Short-term low-intensity aerobic exercise programs, such as aerobic dance, walking, or Tai Chi, have produced improvements in physical functioning, strength, and cardiorespiratory fitness [134]. Strength training also can improve the physical functioning of frail older women. In a study of 100 nursing home residents [135], subjects randomized to a high-intensity strength training condition experienced significant gains in strength and functional status relative to controls, and they increased their spontaneous activity as well. Even in the oldest old, strength training can improve strength and walking speed. In a study of frail, institutionalized older patients with a mean age of 90 years [136], an eight-week strength training program produced significant gains in muscle strength, size, and functional mobility.

Moreover, exercise programs need not be expensive to be effective. In the Community Healthy Activities Model Program for Seniors (CHAMPS), older adults were encouraged to use existing exercise programs and classes already available in the community [137]. This program resulted in increased activity levels, improved self-esteem, reduced anxiety, reduced depression, and increased psychological well-being. Effective programs to increase physical activity may be especially important for older women. Older women are more likely to be sedentary than are older men, and they report more barriers to physical activity [77]. Programs that can remove these barriers and encourage older women to exercise have the potential to contribute significantly to successful aging.

C. Psychosocial Interventions

Interventions to provide or increase social support may enhance successful aging among older women. A randomized controlled trial of a social support intervention [138] produced improvements in self-reported health, although it did not affect physical functioning or mortality. Unfortunately, social support provided as part of an experimental intervention may not be an acceptable substitute for social support that develops as a natural part of a relationship. In one study [139], older women were paired to form dyads that could provide reciprocal social support by telephone. This intervention failed to show effects on physical or psychological functioning, perhaps because many of the women still reported feelings of dissatisfaction from their naturally occurring social networks and families. Although overly contrived programs to provide social support may not be effective, it may be possible to encourage or enable older women to enhance their own support networks by providing them with locations and community activities in which they can interact with other people.

XI. Epidemiologic Issues

As the Baby Boom Generation ages, the older population in the United States will increase dramatically. In the year 2040, the proportion of the U.S. population that is 65 years of age or older is expected to reach a peak of 20.7%, as compared with 12.5% in 1990 and 12.8% in 2000 [140]. This will result in growth of the older population from 31,079,000 in 1990 to 77,014,000 in 2040—over a twofold increase in 50 years. Among the older population, those 85 years of age and older are the most rapidly growing group.

The majority of these older people in the United States, especially the oldest old, are (and will be) women. In 1990, there were only 38.6 males for every 100 females 85 years of age and older in the United States [140]. If these older women experience physical and cognitive impairment in old age, they will create an enormous public health burden. If, however, they are able to maintain high levels of functioning for most of this period, they will be able to maintain their independence, enjoy their old age, and provide wisdom and inspiration to younger generations.

XII. Summary and Conclusions/Future Directions

A. The Need for Research in Ethnic Minority Populations

Approximately 89% of women age 65 years and older in the United States are white [1]. As the nonwhite population in the United States increases, ethnic minority women will comprise a larger proportion of the older population of the United States. Population projections indicate that by the year 2050, 10% of the U.S. population aged 65 or older will be African-American, 16% will be Hispanic, and 8% will be Asian, Pacific Islander, Native American, Eskimo, or Aleut [1]. Because the prevalence of physical impairment is higher among minority older women than among white older women, disabled minority older women represent a significant public health burden [40]. In addition, the gender differences in life expectancy and physical impairment among ethnic minority older adults are similar to those of whites, suggesting that ethnic minority women are at high risk for spending a large portion of their later life with physical disabilities [38]. Several studies of aging (e.g., the MacArthur Study of Successful Aging) have compared African-American women with white women. Analyses of the older respondents in the Hispanic Health and Nutrition Examination Survey (Hispanic HANES) have provided some information about the health status of older Hispanics [38,39]. However, no large-scale population-based longitudinal studies of aging in the United States have focused on ethnic differences in successful aging. Research is needed to determine the predictors of successful aging among older minority women, and community programs are needed to decrease disability among these women.

B. The Need for Research on HIV-Risk Sexual Behaviors among Older Women

Although a great deal of research has focused on the physical, cognitive, and social functioning of older women, little research has focused on their sexual functioning. Despite this dearth of research, evidence suggests that many older women do indeed remain involved in sexual activity throughout old age [141]. Despite public opinion that unsafe sexual behavior is an activity of the young, research indicates that many older adults also engage in sexual practices that may place them at risk for HIV infection. In a national sample of adults over 50 years of age [142], 9% reported at least one HIV-risk factor (multiple sexual partners, risky sexual partners, blood transfusion, intravenous drug use, and/or hemophilia). In addition, approximately 95% of postmenopausal women fail to use condoms consistently [141]. As the population of healthy, active older women increases so too will the population of older women who are at risk for HIV infection through sexual contact.

HIV-prevention programs have not yet focused on older women. Interventions to promote successful aging should emphasize safer sex along with other health behaviors, such as physical activity, nutrition, and avoidance of smoking.

C. How Can Disabled Older Women Age Successfully?

Despite the potential for improvement in physical functioning among older women, it is inevitable that a proportion of older women will experience chronic disease and disability. However, chronic disease need not prevent a woman from enjoying successful aging. In the Women's Health and Aging Study [25], 35% of older women with moderate to severe disabilities reported high emotional vitality, defined as a high sense of personal mastery and happiness and low levels of depression and anxiety. Further research is needed to determine the personality, social network, and environmental characteristics that enable disabled older women to overcome their physical limitations and enjoy an emotionally vital old age. Instead of measuring how many women are aging "successfully," it may be more useful to measure how many are aging "effectively"—that is, how many are able to maximize their functional abilities and maintain a high quality of life even if they are faced with chronic illness or disability [24].

D. Conclusions

Successful aging is a complex concept that involves physical, cognitive, and emotional health. As the population of older women in the United States continues to grow, successful aging will become an increasingly important women's health issue. Continued research is needed to determine the predictors of successful aging among women and to develop interventions to promote successful aging.

References

1. U.S. Bureau of the Census (1996). 65+ in the United States. *Current Population Reports, Special Studies, P23-190.* U.S. Gov. Printing Office, Washington, DC.
2. Sheehy, G. (1996). "New Passages: Mapping Your Life Across Time." Ballantine Books, New York.
3. Crimmins, E. M., Hayward, M. D., and Saito, Y. (1996). Differentials in active life expectancy in the older population of the United States. *J. Gerontol. Soc. Sci.* **51B,** S111–S120.
4. Katz, S., Branch, L. G., and Branson, M. H. (1983). Active life expectancy. *N. Engl. J. Med.* **309,** 1218–1224.
5. Rowe, J. W., and Kahn, R. L. (1998). "Successful Aging." Pantheon Books, New York.
6. Baltes, P. B., and Baltes, M. M. (1990). Psychological perspectives on successful aging: The model of selective optimization with compensation. *In* "Successful Aging: Perspectives from the Social Sciences" (P. B. Baltes and M. M. Baltes, eds.), pp. 1–34. Cambridge University Press, Cambridge, UK.
7. Fries, J. F. (1990). Medical perspectives upon successful aging. *In* "Successful Aging: Perspectives from the Social Sciences" (P. B. Baltes and M. M. Baltes, eds.), pp. 35–49. Cambridge University Press, Cambridge, UK.
8. Seeman, T. E., Berkman, L. F., Charpentier, P. A., Blazer, D. G., Albert, M. S., and Tinetti, M. E. (1995). Behavioral and psycho-social predictors of physical performance: MacArthur Studies of Successful Aging. *J. Gerontol. Med. Sci.* **50,** M177–M183.
9. Katz, S., Downs, T. D., Cash, H. R., and Grotz, R. C. (1970). Progress in development of the index of ADL. *Gerontologist* **10,** 20–30.
10. Fillenbaum, G. (1985). Screening the elderly—a brief instrumental activities of daily living measure. *J. Am. Geriatr. Soc.* **33,** 698–706.
11. Rosow, I., and Breslau, N. (1969). A Guttman Health Scale for the aged. *J. Gerontol.* **21,** 557–560.
12. Nagi, S. Z. (1976). An epidemiology of disability among adults in the United States. *Milbank Mem. Fund Q.* **54,** 439–468.
13. Guralnik, J. M., Simonsick, E. M., Ferrucci, L., and Glynn, R. J. (1994). A short physical performance battery assessing lower extremity function: Association with self-reported disability and prediction of mortality and nursing home admission. *J. Gerontol. Med. Sci.* **49,** M85–M94.
14. Merrill, S. S., Seeman, T. E., Kasl, S. V., and Berkman, L. F. (1997). Gender differences in the comparison of self-reported disability and performance measures. *J. Gerontol. Med. Sci.* **52A,** M19–M26.
15. Rathouz, P. J., Kasper, J. D., Zeger, S. L., Ferrucci, L., Bandeen-Roche, K., Miglioretti, D. L., and Fried, L. P. (1998). Short-term consistency in self-reported physical functioning among elderly women: The Women's Health and Aging Study. *Am. J. Epidemiol.* **147,** 764–773.
16. Perlmutter, M. (1988). Cognitive potential throughout life. *In* "Emergent Theories of Aging" (V. L. Bengtson and J. E. Birren, eds.), pp. 247–268. Springer, New York.
17. Herzog, A. R., and Wallace, R. B. (1997). Measures of cognitive functioning in the AHEAD study. *J. Gerontol. Psychol. Soc. Sci., Spe. Issue* **52B,** 37–48.
18. Folstein, M. F., Folstein, S. E., and McHugh, P. R. (1975). Mini-mental state: A practical method for grading the cognitive state of patients for the clinician. *J. Psychiatr. Res.* **12,** 189–198.
19. Tombaugh, T. N., and McIntyre, N. J. (1992). The Mini-Mental State Examination: A comprehensive review. *J. Am. Geriatr. Soc.* **40,** 922–935.
20. Aguero-Torres, H., Fratiglioni, L., Guo, Z., Viitanen, M., von Strauss, E., and Winblad, B. (1998). Dementia is the major cause of functional dependence in the elderly: 3-year follow-up data from a population-based study. *Am. J. Public Health* **88,** 1452–1456.
21. Cumming, E., and Henry, W. H. (1961). "Growing Old." Basic Books, New York.
22. Havighurst, R. J. (1963). Successful aging. *In* "Processes of Aging: Social and Psychological Perspectives" (R. H. Williams, C. Tibbits, and W. Donahue, eds.), p. 299. Atherton, New York.
23. Havighurst, R. J. (1968). A social-psychological perspective on aging. *Gerontologist* **8,** 67–71.
24. Curb, J. D., Guralnik, J. M., LaCroix, A. Z., Korper, S. P., Deeg, D., Miles, T., and White, L. (1990). Effective aging: Meeting the challenge of growing older. *J. Am. Geriatr. Soc.* **38,** 827–828.
25. Penninx, B. W., Guralnik, J. M., Simonsick, E. M., Kasper, J. D., Ferrucci, L., and Fried, L. P. (1998). Emotional vitality among disabled older women: The Women's Health and Aging Study. *J. Am. Geriatr. Soc.* **46,** 807–815.
26. Kane, R. L., Ouslander, J. G., and Abrass, I. B. (1994). "Essentials of Clinical Geriatrics," 3rd ed. McGraw-Hill, New York.
27. Mendes de Leon, C. F., Fillenbaum, G. G., Williams, C. S., Brock, D. B., Beckett, L. A., and Berkman, L. F. (1995). Functional disability among elderly blacks and whites in two diverse areas: The New Haven and North Carolina EPESE. Established Populations for the Epidemiologic Studies of the Elderly. *Am. J. Public Health* **85,** 994–998.

28. Rosenberg, M. W., and Moore, E. G. (1997). The health of Canada's elderly population: Current status and future implications. *Can. Med. Assoc. J.* **157**, 1025–1032.

29. Crimmins, E. M., Saito, Y., and Reynolds, S. L. (1997). Further evidence on recent trends in the prevalence and incidence of disability among older Americans from two sources: The LSOA and the NHIS. *J. Gerontol. Soc. Sci.* **52B**, S59–S71.

30. Guralnik, J. M., Leveille, S. G., Hirsch, R., Ferrucci, L., and Fried, L. P. (1997). The impact of disability in older women. *J. Am. Med. Women's Assoc.* **52**, 113–20.

31. Strawbridge, W. J., Kaplan, G. A., Camacho, T., and Cohen, R. D. (1992). The dynamics of disability and functional change in an elderly cohort: Results from the Alameda County Study. *J. Am. Geriatr. Soc.* **40**, 799–806.

32. National Center for Health Statistics (1998). ''Health, United States, 1998. With Socioeconomic Status and Health Chartbook'' DHHS Publ. No. (PHS) 98-1232. National Center for Health Statistics, Hyattsville, MD.

33. Henderson, A. S., and Kay, D. W. K. (1997). The epidemiology of functional psychoses of late onset. *Eur. Arch. Psychiatry Clin. Neurosci.* **247**, 176–189.

34. Brayne, C., Gill, C., Paykel, E. S., Huppert, F. *et al.* (1995). Cognitive decline in an elderly population: A two wave study of change. *Psychol. Med.* **25**, 673–683.

35. National Institutes of Health Consensus Development Conference (1987). Differential diagnosis of dementing diseases. *JAMA, J. Am. Med. Assoc.* **258**, 3411–3416.

36. Jorm, A. F., Korten, A. E., and Henderson, A. S. (1987). The prevalence of dementia: A quantitative integration of the literature. *Acta Psychiatr. Scand.* **76**, 456–479.

37. U.S. Preventive Services Task Force (1996). ''Guide to Clinical Preventive Services,'' 2nd ed. Williams & Wilkins, Baltimore, MD.

38. Markides, K. S. (1989). Consequences of gender differentials in life expectancy for black and Hispanic Americans. *Int. J. Aging Hum. Dev.* **29**, 95–102.

39. Markides, K. S., and Lee, D. J. (1991). Predictors of health status in middle-aged and older Mexican Americans. *J. Gerontol. Soc. Sci.* **46**, S243–S249.

40. Seeman, T. E., Charpentier, P. A., Berkman, L. F., Tinetti, M. E., Guralnik, J. M., Albert, M., Blazer, D., and Rowe, J. W. (1994). Predicting changes in physical performance in a high-functioning elderly cohort: MacArthur Studies of Successful Aging. *J. Gerontol. Med. Sci.* **49**, M97–M108.

41. Heyman, A., Fillenbaum, G., Prosnitz, B., and Raiford, K. (1991). Estimated prevalence of dementia among elderly Black and White community residents. *Arch. Neurol. (Chicago)* **48**, 594–598.

42. Guralnik, J. M., Land, K. C., Blazer, D., Fillenbaum, G. G., and Branch, L. G. (1993). Educational status and active life expectancy among older blacks and whites. *N. Engl. J. Med.* **329**(2), 110–116.

43. Parker, S. L., Davis, K. J., Wingo, P. A., Ries, L. A., and Heath, C. W., Jr. (1998). Cancer statistics by race and ethnicity. *Ca: Cancer J. Clin.* **48**, 31–48.

44. Savage, P. J., and Harlan, W. R. (1991). Racial and ethnic diversity in obesity and other risk factors for cardiovascular disease: Implications for studies and treatment. *Ethnicity Dis.* **1**, 200–211.

45. Gasperino, J. (1996). Ethnic differences in body composition and their relation to health and disease in women. *Ethnicity Health,* **1**, 337–347.

46. Bengtson, V. L., Burton, L. M., and Rosenthal, C. (1993). Families and aging: Diversity and heterogeneity. In ''Life-span Development: A Diversity Reader'' (M. A. Black and R. A. Pierce, eds.), pp. 226–246. Kendall/Hunt, Dubuque, IA.

47. Seeman, T. E., Bruce, M. L., and McAvay, G. J. (1996). Social network characteristics and onset of ADL disability: MacArthur Studies of Successful Aging. *J. Gerontol. Soc. Sci.* **51B**, S191–S200.

48. Silverstein, M., and Waite, L. J. (1993). Are Blacks more likely than Whites to receive and provide social support in middle and old age? Yes, no, and maybe so. *J. Gerontol. Soc. Sci.* **48**, S212–S222.

49. Guralnik, J. M., LaCroix, A. Z., Abbott, R. D., Berkman, L. F., Satterfield, S., Evans, D. A., and Wallace, R. B. (1993). Maintaining mobility in late life. I. Demographic characteristics and chronic conditions. *Am. J. Epidemiol.* **137**, 845–857.

50. Tran, T. V., and Williams, L. F. (1998). Poverty and impairment in activities of living among elderly Hispanics. *Soc. Work Health Care* **26**, 59–78.

51. Seeman, T. E., and Adler, N. (1998). Older Americans: Who will they be? Socio-demographic changes in the population of older Americans and their potential health consequences. *Natl. Forum* **78**, 22–25.

52. House, J. S., Lepkowski, J. M., Kinney, A. M., and Mero, R. P. (1994). The social stratification of aging and health. *J. Health Soc. Behav.* **35**, 213–234.

53. Hubert, H. B., Bloch, D. A., and Fries, J. F. (1993). Risk factors for physical disability in an aging cohort: The NHANES I Epidemiologic Followup Study. *J. Rheumatol.* **20**, 480–488.

54. Ravaglia, G., Forti, P., Maioli, F., Boschi, F., Cicognani, A., Bernardi, M., Pratelli, L., Pizzoferrato, A., Porcu, S., and Gasbarrini, G. (1997). Determinants of functional status in healthy Italian nonagenarians and centenarians: A comprehensive functional assessment by the instruments of geriatric practice. *J. Am. Geriatr. Soc.* **45**, 1196–1202.

55. Albert, M. S., Jones, K., Savage, C. R., Berkman, L., Seeman, T., Blazer, D., and Rowe, J. W. (1995). Predictors of cognitive change in older persons: MacArthur Studies of Successful Aging. *Psychol. Aging* **10**, 578–589.

56. Timiras, P. S. (1995). Education, homeostasis, and longevity. *Exp. Gerontol.* **30**(3–4), 189–198.

57. De Ronchi, D., Fratiglioni, L., Rucci, P., Paternico, A., Graziani, S., and Dalmonte, E. (1998). The effect of education on dementia occurrence in an Italian population with middle to high socioeconomic status. *Neurology* **50**, 1231–1238.

58. Katzman, R. (1993). Education and the prevalence of dementia and Alzheimer's disease. *Neurology* **43**, 13–20.

59. Mortimer, J. A., and Graves, A. B. (1993). Education and other socioeconomic determinants of dementia and Alzheimer's disease. *Neurology* **43**(Suppl. 4), S39–S44.

60. Schmand, B., Smit, J. H., Geerlings, M. I., and Lindeboom, J. (1997). The effects of intelligence and education on the development of dementia. A test of the brain reserve hypothesis. *Psychol. Med.* **27**, 1337–1344.

61. Beekman, A. T. F., Deeg, D. J. H., van Tilburg, T., Smit, J. H., Hooijer, C., and van Tilburg, W. (1995). Major and minor depression in later life: A study of prevalence and risk factors. *J. Affect. Disord.* **36**, 65–75.

62. Kiljunen, M., Sulkava, R., Niinistoe, L., Polvikoski, T., Verkkoniemi, A., and Halonen, P. (1997). Depression measured by the Zung depression status inventory is very rare in a Finnish population aged 85 years and over. *Intl. Psychogeriatr.* **9**, 359–368.

63. Zunzunegui, M. V., Beland, F., Llacer, A., and Leon, V. (1998). Gender differences in depressive symptoms among Spanish elderly. *Soc. Psychiatry Psychiatr. Epidemiol.* **33**, 195–205.

64. Bruce, M. L., Seeman, T. E., Merrill, S. S., and Blazer, D. G. (1994). The impact of depressive symptomatology on physical disability: MacArthur Studies of Successful Aging. *Am. J. Public Health* **84**, 1796–1799.

65. Penninx, B. W., Guralnik, J. M., Ferrucci, L., Simonsick, E. M., Deeg, D. J., and Wallace, R. B. (1998). Depressive symptoms and physical decline in community-dwelling older persons. *JAMA, J. Am. Med. Assoc.* **279**, 1720–1726.

66. Heidrich, S. M. (1998). Older women's lives through time. *Adv. Nurs. Sci.* **20**, 65–75.

67. Kivelae, S., Longaes-Saviaro, P., Kimmo, P., and Kesti, E. (1996). Health, health behaviour and functional ability predicting depression in old age: A longitudinal study. *Int. J. Geriatr. Psychiatry* **11**, 871–877.

68. Cipolli, C., Neri, M., De Vreese, L. P., and Pinelli, M. (1996). The influence of depression on memory and metamemory in the elderly. *Arch. Gerontol. Geriatr.* **23**, 111–127.

69. LaCroix, A. Z., Guralnik, J. M., Berkman, L. F., Wallace, R. B., and Satterfield, S. (1993). Maintaining mobility in late life. II. Smoking, alcohol consumption, physical activity, and body mass index. *Am. J. Epidemiol.* **137**, 858–869.

70. Shephard, R. J. (1993). Exercise and aging: Extending independence in older adults. *Geriatrics* **48**, 61–64.

71. Wagner, E. H., LaCroix, A. Z., Buchner, D. M., and Larson, E. B. (1992). Effects of physical activity on health status in older adults. I: Observational studies. *Annu. Rev. Public Health* **13**, 451–468.

72. Unger, J. B., Johnson, C. A., and Marks, G. (1997). Functional decline in the elderly: Evidence for direct and stress-buffering protective effects of social interactions and physical activity. *Ann. Behav. Med.* **19**, 152–160.

73. Mor, V., Murphy, J., Masterson-Allen, S., Willey, C., Razmpour, A., Jackson, M. E., Greer, D., and Katz, S. (1989). Risk of functional decline among well elders. *J. Clin. Epidemiol.* **42**, 895–904.

74. Simonsick, E. M., Lafferty, M. E., Phillips, C. L., Mendes de Leon, C. F., Kasl, S. V., Seeman, T. E., Fillenbaum, G., Hebert, P., and Lemke, J. H. (1993). Risk due to inactivity in physically capable older adults. *Am. J. Public Health* **83**, 1443–1450.

75. Unger, J. B. (1995). Sedentary lifestyle as a risk factor for self-reported poor physical and mental health. *Am. J. Health Promot.* **9**, 15–17.

76. Leigh, J. P., and Fries, J. F. (1992). Health habits, health care use and costs in a sample of retirees. *Inquiry* **29**, 44–54.

77. Lee, C. (1993). Factors related to the adoption of exercise among older women. *J. Behav. Med.* **16**, 323–334.

78. Gillis, K. J., and Hirdes, J. P. (1996). The quality of life implications of health practices among older adults: Evidence from the 1991 Canadian General Social Survey. *Can. J. Aging* **15**, 299–314.

79. Ranieri, P., Bertozzi, B., Frisoni, G. B., Rozzini, R., and Trabucchi, M. (1996). Determinants of malnutrition in a geriatric ward: Role of comorbidity and functional status. *J. Nut. Elderly* **16**, 11–22.

80. Ribot, C., Tremollieres, F., and Pouilles, J. M. (1994). The effect of obesity on postmenopausal bone loss and the risk of osteoporosis. *Adv. Nutr. Res.* **9**, 257–271.

81. Stoddard, C. E., and Thompson, D. L. (1996). Alcohol and the elderly: Special concerns for counseling professionals. *Alcohol. Treat. Q.* **14**, 59–69.

82. Smith, J. W. (1995). Medical manifestations of alcoholism in the elderly. *Int. J. Addict.* **30**, 1749–1798.

83. Serdula, M. K., Koong, S. L., Williamson, D. F., Anda, R. F., Madans, J. H., Kleinman, J. C., and Byers, T. (1995). Alcohol intake and subsequent mortality: Findings from the NHANES I Follow-up Study. *J. Stud. Alcohol* **56**, 233–239.

84. Baldini, E. H., and Strauss, G. M. (1997). Women and lung cancer: Waiting to exhale. *Chest* **112**, 229S–234S.

85. Hennekens, C. H. (1998). Risk factors for coronary heart disease in women. *Cardiol. Clin.* **16**, 1–8.

86. Kendrick, J. S., and Merritt, R. K. (1996). Women and smoking: An update for the 1990s. *Am. J. Obstet. Gynecol.* **175**, 528–535.

87. Maxwell, C. J., and Hirdes, J. P. (1993). The prevalence of smoking and implications for quality of life among the community-based elderly. *Am. J. Prev. Med.* **9**, 338–345.

88. LaCroix, A. Z., and Omenn, G. S. (1992). Older adults and smoking. *Clin. Geriatr. Med.* **8**, 69–87.

89. Pierce, J. P., and Gilpin, E. A. (1995). A historical analysis of tobacco marketing and the uptake of smoking by youth in the United States: 1890–1977. *Health Psychol.* **14**, 500–508.

90. Bandura A. (1977). Self-efficacy: Toward a unifying theory of behavioral change. *Psychol. Rev.* **84**, 191–215.

91. Seeman, T. E., Unger, J., McAvay, G., and Mendes de Leon, C. (1999). The role of self-efficacy beliefs in perceptions of functional disability: MacArthur Studies of Successful Aging.

92. Seeman, T., McAvay, G., Merrill, S., Albert, M., and Rodin, J. (1996). Self-efficacy beliefs and change in cognitive performance: MacArthur Studies on Successful Aging. *Psychol. Aging* **11**, 538–551.

93. Mendes de Leon, C. F., Seeman, T. E., Baker, D. I., Richardson, E. D., and Tinetti, M. E. (1996). Self-efficacy, physical decline, and change in functioning in community-living elders: A prospective study. *J. Gerontol. Soc. Sci.* **51B**, S183–S190.

94. Grembowski, D., Patrick, D., Diehr, P., and Durham, M. (1993). Self-efficacy and health behavior among older adults. *J. Health Soc. Behav.* **34**, 89–104.

95. Heaney, C. A., and Israel, B. A. (1997). Social networks and social support. *In* "Health Behavior and Health Education: Theory, Research, and Practice" (K. Glanz, F. M. Lewis, and B. K. Rimer, eds.), 2nd ed., pp. 179–205. Jossey-Bass, San Francisco.

96. Bowling, A., and Browne, P. D. (1991). Social networks, health, and emotional well-being among the oldest old in London. *J. Gerontol. Soc. Sci.* **46**, S20–S22.

97. Liu, X., Liang, J., Muramatsu, N., and Sugisawa, H. (1995). Transitions in functional status and active life expectancy among older people in Japan. *J. Gerontol. Soc. Sci.* **50B**, S383–S394.

98. Mor, V., Wilcox, V., Rakowski, W., and Hiris, J. (1994). Functional transitions among the elderly: Patterns, predictors, and related hospital use. *Am. J. Public Health* **84**, 1274–1280.

99. Weinberger, M., Tierney, W. M., Booher, P., and Hiner, S. L. (1990). Social support, stress and functional status in patients with osteoarthritis. *Soc. Sci. Med.* **30**, 503–508.

100. Wolinsky, F. D., and Johnson, R. J. (1992). Widowhood, health status, and the use of health services by older adults: A cross-sectional and prospective approach. *J. Gerontol. Soc. Sci.* **47**, S8–S16.

101. Strawbridge, W. J., Cohen, R. D., Shema, S. J., and Kaplan, G. A. (1996). Successful aging: Predictors and associated activities. *Am. J. Epidemiol.* **144**(2), 135–141.

102. Silverstein, M., Chen, X., and Heller, K. (1996). Too much of a good thing? Intergenerational social support and the psychological well-being of older parents. *J. Marriage Fam.* **58**, 970–982.

103. Ikkink, K. K., and van Tilburg, T. (1998). Do older adults' network members continue to provide instrumental support in unbalanced relationships? *J. Soc. Pers. Relat.* **15**, 59–75.

104. Glass, T. A., and Maddox, G. L. (1992). The quality and quantity of social support: Stroke recovery as psycho-social transition. *Soc. Sci. Med.* **34**, 1249–1261.

105. Helgeson, V. S., and Cohen, S. (1996). Social support and adjustment to cancer: Reconciling descriptive, correlational, and intervention research. *Health Psychol.* **15**, 135–148.

106. Uchino, B. N., Cacioppo, J. T., Malarkey, W., and Glaser, R. (1995). Appraisal support predicts age-related differences in cardiovascular function in women. *Health Psychol.* **14**, 556–562.

107. Ward, M. M., and Leigh, J. P. (1993). Marital status and the progression of functional disability in patients with rheumatoid arthritis. *Arthritis Rheum.* **36**, 581–588.

108. Barry, K. L., and Fleming, M. F. (1988). Widowhood: A review of the social and medical implications. *Fam. Med.* **20**, 413–417.

109. Martikainen, P., and Valkonen, T. (1996). Mortality after death of spouse in relation to duration of bereavement in Finland. *J. Epidemiol. Commun. Health* **50**, 264–268.

110. Stroebe, M. S. (1998). New directions in bereavement research: Exploration of gender differences. *Palliative Med.* **12**, 5–12.

111. Bennett, K. M. (1997). A longitudinal study of wellbeing in widowed women. *Int. J. Geriatr. Psychiatry* **12**, 61–66.

112. Dorfman, L. T., and Rubenstein, L. M. (1993). Paid and unpaid activities and retirement satisfaction among rural seniors. *Phys. Occup. Ther. Geriatr.* **12**, 45–63.

113. Aquino, J. A., Russell, D. W., Cutrona, C. E., and Altmaier, E. M. (1996). Employment status, social support, and life satisfaction among the elderly. *J. Coun. Psychol.* **43**, 480–489.

114. Fischer, L. R., Mueller, D. P., and Cooper, P. W. (1991). Older volunteers: A Discussion of the Minnesota Senior Study. *Gerontologist* **31**, 183–194.

115. Weinstein, L., Xie, X., and Cleanthous, C. C. (1995). Purpose in life, boredom, and volunteerism in a group of retirees. *Psychol. Rep.* **76**, 482.

116. Mui, A. C., and Burnette, J. D. (1994). A comparative profile of frail elderly persons living alone and those living with others. *J. Gerontol. Soc. Work* **21**, 5–26.

117. Freedman, V. A. (1996). Family structure and the risk of nursing home admission. *J. Gerontol. Soc. Sci.* **51**, S61–S69.

118. Freedman, V. A., Berkman, L. F., Rapp, S. R., and Ostfeld, A. M. (1994). Family networks: Predictors of nursing home entry. *Am. J. Public Health* **84**(5), 843–845.

119. Deevey, S. (1990). Older lesbian women: An invisible minority. *J. Gerontol. Nurs.* **16**, 35–37, 39.

120. Kehoe M. (1988). Lesbians over 60 speak for themselves. *J. Homosex.* **16**, 1–111.

121. Friend, R. A. (1989). Older lesbian and gay people: Responding to homophobia. *Marriage Fam. Rev.* **14**, 241–263.

122. McEwen, B. S. (1998). Stress, adaptation, and disease. Allostasis and allostatic load. *Ann. N. Y. Acad. Sci.* **840**, 33–44.

123. McEwen, B. S., and Stellar, E. (1993). Stress and the individual: Mechanisms leading to disease. *Arch. Intern. Med.* **153**, 2093–2101.

124. Seeman, T. E., Singer, B. H., Rowe, J. W., Horwitz, R. I., and McEwen, B. S. (1997). Price of adaptation-allostatic load and its health consequences: MacArthur Studies of Successful Aging. *Arch. Intern. Med.* **157**, 2259–2268.

125. Greendale, G. A., Unger, J. B., Rowe, J. W., and Seeman, T. E. (1999). The relation between cortisol excretion and fractures in healthy elderly: Results from the MacArthur Studies. *J. Am. Geriatr. Soc.*, in press.

126. Seeman, T. E., McEwen, B. S., Singer, B. H., Albert, M. S., and Rowe, J. W. (1997). Increase in urinary cortisol excretion and memory decline: MacArthur Studies of Successful Aging. *J. Clin. Endocrinol. Metab.* **82**, 2458–2465.

127. Yaffe, K., Sawaya, G., Leiberburg, I., and Grady, D. (1998). Estrogen therapy in postmenopausal women. *JAMA, J. Am. Med. Assoc.* **279**, 688–695.

128. Barrett-Connor, E., Slone, S., Greendale, G., Kritz-Silverstein, D., Espeland, M., Johnson, S. R., Waclawiw, M., and Fineberg, S. E. (1997). The Postmenopausal Estrogen/Progestin Interventions Study: Primary outcomes in adherent women. *Maturitas* **27**, 261–274.

129. The Writing Group for the PEPI Trial (1996). Effects of hormone therapy on bone mineral density: Results from the postmenopausal estrogen/progestin interventions (PEPI) trial. *JAMA, J. Am. Med. Assoc.* **276**, 1389–1396.

130. Bonjour, J. P., Schurch, M. A., and Rizzoli, R. (1996). Nutritional aspects of hip fractures. *Bone* **18**, 139S–144S.

131. Reid, I. R. (1996). Therapy of osteoporosis: Calcium, vitamin D, and exercise. *Am. J. Med. Sci.* **312**, 278–286.

132. Sankaran, S. K. (1996). Osteoporosis prevention and treatment. Pharmacological management and treatment implications. *Drugs Aging* **9**, 472–477.

133. Wark, J. D. (1996). Osteoporotic fractures: Background and prevention strategies. *Maturitas* **23**, 193–207.

134. Chandler, J. M., and Hadley, E. C. (1996). Exercise to improve physiologic and functional performance in old age. *Clin. Geriatr. Med.* **12**(4), 761–784.

135. Evans, W. J. (1995). Effects of exercise on body composition and functional capacity of the elderly. *J. Gerontol. Biol. Med. Spec. Issue Sci.* **50**, 147–150.

136. Fiatarone, M. A., O'Neill, E. F., Ryan, N. D., Clements, K. M., Solares, G. R., Nelson, M. E., Roberts, S. B., Kehayias, J. J., Lipsitz, L. A., and Evans, W. J. (1994). Exercise training and nutritional supplementation for physical frailty in very elderly people. *N. Engl. J. Med.* **330**, 1769–1775.

137. Stewart, A. L., Mills, K. M., Sepsis, P. G., King, A. C., McLellan, B. Y., Roitz, K., and Ritter, P. L. (1997). Evaluation of CHAMPS, a physical activity promotion program for older adults. *Ann. Behav. Med.* **19**, 353–361.

138. Clarke, M., Clarke, S. J., and Jagger, C. (1992). Social intervention and the elderly: A Randomized Controlled Trial. *Am. J. Epidemiol.* **136**, 1517–1523.

139. Heller, K., Thompson, M. G., Trueba, P. E., Hogg, J. R., and Vlachos-Weber, I. (1991). Peer support telephone dyads for elderly women: Was this the wrong intervention? *Am. J. Commun. Psychol.* **19**, 53–74.

140. Hobbs, F. B., and Damon, B. L. (1996). "Current Population Reports, Special Study P23-190, 65+ in the United States." U.S. Department of Health and Human Services, Washington, DC.

141. Leigh, B. C., Temple, M. T., and Trocki, K. F. (1993). The sexual behavior of U.S. adults: Results from a national survey. *Am. J. Public Health* **83**, 1400–1408.

142. Stall, R., and Catania, J. (1994). AIDS risk behaviors among late middle-aged and elderly Americans: The National AIDS Behavioral Surveys. *Arch. Intern. Med.* **154**, 57–63.

100

Epilogue

JENNIFER L. KELSEY
Department of Health Research and Policy
Stanford University School of Medicine
Stanford, California

The last part of the twentieth century has been an exciting era for the study of women's health. One major accomplishment has been the recognition by scientists, health care providers, politicians, and others that there are many important health-related issues of specific concern to women or of much greater concern to women than men. Consequently, most medical research now includes women, and, in some instances, is focused on women. It is now recognized that women do not always react in the same way as men to preventive efforts, diagnostic procedures, and treatments. Issues that were previously taboo are now openly discussed. New technologies are providing many more choices to women as well as men. The scope of *Women and Health* illustrates well the diversity of characteristics that are now recognized as affecting women's health, including social, psychological, cultural, economic, and political as well as biomedical factors.

The causes of death of greatest importance among women in industrialized countries are cardiovascular diseases and cancers, especially cancers of the breast, colon, and lung. Also of importance are diseases whose primary effect is on the quality of life of older individuals, such as osteoarthritis, osteoporosis, and the dementias. Although these diseases are becoming important in developing countries as well, other conditions that mainly affect younger populations, such as human immunodeficiency virus (HIV), human papilloma virus (HPV), cervical cancer, diarrheal diseases of children, and diseases of pregnancy and childbirth, are currently of higher priority for women in these areas. Improved preventive and therapeutic measures will be important in reducing the impact of these diseases in developing countries, but it is also clear that political and cultural changes that make the environment less repressive to women are also needed if there is to be a significant impact on the occurrence of some of these diseases.

The health issues of greatest concern vary considerably from one woman to another. These concerns depend in part on a woman's own life stage and that of her family and friends. At different times she may be concerned about the health of fetuses, infants, children, adolescents, and health during the reproductive years, the menopausal years, or early and late old age. A woman at one life stage may not appreciate that what she does at that time may have a significant effect on her later health. For instance, it may be difficult to persuade a teenager that she should consume more calcium in order to reduce her risk for hip fracture when she reaches her eighties and nineties, or to eat more fruits and vegetables to reduce her risk of cancer or heart disease in later life. A woman in her eighties or nineties who has experienced hard economic times may have difficulty understanding why her teenage granddaughter refuses to eat because she is so concerned about her body image, or why she uses drugs or becomes pregnant. Memory problems of the elderly may not be well tolerated by those in younger generations. Menopausal symptoms are often poorly appreciated by men and by young women who have not experienced them and who know little about them.

Other life events are not anticipated until they are suddenly thrust upon a person. A woman in her fifties may suddenly have care-giving responsibilities for elderly parents just at a time when her children were grown and she thought her family responsibilities were largely over. A grandmother may suddenly find herself caring for her grandchildren because her daughter is unable or unwilling to do so herself. As knowledge about the different life stages gradually evolves and is disseminated, it is hoped that the problems experienced by women in each life stage will be better understood by those at other stages in life as well as by men and by health care providers. Similarly, women from different cultural or socioeconomic backgrounds may have different health concerns that need to be appreciated by those with other backgrounds.

A woman's health concerns also depend to a great extent on the part of the world in which she lives. While in industrialized countries a young woman may become so obsessed with being thin that she develops an eating disorder and refuses to eat, in other parts of the world starvation is a major concern and refusing to eat food that is offered is unimaginable. In developed countries, there is a great deal of interest in reproductive technologies to assist otherwise infertile couples to have children, while in many parts of the world overpopulation remains an enormous societal problem. While the feminist movement has achieved many gains for women in industrialized countries, in large parts of the world women still have few rights. Health problems resulting in part from this low status of women range from female genital mutilation to HIV infection to high maternal morbidity and mortality. Several of the diseases of greatest concern to women in developing countries, such as AIDS, cervical cancer, diarrheal diseases of children, and diseases of pregnancy and delivery, are not of as much concern to women in developed countries. Violence to women is a problem throughout the world, but the form that violence takes varies greatly depending on culture and many other factors.

Lower salaries for women compared to men, substandard working conditions, differential access to health care, and lower status remain major problems for women throughout the world at the turn of this century, but these problems are particularly severe in developing countries. The globalization of the economy is giving women in developed countries a greater awareness of the enormous problems faced on a daily basis by many women in developing countries.

Looking at women's health from a global perspective, it seems unlikely that the advances being made in biomedical technology will have a major impact on much of the world or on many women in industrialized nations. Thus, higher priority should be given to developing less expensive technologies, medications, and procedures that can help people throughout the world. Many new technologies are also not an option for people of low socioeconomic status in developed countries. There will have to be financial incentives for the development of less expensive technologies. As somewhat effective treatments for HIV infection become available to those in developed countries, it is most unfortunate that they are beyond the financial reach of people living in parts of the world where HIV infection is most prevalent. Development of a vaccine for HPV could be of great benefit to women, but its impact will be limited if it is not economically feasible for use by the women who need it most—that is, women in developing countries.

In industrialized countries, new biomedical technologies are in fact affecting women in many ways. Babies who are born with birth defects can be kept alive much longer, often with the need for a great deal of care that is most often provided by the mother and with a huge financial cost to society. Genetic testing can provide women with guidance as to whether they should try to have children, and if they do become pregnant, whether the pregnancy should be terminated. This, of course, raises ethical issues. To some extent at present, and certainly to a greater extent in the not-too-distant future, genetic testing may be done that will indicate a woman's susceptibility to certain diseases. This, in turn, presents many more decisions. If a woman is BRCA1- or BRCA2-positive, should she have her breasts removed, her ovaries removed, should she take tamoxifen, or should she just have frequent mammograms? It is not known which, if any, of these options will be most effective in reducing risk for breast cancer in the genetically susceptible woman.

For a woman entering the menopausal years, many choices of drugs and combinations of drugs are already available to treat menopausal symptoms and also to affect the subsequent likelihood of developing such important diseases as osteoporosis, heart disease, and breast cancer. With the advent of the synthetic selective estrogen receptor modulators (SERMs), also called "designer estrogens," there is the prospect of reducing the likelihood of occurrence of several major diseases, but long-term effects of these agents are not known. More options are going to become available in the near future, and continued monitoring of the various risks and benefits is essential. Prompt dissemination of scientific information such as that included in this book will be important if informed decisions are to be made.

There is increasing awareness that women and men differ biologically and behaviorally. The diagnosis, prognosis, and optimal treatment of some types of coronary heart disease and the prognosis and optimal treatment of some forms of stroke differ in women and men. Many autoimmune diseases are more common in women than men for reasons that are not clear. In the workplace women tend to have different goals and priorities from men and react differently in stressful situations. They tend to gravitate to different types of jobs. All of these workplace factors can affect the mental and physical health of women.

It is recognized that results from randomized trials in men may not necessarily be applicable to women. Consequently, several large randomized trials in women, the largest of which is the Women's Health Initiative in the United States, have been launched. It is important, nevertheless, that women do not become so enthusiastic about being participants in randomized trials that they fail to realize that participation may involve potential risks as well as benefits.

As women in any society have more choices and as different behaviors become more socially acceptable, the choices women make are not always in their best interest. The huge increase in the number of women who smoke cigarettes worldwide is already taking a toll on their health, ranging from lung cancer to delivery of low birth weight infants to coronary heart disease. Obesity is a problem for women in many parts of the world and is likely to become even more prevalent in the future. The increase in the frequency of sexually transmitted diseases is also in part a consequence of changing societal mores.

Many areas are now the object of serious study that were not considered appropriate topics for casual conversation in previous decades. Examples are menopause, lesbian health, premenstrual syndrome, posttraumatic stress syndrome following interpersonal violence, and urinary incontinence. Increased research into some of these topics has led to new methods of prevention and treatment, with a resulting beneficial effect on the quality of life for certain subgroups of women and, to some extent, of men.

In developed countries, approximately one-third of a woman's life is now spent in the postmenopausal years. Conditions that affect older women, such as arthritis, osteoporosis, cardiovascular disease, vision impairment, hearing impairment, and dementia, can greatly affect quality of life. These and other diseases of the elderly will become increasingly important problems in both developed and developing countries because of the sheer number of elderly who will be residing in most countries, even if the proportion of elderly in many populations is still relatively small. In the developing world, the number of people aged 65 years or older is expected to increase by 200–400% between 1990 and 2025. Many of these elderly will be women. Treating diseases of the elderly is expensive and will be a burden borne by all countries. Thus, throughout the world, improving the quality of life of the elderly in a cost-effective manner is of high priority. As an increasing number of women reach very old ages, the issue of what makes some women age in a relatively healthy manner while others have a very poor quality of life is of great interest and importance. Biological, lifestyle, and psychological factors undoubtedly contribute, and multidisciplinary research is needed to identify the important contributing factors.

It is apparent that many areas in women's health are of high priority for further research and disease control efforts. As we enter the twenty-first century, we of course have no way of knowing all of the changes that are in store for us, but the progress that has been made in recent decades in recognizing and understanding many health issues of particular concern to women will provide a good foundation for future research, teaching, prevention, and clinical care.

Epidemiology Glossary

Analytic epidemiology, Epidemiologic investigations that are designed to test hypotheses using comparison groups to provide baseline data. Data from analytic studies are used to quantify the association between exposures and outcomes. Common types of analytic designs are cohort and case-control studies.

Association, The statistical relationship between two or more events, characteristics, or other variables.

Attack rate, A synonym for cumulative incidence used in studies of infectious disease. The period of observation corresponds to the duration of an epidemic.

Attributable rate percent (AR%), The rate difference (attributable rate) divided by the disease rate among the exposed, expressed as a percentage. This estimates the proportion of disease among the exposed that is attributable to the exposure if the association is causal. *(Rate difference percent, Attributable proportion, Etiologic fraction, Exposed attributable fraction)*

Attributable risk percent (AR%), The risk difference divided by the cumulative incidence of disease among the exposed, expressed as a percentage.

Bias, Deviation of results from the truth. Any trend in the collection, analysis, interpretation, publication, or review of data that can lead to conclusions that are systematically different from the truth.

Case, A countable instance in a population or study group of a particular disease, health disorder, or condition under investigation. Also, an individual with the particular disease or condition under study.

Case-control study, A type of observational analytic study where groups of individuals are enrolled into the study based on the presence ("case") or absence ("control") of a given disease or condition. Exposure histories are then compared between cases and controls. *(Case-referent study)*

Case definition, A set of standard criteria for determining whether a person has a particular disease or condition, by specifying clinical criteria and limitations on time, place, and person.

Case fatality rate, The proportion of persons with a particular condition (cases) who die from that condition. The numerator is the number of cause-specific deaths and the denominator is the number of incident cases with the condition. *(Case fatality ratio)*

Cause of disease, A factor, characteristic, behavior, event, etc. that directly influences the occurrence of disease. A reduction of the factor in the population leads to a reduction in the occurrence of disease.

Cohort, A defined group of people who have had a common experience or exposure who are then followed to quantify the incidence of new diseases or events. The term birth cohort is used to refer to a group of people born during a particular period or year.

Cohort study, See Follow-up study.

Confidence interval (CI), A range, estimated from the degree of presumed random variability of the data, such that the true value of a parameter is expected to be included in x% of repeated sampling estimations. The specified probability of the range is called the confidence level and the end points of the confidence interval are called the confidence limits.

Confounding, An admixture of effects in which the crude measure of association between an exposure and disease is influenced by an association between the exposure of interest and another risk factor. A factor confounds an association between an exposure and a disease when it is independently associated with both the exposure and the disease. This occurs when the confounding factor is associated with the disease among nonexposed individuals, is associated with the exposure among nondiseased persons, and is not an intermediate step in the causal chain. *(Confounder)*

Control, Person without the disease or condition of interest in a case-control study. Unexposed persons in a follow-up study are not controls; they are comparison subjects.

Cross-sectional study, A descriptive study in which the study population is selected independent of disease and exposure status, which are then determined at the time of the study.

Cumulative incidence (CI), The proportion of initially healthy people under observation who become cases by the end of a specified period of observation. *(Attack rate, Risk)*

Cumulative incidence ratio (CIR), A measure of association that is the ratio of the cumulative incidence in an exposed population to that in an unexposed population. *(Risk ratio)*

Dependent variable, The outcome variable in a statistical analysis or the variable whose values are a function of the independent variables in the relationship under study.

Descriptive epidemiology, Studies of the distribution of disease that are helpful in generating hypotheses. These studies often organize and summarize existing health-related data. Types of descriptive studies include case reports, correlational studies, and cross-sectional studies.

Determinant, Any factor, event, characteristic, or other entity that brings about change in a health condition.

Differential misclassification, Bias in measuring exposure or outcome that results in differential quality (accuracy) of information among subjects conditional on the presence of the other factor (outcome or exposure). *(Information bias, Nonrandom misclassification, Observation bias)*

Distribution, The frequency and pattern of health-related characteristics and events in a population. Statistically, the observed or theoretical frequency of values of a variable.

Endemic disease, The constant presence of a disease or infectious agent within a given geographic area or population group.

Environmental factor, An extrinsic factor such as climate, sanitation, geology, air quality, etc. that affects the population under study and the opportunity for exposure.

Epidemic, The occurrence of more cases of disease than expected in a given area or among a specific group of people over a particular period of time. *(Outbreak)*

Epidemiology, The study of the distribution and determinants of disease frequency in human populations.

Follow-up study, A type of observational analytic study in which groups of individuals defined on the basis of their exposure status are followed to assess the occurrence of disease. *(Cohort study)*

Health, A state of physical, mental, and social well-being; not merely the absence of disease or infirmity.

Health indicator, A measure that reflects, or indicates, the state of health of persons in a defined population (*e.g.,* mortality rate).

Health information system, A combination of health statistics from various sources, used to derive information about health status, health care, provision and use of services, and impact on health.

Host factor, An intrinsic factor such as age, race, sex, behavior, etc. that influences an individual's exposure, susceptibility, or response to a causative agent.

Hypothesis, A supposition, arrived at from observation or reflection, that leads to refutable predictions. Any conjecture cast in a form that will allow it to be tested and refuted.

Hypothesis, null, The first step in testing for statistical significance in which it is assumed that the exposure is not related to the disease under study.

Hypothesis, alternative, The hypothesis to be adopted if the null hypothesis proves implausible, in which the exposure is associated with the disease under study.

Immunity, active, Resistance developed in response to stimulus by an antigen (infecting agent or vaccine) and usually characterized by the presence of antibody produced by the host.

Immunity, herd, The resistance of a group to invasion and spread of an infectious agent based on the resistance to infection of a high proportion of individual members of the group. The resistance is a product of the number susceptible and the probability that those who are susceptible will come into contact with an infected person.

Immunity, passive, Immunity conferred by an antibody produced in another host and acquired naturally by an infant from its mother or artificially by administration of an antibody-containing preparation (antiserum or immune globulin).

Incidence rate (IR), The number of new cases divided by the person-time of observation. *(Incidence, Incidence density)*

Incubation period, A period of subclinical or inapparent pathologic changes following exposure, ending with the onset of symptoms of infectious disease.

Independent variable, An exposure, risk factor, or other characteristic being observed or measured that is hypothesized to influence an event or manifestation (the dependent variable).

Inference, statistical, The development of generalizations from sample data, usually with calculated degrees of uncertainty.

Information bias, Bias in measuring an association that results from some level of misclassification of subjects in exposure or disease status.

Intervention study, A follow-up study in which the subjects are allocated to a study group by a predetermined randomizing procedure to receive one, two, or more exposures, the effects of which are being evaluated. *(Experimental study, Randomized trial)*

Latency period, A period of subclinical or inapparent pathologic changes following exposure, ending with the onset of symptoms of chronic disease.

Mean, arithmetic, A measure of central tendency that is calculated by adding together all of the individual values in a group of measurements and dividing by the number of values in the group. *(Average)*

Mean, geometric, The mean or average of a set of data measured on a logarithmic scale.

Measure of association, A quantified relationship between an exposure and disease. Includes the rate ratio, risk ratio, odds ratio, standardized mortality ratio, etc. and often is simply called the relative risk and abbreviated RR.

Measure of central tendency, A central value that best represents a distribution of data. Measures of central tendency (or location) include the mean, median, and mode.

Measure of dispersion, A measure of the spread of a distribution from its central value. Measures used in epidemiology include the variance and standard deviation.

Median, The measure of central tendency that divides a set of data into two equal parts.

Medical surveillance, The monitoring of potentially exposed individuals to detect early symptoms of disease.

Mode, The most frequently occurring value in a set of observations.

Morbidity, Any departure from a state of physiological or psychological well-being.

Mortality rate (MR), The measure of the frequency of occurrence of death in a defined population during a specified interval of time. The numerator is the number of deaths and the denominator is the person-time of observation.

Natural history of disease, The temporal course of disease from onset to resolution.

Necessary cause, A causal factor whose presence is required for the occurrence of the disease or outcome.

Non-differential misclassification, Inaccuracy in categorization of subjects by exposure or disease status that is independent of their other status (disease or exposure). *(Random misclassification, Error)*

Null value, The expected value of a measure of association when there is no association between the exposure and outcome. For a relative risk, the null value is 1.0; for a rate difference it is 0.

Odds, The ratio of the frequency of occurrence to that of nonoccurrence.

Odds ratio (OR), The ratio of two odds. In a case-control study, a measure of association that is the ratio of the odds of exposure among the cases to that among the controls. With newly incident cases (and in most other situations), the odds ratio is an appropriate estimate of the rate ratio. *(Relative odds, Exposure odds ratio, Relative risk)*

Pandemic, An epidemic occurring over a very wide area (several countries or continents) and usually affecting a large proportion of the population.

Period prevalence, The number of existing cases of a disease divided by the number of persons in a population over a specified period of time.

Person-time, The number of persons under observation multiplied by the duration of observation. Commonly given as person-years and used as the denominator in calculation of a rate. For example, if ten persons are under study for ten years there are 100 person-years of observation. The number of person-years would be the same if 100 persons were under observation for one year or 200 persons for six months.

Population, The total number of inhabitants of a given geographic location, identified resource, or other defined cohort. *(Source population, Study base, Base population)*

Population attributable rate (PAR), A measure of the amount of disease associated with an exposure within a population. It is the difference between the incidence rate of disease in the entire population and that among the nonexposed.

Population attributable risk (PAR), The difference between the cumulative incidence of disease in the entire population and that among the nonexposed.

Population attributable rate percent (PAR%), The population attributable rate divided by the disease rate in the study population, expressed as a percentage. This estimates the proportion of disease in the study population that is attributable to the exposure. The PAR and PAR% are meaningful only when the proportion exposed in the study population is the same as that exposed in the source population. *(Compare with the Population attributable risk percent)*

Predictive value positive (PV+), The probability that a person with a positive screening test truly has the disease.

Prevalence (P), The number of existing cases divided by the number of persons (both ill and well) in a population at a specified point in time. A proportion rather than a rate. *(Prevalence proportion)*

Prevalence ratio (PR), A measure of association calculated as the ratio of the prevalence of a disease among the exposed to that among the nonexposed.

Proportion, A type of ratio in which the numerator is wholly included in the denominator. The ratio of a part to the whole, expressed as a decimal (0.2), a fraction (1/5), or, loosely, as a percentage (20%).

Prospective, Describes a study in which the disease of interest has not occurred among subjects at the start of the study.

P value, A measure of the consistency of the data with the null hypothesis.

Rate, The number of new events divided by the person-time of observation. Note that time is an integral component of a rate. The term rate is often misused; for example, case fatality rate, maternal mortality rate, and attack rate are actually proportions.

The word risk is also often incorrectly used synonymously with rate. A risk is the accumulated effect of a rate operating during a specified period of time.

Rate difference (RD), A measure of association that is the difference in incidence rates of disease between exposed and unexposed populations. *(Incidence rate difference, Attributable rate)*

Rate ratio (RR), A measure of association that is the ratio of a rate of disease in an exposed population to that in an unexposed population. *(Incidence density ratio, Incidence rate ratio, Relative risk)*

Relative risk (RR), A general term often used to mean a rate ratio, risk ratio, odds ratio, or prevalence ratio, etc.

Retrospective, Describes a study in which the disease of interest has already occurred among subjects at the start of the study.

Risk, A probability that an individual will become ill within a stated period of time or at a certain age. Also, a nontechnical term encompassing a variety of measures of the probability of a generally unfavorable outcome.

Risk difference, A measure of association that is the difference between the risk of disease in an exposed population and that in one not exposed. *(Attributable risk)*

Risk factor, An aspect of personal behavior or lifestyle, an environmental exposure, or an inherited characteristic that is associated with an increased occurrence of disease or other health-related event.

Risk ratio (RR), A measure of association that is the ratio of the risk of disease in an exposed population to that in a nonexposed population. *(Cumulative incidence ratio)*

Secular trend, Changes over a period of time, often years or decades.

Selection bias, Bias in measuring an association that results when the selection criteria for subjects differ between compared groups in relation to the exposure or outcome of interest.

Sensitivity, The proportion of persons with the disease who are correctly identified by a screening test or case definition as having the disease.

Specificity, The proportion of persons who do not have a disease who are correctly identified as nondiseased by a screening test or case definition.

Standardized mortality ratio (SMR), The ratio of observed to expected numbers of deaths, commonly multiplied by 100.

Standardized rate ratio (SRR), A rate ratio in which the rates have been standardized to the same (standard) set of weights.

Sufficient cause, A causal factor or collection of factors whose presence is always followed by the occurrence of the disease.

Vital statistics, Systematically tabulated information about births, marriages, divorces, and deaths.

Note: In this glossary, "exposure" means the hypothetical cause and "disease" means the outcome or effect. Adapted from glossaries developed by the Department of Epidemiology, Harvard School of Public Health, and the Centers for Disease Control and Prevention (http://www.cdc.gov/nccdphp/drh/eip_gloss.htm). Further explanation of epidemiologic terms and concepts may be found in the following texts:

Gordis, L. (1996). "Epidemiology." Saunders, Philadelphia.
Hennekens, C. H., and Buring, J. E. (1987). "Epidemiology in Medicine." Little, Brown, Boston.

Last, J. M. (1995). "A Dictionary of Epidemiology," 3rd ed.
    Oxford University Press, New York.
MacMahon, B., and Trichopoulos, D. (1996). "Epidemiology:
    Principles and Methods," 2nd ed. Little, Brown, Boston.

Index

Rate per 100,000 population

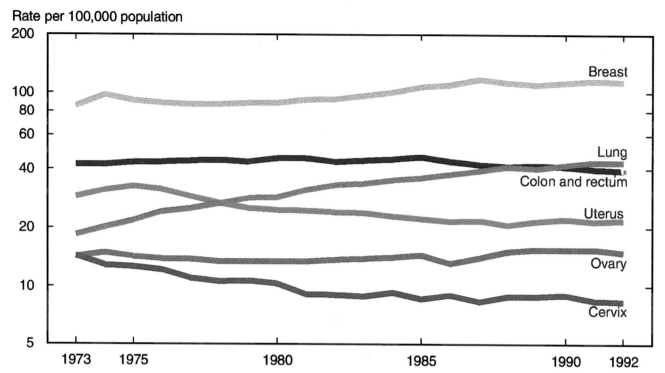

FIGURE 1.9 Incidence rates for selected cancer sites among women: Selected geographic areas of the United States, 1973–1992. Source: National Institutes of Health, National Cancer Institute, Surveillance, Epidemiology, and End Results (SEER) Program [33].

Deaths per 100,000 population

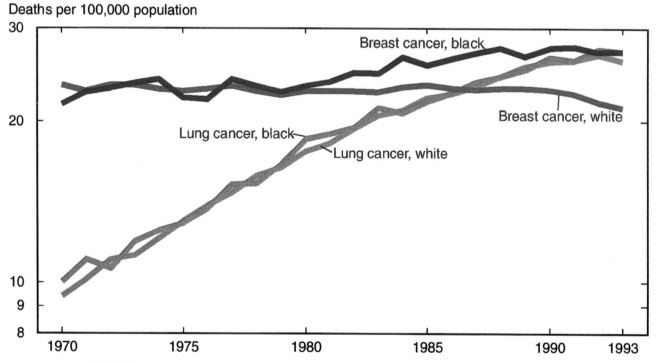

FIGURE 1.10 Death rates for lung and breast cancer among women by race: United States, 1970–1993. Source: Centers for Disease Control and Prevention, National Center for Health Statistics, National Vital Statistics System [33].

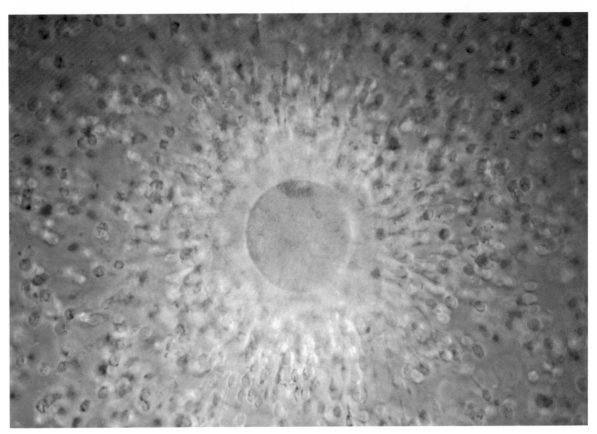

FIGURE 17.4 Photomicrograph of a mature oocyte with surrounding cumulus.

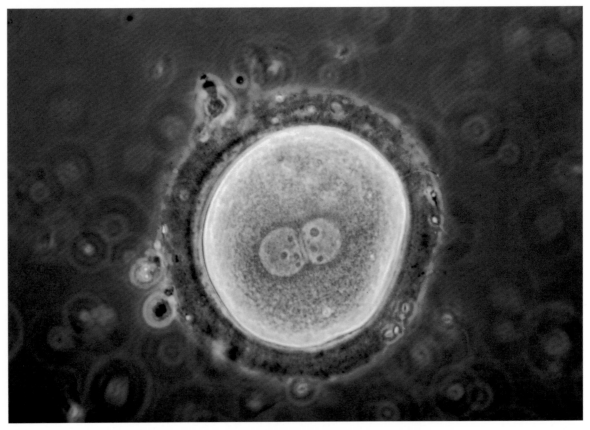

FIGURE 17.5 Demonstration of two pronuclei at 16 hours post fertilization.

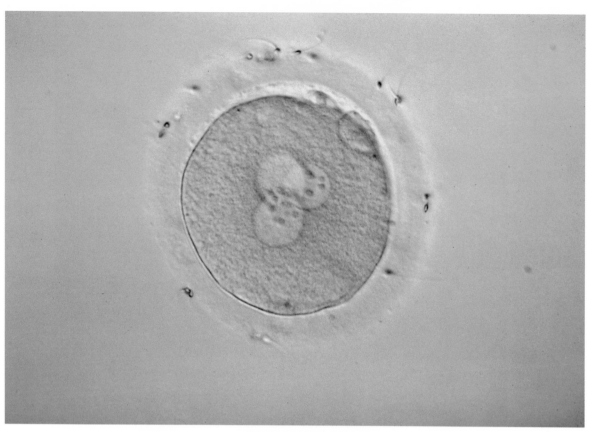

FIGURE 17.6 Demonstrating three pronuclei (polyspermic fertilization) at 16 hours post fertilization.

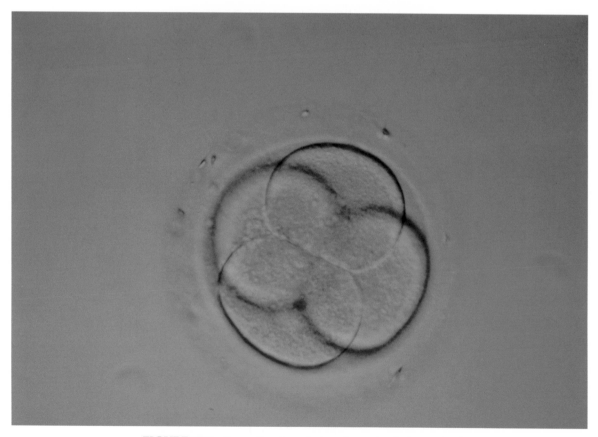

FIGURE 17.7 Four cell embryo for transfer on day 2 after retrieval.

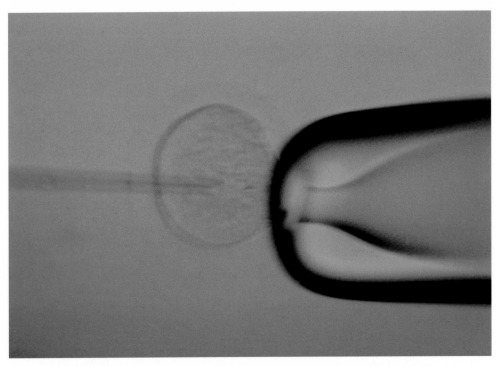

FIGURE 17.8 Mature oocyte undergoing intracytoplasmic sperm injection (ICSI).

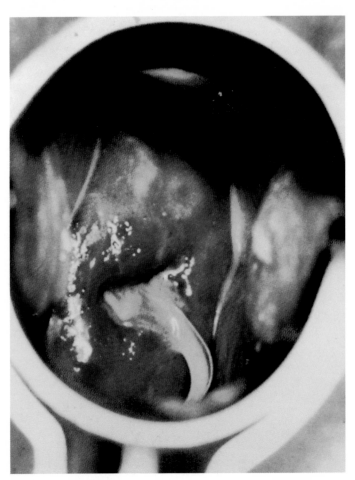

FIGURE 21.3 Mucopurulent cervicitis (from Holmes, K.K., *et al.*, eds. (1999). *Sexually Transmitted Diseases, 3rd Ed.* McGraw-Hill, New York).

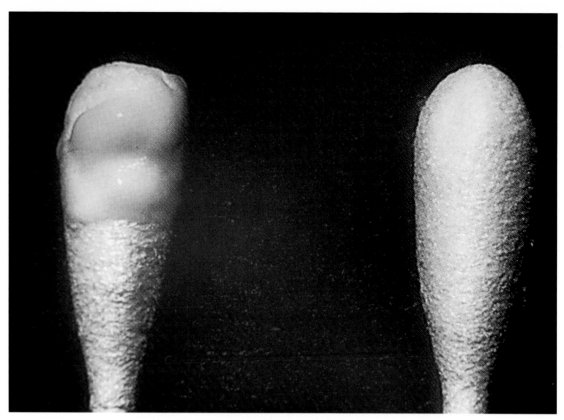

FIGURE 21.4 Positive swab test

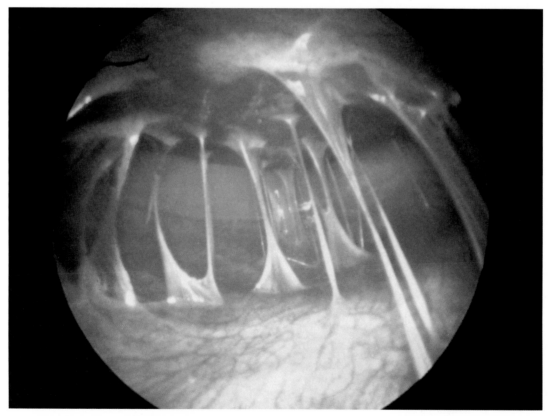

FIGURE 21.5 Adhesions between liver capsule and anterior abdominal wall in salpingitis associated peri-hepatitis (from Holmes, K.K., *et al.*, eds. (1999). *Sexually Transmitted Diseases, 3rd Ed.* McGraw-Hill, New York).

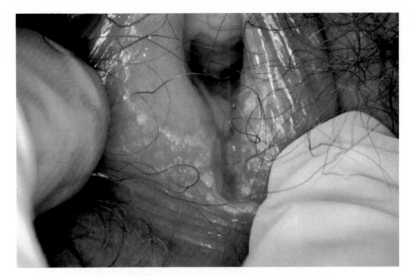

FIGURE 25.1   Vulvar condyloma acuminata.

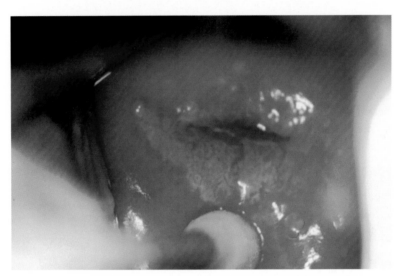

FIGURE 25.2   This woman's biopsy was LSIL. Note the acetowhite epithelium with well-demarcated edges and coarse mosaic vascular pattern typical of SIL.

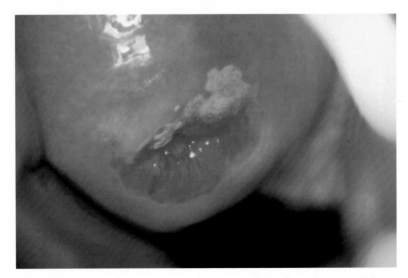

FIGURE 25.3   Colpophotograph of a woman with an atypical transformation zone. Colposcopically this may be indistinguishable from SIL. This young woman had no cytologic or histologic evidence of SIL.